Electricity

$$F = \frac{kq_1q_2}{r^2}$$

$$R = \frac{\rho L}{A}$$

$$I = \frac{V}{R}$$

Series Circuits

(a) $I = I_1 = I_2 = I_3 = \cdots$

(b) $R = R_1 + R_2 + R_3 + \cdots$

(c) $E = V_1 + V_2 + V_3 + \cdots$

Parallel Circuits

(a) $I = I_1 + I_2 + I_3 + \cdots$

(b) $\frac{1}{R} = \frac{1}{R_1} + \frac{1}{R_2} + \frac{1}{R_3} + \cdots$

(c) $E = V_1 = V_2 = V_3 = \cdots$

Cells in Series

(a) $I = I_1 = I_2 = I_3 = \cdots$

(b) $r = r_1 + r_2 + r_3 + \cdots$

(c) $E = E_1 + E_2 + E_3 + \cdots$

Cells in Parallel

(a) $I = I_1 + I_2 + I_3 + \cdots$

(b) $r = \frac{r \text{ of one cell}}{\text{number of like cells}}$

(c) $E = E_1 = E_2 = E_3 = \cdots$

$$V = E - Ir$$

$$P = VI = I^2R = \frac{V^2}{R}$$

Magnetism

$$B = \frac{\mu_0 I}{2\pi R}$$

$$B = \pi_0 In$$

Transformers

$$\frac{V_P}{V_S} = \frac{N_P}{N_S}$$

$$\frac{I_S}{I_P} = \frac{N_P}{N_S}$$

ac Circuits

$$X_L = 2\pi fL$$

$$I = \frac{E}{X_L}$$

$$I = \frac{E}{Z}$$

$$Z = \sqrt{R^2 + X_L^2}$$

$$\tan\phi = \frac{X_L}{R}$$

$$X_C = \frac{1}{2\pi fC}$$

$$Z = \sqrt{R^2 + X_C^2}$$

$$\tan\phi = \frac{X_C}{R}$$

$$Z = \sqrt{R^2 + (X_L - X_C)^2}$$

$$\tan\phi = \frac{X_L - X_C}{R}$$

$$f = \frac{1}{2\pi\sqrt{LC}}$$

Light

$$c = \lambda f$$

$$E = hf$$

$$E = \frac{I}{4\pi r^2}$$

$$\frac{1}{f} = \frac{1}{s_o} + \frac{1}{s_i}$$

$$M = \frac{h_i}{h_o} = \frac{-s_i}{s_o}$$

$$n = \frac{\sin i}{\sin r} = \frac{\text{speed of light in vacuum}}{\text{speed of light in substance}}$$

$$\sin i_c = \frac{1}{n}$$

Modern Physics

$$E = -\frac{kZ^2}{n^2}$$

$$E = \Delta mc^2$$

$$Q = (M_p - M_d - m_\alpha)c^2$$

$$N = N_0e^{-\lambda t}$$

$$T_{1/2} = \frac{0.693}{\lambda}$$

$$A = \lambda N = \lambda N_0e^{-\lambda t} = A_0e^{-\lambda t}$$

APPLIED PHYSICS

EIGHTH EDITION

DALE EWEN

Parkland Community College
Champaign, Illinois

NEILL SCHURTER

P. ERIK GUNDERSEN

Pascack Valley Regional High School District
Montvale, New Jersey

PEARSON
Prentice Hall

Upper Saddle River, New Jersey
Columbus, Ohio

Library of Congress Cataloging-in-Publication Data

Ewen, Dale
Applied physics, — 8th ed. / Dale Ewen, Neill Schurter, P. Erik Gundersen.
p. cm.
Includes index
ISBN 0-13-110169-2
1. Physics. I. Schurter, Neill. II. Gundersen, P. Erik. III. Title.
QC23.2.E88 2005
530—dc22
2003021948

Editor in Chief: Stephen Helba
Senior Acquisitions Editor: Gary Bauer
Editorial Assistant: Natasha Holden
Development Editor: Michelle Churma
Production Editor: Louise N. Sette
Production Supervision: *The GTS Companies*/York, PA Campus
Design Coordinator: Diane Ernsberger
Cover Designer: Jason Moore
Cover art: Corbis
Production Manager: Pat Tonneman
Marketing Manager: Leigh Ann Sims

This book was set in Times by *The GTS Companies*/York, PA Campus. It was printed and bound by Courier Kendallville, Inc. The cover was printed by The Lehigh Press, Inc.

Previous editions entitled *Physics for Career Education.*

Photo Credits: Corbis Corporation, p.114; Dorling Kindersley, p.582; Getty Images, p.244, 400, 630; Greg Probst/Getty Images, Inc., p.504; Hank Morgan/Photo Researchers, Inc., p.556; Image Stock Imagery, p.84; Insurance Institute for Highway Safety, p.134; Jim Steinberg/Photo Researchers, Inc., p.386; Jonathan Nourok/PhotoEdit, Inc., p.244; Jurgen Vogt/Getty Images, Inc., p.150; Mark D. Phillips/Photo Researchers, Inc., p.56; Michael Rosenfeld/Getty Images, Inc., p.12; Railway Technical Research Institute, p.482; Richard Megna/Fundamental Photographs, p.244; Robert Brenner/PhotoEdit, Inc., p.244; Ron Thomas/Getty Images, Inc., p.538; St. Meyers/Okapia/Photo Researchers, Inc., p.344

Pearson Education Ltd.
Pearson Education Singapore Pte. Ltd.
Pearson Education Canada, Ltd.
Pearson Education—Japan
Pearson Education Australia Pty. Limited
Pearson Education North Asia Ltd.
Pearson Educación de Mexico, S.A. de C.V.
Pearson Education Malaysia Pte. Ltd.

PEARSON
Prentice Hall

10 9 8 7 6 5 4 3
ISBN 0-13-110169-2

CONTENTS

PREFACE ix

CHAPTER 0 **An Introduction to Physics** 3

0.1 Why Study Physics? 4
0.2 Physics and Its Role in Technology 5
0.3 Physics and Its Connections to Other Fields and Sciences 7
0.4 Theories, Laws, and the Problem-Solving Method 8

CHAPTER 1 **The Physics Tool Kit** 13

1.1 Standards of Measure 14
1.2 Introduction to the Metric System 16
1.3 Length 18
1.4 Area and Volume 22
1.5 Other Units 32
1.6 Measurement: Significant Digits and Accuracy 36
1.7 Measurement: Precision 38
1.8 Calculations with Measurements 40
1.9 Problem-Solving Method 45

CHAPTER 2 **Vectors** 57

2.1 Vectors and Scalars 58
2.2 Components of a Vector 65
2.3 Vectors in Standard Position 73

CHAPTER 3 **Motion** 85

3.1 Speed Versus Velocity 86
3.2 Acceleration 93
3.3 Uniformly Accelerated Motion and Free Fall 97
3.4 Projectile Motion 106

CHAPTER 4 **Force** 115

4.1 Force and the Law of Inertia 116
4.2 Force and the Law of Acceleration 118
4.3 Friction 121
4.4 Total Forces in One Dimension 125
4.5 Gravity and Weight 127
4.6 Law of Action and Reaction 129

CHAPTER 5 **Momentum 135**

5.1 Impulse and Momentum 136
5.2 Collisions 144

CHAPTER 6 **Concurrent and Parallel Forces 151**

6.1 Forces in Two Dimensions 152
6.2 Concurrent Forces in Equilibrium 157
6.3 Torque 166
6.4 Parallel Force Problems 169
6.5 Center of Gravity 174

CHAPTER 7 **Work and Energy 185**

7.1 Work 186
7.2 Power 191
7.3 Energy 197
7.4 Conservation of Mechanical Energy 202

CHAPTER 8 **Rotational Motion 211**

8.1 Measurement of Rotational Motion 212
8.2 Angular Momentum 219
8.3 Centripetal Force 221
8.4 Power in Rotational Systems 224
8.5 Transferring Rotational Motion 227
8.6 Gears 229
8.7 Pulleys Connected with a Belt 237

CHAPTER 9 **Simple Machines 245**

9.1 Machines and Energy Transfer 246
9.2 The Lever 248
9.3 The Wheel-and-Axle 253
9.4 The Pulley 255
9.5 The Inclined Plane 259
9.6 The Screw 262
9.7 The Wedge 264
9.8 Compound Machines 265
9.9 The Effect of Friction on Simple Machines 267

CHAPTER 10 **Universal Gravitation and Satellite Motion 275**

10.1 Universal Gravitation 276
10.2 Gravitational Fields 279
10.3 Satellite Motion 280

CHAPTER 11 **Matter 287**

11.1 Properties of Matter 288
11.2 Properties of Solids 290
11.3 Properties of Liquids 303

11.4 Properties of Gases 306
11.5 Density 306

CHAPTER 12 **Fluids 319**

12.1 Hydrostatic Pressure 320
12.2 Hydraulic Principle 326
12.3 Air Pressure 330
12.4 Buoyancy 333
12.5 Fluid Flow 336

CHAPTER 13 **Temperature and Heat Transfer 345**

13.1 Temperature 346
13.2 Heat 350
13.3 Heat Transfer 352
13.4 Specific Heat 357
13.5 Method of Mixtures 360
13.6 Expansion of Solids 363
13.7 Expansion of Liquids 368
13.8 Change of Phase 370

CHAPTER 14 **Properties of Gases 387**

14.1 Charles' Law 388
14.2 Boyle's Law 390
14.3 Charles' and Boyle's Laws Combined 393

CHAPTER 15 **Wave Motion and Sound 401**

15.1 Characteristics of Waves 402
15.2 Electromagnetic Waves 409
15.3 Sound Waves 412
15.4 The Doppler Effect 415
15.5 Resonance 418
15.6 Simple Harmonic Motion 420

CHAPTER 16 **Basic Electricity 429**

16.1 Electric Charges 430
16.2 Induction 431
16.3 Coulomb's Law 434
16.4 Electric Fields 436
16.5 Simple Circuits 439
16.6 Ohm's Law 444
16.7 Series Circuits 447
16.8 Parallel Circuits 451
16.9 Compound Circuits 457
16.10 Electric Instruments 462
16.11 Voltage Sources 466
16.12 Cells in Series and Parallel 467
16.13 Electric Power 470

CHAPTER 17 Magnetism 483

17.1 Introduction to Magnetism 484
17.2 Magnetic Effects of Currents 485
17.3 Induced Magnetism and Electromagnets 492
17.4 Induced Current 493
17.5 Generators 494
17.6 The Motor Principle 497
17.7 Magnetic Forces on Moving Charged Particles 500

CHAPTER 18 Alternating Current Electricity 505

18.1 What Is Alternating Current? 506
18.2 ac Power 509
18.3 Inductance 517
18.4 Inductance and Resistance in Series 520
18.5 Capacitance 523
18.6 Capacitance and Resistance in Series 525
18.7 Capacitance, Inductance, and Resistance in Series 527
18.8 Resonance 529
18.9 Rectification and Amplification 531
18.10 Commercial Generator Power Output 532

CHAPTER 19 Light 539

19.1 Nature of Light 540
19.2 The Speed of Light 543
19.3 Light as a Wave 545
19.4 Light as a Particle 547
19.5 Photometry 549

CHAPTER 20 Reflection and Refraction 557

20.1 Mirrors and Images 558
20.2 Images Formed by Plane Mirrors 560
20.3 Images Formed by Concave Mirrors 561
20.4 Images Formed by Convex Mirrors 563
20.5 The Mirror Formula 564
20.6 The Law of Refraction 566
20.7 Total Internal Reflection 570
20.8 Types of Lenses 573
20.9 Images Formed by Converging Lenses 574
20.10 Images Formed by Diverging Lenses 575

CHAPTER 21 Color 583

21.1 The Color of Light 584
21.2 Diffraction of Light 592
21.3 Interference 593
21.4 Polarization of Light 595

CHAPTER 22 **Survey of Modern Physics** **603**

22.1 Quantum Theory 604
22.2 The Atom 604
22.3 Atomic Structure and Atomic Spectra 605
22.4 Quantum Mechanics and Atomic Properties 608
22.5 The Nucleus—Structure and Properties 610
22.6 Nuclear Mass and Binding Energy 613
22.7 Radioactive Decay 616
22.8 Nuclear Reactions—Fission and Fusion 621
22.9 Detection and Measurement of Radiation 622
22.10 Radiation Penetrating Power 624

CHAPTER 23 **Special and General Relativity** **631**

23.1 Albert Einstein 632
23.2 Special Theory of Relativity 632
23.3 Space-Time 635
23.4 General Theory of Relativity 636

APPENDIX A **Mathematics Review** **639**

A.1 Signed Numbers 639
A.2 Powers of 10 642
A.3 Scientific Notation 644
A.4 Solving Linear Equations 647
A.5 Solving Quadratic Equations 650
A.6 Formulas 653
A.7 Substituting Data into Formulas 656
A.8 Right-Triangle Trigonometry 659
A.9 Law of Sines and Law of Cosines 666

APPENDIX B **Scientific Calculator** **677**

B.1 Scientific Notation 677
B.2 Squares and Square Roots 678
B.3 Trigonometric Operations 679
B.4 Finding a Power 681

APPENDIX C **Tables** **683**

Table 1 U.S. Weights and Measures 683
Table 2 Conversion Table for Length 683
Table 3 Conversion Table for Area 683
Table 4 Conversion Table for Volume 684
Table 5 Conversion Table for Mass 684
Table 6 Conversion Table for Density 684
Table 7 Conversion Table for Time 685
Table 8 Conversion Table for Speed 685
Table 9 Conversion Table for Force 685
Table 10 Conversion Table for Power 685
Table 11 Conversion Table for Pressure 686

Table 12 Mass and Weight Density 686
Table 13 Specific Gravity of Certain Liquids 687
Table 14 Conversion Table for Energy, Work, and Heat 687
Table 15 Heat Constants 687
Table 16 Coefficient of Linear Expansion 688
Table 17 Coefficient of Volume Expansion 688
Table 18 Charge 688
Table 19 Relationships of Metric SI Base and Derived Units 689
Table 20 Electric Symbols 690
Table 21 Periodic Table 691
Table 22 The Greek Alphabet 692

APPENDIX D **Glossary 693**

APPENDIX E **Answers to Odd-Numbered Problems and to Chapter Review Questions and Problems 707**

INDEX *743*

PREFACE

Applied Physics, eighth edition, formerly *Physics for Career Education*, provides comprehensive and practical coverage of physics for students needing an applied physics approach or considering a vocational–technical career. It emphasizes physical concepts as applied to industrial–technical fields and uses common applications to improve the physics and mathematics competence of the student. This eighth edition has been carefully reviewed and special efforts have been taken to emphasize clarity and accuracy of presentation.

This text is divided into five major areas: mechanics, matter and heat, wave motion and sound, electricity and magnetism, and light and modern physics.

Key Features

- Real-world applications to motivate students
- Clear to-the-point topic coverage
- Over 3900 problems and questions to assist student learning
- Unique problem-solving format is consistently used throughout the text
- Numerous drawings, diagrams, photographs, and examples illustrate the application of physics in the real world
- Extensive problem sets at the end of each section provide students with ample opportunity for practice
- Comprehensive discussion and consistent use of the results of measurements and significant digits
- Biographical sketches of important scientists appear in most chapters
- Answers to odd-numbered problems and chapter review questions and problems in Appendix E
- Comprehensive Glossary—one-stop reference in Appendix D
- Basic Scientific Calculator Instructions are presented in Appendix B
- Basic Math Review—provides students with a refresher of math needed for the course in Appendix A

Special New Features

- **NEW—Four-Color Format**—photos, illustrations, and diagrams are now in color, improving student interest and comprehension.
- **NEW—Try this Activity**—provide students with opportunities to experiment with physics concepts. Activities involve a demonstration or mini-activity that can be performed by students on their own to experience a physics concept, allowing for active vs. passive learning.

- **NEW—Physics Connections**—apply physics to familiar real-world situations and events. These brief readings help students bridge the gap between what is taught in the chapter and "real-world" technical applications.
- **NEW—Applied Concepts**—application-based questions at the end of each chapter that develop problem-solving skills in real-life physics applications
- **NEW—3 new chapters:**
  Chapter 10—*Universal Gravitation and Satellite Motion*
  Chapter 21—*Color*
  Chapter 23—*Special and General Relativity*
  These chapters have been added to broaden the coverage in the text for use in general education courses as well as technical courses.
- **NEW—Comprehensive Glossary**—Appendix D provides a one-stop student reference.

Examples of Key Features

Unique Problem-Solving Method

This textbook teaches students to use a proven effective problem-solving methodology. The consistent use of this method trains students to make a sketch, identify the data elements, select the appropriate equation, solve for the unknown quantity, and substitute the data in the working equation. An icon that outlines the method is placed in the margin of most problem sets as a reminder to students.

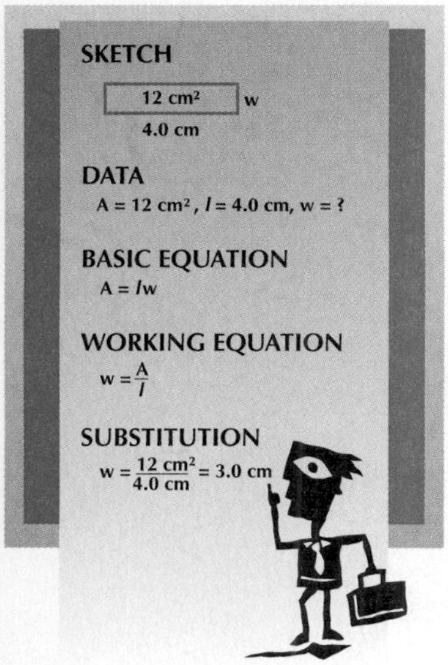

Figure P. 1 shows examples illustrating how the problem-solving method is used in the text. See pages 45–46 for the detailed presentation of the problem-solving method.

Figure P.1

EXAMPLE 1

A ball rolls at a constant speed of 0.700 m/s as it reaches the end of a 1.30-m-high table (Fig. 3.22). How far from the edge of the table does the ball land?

Sketch:

Figure 3.22

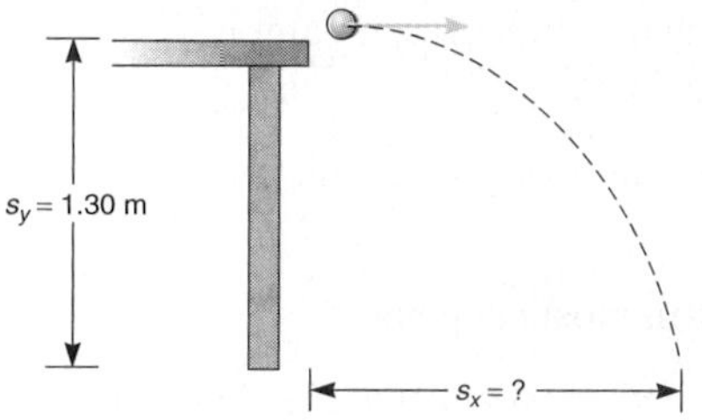

Data:

$$v_{iy} = 0 \text{ m/s} \qquad v_x = 0.700 \text{ m/s}$$
$$s_y = 1.30 \text{ m} \qquad s_x = ?$$

Basic Equations:

$$s_y = v_{iy}t + \tfrac{1}{2}a_y t^2 \qquad s_x = v_x t$$

Working Equations (with $v_{iy} = 0$):

$$t = \sqrt{\frac{2s_y}{a}} \qquad s_x = v_x t$$

Substitution:

$$t = \sqrt{\frac{2(1.30 \text{ m})}{9.80 \text{ m/s}^2}}$$
$$t = 0.515 \text{ s}$$
$$s_x = (0.700 \text{ m/s})(0.515 \text{ s})$$
$$s_x = 0.361 \text{ m}$$

Figure P.2

EXAMPLE 4

Juan and Sonja use a push mower to mow a lawn. Juan, who is taller, pushes at a constant force of 33.1 N on the handle at an angle of 55.0° with the ground. Sonja, who is shorter, pushes at a constant force of 23.2 N on the handle at an angle of 35.0° with the ground. Assume they each push the mower 3000 m. Who does more work and by how much?

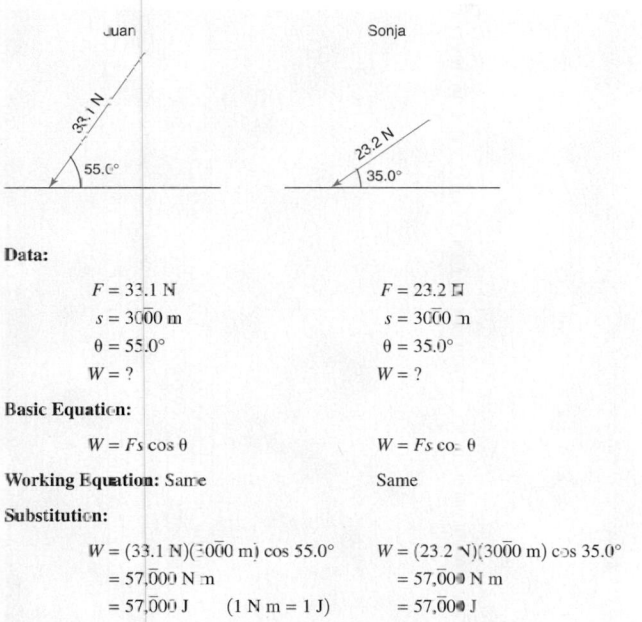

Sketch:

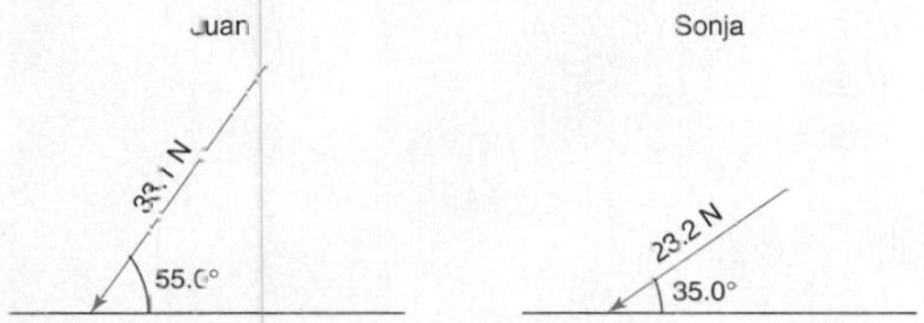

Data:

Juan	Sonja
$F = 33.1\text{ N}$	$F = 23.2\text{ N}$
$s = 3\overline{0}00\text{ m}$	$s = 3\overline{0}00\text{ m}$
$\theta = 55.0°$	$\theta = 35.0°$
$W = ?$	$W = ?$

Basic Equation:

$W = Fs\cos\theta$ $\qquad$ $W = Fs\cos\theta$

Working Equation: Same $\qquad$ Same

Substitution:

$$\begin{aligned} W &= (33.1\text{ N})(3\overline{0}00\text{ m})\cos 55.0° \\ &= 57,\overline{0}00\text{ N m} \\ &= 57,\overline{0}00\text{ J} \quad (1\text{ N m} = 1\text{ J}) \end{aligned}$$

$$\begin{aligned} W &= (23.2\text{ N})(3\overline{0}00\text{ m})\cos 35.0° \\ &= 57,\overline{0}00\text{ N m} \\ &= 57,\overline{0}00\text{ J} \end{aligned}$$

They do the same amount of work. However, Juan must exert more energy because he pushes into the ground more than Sonja, who pushes more in the direction of the motion.

Worked Examples

Worked examples are consistently displayed in the problem-solving format and used to illustrate and clarify basic concepts and problems. Since many students learn by example, a large number of examples are provided. The example in Figure P.3 shows how conversion factors are displayed and used.

Figure P.3

EXAMPLE 2

Find the depth in a lake at which the pressure is 105 lb/in^2.

Data:

$$P = 105\text{ lb/in}^2$$
$$D_w = 62.4\text{ lb/ft}^3$$
$$h = ?$$

Basic Equation

$$P = hD_w$$

Working Equation:

$$h = \frac{P}{D_w}$$

Substitution:

$$h = \frac{105\text{ lb/in}^2}{62.4\text{ lb/ft}^3}$$

$$= 1.68\frac{\text{ft}^3}{\text{in}^2}$$

$$\boxed{\frac{\text{lb/in}^2}{\text{lb/ft}^3} = \frac{\text{lb}}{\text{in}^2} \div \frac{\text{lb}}{\text{ft}^3} = \frac{\text{lb}}{\text{in}^2} \times \frac{\text{ft}^3}{\text{lb}} = \frac{\text{ft}^3}{\text{in}^2}}$$

$$= 1.68\frac{\text{ft}^3}{\text{in}^2} \times \left(\frac{12\text{ in.}}{1\text{ ft}}\right)^2$$

$$= 242\text{ ft}$$

New High-Interest Chapter Openers

Chapter opening photos feature topics of interest to students with hand-written formula notes relating the action in the photo to a physical principle discussed in the chapter.

Figure P.4

Try this Activity

These activities provide students with opportunities to experiment with physics concepts. Activities involve a demonstration or mini-activity that can be performed by students on their own to experience a physics concept, allowing for active vs. passive learning.

Figure P.5

TRY THIS ACTIVITY

Free Fall in a Vacuum

Drop a piece of paper and a book at the same time and clock the time it takes for each to hit the floor. Now place that paper on top of the book as shown in Fig. 3.14 (***Note:*** The top surface area of the book must be larger than that of the paper.) What happens to the time it takes the book and paper to fall? What does this show about objects falling in a vacuum? (A vacuum is a space in which there is no air resistance present.)

Figure 3.14 Place the paper on top of the book. The book must be larger than the paper.

Physics Connections

These features apply physics to familiar real-world situations and events. These brief readings help students bridge the gap between what is taught in the chapter and "real-world" technical applications.

Figure P.6

PHYSICS CONNECTIONS

Fiber Optic Cables

Most transmission of information travels as electric impulses through electric and telephone lines and fiber optic cables. Electric signals travel relatively slowly, cause wires to heat up, and need transformers to boost the voltage of the signal traveling over long distances. Electric signals and wires are being replaced with light signals traveling through flexible, low-cost strands of glass. Because light travels through glass optical fibers, there is no resistance to weaken the light signal, and the signal travels at the speed of light, which is much faster than the speed of conventional electric signals. Such advances in fiber optics communications are revolutionizing the way we communicate.

Light traveling in the same medium travels in a straight line, whereas fiber optic cables can transmit a signal while twisting and turning, because of total internal reflection. The angle at which the light strikes the cladding of the fiber is always greater than the critical angle of the cladding and the core. The low critical angle allows the light to continually reflect and travel great distances without needing to be reamplified. In order for the cable to maintain a low critical angle, the glass must contain no imperfections or bubbles that would cause the light to be directed out or backward through the cable (see Fig. 20.24).

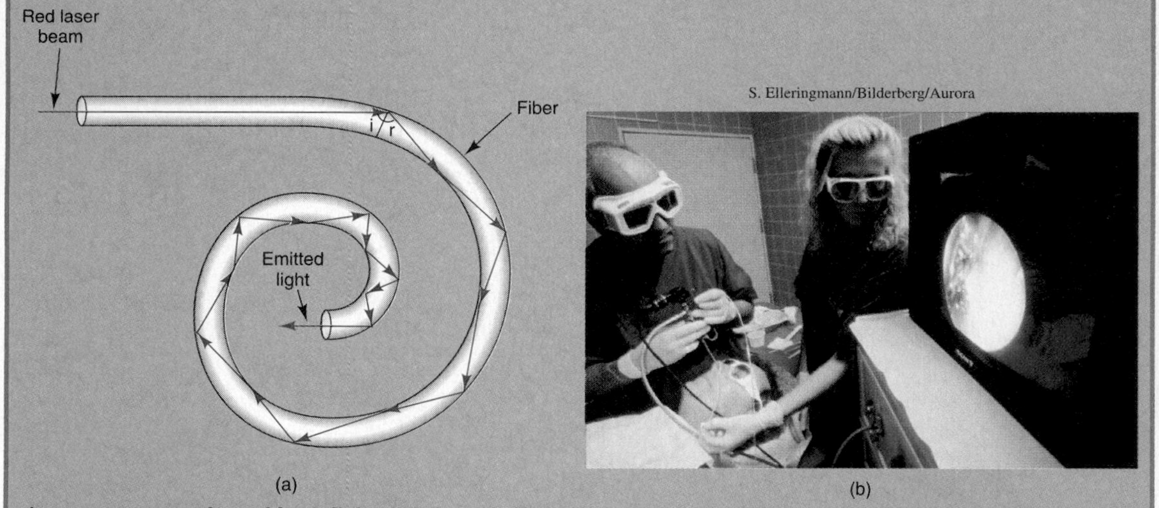

Figure 20.24 (a) The red laser light entering the fiber optic cable is totally internally reflected, which results in the light emerging at the end of the cable. (b) An endoscope is a bundle of fiber optic cables used in many minimally invasive surgeries. Here an endoscope is used in the removal of nose adenoids with a laser therapy procedure.

Fiber optic cables are used in telecommunications, computer networks, and medicine. A few strands of glass fiber can carry thousands of separate digital telephone conversations by slightly altering the frequency of the light for each phone conversation. A digital signal transmitted at one frequency cannot be confused by a signal carried at another frequency. Many computer networks and internal components in computers use fiber optic cables to carry data. By eliminating electric wiring, the fiber optic cable helps to reduce the temperature inside computers and servers. Finally, physicians use fiber optic bundles to perform minimally invasive procedures. A tool called an endoscope, composed of a bundle of fiber optic cables, transmits light into a patient's body while another bundle of fibers on the endoscope functions as a digital camera. The camera picks up the image and sends it back through the fiber optic cable to a monitor in the procedure room.

Applied Concepts

Application-based questions at the end of each chapter that develop problem-solving skills in real-life physics applications

Figure P.7

APPLIED CONCEPTS

1. Rosita needs to purchase a sump pump for her basement. (a) If the pump must carry 10.0 kg of water to a height of 2.75 m each minute, what minimum wattage pump is needed? (b) What three main factors determine power for a sump pump?
2. As a roller coaster designer, you must carefully balance the desire for excitement and the need for safety. Your most recent design is shown in Fig. 7.17. (a) If a 355-kg roller coaster car has zero velocity on the top of the first hill, determine its potential energy. (b) What is the velocity of the roller coaster car at the specified locations in the design? (c) Explain the relationship between velocity and the position on the track throughout the ride. (Consider the track to be frictionless.)

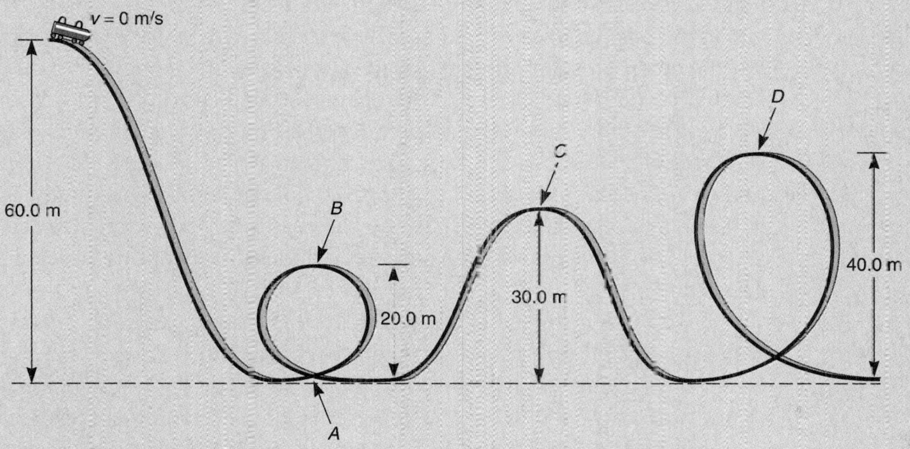

Figure 7.17

3. A 22,500-kg Navy fighter jet flying 235 km/h must catch an arresting cable to land safely on the runway strip of an aircraft carrier. (a) How much energy must the cable absorb to stop the fighter jet? (b) If the cable allows the jet to move 115 m before coming to rest, what is the average force that the cable exerts on the jet? (c) If the jet were given more than 115 m to stop, how would the force applied by the cable change?
4. The hydroelectric plant at the Itaipu Dam, located on the Parana River between Paraguay and Brazil, uses the transfer of potential to kinetic energy of water to generate electricity. (a) If 1.00×10^6 gallons of water (3.79×10^6 kg) flows down 142 m into the turbines each second, how much power does the hydroelectric power plant generate? (For comparison purposes, the Hoover Dam generates 1.57×10^6 W of power.) (b) How much power could the plant produce if the Itaipu Dam were twice its actual height? (c) Explain why the height of a dam is important for hydroelectric power plants.
5. A 1250-kg wrecking ball is lifted to a height of 12.7 m above its resting point. When the wrecking ball is released, it swings toward an abandoned building and makes an indentation of 43.7 cm in the wall. (a) What is the potential energy of the wrecking ball at a height of 12.7 m? (b) What is its kinetic energy as it strikes the wall? (c) If the wrecking ball transfers all of its kinetic energy to the wall, how much force does the wrecking ball apply to the wall? (d) Why should a wrecking ball strike a wall at the lowest point in its swing?

Ancillaries

- Companion Laboratory Manual (0-13-110353-9)
- Instructor's Resource Manual with Complete Solutions
- PowerPoint transparencies
- Test Item File
- The Prentice Hall TestGen, which provides the Test Item File on CD-ROM.

To the Faculty

This text is written at a language level and at a mathematics level that is cognizant of and beneficial to *most* students in programs that do not require a high level of mathematics. The authors have assumed that the student has successfully completed one year of high school algebra or its equivalent. Simple equations and formulas are reviewed and any mathematics beyond this level is developed in the text or in an appendix. The manner in which the mathematics is used in the text displays the need for mathematics in technology. For the better prepared student, the mathematics sections may be omitted with no loss in continuity. This text is designed so that faculty have flexibility in selecting the topics, as well as the order of topics, that meet the needs of their students and programs of study.

Sections are short, and each deals with only one concept. The need for the investigation of a physical principle is developed before undertaking its study, and many diagrams are used to aid students in visualizing the concept. Many examples and problems are given to help students develop and check their mastery of one concept before moving to another.

This text is designed to be used in a vocational–technical program in a community college, a technical institute, or a high school for students who plan to pursue a technical career or in a general physics course where an applied physics approach is preferred. The topics were chosen with the assistance of technicians and management in several industries and faculty consultants. Suggestions from users and reviewers of the previous edition were used extensively in this edition.

A general introduction to physics is presented in Chapter 0. Chapter 1 introduces students to basic units of measurement. For students who lack a metric background or who need a review, an extensive discussion of the metric system is given in Chapter 1, where it is shown how the results of measurements are approximate numbers, which are then used consistently throughout the text. Those who need to review some mathematical skills are referred to the appendices as necessary. Chapter 1 also introduces students to a problem-solving method that is consistently used in the rest of the text. Vectors are developed in Chapter 2, followed by a comprehensive study of motion, force, work and energy, rotational energy, simple machines, and universal gravitation and satellite motion.

The treatment of matter includes a discussion of the three states of matter, density, fluids, pressure, and Pascal's principle. The treatment of heat includes temperature, specific heat, thermal expansion, change of state, and ideal gas laws.

The section on wave motion and sound deals with basic wave characteristics, the nature and speed of sound, the Doppler effect, and resonance.

The section on electricity and magnetism begins with a brief discussion of static electricity, followed by an extensive treatment of dc circuits and sources, Ohm's law, and series and parallel circuits. The chapter on magnetism, generators, and motors is largely descriptive, but it allows for a more in-depth study if desired. Then ac circuits and transformers are treated extensively.

The chapter on light briefly discusses the wave and particle nature of light, but deals primarily with illumination. The chapter on reflection and refraction treats the images formed by mirrors and lenses. A brief introduction to color includes diffraction, interference, and polarization of light.

The section on modern physics provides an introduction to the structure and properties of the atomic nucleus, radioactive decay, nuclear reactions, and radioactivity followed by a very brief introduction to relativity.

A companion laboratory manual is available. An *Instructor's Resource Manual* that includes Complete Solutions, Transparency Masters, and a Test Item File is available at no charge to instructors using this text.

To the Student: Why Study Physics?

Physics is useful. Architects, mechanics, builders, carpenters, electricians, plumbers, and engineers are only some of the people who use physics every day in their jobs or professions. In fact, every person uses physics principles every hour of every day. The movement of an arm can be described using principles of the lever. All building trades, as well as the entire electronics industry, also use physics.

Physics is often defined as the study of matter, energy, and their transformations. The physicist uses scientific methods to observe, measure, and predict physical events and behaviors. However, gathered data left in someone's notebook in a laboratory are of little use to society.

Physics provides a universal means of describing and communicating about physical phenomena in the language of mathematics. Mechanics is the base on which almost all other areas of physics are built. Motion, force, work, electricity, and light are topics confronted daily in industry and technology. The basic laws of conservation of energy are needed to understand heat, sound, wave motion, electricity, and electromagnetic radiation.

Physics is always changing as new frontiers are being established in the study of the nature of matter. The topics studied in this course, however, will probably not greatly change with new research and will remain a classical foundation for work in many, many fields. We begin our study with the rules of the road—measurement, followed by a systematic problem-solving method. The end result should be a firm base on which to build a career in almost any field.

ACKNOWLEDGMENTS

The authors thank the many faculty and students who have used the previous editions, and especially those who have offered suggestions. If anyone wishes to correspond with us regarding suggestions, criticisms, questions, or errors, please contact Dale Ewen through Prentice Hall or through the web address http://www.prenhall.com.

We thank the following reviewers: Dr. A. M. Bloom, Hallmark Institute (TX); Robert H. Hadley, DeVry Institute of Technology (NJ); and Grace Wong, Heald College (CA).

We extend our sincere and special thanks to our Prentice Hall editor, Gary Bauer; development editor, Michelle Churma; senior production editor, Louise Sette; project manager, Wendy Druck at GTS Companies; and copyeditor, Philip Koplin.

Finally, we are especially grateful to Joyce Ewen for her excellent proofreading assistance, Amy Gundersen and Anette Brouhle for their assistance with the solutions manual, and to our families for their encouragement.

Dale Ewen
Neill Schurter
P. Erik Gundersen

CHAPTER 0

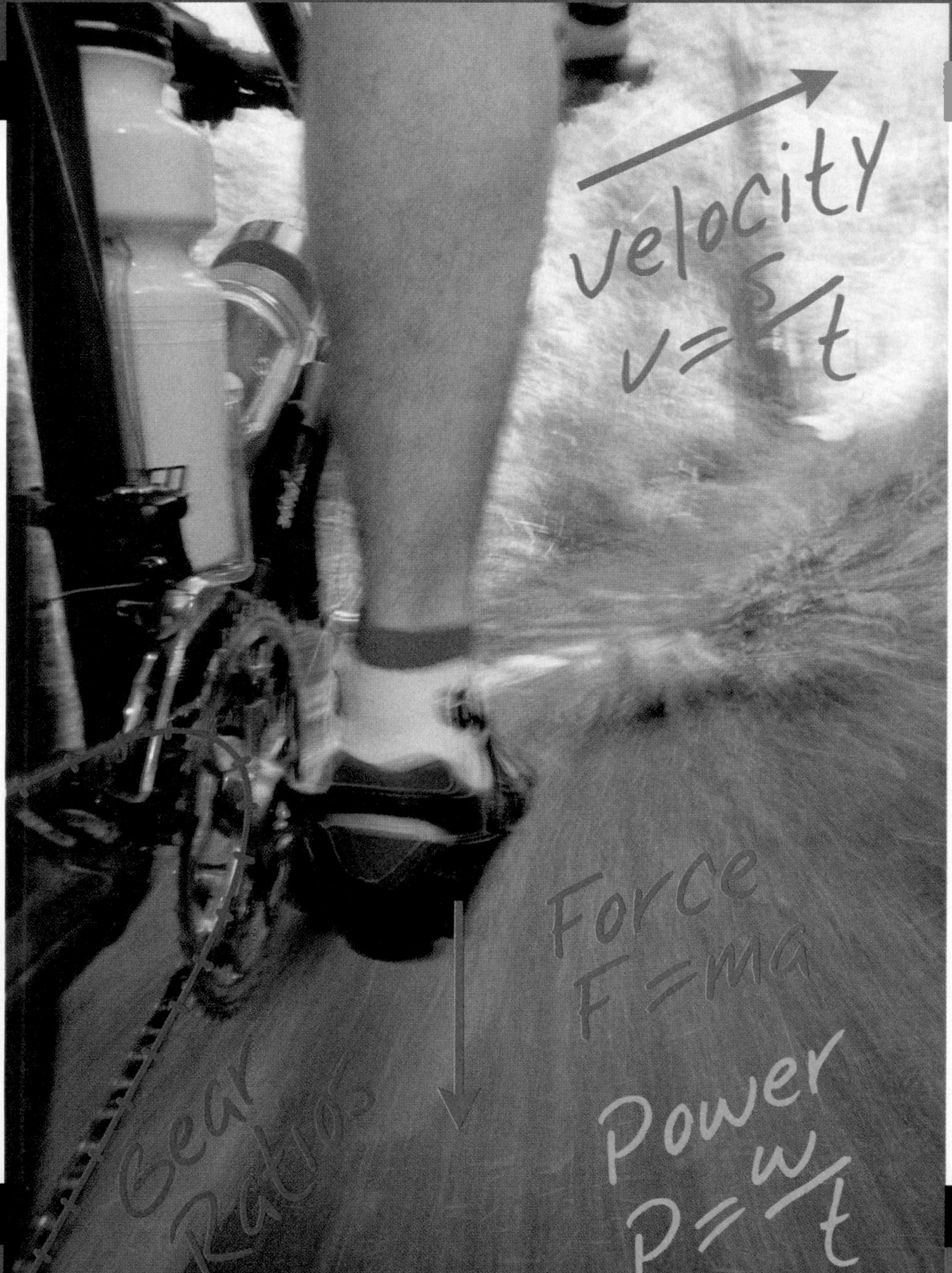

AN INTRODUCTION TO PHYSICS

Physics plays an important role in all aspects of our lives. For this mountain biker, the forces exerted by his legs, the air pressure in the tires, and the correct gearing combination will help determine the outcome of the race.

Before enrolling in a physics course, you may have taken physics for granted. In this chapter we will introduce physics to you and help you appreciate the impact that physics will have on your life and career.

Objectives

The major goals of this chapter are to enable you to:

1. Determine what physics governs and controls.
2. Conclude that physics is a building block of all the sciences.
3. Identify areas in your life that will be impacted by studying physics.
4. Differentiate between laws and theories.
5. Provide reasons why problem-solving techniques are vital in the study of physics.

0.1 Why Study Physics?

What do flying birds, automobiles, blue skies, and cell phones have in common? They all involve physics. **Physics** is the branch of science that describes the motion and energy of all matter throughout the universe. Birds, for example, use the difference in air pressures above and below their wings to keep themselves aloft. Automobiles use the principles of mechanics and thermodynamics to transfer stored chemical energy in gasoline to moving energy in rolling tires. The sky is blue when sunlight strikes and scatters off nitrogen and oxygen molecules in our atmosphere. Finally, cell phones use electronic components and the principles of electromagnetic waves to transfer energy and information from one cell phone to another (Fig. 0.1).

Physics is often considered to be the most fundamental of all the sciences. In order to study biology, chemistry, or any other natural science, one should have a firm understanding of the principles of physics. For example, **biology,** the branch of science that studies living organisms, uses the principles of fluid movement to understand how the blood moves through the heart, arteries, and veins. **Chemistry,** the branch of science that studies the composition, structure, properties, and reactions of matter, relies on the physics of subatomic particles to understand why chemical reactions take place. **Geology,** the branch of science that studies the origin, history, and structure of the earth, uses the physics of mechanical waves and energy transfer to determine the magnitude and location of earthquakes. Finally, **astronomy,** the field of science that studies everything that takes place outside the earth's atmosphere, relies on the laws of gravity and theory of relativity to describe the workings of the universe.

Students often wonder, "Why should I study physics? What is it going to do for me?" The answer is that physics plays an important role in everyday life and in the careers of many people. Choosing the right bat, golf club, or ski can be made easier with a bit of physics knowledge (Fig. 0.2). While on the job, architects, engineers, electricians, medical technicians, surveyors, and others use the principles of physics every day. When understood, physics can help us solve difficult physical problems and make us better decision makers.

Figure 0.1 Physics is involved in all aspects of cellular phone technology. It controls everything from the electrical circuits in the phone to the transmission of radio waves between phones.

Figure 0.2 A baseball player's understanding of physics can help improve all aspects of his game, including pitching, batting, and fielding. Nolan Ryan, Texas Rangers, May 1, 1991.

Photo courtesy of AP/Worldwide Photos. Reprinted with permission

Figure 0.3 Albert Einstein (1879–1955) is often considered one of the most influential scientists of the twentieth century. His work on relativity, as made famous through the equation $E = mc^2$, and the photoelectric effect changed the way the world viewed physics.

Photo courtesy of the National Archives and Records Administration

TRY THIS ACTIVITY

Physics All Around Us

Look around and find something that may have to do with physics. Although you may not yet have studied many physics principles, you should know that physics governs things that move and transfer energy. Be as general as you need to be in your observations. The point is for you to see that physics plays a role in most everything.

A **physicist** is a person who is an expert in or who studies physics. It is a physicist's job to seek an understanding of how the physical universe behaves. Albert Einstein, perhaps one of the most famous physicists of all time, once said, "I am like a child, I always ask the simplest questions" (Fig. 0.3). Such **theoretical physicists** often spend their professional lives researching previous theories and mathematical models to form new theories in physics. **Experimental physicists,** however, focus on performing experiments to develop and confirm physical theories.

It is generally accepted that physics evolved from ancient Greek philosophers like Plato (c. 428–347 BC) and Aristotle (384–322 BC). Aristotle believed that there were two types of motion: Natural motion occurred because objects wanted to seek their "natural" resting place (smoke rising or rocks falling), whereas violent motion occurred when objects were unnaturally pulled or dragged from place to place (person dragging a crate). Although he was not correct in his analysis, it was the beginning of observing and documenting physical phenomena. Plato, Aristotle, and others like them can be considered theoretical physicists.

It was not until the days of Archimedes (287–212 BC) that experiments were conducted to document and prove physical theories (Fig. 0.4). Since then, thousands of physicists have built and improved upon the knowledge base developed by those before them. It is now your turn to use the physics that you will learn to help you understand, improve, and make advances in our technological world. You will see that physics has use!

Figure 0.4 Archimedes is best known for observing that water was displaced when he stepped into his bath. He proceeded to conduct experiments where he measured the amount of water that overflowed when objects were placed into a tub full of water. He established a principle that states that an object immersed in a fluid will experience a buoyant force equal to the weight of the displaced fluid. It was said that following his great experiment and conclusion, he was seen running around his town in Sicily, exclaiming "Eureka!" Archimedes is also recognized for his work with simple machines like the screw, lever, and pulley. Legend has it that Archimedes said, "Give me a firm spot on which to stand and I will move the earth" (referring to the use of a lever).

Photo courtesy of Dorling Kindersley

0.2 Physics and Its Role in Technology

Although often discussed as though they are the same thing, science and technology are quite different. **Science** is a system of knowledge that is concerned with establishing accurate conclusions about the behavior of everything in the universe. It is a field in which **hypotheses** (scientifically based predictions) are made, information is gathered, and experiments are performed to determine how something in our natural world works or behaves. **Technology,** on the other hand, is a field that uses scientific knowledge to develop material products or processes that satisfy human needs and desires. Technology and science rely closely on one another to make further advances in their respective fields.

Thomas A. Edison (1847–1931) used scientific information and the discoveries of other scientists to create over 1000 inventions (Fig. 0.5). Edison's development of the first practical lighting system was made possible by applying the science of electricity and the science of materials and then putting that knowledge to use to satisfy his technological need. The following illustrate how science has played a role in improving technology.

Robotics: Due to advances in electronics, materials, and machines, robots commonly perform a variety of tasks from assembling cars on a production line to exploring the surface of Mars. NASA's Sojourner was the first robotic device to explore the surface of another planet (Fig. 0.6).

Bridges: Work in materials science and structural engineering has paved the way for advances in bridge design and construction. The New Clark Bridge in Alton, Illinois, is just one example of a cable-stayed bridge that has used scientific

Figure 0.5 Thomas A. Edison.

Photo courtesy of National Park Service, Edison National Historic Site

breakthroughs in materials science and physics to increase the structural integrity of the bridge and cut costs (Fig. 0.7).

Superconductors: Superconductors allow electric current to travel through materials with virtually no resistance. Materials such as aluminum, lead, and niobium are cooled by liquid helium to bring the temperature down to the low critical point. At the low critical point temperatures, the materials achieve zero electric resistance (Fig. 0.8). Scientific research is under way to develop superconducting materials that can operate closer to room temperature; this would bring about tremendous improvements in energy efficiency.

Active Noise Cancellation: Audiologists will tell you that noise increases stress levels. Acoustic and electrical engineers are now able to produce inverted noise patterns that cancel out disturbing noise (Fig. 0.9). Helicopter pilots, factory workers, and business travelers are using this technology to reduce stressful noise levels in their environment.

Liquid Crystal Displays: With advances in optics and electronics, physicists and chemists have created more advanced liquid crystal displays (LCDs), which are used as screens on laptop computers, personal digital assistants (PDAs), watches, and televisions (Fig. 0.10).

Magnetic Levitation: The speed limitations of traditional trains have created a need for super-fast, magnetic levitation (maglev) trains (Fig. 0.11). Such a vehicle is levitated off a monorail by virtue of the repulsion between the train and the rail, which have like magnetic poles, which repel one another. Electromagnets are used to propel the train forward as it glides above the rail. Such improvements will greatly reduce frictional resistance and allow trains to travel at twice the speed of conventional trains—up to 250 miles per hour.

Gyroscope: A gyroscope is a heavy wheel that uses rotational inertia to prevent tilting and is used to steady compasses, ships, airplanes, and rockets (Fig. 0.12). Advances in gyroscopes and electronic sensors have made it possible to create gyrostabilizers for ships. Such devices send signals to the ship's computer specifying how its fins should be positioned to prevent significant rolling motions.

Figure 0.7 The New Clark Bridge, a cable-stayed bridge in Alton, Illinois.

Figure 0.6 NASA's robotic Sojourner Rover on Mars in 1997.

Photo courtesy of NASA Headquarters

Figure 0.8 Overview of the Relativistic Heavy Ion Collider (RHIC) superconductor at the Brookhaven National Laboratory.

Photo courtesy of Brookhaven National Laboratory

Figure 0.9 Active noise cancellation technology for the consumer can be found in small, noise-reduction headsets.

Figure 0.10 Advances in LCD panels allow small color PDAs to have color screens.

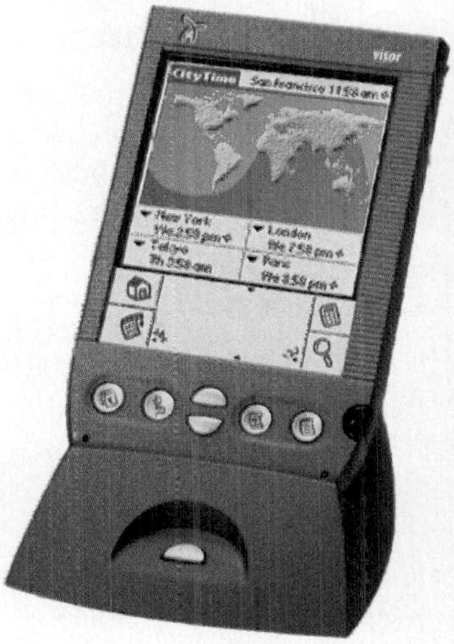

Copyright of PalmOne, Inc. Photo reprinted with permission

Figure 0.11 Maglev train technology has high-speed trains competing with airplane service.

Copyright of Apollo/PhotoEdit, Inc. Photo reprinted with permission

Figure 0.12 Gyroscope

Photo courtesy of Dorling Kindersley

0.3 Physics and Its Connection to Other Fields and Sciences

Ancient Greeks like Plato and Aristotle did not specialize in physics. In fact, it was not until the 1800s that physics was considered a science. Prior to the 1800s, Plato, Aristotle, Copernicus, and Galileo were considered natural philosophers, not physicists. Today, every physicist specializes in a subdivision of physics. There is simply too much information to allow someone to study every type of physics.

The following is a listing of the 18 subdivisions of physics:

Mechanics: Study of forces, motion, and energy.
Thermodynamics: Study of heat energy transfer.
Cryogenics: Study of matter at extremely low temperatures.
Plasma Physics: Study of electrically charged, ionized gas.

PHYSICS CONNECTIONS

Physics, Technology, and Sports

Physics plays a major role in sports. From the padding in a baseball glove to the stance of a wrestler, a good working knowledge of physics helps athletes and sports equipment companies achieve greater successes. Ski companies employ engineers who focus solely on the physics and engineering of improving a skier's time down the mountain. At the Winter Olympics in Salt Lake City, adjustments made to the length, shape, and composition of the skis played an important role in the success of the skiers. Such variables determine the amount of pressure the skier places on the snow and the friction of the ski. There are tradeoffs as well. Whereas a wider ski front increases its turning abilities, it also creates large vibrations that can slow down the skier. The use of titanium and various fibers and adhesives decreases those vibrations and results in lighter, stiffer skis. The application of physics has taken the once-simple wooden ski and has created a complex, high-performance device (Fig. 0.13).

Figure 0.13 Each ski design is created for use based on scientific knowledge.

Solid State Physics: Study of the physical properties of solid materials, also known as condensed matter physics.

Geophysics: Study of the interaction of forces and energy found within the earth; closely related to geology.

Astrophysics: Study of the interaction of forces and energy between interstellar objects; closely related to astronomy.

Acoustics: Study of the creation and transmission of sound under various conditions.

Optics: Study of the behavior of light in a variety of conditions.

Electromagnetism: Study of the relationship between electricity and magnetism.

Fluid Dynamics: Study of how liquids and gases move from one location to another.

Mathematical Physics: Mathematics of physics and its related fields.

Statistical Mechanics: Study of the development of statistical models that simulate the effects of systems composed of many particles.

High-Energy Physics: Study of new fundamental, subatomic particles using high-energy machines that send known subatomic particles colliding into one another; simulation of what the universe was like close to the time of the "big bang."

Atomic Physics: Study of the structure of the atom based on the knowledge gained in the field of high-energy physics.

Molecular Physics: Study of the structure of molecules based on the knowledge gained in atomic physics.

Nuclear Physics: Study of nuclear interactions.

Quantum Physics: Study of small particles and their energy.

0.4 Theories, Laws, and the Problem-Solving Method

Physics is constantly being refined. Although the major principles of physics do not change drastically over time, newer theories requiring a tremendous amount of experimentation can slightly change our understanding of physics. A **theory** is a scientific conclusion that attempts to explain natural occurrences. Typically it has been tested in the laboratory but has not been proven with absolute certainty. A **principle** is a step closer to a law in physics.

Figure 0.14 These students are using a problem-solving method to help them arrive at a solution.

Principles have been experimentally proven in the laboratory, have stood the test of various conditions, and continue to hold true. Laws are the final degree of scientific certainty. **Laws** are often defined using formulas. For example, Newton's second law of motion, $F = ma$, has been proven to be true and is considered a law of physics.

The **scientific method** is an orderly procedure used by scientists in collecting, organizing, and analyzing new information which refutes or supports a scientific hypothesis. The constant use of the scientific method and the development of theories, principles, and laws is similar to the problem-solving method discussed in detail in Chapter 1. The **problem-solving method** is an orderly procedure that aids in understanding questions and solving problems. Nonscientists use the problem-solving method more often than the scientific method. The problem-solving method is helpful when a problem arises either in this text or on the job. An individual or a team must develop the skills needed to collect data, analyze a problem, and work toward finding its solution in a logical and orderly fashion (Fig. 0.14). In order to find solutions to problems, tools are needed to make the job easier. In the next chapter, we will familiarize ourselves with two important tools of physics: measurement and mathematics.

Glossary

Astronomy The branch of science that studies everything that takes place outside of the earth's atmosphere. (p. 4)

Biology The branch of science that studies living organisms. (p. 4)

Chemistry The branch of science that studies the composition, structure, properties, and reactions of matter. (p. 4)

Experimental Physicist A physicist who performs experiments to develop and confirm physical theories. (p. 5)

Geology The branch of science that studies the origin, history, and structure of the earth. (p. 4)

Hypothesis A scientifically based prediction that needs testing to verify its validity. (p. 5)

Law The highest level of certainty for an explanation of physical occurrences. A law is often accompanied by a formula. (p. 9)

Physics The branch of science that describes the motion and energy of all matter throughout the universe. (p. 4)
Physicist A person who is an expert in or who studies physics. (p. 5)
Principle A rule or fundamental assumption that has been proven in the laboratory. (p. 8)
Problem-Solving Method An orderly procedure that aids in understanding questions and solving problems. (p. 9)
Science A system of knowledge that is concerned with establishing accurate conclusions about the behavior of everything in the universe. (p. 5)
Scientific Method An orderly procedure used by scientists in collecting, organizing, and analyzing new information which refutes or supports a scientific hypothesis. (p. 9)
Technology The field that uses scientific knowledge to develop material products or processes that satisfy human needs and desires. (p. 5)
Theoretical Physicist A physicist who predominantly uses previous theories and mathematical models to form new theories in physics. (p. 5)
Theory A scientifically accepted principle that attempts to explain natural occurrences. (p. 8)

Review Questions

1. Physics is a field of study that governs
   (a) how the planets orbit the sun.
   (b) the rate at which blood flows through a person's veins.
   (c) how quickly a helium balloon will rise into the air.
   (d) all of the above.
2. Who among the following is an example of a theoretical physicist?
   (a) Archimedes, who measured the volume of water that was displaced after placing objects in a tub of water.
   (b) Albert Einstein, who performed various thought experiments in his mind to arrive at his theories of relativity.
   (c) Marie Curie, who, along with her husband, was credited with discovering radioactivity through a series of laboratory experiments.
   (d) Benjamin Franklin, who through various laboratory experiments determined that electricity is the flow of microscopic charged particles.
3. Why are Isaac Newton's conclusions on motion considered laws of physics?
   (a) Newton himself declared them laws.
   (b) Newton performed various thought experiments on motion.
   (c) The formulas accompanying Newton's laws have proved correct in experiments for years.
   (d) Newton's reputation alone made his scientific conclusions laws.
4. Which of the following is not considered a branch of physics?
   (a) Thermodynamics (b) Astronomy
   (c) Geophysics (d) Atomic physics
5. Analyzing the braking distance of a sports car would most likely utilize which field of physics?
   (a) Molecular physics (b) Quantum physics
   (c) Fluid dynamics (d) Mechanics
6. Who is considered to be the first true physicist and what did he do to deserve this recognition in scientific history?
7. Explain the difference between science and technology. Are the two fields related?
8. Provide two examples of scientific knowledge and a technological development that relies on that scientific knowledge.
9. What is the difference between the scientific method and the problem-solving method?
10. Why is it important to study physics? Provide a few examples of what an understanding of the physical world can do for you today and in your future.

CHAPTER 1

Problem-Solving method
Conversion Factor
650 ft (0.305 m / ft) = 1.98 m
mass vs. Weight = 9.03 × 10³ N
Accuracy = 3
Precision = 10 N

THE PHYSICS TOOL KIT

A good mechanic needs not only the right tools for the job but also to be proficient in using those tools. The same applies to physics. Analyzing the problem, choosing the correct formula, and manipulating the equation will help you become a good physics student. In this chapter, we discuss mathematical techniques, significant digits, accuracy, precision, and the problem-solving method. These will be your basic tools for physics.

Objectives

The major goals of this chapter are to enable you to:

1. Explain the need for standardization of measurement.
2. Use the metric system of measurement.
3. Convert measurements from one system to another.
4. Solve problems involving length, area, and volume.
5. Distinguish between mass and weight.
6. Use significant digits to determine the accuracy of measurements.
7. Differentiate between accuracy and precision.
8. Solve problems with measurements and consistently express the results with the correct significant digits.
9. Use a systematic approach to solving physics problems.
10. Analyze problems using the problem-solving method.

1.1 Standards of Measure

When two people work together on the same job, they should both use the same standards of measure. If not, the result can be a problem (Fig. 1.1).

Figure 1.1 The trouble with inconsistent systems of measurement

Standards of measure are sets of units of measurement for length, weight, and other quantities defined in a way that is useful to a large number of people. Throughout history, there have been many standards by which measurements have been made:

- *Chain:* A measuring instrument of 100 links used in surveying. One chain has a length of 66 feet.
- *Rod:* A length determined by having each of 16 men put one foot behind the foot of the man before him in a straight line [Fig. 1.2(a)]. The rod is now standardized as $16\frac{1}{2}$ feet.
- *Yard:* The distance from the tip of the king's nose to the fingertips of his outstretched hand [Fig. 1.2(b)].
- *Foot:* The rod divided by 16; it was also common to use the length of one's own foot as the unit foot.
- *Inch:* The length of three barley corns, round and dry, taken from the center of the ear, and laid end to end [Fig. 1.2(c)].

Figure 1.2 Definitions of some old units

(a) A rod used to be 16 "people feet." Distance divided by 16 equals one foot.

(b) The "old" yard.

(c) At one time, three barley corns were used to define one inch.

The U.S. system of measure, which is derived from and sometimes called the English system, is a combination of makeshift units of Anglo-Saxon, Roman, and French-Norman weights and measures.

After the standards based on parts of the human body and on other gimmicks, basic standards were accepted by world governments. They also agreed to construct and distribute accurate standard copies of all the standard units. During the 1790s, a decimal system based on our number system, the metric system, was developed in France. Its acceptance was gained mostly because it was easy to use and easy to remember. Many nations began adopting it as their official system of measurement. By 1900, most of Europe and South America were metric. In 1866, metric measurements for official use were legalized in the

TRY THIS ACTIVITY

Stepping Off

When a short distance needs to be measured and a tape measure is not available, some people measure the approximate length by using the length of their own foot as a unit. Measure the distance between two points approximately 15 to 25 ft apart by placing one foot in front of the other and counting the steps. Then, measure the same distance with a tape measure. How close to the standard foot is the length of your own foot? How much error did you generate?

For longer distances, some measure the approximate length by pacing off the distance using one stride as approximately 1 yd or 3 ft. Measure the distance between two points approximately 50 to 75 ft apart by pacing off the distance and counting the strides. Then, measure the same distance with a tape measure How close to the standard yard is the length of your own stride? How much error did you generate?

United States. In 1893, the Secretary of the Treasury, by administrative order, declared the new metric standards to be the nation's "fundamental standards" of mass and length. Thus, indirectly, the United States *officially* became a metric nation. Even today, the U.S. units are officially defined in terms of the standard metric units.

Throughout U.S. history, several attempts have been made to convert the nation to the metric system. By the 1970s, the United States found itself to be the only nonmetric industrialized country left in the world. However government inaction to implement the metric system resulted in the United States regularly using a greater variety of confusing units than any other country. Industry and business, however, found their foreign markets drying up because metric products were preferred Now many segments of American industry and business have independently gone metric because world trade is geared toward the metric

PHYSICS CONNECTIONS

Lost in Space

Even professional engineers and scientists sometimes forget to include units or make mistakes when converting from one system of measurement to another. NASA's Mars Climate Orbiter (Fig. 1.3) was lost in space on September 23, 1999, after engineers from Lockheed Martin used U.S. measurements when calculating rocket thrusts and did not convert those measurements to the metric units used by NASA engineers. Such a simple mistake, common to students on physics exams, cost Lockheed Martin and NASA a $125 million space probe and a great deal of embarrassment.

According to officials, NASA assumed that the twice-daily rocket thrust calculations were based on metric units, whereas Lockheed Martin had neither converted the numbers to metric nor labeled them in U.S. units. As a result, the Mars Climate Orbiter which was to orbit Mars, was inadvertently sent off course and lost. The orbiter was intended to collect atmospheric and surface data while also serving as a communications link for the Mars Lander.

Figure 1.3 NASA's Mars Climate Orbiter—artist's conception

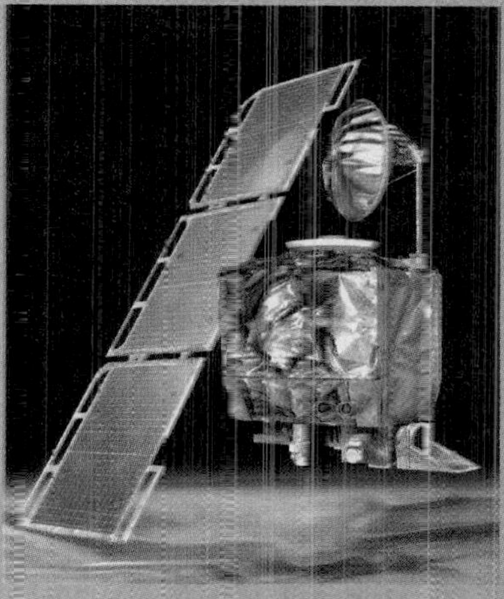

Photo courtesy of NASA Headquarters/Jet Propulsion Laboratory

Note: Dependent upon the skills of the class or individual students, the class or individual students may need to review or study sections in Appendix A before proceeding with given sections in this text.

system of measurement. The inherent simplicity of the metric system of measurement and standardization of weights and measures has led to major cost savings in industries that have converted to it.

Most major U.S. industries, such as the automotive, aviation, and farm implement industries, as well as the Department of Defense and other federal agencies have effectively converted to the metric system. In this text approximately 70% of our examples and exercises are metric in the sections where both U.S. and metric systems are still commonly used. In some industries, you—the student and worker—will need to know and use both systems.

1.2 Introduction to the Metric System

Gabriel Mouton (1618–1694),

a French vicar who spent much of his time studying mathematics and astronomy, is credited by many for originating the metric system. French scientists in the late 18th century are credited with replacing the chaotic collection of systems then in use with the metric system.

The modern metric system is identified in all languages by the abbreviation **SI** (for Système International d'Unités—the international system of units of measurement written in French). The SI metric system has seven *basic units* [Table 1.1(a)]. All other SI units are called *derived units*; that is, they can be defined in terms of these seven basic units (see Appendix C, Table 19). For example, the newton (N) is defined as 1 kg m/s^2 (kilogram metre per second per second). Many derived units will be presented and discussed in this text. Some derived SI units are given in Table 1.1(b). **Gabriel Mouton** is often credited for originating the metric system.

Table 1.1 SI Units of Measure

(a) Basic Unit	SI Abbreviation	Used for Measuring
Metre*	m	Length
Kilogram	kg	Mass
Second	s	Time
Ampere	A	Electric current
Kelvin	K	Temperature
Candela	cd	Light intensity
Mole	mol	Molecular substance
(b) Derived Unit	**SI Abbreviation**	**Used for Measuring**
Litre*	L or ℓ	Volume
Cubic metre	m^3	Volume
Square metre	m^2	Area
Newton	N	Force
Metre per second	m/s	Speed
Joule	J	Energy
Watt	W	Power

*At present, there is some difference of opinion in the United States on the spelling of metre and litre. We have chosen the "re" spellings for two reasons. First, this is the internationally accepted spelling for all English-speaking countries. Second, the word "meter" already has many different meanings—parking meter, electric meter, odometer, and so on. Many feel that the metric units of length and volume should be distinctive and readily recognizable—thus the spellings "metre" and "litre."

Because the metric system is a decimal or base 10 system, it is very similar to our decimal number system and any decimal money system. It is an easy system to use because calculations are based on the number 10 and its multiples. Special prefixes are used to name these multiples and submultiples, which may be used with most all SI units. Because the same prefixes are used repeatedly, the memorization of many conversions has been significantly reduced. Table 1.2 shows these prefixes and the corresponding symbols.

Table 1.2 Prefixes for SI Units

Multiple or Submultiple[a] Decimal Form	Power of 10	Prefix[b]	Prefix Symbol	Pronunciation	Meaning
1,000,000,000,000	10^{12}	tera	T	tĕr'ă	One trillion times
1,000,000,000	10^{9}	giga	G	jĭg'ă	One billion times
1,000,000	10^{6}	mega	M	mĕg'ă	One million times
1,000	10^{3}	kilo	k	kĭl'ō	One thousand times
100	10^{2}	hecto	h	hĕk'tō	One hundred times
10	10^{1}	deka	da	dĕk'ă	Ten times
0.1	10^{-1}	deci	d	dĕs'ĭ	One tenth of
0.01	10^{-2}	centi	c	sĕnt'ĭ	One hundredth of
0.001	10^{-3}	milli	m	mĭl'ĭ	One thousandth of
0.000001	10^{-6}	micro	μ	mī'krō	One millionth of
0.000000001	10^{-9}	nano	n	năn'ō	One billionth of
0.000000000001	10^{-12}	pico	p	pē'kō	One trillionth of

[a]Factor by which the unit is multiplied.
[b]The same prefixes are used with all SI metric units.

EXAMPLE 1

Write the SI abbreviation for 36 centimetres.

The symbol for the prefix *centi* is c.
The symbol for the unit *metre* is m.

Thus, 36 cm is the SI abbreviation for 36 centimetres.

EXAMPLE 2

Write the SI metric unit for the abbreviation 45 kg.

The prefix for k is *kilo*; the unit for g is *gram*.
Thus, 45 kilograms is the SI metric unit for 45 kg.

PROBLEMS 1.2

Give the metric prefix for each value.

1. 1000 2. 0.01 3. 100 4. 0.1
5. 0.001 6. 10 7. 1,000,000 8. 0.000001

Give the metric symbol, or abbreviation, for each prefix.

9. hecto 10. kilo 11. milli 12. deci
13. mega 14. deka 15. centi 16. micro

Write the abbreviation for each quantity.

17. 135 millimetres 18. 83 dekagrams 19. 28 kilolitres
20. 52 centimetres 21. 49 centigrams 22. 85 milligrams
23. 75 hectometres 24. 15 decilitres

Write the SI unit for each abbreviation.

25. 24 m 26. 185 L 27. 59 g 28. 125 kg
29. 27 mm 30. 25 dL 31. 45 dam 32. 27 mg
33. 26 Mm 34. 275 μg
35. The basic metric unit of length is _____.
36. The basic unit of mass is _____.
37. Two common metric units of volume are _____ and _____.
38. The basic unit for electric current is _____.
39. The basic metric unit for time is _____.
40. The common metric unit for power is _____.

1.3 Length

In most sections that introduce units of measure, we present the units in subsections as follows: metric units, U.S. units, and conversions between metric and U.S. units.

Metric Length

The basic SI unit of length is the **metre** (m) (Fig. 1.4). The first standard metre was chosen in the 1790s to be one ten-millionth of the distance from the earth's equator to either pole. Modern measurements of the earth's circumference show that the first length is off by about 0.02% from this initial standard. The current definition adopted in 1983 is based on the speed of light in a vacuum and reads "The metre is the length of path traveled by light in a vacuum during a time interval of 1/299,792,458 of a second." Long distances are measured in kilometres (km) (Fig. 1.5). We use the centimetre (cm) to measure short distances, such as the length of this book or the width of a board [Fig. 1.6(a)]. The millimetre (mm) is used to measure very small lengths, such as the thickness of this book or the depth of a tire tread [Fig. 1.6(b)]. A metric ruler is shown in Fig. 1.6(c).

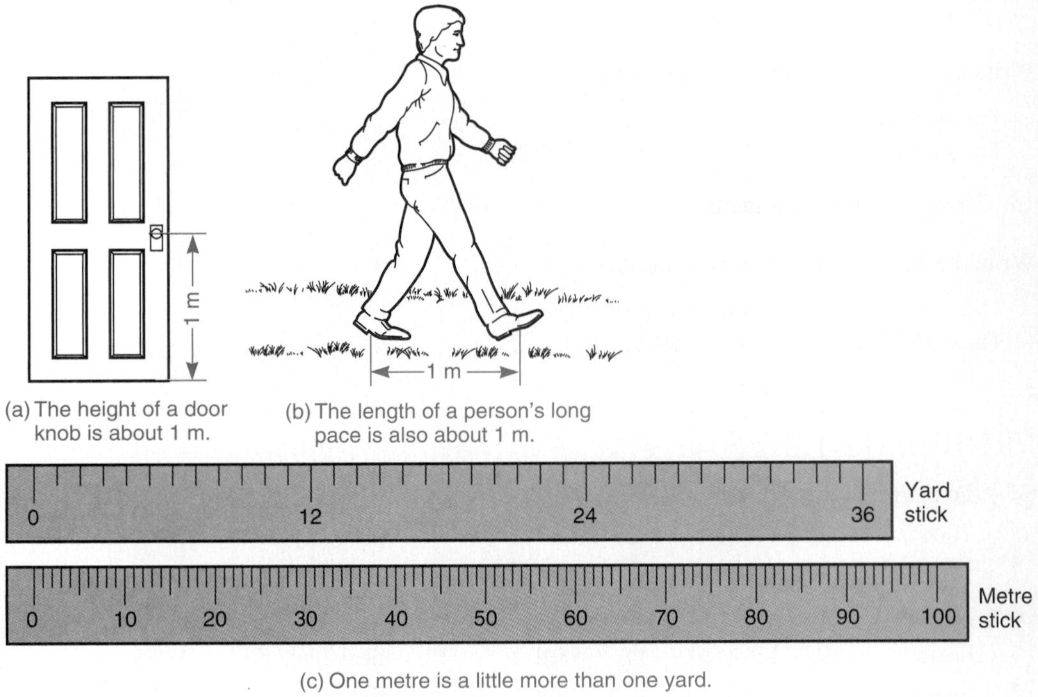

Figure 1.4 One metre

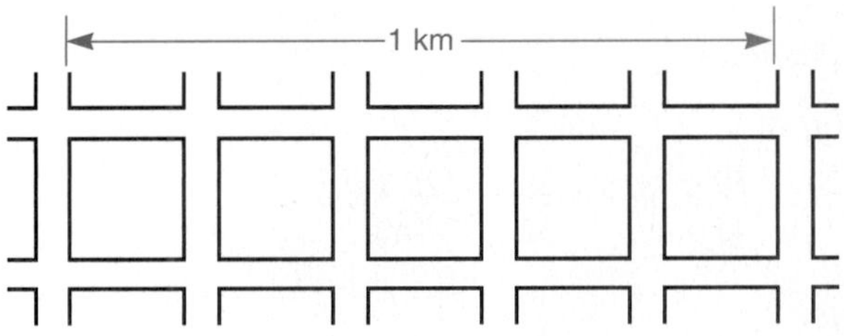

Figure 1.5 The length of five city blocks is about 1 km.

A **conversion factor** is an expression used to change from one unit or set of units to another. We know that we can multiply any number or quantity by 1 without changing the value of the original quantity. We also know that any fraction equals 1 when its numerator

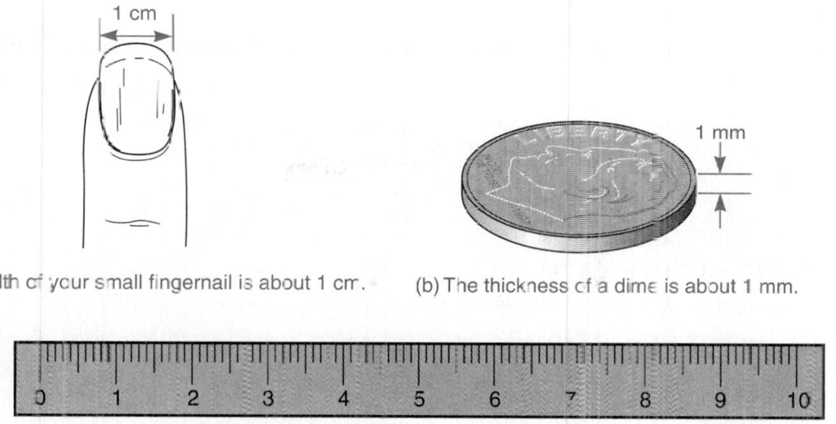

(a) The width of your small fingernail is about 1 cm.

(b) The thickness of a dime is about 1 mm.

(c) The large numbered divisions are centimetres. Each centimetre is divided into 10 equal parts, called millimetres.

Figure 1.6 Small metric length units

and denominator are equal. For example, $\frac{5}{5} = 1$, $\frac{12\text{ m}}{12\text{ m}} = 1$, and $\frac{6.5\text{ kg}}{6.5\text{ kg}} = 1$. In addition, since 1 m = 100 cm, $\frac{1\text{ m}}{100\text{ cm}} = 1$. Similarly, $\frac{100\text{ cm}}{1\text{ m}} = 1$, because the numerator equals the denominator. We call such names for 1 *conversion factors*. The information necessary for forming a conversion factor is usually found in tables. As in the case 1 m = 100 cm, there are two conversion factors for each set of data:

$$\frac{1\text{ m}}{100\text{ cm}} \quad \text{and} \quad \frac{100\text{ cm}}{1\text{ m}}$$

CONVERSION FACTORS

Choose a conversion factor in which the old units are in the numerator of the original expression and in the denominator of the conversion factor, or in the denominator of the original expression and in the numerator of the conversion factor. That is, we want the old units to cancel each other.

EXAMPLE 1

Change 215 cm to metres.

As we saw before, the two possible conversion factors are

$$\frac{1\text{ m}}{100\text{ cm}} \quad \text{and} \quad \frac{100\text{ cm}}{1\text{ m}}$$

We choose the conversion factor with centimetres in the *denominator* so that the cm units cancel each other.

$$215\ \cancel{\text{cm}} \times \frac{1\text{ m}}{100\ \cancel{\text{cm}}} = 2.15\text{ m}$$

Note: Conversions *within* the metric system involve only moving the decimal point.

EXAMPLE 2

Change 4 m to centimetres.

$$4\ \cancel{\text{m}} \times \frac{100\text{ cm}}{1\ \cancel{\text{m}}} = 400\text{ cm}$$

EXAMPLE 3

Change 39.5 mm to centimetres.

Choose the conversion factor with millimetres in the denominator so that the mm units cancel each other.

$$39.5\ \cancel{\text{mm}} \times \frac{1\text{ cm}}{10\ \cancel{\text{mm}}} = 3.95\text{ cm}$$

EXAMPLE 4

Change 0.05 km to centimetres.

First, change to metres and then to centimetres.

$$0.05\ \cancel{\text{km}} \times \frac{1000\text{ m}}{1\ \cancel{\text{km}}} = 50\text{ m}$$

$$50\ \cancel{\text{m}} \times \frac{100\text{ cm}}{1\ \cancel{\text{m}}} = 5000\text{ cm}$$

Or,

$$0.05\ \cancel{\text{km}} \times \frac{1000\ \cancel{\text{m}}}{1\ \cancel{\text{km}}} \times \frac{100\text{ cm}}{1\ \cancel{\text{m}}} = 5000\text{ cm}$$

U.S. Length

The basic units of the U.S. system are the foot, the pound, and the second. The foot is the basic unit of length and may be divided into 12 equal parts or inches. Common U.S. length conversions include

1 foot (ft) = 12 inches (in.)
1 yard (yd) = 3 ft
1 mile (mi) = 5280 ft

See Table 1 of Appendix C for U.S. weights and measures.

We also use a conversion factor to change from one U.S. length unit to another.

EXAMPLE 5

Change 84 in. to feet.

Choose the conversion factor with inches in the denominator and feet in the numerator.

$$84\ \cancel{\text{in.}} \times \frac{1\text{ ft}}{12\ \cancel{\text{in.}}} = 7\text{ ft}$$

Metric–U.S. Conversions

To change from a U.S. unit to a metric unit or from a metric unit to a U.S. unit, again use a conversion factor, such as 1 in. = 2.54 cm.

EXAMPLE 6

Express 10 inches in centimetres.

$$1\text{ in.} = 2.54\text{ cm} \quad \text{so} \quad 10\ \cancel{\text{in.}} \times \frac{2.54\text{ cm}}{1\ \cancel{\text{in.}}} = 25.4\text{ cm}$$

The conversion factors you will need are given in Appendix C. The following examples show you how to use these tables.

EXAMPLE 7

Change 15 miles to kilometres.

From Table 2 in Appendix C, we find 1 mile listed in the left-hand column. Moving over to the fourth column, under the heading "km," we see that 1 mile (mi) = 1.61 km. Then we have

$$15 \cancel{\text{mi}} \times \frac{1.61 \text{ km}}{1 \cancel{\text{mi}}} = 24.15 \text{ km}$$

EXAMPLE 8

Change 220 centimetres to inches.

Find 1 centimetre in the left-hand column and move to the fifth column under the heading "in." We find that 1 centimetre = 0.394 in. Then

$$220 \cancel{\text{cm}} \times \frac{0.394 \text{ in.}}{1 \cancel{\text{cm}}} = 86.68 \text{ in.}$$

EXAMPLE 9

Change 3 yards to centimetres.

Since there is no direct conversion from yards to centimetres in the tables, we must first change yards to inches and then inches to centimetres:

$$3 \cancel{\text{yd}} \times \frac{36 \cancel{\text{in.}}}{1 \cancel{\text{yd}}} \times \frac{2.54 \text{ cm}}{1 \cancel{\text{in.}}} = 274.32 \text{ cm}$$

PROBLEMS 1.3

Which unit is longer?

1. 1 metre or 1 centimetre
2. 1 metre or 1 millimetre
3. 1 metre or 1 kilometre
4. 1 centimetre or 1 millimetre
5. 1 centimetre or 1 kilometre
6. 1 millimetre or 1 kilometre

Which metric unit (km, m, cm, or mm) would you use to measure the following?

7. Length of a wrench
8. Thickness of a saw blade
9. Height of a barn
10. Width of a table
11. Thickness of a hypodermic needle
12. Distance around an automobile racing track
13. Distance between New York and Miami
14. Length of a hurdle race
15. Thread size on a pipe
16. Width of a house lot

Fill in each blank with the most reasonable metric unit (km, m, cm, or mm).

17. Your car is about 6 _____ long.
18. Your pencil is about 20 _____ long.
19. The distance between New York and San Francisco is about 4200 _____.
20. Your pencil is about 7 _____ thick.
21. The ceiling in my bedroom is about 240 _____ high.
22. The length of a football field is about 90 _____.
23. A jet plane usually cruises at an altitude of 9 _____.
24. A standard film size for cameras is 35 _____.
25. The diameter of my car tire is about 60 _____.
26. The zipper on my jacket is about 70 _____ long.
27. Juan drives 9 _____ to school each day.
28. Jacob, our basketball center, is 203 _____ tall.
29. The width of your hand is about 80 _____.
30. A handsaw is about 70 _____ long.
31. A newborn baby is usually about 45 _____ long.
32. The standard metric piece of plywood is 120 _____ wide and 240 _____ long.

Fill in each blank.

33. 1 km = _____ m
34. 1 mm = _____ m
35. 1 m = _____ cm
36. 1 m = _____ hm
37. 1 dm = _____ m
38. 1 dam = _____ m
39. 1 m = _____ mm
40. 1 m = _____ dm
41. 1 hm = _____ m
42. 1 cm = _____ m
43. 1 m = _____ km
44. 1 m = _____ dam
45. 1 cm = _____ mm
46. Change 250 m to cm.
47. Change 250 m to km.
48. Change 546 mm to cm.
49. Change 178 km to m.
50. Change 35 dm to dam.
51. Change 830 cm to m.
52. Change 75 hm to km.
53. Change 375 cm to mm.
54. Change 7.5 mm to μm.
55. Change 4 m to μm.
56. State your height in centimetres and in metres.
57. The wheelbase of a certain automobile is 108 in. long. Find its length (a) in feet. (b) in yards.
58. Change 43,296 ft (a) to miles. (b) to yards.
59. Change 6.25 mi (a) to yards. (b) to feet.
60. The length of a connecting rod is 7 in. What is its length in centimetres?
61. The distance between two cities is 256 mi. Find this distance in kilometres.
62. Change 5.94 m to feet.
63. Change 7.1 cm to inches.
64. Change 1.2 in. to centimetres.
65. The turning radius of an auto is 20 ft. What is this distance in metres?
66. Would a wrench with an opening of 25 mm be larger or smaller than a 1-in. wrench?
67. How many reamers each 20 cm long can be cut from a bar 6 ft long, allowing 3 mm for each saw cut?
68. If 214 pieces each 47 cm long are to be turned from $\frac{1}{4}$-in. round steel stock with $\frac{1}{8}$ in. of waste allowed on each piece, what length (in metres) of stock is required?

Figure 1.7

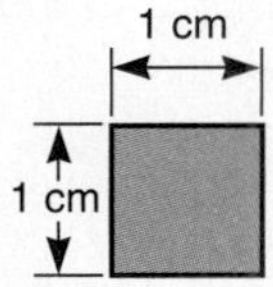

One square centimetre (cm^2)

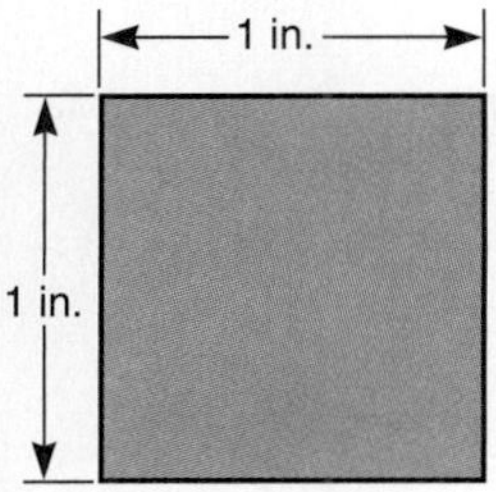

One square inch (in^2)

1.4 Area and Volume

Area

The **area** of a figure is the number of square units that it contains. To measure a surface area of an object, you must first decide on a standard unit of area. Standard units of area are based on the square and are called *square inches*, *square centimetres*, *square miles*, or some other square unit of measure. An area of 1 square centimetre (cm^2) is the amount of area found within a square 1 cm on each side. An area of 1 square inch (in^2) is the amount of area found within a square of 1 in. on each side (Fig. 1.7).

In general, when multiplying measurements of like units, multiply the numbers and then multiply the units as follows:

$$2 \text{ cm} \times 4 \text{ cm} = (2 \times 4)(\text{cm} \times \text{cm}) = 8 \text{ cm}^2$$
$$3 \text{ in.} \times 5 \text{ in.} = (3 \times 5)(\text{in.} \times \text{in.}) = 15 \text{ in}^2$$
$$1.4 \text{ m} \times 6.7 \text{ m} = (1.4 \times 6.7)(\text{m} \times \text{m}) = 9.38 \text{ m}^2$$

Figure 1.8

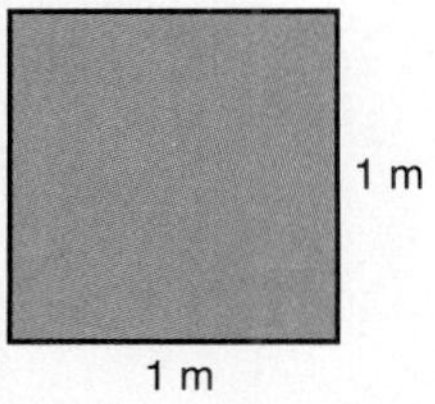

One square metre (m^2)

Metric Area

The basic unit of area in the metric system is the *square metre* (m^2), the area in a square whose sides are 1 m long (Fig. 1.8). The square centimetre (cm^2) and the square millimetre (mm^2) are smaller units of area. Larger units of area are the square kilometre (km^2) and the hectare (ha).

EXAMPLE 1

Find the area of a rectangle 5 m long and 3 m wide.

Each square in Fig. 1.9 represents 1 m^2. By simply counting the number of squares (square metres), we find that the area of the rectangle is 15 m^2. We can also find the area of the rectangle by using the formula

$$A = lw = (5 \text{ m})(3 \text{ m}) = 15 \text{ m}^2 \qquad (\textbf{\textit{Note:}}\ \text{m} \times \text{m} = \text{m}^2)$$

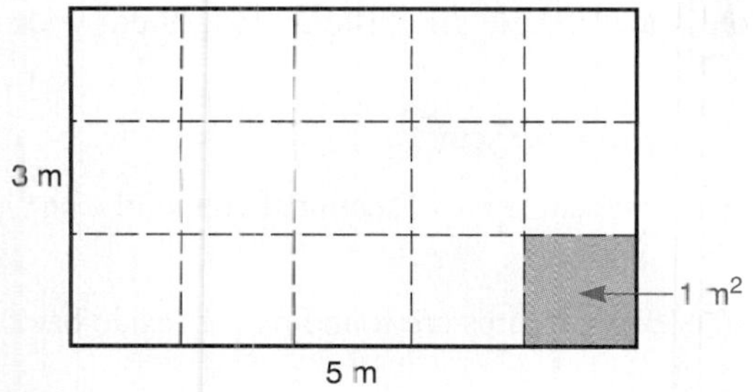

Figure 1.9

EXAMPLE 2

Find the area of the metal plate shown in Fig. 1.10.

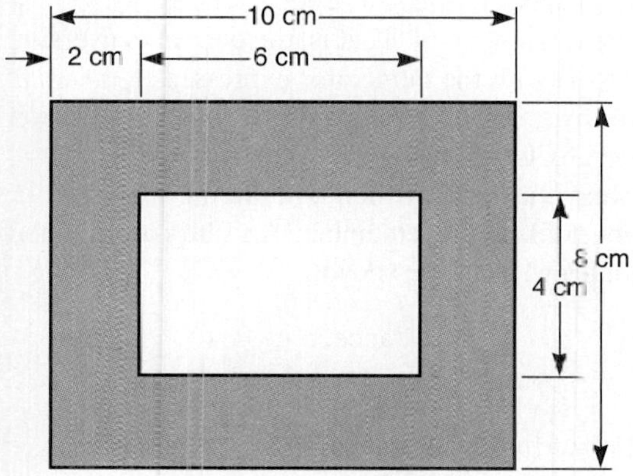

Figure 1.10

To find the area of the metal plate, find the area of each of the two rectangles and then find the difference of their areas. The large rectangle is 10 cm long and 8 cm wide. The small rectangle is 6 cm long and 4 cm wide. Thus,

$$\begin{aligned} &\text{area of large rectangle: } A = lw = (10 \text{ cm})(8 \text{ cm}) = 80 \text{ cm}^2 \\ &\text{area of small rectangle: } A = lw = (6 \text{ cm})(4 \text{ cm}) = \underline{24 \text{ cm}^2} \\ &\text{area of metal plate: } \qquad 56 \text{ cm}^2 \end{aligned}$$

The surface that would be seen by cutting a geometric solid with a thin plate parallel to one side of the solid represents the *cross-sectional area* of the solid.

EXAMPLE 3

Find the smallest cross-sectional area of the box shown in Fig. 1.11(a).

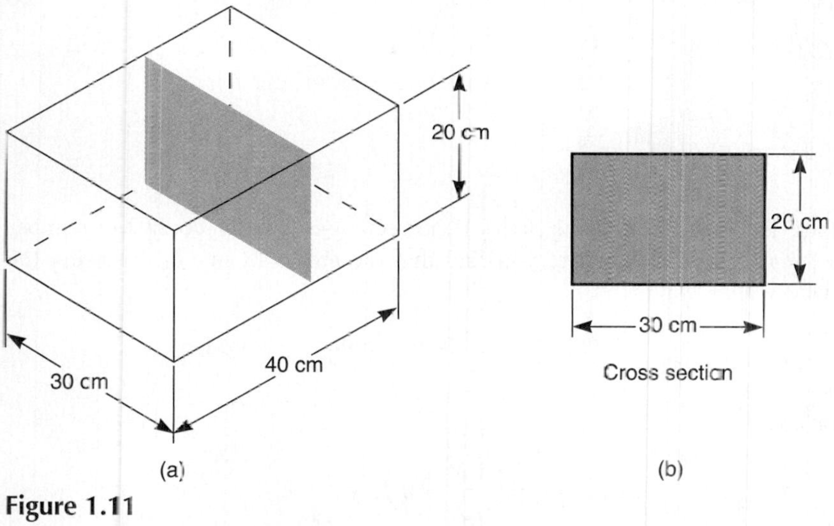

Figure 1.11

The indicated cross section of this box is a rectangle 30 cm long and 20 cm wide [Fig. 1.11(b)]. Thus

$$A = lw = (30 \text{ cm})(20 \text{ cm}) = 600 \text{ cm}^2$$

The area of this rectangle is 600 cm^2, which represents the cross-sectional area of the box.

The formulas for finding the areas of other plane figures are found on the inside back cover.

Figure 1.12 One hectare

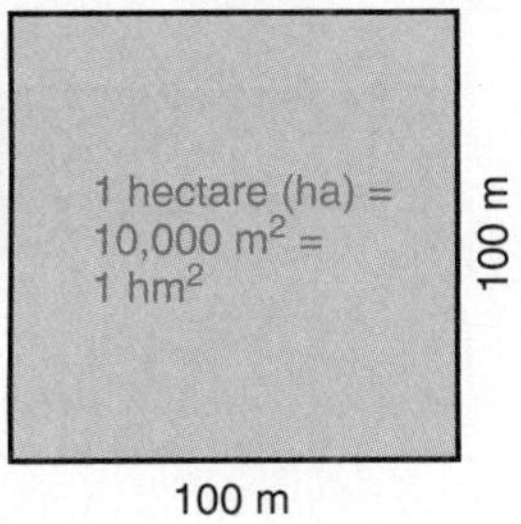

The *hectare* is the fundamental SI unit for land area. An area of 1 hectare equals the area of a square 100 m on a side (Fig. 1.12). The hectare is used because it is more convenient to say and use than square hectometre. The metric prefixes are *not* used with the hectare unit. That is, instead of saying "2 kilohectares," we say "2000 hectares."

To convert area or square units, use a conversion factor. That is, the correct conversion factor will be in fractional form and equal to 1, with the numerator expressed in the units you wish to convert to and the denominator expressed in the units given. The conversion table for area is provided as Table 3 of Appendix C.

The conversion of area units will be shown using a method of squaring the linear or length conversion factor which you are most likely to remember. An alternate method emphasizing direct use of the conversion tables will also be shown.

EXAMPLE 4

Change 258 cm^2 to m^2.

$$258 \text{ cm}^2 \times \left(\frac{1 \text{ m}}{100 \text{ cm}}\right)^2 = 258 \cancel{\text{cm}^2} \times \frac{1^2 \text{ m}^2}{100^2 \cancel{\text{cm}^2}} = 0.0258 \text{ m}^2$$

Note: The intermediate step is usually not shown.

Alternate Method:

$$258 \cancel{\text{cm}^2} \times \frac{1 \text{ m}^2}{10{,}000 \cancel{\text{cm}^2}} = 0.0258 \text{ m}^2$$

U.S. Area

EXAMPLE 5

Find the area of a rectangle that is 6 in. long and 4 in. wide (Fig. 1.13).

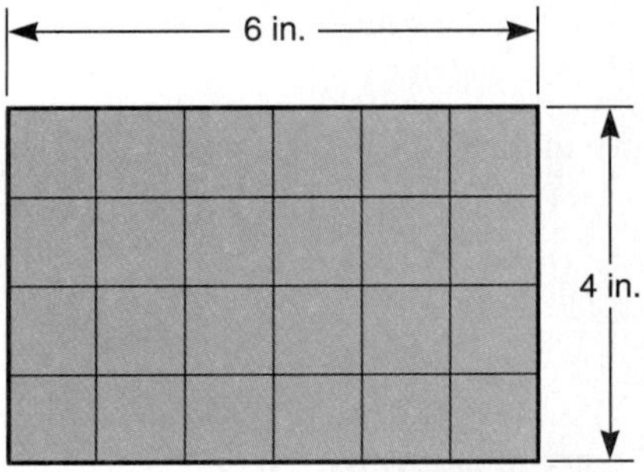

Figure 1.13

Each square is 1 in^2. To find the area of the rectangle, simply count the number of squares in the rectangle. Therefore, you find that the area = 24 in^2, or, by using the formula,

$$A = lw = (6 \text{ in.})(4 \text{ in.}) = 24 \text{ in}^2$$

EXAMPLE 6

Change 324 in^2 to yd^2.

$$324 \text{ in}^2 \times \left(\frac{1 \text{ yd}}{36 \text{ in.}}\right)^2 = 0.25 \text{ yd}^2$$

Alternate Method:

$$324 \text{ in}^2 \times \frac{1 \text{ yd}^2}{1296 \text{ in}^2} = \frac{324}{1296} \text{ yd}^2 = 0.25 \text{ yd}^2$$

Metric–U.S. Area Conversions

EXAMPLE 7

Change 25 cm^2 to in^2.

$$25 \text{ cm}^2 \times \left(\frac{1 \text{ in.}}{2.54 \text{ cm}}\right)^2 = 3.875 \text{ in}^2$$

Alternate Method:

$$25 \text{ cm}^2 \times \frac{0.155 \text{ in}^2}{1 \text{ cm}^2} = 3.875 \text{ in}^2$$

Conversion factors found in tables are usually rounded. There are many rounding procedures in general use. We will use one of the simplest methods, stated as follows:

ROUNDING NUMBERS

To round a number to a particular place value:

1. If the digit in the next place to the right is less than 5, drop that digit and all other following digits. Replace any whole number places dropped with zeros.
2. If the digit in the next place to the right is 5 or greater, add 1 to the digit in the place to which you are rounding. Drop all other following digits. Replace any whole number places dropped with zeros.

EXAMPLE 8

Change 28.5 m^2 to in^2.

$$28.5 \text{ m}^2 \times \left(\frac{39.4 \text{ in.}}{1 \text{ m}}\right)^2 = 44{,}242.26 \text{ in}^2$$

Alternate Method:

$$28.5 \text{ m}^2 \times \frac{1550 \text{ in}^2}{1 \text{ m}^2} = 44{,}175 \text{ in}^2$$

Note: The choice of rounded conversion factors will often lead to results that differ slightly. When checking your answers, you must allow for such rounding differences.

To convert between metric and U.S. land area units, use the relationship

$$1 \text{ hectare} = 2.47 \text{ acres}$$

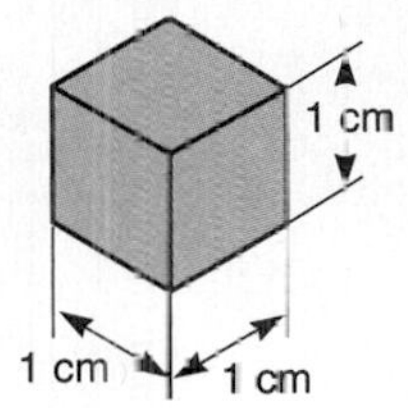

One cubic centimetre (cm^3)

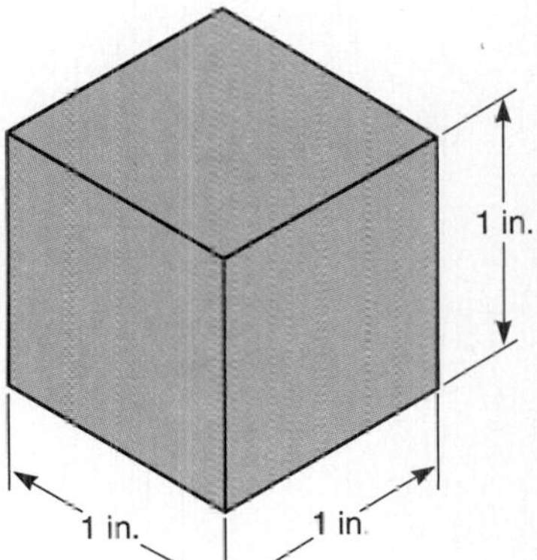

One cubic inch (in^3)

Figure 1.14

Volume

The **volume** of a figure is the number of cubic units that it contains. Standard units of volume are based on the cube and are called *cubic centimetres, cubic inches, cubic yards,* or some other cubic unit of measure. A volume of 1 cubic centimetre (cm^3) is the same as the amount of volume contained in a cube 1 cm on each side. One cubic inch (in^3) is the volume contained in a cube 1 in. on each side (Fig. 1.14).

Note: When multiplying measurements of like units, multiply the numbers and then multiply the units as follows:

$$3 \text{ in.} \times 5 \text{ in.} \times 4 \text{ in.} = (3 \times 5 \times 4)(\text{in.} \times \text{in.} \times \text{in.}) = 60 \text{ in}^3$$
$$2 \text{ cm} \times 4 \text{ cm} \times 1 \text{ cm} = (2 \times 4 \times 1)(\text{cm} \times \text{cm} \times \text{cm}) = 8 \text{ cm}^3$$
$$1.5 \text{ ft} \times 8.7 \text{ ft} \times 6 \text{ ft} = (1.5 \times 8.7 \times 6)(\text{ft} \times \text{ft} \times \text{ft}) = 78.3 \text{ ft}^3$$

EXAMPLE 9

Metric Volume

Find the volume of a rectangular prism 6 cm long, 4 cm wide, and 5 cm high.

Each cube shown in Fig. 1.15 is 1 cm^3. To find the volume of the rectangular solid, count the number of cubes in the bottom layer of the rectangular solid and then multiply that number by the number of layers that the solid can hold. Therefore, there are 5 layers of 24 cubes, which is 120 cubes or 120 cubic centimetres.

Figure 1.15

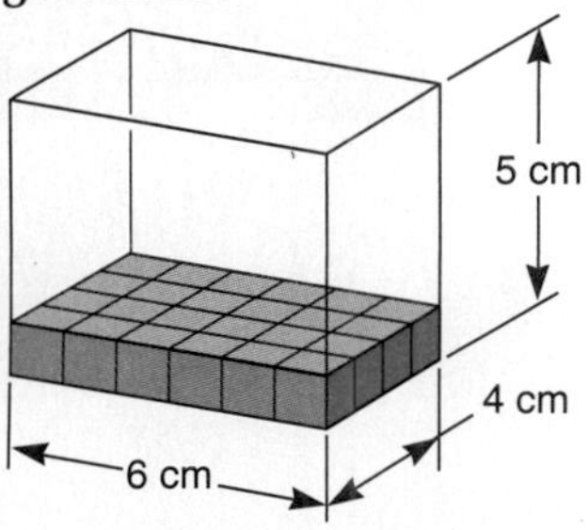

Or, by formula, $V = Bh$, where B is the area of the base and h is the height. However, the area of the base is found by lw, where l is the length and w is the width of the rectangle. Therefore, the volume of a rectangular solid can be found by the formula

$$V = lwh = (6 \text{ cm})(4 \text{ cm})(5 \text{ cm}) = 120 \text{ cm}^3$$

Note: $\text{cm} \times \text{cm} \times \text{cm} = \text{cm}^3$.

A common unit of volume in the metric system is the *litre* (L) (Fig. 1.16). The litre is commonly used for liquid volumes.

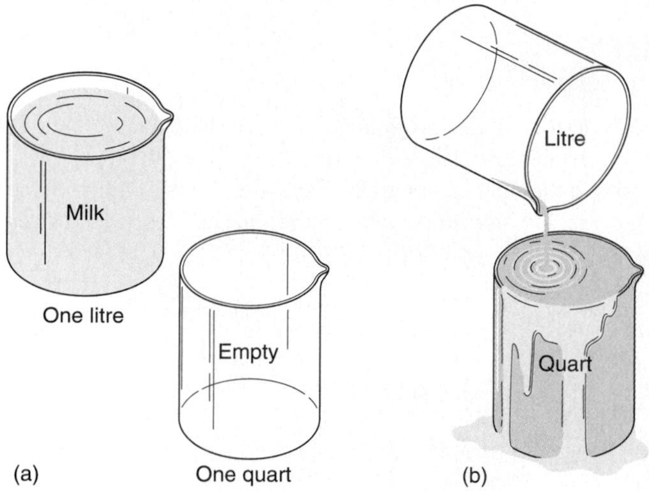

Figure 1.16 One litre of milk is a little more than 1 quart of milk.

The cubic metre (m^3) is used to measure large volumes. The cubic metre is the volume in a cube 1 m on an edge. For example, the usual teacher's desk could be boxed into 2 cubic metres side by side.

The relationship between the litre and the cubic centimetre deserves special mention. The litre is defined as the volume in 1 cubic decimetre (dm^3). That is, 1 litre of liquid fills a cube 1 dm (10 cm) on an edge (Fig. 1.17). The volume of this cube can be found by using the formula

$$V = lwh = (10 \text{ cm})(10 \text{ cm})(10 \text{ cm}) = 1000 \text{ cm}^3$$

That is,

$$1 \text{ L} = 1000 \text{ cm}^3$$

Then

$$\frac{1}{1000} \text{ L} = 1 \text{ cm}^3$$

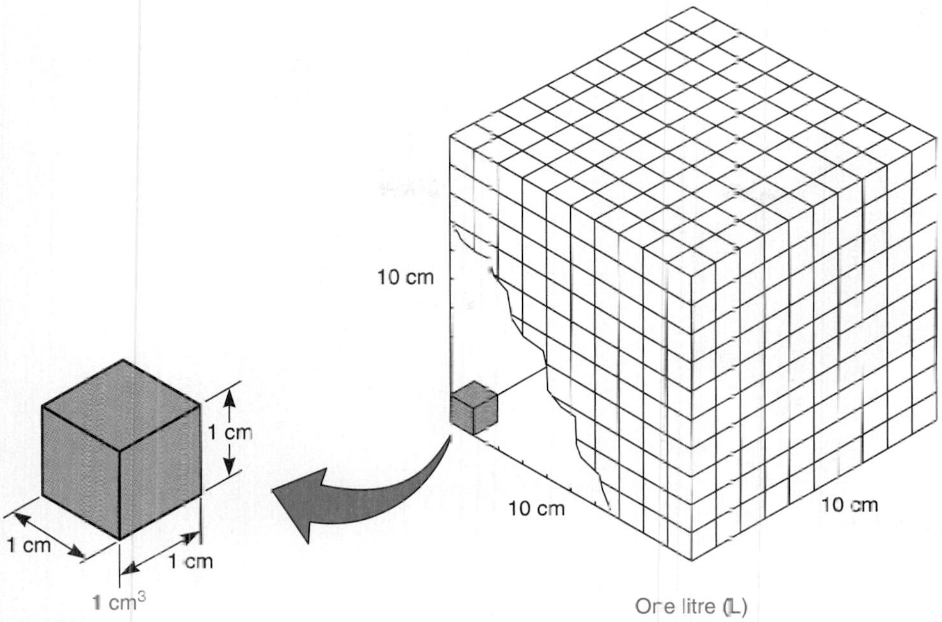

Figure 1.17
One litre contains 1000 cm³.

But

$$\frac{1}{1000}\text{ L} = 1\text{ mL}$$

Therefore,

$$1\text{ mL} = 1\text{ cm}^3$$

Milk, soda, and gasoline are usually sold by the litre in countries using the metric system. Liquid medicine, vanilla extract, and lighter fluid are usually sold by the millilitre. Many metric cooking recipes are given in millilitres. Very large quantities of oil are sold by the kilolitre (1000 L).

EXAMPLE 10

Change 0.75 L to millilitres.

$$0.75\ \cancel{\text{L}} \times \frac{1000\text{ mL}}{1\ \cancel{\text{L}}} = 750\text{ mL}$$

Similarly, the conversion of volume cubic units will be shown using a method of cubing the linear, or length, conversion factor that you are most likely to remember. An alternate method emphasizing direct use of the conversion tables will also be shown.

EXAMPLE 11

Change 0.65 cm³ to cubic millimetres.

$$0.65\text{ cm}^3 \times \left(\frac{10\text{ mm}}{1\text{ cm}}\right)^3 = 0.65\ \cancel{\text{cm}^3} \times \frac{10^3\text{ mm}^3}{1^3\ \cancel{\text{cm}^3}} = 650\text{ mm}^3$$

Note: The intermediate step is usually not shown.

Alternate Method:

$$0.65\ \cancel{\text{cm}^3} \times \frac{1000\text{ mm}^3}{1\ \cancel{\text{cm}^3}} = 650\text{ mm}^3$$

EXAMPLE 12

U.S. Volume

Find the volume of the prism shown in Fig. 1.18.

$$V = lwh = (8\text{ in.})(4\text{ in.})(5\text{ in.}) = 160\text{ in}^3$$

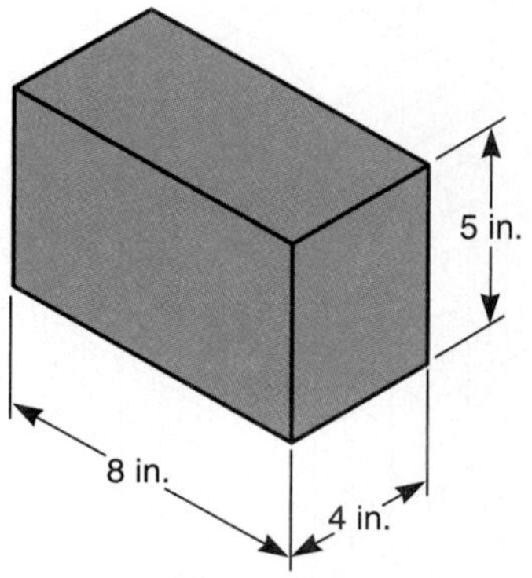

Figure 1.18

EXAMPLE 13

Change 24 ft^3 to in^3.

$$24\text{ ft}^3 \times \left(\frac{12\text{ in.}}{1\text{ ft}}\right)^3 = 41{,}472\text{ in}^3$$

Alternate Method:

$$24\text{ }\cancel{\text{ft}^3} \times \frac{1728\text{ in}^3}{1\text{ }\cancel{\text{ft}^3}} = 41{,}472\text{ in}^3$$

Metric–U.S. Volume Conversions

EXAMPLE 14

Change 56 in^3 to cm^3.

$$56\text{ in}^3 \times \left(\frac{2.54\text{ cm}}{1\text{ in.}}\right)^3 = 917.68\text{ cm}^3$$

Alternate Method:

$$56\text{ }\cancel{\text{in}^3} \times \frac{16.4\text{ cm}^3}{1\text{ }\cancel{\text{in}^3}} = 918.4\text{ cm}^3$$

EXAMPLE 15

Change 28 m^3 to ft^3.

$$28\text{ m}^3 \times \left(\frac{3.28\text{ ft}}{1\text{ m}}\right)^3 = 988.1\text{ ft}^3$$

Alternate Method:

$$28\text{ }\cancel{\text{m}^3} \times \frac{35.3\text{ ft}^3}{1\text{ }\cancel{\text{m}^3}} = 988.4\text{ ft}^3$$

Surface Area

The **lateral** (side) **surface area** of any geometric solid is the area of all the lateral faces. The **total surface area** of any geometric solid is the lateral surface area plus the area of the bases.

EXAMPLE 16

Find the lateral surface area of the prism shown in Fig. 1.19.

$$\begin{aligned}
\text{area of lateral face 1} &= (6 \text{ in.})(5 \text{ in.}) = 30 \text{ in}^2 \\
\text{area of lateral face 2} &= (5 \text{ in.})(4 \text{ in.}) = 20 \text{ in}^2 \\
\text{area of lateral face 3} &= (6 \text{ in.})(5 \text{ in.}) = 30 \text{ in}^2 \\
\text{area of lateral face 4} &= (5 \text{ in.})(4 \text{ in.}) = \underline{20 \text{ in}^2} \\
\text{lateral surface area} &= 100 \text{ in}^2
\end{aligned}$$

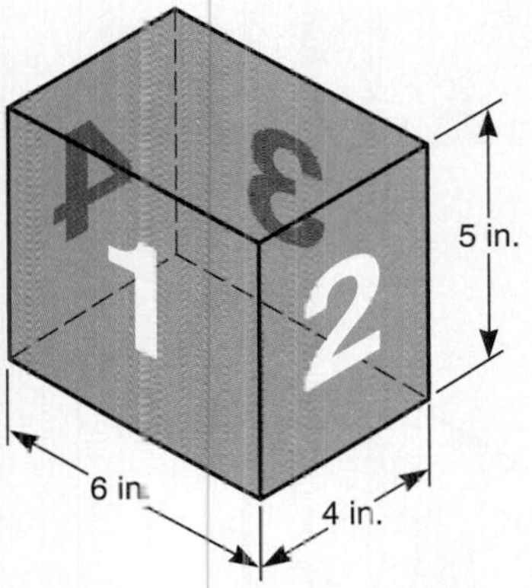

Figure 1.19

EXAMPLE 17

Find the total surface area of the prism shown in Fig. 1.19.

$$\begin{aligned}
\text{total surface area} &= \text{lateral surface area} + \text{area of the bases} \\
\text{area of base} &= (6 \text{ in.})(4 \text{ in.}) = 24 \text{ in}^2 \\
\text{area of both bases} &= 2(24 \text{ in}^2) = 48 \text{ in}^2 \\
\text{total surface area} &= 100 \text{ in}^2 + 48 \text{ in}^2 = 148 \text{ in}^2
\end{aligned}$$

Area formulas, volume formulas, and lateral surface area formulas are provided on the inside back cover.

PROBLEMS 1.4

Find the area of each figure.

1.

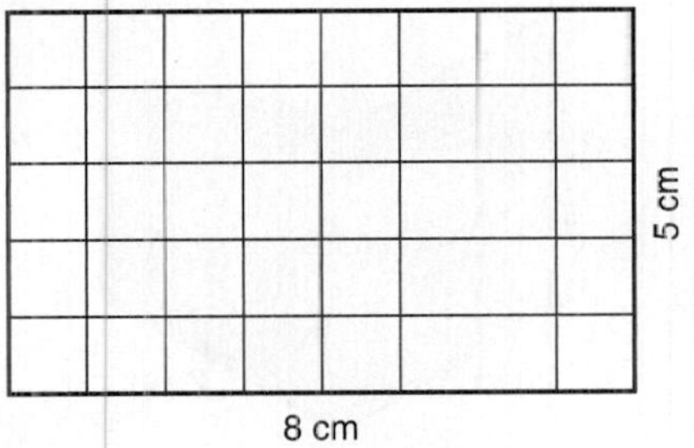

2.

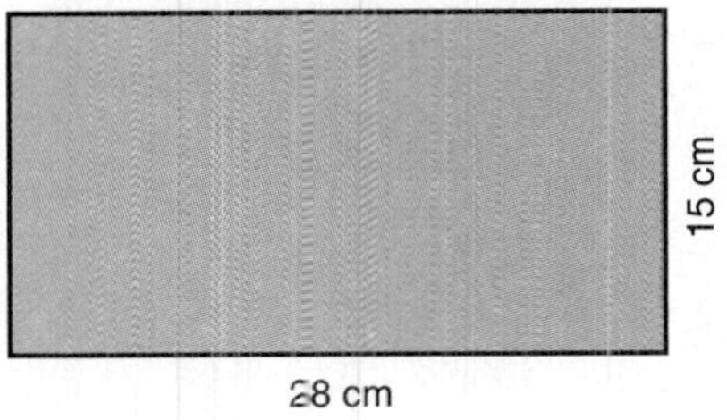

3.

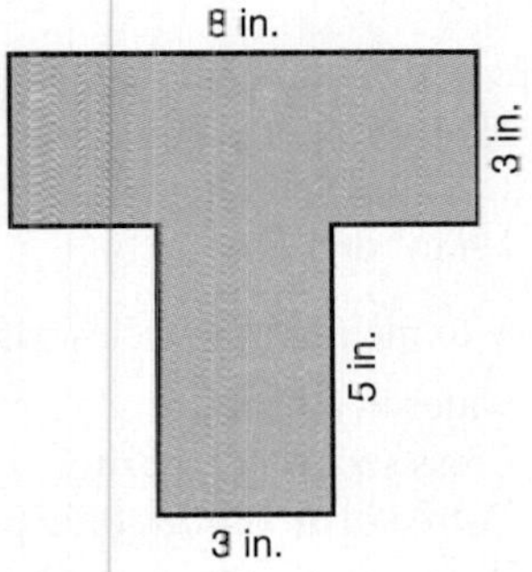

4.

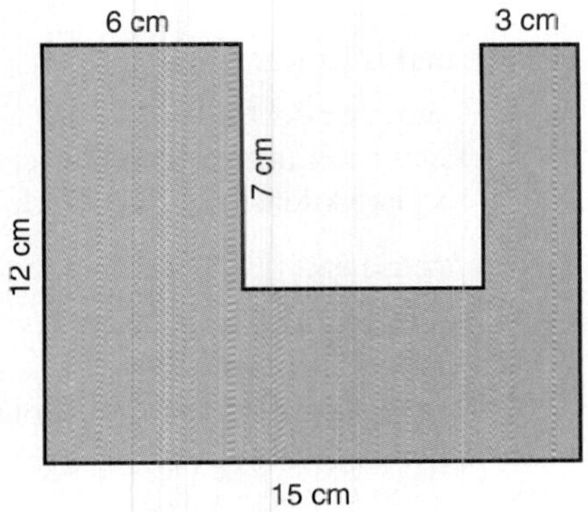

5. Find the cross-sectional area of the I-beam.

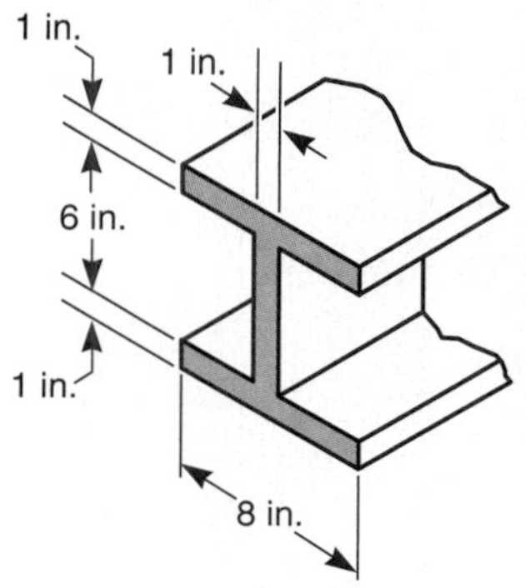

6. Find the largest cross-sectional area of the figure.

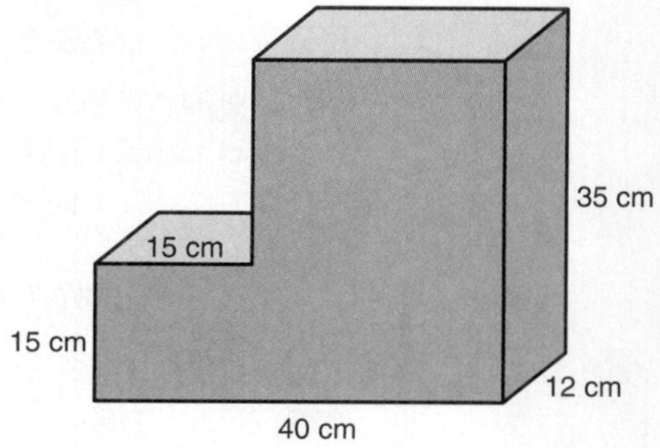

Find the volume in each figure.

7.

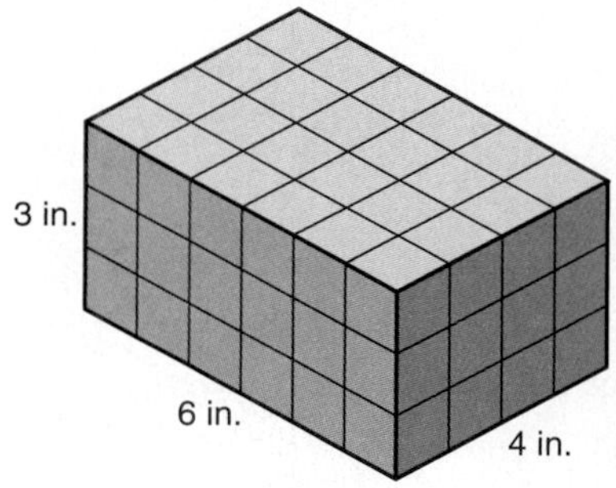

8.

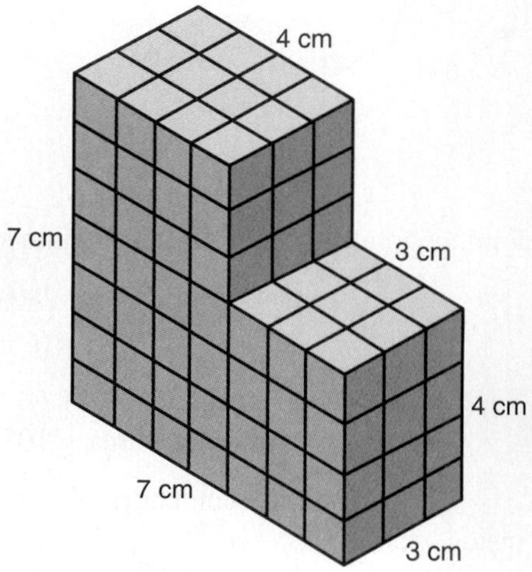

9.

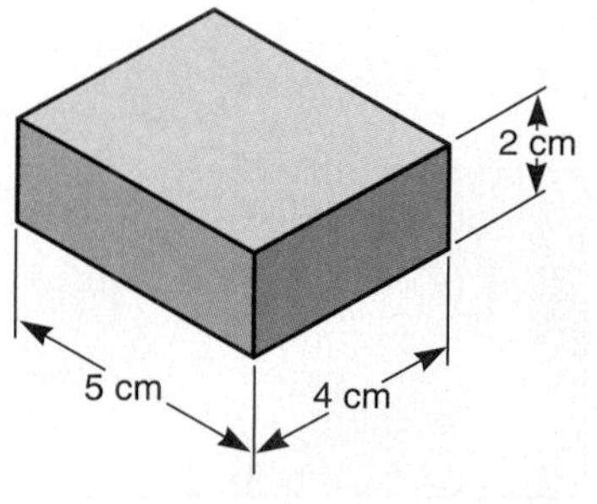

10.

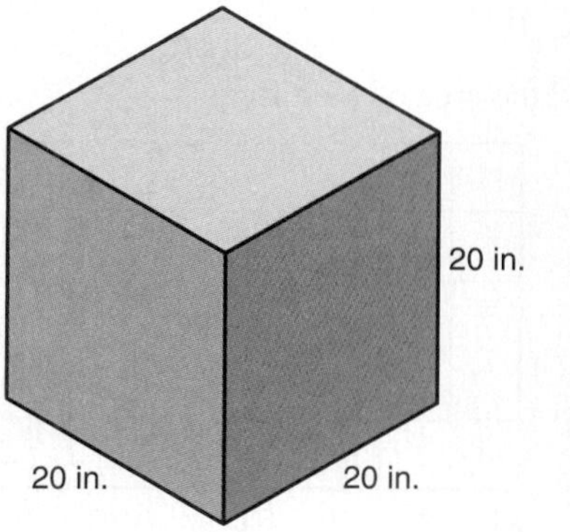

Which unit is larger?

11. 1 litre or 1 centilitre
12. 1 millilitre or 1 kilolitre
13. 1 cubic millimetre or 1 cubic centimetre
14. 1 cm^3 or 1 m^3
15. 1 square kilometre or 1 hectare
16. 1 mm^2 or 1 dm^2

Which metric unit (m^3, L, mL, m^2, cm^2, ha) would you use to measure the following?

17. Oil in your car's crankcase
18. Water in a bathtub
19. Floor space in a house
20. Cross section of a piston
21. Storage space in a miniwarehouse
22. Coffee in an office coffeepot
23. Size of a field of corn
24. Page size of a newspaper
25. A dose of cough syrup
26. Size of a cattle ranch

27. Cargo space in a truck
28. Gasoline in your car's gas tank
29. Piston displacement of an engine
30. Paint needed to paint a house
31. Dose of eye drops
32. Size of a plot of timber

Fill in the blank with the most reasonable metric unit (m^3, L, mL, m^2, cm^2, ha).

33. Go to the store and buy 4 _____ of root beer for the party.
34. I drank 200 _____ of orange juice for breakfast.
35. Craig bought a 30-_____ tarpaulin for his truck.
36. The cross section of a log is 3200 _____.
37. A farmer's gasoline storage tank holds 4000 _____.
38. Our city water tower holds 500 _____ of water.
39. Brian planted 60 _____ of soybeans this year.
40. David needs some copper tubing with a cross section of 3 _____.
41. Paula ordered 15 _____ of concrete for her new driveway.
42. Barbara heats 420 _____ of living space in her house.
43. Joyce's house has 210 _____ of floor space.
44. Kurt mows 5 _____ of grass each week.
45. Amy is told by her doctor to drink 2 _____ of water each day.
46. My favorite coffee cup holds 225 _____ of coffee.

Fill in each blank.

47. 1 L = _____ mL
48. 1 kL = _____ L
49. 1 L = _____ daL
50. 1 L = _____ kL
51. 1 L = _____ hL
52. 1 L = _____ dL
53. 1 daL = _____ L
54. 1 mL = _____ L
55. 1 mL = _____ cm^3
56. 1 L = _____ cm^3
57. 1 m^3 = _____ cm^3
58. 1 cm^3 = _____ mL
59. 1 cm^3 = _____ L
60. 1 dm^3 = _____ L
61. 1 m^2 = _____ cm^2
62. 1 km^2 = _____ m^2
63. 1 cm^2 = _____ mm^2
64. 1 mm^2 = _____ m^2
65. 1 dm^2 = _____ m^2
66. 1 ha = _____ m^2
67. 1 km^2 = _____ ha
68. 1 ha = _____ km^2
69. Change 7500 mL to L.
70. Change 0.85 L to mL.
71. Change 1.6 L to mL.
72. Change 9 mL to L.
73. Change 275 cm^3 to mm^3.
74. Change 5 m^3 to cm^3.
75. Change 4 m^3 to mm^3.
76. Change 520 mm^3 to cm^3.
77. Change 275 cm^3 to mL.
78. Change 125 cm^3 to L.
79. Change 1 m^3 to L.
80. Change 150 mm^3 to L.
81. Change 7.5 L to cm^3.
82. Change 450 L to m^3.
83. Change 5000 mm^2 to cm^2.
84. Change 1.75 km^2 to m^2.
85. Change 5 m^2 to cm^2.
86. Change 250 cm^2 to mm^2.
87. Change 4×10^8 m^2 to km^2.
88. Change 5×10^7 cm^2 to m^2.
89. Change 5 yd^2 to ft^2.
90. How many m^2 are in 225 ft^2?
91. Change 15 ft^2 to cm^2.
92. How many ft^2 are in a rectangle 15 m long and 12 m wide?
93. Change 108 in^2 to ft^2.
94. How many in^2 are in 51 cm^2?
95. How many in^2 are in a square 11 yd on a side?
96. How many m^2 are in a doorway whose area is 20 ft^2?
97. Change 19 yd^3 to ft^3.
98. How many in^3 are in 29 cm^3?
99. How many yd^3 are in 23 m^3?
100. How many cm^3 are in 88 in^3?
101. Change 8 ft^3 to in^3.
102. How many in^3 are in 12 m^3?
103. The volume of a casting is 38 in^3. What is its volume in cm^3?
104. How many castings of 14 cm^3 can be made from a 12-ft^3 block of steel?
105. Find the lateral surface area of the figure in Problem 9.
106. Find the lateral surface area of the figure in Problem 10.
107. Find the total surface area of the figure in Problem 9.
108. Find the total surface area of the figure in Problem 10.
109. How many mL of water would the figure in Problem 9 hold?
110. How many mL of water would the figure in Problem 8 hold?

1.5 Other Units

Mass and Weight

The **mass** of an object is the quantity of material making up the object. One unit of mass in the metric system is the *gram* (g) (Fig. 1.20). The gram is defined as the mass of 1 cubic centimetre (cm^3) of water at its maximum density. Since the gram is so small, the **kilogram** (kg) is the basic unit of mass in the metric system. One kilogram is defined as the mass of 1 cubic decimetre (dm^3) of water at its maximum density. The standard kilogram is a special platinum–iridium cylinder at the International Bureau of Weights and Measures near Paris, France. Since 1 dm^3 = 1 L, 1 litre of water has a mass of 1 kilogram.

(a) A common paper clip has a mass of about 1 g. (b) Three aspirin have a mass of about 1 g.

Figure 1.20

For very, very small masses, such as medicine dosages, we use the *milligram* (mg). One grain of salt has a mass of about 1 mg. The *metric ton* (1000 kg) is used to measure the mass of very large quantities, such as the coal on a barge, a trainload of grain, or a shipload of ore.

EXAMPLE 1

Change 74 kg to grams.

Choose the conversion factor with kilograms in the denominator so that the kg units cancel each other.

$$74\ \cancel{\text{kg}} \times \frac{1000\text{ g}}{1\ \cancel{\text{kg}}} = 74{,}000\text{ g}$$

EXAMPLE 2

Change 600 mg to grams.

$$600\ \cancel{\text{mg}} \times \frac{1\text{ g}}{1000\ \cancel{\text{mg}}} = 0.6\text{ g}$$

Figure 1.21

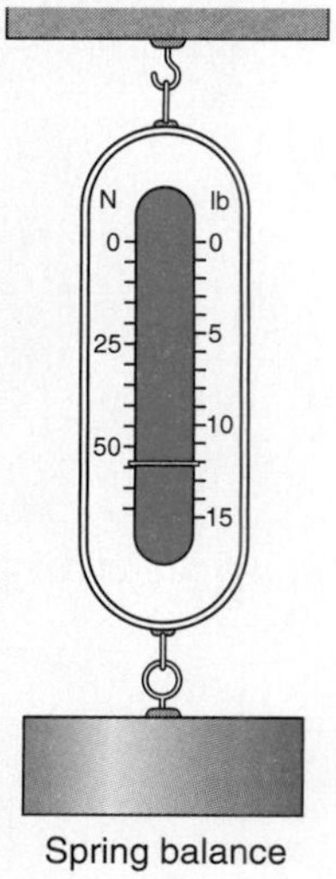

Spring balance

The **weight** of an object is a measure of the gravitational force or pull acting on it. The weight unit in the metric system is the *newton* (N). A small apple weighs about one newton.

The *pound* (lb), a unit of force, is one of the basic U.S. system units. It is defined as the pull of the earth on a cylinder of a platinum–iridium alloy that is stored in a vault at the U.S. Bureau of Standards. The *ounce* (oz) is another common unit of weight in the U.S. system. The relationship between pounds and ounces is

$$1\text{ lb} = 16\text{ oz}$$

The following relationships can be used for conversion between systems of units:

$$1\text{ N} = 0.225\text{ lb} \quad \text{or} \quad 1\text{ lb} = 4.45\text{ N}$$

The mass of an object remains constant, but its weight changes according to its distance from the earth or another planet. Mass and weight and their units of measure are discussed in more detail in Section 4.5.

A **spring balance** (Fig. 1.21) is an instrument containing a spring, which stretches in proportion to the weight of the body, and a pointer attached to the spring with a calibrated scale read directly in pounds or newtons. The common bathroom scale uses this principle to measure weight.

A **platform balance** (Fig. 1.22) consists of two platforms connected by a horizontal rod that balances on a knife edge. This device compares the pull of gravity on objects that are on the two platforms. The platforms are at the same height only when the unknown mass of the object on the left is equal to the known mass placed on the right. It is also possible to use one platform and a mass that slides along a calibrated scale. Variations of this basic design are found in some meat market and truck scales.

Figure 1.22 Platform balance

Photo courtesy of Dorling Kindersley

EXAMPLE 3

The weight of the intake valve of an auto engine is 0.18 lb. What is its weight in ounces and in newtons?

To find the weight in ounces, we simply use a conversion factor as follows:

$$0.18\ \cancel{\text{lb}} \times \frac{16\ \text{oz}}{1\ \cancel{\text{lb}}} = 2.88\ \text{oz}$$

To find the weight in newtons, we again use a conversion factor:

$$0.18\ \cancel{\text{lb}} \times \frac{4.45\ \text{N}}{1\ \cancel{\text{lb}}} = 0.801\ \text{N}$$

Time

Airlines and other transportation systems run on time schedules that would be meaningless if we did not have a common unit for time measurement. All the common units for time measurement are the same in both systems. These units are based on the motion of the earth and the moon (Fig. 1.23). The year is approximately the time required for one complete revolution of the earth about the sun. The month is approximately the time for one complete revolution of the moon about the earth. The day is the time for one rotation of the earth about its axis.

Figure 1.23

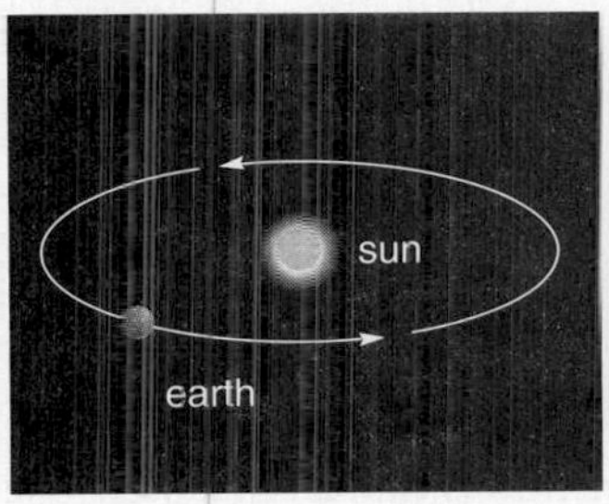

Revolution of earth about the sun

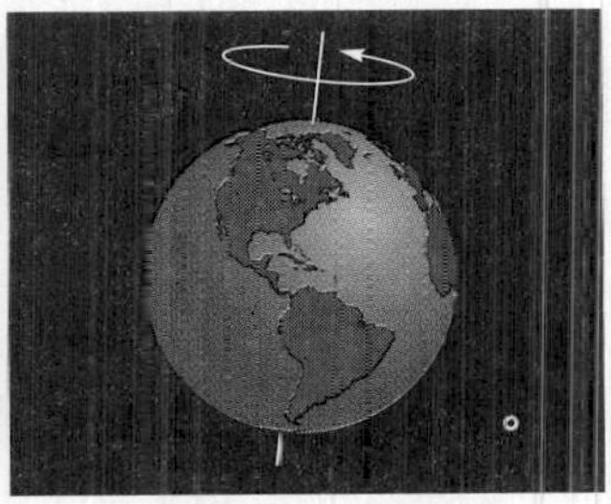

Rotation of earth about its axis

The basic time unit is the **second** (s). For many years, the second was defined as $\frac{1}{86,400}$ of a mean solar day. The standard second adopted in 1967 is defined more precisely in terms of the frequency of radiation emitted by cesium atoms when they pass between two particular states; that is, the time required for 9,192,631,770 periods of this radiation. The second is not always convenient to use, so other units are necessary. The *minute* (min) is 60 seconds, the *hour* (h) is 60 minutes, and the *day* is 24 hours. The *year* is 365 days in length except for every fourth year, when it is 366 days long. This difference is necessary to keep the seasons at the same time each year, since one revolution of the earth about the sun takes approximately $365\frac{1}{4}$ days.

The Julian calendar introduced by Julius Caesar in 46 BC provided for an ordinary year of 365 days and a leap year of 366 days every fourth year. Astronomers have found that the length of a year has varied from 365.24253 days in 5000 BC to 365.24219 days in 2000 AD. The Gregorian calendar now used in most countries of the world was introduced by Pope Gregory XIII in 1582 to correct the Julian calendar discrepancies. The Gregorian calendar provides for an ordinary year of 365 days and a leap year of 366 days in years divisible by four except in century years not divisible by 400. Thus, years 1600 and 2000 are leap years but years 1700, 1800, 1900, and 2100 are not. In 1582, 10 days were omitted from the calendar to adjust for the accumulated difference of the Julian calendar since 46 BC. By decree, the day following October 4, 1582, became October 15, 1582. Now, small fractions-of-a-second adjustments are made in the calendar annually by international agreement to compensate for the variation of the earth's orbit around the sun.

Figure 1.24

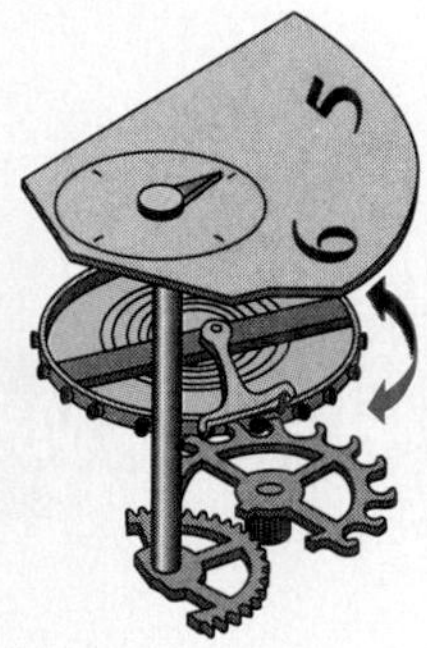

Oscillation of balance wheel

Common devices for time measurement are the electric clock, the mechanical watch, and the quartz crystal watch. The accuracy of an electric clock depends on how accurately the 60-Hz (hertz = cycles per second) line voltage is controlled. In the United States this is controlled very accurately. Most mechanical watches have a balance wheel that oscillates near a given frequency, usually 18,000 to 36,000 vibrations per hour, and drives the hands of the watch (Fig. 1.24). The quartz crystal in a watch is excited by a small power cell and vibrates 32,768 times per second. The accuracy of the watch depends on how well the frequency of oscillation is controlled.

EXAMPLE 4

Change 2 h 15 min to seconds.

First,

$$2\,\cancel{h} \times \frac{60\text{ min}}{1\,\cancel{h}} = 120\text{ min}$$

Then

$$2\text{ h }15\text{ min} = 120\text{ min} + 15\text{ min} = 135\text{ min}$$

and

$$135\,\cancel{\text{min}} \times \frac{60\text{ s}}{1\,\cancel{\text{min}}} = 8100\text{ s}$$

Very short periods of time are measured in parts of a second, given with the appropriate metric prefix. Such units are commonly used in electronics.

EXAMPLE 5

What is the meaning of each unit?

(a) 1 ms = 1 millisecond = 10^{-3} s and means one one-thousandth of a second.
(b) 1 μs = 1 microsecond = 10^{-6} s and means one one-millionth of a second.
(c) 1 ns = 1 nanosecond = 10^{-9} s and means one one-billionth of a second.
(d) 1 ps = 1 picosecond = 10^{-12} s and means one one-trillionth of a second.

Note: The Greek letter μ is pronounced "mu." However, 1 μs is stated or read as "one microsecond."

EXAMPLE 6

Change 45 ms to seconds.
Since 1 ms = 10^{-3} s,

$$45\ \cancel{ms} \times \frac{10^{-3}\ s}{1\ \cancel{ms}} = 45 \times 10^{-3}\ s = 0.045\ s$$

EXAMPLE 7

Change 0.000000025 s to nanoseconds.
Since 1 ns = 10^{-9} s,

$$0.000000025\ \cancel{s} \times \frac{1\ ns}{10^{-9}\ \cancel{s}} = 25\ ns$$

PROBLEMS 1.5

Which unit is larger?

1. 1 gram or 1 centigram
2. 1 gram or 1 milligram
3. 1 gram or 1 kilogram
4. 1 centigram or 1 milligram
5. 1 centigram or 1 kilogram
6. 1 milligram or 1 kilogram

Which metric unit (kg, g, mg, or metric ton) would you use to measure the following?

7. Your mass
8. An aspirin
9. A bag of lawn fertilizer
10. A bar of hand soap
11. A trainload of grain
12. A sewing needle
13. A small can of corn
14. A channel catfish
15. A vitamin capsule
16. A car

Fill in each blank with the most reasonable metric unit (kg, g, mg, or metric ton).

17. A newborn's mass is about 3 _____.
18. An elevator in a local department store has a load limit of 2000 _____.
19. Margie's diet calls for 250 _____ of meat.
20. A 200-car train carries 11,000 _____ of soybeans.
21. A truckload shipment of copper pipe has a mass of 900 _____.
22. A carrot has a mass of 75 _____.
23. A candy recipe calls for 175 _____ of chocolate.
24. My father has a mass of 70 _____.
25. A pencil has a mass of 10 _____.
26. Postage rates for letters would be based on the _____.
27. A heavyweight boxing champion has a mass of 93 _____.
28. A nickel has a mass of 5 _____.
29. My favorite spaghetti recipe calls for 1 _____ of ground beef.
30. My favorite spaghetti recipe calls for 150 _____ of tomato paste.
31. Our local grain elevator shipped 10,000 _____ of wheat last year.
32. A slice of bread has a mass of about 25 _____.
33. I bought a 5-_____ bag of potatoes at the store today.
34. My grandmother takes 250-_____ capsules for her arthritis.

Fill in each blank.

35. 1 kg = _____ g
36. 1 mg = _____ g
37. 1 g = _____ cg
38. 1 g = _____ hg
39. 1 dg = _____ g
40. 1 dag = _____ g
41. 1 g = _____ mg
42. 1 g = _____ dg
43. 1 hg = _____ g
44. 1 cg = _____ g
45. 1 g = _____ kg
46. 1 g = _____ dag
47. 1 g = _____ μg
48. 1 mg = _____ μg
49. Change 575 g to mg.
50. Change 575 g to kg.
51. Change 650 mg to g.
52. Change 375 kg to g.
53. Change 50 dg to g.
54. Change 485 dag to dg.
55. Change 30 kg to mg.
56. Change 4 metric tons to kg.

57. Change 25 hg to kg.
58. Change 58 μg to g.
59. Change 400 μg to mg.
60. Change 30,000 kg to metric tons.
61. What is the mass of 750 mL of water?
62. What is the mass of 1 m^3 of water?
63. The weight of a car is 3500 lb. Find its weight in newtons.
64. A certain bridge is designed to support 150,000 lb. Find the maximum weight that it will support in newtons.
65. Jose weighs 200 lb. What is his weight in newtons?
66. Change 80 lb to newtons.
67. Change 2000 N to pounds.
68. Change 2000 lb to newtons.
69. Change 120 oz to pounds.
70. Change 3.5 lb to ounces.
71. Change 10 N to ounces.
72. Change 25 oz to newtons.
73. Find the metric weight of a 94-lb bag of cement.
74. What is the weight in newtons of 500 blocks if each weighs 3 lb?

Fill in each blank.

75. The basic metric unit of time is _____. Its abbreviation is _____.
76. The basic metric unit of mass is _____. Its abbreviation is _____.
77. The common metric unit of weight is _____. Its abbreviation is _____.

Which is larger?

78. 1 second or 1 millisecond
79. 1 millisecond or 1 nanosecond
80. 1 ps or 1 μs
81. 1 ms or 1 μs

Write the abbreviation for each unit.

82. 8.6 microseconds
83. 45 nanoseconds
84. 75 picoseconds
85. Change 345 μs to s.
86. Change 1 h 25 min to min.
87. Change 4 h 25 min 15 s to s.
88. Change 7×10^6 s to h.
89. Change 4 s to ns.
90. Change 1 h to ps.

1.6 Measurement: Significant Digits and Accuracy

Up to this time in your studies, probably all numbers and all measurements have been treated as exact numbers. An **exact number** is a number that has been determined as a result of counting, such as 24 students are enrolled in this class, or by some definition, such as 1 h = 60 min or 1 in. = 2.54 cm, a conversion definition agreed to by the world governments' bureaus of standards. Generally, the treatment of the addition, subtraction, multiplication, and division of exact numbers is the emphasis or main content of elementary mathematics.

However, nearly all data of a technical nature involve **approximate numbers;** that is, numbers determined as a result of some measurement process—some direct, as with a ruler, and some indirect, as with a surveying transit or reading an electric meter. First, realize that no measurement can be found exactly. The length of the cover of this book can be found using many instruments. The better the measuring device used, the better is the measurement.

A measurement may be expressed in terms of its accuracy or its precision. The **accuracy** of a measurement refers to the number of digits, called **significant digits,** which indicates the number of units that we are reasonably sure of having counted when making a measurement. The greater the number of significant digits given in a measurement, the better is the accuracy and vice versa.

EXAMPLE 1

The average distance between the moon and the earth is 385,000 km. This measurement indicates measuring 385 thousands of kilometres; its accuracy is indicated by three significant digits.

EXAMPLE 2

A measurement of 0.025 cm indicates measuring 25 thousandths of a centimetre; its accuracy is indicated by two significant digits.

EXAMPLE 3

A measurement of 0.0500 s indicates measuring $5\bar{0}0$ ten-thousandths of a second; its accuracy is indicated by three significant digits.

Notice that sometimes a zero is significant and sometimes it is not. To clarify this, we use the following rules for significant digits:

SIGNIFICANT DIGITS

1. All nonzero digits are significant: 156.4 m has four significant digits (this measurement indicates 1564 tenths of metres).
2. All zeros between significant digits are significant: 306.02 km has five significant digits (this measurement indicates 30,602 hundredths of kilometres).
3. In a number greater than 1, a zero that is specially tagged, such as by a bar above it, is significant: $23\bar{0}$,000 km has three significant digits (this measurement indicates $23\bar{0}$ thousands of kilometres).
4. All zeros to the right of a significant digit *and* a decimal point are significant: 86.10 cm has four significant digits (this measurement indicates $861\bar{0}$ hundredths of centimetres).
5. In whole-number measurements, zeros at the right that are not tagged are *not* significant: 2500 m has two significant digits (25 hundreds of metres).
6. In measurements of less than 1, zeros at the left are *not* significant: 0.00752 m has three significant digits (752 hundred-thousandths of a metre).

When a number is written in scientific notation, the decimal part indicates the number of significant digits. For example, $20\bar{0}$,000 m would be written in scientific notation as 2.00×10^5 m.

In summary:

To find the number of significant digits:

1. All nonzero digits are significant.
2. Zeros are significant when they
   (a) are between significant digits;
   (b) follow the decimal point and a significant digit; or
   (c) are in a whole number and a bar is placed over the zero.

EXAMPLE 4

Determine the accuracy (the number of significant digits) of each measurement.

Measurement	Accuracy (significant digits)
(a) 2642 ft	4
(b) 2005 m	4 (Both zeros are significant.)
(c) 2050 m	3 (Only the first zero is significant.)
(d) 2500 m	2 (No zero is significant.)
(e) $25\bar{0}0$ m	3 (Only the first zero is significant.)
(f) $250\bar{0}$ m	4 (Both zeros are significant.)

(*continued*)

Measurement	Accuracy (significant digits)
(g) 34,000 mi	2 (No zeros are significant.)
(h) 15,670,000 lb	4 (No zeros are significant.)
(i) 203.05 km	5 (Both zeros are significant.)
(j) 0.000345 kg	3 (No zeros are significant.)
(k) 75 N	2
(l) 2.3 s	2
(m) 0.02700 g	4 (Only the right two zeros are significant.)
(n) 2.40 cm	3 (The zero is significant.)
(o) 4.050 μs	4 (All zeros are significant.)
(p) 100.050 km	6 (All zeros are significant.)
(q) 0.004 s	1 (No zeros are significant.)
(r) 2.03×10^4 m^2	3 (The zero is significant.)
(s) 1.0×10^{-3} N	2 (The zero is significant.)
(t) 5×10^6 kg	1
(u) 3.060×10^8 m^3	4 (Both zeros are significant.)

PROBLEMS 1.6

Determine the accuracy (the number of significant digits) of each measurement.

1. 536 ft	2. 307.3 mi	3. 5007 m
4. 5.00 cm	5. 0.0070 in.	6. 6.010 cm
7. $84\bar{0}0$ km	8. $30\bar{0}0$ ft	9. 187.40 m
10. $5\bar{0}0$ g	11. 0.00700 in.	12. 10.30 cm
13. 376.52 m	14. 3.05 mi	15. 4087 kg
16. 35.00 mm	17. 0.0160 in.	18. $37\bar{0}$ lb
19. $4\bar{0}00$ N	20. 5010 ft^3	21. 7 N
22. 32,000 tons	23. 70.00 m^2	24. 0.007 m
25. 2.4×10^3 kg	26. 1.20×10^{-5} ms	27. 3.00×10^{-4} kg
28. 4.0×10^6 ft	29. 5.106×10^7 kg	30. 1×10^{-9} m

1.7 Measurement: Precision

The **precision** of a measurement refers to the smallest unit with which a measurement is made, that is, the position of the last significant digit.

EXAMPLE 1

The precision of the measurement 385,000 km is 1000 km. (The position of the last significant digit is in the thousands place.)

EXAMPLE 2

The precision of the measurement 0.025 cm is 0.001 cm. (The position of the last significant digit is in the thousandths place.)

EXAMPLE 3

The precision of the measurement 0.0500 s is 0.0001 s. (The position of the last significant digit is in the ten-thousandths place.)

Unfortunately, the terms *accuracy* and *precision* have several different common meanings. Here we will use each term consistently as we have defined them. A measurement of 0.0004 cm has good precision and poor accuracy when compared with the measurement 378.0 cm

Measurement	Precision	Accuracy
0.0004 cm	0.0001 cm	1 significant digit
378.0 cm	0.1 cm	4 significant digits

EXAMPLE 4

Determine the precision of each measurement given in Example 4 of Section 1.6.

Measurement	Precision	Accuracy (significant digits)
(a) 2642 ft	1 ft	4
(b) 2005 m	1 m	4
(c) 2050 m	10 m	3
(d) 2500 m	100 m	2
(e) $25\bar{0}0$ m	10 m	3
(f) $250\bar{0}$ m	1 m	4
(g) 34,000 mi	1000 mi	2
(h) 15,670,000 lb	10,000 lb	4
(i) 203.05 km	0.01 km	5
(j) 0.000345 kg	0.000001 kg	3
(k) 75 N	1 N	2
(l) 2.3 s	0.1 s	2
(m) 0.02700 g	0.00001 g	4
(n) 2.40 cm	0.01 cm	3
(o) 4.050 μs	0.001 μs	4
(p) 100.050 km	0.001 km	6
(q) 0.004 s	0.001 s	1
(r) 2.03×10^4 m^2	0.01×10^4 m^2 or 100 m^2	3
(s) 1.0×10^{-3} N	0.1×10^{-3} N or 0.0001 N	2
(t) 5×10^6 kg	1×10^6 kg or 1,000,000 kg	1
(u) 3.060×10^8 m^3	0.001×10^8 m^3 or 1×10^5 m^3 or 100,000 m^3	4

PHYSICS CONNECTIONS

Precision and the New Clark Bridge

Bridges are usually not built from one end to the other. Construction typically begins in the middle of the bridge or on each end and meets in the middle. In doing so, it becomes extremely important that every section is in precise alignment so that the bridge will meet at the critical connection points.

For example, the New Clark Bridge crossing the Mississippi at Alton, Illinois, spans over 4600 ft and was designed to high precision so that each member of the bridge would be no more than $\frac{1}{8}$ in out of alignment. As frames and towers were built in the flowing river, teams of surveyors used fixed points of reference and laser beams to survey the placement of each tower and pier. Ignoring the importance of accuracy and precision would have caused serious problems.

TRY THIS ACTIVITY

Accuracy and Precision

Measure the time it takes you to run the 100-yard dash. Use a digital stopwatch and a regular wristwatch with a second hand. How accurate is each of the measurements? How precise is each of the measurements?

PROBLEMS 1.7

Determine the precision of each measurement.

1. 536 ft	2. 307.3 mi	3. 5007 m
4. 5.00 cm	5. 0.0070 in.	6. 6.010 cm
7. $84\bar{0}0$ km	8. $30\bar{0}0$ ft	9. 187.40 m
10. $5\bar{0}0$ g	11. 0.00700 in.	12. 10.30 cm
13. 376.52 m	14. 3.05 mi	15. 4087 kg
16. 35.00 mm	17. 0.0160 in.	18. $37\bar{0}$ lb
19. $4\bar{0}00$ N	20. 5010 ft^3	21. 7 N
22. 32,000 tons	23. 70.00 m^2	24. 0.007 m
25. 2.4×10^3 kg	26. 1.20×10^{-5} ms	27. 3.00×10^{-4} kg
28. 4.0×10^6 ft	29. 5.106×10^7 kg	30. 1×10^{-9} m

In each set of measurements, find the measurement that is (a) the most accurate and (b) the most precise.

31. 15.7 in.; 0.018 in.; 0.07 in.	32. 368 ft; 600 ft; 180 ft
33. 0.734 cm; 0.65 cm; 16.01 cm	34. 3.85 m; 8.90 m; 7.00 m
35. 0.0350 s; 0.025 s; 0.00040 s; 0.051 s	36. 125.00 g; 8.50 g; 9.000 g; 0.05 g
37. $27{,}0\bar{0}0$ L; 350 L; 27.6 L; 4.75 L	38. 8.4 m; 15 m; 180 m; 0.40 m
39. 500 N; 10,000 N; 500,000 N; 50 N	40. 7.5 ms; 14.2 ms; 10.5 ms; 120.0 ms

In each set of measurements, find the measurement that is (a) the least accurate and (b) the least precise.

41. 16.4 in.; 0.075 in.; 0.05 in.	42. 475 ft; 300 ft; 360 ft
43. 27.5 m; 0.65 m; 12.02 m	44. 5.7 kg; 120 kg; 0.025 kg
45. 0.0250 g; 0.015 g; 0.00005 g; 0.75 g	46. 185.0 m; 6.75 m; 5.000 m; 0.09 m
47. 45,000 N; 250 N; 16.8 N; 0.25 N; 3 N	48. 2.50 kg; 42.0 kg; $15\bar{0}$ kg; 0.500 kg
49. $20\bar{0}0$ kg; $10{,}\bar{0}00$ kg; $40\bar{0}{,}000$ kg; 20 kg	50. 80 ft; 250 ft; 12,550 ft; $26\bar{0}0$ ft

Figure 1.25 Micrometer with precision 0.01 mm

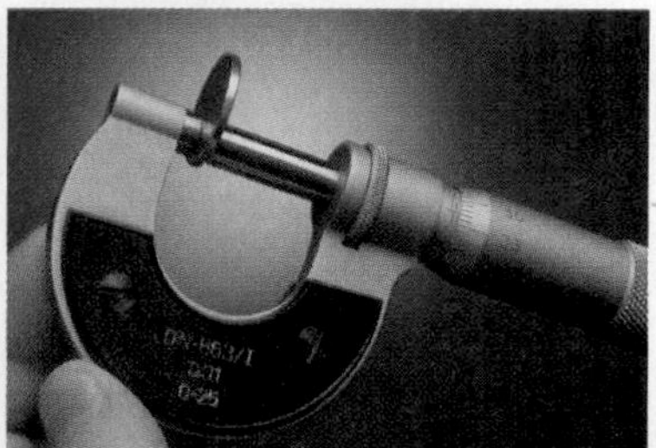

1.8 Calculations with Measurements

If one person measures the length of one of two parts of a shaft with a micrometer calibrated in 0.01 mm as 42.28 mm and another person measures the second part with a ruler calibrated in mm as 54 mm, would the total length be 96.28 mm? Note that the sum 96.28 mm indicates a precision of 0.01 mm. The precision of the ruler is 1 mm, which means that the measurement 54 mm with the ruler could actually be anywhere between 53.50 mm and 54.50 mm using the micrometer (which has a precision of 0.01 mm and is shown in Fig. 1.25). That is, using the ruler, any measurement between 53.50 mm and 54.50 mm can only be read as 54 mm. Of course, this means that the tenths and hundredths digits in the sum 96.28 mm are really meaningless. In other words, *the sum or difference of measurements can be no more precise than the least precise measurement.* That is,

To add or subtract measurements:

1. Make certain that all the measurements are expressed in the same unit. If they are not, convert them all to the same unit.
2. Add or subtract.
3. Round the results to the same precision as the least precise measurement.

EXAMPLE 1

Add the measurements 16.6 mi, 124 mi, 3.05 mi, and 0.837 mi.

All measurements are in the same unit, so add.

$$\begin{array}{r} 16.6\ \text{mi} \\ 124\ \text{mi} \\ 3.05\ \text{mi} \\ \underline{0.837\ \text{mi}} \\ 144.487\ \text{mi} \end{array} \rightarrow 144\ \text{mi}$$

Then, round this sum to the same precision as the least precise measurement, which is 124 mi. Thus, the sum is 144 mi.

EXAMPLE 2

Add the measurements 1370 cm, 1575 mm, 2.374 m, and 8.63 m.

First, convert all measurements to the same unit, say m.

$$1370\ \text{cm} = 13.7\ \text{m}$$
$$1575\ \text{mm} = 1.575\ \text{m}$$

Then add.

$$\begin{array}{r} 13.7\ \text{m} \\ 1.575\ \text{m} \\ 2.374\ \text{m} \\ \underline{8.63\ \text{m}} \\ 26.279\ \text{m} \end{array} \rightarrow 26.3\ \text{m}$$

Then, round this sum to the same precision as the least precise measurement, which is 13.7 m. Thus, the sum is 26.3 m.

EXAMPLE 3

Subtract the measurements 3457.8 g − 2.80 kg.

First, convert both measurements to the same unit, say g.

$$2.80\ \text{kg} = 28\bar{0}0\ \text{g}$$

Then subtract.

$$\begin{array}{r} 3457.8\ \text{g} \\ \underline{28\bar{0}0\ \text{g}} \\ 657.8\ \text{g} \end{array} \rightarrow 660\ \text{g}$$

Then, round this difference to the same precision as the least precise measurement, which is $28\bar{0}0$ g. Thus, the difference is 660 g.

Now suppose that you wish to find the area of the base of a rectangular building. You measure its length as 54.7 m and its width as 21.5 m. Its area is then

$$A = lw$$
$$A = (54.7\ \text{m})(21.5\ \text{m})$$
$$= 1176.05\ \text{m}^2$$

Note that the result contains six significant digits, whereas each of the original measurements contains only three significant digits. To rectify this inconsistency, we say that the product or quotient of measurements can be no more accurate than the least accurate measurement. That is,

To multiply or divide measurements:

1. Multiply or divide the measurements as given.
2. Round the result to the same number of significant digits as the measurement with the least number of significant digits.

Using the preceding rules, we find that the area of the base of the rectangular building is 1180 m^2.

Note: We assume throughout that you are using a calculator to do all calculations.

EXAMPLE 4

Multiply the measurements (124 ft)(187 ft).

$$(124 \text{ ft})(187 \text{ ft}) = 23{,}188 \text{ ft}^2$$

Round this product to three significant digits, which is the accuracy of the least accurate measurement (and also the accuracy of each measurement in the example). That is,

$$(124 \text{ ft})(187 \text{ ft}) = 23{,}200 \text{ ft}^2$$

EXAMPLE 5

Multiply the measurements (2.75 m)(1.25 m)(0.75 m).

$$(2.75 \text{ m})(1.25 \text{ m})(0.75 \text{ m}) = 2.578125 \text{ m}^3$$

Round this product to two significant digits, which is the accuracy of the least accurate measurement (0.75 m). That is,

$$(2.75 \text{ m})(1.25 \text{ m})(0.75 \text{ m}) = 2.6 \text{ m}^3$$

EXAMPLE 6

Divide the measurements 144,000 $ft^3 \div$ 108 ft.

$$144{,}000 \text{ ft}^3 \div 108 \text{ ft} = 1333.333\ldots \text{ ft}^2$$

Round this quotient to three significant digits, which is the accuracy of the least accurate measurement (the accuracy of both measurements in this example). That is,

$$144{,}000 \text{ ft}^3 \div 108 \text{ ft} = 1330 \text{ ft}^2$$

EXAMPLE 7

Find the value of $\dfrac{(68 \text{ ft})(10{,}\bar{0}00 \text{ lb})}{95.6 \text{ s}}$.

$$\frac{(68 \text{ ft})(10{,}\bar{0}00 \text{ lb})}{95.6 \text{ s}} = 7112.9707\ldots \frac{\text{ft lb}}{\text{s}}$$

Round this result to two significant digits, which is the accuracy of the least accurate measurement (68 ft). That is,

$$\frac{(68 \text{ ft})(10{,}\bar{0}00 \text{ lb})}{95.6 \text{ s}} = 7100 \text{ ft lb/s}$$

EXAMPLE 8

Find the value of $\dfrac{(58.0\text{ kg})(2.40\text{ m/s})^2}{5.40\text{ m}}$.

$$\frac{(58.0\text{ kg})(2.40\text{ m/s})^2}{5.40\text{ m}} = 61.8666\ldots\frac{\text{kg m}}{\text{s}^2}$$

Carefully simplify the units:

$$\frac{(\text{kg})(\text{m/s})^2}{\text{m}} = \frac{(\text{kg})(\text{m}^2/\text{s}^2)}{\text{m}} = \frac{\text{kg m}}{\text{s}^2}$$

Round this result to three significant digits, which is the accuracy of the least accurate measurement (the accuracy of all measurements in this example). That is,

$$\frac{(58.0\text{ kg})(2.40\text{ m/s})^2}{5.40\text{ m}} = 61.9\text{ kg m/s}^2$$

Note: To multiply or divide measurements, the units do not need to be the same. The units must be the same to add or subtract measurements. Also, the units are multiplied and/or divided in the same manner as the corresponding numbers.

Any power or root of a measurement should be rounded to the same accuracy as the given measurement.

COMBINATIONS OF OPERATIONS WITH MEASUREMENTS

For combinations of additions, subtractions, multiplications, divisions, and powers involving measurements, follow the usual order of operations used in mathematics:

1. Perform all operations inside parentheses first.
2. Evaluate all powers.
3. Perform any multiplications or divisions, in order, from left to right; then express each product or quotient using its correct accuracy.
4. Perform any additions or subtractions, in order, from left to right; then express the final result using the correct precision.

EXAMPLE 9

Find the value of $(4.00\text{ m})(12.65\text{ m}) + (24.6\text{ m})^2 + \dfrac{235.0\text{ m}^3}{16.00\text{ m}}$.

$$(4.00\text{ m})(12.65\text{ m}) + (24.6\text{ m})^2 + \frac{235.0\text{ m}^3}{16.00\text{ m}} =$$

$$50.6\text{ m}^2 + 605\text{ m}^2 \quad + 14.69\text{ m}^2 = 67\bar{0}\text{ m}^2$$

Obviously, such calculations with measurements should be done with a calculator. When no calculator is available, you may round the original measurements or any intermediate results to one more digit than the required accuracy or precision as required in the final result.

If both exact numbers and approximate numbers (measurements) occur in the same calculation, only the approximate numbers are used to determine the accuracy or precision of the result.

The procedures for operations with measurements shown here are based on methods followed and presented by the American Society for Testing and Materials. There are even more sophisticated methods for dealing with the calculations of measurements. The method one uses, and indeed whether one should even follow any given procedure, depends on the number of measurements and the sophistication needed for a particular situation.

In this book, we generally follow the customary practice of expressing measurements in terms of three significant digits, which is the accuracy used in most engineering and design work.

PROBLEMS 1.8

Use the rules for addition of measurements to add each set of measurements.

1.	2.	3.	4.
3847 ft	8,560 m	42.8 cm	0.456 g
5800 ft	84,000 m	16.48 cm	0.93 g
4520 ft	18,476 m	1.497 cm	0.402 g
	12,500 m	12.8 cm	0.079 g
		9.69 cm	0.964 g

5. 39,000 N; 19,600 N; 8470 N; 2500 N
6. 6800 ft; 2760 ft; $4\bar{0}00$ ft; $20\bar{0}0$ ft
7. 467 m; 970 cm; $12\bar{0}0$ cm; 1352 cm; $30\bar{0}$ m
8. 36.8 m; 147.5 cm; 1.967 m; 125.0 m; 98.3 cm
9. 12 s; 1.004 s; 0.040 s; 3.9 s; 0.87 s
10. 160,000 N; 84,200 N; 4300 N; 239,000 N; 17,450 N

Use the rules for subtraction of measurements to subtract each second measurement from the first.

11.	12.	13.	14.
2876 kg	14.73 m	45.585 g	34,500 kg
2400 kg	9.378 m	4.6 g	$9{,}5\bar{0}0$ kg

15. 4200 km – 975 km
16. 64.73 g – 9.4936 g
17. 1,600,000 kg – 685,000 kg
18. 170 mm – 10.2 cm
19. 3.00 m – $26\bar{0}$ cm
20. 1.40 ms – 0.708 ms

Use the rules for multiplication of measurements to multiply each set of measurements.

21. (125 m)(39 m)
22. (470 ft)(1200 ft)
23. (1637 km)(857 km)
24. (9100 m)($6\bar{0}0$ m)
25. (18.70 m)(39.45 m)
26. (565 cm)(180 cm)
27. (14.5 cm)(18.7 cm)(20.5 cm)
28. (0.046 m)(0.0317 m)(0.0437 m)
29. ($45\bar{0}$ in.)(315 in.)(205 in.)
30. (18.7 kg)(217 m)

Use the rules for division of measurements to divide.

31. $360 \text{ ft}^3 \div 12 \text{ ft}^2$
32. $125 \text{ m}^2 \div 3.0 \text{ m}$
33. $275 \text{ cm}^2 \div 90.0 \text{ cm}$
34. $185 \text{ mi} \div 4.5 \text{ h}$
35. $\dfrac{347 \text{ km}}{4.6 \text{ h}}$
36. $\dfrac{2700 \text{ m}^3}{9\bar{0}0 \text{ m}^2}$
37. $\dfrac{8800 \text{ mi}}{8.5 \text{ h}}$
38. $\dfrac{4960 \text{ ft}}{2.95 \text{ s}}$

Use the rules for multiplication and division of measurements to find the value of each of the following.

39. $\dfrac{(18 \text{ ft})(290 \text{ lb})}{4.6 \text{ s}}$
40. $\dfrac{(18.5 \text{ kg})(4.65 \text{ m})}{19.5 \text{ s}}$
41. $\dfrac{4500 \text{ mi}}{12.3 \text{ h}}$
42. $\dfrac{48.9 \text{ kg}}{(1.5 \text{ m})(3.25 \text{ m})}$
43. $\dfrac{(48.7 \text{ m})(68.5 \text{ m})(18.4 \text{ m})}{(35.5 \text{ m})(40.0 \text{ m})}$
44. $\frac{1}{2}(270 \text{ kg})(16.4 \text{ m/s})^2$

45. $\dfrac{(85.7\text{ kg})(25.7\text{ m/s})^2}{12.5\text{ m}}$

46. $\dfrac{(45.2\text{ kg})(13.7\text{ m})}{(2.65\text{ s})^2}$

47. $\frac{4}{3}\pi(13.5\text{ m})^3$

48. $\dfrac{140\text{ g}}{(3.4\text{ cm})(2.8\text{ cm})(5.6\text{ cm})}$

49. $(213\text{ m})(65.3\text{ m}) - (175\text{ m})(44.5\text{ m})$

50. $(4.5\text{ ft})(7.2\text{ ft})(12.4\text{ ft}) + (5.42\text{ ft})^3$

51. $\dfrac{(125\text{ ft})(295\text{ ft})}{44.7\text{ ft}} + \dfrac{(215\text{ ft})^3}{(68.8\text{ ft})(12.4\text{ ft})} + \dfrac{(454\text{ ft})^3}{(75.5\text{ ft})^2}$

52. $(12.5\text{ m})(46.75\text{ m}) + \dfrac{(6.76\text{ m})^3}{4910\text{ m}} - \dfrac{(41.5\text{ m})(21\text{ m})(28.8\text{ m})}{31.7\text{ m}}$

1.9 Problem-Solving Method

Problem solving in technical fields is more than substituting numbers and units into formulas. You must develop skill in taking data, analyzing the problem, and finding the solution in an orderly manner. Understanding the principle involved in solving a problem is more important than blindly substituting into a formula. By following an orderly procedure for problem solving, we develop an approach to problem solving that you can use in your studies and on the job.

The following **problem-solving method** aids in understanding and solving problems and will be applied to all problems in this book where appropriate.

1. ***Read the problem carefully.*** This might appear obvious, but it is the most important step in solving a problem. As a matter of habit, you should read the problem at least twice.
   (a) The first time you should read the problem straight through from beginning to end. Do not stop to think about setting up an equation or formula. You are only trying to get a general overview of the problem during this first reading.
   (b) Read through a second time slowly and *completely*, beginning to think ahead to the following steps.
2. ***Make a sketch.*** Some problems may not lend themselves to a sketch. However, make a sketch whenever possible. Many times, seeing a sketch will show if you have forgotten important parts of the problem and may suggest the solution. This is a *very important* part of problem solving and is often overlooked.
3. ***Write all given information including units.*** This is necessary to have all essential facts in mind before looking for the solution. There are some common phrases that have understood physical meanings. For example, the term *from rest* means the initial velocity equals zero or $v_i = 0$; the term *smooth surface* means assume that no friction is present.
4. ***Write the unknown or quantity asked for in the problem.*** Many students have difficulty solving problems because they don't know what they are looking for and solve for the wrong quantity.
5. ***Write the basic equation or formula that relates the known and unknown quantities.*** Find the basic formula or equation to use by studying what is given and what you are asked to find. Then look for a formula or equation that relates these quantities. Sometimes, you may need to use more than one equation or formula in a problem.
6. ***Find a working equation by solving the basic equation or formula for the unknown quantity.***
7. ***Substitute the data in the working equation, including the appropriate units.*** It is important that you *carry the units all the way through the problem* as a check that you have solved the problem correctly. For example, if you are asked to find the weight of an object in newtons and the units of your answer work out to be metres, you need to review your solution for the error. (When the unit analysis is not obvious, we will go through it step by step in a box nearby.)

Figure 1.26

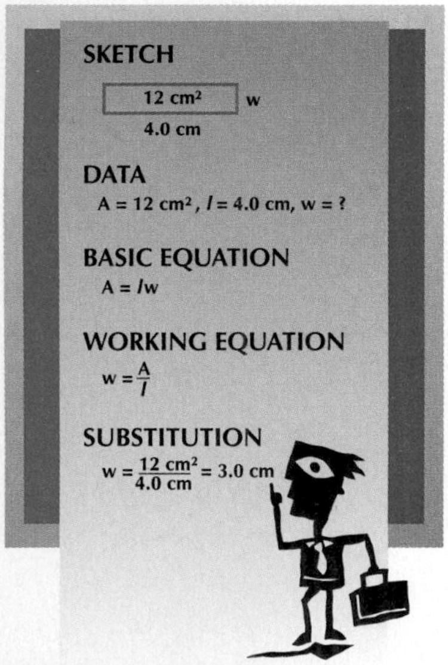

8. ***Perform the indicated operations and work out the solution.*** Although this will be your final written step, you should always ask yourself, "Is my answer reasonable?" Here and on the job you will be dealing with practical problems. A quick estimate will often reveal an error in your calculations.
9. ***Check your answer.*** Ask yourself, "Did I answer the questions?"

To help you recall this procedure, with almost every problem set that follows, you will find Fig. 1.26 as shown here. This figure is not meant to be complete, and is only an outline to assist you in remembering and following the procedure for solving problems. *You should follow this outline in solving all problems in this course.*

This problem-solving method will now be demonstrated in terms of relationships and formulas with which you are probably familiar. The formulas for finding area and volume can be found on the inside back cover.

You may find it helpful to review Appendix A.6 Formulas and A.7 Substituting Data into Formulas now.

EXAMPLE 1

Find the volume of concrete required to fill a rectangular bridge abutment whose dimensions are 6.00 m × 3.00 m × 15.0 m.

Sketch:

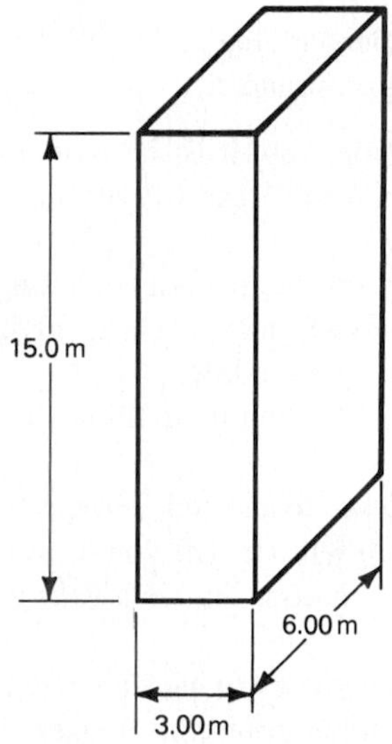

Data:

$l = 6.00$ m
$w = 3.00$ m
$h = 15.0$ m
} This is a listing of the information that is known.

$V = ?$ This identifies the unknown.

Basic Equation:

$$V = lwh$$

Working Equation: Same

Substitution:

$$V = (6.00 \text{ m})(3.00 \text{ m})(15.0 \text{ m})$$
$$= 27\bar{0} \text{ m}^3$$

Note: $\text{m} \times \text{m} \times \text{m} = \text{m}^3$

EXAMPLE 2

A rectangular holding tank 24.0 m in length and 15.0 m in width is used to store water for short periods of time in an industrial plant. If 2880 m^3 of water is pumped into the tank, what is the depth of the water?

Sketch:

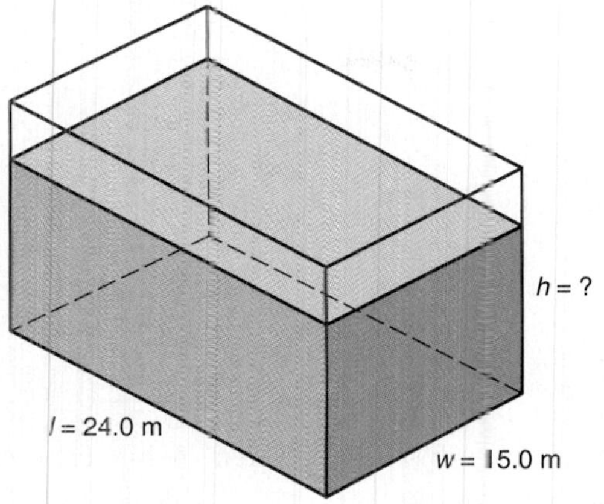

Data:

$$V = 2880 \text{ m}^3$$
$$l = 24.0 \text{ m}$$
$$w = 15.0 \text{ m}$$
$$h = ?$$

Basic Equation:

$$V = lwh$$

Working Equation:

$$h = \frac{V}{lw}$$

Substitution:

$$h = \frac{2880 \text{ m}^3}{(24.0 \text{ m})(15.0 \text{ m})}$$
$$= 8.00 \text{ m}$$

$$\frac{\text{m}^3}{\text{m} \times \text{m}} = \text{m}$$

EXAMPLE 3

A storage bin in the shape of a cylinder contains 814 m^3 of storage space. If its radius is 6.00 m, find its height.

Sketch:

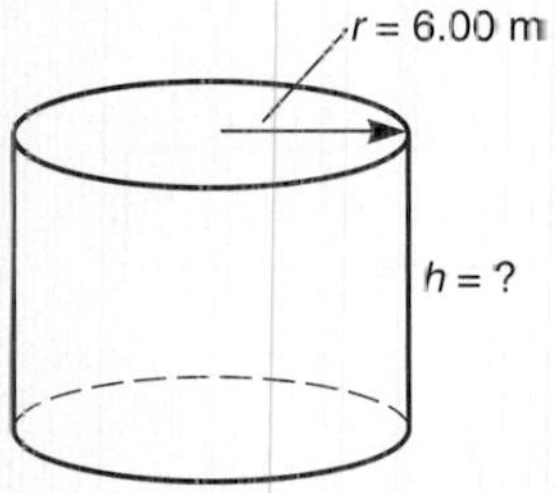

Data:

$$V = 814 \text{ m}^3$$
$$r = 6.00 \text{ m}$$
$$h = ?$$

Basic Equation:

$$V = \pi r^2 h$$

Working Equation:

$$h = \frac{V}{\pi r^2}$$

Substitution:

$$h = \frac{814 \text{ m}^3}{\pi(6.00 \text{ m})^2}$$

$$= 7.20 \text{ m}$$

$$\boxed{\frac{\text{m}^3}{\text{m}^2} = \text{m}}$$

EXAMPLE 4

A rectangular piece of sheet metal measures 45.0 cm by 75.0 cm. A 10.0-cm square is then cut from each corner. The metal is then folded to form a box-like container without a top. Find the volume of the container.

Sketch:

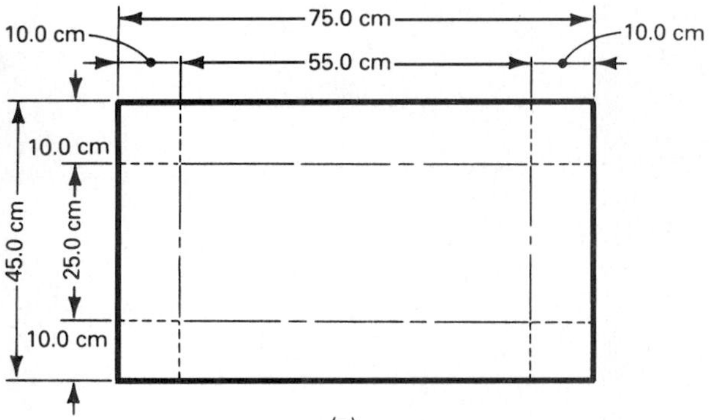

(a)

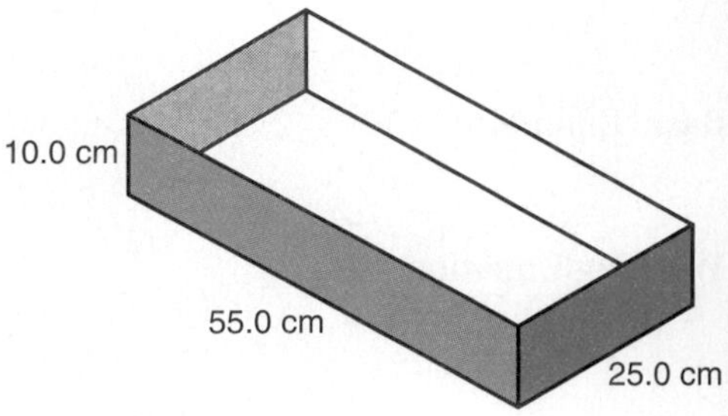

(b)

Data:

$$l = 55.0 \text{ cm}$$
$$w = 25.0 \text{ cm}$$
$$h = 10.0 \text{ cm}$$
$$V = ?$$

Basic Equation:

$$V = lwh$$

Working Equation: Same

Substitution:

$$V = (55.0 \text{ cm})(25.0 \text{ cm})(10.0 \text{ cm})$$
$$= 13{,}800 \text{ cm}^3$$

$$\boxed{\text{cm} \times \text{cm} \times \text{cm} = \text{cm}^3}$$

EXAMPLE 5

The cross-sectional area of a hole is 725 cm^2. Find its radius.

Sketch:

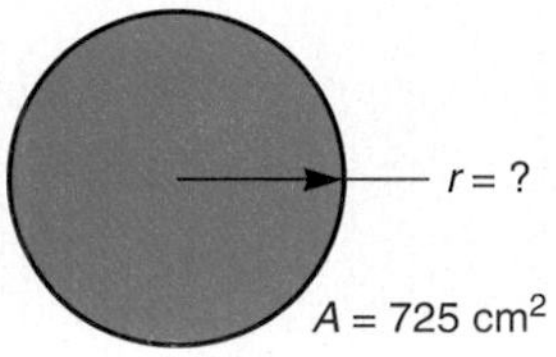

Data:

$$A = 725 \text{ cm}^2$$
$$r = ?$$

Basic Equation:

$$A = \pi r^2$$

Working Equation:

$$r = \sqrt{\frac{A}{\pi}}$$

Substitution:

$$r = \sqrt{\frac{725 \text{ cm}^2}{\pi}}$$
$$= 15.2 \text{ cm}$$

$\sqrt{\text{cm}^2} = \text{cm}$

PHYSICS CONNECTIONS

Eratosthenes, a third-century Egyptian, used a problem-solving method to determine that the earth was not flat, but round, a fact that Columbus has been credited with discovering more than 1000 years later. Eratosthenes wondered why it was that at noon of the summer solstice, towers in Syene, Egypt, made no shadows, whereas documentation showed that towers in Alexandria, Egypt, did make distinct shadows. Eratosthenes decided to determine why towers in one city would cast shadows while towers in another city would not.

Eratosthenes sketched the problem, gathered data, and collected geometrical equations to solve this complex problem. He hired a person to pace the distance between the two cities (800 km) and used geometry to solve the problem (Fig. 1.27). After calculating the difference in the positions of the two cities to be approximately 7° out of a 360° sphere, he concluded that the earth's circumference was 40,000 km—remarkably accurate when compared to today's calculations. Eratosthenes had a problem to solve, so he, like all good scientists and problem solvers, followed several steps that included analyzing the problem, collecting data, selecting appropriate equations, and making the calculations.

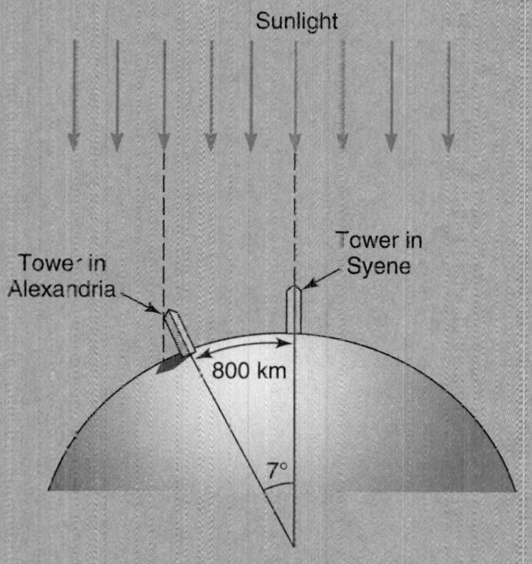

Figure 1.27 Towers in Alexandria and Syene, the shadow cast by the tower in Alexandria, and the curvature of the earth with the angle and the distance between the two cities

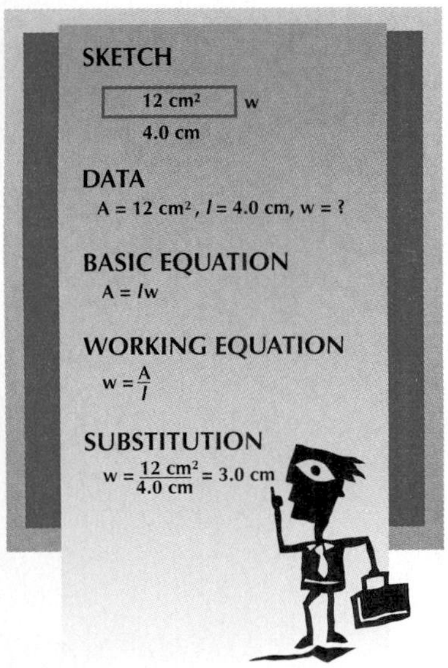

PROBLEMS 1.9

Use the problem-solving method to work each problem. (Here, as throughout the text, follow the rules for calculations with measurements.)

1. Find the volume of the box in Fig. 1.28.
2. Find the volume of a cylinder whose height is 7.50 in. and diameter is 4.20 in. (Fig. 1.29).
3. Find the volume of a cone whose height is 9.30 cm if the radius of the base is 5.40 cm (Fig. 1.30).

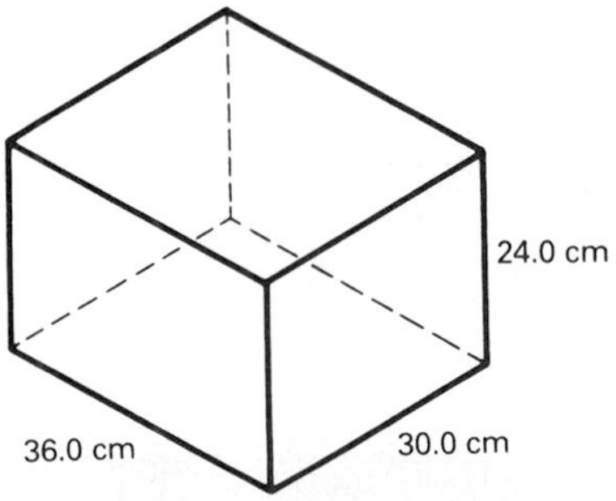

Figure 1.28

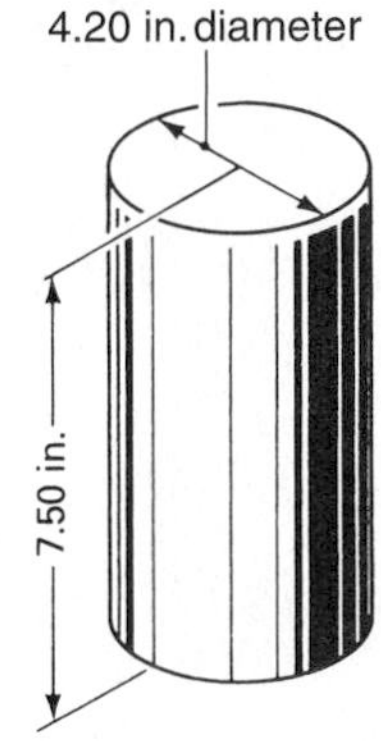

Figure 1.29

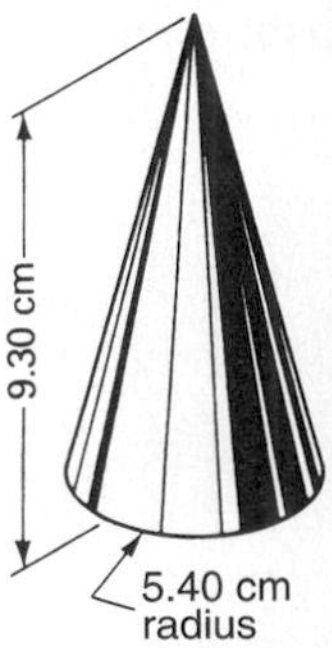

Figure 1.30

The cylinder in an engine of a road grader as shown in Fig. 1.31 is 11.40 cm in diameter and 24.00 cm high. Use Fig. 1.31 for Problems 4 through 6.

4. Find the volume of the cylinder.
5. Find the cross-sectional area of the cylinder.
6. Find the lateral surface area of the cylinder.

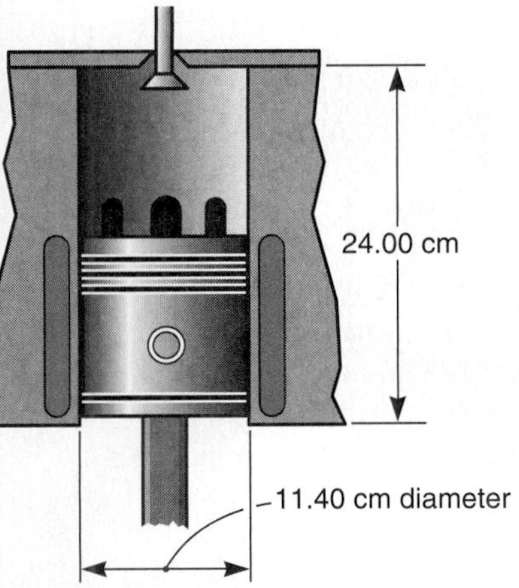

Figure 1.31

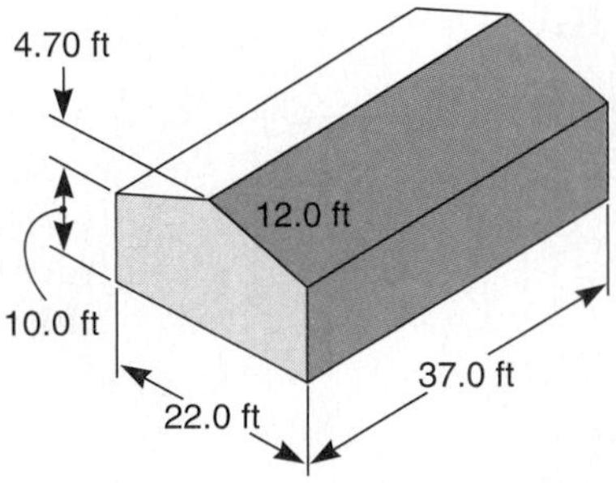

Figure 1.32

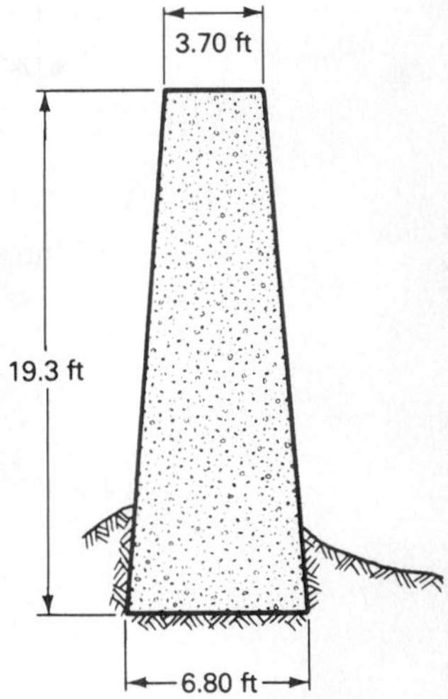

Figure 1.33

7. Find the total volume of the building shown in Fig. 1.32.
8. Find the cross-sectional area of the concrete retaining wall shown in Fig. 1.33.
9. Find the volume of a rectangular storage facility 9.00 ft by 12.0 ft by 8.00 ft.
10. Find the cross-sectional area of a piston head with a diameter of 3.25 cm.
11. Find the area of a right triangle that has legs of 4.00 cm and 6.00 cm.
12. Find the length of the hypotenuse of the right triangle in Problem 11.
13. Find the cross-sectional area of a pipe with outer diameter 3.50 cm and inner diameter 3.20 cm.
14. Find the volume of a spherical water tank with radius 8.00 m.

15. The area of a rectangular parking lot is $90\bar{0}$ m^2. If the length is 25.0 m, what is the width?
16. The volume of a rectangular crate is 192 ft^3. If the length is 8.00 ft and the width is 4.00 ft, what is the height?
17. A cylindrical silo has a circumference of 29.5 m. Find its diameter.
18. If the silo in Problem 17 has a capacity of $100\bar{0}$ m^3, what is its height?
19. A wheel 30.0 cm in diameter moving along level ground made 145 complete rotations. How many metres did the wheel travel?
20. The side of the silo in Problems 17 and 18 needs to be painted. If each litre of paint covers 5.0 m^2, how many litres of paint will be needed? (Round up to the nearest litre.)
21. You are asked to design a cylindrical water tank that holds $50\bar{0},000$ gal with radius 18.0 ft. Find its height. (1 ft^3 = 7.50 gal)
22. If the height of the water tank in Problem 21 were 42.0 ft, what would be its radius?
23. A ceiling is 12.0 ft by 15.0 ft. How many suspension panels 1.00 ft by 3.00 ft are needed to cover the ceiling?
24. Find the cross-sectional area of the dovetail slide shown in Fig. 1.34.

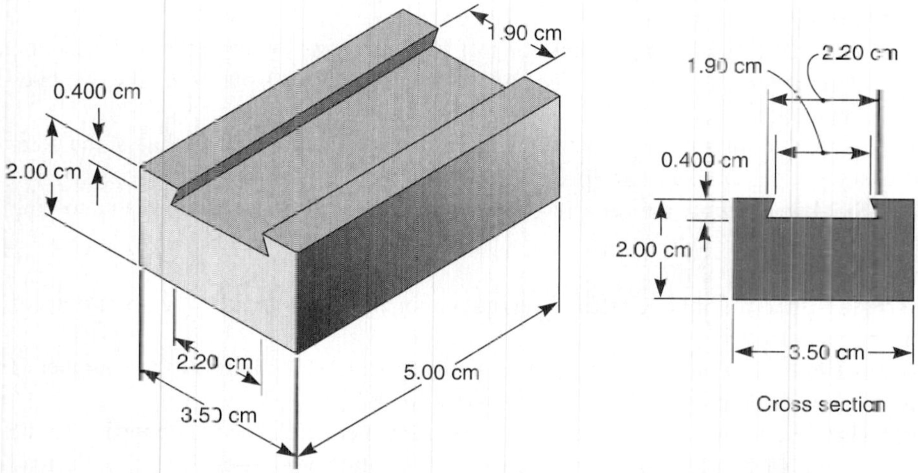

Figure 1.34

25. Find the volume of the storage bin shown in Fig. 1.35.
26. The maximum cross-sectional area of a spherical propane storage tank is 3.05 m^2. Will it fit into a 2.00-m-wide trailer?
27. How many cubic yards of concrete are needed to pour a patio 12.0 ft × 20.0 ft and 6.00 in. thick?
28. What length of sidewalk 4.00 in. thick and 4.00 ft wide could be poured with 2.00 yd^3 of concrete?

Figure 1.35

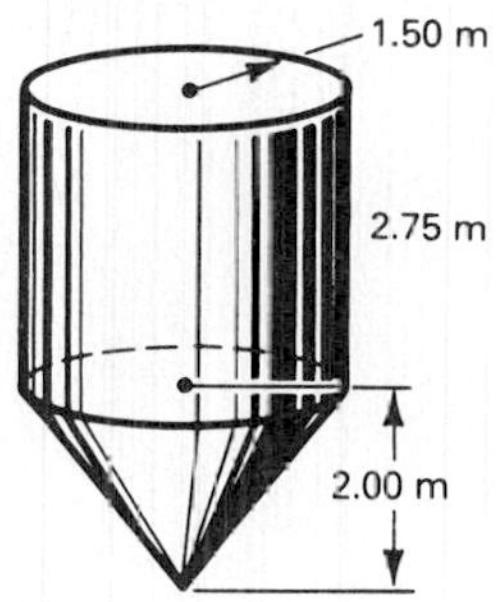

Find the volume of each figure.

29.

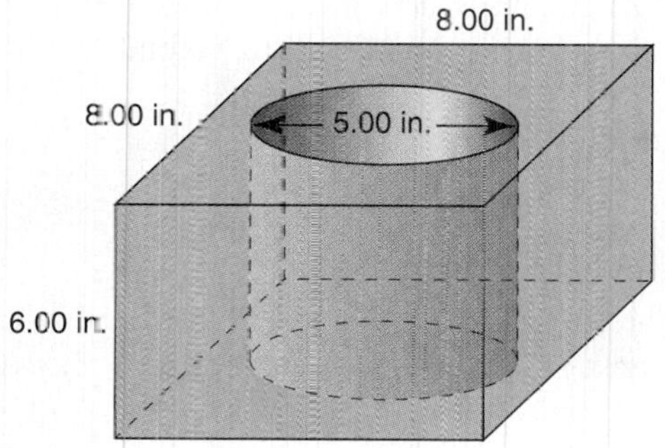

30.

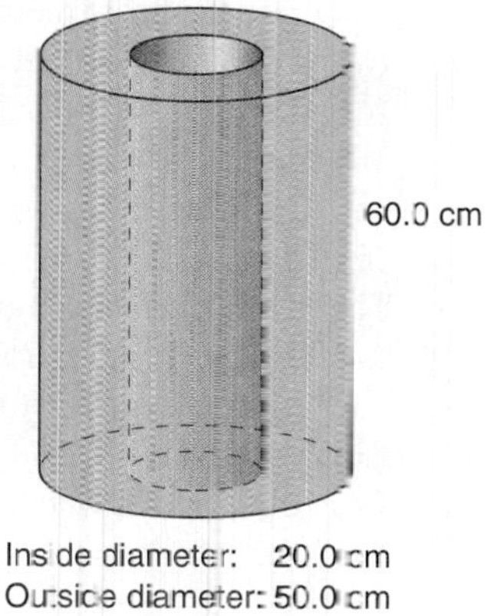

Glossary

Accuracy The number of digits, called significant digits, in a measurement, which indicates the number of units that we are reasonably sure of having counted. The greater the number of significant digits, the better is the accuracy. (p. 36)

Approximate Number A number that has been determined by some measurement or estimation process. (p. 36)

Area The number of square units contained in a figure. (p. 22)

Conversion Factor An expression used to convert from one set of units to another. Often expressed as a fraction whose numerator and denominator are equal to each other although in different units. (p. 18)

Exact Number A number that has been determined as a result of counting, such as 21 students enrolled in a class, or by some definition, such as 1 h = 60 min. (p. 36)

Kilogram The basic metric unit of mass. (p. 32)

Lateral Surface Area The area of all the lateral (side) faces of a geometric solid. (p. 28)

Mass A measure of the quantity of material making up an object. (p. 32)

Metre The basic metric unit of length. (p. 18)

Platform Balance An instrument consisting of two platforms connected by a horizontal rod that balances on a knife edge. The pull of gravity on objects placed on the two platforms is compared. (p. 33)

Precision Refers to the smallest unit with which a measurement is made, that is, the position of the last significant digit. (p. 38)

Problem-Solving Method An orderly procedure that aids in understanding and solving problems. (p. 45)

Second The basic unit of time. (p. 34)

SI (Système International d'Unités) The international modern metric system of units of measurement. (p. 16)

Significant Digits The number of digits in a measurement, which indicates the number of units we are reasonably sure of having counted. (p. 36)

Spring Balance An instrument containing a spring, which stretches in proportion to the force applied to it, and a pointer attached to the spring with a calibrated scale read directly in given units. (p. 32)

Standards of Measure A set of units of measurement for length, weight, and other quantities defined in such a way as to be useful to a large number of people. (p. 14)

Total Surface Area The total area of all the surfaces of a geometric solid; that is, the lateral surface area plus the area of the bases. (p. 28)

Volume The number of cubic units contained in a figure. (p. 25)

Weight A measure of the gravitational force or pull acting on an object. (p. 32)

Review Questions

1. What are the basic metric units for length, mass, and time?
   (a) Foot, pound, hour (b) Newton, litre, second
   (c) Metre, kilogram, second (d) Mile, ton, day
2. When a value is multiplied or divided by 1, the value is
   (a) increased. (b) unchanged.
   (c) decreased. (d) none of the above.
3. The lateral surface area of a solid is
   (a) always equal to total surface area. (b) never equal to total surface area.
   (c) usually equal to total surface area. (d) rarely equal to total surface area.
4. Accuracy is
   (a) the same as precision.
   (b) the smallest unit with which a measurement is made.

(c) the number of significant digits.
(d) all of the above.

5. When multiplying or dividing two or more measurements, the units
(a) must be the same. (b) must be different. (c) can be different.
6. Cite three examples of problems that would arise in the construction of a home by workers using different systems of measurement.
7. Why is the metric system preferred worldwide to the U.S. system of measurement?
8. List a very large and a very small measurement that could be usefully written in scientific notation.
9. When using conversion factors, can units be treated like other algebraic quantities?
10. What is the meaning of cross-sectional area?
11. Can a brick have more than one cross-sectional area?
12. What is the fundamental metric unit for land area?
13. Which is larger, a litre or a quart?
14. List three things that might conveniently be measured in millilitres.
15. How do weight and mass differ?
16. What is the basic metric unit of weight?
17. A microsecond is one-_____ of a second.
18. Why must we concern ourselves with significant digits?
19. Can the sum or difference of two measurements ever be more precise than the least precise measurement?
20. When rounding the product or quotient of two measurements, is it necessary to consider significant digits?
21. Why is reading the problem carefully the most important step in problem solving?
22. How can making a sketch help in problem solving?
23. What do we call the relationship between data that are given and what we are asked to find?
24. How is a working equation different from a basic equation?
25. How can analysis of the units in a problem assist in solving the problem?
26. How can making an estimate of your answer assist in the correct solution of problems?

Review Problems

Give the metric prefix for each value:

1. 1000
2. 0.001

Give the metric symbol, or abbreviation, for each prefix:

3. micro
4. mega

Write the abbreviation for each quantity:

5. 45 milligrams
6. 138 centimetres

Which is larger?

7. 1 L or 1 mL
8. 1 kg or 1 mg
9. 1 L or 1 m^3

Fill in each blank (round to three significant digits when necessary):

10. 250 m = _____ km
11. 850 mL = _____ L
12. 5.4 kg = _____ g
13. 0.55 s = _____ μs
14. 25 kg = _____ g
15. 75 μs = _____ ns
16. 275 cm^2 = _____ mm^2
17. 350 cm^2 = _____ m^2
18. 0.15 m^3 = _____ cm^3
19. 500 cm^3 = _____ mL
20. 150 lb = _____ kg
21. 36 ft = _____ m
22. 250 cm = _____ in.
23. 150 in^2 = _____ cm^2
24. 24 yd^2 = _____ ft^2
25. 6 m^3 = _____ ft^3
26. 16 lb = _____ N
27. 15,600 s = _____ h _____ min

Determine the accuracy (the number of significant digits) in each measurement:

28. 5.08 kg   29. 20,570 lb   30. 0.060 cm   31. 2.00×10^{-4} s

Determine the precision of each measurement:

32. 30.6 ft   33. 0.0500 s   34. 18,000 mi   35. 4×10^5 N

For each set of measurements, find the measurement that is

(a) the most accurate.   (b) the least accurate.
(c) the most precise.   (d) the least precise.

36. 12.00 m; 0.150 m; 2600 m; 0.008 m
37. 208 L; 18,050 L; 21.5 L; 0.75 L

Use the rules of measurements to add the following measurements:

38. 0.0250 s; 0.075 s; 0.00080 s; 0.024 s
39. 2100 N; 36,800 N; 24,000 N; 14.5 N; 470 N

Use the rules for multiplication and division of measurements to find the value of each of the following:

40. (450 cm)(18.5 cm)(215 cm)   41. $\dfrac{1480\ \text{m}^3}{9.6\ \text{m}}$

42. $\dfrac{(25.0\ \text{kg})(1.20\ \text{m/s})^2}{3.70\ \text{m}}$

43. Find the area of a rectangle 4.50 m long and 2.20 m wide.
44. Find the volume of a rectangular box 9.0 cm long, 6.0 cm wide, and 13 cm high.
45. A cone has a volume of 314 cm^3 and radius of 5.00 cm. What is its height?
46. A right triangle has a side of 41.2 mm and a side of 9.80 mm. Find the length of the hypotenuse.
47. Given a cylinder with a radius of 7.20 cm and a height of 13.4 cm, find the lateral surface area.
48. A rectangle has a perimeter of 40.0 cm. One side has a length of 14.0 cm. What is the length of an adjacent side?
49. The formula for the volume of a cylinder is $V = \pi r^2 h$. If $V = 21\bar{0}0\ \text{m}^3$ and $h = 17.0$ m, find r.
50. The formula for the area of a triangle is $A = \frac{1}{2}bh$. If $b = 12.3$ m and $A = 88.6\ \text{m}^2$, find h.
51. Find the volume of the lead sleeve with the cored hole in Fig. 1.36.
52. A rectangular plot of land measures 40.0 m by $12\bar{0}$ m with a parcel 10.0 m by 12.0 m out of one corner for an electrical transformer. What is the area of the remaining plot?

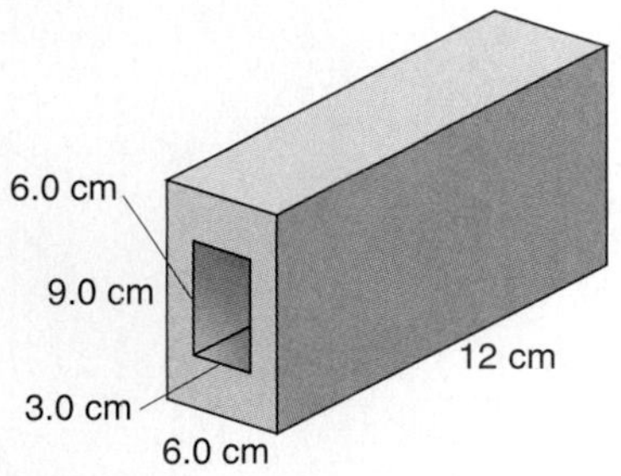

Figure 1.36

APPLIED CONCEPTS

1. You run a landscaping business and know that you want to charge \$50.00 to mow a person's lawn whose property is 10$\overline{0}$ ft × 20$\overline{0}$ ft. If the house dimensions take up a 35.0 ft × 80.0 ft area, how much are you charging per square yard? (Use the problem-solving method outlined in Section 1.9 to solve the problem.)
2. A room that measures 10.0 ft wide, 32.0 ft long, and 8.00 ft high needs a certain amount of air pumped into it per minute to keep the air quality up to regulations. If the room needs completely new air every 20.0 minutes, what is the volume of air per second that is being pumped into the room? (Use the problem-solving method outlined in Section 1.9 to solve the problem.)
3. Instead of using a solid iron beam, structural engineers and contractors use I-beams to save materials and money. How many I-beams can be molded from the same amount of iron contained in the solid iron beam as shown in Fig. 1.37? (Use the problem-solving method outlined in Section 1.9 to solve the problem.)

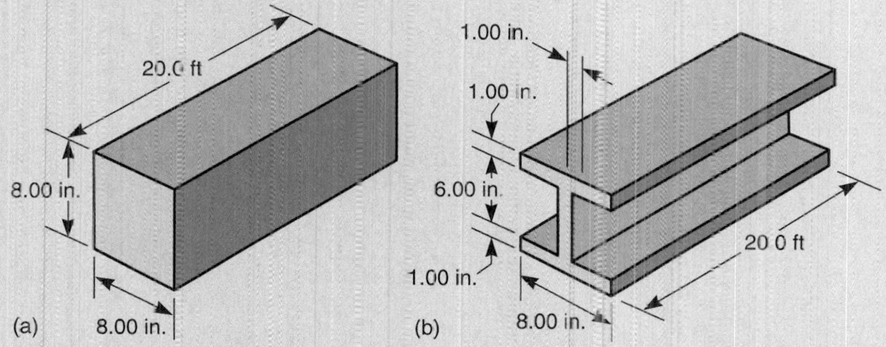

Figure 1.37

4. A shipping specialist at a craft store needs to pack Styrofoam balls of radius 4.00 in. into a 1.40 ft × 2.80 ft × 1.40 ft rectangular cardboard container. What is the maximum number of balls that can fit in the container? Hint: Spherical balls have spaces around them when packed in rectangular containers.
5. A crane needs to lift a spool of fine steel cable to the top of a bridge deck. The type of steel in the cable has a density of 7750 kg/m^3. The maximum lifting mass of the crane is 43,400 kg. (a) Given the dimensions of the spool in Fig. 1.38, find the volume of the spool. (b) Can the crane safely lift the spool?

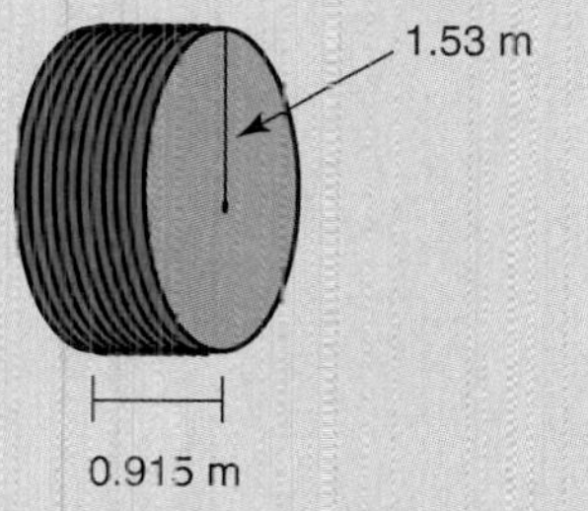

Figure 1.38

CHAPTER 2

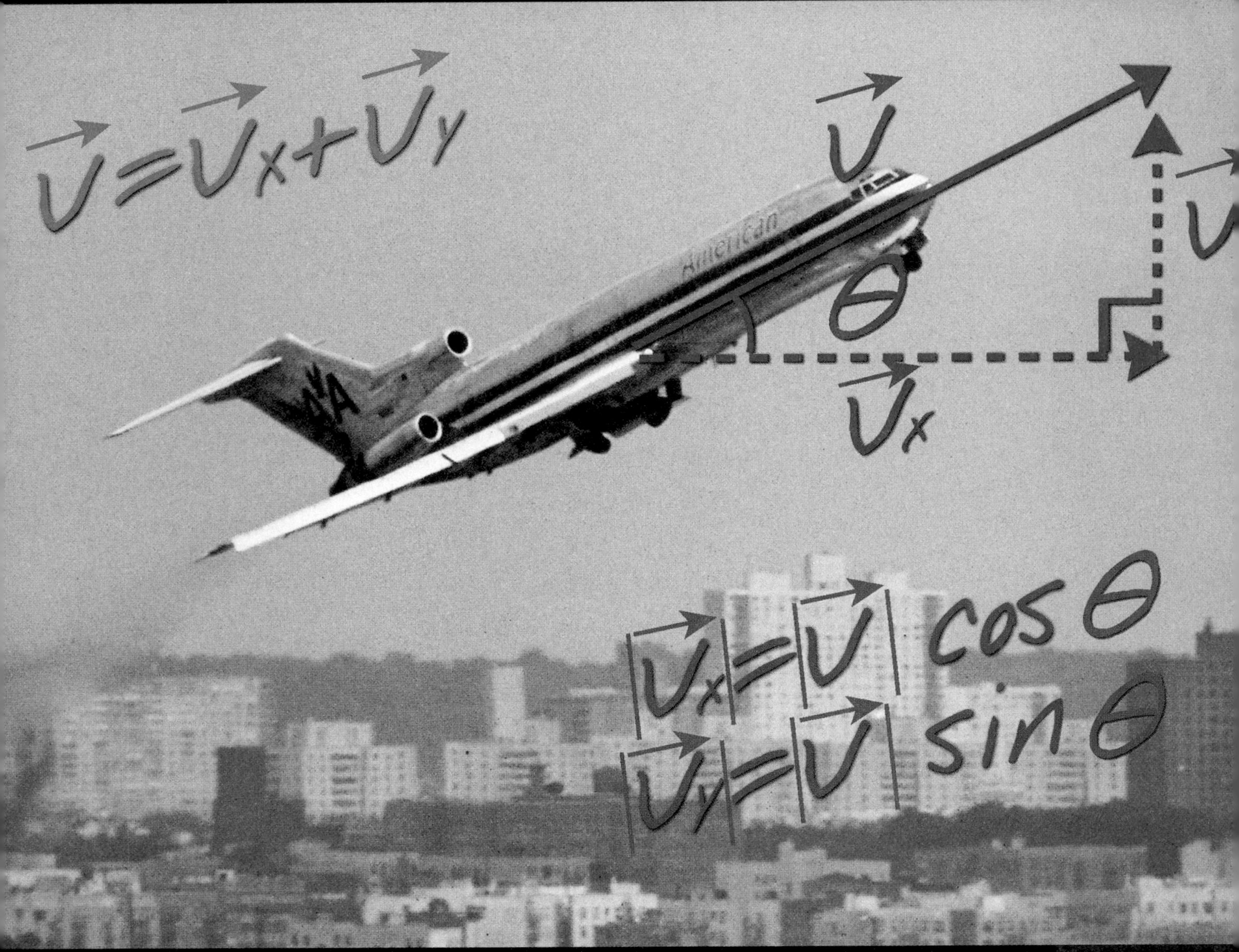

$\vec{V} = \vec{V}_x + \vec{V}_y$
$\vec{V}$
θ
$\vec{V}_x$
$|\vec{V}_x| = |\vec{V}| \cos\theta$
$|\vec{V}_y| = |\vec{V}| \sin\theta$

VECTORS

Some physical quantities, called *scalars,* may be described by and involve calculations with numerical quantities alone. Other physical quantities, called *vectors,* require both a numerical quantity and a direction to be completely described and often involve calculations using trigonometry. Vectors are developed in this chapter prior to their use in the following chapters.

Objectives

The major goals of this chapter are to enable you to:

1. Distinguish between a vector and a scalar quantity.
2. Add vectors graphically.
3. Find the components of a vector.
4. Work with vectors in standard position.
5. Apply the basic concepts of right-triangle trigonometry using displacement vectors.

2.1 Vectors and Scalars*

Every physical quantity can be classified as either a scalar or a vector quantity. A **scalar** is a quantity that can be completely described by a number (called its magnitude) and a unit. Examples of scalars are length, temperature, and volume. All these quantities can be expressed by a number with the appropriate units. For example, the length of a steel beam can be expressed as 18 ft; the temperature at 11:00 A.M. is 15°C; and the volume of a room is 300 m^3.

Figure 2.1 Displacement

A **vector** is a quantity that requires both *magnitude* (size) and *direction* to be completely described. Examples of vectors are force, displacement, and velocity. To completely describe a force, you must give not only its magnitude (size or amount), but also its direction.

To describe the change of position of an object, such as an airplane flying from one city to another, we use the term *displacement.* **Displacement** is the net change in position of an object, or the direct distance and direction it moves. For example, to completely describe the flight of a plane between two cities requires both the *distance* between them and the *direction from* the first city *to* the second (Fig. 2.1).

Suppose that a friend asks you how to reach your home from school. If you replied that he should walk four blocks, you would not have given him enough information [Fig. 2.2(a)]. Obviously, you would need to tell him which direction to go. If you had replied, "Four blocks north," your friend could then find your home [Fig. 2.2(b)].

Figure 2.2 Displacement involves both a distance and a direction.

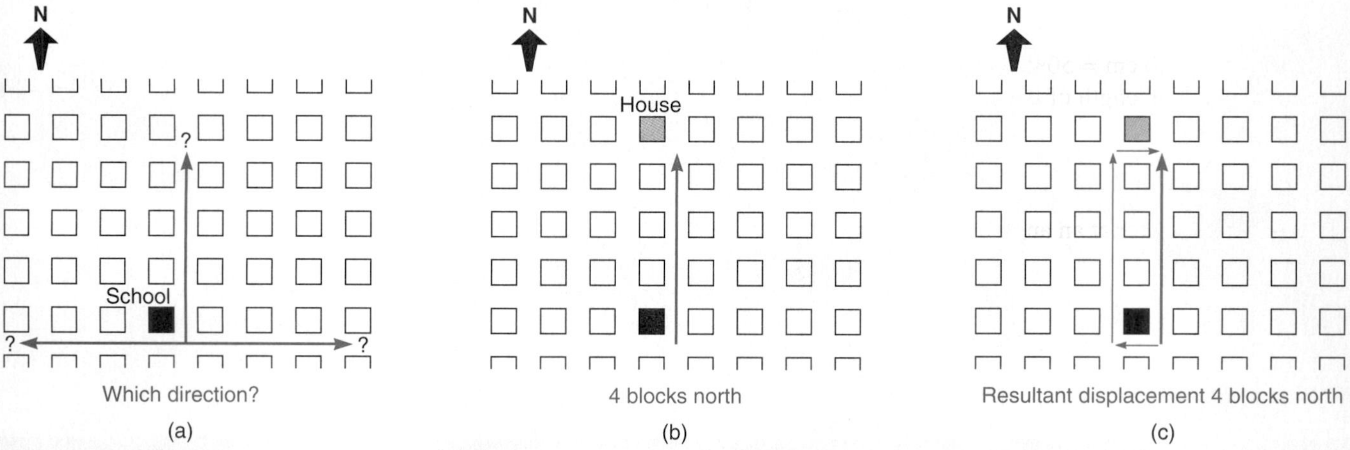

Figure 2.3

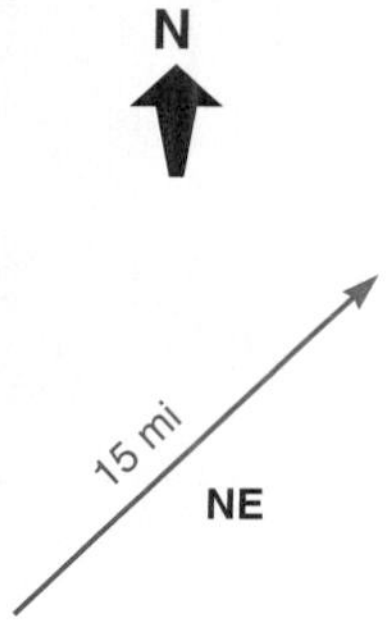

Displacement involves all the necessary information about a change in position; that is, it includes both *distance* and *direction.* It does not contain any information about the path that has been followed. *The units of displacement are length units,* such as metres, feet, or miles. If your friend decides to walk one block west, four blocks north, and then one block east, he will still arrive at your house. This resultant displacement is the same as if he had walked four blocks north [Fig. 2.2(c)].

The magnitude of the displacement vector "15 miles NE" is 15 miles (Fig. 2.3). *Thus, a vector has both magnitude and direction.*

To represent a vector in our diagrams, we draw an arrow that points in the correct direction. The magnitude of the vector is indicated by the length of the arrow. We usually

*Sections in this chapter may be omitted for those who have mastered these skills; others may need to review them. Begin with Section A.8 in the appendix if you need to study or review right-triangle trigonometry.

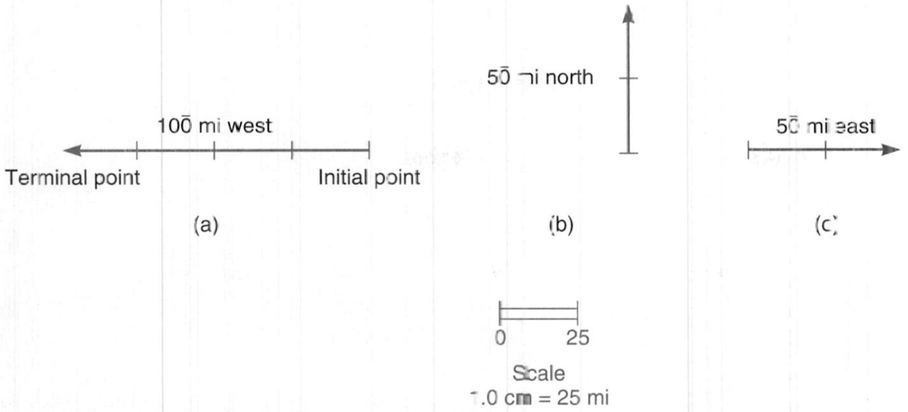

Figure 2.4 Use a scale to draw the proper length of a given vector.

choose a scale, such as 1.0 cm = 25 mi, for this purpose (Fig. 2.4). Thus, a displacement of $10\bar{0}$ mi west is drawn as an arrow (pointing west) 4.0 cm long [Fig. 2.4(a)] since

$$10\bar{0}\ \text{mi} \times \frac{1.0\ \text{cm}}{25\ \text{mi}} = 4.0\ \text{cm}$$

One end of the vector is called the initial point and the other is called the terminal point, as shown in Fig. 2.4(a). Displacements of $5\bar{0}$ mi north [Fig. 2.4(b)] and $5\bar{0}$ mi east [Fig. 2.4(c)] using the same scale are also shown.

EXAMPLE 1

Using the scale 1.0 cm = $5\bar{0}$ km, draw the displacement vector 275 km at 45° north of west.

First, find the length of the vector.

$$275\ \text{km} \times \frac{1.0\ \text{cm}}{5\bar{0}\ \text{km}} = 5.5\ \text{cm}$$

Then draw the vector at an angle 45° north of west (Fig. 2.5).

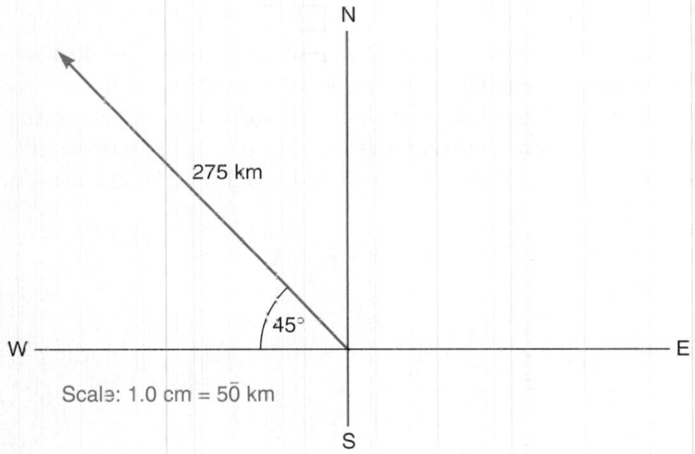

Figure 2.5

EXAMPLE 2

Using the scale $\frac{1}{4}$ in. = $2\bar{0}$ mi, draw the displacement vector $15\bar{0}$ mi at 22° east of south.

First, find the length of the vector.

$$15\bar{0}\ \text{mi} \times \frac{\frac{1}{4}\ \text{in.}}{2\bar{0}\ \text{mi}} = 1\tfrac{7}{8}\ \text{in.}$$

Then draw the vector at 22° east of south (Fig. 2.6).

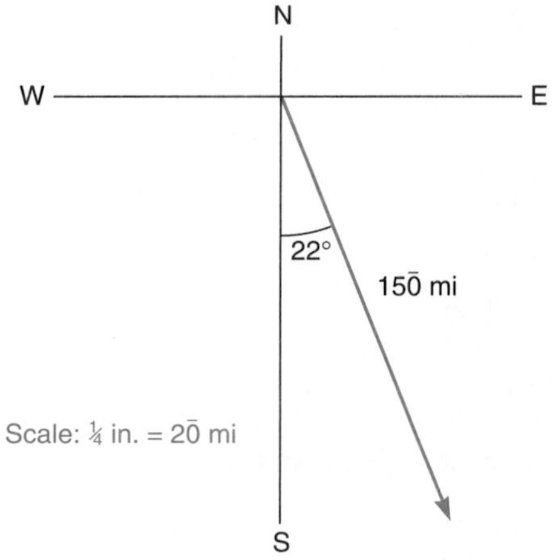

Figure 2.6

Graphical Addition of Vectors

Vectors may be denoted by a single letter with a small arrow above, such as $\vec{A}$, $\vec{v}$, or $\vec{R}$ [Fig. 2.7(a)]. This notation is especially useful when writing vectors on paper or on a chalkboard. In this book we use the traditional boldface type to denote vectors, such as **A**, **v**, or **R** [Fig. 2.7(b)]. The length of vector $\vec{A}$ is written $|\vec{A}|$; the length of vector **A** is written $|\mathbf{A}|$.

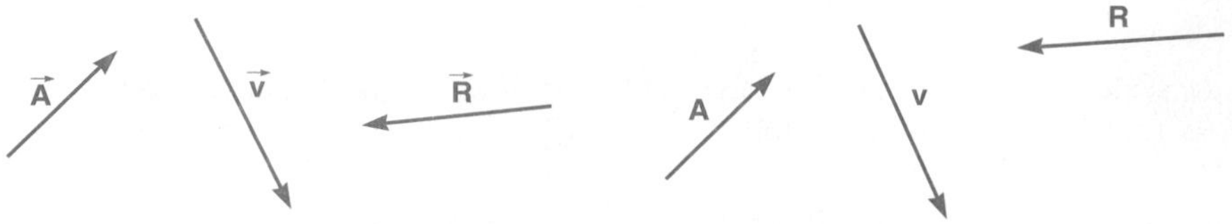

(a) Vector quantities $\vec{A}$, $\vec{v}$, and $\vec{R}$ usually have arrows when writing them on paper or on a chalkboard.

(b) Vector quantities **A**, **v**, and **R** usually are written in boldface type in textbooks.

Figure 2.7

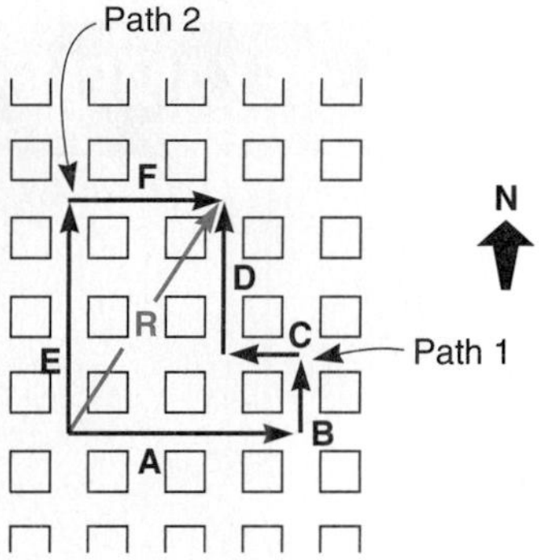

The resultant vector **R** is the graphic sum of the component sets of vectors **A**, **B**, **C**, and **D**, and **E** and **F**. That is, **A** + **B** + **C** + **D** = **R** and **E** + **F** = **R**.

Figure 2.8

Any given displacement can be the result of many different combinations of displacements. In Fig. 2.8, the displacement represented by the arrow labeled **R** for resultant is the result of either of the two paths shown. This vector is called the **resultant** of the vectors that make up either path 1 or path 2. *The resultant vector is the sum of a set of vectors.* The resultant vector, **R**, in Fig. 2.8 is the sum of the vectors **A**, **B**, **C**, and **D**. It is also the sum of vectors **E** and **F**. That is,

$$\mathbf{A} + \mathbf{B} + \mathbf{C} + \mathbf{D} = \mathbf{R} \quad \text{and} \quad \mathbf{E} + \mathbf{F} = \mathbf{R}$$

To solve a vector addition problem such as displacement:

1. Choose a suitable scale and calculate the length of each vector.
2. Draw the north–south reference line. Graph paper should be used.
3. Using a ruler and protractor, draw the first vector and then draw the other vectors so that the initial end of each vector is placed at the terminal end of the previous vector.
4. Draw the sum or resultant vector from the initial end of the first vector to the terminal end of the last vector.
5. Measure the length of the resultant and use the scale to find the magnitude of the vector. Use a protractor to measure the angle of the resultant.

EXAMPLE 3

Find the resultant displacement of an airplane that flies $2\bar{0}$ mi due east, then $3\bar{0}$ mi due north, and then $1\bar{0}$ mi at $6\bar{0}°$ west of south.

We choose a scale of 1.0 cm = 5.0 mi so that the vectors are large enough to be accurate and small enough to fit on the paper. (Here each block represents 0.5 cm.) The length of the first vector is

$$|\mathbf{A}| = 2\bar{0}\ \cancel{\text{mi}} \times \frac{1.0\ \text{cm}}{5.0\ \cancel{\text{mi}}} = 4.0\ \text{cm}$$

The length of the second vector is

$$|\mathbf{B}| = 3\bar{0}\ \cancel{\text{mi}} \times \frac{1.0\ \text{cm}}{5.0\ \cancel{\text{mi}}} = 6.0\ \text{cm}$$

The length of the third vector is

$$|\mathbf{C}| = 1\bar{0}\ \cancel{\text{mi}} \times \frac{1.0\ \text{cm}}{5.0\ \cancel{\text{mi}}} = 2.0\ \text{cm}$$

Draw the north–south reference line, and draw the first vector as shown in Fig. 2.9(a). The second and third vectors are then drawn as shown in Fig. 2.9(b) and 2.9(c).

Using a ruler, we find that the length of the resultant measures 5.5 cm [Fig. 2.9(d)]. Since 1.0 cm = 5.0 mi, this represents a displacement with magnitude

$$|\mathbf{R}| = 5.5\ \cancel{\text{cm}} \times \frac{5.0\ \text{mi}}{1.0\ \cancel{\text{cm}}} = 28\ \text{mi}$$

The angle measures 24°, so the resultant is 28 mi at 24° east of north.

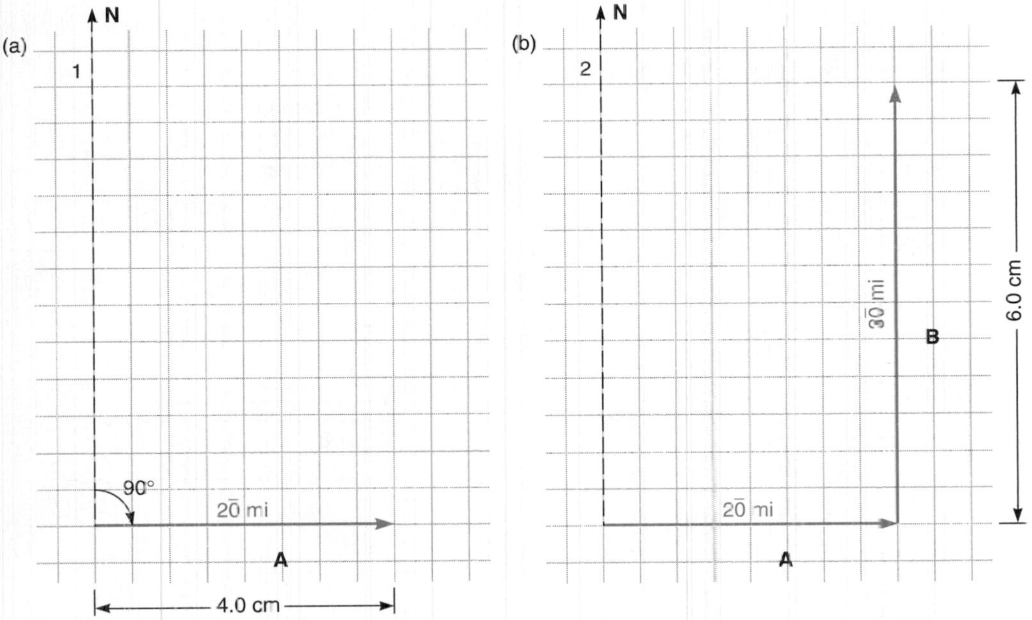

1. Draw the north–south reference line and the first vector: $2\bar{0}$ mi due east.

2. Draw the second vector: $3\bar{0}$ mi due north.

Figure 2.9 ***(Continued)***

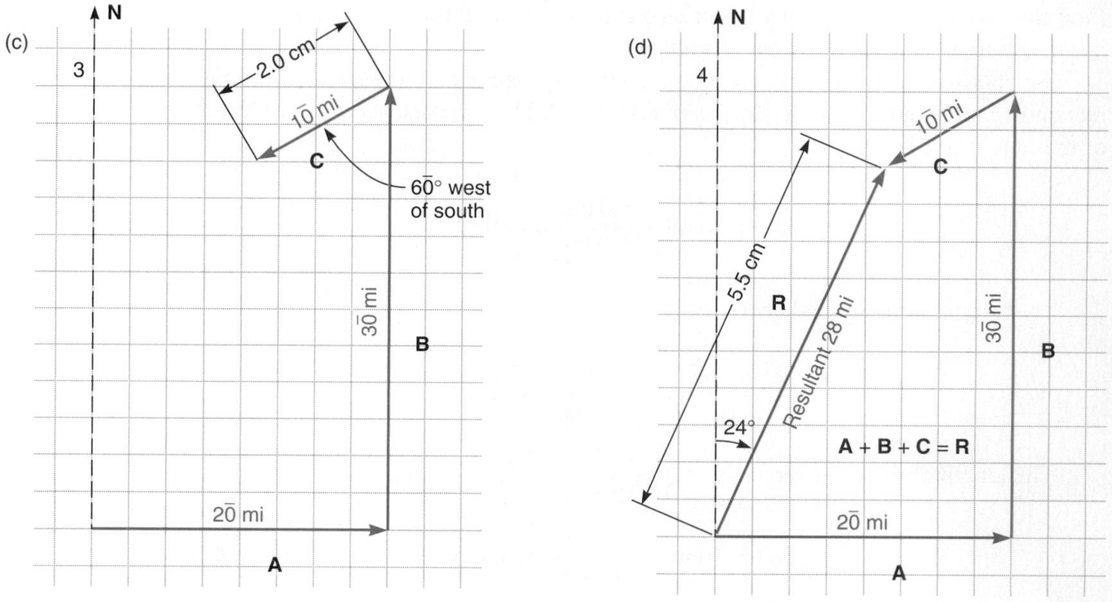

3. Draw the third vector: 1$\bar{0}$ mi at 6$\bar{0}$° west of south.

4. Draw the resultant vector, which is 28 mi at 24° east of north.

Scale: 1.0 cm = 5.0 mi

EXAMPLE 4

Find the resultant of the displacements 15$\bar{0}$ km due west, then 20$\bar{0}$ km due east, and then 125 km due south.

Choose a scale of 1.0 cm = 5$\bar{0}$ km. Follow the procedure in Fig. 2.10. The length of the resultant measures 2.6 cm. Since 1.0 cm = 5$\bar{0}$ km,

$$|\mathbf{R}| = 2.6\ \cancel{\text{cm}} \times \frac{5\bar{0}\ \text{km}}{1.0\ \cancel{\text{cm}}} = 130\ \text{km}$$

The angle measures 22°, so the resultant is 130 km at 22° east of south.

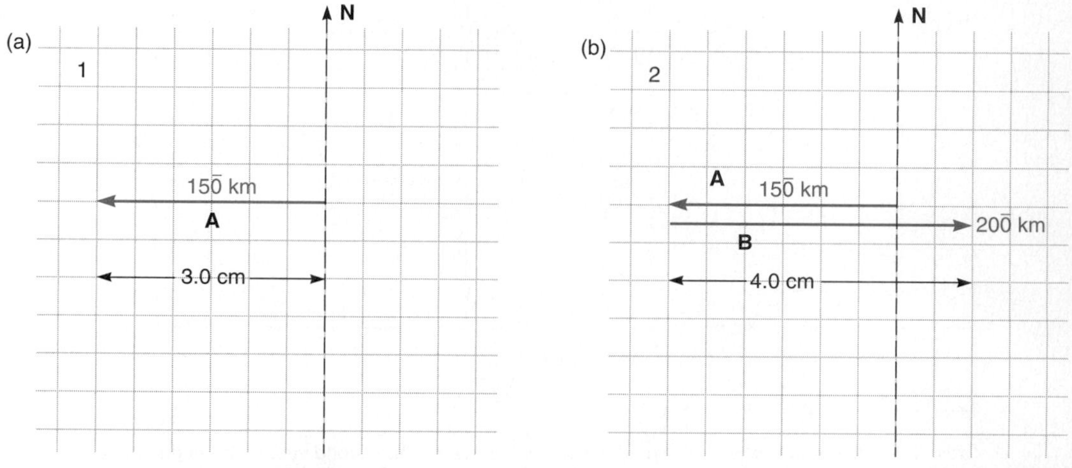

1. Draw the north–south reference line and the first vector: 15$\bar{0}$ km due west.

2. Draw the vector: 20$\bar{0}$ km due east.

Figure 2.10 ***(Continued)***

(c)

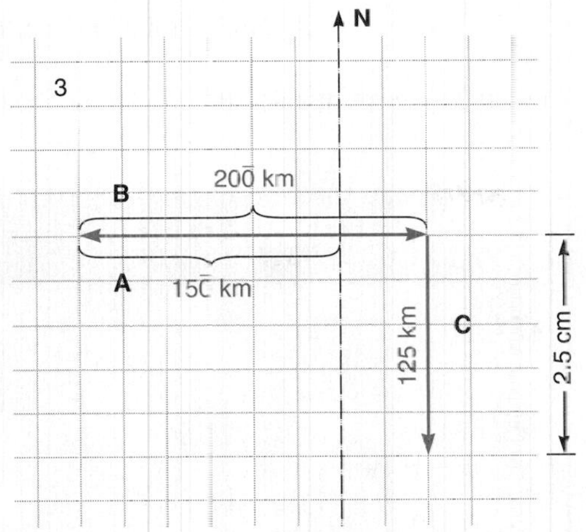

3. Draw the vector: 125 mi due south.

(d)

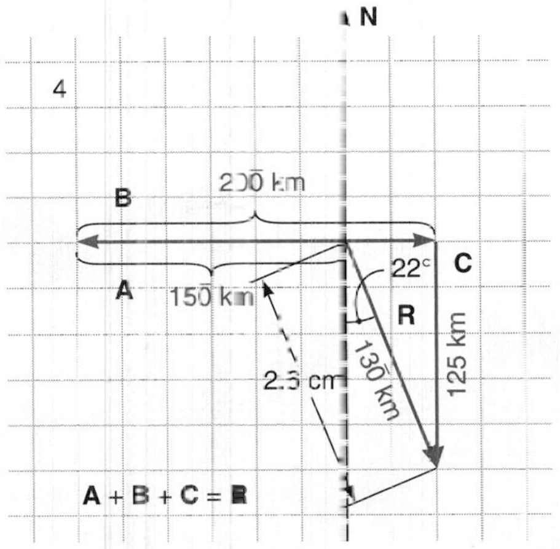

4. The length of the resultant is 2.6 cm, which represents $13\bar{0}$ km at 22° east of south.

Scale: 1.0 cm = $5\bar{0}$ km

PROBLEMS 2.1

Using the scale 1.0 cm = $5\bar{0}$ km, find the length of the vector that represents each displacement.

1.	Displacement $1\bar{0}0$ km east	length = ______ cm
2.	Displacement 125 km south	length = ______ cm
3.	Displacement $14\bar{0}$ km at 45° east of south	length = ______ cm
4.	Displacement $26\bar{0}$ km at $3\bar{0}$° south of west	length = ______ cm
5.	Displacement 315 km at 65° north of east	length = ______ cm
6.	Displacement 187 km at 17° north of west	length = ______ cm
7–12.	Draw the vectors in Problems 1 through 6 using the scale indicated.	

Using the scale $\frac{1}{4}$ in = $2\bar{0}$ mi, find the length of the vector that represents each displacement.

13.	Displacement $1\bar{0}0$ mi west	length = ______ in.
14.	Displacement $17\bar{0}$ mi north	length = ______ in.
15.	Displacement $21\bar{0}$ mi at 45° south of west	length = ______ in.
16.	Displacement 145 mi at $6\bar{0}$° north of east	length = ______ in.
17.	Displacement 75 mi at 25° west of north	length = ______ in.
18.	Displacement $16\bar{0}$ mi at 72° west of south	length = ______ in.
19–24.	Draw the vectors in Problems 13 through 18 using the scale indicated.	

Use graph paper to find the resultant of each displacement pair.

25. 35 km due east, then $5\bar{0}$ km due north
26. $6\bar{0}$ km due west, then $9\bar{0}$ km due south
27. $5\bar{0}0$ mi at 75° east of north, then $15\bar{0}0$ mi at $2\bar{0}$° west of south
28. $2\bar{0}$ mi at 3° north of east, then 17 mi at 9° west of south
29. 67 km at 55° north of west, then 46 km at 25° south of east
30. 4.0 km at 25° west of south, then 2.0 km at 15° north of east

Use graph paper to find the resultant of each set of displacements.

31. $6\bar{0}$ km due south, then $9\bar{0}$ km at 15° north of west, and then 75 km at 45° north of east
32. 110 km at $5\bar{0}$° north of east, then 170 km at $3\bar{0}$° east of south, and then 145 km at $2\bar{0}$° north of east
33. 1700 mi due north, then 2400 mi at $1\bar{0}$° north of east, and then $2\bar{0}00$ mi at $2\bar{0}$° south of west
34. $9\bar{0}$ mi at $1\bar{0}$° west of north, then 75 mi at $3\bar{0}$° west of south, and then 55 mi at $2\bar{0}$° east of south
35. 75 km at 25° north of east, then 75 km at 65° south of west, and then 75 km due south
36. 17 km due north, then $1\bar{0}$ km at 7° south of east, and then 15 km at $1\bar{0}$° west of south

37. 12 mi at 58° north of east, then 16 mi at 78° north of east, then $1\bar{0}$ mi at 45° north of east, and then 14 mi at $1\bar{0}$° north of east
38. $1\bar{0}$ km at 15° west of south, then 27 km at 35° north of east, then 31 km at 5° north of east, and then 22 km at $2\bar{0}$° west of north

PHYSICS CONNECTIONS

Global Positioning Satellites

Navigators continually struggle to find better tools to help them determine their location. The first explorers used the sun and stars to help them steer a straight course, but this method of navigation only worked under clear skies. Magnetic compasses were developed, yet could only be used to determine longitude, not latitude. Finally, the mechanical clock, in conjunction with the compass, provided navigators with the most accurate method of determining location. Today, most navigators use a hand-held device that functions in concert with a series of 24 orbiting satellites. This network, the Global Positioning System (GPS), can determine your position and altitude anywhere on earth.

The GPS pinpoints your location by sending out radio signals to locate any 4 of the 24 orbiting GPS satellites. Once the satellites are found, the GPS measures the length of time it takes for a radio signal to reach the hand-held receiver. When the time is determined for each of 4 satellites, the distance is calculated, and the longitude, latitude, and altitude are displayed on the screen [Fig. 2.11(a)].

GPS was first developed solely for military use. Eventually, the GPS system was made available for civilian businesses. Shipping, airline, farming, surveying, and geological companies made use of the technology. Today, GPS receivers are affordable and are used by the general public [Fig. 2.11(b)]. More sophisticated receivers not only locate a position, but can also guide the navigator to a predetermined location. Several automobile manufacturers have included GPS receivers as an option in their cars. Such receivers come complete with voice commands such as, "Turn left at the next traffic light," as part of their option packages.

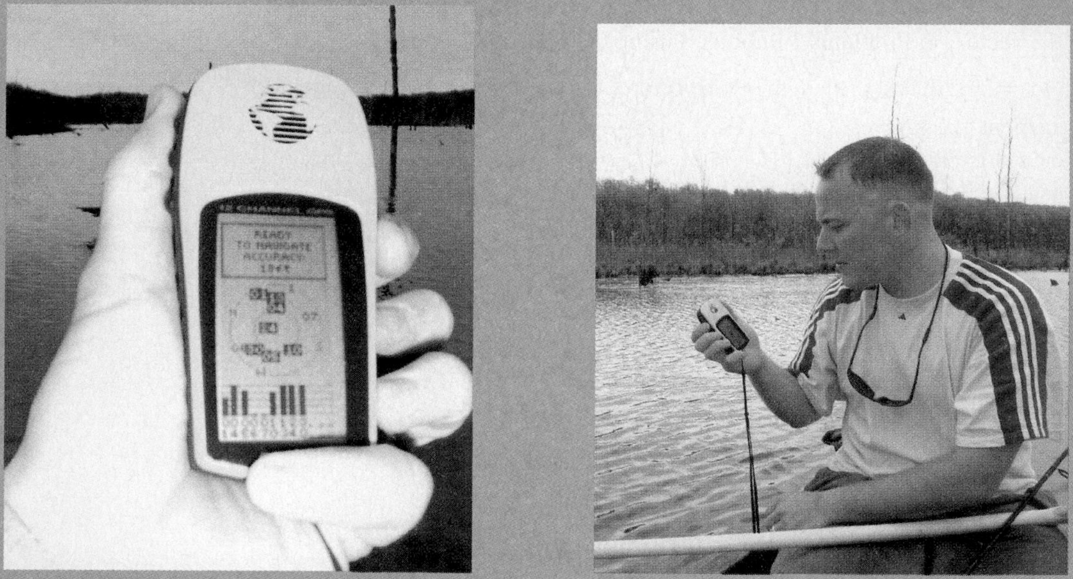

(a) (b)

Figure 2.11 **(a) The screen on the GPS receiver shows the position and strength of the signal between the receiver and the various satellites. At the time this photograph was taken, the receiver picked up 7 of the 12 overhead satellites, bringing the precision to within 20 ft of the actual location. (Photo by William Brouhle.) (b) Global Positioning Systems have allowed for an enormous step forward in navigation. The GPS receiver shown has monitored and recorded precisely where the person has traveled and is now helping the user find his way back to camp.**

2.2 Components of a Vector

René Descartes (1596–1650), *mathematician and philosopher, was born in France. He founded analytic or coordinate geometry, often called Cartesian geometry, and made major contributions in optics.*

Before our further study of vectors, we need to discuss components of vectors. This requires using the number plane. The **number plane** (sometimes called the *Cartesian coordinate system* after **René Descartes**) is determined by a horizontal line called the x-axis and a vertical line called the y-axis intersecting at right angles as shown in Fig. 2.12. These two lines divide the number plane into four quadrants, which we label as quadrants I, II, III, and IV. Each axis has a scale, and the intersection of the two axes is called the *origin.* The x-axis contains positive numbers to the right of the origin and negative numbers to the left of the origin. The y-axis contains positive numbers above the origin and negative numbers below the origin.

Graphically, a vector is represented by a directed line segment. The length of the line segment indicates the magnitude of the quantity. An arrowhead indicates the direction. If A and B are the end points of a line segment as in Fig. 2.13, the symbol **AB** denotes the *vector from A to B.* Point A is called the *initial point.* Point B is called the *terminal point* or *end point* of the vector. Vector **BA** has the same length as vector **AB** but has the opposite direction. Vectors may also be denoted by a single letter, such as **u**, **v**, or **R**.

The sum of two or more vectors is called the **resultant vector**. When two or more vectors are added, each of these vectors is called a **component** of the resultant, or sum, vector. The components of vector **R** in Fig. 2.14(a) are vectors **A**, **B**, and **C**. *Note:* A vector may have more than one set of component vectors. The components of vector **R** in Fig. 2.14(b) are vectors **E** and **F**.

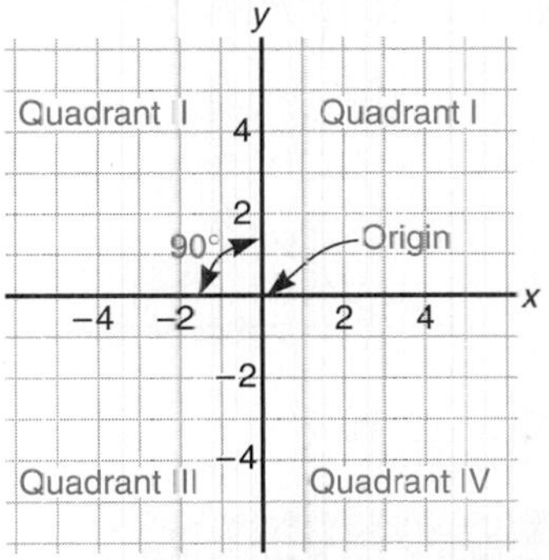

Figure 2.12 Number plane

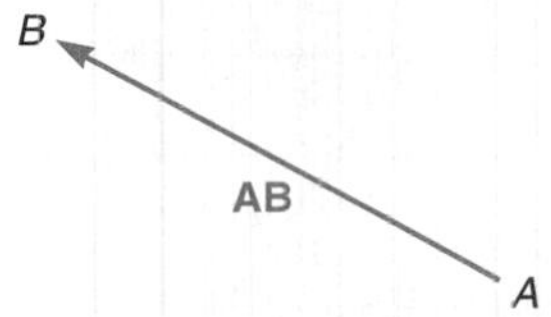

Figure 2.13 Vector from A to B

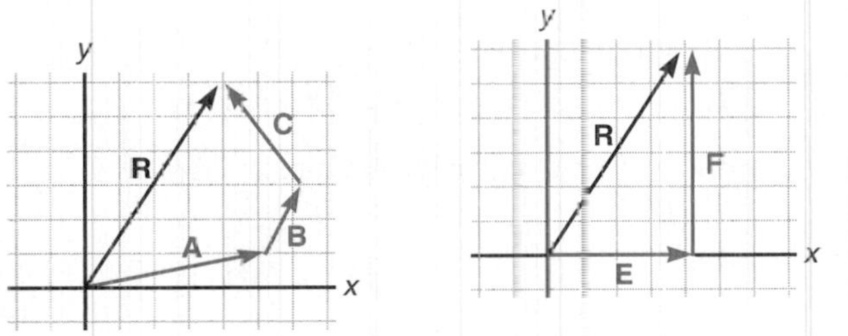

(a) Vectors **A**, **B**, and **C** are components of the resultant vector **R**.

(b) Vector **E** is a horizontal component and vector **F** is a vertical component of the resultant vector **R**.

Figure 2.14

We are most interested in the components of a vector that are perpendicular to each other and that are on or parallel to the x- and y-axes. In particular, we are interested in the type of component vectors shown in Fig. 2.14(b) (component vectors **E** and **F**). The horizontal component vector that lies on or is parallel to the x-axis is called the **x-component.** The vertical component vector that lies on or is parallel to the y-axis is called the **y-component.** Three examples are shown in Fig. 2.15.

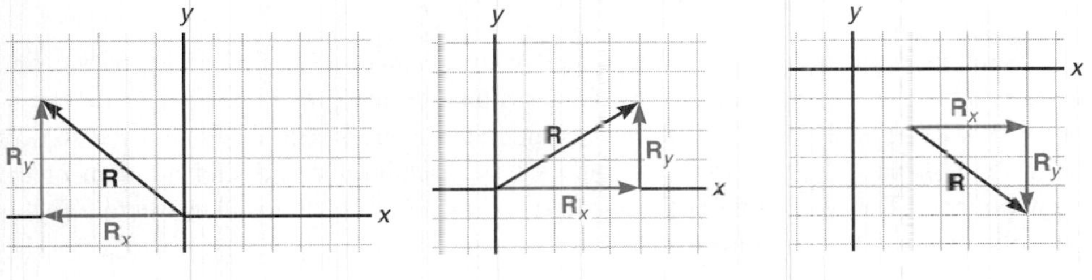

(a) (b) (c)

$\mathbf{R}_x$ = the x-component of vector **R**
$\mathbf{R}_y$ = the y-component of vector **R**

Figure 2.15

The x- and y-components of vectors can also be expressed as signed numbers. The absolute value of the signed number corresponds to the magnitude of the vector. The sign of the number corresponds to the direction of the component as follows:

x-component	*y*-component
+, if right	+, if up
−, if left	−, if down

EXAMPLE 1

Find the x- and y-components of vector **R** in Fig. 2.16.

$$\mathbf{R}_x = x\text{-component of } \mathbf{R} = +4$$
$$\mathbf{R}_y = y\text{-component of } \mathbf{R} = +3$$

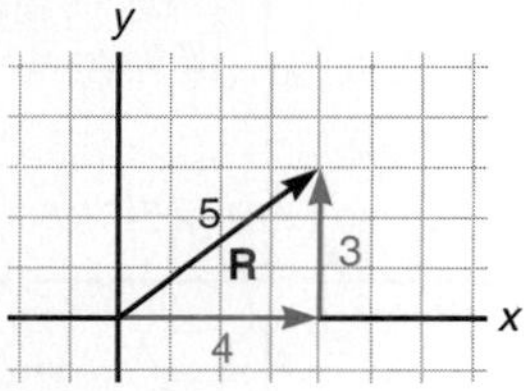

Figure 2.16

EXAMPLE 2

Find the x- and y-components of vector **R** in Fig. 2.17.

$$\mathbf{R}_x = x\text{-component of } \mathbf{R} = +6$$
$$\mathbf{R}_y = y\text{-component of } \mathbf{R} = -8$$

(y-component points in a negative direction)

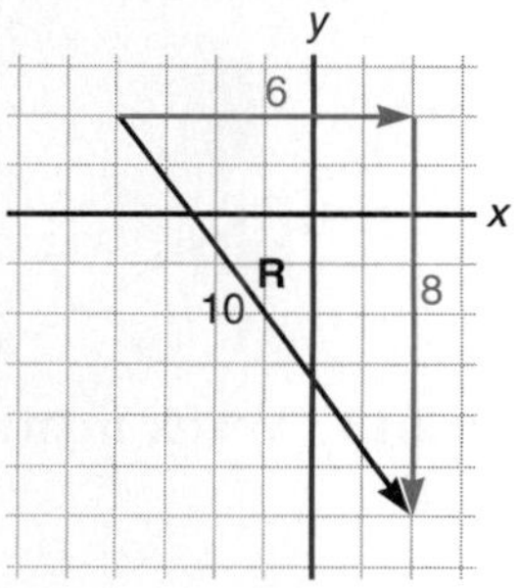

Figure 2.17

EXAMPLE 3

Find the x- and y-components of vector **R** in Fig. 2.18.

$$\mathbf{R}_x = -12$$
$$\mathbf{R}_y = -9$$

(both x- and y-components point in a negative direction)

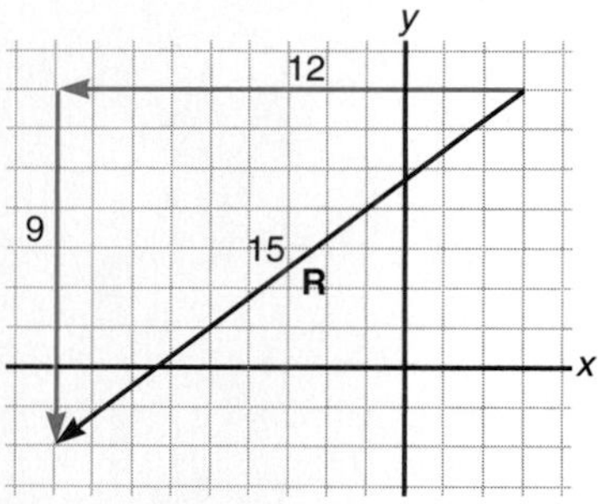

Figure 2.18

Now that we have expressed the x- and y-components as signed numbers, we find the resultant vector of several vectors using arithmetic and graphing. To find the resultant vector of several vectors, find the x-component of each vector and find the sum of the x-components. Then find the y-component of each vector and find the sum of the y-components. The two sums are the x- and y-components of the resultant vector. This is shown in the following examples.

EXAMPLE 4

Given vectors **A** and **B** in Fig. 2.19, graph and find the x- and y-components of the resultant vector **R**.

Graph resultant vector **R** by connecting the initial point of vector **A** to the end point of vector **B** [Fig. 2.20(a)]. The resultant vector **R** is shown in Fig. 2.20(b).

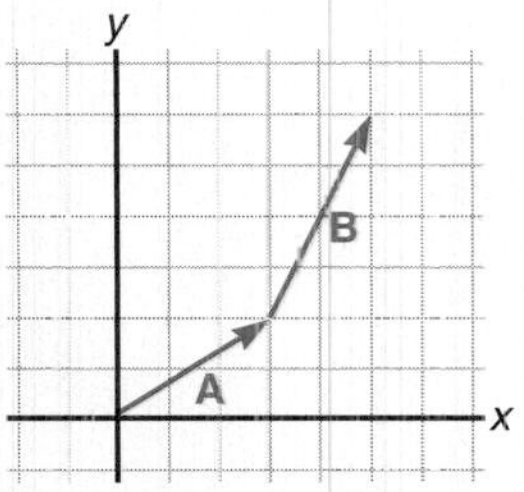

Figure 2.19

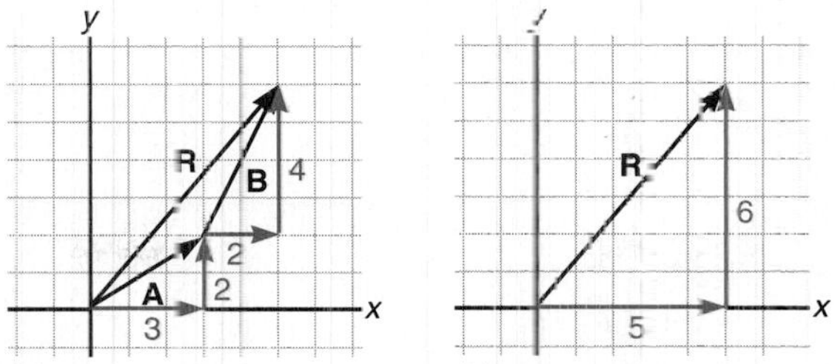

(a) A + B = R (b)

Figure 2.20

Find the x-component of **R** by adding the x-components of **A** and **B**.

$$\mathbf{A}_x = +3$$
$$\mathbf{B}_x = \underline{+2}$$
$$\mathbf{R}_x = +5$$

Find the y-component of **R** by adding the y-components of **A** and **B**.

$$\mathbf{A}_y = +2$$
$$\mathbf{B}_y = \underline{+4}$$
$$\mathbf{R}_y = +6$$

EXAMPLE 5

Given vectors **A**, **B**, and **C** in Fig. 2.21, graph and find the x- and y-components of the resultant vector **R**.

Graph resultant vector **R** by connecting the initial point of vector **A** to the end point of vector **C** [Fig. 2.22(a)]. The resultant vector **R** is shown in Fig. 2.22(b).

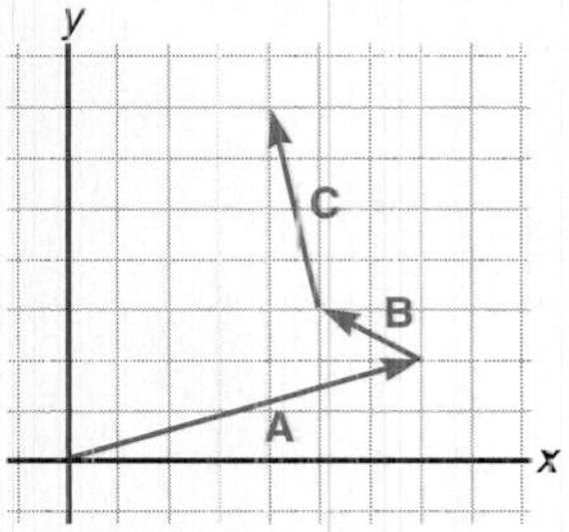

Figure 2.21

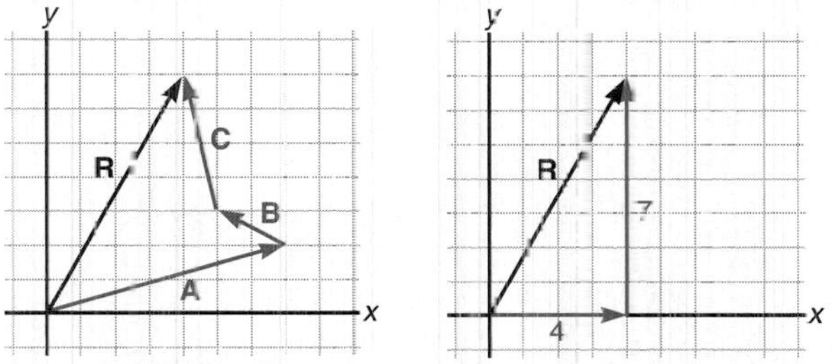

(a) A − B + C = R (b)

Figure 2.22

Find the x-component of **R** by adding the x-components of **A**, **B**, and **C**. Find the y-component of **R** by adding the y-components of **A**, **B**, and **C**.

Vector	x-component	y-component
A	+7	+2
B	−2	+1
C	−1	+4
R	+4	+7

EXAMPLE 6

Given vectors **A**, **B**, **C**, and **D** in Fig. 2.23, graph and find the x- and y-components of the resultant vector **R**.

Graph resultant vector **R** by connecting the initial point of vector **A** to the end point of vector **D** [Fig. 2.24(a)]. The resultant vector **R** is shown in Fig. 2.24(b).

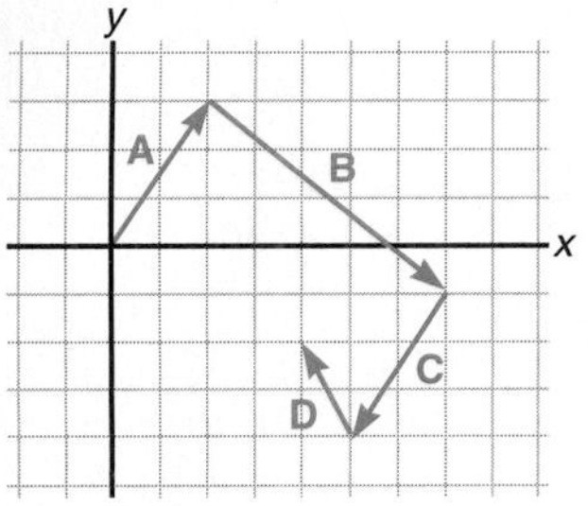

Figure 2.23

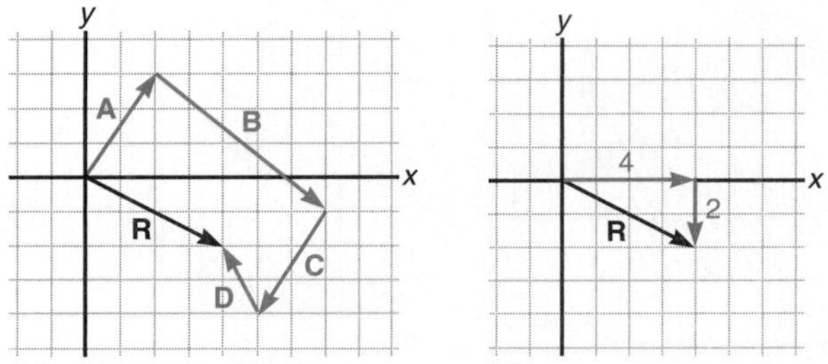

(a) A + B + C + D = R (b)

Figure 2.24

Find the x-component of **R** by adding the x-components of **A**, **B**, **C**, and **D**. Find the y-component of **R** by adding the y-components of **A**, **B**, **C**, and **D**.

Vector	x-component	y-component
A	+2	+3
B	+5	−4
C	−2	−3
D	−1	+2
R	+4	−2

Two vectors are equal when they have the same magnitude and the same direction [Fig. 2.25(a)]. Two vectors are opposites or negatives of each other when they have the same magnitude but opposite directions [Fig. 2.25(b)].

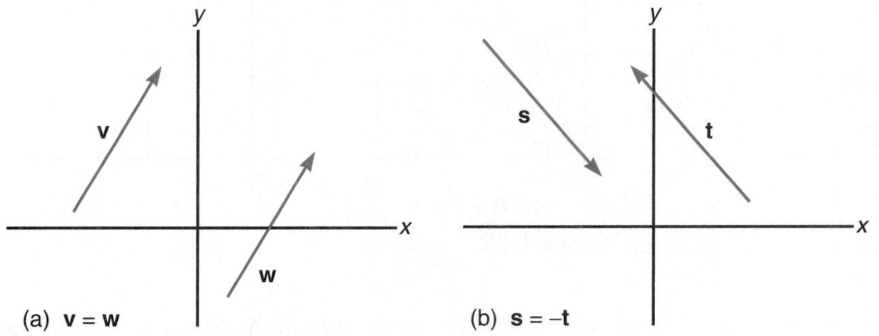

Figure 2.25

A vector may be placed in any position in the number plane as long as its magnitude and direction are not changed [Fig. 2.26(a)]. Note that the vectors in Fig. 2.26(b) are equal because they have the same magnitude (length) and the same direction.

To add two or more vectors graphically, construct the first vector with its initial point at the origin and parallel to its given position. Then, construct the second vector with its initial point on the end point of the first vector and parallel to its given position. Then, construct the third vector with its initial point on the end point of the second vector and parallel to its given position. Continue this process until all vectors have been so constructed. The sum or resultant vector is the vector joining the initial point of the first vector (origin) to the end point of the last vector. (The order of adding or constructing the given vectors does not matter.)

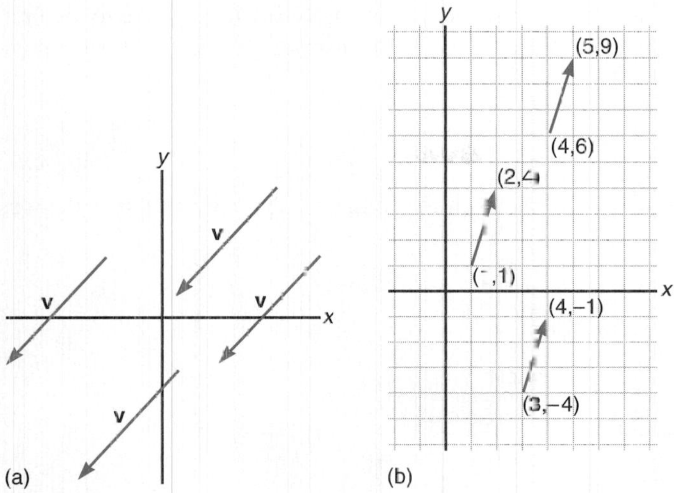

Figure 2.26

EXAMPLE 7

Given vectors **A**, **B**, and **C** in Fig. 2.27(a), graph and find the x- and y-components of the resultant vector **R**.

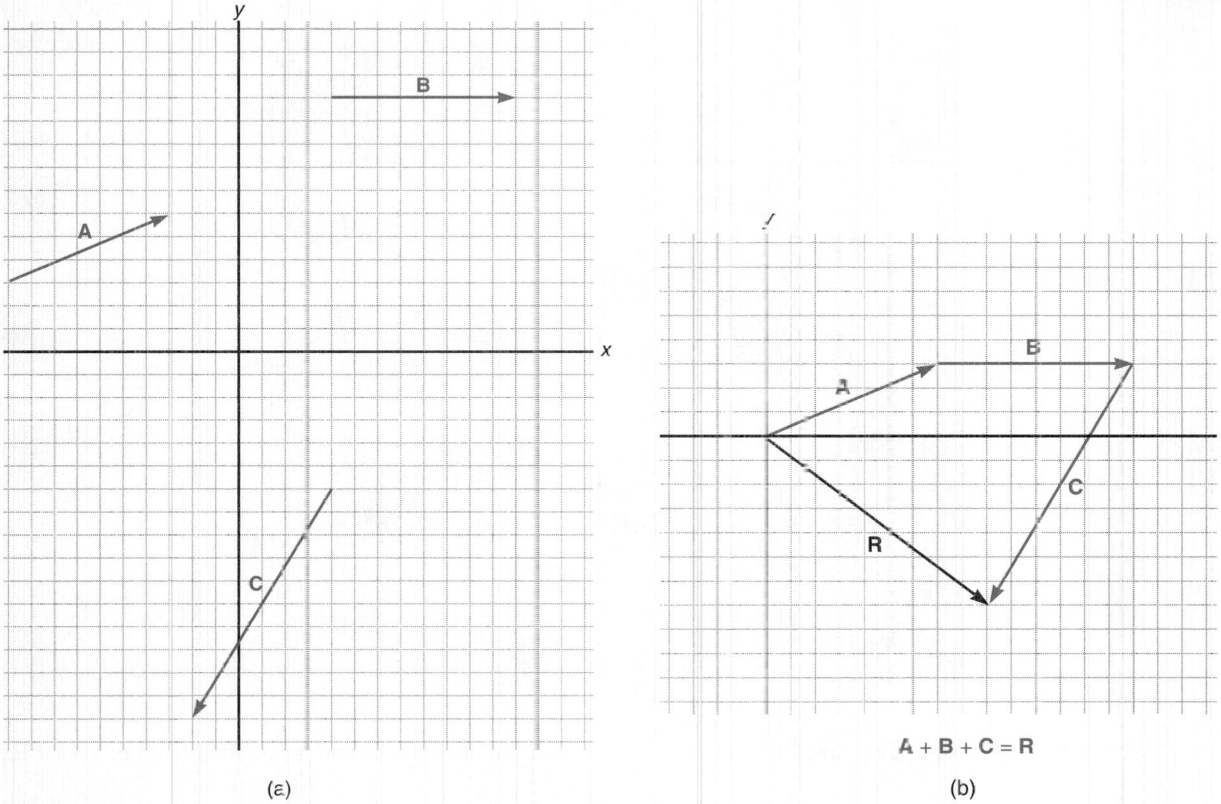

Figure 2.27

Construct vector **A** with its initial point at the origin and parallel to its given position as in Fig. 2.27(b). Next, construct vector **B** with its initial point on the end point of vector **A** and parallel to its given position. Then, construct vector **C** with its initial point on the end point of vector **B** and parallel to its given position. The resultant vector **R** is the vector with its initial point at the origin and its end point at the end point of vector **C**.

From the graph in Fig. 2.27(b), we read the x-component of **R** as 9 and the y-component of **R** as -7.

Figure 2.28
$\mathbf{R} = \mathbf{v} - \mathbf{w} = \mathbf{v} + (-\mathbf{w})$

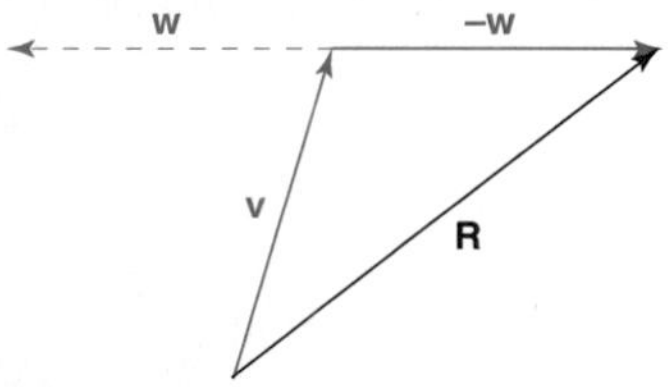

One vector may be subtracted from a second vector by adding its negative to the first. That is, $\mathbf{v} - \mathbf{w} = \mathbf{v} + (-\mathbf{w})$. Construct **v** as usual. Then construct −**w** and find the resultant **R** as shown in Fig. 2.28.

PROBLEMS 2.2

Find the *x*- and *y*-components of each vector in the following diagram. (Express them as signed numbers and then graph them as vectors.)

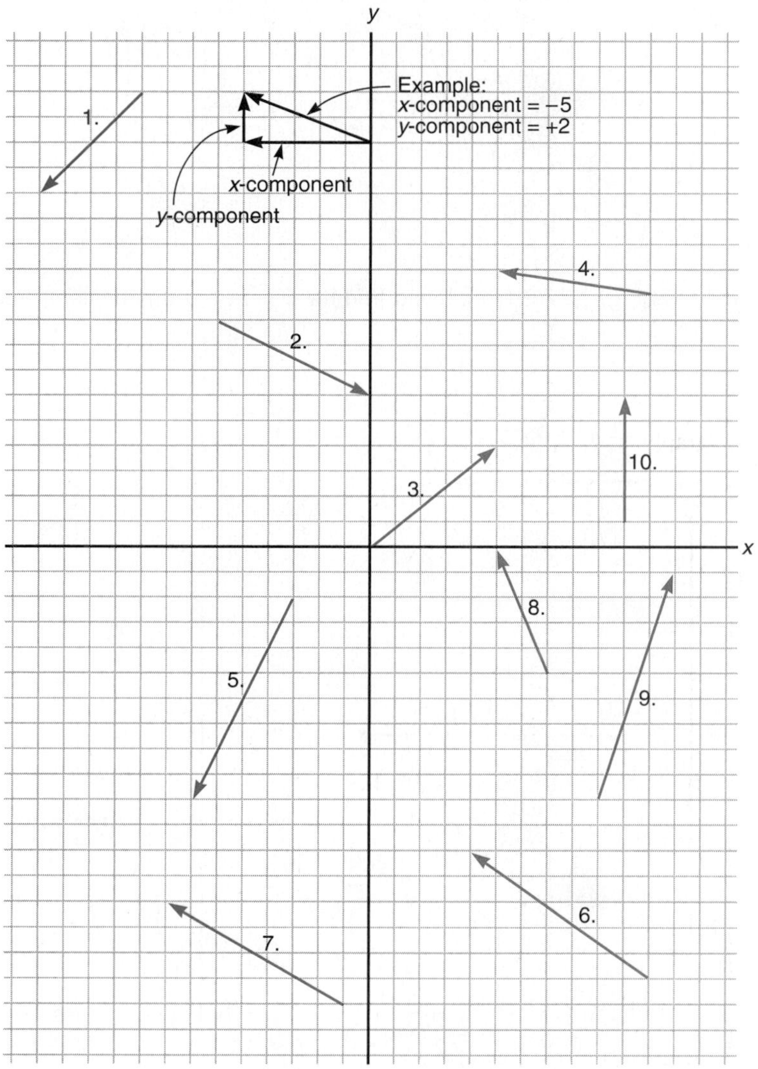

Find the *x*- and *y*-components of each resultant vector **R** and graph the resultant vector **R**.

Vector	*x*-component	*y*-component	Vector	*x*-component	*y*-component
11. **A**	+2	+3	12. **A**	+9	−5
B	+7	+2	**B**	−4	−6
R			**R**		
13. **A**	−2	+13	14. **A**	+10	−5
B	−11	+1	**B**	−13	−9
C	+3	−4	**C**	+4	+3
R			**R**		

Vector	x-component	y-component	Vector	x-component	y-component
15. **A**	+17	+7	16. **A**	+1	+7
B	−14	+11	**B**	+9	−4
C	+7	+9	**C**	−4	+13
D	−6	−15	**D**	−11	−4
R			**R**		
17. **A**	+1.5	−1.5	18. **A**	+1	−1
B	−3	−2	**B**	−4	−2
C	+7.5	−3	**C**	+2	+4
D	+2	+2.5	**D**	+5	−3
R			**E**	+3	+5
			R		
19. **A**	+1.5	+2.5	20. **A**	−7	+15
B	−2	−3	**B**	+13.5	−17.5
C	+3.5	−7.5	**C**	−7.5	−20
D	−4	+6	**D**	+6	+13.5
E	−5.5	+2	**E**	+2.5	+2.5
R			**F**	−11	+1[illegible].5
			R		

For each set of vectors, graph and find the x- and y-components of the resultant vector **R**.

21.

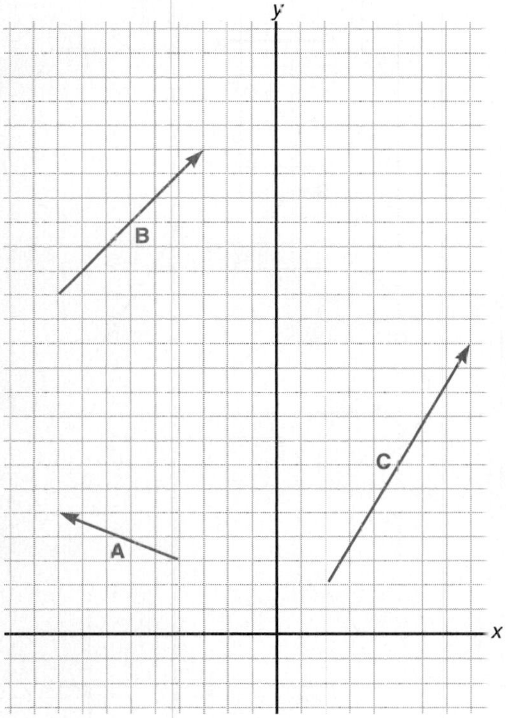

22.

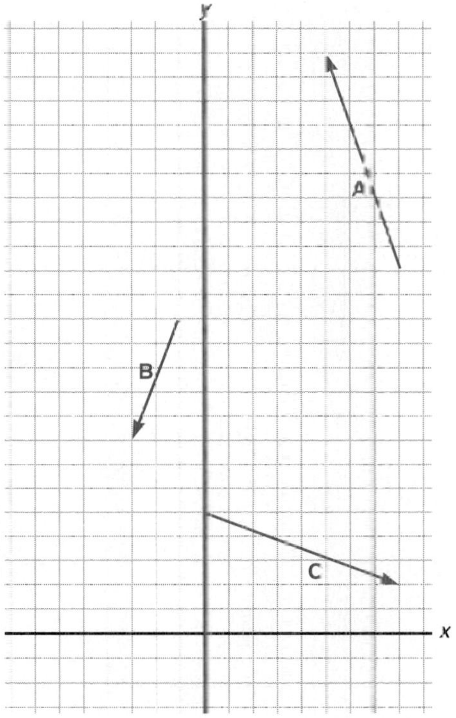

23.

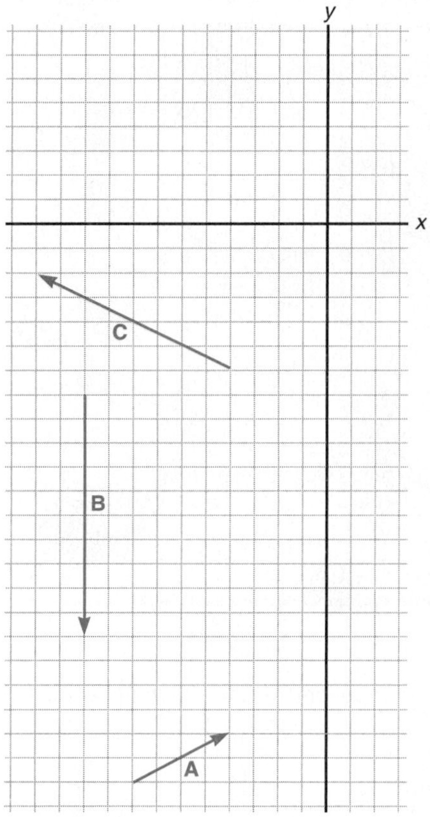

24.

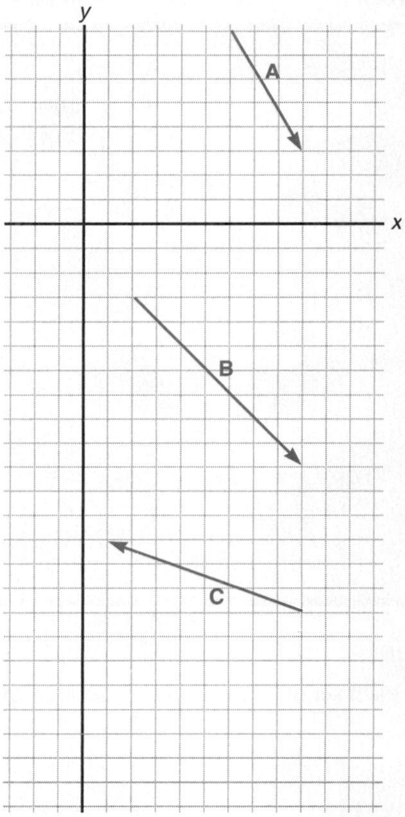

25.

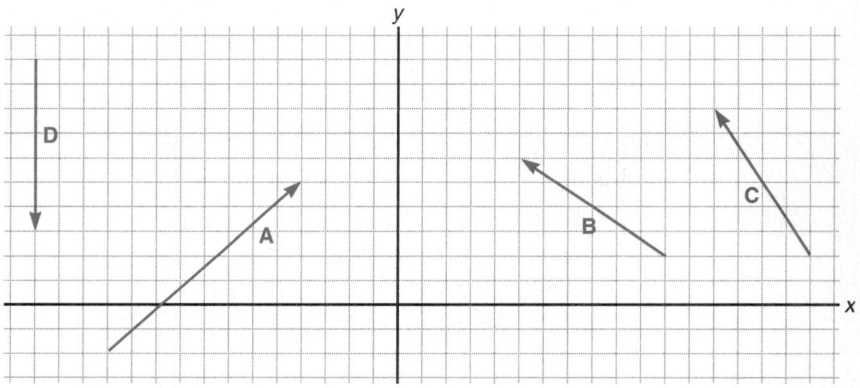

26.

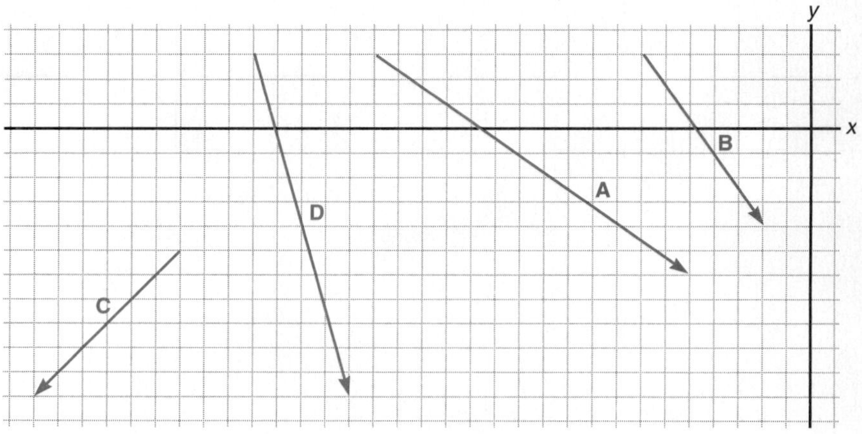

2.3 Vectors in Standard Position*

A vector may be placed in any position in the number plane as long as its magnitude and direction are not changed. A vector is in **standard position** when its initial point is at the origin of the number plane. A vector in standard position is expressed in terms of its length and its angle θ, where θ *is measured counterclockwise from the positive x-axis to the vector.* The vectors shown in Fig. 2.29 are in standard position.

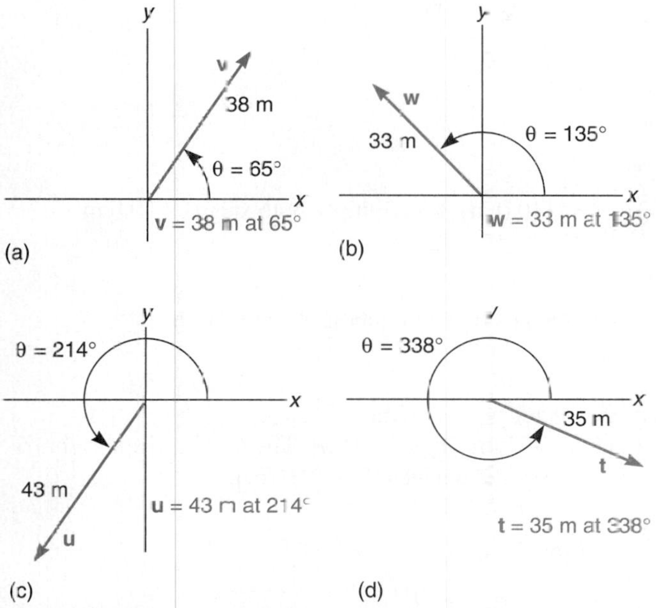

Figure 2.29 **Vectors in standard position**

Finding the Components of a Vector

EXAMPLE 1

Find the x- and y-components of the vector $\mathbf{v} = 10.0$ m at $60.0°$.

First, draw the vector in standard position [Fig. 2.30(a)]. Then, draw a right triangle where the legs represent the x- and y-components [Fig. 2.30(b)]. The absolute value of the x-component of the vector is the length of the side adjacent to the $60.0°$ angle. Therefore, to find the x-component,

$$\cos 60.0° = \frac{\text{side adjacent to } 60.0°}{\text{hypotenuse}} = \frac{|\mathbf{v}_x|}{10.0 \text{ m}}$$

$$\cos 60.0° = \frac{|\mathbf{v}_x|}{10.0 \text{ m}}$$

$$(\cos 60.0°)(10.0 \text{ m}) = \left(\frac{|\mathbf{v}_x|}{10.0 \text{ m}}\right)(10.0 \text{ m}) \qquad \text{Multiply both sides by 10.0 m.}$$

$$5.00 \text{ m} = |\mathbf{v}_x|$$

Since the x-component is pointing in the positive x-direction, $\mathbf{v}_x = +5.00$ m.

The absolute value of the y-component of the vector is the length of the side opposite the $60.0°$ angle. Therefore, to find the y-component,

$$\sin 60.0° = \frac{\text{side opposite } 60.0°}{\text{hypotenuse}} = \frac{|\mathbf{v}_y|}{10.0 \text{ m}}$$

$$\sin 60.0° = \frac{|\mathbf{v}_y|}{10.0 \text{ m}}$$

*You may want to review or study Section A.8 of Appendix A (Right-Triangle Trigonometry) and Section B.3 of Appendix B (Trigonometric Operations) for using a scientific calculator.

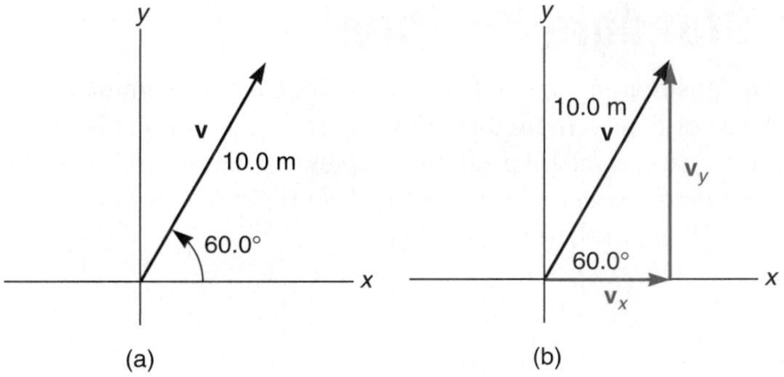

Figure 2.30

$$(\sin 60.0°)(10.0 \text{ m}) = \left(\frac{|\mathbf{v}_y|}{10.0 \text{ m}}\right)(10.0 \text{ m}) \qquad \text{Multiply both sides by 10.0 m.}$$

$$8.66 \text{ m} = |\mathbf{v}_y|$$

Since the y-component is pointing in the positive y-direction, $\mathbf{v}_y = +8.66$ m.

EXAMPLE 2

Find the x- and y-components of the vector $\mathbf{w} = 13.0$ km at 220.0°.

First, draw the vector in standard position [Fig. 2.31(a)]. Then, complete a right triangle with the x- and y-components being the two legs [Fig. 2.31(b)].

Find angle A as follows:

$$180° + A = 220.0°$$
$$A = 40.0°$$

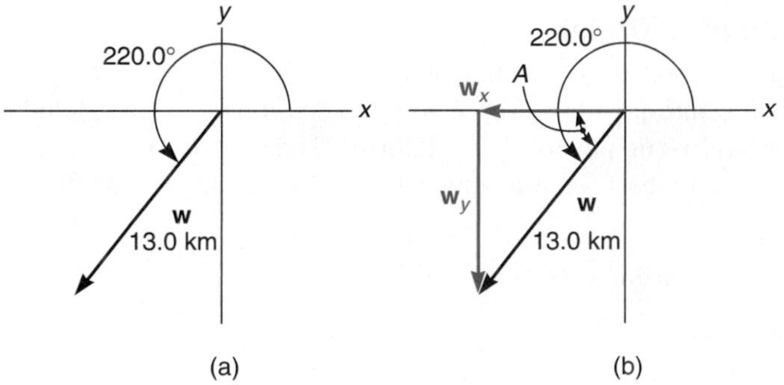

Figure 2.31

The absolute value of the x-component is the length of the side adjacent to angle A. Therefore, to find the x-component,

$$\cos A = \frac{\text{side adjacent to } A}{\text{hypotenuse}}$$

$$\cos 40.0° = \frac{|\mathbf{w}_x|}{13.0 \text{ km}}$$

$$(\cos 40.0°)(13.0 \text{ km}) = \left(\frac{|\mathbf{w}_x|}{13.0 \text{ km}}\right)(13.0 \text{ km}) \qquad \text{Multiply both sides by 13.0 km.}$$

$$9.96 \text{ km} = |\mathbf{w}_x|$$

Since the x-component is pointing in the negative x-direction, $\mathbf{w}_x = -9.96$ km.

The absolute value of the y-component of the vector is the length of the side opposite angle A. Therefore, to find the y-component,

$$\sin A = \frac{\text{side opposite } A}{\text{hypotenuse}}$$

$$\sin 40.0° = \frac{|\mathbf{w}_y|}{13.0 \text{ km}}$$

$$(\sin 40.0°)(13.0 \text{ km}) = \left(\frac{|\mathbf{w}_y|}{13.0 \text{ km}}\right)(13.0 \text{ km}) \qquad \text{Multiply both sides by 13.0 km.}$$

$$8.36 \text{ km} = |\mathbf{w}_y|$$

Since the y-component is pointing in the negative y-direction, $\mathbf{w}_y = -8.36$ km.

In general, find the x- and y-components of a vector as follows. First, draw any vector $\mathbf{v}$ in standard position; then, draw its x- and y-components as shown in Fig. 2.32. Use the right triangle to find the x-component as follows:

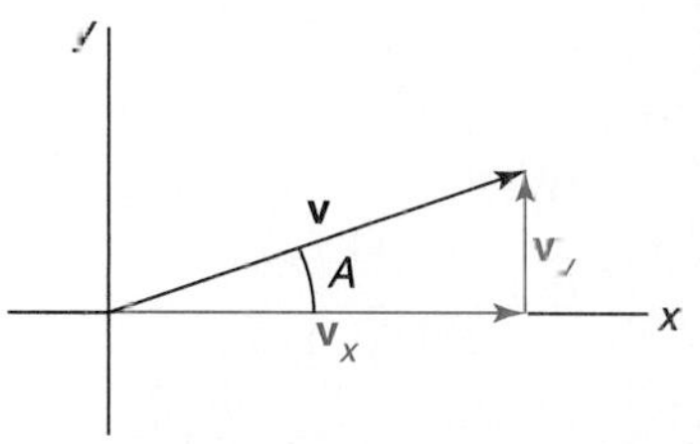

Vector $\mathbf{v}$ in standard position with its horizontal component $\mathbf{v}_x$ and its vertical component $\mathbf{v}_y$

Figure 2.32

$$\cos A = \frac{\text{side adjacent to } A}{\text{hypotenuse}}$$

$$\cos A = \frac{|\mathbf{v}_x|}{|\mathbf{v}|}$$

$$|\mathbf{v}|(\cos A) = \left(\frac{|\mathbf{v}_x|}{|\mathbf{v}|}\right)|\mathbf{v}| \qquad \text{Multiply both sides by } |\mathbf{v}|.$$

$$|\mathbf{v}|(\cos A) = |\mathbf{v}_x|$$

Similarly, we use the right triangle to find the y-component as follows:

$$\sin A = \frac{\text{side opposite } A}{\text{hypotenuse}}$$

$$\sin A = \frac{|\mathbf{v}_y|}{|\mathbf{v}|}$$

$$|\mathbf{v}|(\sin A) = \left(\frac{|\mathbf{v}_y|}{|\mathbf{v}|}\right)|\mathbf{v}| \qquad \text{Multiply both sides by } |\mathbf{v}|.$$

$$|\mathbf{v}|(\sin A) = |\mathbf{v}_y|$$

The signs of the x- and y-components are determined by the quadrants in which the vector in standard position lies.

In general:

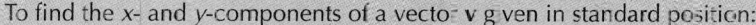

To find the x- and y-components of a vector $\mathbf{v}$ given in standard position:

1. Complete the right triangle with the legs being the x- and y-components of the vector.
2. Find the lengths of the legs of the right triangle as follows:

$$|\mathbf{v}_x| = |\mathbf{v}|(\cos A)$$

$$|\mathbf{v}_y| = |\mathbf{v}|(\sin A)$$

3. Determine the signs of the x- and y-components.

EXAMPLE 3

Find the x- and y-components of the vector $\mathbf{v} = 27.0$ ft at $125.0°$.

First, draw the vector in standard position [Fig. 2.33(a)]. Then, complete a right triangle with the x- and y-components being the two legs [Fig. 2.33(b)]. Find angle A as follows:

$$A + 125.0° = 180°$$
$$A = 55.0°$$

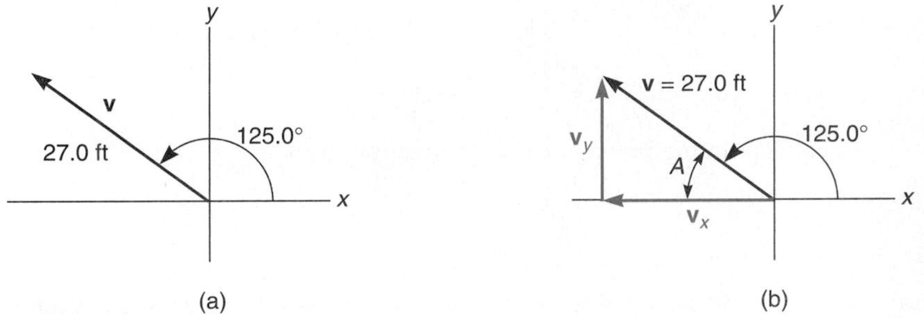

Figure 2.33

Next, find the x-component as follows:

$$|\mathbf{v}_x| = |\mathbf{v}|(\cos A)$$
$$|\mathbf{v}_x| = (27.0 \text{ ft})(\cos 55.0°)$$
$$= 15.5 \text{ ft}$$

Since the x-component is pointing in the negative x-direction,

$$\mathbf{v}_x = -15.5 \text{ ft}$$

Then, find the y-component as follows:

$$|\mathbf{v}_y| = |\mathbf{v}|(\sin A)$$
$$|\mathbf{v}_y| = (27.0 \text{ ft})(\sin 55.0°)$$
$$= 22.1 \text{ ft}$$

Since the y-component is pointing in the positive y-direction,

$$\mathbf{v}_y = +22.1 \text{ ft}$$

Finding a Vector from Its Components

EXAMPLE 4

Find vector $\mathbf{R}$ in standard position with $\mathbf{R}_x = +3.00$ m and $\mathbf{R}_y = +4.00$ m.

First, graph the x- and y-components (Fig. 2.34) and complete the right triangle. The hypotenuse is the resultant vector. Find angle A as follows:

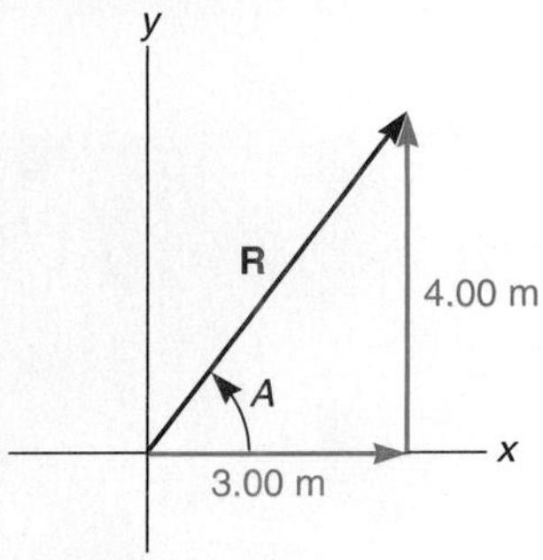

Figure 2.34

$$\tan A = \frac{\text{side opposite } A}{\text{side adjacent to } A}$$
$$\tan A = \frac{4.00 \text{ m}}{3.00 \text{ m}} = 1.333$$
$$A = 53.1° \qquad \text{(see Appendix B, Section B.3)}$$

Find the magnitude of $\mathbf{R}$ using the Pythagorean theorem:

$$|\mathbf{R}| = \sqrt{|\mathbf{R}_x|^2 + |\mathbf{R}_y|^2}$$
$$|\mathbf{R}| = \sqrt{(3.00 \text{ m})^2 + (4.00 \text{ m})^2}$$
$$= 5.00 \text{ m}$$

That is, $\mathbf{R} = 5.00$ m at $53.1°$.

In general:

To find vector **R** in standard position when the *x*- and *y*-components are given:

1. Complete the right triangle with the legs being the *x*- and *y*-components of the vector.
2. Find the acute angle *A* of the right triangle whose vertex is at the origin by using tan *A*.
3. Find angle θ in standard position as follows:

$\theta = A$ (θ in first quadrant)

$\theta = 180° - A$ (θ in second quadrant)

$\theta = 180° + A$ (θ in third quadrant)

$\theta = 360° - A$ (θ in fourth quadrant)

4. Find the magnitude of the vector using the Pythagorean theorem:

$$\mathbf{R} = \sqrt{|\mathbf{R}_x|^2 + |\mathbf{R}_y|^2}$$

The Greek letter θ (theta) is often used to represent the measure of an angle.

EXAMPLE 5

Find vector **R** in standard position whose *x*-component is +7.00 mi and *y*-component is −5.00 mi.

First, graph the *x*- and *y*-components (Fig. 2.35) and complete the right triangle. The hypotenuse is the resultant vector. Find angle *A* as follows:

$$\tan A = \frac{\text{side opposite } A}{\text{side adjacent to } A}$$

$$\tan A = \frac{5.00 \text{ mi}}{7.00 \text{ mi}} = 0.7143$$

$$A = 35.5°$$

Figure 2.35

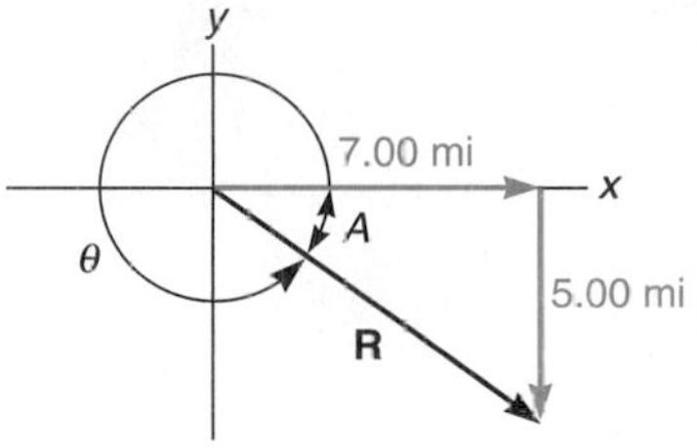

Then

$$\theta = 360° - A$$
$$= 360° - 35.5°$$
$$= 324.5°$$

Find the magnitude of **R** using the Pythagorean theorem:

$$|\mathbf{R}| = \sqrt{|\mathbf{R}_x|^2 + |\mathbf{R}_y|^2}$$
$$|\mathbf{R}| = \sqrt{(7.00 \text{ mi})^2 + (5.00 \text{ mi})^2}$$
$$= 8.60 \text{ mi}$$

That is, **R** = 8.60 mi at 324.5°.

EXAMPLE 6

Find vector **R** in standard position with $\mathbf{R}_x = -115$ km and $\mathbf{R}_y = +175$ km.

First, graph the *x*- and *y*-components (Fig. 2.36) and complete the right triangle. The hypotenuse is the resultant vector. Find angle *A* as follows:

$$\tan A = \frac{\text{side opposite } A}{\text{side adjacent to } A}$$

$$\tan A = \frac{175 \text{ km}}{115 \text{ km}} = 1.522$$

$$A = 56.7°$$

Figure 2.36

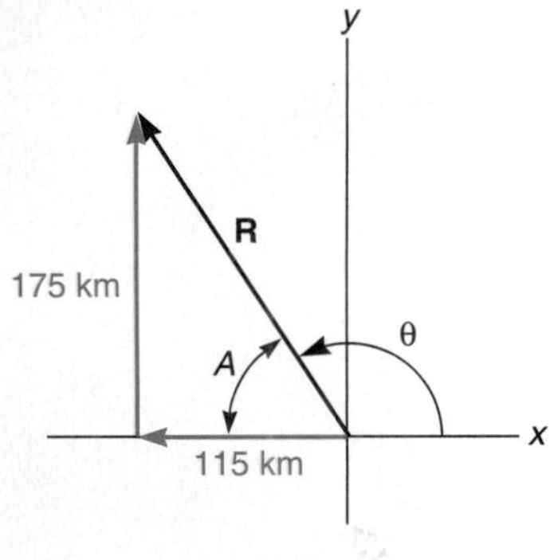

Then

$$\theta = 180° - A$$
$$= 180° - 56.7°$$
$$= 123.3°$$

Find the magnitude of **R** using the Pythagorean theorem:

$$|\mathbf{R}| = \sqrt{|\mathbf{R}_x|^2 + |\mathbf{R}_y|^2}$$
$$|\mathbf{R}| = \sqrt{(115 \text{ km})^2 + (175 \text{ km})^2}$$
$$= 209 \text{ km}$$

That is, **R** = 209 km at 123.3°.

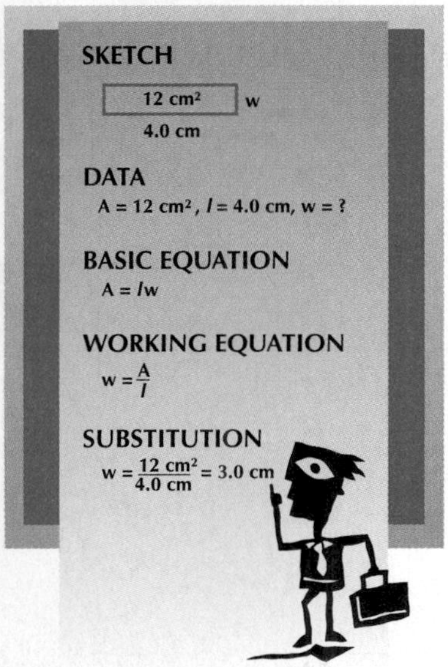

PROBLEMS 2.3

Make a sketch of each vector in standard position. Use the scale 1.0 cm = $1\bar{0}$ m.

1. **v** = $2\bar{0}$ m at 25°
2. **w** = 25 m at 125°
3. **u** = 25 m at 245°
4. **s** = $2\bar{0}$ m at 345°
5. **t** = 15 m at 105°
6. **r** = 35 m at 291°
7. **m** = $3\bar{0}$ m at 405°
8. **n** = 25 m at 525°

Find the *x*- and *y*-components of each vector.

9.

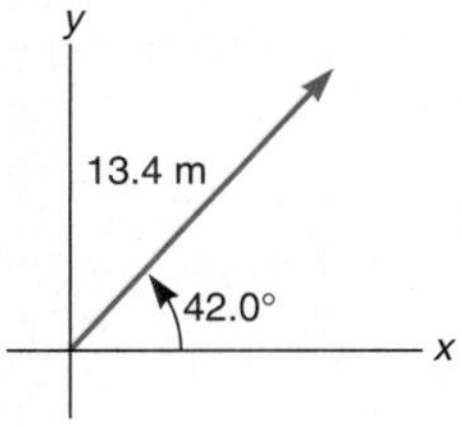

10.

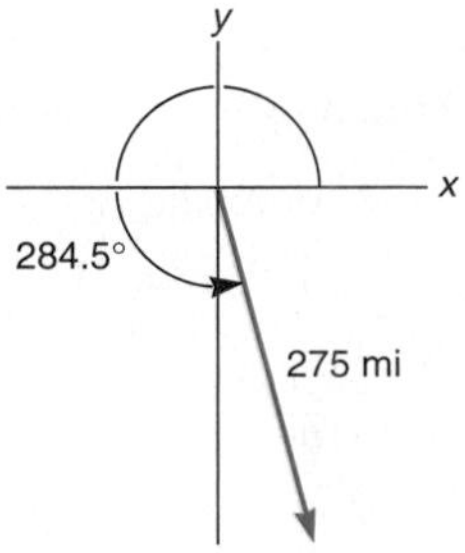

11.

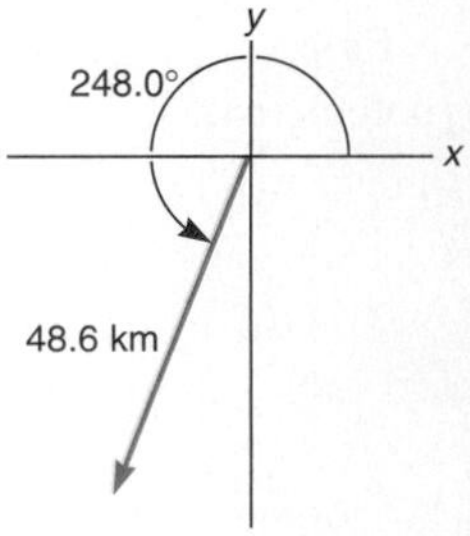

12.

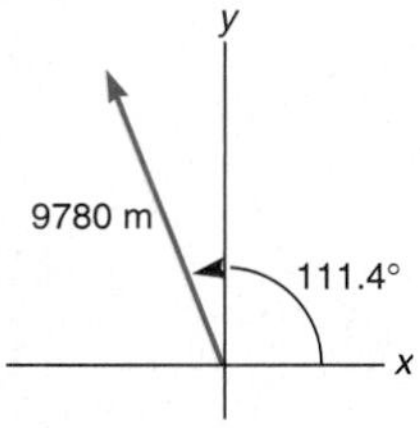

13.

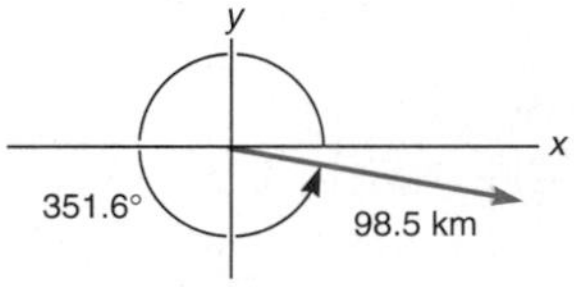

14.

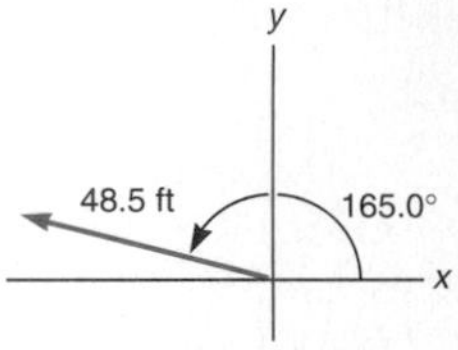

Find the *x*- and *y*-components of each vector given in standard position.

15. **v** = 38.9 m at 10.5°
16. **u** = 478 ft at 195.0°
17. **w** = 9.60 km at 310.0°
18. **s** = 5430 mi at 153.7°
19. **t** = 29.5 m at 101.5°
20. **m** = 154 mi at 273.2°

In Problems 21 through 32, find each resultant vector **R**. Give **R** in standard position.

21.

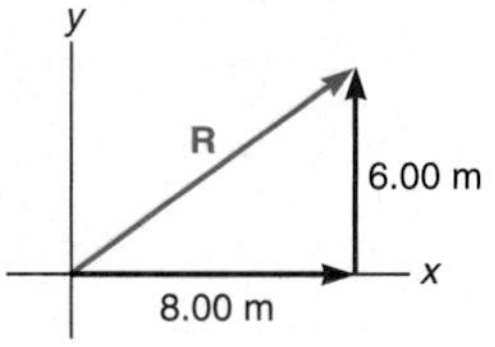

22.

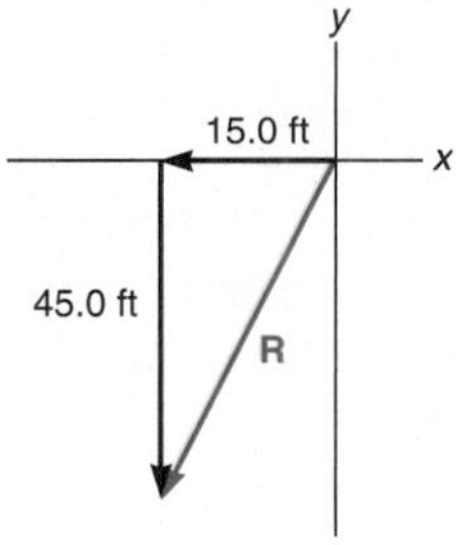

23.

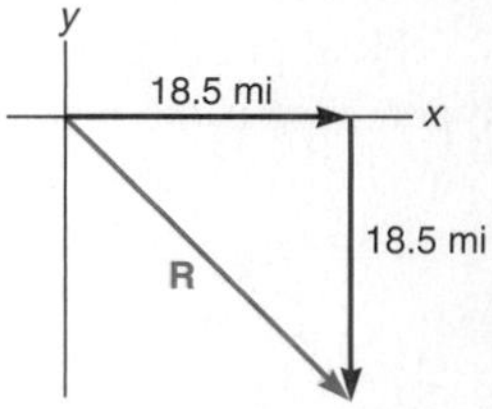

24.

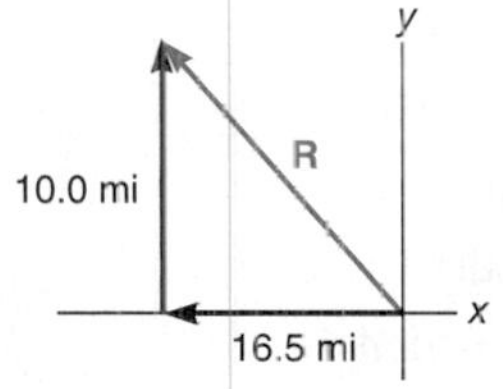

25.

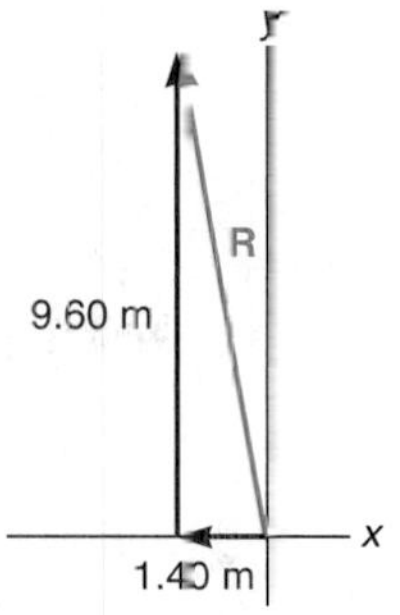

26.

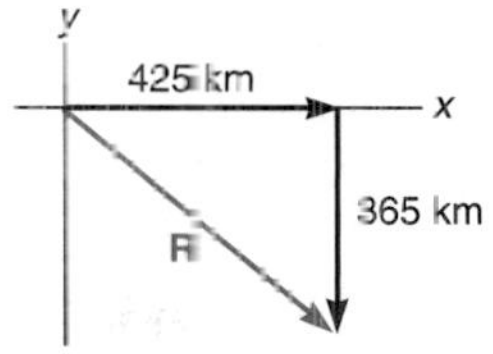

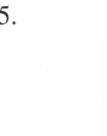

	x-component	*y-component*		*x-component*	*y-component*
27.	+19.5 m	−49.6 m	28.	−158 km	+236 km
29.	+14.7 mi	+16.8 mi	30.	−3240 ft	−1890 ft
31.	−9.65 m	+4.36 m	32.	+375 km	−408 km

Glossary

Component Vector When two or more vectors are added, each of the vectors is called a component of the resultant, or sum, vector. (p. 65)

Displacement The net change in position of an object, or the direct distance and direction it moves; a vector. (p. 58)

Number Plane A plane determined by the horizontal line called the *x*-axis and a vertical line called the *y*-axis intersecting at right angles. These two lines divide the number plane into four quadrants. The *x*-axis contains positive numbers to the right of the origin and negative numbers to the left of the origin. The *y*-axis contains positive numbers above the origin and negative numbers below the origin. (p. 65)

Resultant Vector The sum of two or more vectors. (pp. 60 and 65)

Scalar A physical quantity that can be completely described by a number (called its magnitude) and a unit. (p. 58)

Standard Position A vector is in standard position when its initial point is at the origin of the number plane. The vector is expressed in terms of its length and its angle θ, where θ is measured counterclockwise from the positive *x*-axis to the vector. (p. 73)

Vector A physical quantity that requires both magnitude (size) and direction to be completely described. (p. 58)

***x*-component** The horizontal component of a vector that lies along the *x*-axis. (p. 65)

***y*-component** The vertical component of a vector that lies along the *y*-axis. (p. 65)

Formulas

2.3 To find the *x*- and *y*-components of a vector **v** given in standard position (Fig. 2.37):

1. Complete the right triangle with the legs being the *x*- and *y*-components of the vector.
2. Find the lengths of the legs of the right triangle as follows:

$$|\mathbf{v}_x| = |\mathbf{v}|(\cos A)$$

$$|\mathbf{v}_y| = |\mathbf{v}|(\sin A)$$

3. Determine the signs of the *x*- and *y*- components.

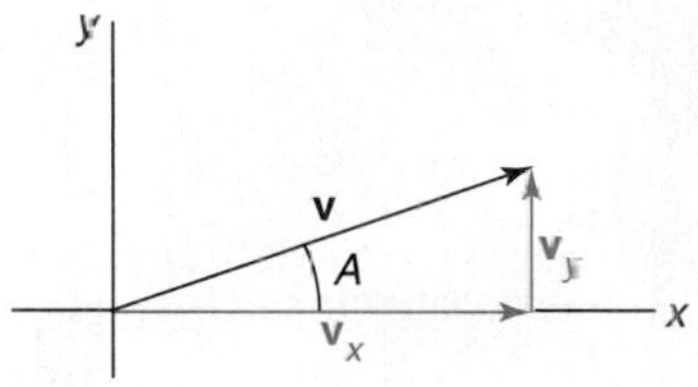

Figure 2.37

To find vector **R** in standard position when the *x*- and *y*-components are given:

1. Complete the right triangle with the legs being the *x*- and *y*-components of the vector.
2. Find the acute angle *A* of the right triangle whose vertex is at the origin by using tan *A*.

3. Find angle θ in standard position as follows:

$$\theta = A \qquad (\theta \text{ in first quadrant})$$
$$\theta = 180° - A \qquad (\theta \text{ in second quadrant})$$
$$\theta = 180° + A \qquad (\theta \text{ in third quadrant})$$
$$\theta = 360° - A \qquad (\theta \text{ in fourth quadrant})$$

4. Find the magnitude of the vector using the Pythagorean theorem:

$$\mathbf{R} = \sqrt{|\mathbf{R}_x|^2 + |\mathbf{R}_y|^2}$$

Review Questions

1. Displacement
   (a) can be interchanged with direction.
   (b) is a measurement of volume.
   (c) can be described only with a number.
   (d) is the net distance an object travels, showing direction and distance.
2. When adding vectors, the order in which they are added
   (a) is not important.
   (b) is important.
   (c) is important only in certain cases.
3. A vector is in standard position when its initial point is
   (a) at the origin.
   (b) along the x-axis.
   (c) along the y-axis.
4. Discuss number plane, origin, and axis in your own words.
5. Can every vector be described in terms of its components?
6. Can a vector have more than one set of component vectors?
7. Describe how to add two or more vectors graphically.
8. Describe how to find a resultant vector if given its x- and y-components.
9. Is a vector limited to a single position in the number plane?
10. Is the angle of a vector in standard position measured clockwise or counterclockwise?
11. What are the limits on the angle measure of a vector in standard position in the third quadrant?
12. Describe how to find the x- and y-components of a vector given in standard position.
13. Describe how to find a vector in standard position when the x- and y-components are given.

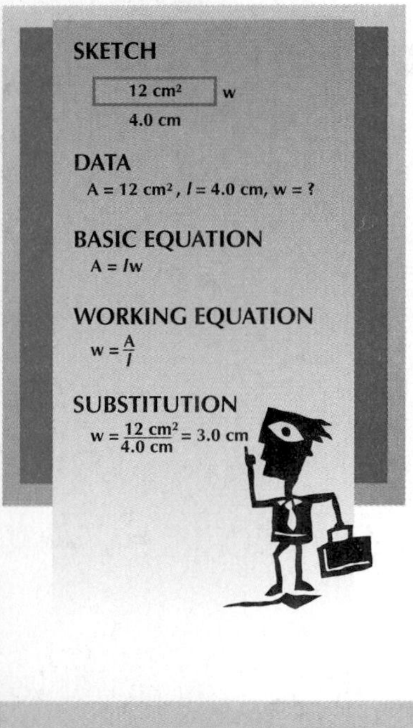

Review Problems

1. A hiker is plotting his course on a map with a scale of 1.00 cm = 3.00 km. If the hiker walks 2.50 cm north, then turns south and walks 1.50 cm, what is the actual displacement of the hiker in km?
2. A hiker is plotting his course on a map with a scale of 1.00 cm = 3.00 km. If the hiker walks 1.50 cm north, then turns south and walks 2.50 cm, what is the actual displacement of the hiker in km?
3. A co-pilot is charting her course on a map with a scale of 1.00 cm = 20.0 km. If the plane is charted to head 13.0 cm west, 9.00 cm north, and 2.00 cm east, what is the actual displacement of the plane in km?
4. A co-pilot is charting her course on a map with a scale of 1.00 cm = 20.0 km. If the plane is charted to head 25.0° north of east for 16.0 cm, north for 6.00 cm, and west for 5.00 cm, what is the actual displacement of the plane in km?
5. Vector **R** has x-component = +14.0 and y-component = +3.00. Find its length.
6. Vector **R** has x-component = –5.00 and y-component = +10.0. Find its length.
7. Vector **R** has x-component = +8.00 and y-component = –2.00. Find its length.
8. Vector **R** has x-component = –3.00 and y-component = –4.00. Find its length.

9. Vectors **A**, **B**, and **C** are given. Vector **A** has x-component = +3.00 and y-component = +4.00. Vector **B** has x-component = +5.00 and y-component = −7.00. Vector **C** has x-component = −2.00 and y-component = +1.00. Find the resultant vector **R**.
10. Vectors **A**, **B**, and **C** are given. Vector **A** has x-component = +5.00 and y-component = +7.00. Vector **B** has x-component = +9.00 and y-component = −3.00. Vector **C** has x-component = −5.00 and y-component = +5.00. Find the x- and y-components of the resultant vector **R**.
11. Vectors **A**, **B**, and **C** are given. Vector **A** has x-component = −3.00 and y-component = −4.00. Vector **B** has x-component = −5.00 and y-component = +7.00. Vector **C** has x-component = +2.00 and y-component = −1.00. Find the x- and y-components of the resultant vector **R**.
12. Vectors **A**, **B**, and **C** are given. Vector **A** has x-component = −5.00 and y-component = −7.00. Vector **B** has x-component = −9.00 and y-component = +3.00. Vector **C** has x-component = +5.00 and y-component = −5.00. Find the x- and y-components of the resultant vector **R**.

Graph and find the x- and y-components of each resultant vector **R**, where **R** = **A** + **B** + **C** + **D**.

13.

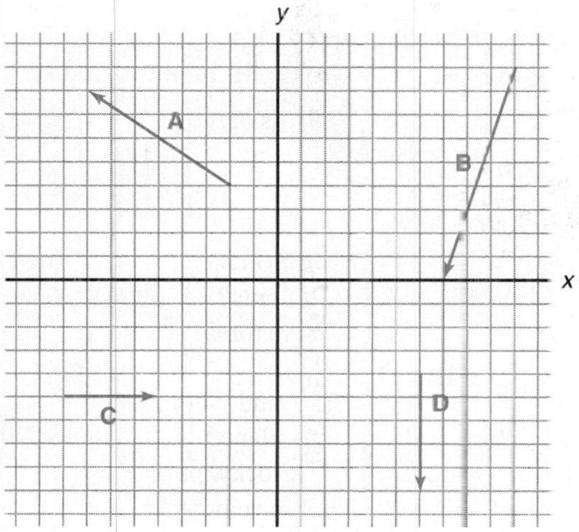

14.

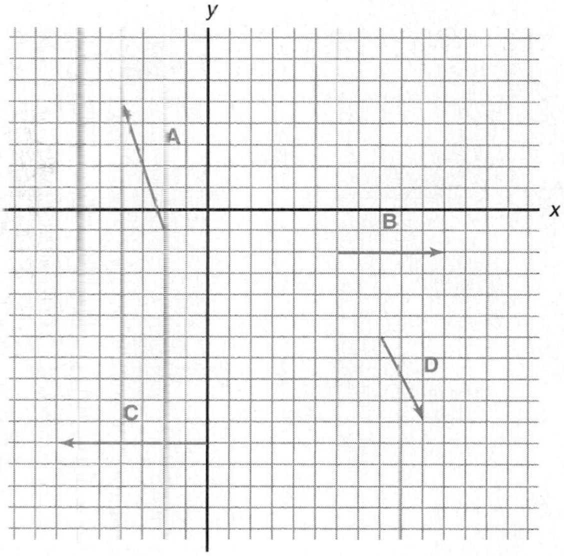

15. Find the x- and y-components of vector **R** which has a length of 13.0 cm at 30.0°.
16. Find the x- and y-components of vector **R**, which has a length of 10.0 cm at 60.0°.
17. Find the x- and y-components of vector **R**, which has a length of 20.0 cm at 50.0°.
18. Vector **R** has length 9.00 cm at 240.0°. Find its x- and y-components.
19. Vector **R** has length 9.00 cm at 40.0°. Find its x- and y-components.
20. Vector **R** has length 18.0 cm at 305.0°. Find its x- and y-components.

APPLIED CONCEPTS

1. The New Clark Bridge is an elegant cable-stayed bridge. Its design requires cables to reach from the road deck up to the tower and back down to the road deck on the other side of the tower as shown in Fig. 2.38. In order to determine the best method for shipping the cables, the shipping company needs to know the lengths of the shortest and longest cables. Given the measurements in the diagram, determine the total lengths of the indicated shortest and longest cables.

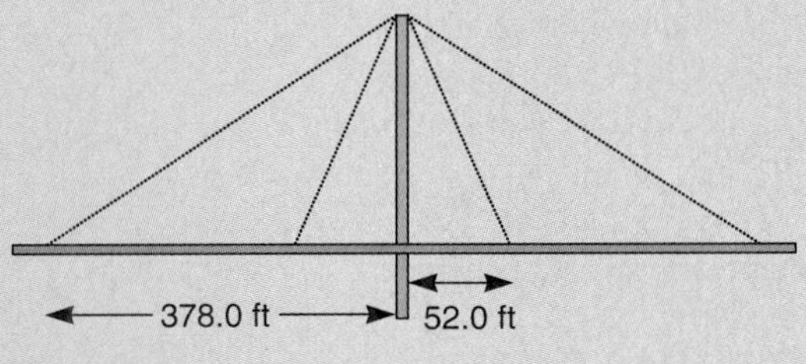

Figure 2.38

2. Frank just learned that the 80$\bar{0}$-m section of Broadway that he uses to get to work will be closed for several days. Given the information from a map of Manhattan (Fig. 2.39), what will be the distance of Frank's next shortest route?

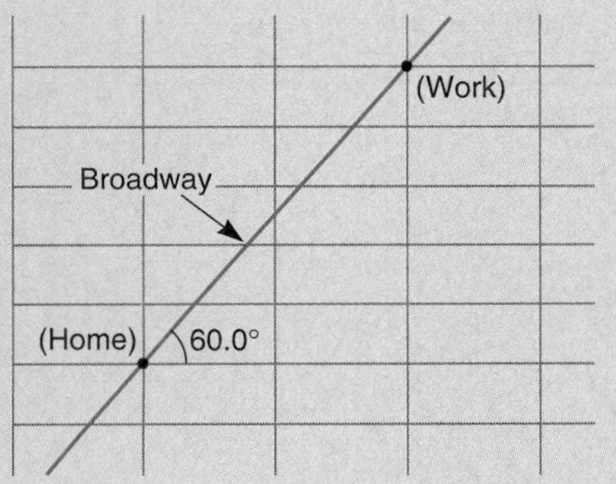

Figure 2.39

3. Power cables need to be suspended by the power company across a river to a new condominium development. Find the distance across the river in Fig. 2.40.

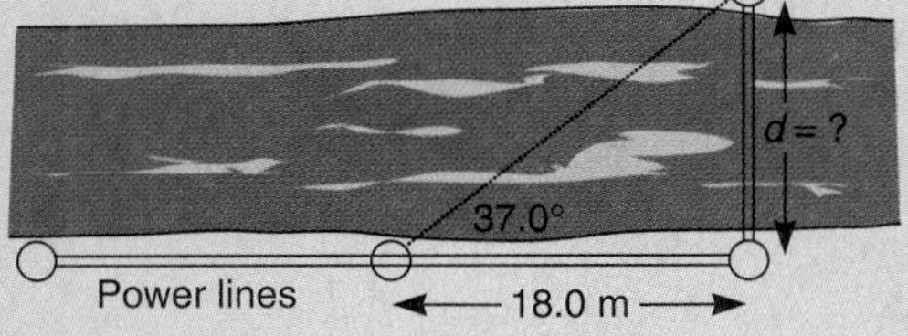

Figure 2.40

4. Bill has set his GPS to track his route. At the conclusion of his hike, the receiver indicates that he walked 3.50 mi north, 1.00 mi northeast, and 1.50 mi south. How far away is Bill from his original position?
5. With the airplane cruising at 30,$\bar{0}$00 ft, the navigator indicates to the captain that the plane should continue traveling north for 50$\bar{0}$ km and then turn to a heading of 45.0° east of north for 20$\bar{0}$ km. What will be the resultant distance traveled?

CHAPTER 3

MOTION

Motion is a change of position. Velocity and acceleration describe important kinds of motion. An analysis of motion helps introduce the real nature of physics—to understand the nature and behavior of the physical world.

Objectives

The major goals of this chapter are to enable you to:

1. Distinguish between speed and velocity.
2. Use vectors to illustrate and solve velocity problems.
3. Evaluate the difference between velocity and acceleration.
4. Utilize vectors to illustrate and solve acceleration problems.
5. Analyze the motion of an object in free fall.
6. Solve two-dimensional motion problems.
7. Calculate the range of projectile motion.

3.1 Speed Versus Velocity

This chapter begins our study of mechanics, the study of motion. **Motion** can be defined as an object's change in position. How quickly the object changes its position is called its speed.

The ability to analyze and determine the speed of an object is important in many areas of science and technology. Automotive engineers are concerned not only with the motion of the entire vehicle, but also with the motion of the pistons, valves, driveshaft, and so on. The particular speeds of all the internal parts have a direct and very important effect on the motion of the vehicle.

Speed, as measured on a speedometer, is the distance traveled per unit of time. The speed of an automobile is represented in either miles per hour or kilometres per hour. These units actually help define the formula for calculating speed (Fig. 3.1):

Figure 3.1 A speedometer measures speed, but not velocity.

Copyright of Image Stock Imagery. Photo reprinted with permission

$$\text{speed} = \frac{\text{distance traveled}}{\text{time to move that distance}}$$

Speed is a scalar value, for it shows only the magnitude of the position change per unit of time and does not indicate a direction. The unit for speed is a distance unit divided by a time unit, such as miles per hour (mi/h), kilometres per hour (km/h), metres per second (m/s), and feet per second (ft/s). For example, if you drive $35\bar{0}$ mi in 7.00 h, your average speed is

$$\frac{35\bar{0}\text{ mi}}{7.00\text{ h}} = 50.0\text{ mi/h}$$

Speed represents how fast something is moving, yet it does not indicate the direction in which it is traveling. Suppose you started driving from Chicago at a speed of 50 mi/h for 6 h. Where did you end your trip? You may have driven 50 mi/h southwest toward St. Louis, 50 mi/h northeast toward Detroit, 50 mi/h southeast toward Louisville, or 50 mi/h in a loop that brought you back to Chicago. Although speed may indicate how fast you are moving, it may not give you all the information you need to solve a problem.

The **velocity** of an object is the rate of motion in a particular direction. Velocity is a vector that not only represents the speed, but also indicates the direction of motion. The relationship may be expressed by the equation

$$v_{avg} = \frac{s}{t}$$

or

$$\boxed{s = v_{avg}t}$$

where

s = displacement or distance
v_{avg} = average velocity or speed
t = time

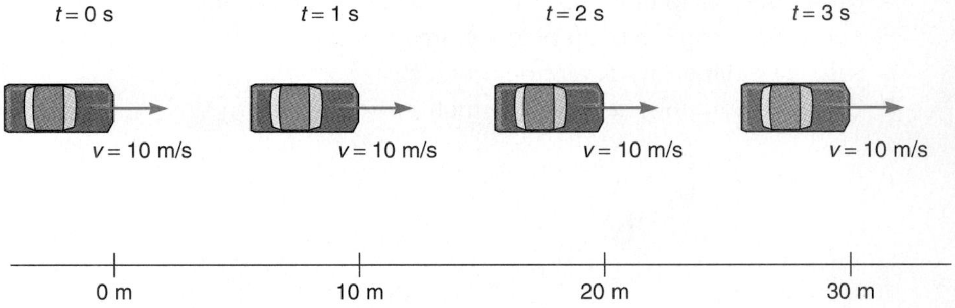

Figure 3.2 The velocity, distance, and time for a car traveling at a constant velocity of 10 m/s to the right is shown in 1-s intervals.

This equation is used to find either average speed (a scalar quantity) or the magnitude of the velocity (a vector quantity). Remember that if indicating velocity, the direction must be included with the speed. Therefore, a speed of 50 mi/h would be written 50 mi/h northeast, 50 mi/h up, or 50 mi/h 30° east of south as a velocity.

Figure 3.2 shows an illustration of a car traveling at constant velocity.

EXAMPLE 1

Find the average speed of an automobile that travels 160 km in 2.0 h.

Sketch: None needed

Data:

$$s = 160 \text{ km}$$
$$t = 2.0 \text{ h}$$
$$v_{avg} = ?$$

Basic Equation:

$$s = v_{avg}t$$

Working Equation:

$$v_{avg} = \frac{s}{t}$$

Substitution:

$$v_{avg} = \frac{160 \text{ km}}{2.0 \text{ h}}$$
$$= 8\bar{0} \text{ km/h}$$

EXAMPLE 2

An airplane flies $35\bar{0}0$ mi in 5.00 h. Find its average speed.

Sketch: None needed

Data:

$$s = 35\bar{0}0 \text{ mi}$$
$$t = 5.00 \text{ h}$$
$$v_{avg} = ?$$

Basic Equation:

$$s = v_{avg}t$$

Working Equation:

$$v_{avg} = \frac{s}{t}$$

Substitution:

$$v_{avg} = \frac{35\bar{0}0 \text{ mi}}{5.00 \text{ h}}$$
$$= 7\bar{0}0 \text{ mi/h}$$

EXAMPLE 3

Find the velocity of a plane that travels $60\bar{0}$ km due north in 3 h 15 min.

Sketch: None needed

Data:

$$s = 6\bar{0}0 \text{ km}$$
$$t = 3 \text{ h } 15 \text{ min} = 3.25 \text{ h}$$
$$v_{avg} = ?$$

Basic Equation:

$$s = v_{avg}t$$

Working Equation:

$$v_{avg} = \frac{s}{t}$$

Substitution:

$$v_{avg} = \frac{6\bar{0}0 \text{ km}}{3.25 \text{ h}} = 185 \text{ km/h}$$

The direction is north. Thus, the velocity is 185 km/h due north.

To find the sum (resultant vector) of velocity vectors, use the component method as outlined in Chapter 2.

EXAMPLE 4

A plane is flying due north (at 90°) at 265 km/h. Suddenly there is a wind from the east (at 180°) at 55.0 km/h. What is the plane's new velocity with respect to the ground in standard position?

First, graph the plane's old velocity as the y-component and the wind velocity as the x-component (Fig. 3.3). The resultant vector is the plane's new velocity with respect to the ground. Find angle A as follows:

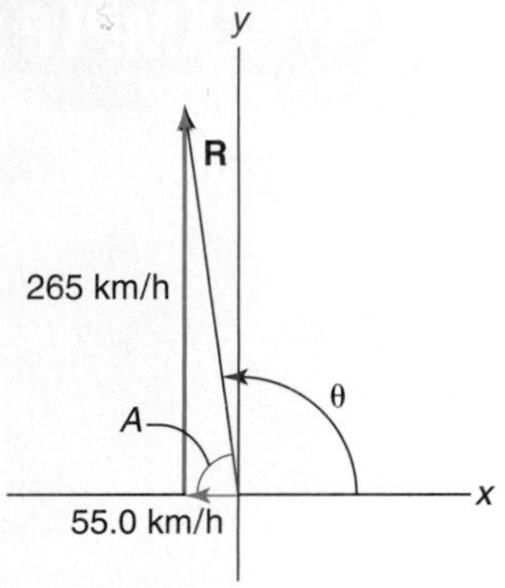

Figure 3.3

$$\tan A = \frac{\text{side opposite } A}{\text{side adjacent to } A}$$
$$\tan A = \frac{265 \text{ km/h}}{55.0 \text{ km/h}} = 4.818$$
$$A = 78.3°$$

then

$$\theta = 180° - 78.3° = 101.7°$$

Find the magnitude of the new velocity (ground speed) using the Pythagorean theorem:

$$|\mathbf{R}| = \sqrt{|\mathbf{R}_x|^2 + |\mathbf{R}_y|^2}$$
$$|\mathbf{R}| = \sqrt{(55.0 \text{ km/h})^2 + (265 \text{ km/h})^2} = 271 \text{ km/h}$$

That is, the new velocity of the plane is 271 km/h at 101.7°.

EXAMPLE 5

A plane is flying northwest (at 135.0°) at 315 km/h. Suddenly there is a wind from 30.0° south of west (at 30.0°) at 65.0 km/h. What is the plane's new velocity with respect to the ground in standard position?

First, graph the plane's old velocity and the wind velocity as vectors in standard position (Fig. 3.4). The resultant vector is the plane's new velocity with respect to the ground.

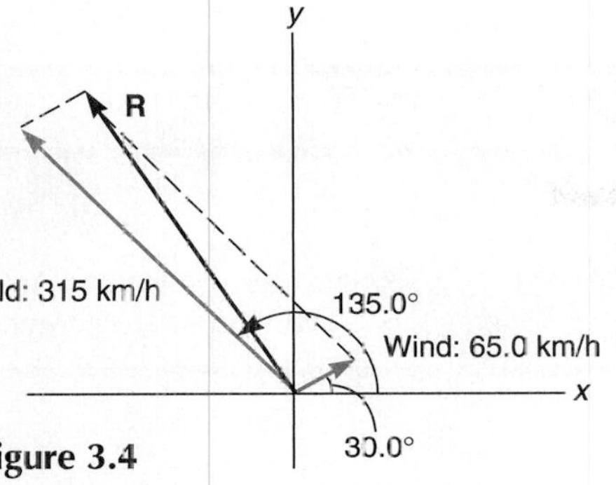

Figure 3.4

Then, find the x- and y-components of the plane's old velocity and the wind velocity using Fig. 3.5.

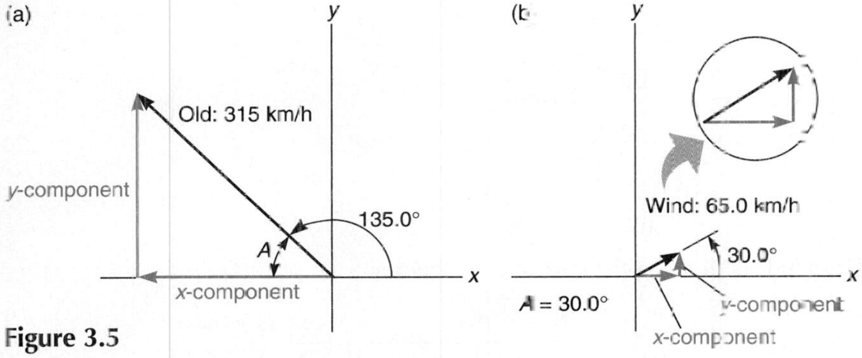

Figure 3.5

Plane: See Fig. 3.5(a). $A = 180° - 135.0° = 45.0°$

***x*-component**	***y*-component**
$\cos A = \dfrac{\text{side adjacent to } A}{\text{hypotenuse}}$	$\sin A = \dfrac{\text{side opposite } A}{\text{hypotenuse}}$
$\cos 45.0° = \dfrac{x\text{-component}}{315 \text{ km/h}}$	$\sin 45.0° = \dfrac{y\text{-component}}{315 \text{ km/h}}$
$(315 \text{ km/h})(\cos 45.0°) = x\text{-component}$	$(315 \text{ km/h})(\sin 45.0°) = y\text{-component}$
$223 \text{ km/h} = x\text{-component}$	$223 \text{ km/h} = y\text{-component}$
Thus, x-component $= -223$ km/h	y-component $= +223$ km/h

Wind: See Fig. 3.5(b).

***x*-component**	***y*-component**
$\cos A = \dfrac{\text{side adjacent to } A}{\text{hypotenuse}}$	$\sin A = \dfrac{\text{side opposite } A}{\text{hypotenuse}}$
$\cos 30.0° = \dfrac{x\text{-component}}{65.0 \text{ km/h}}$	$\sin 30.0° = \dfrac{y\text{-component}}{65.0 \text{ km/h}}$
$(65.0 \text{ km/h})(\cos 30.0°) = x\text{-component}$	$(65.0 \text{ km/h})(\sin 30.0°) = y\text{-component}$
$56.3 \text{ km/h} = x\text{-component}$	$32.5 \text{ km/h} = y\text{-component}$
Thus, x-component $= +56.3$ km/h	y-component $= -32.5$ km/h

To find **R**:

	x-component	*y*-component	
Plane:	−223 km/h	+223 km/h	
Wind:	+56.3 km/h	+32.5 km/h	
Sum:	−167 km/h	+256 km/h	(Round each component sum to its least precise component.)

Find angle *A* from Fig. 3.6 as follows:

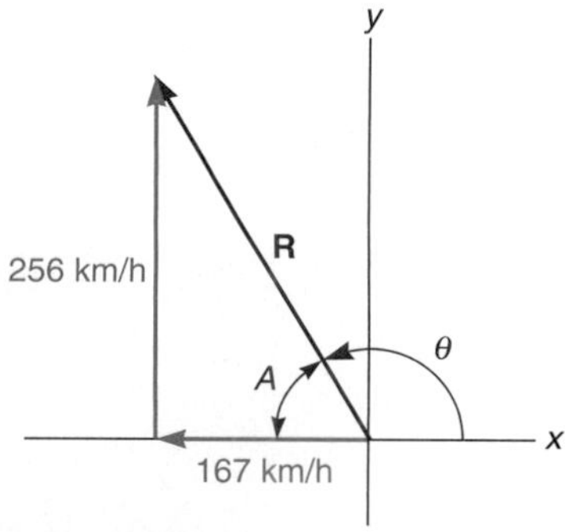

Figure 3.6

$$\tan A = \frac{\text{side opposite } A}{\text{side adjacent to } A}$$

$$\tan A = \frac{256 \text{ km/h}}{167 \text{ km/h}} = 1.533$$

$$A = 56.9°$$

and

$$\theta = 180° - 56.9° = 123.1°$$

Find the magnitude of **R** using the Pythagorean theorem:

$$|\mathbf{R}| = \sqrt{|\mathbf{R}_x|^2 + |\mathbf{R}_y|^2}$$

$$|\mathbf{R}| = \sqrt{(167 \text{ km/h})^2 + (256 \text{ km/h})^2}$$

$$= 306 \text{ km/h}$$

That is, the new velocity of the plane is 306 km/h at 123.1°.

PHYSICS CONNECTIONS

Vectors Across Rivers

Crossing the Hudson River, which separates New York City from New Jersey, can be challenging. A working knowledge of velocity and vectors is absolutely essential, especially when attempting to cross the river in strong currents, brisk winds, driving rain, and dense fog. In addition, maneuvering between barges, cruise ships, recreational boaters, and driftwood can make the job even more difficult.

Ferry captains like Mike and John combine vectors every time they cross the river. Captain Mike said, "At times crossing the river can be quite tricky. Your ferry might be pointing directly across the river, but the current is pushing you farther down river. In order to combat this, you need to change the boat's heading so when your velocity and the current's velocity combine, you arrive at your planned destination" (Fig. 3.7).

Sometimes, when the current and wind are strong and headed in the same direction, ferryboats can appear to be heading up to 45° away from their destination, yet still travel directly across the river. In such situations, docking can be nerve-racking (Fig. 3.8).

The two experienced sea captains say vectors are even more important in the open seas. Captain Mike said, "Combining your boat's velocity vector with the current and wind vectors can mean the difference between arriving at your home port or at a port 100 miles away." These days the process is easier with the use of radar and computer navigation equipment and software. Looking at the monitor, Captain Mike said, "We can see the velocity of the current and our boat's intended velocity directly on the monitor. Instead of our combining the vectors, the computer can instantaneously combine them and provide the resultant velocity" (Fig. 3.9).

Figure 3.7 An example of how the velocity of a boat and the velocity of the current are combined so the resultant velocity is directed toward the desired location.

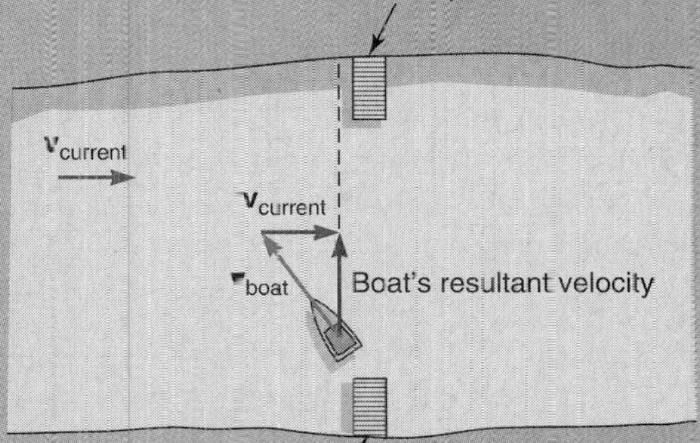

Figure 3.8 Although the boat is not pointed toward the dock, the combination of the boat's velocity (green vector) plus the current's velocity (blue vector) results in a perfect docking (red vector).

Figure 3.9(a) Captain Mike combines velocity vectors every time he navigates across the Hudson River.

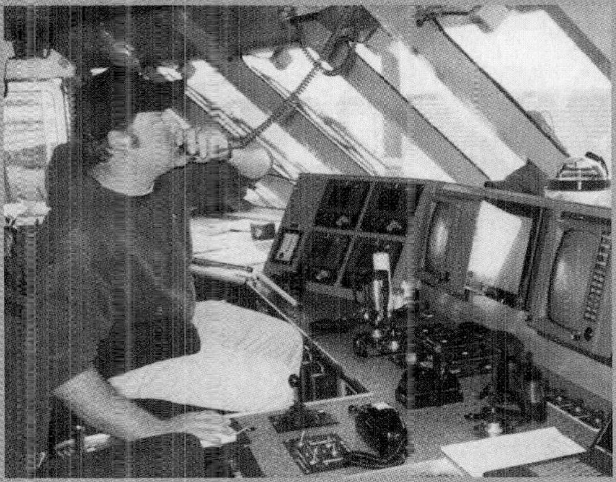

Figure 3.9(b) Captain John's high-speed ferry is equipped with computers that automatically combine and display the boat's resultant velocity vector.

SKETCH

12 cm² w
4.0 cm

DATA

$A = 12\ cm^2$, $l = 4.0$ cm, $w = ?$

BASIC EQUATION

$A = lw$

WORKING EQUATION

$w = \frac{A}{l}$

SUBSTITUTION

$w = \frac{12\ cm^2}{4.0\ cm} = 3.0$ cm

PROBLEMS 3.1

Find the average speed (in the given units) of an auto that travels each distance in the given time.

1. Distance of 150 mi in 3.0 h (in mi/h)
2. Distance of 190 m in 8.5 s (in m/s)
3. Distance of 8550 m in 6 min 35 s (in m/s)
4. Distance of 45 km in 0.50 h (in km/h)
5. Distance of 785 ft in 11.5 s (in ft/s)
6. Find the average speed (in mi/h) of a racing car that turns a lap on a 1.00-mi oval track in 30.0 s.
7. While driving at $9\overline{0}$ km/h, how far can you travel in 3.5 h?
8. While driving at $9\overline{0}$ km/h, how far (in metres) do you travel in 1.0 s?
9. An automobile is traveling at 55 mi/h. Find its speed
   (a) in ft/s. (b) in m/s. (c) in km/h.
10. An automobile is traveling at 22.0 m/s. Find its speed
    (a) in km/h. (b) in mi/h. (c) in ft/s.

Find the velocity for each displacement and time.

11. 160 km east in 2.0 h
12. $1\overline{0}0$ km north in 3.0 h
13. $10\overline{0}0$ mi south in 8.00 h
14. 31.0 mi west in 0.500 h
15. 275 km at $3\overline{0}°$ south of east in 4.50 h
16. 426 km at 45° north of west in 2.75 h
17. Milwaukee is 121 mi (air miles) due west of Grand Rapids. Maria drives 255 mi in 4.75 h from Grand Rapids to Milwaukee around Lake Michigan. Find (a) her average driving speed and (b) her average travel velocity.
18. Telluride, Colorado, is 45 air miles at 11° east of north of Durango. On a winter day, Chuck drove 120 mi from Durango to Telluride around a mountain in $4\frac{1}{4}$ h including a traffic delay. Find (a) his average driving speed and (b) his average travel velocity.
19. A plane is flying due north at 325 km/h. Suddenly there is a wind from the south at 45 km/h. What is the plane's new velocity with respect to the ground in standard position?
20. A plane is flying due west at 275 km/h. Suddenly there is a wind from the west at $8\overline{0}$ km/h. What is the plane's new velocity with respect to the ground in standard position?
21. A plane is flying due west at 235 km/h. Suddenly there is a wind from the north at 45.0 km/h. What is the plane's new velocity with respect to the ground in standard position?
22. A plane is flying due north at 185 mi/h. Suddenly there is a wind from the west at 35.0 mi/h. What is the plane's new velocity with respect to the ground in standard position?
23. A plane is flying southwest at 155 mi/h. Suddenly there is a wind from the west at 45.0 mi/h. What is the plane's new velocity with respect to the ground in standard position?
24. A plane is flying southeast at 215 km/h. Suddenly there is a wind from the north at 75.0 km/h. What is the plane's new velocity with respect to the ground in standard position?
25. A plane is flying at 25.0° north of west at $19\overline{0}$ km/h. Suddenly there is a wind from 15.0° north of east at 45.0 km/h. What is the plane's new velocity with respect to the ground in standard position?
26. A plane is flying at 36.0° south of west at $15\overline{0}$ mi/h. Suddenly there is a wind from 75.0° north of east at 55.0 mi/h. What is the plane's new velocity with respect to the ground in standard position?

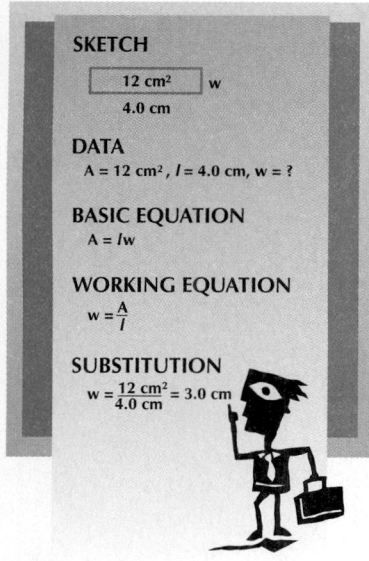

3.2 Acceleration

When the dragster shown in Fig. 3.10 travels down a quarter-mile track, its velocity changes. Its velocity at the end of the race is much greater than its velocity near the start. The faster the velocity of the dragster changes, the less its travel time will be.

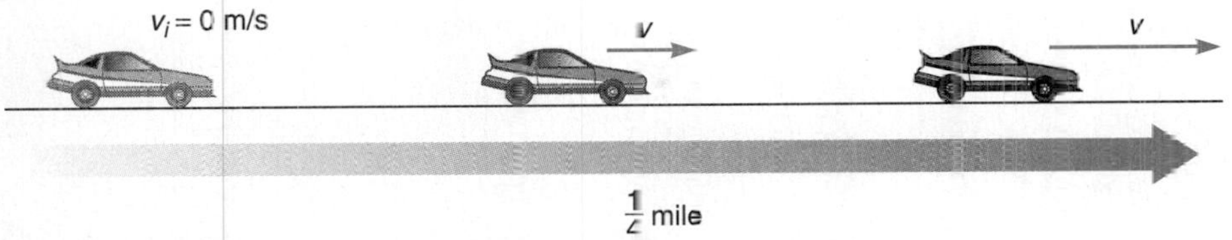

Figure 3.10 The velocity of the dragster changes in magnitude from zero at the start to its final velocity at the finish.

And the faster its velocity changes, the larger its acceleration will be. **Acceleration** *is the change in velocity per unit time.* That is,

$$\text{average acceleration} = \frac{\text{change in velocity (or speed)}}{\text{elapsed time}}$$
$$= \frac{\text{final velocity} - \text{initial velocity}}{\text{time}}$$

This relationship can be expressed by the equation

$$a = \frac{\Delta v}{t} = \frac{v_f - v_i}{t}$$

or

$$\boxed{\Delta v = at}$$

where

Δv = change in velocity (or speed)
a = acceleration
t = time

The Greek letter Δ (capital delta) is used to mean "change in."

EXAMPLE 1

A dragster starts from rest (velocity = 0 ft/s) and attains a speed of $15\bar{0}$ ft/s in 10.0 s. Find its acceleration.

Sketch: None needed

Data:

$$\Delta v = 15\bar{0}\text{ ft/s} - 0\text{ ft/s} = 15\bar{0}\text{ ft/s}$$
$$t = 10.0\text{ s}$$
$$a = ?$$

Basic Equation:

$$\Delta v = at$$

Working Equation:

$$a = \frac{\Delta v}{t}$$

Substitution:

$$a = \frac{15\bar{0}\ \text{ft/s}}{10.0\ \text{s}}$$

$$= 15.0\ \frac{\text{ft/s}}{\text{s}} \text{ or 15.0 feet per second per second}$$

Recall from arithmetic that to simplify fractions in the form

$$\frac{\frac{a}{b}}{\frac{c}{d}}$$

we divide by the denominator; that is, invert and multiply:

$$\frac{\frac{a}{b}}{\frac{c}{d}} = \frac{a}{b} \div \frac{c}{d} = \frac{a}{b} \cdot \frac{d}{c} = \frac{ad}{bc}$$

Use this idea to simplify the units 15.0 feet per second per second:

$$\frac{\frac{15.0\ \text{ft}}{\text{s}}}{\frac{\text{s}}{1}} = \frac{15.0\ \text{ft}}{\text{s}} \div \frac{\text{s}}{1} = \frac{15.0\ \text{ft}}{\text{s}} \cdot \frac{1}{\text{s}} = \frac{15.0\ \text{ft}}{\text{s}^2} \text{ or } 15.0\ \text{ft/s}^2$$

The units of acceleration are usually ft/s^2 or m/s^2.

When the speed of an automobile increases from rest to 5 mi/h in the first second, to 10 mi/h in the next second, and to 15 mi/h in the third second, its acceleration is $5\ \frac{\text{mi/h}}{\text{s}}$. That is, its increase in speed is 5 mi/h during each second. If an automobile increases in speed from 6 m/s to 9 m/s in the first second, to 12 m/s in the next second, and to 15 m/s in the third second, its acceleration is $3\ \frac{\text{m/s}}{\text{s}}$, usually written $3\ \text{m/s}^2$ (Fig. 3.11). This means that the speed of the automobile increases 3 m/s during each second.

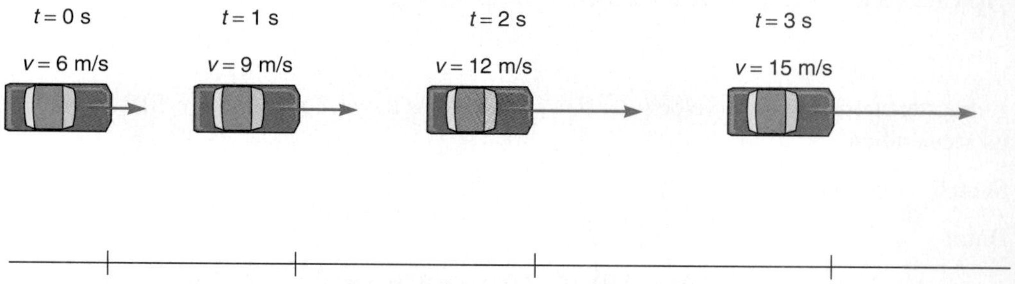

Figure 3.11 This car is speeding up with a constant acceleration. Note how the distance covered and the velocity change during each time interval.

EXAMPLE 2

A car accelerates from 45 km/h to $8\bar{0}$ km/h in 3.00 s. Find its acceleration (in m/s^2).

Sketch: None needed

Data:

$$\Delta v = 8\bar{0}\ \text{km/h} - 45\ \text{km/h} = 35\ \text{km/h}$$

$$t = 3.00\ \text{s}$$

$$a = ?$$

Basic Equation:

$$\Delta v = at$$

Working Equation:

$$a = \frac{\Delta v}{t}$$

Substitution:

$$a = \frac{35\ \text{km/h}}{3.00\ \text{s}} \times \frac{1000\ \text{m}}{1\ \text{km}} \times \frac{1\ \text{h}}{3600\ \text{s}}$$
$$= 3.2\ \text{m/s}^2$$

Note the use of the conversion factors to change the units km/h/s to m/s^2.

EXAMPLE 3

A plane accelerates at $8.5\ \text{m/s}^2$ for 4.5 s. Find its increase in speed (in m/s).

Sketch: None needed

Data:

$$a = 8.5\ \text{m/s}^2$$
$$t = 4.5\ \text{s}$$
$$\Delta v = ?$$

Basic Equation:

$$\Delta v = at$$

Working Equation: Same

Substitution:

$$\Delta v = (8.5\ \text{m/s}^2)(4.5\ \text{s})$$
$$= 38\ \text{m/s}$$

$$\boxed{\frac{\text{m}}{\text{s}^2} \times \text{s} = \frac{\text{m}}{\text{s}}}$$

Acceleration means more than just an increase in speed. In fact, since velocity is the speed of an object and its direction of motion, acceleration can mean speeding up, slowing down, or changing direction. The next example will illustrate negative acceleration (sometimes called deceleration). **Deceleration** is an acceleration that usually indicates that an object is slowing down (Fig. 3.12). Acceleration when an object changes direction will be discussed in chapter 8.

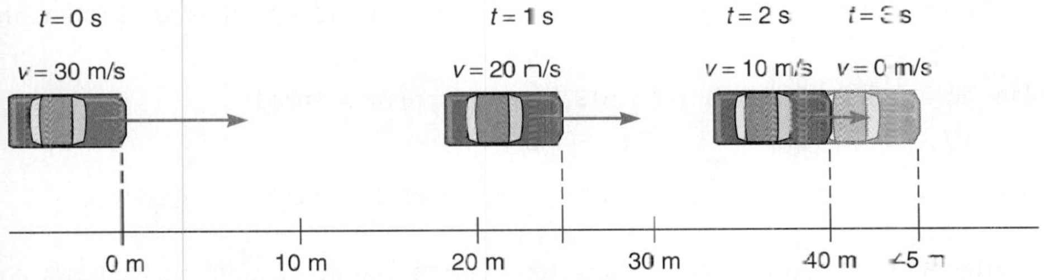

Figure 3.12 This car is slowing down with a constant acceleration of $-10\ \text{m/s}^2$. Note how the distance covered and the velocity change during each unit of time interval.

EXAMPLE 4

A driver steps off the gas pedal and coasts at a rate of $-3.00\ \text{m/s}^2$ for 5.00 s. Find the driver's new speed if she was originally traveling at a velocity of 20.0 m/s. (The negative acceleration indicates that the acceleration is pointed in the opposite direction of the velocity; that is, the object is slowing down.)

Sketch: None needed

Data:

$$a = -3.00 \text{ m/s}^2$$
$$t = 5.00 \text{ s}$$
$$v_i = 20.0 \text{ m/s}$$

Basic Equation:

$$\Delta v = at$$
$$v_f - v_i = at$$

Working Equation:

$$v_f = v_i + at$$

Substitution:

$$v_f = 20.0 \text{ m/s} + (-3.00 \text{ m/s}^2)(5.00 \text{ s})$$
$$v_f = 5.0 \text{ m/s}$$

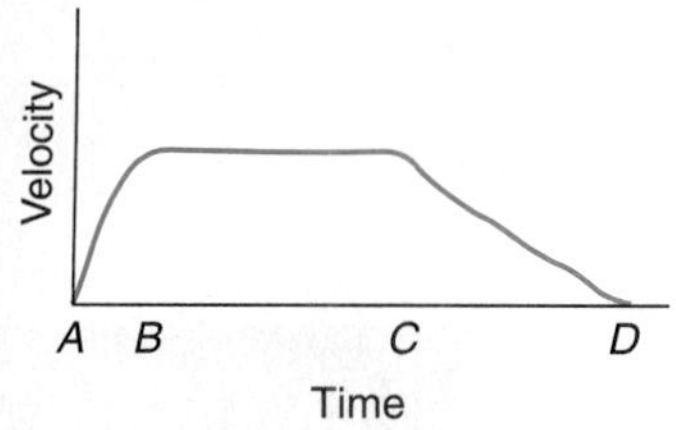

Figure 3.13 Motion of a high-speed train going from one station to another. When the speed increases, the acceleration is positive. When the speed is constant, the acceleration is zero. When the speed decreases, the acceleration is negative.

Acceleration has both magnitude and direction. Consider a high-speed train moving out of the station due east (positive direction). Its velocity increases in magnitude until it reaches its cruising speed (Fig. 3.13). Its acceleration is greatest at the start (from *A* to *B*), when its increase in velocity is largest. When the train is moving at relatively constant velocity (from *B* to *C*), its acceleration is near zero. Then as the train approaches the next station (from *C* to *D*), its speed decreases as it decelerates.

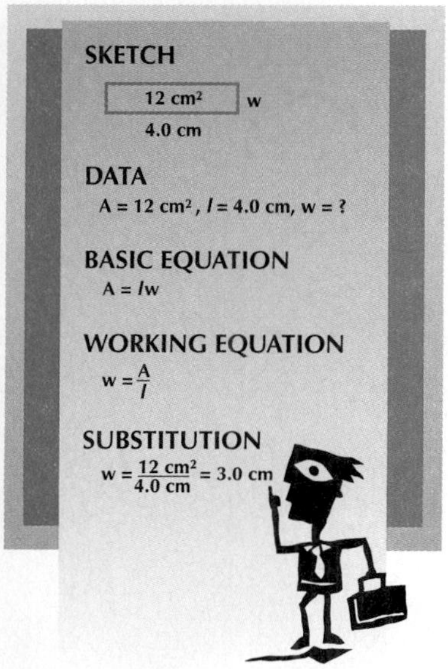

PROBLEMS 3.2

An automobile changes speed as shown. Find its acceleration.

	Speed Change	*Time Interval*	*Find a*
1.	From 0 to 15 m/s	1.0 s	in m/s^2
2.	From 0 to 18 m/s	3.0 s	in m/s^2
3.	From $6\bar{0}$ ft/s to $7\bar{0}$ ft/s	1.0 s	in ft/s^2
4.	From 45 m/s to 65 m/s	2.0 s	in m/s^2
5.	From 25 km/h to $9\bar{0}$ km/h	5.6 s	in m/s^2
6.	From $1\bar{0}$ mi/h to $5\bar{0}$ mi/h	3.5 s	in ft/s^2

7. A dragster starts from rest and reaches a speed of 62.5 m/s in 10.0 s. Find its acceleration (in m/s^2).
8. A car accelerates from 25 mi/h to 55 mi/h in 4.5 s. Find its acceleration (in ft/s^2).
9. A train accelerates from $1\bar{0}$ km/h to $11\bar{0}$ km/h in 2 min 15 s. Find its acceleration (in m/s^2).

A plane accelerates at 30.0 ft/s^2 for 3.30 s. Find its increase in speed

10. in ft/s.
11. in mi/h.

A rocket accelerates at 10.0 m/s^2 from rest for 20.0 s. Find its increase in speed

12. in m/s.
13. in km/h.
14. How long (in seconds) does it take for a rocket sled accelerating at 15.0 m/s^2 to change its speed from 20.0 m/s to 65.0 m/s?
15. How long (in seconds) does it take for a truck accelerating at 1.50 m/s^2 to go from rest to 90.0 km/h?
16. How long (in seconds) does it take for a car accelerating at 3.50 m/s^2 to go from rest to $12\bar{0}$ km/h?
17. The speed of a delivery van increases from 2.00 m/s at 1.00 s to 16.0 m/s at 4.50 s. What is its average speed?

18. A go-cart rolls backward down a driveway. We define forward speed as positive and backward speed as negative. The cart's speed changes from –2.00 m/s to –9.00 m/s in 2.00 s. What is its acceleration?
19. A stock car is moving at 25.0 m/s when the driver applies the brakes. If it stops in 3.00 s, what is its average acceleration?
20. If the car in Problem 19 took twice as long to stop, what would its acceleration be?
21. If the car in Problem 19 was going twice as fast but was able to stop in the same time, what would its acceleration be?
22. If the car in Problem 19 was going twice the speed and stopped in twice the time, what would its acceleration be?

3.3 Uniformly Accelerated Motion and Free Fall

The following equations apply to freely falling bodies and uniformly accelerated motion in general.

ACCELERATED MOTION

$$v_{avg} = \frac{v_f + v_i}{2} \qquad s = v_i t + \tfrac{1}{2} a_{avg} t^2$$

$$a_{avg} = \frac{v_f - v_i}{t} \qquad s = \tfrac{1}{2}(v_f + v_i)t$$

$$v_f = v_i + a_{avg} t \qquad 2a_{avg} s = v_f^2 - v_i^2$$

where

s = displacement v_{avg} = average velocity
v_f = final velocity a_{avg} = average acceleration
v_i = initial velocity t = time

Now consider some problems using these equations, applying our problem-solving method.

EXAMPLE 1

The average velocity of a rolling freight car is 2.00 m/s. How long does it take for the car to roll 15.0 m?

Sketch: None needed

Data:

$$s = 15.0 \text{ m}$$
$$v_{avg} = 2.00 \text{ m/s}$$
$$t = ?$$

Basic Equation:

$$s = v_{avg} t$$

Working Equation:

$$t = \frac{s}{v_{avg}}$$

Substitution:

$$t = \frac{15.0\text{ m}}{2.00\text{ m/s}}$$
$$= 7.50\text{ s}$$

$$\boxed{\frac{\text{m}}{\text{m/s}} = \text{m} \div \frac{\text{m}}{\text{s}} = \text{m} \cdot \frac{\text{s}}{\text{m}} = \text{s}}$$

EXAMPLE 2

A dragster starting from rest reaches a final velocity of 318 km/h. Find its average velocity.

Sketch: None needed

Data:

$$v_i = 0$$
$$v_f = 318\text{ km/h}$$
$$v_{avg} = ?$$

Basic Equation:

$$v_{avg} = \frac{v_f + v_i}{2}$$

Working Equation: Same

Substitution:

$$v_{avg} = \frac{318\text{ km/h} + 0\text{ km/h}}{2}$$
$$= 159\text{ km/h}$$

EXAMPLE 3

A train slowing to a stop has an average acceleration of -3.00 m/s^2. [Note that a minus (–) acceleration is commonly called *deceleration*, meaning that the train is slowing down.] If its initial velocity is 30.0 m/s, how far does it travel in 4.00 s?

Sketch: None needed

Data:

$$a_{avg} = -3.00\text{ m/s}^2$$
$$v_i = 30.0\text{ m/s}$$
$$t = 4.00\text{ s}$$
$$s = ?$$

Basic Equation:

$$s = v_i t + \tfrac{1}{2}a_{avg}t^2$$

Working Equation: Same

Substitution:

$$s = (30.0\text{ m/s})(4.00\text{ s}) + \tfrac{1}{2}(-3.00\text{ m/s}^2)(4.00\text{ s})^2$$
$$= 12\bar{0}\text{ m} - 24.0\text{ m}$$
$$= 96\text{ m}$$

EXAMPLE 4

An automobile accelerates from 67.0 km/h to 96.0 km/h in 7.80 s. What is its acceleration (in m/s^2)?

Sketch: None needed

Data:

$$v_f = 96.0 \text{ km/h}$$
$$v_i = 67.0 \text{ km/h}$$
$$t = 7.80 \text{ s}$$
$$a = ?$$

Basic Equation:

$$a_{\text{avg}} = \frac{v_f - v_i}{t}$$

Working Equation: Same

Substitution:

$$a_{\text{avg}} = \frac{96.0 \text{ km/h} - 67.0 \text{ km/h}}{7.80 \text{ s}}$$
$$= \frac{29.0 \text{ km/h}}{7.80 \text{ s}}$$
$$= \frac{29.0 \dfrac{\cancel{\text{km}}}{\cancel{\text{h}}} \times \dfrac{1000 \text{ m}}{1 \cancel{\text{km}}} \times \dfrac{1 \cancel{\text{h}}}{3600 \text{ s}}}{7.80 \text{ s}}$$
$$= 1.03 \text{ m/s}^2$$

Freely falling bodies undergo constant acceleration. In a vacuum (in the absence of air resistance) a ball and a feather fall at the same rate. In 1971, astronaut David Scott dropped a hammer and a feather simultaneously. Both objects fell at the same rate and landed at the same time on the surface of the moon where there is no atmosphere.

TRY THIS ACTIVITY

Free Fall in a Vacuum

Drop a piece of paper and a book at the same time and clock the time it takes for each to hit the floor. Now place that paper on top of the book as shown in Fig. 3.14. (***Note:*** The top surface area of the book must be larger than that of the paper.) What happens to the time it takes the book and paper to fall? What does this show about objects falling in a vacuum? (A vacuum is a space in which there is no air resistance present.)

Figure 3.14 **Place the paper on top of the book. The book must be larger than the paper.**

The acceleration of a freely falling body, also called the **acceleration due to gravity,** is denoted by the symbol g, where $a = g = 9.80\ \text{m/s}^2$ (metric system) or $a = g = 32.2\ \text{ft/s}^2$ (U.S. system) on the earth's surface.

What does $a = 9.80\ \text{m/s}^2$ mean? When a ball is dropped from a building, the speed of the ball increases by 9.80 m/s during each second. Figure 3.15 shows the total distance traveled and the speed at 1-s intervals.

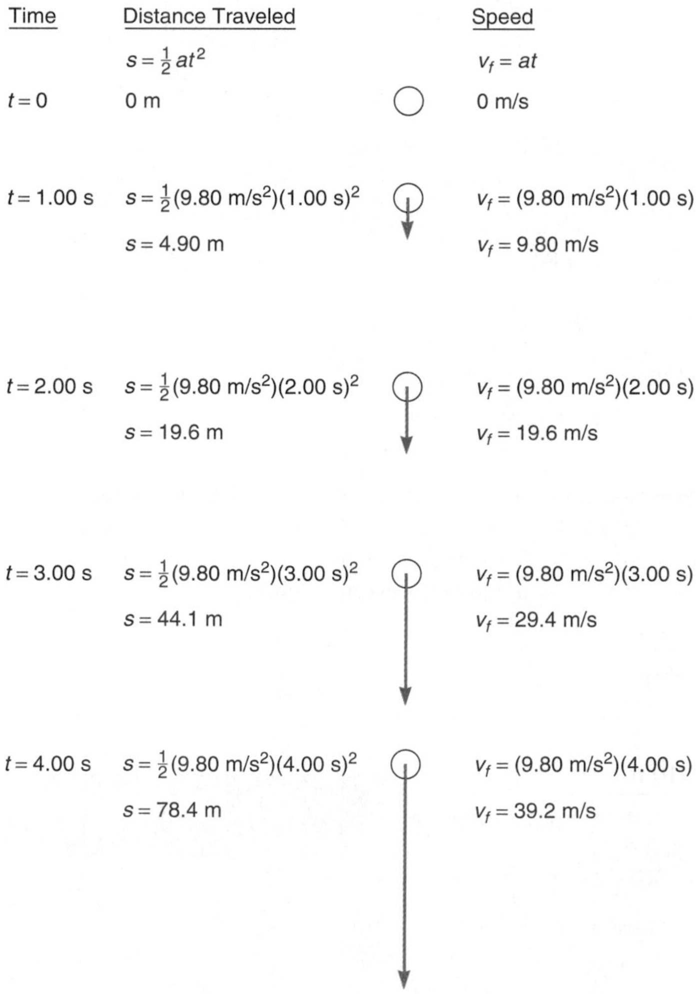
Time
Distance Traveled
Speed
$s = \frac{1}{2}at^2$
$v_f = at$
$t = 0$
0 m
0 m/s
$t = 1.00$ s
$s = \frac{1}{2}(9.80\ \text{m/s}^2)(1.00\ \text{s})^2$
$s = 4.90$ m
$v_f = (9.80\ \text{m/s}^2)(1.00\ \text{s})$
$v_f = 9.80$ m/s
$t = 2.00$ s
$s = \frac{1}{2}(9.80\ \text{m/s}^2)(2.00\ \text{s})^2$
$s = 19.6$ m
$v_f = (9.80\ \text{m/s}^2)(2.00\ \text{s})$
$v_f = 19.6$ m/s
$t = 3.00$ s
$s = \frac{1}{2}(9.80\ \text{m/s}^2)(3.00\ \text{s})^2$
$s = 44.1$ m
$v_f = (9.80\ \text{m/s}^2)(3.00\ \text{s})$
$v_f = 29.4$ m/s
$t = 4.00$ s
$s = \frac{1}{2}(9.80\ \text{m/s}^2)(4.00\ \text{s})^2$
$s = 78.4$ m
$v_f = (9.80\ \text{m/s}^2)(4.00\ \text{s})$
$v_f = 39.2$ m/s

Figure 3.15 Depiction of a ball falling with constant acceleration $a = g = 9.80\ \text{m/s}^2$ with the speed and the distance traveled calculated at the given times. Because the ball was dropped, $v_i = 0$ and the formulas for v_f and s are shown simplified. Note how the velocity and the distance traveled increase during each successive time interval.

TRY THIS ACTIVITY

Calculating Height

Drop a ball from a window that is at least two stories high (or from some other height) and time how long it takes the ball to fall to the ground. Use the formulas for accelerated motion to find the height. Measure the actual height with a tape measure. How does your calculated height compare with the actual height? What factors may have made your calculated height different from the actual height? Remember to use the problem-solving method to help you solve this problem.

We limit our discussion to uniformly accelerated motion because the mathematical tools needed to study other kinds of motion are beyond the scope of this book. We also need to assume that air resistance of a freely falling body is negligible for our calculations. However, air resistance is, in fact, an important factor in the design of machines that move through the atmosphere.

When the air resistance of a falling object equals its weight, the net force is zero and no further acceleration occurs. That is, a falling object reaches its **terminal speed.**

EXAMPLE 5

A rock is thrown straight down from a cliff with an initial velocity of 10.0 ft/s. Its final velocity when it strikes the water below is $31\bar{0}$ ft/s. The acceleration due to gravity is 32.2 ft/s². How long is the rock in flight?

Sketch: None needed

Data:

$$v_i = 10.0 \text{ ft/s}$$
$$a = 32.2 \text{ ft/s}^2$$
$$v_f = 31\bar{0} \text{ ft/s}$$
$$t = ?$$

Note the importance of listing all the data as an aid to finding the basic equation.

Basic Equation:

$$v_f = v_i + a_{\text{avg}}t \quad \text{or} \quad a_{\text{avg}} = \frac{v_f - v_i}{t}$$ (two forms of the same equation)

Working Equation:

$$t = \frac{v_f - v_i}{a_{\text{avg}}}$$

Substitution:

$$t = \frac{31\bar{0} \text{ ft/s} - 10.0 \text{ ft/s}}{32.2 \text{ ft/s}^2}$$
$$= \frac{30\bar{0} \text{ ft/s}}{32.2 \text{ ft/s}^2}$$
$$= 9.32 \text{ s}$$

$$\frac{\text{ft/s}}{\text{ft/s}^2} = \frac{\text{ft}}{\text{s}} \div \frac{\text{ft}}{\text{s}^2} = \frac{\text{ft}}{\text{s}} \cdot \frac{\text{s}^2}{\text{ft}} = \text{s}$$

When any object is thrown or hurled vertically upward, its upward speed is uniformly decreased by the force of gravity until it stops for an instant at its peak before falling back to the ground. As it is falling to the ground, it is uniformly accelerated by gravity the same as it would have been if dropped from its peak height. If an object is thrown vertically upward and if the initial velocity is known, the previous acceleration/gravity formulas may be used to find how high the object rises, how long it is in flight, and so on.

Note: When we consider a problem involving an object being thrown upward, we will consider an upward direction to be negative and the opposing gravity in its normal downward direction to be positive.

EXAMPLE 6

A baseball is thrown vertically upward with an initial velocity of 25.0 m/s (see Fig. 3.16).

(a) How high does it go?
(b) How long does it take to reach its maximum height?
(c) How long is it in flight?

Figure 3.16

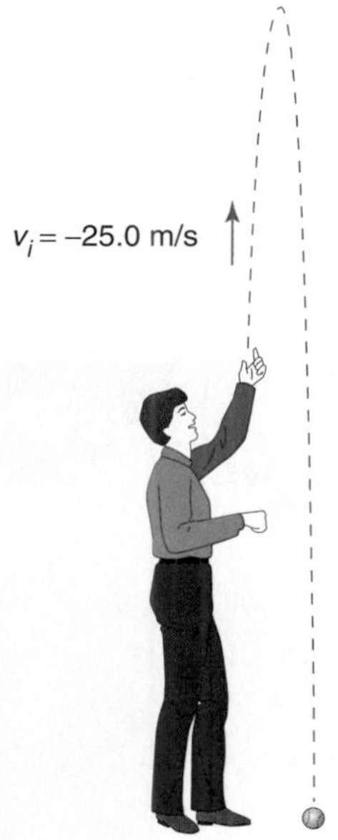

(a) Data:

$v_i = -25.0$ m/s (v_i is negative because the initial velocity is directed opposite gravity, g.)

$v_f = 0$ (At the instant of the ball's maximum height, its velocity is zero.)

$a_{avg} = g = 9.80 \text{ m/s}^2$

$s = ?$

Basic Equation:

$$2a_{avg}s = v_f^2 - v_i^2$$

Working Equation:

$$s = \frac{v_f^2 - v_i^2}{2a_{avg}}$$

Substitution:

$$s = \frac{0^2 - (-25.0 \text{ m/s})^2}{2(9.80 \text{ m/s}^2)}$$

$$\frac{(\text{m/s})^2}{\text{m/s}^2} = \frac{\text{m}^2/\text{s}^2}{\text{m/s}^2} = \frac{\text{m}^2}{\text{s}^2} \div \frac{\text{m}}{\text{s}^2} = \frac{\overset{\text{m}}{\cancel{\text{m}^2}}}{\cancel{\text{s}^2}} \times \frac{\cancel{\text{s}^2}}{\cancel{\text{m}}} = \text{m}$$

$s = -31.9$ m (s being negative indicates an upward displacement.)

(b) Data:

$v_i = -25.0$ m/s

$v_f = 0$

$a_{avg} = g = 9.80 \text{ m/s}^2$

$t = ?$

Basic Equation:

$$v_f = v_i + a_{avg}t$$

Working Equation:

$$t = \frac{v_f - v_i}{a_{avg}}$$

Substitution:

$$t = \frac{0 - (-25.0 \text{ m/s})}{9.80 \text{ m/s}^2} = 2.55 \text{ s}$$

$$\frac{\text{m/s}}{\text{m/s}^2} = \frac{\text{m}}{\text{s}} \div \frac{\text{m}}{\text{s}^2} = \frac{\cancel{\text{m}}}{\cancel{\text{s}}} \times \frac{\overset{\text{s}}{\cancel{\text{s}^2}}}{\cancel{\text{m}}} = \text{s}$$

(c) The ball decelerates on the way up and accelerates on the way down at the same rate because the acceleration due to gravity is constant (9.80 m/s^2). Therefore, the time for the ball to reach its peak is the same as the time for it to fall to the ground. The total time in flight is 2(2.55 s) = 5.10 s.

With what speed does the ball in Example 6 hit the ground? The answer is 25.0 m/s. Can you explain why?

Earlier in this section, we assumed no air resistance. We know that two dense and compact objects, such a bowling ball and a marble, will fall at the same rate in air. We also know that two unlike objects, such as a marble and a feather, fall at different rates because of air resistance.

A parachute takes advantage of air resistance to slow a sky diver's descent [Fig. 3.17(a)]. What would happen if a parachute does not open [Fig. 3.17(b)]? Does this

(a) (b)

Figure 3.17 The sky divers in (a) have a different acceleration toward the ground than those in (b) as a result of different air resistance.

mean that the sky diver's velocity would increase constantly until he or she hits the ground? As the velocity increases, the air resistance also increases. Since the gravitational pull and the air resistance are directed opposite each other, they tend to oppose or equalize each other. (Here, the velocity and acceleration are both directed downward, while the air resistance is directed upward.) This equalization occurs when the friction of the air's resistance equals the force of gravity. When this equalization occurs, the falling object stops accelerating and continues to fall at a constant velocity called *terminal velocity*.

The terminal speed of a sky diver varies from 150 to 200 km/h, depending on the person's weight and position. A heavier person will attain a greater terminal speed than a lighter person because the larger weight results in a larger acceleration before the air resistance equals the weight. Body position also makes a difference. When a body is spread out like that of a bird gliding with outstretched wings, its surface area increases, which results in more air resistance. Terminal speed can be controlled by varying the body position. A light sky diver and a heavy sky diver can remain in close proximity to each other if the light person decreases his or her air resistance by falling head or feet first while the heavy person spreads out and increases his or her air resistance. A parachute greatly increases air resistance and reduces the terminal speed to approximately 15 to 25 km/h, which is slow enough for a safe landing.

In general, the terminal velocity of an object varies with its weight and its aerodynamic features, which include the following:

1. The shape of the object. (A symmetrical object is more aerodynamic than a nonsymmetrical one.)
2. The orientation of the object as it is traveling. (A sky diver slows the fall by spreading out his or her arms and legs parallel to the ground while falling. The speed of the fall increases if the person falls head or feet first.)
3. The smoothness of the surface. (A body with a smooth surface provides less air resistance than a body with a rough surface and falls or flies faster as a result.)

TRY THIS ACTIVITY

Air Resistance and Acceleration

Drop several objects of varying sizes, weights, and shapes outside a second-floor window. Which objects hit the ground faster than others? If gravity acts equally on all objects, then what factors might cause the objects to fall at different rates?

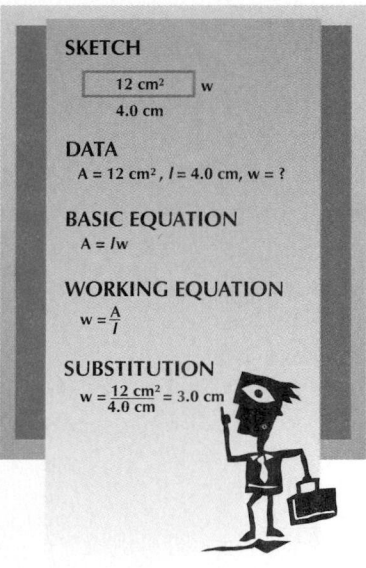

SKETCH

12 cm² w
4.0 cm

DATA

$A = 12\text{ cm}^2,\ l = 4.0\text{ cm},\ w = ?$

BASIC EQUATION

$A = lw$

WORKING EQUATION

$w = \frac{A}{l}$

SUBSTITUTION

$w = \frac{12\text{ cm}^2}{4.0\text{ cm}} = 3.0\text{ cm}$

PROBLEMS 3.3

Substitute in the given equation and find the unknown quantity.

1. Given: $v_{avg} = \dfrac{v_f + v_i}{2}$

   $v_f = 6.20\text{ m/s}$

   $v_i = 3.90\text{ m/s}$

   $v_{avg} = ?$

2. Given: $a_{avg} = \dfrac{v_f - v_i}{t}$

   $a_{avg} = 3.07\text{ m/s}^2$

   $v_f = 16.8\text{ m/s}$

   $t = 4.10\text{ s}$

   $v_i = ?$

3. Given: $s = v_i t + \frac{1}{2}a_{avg}t^2$

   $t = 3.00\text{ s}$

   $a_{avg} = 6.40\text{ m/s}^2$

   $v_i = 33.0\text{ m/s}$

   $s = ?$

4. Given: $2a_{avg}s = v_f^2 - v_i^2$

   $a_{avg} = 8.41\text{ m/s}^2$

   $s = 4.81\text{ m}$

   $v_i = 1.24\text{ m/s}$

   $v_f = ?$

5. Given: $v_f = v_i + a_{avg}t$

   $v_f = 10.40\text{ ft/s}$

   $v_i = 4.01\text{ ft/s}$

   $t = 3.00\text{ s}$

   $a_{avg} = ?$

6. The average velocity of a mini-bike is 15.0 km/h. How long does it take for the bike to go 35.0 m?
7. A sprinter starting from rest reaches a final velocity of 18.0 mi/h. What is her average velocity?
8. A coin is dropped with no initial velocity. Its final velocity when it strikes the earth is 50.0 ft/s. The acceleration due to gravity is 32.2 ft/s^2. How long does it take to strike the earth?
9. A rocket lifting off from earth has an average acceleration of 44.0 ft/s^2. Its initial velocity is zero. How far into the atmosphere does it travel during the first 5.00 s, assuming that it goes straight up?
10. The final velocity of a truck is 74.0 ft/s. If it accelerates at a rate of 2.00 ft/s^2 from an initial velocity of 5.00 ft/s, how long will it take for it to attain its final velocity?
11. A truck accelerates from 85 km/h to $12\bar{0}$ km/h in 9.2 s. Find its acceleration in m/s^2.
12. How long does it take a rock to drop 95.0 m from rest? Find the final speed of the rock.
13. An aircraft with a landing speed of 295 km/h lands on an aircraft carrier with a landing area 205 m long. Find the minimum constant deceleration required for a safe landing.
14. A ball is thrown downward from the top of a 43.0-ft building with an initial speed of 62.0 ft/s. Find its final speed as it strikes the ground.
15. A car is traveling at $7\bar{0}$ km/h. It then uniformly decelerates to a complete stop in 12 s. Find its acceleration (in m/s^2).
16. A car is traveling at $6\bar{0}$ km/h. It then accelerates at 3.6 m/s^2 to $9\bar{0}$ km/h.
    (a) How long does it take to reach the new speed?
    (b) How far does it travel while accelerating?

17. A rock is dropped from a bridge to the water below. It takes 2.40 s for the rock to hit the water.
(a) Find the speed (in m/s) of the rock as it hits the water.
(b) How high (in metres) is the bridge above the water?
18. A bullet is fired vertically upward from a gun and reaches a height of $70\bar{0}0$ ft.
(a) Find its initial velocity.
(b) How long does it take to reach its maximum height?
(c) How long is it in flight?
19. A bullet is fired vertically upward from a gun with an initial velocity of $25\bar{0}$ m/s.
(a) How high does it go?
(b) How long does it take to reach its maximum height?
(c) How long is it in flight?
20. A rock is thrown down with an initial speed of 30.0 ft/s from a bridge to the water below. It takes 3.50 s for the rock to hit the water.
(a) Find the speed (in ft/s) of the rock as it hits the water.
(b) How high is the bridge above the water?
21. A rock is thrown straight up with an initial speed of 10.0 m/s from a deck that is 25.0 m above the ground.
(a) How long does it take to reach its maximum height?
(b) What maximum height above the deck does it reach?
(c) At what speed does it hit the ground?
(d) What total length of time is the rock in the air?
22. A rock is thrown straight up and reaches a height of 15.0 m above a deck that is 40.0 m above the ground.
(a) What is the initial speed of the rock?
(b) How long does it take to reach its maximum height?
(c) At what speed does it hit the ground?
(d) What total length of time is the rock in the air?
23. John is standing on a steel beam 255.0 ft above the ground. Linda is standing 30.0 ft directly above John.
(a) For John to throw a hammer up to Linda, at what initial speed must John throw the hammer for it to just reach Linda?
(b) Suppose the hammer reaches the correct height, but Linda just misses catching it. How long does someone on the ground have to move out of the way from the time the hammer reaches its maximum height?
(c) At what speed does the hammer hit the ground?
24. Kurt is standing on a steel beam 275.0 ft above the ground and throws a hammer straight up at an initial speed of 40.0 ft/s. At the instant he releases the hammer, he also drops a wrench from his pocket. Assume that neither the hammer nor the wrench hits anything while in flight.
(a) Find the time difference between when the wrench and the hammer hit the ground.
(b) Find the speed at which the wrench hits the ground.
(c) Find the speed at which the hammer hits the ground.
(d) How long does it take for the hammer to reach its maximum height?
(e) How high above the ground is the wrench at the time the hammer reaches its maximum height?
25. One ball is dropped from a cliff. A second ball is thrown down 1.00 s later with an initial speed of 40.0 ft/s. How long after the second ball is thrown will the second ball overtake the first?
26. A car with velocity 2.00 m/s at $t = 0$ accelerates at 4.00 m/s^2 for 2.50 s. What is its velocity at $t = 2.50$ s?
27. A truck moving at 30.0 km/h accelerates at a constant rate of 3.50 m/s^2 for 6.80 s. Find its final velocity in km/h.
28. A bus accelerates from rest at a constant 5.50 m/s^2. How long will it take to reach 28.0 m/s?
29. A motorcycle slows from 22.0 m/s to 3.00 m/s with constant acceleration -2.10 m/s^2. How much time is required to slow down?

3.4 Projectile Motion

A **projectile** is a propelled object that travels through the air but has no capacity to propel itself. Up to this point, we have discussed projectiles either being thrown straight up or dropped straight down. We now discuss objects being thrown at an angle. This type of motion uses the same principles and formulas as the previous section on uniformly accelerated motion. However, two dimensions are needed to analyze projectile motion; x will represent horizontal and y will represent vertical.

A baseball pop fly to right field, a car driving off a cliff, and water flowing out of a hose are all examples of projectile motion. **Projectile motion** is the movement of a projectile as it travels through the air influenced only by its original velocity and gravitational acceleration.

Consider the example of a ball being rolled across a table at a constant 2.00 m/s. The ball moves 2.00 m for every second that it travels. Assuming no friction, the ball would continue to roll at that same speed (Fig. 3.18). As the ball approaches and rolls off the edge of the table, the ball continues to move horizontally at 2.00 m/s until it strikes the floor. Horizontally, there is nothing causing the ball to speed up or slow down, so it continues with that same horizontal motion.

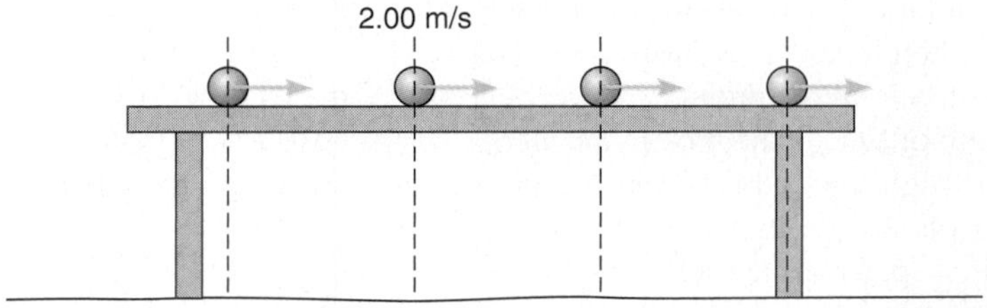

Figure 3.18 A ball rolling across a table with a constant velocity

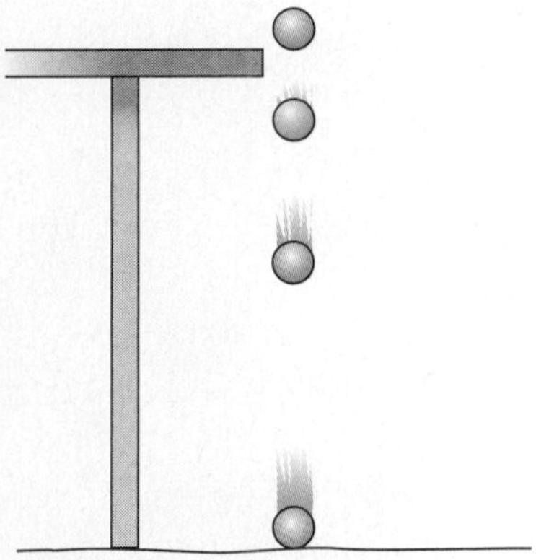

Figure 3.19 If a ball is simply dropped from the edge of a table, it will accelerate toward the ground, picking up speed and covering a greater and greater distance as time passes.

As the ball rolls off the table, gravity is accelerating the ball downward at a rate of $g = 9.80\text{ m/s}^2$ (Fig. 3.19). As a result, the ball increases its vertical speed by 9.80 m/s for every second it falls. Vertically, the problem can be treated as any other uniformly accelerated motion problem as seen in the previous section.

A projectile motion problem may be separated into the horizontal frame, the nongravitational acceleration component, and into the vertical frame, the gravitational acceleration component. Using formulas in both the horizontal and vertical frames allows you to solve problems such as finding the speed the ball was traveling as it came off the table, the distance the ball landed from the edge of the table, and the time the ball was in the air (Fig. 3.20).

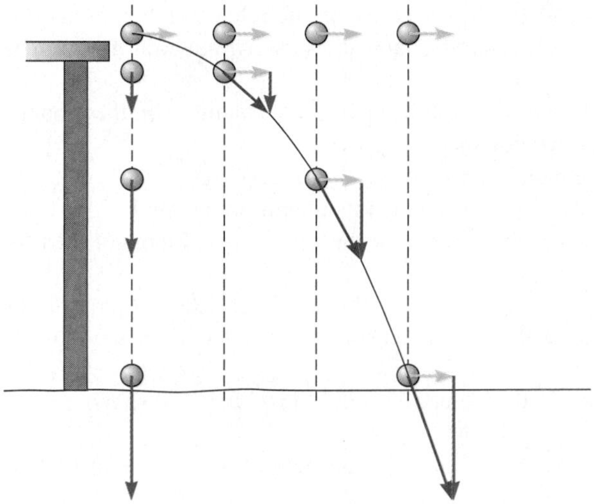

Figure 3.20 As the ball rolls off the table, it continues its horizontal motion, yet also accelerates vertically. The result is the curved motion shown.

TRY THIS ACTIVITY

Quarter and the Ruler Trick

Set up two coins and a ruler as shown in Fig. 3.21. Flick the ruler so the two coins are launched off the table at the same time. The coin closer to your stationary hand will travel more slowly than the coin farther away from your stationary hand. Which coin hits the ground first? Why?

Figure 3.21

EXAMPLE 1

A ball rolls at a constant speed of 0.700 m/s as it reaches the end of a 1.30-m-high table (Fig. 3.22). How far from the edge of the table does the ball land?

Sketch:

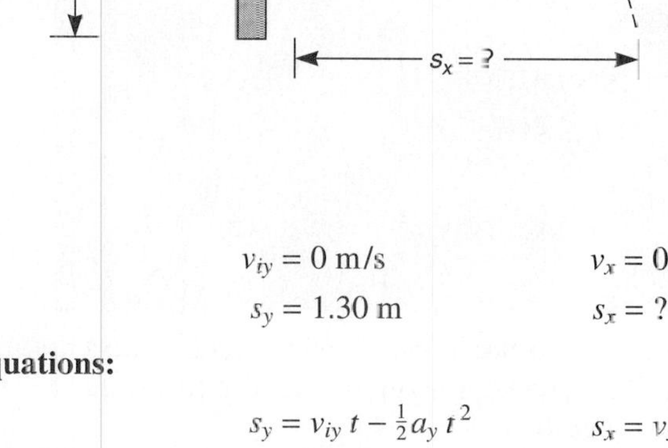

Figure 3.22

Data:

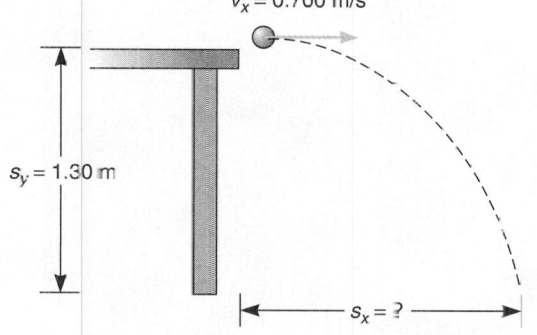

$$v_{iy} = 0 \text{ m/s} \qquad v_x = 0.700 \text{ m/s}$$
$$s_y = 1.30 \text{ m} \qquad s_x = ?$$

Basic Equations:

$$s_y = v_{iy}\,t - \tfrac{1}{2}a_y\,t^2 \qquad s_x = v_x\,t$$

Working Equations (with $v_{iy} = 0$):

$$t = \sqrt{\frac{2s_y}{a}} \qquad s_x = v_x t$$

Substitution:

$$t = \sqrt{\frac{2(1.30\text{ m})}{9.80\text{ m/s}^2}}$$

$$t = 0.515\text{ s}$$

$$s_x = (0.700\text{ m/s})(0.515\text{ s})$$

$$s_x = 0.361\text{ m}$$

Projectiles launched at an angle need to be treated the same as a ball rolled off a table. However, unlike such a ball, which has an initial vertical velocity of zero, a projectile launched at an angle has an initial vertical velocity. The **range** is the horizontal distance that a projectile will travel before striking the ground. The range and the flight time may be found as follows.

1. Separate the original speed of the projectile into *horizontal* (x-component of the velocity) and *vertical components* (y-component of the velocity) using vectors and trigonometry (Fig. 3.23).
2. Use the equation $s_x = v_x t$ to find the range. The horizontal component of the velocity does not change because gravitational acceleration does not act in the horizontal direction (Fig. 3.24). (Gravity pulls objects down in the vertical direction.)
3. Use the vertical component of the velocity and uniformly accelerated motion equations to determine how long the projectile will be in the air. Since gravity accelerates the projectile, use the following formula and solve for time: $v_f = v_i + at$.

***Note*:** Since the vertical components of v_i and v_f are equal but opposite in direction and therefore have opposite signs, the time for the projectile going up is the same as the time for it going down.

Figure 3.23 The ball is kicked with a horizontal (yellow vector) and vertical (blue vector) velocity. The red vector represents the resultant of the horizontal and vertical velocities.

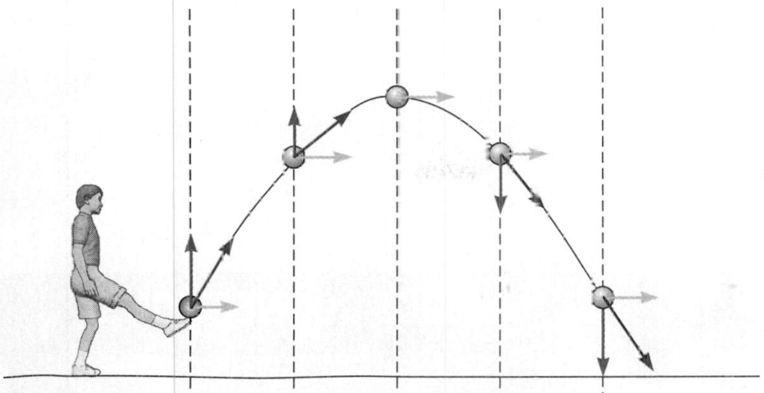

Figure 3.24 Velocity vectors as a ball kicked into the air travels throughout its trajectory

EXAMPLE 2

A baseball is hit and moves initially at an angle of 35.0° above the horizontal ground with a velocity of 25.0 m/s as shown in Fig. 3.25. (a) What are the vertical and horizontal components of the initial velocity of the ball? (b) How long will the ball be in the air? (c) What will be the range for this projectile?

Sketch:

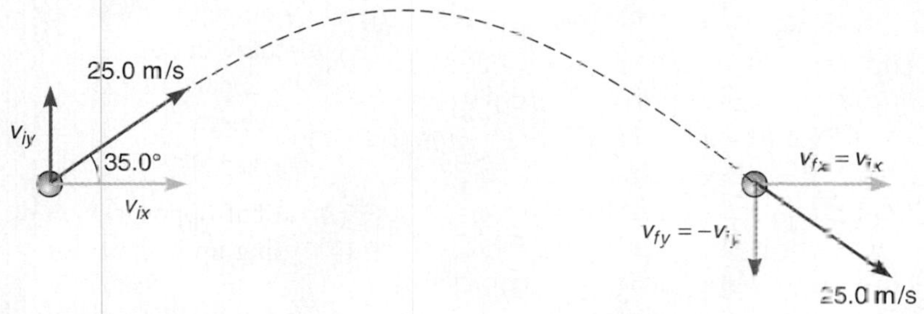

Figure 3.25

Data:

$$v = 25.0 \text{ m/s}$$
$$A = 35.0°$$

Basic Equations:

$$|\mathbf{v}_x| = |\mathbf{v}| \cos A$$
$$|\mathbf{v}_y| = |\mathbf{v}| \sin A$$
$$v_{fy} = v_{iy} + a_y t$$
$$s_x = v_x t$$

Working Equations: Same except for

$$t = \frac{v_{fy} - v_{iy}}{a_y}$$

Substitutions:

(a)

$$|\mathbf{v}_x| = (25.0 \text{ m/s})(\cos 35.0°)$$
$$|\mathbf{v}_x| = 20.5 \text{ m/s}$$
$$|\mathbf{v}_y| = (25.0 \text{ m/s})(\sin 35.0°)$$
$$|\mathbf{v}_y| = 14.3 \text{ m/s}$$

(b) $$t = \frac{(14.3\text{ m/s}) - (-14.3\text{ m/s})}{9.80\text{ m/s}^2}$$

$$t = 2.92\text{ s}$$

(c) $$s_x = (20.5\text{ m/s})(2.92\text{ s})$$

$$s_x = 59.9\text{ m}$$

PHYSICS CONNECTIONS

Orbiting Cannonballs?

Isaac Newton once said that if he could fire a cannonball with enough velocity, he could get it to circle the globe. Is this true? Can you really put something into orbit by launching it fast enough? In reality, there is too much air resistance and too many obstacles to allow this to happen. Theoretically, though, if we launch something with enough horizontal velocity, the earth itself would curve away before the cannonball strikes it. Figure 3.26 shows a cannonball launched with varying initial velocities. As the horizontal velocity of the cannonball increases, its range increases as well. In addition, the ball appears to fall farther due to the curvature of the earth. Finally, when the ball is launched with a large enough velocity, it completely misses the earth and achieves orbit.

Figure 3.26 Newton's diagram of a cannonball orbiting the earth

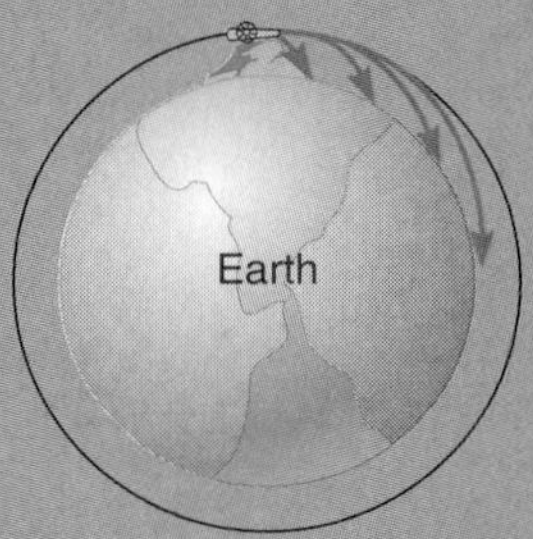

To place a satellite into orbit, a rocket or space shuttle must bring the satellite to a point above the earth's atmosphere where it will not experience any air resistance. The object is then given enough horizontal velocity so that although it is falling toward the earth, its horizontal velocity will prevent it from getting any closer to the surface. The concept of a falling object continuously missing the earth was a revolutionary concept developed by Isaac Newton over 300 years ago.

Communications satellites, the moon, and the astronauts in a space shuttle all are in a constant state of free fall but have the correct amount of horizontal velocity to prevent them from striking the earth. The speed required to keep a cannonball in orbit around the earth is approximately 17,700 mi/h. A space shuttle travels at approximately 17,400 mi/h because it orbits farther away from the earth.

PROBLEMS 3.4

Find the horizontal range for projectiles with the following speeds and angles.

	Angle	*Initial Speed*
1.	15.0°	35.0 m/s
2.	75.0°	35.0 m/s
3.	35.0°	35.0 m/s
4.	55.0°	35.0 m/s
5.	45.0°	35.0 m/s

6. Draw a conclusion about range and angles based on the answers to 1 through 5.
7. Part of military training involves aiming and shooting a cannon. If a soldier sets the cannon at an angle of 62.0° and a launch velocity of 67.0 m/s, how far will the projectile travel horizontally?
8. A faulty fireworks rocket launches but never discharges. If the rocket launches with an initial velocity of 33.0 ft/s at an angle of 85.0°, how far away from the launch site does the fireworks land?
9. An outfielder throws a baseball at a speed of $11\bar{0}$ ft/s at an angle of 25.0° above the horizontal to home plate $29\bar{0}$ ft away. Will the ball reach the catcher on the fly or will it bounce first?

Glossary

Acceleration Change in velocity per unit time. (p. 93)
Acceleration Due to Gravity The acceleration of a freely falling object. On the earth's surface the acceleration due to gravity is 9.80 m/s^2 (metric) or 32.2 ft/s^2 (U.S.) (p. 100)
Deceleration An acceleration that indicates an object is slowing down. (p. 95)
Motion A change of position. (p. 86)
Projectile A propelled object that travels through the air but has no capacity to propel itself. (p. 106)
Projectile Motion The motion of a projectile as it travels through the air influenced only by its original velocity and gravitational acceleration. (p. 106)
Range The horizontal distance that a projectile will travel before striking the ground. (p. 108)
Speed The distance traveled per unit of time. A scalar because it is described by a number and a unit, not a direction. (p. 86)
Terminal Speed The speed attained by a freely falling body when the air resistance equals its weight and no further acceleration occurs. (p. 101)
Velocity The rate of motion in a particular direction. The time rate of change of an object's displacement. Velocity is a vector that gives the direction of travel and the distance traveled per unit of time. (p. 86)

Formulas

3.1 $s = v_{avg}t$

3.2 $\Delta v = at$

3.3 $v_{avg} = \dfrac{v_f + v_i}{2}$ $\quad s = v_it + \frac{1}{2}a_{avg}t^2$

$a_{avg} = \dfrac{v_f - v_i}{t}$ $\quad s = \frac{1}{2}(v_f + v_i)t$

$v_f = v_i + a_{avg}t$ $\quad 2a_{avg}s = v_f^2 - v_i^2$

Review Questions

1. Velocity is
   (a) the distance traveled per unit of time.
   (b) the same as speed.
   (c) direction of travel and distance traveled per unit of time.
   (d) only the direction of travel.
2. A large heavy rock and a small marble are dropped at the same time from the roof of a three-story building. Neglecting air resistance, which object will strike the ground first?
   (a) The marble. (b) The rock.
   (c) They both strike the ground at the same time.
3. One ball is thrown horizontally while another is dropped vertically. Which ball will strike the ground first?
   (a) Both strike the ground at the same time.
   (b) The horizontally thrown ball strikes first.
   (c) The vertically dropped ball strikes first.
4. At what launch angle with the ground does a projectile have the greatest horizontal range?
   (a) 0° (b) 45° (c) 60° (d) 90°
5. Where in a projectile's path would its speed be the least?
   (a) Just as it is launched.

(b) Just before it lands.
(c) It has the same speed throughout its entire motion.
(d) At the top of its path.

6. Explain your answer to Question 2.
7. Explain your answer to Question 3.
8. Distinguish between velocity and speed.
9. Is velocity always constant?
10. Why are vectors important in measuring motion? Provide two examples where vectors are used to help measure motion.
11. Give three familiar examples of acceleration.
12. Distinguish among acceleration, deceleration, and average acceleration.
13. State the values of the acceleration due to gravity for freely falling bodies in both the metric and U.S. systems.

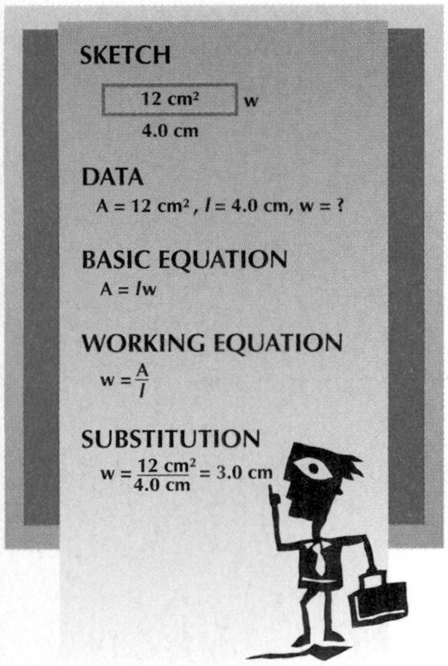

Review Problems

1. A boat travels at 17.0 mi/h for 1.50 h. How far does the boat travel?
2. A jet flies at $11\bar{0}0$ mi/h for $30\bar{0}0$ mi. How long is the jet flying?
3. A plane flies north at 215 km/h. A wind from the east blows at 69 km/h. What is the plane's new velocity with respect to the ground in standard position?
4. A glider flies southeast (at 320.0°) at 25.0 km/h. A wind picks up to 12.0 km/h from 15.0° south of west. What is the new velocity of the glider with respect to the ground in standard position?
5. A runner starts from rest and attains a speed of 8.00 ft/s after 2.00 s. What is the runner's acceleration?
6. A race car goes from rest to 150 km/h with an acceleration of 6.0 m/s^2. How many seconds does it take?
7. A sailboat has an initial velocity of 10.0 km/h and accelerates to 20.0 km/h. Find its average velocity.
8. A skateboarder starts from rest and accelerates at a rate of 1.30 m/s^2 for 3.00 s. What is his final velocity?
9. A plane has an average velocity of $5\bar{0}0$ km/h. How long does it take to travel 1.5×10^4 km?
10. A train has a final velocity of 110 km/h. It accelerated for 36 s at 0.50 m/s^2. What was its initial velocity?
11. A boulder is rolling down a hill at 8.00 m/s before it comes to rest 17.0 s later. What is its average velocity?
12. A truck accelerates from rest to 120 km/h in 13 s. Find its acceleration.
13. An airplane reaches a velocity of 71.0 m/s when it takes off. What must its acceleration be if the runway is 1.00 km long?
14. An airplane accelerates at 3.00 m/s^2 from a velocity of 21.0 m/s over a distance of 535 m. What is its final velocity?
15. A bullet is fired vertically upward and reaches a height of 2150 m.
    (a) Find its initial velocity.
    (b) How long does it take to reach its maximum height?
    (c) How long is it in flight?
16. A rock is thrown down with an initial speed of 10.0 m/s from a bridge to the water below. It takes 2.75 s for the rock to hit the water.
    (a) Find the speed of the rock as it hits the water.
    (b) How high is the bridge above the water?
17. A shot put is hurled at 9.43 m/s at an angle of 55.0°. Ignoring the height of the shot putter, what will be the range of the shot?
18. An archer needs to hit a bull's eye on a target at eye level 60.0 ft away. If the archer fires the arrow from eye level with a speed of 47.0 ft/s at an angle of 25.0° above the horizontal, will the arrow hit the target?

APPLIED CONCEPTS

1. Amy walks at an average speed of 1.75 m/s toward her airport gate. When she comes within 57.5 m of her gate she gets onto and continues to walk on a people mover at the same rate. (a) If she arrives at the gate 15.3 s after getting on the people mover, how fast was she moving relative to the ground? (b) What was the speed of the people mover?
2. A novice captain is pointing his ferryboat directly across the river at a speed of 15.7 mi/h. If he does not pay attention to the current that is headed downriver at 5.35 mi/h, what will be his resultant speed and direction? (Consider that the current and the boat's initial heading are perpendicular to each other.)
3. Anette is a civil engineer and needs to determine the length of a highway on-ramp before construction begins. If the average vehicle takes 10.8 s to go from 20.0 mi/h to 60.0 mi/h, how long should the separate merge lane be so a car can be up to speed to 60.0 mi/h before merging?
4. As a movie stunt coordinator, you need to be sure a stunt will be safe before it is performed. If a stuntwoman is to run horizontally off the roof of a three-story building and land on a foam pad 7.35 m away from the base of the building, how fast should she run as she leaves the roof? (Each story is 4.00 m.)
5. As a newspaper delivery boy, Jason needs to know his projectile motion to throw a paper horizontally from a height of 1.40 m to a door that is 13.5 m away. What must the paper's velocity be for it to reach the door?

CHAPTER 4

FORCE

Classical physics is sometimes called Newtonian physics in honor of Sir Isaac Newton, who lived from 1642 to 1727 and formulated three laws of motion that summarize much of the behavior of moving bodies.

Forces may cause motion. Inertia tends to resist the influence of an applied force. Forces, inertia, friction, and how they relate to motion are considered now.

Objectives

The major goals of this chapter are to enable you to:

1. Relate force and the law of inertia.
2. Apply the law of acceleration.
3. Identify components of friction.
4. Analyze forces in one dimension.
5. Distinguish among weight, mass, and gravity.
6. Analyze how the law of action and reaction is used.

4.1 Force and the Law of Inertia

As discussed in the previous chapter, if an object changes its velocity, we say it accelerates. But what causes an object to accelerate? Let's take an example of a soccer ball at rest on a field. What must you do to accelerate the ball? Similarly, if your car is approaching a red stoplight, what must you do to make the car accelerate to rest? The answer in both instances is to apply a force.

A **force** is any push or pull. Forces tend to either change the motion of an object or prevent the object from changing its motion. Force is a vector quantity and therefore has both magnitude and direction. The force tends to produce acceleration in the direction of its application. Therefore, if you want to accelerate the soccer ball, you need to kick it in the direction you want the ball to move. To get the car to slow down, you apply the brakes, which creates a force in the opposite direction of the car's motion and causes a negative acceleration.

Not all forces result in acceleration. For example, if you attempt to push a file cabinet, it may not accelerate because another force, in this case a frictional force, matches your pushing force, preventing you from accelerating the cabinet (Fig. 4.1). This concept of concurrent forces will be covered in Chapter 6.

The units for measuring force are the newton (N) in the metric system and the pound (lb) in the U.S. system. The conversion factor is

$$4.45 \text{ N} = 1 \text{ lb}$$

Isaac Newton (1642–1727),

physicist, mathematician, and astronomer, was born in England. He is credited with discovering the laws of motion and gravitation, studying the nature of light and finding that white light is a mixture of colors which can be separated by refraction, inventing the calculus, and devising the first reflecting telescope. The metric unit of force, the newton, is named after him.

Let's examine the relationship between forces and motion. There are three relationships or laws that were discovered by **Isaac Newton** during the late seventeenth century. The three laws are often called Newton's laws. The first law is as follows:

LAW OF INERTIA: NEWTON'S FIRST LAW

A body that is in motion continues in motion with the same velocity (at constant speed and in a straight line) and a body at rest continues at rest unless an unbalanced (outside) force acts upon it.

Figure 4.1 The pushing force and the frictional force cancel each other out, resulting in zero acceleration.

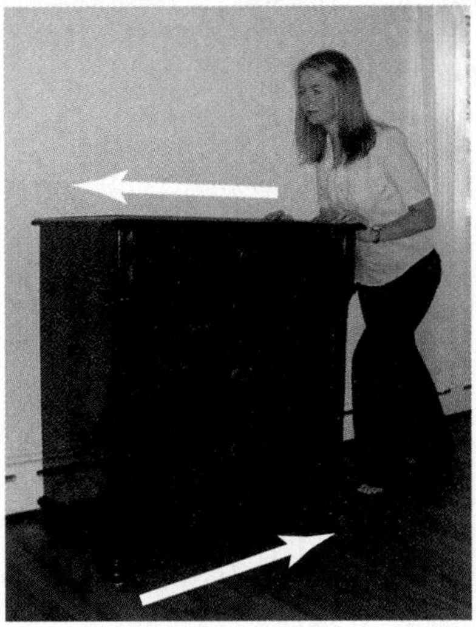

If an automobile is stopped (at rest) on level ground, it resists being moved. That is, a person is required to exert a tremendous push to get it moving. Similarly, if an automobile is moving—even slowly—it takes a large force to stop it. This property of resisting a change in motion is called inertia. **Inertia** is the property of a body that causes it to remain at rest if it is at rest or to continue moving with a constant velocity unless an unbalanced force acts upon it. When the accelerating force of an automobile engine is no longer applied to a moving car, the car will slow down. This is not a violation of the law of inertia because there are forces being applied to the car through air resistance, friction in the bearings, and the rolling resistance of the tires [Fig. 4.2(a)]. If these forces could be removed, the auto would continue moving with a constant velocity. Anyone who has tried to stop quickly on ice knows the effect of the law of inertia when frictional forces are small [Fig. 4.2(b)].

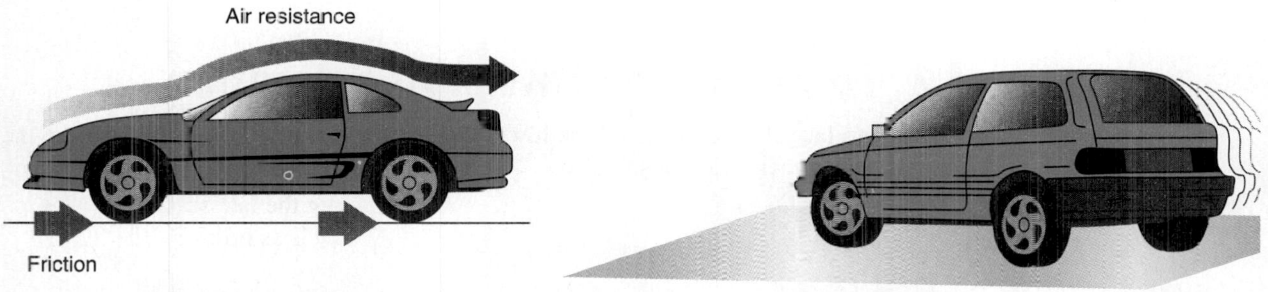

(a) Air resistance and friction slow the car.

(b) Inertia makes it hard to stop a car on ice.

Figure 4.2

TRY THIS ACTIVITY

Inertia Tricks

Place a card on top of a glass. Then place a quarter on top of the card as in Fig. 4.3(a). Quickly flick the card horizontally. The inertia of the coin tends to keep it at rest horizontally. The force that causes the coin to move is the vertical force of gravity, which pulls it straight down into the glass as in Fig. 4.3(b).

How could you remove a tablecloth from under a set of dishes without removing the dishes from the table? The trick to keeping the dishes on the table is to use a very smooth tablecloth without any hem and very quickly jerk the tablecloth backward in a slightly downward direction. We suggest that you practice with heavy objects on plastic plates. Remember that the more mass placed on the plates, the greater is the inertia.

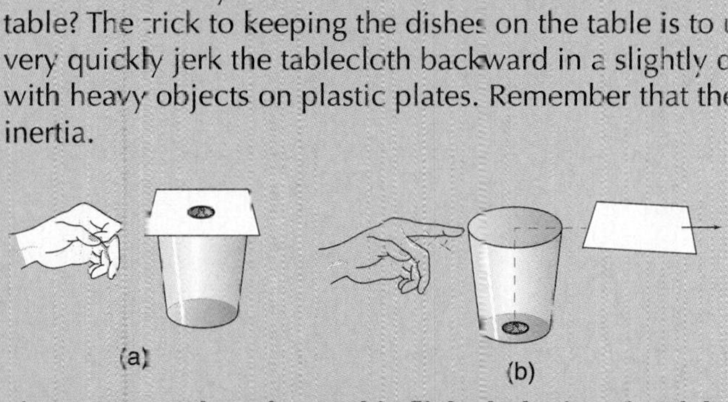

Figure 4.3 When the card is flicked, the inertia of the quarter keeps it at rest until it falls into the glass.

Some objects more than others tend to resist changes in their motion. It is much easier to push a small automobile than to push a large truck into motion (Fig. 4.4). **Mass** *is a measure of the inertia of a body;* that is, a measure of the resistance a body has to changing its motion. The common units of mass are the kilogram (kg) in the metric system and the slug in the U.S. system. The conversion factor is 1 kg = 0.0685 slug.

Figure 4.4 A larger body (in mass) has a greater resistance than does a smaller one to a change in its motion.

4.2 Force and the Law of Acceleration

The second law of motion, called the **law of acceleration**, relates the applied force, the mass, and the acceleration of an object:

LAW OF ACCELERATION: NEWTON'S SECOND LAW
The total force acting on a body is equal to the mass of the body times its acceleration.

In equation form this law is

$$F = ma$$

where

F = total force
m = mass
a = acceleration

The formula states that when a force is applied to an object, the force causes the object to accelerate. The stronger the force, the larger is the acceleration. The weaker the force, the smaller is the acceleration. In addition, when pushed with the same force, a more massive object will accelerate less and a less massive object will accelerate more.

If the mass is kept constant,

$F \sim a$ (The force is directly proportional to the acceleration.)

If the force is kept constant,

$m \sim 1/a$ (The mass is inversely proportional to the acceleration.)

In SI units, the mass unit is the kilogram (kg) and the acceleration unit is metre/second/second (m/s^2). The force required to accelerate 1 kg of mass at a rate of 1 m/s^2 is

$$\begin{aligned} F &= ma \\ &= (1\text{ kg})(1\text{ m/s}^2) \\ &= 1\text{ kg m/s}^2 \end{aligned}$$

The SI force unit is the newton (N), named in honor of Isaac Newton, and is defined as

$$1 \text{ N} = 1 \text{ kg m/s}^2$$

In the U.S. system, the mass unit is the slug and the acceleration unit is foot/second/second (ft/s^2). The force required to accelerate 1 slug of mass at the rate of 1 ft/s^2 is

$$\begin{aligned} F &= ma \\ &= (1 \text{ slug})(1 \text{ ft/s}^2) \\ &= 1 \text{ slug ft/s}^2 \end{aligned}$$

The U.S. force unit is the pound (lb) and is defined as

$$1 \text{ lb} = 1 \text{ slug ft/s}^2$$

Note: The other metric unit of force is the dyne. One dyne is the force required to accelerate 1 g of mass at the rate of 1 cm/s^2. The dyne is not an SI unit, and its use is less common.

EXAMPLE 1

What force is necessary to produce an acceleration of 6.00 m/s^2 on a mass of 5.00 kg?

Data:

$$\begin{aligned} m &= 5.00 \text{ kg} \\ a &= 6.00 \text{ m/s}^2 \\ F &= ? \end{aligned}$$

Basic Equation:

$$F = ma$$

Working Equation: Same

Substitution:

$$\begin{aligned} F &= (5.00 \text{ kg})(6.00 \text{ m/s}^2) \\ &= 30.0 \text{ kg m/s}^2 \\ &= 30.0 \text{ N} \qquad (1 \text{ N} = 1 \text{ kg m/s}^2) \end{aligned}$$

EXAMPLE 2

What force is necessary to produce an acceleration of 2.00 ft/s^2 on a mass of 3.00 slugs?

Data:

$$\begin{aligned} m &= 3.00 \text{ slugs} \\ a &= 2.00 \text{ ft/s}^2 \\ F &= ? \end{aligned}$$

Basic Equation:

$$F = ma$$

Working Equation: Same

Substitution:

$$\begin{aligned} F &= (3.00 \text{ slugs})(2.00 \text{ ft/s}^2) \\ &= 6.00 \text{ slug ft/s}^2 \\ &= 6.00 \text{ lb} \qquad (1 \text{ lb} = 1 \text{ slug ft/s}^2) \end{aligned}$$

EXAMPLE 3

Find the acceleration produced by a force of $50\bar{0}$ N applied to a mass of 20.0 kg.

Sketch: None needed

Data:

$$F = 50\bar{0}\text{ N}$$
$$m = 20.0\text{ kg}$$
$$a = ?$$

Basic Equation:

$$F = ma$$

Working Equation:

$$a = \frac{F}{m}$$

Substitution:

$$a = \frac{50\bar{0}\text{ N}}{20.0\text{ kg}}$$
$$= 25.0\frac{\text{N}}{\text{kg}}$$
$$= 25.0\frac{\cancel{\text{N}}}{\cancel{\text{kg}}} \times \frac{1\ \cancel{\text{kg}}\text{ m/s}^2}{1\ \cancel{\text{N}}}$$
$$= 25.0\text{ m/s}^2$$

Note: We use a conversion factor to obtain acceleration units.

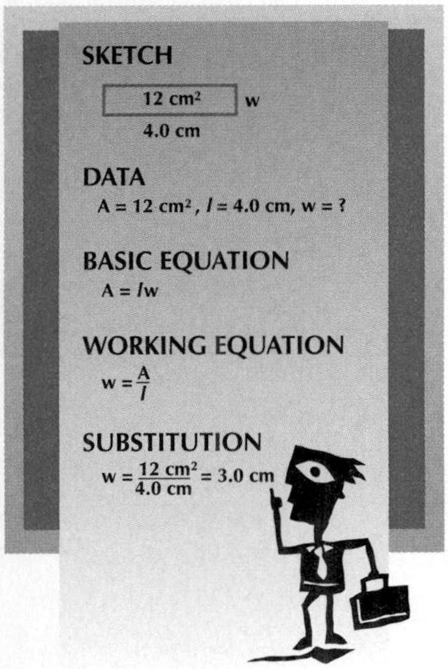

PROBLEMS 4.2

Find the total force necessary to give each mass the given acceleration.

1. $m = 15.0$ kg, $a = 2.00$ m/s^2
2. $m = 4.00$ kg, $a = 0.500$ m/s^2
3. $m = 111$ slugs, $a = 6.70$ ft/s^2
4. $m = 91.0$ kg, $a = 6.00$ m/s^2
5. $m = 28.0$ slugs, $a = 9.00$ ft/s^2
6. $m = 42.0$ kg, $a = 3.00$ m/s^2
7. $m = 59.0$ kg, $a = 3.90$ m/s^2
8. $m = 2.20$ slugs, $a = 1.53$ ft/s^2

Find the acceleration of each mass with the given total force.

9. $m = 19\bar{0}$ kg, $F = 76\bar{0}0$ N
10. $m = 7.00$ slugs, $F = 12.0$ lb
11. $m = 3.60$ kg, $F = 42.0$ N
12. $m = 0.790$ kg, $F = 13.0$ N
13. $m = 11\bar{0}$ kg, $F = 57.0$ N
14. $m = 84.0$ kg, $F = 33.0$ N
15. $m = 9.97$ slugs, $F = 13.9$ lb
16. $m = 21\bar{0}$ kg, $F = 41.0$ N
17. Find the total force necessary to give an automobile of mass 1750 kg an acceleration of 3.00 m/s^2.
18. Find the acceleration produced by a total force of 93.0 N on a mass of 6.00 kg.
19. Find the total force necessary to give an automobile of mass $12\bar{0}$ slugs an acceleration of 11.0 ft/s^2.
20. Find the total force necessary to give a rocket of mass $25,\bar{0}00$ slugs an acceleration of 28.0 ft/s^2.
21. Find the total force necessary to give a $14\bar{0}$ kg mass an acceleration of 41.0 m/s^2.
22. Find the acceleration produced by a total force of $30\bar{0}$ N on a mass of 0.750 kg.
23. Find the mass of an object with acceleration 15.0 m/s^2 when an unbalanced force of 90.0 N acts on it.

24. An automobile has a mass of $10\bar{0}$ slugs. The passengers it carries have a mass of 5.00 slugs each.
    (a) Find the acceleration of the auto and one passenger if the total force acting on them is $15\bar{0}0$ lb.
    (b) Find the acceleration of the auto and six passengers if the total force is again $15\bar{0}0$ lb.
25. Find the acceleration produced by a force of 6.75×10^6 N on a rocket of mass 5.27×10^5 kg.
26. An astronaut has a mass of 80.0 kg. His space suit has a mass of 15.5 kg. Find the acceleration of the astronaut during his space walk when his backpack propulsion unit applies a force to him (and his suit) of 85.0 N.
27. A discus thrower exerts a force of $14\bar{0}$ N on the disc. (a) If the disc has an acceleration of 19.0 m/s^2, what is the mass of the disc? (b) If the mass of the disc is doubled and the applied force remains the same what is the new acceleration? (c) If the force is doubled and the mass is unchanged, how is the acceleration affected?
28. A scooter and rider together have a mass of 275 kg. (a) If the scooter slows with an acceleration of -4.50 m/s^2, what is the net force on the scooter and rider? (b) Describe the direction of the force and state the meaning of the (−) sign.
29. A pickup truck with mass of 1230 kg moving at 105 km/h stops within a distance of 53.0 m. (a) What is the direction and size of the force that acts on the truck? (b) How much more slowing force would be required to stop the truck in half the distance?

4.3 Friction

Friction is a force that resists the relative motion of two objects in contact caused by the irregularities of the two surfaces sliding or rolling across each other, which tend to catch on each other (Fig. 4.5). In general, it is found that if two rough surfaces are polished, the frictional force between them is lessened. However, there is a point beyond which this decrease in friction is not observed. If two objects are polished such that the surfaces are very smooth, then the frictional force actually increases. (Two panes of glass is an example.) This is related to the fact that friction can also be caused by the adhesion of molecules of one surface to the molecules of the other surface. This adhesive force is similar to the electric forces that hold atoms together in solids.

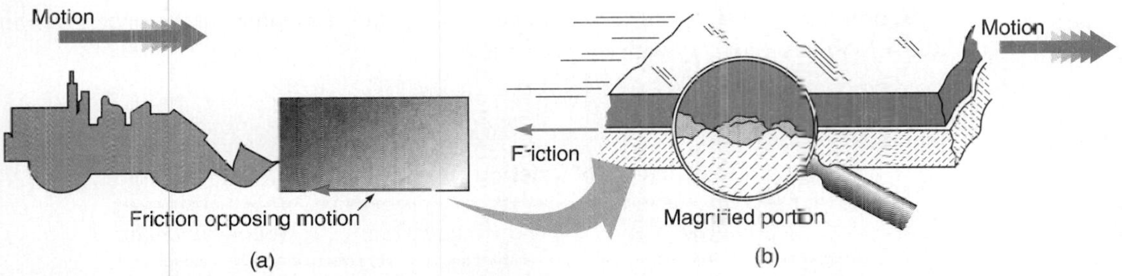

Figure 4.5 **Friction resists motion of objects in contact with each other.**

Friction is both a necessity and a hindrance to our everyday living. Experiments with frictional forces indicate the following general characteristics:

1. ***Friction is a force that always acts parallel to the surface in contact and opposite to the direction of motion.*** If there is no motion, friction acts in the direction opposite any force that tends to produce motion [Fig. 4.6(a)].
2. ***Starting friction is greater than sliding friction.*** When you push a large box across the floor, you probably notice that it takes more force to start it moving than to keep it

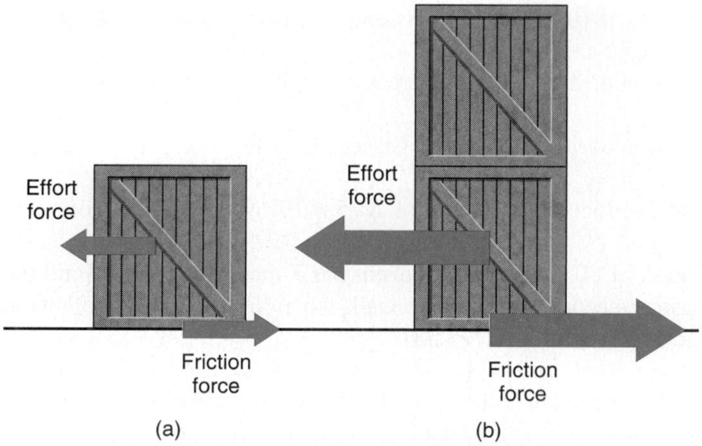

Figure 4.6 Friction increases as the force between the surfaces increases.

moving. This is due to inertia. A box at rest tends to remain at rest, whereas a moving box tends to continue moving.

3. ***Friction increases as the force between the surfaces increases.*** It is much easier to slide a light crate than a heavy one across the floor [Fig. 4.6(b)]. *The area of contact is not relevant.*

The characteristics of friction can be described by the following equation:

$$F_f = \mu F_N$$

where

F_f = frictional force
F_N = **normal force** (force perpendicular to the contact surface)
μ = coefficient of friction

The **coefficient of friction** is the ratio between the frictional force and the normal force of the object. The coefficient describes how rough or smooth the two surfaces are when they are in contact with each other. A higher coefficient of friction indicates two rough surfaces, whereas a lower number indicates two smooth surfaces. The Greek letter μ is pronounced "myu."

Representative values for the coefficients of friction for some surfaces are given in Table 4.1. Values may vary with surface conditions.

Table 4.1 Coefficients of Friction (μ)

Material	Starting Friction	Sliding Friction
Steel on steel	0.58	0.20
Steel on steel (lubricated)	0.13	0.13
Glass on glass	0.95	0.40
Hardwood on hardwood	0.40	0.25
Steel on concrete		0.30
Aluminum on aluminum	1.9	
Rubber on dry concrete	2.0	1.0
Rubber on wet concrete	1.5	0.97
Aluminum on wet snow	0.4	0.02
Steel on Teflon	0.04	0.04

TRY THIS ACTIVITY

How to Find μ

Establish your own chart of coefficients of friction for various pairs of surfaces. Take any fairly massive object, perhaps a textbook, and measure its normal force by weighing it with a spring scale. (We will only be doing this activity on horizontal surfaces so the normal force is the same as the weight of the object.) To determine the coefficient of friction between the two surfaces, slowly pull horizontally on the spring scale until the object moves at a constant velocity. As the object moves, the reading on the scale gives the applied pulling force, which is also the force of friction. Record your data in a table and compare. The rougher materials should have a larger coefficient of friction than the smoother materials.

In general:

To reduce sliding friction:

1. Use smoother surfaces.
2. Use lubrication to provide a thin film between surfaces.
3. Use Teflon to greatly reduce friction between surfaces when an oil lubricant is not desirable, such as in electric motors.
4. Substitute rolling friction for sliding friction. Using ball bearings and roller bearings greatly reduces friction.

EXAMPLE 1

A force of 170 N is needed to keep a 530-N wooden box sliding on a wooden floor. What is the coefficient of sliding friction?

Sketch:

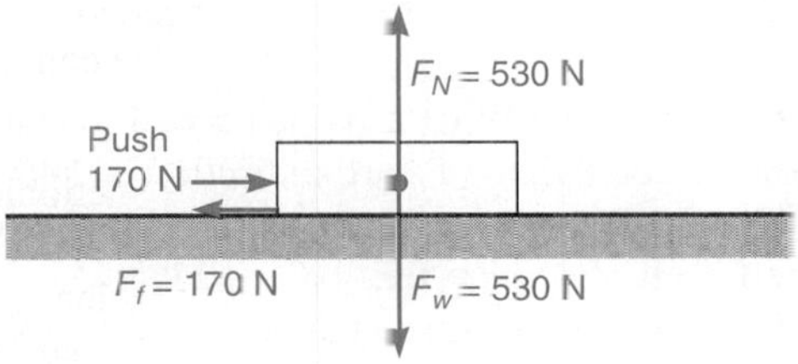

Data:

$$F_f = 170 \text{ N}$$
$$F_N = 530 \text{ N}$$
$$\mu = ?$$

Basic Equation:

$$F_f = \mu F_N$$

Working Equation:

$$\mu = \frac{F_f}{F_N}$$

Substitution:

$$\mu = \frac{170 \text{ N}}{530 \text{ N}}$$
$$= 0.32$$

Note that μ does not have a unit because the force units always cancel.

PHYSICS CONNECTIONS

Friction and Antilock Brakes

Since the invention of the automobile, engineers have worked to design brakes that reduce the stopping distance for automobiles. The most impressive advance in automobile braking technology is the antilock braking system (ABS). ABS technology is found on most new automobiles and works to prevent changing the friction between the tires and the road to sliding friction. The ratio of the coefficients of starting friction to sliding friction of rubber on dry cement is 2.0 to 1.0.

Automobiles stop by using friction to slow down during normal braking situations. This friction takes place while the tires are in contact with the pavement. Locking the wheels or braking on ice or a film of water can quickly change normal friction between the road and the tires to sliding friction between the tires and the road. Sliding tires cause cars to go out of control and result in longer stopping distances.

Antilock braking systems use computerized sensors to monitor the rotational speed of the tires while braking. If the sensors detect a sudden decrease in the turning speed of the tires (indicating a skid), the control unit repeatedly releases and then restores the pressure to the brakes at a rate of up to 15 times per second. In essence, the ABS pulses the brakes on and off to keep the tires from sliding. Maintaining the normal frictional force throughout braking keeps the car in control and reduces its stopping distance.

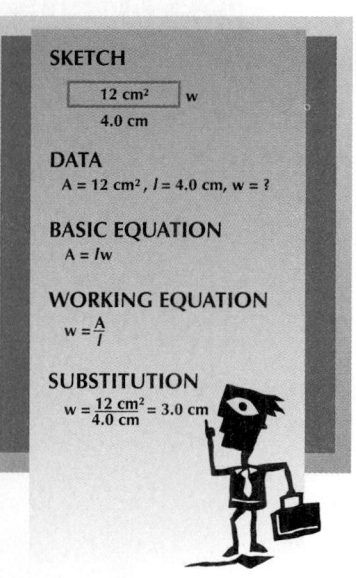

SKETCH

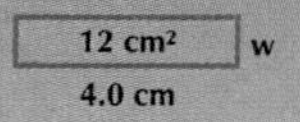

DATA

$A = 12\text{ cm}^2, l = 4.0\text{ cm}, w = ?$

BASIC EQUATION

$A = lw$

WORKING EQUATION

$w = \frac{A}{l}$

SUBSTITUTION

$w = \frac{12\text{ cm}^2}{4.0\text{ cm}} = 3.0\text{ cm}$

PROBLEMS 4.3

1. A cart on wheels weighs 2400 N. The coefficient of rolling friction between the wheels and floor is 0.16. What force is needed to keep the cart rolling uniformly?
2. A wooden crate weighs 780 lb. What force is needed to start the crate sliding on a wooden floor when the coefficient of starting friction is 0.40?
3. A piano weighs 4700 N. What force is needed to start the piano rolling across the floor when the coefficient of starting friction is 0.23?
4. A force of 850 N is needed to keep the piano in Problem 3 rolling uniformly. What is the coefficient of rolling friction?
5. A dog sled weighing 750 lb is pulled over level snow at a uniform speed by a dog team exerting a force of $6\bar{0}$ lb. Find the coefficient of friction.
6. A horizontal conveyor belt system has a coefficient of moving friction of 0.65. The motor driving the system can deliver a maximum force of 2.5×10^6 N. What maximum total weight can be placed on the conveyor system?
7. A tow truck can deliver 2500 lb of pulling force. What is the maximum-weight truck that can be pulled by the tow truck if the coefficient of rolling friction of the truck is 0.10?
8. A snowmobile is pulling a large sled across a snow-covered field. The weight of the sled is 3560 N. If the coefficient of friction of the sled is 0.12, what is the pulling force supplied by the snowmobile?
9. An automobile weighs 12,000 N and has a coefficient of starting friction of 0.13. What force is required to start the auto rolling?
10. If the coefficient of rolling friction of the auto in Problem 9 is 0.080, what force is required to keep it moving once it is in motion?
11. An alloy block is placed on a smooth composite table. If a force of 14.0 N is required to keep the 40.0-N block moving at a constant velocity, what is the coefficient of sliding friction for the table and block?

12. If a 20.0-N casting is placed on the block in Problem 11, what force is necessary to keep the block and casting moving at constant velocity?
13. Rubber tires and wet blacktop have a coefficient of sliding friction of 0.500. A pickup truck with mass $75\overline{0}$ kg traveling 30.0 m/s skids to a stop. (a) What is the size and direction of the frictional force that the road exerts on the truck? (b) Find the acceleration of the truck. (c) How far would the truck travel before coming to rest?
14. The coefficient of friction in Problem 13 is 0.700 if the tires do not skid. (a) Would the force of friction be greater than, less than, or equal to that in Problem 13? (b) Would the truck in (a) stop in a shorter, a longer, or an equal distance compared to the truck in Problem 13?

4.4 Total Forces in One Dimension

In the examples used to illustrate the law of acceleration, we discussed only total forces. We need to remember that forces are vectors and have magnitude and direction. The total force acting on an object is the resultant of the separate forces. *When forces act in the same or opposite directions (in one dimension), the total, or net, force can be found by adding the forces that act in one direction and subtracting the forces that act in the opposite direction.* It is often useful to draw the forces as vectors.

EXAMPLE 1

Two workers push in the same direction (to the right) on a crate. The force exerted by one worker is $15\overline{0}$ lb. The force exerted by the other is 175 lb. Find the net force exerted.

Sketch:

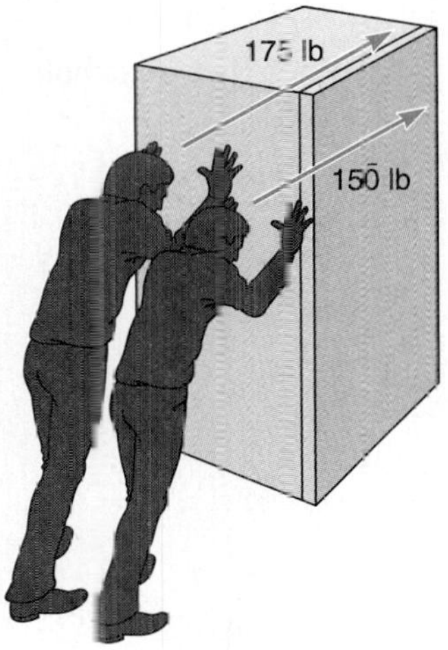

Both forces act in the same direction, so the total force is the sum of the two.

Note: The Greek letter Σ (sigma) means "sum of."

$$\Sigma \mathbf{F} = 175 \text{ lb} + 15\overline{0} \text{ lb}$$
$$= 325 \text{ lb to the right}$$

EXAMPLE 2

The same two workers push the crate to the right, and the motion is opposed by a frictional force of $30\overline{0}$ lb. Find the net force.

Sketch:

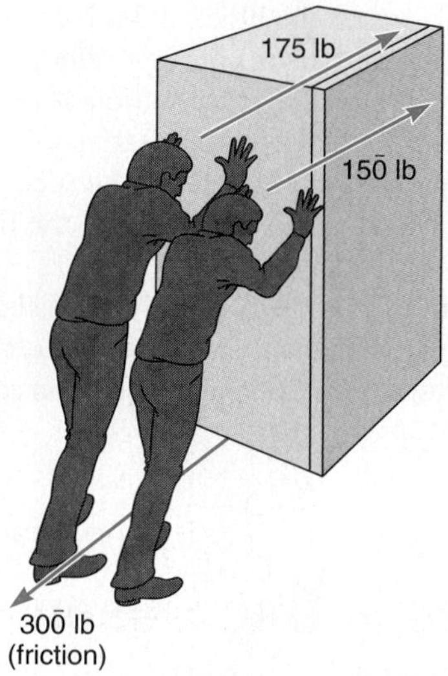

The workers push in one direction and friction pushes in the opposite direction, so we add the forces exerted by the workers and subtract the frictional force.

$$\Sigma \mathbf{F} = 175 \text{ lb} + 15\bar{0} \text{ lb} - 30\bar{0} \text{ lb}$$
$$= 25 \text{ lb to the right}$$

EXAMPLE 3

The crate in Example 2 has a mass of 5.00 slugs. What is its acceleration when the workers are pushing against the frictional force?

Sketch: None needed

Data:

$$F = 25 \text{ lb (from Example 2)}$$
$$m = 5.00 \text{ slugs}$$
$$a = ?$$

Basic Equation:

$$F = ma$$

Working Equation:

$$a = \frac{F}{m}$$

Substitution:

$$a = \frac{25 \text{ lb}}{5.00 \text{ slugs}}$$
$$= 5.0 \frac{\cancel{\text{lb}}}{\cancel{\text{slugs}}} \times \frac{1 \cancel{\text{slug}} \text{ ft/s}^2}{1 \cancel{\text{lb}}}$$
$$= 5.0 \text{ ft/s}^2$$

Note: We use a conversion factor to obtain acceleration units.

EXAMPLE 4

Two workers push in the same direction on a large pallet. The force exerted by one worker is 645 N. The force exerted by the other worker is 755 N. The motion is opposed by a frictional force of 1175 N. Find the net force.

$$\Sigma \mathbf{F} = 645 \text{ N} + 755 \text{ N} - 1175 \text{ N}$$
$$= 225 \text{ N}$$

PROBLEMS 4.4

Find the net force including its direction when each force acts in the direction indicated.

1. 17.0 N to the left, 20.0 N to the right.
2. 265 N to the left, 85 N to the right.
3. 100.0 N to the left, 75.0 N to the right, and 10.0 N to the right.
4. $19\bar{0}$ lb to the left, 87 lb to the right, and 49 lb to the right.
5. 346 N to the right, 247 N to the left, and 103 N to the left.
6. 37 N to the right, 24 N to the left, 65 N to the right, and 85 N to the right.
7. Find the acceleration of an automobile of mass $10\bar{0}$ slugs acted upon by a driving force of $50\bar{0}$ lb that is opposed by a frictional force of $10\bar{0}$ lb (Fig. 4.7).
8. Find the acceleration of an automobile of mass $15\bar{0}0$ kg acted upon by a driving force of $22\bar{0}0$ N that is opposed by a frictional force of 450 N.
9. A truck of mass 13,100 kg is acted upon by a driving force of $89\bar{0}0$ N. The motion is opposed by a frictional force of 2230 N. Find the acceleration.
10. A speedboat of mass 30.0 slugs has a $30\bar{0}$-lb force applied by the propellers. The friction of the water on the hull is a force of $10\bar{0}$ lb. Find the acceleration.

SKETCH

12 cm² | w

4.0 cm

DATA

$A = 12\ \text{cm}^2$, $l = 4.0$ cm, $w = ?$

BASIC EQUATION

$A = lw$

WORKING EQUATION

$w = \frac{A}{l}$

SUBSTITUTION

$w = \frac{12\ \text{cm}^2}{4.0\ \text{cm}} = 3.0$ cm

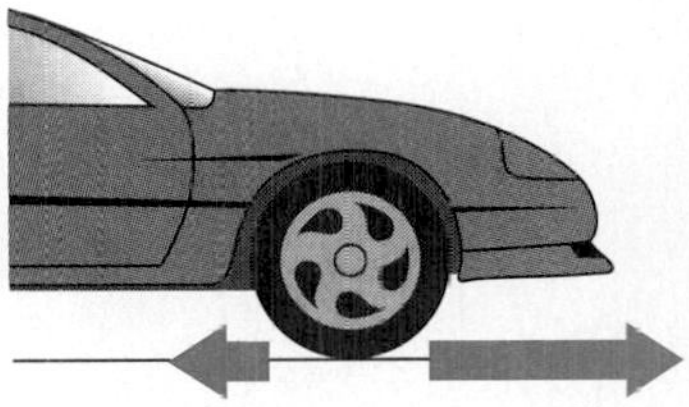

Figure 4.7

4.5 Gravity and Weight

The **weight** of an object is the amount of *gravitational pull* exerted on an object by the earth. If this force is not balanced by other forces, an acceleration is produced. When you hold a brick in your hand as in Fig. 4.8, you exert an upward force on the brick that balances the downward force (weight). If you remove your hand, the brick moves downward due to the unbalanced force. The velocity of the falling brick increases, but the acceleration (rate of change of the velocity) is constant.

The acceleration of all objects near the surface of the earth is the same if air resistance is ignored. We call this acceleration due to the gravitational pull of the earth g. Its value is 9.80 m/s² in the metric system and 32.2 ft/s² in the U.S. system.

The *weight* of an object is the force exerted by the earth (or by another large body) and gives the object an acceleration g. This force can be found using $F = ma$, where $a = g$. If we abbreviate weight by F_w, the equation for weight is

$$F_w = mg$$

where

F_w = weight
m = mass
g = acceleration due to gravity
$g = 9.80\ \text{m/s}^2$ (metric)
$g = 32.2\ \text{ft/s}^2$ (U.S.)

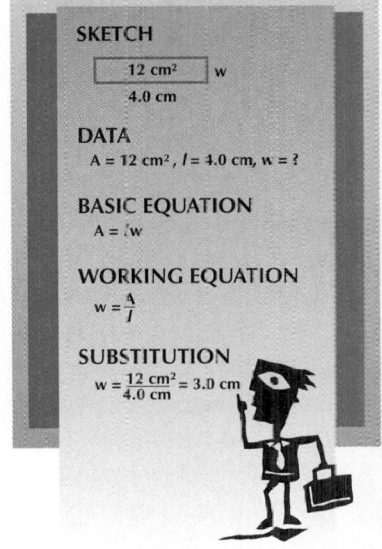

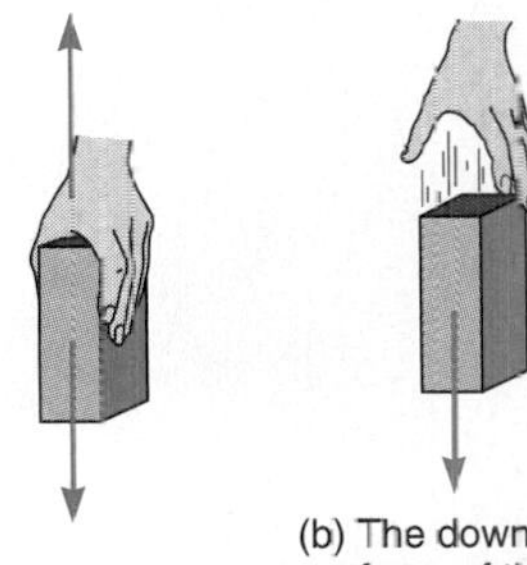

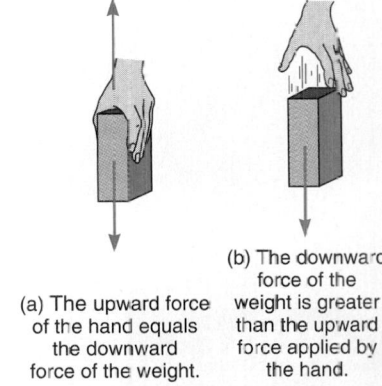

(a) The upward force of the hand equals the downward force of the weight.

(b) The downward force of the weight is greater than the upward force applied by the hand.

Figure 4.8 Balancing gravity in holding a brick

EXAMPLE 1

Find the weight of 5.00 kg.

Data:

$m = 5.00$ kg
$g = 9.80\ \text{m/s}^2$
$F_w = ?$

Basic Equation:

$$F_w = mg$$

Working Equation: Same

Substitution:

$$\begin{aligned} F_w &= (5.00 \text{ kg})(9.80 \text{ m/s}^2) \\ &= 49.0 \text{ kg m/s}^2 \\ &= 49.0 \text{ N} \qquad (1 \text{ N} = 1 \text{ kg m/s}^2) \end{aligned}$$

EXAMPLE 2

Find the weight of 12.0 slugs.

Data:

$$\begin{aligned} m &= 12.0 \text{ slugs} \\ g &= 32.2 \text{ ft/s}^2 \\ F_w &= ? \end{aligned}$$

Basic Equation:

$$F_w = mg$$

Working Equation: Same

Substitution:

$$\begin{aligned} F_w &= (12.0 \text{ slugs})(32.2 \text{ ft/s}^2) \\ &= 386 \text{ slug ft/s}^2 \\ &= 386 \text{ lb} \qquad (1 \text{ lb} = 1 \text{ slug ft/s}^2) \end{aligned}$$

Mass Versus Weight

Note that the mass of an object remains the same, but its weight varies according to the gravitational pull. For example, an astronaut of mass 75.0 kg has a weight of

$$F_w = mg = (75.0 \text{ kg})(9.80 \text{ m/s}^2) = 735 \text{ N}$$

on the earth. If that astronaut lands on the moon, where the acceleration due to gravity is less than it is on the earth, the astronaut will still have the same 75.0 kg mass but will weigh significantly less than 735 N. (The moon has less acceleration due to gravity than the earth in part because the moon is less massive than the earth. A full explanation will be given in Chapter 10.) The following example shows the astronaut's weight on the moon.

EXAMPLE 3

Find the 75.0-kg astronaut's weight on the moon, where $g = 1.63 \text{ m/s}^2$.

Sketch: None needed

Data:

$$\begin{aligned} m &= 75.0 \text{ kg} \\ g &= 1.63 \text{ m/s}^2 \\ F_w &= ? \end{aligned}$$

Basic Equation:

$$F_w = mg$$

Working Equation: Same

Substitution:

$$\begin{aligned} F_w &= (75.0 \text{ kg})(1.63 \text{ m/s}^2) \\ &= 122 \text{ N} \qquad (1 \text{ N} = 1 \text{ kg m/s}^2) \end{aligned}$$

Note that the astronaut's *mass* on the moon remains 75.0 kg.

Mass Versus Volume

Do not confuse mass and volume. The volume of an object is the measure of the space it occupies. Volume is measured in cubic units such as cm^3, ft^3, or L. The mass of an object is the amount of inertia or the amount of material it contains. The more mass contained in an object, the greater its inertia and the more force it takes to move it or change its motion. Compare the masses of two boxes of identical size, one filled with books and one empty. The box filled with books has more mass and requires more force to move it.

As an astronaut goes from the earth to the moon, his or her weight changes but mass and volume remain the same. Weight and mass are directly proportional in a given place as we saw earlier.

PROBLEMS 4.5

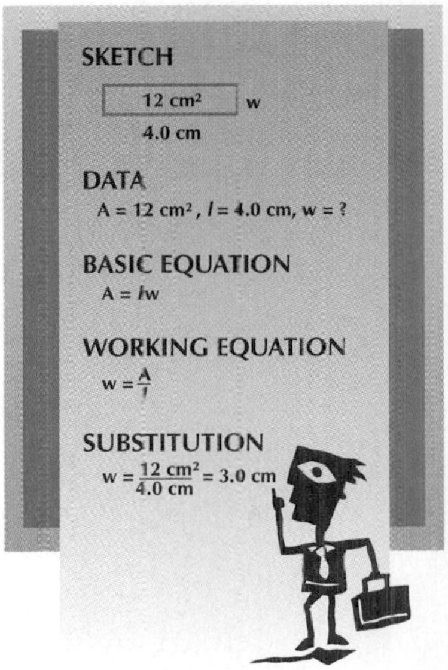

Find the weight for each mass.

1. $m = 30.0$ kg
2. $m = 60.0$ kg
3. $m = 10.0$ slugs
4. $m = 9.00$ kg

Find the mass for each weight.

5. $F_w = 17.0$ N
6. $F_w = 21.0$ lb
7. $F_w = 12,\bar{0}00$ N
8. $F_w = 25,\bar{0}00$ N
9. $F_w = 6.7 \times 10^{12}$ N
10. $F_w = 5.5 \times 10^6$ lb
11. Find the weight of an 1150-kg automobile.
12. Find the weight of an 81.5-slug automobile.
13. Find the mass of a 2750-lb automobile.
14. Find the mass of an 11,500-N automobile.
15. Find the weight of a 1350-kg automobile (a) on the earth and (b) on the moon.
16. Maria weighs 115 lb on the earth. What are her (a) mass and (b) weight on the moon?
17. John's mass is 65.0 kg on the earth. What are his (a) mass and (b) weight on the moon?
18. What is your weight in newtons and in pounds?
19. What is your mass in kilograms and in slugs?
20. What are your U.S. mass and weight on the moon?
21. What are your metric mass and weight on the moon?
22. John's mass is 65.0 kg on the earth. What are his U.S. (a) mass and (b) weight $40\bar{0}0$ mi above the surface of the earth, where $g = 7.85$ ft/s^2?
23. Maria weighs 115 lb on the earth. What are her U.S. (a) mass and (b) weight on Jupiter, where $g = 85.0$ ft/s^2?
24. John's mass is 65.0 kg on the earth. What are his metric (a) mass and (b) weight on Mars, where $g = 3.72$ m/s^2?
25. What are your metric mass and weight on Jupiter, where $g = 25.9$ m/s^2?
26. What are your metric mass and weight on Mars, where $g = 3.72$ m/s^2?

4.6 Law of Action and Reaction

When an automobile accelerates, we know that a force is being applied to it. What applies this force? You may think that the tires exert this force on the auto. This is not correct, because the tires move along with the auto and there must be a force applied to them also. The ground below the tires actually supplies the force that accelerates the car (Fig. 4.9). This force is called a *reaction* to the force exerted by the tires on the ground, which is called the *action force*. The third law of motion, the **law of action and reaction,** can be stated as follows:

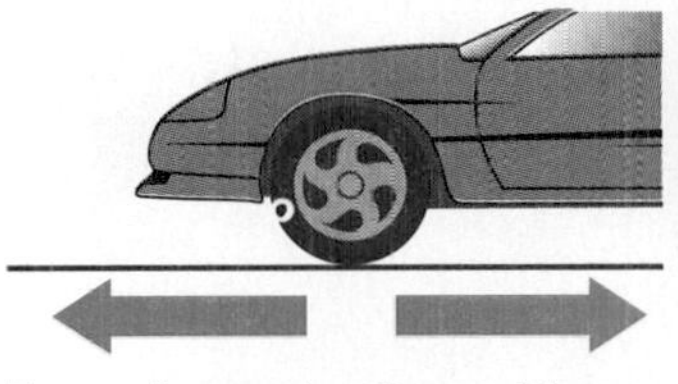

$F_1 = -F_2$

Figure 4.9 The ground supplies the force to accelerate the car.

LAW OF ACTION AND REACTION: NEWTON'S THIRD LAW

For every force applied by object *A* to object *B* (action), there is an equal but opposite force exerted by object *B* to object *A* (reaction).

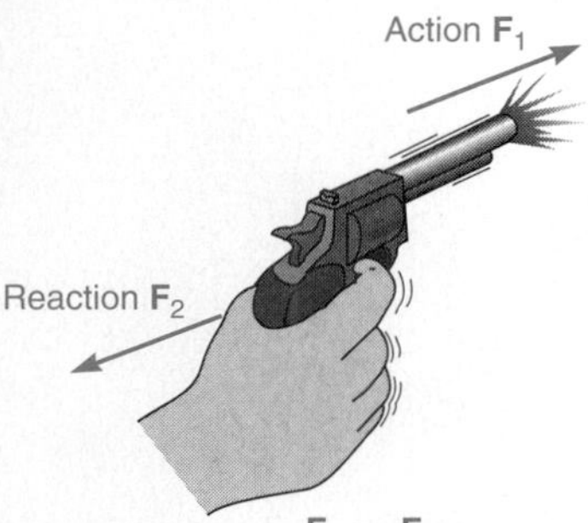

Figure 4.10 The magnitude of the reaction force always equals the magnitude of the action force.

When a bullet is fired from a handgun (action), the recoil felt is the reaction. These forces are shown in Fig. 4.10. Note that the action and reaction forces *never* act on the same object.

For every interaction, the forces always occur in pairs and are equal and opposite. When you sit on a chair, your weight pushes *down* on the chair; the chair pushes *up* with a force equal to your weight. If the chair pushed up with a force less than your weight, you would fall through it. If the chair pushed up with a force greater than your weight, you would be pushed up above the seat.

Most interactions depend on force. What would happen if you were standing in a small boat and tried to jump across to the nearby boat dock? You would fall into the water. Why? The force you would exert against the boat as you jumped would push it away, and the equal and opposite force the boat would exert on you would not result in much forward motion for you.

Glossary

Coefficient of Friction The ratio between the frictional force and the normal force of an object. The number represents how rough or smooth two surfaces are when moving across one another. (p. 122)

Force A push or a pull that tends to change the motion of an object or prevent an object from changing motion. Force is a vector quantity and thus has both magnitude and direction. (p. 116)

Friction A force that resists the relative motion of two objects in contact caused by the irregularities of two surfaces sliding or rolling across each other. (p. 121)

Inertia The property of a body that causes it to remain at rest if it is at rest or to continue moving with a constant velocity unless an unbalanced force acts upon it. (p. 117)

Law of Acceleration The total force acting on a body is equal to the mass of the body times its acceleration. (Newton's second law) (p. 118)

Law of Action and Reaction For every force applied by object *A* to object *B* (action), there is an equal but opposite force exerted by object *B* on object *A* (reaction). (Newton's third law). (p. 129)

Law of Inertia A body that is in motion continues in motion with the same velocity (at constant speed and in a straight line) and a body at rest continues at rest unless an unbalanced (outside) force acts upon it (Newton's first law). (p. 116)

Mass A measure of the inertia of a body. (p. 118)

Normal Force Force perpendicular to the contact surface. (p. 122)

Weight The amount of gravitational pull exerted on an object by the earth or by another large body. (p. 127)

Formulas

4.2 $F = ma$

4.3 $F_f = \mu F_N$

4.5 $F_w = mg$

where $g = 9.80\ \text{m/s}^2$ (metric)
$g = 32.2\ \text{ft/s}^2$ (U.S.)

Review Questions

1. Force
   (a) is a vector quantity. (b) may be different from weight.
   (c) does not always cause motion. (d) all of the above.

2. The metric weight of a 10-lb bag of sugar is approximately
   (a) 4.45 N. (b) 44.5 N.
   (c) 445 N. (d) none of the above.
3. Mass and weight
   (a) are the same. (b) are different.
   (c) do not change wherever you are.
4. According to Newton's second law, the law of acceleration,
   (a) acceleration is equal to mass times force.
   (b) mass is equal to mass times acceleration.
   (c) force is equal to mass times acceleration.
   (d) none of the above.
5. Friction
   (a) always acts parallel to the surface of contact and opposite to the direction of motion.
   (b) acts in the direction of motion.
   (c) is smaller when starting than moving.
   (d) is an imaginary force.
6. Cite three examples of forces acting without motion being produced.
7. (a) Does a pound of feathers have more inertia than a pound of lead?
   (b) Does the pound of feathers have more mass than the pound of lead?
8. How is inertia a factor in multicar pileups?
9. Using your own words, state Newton's first law the law of inertia.
10. Distinguish between velocity and acceleration.
11. When the same force is applied to two different masses, which will have a greater acceleration?
12. Is 3 pounds heavier than 10 newtons?
13. Explain how life would be easier or more difficult without friction.
14. Explain how the weight of an astronaut is different on the moon than on the earth. Would the astronaut's mass be different?
15. Explain the difference between action and reaction forces.
16. State Newton's third law of motion, the law of action and reaction, in your own words.

Review Problems

1. A crate of mass 6.00 kg is moved by a force of 18.0 N. What is its acceleration?
2. A 825-N force is required to pedal a bike with an acceleration of 11.0 m/s^2. What is the mass of the bike and person?
3. A block of mass 0.89 slug moves with a force of 17.0 lb. Find the block's acceleration.
4. What is the force necessary for a 2400-kg truck to accelerate at a rate of 8.0 m/s^2?
5. Two movers push a piano across a frictionless surface. One pushes with 29.0 N of force and the other mover exerts 35.0 N. What is the total force?
6. A 340-N box has a frictional force of 57 N. Find the coefficient of sliding friction.
7. A truck pulls a trailer with a frictional force of 870 N and a coefficient of friction of 0.23. What is the trailer's normal force?
8. A steel box is slid along a steel surface. It has a normal force of 57 N. What is the frictional force?
9. A rock of a mass 13.0 kg is dropped from a cliff. Find its weight.
10. A projectile has a mass of 0.37 slug. Find its weight.

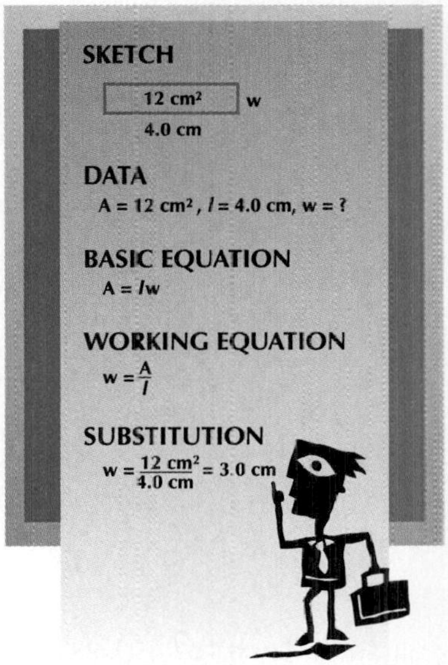

APPLIED CONCEPTS

1. Engineers at Boeing developing specs for their "next-generation" 737 aircraft need to know the acceleration of the 737-900 during a typical take-off. (a) What acceleration would they calculate given the plane's 78,200-kg mass and its maximum engine force of 121,000 N? (b) How fast would the plane be traveling after the first $5\overline{0}0$ m of runway? (c) How fast would it travel after the first $15\overline{0}0$ m of runway?
2. The Apollo spacecrafts were launched toward the moon using the Saturn rocket, the most powerful rocket available. Each rocket had five engines producing a total of 33.4×10^6 N of force to launch the 2.77×10^6-kg spacecraft toward the moon. (a) Find the average acceleration of the spacecraft. (b) Calculate the altitude of the rocket 2.50 min after launch—the point when the spacecraft loses its first stage.
3. Kirsten's mass is 3.73 slugs. Being the physics fan that she is, she decides to see what her apparent weight will be during an elevator ride. Beginning at rest, the elevator accelerates upward at 4.50 ft/s^2 for 3.00 s and then continues at a constant upward velocity. Finally, as the elevator comes to a stop at the top floor, the elevator slows down (accelerates downward but continues to move upward) at a rate of −5.5 ft/s^2 (the negative sign represents the downward direction). Find Kirsten's weight while the elevator is (a) at rest, (b) speeding up, (c) moving at a constant velocity, and (d) slowing down. The next time you ride in an elevator, concentrate on when you feel heavier and when you feel lighter.
4. A motorcycle racer traveling at 145 km/h loses control in a corner of the track and slides across the concrete surface. The combined mass of the rider and bike is 243 kg. The steel of the motorcycle rubs against the concrete road surface. (a) What is the frictional force between the road and the motorcycle and rider? (b) What would be the acceleration of the motorcycle and rider during the wipeout? (c) Assuming there were no barriers to stop the motorcycle and rider, how long would it take the bike and the rider to slow to a stop?
5. The motorcycle and rider are sliding with the same acceleration as found in Problem 4. If the motorcycle and rider have been sliding for 4.55 s, what will be the force applied to the motorcycle and the rider when they strike the side barrier and come to rest in 0.530 s?

CHAPTER 5

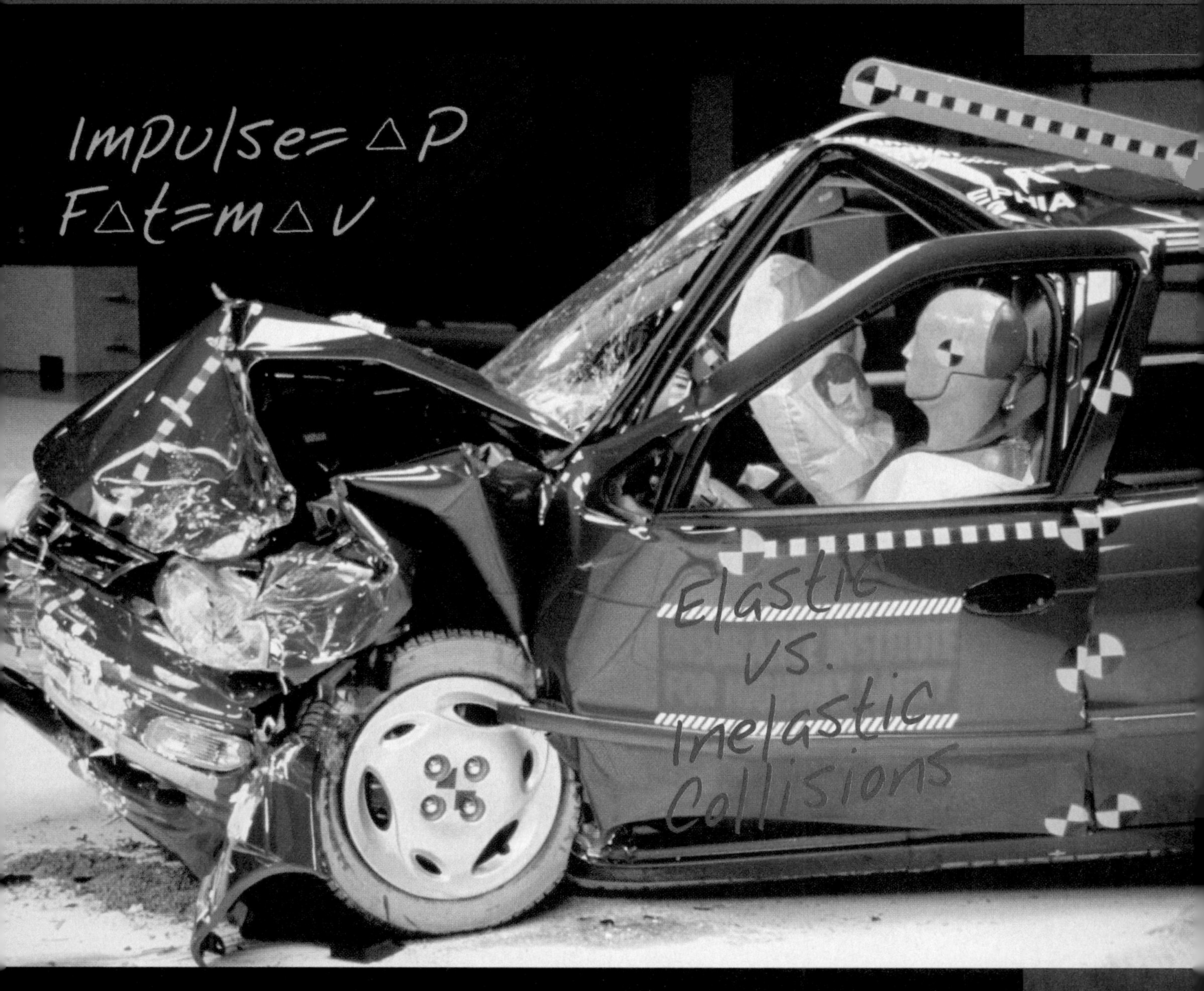

MOMENTUM

Impulse and momentum are important concepts in describing and understanding the motion of objects and the related effects on those objects. The law of conservation of momentum is an important law of physics, which helps us analyze how two objects interact with each other when they are in contact with each other and when they collide.

Objectives

The major goals of this chapter are to enable you to:

1. Use impulse and momentum in describing motion.
2. State the law of conservation of momentum and apply it to physical problems.
3. Analyze elastic and inelastic collisions of two objects.

5.1 Impulse and Momentum

We know that it is much more difficult to stop a large truck than a small car traveling at the same speed. The truck has more *inertia* and is more difficult to bring to a stop or to begin moving than the car. Momentum is a measure of the amount of inertia and motion an object has or of the difficulty in bringing a moving object to rest. **Momentum** *equals the product of the mass times the velocity of an object.*

$$p = mv$$

where

p = momentum
m = mass
v = velocity

The momentum of a train makes it impossible to stop within a short distance and explains why it cannot stop at a railroad crossing when the engineer sees someone stopped or stalled at it.

The units of momentum are kg m/s in the metric system and slug ft/s in the U.S. system. Momentum is a vector quantity whose direction is the same as the velocity.

EXAMPLE 1

Find the momentum of an auto with mass 105 slugs traveling 60.0 mi/h.

Sketch: None needed

Data:

$$m = 105 \text{ slugs}$$
$$v = 60.0 \text{ mi/h} = 88.0 \text{ ft/s}$$
$$p = ?$$

Basic Equation:

$$p = mv$$

Working Equation: Same

Substitution:

$$p = (105 \text{ slugs})(88.0 \text{ ft/s})$$
$$= 9240 \text{ slugs ft/s}$$

EXAMPLE 2

Find the momentum of an auto with mass 1350 kg traveling 75.0 km/h.

Sketch: None needed

Data:

$$m = 1350 \text{ kg}$$
$$v = 75.0 \frac{\cancel{\text{km}}}{\cancel{\text{h}}} \times \frac{1000 \text{ m}}{1 \cancel{\text{km}}} \times \frac{1 \cancel{\text{h}}}{3600 \text{ s}} = 20.8 \text{ m/s}$$
$$p = ?$$

Basic Equation:

$$p = mv$$

Working Equation: Same

Substitution:

$$p = (1350 \text{ kg})(20.8 \text{ m/s})$$
$$= 28{,}100 \text{ kg m/s}$$

EXAMPLE 3

Find the velocity a bullet of mass 1.00×10^{-2} kg would have to have so that it has the same momentum as a lighter bullet of mass 1.80×10^{-3} kg and velocity 325 m/s.

Sketch:

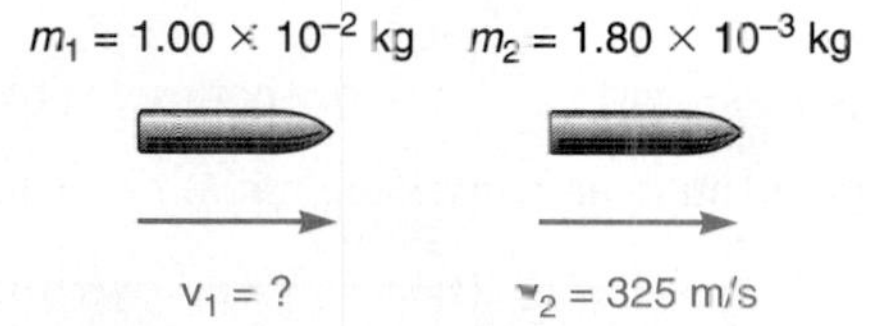

Data:

Heavier Bullet	Lighter Bullet
$m_1 = 1.00 \times 10^{-2}$ kg	$m_2 = 1.80 \times 10^{-3}$ kg
$v_1 = ?$	$v_2 = 325$ m/s
$p_1 = ?$	$p_2 = ?$

Basic Equations:

$$p_1 = m_1 v_1$$
$$p_2 = m_2 v_2$$

We want

$$p_1 = p_2$$

or

$$m_1 v_1 = m_2 v_2$$

Working Equation:

$$v_1 = \frac{m_2 v_2}{m_1}$$

Substitution:

$$v_1 = \frac{(1.80 \times 10^{-3} \text{ kg})(325 \text{ m/s})}{1.00 \times 10^{-2} \text{ kg}}$$
$$= 58.5 \text{ m/s}$$

The **impulse** on an object is the product of the force and the time interval during which the force acts on the object. That is,

$$\text{impulse} = Ft$$

where

F = force

t = time interval during which the force acts

How are impulse and momentum related? Recall that

$$a = \frac{v_f - v_i}{t}$$

If we substitute this equation into Newton's second law of motion, we have

$$F = ma$$

$$F = m\left(\frac{v_f - v_i}{t}\right)$$

$$F = \frac{mv_f - mv_i}{t} \qquad \text{Remove parentheses.}$$

$$Ft = mv_f - mv_i \qquad \text{Multiply both sides by } t.$$

Note that mv_f is the final momentum and mv_i is the initial momentum. That is,

$$\text{impulse} = \Delta p \text{ (change in momentum)} = Ft = mv_f - mv_i$$

Note: The Greek letter Δ ("delta") is used to designate "change in."

The impulse is the measure of the change in momentum of an object in response to an exerted force. To change an object's momentum or motion, a force must be applied to the object for a given period of time. The amount of force and the length of time the force is applied will determine the change in momentum. A common example that illustrates this relationship is a golf club hitting a golf ball (Fig. 5.1). When a golf ball is on the tee, it has zero momentum because its velocity is zero. To give it momentum (impulse), you apply a force for a given period of time. During the time that the club and ball are in contact, the force of the swinging club is transferring most of its momentum to the ball. The impulse given to the ball is the product of the *force* with which the ball is hit and the length of *time* that the club and ball are in direct contact. You can give it more momentum (impulse) by increasing the *force* (by swinging the golf club faster) or increasing the *time* (by keeping the golf club in contact with the ball longer, which shows the importance of "followthrough").

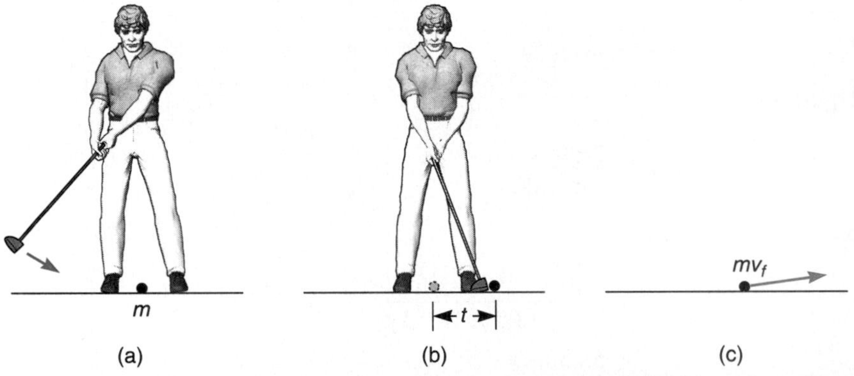

Figure 5.1 When a person hits a golf ball with a golf club, the club applies a force *F* during the time *t* that the club is in contact with the ball (b). The impulse (change in momentum) is $Ft = mv_f - mv_i = mv_f$ because $v_i = 0$ (c).

EXAMPLE 4

A 17.5-g bullet is fired at a muzzle velocity of 582 m/s from a gun with a mass of 8.00 kg and a barrel length of 75.0 cm.

(a) How long is the bullet in the barrel?
(b) What is the force on the bullet while it is in the barrel?
(c) Find the impulse exerted on the bullet while it is in the barrel.
(d) Find the bullet's momentum as it leaves the barrel.

Sketch:

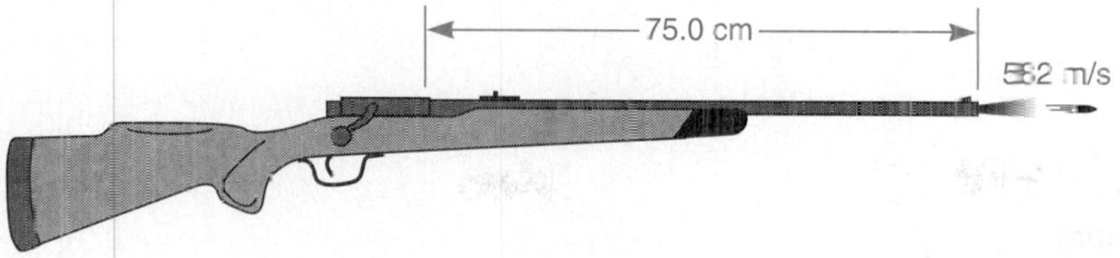

(a) Data:

$$s = 75.0 \text{ cm} = 0.750 \text{ m}$$
$$v_f = 582 \text{ m/s}$$
$$v_i = 0 \text{ m/s}$$
$$v_{avg} = \frac{v_f + v_i}{2} = \frac{582 \text{ m/s} + 0 \text{ m/s}}{2} = 291 \text{ m/s}$$
$$t = ?$$

Basic Equation:

$$s = v_{avg}t$$

Working Equation:

$$t = \frac{s}{v_{avg}}$$

Substitution:

$$t = \frac{0.750 \text{ m}}{291 \text{ m/s}}$$
$$= 0.00258 \text{ s}$$

Note: This is the length of time that the force is applied to the bullet.

(b) Data:

$$t = 0.00258 \text{ s}$$
$$m = 17.5 \text{ g} = 0.0175 \text{ kg}$$
$$v_f = 582 \text{ m/s}$$
$$v_i = 0 \text{ m/s}$$
$$F = ?$$

Basic Equation:

$$Ft = mv_f - mv_i$$

Working Equation:

$$F = \frac{mv_f - mv_i}{t}$$

Substitution:

$$F = \frac{(0.0175 \text{ kg})(582 \text{ m/s}) - (0.0175 \text{ kg})(0 \text{ m/s})}{0.00258 \text{ s}}$$
$$= 3950 \text{ kg m/s}^2$$
$$= 3950 \text{ N} \qquad (1 \text{ N} = 1 \text{ kg m/s}^2)$$

(c) Data:

$$t = 0.00258 \text{ s}$$
$$F = 3950 \text{ N}$$
$$\text{impulse} = ?$$

Basic Equation:

$$\text{impulse} = Ft$$

Working Equation: Same

Substitution:

$$\begin{aligned} \text{impulse} &= (3950 \text{ N})(0.00258 \text{ s}) \\ &= 10.2 \text{ N s} \\ &= 10.2 \text{ (kg m/s}^2\text{)(s)} \qquad (1 \text{ N} = 1 \text{ kg m/s}^2) \\ &= 10.2 \text{ kg m/s} \end{aligned}$$

(d) Data:

$$m = 17.5 \text{ g} = 0.0175 \text{ kg}$$
$$v = 582 \text{ m/s}$$
$$p = ?$$

Basic Equation:

$$p = mv$$

Working Equation: Same

Substitution:

$$\begin{aligned} p &= (0.0175 \text{ kg})(582 \text{ m/s}) \\ &= 10.2 \text{ kg m/s} \end{aligned}$$

Note: The impulse equals the change in momentum.

TRY THIS ACTIVITY

Scrambled Eggs

Drop a raw egg from a height of a few feet onto a surface that can be cleaned. Observe the motion of the egg as it hits the surface and note the time the egg takes to come to rest. Drop another raw egg from the same height into a suspended bed sheet. Again observe the motion of the egg as it hits the bed sheet and note the time the egg takes to come to rest (Fig. 5.2). Explain the connection between what happened to these eggs and how airbags in automobiles work.

Figure 5.2

One of the most important laws of physics is the following:

LAW OF CONSERVATION OF MOMENTUM

When no outside forces are acting on a system of moving objects, the total momentum of the system remains constant.

For example, consider a 35-kg boy and a 75-kg man standing next to each other on ice skates on "frictionless" ice (Fig. 5.3). The man pushes on the boy, which gives the boy a velocity of 0.40 m/s. What happens to the man? Initially, the total momentum was zero because the initial velocity of each was zero. According to the law of conservation of momentum, the total momentum must still be zero. That is,

$$m_{boy}v_{boy} + m_{man}v_{man} = 0$$
$$(35 \text{ kg})(0.40 \text{ m/s}) + (75 \text{ kg})v_{man} = 0$$
$$v_{man} = -0.19 \text{ m/s}$$

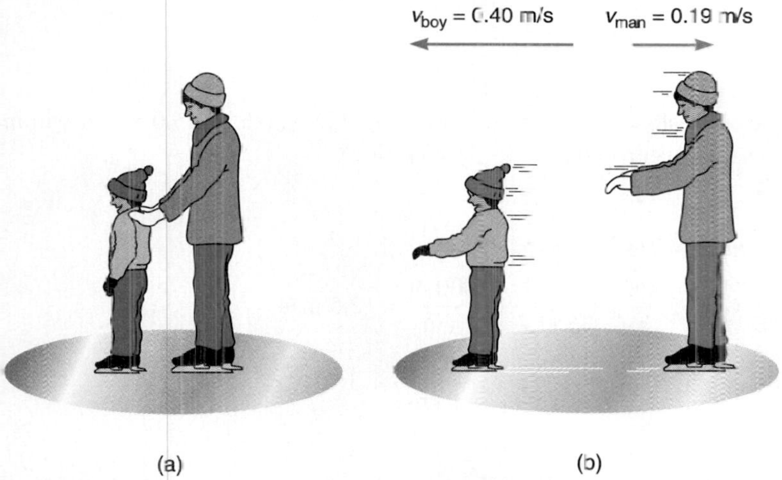

Figure 5.3 Momentum is conserved by the lighter boy moving faster than the heavier man.

Note: The minus sign indicates that the man's velocity and the boy's velocity are in opposite directions.

Rocket propulsion is another illustration of conservation of momentum. **Wernher von Braun** was a pioneering rocket scientist. As in the example of the skaters, the total momentum of a rocket on the launch pad is zero. When the rocket engines are fired, hot exhaust gases (actually gas molecules) are expelled downward through the rocket nozzle at tremendous speeds. As the rocket takes off, the sum of the total momentums of the rocket and the gas particles must remain zero. The total momentum of the gas particles is the sum of the products of each mass and its corresponding velocity and is directed down. The momentum of the rocket is the product of its mass and its velocity and is directed up.

When the rocket is in space, its propulsion works in the same manner The conservation of momentum is still valid except that when the rocket engines are fired, the total momentum is a nonzero constant. This is because the rocket has velocity.

Actually, repair work is more difficult in space than it is on the earth because of the conservation of momentum and the "weightlessness" of objects in orbit. On the earth, when a hammer is swung, the person is coupled to the earth by frictional forces, so that the person's mass includes that of the earth. In space orbit, because the person is weightless there is no friction to couple him or her to the spaceship. A person in space has roughly the same problem driving a nail as a person on the earth would have wearing a pair of "frictionless" roller skates.

A change in momentum takes force and time because

$$\text{change in momentum} = \text{impulse} = Ft$$

Wernher von Braun (1912–1977), *engineer and rocket expert, was born in Germany. He was chiefly responsible for the manufacture and launching of the first American artificial earth satellite, Explorer I, in 1958. As director of the Marshall Space Flight Center from 1960 to 1970, he developed the Saturn rocket for the Apollo 8 moon landing in 1969.*

As we noted earlier in this section, it is more difficult to stop a large truck than a small car traveling at the same speed and impossible to stop a rapidly moving train within a short distance. These events can be explained in terms of the **impulse–momentum theorem** as follows.

If the mass of an object is constant, then a change in its velocity results in a change in its momentum. That is,

$$\Delta p = m\Delta v$$

The impulse of an object equals its change in momentum. That is,

$$F\Delta t = \Delta p$$

Then,

IMPULSE–MOMENTUM THEOREM

$$F\Delta t = \Delta p = m\Delta v = mv_f - mv_i$$

EXAMPLE 5

What force is required to slow a 1450-kg car traveling 115 km/h to 45.0 km/h within 3.00 s? How far does the car travel during its deceleration?

Data:

$$m = 1450 \text{ kg}$$

$$v_f = 45\,\frac{\text{km}}{\text{h}} \times \frac{1 \text{ h}}{3600 \text{ s}} \times \frac{1000 \text{ m}}{1 \text{ km}} = 12.5 \text{ m/s}$$

$$v_i = 115\,\frac{\text{km}}{\text{h}} \times \frac{1 \text{ h}}{3600 \text{ s}} \times \frac{1000 \text{ m}}{1 \text{ km}} = 31.9 \text{ m/s}$$

$$\Delta t = 3.00 \text{ s}$$

$$F = ?$$

Basic Equation:

$$F\Delta t = mv_f - mv_i$$

Working Equation:

$$F = \frac{mv_f - mv_i}{\Delta t}$$

Substitution:

$$F = \frac{(1450 \text{ kg})(12.5 \text{ m/s}) - (1450 \text{ kg})(31.9 \text{ m/s})}{3.00 \text{ s}}$$

$$= -9380 \text{ kg m/s}^2 = -9380 \text{ N}$$

Note: The negative sign indicates a deceleration force.

Basic Equation:

$$s = \tfrac{1}{2}(v_f + v_i)t$$

Working Equation: Same

Substitution:

$$s = \tfrac{1}{2}(12.5 \text{ m/s} + 31.9 \text{ m/s})(3.00 \text{ s})$$

$$= 66.6 \text{ m}$$

PHYSICS CONNECTIONS

Airbags

During an automobile front-end collision, passengers will continue to travel forward until the dashboard, seat belt, or airbag applies a force on them to stop them. Airbags are designed to provide a cushion-like effect to gradually bring passengers to rest. Airbags increase the time it takes to bring passengers to a stop and reduce the force of the impact (Fig. 5.4). Airbags used in conjunction with seat belts help prevent death and serious injury.

Airbags expand from the steering wheel or dashboard when a sudden impulse or a change in momentum of the vehicle triggers a sensor that is connected to a heating element. The heating element causes a chemical reaction with a propellant that fills the airbag with nitrogen gas within $\frac{1}{20}$ s. This short inflation time gives the airbag enough time to inflate before the passenger strikes it. Within $\frac{1}{2}$ s, the collision is completed and the airbag deflates.

Airbags are designed to strike the average seat-belted male in the midsection of the body. An airbag, which expands at a rate of 150 mi/h, can be quite dangerous if the bag strikes short individuals in the face. Injuries to women and children caused by airbags are a serious problem. Efforts are being made to automatically adjust airbag deployment to make airbags safer for all passengers. Airbags are also used for side-impact collisions.

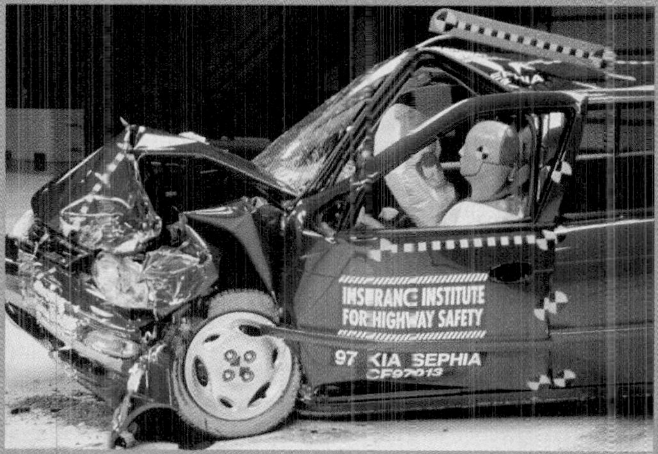

Figure 5.4 An airbag increases the time it takes to bring a passenger to a stop in a collision by reducing the force of the impact applied to the passenger.

PROBLEMS 5.1

Find the momentum of each object.

1. $m = 2.00$ kg, $v = 40.0$ m/s
2. $m = 5.00$ kg, $v = 90.0$ m/s
3. $m = 17.0$ slugs, $v = 45.0$ ft/s
4. $m = 38.0$ kg, $v = 97.0$ m/s
5. $m = 3.8 \times 10^5$ kg, $v = 2.5 \times 10^2$ m/s
6. $m = 3.84$ kg, $v = 1.6 \times 10^5$ m/s
7. $F_w = 1.50 \times 10^5$ N, $v = 4.50 \times 10^4$ m/s
8. $F_w = 3200$ lb, $v = 6\bar{0}$ mi/h (change to ft/s)
9. (a) Find the momentum of a heavy automobile of mass $18\bar{0}$ slugs traveling 70.0 ft/s.
   (b) Find the velocity of a light auto of mass 80.0 slugs so that it has the same momentum as the auto in part (a).
   (c) Find the weight (in lb) of each auto in parts (a) and (b).
10. (a) Find the momentum of a bullet of mass 1.00×10^{-3} slug traveling $70\bar{0}$ ft/s.
   (b) Find the velocity of a bullet of mass 5.00×10^{-4} slug so that it has the same momentum as the bullet in part (a).

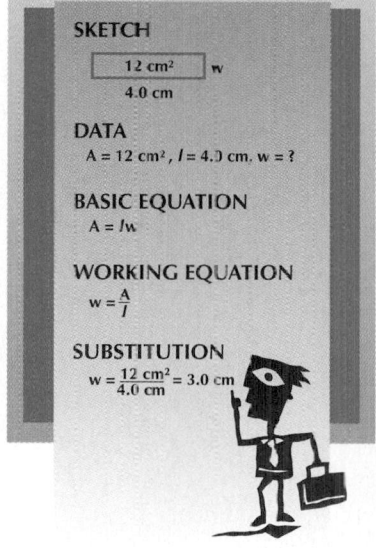

SKETCH

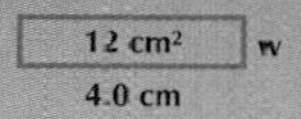

DATA

$A = 12\ \text{cm}^2,\ l = 4.0\ \text{cm},\ w = ?$

BASIC EQUATION

$A = lw$

WORKING EQUATION

$w = \frac{A}{l}$

SUBSTITUTION

$w = \frac{12\ \text{cm}^2}{4.0\ \text{cm}} = 3.0\ \text{cm}$

11. (a) Find the momentum of an automobile of mass 2630 kg traveling 21.0 m/s.
    (b) Find the velocity (in km/h) of a light auto of mass 1170 kg so that it has the same momentum as the auto in part (a).
12. A ball of mass 0.50 kg is thrown straight up at 6.0 m/s.
    (a) What is the initial momentum of the ball?
    (b) What is the momentum of the ball at its peak?
    (c) What is the momentum of the ball as it hits the ground?
13. A bullet with mass 60.0 g is fired with an initial velocity of 575 m/s from a gun with mass 4.50 kg. What is the speed of the recoil of the gun?
14. A cannon is mounted on a railroad car. The cannon shoots a 1.75-kg ball with a muzzle velocity of $30\bar{0}$ m/s. The cannon and the railroad car together have a mass of $45\bar{0}0$ kg. If the ball, cannon, and railroad car are initially at rest, what is the recoil velocity of the car and cannon?
15. A 125-kg pile driver falls from a height of 10.0 m to hit a piling.
    (a) What is its speed as it hits the piling?
    (b) With what momentum does it hit the piling?
16. A person is traveling 75.0 km/h in an automobile and throws a bottle of mass 0.500 kg out the window.
    (a) With what momentum does the bottle hit a roadway sign?
    (b) With what momentum does the bottle hit an oncoming automobile traveling 85.0 km/h in the opposite direction?
    (c) With what momentum does the bottle hit an automobile passing and traveling 85.0 km/h in the same direction?
17. A 75.0-g bullet is fired with a muzzle velocity of $46\bar{0}$ m/s from a gun with mass 3.75 kg and barrel length of 66.0 cm.
    (a) How long is the bullet in the barrel?
    (b) What is the force on the bullet while it is in the barrel?
    (c) Find the impulse exerted on the bullet while it is in the barrel.
    (d) Find the bullet's momentum as it leaves the barrel.
18. A 60.0-g bullet is fired at a muzzle velocity of 525 m/s from a gun with mass 4.50 kg and a barrel length of 55.0 cm.
    (a) How long is the bullet in the barrel?
    (b) What is the force on the bullet while it is in the barrel?
    (c) Find the impulse exerted on the bullet while it is in the barrel.
    (d) Find the bullet's momentum as it leaves the barrel.
19. What force is required to stop a 1250-kg car traveling 95.0 km/h within 4.00 s? How far does the car travel during its deceleration?
20. What force is required to slow a 1350-kg car traveling 90.0 km/h to 25.0 km/h within 4.00 s? How far does the car travel during its deceleration? How long does it take for the car to come to a complete stop at this same rate of deceleration?

5.2 Collisions

The collision of two objects is an excellent example that demonstrates the law of conservation of momentum. Whenever objects collide in the absence of any external forces, the total momentum of the objects *before* the collision equals the total momentum *after* the collision. That is,

$$\text{total momentum}_{\text{before collision}} = \text{total momentum}_{\text{after collision}}$$

In the remainder of this section, we will study collisions by discussing the two extreme types, perfectly elastic and inelastic.

Elastic Collisions

In an **elastic collision,** two objects collide and return to their original shape without being permanently deformed. This happens when two billiard balls collide.

EXAMPLE 1

One ball of mass 0.600 kg traveling 9.00 m/s to the right collides with a second ball of mass 0.300 kg traveling 8.00 m/s to the left. After the collision, the heavier ball is traveling 2.33 m/s to the left. What is the velocity of the lighter ball after the collision?

Sketch:

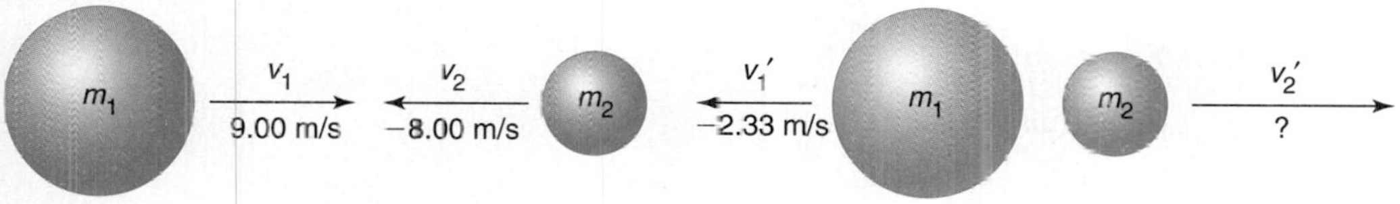

(a) Before impact (b) After impact

Data:

$$m_1 = 0.600 \text{ kg}$$
$$m_2 = 0.300 \text{ kg}$$
$$v_1 = 9.00 \text{ m/s} \quad \text{(right is positive direction)}$$
$$v_2 = -8.00 \text{ m/s} \quad \text{(left is negative direction)}$$
$$v_1' = -2.33 \text{ m/s}$$
$$v_2' = ?$$

Basic Equation:

$$m_1v_1 + m_2v_2 = m_1v_1' + m_2v_2'$$

Working Equation:

$$v_2' = \frac{m_1v_1 + m_2v_2 - m_1v_1'}{m_2}$$

Substitution:

$$v_2' = \frac{(0.600 \text{ kg})(9.00 \text{ m/s}) + (0.300 \text{ kg})(-8.00 \text{ m/s}) - (0.600 \text{ kg})(-2.33 \text{ m/s})}{0.300 \text{ kg}}$$
$$= 14.66 \text{ m/s} \quad \text{(to the right)}$$

Inelastic Collisions

In an **inelastic collision,** two objects collide and couple together. This happens when two railroad cars collide and couple together and move along the tracks.

EXAMPLE 2

A 1.75×10^4-kg railroad car traveling 8.00 m/s to the east collides and couples with a stopped 2.25×10^4-kg railroad car. What is the velocity of the joined railroad cars after the collision?

Sketch:

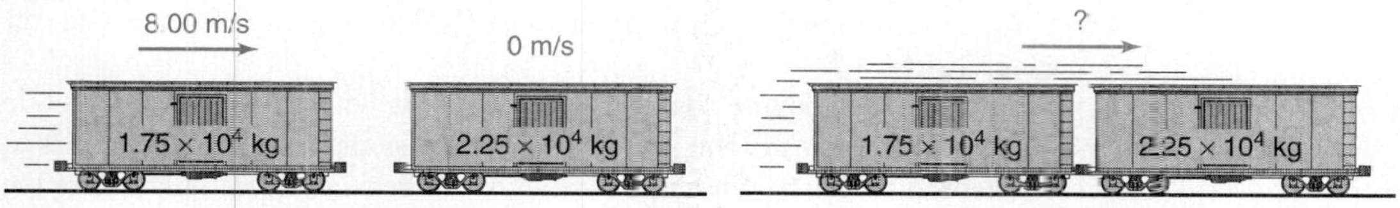

(a) Before impact (b) After impact

Data:

$$m_1 = 1.75 \times 10^4 \text{ kg}$$
$$m_2 = 2.25 \times 10^4 \text{ kg}$$
$$v_1 = 8.00 \text{ m/s}$$
$$v_2 = 0 \text{ m/s}$$
$$v' = ?$$

Basic Equation:

$$m_1v_1 + m_2v_2 = (m_1 + m_2)v'$$

Working Equation:

$$v' = \frac{m_1v_1 + m_2v_2}{m_1 + m_2}$$

Substitution:

$$v' = \frac{(1.75 \times 10^4 \text{ kg})(8.00 \text{ m/s}) + (2.25 \times 10^4 \text{ kg})(0 \text{ m/s})}{1.75 \times 10^4 \text{ kg} + 2.25 \times 10^4 \text{ kg}}$$
$$= 3.50 \text{ m/s (east)}$$

Collisions that demonstrate both elastic and inelastic characteristics as well as collisions in two-dimensional and three-dimensional space will not be discussed in this text.

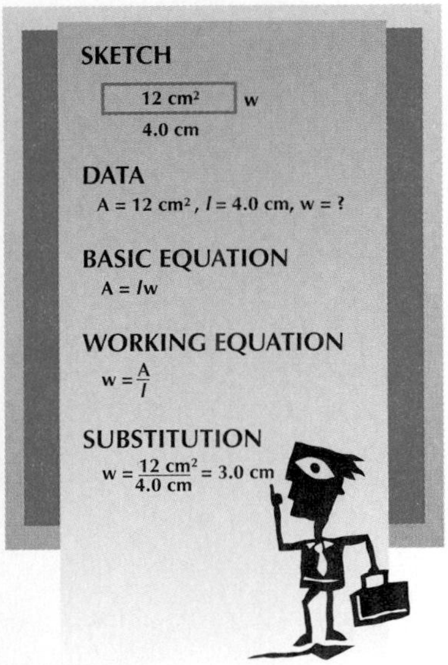

PROBLEMS 5.2

Assume a frictionless surface in each of the following problems.

1. One ball of mass 0.500 kg traveling 6.00 m/s to the right collides with a ball of mass 0.200 kg initially at rest. After the collision, the heavier ball is traveling 2.57 m/s to the right. What is the velocity of the lighter ball after the collision?
2. A ball of mass 625 g traveling 4.00 m/s to the right collides with another ball of equal mass that is initially at rest. The first ball is at rest after the collision. (a) What is the velocity of the second ball after the collision? (b) What is the velocity of the second ball if the size of the two balls of equal mass changes?
3. A 0.600-kg ball traveling 4.00 m/s to the right collides with a 1.00-kg ball traveling 5.00 m/s to the left. After the collision, the lighter ball is traveling 7.25 m/s to the left. What is the velocity of the heavier ball after the collision?
4. A 90.0-g disk traveling 3.00 m/s to the right collides with a 75.0-g disk traveling 8.00 m/s to the left. After the collision, the heavier disk is traveling 7.00 m/s to the left. What is the velocity of the lighter disk after the collision?
5. A railroad car of mass 2.00×10^4 kg is traveling north 6.00 m/s and collides with a railroad car of mass 1.50×10^4 kg traveling south 4.00 m/s. Find the velocity of the railroad cars that become coupled after the collision.
6. Find the velocity of the railroad cars in Problem 5 if the lighter car is initially at rest.
7. One cart of mass 12.0 kg is moving 6.00 m/s to the right on a frictionless track and collides with a cart of mass 4.00 kg moving in the opposite direction 3.00 m/s. Find the final velocity of the carts that become stuck together after the collision.
8. One cart of mass 15.0 kg is moving 5.00 m/s to the right on a frictionless track and collides with a cart of mass 3.00 kg. The final velocity of the carts that become stuck together after the collision is 1.50 m/s to the right. Find the velocity of the second cart before the collision.
9. A 1650-kg automobile moving south 12.0 m/s collides with a 2450-kg automobile moving north on an icy road. The automobiles stick together and move 3.00 m/s to the north after the collision. What is the speed of the heavier automobile before the collision?

10. A 16.0-g bullet is shot into a wooden block at rest with mass 4550 g on a frictionless surface. The block moves 1.20 m/s after the bullet strikes and becomes lodged in the block. Find the speed of the bullet before striking the block.
11. A 2450-kg automobile moving north 12.0 m/s collides with a 1650-kg automobile moving 8.00 m/s on an icy road. The automobiles stick together and move after the collision. Find the velocity of the automobiles after the collision if the automobiles were traveling in (a) opposite directions and (b) the same direction before the collision.

Glossary

Elastic Collision A collision in which two objects return to their original shape without being permanently deformed. (p. 145)

Impulse The product of the force exerted and the time interval during which the force acts on the object. Impulse equals the change in momentum of an object in response to the exerted force. (p. 137)

Impulse–Momentum Theorem If the mass of an object is constant, then a change in its velocity results in a change of its momentum.
That is, $F\Delta t = \Delta p = m\Delta v = mv_f - mv_i$. (p. 142)

Inelastic Collision A collision in which two objects couple together. (p. 145)

Law of Conservation of Momentum When no outside forces are acting on a system of moving objects, the total momentum of the system remains constant. (p. 141)

Momentum A measure of the amount of inertia and motion an object has or the difficulty in bringing a moving object to rest. Momentum equals the mass times the velocity of an object. (p. 136)

Formulas

5.1 $p = mv$
impulse = Ft
impulse = change in momentum
impulse–momentum theorem

$$F\Delta t = \Delta p = m\Delta v = mv_f - mv_i$$

5.2 total momentum$_{\text{before collision}}$ = total momentum$_{\text{after collision}}$

Review Questions

1. Momentum is
   (a) equal to speed times weight.
   (b) equal to mass times velocity.
   (c) the same as force.
2. Impulse is
   (a) a force applied to an object.
   (b) the initial force applied to an object.
   (c) the initial momentum applied to an object.
   (d) the change in momentum due to a force being applied to an object during a given time.
3. Why do a slow-moving loaded truck and a speeding rifle bullet each have a large momentum?
4. How are impulse and change in momentum related?
5. Why is "followthrough" important in hitting a baseball or a golf ball?
6. Describe in your own words the law of conservation of momentum.

7. Describe conservation of momentum in terms of a rocket being fired.
8. One billiard ball striking another is an example of a(n) _____ collision.
9. One moving loaded railroad car striking and coupling with a parked empty railroad car and then both moving on down the track is an example of a(n) _____ collision.
10. A father and 8-year-old son are standing on ice skates in an ice arena. The father then pushes the son on the back to give him a quick start. What do we know about the momentum of each person?

Review Problems

1. A truck with mass 1475 slugs travels 57.0 mi/h. Find its momentum.
2. A projectile with mass 27.0 kg is fired with a momentum of 5.50 kg m/s. Find its velocity.
3. A box is pushed with a force of 125 N for 2.00 min. What is the impulse?
4. What is the momentum of a bullet of mass 0.034 kg traveling at 250 m/s?
5. A 4.00-g bullet is fired from a 4.50-kg gun with a muzzle velocity of 625 m/s. What is the speed of the recoil of the gun?
6. A 150-kg pile driver falls from a height of 7.5 m to hit a piling.
   (a) What is its speed as it hits the piling?
   (b) With what momentum does it hit the piling?
7. A 15.0-g bullet is fired at a muzzle velocity of 3250 m/s from a high-powered rifle with a mass of 4.75 kg and barrel of length 75.0 cm.
   (a) How long is the bullet in the barrel?
   (b) What is the force on the bullet while it is in the barrel?
   (c) Find the impulse exerted on the bullet while it is in the barrel.
   (d) Find the bullet's momentum as it leaves the barrel.
8. What force is required to slow a 1250-kg car traveling 115 km/h to 30.0 km/h within 3.50 s? (a) How far does the car travel during its deceleration? (b) How long does it take for the car to come to a complete stop at this same rate of deceleration?
9. One ball of mass 575 g traveling 3.50 m/s to the right collides with another ball of mass 425 g that is initially at rest. After the collision, the lighter ball is traveling 4.03 m/s. What is the velocity of the heavier ball after the collision?
10. A railroad car of mass 2.25×10^4 kg is traveling east 5.50 m/s and collides with a railroad car of mass 3.00×10^4 kg traveling west 1.50 m/s. Find the velocity of the railroad cars that become coupled after the collision.
11. A 195-g ball traveling 4.50 m/s to the right collides with a 125-g ball traveling 12.0 m/s to the left. After the collision, the heavier ball is traveling 8.40 m/s to the left. What is the velocity of the lighter ball after the collision?

APPLIED CONCEPTS

1. A coach knows it is vital that the volleyballs be fully inflated before a match. (a) Calculate the impulse on a spiked 0.123-slug volleyball when the incoming velocity of the ball is −11.5 ft/s and the outgoing velocity is 57.3 ft/s. (b) Using physics terms, explain what would happen to the outgoing velocity and the impulse on the ball if it were not fully inflated.
2. An automobile accident causes both the driver and passenger front airbags to deploy. (a) If the vehicle was traveling at a speed of 38.6 km/h and is now at rest, find the change in momentum for both the 68.4-kg adult driver and the 34.2-kg child passenger. (b) The adult took 0.564 s and the child took 0.260 s to come to rest. Find the force that the airbag exerted on each individual. Explain why airbags tend to be dangerous for children.
3. Several African tribes engage in a ritual much like bungee jumping, in which a tree vine is used instead of a bungee cord. (a) A 70.8-kg person falls from a cliff with such a vine attached to his ankles. Find the force applied to his ankles if it takes 0.355 s to change his velocity from −18.5 m/s to rest (the negative sign represents the downward direction). (b) Find the force applied to the person's ankles when he takes 1.98 s to change his velocity from −18.5 m/s to +9.75 m/s using a manufactured bungee cord (remember that a bungee cord causes the jumper to bounce back upward). (c) Using physics terms, describe why it is safer to use a bungee cord than a tree vine.
4. Sally, who weighs 125 lb, knows that getting out of a 65.5-lb canoe can be a difficult experience. (a) What happens to the canoe's velocity if she attempts to step out of the canoe and onto the dock with a velocity of 3.50 ft/s? (b) If the canoe were heavier, would it be easier or harder to step out of it?
5. An automobile accident investigator needs to determine the initial westerly velocity of a Jeep (m = 1720 kg) that may have been speeding before colliding head-on with a Volkswagen (m = 1510 kg) that was moving with a velocity of 75.7 km/h east. The speed limit on this road was 90 km/h. After the collision, the Jeep and the Volkswagen stuck together and continued to travel with a velocity of 15.5 km/h west. (a) Find the initial westerly velocity of the Jeep. (b) Was the Jeep speeding?

CHAPTER 6

F_2
S_{t_3}
F_3
S_{t_1}
F_1
$\Sigma F = 0$
$\Sigma \tau$ clockwise =
$\Sigma \tau$ counter clockwise

CONCURRENT AND PARALLEL FORCES

Not all forces cause motion. A body that is static, or not moving, may be in equilibrium. Huge forces may be acting on a bridge without producing motion. We will consider concurrent forces (forces acting at the same point) in equilibrium.

When forces act nonconcurrently (that is, not at the same point), they may tend to produce rotational motion. We will consider torque (an applied force that causes a rotation), parallel force problems, equilibrium, and the concept of center of gravity.

Objectives

The major goals of this chapter are to enable you to:

1. Find the vector sum of concurrent forces.
2. Analyze equilibrium in one dimension.
3. Analyze concurrent force situations using force diagrams.
4. Distinguish compression and tension.
5. Apply the torque equation to rotational problems.
6. Solve parallel force problems.
7. Express the conditions of equilibrium using torque concepts.
8. Use the center of gravity to solve parallel force problems.

6.1 Forces in Two Dimensions

Concurrent forces are those forces that are applied to or act at the same point, as in Fig. 6.1. When two or more forces act at the same point, the **resultant force** is the sum of the forces applied at that point. The resultant force is the single force that has the same effect as the two or more forces acting together.

As we saw in Section 4.4, when forces act in the same or opposite directions (in one dimension), the total, or net, force can be found by adding the forces that act in one direction and subtracting the forces that act in the opposite direction. What is the result when the forces are acting in two dimensions? In Section 2.2, addition of two vectors was shown by connecting the end point of the first vector to the initial point of the second vector as shown in Fig. 6.2. This is often called the *vector triangle method.*

This sum may also be obtained by constructing a parallelogram using the two vectors as adjacent sides and then constructing the opposite sides parallel as shown in Fig. 6.3. The diagonal of the parallelogram is the resultant, or sum, of the two vectors. This is often called the *parallelogram method.*

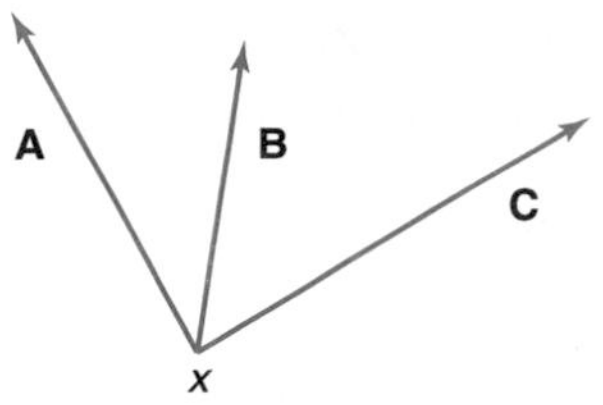

Figure 6.1 Concurrent forces are applied to or act at the same point.

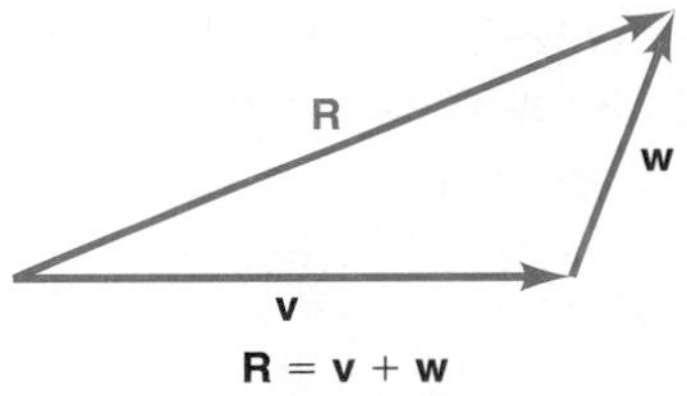

Figure 6.2 Vector triangle method of adding two vectors

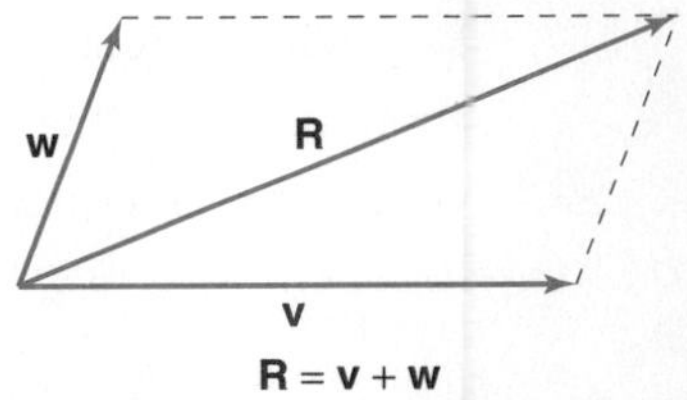

Figure 6.3 Parallelogram method of adding two vectors

In Fig. 6.4, two people on opposite banks are pulling a boat up a small river. Note that the effort forces must be equal to keep the boat in the middle of the river, the direction of the equal effort forces is along the ropes, and the direction of the resultant force is upriver. What would happen if the effort forces were not equal?

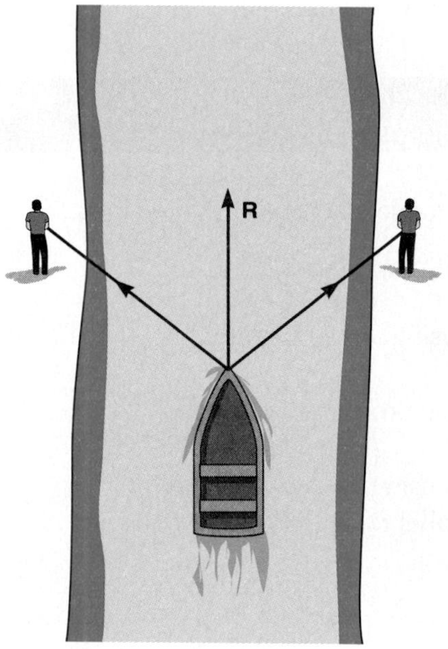

Figure 6.4

EXAMPLE 1

Two workers move a large crate by applying two ropes at the same point. The first worker applies a force of 525 N while the second worker applies a force of 763 N at the same point at right angles as shown in Fig. 6.5. Find the resultant force.

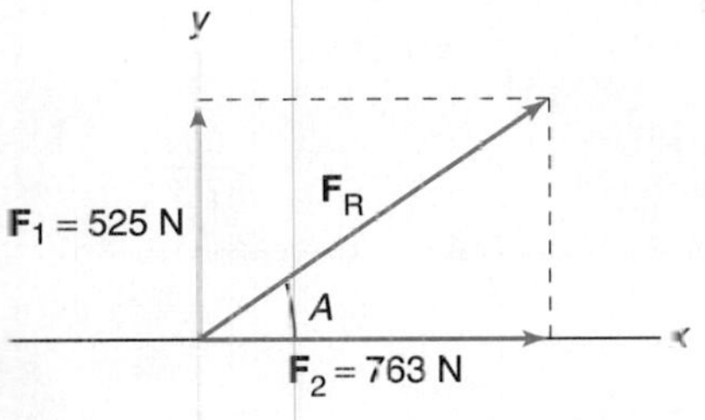

Figure 6.5

To find $\mathbf{F}_R$, find the x- and y-components of each vector and add the components as follows:

Vector	x-component	y-component
$\mathbf{F}_1$	0 N	525 N
$\mathbf{F}_2$	763 N	0 N
$\mathbf{F}_R$	763 N	525 N

Find angle A as follows:

$$\tan A = \frac{\text{side opposite } A}{\text{side adjacent to } A} = \frac{|\mathbf{F}_{Ry}|}{|\mathbf{F}_{Rx}|}$$

$$\tan A = \frac{525 \text{ N}}{763 \text{ N}} \qquad \text{(In a parallelogram, opposite sides are equal.)}$$

$$= 0.6881$$

$$A = 34.5°$$

Find the magnitude of $\mathbf{F}_R$ using the Pythagorean theorem:

$$|\mathbf{F}_R| = \sqrt{|\mathbf{F}_{Rx}|^2 + |\mathbf{F}_{Ry}|^2}$$

$$|\mathbf{F}_R| = \sqrt{(763 \text{ N})^2 + (525 \text{ N})^2}$$

$$= 926 \text{ N}$$

That is, $\mathbf{F}_R$ = 926 N at 34.5°.

EXAMPLE 2

Two workers move a large crate by applying two ropes at the same point. The first worker applies a force of 525 N while the second worker applies a force of 763 N at the same point as shown in Fig. 6.6. Find the resultant force.

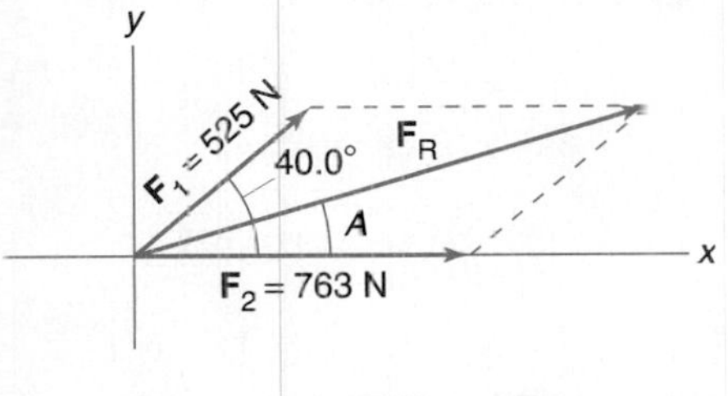

Figure 6.6

To find $\mathbf{F}_R$, find the x- and y-components of each vector and add the components as follows:

Vector	x-component	y-component
$\mathbf{F}_1$	$\|\mathbf{F}_1\| \cos\theta =$ (525 N) cos 40.0° = 402 N	$\|\mathbf{F}_1\| \sin\theta =$ (525 N) sin 40.0° = 337 N
$\mathbf{F}_2$	763 N	0 N
$\mathbf{F}_R$	1165 N	337 N

Find angle A as follows:

$$\tan A = \frac{|\mathbf{F}_{Ry}|}{|\mathbf{F}_{Rx}|} = \frac{337\text{ N}}{1165\text{ N}} = 0.2893$$

$$A = 16.1°$$

Find the magnitude of $\mathbf{F}_R$ using the Pythagorean theorem:

$$|\mathbf{F}_R| = \sqrt{|\mathbf{F}_{Rx}|^2 + |\mathbf{F}_{Ry}|^2}$$

$$|\mathbf{F}_R| = \sqrt{(1165\text{ N})^2 + (337\text{ N})^2} = 1210\text{ N}$$

That is, $\mathbf{F}_R = 1210$ N at 16.1°.

EXAMPLE 3

Forces of $\mathbf{F}_1 = 375$ N, $\mathbf{F}_2 = 575$ N, and $\mathbf{F}_3 = 975$ N are applied at the same point. The angle between $\mathbf{F}_1$ and $\mathbf{F}_2$ is 60.0° and the angle between $\mathbf{F}_2$ and $\mathbf{F}_3$ is 80.0°. $\mathbf{F}_2$ is between $\mathbf{F}_1$ and $\mathbf{F}_3$. Find the resultant force.

First, draw a force diagram as in Fig. 6.7. Place the point of application at the origin and one of the forces on the x-axis for ease in computing the components.

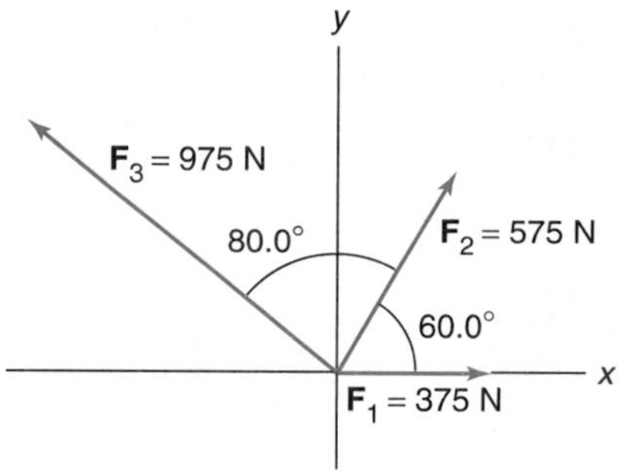

Figure 6.7

To find $\mathbf{F}_R$, find the x- and y-components of each vector and add the components as follows:

Vector	x-component	y-component
$\mathbf{F}_1$	375N	0 N
$\mathbf{F}_2$	$\|\mathbf{F}_2\| \cos A =$ (575 N) cos 60.0° = 288 N	$\|\mathbf{F}_2\| \sin A =$ (575 N) sin 60.0° = 498 N
$\mathbf{F}_3$	$\|\mathbf{F}_3\| \cos A =$ −(975 N) cos 40.0° = −747 N	$\|\mathbf{F}_3\| \sin A =$ (975 N) sin 40.0° = 627 N
Note: $A = 180° - 140.0° = 40.0°$		
$\mathbf{F}_R$	−84 N	1125 N

Find angle A of the resultant vector as follows:

$$\tan A = \frac{|\mathbf{F}_{Ry}|}{|\mathbf{F}_{Rx}|} = \frac{1125\text{ N}}{84\text{ N}} = 13.39$$

$$A = 85.7°$$

Note: The x-component of $\mathbf{F}_R$ is negative and its y-component is positive; this means that $\mathbf{F}_R$ is in the second quadrant. Its angle in standard position is $180° - 85.7° = 94.3°$.

Find the magnitude of $\mathbf{F}_R$ using the Pythagorean theorem:

$$|\mathbf{F}_R| = \sqrt{|\mathbf{F}_{Rx}|^2 + |\mathbf{F}_{Ry}|^2}$$

$$|\mathbf{F}_R| = \sqrt{(84\text{ N})^2 + (1125\text{ N})^2}$$

$$= 1130\text{ N}$$

That is, $\mathbf{F}_R = 1130$ N at 94.3°, or 94.3° from $\mathbf{F}_1$.

The resultant vector is shown in Fig. 6.8.

Figure 6.8

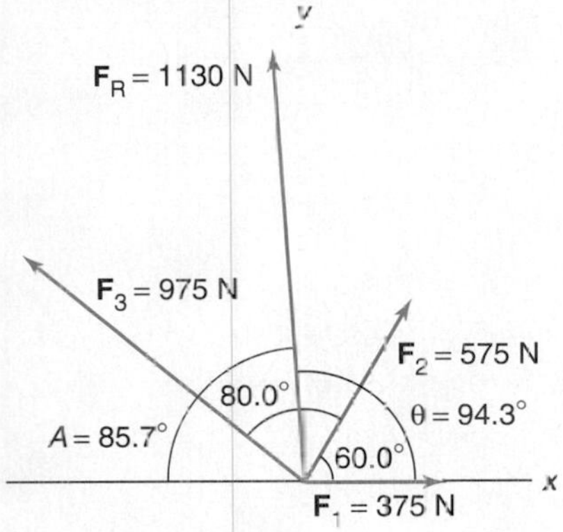

PROBLEMS 6.1

Find the sum of each set of forces acting at the same point in a straight line.

1. 355 N (right); 475 N (right); 245 N (left); 555 N (left)
2. 703 N (right); 829 N (left); 125 N (left); 484 N (left)
3. Forces of 225 N and 175 N act at the same point.
   (a) What is the magnitude of the maximum net force the two forces can exert together?
   (b) What is the magnitude of the minimum net force the two forces can exert together?
4. Three forces with magnitudes of 225 N, 175 N, and 125 N act at the same point.
   (a) What is the magnitude of the maximum net force the three forces can exert together?
   (b) What is the magnitude of the minimum net force the three forces can exert together?

SKETCH

12 cm² | w
4.0 cm

DATA

$A = 12\text{ cm}^2$, $l = 4.0$ cm, $w = ?$

BASIC EQUATION

$A = lw$

WORKING EQUATION

$w = \frac{A}{l}$

SUBSTITUTION

$w = \frac{12\text{ cm}^2}{4.0\text{ cm}} = 3.0$ cm

Find the sum of each set of vectors. Give angles in standard position.

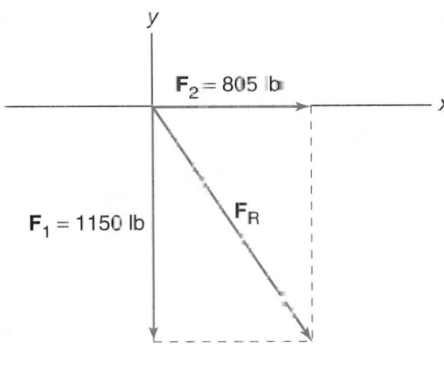

5.

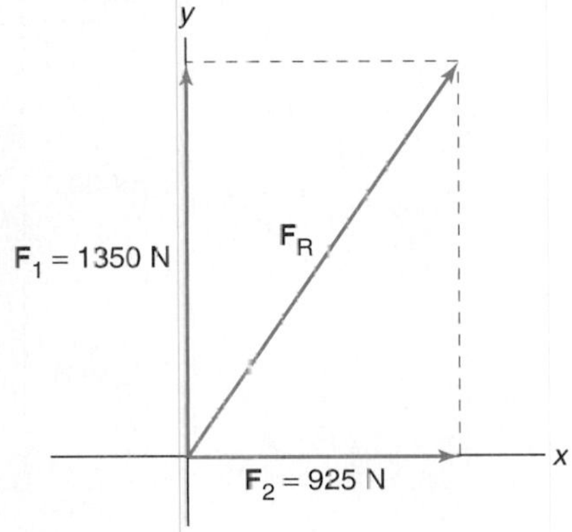

6.

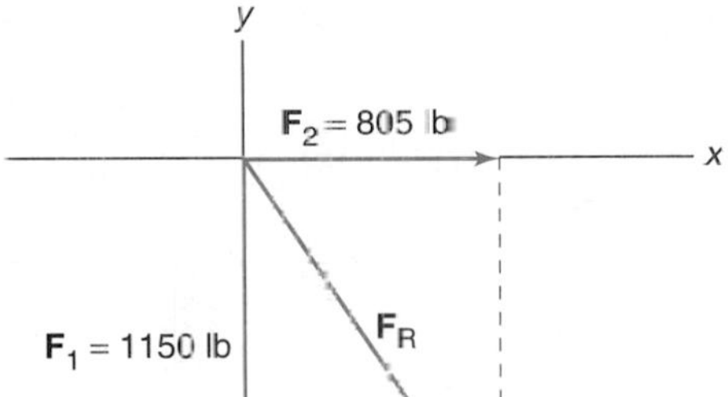

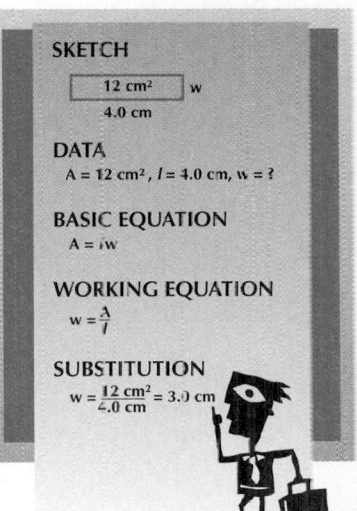

PHYSICS CONNECTIONS

The Cable-Stayed Bridge

All bridges are designed and constructed according to the needs of the community, the desired aesthetics, the costs, and the geographic and geological conditions around the bridge site. One of the most popular, attractive, and cost-effective designs is the cable-stayed bridge. The physical strength and relatively low cost of the design made the cable-stayed bridge ideal for the midlength span across the Mississippi River at Alton, Illinois (Fig. 6.9).

Cable-stayed bridges support the roadbed by attaching one end of multiple cables directly to the deck, passing them through a vertical tower, and attaching them to the deck on the opposite side of the tower. Through the use of lighter, stronger materials, engineers are able to avoid the need for the heavy and expensive steel and massive anchorages that are needed to support more traditional suspension bridges.

The combination of compression, tension, shear, and bending forces keeps the cable-stayed New Clark Bridge static. This particular cable-stayed bridge was designed to replace a deteriorating truss bridge that the community had outgrown. The New Clark Bridge meets the needs of the growing community and a busy shipping channel, is aesthetically pleasing, and is economically viable. The bridge also met the geographic and geological conditions as dictated by the Mississippi River and surrounding landscape.

Figure 6.9 The new cable-stayed design of the New Clark Bridge at Alton, Illinois

7.

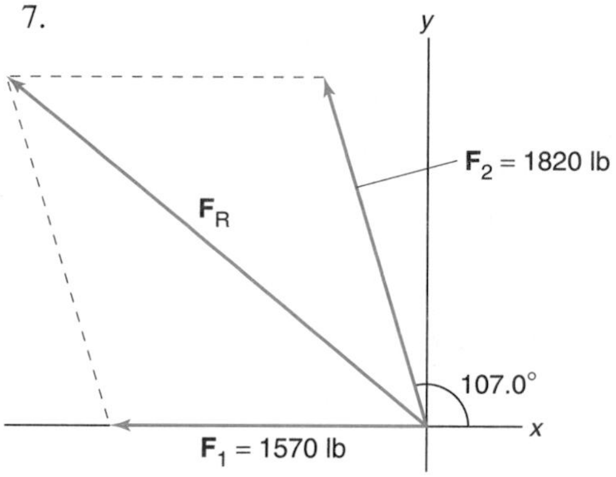

8.

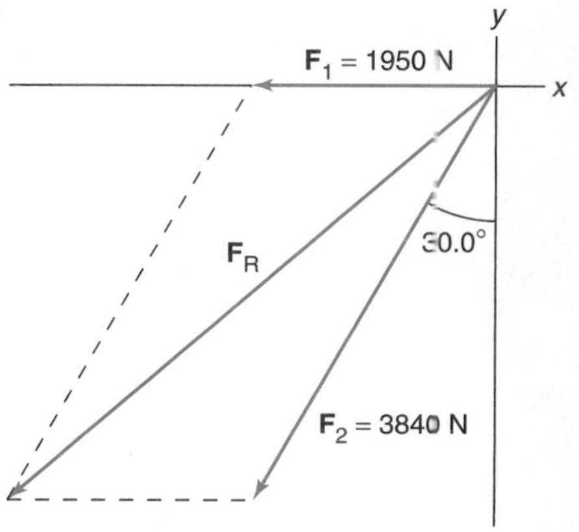

9.

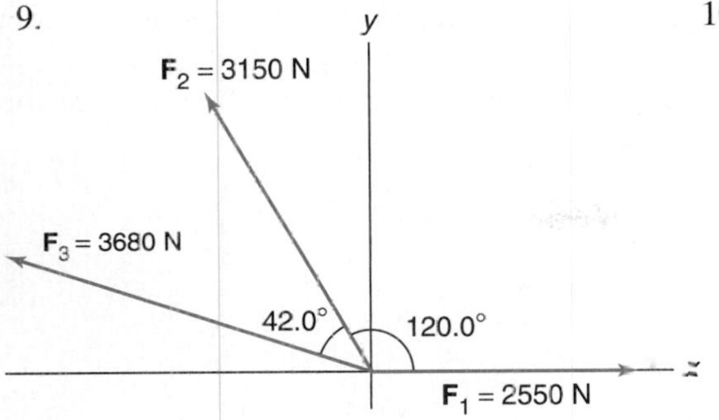

10.

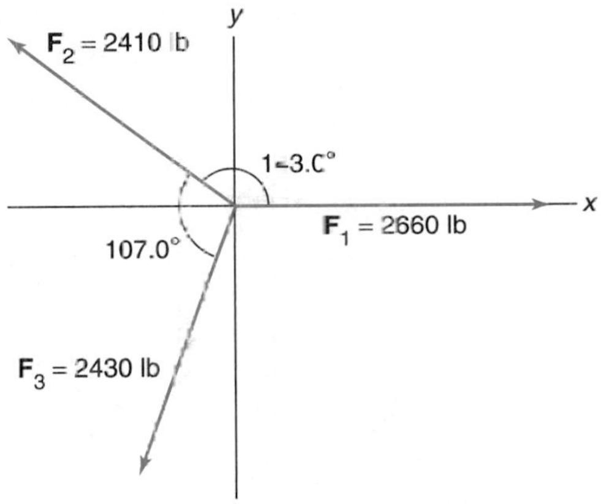

11. Forces of $\mathbf{F}_1$ = 1150 N, $\mathbf{F}_2$ = 875 N, and $\mathbf{F}_3$ = 1450 N are applied at the same point. The angle between $\mathbf{F}_1$ and $\mathbf{F}_2$ is 90.0° and the angle between $\mathbf{F}_2$ and $\mathbf{F}_3$ is 120.0°. $\mathbf{F}_2$ is between $\mathbf{F}_1$ and $\mathbf{F}_3$. Find the resultant force.
12. Four forces, each of magnitude 2750 lb, act at the same point. The angle between adjacent forces is 30.0°. Find the resultant force.

6.2 Concurrent Forces in Equilibrium

Equilibrium in One Dimension

Equilibrium *is the state of a body in which there is no change in its motion.* A body is in equilibrium when the net force acting on it is zero. That is, it is not accelerating; it is either at rest or moving at a constant velocity. The study of objects in equilibrium is called **statics.**

The forces applied to an object in one dimension act in the same direction or in opposite directions. For the net force to be zero, the forces in one direction must equal the forces in the opposite direction. We can write the equation for equilibrium in one dimension as

$$\mathbf{F}_+ = \mathbf{F}_-$$

where

$\mathbf{F}_+$ = the sum of all forces acting in one direction (call it the positive direction)
$\mathbf{F}_-$ = the sum of all the forces acting in the opposite (negative) direction.

Note in Fig. 6.10 that the downward force (weight of the bridge) must equal the sum of the upward forces produced by the two bridge supports for the bridge to be in equilibrium.

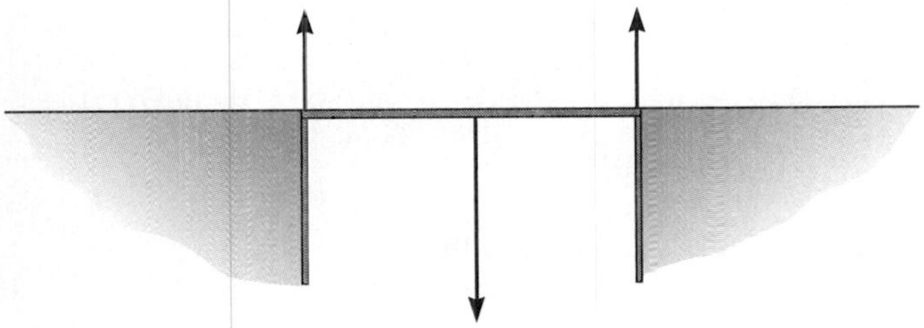

Figure 6.10

EXAMPLE 1

A cable supports a large crate of weight 1250 N (Fig. 6.11). What is the upward force on the crate if it is in equilibrium?

Sketch: Draw a force diagram of the crate in equilibrium, and show the forces that act on it. Note that we call the upward direction positive as indicated by the arrow.

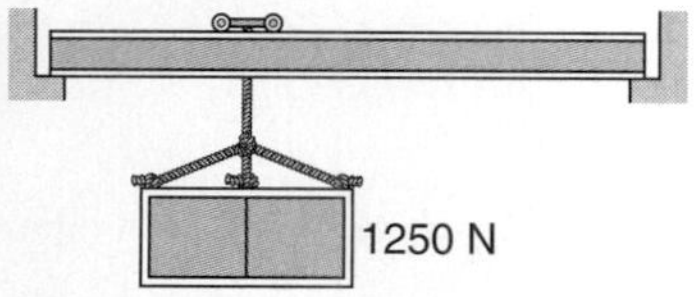

Figure 6.11

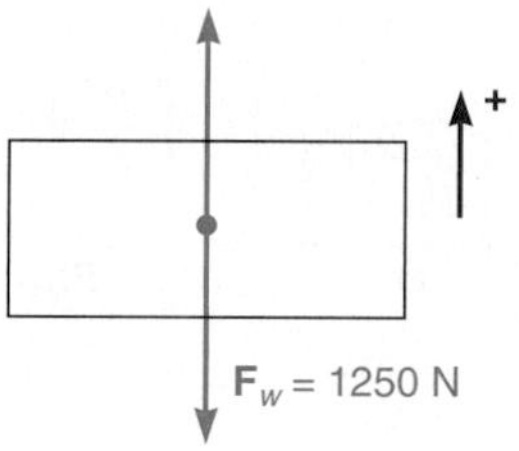

Data:

$$\mathbf{F}_w = 1250 \text{ N}$$
$$\mathbf{S} = ?$$

Basic Equation:

$$\mathbf{F}_+ = \mathbf{F}_-$$

Working Equation:

$$\mathbf{S} = \mathbf{F}_w$$

Substitution:

$$\mathbf{S} = 1250 \text{ N}$$

EXAMPLE 2

Four persons are having a tug-of-war with a rope. Harry and Mary are on the left; Bill and Jill are on the right. Mary pulls with a force of 105 lb, Harry pulls with a force of 255 lb, and Jill pulls with a force of 165 lb. With what force must Bill pull to produce equilibrium?

Sketch:

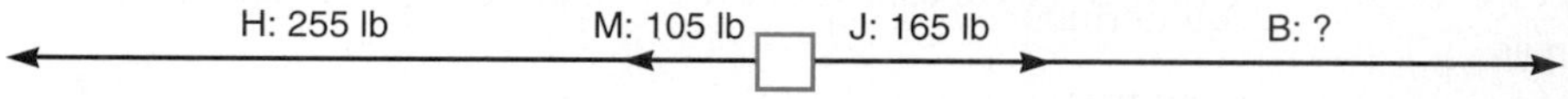

Data:

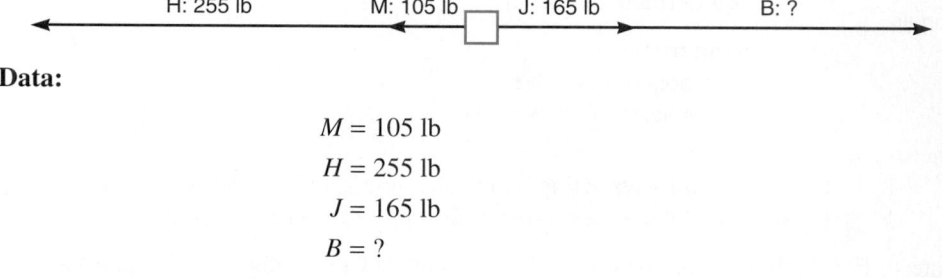

$$M = 105 \text{ lb}$$
$$H = 255 \text{ lb}$$
$$J = 165 \text{ lb}$$
$$B = ?$$

Basic Equation:

$$\mathbf{F}_+ = \mathbf{F}_- \quad \text{or}$$
$$M + H = J + B$$

Working Equation:

$$B = M + H - J$$

Substitution:

$$B = 105 \text{ lb} + 255 \text{ lb} - 165 \text{ lb}$$
$$= 195 \text{ lb}$$

Equilibrium in Two Dimensions

A body is in equilibrium when it is either at rest or moving at a constant speed in a straight line. Figure 6.12(a) shows the resultant force of the sum of two forces from Example 1 in Section 6.1. When two or more forces act at a point, the **equilibrant force** is the force that when applied at that same point produces equilibrium. *The equilibrant force is equal in magnitude to that of the resultant force but it acts in the opposite direction* [see Fig. 6.12(b)]. In this case, the equilibrant force is 926 N at 214.5° (180° + 34.5°).

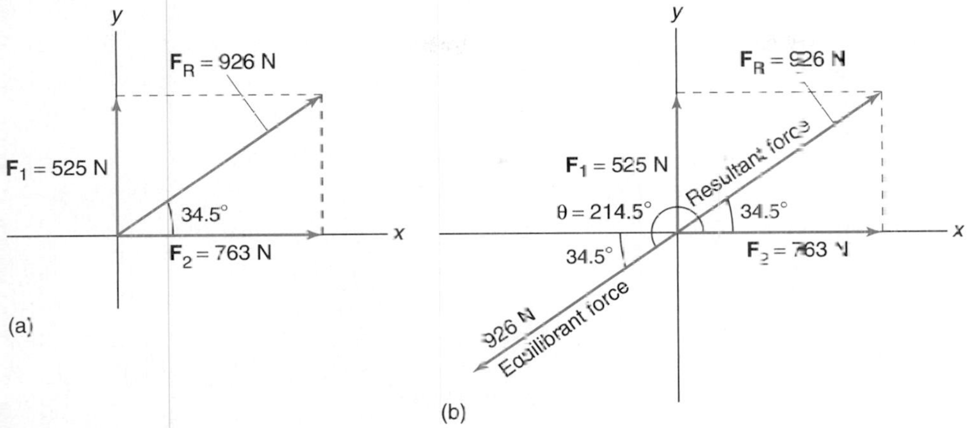

Figure 6.12

If an object is in equilibrium in two dimensions, the net force acting on it must be zero. For the net force to be zero, the sum of the x-components must be zero and the sum of the y-components must be zero. For forces **A**, **B**, and **C** with x-components $\mathbf{A}_x$, $\mathbf{B}_x$, and $\mathbf{C}_x$, respectively, and with y-components $\mathbf{A}_y$, $\mathbf{B}_y$, and $\mathbf{C}_y$, respectively, to be in equilibrium, both of the following conditions must hold:

CONDITIONS FOR EQUILIBRIUM

1. The sum of x-components = 0; that is, $\mathbf{A}_x + \mathbf{B}_x + \mathbf{C}_x = 0$.
2. The sum of y-components = 0; that is, $\mathbf{A}_y + \mathbf{B}_y + \mathbf{C}_y = 0$.

In general, to solve equilibrium problems:

1. Draw a force diagram from the point at which the unknown forces act.
2. Find the x- and y-component of each force.
3. Substitute the components in the equations

$$\text{sum of } x\text{-components} = 0$$

$$\text{sum of } y\text{-components} = 0$$

4. Solve for the unknowns. This may involve two simultaneous equations.

Figure 6.13 Tension and compression forces

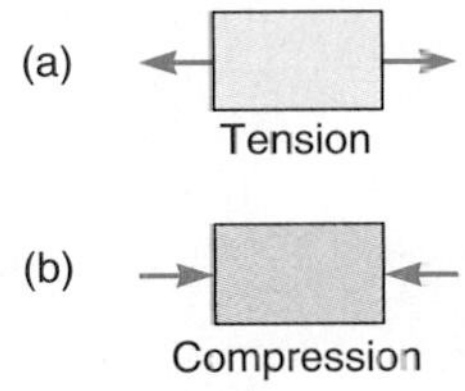

We may need to find the tension or compression in part of a structure, such as in a beam or a cable. **Tension** is a stretching force produced by forces pulling outward on the ends of an object [Fig. 6.13(a)]. **Compression** is a force produced by forces pushing inward on the ends of an object [Fig. 6.13(b)]. A rubber band being stretched is an example of tension [Fig. 6.14(a)]. A valve spring whose ends are pushed together is an example of compression [Fig. 6.14(b)].

Figure 6.14

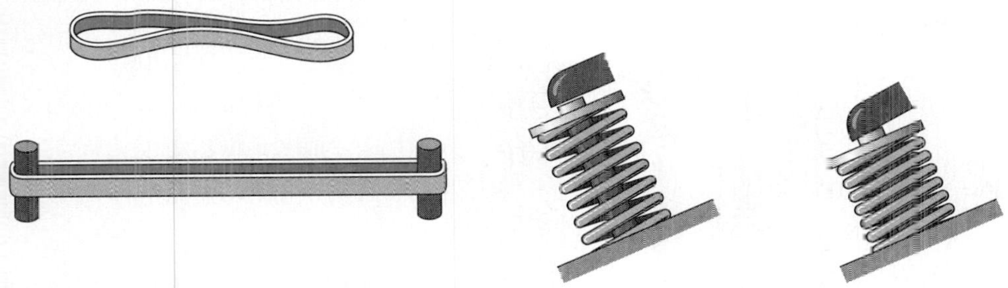

(a) Tension in a rubber band. (b) Compression in a valve spring

EXAMPLE 3

Find the forces **F** and **F′** necessary to produce equilibrium in the force diagram shown in Fig. 6.15.

1.

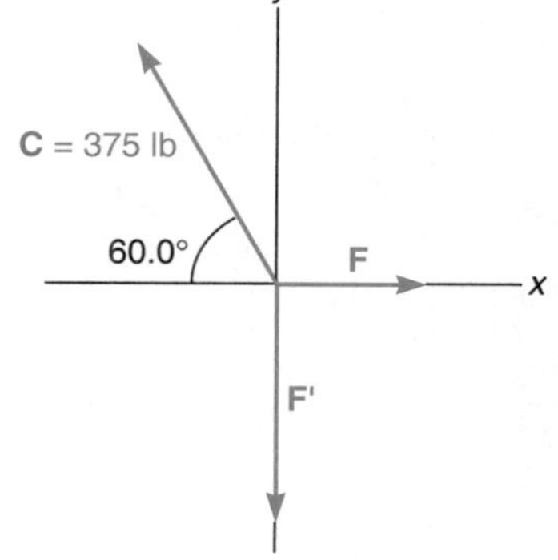

Figure 6.15

2. *x-components*
   $\mathbf{F}_x = \mathbf{F}$
   $\mathbf{F}'_x = 0$
   $\mathbf{C}_x = -(375 \text{ lb})(\cos 60.0°)$
   $= -188 \text{ lb}$

   y-components
   $\mathbf{F}_y = 0$
   $\mathbf{F}'_y = -\mathbf{F}'$
   $\mathbf{C}_y = (375 \text{ lb})(\sin 60.0°)$
   $= 325 \text{ lb}$

3. Sum of *x*-components = 0
   $\mathbf{F} + 0 + (-188 \text{ lb}) = 0$

   Sum of *y*-components = 0
   $0 + (-\mathbf{F}') + 325 \text{ lb} = 0$

4. $\mathbf{F} = 188 \text{ lb}$ $\qquad$ $\mathbf{F}' = 325 \text{ lb}$

EXAMPLE 4

Find the forces **F** and **F′** necessary to produce equilibrium in the force diagram shown in Fig. 6.16.

1.

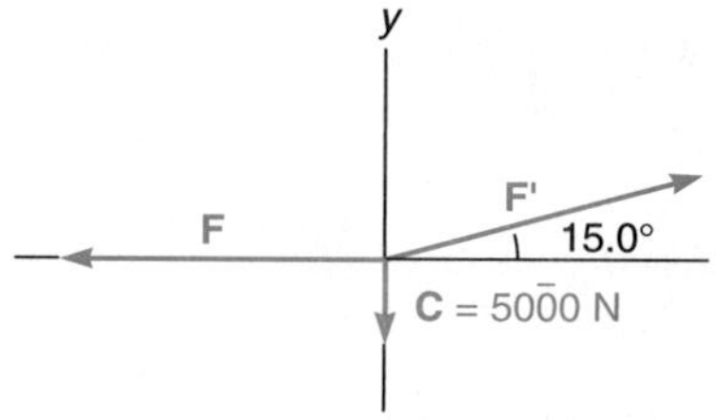

Figure 6.16

2. *x-components*
   $\mathbf{F}_x = -\mathbf{F}$
   $\mathbf{F}'_x = \mathbf{F}' \cos 15.0°$
   $\mathbf{C}_x = 0$

   y-components
   $\mathbf{F}_y = 0$
   $\mathbf{F}'_y = -\mathbf{F}' \sin 15.0°$
   $\mathbf{C}_y = -50\bar{0}0 \text{ N}$

3. Sum of *x*-components = 0
   $(-\mathbf{F}) + \mathbf{F}' \cos 15.0° + 0 = 0$

   Sum of *y*-components = 0
   $0 + \mathbf{F}' \sin 15.0° + (-50\bar{0}0 \text{ N}) = 0$

4. ***Note:*** Solve for **F′** in the right-hand equation first. Then substitute this value in the left-hand equation to solve for **F**:

$$\mathbf{F}' = \frac{50\bar{0}0 \text{ N}}{\sin 15.0°}$$
$$= 19{,}300 \text{ N}$$

$$\mathbf{F} = \mathbf{F}' \cos 15.0°$$
$$= (19{,}300 \text{ N})(\cos 15.0°)$$
$$= 18{,}600 \text{ N}$$

EXAMPLE 5

The crane shown in Fig. 6.17 is supporting a beam that weighs 6000 N. Find the tension in the horizontal supporting cable and the compression in the boom.

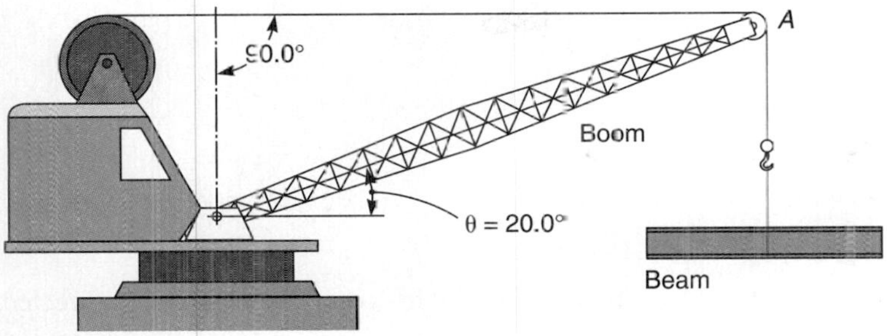

Figure 6.17

1. Draw the force diagram showing the forces acting at point A.

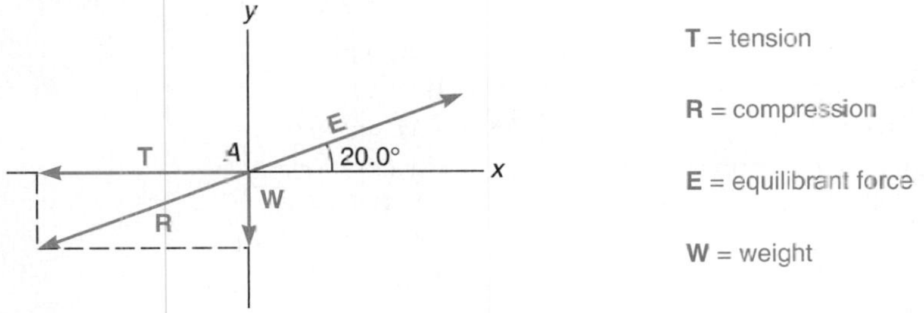

T is the force exerted at A by the horizontal supporting cable.
E is the force exerted by the boom at A.
W is the force (weight of the beam) pulling straight down at A.
R is the sum of forces **W** and **T**, which is equal in magnitude but opposite in direction to force **E** ($\mathbf{R} = -\mathbf{E}$).

2. *x-components*
   $\mathbf{E}_x = \mathbf{E} \cos 20.0°$
   $\mathbf{T}_x = -\mathbf{T}$
   $\mathbf{W}_x = 0$

   y-components
   $\mathbf{E}_y = \mathbf{E} \sin 20.0°$
   $\mathbf{T}_y = 0$
   $\mathbf{W}_y = -6000 \text{ N}$
3. Sum of x-components = 0
   $\mathbf{E} \cos 20.0° + (-\mathbf{T}) = 0$

   Sum of y-components = 0
   $\mathbf{E} \sin 20.0° + (-6000 \text{ N}) = 0$
4. $\mathbf{T} = \mathbf{E} \cos 20.0°$

   $$\mathbf{E} = \frac{6000 \text{ N}}{\sin 20.0°} = 17{,}500 \text{ N}$$

   $$\mathbf{T} = (17{,}500 \text{ N})(\cos 20.0°) = 16{,}400 \text{ N}$$

EXAMPLE 6

A homeowner pushes a 40.0-lb lawn mower at a constant velocity (Fig. 6.18). The frictional force on the mower is 20.0 lb. What force must the person exert on the handle, which makes an angle of 30.0° with the ground? Also, find the normal (perpendicular to ground) force.

This is an equilibrium problem because the mower is not accelerating and the net force is zero.

1. Draw the force diagram.

Figure 6.18

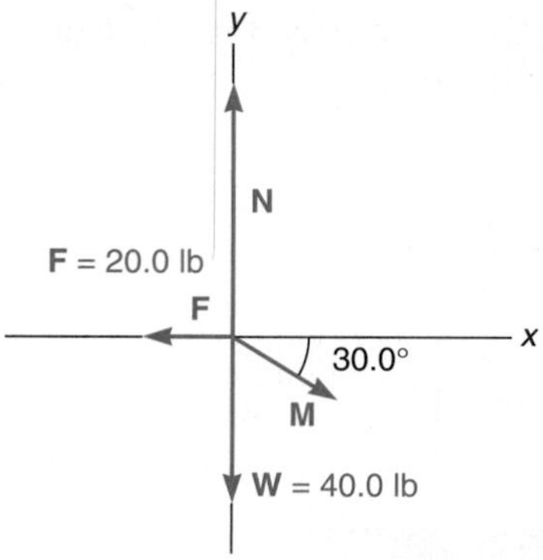

M is the force exerted on the mower by the person; this compression force is directed down along the handle.
W is the weight of the mower directed straight down.
N is the force exerted upward on the mower by the ground, which keeps the mower from falling through the ground.
F is the frictional force that opposes the motion.

2. *x-components*

$\mathbf{N}_x = 0$
$\mathbf{W}_x = 0$
$\mathbf{F}_x = -20.0 \text{ lb}$
$\mathbf{M}_x = \mathbf{M} \cos 30.0°$

y-components

$\mathbf{N}_y = \mathbf{N}$
$\mathbf{W}_y = -40.0 \text{ lb}$
$\mathbf{F}_y = 0$
$\mathbf{M}_y = -\mathbf{M} \sin 30.0°$

3. Sum of x-components = 0

$0 + 0 + (-20.0 \text{ lb}) + \mathbf{M} \cos 30.0° = 0$

Sum of y-components = 0

$\mathbf{N} + (-40.0 \text{ lb}) + 0 + (-\mathbf{M} \sin 30.0°) = 0$

$\mathbf{N} = \mathbf{M} \sin 30.0 + 40.0 \text{ lb}$

4. $\mathbf{M} = \dfrac{20.0 \text{ lb}}{\cos 30.0°} = 23.1 \text{ lb}$

$\mathbf{N} = (23.1 \text{ lb})(\sin 30.0°) + 40.0 \text{ lb} = 51.6 \text{ lb}$

EXAMPLE 7

The crane shown in Fig. 6.19 is supporting a beam that weighs $60\bar{0}0$ N. Find the tension in the supporting cable and the compression in the boom.

1. Draw the force diagram showing the forces acting at point A.

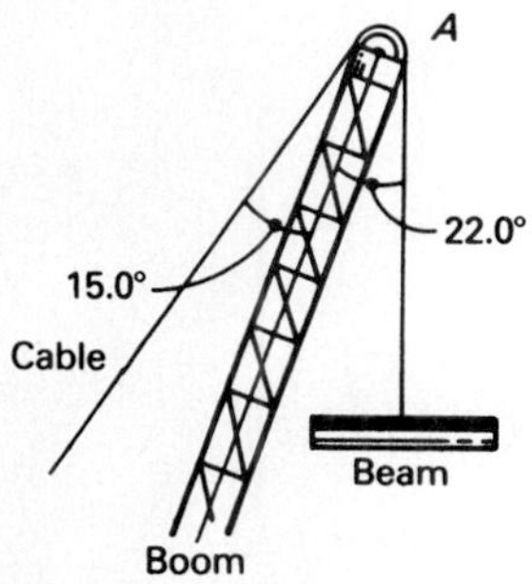

Figure 6.19

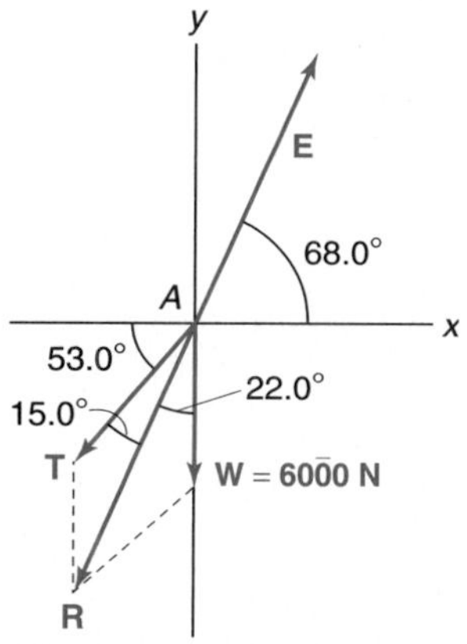

W is the weight of the beam, which pulls straight down.
T is the force exerted at *A* by the supporting cable.
E is the force exerted by the boom at *A*.
R is the sum of forces **W** and **T**, which is equal in magnitude but opposite in direction to force **E** (**R** = –**E**).

2. *x-components* / *y-components*

$\mathbf{E}_x = \mathbf{E}\cos 68.0°$ $\qquad$ $\mathbf{E}_y = \mathbf{E}\sin 68.0°$

$\mathbf{T}_x = -\mathbf{T}\cos 53.0°$ $\qquad$ $\mathbf{T}_y = -\mathbf{T}\sin 53.0°$

$\mathbf{W}_x = 0$ $\qquad$ $\mathbf{W}_y = -6\bar{0}00\text{ N}$

3. Sum of *x*-components = 0 $\qquad$ Sum of *y*-components = 0

$\mathbf{E}\cos 68.0° + (-\mathbf{T}\cos 53.0°) + 0 = 0$ $\qquad$ $\mathbf{E}\sin 68.0° + (-\mathbf{T}\sin 53.0°) + (-6\bar{0}00\text{ N}) = 0$

4. ***Note:*** Solve the left equation for **E**. Then substitute this quantity in the right equation and solve for **T**:

$$\mathbf{E} = \frac{\mathbf{T}\cos 53.0°}{\cos 68.0°} \qquad \left(\frac{\mathbf{T}\cos 53.0°}{\cos 68.0°}\right)(\sin 68.0°) - \mathbf{T}\sin 53.0° = 6\bar{0}00\text{ N}$$

$$1.490\mathbf{T} - 0.799\mathbf{T} = 6\bar{0}00\text{ N}$$

$$0.691\mathbf{T} = 6\bar{0}00\text{ N}$$

$$\mathbf{T} = \frac{6\bar{0}00\text{ N}}{0.691} = 8680\text{ N}$$

$$\mathbf{E} = \frac{(8680\text{ N})(\cos 53.0°)}{\cos 68.0°} = 13{,}900\text{ N}$$

Alternate Method. You can orient a force diagram any way you want on the *x*–*y* axes. You should orient it so that as many of the vectors as possible are on an *x*- or a *y*-axis. The result will be the same. Let's rework Example 7 as follows:

1. Draw the force diagram showing the forces acting at point *A* using the same notation as follows:

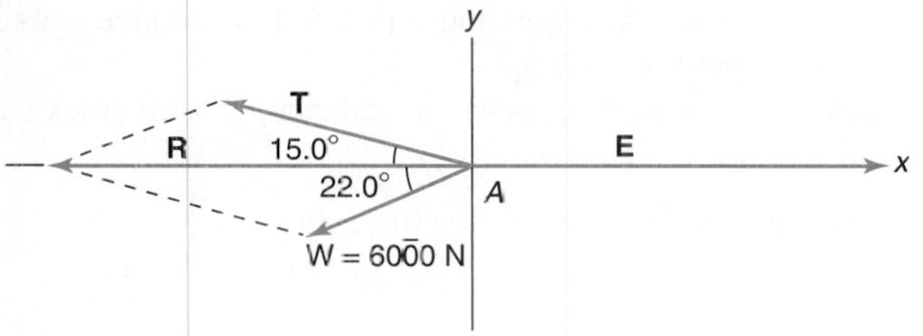

2. *x-components* / *y-components*

$\mathbf{E}_x = \mathbf{E}$ $\qquad$ $\mathbf{E}_y = 0$

$\mathbf{T}_x = -\mathbf{T}\cos 15.0°$ $\qquad$ $\mathbf{T}_y = \mathbf{T}\sin 15.0°$

$\mathbf{W}_x = -(6\bar{0}00\text{ N})(\cos 22.0°)$ $\qquad$ $\mathbf{W}_y = -(6\bar{0}00\text{ N})(\sin 22.0°)$

3. Sum of *x*-components = 0 $\qquad$ Sum of *y*-components = 0

$\mathbf{E} + (-\mathbf{T}\cos 15.0°) + (-6\bar{0}00\text{ N})(\cos 22.0°) = 0$ $\qquad$ $0 + \mathbf{T}\sin 15.0° + (-6\bar{0}00\text{ N})(\sin 22.0°) = 0$

4. Solve the right equation for **T** (since it has only one variable). Then solve the left equation for **E** and substitute this quantity:

$$\mathbf{T} = \frac{(6\bar{0}00\text{ N})(\sin 22.0°)}{\sin 15.0°} = 8680\text{ N}$$

$$\mathbf{E} = \mathbf{T}\cos 15.0° + (6\bar{0}00\text{ N})(\cos 22.0°)$$

$$\mathbf{E} = (8680\text{ N})(\cos 15.0°) + (6\bar{0}00\text{ N})(\cos 22.0°) = 13{,}900\text{ N}$$

SKETCH

12 cm² w 4.0 cm

DATA

$A = 12\text{ cm}^2, l = 4.0\text{ cm}, w = ?$

BASIC EQUATION

$A = lw$

WORKING EQUATION

$w = \frac{A}{l}$

SUBSTITUTION

$w = \frac{12\text{ cm}^2}{4.0\text{ cm}} = 3.0\text{ cm}$

PROBLEMS 6.2

Find the force **F** that will produce equilibrium in each force diagram.

1. 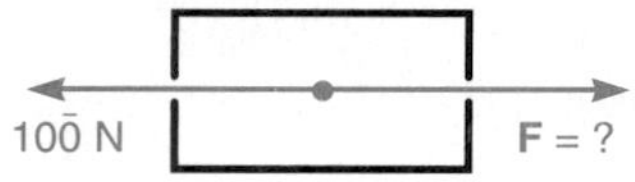

2. 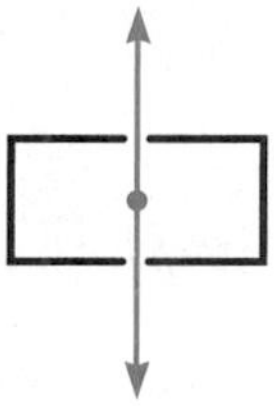

3. 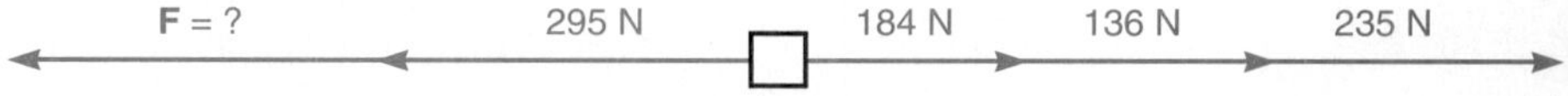

4. 

5. 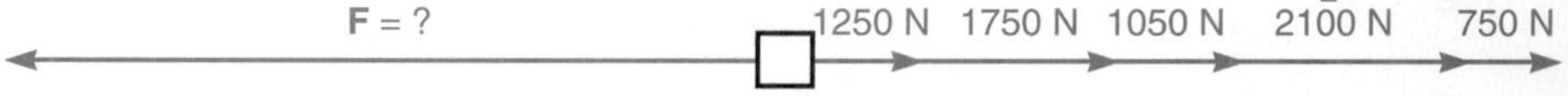

6. 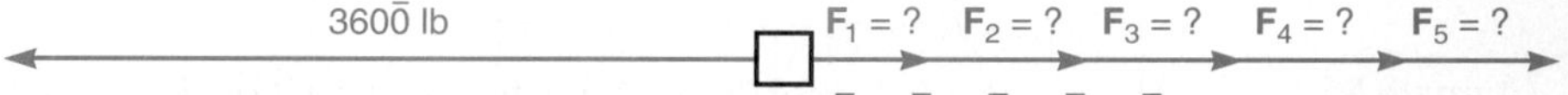

$F_1 = F_2 = F_3 = F_4 = F_5$

7. Five persons are having a tug-of-war. Kurt and Brian are on the left; Amy, Barbara, and Joyce are on the right. Amy pulls with a force of 225 N, Barbara pulls with a force of 495 N, Joyce pulls with a force of 455 N, and Kurt pulls with a force of 605 N. With what force must Brian pull to produce equilibrium?
8. A certain wire can support 6450 lb before it breaks. Seven 820-lb weights are suspended from the wire. Can the wire support an eighth weight of 820 lb?
9. The frictional force of a loaded pallet in a warehouse is 385 lb. Can three workers, each exerting a force of 135 lb, push it to the side?
10. A bridge has a weight limit of 7.0 tons. How heavy a load can a 2.5-ton truck carry across?

Find the forces $\mathbf{F}_1$ and $\mathbf{F}_2$ that produce equilibrium in each force diagram.

11. 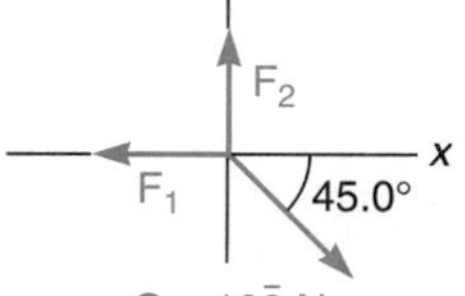

12. 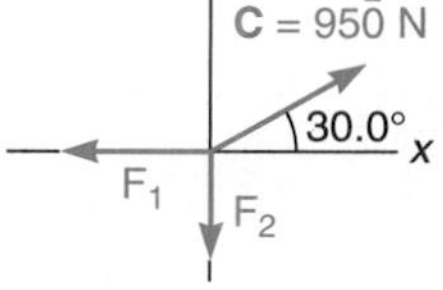

13. 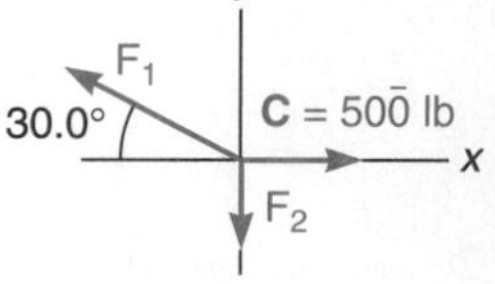

14. 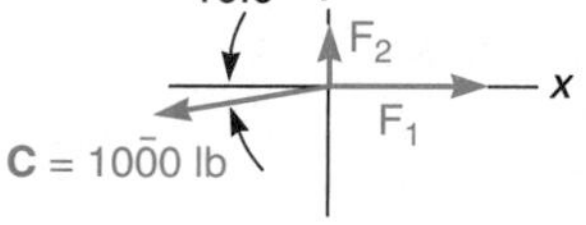

15. 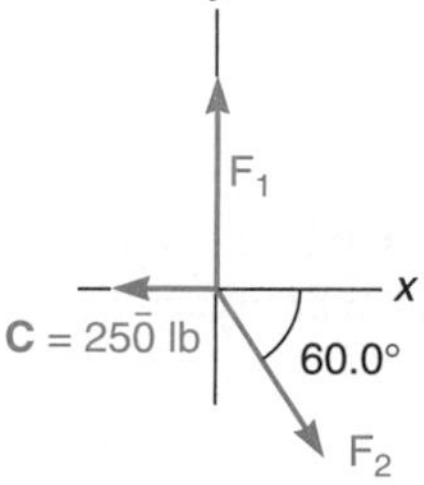

16. 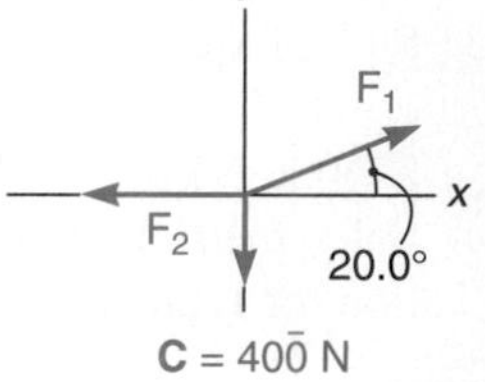

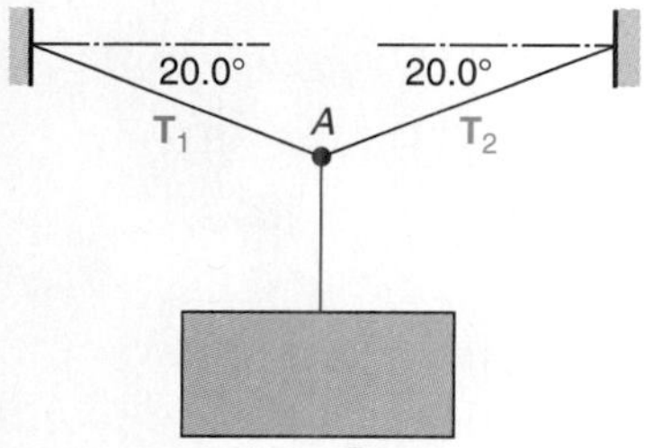

Figure 6.20

17. A rope is attached to two buildings and supports a $50\bar{0}$-lb sign (Fig. 6.20). Find the tensions in the two ropes $\mathbf{T}_1$ and $\mathbf{T}_2$. (*Hint:* Draw the force diagram of the forces acting at the point labeled *A*.)
18. If the angle between the horizontal and the ropes in Problem 17 is changed to 10.0°, what are the tensions in the two ropes $\mathbf{T}_1$ and $\mathbf{T}_2$?

19. If the angles between the horizontal and the ropes in Problem 17 are changed to 20.0° and 30.0°, find the tension in each rope.
20. Find the tension in the horizontal supporting cable and the compression in the boom of the crane shown in Fig. 6.21, which supports an $89\bar{0}0$-N beam.

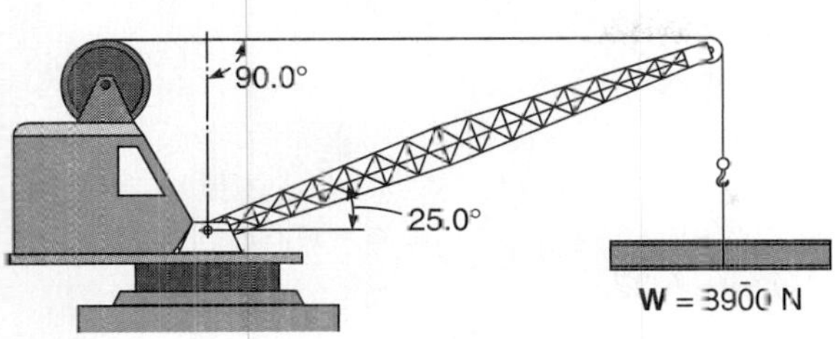

Figure 6.21

21. Find the tension in the horizontal supporting cable and the compression in the boom of the crane shown in Fig. 6.22, which supports a $15\bar{0}0$-lb beam.
22. The frictional force of the mower shown in Fig. 6.23 is $2\bar{0}$ lb. What force must the man exert along the handle to push it at a constant velocity?

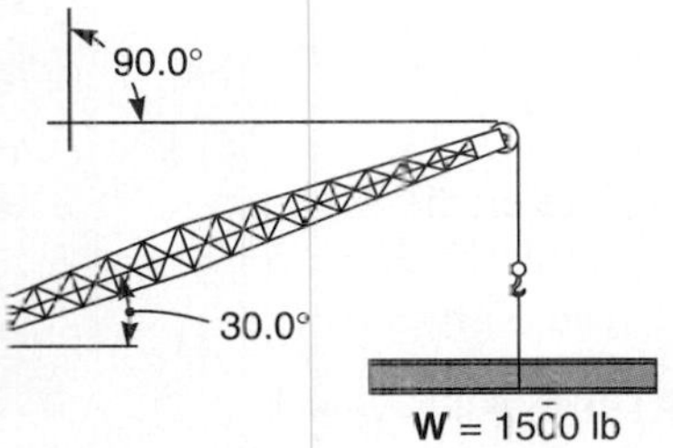

Figure 6.22

Figure 6.23

23. A vehicle that weighs 16,200 N is parked on a 20.0° hill (Fig. 6.24). What braking force is necessary to keep it from rolling? Neglect frictional forces. (*Hint:* When you draw the force diagram, tilt the x- and y-axes as shown. **B** is the braking force directed up the hill and along the x-axis.)

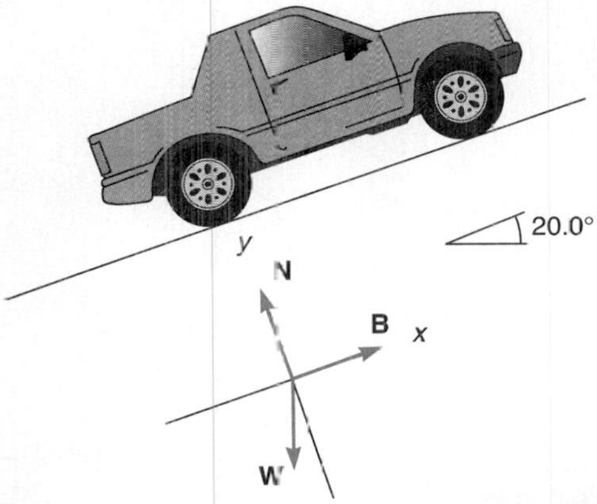

Figure 6.24

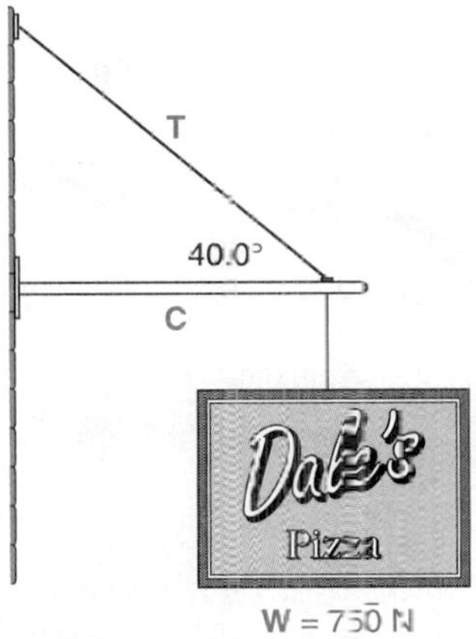

Figure 6.25

24. Find the tension in the cable and the compression in the support of the sign shown in Fig. 6.25.
25. The crane shown in Fig. 6.26 is supporting a load of 1850 lb. Find the tension in the supporting cable and the compression in the boom.
26. The crane shown in Fig. 6.27 is supporting a load of 11,500 N. Find the tension in the supporting cable and the compression in the boom.

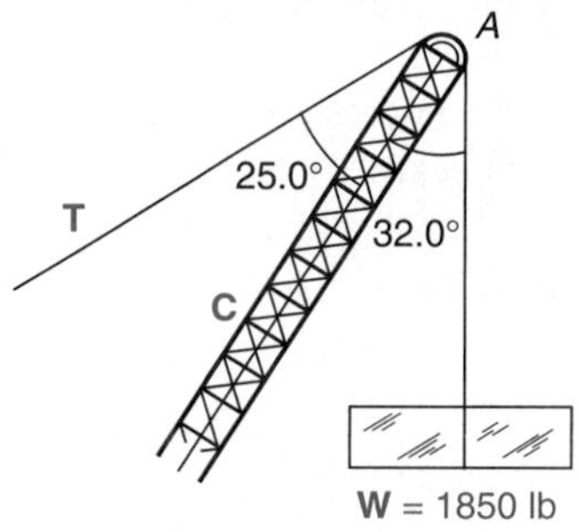

Figure 6.26

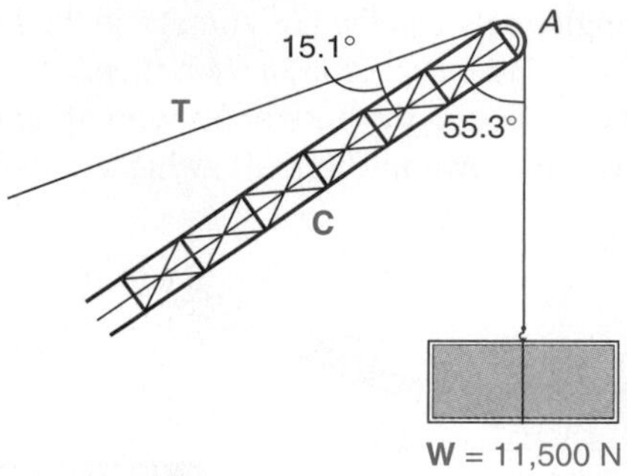

Figure 6.27

6.3 Torque

A *torque* is produced when a force is applied to produce a rotation, as, for example, when a wrench is used to turn a bolt or a claw hammer is used to pull a nail from wood. **Torque** is the tendency to produce change in rotational motion.

The torque developed depends on two factors:

1. The amount of force applied
2. How far from the point of rotation the force is applied

Torque is expressed by the equation

$$\tau = Fs_t$$

where

τ = torque (N m or lb ft) (τ is the lowercase Greek letter "tau.")
F = applied force (N or lb)
s_t = length of torque arm (m or ft)

Note that s_t, the length of the torque arm, is different from s in the equation defining work ($W = Fs$). Recall that s in the work equation is the linear distance over which the force acts.

When you use a wrench to turn a bolt, less effort is used (greater torque is produced) as the distance you place your hand from the bolt increases (Fig. 6.28). Plumbers often use a wrench with a long torque arm to loosen or tighten large bolts and fittings.

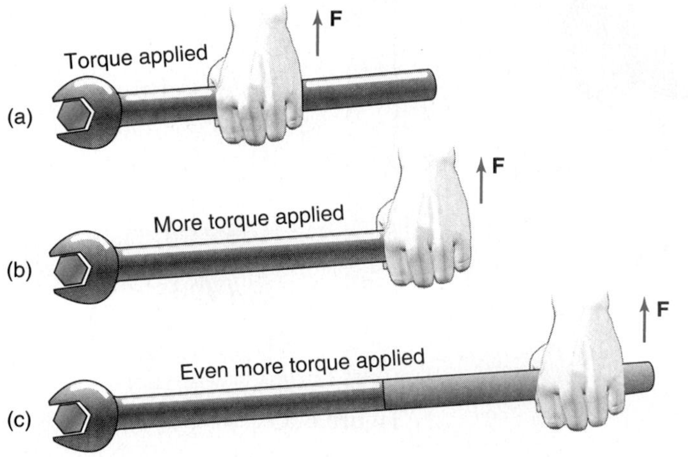

Figure 6.28 Even though the same force is used, the torque applied to the bolt increases as the distance from your hand to the bolt increases. In (b), you produce more torque by placing your hand on the end of the wrench handle. In (c), you produce even more torque by using an extender sleeve.

In all torque problems, we are concerned with motion about a point or axis of rotation as in pedaling a bicycle (Fig. 6.29). In pedaling, we apply a force to the pedal, causing the sprocket to rotate. The torque arm is the *perpendicular* distance from the point of rotation to the applied force [Fig. 6.29(a)]. In torque problems, s_t is always perpendicular to the force [Fig. 6.29(a)]. Note that s_t is the distance from the pedal to the axle. The units of torque look similar to those of work, but note the difference between s and s_t.

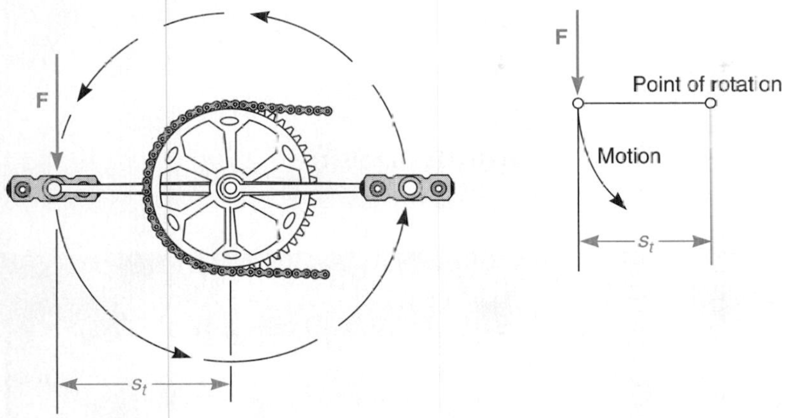

Figure 6.29 Torque produced in pedaling a bicycle

If the force is not exerted tangent to the circle made by the pedal (Fig. 6.30), the length of the torque arm is *not* the length of the pedal arm. The torque arm, s_t, is measured as the perpendicular distance to the force. Since s_t is therefore shorter, the product $F \cdot s_t$ is smaller, and the turning effect, the torque, is less in the pedal position shown in Fig. 6.30. Maximum torque is produced when the pedals are horizontal and the force applied is straight down.

Torque is a vector quantity that acts along the axis of rotation (not along the force) and points in the direction in which a right-handed screw would advance if turned by the torque as in Fig. 6.31(a). The *right-hand rule* is often used to determine the direction of the torque as follows: Grasp the axis of rotation with your right hand so that your fingers circle it in the direction that the torque tends to induce rotation. Your thumb will point in the direction of the torque vector [Fig. 6.31(b)]. Thus, the torque vector in Fig. 6.31(a) is perpendicular to and points out of the page.

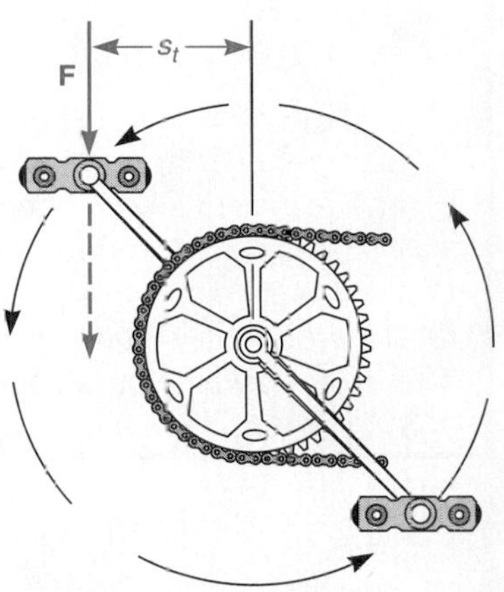

Figure 6.30 Maximum torque is only produced when the pedal reaches a position perpendicular to the applied force.

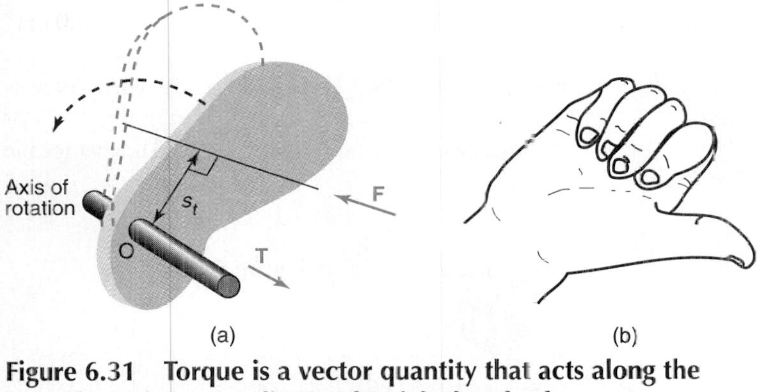

Figure 6.31 Torque is a vector quantity that acts along the axis of rotation according to the right-hand rule.

EXAMPLE

A force of 10.0 lb is applied to a bicycle pedal. If the length of the pedal arm is 0.850 ft, what torque is applied to the shaft?

Sketch:

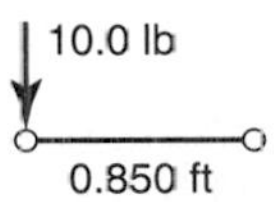

Data:

$$F = 10.0 \text{ lb}$$
$$s_t = 0.850 \text{ ft}$$
$$\tau = ?$$

Basic Equation:

$$\tau = Fs_t$$

Working Equation: Same

Substitution:

$$\tau = (10.0 \text{ lb})(0.850 \text{ ft})$$
$$= 8.50 \text{ lb ft}$$

TRY THIS ACTIVITY

Hammers and Screwdrivers

Torque is an essential component of most hand tools. Drive a nail into a piece of wood by holding a hammer near its head. Count the number of hits it takes to drive the nail into the wood. Then, drive a like nail into the wood by holding the hammer near the end of the handle. Count the number of hits it takes. Using physics terminology, explain which handle grip is better for driving nails into wood.

Use two screwdrivers with different diameter handles to screw two similar screws into a board. Which screwdriver is able to apply more torque to a screw? If one screwdriver applies more torque to a screw than the other, why would anyone want to use a screwdriver that cannot exert the maximum torque?

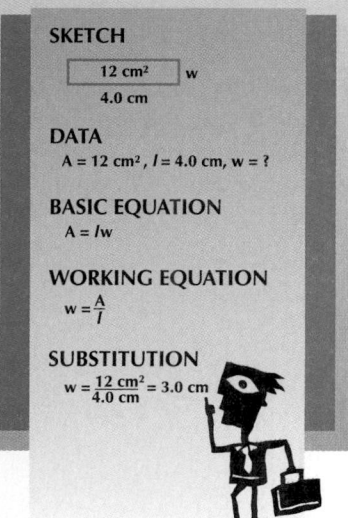

SKETCH

12 cm² w

4.0 cm

DATA

$A = 12 \text{ cm}^2$, $l = 4.0$ cm, $w = ?$

BASIC EQUATION

$A = lw$

WORKING EQUATION

$w = \frac{A}{l}$

SUBSTITUTION

$w = \frac{12 \text{ cm}^2}{4.0 \text{ cm}} = 3.0$ cm

PROBLEMS 6.3

Assume that each force is applied perpendicular to the torque arm.

1. Given: $F = 16.0$ lb, $s_t = 6.00$ ft, $\tau = ?$
2. Given: $F = 10\bar{0}$ N, $s_t = 0.420$ m, $\tau = ?$
3. Given: $\tau = 60.0$ N m, $F = 30.0$ N, $s_t = ?$
4. Given: $\tau = 35.7$ lb ft, $s_t = 0.0240$ ft, $F = ?$
5. Given: $\tau = 65.4$ N m, $s_t = 35.0$ cm, $F = ?$
6. Given: $F = 63\bar{0}$ N, $s_t = 74.0$ cm, $\tau = ?$
7. If the torque on a shaft of radius 2.37 cm is 38.0 N m (Fig. 6.32), what force is applied to the shaft?
8. If a force of 56.2 lb is applied to a torque wrench 1.50 ft long (Fig. 6.33), what torque is indicated by the wrench?

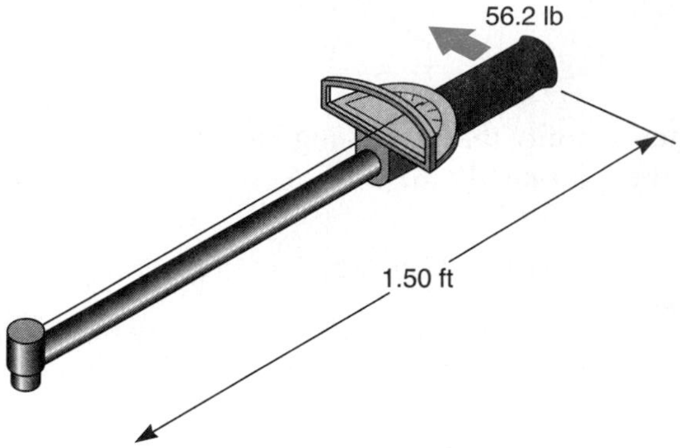

Figure 6.33

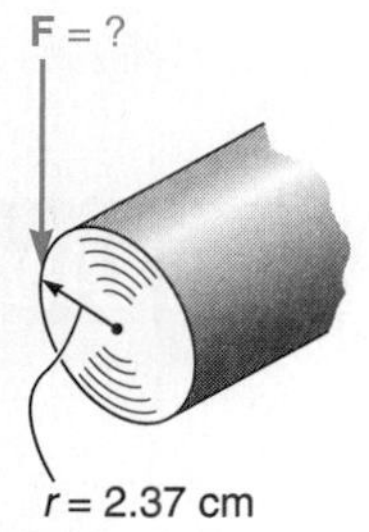

Figure 6.32

9. A motorcycle head bolt is torqued to 25.0 N m. What length shaft do we need on a wrench to exert a maximum force of 70.0 N?
10. A force of 112 N is applied to a shaft of radius 3.50 cm. What is the torque on the shaft?
11. A torque of 175 lb ft is needed to free a large rusted-on nut. The length of the wrench is 1.10 ft. What force must be applied to free it?
12. A torque wrench reads 14.5 N m. If its length is 25.0 cm, what force is being applied to the handle?
13. The torque on a shaft of radius 3.00 cm is 12.0 N m. What force is being applied to the shaft?
14. An automobile bolt is torqued to 27.0 N m. If the length of the wrench is 30.0 cm, what force is applied to the wrench?
15. A torque wrench reads 25 lb ft. (a) If its length is 1.0 ft, what force is being applied to the wrench? (b) What is the force if the length is doubled? Explain the results.
16. If 13 N m of torque is applied to a bolt with an applied force of 25 N, what is the length of the wrench?
17. If the torque required to loosen a nut on the wheel of a pickup truck is 40.0 N m, what minimum force must be applied to the end of a wrench 30.0 cm long to loosen the nut?
18. How is the required force to loosen the nut in Problem 17 affected if the length of the wrench is doubled?

6.4 Parallel Force Problems

A painter stands 2.00 ft from one end of a 6.00-ft plank that is supported at each end by a scaffold [Fig. 6.34(a)]. How much of the painter's weight must each end of the scaffold support? Problems of this kind are often faced in the construction industry, particularly in the design of bridges and buildings. Using some things we learned about torques and equilibrium, we can now solve problems of this type.

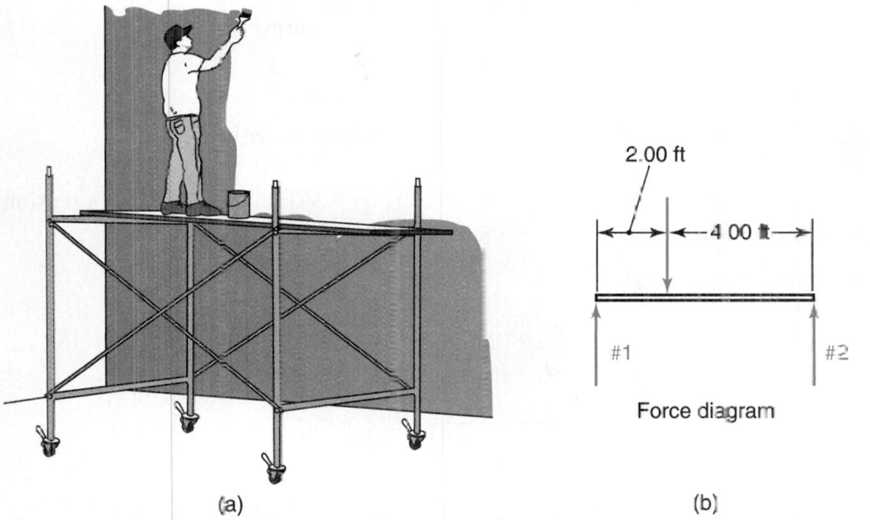

Figure 6.34 Parallel forces shown by the example of a painter on a scaffold

Let's look more closely at the painter problem. The force diagram [Fig. 6.34(b)] shows the forces and distances involved. The arrow pointing down represents the weight of the person. The arrows pointing up represent the forces exerted by each end of the scaffold in supporting the plank and painter. (For now, we will neglect the weight of the plank.) We have a condition of equilibrium. The plank and painter are not moving. The sum of the forces exerted by the ends of the scaffold is equal to the weight of the painter (Fig. 6.35).

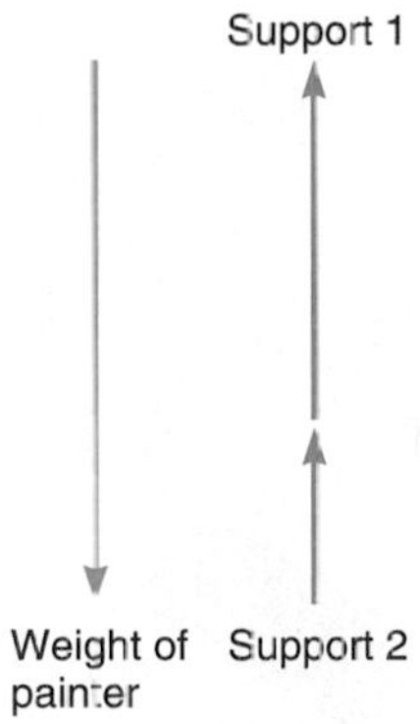

Figure 6.35 In equilibrium, the sum of the forces is zero.

Since these forces are vectors and are parallel, we can show that their sum is zero. Using engineering notation, we write

$$\Sigma \mathbf{F} = 0$$

where Σ (Greek capital letter sigma) means summation or "the sum of" and **F** is force, a vector quantity. So $\Sigma\mathbf{F}$ means "the sum of forces," in this case the sum of parallel forces.

FIRST CONDITION OF EQUILIBRIUM

The sum of all parallel forces on a body in equilibrium must be zero.

If the vector sum is not zero (forces up unequal to forces down), we have an unbalanced force tending to cause motion.

Now consider this situation: One end of the scaffold remains firmly in place, supporting the man, and the other is removed. What happens to the painter? The plank, supported only on one end, falls (Fig. 6.36), and the painter has a mess to clean up!

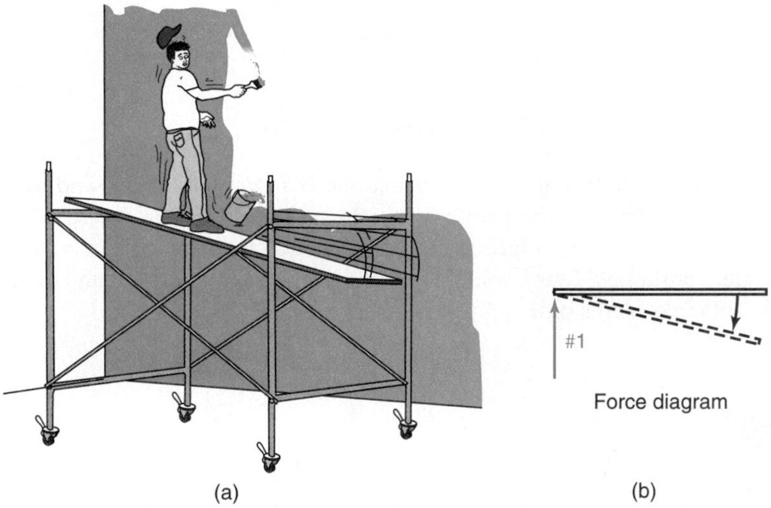

Figure 6.36 The position of the supporting force is important!

EXAMPLE 1

A sign of weight 150$\bar{0}$ lb is supported by two cables (Fig. 6.37). If one cable has a tension of 60$\bar{0}$ lb, what is the tension in the other cable?

Sketch: Draw the force diagram.

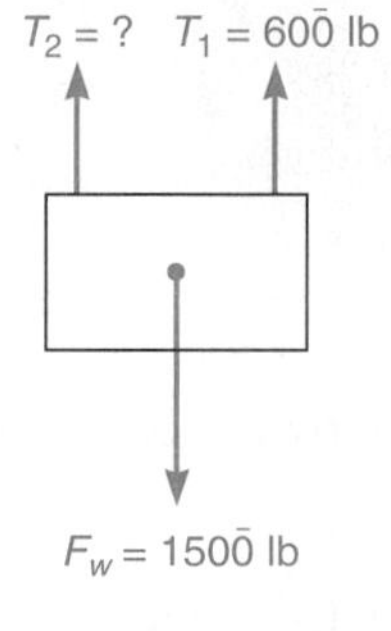

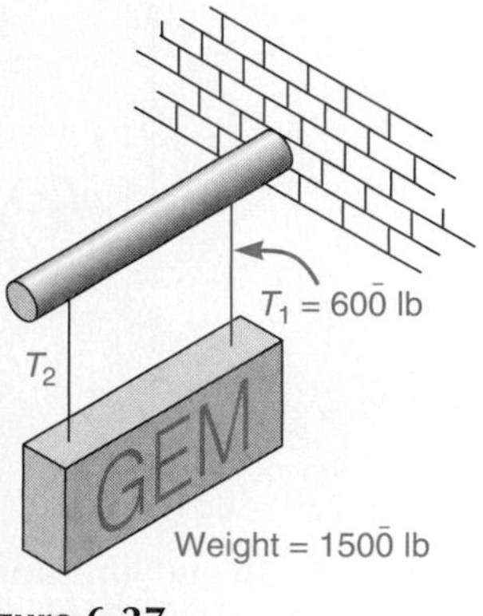

Figure 6.37

Data:

$$F_w = 150\bar{0} \text{ lb}$$
$$T_1 = 60\bar{0} \text{ lb}$$
$$T_2 = ?$$

Basic Equation:

$$F_+ = F_-$$

Working Equation:

$$T_1 + T_2 = F_w$$
$$T_2 = F_w - T_1$$

Substitution:

$$T_2 = 150\bar{0}\ \text{lb} - 60\bar{0}\ \text{lb}$$
$$= 90\bar{0}\ \text{lb}$$

Not only must the forces balance each other (vector sum $= 0$), but they must also be positioned so that there is no rotation in the system. To avoid rotation, we can have no unbalanced torques.

Sometimes there will be a natural point of rotation, as in our painter problem. We can, however, choose any point as our center of rotation as we consider the torques present. We will soon see that one of any number of points could be selected. What is necessary, though, is that there be no rotation (no unbalanced torques).

Again, using engineering notation, we write

$$\Sigma\tau_{\text{any point}} = 0$$

where $\Sigma\tau_{\text{any point}}$ is the sum of the torques about any chosen point or,

SECOND CONDITION OF EQUILIBRIUM

The sum of the clockwise torques on a body in equilibrium must equal the sum of the counterclockwise torques about any point.

$$\Sigma\tau_{\text{clockwise (cw)}} = \Sigma\tau_{\text{counterclockwise (ccw)}}$$

EXAMPLE 2

To illustrate these principles, we will find how much weight each end of the scaffold must support if our painter weighs $15\bar{0}$ lb.

Sketch:

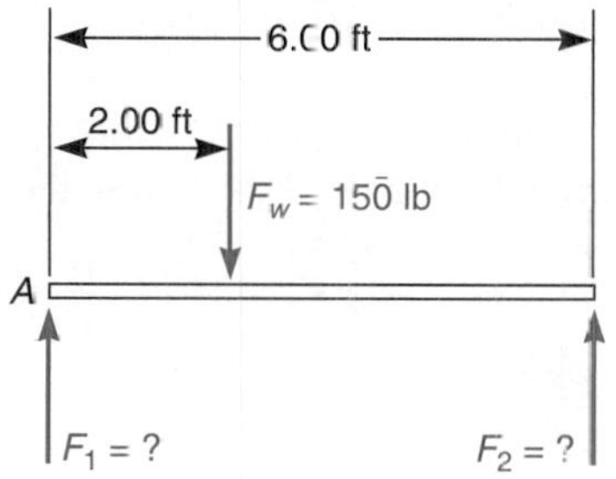

Data:

$$F_w = 15\bar{0}\ \text{lb}$$
$$\text{plank} = 6.00\ \text{ft}$$
$$F_w \text{ is } 2.00 \text{ ft from one end}$$

Basic Equations:

1. $\Sigma F = 0$
   sum of forces = 0

$$F_1 + F_2 - F_w = 0 \qquad \text{(\textbf{Note:} } F_w \text{ is negative because its direction is opposite } F_1 \text{ and } F_2.)$$
$$\text{or } F_1 + F_2 = F_w$$
$$F_1 + F_2 = 15\bar{0} \text{ lb}$$

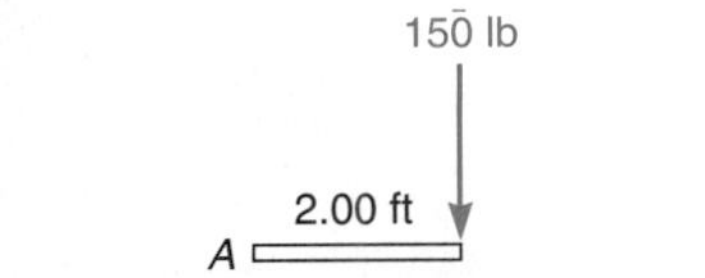

Figure 6.38 Torque arm of painter about point *A*

2. $\Sigma\tau_{\text{clockwise}} = \Sigma\tau_{\text{counterclockwise}}$

First, select a point of rotation. Choosing an end is usually helpful in simplifying the calculations. Choose the left end (point A) where F_1 acts. What are the clockwise torques about this point?

The force due to the weight of the painter tends to cause clockwise motion. The torque arm is 2.00 ft (Fig. 6.38). Then $\tau = (15\bar{0}\text{ lb})(2.00\text{ ft})$. This is the only clockwise torque.

The only counterclockwise torque is F_2 times its torque arm, 6.00 ft (Fig. 6.39). $\tau = (F_2)(6.00\text{ ft})$. There is no torque involving F_1 because its torque arm is zero. Setting $\Sigma\tau_{\text{clockwise}} = \Sigma\tau_{\text{counterclockwise}}$ we have the equation:

$$(15\bar{0}\text{ lb})(2.00\text{ ft}) = (F_2)(6.00\text{ ft})$$

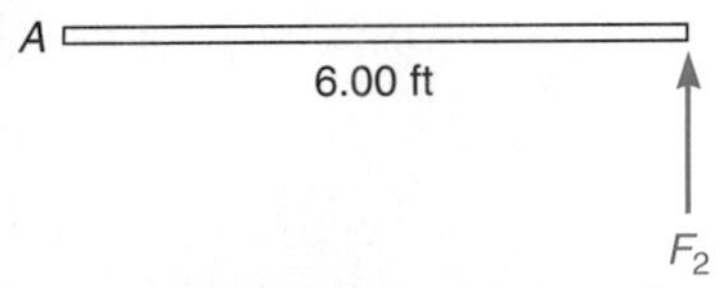

Figure 6.39 Torque arm of F_2 about point *A*

Note that by selecting an end as the point of rotation, we were able to have an equation with just one variable (F_2). Solving for F_2 gives the working equation:

$$F_2 = \frac{(15\bar{0}\text{ lb})(2.00\text{ ft})}{6.00\text{ ft}} = 50.0\text{ lb}$$

Since $\Sigma F = F_1 + F_2 = F_w$, substitute for F_2 and F_w to find F_1:

$$F_1 + 50.0\text{ lb} = 15\bar{0}\text{ lb}$$
$$F_1 = 15\bar{0}\text{ lb} - 50.0\text{ lb}$$
$$= 10\bar{0}\text{ lb}$$

To solve parallel force problems:

1. Sketch the problem.
2. Write an equation setting the sums of the opposite forces equal to each other.
3. Choose a point of rotation. Eliminate a variable, if possible (by making its torque arm zero).
4. Write the sum of all clockwise torques.
5. Write the sum of all counterclockwise torques.
6. Set $\Sigma\tau_{\text{clockwise}} = \Sigma\tau_{\text{counterclockwise}}$.
7. Solve the equation $\Sigma\tau_{\text{clockwise}} = \Sigma\tau_{\text{counterclockwise}}$ for the unknown quantity.
8. Substitute the value found in step 7 into the equation in step 2 to find the other unknown quantity.

EXAMPLE 3

A bricklayer weighing 175 lb stands on an 8.00-ft scaffold 3.00 ft from one end (Fig. 6.40). He has a pile of bricks, which weighs 40.0 lb, 3.00 ft from the other end. How much weight must each end support?

Figure 6.40

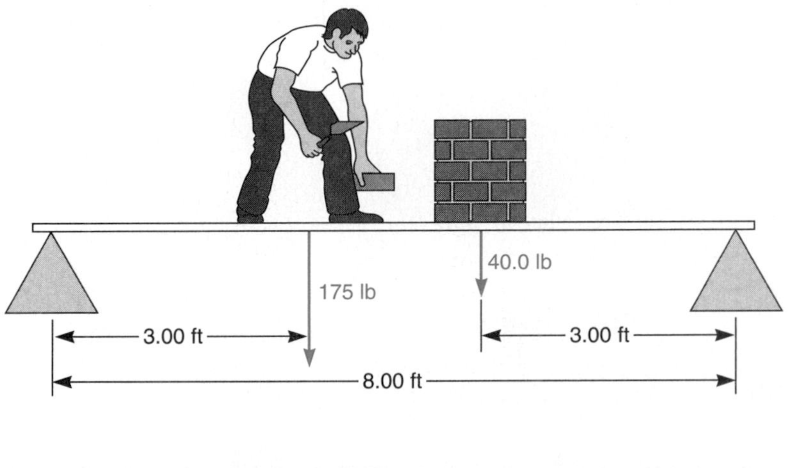

1. 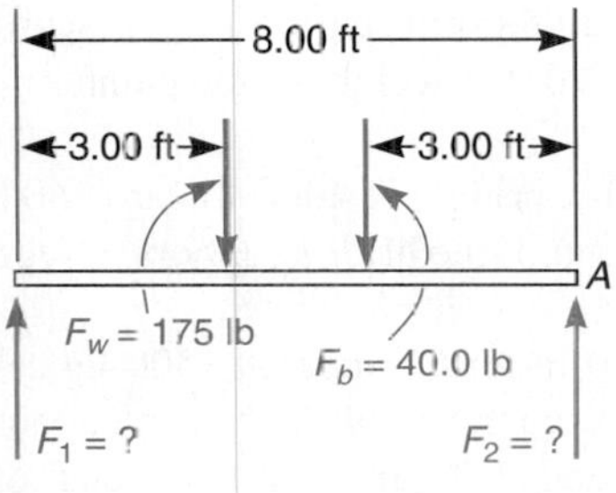

2. $\Sigma F = F_1 + F_2 = 175 \text{ lb} + 40.0 \text{ lb}$
3. Choose a point of rotation. Choose either end to eliminate one of the variables F_1 or F_2. Let us choose the right end and label it A.
4. $\Sigma\tau_{\text{clockwise}} = (F_1)(8.00 \text{ ft})$
5. $\Sigma\tau_{\text{counterclockwise}} = (40.0 \text{ lb})(3.00 \text{ ft}) + (175 \text{ lb})(5.00 \text{ ft})$
   Note that there are two counterclockwise torques.
6. $\Sigma\tau_{\text{clockwise}} = \Sigma\tau_{\text{counterclockwise}}$
   $F_1(8.00 \text{ ft}) = (40.0 \text{ lb})(3.00 \text{ ft}) + (175 \text{ lb})(5.00 \text{ ft})$
7. $F_1 = \dfrac{(40.0 \text{ lb})(3.00 \text{ ft}) + (175 \text{ lb})(5.00 \text{ ft})}{8.00 \text{ ft}}$
   $= \dfrac{12\bar{0} \text{ lb ft} + 875 \text{ lb ft}}{8.00 \text{ ft}} = \dfrac{995 \text{ lb ft}}{8.00 \text{ ft}} = 124 \text{ lb}$
8. $F_1 + F_2 = 175 \text{ lb} + 40.0 \text{ lb}$
   $124 \text{ lb} + F_2 = 215 \text{ lb}$ $\quad (F_1 = 124 \text{ lb})$
   $F_2 = 91 \text{ lb}$

PROBLEMS 6.4

Find the force F that will produce equilibrium for each force diagram. Use the same procedure as in Example 1.

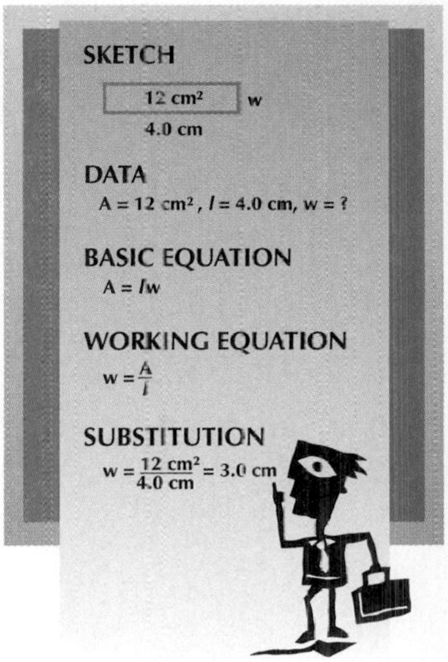

1. 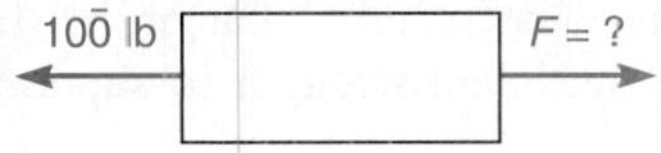

2. 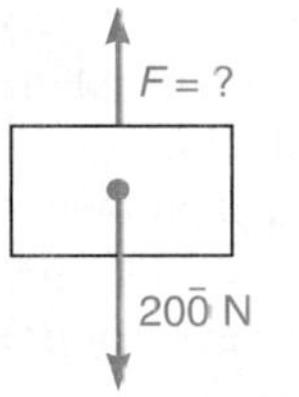

3. 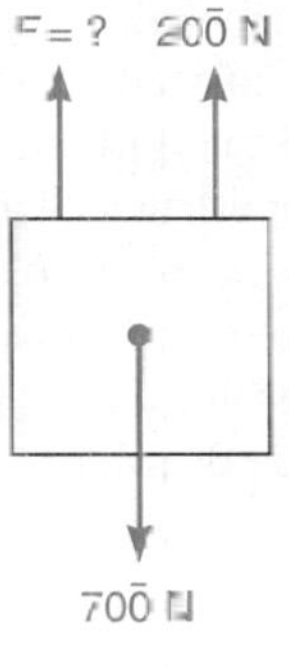

4. 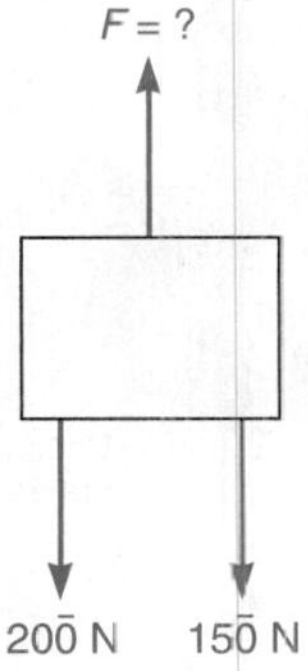

5. 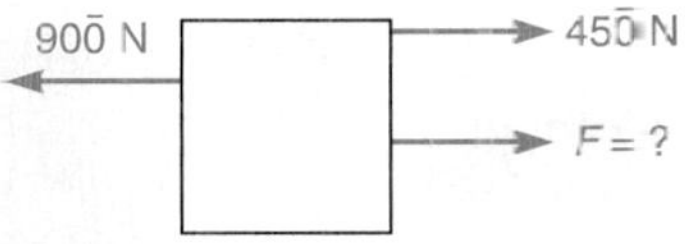

6. 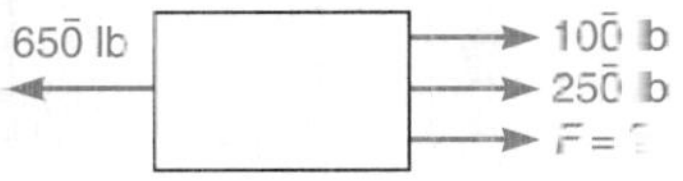

7. 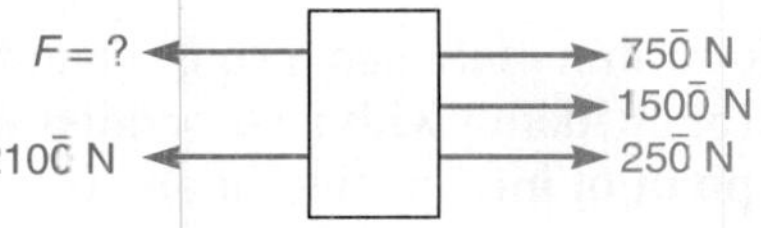

8. 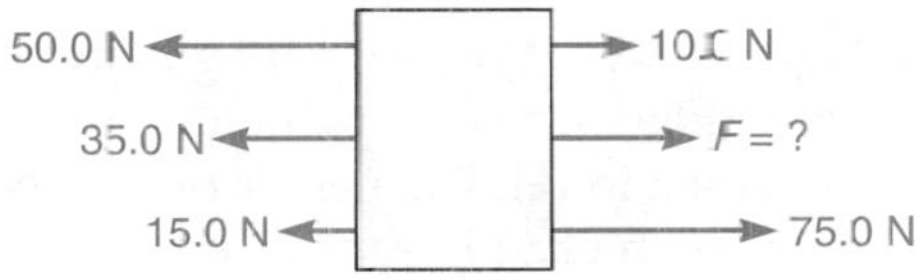

9. A 90.0-kg painter stands 3.00 m from one end of an 8.00-m scaffold. If the scaffold is supported at each end by a stepladder, how much of the weight of the painter must each ladder support?
10. A $50\bar{0}0$-lb truck is 20.0 ft from one end of a 50.0-ft bridge. A $40\bar{0}0$-lb car is 40.0 ft from the same end. How much weight must each end of the bridge support? (Neglect the weight of the bridge.)
11. A $24\bar{0}0$-kg truck is 6.00 m from one end of a 27.0-m-long bridge. A $15\bar{0}0$-kg car is 10.0 m from the same end. How much weight must each end of the bridge support?
12. An auto transmission of mass 165 kg is located 1.00 m from one end of a 2.50-m bench. What weight must each end of the bench support?
13. A bar 8.00 m long supports masses of 20.0 kg on the left end and 40.0 kg on the right end. At what distance from the 40.0-kg mass must the bar be supported for the bar to balance?
14. Two painters, each of mass 75.0 kg, stand on a 12.0 m scaffold, 6.00 m apart and 3.00 m from each end. They share a paint container of mass 21.0 kg in the middle of the scaffold. What weight must be supported by each of the ropes secured to the ends of the scaffold?
15. Two painters, one of mass 75.0 kg and the other 90.0 kg, stand on a 12.00-m scaffold, 6.00 m apart and 3.00 m from each end. They share a paint container of mass 21.0 kg in the middle of the scaffold. What weight must be supported by each of the ropes secured at the end of the scaffold?
16. Two painters stand on a 10.00-m scaffold. One, of mass 65.0 kg, stands 2.00 m from one end. The other, of mass 95.0 kg, stands 4.00 m from the other end. They share a paint container of mass 18.0 kg located between the two and 2.50 m from the larger person. What weight must be supported by each of the ropes secured at the ends of the scaffold?

6.5 Center of Gravity

In Section 6.4 we neglected the weight of the plank in the painter example. In practice, the weight of the plank or bridge is extremely important. The weight of a bridge being designed must be known in order to use materials of sufficient strength to support the bridge and the traffic and not collapse.

An important idea in this kind of problem is center of gravity. *The* **center of gravity** *of any body is that point at which all of its weight can be considered to be concentrated.* A body such as a brick or a uniform rod has its center of gravity at its middle or center. The center of gravity of something like an automobile, however, is not at its center or middle because its weight is not evenly distributed throughout. Its center of gravity is located nearer the heavy engine.

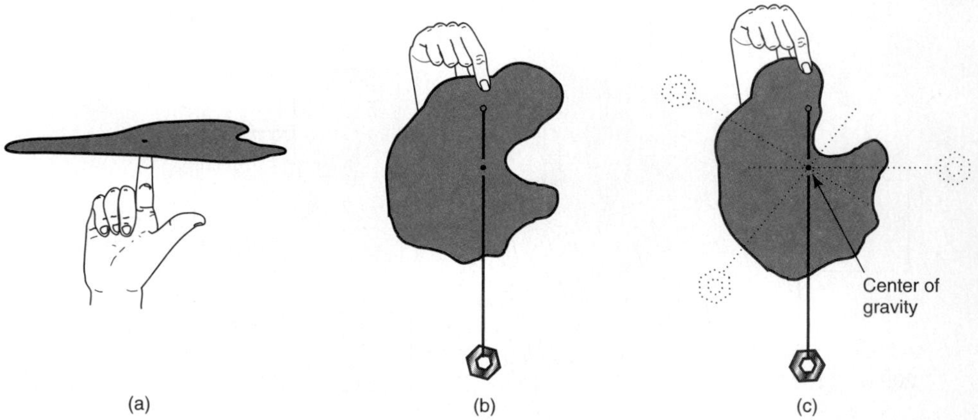

Figure 6.41 The center of gravity of a uniform thin plate can also be found by suspending it from a point and using a vertical chalkline with a suspended weight as shown in (b). The center of gravity is the point of intersection of any two or more such chalklines as in (c).

The center of gravity of an irregularly shaped uniform thin plate is the point at which it can be supported as in Fig. 6.41(a). The center of gravity of a uniform thin plate can also be found by suspending it from a point and using a vertical chalkline with a suspended weight as shown in Fig. 6.41(b). The center of gravity is the point of intersection of any two or more such chalklines as in Fig. 6.41(c).

You have probably had the experience of carrying a long board by yourself. If the board was not too heavy, you could carry it yourself by balancing it at its middle (Fig. 6.42). You didn't have to hold up both ends. You applied the principle of center of gravity and balanced the board at that point.

Figure 6.42 Support of the board at its center of gravity

We shall represent the weight of a body by a vector through its center of gravity. We use a vector to show the weight (force due to gravity) of the body (Fig. 6.43). It is placed through the center of gravity to show that all the weight may be considered concentrated at that point. If the center of gravity is not at the middle of the body, its location will be given (Fig. 6.43). In solving problems, the weight of the plank or bridge is represented like the other forces by a vector, which in the case of weight is through the center of gravity of the object.

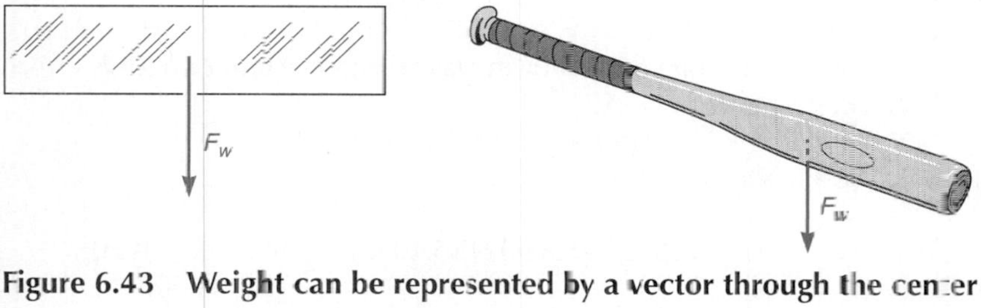

Figure 6.43 Weight can be represented by a vector through the center of gravity.

EXAMPLE 1

A carpenter stands 2.00 ft from one end of a 6.00-ft scaffold that is uniform and weighs 20.0 lb. If the carpenter weighs 165 lb, how much weight must each end support?

1. **Sketch:**

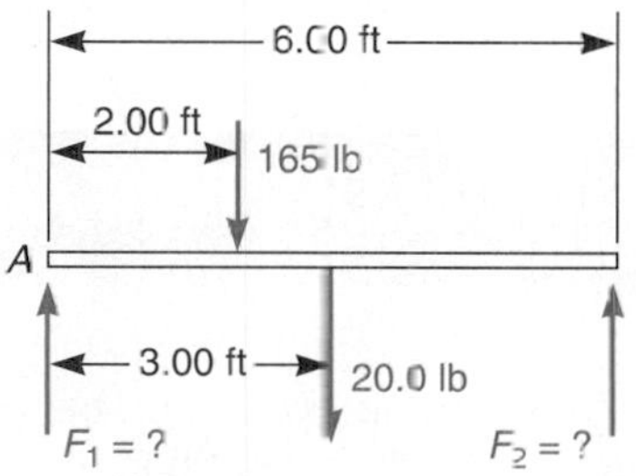

Since the plank is uniform, its center of gravity is at the middle.

2. $\Sigma F = F_1 + F_2 = 165 \text{ lb} + 20.0 \text{ lb}$
3. Choose the left end as the point of rotation and label it A.
4. $\Sigma\tau_{\text{clockwise}} = (165 \text{ lb})(2.00 \text{ ft}) + (20.0 \text{ lb})(3.00 \text{ ft})$
5. $\Sigma\tau_{\text{counterclockwise}} = (F_2)(6.00 \text{ ft})$
6. $(165 \text{ lb})(2.00 \text{ ft}) + (20.0 \text{ lb})(3.00 \text{ ft}) = (F_2)(6.00 \text{ ft})$
7. $F_2 = \dfrac{33\bar{0} \text{ lb ft} + 60.0 \text{ lb ft}}{6.00 \text{ ft}} = \dfrac{39\bar{0} \text{ lb ft}}{6.00 \text{ ft}} = 65.0 \text{ lb}$
8. $F_1 + 65.0 \text{ lb} = 165 \text{ lb} + 20.0 \text{ lb}$

$$F_1 = 165 \text{ lb} + 20.0 \text{ lb} - 65.0 \text{ lb} = 12\bar{0} \text{ lb}$$

We can also use this method to find the magnitude and position of a parallel force vector that produces equilibrium.

EXAMPLE 2

Find the magnitude, direction, and placement (from point A) of a parallel vector F_6 that will produce equilibrium in the parallel force diagram in Fig. 6.44.

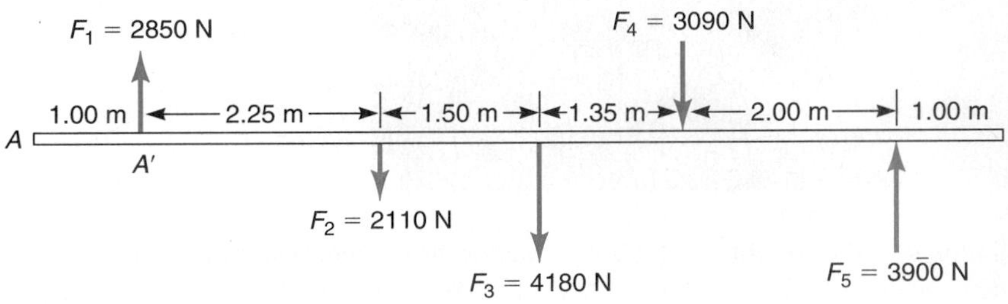

Figure 6.44

1. $\Sigma F = F_1 + F_5 + F_6 = F_2 + F_3 + F_4$

$$2850 \text{ N} + 39\bar{0}0 \text{ N} + F_6 = 2110 \text{ N} + 4180 \text{ N} + 3090 \text{ N}$$
$$F_6 = 2630 \text{ N (up)}$$

2. Choose A' instead of A as the point of rotation to make the torque arm zero for F_1. Also, let x be the distance of F_6 from point A'.
3. $\Sigma\tau_{\text{clockwise}} = (2110 \text{ N})(2.25 \text{ m}) + (4180 \text{ N})(3.75 \text{ m}) + (3090 \text{ N})(5.10 \text{ m})$
4. $\Sigma\tau_{\text{counterclockwise}} = (39\bar{0}0 \text{ N})(7.10 \text{ m}) + (2630 \text{ N})(x)$
5. $\Sigma\tau_{\text{clockwise}} = \Sigma\tau_{\text{counterclockwise}}$
6. $(2110 \text{ N})(2.25 \text{ m}) + (4180 \text{ N})(3.75 \text{ m}) + (3090 \text{ N})(5.10 \text{ m}) = (39\bar{0}0 \text{ N})(7.10 \text{ m}) + (2630 \text{ N})(x)$

$$3.23 \text{ m} = x \text{ (from } A')$$
$$\text{or } 4.23 \text{ m} = x \text{ (from } A)$$

Thus, the equilibrium vector is 2630 N (up) placed at 4.23 m from point A.

TRY THIS ACTIVITY

Center of Mass

Stand with your back and heels touching a wall. Without moving your feet, bend over and touch your legs below your knees. Using the concept of center of gravity, explain what happened.

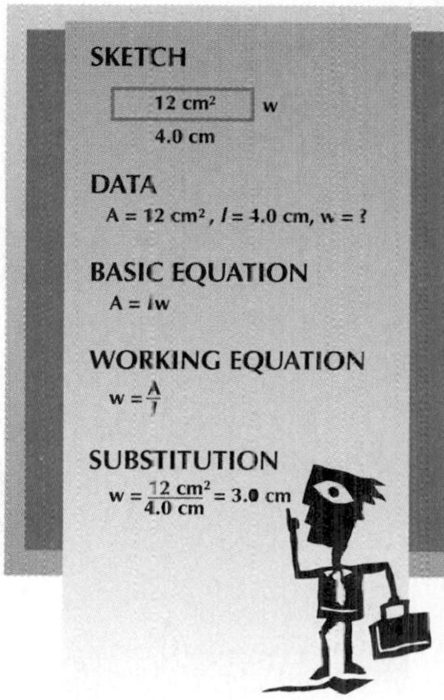

PROBLEMS 6.5

Solve each problem using the methods outlined in this chapter.

1. Solve for F_1: $30.0F_1 = (14.0)(18.0) + (25.0)(17.0)$
2. Solve for F_w: $(12.0)(15.0) + 45.0F_w = (21.0)(65.0) + (22.0)(32.0)$
3. Two workers carry a uniform 15.0-ft plank that weighs 22.0 lb (Fig. 6.45). A load of blocks weighing 165 lb is located 7.00 ft from the first worker. What force must each worker exert to hold up the plank and load?

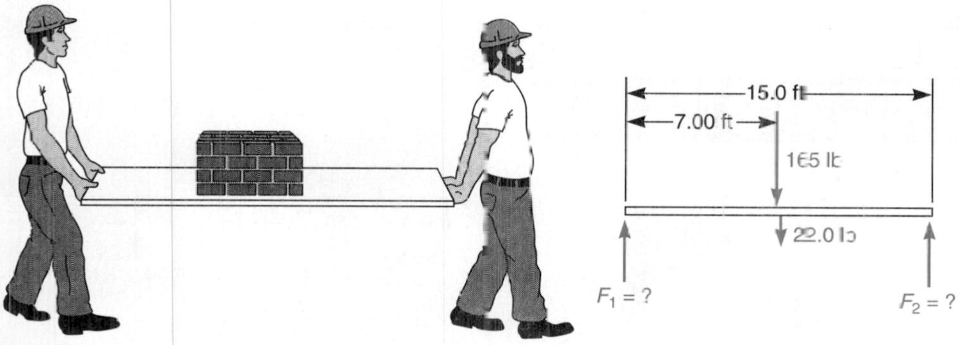

Figure 6.45

4. Juan and Pablo carry a load weighing 720 N on a pole between them. (a) If the pole is 2.0 m long and the load is 0.50 m from Pablo, what force does each person support? Neglect the weight of the pole. (b) If the weight of the 120-N pole is considered, what force does each person support?
5. A wooden beam is 3.30 m long and has its center of gravity 1.30 m from one end. If the beam weighs 2.50×10^4 N what force is needed to support each end?
6. An auto engine weighs 650 lb and is located 4.00 ft from one end of a 10.0-ft workbench. If the bench is uniform and weighs 75.0 lb, what weight must each end of the bench support?
7. A bridge across a country stream weighs 89,200 N. A large truck stalls 4.00 m from one end of the 9.00-m bridge. What weight must each of the piers support if the truck weighs 98,000 N?
8. A window washer's scaffold 12.0 ft long and weighing 75.0 lb is suspended from each end. One washer weighs 155 lb and is 3.00 ft from one end. The other washer is 4.00 ft from the other end. If the force supported by the end near the first washer is 200 lb, how much does the second washer weigh?
9. A porch swing weighs 29.0 lb. It is 4.40 ft long and has a dog weighing 14.0 lb sleeping on it 1.90 ft from one end and a 125-lb person sitting 1.00 ft from the other end. What weight must the support ropes on each end hold up?
10. A wooden plank is 5.00 m long and supports a 75.0-kg block 2.00 m from one end. If the plank is uniform with mass 30.0 kg, how much force is needed to support each end?
11. A bridge has a mass of 1.60×10^4 kg, is 21.0 m long, and has a 3500-kg truck 7.00 m from one end. What force must each end of the bridge support?
12. A uniform steel beam is 5.00 m long and weighs 3.60×10^5 N. What force is needed to lift one end?
13. A wooden pole is 4.00 m long, weighs 315 N, and has its center of gravity 1.50 m from one end. What force is needed to lift each end?
14. A bridge has a mass of 2.60×10^4 kg, is 32.0 m long, and has a 3500-kg truck 15.0 m from one end. What force must each end of the bridge support?
15. An auto engine of mass 295 kg is located 1.00 m from one end of a 4.00-m workbench. If the uniform bench has a mass of 45.0 kg, what weight must each end of the bench support?

16. A 125-kg horizontal beam is supported at each end. A 325-kg mass rests one-fourth of the way from one end. What weight must be supported at each end?
17. The sign shown in Fig. 6.46 is 4.00 m long, weighs $155\bar{0}$ N, and is made of uniform material. A weight of 245 N hangs 1.00 m from the end. Find the tension in each support cable.
18. The uniform bar in Fig. 6.47 is 5.00 m long and weighs 975 N. A weight of 255 N is attached to one end while a weight of 375 N is attached 1.50 m from the other end. (a) Find the tension in the cable. (b) Where should the cable be tied to lift the bar and its weights so that the bar hangs in a horizontal equilibrium position?

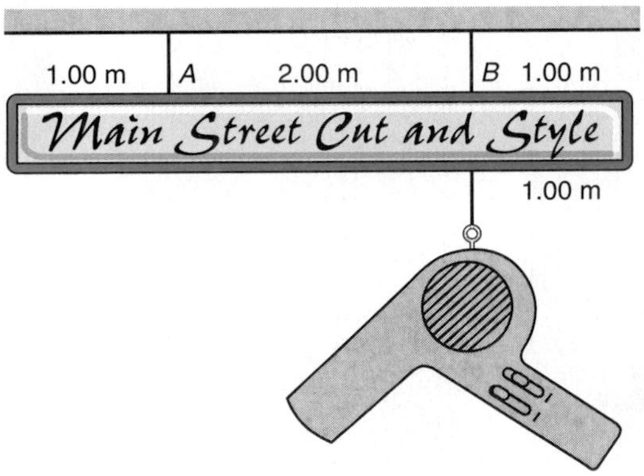

Figure 6.46

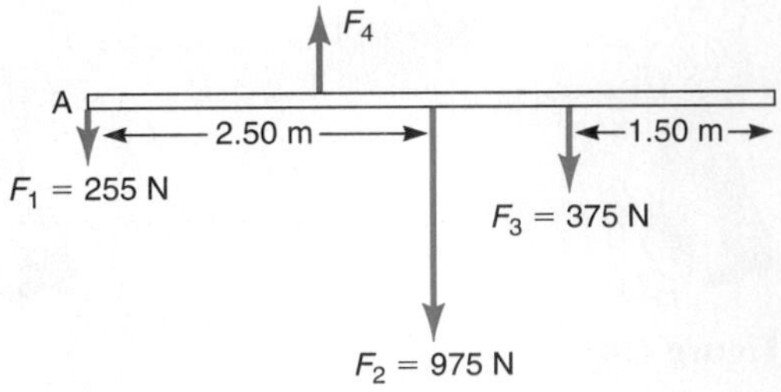

Figure 6.47

Find the magnitude, direction, and placement (from point A) of a parallel vector F_6 that will produce equilibrium in each force diagram.

19.

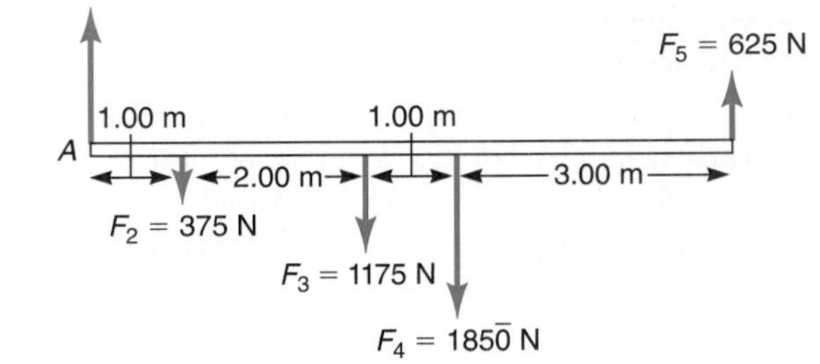

20.

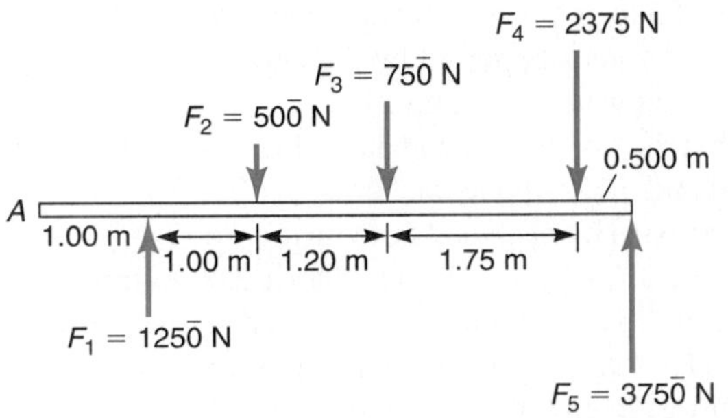

Glossary

Center of Gravity The point of any body at which all of its weight can be considered to be concentrated. (p. 174)

Compression A force produced by forces pushing inward on the ends of an object (p. 159)

Concurrent Forces Two or more forces applied to, or acting at, the same point. (p. 152)

Equilibrant Force The force that produces equilibrium. (p. 158)

Equilibrium An object is said to be in equilibrium when the net force acting on it is zero. A body that is in equilibrium is either at rest or moving at a constant velocity. (p. 157)

First Condition of Equilibrium The sum of all parallel forces on a body in equilibrium must be zero. (p. 170)

Resultant Force The sum of the forces applied at the same point, the single force that has the same effect as the two or more forces acting together. (p. 152)

Second Condition of Equilibrium The sum of the clockwise torques on a body in equilibrium must be equal to the sum of the counterclockwise torques about any point. (p. 171)

Statics The study of objects that are in equilibrium. (p. 157)

Tension A stretching force produced by forces pulling outward on the ends of an object. (p. 159)

Torque The tendency to produce change in rotational motion. Equal to the applied force times the length of the torque arm. (p. 166)

Formulas

6.1 To find the resultant vector $\mathbf{F}_R$ of two or more vectors:

(a) find the x- and y-components of each vector and add the components.

(b) find angle A as follows:

$$\tan A = \frac{|\text{sum of } y\text{-components}|}{|\text{sum of } x\text{-components}|} = \frac{|\mathbf{F}_{Ry}|}{|\mathbf{F}_{Rx}|}$$

Determine the quadrant of the angle from the signs of the sum of the x- and y-components.

(c) find the magnitude of $\mathbf{F}_R$ using the Pythagorean theorem:

$$|\mathbf{F}_R| = \sqrt{|\mathbf{F}_{Rx}|^2 + |\mathbf{F}_{Ry}|^2}$$

6.2 Condition for equilibrium in one dimension:

$$\mathbf{F}_+ = \mathbf{F}_-$$

where $\mathbf{F}_+$ is the sum of the forces acting in one direction (call it the positive direction) and $\mathbf{F}_-$ is the sum of the forces acting in the opposite (negative) direction.

Conditions for equilibrium in two dimensions:

(a) The sum of x-components = 0; that is, $\mathbf{A}_x + \mathbf{B}_x + \mathbf{C}_x = 0$; and

(b) The sum of y-components = 0; that is, $\mathbf{A}_y + \mathbf{B}_y + \mathbf{C}_y = 0$.

To solve equilibrium problems:

1. Draw a force diagram from the point at which the unknown forces act.
2. Find the x- and y-components of each force.
3. Substitute the components in the equations

$$\text{sum of } x\text{-components} = 0$$
$$\text{sum of } y\text{-components} = 0$$

4. Solve for the unknowns. This may involve two simultaneous equations.

6.3 $\tau = Fs_t$

6.4 *First condition of equilibrium:* The sum of all parallel forces on an object must be zero.

$$\Sigma F = 0$$

Second condition of equilibrium: The sum of the clockwise torques on an object must equal the sum of the counterclockwise torques.

$$\Sigma\tau_{\text{clockwise}} = \Sigma\tau_{\text{counterclockwise}}$$

Review Questions

1. Concurrent forces act at
   (a) two or more different points. (b) the same point.
   (c) the origin.
2. The resultant force is
   (a) the last force applied.
   (b) the single force that has the same effect as the two or more forces acting together.
   (c) equal to either diagonal when using the parallelogram method to add vectors.
3. A moving object
   (a) can be in equilibrium. (b) is never in equilibrium.
   (c) has no force being applied.
4. The study of an object in equilibrium is called
   (a) dynamics. (b) astronomy. (c) statics. (d) biology.
5. Torque is
   (a) applied force in rotational motion.
   (b) the length of the torque arm.
   (c) applied force times the length of the torque arm.
   (d) none of the above.
6. The first condition of equilibrium states that
   (a) all parallel forces must be zero.
   (b) all perpendicular forces must be zero.
   (c) all frictional forces must be zero.
7. In the second condition of equilibrium,
   (a) clockwise and counterclockwise torques are unequal.
   (b) clockwise and counterclockwise torques are equal.
   (c) there are no torques.
8. The center of gravity of an object
   (a) is always at its geometric center.
   (b) does not have to be at the geometric center.
   (c) exists only in symmetrical objects.
9. Is motion produced every time a force is applied to an object?
10. What is the relationship between opposing forces on a body that is in equilibrium?
11. Define *equilibrium*.
12. In what direction does the force due to gravity always act?
13. What may be said about concurrent forces whose sum of x-components equals zero and whose sum of y-components equals zero?
14. What is a force diagram?
15. Is the length of the pedal necessarily the true length of the torque arm in pedaling a bicycle?
16. In your own words, explain the second condition of equilibrium.
17. What is the primary consideration in the selection of a point of rotation in an equilibrium problem?
18. List three examples from daily life in which you use the concept of center of gravity.
19. Is the center of gravity of an object always at its geometric center?
20. On a 3.00-m scaffold of uniform mass, with supports at each end, there is a pile of bricks 0.500 m from one end. Which support will exert a greater force: the one closer to the bricks or the one farther away?

Review Problems

1. Find the sum of the following forces acting at the same point in a straight line: 345 N (right); 108 N (right); 481 N (left); 238 N (left); 303 N (left).

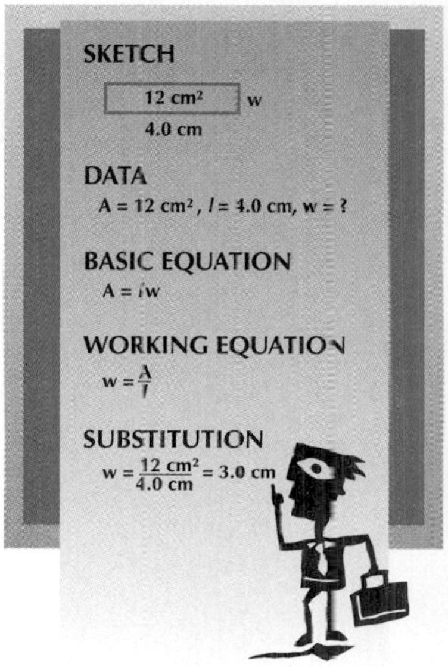
SKETCH
12 cm²
w
4.0 cm
DATA
A = 12 cm², l = 4.0 cm, w = ?
BASIC EQUATION
A = lw
WORKING EQUATION
w = A/l
SUBSTITUTION
w = 12 cm²/4.0 cm = 3.0 cm

2. Forces of 275 lb and 225 lb act at the same point.
   (a) What is the magnitude of the maximum net force the two forces can exert together?
   (b) What is the magnitude of the minimum net force the two forces can exert together?

Find the sum of each set of vectors. Give angles in standard position.

3.

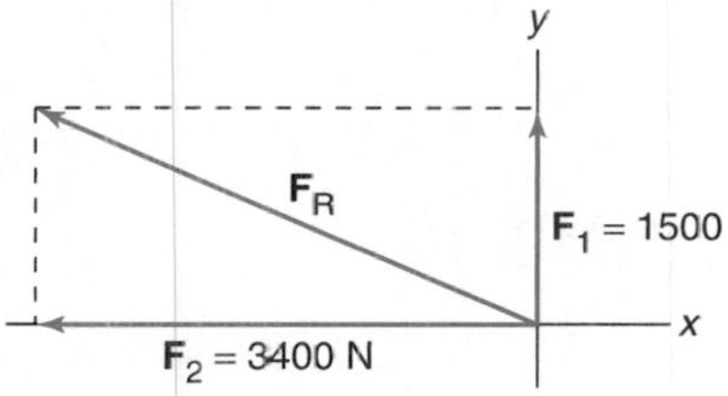

4.

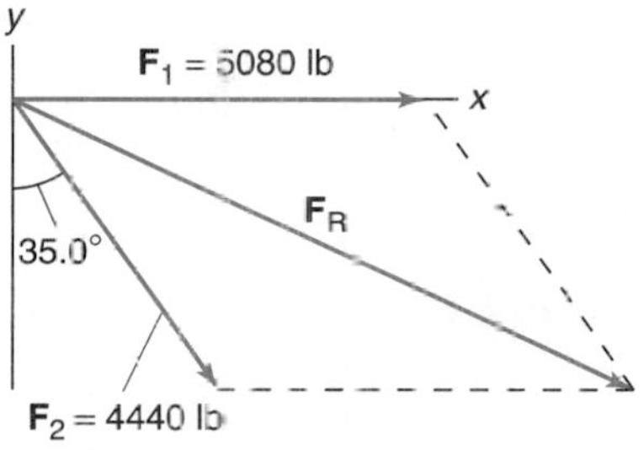

5.

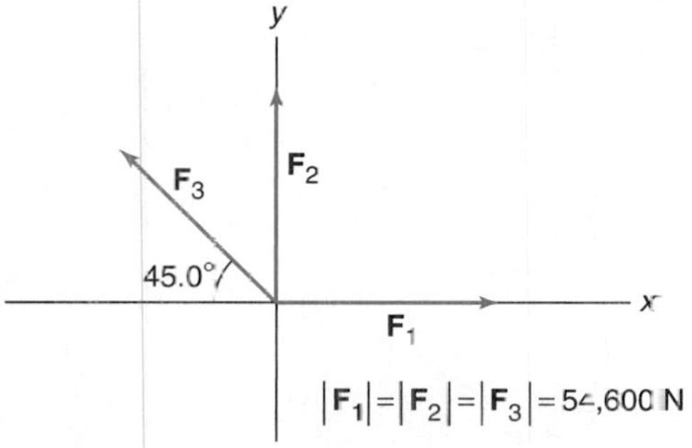

6. Forces of $\mathbf{F}_1 = 1250$ N, $\mathbf{F}_2 = 625$ N, and $\mathbf{F}_3 = 1850$ N are applied at the same point. The angle between $\mathbf{F}_1$ and $\mathbf{F}_2$ is 120.0° and the angle between $\mathbf{F}_2$ and $\mathbf{F}_3$ is 30.0°. $\mathbf{F}_2$ is between $\mathbf{F}_1$ and $\mathbf{F}_3$. Find the resultant force.
7. Eight people are involved in a tug-of-war. The blue team members pull with forces of 220 N, 340 N, 180 N, and 560 N. Three members of the red team pull with forces of 250 N, 160 N, and 420 N. With what force must the fourth person pull to maintain equilibrium?
8. A bridge has a weight limit of 14.0 tons. What is the maximum weight an 8.0-ton truck can carry across and still maintain equilibrium?
9. The x-components of three vectors are $\mathbf{F}_x$, 375 units, and 150 units. If their sum is equal to zero, what is $\mathbf{F}_x$?
10. If $\mathbf{W}_y = 60\bar{0}$ N and $\mathbf{W}_x = 90\bar{0}$ N, what are the magnitude and direction of the resultant $\mathbf{W}$?

Find forces $\mathbf{F}_1$ and $\mathbf{F}_2$ that produce equlibrium in each force diagram.

11.

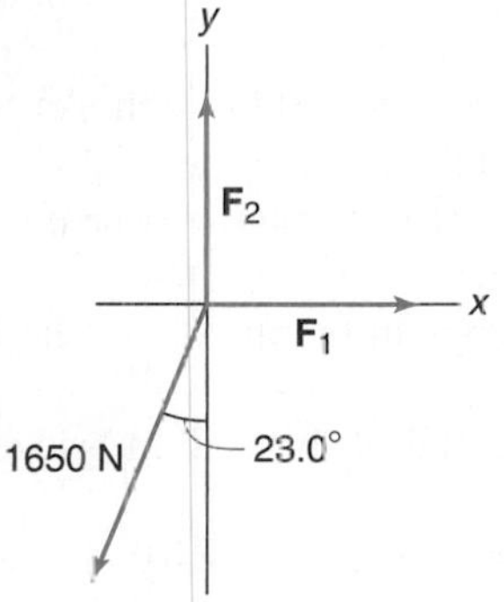

12.

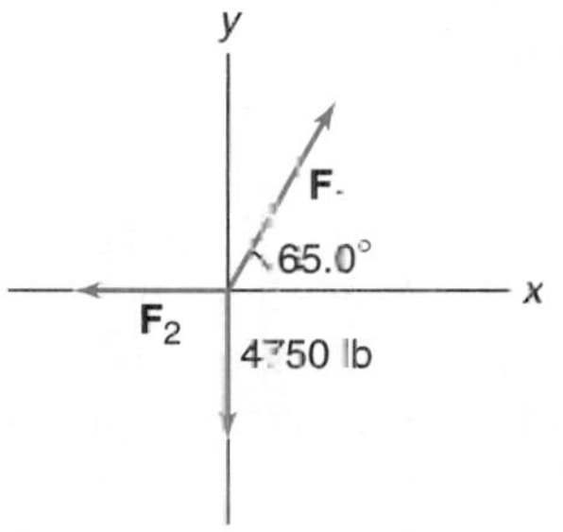

13. Find the tension in the cable and the compression in the support of the sign shown in Fig. 6.48.

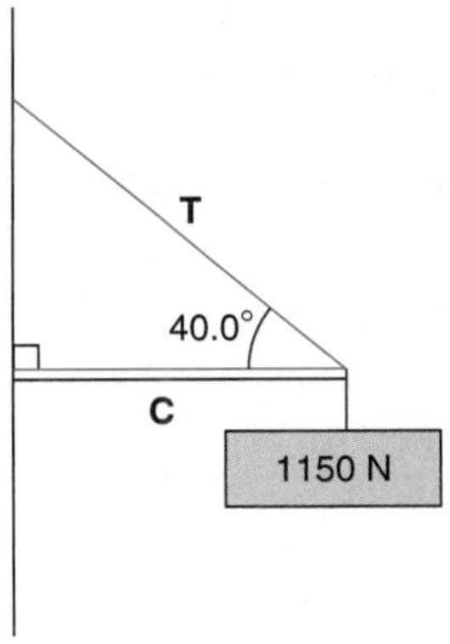

Figure 6.48

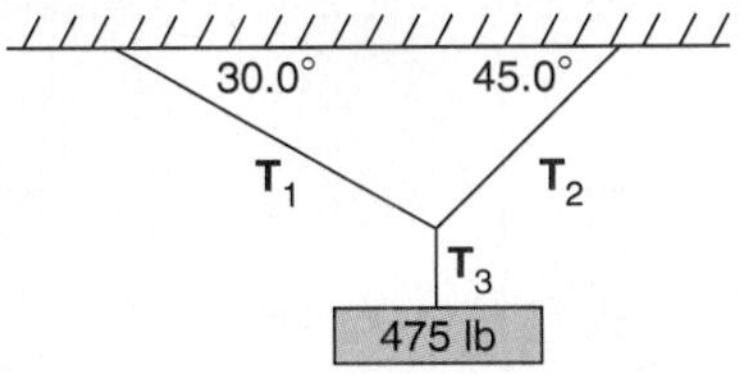

Figure 6.49

14. Find the tension in each cable in Fig. 6.49.
15. Find the tension in each cable in Fig. 6.50.

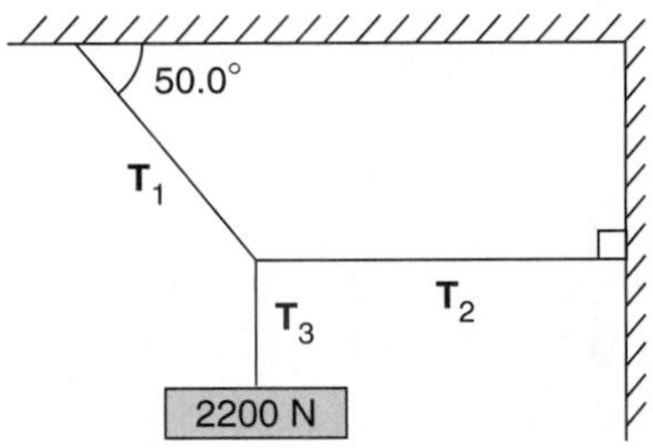

Figure 6.50

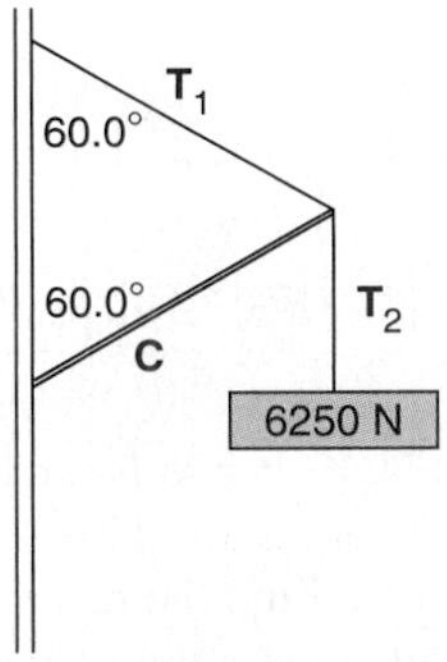

Figure 6.51

16. Find the tension and the compression in Fig. 6.51.
17. A man is changing a flat tire using a tire iron that is 50.0 cm long. If he exerts a force of 53.0 N, how much torque (in N m) does he produce?
18. A torque of 81.0 lb ft is produced by a torque arm of 3.00 ft. What force is being applied?
19. A hanging sign has mass $20\bar{0}$ kg. If the tension in one support cable is 1080 N, what is the tension in the other support cable?
20. A scaffold supports a bricklayer and bricks weighing 450 lb. If the force in one end support is 290 lb, what is the supporting force in the other?
21. Two ladders at the ends of a scaffold support a mass of 90.0 kg each. An 80.0-kg worker is on the scaffold with a pile of bricks. Find the mass of the bricks.
22. How far from the light end of a 68.0-cm bat would its center of gravity be if it is one-fourth of the length of the bat from the heavy end?
23. A bridge has mass $80\bar{0}0$ kg. If a $32\bar{0}0$-kg truck stops in the middle of the bridge, what mass must each pier support?
24. If the truck in Problem 23 stops 7.00 m from one end of the 26.0-m bridge, what weight must each end support?
25. A uniform 2.20-kg steel bar with length 2.70 m is suspended on each end by a chain. A 40.0-kg person hangs 70.0 cm from one end while a 55.0-kg person hangs 50.0 cm from the other end. Another person pushes up halfway between the two persons with a force of 127 N. How much weight does each chain support?

26. Find the vertical force needed to support the 4.00-m-long uniform beam in Fig. 6.52, which weighs 3475 N.

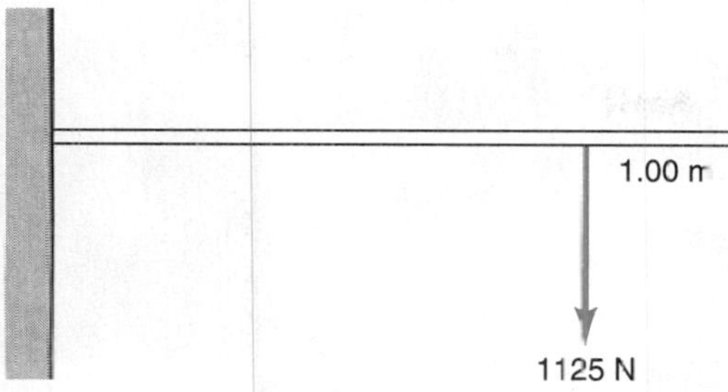

Figure 6.52

APPLIED CONCEPTS

1. Archeologists in Egypt are attempting to open a 2.57-m-wide door in the tomb of Tutankhamen. Miles, one of the rookie archeologists, pushes perpendicularly on the door with a force of 894 N. (a) How much torque does Miles exert on the door? (b) As the door opens, Miles continues to push with the same force, but is at an angle of 30.0° to the face of the door. How much torque does Miles apply to the door? (c) What could Miles do to continually exert the same amount of torque on the door?
2. Sean and Greg are on a job site standing on two beams 11.5 ft apart. They need to lift their crate of tools midway between them with ropes up 33.5 ft to where they are working. (a) What is the angle between the ropes when the crate is on the ground? (b) How much force do Sean and Greg need to exert on the ropes when lifting the 115-lb crate off the ground? (c) How much force do both Sean and Greg need to exert when the crate is 5.75 ft below them? (d) Explain why the force to lift the crate changes as it moves closer to them.
3. Maria has severe arthritis and can only apply a maximum force of 25.5 N when opening her front door. (a) How much more torque would Maria be able to apply if she purchased a lever-style door opener with a handle 12.7 cm long compared to her conventional doorknob with a radius of 3.74 cm? (b) How much less force would she need to apply to the new door handle in order to maintain the original amount of torque she had with her old doorknob?
4. Krista's flagpole bracket is mounted at an angle of 45.0° to the wall and is designed to support a maximum torque of 40.0 N m. (a) If the new flag and pole that Krista bought have a mass of 8.75 kg and their center of mass is located 0.750 m from the bracket, will the bracket support the flag and pole? (b) What could Krista do to the angle of the bracket to reduce the torque?
5. Luisa, whose mass is 45.0 kg, is standing at the end of a 5.50-m diving board, which has a mass of 35.7 kg (Fig. 6.53). (a) What force must the bracket and fulcrum exert on the diving board? (b) Where would Luisa stand to have both the fulcrum and the bracket apply an upward force?

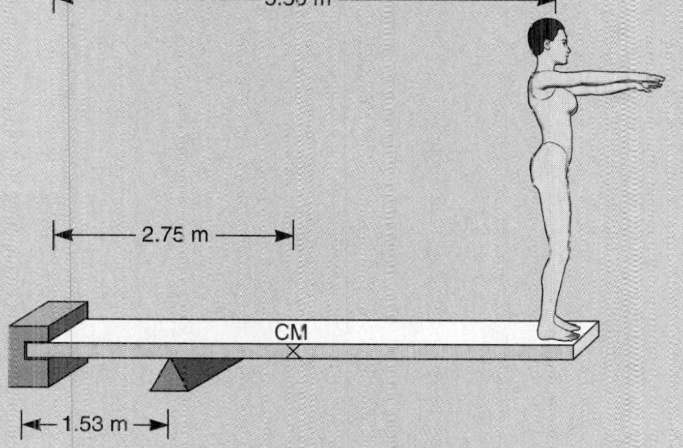

Figure 6.53

CHAPTER 7

WORK AND ENERGY

Work, power, and energy are common terms used to describe changes in physical activity. In science each of these terms has a limited definition. Work, for example, is accomplished only where there is movement of the object on which the applied force acts. Effort alone is insufficient.

We will now study how the scientist and engineer use work, power, and energy; how they are related; and how they differ from their everyday meanings.

Objectives

The major goals of this chapter are to enable you to:

1. Distinguish the common and technical definitions of work.
2. Analyze how power is used and described in technical applications.
3. Relate kinetic and potential energy to the law of conservation of mechanical energy.

7.1 Work

What is work? The common idea of the definition of work is quite different from the technical definition. We often associate work with physical or mental effort that leads to fatigue. The technical definition of work is more limited. If we try to lift a heavy crate that doesn't budge [Fig. 7.1(a)], we would probably say that we have done work because we strained our muscles and feel tired, but in a technical sense, no work was done, because the crate did not move. Work would be done *on* the crate if we were to move it across the floor [Fig. 7.1(b)]. In this case, work was done *by* us, and work was done *on* the crate.

(a) Crate is not moved. (b) Crate is moved.

Figure 7.1 Work is done only in the case shown in part (b).

The technical meaning of work requires that work must be done *by* one object *on* another object. When a stake is driven into the ground (Fig. 7.2), work is done *by* the moving sledgehammer and work is done *on* the stake. When a bulldozer pushes a boulder (Fig. 7.3), work is done *by* the bulldozer and work is done *on* the boulder.

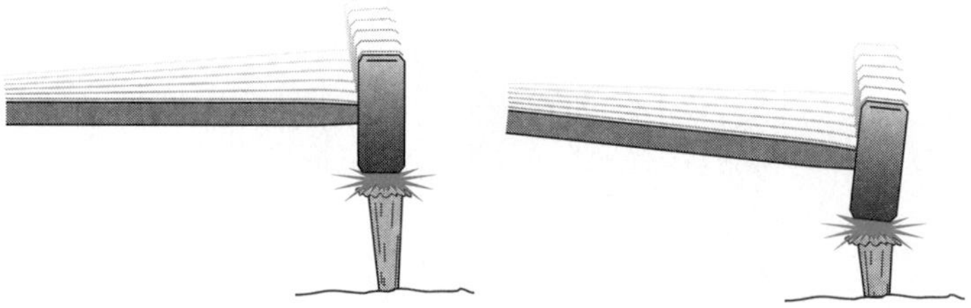

(a) (b)

Figure 7.2 Work is done by one object on another.

Figure 7.3 Work is done *by* the bulldozer *on* the boulder.

The previous examples show a limited meaning of work: that work is done when a force acts through a distance. The physical definition of **work** is even narrower: *Work is the product of the force in the direction of the motion and the displacement.*

$$W = Fs$$

where

W = work
F = force applied *in the direction of the motion*
s = displacement

Now, let us apply our technical definition of work to our unsuccessful effort to lift the crate. We applied a force by lifting on the crate but were unable to move it. Therefore, the displacement was zero, and the product of the force and the displacement must also be zero. Therefore, no work was done.

From the equation for work, we can determine the units for work. In the metric system, force is expressed in newtons and the displacement in metres:

$$\text{work} = \text{force} \times \text{displacement} = \text{newton} \times \text{metre} = \text{N m}$$

This unit (N m) has a special name in honor of **James P. Joule.** It is the joule (J) [pronounced jōōl]:

$$1 \text{ N m} = 1 \text{ joule} = 1 \text{ J}$$

In the U.S. system, force is expressed in pounds (lb) and the displacement in feet (ft):

$$\text{work} = \text{force} \times \text{displacement} = \text{pounds} \times \text{feet} = \text{ft lb}$$

The U.S. unit of work is called the foot-pound.

Work is not a vector quantity because it has no particular direction. It is a scalar and has only magnitude.

James P. Joule (1818–1889),

physicist, was born in England. In a series of experiments from 1843 to 1878, he showed that heat is a form of energy and established the mechanical equivalent of heat, which formed the basis of the theory of the conservation of energy. He also worked with Lord Kelvin on temperature changes in gases, which led to the founding of the refrigeration industry. The unit of work, the joule, is named after him.

EXAMPLE 1

Find the amount of work done by a worker lifting 225 N of bricks to a height of 1.75 m as shown in Fig. 7.4.

Data:

$$F = 225 \text{ N}$$
$$s = 1.75 \text{ m}$$
$$W = ?$$

Basic Equation:

$$W = Fs$$

Working Equation: Same

Substitution:

$$W = (225 \text{ N})(1.75 \text{ m})$$
$$= 394 \text{ N m} \quad \text{or} \quad 394 \text{ J}$$

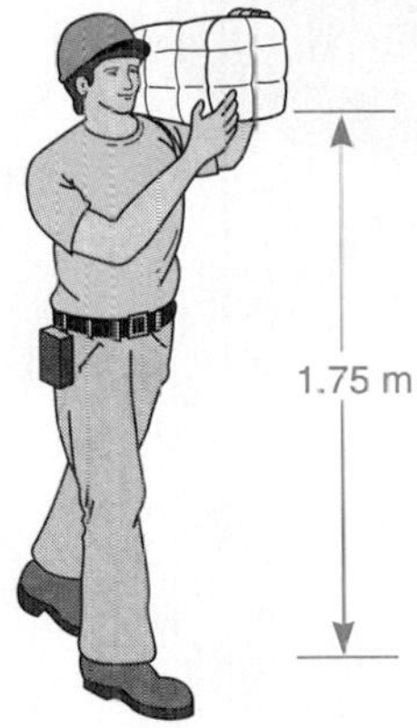

Figure 7.4

EXAMPLE 2

A worker pushes a 350-lb cart a distance of 30̄ ft by exerting a constant force of 40̄ lb as shown in Fig. 7.5. How much work does the person do?

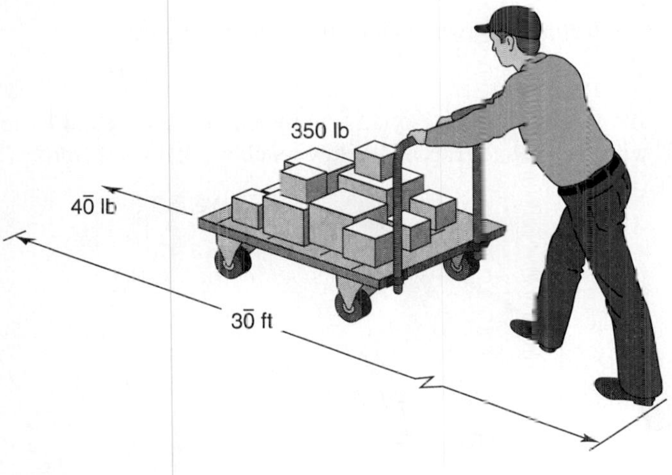

Figure 7.5

Data:

$$F = 4\bar{0} \text{ lb}$$
$$s = 3\bar{0} \text{ ft}$$
$$W = ?$$

Basic Equation:

$$W = Fs$$

Working Equation: Same

Substitution:

$$W = (4\bar{0} \text{ lb})(3\bar{0} \text{ ft})$$
$$= 1200 \text{ ft lb}$$

Note: In Example 2 the cart weighs 350 lb but $F = 4\bar{0}$ lb. (Recall that the weight of an object is the measure of its gravitational attraction to the earth and is represented by a vertical vector pointing down to the center of the earth.) There is no vertical motion in the direction of this gravitational force. Therefore, the weight of the box is not the force used to determine the amount of work being done.

Work is being done by the worker pushing the pallet. Exerting a force of $4\bar{0}$ lb results in a displacement in the direction of the applied force. The work done is the product of this force ($4\bar{0}$ lb) and the displacement ($3\bar{0}$ ft) in the direction the force is applied.

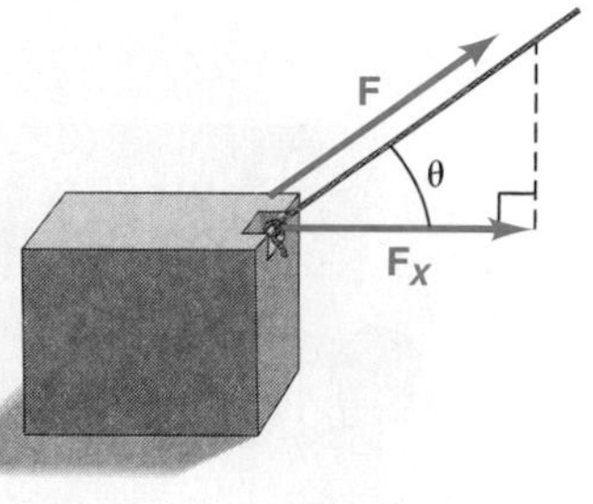

Figure 7.6

Recall that the definition of work states that work is the product of the *force in the direction of the motion* and the displacement. To determine the work when the force is not applied in the direction of the motion, consider a block being pulled by a rope with a force $\mathbf{F}$ that makes an angle θ with level ground as shown in Fig. 7.6. First, draw the horizontal component $\mathbf{F}_x$ and complete the right triangle. Note that $\mathbf{F}_x$ is the force in the direction of the motion. From the right triangle we have

$$\cos\theta = \frac{\text{side adjacent to } \theta}{\text{hypotenuse}} = \frac{|\mathbf{F}_x|}{|\mathbf{F}|}$$

Or,

$$|\mathbf{F}_x| = |\mathbf{F}| \cos\theta$$

That is, when the applied force is not in the direction of the motion, the work done is

$$\boxed{W = Fs \cos\theta}$$

where

W = the work done
F = the applied force
s = the displacement
θ = the angle between the applied force and the direction of the motion

EXAMPLE 3

A person pulls a sled along level ground a distance of 15.0 m by exerting a constant force of 215 N at an angle of 30.0° with the ground (Fig. 7.7). How much work does she do?

Figure 7.7

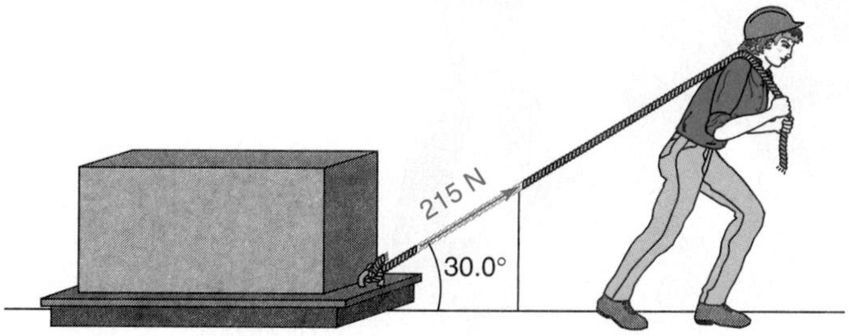

Data:

$$F = 215\text{ N}$$
$$s = 15.0\text{ m}$$
$$\theta = 30.0°$$
$$W = ?$$

Basic Equation:

$$W = Fs\cos\theta$$

Working Equation: Same

Substitution:

$$W = (215\text{ N})(15.0\text{ m})\cos 30.0°$$
$$= 2790\text{ N m}$$
$$= 2790\text{ J} \qquad (1\text{ N m} = 1\text{ J})$$

EXAMPLE 4

Juan and Sonja use a push mower to mow a lawn. Juan, who is taller, pushes at a constant force of 33.1 N on the handle at an angle of 55.0° with the ground. Sonja, who is shorter, pushes at a constant force of 23.2 N on the handle at an angle of 35.0° with the ground. Assume they each push the mower $3\bar{0}00$ m. Who does more work and by how much?

Sketch:

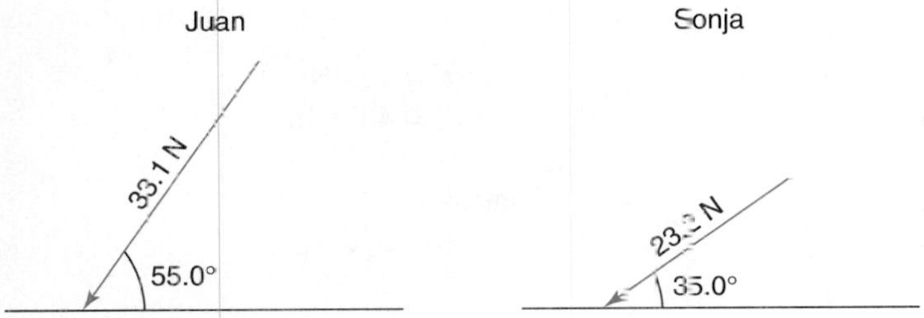

Data:

$F = 33.1\text{ N}$	$F = 23.2\text{ N}$
$s = 3\bar{0}00\text{ m}$	$s = 3\bar{0}00\text{ m}$
$\theta = 55.0°$	$\theta = 35.0°$
$W = ?$	$W = ?$

Basic Equation:

$W = Fs\cos\theta$	$W = Fs\cos\theta$

Working Equation: Same | Same

Substitution:

$W = (33.1\text{ N})(3\bar{0}00\text{ m})\cos 55.0°$	$W = (23.2\text{ N})(3\bar{0}00\text{ m})\cos 35.0°$
$= 57{,}\bar{0}00\text{ N m}$	$= 57{,}\bar{0}00\text{ N m}$
$= 57{,}\bar{0}00\text{ J} \quad (1\text{ N m} = 1\text{ J})$	$= 57{,}\bar{0}00\text{ J}$

They do the same amount of work. However, Juan must exert more energy because he pushes into the ground more than Sonja, who pushes more in the direction of the motion.

EXAMPLE 5

Find the amount of work done in vertically lifting a steel beam with mass 750 kg at uniform speed a distance of 45 m.

Here the force is the weight of the beam.

Data:

$$F = mg$$
$$m = 750 \text{ kg}$$
$$g = 9.80 \text{ m/s}^2$$
$$s = 45 \text{ m}$$
$$W = ?$$

Basic Equation:

$$W = Fs = mgs$$

Working Equation: Same

Substitution:

$$\begin{aligned} W &= (750 \text{ kg})(9.80 \text{ m/s}^2)(45 \text{ m}) \\ &= 3.3 \times 10^5 \text{ (kg m/s}^2\text{)(m)} \\ &= 3.3 \times 10^5 \text{ N m} \qquad (1 \text{ N} = 1 \text{ kg m/s}^2) \\ &= 3.3 \times 10^5 \text{ J} \qquad (1 \text{ J} = 1 \text{ N m}) \end{aligned}$$

Do you see that 330 kJ would also be an acceptable answer?

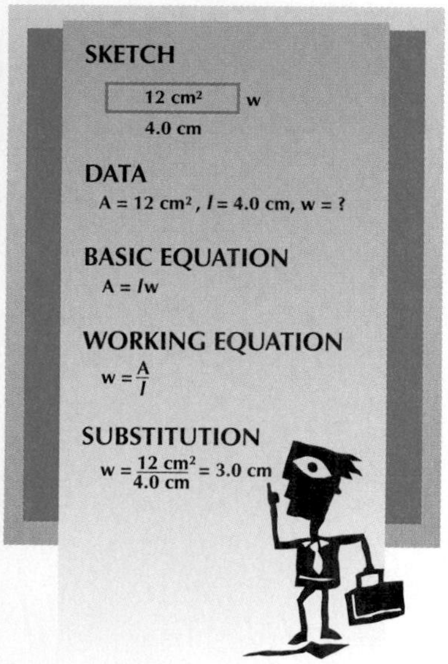

PROBLEMS 7.1

1. Given: $F = 10.0$ N, $s = 3.43$ m, $W = ?$
2. Given: $F = 125$ N, $s = 4875$ m, $W = ?$
3. Given: $F = 1850$ N, $s = 625$ m, $\theta = 37.5°$, $W = ?$
4. Given: $W = 697$ ft lb, $s = 976$ ft, $F = ?$
5. Given: $F = 25{,}700$ N, $s = 238$ m, $W = 5.57 \times 10^6$ J, $\theta = ?$
6. Given: $F = ma$, $m = 16.0$ kg, $a = 9.80 \text{ m/s}^2$, $s = 13.0$ m, $W = ?$
7. How much work is required for a mechanical hoist to lift a $900\bar{0}$-N automobile to a height of 1.80 m for repairs?
8. A hay wagon is used to move bales from the field to the barn. The tractor pulling the wagon exerts a constant force of 350 lb. The distance from field to barn is $\frac{1}{2}$ mi. How much work (ft lb) is done in moving one load of hay to the barn?
9. A worker lifts 75 concrete blocks a distance of 1.50 m to the bed of a truck. Each block has a mass of 4.00 kg. How much work is done to lift all the blocks to the truck bed?
10. The work required to lift eleven 94.0-lb bags of cement from the ground to the back of a truck is 4340 ft lb. What is the distance from the ground to the bed of the truck?
11. How much work is done in lifting 450 lb of cement 75 ft above the ground?
12. How much work is done lifting a $20\bar{0}$-kg wrecking ball 6.50 m above the ground?
13. A gardener pushes a mower a distance of $90\bar{0}$ m in mowing a yard. The handle of the mower makes an angle of 40.0° with the ground. The gardener exerts a force of 35.0 N along the handle of the mower (Fig. 7.8). How much work does the gardener do in mowing the lawn?

Figure 7.8

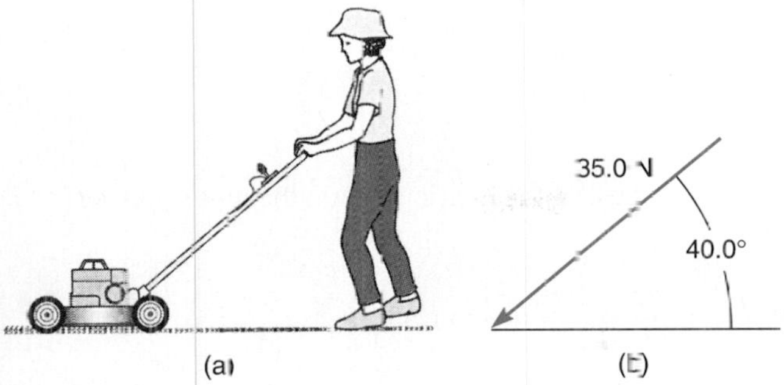

14. The handle of a vegetable wagon makes an angle of 25.0° with the horizontal (Fig. 7.9). If the peddler exerts a force of 35.0 lb along the handle, how much work does the peddler do in pulling the cart 1.00 mi?

Figure 7.9

15. A crate is pulled 675 ft across a warehouse floor by a worker using a rope that makes an angle of 50.0° with the floor. If 375 lb is exerted on the rope, how much work is done in pulling the crate across the floor?
16. A man pulls a sled a distance of 231 m. The rope attached to the sled makes an angle of 30.0° with the ground. The man exerts a force of 775 N on the rope. How much work does the man do in pulling the sled?
17. A tractor tows a barge through a canal with a towrope that makes an angle of 21° with the bank of the canal. If the tension in the rope is 12,000 N, how much work is done in moving the barge 550 m?
18. Two tractors tow a barge through a canal; each tractor uses a towrope that makes an angle of 21° with the bank of the canal. If the tension in each rope is 12,000 N, how much work is done in moving the barge 550 m?
19. Two students push a dune buggy 35.0 m across a lot. The force required is 825 N. How much work is done?
20. After a rain, the force necessary to push the dune buggy in Problem 19 through the mud is doubled. How does the amount of work done by the students change?
21. A delivery person carries a 215-N box up stairs 4.20 m vertically and 6.80 m horizontally.
    (a) How much work does the delivery person do?
    (b) How much work does the delivery person do in carrying the box down the stairs?
22. A crate is pulled by a force of 628 N across the floor by a worker using a rope making an angle of 46.0° with the floor. If the crate is pulled 15.0 m, how much work does the force on the rope do?

7.2 Power

Power is the rate of doing work; that is,

$$P = \frac{W}{t}$$

where

$$P = \text{power}$$
$$W = \text{work}$$
$$t = \text{time}$$

The units of power are familiar to most of us. In the metric system, the unit of power is the *watt:*

$$P = \frac{W}{t} = \frac{Fs}{t} = \frac{\text{N m}}{\text{s}} = \frac{\text{J}}{\text{s}} = \text{watt}$$

Power is often expressed in kilowatts and megawatts.

$$1000 \text{ watts (W)} = 1 \text{ kilowatt (kW)}$$
$$1{,}000{,}000 \text{ watts} = 1 \text{ megawatt (MW)}$$

In the U.S. system, the unit of power is either ft lb/s or horsepower:

$$P = \frac{W}{t} = \frac{Fs}{t} = \frac{\text{ft lb}}{\text{s}}$$

Horsepower (hp) is a unit defined by **James Watt:**

James Watt (1736–1819), *engineer and inventor, was born in Scotland. He made fundamental improvements to steam engines and is credited for several related inventions. The term horsepower was first used by him; the SI unit of power, the watt, is named after him.*

1 horsepower (hp) = 550 ft lb/s = 33,000 ft lb/min

One horsepower is equivalent to moving a force of 550 lb a distance of 1 ft in 1 s or moving 1 lb a distance of 550 ft in 1 s.

Note: Since the above is a definition, treat any conversion factor as an exact number, which does not affect the number of significant digits in any calculation.

TRY THIS ACTIVITY

Human Horsepower

How much equivalent horsepower do you possess? Using a tape measure, a stopwatch, and a flight of stairs, determine the horsepower in your legs when climbing a flight of stairs. First, measure the vertical height of the stairs. Use this distance and your weight to determine the work required to move your body's weight up the stairs. Clock the time it takes to walk and to run up the stairs. Then, find and compare the power. Convert your power to horsepower. How does your running horsepower compare to the horsepower of a lawnmower or a car?

EXAMPLE 1

A freight elevator with operator weighs $50\bar{0}0$ N. If it is raised to a height of 15.0 m in 10.0 s, how much power is developed?

Data:

$$F = 50\bar{0}0 \text{ N}$$
$$s = 15.0 \text{ m}$$
$$t = 10.0 \text{ s}$$
$$P = ?$$

Basic Equations:

$$P = \frac{W}{t} \quad \text{and} \quad W = Fs$$

Working Equation:

$$P = \frac{Fs}{t}$$

Substitution:

$$P = \frac{(50\bar{0}0\text{ N})(15.0\text{ m})}{10.0\text{ s}}$$
$$= 75\bar{0}0\text{ N m/s}$$

EXAMPLE 2

The power expended in lifting an 825-lb girder to the top of a building $10\bar{0}$ ft high is 10.0 hp. How much time is required to raise the girder?

Data:

$$F = 825\text{ lb}$$
$$s = 10\bar{0}\text{ ft}$$
$$P = 10.0\text{ hp}$$
$$t = ?$$

Basic Equations:

$$P = \frac{W}{t} \quad \text{and} \quad W = Fs$$

Working Equation:

$$t = \frac{W}{P} = \frac{Fs}{P}$$

Substitution:

$$t = \frac{(825\text{ lb})(10\bar{0}\text{ ft})}{10.0\text{ hp}}$$
$$= \frac{(825\text{ lb})(10\bar{0}\text{ ft})}{10.0\text{ hp}} \times \frac{1\text{ hp}}{550\,\frac{\text{ft lb}}{\text{s}}}$$
$$= 15.0\text{ s}$$

$$\frac{\text{lb ft}}{\text{hp}} \times \frac{\text{hp}}{\frac{\text{ft lb}}{\text{s}}} = \frac{\text{lb ft}}{\text{hp}} \times \left(\text{hp} \div \frac{\text{ft lb}}{\text{s}}\right) = \frac{\cancel{\text{lb ft}}}{\cancel{\text{hp}}} \times \left(\cancel{\text{hp}} \times \frac{\text{s}}{\cancel{\text{ft lb}}}\right) = \text{s}$$

Note: We use a conversion factor to obtain time units.

EXAMPLE 3

The mass of a large steel wrecking ball is $20\bar{0}0$ kg. What power is used to raise it to a height of 40.0 m if the work is done in 20.0 s?

Data:

$$m = 20\bar{0}0\text{ kg}$$
$$s = 40.0\text{ m}$$
$$t = 20.0\text{ s}$$
$$P = ?$$

Basic Equations:

$$P = \frac{W}{t} \quad \text{and} \quad W = Fs$$

Working Equation:

$$P = \frac{Fs}{t}$$

Substitution: Note that we cannot directly substitute into the working equation because our data are given in terms of *mass* and we must find *force* to substitute in $P = Fs/t$. The force is the weight of the ball:

$$F = mg = (20\bar{0}0 \text{ kg})(9.80 \text{ m/s}^2) = 19{,}600 \text{ kg m/s}^2 = 19{,}600 \text{ N}$$

Then

$$\begin{aligned} P = \frac{Fs}{t} &= \frac{(19{,}600 \text{ N})(40.0 \text{ m})}{20.0 \text{ s}} \\ &= 39{,}200 \text{ N m/s} \\ &= 39{,}200 \text{ W} \quad \text{or} \quad 39.2 \text{ kW} \end{aligned}$$

EXAMPLE 4

A machine is designed to perform a given amount of work in a given amount of time. A second machine does the same amount of work in half the time. Find the power of the second machine compared with the first.

Data (for the second machine given in terms of the first):

$$W = W$$

$$t = \tfrac{1}{2}t = \frac{t}{2}$$

$$P = ?$$

Basic Equation:

$$P = \frac{W}{t}$$

Working Equation: Same

Substitution:

$$P = \frac{W}{\frac{t}{2}} = W \div \frac{t}{2} = W \times \frac{2}{t} = 2\left(\frac{W}{t}\right) = 2P$$

Thus, the power is doubled when the time is halved.

EXAMPLE 5

A motor is capable of developing 10.0 kW of power. How large a mass can it lift 75.0 m in 20.0 s?

Data:

$$P = 10.0 \text{ kW} = 10{,}\bar{0}00 \text{ W}$$

$$s = 75.0 \text{ m}$$

$$t = 20.0 \text{ s}$$

$$F = ?$$

Basic Equations:

$$P = \frac{W}{t} \quad \text{and} \quad W = Fs \quad \text{or} \quad P = \frac{Fs}{t}$$

Working Equation:

$$F = \frac{Pt}{s}$$

Substitution:

$$F = \frac{(10,\bar{0}00 \text{ W})(20.0 \text{ s})}{75.0 \text{ m}}$$

$$= 2670 \frac{\cancel{\text{W}}\ \cancel{\text{s}}}{\cancel{\text{m}}} \times \frac{1 \text{ N } \cancel{\text{m}}/\cancel{\text{s}}}{1\ \cancel{\text{W}}} \qquad (1 \text{ W} = 1 \text{ J/s} = 1 \text{ N m/s})$$

$$= 2670 \text{ N}$$

Next, change the weight to mass as follows:

Data:

$F = 2670$ N

$g = 9.80$ m/s^2

$m = ?$

Basic Equation:

$$F = mg$$

Working Equation:

$$m = \frac{F}{g}$$

Substitution:

$$m = \frac{2670\ \cancel{\text{N}}}{9.80\ \cancel{\text{m/s}^2}} \times \frac{1 \text{ kg } \cancel{\text{m/s}^2}}{1\ \cancel{\text{N}}} \qquad (1 \text{ N} = 1 \text{ kg m/s}^2)$$

$$= 272 \text{ kg}$$

EXAMPLE 6

A pump is needed to lift $15\bar{0}0$ L of water per minute a distance of 45.0 m. What power, in kW, must the pump be able to deliver? (1 L of water has a mass of 1 kg.)

Data:

$$m = 15\bar{0}0 \text{ L} \times \frac{1 \text{ kg}}{1 \text{ L}} = 15\bar{0}0 \text{ kg}$$

$s = 45.0$ m

$t = 1$ min $= 60.0$ s

$g = 9.80$ m/s^2

$P = ?$

Basic Equations:

$$P = \frac{W}{t}, \quad W = Fs, \quad \text{and} \quad F = mg, \quad \text{or} \quad P = \frac{mgs}{t}$$

Working Equation:

$$P = \frac{mgs}{t}$$

Substitution:

$$P = \frac{(1\bar{5}00\text{ kg})(9.80\text{ m/s}^2)(45.0\text{ m})}{60.0\text{ s}}$$

$$= 1.10 \times 10^4\text{ kg m}^2\text{/s} \qquad \left(1\text{ W} = \frac{1\text{ J}}{\text{s}} = \frac{1\text{ N m}}{\text{s}} = \frac{1\text{ (kg m/s}^2\text{)(m)}}{\text{s}} = 1\text{ kg m}^2\text{/s}\right)$$

$$= 1.10 \times 10^4\ \cancel{\text{W}} \times \frac{1\text{ kW}}{10^3\ \cancel{\text{W}}}$$

$$= 11.0\text{ kW}$$

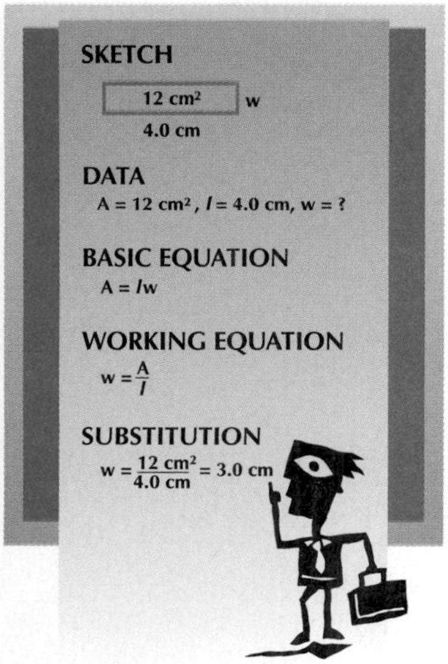

PROBLEMS 7.2

1. Given: $W = 132$ J
   $t = 7.00$ s
   $P = ?$
2. Given: $P = 231$ ft lb/s
   $t = 14.3$ s
   $W = ?$
3. Given: $P = 75.0$ W
   $W = 40.0$ J
   $t = ?$
4. Given: $W = 55.0$ J
   $t = 11.0$ s
   $P = ?$
5. The work required to lift a crate is $31\bar{0}$ J. If the crate is lifted in 25.0 s, what power is developed?
6. When a $36\bar{0}0$-lb automobile runs out of gas, it is pushed by its unhappy driver and a friend a quarter of a mile (0.250 mi). To keep the car rolling, they must exert a constant force of 175 lb.
   (a) How much work do they do?
   (b) If it takes them 15.0 min, how much power do they develop?
   (c) Expressed in horsepower, how much power do they develop?
7. An electric golf cart develops 1.25 kW of power while moving at a constant speed.
   (a) Express its power in horsepower.
   (b) If the cart travels $20\bar{0}$ m in 35.0 s, what force is exerted by the cart?
8. How many seconds would it take a 7.00-hp motor to raise a 475-lb boiler to a platform 38.0 ft high?
9. How long would it take a $95\bar{0}$-W motor to raise a $36\bar{0}$-kg mass to a height of 16.0 m?
10. A $15\bar{0}0$-lb casting is raised 22.0 ft in 2.50 min. Find the required horsepower.
11. What is the rating in kW of a 2.00-hp motor?
12. A wattmeter shows that a motor is drawing $22\bar{0}0$ W. What horsepower is being delivered?
13. A 525-kg steel beam is raised 30.0 m in 25.0 s. How many kilowatts of power are needed?
14. How long would it take a 4.50-kW motor to raise a 175-kg boiler to a platform 15.0 m above the floor?
15. An escalator is needed to carry 75 passengers per minute a vertical distance of 8.0 m. Assume that the mass of each passenger is $7\bar{0}$ kg.
    (a) What is the power (in kW) of the motor needed?
    (b) Express this power in horsepower.
    (c) What is the power (in kW) of the motor needed if 35% of the power is lost to friction and other losses?
16. A pump is needed to lift $75\bar{0}$ L of water per minute a distance of 25.0 m. What power (in kW) must the pump be able to deliver? (1 L of water has a mass of 1 kg.)
17. A machine is designed to perform a given amount of work in a given amount of time. A second machine does twice the same amount of work in half the time. Find the power of the second machine compared with the first.

18. A machine is designed to perform a given amount of work in a given amount of time. A second machine does 2.5 times the same amount of work in one-third the time. Find the power of the second machine compared with the first.
19. A motor on an escalator is capable of developing 12 kW of power.
    (a) How many passengers of mass 75 kg each can it lift a vertical distance of 9.0 m per min, assuming no power loss?
    (b) What power, in kW, motor is needed to move the same number of passengers at the same rate if 45% of the actual power developed by the motor is lost to friction and heat loss?
20. A pump is capable of developing 4.00 kW of power. How many litres of water per minute can be lifted a distance of 35.0 m? (1 L of water has a mass of 1 kg.)
21. A pallet weighing 575 N is lifted a distance of 20.0 m vertically in 10.0 s. What power is developed in kilowatts?
22. An ironworker carries a 7.50-kg toolbag up a vertical ladder on a high-rise building under construction.
    (a) After 30.0 s, he is 8.20 m above his starting point. How much work does the worker do on the toolbag?
    (b) If the worker weighs 645 N, how much work does he do in lifting himself and the toolbag?
    (c) What is the average power developed by the worker?

7.3 Energy

Energy is defined as the ability to do work. There are many forms of energy, such as mechanical, electrical, thermal, fluid, chemical, atomic, and sound.

The mechanical energy of a body or a system is due to its position, its motion, or its internal structure. There are two kinds of mechanical energy: potential energy and kinetic energy. **Potential energy** is the stored energy of a body due to its internal characteristics or its position. **Kinetic energy** is the energy due to the mass and the velocity of a moving object.

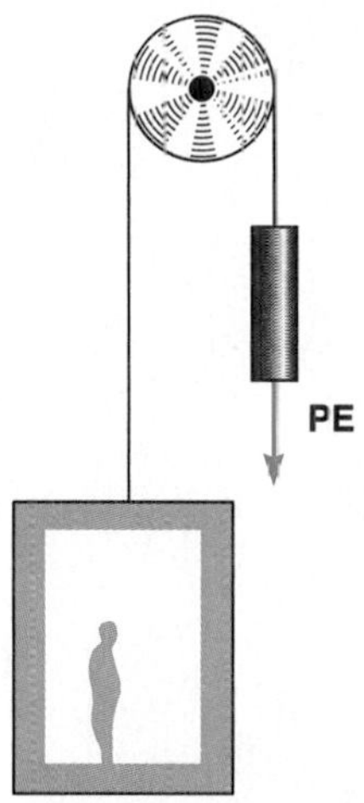

Figure 7.10 Potential energy of position

Internal potential energy is determined by the nature or condition of the substance; for example, gasoline, a compressed spring, or a stretched rubber band has internal potential energy due to its internal characteristics. **Gravitational potential energy** is determined by the position of an object relative to a particular reference level; for example, a rock lying on the edge of a cliff, the raised counterweight on an elevator (Fig. 7.10), or a raised pile driver has potential energy due to its position. Each weight has the ability to do work because of the pull of gravity on it. The unit of energy is the joule (J) in the metric system and the foot-pound (ft lb) in the U.S. system.

The formula for gravitational potential energy is

$$\text{PE} = mgh$$

where

PE = potential energy
m = mass
$g = 9.80\ \text{m/s}^2$ or $32.2\ \text{ft/s}^2$
h = height above reference level

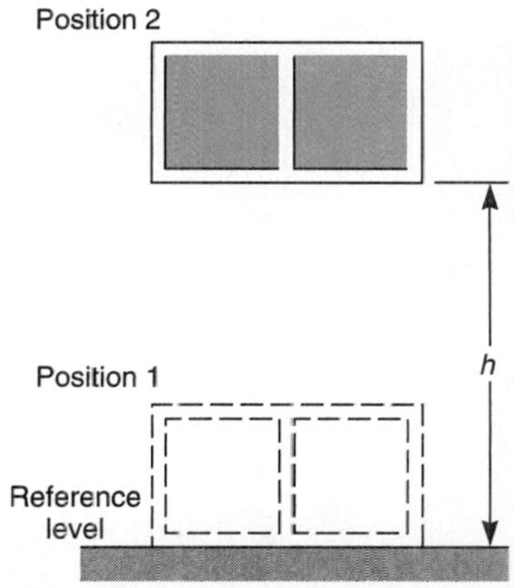

Figure 7.11 Work done in raising the crate gives it potential energy.

In position 1 in Fig. 7.11, the crate is at rest on the floor. It has no ability to do work because it is in its lowest position. To raise the crate to position 2, work must be done to lift it. In the raised position, however, it now has stored ability to do work (by falling to the floor). Its PE (potential energy) can be calculated by multiplying the mass of the crate times acceleration of gravity (g) times height above reference level (h). Note that we can calculate the potential energy of the crate with respect to any level we choose. Here we have chosen the floor as the zero or lowest reference level.

EXAMPLE 1

A wrecking ball of mass $20\bar{0}$ kg is poised 4.00 m above a concrete platform whose top is 2.00 m above the ground.

(a) With respect to the platform, what is the potential energy of the ball?
(b) With respect to the ground, what is the potential energy of the ball?

Sketch:

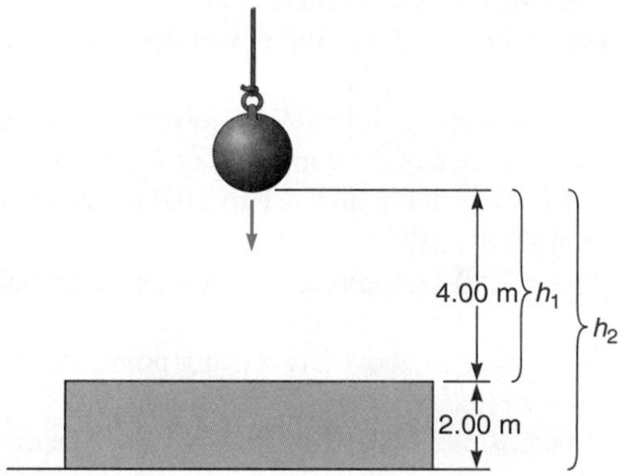

Data:

$m = 20\bar{0}$ kg
$h_1 = 4.00$ m
$h_2 = 6.00$ m
PE = ?

Basic Equation:

$\text{PE} = mgh$

Working Equation: Same

(a) Substitution:

$$\text{PE} = (20\bar{0}\text{ kg})(9.80\text{ m/s}^2)(4.00\text{ m})$$
$$= 7840\,\frac{\text{kg m}^2}{\text{s}^2} \times \frac{1\text{ J}}{\text{kg m}^2/\text{s}^2} \qquad [1\text{ J} = 1\text{ N m} = 1\text{ (kg m/s}^2)(\text{m}) = 1\text{ kg m}^2/\text{s}^2]$$
$$= 7840\text{ J}$$

(b) Substitution:

$$\text{PE} = (20\bar{0}\text{ kg})(9.80\text{ m/s}^2)(6.00\text{ m})$$
$$= 11{,}800\,\frac{\text{kg m}^2}{\text{s}^2} \times \frac{1\text{ J}}{\text{kg m}^2/\text{s}^2}$$
$$= 11{,}800\text{ J}$$

Kinetic energy is due to the mass and the velocity of a moving object and is given by the formula

$$\boxed{\text{KE} = \tfrac{1}{2}mv^2}$$

where

KE = kinetic energy
m = mass of moving object
v = velocity of moving object

A pile driver (Fig. 7.12) shows the relation of energy of motion to useful work. The energy of the driver is its kinetic energy as it hits. When the driver strikes the pile, work is done on the pile, and it is forced into the ground. The depth it goes into the ground is determined by the force applied to it. The force applied is determined by the energy of the driver. If all the kinetic energy of the driver is converted to useful work, then

$$\frac{1}{2}mv^2 = Fs$$

Figure 7.12 Energy of motion becomes useful work in the pile driver.

EXAMPLE 2

A pile driver with mass 10,000 kg strikes a pile with velocity 10.0 m/s.

(a) What is the kinetic energy of the driver as it strikes the pile?
(b) If the pile is driven 20.0 cm into the ground, what force is applied to the pile by the driver as it strikes the pile? Assume that all the kinetic energy of the driver is converted to work.

Sketch:

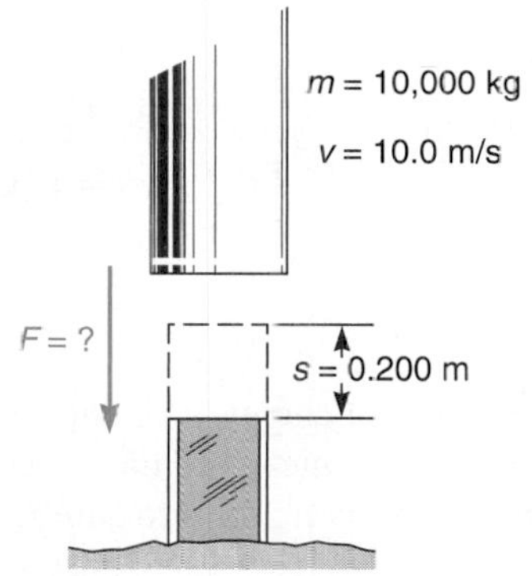

Data:

$m = 1.00 \times 10^4$ kg
$v = 10.0$ m/s
$s = 20.0$ cm $= 0.200$ m
$F = ?$

(a) Basic Equation:

$$\text{KE} = \tfrac{1}{2}mv^2$$

Working Equation: Same

Substitution:

$$\text{KE} = \tfrac{1}{2}(1.00 \times 10^4 \text{ kg})(10.0 \text{ m/s})^2$$

$$= 5.00 \times 10^5 \frac{\cancel{\text{kg}}\,\cancel{\text{m}^2}}{\cancel{\text{s}^2}} \times \frac{1 \text{ J}}{\cancel{\text{kg}}\,\cancel{\text{m}^2/\text{s}^2}} \qquad [1 \text{ J} = 1 \text{ N m} = 1 \text{ (kg m/s}^2)(\text{m}) = 1 \text{ kg m}^2/\text{s}^2]$$

$$= 5.00 \times 10^5 \text{ J} \quad \text{or} \quad 50\bar{0} \text{ kJ}$$

(b) Basic Equation:

$$\text{KE} = W = Fs$$

Working Equation:

$$F = \frac{\text{KE}}{s} \qquad [\text{Use KE from part (a).}]$$

Substitution:

$$F = \frac{5.00 \times 10^5 \cancel{\text{J}}}{0.200 \cancel{\text{m}}} \times \frac{1 \text{ N} \cancel{\text{m}}}{1 \cancel{\text{J}}} \qquad (1 \text{ J} = 1 \text{ N m})$$

$$= 2.50 \times 10^6 \text{ N}$$

EXAMPLE 3

A 60.0-g bullet is fired from a gun with 3150 J of kinetic energy. Find its velocity.

Data:

$$\text{KE} = 3150 \text{ J}$$
$$m = 60.0 \text{ g} = 0.0600 \text{ kg}$$
$$v = ?$$

Basic Equation:

$$\text{KE} = \tfrac{1}{2}mv^2$$

Working Equation:

$$v = \sqrt{\frac{2(\text{KE})}{m}}$$

Substitution:

$$v = \sqrt{\frac{2(3150 \cancel{\text{J}})}{0.0600 \cancel{\text{kg}}} \times \frac{1 \cancel{\text{kg}} \text{ m}^2/\text{s}^2}{1 \cancel{\text{J}}}} \qquad [1 \text{ J} = 1 \text{ N m} = 1 \text{ (kg m/s}^2)(\text{m}) = 1 \text{ kg m}^2/\text{s}^2]$$

$$v = 324 \text{ m/s}$$

SKETCH

12 cm² (w; 4.0 cm)

DATA

$A = 12 \text{ cm}^2, l = 4.0 \text{ cm}, w = ?$

BASIC EQUATION

$A = lw$

WORKING EQUATION

$w = \frac{A}{l}$

SUBSTITUTION

$w = \frac{12 \text{ cm}^2}{4.0 \text{ cm}} = 3.0 \text{ cm}$

We have discussed only two types of energy—kinetic and potential. Keep in mind that energy exists in many other forms—chemical, atomic, electrical, sound, and heat. These forms and the conversion of energy from one form to another will be studied later.

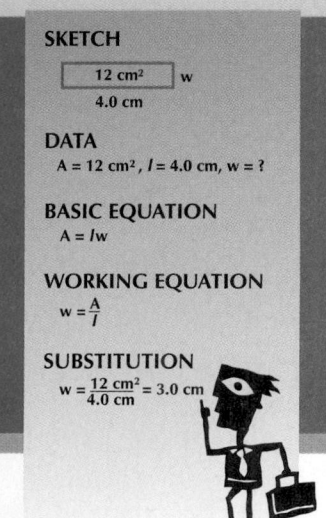

PROBLEMS 7.3

1. Given: $m = 11.4$ kg
   $g = 9.80$ m/s²
   $h = 22.0$ m
   PE = ?

2. Given: $m = 3.50$ kg
   $g = 9.80$ m/s²
   $h = 15.0$ m
   PE = ?

3. Given: $m = 4.70$ kg
$v = 9.60$ m/s
KE = ?

4. Given: PE = 93.6 J
$g = 9.80$ m/s^2
$m = 2.30$ kg
$h = ?$

5. A truck with mass $95\overline{0}$ slugs is driven 55.0 mi/h.
   (a) What is its velocity in ft/s?
   (b) What is its kinetic energy?
6. A bullet with mass 12.0 g travels 415 m/s. Find its kinetic energy (*Hint:* Convert 12.0 g to kg.)
7. A bicycle and rider together have a mass of 7.40 slugs. If the kinetic energy is 742 ft lb, find the velocity.
8. A crate of mass 475 kg is raised to a height 17.0 m above the floor. What potential energy has it acquired with respect to the floor?
9. A tank of water containing $25\overline{0}0$ L of water is stored on the roof of a building.
   (a) Find its potential energy with respect to the floor, which is 12.0 m below the roof.
   (b) Find its potential energy with respect to the basement, which is 4.0 m below the first floor.
10. The potential energy of a girder, after being lifted to the top of a building, is 5.17×10^5 ft lb. If its mass is 173 slugs, how high is the girder?
11. A 30.0-g bullet is fired from a gun and possesses 1750 J of kinetic energy. Find its velocity.
12. The Hoover Dam is 726 ft high. Find the potential energy of 1.00 million ft^3 of water at the top of the dam. (1 ft^3 of water weighs 62.4 lb.)
13. A 250-kg part falls from a plane and hits the ground at 150 km/h. Find its kinetic energy.
14. A meteorite is a solid composed of stone and/or metal material from outer space that passes through the atmosphere and hits the earth's surface. Find the kinetic energy of a meteorite with mass 250 kg that hits the earth at 25 km/s.
15. Water is pumped at $25\overline{0}$ m^3/min from a lake into a tank 65.0 m above the lake.
   (a) What power, in kW, must be delivered by the pump?
   (b) What horsepower rating does this pump motor have?
   (c) What is the increase in potential energy of the water each minute?
16. Oil is pumped at 25.0 m^3/min into a tank 10.0 m above the ground. (1 L of oil has a mass of 0.68 kg.)
   (a) What power, in kW, must be delivered by the pump?
   (b) What is the increase in potential energy of the oil after 10.0 min?
   (c) Find the increase in potential energy of the oil after 10.0 min if the tank is 5.00 m above the ground.
17. If the velocity of an object is doubled, by what factor is its kinetic energy increased?
18. If the kinetic energy of an object is doubled, by what factor is its velocity increased?
19. A 4.20-g slug is shot from a rifle at 965 m/s.
   (a) What is the kinetic energy of the slug?
   (b) How much work is done on the slug if it starts from rest?
   (c) Find the average force on the slug if the work is done over a distance of 0.750 m.
   (d) If the slug comes to rest after penetrating 1.50 cm into metal, what is the magnitude of the average force it exerts?
20. A window washer with mass 90.0 kg first climbs 45.0 m upward to the top of a building, then from the top goes down 85.0 m to the ground.
   (a) What is the potential energy of the window washer at the top of the building, using his initial position as the reference level?
   (b) Find the potential energy of the window washer at ground level with respect to his initial position.
21. A painter weighing $63\overline{0}$ N climbs to a height of 5.00 m on a ladder.
   (a) How much work does he do in climbing the ladder?
   (b) What is the increase in gravitational potential energy of the painter?
   (c) Where does the energy come from to cause this increase in potential energy?

7.4 Conservation of Mechanical Energy

Kinetic and potential energy are related by the **law of conservation of mechanical energy.**

LAW OF CONSERVATION OF MECHANICAL ENERGY

The sum of the kinetic energy and the potential energy in a system is constant if no resistant forces do work.

A pile driver shows this energy conservation. When the driver is at its highest position, the potential energy is maximum and the kinetic energy is zero [Fig. 7.13(a)]. Its potential energy is

$$\text{PE} = mgh$$

and its kinetic energy is

$$\text{KE} = \tfrac{1}{2}mv^2 = \tfrac{1}{2}m(0)^2 = 0$$

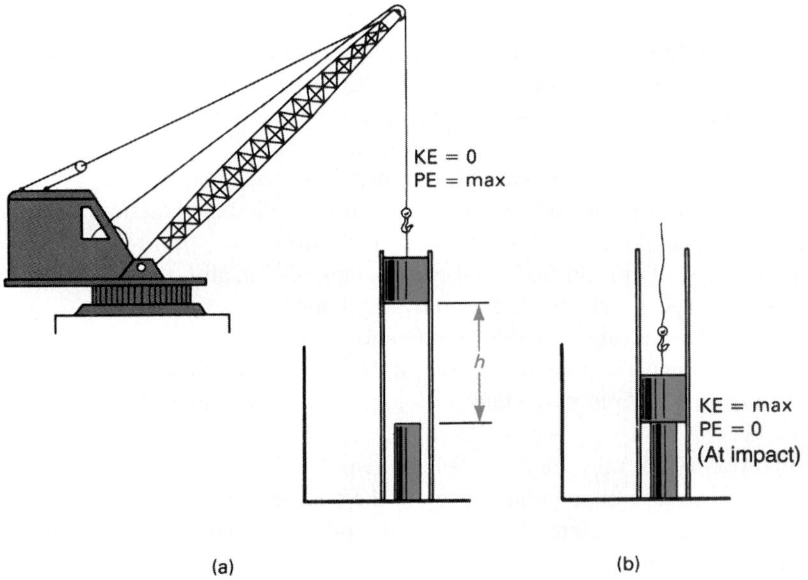

Figure 7.13

When the driver hits the top of the pile [Fig. 7.13(b)], it has its maximum kinetic energy; its potential energy is

$$\text{PE} = mgh = mg(0) = 0$$

Since the total energy in the system must remain constant, the maximum potential energy must equal the maximum kinetic energy:

$$\text{PE}_{\text{max}} = \text{KE}_{\text{max}}$$
$$mgh = \tfrac{1}{2}mv^2$$

Solving for the velocity of the driver as it hits the pile gives

$$v = \sqrt{2gh}$$

when the initial velocity of the driver is zero.

EXAMPLE 1

A pile driver falls freely from a height of 3.50 m above a pile. What is its velocity as it hits the pile?

Data:

$$h = 3.50 \text{ m}$$
$$g = 9.80 \text{ m/s}^2$$
$$v = ?$$

Basic Equation:

$$v = \sqrt{2gh}$$

Working Equation: Same

Substitution:

$$v = \sqrt{2(9.80 \text{ m/s}^2)(3.50 \text{ m})}$$
$$= 8.28 \text{ m/s}$$

$\sqrt{m^2/s^2} = m/s$

The conservation of mechanical energy can also be illustrated by considering a swinging pendulum bob where there is no resistance involved. Pull the bob over to the right side so that the string makes an angle of 65° with the vertical [Fig. 7.14(a)]. At this point, the bob contains its maximum potential energy and its minimum kinetic energy (zero). Note that a larger maximum potential energy is possible when an initial deflection of greater than 65° is made.

An instant later, the bob has lost some of its potential or stored energy, but it has gained in kinetic energy due to its motion [Fig. 7.14(b)]. At the bottom of its arc of swing [Fig. 7.14(c)], its potential energy is zero and its kinetic energy is maximum (its velocity is maximum). The kinetic energy of the bob then causes the bob to swing upward to the left. As it completes its swing [Fig. 7.14(d)], its kinetic energy is decreasing and its potential energy is increasing. That is, its kinetic energy is changing to potential energy.

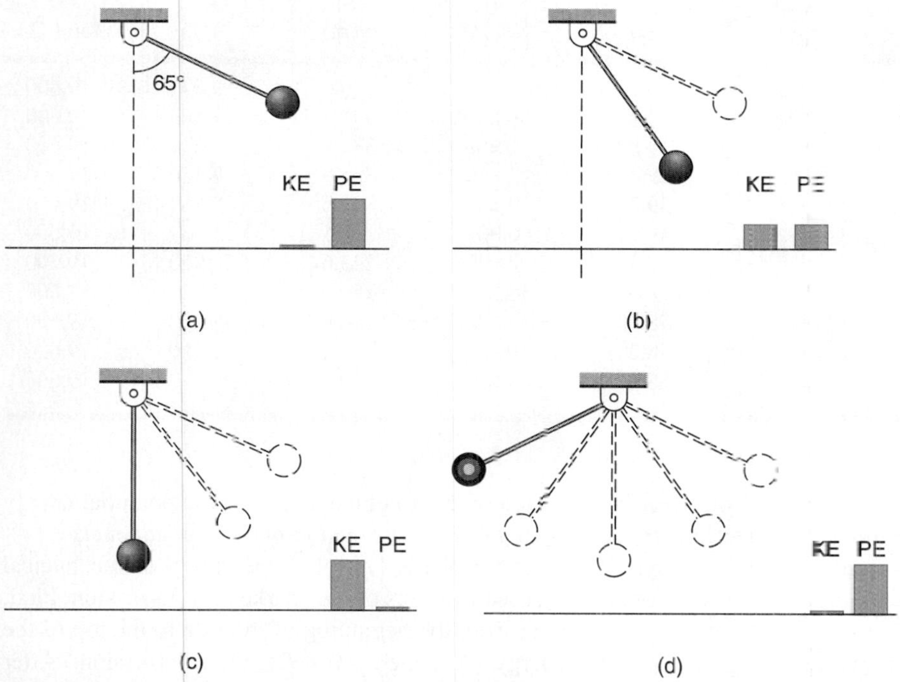

Figure 7.14 Kinetic and potential energy changes in a pendulum

According to the law of conservation of mechanical energy, the sum of the kinetic energy and the potential energy of the bob at any instant is a constant. Assuming no resistant forces, such as friction or air resistance, the bob would swing uniformly "forever."

EXAMPLE 2

Drop a 5.000-kg mass from a hot air balloon 400.0 m above the ground. Find its kinetic energy, its potential energy, and the sum of the kinetic energy and the potential energy in 1-s intervals until the mass hits the ground. (Assume no air resistance.)

Data:

$$m = 5.000 \text{ kg}$$
$$g = 9.80 \text{ m/s}^2$$

In the accompanying table, fill in each column as follows:

In column (1), list the times, t, in 1.000-s increments until the mass hits the ground.

In column (2), use $s = \frac{1}{2}gt^2$ to find the distance, s, the mass has fallen from the balloon at each time, t, rounded to the nearest 0.1 m.

In column (3), use $v = \sqrt{2gh} = \sqrt{2gs}$ to find its velocity at each time, t, rounded to the nearest 0.01 m/s.

In column (4), use $\text{KE} = \frac{1}{2}mv^2$ to find the kinetic energy at each time, t, rounded to the nearest 10 J.

In column (5), use $h = 400.0 \text{ m} - s$ to find the height above the ground at each time, t, rounded to the nearest 0.1 m.

In column (6), use $\text{PE} = mgh$ to find the potential energy at each time, t, rounded to the nearest 10 J.

In column (7), find the sum of the KE and PE columns at each time, t.

(1) t (s)	(2) s (m)	(3) v (m/s)	(4) KE (J)	(5) h (m)	(6) PE (J)	(7) Total (J)
0.000	0.0	0.00	0	400.0	19,6$\bar{0}$0	19,6$\bar{0}$0
1.000	4.9	9.80	240	395.1	19,360	19,6$\bar{0}$0
2.000	19.6	19.60	960	380.4	18,640	19,6$\bar{0}$0
3.000	44.1	29.40	2,160	355.9	17,440	19,6$\bar{0}$0
4.000	78.4	39.20	3,840	321.6	15,760	19,6$\bar{0}$0
5.000	122.5	49.00	6,0$\bar{0}$0	277.5	13,6$\bar{0}$0	19,6$\bar{0}$0
6.000	176.4	58.80	8,640	223.6	10,960	19,6$\bar{0}$0
7.000	240.1	68.60	11,760	159.9	7,840	19,6$\bar{0}$0
8.000	313.6	78.40	15,370	86.4	4,230	19,6$\bar{0}$0
9.000	396.9	88.20	19,450	3.1	150	19,6$\bar{0}$0
9.035	400.0	88.54	19,6$\bar{0}$0	0.0	0	19,6$\bar{0}$0

As you can see from the table, the sum of the kinetic energy and the potential energy at each time, t, is constant according to the *law of conservation of mechanical energy*.

A roller coaster is an excellent example of the law of conservation of mechanical energy. Figure 7.15 shows a roller coaster with various points marked for discussion. First, work is done to take the roller coaster car from the beginning of the ride to the top of the structure (point P) to give it potential energy. (Assume $v = 0$ at P for this discussion.) After

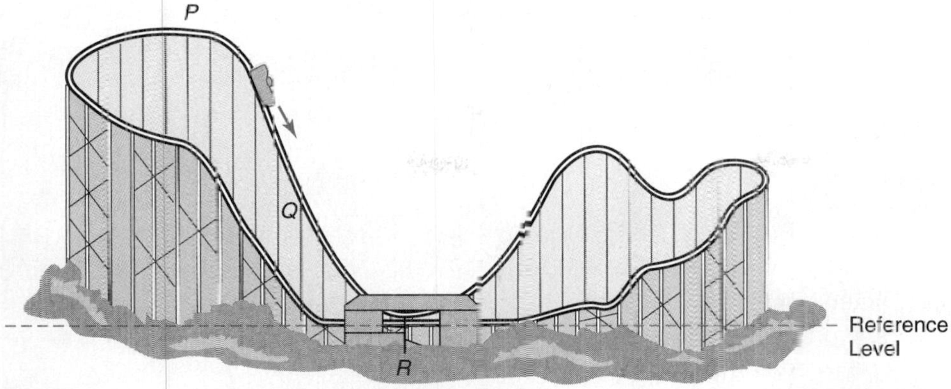

Figure 7.15 Work is done to take the roller coaster car from the beginning of the ride to the top of the structure to give it potential energy. After it leaves the peak of the ride, the sum of its potential energy and its kinetic energy is equal at each point of the ride.

it reaches the peak of the ride, the sum of its potential energy and its kinetic energy is equal at each point of the ride. That is,

Point P			Point Q			Point R	
Kinetic	Potential		Kinetic	Potential		Kinetic	Potential
0 +	mgh_P	=	$\frac{1}{2}mv_Q^2$ +	mgh_Q	=	$\frac{1}{2}mv_R^2$ +	0

In studying collisions (Section 5.2), we learned that whenever objects collide in the absence of any external forces, the total momentum of the objects before the collision equals the total momentum after the collision. This was an example of the law of conservation of momentum.

> In an *elastic* collision, the total kinetic energy in the system *before* the collision equals the total kinetic energy in the system *after* the collision.

That is,

$$\text{total kinetic energy}_{\text{before collision}} = \text{total kinetic energy}_{\text{after collision}}$$

Let's check this principle in Example 1 in Section 5.2.

Kinetic energy before collision:

$$\tfrac{1}{2}m_1v_1^2 + \tfrac{1}{2}m_2v_2^2 = \tfrac{1}{2}(0.600\text{ kg})(9.00\text{ m/s})^2 + \tfrac{1}{2}(0.300\text{ kg})(-8.00\text{ m/s})^2$$
$$= 33.9\text{ kg m}^2/\text{s}^2 = 33.9\text{ N m} = 33.9\text{ J}$$

Kinetic energy after collision:

$$\tfrac{1}{2}m_1(v_1')^2 + \tfrac{1}{2}m_2(v_2')^2 = \tfrac{1}{2}(0.600\text{ kg})(-2.33\text{ m/s})^2 + \tfrac{1}{2}(0.300\text{ kg})(14.66\text{ m/s})^2$$
$$= 33.9\text{ kg m}^2/\text{s}^2 = 33.9\text{ N m} = 33.9\text{ J}$$

SKETCH

12 cm² w
4.0 cm

DATA

$A = 12\text{ cm}^2$, $l = 4.0\text{ cm}$, $w = ?$

BASIC EQUATION

$A = lw$

WORKING EQUATION

$w = \frac{A}{l}$

SUBSTITUTION

$w = \frac{12\text{ cm}^2}{4.0\text{ cm}} = 3.0\text{ cm}$

PROBLEMS 7.4

1. A pile driver falls a distance of 2.50 m before hitting a pile. Find its velocity as it hits the pile.
2. A sky diver jumps out of a plane at a height of 50$\bar{0}$0 ft. If her parachute does not open until she reaches 10$\bar{0}$0 ft, what is her velocity at that point if air resistance is neglected?

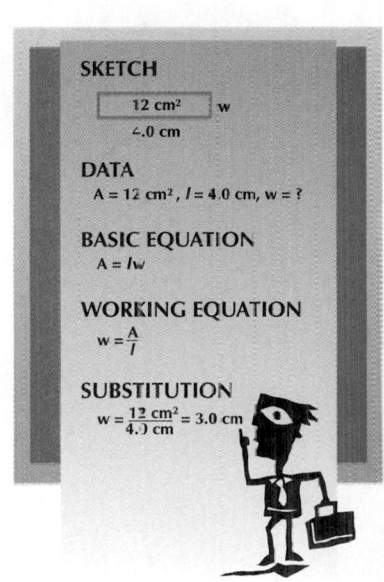

PHYSICS CONNECTIONS

Landing on an Aircraft Carrier

Landing any airplane is a complicated procedure. Furthermore, placing a fast-moving jet on the deck of a ship in the middle of the ocean is one of the most complicated and dangerous tasks for a Navy pilot. Although the length of an aircraft carrier is over 300 m (approximately three football fields in length), its landing strip for the jets is only half that length. Even if the pilot can place the jet on the landing strip, he or she is not able to stop the jet in such a short distance without applying an external force on it.

Upon approaching the carrier, the pilot extends a mechanism called a tail-hook from the rear of the jet. This tail-hook is used to catch one of the four hydraulically controlled arresting cables on the deck of the aircraft carrier. Once the tail-hook catches one of the cables, the hydraulic cabling system extends and applies a force to slow the aircraft (Fig. 7.16). The force from the cable, performed over the distance that the cable extends, works to bring the aircraft to a quick, safe stop. The hydraulic cabling system absorbs the kinetic energy of the moving plane. If the pilot misses all four arresting cables, he or she must increase the throttle to remain safely airborne after leaving the deck of the aircraft carrier and try again.

Figure 7.16 The cable must absorb the kinetic energy of the landing jet.

3. A piece of shattered glass falls from the 82nd floor of a building, $27\bar{0}$ m above the ground. What is the velocity of the glass when it hits the ground, if air resistance is neglected?
4. A 10.0-kg mass is dropped from a hot air balloon at a height of 325 m above the ground. Find its speed at points $30\bar{0}$ m, $20\bar{0}$ m, and $10\bar{0}$ m above the ground and as it hits the ground.
5. A 0.175-lb ball is thrown upward with an initial velocity of 75.0 ft/s. What is the maximum height reached by the ball?
6. A pile driver falls a distance of 1.75 m before hitting a pile. Find its velocity as it hits the pile.
7. A sandbag is dropped from a hot air balloon at a height of 125 m above the ground. Find its velocity as it hits the ground. Ignore air resistance.

8. A piece of broken glass with mass 15.0 kg falls from the side of a building 8.00 m above the street.
   (a) What is the kinetic energy of the glass as it hits the street?
   (b) What is the speed of the glass as it hits the street?
9. A ball is thrown downward from the top of a building at a speed of 75 ft/s. Find its velocity as it hits the ground 475 ft below. Ignore air resistance.
10. Find the maximum height reached by a ball thrown upward at a velocity of 95 ft/s.
11. Drop a 4.000-kg mass from a hot air balloon 300.0 m above the ground. Find its kinetic energy, its potential energy, and the sum of the kinetic energy and the potential energy in 1-s intervals until the mass hits the ground. (Assume no air resistance.)
12. A 2.00-kg projectile is fired vertically upward with an initial velocity of 98.0 m/s. Find its kinetic energy, its potential energy, and the sum of its kinetic and potential energies at each of the following times
   (a) the instant of its being fired
   (b) $t = 1.00$ s
   (c) $t = 2.00$ s
   (d) $t = 5.00$ s
   (e) $t = 10.00$ s
   (f) $t = 12.00$ s
   (g) $t = 15.00$ s
   (h) $t = 20.00$ s

Glossary

Energy The ability to do work. There are many forms of energy, such as mechanical, electrical, thermal, fluid, chemical, atomic, and sound. (p. 197)

Gravitational Potential Energy The energy determined by the position of an object relative to a particular reference level. (p. 197)

Internal Potential Energy The energy determined by the nature or condition of a substance. (p. 197)

Kinetic Energy The energy due to the mass and the velocity of a moving object. (p. 197)

Law of Conservation of Mechanical Energy The sum of the kinetic energy and the potential energy in a system is constant if no resistant forces do work. (p. 202)

Potential Energy The stored energy of a body due to its internal characteristics or its position. (p. 197)

Power The rate of doing work (work divided by time). (p. 191)

Work The product of the force in the direction of motion and the displacement. (p. 186)

Formulas

7.1. $W = Fs$
$W = Fs \cos \theta$

7.2. $P = \dfrac{W}{t}$

7.3. $\text{PE} = mgh$
$\text{KE} = \frac{1}{2}mv^2$

7.4. $v = \sqrt{2gh}$

Review Questions

1. Work is done when
   (a) a force is applied.
   (b) a person tries unsuccessfully to move a crate.
   (c) force is applied and an object is moved.

2. Power
   (a) is work divided by time. (b) is measured in newtons.
   (c) is time divided by work. (d) none of the above.
3. A large boulder at rest possesses
   (a) potential energy. (b) kinetic energy. (c) no energy.
4. A large boulder rolling down a hill possesses
   (a) potential energy. (b) kinetic energy. (c) no energy.
   (d) both kinetic and potential energy.
5. With no air resistance and no friction, a pendulum would
   (a) not swing. (b) swing for a short time. (c) swing forever.
6. Can work be done by a moving object on itself?
7. Has a man swinging a sledgehammer done work if he misses the stake at which he is swinging?
8. Develop the units associated with work from the components of the definition: work = force × displacement.
9. Is work a vector quantity?
10. Is work being done on a boulder by gravity?
11. Is work being done by the weight of a grandfather clock?
12. How could the power developed by a man pushing a stalled car be measured?
13. How does water above a waterfall possess potential energy?
14. What are two devices possessing gravitational potential energy?
15. Is kinetic energy dependent on time?
16. At what point is the kinetic energy of a swinging pendulum bob at a maximum?
17. At what point is the potential energy of a swinging pendulum bob at a maximum?
18. Is either kinetic or potential energy a vector quantity?
19. Can an object possess both kinetic and potential energy at the same time?
20. Why is a person more likely to be severely injured by a bolt falling from the fourth floor of a job site than by one falling from the second floor?

SKETCH

12 cm² w

4.0 cm

DATA

$A = 12\text{ cm}^2,\ l = 4.0\text{ cm},\ w = ?$

BASIC EQUATION

$A = lw$

WORKING EQUATION

$w = \frac{A}{l}$

SUBSTITUTION

$w = \frac{12\text{ cm}^2}{4.0\text{ cm}} = 3.0\text{ cm}$

Review Problems

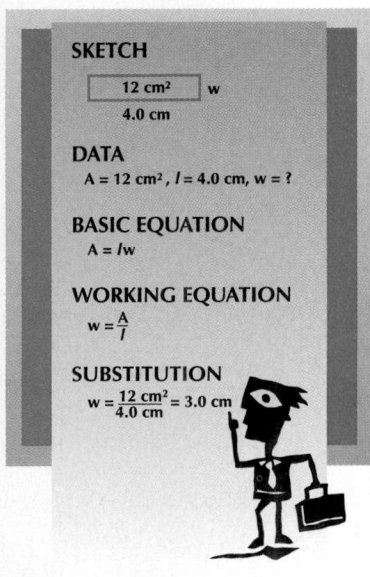

1. How many joules are in one kilowatt-hour?
2. An endloader holds $1\bar{5}00$ kg of sand 2.00 m off the ground for 3.00 min. How much work does it do?
3. How high can a 10.0-kg mass be lifted by $1\bar{0}00$ J of work?
4. A 40.0-kg pack is carried up a $2\bar{5}00$-m-high mountain in 10.0 h. How much work is done?
5. Find the average power output in Problem 4 in (a) watts; (b) horsepower.
6. A 10.0-kg mass has a potential energy of 10.0 J when it is at what height?
7. A 10.0-lb weight has a potential energy of 20.0 ft lb at what height?
8. At what speed does a 1.00-kg mass have a kinetic energy of 1.00 J?
9. At what speed does a 10.0-N weight have a kinetic energy of 1.00 J?
10. What is the kinetic energy of a $3\bar{0}00$-lb automobile moving at 55.0 mi/h?
11. What is the potential energy of an 80.0-kg diver standing 3.00 m above the water?
12. What is the kinetic energy of a 0.020-kg bullet having a velocity of 550 m/s?
13. What is the potential energy of an 85.0-kg high jumper clearing a 2.00-m bar?
14. A worker pulls a crate 10.0 m by exerting a force of $3\bar{0}0$ N.
    (a) How much work does the worker do?
    (b) How much work does the worker do pulling the crate a distance of 10.0 m by exerting the same force at an angle of 20.0° with the horizontal?
15. A hammer falls from a scaffold on a building 50.0 m above the ground. Find its speed as it hits the ground.

APPLIED CONCEPTS

1. Rosita needs to purchase a sump pump for her basement. (a) If the pump must carry 10.0 kg of water to a height of 2.75 m each minute, what minimum wattage pump is needed? (b) What three main factors determine power for a sump pump?
2. As a roller coaster designer, you must carefully balance the desire for excitement and the need for safety. Your most recent design is shown in Fig. 7.17. (a) If a 355-kg roller coaster car has zero velocity on the top of the first hill, determine its potential energy. (b) What is the velocity of the roller coaster car at the specified locations in the design? (c) Explain the relationship between velocity and the position on the track throughout the ride. (Consider the track to be frictionless.)

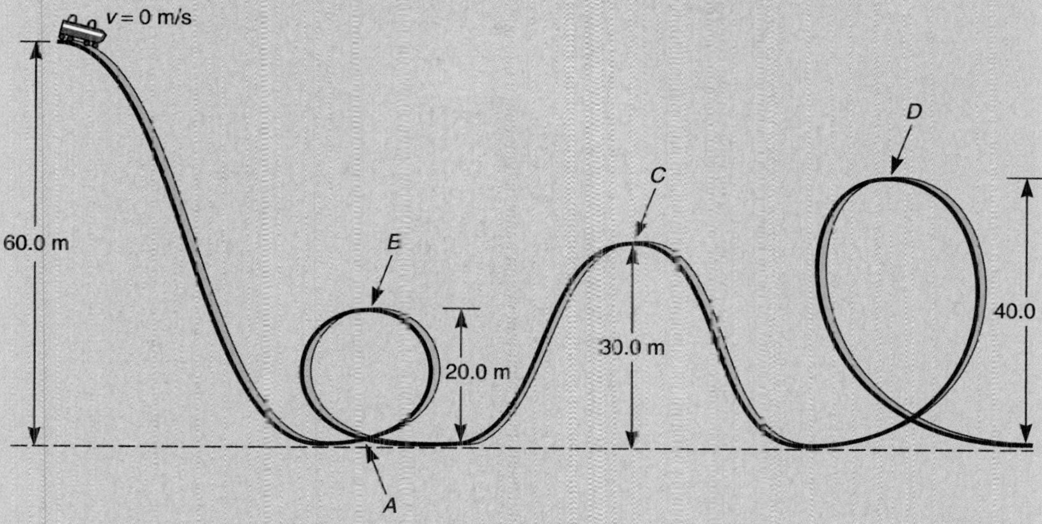

Figure 7.17

3. A 22,500-kg Navy fighter jet flying 235 km/h must catch an arresting cable to land safely on the runway strip of an aircraft carrier. (a) How much energy must the cable absorb to stop the fighter jet? (b) If the cable allows the jet to move 115 m before coming to rest, what is the average force that the cable exerts on the jet? (c) If the jet were given more than 115 m to stop, how would the force applied by the cable change?
4. The hydroelectric plant at the Itaipu Dam, located on the Parana River between Paraguay and Brazil, uses the transfer of potential to kinetic energy of water to generate electricity. (a) If 1.00×10^6 gallons of water (3.79×10^6 kg) flows down 142 m into the turbines each second, how much power does the hydroelectric power plant generate? (For comparison purposes, the Hoover Dam generates 1.57×10^6 W of power.) (b) How much power could the plant produce if the Itaipu Dam were twice its actual height? (c) Explain why the height of a dam is important for hydroelectric power plants.
5. A 1250-kg wrecking ball is lifted to a height of 12.7 m above its resting point. When the wrecking ball is released, it swings toward an abandoned building and makes an indentation of 43.7 cm in the wall. (a) What is the potential energy of the wrecking ball at a height of 12.7 m? (b) What is its kinetic energy as it strikes the wall? (c) If the wrecking ball transfers all of its kinetic energy to the wall, how much force does the wrecking ball apply to the wall? (d) Why should a wrecking ball strike a wall at the lowest point in its swing?

CHAPTER 8

ROTATIONAL MOTION

The concepts of displacement, velocity, acceleration, vectors, and forces in a straight line also apply to motion in a curved path and rotational motion.

Objectives

The major goals of this chapter are to enable you to:

1. Distinguish between rectilinear curvilinear, and rotational motion.
2. Find angular displacement, velocity, and acceleration.
3. Use conservation of angular momentum to describe rotational motion.
4. Find centripetal force.
5. Find power in rotational systems.
6. Analyze how gears, gear trains, and pulleys are used to transfer rotational motion.

8.1 Measurement of Rotational Motion

Until now we have considered only motion in a straight line, called **rectilinear motion.** Technicians are often faced with many problems with motion along a curved path or with objects rotating about an axis. Although these kinds of motion are similar, we must distinguish between them.

Motion along a curved path is called **curvilinear motion.** A satellite in orbit around the earth is an example of curvilinear motion [Fig. 8.1(a)].

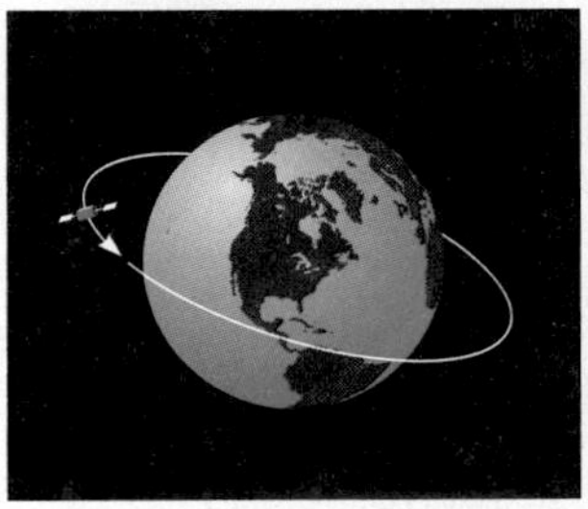

(a) Curvilinear motion of an orbiting satellite

(b) Rotational motion of earth spinning on its axis

(c) Rotational motion of a wheel spinning on its axle

Figure 8.1

Rotational motion occurs when the body itself is spinning. Examples of rotational motion are the earth spinning on its axis, a turning wheel, a turning driveshaft, and the turning shaft of an electric motor [Fig. 8.1(b) and (c)].

We can see a wheel turn, but to gather useful information about its motion, we need a system of measurement. There are three basic systems of defining angle measurement. The **revolution** is one complete rotation of a body. This unit of measurement of rotational motion is the number of rotations—how many times the object goes around. The unit of rotation (most often used in industry) is the revolution (rev). A second system of angular measurement divides the circle of rotation into 360 degrees ($360° = 1$ rev). One **degree** is 1/360 of a complete revolution.

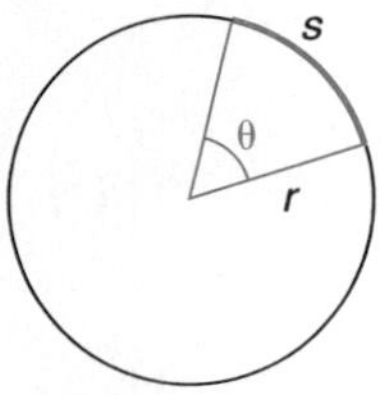

When $s = r$, $\theta = 1$ rad

Figure 8.2

The **radian** (rad), which is approximately 57.3° or exactly $\left(\frac{360}{2\pi}\right)^{\circ}$, is a third angular unit of measurement. A radian is defined as that angle with its vertex at the center of a circle whose sides cut off an arc on the circle equal to its radius (Fig. 8.2), where $s = r$ and θ (Greek lowercase theta, used as a variable for the angle) $= 1$ rad.

Stated as a formula,

$$\theta = \frac{s}{r}$$

where

θ = angle determined by s and r
s = length of the arc of the circle
r = radius of the circle

Technically, angle θ measured in radians is defined as the ratio of two lengths: the lengths of the arc and the radius of a circle. Since the length units in the ratio cancel, the radian is a unitless dimension. As a matter of convenience, "rad" is often used to show radian measurement. A useful relationship is 2π rad equals one revolution. Therefore,

$$1 \text{ rev} = 360° = 2\pi \text{ rad}$$

You may need to use this conversion between systems of measurement.

EXAMPLE 1

Convert the angle 10π rad (a) to rev and (b) to degrees.

Using 1 rev = 360° = 2π rad, form conversion factors so that the old units are in the denominator and the new units are in the numerator.

(a) $\theta = (10\pi \text{ rad})\left(\frac{1 \text{ rev}}{2\pi \text{ rad}}\right)$ (b) $\theta = (10\pi \text{ rad})\left(\frac{360^\circ}{2\pi \text{ rad}}\right)$

$= 5$ rev $= 1800^\circ$

Angular displacement is the angle through which any point on a rotating body moves. Note that on any rotating body, all points on that body move through the same angle in any given amount of time—even though each may travel different linear distances. Point *A* on the flywheel shown in Fig. 8.3 travels much farther than point *B* (along a curved line), but during one revolution both travel through the same angle (have equal angular displacements).

In the automobile industry, technicians are concerned with the *rate* of rotational motion. Recall that in the linear system, velocity is the rate of motion (displacement/time). Similarly, **angular velocity** in the rotational system is the rate of angular displacement. Angular velocity (designated ω, the Greek lowercase letter omega) is usually measured in rev/min (rpm) for relatively slow rotations (e.g., automobile engines) and rev/s or rad/s for high-speed instruments. We use the term *angular velocity* when referring to a vector that includes the direction of rotation. We use the term *angular speed* in referring to a magnitude when the direction of rotation is either not known or not important.

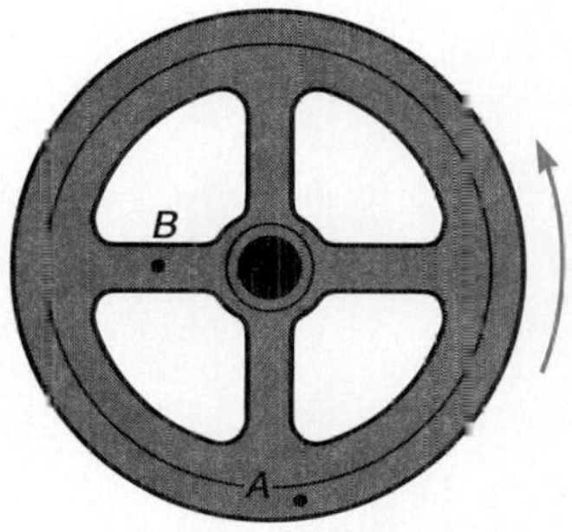

Figure 8.3 The angular displacements of points *A* and *B* on the flywheel are always the same.

$$\omega = \text{angular velocity} = \frac{\text{number of revolutions}}{\text{time}} = \frac{\text{angular displacement}}{\text{time}}$$

Written as a formula,

$$\omega = \frac{\theta}{t}$$

where

ω = angular velocity or speed (rad/s) or ω = angular velocity or speed (rev/min)
θ = angle (in radians) θ = angle (in revolutions)
t = time (in seconds) t = time (in minutes)

EXAMPLE 2

A motorcycle wheel turns $36\bar{0}0$ times while being ridden for 6.40 min. What is the angular speed in rev/min?

Data:

$$t = 6.40 \text{ min}$$
$$\text{number of revolutions} = 36\bar{0}0 \text{ rev}$$
$$\omega = ?$$

Basic Equation:

$$\omega = \frac{\theta}{t}$$

Working Equation: Same

Substitution:

$$\omega = \frac{36\bar{0}0 \text{ rev}}{6.40 \text{ min}}$$
$$= 563 \text{ rev/min or } 563 \text{ rpm}$$

Formulas for linear speed of a rotating point on a circle and angular speed are related as follows. We know

$$(1)\quad \theta = \frac{s}{r} \qquad (2)\quad v = \frac{s}{t} \qquad (3)\quad \omega = \frac{\theta}{t}$$

Therefore, combining and substituting s/r for θ in (3), we obtain

$$\omega = \frac{s/r}{t}$$

$$\omega(r) = \frac{(s/r)(r)}{t} \qquad \text{Multiply both sides by } r.$$

$$\omega r = \frac{s}{t}$$

$$\omega r = v \qquad \text{Recall that } v = s/t.$$

Thus,

$$\boxed{v = \omega r}$$

where

v = linear velocity of a point on the circle
ω = angular speed
r = radius

EXAMPLE 3

A wheel of 1.00 m radius turns at $10\bar{0}0$ rpm.

(a) Express the angular speed in rad/s.
(b) Find the angular displacement in 2.00 s.
(c) Find the linear speed of a point on the rim of the wheel.

Sketch:

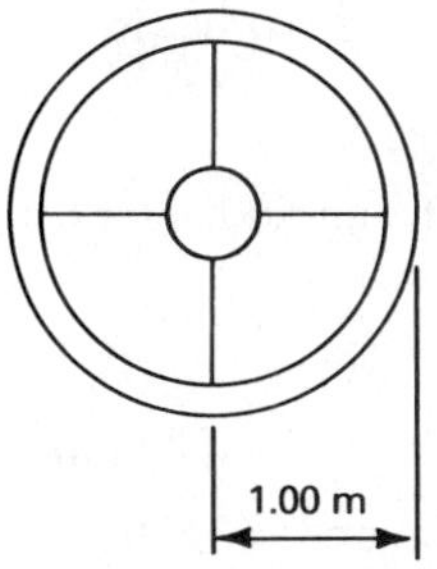

(a) Data:

$$\omega = 10\bar{0}0 \text{ rpm} \qquad \text{(change to rad/s)}$$

$$\omega = 10\bar{0}0 \frac{\cancel{\text{rev}}}{\cancel{\text{min}}} \times \frac{2\pi \text{ rad}}{1 \cancel{\text{rev}}} \times \frac{1 \cancel{\text{min}}}{60 \text{ s}} = 105 \text{ rad/s}$$

(b) Data:

$$\omega = 105 \text{ rad/s}$$
$$t = 2.00 \text{ s}$$
$$\theta = ?$$

Basic Equation:

$$\omega = \frac{\theta}{t}$$

Working Equation:

$$\theta = \omega t$$

Substitution:

$$\theta = (105 \text{ rad/s})(2.00 \text{ s})$$
$$= 21\bar{0} \text{ rad}$$

(c) Data:

$$\omega = 105 \text{ rad/s}$$
$$r = 1.00 \text{ m}$$
$$v = ?$$

Basic Equation:

$$v = \omega r$$

Working Equation: Same

Substitution:

$$v = (105 \text{ rad/s})(1.00 \text{ m})$$
$$= 105 \text{ m/s}$$

(rad/s)(m) = m/s because the rad is a unitless dimension.

A device called a *stroboscope* or strobe light may be used to measure or check the speed of rotation of a shaft or other machinery part. Repeating motion is "slowed down" so it can be observed more easily. The light flashes rapidly, and the rate of flash can be adjusted to coincide with the rotation of a point or points on the rotating object. Knowing the rate of flashing will also then reveal the rate of rotation. A slight variation in the rate of rotation and flash will cause the observed point to appear to move either forward or backward as the stagecoach wheels in old western movies sometimes appear to do. Figure 8.4 shows a stroboscopic linear motion time-lapse photo of a woman running.

Figure 8.4

In linear motion, we found a change in velocity results in an acceleration. Similarly, in rotational motion, changing the rate of rotation involves a change in angular velocity and results in an angular acceleration. For uniformly accelerated rotational motion, **angular acceleration** is the rate of change of angular velocity. That is,

$$\alpha = \frac{\Delta\omega}{t}$$

where

α = angular acceleration
$\Delta\omega$ = change in angular velocity
t = time

The equations for uniformly accelerated linear motion in Section 3.3 may easily be transformed into the corresponding equations for uniformly accelerated rotational motion by substituting θ for s, ω for v, and α for a.

Linear Motion	Rotational Motion
$s = v_{avg}t$	$\theta = \omega_{avg}t$
$s = v_it + \frac{1}{2} a_{avg}t^2$	$\theta = \omega_it + \frac{1}{2} \alpha_{avg}t^2$
$v_{avg} = \frac{v_f + v_i}{2}$	$\omega_{avg} = \frac{\omega_f + \omega_i}{2}$
$v_f = v_i + a_{avg}t$	$\omega_f = \omega_i + \alpha_{avg}t$
$a_{avg} = \frac{v_f - v_i}{t}$	$\alpha_{avg} = \frac{\omega_f - \omega_i}{t}$
$s = \frac{1}{2}(v_f + v_i)t$	$\theta = \frac{1}{2}(\omega_f + \omega_i)t$
$2a_{avg}s = v_f^2 - v_i^2$	$2\alpha_{avg}\theta = \omega_f^2 - \omega_i^2$

where

s = linear displacement
v_f = final linear velocity
v_i = initial linear velocity
v_{avg} = average linear velocity
a_{avg} = average linear acceleration
t = time

where

θ = angular displacement
ω_f = final angular velocity
ω_i = initial angular velocity
ω_{avg} = average angular velocity
α_{avg} = average angular acceleration
t = time

EXAMPLE 4

A rotating pulley 24.0 cm in diameter is rotating at an initial angular speed of 30.5 rad/s. The speed is increased to 41.5 rad/s within 6.30 s. (a) Find the pulley's average angular acceleration. (b) Find the final linear speed of a point on its rim.

(a) Data:

$\omega_i = 30.5$ rad/s
$\omega_f = 41.5$ rad/s
$t = 6.30$ s
$\alpha_{avg} = ?$

Basic Equation:

$$\omega_f = \omega_i + \alpha_{avg}t$$

Working Equation:

$$\alpha_{avg} = \frac{\omega_f - \omega_i}{t}$$

Substitution:

$$\alpha_{avg} = \frac{41.5 \text{ rad/s} - 30.5 \text{ rad/s}}{6.30 \text{ s}}$$
$$= 1.75 \text{ rad/s}^2$$

(b) Data:

$$\omega = 41.5 \text{ rad/s}$$
$$r = 12.0 \text{ cm}$$
$$v = ?$$

Basic Equation:

$$v = \omega r$$

Working Equation: Same

Substitution:

$$v = (41.5 \text{ rad/s})(12.0 \text{ cm})$$
$$= 498 \text{ cm/s}$$

PROBLEMS 8.1

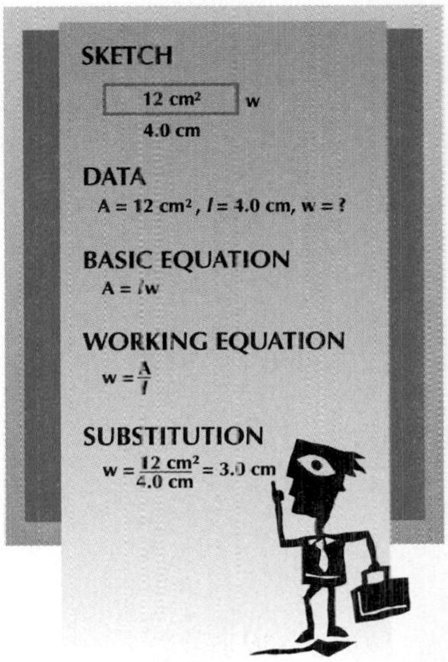

1. Convert $6\frac{1}{2}$ revolutions
   (a) to radians.
   (b) to degrees.
2. Convert 2880°
   (a) to revolutions.
   (b) to radians.
3. Convert 25π rad
   (a) to revolutions.
   (b) to degrees.
4. Convert 12.0 revolutions
   (a) to radians.
   (b) to degrees.

Find the angular speed in Problems 5–10.

5. Number of revolutions = 525
   $t = 3.42$ min
   $\omega =$ _____ rpm
6. Number of revolutions = 7360
   $t = 37.0$ s
   $\omega =$ _____ rev/s
7. Number of revolutions = 4.00
   $t = 3.00$ s
   $\omega =$ _____ rad/s
8. Number of revolutions = 325
   $t = 5.00$ min
   $\omega =$ _____ rpm
9. Number of revolutions = 6370
   $t = 18.0$ s
   $\omega =$ _____ rev/s
10. Number of revolutions = 625
    $t = 5.05$ s
    $\omega =$ _____ rad/s
11. Convert 675 rad/s to rpm.
12. Convert 285 rpm to rad/s.
13. Convert 136 rpm to rad/s.
14. Convert 88.4 rad/s to rpm.
15. A motor turns at a rate of 11.0 rev/s. Find its angular speed in rpm.
16. A rotor turns at a rate of 180 rpm. Find its angular speed in rev/s.
17. A rotating wheel completes one revolution in 0.150 s. Find its angular speed
    (a) in rev/s. (b) in rpm. (c) in rad/s.
18. A rotor completes 50.0 revolutions in 3.25 s. Find its angular speed
    (a) in rev/s. (b) in rpm. (c) in rad/s.
19. A flywheel rotates at 1050 rpm.
    (a) How long (in s) does it take to complete one revolution?
    (b) How many revolutions does it complete in 5.00 s?

20. A wheel rotates at 36.0 rad/s.
    (a) How long (in s) does it take to complete one revolution?
    (b) How many revolutions does it complete in 8.00 s?
21. A shaft of radius 8.50 cm rotates 7.00 rad/s. Find its angular displacement (in rad) in 1.20 s.
22. A wheel of radius 0.240 m turns at 4.00 rev/s. Find its angular displacement (in rev) in 13.0 s.
23. A pendulum of length 1.50 m swings through an arc of 5.0°. Find the length of the arc through which the pendulum swings.
24. An airplane circles an airport twice while 5.00 mi from the control tower. Find the length of the arc through which the plane travels.
25. A wheel of radius 27.0 cm has an angular speed of 47.0 rpm. Find the linear speed (in m/s) of a point on its rim.
26. A belt is placed around a pulley that is 30.0 cm in diameter and rotating at 275 rpm. Find the linear speed (in m/s) of the belt. (Assume no belt slippage on the pulley.)
27. A flywheel of radius 25.0 cm is rotating at 655 rpm.
    (a) Express its angular speed in rad/s.
    (b) Find its angular displacement (in rad) in 3.00 min.
    (c) Find the linear distance traveled (in cm) by a point on the rim in one complete revolution.
    (d) Find the linear distance traveled (in m) by a point on the rim in 3.00 min.
    (e) Find the linear speed (in m/s) of a point on the rim.
28. An airplane propeller with blades 2.00 m long is rotating at 1150 rpm.
    (a) Express its angular speed in rad/s.
    (b) Find its angular displacement in 4.00 s.
    (c) Find the linear speed (in m/s) of a point on the end of the blade.
    (d) Find the linear speed (in m/s) of a point 1.00 m from the end of the blade.
29. An automobile is traveling at 60.0 km/h. Its tires have a radius of 33.0 cm.
    (a) Find the angular speed of the tires (in rad/s).
    (b) Find the angular displacement of the tires in 30.0 s.
    (c) Find the linear distance traveled by a point on the tread in 30.0 s.
    (d) Find the linear distance traveled by the automobile in 30.0 s.
30. Find the angular speed (in rad/s) of the following hands on a clock.
    (a) Second hand (b) Minute hand (c) Hour hand
31. A bicycle wheel of diameter 30.0 in. rotates twice each second. Find the linear velocity of a point on the wheel.
32. A flywheel with radius 1.50 ft has a linear velocity of 30.0 ft/s. Find the time for it to complete 4π rad.
33. The earth rotates on its axis at an angular speed of 1 rev/24 h. Find the linear speed (in km/h)
    (a) of Singapore, which is nearly on the equator.
    (b) of Houston, which is approximately 30.0° north latitude.
    (c) of Minneapolis, which is approximately 45.0° north latitude.
    (d) of Anchorage, which is approximately 60.0° north latitude.
34. A truck tire rotates at an initial angular speed of 21.5 rad/s. The driver accelerates, and after 3.50 s the tire's angular speed is 28.0 rad/s. What is the tire's average angular acceleration during its linear acceleration?
35. Find the angular acceleration of a radiator fan blade as its angular speed increases from 8.50 rad/s to 15.4 rad/s in 5.20 s.
36. A wheel of radius 20.0 cm starts from rest and makes 6.00 revolutions in 2.50 s. (a) Find its average angular velocity in rad/s. (b) Find its average angular acceleration. (c) Find the final linear speed of a point on the rim of the wheel. (d) Find the linear acceleration of a point on the rim of the wheel.
37. A circular disk 30.0 cm in diameter is rotating at 275 rpm and then uniformly stopped within 8.00 s. (a) Find its average angular acceleration. (b) Find the initial linear speed of a point on its rim. (c) How many revolutions does the disk make before it stops?

38. A rotating flywheel of diameter 40.0 cm accelerates from rest to 250 rad/s in 15.0 s. (a) Find its average angular acceleration. (b) Find the linear velocity of a point on the rim of the wheel after 15.0 s. (c) How many revolutions does the wheel make during the 15.0 s?

8.2 Angular Momentum

Recall that in Section 4.1 we saw that *inertia* is the property of a body that causes it to remain at rest if it is at rest or to continue moving with a constant velocity unless an unbalanced force acts upon it. Similarly, *rotational inertia,* called the **moment of inertia,** is the property of a rotating body that causes it to continue to turn until a torque causes it to change its rotational motion. A freely spinning wheel on an upside-down bicycle continues to spin after you stop hand cranking the pedals because of its rotational inertia.

In Section 6.3, we saw that a *torque* is produced when a force is applied to produce a rotation ($\tau = Fs_t$). To increase the speed of the spinning bicycle wheel above, additional torque must be applied by increasing the applied force, F. The angular acceleration of a rotating body is found to be directly proportional to the torque applied to it. This applied torque can be expressed as follows:

$$\tau = I\alpha$$

where

τ = applied torque
I = moment of inertia (rotational inertia)
α = angular acceleration

This is the rotational equivalent of Newton's second law of motion and applies to a rigid body rotating about a fixed axis.

The moment of inertia, I, is a measure of the rotational inertia of a body. The rotational inertia is determined by the mass of the rotating object and how far away that mass is from its axis of rotation. Figure 8.5 shows two cylinders of equal mass, one solid and one hollow. The hollow cylinder has more rotational inertia because its mass is concentrated farther from its axis of rotation. A flywheel is a mechanical device, which is usually a heavy metal rotating wheel attached to a drive shaft with most of its weight concentrated at its circumference. A small motor can slowly increase the speed of a flywheel to store up kinetic energy; then, the small motor uses this inertia to perform a task for which it is ordinarily too small. A flywheel is also used to minimize rotational variations due to fluctuations in load and applied torque. Figure 8.6 shows a flywheel that is used to produce a steady rotation where the applied force of the piston is intermittent in this one-cylinder engine.

In Section 5.1 we saw that (linear) *momentum* is a measure of the amount of inertia and motion an object has or of the difficulty in bringing a moving body to rest. The formula for linear momentum is $p = mv$. There we studied applications involving linear motion and

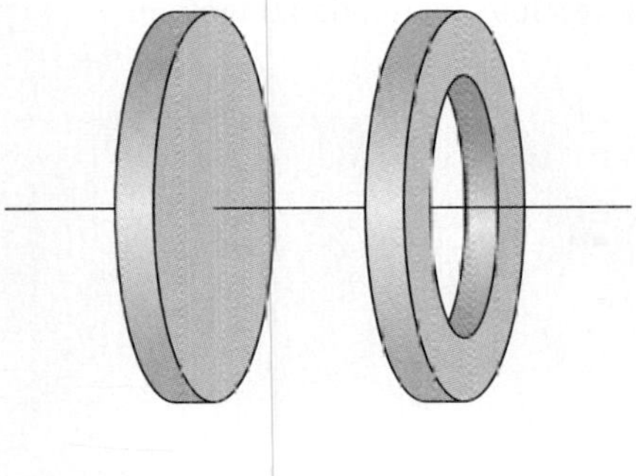

Figure 8.5 Even though the two cylinders are of equal mass, the hollow cylinder has more rotational inertia because its mass is concentrated farther from its axis of rotation.

Figure 8.6 The large flywheel on this 1913 Case 30–60 tractor produces steady rotation of the crankshaft between fuel ignition cycles in its two-cylinder engine.

found the impulse is the change in linear momentum, $Ft = mv_f - mv_i$. Similarly, **angular momentum** for a rotating body about a fixed axis is defined as

$$L = I\omega$$

where

L = angular momentum
I = moment of inertia (rotational inertia)
ω = angular velocity

Note the comparison with linear dynamics:

angular momentum = (moment of inertia) × (angular velocity)
linear momentum = (mass, a measure of inertia) × (linear velocity)

Furthermore, the *angular impulse* is the change in angular momentum.

$$\tau t = I\omega_f - I\omega_i$$

where

τ = torque
t = time
I = moment of inertia
ω_f = final angular velocity
ω_i = initial angular velocity

Compare the following pairs of equations for linear motion and rotational motion:

Linear Motion	Rotational Motion
$F = ma$	$\tau = I\alpha$
$p = mv$	$L = I\omega$
$Ft = mv_f - mv_i$	$\tau t = I\omega_f - I\omega_i$

where

F = applied force
m = mass (inertia)
a = linear acceleration
p = linear momentum
v = linear velocity
t = time
v_f = final linear velocity
v_i = initial linear velocity

where

τ = applied torque
I = moment of inertia
α = angular acceleration
L = angular momentum
ω = angular velocity
t = time
ω_f = final angular velocity
ω_i = initial angular velocity

Conservation of Angular Momentum

In Section 5.1 we learned from the law of conservation of momentum that the total linear momentum ($p = mv$) of a system remains unchanged unless an external force acts on it. Similarly, the **law of conservation of angular momentum** states that the total angular momentum ($L = I\omega$) of a system remains unchanged unless an external torque acts on it.

LAW OF CONSERVATION OF ANGULAR MOMENTUM

The angular momentum of a system remains unchanged unless an external torque acts on it.

A spinning ice skater is an interesting example. When the skater's arms are extended, the rotational inertia, I, is relatively large and the angular velocity, ω, is relatively small. Often at the end of a spin, the skater pulls his or her arms tight to the body resulting in a much faster spin (larger angular velocity) because of a much smaller rotational inertia, I. When a rotating body contracts, its angular velocity, ω, increases; and when a rotating body expands, its angular velocity decreases. This phenomenon is the result of the conservation of angular momentum.

Similarly, gymnasts and divers generate their spins (torque) from a solid base or a diving board after which the angular momentum remains unchanged. The usual somersaults and twists result from making variations in their rotational inertia. Astronauts must also learn to control their spins as they maneuver their bodies to work in space.

TRY THIS ACTIVITY

Spinning Non-Ice Skaters

Ice skaters are able to control their rotational speed by changing their rotational inertia. To do this without a pair of skates, sit in an office chair that rotates freely and hold a couple of weights or heavy books close to your chest. Have someone spin the chair and gradually increase your rotational speed. After reaching your optimum rotational speed, stretch out your arms holding the weights in your hands. What happens to your rotational speed as you move your arms from close to your chest to an outstretched position? Try this a few times. Explain how you can change your rotational speed like ice skaters do.

8.3 Centripetal Force

Newton's laws of motion apply to motion along a curved path as well as in a straight line. Recall that a moving body tends to continue in a straight line because of inertia. If we are to cause the body to move in a circle, we must constantly apply a force perpendicular to

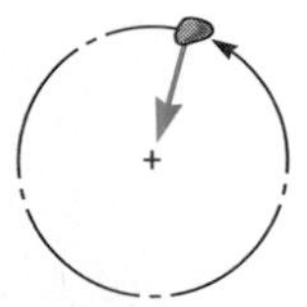

Figure 8.7 Rock on a string being swung in a circle. The centripetal force is directed toward the center.

the line of motion of the body. A simple example is a rock on the end of a string being swung in a circle (Fig. 8.7). By Newton's first law, the rock tends to go in a straight line but the string exerts a constant force on the rock perpendicular to this line of travel. The resulting path of the rock is a circle [Fig. 8.8(a)]. The force of the string on the rock is the *centripetal* (toward the center) force. The **centripetal force** acting on a body in circular motion causes it to move in a circular path. This force is exerted toward the center of the circle. If the string should break, however, there would no longer be a centripetal force acting on the rock and it would fly off tangent to the circle [Fig. 8.8(b)].

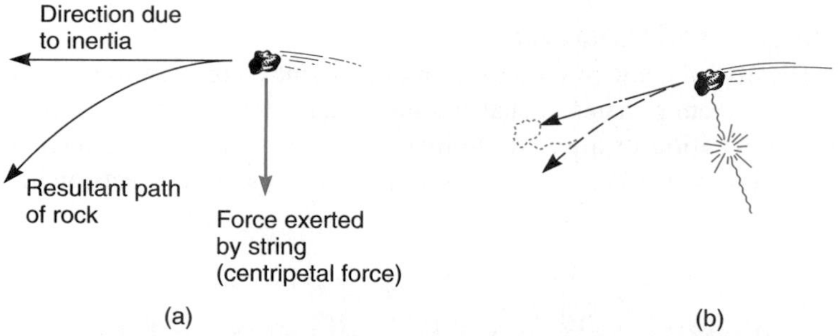

Figure 8.8 Centripetal force on a rock being swung in a circle

The equation for finding the centripetal force on any body moving along a curved path is

$$F = \frac{mv^2}{r}$$

where

F = centripetal force
m = mass of the body
v = velocity of the body
r = radius of curvature of the path of the body

TRY THIS ACTIVITY

Whirling a Bucket of Water

Hold a bucket with some water in your hand and quickly swing it vertically in a circular motion. Why does the water not fall out of the bucket when the bucket is upside down? Gradually decrease the speed of the whirling bucket while measuring the approximate rotational speed of the bucket. Measure the radius of the circular arc from your shoulder to the bottom of the bucket, the mass of the bucket and water, and the rotational speed of the bucket to find the minimum centripetal force needed to keep the water in the bucket. Next, double the amount of water in the bucket and repeat the experiment. How does the amount of water influence the rotational speed needed to keep the water in the bucket?

EXAMPLE

An automobile of mass 1640 kg rounds a curve of radius 25.0 m with a velocity of 15.0 m/s (54.0 km/h). What centripetal force is exerted on the automobile while rounding the curve?

Sketch:

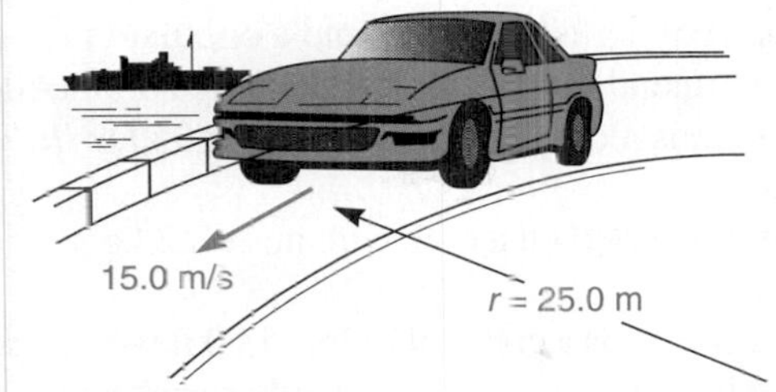

Data:

$$m = 1640 \text{ kg}$$
$$v = 15.0 \text{ m/s}$$
$$r = 25.0 \text{ m}$$
$$F = ?$$

Basic Equation:

$$F = \frac{mv^2}{r}$$

Working Equation: Same

Substitution:

$$F = \frac{(1640 \text{ kg})(15.0 \text{ m/s})^2}{25.0 \text{ m}}$$
$$= 14{,}800 \text{ kg m/s}^2 \qquad (\text{Recall: } 1 \text{ N} = 1 \text{ kg m/s}^2)$$
$$= 14{,}800 \text{ N}$$

PROBLEMS 8.3

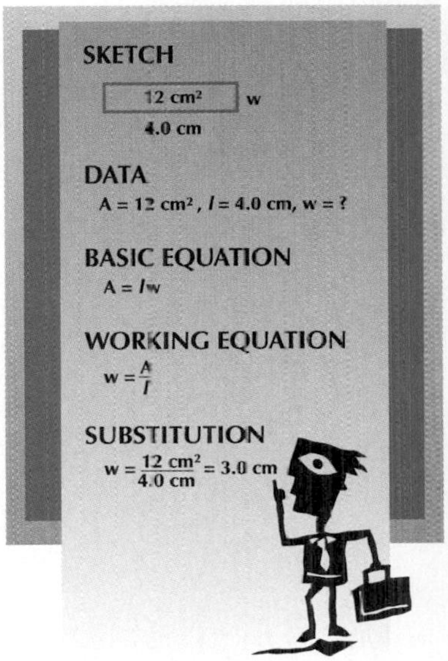

1. Given: $m = 64.0$ kg, $v = 34.0$ m/s, $r = 17.0$ m, $F =$ _____ N
2. Given: $m = 11.3$ slugs, $v = 3.00$ ft/s, $r = 3.24$ ft, $F =$ _____ lb
3. Given: $F = 25\bar{0}0$ lb, $v = 47.6$ ft/s, $r = 72.0$ ft, $m =$ _____ slugs
4. Given: $F = 587$ N, $v = 0.780$ m/s, $m = 67.0$ kg, $r =$ _____ m
5. Given: $F = 602$ N, $m = 63.0$ kg, $r = 3.20$ m, $v =$ _____ m/s
6. Given: $m = 37.5$ kg, $v = 17.0$ m/s, $r = 3.75$ m, $F =$ _____ N
7. Given: $F = 75.0$ N, $v = 1.20$ m/s, $m = 10\bar{0}$ kg, $r =$ _____ m
8. Given: $F = 80.0$ N, $m = 43.0$ kg, $r = 17.5$ m, $v =$ _____ m/s
9. An automobile of mass 117 slugs follows a curve of radius 79.0 ft with a speed of 49.3 ft/s. What centripetal force is exerted on the automobile while it is rounding the curve?
10. Find the centripetal force exerted on a 7.12-kg mass moving at a speed of 2.98 m/s in a circle of radius 2.72 m.
11. The centripetal force on a car of mass $8\bar{0}0$ kg rounding a curve is 6250 N. If its speed is 15.0 m/s, what is the radius of the curve?

12. The centripetal force on a runner is 17.0 lb. If the runner weighs 175 lb and his speed is 14.0 mi/h, find the radius of the curve.
13. An automobile with mass 1650 kg is driven around a circular curve of radius $15\bar{0}$ m at 80.0 km/h. Find the centripetal force of the road on the automobile.
14. A cycle of mass 510 kg rounds a curve of radius $4\bar{0}$ m at 95 km/h. What is the centripetal force on the cycle?
15. What is the centripetal force exerted on a rock with mass 3.2 kg moving at 3.5 m/s in a circle of radius 2.1 m?
16. A truck with mass 215 slugs rounds a curve of radius 53.0 ft with a speed of 62.5 ft/s. (a) What centripetal force is exerted on the truck while rounding the curve? (b) How does the centripetal force change when the velocity is doubled? (c) What is the new force?
17. A 225-kg dirt bike is rounding a curve with linear velocity of 35 m/s and an angular speed of 0.25 rad/s. Find the centripetal force exerted on the bike.

8.4 Power in Rotational Systems

One of the most important aspects of rotational motion to the technician is the power developed. Recall that torque was discussed in Section 6.3. Power, however, must be considered whenever an engine or motor is used to turn a shaft. Some common examples are the use of winches and drive trains (Fig. 8.9).

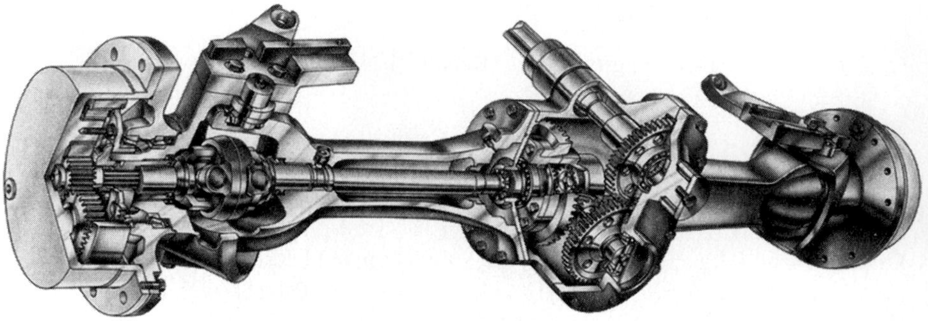

Figure 8.9 This driveshaft connects the engine transmission with the axle to supply power to the wheels and other components of this tractor. (Courtesy of Deere & Company.)

Earlier we learned that

$$\text{power} = \frac{\text{force} \times \text{displacement}}{\text{time}} = \frac{\text{work}}{\text{time}}$$

in the linear system. In the rotational system

$$\begin{aligned} P &= \frac{(\text{torque})(\text{angular displacement})}{\text{time}} \\ &= (\text{torque})(\text{angular velocity}) \\ &= \tau\omega \end{aligned}$$

Recall that angular displacement is the angle through which a shaft is turned. In the metric system, angular displacement must be expressed in radians (1 rev = 2π radians).

Substituting symbols and units, we have, in watts (W):

$$P = \tau\omega$$
$$= (\text{N m})\left(\frac{1}{\text{s}}\right) = \frac{\text{N m}}{\text{s}} = \frac{\text{J}}{\text{s}} = \text{W}$$

To find the power in kilowatts (kW), multiply the number of watts by the conversion factor

$$\frac{1\ \text{kW}}{1000\ \text{W}}$$

Note: In problem solving, the radian unit is a unitless dimension; ω is expressed with the unit/s.

EXAMPLE 1

How many watts of power are developed by a mechanic tightening bolts using 50.0 N m of torque at a rate of 2.50 rad/s? How many kW?

Data:

$$\tau = 50.0\ \text{N m}$$
$$\omega = 2.50/\text{s}$$
$$P = ?$$

Basic Equation:

$$P = \tau\omega$$

Working Equation: Same

Substitution:

$$P = (50.0\ \text{N m})(2.50/\text{s})$$
$$= 125\ \text{N m/s}$$
$$= 125\ \text{W} \qquad (1\ \text{W} = 1\ \text{N m/s})$$

To find the power in kW:

$$125\ \text{W} \times \frac{1\ \text{kW}}{1000\ \text{W}} = 0.125\ \text{kW}$$

In the U.S. system, we measure angular displacement by multiplying the number of revolutions by 2π:

$$\text{angular displacement} = (\text{number of revolutions})(2\pi)$$

For the rotational system

$$\text{power} = \frac{(\text{torque})(2\pi\ \text{revolutions})}{\text{time}}$$

When time is in minutes

$$\text{power} = \text{torque} \times 2\pi \times \frac{\text{rev}}{\text{min}} \times \frac{1\ \text{min}}{60\ \text{s}}$$

$$\boxed{\text{power in } \frac{\text{ft lb}}{\text{s}} = \text{torque in lb ft} \times \frac{\text{number of revolutions}}{\text{min}} \times 0.105 \frac{\text{min}}{\text{rev s}}}$$

$$\underbrace{0.105 \frac{\text{min}}{\text{rev s}}}_{2\pi \times \frac{1\ \text{min}}{60\ \text{s}}}$$

Another common unit of power is the horsepower (hp). The conversion factor between $\frac{\text{ft lb}}{\text{s}}$ and hp is

$$\text{power in hp} = \text{power in } \frac{\text{ft lb}}{\text{s}} \times \frac{\text{hp}}{550 \text{ ft lb/s}}$$

EXAMPLE 2

What power (in ft lb/s) is developed by an electric motor with torque 5.70 lb ft and speed 425 rpm?

Data:

$$\tau = 5.70 \text{ lb ft}$$
$$\omega = 425 \text{ rpm}$$
$$P = ?$$

Basic Equation:

$$P = \text{torque} \times \frac{\text{rev}}{\text{min}} \times 0.105 \frac{\text{min}}{\text{rev s}}$$

Working Equation: Same

Substitution:

$$P = (5.70 \text{ lb ft})\left(425 \frac{\cancel{\text{rev}}}{\cancel{\text{min}}}\right)\left(0.105 \frac{\cancel{\text{min}}}{\cancel{\text{rev}} \text{ s}}\right)$$
$$= 254 \text{ ft lb/s}$$

EXAMPLE 3

What horsepower is developed by a racing engine with torque 545 lb ft at $65\bar{0}0$ rpm?

First, find power in ft lb/s and then convert to hp.

Sketch: None needed

Data:

$$\tau = 545 \text{ lb ft}$$
$$\omega = 65\bar{0}0 \text{ rpm}$$
$$P = ?$$

Basic Equation:

$$P = \text{torque} \times \frac{\text{rev}}{\text{min}} \times 0.105 \frac{\text{min}}{\text{rev s}}$$

Working Equation: Same

Substitution:

$$P = (545 \text{ lb ft})\left(65\bar{0}0 \frac{\cancel{\text{rev}}}{\cancel{\text{min}}}\right)\left(0.105 \frac{\cancel{\text{min}}}{\cancel{\text{rev}} \text{ s}}\right)$$
$$= 372{,}000 \frac{\cancel{\text{ft lb}}}{\cancel{\text{s}}} \times \frac{1 \text{ hp}}{550 \frac{\cancel{\text{ft lb}}}{\cancel{\text{s}}}}$$
$$= 676 \text{ hp}$$

PROBLEMS 8.4

1. Given: $\tau = 125$ lb ft
   $\omega = 555$ rpm
   $P =$ ____ ft lb/s
2. Given: $\tau = 39.4$ N m
   $\omega = 6.70$/s
   $P =$ ____ W
3. Given: $\tau = 372$ lb ft
   $\omega = 264$ rpm
   $P =$ ____ hp
4. Given: $\tau = 65\bar{0}$ N m
   $\omega = 45.0$/s
   $P =$ ____ kW
5. Given: $P = 8950$ W
   $\omega = 4.80$/s
   $\tau =$ ____
6. Given: $P = 650$ W
   $\tau = 540$ N m
   $\omega =$ ____
7. What horsepower is developed by an engine with torque $40\bar{0}$ lb ft at $450\bar{0}$ rpm?
8. What torque must be applied to develop 175 ft lb/s of power in a motor if $\omega = 394$ rpm?
9. Find the angular velocity of a motor developing 649 W of power with torque 131 N m.
10. A high-speed industrial drill develops 0.500 hp at $160\bar{0}$ rpm. What torque is applied to the drill bit?
11. An engine has torque of 550 N m at 8.3 rad/s. What power in watts does it develop?
12. Find the angular velocity of a motor developing 33.0 N m/s of power with a torque of 6.0 N m.
13. What power (in hp) is developed by an engine with torque 524 lb ft
    (a) at $300\bar{0}$ rpm? (b) at $600\bar{0}$ rpm?
14. Find the angular velocity of a motor developing 650 W of power with a torque of 130 N m.
15. A drill develops 0.500 kW of power at $180\bar{0}$ rpm. What torque is applied to the drill bit?
16. What power is developed by an engine with torque $75\bar{0}$ N m applied at $450\bar{0}$ rpm?
17. A tangential force of 150 N is applied to a flywheel of diameter 45 cm to maintain a constant angular velocity of 175 rpm. How much work is done per minute?
18. Find the power developed by an engine with torque 1250 N m applied at $50\bar{0}0$ rpm.
19. Find the angular velocity of a motor developing $10\bar{0}0$ W of power with a torque of $15\bar{0}$ N m.
20. A motor develops 0.75 kW of power at 2000 revolutions per $1\bar{0}$ min. What torque is applied to the motor shaft?
21. What power is developed when a tangential force of 175 N is applied to a flywheel of diameter 86 cm, causing it to have an angular velocity of 36 revolutions per 6.0 s?
22. What power is developed when a tangential force of $25\bar{0}$ N is applied to a wheel 57.0 cm in diameter with an angular velocity of 25.0 revolutions in 13.0 s?
23. An engine develops 1.50 kW of power at $10,\bar{0}00$ revolutions per 5.00 min. What torque is applied to the engine's crankshaft?

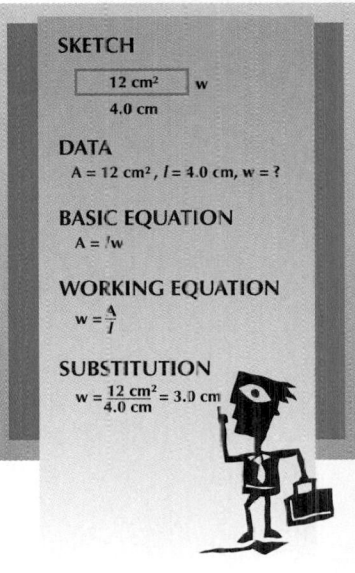

SKETCH

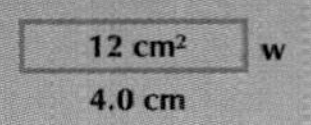

DATA
$A = 12\text{ cm}^2,\ l = 4.0\text{ cm},\ w = ?$

BASIC EQUATION
$A = lw$

WORKING EQUATION
$w = \frac{A}{l}$

SUBSTITUTION
$w = \frac{12\text{ cm}^2}{4.0\text{ cm}} = 3.0\text{ cm}$

8.5 Transferring Rotational Motion

Suppose that two disks are touching each other as in Fig. 8.10. Disk A is driven by a motor and turns disk B (wheel) by making use of the friction between them. The relationship between the diameters of the two disks and their number of revolutions is

$$D \cdot N = d \cdot n$$

where

D = diameter of the driver disk
d = diameter of the driven disk
N = number of revolutions of the driver disk
n = number of revolutions of the driven disk

However, using two disks to transfer rotational motion is not very efficient due to slippage that may occur between them. The most common ways to prevent disk slippage are

Figure 8.10 In the self-propelled lawn mower, disk *A*, driven by the motor, turns disk *B*, the wheel, which results in the mower moving along the ground.

placing the teeth on the edge of the disk and connecting the disks with a belt. Therefore, instead of using disks, we use gears or belt-driven pulleys to transfer this motion. The teeth on the gears eliminate the slippage; the belt connecting the pulleys helps reduce the slippage and provides for distance between rotating centers (Fig. 8.11).

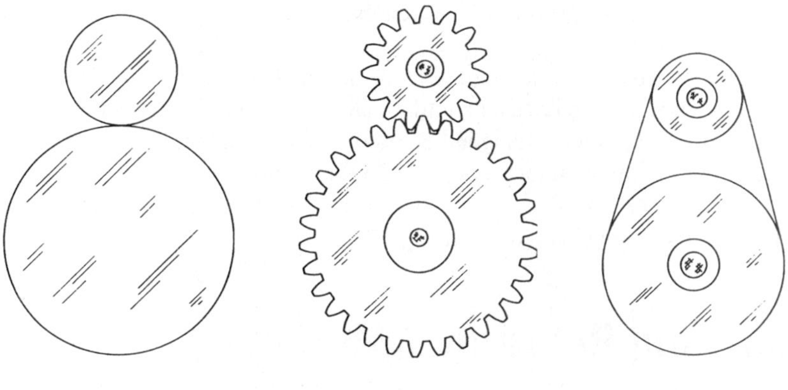

Figure 8.11 Gears and pulleys are used to reduce slippage in transferring rotational motion.

We can change the equation $D \cdot N = d \cdot n$ to the form $D/d = n/N$ by dividing both sides by dN. The left side indicates the ratio of the diameters of the disks. If the ratio is 2, this means that the larger disk must have a diameter two times the diameter of the smaller disk. The same ratio would apply to gears and pulleys. The ratio of the diameters of the gears must be 2 to 1, and the ratio of the diameters of the pulleys must be 2 to 1. In fact, the ratio of the number of teeth on the gears must be 2 to 1.

The right side of the equation indicates the ratio of the number of revolutions of the two disks. If the ratio is 2, this means that the smaller disk makes two revolutions while the larger disk makes one revolution. The same would be true for gears and for pulleys connected by a belt.

Gears and pulleys are used to increase or reduce the angular velocity of a rotating shaft or wheel. When two gears or pulleys are connected, the speed at which each turns compared to the other is inversely proportional to the diameter of that gear or pulley. The larger the diameter of a pulley or gear, the slower it turns. The smaller the diameter of a pulley or gear, the faster it will turn when connected to a larger one.

8.6 Gears

Gears are used to transfer rotational motion from one gear to another. The gear that causes the motion is called the *driver gear.* The gear to which the motion is transferred is called the *driven gear*.

There are many different sizes, shapes, and types of gears. Some examples are shown in Fig. 8.12. For any type of gear, we use one basic formula:

$$\boxed{T \cdot N = t \cdot n}$$

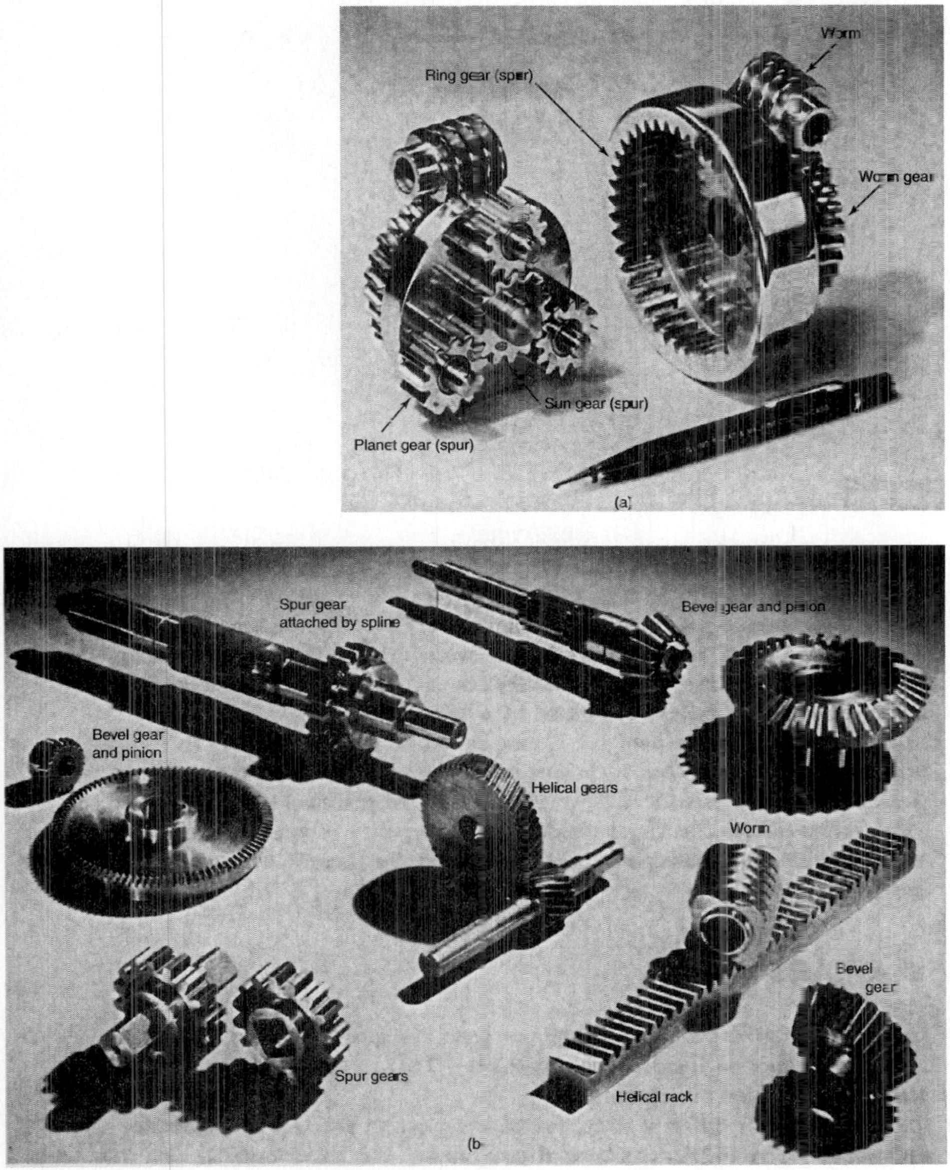

Figure 8.12 Examples of different types of gears. (Courtesy of Foote-Jones/ Illinois Gear, Chicago, IL.)

where

T = number of teeth on the driver gear
N = number of revolutions of the driver gear
t = number of teeth on the driven gear
n = number of revolutions of the driven gear

EXAMPLE 1

A driver gear has 30 teeth. How many revolutions does the driven gear with 20 teeth make while the driver makes one revolution?

Data:

$$T = 30 \text{ teeth} \qquad t = 20 \text{ teeth}$$
$$N = 1 \text{ revolution} \qquad n = ?$$

Basic Equation:

$$T \cdot N = t \cdot n$$

Working Equation:

$$n = \frac{T \cdot N}{t}$$

Substitution:

$$n = \frac{(30 \text{ teeth})(1 \text{ rev})}{20 \text{ teeth}}$$
$$= 1.5 \text{ rev}$$

EXAMPLE 2

A driven gear of 70 teeth makes 63.0 revolutions per minute (rpm). The driver gear makes 90.0 rpm. What is the number of teeth required for the driver gear?

Data:

$$N = 90.0 \text{ rpm}$$
$$t = 70 \text{ teeth}$$
$$n = 63.0 \text{ rpm}$$
$$T = ?$$

Basic Equation:

$$T \cdot N = t \cdot n$$

Working Equation:

$$T = \frac{t \cdot n}{N}$$

Substitution:

$$T = \frac{(70 \text{ teeth})(63.0 \text{ rpm})}{90.0 \text{ rpm}}$$
$$= 49 \text{ teeth}$$

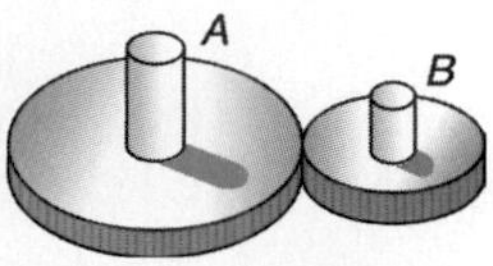

Figure 8.13 Meshed gears shown as cylinders for simplicity

Gear Trains

When two gears mesh (Fig. 8.13),* they turn in opposite directions. If gear A turns clockwise, gear B turns counterclockwise. If gear A turns counterclockwise, gear B turns clockwise. If a

*Although gears have teeth, in technical work they are often shown as cylinders.

third gear is inserted between the two (Fig. 8.14), then gears A and B are rotating in the same direction. This third gear is called an *idler*. A **gear train** is a series of gears that transfers rotational motion from one gear to another.

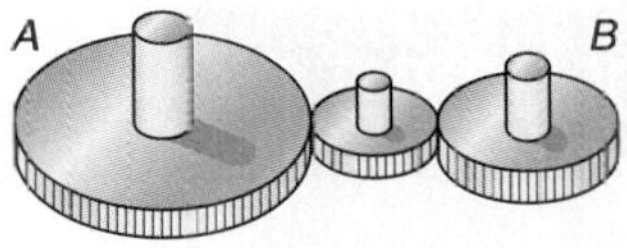

Figure 8.14 Gear train of three gears

When the number of shafts in a gear train is odd (such as 1, 3, 5, . . .), the first gear and the last gear rotate in the same direction. When the number of shafts is even, the gears rotate in opposite directions.

When a complex gear train is considered, the relationship between revolutions and number of teeth is still present. This relationship is: The number of revolutions of the first driver times the product of the number of teeth of all the driver gears equals the number of revolutions of the final driven gear times the product of the number of teeth on all the driven gears. That is,

$$NT_1T_2T_3T_4\cdots = nt_1t_2t_3t_4\cdots$$

where
N = number of revolutions of first driver gear
T_1 = teeth on first driver gear
T_2 = teeth on second driver gear
T_3 = teeth on third driver gear
T_4 = teeth on fourth driver gear
n = number of revolutions of last driven gear
t_1 = teeth on first driven gear
t_2 = teeth on second driven gear
t_3 = teeth on third driven gear
t_4 = teeth on fourth driven gear

EXAMPLE 3

Determine the relative motion of gears A and B in Fig. 8.15.

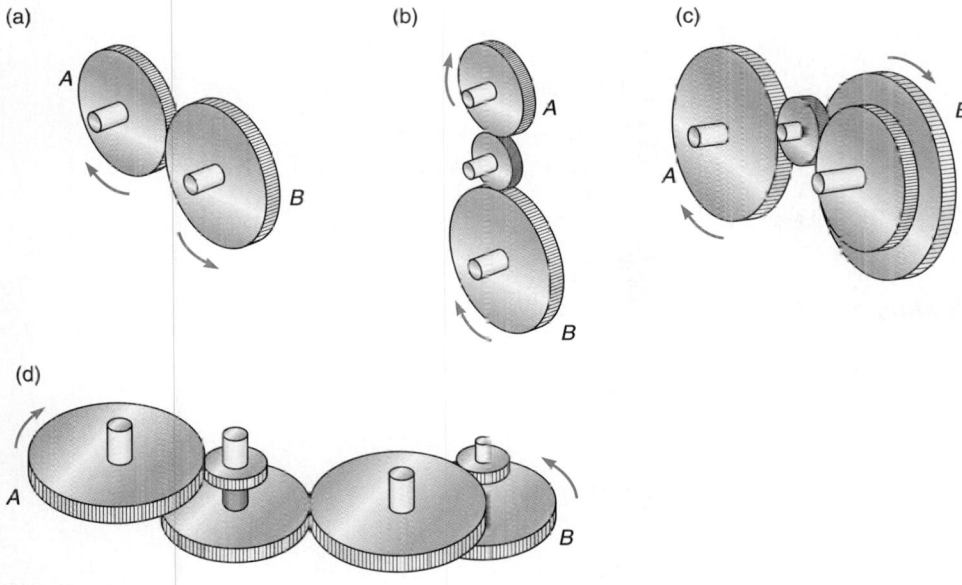

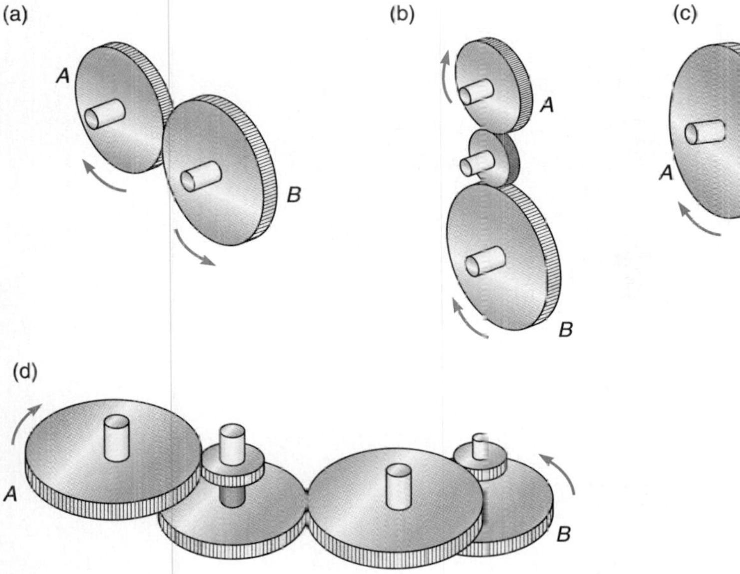

Figure 8.15

EXAMPLE 4

Find the number of revolutions per minute of gear D in Fig. 8.16 if gear A rotates at 20.0 rpm. Gears A and C are drivers and gears B and D are driven.

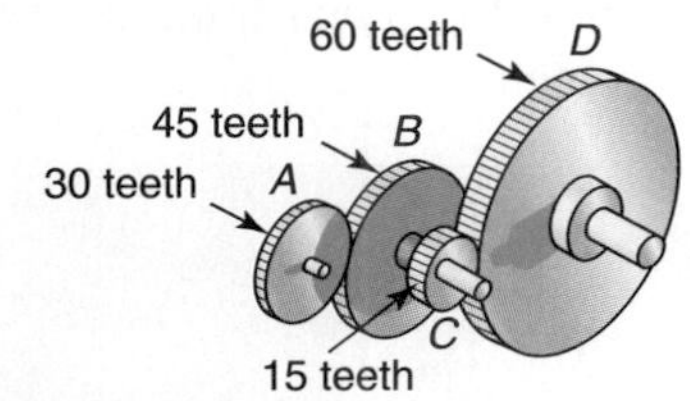

Figure 8.16

Data:

$$N = 20.0 \text{ rpm} \qquad t_1 = 45 \text{ teeth}$$
$$T_1 = 30 \text{ teeth} \qquad t_2 = 60 \text{ teeth}$$
$$T_2 = 15 \text{ teeth} \qquad n = ?$$

Basic Equation:

$$NT_1T_2 = nt_1t_2$$

Working Equation:

$$n = \frac{NT_1T_2}{t_1t_2}$$

Substitution:

$$n = \frac{(20.0 \text{ rpm})(30 \text{ teeth})(15 \text{ teeth})}{(45 \text{ teeth})(60 \text{ teeth})}$$
$$= 3.33 \text{ rpm}$$

EXAMPLE 5

Find the rpm of gear D in the train shown in Fig. 8.17. Gears A and C are drivers and gears B and D are driven.

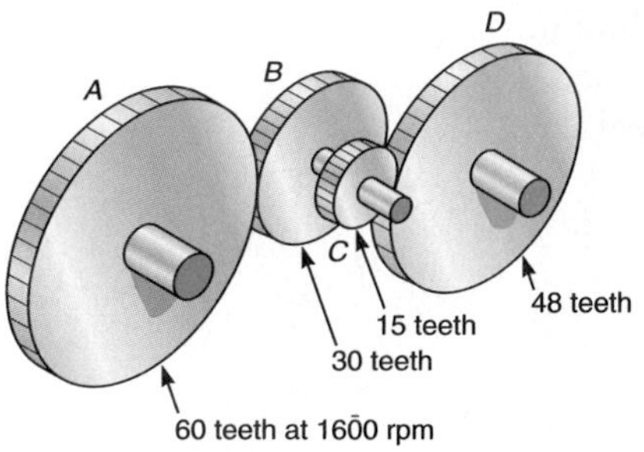

Figure 8.17

Data:

$$N = 16\bar{0}0 \text{ rpm} \qquad t_1 = 30 \text{ teeth}$$
$$T_1 = 60 \text{ teeth} \qquad t_2 = 48 \text{ teeth}$$
$$T_2 = 15 \text{ teeth} \qquad n = ?$$

Basic Equation:

$$NT_1T_2 = nt_1t_2$$

Working Equation:

$$n = \frac{NT_1T_2}{t_1t_2}$$

Substitution:

$$n = \frac{(16\bar{0}0 \text{ rpm})(60 \text{ teeth})(15 \text{ teeth})}{(30 \text{ teeth})(48 \text{ teeth})}$$
$$= 10\bar{0}0 \text{ rpm}$$

EXAMPLE 6

In the gear train shown in Fig. 8.18, find the speed in rpm of gear A.

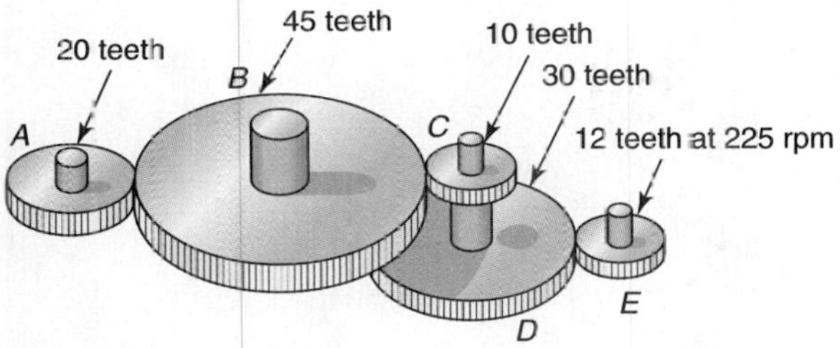

Figure 8.18

Data:

$$t_1 = 45 \text{ teeth} \qquad T_1 = 20 \text{ teeth}$$
$$t_2 = 10 \text{ teeth} \qquad T_2 = 45 \text{ teeth}$$
$$t_3 = 12 \text{ teeth} \qquad T_3 = 30 \text{ teeth}$$
$$n = 225 \text{ rpm} \qquad N = ?$$

Gear B is both a driver and a driven gear.

Basic Equation:

$$NT_1T_2T_3 = nt_1t_2t_3$$

Working Equation:

$$N = \frac{nt_1t_2t_3}{T_1T_2T_3}$$

Substitution:

$$N = \frac{(225 \text{ rpm})(45 \text{ teeth})(10 \text{ teeth})(12 \text{ teeth})}{(20 \text{ teeth})(45 \text{ teeth})(30 \text{ teeth})}$$
$$= 45.0 \text{ rpm}$$

In a gear train, when a gear is both a driver gear and a driven gear, it may be omitted from the computation.

EXAMPLE 7

The problem in Example 6 could have been worked as follows because gear B is both a driver and a driven.

Basic Equation:

$$NT_1T_3 = nt_2t_3$$

Working Equation:

$$N = \frac{nt_2t_3}{T_1T_3}$$

Substitution:

$$N = \frac{(225 \text{ rpm})(10 \text{ teeth})(12 \text{ teeth})}{(20 \text{ teeth})(30 \text{ teeth})}$$
$$= 45.0 \text{ rpm}$$

PHYSICS CONNECTIONS

Bicycle Gears

The gearing system on a bicycle allows a cyclist to choose how much force he or she would like to exert when riding a bicycle. What gear ratio should be used when riding uphill, downhill, or on level ground? Prior to the use of gears on bicycles, a rider needed to sit directly above the front wheel in order to pedal. Gears, chains, and other advances have made the modern bicycle much more comfortable and efficient.

To find the gear ratio of a bicycle's gearing system, divide the number of teeth on the rear, driven gear by the number of teeth on the front, driver gear. When the number of teeth on the front gear is larger, the gear ratio is less than one and the rear wheel turns faster than the pedals. This results in high speeds for going down slight inclines or for level ground and allows the cyclist to pedal fewer revolutions while traveling a greater distance. When the number of teeth in the rear gear is larger, the gear ratio is greater than one. Here, the pedals turn faster than the rear wheel. This results in low speeds for going up large hills and allows the cyclist to pedal more revolutions while traveling a shorter distance but exerting a more manageable leg force. The following table illustrates the differences:

Number of Teeth on Front Gear (Driver)	Number of Teeth on Rear Gear (Driven)	Gear Ratio (Driven/Driver)	Number of Teeth on Front Gear (Driver)	Number of Teeth on Rear Gear (Driven)	Gear Ratio (Driven/Driver)
44	11	1/4	15	30	2/1
When the front gear turns once, the back gear turns four times (good for traveling at high speeds).			When the front gear turns once, the back gear turns only ½ a rotation (good for reducing the force needed to pedal while going uphill).		

Most beginner bicycles simply connect the chain on the front gear to the rear gear with no option for changing the gear ratios. Children must stand up to pedal with more force. Geared bicycles, including road and mountain bikes, are engineered with a variety of gear ratios to allow a cyclist to travel more easily over many terrains (Fig. 8.19).

Figure 8.19 (a) Beginner bicycle (b) Mountain bike

PROBLEMS 8.6

Fill in the blanks.

	Number of Teeth		*rpm*	
	Driver	*Driven*	*Driver*	*Driven*
1.	16	48	156	_____
2.	36	24	_____	225
3.	18	_____	72.0	54.0
4.	_____	64	148	55.5
5.	48	36	_____	276
6.	16	12	144	_____

7. A driver gear has 36 teeth and makes 85.0 rpm. Find the rpm of the driven gear with 72 teeth.
8. A motor turning at 1250 rpm is fitted with a gear having 54 teeth. Find the speed of the driven gear if it has 45 teeth.
9. A gear running at $25\bar{0}$ rpm meshes with another revolving at $10\bar{0}$ rpm. If the smaller gear has 30 teeth, how many teeth does the larger gear have?
10. A driver gear with 40 teeth makes 154 rpm. How many teeth must the driven gear have if it makes $22\bar{0}$ rpm?
11. Two gears have a speed ratio of 4.2 to 1. If the smaller gear has 15 teeth, how many teeth does the larger gear have?
12. What size gear should be meshed with a 15-tooth pinion to achieve a speed reduction of 10 to 3?
13. A driver gear has 72 teeth and makes 162 rpm. Find the rpm of the driven gear with 81 teeth.
14. A driver gear with 60 teeth makes 1600 rpm. How many teeth must the driven gear have if it makes 480 rpm?
15. What size gear should be meshed with a 20-tooth pinion to achieve a speed reduction of 3 to 1?
16. A motor turning at $15\bar{0}0$ rpm is fitted with a gear having 60 teeth. Find the speed of the driven gear if it has 40 teeth
17. The larger of two gears in a clock has 36 teeth and turns at a rate of 0.50 rpm. How many teeth does the smaller gear have if it rotates at 1/30 rev/s?
18. How many revolutions does an 88-tooth gear make in 10.0 min when it is meshed with a 22-tooth pinion rotating at 44 rpm?

If gear *A* turns in a clockwise motion, determine the motion of gear *B* in each gear train.

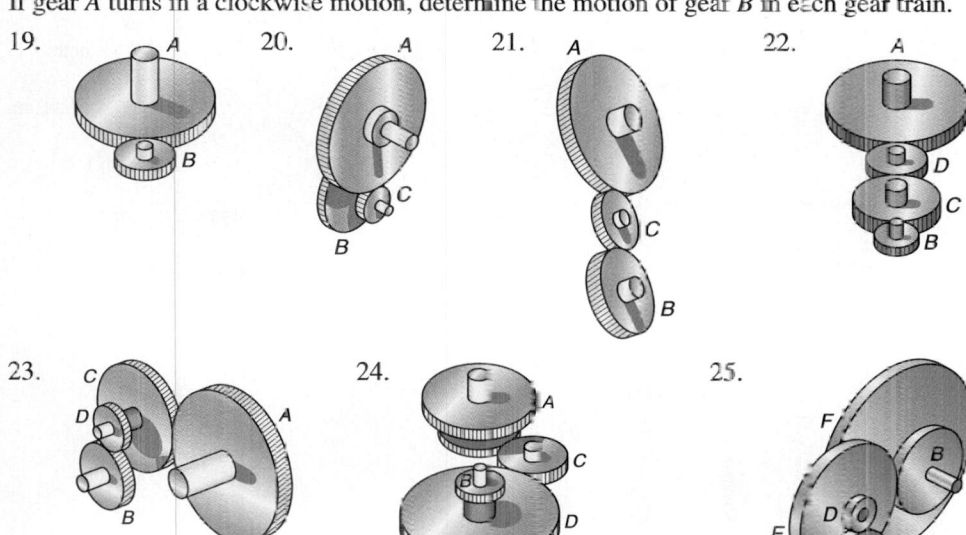

19.
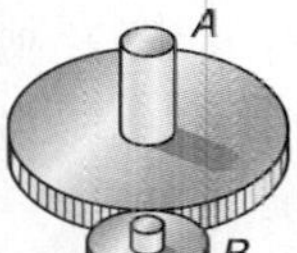

20.
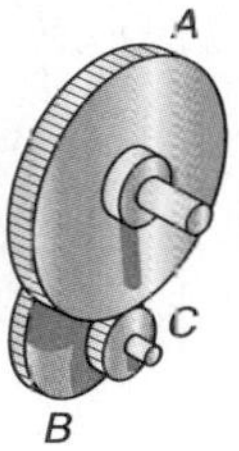

21.
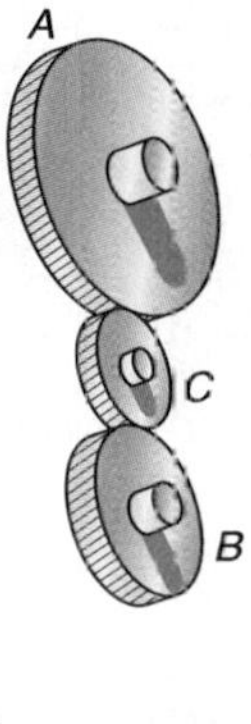

22.
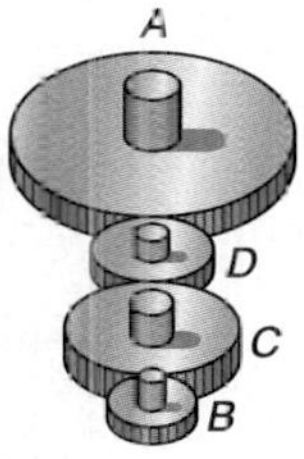

23.
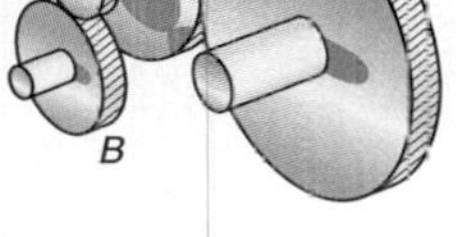

24.
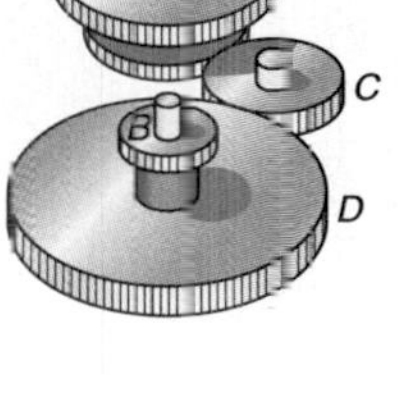

25.
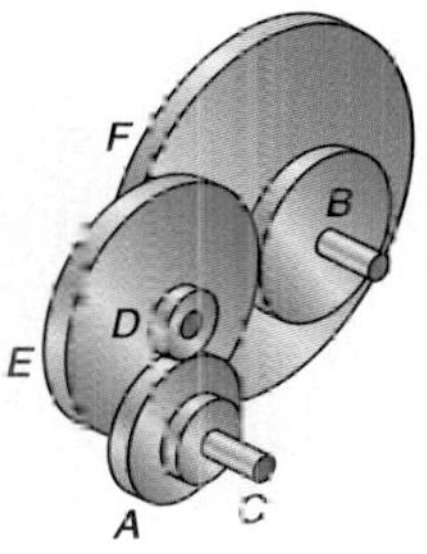

SKETCH

12 cm² w
4.0 cm

DATA

$A = 12\ cm^2$, $l = 4.0$ cm, $w = ?$

BASIC EQUATION

$A = lw$

WORKING EQUATION

$w = \frac{A}{l}$

SUBSTITUTION

$w = \frac{12\ cm^2}{4.0\ cm} = 3.0\ cm$

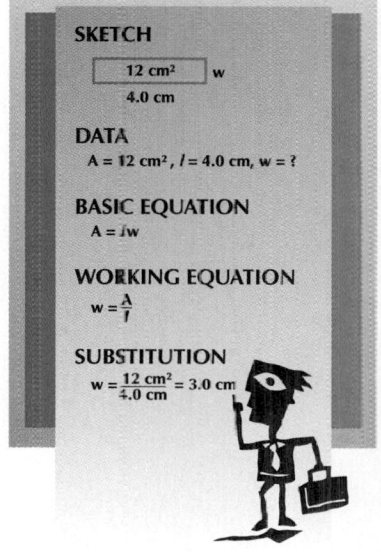

26.

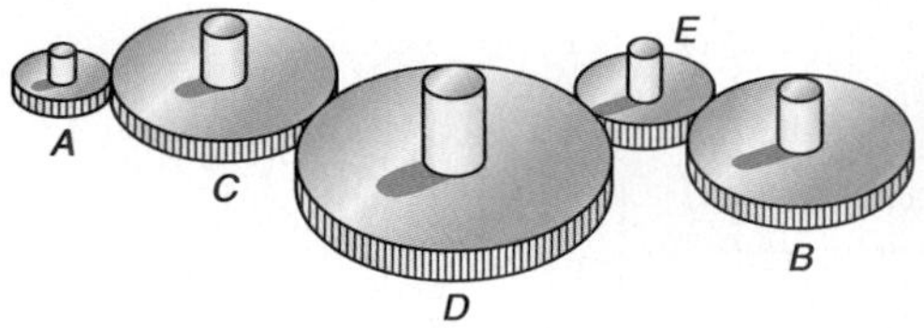

27.

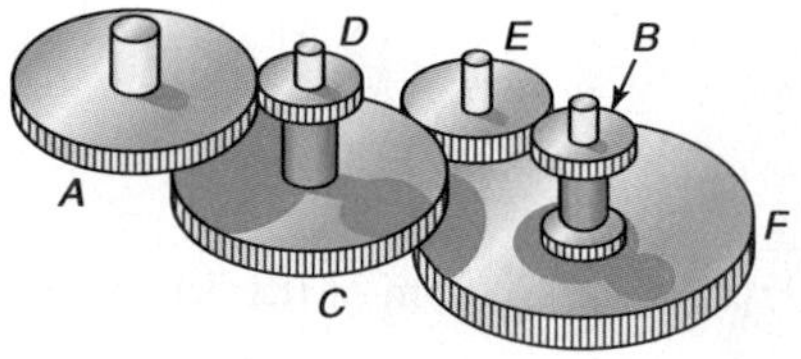

28.

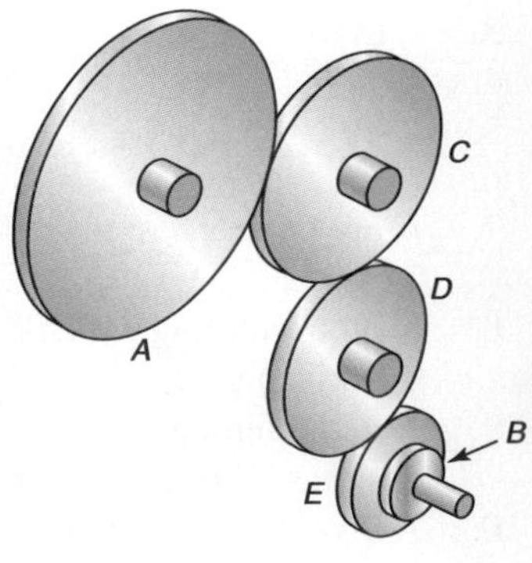

Find the speed in rpm of gear D in each gear train.

29.

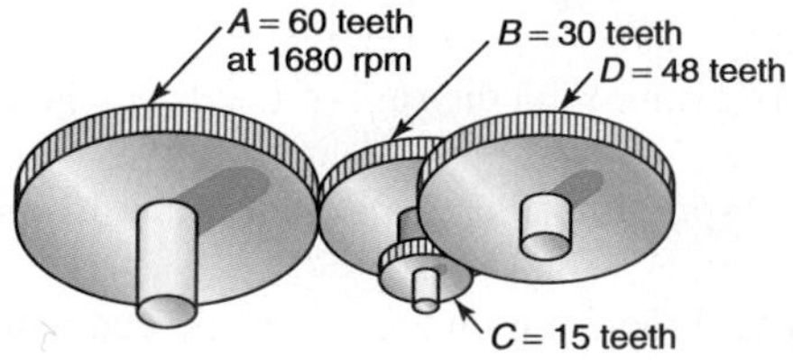

30.

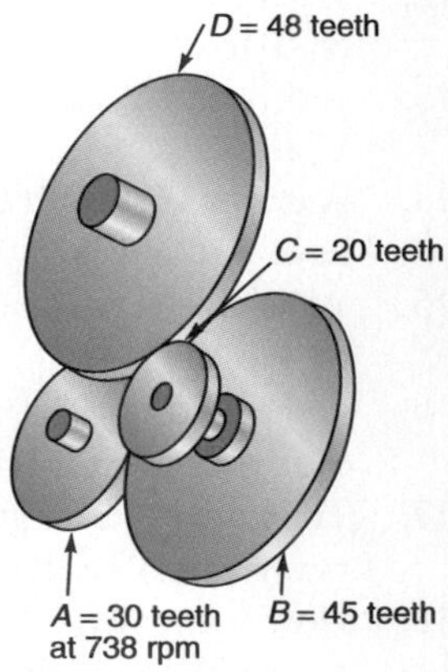

31.

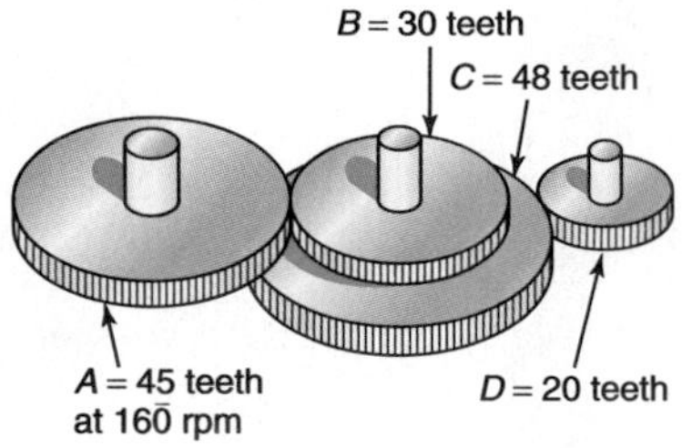

32.

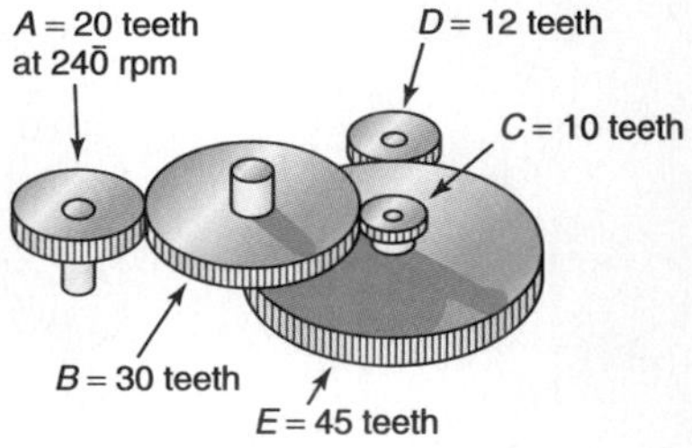

33.

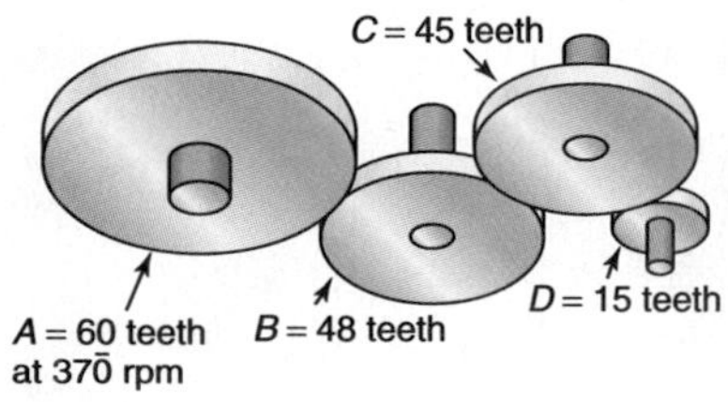

Find the number of teeth for gear D in each gear train.

34.

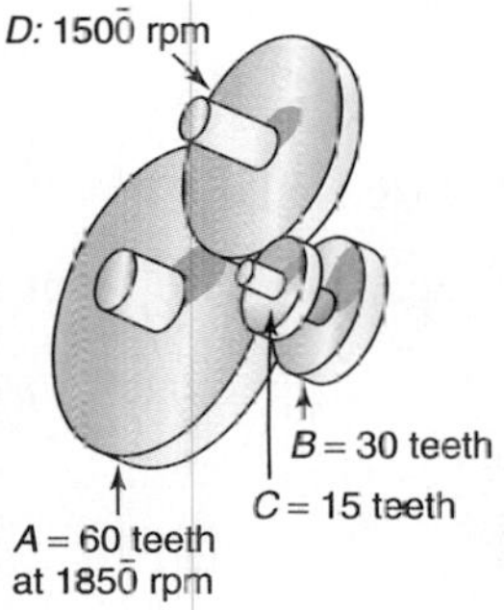

35.

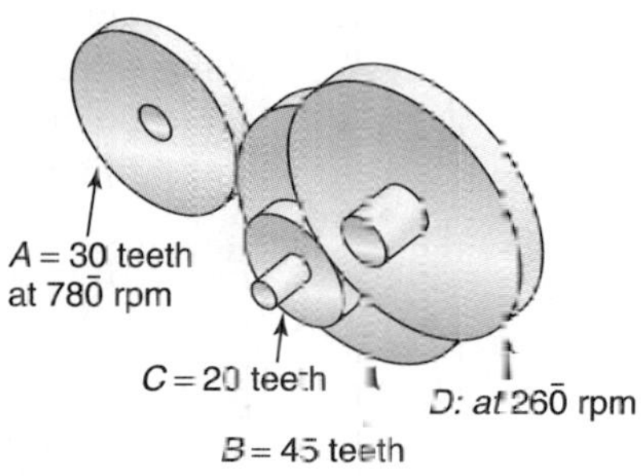

36.

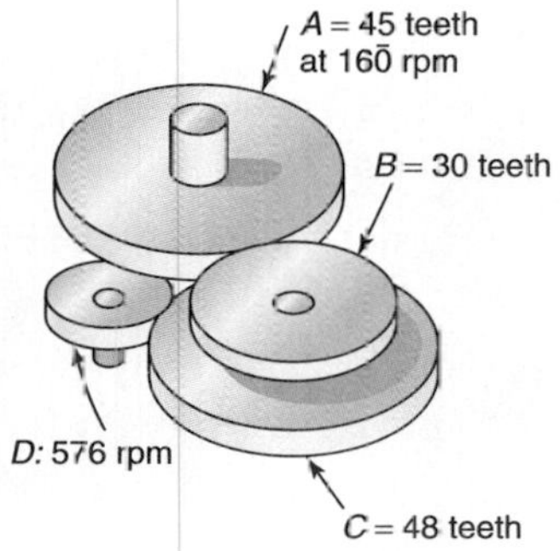

37.

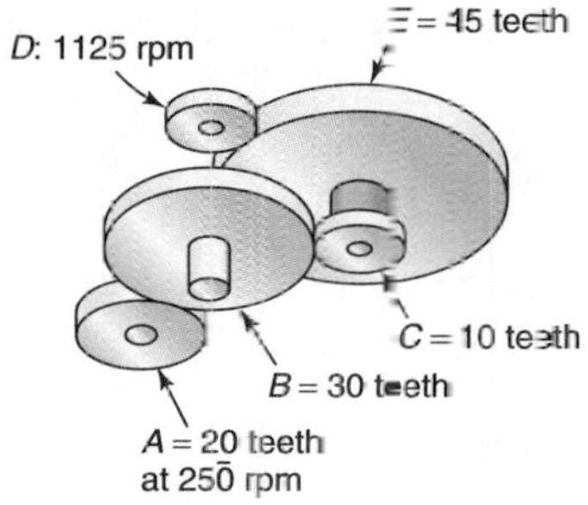

38.

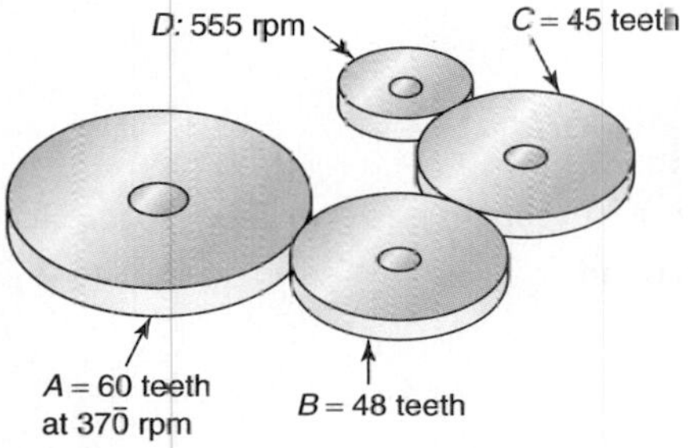

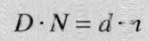

39. Find the direction of rotation of gear B if gear A is turned counterclockwise in Problems 22 through 28.
40. Find the effect of doubling the number of teeth on gear A in Problem 38.

8.7 Pulleys Connected with a Belt

Pulleys connected with a belt are used to transfer rotational motion from one shaft to another (Fig. 8.20). Assuming no slippage, the linear speed of any point on the belt equals the linear speed of any point on the rim of each pulley as the belt travels around each pulley. The larger the pulley, the larger is its circumference: $C = \pi d$. The larger the circumference of the pulley, the longer a point on the belt stays in contact with the pulley. The smaller the circumference of the pulley, the shorter a point on the belt stays in contact with the pulley, which causes the smaller pulley to rotate faster than the larger pulley. Two pulleys connected with a belt have a relationship similar to gears. Assuming no slippage, when two pulleys are connected

$$D \cdot N = d \cdot n$$

Figure 8.20 A single belt drives several components from this engine. (Courtesy of Deere & Company.)

where

D = diameter of the driver pulley
N = number of revolutions per minute of the driver pulley
d = diameter of the driven pulley
n = number of revolutions per minute of the driven pulley

The preceding equation may be generalized in the same manner as for gear trains as follows:

$$ND_1D_2D_3\cdots = nd_1d_2d_3\cdots$$

EXAMPLE

Find the speed in rpm of pulley A shown in Fig. 8.21.

Data:

$$D = 6.00 \text{ in.}$$
$$d = 30.0 \text{ in.}$$
$$n = 35\bar{0} \text{ rpm}$$
$$N = ?$$

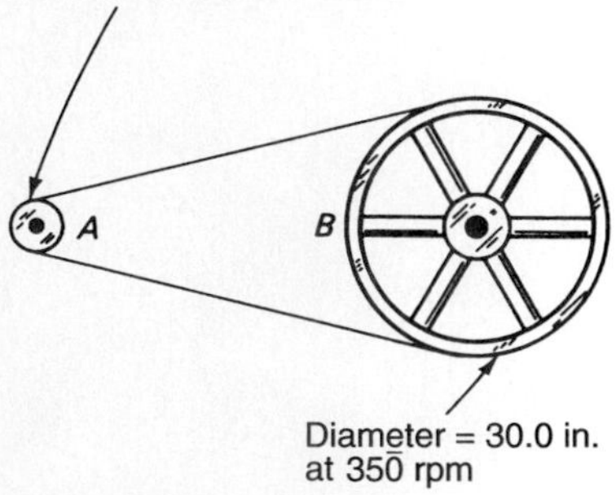

Figure 8.21

Basic Equation:

$$D \cdot N = d \cdot n$$

Working Equation:

$$N = \frac{dn}{D}$$

Substitution:

$$N = \frac{(30.0 \not{\text{in.}})(35\bar{0} \text{ rpm})}{6.00 \not{\text{in.}}}$$
$$= 1750 \text{ rpm}$$

When two pulleys are connected with an open-type belt, the pulleys turn in the same direction. When two pulleys are connected with a cross-type belt, the pulleys turn in opposite directions. See Fig. 8.22.

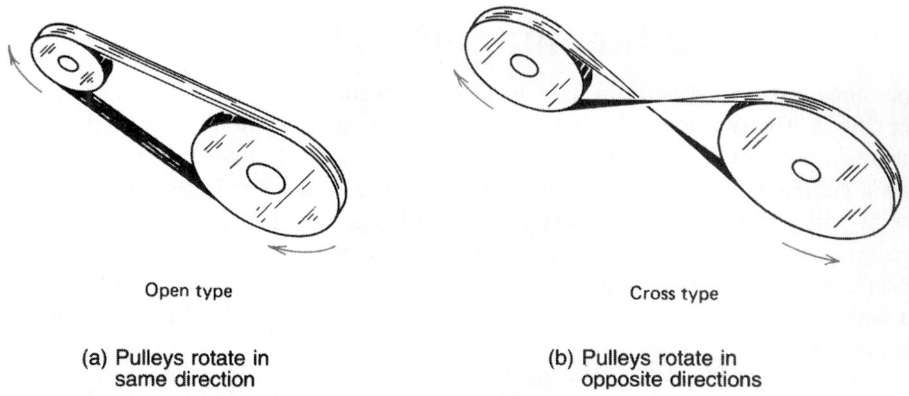

Figure 8.22 Crossing a belt reverses direction.

PROBLEMS 8.7

Find each missing quantity using $D \cdot N = d \cdot n$.

	D	N	d	n
1.	18.0	$150\bar{0}$	12.0	____
2.	36.0	____	9.00	972
3.	12.0	$18\bar{0}0$	6.00	____
4.	____	2250	9.00	1125
5.	49.0	1860	____	$62\bar{0}$

6. A driver pulley of diameter 6.50 in. revolves at 1650 rpm. Find the speed of the driven pulley if its diameter is 26.0 in.
7. A driver pulley of diameter 25.0 cm revolves at $12\bar{0}$ rpm. At what speed will the driven pulley turn if its diameter is 48.0 cm?
8. One pulley of diameter 36.0 cm revolves at $60\bar{0}$ rpm. Find the diameter of the second pulley if it rotates at $36\bar{0}$ rpm.
9. One pulley rotates at $45\bar{0}$ rpm. The diameter of the second pulley is 15.0 in. and rotates at 675 rpm. Find the diameter of the first pulley.
10. The radius of a pulley is 10.0 cm and rotates at $12\bar{0}$ rpm. The radius of the second pulley is 15.0 cm; find its rpm.

Determine the direction of pulley B in each pulley system.

11.

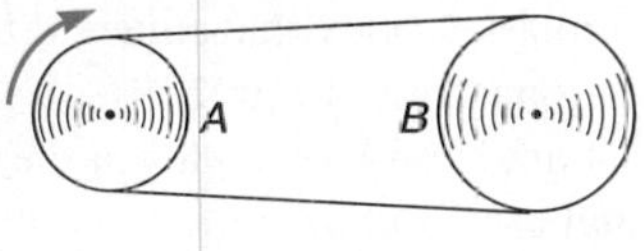

12.

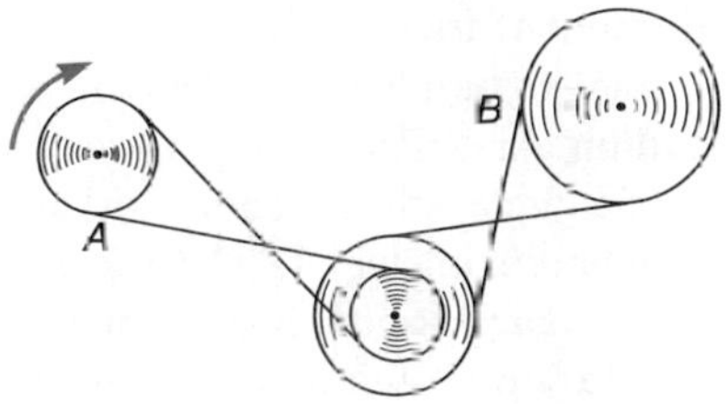

13.

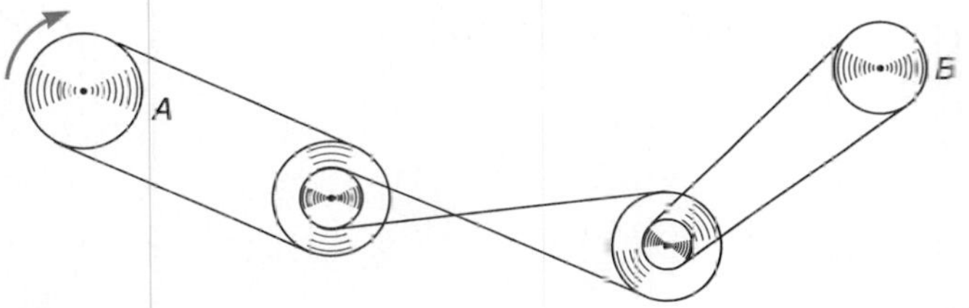

14.

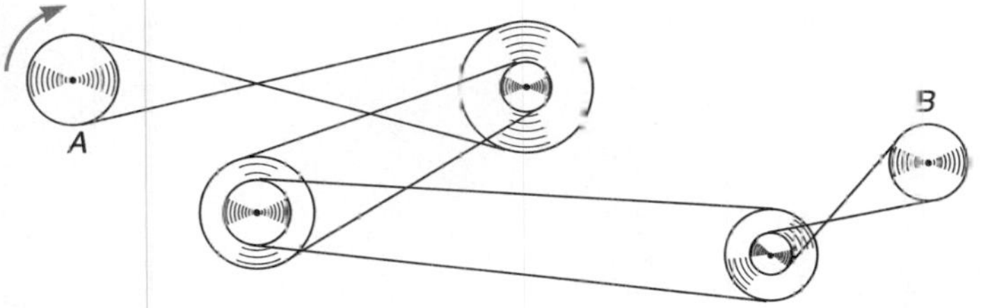

15.

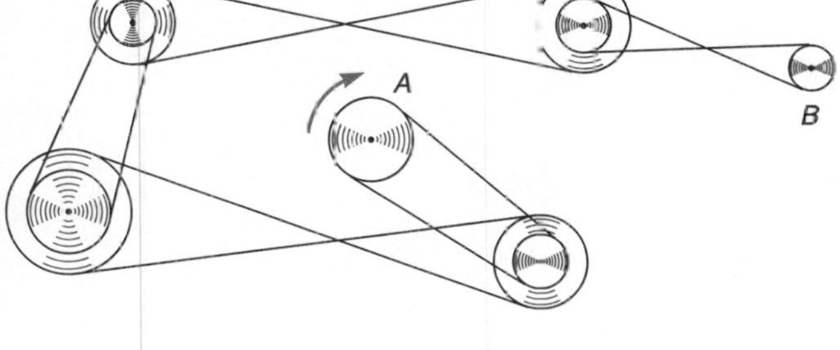

SKETCH

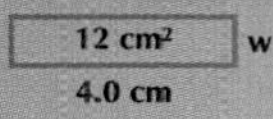

DATA

$A = 12\ \text{cm}^2,\ l = 4.0\ \text{cm},\ w = ?$

BASIC EQUATION

$A = lw$

WORKING EQUATION

$w = \frac{A}{l}$

SUBSTITUTION

$w = \frac{12\ \text{cm}^2}{4.0\ \text{cm}} = 3.0\ \text{cm}$

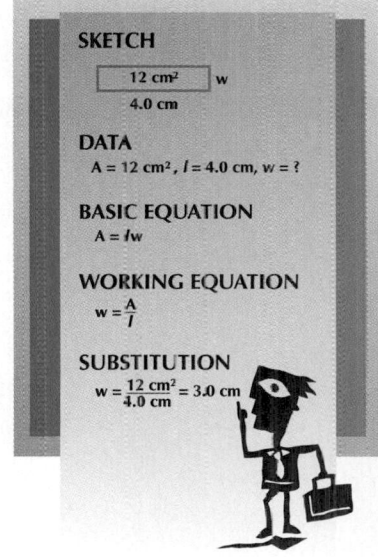

16. What size pulley should be placed on a countershaft turning $15\bar{0}$ rpm to drive a grinder with a 12.0-cm pulley that is to turn at $120\bar{0}$ rpm?

Glossary

Angular Acceleration The rate of change of angular velocity (change in angular velocity/time). (p. 216)

Angular Displacement The angle through which any point on a rotating body moves. (p. 213)

Angular Momentum For a rotating body about a fixed axis, the angular momentum is the product of the moment of inertia and the angular velocity of the body. (p. 220)

Angular Velocity The rate of angular displacement (angular displacement/time). (p. 213)

Centripetal Force The force acting on a body in circular motion that causes it to move in a circular path. This force is exerted toward the center of the circle. (p. 222)

Curvilinear Motion Motion along a curved path. (p. 212)

Degree An angular unit of measure. Defined as 1/360 of one complete revolution. (p. 212)

Gear Train A series of gears that transfers rotational motion from one gear to another. (p. 231)

Law of Conservation of Angular Momentum The angular momentum of a system remains unchanged unless an outside torque acts on it. (p. 221)

Moment of Inertia Rotational inertia; the property of a rotating body that causes it to continue to turn until a torque causes it to change its rotational motion. (p. 219)

Radian An angular unit of measurement. Defined as that angle with its vertex at the center of a circle whose sides cut off an arc on the circle equal to its radius. Equal to ($360°/2\pi$) or approximately 57.3°. (p. 212)

Rectilinear Motion Motion in a straight line. (p. 212)

Revolution A unit of measurement in rotational motion. One complete rotation of a body. (p. 212)

Rotational Motion Spinning motion of a body. (p. 212)

Formulas

8.1 $\theta = \frac{s}{r}$

$1 \text{ rev} = 360° = 2\pi \text{ rad}$

$\omega = \frac{\theta}{t}$

$v = \omega r$

$\alpha = \frac{\Delta\omega}{t}$

$\theta = \omega_{avg} t$

$\theta = \omega_i t + \frac{1}{2}\alpha_{avg} t^2$

$\omega_{avg} = \frac{\omega_f + \omega_i}{2}$

$\omega_f = \omega_i + \alpha_{avg} t$

$\alpha_{avg} = \frac{\omega_f - \omega_i}{t}$

$\theta = \frac{1}{2}(\omega_f + \omega_i)t$

$2\alpha_{avg}\,\theta = \omega_f^2 - \omega_i^2$

8.2 $\tau = I\alpha$

$L = I\omega$

$\tau t = I\omega_f - I\omega_i$

8.3 $F = \dfrac{mv^2}{r}$

8.4 $P = \tau\omega$

$$\text{power in } \frac{\text{ft lb}}{\text{s}} = \text{torque in lb ft} \times \frac{\text{number of revolutions}}{\text{min}} \times 0.105 \frac{\text{min}}{\text{rev s}}$$

$$\text{power in hp} = \text{power in} \frac{\text{ft lb}}{\text{s}} \times \frac{\text{hp}}{550 \text{ ft lb/s}}$$

8.5 $D \cdot N = d \cdot n$

8.6 $T \cdot N = t \cdot n$

$NT_1T_2T_3T_4 \cdots = nt_1t_2t_3t_4 \cdots$

8.7 $D \cdot N = d \cdot n$

$ND_1D_2D_3 \cdots = nd_1d_2d_3 \cdots$

Review Questions

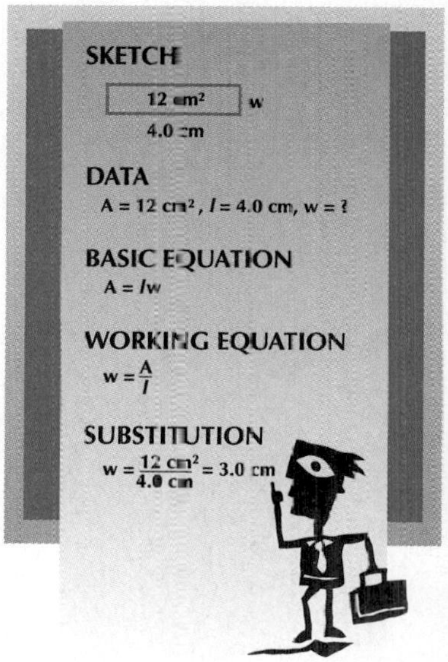

1. Angular velocity is measured in
   (a) revolutions/minute. (b) radians/second.
   (c) revolutions/second. (d) all of the above.
2. Power in the rotational system
   (a) is found in the same way as in the linear system.
   (b) is found differently than it is in the linear system.
   (c) cannot be determined.
   (d) is always a constant.
3. A gear train has 13 directly connected gears. The first and last gears will
   (a) rotate in opposite directions. (b) rotate in the same direction.
   (c) not rotate.
4. Distinguish between curvilinear motion and rotational motion.
5. Name the two types of measurement of rotation.
6. In your own words, define *radian*.
7. What is angular displacement? In what units is it measured?
8. How is linear velocity of a point on a circle related to angular velocity?
9. How do equations for uniformly accelerated rotational motion compare with those for uniformly accelerated linear motion?
10. A girl jumping from a high platform into a pool tucks her body into a tight ball to complete two somersaults before extending her body as she enters the water. Her body rotates much more quickly during the somersaults than during her extension into the water. This is an example of what law?
11. Is the tangent to a circle always perpendicular to the radius?
12. Will inertia tend to keep a moving body following a curved path?
13. Explain the relationship between the number of teeth on two interlocking gears and their relative number of revolutions.
14. How does the presence of an idler gear affect the relationship between a driver gear and a driven gear in a gear train?
15. When the number of directly connected gears in a gear train is four, do the first and last gears in the train rotate in the same or in opposite directions?
16. Why can a gear that is both a driver gear and a driven gear be omitted from a computation?
17. How do pulley combination equations compare to gear train equations?
18. If a large pulley and a small pulley are connected with a belt, which will turn faster?
19. How do we know the belt connecting two pulleys travels at the same rate while in contact with the different-size pulleys?

SKETCH

12 cm² | w
4.0 cm

DATA
$A = 12\text{ cm}^2$, $l = 4.0$ cm, $w = ?$

BASIC EQUATION
$A = lw$

WORKING EQUATION
$w = \frac{A}{l}$

SUBSTITUTION
$w = \frac{12\text{ cm}^2}{4.0\text{ cm}} = 3.0\text{ cm}$

Review Problems

1. Convert 13 revolutions to (a) radians and (b) degrees.
2. A bicycle wheel turns 25π rad during 45 s. Find the angular velocity of the wheel.
3. A lawn tractor tire turns at 65.0 rpm and has a radius of 13.0 cm. Find the linear speed of the tractor in m/s.
4. A model plane pulls into a tight curve of a radius of 25.0 m. The 3$\bar{0}$0-g plane is traveling at 90.0 km/h. What is the plane's centripetal force?
5. A 0.950-kg mass is spun in a circle with a centripetal force of 12.0 N on a string of radius 60.0 cm. At what velocity does it travel?
6. A girl riding her bike creates a torque of 1.20 lb ft with an angular speed of 45.0 rpm. How much power does she produce?
7. A motor generates 30$\bar{0}$ W of power. The torque necessary is 50.0 N m. Find the angular velocity.
8. Two rollers are side by side, with the large one turning the small one. The diameter of the small one is 2.00 cm and it turns at 15.0 rpm. The large roller has a diameter of 5.00 cm. How many revolutions does it make in one minute?
9. A clock is driven by a series of gears. The first gear has 30 teeth and rotates 60.0 times a minute. The second gear rotates at 90.0 rpm. How many teeth does it have?
10. Two gears have 13 and 26 teeth, respectively. The first gear turns at 115 rpm. How many times per minute does the second gear rotate?
11. A gear train has 17 directly connected gears. Do the first and last gears rotate in the same direction?
12. A pulley of diameter 14.0 cm is driven by an electric motor to revolve 75.0 rpm. The pulley drives a second one of diameter 10.0 cm. How many revolutions does the second pulley make in 1.00 min?
13. A pulley of diameter 5.00 cm is driven at 10$\bar{0}$ rpm. Find the diameter of a second pulley if it is driven by the first at 25$\bar{0}$ rpm.
14. If gear C turns counterclockwise, in what direction does gear F turn?

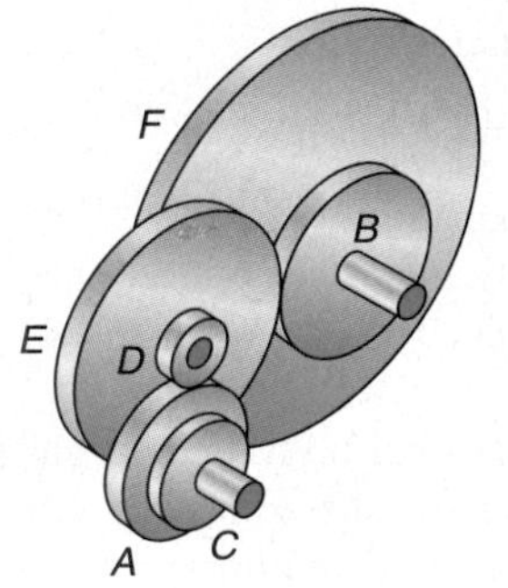

15. Find the speed in rpm of gear D.

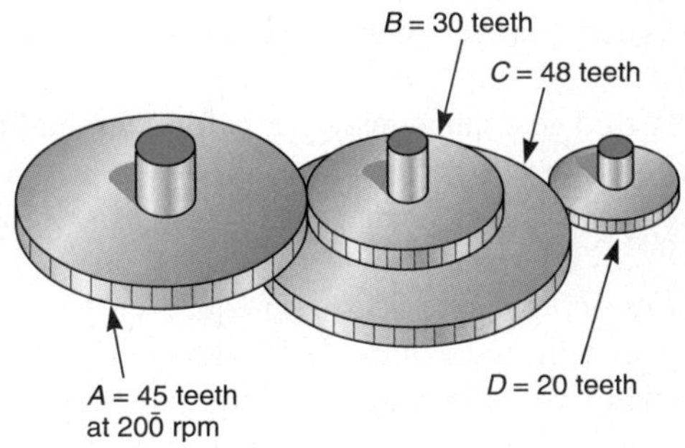

16. Find the number of teeth in gear D.

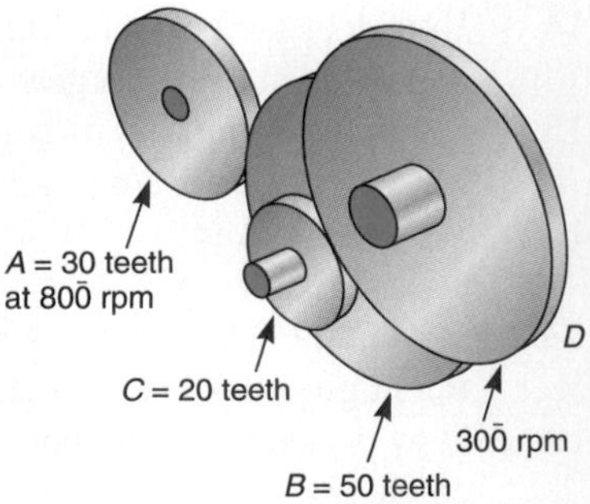

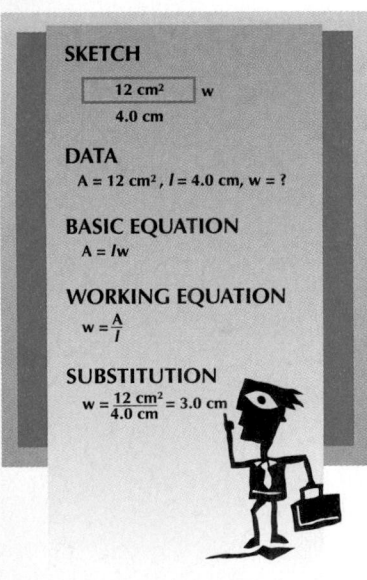

APPLIED CONCEPTS

1. As part of their training, NASA astronauts are placed in large rotating machines to see how well their bodies can withstand various "g" forces. (One g equals the gravitational force of gravity at the earth's surface.) (a) What rotational speed is needed in a device with radius 6.25 m to allow a 75.8-kg astronaut to experience a force that is twice his normal weight or "2 g"? (b) What rotational speed is needed to achieve a "4 g" effect? (c) Would the rotational speeds found in (a) and (b) change if the astronaut had a different mass?
2. Waterwheels are used to convert kinetic energy from falling water into useful mechanical energy. (a) If an average of 10.0 kg of water flows onto a waterwheel every second, how much torque does the water exert on a waterwheel of radius 2.45 m? (b) The torque exerted by the water causes the wheel to move with an angular acceleration of 0.593 rad/s^2. Using this information, find the rotational inertia of the water wheel. (c) If you were the designer of this waterwheel, what are two things you could do to increase the rotational inertia of the waterwheel?
3. In Chapter 7 (Fig. 7.17), you were asked to calculate the velocity of a frictionless roller coaster car at various locations around the track. To make sure the loops are safe, you must make sure the track can support between two and five times the normal weight (that is, between 2 g and 5 g) of a car. (a) How many g's does the 355-kg roller coaster car experience in each loop? (b) Are the loops safe? If not, explain what can be done to make the loops safe.
4. A hairpin turn on a concrete racetrack has a radius of 117 ft. (a) Since the frictional force between the road and the tires acts as the centripetal force, find the vehicle's maximum possible speed around the turn. (The coefficient of starting friction between rubber and concrete is 2.00.) (b) What would be the maximum velocity around the next turn if it has double the radius of the first turn? (c) Explain how the radius of a turn influences the centripetal force and the maximum velocity for a vehicle.
5. (a) How much power does a motorcycle need to produce 72.0 lb ft of torque while the engine crankshaft rotates at $30\overline{0}0$ rpm? (b) When shifted into a higher gear, the same amount of power rotates the engine crankshaft at $50\overline{0}0$ rpm. How much torque does the engine produce in this higher gear? (c) Using your knowledge of power, torque, and angular speed, explain why a motorcycle should be in a low gear when climbing a steep hill.

CHAPTER 9

Inclined Plane
M.A.= $\frac{S_E}{S_R}$
S_R
S_E

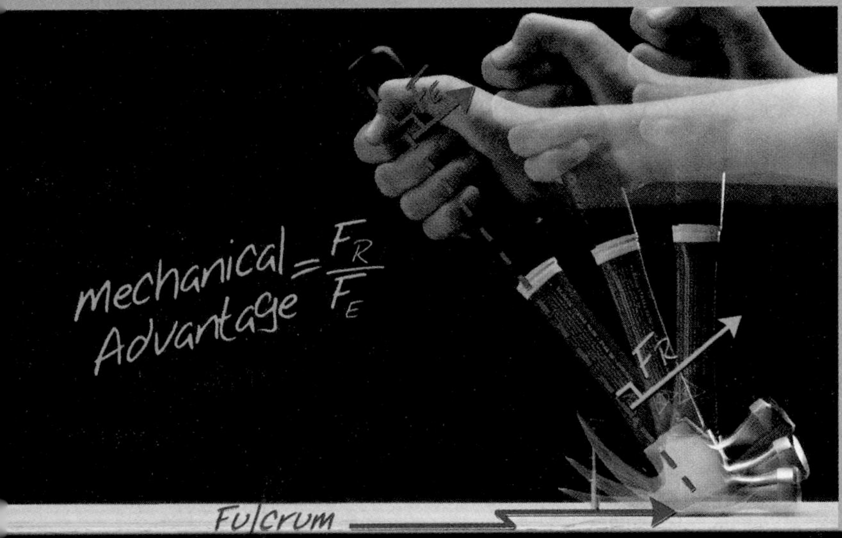
mechanical Advantage = $\frac{F_R}{F_E}$
F_R
Fulcrum

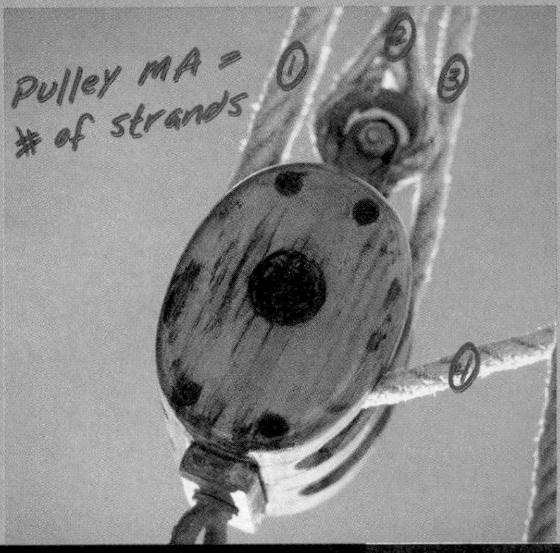
Pulley MA =
# of strands
1
2
3
4

SIMPLE MACHINES

Machines may be used to transfer energy, multiply force, or multiply speed. We use them to obtain a mechanical advantage—to do something more efficiently that would be difficult or impossible without mechanical help. In this chapter we examine simple machines and their usefulness in technology.

Objectives

The major goals of this chapter are to enable you to:

1. Determine how energy is transferred using simple machines.
2. Analyze the efficiency and mechanical advantages of simple machines.
3. Distinguish the three types of levers and the mechanical advantage of each.
4. Analyze the mechanical advantage of the wheel-and-axle, the pulley, the inclined plane, and the screw.

9.1 Machines and Energy Transfer

A **machine** is used to transfer energy from one place to another and allows work to be done that could not otherwise be done or could not be done as easily. By using a pulley system [Fig. 9.1(a)], one person can easily lift an engine from an automobile. Pliers [Fig. 9.1(b)] allow a person to cut a wire or turn a nut with the strength of his or her hand.

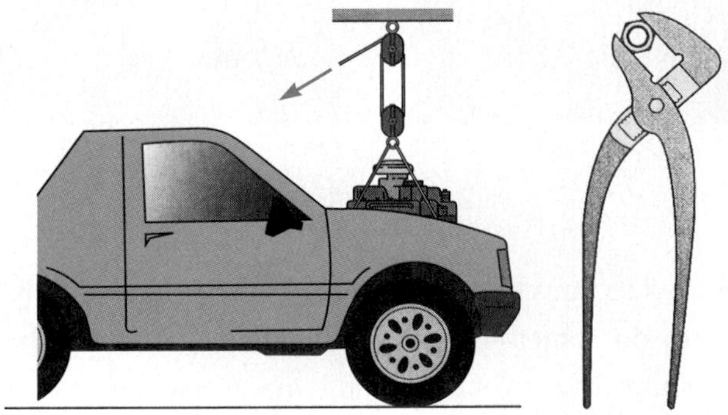

(a) (b)

Figure 9.1 Simple machines used to transfer energy

Machines are sometimes used to multiply force. By applying a small force, we can use a machine to jack up an automobile [Fig. 9.2(a)]. Machines are sometimes used to multiply speed, as with the gears on a bicycle [Fig. 9.2(b)]. Machines are used to change direction. When we use a single fixed pulley on a flag pole to raise a flag [Fig. 9.2(c)], the only advantage we get is the change in direction. (We pull the rope down, and the flag goes up.)

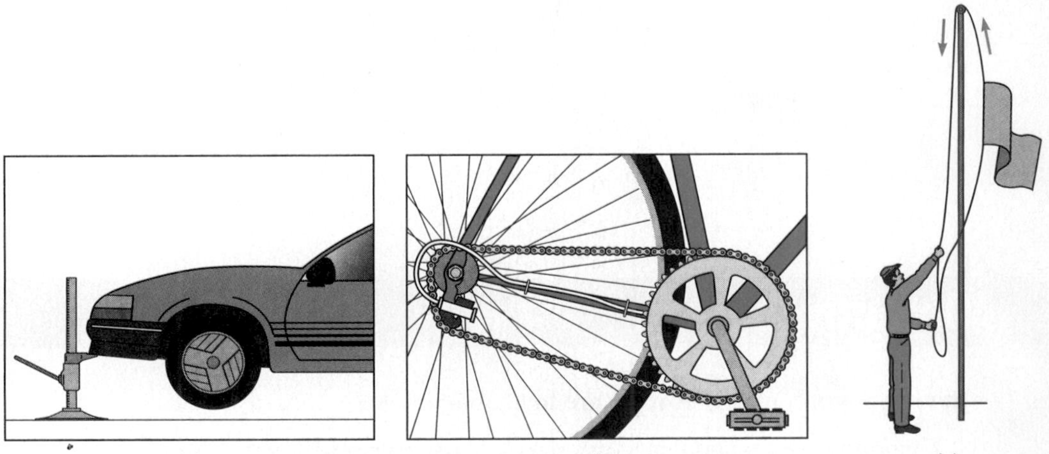

(a) A jack is used to multiply force.

(b) Gears are used to multiply speed.

(c) A pulley is used to change direction.

Figure 9.2 Simple machines may be used to multiply force or speed or to change direction.

A **simple machine** is any one of six mechanical devices in which an applied force results in useful work (Fig. 9.3). All other machines—no matter how complex—are combinations of two or more of these simple machines.

In every machine we are concerned with two forces—effort and resistance. The **effort** is the force applied *to* the machine. The **resistance** is the force overcome *by* the machine. A person applies $3\bar{0}$ lb on the jack handle in Fig. 9.4 to produce a lifting force of $60\bar{0}$ lb on the car. The effort force is $3\bar{0}$ lb. The resistance force is $60\bar{0}$ lb.

Figure 9.3 Six simple machines

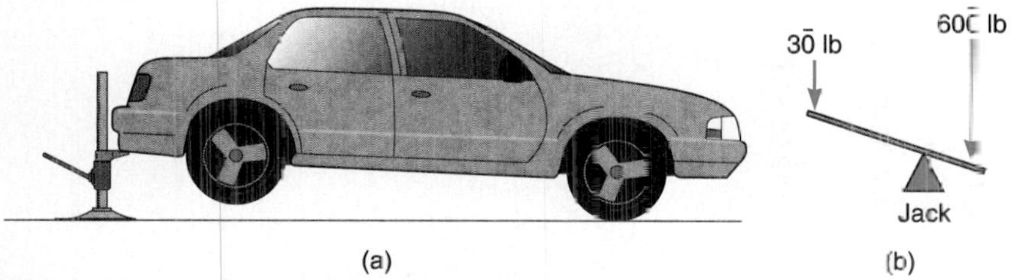

Figure 9.4 This lever multiplies force.

> **LAW OF SIMPLE MACHINES**
> resistance force × resistance distance = effort force × effort distance

Mechanical Advantage and Efficiency

The **mechanical advantage** (MA) is the ratio of the resistance force to the effort force. By formula,

$$\text{MA} = \frac{\text{resistance force}}{\text{effort force}}$$

The MA of the jack in Fig. 9.4 is found as follows:

$$\text{MA} = \frac{\text{resistance force}}{\text{effort force}} = \frac{60\bar{0}\ \cancel{\text{lb}}}{3\bar{0}\ \cancel{\text{lb}}} = \frac{2\bar{0}}{1}$$

This MA means that, for each pound applied by the person, he or she lifts $2\bar{0}$ pounds. Note that MA has no units. Why?

Each time a machine is used, part of the energy or effort applied to the machine is lost due to friction (Fig. 9.5). The **efficiency** of a machine is the ratio of the work output to the work input. By formula,

$$\text{efficiency} = \frac{\text{work output}}{\text{work input}} \times 100\% = \frac{F_{\text{output}} \times s_{\text{output}}}{F_{\text{input}} \times s_{\text{input}}} \times 100\%$$

Figure 9.5 Some work is always lost to friction.

9.2 The Lever

A **lever** consists of a rigid bar free to turn on a pivot called a **fulcrum** (Fig. 9.6). The mechanical advantage (MA) is the ratio of the effort arm (s_E) to the resistance arm (s_R):

$$\text{MA}_{\text{lever}} = \frac{\text{effort arm}}{\text{resistance arm}} = \frac{s_E}{s_R}$$

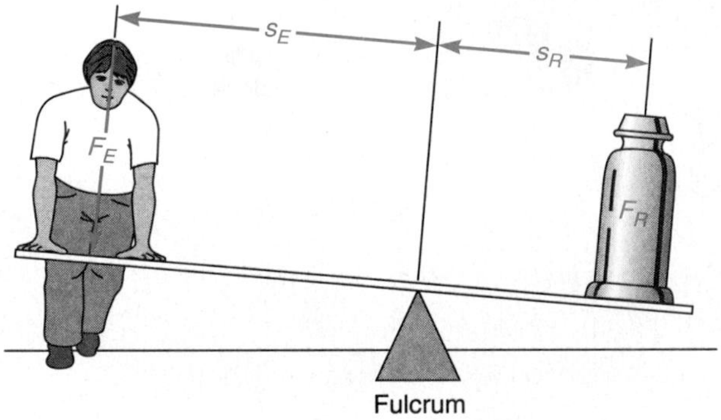

Figure 9.6 Mechanical advantage of the lever: $MA = \frac{s_E}{s_R}$.

The **effort arm** is the distance from the effort to the fulcrum. The **resistance arm** is the distance from the fulcrum to the resistance. The three types or classes of levers are shown in Fig. 9.7. The law of simple machines as applied to levers (basic equation) is

$$F_R \cdot s_R = F_E \cdot s_E$$

where

$$F_R = \text{resistance force}$$
$$s_R = \text{length of resistance arm}$$
$$F_E = \text{effort force}$$
$$s_E = \text{length of effort arm}$$

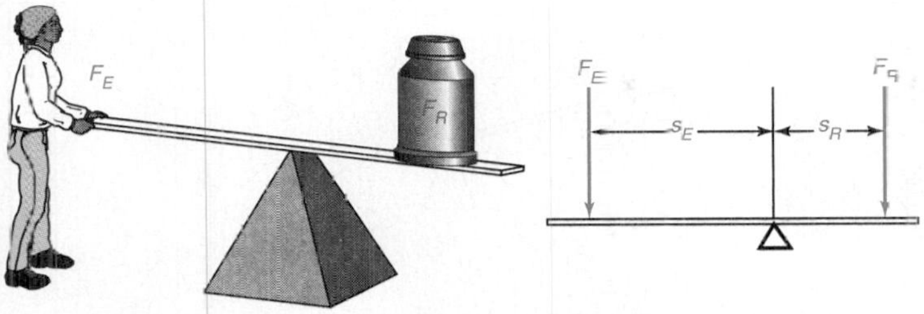

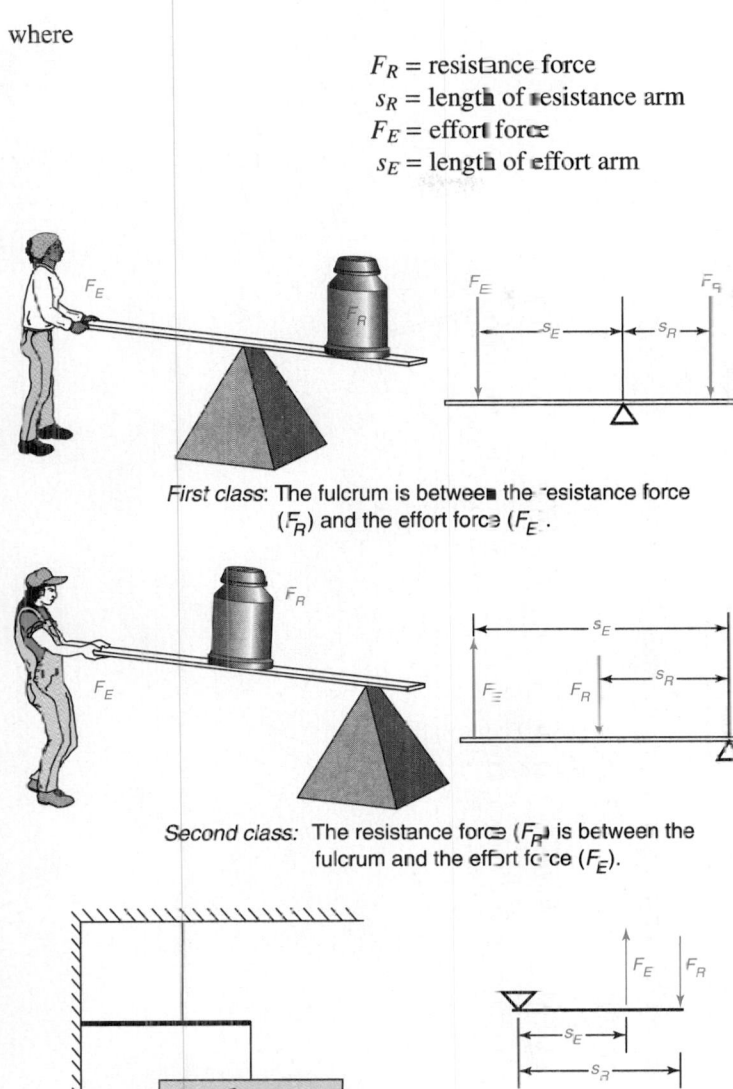

First class: The fulcrum is between the resistance force (F_R) and the effort force (F_E).

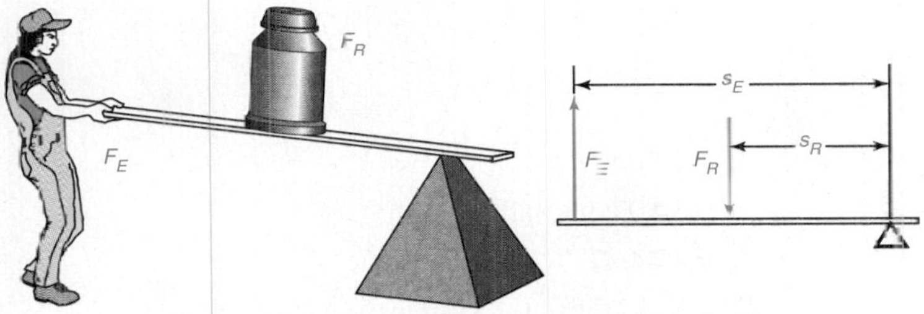

Second class: The resistance force (F_R) is between the fulcrum and the effort force (F_E).

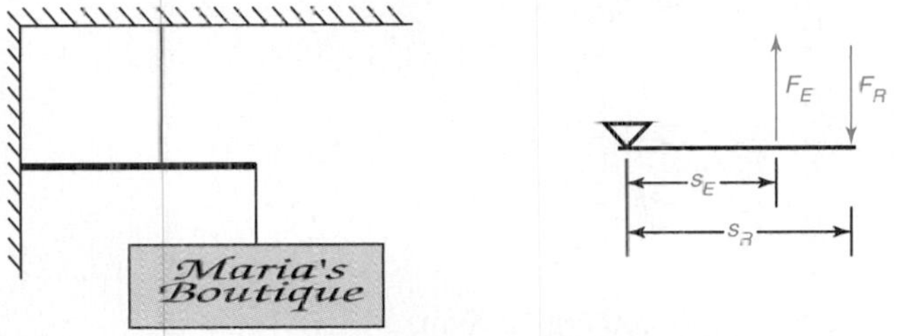

Third class: The effort force (F_E) is between the fulcrum and the resistance force (F_R).

Figure 9.7 Three classes of levers

TRY THIS ACTIVITY

Pulling Nails

Hammer two identical nails into a piece of wood so that the heads are slightly above the wood. Wedge the jaws of the hammer beneath the head of the first nail and hold the handle close to the head of the hammer. Before removing the nail, note the resistance distance, the position of the fulcrum, and the effort distance. Note the amount of effort force necessary to remove the first nail. Next, hold the end of the handle as you remove the second nail. Explain why simple machines, such as hammers pulling nails, are able to reduce the effort force.

EXAMPLE 1

A bar is used to raise a $12\bar{0}0$-N stone. The pivot is placed 30.0 cm from the stone. The worker pushes 2.50 m from the pivot. What is the mechanical advantage? What force is exerted?

Sketch:

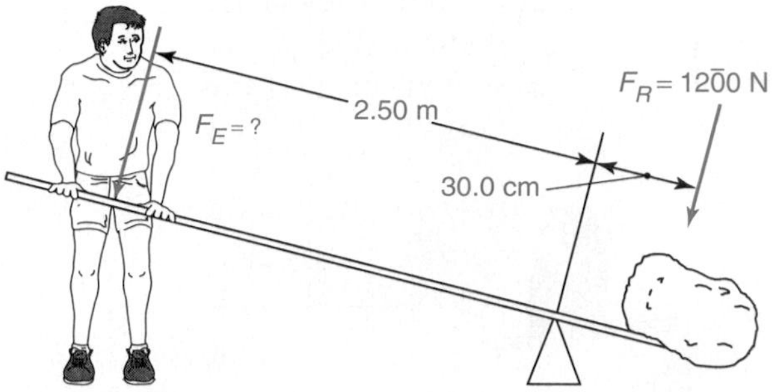

First, find MA:

$$\text{MA}_{\text{lever}} = \frac{s_E}{s_R} = \frac{2.50\ \cancel{m}}{0.300\ \cancel{m}} = \frac{8.33}{1}$$

To find the force:

Data:

$$s_E = 2.50\text{ m}$$
$$s_R = 30.0\text{ cm} = 0.300\text{ m}$$
$$F_R = 12\bar{0}0\text{ N}$$
$$F_E = ?$$

Basic Equation:

$$F_R \cdot s_R = F_E \cdot s_E$$

Working Equation:

$$F_E = \frac{F_R \cdot s_R}{s_E}$$

Substitution:

$$F_E = \frac{(12\bar{0}0\text{ N})(0.300\ \cancel{m})}{2.50\ \cancel{m}}$$
$$= 144\text{ N}$$

EXAMPLE 2

A wheelbarrow 2.00 m long has a $9\bar{0}0$-N load 50.0 cm from the axle. What is the MA? What force is needed to lift the wheelbarrow?

Sketch:

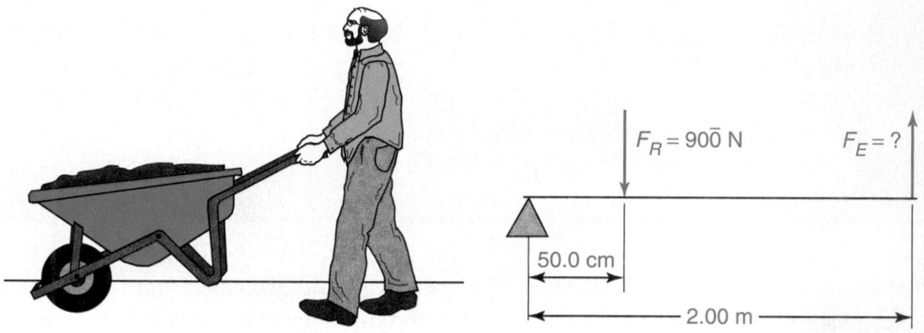

First, find MA:

$$\text{MA} = \frac{s_E}{s_R} = \frac{2.00 \text{ m}}{0.500 \text{ m}} = \frac{4.00}{1}$$

To find the force:

Data:

$$s_E = 2.00 \text{ m}$$
$$s_R = 50.0 \text{ cm} = 0.500 \text{ m}$$
$$F_R = 90\bar{0} \text{ N}$$
$$F_E = ?$$

Basic Equation:

$$F_R \cdot s_R = F_E \cdot s_E$$

Working Equation:

$$F_E = \frac{F_R \cdot s_R}{s_E}$$

Substitution:

$$F_E = \frac{(90\bar{0} \text{ N})(0.500 \text{ m})}{2.00 \text{ m}}$$
$$= 225 \text{ N}$$

EXAMPLE 3

The MA of a pair of pliers is 6.0/1. A force of 8.0 lb is exerted on the handle. What force is exerted on a wire in the pliers?

MA = 6.0/1 means that for each pound of force applied on the handle, 6.0 lb is exerted on the wire. Therefore, if a force of 8.0 lb is applied on the handle, a force of (6.0)(8.0 lb) or 48 lb is exerted on the wire.

PROBLEMS 9.2

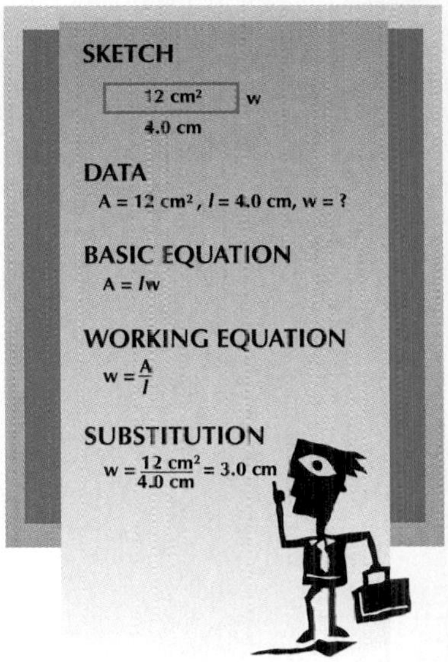

Given $F_R \cdot s_R = F_E \cdot s_E$, find each missing quantity.

	F_R	F_E	s_R	s_E
1.	20.0 N	5.00 N	3.70 cm	____ cm
2.	____ N	176 N	49.2 cm	76.3 cm
3.	37.0 N	12.0 N	____ cm	112 cm
4.	23.4 lb	9.80 lb	____ in.	53.9 in.
5.	119 N	____ N	29.7 cm	67.4 cm

Given $\text{MA}_{\text{lever}} = \frac{F_R}{F_E}$, find each missing quantity.

	MA	F_R	F_E
6.	____	20.0 N	5.00 N
7.	____	23.4 lb	9.80 lb
8.	7.00	119 N	____ N
9.	4.00	____ lb	12.2 lb
10.	____	37.0 N	12.0 N

PHYSICS CONNECTIONS

The Human Body—A Complex Machine

The human arm is a classic example of a human simple machine. Figure 9.8(a) illustrates the forces and lever arms that are in place while lifting a weight. The elbow's hinge joint acts as a relatively low-friction fulcrum. The bicep is the muscle that exerts the effort force, and the barbell is the resistance force. According to the equation for mechanical advantage, when the resistance force is farther from the fulcrum than the effort force, the mechanical advantage is less than one. Instead of reducing the force needed to lift the object, the bicep actually exerts more force than the weight of the barbell.

Machines with mechanical advantages less than one, such as the human arm, are useful because the effort force does not have to move far in order for the weight to move large distances. The bicep muscle contracts a relatively small distance, whereas the barbell moves a larger distance. When the arm is used to throw a ball, the muscles exert a large force over a small distance, whereas the end of the forearm moves a relatively large distance with a high velocity. As seen in Fig. 9.8(b), the arm acts like several levers when throwing a ball. Although the force needed to throw the ball is greater than the weight of the ball, the great distance covered by the ball in a short period of time translates into a high velocity of the ball.

(a)

(b)

Figure 9.8 (a) The bicep moves very little while exerting a large force to lift the heavy barbell. (b) The parts of the arm act as several simple machines to throw a ball with a high velocity.

Given $MA_{lever} = \dfrac{s_E}{s_R}$, find each missing quantity.

	MA	s_R	s_E
11.	____	49.2 cm	76.3 cm
12.	7.00	29.7 in.	____ in.
13.	____	29.7 cm	67.4 cm
14.	4.00	____ cm	67.4 cm

15. A pole is used to lift a car that fell off a jack (Fig. 9.9). The pivot is 2.00 ft from the car. Two people together exert 275 lb of force 8.00 ft from the pivot. (a) What force is applied to the car? (Ignore the weight of the pole.) (b) Find the MA.
16. A bar is used to lift a $10\bar{0}$-kg block of concrete. The pivot is 1.00 m from the block. (a) If the worker pushes down on the other end of the bar a distance of 2.50 m from the pivot, what force (in N) must the worker apply? (b) Find the MA.

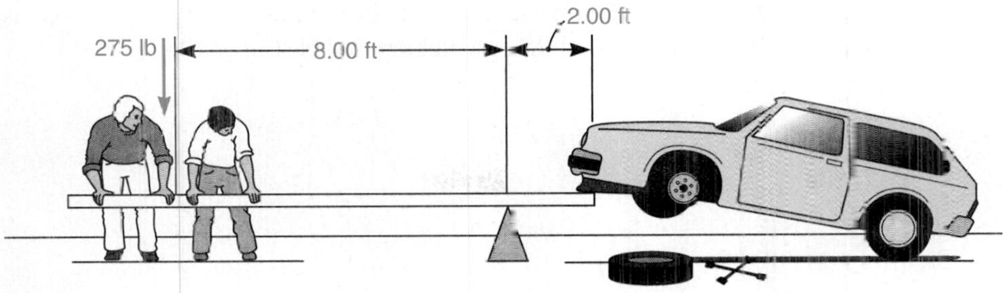

Figure 9.9

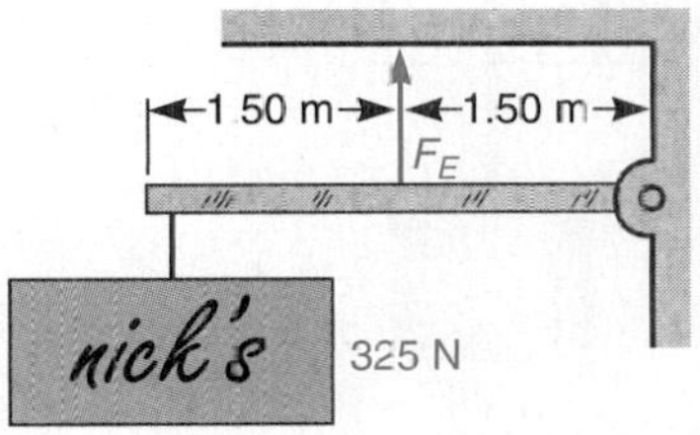

Figure 9.10

17. A wheelbarrow 6.00 ft long is used to haul a $18\bar{0}$-lb load. (a) How far from the wheel is the load placed so that a person can lift the load with a force of 45.0 lb? (b) Find the MA.
18. (a) Find the force, F_E, pulling up on the beam holding the sign shown in Fig. 9.10. (b) Find the MA.

9.3 The Wheel-and-Axle

The **wheel-and-axle** consists of a large wheel attached to an axle so that both turn together (Fig. 9.11). Other examples include a doorknob and a screwdriver with a thick handle.

The law of simple machines as applied to the wheel-and-axle (basic equation) is

$$F_R \cdot r_R = F_E \cdot r_E$$

where

F_R = resistance force
r_R = radius of resistance force
F_E = effort force
r_E = radius of effort force

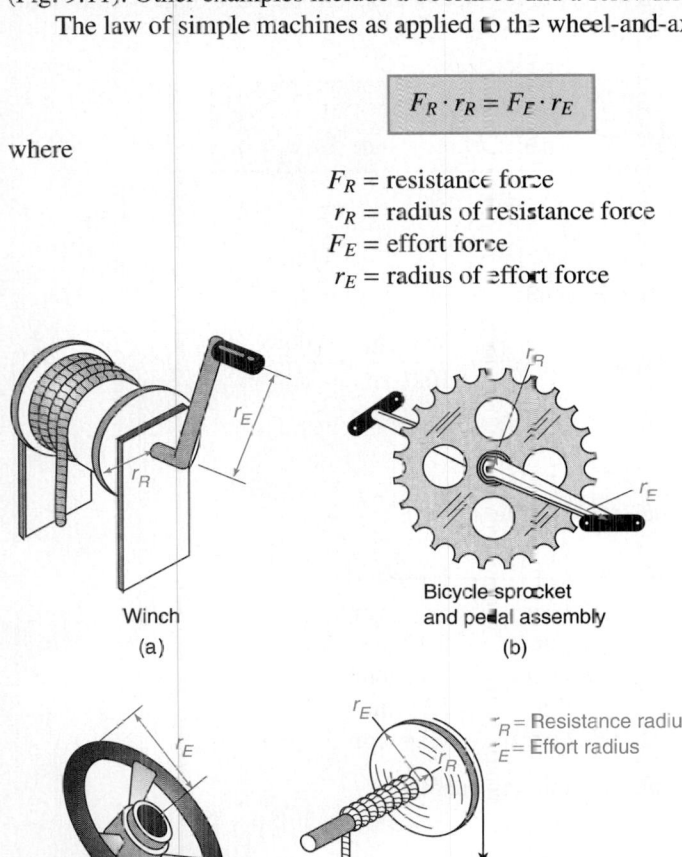

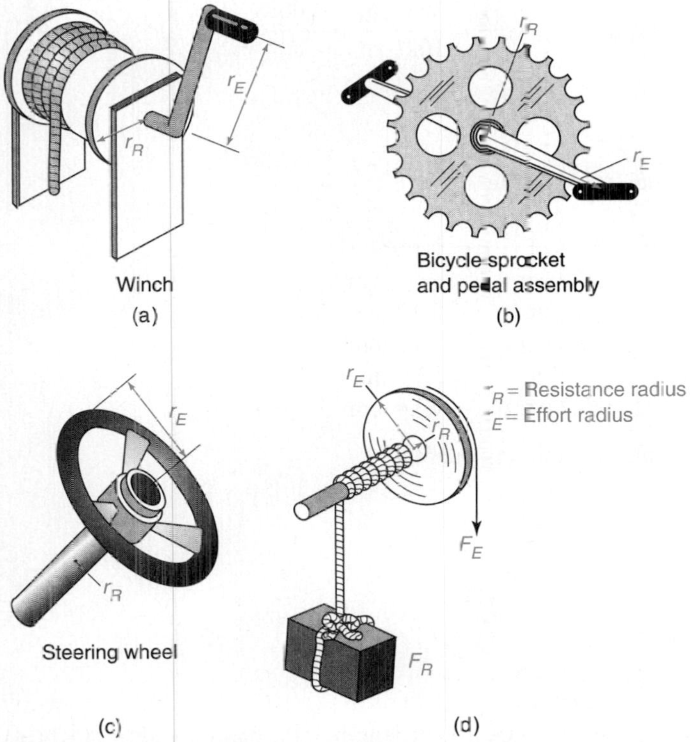

Figure 9.11 Examples of the wheel-and-axle

EXAMPLE 1

A winch has a handle that turns in a radius of 30.0 cm. The radius of the drum or axle is 10.0 cm. Find the force required to lift a bucket weighing $50\bar{0}$ N (Fig. 9.12).

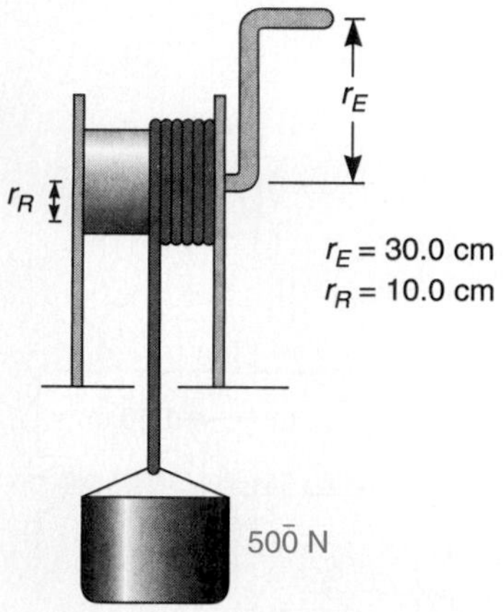

Figure 9.12

Data:

$$F_R = 50\bar{0}\text{ N}$$
$$r_E = 30.0\text{ cm}$$
$$r_R = 10.0\text{ cm}$$
$$F_E = ?$$

Basic Equation:

$$F_R \cdot r_R = F_E \cdot r_E$$

Working Equation:

$$F_E = \frac{F_R \cdot r_R}{r_E}$$

Substitution:

$$F_E = \frac{(50\bar{0}\text{ N})(10.0\text{ cm})}{30.0\text{ cm}}$$
$$= 167\text{ N}$$

The mechanical advantage (MA) of the wheel-and-axle is the ratio of the radius of the effort force to the radius of the resistance force.

$$\text{MA}_{\text{wheel-and-axle}} = \frac{\text{radius of effort force}}{\text{radius of resistance force}} = \frac{r_E}{r_R}$$

EXAMPLE 2

Calculate the MA of the winch in Example 1.

$$\text{MA}_{\text{wheel-and-axle}} = \frac{r_E}{r_R} = \frac{30.0\text{ cm}}{10.0\text{ cm}} = \frac{3.00}{1}$$

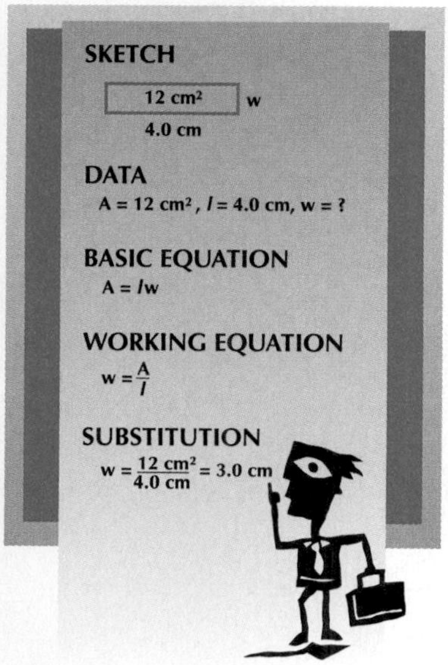

PROBLEMS 9.3

Given $F_R \cdot r_R = F_E \cdot r_E$, find each missing quantity.

	F_R	F_E	r_R	r_E
1.	20.0 N	5.30 N	3.70 cm	____ cm
2.	$37\bar{0}$ N	$12\bar{0}$ N	____ m	1.12 m
3.	____ N	175 N	49.2 cm	76.3 cm
4.	23.4 lb	9.80 lb	____ in.	53.9 in.
5.	1190 N	____ N	29.7 cm	67.4 cm

Given $\text{MA}_{\text{wheel-and-axle}} = \frac{r_E}{r_R}$, find each missing quantity.

	MA	r_E	r_R
6.	7.00	119 mm	____ mm
7.	4.00	____ in.	12.2 in.
8.	____	49.2 cm	31.7 cm
9.	3.00	61.3 cm	____ cm
10.	____	67.4 mm	29.7 mm

11. The radius of the axle of a winch is 3.00 in. The length of the handle (radius of wheel) is 1.50 ft. (a) What weight will be lifted by an effort of 73.0 lb? (b) Find the MA.

12. A wheel having a radius of 70.0 cm is attached to an axle of radius 20.0 cm. (a) What force must be applied to the rim of the wheel to raise a weight of 1500 N? (b) Find the MA. (c) What weight can be lifted if a force of 575 N is applied?
13. The diameter of the wheel of a wheel-and-axle is 10.0 cm. (a) If a force of 475 N is raised by applying a force of 142 N, find the diameter of the axle. (b) Find the MA.
14. Two persons use a windlass to raise a mass of 470 kg. The radius of the wheel is 48 cm and the radius of the axle is 4.0 cm. (a) What force is required to lift the load? (b) Find the MA of the windlass. (c) If the efficiency of the windlass is 60% and each person exerts the same force, how much force must each apply?

9.4 The Pulley

A **pulley** is a grooved wheel that turns readily on an axle and is supported in a frame. It can be fastened to a fixed object or to the resistance that is to be moved. A **fixed pulley** is a pulley fastened to a fixed object [Fig. 9.13(a)]. A **movable pulley** is fastened to the object to be moved [Fig. 9.13(b)]. A pulley system consists of combinations of fixed and movable pulleys [Fig. 9.13(c)–(e)].

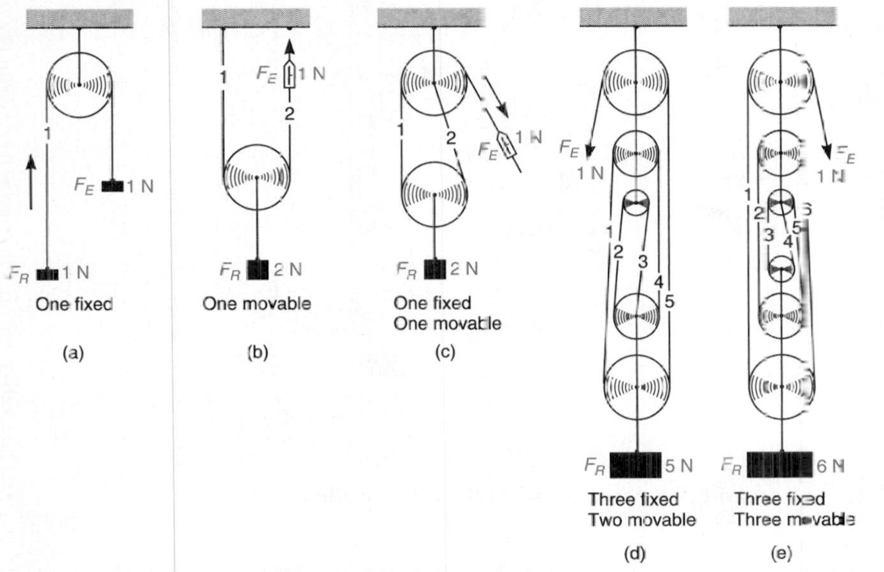

Figure 9.13 Pulleys and pulley systems

The law of simple machines as applied to pulleys (Fig. 9.14) is

$$F_R \cdot s_R = F_E \cdot s_E$$

Here, s refers to the distance moved. From the preceding equation,

$$\frac{F_R}{F_E} = \frac{s_E}{s_R} = \text{MA}_{\text{pulley}}$$

However, when one continuous cord is used, this ratio reduces to the number of strands holding the resistance in the pulley system. Therefore,

$$\text{MA}_{\text{pulley}} = \text{number of strands holding the resistance}$$

This result may be explained as follows: When a weight is supported by two strands, each individual strand supports one-half of the total weight. Thus, the MA = 2. If a weight is

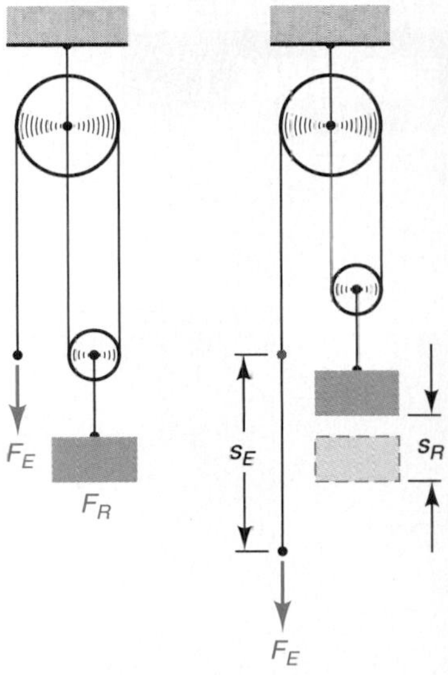

Figure 9.14 Law of simple machines applied to pulleys: $F_R \cdot s_R = F_E \cdot s_E$

supported by three strands, each individual strand supports one-third of the total weight. Thus, the MA = 3. If a weight is supported by four strands, each individual strand supports one-fourth of the total weight. In general, when a weight is supported by n strands, each individual strand supports $\frac{1}{n}$ of the total weight. Thus, the MA = n.

Stated another way, the resistance force, F_R, is spread equally among the supporting strands. Thus, $F_R = nT$, where n is the number of strands holding the resistance and T is the tension in each supporting strand.

The effort force, F_E, equals the tension, T, in each supporting strand. The equation may then be written

$$\text{MA}_{\text{pulley}} = \frac{F_R}{F_E} = \frac{nT}{T} = n$$

Note: The mechanical advantage of the pulley does not depend on the diameter of the pulley.

A number of examples are shown in Fig. 9.15.

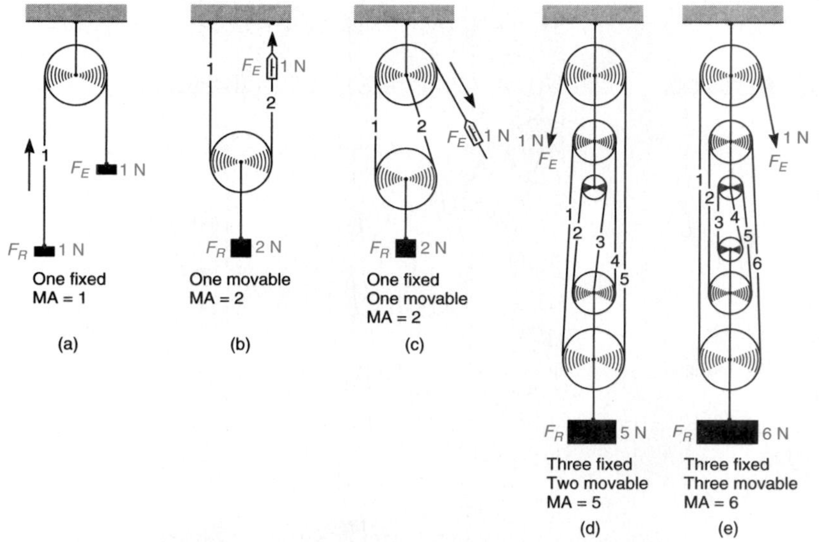

Figure 9.15 Mechanical advantage of pulleys and pulley systems

EXAMPLE 1

Draw two different sets of pulleys, each with an MA of 4.

Sketch:

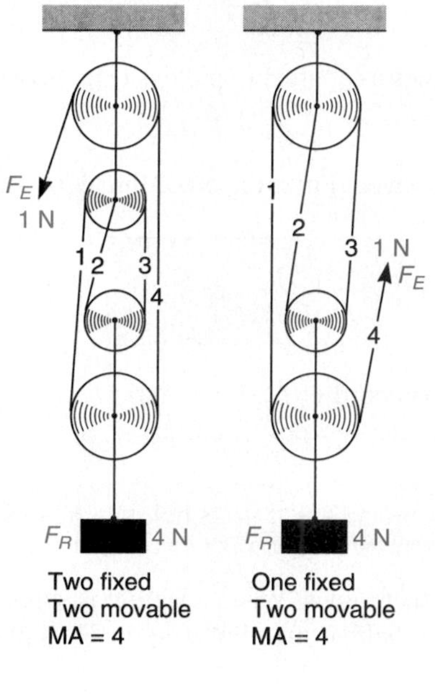

EXAMPLE 2

What effort will lift a resistance of 480 N in the pulley systems in Example 1?

Data:

$$MA_{\text{pulley}} = 4$$
$$F_R = 480 \text{ N}$$
$$F_E = ?$$

Basic Equation:

$$MA_{\text{pulley}} = \frac{F_R}{F_E}$$

Working Equation:

$$F_E = \frac{F_R}{MA_{\text{pulley}}}$$

Substitution:

$$F_E = \frac{480 \text{ N}}{4}$$
$$= 120 \text{ N}$$

EXAMPLE 3

If the resistance moves 7.00 ft, what is the effort distance of the pulley system in Example 1?

Data:

$$MA_{\text{pulley}} = 4$$
$$s_R = 7.00 \text{ ft}$$
$$s_E = ?$$

Basic Equation:

$$MA_{\text{pulley}} = \frac{s_E}{s_R}$$

Working Equation:

$$s_E = s_R(MA_{\text{pulley}})$$

Substitution:

$$s_E = (7.00 \text{ ft})(4)$$
$$= 28.0 \text{ ft}$$

EXAMPLE 4

The pulley system in Fig. 9.16 is used to raise a 650-lb object 25 ft. What is the mechanical advantage? What force is exerted?

$$MA_{\text{pulley}} = \text{number of strands holding the resistance}$$
$$= 5$$

To find the force exerted:

Data:

$$MA_{\text{pulley}} = 5$$
$$F_R = 650 \text{ lb}$$
$$F_E = ?$$

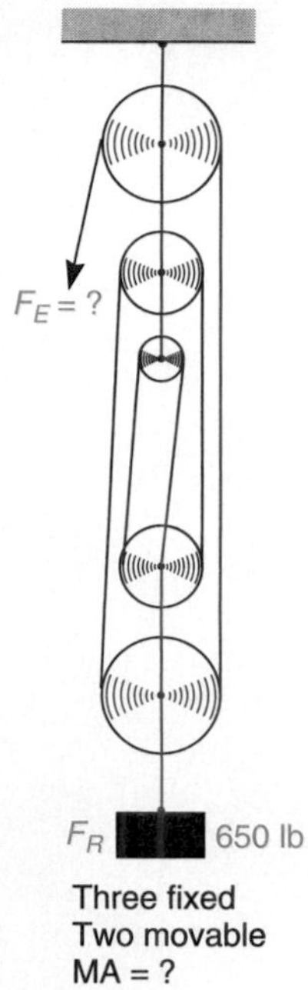

Figure 9.16

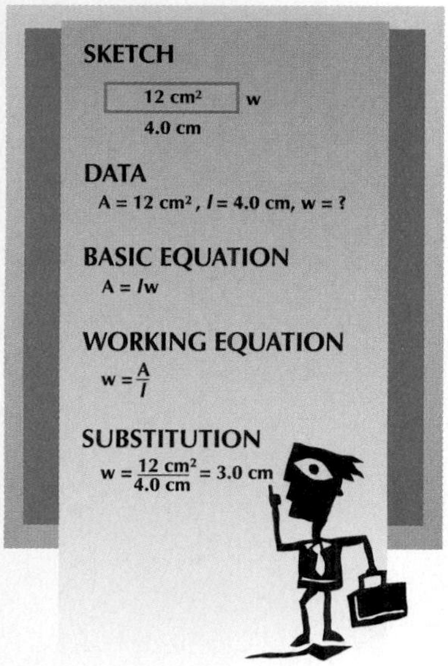

Basic Equation:

$$\text{MA}_{\text{pulley}} = \frac{F_R}{F_E}$$

Working Equation:

$$F_E = \frac{F_R}{\text{MA}_{\text{pulley}}}$$

Substitution:

$$F_E = \frac{650 \text{ lb}}{5}$$

$$= 130 \text{ lb}$$

PROBLEMS 9.4

Find the mechanical advantage of each pulley system.

1. 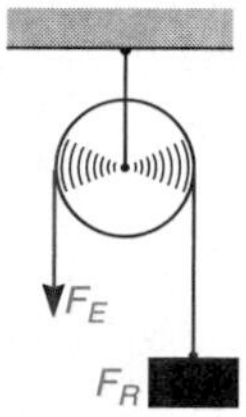

2. 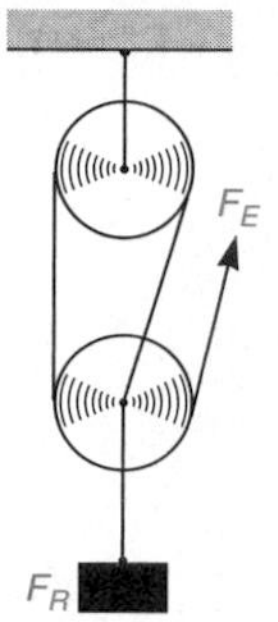

3. 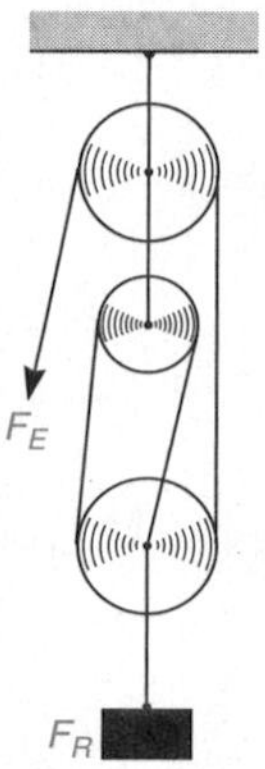

4. 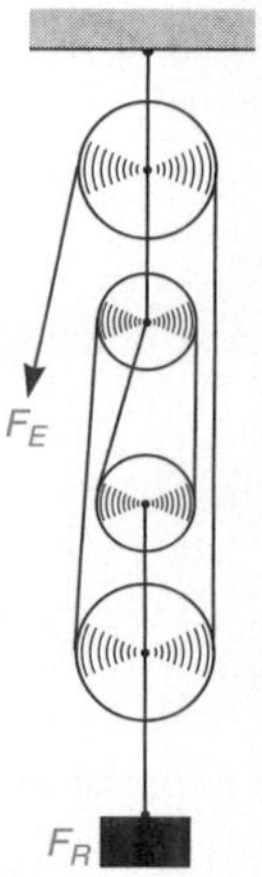

5. 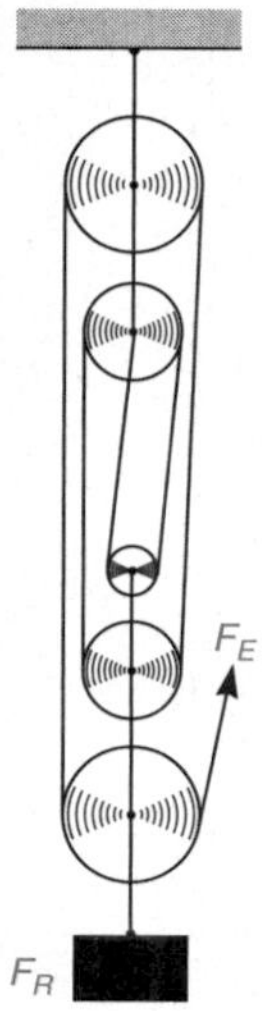

6. 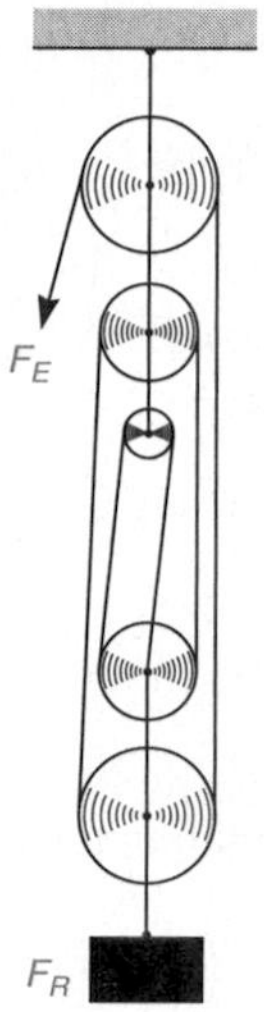

7. 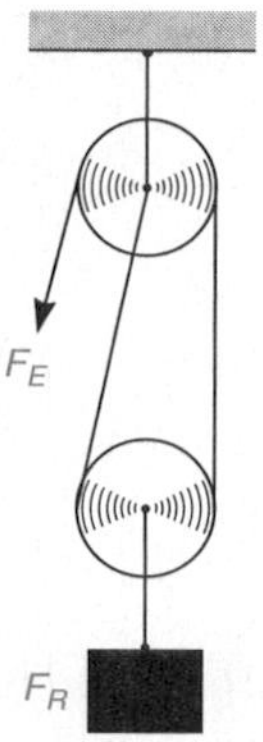

8. 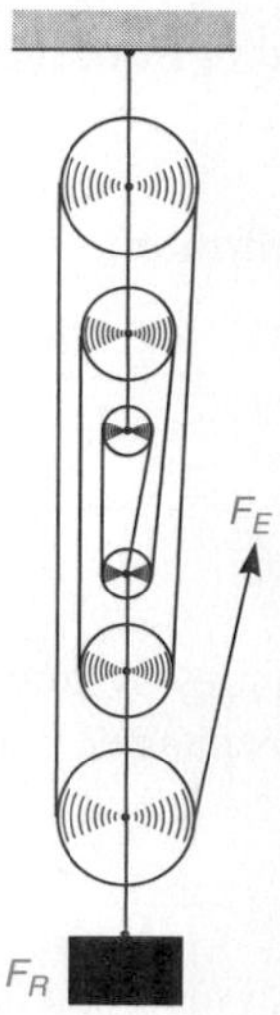

Draw each pulley system for Problems 9–14.

9. One fixed and two movable. Find the system's MA.
10. Two fixed and two movable with an MA of 5.
11. Three fixed and three movable with an MA of 6.
12. Four fixed and three movable. Find the system's MA.

13. Four fixed and four movable with an MA of 8.
14. Three fixed and four movable with an MA of 8.
15. What is the MA of a single movable pulley?
16. (a) What effort will lift a 250-lb weight by using a single movable pulley? (b) If the weight is moved 15.0 ft, how many feet of rope are pulled by the person exerting the effort?
17. A system consisting of two fixed pulleys and two movable pulleys has a mechanical advantage of 4. (a) If a force of 97.0 N is exerted, what weight is raised? (b) If the weight is raised 20.5 m, what length of rope is pulled?
18. A 400-lb weight is lifted 30.0 ft. (a) Using a system of one fixed and two movable pulleys, find the effort force and effort distance. (b) If an effort force of 65.0 N is applied through an effort distance of 13.0 m, find the weight of the resistance and the distance it is moved.
19. Can an effort force of 75.0 N lift a 275-N weight using the pulley system in Problem 3?
20. (a) What effort will lift a 1950-N weight using the pulley system in Problem 5? (b) If the weight is moved 3.00 m, how much rope must be pulled through the pulley system by the person exerting the force?
21. Complete the following pulley system mechanical advantage chart, which lists two possible arrangements of fixed and movable pulleys for each given mechanical advantage.

	Mechanical Advantage (MA)							
Pulleys	**1**	**2**	**3**	**4**	**5**	**6**	**7**	**8**
Fixed	1	1	2					
Movable	0	1	1					
Fixed		0	1					
Movable		1	1					

22. Can you arrange a pulley system containing 10 pulleys and obtain a mechanical advantage of 12? Why or why not?

9.5 The Inclined Plane

An **inclined plane** is a plane surface set at an angle from the horizontal used to raise objects that are too heavy to lift vertically. Gangplanks, chutes, and ramps are all examples of the inclined plane (Fig. 9.17). The work done in raising a resistance using the inclined plane equals the resistance times the height. This must also equal the work input, which can be found by multiplying the effort times the length of the plane.

$$F_R \cdot s_R = F_E \cdot s_E \quad \text{(law of machines)}$$

$$F_R \cdot \text{height of plane} = F_E \cdot \text{length of plane}$$

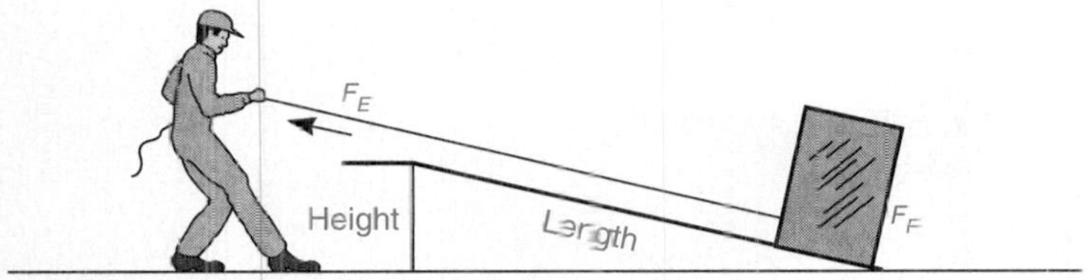

Figure 9.17 Inclined plane

From the preceding equation,

$$\frac{F_R}{F_E} = \frac{\text{length of plane}}{\text{height of plane}} = \text{MA}_{\text{inclined plane}}$$

EXAMPLE 1

A worker is pushing a box weighing $15\bar{0}0$ N up a ramp 6.00 m long onto a platform 1.50 m above the ground. What is the mechanical advantage? What effort is applied?

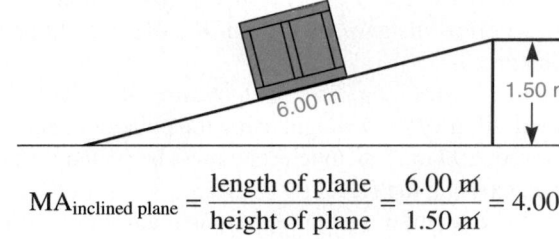

Sketch:

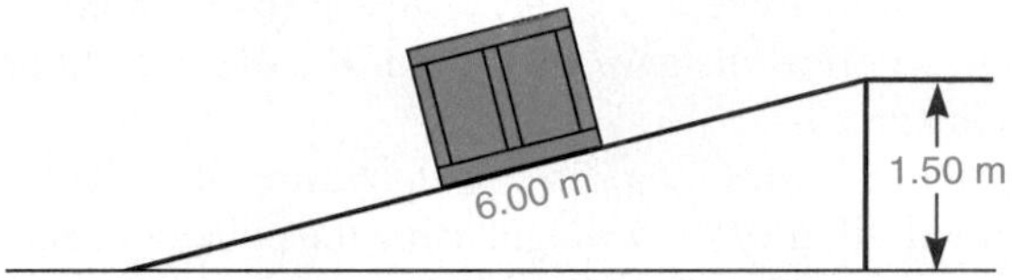

$$\text{MA}_{\text{inclined plane}} = \frac{\text{length of plane}}{\text{height of plane}} = \frac{6.00\ \cancel{\text{m}}}{1.50\ \cancel{\text{m}}} = 4.00$$

To find the effort force:

Data:

$$F_R = 15\bar{0}0 \text{ N}$$
$$\text{MA}_{\text{inclined plane}} = 4.00$$
$$F_E = ?$$

Basic Equation:

$$\text{MA}_{\text{inclined plane}} = \frac{F_R}{F_E}$$

Working Equation:

$$F_E = \frac{F_R}{\text{MA}_{\text{inclined plane}}}$$

Substitution:

$$F_E = \frac{15\bar{0}0 \text{ N}}{4.00}$$
$$= 375 \text{ N}$$

EXAMPLE 2

Find the length of the shortest ramp that can be used to push a $60\bar{0}$-lb resistance onto a platform 3.50 ft high by exerting a force of 72.0 lb.

Data:

$$F_R = 60\bar{0} \text{ lb}$$
$$F_E = 72.0 \text{ lb}$$
$$\text{height} = 3.50 \text{ ft}$$
$$\text{length} = ?$$

Basic Equation:

$$F_R \cdot \text{height} = F_E \cdot \text{length}$$

Working Equation:

$$\text{length} = \frac{F_R \cdot \text{height}}{F_E}$$

Substitution:

$$\text{length} = \frac{(60\bar{0}\ \cancel{\text{lb}})(3.50\ \text{ft})}{72.0\ \cancel{\text{lb}}} = 29.2\ \text{ft}$$

EXAMPLE 3

An inclined plane is 13.0 m long and 5.00 m high. What is its mechanical advantage and what weight can be raised by exerting a force of 375 N?

$$\text{MA}_{\text{inclined plane}} = \frac{\text{length of plane}}{\text{height of plane}} = \frac{13.0\ \cancel{\text{m}}}{5.00\ \cancel{\text{m}}} = 2.60$$

To find the weight of the resistance:

Data:

$$\text{MA}_{\text{inclined plane}} = 2.60$$
$$F_E = 375\ \text{N}$$
$$F_R = ?$$

Basic Equation:

$$\text{MA}_{\text{inclined plane}} = \frac{F_R}{F_E}$$

Working Equation:

$$F_R = (F_E)(\text{MA}_{\text{inclined plane}})$$

Substitution:

$$F_R = (375\ \text{N})(2.60) = 975\ \text{N}$$

PROBLEMS 9.5

Given $F_R \cdot \text{height} = F_E \cdot \text{length}$, find each missing quantity.

	F_R	F_E	*Height of Plane*	*Length of Plane*
1.	20.0 N	5.30 N	3.40 cm	____ cm
2.	$98\bar{0}0$ N	2340 N	____ m	3.79 m
3.	119 lb	____ lb	13.2 in.	74.0 in.
4.	____ N	1760 N	82.1 cm	3.79 m
5.	$37\bar{0}0$ N	$12\bar{0}0$ N	____ cm	112 cm

Given $\text{MA}_{\text{inclined plane}} = \dfrac{\text{length of plane}}{\text{height of plane}}$, find each missing quantity.

	MA	*Length of Plane*	*Height of Plane*
6.	9.00	3.40 ft	____ ft
7.	____	3.79 m	0.821 m
8.	1.30	____ ft	9.72 ft
9.	____	74.0 cm	13.2 cm
10.	17.4	____ in.	13.4 in.

11. An inclined plane is 10.0 m long and 2.50 m high. (a) Find its mechanical advantage. (b) A resistance of 727 N is pushed up the plane. What effort is needed? (c) An effort

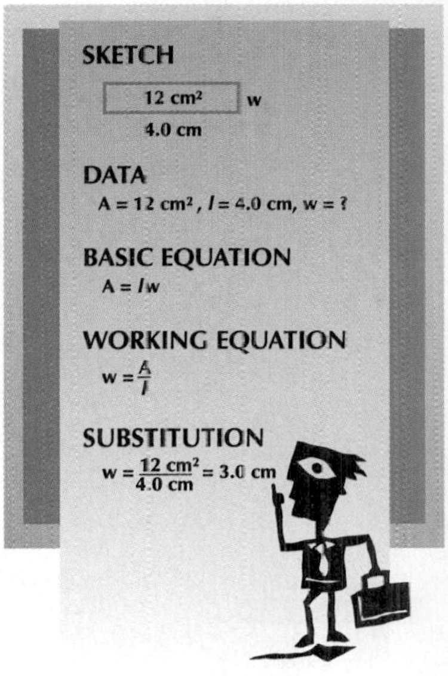

of 20$\bar{0}$ N is applied to push an 815-N resistance up the inclined plane. Is the effort enough?

12. A safe is loaded onto a truck whose bed is 5.50 ft above the ground. The safe weighs 538 lb. (a) If the effort applied is 14$\bar{0}$ lb, what length of ramp is needed? (b) What is the MA of the inclined plane? (c) Another safe weighing 257 lb is loaded onto the same truck. If the ramp is 21.1 ft long, what effort is needed?
13. A resistance of 325 N is raised by using a ramp 5.76 m long and by applying a force of 75.0 N. (a) How high can it be raised? (b) Find the MA of the ramp.
14. A plank 12 ft long is used as an inclined plane to a platform 3.0 ft high. (a) What force must be used to push a load weighing 480 lb up the plank? (b) Find the MA of the inclined plane.

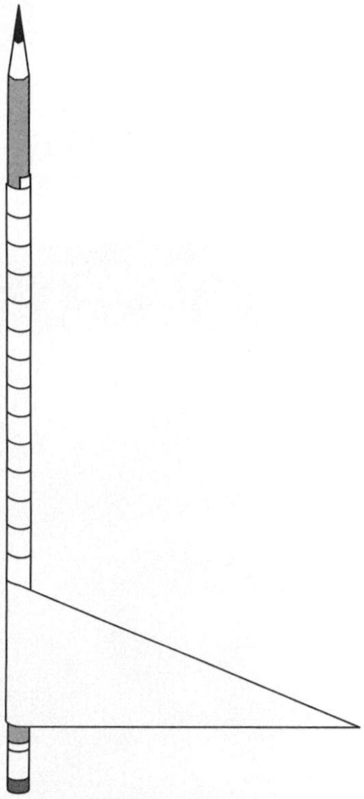
Figure 9.18 The screw is an inclined plane wound around a cylinder. The hypotenuse of the triangular section of paper corresponds to the inclined plane (threads) of a screw as it is wound around the pencil.

9.6 The Screw

A **screw** is an inclined plane wrapped around a cylinder. To illustrate, cut a sheet of paper in the shape of a right triangle and wind it around a pencil as shown in Fig. 9.18. The jackscrew, wood screw, and auger are examples of this simple machine (Fig. 9.19). The distance a beam rises or the distance the wood screw advances into a piece of wood in one revolution is called the **pitch** of the screw. Therefore, the pitch of a screw is also the distance between two successive threads.

From the law of machines,

$$F_R \cdot s_R = F_E \cdot s_E$$

However, for advancing a screw with a screwdriver

$$s_R = \text{pitch of screw}$$
$$s_E = \text{circumference of the handle of the screwdriver}$$

or

$$s_E = 2\pi r$$

where r is the radius of the handle of the screwdriver. Therefore,

$$\boxed{F_R \cdot \text{pitch} = F_E \cdot 2\pi r}$$

so

$$\boxed{\frac{F_R}{F_E} = \frac{2\pi r}{\text{pitch}} = \text{MA}_{\text{screw}}}$$

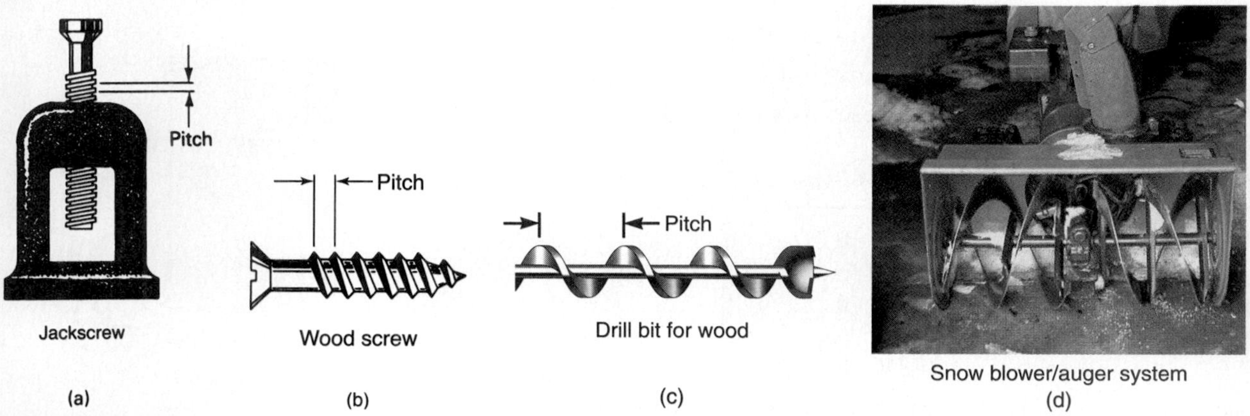

Figure 9.19

In the case of a jackscrew, *r is the length of the handle turning the screw and not the radius of the screw.*

EXAMPLE 1

Find the mechanical advantage of a jackscrew having a pitch of 25.0 mm and a handle radius of 35.0 cm.

Data:

$$\text{pitch} = 25.0 \text{ mm} = 2.50 \text{ cm}$$
$$r = 35.0 \text{ cm}$$
$$\text{MA}_{\text{screw}} = ?$$

Basic Equation:

$$\text{MA}_{\text{screw}} = \frac{2\pi r}{\text{pitch}}$$

Working Equation: Same

Substitution:

$$\text{MA}_{\text{screw}} = \frac{2\pi(35.0 \text{ cm})}{2.50 \text{ cm}}$$
$$= 88.0$$

EXAMPLE 2

What resistance can be lifted using the jackscrew in Example 1 if an effort of 203 N is exerted?

Data:

$$\text{MA}_{\text{screw}} = 88.0$$
$$F_E = 203 \text{ N}$$
$$F_R = ?$$

Basic Equation:

$$\text{MA}_{\text{screw}} = \frac{F_R}{F_E}$$

Working Equation:

$$F_R = (F_E)(\text{MA}_{\text{screw}})$$

Substitution:

$$F_R = (203 \text{ N})(88.0)$$
$$= 17{,}900 \text{ N}$$

EXAMPLE 3

A 19,400-N weight is raised using a jackscrew having a pitch of 5.00 mm and a handle length of 255 mm. What force must be applied?

Data:

$$\text{pitch} = 5.00 \text{ mm}$$
$$r = 255 \text{ mm}$$
$$F_R = 19{,}400 \text{ N}$$
$$F_E = ?$$

Basic Equation:

$$F_R \cdot \text{pitch} = F_E \cdot 2\pi r$$

Working Equation:

$$F_E = \frac{F_R\,(\text{pitch})}{2\pi r}$$

Substitution:

$$F_E = \frac{(19{,}400\text{ N})(5.00\text{ mm})}{2\pi\,(255\text{ mm})}$$
$$= 60.5\text{ N}$$

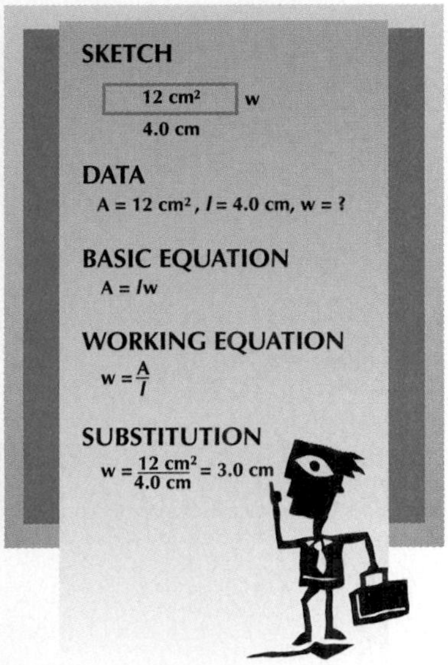

PROBLEMS 9.6

Given $F_R \cdot \text{pitch} = F_E \cdot 2\pi r$, find each missing quantity.

	F_R	F_E	*Pitch*	r
1.	20.7 N	5.30 N	3.70 mm	____ mm
2.	____ lb	17.6 lb	0.130 in.	24.5 in.
3.	234 N	9.80 N	____ mm	53.9 mm
4.	1190 N	____ N	2.97 mm	67.4 mm
5.	$37\overline{0}$ lb	12.0 lb	____ in.	11.2 in.

Given $\text{MA}_{\text{screw}} = \dfrac{2\pi r}{\text{pitch}}$, find each missing quantity.

	MA	r	*Pitch*
6.	7.00	34.0 mm	____ mm
7.	____	3.79 in.	0.812 in.
8.	9.00	____ in.	0.970 in.
9.	____	7.40 mm	1.32 mm
10.	13.0	____ mm	2.10 mm

11. A 3650-lb car is raised using a jackscrew having eight threads to the inch and a handle 15.0 in. long. (a) What effort must be applied? (b) What is the MA?
12. The mechanical advantage of a jackscrew is 97.0. (a) If the handle is 34.5 cm long, what is the pitch? (b) How much weight can be raised by applying an effort of 405 N to the jackscrew?
13. A wood screw with pitch 0.125 in. is advanced into wood using a screwdriver whose handle is 1.50 in. in diameter. (a) What is the mechanical advantage of the screw? (b) What is the resistance of the wood if 15.0 lb of effort is applied on the wood screw? (c) What is the resistance of the wood if 15.0 lb of effort is applied to the wood screw using a screwdriver whose handle is 0.500 in. in diameter?
14. The handle of a jackscrew is 60.0 cm long. (a) If the mechanical advantage is 78.0, what is the pitch? (b) How much weight can be raised by applying a force of $43\overline{0}$ N to the jackscrew handle?

9.7 The Wedge

A **wedge** is an inclined plane in which the plane is moved instead of the resistance. Examples are shown in Fig. 9.20.

Finding the mechanical advantage of a wedge is not practical because of the large amount of friction. A narrow wedge is easier to drive than a thick wedge. Therefore, the mechanical advantage depends on the ratio of its length to its thickness.

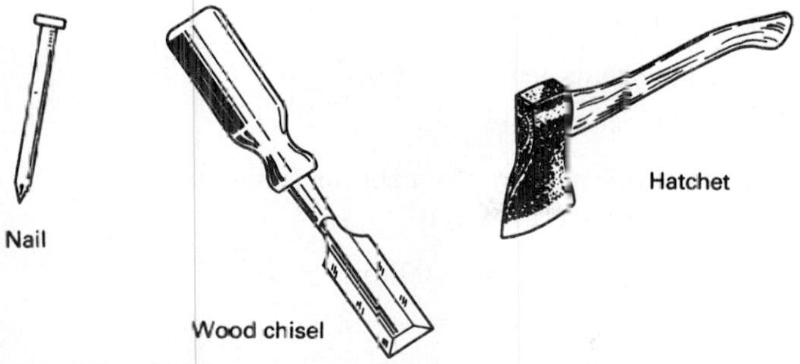

Figure 9.20 Inclined planes where the plane moves instead of the resistance

9.8 Compound Machines

A **compound machine** is a combination of simple machines. Examples are shown in Fig. 9.21. In most compound machines, *the total mechanical advantage is the product of the mechanical advantage of each simple machine.*

$$MA_{\text{compound machine}} = (MA_1)(MA_2)(MA_3)\cdots$$

Figure 9.21 Compound machines multiply mechanical advantage. (Reprinted courtesy of Caterpillar Inc.)

EXAMPLE

A crate weighing 95$\bar{0}$0 N is pulled up the inclined plane using the pulley system shown in Fig. 9.22.

(a) Find the mechanical advantage of the total system.
(b) What effort force (F_E) is needed?

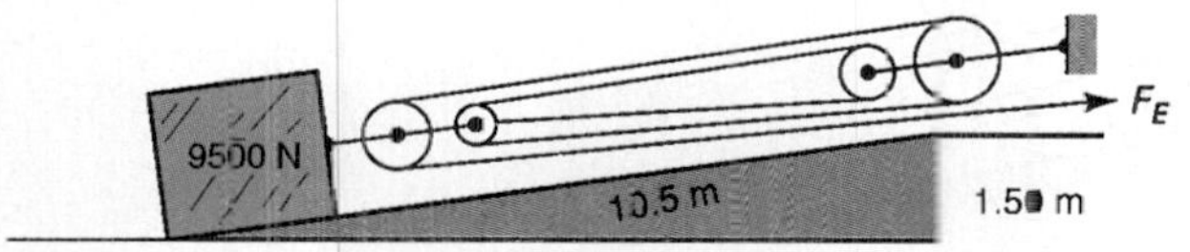

Figure 9.22

(a) First, find the MA of the inclined plane.

$$MA_{\text{inclined plane}} = \frac{\text{length of plane}}{\text{height of plane}} = \frac{10.5\ \cancel{m}}{1.50\ \cancel{m}} = 7.00$$

The MA of the pulley system = 5 (the number of supporting strands).
The MA of the total system (compound machine) is

$$(MA_{\text{inclined plane}})(MA_{\text{pulley system}}) = (7.00)(5) = 35.0$$

(b) Data:

$$MA_{\text{compound machine}} = 35.0$$
$$F_R = 95\bar{0}0\text{ N}$$
$$F_E = ?$$

Basic Equation:

$$MA_{\text{compound machine}} = \frac{F_R}{F_E}$$

Working Equation:

$$F_E = \frac{F_R}{MA_{\text{compound machine}}}$$

Substitution:

$$F_E = \frac{95\bar{0}0\text{ N}}{35.0}$$
$$= 271\text{ N}$$

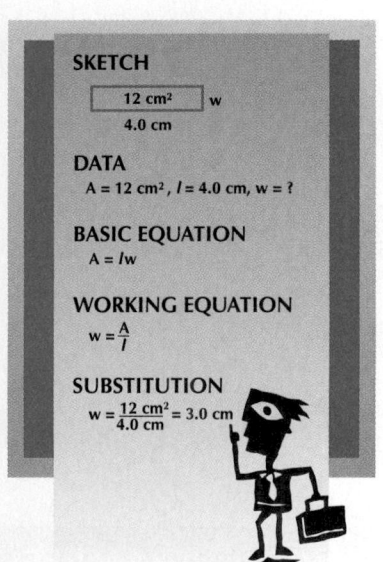

SKETCH

12 cm² w
4.0 cm

DATA
$A = 12\text{ cm}^2, l = 4.0\text{ cm}, w = ?$

BASIC EQUATION
$A = lw$

WORKING EQUATION
$w = \frac{A}{l}$

SUBSTITUTION
$w = \frac{12\text{ cm}^2}{4.0\text{ cm}} = 3.0\text{ cm}$

PROBLEMS 9.8

1. The box shown in Fig. 9.23 being pulled up an inclined plane using the indicated pulley system (called a block and tackle) weighs 9790 N. If the inclined plane is 6.00 m long and the height of the platform is 2.00 m, find the mechanical advantage of this compound machine.

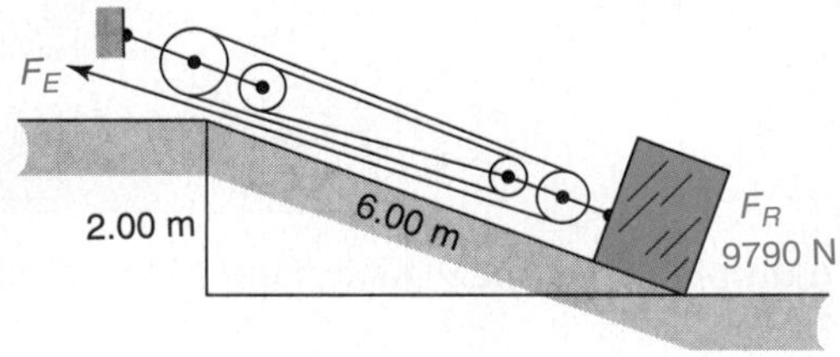

Figure 9.23

2. What effort force must be exerted to move the box to the platform in Problem 1?
3. Find the mechanical advantage of the compound machine shown in Fig. 9.24. The radius of the crank is 1.00 ft and the radius of the axle is 0.500 ft.

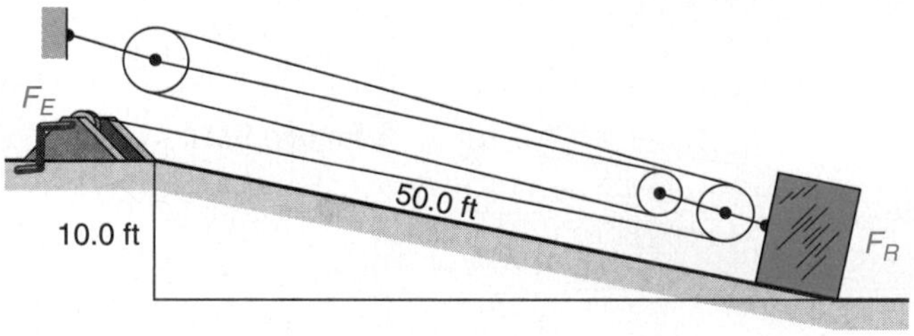

Figure 9.24

4. If an effort of $30\bar{0}$ lb is exerted, what weight can be moved up the inclined plane using the compound machine in Problem 3?
5. What effort is required to move a load of 1.50 tons up the inclined plane using the compound machine in Problem 3? (1 ton = $200\bar{0}$ lb)
6. Find the mechanical advantage of the compound machine in Problem 1 if the inclined plane is 8.00 m long and 2.00 m high.
7. What effort force is needed to move a box of weight $250\bar{0}$ N to the platform in Problem 6?
8. Find the mechanical advantage of the compound machine in Problem 3 if the radius of the crank is 40.0 cm, the radius of the axle is 12.0 cm, the length of the inclined plane is 12.0 m, and the height of the inclined plane is 50.0 cm.
9. If an effort of $45\bar{0}$ N is exerted in Problem 8, what weight can be moved up the inclined plane?
10. What effort force (in N) is needed to move 2.50 metric tons up the inclined plane in Problem 8?

9.9 The Effect of Friction on Simple Machines

The *law of simple machines* has been stated for a particular machine and in the general case in the previous sections. Each has been stated in terms of what is called *ideal mechanical advantage* (IMA), in which we have 100% efficiency. Actually, in every machine energy is lost through heat to overcome friction. This lost energy decreases the efficiency of the machine; that is, more work must be put into a machine than is gotten out of the machine. This lost energy is heat energy, which results in machine wear or even burning out of certain parts of the machine.

Throughout this chapter we have been discussing simple machines, mechanical advantage, resistance force, effort force, resistance distance, and effort distance in the ideal case while ignoring friction. For example, in the pulley system in Fig. 9.25, we find the IMA is 4 to 1. Ideally, it takes 300 N of effort force to lift the resistant force of 1200 N; that is, it ideally takes 1 N of force to raise 4 N of weight. However, it actually takes 400 N of effort to lift the 1200-N weight; the *actual mechanical advantage* (AMA) is then 3 to 1; that is, it actually takes 1 N of force to raise 3 N of weight.

In general, the actual mechanical advantage is found by the following formula:

$$\text{AMA} = \frac{F_R}{F_E} = \frac{\text{resistance force}}{\text{effort force}}$$

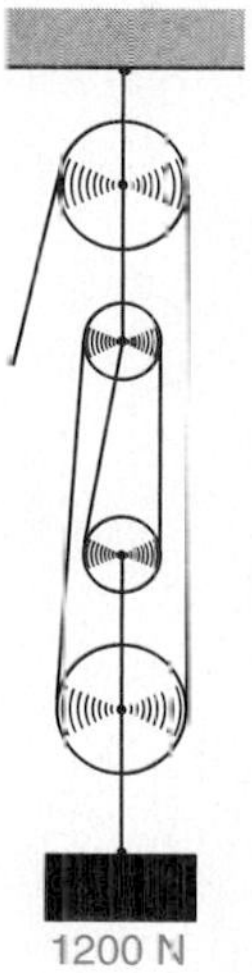

Figure 9.25 Friction has been ignored in our study of pulleys.

In Section 4.3, we studied the effects of sliding friction. The actual effects of sliding friction are substantial in inclined plane problems. In Example 1 of Section 9.5, the inclined plane (repeated in Fig. 9.26) has an IMA of 4 to 1. Actually, it takes 545 N of effort to move the $150\bar{0}$-N box up the ramp. Therefore, the AMA is

$$\text{AMA} = \frac{F_R}{F_E} = \frac{150\bar{0}\text{ N}}{545\text{ N}} = 2.75$$

That is, it actually takes 1.00 N of force to push 2.75 N up the ramp.

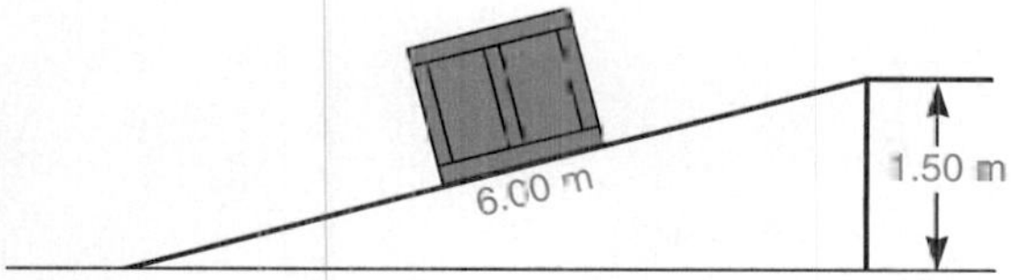

Figure 9.26

TRY THIS ACTIVITY

Reducing Friction

Observe the amount of effort force required to drag a heavy book up an inclined plane. Before repeating this process, place several small wooden dowels or marbles under the book. Drag the book up the incline again and observe the amount of effort force required. How do the wooden dowels or marbles reduce the effort force required to move the book up the incline? The ancient Egyptians may have used large dowels to reduce the effort force to drag massive stone blocks to the tops of pyramids.

Glossary

Compound Machine A combination of simple machines. Its total mechanical advantage is the product of the mechanical advantage of each simple machine. (p. 265)
Efficiency The ratio of the work output to the work input of a machine. (p. 248)
Effort The force applied to a machine. (p. 246)
Effort Arm The distance from the effort force to the fulcrum of a lever. (p. 248)
Fixed Pulley A pulley that is fastened to a fixed object. (p. 255)
Fulcrum A pivot about which a lever is free to turn. (p. 248)
Inclined Plane A plane surface set at an angle from the horizontal used to raise objects that are too heavy to lift vertically. (p. 259)
Law of Simple Machines Resistance force × resistance distance = effort force × effort distance. (p. 247)
Lever A rigid bar free to turn on a pivot called a *fulcrum*. (p. 248)
Machine An object or system that is used to transfer energy from one place to another and allows work to be done that could not otherwise be done or could not be done as easily. (p. 246)
Mechanical Advantage The ratio of the resistance force to the effort force. (p. 247)
Movable Pulley A pulley that is fastened to the object to be moved. (p. 255)
Pitch The distance a screw advances in one revolution of the screw. Also the distance between two successive threads. (p. 262)
Pulley A grooved wheel that turns readily on an axle and is supported in a frame. (p. 255)
Resistance The force overcome by a machine. (p. 246)
Resistance Arm The distance from the resistance force to the fulcrum of a lever. (p. 248)
Screw An inclined plane wrapped around a cylinder. (p. 262)
Simple Machine Any one of six mechanical devices in which an applied force results in useful work. The six simple machines are the lever, the wheel and axle, the pulley, the inclined plane, the screw, and the wedge. (p. 246)
Wedge An inclined plane in which the plane is moved instead of the resistance. (p. 264)
Wheel-and-Axle A large wheel attached to an axle so that both turn together. (p. 253)

Formulas

9.1 resistance force × resistance distance = effort force × effort distance

$$\text{MA} = \frac{\text{resistance force}}{\text{effort force}}$$

$$\text{efficiency} = \frac{\text{work output}}{\text{work input}} \times 100\% = \frac{F_{\text{output}} \times s_{\text{output}}}{F_{\text{input}} \times s_{\text{input}}} \times 100\%$$

9.2 $MA_{lever} = \dfrac{\text{effort arm}}{\text{resistance arm}} = \dfrac{s_E}{s_R}$

$F_R \cdot s_R = F_E \cdot s_E$

9.3 $MA_{wheel\text{-}and\text{-}axle} = \dfrac{\text{radius of effort force}}{\text{radius of resistance force}} = \dfrac{r_E}{r_R}$

$F_R \cdot r_R = F_E \cdot r_E$

9.4 MA_{pulley} = number of strands holding the resistance

$MA_{pulley} = \dfrac{s_E}{s_R}$

9.5 $MA_{inclined\ plane} = \dfrac{\text{length of plane}}{\text{height of plane}}$

$F_R \cdot \text{height} = F_E \cdot \text{length}$

9.6 $MA_{screw} = \dfrac{2\pi r}{\text{pitch}}$

$F_R \cdot \text{pitch} = F_E \cdot 2\pi r$

9.8 $MA_{compound\ machine} = (MA_1)(MA_2)(MA_3)\cdots$

9.9 $AMA = \dfrac{F_R}{F_E} = \dfrac{\text{resistance force}}{\text{effort force}}$

Review Questions

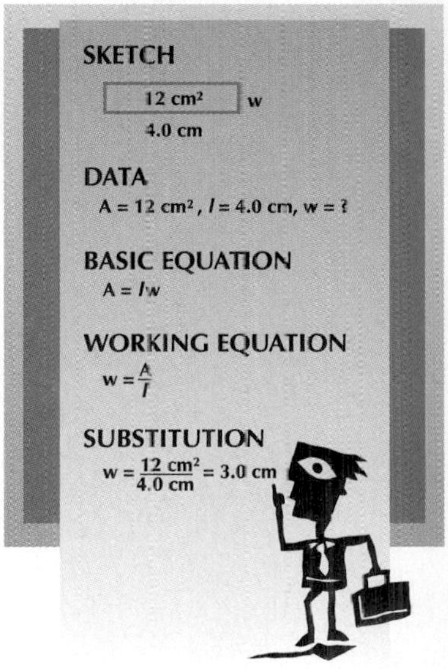

1. Which of the following is not a simple machine?
   (a) Pulley (b) Lever (c) Wedge (d) Automobile
2. The force applied to the machine is the
   (a) effort. (b) frictional. (c) horizontal. (d) resistance.
3. Efficiency is
   (a) the same as mechanical advantage. (b) a percentage.
   (c) impossible to determine.
4. A second-class lever has
   (a) two fulcrums. (b) two effort arms.
   (c) two resistance arms. (d) a resistance arm shorter than the effort arm.
5. A pulley has eight strands holding the resistance. The mechanical advantage is
   (a) 4. (b) 8. (c) 16. (d) 64.
6. The mechanical advantage of a compound machine
   (a) is the sum of the MA of each simple machine.
   (b) is the product of the MA of each simple machine.
   (c) cannot be found.
   (d) is none of the above.
7. Cite three examples of machines used to multiply speed.
8. What name is given to the force overcome by the machine?
9. State the law of simple machines in your own words.
10. What is the term used for the ratio of the resistance force to the effort force?
11. What is the term used for the ratio of the amount of work obtained from a machine to the amount of work put into the machine?
12. Does a friction-free machine exist?
13. What is the pivot point of a lever called?
14. In your own words, state how to find the MA of a lever.
15. Which type of lever do you think would be most efficient?

16. State the law of simple machines as it is applied to levers.
17. Where is the fulcrum located in a third-class lever?
18. In your own words, explain the law of simple machines as applied to the wheel-and-axle.
19. Does the MA of a wheel-and-axle depend on the force applied?
20. Describe the difference between a fixed pulley and a movable pulley.
21. Does the MA of a pulley depend on the radius of the pulley?
22. How can you find the MA of an inclined plane?
23. In your own words, describe the pitch of a screw.
24. How does the MA of a jackscrew differ from the MA of a screwdriver?

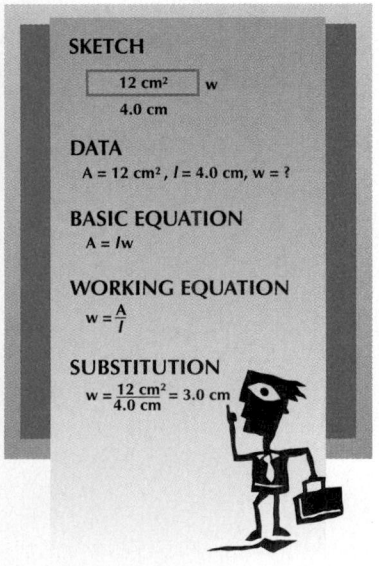

SKETCH

12 cm² w
4.0 cm

DATA

$A = 12\ \text{cm}^2$, $l = 4.0$ cm, $w = ?$

BASIC EQUATION

$A = lw$

WORKING EQUATION

$w = \frac{A}{l}$

SUBSTITUTION

$w = \frac{12\ \text{cm}^2}{4.0\ \text{cm}} = 3.0$ cm

Review Problems

1. A girl uses a lever to lift a box. The box has a resistance force of $25\bar{0}$ N while she exerts an effort force of 125 N. What is the mechanical advantage of the lever?
2. A bicycle requires 1575 N m of input but only puts out 1150 N m of work. What is the bicycle's efficiency?
3. A lever uses an effort arm of 2.75 m and has a resistance arm of 72.0 cm. What is the lever's mechanical advantage?
4. Two people are on a teeter-totter. One person exerts a force of 540 N and is 2.00 m from the fulcrum. If they are to remain balanced, how much force does the other person exert if she is (a) also 2.00 m from the fulcrum? (b) 3.00 m from the fulcrum?
5. A wheel-and-axle has an effort force of 125 N and an effort radius of 17.0 cm. (a) If the resistance force is 325 N, what is the resistance radius? (b) Find the mechanical advantage.
6. What is the mechanical advantage of a pulley system having 12 strands holding the resistance?
7. A pulley system has a mechanical advantage of 5. What is the resistance force if an effort of 135 N is exerted?
8. An inclined plane has a height of 1.50 m and a length of 4.50 m. (a) What effort must be exerted to pull up an 875-N box? (b) What is the mechanical advantage?
9. What height must a 10.0-ft-long inclined plane be to lift a $10\bar{0}0$-lb crate with $23\bar{0}$ lb of effort?
10. A screw has a pitch of 0.0200 cm. An effort force of 29.0 N is used to turn a screwdriver whose handle diameter is 36.0 mm. What is the maximum resistance force?
11. A 945-N resistance force is overcome with a 13.5-N effort using a screwdriver whose handle is 24.0 mm in diameter. What is the pitch of the screw?
12. Find the mechanical advantage of a jackscrew with a 1.50-cm pitch and a handle 36.0 cm long.
13. A courier uses a bicycle with rear wheel radius 35.6 cm and gear radius 4.00 cm. If a force of 155 N is applied to the chain, the wheel rim moves 14.0 cm. (a) If the efficiency is 95.0%, what is the ideal mechanical advantage of the wheel and gear? (b) What is the actual mechanical advantage of the wheel and gear? (c) What is the force on the pavement applied by the wheel?
14. (a) If the gear radius is doubled on the courier's bicycle in Problem 13, how does the mechanical advantage change? (b) How far did the courier move the chain to produce the 14.0-cm linear movement of the rim?
15. A farmer uses a pulley system to raise a 225-N bale 16.5 m. A 129-N force is applied by pulling a rope 33.0 m. What is the mechanical advantage of the pulley system?
16. A laborer uses a lever to raise a 1250-N rock a distance of 13.0 cm by applying a force of 225 N. If the efficiency of the lever is 88.7%, how far did the laborer have to move his end of the lever?

Find the mechanical advantage of each pulley system.

17.

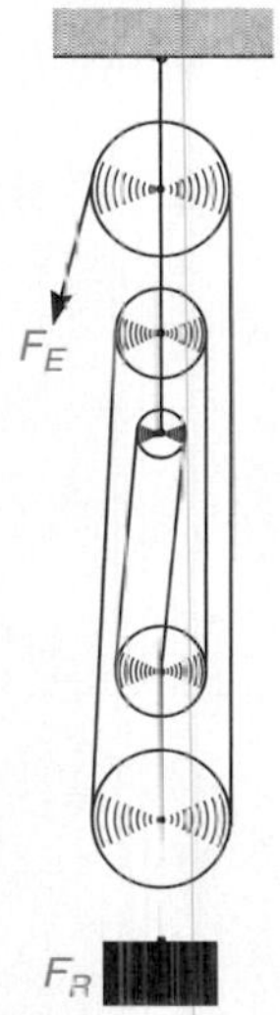

18.

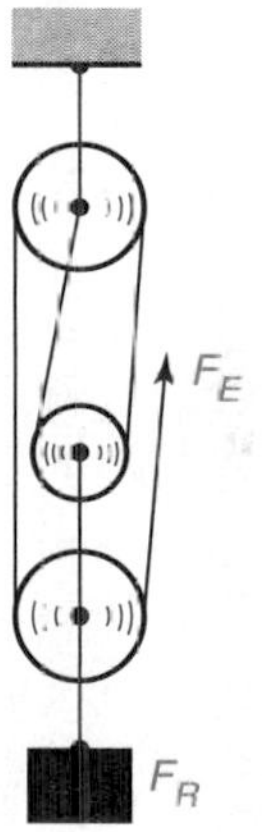

19. (a) Find the mechanical advantage of the compound machine in Fig. 9.27. The radius of the crank is 32.0 cm and the radius of the axle is 8.00 cm. (b) If an effort force of 75 N is applied to the handle of the crank, what force can be moved up the inclined plane?

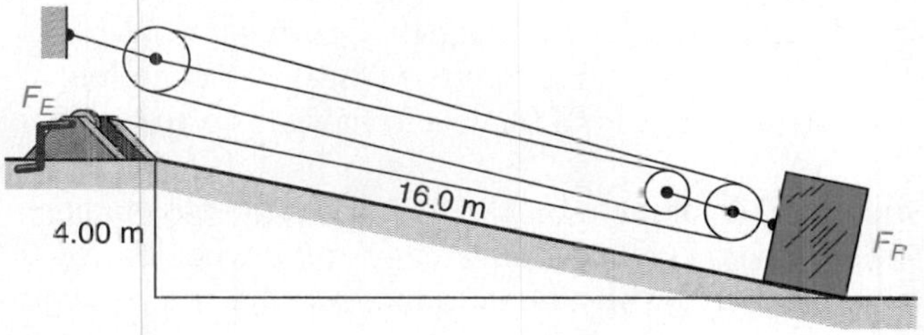

Figure 9.27

20. If an effort force of 45 N is applied to a simple machine and moves a resistance of 270 N, what is the actual mechanical advantage?

APPLIED CONCEPTS

1. In the third century BC, Archimedes said, "Give me a long enough lever and a firm fulcrum and I will lift the earth." (a) Using the moon as a fixed fulcrum, how long a lever would Archimedes have needed to lift the earth? (b) What would be the mechanical advantage of such a simple machine? (Disregard the mass of the lever. Assume the weight of the earth is 5.85×10^{25} N, Archimedes' weight is 858 N, and the distance from the earth to the moon is 3.84×10^{8} m.)
2. (a) What is the mechanical advantage of the fishing pole in Fig. 9.28? (b) If the fisherman's hand moves forward 30.5 cm, how far will the tip of the fishing pole move? (c) How fast will the tip of the fishing pole move if the forward motion is completed in 0.554 s? (d) In general, what is the benefit of using a simple machine that has a mechanical advantage less than 1.0?
3. A snowblower auger has a radius of 7.75 in. and a pitch of 5.80 in. (a) What is the mechanical advantage of this auger system? (b) How much effort force is needed for the auger to throw 23.4 lb of snow? (c) What happens to the mechanical advantage if the pitch is reduced?

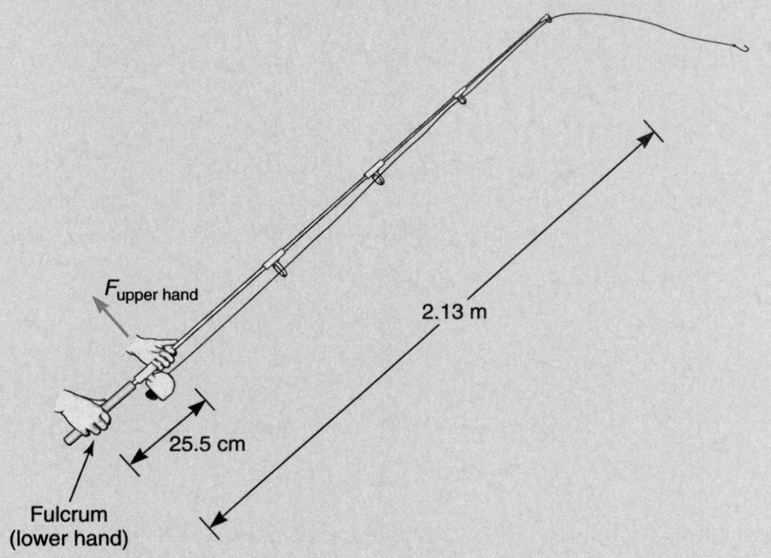

Figure 9.28

4. Aaron, a bicycle mechanic, is studying the mechanical advantage of the rear wheel. The rear wheel's radius is 14.0 in. and the chain is currently on a gear with radius 1.65 in. (Fig. 9.29). (a) If the chain applies a force of 54.5 lb on the wheel, what is the mechanical advantage of the wheel? (b) What is the resistance force? (c) How many inches of chain must be used to turn the wheel one complete rotation? (d) Finally, under what circumstances is it beneficial for a bicycle to have a mechanical advantage less than 1.0?
5. Willie is using a wheelbarrow (Fig. 9.30) to move 345 lb of patio block. (a) What is the mechanical advantage of the wheelbarrow handles? (b) What is the mechanical advantage of the wheelbarrow's wheel-and-axle? (c) How can the design of the wheelbarrow be altered to increase the mechanical advantage?

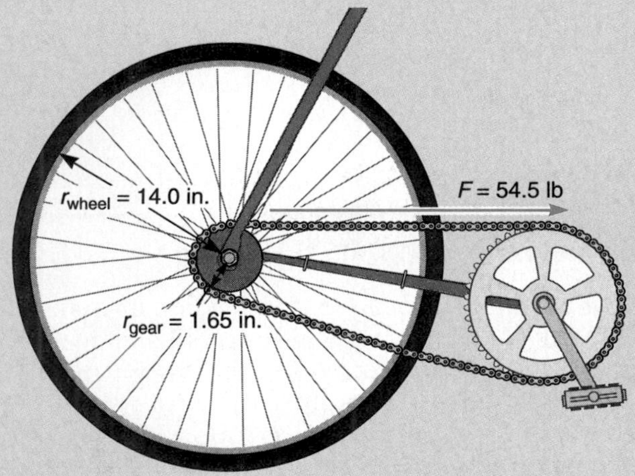

Figure 9.29

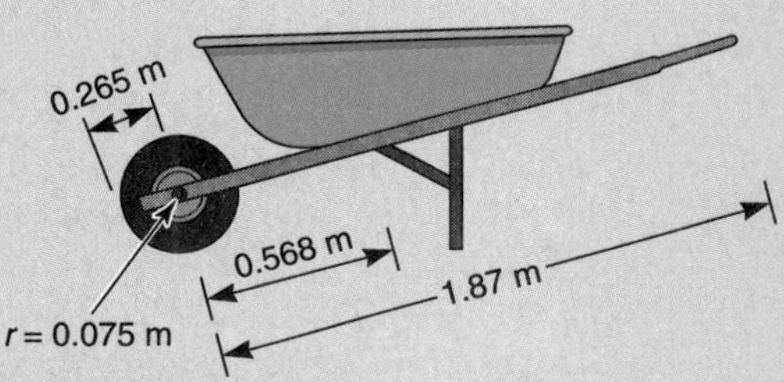

Figure 9.30

CHAPTER 10

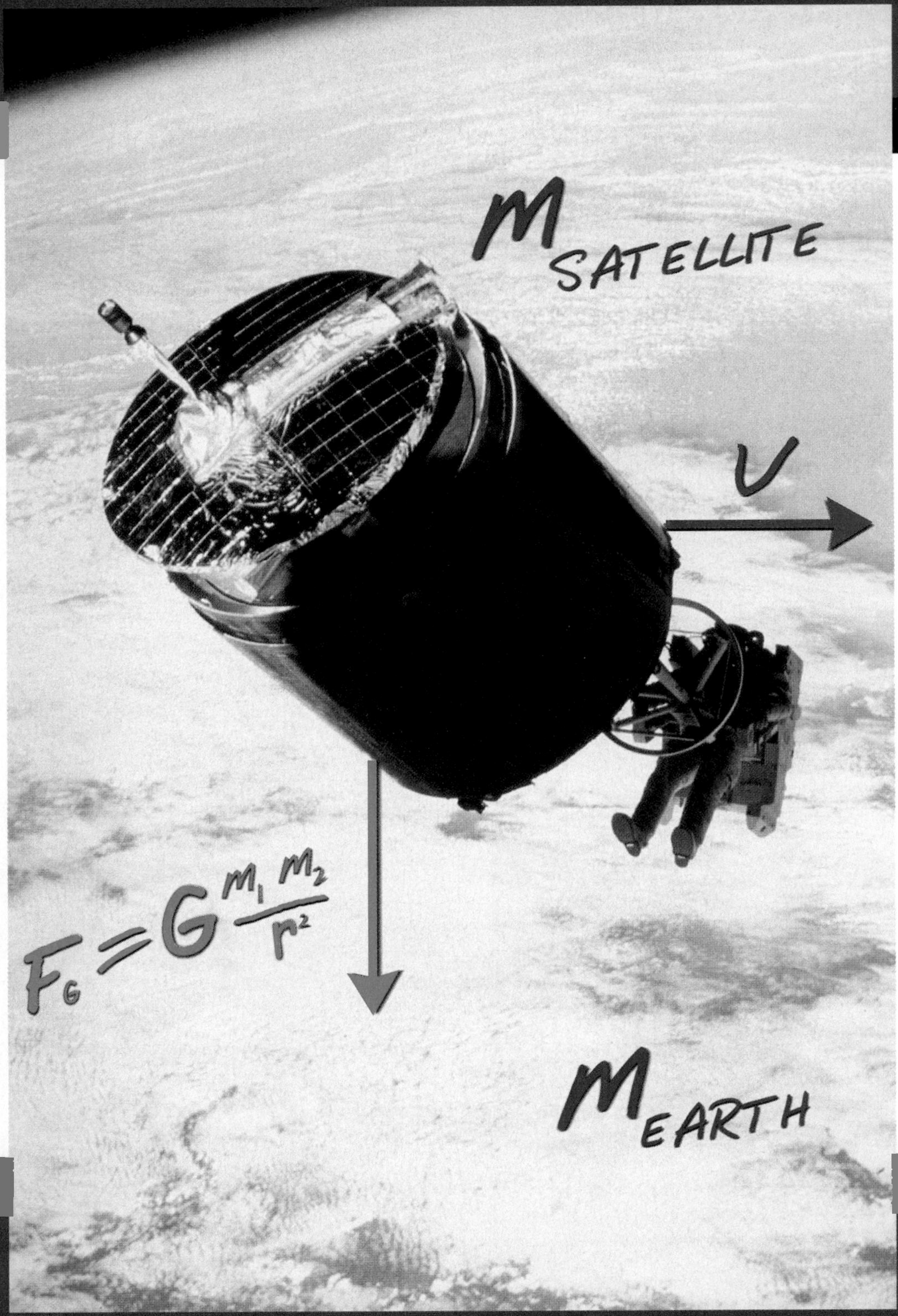

Photo courtesy of NASA Headquarters

UNIVERSAL GRAVITATION AND SATELLITE MOTION

What allows thousands of satellites to orbit the earth? Why does vertical motion on the moon appear to take place in slow motion? In this chapter we will discover how we can send massive objects into orbit and why a person weighs less on the moon than on the earth. In this chapter we appreciate the importance of understanding the difference between mass and weight.

Objectives

The major goals of this chapter are to enable you to:

1. Describe how gravitation acts between all objects with mass.
2. Calculate what your weight would be on various planets.
3. Explain the concept of a gravitational field.
4. Express how objects are able to orbit the earth.
5. Determine the connection between orbiting the earth and being in free fall.

10.1 Universal Gravitation

What makes an astronaut weigh less on the moon than on the earth? **Newton's law of universal gravitation** states that all objects that have mass are attracted to one another by a gravitational force. Newton determined that the greater the mass of two objects, the stronger is the attractive gravitational force between them. He also discovered that as objects move away from each other, the attraction between them diminishes dramatically. Newton's law of universal gravitation is

$$F_G = G\frac{m_1 m_2}{r^2}$$

where

F_G = gravitational force between the two objects
$G = 6.67 \times 10^{-11}\ \text{N m}^2/\text{kg}^2 = 3.44 \times 10^{-8}\ \text{lb ft}^2/\text{slug}^2$ (universal gravitational constant)
m_1 = mass of the first object
m_2 = mass of the second object
r = distance between the centers of mass of the two objects.

The law of universal gravitation makes it possible to calculate an object's weight on another planet, the gravitational force exerted between the person sitting next to you in class and you, and the gravitational forces between planets, moons, and stars.

Table 10.1 Table of Planetary Data

Object	Average Radius (m)	Mass (kg)	Mean Sun to Planet Distance (m)
Sun	6.96×10^{8}	1.99×10^{30}	—
Mercury	2.44×10^{6}	3.30×10^{23}	5.79×10^{10}
Venus	6.05×10^{6}	4.87×10^{24}	1.08×10^{11}
Earth	6.38×10^{6}	5.97×10^{24}	1.50×10^{11}
Mars	3.40×10^{6}	6.42×10^{23}	2.28×10^{11}
Jupiter	7.15×10^{7}	1.90×10^{27}	7.78×10^{11}
Saturn	6.03×10^{7}	5.69×10^{26}	1.43×10^{12}
Uranus	2.56×10^{7}	8.66×10^{25}	2.87×10^{12}
Neptune	2.48×10^{7}	1.03×10^{26}	4.50×10^{12}
Pluto	1.15×10^{6}	1.5×10^{22}	5.91×10^{12}

Other information:

earth to moon distance $= 3.84 \times 10^{8}$ m
moon's radius $= 1.74 \times 10^{6}$ m
moon's mass $= 7.35 \times 10^{22}$ kg

EXAMPLE 1

Compare (a) the gravitational force between Vince and the earth to (b) the gravitational force between Vince and Matt, who is sitting 2.34 m away from Vince. Both Matt and Vince have a mass of 85.5 kg.

Data:

$$m_{\text{earth}} = 5.97 \times 10^{24}\ \text{kg}$$
$$m_{\text{matt}} = 85.5\ \text{kg}$$
$$m_{\text{vince}} = 85.5\ \text{kg}$$
$$r_{\text{vince-earth}} = 6.38 \times 10^{6}\ \text{m}$$
$$r_{\text{vince-matt}} = 2.34\ \text{m}$$

Basic Equation:

$$F_G = G\frac{m_1 m_2}{r^2}$$

Working Equation: Same

Substitution:

(a) $$F_G = 6.67 \times 10^{-11} \frac{\text{N m}^2}{\text{kg}^2} \frac{(85.5\text{ kg})(5.97 \times 10^{24}\text{ kg})}{(6.38 \times 10^6\text{ m})^2}$$

$$= 836\text{ N}$$

(b) $$F_G = 6.67 \times 10^{-11} \frac{\text{N m}^2}{\text{kg}^2} \frac{(85.5\text{ kg})(85.5\text{ kg})}{(2.34\text{ m})^2}$$

$$= 8.90 \times 10^{-8}\text{ N}$$

The gravitational force between Vince and Matt is virtually undetectable compared to the gravitational force between Vince and the earth.

Henry Cavendish (1731–1810),

English physicist and chemist, was born in France. He was the first person to experimentally determine the universal gravitational constant used in Isaac Newton's law of universal gravitation (Fig. 10.1). He also found that water was composed of hydrogen and oxygen and estimated the density of the earth.

Photo courtesy of Dorling Kindersley

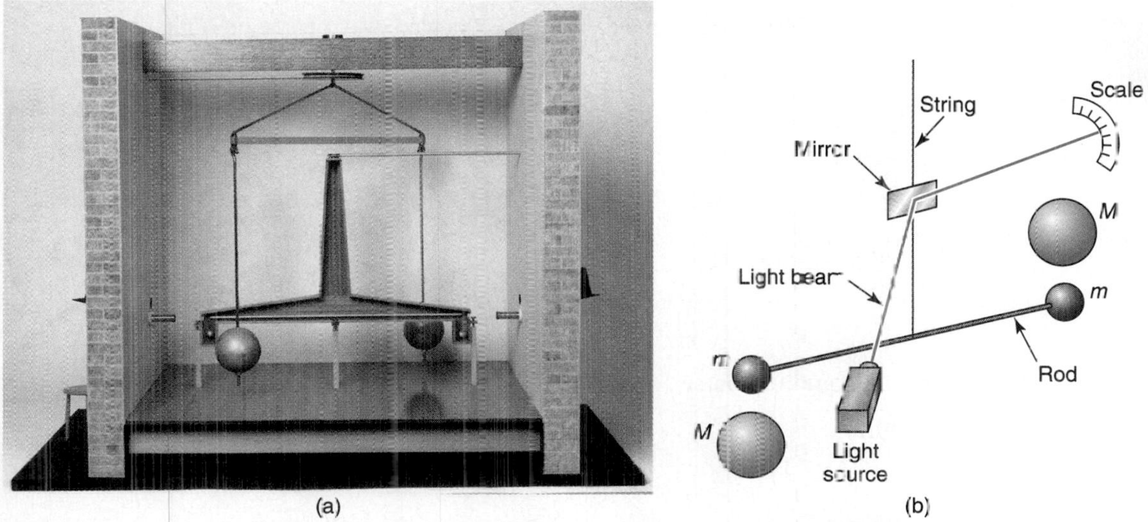

Figure 10.1 The Cavendish experiment. In 1798, Henry Cavendish attached a pair of 2-in. lead balls to the ends of a light horizontal rod, which was suspended by a string onto which a mirror was mounted. When this pair of lead balls was brought near two more massive 12-in. balls, the gravitational force caused the smaller balls to move toward the more massive balls, which caused the string and mirror to twist. The tiny movement was measured by reflecting a narrow light beam onto a scale. By carefully measuring the very small twisting motion of the suspended balls, Cavendish was able to determine *G* within 1% of today's accepted value.

EXAMPLE 2

One of Newton's preliminary studies of universal gravitation involved the moon and a falling apple. (a) Find the force between an apple ($m = 0.153$ kg) and the earth. (b) Compare this to the force between the moon and the earth, two objects that are quite massive yet far away from each other. See Table 10.1.

Data:

$$m_{apple} = 0.153\text{ kg}$$
$$m_{moon} = 7.35 \times 10^{22}\text{ kg}$$
$$m_{earth} = 5.97 \times 10^{24}\text{ kg}$$
$$r_{apple-earth} = 6.38 \times 10^6\text{ m}$$
$$r_{moon-earth} = 3.84 \times 10^8\text{ m}$$

Basic Equation:

$$F_G = G\frac{m_1 m_2}{r^2}$$

Working Equation: Same

Substitutions:

(a) $F_G = 6.67 \times 10^{-11}\,\frac{\text{N m}^2}{\text{kg}^2}\,\frac{(0.153\text{ kg})(5.97 \times 10^{24}\text{ kg})}{(6.38 \times 10^6\text{ m})^2}$

$= 1.50\text{ N}$

(b) $F_G = 6.67 \times 10^{-11}\,\frac{\text{N m}^2}{\text{kg}^2}\,\frac{(7.35 \times 10^{22}\text{ kg})(5.97 \times 10^{24}\text{ kg})}{(3.84 \times 10^8\text{ m})^2}$

$= 1.98 \times 10^{20}\text{ N}$

The force between the earth and the moon is much greater than the force between the earth and the apple. Although the moon is much farther away from the earth than the apple is, the moon's much larger mass more than makes up for the difference in the distances.

EXAMPLE 3

What would be the gravitational force exerted on a 65.0-kg person on Jupiter?

Data:

$$m_{\text{person}} = 65.0\text{ kg}$$
$$m_{\text{Jupiter}} = 1.90 \times 10^{27}\text{ kg}$$
$$r_{\text{Jupiter}} = 7.15 \times 10^7\text{ m}$$

Basic Equation:

$$F_G = G\frac{m_1 m_2}{r^2}$$

Working Equation: Same

Substitution:

$$F_G = 6.67 \times 10^{-11}\,\frac{\text{N m}^2}{\text{kg}^2}\,\frac{(65.0\text{ kg})(1.90 \times 10^{27}\text{ kg})}{(7.15 \times 10^7\text{ m})^2}$$

$$= 1610\text{ N}$$

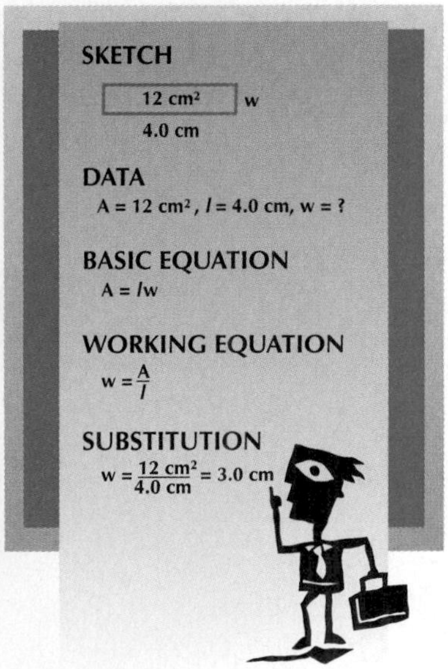

As these examples show, the gravitational force becomes stronger as the mass of the objects increases and as the distance between them decreases.

PROBLEMS 10.1

1. Compare the gravitational force that (a) the earth exerts on an 84.3-kg person and (b) the force that the sun exerts on the same person.
2. Find the gravitational force between the sun and the earth.
3. Find the gravitational force between the sun and Mercury.
4. Find the gravitational force between the sun and Jupiter.
5. Find the gravitational force between the sun and Pluto.
6. Explain why the gravitational force between the sun and Jupiter is greater than the gravitational force between the sun and the earth even though the sun and the earth are much closer to one another than are the sun and Jupiter.
7. A satellite is orbiting 3.22×10^5 m above the surface of the earth. If the mass of the satellite is 3.80×10^4 kg, what is the weight or gravitational force exerted on the satellite by the earth?

8. If the satellite in Problem 7 is orbiting at twice its original distance from the earth, what would be the weight or gravitational force exerted on the satellite by the earth?
9. What is the gravitational force exerted between an electron ($m = 9.11 \times 10^{-31}$ kg) and a proton ($m = 1.67 \times 10^{-27}$ kg) in a hydrogen atom where the distance between the electron and proton is 5.3×10^{-9} m?
10. The Apollo 16 lunar module had a mass of 4240 kg. Using Newton's law of universal gravitation, find its weight (a) on the earth and (b) on the moon.

10.2 Gravitational Fields

Isaac Newton's contemporaries did not welcome Newton's laws. The thought that the earth could exert a force on the moon almost 240,000 mi away without touching it seemed outrageous to many at the time. Until then, it was thought that most objects needed to be physically touching in order to have an effect on one another.

The area around a massive body where an object experiences a gravitational force is called a **gravitational field.** By introducing the concept of a gravitational field, physicists have been able to determine that an object does not need to be in physical contact with another object to exert a force on it (Fig. 10.2). Movement caused by the gravitational field is known as acceleration due to gravity.

The gravitational field around the surface of the earth is quite strong. However, if one travels out into space a bit, the distance alone affects the intensity of the field. That is, the acceleration due to gravity decreases significantly as shown in Fig. 10.3.

The concept of a gravitational field inspired Albert Einstein to establish a theory that would describe a warped space-time field in our universe. This fascinating, yet conceptually challenging, theory will be covered in Chapter 23.

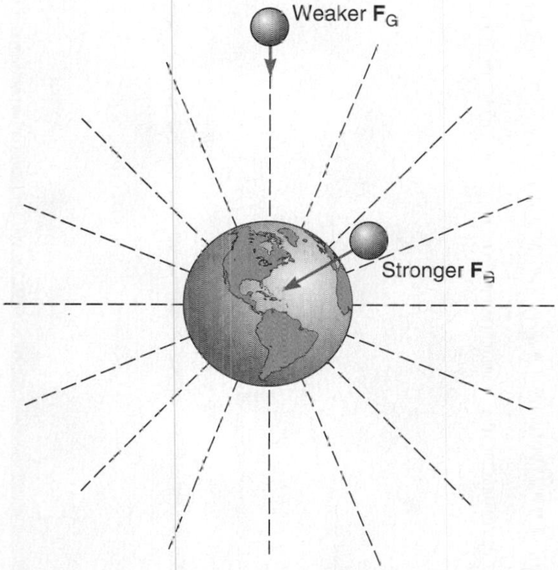

Figure 10.2 Gravitational field lines representing the strength of gravity around the earth. The closer the lines, the stronger is the gravitational attraction as shown by the force vectors.

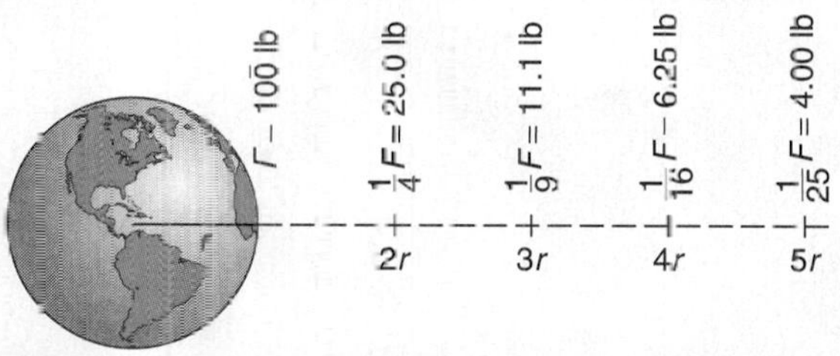

Figure 10.3 The gravitational force depends on the object's distance from the center of the earth. For example, if an object weighs 100 lb on the earth's surface, its weight becomes dramatically less the farther away the object is placed.

Photo courtesy of NASA/Johnson Space Center

Figure 10.4 Although an astronaut in orbit feels weightless, there is still a gravitational force acting on the astronaut. Here astronaut Joseph R. Tanner (right) stands fixed on the end of Discovery's remote manipulator system arm and aims a camera at the solar array panels of the Hubble Space Telescope as astronaut Gregory J. Harbaugh assists.

10.3 Satellite Motion

An **orbit** is the path taken by an object during its revolution around another object, such as the path of the moon or a satellite about the earth or of a planet about the sun. We have seen pictures of astronauts weightless and appearing to "float" in space as they orbit the earth (Fig. 10.4). The reason is because of gravity, not the lack of gravity, at that point in space.

In Chapter 3, we learned that when an object is dropped or falls from an elevated point on earth, nothing is pushing back up on the object so it appears weightless (Fig. 10.5). Gravity may be pulling the object down, but there is no normal force to support the object. In essence, when astronauts orbit the earth, they are in a constant state of free fall.

A person in free fall feels weightless, just like the astronauts. However, as the person continues to fall, he or she will crash to the earth. Newton realized that the moon and other orbiting objects must sustain a large enough horizontal velocity to remain in orbit and avoid crashing to the earth.

(a)

(b)

Figure 10.5 (a) When supported, the 0.50-kg mass weighs 4.90 N. (b) When no longer supported, the object appears weightless and the scale reads 0 N.

TRY THIS ACTIVITY

Weightless Water

Place two small holes on opposite sides near the bottom of a Styrofoam cup. Cover the holes with your thumb and finger and fill the cup with water. While standing on a stepladder or a table, grasp the cup near the top with your other hand. Uncover the holes just long enough to see the water streaming out. Then drop the cup. What happens to the streams of water as the cup and water fall? How is this activity similar to what an astronaut orbiting the earth experiences? How is it different?

An object launched beyond the earth's atmosphere and given a horizontal velocity moves sideways while falling at the same time. Moving horizontally at just the right velocity while falling due to gravity allows the space shuttle, astronauts, and other objects to continually miss the surface of the earth and achieve orbit (Fig. 10.6).

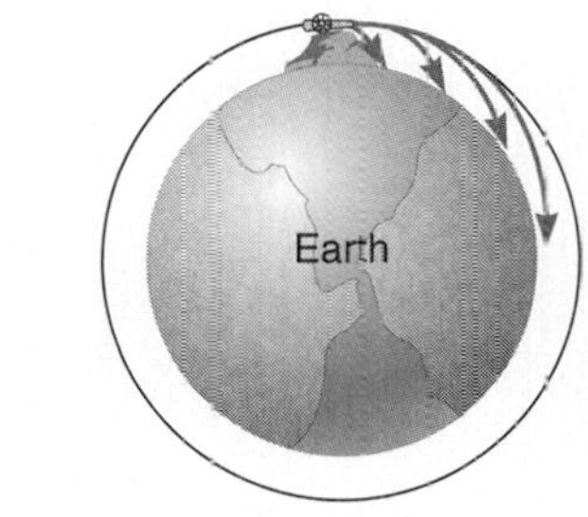

Figure 10.6

Isaac Newton conceived this concept several centuries before the development of rockets and artificial satellites and predicted that orbiting the earth could theoretically be done. Newton made the connection between satellite motion and the orbit of the moon around the earth. The velocity of satellites orbiting the earth or any other planet can be calculated by combining centripetal motion and the law of universal gravitation:

$$F_c = \frac{mv^2}{r}, \qquad F_G = G\frac{m_1 m_2}{r^2}$$

Solve the centripetal force equation for velocity,

$$v = \sqrt{\frac{F_c r}{m}}$$

Since the centripetal force holding a satellite in orbit is actually the gravitational force between the earth and the satellite, F_G can be substituted for F_c:

$$v = \sqrt{\frac{Gm_1 m_2 r}{r^2 m_1}} \qquad \text{(Cancel like terms.)}$$

$$\boxed{v = \sqrt{\frac{Gm_2}{r}}}$$

where

v = velocity of the satellite
$G = 6.67 \times 10^{-11}\ \text{N m}^2/\text{kg}^2 = 3.44 \times 10^{-8}\ \text{lb ft}^2/\text{slug}^2$
m_1 = mass of the satellite
m_2 = mass of the object being orbited
r = distance from the center of the earth to the satellite

The time or period, T, to orbit the earth or other celestial body is

$$\boxed{T = 2\pi\sqrt{\frac{r^3}{Gm_2}}}$$

(The derivation of this equation is beyond the scope of this text.)

Note that just as acceleration due to gravity is independent of the mass of falling objects, the velocity and the period of the object are independent of the mass of the satellite as well.

EXAMPLE 1

If a satellite orbits the earth at 20$\bar{0}$0 km above sea level, (a) how fast will the orbiting satellite travel and (b) how long will it take to orbit the earth once?

Data:

$$r_{\text{earth–satellite}} = 6.38 \times 10^6 \text{ m} + 2.00 \times 10^6 \text{ m} = 8.38 \times 10^6 \text{ m}$$
$$m_{\text{earth}} = 5.97 \times 10^{24} \text{ kg}$$

Basic Equations:

$$v = \sqrt{\frac{Gm}{r}}$$

$$T = 2\pi\sqrt{\frac{r^3}{Gm}}$$

Working Equations: Same

Substitutions:

(a) $$v = \sqrt{\frac{(6.67 \times 10^{-11} \text{ N m}^2/\text{kg}^2)(5.97 \times 10^{24} \text{ kg})}{8.38 \times 10^6 \text{ m}}}$$
$$= 6.89 \times 10^3 \text{ m/s}$$

$$\boxed{\sqrt{\frac{\text{N m}^2/\text{kg}^2 \text{ kg}}{\text{m}}} = \sqrt{\frac{(\text{kg m/s}^2)\text{ m}^2/\text{kg}^2 \text{ kg}}{\text{m}}} = \sqrt{\text{m}^2/\text{s}^2} = \text{m/s}}$$

(b) $$T = 2\pi\sqrt{\frac{(8.38 \times 10^6 \text{ m})^3}{(6.67 \times 10^{-11} \text{ N m}^2/\text{kg}^2)(5.97 \times 10^{24} \text{ kg})}}$$
$$= 7.64 \times 10^3 \text{ s} = 2.12 \text{ h}$$

$$\boxed{\sqrt{\frac{\text{m}^3}{\text{N m}^2/\text{kg}^2 \text{ kg}}} = \sqrt{\frac{\text{m}^3}{(\text{kg m/s}^2)\text{ m}^2/\text{kg}^2 \text{ kg}}} = \sqrt{\text{s}^2} = \text{s}}$$

EXAMPLE 2

An asteroid orbits the sun 8.35×10^{11} m from the sun. (a) How fast must the asteroid travel to maintain its orbit around the sun? (b) How long will it take the asteroid to orbit the sun?

Data:

$$r_{\text{sun–asteroid}} = 8.35 \times 10^{11} \text{ m}$$
$$m_{\text{sun}} = 1.99 \times 10^{30} \text{ kg}$$

Basic Equations:

$$v = \sqrt{\frac{Gm}{r}}$$

$$T = 2\pi\sqrt{\frac{r^3}{Gm}}$$

Working Equations: Same

Substitutions:

(a) $$v = \sqrt{\frac{(6.67 \times 10^{-11} \text{ N m}^2/\text{kg}^2)(1.99 \times 10^{30} \text{ kg})}{8.35 \times 10^{11} \text{ m}}}$$
$$= 1.26 \times 10^4 \text{ m/s}$$

(b) $$T = 2\pi\sqrt{\frac{(8.35 \times 10^{11} \text{ m})^3}{(6.67 \times 10^{-11} \text{ N m}^2/\text{kg}^2)(1.99 \times 10^{30} \text{ kg})}}$$
$$= 4.16 \times 10^8 \text{ s} = 13.2 \text{ yr}$$

PHYSICS CONNECTIONS

Satellite Orbits

Hundreds of functioning artificial satellites currently orbit the earth. These satellites have many functions such as tracking weather patterns, collecting scientific data about the earth and space, and transmitting telephone, television, and Internet communications. Our society has become dependent upon satellites over the last 45 years; much of our use of technology today would not be possible without satellites orbiting the earth.

Several types of orbits allow artificial satellites to maintain a stable orbit (Fig. 10.7). Communication and weather satellites need to remain over a fixed point on the earth. A television satellite dish on a house always points at a particular satellite in a stationary orbit. Similarly, common weather images are taken from satellites in a stationary orbit, called geosynchronous. Geosynchronous satellites orbit 22,400 mi (36,000 km) above the equator.

Imaging and intelligence satellites need to take clear and detailed pictures of the entire earth. Such satellites are placed in low-altitude, north-south polar orbits. The earth rotates underneath the satellites, allowing pictures and images to be collected for most of the earth.

Finally, the most versatile orbit for satellites is the asynchronous orbit. The space shuttle, scientific research satellites, and global positioning satellites need to be in an orbit that allows them to pass over various locations on earth at different times. These satellites typically orbit at varying altitudes ranging from 200 to 12,000 mi above the surface of the earth.

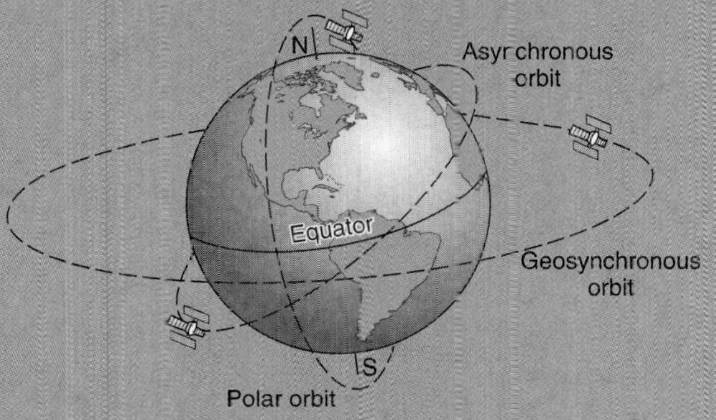

Figure 10.7 Three types of orbits for artificial satellites.

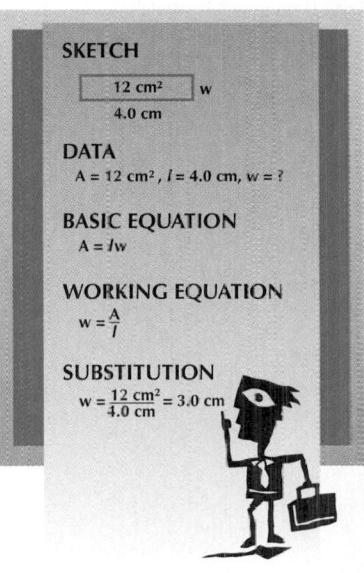

SKETCH

12 cm² w

4.0 cm

DATA

A = 12 cm², l = 4.0 cm, w = ?

BASIC EQUATION

$A = lw$

WORKING EQUATION

$w = \frac{A}{l}$

SUBSTITUTION

$w = \frac{12\text{ cm}^2}{4.0\text{ cm}} = 3.0\text{ cm}$

PROBLEMS 10.3

1. The moon orbits 3.84×10^8 m from the earth. How fast does the moon travel?
2. The moon orbits 3.84×10^8 m from the earth. How long does it take to orbit the earth?
3. Find the orbital velocity for Mercury as it orbits the sun.
4. Find the orbital velocity for the earth as it orbits the sun.
5. Find the orbital velocity for Saturn as it orbits the sun.
6. Find the orbital velocity for Uranus as it orbits the sun.
7. Find the time Mercury takes to orbit the sun.
8. Find the time the earth takes to orbit the sun.
9. Find the time Saturn takes to orbit the sun.
10. Find the time Uranus takes to orbit the sun.

Glossary

Gravitational Field The area around a massive body in which an object experiences a gravitational force. The more massive and closer an object is to that body, the stronger is the gravitational field. (p. 279)

Newton's Law of Universal Gravitation All objects that have mass are attracted to one another by a gravitational force. (p. 276)

Orbit The path taken by an object during its revolution around another object, such as the path of the moon or a satellite about the earth or of a planet about the sun. (p. 280)

Formulas

10.1 $F_G = G\frac{m_1 m_2}{r^2}$

10.3 $v = \sqrt{\frac{Gm}{r}}$

$T = 2\pi\sqrt{\frac{r^3}{Gm}}$

Review Questions

1. What type of force is related to the mass of objects?
   (a) Electric force (b) Strong force
   (c) Magnetic force (d) Gravitational force
2. As the distance increases between two objects, the gravitational force between the objects
   (a) increases. (b) decreases. (c) remains constant.
3. As the mass of two objects increases, the gravitational force between the objects
   (a) increases. (b) decreases. (c) remains constant.
4. The mass of a satellite is increased. In order to maintain the same orbital period, its distance from the earth must
   (a) increase. (b) decrease. (c) remain constant.
5. As the distance increases between a satellite and the earth, what happens to the time it takes to complete an orbit?
   (a) Increases (b) Decreases (c) Remains constant
6. Explain why the gravitational force that exists between the person sitting next to you in class and you is much less than the gravitational force that exists between the earth and you.
7. What would happen to your weight on earth if the radius of the earth doubled, but its mass stayed the same?
8. What would happen to your weight on earth if the mass of the earth doubled, but its radius stayed the same?
9. Explain how a satellite in orbit is in a constant state of free fall, yet does not crash to the earth.
10. According to Isaac Newton, how is the motion of a falling apple different from the motion of the moon orbiting the earth?
11. Most planets actually have slightly elliptical orbits around the sun. What is the force exerted on a planet at its perigee (the point closest to the sun) compared to the force at its apogee (the point farthest from the sun)?
12. Does the mass of a satellite influence the time it takes to orbit the earth?

Review Problems

1. Two 0.300-kg apples are 25.0 cm apart from one center to the other. Find the gravitational attraction between the two apples.
2. Two 65.0-kg people are standing 1.00 m apart. What is the attractive gravitational force between them?
3. Find the weight of a 65.0-kg person on the earth (in newtons and pounds).
4. Find the weight of a 65.0-kg person on Jupiter (in newtons and pounds).
5. Find the weight of a 65.0-kg person on Pluto (in newtons and pounds).
6. If the moon orbited at one-half the present distance to the earth, what would be the orbiting time for the moon?
7. If the moon orbited at twice the present distance to the earth, what would be the orbiting time for the moon?
8. If the moon orbited at four times the present distance to the earth, what would be the orbiting time for the moon?
9. Using Newton's law of universal gravitation, find the amount of gravitational force on an 85.0-kg astronaut on the launch pad.
10. If an 85.0-kg astronaut in a space shuttle orbits the earth 362 km above sea level, what is the amount of gravitational force between the astronaut and the earth?

APPLIED CONCEPTS

1. The gravitational differences between the earth and Mars is a factor that engineers and scientists must consider before sending an astronaut to the "Red Planet." (a) What is the acceleration due to gravity on Mars? (b) If an 85.0-kg astronaut landed on Mars, how much would the astronaut weigh there compared to his weight on the earth? (c) Based on the acceleration due to gravity on Mars, what might happen to the strength of the astronaut's muscles while the astronaut is away from earth?
2. (a) How far from the center of the earth must a person be located in space so that his or her weight would be the same as when he or she is standing on the moon? (b) How many earth radii would this location be from the center of the earth?
3. A geosynchronous communication satellite orbits at a fixed point above the earth's surface. The earth takes 24.0 h to rotate on its axis. (a) Find the communication satellite's orbiting altitude from the center of the earth and from the surface of the earth. (b) Find its linear speed.
4. Flight engineers for the Apollo Lunar Orbiter placed the orbiter $15\bar{0}$ km above the surface of the moon. What was the Orbiter's (a) linear velocity and (b) period as it orbited the moon? (c) How would increasing the altitude affect the Orbiter's velocity and period?
5. (a) What is the gravitational force on a 65.7-kg space shuttle astronaut orbiting the earth 427 km above the surface of the earth? (b) Explain why astronauts orbiting the earth experience weightlessness even though the earth continually applies a force on their bodies.

CHAPTER 11

MATTER

Technology is used to take raw materials and shape, refine, mold, and transform them into products useful to our society. To do this, it is necessary to have some understanding of the nature and properties of matter and basic characteristics of its various forms of solids, liquids, and gases. Once these materials take shape and become useful products, they are subjected to various stresses and strains that can ultimately cause failure. We will show how to analyze these stresses and strains so that materials can be evaluated before being put to use.

Objectives

The major goals of this chapter are to enable you to:

1. Describe the properties of matter.
2. Apply Hooke's law.
3. Describe the properties of solids, liquids, and gases.
4. Solve density and specific gravity problems.
5. Calculate the amount of stress on objects.

11.1 Properties of Matter

What are the building blocks of matter? First, **matter** is anything that occupies space and has mass. Suppose that we take a cube of sugar and divide it into two pieces. Then we divide a resulting piece into two pieces. Can we continue this process indefinitely and get smaller and smaller particles of sugar each time? No, at some point the subdivision will result in something different from sugar.

An **element** is a substance that cannot be separated into simpler substances. A **compound** is a substance containing two or more elements.

A **molecule** is the smallest particle of an element that can exist in a free state and still retain the characteristics of that element or compound. Most simple molecules are about 3×10^{-10} m in diameter. An **atom** is the smallest particle of an element that can exist in a stable or independent state. The molecules of elements consist of one atom or two or more similar atoms; the molecules of compounds consist of two or more different atoms.

What do we get if we divide the sugar molecule? The resulting particles are carbon, hydrogen, and oxygen atoms. Models of water and sugar molecules are shown in Fig. 11.1. Not all atoms are the same size. The hydrogen atom is the smallest, with diameter 6×10^{-11} m and mass 1.67×10^{-27} kg. Uranium is one of the heaviest atoms, 3.96×10^{-25} kg.

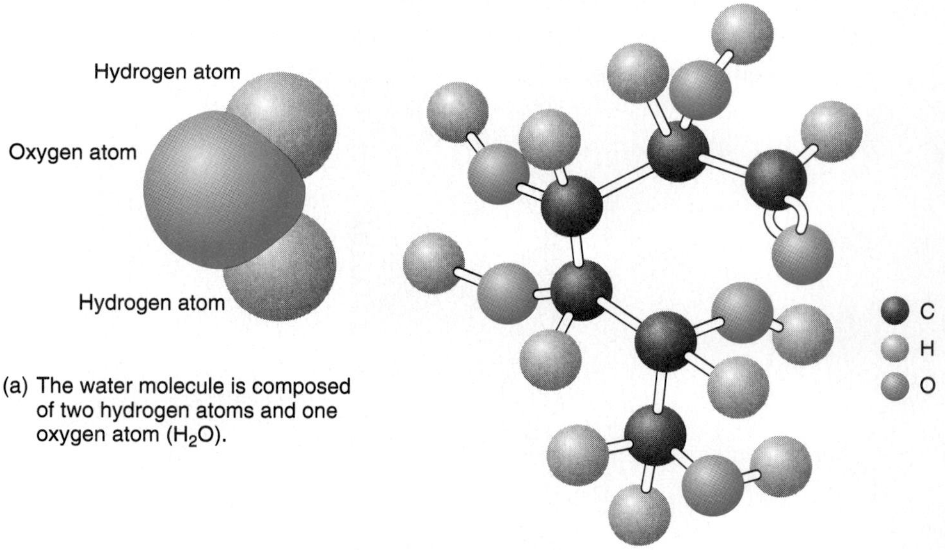

(a) The water molecule is composed of two hydrogen atoms and one oxygen atom (H_2O).

(b) The sugar (glucose) molecule is composed of six carbon atoms, twelve hydrogen atoms, and six oxygen atoms ($C_6H_{12}O_6$).

Figure 11.1

What happens if an atom is subdivided? Several constituent particles of the atom have been discovered. Of these, the three most important are (a) the **proton**, a particle with a positive charge, (b) the **electron**, a particle with a negative charge, and (c) the **neutron**, a particle that does not have an electric charge. Although there are smaller particles constituting the atom, we limit our discussion here to these three. Table 11.1 provides some basic information about these three particles.

Table 11.1 Properties of Atomic Particles

Particle	Mass	Diameter	Charge
Proton	1.673×10^{-27} kg	8.2×10^{-16} m	+1
Electron	9.109×10^{-31} kg	0*	−1
Neutron	1.675×10^{-27} kg	8.2×10^{-16} m	0

*The zero size of an electron is due to its lack of any internal structure.

Models of the hydrogen atom and the carbon atom are shown in Fig. 11.2. The **nucleus** is the center part of an atom made up of protons and neutrons, while the electrons surround the nucleus. Electrons do not really orbit the nuclei like planets around the sun; they are more like a cloud around a given nucleus. The atoms of the nuclei are held together by strong nuclear forces. The molecules are held together by electrical forces.

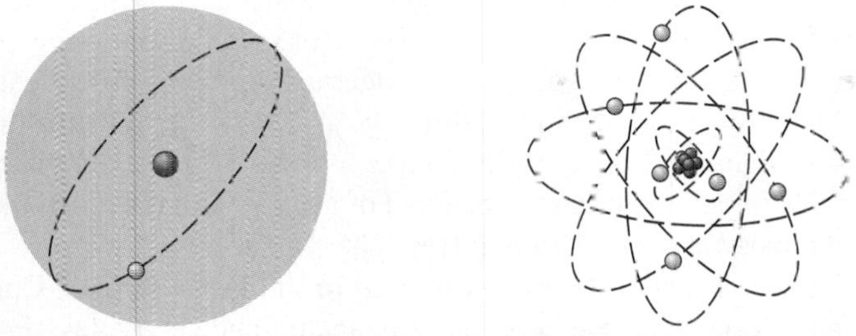

(a) The hydrogen atom is composed of a nucleus which contains one proton. Its one electron surrounds the nucleus.

(b) The carbon atom is composed of a nucleus which contains six protons and six neutrons. Its six electrons surround the nucleus.

Figure 11.2

Protons, electrons, and neutrons, formed in various combinations, give us more than 100 known atoms or chemical elements. The atoms, formed in various combinations, give us the very long list of known molecules. **Ernest Rutherford** developed the concept of the atomic nucleus.

Ernest Rutherford (1871–1937),

physicist, was born in New Zealand. He developed the concept of the atomic nucleus, proposed in 1903, that radioactivity results from the disintegration of atoms, and received the Nobel Prize for Chemistry in 1908.

Matter exists in three states: solid, liquid, and gas. A **solid** is a substance that has a definite shape and a definite volume. A **liquid** is a substance that takes the shape of its container and has a definite volume. A **gas** is a substance that takes the shape of its container and has the same volume as its container.

The molecules of a solid are fixed in relation to each other [Fig. 11.3(a)]. They vibrate in a back-and-forth motion. They are so close that a solid can be compressed only slightly. Solids are usually crystalline substances, meaning that their molecules are arranged in a definite pattern and in fixed positions. This is why a solid tends to hold its shape and has a definite volume.

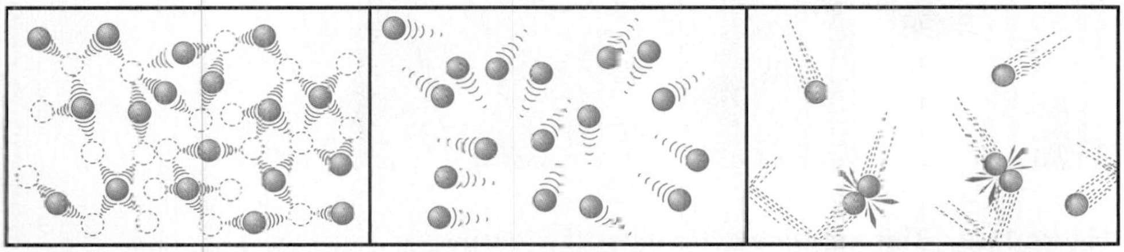

(a) Solid molecules vibrate in fixed positions.

(b) Liquid molecules flow over each other.

(c) Gas molecules move rapidly in all directions and collide.

Figure 11.3

The molecules of a liquid are not fixed in relation to each other [Fig. 11.3(b)]. They normally move in a flowing type of motion but are so close together that they are practically incompressible and have a definite volume. Because the molecules move in a smooth flowing motion and not in any fixed manner, a liquid takes the shape of its container.

The molecules of a gas are not fixed in relation to each other and move rapidly in all directions, colliding with each other [Fig. 11.3(c)]. They are much farther apart than molecules in a liquid, and they are extremely far apart when compared to the distance between molecules in solids. The movement of the molecules is limited only by the container. Therefore, a gas takes the shape of its container. Because the molecules are far apart, a gas can easily be compressed, and it has the same volume as its container.

11.2 Properties of Solids

Solids have a definite shape and a definite volume and have molecules that are usually arranged in a definite pattern. The following properties are common to most solids.

Cohesion and Adhesion

The molecules of a solid are held together by large internal molecular forces. **Cohesion** is the force of attraction between like molecules. The cohesive forces hold the closely packed molecules of a solid together and keep its shape and volume from being easily changed. Cohesion can also be shown by grinding and polishing the surfaces of two like solids and then sliding their surfaces together. For example, take two pieces of polished plate glass and slide them together. Try to pull them apart. The force of attraction of the like molecules of the two pieces of glass makes it difficult to pull them apart.

Adhesion is the force of attraction between different or unlike molecules. Common examples include glue and wood, adhesive tape and skin, and tar and shoe soles.

Tensile Strength

The **tensile strength** of a solid is a measure of its resistance to being pulled apart. That is, the tensile strength of a solid is a measure of its cohesive forces between adjacent molecules. The tensile strength of a rod or wire is found by putting it in a machine that pulls the rod or wire until it breaks (Fig. 11.4). The tensile strength is the ratio

$$\frac{\text{force required to break the rod or wire}}{\text{cross-sectional area of the rod or wire}}$$

Figure 11.4 This machine determines the tensile strength of a metal rod by finding the force needed to pull the rod until it breaks.

Johan August Brinell (1849–1925), *engineer and metallurgist, was born in Sweden. He invented the Brinell machine for measuring hardness of alloys and metals.*

Hardness

The **hardness** of a solid is a measure of the internal resistance of its molecules to being forced farther apart or closer together. More commonly, we classify the hardness of a solid in terms of its difficulty in being scratched using a "scratch test." The given material is scratched in a certain way. Its scratch is then compared with a series of standard scratches of materials that form an arbitrary hardness scale from very soft solids to the hardest known substance, diamond.

The **Brinell method,** named after **Johan Brinell,** is a common industrial method used to measure the hardness of a metal. A machine is used to press a 10-mm hardened chrome-steel

Photo courtesy of Dorling Kindersley

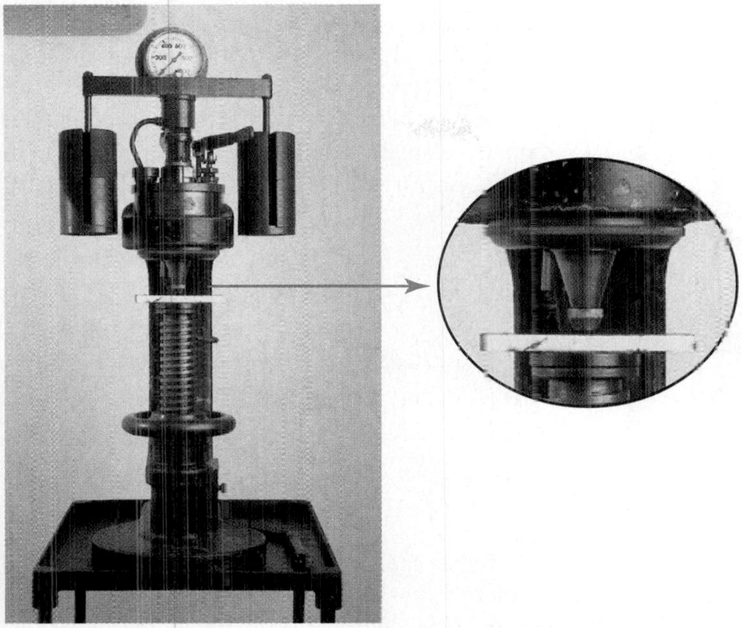

Figure 11.5 The Brinell hardness testing machine determines the hardness of a metal.

ball with the same force as an equivalent mass of 3000 kg into the metal being tested (Fig. 11.5). The diameter of the resulting impression is used as a measure of the metal's hardness. The Brinell value or number is the ratio

$$\frac{\text{mass placed on the object (in kg)}}{\text{surface area of the impression (in mm}^2\text{)}}$$

This value is compared with a scale of the accepted hardnesses of given metals. The larger the Brinell number, the harder is the metal.

Steel can be hardened by heating it to a very high temperature, then suddenly cooling it by putting it in water. However, it then becomes brittle. This cooled steel can then be tempered (toughened) by reheating it and allowing it to cool slowly. As the steel cools, it loses hardness and gains toughness. If the steel cools down slowly and completely, we say that it is *annealed*. Annealed steel is soft and tough but not brittle.

Ductility

Ductility is the property of a metal that enables it to be drawn through a die to produce a wire. As the rod is pulled through the die, its diameter is decreased and its length is increased as it becomes a wire (Fig. 11.6).

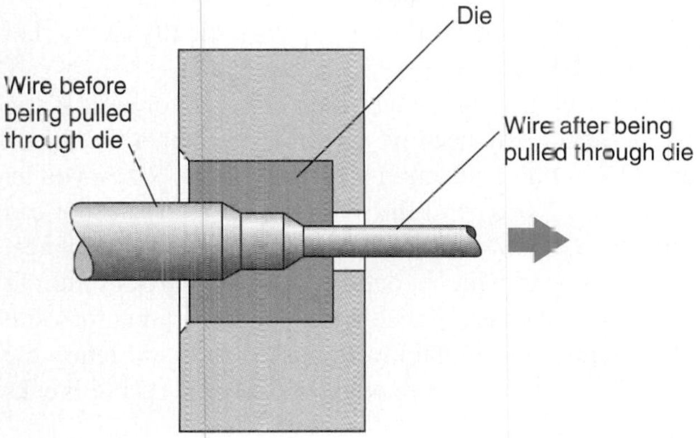

Figure 11.6 Ductility: the ability of a metal to be drawn into a wire

Malleability

Malleability is the property of a metal that enables it to be hammered and rolled into a sheet. As the metal is hammered or rolled as in Fig. 11.7, its shape or thickness is changed. During this process, the atoms slide over each other and change positions. The cohesive forces are relatively strong; thus, the atoms do not become widely separated during their rearrangement and the resulting shape remains relatively stable.

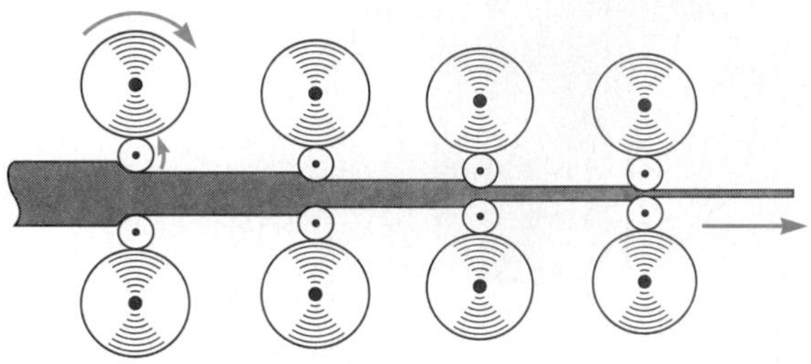

Figure 11.7 Malleability: the ability of a metal to be rolled into a sheet

Elasticity

An object becomes deformed when outside forces change its shape or size. **Elasticity** is a measure of a deformed object's ability to return to its original size and shape once the outside forces are removed. When the solid is being deformed, sometimes the molecules attract each other and sometimes they repel each other. For instance, try to pull a rubber ball apart (Fig. 11.8). You notice that the ball stretches out of shape.

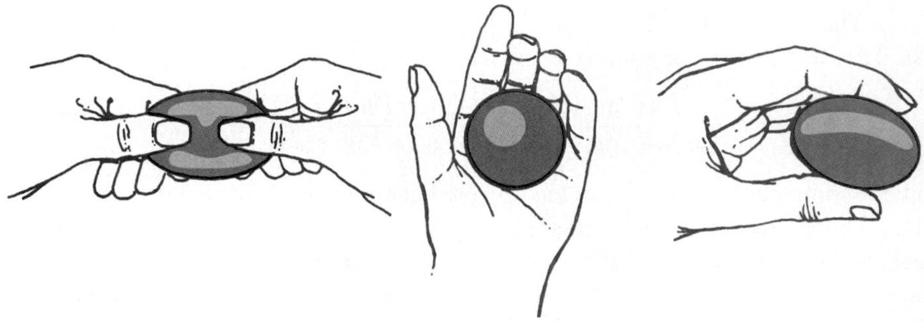

(a) (b) (c)

Figure 11.8 Elasticity in a rubber ball

However, when you release the pulling force, the ball returns to its original shape because the molecules, being farther apart than normal, attract each other. If you squeeze the ball, it will again become out of shape. Now release the pressure and the ball will again return to its original shape because the molecules, being too close together, repel each other. Therefore, we can see that when molecules are slightly pulled out of position, they attract each other. When they are pressed too close together, they repel each other.

Most solids have the property of elasticity; however, some are only slightly elastic. For example, wood and Styrofoam are two solids whose elasticity is small.

Not every elastic object returns to its original shape after being deformed. If too large a deforming force is applied, an object may become deformed permanently. Take a spring [Fig. 11.9(a)] and pull it apart by a moderate amount [Fig. 11.9(b)]. When you let it go, it should return to its original shape. Next, pull the spring apart as far as you can [Fig. 11.9(c)]. When you let it go this time, it will probably not return to its original shape. The **elastic limit** of a solid is the point beyond which a deformed object cannot return to its original shape. The spring's molecules have been pulled far enough apart that they slid past one another beyond the point at which the original molecular forces could return the spring to its original shape. If the deforming force is enough greater, the spring breaks apart [Fig. 11.9(d)].

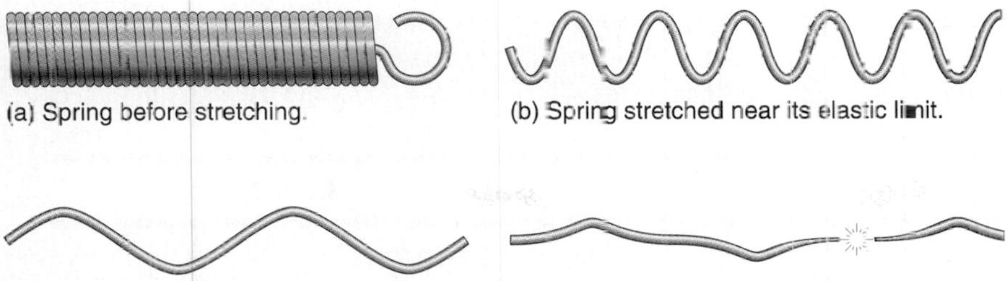

(a) Spring before stretching. (b) Spring stretched near its elastic limit.

(c) Spring stretched beyond its elastic limit. (d) Spring stretched much beyond its elastic limit . . . break occurs!

Figure 11.9

Stress is the ratio of the outside applied force, which tends to cause a distortion, to the area over which the force acts. In other words,

$$\text{stress} = \frac{\text{applied force}}{\text{area over which the force acts}}$$

or

$$S = \frac{F}{A}$$

where

$S =$ stress, usually in N/m^2 (Pa) or lb/in^2 (psi)
$F =$ force applied, N or lb, perpendicular to the surface to which it is applied
$A =$ area, m^2 or in^2

Since the SI metric unit for force is the newton (N) and that for area is the square metre (m^2), the corresponding pressure unit is N/m^2. This unit is given the special name *pascal* (Pa), named after **Blaise Pascal,** who made important discoveries in science and mathematics.

Blaise Pascal (1623–1662),

mathematician, physicist, and theologian, was born in France. He invented the calculating machine in 1647 and later the barometer, the hydraulic press, and the syringe. He also formulated the modern theory of probability.

$$1\ N/m^2 = 1\ Pa$$

Imagine a brick weighing 12.0 N first lying on its side on a table and then standing on one end (Fig. 11.10). The weight of the brick is the same no matter what its position, so the total force (the weight of the brick) on the table is the same in both cases. However, the position of

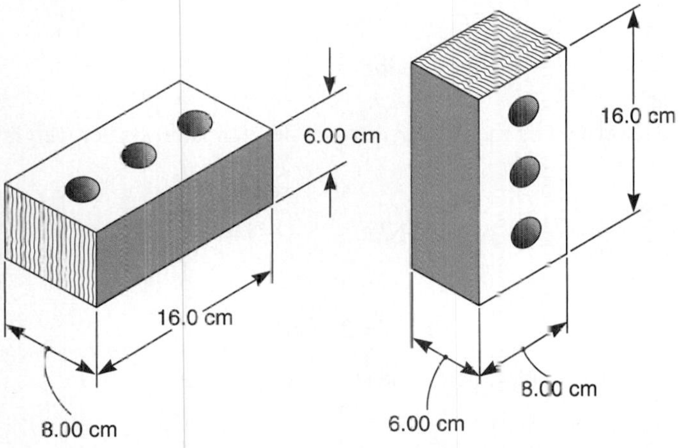

Figure 11.10 The weight of the brick is constant, but the stress in part (b) is greater.

the brick does make a difference in the stress exerted on the table. In which case is the stress greater? When standing on end, the brick exerts a greater stress on the table because the area of contact on the end is *smaller* than on the side. Using $S = F/A$, find the stress in each case:

Case 1	Case 2
$F = 12.0\text{ N}$	$F = 12.0\text{ N}$
$A = 8.00\text{ cm} \times 16.0\text{ cm} = 128\text{ cm}^2$	$A = 6.00\text{ cm} \times 8.00\text{ cm} = 48.0\text{ cm}^2$
$S = \frac{F}{A} = \frac{12.0\text{ N}}{128\text{ cm}^2} \times \left(\frac{100\text{ cm}}{1\text{ m}}\right)^2$	$S = \frac{F}{A} = \frac{12.0\text{ N}}{48.0\text{ cm}^2} \times \left(\frac{100\text{ cm}}{1\text{ m}}\right)^2$
$= 938\text{ N/m}^2 = 938\text{ Pa}$	$= 25\bar{0}0\text{ N/m}^2 = 25\bar{0}0\text{ Pa}$

This shows that when the same force is applied to a smaller area, the stress is greater. From the discussion so far, would you rather a woman step on your foot with a pointed-heel shoe or with a flat-heel shoe? (See Fig. 11.11.) This is a serious issue for the aircraft industry. They must design and construct airplane floors light in weight but strong enough to withstand the stress of pointed-heel shoes. For example, if a 160-lb woman rests her weight on a 4.0-in^2 heel, the stress is

$$S = \frac{F}{A} = \frac{160\text{ lb}}{4.0\text{ in}^2} = 4\bar{0}\text{ lb/in}^2$$

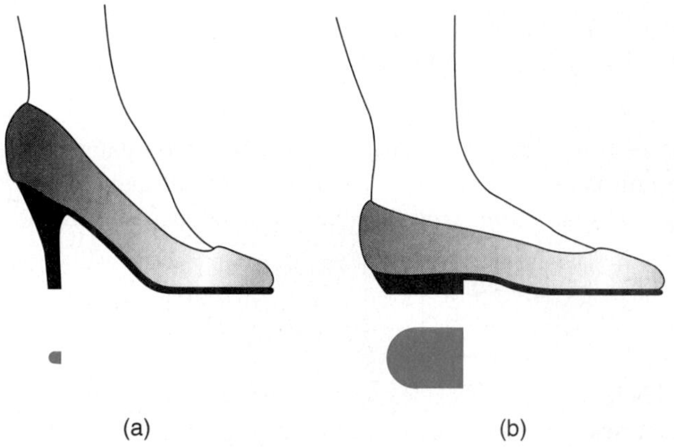

Figure 11.11 The stress of the heel in part (a) is greater because the weight of the woman rests on a smaller area.

But if she rests her weight on a pointed heel of $\frac{1}{4}$ in^2, which is a common area of a pointed heel, the stress is

$$S = \frac{F}{A} = \frac{160\text{ lb}}{\frac{1}{4}\text{ in}^2} = 640\text{ lb/in}^2$$

A similar comparison may be shown using metric units. If a 65-kg woman rests her weight on a 25-cm^2 heel, the stress is

$$S = \frac{F}{A} = \frac{(65\text{ kg})(9.80\text{ m/s}^2)}{25\text{ cm}^2 \times \left(\frac{1\text{ m}}{100\text{ cm}}\right)^2} = 2.5 \times 10^5\text{ N/m}^2 \qquad (1\text{ N} = 1\text{ kg m/s}^2)$$

$$= 2.5 \times 10^5\text{ Pa} = 250\text{ kPa}$$

If she rests her weight on a pointed heel of 1.0 cm^2, the stress is

$$S = \frac{F}{A} = \frac{(65\text{ kg})(9.80\text{ m/s}^2)}{1.0\text{ cm}^2 \times \left(\frac{1\text{ m}}{100\text{ cm}}\right)^2} = 6.4 \times 10^6\text{ N/m}^2$$

$$= 6.4 \times 10^6\text{ Pa} = 6400\text{ kPa}$$

Since the pascal is a relatively small unit, the kilopascal (kPa) is a commonly used unit of pressure.

Five basic types of stresses are as follows:

Tension is a stress caused by two forces acting directly opposite each other. This stress tends to cause objects to become longer and thinner. An example of such a stress is that on the rope in "tug-of-war" (Fig. 11.12). The rope has one team's force pulling one way and another team's force pulling in the opposite direction. If the rope is not strong enough to withstand the tension, it could ultimately stretch beyond its elastic limit and break.

Figure 11.12 The rope in a tug-of-war competition is in constant tension.

Compression is a stress caused by two forces acting directly toward each other. This stress tends to cause objects to become shorter and thicker. An example of compression is the stress present in a supporting column (Fig. 11.13). A load is pushing down on the column, while the ground is applying a force pushing up on the column. As a result, the pillar compresses.

Figure 11.13 A column under the New Clark Bridge crossing the Mississippi River is in compression.

Shearing is a stress caused by two forces applied in parallel, opposite directions. In Fig. 11.14, the table pushing the book to the left counteracts the force of the hand pushing

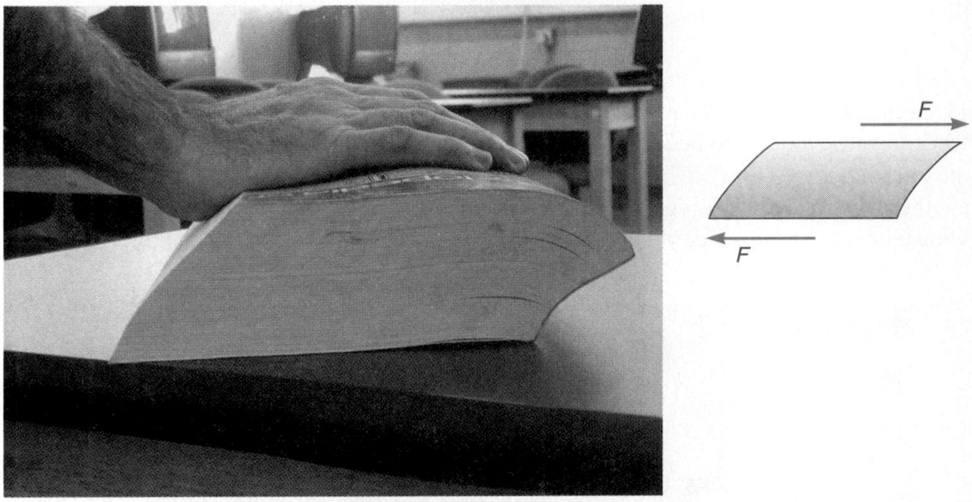

Figure 11.14 A book being pushed in this way is undergoing shear.

the book to the right. The normally rectangular shape of the book is altered. Scissors use shearing to cut paper.

Torsion is a stress related to a twisting motion. Torsion occurs when two torques act in opposite directions. This type of stress severely compromises the strength of most materials. An example of torsion is the stress on a bolt or a screw as it is being tightened (Fig. 11.15).

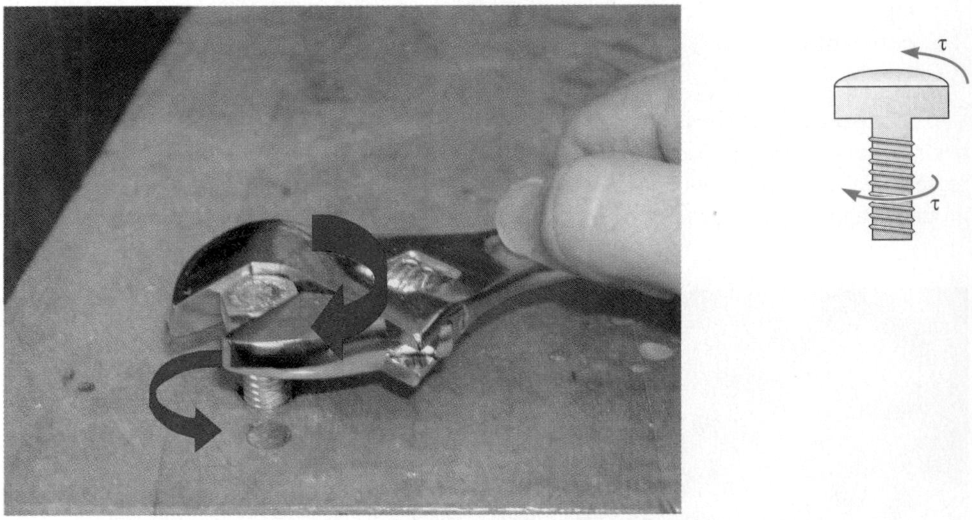

Figure 11.15 The twisting of the bolt in one direction is counteracted by the force of the wood resisting the turning motion.

Bending consists of both tension and compression stresses. It occurs when a force is placed on an object causing it to sag. An example of bending is caused by a person sitting on a board (Fig. 11.16). The top section of the board is being pushed together, in compression, while the bottom section of the board is being pulled apart, in tension.

Whenever a stress is applied to an object, the object is changed minutely, at least. If you stand on a steel beam, it bends—at least slightly. **Strain** is the deformation of an object due to an applied force. That is, strain is the relative amount of deformation of a body that is under stress. Or, strain is *change in length per unit of length,* change in volume per unit of volume, and so on. Strain is a direct and necessary consequence of stress.

Figure 11.16 A beam that is bending

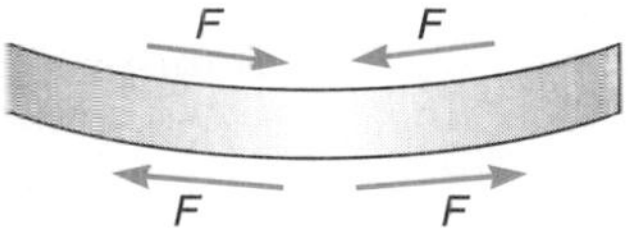

TRY THIS ACTIVITY

Stresses

Foam is a flexible material that can easily demonstrate the various types of stresses on solid materials. Using a permanent marker, draw lines at 1.0-in. intervals along the top and bottom sides of a rectangular piece of foam (Fig. 11.17). Using Figures 11.12–11.16 as a guide, apply the appropriate forces to the foam to simulate the five types of stress. Observe how the stresses affect the spacing of the drawn lines. Describe how the strength of building materials such as concrete, wood, or steel is affected when the materials are subjected to the five types of stress.

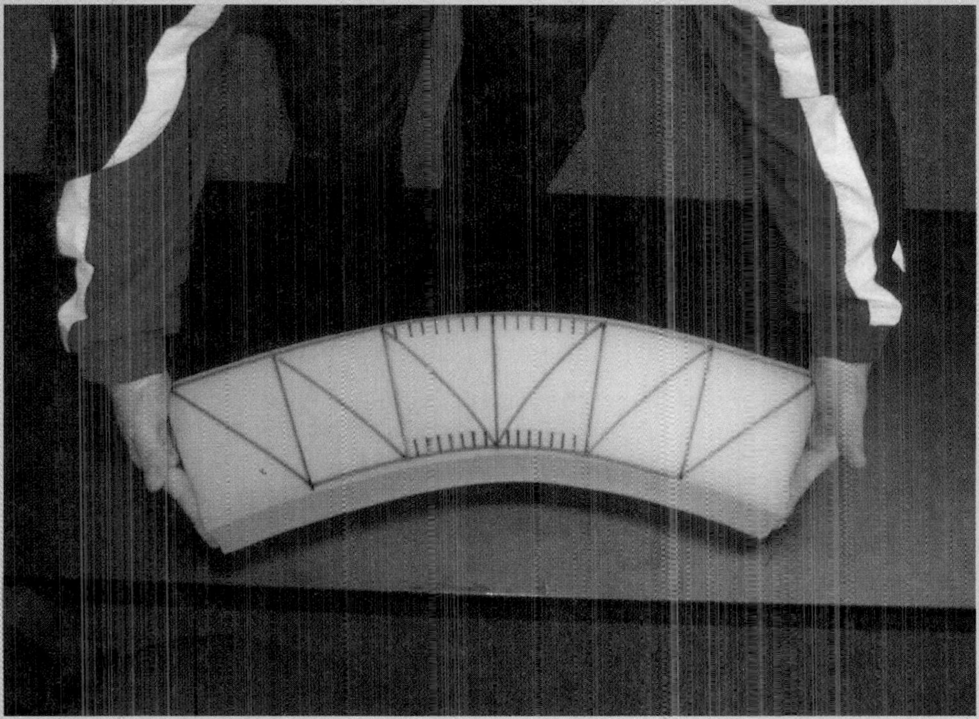

Figure 11.17

EXAMPLE 1

A steel column in a building has a cross-sectional area of $25\bar{0}0$ cm^2 and supports a weight of 1.50×10^5 N. Find the stress on the column.

Sketch: None needed

Data:

$$A = 25\bar{0}0 \text{ cm}^2 \times \left(\frac{1 \text{ m}}{100 \text{ cm}}\right)^2 = 0.250 \text{ m}^2$$

$$F = 1.50 \times 10^5 \text{ N}$$

$$S = ?$$

Basic Equation:

$$S = \frac{F}{A}$$

Working Equation: Same

Substitution:

$$S = \frac{1.50 \times 10^5 \text{ N}}{0.250 \text{ m}^2}$$
$$= 6.00 \times 10^5 \text{ N/m}^2$$
$$= 6.00 \times 10^5 \text{ Pa} \quad \text{or} \quad 6\bar{0}0 \text{ kPa}$$

Robert Hooke (1635–1703), *chemist and physicist, was born in England. He formulated the law governing elasticity (Hooke's law), invented the balance spring for watches, worked with and made important observations with the telescope and the microscope, and formulated the theory of planetary movement.*

Hooke's Law

One of the most basic principles related to the elasticity of solids is **Hooke's law,** named after **Robert Hooke.**

HOOKE'S LAW

The ratio of the force applied to an object to its change in length (resulting in its being stretched or compressed by the applied force) is constant as long as the elastic limit has not been exceeded.

Or, stated another way: Stress is directly proportional to strain as long as the elastic limit has not been exceeded. (See Fig. 11.18.) In equation form,

$$\frac{F}{\Delta l} = k$$

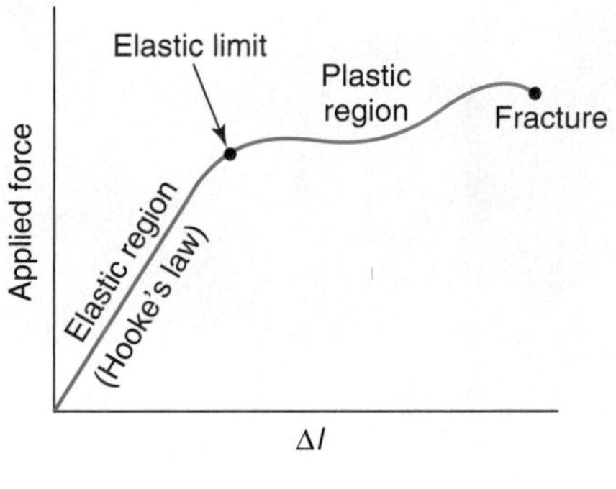

Figure 11.18 Graph of Hooke's law showing behavior within and beyond the elastic limit

where

$$F = \text{applied force}$$
$$\Delta l = \text{change in length}$$
$$k = \text{elastic constant}$$

Note: Δ (the Greek letter delta) is often used in mathematics and science to mean "change in."

EXAMPLE 2

A force of 5.00 N is applied to a spring whose elastic constant is 0.250 N/cm. Find its change in length.

Sketch:

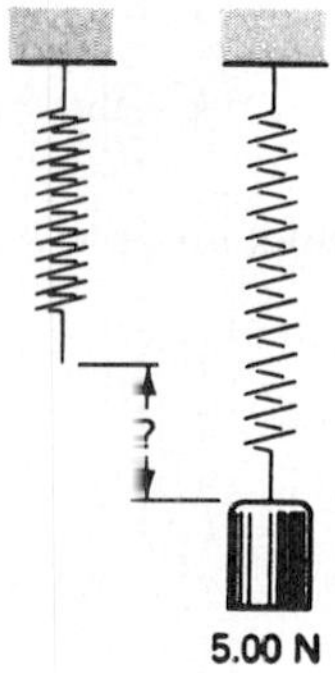

Data:

$$F = 5.00 \text{ N}$$
$$k = 0.250 \text{ N/cm}$$
$$\Delta l = ?$$

Basic Equation:

$$\frac{F}{\Delta l} = k$$

Working Equation:

$$\Delta l = \frac{F}{k}$$

Substitution:

$$\Delta l = \frac{5.00 \text{ N}}{0.250 \text{ N/cm}} = 20.0 \text{ cm}$$

$$\frac{\text{N}}{\text{N/cm}} = \text{N} \div \frac{\text{N}}{\text{cm}} = \cancel{\text{N}} \cdot \frac{\text{cm}}{\cancel{\text{N}}} = \text{cm}$$

EXAMPLE 3

A force of 3.00 lb stretches a spring 12.0 in. What force is required to stretch the spring 15.0 in.?

Sketch:

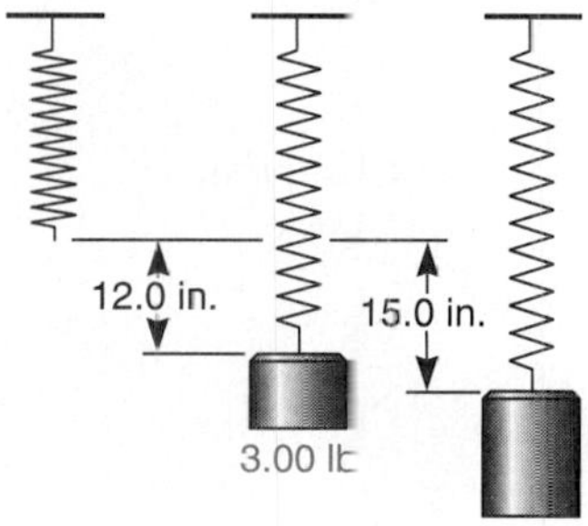

Data:

$$F_1 = 3.00 \text{ lb}$$
$$l_1 = 12.0 \text{ in.}$$
$$l_2 = 15.0 \text{ in.}$$
$$F_2 = ?$$

Basic Equation:

$$\frac{F}{\Delta l} = k$$

Working Equations:

$$\frac{F}{\Delta l} = k \quad \text{and} \quad F = k(\Delta l)$$

Substitution: There are two substitutions, one to find k and one to find the second force F_2:

$$\frac{3.00 \text{ lb}}{12.0 \text{ in.}} = k$$
$$0.250 \text{ lb/in.} = k$$

$$F_2 = (0.250 \text{ lb/}\cancel{\text{in.}})(15.0 \cancel{\text{in.}})$$
$$= 3.75 \text{ lb}$$

EXAMPLE 4

A support column is compressed 3.46×10^{-4} m under a weight of 6.42×10^5 N. How much is the column compressed under a weight of 5.80×10^6 N?

Sketch: None needed

First find k:

Data:

$$F_2 = 6.42 \times 10^5 \text{ N}$$
$$\Delta l_2 = 3.46 \times 10^{-4} \text{ m}$$
$$k = ?$$

Basic Equation:

$$\frac{F_2}{\Delta l_2} = k$$

Working Equation: Same

Substitution:

$$k = \frac{6.42 \times 10^5 \text{ N}}{3.46 \times 10^{-4} \text{ m}}$$
$$= 1.86 \times 10^9 \text{ N/m}$$

Then:

Data:

$$k = 1.86 \times 10^9 \text{ N/m}$$
$$F_1 = 5.80 \times 10^6 \text{ N}$$
$$\Delta l_1 = ?$$

Basic Equation:

$$\frac{F_1}{\Delta l_1} = k$$

Working Equation:

$$\Delta l_1 = \frac{F_1}{k}$$

Substitution:

$$\Delta l_1 = \frac{5.80 \times 10^6 \text{ N}}{1.86 \times 10^9 \text{ N/m}}$$
$$= 3.12 \times 10^{-3} \text{ m or } 3.12 \text{ mm}$$

PROBLEMS 11.2

1. A packing crate 2.50 m × 0.80 m × 0.45 m weighs 1.41×10^5 N. Find the stress (in kPa) exerted by the crate on the floor in each of its three possible positions.
2. A packing crate 2.50 m × 20.0 cm × 30.0 cm has a mass of 975 kg. Find the stress (in kPa) exerted by the crate on the floor in each of its three possible positions.
3. A spring is stretched 24.0 in. by a force of 54.0 lb.
   (a) How far will it stretch if a force of 104 lb is applied?
   (b) What weight will stretch the spring 9.00 in.?
4. A 17.0-N force stretches a wire 0.650 cm.
   (a) What force will stretch a similar piece of wire 1.87 cm?
   (b) A force of 21.3 N is applied to a similar piece of wire. How far will it stretch?
5. A force of 36.0 N stretches a spring 18.0 cm. Find the spring constant (in N/m).
6. A force of 5.00 N is applied to a spring whose spring constant is 0.250 N/cm. Find its change in length (in cm).
7. Each vertical steel column of an office building supports 1.30×10^5 N and is compressed 5.90×10^{-3} cm.
   (a) Find the compression in each column if a weight of 5.50×10^5 N is supported.
   (b) If the compression of each steel column is 0.0710 cm, what weight is supported by each column?

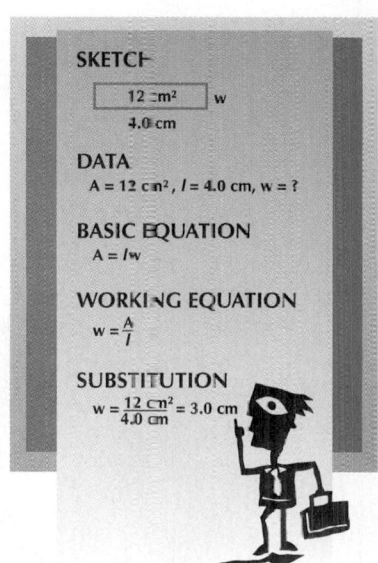

SKETCH

12 cm² | w

4.0 cm

DATA

$A = 12 \text{ cm}^2$, $l = 4.0$ cm, $w = ?$

BASIC EQUATION

$A = lw$

WORKING EQUATION

$w = \frac{A}{l}$

SUBSTITUTION

$w = \frac{12 \text{ cm}^2}{4.0 \text{ cm}} = 3.0 \text{ cm}$

PHYSICS CONNECTIONS

Development of Materials

Engineers choose materials based on two main factors: cost and strength. The strength of a material used depends on its purpose. For example, will the material be used underneath a roadway to support cars and trucks, or will it be used to suspend a roadway? Advances in the science and engineering of materials have helped in the development of more economical bridges that span greater and greater distances.

Ancient Roman, Mesopotamian, and Chinese bridge builders used stone to create their arch bridges. Stone was abundant and very strong under compression forces. However, the need for lighter and stronger materials sparked technological improvements in building materials during the Industrial Revolution of the 18th and 19th centuries. During that time, an English engineer, Abraham Darby, built the first cast-iron arch bridge (Fig. 11.19). The low cost and high compression strength of iron made it a much better alternative to the expensive and heavy stone arch. In later years, however, problems with cast-iron bridges became evident, as the brittle nature of iron under tension and shear stresses caused fractures in cast-iron structures.

In the late 19th century, an American engineer, James Eads, constructed the first steel arch bridge over the Mississippi River. The steel used in this bridge was not much stronger under compression than iron but was more than twice as strong as iron under tension and shearing stresses. Steel is used for beams and cables in almost all bridges today.

Although concrete is extremely weak under shearing and tension stresses, it is used to combat compression stresses. To improve concrete's ability to withstand tension and shearing stresses, stretched steel wire mesh or cables are embedded in the concrete. After the concrete dries, the tension in the wire mesh or cables is released, placing the concrete in a permanent state of compression. Therefore, when a tension force is placed on the prestressed concrete, the concrete itself remains under compression from the steel mesh or cables.

Photo courtesy of Dorling Kindersley

(a)

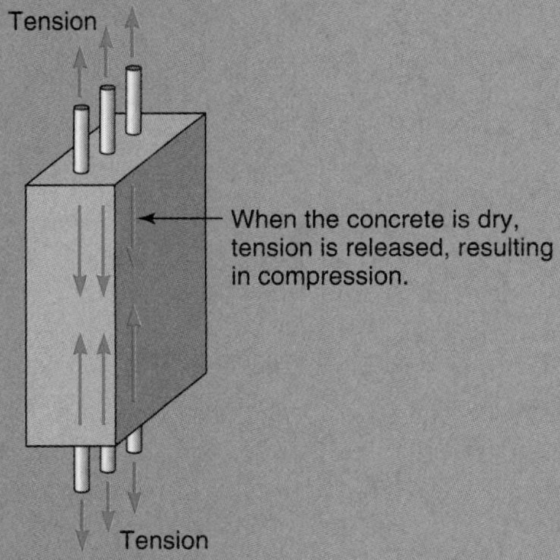
Tension
When the concrete is dry, tension is released, resulting in compression.
Tension

(b)

Figure 11.19 (a) Darby's cast iron bridge, constructed at Coalbrookdale, England, in 1779 (b) Pre-stressed concrete requires tension on the cables while the concrete dries. After the concrete dries, the tension is released, placing the concrete in compression.

8. Each vertical steel column of an office building supports 30,000 lb and is compressed 0.00234 in. Find the compression of each column if a weight of 125,000 lb is supported.
9. The compression of each steel column in Problem 8 is 0.0279 in. What weight is supported by each column?
10. A coiled spring is stretched 40.0 cm by a 5.00-N weight.
    (a) How far is it stretched by a 15.0-N weight?
    (b) What weight will stretch the spring 60.0 cm?
11. A 12,000-N load is hanging from a steel cable that is 10.0 m long and 16.0 mm in diameter. Find the stress.
12. A rectangular cast-iron column 25.0 cm × 25.0 cm × 5.00 m supports a weight of 6.80×10^6 N. Find the stress in pascals on the column.
13. In a Hooke's law experiment, the following weights were attached to a spring, resulting in the following elongations:

Weight (N)	Elongation (cm)
50	2.0
75	3.3
105	4.2
125	5.0
150	6.0
175	7.4
225	9.5
275	11.1

    (a) Plot the graph of weight versus elongation and draw the best straight line through the data.
    (b) From the graph, what weight corresponds to an elongation of 7.5 cm?
    (c) From the graph, what elongation corresponds to a weight of 220 N?
    (d) From the graph, determine the spring constant.
14. What was the original length of a spring with spring constant 96.0 N/m that is stretched to 28.0 cm by a 15.0-N weight?
15. Two hanging springs, each 15.0 cm long, with spring constants 0.970 N/cm and 1.45 N/cm, respectively, are stretched by a bar weighing 26.0 N that connects them. The bar is 6.00 m long and has a center of mass 2.00 m from the spring with constant 0.970 N/cm. How far does each spring stretch?
16. A firefighter weighs 725 N. She wears shoes that each cover an area of 206 cm^2.
    (a) What is the average stress she applies to the ground on which she is standing?
    (b) How does the stress change if she stands on only one foot?
17. Two identical wires are 125 cm and 375 cm long, respectively. The first wire is broken by a force of 489 N. What force is needed to break the other?
18. The cross-sectional area of a wire is 2.50×10^{-3} cm^2 and its tensile strength is 1.00×10^5 N/cm^2. What force will break the wire?
19. A spring having a force constant of 1.25 N/cm is stretched through a distance of 11.5 cm. How much work is required to stretch the spring?

11.3 Properties of Liquids

As noted previously, a liquid is a substance that has a definite volume and takes the shape of its container. The molecules move in a flowing motion, yet are so close together that it is very difficult to compress a liquid. Most liquids share the following common properties.

Cohesion and Adhesion

Cohesion, the force of attraction between like molecules, causes a liquid like molasses to be sticky. Adhesion, the force of attraction between unlike molecules, causes the molasses to also stick to your finger. In the case of water, its adhesive forces are greater than its cohesive forces. Put a glass in water and pull it out. Some water remains on the glass.

In the case of mercury, the opposite is true. Mercury's cohesive forces are greater than its adhesive forces. If glass is submerged in mercury and then pulled out, virtually no mercury remains on the glass.

A liquid whose adhesive forces are greater than its cohesive forces tends to wet any surface that comes in contact with it. A liquid whose cohesive forces are greater than its adhesive forces tends to leave dry or not wet any surface that comes in contact with it.

Surface Tension

Surface tension is the ability of the surface of a liquid to act like a thin, flexible film. The ability of the surface of water to support a needle is an example. The water's surface acts like a thin, flexible surface film. The surface tension of water can be reduced by adding soap to the water (Fig. 11.20). Soaps are added to laundry water to decrease the surface tension of water so that the water more easily penetrates the fibers of the clothes being washed.

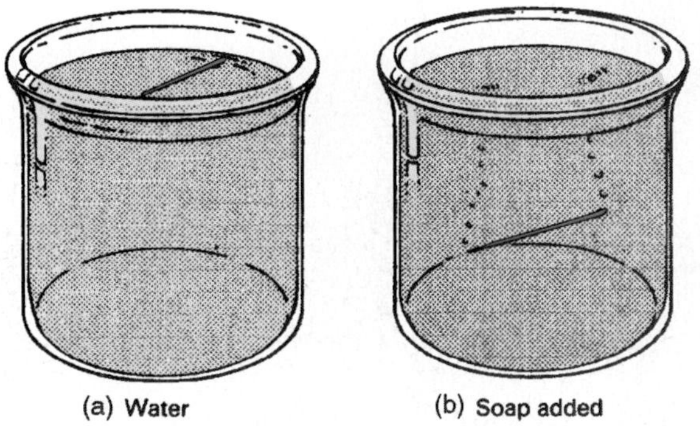

Figure 11.20 (a) The surface tension of water will support a needle. (b) Adding soap reduces surface tension.

Surface tension causes a raindrop to hold together and a small drop of mercury to keep an almost spherical shape. A liquid drop suspended in space is spherical. A falling raindrop's shape is due to the friction with the air (Fig. 11.21).

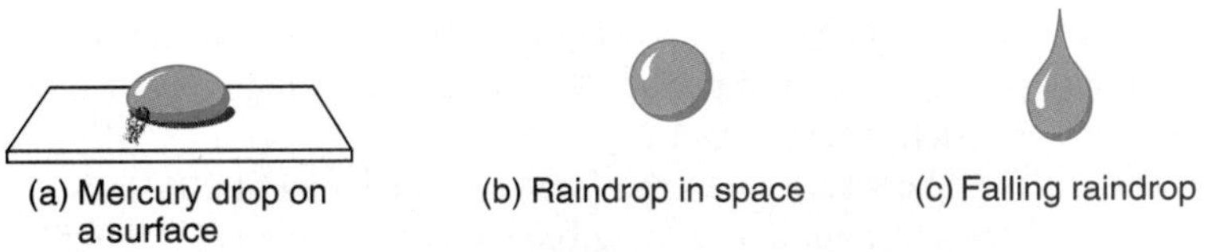

Figure 11.21 (a) Surface tension causes a drop of mercury to be more spherical. (b) Surface tension causes a raindrop to hold together. (c) The shape of a falling raindrop is due to the friction with air.

Viscosity

Viscosity is the internal friction of a fluid caused by molecular attraction, which makes it resist a tendency to flow. The greater the molecular attraction, the more is the friction and the greater is the viscosity. For example, it takes more force to move a block of wood through oil than through water. This is because oil is more viscous than water.

If a liquid's temperature is increased, its viscosity decreases. For example, the viscosity of oil in a car engine before it is started on a winter morning at −10.0°C is greater than after the engine has been running for an hour (Fig. 11.22).

One common misunderstanding is that higher viscosity means higher density. For example, oil is more viscous, but water is denser. Therefore, oil floats on water.

Figure 11.22 Cold oil is more viscous than hot oil.

Capillary Action

Capillary action is the behavior of liquids that causes the liquid level in small-diameter tubes to be different than that in larger tubes. A liquid keeps the same level in connected tubes filled with it if the tubes have a large enough diameter (Fig. 11.23). In tubes of different small diameters, water does not stand at the same level. The smaller the diameter, the higher the water rises. In the case of mercury, it does not stand at the same level either, but instead of rising up the narrower tube, the mercury level falls in the smaller diameter tube or its level is depressed. The smaller the diameter, the lower its level is depressed (Fig. 11.24).

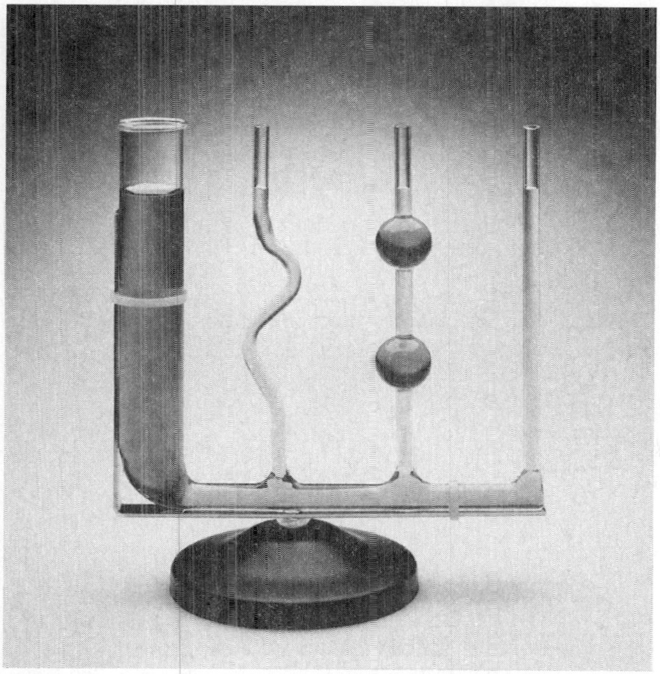

Figure 11.23 A liquid keeps the same level in tubes of large enough diameter that are connected. Neither the shape nor the size of the containers makes a difference.

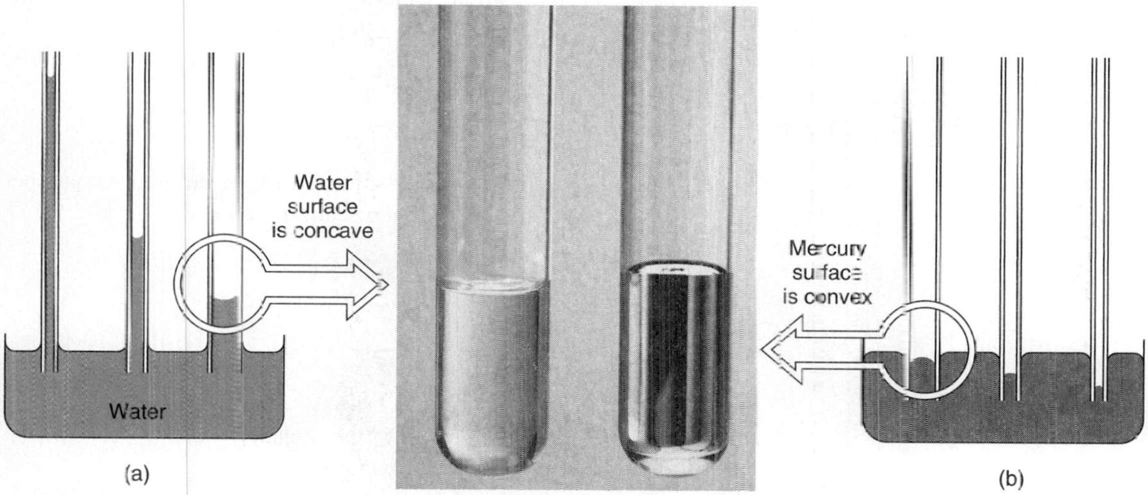

Figure 11.24 Capillary action. In small tubes, water stands at higher levels for smaller diameter tubes. The water surface is concave. In small tubes, mercury stands at lower levels for smaller diameters. The mercury surface is convex. [The tube diameters and differences in liquid levels are exaggerated in (a) and (b).]

Capillary action is due to both the adhesion of the liquid molecules with the tube and the surface tension of the liquid. For water, the adhesive forces are greater than the cohesive molecular forces. Thus, water creeps up the sides of the tube and produces a concave water surface. The surface tension of the water tends to flatten the concave surface. Together, these two forces raise the water up the tube until it is counterbalanced by the weight of the water column itself.

In the case of mercury, the cohesive molecular forces are greater than the adhesive forces and produce a convex mercury surface. The surface tension of the mercury tends to further hold down its level. This crescent-shaped surface of a liquid column in a tube, whether concave or convex, is called a **meniscus.** To measure the height of a liquid in a tube, measure to the lowest point of a concave meniscus or to the highest point of a convex meniscus.

Experimentally, scientists have found that:

1. Liquids rise in capillary tubes they tend to wet and are depressed in tubes they tend not to wet.
2. Elevation or depression in the tube is inversely proportional to the diameter of the tube.
3. The elevation or depression decreases as the temperature increases.

Capillary action causes the rise of oil (or kerosene) in the wick of an oil lamp. Towels also absorb water because of capillary action.

11.4 Properties of Gases

Expansion is a property of a gas in which the rapid random movement of its molecules causes the gas to completely occupy the volume of its container.

Diffusion is the process by which molecules of a gas mix with the molecules of a solid, a liquid, or another gas. If you remove the cap from a can of gasoline, you soon smell the fumes. The air molecules and the gasoline molecules mix throughout the room because of diffusion.

A balloon inflates due to the pressure of the air molecules on its inside surface. This pressure is caused by the bombardment on the walls by the moving molecules. The pressure may be increased by increasing the number of molecules by blowing more air into the balloon. Pressure may also be increased by heating the air molecules already in the balloon. Heat increases the velocity of the molecules.

The behavior of liquids and gases is often very similar. A **fluid** is a substance that takes the shape of its container. The term is used when discussing principles and behaviors common to both liquids and gases.

11.5 Density

Density is a property of all three states of matter. **Mass density,** D_m, is defined as mass per unit volume. **Weight density,** D_w, is defined as weight per unit volume, or,

$$D_m = \frac{m}{V} \qquad D_w = \frac{F_w}{V}$$

where

D_m = mass density
m = mass
V = volume

D_w = weight density
F_w = weight
V = volume

Although mass density and weight density can be expressed in both the metric system and the U.S. system, mass density is usually given in the metric units kg/m^3 and weight density is usually given in the U.S. units lb/ft^3 (Table 11.2).

Table 11.2 Densities for Various Substances

Substance	Mass Density (kg/m^3)	Weight Density (lb/ft^3)
Solids		
Copper	8,890	555
Iron	7,800	490
Lead	11,300	708
Aluminum	2,700	169
Ice	917	57
Wood, white pine	420	26
Concrete	2,300	140
Cork	240	15
Liquids		
Water	1,000	62.4
Seawater	1,025	64.0
Oil	870	54.2
Mercury	13 600	846
Alcohol	790	49.4
Gasoline	680	42.0
Gases*	At 0°C and 1 atm pressure	At 32°F and 1 atm pressure
Air	1.29	0.081
Carbon dioxide	1.96	0.123
Carbon monoxide	1.25	0.078
Helium	0.178	0.011
Hydrogen	0.0899	0.0056
Oxygen	1.43	0.089
Nitrogen	1.25	0.078
Ammonia	0.760	0.047
Propane	2.02	0.126

*The density of a gas is found by pumping the gas into a container, measuring its volume and mass or weight, and then using the appropriate density formula.

The mass density of water is $1\bar{0}00\ kg/m^3$; that is, 1 cubic metre of water has a mass of $1\bar{0}00$ kg. The weight density of water is 62.4 lb/ft^3; that is, 1 cubic foot of water weighs 62.4 lb. (A suggested project is to take a container 1 cubic foot in volume, pour it full of water, and find the weight of the water. If you fill the container with a gallon container, you will also find that 1 ft^3 is approximately 7.5 gal.)

In nearly all forms of matter, the density usually decreases as the temperature increases and increases as the temperature decreases. Water does not follow the usual pattern of increasing density at lower temperatures; ice is actually less dense than liquid water. This phenomenon is discussed more fully in Section 13.7.

Note: Conversion factors must often be used to obtain the desired units.

EXAMPLE 1

Find the weight density of a block of wood 3.00 in. × 4.00 in. × 5.00 in. with weight 0.700 lb.

Sketch:

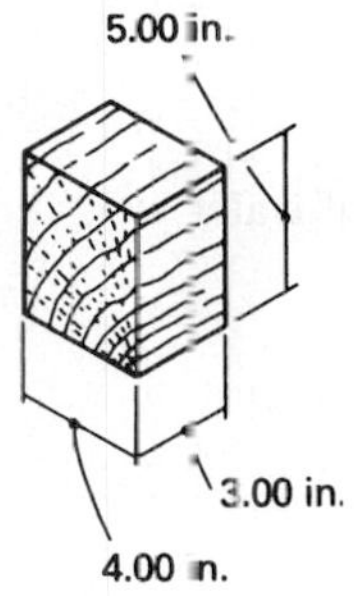

Data:

$$l = 4.00 \text{ in.}$$
$$w = 3.00 \text{ in.}$$
$$h = 5.00 \text{ in.}$$
$$F_w = 0.700 \text{ lb}$$
$$D_w = ?$$

Basic Equations:

$$V = lwh \quad \text{and} \quad D_w = \frac{F_w}{V}$$

Working Equations: Same

Substitutions:

$$V = (4.00 \text{ in.})(3.00 \text{ in.})(5.00 \text{ in.})$$
$$= 60.0 \text{ in}^3$$

$$D_w = \frac{0.700 \text{ lb}}{60.0 \text{ in}^3}$$
$$= 0.0117 \frac{\text{lb}}{\cancel{\text{in}^3}} \times \left(\frac{12 \cancel{\text{in.}}}{1 \text{ ft}}\right)^3$$
$$= 20.2 \text{ lb/ft}^3$$

EXAMPLE 2

Find the mass density of a ball bearing with mass 22.0 g and radius 0.875 cm.

Data:

$$r = 0.875 \text{ cm}$$
$$m = 22.0 \text{ g}$$
$$D_m = ?$$

Basic Equations:

$$V = \tfrac{4}{3}\pi r^3 \qquad \text{and} \qquad D_m = \frac{m}{V}$$

Working Equations: Same

Substitutions:

$$V = \tfrac{4}{3}\pi (0.875 \text{ cm})^3$$
$$= 2.81 \text{ cm}^3$$

$$D_m = \frac{22.0 \text{ g}}{2.81 \text{ cm}^3}$$
$$= 7.83 \text{ g/cm}^3$$
$$= 7.83 \frac{\cancel{\text{g}}}{\cancel{\text{cm}^3}} \times \left(\frac{100 \cancel{\text{cm}}}{1 \text{ m}}\right)^3 \times \frac{1 \text{ kg}}{10^3 \cancel{\text{g}}} = 7830 \text{ kg/m}^3$$

EXAMPLE 3

Find the weight density of a gallon of water weighing 8.34 lb.

Data:

$$F_w = 8.34 \text{ lb}$$
$$V = 1 \text{ gal} = 231 \text{ in}^3$$
$$D_w = ?$$

Basic Equation:

$$D_w = \frac{F_w}{V}$$

Working Equation: Same

Substitution:

$$\begin{aligned} D_w &= \frac{8.34\text{ lb}}{231\text{ in}^3} \\ &= 0.0361\frac{\text{lb}}{\text{in}^3} \times \left(\frac{12\text{ in.}}{1\text{ ft}}\right)^3 \\ &= 62.4\text{ lb/ft}^3 \end{aligned}$$

EXAMPLE 4

Find the weight density of a can of oil (1 quart) weighing 1.90 lb.

Data:

$$\begin{aligned} V &= 1\text{ qt} = \tfrac{1}{4}\text{ gal} = \tfrac{1}{4}(231\text{ in}^3) = 57.8\text{ in}^3 \\ F_w &= 1.90\text{ lb} \\ D_w &= ? \end{aligned}$$

Basic Equation:

$$D_w = \frac{F_w}{V}$$

Working Equation: Same

Substitution:

$$\begin{aligned} D_w &= \frac{1.90\text{ lb}}{57.8\text{ in}^3} \\ &= 0.0329\frac{\text{lb}}{\text{in}^3} \times \left(\frac{12\text{ in.}}{1\text{ ft}}\right)^3 \\ &= 56.9\text{ lb/ft}^3 \end{aligned}$$

EXAMPLE 5

A quantity of gasoline weighs 5.50 lb with weight density 42.0 lb/ft^3. Find its volume.

Data:

$$\begin{aligned} D_w &= 42.0\text{ lb/ft}^3 \\ F_w &= 5.50\text{ lb} \\ V &= ? \end{aligned}$$

Basic Equation:

$$D_w = \frac{F_w}{V}$$

Working Equation:

$$V = \frac{F_w}{D_w}$$

Substitution:

$$\begin{aligned} V &= \frac{5.50\text{ lb}}{42.0\text{ lb/ft}^3} \\ &= 0.131\text{ ft}^3 \end{aligned}$$

The density of an irregular solid (rock) cannot be found directly because of the difficulty of finding its volume. However, we could find the amount of water the solid displaces, which is the same as the volume of the irregular solid (Fig. 11.25). The volume of water in the small beaker equals the volume of the rock.

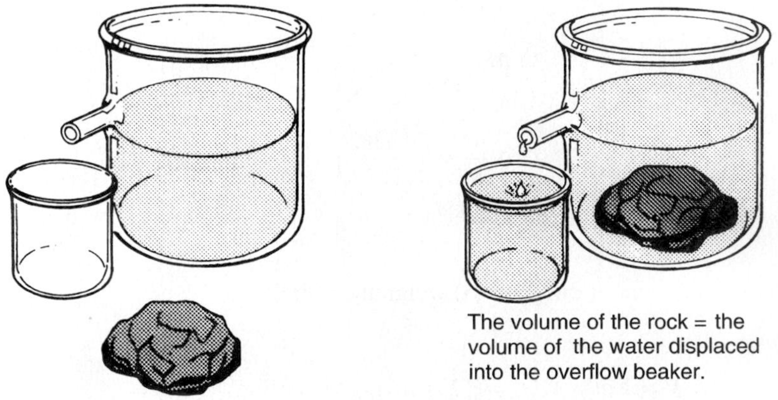

The volume of the rock = the volume of the water displaced into the overflow beaker.

Figure 11.25 The volume of this rock can be found by measuring the volume of the liquid it displaces.

EXAMPLE 6

A rock of mass 10.8 kg displaces $3\bar{2}00$ cm^3 of water. What is the mass density of the rock?

Data:

$$m = 10.8 \text{ kg}$$
$$V = 3\bar{2}00 \text{ cm}^3$$
$$D_m = ?$$

Basic Equation:

$$D_m = \frac{m}{V}$$

Working Equation: Same

Substitution:

$$D_m = \frac{10.8 \text{ kg}}{3\bar{2}00 \text{ cm}^3} \times \left(\frac{100 \text{ cm}}{1 \text{ m}}\right)^3$$
$$= 3380 \text{ kg/m}^3$$

EXAMPLE 7

A rock displaces 3.00 gal of water and has a weight density of 156 lb/ft^3. What is its weight?

Data:

$$D_w = 156 \text{ lb/ft}^3$$
$$V = 3.00 \text{ gal}$$
$$F_w = ?$$

Basic Equation:

$$D_w = \frac{F_w}{V}$$

Working Equation:

$$F_w = D_w V$$

Substitution:

$$F_w = 156\frac{\text{lb}}{\text{ft}^3} \times 3.00 \text{ gal} \times \frac{231 \text{ in}^3}{1 \text{ gal}} \times \left(\frac{1 \text{ ft}}{12 \text{ in.}}\right)^3$$
$$= 62.6 \text{ lb}$$

To compare the densities of two materials, we compare each with the density of water. The **specific gravity** of any material is the ratio of the density of the material to the density of water. That is,

$$\text{specific gravity (sp gr)} = \frac{D_{\text{material}}}{D_{\text{water}}}$$

Note that specific gravity is a unitless quantity.

EXAMPLE 8

The density of iron is 7830 kg/m^3. Find its specific gravity.

Data:

$$D_{\text{material}} = 7830 \text{ kg/m}^3$$
$$D_{\text{water}} = 1\bar{0}00 \text{ kg/m}^3$$
$$\text{sp gr} = ?$$

Basic Equation:

$$\text{sp gr} = \frac{D_{\text{material}}}{D_{\text{water}}}$$

Working Equation: Same

Substitution:

$$\text{sp gr} = \frac{7830 \text{ kg/m}^3}{1\bar{0}00 \text{ kg/m}^3}$$
$$= 7.83$$

This means that iron is 7.83 times as dense as water, and thus it sinks in water.

EXAMPLE 9

The density of oil is 54.2 lb/ft^3. Find its specific gravity.

$$D_{\text{material}} = 54.2 \text{ lb/ft}^3$$
$$D_{\text{water}} = 62.4 \text{ lb/ft}^3$$
$$\text{sp gr} = ?$$

Basic Equation:

$$\text{sp gr} = \frac{D_{\text{material}}}{D_{\text{water}}}$$

Working Equation: Same

Substitution:

$$\text{sp gr} = \frac{54.2 \text{ lb/ft}^3}{62.4 \text{ lb/ft}^3}$$
$$= 0.869$$

This means oil is 0.869 times as dense as water and thus it floats on water.

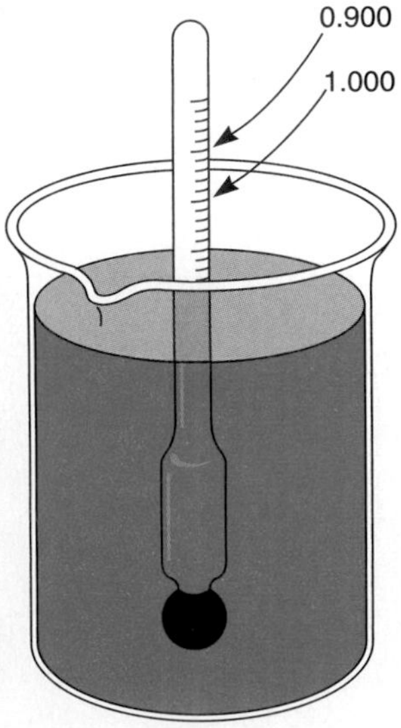

Figure 11.26 The hydrometer measures density of a liquid.

In general, the specific gravity of

water = 1
a material denser than water > 1
a material less dense than water < 1

When we check the antifreeze in a radiator in winter, we are really finding the specific gravity of the liquid. Specific gravity is a comparison of the density of a substance to that of water. Because the density of antifreeze is different from the density of water, we find the concentration of antifreeze (and thus the amount of protection from freezing) by measuring the specific gravity of the solution in the radiator.

A **hydrometer** is a sealed glass tube weighted at one end so that it floats vertically in a liquid (Fig. 11.26). It sinks in the liquid until it displaces an amount of liquid equal to its own weight. The densities of the displaced liquids are inversely proportional to the depths to which the tube sinks. That is, the greater the density of the liquid, the less the tube sinks; the less the density of the liquid, the more the tube sinks. A hydrometer usually has a scale inside the tube and is calibrated so that it floats in water at the 1.000 mark. Anything with a specific gravity greater than 1 sinks in water. A substance with a specific gravity less than 1 floats in water; its specific gravity indicates the fractional volume that is under water.

Hydrometers are commonly used to measure the specific gravities of battery acid and antifreeze in radiators (Fig. 11.27). In a lead storage battery, the electrolyte is a solution of sulfuric acid and water, and the specific gravity of the solution varies with the amount of charge of the battery. Table 11.3 gives common specific gravities of conditions of a lead storage battery. Table 11.4 gives various specific gravities and the corresponding temperatures below which the antifreeze and water solution will freeze.

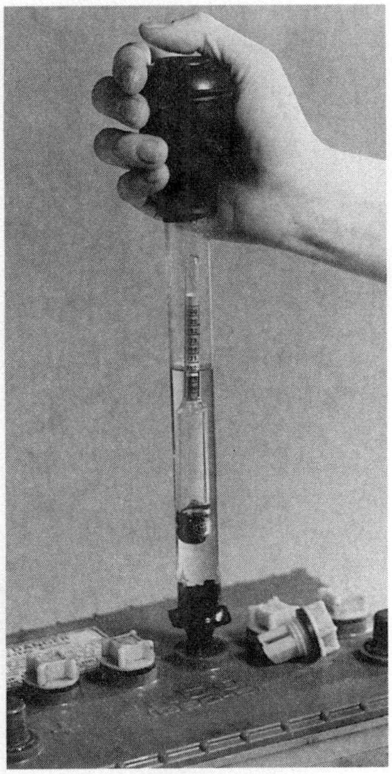

Figure 11.27 A common hydrometer

Table 11.3 Specific Gravities for a Lead Storage Battery

Condition	Specific Gravity
New (fully charged)	1.30
Old (discharged)	1.15

Table 11.4 Specific Gravities for Antifreeze and Water Solution

Temperature (°C)	Specific Gravity
−1.24	1.00
−2.99	1.01
−6.89	1.02
−19.82	1.05
−44.83	1.07
−51.23	1.08

One other factor must be considered in the use of the hydrometer—that of temperature. Significant differences in readings will occur over a range of temperatures. Specific gravities of some common liquids at room temperature are given in Table 11.5.

Table 11.5 Specific Gravities of Common Liquids at Room Temperature (20°C or 68°F)

Liquid	Specific Gravity
Benzene	0.9
Ethyl alcohol	0.79
Gasoline	0.68
Kerosene	0.82
Mercury	13.6
Seawater	1.025
Sulfuric acid	1.84
Turpentine	0.87
Water	1.000

PROBLEMS 11.5

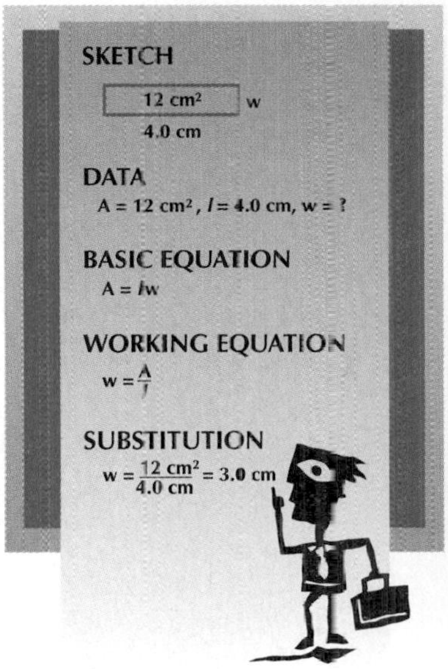

Express mass density in kg/m^3 and weight density in lb/ft^3.

1. Find the mass density of a chunk of rock of mass 215 g that displaces a volume of 75.0 cm^3 of water.
2. A block of wood is 55.9 in. × 71.1 in. × 25.4 in. and weighs 1810 lb. Find its weight density.
3. If a block of wood of the size in Problem 2 has a weight density of 30.0 lb/ft^3, what does it weigh?
4. Find the volume (in cm^3) of 1350 g of mercury.
5. Find the volume (in cm^3) of 1350 g of cork
6. Find the volume (in m^3) of 1350 g of nitrogen at 0°C and 1 atm pressure.
7. A block of gold 9.00 in. × 8.00 in. × 6.00 in. weighs 302 lb. Find its weight density.
8. A cylindrical piece of copper is 9.00 in. tall and 1.40 in. in radius. How much does it weigh?
9. A piece of aluminum of mass 6.24 kg displaces water that fills a container 12.0 cm × 12.0 cm × 16.0 cm. Find its mass density.
10. If 1.00 pint of turpentine weighs 0.907 lb, what is its weight density?
11. Find the mass density of gasoline if 105 g occupies 155 cm^3.
12. How much does 1.00 gal of gasoline weigh?
13. Determine the volume (in m^3) of 3045 kg of oil.
14. How many ft^3 will 573 lb of water occupy?
15. If 20.4 in^3 of linseed oil weighs 0.694 lb, what is its weight density?
16. If 108 in^3 of ammonia gas weighs 0.00301 lb, what is its weight density?
17. Find the volume of 3.00 kg of propane at 0°C and 1 atm pressure.
18. Granite has a mass density of 2650 kg/m^3. Find its weight density in lb/ft^3.
19. Find the mass density of a metal block 18.0 cm × 24.0 cm × 8.00 cm with mass 9.76 kg.
20. Find the mass (in kg) of 1.00 m^3 of
    (a) water. (b) gasoline. (c) copper.
    (d) mercury. (e) air at 0°C and 1 atm pressure.
21. What size tank (in litres) is needed for 1000 kg of
    (a) water? (b) gasoline? (c) mercury?
22. Copper has a mass density of 8890 kg/m^3. Find its mass density in g/cm^3.

Use Table 11.2 to find the specific gravity of each material.

23.	Ice	24.	Concrete	25.	Iron
26.	Air	27.	Gasoline	28.	Cork

29. The specific gravity of material X is 0.82. Does it sink in or float on water?
30. The specific gravity of material Y is 1.7. Does it sink in or float on water?
31. The specific gravity of material Z is 0.52. Does it sink in or float on gasoline?
32. The specific gravity of material W is 11.5. Does it sink in or float on mercury?
33. A proton has mass 1.67×10^{-27} kg and diameter 8.2×10^{-16} m. Find its specific gravity.
34. Find the mass density of a 315-g object that displaces 0.275 m^3 of water.
35. What is the mass density of a 500-g block that displaces 215 cm^3 of water?

Glossary

Adhesion The force of attraction between different or unlike molecules. (p. 290)

Atom The smallest particle of an element that can exist in a stable or independent state. (p. 288)

Bending Consists of both tension and compression stresses. It occurs when a force is placed on a beam causing it to sag. (p. 296)

Brinell Method Common industrial method used to measure the hardness of a metal. (p. 290)

Capillary Action The behavior of liquids that causes the liquid level in small tubes to be different than in larger tubes. This behavior is due both to adhesion of the liquid molecules to the tube and to the surface tension of the liquid. (p. 305)

Cohesion The force of attraction between like molecules. Holds the closely packed molecules of a solid together. (p. 290)

Compound A substance containing two or more elements. (p. 288)

Compression A stress caused by two forces acting directly toward each other. This stress tends to cause objects to become shorter and thicker. (p. 295)

Diffusion The process by which molecules of a gas mix with the molecules of a solid, a liquid, or another gas. (p. 306)

Ductility A property of a metal that enables it to be drawn through a die to produce a wire. (p. 291)

Elastic Limit The point beyond which a deformed object cannot return to its original shape. (p. 292)

Elasticity A measure of a deformed object's ability to return to its original size and shape once the deforming force is removed. (p. 292)

Electron One of the particles that makes up atoms. Has a negative charge. (p. 288)

Element A substance that cannot be separated into simpler substances. (p. 288)

Expansion Property of a gas in which the rapid random movement of its molecules causes the gas to completely occupy the volume of its container. (p. 306)

Fluid A substance that takes the shape of its container. Either a liquid or a gas. (p. 306)

Gas A substance that takes the shape of its container and has the same volume as its container. (p. 289)

Hardness A measure of the internal resistance of the molecules of a solid being forced farther apart or closer together. (p. 290)

Hooke's Law A principle of elasticity in solids: The ratio of the force applied to an object to its change in length (resulting in its being stretched or compressed by the applied force) is constant as long as the elastic limit has not been exceeded. (p. 298)

Hydrometer A sealed glass tube weighted at one end so that it floats vertically in a liquid; an instrument used to determine specific gravity. (p. 312)

Liquid A substance that takes the shape of its container and has a definite volume. (p. 289)

Malleability A property of a metal that enables it to be hammered and rolled into a sheet. (p. 292)

Mass Density The mass per unit volume of a substance. (p. 306)

Matter Anything that occupies space and has mass. (p. 288)

Meniscus The crescent-shaped surface of a liquid column in a tube. (p. 306)

Molecule The smallest particle of a substance that exists in a stable and independent state. (p. 288)

Neutron One of the particles that makes up atoms. Does not carry an electric charge. (p. 288)

Nucleus The center part of an atom made up of protons and neutrons. (p. 289)

Proton One of the particles that makes up atoms. Has a positive charge. (p. 288)

Shearing A stress caused by two forces applied in parallel, opposite directions. (p. 295)

Solid A substance that has a definite shape and a definite volume. (p. 289)

Specific Gravity The ratio of the density of any material to the density of water. (p. 311)

Strain The deformation of an object due to an applied force. (p. 296)

Stress The ratio of an outside applied distorting force to the area over which the force acts. (p. 293)

Surface Tension The ability of the surface of a liquid to act like a thin, flexible film. (p. 304)

Tensile Strength A measure of a solid's resistance to being pulled apart. (p. 290)

Tension A stress caused by two forces acting directly opposite each other. This stress tends to cause objects to become longer and thinner. (p. 295)

Torsion A stress related to a twisting motion. This type of stress severely compromises the strength of most materials. (p. 296)

Viscosity The internal friction of a fluid caused by molecular attraction, which makes it resist a tendency to flow. (p. 304)
Weight Density The weight per unit volume of a substance. (p. 306)

Formulas

11.2 $S = \dfrac{F}{A}$

$\dfrac{F}{\Delta l} = k$

11.5 $D_m = \dfrac{m}{V}$

$D_w = \dfrac{F_w}{V}$

$\text{specific gravity (sp gr)} = \dfrac{D_{\text{material}}}{D_{\text{water}}}$

Review Questions

1. The most important particles that make up atoms include which of the following?
   (a) Neutron (b) Molecule (c) Electron
   (d) Hydrogen (e) Proton
2. Matter exists in which of the following?
   (a) Gas (b) Neutrons (c) Electrons
   (d) Solid (e) Liquid
3. The common industrial method used to measure the hardness of a metal is
   (a) the Bernoulli method. (b) Hooke's method.
   (c) the capillary method. (d) the Brinell method.
   (e) none of the above.
4. Density is a property of
   (a) gases. (b) liquids.
   (c) solids. (d) all of the above.
   (e) none of the above.
5. The process by which molecules of a gas mix with the molecules of a solid, a liquid, or a gas is called
   (a) expansion. (b) contraction.
   (c) capillary action. (d) diffusion.
   (e) none of the above.
6. Capillary action refers to
   (a) the mixing of molecules of different types.
   (b) the behavior of liquids in small tubes.
   (c) the attractive force between molecules.
   (d) stretching beyond the elastic limit.
7. The relationship of the change in length of a stretched or compressed object to the force causing the change is given by
   (a) Pascal's law. (b) Brinell's law. (c) the elastic limit.
   (d) Hooke's law. (e) none of the above.
8. The ability of the surface of water to support a needle is an example of
   (a) mass density. (b) Hooke's law. (c) diffusion.
   (d) stress. (e) surface tension.
9. In your own words, describe the difference between mass density and weight density.
10. Would the mass density of an object be the same if the object were on the moon rather than on the earth? Would the weight density be the same?
11. In your own words, describe capillary action.

12. What is the difference between adhesion and cohesion?
13. Give one example of the effect of surface tension that is not described in this book.
14. The mass of a proton is approximately _____ times heavier than the mass of an electron.
15. The applied force divided by the area over which the force acts is called _____.
16. In your own words, state Hooke's law.
17. The commonly used unit of stress in the metric system is the _____.
18. Describe how to find the specific gravity of an object.
19. What is the ratio of mass to volume called?
20. What is friction in liquids called?
21. A spring that has been permanently deformed is said to have been deformed past its_____ _____.
22. List the three states of matter.
23. Distinguish between a molecule and an atom.
24. Distinguish between a neutron and a proton.
25. List the five basic stresses.
26. Explain how a hydrometer measures the charge in a lead storage battery. Does the temperature affect the measurement?

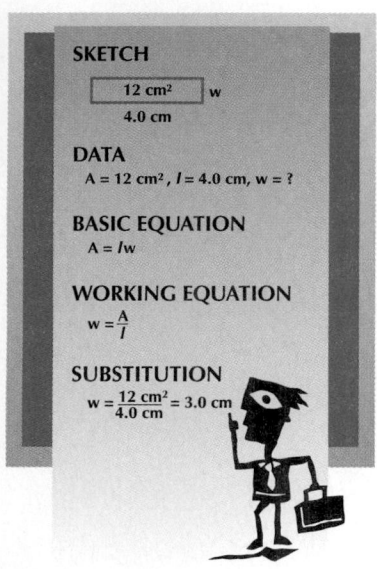

SKETCH

12 cm² w
4.0 cm

DATA

$A = 12\text{ cm}^2$, $l = 4.0\text{ cm}$, $w = ?$

BASIC EQUATION

$A = lw$

WORKING EQUATION

$w = \frac{A}{l}$

SUBSTITUTION

$w = \frac{12\text{ cm}^2}{4.0\text{ cm}} = 3.0\text{ cm}$

Review Problems

1. A force of 32.5 N stretches a wire 0.470 cm. What force will stretch a similar piece of wire 2.39 cm?
2. A force of 7.33 N is applied to a spring whose spring constant is 0.298 N/cm. Find its change in length.
3. Each vertical steel column of an office building supports 42,100 lb and is compressed 0.0258 in. What is the spring constant of the steel? Find the compression if the weight were 51,700 lb.
4. A rectangular cast-iron column 16.0 cm × 16.0 cm × 4.50 m supports a weight of 7.95×10^6 N. Find the stress on the top of the column.
5. Find the weight density of a block of metal 7.00 in. × 6.50 in. × 8.00 in. that weighs 425 lb.
6. A cylindrical piece of aluminum is 4.25 cm tall and 1.95 cm in radius. How much does it weigh?
7. A piece of metal has a mass of 8.36 kg. If it displaces water that fills a container 9.34 cm × 10.0 cm × 10.0 cm, what is the mass density of the metal?
8. A block of wood is 27.7 in. × 36.3 in. × 12.4 in. and weighs 602 lb. Find its weight density.
9. Find the volume (in cm^3) of 759 g of mercury.
10. Find the volume (in m^3) of 1970 g of hydrogen at 0°C and 1 atm.
11. Find the mass of 1510 m^3 of oxygen at 0°C and 1 atm.
12. Find the weight of 951 ft^3 of water.
13. Find the weight density of a block of material 4.27 in. × 3.87 in. × 5.44 in. that weighs 0.982 lb.
14. Find the weight density of 2.00 quarts of liquid weighing 3.67 lb.
15. A quantity of liquid weighs 4.65 lb with a weight density of 39.8 lb/ft^3. What is its volume?
16. The density of a metal is 694 kg/m^3. Find its specific gravity.
17. A solid displaces 4.30 gal of water and has a weight density of 135 lb/ft^3. What is its weight?
18. Find the mass of a rectangular gold bar 4.00 cm × 6.00 cm × 20.00 cm. The mass density of gold is 19,300 kg/m^3.
19. Find the mass density of a chunk of rock using only a scale knowing the following information: mass of rock is 225 g; mass of water the rock displaces is 75.9 g.
20. The specific gravity of an unknown substance is 0.80. Will it float on or sink in gasoline?

APPLIED CONCEPTS

1. Instead of carrying a full-size spare tire, many automobiles are equipped with a significantly smaller spare tire with a warning that the car should not exceed a speed of 50 mi/h. (a) How much pressure does the smaller tire withstand if it must support one-fourth of a 1650-kg minivan and its area of contact with the road measures 8.26 cm × 15.9 cm? (b) Find the pressure on a full-size spare tire if its area of contact with the road measures 15.2 cm × 21.0 cm. (c) Why is it important to limit the speed of the smaller spare tire?
2. Observe the warped lines on asphalt pavement in front of a stoplight (Fig. 11.28). (a) What type of stress causes such curved lines? (b) Why are the warped lines, as seen in the photograph, more noticeable at stop signs and stoplights than on other sections of road?

Figure 11.28

3. Raul weighs 235 lb and is able to float in seawater but not in fresh water. Find his volume assuming that his density is the average density of seawater and fresh water.
4. A tanker truck with a cylindrical container 11.3 m long and 2.35 m in diameter transports several types of liquids. How much mass does it carry if it is transporting a full load of (a) water, (b) oil, and (c) gasoline?
5. Every morning Shakira weighs herself on a bathroom spring scale. Today, the scale reads 134 lb. (a) If the spring in the bathroom scale compresses 0.752 in., find its spring constant. (b) If her husband steps on the scale and the spring compresses 1.13 in., how much does he weigh? (c) Why is there a limit to how much weight a spring scale can measure?

CHAPTER 12

FLUIDS

Substances that flow are called fluids. Because liquids and gases behave in much the same manner, they will be presented together. From ships to airplanes to automobile brake systems to balloons, the utilization of fluids and their properties is important to technology and everyday life. We will now study hydraulics, hydrostatic pressure, buoyancy, and flow.

Objectives

The major goals of this chapter are to enable you to:

1. Describe the behavior of fluids.
2. Determine pressure using the hydraulic principle.
3. Distinguish between gauge pressure and absolute pressure.
4. Calculate buoyancy using Archimedes' principle.
5. Analyze fluid flow and Bernoulli's principle.

12.1 Hydrostatic Pressure

Since liquids and gases behave in much the same manner, they are often studied together as fluids. The gas and water piped to your home are fluids having several common characteristics.

Pressure is the force applied per unit area. **Hydrostatic pressure** is the pressure a liquid at rest exerts on a submerged object (Fig. 12.1). As you probably know, the pressure on an object increases as the water depth increases. Liquids differ from solids in that they exert force in all directions, whereas solids exert only a downward force due to gravity.

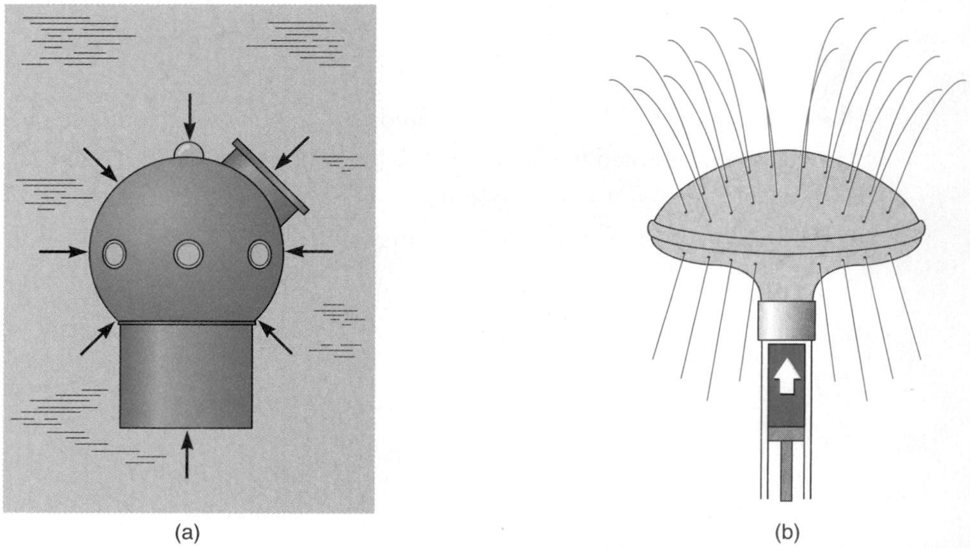

Figure 12.1 (a) The pressure on a submerged object is exerted by the liquid in all directions. The pressure at ocean-floor level is so great that very strong containers must be built to withstand the enormous pressures. (b) When a force is exerted on a liquid in a container, the force is exerted uniformly in all directions. Note that the liquid squirts out of the holes uniformly in all directions.

The pressure in a liquid depends *only* on the depth and weight density of the liquid and *not* on the surface area. Because the pressure exerted by water increases with depth, dams are built much thicker at the base than at the top (Fig. 12.2).

TRY THIS ACTIVITY

Water Pressure

The next time you go swimming in a pool, swim toward the bottom. As you swim deeper, notice the gradual change in pressure on your eardrums. Explain how the change in water pressure on your eardrums can be used to describe hydrostatic pressure. How would the pressure change if the pool were full of salt water?

This general pressure principle may also be illustrated with a pile of bricks. The weight of the pile and the pressure on the ground clearly increase as the pile is made taller. The force (weight) applied to the same area at the bottom of the pile increases the pressure at the bottom of the pile with the addition of each brick. Likewise, the pressure on a submerged object increases the farther it is pushed down due to the weight of the additional water above it.

Figure 12.2 Dams must be built much thicker at the base because the pressure exerted by the water increases with depth.

To find the pressure at a given depth in a liquid, use the formula

$$P = hD_w$$

where

P = pressure
h = height (or depth)
D_w = weight density of the liquid

One way to increase water pressure is to raise a water storage tank above the ground. This same formula applies except h is the sum of the depth of the water in the tank and the distance the bottom of the water tank is above the ground or some other reference such as a building. This is why community water storage tanks are placed on towers or on the tops of large hills.

TRY THIS ACTIVITY

Water Pressure and Depth

Punch three small holes near the top, the middle, and the bottom on the side of a large juice can with its lid removed. Next, fill the can with water and compare the streams of water spurting out each hole (Fig. 12.3). Pressure in a liquid also depends on the weight density of the liquid. Pressure on an object 50 m below the surface of fresh water is less than the pressure at the same depth in salt water because fresh water is less dense than salt water.

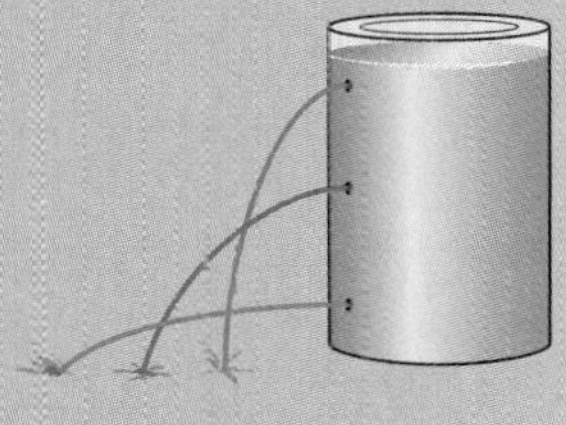

Figure 12.3

EXAMPLE 1

Find the pressure at the bottom of a water-filled drum 4.00 ft high.

Sketch:

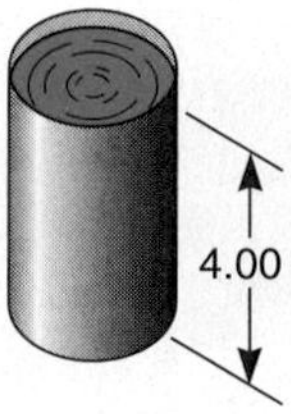

Data:

$$h = 4.00 \text{ ft}$$
$$D_w = 62.4 \text{ lb/ft}^3$$
$$P = ?$$

Basic Equation:

$$P = hD_w$$

Working Equation: Same

Substitution:

$$P = (4.00 \text{ ft})\left(62.4 \frac{\text{lb}}{\text{ft}^3}\right)$$
$$= 25\bar{0} \frac{\text{lb}}{\cancel{\text{ft}^2}} \times \left(\frac{1 \cancel{\text{ft}}}{12 \text{ in.}}\right)^2$$
$$= 1.74 \text{ lb/in}^2$$

Note: The pressure depends only on the height, not the width or area of the container.

EXAMPLE 2

Find the depth in a lake at which the pressure is 105 lb/in^2.

Data:

$$P = 105 \text{ lb/in}^2$$
$$D_w = 62.4 \text{ lb/ft}^3$$
$$h = ?$$

Basic Equation:

$$P = hD_w$$

Working Equation:

$$h = \frac{P}{D_w}$$

Substitution:

$$h = \frac{105 \text{ lb/in}^2}{62.4 \text{ lb/ft}^3}$$
$$= 1.68 \frac{\text{ft}^3}{\text{in}^2}$$

$$\frac{\text{lb/in}^2}{\text{lb/ft}^3} = \frac{\text{lb}}{\text{in}^2} \div \frac{\text{lb}}{\text{ft}^3} = \frac{\cancel{\text{lb}}}{\text{in}^2} \times \frac{\text{ft}^3}{\cancel{\text{lb}}} = \frac{\text{ft}^3}{\text{in}^2}$$

$$= 1.68 \frac{\overset{\text{ft}}{\cancel{\text{ft}^3}}}{\cancel{\text{in}^2}} \times \left(\frac{12 \cancel{\text{in.}}}{1 \cancel{\text{ft}}}\right)^2$$
$$= 242 \text{ ft}$$

EXAMPLE 3

Find the height of a water column where the pressure at the bottom of the column is $40\bar{0}$ kPa and the weight density of water is $98\bar{0}0$ N/m^3.

Sketch: None needed

Data:

$$P = 40\bar{0} \text{ kPa}$$
$$D_w = 98\bar{0}0 \text{ N/m}^3$$
$$h = ?$$

Basic Equation:

$$P = hD_w$$

Working Equation:

$$h = \frac{P}{D_w}$$

Substitution:

$$h = \frac{40\bar{0} \text{ kPa}}{98\bar{0}0 \text{ N/m}^3}$$
$$= \frac{40\bar{0} \text{ kPa}}{98\bar{0}0 \text{ N/m}^3} \times \frac{10^3 \text{ N/m}^2}{1 \text{ kPa}} \qquad (\textit{Recall: } 1 \text{ kPa} = 10^3 \text{ N/m}^2)$$
$$= 40.8 \text{ m}$$

$$\frac{\cancel{\text{kPa}}}{\text{N/m}^3} \times \frac{\text{N/m}^2}{\cancel{\text{kPa}}} = \text{N/m}^2 \div \text{N/m}^3 = \frac{\cancel{\text{N}}}{\text{m}^2} \times \frac{\text{m}^3}{\cancel{\text{N}}} = \text{m}$$

Total Force Exerted by Liquids

The *total force* exerted by a liquid on a horizontal surface (such as the bottom of a barrel) depends on the area of the surface, the depth of the liquid, and the weight density of the liquid. By formula,

$$F_t = AhD_w$$

where

F_t = total force
A = area of bottom or horizontal surface
h = height or depth of the liquid
D_w = weight density

EXAMPLE 4

Find the total force on the bottom of a rectangular tank 10.0 ft by 5.00 ft by 4.00 ft deep filled with water.

Sketch:

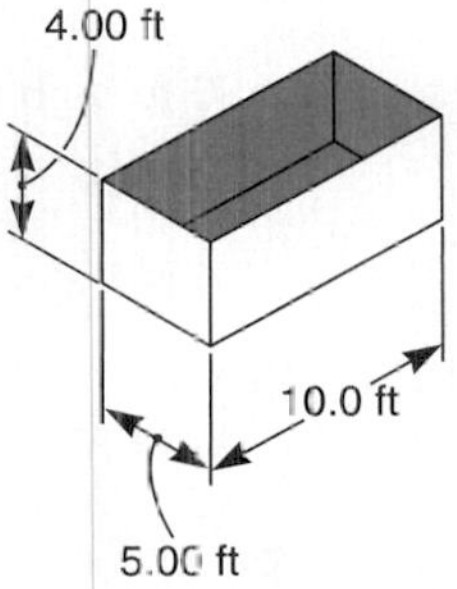

Data:

$$A = lw = (10.0\text{ ft})(5.00\text{ ft}) = 50.0\text{ ft}^2$$
$$h = 4.00\text{ ft}$$
$$D_w = 62.4\text{ lb/ft}^3$$
$$F_t = ?$$

Basic Equation:

$$F_t = AhD_w$$

Working Equation: Same

Substitution:

$$F_t = (50.0\text{ ft}^2)(4.00\text{ ft})\left(62.4\frac{\text{lb}}{\text{ft}^3}\right)$$
$$= 12{,}500\text{ lb}$$

The total force on a vertical surface F_s (such as the *side* of a tank) is found by using half the vertical height (average height):

$$\boxed{F_s = \tfrac{1}{2}AhD_w}$$

where A is the area of the side or vertical surface.

EXAMPLE 5

Find the total force on the small side of the rectangular tank in Example 4.

Data:

$$A = lw = (5.00\text{ ft})(4.00\text{ ft}) = 20.0\text{ ft}^2$$
$$h = 4.00\text{ ft}$$
$$D_w = 62.4\text{ lb/ft}^3$$
$$F_s = ?$$

Basic Equation:

$$F_s = \tfrac{1}{2}AhD_w$$

Working Equation: Same

Substitution:

$$F_s = \tfrac{1}{2}(20.0\text{ ft}^2)(4.00\text{ ft})(62.4\text{ lb/ft}^3)$$
$$= 25\bar{0}0\text{ lb}$$

EXAMPLE 6

Find the total force on the side of a water-filled cylindrical tank 3.00 m high with radius 5.00 m. The weight density of water is $98\bar{0}0\text{ N/m}^3$.

Sketch:

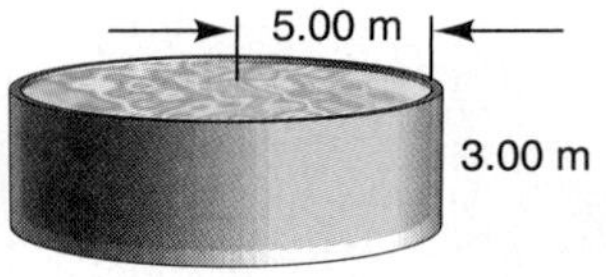

Data:

$A = 2\pi rh = 2\pi(5.00\text{ m})(3.00\text{ m})$ *Note:* The area of the vertical surface is the lateral surface of the cylinder.

$h = 3.00\text{ m}$

$D_w = 98\bar{0}0\text{ N/m}^3$

$F_s = ?$

Basic Equation:

$$F_s = \tfrac{1}{2}AhD_w$$

Working Equation: Same

Substitution:

$$F_s = \tfrac{1}{2}[2\pi(5.00\text{ m})(3.00\text{ m})](3.00\text{ m})(98\bar{0}0\text{ N/m}^3)$$
$$= 1.39 \times 10^6\text{ N}$$

PROBLEMS 12.1

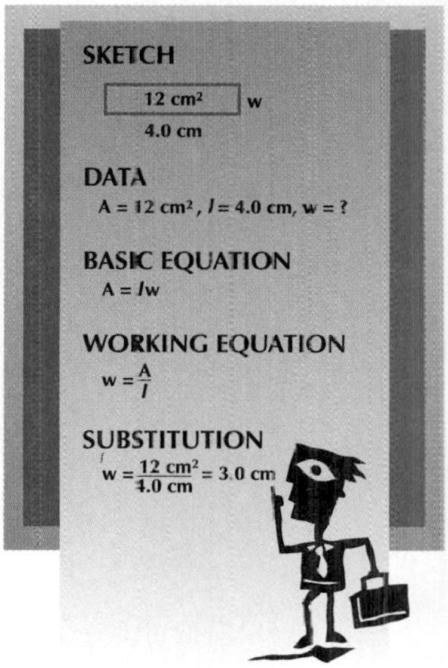

1. Find the pressure (in lb/in^2) at the bottom of a tower with water 50.0 ft deep.
2. Find the height of a column of water where the pressure at the bottom of the column is 20.0 lb/in^2.
3. Find the density of a liquid that exerts a pressure of 0.400 lb/in^2 at a depth of 42.0 in.
4. (a) Find the total force on the bottom of a water-filled circular cattle tank 0.750 m high with radius 1.30 m where the weight density of water is $98\bar{0}0\text{ N/m}^3$. (b) Find the total force on the side of the tank.
5. What must the water pressure be to supply water to the third floor of a building (35.0 ft up) with a pressure of 40.0 lb/in^2 at that level?
6. A small rectangular tank 5.00 in. by 9.00 in. is filled with mercury. (a) If the total force on the bottom of the tank is 165 lb, how deep is the mercury? (Weight density of mercury = 0.490 lb/in^3.) (b) Find the total force on the larger side of the tank.
7. Find the water pressure (in kPa) at the 25.0-m level of a water tower containing water 50.0 m deep.
8. Find the height of a column of water where the pressure at the bottom is 115 kPa.
9. What is the mass density of a liquid that exerts a pressure of 178 kPa at a depth of 24.0 m?
10. (a) Find the total force on the bottom of a cylindrical gasoline storage tank 15.0 m high with radius 23.0 m. (b) Find the total force on the side of the tank.
11. What must the water pressure be to supply the second floor (18.0 ft up) with a pressure of 50.0 lb/in^2 at that level?
12. Find the water pressure at ground level to supply water to the third floor of a building 8.00 m high with a pressure of 325 kPa at the third-floor level.
13. What pressure must a pump supply to pump water up to the thirtieth floor of a skyscraper with a pressure of 25 lb/in^2? Assume that the pump is located on the first floor and that there are 16.0 ft between floors.
14. A submarine is submerged to a depth of 3550 m in the Pacific Ocean. What air pressure (in kPa) is needed to blow water out of the ballast tanks?

A filled water tower sits on the top of the highest hill in a town (use Fig. 12.4 for Problems 15–19). The cylindrical tower has a radius of 12.0 m and a height of 50.0 m.

15. Find the total force on the bottom of the water tower.
16. Find the total force on the sides of the water tower.
17. Find the pressure (in kPa) on the bottom of the water tower.
18. What is the water pressure (in kPa) at the fire station?
19. What is the water pressure (in kPa) at the school?

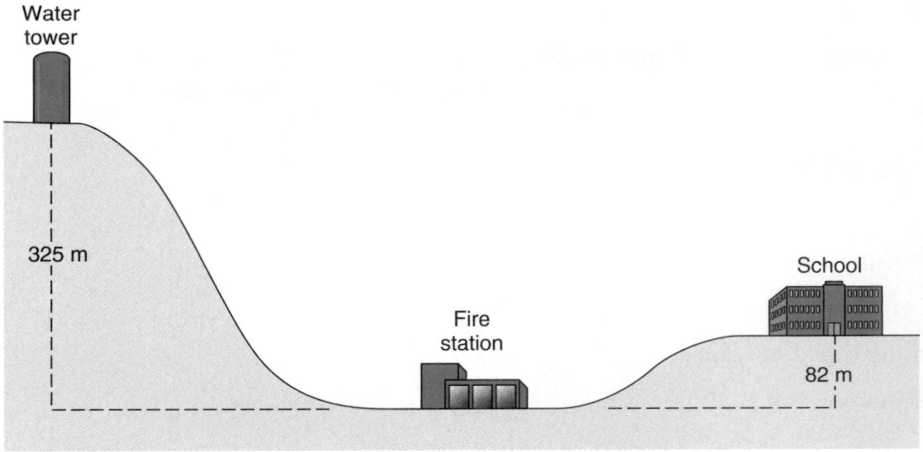

Figure 12.4

20. A cylindrical grain bin 24.0 ft in diameter is filled with corn whose weight density is 45.1 lb/ft^3. How tall can the bin be for the floor to support 94.0 lb/in^2 of pressure?

12.2 Hydraulic Principle

Put a stopper in one end of a metal pipe. Fill it with a fluid such as water. Then put a second stopper into the open end. Put the pipe in a horizontal position as in Fig. 12.5 and push on stopper A. What happens? Stopper B is pushed out. This illustrates a basic principle of hydraulics: The liquid in the pipe transmits the pressure from one stopper or piston to another without measurable loss.

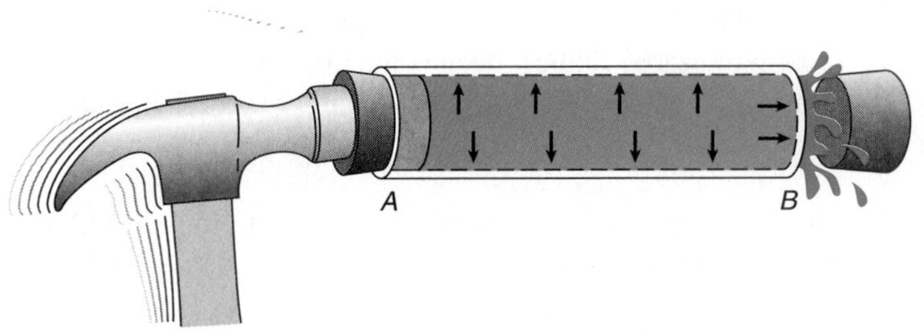

Figure 12.5 Pressure on a confined liquid is transmitted in all directions without measurable loss.

HYDRAULIC PRINCIPLE (PASCAL'S PRINCIPLE)

The pressure applied to a confined liquid is transmitted without measurable loss throughout the entire liquid to all inner surfaces of the container.

The rear-wheel hydraulic brake system of a front-wheel-drive automobile (Fig. 12.6) is an application of Pascal's principle. When the driver pushes the brake pedal, the pressure on the piston in the master cylinder is transmitted through the brake fluid to the two pistons in the brake cylinder. This transmitted pressure then forces the brake-cylinder pistons to push the brake shoes against the brake drum and stop the automobile. Releasing the brake pedal releases the pressure on the pistons in the brake cylinder. The spring pulls the brake shoes away from the brake drum, which allows the wheels to turn freely again.

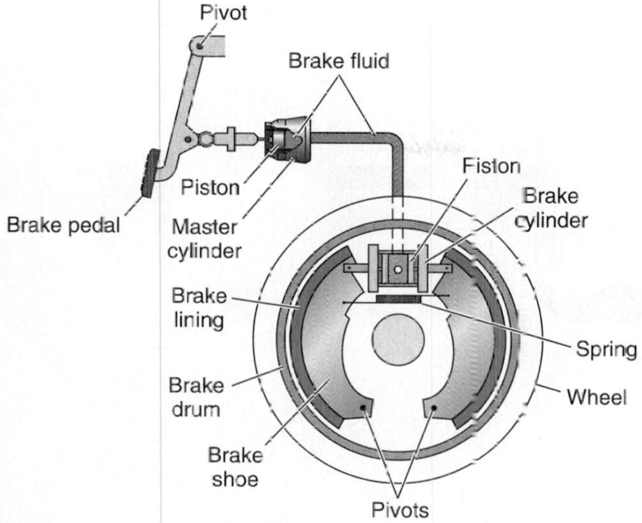

Figure 12.6 Hydraulic brake system of an automobile

The hydraulic jack, lift, and press are applications of hydraulics being used as a simple machine to multiply force. If we apply a force to the small piston of the hydraulic lift in Fig. 12.7, the pressure is transmitted without measurable loss in all directions. The reason for this is the virtual noncompressibility of liquids. The *pressure* on the large piston is the same as the pressure on the small piston; however, the *total force* on the large piston is greater because of its larger surface area.

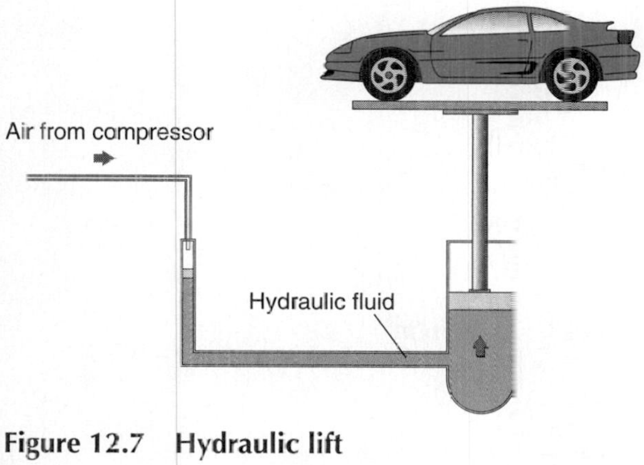

Figure 12.7 Hydraulic lift

Although it may seem we are acquiring a "free" increase in force by being able to lift an automobile using a very small applied force, that is really not the case. The principle is similar to the mechanical advantage of simple machines; we can multiply force but at the expense of distance. The lift, although able to move a large weight, can only move it a relatively short distance.

EXAMPLE 1

From the diagram of the hydraulic jack in Fig. 12.8, find

(a) the pressure on the small piston.
(b) the pressure on the large piston.
(c) the total force on the large piston.
(d) the mechanical advantage of the jack.

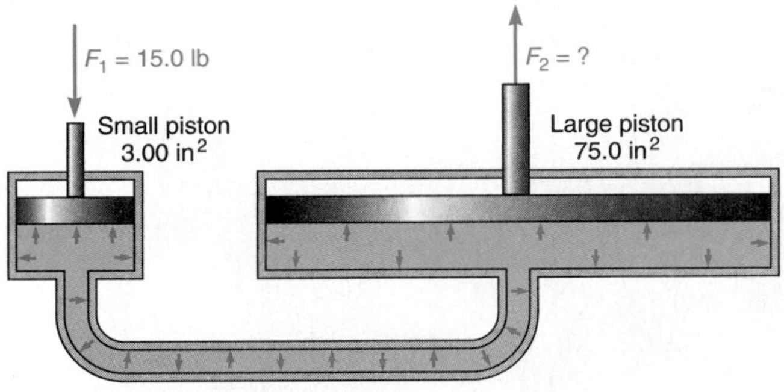

Figure 12.8 Hydraulic jack

Data:

$$F_1 = 15.0 \text{ lb}$$
$$A_{\text{small piston}} = A_1 = 3.00 \text{ in}^2$$
$$A_{\text{large piston}} = A_2 = 75.0 \text{ in}^2$$
$$P_1 = ?$$
$$P_2 = ?$$
$$F_2 = ?$$
$$\text{MA} = ?$$

(a) Basic Equation:

$$P_1 = \frac{F_1}{A_1}$$

Working Equation: Same

Substitution:

$$P_1 = \frac{15.0 \text{ lb}}{3.00 \text{ in}^2}$$
$$= 5.00 \text{ lb/in}^2$$

(b) Applying Pascal's Principle:

$$P_2 = P_1 = 5.00 \text{ lb/in}^2$$

(c) Basic Equation:

$$P_2 = \frac{F_2}{A_2}$$

Working Equation:

$$F_2 = P_2 A_2$$

Substitution:

$$F_2 = \left(5.00 \frac{\text{lb}}{\cancel{\text{in}^2}}\right)(75.0 \cancel{\text{in}^2})$$
$$= 375 \text{ lb}$$

(d) Basic Equation:

$$\text{MA} = \frac{F_R}{F_E}$$

Working Equation: Same

Substitution:

$$\text{MA} = \frac{375 \text{ lb}}{15.0 \text{ lb}}$$
$$= 25.0$$

EXAMPLE 2

The small piston of a hydraulic press has an area of 10.0 cm². If the applied force is 50.0 N, what must the area of the large piston be to exert a pressing force of 4800 N?

Sketch: None needed

Data:

$$A_1 = 10.0 \text{ cm}^2$$
$$F_1 = 50.0 \text{ N}$$
$$F_2 = 4800 \text{ N}$$
$$A_2 = ?$$

Basic Equations:

$$P_1 = \frac{F_1}{A_1}, \quad P_2 = \frac{F_2}{A_2}, \quad \text{and since } P_1 = P_2, \quad \frac{F_1}{A_1} = \frac{F_2}{A_2}$$

Working Equation:

$$A_2 = \frac{A_1 F_2}{F_1}$$

Substitution:

$$A_2 = \frac{(10.0 \text{ cm}^2)(4800 \text{ N})}{50.0 \text{ N}}$$
$$= 960 \text{ cm}^2$$

PROBLEMS 12.2

1. The area of the small piston in a hydraulic jack is 0.750 in². The area of the large piston is 3.00 in². If a force of 15.0 lb is applied to the small piston, what weight can be lifted by the large one?
2. The mechanical advantage of a hydraulic press is 25. What applied force is necessary to produce a pressing force of 2400 N?
3. Find the mechanical advantage of a hydraulic press that produces a pressing force of 8250 N when the applied force is 375 N.
4. The mechanical advantage of a hydraulic press is 13. What applied force is necessary to produce a pressing force of 990 lb?
5. Find the mechanical advantage of a hydraulic press that produces a pressing force of 1320 N when the applied force is 55.0 N.
6. The small piston of a hydraulic press has an area of 8.00 cm². If the applied force is 25.0 N, find the area of the large piston to exert a pressing force of 3500 N.
7. The MA of a hydraulic jack is 250. What force must be applied to lift an automobile weighing 12,000 N?
8. The small piston of a hydraulic press has an area of 4.00 in². If the applied force is 10.0 lb, what must the area of the large piston be to exert a pressing force of 865 lb?
9. The MA of a hydraulic jack is 420. Find the weight of the heaviest automobile that can be lifted by an applied force of 55 N.
10. The mechanical advantage of a hydraulic jack is 450. Find the weight of the heaviest automobile that can be lifted by an applied force of 60.0 N.

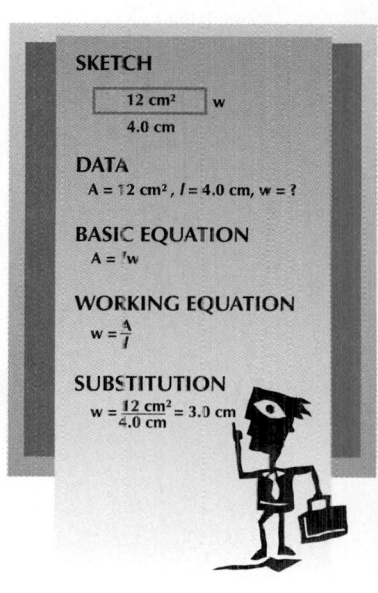

SKETCH

12 cm² w

4.0 cm

DATA

$A = 12 \text{ cm}^2$, $l = 4.0 \text{ cm}$, $w = ?$

BASIC EQUATION

$A = lw$

WORKING EQUATION

$w = \frac{A}{l}$

SUBSTITUTION

$w = \frac{12 \text{ cm}^2}{4.0 \text{ cm}} = 3.0 \text{ cm}$

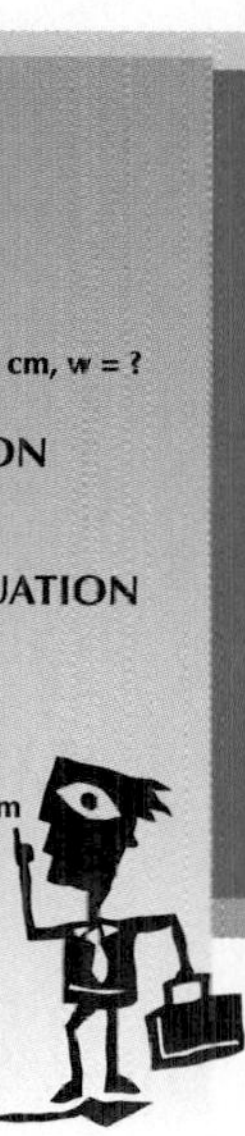

11. The pistons of a hydraulic press have radii of 2.00 cm and 12.0 cm, respectively.
    (a) What force must be applied to the smaller piston to exert a force of 5250 N on the larger?
    (b) What is the pressure (in N/cm^2) on each piston?
    (c) What is the mechanical advantage of the press?
12. The small circular piston of a hydraulic press has an area of 8.00 cm^2. If the applied force is 25.0 N, what must the area of the large piston be to exert a pressing force of 3650 N?
13. The large piston on a hydraulic lift has radius 40.0 cm. The small piston has radius 5.00 cm to which a force of 75.0 N is applied. Find
    (a) the force exerted by the large piston.
    (b) the pressure on the large piston.
    (c) the pressure on the small piston.
    (d) the mechanical advantage of the lift.
    (e) What happens when the area of the small piston is half as large?
    (f) What happens when the radius of the small piston is half as large?
14. In a hydraulic system a 20.0-N force is applied to the small piston with cross-sectional area 25.0 cm^2. What weight can be lifted by the large piston with cross-sectional area 50.0 cm^2?
15. If the diameter of the larger piston in Problem 14 is doubled, how is the weight able to be lifted changed?
16. If a dentist's chair weighs $16\bar{0}0$ N and is raised by a large piston with cross-sectional area of 75.0 cm^2, what force must be exerted on a small piston of cross-sectional area 3.75 cm^2 to lift the chair?
17. A hydraulic jack whose piston has a cross-sectional area of 115 cm^2 supports a pickup truck weighing 1.20×10^4 N. Compressed air is used to apply a force on the second piston with cross-sectional area 25.0 cm^2. How large must this force be to support the truck?
18. Compressed air in a car lift applies a force to a piston with radius 5.00 cm. This pressure is transmitted through a hydraulic system to a second piston with radius 15.0 cm. (a) How much force must the compressed air exert to lift a vehicle weighing 1.33×10^4 N? (b) What pressure produces the lift?

12.3 Air Pressure

Since air has weight, as does any fluid, it exerts pressure. The atmosphere exerts pressure on objects on the surface of the earth. This atmospheric pressure can be illustrated by using a bell jar with a hole in the top over which a thin rubber membrane is stretched as in Fig. 12.9(a). As air is pumped out of the bell jar, the inside air pressure is reduced by removing a number of air molecules. Thus, there are fewer molecular bombardments on the inside surface of the rubber membrane than there are on its outside surface. The outside air pressure, now greater than the inside air pressure, pushes the rubber membrane down into the bell jar. When a straw is used to drink, the air pressure inside the straw is reduced [Fig. 12.9(b)]. As a result, the outside air pressure is higher than the pressure in the straw, which forces the fluid up the straw.

In Section 12.1 we saw that the pressure on a submerged body increases as the body goes deeper into the liquid. Some creatures live near the bottom of the ocean, where the pressure of the water is so great that it would collapse any human body and most submarines, but through the process of evolution, such creatures have adapted to this tremendous pressure. Similarly, we on the earth live at the bottom of a fluid, air, that is several miles deep. The pressure from this fluid is normally 14.7 lb/in^2 or 101.32 kPa at sea level. We do not feel this pressure because it normally is almost the same from all directions and also because living bodies maintain an internal pressure that balances the external pressure. **Atmospheric pressure** is the pressure caused by the weight of the air in the atmosphere. Air pressure and the amount of air decrease with altitude. Mountain climbers often must use oxygen tanks to help them breathe at high altitudes. Aircraft must have pressurized

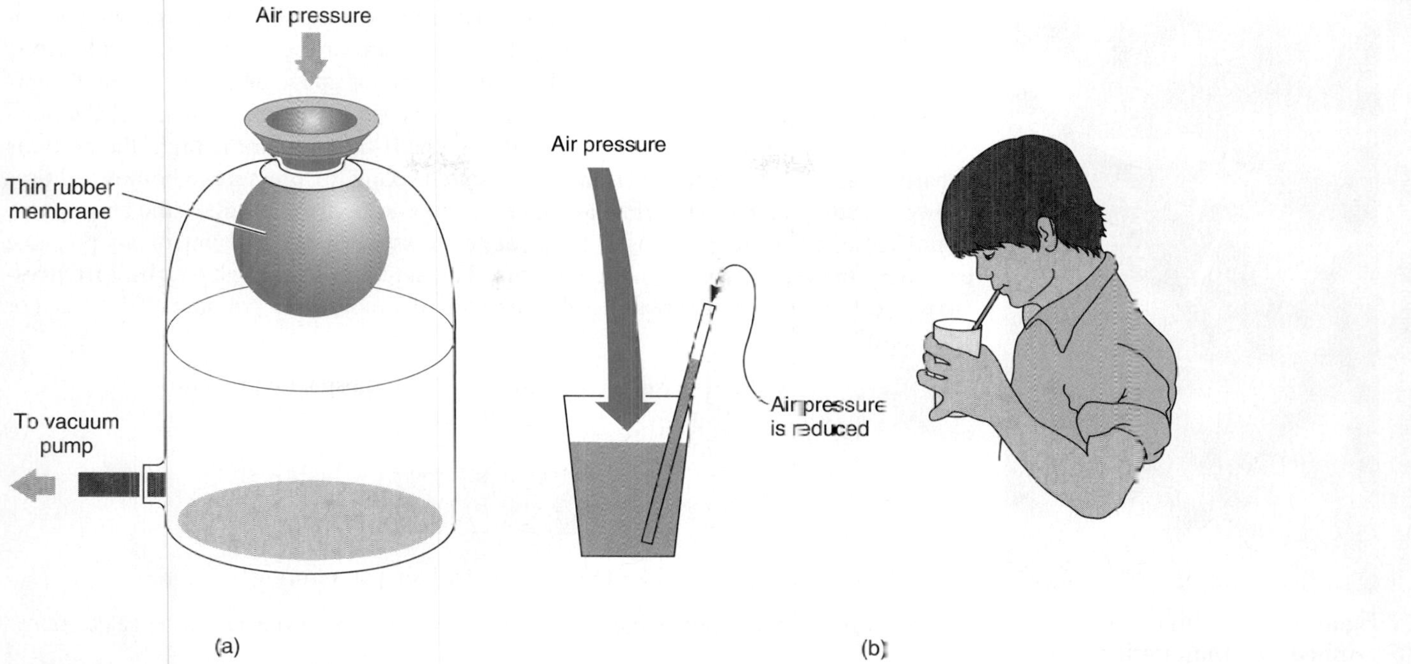

Figure 12.9 Effects of air pressure

cabins for passengers to be able to breathe and function. When the air pressure becomes unequal, its force becomes quite evident in the form of wind. This wind may be a cool summer breeze or the tremendous concentrated force of a tornado.

What is the *pressure of our atmosphere* equivalent to? Experiments have shown that the atmosphere supports a column of water 33.9 ft or 10.3 m high in a tube from which the air has been removed. The atmosphere supports 29.9 in. or 76.0 cm of mercury in a similar tube (Fig. 12.10). This is not surprising; mercury is 13.6 times as dense as water and

$$\frac{1}{13.6} \times 33.9 \text{ ft} \times \frac{12 \text{ in.}}{1 \text{ ft}} = 29.9 \text{ in.}$$

The height of the mercury column in a barometer is independent of the width (or diameter or cross-sectional area) of the barometer tube. This "inches of mercury" measurement has been standard for many years on TV weather programs but is increasingly being replaced by the metric standard measurement in kilopascals (kPa).

The pressure of the atmosphere can be expressed in terms of the pressure of an equivalent column of mercury. Air pressure at sea level is normally 29.9 in. or 76.0 cm or 760 mm of mercury. How do we arrive at the 14.7 lb/in^2 measurement? In the case of mercury, the height of the column is 29.9 in. or 2.49 ft. Its density is 13.6 × 62.4 lb/ft^3 or 849 lb/ft^3. Therefore,

$$P = hD_w$$

$$P = 2.49 \text{ ft} \times 849 \frac{\text{lb}}{\text{ft}^3} \times \left(\frac{1 \text{ ft}}{12 \text{ in.}}\right)^2$$

$$= 14.7 \text{ lb/in}^2 \text{ at standard temperature}$$

A pressure of 2 atm is equivalent to 29.4 lb/in^2 or 202.64 kPa. If the pressure is $\frac{1}{2}$ atm at a given point in the atmosphere, it is 7.35 lb/in^2 or 50.66 kPa.

An interesting demonstration illustrates atmospheric pressure by using it to collapse a metal can. A can that can be tightly capped is placed on a burner with its cap off. A small amount of water is placed in the can and heated until it boils. The cap is then secured, and the can is removed from the heat source. The steam has pushed air from the can, and any remaining air has expanded from being heated. When the sealed can cools and the steam inside condenses, the pressure in the can is reduced. The greater pressure of the atmosphere

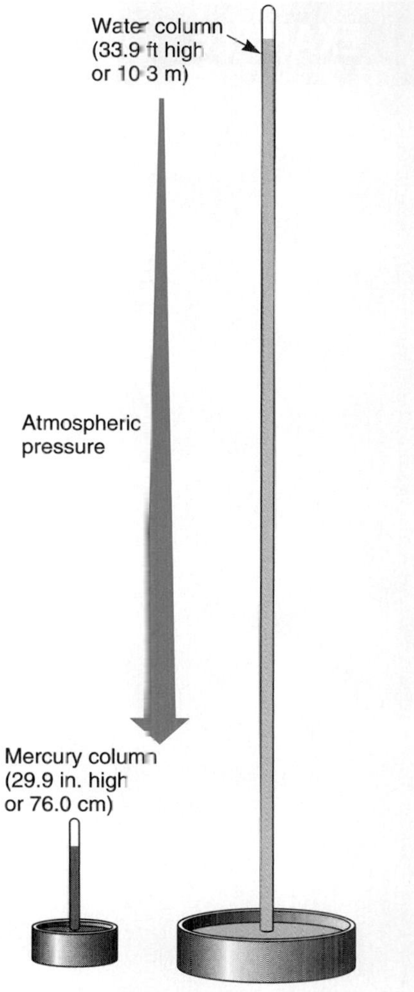

Figure 12.10 The barometer measures air pressure.

Figure 12.11 This can was crushed by atmospheric pressure.

outside the can crushes the can (Fig. 12.11). The same concept has been used for generations to seal jars of food in home canning, but the jars are strong enough to not collapse.

Different types of gauges read either atmospheric or gauge pressure. When we purchase bottled gas, the amount of gas and its density vary with the pressure. If the pressure is low, the amount of gas in the bottle is low. If the pressure is high, the bottle is "nearly full." The *gauge* that is usually used for checking the pressure in bottles and tires shows a reading of zero at normal atmospheric pressure. The pressure of the atmosphere is not included in this reading. Thus, **gauge pressure** is the amount of air pressure excluding the normal atmospheric pressure. The actual pressure, called **absolute pressure,** is the gauge pressure reading plus the normal atmospheric pressure, 101.32 kPa or 14.7 lb/in^2. That is,

$$\text{absolute pressure} = \text{gauge pressure} + \text{atmospheric pressure}$$

or

$$\boxed{P_{abs} = P_{ga} + P_{atm}}$$

where $P_{atm} = 101.32$ kPa or 14.7 lb/in^2 at standard temperature.

Note: Atmospheric pressure has not been included in our previous pressure calculations.

EXAMPLE

What is the absolute pressure in a tire inflated to 32.0 lb/in^2

(a) in lb/in^2?
(b) in kPa?

(a) Sketch: None needed

Data:

$$P_{ga} = 32.0 \text{ lb/in}^2$$
$$P_{atm} = 14.7 \text{ lb/in}^2$$
$$P_{abs} = ?$$

Basic Equation:

$$P_{abs} = P_{ga} + P_{atm}$$

Working Equation: Same

Substitution:

$$P_{abs} = 32.0 \text{ lb/in}^2 + 14.7 \text{ lb/in}^2$$
$$= 46.7 \text{ lb/in}^2$$

(b) We use the conversion factor:

$$101.32 \text{ kPa} = 14.7 \text{ lb/in}^2$$

Therefore,

$$P_{abs} = 46.7 \cancel{\text{lb/in}^2} \times \frac{101.32 \text{ kPa}}{14.7 \cancel{\text{lb/in}^2}}$$
$$= 322 \text{ kPa}$$

PROBLEMS 12.3

1. Change 815 kPa to lb/in^2.
2. Change 64.3 lb/in^2 to kPa.
3. Change 42.5 lb/in^2 to kPa.
4. Change 215 kPa to lb/in^2.

5. Find the pressure of
   (a) 3 atm (in kPa). (b) 2 atm (in kPa).
   (c) 6 atm (in lb/in^2). (d) 5 atm (in kPa).
   (e) $\frac{1}{3}$ atm (in kPa). (f) $\frac{1}{4}$ atm (in kPa).
6. A barometer in the Rocky Mountains reads 516 mm of mercury. Find this pressure (a) in kPa and (b) in lb/in^2.
7. Find the absolute pressure in a bicycle tire with a gauge pressure of 485 kPa.
8. Find the absolute pressure of a motorcycle tire with a gauge pressure of 255 kPa.
9. Find the gauge pressure of a tire with an absolute pressure of 45.0 lb/in^2.
10. Find the gauge pressure of a tire with an absolute pressure of 425 kPa.
11. Find the absolute pressure of a tire gauge that reads 205 kPa.
12. Find the absolute pressure of a tank whose gauge pressure reads 362 lb/in^2.
13. Find the gauge pressure of a tank whose absolute pressure is 1275 kPa.
14. Find the gauge pressure of a tank whose absolute pressure is 218 lb/in^2.
15. Find the absolute pressure of a cycle tire with gauge pressure 3.00×10^5 Pa
16. Find the absolute pressure in a hydraulic jack with a small piston of area 23.0 cm^2 when a force of 125 N is applied. The area of the large piston is 46.0 cm^2.

SKETCH

12 cm² | w
4.0 cm

DATA

$A = 12\ cm^2,\ l = 4.0\ cm,\ w = ?$

BASIC EQUATION

$A = lw$

WORKING EQUATION

$w = \frac{A}{l}$

SUBSTITUTION

$w = \frac{12\ cm^2}{4.0\ cm} = 3.0\ cm$

12.4 Buoyancy

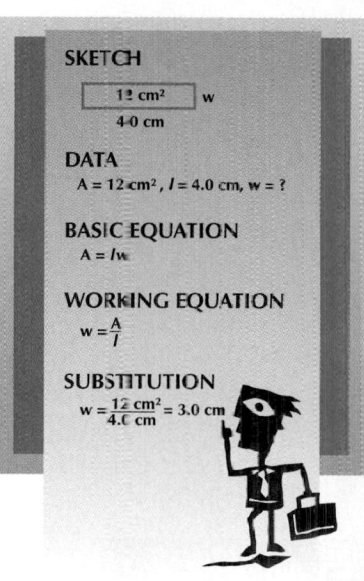

Archimedes was one of the first to study fluids and formulated what is now called **Archimedes' principle.**

ARCHIMEDES' PRINCIPLE

Any object placed in a fluid apparently loses weight equal to the weight of the displaced fluid.

In Fig. 12.12, the toy boat floats on water. Note that the weight of the boat equals the weight of the water it displaces. Three people in a boat displace more water than only one person in the boat. The boat rides lower due to the increased weight. Similarly, a loaded ship displaces more water than an empty ship and rides lower due to its increased weight (Fig. 12.13).

What happens to the weight of a brick or some other object that sinks in water? First, weigh the brick in air. Then, lower the brick under the water and weigh it again. It weighs less. The difference between the two weights is the buoyant (upward) force of the water. That is, the **buoyant force** is the upward force exerted on a submerged or partially submerged object. This is illustrated in the following example.

Archimedes (287 B.C.–212 B.C.),

Greek mathematician, was born in Syracuse. He is remembered for the construction of siege-engines against the Romans, Archimedes' screw (still used for raising water), discovering the principle for the buoyant force of a floating body, and founding the science of hydrostatics. In mathematics, he discovered the formulas for the areas and volumes of spheres, cylinders, paraboias, and other plane and solid figures.

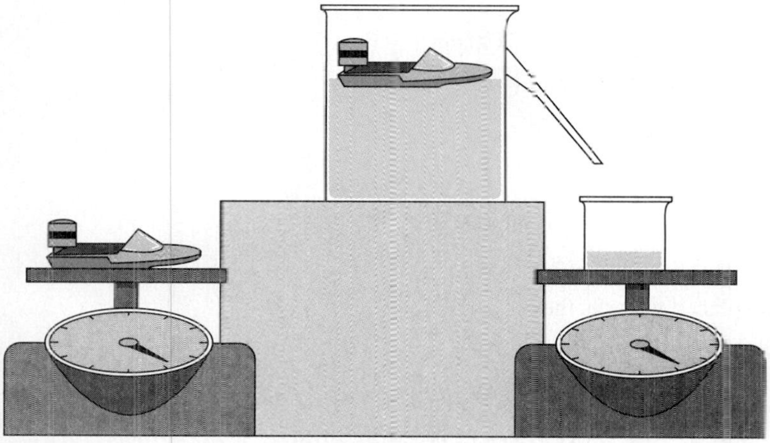

Figure 12.12 Weight of object = weight of displaced water

Figure 12.13 A floating ship or boat displaces an amount of water equal to its total weight. This is why a loaded ship sinks lower in the water and displaces more water than an empty ship.

EXAMPLE 1

A solid concrete block 15.0 cm × 20.0 cm × 10.0 cm weighs 67.6 N in air (Fig. 12.14). When lowered into water, it weighs 38.2 N. The buoyant force is 67.6 N − 38.2 N = 29.4 N.

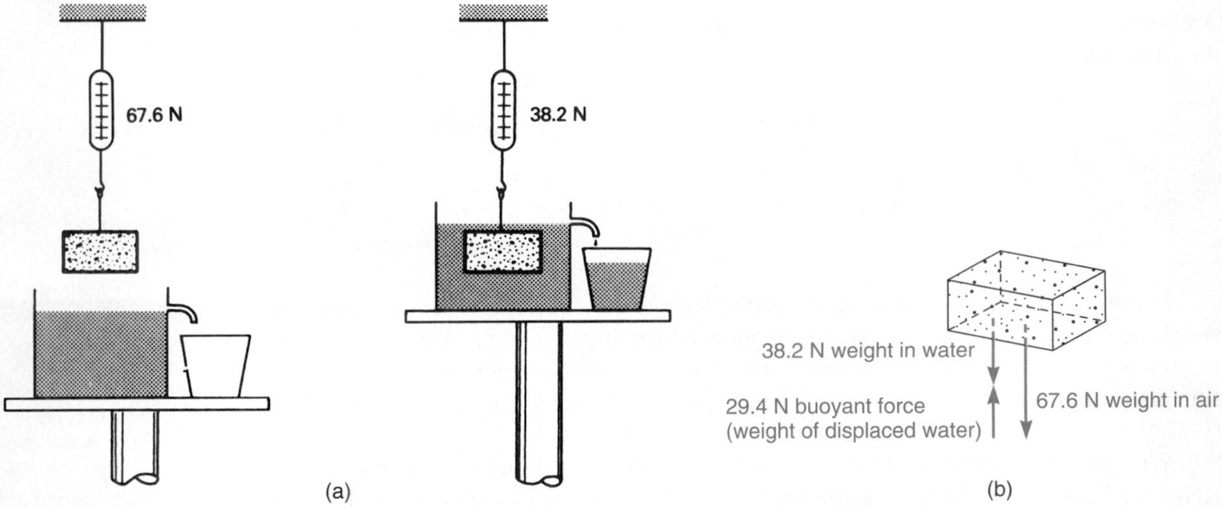

Figure 12.14 Weight in water = weight in air – buoyant force (weight of displaced water)

The volume of the displaced water is

$$V = lwh$$
$$V = (15.0 \text{ cm})(20.0 \text{ cm})(10.0 \text{ cm})$$
$$= 3\bar{0}00 \text{ cm}^3 = 3.00 \times 10^{-3} \text{ m}^3$$

The mass of the displaced water is

$$m = D_m V$$
$$m = (100\bar{0} \text{ kg/}\cancel{\text{m}^3})(3.00 \times 10^{-3} \cancel{\text{m}^3})$$
$$= 3.00 \text{ kg}$$

The weight of the displaced water is then

$$F_w = mg$$
$$F_w = (3.00 \text{ kg})(9.80 \text{ m/s}^2)$$
$$= 29.4 \text{ N} \qquad (1 \text{ N} = 1 \text{ kg m/s}^2)$$

which equals the buoyant force.

A submarine uses these same principles to dive and rise in the ocean. A submarine has ballast tanks that are filled with seawater, making it heavier than the volume of water it displaces, and so it descends. For the submarine to rise, the water in the ballast tanks is blown out by compressed air. The submarine then becomes lighter than the volume of water it displaces and so it rises.

Primitive peoples made their boats of wood, which has densities less than water. How can a barge made of steel, which is approximately eight times as dense as water, float? The barge must be built so that it displaces at least its own weight plus its cargo load equivalent of water. The wider the barge and the deeper it is immersed, the more water is displaced and the greater is the buoyant force exerted on the barge by the water.

Imagine a loaded fishing boat coming in with the day's catch and just barely clearing a bridge on its way to the dock. After it is emptied of its cargo and heads back to sea, will it have more or less clearance under the bridge? Use Fig. 12.13 to explain your answer.

EXAMPLE 2

A rectangular boat is 4.00 m wide, 8.00 m long, and 3.00 m deep.

(a) How many m^3 of water will it displace if the top stays 1.00 m above the water?
(b) What load (in newtons) will the boat contain under these conditions if the empty boat weighs 8.60×10^4 N in dry dock?

(a) The volume of water displaced by the boat is

$$V = lwh$$
$$V = (8.00\text{ m})(4.00\text{ m})(2.00\text{ m})$$
$$= 64.0\text{ m}^3$$

(b) The load of the boat is the buoyant force of the displaced water (D_wV) minus the weight of the boat in dry dock.

$$(98\bar{0}0\text{ N/m}^3)(64.0\text{ m}^3) - (8.60 \times 10^4\text{ N}) = 5.41 \times 10^5\text{ N}$$

Note: The weight density of water is $98\bar{0}0$ N/m^3.

Archimedes' principle applies to gases as well as liquids. Lighter-than-air craft (such as the Goodyear blimps) operate on this principle. Since they are filled with helium, which is lighter than air, the buoyant force on them causes them to be supported by the air. Being "submerged" in the air, a blimp is buoyed up by the weight of the air it displaces, which equals the buoyant force of the air on the balloon.

PROBLEMS 12.4

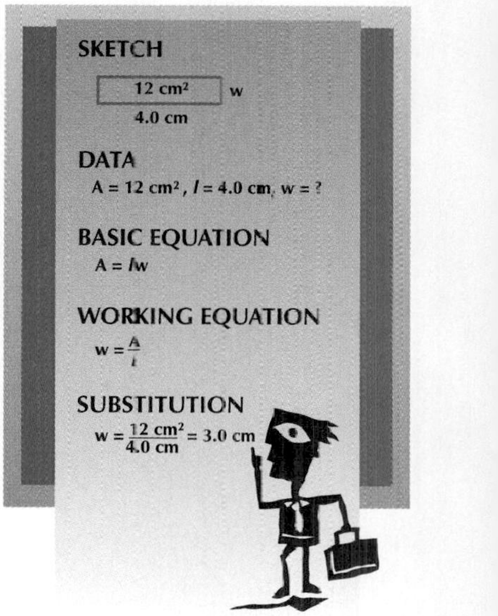

1. A metal alloy weighs 81.0 lb in air and 68.0 lb when under water. Find the buoyant force of the water.
2. A piece of metal weighs 67.0 N in air and 62.0 N in water. Find the buoyant force of the water.
3. A rock weighs 25.7 N in air and 21.8 N in water. What is the buoyant force of the water?
4. A metal bar weighs 455 N in air and 437 N in water. What is the buoyant force of the water?
5. A rock displaces 1.21 ft^3 of water. What is the buoyant force of the water?
6. A metal displaces 16.8 m^3 of water. Find the buoyant force of the water.
7. A metal casting displaces 327 cm^3 of water. Find the buoyant force of the water.
8. A piece of metal displaces 657 cm^3 of water. Find the buoyant force of the water.
9. A metal casting displaces 2.12 ft^3 of alcohol. Find the buoyant force of the alcohol.
10. A metal cylinder displaces 515 cm^3 of gasoline. Find the buoyant force of the gasoline.
11. A 75.0-kg rock lies at the bottom of a pond. Its volume is 3.10×10^4 cm^3. How much force is needed to lift the rock?

12. A 125-lb rock lies at the bottom of a pond. Its volume is 0.800 ft^3. How much force is needed to lift the rock?
13. A flat-bottom river barge is 30.0 ft wide, 85.0 ft long, and 15.0 ft deep.
    (a) How many ft^3 of water will it displace while the top stays 3.00 ft above the water?
    (b) What load in tons will the barge contain under these conditions if the empty barge weighs $16\bar{0}$ tons in dry dock?
14. A flat-bottom river barge is 12.0 m wide, 30.0 m long, and 6.00 m deep.
    (a) How many m^3 of water will it displace while the top stays 1.00 m above the water?
    (b) What load (in newtons) will the barge contain under these conditions if the empty barge weighs 3.55×10^6 N in dry dock?
15. What is the volume (in m^3) of the water displaced by a submerged air tank that is acted on by a buoyant force of 7.50×10^4 N?
16. A lifeguard swims with her head just above the water. What is the volume of the submerged part of her body if she weighs $60\bar{0}$ N?
17. An underwater camera weighing 1250 N in air is submerged and supported by a tether line. If the volume of the camera is 8.30×10^{-2} m^3, what is the tension in the line?

12.5 Fluid Flow

Think for a minute about the motion of water flowing down a fast-moving mountain stream that contains boulders and rapids and about the motion of the air during a thunderstorm or during a tornado. These types of motion are complex, indeed. Our discussion will focus on the simpler examples of fluid flow.

Streamline flow, also known as *laminar flow,* is the smooth flow of a fluid through a tube (Fig. 12.15). By smooth flow we mean that all particles of the fluid follow the same uniform path. **Turbulent flow,** also known as *nonlaminar flow,* is the erratic, unpredictable flow of a fluid resulting from excessive speed of the flow or sudden changes in direction or size of the tube.

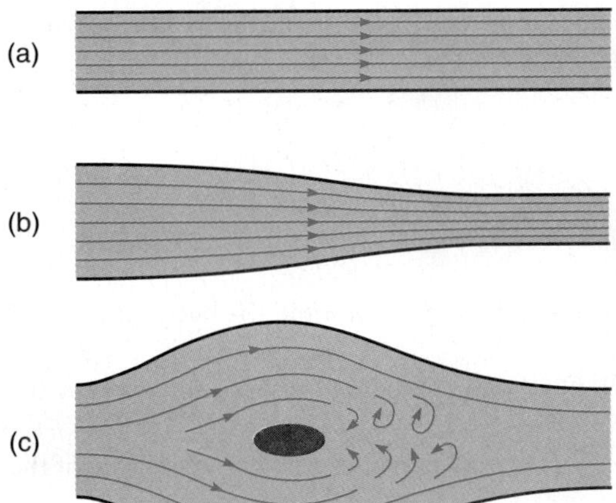

Figure 12.15 Streamline flow of a fluid through a smooth tube or pipe is shown in (a) and (b). Water flowing in a creek or a mountain stream over and around rocks, resulting in sudden changes in direction and speed, is a common example of turbulent flow as in (c).

The **flow rate** of a fluid is the volume of fluid flowing past a given point in a pipe per unit time. Assume that we have a streamline flow through a straight section of pipe at speed v. During a time interval of t, each particle of fluid travels a distance vt. If A is the cross-sectional area of the pipe, the volume of fluid passing a given point during the time interval t is vtA. Thus, the flow rate, Q, is given by

$$Q = \frac{vtA}{t}$$

or

$$\boxed{Q = vA}$$

where

Q = flow rate
v = speed of the fluid through the tube or pipe
A = cross-sectional area of the tube or pipe

EXAMPLE

Water flows through a fire hose of diameter 6.40 cm at a speed of 5.90 m/s. Find the flow rate of the fire hose in L/min.

Data:

$$v = 5.90 \text{ m/s}$$
$$r = 3.20 \text{ cm} = 0.0320 \text{ m}$$
$$A = \pi r^2 = \pi(0.0320 \text{ m})^2$$
$$Q = ?$$

Basic Equation:

$$Q = vA$$

Working Equation: Same

Substitution:

$$Q = \left(5.90\frac{\text{m}}{\text{s}}\right)\pi(0.0320 \text{ m})^2 \times \frac{10^3 \text{ L}}{1 \text{ m}^3} \times \frac{60 \text{ s}}{1 \text{ min}}$$
$$= 1140 \text{ L/min}$$

The overall volume rate of flow of water in a stream or creek is relatively constant; that is, the volume of water passing through the various sections is constant. Thus, the speed of the water increases within sections where the stream is narrow and decreases within sections where the stream is broad.

For an incompressible fluid, the flow rate is constant throughout the pipe. If the cross-sectional area of the pipe changes and streamline flow is maintained, the flow rate is the same all along the pipe. That is, as the cross-sectional area increases, the velocity decreases, and vice versa (Fig. 12.16).

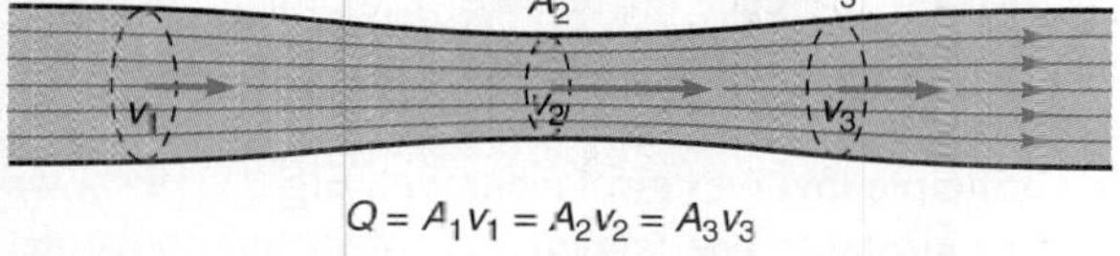

Figure 12.16 For an incompressible fluid, the flow rate is constant throughout.

TRY THIS ACTIVITY

The Venturi Hose

Turn on the water to a garden hose that does not have a nozzle. Note the speed of the water as it flows out of the hose. Without turning the faucet, pinch the hose just behind the metal coupling allowing only a small space for the water to flow. Describe what happens to the speed of the water as it now comes out of the hose. Explain why the speed changes although the faucet remained the same.

Giovanni Battista Venturi (1746–1822),

physicist, was born in Italy. He worked on the flow of fluids and is remembered for his discovery of the Venturi effect, the decrease of the pressure in a fluid in a pipe as the diameter is gradually reduced. The effect has many applications, such as in the carburetor and fluid-flow measuring instruments.

Daniel Bernoulli (1700–1782),

mathematician, was born in The Netherlands. He worked on trigonometry, mechanics, vibrating systems, and hydrodynamics (leading to the kinetic theory of gases), and developed Bernoulli's principle.

What happens to the pressure as the cross-sectional area of the pipe changes? This concept can be illustrated by use of a Venturi meter, named after **Giovanni Battista Venturi** (Fig. 12.17). Here the vertical tubes act like pressure gauges; the higher the column, the higher the pressure. As you can see, the higher the speed, the lower the pressure, and vice versa. This change in pressure of a fluid in streamline flow was first explained by **Daniel Bernoulli.**

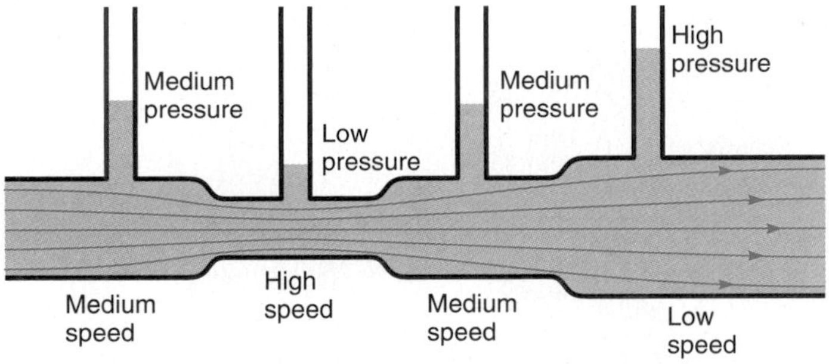

Figure 12.17 A Venturi meter shows that the higher the speed of a fluid through a tube, the lower the pressure; the lower the speed of a fluid, the higher the pressure.

Do you recall walking into the wind in winter on a city street lined with several tall buildings when the wind seemed stronger than usual? The wind actually was stronger because it was acting as a fluid as illustrated using the Venturi meter principle in Fig. 12.17. That is, the speed of the wind increased as it was forced to flow between the tall buildings along the street.

BERNOULLI'S PRINCIPLE

For the horizontal flow of a fluid through a tube, the sum of the pressure and energy of motion (kinetic energy) per unit volume of the fluid is constant.

One application of **Bernoulli's principle** involves a small engine carburetor (Fig. 12.18). The volume of airflow is determined by the position of the butterfly valve. As the air flows through the throat, the air gains speed and loses pressure. The pressure in the fuel bowl equals the pressure above the throat. Due to the difference in pressure between the fuel bowl and the throat, gasoline is drawn into and is mixed with the airstream. The reduced pressure in the throat also helps the gasoline to vaporize.

Another application of Bernoulli's principle involves airplane travel. Fig. 12.19 shows the flow of air rushing past the wing of an airplane. The velocity, v_1, of the air above the wing is greater than the velocity of the air below, v_2, because it has farther to travel in a given time. Thus, the pressure at point 2 is greater, which causes lift on the wing.

An airplane requires a longer distance for takeoff in summer than in winter. Why? Hot air is less dense and has fewer air molecules to lift the plane as it moves down the runway.

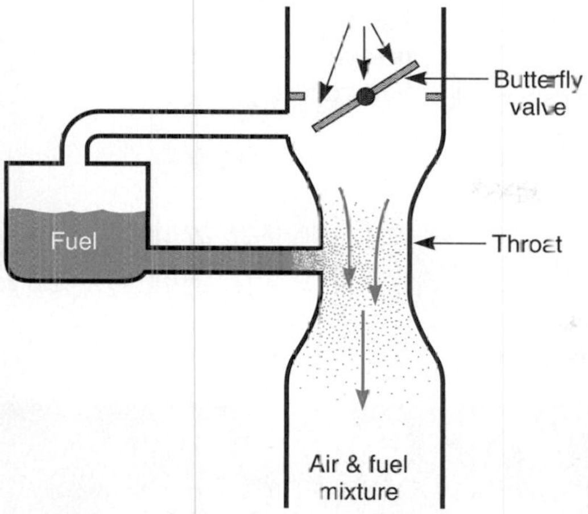

Figure 12.18 Small engine carburetor

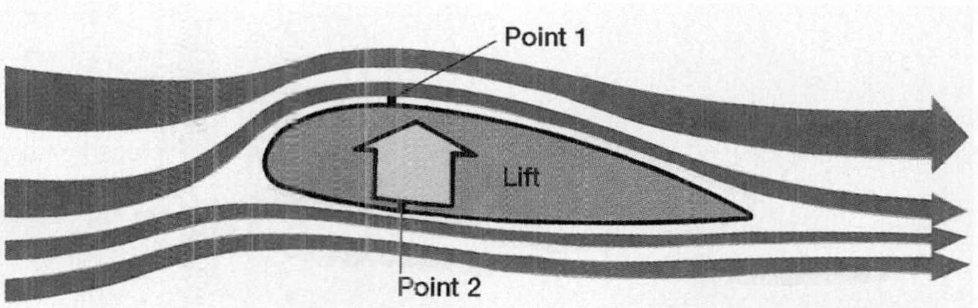

Figure 12.19 Air flowing past an airplane wing creates lift.

Two other examples that illustrate Bernoulli's principle are a curving baseball and a paint spray gun. The rotating baseball in Fig. 12.21 drags a layer of air immediately next to its surface. This reduces the speed of air at the bottom of the ball and increases the speed at the top of the ball causing the pressure to drop more at the top than at the bottom so that the ball curves as shown.

When the air in a spray gun is accelerated through a narrowing in the line, the pressure is reduced, and paint is drawn into the airstream and is then forced from the gun.

TRY THIS ACTIVITY

Low Pressure

Tear a sheet of paper in half and position each half as shown in Fig. 12.20. What do you think will happen to the positions of the papers when a continuous stream of air is blown between them? Will the sheets separate or move closer to one another? Test your prediction by blowing a continuous stream of air between the two sheets of paper. Was your prediction correct? Use Bernoulli's principle to explain the behavior of the paper.

Figure 12.20

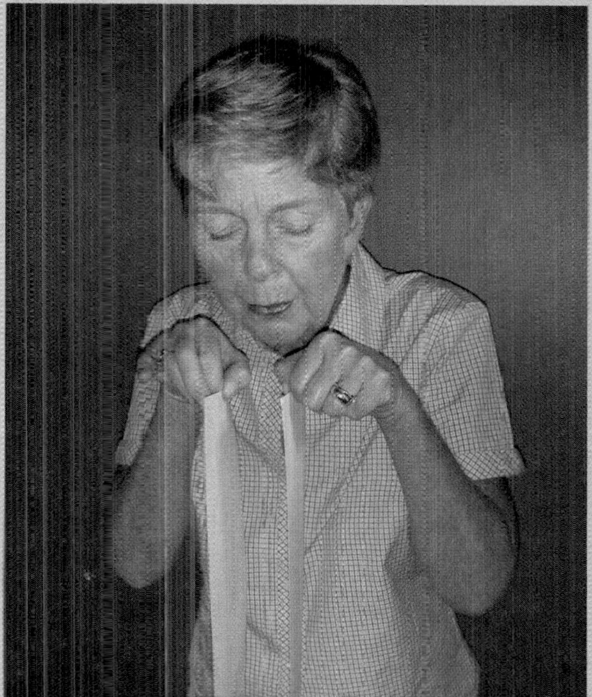

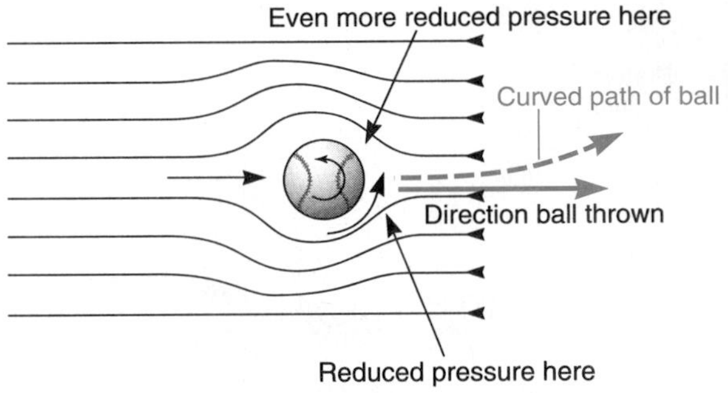

Figure 12.21 Bernoulli's principle explains why a baseball curves.

PHYSICS CONNECTIONS

Blood Pressure

A fluid exerts pressure on the walls of its closed container. As a fluid, blood exerts pressure on the walls of the heart, arteries, vessels, and capillaries that make up the circulatory system. When the diameter of the arteries and vessels is narrowed, an increased pressure is exerted on the walls of the blood vessels.

A sphygmomanometer (Fig. 12.22) is a device used to measure a person's blood pressure. The sphygmomanometer cuff is placed around the upper arm, inflated, and then deflated while the meter measures the pressure of blood passing through that section of the arm. As seen in this chapter, liquid pressure is dependent on the depth of the fluid. Since a sphygmomanometer cannot be placed around the heart and the depth of the fluid must be at the same level as the heart, the upper arm becomes a convenient location to measure blood pressure. When a person is lying down, the depth of the fluid throughout the body is roughly the same so that the pressure could be measured on any limb of the body.

When standing on one's head, one notices that the blood vessels in the head experience a great deal of pressure. As a result, the blood vessels on the sides of the temple protrude. The vessels in the feet and legs typically withstand much higher pressures than the vessels in the head.

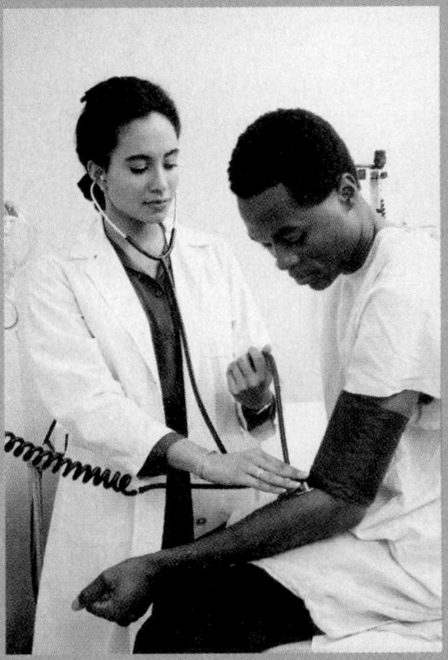

Figure 12.22 A person having his blood pressure measured. Notice the sphygmomanometer is at the same level as the heart.

SKETCH

12 cm² (rectangle, length 4.0 cm, width w)

DATA

$A = 12\ \text{cm}^2$, $l = 4.0$ cm, $w = ?$

BASIC EQUATION

$A = lw$

WORKING EQUATION

$w = \frac{A}{l}$

SUBSTITUTION

$w = \frac{12\ \text{cm}^2}{4.0\ \text{cm}} = 3.0$ cm

PROBLEMS 12.5

1. Water flows through a hose of diameter 3.90 cm at a velocity of 5.00 m/s. Find the flow rate of the hose in L/min.
2. Water flows through a 15.0-cm fire hose at a rate of 5.00 m/s.
   (a) Find the rate of flow through the hose in L/min.
   (b) How many litres pass through the hose in 30.0 min?
3. Water flows from a pipe at 650 L/min.
   (a) What is the diameter (in cm) of the pipe if the velocity of the water is 1.5 m/s?
   (b) Find the velocity (in m/s) of the water if the diameter of the pipe is 20.0 cm.
4. Water flows through a pipe of diameter 8.00 cm at 45.0 m/min. Find the flow rate (a) in m^3/min and (b) in L/s.
5. A pump is rated to deliver 50.0 gal/min. Find the velocity of water in (a) a 6.00-in.-diameter pipe and (b) a 3.00-in.-diameter pipe.
6. What size pipe needs to be attached to a pump rated at 36.0 gal/min if the desired velocity is 10.0 ft/min? Give the inner diameter in inches.

7. What is the diameter of a pipe in which water travels 32.0 m in 15.0 s and has a flow rate of 2620 L/min?
8. A garden hose is used to fill a bucket in 30.0 s. If you cover part of the hose nozzle so the speed of the water leaving the hose doubles, how long does it take to fill the bucket?

Glossary

Absolute Pressure The actual air pressure given by the gauge reading plus the normal atmospheric pressure. (p. 332)

Archimedes' Principle Any object placed in a fluid apparently loses weight equal to the weight of the displaced fluid. (p. 333)

Atmospheric Pressure The pressure caused by the weight of the air in the atmosphere. (p. 330)

Bernoulli's Principle For the horizontal flow of a fluid through a tube, the sum of the pressure and energy of motion (kinetic energy) per unit volume of the fluid is constant. (p. 338)

Buoyant Force The upward force exerted on a submerged or partially submerged object. (p. 333)

Flow Rate The volume of fluid flowing past a given point in a pipe per unit time. (p. 336)

Gauge Pressure The amount of air pressure excluding the normal atmospheric pressure. (p. 332)

Hydraulic Principle (Pascal's Principle) The pressure applied to a confined liquid is transmitted without measurable loss throughout the entire liquid to all inner surfaces of the container. (p. 326)

Hydrostatic Pressure The pressure a liquid at rest exerts on a submerged object. (p. 320)

Pressure The force applied per unit area. (p. 320)

Streamline Flow The smooth flow of a fluid through a tube. (p. 336)

Turbulent Flow The erratic, unpredictable flow of a fluid resulting from excessive speed of the flow or sudden changes in direction or size of the tube. (p. 336)

Formulas

12.1 $P = hD_w$

$F_t = AhD_w$

$F_s = \frac{1}{2}AhD_w$

12.3 $P_{abs} = P_{ga} + P_{atm}$

12.5 $Q = vA$

Review Questions

1. The force applied to a unit area is called
   (a) strain. (b) total force.
   (c) pressure. (d) none of the above.
2. The statement that the pressure applied to a confined liquid is transmitted without measurable loss throughout the entire liquid to all inner surfaces of the container is called
   (a) Hooke's law. (b) Pascal's principle.
   (c) Archimedes' principle. (d) none of the above.
3. For an incompressible fluid, the flow rate is
   (a) equal for all surfaces. (b) constant throughout the pipe.
   (c) greater for the larger parts of the pipe. (d) none of the above.

4. Bernoulli's principle states that for horizontal flow of a fluid through a tube, the sum of the pressure and energy of motion per unit volume is
   (a) increasing with time. (b) decreasing with time.
   (c) constant. (d) none of the above.
5. Bernoulli's principle explains
   (a) curving baseballs. (b) the hydraulic principle.
   (c) absolute pressure. (d) buoyant forces.
   (e) none of the above.
6. What is the metric unit for pressure?
7. In your own words, define *pressure*.
8. In your own words, state how to find the force exerted on the vertical side of a rectangular water tank.
9. In your own words, state the hydraulic principle.
10. Describe why a ship floats.
11. Describe how a rotating baseball follows a curved path.
12. How does an airplane wing provide lift?
13. What is the difference between streamline and turbulent flow?
14. Give an example of how Archimedes' principle applies to gases.
15. Describe the difference between absolute and gauge pressure. Which do you use when you measure the pressure in your automobile tires?
16. Is the pressure on a small piston different from the pressure on a large piston in the same hydraulic system? Are the forces on the two pistons the same?
17. On what does the total force exerted by a liquid on a horizontal surface depend?
18. Why must the thickness of a dam be greater at the bottom than at the top?
19. Is the hydraulic piston in the master brake cylinder in an automobile larger or smaller than the piston in the brake cylinder at the wheels? Why?
20. Would a drinking straw work in space where there is no gravity? Explain.

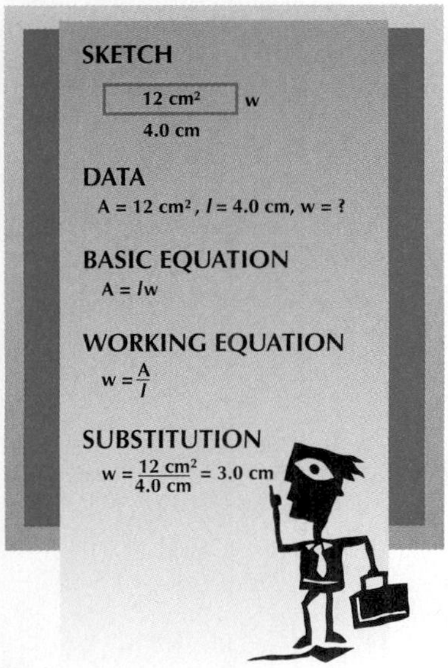

Review Problems

1. Find the pressure (in kPa) at the bottom of a water-filled drum 3.24 m high.
2. Find the depth in a lake at which the pressure is 197 lb/in^2.
3. Find the height of a water column when the pressure at the bottom of the column is 297 kPa.
4. What is the total force exerted on the bottom of a rectangular tank 8.67 ft by 4.83 ft by 3.56 ft deep?
5. Find the water pressure (in kPa) at a point 35.0 m from the bottom of a 55.0-m-tall full water tower.
6. Find the total force on the bottom of a cylindrical water tower 55.0 m high and 7.53 m in radius.
7. Find the total force on the side of a cylindrical water tower 55.0 m high and 7.53 m in radius.
8. Find the total force on the side of a rectangular water trough 1.25 m high by 1.55 m by 2.95 m.
9. What must the water pressure (in kPa) be on the ground to supply a water pressure of 252 N/cm^2 on the third floor, which is 9.00 m above the ground?
10. What water pressure must a pump that is located on the first floor supply to have water on the twenty-fifth floor of a building with a pressure of 26 lb/in^2? Assume that the distance between floors is 16.0 ft.
11. A submarine is submerged to a depth of 3150 ft in the Atlantic Ocean. What air pressure (in kPa) is needed to blow water out of the ballast tanks?
12. The area of the large piston in a hydraulic jack is 4.75 in^2. The area of the small piston is 0.564 in^2. (a) What force must be applied to the small piston if a weight of $65\bar{0}$ lb is to be lifted? (b) What is the mechanical advantage of the hydraulic jack?
13. The MA of a hydraulic jack is 324. What force must be applied to lift an automobile weighing 11,500 N?

14. The pistons of a hydraulic press have radii of 0.543 cm and 3.53 cm, respectively. (a) What force must be applied to the smaller piston to exert a force of 4350 N on the larger? (b) What is the pressure (in N/cm^2) on each piston? (c) What is the mechanical advantage of the press?
15. Find the absolute pressure in a bicycle tire with a gauge pressure of 202 kPa.
16. Find the gauge pressure of a tire with an absolute pressure of 655 kPa.
17. Find the gauge pressure of a tank whose absolute pressure is 314 lb/in^2.
18. A rock weighs 55.4 N in air and 52.1 N in water. Find the buoyant force on the rock.
19. A metal displaces 643 cm^3 of water. Find the buoyant force of the water.
20. A rock displaces 314 cm^3 of alcohol. Find the buoyant force on the rock.
21. A flat-bottom barge is 22.3 ft wide, 87.5 ft long, and 16.5 ft deep. How many ft^3 of water will it displace while the top stays 3.20 ft above the water? What load in tons will the barge contain if the barge weighs 157 tons in dry dock?
22. Water flows through a hose of diameter 3.00 cm at a velocity of 4.43 m/s. Find the flow rate of the hose in L/min.
23. Water flows through a 13.0-cm-diameter fire hose at a rate of 4.53 m/s. What is the rate of flow through the hose in L/min? How many litres pass through the hose in 25.0 min?
24. (a) What is the weight density of a liquid that exerts a total force of 433 N on the sides of a 3.00-m-tall cylindrical tank? The radius of the tank is 0.913 m, and it is filled to 1.75 m. (b) What liquid might this be?

APPLIED CONCEPTS

1. An aquarium's main tank holds 200,000 gal or 758 m^3 of salt water. (a) What is the lateral surface area of the glass wall if the cylindrical saltwater tank is 14.5 m tall? (b) What is the force applied to the vertical glass surface? (c) Steel bands are often placed around aquariums to reinforce the glass walls. Explain how the steel bands should be spaced toward the bottom of the tank.
2. The piston in a master cylinder has a radius of 0.570 in. and the radius of each of the two brake cylinder pistons is 1.75 in. (a) How much pressure is created in the master cylinder if a driver quickly applies 45.5 lb of force to the automobile's brake pedal? (b) Given the area of the two brake cylinder pistons, what is the force applied to each of the brake drums? (See Fig. 12.6.)
3. A crane that can lift a maximum of 9000 N is preparing to lift and move an underwater concrete mooring. The 1.25 m × 1.25 m × 0.450 m concrete block is located in seawater. Verify that the crane is strong enough to lift the mooring by finding (a) its volume, (b) its dry-dock weight, (c) the water's buoyant force, and (d) the force needed to lift the mooring while it is under water. (Refer to Table 11.2 for the mass density of concrete and of water.)
4. Wind tunnels are used to measure the aerodynamic properties of prototype models. (a) If a fan generates a wind speed of 25.0 mi/h inside an 8.75-ft^2 section of a wind tunnel, what is the wind speed as the air enters the narrower, 4.35-ft^2 section of the wind tunnel? (b) Explain why it is often windier on city streets surrounded by tall buildings than in more open areas.
5. A flexible hose with inside radius 0.250 in. leads to a shower fixture with 15 holes, each with radius 3.13×10^{-2} in. (a) If a faucet is opened to allow the water to move through the pipes at 3.94 ft/s, what is the speed of the water as it comes out of the holes? (b) What are two ways to increase the speed of the water from a shower fixture without opening the faucet farther? Explain.

CHAPTER 13

TEMPERATURE AND HEAT TRANSFER

Almost all forms of technology have concerns about temperature and heat transfer. The concern may be direct, as in refrigeration, or indirect, as in the thermal expansion of highways. Being able to measure heat transfer can mean the difference between success and failure of many things, from steam heat to space travel.

Objectives

The major goals of this chapter are to enable you to:

1. Distinguish between temperature and heat.
2. Express temperature using different scales.
3. Analyze heat transfer applications.
4. Determine final temperature using the method of mixtures.
5. Relate heat transfer to the expansion of solids and liquids.
6. Find the heat required for change of phase of solids, liquids, and gases.

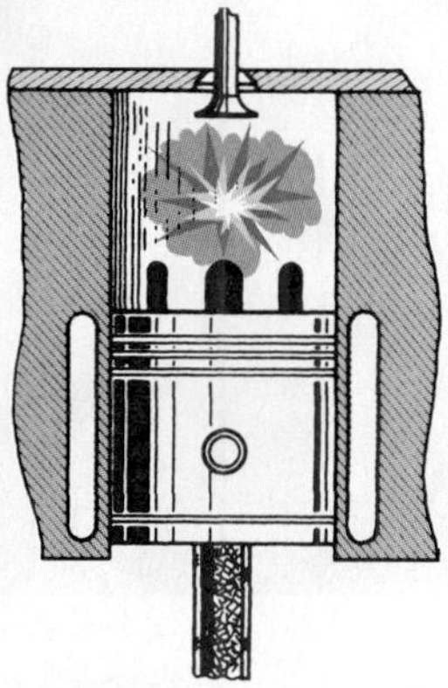

Figure 13.1 Force on piston produced by hot expanding gas

13.1 Temperature

An understanding of temperature and heat is very important. An automotive technician is concerned with the heat energy released by a fuel mixture in a combustion chamber (Fig. 13.1). The excess heat produced by an engine must be transferred to the atmosphere. Engineers are concerned with the expansion and contraction of bridges and roads when the temperature changes.

Basically, **temperature** is a measure of the hotness or coldness of an object. Temperature could be measured in a simple way by using your hand to sense the hotness or coldness of an object. However, the range of temperatures that your hand can withstand is too small, and your hand is not precise enough to measure temperature adequately. Therefore, other methods are used for measuring temperature.

	Colors	*°C*	*°F*	*Processes*
Heat colors	White	1371	2500	Welding
		1315	2400	High-speed steel hardening (2150–2450°F)
	Yellow white	1259	2300	
		1204	2200	
		1149	2100	
	Yellow	1093	2000	
		1036	1900	Alloy tool steel hardening (1500–1950°F)
	Orange red	981	1800	
		926	1700	
		871	1600	
	Light cherry red	815	1500	Carbon tool steel hardening (1350–1550°F)
	Cherry red	760	1400	
		704	1300	
	Dark red	648	1200	
		593	1100	High-speed steel tempering (1000–1100°F)
	Very dark red	538	1000	Carbon tool steel tempering (300–1050°F)
		482	900	
	Black red in dull light or blackness	426	800	
		371	700	
Temper colors	Pale blue (590°F) Violet (545°F) Purple (525°F) Yellowish brown (490°F) Dark straw (465°F) Light straw (425°F)	315	600	
		260	500	
		204	400	
		149	300	
		93	200	
		38	100	
		18	0	

Figure 13.2 Metallurgy and heat treatment: temperatures, steel colors, and related processes. (Courtesy of Allegheny Ludlum Steel Corp. Reprinted by permission.)

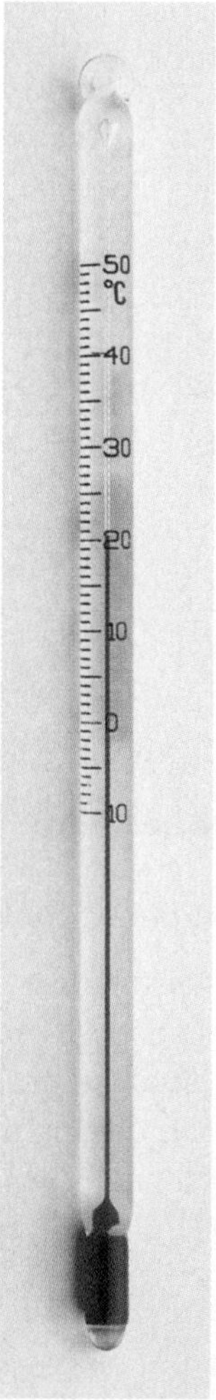

Photo courtesy of Dorling Kindersley

Figure 13.3 Common thermometer

Certain properties of matter vary with their temperature. For example, when objects are heated, they give off light of different colors. When an object is heated, in the absence of chemical reactions, it first gives off red light. As it is heated more, it appears white.

Chemical reactions sometimes cause different colors. When carbon steel is heated and exposed to air, several colors are observed before the rod appears red (see Fig. 13.2). This is due to a chemical reaction involving the carbon. If we could measure the color of the light, we could then determine the temperature. Although this works only for high temperatures, it is used in the production of metal alloys. The temperature of hot molten metals is determined this way.

Another property of matter that we use to find temperature is the change in volume of a liquid or a solid as its temperature changes. The liquid in glass thermometers is an example. This type of thermometer (Fig. 13.3) consists of a hollow glass bulb and a hollow glass tube joined together. A small amount of liquid such as alcohol is placed in the bulb. The air is removed from the tube. When the liquid is heated, it expands and rises up the glass tube. The height to which the liquid rises indicates the temperature.

The thermometer is standardized by marking two points on the glass that indicate the liquid level at two known temperatures. The temperatures used are the *freezing point* of water and the *boiling point* of water at sea level. The distance between these marks is then divided up into equal segments called *degrees*.

We will study the four temperature scales shown in Fig. 13.4. The common metric temperature scale is the **Celsius scale** with freezing point 0°C and boiling point 100°C. To write a temperature, we write the number followed by the degree symbol (°) followed by the capital letter of the scale used. Temperatures below zero on a scale are written as negative numbers. Thus, 20° below zero on the Celsius scale is written as −20°C.

The U.S. temperature scale is the **Fahrenheit scale** with freezing point 32°F and boiling point 212°F. The relationship between Fahrenheit temperatures (T_F) and Celsius temperatures (T_C) is given by

$$T_C = \frac{5}{9}(T_F - 32°)$$

$$T_F = \frac{9}{5}T_C + 32°$$

where

T_C = Celsius temperature

T_F = Fahrenheit temperature

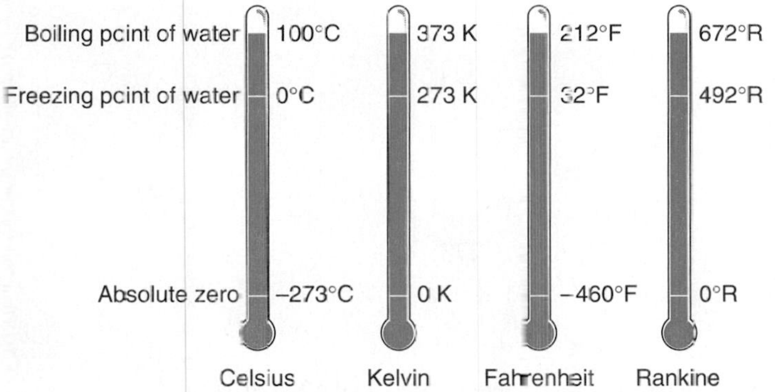

Figure 13.4 Four basic temperature scales

Anders Celsius (1701–1744),

astronomer, was born in Sweden. He devised the centigrade scale of temperature in 1742. The Celsius scale (formerly the centigrade scale) is named after him.

Gabriel Daniel Fahrenheit (1686–1736),

physicist, was born in Poland. He invented the alcohol thermometer in 1709 and the mercury thermometer in 1714.

EXAMPLE 1

The human body average temperature is 98.6°F. What is it in degrees Celsius?

Data:

$$T_F = 98.6°F$$
$$T_C = ?$$

Basic Equation:

$$T_C = \frac{5}{9}(T_F - 32°)$$

Working Equation: Same

Substitution:

$$\begin{aligned} T_C &= \frac{5}{9}(98.6° - 32°) \\ &= \frac{5}{9}(66.6°) \\ &= 37.0°C \end{aligned}$$

Lord Kelvin (Sir William Thomson) (1824–1907),

mathematician and physicist, was born in Belfast. He helped develop the law of conservation of energy and the absolute temperature scale (now named the Kelvin scale), did fundamental research in thermodynamics, presented the dynamic theory of heat, developed theorems for the mathematical analysis of electricity and magnetism, and designed several kinds of electrometers.

William Rankine (1820–1872),

engineer and scientist, was born in Scotland. He is noted for his work on the steam engine, machinery, shipbuilding, applied mechanics, the new science of thermodynamics, and the theories of elasticity and of waves.

Sometimes it is necessary to use the *absolute temperature scales,* which are the Kelvin scale and the Rankine scale. These are called absolute scales because 0 on either scale refers to the lowest limit of temperature, called *absolute zero.*

The **Kelvin scale** is the metric absolute temperature scale on which absolute zero is 0 K and is closely related to the Celsius scale. The relationship is*

$$\boxed{T_K = T_C + 273}$$

The **Rankine scale** is the U.S. absolute temperature scale on which absolute zero is 0°R and is closely related to the Fahrenheit scale. The relationship is

$$\boxed{T_R = T_F + 46\bar{0}°}$$

EXAMPLE 2

Change 18°C to Kelvin.

Data:

$$T_C = 18°C$$
$$T_K = ?$$

Basic Equation:

$$T_K = T_C + 273$$

Working Equation: Same

Substitution:

$$\begin{aligned} T_K &= 18 + 273 \\ &= 291 \text{ K} \end{aligned}$$

*The degree symbol (°) is not used when writing a temperature on the Kelvin scale.

EXAMPLE 3

Change 535°R to degrees Fahrenheit.

Data:

$$T_R = 535°\text{R}$$
$$T_F = ?$$

Basic Equation:

$$T_R = T_F + 46\bar{0}°$$

Working Equation:

$$T_F = T_R - 46\bar{0}°$$

Substitution:

$$T_F = 535° - 46\bar{0}°$$
$$= 75°\text{F}$$

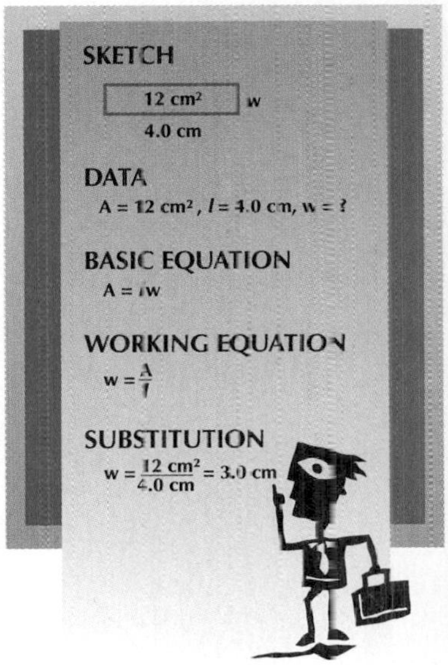

PROBLEMS 13.1

Find each temperature as indicated.

1. $T_F = 77°\text{F}$, $T_C =$ _____
2. $T_F = 113°\text{F}$, $T_C =$ _____
3. $T_F = 257°\text{F}$, $T_C =$ _____
4. $T_C = 15°\text{C}$, $T_F =$ _____
5. $T_C = 145°\text{C}$, $T_F =$ _____
6. $T_C = 35°\text{C}$, $T_F =$ _____
7. $T_F = 1\bar{0}°\text{F}$, $T_C =$ _____
8. $T_F = 2\bar{0}°\text{F}$, $T_C =$ _____
9. $T_C = 95°\text{C}$, $T_F =$ _____
10. $T_F = -5\bar{0}°\text{F}$, $T_C =$ _____
11. $T_C = 25°\text{C}$, $T_K =$ _____
12. $T_F = -45°\text{F}$, $T_R =$ _____
13. $T_K = 406\text{ K}$, $T_C =$ _____
14. $T_C = 75°\text{C}$, $T_K =$ _____
15. $T_C = -5\bar{0}°\text{C}$, $T_K =$ _____
16. $T_K = 175\text{ K}$, $T_C =$ _____
17. $T_K = 600\bar{0}\text{ K}$, $T_C =$ _____
18. The melting point of pure iron is 1505°C. What Fahrenheit temperature is this?
19. The melting point of mercury is −38.0°F. What Celsius temperature is this?
20. A welding white heat is approximately 14$\bar{0}$0°C. Find this temperature expressed in degrees Fahrenheit.
21. The temperature in a crowded room is 85°F. What is the Celsius reading?
22. The temperature of an iced tea drink is 5°C. What is the Fahrenheit reading?
23. The boiling point of liquid nitrogen is −196°C. What is the Fahrenheit reading?
24. The melting point of ethyl alcohol is −179°F. What is the Celsius reading?

During the forging and heat-treating of steel, the color of heated steel is used to determine its temperature. Complete the following table, which shows the color of heat-treated steel and the corresponding approximate temperatures in degrees Celsius and Fahrenheit. (Round to three significant digits.)

	Color	°*C*	°*F*
25.	White	——	22$\bar{0}$0
26.	Yellow	11$\bar{0}$0	——
27.	Orange	——	1725
28.	Cherry red	718	——
29.	Dark red	635	——
30.	Faint red	——	90$\bar{0}$
31.	Pale blue	31$\bar{0}$	——

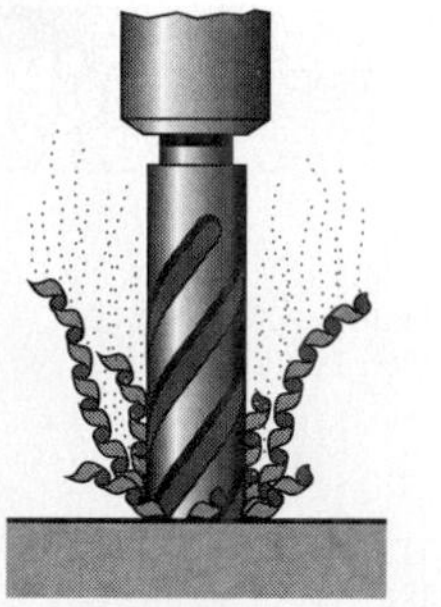

Figure 13.5 Friction causes a rise in temperature of the drill and plate.

13.2 Heat

When a hole is drilled in a metal block (Fig. 13.5), it becomes very hot. As the drill does mechanical work on the metal, the temperature of the metal increases. How can we explain this? Note the difference between the metal at low temperatures and at high temperatures. At high temperatures, the atoms in the metal vibrate more rapidly than at low temperatures. Their velocity is higher at high temperatures, and thus their kinetic energy ($KE = \frac{1}{2}mv^2$) is greater. To raise the temperature of a material, we must speed up the atoms; that is, we must add energy to them. **Heat** is a form of internal kinetic and potential energy contained in an object associated with the motion of its atoms or molecules and may be transferred from an object at a higher temperature to one at a lower temperature.

Drilling a hole in a metal block causes a temperature increase. As the drill turns, it collides with atoms of the metal, causing them to speed up. This mechanical work done on the metal has caused an increase in the energy of the atoms. For this reason, any friction between two surfaces results in a temperature rise of the materials.

Since heat is a form of energy, we could measure it in joules or ft lb, which are energy units. However, before it was known that heat is a form of energy, special units for heat were developed, which are still in use. These units are the calorie and the kilocalorie in the metric system and the Btu (British thermal unit) in the U.S. system. The **kilocalorie** (kcal) is the amount of heat necessary to raise the temperature of 1 kg of water 1°C. ***Note:*** The precise definition is based on the amount of heat needed to raise the temperature of 1 kg of water from 14.5°C to 15.5°C; however, the variation for each 1°C change in temperature is so minimal that it can be ignored for all practical purposes. The **Btu** is the amount of heat (energy) necessary to raise the temperature of 1 lb of water 1°F. The **calorie** (cal) is the amount of heat (energy) necessary to raise the temperature of 1 g of water 1°C. ***Note:*** One food calorie is the same as 1 kcal.

To lower the temperature of a substance, we need to remove some of the heat, the energy of motion of the molecules. When we have removed all the heat possible (when the molecules are moving as slowly as possible), we have reached **absolute zero,** the lowest possible temperature. Lower temperatures cannot be reached because all the heat has been removed. However, there is no upper limit on temperature because we can always add more heat (energy) to a substance to increase its temperature.

As mentioned before, heat and work are somehow related. James Prescott Joule determined by experiments the relationship between heat and work, called the **mechanical equivalent of heat.** He found that

1. 1 cal of heat is produced by 4.19 J of work.
2. 1 kcal of heat is produced by 4190 J of work.
3. 1 Btu of heat is produced by 778 ft lb of work.

The following are some examples in which heat is converted into useful work:

1. ***In our bodies.*** When food is oxidized, heat energy is produced, which can be converted into muscular energy, which in turn can be turned into work. Experiments have shown that only about 25% of the heat energy from our food is converted into muscular energy. That is, our bodies are about 25% efficient.
2. ***By burning gases.*** When a gas is burned, the gas expands and builds up a tremendous pressure that may convert heat to work by exerting a force to move a piston in an engine or turn the blades of a turbine. Since the burning of the fuel occurs within the cylinder or turbine, such engines are called *internal combustion engines*.
3. ***By steam.*** Heat from burning oil, coal, or wood may be used to generate steam. When water changes to steam under normal atmospheric pressure, it expands about 1700 times. When confined to a boiler, the pressure exerts a force against the piston in a steam engine or against the blades of a steam turbine. Since the fuel burns outside the engine, most steam engines or steam turbines are *external combustion engines*.

Technically, what is the difference between temperature and heat? *Temperature* is a measure of the hotness or coldness of an object. *Heat* is the total thermal energy (kinetic and potential) that can be transferred from an object at a higher temperature to one at a lower temperature. There are two basic ways of changing the temperature of an object:

1. By doing work *on* the object, such as the work done by the drill on the metal block in Fig. 13.5
2. By supplying energy *to* the object, such as mechanical, chemical, or electrical energy.

EXAMPLE 1

Find the amount of work (in J) that is equivalent to 4850 cal of heat.

$$4850 \text{ cal} \times \frac{4.19 \text{ J}}{1 \text{ cal}} = 20{,}300 \text{ J} \quad \text{or} \quad 20.3 \text{ kJ}$$

EXAMPLE 2

How much work must a person do to offset eating a 775-calorie breakfast?

First, note that one food calorie equals one kilocalorie.

$$775 \text{ kcal} \times \frac{4190 \text{ J}}{1 \text{ kcal}} = 3.25 \times 10^6 \text{ J} \quad \text{or} \quad 3.25 \text{ MJ}$$

EXAMPLE 3

A given coal gives off 7150 kcal/kg of heat when burned. How many joules of work result from burning one metric ton, assuming that 35.0% of the heat is lost?

First, note that one metric ton equals 1000 kg.

$$7150 \frac{\text{kcal}}{\text{kg}} \times \frac{4190 \text{ J}}{\text{kcal}} \times 1000 \text{ kg} \times 0.350 = 1.05 \times 10^{10} \text{ J}$$

PROBLEMS 13.2

1. Find the amount of heat in cal generated by 95 J of work.
2. Find the amount of heat in kcal generated by 7510 J of work.
3. Find the amount of work that is equivalent to 1550 Btu.
4. Find the amount of work that is equivalent to 3850 kcal.
5. Find the mechanical work equivalent (in J) of 765 kcal of heat.
6. Find the mechanical work equivalent (in J) of 8550 cal of heat.
7. Find the heat equivalent (in Btu) of 3.46×10^6 ft lb of work.
8. Find the heat equivalent (in kcal) of 7.53×10^5 J of work.
9. How much work must a person do to offset eating a piece of cake containing 625 cal?
10. How much work must a person do to offset eating a $\bar{2}00$-g bag of potato chips if 28 g of chips contain 150 cal?
11. A given gasoline yields 1.15×10^4 cal/g when burned. How many joules of work are obtained by burning 875 g of gasoline?
12. A coal sample yields 1.25×10^4 Btu/lb. How many foot-pounds of work result from burning 1.00 ton of this coal?
13. Natural gas burned in a gas turbine has a heating value of 1.10×10^5 cal/g. If the turbine is 24.0% efficient and 2.50 g of gas is burned each second, find (a) how many joules of work are obtained and (b) the power output in kilowatts.
14. Find the amount of heat energy that must be produced by the body to be converted into muscular energy and then into $1\bar{0}00$ ft lb of work. Assume that the body is 25% efficient.

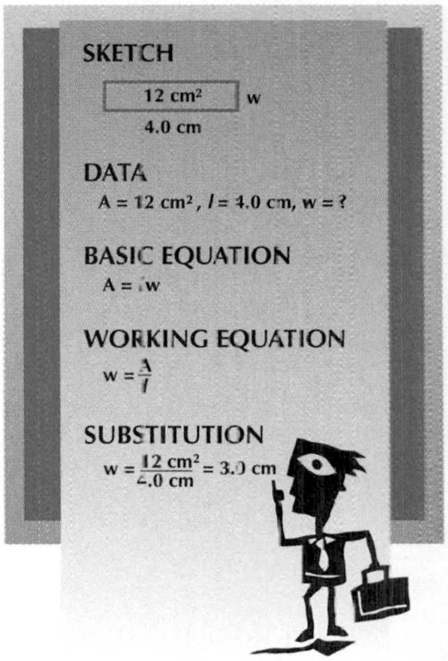

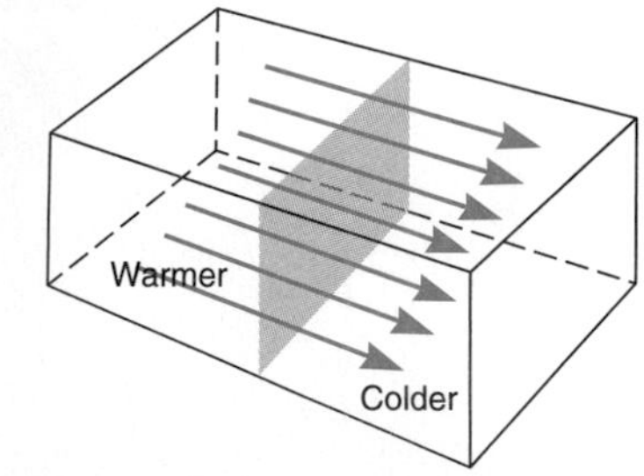

Figure 13.6 Transfer of heat from a warmer to a colder area

13.3 Heat Transfer

The movement of heat from a hot engine to the air is necessary to keep the engine from overheating. The heat produced by a furnace must be transferred to the various rooms in a house. The movement of heat is a major technical application.

The transfer of heat from one object to another is always from the warmer object to the colder one or from the warmer part of an object to a colder part (Fig. 13.6). There are three methods of heat transfer: *conduction, convection,* and *radiation*. **Conduction** is the heat transfer from a warmer part of a substance to a cooler part as a result of molecular collisions, which cause the slower-moving molecules to move faster. Conduction is the usual method of heat transfer in solids. When one end of a metal rod is heated, the molecules in that end move faster than before. These molecules collide with other molecules and cause them to move faster also. In this way, the heat is transferred from one end of the metal to the other (Fig. 13.7). Another example of conduction is the transfer of the excess heat produced in the combustion chamber of an engine through the engine block into the water coolant (Fig. 13.8).

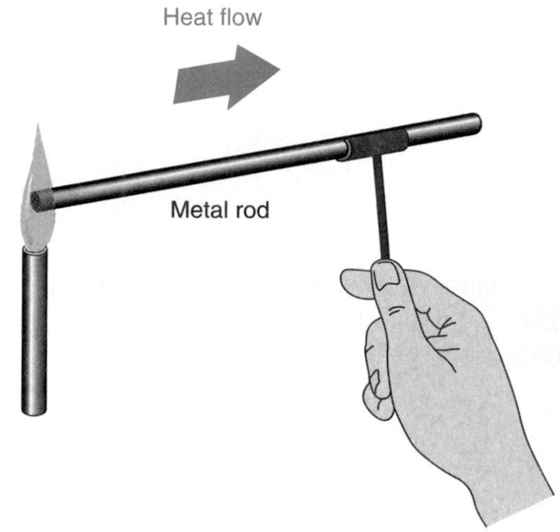

Figure 13.7 Heat flows by conduction in the metal rod.

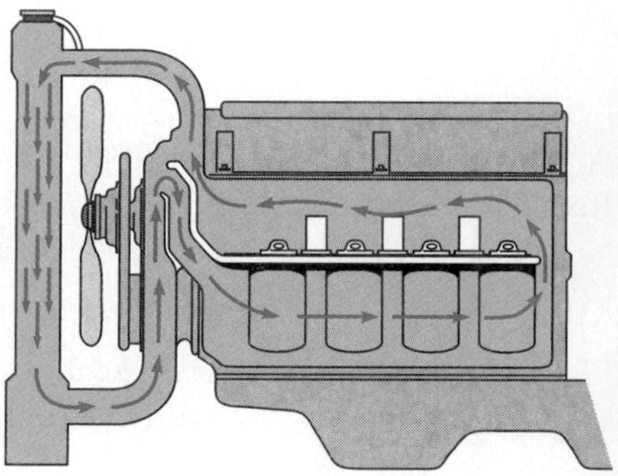

Figure 13.8 Heat conduction in an auto engine cooling system

The conduction of heat through some materials is better than through others. A poor conductor of heat is called an *insulator*. A list of good conductors and poor conductors is given in Table 13.1.

Table 13.1 Good and Poor Heat Conductors

Good Heat Conductors	Poor Heat Conductors
Copper	Asbestos
Aluminum	Glass
Steel	Wood
	Air
	Snow

Sitting on aluminum bleachers on a cold day feels colder than sitting on wood bleachers at the same temperature because aluminum is a much better heat conductor than wood. Wood is a good insulator and is often used as handles for pots and pans. In winter, animals find shelter in snow banks by making snow holes because snow is a poor heat conductor. Snow causes not only the ground heat, but also the animals' body heat to be retained. Before health concerns were realized, asbestos was once widely used in insulating buildings because it is such a poor heat conductor.

Convection is the heat transfer by the movement of warm molecules from one region of a gas or a liquid to another. The wind carries heat along with it. The water coolant in an engine carries hot water from the engine block to the radiator by a convection process. Heat transfer by the wind is a natural convection process. Heat transfer by the engine coolant is a forced convection process because it depends on a pump.

A dramatic illustration of the difference between convection and conduction can be shown by restraining some ice at the bottom of a test tube with some steel wool and filling the tube with water. When the water in the top of the tube is heated with a flame, it will boil without melting the ice. The poor conductivity of water keeps the less dense boiling water at the top and any heat transfer to the ice must be by conduction.

Convection currents are caused by the expansion of liquids or gases as they are heated or cooled. This expansion makes the hot gas or liquid less dense than the surrounding fluid. The lighter fluid is then forced upward by the heavier, surrounding fluid, which then flows in to replace it (Fig. 13.9). This type of behavior occurs in a fireplace as hot air goes up the chimney and is replaced by cool air from the adjacent room. The cool air draft, as this is called, is eventually supplied from outside air. This is why a fireplace is not very effective in heating a house. An airtight woodburning stove, however, draws little air from the inside of a house and is therefore much more efficient at heating the house.

Figure 13.9 A fireplace draws room air up into the chimney.

Conduction and convection require the presence of matter. All life on earth depends on the transfer of energy from the sun; this energy is transferred through nearly empty space. **Radiation** is heat transfer through energy being transmitted in the forms of rays, waves, or particles. Put your hand several inches from a hot iron (Fig. 13.10). The heat you feel is not transferred by conduction, because air is a poor conductor. It is not transferred by convection, because the hot air rises. This heat transfer is through radiation. This radiant heat is similar to light and passes through air, glass, and the vacuum of space. The energy that comes to us from the sun is in the form of radiant energy. At night, heat in the ground is radiated into the air.

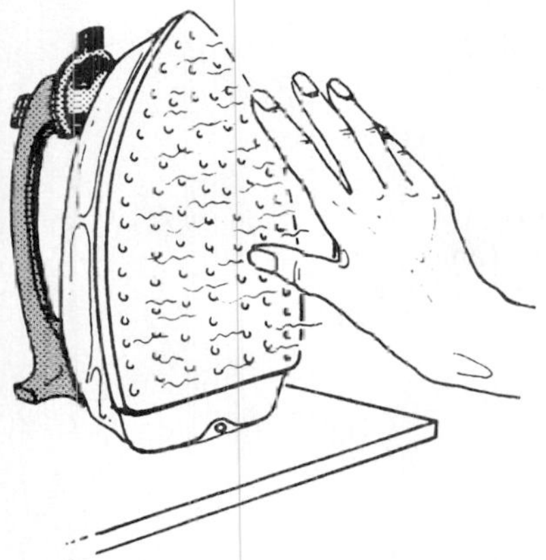

Figure 13.10 Heat radiation

Figure 13.11 Dark objects absorb more radiant heat than light objects.

Dark objects absorb radiant heat and light objects reflect radiant heat. This is why we feel cooler on a hot day in light-colored clothing than in dark clothing (Fig. 13.11).

Heat flow calculations are important because of concern about energy conservation. All three of the heat-transfer mechanisms discussed here must be considered in any estimation of heat loss from a building. In addition, infiltration losses that arise from leakage through cracks and openings near doors, windows, and other such areas must also be considered. Infiltration losses are, of course, a form of convective transfer but are often treated separately. We will discuss methods for calculating heat loss by conduction.

The equations describing the flow of heat through an object are very similar to those for the flow of electricity, which are developed in later chapters. The driving potential for heat flow is the temperature difference between the hot and cold sides of the object. Heat flow is similar to the flow of electrical charge.

The ability of a material to transfer heat by conduction is called its **thermal conductivity.** Metals are good conductors of heat. Glass and air are poor conductors. The rate at which heat is transferred through an object depends on the following factors:

1. The thermal conductivity
2. The cross-sectional area through which the heat flows
3. The thickness of the material
4. The temperature difference between the two sides of the material

The total amount of heat transferred is given by the equation

$$Q = \frac{KAt(T_2 - T_1)}{L}$$

where

Q = heat transferred in J or Btu
K = thermal conductivity (from Table 13.2)
A = cross-sectional area
t = total time
T_2 = temperature of the hot side
T_1 = temperature of the cool side
L = thickness of the material

Table 13.2 gives the thermal conductivities of some common materials.

Table 13.2 Thermal Conductivities

Substance	J/(s m °C)	Btu/(ft °F h)
Air	0.025	0.015
Aluminum	230	140
Brass	120	68
Brick/concrete	0.84	0.48
Cellulose fiber (loose fill)	0.039	0.023
Copper	380	220
Corkboard	0.042	0.024
Glass	0.75	0.50
Gypsum board (sheetrock)	0.16	0.092
Mineral wool	0.045	0.026
Plaster	0.14	0.083
Polystyrene foam	0.035	0.020
Polyurethane (expanded)	0.024	0.014
Steel	45	26
Water	0.56	0.32

EXAMPLE 1

Find the heat flow in an 8.0-h period through a 36 in. × 36 in. pane of glass (0.125 in. thick) if the temperature of the inner surface of the glass is 65°F and the temperature of the outer surface is 15°F.

Data:

$$K = 0.50 \text{ Btu/(ft °F h)}$$
$$A = 36 \text{ in.} \times 36 \text{ in.} = 3.0 \text{ ft} \times 3.0 \text{ ft} = 9.0 \text{ ft}^2$$
$$t = 8.0 \text{ h}$$
$$T_2 = 65°\text{F}$$
$$T_1 = 15°\text{F}$$
$$L = 0.125 \text{ in.} \times \left(\frac{1 \text{ ft}}{12 \text{ in.}}\right) = 0.0104 \text{ ft}$$
$$Q = ?$$

Basic Equation:

$$Q = \frac{KAt(T_2 - T_1)}{L}$$

Working Equation: Same

Substitution:

$$Q = \frac{[0.50 \text{ Btu/(ft °F h)}](9.0 \text{ ft}^2)(8.0 \text{ h})(65°\text{F} - 15°\text{F})}{0.0104 \text{ ft}}$$
$$= 1.7 \times 10^5 \text{ Btu}$$

The insulation value of construction material is often expressed in terms of the *R value,* which indicates the ability of the material to resist the flow of heat and uses U.S. units. The *R* value is inversely proportional to the thermal conductivity and directly proportional to the thickness. Low thermal conductivity is characteristic of good insulators. This is described by the equation

$$R = \frac{L}{K}$$

where

$$R = R \text{ value (in ft}^2\ ^\circ\text{F/Btu/h)}$$
$$K = \text{thermal conductivity}$$
$$L = \text{thickness of the material (in ft)}$$

EXAMPLE 2

Calculate the R value of 6.0 in. of mineral wool insulation.

Data:

$$L = 6.0 \text{ in.} = 0.50 \text{ ft}$$
$$K = 0.026 \text{ Btu/(ft }^\circ\text{F h)}$$
$$R = ?$$

Basic Equation:

$$R = \frac{L}{K}$$

Working Equation: Same

Substitution:

$$R = \frac{0.50 \text{ ft}}{0.026 \text{ Btu/(ft }^\circ\text{F h)}}$$
$$= 19 \text{ ft}^2\ ^\circ\text{F/Btu/h}$$

$$\frac{\text{ft}}{\frac{\text{Btu}}{\text{ft }^\circ\text{F h}}} = \text{ft} \div \frac{\text{Btu}}{\text{ft }^\circ\text{F h}} = \text{ft} \cdot \frac{\text{ft }^\circ\text{F h}}{\text{Btu}} = \frac{\text{ft}^2\ ^\circ\text{F h}}{\text{Btu}} = \frac{\text{ft}^2\ ^\circ\text{F}}{\text{Btu/h}}$$

This result could also have been written R-19. There is no equivalent in the metric system.

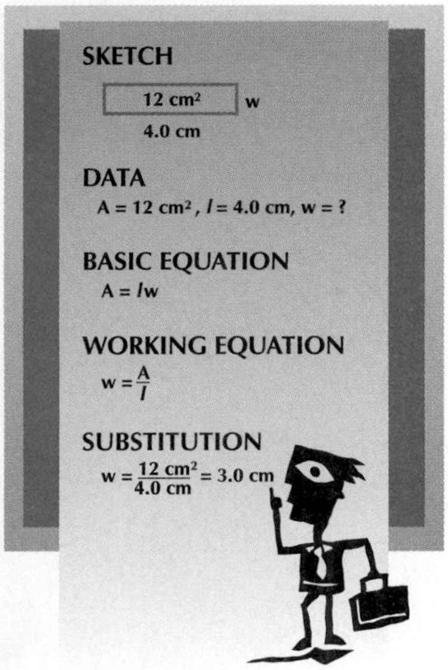

PROBLEMS 13.3

1. Find the R value of a pane of 0.125-in.-thick glass.
2. Find the R value of a brick wall 4.0 in. thick.
3. Find the R value of 0.50-in.-thick sheetrock.
4. Find the thermal conductivity of a piece of building material 0.25 in. thick that has an R value of 1.6 ft^2 °F/Btu/h.
5. Find the R value of 0.50-in.-thick corkboard.
6. The dimensions of a rectangular building are $2\bar{0}$ ft × $10\bar{0}$ ft. The average outer wall temperature is $2\bar{0}$°F and the average inner wall temperature is 55°F. Find the amount of heat conducted through the walls of the building in 24 h if the R value of the walls is 11 ft^2 °F/Btu/h.
7. Find the heat flow during 30.0 days through a glass window of thickness 0.20 in. with area 15 ft^2 if the average outer surface temperature is 25°F and the average inner glass surface temperature is $5\bar{0}$°F.
8. Find the heat flow in 30.0 days through a 0.25-cm-thick steel plate with cross section 45 cm × 75 cm. Assume a temperature differential of 95°C.
9. Find the heat flow in 75 s through a steel rod of length 85 cm and diameter 0.50 cm if the temperature of the hot end of the rod is $11\bar{0}$°C and the temperature of the cool end is −25°C.
10. Find the heat flow in 15 min through a 0.10-cm-thick copper plate with cross-sectional area 150 cm^2 if the temperature of the hot side is $99\bar{0}$°C and the temperature of the cool side is 5°C.

PHYSICS CONNECTIONS

Insulation

The purpose of insulating a home is to limit the amount of heat transfer through the walls and ceiling. Insulation limits heat transfer via conduction, convection, and radiation. Different insulating materials and methods are used. Less dense materials are often more effective in preventing heat transfer. Attic insulation must be fluffy and not compressed to be less dense. Molecules farther apart make fewer collisions, which results in less transfer of heat energy to a cooler environment. Other examples of less dense materials limiting heat transfer via conduction are argon gas placed between double-pane windows and evacuated linings in vacuum bottles. Argon has very low density; evacuated linings contain almost no air and therefore few molecules to transfer heat.

Fiberglass is one of the most popular types of insulation. The glass fiber is a poor conductor of heat. In addition, the air pockets in the fluffy fiberglass insulation prevent convection currents from transferring energy between the molecules and transferring heat (Fig. 13.12).

Phil Degginger/Color-Pic, Inc.

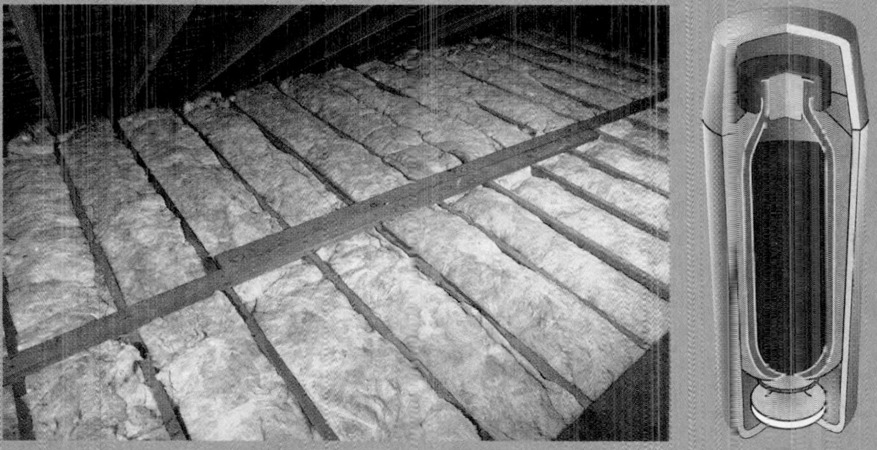

(a) (b)

Figure 13.12 **(a) Insulation is made of fiberglass, a poor conductor of heat. The fluffiness of the insulation creates air pockets to also eliminate convection currents. (b) The evacuated chamber inside a vacuum bottle prevents heat from transferring through the sides of the bottle.**

11. Find the heat flow in 24 h through a refrigerator door 30.0 in. × 58.0 in. insulated with cellulose fiber 2.0 in. thick. The temperature inside the refrigerator is 38°F. Room temperature is 72°F.
12. Find the heat flow in 30.0 days through a freezer door 30.0 in. × 58.0 in. insulated with cellulose fiber 2.0 in. thick. The temperature inside the freezer is −10°F. Room temperature is 72°F.
13. Find the heat flow in 24 h through a refrigerator door 76.0 cm × 155.0 cm insulated with cellulose fiber 5.0 cm thick. The temperature inside the refrigerator is 3°C. Room temperature is 21°C.
14. Find the heat flow in 30.0 days through a freezer door 76.0 cm × 155.0 cm insulated with cellulose fiber 5.0 cm thick. The temperature inside the freezer is −18°C. Room temperature is 21°C.
15. Find the heat flow through the sides of an 18-cm-tall glass of ice water in 45 s. The glass is 6.00 mm thick; the temperature inside is 28.0°C. The temperature outside is 43.3°C and the radius is 7.0 cm.

13.4 Specific Heat

If we placed a piece of steel and a pan of water in the direct summer sunlight, we would find that the water becomes only slightly warmer whereas the steel gets quite hot. Why should one get so much hotter than the other? If equal masses of steel and water were placed over the same flame for 1 min, the temperature of the steel would increase almost 10 times more than that of the water. The water has a greater capacity to absorb heat.

Because water has a much higher capacity for storing energy than most common materials, it is very useful in cooling systems in engines and power plants. This property of water affects the climate of many places. Cities on large lakes, like Chicago, are warmer in the winter and cooler in the summer near the lake because of the high heat capacity of water. High temperatures in summer and low temperatures in winter in the middle of large continents are largely due to the absence of large bodies of water. Europe is warmer in the winter than mid-Canada because warm air from over the Atlantic Ocean is blown by prevailing westerly winds over the land.

The specific heat of a substance is a measure of its capacity to absorb or give off heat per degree change in temperature. This property of water to absorb or give off large amounts of heat makes it an effective substance for transferring heat in industrial processes.

The **specific heat** of a substance is the amount of heat necessary to change 1 kg of it 1°C (1 lb of it 1°F in the U.S. system). By formula,

$$c = \frac{Q}{m\Delta T} \quad \text{(metric)} \qquad c = \frac{Q}{w\Delta T} \quad \text{(U.S.)}$$

To find the amount of heat added or taken away from a substance to produce a certain temperature change, we use

$$Q = cm\Delta T \quad \text{(metric)} \qquad Q = cw\Delta T \quad \text{(U.S.)}$$

where

c = specific heat
Q = heat
m = mass
w = weight
ΔT = change in temperature

A list of specific heats is given in Table 15 of Appendix C.

TRY THIS ACTIVITY

Cool Floors

A dramatic example of heat conduction is often experienced on cold winter mornings. While standing with your bare feet on a cold tile floor, note how quickly heat is transferred from your feet to the tile. Then, stand in a doorway with one bare foot on tile and one bare foot on wood and note the difference in the rate at which heat is transferred. What are the general characteristics that determine the heat capacity for your floors? Why are mats commonly placed on bathroom floors?

EXAMPLE 1

How many kilocalories of heat must be added to 10.0 kg of steel to raise its temperature $15\overline{0}$°C?

Data:

$m = 10.0$ kg
$\Delta T = 15\overline{0}$°C
$c = 0.115$ kcal/kg °C (from Table 15 of Appendix C)
$Q = ?$

Basic Equation:

$$Q = cm\Delta T$$

Working Equation: Same

Substitution:

$$Q = \left(0.115 \frac{\text{kcal}}{\text{kg °C}}\right)(10.0 \text{ kg})(15\overline{0} \text{°C})$$

$$= 173 \text{ kcal}$$

EXAMPLE 2

How many joules of heat must be absorbed to cool 5.00 kg of water from 75.0°C to 10.0°C?

Data:

$$m = 5.00 \text{ kg}$$

$$\Delta T = 75.0\text{°C} - 10.0\text{°C} = 65.0\text{°C}$$

$$c = 4190 \text{ J/kg °C} \quad \text{(from Table 15 of Appendix C)}$$

$$Q = ?$$

Basic Equation:

$$Q = cm\Delta T$$

Working Equation: Same

Substitution:

$$Q = \left(4190 \frac{\text{J}}{\text{kg °C}}\right)(5.00 \text{ kg})(65.0\text{°C})$$

$$= 1.36 \times 10^6 \text{ J} \quad \text{or} \quad 1.36 \text{ MJ}$$

PROBLEMS 13.4

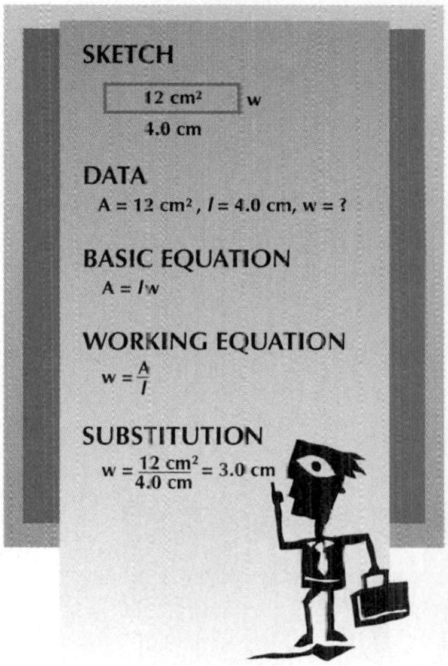

Find Q for each material.

1. Steel, $w = 3.00$ lb, $\Delta T = 50\overline{0}$°F, $Q =$ _____ Btu
2. Copper, $m = 155$ kg, $\Delta T = 170$°C, $Q =$ _____ kcal
3. Water, $w = 19.0$ lb, $\Delta T = 20\overline{0}$°F, $Q =$ _____ Btu
4. Water, $m = 25\overline{0}$ g, $\Delta T = 17.0$°C, $Q =$ _____ cal
5. Ice, $m = 5.00$ kg, $\Delta T = 2\overline{0}$°C, $Q =$ _____ J
6. Steam, $w = 5.00$ lb, $\Delta T = 4\overline{0}$°F, $Q =$ _____ Btu
7. Aluminum, $m = 79.0$ g, $\Delta T = 16$°C, $Q =$ _____ cal
8. Brass, $m = 750$ kg, $\Delta T = 125$°C, $Q =$ _____ J
9. Steel, $m = 1250$ g, $\Delta T = 50.0$°C, $Q =$ _____ J
10. Aluminum, $m = 85\overline{0}$ g, $\Delta T = 115$°C, $Q =$ _____ kcal
11. Water, $m = 80\overline{0}$ g, $\Delta T = 80.0$°C, $Q =$ _____ kcal
12. Lead, $m = 475$ kg, $\Delta T = 245$°C, $Q =$ _____ J
13. How many Btu of heat must be added to 1200 lb of copper to raise its temperature from 10$\overline{0}$°F to 45$\overline{0}$°F?
14. How many Btu of heat are given off by 50$\overline{0}$ lb of aluminum when it cools from 65$\overline{0}$°F to 75°F?
15. How many kcal of heat must be added to 1250 kg of copper to raise its temperature from 25°C to 275°C?
16. How many joules of heat are absorbed by an electric freezer in lowering the temperature of 1850 g of water from 80.0°C to 10.0°C?
17. How many joules of heat are required to raise the temperature of 75$\overline{0}$ kg of water from 15.0°C to 75.0°C?

18. How many kilocalories of heat must be added to $75\bar{0}$ kg of steel to raise its temperature from 75°C to $30\bar{0}$°C?
19. How many joules of heat are given off when 125 kg of steel cools from 1425°C to 82°C?
20. A 525-kg steam boiler is made of steel and contains 315 kg of water at 40.0°C. Assuming that 75% of the heat is delivered to the boiler and water, how many kilocalories are required to raise the temperature of both the boiler and water to 100.0°C?
21. Find the initial temperature of a 49.0-N cube of zinc, 16.0 cm on a side, that gives off 3.36×10^5 J of heat while cooling to 80.0°C.
22. A coolant lowers the temperature 13°C in a steel engine weighing 16,250 N and running at a temperature of $11\bar{0}$°C. What is the heat reduction (in joules) in the steel engine?
23. A block of iron with mass 0.400 kg is heated to 325°C from 295°C. How much heat is absorbed by the iron?
24. A block of copper is heated from 20.0°C to 80.0°C. How much heat is absorbed by the copper if its mass is 60.0 g?
25. The cooling system of a truck engine contains 20.0 L of water. (1 L of water has a mass of 1 kg.) (a) If the engine is run until 845 kJ of heat is added, what is the change in temperature of the water? (b) In the winter, the system was filled with 20.0 L of ethyl alcohol with density 0.800 g/cm^3. If the ethyl alcohol absorbed the same 845 kJ of heat, what would be the increase in temperature of the alcohol? (c) Would ethyl alcohol or water be a better coolant? Why?

13.5 Method of Mixtures

When two substances at different temperatures are mixed together, heat flows from the warmer body to the cooler body until they reach the same temperature (Fig. 13.13). This is known as thermal equilibrium, achieved in this case by the **method of mixtures.** Part of the heat lost by the warmer body is transferred to the cooler body and part is lost to the surrounding objects or the air. In most cases almost all the heat is transferred to the cooler body. We assume here that all the heat lost by the warmer body equals the heat gained by the cooler body. The amount of heat lost or gained by a body is

Figure 13.13 Heat flows from the warmer substance to the cooler.

$$Q = cm\Delta T \quad \text{or} \quad Q = cw\Delta T$$

By formula,

$$Q_{\text{lost}} = Q_{\text{gained}}$$
$$c_l m_l (T_l - T_f) = c_g m_g (T_f - T_g)$$

where the subscript l refers to the warmer body, which *loses* heat, the subscript g refers to the cooler body, which *gains* heat, and T_f is the final temperature of the mixture.

EXAMPLE 1

A 10.0-lb piece of hot copper is dropped into 30.0 lb of water at $5\bar{0}$°F. If the final temperature of the mixture is 65°F, what was the initial temperature of the copper?

Data:

$w_l = 10.0$ lb $\qquad w_g = 30.0$ lb

$c_l = 0.093$ Btu/lb °F $\qquad c_g = 1.00$ Btu/lb °F

$T_l = ?$ $\qquad T_g = 5\bar{0}$°F

$T_f = 65$°F

Basic Equation:

$$c_l w_l (T_l - T_f) = c_g w_g (T_f - T_g)$$

Working Equation:

$$T_l = \frac{c_g w_g}{c_l w_l}(T_f - T_g) + T_f$$

Substitution:

$$T_l = \frac{(1.00 \text{ Btu/lb °F})(30.0 \text{ lb})}{(0.093 \text{ Btu/lb °F})(10.0 \text{ lb})}(65°\text{F} - 5\bar{0}°\text{F}) + 65°\text{F}$$
$$= 550°\text{F}$$

Some find it easier to find T_l using a second method. Substitute the data directly into the basic equation. Then solve for T_l as follows:

$$\left(0.093\frac{\text{Btu}}{\text{lb °F}}\right)(10.0 \text{ lb})(T_l - 65°\text{F}) = \left(1.00\frac{\text{Btu}}{\text{lb °F}}\right)(30.0 \text{ lb})(65°\text{F} - 5\bar{0}°\text{F})$$

$$0.93T_l \text{ Btu/°F} - 6\bar{0} \text{ Btu} = 450 \text{ Btu}$$

$$0.93T_l \text{ Btu/°F} = 510 \text{ Btu}$$

$$T_l = \frac{510 \text{ Btu}}{0.93 \text{ Btu/°F}}$$

$$T_l = 550°\text{F}$$

EXAMPLE 2

If $20\bar{0}$ g of steel at $22\bar{0}$°C is added to $50\bar{0}$ g of water at 10.0°C, find the final temperature of this mixture.

Data:

$$c_l = 0.115 \text{ cal/g °C} \qquad c_g = 1.00 \text{ cal/g °C}$$
$$m_l = 20\bar{0} \text{ g} \qquad m_g = 50\bar{0} \text{ g}$$
$$T_l = 22\bar{0}°\text{C} \qquad T_g = 10.0°\text{C}$$
$$T_f = ?$$

Basic Equation:

$$c_l m_l (T_l - T_f) = c_g m_g (T_f - T_g)$$

Working Equation:

$$T_f = \frac{c_l m_l T_l + c_g m_g T_g}{c_l m_l + c_g m_g}$$

Substitution:

$$T_f = \frac{(0.115 \text{ cal/g °C})(20\bar{0} \text{ g})(22\bar{0}°\text{C}) + (1.00 \text{ cal/g °C})(50\bar{0} \text{ g})(10.0°\text{C})}{(0.115 \text{ cal/g °C})(20\bar{0} \text{ g}) + (1.00 \text{ cal/g °C})(50\bar{0} \text{ g})}$$
$$= 19.2°\text{C}$$

To find T_f by the second method, substitute the data directly into the basic equation. Then, solve for T_f as follows:

$$\left(0.115\frac{\text{cal}}{\text{g °C}}\right)(20\bar{0} \text{ g})(22\bar{0}°\text{C} - T_f) = \left(1.00\frac{\text{cal}}{\text{g °C}}\right)(50\bar{0} \text{ g})(T_f - 10.0°\text{C})$$

$$5060 \text{ cal} - 23.0\frac{\text{cal}}{°\text{C}}T_f = 50\bar{0}\frac{\text{cal}}{°\text{C}}T_f - 50\bar{0}0 \text{ cal}$$

$$10{,}060 \text{ cal} = 523\frac{\text{cal}}{°\text{C}}T_f$$

$$\frac{10{,}060 \text{ cal}}{523 \text{ cal/°C}} = T_f$$

$$19.2°\text{C} = T_f$$

SKETCH

12 cm² w
4.0 cm

DATA

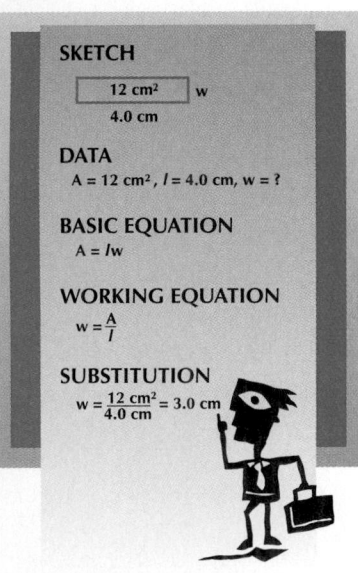

$A = 12\text{ cm}^2,\ l = 4.0\text{ cm},\ w = ?$

BASIC EQUATION

$A = lw$

WORKING EQUATION

$w = \frac{A}{l}$

SUBSTITUTION

$w = \frac{12\text{ cm}^2}{4.0\text{ cm}} = 3.0\text{ cm}$

PROBLEMS 13.5

Refer to Table 15 of Appendix C.

1. A 2.50-lb piece of steel is dropped into 11.0 lb of water at 75.0°F. The final temperature is 84.0°F. What was the initial temperature of the steel?
2. Mary mixes 5.00 lb of water at $20\bar{0}$°F with 7.00 lb of water at 65.0°F. Find the final temperature of the mixture.
3. A $25\bar{0}$-g piece of tin at 99°C is dropped in $10\bar{0}$ g of water at $1\bar{0}$°C. If the final temperature of the mixture is $2\bar{0}$°C, what is the specific heat of the tin?
4. How many grams of water at $2\bar{0}$°C are necessary to change $80\bar{0}$ g of water at $9\bar{0}$°C to $5\bar{0}$°C?
5. A 159-lb piece of aluminum at $50\bar{0}$°F is dropped into $40\bar{0}$ lb of water at $6\bar{0}$°F. What is the final temperature?
6. A 42.0-lb piece of steel at $67\bar{0}$°F is dropped into $10\bar{0}$ lb of water at 75.0°F. What is the final temperature of the mixture?
7. If 1250 g of copper at 20.0°C is mixed with $50\bar{0}$ g of water at 95.0°C, find the final temperature of the mixture.
8. If $50\bar{0}$ g of brass at $20\bar{0}$°C and $30\bar{0}$ g of steel at $15\bar{0}$°C are added to $90\bar{0}$ g of water in an aluminum pan of mass $15\bar{0}$ g both at 20.0°C, find the final temperature, assuming no loss of heat to the surroundings.
9. The following data were collected in the laboratory to determine the specific heat of an unknown metal:

Mass of copper calorimeter	153 g
Specific heat of calorimeter	0.092 cal/g °C
Mass of water	275 g
Specific heat of water	1.00 cal/g °C
Mass of metal	236 g
Initial temperature of water and calorimeter	16.2°C
Initial temperature of metal	99.6°C
Final temperature of calorimeter, water, and metal	22.7°C

Find the specific heat of the unknown metal. ***Note:*** A calorimeter is usually a metal cup inside another metal cup that is insulated by the air between them (Fig. 13.14).

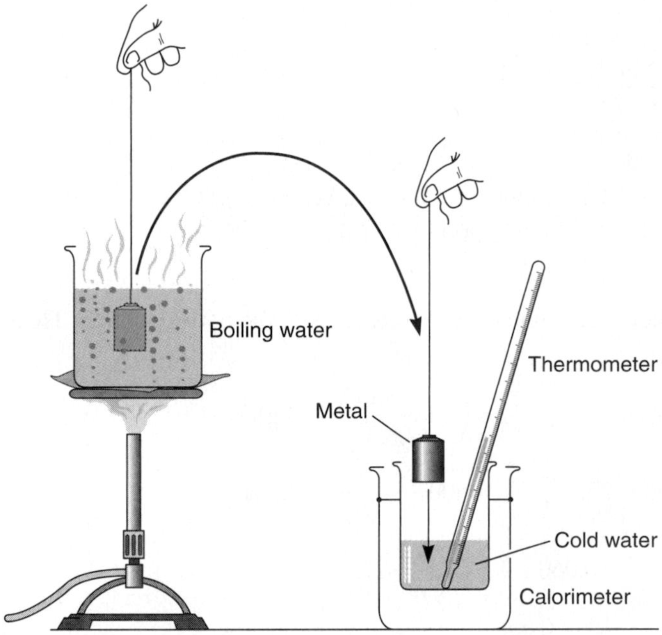

Figure 13.14 Apparatus for measuring the specific heat of a metal by the method of mixtures

10. The following data were collected in the laboratory to determine the specific heat of an unknown metal:

Mass of aluminum calorimeter	132 g
Specific heat of calorimeter	$92\bar{0}$ J/kg °C
Mass of water	285 g
Specific heat of water	4190 J/kg °C
Mass of metal	215 g
Initial temperature of water and calorimeter	12.6°C
Initial temperature of metal	99.1°C
Final temperature of calorimeter, water, and metal	18.6°C

Find the specific heat of the unknown metal.

11. Determine the original temperature of a $56\bar{0}$-g piece of lead placed in a 165-g brass calorimeter that contains 325 g of water. The initial temperature of the water and calorimeter was 18.0°C. The final temperature of the lead, calorimeter, and water is 31.0°C.
12. How much heat must be absorbed by its surroundings to cool a 565-g cube of iron from 100.0°C to 20.0°C?
13. How much water at 0°C would be needed to cool the iron in Problem 12 to 20.0°C?
14. The specific heat of water is 4190 J/kg °C. The specific heat of steel is 481 J/kg °C. Why may you burn your tongue on hot coffee but not on the spoon when both are at the same temperature?

13.6 Expansion of Solids

Most solids expand when heated and contract when cooled. They expand or contract in all three dimensions—length, width, and thickness. When a solid is heated, the expansion is due to the increased length of the vibrations of the atoms and molecules. This results in the solid expanding in all directions. This increase in volume results in a decrease in weight density, which was discussed in Chapter 11. Engineers, technicians, and designers must know the effects of thermal expansion. You have no doubt heard of highway pavements buckling on a hot summer day (Fig. 13.15). Bridges are built with special joints that allow for

Figure 13.15 **Thermal expansion causes pavement to sometimes break up (buckle) in the summer. Air temperatures in the mid-90s (°F) or low-30s (°C) can easily translate into pavement temperatures in the 125°F or 52°C range. As the pavement absorbs heat, it expands and can reach the point where the concrete buckles across the traffic lanes.**

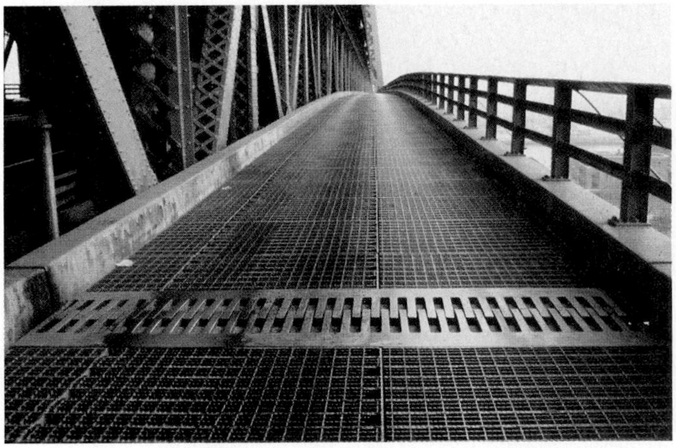

Figure 13.16 Thermal expansion joint on a bridge

expansion and contraction of the bridge deck (Fig. 13.16). Similarly, TV towers, pipelines, and buildings must be designed and built to allow for this expansion and contraction.

There are some advantages to solids expanding. A bimetallic strip is made by fusing two different metals together side by side as illustrated in Fig. 13.17. When heated, the brass expands more than the steel, which makes the strip curve. If the bimetallic strip is cooled below room temperature, the brass will contract more than the steel, forcing the strip to curve in the opposite direction. The thermostat operates on this principle. As shown in Fig. 13.18, the basic parts of a thermostat are a bimetallic strip on the right and a regular metal strip on the left. The bimetallic strip of brass and steel bends with the temperature. The regular metal strip is moved by hand to set the temperature desired. This particular bimetallic strip is made and placed so that it bends to the left when cooled. As a result, when it comes in contact with the strip on the left, it completes a circuit, which turns on the furnace. When the room warms to the desired temperature, the bimetallic strip moves back to the right, which opens the contacts and shuts off the heat. Bimetallic strips are in spiral form in some thermostats (Fig. 13.19).

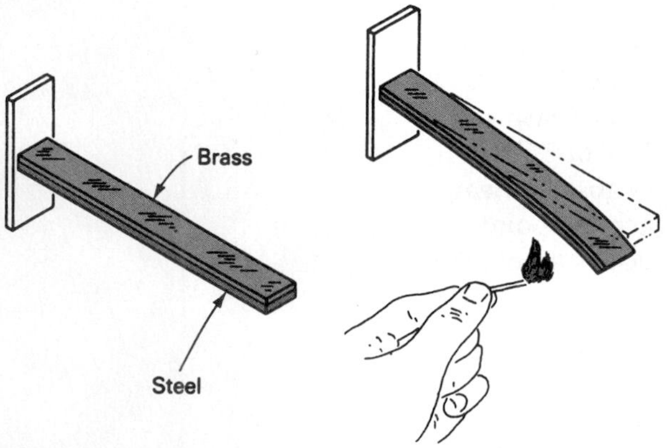

Figure 13.17 Thermal expansion of a bimetallic strip

(a) Circuit open (b) Circuit closed

Figure 13.18 Simple thermostat

Linear Expansion

The amount that a solid expands depends on the following:

1. ***Material.*** Different materials expand at different rates. Steel expands at a rate less than that of brass.
2. ***Length of the solid.*** The longer the solid, the larger is the expansion. A 20.0-cm steel rod will expand twice as much as a 10.0-cm steel rod.

3. ***Amount of change in temperature.*** The greater the change in temperature, the greater is the expansion.

This can be written as a formula:

$$\Delta l = \alpha l \Delta T$$

where

Δl = change in length
α = a constant called the **coefficient of linear expansion***
l = original length
ΔT = change in temperature

Table 13.3 lists the coefficients of linear expansion for some common solids.

Figure 13.19 Thermostat using spiral bimetallic nut

Table 13.3 Coefficients of Linear Expansion

Material	α (metric)	α (U.S.)
Aluminum	2.3×10^{-5}/C°	1.3×10^{-5}/F°
Brass	1.9×10^{-5}/C°	1.0×10^{-5}/F°
Concrete	1.1×10^{-5}/C°	6.0×10^{-6}/F°
Copper	1.7×10^{-5}/C°	9.5×10^{-6}/F°
Glass	9.0×10^{-6}/C°	5.1×10^{-6}/F°
Pyrex	3.0×10^{-6}/C°	1.7×10^{-6}/F°
Steel	1.3×10^{-5}/C°	6.5×10^{-6}/F°
Zinc	2.6×10^{-5}/C°	1.5×10^{-5}/F°

Comparing the coefficients of linear expansion of common glass and Pyrex, we can see that Pyrex expands and contracts approximately one-third as much as glass. This is why it is used in cooking and chemical laboratories.

EXAMPLE 1

A steel railroad rail is 40.0 ft long at 0°F. How much will it expand when heated to $10\bar{0}$°F?

Data:

$$l = 40.0 \text{ ft}$$
$$\Delta T = 10\bar{0}°\text{F} - 0°\text{F} = 10\bar{0}°\text{F}$$
$$\alpha = 6.5 \times 10^{-6}/\text{F}°$$
$$\Delta l = ?$$

Basic Equation:

$$\Delta l = \alpha l \Delta T$$

Working Equation: Same

Substitution:

$$\Delta l = (6.5 \times 10^{-6}/\text{F}°)(40.0 \text{ ft})(10\bar{0}°\text{F})$$
$$= 0.026 \text{ ft} \quad \text{or} \quad 0.31 \text{ in.}$$

Pipes that undergo large temperature changes are installed to allow for expansion and contraction (Fig. 13.20).

*Defined as change in unit length of a solid when its temperature is changed 1 degree.

Figure 13.20 Thermal expansion joints in pipes

EXAMPLE 2

What allowance for expansion must be made for a steel pipe $12\bar{0}$ m long that handles coolants and must undergo temperature changes of $20\bar{0}$°C?

Data:

$$\alpha = 1.3 \times 10^{-5}/\text{C}^\circ$$
$$l = 12\bar{0}\ \text{m}$$
$$\Delta T = 20\bar{0}^\circ\text{C}$$
$$\Delta l = ?$$

Basic Equation:

$$\Delta l = \alpha l \Delta T$$

Working Equation: Same

Substitution:

$$\Delta l = (1.3 \times 10^{-5}/\cancel{\text{C}}^\circ)(12\bar{0}\ \text{m})(20\bar{0}\ {}^\circ\cancel{\text{C}})$$
$$= 0.31\ \text{m} \quad \text{or} \quad 31\ \text{cm}$$

Area and Volume Expansion of Solids

Solids expand in width and thickness as well as in length when heated. The area of a hole cut out of a metal sheet will expand in the same way as the surrounding material. To allow for this expansion the following formulas are used:

Area expansion:	$\Delta A = 2\alpha A\ \Delta T$
Volume expansion:	$\Delta V = 3\alpha V\ \Delta T$

where

$$A = \text{original area}$$
$$V = \text{original volume}$$

EXAMPLE 3

The top of a circular copper disk has an area of 64.2 in^2 at 20̄°F. What is the change in area when the temperature is increased to 150̄°F?

Data:

$$\alpha = 9.5 \times 10^{-6}/\text{F}^\circ$$
$$A = 64.2 \text{ in}^2$$
$$\Delta T = 15\bar{0}^\circ\text{F} - 2\bar{0}^\circ\text{F} = 13\bar{0}^\circ\text{F}$$
$$\Delta A = ?$$

Basic Equation:

$$\Delta A = 2\alpha A \, \Delta T$$

Working Equation: Same

Substitution:

$$\Delta A = 2(9.5 \times 10^{-6}/\text{F}^\circ)(64.2 \text{ in}^2)(13\bar{0}^\circ\text{F})$$
$$= 0.16 \text{ in}^2$$

EXAMPLE 4

A section of concrete measures 6.00 m × 12.0 m × 30.0 m at 38°C. What allowance for change in volume is necessary for a temperature of −15°C?

Data:

$$V = (6.00 \text{ m})(12.0 \text{ m})(30.0 \text{ m}) = 2160 \text{ m}^3$$
$$\alpha = 1.1 \times 10^{-5}/\text{C}^\circ$$
$$\Delta T = 38^\circ\text{C} - (-15^\circ\text{C}) = 53^\circ\text{C}$$

Basic Equation:

$$\Delta V = 3\alpha V \, \Delta T$$

Working Equation: Same

Substitution:

$$\Delta V = 3(1.1 \times 10^{-5}/\text{C}^\circ)(2160 \text{ m}^3)(53^\circ\text{C})$$
$$= 3.8 \text{ m}^3$$

PROBLEMS 13.6

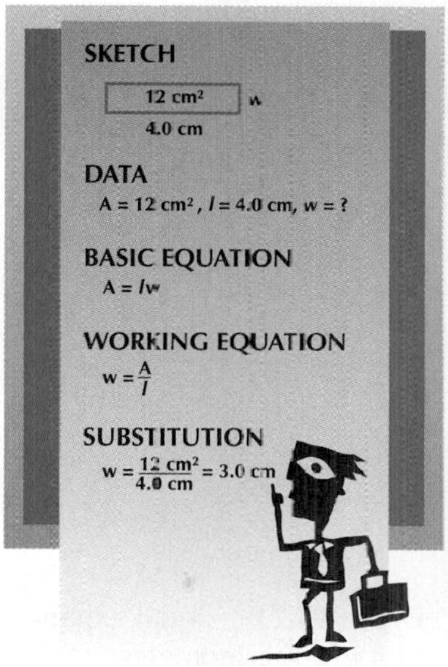

1. Find the increase in length of copper tubing 200.0 ft long at 40.0°F when it is heated to 200.0°F.
2. Find the increase in length of a zinc rod 50.0 m long at 15.0°C when it is heated to 130.0°C.
3. Find the increase in length of 300.00 m of copper wire when its temperature changes from 14°C to 34°C.
4. A steel pipe 8.25 m long is installed at 45°C. Find the decrease in length when coolants at −60°C pass through the pipe.
5. A steel tape measures 200.00 m at 15°C. What is its length at 55°C?
6. A brass rod 1.020 m long expands 3.0 mm when it is heated. Find the temperature change.
7. The road bed on a bridge 500.0 ft long is made of concrete. What allowance is needed for temperatures of −40̄°F in winter and 140̄°F in summer?
8. An aluminum plug has a diameter of 10.003 cm at 40.0°C. At what temperature will it fit precisely into a hole of constant diameter 10.000 cm?

9. The diameter of a steel drill at 45°F is 0.750 in. Find its diameter at 375°F.
10. A brass ball with diameter 12.000 cm is 0.011 cm too large to pass through a hole in a copper plate when the ball and plate are at a temperature of 20.0°C. What is the temperature of the ball when it will just pass through the plate, assuming that the temperature of the plate does not change? What is the temperature of the plate when the ball will just pass through, assuming that the temperature of the ball does not change?
11. A brass cylinder has a cross-sectional area of 482 cm^2 at –5°C. Find its change in area when heated to 95°C.
12. The volume of the cylinder in Problem 11 is 4820 cm^3 at 240.0°C. Find its change in volume when cooled to –75.0°C.
13. An aluminum pipe has a cross-sectional area of 88.40 cm^2 at 15°C. What is its cross-sectional area when the pipe is heated to 155°C?
14. A steel pipe has a cross-sectional area of 127.20 in^2 at 25°F. What is its cross-sectional area when the pipe is heated to 175°F?
15. A glass plug has a volume of 60.00 cm^3 at 12°C. What is its volume at 76°C?
16. The diameter of a hole drilled through brass at 21°C measures 6.500 cm. Find the diameter and area of the hole when the brass is heated to 175°C.
17. Steel rails 15.000 m long are laid at 10.0°C. How much space should be left between them if they are to just touch at 35.0°C?
18. Steel beams 120.000 ft long are placed in a highway overpass to allow for expansion and contraction. The temperature range allowance is $-3\bar{0}$°F to $13\bar{0}$°F.
    (a) Find the space allowance (in inches) between the beams at $-3\bar{0}$°F if the beams touch at $13\bar{0}$°F.
    (b) Find the space allowance between the beams if placed at 75°F and touch at $13\bar{0}$°F.
19. The spaces between 13.00-m steel rails are 0.711 cm at –15°C. If the rails touch at 35.5°C, what is the coefficient of linear expansion?
20. A section of concrete dam is a rectangular solid 20.0 ft by 50.0 ft by 80.0 ft at 115°F. What allowance for change in volume is necessary for a temperature of –15°F?
21. A glass ball has a radius of 12.000 cm at 6.0°C. Find its change in volume when the temperature is increased to 81.0°C.
22. Find the final height of a concrete column that is 1.250 m × 1.250 m × 4.250 m at 0.0°C when the column is heated to 45.0°C.
23. What is the final volume of a glass right circular cylinder with original height 1.200 m and radius 30.00 cm that is heated from 13.0°C to 56.0°C?
24. A metal bar at 21.0°C is 2.6000 m long. If the bar is heated to 93.0°C, its change in length is 3.40 mm. What is the coefficient of linear expansion of the bar?

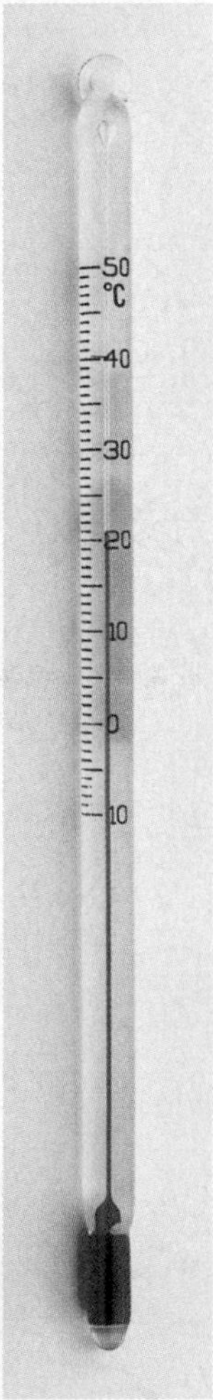

Photo courtesy of Dorling Kindersley

Figure 13.21 Liquid expansion in a thermometer

13.7 Expansion of Liquids

Liquids also generally expand when heated and contract when cooled. The thermometer is made using this principle. When a thermometer is placed under your tongue, the heat from your mouth causes the liquid in the bottom of the thermometer to expand. The liquid is then forced to rise up the thin calibrated tube (Fig. 13.21). Similarly, when the gasoline tank on a car is filled to capacity on a hot summer day and then parked in a hot parking lot, it overflows. The cold gas from the underground storage tanks expands as it warms up in the car's tank. An automobile radiator filled to the brim with cold water would likewise overflow as the engine heats the water. The formula for *volume expansion of liquids* is

$$\Delta V = \beta V \Delta T$$

where

β = coefficient of volume expansion for liquids
V = original volume

Table 13.4 lists the coefficients of volume expansion for some common liquids.

Table 13.4 Coefficients of Volume Expansion

Liquid	β (metric)	β (U.S.)
Acetone	$1.49 \times 10^{-3}/C°$	$8.28 \times 10^{-4}/F°$
Alcohol, ethyl	$1.12 \times 10^{-3}/C°$	$6.62 \times 10^{-4}/F°$
Carbon tetrachloride	$1.24 \times 10^{-3}/C°$	$6.89 \times 10^{-4}/F°$
Mercury	$1.8 \times 10^{-4}/C°$	$1.0 \times 10^{-4}/F°$
Petroleum	$9.6 \times 10^{-4}/C°$	$5.33 \times 10^{-4}/F°$
Turpentine	$9.7 \times 10^{-4}/C°$	$5.39 \times 10^{-4}/F°$
Water	$2.1 \times 10^{-4}/C°$	$1.17 \times 10^{-4}/F°$

EXAMPLE 1

If petroleum at 0°C occupies $25\bar{0}$ L, what is its volume at $5\bar{0}$°C?

Data:

$$\beta = 9.6 \times 10^{-4}/C°$$
$$V = 25\bar{0}\ L$$
$$\Delta T = 5\bar{0}°C$$
$$\Delta V = ?$$

Basic Equation:

$$\Delta V = \beta V \Delta T$$

Working Equation: Same

Substitution:

$$\Delta V = (9.6 \times 10^{-4}/\cancel{C°})(25\bar{0}\ L)(5\bar{0}\cancel{°C})$$
$$= 12\ L$$
$$\text{volume at } 5\bar{0}°C = V + \Delta V$$
$$= 25\bar{0}\ L + 12\ L = 262\ L$$

EXAMPLE 2

Find the increase in volume of 18.2 in^3 of water when the water is heated from $4\bar{0}$°F to $18\bar{0}$°F.

Data:

$$\beta = 1.17 \times 10^{-4}/F°$$
$$V = 18.2\ in^3$$
$$\Delta T = 18\bar{0}°F - 4\bar{0}°F = 14\bar{0}°F$$
$$\Delta V = ?$$

Basic Equation:

$$\Delta V = \beta V \Delta T$$

Working Equation: Same

Substitution:

$$\Delta V = (1.17 \times 10^{-4}/\cancel{F°})(18.2\ in^3)(14\bar{0}\cancel{°F})$$
$$= 0.298\ in^3$$

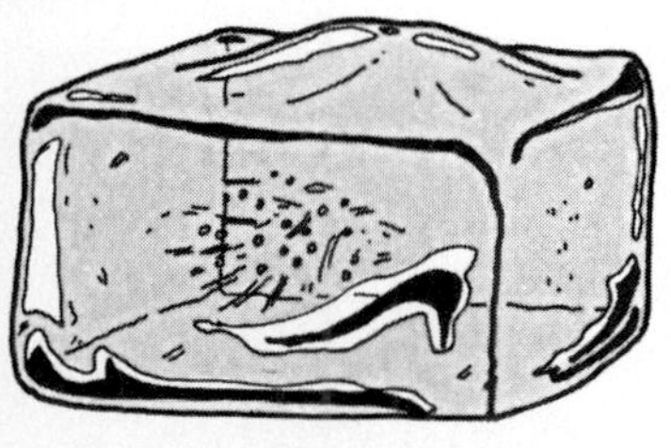

Figure 13.22 Expansion of water in change from liquid to solid

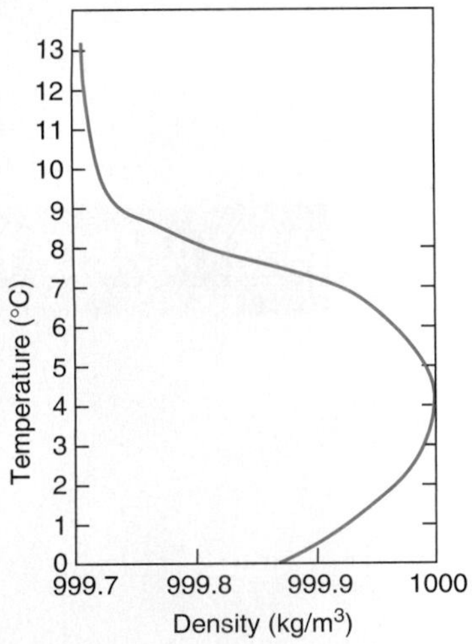

Figure 13.23 Change in density of water with change in temperature

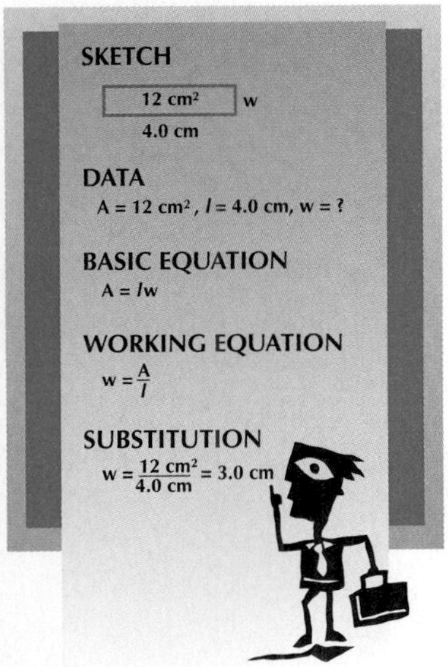

Expansion of Water

Water is unusual in its expansion characteristics. Recall the mound in the middle of each ice cube in an ice cube tray (Fig. 13.22). This is evidence of the expansion of water during its change of state from liquid to solid form.

Nearly all liquids are the most dense at their lowest temperature before a change of phase to become solids. As the temperature drops, the molecular motion slows and the substance becomes denser. Water does not follow this general rule. Because of its unusual structural characteristics, water is densest at 4°C or 39.2°F instead of 0°C or 32°F. A graph of its change in density with increase in temperature is shown in Fig. 13.23. As ice melts and the water temperature is slightly increased, there are still groups of molecules that have the open crystallographic structure of ice, which is less dense than water. As the water is heated to 4°C, these groupings disappear and the water becomes denser. Above 4°C, water then expands normally as the temperature is raised.

This unique behavior of water is critically important in lakes that freeze in winter. If, as in most liquids, water were most dense at its freezing point, the coldest water would settle to the bottom and the lake would freeze from the bottom up. Any living creatures would be killed. Fortunately, the most dense water at 4°C is at the bottom and the less dense water at 0°C is above it, so ice forms at the surface, the water below remains liquid, and the lake freezes from the top down.

When ice melts at 0°C or 32°F, the water formed *contracts* as the temperature is raised to 4°C or 39.2°F. Then it begins to *expand,* as do most other liquids.

PROBLEMS 13.7

1. A quantity of carbon tetrachloride occupies 625 L at 12°C. Find its volume at 48°C.
2. Some mercury occupies 157 in^3 at $-3\bar{0}$°F. What is its change in volume when heated to $9\bar{0}$°F?
3. Some petroleum occupies 11.7 m^3 at −17°C. Find its volume at 28°C.
4. Find the increase in volume of 35 L of acetone heated from 28°C to 38°C.
5. Some water at $18\bar{0}$°F occupies $378\bar{0}$ ft^3. What is its volume at 122°F?
6. A $120\bar{0}$-L tank of petroleum is completely filled at 9°C. How much spills over if the temperature rises to 45°C?
7. Find the increase in volume of 215 cm^3 of mercury when its temperature increases from $1\bar{0}$°C to 25°C.
8. Find the decrease in volume of $200\bar{0}$ ft^3 of alcohol in a railroad tank car if the temperature drops from 75°F to 54°F.
9. A gasoline service station owner receives a truckload of $33{,}00\bar{0}$ L of gasoline at 32°C. It cools to 15°C in the underground tank. At 40 cents/L, how much money is lost as a result of the contraction of the gasoline?
10. A Pyrex container is completely filled with 275 cm^3 of mercury at 10.0°C. How much mercury spills over when heated to 75.0°C?
11. What was the temperature of $18\bar{0}$ mL of acetone before it was heated to 98°C and increased to a volume of $20\bar{0}$ mL?

13.8 Change of Phase

Many industries are concerned with a change of phase in the materials they use. In foundries the principal activity is to change solid metals to liquid, pour the liquid metal into molds, and allow it to become solid again (Fig. 13.24). **Change of phase** (sometimes called *change of state*) is a change in a substance from one form of matter (solid, liquid, or gas) to another.

Fusion

The change of phase from solid to liquid is called **melting** or **fusion.** The change from liquid to solid is called **freezing** or **solidification.** Most solids have a crystalline structure and a definite melting point at any given pressure. Melting and solidification of these substances occur at the same temperature. For example, water at 0°C (32°F) changes to ice and ice

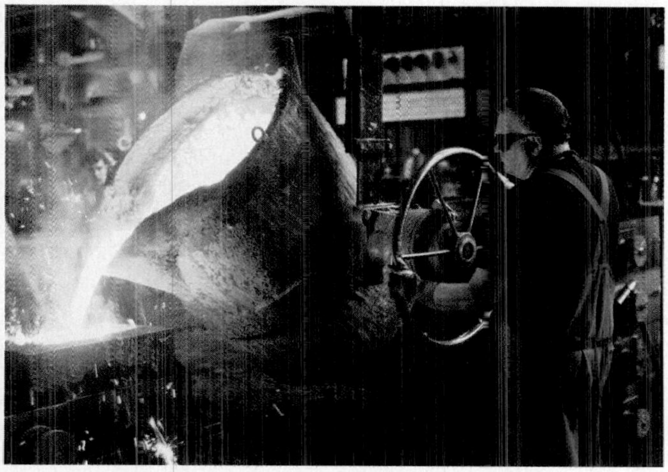

Figure 13.24 Molten pig iron from the blast furnace is poured into an open hearth furnace, refined, and purified into steel at temperatures about 2900°F.

changes to water at the same temperature. There is no temperature change during change of phase. Ice at 0°C changes to water at 0°C. Only a few substances, such as butter and glass, have no particular melting temperature but change phase gradually.

Although there is no temperature change during a change of phase, *there is a transfer of heat.* A melting solid *absorbs* heat and a solidifying liquid *gives off* heat. When 1 g of ice at 0°C melts, it absorbs $8\bar{0}$ cal of heat. Similarly, when 1 g of water freezes at 0°C, ice at 0°C is produced, and $8\bar{0}$ cal of heat is released.

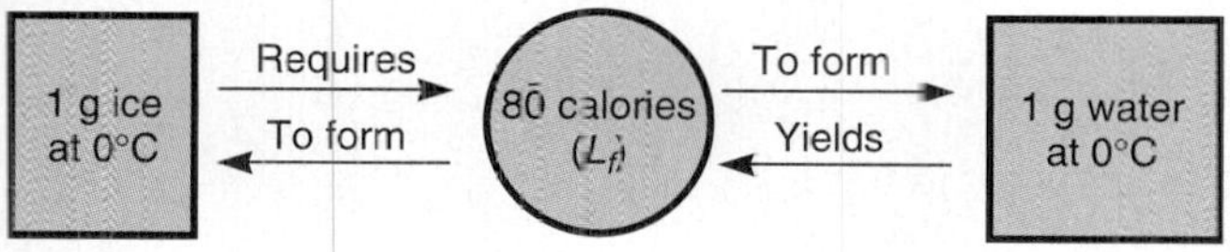

When 1 kg of ice at 0°C melts, it absorbs $8\bar{0}$ kcal of heat. Similarly, when 1 kg of water freezes at 0°C, ice at 0°C is produced and $8\bar{0}$ kcal of heat is released.

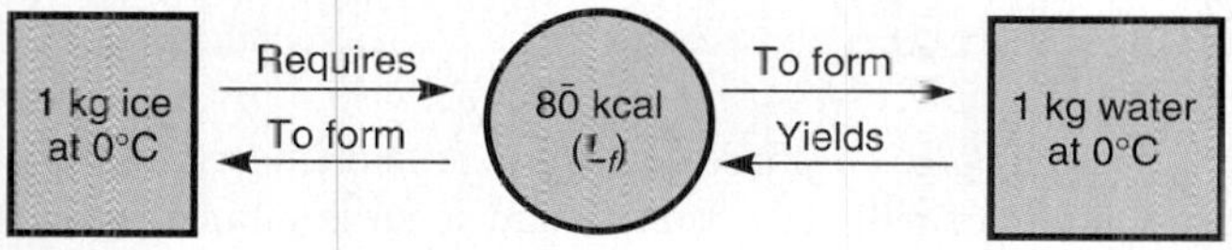

Or, when 1 kg of ice at 0°C melts, it absorbs 335 kilojoules (kJ) of heat. Then, when 1 kg of water freezes at 0°C, ice at 0°C is produced and 335 kJ of heat is released.

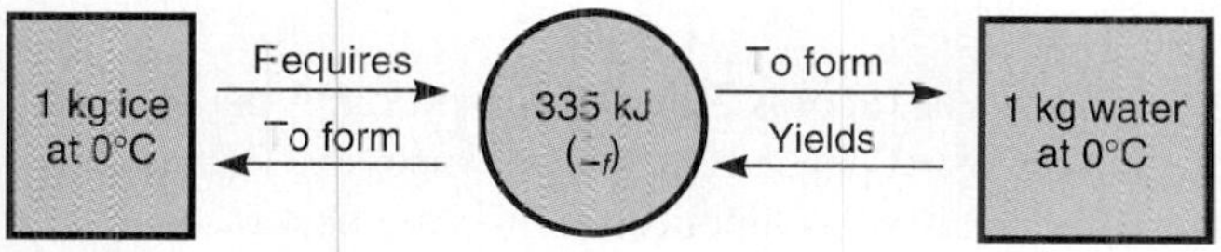

When 1 lb of ice at 32°F melts, it absorbs 144 Btu of heat. Similarly when 1 lb of water freezes at 32°F, ice at 32°F is produced and 144 Btu of heat is released.

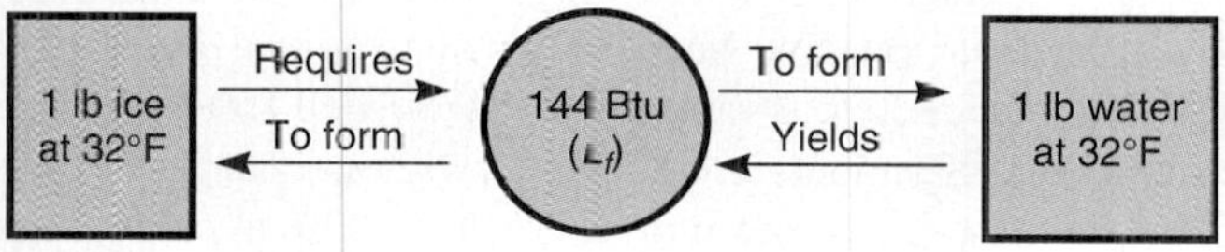

The amount of heat required to melt 1 g or 1 kg or 1 lb of a liquid is called its **heat of fusion,** designated L_f.

$$L_f = \frac{Q}{m} \quad \text{(metric)} \qquad L_f = \frac{Q}{w} \quad \text{(U.S.)}$$

where

L_f = heat of fusion (see Table 15 in Appendix C)
Q = quantity of heat
m = mass of substance (metric system)
w = weight of substance (U.S. system)

EXAMPLE 1

If 1340 kJ of heat is required to melt 4.00 kg of ice at 0°C into water at 0°C, what is the heat of fusion of water?

Data:

$$Q = 1340 \text{ kJ}$$
$$m = 4.00 \text{ kg}$$
$$L_f = ?$$

Basic Equation:

$$L_f = \frac{Q}{m}$$

Working Equation: Same

Substitution:

$$L_f = \frac{1340 \text{ kJ}}{4.00 \text{ kg}}$$
$$= 335 \text{ kJ/kg}$$

heat of fusion (water) = $8\bar{0}$ cal/g, or $8\bar{0}$ kcal/kg, or 335 kJ/kg, or 144 Btu/lb

A very interesting (and delicious) change-of-phase activity is to make homemade ice cream. A sealed container with a mixture of milk, egg, vanilla, and sugar is submerged in a mixture of rock salt and crushed ice. The salt causes the ice to rapidly melt, which requires heat while the ice changes phase from solid to liquid. Most of this heat is transferred from the ice cream mixture, which hardens into ice cream.

Vaporization

The change of phase from a liquid to a gas or vapor is called **vaporization.** A pot of boiling water (Fig. 13.25) vividly shows this change of phase as the steam evaporates and leaves the liquid. Note that vaporization requires that heat be supplied; in this case heat is required to boil the water. The reverse process (change from a gas to a liquid) is called **condensation.** As steam condenses in radiators (Fig. 13.26), large amounts of heat are released.

At the point of condensation, the vapor becomes *saturated;* that is, the vapor cannot hold any more moisture. For example, water vapor is always present in some amount in the earth's atmosphere. The weather term **relative humidity** is the ratio of the actual amount of vapor in the atmosphere to the amount of vapor required to reach 100% of saturation at the existing temperature. As the air temperature decreases without change in pressure or vapor content, the relative humidity increases until it reaches 100% at saturation.

Figure 13.25 Heat supplied to boiling water changes liquid water into steam—the gas form of water.

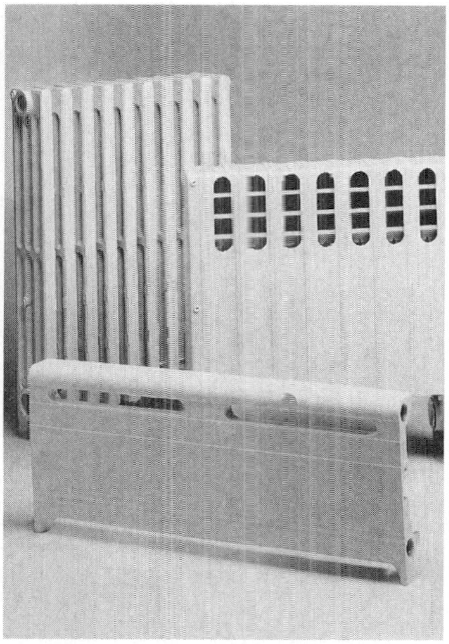

Figure 13.26 A large amount of heat is released by condensation of steam in a radiator.

The temperature at which saturation is reached is called the **dew point.** Once saturation is reached and the temperature continues to decrease, condensation occurs in the form of dew, fog, mist, clouds, and rain or other forms of precipitation.

While a liquid is boiling, the temperature of the liquid does not change. However, there is a transfer of heat. A liquid being vaporized (boiled) *absorbs* heat. As a vapor condenses, heat is given off.

The amount of heat required to vaporize 1 g or 1 kg or 1 lb of a liquid is called its **heat of vaporization,** designated L_v. So, when 1 g of water at 10$\bar{0}$°C changes to steam at 10$\bar{0}$°C, it absorbs 54$\bar{0}$ cal; when 1 g of steam at 10$\bar{0}$°C condenses to water at 10$\bar{0}$°C, 54$\bar{0}$ cal of heat is given off.

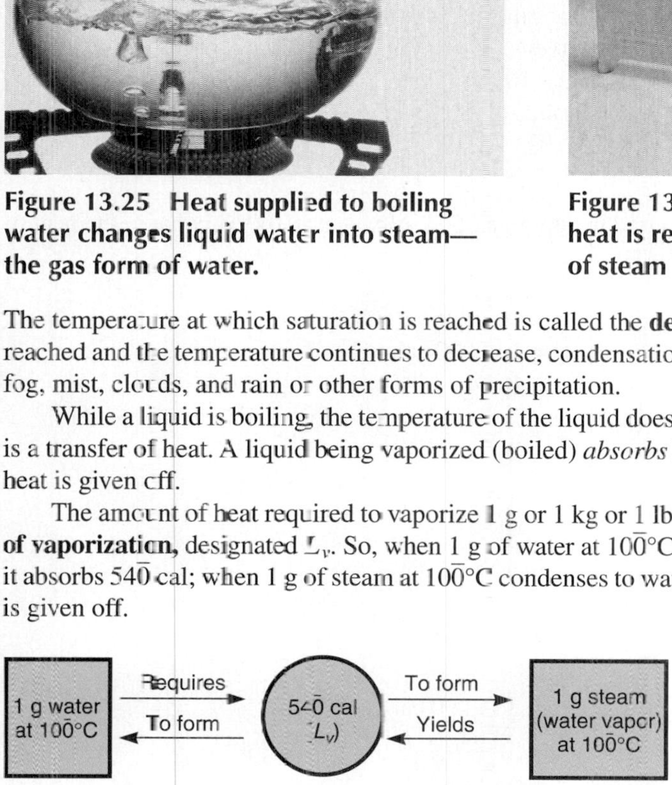

When 1 kg of water at 10$\bar{0}$°C changes to steam at 10$\bar{0}$°C, it absorbs 54$\bar{0}$ kcal of heat. Similarly, when 1 kg of steam at 10$\bar{0}$°C condenses to water at 10$\bar{0}$°C, 54$\bar{0}$ kcal of heat is given off.

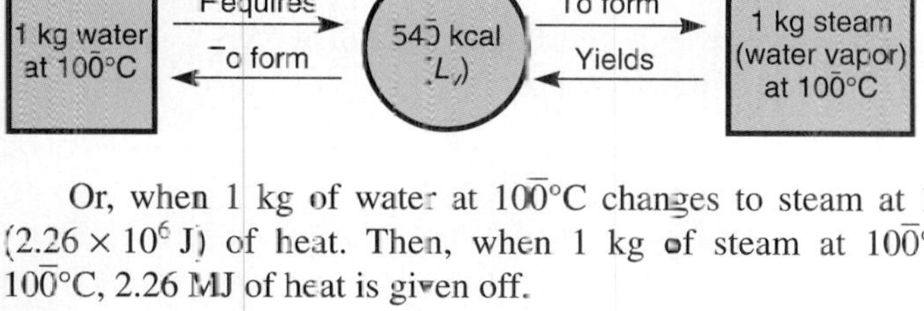

Or, when 1 kg of water at 10$\bar{0}$°C changes to steam at 10$\bar{0}$°C, it absorbs 2.26 MJ (2.26×10^6 J) of heat. Then, when 1 kg of steam at 10$\bar{0}$°C condenses to water at 10$\bar{0}$°C, 2.26 MJ of heat is given off.

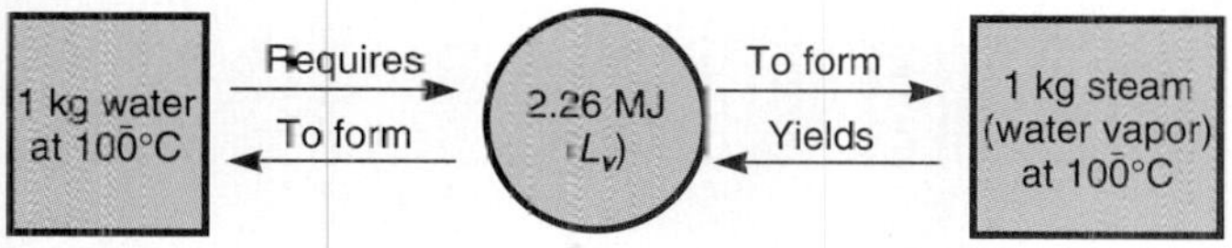

When 1 lb of water at 212°F changes to steam at 212°F, $97\bar{0}$ Btu of heat is absorbed; when 1 lb of steam at 212°F condenses to water at 212°F, $97\bar{0}$ Btu of heat is given off.

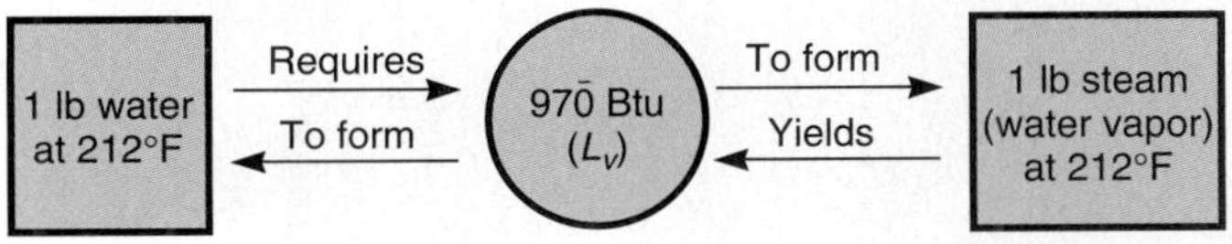

$$L_v = \frac{Q}{m} \quad \text{(metric)} \qquad L_v = \frac{Q}{w} \quad \text{(U.S.)}$$

where

L_v = heat of vaporization (see Table 15 in Appendix C)
Q = quantity of heat
m = mass of substance (metric system)
w = weight of substance (U.S. system)

EXAMPLE 2

If 135,000 cal of heat is required to vaporize $25\bar{0}$ g of water at $1\bar{0}0$°C, what is the heat of vaporization of water?

Data:

$$Q = 135{,}000 \text{ cal}$$
$$m = 25\bar{0} \text{ g}$$
$$L_v = ?$$

Basic Equation:

$$L_v = \frac{Q}{m}$$

Working Equation: Same

Substitution:

$$L_v = \frac{135{,}000 \text{ cal}}{25\bar{0} \text{ g}} = 54\bar{0} \text{ cal/g}$$

heat of vaporization (water) = $54\bar{0}$ cal/g, or $54\bar{0}$ kcal/kg, or 2.26 MJ/kg, or $97\bar{0}$ Btu/lb

EXAMPLE 3

If 15.8 MJ of heat is required to vaporize 18.5 kg of ethyl alcohol at 78.5°C (its boiling point), what is the heat of vaporization of ethyl alcohol?

Data:

$$Q = 15.8 \text{ MJ}$$
$$m = 18.5 \text{ kg}$$
$$L_v = ?$$

Basic Equation:

$$L_v = \frac{Q}{m}$$

Working Equation: Same

Substitution:

$$L_v = \frac{15.8 \text{ MJ}}{18.5 \text{ kg}}$$

$$= 0.854 \text{ MJ/kg} \quad \text{or} \quad 854 \text{ kJ/kg} \quad \text{or} \quad 8.54 \times 10^5 \text{ J/kg}$$

Figures 13.27 through 13.29 show the heat gained by one unit of ice at a temperature below its melting point as it warms to its melting point, changes to water, warms to its boiling point, changes to steam and then is heated above its boiling point in joules, Btu, and calories. Note that during changes of phase there are no temperature changes. Recall the basic shape of these graphs because we will use it to find the amount of heat gained or lost when a quantity of material goes through one or both changes of phase. Refer to Fig. 13.30 to do such problems. See Table 15 of Appendix C for heat constants of some common substances.

Figure 13.27

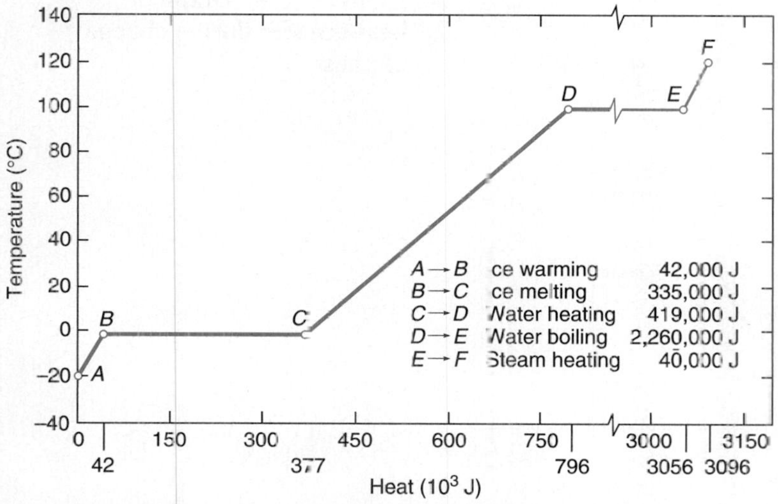

Heat gained by one kilogram of ice at –20°C as it is converted to steam at 120°C.

Figure 13.28

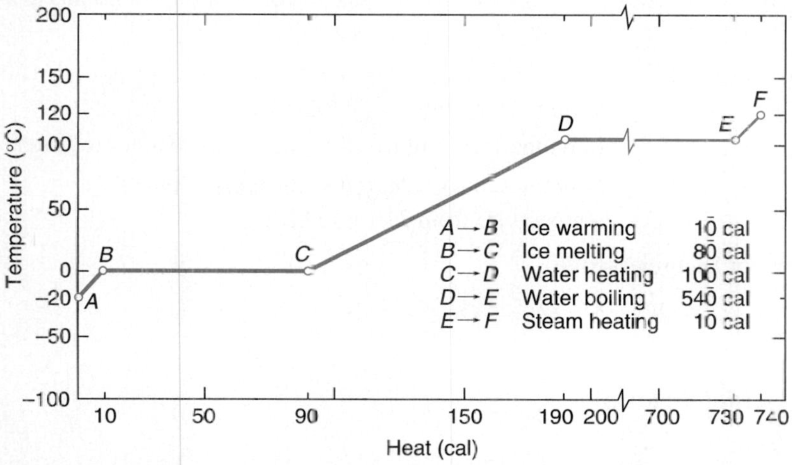

Heat gained by one gram of ice at –20°C as it is converted to steam at 120°C.

Figure 13.29

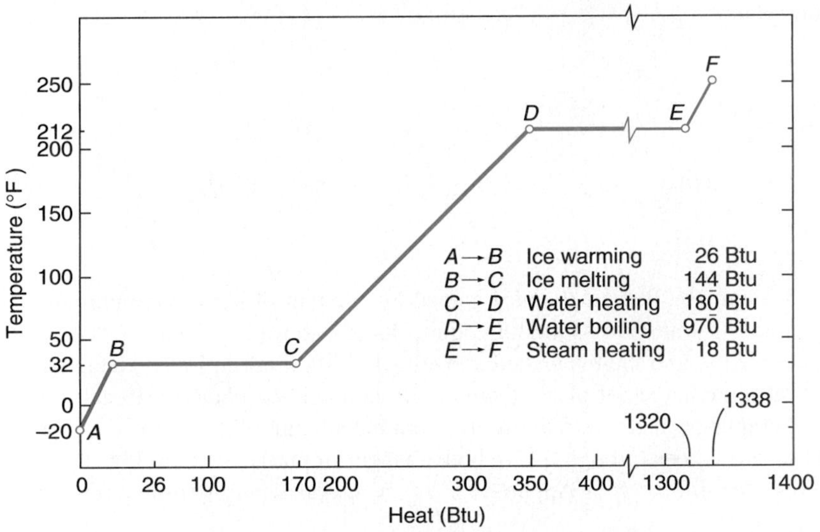

Heat gained by one pound of ice at –20°F as it is converted to steam at 250°F.

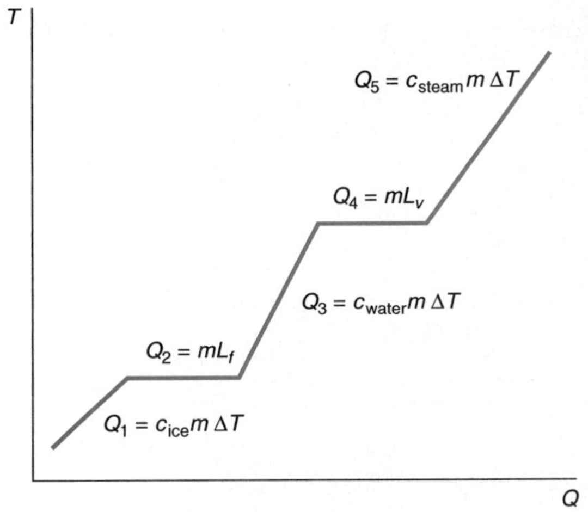

Figure 13.30 Graph of heat transfer during change of phase

EXAMPLE 4

How many Btu of heat are released when 4.00 lb of steam at 222°F is cooled to water at 82°F?

To find the amount of heat released when steam at a temperature above its vaporization point is cooled to water below its boiling point, we need to consider three amounts (see Fig. 13.31):

$Q_5 = c_{steam} w \Delta T$	(amount of heat released as the steam changes temperature from 222°F to 212°F)
$Q_4 = wL_v$	(amount of heat released as the steam changes to water)
$Q_3 = c_{water} w \Delta T$	(amount of heat released as the water changes temperature from 212°F to 82°F)

So the total amount of heat released is

$$Q = Q_5 + Q_4 + Q_3$$

Data:

$$w = 4.00 \text{ lb}$$
$$T_i \text{ of steam} = 222°\text{F}$$
$$T_f \text{ of water} = 82°\text{F}$$
$$Q = ?$$

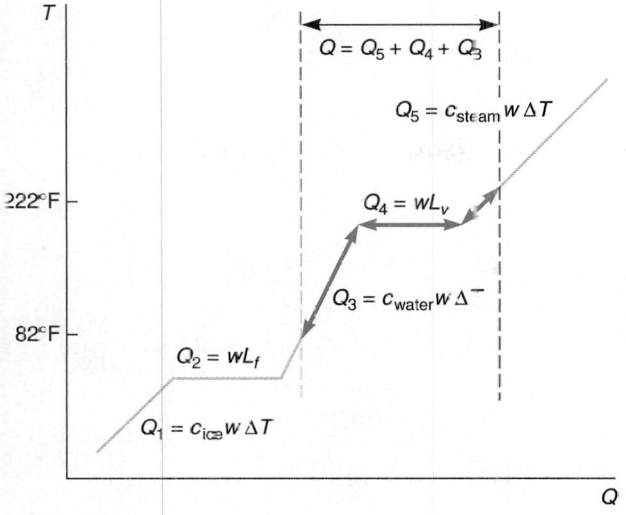

Figure 13.31

Basic Equation:

$$Q = Q_5 + Q_4 + Q_3$$

Working Equation:

$$Q = c_{\text{steam}} w \Delta T + wL_v + c_{\text{water}} w \Delta T$$

Substitution:

$$Q = \left(0.48 \frac{\text{Btu}}{\text{lb °F}}\right)(4.00 \text{ lb})(10°\text{F}) + (4.00 \text{ lb})\left(970 \frac{\text{Btu}}{\text{lb}}\right)$$
$$+ \left(1.00 \frac{\text{Btu}}{\text{lb °F}}\right)(4.00 \text{ lb})(130°\text{F})$$
$$= 4420 \text{ Btu}$$

EXAMPLE 5

How many joules of heat are needed to change 3.50 kg of ice at −15.0°C to steam at 120.0°C?

Sketch:

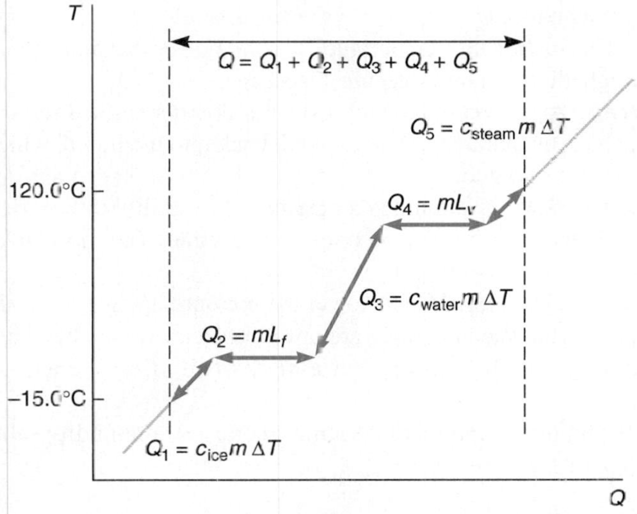

Data:

$$m = 3.50 \text{ kg}$$
$$T_i \text{ of ice} = -15.0°\text{C}$$
$$T_f \text{ of steam} = 120.0°\text{C}$$
$$Q = ?$$

Basic Equation:

$$Q = Q_1 + Q_2 + Q_3 + Q_4 + Q_5$$

Working Equation:

$$Q = c_{\text{ice}} m\Delta T + mL_f + c_{\text{water}} m\Delta T + mL_v + c_{\text{steam}} m\Delta T$$

Substitution:

$$\begin{aligned} Q &= \left(2100 \frac{\text{J}}{\text{kg °C}}\right)(3.50\ \text{kg})(15.0\text{°C}) + (3.50\ \text{kg})\left(335 \frac{\text{kJ}}{\text{kg}}\right) \\ &\quad + \left(4190 \frac{\text{J}}{\text{kg °C}}\right)(3.50\ \text{kg})(100.0\text{°C}) + (3.50\ \text{kg})\left(2.26 \frac{\text{MJ}}{\text{kg}}\right) \\ &\quad + \left(2\bar{0}00 \frac{\text{J}}{\text{kg °C}}\right)(3.50\ \text{kg})(20.0\text{°C}) \\ &= 1.080 \times 10^7\ \text{J} \quad \text{or} \quad 10.80\ \text{MJ} \end{aligned}$$

Evaporation as a Cooling Process

Evaporation is the process by which high-energy molecules of a liquid continually leave its surface. This change of a liquid to a gas helps keep your body cool. When you become too warm, your sweat glands produce water, which evaporates from your skin. As the water evaporates, your body loses heat at the rate of $\frac{Q}{m} = L_v$. In a cool summer breeze, the perspiration evaporates more rapidly, which cools you faster. On a hot, humid day, you tend to remain hot because the perspiration does not evaporate as quickly.

Years ago when a person, especially a child, had a very high body temperature, the common medical practice was to rub the body with rubbing alcohol because it quickly evaporates from the skin. As it evaporates, it removes heat from the body. This practice is not recommended now because we know the alcohol is also absorbed in the body.

During evaporation the molecules of a liquid continually leave the surface of the liquid. Some molecules have enough energy to leave, freeing themselves from the liquid's surface. The rest, not having enough energy to leave, fall back and remain as part of the liquid. The rate of evaporation of a liquid depends on the following:

1. ***Amount of surface area.*** The larger the surface area, the greater is the number of molecules that have a chance to escape from the surface.
2. ***Temperature.*** The higher the temperature, the higher is the molecular energy of the molecules, which allows more molecules to escape.
3. ***Surface currents.*** Air currents blowing over the liquid's surface remove many of the molecules that have evaporated before they fall back into the liquid, which is why a cool summer breeze "feels so good."
4. ***Volatility.*** The **volatility** of a liquid is a measure of its ability to vaporize. Examples of highly volatile liquids are rubbing alcohol and gasoline. The more volatile the liquid, the greater is its rate of evaporation.
5. ***Pressure on or above the liquid.*** The lower the pressure, the greater is the rate of evaporation. Under a partial vacuum, there are fewer molecules available with which the liquid molecules may collide, allowing for a higher rate of escape and a higher rate of evaporation.
6. ***Humidity.*** If the liquid is exposed to the atmosphere, lower humidity values will provide for greater evaporation.

Heat Pump

A **heat pump** transfers heat from a lower-temperature source to a higher-temperature source or vice versa. It is often used for heating in the winter and cooling in the summer.

A heat pump contains a vapor, usually called a *refrigerant,* that is easily condensed to a liquid when under pressure. The liquid refrigerant gives up the heat gained during compression to the higher-temperature source. The liquid is then released to a low-pressure part of the heat pump, where it quickly evaporates and takes its heat of vaporization from the lower-temperature heat source. The vapor is then compressed again, and the cycle repeats. Work is done on the vapor for the heat pump to transfer the heat from a lower-temperature source to a higher-temperature one.

In winter, a house may be heated by a heat pump that extracts heat from the outside air and transfers it into the house. In summer, the heat pump is reversed and extracts heat from the inside air and transfers it to the outdoors (Fig. 13.32).

Refrigerators and freezers are forms of heat pumps. Heat is transferred from the appliance to the room.

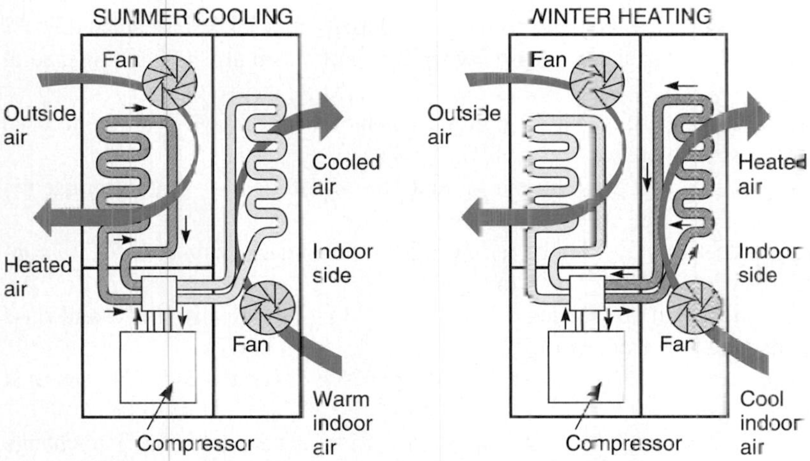

Figure 13.32 The heat pump is used to heat a home in winter and cool it in summer.

Effects of Pressure and Impurities on Change of Phase

Automobile cooling systems present important problems concerning change of phase. Most substances contract on solidifying. However, water and a few other substances expand. The tremendous force exerted by this expansion is shown by the number of cracked automobile blocks and burst radiators suffered by careless motorists every winter.

Impurities in water tend to *lower* the freezing point. Alcohol has a lower freezing point than water and is used in some types of antifreeze. By mixing antifreeze with water in the cooling system, one can lower the freezing point of the water to avoid freezing in winter. Automobile engines may also be ruined in winter by overheating if the water in the radiator is frozen, preventing the engine from being cooled by circulation in the system.

An increase in the pressure on a liquid *raises* the boiling point. Automobile manufacturers utilize this fact by pressurizing their cooling systems and thereby raising the boiling point of the coolant used.

A decrease in the pressure on a liquid *lowers* the boiling point. Frozen concentrated orange juice is produced by subjecting the pure juice to very low pressures at which the water in the juice is evaporated. Then the consumer must restore the lost water before serving the juice.

SKETCH

12 cm² | w
4.0 cm

DATA

$A = 12\text{ cm}^2$, $l = 4.0\text{ cm}$, $w = ?$

BASIC EQUATION

$A = lw$

WORKING EQUATION

$w = \frac{A}{l}$

SUBSTITUTION

$w = \frac{12\text{ cm}^2}{4.0\text{ cm}} = 3.0\text{ cm}$

PROBLEMS 13.8

1. How many calories of heat are required to melt 14.0 g of ice at 0°C?
2. How many pounds of ice at 32°F can be melted by the addition of 635 Btu of heat?
3. How many Btu of heat are required to vaporize 1_.0 lb of water at 212°F?
4. How many grams of steam in a boiler at 100°C can be condensed to water at 100°C by the removal of 1520 cal of heat?

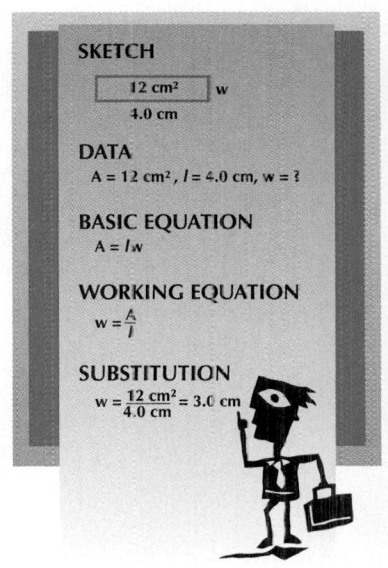

5. How many calories of heat are required to melt 320 g of ice at 0°C?
6. How many calories of heat are given off when 3250 g of steam is condensed to water at $10\bar{0}$°C?
7. How many joules of heat are required to melt 20.0 kg of ice at 0°C?
8. How many kilocalories of heat are required to melt 20.0 kg of ice at 0°C?
9. How many joules of heat need to be removed to condense 1.50 kg of steam at $10\bar{0}$°C?
10. How many litres of water at $10\bar{0}$°C are vaporized by the addition of 5.00 MJ of heat?
11. How many Btu of heat are required to melt 33.0 lb of ice at 32°F and to raise the temperature of the melted ice to 72°F?
12. How many Btu of heat are released when 20.0 lb of water at $8\bar{0}$°F is cooled to 32°F and then frozen in an ice plant?
13. How many Btu of heat are required to change 9.00 lb of ice at $1\bar{0}$°F to steam at 232°F?
14. How many calories of heat are released when $20\bar{0}$ g of steam at $12\bar{0}$°C is changed to ice at −12°C?
15. How many kilocalories of heat are required to melt 50.0 kg of ice at 0°C and to raise the temperature of the melted ice to $2\bar{0}$°C?
16. How many joules of heat are required to melt 15.0 kg of ice at 0°C and to raise the temperature of the melted ice to 75°C?
17. How many joules of heat need to be removed from 1.25 kg of steam at 115°C to condense it to water and cool the water to $5\bar{0}$°C?
18. How many kcal of heat are needed to vaporize 5.00 kg of water at $10\bar{0}$°C and raise the temperature of the steam to 145°C?
19. How many calories of heat are needed to change 625 g of ice at −24.0°C to steam at 132.0°C?
20. How many Btu of heat must be withdrawn from 5.65 lb of steam at 236.0°F to change it to ice at 12.0°F?
21. How many kilocalories of heat are needed to change 143 N of ethyl alcohol at 65.0°C to vapor?
22. How many joules of heat does 620 g of mercury require to go from a solid at −38.9°C to vapor?

Glossary

Absolute Zero The lowest possible temperature. (p. 350)

Btu (British thermal unit) The amount of heat (energy) necessary to raise the temperature of 1 lb of water 1°F. (p. 350)

Calorie The amount of heat necessary to raise the temperature of 1 g of water 1°C. (p. 350)

Celsius Scale The metric temperature scale on which ice melts at 0° and water boils at 100°. (p. 347)

Change of Phase (sometimes called *change of state*) A change in a substance from one form of matter (solid, liquid, or gas) to another. (p. 370)

Coefficient of Linear Expansion A constant that indicates the amount by which a solid expands or contracts when its temperature is changed 1 degree. (p. 365)

Condensation The change of phase from gas or vapor to a liquid. (p. 372)

Conduction A form of heat transfer from a warmer part of a substance to a cooler part as a result of molecular collisions, which cause the slower-moving molecules to move faster. (p. 352)

Convection A form of heat transfer by the movement of warm molecules from one region of a gas or a liquid to another. (p. 353)

Dew Point The temperature at which air becomes saturated with water vapor and condensation occurs. (p. 373)

Evaporation The process by which high-energy molecules of a liquid continually leave its surface. (p. 378)

Fahrenheit Scale The U.S. temperature scale on which ice melts at 32° and water boils at 212°. (p. 347)
Freezing The change of phase from liquid to solid. Also called *solidification*. (p. 370)
Fusion The change of phase from solid to liquid. Also called *melting*. (p. 370)
Heat A form of internal kinetic and potential energy contained in an object associated with the motion of its atoms or molecules and which may be transferred from an object at a higher temperature to one at a lower temperature. (p. 350)
Heat of Fusion The heat required to melt 1 g or 1 kg or 1 lb of a liquid. (p. 372)
Heat of Vaporization The amount of heat required to vaporize 1 g or 1 kg or 1 lb of a liquid. (p. 373)
Heat Pump A pump containing a vapor (refrigerant) that is easily condensed to a liquid when under pressure. Produces heat during compression and cooling during vaporization. (p. 378)
Kelvin Scale The metric absolute temperature scale on which absolute zero is 0 K and the units are the same as on the Celsius scale (p. 348)
Kilocalorie The amount of heat necessary to raise the temperature of 1 kg of water 1°C. (p. 350)
Mechanical Equivalent of Heat The relationship between heat and mechanical work. (p. 350)
Melting The change of phase from solid to liquid. Also called *fusion*. (p. 370)
Method of Mixtures When two substances at different temperatures are "mixed together," heat flows from the warmer body to the cooler body until they reach the same temperature. Part of the heat lost by the warmer body is transferred to the cooler body and to surrounding objects. If the two substances are well insulated from surrounding objects, the heat lost by the warmer body is equal to the heat gained by the cooler body. (p. 360)
Radiation A form of heat transfer through energy being radiated or transmitted in the forms of rays, waves, or particles. (p. 353)
Rankine Scale The U.S. absolute temperature scale on which absolute zero is 0° R and the degree units are the same as on the Fahrenheit scale. (p. 348)
Relative Humidity Ratio of the actual amount of vapor in the atmosphere to the amount of vapor required to reach 100% of saturation at the existing temperature. (p. 372)
Solidification The change of phase from liquid to solid. Also called *freezing*. (p. 370)
Specific Heat The amount of heat necessary to change the temperature of 1 kg of a substance by 1°C in the metric system or 1 lb of a substance by 1°F in the U.S. system. (p. 358).
Temperature A measure of the hotness or coldness of an object. (p. 346)
Thermal Conductivity The ability of a material to transfer heat by conduction. (p. 354)
Vaporization The change of phase from liquid to a gas or vapor. (p. 372)
Volatility A measure of a liquid's ability to vaporize. The more volatile the liquid, the greater is its rate of evaporation. (p. 378)

Formulas

13.1 $T_C = \frac{5}{9}(T_F - 32°)$

$T_F = \frac{9}{5}T_C + 32°$

$T_K = T_C + 273$

$T_R = T_F + 460°$

13.3 $Q = \dfrac{KAt(T_2 - T_1)}{L}$

$R = \dfrac{L}{K}$

13.4 $Q = cm\Delta T$

$Q = cw\Delta T$

13.5 $Q_{\text{lost}} = Q_{\text{gained}}$

$$c_l m_l (T_l - T_f) = c_g m_g (T_f - T_g)$$

13.6 $\Delta l = \alpha l \Delta T$

$$\Delta A = 2\alpha A \Delta T$$
$$\Delta V = 3\alpha V \Delta T$$

13.7 $\Delta V = \beta V \Delta T$

13.8 $L_f = \frac{Q}{m} \quad L_f = \frac{Q}{w}$

$$L_v = \frac{Q}{m} \quad L_v = \frac{Q}{w}$$

Review Questions

1. Which of the following are methods of heat transfer?
   (a) Convection (b) Conduction
   (c) Temperature (d) Radiation
   (e) Potential energy
2. Which of the following are good conductors of heat?
   (a) Air (b) Copper (c) Steel
   (d) Aluminum (e) Brick (f) Mineral wool
3. The amount that a solid expands when heated depends on
   (a) the type of material. (b) the length of the solid.
   (c) the density of the solid. (d) the amount of temperature change.
   (e) all of the above.
4. The rate of evaporation from the surface of a liquid depends on
   (a) the temperature. (b) the volatility of the liquid.
   (c) the mass density of the liquid. (d) the air pressure above the liquid.
5. The amount of heat required to melt 1 kg of a solid is called its
   (a) heat of vaporization. (b) mass density.
   (c) weight density. (d) heat of fusion.
   (e) volume expansion.
6. The operation of a simple thermostat depends on
   (a) the mechanical equivalent of heat. (b) specific heat.
   (c) thermal expansion. (d) the *R* value.
7. In your own words, describe the method of mixtures.
8. What is the mechanical equivalent of heat in the U.S. system?
9. Which other temperature scale is closely related to the Fahrenheit scale?
10. Which other temperature scale is closely related to the Celsius scale?
11. Distinguish between the Celsius and Fahrenheit temperature scales.
12. Distinguish between heat and temperature.
13. Give three examples of the conversion of heat into useful work.
14. Give three examples of the conversion of work into heat.
15. Should you wear light- or dark-colored clothing on a hot sunny summer day? Explain.
16. Does the area of a hole cut out of a metal block increase or decrease as the metal is heated? Explain.
17. (a) At what temperature does water have its highest density? (b) How does water differ from other liquids in this regard? (c) What is the impact of this unique characteristic of water?
18. Which would cool a hot object better: 10 kg of water at 0°C or 10 kg of ice at 0°C? Explain.
19. Steam can cause much more severe burns than hot water. Explain.
20. Why are ice cubes often observed to have a slight mound on the top of the cube?

21. In your own words, describe each method of heat transfer.
22. Describe why automotive cooling systems are designed to operate at elevated pressures.
23. Explain how a heat pump works to heat in the winter and cool in the summer.

Review Problems

SKETCH

12 cm² w
4.0 cm

DATA

$A = 12\text{ cm}^2$, $l = 4.0\text{ cm}$, $w = ?$

BASIC EQUATION

$A = lw$

WORKING EQUATION

$w = \frac{A}{l}$

SUBSTITUTION

$w = \frac{12\text{ cm}^2}{4.0\text{ cm}} = 3.0\text{ cm}$

1. Change 344 K to degrees Celsius.
2. Change 24°C to Kelvin.
3. Change 5110°C to degrees Fahrenheit.
4. Change 635°F to degrees Celsius.
5. Find the amount of heat in cal generated by 43.0 J of work.
6. Find the amount of heat in kcal generated by 6530 J of work.
7. Find the amount of work equivalent to 435 Btu.
8. Find the heat flow during 4.10 h through a glass window of thickness 0.15 in. with an area of 33 ft^2 if the average outer surface temperature is 22°F and the average inner glass surface is 48°F.
9. Find the heat flow in 25.0 days through a freezer door 80.0 cm × 144 cm insulated with cellulose fiber 4.0 cm thick. The temperature inside the freezer is −14°C. Room temperature is 22°C.
10. How many Btu of heat must be added to 835 lb of steel to raise its temperature from 20.0°F to 455°F?
11. How many kcal of heat must be added to 148 kg of aluminum to raise its temperature from 21.5°C to 485°C?
12. A 161-kg steam boiler is made of steel and contains 1127 N of water at 8.9°C. How much heat is required to raise the temperature of both the boiler and the water to 10$\bar{0}$°C?
13. A 3.80-lb piece of copper is dropped into 3.35 lb of water at 48.0°F. The final temperature is 98.2°F. What was the initial temperature of the copper?
14. A 355-g piece of metal at 48.0°C is dropped into 111 g of water at 15.0°C. If the final temperature of the mixture is 32.5°C, what is the specific heat of the metal?
15. A brass rod 45.2 cm long expands 0.734 mm when heated. Find the temperature change.
16. The length of a steel rod at 5°C is 12.500 m. What is its length when heated to 154°C?
17. The diameter of a hole drilled through aluminum at 22°C is 7.50 mm. Find the diameter and the area of the hole at 89°C.
18. A steel ball has a radius of 1.54 cm at 35°C. Find its change in volume when the temperature is increased to 84.5°C.
19. Find the increase in volume of 44.8 L of acetone when it is heated from 37.0°C to 75.5°C.
20. What is the decrease in volume of 3450 ft^3 of alcohol in a railroad tank car if the temperature drops from 87.0°F to 33.0°F?
21. How many kcal of heat are required to vaporize 21.5 kg of water at 10$\bar{0}$°C?
22. How many Btu of heat are required to melt 8.35 lb of ice at 32°F?
23. How many kcal of heat must be withdrawn from 4.56 kg of steam at 125°C to change it to ice at −44.5°C?
24. How many joules of heat are required to change 336 g of ethyl alcohol from a solid at −117°C to a vapor at 78.5°C?

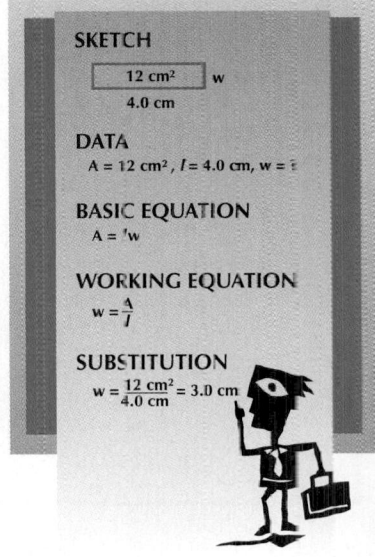

APPLIED CONCEPTS

1. A polystyrene foam cover prevents an ice–water mixture from absorbing heat from 25.3°C air outside a cooler. If the cover is 54.5 cm × 37.8 cm in cross section and 5.25 cm thick, how much heat will transfer through the foam in 60.0 min?
2. Every winter a local recreation department fills a parking lot with water so that the water will freeze and people can ice skate on it. Although the weather does not get cold enough for the water to freeze, the water temperature drops from 4.30°C to 0.00°C. If the water is 12.7 cm deep and the parking lot is 24.5 m × 33.8 m, how much heat energy does the water release into the air?
3. (a) What is the heat of fusion for the water in Problem 2 to freeze into a solid block of ice? (b) What is the total energy needed for the water at 4.30°C to change to ice at 0.00°C? (c) How much additional heat is removed from the ice to lower its temperature to −3.43°C?
4. Pedro, a contractor, is trying to choose between purchasing a steel or an aluminum 50-ft tape measure. (a) If he wants the tape measure that expands and contracts less, which tape measure should he purchase? Consider that the tape measures were calibrated at 68.0°F in the factory. (b) How much can he expect his new tape measure to contract if the temperature outside is 18.5°F?
5. In anticipation of winter snowstorms, Jamal fills his 2.50-gal gas can at the local gas station. (a) If the temperature is 65.3°F on the day he fills the gas can, what volume of gas will Jamal have when the temperature drops to 10.5°F? (b) If the gas cost $1.97/gal, how much money does Jamal lose?

CHAPTER 14

PROPERTIES OF GASES

Properties of gases are related to the temperature and the pressure under which the gas is contained. Gas laws concern the behavior of "ideal" gases at standard conditions of temperature and pressure. The focus of this chapter is on the behavior of gases as temperature and pressure are varied.

Objectives

The major goals of this chapter are to enable you to:

1. Use Charles' law to determine thermal expansion.
2. Apply Boyle's law to calculate volume changes of gases.
3. Relate gas density to pressure and temperature.

14.1 Charles' Law

Before making a long summer trip, you notice that the tires are low. You stop at a gas station around the corner and add air to 28 lb/in^2 gauge pressure. Later in the afternoon you stop for gas. Since your tires were low that morning, you decide to check them again. Now you notice that they look a bit larger, and the gauge pressure is 40 lb/in^2. What happened? When a gas is heated, the increased kinetic energy causes the volume to increase, the pressure to increase, or both to increase.

To study these concepts we will use the idea of an "ideal gas." When the density of a gas is sufficiently low, the pressure, volume, and temperature of the gas tend to be related in a rather simple way. These relationships are fairly accurate over a broad range of temperatures and pressures in real gases. They also apply over a broad range of different gases. We will therefore consider the wide variety of gases and their behaviors by using a model which we will call an ideal gas. Charles' and Boyle's laws mathematically describe these relationships.

CHARLES' LAW

If the pressure on a gas is constant, the volume is directly proportional* to its absolute (Kelvin or Rankine) temperature.

Jacques Charles (1746–1823), *physicist, was born in France. He became famous for making the first manned ascent by hydrogen balloon in 1783. His interest in gases led him to formulate Charles' law.*

By formula,

$$\frac{V}{T} = \frac{V'}{T'} \quad \text{or} \quad VT' = V'T$$

where

V = original volume
T = original temperature
V' = final volume
T' = final temperature

EXAMPLE 1

A gas occupies $45\bar{0}$ cm^3 at $3\bar{0}$°C. At what temperature will the gas occupy $48\bar{0}$ cm^3?

Data:

$V = 45\bar{0}\ \text{cm}^3$
$T = 3\bar{0}° + 273 = 303\ \text{K}$ (***Note:*** We must use Kelvin temperature.)
$V' = 48\bar{0}\ \text{cm}^3$
$T' = ?$

Basic Equation:

$$\frac{V}{T} = \frac{V'}{T'}$$

Working Equation:

$$T' = \frac{TV'}{V}$$

Substitution:

$$T' = \frac{(303\ \text{K})(48\bar{0}\ \cancel{\text{cm}^3})}{45\bar{0}\ \cancel{\text{cm}^3}}$$

$$= 323\ \text{K} \quad \text{or} \quad 5\bar{0}°\text{C}$$

---

*Directly proportional means that as temperature increases, volume increases, and as temperature decreases, volume decreases.

EXAMPLE 2

At $4\bar{0}$°F, some helium occupies 15.0 ft³. What will be its volume at $9\bar{0}$°F?

Data:

$$V = 15.0 \text{ ft}^3$$
$$T = 4\bar{0}° + 46\bar{0}° = 50\bar{0}°\text{R} \qquad (\textit{Note:} \text{ We must use Rankine temperature.})$$
$$T' = 9\bar{0}° + 46\bar{0}° = 55\bar{0}°\text{R}$$
$$V' = ?$$

Basic Equation:

$$\frac{V}{T} = \frac{V'}{T'}$$

Working Equation:

$$V' = \frac{VT'}{T}$$

Substitution:

$$V' = \frac{(15.0 \text{ ft}^3)(55\bar{0}°\cancel{\text{R}})}{50\bar{0}°\cancel{\text{R}}}$$
$$= 16.5 \text{ ft}^3$$

TRY THIS ACTIVITY

Mylar® Balloons

Take a room-temperature, helium-filled Mylar® balloon and place it in a significantly colder environment, such as outside in the cold winter or in a freezer. As the temperature of the helium decreases, observe what happens to the volume of the balloon. After it has been outside several minutes, bring the balloon back into the room-temperature environment. Observe the change that occurs. Then, place the balloon in a warmer area such as near a hot radiator. Why does the volume of the balloon change in different temperature environments?

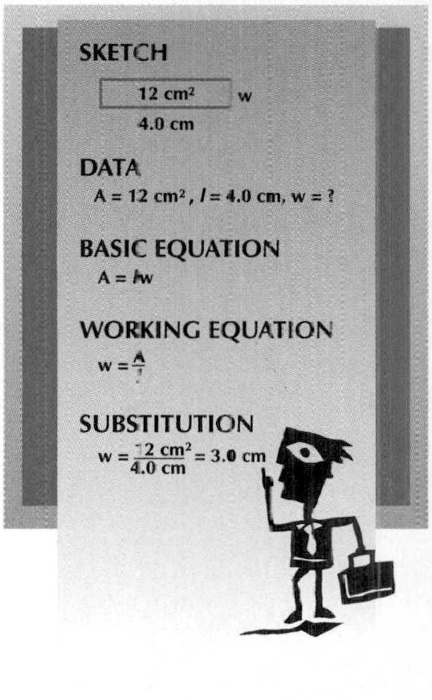

PROBLEMS 14.1

1. Change 15°C to K.
2. Change −14°C to K.
3. Change 317 K to °C
4. Change 235 K to °C.
5. Change 72°F to °R.
6. Change −55°F to °R.
7. Change $55\bar{0}$°R to °F.
8. Change 375°R to °F.

Use $\frac{V}{T} = \frac{V'}{T'}$ to find each quantity:

9. $T = 315$ K, $V' = 225$ cm³, $T' = 275$ K, find V.
10. $T = 615$°R, $V = 60.3$ in³, $T' = 455$°R, find V'.
11. $V = 2\bar{0}0$ ft³, $T = 95$°F, $V' = 25\bar{0}$ ft³, find T'
12. $V = 19.7$ L, $T = 51$°C, $V' = 25.2$ L, find T'.
13. Some gas occupies a volume of 325 m³ at 41°C. What is its volume at 94°C?
14. Some oxygen occupies 275 in³ at 35°F. Find its volume at 95°F.
15. Some methane occupies 1575 L at 45°C. Find its volume at 15°C.
16. Some helium occupies $12\bar{0}0$ ft³ at 70°F. At what temperature will its volume be $6\bar{0}0$ ft³?

17. Some nitrogen occupies 14,300 cm^3 at 25.6°C. What is the temperature when its volume is 10,250 cm^3?
18. Some propane occupies 1270 cm^3 at 18.0°C. What is the temperature when its volume is 1530 cm^3?
19. Some carbon dioxide occupies 34.5 L at 49.0°C. Find its volume at 12.0°C.
20. Some oxygen occupies 28.7 ft^3 at 11.0°F. Find its temperature when its volume is 18.5 ft^3.
21. A balloon contains 26.0 L of hydrogen at 40.0°F. What is the Kelvin temperature change needed to make the balloon expand to 36.0 L?
22. Using Charles' law, determine the effect
    (a) on the temperature of a gas when the volume is doubled.
    (b) on the temperature of a gas when the volume is tripled.
    (c) on the volume when temperature is doubled.
    (d) Explain the relationship between the volume and the temperature.
23. If 38.0 L of hydrogen is heated to $11\bar{0}$°C and expands to 90.0 L, what was the original temperature?
24. A tank contains 3.00 L of acetylene at 4.00°C. Find the volume at 12.0°C.
25. A hot air balloon contains 147 m^3 of air at 19.0°C. What is the volume of air at 32.0°C?
26. A tank with 139 L of propane is cooled from 91.0°C to 37.0°C. Find the original volume.

14.2 Boyle's Law

BOYLE'S LAW

If the temperature of a gas is constant, the volume is inversely proportional* to the absolute pressure.

Robert Boyle (1627–1691),

chemist, was born in Ireland. Working with Robert Hooke as his assistant, he performed experiments on air, vacuums, combustion, and respiration. He developed Boyle's law in 1662. He also did research on properties of acids and alkalis, specific gravity, crystallography, and refraction.

By formula,

$$\frac{V}{V'} = \frac{P'}{P} \quad \text{or} \quad VP = V'P'$$

where

V = original volume
V' = final volume
P = original pressure
P' = final pressure

Note: The pressure must be expressed in terms of *absolute pressure*.

EXAMPLE 1

Some oxygen occupies $50\bar{0}$ in^3 at an absolute pressure of 40.0 lb/in^2 (psi). What is its volume at an absolute pressure of $10\bar{0}$ psi?

Data:

$V = 50\bar{0}\ in^3$
$P = 40.0$ psi
$P' = 10\bar{0}$ psi
$V' = ?$

*Inversely proportional means that as volume increases, pressure decreases, and as volume decreases, pressure increases.

Basic Equation:

$$\frac{V}{V'} = \frac{P'}{P}$$

Working Equation:

$$V' = \frac{VP}{P'}$$

Substitution:

$$V' = \frac{(50\bar{0}\ \text{in}^3)(40.0\ \text{psi})}{10\bar{0}\ \text{psi}}$$
$$= 20\bar{0}\ \text{in}^3$$

EXAMPLE 2

Some nitrogen occupies 20.0 m^3 at a gauge pressure of 274 kPa. Find the absolute pressure when its volume is 30.0 m^3.

Data:

$$V = 20.0\ \text{m}^3$$
$$P_{\text{abs}} = 274\ \text{kPa} + 101\ \text{kPa} = 375\ \text{kPa}$$
$$V' = 30.0\ \text{m}^3$$
$$P' = ?$$

Basic Equation:

$$\frac{V}{V'} = \frac{P'}{P}$$

Working Equation:

$$P' = \frac{VP}{V'}$$

Substitution:

$$P' = \frac{(20.0\ \text{m}^3)(375\ \text{kPa})}{30.0\ \text{m}^3}$$
$$= 25\bar{0}\ \text{kPa}$$

Density and Pressure

If the pressure of a given amount (constant volume) of gas is increased, its density increases as the gas molecules are forced closer together. (Recall that density is discussed in Section 11.5.) Also, if the pressure is decreased, the density decreases. That is, the *density of a gas is directly proportional to its pressure* as long as there is no change in state. In equation form,

$$\frac{D}{D'} = \frac{P}{P'} \quad \text{or} \quad DP' = D'P$$

where

D = original density
D' = final density
P = original pressure (absolute)
P' = final pressure (absolute)

EXAMPLE 3

Some carbon dioxide has a density of 1.60 kg/m^3 at an absolute pressure of 95.0 kPa. What is the density when the pressure is decreased to 80.0 kPa?

Data:

$$D = 1.60 \text{ kg/m}^3$$
$$P = 95.0 \text{ kPa}$$
$$P' = 80.0 \text{ kPa}$$
$$D' = ?$$

Basic Equation:

$$\frac{D}{D'} = \frac{P}{P'}$$

Working Equation:

$$D' = \frac{DP'}{P}$$

Substitution:

$$D' = \frac{(1.60 \text{ kg/m}^3)(80.0 \cancel{\text{kPa}})}{95.0 \cancel{\text{kPa}}}$$
$$= 1.35 \text{ kg/m}^3$$

EXAMPLE 4

A gas has a density of 2.00 kg/m^3 at a gauge pressure of $16\bar{0}$ kPa. What is the density at a gauge pressure of $30\bar{0}$ kPa?

Data:

$$D = 2.00 \text{ kg/m}^3$$
$$P = 16\bar{0} \text{ kPa} + 101 \text{ kPa} = 261 \text{ kPa}$$
$$P' = 30\bar{0} \text{ kPa} + 101 \text{ kPa} = 401 \text{ kPa}$$
$$D' = ?$$

Basic Equation:

$$\frac{D}{D'} = \frac{P}{P'}$$

Working Equation:

$$D' = \frac{DP'}{P}$$

Substitution:

$$D' = \frac{(2.00 \text{ kg/m}^3)(401 \cancel{\text{kPa}})}{261 \cancel{\text{kPa}}}$$
$$= 3.07 \text{ kg/m}^3$$

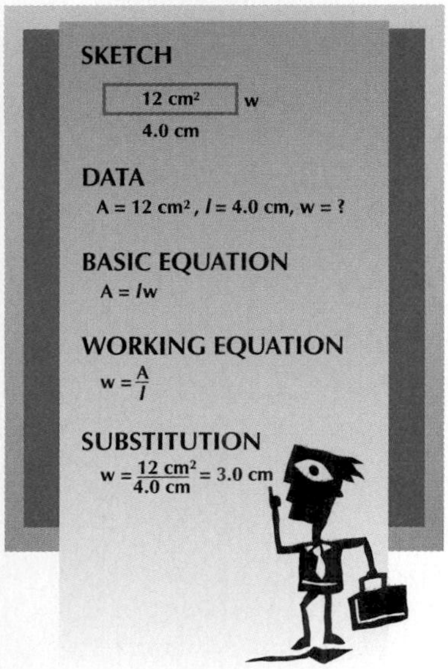

PROBLEMS 14.2

Use $\frac{V}{V'} = \frac{P'}{P}$ or $\frac{D}{D'} = \frac{P}{P'}$ to find each quantity. (All pressures are absolute unless otherwise stated.)

1. $V' = 315 \text{ cm}^3$, $P = 101$ kPa, $P' = 85.0$ kPa; find V.
2. $V = 45\bar{0}$ L, $V' = 70\bar{0}$ L, $P = 75\bar{0}$ kPa; find P'.
3. $V = 76.0 \text{ m}^3$, $V' = 139 \text{ m}^3$, $P' = 41.0$ kPa; find P.

4. $V = 439 \text{ in}^3$, $P' = 38.7$ psi, $P = 47.1$ psi; find V'.
5. $D = 1.80 \text{ kg/m}^3$, $P = 108$ kPa, $P' = 125$ kPa; find D'.
6. $D = 1.65 \text{ kg/m}^3$, $P = 87.0$ kPa, $D' = 1.85 \text{ kg/m}^3$; find P'.
7. $P = 51.0$ psi, $P' = 65.3$ psi, $D' = 0.231 \text{ lb/ft}^3$; find D.
8. Some air at 22.5 psi occupies $14\bar{0}0 \text{ in}^3$. What is its volume at 18.0 psi?
9. Some nitrogen at a pressure of 110.0 kPa occupies 185 m^3. Find its pressure if its volume is changed to 225 m^3.
10. Some methane at 185.0 kPa occupies 65.0 L. What is its volume at a pressure of 95.0 kPa?
11. Some carbon dioxide has a density of 3.75 kg/m^3 at 815 kPa. What is its density if the pressure is decreased to 725 kPa?
12. Some oxygen has a density of 1.75 kg/m^3 at normal atmospheric pressure. What is its pressure (in kPa) when the density is changed to 1.45 kg/m^3?
13. Some methane at $50\bar{0}$ kPa gauge pressure occupies $75\bar{0} \text{ m}^3$. What is its gauge pressure if its volume is $50\bar{0} \text{ m}^3$?
14. Some helium at 15.0 psi gauge pressure occupies 20.0 ft^3. Find its volume at 20.0 psi gauge pressure.
15. Some nitrogen at 80.0 psi gauge pressure occupies 13.0 ft^3. Find its volume at 50.0 psi gauge pressure.
16. Some carbon dioxide has a density of 6.35 kg/m^3 at 685 kPa gauge pressure. What is the density when the gauge pressure is 455 kPa?
17. Some propane has a density of 48.5 oz/ft^3 at 265 psi gauge pressure. What is the gauge pressure when the density is 30.6 oz/ft^3?
18. Some air occupies 4.5 m^3 at a gauge pressure of 46 kPa. What is the volume at a gauge pressure of 13 kPa?
19. Some oxygen at 87.6 psi (absolute) occupies 75.0 in^3. Find its volume if its absolute pressure is (a) doubled, (b) tripled, (c) halved.
20. A gas at $30\bar{0}$ kPa (absolute) occupies 40.0 m^3. Find its absolute pressure if its volume is (a) doubled, (b) tripled, (c) halved.
21. A volume of 58.0 L of hydrogen is heated from 33°C to 68°C. If its original density is 4.85 kg/m^3 and its original absolute pressure is $12\bar{0}$ kPa, what is the resulting density?
22. Some argon gas is in a 42.0-L container at a pressure of $32\bar{0}$ kPa. What is the pressure if it is transferred to a container with a volume of 51.0 L?
23. A 2.00-L plastic bottle contains air at a pressure of 33.0 kPa. A person squeezes the bottle resulting in a pressure of 57.0 kPa. What is the new volume of the container?
24. The 3.25-cm^3 volume of air remaining in a water balloon is reduced to 2.75 cm^3 upon impact when the air pressure is 48.3 kPa. (a) Is the original pressure more or less than that upon impact? (b) Find the original pressure.
25. A mass of 1.31 kg of neon is in a 3.00-m^3 container at a pressure of 121 kPa. When the pressure is reduced to 97.4 kPa, what is the final density?
26. The air density in a tractor tire is 1.40 kg/m^3 at a pressure of 314 kPa. (a) As the air pressure increases to $7\bar{0}0$ kPa, does the final density of the air increase or decrease? (b) What is the resulting density of the air?
27. An unknown gas is in a tank at 13.3 kPa. (a) If the density of the gas changes from 1.45 kg/m^3 to 1.35 kg/m^3, will the resulting pressure be higher or lower than the original pressure? (b) Find the resulting pressure.

14.3 Charles' and Boyle's Laws Combined

Most of the time it is very difficult to keep the pressure constant or the temperature constant. In this case we combine Charles' law and Boyle's law as follows:

$$\frac{VP}{T} = \frac{V'P'}{T'} \quad \text{or} \quad VPT' = V'P'T$$

Note: Both pressure and temperature must be absolute.

EXAMPLE

We have 5.00 m^3 of acetylene at 4.00°C at $16\bar{0}$ kPa (absolute). What is the pressure if its volume is changed to 4.00 m^3 at 30.0°C?

Data:

$$V = 5.00 \text{ m}^3$$
$$P = 16\bar{0} \text{ kPa}$$
$$T = 4.00° + 273° = 277 \text{ K}$$
$$V' = 4.00 \text{ m}^3$$
$$T' = 30.0° + 273° = 303 \text{ K}$$
$$P' = ?$$

Basic Equation:

$$\frac{VP}{T} = \frac{V'P'}{T'}$$

Working Equation:

$$P' = \frac{VPT'}{TV'}$$

Substitution:

$$P' = \frac{(5.00 \cancel{\text{m}^3})(16\bar{0} \text{ kPa})(303 \cancel{\text{K}})}{(277 \cancel{\text{K}})(4.00 \cancel{\text{m}^3})}$$
$$= 219 \text{ kPa}$$

The gas laws are reasonably accurate except at very low temperatures and under extreme pressures. A commonly used reference in gas laws is called **standard temperature and pressure** (STP). Standard temperature is the freezing point of water, 0°C or 32°F. Standard pressure is equivalent to atmospheric pressure, 101.32 kPa or 14.7 lb/in^2.

Vapor Pressure and Humidity

When a liquid evaporates, molecules of the liquid pass from its surface into the air above the liquid. This increase in the number of molecules in the air causes an increase in the pressure above the liquid. This increase in pressure is called *vapor pressure*.

Water boiling is an example of the change of phase of a liquid to a gas. Bubbles of gas form below the surface of the water, rise to the surface, and escape into the air. Recall that boiling water remains at a constant 100°C temperature at standard pressure. Water can, however, have its temperature raised above 100°C and be kept from boiling in a pressure cooker. This pot has a heavy, tight-fitting lid with a steam escape valve. As the water in the pot begins to boil, the pressure above the water builds up and interferes with further boiling. This way, the boiling temperature of the water can be raised above 100°C, allowing for fast cooking of a roast or other foods. Likewise, at a higher altitude, water boils at a lower temperature. In the mile-high city of Denver, water boils at 95°C, making for longer cooking times. Many premixed baking foods provide alternate baking times for high altitudes.

Evaporation of water molecules into the air from a lake followed by cooling as the air passes over cooler land may create a fog. Fog is basically a cloud that forms near the ground. The returning of some water vapor molecules in the air to a liquid is an example of a process called *condensation*. Condensation occurs at a point in temperature called the *dew point*. Dew forms when moist air is cooled by the earth's surface. It can be closely reproduced in the laboratory by taking a glass container partly filled with water and adding pieces of ice. When the water and glass reach the dew point in temperature, water will condense from the air on the outside of the glass. The condensation on a can of cold soda is a familiar example, in which water vapor in the air is chilled upon contact with the cold can. The water molecules slow when cooled and change phase to liquid.

Condensation of steam can be very dangerous. Since steam gives up considerable energy when it changes from vapor to liquid (540 cal/g °C), getting burned by steam can be much worse than getting burned by water at the same temperature. This same concept explains why a steam heating system produces much more heat than a hot water system.

The maximum amount of water vapor that air will hold at a given temperature is *absolute humidity*. Any increase becomes rainfall. *Relative humidity*, in contrast, is the ratio of the amount of water vapor a sample of air holds to the maximum amount it can hold if it is saturated.

Another example of the effect of condensation on people can be seen by comparing the "dry heat" in Arizona with the humid air at the same temperature on the coast of the Gulf of Mexico. In Phoenix, where the humidity is very low, evaporation is substantially greater than condensation. In New Orleans, however, where the humidity is much higher, condensation is much greater and offsets evaporation so people are more uncomfortable as condensation limits the cooling effect of the evaporation of a person's perspiration. For most people, a relative humidity of 50% to 60% at about 20°C is comfortable. When the relative humidity is higher, the air feels heavy and uncomfortable.

PHYSICS CONNECTIONS

Decompression Sickness: "The Bends"

When a scuba diver dives in deep water, he or she experiences a greater water pressure than when closer to the surface. According to Boyle's law, when the pressure is increased on an object, its volume decreases. In the case of a scuba diver, the volume of air inside the diver's lungs decreases as well. In order for the diver to breathe correctly, the air in the diver's tank must be highly pressurized and comprise 10% to 20% oxygen and 80% to 90% nitrogen.

Under normal atmospheric pressure, nitrogen does not dissolve in blood. Under pressure, nitrogen's volume decreases, thereby increasing its solubility in blood. As the diver swims deeper, the pressure in the lungs increases, and the blood accepts the smaller nitrogen bubbles. If the scuba diver returns too quickly to the surface, he or she experiences a quickly reduced pressure on his or her body, which results in enlarged nitrogen bubbles in the blood stream. The presence of nitrogen bubbles in the blood can cause generalized barotrauma, otherwise known as decompression sickness or "the bends."

Common symptoms of "the bends" include joint pain, dizziness, and numbness. The only way to prevent "the bends" is to slowly rise to the surface. This gradually reduces the pressure on the diver and the nitrogen and gives the nitrogen bubbles time to dissolve and prevents them from forming large bubbles in the blood.

Figure 14.1 A hyperbaric chamber

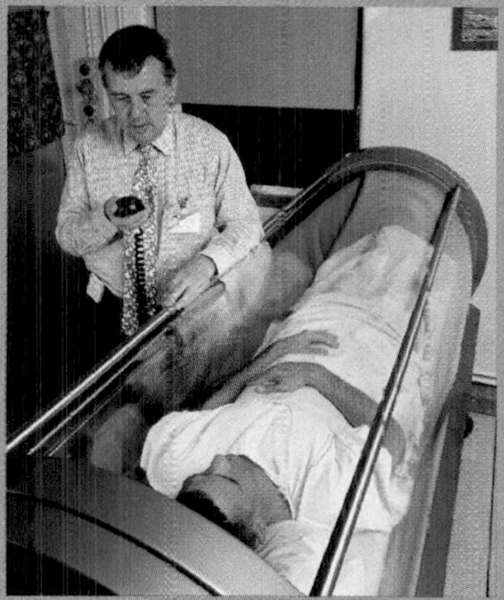

In the mid-1800s, workers on the Brooklyn Bridge were the first individuals to be documented with "the bends." They worked under the Hudson River in high-pressure compartments where they would enter and exit several times per day without slowly decompressing. Symptoms were documented and the cause of this extremely painful condition was eventually discovered. Today, someone suffering from "the bends" is taken to decompression chambers, called hyperbaric chambers, where the patient breathes highly pressurized oxygen that is gradually decompressed to normal atmospheric pressure (Fig. 14.1).

SKETCH

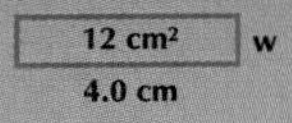

DATA

$A = 12\text{ cm}^2$, $l = 4.0$ cm, $w = ?$

BASIC EQUATION

$A = lw$

WORKING EQUATION

$w = \frac{A}{l}$

SUBSTITUTION

$w = \frac{12\text{ cm}^2}{4.0\text{ cm}} = 3.0$ cm

PROBLEMS 14.3

Use $\frac{VP}{T} = \frac{V'P'}{T'}$ to find each quantity. (All pressures are absolute unless otherwise stated.)

1. $P = 825$ psi, $T = 575°\text{R}$, $V' = 1550\text{ in}^3$, $P' = 615$ psi, $T' = 525°\text{R}$; find V.
2. $V = 50\bar{0}\text{ in}^3$, $T = 50\bar{0}°\text{R}$, $V' = 80\bar{0}\text{ in}^3$, $P' = 80\bar{0}$ psi, $T' = 45\bar{0}°\text{R}$; find P.
3. $V = 90\bar{0}\text{ m}^3$, $P = 105$ kPa, $T = 30\bar{0}$ K, $P' = 165$ kPa, $T' = 265$ K; find V'.
4. $V = 18.0\text{ m}^3$, $P = 112$ kPa, $V' = 15.0\text{ m}^3$, $P' = 135$ kPa, $T' = 235$ K; find T.
5. $V = 532\text{ m}^3$, $P = 135$ kPa, $T = 87°\text{C}$, $V' = 379\text{ m}^3$, $P' = 123$ kPa; find T'.
6. We have $60\bar{0}\text{ in}^3$ of oxygen at $150\bar{0}$ psi at 65°F. What is the volume at $120\bar{0}$ psi at $9\bar{0}$°F?
7. We have $80\bar{0}\text{ m}^3$ of natural gas at 235 kPa at $3\bar{0}$°C. What is the temperature if the volume is changed to $120\bar{0}\text{ m}^3$ at 215 kPa?
8. We have $140\bar{0}$ L of nitrogen at 135 kPa at 54°C. What is the temperature if the volume is changed to $80\bar{0}$ L at 275 kPa?
9. An acetylene welding tank has a pressure of $20\bar{0}0$ psi at $4\bar{0}$°F. If the temperature rises to $9\bar{0}$°F, what is the new pressure?
10. What is the new pressure in Problem 9 if the temperature falls to $-3\bar{0}$°F?
11. An ideal gas occupies a volume of 5.00 L at STP. What is its gauge pressure (in kPa) if the volume is halved and its temperature increases to $40\bar{0}$°C?
12. An ideal gas occupies a volume of 5.00 L at STP.
    (a) What is its temperature if its volume is halved and its absolute pressure is doubled?
    (b) What is its temperature if its volume is doubled and its absolute pressure is tripled?
13. Some propane occupies 2.00 m^3 at 18.0°C at an absolute pressure of $3.50 \times 10^5\text{ N/m}^2$. (a) Find the absolute pressure (in kPa) at the same temperature when the volume is halved. (b) Find the new temperature when the absolute pressure is doubled and the volume is doubled. (c) Find the new volume when the absolute pressure is halved and the temperature is decreased to −12.0°C. (d) Find the new volume if the absolute pressure is $1.30 \times 10^6\text{ N/m}^2$ and the temperature is 31.0°C.
14. A balloon with volume $32\bar{0}0$ mL of xenon gas is at a gauge pressure of 122 kPa and a temperature of 27°C. What is the volume when the balloon is heated to 65°C and the gauge pressure is decreased to 112 kPa?
15. A 7.85-L helium-filled balloon experiences a change in both pressure from normal atmospheric pressure to 60.5 kPa and temperature from 24.0°C to 6.00°C. What is the resulting volume?

Glossary

Boyle's Law If the temperature of a gas is constant, the volume is inversely proportional to the absolute pressure, $V/V' = P'/P$. (p. 390)

Charles' Law If the pressure on a gas is constant, the volume is directly proportional to its Kelvin or Rankine temperature, $V/T = V'/T'$. (p. 388)

Standard Temperature and Pressure (STP) A commonly used reference in gas laws. Standard temperature is the freezing point of water. Standard pressure is equivalent to atmospheric pressure. (p. 394)

Formulas

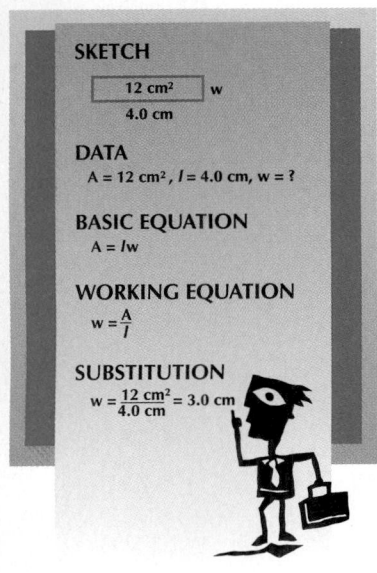

14.1 Charles' Law: $\frac{V}{T} = \frac{V'}{T'}$

14.2 Boyle's law: $\dfrac{V}{V'} = \dfrac{P'}{P}$

$$\frac{D}{D'} = \frac{P}{P'}$$

14.3 $\dfrac{VP}{T} = \dfrac{V'P'}{T'}$

Review Questions

1. The gas law that relates volume and temperature is called
   (a) Boyle's law. (b) Hooke's law.
   (c) Charles' law. (d) none of the above.
2. The gas law that relates volume and pressure is called
   (a) Boyle's law. (b) Hooke's law.
   (c) Charles' law. (d) none of the above.
3. If the temperature of a gas is constant and the volume is decreased, the pressure will
   (a) stay the same. (b) decrease.
   (c) increase. (d) increase or decrease, depending on the gas.
4. If the temperature of a gas is constant and the pressure is decreased, the volume will
   (a) stay the same. (b) decrease.
   (c) increase. (d) increase or decrease, depending on the gas.
5. If the pressure on a gas is constant and the temperature is decreased, the volume will
   (a) stay the same. (b) decrease.
   (c) increase. (d) increase or decrease, depending on the gas.
6. If the pressure on a gas is constant and the volume is decreased, the temperature will
   (a) stay the same. (b) decrease.
   (c) increase. (d) increase or decrease, depending on the amount of gas.
7. Describe the conditions of standard temperature and pressure.
8. Describe what happens to the volume of a gas if its temperature and pressure increase.
9. Describe what happens to the temperature of a gas if its volume and pressure increase.
10. What causes the tendency of the volume and pressure of a gas to increase when it is heated?
11. What causes the tendency of the temperature of a gas to increase when it is compressed?
12. What causes the tendency of the pressure of a gas to decrease when the volume is increased?

Review Problems

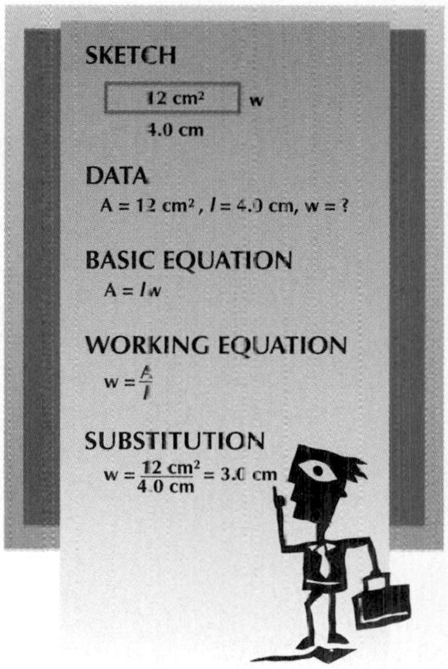

All pressures are absolute unless otherwise stated.

1. A gas occupies 13.5 ft^3 at 35.8°F. What will the volume of this gas be at 88.6°F if the pressure is constant?
2. A gas occupies 3.45 m^3 at 18.5°C. What will the volume of this gas be at 98.5°C if the pressure is constant?
3. Some hydrogen occupies 115 ft^3 at 54.5°F. What is the temperature when the volume is 132 ft^3 if the pressure is constant?
4. Some carbon dioxide occupies 45.3 L at 38.5°C. What is the temperature when the volume is 44.2 L if the pressure is constant?
5. Some propane occupies 145 cm^3 at 12.4°C. What is the temperature when the volume is 156 cm^3 if the pressure is constant?
6. Some air at 276 kPa occupies 32.4 m^3. What is its absolute pressure if its volume is doubled at constant temperature?
7. Some helium at 17.5 psi gauge pressure occupies 35.0 ft^3. What is the volume at 32.4 psi if the temperature is constant?

8. Some carbon dioxide has a density of 6.35 kg/m^3 at 685 kPa gauge pressure. What is the density when the gauge pressure is 355 kPa if the temperature is constant?
9. We have 435 in^3 of nitrogen at 1340 psi gauge pressure at 75°F. What is the volume at 1150 psi gauge pressure at 45°F?
10. We have 755 m^3 of carbon dioxide at 344 kPa at 25°C. Find the temperature if the volume is changed to 1330 m^3 at 197 kPa.
11. A welding tank has a gas pressure of 1950 psi at 38°F. (a) What is the new pressure if the temperature rises to 98°F? (b) What is the temperature if the gas pressure falls to 1870 psi?
12. An ideal gas occupies a volume of 4.50 L at STP. What is its gauge pressure (in kPa) if the volume is halved and the absolute temperature is doubled?
13. An ideal gas occupies a volume of 5.35 L at STP. What is its gauge pressure (in kPa) if the volume is halved and the temperature is increased by 45.5°C?
14. A volume of 1120 L of helium at $40\bar{0}0$ Pa is heated from 45°C to 77°C, increasing the pressure to $60\bar{0}0$ Pa. What is the resulting volume of the gas?
15. In a 47-cm-tall cylinder of radius 7.0 cm, hydrogen of density 2.50 kg/m^3 is at a gauge pressure of 327 kPa. What is the density when the absolute pressure is changed to 525 kPa?

APPLIED CONCEPTS

1. Fran purchases a 1.85-ft^3, helium-filled Mylar® balloon in a store with an inside temperature of 65.5°F. When she walks outside, the air temperature is only 11.0°F. (a) What happens to the balloon as Fran walks toward her car? (b) What is the volume of the balloon when the temperature of the helium reaches the air temperature? (c) What would have been the volume of the balloon if she had purchased it in the summer with an outside temperature of 101°F? (d) Depending on the season, what can the store clerk do to compensate for the outside temperature and its impact on the volume of the balloon?
2. An automobile tire is filled to an air pressure of 32 lb/in^2 at an air temperature of $4\bar{0}$°F. (a) After the car is driven for a period of time, the temperature of the air inside the tire increases to 145°F. If the volume of the tire remains constant, what is the pressure inside the tire? (b) If a person wants to maintain an air pressure of 32 lb/in^2 when the tire is warm, what initial air pressure should be put in the tire?
3. A 15.0-cm-long cylinder has a movable piston with radius 2.00 cm. (a) What is the volume of the cylinder? (b) If the initial air pressure is 101 kPa at 0.0°C, what is the air pressure inside the cylinder when the piston is pushed down 4.00 cm? (c) If the air temperature inside the compressed cylinder is heated to 20.0°C, what is the new air pressure?
4. A 0.0300-m^3 steel tank containing helium is stored at 20.0°C under a pressure of 22.5×10^3 kPa. (a) If the helium is released into a standard 20.0°C, 1010-kPa atmosphere, how much volume will the helium occupy? (b) How much volume will the released helium occupy if released into a −28.5°C, 285-Pa stratosphere?
5. A lightweight weather-collecting sensor is attached to a 2.50-L balloon. The balloon is released on a day when the temperature is 23.4°C and the atmospheric pressure is 1.35×10^5 Pa. The balloon will collect data until it reaches the stratosphere. (a) The weather sensor measures a temperature of −28.2°C and an atmospheric pressure of 285 Pa. What is the volume of the balloon? (b) Given the change in the volume of the balloon as it reaches the stratosphere, should the balloon be fully inflated when released from the ground?

CHAPTER 15

WAVE MOTION AND SOUND

The importance of radio and television signals and other forms of electromagnetic waves cannot be overstated. Communication using these phenomena is the backbone of modern civilization. We will first study mechanical waves to see how waves in general behave and then study electromagnetic radiation and sound.

Objectives

The major goals of this chapter are to enable you to:

1. Describe characteristics of mechanical waves.
2. Describe electromagnetic waves and the electromagnetic spectrum.
3. Analyze sound waves and explain the Doppler effect.

15.1 Characteristics of Waves

Energy may be transferred by the motion of particles. Electricity, for example, is conducted along a wire by the motion of electrons. Heat is conducted by the motion of atoms and molecules. Tides and winds are examples of transfer of energy by the motion of fluids.

Energy may also be transferred by *wave motion*. The sun's energy is transported to the earth by light waves. Radio waves are an illustration of energy transfer for communications. Light waves produced by lasers are being used for voice and data transmission in optical fibers. Sound waves are yet another method by which energy may be transferred. The source of all waves is something that vibrates. The frequency of the vibrating source and the frequency of the wave it produces are the same.

A **wave** is a disturbance that moves through a medium or through space. This disturbance may be a displacement of atoms away from their equilibrium positions in an elastic medium, a pulse in a spring, a change in pressure of a gas, or a variation in light intensity. There is a transfer of energy in the direction of propagation of the disturbance for each type of wave. A leaf floating in a pond as a wave passes will bob up and down with the water but will not move along with the wave.

The elastic medium through which a wave travels or propagates is in many respects similar to a chain of particles connected by a series of springs like those shown in Fig. 15.1. If particle *A* is pulled to the left away from its equilibrium position, the neighboring spring exerts a force that tends to return *A* to its equilibrium position. The same spring exerts an equal but opposite force on particle *B*, which also tends to displace *B* to the left. As particle *B* moves to the left, the next particle experiences a force to the left, and so on until each particle experiences a displacement. If particle *A* is returned to its equilibrium position, the other particles will return to their equilibrium positions at a later time.

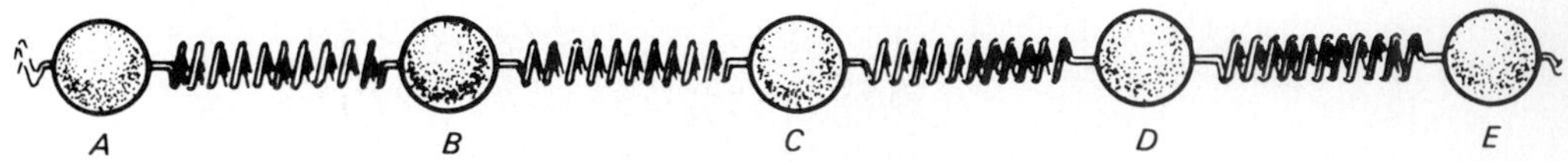

Figure 15.1 Springs in series are a form of elastic medium.

If particle *A* is forced to oscillate about its equilibrium position, all the other particles will also oscillate about their equilibrium positions. The kinetic energy given to the first particle is transmitted to each successive particle in the system. Although energy is transferred through the connecting springs, there is no transfer of particles from position *A* to *E*. This energy transfer without matter transfer is typical of all types of wave motion.

Another type of wave motion is shown in Fig. 15.2. In this case the elastic medium is a long spring. If the left end of the spring is rapidly lifted up and then returned to its starting position, a crest is formed that travels to the right [Fig. 15.2(a)]. If the left end is displaced downward and rapidly returned to its original position, a trough is formed that travels to the right [Fig. 15.2(b)].

A **pulse** is a nonrepeated disturbance that carries energy through a medium or through space. If the pulse is repeated periodically, then a series of crests and troughs will travel through the medium, creating a traveling wave. A **transverse wave** is a disturbance in a medium in which the motion of the particles is perpendicular to the direction of the wave motion (Fig. 15.3). Water waves are another example of transverse waves. The **amplitude** of a wave is the maximum displacement of any part of the wave from its equilibrium, or rest, position.

If a spring is compressed at the left end as shown in Fig. 15.4 and then released, the compression will travel to the right. Similarly, if the spring is stretched, a rarefaction (the stretched portion of the spring) is formed that will propagate or travel to the right. In this case the particle motion is along the direction of the wave travel. A **longitudinal wave** is a disturbance in a medium in which the motion of the particles is along the direction of the wave travel. Sound is another example.

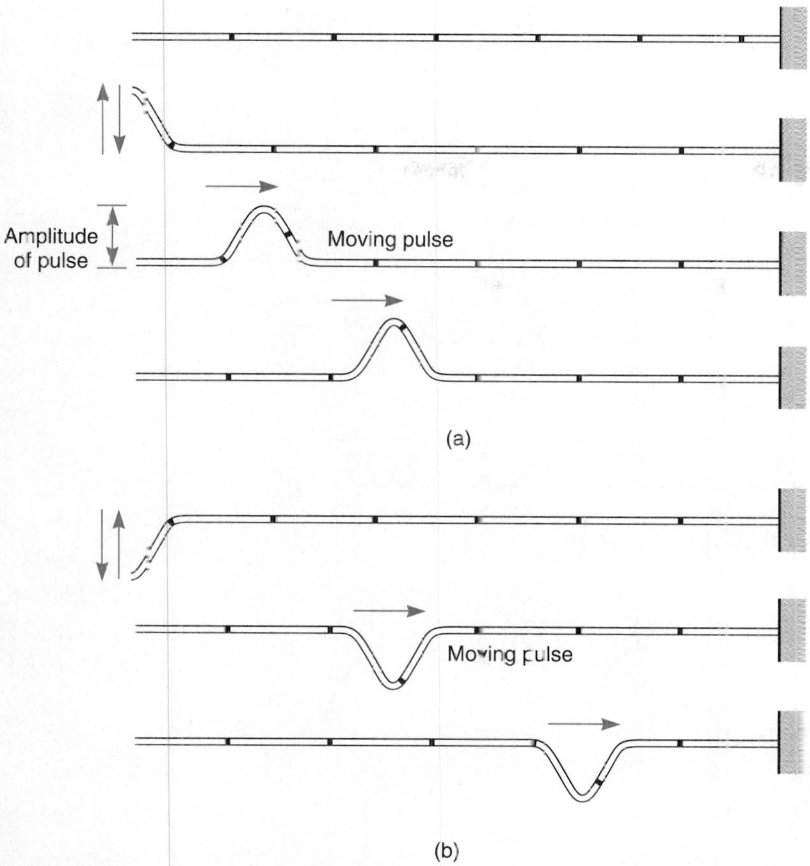

Figure 15.2 Pulses in a long spring

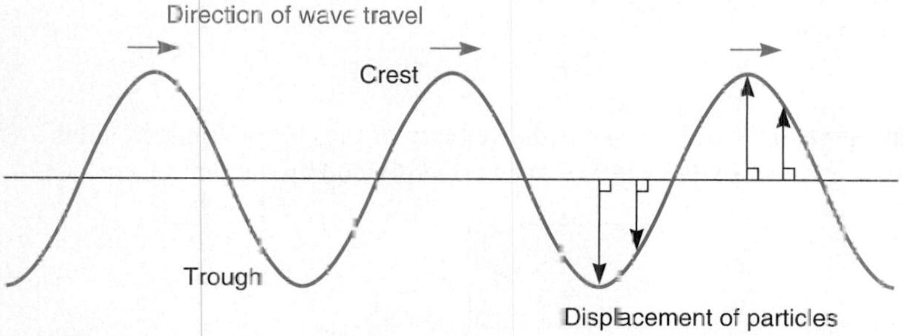

Figure 15.3 Transverse waves

The **wavelength** λ is the minimum distance between particles that have the same displacement and are moving in the same direction (Fig. 15.5).

The **period** is the time required for a single wave to pass a given point. The **frequency** is the number of complete waves passing a given point per unit time. The common unit for frequency is the hertz (Hz) (named after **Heinrich Hertz**), where one oscillation per second is equal to 1 hertz (1 Hz = 1/s). Higher frequencies are measured in kilohertz (kHz), megahertz (MHz), and gigahertz (GHz). Radar and microwaves are measured in GHz, frequency-modulated (FM) radio waves are measured in MHz, and amplitude-modulated (AM) radio waves are measured in kHz. The period and the frequency are related by

$$f = \frac{1}{T}$$

where f = frequency
T = period

Heinrich Hertz (1857–1894), *physicist, was born in Germany. His main work was on electromagnetic waves; he was the first to broadcast and receive radio waves. The unit of frequency, the hertz, is named after him.*

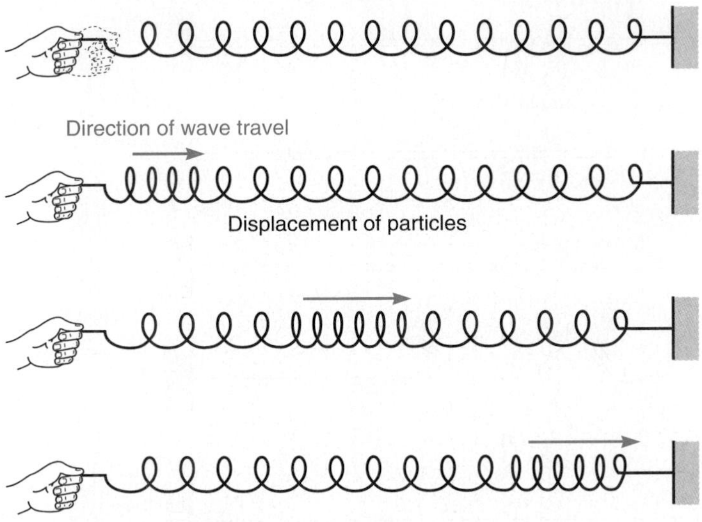

Figure 15.4 Longitudinal wave in a spring

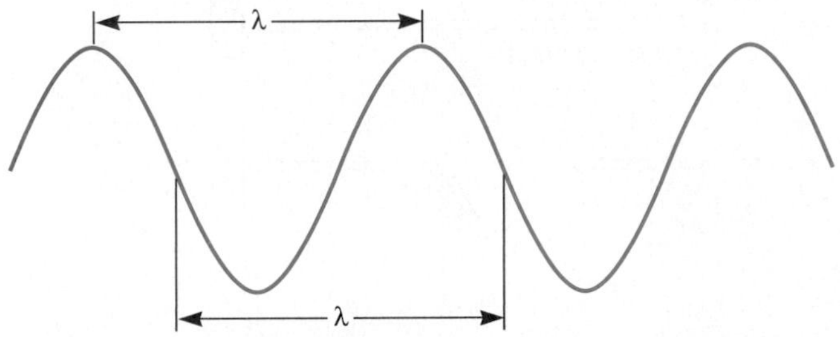

Figure 15.5 The wavelength of a wave is the distance between successive corresponding points on a uniformly repeated wave.

The **propagation velocity** v of a wave is the velocity of the energy transfer and is given by the distance traveled by the wave in one period divided by the period, or

$$v = \frac{\lambda}{T} = \lambda f$$

where v = velocity
λ = wavelength
T = period
f = frequency

These relationships apply to sound, water, light, and all other waves.

EXAMPLE

Find the velocity of a wave with wavelength 2.5 m and frequency 44 Hz.

Data:

$$\lambda = 2.5 \text{ m}$$
$$f = 44 \text{ Hz} = 44/\text{s}$$
$$v = ?$$

Basic Equation:

$$v = \lambda f$$

Working Equation: Same

Substitution:

$$v = (2.5\text{ m})(44/\text{s})$$
$$= 110\text{ m/s}$$

Superposition of Waves

When two waves of a similar type pass through the same medium, a new wave is created by the **superposition of waves.** This new wave is the algebraic sum of the separate displacements of the individual waves (Fig. 15.6). **Constructive interference** occurs where the waves add together to form a larger displacement as at point *A*. **Destructive interference** occurs where the waves add together to form a smaller displacement as at point *B*.

The addition is algebraic, so two positive amplitudes (that is, two amplitudes in the same direction) add together to make a larger amplitude in the same direction (constructive interference). Where the amplitudes are in opposite directions, the smaller counteracts the larger with a net amplitude in the direction of the larger (destructive interference). See location (1) of Fig. 15.6(c) for an illustration of constructive interference and location (2) for an illustration of destructive interference.

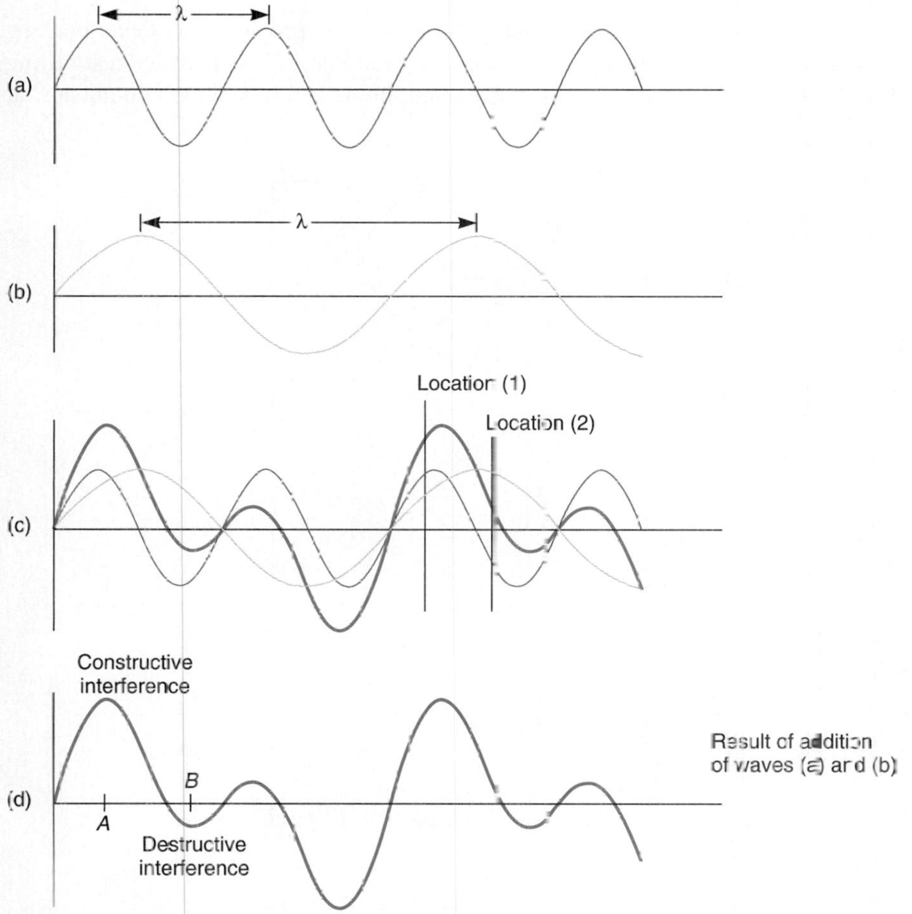

Figure 15.6 Superposition of two waves to form a new wave by adding their displacements

Standing Waves

When a transverse pulse reaches the end of a spring that is fastened to a rigid support (Fig. 15.7), the pulse is reflected along the spring with the displacement of the reflected wave opposite in direction to that of the incident wave. A traveling wave is also reflected at the rigid end of a spring, producing two waves moving in opposite directions. Reflections from a free end are not inverted.

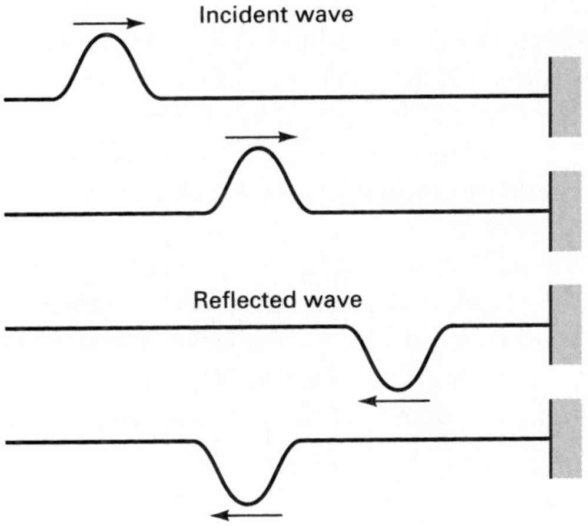

Figure 15.7 The reflected wave in the spring has the same amplitude as the incident wave.

In one special case two waves combine so that there is no propagation of energy along the wave. The wave displacements are constant and motionless. This is called a **standing wave** (Fig. 15.8) because the two waves of equal amplitude and wavelength do not appear

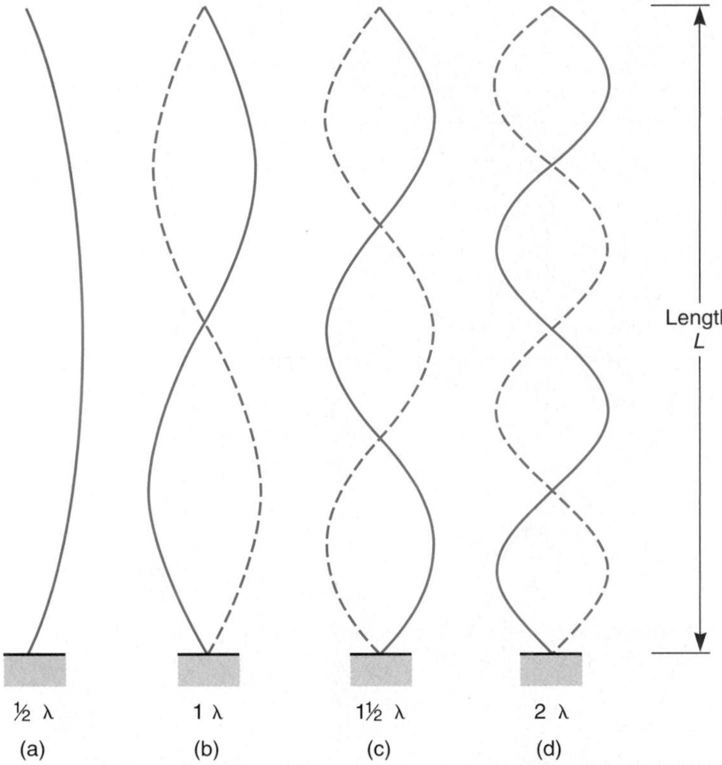

Figure 15.8 Standing waves in a string generated by increasing the frequency of vibration

to be traveling. The points of destructive interference and constructive interference remain in fixed positions. Figure 15.8 shows an example of a standing wave on a string that could produce sound on a musical instrument. Note that there is no motion of the string at the end points. Although there is no propagation of energy along the string, there may be energy transfer from the string to the air surrounding it, producing sound waves at the same frequency of oscillation as that of the vibrating string.

Standing waves are formed in strings of musical instruments and in the air in an organ pipe, a flute, and other wind instruments.

Interference and Diffraction

If two rocks are simultaneously dropped into a pool, each will produce a set of waves or ripples (Fig. 15.9). Wherever two wave crests cross each other, the water height is higher than for either crest alone. Where two troughs cross, the water level is lower than for one alone. If a trough crosses a crest, the water level is nearly undisturbed. This is an example of wave **interference.** Constructive interference occurs when two crests or troughs meet, giving a larger disturbance than for either wave alone. Destructive interference occurs when a wave and a trough meet and cancel each other out. Those areas where waves cancel each other out are called *nodes* [Fig. 15.9(b)].

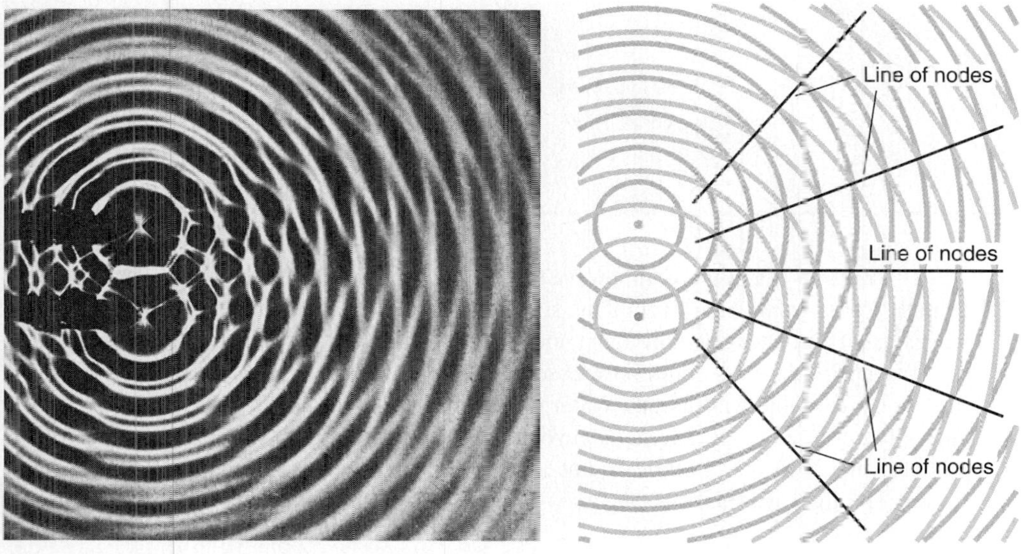

(a) (b)

Figure 15.9 Interference of waves from two sources. The waves combine to form larger waves and cancel each other out as the wave fronts meet.

Interference of waves can occur for any type of wave, including sound, light, and radio waves. Some directional radio broadcast antennas rely on constructive interference between waves from different parts of the antenna to direct the signal in the desired direction. Destructive interference prevents the signal from propagating in undesired directions. "Dead spots" in an auditorium are caused by destructive interference of waves coming directly from the sound source with sound waves reflected from the walls, ceiling, or floor of the room. Proper choice of room shape and proper placement of materials that absorb sound well can lead to a room with good acoustical characteristics.

TRY THIS ACTIVITY

Interfering Waves

Place either a long Slinky® toy or a length of rope on a smooth, low-friction floor. With one person at each end of the rope, create a "wave pulse" toward each other. Observe what happens when the two waves interfere with each other. Explain what happens when two positive pulses interfere, two negative pulses interfere, and a positive and a negative pulse interfere with one another (Fig. 15.10).

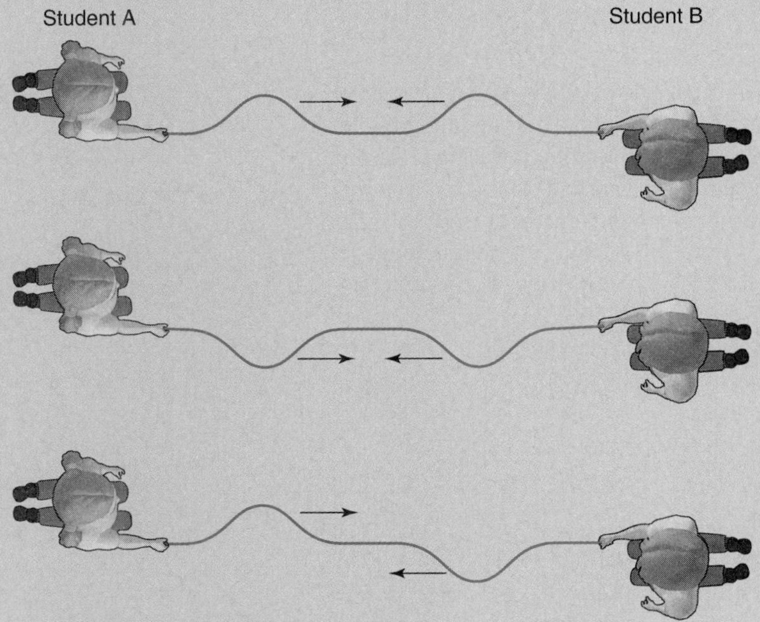

Figure 15.10

Diffraction is a property of a wave that describes its ability to bend around obstacles in its path. Water waves bend around the supports of a pier or a large rock (Fig. 15.11). Sound waves pass from one room through a door and spread into a second room (Fig. 15.12).

Wave diffraction is commonly observed only when the obstacle or opening is nearly the same size as the wavelength. Water waves and sound waves often have wavelengths in an easily observed range. Light, however, has a wavelength approximately 5×10^{-7} m. For this reason diffraction of light is not as easily recognized as diffraction of some other waves. The light waves from the sun when they encounter obstacles (air molecules in the

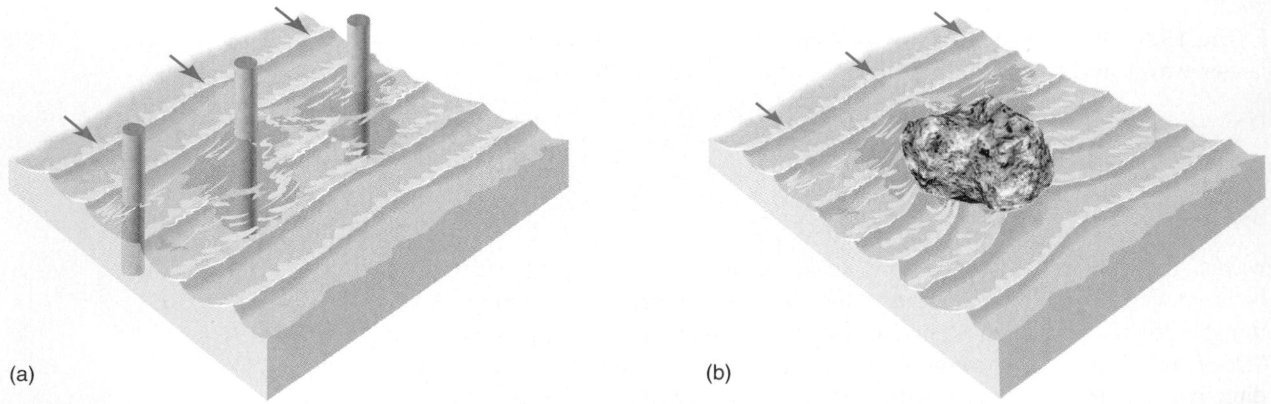

Figure 15.11 Water waves bend around obstacles and pass into the region behind (a) the pier supports or (b) a large rock. This property is called diffraction.

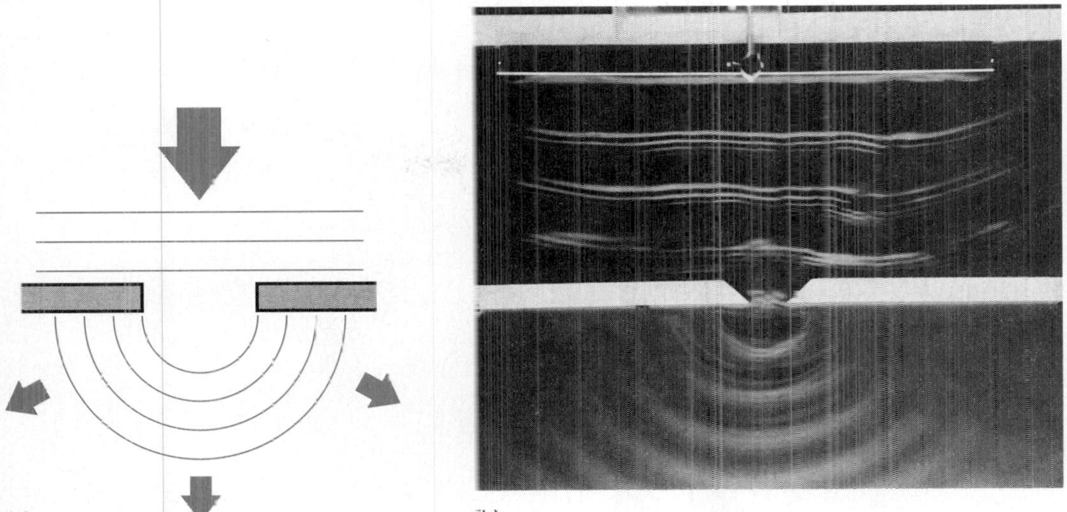

Figure 15.12 Diffraction of (a) sound waves entering a room through a door and (b) water waves passing through a small opening

atmosphere) are diffracted around them. The diffracted wave fronts are more or less spherical and are spread out or *scattered*.

Blue sky and a red setting sun are a result of scattering. When we look up at the sky, we see scattered light. Those colors with the shortest wavelengths (blue) are scattered the most; the longer wavelengths (red and yellow) are transmitted with very little scattering. (These topics are covered in Chapter 21.)

15.2 Electromagnetic Waves

An **electromagnetic wave** is a transverse wave resulting from a periodic disturbance in an electromagnetic field having an electric component and a magnetic component, each being perpendicular to the other and both perpendicular to the direction of travel. All electromagnetic waves travel with velocity $v = c =$ **speed of light** $= 3.00 \times 10^8$ m/s. So, for electromagnetic waves

$$\boxed{c = \lambda f}$$

where

$c =$ speed of light (3.00×10^8 m/s)
$\lambda =$ wavelength
$f =$ frequency

Note that λ and f are inversely proportional. That is, when the frequency increases, the wavelength decreases; and when the frequency decreases, the wavelength increases.

EXAMPLE

The FM band of a radio is centered around a frequency of $10\bar{0}$ megahertz (MHz). Find the length of an FM antenna if each arm must be a quarter-wavelength.

First, find the wavelength, λ.

Data:

$$f = 10\bar{0}\text{ MHz} = 10\bar{0} \times 10^6\text{ Hz} = 1.00 \times 10^8/\text{s} \qquad (1\text{ Hz} = 1/\text{s})$$
$$c = 3.00 \times 10^8\text{ m/s}$$
$$\lambda = ?$$

Basic Equation:

$$c = \lambda f$$

Working Equation:

$$\lambda = \frac{c}{f}$$

Substitution:

$$\lambda = \frac{3.00 \times 10^8\text{ m/s}}{1.00 \times 10^8/\text{s}}$$
$$= 3.00\text{ m}$$

Therefore,

$$\frac{\lambda}{4} = \frac{3.00\text{ m}}{4} = 0.750\text{ m}$$

The **electromagnetic spectrum** is the entire range of electromagnetic waves classified according to frequency (Fig. 15.13). All electromagnetic waves travel at the same speed. The radio broadcast band is in the region of 1 MHz. The very high frequency (VHF) television band starts in the region of 50 MHz; the ultra high frequency (UHF) band is even higher. The highest-frequency waves generated by electronic oscillators are microwaves. Microwaves may be used to relay TV signals from a remote location to the studio transmitter.

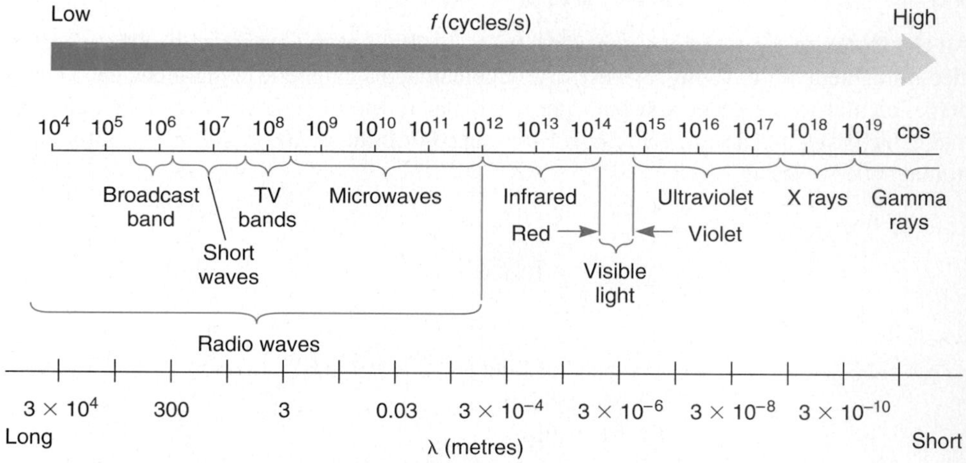

Figure 15.13 The electromagnetic spectrum

Molecular and atomic oscillations create waves of even higher frequencies, including infrared, visible light, ultraviolet, and X-ray waves. Visible light is electromagnetic radiation in the range 4.3×10^{14} to 7×10^{14} Hz.

We are surrounded by and bombarded by a sea of electromagnetic radiation. The light we see is only a very tiny part of this radiation.

PHYSICS CONNECTIONS

Evading Radar

Radar (RAdio Detecting And Ranging) is used as a navigation, early-warning, and detection device. Radar transmitters send electromagnetic radio waves into the atmosphere to measure the time and the frequency of radio waves after they have reflected off objects. (See Section 15.4 for more information on the "Doppler effect.") The reflected radio waves provide the observer with the object's position, size, and velocity.

A futuristic French fighter plane known as Rafale uses a device to help the plane evade radar. The plane uses a technology called active cancellation. As radio waves from radar strike the Rafale, an inverted version of the reflected radio wave is emitted. When two identical but opposite waves meet, they cancel each other, which results in destructive interference. Through destructive interference, the producer of a radar signal will not see any reflected radio waves from the Rafale.

Figure 15.14 Stealth aircraft is designed to be invisible to radar.

Stealth aircraft avoid radar detection because of their body shape and design. The odd shapes and angles of the stealth body are designed to deflect the radio waves around the plane instead of reflecting them to the producer of the radar signal. Some of the stealth material is also designed to absorb the radio waves (Fig. 15.14).

PROBLEMS 15.2

1. Find the period of a wave whose frequency is $50\bar{0}$ Hz.
2. Find the frequency of a wave whose period is 0.550 s.
3. Find the velocity of a wave with wavelength 2.00 m and frequency $40\bar{0}$ Hz.
4. (a) What is the frequency of a light wave with wavelength 5.00×10^{-7} m and velocity 3.00×10^{8} m/s? (b) Find the period of the wave.
5. What is the speed of a wave with frequency 3.50 Hz and wavelength 0.550 m?
6. Find the wavelength of water waves with frequency 0.650 Hz and velocity 1.50 m/s.
7. What is the wavelength of longitudinal waves in a coil spring with frequency 7.50 Hz and velocity 6.10 m/s?
8. A wave generator produces 20 pulses in 3.50 s. (a) What is its period? (b) What is its frequency?
9. Find the frequency of a wave produced by a generator that emits 30 pulses in 2.50 s.
10. What is the wavelength of an electromagnetic wave with frequency 50.0 MHz?
11. Find the frequency of an electromagnetic wave with wavelength 1.50 m.
12. Find the wavelength of a wave traveling at 2.68×10^{6} m/s with a period of 0.0125 s.
13. Find the wavelength of a wave traveling twice the speed of sound (speed of sound = 331 m/s) that is produced by an oscillator emitting 63 pulses every 8.3×10^{-6} min.
14. (a) Find the velocity of X rays emitted with wavelength 3.00×10^{-9} m and frequency 3.00×10^{18} Hz. (b) Find the period of the waves.
15. Find the velocity of microwaves having wavelength 0.750 m and frequency 2.75×10^{10} Hz.

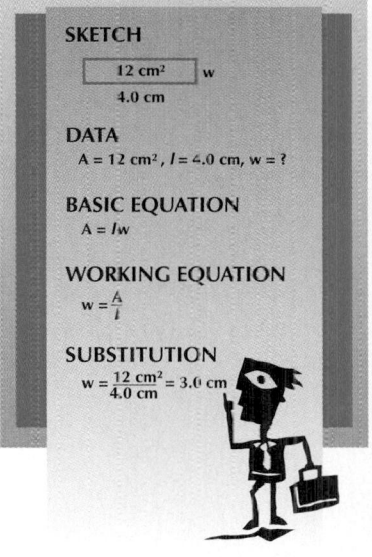

SKETCH

12 cm² w

4.0 cm

DATA

$A = 12\text{ cm}^2$, $l = 4.0$ cm, $w = ?$

BASIC EQUATION

$A = lw$

WORKING EQUATION

$w = \frac{A}{l}$

SUBSTITUTION

$w = \frac{12\text{ cm}^2}{4.0\text{ cm}} = 3.0\text{ cm}$

15.3 Sound Waves

Daily we are exposed to many sounds and communicate with other people using the medium of sound. Whether the sound is pleasant music, a voice, or a loud siren, there are three requirements for the detection of sound in a physiological sense. There must be a source of sound, a medium (such as air) for transmitting it, and an ear to receive it. In a physical sense, sound is a vibratory disturbance in an elastic medium which may produce the sensation of sound. The frequency range over which the human ear responds is approximately 20 to 20,000 Hz. Ultrasonic waves have a frequency higher than 20,000 Hz.

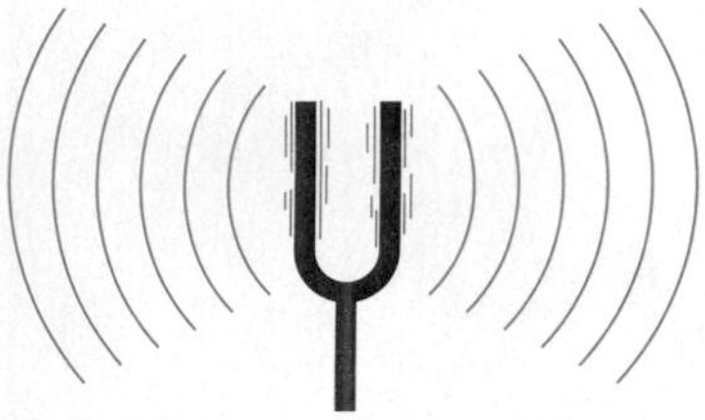

Figure 15.15 Sound waves produced by a vibrating tuning fork

Sound refers to those waves transmitted through a medium with frequencies capable of being detected by the human ear and is produced by a vibrating source. A ringing bell, the vibrating head of a drum, and a tuning fork (Fig. 15.15) are common examples of vibrating sources of sound. Other vibrating sources are not as easily recognized. Vibrating vocal cords produce speech. The notes of a clarinet originate with a vibrating reed. An auto horn uses an electrically driven vibrating diaphragm.

The most common medium for the propagation of sound is air. Sound will also propagate in solids or liquids. A vacuum will not carry sound. A mechanic may listen to the sounds of a running engine by placing one end of a metal rod against the engine and the other end against his or her ear. Sounds transmitted through water are utilized by passive sonar receivers aboard ships or submarines to identify other ships nearby.

Sound waves transmitted through the earth may be detected by an instrument called a *seismograph,* which can detect small motions of the earth's crust. An earthquake produces both longitudinal and transverse waves, which propagate with different velocities through the earth's crust. The distance to the source can be determined by measuring the time interval between the arrival of the two types of waves. Comparison of such data from seismographs at several points on the surface of the earth helps locate the epicenter of the earthquake. Waves can be intentionally set off by buried explosives. Reflections of these sound waves from different rock formations are recorded by seismographs. These recordings allow geologists to determine the underlying structure of the earth and predict the location of possible oil- or gas-producing regions.

Musical instruments utilize vibrating strings and vibrating columns of air to produce regular sounds. Wind instruments use an enclosed or partially enclosed column of air where waves are produced in the confined air by a pressure change (usually, blowing). Standing waves can then be produced in a column of proper length.

You may have watched distant lightning and noticed the time lapse before the sound of thunder reaches you. This is an example of the relatively slow speed of sound compared to the **speed of light** (3.00×10^8 m/s). The **speed of sound** in dry air at 1 atm pressure and 0°C is 331 m/s. Changes in humidity and temperature cause a variation in the speed of sound. The speed of sound increases with temperature at the rate of 0.61 m/s/°C. The speed of sound in dry air at 1 atm pressure is then given by

$$v = 331\text{ m/s} + \left(0.61\frac{\text{m/s}}{°\text{C}}\right)T$$

$$v = 1087\text{ ft/s} + \left(1.1\frac{\text{ft/s}}{°\text{F}}\right)(T - 32°\text{F})$$

where

v = speed of sound in air
T = air temperature

EXAMPLE 1

Find the speed of sound in dry air at 1 atm pressure if the temperature is 23°C.

Data:

$$T = 23°\text{C}$$
$$v = ?$$

Basic Equation:

$$v = 331 \text{ m/s} + \left(0.61 \frac{\text{m/s}}{°\text{C}}\right)T$$

Working Equation: Same

Substitution:

$$v = 331 \text{ m/s} + \left(0.61 \frac{\text{m/s}}{°\cancel{\text{C}}}\right)(23°\cancel{\text{C}})$$

$$= 345 \text{ m/s}$$

EXAMPLE 2

What is the time required for the sound from an explosion to reach an observer $190\bar{0}$ m away for the conditions of Example 1?

Data:

$$v = 345 \text{ m/s}$$
$$s = 190\bar{0} \text{ m}$$
$$t = ?$$

Basic Equation:

$$s = vt \qquad \text{(from Section 3.1)}$$

Working Equation:

$$t = \frac{s}{v}$$

Substitution:

$$t = \frac{190\bar{0} \text{ m}}{345 \text{ m/s}}$$

$$= 5.51 \text{ s}$$

$$\boxed{\frac{\text{m}}{\text{m/s}} = \text{m} \div \frac{\text{m}}{\text{s}} = \cancel{\text{m}} \cdot \frac{\text{s}}{\cancel{\text{m}}} = \text{s}}$$

Sound propagates faster in a dense medium such as water than it does in a less dense medium such as air. A list of the speed of sound in various media is given in Table 15.1.

Table 15.1 Speed of Sound in Various Media

Medium	Speed	
	m/s	ft/s
Aluminum	6,420	21,100
Brass	4,700	15,400
Steel	5,960	19,500
Granite	6,000	19,700
Alcohol	1,210	3,970
Water (25°C)	1,500	4,920
Air, dry (0°C)	331	1,090
Vacuum	0	0

All sounds have characteristics that we associate only with that sound. A siren is loud. A whisper is soft. Music can be loud or soft. The physical properties that differ for these

Alexander Graham Bell (1847–1922),

inventor, was born in Scotland. He invented the telephone and the telegraph in 1876 and established the Bell Telephone Company in 1877.

sounds are intensity and frequency. The physiological characteristics of these sounds are loudness and pitch. Sound quality is related to the number of frequencies present. Pitch and frequency are closely related terms.

Intensity is the energy transferred by sound per unit time through unit area, thus $\frac{\text{power}}{\text{area}}$. **Loudness** refers to the strength of the sensation of sound heard by an observer and describes how strong or faint the sensation of sound seems. The ear does not respond equally to all frequencies. Sound must reach a certain intensity before it can be heard. The human ear normally detects sounds ranging in intensity from 10^{-12} W/m^2 (the threshold of hearing) to 10^0 W/m^2 or 1 (the threshold of pain). Levels of intensity are also measured on a logarithmic scale in decibels (dB); the unit "bel" is named after **Alexander Graham Bell.** Table 15.2 shows a range of familiar sounds.

The **pitch** of a sound is the effect of the frequency of sound waves on the ear. Higher pitched sounds have a higher frequency than lower ones. The quality of sound can easily be determined by a casual listener. Irregular vibrations tend to produce noise, whereas regular vibrations in multiples of the fundamental vibration rate of an object produce sounds more pleasing to the human ear. These multiples of the fundamental tone are called harmonics. The quality of a sound depends on the number of harmonics produced and their relative intensities.

Sound waves from two sources reaching an observer at the same time will constructively or destructively interfere with each other. Identical waves from two separate sources such as stereo loudspeakers will enhance or reduce each other at different points in space. Sound waves of slightly different frequencies will also enhance or reduce each other at different points in time. The alternating periods of increasing and decreasing volume are called *beats*. The number of times per second the sound reaches a maximum is known as the *beat frequency* and is equal to the difference in frequencies of the two sound waves. This phenomenon is used for tuning musical instruments against a set of standard-frequency tuning forks. When the beat frequency is small, the instrument is well matched with the tuning fork.

Table 15.2 Range of Sound Levels and Intensities

Situation	Sound Level (dB)	Intensity (W/m^2)	Sound
Threshold of hearing	0	10^{-12}	Scarcely audible
Minimal sounds	10	10^{-11}	
Ticking watch	20	10^{-10}	
Whisper	30	10^{-9}	Faint
Leaves rustling, refrigerator	40	10^{-8}	
Average home, neighborhood street	50	10^{-7}	Moderate
Normal conversation, microwave	60	10^{-6}	
Car, alarm clock, city traffic	70	10^{-5}	Loud
Garbage disposal, vacuum cleaner, outboard motor, noisy restaurant	80	10^{-4}	Very loud
Factory, electric shaver, screaming child	85		
Lawn mower, passing motorcycle, convertible ride on a freeway	90	10^{-3}	
Blow dryer, diesel truck, chain saw, helicopter, subway train	100	10^{-2}	
Car horn, snowblower	110	10^{-1}	Deafening
Rock concert, prop plane	120	10^{0}	Painful
100 ft from jet engine, air raid siren	130	10^{1}	
Shotgun blast	140	10^{2}	

Source: Better Hearing Institute, *Self-Help for Hard of Hearing People.* Each increase of 10 dB is a tenfold increase in sound intensity; that is, 90 dB is ten times noisier than 80 dB. The U.S. government advises wearing earplugs or other hearing protection for anyone exposed to 85 dB for a period of more than a few hours. Some hearing experts have found that hearing damage is done at sound levels as low as 70 dB. Federal regulations require a hearing conservation program in the workplace where employees are exposed to 85 dB or more during an 8-h work period.

The ear is the human organ used to detect sound. It consists of three sections: the outer ear, the middle ear, and the inner ear (Fig. 15.16). Sound travels through the ear canal, which ends at the eardrum. The sound waves cause the eardrum to vibrate. The sound energy is transferred through the three smallest bones in the body to a fluid in the inner ear, which excites tiny hairs called cilia. Cilia resonate in the fluid, sending nerve impulses to the brain in the form of electrical waves. The brain then interprets the sound and we are able to hear.

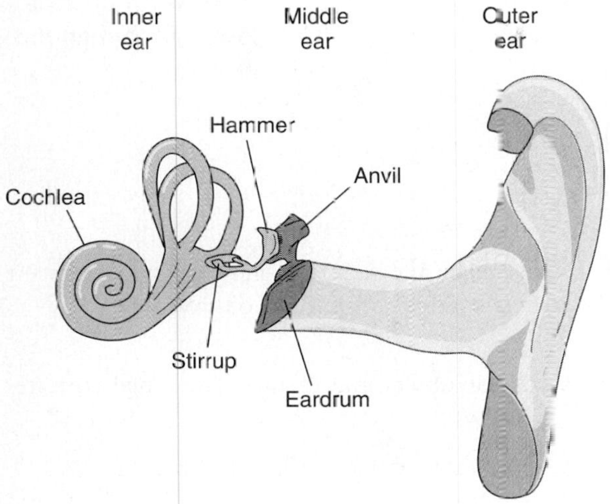

Figure 15.16 Sound waves move through all three regions of the ear and are sent to the brain in the form of electrical waves.

15.4 The Doppler Effect

The **Doppler effect** refers to the variation of the frequency heard when a source of sound and the ear are moving relative to each other. As an automobile or a train passes by at a high speed sounding its horn or whistle, the frequency or pitch of the sound drops noticeably as it passes the observer. This variation in pitch is called the Doppler effect, named after **Christian Johann Doppler.**

Water waves spread over the flat surface of the water. Sound and light waves, though, travel in three-dimensional space in all directions like an expanding balloon. Just as circular wave crests are closer together in front of a swimming duck, spherical sound or light wave crests ahead of a moving source are closer together than those behind the source and reach a receiver more frequently (Fig. 15.17).

Motion of a source of sound toward an observer increases the rate at which he or she receives the vibrations. The velocity of each vibration is the speed of sound whether the source is moving or not. Each vibration from an approaching source has a shorter distance to travel. The wavelength is shortened when the source is moving toward the observer and is lengthened when the source is moving away from the observer. The vibrations are therefore received at a higher frequency than they are sent. Similarly, sound waves from a receding source are received at a lower frequency than that at which they are sent (Fig. 15.18).

The apparent Doppler-shifted frequency for sound is given by the equation

$$f' = f\left(\frac{v}{v \pm v_s}\right)$$

where

f' = Doppler-shifted frequency
f = actual source frequency
v = speed of sound
v_s = speed of the source

Christian Johann Doppler (1803–1853),

physicist and mathematician, was born in Austria. He is best known for his explanation of the perceived frequency variation of sound and light waves relative to the motion of the source and detector, commonly known as the Doppler effect.

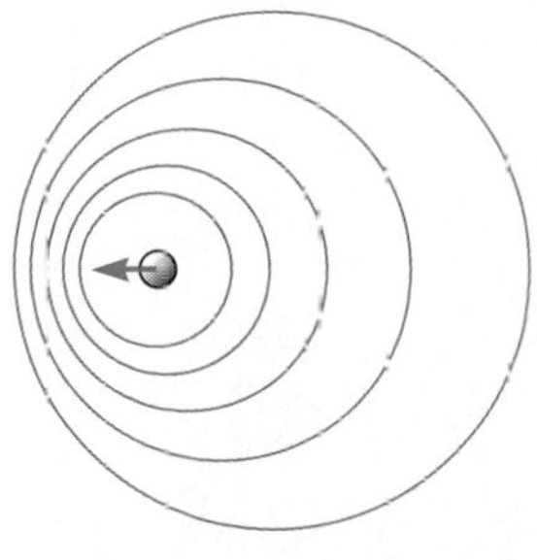

Figure 15.17 A source moving to the left creates higher frequency sound waves ahead and lower frequency sound waves behind.

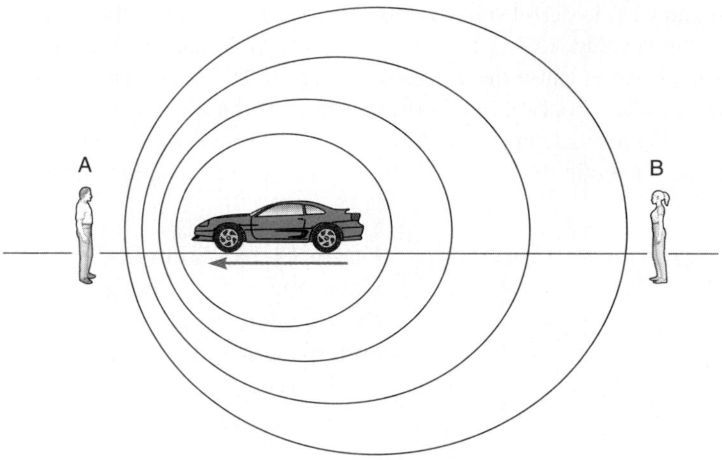

Figure 15.18 **As this car moves to the left, Observer A in front of the car hears the car horn at a higher pitch than the driver, whereas Observer B hears a lower pitch than the driver.**

The + sign in the denominator is used when the source is moving away from the observer. The − sign is used when the source is moving toward the observer.

EXAMPLE

An automobile sounds its horn while passing an observer at 25 m/s. The actual horn frequency is $40\bar{0}$ Hz.

(a) What is the frequency heard by the observer while the car is approaching?
(b) What is the frequency heard when the car is leaving?
Assume that the speed of sound is 345 m/s.

(a) Data:

$$f = 40\bar{0}\text{ Hz}$$
$$v = 345\text{ m/s}$$
$$v_s = 25\text{ m/s toward observer}$$
$$f' = ?$$

Basic Equation:

$$f' = f\left(\frac{v}{v - v_s}\right)$$

Working Equation: Same

Substitution:

$$f' = (40\bar{0}\text{ Hz})\left(\frac{345\text{ m/s}}{345\text{ m/s} - 25\text{ m/s}}\right)$$
$$= 431\text{ Hz}$$

(b) We simply change the sign from − to + in the basic equation of part (a). All other data remain the same. We then find

$$f' = (40\bar{0}\text{ Hz})\left(\frac{345\text{ m/s}}{345\text{ m/s} + 25\text{ m/s}}\right)$$
$$= 373\text{ Hz}$$

The Doppler effect can be easily demonstrated by two students and a toy horn. One is the observer and remains stationary. The other, with the horn, blows it while turning around. The change in pitch heard by the observer is the result of the Doppler effect.

The Doppler effect is usually experienced with sound waves. However, it is common to all waves. For visible light, the Doppler effect is seen as a change in color since the color of light is determined by its frequency. Astronomers use this principle in the study of the universe to determine whether heavenly bodies are approaching or moving away from us.

PHYSICS CONNECTIONS

Ultrasound

Ultrasound is used to examine tissue and liquid-based internal organs and systems without subjecting the patient to invasive procedures. Ultrasound is an extremely high-frequency sound wave ranging from 20 kHz to 5 MHz. A device called a transducer sends sound waves of a particular frequency and measures the frequency as the waves reflect off various media or organs. Since the speed of sound is dependent on the medium through which it travels, the ultrasound processor that takes the information from the device can determine the position and density of the tissue.

Ultrasound is often used for diagnostic purposes in place of X rays because it does not use radiation and is safer for the person and/or the fetus being examined. Along with monitoring the status of a fetus and generating deep heat for therapeutic purposes, ultrasound is also used to observe various human systems such as the nervous, urinary, reproductive, and circulatory systems. Ultrasound is not used to view bone structures because the high-frequency sound waves reflect well off only liquid-based objects. Recent technological improvements in ultrasound have led researchers to advances in three-dimensional imaging and Doppler ultrasound, where the movement of blood can be monitored using sound waves (Fig. 15.19).

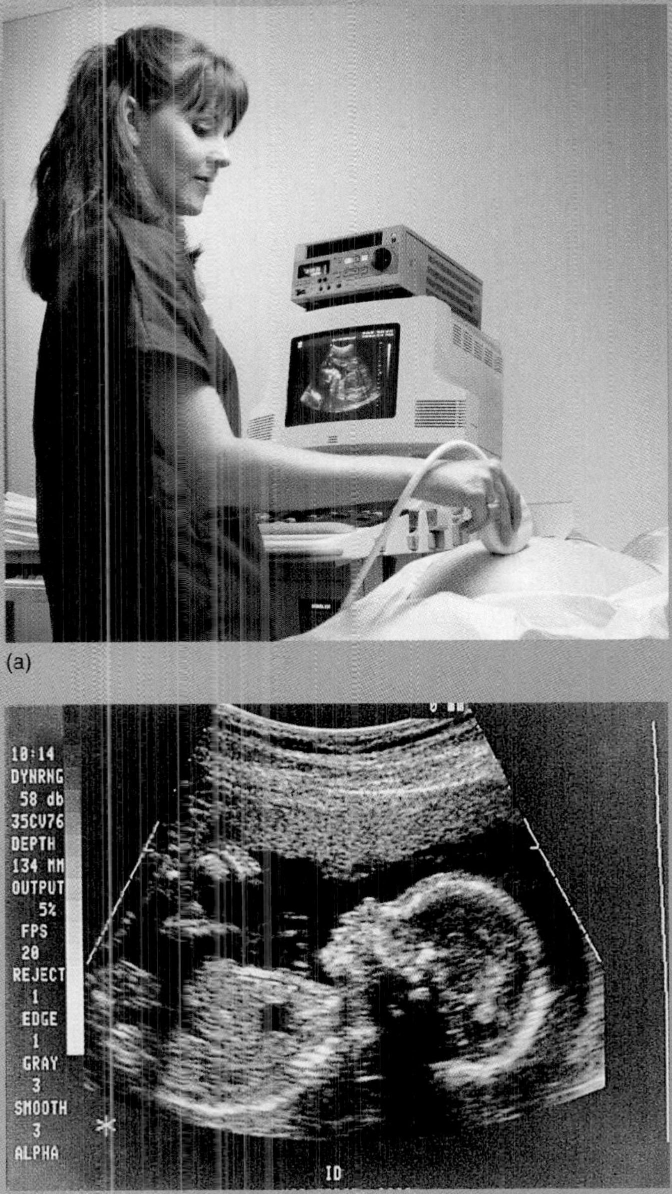

Figure 15.19 (a) Ultrasound technology is used to visualize tissue inside the human body without subjecting the person to invasive procedures or harmful radiation. (b) An ultrasound of a human fetus.

PROBLEMS 15.4

1. Find the speed of sound in m/s at $1\bar{0}$°C at 1 atm pressure in dry air.
2. Find the speed of sound in m/s at 35°C at 1 atm pressure in dry air.
3. Find the speed of sound in m/s at −23°C at 1 atm pressure in dry air.
4. How long will it take a sound to travel 21.0 m for the conditions of Problem 1?

SKETCH

12 cm² | w
4.0 cm

DATA

$A = 12\text{ cm}^2$, $l = 4.0\text{ cm}$, $w = ?$

BASIC EQUATION

$A = lw$

WORKING EQUATION

$w = \frac{A}{l}$

SUBSTITUTION

$w = \frac{12\text{ cm}^2}{4.0\text{ cm}} = 3.0\text{ cm}$

5. How long will it take a sound to travel through 750$\bar{0}$ m of water at 25°C?
6. A sound wave is transmitted through water from one submarine, is reflected off another submarine 15 km away, and returns to the sonar receiver on the first submarine. What is the round-trip transit time for the sound wave? Assume that the water temperature is 25°C.
7. A sonar receiver detects a reflected sound wave from another ship 3.52 s after the wave was transmitted. How far away is the other ship? Assume that the water temperature is 25°C.
8. A woman is swimming when she hears the underwater sound wave from an exploding ship across the harbor. She immediately lifts her head out of the water. The sound wave from the explosion propagating through the air reaches her 4.00 s later. How far away is the ship? Assume that the water temperature is 25°C and the air temperature is 23°C.
9. A train traveling at a speed of 4$\bar{0}$ m/s approaches an observer at a station and sounds a 55$\bar{0}$-Hz whistle. What frequency will be heard by the observer? Assume that the sound velocity in air is 345 m/s.
10. What frequency is heard by an observer who hears the 45$\bar{0}$-Hz siren on a police car traveling at 35 m/s away from her? Assume that the velocity of sound in air is 345 m/s.
11. A car is traveling toward you at 40.0 mi/h. The car horn produces a sound at a frequency of 48$\bar{0}$ Hz. What frequency do you hear? Assume that the sound velocity in air is 1090 ft/s.
12. A car is traveling away from you at 40.0 mi/h. The car horn produces a sound at a frequency of 48$\bar{0}$ Hz. What frequency do you hear? Assume that the sound velocity in air is 1090 ft/s.
13. A jet airplane taxiing on the runway at 13.0 km/h is moving away from you. The engine produces a frequency of 66$\bar{0}$ Hz in −6.0°C air. What frequency do you hear?
14. While snorkeling you hear a dolphin's sound as it approaches at 5.00 m/s. If the perceived frequency is 85$\bar{0}$ Hz, what is the actual frequency being emitted?
15. Two construction workers stand 112 m apart. One strikes a steel beam with a hammer. How long does it take for the other to hear the sound? Assume the velocity of sound in air is 345 m/s.
16. What is the length of a brass pipe through which a sound wave is transmitted in 0.136 s?
17. A crop duster airplane flies overhead at 44.7 m/s. The frequency of the sound is 605 Hz. (a) What do you perceive as the frequency as it approaches? Assume the velocity of sound in air is 345 m/s. (b) What frequency do you hear as the plane flies away and has accelerated to 55.0 m/s?

15.5 Resonance

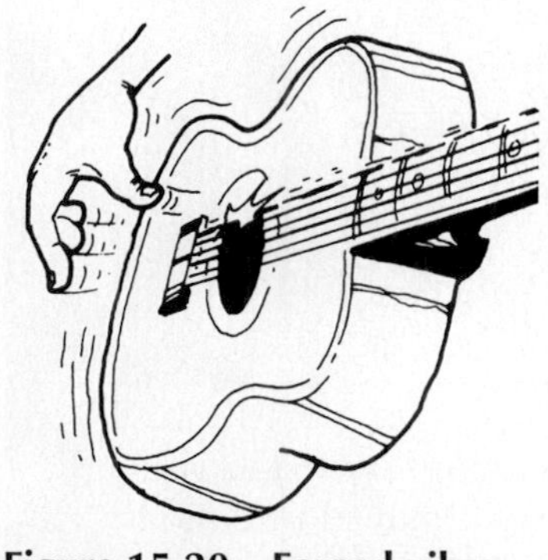

Figure 15.20 Forced vibration of a guitar sounding board

The **natural frequency** of an object, such as a tuning fork, is the frequency at which it vibrates when struck by another object, such as a rubber hammer. This frequency depends on the length and thickness of the tuning fork and the material from which it is made. Strings on a guitar also vibrate at a natural frequency. The sounding board of a guitar is forced to vibrate at the same frequency as the strings because of energy transfer from the strings to the sounding board (Fig. 15.20). This is an example of *forced vibration*. The natural frequency of the board is typically different from that of the strings or tuning fork. Because the area of the sounding board is large, energy transfer into sound waves is very efficient. Therefore, the vibrating string or tuning fork loses its energy or dies out more rapidly if in contact with a sounding board.

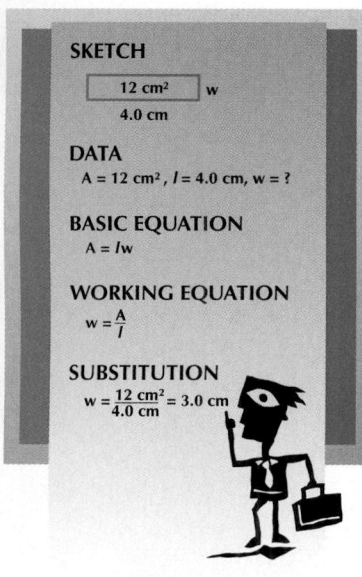

Consider two objects such as tuning forks with the same natural frequency that are set close together (Fig. 15.21). One is set into vibration and then stopped after a few seconds. It is found that the other tuning fork is weakly vibrating. The sound waves of the first fork cause the second to vibrate. This is called *sympathetic vibration* or *resonance*. **Resonance** occurs when the natural vibration rates of two objects are the same. Energy transfer into vibrations of the second fork is found to be much more efficient when both forks have the same frequency than when they have different frequencies. Large vibrations can be set up if the

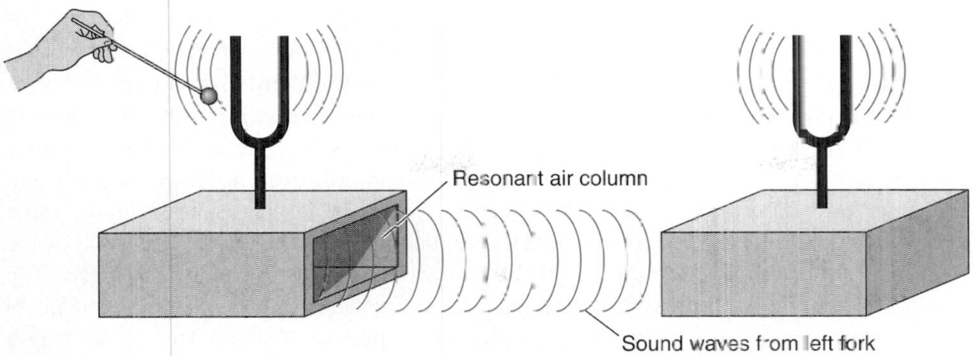

Figure 15.21 Identical tuning forks with resonant air column. Sound waves of the left fork cause the right fork to vibrate.

TRY THIS ACTIVITY

Singing Wineglasses

Moisten the tip of your finger and gently rub it around the rim of a quality wineglass. At just the right speed, the friction between your finger and the wineglass causes the glass to vibrate at its natural frequency (Fig. 15.22). The glass forms a standing wave, begins to resonate, and produces the pure ringing sound of the wineglass. What can be done to the glass to change the frequency of the resonating sound?

Figure 15.22

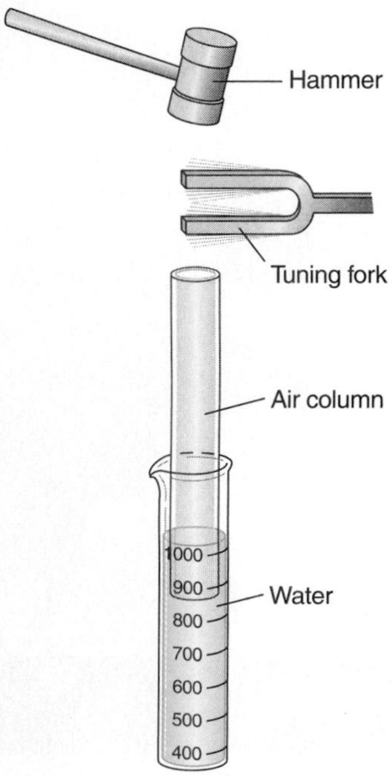

Figure 15.23 Resonance of an air column

driving force is at the natural frequency of a system. Auto body rattles sometimes occur at certain speeds and disappear for small speed changes. Radio receivers operate on the principle of resonance. The natural frequency of vibration of electrical currents in a circuit may be tuned to that of an incoming radio signal, which is then amplified and converted into sound.

The playing of a musical instrument demonstrates resonance. In wind instruments either the lips or a reed vibrates. Without an air column, however, no music is produced. The tube that makes up the instrument is necessary. Air in the tube vibrates at the same frequency as the lips or reed (resonance) to produce the musical sound. The pitch of the sound is varied by changing the length of the resonating column of vibrating air. The length of the air column determines the resonant frequencies of the vibrating air. This is easily demonstrated by holding a tuning fork above a hollow tube and varying the length of the air column (Fig. 15.23). The sound is louder when the air column is in resonance with the tuning fork.

The tuning fork produces alternating high- and low-pressure variations in the air as it vibrates. Because of the movement of waves up and down the tube with accompanying constructive and destructive interference between the waves, resonance is found at one-fourth wavelengths in a closed tube. Resonance is found at multiples of $\lambda/4$. Open pipes also resonate. However, resonance in open pipes is at half-wavelength multiples ($\lambda/2$, λ, $3\lambda/2$, etc.).

15.6 Simple Harmonic Motion

Periodic motion occurs when an object moves repeatedly over the same path in equal time intervals. Attach a mass m to a spring suspended from a support [Fig. 15.24(a)]. Pull the mass down (b) and release it (c). The mass moves up and down in periodic motion.

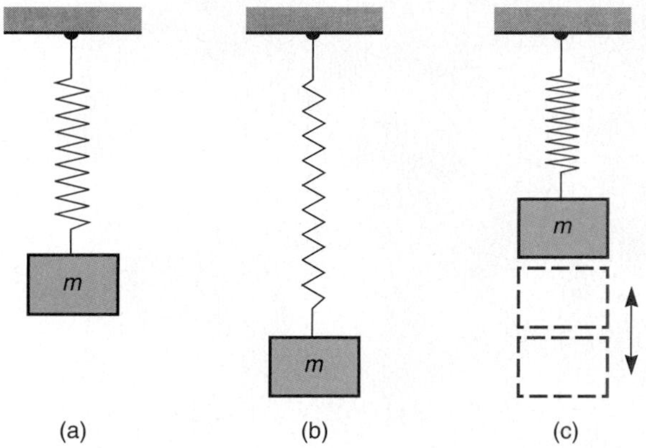

Figure 15.24 Mass suspended by a spring from a support in simple harmonic motion

Simple harmonic motion is a type of linear motion in which the acceleration of an object is directly proportional to its displacement from its equilibrium position; the motion is always directed toward the equilibrium position. That is, the farther the spring is pulled down, the faster the spring moves when it is released, and the motion is always directed toward the equilibrium (rest) position. The mass on the spring in Fig. 15.24 is an example of an object in simple harmonic motion.

Next, we will compare simple harmonic motion and circular motion and discuss some of the corresponding terms. Assume that a Ferris wheel is rotating uniformly, with you in the only seat and with the sun directly overhead (Fig. 15.25). Now compare your position with the position of your shadow. When you are in position a, your shadow is in position a'; when you are in position b, your shadow is in position b'; and so on. As you complete one revolution on the Ferris wheel, your shadow makes one complete vibration (cycle) on the ground in simple harmonic motion. When your shadow is at b', the *displacement* of your shadow is the distance $b'O'$, which is the distance from your shadow to the midpoint of its vibration, O'. In general, the **displacement** of an object in simple harmonic motion is its distance from its equilibrium, or rest, position. The **amplitude** of the vibration is the maximum displacement $O'P$ or $O'Q$, which is also the radius of the Ferris wheel. The **period** is the time

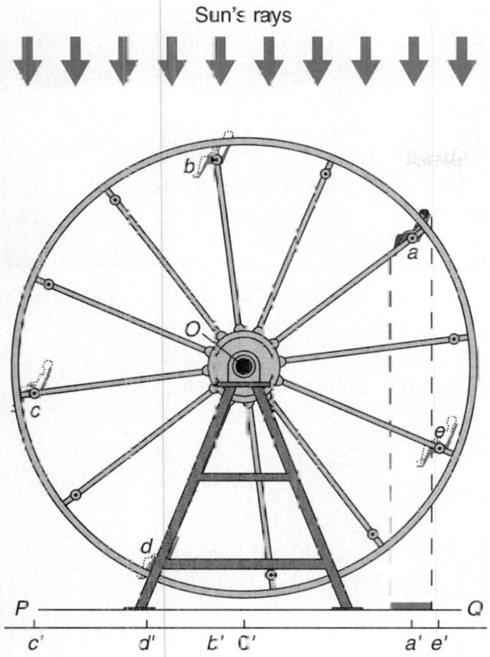

Figure 15.25 Simple harmonic motion of the shadow of a person on line *PQ*, where the person is rotating at uniform speed in a circle on a Ferris wheel

required for one complete vibration—the time required for you to make one complete revolution on the Ferris wheel and the time required for your shadow to make one complete vibration on the ground. The **frequency** is the number of complete vibrations per unit of time or the number of complete revolutions that you make on the Ferris wheel per unit of time. The motion of the rider when graphed over time produces a special curve called a sine wave as shown in Fig. 15.26. The frequency f equals the reciprocal of the period T. That is, $f = \frac{1}{T}$. The equilibrium position of the shadow is the midpoint of its path, O'.

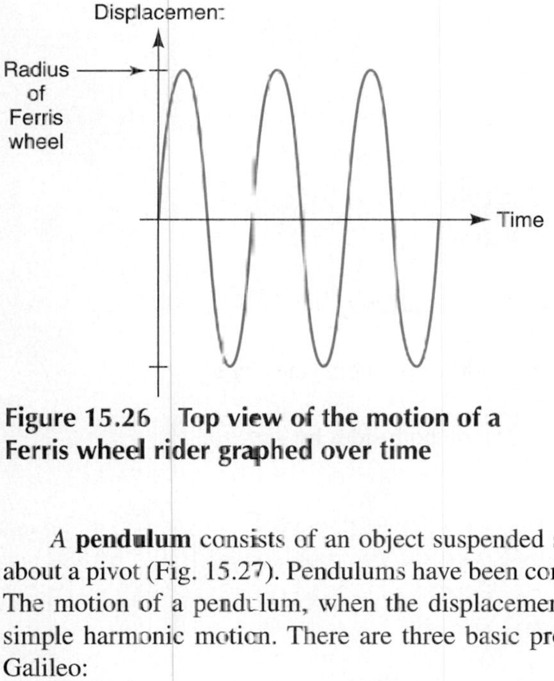

Figure 15.26 Top view of the motion of a Ferris wheel rider graphed over time

A **pendulum** consists of an object suspended so that it swings freely back and forth about a pivot (Fig. 15.27). Pendulums have been commonly used in clocks for many years. The motion of a pendulum, when the displacement is small, very closely approximates simple harmonic motion. There are three basic properties of a pendulum discovered by Galileo:

Figure 15.27 Free-swinging pendulum of length *l*

1. Its period is independent of its mass. (Air resistance is more affected by the size and shape of the bob than by its mass.)

TRY THIS ACTIVITY

Swing Set Physics

A swing often serves as a good pendulum. Using a stopwatch, measure the period of a swing as it moves forward and backward. Vary the amplitude of the swing, the mass of the person sitting on it, and the length of the swing's chain. Which variable alters the period?

2. Its period is independent of the amplitude when the arc is small, that is, when its arc is less than 10°.
3. Subject to these conditions, its period is given by

$$T = 2\pi\sqrt{\frac{l}{g}}$$

where

T = period (usually, in seconds)
l = length of pendulum (m or ft)
$g = 9.80\ \text{m/s}^2$ or $32.2\ \text{ft/s}^2$

The period of any pendulum depends only on its length and the acceleration of gravity. The longer the pendulum, the longer is the time for each complete swing or period.

EXAMPLE

Find the length (in cm) of a pendulum with a period of 1.50 s.

Data:

$$T = 1.50\ \text{s}$$
$$g = 9.80\ \text{m/s}^2$$
$$l = ?$$

Basic Equation:

$$T = 2\pi\sqrt{\frac{l}{g}}$$

$$T^2 = 4\pi^2\frac{l}{g} \quad \text{Square both sides.}$$

$$gT^2 = 4\pi^2 l \quad \text{Multiply both sides by } g.$$

$$\frac{gT^2}{4\pi^2} = l \quad \text{Divide both sides by } 4\pi^2.$$

Working Equation:

$$l = \frac{gT^2}{4\pi^2}$$

Substitution:

$$l = \frac{(9.80\ \text{m/s}^2)(1.50\ \text{s})^2}{4\pi^2}$$
$$= 0.559\ \text{m} = 55.9\ \text{cm}$$

PROBLEMS 15.6

1. Find the length (in cm) of a pendulum with a period of 1.50 s.
2. Find the length (in ft) of a pendulum with a period of 3.00 s.
3. Find the period of a pendulum 1.25 m long.
4. Find the period of a pendulum 2.00 ft long
5. Find the length (in in.) of a pendulum with a period of 2.25 s.
6. Find the length (in m) of a pendulum with a period of 0.700 s.
7. Find the period of a pendulum 18.0 in. long.
8. Find the period of a pendulum 35.0 cm long.
9. If you double the length of a pendulum, what happens to its period?
10. If you double the period of a pendulum, what happens to its length?
11. A grandfather clock has a 0.750-m pendulum. What is its period?
12. A grandfather clock has a pendulum with period 2.40 s. (a) Find the length of this pendulum. (b) Is the length of this pendulum longer or shorter than the length of the pendulum of the clock in Problem 11?

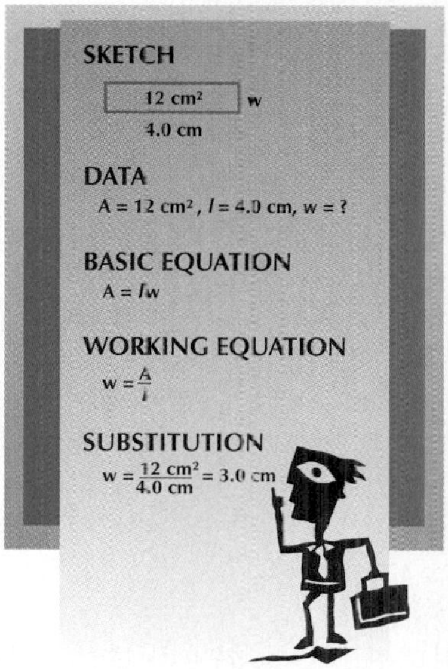

Glossary

Amplitude The maximum displacement of any part of a wave or a vibration from its equilibrium, or rest, position. (pp. 402, 420)

Constructive Interference The superposition of waves to form a larger disturbance (wave) in a medium. Occurs when two crests or troughs of superimposed waves meet. (p. 405)

Destructive Interference The superposition of waves to form a smaller disturbance (wave) in a medium. (p. 405)

Diffraction The property of a wave that describes its ability to bend around obstacles in its path. (p. 407)

Displacement The distance of an object in simple harmonic motion from its equilibrium, or rest, position. (p. 420)

Doppler Effect The variation of the frequency heard when a source of sound and the ear are moving relative to each other. (p. 415)

Electromagnetic Spectrum The entire range of electromagnetic waves classified according to frequency. (p. 410)

Electromagnetic Wave A transverse wave resulting from a periodic disturbance in an electromagnetic field having an electric component and a magnetic component, each being perpendicular to the other and both perpendicular to the direction of travel. (p. 409)

Frequency The number of complete waves passing a given point per unit time; the number of complete vibrations per unit time in simple harmonic motion. (pp. 403, 421)

Intensity The energy transferred by sound per unit time through unit area. (p. 414)

Interference The effect of two intersecting waves resulting in a loss of displacement in certain areas and an increase in displacement in others. (p. 407)

Longitudinal Wave A disturbance in a medium in which the motion of the particles is along the direction of the wave travel. (p. 402)

Loudness The strength of the sensation of sound to an observer. (p. 414)

Natural Frequency The frequency at which an object vibrates when struck by another object, such as a rubber hammer. (p. 418)

Pendulum An object suspended so that it swings freely back and forth about a pivot. (p. 421)

Period The time required for a single wave to pass a given point or the time required for one complete vibration of an object in simple harmonic motion. (pp. 403, 420)

Pitch The effect of the frequency of sound waves on the ear. (p. 414)

Propagation Velocity The velocity of energy transfer of a wave, given by the distance traveled by the wave in one period divided by the period. (p. 404)

Pulse Nonrepeated disturbance that carries energy through a medium or through space. (p. 402)

Resonance A sympathetic vibration of an object caused by the transfer of energy from another object vibrating at the natural frequency of vibration of the first object. (p. 418)

Simple Harmonic Motion A type of linear motion of an object in which the acceleration is directly proportional to its displacement from its equilibrium position and the motion is always directed to the equilibrium position. (p. 420)

Sound Those waves transmitted through a medium with frequencies capable of being detected by the human ear. (p. 412)

Speed of Light The speed at which light and other forms of electromagnetic radiation travel: 3.00×10^8 m/s in a vacuum. (pp. 409, 412)

Speed of Sound The speed at which sound waves travel in a medium: 331 m/s in dry air at 1 atm pressure and 0°C. (p. 412)

Standing Waves A special case of superposition of two waves when no energy propagation occurs along the wave. The two waves of equal amplitude and wavelength do not appear to be traveling. (p. 406)

Superposition of Waves The algebraic sum of the separate displacements of two or more individual waves passing through a medium. (p. 405)

Transverse Wave A disturbance in a medium in which the motion of the particles is perpendicular to the direction of the wave motion. (p. 402)

Wave A disturbance that moves through a medium or through space. (p. 402)

Wavelength The minimum distance between particles in a wave that have the same displacement and are moving in the same direction. (p. 403)

Formulas

15.1 $f = \frac{1}{T}$

$v = \lambda f$

15.2 $c = \lambda f$

15.3 $v = 331 \text{ m/s} + \left(0.61 \frac{\text{m/s}}{°\text{C}}\right)T$

$v = 1087 \text{ ft/s} + \left(1.1 \frac{\text{ft/s}}{°\text{F}}\right)(T - 32°\text{F})$

15.4 $f' = f\left(\frac{v}{v \pm v_s}\right)$

15.6 $T = 2\pi\sqrt{\frac{l}{g}}$

Review Questions

1. Which of the following are methods of energy transfer?
   (a) Conduction (b) Radiation
   (c) Wave motion (d) None of the above
2. The minimum distance between particles in a wave that have the same displacement and are moving in the same direction is called
   (a) the period. (b) the frequency.
   (c) the wavelength. (d) none of the above.
3. Which of the following refers to the time required for a single wave to pass a given point?
   (a) The period (b) The frequency
   (c) The wavelength (d) None of the above

4. Which of the following refers to the number of complete waves passing a given point per unit time?
   (a) The period (b) The frequency
   (c) The wavelength (d) None of the above
5. An example of a transverse wave is
   (a) a sound wave. (b) a water wave.
   (c) interference. (d) none of the above.
6. Which of the following is an example of longitudinal waves?
   (a) Sound waves (b) Water waves
   (c) Interference (d) None of the above
7. Which of the following are electromagnetic waves?
   (a) Sound (b) Water waves
   (c) Radar waves (d) X rays
   (e) All of the above
8. Explain the difference between interference and diffraction.
9. Explain the difference between constructive and destructive interference.
10. If waves did not exhibit the property of diffraction, under what conditions would your stereo system sound different?
11. Give an example of diffraction of water waves.
12. What happens to the frequency of a vibrating string on a guitar if the length of the string is decreased?
13. Explain the difference between a wave and a pulse.
14. Give an example of a pulse.
15. Why does the setting sun appear reddish in color?
16. Why does the sky appear blue?
17. What happens to the speed of sound when the temperature increases? Explain why this might happen.
18. Explain how a seismograph works.
19. How does the speed of sound differ in water and air? Explain the reason for this difference.
20. In your own words, explain the Doppler effect.
21. Distinguish between sympathetic and forced vibration.
22. In your own words, explain resonance.
23. State a reason that might explain why many stars appear to have their light shifted to the red (longer wavelength) part of the electromagnetic spectrum when viewed from the earth.
24. Distinguish between amplitude and displacement.
25. Distinguish between period and frequency.
26. Does the period of a pendulum depend on its mass, and if so, how?

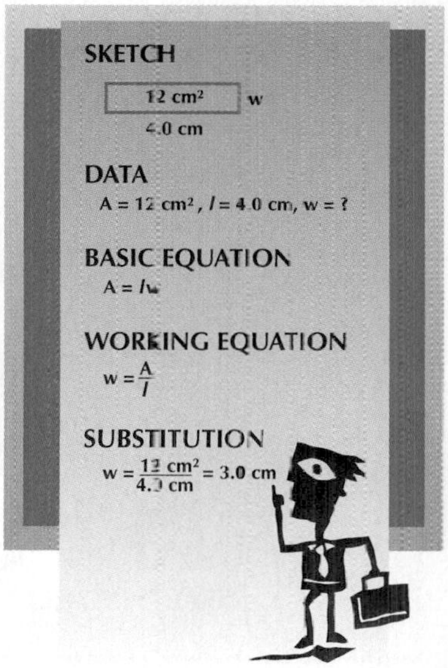

Review Problems

1. Find the period of a wave with frequency 355 kHz.
2. Find the frequency of a wave with period 0.320 s.
3. (a) What is the frequency of a light wave with wavelength 4.50×10^{-7} m and velocity 3.00×10^{8} m/s? (b) Find the period of the wave.
4. Find the speed of a wave with frequency 8.97 Hz and wavelength 0.654 m.
5. What is the wavelength of longitudinal waves in a coil spring with frequency 4.65 Hz and velocity 5.78 m/s?
6. Find the frequency of a wave produced by a generator that emits 85 pulses in 1.3 s.
7. What is the wavelength of an electromagnetic wave with frequency 65.5 MHz?
8. Find the speed of sound in m/s at 85°C at 1 atm pressure in dry air.
9. Find the speed of sound in m/s at −35°C at 1 atm pressure in dry air.
10. How long will it take a sound wave to travel through 1450 m of water at 25°C?

11. A sound wave is transmitted through water from one ship, is reflected off another ship 22 km away, and returns to the sonar receiver on the first ship. What is the round-trip transit time for the sound wave if the water temperature is 23°C?
12. A train traveling at a speed of 95 mi/h approaches an observer at a station and sounds a 525-Hz whistle. What frequency will be heard by the observer? Assume that the sound velocity in air is 1090 ft/s.
13. A car is traveling toward you at 95 km/h. The car horn produces a sound at frequency 4950 Hz. (a) What frequency do you hear? Assume that the sound velocity in air is 345 m/s. (b) What frequency do you hear if the car is traveling away from you?
14. What is the frequency of the sound waves being emitted from a train whistle while approaching at 45 m/s in air that is 11°C? The perceived frequency is 425 Hz.
15. The tail light on a car produces light with wavelength 5.00×10^{-7} m. What frequency do you observe when the car is departing at 24 m/s at 0°C?
16. A pendulum has a length of 0.450 m. What is its period?
17. A pendulum has a period of 0.700 s. Find the length of the pendulum in inches.

APPLIED CONCEPTS

1. The pendulum on a grandfather clock is calibrated so its period equals 1.00 s. (a) What is the length of the pendulum cable? (b) If the grandfather clock is moved to the moon where the acceleration due to gravity is 1.62 m/s^2, will the clock keep the correct time? (c) If not, what can be done to the pendulum to correct the problem?
2. The Tacoma Narrows Bridge, built across Puget Sound in Washington, formed a standing wave before it collapsed on November 7, 1940. As seen in the photo in Fig. 15.28, (a) how many full wavelengths were between the two towers, which were spaced 2800 ft apart? (b) If the frequency of the vibrations was 0.20 Hz, find the wave speed for this mechanical wave. (c) Find the wavelength of the bridge's standing wave if the frequency is doubled.

Figure 15.28

3. Maintaining strong AM radio reception when driving under overpasses or through tunnels is often difficult. FM radio frequencies typically are not as affected as AM frequencies. (a) Using the frequency range for AM radio, 550 kHz to 1650 kHz, and FM radio, 88 MHz to 108 MHz, find the range of wavelengths for both AM and FM bands. (b) In terms of wave diffraction, explain why AM radio waves have a more difficult time passing under overpasses and through tunnels.
4. Dave, a jet engine technician, is exposed to sound intensities of 10.3 W/m^2 (12$\bar{0}$ dB) at 30.4 m from a jet (Fig. 15.29). (a) What is the audible power produced by the jet engine? (b) Since the sound power produced by the engine remains constant and the sound propagates in all directions (spherical area), what is the intensity of the sound 60.8 m from the engine? (c) Compare the intensity of the sound at 30.4 m to the intensity of the sound at 60.8 m from the engine.

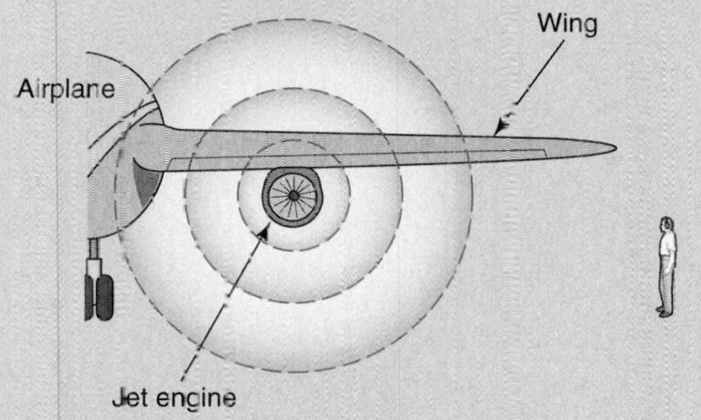

Figure 15.29

5. The speed of an automobile can be determined using the Doppler effect by sounding the horn. (a) An automobile's horn produces a frequency of 765 Hz. How fast is the car traveling if a stationary microphone measures the horn's frequency as 836 Hz at a temperature of 23.3°C? (b) What is the frequency of the sound wave as the car moves away from the microphone?

CHAPTER 16

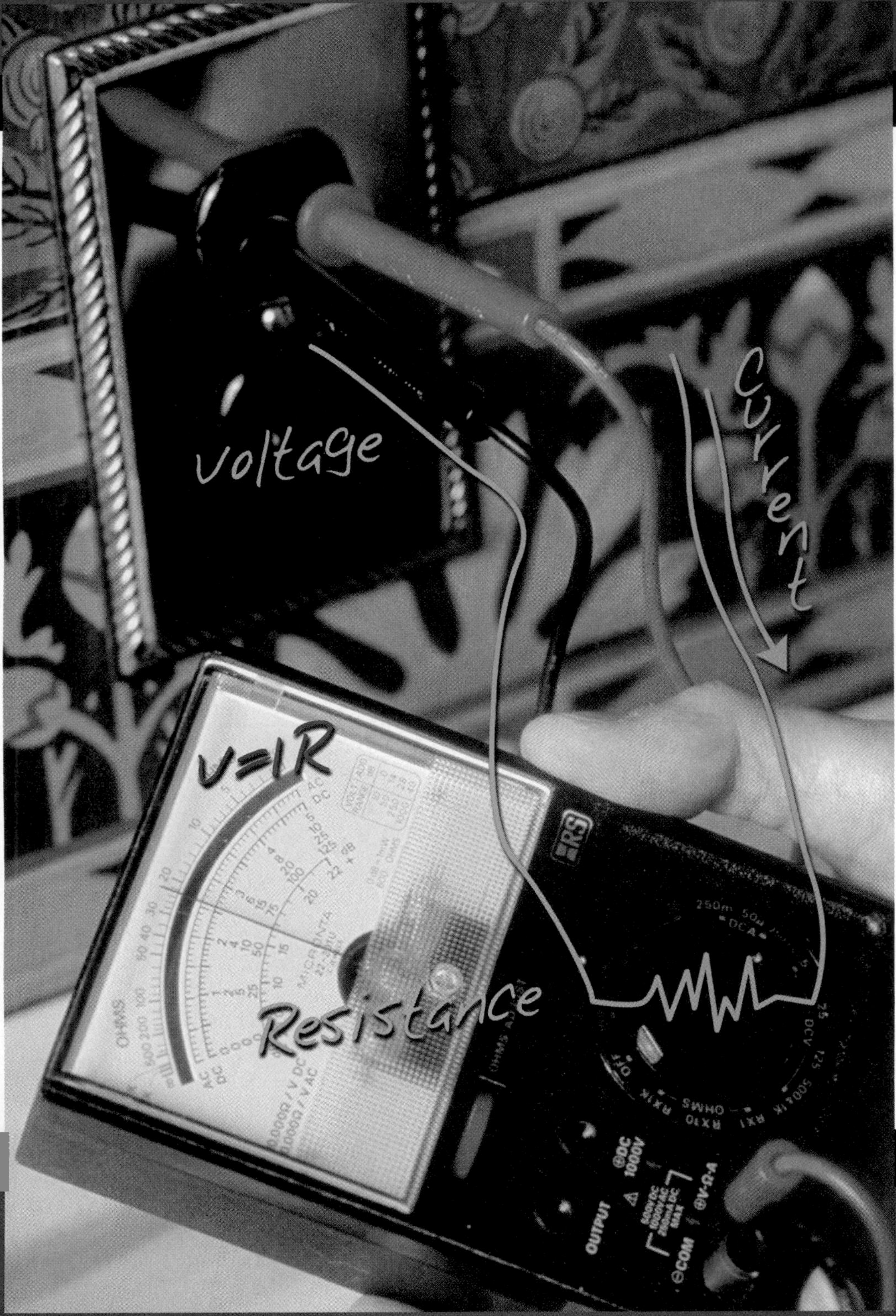

BASIC ELECTRICITY

Electricity is the best means of transmitting energy for many purposes. We consider the basics of electricity including electric charge, electric fields, static electricity, electric current and circuits, batteries, and electric power.

Objectives

The major goals of this chapter are to enable you to:

1. Describe the nature of electric charges.
2. Distinguish conduction and induction.
3. Use Coulomb's law to find the force between charges.
4. Describe the characteristics of electricity.
5. Use Ohm's law to solve electric flow problems.
6. Use electrical symbols to describe circuits.
7. Find current, voltage, and resistance in simple circuits.
8. Describe the nature of cells and batteries.
9. Analyze circuits with cells in series and parallel.
10. Find electric power.

16.1 Electric Charges

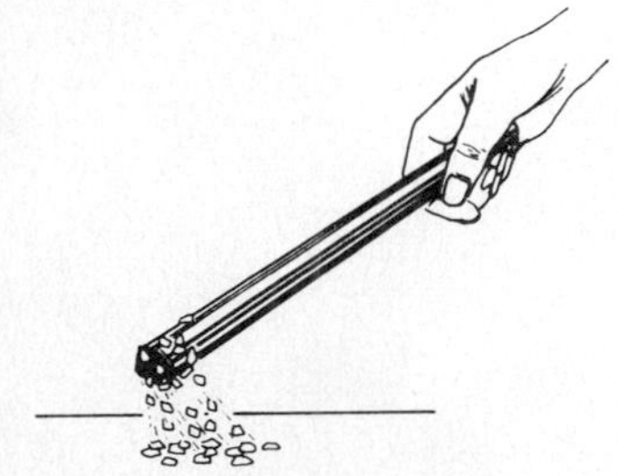

Figure 16.1 Amber rod attracting bits of paper after being rubbed

Electrification was first studied 2500 years ago in ancient Greece. It was found that when an amber rod was rubbed with a wool cloth, the rod attracted small objects (Fig. 16.1). When two objects are rubbed together, they become charged.

When you slide rubber-soled shoes on a wool rug on a dry day, you become electrified. That is, you have acquired a static charge. This static charge is usually lost when you touch an object at a different potential as in Fig. 16.2. Part of the static charge may be lost when you touch an object, such as another person, to which charge is transferred.

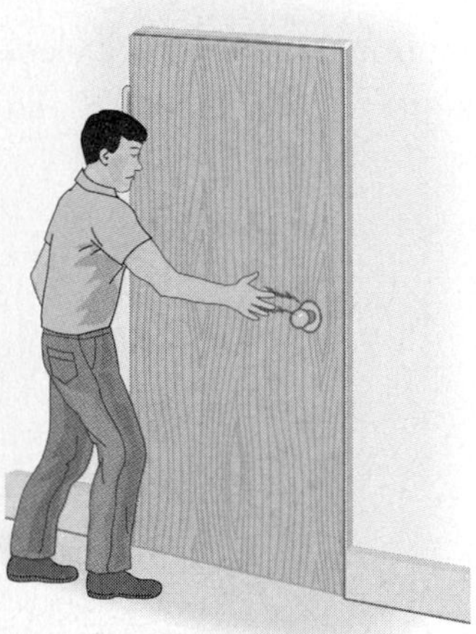

Figure 16.2 Stored charge being transferred

To understand electricity, we need to know more about the structure of matter. We have seen that all matter is made up of atoms. These atoms are made of electrons, protons, and neutrons. Each **proton** has one unit of positive charge and each **electron** has one unit of negative charge. The **neutron** has no charge. The protons and neutrons are tightly packed into what is called the *nucleus*. Electrons may be thought of as small charged clouds that surround the nucleus of atoms (Fig. 16.3). An atom normally has the same number of electrons as protons and thus is uncharged. If an electron is removed, the atom is left with a *positive charge* (+), that is, an excess of protons. If an extra electron is added, the atom has a *negative charge* (−), that is, an excess of electrons. **Benjamin Franklin** first referred to electric charges as being positive and negative. The study of electric charges and the forces between them is referred to as *electrostatics*. As the name implies, *static electricity* is electricity that does not flow. *Current electricity* is the flow of electric charge through a conductor.

When two materials are rubbed together, the atoms on the two surfaces move across each other and brush off electrons. The electrons are transferred from one surface to the other. One surface is then left with a *positive charge* and the other is *negative*. This process is called *electrification*.

The two types of electric charges can be observed indirectly by using an electroscope. A very simple electroscope is a ball of wood pith on a silk thread [Fig. 16.4(a)]. We can produce a charge on a hard rubber rod by rubbing it with a wool cloth. The rubber rod acquires a negative charge and the wool acquires a positive charge. The universal acceptance of the description of these charges establishes the convention of positive and negative charge in electric circuits. Now transfer some of this negative charge from the rubber rod to the pith ball [Fig. 16.4(b)]. The pith ball becomes negatively charged by **conduction,** a transfer of charge from one place to another. Another pith ball charged in the same way is repelled by the other pith ball [Fig. 16.4(c)]. This charge is *negative* (−).

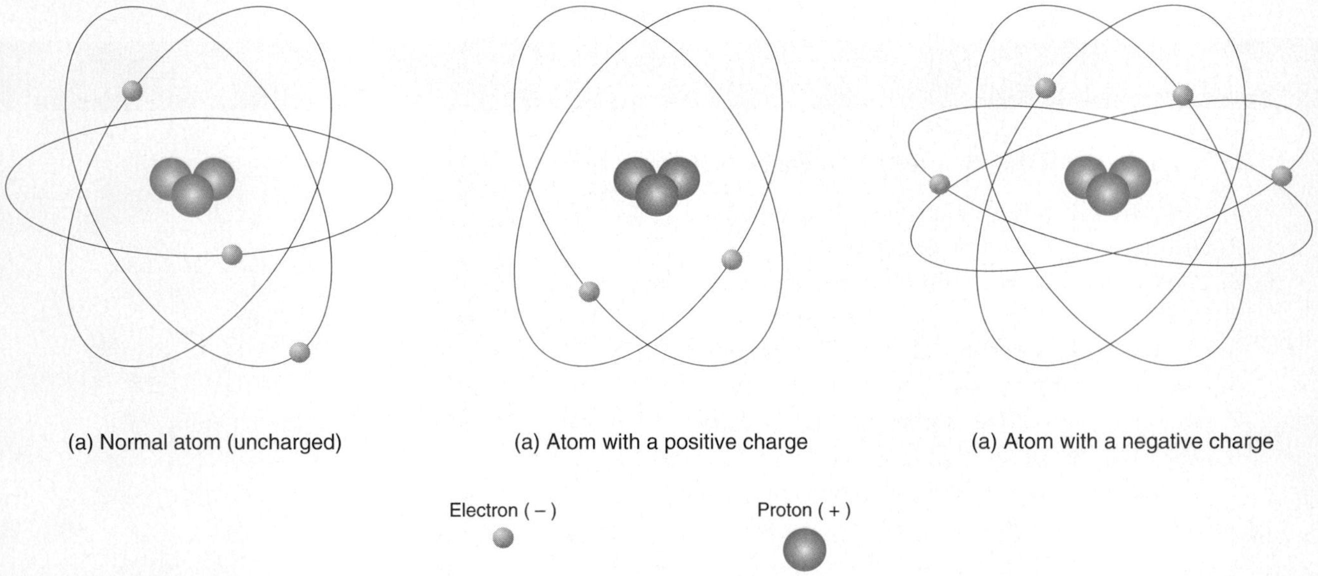

Figure 16.3

TRY THIS ACTIVITY

Picking Up Dust

Rub a furry or wool fabric against a plastic rod for approximately 15 to 20 s. Bring the rod into a dusty area or near a pile of sawdust. What happens to some of the dust or sawdust? Explain what is taking place in relation to the movement of charge.

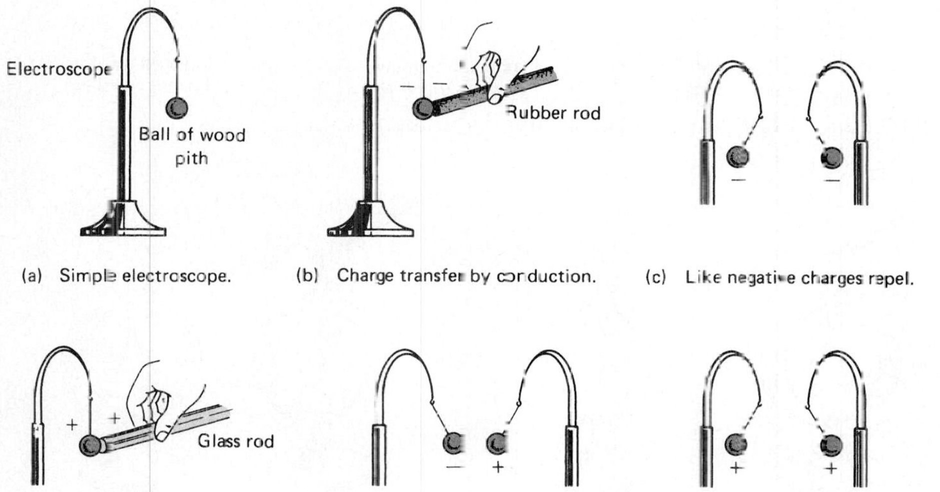

(a) Simple electroscope. (b) Charge transfer by conduction. (c) Like negative charges repel.

(d) Positive charge by conduction. (e) Unlike charges attract. (f) Like positive charges repel.

Figure 16.4 Charging a simple electroscope

Now rub a glass rod with silk. The glass rod acquires a positive charge and the silk acquires a negative charge. Transfer some of the positive charge from the glass rod to a pith ball [Fig. 16.4(d)]. This pith ball is attracted to the negatively charged pith ball [Fig. 16.4(e)]. The charge produced by glass and silk is called a *positive charge* (+). Two pith balls that are positively charged will repel each other [Fig. 16.4(f)].

Benjamin Franklin (1706–1790),

printer, writer, scientist, and statesman, was born in Boston, Massachusetts. He first became a skilled printer and successful businessman. He retired from business in 1748 to devote more time to his scientific interests. He developed the fuel-efficient Franklin stove and conducted a series of experiments in electricity, which brought him international recognition as a scientist. Through his famous kite experiment, he demonstrated that lightning is an electric discharge. He invented the lightning rod and the bifocal lens. Later in his life, he became a well-known statesman. The two types of electric charge were referred to as positive and negative by Franklin. He arbitrarily called the charge on the rubbed glass rod "positive" and the charge on the rubber rod "negative." We still follow his convention today.

16.2 Induction

Induction is a method of charging one object by bringing a charged object near to, but not touching, it. The leaf electroscope is more sensitive than the pith ball type and can also be used to show electrification or charging by induction. The leaf electroscope [Fig. 16.5(a)] usually consists of a metal rod with a metal ball on one end and two thin strips of gold foil leaf hanging from the other end. The delicate leaves are enclosed to protect them and the rod is insulated from the enclosure. When charge is placed on the leaves, they diverge because of the force of repulsion of their similar charge.

Electroscopes may be charged by *conduction* by touching a charged object to the metal ball. Electroscopes are charged by *induction* by bringing a charged object near to, but not touching, the metal ball. Charges of the same sign as the charged object are repelled to the leaves, which repel each other and separate [Fig. 16.5(b) and (c)].

When the charged object is removed, the leaves close as the free electrons redistribute themselves over the ball, rod, and leaves. The electroscope has been only temporarily charged by induction. A residual charge by induction may be obtained by charging the electroscope temporarily as in Fig. 16.6(a) and then touching the electroscope with a neutral object (like your finger) while the charged object is still held close [Fig. 16.6(b)]. The neutral object provides a path for some of the induced charge on the electroscope to escape. When everything is removed, a residual charge by induction remains on the electroscope [Fig. 16.6(d)].

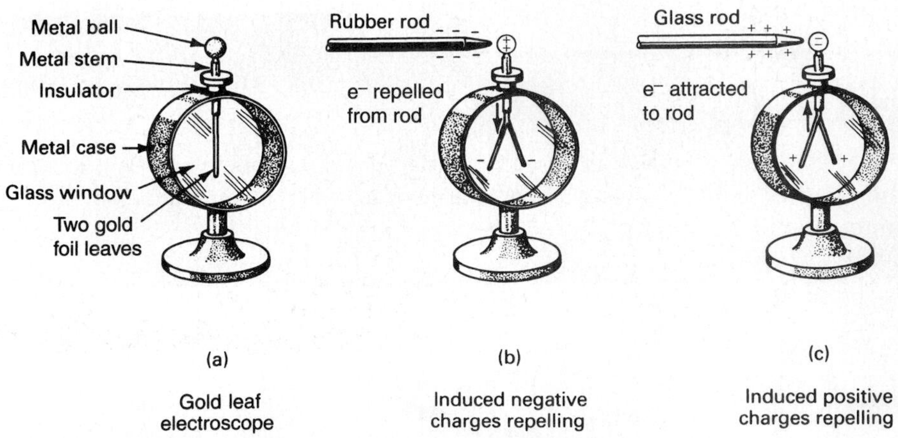

Figure 16.5 Charging an electroscope by induction

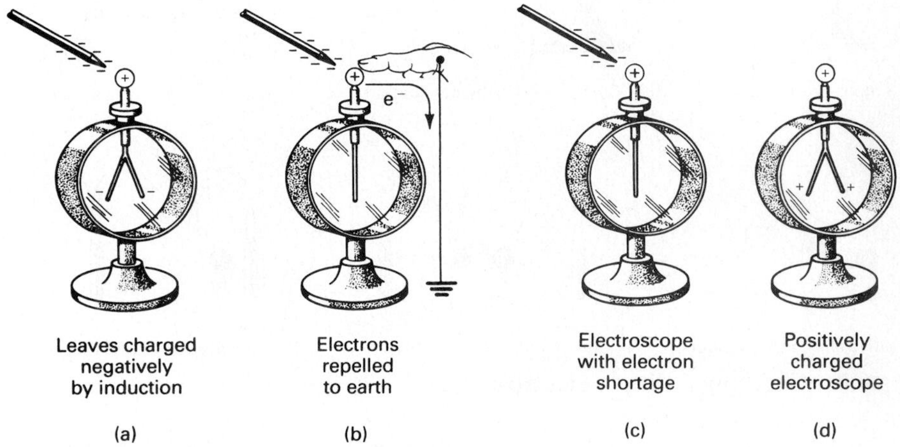

Figure 16.6 Residual charge by induction

In summary,

> Like charges repel each other and unlike charges attract each other.

TRY THIS ACTIVITY

Repelled Balloons

Inflate two latex balloons, tie a piece of light string to each balloon, and suspend them from the ceiling or a horizontal rod so they are 1 ft or so apart. Rub each of the balloons with a furry fabric or your hair. Both balloons are now charged. Slowly reduce the distance between the two balloons. Look for evidence of an electric force. If you see such evidence, explain why the force is either attractive or repulsive. In addition, what are the two main factors that play a role in the strength of the electric force?

Lightning is simply a huge static electricity spark produced in the atmosphere by moving air masses that results in a tremendous discharge (Fig. 16.7). During a thunderstorm the negative charge build-up at the bottom of a layer of clouds induces a positive charge on the ground below. The ground is charged by induction, and lightning is the resulting electric

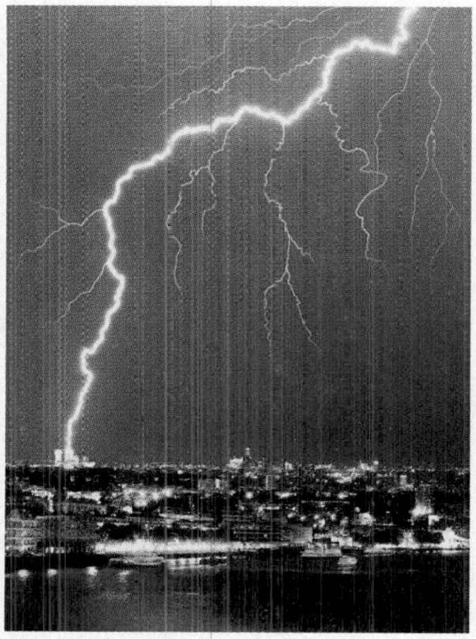

Courtesy of Pearson Education-Asia

Figure 16.7 Lightning: static electric discharge

discharge between the negatively charged clouds and the positively charged ground. Lightning also occurs as electric discharges between oppositely charged parts of clouds.

The lightning rod was invented by Benjamin Franklin, who observed that electricity from the air tended to accumulate on pointed objects. He found that a build-up of charge is decreased by installing a pointed rod or rods on the roof of a building and connecting the rod to the ground with a heavy conducting wire. A lightning rod not only prevents a large build-up of charge by induction, but also provides a conducting path to the ground for any sudden tremendous discharge of a lightning strike. This often results in protecting the building from damage.

The *van de Graaff generator*, named after **Robert van de Graaff,** is a laboratory machine that is used to produce static electricity and transfer it to a metal sphere by conduction. The van de Graaff consists of an electron source, a rubber belt driven by a motor, and a metal sphere supported by an insulating stand (Fig. 16.8). The electron source

Robert van de Graaff (1901–1967),

physicist, was born in Tuscaloosa, Alabama. He invented a constant-potential electrostatic generator (later known as the van de Graaff generator) and later developed this generator for use as a particle accelerator. He invented the insulating-core transformer in the late 1950s.

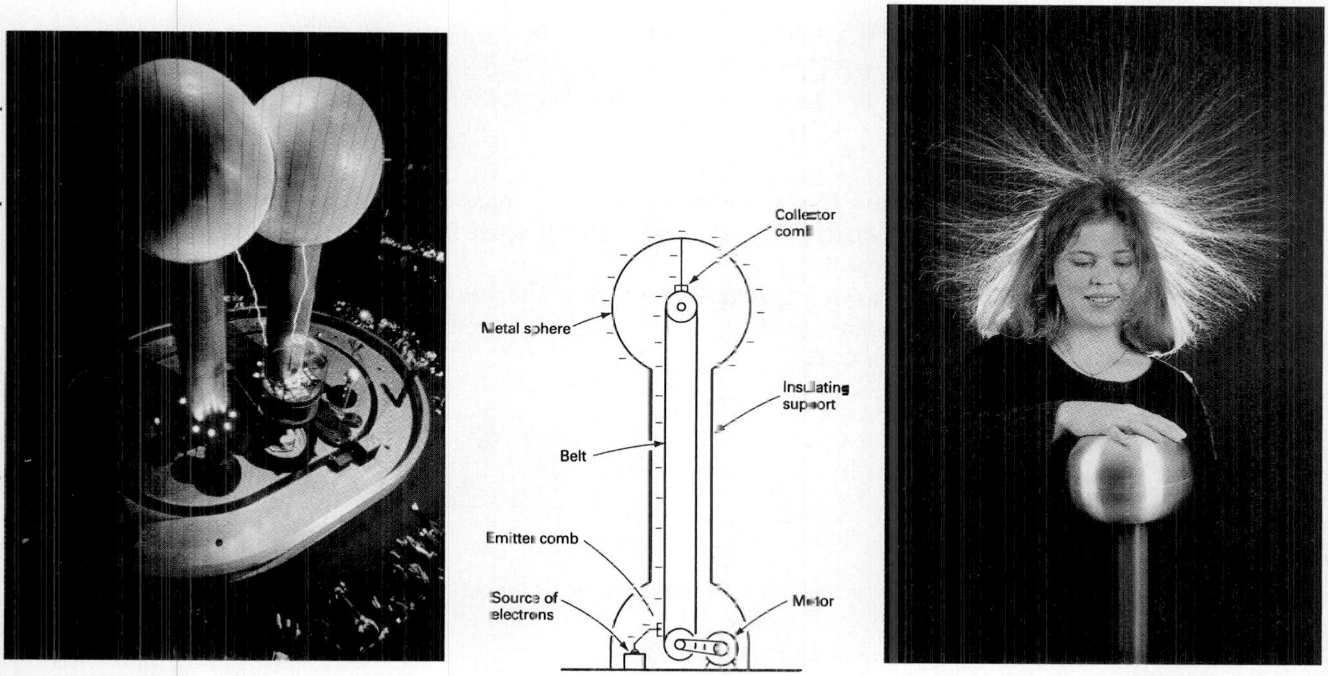

Copyright of Hank Morgan/Photo Researchers, Inc. Photo reprinted with permission

Copyright of Richard Megna/Fundamental Photographs

(a) Laboratory model (b) Schematic (c) Student model

Figure 16.8 Van de Graaff generator

charges the rubber belt as it passes by, which carries the electrons up to the sphere and deposits them there. This builds up a high potential difference (several hundred thousand volts) between the sphere and the ground. Note that there is no charge on the inside surface of the sphere.

16.3 Coulomb's Law

Charles Coulomb (1736–1806),

physicist, was born in France. He pioneered research in electricity and magnetism. The unit of measuring electricity, the coulomb, is named after him.

The charge on a proton is denoted by the symbol e^+ and the electron's charge by e^-. In the study of electricity, a much larger unit of charge is required and is called the *coulomb*, C, named after **Charles Coulomb.** The measurement of the unit charges gives $e^+ = +1.60 \times 10^{-19}$ C and $e^- = -1.60 \times 10^{-19}$ C. Thus, a collection of 6.25×10^{18} electrons, which is defined by

$$\frac{1}{1.60 \times 10^{-19}} = 6.25 \times 10^{18}$$

has a total charge of 1.00 C. Although this seems like a lot of electrons, it represents only the amount of charge that passes through a 100-W light bulb in about 1 s. We use q to denote amount of electric charge.

In 1789, Coulomb made a scientific study of the forces of attraction and repulsion between charged objects using a very sensitive torsion balance. From his experiments he determined the existence of an *inverse square law* for charged particles that can be used to calculate the forces of attraction or repulsion between charged objects. The inverse-square-law behavior was also found to apply to the much weaker gravitational force studied in Chapter 4. While the gravitational force is always attractive, the electric force can be either attractive or repulsive. These two forces are important because one holds the solar system together (gravity) and the other holds atoms and molecules together (electricity). Electric forces, however, are billions of times greater than the earth's gravitational force.

COULOMB'S LAW OF ELECTROSTATICS

The force between two point charges q_1 and q_2 is directly proportional to the product of their magnitudes and inversely proportional to the square of the distance separating them, r.

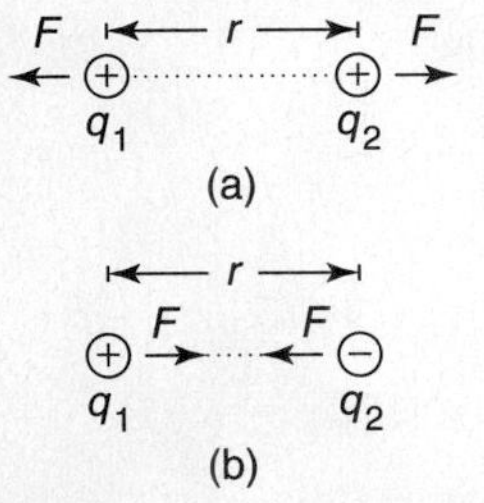

Figure 16.9 (a) Two like charges at distance *r* apart repel each other with force *F*. (b) Two unlike charges attract.

Figure 16.9(a) shows the repulsive force between two like (positive in this case) charges separated by a distance r. The attractive force between two unlike charges is shown in Fig. 16.9(b).

We use a *proportionality constant* k in writing Coulomb's law as an equation to take into account the air or other medium between the charges. Written in equation form, **Coulomb's law** becomes

$$\boxed{F = \frac{kq_1q_2}{r^2}}$$

where

$F =$	force of attraction or repulsion	(in newtons)
$k = 9.00 \times 10^9$ N m^2/C^2		(k was found by experiment)
$q_1, q_2 =$	electric charges	(in coulombs)
$r =$	distance between the charges	(in metres)

The force between the charges is a vector quantity that acts on each charge.

TRY THIS ACTIVITY

Attracting Water

Turn on a faucet so only a very thin stream of water comes out of the tap. Charge a plastic rod, latex balloon, or plastic comb by rubbing it with a piece of wool or fur. Slowly bring the charged object toward the stream of water. Explain what happens to the stream of water as it flows from the tap.

EXAMPLE

Two charges, each with magnitude +6.50 μC, are separated by a distance of 0.200 cm. Find the force of repulsion between them.

Data:

$$q_1 = q_2 = +6.50\ \mu\text{C} = +6.50 \times 10^{-6}\ \text{C}$$
$$r = 0.200\ \text{cm} = 0.00200\ \text{m} = 2.00 \times 10^{-3}\ \text{m}$$
$$k = 9.00 \times 10^{9}\ \text{N m}^2/\text{C}^2$$
$$F = ?$$

Basic Equation:

$$F = \frac{kq_1q_2}{r^2}$$

Working Equation: Same

Substitution:

$$F = \frac{(9.00 \times 10^{9}\ \text{N m}^2/\text{C}^2)(6.50 \times 10^{-6}\ \text{C})(6.50 \times 10^{-6}\ \text{C})}{(2.00 \times 10^{-3}\ \text{m})^2}$$
$$= 9.51 \times 10^{4}\ \text{N}$$

PROBLEMS 16.3

1. Two identical charges, each -8.00×10^{-5} C, are separated by a distance of 25.0 cm. What is the force of repulsion?
2. The force of repulsion between two identical positive charges is 0.800 N when the charges are 0.100 m apart. Find the value of each charge.
3. A charge of $+3.0 \times 10^{-6}$ C exerts a force of 940 N on a charge of -6.0×10^{-6} C. How far apart are the charges?
4. A charge of -3.0×10^{-8} C exerts a force of 0.045 N on a charge of $+5.0 \times 10^{-7}$ C. How far apart are the charges?
5. When a −9.0-μC charge is placed 0.12 cm from a charge q in a vacuum, the force between the two charges is 850 N. What is the value of q?
6. How far apart are two identical charges of +6.00 μC if the force between them is 25.0 N?
7. Three charges are located along the x-axis. Charge A (+3.00 μC) is located at the origin. Charge B (+5.50 μC) is located at $x = +0.400$ m. Charge C (−4.60 μC) is located at $x = +0.750$ m. (a) Find the total force (and direction) on charge B. (b) Find the total force (and direction) on charge A. (c) Find the total force (and direction) on charge C.
8. Three charges are located along the x-axis. Charge A (+5.00 μC) is located at the origin. Charge B (+4.50 μC) is located at $x = +0.250$ m. Charge C (−4.20 μC) is located at $x = +0.650$ m. Find the total force (and direction) on charge B.

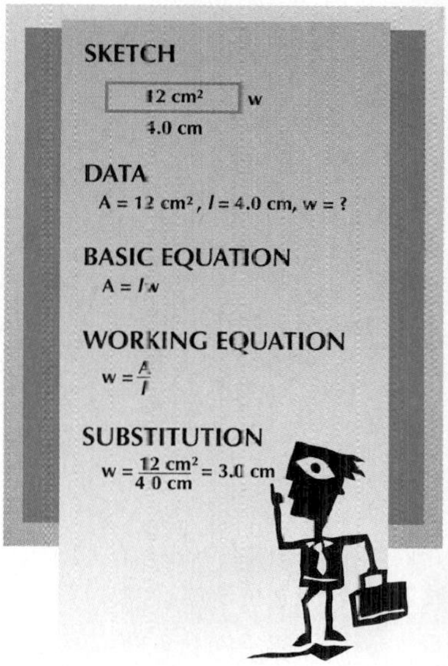

16.4 Electric Fields

Courtsey of Pearson Learning

Figure 16.10 Common electric field. These two girls are using static electricity to adhere balloons to the wall as shown.

So far, we have discussed electrification due to the brushing of electrons from a surface. The concept of the electric field is also an important part of the study of static electricity. Two magnets may either attract or repel each other even though they may not be touching each other. This illustrates the idea of the "field" in that even though they are not touching each other, there is an invisible region around each one that affects the other magnet if that magnet is placed in the region.

In terms of static electricity, an **electric field** exists where an electric force (of attraction or repulsion) acts on a charge brought into the area. A charged balloon put on a wall illustrates this principle (Fig. 16.10). Note that the balloon attracts the wall even without physical contact. The balloon has acquired a negative charge through friction, but the wall surface acquires a positive charge produced by the electric field of the charged balloon. Such an invisible electric field is present around every charged object.

Static electricity can be a real hazard in industry as well as a curiosity and sometimes a nuisance in daily life. The electric spark from static electricity, particularly in synthetic fiber textile mills, is extremely dangerous. Also, some workers in cosmetic factories in which aerosol (spray) products are made with hydrocarbon (petroleum type) propellants are required to wear cotton clothes rather than those made with synthetic fibers.

Electric fields can be measured by using test charges. Fields are represented using lines to show the direction and intensity of the field. Field lines do not really exist. They are just a means of providing a model of the field to visualize how the field is stronger where the lines are closer together and also to represent the direction of the field. Electric fields really do exist, though we study them by observing the effects they produce. Keep in mind electric fields exist in three dimensions though our drawings are only two-dimensional models (Fig. 16.11).

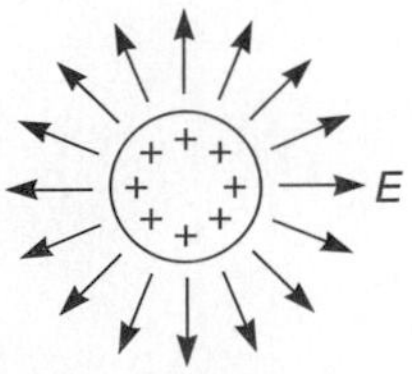

Figure 16.11 An electric field is created in the space around a charged object.

We can use the test charge to calculate the strength of a field but we do not yet know why charged bodies exert forces on each other. Still, we can detect and measure fields because fields produce forces on charges placed in the field.

A test charge creates an electric field about it in all directions. If a second charge is placed at some point in the field of the first charge, it interacts with the first charge. Gravitational fields exist around bodies like the earth and exert a force on nearby bodies, like people. We recall that $F_w = mg$, and we can find the gravitational field to be $g = F_w/m$. Similarly an electric field can be described by

$$E = \frac{F}{q'}$$

where

E = electric field
F = force on a test charge placed in the field
q' = test charge, measured in coulombs

The result is a vector quantity that is the magnitude of the electric field. By measuring the force on the test charge at different locations in the field, we can map the field and then represent it by using field lines for a model.

EXAMPLE

A positive test charge of 30.0 μC is placed in an electric field. The force on it is 0.600 N. What is the magnitude of the electric field at the location of the test charge?

Data:

$q' = 30.0\ \mu\text{C} = 3.00 \times 10^{-5}\ \text{C}$
$F = 0.600\ \text{N}$
$E = ?$

Basic Equation:

$$E = \frac{F}{q'}$$

Working Equation: Same

Substitution:

$$E = \frac{0.500\ \text{N}}{3.00 \times 10^{-5}\ \text{C}}$$
$$= 2.00 \times 10^{4}\ \text{N/C}$$

Electric fields are used for many applications in electronics and elsewhere. An ink-jet printer uses the deflection caused by an electric field on charged ink droplets to direct ink to the appropriate spot on the paper. In a similar way, the electron beam in a TV picture tube is deflected to the correct spots on the screen to produce a picture (Fig. 16.12).

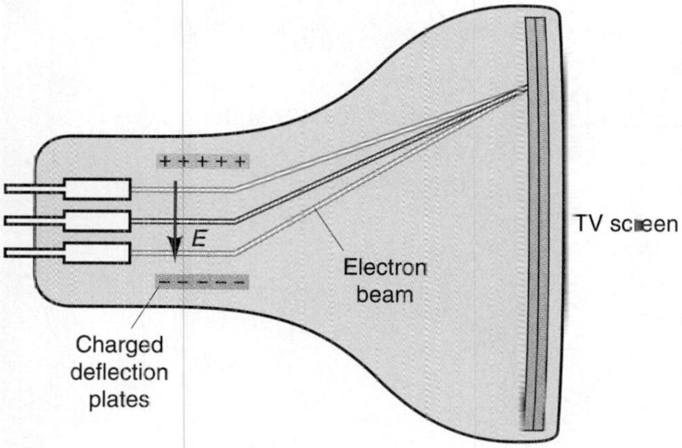

Figure 16.12 Electron beam deflection in a television

Energy can be transmitted by means of an electric field through empty space. When charges move in an electric field, they cause a disturbance that produces changes near the moving charges and results in the transfer of energy.

PHYSICS CONNECTIONS

Lightning Safety

The earth experiences about 25 lightning ground strikes every second. Each year in the United States approximately 400 people are struck by lightning, which results in 100 deaths. Many of these strikes take place on ball fields, golf courses, and other open areas.

The safest place to be during a lightning storm is inside a building or a car. The charge from the lightning strike gathers on the outside of the building or car because like electric charges repel and remain as far away from each other as possible. After accumulating on the exterior of the building or car, the charges move quickly to a ground. Although you may be safest inside a building or a car, you should stay away from electric appliances and plumbing fixtures that are connected to the ground and will act as conductors of electricity if the building is struck by lightning.

If you are not able to get inside a building or a car, the following precautions should be taken during a lightning storm:

1. Move to the lowest section of ground and crouch down [Fig. 16.13]. By reducing your height, you decrease the likelihood of becoming a lightning target. Do not let your hands or other parts of your unprotected body touch the ground. When lightning strikes the ground, it typically spreads out along the ground and can still reach you. If only your shoes are on the ground, the amount of current passing through your body will be reduced. If you are lying on the ground, the electric current could easily pass through your entire body, including your heart.
2. Remove all metal objects from your body. Metal on your body acts as a conductor of electricity and will attract lightning.
3. Avoid tall objects such as trees and hilltops and open areas like water and fields. Being the tallest object or standing under tall objects makes a person part of a giant lightning rod. Tall objects, like trees and masts, attract the charges from a thundercloud and lightning strikes.
4. If you are on a body of water, get back to shore as quickly as possible. If this is not possible, crouch down and move away from tall metal masts.

Figure 16.13 The proper position to take if one is not able to get inside during a thunderstorm.

New advances in measuring electric fields are being used at ball fields. These devices measure the electric field intensity in the atmosphere and warn of possible lightning strikes. Since many lightning strikes occur at ball fields, such devices are becoming standard equipment at outdoor sporting events.

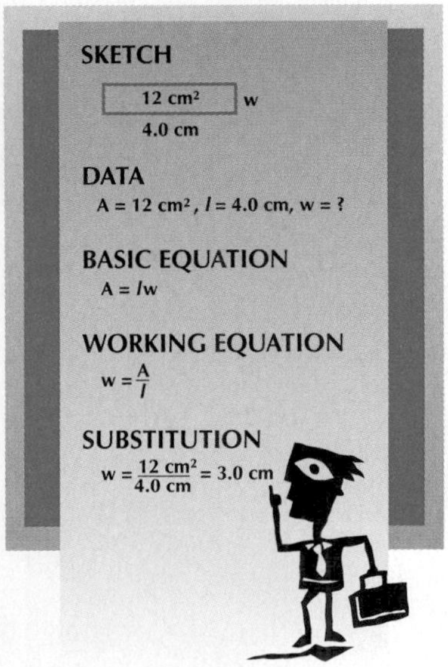

PROBLEMS 16.4

1. An electric field has a positive test charge of 4.00×10^{-5} C placed on it. The force on it is 0.600 N. What is the magnitude of the electric field at the test charge location?
2. What is the field magnitude of an electric field in which a negative charge of 2.00×10^{-8} C experiences a force of 0.0600 N?
3. An electric field exerts a force of 2.50×10^{-4} N on a positive test charge of 5.00×10^{-4} C. Find the magnitude of the field at the charge location.
4. An electric field exerts a force of 3.00×10^{-4} N on a positive test charge of 7.50×10^{-4} C. Find the magnitude of the field at the charge location.
5. An electric field of magnitude 0.450 N/C exerts a force of 8.00×10^{-4} N on a test charge placed in the field. What is the magnitude of the test charge?
6. An electric field of magnitude 0.370 N/C exerts a force of 6.20×10^{-4} N on a test charge placed in the field. What is the magnitude of the test charge?
7. What force is exerted on a test charge of 3.86×10^{-5} C if it is placed in an electric field of magnitude 1.75×10^{4} N/C?
8. What force is exerted on a test charge of 4.00×10^{-5} C if it is placed in an electric field of magnitude 3.00×10^{6} N/C?

16.5 Simple Circuits

Electrons moving in a wire produce a current in the wire. When the electron current flows in only one direction (Fig. 16.14), it is called **direct current** (dc). Current that changes direction is called **alternating current** (ac). Alternating current will be considered in Chapter 18.

An electric current is a convenient and cost-effective means of transmitting energy. We all face many daily situations that require energy to do a particular task. To drill a hole in a metal stud (Fig. 16.15), energy must be supplied and transformed into mechanical energy to turn the drill bit. The problem is how to supply energy to the machine being used in a form that the machine can turn into useful work. Electricity is often the most satisfactory means of transmitting energy.

We begin our study of the use of electricity in transferring energy with a circuit of a simple flashlight (Fig. 16.16). An **electric circuit** is a conducting loop in which electrons carrying electric energy may be transferred from a suitable source to do useful work and returned to the source. Energy is stored in the battery. When the switch is closed, energy is transmitted to the light, and the light glows.

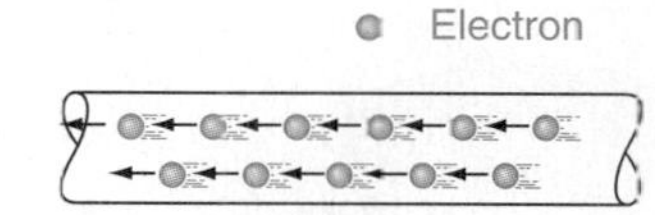

Figure 16.14 Current flowing in only one direction is direct current.

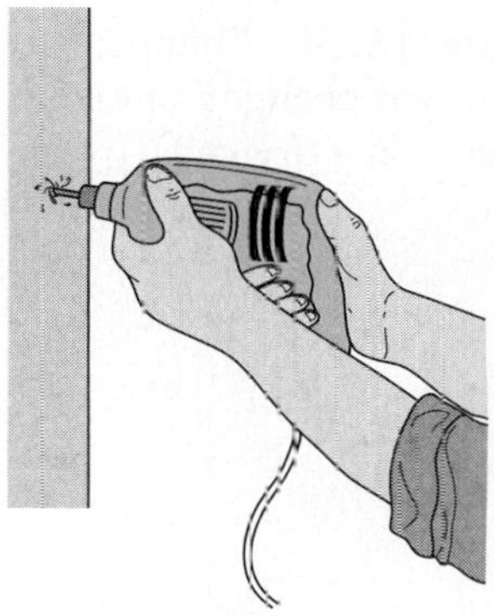

Figure 16.15 Changing electric energy to mechanical energy in drilling

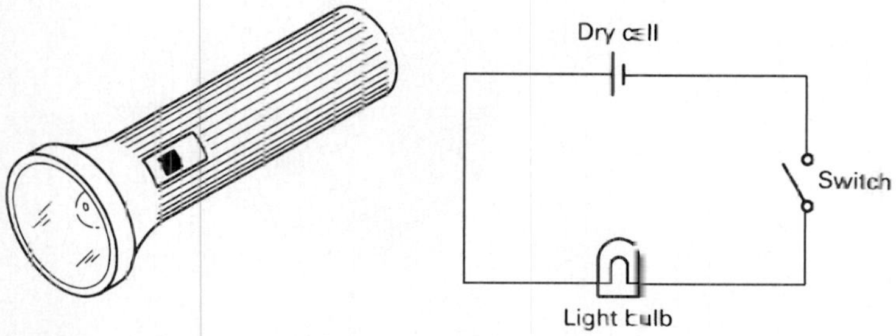

Figure 16.16 Simple electric circuit

Current electricity is the flow of energized electrons through an electron carrier called a conductor (Fig. 16.17). The electrons move from the energy *source* (the battery, here) to the *load* (where the transmitted energy is turned into useful work). There they lose energy picked up in the source. We now consider each part of the circuit and determine its function.

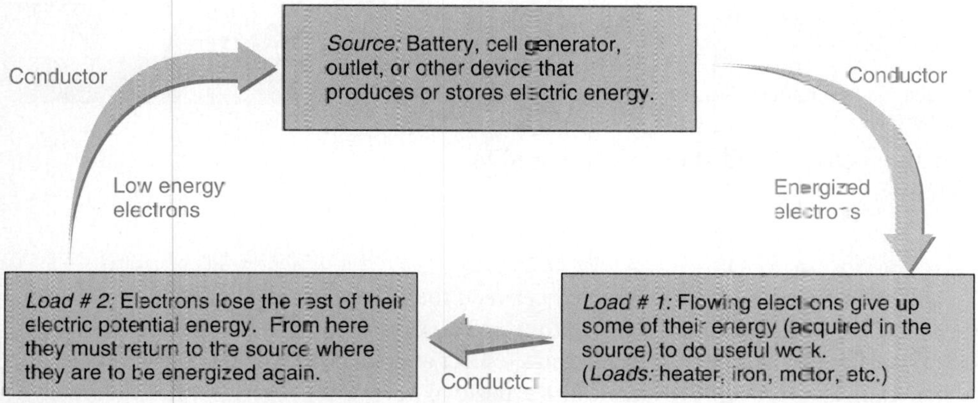

Figure 16.17 Flow of energized electrons through a conductor

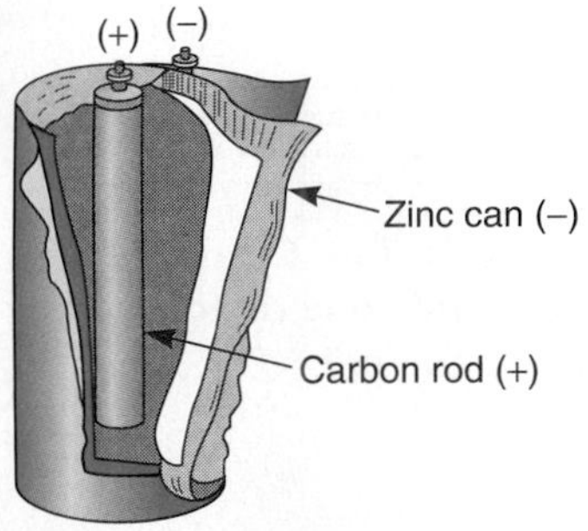

Figure 16.18 Chemical energy is changed to electric energy in a dry cell.

The Source of Energy

The **source** is the object that supplies electric energy for the flow of electric charge (electrons) in a circuit. The dry cell (Fig. 16.18) is a device that converts chemical energy to electric energy. How the cell does this will be studied later in this chapter. Here, we simply state that, by chemical action, electrons are given energy in the cell. When energy is given to electrons in this manner, their electric potential energy is raised.

What does "electric potential energy" mean? The flow of charge in an electric circuit is often compared to the flow of water in a hydraulic system, as shown in Fig. 16.19(b). Water naturally flows from a position of high potential to a position of low potential and performs work in the process, such as turning a waterwheel or turbine. A pump is needed to return the water from its low-potential position to its high-potential position. There is a *difference in potential* due to its position. Work has been done in lifting it against gravity to the higher position. In a source of electric energy something similar happens. In the source (the battery), work is done on electrons that gives them potential energy. This potential difference between the high-potential energy energized electrons [at the negative (–) pole] and the low-potential energy electrons [at the positive (+) pole] causes the electrons to flow from one point (–) to the other (+) when connected [Fig. 16.19(a)].

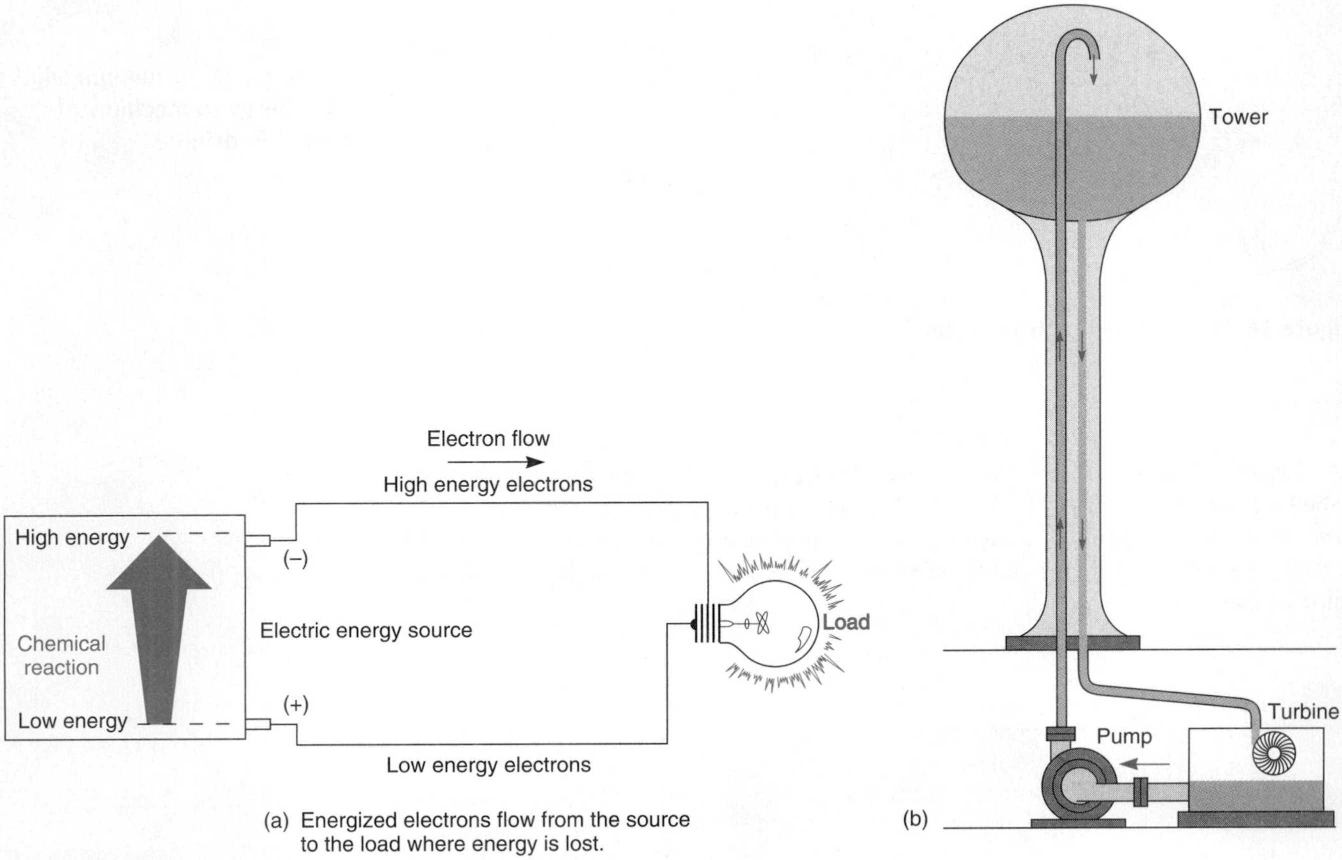

Figure 16.19 The flow of charge in an electric circuit is often compared to the flow of water in a hydraulic system.

Think of this potential difference between the two points as an electric field set up between the poles that drives the electrons in the external circuit from the negative pole to the positive pole. The energized electrons collect at the source's negative pole, repel each other, and flow through the circuit to the positive pole. They lose their potential electric energy to the load.

The Conductor

A conductor carries or transfers the electric charge to the load [Fig. 16.19(a), the light bulb]. A **conductor** is a material (such as copper) through which an electric charge is readily transferred. Such materials have large numbers of free electrons (electrons that are free to move throughout the conductor). As high-energy electrons from the source pass through the conductor, they collide with other electrons in the conductor. These electrons then carry the energy farther along the wire until they collide again and transfer energy on through the wire.

Silver, copper, and aluminum are metals that allow electrons to pass freely through and thus are good conductors. Other metals offer more opposition to the flow of electrons and are poorer conductors (Fig. 16.20). Substances that do not allow electrons to pass readily are called **insulators**. Common insulators are rubber, wool, silk, glass, wood, distilled water, and dry air.

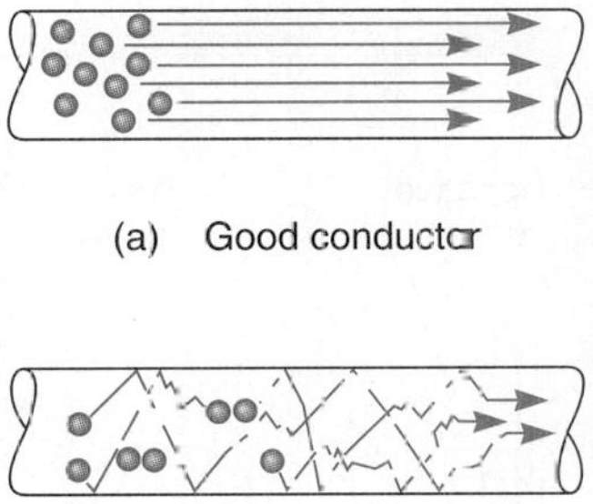

(a) Good conductor

(b) Poor conductor

Figure 16.20

A small number of materials, called **semiconductors,** fall between conductors and insulators in their ability to conduct electric current. Their importance is due to the fact that these materials under certain conditions allow current to flow in one direction only. Silicon is a semiconductor used in transistors and integrated circuits (ICs). Semiconductors are neither good conductors nor good insulators in their pure form. However, they become excellent conductors or insulators when an impurity is added. Transistors are made by layering semiconductor materials together.

Selenium is a semiconductor used in the process of making photocopies. Charge built up on its surface will remain there in the dark. When a certain colored light shines on it, the charge will dissipate very quickly from the areas exposed to the light. A powder sticking only to the charged areas is transferred to a piece of paper called a photocopy.

A **superconductor** is a material that continuously conducts electric current without resistance when cooled to typically very low temperatures, usually near absolute zero. H. Kamerlingh-Onnes discovered superconductivity in 1911 shortly after he discovered how to liquefy helium gas. He determined that mercury metal lost its resistance to the flow of electricity at temperatures just below 4.2 K, the boiling point of helium. Scientists are currently finding materials in which superconductivity exists at higher temperatures. The ultimate goal is to find a material in which superconductivity exists at room temperature.

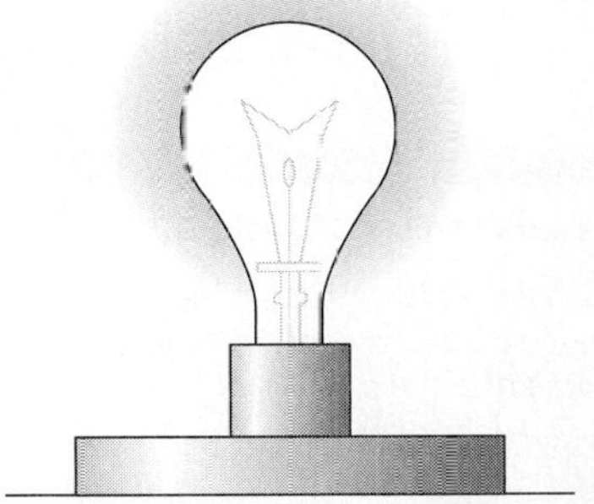

Figure 16.21 Electric energy is changed to light and heat in the load.

The Load

In the load, electrons lose their energy. The **load** in a circuit converts the electric energy into other forms of energy or work. In a light bulb, electric energy is changed to light and heat (Fig. 16.21). An electric motor changes electric energy to mechanical energy. The load may be a complex motor or only a simple resistor with heat the only new form of energy. The electrons do not collect and remain in the load, but continue back to the low-energy side of the battery (+). There they are energized again for another trip through the circuit.

Current

The flow of electrons through a conductor is called **current.** We could count the electrons passing a point during a certain time to get the rate of flow. This is impractical because the flow of electrons is so large (about 10^{18}/s). To have a workable unit of electric charge, we define a charge of 6.25×10^{18} electrons as 1 *coulomb* (C). The *ampere* (A) is the rate of flow of 1 C of charge passing a point in 1 s. We define a unit for the rate of flow of charge as follows:

$$\boxed{1 \text{ ampere (A)} = \frac{1 \text{ coulomb (C)}}{1 \text{ second (s)}}}$$

As mentioned earlier, the charge carriers in metals are electrons. In some other conductors, such as electrolytes (conducting liquid solutions), the charge carriers may be positive or negative or both. An agreement must be made to determine which charge carriers should be assumed in our following discussions.

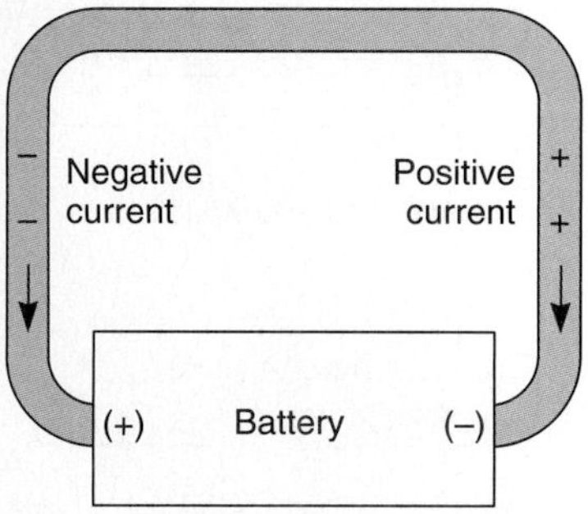

Figure 16.22 Current flow as positive current or negative current

Note that positive charges flow in the opposite direction (toward the negative terminal) from that of negative charges (toward the positive terminal) when a battery is connected to a circuit (Fig. 16.22). A positive current moving in one direction is equivalent for almost all measurements to a current of negative charges flowing in the opposite direction.

In this book we assume that the charge carriers are positive, and we draw our current arrows in the direction that a positive charge would flow. This is the practice of the majority of engineers and technicians. If you encounter the negative-current convention, you should remember that a negative current flows in the opposite direction from that of a positive current. Regardless of the method used, the analysis of a situation by either method will give the correct result. Some of the rules discussed later, such as the right-hand rule for finding the direction of the magnetic field, will be different if the negative-current convention is used.

Voltage

We have seen that current flows in a circuit because of the difference in potential of the different points in the circuit. Work is done as a charge moves from one point to another in an electric field. Work is required to move a charge from one point to another when such points differ in electric potential. The *potential difference* between two points in an electric field is the work done per unit of charge as the charge is moved between two points. That is,

$$\text{potential difference} = \frac{\text{work}}{\text{charge}}$$

In *sources,* the raising of the potential energy of electrons that results in a potential difference across a source is called **emf** (E). In *circuits,* the lowering of the potential difference across a load is called **voltage drop.**

The *volt* (V), named after **Allessandro Volta,** is the unit of both emf and voltage drop. We define the volt as the potential difference between two points if 1 J of work is produced or used in moving 1 C of charge from one point to another:

Allessandro Volta (1745–1827), *physicist, was born in Italy. He invented the first electric battery. The unit of electric potential, the volt, is named after him.*

$$\boxed{1 \text{ volt (V)} = \frac{1 \text{ joule (J)}}{1 \text{ coulomb (C)}}}$$

Resistance

Not all substances and not even all metals are good conductors of electricity. Those with few free electrons tend to have greater opposition to the flow of charge. This opposition to current flow is called **resistance.** The unit of resistance is the *ohm* (Ω). It is not a fundamental unit and is discussed in Section 16.6.

The resistance of a wire is determined by several factors. Among these are:

1. ***Temperature.*** An increase in temperature results in an increase in resistance in a wire, for most metals. Other materials, such as semiconductors, show a decrease in resistance with increasing temperature.
2. ***Length.*** Resistance varies directly with length. If we double the length of a given wire, the resistance is doubled [Fig. 16.23(a)].

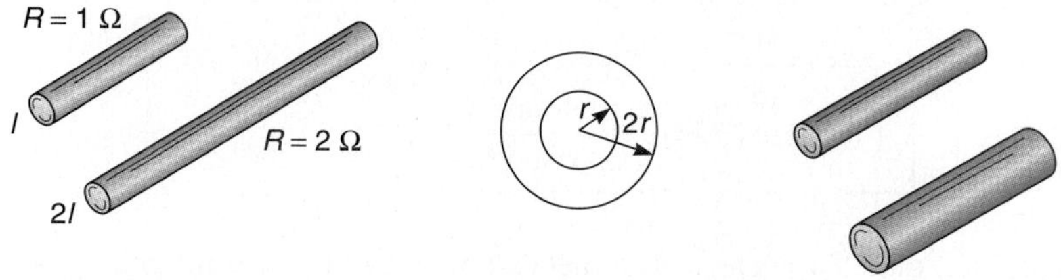

(a) Resistance varies directly with length.

(b) Doubling the radius more than doubles the cross-sectional area.

Figure 16.23

3. ***Cross-sectional area.*** Resistance varies inversely with cross-sectional area. If we double the cross section of a wire, the resistance is *halved*. This is similar to water flowing through two pipes. It flows more easily through the larger pipe. (Note that doubling the radius of a wire [Fig. 16.23(b)] *more than* doubles the cross-sectional area: $A = \pi r^2$.)
4. ***Material.*** Resistance depends on the nature of the material. For example, copper is a better conductor than steel. The conducting characteristic of various materials is described by resistivity. **Resistivity** (ρ) is the resistance per unit length of a material with uniform cross section.

These factors are related by the equation

$$R = \frac{\rho l}{A}$$

where

R = resistance
ρ = resistivity
l = length
A = cross-sectional area

EXAMPLE

Find the resistance of a copper wire 20.0 m long with cross-sectional area of 6.56×10^{-3} cm^2 at 20°C. The resistivity of copper at 20°C is $1.72 \times 10^{-6}\ \Omega$ cm.

Data:

$$l = 20.0 \text{ m} = 2.00 \times 10^3 \text{ cm}$$
$$A = 6.56 \times 10^{-3} \text{ cm}^2$$
$$\rho = 1.72 \times 10^{-6}\ \Omega \text{ cm}$$
$$R = ?$$

Basic Equation:

$$R = \frac{\rho l}{A}$$

Working Equation: Same

Substitution:

$$R = \frac{(1.72 \times 10^{-6}\ \Omega\ \text{cm})(2.00 \times 10^3\ \text{cm})}{6.56 \times 10^{-3}\ \text{cm}^2}$$
$$= 0.524\ \Omega$$

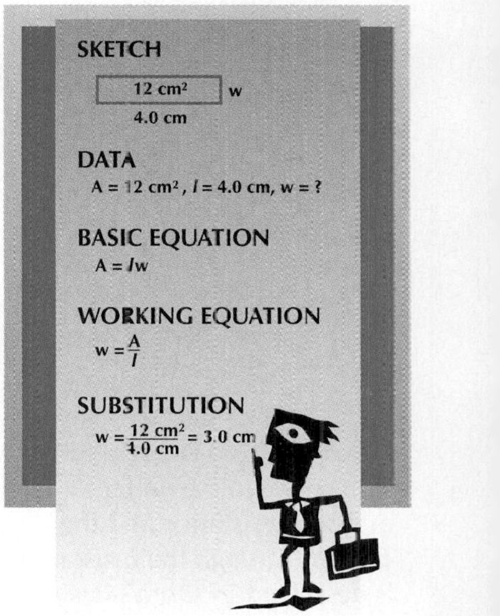

PROBLEMS 16.5

1. Find the resistance of 78.0 m of No. 20 aluminum wire at 20°C. ($\rho = 2.83 \times 10^{-6}\ \Omega$ cm, $A = 2.07 \times 10^{-2}$ cm^2.)
2. Find the resistance of 315 ft of No. 24 copper wire with resistance 0.0262 Ω/ft.
3. Find the resistance per foot of No. 22 copper wire if $58\bar{0}$ ft has a resistance of 9.57 Ω.
4. At 77°F, $10\bar{0}$ ft of No. 18 copper wire has a resistance of 0.651 Ω. Find the resistance of $50\bar{0}$ ft of this wire.
5. Find the resistance of 475 m of No. 20 copper wire at 20°C. ($\rho = 1.72 \times 10^{-6}\ \Omega$ cm, $A = 2.07 \times 10^{-2}$ cm^2.)
6. Find the resistance of $10\bar{0}$ m of No. 20 copper wire at 20°C. ($\rho = 1.72 \times 10^{-6}\ \Omega$ cm, $A = 2.07 \times 10^{-2}$ cm^2.)

7. Find the resistance of 50.0 m of No. 20 aluminum wire at 20°C. ($\rho = 2.83 \times 10^{-6}\ \Omega$ cm, $A = 2.07 \times 10^{-2}$ cm^2.)
8. Find the length of copper wire with resistance 0.0262 Ω/ft and total resistance 3.00 Ω.
9. Find the cross-sectional area of copper wire at 20°C that is 60.0 m long and has resistivity $\rho = 1.72 \times 10^{-6}\ \Omega$ cm and resistance 0.788 Ω.
10. Find the length of a copper wire with resistance 0.0262 Ω/ft and total resistance 5.62 Ω.

Georg Simon Ohm (1787–1854), *physicist, was born in Bavaria. Ohm's law resulted from his work and research in electricity. The unit of the measure of electrical resistance, the ohm, is named after him.*

16.6 Ohm's Law

When a voltage is applied *across* a material that conducts electrical current, the relationship between current *through* the material and voltage across it depends upon the type of material as shown in Fig. 16.24. The straight-line relationship shown in Fig. 16.24(a) is typical of many materials, including metal conductors. Other materials, such as semiconductors shown in Fig. 16.24(b), show a nonlinear relationship between I and V. For the materials with a straight-line relationship, the equation relating I and V was determined by **Georg Simon Ohm.** The relationship is called **Ohm's law** (see Fig. 16.25).

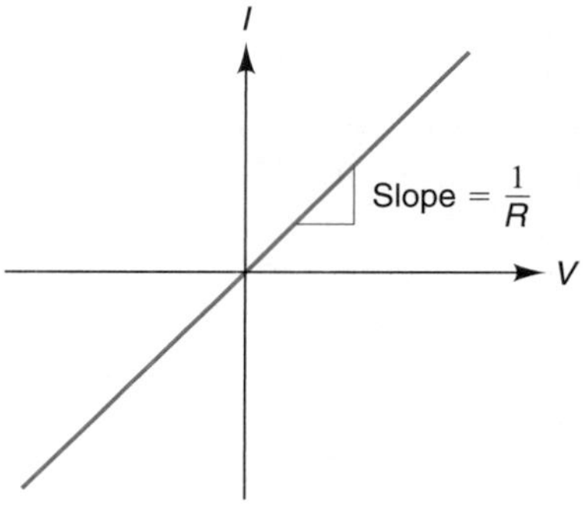

(a) Straight-line (linear) I–V characteristic (typical of resistors)

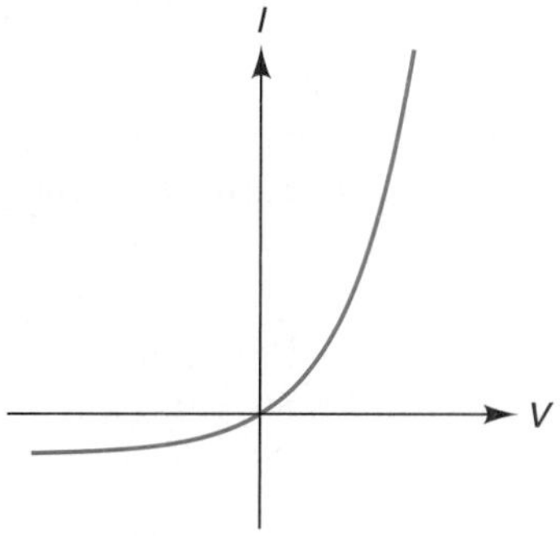

(b) Nonlinear I–V characteristic (typical of semiconductor diodes)

Figure 16.24

Ohm's law

$$I = \frac{V}{R}$$

where

I = current *through* the resistance
V = voltage drop *across* the resistance
R = resistance

Ohm's law can also be written

$$I = \frac{E}{R}$$

where

E = emf of the source of electrical energy

Ohm's law applies to dc circuits containing linear resistors and those ac circuits containing only resistance. It may be applied to the whole circuit or to any part of it.

Ohm's law can aid us in understanding resistance. As we mentioned earlier, the ohm (Ω) is a derived unit. From Ohm's law,

$$I = \frac{V}{R}$$

Solving for R, we obtain

$$R = \frac{V}{I}$$

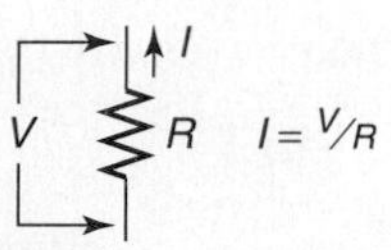

Figure 16.25 The relationship between the voltage *across* a resistance and the current *through* the resistance is described by Ohm's law.

Substituting units, we obtain

$$\Omega = \frac{V}{A}$$

TRY THIS ACTIVITY

String of Lights

Attach a D-cell battery to a small 2.5-V or 3.5-V light bulb and observe the brightness of the light. Disconnect the circuit and attach another light bulb between the first bulb and the battery. Observe the relative brightness of the bulbs compared to when only one bulb was lit. Repeat the process with two or three additional bulbs. Using Ohm's law, explain what happened to the brightness of each bulb.

EXAMPLE 1

A heating element on an electric range operating on 240 V has a resistance of 30.0 Ω. What current does it draw?

Data:

$$E = 240 \text{ V}$$
$$R = 30.0\ \Omega$$
$$I = ?$$

Basic Equation:

$$I = \frac{E}{R}$$

Working Equation: Same

Substitution:

$$I = \frac{240 \text{ V}}{30.0\ \Omega}$$
$$= 8.0 \text{ V}/\Omega$$
$$= 8.0 \text{ A} \qquad \boxed{\frac{V}{\Omega} = A}$$

EXAMPLE 2

A flashlight bulb is connected to two dry cells with an equivalent voltage of 3.0 V. If it draws 15 mA, what is its resistance?

Sketch:

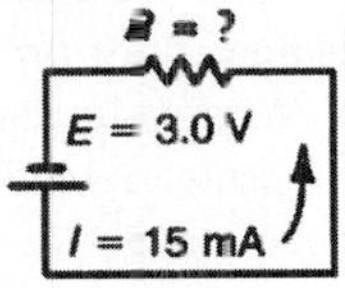

Data:

$$E = 3.0 \text{ V}$$
$$I = 15 \text{ mA} = 15 \times 10^{-3} \text{ A} = 0.015 \text{ A}$$
$$R = ?$$

Basic Equation:

$$I = \frac{E}{R}$$

Working Equation:

$$R = \frac{E}{I}$$

Substitution:

$$R = \frac{3.0 \text{ V}}{0.015 \text{ A}}$$
$$= 2\bar{0}0 \text{ V/A}$$
$$= 2\bar{0}0\ \Omega$$

Electric shock is a very real hazard to electricians. High voltage does not carry the danger that a large current does. The resistance of the human body dry would normally be in the range of 100,000 to 500,000 Ω. If a person is standing in a bathtub, however, his or her resistance is greatly lowered by the water, and any electric appliance falling into the water could deliver sufficient current to cause death. A current as small as 0.070 A can cause serious damage to the nervous system and be fatal. Electricians can protect themselves by using only one hand in some circumstances where a wire must be grasped so that no circuit is completed. Another technique is to touch any questionable wire with the back of one's hand so that if there is any current present, an unexpected shock will not cause a muscular contraction that will keep the hand gripping the wire. In any attempted rescue of a shock victim, the first task is to clear the person from the electric supply with a piece of wood or other nonconductor to avoid the rescuer becoming a second victim.

Appliances today are usually supplied with a third prong on the electric plug to conduct any charge buildup on the appliance to the ground. Sometimes appliances are also made with insulating cases to achieve the same goal—to provide a path not through the human body for electricity to flow.

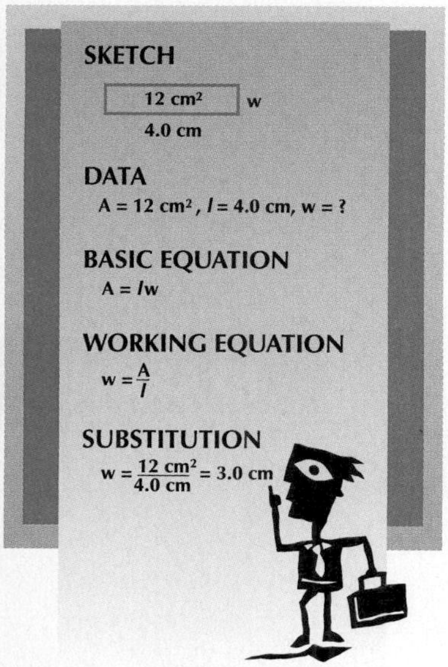

PROBLEMS 16.6

1. A heating element operates on 115 V. If it has a resistance of 24.0 Ω, what current does it draw?
2. A coffeepot operates on 12.0 V. If it draws 2.50 A, find its resistance.
3. An electric heater draws a maximum of 14.0 A. If its resistance is 15.7 Ω, on what voltage is it operating?
4. A heating coil operates on $22\bar{0}$ V. If it draws 15.0 A, find its resistance.
5. Find the resistance that draws 0.750 A on 115 V.
6. What current does a 75.0-Ω resistance draw on 115 V?
7. A heater operates on $22\bar{0}$ V. If it draws 12.5 A, what is its resistance?
8. What current does a 50.0-Ω resistance draw on 115 V?
9. What current does a 175-Ω resistance draw on $22\bar{0}$ V?
10. A heater draws 3.50 A on 115 V. What is its resistance?
11. (a) What current does a 150-Ω resistance draw on a $1\bar{0}$-V battery? (b) What voltage battery would produce 3 times the current in (a)? (c) What current would a 75-Ω resistor draw on the $1\bar{0}$-V battery?
12. A heater draws 4.25 A on 32.0 V. (a) What is the resistance of the heater? (b) What resistance heater would draw 8.50 A on 32.0 V?
13. Electric characteristics of all consumer electric devices must be shown on an attached plate. What is the resistance of an iron that discloses 6.40 A of current used on a $12\bar{0}$-V line?

14. What is the effective resistance of a television that draws 2.50 A on a 115-V line?
15. Find the current used by a stereo with resistance 65.0 Ω in a $12\bar{0}$-V system.
16. What is the current used by a microwave oven with resistance 20.0 Ω in a $12\bar{0}$-V system?

16.7 Series Circuits

In order to communicate about problems in electricity, technicians have developed a "picture language" of their own using symbols and diagrams. The circuit diagram is the most common and useful way to show a circuit. Note how each component (part) of the picture in Fig. 16.26(a) is represented by its symbol in the symbol diagram in its relative position in Fig. 16.26(b). The light bulb can be represented as a resistance. Then the circuit diagram appears as in Fig. 16.26(c). Some of the symbols used most often appear in Appendix C, Table 20.

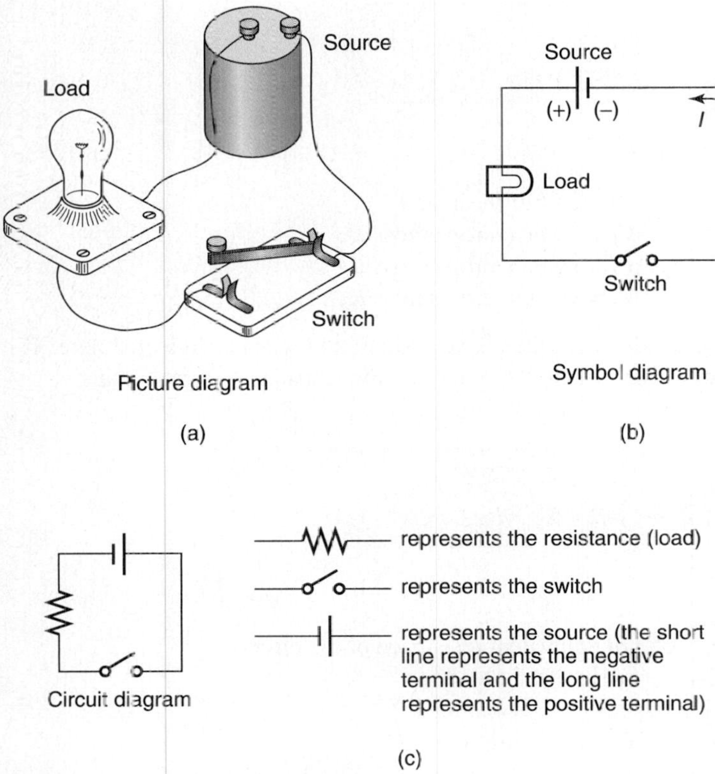

Figure 16.26 **A series circuit**

There are two basic types of circuits: series and parallel. A fuse in a house is wired in series with the outlets. The outlets themselves are wired in parallel. A study of series and parallel circuits is basic to a study of electricity.

An electric circuit with only one path for the current to flow (Fig. 16.27) is called a **series circuit.** The current in a series circuit is the same throughout. That is, the current flows out of one resistance and into the next resistance. Therefore, the total current is the same as the current flowing through each resistance in the circuit.

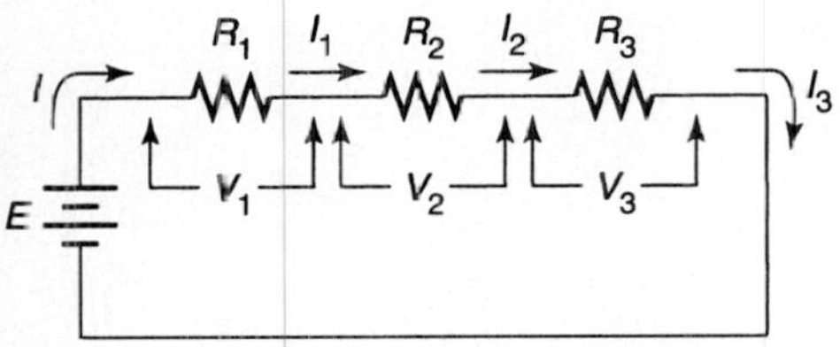

Figure 16.27 **Series circuit**

$$\textit{SERIES}$$
$$I = I_1 = I_2 = I_3 = \cdots$$

where

I = total current
I_1 = current through R_1
I_2 = current through R_2
I_3 = current through R_3

In a series circuit, the emf of the source equals the sum of the separate voltage drops in the circuit (Fig. 16.27):

$$\textit{SERIES}$$
$$E = V_1 + V_2 + V_3 + \cdots$$

where

E = emf of the source
V_1 = voltage drop across R_1
V_2 = voltage drop across R_2
V_3 = voltage drop across R_3

The resistance of the conducting wires is very small and will be neglected here. The total resistance of a series circuit equals the sum of all the resistances in the circuit:

$$\textit{SERIES}$$
$$R = R_1 + R_2 + R_3 + \cdots$$

where

R = total or equivalent resistance of the circuit
R_1 = resistance of first load
R_2 = resistance of second load
R_3 = resistance of third load

The **equivalent resistance** is the single resistance that can replace a series and/or parallel combination of resistances in a circuit and provide the same current flow and voltage drop. The equivalent resistance of a series combination is larger than the resistance of any one of the resistances in series.

EXAMPLE 1

Find the total resistance of the circuit shown in Fig. 16.28.

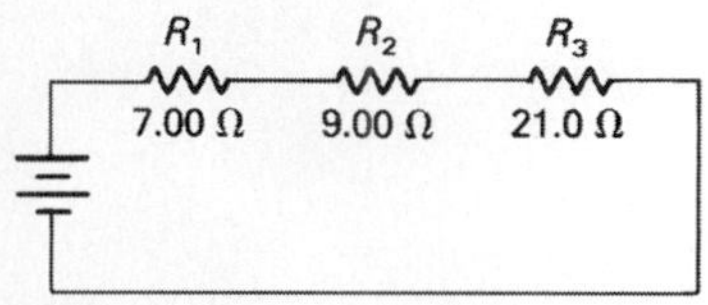

Figure 16.28

Data:

$R_1 = 7.00\ \Omega$

$R_2 = 9.00\ \Omega$

$R_3 = 21.0\ \Omega$

$R = ?$

Basic Equation:

$$R = R_1 + R_2 + R_3$$

Working Equation: Same

Substitution:

$$R = 7.00\ \Omega + 9.00\ \Omega + 21.0\ \Omega$$
$$= 37.0\ \Omega$$

EXAMPLE 2

Find the current in the circuit shown in Fig. 16.29.

Data:

$$R_1 = 5.00\ \Omega$$
$$R_2 = 13.0\ \Omega$$
$$R_3 = 12.0\ \Omega$$
$$R_4 = 96.0\ \Omega$$
$$E = 90.0\ \text{V}$$
$$I = ?$$

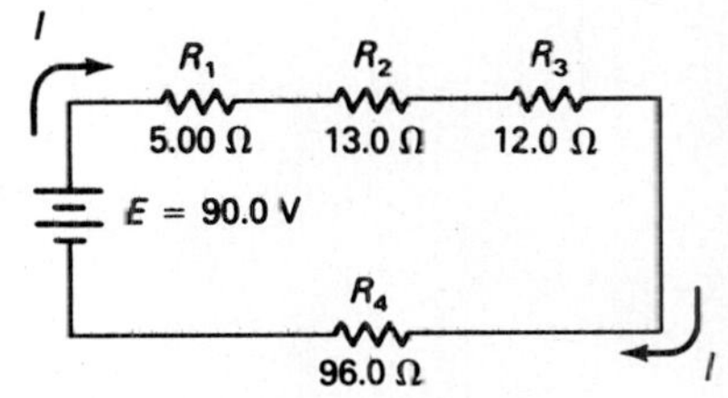

Figure 16.29

Basic Equations:

$$R = R_1 + R_2 + R_3 + R_4 \quad \text{and} \quad I = \frac{E}{R}$$

Working Equations: Same

Substitutions:

$$R = 5.00\ \Omega + 13.0\ \Omega + 12.0\ \Omega + 96.0\ \Omega$$
$$= 126.0\ \Omega$$

$$I = \frac{90.0\ \text{V}}{126.0\ \Omega}$$
$$= 0.714\ \text{A}$$

EXAMPLE 3

Find the value of R_3 in the circuit shown in Fig. 16.30.

Data:

$$I = 3.00\ \text{A}$$
$$E = 115\ \text{V}$$
$$R_1 = 23.0\ \Omega$$
$$R_2 = 14.0\ \Omega$$
$$R_3 = ?$$

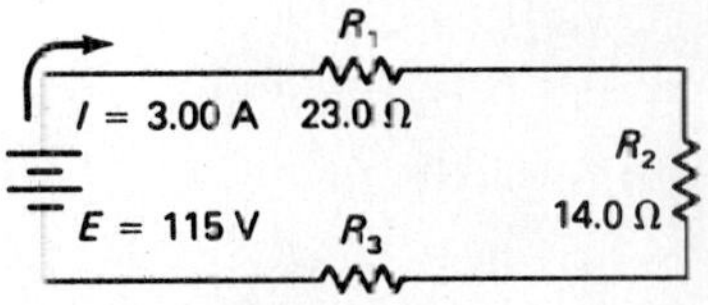

Figure 16.30

Basic Equations:

$$I = \frac{E}{R} \quad \text{and} \quad R = R_1 + R_2 + R_3$$

Working Equations:

$$R = \frac{E}{I} \quad \text{and} \quad R_3 = R - R_1 - R_2$$

Substitutions:

$$R = \frac{115\ \text{V}}{3.00\ \text{A}}$$
$$= 38.3\ \Omega$$

$$R_3 = 38.3\ \Omega - 23.0\ \Omega - 14.0\ \Omega$$
$$= 1.3\ \Omega$$

EXAMPLE 4

Find the voltage drop across R_3 in Example 3.

Data:

$$I = I_3 = 3.00 \text{ A}$$
$$R_3 = 1.3 \text{ }\Omega$$
$$V_3 = ?$$

Basic Equation:

$$I_3 = \frac{V_3}{R_3}$$

Working Equation:

$$V_3 = I_3R_3$$

Substitution:

$$\begin{aligned} V_3 &= (3.00 \text{ A})(1.3 \text{ }\Omega) \\ &= 3.9 \text{ V} \end{aligned}$$

Table 16.1 summarizes the characteristics of series circuits.

Table 16.1 Characteristics of Series Circuits

	Series
Current	$I = I_1 = I_2 = I_3 = \cdots$
Equivalent Resistance	$R = R_1 + R_2 + R_3 + \cdots$
Voltage	$E = V_1 + V_2 + V_3 + \cdots$

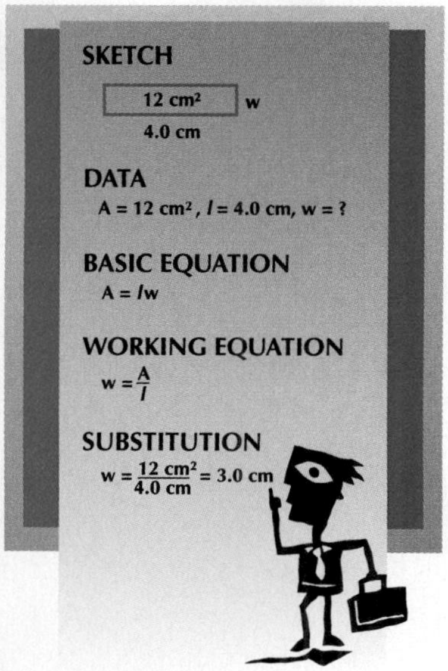

PROBLEMS 16.7

1. Three resistors of 2.00 Ω, 5.00 Ω, and 6.50 Ω are connected in series with a 24.0-V battery. Find the total resistance of the circuit.
2. Find the current in Problem 1.
3. Find the equivalent resistance in the circuit shown in Fig. 16.31.
4. Find the current through R_2 in Problem 3.
5. Find the current in the circuit shown in Fig. 16.32.
6. Find the voltage drop across R_1 in Problem 5.
7. What emf is needed for the circuit shown in Fig. 16.33?
8. Find the voltage drop across R_3 in Problem 7.
9. Find the equivalent resistance in the circuit shown in Fig. 16.34.
10. Find R_3 in the circuit in Problem 9.
11. Find the values of R_1, R_2, and R_3 in Fig. 16.35.
12. Find the values of V_1, R_2, and V_3 in Fig. 16.36.
13. Find the values of R_1, V_2, and R_3 in Fig. 16.37.

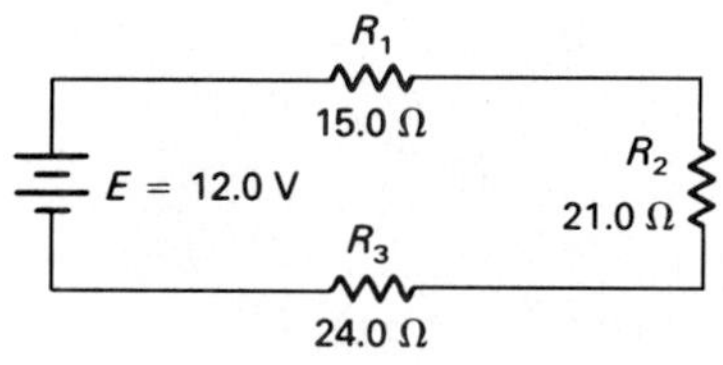

Figure 16.31

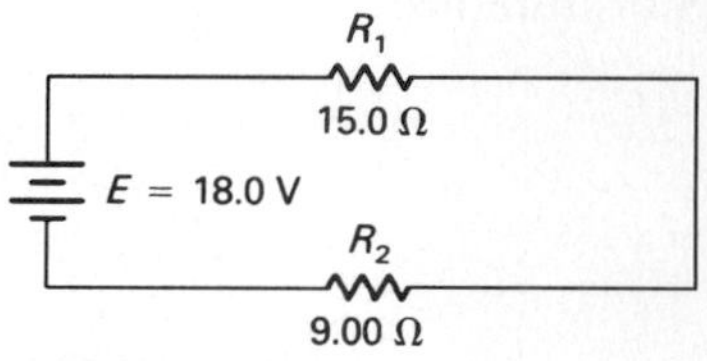

Figure 16.32

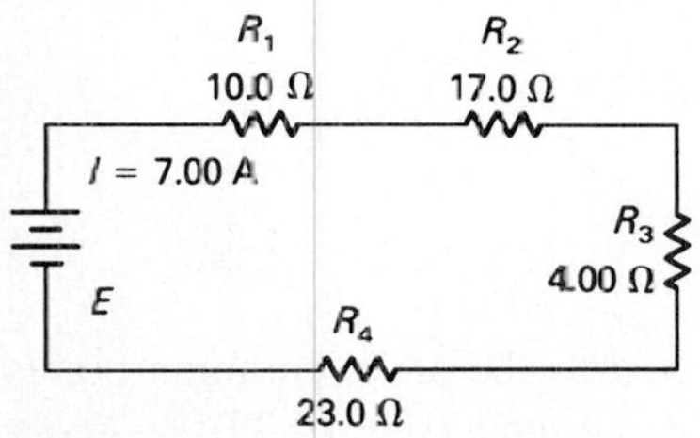

Figure 16.33

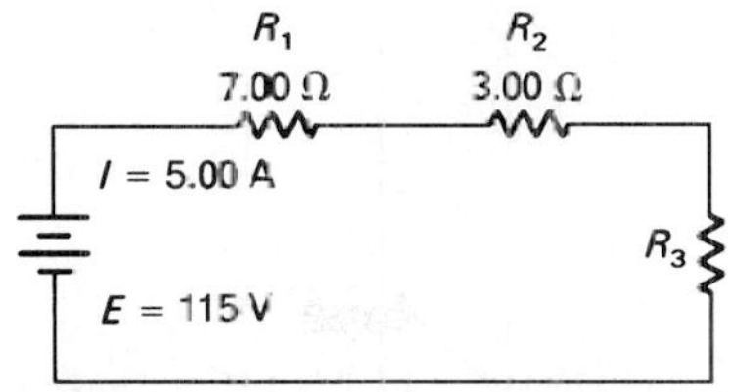

Figure 16.34

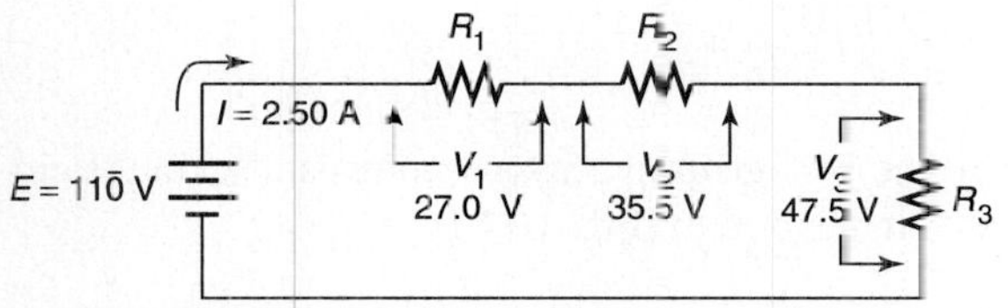

Figure 16.35

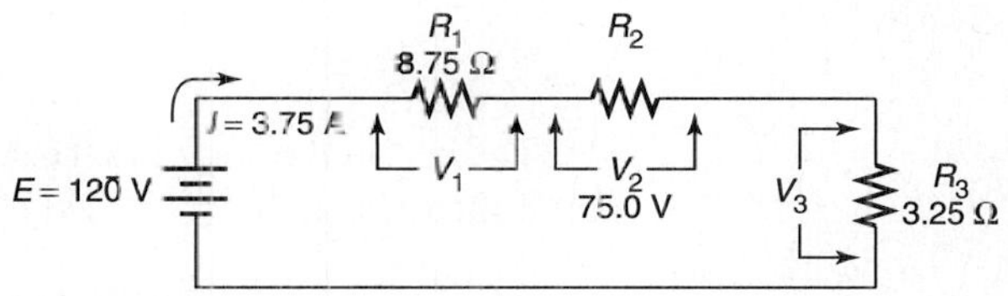

Figure 16.36

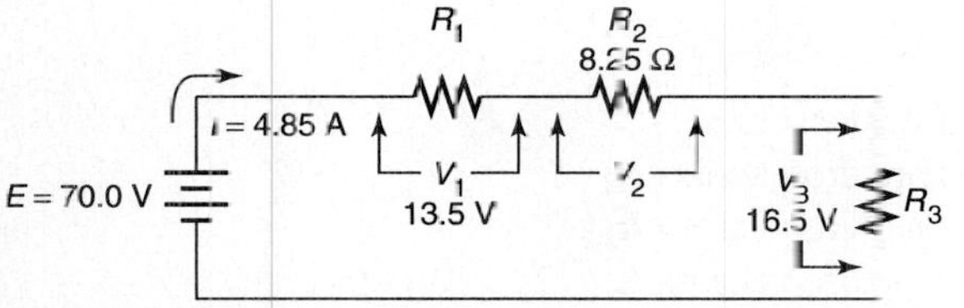

Figure 16.37

16.8 Parallel Circuits

An electric circuit with more than one path for the current to flow (Fig. 16.38) is called a **parallel circuit.** All resistances connected in parallel have their ends connected to two common points (nodes) in the circuit (points A and B in Fig. 16.38). The current in a parallel circuit is divided among the branches of the circuit (Fig. 16.39). How it is divided depends on the resistance of each branch. The paths with the least resistance allow the

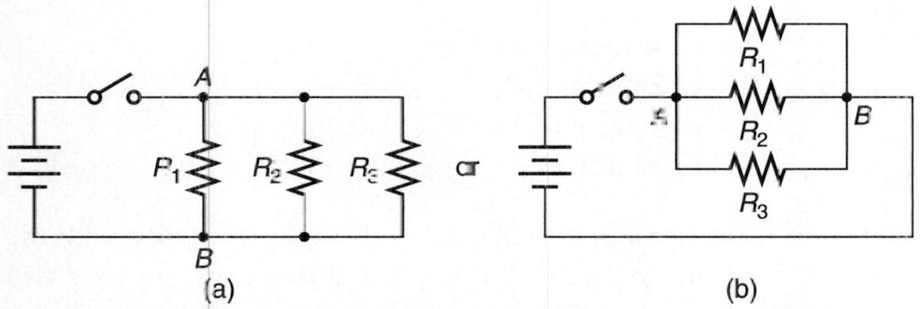

Figure 16.38 Different ways to represent a parallel circuit

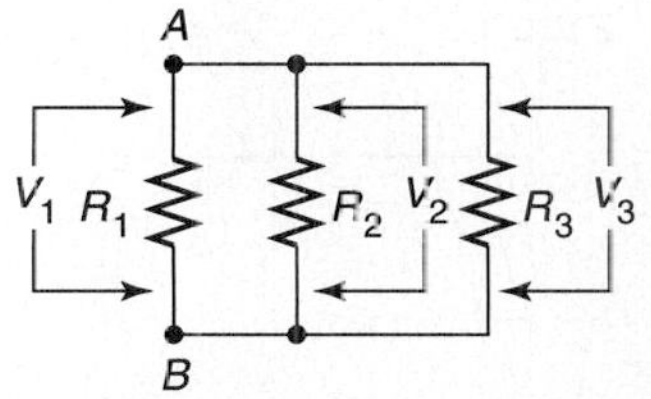

Figure 16.39 $I = I_1 + I_2 + I_3$

largest currents to flow. Since the current divides, the current from the source equals the sum of the currents through each of the branches

PARALLEL

$$I = I_1 + I_2 + I_3 + \cdots$$

where

I = total current in the circuit
I_1 = current through R_1
I_2 = current through R_2
I_3 = current through R_3

Since the ends of all resistances in parallel are connected to the same common points (nodes) in the circuit, the voltage across each resistance is the same (Fig. 16.39):

PARALLEL

$$V_1 = V_2 = V_3 = \cdots$$

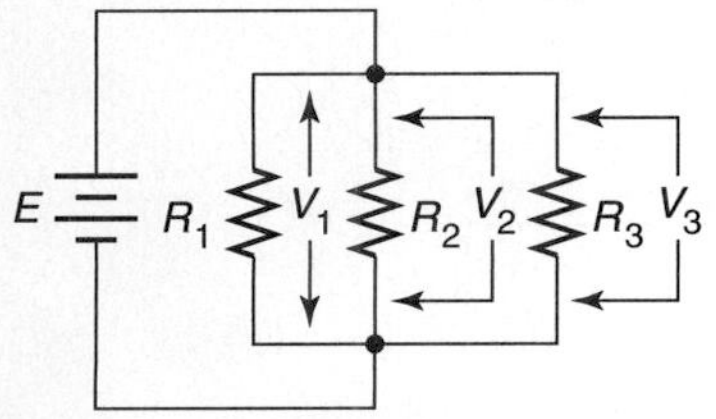

Figure 16.40
$E = V_1 = V_2 = V_3$

The emf of the source is the same as the voltage drop across each resistance in the circuit if there are no other (series) elements in the circuit (Fig. 16.40):

PARALLEL WITH VOLTAGE SOURCE

$$E = V_1 = V_2 = V_3 = \cdots$$

where

E = emf of the source
V_1 = voltage drop across R_1
V_2 = voltage drop across R_2
V_3 = voltage drop across R_3

Therefore, several different loads requiring the same voltage are connected in parallel.

The single resistance that would result in the same current flow and voltage drop as the combination of resistances is called the *equivalent resistance*. The equivalent resistance of a parallel circuit is less than the resistance of any single branch of the circuit. To find the equivalent resistance, use the formula

PARALLEL

$$\frac{1}{R} = \frac{1}{R_1} + \frac{1}{R_2} + \frac{1}{R_3} + \cdots$$

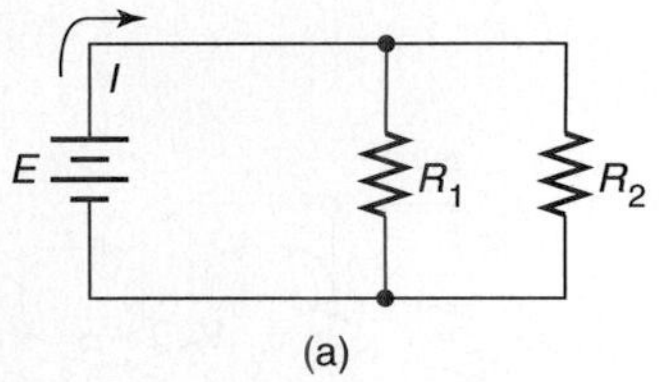

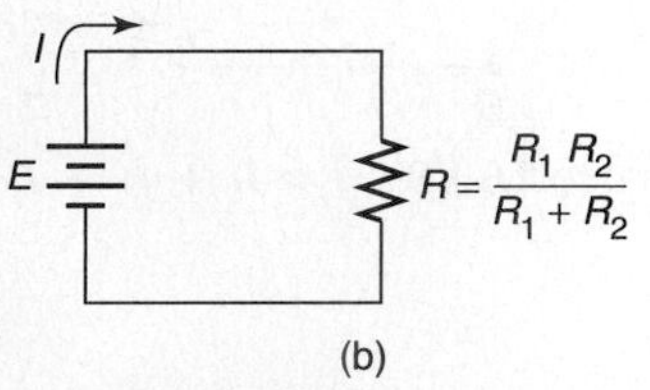

Figure 16.41 Resistor R in part (b) is equivalent to the pair of resistances R_1 and R_2 connected in parallel in part (a).

where

R = equivalent resistance
R_1 = resistance of R_1
R_2 = resistance of R_2
R_3 = resistance of R_3

If the parallel combination of resistances is replaced by a single resistance with the resistance R, the same current flows in the circuit. In the case where there are only two resistances in parallel, then

$$\frac{1}{R} = \frac{1}{R_1} + \frac{1}{R_2}$$

$$R = \frac{R_1 R_2}{R_1 + R_2}$$

(See Fig. 16.41.)

For comparison to parallel circuits, consider the water system shown in Fig. 16.42(a).

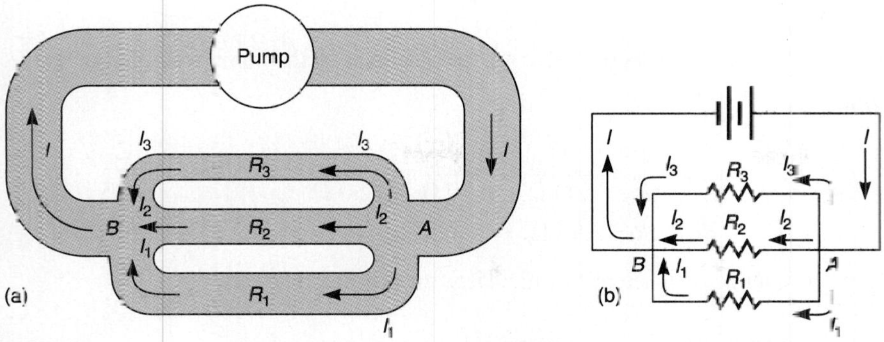

Figure 16.42 A water system may be compared to a parallel electric circuit.

1. The total amount of water flowing through $R_1 + R_2 + R_3$ equals the amount flowing through A or B.
2. The water flowing past point A divides into the three branches R_1, R_2, and R_3.
3. The larger pipes have *less* opposition to water flow than do the smaller pipes. Because R_1 has a larger cross-sectional area than R_2 or R_3, it has less opposition to the flow of water and therefore carries more water than R_2 or R_3.

Similarly, in a parallel electric circuit as in Fig. 16.42(b):

1. The total amount of current flowing through $R_1 + R_2 + R_3$ equals the amount flowing through A or B.
2. The current flowing past point A divides into the three branches R_1, R_2, and R_3.
3. The smaller resistances have *less* opposition to current flow and therefore carry larger currents.

TRY THIS ACTIVITY

Parallel Bulbs

Attach a D-cell battery to a small 2.5-V or 3.5-V light bulb and observe the brightness of the light. Attach a second light bulb in parallel with the first. After adding a third bulb in parallel with the others, note the brightness of the bulbs. Why, when using the same battery, wires, and bulbs as in the "String of Lights" Try This Activity on page 445, does the brightness of the bulbs differ from the bulbs in the series circuit?

EXAMPLE 1

Find the equivalent resistance of the circuit shown in Fig. 16.43.

Data:

$$R_1 = 7.00\ \Omega$$
$$R_2 = 9.00\ \Omega$$
$$R_3 = 12.0\ \Omega$$
$$R = ?$$

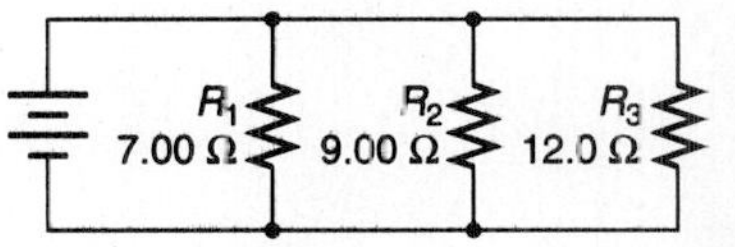

Figure 16.43

Basic Equation:

$$\frac{1}{R} = \frac{1}{R_1} + \frac{1}{R_2} + \frac{1}{R_3}$$

Working Equation:

When using this formula, you should solve for the reciprocal of the unknown, then substitute.

Substitution:

$$\frac{1}{R} = \frac{1}{7.00\ \Omega} + \frac{1}{9.00\ \Omega} + \frac{1}{12.0\ \Omega}$$

$$R = 2.96\ \Omega$$

The key entry sequence on a scientific calculator for this calculation is

7 [x^{-1}] [+] 9 [x^{-1}] [+] 12 [x^{-1}] [=] [x^{-1}] [=]

2.964705882

EXAMPLE 2

Find the total current in the circuit shown in Fig. 16.44.

Data:

$$R_1 = 23.0\ \Omega$$
$$R_2 = 14.0\ \Omega$$
$$R_3 = 5.00\ \Omega$$
$$E = 90.0\ \text{V}$$
$$I = ?$$

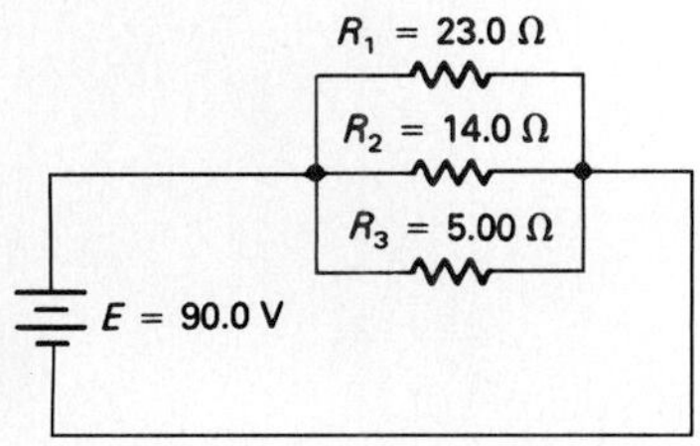

Figure 16.44

First, find the equivalent resistance, R. Second, find the total current, I. To find R:

Basic Equation:

$$\frac{1}{R} = \frac{1}{R_1} + \frac{1}{R_2} + \frac{1}{R_3}$$

Working Equation: Same

Substitution:

$$\frac{1}{R} = \frac{1}{23.0\ \Omega} + \frac{1}{14.0\ \Omega} + \frac{1}{5.00\ \Omega}$$

Using a calculator sequence as in Example 1, we find

$$R = 3.18\ \Omega$$

To find I:

Basic Equation:

$$I = \frac{E}{R}$$

Working Equation: Same

Substitution:

$$I = \frac{90.0\ \text{V}}{3.18\ \Omega}$$
$$= 28.3\ \text{A}$$

EXAMPLE 3

Find the current through R_2 in Fig. 16.44 from Example 2.

Data:

$$R_2 = 14.0\ \Omega$$
$$E = 90.0\text{ V} = V_2$$
$$I_2 = ?$$

Basic Equation:

$$I_2 = \frac{V_2}{R_2}$$

Working Equation: Same

Substitution:

$$I_2 = \frac{90.0\text{ V}}{14.0\ \Omega} = 6.43\text{A}$$

EXAMPLE 4

Find the equivalent resistance and the value of R_3 in the circuit shown in Fig. 16.45.

Data:

$$E = 115\text{ V}$$
$$I = 7.00\text{ A}$$
$$R_1 = 38.0\ \Omega$$
$$R_2 = 49.0\ \Omega$$
$$R_3 = ?$$

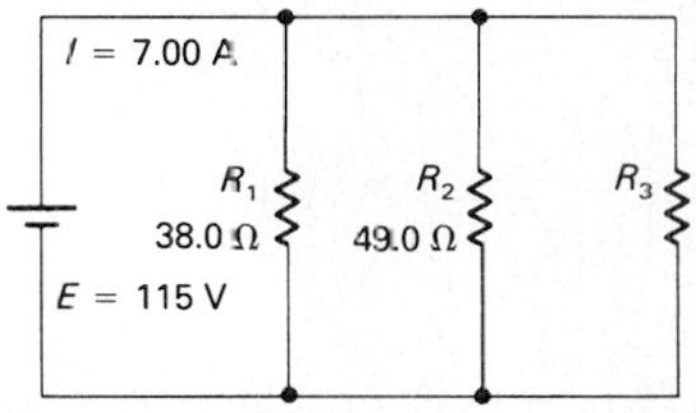

Figure 16.45

First find R:

Basic Equation:

$$I = \frac{E}{R}$$

Working Equation:

$$R = \frac{E}{I}$$

Substitution:

$$R = \frac{115\text{ V}}{7.00\text{ A}} = 16.4\ \Omega$$

To find R_3:

Basic Equation:

$$\frac{1}{R} = \frac{1}{R_1} + \frac{1}{R_2} + \frac{1}{R_3}$$

Working Equation:

$$\frac{1}{R_3} = \frac{1}{R} - \frac{1}{R_1} - \frac{1}{R_2}$$

Substitution:

$$\frac{1}{R_3} = \frac{1}{16.4\ \Omega} - \frac{1}{38.0\ \Omega} - \frac{1}{49.0\ \Omega}$$

$$R_3 = 70.2\ \Omega$$

The key entry sequence on a scientific calculator for this calculation is

16.4 [x^{-1}] [−] 38 [x^{-1}] [−] 49 [x^{-1}] [=] [x^{-1}] [=]

70.16727941

The characteristics of parallel circuits are summarized in Table 16.2.

Table 16.2 Characteristics of Parallel Circuits

	Parallel
Current	$I = I_1 + I_2 + I_3 + \cdots$
Resistance	$\frac{1}{R} = \frac{1}{R_1} + \frac{1}{R_2} + \frac{1}{R_3} + \cdots$
Voltage	$E = V_1 = V_2 = V_3 = \cdots$

Overloading is a concern with parallel circuits like those used in household wiring. As more load (bulbs, appliances, etc.) is added to a parallel circuit, the circuit draws more current. Although no change of current occurs in any individual branch, the current in the circuit as a whole increases. As the current increases, more and more heat is produced, which may cause a fire.

Circuit breakers or fuses (Fig. 16.46) are used to prevent overloading in circuits. Circuit breakers have magnets or bimetallic strips that open a switch when the current in the circuit becomes too large. A fuse has a metal strip that melts when the heat from a given amount of current passes through it. All fuses have a rating that describes the amount of current that can pass through the fuse before it melts or "blows" and breaks the circuit.

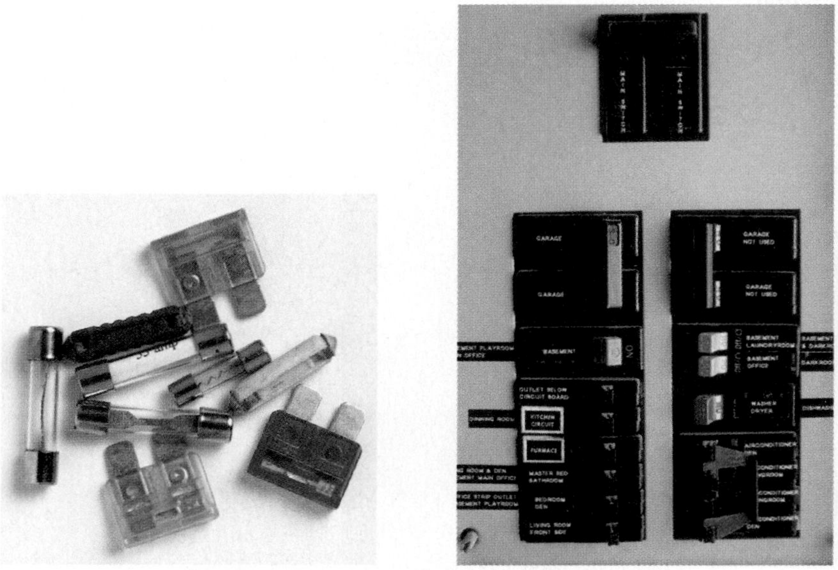

Photo courtesy of Dorling Kindersley

Courtesy of Pearson Education/PH College

Figure 16.46 (a) Assorted fuses (b) Circuit breakers

PROBLEMS 16.8

1. (a) Find the equivalent resistance in the circuit shown in Fig. 16.47.
   (b) What is the total current in the circuit?
   (c) What is the current through R_1?
   (d) What is the current through R_2?
2. (a) Find I_2 (current through R_2) in the circuit shown in Fig. 16.48.
   (b) Find I_3.
   (c) Find I_1.
   (d) Find the total current in the circuit.
   (e) Find the equivalent resistance in the circuit.

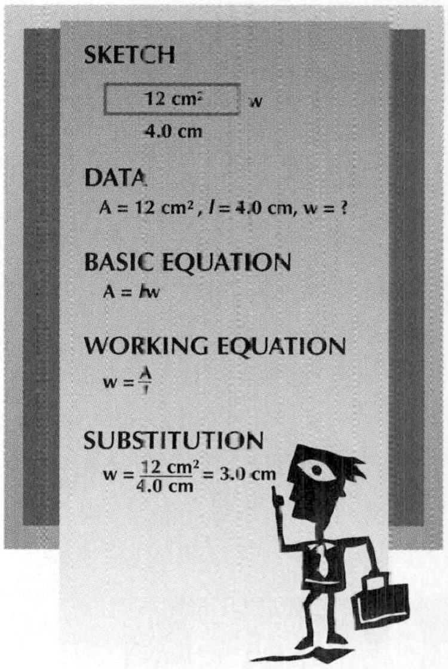

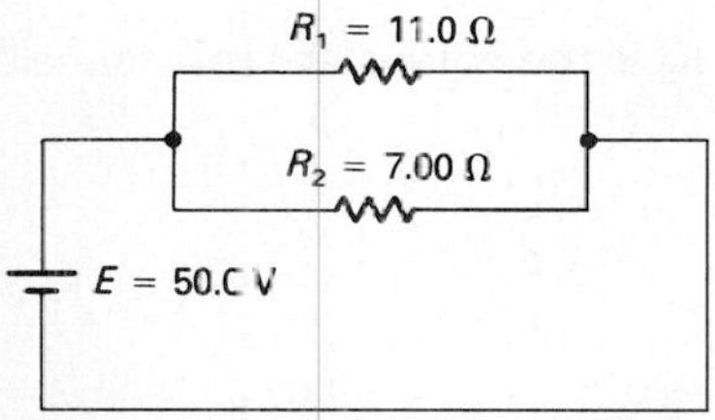

Figure 16.47

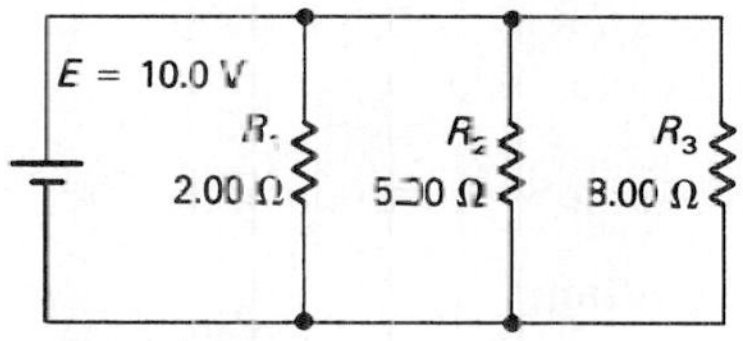

Figure 16.48

3. (a) Find the resistance of R_3 in the circuit in Fig. 16.49.
   (b) What is the current through R_1?
   (c) What is the current through R_3?
4. (a) What is the equivalent resistance in the circuit shown in Fig. 16.50?
   (b) What emf is required for the circuit?
   (c) What is the voltage drop across each resistance?
   (d) What is the current through each resistance?

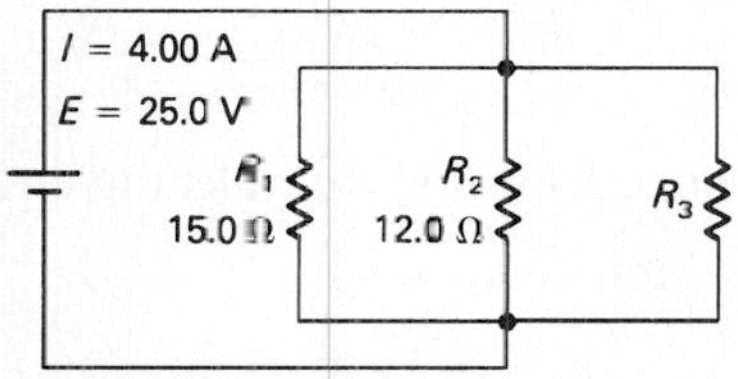

Figure 16.49

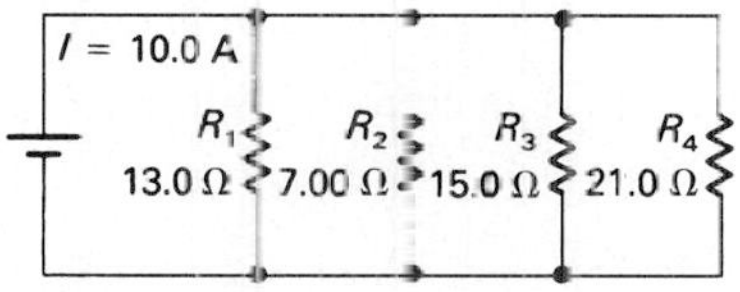

Figure 16.50

16.9 Compound Circuits

A *compound circuit* contains a combination of resistances in series and parallel arrangements. To simplify solving this kind of circuit, we apply the rules for series and parallel circuits to find an equivalent circuit that reduces to a circuit with one resistance.

EXAMPLE 1

Circuit B in Fig. 16.51 is equivalent to circuit A, where $R_4 = R_1 + R_2$. Then, circuit C is equivalent to circuit B, where $\frac{1}{R_5} = \frac{1}{R_3} + \frac{1}{R_4}$.

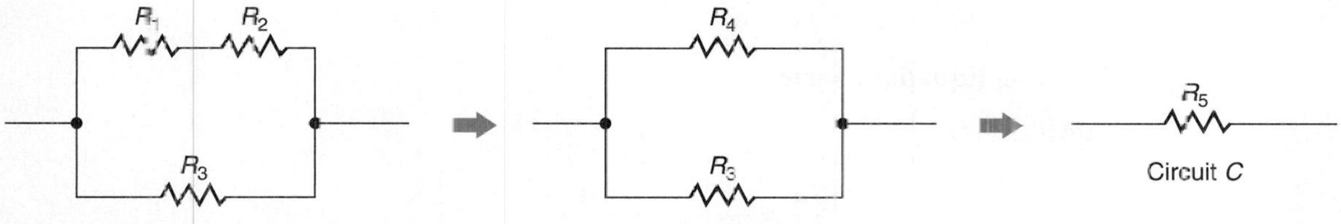

Figure 16.51 **Circuit *A* can be replaced by circuit *C*, where R_5 is the equivalent resistance.**

EXAMPLE 2

Circuit B in Fig. 16.52 is equivalent to circuit A, where $\frac{1}{R_4} = \frac{1}{R_2} + \frac{1}{R_3}$. Then, circuit C is equivalent to circuit B, where $R_5 = R_1 + R_4$.

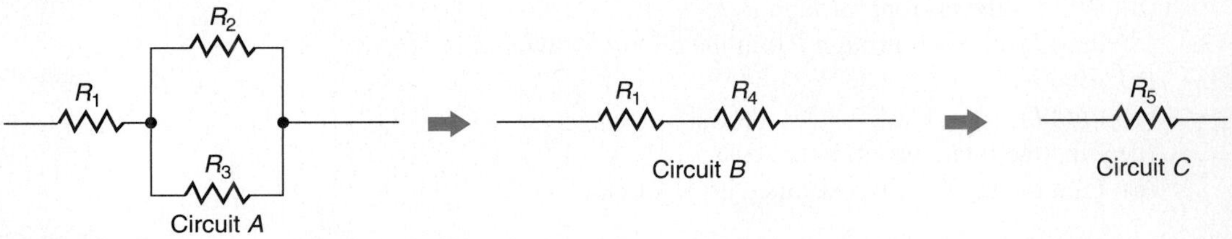

Figure 16.52 Circuit A can be replaced by circuit C, where R_5 is the equivalent resistance.

EXAMPLE 3

Find the total current in the circuit shown in Fig. 16.53.

Solution:

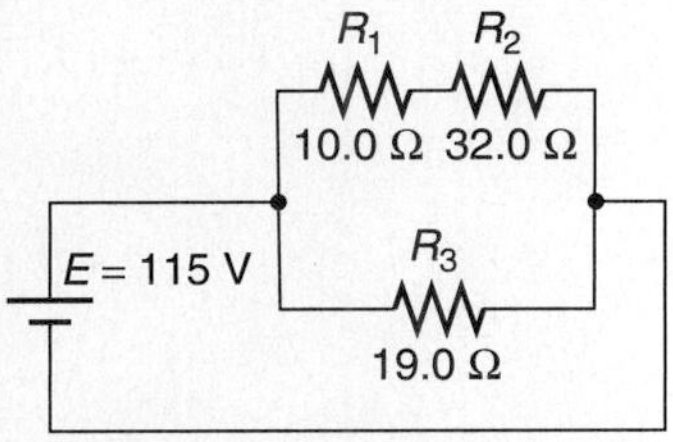

Figure 16.53

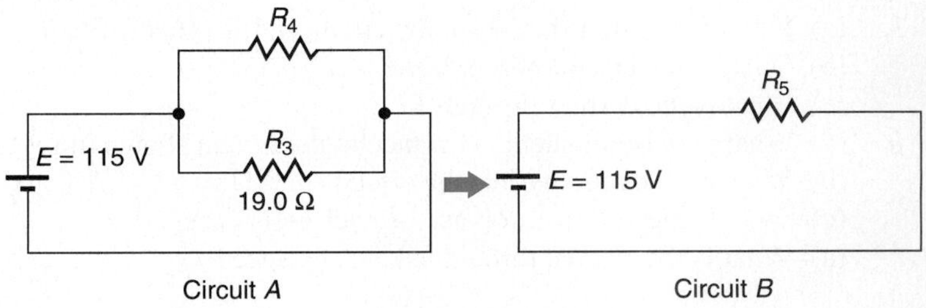

Circuit A is equivalent to the circuit in Fig. 16.53. Then, circuit B is equivalent to circuit A.

Data:

$$E = 115 \text{ V}$$
$$R_1 = 10.0\ \Omega$$
$$R_2 = 32.0\ \Omega$$
$$R_3 = 19.0\ \Omega$$
$$R_4 = R_1 + R_2 = 10.0\ \Omega + 32.0\ \Omega = 42.0\ \Omega$$
$$I = ?$$

First, find the equivalent resistance, R_5. Second, find the total current, I.
To find R_5:

Basic Equation:

$$\frac{1}{R_5} = \frac{1}{R_3} + \frac{1}{R_4}$$

Working Equation: Same

Substitution:

$$\frac{1}{R_5} = \frac{1}{19.0\ \Omega} + \frac{1}{42.0\ \Omega}$$
$$R_5 = 13.1\ \Omega$$

The key entry sequence on a scientific calculator for this calculation is

19 x^{-1} + 42 x^{-1} = x^{-1} =

13.08196721

To find I:

Basic Equation:

$$I = \frac{E}{R_5}$$

Working Equation: Same

Substitution:

$$I = \frac{115 \text{ V}}{13.1 \ \Omega}$$
$$= 8.78 \text{ A}$$

EXAMPLE 4

Find the equivalent resistance in the circuit shown in Fig. 16.54.

Solution:

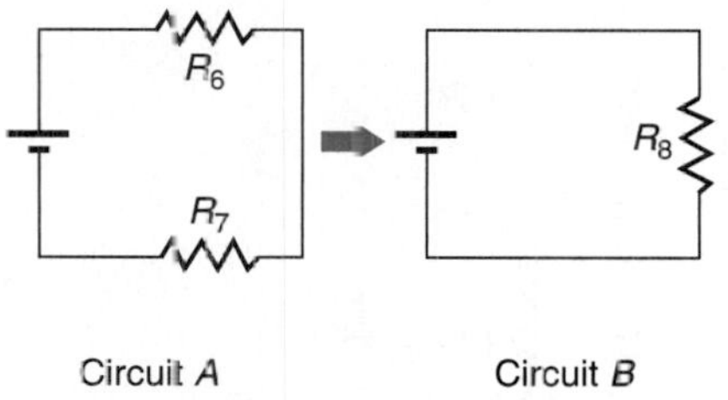

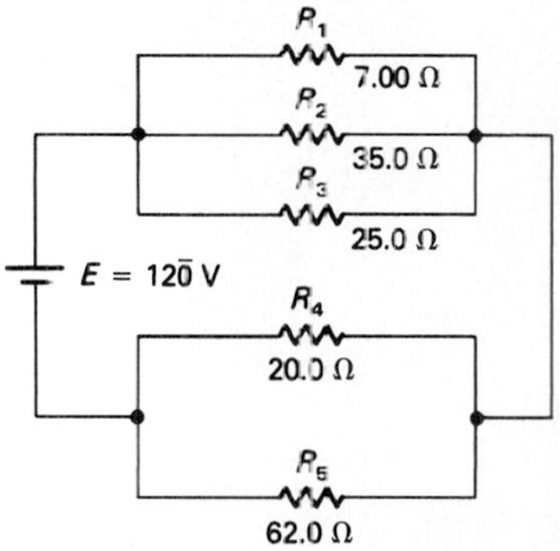

Figure 16.54

Circuit A is equivalent to the circuit in Fig. 16.54, and circuit B is equivalent to circuit A.

Data:

$$R_1 = 7.00 \ \Omega$$
$$R_2 = 35.0 \ \Omega$$
$$R_3 = 25.0 \ \Omega$$
$$R_4 = 20.0 \ \Omega$$
$$R_5 = 62.0 \ \Omega$$
$$E = 12\bar{0} \text{ V}$$
$$R_8 = ?$$

First, find R_6. Second, find R_7. Third, find the equivalent resistance, R_8.
To find R_6:

Basic Equation:

$$\frac{1}{R_6} = \frac{1}{R_1} + \frac{1}{R_2} + \frac{1}{R_3}$$

Working Equation: Same

Substitution:

$$\frac{1}{R_6} = \frac{1}{7.00\ \Omega} + \frac{1}{35.0\ \Omega} + \frac{1}{25.0\ \Omega}$$

$$R_6 = 4.73\ \Omega$$

The key entry sequence on a scientific calculator for this calculation is

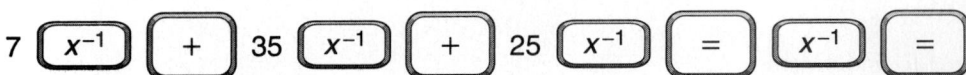

7 [x^{-1}] [+] 35 [x^{-1}] [+] 25 [x^{-1}] [=] [x^{-1}] [=]

4.72972973

To find R_7:

Basic Equation:

$$\frac{1}{R_7} = \frac{1}{R_4} + \frac{1}{R_5}$$

Working Equation: Same

Substitution:

$$\frac{1}{R_7} = \frac{1}{20.0\ \Omega} + \frac{1}{62.0\ \Omega}$$

$$R_7 = 15.1\ \Omega$$

To find R_8:

Basic Equation:

$$R_8 = R_6 + R_7$$

Working Equation: Same

Substitution:

$$\begin{aligned} R_8 &= 4.73\ \Omega + 15.1\ \Omega \\ &= 19.83\ \Omega \text{ or } 19.8\ \Omega \end{aligned}$$

EXAMPLE 5

Find the total current in Example 4.

Data:

$$\begin{aligned} E &= 12\bar{0}\ \text{V} \\ R_8 &= 19.8\ \Omega \\ I &= ? \end{aligned}$$

Basic Equation:

$$I = \frac{E}{R_8}$$

Working Equation: Same

Substitution:

$$\begin{aligned} I &= \frac{12\bar{0}\ \text{V}}{19.8\ \Omega} \\ &= 6.06\ \text{A} \end{aligned}$$

Table 16.3 summarizes the characteristics of series and parallel circuits.

Table 16.3 Characteristics of Series and Parallel Circuits

	Series	Parallel
Current	$I = I_1 = I_2 = I_3 = \cdots$	$I = I_1 + I_2 + I_3 + \cdots$
Resistance	$R = R_1 + R_2 + R_3 + \cdots$	$\frac{1}{R} = \frac{1}{R_1} + \frac{1}{R_2} + \frac{1}{R_3} + \cdots$
Voltage	$E = V_1 + V_2 + V_3 + \cdots$	$E = V_1 = V_2 = V_3 = \cdots$

PROBLEMS 16.9

Use Fig. 16.55 in Problems 1 through 5.

1. (a) Which resistances are connected in parallel?
   (b) What is the equivalent resistance of the resistances connected in parallel?
2. Find the equivalent resistance of the entire circuit.
3. Find the current in R_1.
4. Find the voltage drop across R_1.
5. (a) Find the current through R_3.
   (b) Find the current through R_2.

Use Fig. 16.56 in Problems 6 through 12.

6. What is the equivalent resistance of the resistances connected in parallel?
7. Find the equivalent resistance of the circuit.
8. Find the current in R_1.
9. What is the voltage drop across the parallel part of the circuit?
10. Find the current through R_3.
11. Find the current through R_5.
12. What is the voltage drop across R_3?

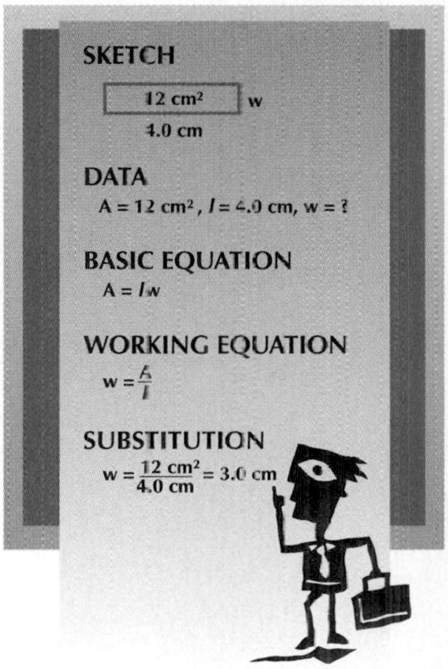

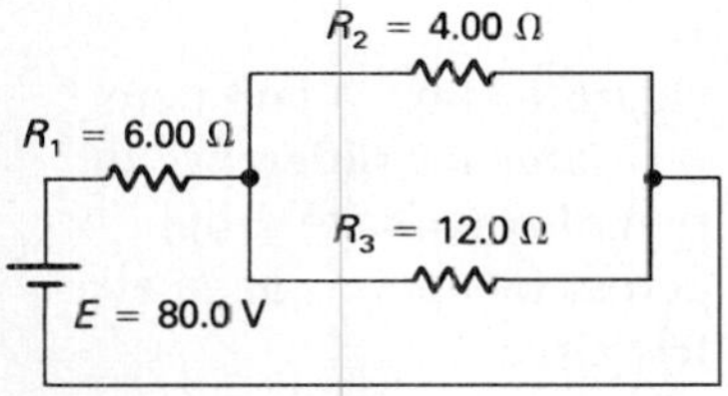

Figure 16.55

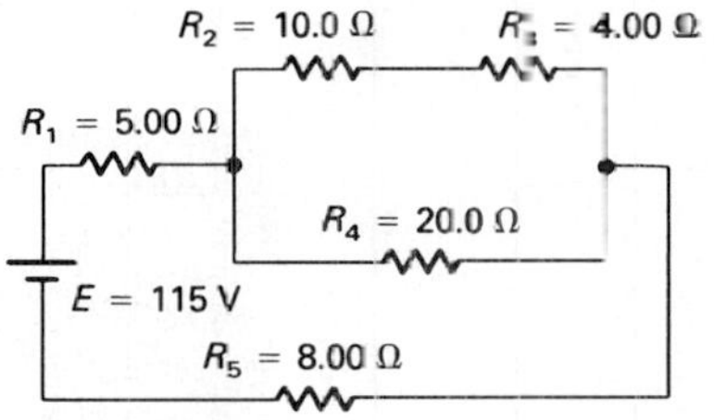

Figure 16.56

Use Fig. 16.57 in Problems 13 through 20.

13. Find the equivalent resistance of the parallel arrangement in the upper branch.
14. Find the equivalent resistance of the parallel arrangement in the lower branch.
15. Find the equivalent resistance of the entire circuit.
16. What emf is required for the given current flow in the circuit?
17. Find the voltage drop across the parallel arrangement in the upper branch.
18. Find the voltage drop across R_4.
19. Find the voltage drop across R_6.
20. Find the current through R_6.

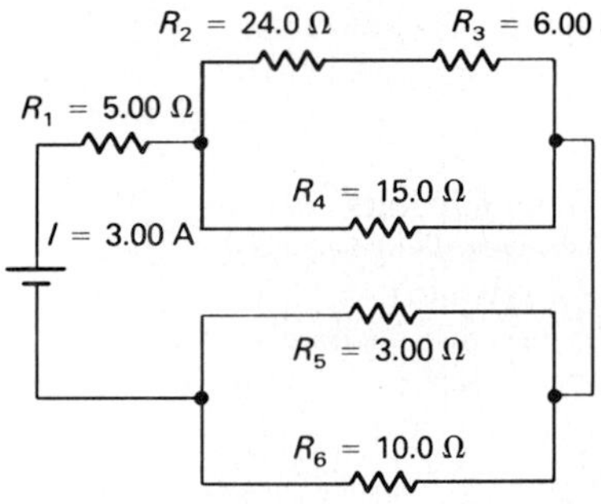

Figure 16.57

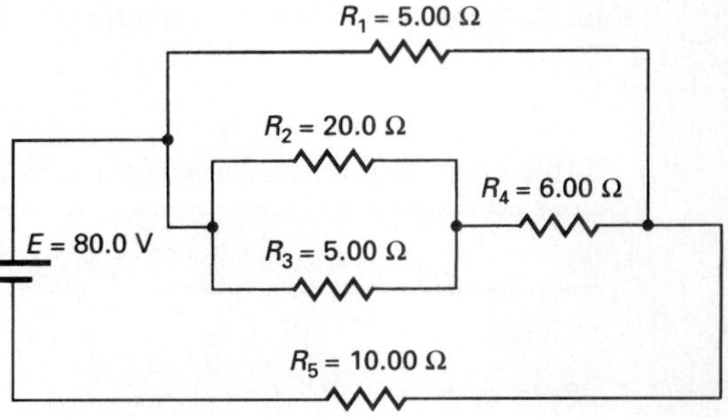

Figure 16.58

Use Fig. 16.58 in Problems 21 through 25.

21. Find the equivalent resistance in the circuit.
22. Find the current through R_5.
23. Find the voltage drop across R_5.
24. Find the voltage drop across R_4.
25. Find the current through R_2.

16.10 Electric Instruments

In the laboratory we use several kinds of electric meters for measurements. Great care must be taken to avoid passing a large current through the meters. Meters are fragile instruments, and abuse will ruin them. A large current will burn out the meter. A **multimeter** is an instrument used to measure current flow, voltage drop, and resistance in ac circuits and dc circuits.

Digital instruments have more than one range on which readings are made (Fig. 16.59). Autorange meters adjust ranges automatically. The reading and use of the different modes of operation will be studied in the laboratory.

1. ***Voltmeter.*** In this mode, the **voltmeter** instrument measures the difference in potential (voltage drop) between two points in a circuit. It should *always* be connected in *parallel* with the part of the circuit across which one wishes to measure the voltage drop (Fig. 16.60). The voltmeter is a high-resistance instrument and draws very little current.
2. ***Ammeter.*** The **ammeter** mode measures the current flowing in a circuit. Therefore, it is connected in *series* in the circuit (Fig. 16.61). Since all the current flows through the

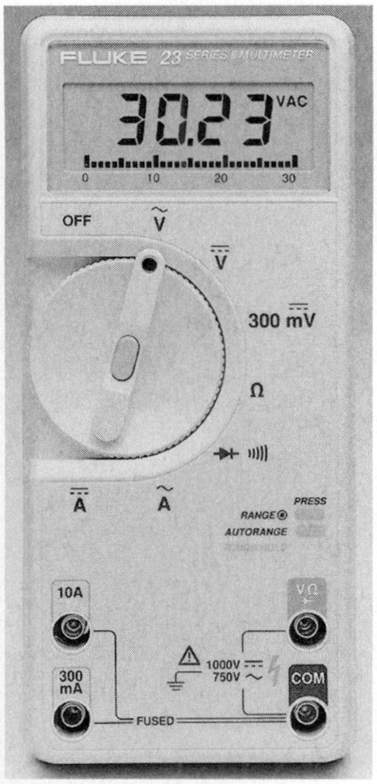

Figure 16.59 Digital multimeter

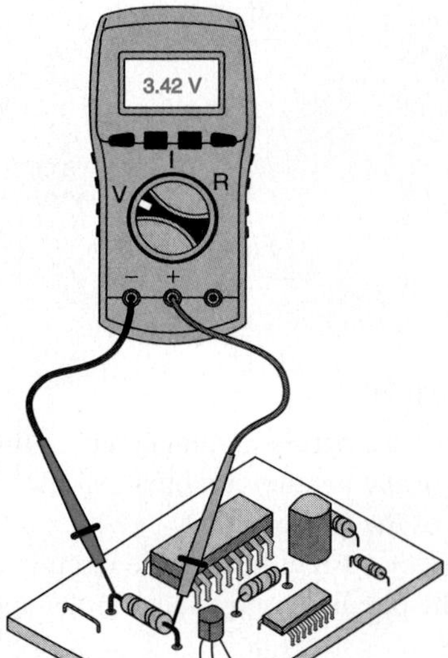

Figure 16.60 A voltmeter measures the difference in potential (voltage drop) across two points in an electric circuit.

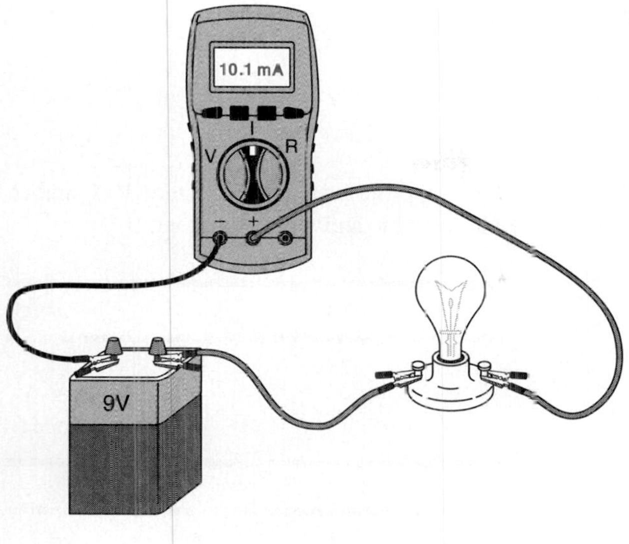

Figure 16.61 An ammeter measures the current flowing in an electric circuit.

meter, it has very low resistance in this mode so that its effect on the circuit will be as small as possible.

3. ***Ohmmeter.*** The **ohmmeter** mode is used to measure the resistance of a circuit component. It should only be used when there is *no current* flowing in the circuit. The ohmmeter has a small battery as a built-in source of energy.

A small current provided by the ohmmeter is caused to flow through the component under test. The presence of another current or a complete circuit connected to the component under test will distort the resistance reading since the rest of the circuit may allow some of this test current to flow "around" the component under test. To avoid this problem, the component should be tested in isolation from any complete circuit (Fig. 16.62).

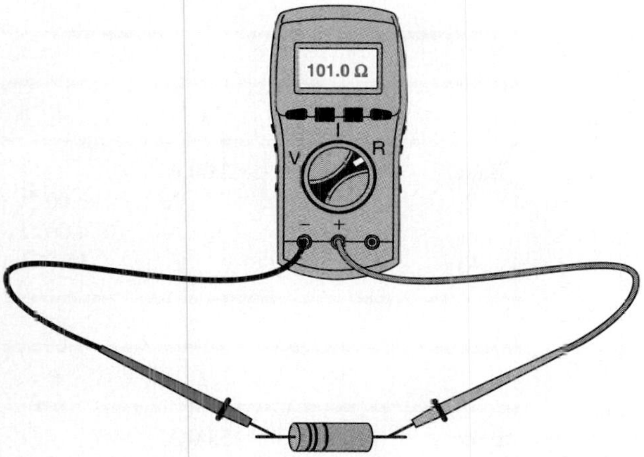

Figure 16.62 An ohmmeter measures the resistance of a component in an electric circuit.

Another instrument is the *galvanometer,* a very sensitive instrument that is used to detect the presence and direction of *very small* currents.

SKETCH

12 cm² w

4.0 cm

DATA

$A = 12\ cm^2$, $l = 4.0\ cm$, $w = ?$

BASIC EQUATION

$A = lw$

WORKING EQUATION

$w = \frac{A}{l}$

SUBSTITUTION

$w = \frac{12\ cm^2}{4.0\ cm} = 3.0\ cm$

PROBLEMS 16.10

Using the formulas for series and parallel circuits, fill in the blanks in the tables shown opposite each circuit. In the blanks across from Battery under

V: Write the emf of the battery.
I: Write the total current in the circuit.
R: Write the equivalent or total resistance of the entire circuit.

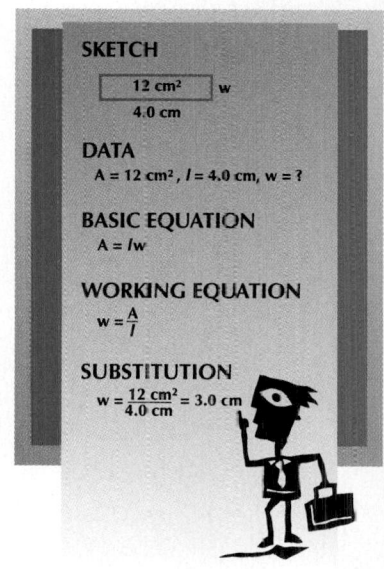

In the blanks across from R_1 under

V: Write the voltage drop across R_1.
I: Write the current flowing through R_1.
R: Write the resistance of R_1.

In the blanks across from R_2, R_3, . . . , fill in the appropriate numbers under V, I, and R. (Begin by looking for key information given in the table and work from there.)

1.

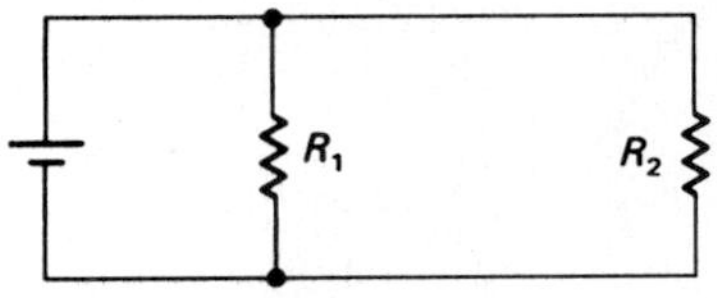

	V	I	R
Battery	12.0 V	A	Ω
R_1	V	A	2.00 Ω
R_2	V	A	4.00 Ω

2.

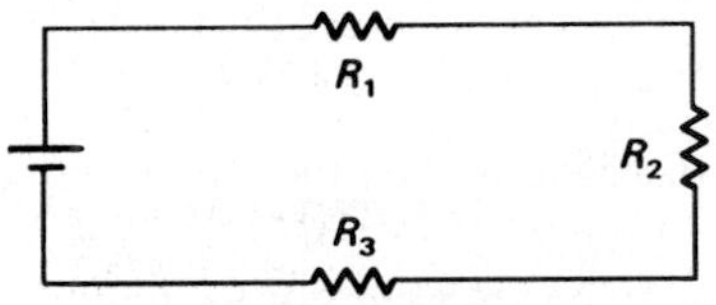

	V	I	R
Battery	V	A	Ω
R_1	V	2.00 A	4.00 Ω
R_2	V	A	6.00 Ω
R_3	V	A	8.00 Ω

3.

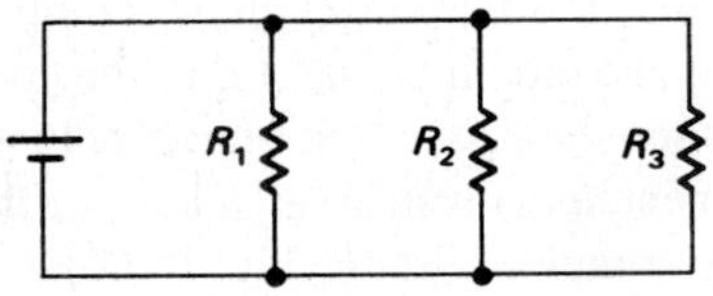

	V	I	R
Battery	V	A	Ω
R_1	V	2.00 A	Ω
R_2	V	3.00 A	12.0 Ω
R_3	V	1.00 A	Ω

4.

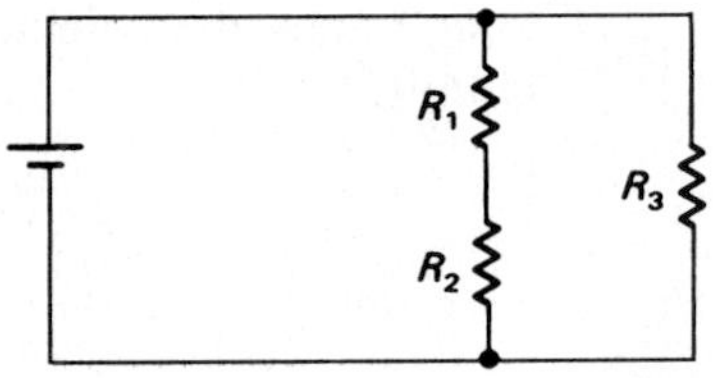

	V	I	R
Battery	12.0 V	2.00 A	Ω
R_1	V	A	6.00 Ω
R_2	V	A	4.00 Ω
R_3	V	A	15.0 Ω

5.

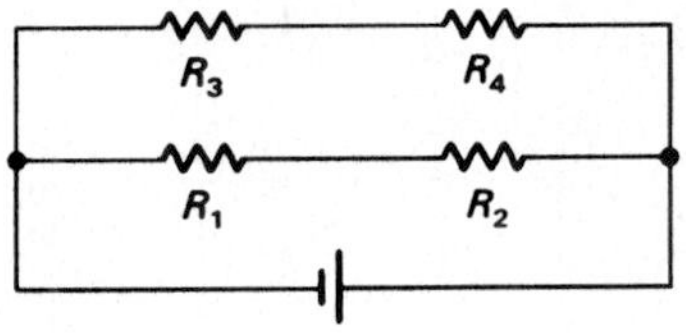

	V	I	R
Battery	50.0 V	5.00 A	Ω
R_1	V	2.00 A	Ω
R_2	25.0 V	A	Ω
R_3	10.0 V	A	Ω
R_4	V	3.00 A	Ω

6.

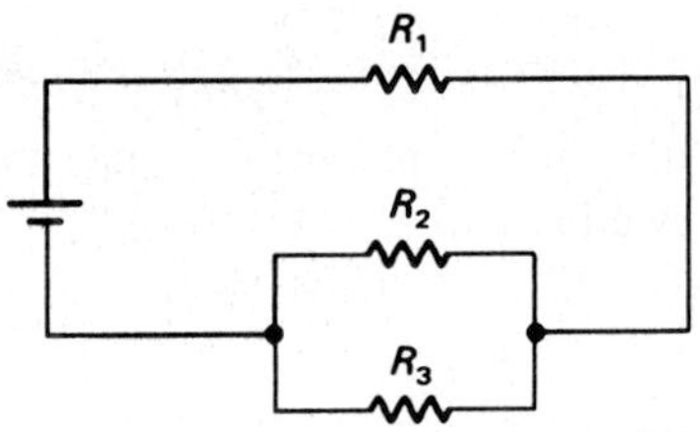

	V	I	R
Battery	24.0 V	A	Ω
R_1	8.00 V	A	Ω
R_2	V	4.00 A	Ω
R_3	V	2.00 A	Ω

7.

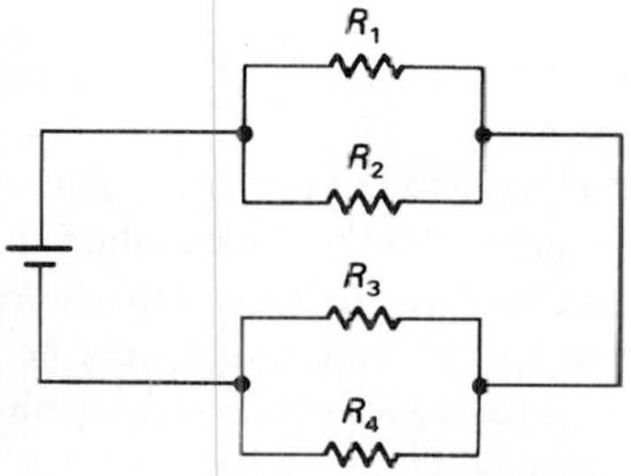

	V	I	R
Battery	V	A	Ω
R_1	12.0 V	A	2.00 Ω
R_2	V	A	4.00 Ω
R_3	24.0 V	A	4.00 Ω
R_4	V	A	8.00 Ω

8.

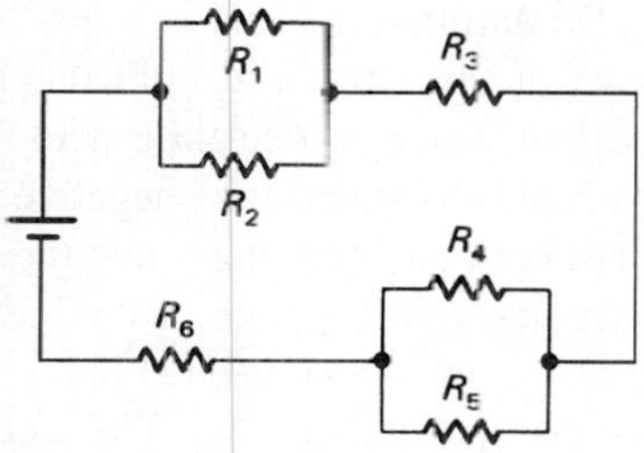

	V	I	R
Battery	30.0 V	A	Ω
R_1	6.00 V	3.00 A	Ω
R_2	V	2.00 A	Ω
R_3	V	A	3.00 Ω
R_4	V	1.00 A	Ω
R_5	8.00 V	A	Ω
R_6	V	A	Ω

9.

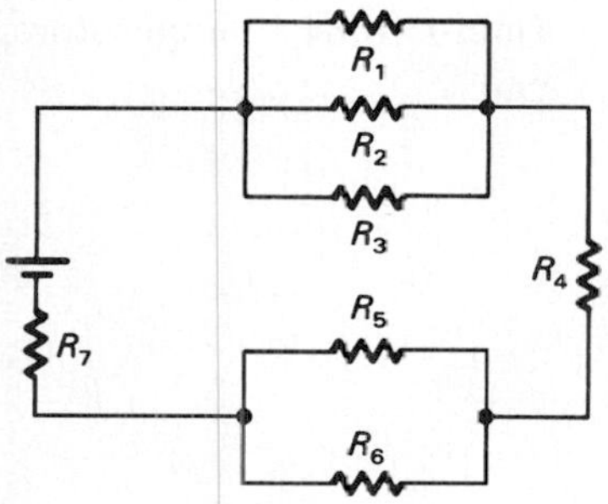

	V	I	R
Battery	V	12.0 A	Ω
R_1	V	A	Ω
R_2	18.0 V	2.00 A	Ω
R_3	V	A	3.00 Ω
R_4	V	A	4.00 Ω
R_5	V	A	2.00 Ω
R_6	V	8.00 A	Ω
R_7	6.00 V	A	Ω

10.

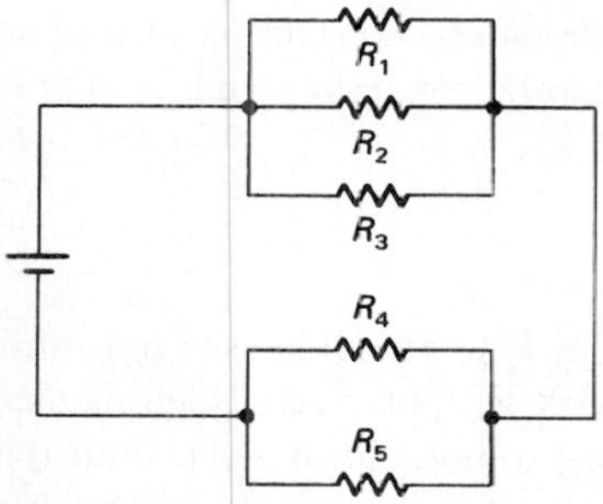

	V	I	R
Battery	46.0 V	A	Ω
R_1	V	3.00 A	Ω
R_2	V	4.00 A	Ω
R_3	V	A	6.00 Ω
R_4	V	3.00 A	Ω
R_5	V	7.00 A	Ω

11.

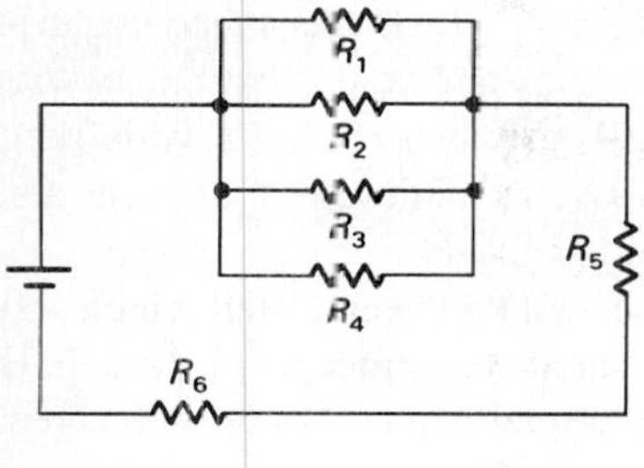

	V	I	R
Battery	V	A	Ω
R_1	V	A	20.0 Ω
R_2	10.0 V	A	Ω
R_3	V	A	4.00 Ω
R_4	V	1.00 A	Ω
R_5	V	5.00 A	5.00 Ω
R_6	V	A	6.00 Ω

16.11 Voltage Sources

Lead Storage Cell

A battery is a group of cells connected together. Each cell consists of a positive plate and a negative plate in a conducting solution. These lead cells are **secondary cells,** which means that they are rechargeable. The passing of an electric current through the cell to restore the original chemicals is called **recharging.** Cells, such as the dry cell, that cannot be efficiently recharged are called **primary cells.** ***Note:*** The cells are all connected together in series (Fig. 16.63). Six storage cells of 2.0 V each connected in series give 12.0 V for an automobile storage battery.

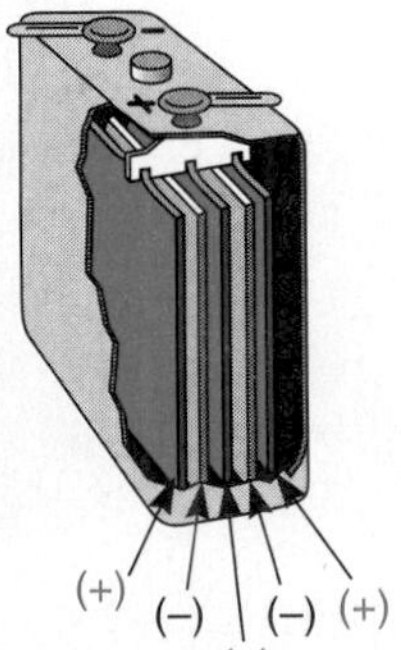

Figure 16.63 Lead storage battery

Lead storage batteries are used to generate voltage in automobiles and in many other types of vehicles and machinery. Lead storage cells are made up of two kinds of lead plates (lead and lead oxide) submerged in a solution of distilled water and sulfuric acid (Fig. 16.64). This acid solution that produces large numbers of free electrons at the negative pole of a cell is called an **electrolyte.** The chemical action between the lead plates and the acid solution produces large numbers of free electrons at the negative (−) pole of the battery. These electrons have a large amount of electric potential energy, which is used in the load in the circuit (for instance, to operate headlights or to turn a starter motor). The work of **Luigi Galvani** led to the production of the electric battery.

Luigi Galvani (1737–1798), *physiologist, was born in Italy. He investigated the effects of electrostatic stimuli applied to muscle fiber. His work inspired his friend, Volta, and led to the production of the electric battery and initiated research into electrophysiology. The galvanometer is named after him.*

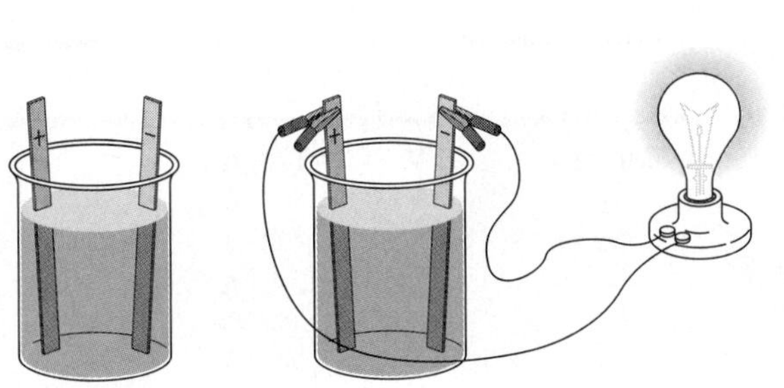

Figure 16.64 Simple storage cell

As the electric energy is used in the load, the battery must be recharged. This is done by a generator or an alternator. Such devices provide an electric current to reverse the chemical reaction taking place in the battery. The recharging process extends the life of the battery, which would otherwise be very short.

Dry Cell

The dry cell is the most widely used primary cell. This kind of cell is used in flashlights and portable products such as cellular phones, notebook computers, drills, etc. A **dry cell** is a voltage-generating cell that consists of an electrolyte in the form of a chemical paste and two electrodes of unlike materials, one of which reacts chemically with the electrolyte to provide energized electrons. The carbon–zinc dry cell is made of a carbon rod, which is the positive (+) terminal or pole, and a zinc can, which acts as the negative (−) terminal (Fig. 16.65). In between is a paste of chemicals and water that reacts with the terminals to provide energized electrons. These cells are available in a wide range of sizes. Common battery voltages range from 1.5 to 9 V. To achieve 9 V requires a series stack of six cells.

Figure 16.65 Dry cell batteries

The dry cell, as well as the lead cell, has resistance within the cell itself which opposes the movement of the electrons. This is called the **internal resistance,** r, of the cell. Every cell has internal resistance. Because current flows in the cell, the emf of the cell is reduced by the voltage drop across the internal resistance (Fig. 16.66). The voltage applied to the external circuit is then

$$V = E - Ir$$

where

V = voltage applied to the circuit
E = emf of the cell
I = current through the cell
r = internal resistance of the cell

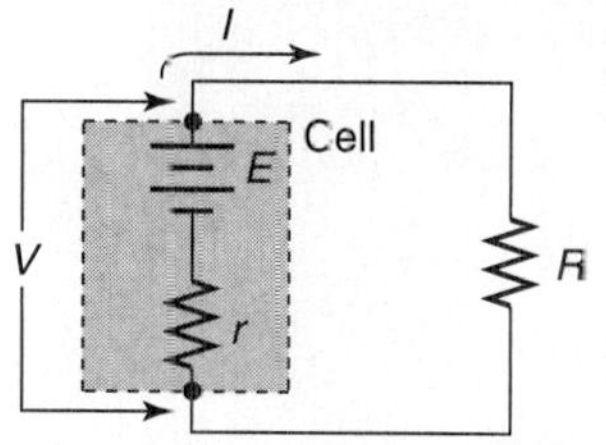

Figure 16.66 Effect of the internal resistance of a dry cell

When the current or the voltage available from a single cell is inadequate for a particular job, we usually connect two or more cells in a parallel or series arrangement.

Alkaline cells resemble carbon–zinc cells but are five to eight times longer lasting. An alkaline cell has a highly porous zinc anode, which oxidizes more readily than does the carbon–zinc cell's anode. Its electrolyte is a strong alkali solution called *potassium hydroxide*. This compound conducts electricity inside the cell very well and enables the alkaline cell to deliver relatively high currents with greater efficiency than that of carbon–zinc cells.

Nickel–cadmium, metal hydride, and lithium batteries are now also in common use. Their advantage over other types of dry cells is that they are rechargeable. Lithium batteries have greater capacity than nickel–cadmium or hydride batteries.

Another type of dc power source is the solar cell. It is commonly made from a semiconductor material (typically amorphous silicon or gallium arsenide). Many small calculators operate with power supplied by solar cells.

16.12 Cells in Series and Parallel

To connect cells in series, the positive terminal of one is connected to the negative terminal of the next cell. This procedure is continued until the desired number of cells is connected (Fig. 16.67). The rules for cells connected in series and parallel are similar to those for resistances.

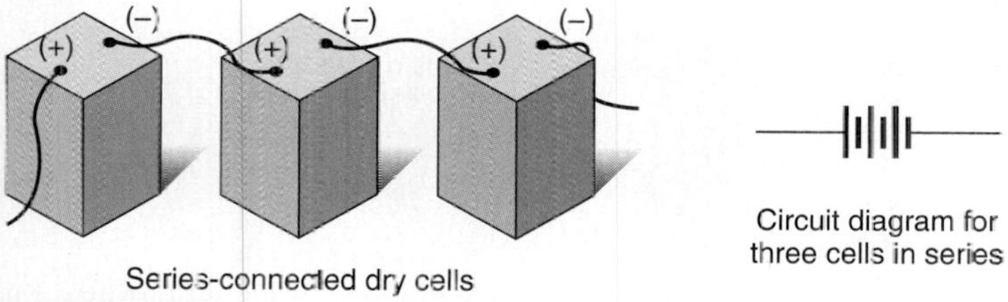

Figure 16.67

CELLS IN SERIES

1. The current in the circuit equals the current in any single cell:

$$I = I_1 = I_2 = I_3 = \cdots$$

2. The internal resistance of the battery equals the sum of the individual internal resistances of the cells:

$$r = r_1 + r_2 + r_3 + \cdots$$

3. The emf of the battery equals the sum of the emf's of the individual cells:

$$E = E_1 + E_2 + E_3 + \cdots$$

EXAMPLE 1

Two 6.00-V cells with internal resistance of 0.100 Ω each are connected in series to form a battery with a current of 0.750 A in each cell (Fig. 16.68).

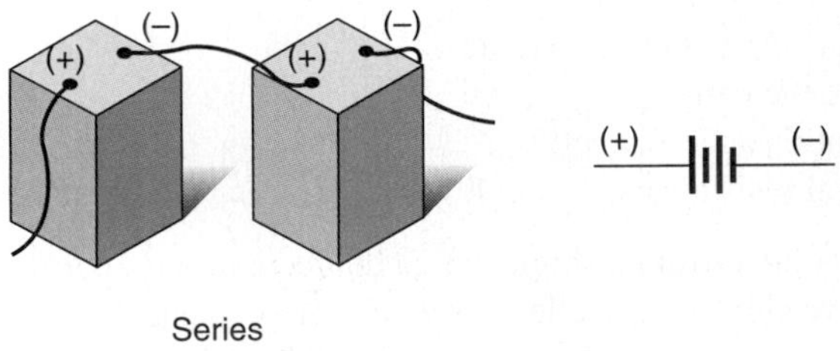

Figure 16.68

(a) What is the emf of the battery?
(b) Find the internal resistance of the battery.
(c) Find the current in the external circuit.

(a) $E = E_1 + E_2 = 6.00\text{ V} + 6.00\text{ V} = 12.00\text{ V}$ (Rule 3)
(b) $r = r_1 + r_2 = 0.100\ \Omega + 0.100\ \Omega = 0.200\ \Omega$ (Rule 2)
(c) $I = 0.750\text{ A}$ (Rule 1)

To connect cells in parallel, the positive terminals of all the cells are connected together and the negative terminals are all connected together (Fig. 16.69).

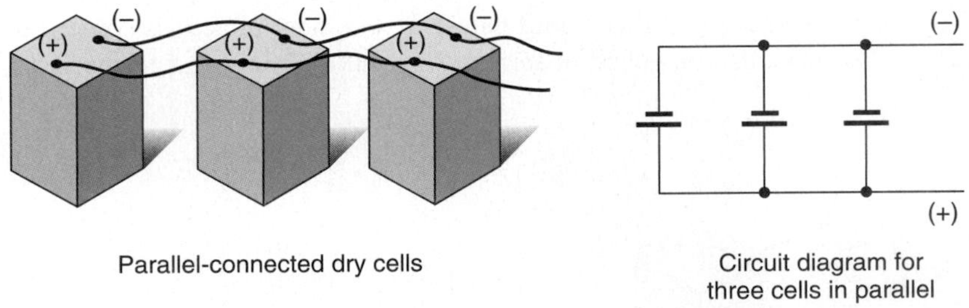

Figure 16.69

A common example of the use of cells in parallel is the practice of jump-starting a car that has a dead battery (Fig. 16.70). It is not common to find cells hooked up in parallel because a mismatch of output voltages could cause problems. The leads from the external circuit may be connected to any positive and negative terminals. (The external circuit is all of the circuit *outside* the battery or cell.)

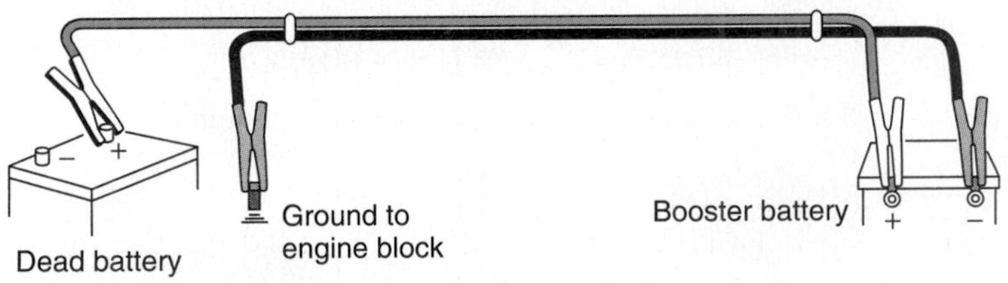

Figure 16.70 How to properly make connections to jump-start a car that has a dead battery

CELLS IN PARALLEL

1. The total current equals the sum of the individual currents in each cell:

$$I = I_1 + I_2 + I_3 + \cdots$$

2. The internal resistance equals the resistance of one cell divided by the number of cells:*

$$r = \frac{r \text{ of one cell}}{\text{number of cells}}$$

3. The emf of the battery equals the emf of any single cell:

$$E = E_1 = E_2 = E_3 = \cdots$$

EXAMPLE 2

Four cells, each 1.50 V and having an internal resistance of 0.0500 Ω, are connected in parallel to form a battery with a current output of 0.250 A in each cell (Fig. 16.71).

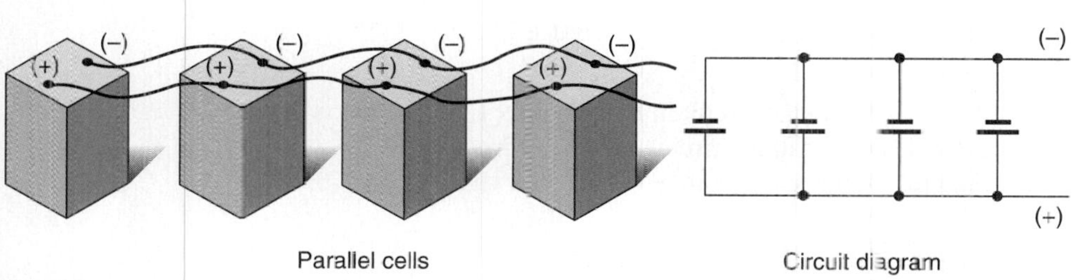

Figure 16.71

(a) What is the emf of the battery?
(b) Find the internal resistance of the battery.
(c) Find the current in the external circuit.

(a) $E = 1.50\text{ V}$ (Rule 3)

(b) $r = \dfrac{r \text{ of one cell}}{\text{number of cells}} = \dfrac{0.0500\ \Omega}{4} = 0.0125\ \Omega$ (Rule 2)

(c) $I = I_1 + I_2 + I_3 + I_4 = 0.250\text{ A} + 0.250\text{ A} + 0.250\text{ A} + 0.250\text{ A} = 1.000\text{ A}$ (Rule 1)

SKETCH

12 cm² w
4.0 cm

DATA

$A = 12\text{ cm}^2$, $l = 4.0\text{ cm}$, $w = ?$

BASIC EQUATION

$A = lw$

WORKING EQUATION

$w = \frac{A}{l}$

SUBSTITUTION

$w = \frac{12\text{ cm}^2}{4.0\text{ cm}} = 3.0\text{ cm}$

PROBLEMS 16.12

1. A cell has an emf of 1.50 V and an internal resistance of 0.0450 Ω. If there is 0.250 A in the cell, what voltage is applied to the external circuit?
2. The voltage applied to a circuit is 11.8 V when the current through the battery is 0.500 A. If the internal resistance of the battery is 0.150 Ω, what is the emf of the battery?
3. The emf of a battery is 12.0 V. If the internal resistance is 0.300 Ω and the voltage applied to the circuit is 11.6 V, what is the current through the battery?

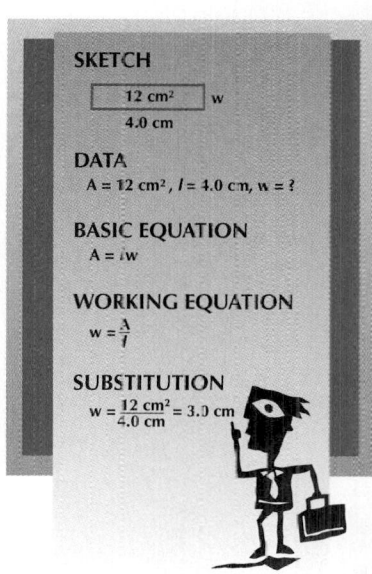

---

*This formula works only when all the cells have the same internal resistance. Otherwise, a formula similar to that for resistors in parallel must be used.

4. Three 1.50-V cells, each with an internal resistance of 0.0500 Ω, are connected in series to form a battery with a current of 0.850 A in each cell.
   (a) Find the current in the external circuit.
   (b) What is the emf of the battery?
   (c) Find the internal resistance of the battery.
5. Five 9.00-V cells, each with internal resistance of 0.100 Ω and current output of 0.750 A, are connected in parallel to form a battery in a certain circuit.
   (a) Find the current in the external circuit.
   (b) What is the emf of the battery?
   (c) Find the internal resistance of the battery.
6. Find the current in the circuit shown in Fig. 16.72.
7. Find the current in the circuit shown in Fig. 16.73.

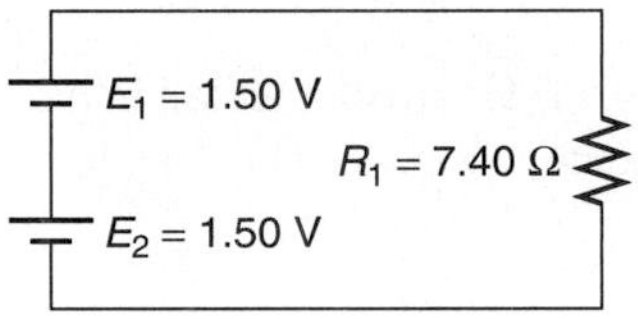

Figure 16.72

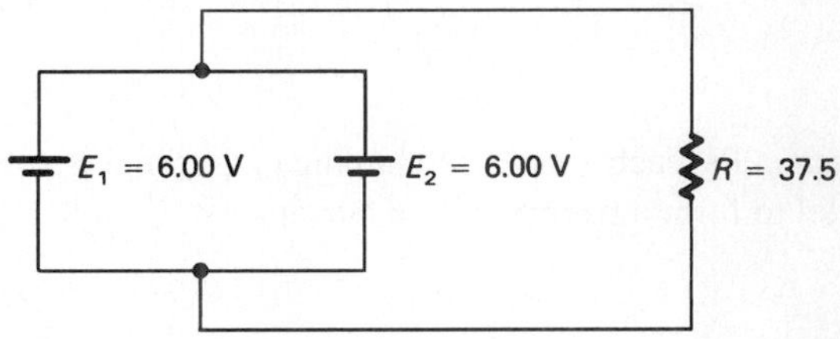

Figure 16.73

8. If the current in the circuit in Fig. 16.74 is 1.20 A, what is the value of R?
9. Find the current in the circuit shown in Fig. 16.75.
10. Find the total resistance in the circuit shown in Fig. 16.75.

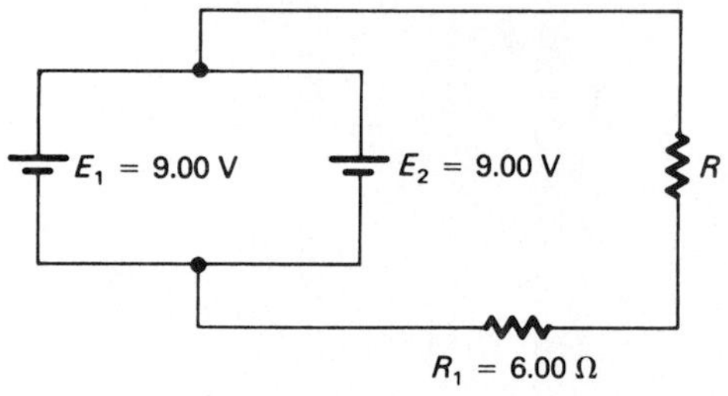

Figure 16.74

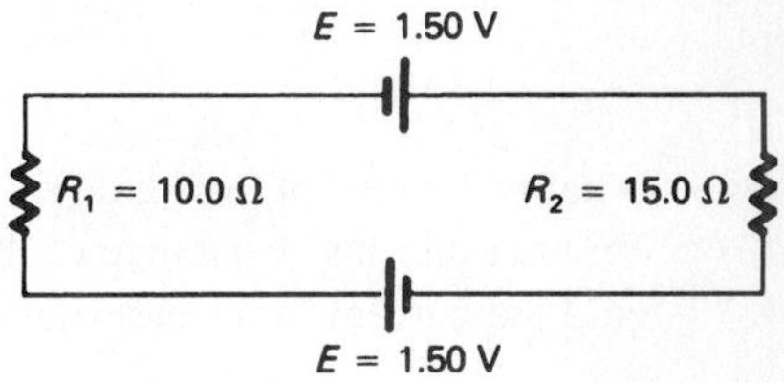

Figure 16.75

16.13 Electric Power

Tremendous quantities of energy are used by industry and sold by power companies. The rate of consuming energy is called **power.** The unit of power is the watt. One *watt* (W) is the power generated by a current of 1 A flowing because of a potential difference of 1 V. A volt is a joule/coulomb (J/C); an ampere is a coulomb/second (C/s). Their product is

$$\text{VA} = \frac{\text{J}}{\text{C}} \cdot \frac{\text{C}}{\text{s}} = \frac{\text{J}}{\text{s}}$$

Thus, 1W = 1 J/s.
Hence, power is

$$P = VI$$

where

$$P = \text{power (watts)}$$
$$V = \text{voltage drop}$$
$$I = \text{current}$$

This equation applies to components of dc circuits and to whole dc circuits as well as to ac circuits with resistance only.

Recalling Ohm's law, $I = V/R$, we find two other equations for power:

Given

$$P = VI$$

substitute for V using $V = IR$ to obtain

$$P = (IR)I = I^2R$$

$$\boxed{P = I^2R}$$

Note from the following unit analysis that amps squared times ohms gives watts:

$$\boxed{\mathrm{A}^2\,\Omega = \mathrm{A}^2 \cdot \frac{\mathrm{V}}{\mathrm{A}} = \mathrm{A\,V} = \frac{\mathrm{C}}{\mathrm{s}} \cdot \frac{\mathrm{J}}{\mathrm{C}} = \frac{\mathrm{J}}{\mathrm{s}} = \mathrm{W}}$$

Also, given

$$P = I^2R$$

substitute

$$I = \frac{V}{R}$$

to get

$$P = \left(\frac{V}{R}\right)^2 R = \frac{V^2}{R^2} \cdot R$$

$$\boxed{P = \frac{V^2}{R}}$$

EXAMPLE 1

A soldering iron draws 7.50 A in a 115-V circuit. What is its wattage rating?

Data:

$$I = 7.50 \text{ A}$$
$$V = 115 \text{ V}$$
$$P = ?$$

Basic Equation:

$$P = VI$$

Working Equation: Same

Substitution:

$$P = (115\text{ V})(7.50\text{ A})$$
$$= 863\text{ W}$$

Therefore, a soldering iron drawing 7.50 A in a 115-V circuit has a rating of 863 W.

EXAMPLE 2

A hand drill draws 4.00 A and has a resistance of 14.6 Ω. What power does it use?

Data:

$$I = 4.00\text{ A}$$
$$R = 14.6\ \Omega$$
$$P = ?$$

Basic Equation:

$$P = I^2R$$

Working Equation: Same

Substitution:

$$P = (4.00\text{ A})^2(14.6\ \Omega)$$
$$= 234\text{ W}$$

Thus, a drill that draws 4.00 A with a resistance of 14.6 Ω has a rating of 234 W.

Since the watt is a relatively small unit, the kilowatt (1 kW = 1000 W) is commonly used in industry.

Although we speak of "paying our power bill," what power companies actually sell is **energy** in a form of work delivered to an electrical component or appliance. Energy is sold in kilowatt-hours (kWh). The amount of energy consumed is equal to the power used times the time it is used. Therefore,

$$\boxed{\text{energy} = \text{power} \times \text{time}}$$

or

$$\text{energy (in kWh)} = (VI)t$$
$$\text{number of kWh} = VIt$$

when V is in volts, I is in amperes, and t is time in hours. Note that electric energy can be expressed in other units (joules), but kilowatt-hours is commonly used. This equation is useful in finding the cost of electric energy. Cost is measured in cents per kilowatt-hour. The cost of operating an electric device may be found as follows:

$$\text{cost} = \text{energy} \times \text{cost per unit energy}$$
$$\text{cost} = (\text{kWh})\left(\frac{\text{cents}}{\text{kWh}}\right)$$

$$\boxed{\text{cost (in cents)} = \text{power (in W)} \times \text{hours} \times \frac{1\text{ kW}}{1000\text{ W}} \times \frac{\text{cents}}{\text{kWh}}}$$

↑ conversion factor

EXAMPLE 3

An iron is rated at 550 W. How much would it cost to operate it for 2.50 h at \$0.08/kWh?

Data:

$$P = 550 \text{ W}$$
$$t = 2.50 \text{ h}$$
$$\text{rate} = \$0.08/\text{kWh}$$
$$\text{cost} = ?$$

Basic Equation:

$$\text{cost} = Pt\left(\frac{\text{kW}}{1000 \text{ W}}\right)\left(\frac{\text{cents}}{\text{kWh}}\right)$$

Working Equation: Same

Substitution:

$$\text{cost} = (550 \cancel{\text{W}})(2.50 \cancel{\text{h}})\left(\frac{\cancel{\text{kW}}}{1000 \cancel{\text{W}}}\right)\left(\frac{\$0.08}{\cancel{\text{kWh}}}\right)$$
$$= \$0.11$$

Remember that the source of electrons in a circuit is the conducting circuit material itself. You may be able to buy an empty water hose but you can't buy an "empty" wire. If you plug in an appliance, *energy* flows from the outlet to the appliance, not electrons. Energy is carried by the electric field and causes motion in the electrons already in the appliance. The power company sells the energy; the appliance supplies the electrons.

PROBLEMS 16.13

1. A heater draws 8.70 A on a 110-V line. What is its wattage rating?
2. What power is needed for a sander that draws 3.50 A and has a resistance of 6.70 Ω?
3. How many amperes will a 75.0-W lamp draw on a 110-V line?
4. Find the resistance of the lamp in Problem 3.
5. How many amperes will a 750-W lamp draw on a 110-V circuit?
6. Find the cost to operate the lamp in Problem 5 for 40.0 h if the cost of energy is \$0.07 per kWh.
7. Six 50.0-W bulbs are operated for 25.0 h on a 115-V circuit. If energy costs \$0.075 per kWh, find the cost of operating them.
8. A small furnace uses 3.00 kW of power. If the cost of operation of the furnace is \$3.84 for a 24.0-h period, what is the cost of energy per kWh?
9. Will a 20.0-A fuse blow if a 1000-W hair dryer, a 1200-W electric skillet, and an 1100-W toaster are all used at once on a 110-V line?
10. How long could you operate a 1000-W soldering iron for \$0.50 if the cost of energy is \$0.075/kWh?
11. Find the cost of operating a 1.50-A motor on a 110-V circuit for 2.00 h at \$0.08/kWh.
12. Find the cost of operating a 2.50-A motor on a 110-V circuit for 3.00 h at \$0.07/kWh.
13. Find the cost of operating a 3.00-A motor on a 110-V circuit for 2.00 h at \$0.07/kWh.
14. How many amperes will a 60-W lamp draw on a 110-V line?
15. Using the following table, list two different combinations of bulbs and appliances that could be used on a 20-A circuit breaker in a 110-V system without causing the circuit breaker to trip.

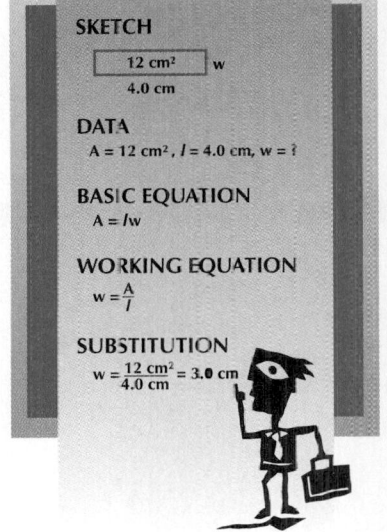

SKETCH

12 cm² | w
4.0 cm

DATA

$A = 12 \text{ cm}^2$, $l = 4.0$ cm, $w = ?$

BASIC EQUATION

$A = lw$

WORKING EQUATION

$w = \frac{A}{l}$

SUBSTITUTION

$w = \frac{12 \text{ cm}^2}{4.0 \text{ cm}} = 3.0 \text{ cm}$

Appliance	Power Rating
Light bulb (60 W)	60 W
Fluorescent bulb (40 W)	40 W
12″ TV	55 W
Projection TV	1500 W
Personal computer	550 W
Hand drill	400 W
Microwave oven	1000 W

16. Using the preceding table, list two different combinations that could be used on a $3\bar{0}$-A circuit breaker in a $11\bar{0}$-V system without causing the circuit breaker to trip.
17. Find the power output of a cell phone charger that delivers $40\bar{0}$ mA of current at 5.90 V dc.
18. A power supply for electronic devices delivers 1.10 A of current at 4.40 V. What is its power output?
19. At what rate does a light bulb convert electric energy to light if it draws 0.500 A of current on a $12\bar{0}$-V outlet?
20. What power is used by a light that draws 2.00 A from a 12.0-V battery?
21. How much electric energy (in joules) is delivered in 5.00 min to an electric motor that draws 0.500 A from a 6.00-V battery?
22. A car has a 12.0-V battery. If the current through the starter is $21\bar{0}$ A, what electric energy (in joules) is delivered to the starter in 10.0 s?
23. (a) How much power does a television use if it draws 2.00 A on a $12\bar{0}$-V line? (b) What energy in kWh does the television use in 30 days if it is used an average of 7.00 h/day? (c) Find the cost of operating the television for 30 days if the cost of energy is \$0.11/kWh.
24. An electric heater is used 5.00 h each day. (a) If it draws 15.0 A on a $12\bar{0}$-V line, how much power does it use? (b) In 30 days, how much energy in kWh does the heater use? (c) At \$0.11/kWh, what does it cost to operate the heater for 30 days?
25. A digital timer is used on a 115-V line. (a) If the resistance of the timer is $12{,}\bar{0}00\ \Omega$, how much current does it draw? (b) How much power does the timer use? (c) What does it cost to operate the timer at \$0.09/kWh for 30 days?

Glossary

Alternating Current Current that changes direction. (p. 439)

Ammeter An instrument that measures the current flowing in a circuit. (p. 462)

Conduction A transfer of charge from one place to another. (p. 430)

Conductor A material through which an electron charge is readily transferred. (p. 441)

Coulomb's Law The force between two point charges is directly proportional to the product of their magnitudes and inversely proportional to the square of the distance between them. (p. 434)

Current The flow of charge that passes through a conductor. (p. 441)

Direct Current Current that flows in one direction. (p. 439)

Dry Cell A voltage-generating cell that consists of a chemical paste and two electrodes of unlike materials, one of which reacts chemically with the electrolyte. (p. 466)

Electric Circuit A conducting loop in which electrons carrying electric energy may be transferred from a suitable source to do useful work and returned to the source. (p. 439)

Electric Field An electric field exists where an electric force acts on a charge brought into the area. (p. 436)

Electrolyte An acid solution that produces large numbers of free electrons at the negative pole of a cell. (p. 466)
Electron A negatively charged particle found in every atom. (p. 430)
emf The potential difference across a source. (p. 442)
Energy Work delivered to an electric component or appliance (power × time). (p. 472)
Equivalent Resistance The single resistance that can replace a series and/or parallel combination of resistances in a circuit and provide the same current flow and voltage drop. (p. 448)
Induction A method of charging one object by bringing a charged object near to, but not touching, it. (p. 431)
Internal Resistance The resistance within a cell that opposes movement of the electrons. (p. 466)
Insulator A substance that does not allow electric current to flow through it readily. (p. 441)
Load The object in a circuit that converts electric energy into other forms of energy or work. (p. 441)
Multimeter An instrument used to measure current flow, voltage drop, and resistance. (p. 462)
Neutron A neutral particle found in the nucleus of most atoms. (p. 430)
Ohm's Law When a voltage is applied across a resistance in an electric circuit, the current equals the voltage drop across the resistance divided by the resistance, $I = V/R$. (p. 444)
Ohmmeter An instrument that measures the resistance of a circuit component. (p. 463)
Parallel Circuit An electric circuit with more than one path for the current to flow. The current is divided among the branches of the circuit. (p. 451)
Power Energy per unit time consumed in a circuit. (p. 470)
Primary Cell A cell that cannot be recharged. (p. 466)
Proton A positively charged particle found in the nucleus of every atom. (p. 430)
Recharging The passing of an electric current through a secondary cell to restore the original chemicals. (p. 466)
Resistance The opposition to current flow. (p. 442)
Resistivity The resistance per unit length of a material with uniform cross section. (p. 443)
Secondary Cell A rechargeable type of cell. (p. 466)
Semiconductors A small number of materials that fall between conductors and insulators in their ability to conduct electric current. (p. 441)
Series Circuit An electric circuit with only one path for the current to flow. The current in a series circuit is the same throughout. (p. 447)
Source The object that supplies electric energy for the flow of electric charge (electrons) in a circuit. (p. 440)
Superconductor A material that continuously conducts electric current without resistance when cooled to typically very low temperatures, often near absolute zero. (p. 441)
Voltage Drop The potential difference across a load in a circuit. (p. 442)
Voltmeter An instrument that measures the difference in potential (voltage drop) between two points in a circuit. (p. 462)

Formulas

16.3 Coulomb's law: $F = \dfrac{kq_1q_2}{r^2}$

16.4 $E = \dfrac{F}{q'}$

16.5 $R = \dfrac{\rho l}{A}$

16.6 Ohm's law: $I = \dfrac{V}{R}$

16.7 *Characteristics of Series Circuits*

Current	$I = I_1 = I_2 = I_3 = \cdots$
Resistance	$R = R_1 + R_2 + R_3 + \cdots$
Voltage	$E = V_1 + V_2 + V_3 + \cdots$

16.8 *Characteristics of Parallel Circuits*

Current	$I = I_1 + I_2 + I_3 + \cdots$
Resistance	$\frac{1}{R} = \frac{1}{R_1} + \frac{1}{R_2} + \frac{1}{R_3} + \cdots$
Voltage	$E = V_1 = V_2 = V_3 = \cdots$

16.11 $V = E - Ir$

16.12 *Cells in series:*

$$I = I_1 = I_2 = I_3 = \cdots$$
$$r = r_1 + r_2 + r_3 + \cdots$$
$$E = E_1 + E_2 + E_3 + \cdots$$

Cells in parallel:

$$I = I_1 + I_2 + I_3 + \cdots$$
$$r = \frac{r \text{ of one cell}}{\text{number of cells}}$$
$$E = E_1 = E_2 = E_3 = \cdots$$

16.13 $P = VI$

$$P = I^2R$$
$$P = \frac{V^2}{R}$$

energy = power × time

$$\text{cost (in cents)} = \text{power (in W)} \times \text{hours} \times \frac{1 \text{ kW}}{1000 \text{ W}} \times \frac{\text{cents}}{\text{kWh}}$$

Review Questions

1. The atomic particle that carries a positive charge is
   (a) the neutron. (b) the proton.
   (c) the electron. (d) none of the above.
2. The atomic particle that carries a negative charge is
   (a) the neutron. (b) the proton.
   (c) the electron. (d) none of the above.
3. The process by which an object becomes charged when it comes in contact with a charged object is called
   (a) induction. (b) electrification.
   (c) conduction. (d) none of the above.
4. The process by which an object becomes permanently charged when it comes near a charged object requires that the first object be
   (a) an insulator. (b) touched by another object.
   (c) a conductor. (d) none of the above.
5. The resistance of a wire is dependent on all of the following except
   (a) temperature. (b) cross-sectional area.
   (c) length. (d) material.
   (e) voltage.
6. Which of the following are good electric conductors?
   (a) Aluminum (b) Wood
   (c) Glass (d) Distilled water
   (e) Silver

7. The total resistance in a circuit containing resistors connected in series is given by
   (a) the sum of the individual resistances.
   (b) the sum of the inverse of the individual resistances.
   (c) the sum of the currents.
   (d) the sum of the voltages.
8. The current in a parallel circuit is given by
   (a) the sum of the inverse currents.
   (b) the sum of the voltages.
   (c) the sum of the currents in the branches.
   (d) none of the above.
9. The emf of a battery with cells connected in series equals the sum of the
   (a) internal resistances of the cells.
   (b) emf of the individual cells.
   (c) current in the individual cells.
10. The current in a battery with cells connected in series equals the
   (a) internal resistances of the cells.
   (b) emf of the individual cells.
   (c) current in the individual cells.
11. The current in a battery with cells connected in parallel equals the
   (a) current in one cell.
   (b) internal resistance.
   (c) sum of the currents in each cell
12. Examples of dry cells include
   (a) lead–zinc cells.
   (b) nickel–cadmium cells.
   (c) carbon–zinc cells.
   (d) fuel cells.
13. In your own words, describe how materials can become charged by electrification.
14. What particles make up an atom?
15. What particles are located in the nucleus (center) of an atom?
16. Where are electrons located in an atom?
17. What are the two types of charge? What atomic particle carries each type of charge?
18. Describe the process of charging an electroscope by conduction.
19. Describe the process of charging an electroscope by induction.
20. In your own words, describe Coulomb's law of electrostatics.
21. Describe an electric field.
22. Describe lightning.
23. The flow of electrons through a conductor is called _____.
24. (a) The unit of current is the _____. (b) The unit of emf is the _____. (c) The unit of resistance is the _____.
25. What effect does doubling the diameter of a wire have on the wire's resistance?
26. In your own words, explain Ohm's law.
27. Differentiate between a series and a parallel circuit.
28. Differentiate between the equivalent resistance in a series circuit and a parallel circuit.
29. In using an electric instrument, with what range should you start when making a measurement?
30. Explain how a parallel water system compares to a parallel electric circuit.
31. How does the current change in a circuit if the resistance increases by a factor of 2?
32. How does the current change in a circuit if the voltage is increased by a factor of 2?
33. How would the resistance of a wire change if the length were to be increased by a factor of 2?
34. Explain the concept of electric potential.
35. Explain the transfer of energy that occurs in a circuit that includes a dry cell and two lamps in series.
36. Distinguish between a primary and a secondary cell.
37. Explain recharging.
38. Describe the function of an electrolyte.
39. In your own words, describe the manner in which a secondary cell produces electric energy.
40. What is the effect of the internal resistance of a cell?
41. The unit of electric power is the _______________.
42. In your own words, explain the relationship between power, voltage, and current.

43. Do we pay the utility company for our power use or our energy use? Explain.
44. Explain the relationship between power, voltage, and resistance.
45. If the current in a circuit is increased by a factor of 2 and the voltage stays constant, how does the power change?
46. If the resistance in a circuit decreases by a factor of 2 and the voltage stays constant, how does the power change?
47. If the voltage and current in a circuit each decrease by a factor of 2, how does the power change?
48. If the current increases in a circuit by a factor of 2 and the voltage stays constant, how would the cost of operating the circuit change?

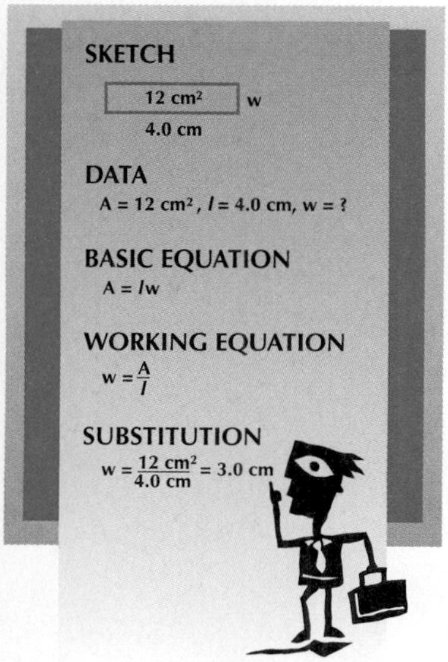

Review Problems

1. Two charges, each $-4.50\ \mu C$, are 0.150 cm apart. Find their repulsive force.
2. The repulsive force between two identical negative charges is 0.750 N when they are 0.100 m apart. Find the amount of each charge.
3. A charge of 2.50×10^{-8} C exerts a force of 0.0250 N on a second charge of 5.00×10^{-7} C. How far apart are the charges?
4. A positive test charge of $2.50\ \mu C$ is placed in an electric field. The force on it is 0.500 N. What is the magnitude of the electric field at the location of the test charge?
5. Find the magnitude of the electric field in which a negative charge of 1.50×10^{-8} C experiences a force of 0.0500 N.
6. What force is exerted on a test charge of 4.25×10^{-5} C if it is placed in an electric field of magnitude 2.50×10^{4} N/C?
7. Find the resistance of 85.5 m of No. 20 aluminum wire ($\rho = 2.83 \times 10^{-6}\ \Omega$ cm, $A = 2.07 \times 10^{-2}\ cm^2$).
8. At 75°F, $12\bar{0}$ ft of wire has a resistance of 0.743 Ω. Find the resistance of $56\bar{0}$ ft of this wire.
9. Find the resistance of 134 m of No. 20 copper wire at 20°C ($\rho = 1.72 \times 10^{-6}\ \Omega$ cm, $A = 2.07 \times 10^{-2}\ cm^2$).
10. Find the length of a copper wire with resistance 0.0273 Ω/ft and total resistance 3.97 Ω.
11. Find the cross-sectional area of copper wire at 20°C that is 55.4 m long and has resistivity $\rho = 1.79 \times 10^{-6}\ \Omega$ cm and resistance 0.943 Ω.
12. A heating element operates on 115 V. If it has a resistance of 15.4 Ω, what current does it draw?
13. A heating coil operates on $22\bar{0}$ V. If it draws 8.75 A, find its resistance.
14. What current does a 234-Ω resistance draw on 115 V?
15. Four resistors of 3.40 Ω, 6.54 Ω, 8.32 Ω, and 1.34 Ω are connected in series with a 12.0-V battery. Find the total resistance of the circuit.
16. Find the current in Problem 15.
17. Find the emf in the circuit shown in Fig. 16.76.
18. Find the equivalent resistance in the circuit shown in Fig. 16.77.

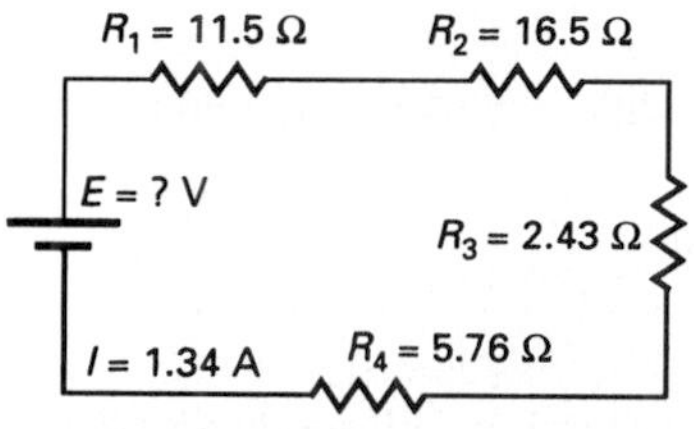

Figure 16.76

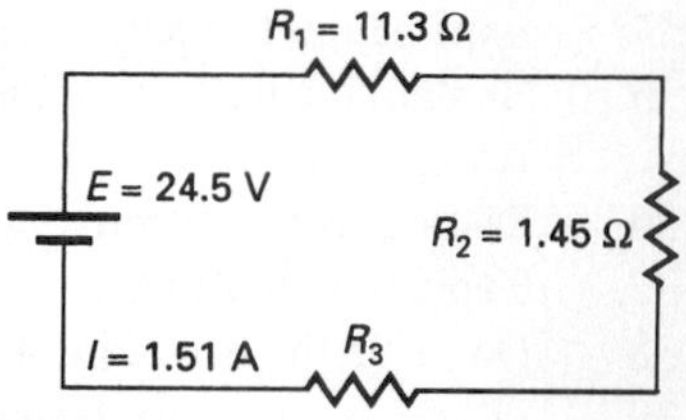

Figure 16.77

19. Find R_3 in the circuit in Problem 18.
20. Find the equivalent resistance in the circuit shown in Fig. 16.78.

21. Find the current in Fig. 16.78.
22. Find the current through R_1 in Fig. 16.78.
23. Find the current through R_2 in Fig. 16.78.
24. Find the equivalent resistance in the circuit of Fig. 16.79.

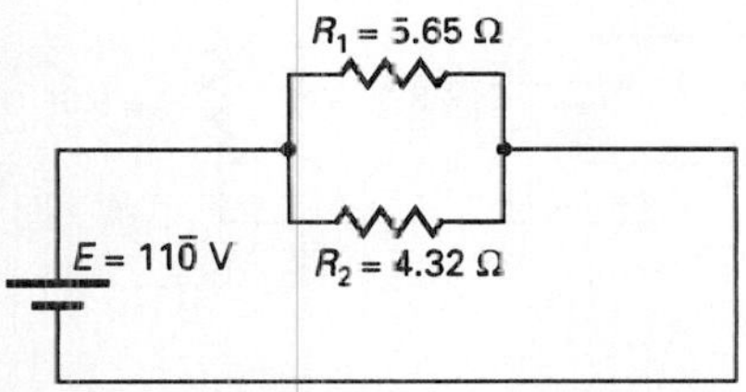

Figure 16.78

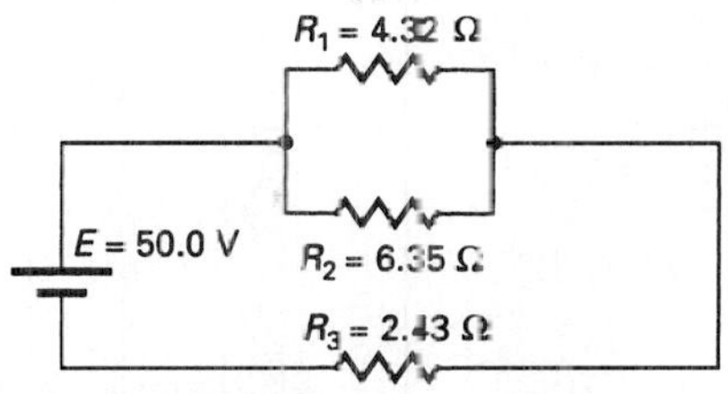

Figure 16.79

25. Find the current through R_3 in Fig. 16.79.
26. Find the current through R_1 in Fig. 16.79; through R_2.
27. Find the equivalent resistance in Fig. 16.80.

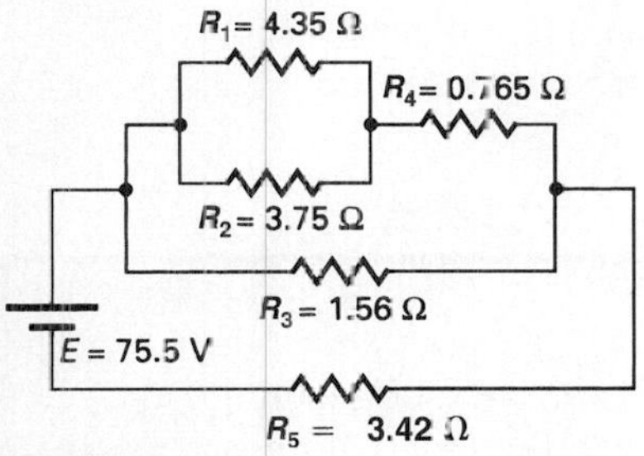

Figure 16.80

28. Find the current in R_5 in Fig. 16.80.
29. Find the voltage drop across R_5 in Fig. 16.80.
30. Find the current in R_1 in Fig. 16.80.
31. Find the voltage drop across R_1 in Fig. 16.80.
32. Using the formulas for series and parallel circuits, fill in the blanks in the table below for the circuit of Fig. 16.81.

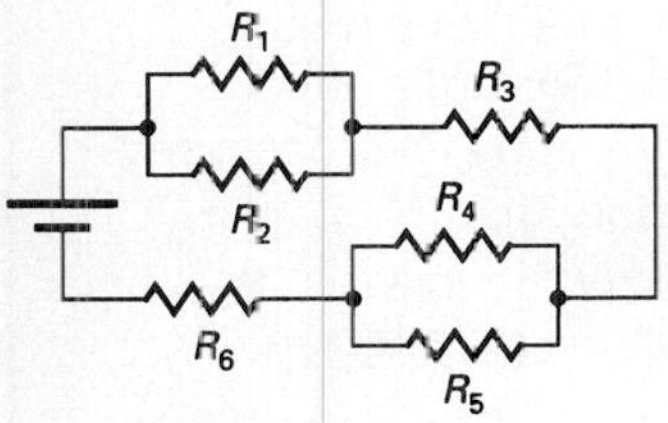

Figure 16.81

	V	I	R
Battery	35.0 V	A	Ω
R_1	5.00 V	2.75 A	Ω
R_2	V	1.95 A	Ω
R_3	V	A	2.80 Ω
R_4	V	0.97 A	Ω
R_5	7.50 V	A	Ω
R_6	V	A	Ω

33. A cell has an emf of 1.44 V and an internal resistance of 0.0550 Ω. If there is 0.135 A in the cell, what voltage is applied to the external circuit?
34. The voltage applied to a circuit is 12.0 V when the current through the battery is 0.858 A. If the internal resistance of the battery is 0.245 Ω, what is the emf?
35. Six 6.00-V cells, each with an internal resistance of 0.0987 Ω and current output of 0.658 A, are connected in parallel to form a battery in a certain circuit. (a) What is the current in the external circuit? (b) What is the emf of the battery? (c) What is the internal resistance of the battery?

36. Find the current in the circuit shown in Fig. 16.82.
37. Find the total resistance in the circuit shown in Fig. 16.83.

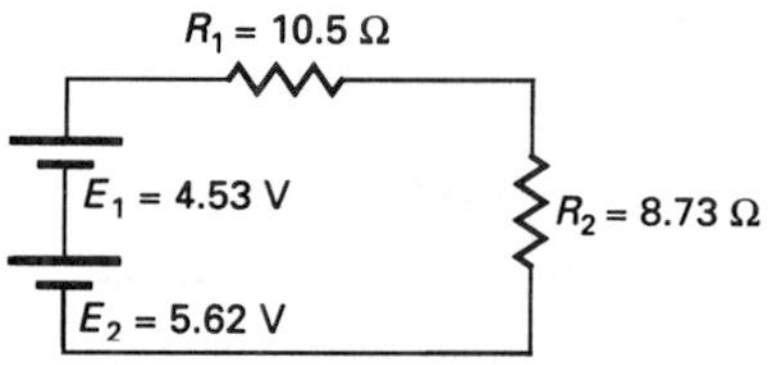

Figure 16.82

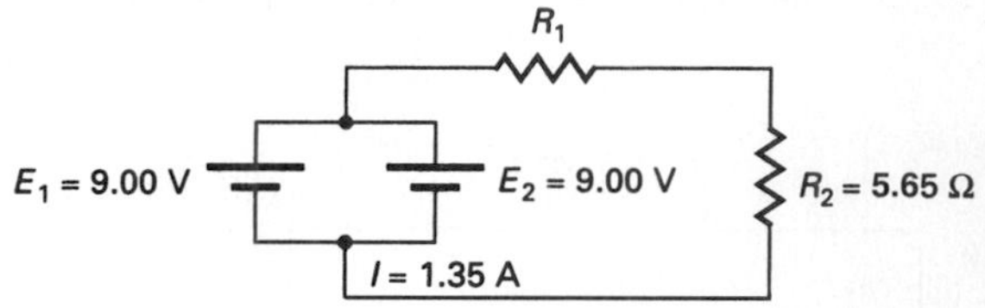

Figure 16.83

38. What power is needed for a drill that draws 2.45 A and has a resistance of 6.55 Ω on a $11\bar{0}$-V circuit?
39. How many amperes will a $15\bar{0}$-W light bulb draw on a $11\bar{0}$-V circuit?
40. What is the cost to operate the lamp in Problem 39 for 135 h if the cost of energy is \$0.05/kWh?
41. If the cost of energy is \$0.043/kWh, how long could you operate a motor that draws 0.40 A on a 110-V line for \$0.45?
42. How many amperes will a $1\bar{0}0$-W lamp draw on a 110-V line?

APPLIED CONCEPTS

1. A hydrogen atom contains one electron and one proton. (a) Find the electric force between the proton and the electron. (The distance between an electron and a proton in a hydrogen atom is 5.29×10^{-11} m. A proton's positive charge has the same magnitude as the negative charge of an electron.) (b) What is the gravitational force between a proton and an electron? (c) If the electric force is stronger than the gravitational force, why don't people accelerate toward charged combs, balloons, and other electrically charged objects?
2. A rod with charge -4.31×10^{-8} C is held 10.3 cm above a piece of sawdust with mass 3.12×10^{-5} kg. (a) How much does the sawdust weigh? (b) How much electrostatic force must the rod apply to lift the sawdust? (c) If the charged rod lifts the sawdust, what is the electric charge of the sawdust? (d) What happens to the strength of the electric field as the sawdust comes closer to the rod?
3. Hairdryers work by blowing heat that is generated by exposed, high-resistance wires. (a) If a woman using a hairdryer comes in direct contact with the $11\bar{0}$-V wires, what is the current that travels through her if she is dry and has a resistance of $50\bar{0}$,000 Ω? (b) If she is wet, the resistance of her body can drop as low as $10\bar{0}$ Ω. What is the current that passes through her body at this resistance? (Death can result when a current of 0.07 A flows through a person's body for more than 1 s.)
4. A $10\bar{0}0$-W microwave, a 40.0-W fluorescent light bulb, and a $55\bar{0}$-W computer are plugged into a $12\bar{0}$-V parallel circuit. (a) What is the current passing through each appliance in the parallel circuit? (b) Find the resistance of each appliance.
5. A $70\bar{0}$-W toaster is plugged into a $11\bar{0}$-V outlet. When turned on, the current flowing through the 6.00-m strand of nichrome wire heats the wire and toasts the bread. As the designer of the toaster, what radius do you choose for the nichrome wire? (Assume the temperature of the wire does not affect the resistance. $\rho_{nichrome} = 1.00 \times 10^{-6}\Omega$ m.)

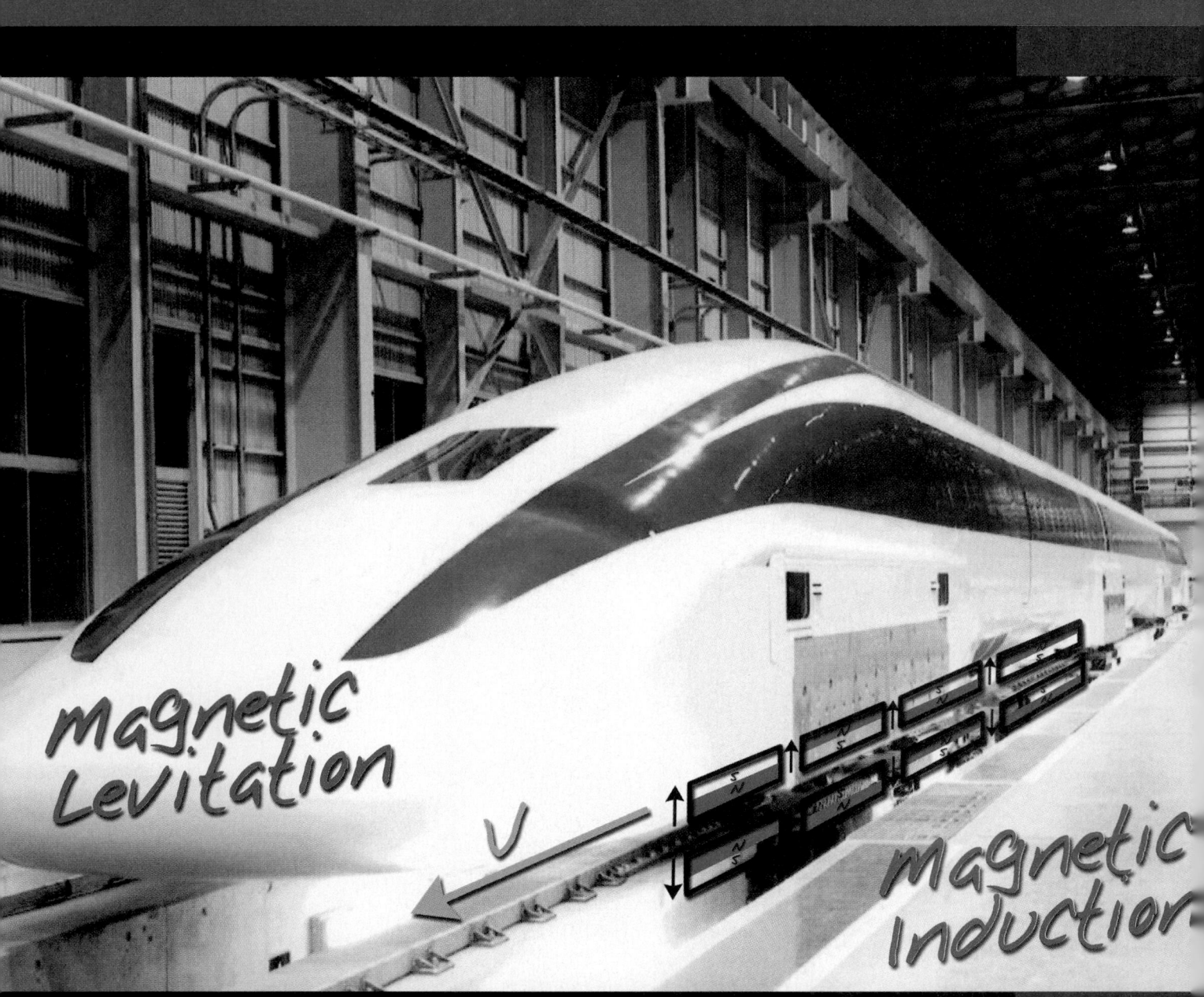

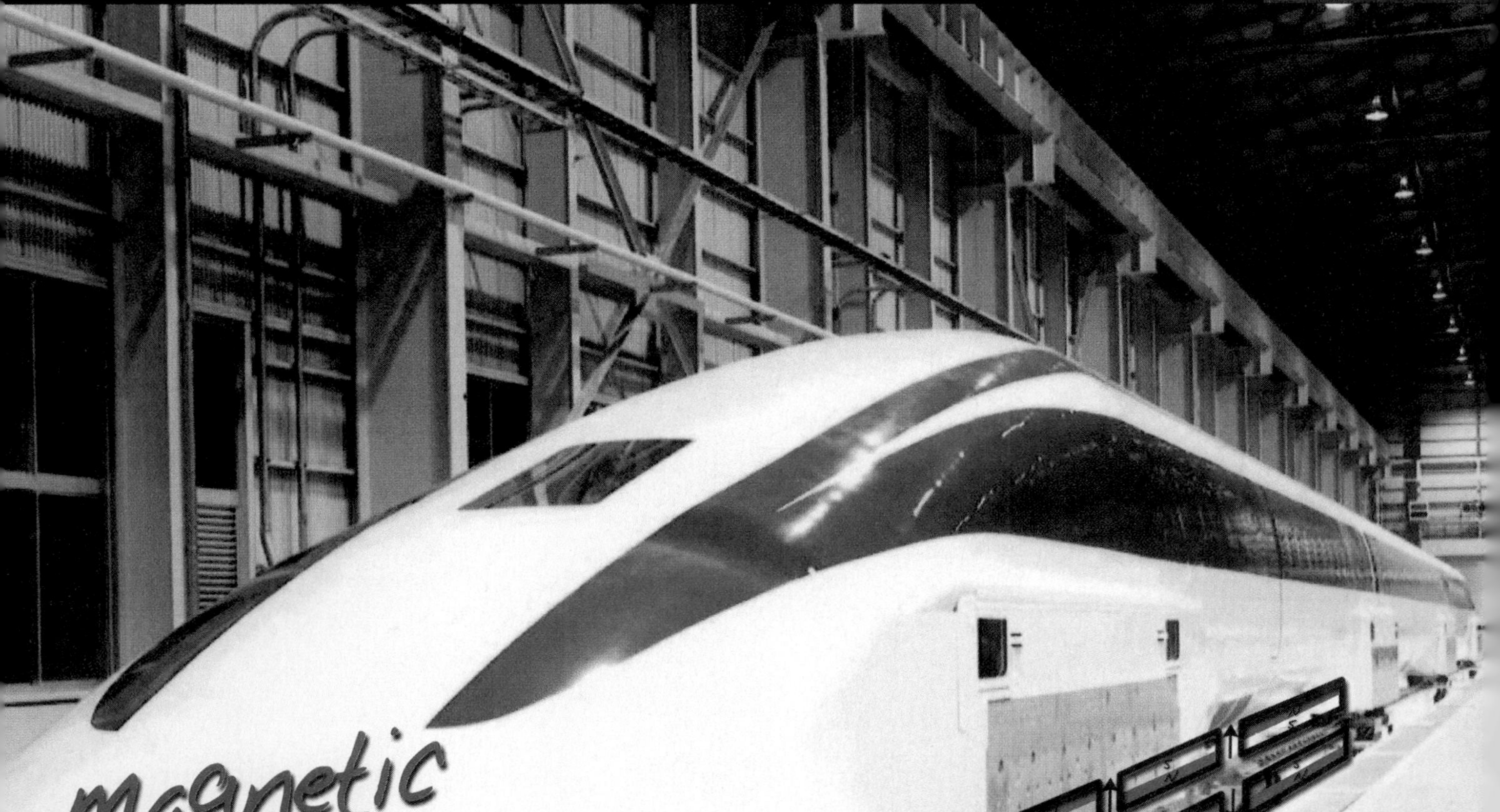
magnetic
Levitation
V
Magnetic
Induction

MAGNETISM

Magnetism and electricity are closely related. Electromagnetism is at the heart of the generation of electric power. We consider how generators and motors use these principles in the production and consumption of electricity.

Objectives

The major goals of this chapter are to enable you to:

1. Describe the nature of magnetism and the magnetic effect of electric currents.
2. Describe how induced magnetism and electric current are related.
3. Distinguish between generators and motors, and describe the principles that apply to both.

17.1 Introduction to Magnetism

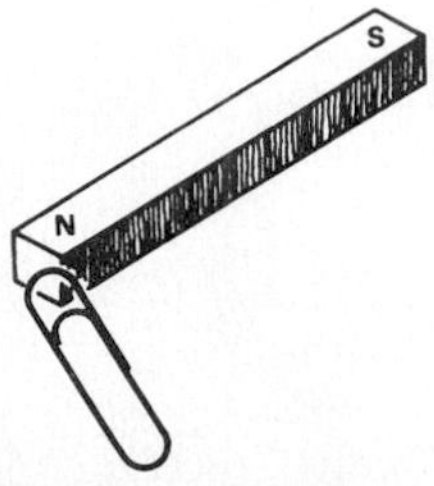

Figure 17.1 Magnetic materials attract iron and steel.

Many devices that use or produce electric energy depend on the relation of magnetism and electric currents. Motors and meters are designed to use the fact that electric currents in wires behave like magnets. Generators produce electric current due to the movement of wires near very large magnets.

In this chapter we investigate the basic properties of magnets and the relation between electric currents and magnetism. In later sections on generators, motors, and transformers, we will use the basic principles of magnetism that are developed here.

Metals with the ability to attract pieces of iron or steel are said to be **magnetic** (Fig. 17.1). Iron ore that is naturally magnetic is called *lodestone*.

Artificial magnets can be made from iron, steel, and several special alloys such as Permalloy™ and alnico. We will discuss the process of creating artificial magnets later. Materials that can be made into magnets are called *magnetic materials*. Most materials are nonmagnetic (examples are wood, aluminum, copper, and zinc).

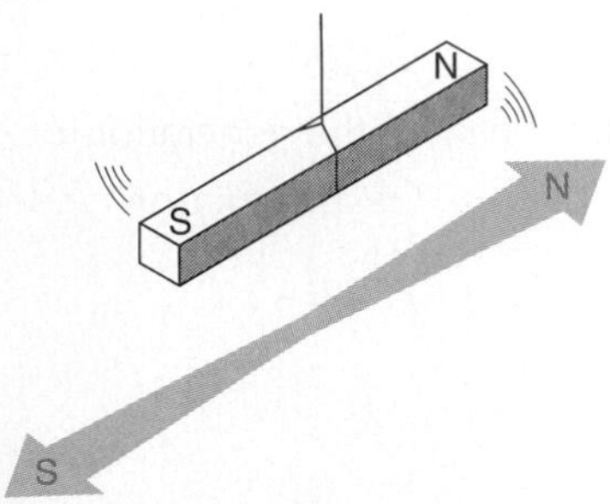

Figure 17.2 A suspended magnet will rotate to line up north and south.

Forces Between Magnets

Suppose that a bar magnet is suspended by a string so that it is free to rotate. It will rotate until one end points north and the other south (Fig. 17.2). The end that points north is called the north-seeking pole, or *north* (N) *pole*. The other end is the south-seeking pole, or *south* (S) *pole*.

If the north pole of another bar magnet is brought near the north pole of this magnet, the two like poles will repel [Fig. 17.3(a)]. The south pole of one magnet will attract the north pole of the other [Fig. 17.3 (b)]. In summary:

> Like magnetic poles repel each other and unlike magnetic poles attract each other.

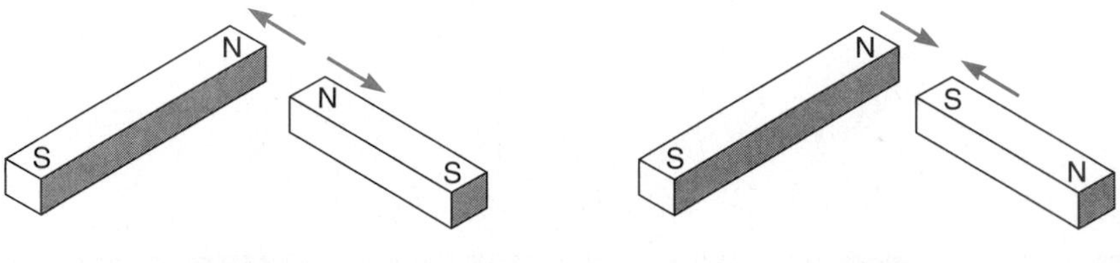

(a) (b)

Figure 17.3 (a) Like poles repel each other. (b) Unlike poles attract each other.

Figure 17.4 Simple compass

A **compass** (Fig. 17.4) is simply a small magnetic needle that is free to rotate on a bearing.

Magnetic Fields of Force

A **magnetic field** is a field of force near a magnetic pole or near an electric current that can be detected using another magnet. We can represent this field of force by drawing lines that indicate the direction of the force exerted on a north pole placed in the field. The field of a bar magnet can be mapped by moving a small compass around the magnet as shown in Fig. 17.5(a). These resulting lines are called **flux lines** (lines of force). The flux lines can also be found by sprinkling iron filings near a magnet [Fig. 17.5(b)].

The fields of combinations of magnets can also be found in this way (Fig. 17.6). Although the iron filing patterns are two-dimensional, the magnetic field around a magnet is actually a three-dimensional field, as shown in Fig. 17.7.

Figure 17.5 (a) Mapping the field of a bar magnet. (b) Iron filings and plotting compasses reveal the flux lines of magnetic force around a bar magnet.

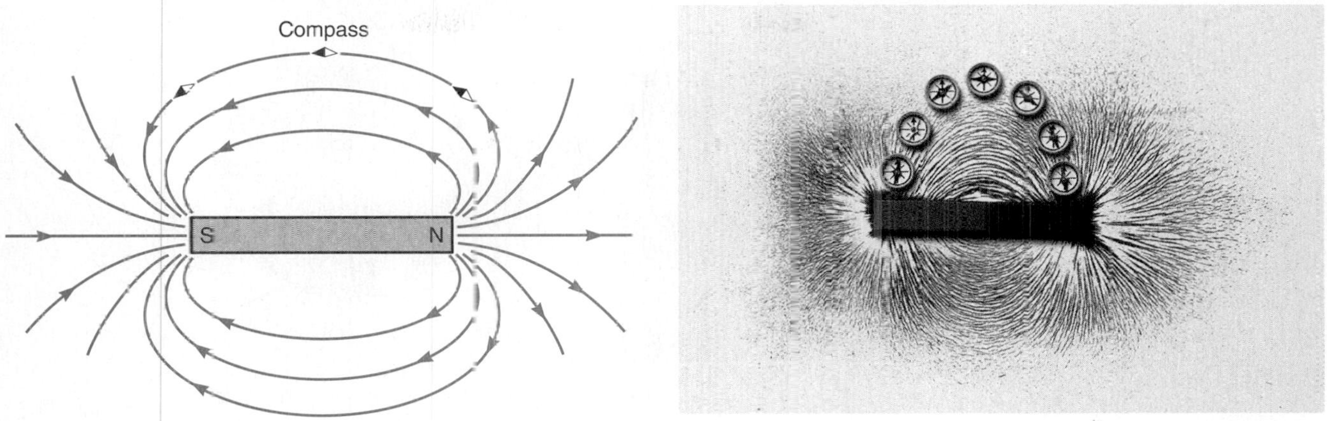

(a) (b)

Photo courtesy of Dorling Kindersley

Photo courtesy of Dorling Kindersley

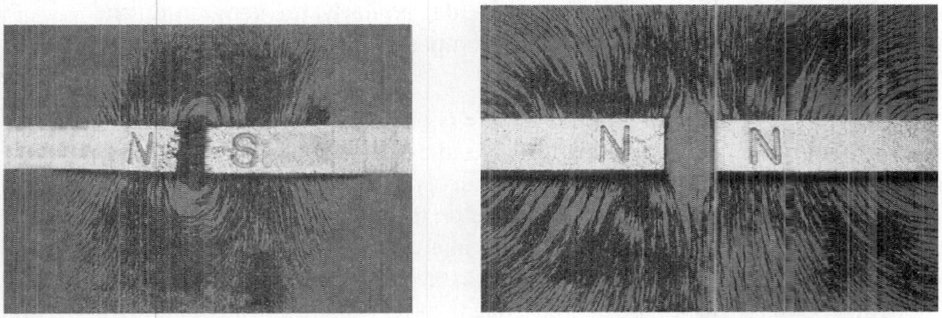

(a) Unlike poles near each other (b) Like poles near each other

Figure 17.6 Flux lines of (a) unlike poles and (b) like poles near each other shown by iron filings.

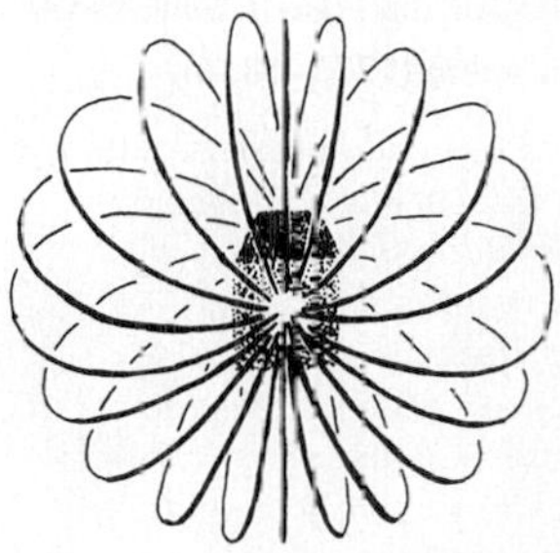

Figure 17.7 A magnetic field is three-dimensional.

Magnetic flux is the total number of magnetic lines of force entering or leaving the pole of a magnet. The symbol for magnetic flux is the Greek letter ϕ. One magnetic line of force is given the unit maxwell (Mx). A more useful unit is 10^8 Mx or one weber (Wb).

The earth's three-dimensional magnetic field is depicted in Fig. 17.8. Many puzzling aspects of the earth's magnetic field have not been resolved. The north magnetic pole and the north geographic pole (sometimes called *true north*) are at different locations. The axis of rotation and the magnetic field axis are slightly different and change approximately 10 min of arc each year. Even more puzzling, scientific evidence indicates that the earth's magnetic field reverses completely every few hundred thousand years without significantly affecting the earth's rotational or orbital motions.

Magnetic poles are similar to the behavior of electric charges in many ways, but there is a very important difference. Electric charges can be isolated, but magnetic poles cannot. A cluster of electrons doesn't need to be accompanied by a cluster of protons. However, a north magnetic pole never exists without the presence of a south magnetic pole. If a magnet is broken in half, each piece still has a north and a south pole.

Hans Christian Oersted (1777–1851),

physicist, was born in Denmark. He was the first to discover the connection between electricity and magnetism in 1820.

17.2 Magnetic Effects of Currents

When a current passes through a conductor, it forms a magnetic field. A compass placed near the current shows the direction of this magnetic field. **Hans Christian Oersted** was the first to discover the connection between electricity and magnetism. We can show this by connecting a

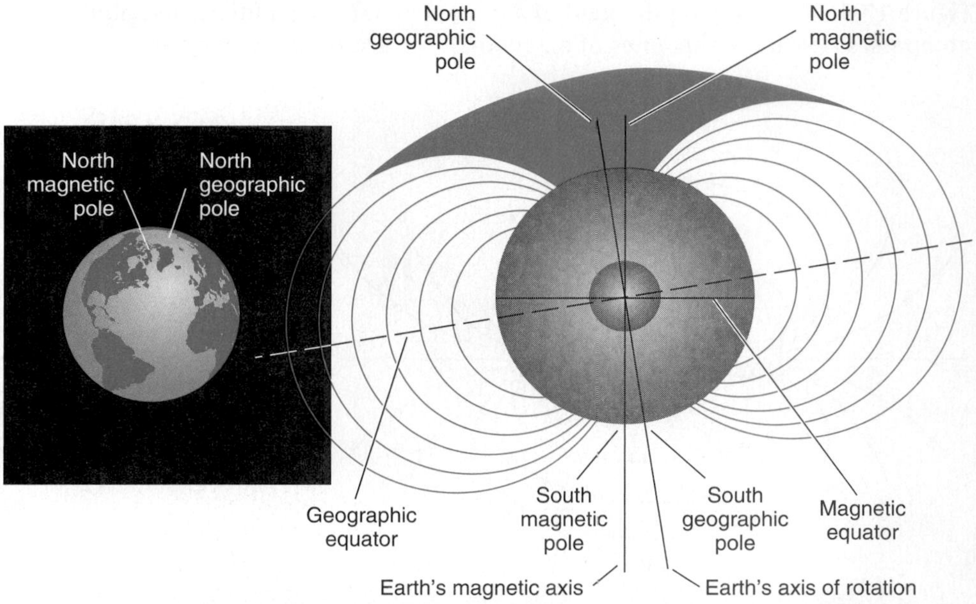

Figure 17.8 The earth's magnetic field is similar to that of other magnets. The geographic and magnetic poles do not coincide; similarly, the geographic and magnetic equators differ. The needle on a compass points to the north magnetic pole

Andre Ampère (1775–1836),

mathematician and physicist, was born in France. His work became the basis of the science of electrodynamics following Oersted's discovery in 1820 of the magnetic effects of electric currents. The SI unit of electric current, the ampere or amp, is named after him.

battery to a wire (Fig. 17.9). A compass needle is placed under the wire as in Fig. 17.9(a). When the switch is closed, the needle is deflected as in Fig. 17.9(b). If the terminals of the battery are reversed, the needle is deflected in the opposite direction [Fig. 17.9(c)]. When the compass needle is placed on top of the conductor, the direction of deflection is reversed in each case [Fig. 17.9(d)–(f)]. When the current in a wire flows in a given direction, the flux lines point in one direction below the wire and in the opposite direction above the wire.

The field actually curves around the straight current-carrying wire [Fig. 17.10(a)]. Iron filings on a sheet of paper perpendicular to a current-carrying wire show the shape of the field. The magnetic field is stronger for large currents than for small currents. The direction of the field near a current in a straight wire is shown in Fig. 17.10(b) and given by **Ampère's rule:**

AMPÈRE'S RULE

Hold the wire in your right hand with your thumb extended in the direction of the current. Your fingers circle the wire in the direction of the flux lines.

The magnetic field near a long current-carrying wire, measured in units of teslas, is circular about the wire and given by Ampère's law:

$$B = \frac{\mu_0 I}{2\pi R}$$

where

B = magnetic field
I = current through the wire
R = perpendicular distance from the center of the wire
$\mu_0 = 4\pi \times 10^{-7}$ T m/A

The magnetic field, B, has the unit *tesla* (T), named after **Nikola Tesla,** and is defined in terms of electric current by the constant μ_0, the permeability constant. The value of μ_0 is

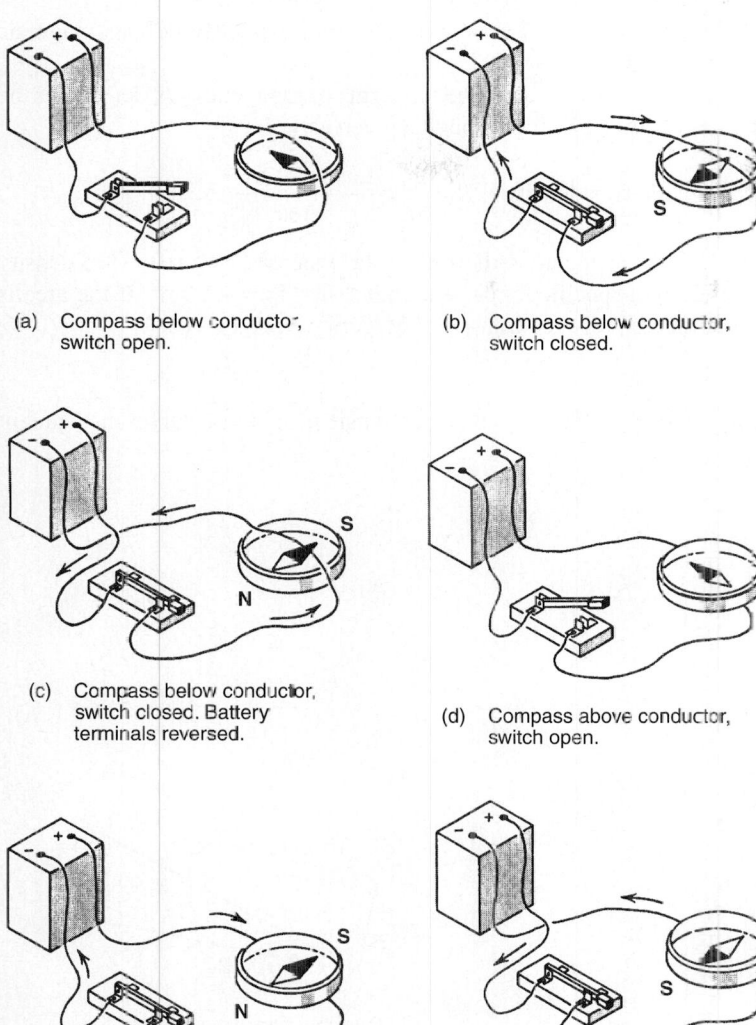

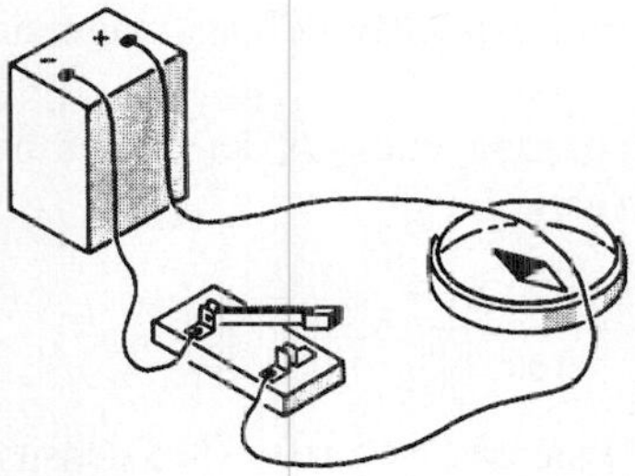

(a) Compass below conductor, switch open.

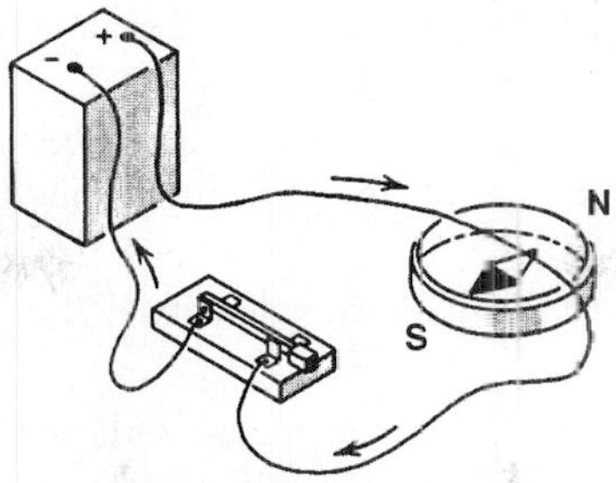

(b) Compass below conductor, switch closed.

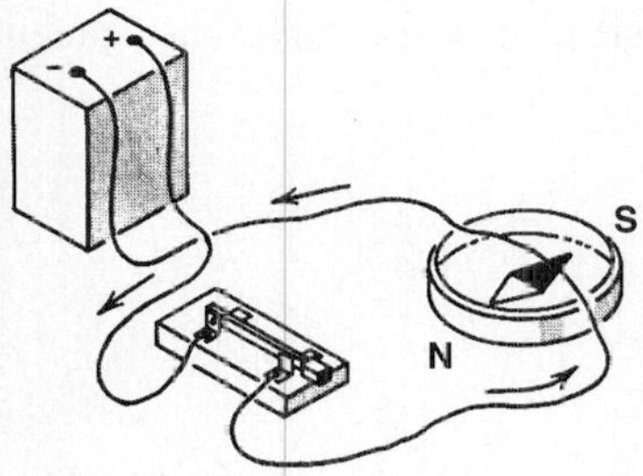

(c) Compass below conductor, switch closed. Battery terminals reversed.

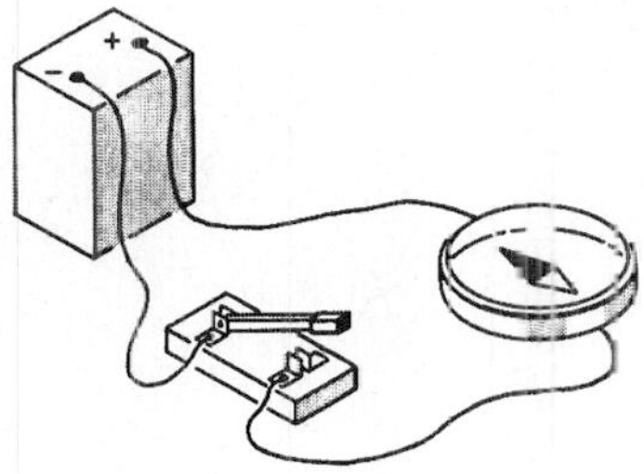

(d) Compass above conductor, switch open.

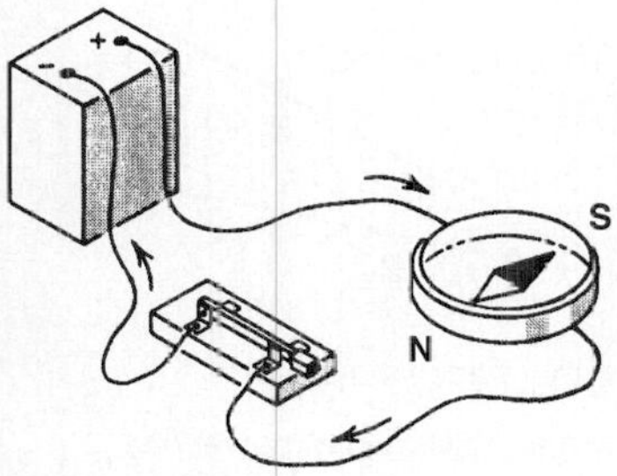

(e) Compass above conductor, switch closed.

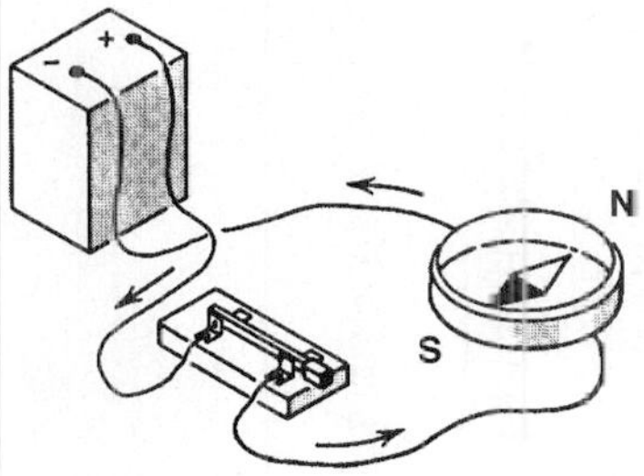

(f) Compass above conductor, switch closed. Battery terminals reversed.

Figure 17.9 Magnetic effects of electric currents

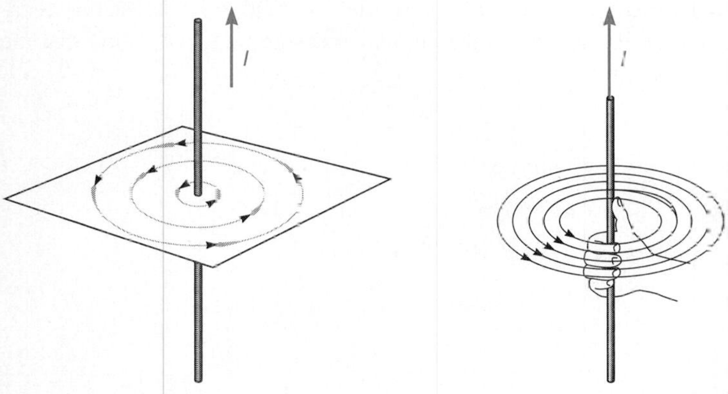

(a) (b)

Figure 17.10 (a) Field around a wire (b) Direction of the field around a current-carrying wire

Nikola Tesla (1856–1943),

electrical engineer and inventor, was born in Croatia. He invented the electromagnetic motor that became the basis for most alternating-current machinery and sold his patents to Westinghouse in 1888. He then produced important inventions involving high-frequency electricity: the Tesla coil and a resonant air-core transformer. His alternating current system illuminated the Chicago World's Fair in 1893 and led to the construction of the Niagara Falls hydroelectric generating plant in 1896. His later inventions included wireless transmission of electricity and radiocontrolled craft. He also worked on pulsed radar, harnessing solar power, and radio communications to other planets.

not experimentally determined but is an assigned value that explicitly defines magnetic field in terms of electric current.

The magnetic field, B, may also be described in terms of magnetic flux density, or the number of magnetic lines of force passing through a given area:

$$\text{magnetic flux density} = B = \frac{\phi}{A} = \frac{\text{magnetic flux}}{\text{area}}$$

Note that magnetic flux, ϕ, includes the total area, whereas magnetic flux density describes a magnetic field in a specified area. The unit tesla (T) = 1 W/m^2. If the area is measured in square centimetres, the unit Gauss = 1 Mx/cm^2 may be used.

EXAMPLE 1

A power line carrying $40\bar{0}$ A is 9.00 m above a transit used by a surveying student [Fig. 17.11(a)].

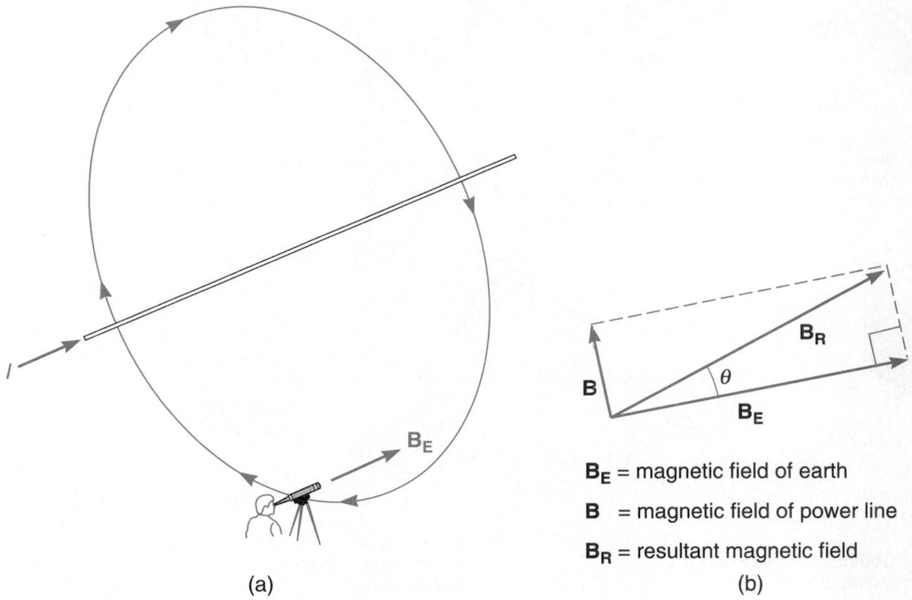

Figure 17.11

(a) Find the magnetic field formed by the power-line current above the transit.
(b) If the earth's horizontal component of magnetic field is 5.20×10^{-5} T at that location, what error could be introduced in the angular measurement? (Assume that the power line runs north–south.)

(a) Data

$$I = 40\bar{0}\ \text{A}$$
$$R = 9.00\ \text{m}$$
$$\mu_0 = 4\pi \times 10^{-7}\ \text{T m/A}$$
$$B = ?$$

Basic Equation:

$$B = \frac{\mu_0 I}{2\pi R}$$

Working Equation: Same

Substitution:

$$B = \frac{(4\pi \times 10^{-7}\ \text{Tm/A})(400\ \text{A})}{2\pi(9.00\ \text{m})}$$
$$= 8.89 \times 10^{-6}\ \text{T}$$

Therefore, the magnetic field from the power line is 8.89×10^{-6} T. With the current from south to north, Ampère's rule shows the direction of B to be east to west.

(b) The angle that the resultant vector [the earth's field plus the wire's field $(B_E + B)$] makes with B_E would be the angular error θ.

Data:

$$B_E = 5.20 \times 10^{-5}\ \text{T} \quad \text{earth's component}$$
$$B = 8.89 \times 10^{-6}\ \text{T} \quad [\text{from part (a)}]$$
$$\theta = ?$$

Basic Equation:

$$\tan\theta = \frac{B}{B_E}$$

Working Equation: Same

Substitution:

$$\tan\theta = \frac{8.89 \times 10^{-6}\ \text{T}}{5.20 \times 10^{-5}\ \text{T}} = 0.171$$
$$\theta = 9.7°$$

The bearing on the surveying student's transit could be in error by 9.7° because of the power line.

Magnetic Field of a Loop

If a wire is bent into a loop, the magnetic field lines become bunched up or concentrated in the center of the loop. A second and subsequent loops increase the intensity of the magnetic field in the loop even more.

To determine the direction of the flux lines of a current in a loop, use Ampère's rule as shown in Fig. 17.12. If several loops are made into a tight spiral as shown in Fig. 17.13, the

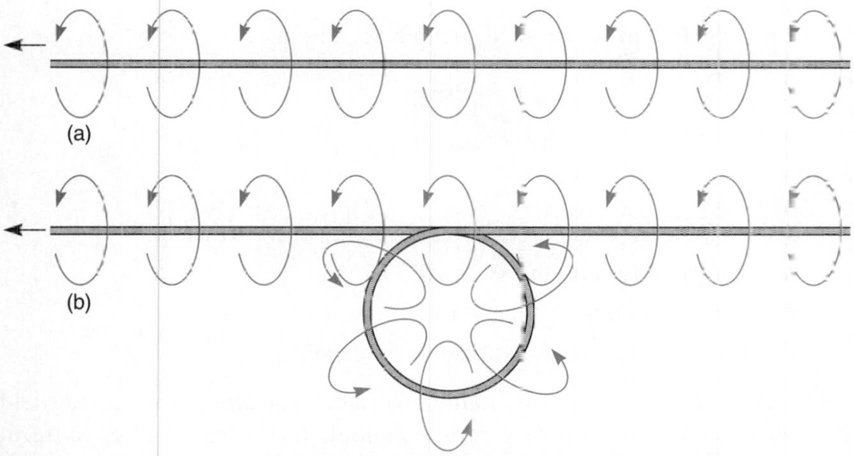

Figure 17.12 **Magnetic field around (a) a straight wire and (b) a loop in a current-carrying wire**

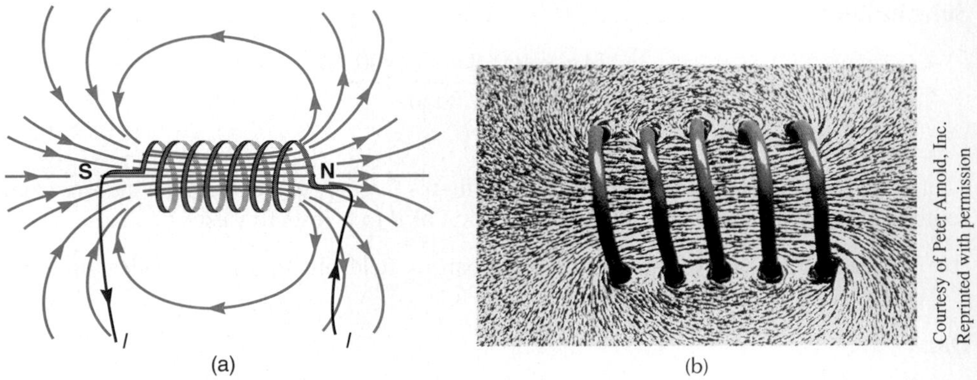

Figure 17.13 (a) Magnetic field of a coil (b) Iron filings show the magnetic lines of force from the magnetic field of a coil of red wire charged with an electric current. The coil of wire passes repeatedly through a white surface.

flux lines add to form the field shown. A coil of tightly wrapped wire is called a **solenoid.** The left side of this solenoid acts like a south magnetic pole. The right side acts like a north magnetic pole. This polarity could be found by using a compass. The rule for finding the polarity of a solenoid is the following:

Hold the solenoid in your right hand so that your fingers circle it in the same direction as the current. Your thumb points to the north pole of the solenoid (Fig. 17.14).

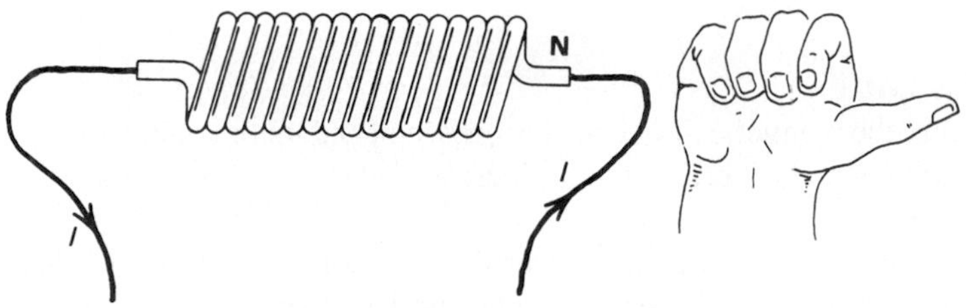

Figure 17.14 Polarity of a solenoid

For a long coil that is tightly turned, the field strength at its center is

$$B = \mu_0 I n$$

where

B = magnetic field in the region at the center of the solenoid
μ_0 = permeability constant, $4\pi \times 10^{-7}$ T m/A
I = current through the solenoid
n = number of turns per unit length of solenoid

The longer the solenoid in relation to its radius, the more uniform the magnetic field is inside the solenoid; for an infinitely long solenoid, the value of B is uniform throughout.

EXAMPLE 2

Find the magnetic field at the center of a solenoid that is 0.425 m long, 0.0750 m in diameter, and has three layers of $85\bar{0}$ turns each, when 0.250 A flows throughout.

Data:

$$I = 0.250 \text{ A}$$

$$n = \frac{3 \times 85\bar{0} \text{ turns}}{0.425 \text{ m}} = 600\bar{0} \text{ turns/m}$$

$$\mu_0 = 4\pi \times 10^{-7} \text{ T m/A}$$

$$B = ?$$

Note: The length, 42.5 cm, can be considered "long" compared with the radius, 3.75 cm (about 11 times longer).

Basic Equation:

$$B = \mu_0 I n$$

Working Equation: Same

Substitution:

$$B = (4\pi \times 10^{-7} \text{ T } \cancel{\text{m}}/\cancel{\text{A}})(0.250 \cancel{\text{A}})(600\bar{0}/\cancel{\text{m}})$$
$$= 1.88 \times 10^{-3} \text{ T}$$

PROBLEMS 17.2

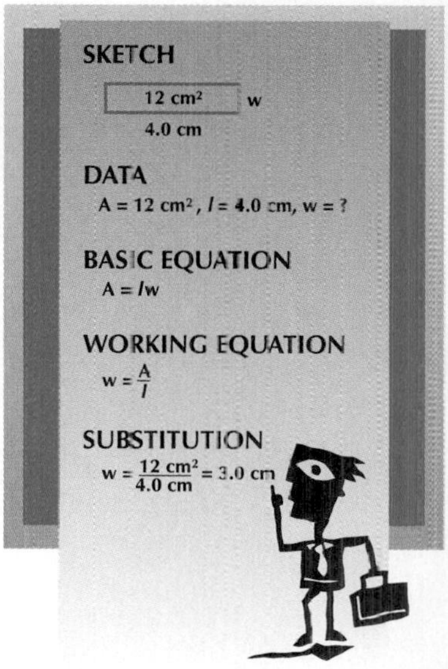

1. Find the magnetic field at 0.250 m from a long wire carrying a current of 15.0 A.
2. Find the magnetic field at 0.500 m from a long wire carrying a current of 7.50 A.
3. What is the current in a wire if the magnetic field is 5.75×10^{-6} T at a distance of 2.00 m from the wire?
4. A power line runs north–south carrying 675 A and is 5.00 m above a transit used by a surveyor.
   (a) What is the magnetic field at the transit because of the power-line current?
   (b) If the earth's horizontal component of magnetic field is 5.20×10^{-5} T, what error is introduced in the surveyor's angular measurement?
5. Find the magnetic field at 0.350 m from a long wire carrying a current of 3.00 A.
6. Find the current in a wire if the magnetic field is 3.50×10^{-6} T at a distance of 2.50 m from the wire.
7. A solenoid has $100\bar{0}$ turns of wire, is 0.320 m long, and carries a current of 5.00 A. What is the magnetic field at the center of the solenoid? Assume that its length is long in comparison with its diameter.
8. A solenoid has $300\bar{0}$ turns of wire and is 0.350 m long. What current is required to produce a magnetic field of 0.100 T at the center of the solenoid? Assume that its length is long in comparison with its diameter.
9. A small solenoid is 0.150 m in length, 0.0150 m in diameter, and has $60\bar{0}$ turns of wire. What current is required to produce a magnetic field of 1.25×10^{-3} T at the center of the solenoid?
10. A solenoid has $250\bar{0}$ turns of wire and is 0.200 m long. What current is required to produce a magnetic field of 0.100 T at the center of the solenoid? Assume that its length is long in comparison with its diameter.
11. A long solenoid has $100\bar{0}$ turns and is 0.250 m long. If the wire carries a current of 2.50 A, what is the strength of the magnetic field at its center?
12. A small solenoid 0.100 m in length has $100\bar{0}$ turns of wire. What current is required to produce a magnetic field of 1.30×10^{-3} T at the center of the solenoid?

17.3 Induced Magnetism and Electromagnets

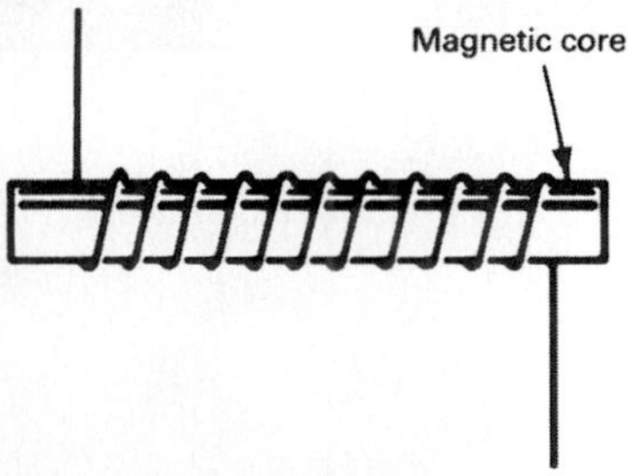

Figure 17.15 Simple electromagnet

Induced magnetism is produced in a magnetic material such as iron when the magnetic material is placed in a magnetic field, such as that produced in the core of a current-carrying solenoid (Fig. 17.15). An **electromagnet** consists of a solenoid and a magnetic core. When a current is passed through the solenoid, the magnetic fields of the atoms in the magnetic material line up to produce a strong magnetic field. When the current through the coil is turned off, the strength of the induced magnet decreases, but some remains. When the core is removed, a magnetic field remains in the core. In materials such as soft iron, very little magnetic field remains in the core after the current flow stops. In other materials, such as steel, alnico, and Permalloy™, a much stronger field remains. The latter materials are used for permanent magnets. However, they are undesirable for use as a core in an induction motor. Soft-iron cores are often used for this application because less energy is required to reverse the polarity of the induced magnetic field.

A magnet can be thought to consist of many atoms, each behaving like a small magnet. In each atom, the electrons orbit about the nucleus and each electron spins about its own axis, producing small current loops that generate magnetic fields. In most materials, these current loops are arranged so that their magnetic fields point in different directions. The result is that the magnetism of one loop (atom) is canceled out by those of its neighbors.

In magnetic materials, the atomic magnetic fields line up with each other in regions called *magnetic domains* when no field is present [Fig. 17.16(a)]. Each domain has a

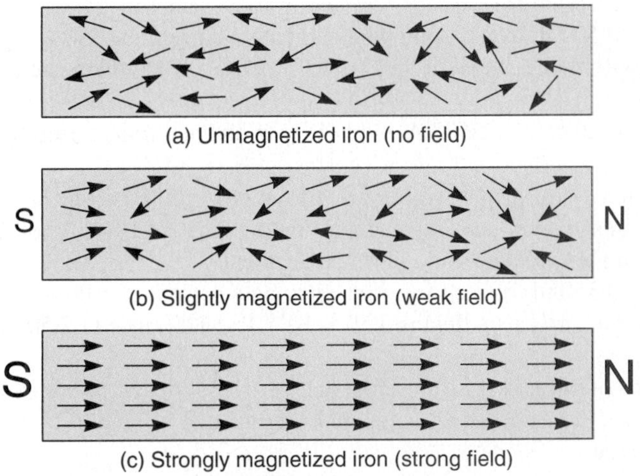

Figure 17.16 Magnetic fields

TRY THIS ACTIVITY

Lifting Paper Clips

Using a large nail, an insulated wire, and a 3.0- to 9.0-V battery, make an electromagnet that will pick up as many paper clips as possible. Without adding more batteries, what can be done to the electromagnet to allow it to pick up more paper clips?

magnetic field direction to which most atoms in the domain are aligned. When no magnetic field is present, the orientations of the domains are random. However, when an external field is present, the domains tend to line up with the field, causing the domain boundaries to shift [Fig. 17.16(b)]. In high electric fields, nearly all the material is aligned [Fig. 17.16(c)]. When the external magnetic field is removed, some materials such as the alloy alnico retain the aligned domains, creating a permanent magnet.

17.4 Induced Current

When a magnet is moved so that its flux lines cut across a wire, an emf is induced in the wire, which is known as **induced current.** The strength of this induced emf depends on the strength of the magnetic field and on the rate at which the flux lines are cut by moving the magnet or wire. Increasing the strength of the field or increasing the rate at which the flux lines are cut also causes the current to increase.

While the magnet shown in Fig. 17.17(a) is moved downward, the galvanometer indicates that a current flows through the wire. If the magnet shown in Fig. 17.17(b) is moved upward, the induced current is in the opposite direction.

A current also flows in the wire if the circuit is closed, the wire is moved, and the magnet is stationary [Fig. 17.17(c)]. The current is produced by the relative motion of the magnet and wire. In commercial generators, magnets are spun inside a set of coils of wire. The induced emf is increased by replacing the single wire with a coil of many turns. For example, tripling the number of turns triples the induced emf [Fig. 17.17(d)].

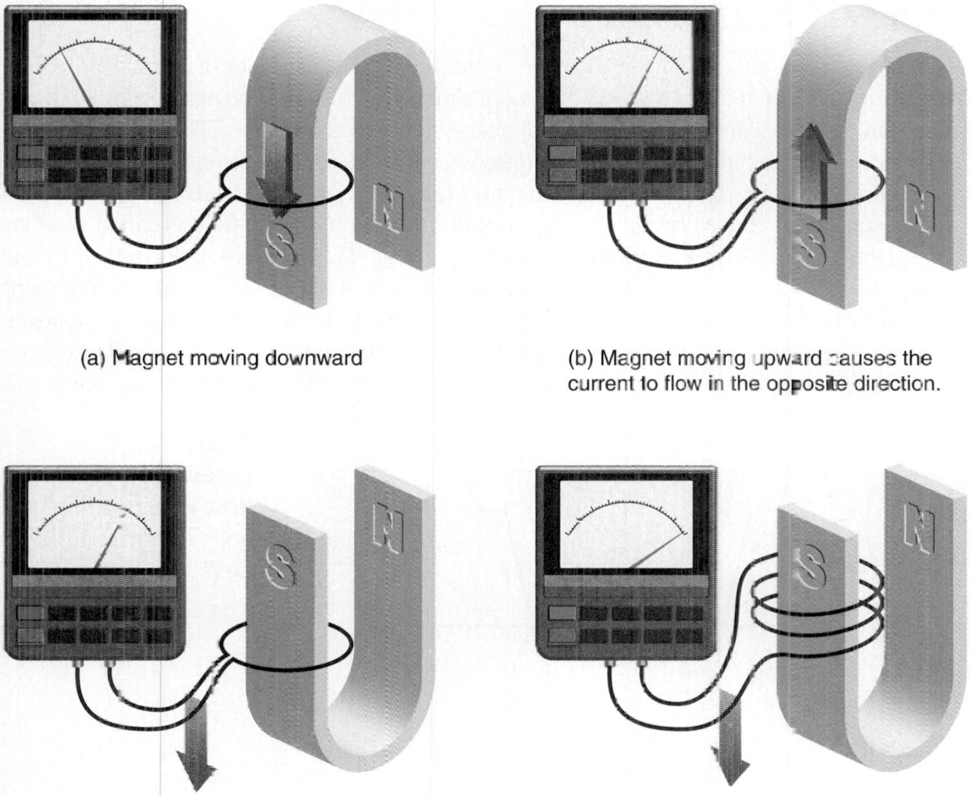

(a) Magnet moving downward

(b) Magnet moving upward causes the current to flow in the opposite direction.

(c) Wire moving downward

(d) Coil moving downward increases the current flow in same direction.

Figure 17.17 Induced current in a wire

PHYSICS CONNECTIONS

Speakers and Electromagnets

A conventional speaker consists of three major components: (a) an amplifier or some other signal source, (b) a cone that vibrates the air to produce a sound wave, and (c) a permanent magnet surrounding a coil of electric wire attached to the back of the cone.

For the speaker to create a sound wave, the amplifier must send an alternating current signal through the wire. As the alternating current travels through the wire, it induces an alternating magnetic field around the coil. The wire's induced magnetic field is repeatedly attracted to and repelled from the magnetic field of the permanent magnet. This repeated motion causes the cone to vibrate at the frequency determined by the amplifier or signal generator (Fig. 17.18).

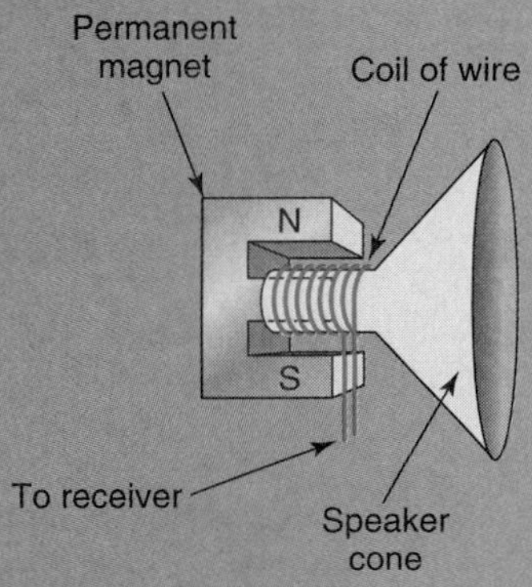

Figure 17.18 The cross section of a conventional speaker

17.5 Generators

The induction of an emf in a coil of wire can be used to produce electrical power. A **generator** consists of a coil of wire rotating in a magnetic field, which produces an induced current in the coil, converting mechanical energy into electric energy. This is the simplest kind of generator to build in the laboratory, so we will study its operation here and compare it to commercial generators, where the magnets (electromagnets) are rotated.

The current produced by rotating the wire through the magnetic field is called an alternating current (ac). As side A of the current loop in Fig. 17.19(a) passes downward by the north pole, the induced current is in one direction. As side A (same side of the rotating loop) passes upward by the south pole as in Fig. 17.19(b), the induced current is in the opposite direction. The result is an alternating current induced in the rotating wire. As side B (the other side of the rotating loop) passes upward and then downward, the current in it also alternates.

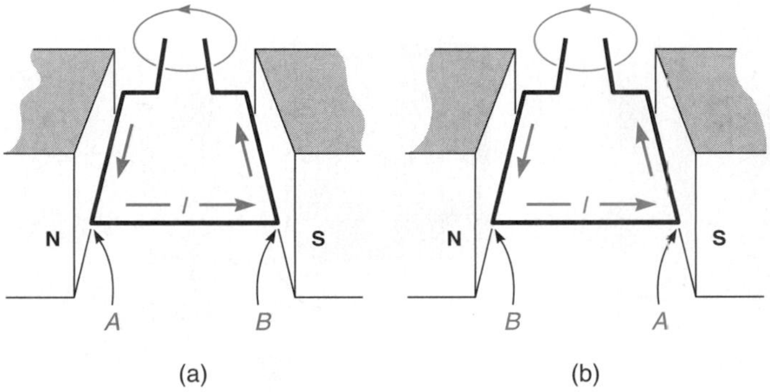

Figure 17.19 Direction of current flow in a rotating wire in a magnetic field

The direction of current flow in a wire as it rotates between the north and south magnet poles is illustrated in Fig. 17.20. In beginning position (1), the current is zero. As the wire passes down through position (2), the current builds and reaches a maximum value in position (3). The current then becomes smaller through position (4) until it is zero again in position (5).

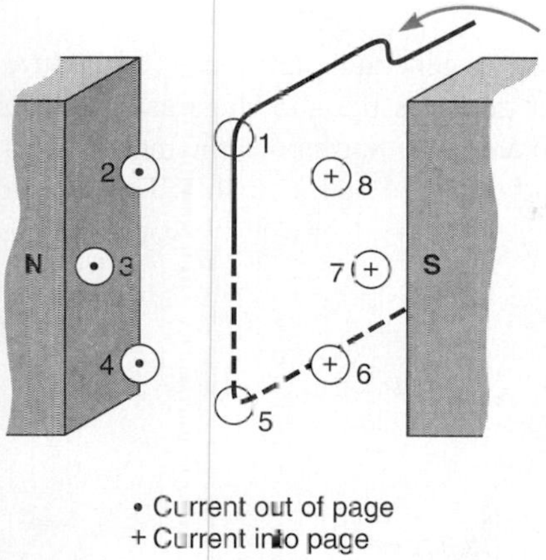

Figure 17.20 Induced current in a wire rotating in a magnetic field

As the wire begins to pass back up through the magnetic field in position (6), the current begins to build and reaches a maximum value in the opposite direction in position (7). It then becomes smaller in position (8) and falls to zero again as it reaches position (1) again. The cycle is then repeated for the next rotation of the wire through the magnetic field.

A graph of the induced current showing the relative magnitude of the current for the changing positions of the wire is shown in Fig. 17.21.

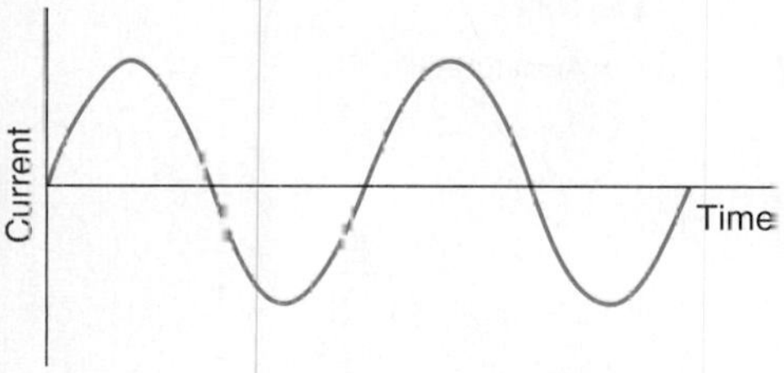

Figure 17.21 Graph of induced current as the wire changes positions in a magnetic field

One cycle is produced by one rotation of the wire. The time required for one cycle depends on the rotational speed of the coil. If the coil rotates 60 times each second, an alternating current of frequency 60 hertz (cycles per second) is produced. The current produced in the coil is conducted by brushes on slip rings to the external circuit as shown in Fig. 17.22. The rotating coil is called the **rotor** or **armature,** and the *field magnets* are called the **stator.**

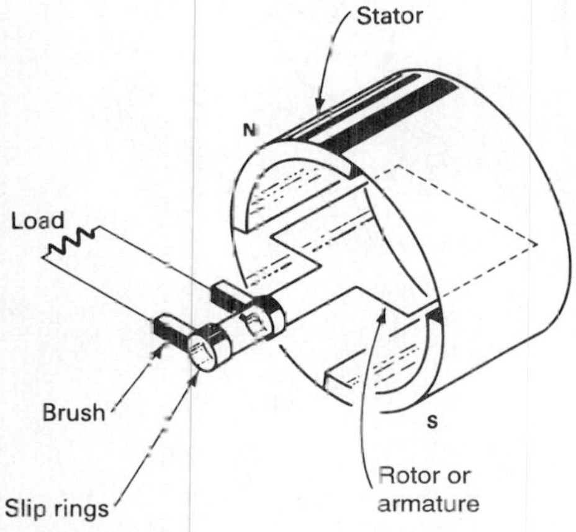

Figure 17.22 Current in a generator

The generator does not actually create electric energy; it changes the mechanical energy of rotation into electric energy. The energy to turn the rotor may be supplied by water falling down a waterfall, a diesel engine, or a steam turbine.

Power companies use large commercial ac generators to produce the current they need to supply to their customers. These generators work similarly to the generator discussed here, but they have many coils and use electromagnets instead of permanent magnets. The large generators used by electric power companies can produce voltages as large as 13,000 V and currents up to 10 A. The alternator used in automobiles is an ac generator that produces voltages about 13 V and currents up to 40 A.

dc Generators

By the use of a special device called a commutator, the ac generator can be used to produce direct current. The **commutator** is a split ring that replaces the slip rings as shown in Fig. 17.23(a). When side *A* of the coil passes upward along the north pole, the induced current flows in the direction shown in Fig. 17.23(b) and is picked up by brush 1. The current in the external circuit is also shown.

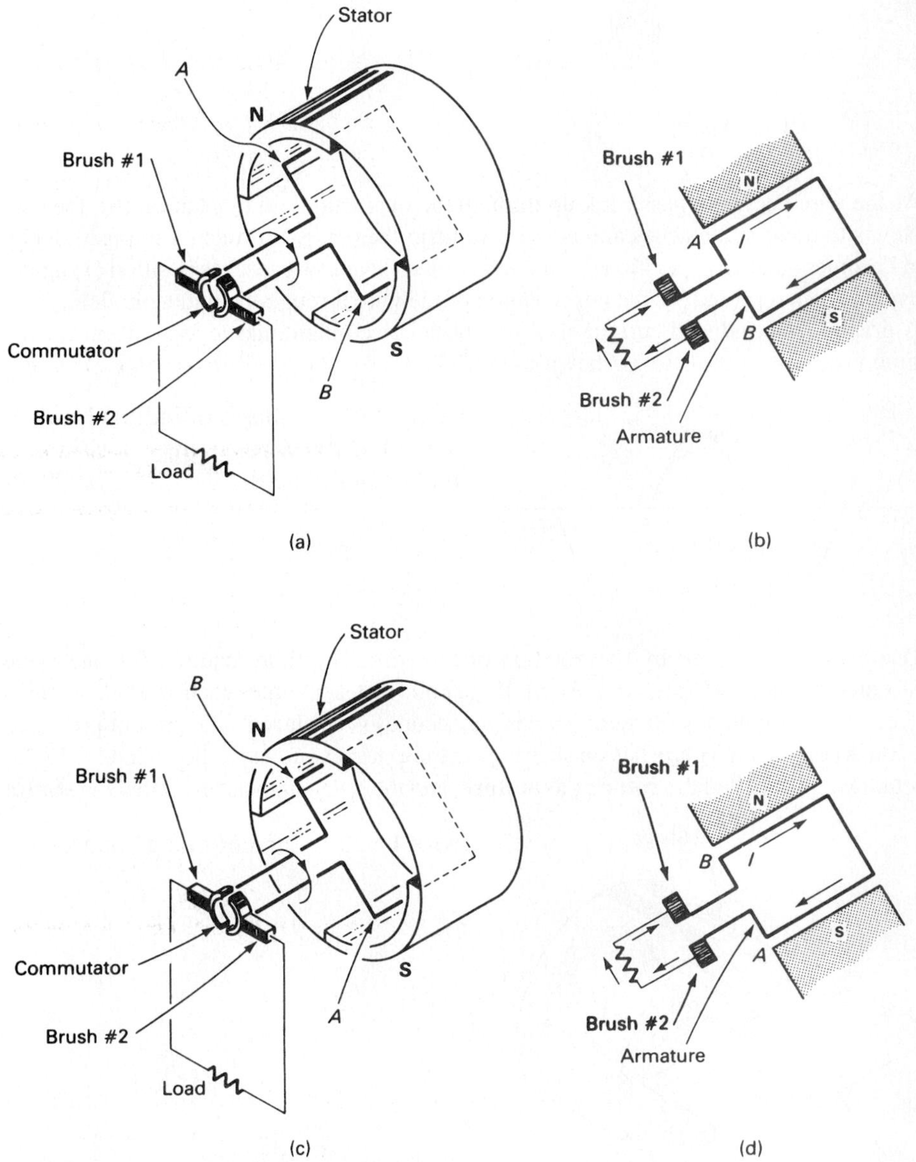

Figure 17.23 Current in a dc generator

When side *B* of the coil passes upward along the north pole, the induced current flows in the direction shown and is picked up by brush 1 [Fig. 17.23(c)]. The current in the external circuit is in the same direction as it was when *A* passed along the north pole [Fig. 17.23(d)]. Thus, this is a direct current.

The current produced by this dc generator does not have the same value at all times. A graph of the induced current is shown in Fig. 17.24(a). Commercial dc generators that are used for industrial purposes contain many coils. The output current has almost the same value at all times due to the large number of coils [Fig. 17.24(b)].

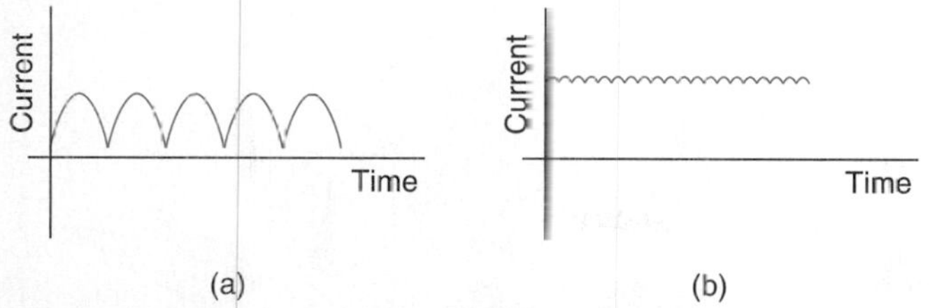

Figure 17.24 Induced current in a dc generator

17.6 The Motor Principle

We have seen that like poles of magnets repel each other. A magnet that is pivoted will spin due to the repulsion of another magnet nearby as shown in Fig. 17.25(a). We can construct an electromagnet by wrapping wire around an iron core and passing a current through the wire [Fig. 17.25(b)]. The north pole of the electromagnet will be repelled by a north pole of another magnet. The electromagnet will turn until its south pole is next to the north pole of the permanent magnet [Fig. 17.25(c)].

If we could suddenly change the polarity of the electromagnet (often called the *armature*), the magnet would repel the north pole and the electromagnet would continue to spin [Fig. 17.25(d)]. If a dc current supply is used [Fig. 17.25(e)], this change can be made by using a commutator (split ring) to change the direction of the current in the electromagnet. Changing the direction of the current flowing through the coil of the electromagnet changes the poles.

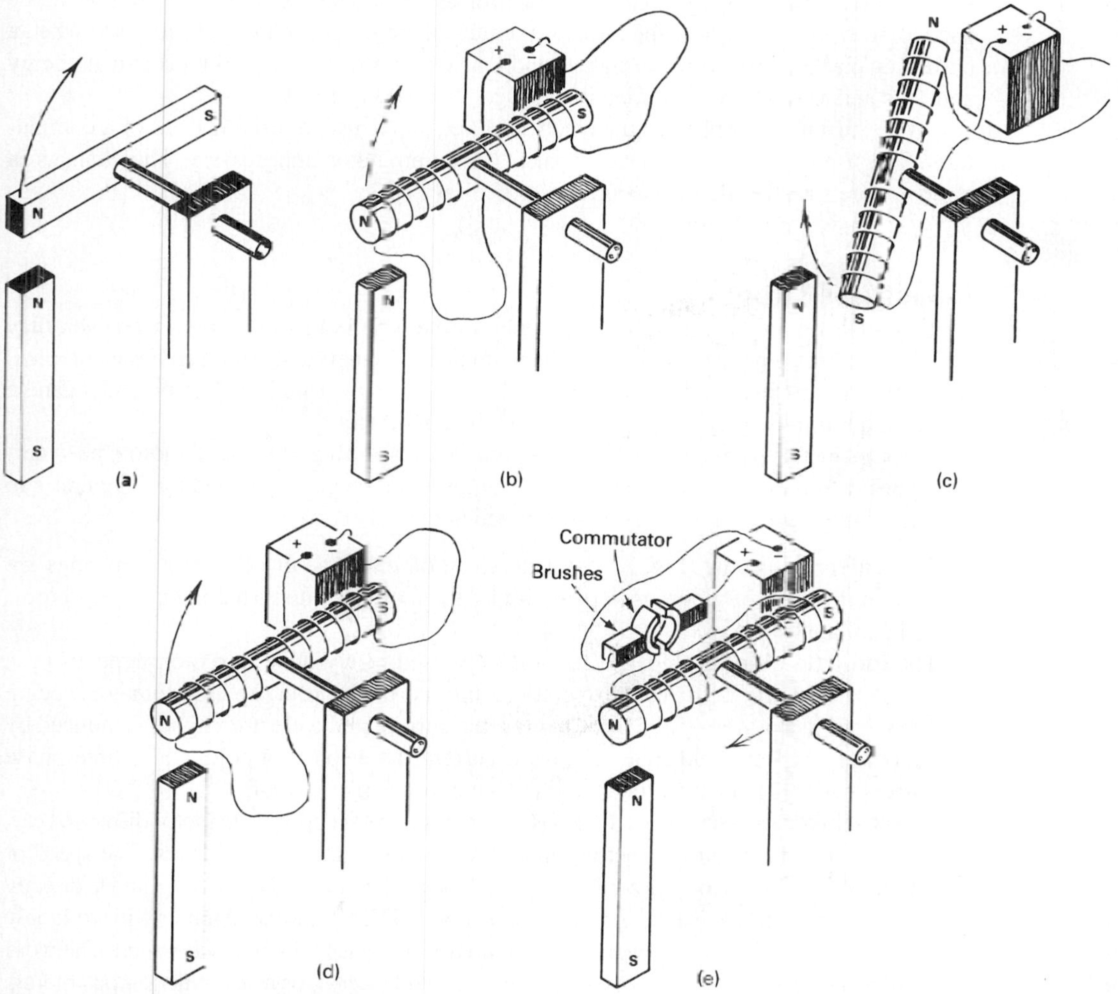

Figure 17.25 Motor principle

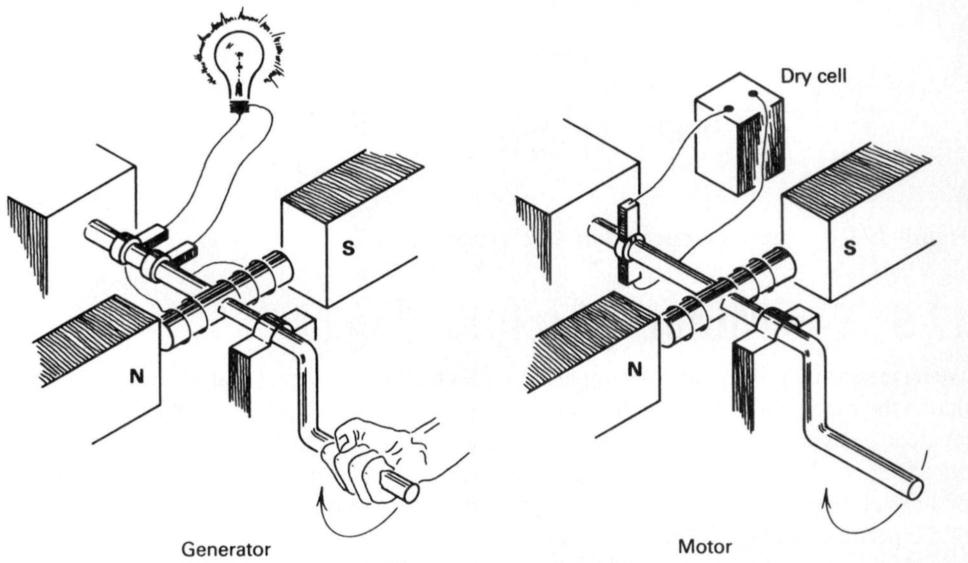

Figure 17.26 A motor performs the reverse function of a generator.

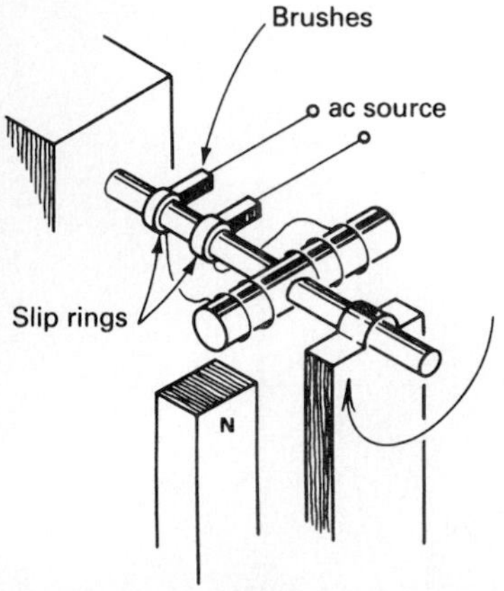

Figure 17.27 Use of ac makes a commutator unnecessary.

As the current changes direction, the electromagnet spins due to the repulsion of like poles. A shaft may be connected to the electromagnet so that the rotational motion can be used to do work. In a **motor** the armature rotates in the magnetic field of the stator when a current is passed through the armature and converts electric energy to mechanical energy and thus performs the reverse function of a generator (Fig. 17.26).

If ac current is supplied to the electromagnet, slip rings are used instead of a commutator. The use of alternating current makes the commutator unnecessary. The changes in direction are supplied by the ac current itself (Fig. 17.27).

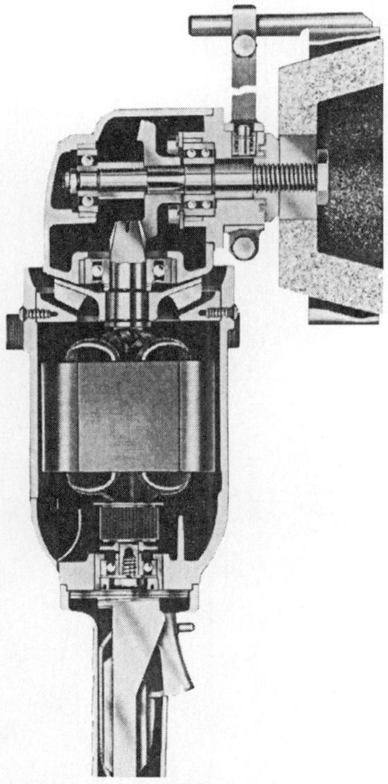

Figure 17.28 Universal motor used in a grinder (Thor Power Tool Company)

Industrial Motors

Commercial motors operate in the same way as the motors just discussed. However, they usually use electromagnets in place of the permanent magnets and are much more complex. Slip rings are not necessary in ac motors. The current in the rotating electromagnet can be induced in the same way a current is induced in a generator.

Motors can be designed for many different purposes. Heavily loaded motors need certain types of starters. The torque and power outputs can be greatly varied by differences in design. Three types of ac motors are discussed here.

1. The **universal motor** (Fig. 17.28) can be run on either ac or dc power. Slip rings are used in this type of motor as in dc motors. This motor is often used in small hand tools and appliances.
2. The **induction motor** (Fig. 17.29), a widely used heavy-duty ac motor, depends on a rotating magnetic field for its operation. Instead of the rotating electromagnet being connected to a power source by slip rings, the current in the electromagnet is induced by a moving magnetic field created by an ac current. Three pairs of poles and a three-phase current cause the stator to rotate and induce currents in the rotor.
3. The **synchronous motor** (Fig. 17.30) is very similar to the slip-ring ac motor discussed earlier. The rotating electromagnet is supplied with current through slip rings. The speed of rotation of a synchronous motor is constant, depends on the number of coils, and is directly proportional to the frequency of its ac power supply. The synchronous motor will work only when operated with an ac power source of the frequency for which it is designed. The word *synchronous* is derived from *syn,* meaning same, and *chrona,* or time. Synchronous motors are used to operate clocks and other devices needing accurate speed control.

Figure 17.29 Induction motor (Courtesy of Bodine Electric Company, Chicago, IL.)

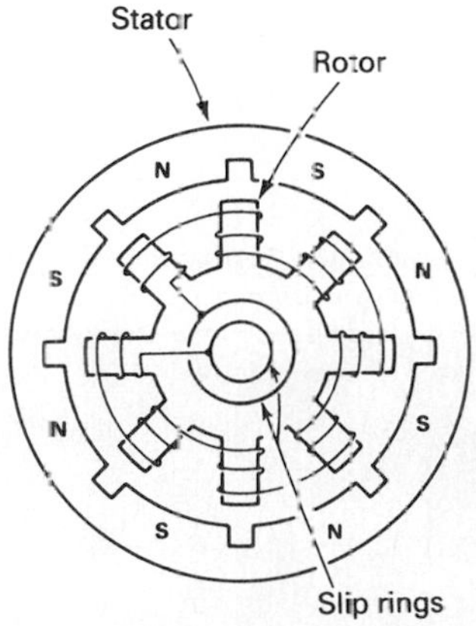

Figure 17.30 Synchronous motor

Figure 17.31 Physicists near the Large Electron–Positron Collider (LEP) accelerator used for particle physics experiments at CERN laboratories in Geneva, Switzerland

PHYSICS CONNECTIONS

Superconducting Electromagnets and Magnetic Resonance Imaging

Superconducting electromagnets are used to accelerate subatomic particles in particle accelerators for nuclear and particle research (Fig. 17.31). The magnetically levitated vehicle, or maglev, now being developed uses electromagnets to elevate the vehicle above a track and virtually eliminates any friction between the vehicle and the track (Fig. 17.32). Air resistance remains a key vehicle design factor.

Magnetic resonance imaging (MRI) is used in medical facilities to provide incredibly clear images of tissues and organs inside the human body without invasive procedures or X-ray exposure. Superconducting electromagnets align hydrogen atoms in the body. After radio waves hit the protons of the hydrogen atoms, the protons return to their previous pattern while emitting electromagnetic signals. Sensors pick up those signals and computers then record and analyze the resulting effects on the different tissues and provide a computer image "slice" of the tissue or organ (Fig. 17.33).

Figure 17.32 Maglev train

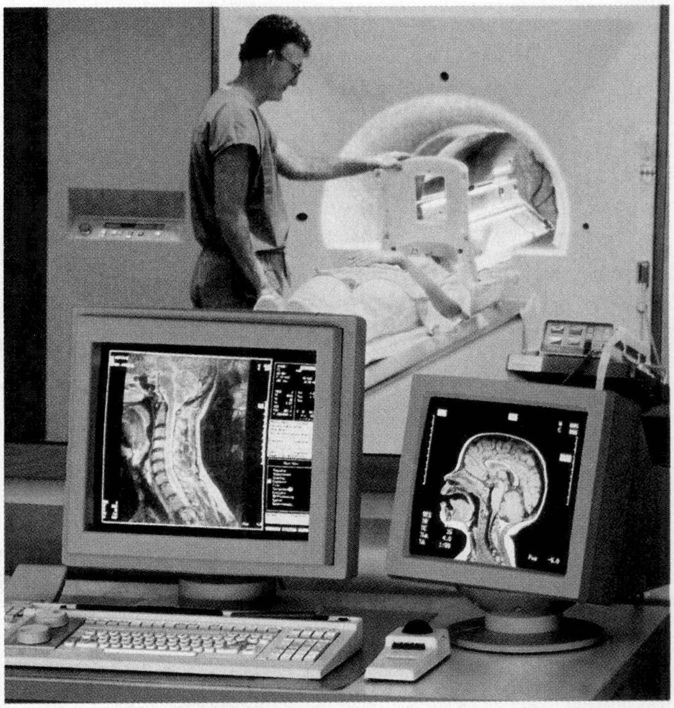

Figure 17.33 Computer screens display a patient's spinal column and brain images during an MRI exam.

17.7 Magnetic Forces on Moving Charged Particles

A charged particle that is not moving will produce no interaction with a magnetic field. However, if the charged particle moves through a magnetic field, it is deflected by the magnetic field. The deflection force is greatest when the direction is perpendicular to the magnetic field lines.

TRY THIS ACTIVITY

Moving Electrons

Take a bar magnet and move it close to the front of an old computer monitor, black and white television, or oscilloscope. (Do not risk using a new television set.) What happens to the image on the screen? Why does the magnet influence the electrons' placement on the screen and appear to warp the image?

The force that acts on a moving charged particle does not act in a direction between the sources of interaction but is perpendicular to both the magnetic field and the direction of the charged particle (Fig. 17.34). This behavior is used to spread electrons on a TV tube to produce a TV picture.

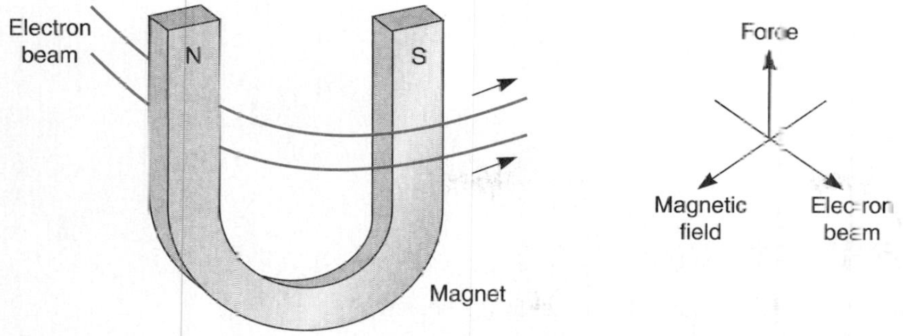

Figure 17.34 **Deflection of an electron beam by a magnetic field**

Glossary

Ampère's Rule To find the direction of a magnetic field near a current and a straight wire, hold the wire in your right hand with your thumb extended in the direction of the current. Your fingers circle the wire in the direction of the flux lines. (p. 486)

Armature The rotating coil or electromagnet in a generator. (p. 495)

Commutator A device in an ac generator that produces a direct current. Composed of a split ring that replaces the slip rings in an ac generator and produces a direct current in the circuit connected to the split ring of the generator. (p. 496)

Compass A small magnetic needle that is free to rotate on a bearing. (p. 484)

Electromagnet A combination of a solenoid and a magnetic material, such as iron, in the core of the solenoid. When a current is passed through the solenoid, the magnetic fields of the atoms in the magnetic material line up to produce a strong magnetic field. (p. 492)

Flux Lines Lines indicating the direction of the magnetic field near a magnetic pole. (p. 484)

Generator An apparatus consisting of a coil of wire rotating in a magnetic field. A current is induced in the coil, converting mechanical energy into electric energy. (p. 494)

Induced Current A current produced in a circuit by the emf produced by motion of the circuit through the flux lines of a magnetic field. (p. 493)

Induced Magnetism Magnetism produced in a magnetic material such as iron when the material is placed in a magnetic field, such as that produced in the core of a current-carrying solenoid. (p. 492)

Induction Motor An ac motor with an electromagnetic current induced by the moving magnetic field of the ac current. (p. 498)

Magnetic Property of metals or other materials that can attract iron or steel. (p. 484)

Magnetic Field A field of force near a magnetic pole or a current that can be detected using a magnet. (p. 484)

Motor A device that is composed of an armature and a stator. When a current is passed through the armature, the armature rotates in the magnetic field of the stator and converts electric energy to mechanical energy. (p. 498)

Rotor The rotating coil in a generator. (p. 495)

Solenoid A coil of tightly wrapped wire. Commonly used to create a strong magnetic field by passing current through the wire. (p. 490)

Stator The field magnets in a generator. (p. 495)

Synchronous Motor An ac motor whose speed of rotation is constant and is directly proportional to the frequency of its ac power supply. (p. 498)

Universal Motor A motor that can be run on either ac or dc power. (p. 498)

Formulas

17.2 $B = \dfrac{\mu_0 I}{2\pi R}$

$B = \mu_0 In$

Review Questions

1. The presence of a magnetic force field may be detected by using
   (a) a compass. (b) iron filings.
   (c) a magnet. (d) all of the above.
2. The deflection of a compass needle placed near a current-carrying wire shows
   (a) the magnetic field of the sun.
   (b) the magnetic field of the wire.
   (c) the electric field.
3. Ampère's rule relates
   (a) the strength of a magnetic field to the magnetic pole.
   (b) the direction of a magnetic field surrounding a current-carrying wire.
   (c) the direction of a magnetic field near a bar magnet.
   (d) none of the above.
   (e) all of the above.
4. The unit used to express the strength of a magnetic field is the ______.
5. Describe how a strong magnetic field can be produced in a solenoid.
6. Describe how to determine the direction of a magnetic field in a solenoid.
7. Describe how a magnetic field is induced by a current-carrying coil surrounding a core of magnetic material.
8. Describe how a generator produces current.
9. Describe the function of a commutator.
10. Describe how a motor works.
11. What is a synchronous motor, and how does it work?
12. Distinguish between a universal motor and an induction motor.
13. Distinguish between an armature and a stator.
14. Describe how an electromagnet works.
15. If the current in a solenoid is increased by a factor of 2, how does the magnetic field change?
16. If the radius of a solenoid decreases by a factor of 2, how does the magnetic field change?
17. If the number of turns per inch in a solenoid were increased by a factor of 4, how would the magnetic field change?
18. Describe how to find the flux lines near a bar magnet.
19. How is alternating current produced by a generator?

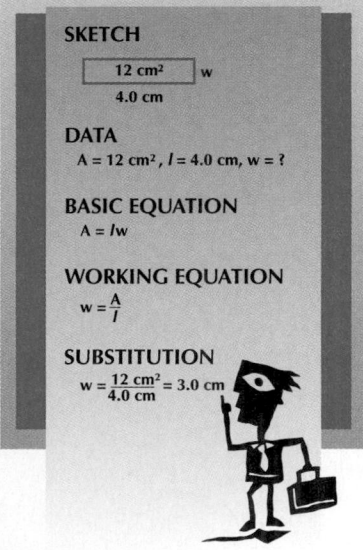

SKETCH

12 cm² w

4.0 cm

DATA

$A = 12\text{ cm}^2, l = 4.0\text{ cm}, w = ?$

BASIC EQUATION

$A = lw$

WORKING EQUATION

$w = \dfrac{A}{l}$

SUBSTITUTION

$w = \dfrac{12\text{ cm}^2}{4.0\text{ cm}} = 3.0\text{ cm}$

Review Problems

1. Find the magnetic field at 0.255 m from a long wire carrying a current of 1.38 A.
2. Find the magnetic field at 0.365 m from a long wire carrying a current of 8.95 A.
3. What is the current in a wire if the magnetic field is 4.75×10^{-6} T at a distance of 1.75 m from the wire?
4. A solenoid has $20\bar{0}0$ turns of wire, is 0.452 m long, and carries a current of 4.55 A. What is the magnetic field at the center of the solenoid?
5. A solenoid has 2750 turns of wire and is 0.182 m long. What current is required to produce a magnetic field of 0.235 T at the center of the solenoid?
6. A power line running north–south carrying $50\bar{0}$ A is 7.00 m above a transit used by a surveyor. What error is induced in the compass used by the surveyor? (Assume the earth's horizontal component of magnetic field is 5.20×10^{-5} T.)

APPLIED CONCEPTS

1. A ship's compass is mistakenly placed 8.35 cm away from a wire carrying a current of 8.25 A. (a) What is the strength of the wire's magnetic field on the compass? (b) The strength of the earth's magnetic field is 5.20×10^{-5} T. How far from the wire must the compass be mounted so that it only experiences a magnetic field of 5.20×10^{-7} T ($\frac{1}{100}$ of the magnetic field of the earth) due to the wire?
2. Figure 17.9 shows a compass near a current-carrying wire. (a) What is the strength of the magnetic field when the compass needle is placed 1.03 cm under the wire? Assume the voltage of the battery is 6.00 V and the resistance of the wire is 15.0 Ω. (b) Compare the strength of the earth's magnetic field to the strength of the magnetic field of the wire.
3. A coaxial cable consists of an inner conducting wire encased in an insulating material surrounded by a conducting metal braid and another insulating sheath. The inner wire carries current in one direction, while the outer braid carries current in the opposite direction. Using Ampère's right-hand rule, explain why there is no significant magnetic field outside the coaxial cable. (Coaxial cables are often used to transmit cable TV and other audiovisual signals.)
4. Figure 17.35 shows a picture tube that uses two solenoids to guide electrons as they move through the magnetic field. (a) What is the orientation of the magnetic field for the two solenoids? (b) As a result of the magnetic field, will the electrons travel toward *a*, *b*, or *c*? (Use Fig. 17.34 as a reference.)

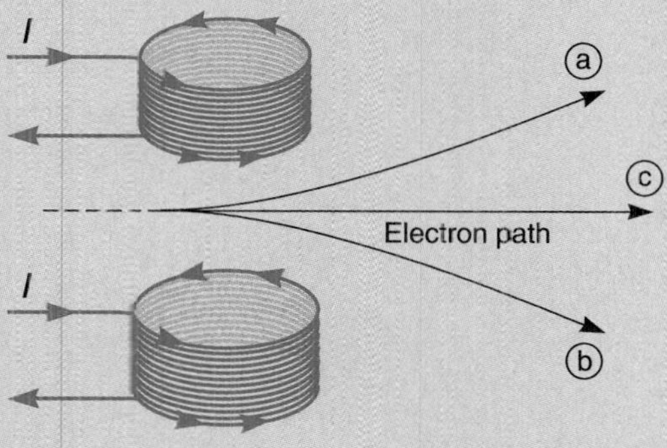

Figure 17.35

5. A copper wire 20.0 m long with radius 4.57×10^{-3} m and $\rho = 1.72 \times 10^{-8}$ Ω cm is used to form a solenoid that will produce a magnetic field. (a) What is the wire's resistance? (b) If the wire is connected to a 4.50-V battery, what is the current that passes through the wire? (c) If the solenoid is designed to have an internal radius of 3.25 cm, how many single-layer coils will be produced? (d) What will be the length of the solenoid if the coils touch one another? (e) Find the strength of the solenoid's magnetic field.

CHAPTER 18

Alternating current
vs.
Time
P=I²R
P=VI
P=V²/R
I
t
1/60 s
Grosvenor House

ALTERNATING CURRENT ELECTRICITY

Ordinary household current is alternating current. The current produced by rotating a loop through a magnetic field is alternating current and constantly changing. We now consider the nature and characteristics of alternating current, transformation of voltage, and various devices used in electric circuits.

Objectives

The major goals of this chapter are to enable you to:

1. Describe the nature and characteristics of alternating current.
2. Describe the use of transformers in changing voltage.
3. Apply inductance and inductive reactance in circuits.
4. Use capacitance and capacitive reactance in circuits.

18.1 What Is Alternating Current?

In Chapter 17 we learned that the current produced by rotating a loop of wire through a magnetic field alternates in direction. This current, called an *alternating current,* is used more frequently in industry and in everyday living than direct current because the transmission of power over long distances is more efficient with alternating current than with direct current.

As its name implies, **alternating current** (ac) is current that flows in one direction in a conductor, then changes direction and flows in the other direction. Electrons in the circuit move first in one direction and then in the opposite direction, alternating back and forth about relatively fixed positions. The direction of flow changes many times in 1 s. Ordinary household current is 60-Hz current. This means that the voltage goes through 60 complete cycles, positive to negative and back again, each second.

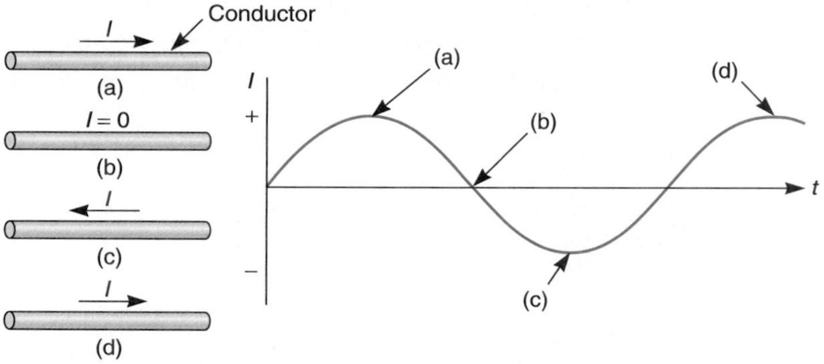

Figure 18.1 Alternating current changes direction as it flows in a conductor.

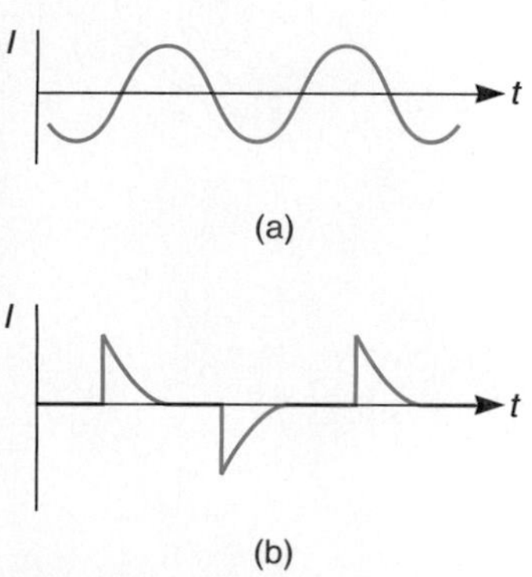

Figure 18.2 Alternating current does not alternate instantly between maximum values in each direction but builds to maximum values as in part (a). Current in a computer digital circuit varies as shown in part (b).

Every time the current repeats itself—flows, changes direction, flows, and changes direction—it goes through one *cycle* (Fig. 18.1). The reason for this alternation is the way current comes from electric generators. The emf and current produced by a generator do not alternate instantly between maximum values in each direction, but they build up to maximum values and then decrease, change direction, and build to maximum values in the other direction (Fig. 18.2).

Direct current is usually a steady flow at a constant value. Graphically, it can be represented as shown in Fig. 18.3. Alternating current, however, is constantly changing. To graphically represent ac, we must show that it builds up and drops off. This can be demonstrated by the curve shown in Fig. 18.4, called a *sine curve*. We form the curve by rotating a vector **V** about a point and plotting the vertical components of **V**. Rotating **V** through 360° graphs one cycle.

The graph in Fig. 18.4 shows one complete cycle of the ac-current curve. A graph of ac voltage is also a sine curve. The current and voltage of ac are constantly changing.

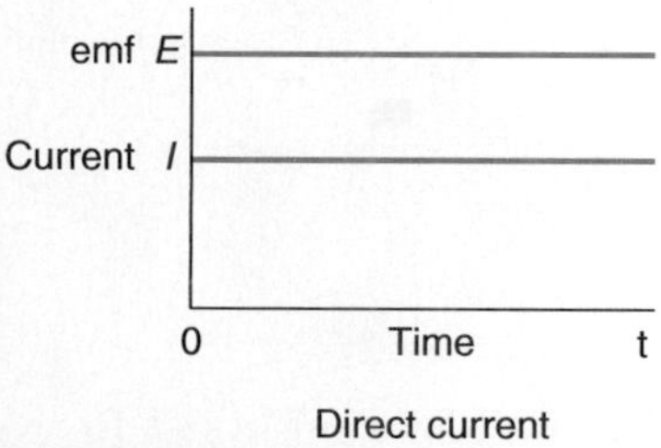

Figure 18.3 Direct current does not alternate but is steady.

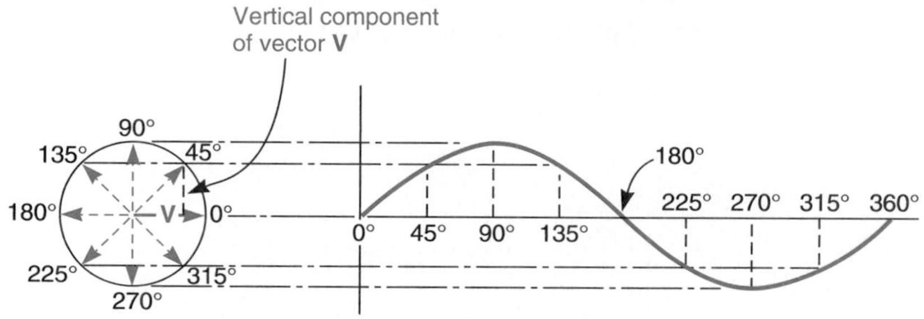

Figure 18.4 Graph of one cycle of alternating current

Instantaneous current is the current at any instant of time; **instantaneous voltage** is the voltage at any instant of time. We can find the value of current or voltage at any instant by using the fact that the graph of each makes a sine curve (Fig. 18.5):

$$i = I_{max} \sin \theta$$
$$e = E_{max} \sin \theta$$

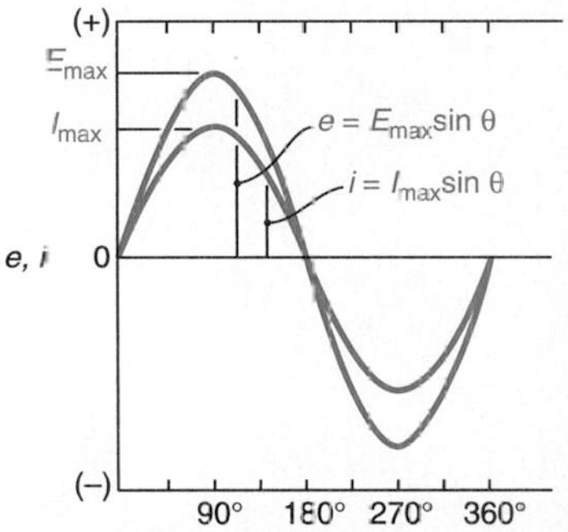

Figure 18.5 **Instantaneous values of current and voltage**

where

i = instantaneous current (current at any instant)
I_{max} = maximum instantaneous current
θ = angle measured from beginning of cycle (see Fig. 18.4)
e = instantaneous voltage
E_{max} = maximum instantaneous voltage

EXAMPLE 1

The maximum voltage in an alternating current is 75 V. Find the instantaneous voltage at $\theta = 35°$.

Data:

$$E_{max} = 75 \text{ V}$$
$$\theta = 35°$$
$$e = ?$$

Basic Equation:

$$e = E_{max} \sin \theta$$

Working Equation: Same

Substitution:

$$e = (75 \text{ V})(\sin 35°)$$
$$= 43 \text{ V}$$

Both curves in Fig. 18.6(a) show e and i reaching a maximum at the same time and falling to zero at the same time. When this occurs, they are "in phase." When there is only resistance in the circuit, e and i are in phase. Later we will study some ac components that will cause e and i to be out of phase [see Fig. 18.6(b)].

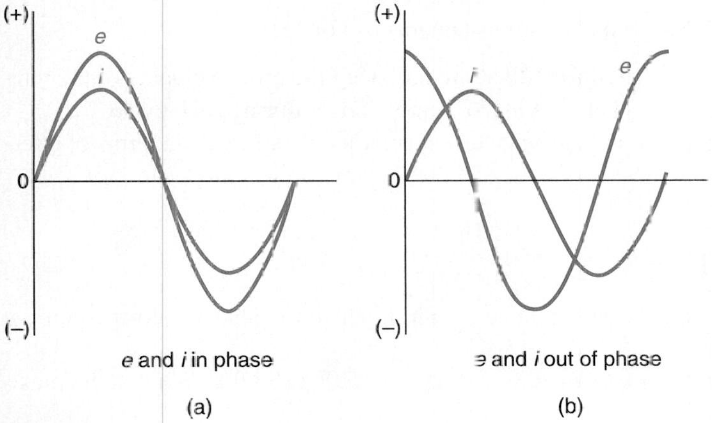

Figure 18.6

Effective Values of ac

A direct measurement of ac is difficult because it is constantly changing. The most useful value of ac is based on its heating effect and is called its effective value. The **effective value** of an alternating current (sometimes called rms, root-mean-square, value) is the number of amperes that produces the same amount of heat in a resistance as an equal number of amperes of a steady direct current.

The numerical factors in the following equations are derived from an average of the sine-wave time variation of the ac current. When this time average is taken, the factors $\sqrt{2} = 1.414$ and $1/\sqrt{2} = 0.707$ are found.

$$I = 0.707\, I_{max}$$
$$I_{max} = 1.41\, I$$

where

$$I = \text{effective value of current}$$
$$I_{max} = \text{maximum instantaneous current}$$

EXAMPLE 2

The current supplied to a woodworking shop is rated at 10.0 A. What is the maximum value of the current supplied?

Data:

$$I = 10.0 \text{ A}$$
$$I_{max} = ?$$

Basic Equation:

$$I_{max} = 1.41\, I$$

Working Equation: Same

Substitution:

$$I_{max} = 1.41\,(10.0 \text{ A}) = 14.1 \text{ A}$$

The effective value for ac voltage may be expressed similarly:

$$E = 0.707\, E_{max}$$
$$E_{max} = 1.41\, E$$

where

$$E = \text{effective value of voltage}$$
$$E_{max} = \text{maximum instantaneous voltage}$$

When we say a house is wired for 120 V, we are using the effective value of the voltage. Actually, the voltage varies between +170 V and −170 V during each cycle.

Unless otherwise stated, ac voltage and current are *always* expressed in terms of effective, or rms, values.

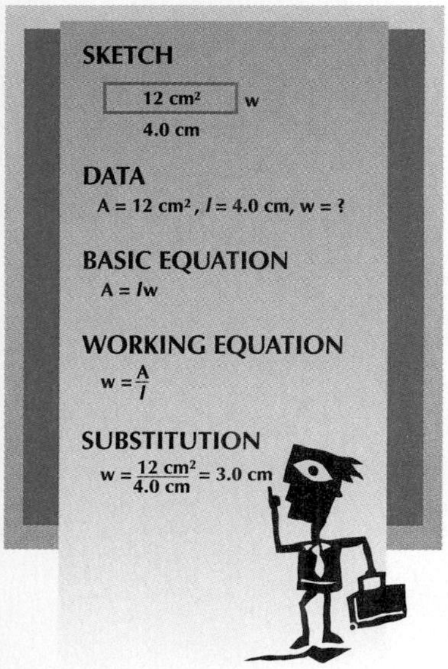

PROBLEMS 18.1

1. What is the maximum voltage in an ac circuit in which the instantaneous voltage at $\theta = 35.0°$ is 27.0 V?
2. The instantaneous voltage in an ac circuit at $\theta = 65.0°$ is 82.0 V. What is the maximum voltage?

3. If the maximum ac voltage on a line is 165 V, what is the instantaneous voltage at $\theta = 45.0°$?
4. The maximum current in an ac circuit is 8.00 A. Find the instantaneous current at $\theta = 60.0°$.
5. The instantaneous current in an ac circuit is 6.50 A at $\theta = 45.0°$. Find the maximum current.
6. What is the maximum voltage in an ac circuit where the instantaneous voltage at $\theta = 51.0°$ is 14.5 V?
7. If the maximum ac voltage on a line is 145 V, what is the instantaneous voltage at $\theta = 35.0°$?
8. The maximum current in an ac circuit is 5.75 A. What is the instantaneous current at $\theta = 80.0°$?
9. Find the maximum current in an ac circuit where the instantaneous current at $\theta = 45.0°$ is 4.00 A.
10. The instantaneous voltage in an ac circuit at $\theta = 55.0°$ is $45\bar{0}$ V. Find the maximum voltage.
11. If $I_{max} = 4.59$ A and $I = 4.32$ A, what is θ?
12. If $I = 1.23$ A and $I_{max} = 3.41$ A, what is θ?
13. Find the effective value of an ac voltage whose maximum voltage is 2250 V.
14. Find the maximum current in an ac circuit with an effective value of 6.00 A.
15. Find the effective value of an ac voltage whose maximum voltage is 165 V.
16. Find the maximum current in an ac circuit with an effective value of 4.00 A.
17. Find the effective value of a current in an ac circuit that reaches a maximum of 17.0 A.
18. Find the effective value of an ac voltage whose maximum voltage is 1150 V.
19. What is the maximum current in a circuit in which an ac ammeter reads 8.50 A?
20. Find the maximum current in a circuit in which an ac ammeter reads 7.00 A.
21. A technician uses an oscilloscope to measure an effective voltage in an ac circuit. Find the effective value if the maximum voltage is 135 V.
22. A technician uses a cathode ray oscilloscope to measure current in an ac circuit which reaches a maximum of 125 A. What is the effective value of the current?
23. A maximum voltage of 34.0 V is developed by an ac generator that delivers a maximum current of 0.170 A to a circuit. (a) What is the effective voltage of the generator? (b) Find the effective current delivered to the circuit by the generator. (c) Find the resistance of the circuit.
24. A power plant generator develops a maximum voltage of $17\bar{0}$ V. What is the effective voltage?
25. A maximum current of 0.700 A flows through a 60.0-W light bulb in a circuit. (a) What is the effective current? (b) Find the resistance of the light bulb when it is lit.
26. If the average power dissipated by an electric light is $15\bar{0}$ W, what is the peak power?
27. If the average power dissipation by a hair dryer is $10\bar{0}0$ W, what is the peak power?

18.2 ac Power

When the load has only resistance, power in ac circuits is found in the same way as in dc circuits:

$$P = I^2R$$

$$P = VI \qquad \text{(using } V = IR\text{)}$$

$$P = \frac{V^2}{R} \qquad \text{(using } I = V/R\text{)}$$

EXAMPLE 1

What power is used in a resistance of 37.0 Ω if it has a current of 0.480 A flowing through it?

Data:

$$R = 37.0\ \Omega$$
$$I = 0.480\ \text{A}$$
$$P = ?$$

Basic Equation:

$$P = I^2R$$

Working Equation: Same

Substitution:

$$P = (0.480\ \text{A})^2(37.0\ \Omega)$$
$$= 8.52\ \text{W}$$

EXAMPLE 2

What power is used in a load of 12.0 Ω resistance if the voltage drop across it is $11\bar{0}$ V?

Data:

$$R = 12.0\ \Omega$$
$$V = 11\bar{0}\ \text{V}$$
$$P = ?$$

Basic Equation:

$$P = \frac{V^2}{R}$$

Working Equation: Same

Substitution:

$$P = \frac{(11\bar{0}\ \text{V})^2}{12.0\ \Omega}$$
$$= 1010\ \text{W} \quad \text{or} \quad 1.01\ \text{kW}$$

The preceding relationships are true only when e and i are in phase. Phase differences produced by capacitance and inductance in an ac circuit are due to reactance. Capacitance, inductance, and reactance will be studied later.

Note that, in the graphs comparing dc and ac power (Fig. 18.7), ac power varies but is always positive (+). The sign indicates only the direction of the current. Even so, p is positive in calculations because the product of $-e$ and $-i$ is positive: $p = (-e)(-i) = ei$.

Transformers and Power

The uses of direct current in industry are somewhat limited. Primary applications are in charging storage batteries, electroplating, generating alternating current, electrolysis, electromagnets, and automobile ignition systems. However, ac can be changed to dc by a simple device called a **rectifier.**

Much more can be done with alternating current. From the kitchen toaster to the largest industrial motors, ac finds wide application. There are very practical reasons for this. The voltage of ac can be easily and efficiently changed in transformers to give almost any desired values. Actually, ac can be used for most purposes just as efficiently as dc. One advantage of ac is that it can be transmitted over long distances with very little heat loss. Heat lost in any electric device is found by the formula

$$\text{heat loss (power) } P = I^2R$$

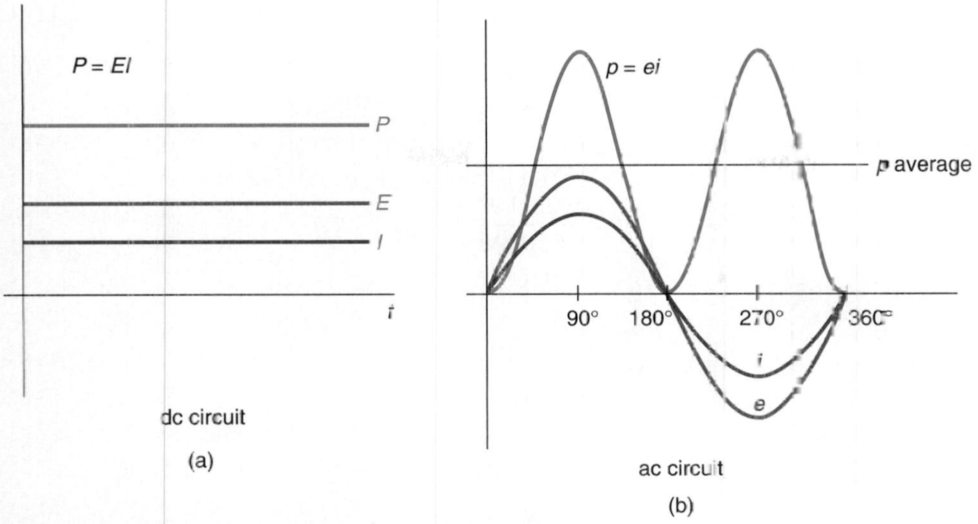

Figure 18.7 dc and ac power compared

The energy wasted as heat can be reduced by making the current smaller. Transformers reduce the current by increasing the voltage since

$$P = \frac{E^2}{R}$$

The major advantage of ac over dc is that ac voltage can easily be changed to meet our needs.

EXAMPLE 3

A plant generates 50.0 kW (50,000 W) of power to be sent to a substation on a line with a resistance of 3.00 Ω. We know that some power will be lost as heat during the transmission. The power lost is $P_{lost} = I^2R$.

(a) How much power is lost if the transmission is at 1150 V?
(b) What percent of the power generated is lost in transmission at 1150 V?
(c) How much power is lost if the transmission is at 11,500 V?
(d) What percent of the power generated is lost in transmission at 11,500 V?
(e) Compare the power losses at the two different transmission voltages.

At 1150 V:

(a)

$$P = VI$$
$$I = \frac{P}{V}$$
$$I = \frac{50,\bar{0}00 \text{ W}}{1150 \text{ V}}$$
$$= 43.5 \text{ A}$$
$$P_{lost} = I^2R$$
$$P_{lost} = (43.5 \text{ A})^2(3.00\ \Omega)$$
$$= 5680 \text{ W}$$
$$= 5.68 \text{ kW}$$

(b)

$$\%_{lost} = \frac{\text{power lost}}{\text{power generated}} \times 100\%$$
$$\%_{lost} = \frac{5.68 \text{ kW}}{50.0 \text{ kW}} \times 100\% = 11.4\%$$

At 11,500 V:

(c)

$$P = VI$$
$$I = \frac{P}{V}$$
$$I = \frac{50,\bar{0}00\ \text{W}}{11,500\ \text{V}}$$
$$= 4.35\ \text{V}$$
$$P_{\text{lost}} = I^2R$$
$$P_{\text{lost}} = (4.35\ \text{A})^2(3.00\ \Omega)$$
$$= 56.8\ \text{W}$$
$$= 0.0568\ \text{kW}$$

(d)

$$\%_{\text{lost}} = \frac{\text{power lost}}{\text{power generated}} \times 100\%$$
$$\%_{\text{lost}} = \frac{0.0568\ \text{kW}}{50.0\ \text{kW}} \times 100\% = 0.114\%$$

(e) This example shows that whereas 11.4% of the power is lost during transmission at 1150 V, only 0.114% is lost at 11,500 V; so by increasing the voltage, the current is correspondingly lowered and the power wasted in transmission is greatly reduced.

Changing Voltage with Transformers

The transformer is a device that is used to change the voltage to reduce the current and thereby lessen the power loss. A **transformer** consists of two coils of wire wrapped on an iron core (Fig. 18.8) and is used to change the voltage. When an alternating current passes through the **primary coil,** an induced current is produced in the **secondary coil** (Fig. 18.9). The magnitude of the voltage induced in the secondary coil depends on

1. The voltage applied to the primary coil.
2. The number of turns in the primary coil.
3. The number of turns in the secondary coil.
4. The power lost between primary and secondary coils.

Figure 18.8 Basic components of a transformer

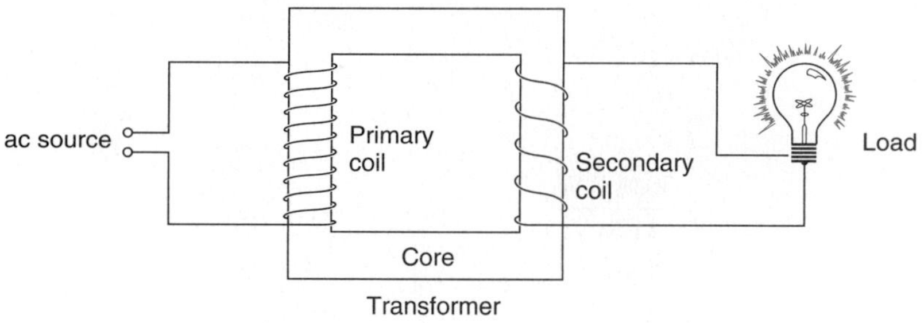

Figure 18.9 An alternating current in the primary coil induces a current in the secondary coil.

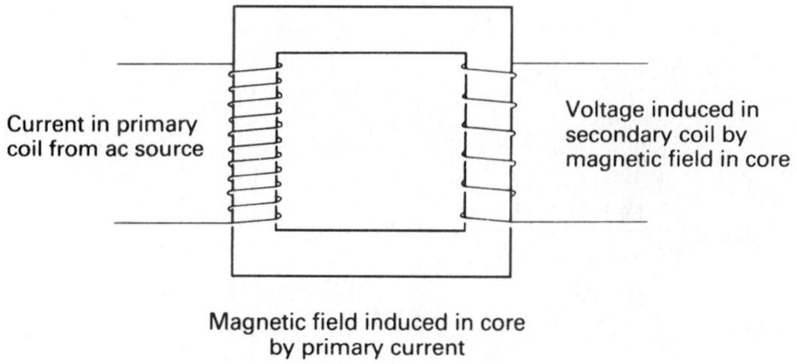

If we assume no power loss between the primary and secondary coils, we have

$$\frac{V_P}{V_S} = \frac{N_P}{N_S}$$

where

V_P = primary voltage
V_S = secondary voltage
N_P = number of primary turns
N_S = number of secondary turns

EXAMPLE 4

A transformer on a neon sign has $1\bar{0}0$ turns in its primary coil and $15{,}\bar{0}00$ turns in its secondary coil. If the voltage applied to the primary coil is $11\bar{0}$ V, what is the secondary voltage?

Data:

$V_P = 11\bar{0}$ V
$N_P = 1\bar{0}0$ turns
$N_S = 15{,}\bar{0}00$ turns
$V_S = ?$

Basic Equation:

$$\frac{V_P}{V_S} = \frac{N_P}{N_S}$$

Working Equation:

$$V_S = \frac{V_P N_S}{N_P}$$

Substitution:

$$V_S = \frac{(11\bar{0}\text{ V})(15{,}\bar{0}00\text{ turns})}{1\bar{0}0\text{ turns}}$$
$$= 16{,}500\text{ V} \quad \text{or} \quad 16.5\text{ kV}$$

Transformers used to raise or lower voltage are called step-up or step-down transformers. *Step-up transformers* are used when a high voltage is needed to operate X-ray tubes or neon signs and to transmit electric power over long distances. A **step-up transformer** increases the voltage by having more turns in the secondary coil than in the primary coil [Fig. 18.10(a)].

A **step-down transformer** lowers the voltage from high-voltage transmission lines to regular 110 V and 220 V for home and industrial use. Voltage is lowered in the step-down transformer because it has more turns in the primary coil than in the secondary coil [Fig. 18.10(b)].

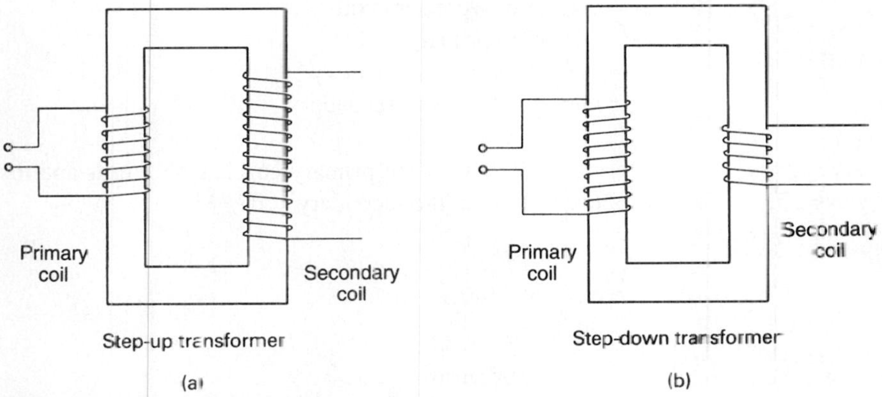

Figure 18.10 Step-up and step-down transformers

Auto transformers are used when a variable output voltage is needed. In this type of transformer, contact can be made across a variable number of the secondary coils using a brush contact. The output voltage is therefore variable from nearly zero to some maximum value. This type of transformer is often used to supply ac power to resistive heater elements to control the heating output.

Transformers do not create energy. In fact, some energy is lost during the change of voltage. Energy losses in transformers are of three types:

1. *Copper losses.* These result from the resistance of the copper wires in the coils and are unavoidable.
2. *Magnetic losses* (called *hysteresis losses*). Some energy is lost (turned into heat) by reversing the magnetism in the core.
3. *Eddy currents.* When a mass of metal (the core) is subjected to a changing magnetic field, *eddy currents* are set up in the metal that do no useful work, waste energy, and produce heat. These losses can be reduced by *laminating* the core. Instead of a solid block of metal for the core [Fig. 18.11(a)], thin sheets of metal with insulated surfaces are used [Fig. 18.11(b)], reducing these induced currents. These ac eddy currents cause the laminations to vibrate, producing the characteristic transformer "hum."

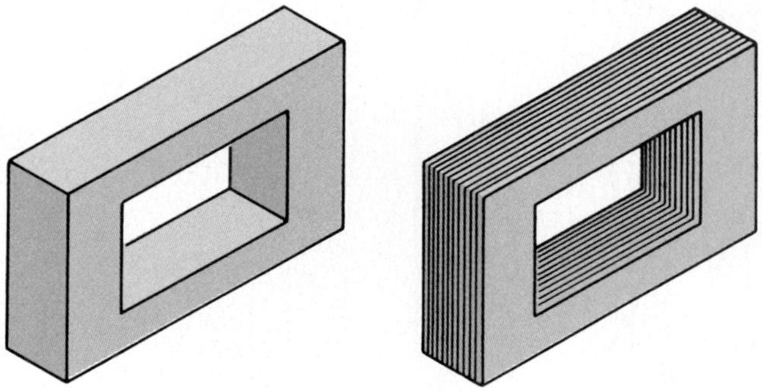

Solid core (a) Laminated core (b)

Figure 18.11 Laminating the core reduces the eddy current losses.

When a transformer steps up the voltage applied to its primary coil, it reduces the current. Energy is conserved—we cannot get any more electrical energy out of a transformer than we put into it. The relationship between primary and secondary currents is

$$\frac{I_S}{I_P} = \frac{N_P}{N_S}$$

where

I_S = current in secondary coil
I_P = current in primary coil
N_P = number of turns in primary
N_S = number of turns in secondary

EXAMPLE 5

The primary current in a transformer is 10.0 A. If the primary coil has $55\overline{0}$ turns and the secondary has $25\overline{0}0$ turns, what current flows in the secondary coil?

Data:

$N_P = 55\overline{0}$ turns
$I_P = 10.0$ A
$N_S = 25\overline{0}0$ turns
$I_S = ?$

Basic Equation:

$$\frac{I_S}{I_P} = \frac{N_P}{N_S}$$

Working Equation:

$$I_S = \frac{I_P N_P}{N_S}$$

Substitution:

$$I_S = \frac{(10.0\text{ A})(55\bar{0}\text{ turns})}{25\bar{0}0\text{ turns}}$$
$$= 2.20\text{ A}$$

It follows from the last two shaded formulas that

$$\frac{V_P}{V_S} = \frac{N_P}{N_S} = \frac{I_S}{I_P}$$

so

$$\frac{V_P}{V_S} = \frac{I_S}{I_P}$$

From this we obtain,

$$I_P V_P = I_S V_S$$

or, using $P = IV$,

$$P_P = P_S$$

which shows that power is conserved between the primary and secondary coils under these assumptions.

EXAMPLE 6

The power in the primary coil of a transformer is 375 W. If the current in the secondary coil is 11.4 A, what is the voltage in the secondary?

Data:

$$P_P = 375\text{ W}$$
$$I_S = 11.4\text{ A}$$
$$V_S = ?$$

Basic Equation:

$$P_P = P_S \quad \text{and} \quad P_S = V_S I_S$$

Working Equations:

$$P_P = V_S I_S \qquad (\textit{Note:}\text{ Substitute for } P_S.)$$

$$V_S = \frac{P_P}{I_S}$$

Substitution:

$$V_S = \frac{375\text{ W}}{11.4\text{ A}}$$
$$= 32.9\text{ V}$$

Good transformers are more than 98% efficient. This is very important in power transmission. It is impractical to generate electricity at high voltage, but high voltage is desirable for transmission. Therefore, transformers are used to step up the voltage for transmission. High voltage is unsuitable, though, for consumer use, so transformers are used to reduce the voltage. A simplified diagram of a power distribution system is shown in Fig. 18.12.

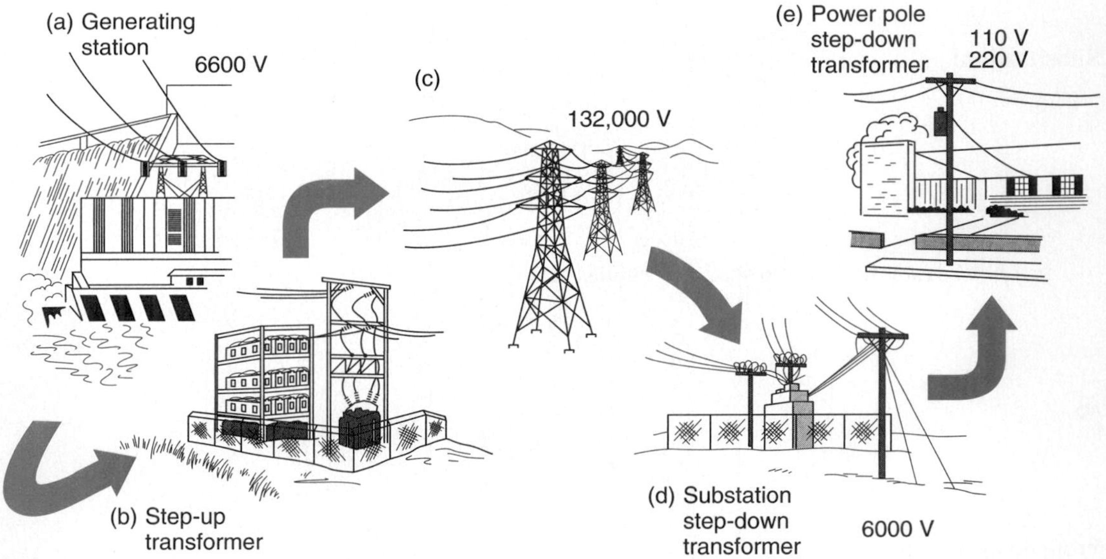

Figure 18.12 Power distribution system

SKETCH

12 cm² | w

4.0 cm

DATA

$A = 12\ \text{cm}^2$, $l = 4.0$ cm, $w = ?$

BASIC EQUATION

$A = lw$

WORKING EQUATION

$w = \frac{A}{l}$

SUBSTITUTION

$w = \frac{12\ \text{cm}^2}{4.0\ \text{cm}} = 3.0\ \text{cm}$

PROBLEMS 18.2

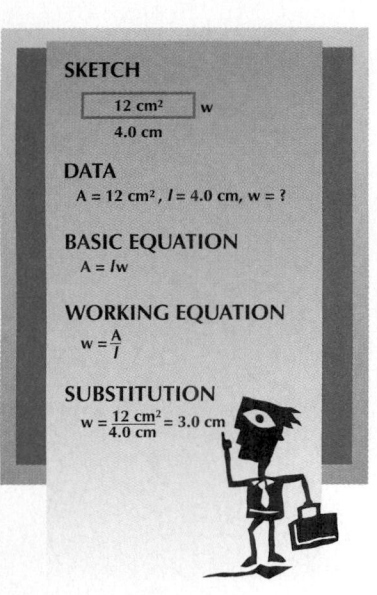

1. A soldering iron is rated at $35\bar{0}$ W. If the current in the iron is 4.00 A, what is the resistance of the iron?
2. What power is developed by a device that draws 6.00 A with resistance 12.0 Ω?
3. Find the output power of a transformer with output voltage $50\bar{0}$ V and current 7.00 A.
4. A heater operates on a $11\bar{0}$-V line and is rated at $45\bar{0}$ W. What is the resistance of the element?
5. A heating element draws 6.00 A on a $22\bar{0}$-V line. What power is used in the element?
6. A 32.0-Ω resistance coil uses 375 W of power. What is the current in the coil?
7. What power is used by a heater that has a resistance of 12.0 Ω and draws a current of 7.00 A?
8. A heater operates on a $11\bar{0}$-V line and is rated at $75\bar{0}$ W. What is the resistance of the heater element?
9. A $11\bar{0}$-Ω resistance coil draws a current of 5.00 A. What power is used?
10. What power is used by a heater with resistance 19.5 Ω and that draws a current of 5.55 A?

11.		12.		13.	
	$V_P = 30.0$ V		$V_P = 25\bar{0}$ V		$I_P = 6.00$ A
	$V_S = 45.0$ V		$N_P = 73\bar{0}$ turns		$I_S = 4.00$ A
	$N_S = 15.0$ turns		$N_S = 275$ turns		$V_P = 39.0$ V
	Find N_P.		Find V_S.		Find V_S.

14. A step-up transformer on a 115-V line provides a voltage of $23\bar{0}0$ V. If the primary coil has 65.0 turns, how many turns does the secondary have?
15. A step-down transformer on a 115-V line provides a voltage of 11.5 V. If the secondary coil has 30.0 turns, how many turns does the primary have?
16. A transformer has 20.0 turns in the primary coil and $22\bar{0}0$ turns in the secondary. If the primary voltage is 12.0 V, what is the secondary voltage?
17. If the current is 9.00 A in the primary coil in Problem 13, find the current in the secondary.

18. If the voltage in the secondary coil of a transformer is $11\bar{0}$ V and the current in it is 15.0 A, what power does it supply?
19. A neon sign has a transformer that changes electricity from $11\bar{0}$ V to $15,\bar{0}00$ V. (a) If the primary current is 8.00 A, find the current in the secondary coil. (b) Find the power in the primary coil.
20. A transformer has an output power of 990 W. If the current in the secondary coil is 0.45 A, what is its voltage?
21. The current in the secondary coil of a transformer is 5.00 A. Find the voltage in the secondary if the power is 775 W.
22. A transformer steps down $66\bar{0}0$ V to $12\bar{0}$ V. (a) If the secondary current is 14.0 A, what is the primary current? (b) Find the power in the primary coil.
23. The primary coil of a step-down transformer has $75\bar{0}0$ turns, and the secondary coil has 125 turns. The voltage across the primary is $72\bar{0}0$ V. (a) Find the voltage across the secondary. (b) If the current in the secondary coil is 36.0 A, find the primary current.
24. A step-up transformer has $30\bar{0}0$ turns in the secondary coil and $2\bar{0}0$ turns in the primary coil. The primary is supplied with alternating current with an effective voltage of 90.0 V. (a) Find the voltage in the secondary coil. (b) If the current in the secondary coil is 2.00 A, what current flows in the primary coil? (c) What power is developed in the primary coil? (d) What power is developed in the secondary coil?

18.3 Inductance

Electronic circuitry in televisions, radios, computers, and electronic instruments has many components other than resistors. These components include capacitors, inductors, diodes, and transistors. With these components weak signals can be amplified, noise can be reduced, and signals at certain frequencies can be detected while signals from other frequencies can be rejected (that is, a circuit can be "tuned in" to a frequency). The analysis of the behavior of circuits with these components can become very complex. As a start toward understanding these circuits, we discuss inductors and capacitors. The operation of diodes and transistors will be discussed only briefly. We begin with a discussion of inductors.

Inductance measures the tendency of a coil of wire to resist a change in the current because the magnetism produced by one part of the coil acts to oppose the change of current in other parts of the coil. Inductance is the property of an electric circuit in which a varying current produces a varying magnetic field that induces voltage in the same circuit or in a nearby circuit. An **inductor** is a circuit component, such as a coil of wire, in which an induced emf opposes any change in the current (Fig. 18.13). The emf is induced in the coil itself as the magnetic field of the coil changes.

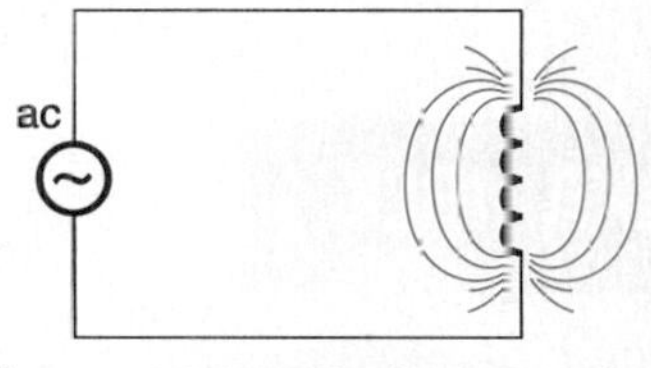

Figure 18.13 **Induced emf opposing current change in an ac circuit**

The unit of inductance, L, is the henry (H), named after **Joseph Henry.** A coil has an inductance of 1 H if an emf of 1 V is induced when the current changes at the rate of 1 A/s. We can express the henry as an ohm second:

$$1\ \text{H} = 1\ \Omega\ \text{s}$$

Joseph Henry (1797–1898),

physicist, was born in Albany, New York. He constructed the first electromagnetic motor, discovered electric induction, and demonstrated the oscillatory nature of electric discharges. He also introduced a system of weather forecasting and investigated the propagation of light and sound waves. The unit of induction, the henry, is named after him.

The henry is a large unit. A more practical unit is the millihenry (mH), which is one one-thousandth of a henry.

Inductance can be illustrated by connecting a coil with a large number of turns and a lamp in series. When connected to a dc source, the lamp burns brightly [Fig. 18.14(a)].

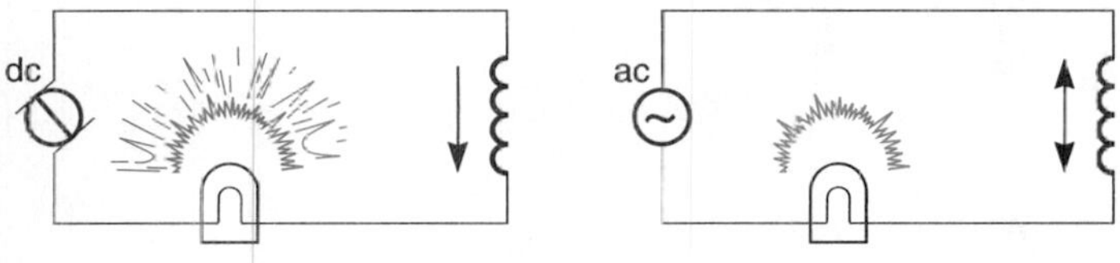

(a) (b)

Figure 18.14 **A coil in an ac circuit produces inductance.**

Figure 18.15 Circuit diagram symbol for inductance

However, when this circuit is connected to an ac power source of the same voltage, the lamp is dimmer because of the inductance of the coil [Fig. 18.14(b)]. The circuit symbol for inductance is shown in Fig. 18.15.

Inductive Reactance

The opposition to ac current flow in an inductor is called **inductive reactance** and is measured in ohms. This is usually represented by X_L. The inductive reactance of a coil is directly proportional to frequency and is found by the following:

$$X_L = 2\pi f L$$

where

X_L = inductive reactance
f = frequency of the ac voltage, expressed in hertz (cycles per second), such as $6\bar{0}$ Hz or $6\bar{0}$/s
L = inductance, in henries

If inductance is given in mH, a conversion must be made to H (henries).

The current in a circuit that has only an ac voltage source and an inductor is given by

$$I = \frac{E}{X_L}$$

where

I = current
E = voltage
X_L = inductive reactance

EXAMPLE

A coil with inductance 1.00 mH is connected to a 60.0-kHz ac power source of $11\bar{0}$ V. What is the current in the circuit?

Sketch:

Data:

$$E = 11\bar{0}\text{ V}$$
$$L = 1.00\text{ mH} = 1.00 \times 10^{-3}\text{ H}$$
$$f = 60.0\text{ kHz} = 60.0 \times 10^{3}\text{ Hz} = 6.00 \times 10^{4}/\text{s}$$
$$I = ?$$

Basic Equations:

$$X_L = 2\pi f L \quad \text{and} \quad I = \frac{E}{X_L}$$

Working Equations: Same

Substitutions:

$$X_L = 2\pi(6.00 \times 10^{4}/\text{s})(1.00 \times 10^{-3}\text{ H})$$
$$= 377\frac{\text{H}}{\text{s}}$$

$$= 377\,\frac{\text{H}}{\text{s}}\left(\frac{1\ \Omega\ \text{s}}{1\ \text{H}}\right) \quad \text{(note conversion factor)}$$

$$= 377\ \Omega$$

$$I = \frac{E}{X_L}$$

$$I = \frac{11\bar{0}\ \text{V}}{377\ \Omega}$$

$$= 0.292\,\frac{\text{V}}{\Omega}\left(\frac{\text{A}\ \Omega}{\text{V}}\right) \quad \text{(note conversion factor)}$$

$$= 0.292\ \text{A}$$

In an inductor, the current lags behind the voltage by one-fourth of a cycle (Fig. 18.16). For example, in a 60-Hz ac circuit, the frequency is 60 cycles/s; that is, the time for one complete cycle is 1/60 s. Thus, the maximum voltage in a 60-Hz circuit occurs

$$\frac{1}{4} \times \frac{1}{60}\ \text{s} = \frac{1}{240}\ \text{s}$$

before the maximum current. The current lag is usually measured in degrees. One-fourth of a cycle is 90°.

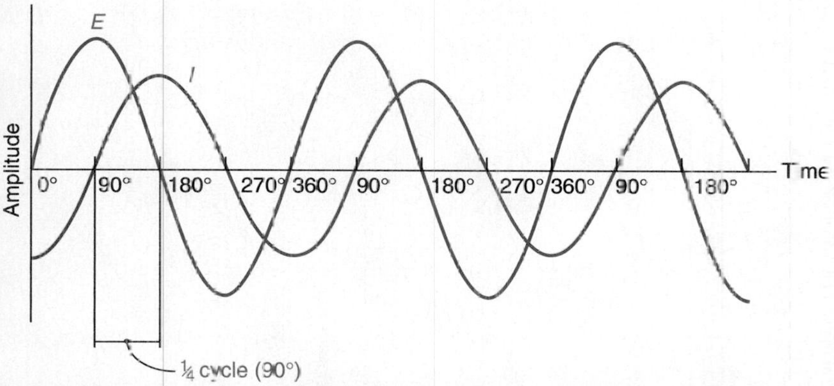

Figure 18.16 In an inductive circuit, the current lags behind the voltage by one-fourth of a cycle.

PROBLEMS 18.3

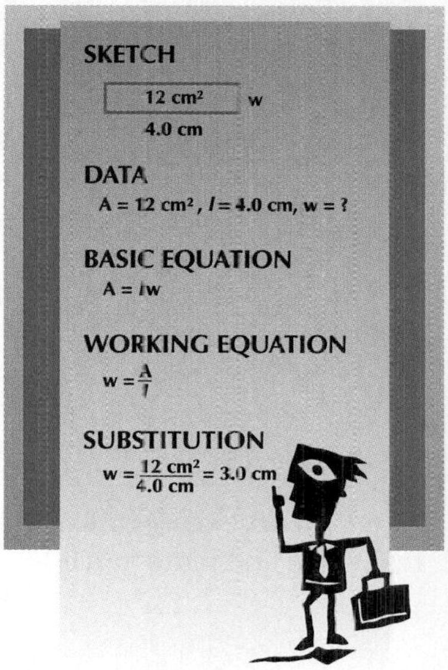

Find the inductive reactance (in ohms) of each inductance at the given frequency.

1. $L = 3.00$ mH, $f = 60.0$ Hz
2. $L = 20.0$ mH, $f = 75.0$ Hz
3. $L = 70.0$ mH, $f = 10.0$ kHz
4. $L = 8.00$ mH, $f = 8.00$ kHz
5. What is the inductive reactance (in ohms) of a 425-μH inductance at a frequency of 15.0 MHz?
6. Find the inductive reactance (in ohms) of a 655-μH inductance at a frequency of 125 MHz.

Find the current (in amperes) in each inductive circuit.

7. $L = 30.0$ mH, $f = 125$ Hz, $E = 14.0$ V
8. $L = 1.00$ mH, $f = 125$ kHz, $E = 145$ V
9. $L = 5.00$ mH, $f = 2.00$ kHz, $E = 50.0$ V
10. $L = 30.0$ mH, $f = 7.00$ MHz, $E = 75.0$ V

11. Find the current (in amperes) in an inductive circuit where $L = 72.0\ \mu H, f = 2.00$ MHz, and $E = 105$ V.
12. Find the current (in amperes) in an inductive circuit where $L = 525\ \mu H, f = 25.0$ MHz, and $E = 65.0$ V.

PHYSICS CONNECTIONS

Induction and the Ground Fault Interrupter

Ground fault interrupt (GFI) outlets (Fig. 18.17) are required by most building codes for electric outlets around sinks or other devices where water can cause electric shocks. The purpose of a GFI is to detect minor losses of current in a circuit. If a working hairdryer were to fall into a sink or a bathtub, the water would cause the hairdryer to short out and send electric charge into the water and metal pipes. In such a situation, the GFI outlet would automatically shut off the current within milliseconds of the short circuit. If someone touched the hairdryer while in the water, the person would be electrocuted if not for the rapid circuit-breaking ability of the GFI.

All ac electric appliances have current moving back and forth at a frequency of 60 Hz. The GFI is designed to measure the changes in the induced magnetic field of that electric current. In other words, the current going into the appliance establishes a magnetic field in the iron loop, while the current leaving the appliance creates an equal and opposite magnetic field in the loop, reversing direction 60 times per second. These two fields, in a normal operating situation, cancel each other out. However, if even a small amount of current does not return through the GFI outlet, the magnetic fields do not cancel out. In that situation, the magnetic field in the loop creates an electric current in the detection coil. Any current passing through the detector coil causes the breaker to trip, which quickly shuts off the circuit and prevents serious injury.

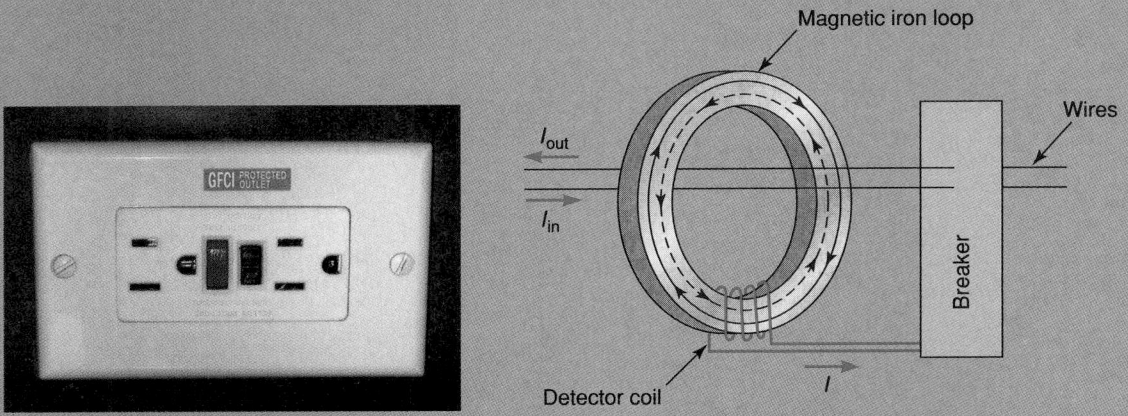

Figure 18.17 (a) A standard ground fault interrupt outlet (b) The basic components of a ground fault interrupt outlet

18.4 Inductance and Resistance in Series

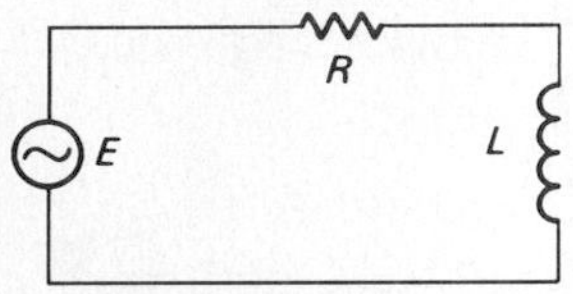

Figure 18.18 ac circuit with resistance and inductance

In addition to inductance, most ac circuits have resistance in the form of lights or other resistors (Fig. 18.18). The current lags behind the voltage by any amount of time greater than zero and as large as one-fourth of a cycle.

Impedance is a measure of the total opposition to current flow in an ac circuit resulting from the effect of both the resistance and the inductive reactance on the circuit. Ohm's law in an ac circuit can be written as

$$I = \frac{E}{Z}$$

where

I = current
E = voltage
Z = impedance

The impedance of a series circuit containing a resistance and an inductance is

$$Z = \sqrt{R^2 + X_L^2}$$
$$Z = \sqrt{R^2 + (2\pi f L)^2}$$

where

Z = impedance
R = resistance
X_L = inductive reactance
f = frequency
L = inductance

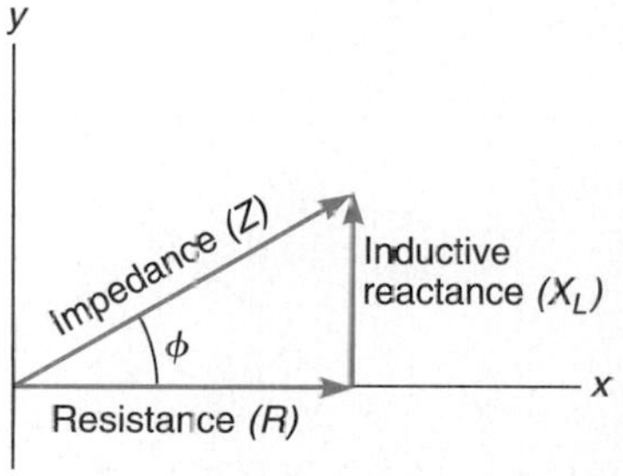

Figure 18.19 Graphic representation of impedance

The impedance can be represented as a vector as the hypotenuse of the right triangle shown in Fig. 18.19. The resistance is always drawn as a vector pointing in the positive x-direction. The inductive reactance is drawn as a vector pointing in the positive y-direction. The angle ϕ shown between the resistance and impedance vectors is the **phase angle** and equals the amount by which the current lags behind the voltage. The phase angle is given by

$$\tan\phi = \frac{X_L}{R}$$

EXAMPLE

A lamp of resistance 40.0 Ω is connected in series with an inductance of 95.0 mH. This circuit is connected to a 115-V, 60.0-Hz power supply. (a) What is the current in the circuit? (b) What is the phase angle?

Sketch:

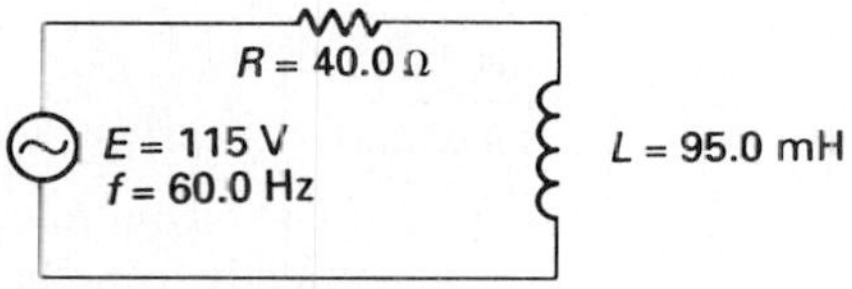

(a) Data:

$E = 115$ V

$f = 60.0$ Hz $= 60.0$/s

$R = 40.0\ \Omega$

$L = 95.0$ mH $= 95.0 \times 10^{-3}$ H $= 0.0950$ H

$Z = ?$

$I = ?$

Basic Equations:

$$Z = \sqrt{R^2 + (2\pi f L)^2} \quad \text{and} \quad I = \frac{E}{Z}$$

Working Equations: Same

Substitutions: First calculate the impedance.

$$Z = \sqrt{(40.0\ \Omega)^2 + [2\pi(60.0/\text{s})(0.0950\ \text{H})]^2}$$
$$= \sqrt{16\bar{0}0\ \Omega^2 + (35.8\ \text{H/s})^2}$$

$$= \sqrt{16\bar{0}0\ \Omega^2 + \left[\left(35.8\frac{\text{H}}{\text{s}}\right)\left(\frac{1\ \Omega\ \text{s}}{1\ \text{H}}\right)\right]^2} \qquad \text{(note conversion factor)}$$

$$= \sqrt{16\bar{0}0\ \Omega^2 + 1280\ \Omega^2}$$

$$= \sqrt{2880\ \Omega^2}$$

$$= 53.7\ \Omega$$

$$I = \frac{E}{Z}$$

$$I = \frac{115\ \text{V}}{53.7\ \Omega}\left(\frac{1\ \text{A}\ \Omega}{1\ \text{V}}\right)$$

$$= 2.14\ \text{A}$$

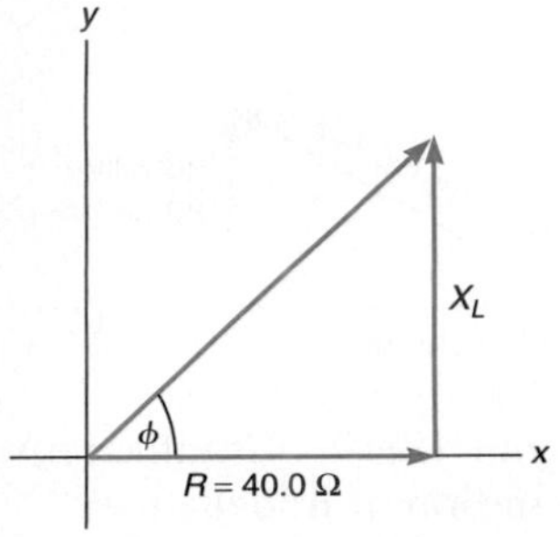

Figure 18.20

(b) To find the phase angle ϕ, first find X_L. Then, construct the vector right triangle as in Fig. 18.20 to find ϕ.

Data:

$$f = 60.0/\text{s}$$
$$L = 0.0950\ \text{H}$$
$$X_L = ?$$
$$\phi = ?$$

Basic Equations:

$$X_L = 2\pi f L \quad \text{and} \quad \tan\phi = \frac{X_L}{R}$$

Working Equations: Same

Substitutions:

$$X_L = 2\pi(60.0/\text{s})(0.0950\ \text{H})$$

$$= 35.8\frac{\text{H}}{\text{s}}$$

$$= 35.8\frac{\text{H}}{\text{s}}\left(\frac{1\ \Omega\ \text{s}}{1\ \text{H}}\right)$$

$$= 35.8\ \Omega$$

[*Note:* This value can also be taken directly from the first working equation in part (a).]

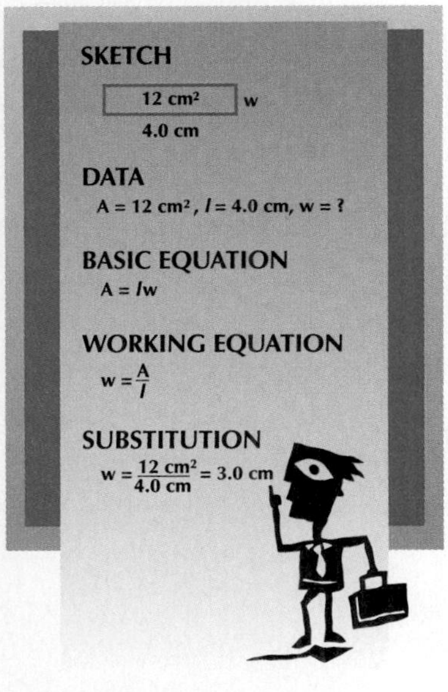

From Fig. 18.20 we have

$$\tan\phi = \frac{35.8\ \Omega}{40.0\ \Omega} = 0.895$$

$$\phi = 41.8°$$

PROBLEMS 18.4

1. For a circuit with $R = 2\bar{0}0\ \Omega$, $L = 10.0$ mH, and $f = 1.25$ kHz:
   (a) find the impedance (in ohms).
   (b) find the phase angle.
   (c) find the current if the voltage is 45.0 V.
2. For a circuit with $R = 12.0\ \Omega$, $L = 1.00$ mH, and $f = 9\bar{0}0$ Hz:
   (a) find the impedance (in ohms).
   (b) find the phase angle.
   (c) find the current if the voltage is 10.0 V.

3. For a circuit with $R = 1.00$ kΩ, $L = 50.0$ mH, and $f = 10.0$ kHz:
   (a) find the impedance (in ohms).
   (b) find the phase angle.
   (c) find the current if the voltage is 15.0 V.
4. For a circuit with resistance 2.00 kΩ, inductance 70.0 mH, and frequency 5.00 kHz:
   (a) find the impedance (in ohms).
   (b) find the phase angle.
   (c) find the current if the voltage is 12.0 V.
5. For a circuit with resistance 300 Ω, inductance 2.00 mH, and frequency 3.00 kHz:
   (a) find the impedance (in ohms).
   (b) find the phase angle.
   (c) find the current if the voltage is 6.00 V.

18.5 Capacitance

An important component of many ac circuits is the capacitor. A **capacitor** is used to store an electric charge, and consists of two conductors that are usually parallel plates separated by a thin insulator. The plates are often made of a metal foil rolled to a convenient size and inserted in a cylinder. Capacitors are represented in circuit diagrams as shown in Fig. 18.21.

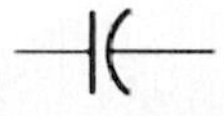

Figure 18.21 Circuit diagram symbol for a capacitor

When a capacitor is connected to a battery, electrons flow from the negative terminal to one capacitor plate as shown in Fig. 18.22(b). When the capacitor is removed from the battery, the charges remain on the capacitor. If the capacitor is then connected to a resistor [Fig. 18.22(c)], electrons will flow through the circuit until the capacitor has lost its charge.

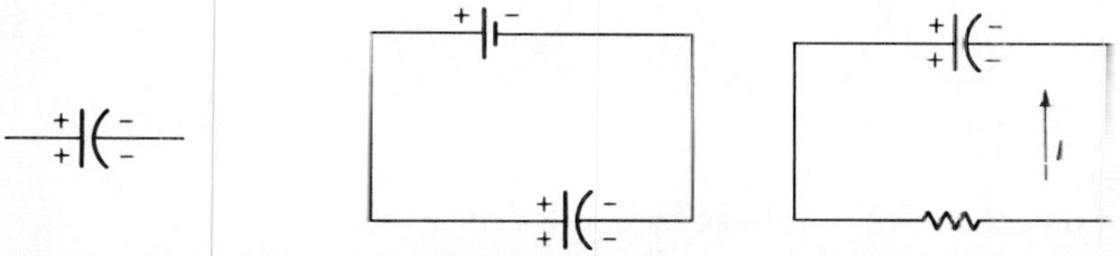

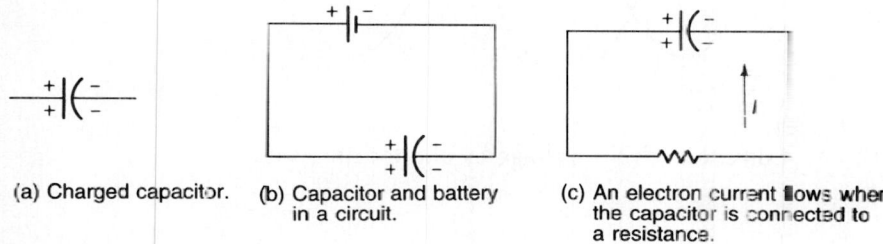

(a) Charged capacitor. (b) Capacitor and battery in a circuit. (c) An electron current flows when the capacitor is connected to a resistance.

Figure 18.22

A capacitor will block a direct current from flowing in a circuit once the capacitor is charged. A low-frequency ac voltage in a capacitive circuit (Fig. 18.23) will cause only a small current to flow because of this blocking nature of a capacitor. A high-frequency ac voltage source will cause a larger current to flow in the capacitive circuit of Fig. 18.23 because a current is required to quickly change the polarity of the capacitor voltage. Capacitors can therefore be used to tune the frequency response of circuits, allowing the blocking of low-frequency electric signals and tuning the resonance frequency of circuits as described in Section 18.8.

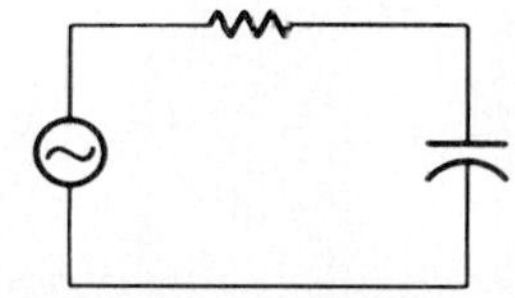

Figure 18.23 Capacitor in an ac circuit

Capacitance is the ratio of the charge on either plate of a capacitor to the potential difference between the plates. That is,

$$C = \frac{Q}{V}$$

where

C = capacitance of a capacitor
Q = amount of charge on either plate of the capacitor
V = potential difference between the two plates of the capacitor

Michael Faraday (1791–1867), *chemist and experimental physicist, was born in England. He introduced the concept of capacitance, discovered electromagnetic induction, and made numerous other contributions to electricity and magnetism. The unit of capacitance, the farad, is named after him.*

The unit of capacitance is the *farad* (F), named after **Michael Faraday.** A more practical unit is the microfarad (μF or 10^{-6} F). **Capacitive reactance** is a measure of the opposition to ac current flow by a capacitor. The effect of a capacitor on a circuit is inversely proportional to frequency and given by

$$X_C = \frac{1}{2\pi f C}$$

where

X_C = capacitive reactance (ohms)
f = frequency
C = capacitance (farads)

$$1 \text{ F} = 1 \text{ s}/\Omega$$

In a circuit that contains only capacitors, the current *leads* the voltage by 90° (one-fourth cycle) as shown in Fig. 18.24.

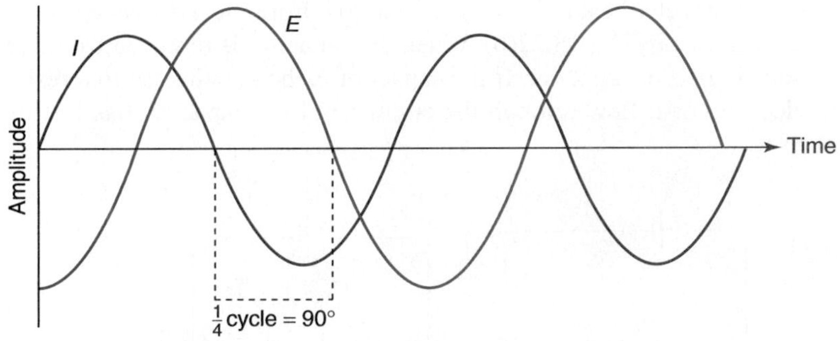

Figure 18.24 Current leads the voltage by one-fourth cycle.

EXAMPLE

Find the capacitive reactance of a 10.0-μF capacitor in a circuit of frequency 1.00 kHz.

Data:

$$C = 10.0\ \mu\text{F} = 10.0 \times 10^{-6}\ \text{F} = 1.00 \times 10^{-5}\ \text{F}$$
$$f = 1.00\ \text{kHz} = 1.00 \times 10^3/\text{s}$$
$$X_C = ?$$

Basic Equation:

$$X_C = \frac{1}{2\pi f C}$$

Working Equation: Same

Substitution:

$$X_C = \frac{1}{2\pi(1.00 \times 10^3/\text{s})(1.00 \times 10^{-5}\ \text{F})}$$

$$= \frac{1}{\left(0.0628\frac{\cancel{\text{F}}}{\cancel{\text{s}}}\right)\left(\frac{1\ \cancel{\text{s}}}{1\ \cancel{\text{F}}\ \Omega}\right)} \quad \text{(note conversion factor)}$$

$$= 15.9\ \Omega$$

PROBLEMS 18.5

Find the capacitive reactance (in ohms) in each ac circuit.

1. $C = 20.0\ \mu\text{F}, f = 1.00\ \text{kHz}$
2. $C = 7.00\ \text{mF}, f = 10\bar{0}\ \text{Hz}$
3. $C = 0.600\ \mu\text{F}, f = 0.100\ \text{kHz}$
4. $C = 30.0\ \text{mF}, f = 2.50\ \text{MHz}$
5. $C = 0.800\ \mu\text{F}, f = 0.250\ \text{MHz}$
6. Find the capacitive reactance of a 15.0-μF capacitor in a circuit of frequency 60.0 Hz.
7. Find the capacitive reactance of a 45.0-μF capacitor in a circuit of frequency 60.0 kHz.
8. Find the capacitive reactance of a 6.00-mF capacitor in a circuit of frequency $10\bar{0}$ Hz.
9. Find the capacitive reactance of a $33\bar{0}$-μF capacitor in a circuit of frequency $30\bar{0}$ Hz.
10. Find the capacitive reactance of a 222-μF capacitor in a circuit of frequency $12\bar{0}$ Hz.

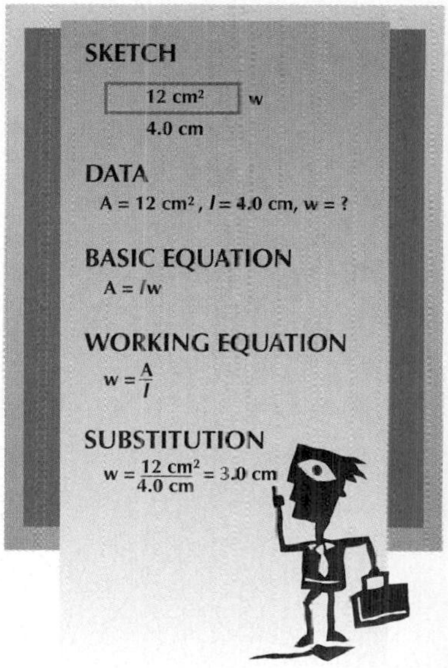

18.6 Capacitance and Resistance in Series

The combined effect of capacitance and resistance in series is measured by the impedance, Z, of the circuit.

$$Z = \sqrt{R^2 + X_C^2}$$

$$Z = \sqrt{R^2 + \left(\frac{1}{2\pi f C}\right)^2}$$

where

Z = impedance
R = resistance
X_C = capacitive reactance
f = frequency
C = capacitance

The current is given by Ohm's law:

$$I = \frac{E}{Z}$$

where

I = current
E = voltage
Z = impedance

The phase angle can be found by drawing the resistance as a vector in the positive x-direction and the capacitive impedance as a vector in the negative y-direction as shown in Fig. 18.25. The phase angle gives the amount by which the voltage lags behind the current.

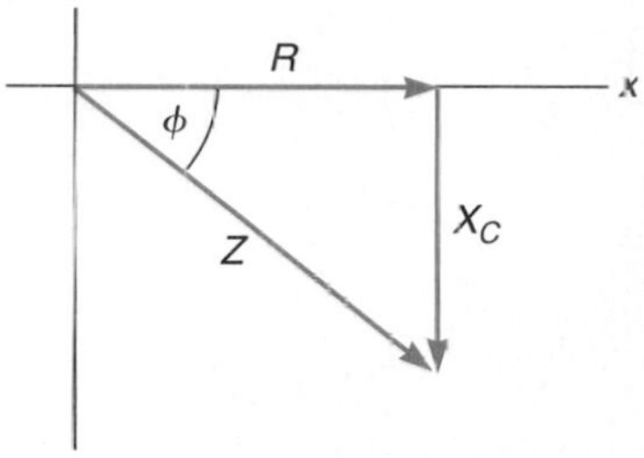

Figure 18.25 Determination of the phase angle in an ac circuit

$$\tan \phi = \frac{X_C}{R}$$

EXAMPLE

What current will flow in a 60.0-Hz ac circuit that includes a $11\bar{0}$-V source, a capacitor of 90.0 μF, and a 16.0-Ω resistance in series? Also find the phase angle.

Sketch:

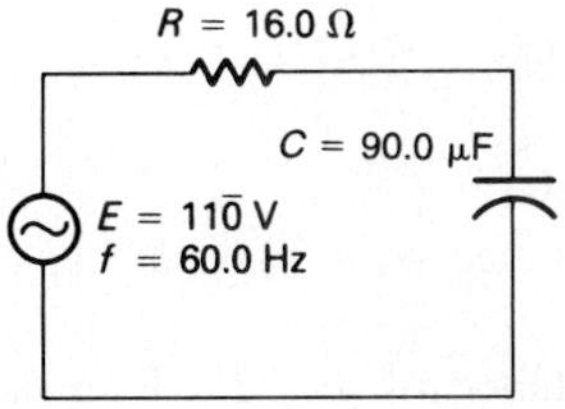

Data:

$$E = 11\bar{0}\text{ V}$$
$$f = 60.0\text{ Hz} = 60.0/\text{s}$$
$$R = 16.0\ \Omega$$
$$C = 90.0\ \mu\text{F} = 90.0 \times 10^{-6}\text{ F} = 9.00 \times 10^{-5}\text{ F}$$
$$Z = ?$$
$$I = ?$$

Basic Equations:

$$Z = \sqrt{R^2 + \left(\frac{1}{2\pi fC}\right)^2} \quad \text{and} \quad I = \frac{E}{Z}$$

Working Equations: Same

Substitutions:

First, find Z:

$$Z = \sqrt{(16.0\ \Omega)^2 + \left(\frac{1}{2\pi(60.0/\text{s})(9.00 \times 10^{-5}\text{ F})}\right)^2}$$
$$= \sqrt{256\ \Omega^2 + \left(\frac{1}{0.0339\text{ F/s}}\right)^2}$$
$$= \sqrt{256\ \Omega^2 + \left(29.5\frac{\cancel{\text{s}}}{\cancel{\text{F}}} \times \frac{1\ \Omega\ \cancel{\text{F}}}{1\ \cancel{\text{s}}}\right)^2}$$
$$= \sqrt{256\ \Omega^2 + 87\bar{0}\ \Omega^2}$$
$$= 33.6\ \Omega$$

Then use Z to find I:

$$I = \frac{E}{Z}$$
$$I = \frac{11\bar{0}\text{ V}}{33.6\ \Omega}$$
$$= 3.27\frac{\cancel{\text{V}}}{\cancel{\Omega}}\left(\frac{1\ \cancel{\Omega}\text{ A}}{1\ \cancel{\text{V}}}\right)$$
$$= 3.27\text{ A}$$

Then find the phase angle:

$$\tan\phi = \frac{X_C}{R}$$
$$\tan\phi = \frac{29.5\ \cancel{\Omega}}{16.0\ \cancel{\Omega}} = 1.84 \qquad (X_C \text{ from above})$$
$$\phi = 61.5°$$

PROBLEMS 18.6

1. For an ac circuit with $R = 1.00$ kΩ, $C = 1.00$ μF, $E = 10\bar{0}$ V, and $f = 10\bar{0}$ Hz:
   (a) find the impedance (in ohms).
   (b) find the phase angle.
   (c) find the current.
2. For an ac circuit with $R = 375$ Ω, $C = 5.00$ μF, $E = 20.0$ V, and $f =$ [illegible].00 kHz:
   (a) find the impedance (in ohms).
   (b) find the phase angle.
   (c) find the current.
3. For an ac circuit with $R = 4.80$ kΩ, $C = 45.0$ μF, $E = 15.0$ V, and $f = 1.75$ kHz:
   (a) find the impedance (in ohms).
   (b) find the phase angle.
   (c) find the current.
4. For an ac circuit with resistance 145 mΩ, capacitance 10.0 μF, frequency 72.5 kHz, and $E = 7.00$ mV:
   (a) find the impedance (in ohms).
   (b) find the phase angle.
   (c) find the current.
5. For an ac circuit with resistance 10.0 mΩ, capacitance 5.00 μF, frequency 10.0 kHz, and $E = 15.0$ mV:
   (a) find the impedance (in ohms).
   (b) find the phase angle.
   (c) find the current.

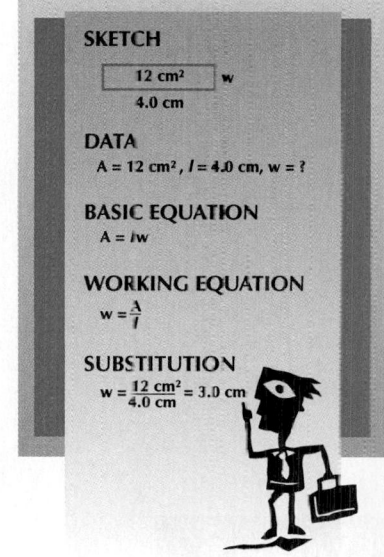

SKETCH

12 cm² w
4.0 cm

DATA

A = 12 cm², l = 4.0 cm, w = ?

BASIC EQUATION

$A = lw$

WORKING EQUATION

$w = \frac{A}{l}$

SUBSTITUTION

$w = \frac{12 \text{ cm}^2}{4.0 \text{ cm}} = 3.0 \text{ cm}$

18.7 Capacitance, Inductance, and Resistance in Series

Many circuits that are important in the design of electronic equipment contain all three types of circuit elements discussed in this chapter. The impedance of a circuit containing resistance, capacitance, and inductance in series can be found from the equation

$$Z = \sqrt{R^2 + (X_L - X_C)^2}$$

where

Z = impedance
R = resistance
X_L = inductive reactance
X_C = capacitive reactance

The vector diagram for this type of circuit is shown in Fig. 18.26. The phase angle is given by

$$\tan \phi = \frac{X_L - X_C}{R}$$

In a circuit containing R, L, and C components, the circuit is *inductive* if $X_L > X_C$ and the current lags behind the voltage. A circuit is *capacitive* if $X_C > X_L$, in which case the voltage lags behind the current. If $X_C = X_L$, the circuit is *resistive;* the voltage and current are *in phase*. If the circuit power is to be maximized, the voltage and current must be kept in phase. The current in this type of circuit is given by

$$I = \frac{E}{Z} = \frac{E}{\sqrt{R^2 + \left(2\pi fL - \frac{1}{2\pi fC}\right)^2}}$$

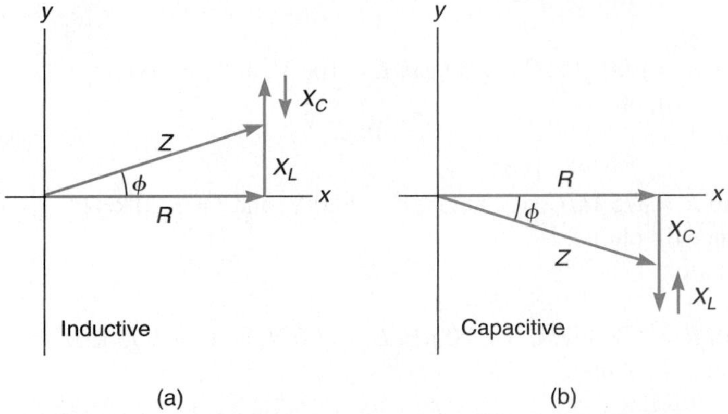

Figure 18.26 Vector diagrams: (a) inductive reactance (b) capacitive reactance

EXAMPLE

An ac circuit contains a $10\bar{0}$-Ω resistance, a 10.0-μF capacitor, and a 10.0-mH inductance in series with a 25.0-V, $20\bar{0}$-Hz source. Find the impedance and the current.

Sketch:

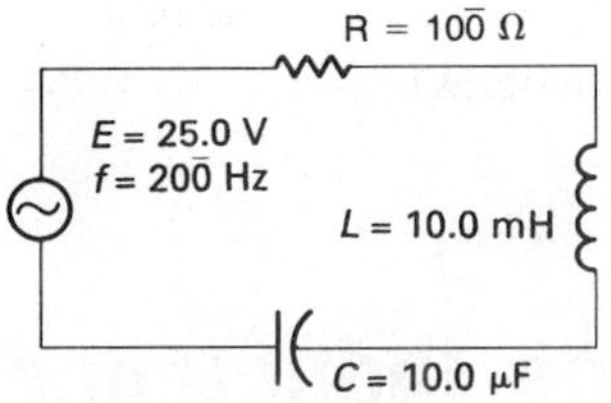

Data:

$$R = 10\bar{0}\ \Omega$$
$$C = 10.0\ \mu\text{F} = 10.0 \times 10^{-6}\ \text{F} = 1.00 \times 10^{-5}\ \text{F}$$
$$L = 10.0\ \text{mH} = 10.0 \times 10^{-3}\ \text{H} = 1.00 \times 10^{-2}\ \text{H}$$
$$E = 25.0\ \text{V}$$
$$f = 20\bar{0}\ \text{Hz} = 20\bar{0}/\text{s}$$
$$Z = ?$$
$$I = ?$$

Basic Equations:

$$X_L = 2\pi f L$$
$$X_C = \frac{1}{2\pi f C}$$
$$Z = \sqrt{R^2 + (X_L - X_C)^2}$$
$$I = \frac{E}{Z}$$

Working Equations: Same

Substitutions:

First find X_L:

$$X_L = 2\pi f L$$
$$X_L = 2\pi(20\bar{0}/\text{s})(1.00 \times 10^{-2}\ \text{H})$$
$$= 12.6 \frac{\cancel{\text{H}}}{\cancel{\text{s}}}\left(\frac{1\ \Omega\ \cancel{\text{s}}}{1\ \cancel{\text{H}}}\right)$$
$$= 12.6\ \Omega$$

Then find X_C:

$$X_C = \frac{1}{2\pi f C}$$

$$X_C = \frac{1}{2\pi(20\bar{0}/\text{s})(1.00 \times 10^{-5}\ \text{F})}$$

$$= 79.6\ \Omega$$

$$\frac{1}{\frac{\text{F}}{\text{s}}} = 1 \div \frac{\text{F}}{\text{s}} = 1 \times \frac{\text{s}}{\text{F}} = \frac{\cancel{\text{s}}}{\cancel{\text{F}}} \times \frac{1\ \Omega\,\cancel{\text{F}}}{1\ \cancel{\text{s}}} = \Omega$$

Then find the impedance:

$$Z = \sqrt{R^2 + (X_L - X_C)^2}$$

$$Z = \sqrt{(10\bar{0}\ \Omega)^2 + (12.6\ \Omega - 79.6\ \Omega)^2}$$

$$= \sqrt{(10\bar{0}\ \Omega)^2 + (-67.0\ \Omega)^2}$$

$$= \sqrt{1.00 \times 10^4\ \Omega^2 + 4490\ \Omega^2}$$

$$= 12\bar{0}\ \Omega$$

Then find the current:

$$I = \frac{E}{Z}$$

$$I = \frac{25.0\ \text{V}}{12\bar{0}\ \Omega}$$

$$= 0.208\frac{\cancel{\text{V}}}{\cancel{\Omega}}\left(\frac{1\ \text{A}\ \cancel{\Omega}}{1\ \cancel{\text{V}}}\right)$$

$$= 0.208\ \text{A}$$

PROBLEMS 18.7

Find the impedance and current in each ac circuit.

1. $R = 25.0\ \Omega$, $L = 50.0$ mH, $C = 50.0\ \mu$F, $f = 60.0$ Hz, $E = 5.00$ V
2. $R = 225\ \Omega$, $L = 10.0$ mH, $C = 0.200\ \mu$F, $f = 1.00$ kHz, $E = 15.0$ V
3. $R = 1.00$ kΩ, $L = 10.0$ mH, $C = 30.0$ mF, $f = 10.0$ kHz, $E = 15.0$ V
4. $R = 1.00$ kΩ, $L = 0.700$ H, $C = 30.0\ \mu$F, $f = 60.0$ Hz, $E = 8.00$ V
5. A circuit contains a $15\bar{0}$-Ω resistance, a 35.0-μF capacitor, and a 0.600-H inductance in series with a 6.00-V, $12\bar{0}$-Hz source. Find the impedance and the current.
6. A circuit contains a 225-Ω resistance, a 5.00-μF capacitor, and a 0.550-H inductance in series with a 7.50-V, 60.0-Hz source. Find the impedance and the current.
7. A circuit contains a 175-Ω resistance, a 4.50-μF capacitor, and a 0.735-H inductance in series with a 5.00-V, $10\bar{0}$-Hz source. Find the impedance and the current.
8. A circuit contains a 575-Ω resistance, a $10\bar{0}$-μF capacitor, and a 0.400-H inductance in series with a $10\bar{0}$-V, $60\bar{0}$-Hz source. Find the impedance and the current.
9. A circuit contains a $45\bar{0}$-Ω resistance, a 35.0-μF capacitor, and a 45.0-mH inductance in series with a 25.0-V, 1.00-kHz source. Find the impedance and the current.
10. A circuit contains a 375-Ω resistance, a $50\bar{0}$-μF capacitor, and a 0.500-H inductance in series with a 55.0-V, $50\bar{0}$-Hz source. Find the impedance and the current.

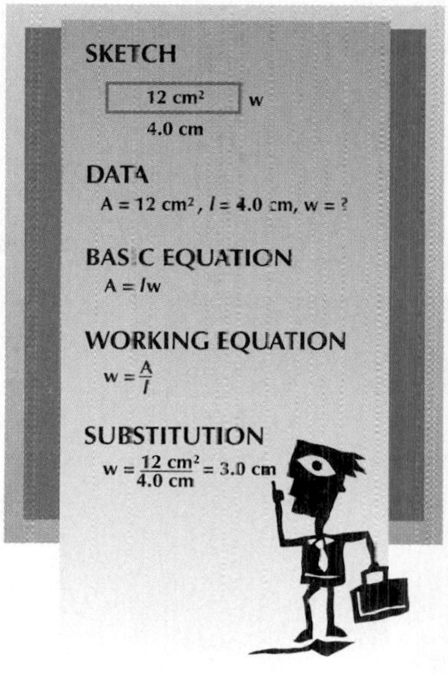

18.8 Resonance

The current in a circuit containing resistance, capacitance, and inductance is given by the equation

$$I = \frac{E}{\sqrt{R^2 + (X_L - X_C)^2}}$$

Courtesy of Pearson Educaton/PH College

Figure 18.27 In this variable air–dielectric capacitor, rotating the movable plates between the fixed plates changes the overlap area and thus the capacitance. Such capacitors were common in tuning circuits in older radios.

When the inductive reactance equals the capacitive reactance, they nullify each other, and the current is given by

$$I = \frac{E}{R}$$

which is its maximum possible value. When this condition exists, the circuit is in **resonance** with the applied voltage. To have resonance, it is essential for the circuit to have both capacitance and inductance.

Resonant circuits are used in radios and televisions. The frequency of a certain station is tuned in when a resonant circuit (antenna circuit) is adjusted to that frequency (Fig. 18.27). This is accomplished by changing the capacitance until the capacitive reactance equals the inductive reactance. The applied voltage is the radio signal picked up by the antenna. A variable capacitor has one set of plates (usually aluminum) mounted on a rotating shaft. The plates have air between them, and the capacitance varies with the amount of overlap of the plates as they are rotated between each other.

The resonant frequency occurs when $X_L = X_C$; power transfer is maximized at resonance as discussed in the preceding section. We find this frequency as follows. We start with

$$X_L = X_C$$

$$2\pi f L = \frac{1}{2\pi f C} \qquad \text{(By substitution)}$$

$$f^2 = \frac{1}{4\pi^2 LC} \qquad \text{(Solve for } f^2\text{.)}$$

$$f = \frac{1}{\sqrt{4\pi^2 LC}} \qquad \text{(Take the square root of both sides.)}$$

$$\boxed{f = \frac{1}{2\pi\sqrt{LC}}}$$

The circuit can be adjusted to any frequency by varying the capacitance or the inductance.

EXAMPLE

Find the resonant frequency of a circuit containing a 5.00-nF capacitor in series with a 2.60-μH inductor.

Sketch:

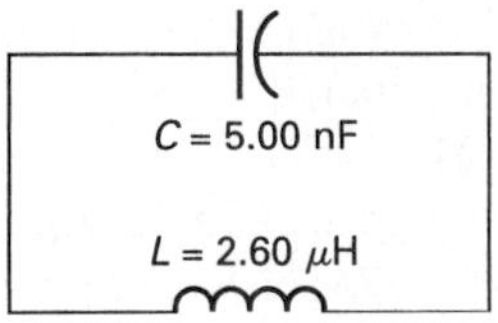

Data:

$$C = 5.00 \text{ nF} = 5.00 \times 10^{-9} \text{ F}$$
$$L = 2.60\ \mu\text{H} = 2.60 \times 10^{-6} \text{ H}$$
$$f = ?$$

Basic Equation:

$$f = \frac{1}{2\pi\sqrt{LC}}$$

Working Equation: Same

Substitution:

$$f = \frac{1}{2\pi\sqrt{(2.60 \times 10^{-6}\text{ H})(5.00 \times 10^{-9}\text{ F})}}$$

$$= \frac{1}{2\pi\sqrt{1.30 \times 10^{-14}\,\text{H F}\left(\frac{1\ \Omega\ \text{s}}{1\ \text{H}}\right)\left(\frac{1\ \text{s}}{1\ \text{F}\ \Omega}\right)}}$$

$$= \frac{1}{7.16 \times 10^{-7}\text{ s}}$$

$$= 1.40 \times 10^{6}\,\frac{\text{cycles}}{\text{s}} \quad \text{or} \quad 14\bar{0}0\,\frac{\text{kilocycles}}{\text{s}} \quad \text{or} \quad 14\bar{0}0\text{ kHz}$$

This frequency is in the AM radio band.

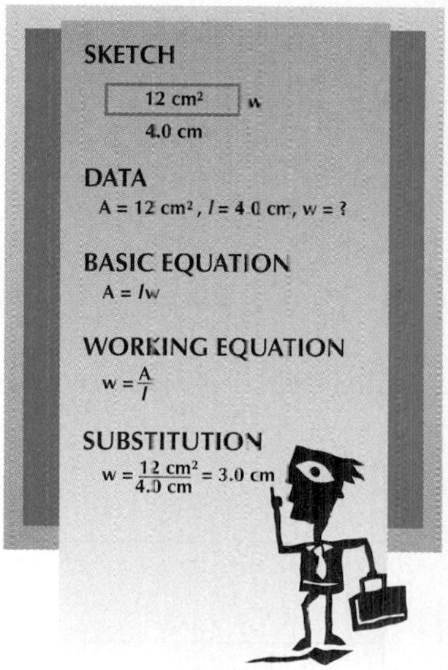

PROBLEMS 18.8

Find the resonant frequency in each ac circuit.

1. $L = 1.00\ \mu\text{H}$ and $C = 4.00\ \mu\text{F}$
2. $L = 2.00\ \mu\text{H}$ and $C = 35.0\ \mu\text{F}$
3. $L = 2.50\ \mu\text{H}$ and $C = 7.00\ \mu\text{F}$
4. $L = 2.65\ \mu\text{H}$ and $C = 35.0\ \mu\text{F}$
5. $L = 42.5\ \mu\text{H}$ and $C = 40.0\ \mu\text{F}$
6. Find the resonant frequency of a circuit containing a 25.0-μF capacitor in series with a 75.0-μH inductor.
7. Find the resonant frequency of a circuit containing a 33.0-μF capacitor in series with a 43.5-μH inductor.
8. Find the resonant frequency of a circuit containing a 10.0-μF capacitor in series with a 37.5-μH inductor.
9. Find the resonant frequency of a circuit containing an 8.00-μF capacitor in series with a 10$\bar{0}$-μH inductor.
10. Find the resonant frequency of a circuit containing a 3.75-μF capacitor in series with a 30$\bar{0}$-μH inductor.

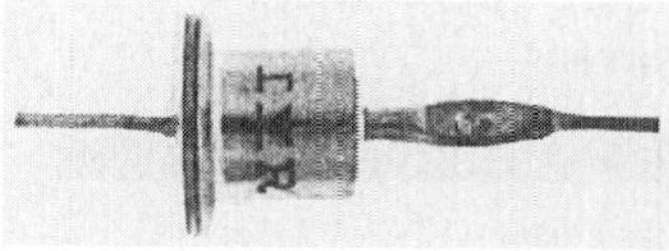

Semiconductor diode (enlarged)

Figure 18.28

18.9 Rectification and Amplification

It is often necessary to change ac into dc to provide dc for charging batteries or to power the integrated circuits (ICs) of computers and other electronic units. This process of changing ac to dc is called **rectification.** A device that accomplishes this is called a *diode.* Early diodes were constructed as vacuum tubes. Modern diodes are made of a semiconductor material and are usually less than 5 mm long (Fig. 18.28).

A **diode** allows current to flow through it in only one direction. It is similar to a turnstile that revolves in only one direction. People can pass the turnstile in one direction but are blocked when they attempt to pass in the opposite direction (Fig. 18.29). A diode allows electrons to pass in only one direction and not in the other (Fig. 18.30).

Thomas Edison found that when a wire filament was heated near a metal plate, an electron charge would begin to flow from the filament across space to the plate. This is called the *Edison effect* and was the beginning of the electronics industry. The electron emitter filament is called the *cathode*. The plate is called the *anode*. The entire device is called a *diode*. The diode allows electrons to flow only in one direction, which is the process of rectification.

A rectifier changes alternating current to direct current by allowing it to pass in only one direction (Fig. 18.31). Additional circuit devices can be added to the rectifier to smooth out the direct current so that it appears as shown in Fig. 18.32. Rectifiers are used in automobiles to change the alternating current produced by the alternator into direct current.

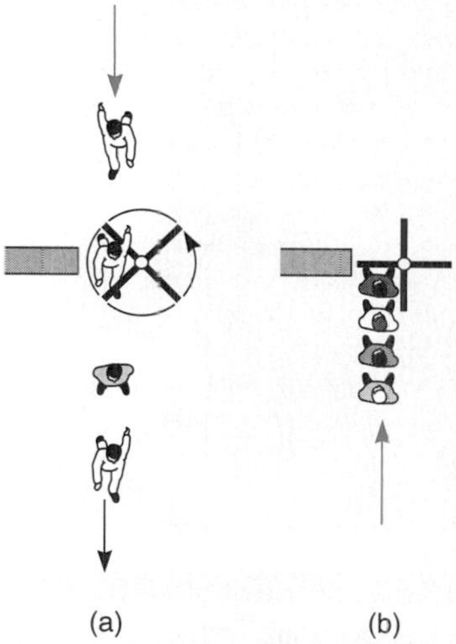

Figure 18.29 A one-way turnstile allowing people to move through in only one direction is like a diode.

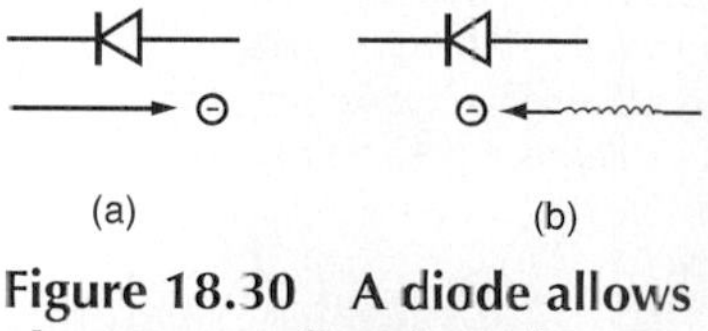

Figure 18.30 A diode allows electrons to flow in only one direction.

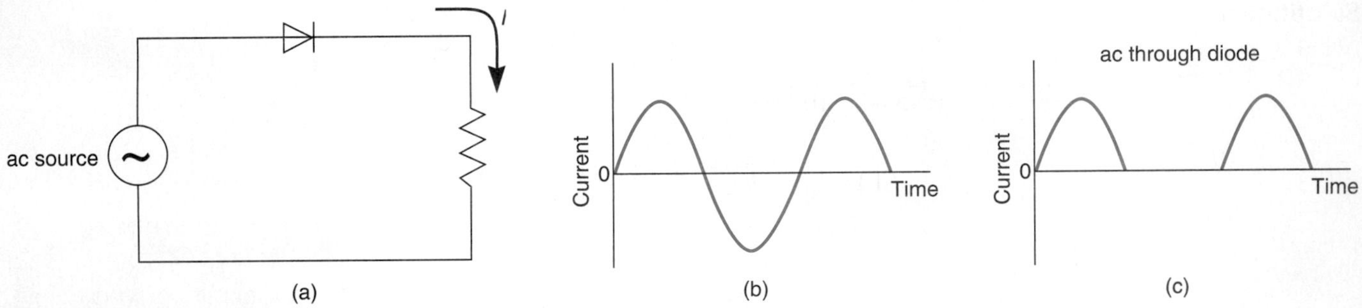

Figure 18.31 A diode can change ac to dc.

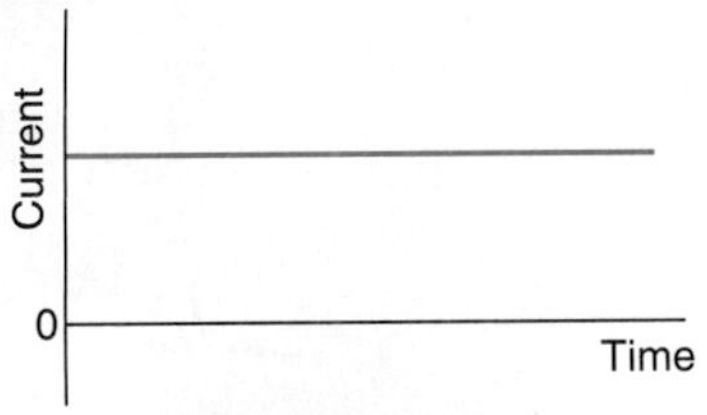

Figure 18.32 Rectified ac current

Thomas Edison (1847–1931),

inventor, was born in Milan, Ohio. He held patents to over 1000 inventions, which included the phonograph, the carbon transmitter that made Alexander Graham Bell's telephone practical, an incandescent lamp, a storage battery, and a system for widespread distribution of electric power from central generating stations. His only true scientific discovery in 1883, called "the Edison effect," for which he could see no commercial application, led to the vacuum tube, which became the basis for the radio. He spent his life inventing many of the devices that shaped the modern world.

John Bardeen (1908–1991),

physicist, was born in Madison, Wisconsin. He developed the transistor in 1947, for which he shared the Nobel Prize in Physics. He shared a second Nobel Prize in Physics in 1972 for the theory of superconductivity, which provided a breakthrough in electromagnetic design. Bardeen became the first person to win the Nobel Prize in Physics twice.

It is often necessary to increase the strength of an electronic signal. This is called **amplification.** Radios, stereos, and many other instruments contain one or more amplifier circuits. Early amplifiers utilized vacuum tubes together with other components. The transistor, developed by **John Bardeen,** has replaced the vacuum tube in most circuitry because of its smaller size and lesser power consumption. A *transistor* amplifier is typically composed of one or more transistors in addition to capacitors, resistors, and possibly inductors to provide the amplifier with the desired gain (amplification), frequency response, and power output. Amplifiers composed of individual transistors, resistors, and capacitors have been replaced for many applications by *integrated circuits* (ICs), which are only slightly larger in size than some transistors. An IC may contain millions of tiny transistors, diodes, resistors, and capacitors on a small chip of silicon less than 1 cm square. In addition to amplifying signals, ICs have been designed to serve as memory or logic units in computers or other applications. These ICs can be programmed to perform arithmetic operations, as in a calculator, or to perform control operations, as in most appliances.

18.10 Commercial Generator Power Output

The power output of the generator is the product of voltage and current. The ac generator converts mechanical energy to electric energy by performing three functions:

1. *Production of voltage:* electric pressure, which pushes the current through the loads
2. *Production of power current:* current converted into heat, light, and mechanical power
3. *Production of magnetizing current:* current transferred back and forth for magnetizing purposes in the generation of electric power, called *reactive kVA* (kilovolt-amperes)

Apparent Power and Reactive kVA

If the current and voltage are not in phase, the product of effective values of alternating current and voltage is the **apparent power** instead of the actual power. Apparent power is measured in kVA (kilovolt-amperes). **Actual power** is a measure of the actual power available to be converted into other forms of energy; it is the product of apparent power and the **power factor:**

$$\text{power factor} = \frac{\text{actual power}}{\text{apparent power}}$$

where the actual power is measured in kW, the apparent power is measured in kVA and is called *reactive kVA,* and the power factor is a unitless ratio less than 1. Note that 1 VA = 1 W.

Mathematically, the power factor is equal to the cosine of the angle by which the current lags behind (or in rare cases leads) the voltage. The power factor is really a correction factor that must be applied to determine actual power produced. The situation is very similar to finding the amount of work done when a force and the motion are not in the same direction.

EXAMPLE

Find the actual power produced by a generating system that produces 13,600 kVA with a power factor of 0.900.

Data:

apparent power = 13,600 kVA
power factor = 0.900
actual power = ?

Basic Equation:

$$\text{power factor} = \frac{\text{actual power}}{\text{apparent power}}$$

Working Equation:

actual power = (apparent power)(power factor)

Substitution:

$$\begin{aligned}\text{actual power} &= (13{,}600\text{ kVA})(0.900)\\ &= 12{,}200\text{ kVA}\\ &= 12{,}200\text{ kW}\end{aligned}$$

PROBLEMS 18.10

1. Find the actual power produced by a generating station that produces 12,600 kVA with a power factor of 0.850.
2. A generating station operates with a power factor of 0.910. What actual power is available on the transmission lines if the apparent power is 12,800 kVA?
3. Find the apparent power produced by a generating station whose actual power is $12\bar{0},000$ kW and whose power factor is 0.900.
4. Find the apparent power produced by a generating station whose actual power is $1{,}9\bar{0}0{,}000$ kW and whose power factor is 0.800.
5. A generating station operates with a power factor of 0.880. What actual power is available on the transmission lines if the apparent power is 11,500 kVA?
6. Find the apparent power produced by a generating station whose actual power is 2,350,000 kW and whose power factor is 0.850.
7. Find the actual power produced by a generating station that produces 23,800 kVA with a power factor of 0.810.
8. A generating station operates with a power factor of 0.840. What actual power is available on the transmission lines if the apparent power is 13,500 kVA?
9. Find the apparent power produced by a generating station whose actual power is $35\bar{0},000$ kW and whose power factor is 0.860.
10. Find the apparent power produced by a generating station whose actual power is 1,250,000 kW and whose power factor is 0.820.
11. Find the power factor of a generating station whose actual power is 55,800 kW and whose apparent power is 63,400 kVA.
12. Find the power factor of a generating station whose apparent power is 645,000 kVA and whose actual power is 587,000 kW.

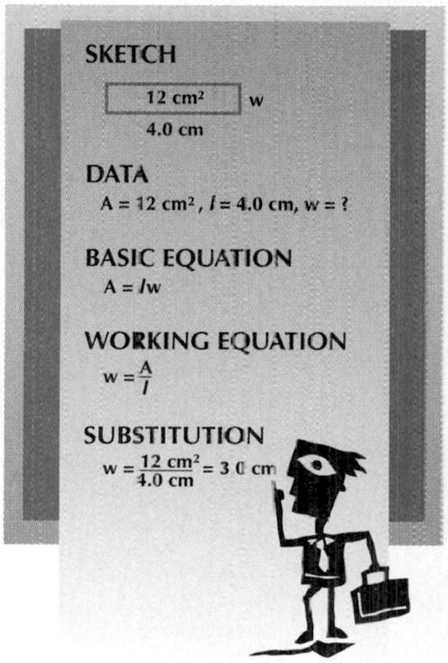

Glossary

Actual Power A measure of the actual power available to be converted into other forms of energy. (p. 532)
Alternating Current A current that flows in one direction in a conductor, changes direction, and then flows in the other direction. (p. 506)
Amplification The process of increasing the strength of an electronic signal. (p. 532)
Apparent Power The product of the effective values of alternating current and voltage. (p. 532)
Capacitance The ratio of the charge on either plate of a capacitor to the potential difference between the plates. (p. 523)
Capacitor A circuit component consisting of two parallel plates separated by a thin insulator used to build up and store charge. (p. 523)
Capacitive Reactance A measure of the opposition to ac current flow by a capacitor. (p. 524)
Diode A device that allows current to flow through it in only one direction. (p. 531)
Effective Value The number of amperes of alternating current that produces the same amount of heat in a resistance as an equal number of amperes of a steady direct current. (p. 508)
Impedance A measure of the total opposition to current flow in an ac circuit resulting from the effect of both the resistance and the inductive reactance on the circuit. (p. 520)
Inductance A measure of the tendency of a coil of wire to resist a change in the current because the magnetism produced by one part of the coil acts to oppose the change of current in other parts of the coil. (p. 517)
Inductive Reactance A measure of the opposition to ac current flow in an inductor. (p. 518)
Inductor A circuit component, such as a coil, in which an induced emf opposes any current change in the circuit. (p. 517)
Instantaneous Current The current at any instant of time. (p. 507)
Instantaneous Voltage The voltage at any instant of time. (p. 507)
Phase Angle The angle between the resistance and impedance vectors in a circuit. (p. 521)
Power Factor The ratio of the actual power to the apparent power. (p. 532)
Primary Coil The coil of a transformer that carries an alternating current and induces a current in the secondary coil. (p. 512)
Rectification The process of changing ac to dc. (p. 531)
Rectifier A device that changes ac to dc. (p. 510)
Resonance A condition in a circuit when the inductive reactance equals the capacitive reactance and they nullify each other. The current that flows in the circuit is then at its maximum value. (p. 530)
Secondary Coil The coil of a transformer in which a current is induced by the current in the primary coil. (p. 512)
Step-Down Transformer A transformer used to lower voltage; it has more turns in the primary coil. (p. 514)
Step-Up Transformer A transformer used to increase voltage; it has more turns in the secondary coil. (p.513)
Transformer A device composed of two coils (primary and secondary) and a magnetic core. Used to step up or step down a voltage. (p. 512)

Formulas

18.1 $i = I_{max} \sin \theta$

$e = E_{max} \sin \theta$

$I = 0.707\ I_{max}$ or $I_{max} = 1.41\ I$

$E = 0.707\ E_{max}$ or $E_{max} = 1.41\ E$

18.2 $P = I^2R = VI = \dfrac{V^2}{R}$

$$\frac{V_P}{V_S} = \frac{N_P}{N_S}$$

$$\frac{I_S}{I_P} = \frac{N_P}{N_S}$$

$$I_P V_P = I_S V_S$$

18.3 $X_L = 2\pi f L$

$$I = \frac{E}{X_L}$$

18.4 $I = \frac{E}{Z}$

$$Z = \sqrt{R^2 + X_L^2}$$

$$Z = \sqrt{R^2 + (2\pi f L)^2}$$

$$\tan\phi = \frac{X_L}{R}$$

18.5 $X_C = \frac{1}{2\pi f C}$

$$C = \frac{Q}{V}$$

18.6 $Z = \sqrt{R^2 + X_C^2}$

$$Z = \sqrt{R^2 + \left(\frac{1}{2\pi f C}\right)^2}$$

$$I = \frac{E}{Z}$$

$$\tan\phi = \frac{X_C}{R}$$

18.7 $Z = \sqrt{R^2 + (X_L - X_C)^2}$

$$\tan\phi = \frac{X_L - X_C}{R}$$

$$I = \frac{E}{Z} = \frac{E}{\sqrt{R^2\left(2\pi f L - \frac{1}{2\pi f C}\right)^2}}$$

18.8 $I = \frac{E}{\sqrt{R^2 + (X_L - X_C)^2}}$

$$f = \frac{1}{2\pi\sqrt{LC}}$$

18.10 $\text{power factor} = \frac{\text{actual power}}{\text{apparent power}}$

Review Questions

1. Which of the following describes alternating current electricity?
   - (a) It can be produced by rotating a loop of wire through a magnetic field.
   - (b) It flows in one direction for a period of time and then reverses direction.
   - (c) It goes through one cycle when it flows in one direction and then reverses direction.
   - (d) All of the above.

2. The voltage, e, and the current, i, in an alternating current circuit are in phase when
   (a) the peak values of both e and i occur at different times.
   (b) the peak values of e and i occur at the same time but their zero values do not.
   (c) the peak values and the zero values of e and i occur simultaneously.
   (d) none of the above.
3. Which of the following affect the voltage induced in the secondary coil of a transformer?
   (a) The current through the primary coil
   (b) The resistance of the primary coil
   (c) The number of turns in the primary coil
   (d) None of the above
4. Which of the following contribute to the energy loss in a transformer?
   (a) Resistance of the copper wires
   (b) Reversing the magnetic field in the core
   (c) Induced currents in the core
   (d) The emf in the outside circuit
   (e) All of the above
5. Explain the difference between maximum current and effective current.
6. Explain the difference between maximum voltage and instantaneous voltage.
7. Explain how power in an ac circuit is related to voltage and current.
8. Explain how power in an ac circuit is related to voltage and resistance.
9. If the number of turns in the secondary coil of a transformer is doubled, how does the output voltage change?
10. The unit of inductance is the ______.
11. Discuss the importance of inductive reactance.
12. How does the inductive reactance depend on frequency?
13. Does the current lead or lag the voltage in an inductive circuit?
14. Describe how energy is stored in a capacitor. How can the stored energy be used?
15. Does the current lead or lag the voltage in a capacitive circuit?
16. How does the reactance of a capacitor depend on frequency?
17. Discuss the condition that leads to resonance.
18. What is the function of a diode in a circuit?
19. Explain the difference between amplification and rectification.
20. Is the phase angle always constant in a circuit containing resistive, capacitive, and inductive elements?

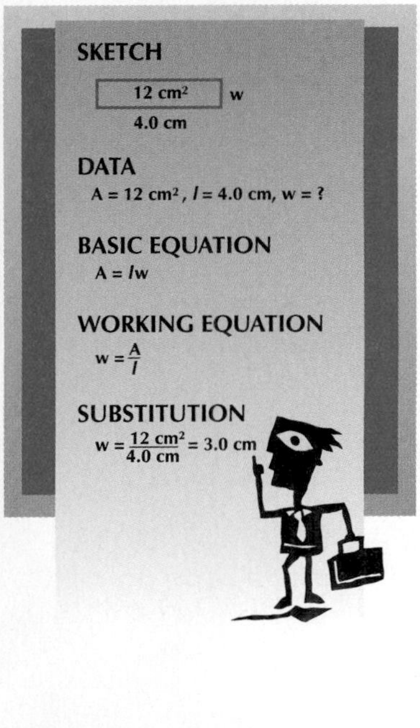

Review Problems

1. What is the maximum voltage in a circuit when the instantaneous value of the voltage is 95.4 V at $\theta = 62°$?
2. If the maximum ac voltage on a line is 185 V, what is the instantaneous voltage at $\theta = 41°$?
3. If the maximum ac voltage on a line is 175 V, what is the instantaneous voltage at $\theta = 23°$?
4. What is the effective value of an ac voltage whose maximum voltage is 135 V?
5. What is the maximum current in a circuit with a current rated at 6.35 A?
6. What power is developed by a device that draws 6.87 A and has a resistance of 15.4 Ω?
7. A heating element draws 4.50 A on a $11\bar{0}$-V line. What power is used in the element?
8. What power is used by a heater with resistance 22.3 Ω and that draws a current of 7.65 A?
9. A step-up transformer on a 115-V line provides a voltage of 2050 V. (a) If the primary coil has 75.0 turns, how many turns does the secondary coil have? (b) If there is a current of 4.55 A in the primary coil, what is the current in the secondary? (c) Find the power in the primary coil.

10. An inductance of 48.0 mH is connected in series with a lamp of resistance 23.0 Ω. This circuit is connected to a 115-V, 60.0-Hz power supply. (a) What is the current in the circuit? (b) What is the phase angle? (c) What is the voltage drop across the inductance?
11. A lamp of resistance 47.5 Ω is connected in series with an inductance of 43.2 mH. This circuit is connected to a 115-V, 60.0-Hz power supply. (a) What is the current in the circuit? (b) What is the phase angle? (c) What is the voltage drop across the resistance?
12. What current will flow in a 60.0-Hz ac series circuit that includes a $11\bar{0}$-V source, a resistor of 19.5 Ω, and a capacitor of 57.4 μF? What is the phase angle? What is the voltage across the resistor? Across the capacitor?
13. A resistor of 21.6 Ω and a capacitor of 38.5 μF are connected in series with a 60.0-Hz ac source with a voltage of $11\bar{0}$ V. What is the current in the circuit? What is the voltage across the capacitor? What is the phase angle?
14. A circuit contains a 175-Ω resistance, a 25.0-μF capacitor, and a 62.0-mH inductance in series with a $11\bar{0}$-V, 60.0-Hz source. Find the impedance and the current.
15. A circuit contains a 115-Ω resistance, a 35.0-μF capacitor, and a 65.0-mH inductance in series with a $11\bar{0}$-V, 60.0-Hz source. Find the impedance and the current.
16. Find the resonant frequency of a circuit containing a 7.50-μF capacitor in series with a 3.70-μH inductance, a 633-Ω resistor, and a $11\bar{0}$-V, 60.0-Hz source. Find the impedance and the current.
17. Find the resonant frequency of a circuit containing a 4.70-μF capacitor in series with a 4.50-μH inductance, a 25.0-Ω resistor, and a $11\bar{0}$-V, 60.0-Hz source. Find the impedance and the current.
18. Find the apparent power produced by a generating station whose actual power is $2{,}9\bar{0}0{,}000$ kW and whose power factor is 0.850.

APPLIED CONCEPTS

1. A microwave oven is designed to draw 11.8 A of current when connected to a $11\bar{0}$-V ac circuit. (a) What is the average power of the microwave? (b) What is the maximum current drawn by the microwave? (c) Provide a rationale for the current's value 180° after the beginning of a cycle.
2. Before converting alternating current to direct current for an electronic video game, a transformer first must reduce the voltage from $12\bar{0}$ V to 9.00 V. (a) If the primary coil has 307 turns, how many turns are in the secondary coil? (b) What is the power of each coil in the transformer? (c) Explain the relationship between the power for the primary coil and the power for the secondary coil.
3. A neighborhood requires 8.50×10^4 W of power from the local substation. (a) What is the current in the wire if the electricity is delivered at $22\bar{0}$ V? (b) What is the current in the wire if the electricity is delivered at $60\bar{0}0$ V? (c) If the resistance of the power line is 0.250 Ω, how much electric power will be converted into heat throughout the $22\bar{0}$-V and $60\bar{0}0$-V power lines? (d) Which power line voltage is the more efficient for transmitting electricity?
4. A 65.5-V, 60.0-Hz ac generator is connected to a radio circuit. (a) If the inductor coil has an inductive reactance of 125 Ω, find the inductance of the coil. (b) What is the effective current in the circuit? (c) Find the power of the circuit.
5. An AM radio tuner circuit has an inductance of 275 mH. (a) What is the capacitance of the variable capacitor when tuned to $88\bar{0}$ kHz? (b) What is the capacitance when tuned to 1010 kHz? (c) Are the capacitor plates larger or smaller when the radio is tuned to a higher-frequency station?

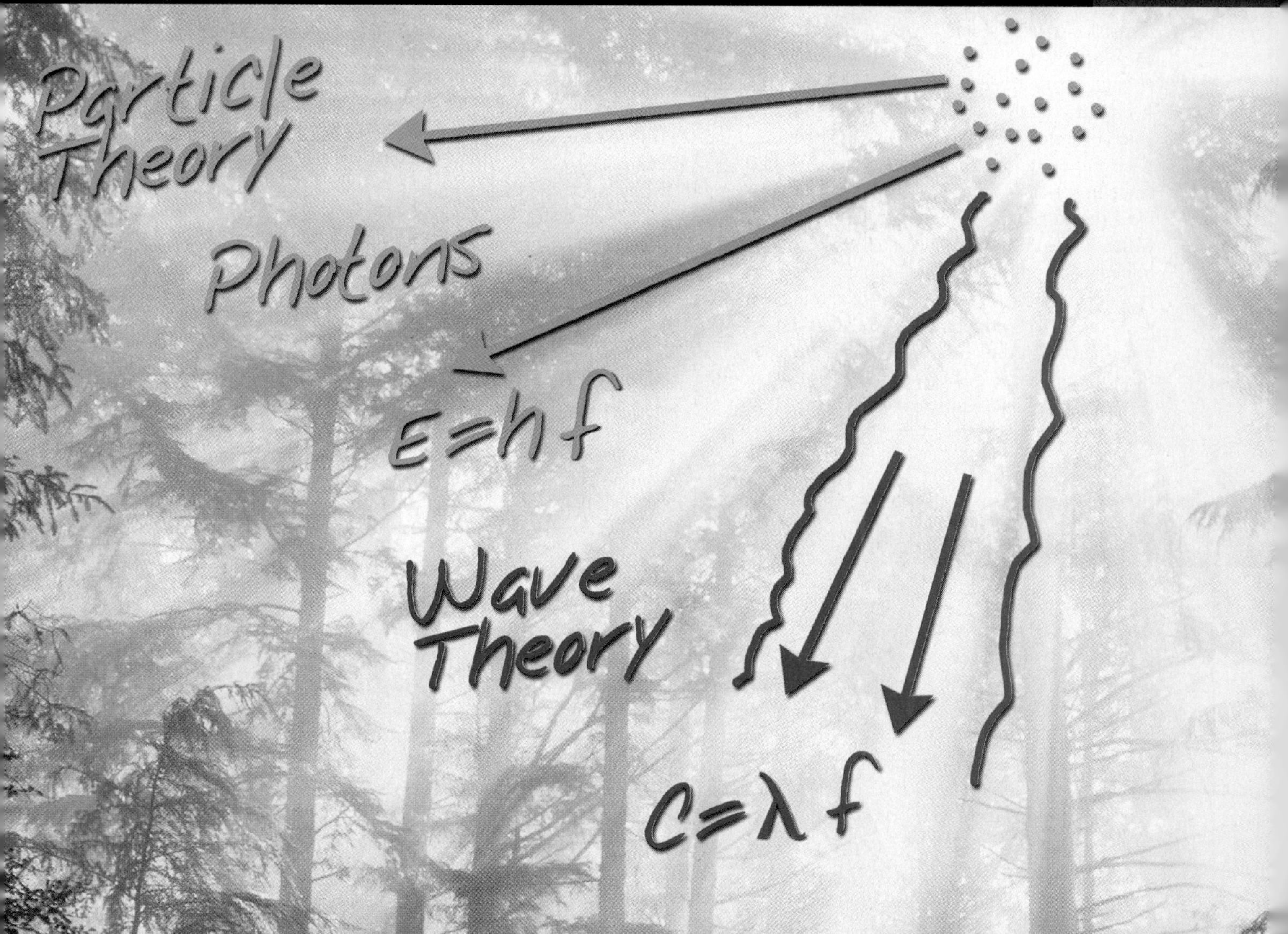
Particle Theory
Photons
E=hf
Wave Theory
c=λf

LIGHT

For centuries scientists have sought to explain the nature of light—why it is reflected and why it is refracted. Even the speed of light has been the subject of study since the seventeenth century.

Light seems to exhibit some characteristics of both particles and waves. The study of the measurement of light is photometry. We begin by examining the nature of light.

Objectives

The major goals of this chapter are to enable you to:

1. Describe the nature of light.
2. Solve problems involving the speed of light.
3. Contrast the wave and particle characteristics of light.
4. Apply principles of photometry to technical problems.

19.1 Nature of Light

Light may be defined as radiant energy that can be seen by the human eye. The search for an explanation for the nature of light has been going on for many centuries. A number of famous scientists, including Isaac Newton, Christiaan Huygens, Albert Einstein, and Louis de Broglie, made major contributions to the current theory of light and its interaction with matter.

In the seventeenth century many of the foundations for modern scientific theories were put forward. Two conflicting theories for the nature of light were proposed. The experimental observations that had to be explained were the following:

1. *Straight-line propagation of light.* An application of this property of light is found in survey work, in which sight lines are commonly used (Fig. 19.1).
2. ***Reflection*** *of light at the boundary between two different media* [Fig. 19.2(a)]. Examples are the reflection of light by a mirror and by the surface of still water.
3. ***Refraction,*** *or bending, of light as it passes through the boundary between two media, such as air and water* [Fig. 19.2(b)]. This bending of light makes objects under water appear to be closer to the surface and farther away from the observer than they really are when viewed from above the surface. It also makes a straight object partially submerged in water appear bent at the surface.

Figure 19.1 Surveyor using the straight-line travel characteristic of light

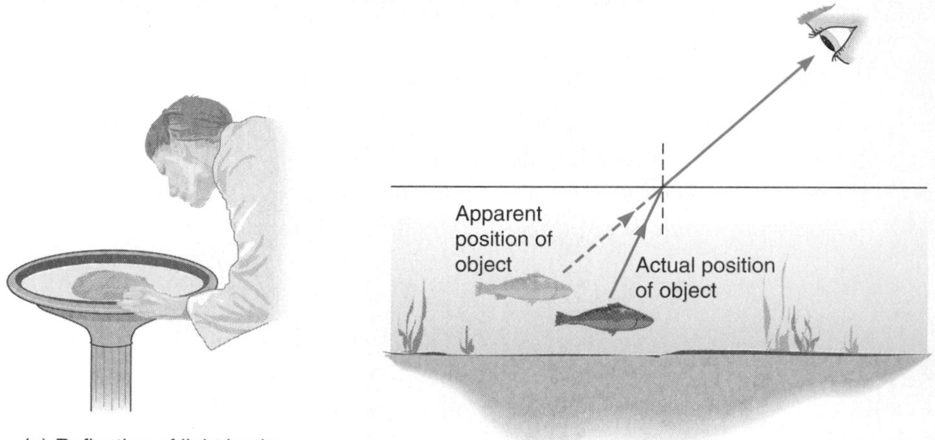

(a) Reflection of light by the surface of the water.

(b) Refraction of light as it passes between media.

Figure 19.2 Reflection and refraction

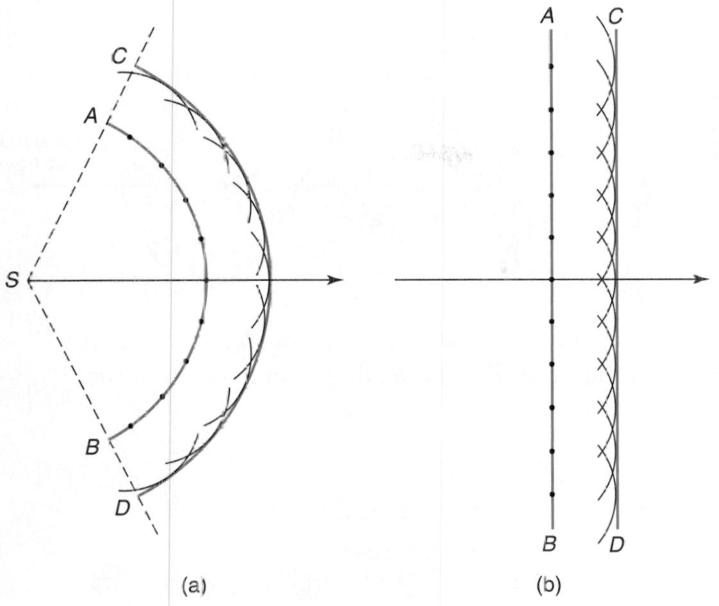

Figure 19.3 Huygens' principle is used to determine the future position of both (a) an advancing spherical wave front *CD* and (b) a plane wave front *CD* when *AB* is given.

The two conflicting theories referred to above are the wave theory and the particle theory of light. **Christiaan Huygens** proposed the **wave theory** of light, according to which light consists of waves traveling out from light sources like water waves traveling out from the point at which a stone is dropped into still water. The waves continue long after the stone drops to the bottom and thus the succeeding waves cannot be caused by the stone's continuing activity. Huygens developed a geometric model as shown in Fig. 19.3 to explain how a wave front advances and reasoned the following important concept named after him:

Christiaan Huygens (1629–1695), *physicist and astronomer, was born in The Netherlands. He made the first pendulum clock, proposed the wave theory of light, discovered the polarization of light, and discovered the ring and fourth satellite of Saturn in 1655.*

Huygens' principle: *Each point on a wave front can be regarded as a new source of small wavelets, which form succeeding waves that spread out uniformly in the forward direction at the same speed.*

The new wave front is the result of all the wavelets. A similar type of wave behavior is sound propagation out from a source of sound. All three types of waves mentioned here—light, water, and sound—travel in straight lines. They also reflect off surfaces or boundaries between media. Refraction of these waves is also observed.

Isaac Newton proposed the **particle theory** of light as an alternative explanation of the experimental observations. Newton thought light was made up of streams of particles. These particles of light, which Newton referred to as *corpuscles,* he felt behaved in a manner similar to particles of matter, which travel in a straight line if the net force acting on them is zero. Particles of matter also rebound or reflect off surfaces. An example of this is a rubber ball bouncing off a wall. A change in the velocity of a particle could produce a change in direction. Newton's particle theory persisted until the early nineteenth century, when diffraction and interference of light were observed. Since these properties could only be explained by the wave theory, the particle theory fell out of favor.

An **electromagnetic wave** consists of two perpendicular transverse waves with one component of the wave being a vibrating electric field and the other being a corresponding magnetic field; the electromagnetic wave moves in a direction perpendicular to both electric and magnetic field components as shown in Fig. 19.4. All such waves travel at the same speed in a vacuum (3.00×10^8 m/s or 186,000 mi/h) but differ in their frequencies and wavelengths. Beginning in the late nineteenth century, experiments confirmed the **electromagnetic theory** and that visible light is only a small portion of the electromagnetic spectrum (Fig. 19.5). Note that as the frequency increases to the right in Fig. 19.5, the wavelength decreases. Electromagnetic waves differ from other transverse and longitudinal waves in that they do not need a medium such as air, water, or a solid through which to travel. As we know, visible light, gamma rays, and X rays travel through space to reach the

Albert Einstein (1879–1955),

theoretical physicist, was born in Germany. His theories of relativity revolutionized modern thought. He won the Nobel Prize in Physics for his work on the photoelectric effect in 1921.

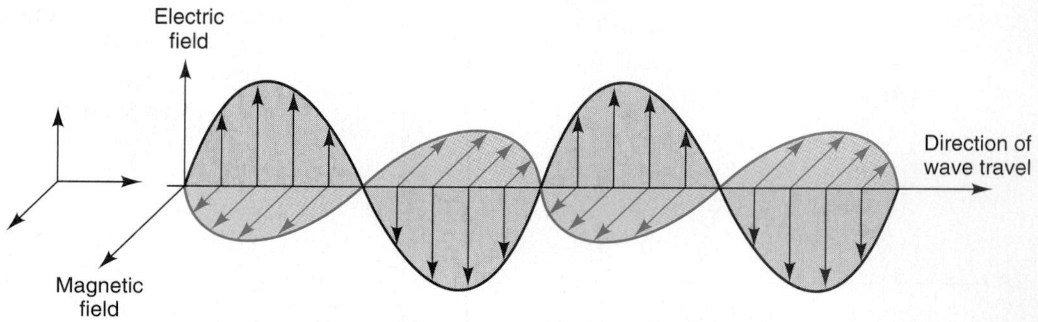

Figure 19.4 The electric and magnetic field components of an electromagnetic wave are perpendicular to each other as well as to the direction of travel of the electromagetic wave.

Guglielmo Marconi (1874–1937),

physicist and inventor, was born in Italy. He experimented on ways to convert electromagnetic waves into electricity. His first successful experiments in wireless telegraphy were made in Bologna in 1895. He formed the Marconi Telegraph Company in London in 1899. In 1898 he transmitted signals across the English Channel and across the Atlantic Ocean in 1901. He later developed short-wave radio equipment and established a worldwide radio telegraph network for the British government. He shared the 1909 Nobel Prize in Physics.

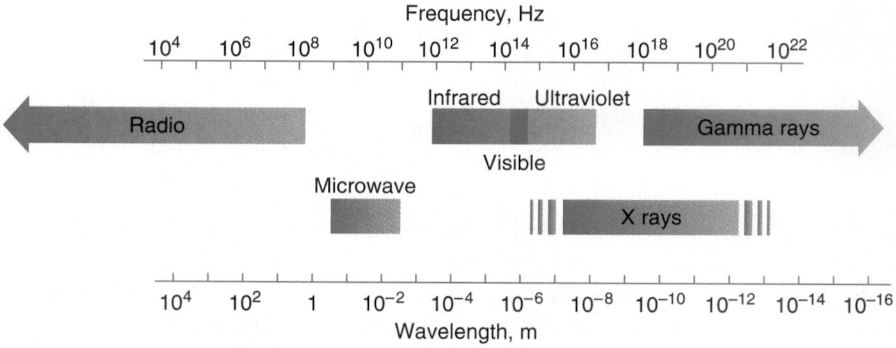

Figure 19.5 Electromagnetic spectrum

earth. Electromagnetic waves are produced by accelerating electric charges, which create an electric field that in turn creates a corresponding magnetic field.

The electromagnetic theory, though, is still not a complete explanation of the nature of light. Early in the twentieth century, the **photoelectric effect** was discovered: the emission (or giving off) of electrons by a substance when struck by electromagnetic radiation. **Albert Einstein** received a Nobel Prize for his work on the photoelectric effect and in the emerging field of theoretical physics. The energy of the emitted electron is related to the energy (frequency) of the incoming light in the same way that the energy of a billiard ball is related to the energy of the cue ball that strikes it. The particle theory of light explains the photoelectric effect, while the wave theory cannot. The electromagnetic wave theory fails to fully explain the nature of these emissions of electrons. **Guglielmo Marconi** experimented on ways to convert electromagnetic waves into electricity.

Max Planck (1858–1947),

theoretical physicist, was born in Germany. His work on the law of thermodynamics and black body radiation led him to question some principles of classical Newtonian physics and introduce the quantum theory in 1900. He received the Nobel Prize in Physics in 1918 for his work in quantum theory.

Later work by **Max Planck** and Albert Einstein developed the idea that light was energy radiated at the speed of light in the form of wave packets of energy, which were called **photons.** Further work showed that in some circumstances light acted like a stream of particles. Based on the work of Planck and Einstein, the **quantum theory** was developed. This theory states that energy, including electromagnetic radiation, is not absorbed or radiated continuously but is radiated in multiples of definite and discontinuous units.

Light therefore appears to have at least a dual character, having properties of both waves and particles (Fig. 19.6). This dual character may be shown by considering how energy may be transported from one point to another.

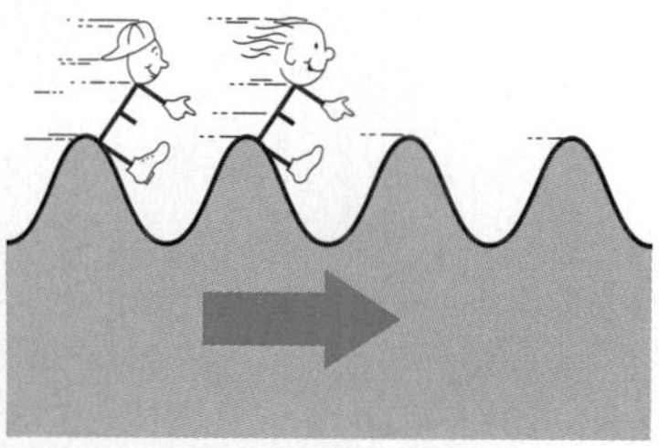

Figure 19.6 Light has properties of both particles and waves.

The first method is to transport particles of matter that carry energy with them. Examples of this method are the conduction of electrons in a wire, the shooting of a bullet, the sending of natural gas through pipelines, and the transport of gasoline.

The second method is the propagation of a wave disturbance through the medium between two points. Sound and water waves are examples of this method of energy transport.

Light is unusual in that it appears to combine characteristics of each method. When light is traveling through a medium, it appears to behave like a wave, with the following characteristics: (1) reflection at the surface of the medium, (2) refraction when passing

TRY THIS ACTIVITY

Flickering Lights

Light emitting diodes (LEDs) are often used to demonstrate the cyclical motion of electrons in an alternating current circuit. Find an extension cord with an LED indicator light in the receptacle end of the cord. Typically the LED is a yellow, green, or red light. After the extension cord is plugged into a 110-V ac outlet, the LED should appear as a constant light. Take the end of the extension cord and spin the light and 2 ft of cord around in circles. Observe the light patterns and explain why the light from the LED no longer appears constant.

from one medium to another, (3) interference (cancellation) when two waves are properly superimposed, and (4) diffraction (bending) when the waves pass the corners of an obstacle. When light interacts with matter, such as when it is absorbed or emitted, it behaves as if it were a massless particle.

19.2 The Speed of Light

One of the most important measured quantities in physics is the **speed of light,** the speed at which light and other forms of electromagnetic radiation travel. At first, the travel of light was assumed to be instantaneous. **Galileo** probably first suggested that time was required for light to travel from place to place. **Ole Roemer** made the first estimate of the speed of light from his study of the time of eclipse of one of the moons of the planet Jupiter as viewed from different places of the orbit of the earth in 1675. Since then, laboratory methods for measuring the speed of light have been developed, most notably by **Albert Michelson.** His measurements were made by using rotating mirrors to reflect a beam of light to a mirror on a mountain 22 miles away and back.

Figure 19.7 shows the mirror arrangement used by Michelson to measure the speed of light. Light only enters the eyepiece of the telescope when the octagonal mirror is at rest or when it is spinning at exactly the right speed. By measuring the time for 1/8 of a rotation of the mirror and the distance to the stationary mirror miles away, Michelson could calculate the speed of light with great accuracy. He also developed an instrument called an *interferometer,* with which in 1920 he made the first accurate measurement of a star's diameter.

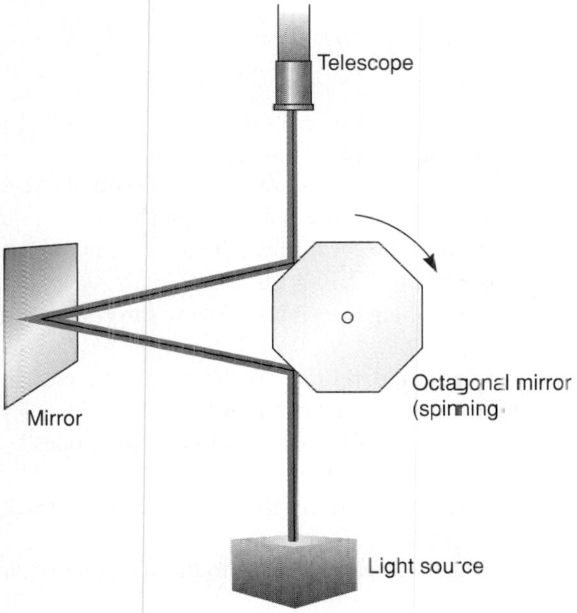

Figure 19.7 Michelson's method for determining the speed of light

Galileo Galilei (1564–1642), *mathematician and astronomer, was born in Italy. He improved the telescope, was the first to use it to study the stars, and supported the Copernican theory that the planets revolved about the sun. His discoveries include the law of uniformly accelerated motion toward earth, the parabolic path of projectiles, and the law that all bodies have weight.*

Ole Roemer (1644–1710), *astronomer, was born in Denmark. He used eclipses of Jupiter's satellites to make the first determination of the speed of light in 1675.*

Albert A. Michelson (1852–1931), *physicist, was born in Germany, but his family moved to the United States when he was 2. He was among the first to accurately measure the speed of light. Michelson received the Nobel Prize in Physics in 1907 for his work and was the first American to receive the Prize.*

Modern laboratory methods have been used to measure accurately the speed of light, which is now defined as 299,792,458 m/s. This is usually rounded to 3.00×10^8 m/s or 186,000 mi/s.

As stated earlier, light is one form of a class of radiation called *electromagnetic radiation,* which also includes radio and television waves, infrared, gamma rays, and X rays. A chart of the entire electromagnetic spectrum is shown in Fig. 19.5.

The distance traveled by any form of electromagnetic radiation can be found by substituting the speed of light c into the equation $s = vt$ as follows:

$$\boxed{s = ct}$$

where

s = distance
c = speed of light, 3.00×10^8 m/s or 186,000 mi/s
t = time

EXAMPLE

Find the distance (in mi) traveled by an X ray in 0.100 s.

Data:

$$c = 186{,}000 \text{ mi/s}$$
$$t = 0.100 \text{ s}$$
$$s = ?$$

Basic Equation:

$$s = ct$$

Working Equation: Same

Substitution:

$$s = (186{,}000 \text{ mi/}\cancel{\text{s}})(0.100\ \cancel{\text{s}})$$
$$= 18{,}600 \text{ mi}$$

Very large distances, such as those between stars, cannot be conveniently expressed in common distance units. Astronomers therefore use the unit light-year to measure such distances. A **light-year** is the distance traveled by light in one earth year, so 1 light-year equals 9.45×10^{15} m or 5.87×10^{12} mi.

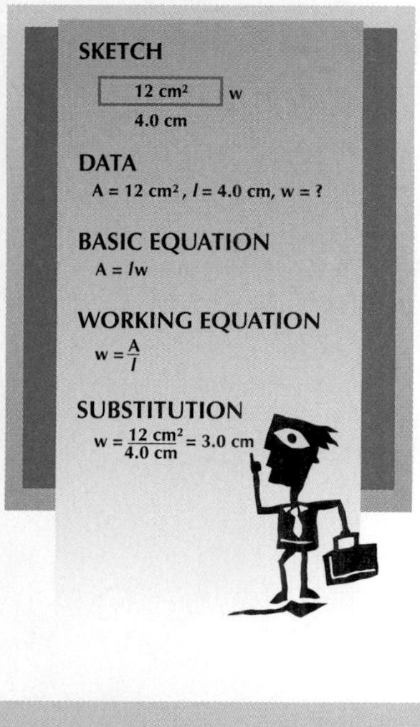

PROBLEMS 19.2

1. Find the distance (in metres) traveled by a radio wave in 5.00 s.
2. Find the distance (in metres) traveled by a light wave in 6.40 s.
3. A television signal is sent to a communications satellite that is $20,\bar{0}00$ mi above a relay station. How long does it take for the signal to reach the satellite?
4. How long does it take for a radio signal from the earth to reach an astronaut on the moon? The distance from the earth to the moon is 2.40×10^5 mi.
5. The sun is 9.30×10^7 mi from the earth. How long does it take light to travel from the sun to the earth?
6. A radar wave is bounced off an airplane and returns to the radar receiver in 2.50×10^{-5} s. How far (in km) is the airplane from the radar receiver?
7. How long does it take for a radio wave to travel $30\bar{0}0$ mi across the United States?
8. How long does it take for a flash of light to travel $10\bar{0}$ m?
9. How long does it take for a police radar beam to travel to a truck and back if the truck is 115 m from the radar unit?
10. How far away (in km) is an airplane if the radar wave returns to the scanning radar unit in 1.24×10^{-3} s?

11. How long does it take for light to reach the earth from Mars when the separation of the two planets is at its smallest? The earth's orbital radius is 143 million kilometres. The orbital radius of Mars is 218 million kilometres. How long does it take when the separation is at its maximum? Assume the planetary orbits are circular. Also make the (nonphysical) assumption that the sun is transparent to the transmission of light between the planets.
12. If it takes 4.31 years for light to reach the earth from Alpha Centauri, the closest star to the earth other than the sun, what is the distance (in miles) to the next nearest neighbor (Barnard's Star), which is 25% farther away?
13. How long does it take light to reach the earth from Jupiter when the separation of the two planets is (a) at its smallest and (b) at its largest? The earth's orbital radius is 143 million kilometres. The orbital radius of Jupiter is 725 million kilometres. Assume the planetary orbits are circular. Also make the (nonphysical) assumption that the sun is transparent to the transmission of light between the planets.
14. Preparing for reentry, astronauts use radar to determine the distance back to the earth. What is their altitude if it takes 0.330 s for the radar wave to travel to the earth and return?
15. The distance to the moon can be calculated by reflecting a ray of light off a mirror left by astronauts. The light travels to the mirror and back in 2.56 s. Find the distance to the moon.
16. Light from the sun travels 1.50×10^8 km to reach the earth. How long does its journey take in minutes?

19.3 Light as a Wave

Light and the other forms of electromagnetic radiation are composed of oscillations in the electric and magnetic fields that exist in space. These oscillations are set up by rapid movement of charged particles such as electrons in radio antennas and electrons in a hot object such as a light bulb filament. A wave is characterized by its **wavelength,** the distance between two successive corresponding points on the wave (Fig. 19.8). This distance is denoted by the Greek lowercase letter lambda, λ. The wavelength of visible light ranges from about 4×10^{-7} m to 7.6×10^{-7} m. The human eye perceives light in the visible spectrum as one or more colors depending upon the frequency or wavelength of the light hitting the retina of the eye. The longest visible wavelengths ($\lambda = \sim 7.5 \times 10^{-7}$ m), which are also the lowest frequencies, are perceived as red. The shortest visible wavelengths ($\lambda = \sim 4 \times 10^{-7}$ m), which are the highest frequencies, are perceived as blue. The wavelengths of other electromagnetic radiations are given in Fig. 19.5.

Another characteristic of waves is the frequency, f. **Frequency** is the number of complete vibrations or cycles per second of a wave. The measurement unit of frequency (cycles/s) is named the hertz (Hz) after Heinrich Hertz, a leader in the study of electromagnetic theory. (1 Hz = 1 cycle per second = 1/s.) Since a "cycle" has no units, it does not appear in the hertz unit.

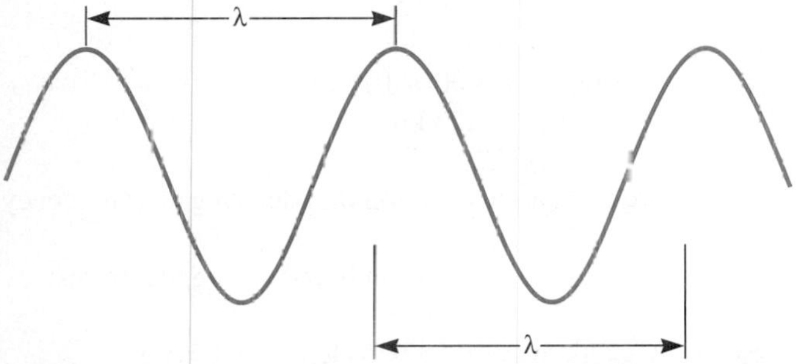

Figure 19.8 The wavelength of a repeating wave is the distance between two successive corresponding points.

The following basic relationship exists for all electromagnetic waves:

$$c = \lambda f$$

where

c = speed of light, 3.00×10^8 m/s or 186,000 mi/s
λ = wavelength
f = frequency

EXAMPLE

Find the frequency of a light wave with a wavelength of 5.00×10^{-7} m.

Data:

$$\lambda = 5.00 \times 10^{-7} \text{ m}$$
$$c = 3.00 \times 10^8 \text{ m/s}$$
$$f = ?$$

Basic Equation:

$$c = \lambda f$$

Working Equation:

$$f = \frac{c}{\lambda}$$

Substitution:

$$f = \frac{3.00 \times 10^8 \text{ m/s}}{5.00 \times 10^{-7} \text{ m}}$$
$$= 6.00 \times 10^{14} \text{ Hz} \qquad \text{(or cycles/s)}$$

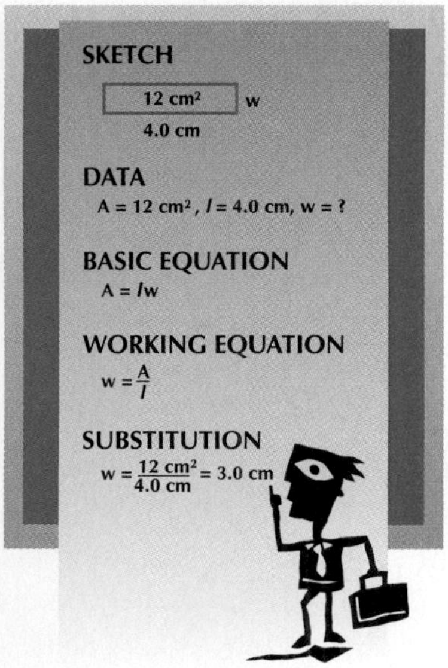

PROBLEMS 19.3

1. $c = 3.00 \times 10^8$ m/s
   $\lambda = 4.55 \times 10^{-5}$ m
   $f = ?$
2. $c = 3.00 \times 10^8$ m/s
   $\lambda = 9.70 \times 10^{-10}$ m
   $f = ?$
3. $c = 3.00 \times 10^8$ m/s
   $f = 9.70 \times 10^{11}$ Hz
   $\lambda = ?$
4. $c = 3.00 \times 10^8$ m/s
   $f = 24.2$ MHz
   $\lambda = ?$
5. $c = 3.00 \times 10^8$ m/s
   $f = 45.6$ MHz
   $\lambda = ?$
6. $c = 3.00 \times 10^8$ m/s
   $f = 415$ Hz
   $\lambda = ?$
7. $c = 3.00 \times 10^8$ m/s
   $\lambda = 6.59 \times 10^{12}$ m
   $f = ?$
8. $c = 3.00 \times 10^8$ m/s
   $\lambda = 9.23$ km
   $f = ?$
9. Find the wavelength of a radio wave from an AM station broadcasting at a frequency of $14\bar{0}0$ kHz.
10. Find the wavelength of a radio wave from an FM station broadcasting at a frequency of $10\bar{0}$ MHz.
11. Find the frequency of an electromagnetic wave if its wavelength is 85.5 m.
12. Find the frequency of an electromagnetic wave if its wavelength is 3.25×10^{-8} m.

13. Find the frequency of blue light.
14. Find the frequency of red light.
15. Find the frequency of yellow light if its wavelength is midway between those of red and blue light.
16. An AM radio station broadcasts a signal with a wavelength of 237 m. Find its frequency in kHz.

19.4 Light as a Particle

As mentioned earlier, light sometimes behaves as if it were a particle. These particles are called *photons,* and each carries a portion of the energy of the wave. This energy is given by

$$E = hf$$

where

f = frequency
$h = 6.626 \times 10^{-34}$ J s   (*Planck's constant*)
E = energy

When Planck introduced his concept of discrete values of energy proportional to the vibration frequencies of electrons, it was initially regarded by many as only a theoretical aid. However, this concept actually triggered the beginning of quantum physics. The only unknown in Planck's equation above was h, which was evaluated to make the mathematical expression fit the experimental data. This value, $h = 6.626 \times 10^{-34}$ J s is known as **Planck's constant.**

EXAMPLE

What is the energy of a photon of electromagnetic radiation with frequency 5.00×10^{12} Hz?

Data:

$h = 6.626 \times 10^{-34}$ J s
$f = 5.00 \times 10^{12}$ Hz $= 5.00 \times 10^{12}$/s
$E = ?$

Basic Equation:

$$E = hf$$

Working Equation: Same

Substitution:

$$E = (6.626 \times 10^{-34} \text{ J s})(5.00 \times 10^{12}/\text{s})$$
$$= 3.31 \times 10^{-21} \text{ J}$$

PHYSICS CONNECTIONS

Light Amplification by Stimulated Emission of Radiation

Light amplification by stimulated emission of radiation is more commonly known by its acronym, LASER. A **laser** is a light source that produces a narrow beam of light with high intensity and is a good example of how light behaves both as a particle and as a wave (Fig. 19.9). A burst of energy from the laser's flash tube causes an increase in the energy level of some of the electrons in the tube. Immediately following the burst of energy, the energy level of the electrons goes back to its original state. When an electron goes to a lower energy level, it emits a particle of light called a photon. When the emitted photon strikes another atom, it causes a similar event for that atom as well. As this process repeats, more and more photons are generated, causing what is known as "stimulated emission of radiation." The photons located inside the flash tube begin to increase their energy as they travel back and forth and repeatedly reflect off the mirrors on each side of the tube. The increase in the photon energy is called "light amplification." As the energy in the tube increases, a small amount of the coherent laser light escapes out of one of the mirrors that is partially transparent. Coherent light is light in which the crests and troughs of the waves line up and are in synchronization with each other.

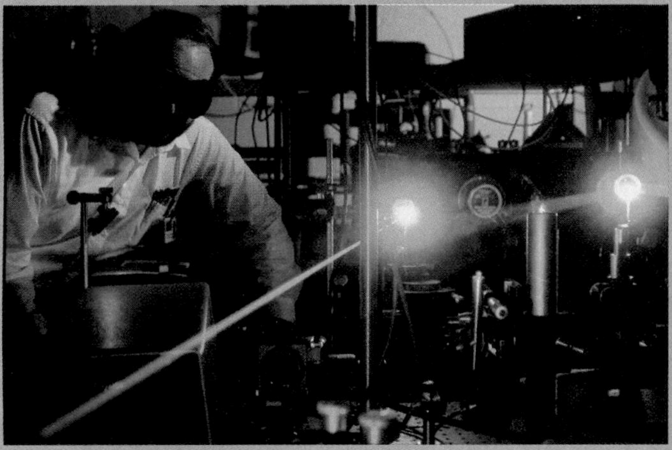

(a)

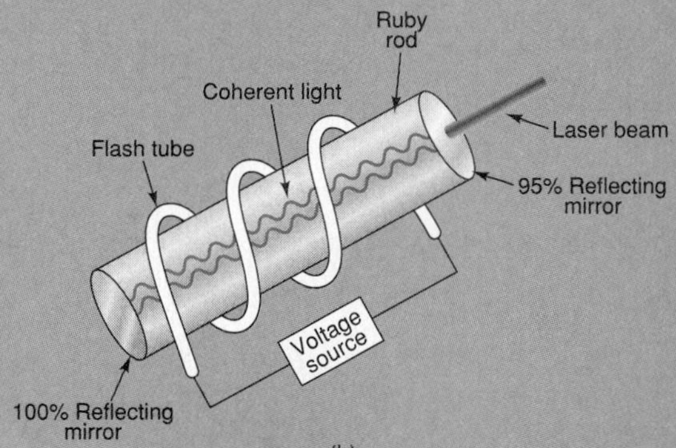

(b)

Figure 19.9 (a) A technician working in a high-tech laser laboratory (b) The inside of a ruby laser

The first laser was constructed by T. H. Maiman in 1960. The type of laser described above is the typical ruby laser. Some other common types of lasers are the red helium–neon laser (λ = 633 nm), the green argon laser (λ = 477 nm, 488 nm, and 515 nm), and the infrared carbon dioxide laser (λ = 10,600 nm).

The laser has become an essential tool in virtually all areas of science and technology. In the medical field, lasers are used to destroy cancerous tumors, correct vision defects, and seal capillaries to prevent excessive bleeding during surgery. In manufacturing, lasers are used to measure or align parts for fabrication, drill accurate holes, and weld metals together. The telecommunications and electronics industries use lasers to transmit digital signals in fiber optic cables, create holograms, read bar-codes, and record or read data from CDs, DVDs, and CD-ROMs. Geologists, astronomers, and physicists use lasers to measure great distances or detect minor changes in position.

PROBLEMS 19.4

1. What is the energy of a photon of electromagnetic radiation with frequency 8.95×10^{10} Hz?
2. What is the frequency of a photon of electromagnetic radiation with energy 3.96×10^{-22} J?
3. What is the energy of a photon of electromagnetic radiation with frequency 4.55×10^{8} Hz?
4. Find the frequency of electromagnetic radiation with energy 2.00×10^{-24} J.
5. Find the frequency of electromagnetic radiation with energy 5.50×10^{-25} J.
6. Find the energy of a photon of electromagnetic radiation with frequency 2.50×10^{12} Hz.
7. Find the frequency of electromagnetic radiation with energy 3.65×10^{-23} J.
8. Find the energy of a photon of electromagnetic radiation with frequency 9.20×10^{16} Hz.
9. Find the energy of a red photon.
10. Find the energy of a blue photon.
11. Find the energy of a yellow photon if the wavelength of its electromagnetic radiation is midway between those of red and blue photons.
12. An AM radio signal has a frequency of 650 kHz. What is the energy of a photon of that electromagnetic radiation?
13. An AM radio station in a nearby town broadcasts a signal with 9.37×10^{-28} J of energy. Find the frequency of this signal in kHz.

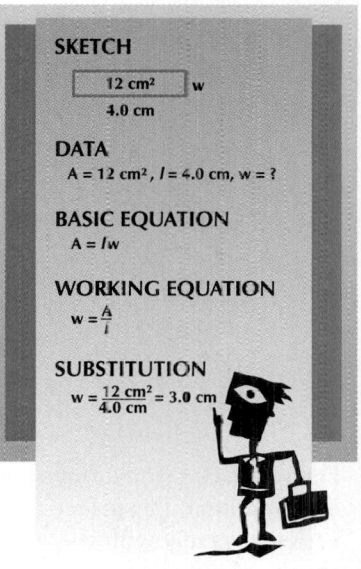

SKETCH

12 cm² | w

4.0 cm

DATA

$A = 12\text{ cm}^2$, $l = 4.0$ cm, $w = ?$

BASIC EQUATION

$A = lw$

WORKING EQUATION

$w = \frac{A}{l}$

SUBSTITUTION

$w = \frac{12\text{ cm}^2}{4.0\text{ cm}} = 3.0\text{ cm}$

19.5 Photometry

Recall that light is produced along with other forms of radiation when substances are heated, like our greatest source of natural light, the sun. Light may be produced in other ways, however, such as by electrons bombarding molecules in a gas such as neon or chemically, as by fireflies. Our most common source of artificial light, however, is the incandescent lamp.

Incandescent lamps (like light bulbs) produce light by the heating of a material (the filament) and the giving off of a wide range of radiation in addition to visible light. Objects that produce light are called *luminous*. On the other hand, objects like the moon, which are not producers of light but only reflect light from another source, are called *illuminated*. When light strikes the surface of most objects, some light is reflected, some is transmitted, and some is absorbed.

The study of the measurement of light is called **photometry.** Two important measurable quantities in photometry are the luminous intensity, I, of a light source, and the illumination, E, of a surface.

Luminous intensity is a measure of the brightness of a light source. The unit for luminous intensity, I, is the candle or *candela* (cd). The early use of certain candles for standards of intensity led to the name of the unit. We now use a platinum source at a certain temperature as the standard for comparison. Another unit, the *lumen*, ℓm, is often used for the measurement of the intensity of a source. One candle produces 4π lumens (ℓm):

$$\boxed{1 \text{ candle} = 4\pi\ \ell\text{m}}$$

To determine the intensity rating in lumens of a 40-W bulb rated at 35 cd, use the following conversion:

$$35 \text{ cd} \times \frac{4\pi\ \ell\text{m}}{1 \text{ cd}} = 440\ \ell\text{m}$$

Thus, a 40-W light bulb rated at 35 cd has an intensity rating of 440 ℓm.

Illumination is the amount of luminous intensity per unit area. The illumination on a surface may be varied by either changing the luminous intensity of the source (using a brighter bulb) or changing the position of the source (moving it closer to or farther from the surface to be illuminated). Of course, the illumination is also less if the surface illuminated is slanted and not directly facing the light source.

The amount of illumination on a surface varies inversely with the square of the distance from the source. For example, if the distance of the illuminated surface from the source is doubled, the illumination is reduced to one–fourth of its former intensity. This can be illustrated by considering a point source of light at the center of concentric (having the same center) spheres (Fig. 19.10). The solid angle is measured in units called *steradians.*

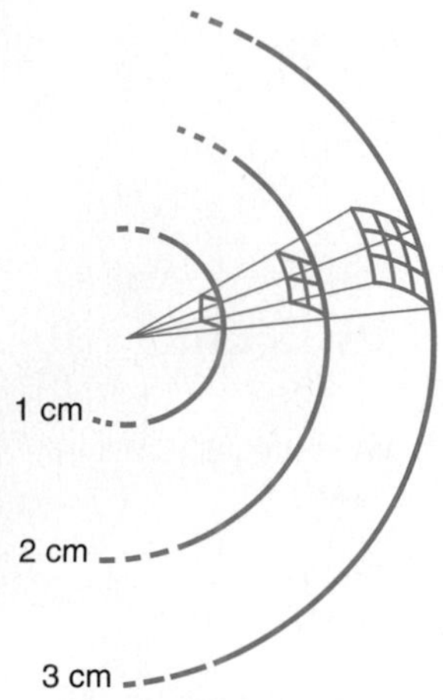

Figure 19.10 The amount of illumination on a surface varies inversely with the square of the distance from the source.

If the source radiates light uniformly in all directions, the light is uniformly distributed over a spherical surface centered at the source. Since the surface area of a sphere is $4\pi r^2$, the illumination, E, at the surface is given by

$$E = \frac{I}{4\pi r^2}$$

where

E = illumination
I = luminous intensity of the source (in lumens)
r = distance between the source and the illuminated surface

The unit of illumination, E, is the *lux:*

$$1 \text{ lux} = 1 \ \ell\text{m/m}^2$$

We assume the surface being illuminated is perpendicular to the source.

Note that the inverse-square-law behavior of light is similar to the two inverse square laws studied earlier, those of gravity and of the electric force between charges described by Coulomb's law.

A common application of the measure of illumination is the rating given to a camcorder, which measures the sensitivity of the instrument.

EXAMPLE 1

Find the illumination E on a surface located 2.00 m from a source with an intensity of $40\bar{0}$ ℓm.

Data:

$$I = 40\bar{0} \ \ell\text{m}$$
$$r = 2.00 \text{ m}$$
$$E = ?$$

Basic Equation:

$$E = \frac{I}{4\pi r^2}$$

Working Equation: Same

Substitution:

$$E = \frac{40\bar{0} \ \ell\text{m}}{4\pi(2.00 \text{ m})^2} = 7.96 \ \ell\text{m/m}^2 \quad \text{or} \quad 7.96 \text{ lux}$$

The unit used for illumination is the lux ($\ell m/m^2$) in the metric system as shown above and the foot-candle ($\ell m/ft^2$) in the U.S system:

$$\boxed{1 \text{ ft-candle} = 1\ \ell\text{m}/\text{ft}^2}$$

EXAMPLE 2

Find the illumination 4.00 ft from a source with an intensity of $60\bar{0}$ ℓm.

Data:

$$I = 60\bar{0}\ \ell\text{m}$$
$$r = 4.00 \text{ ft}$$
$$E = ?$$

Basic Equation:

$$E = \frac{I}{4\pi r^2}$$

Working Equation: Same

Substitution:

$$E = \frac{60\bar{0}\ \ell\text{m}}{4\pi(4.00 \text{ ft})^2}$$

$$E = 2.98 \frac{\ell\text{m}}{\text{ft}^2} \times \frac{1 \text{ ft-candle}}{1\ \ell\text{m}/\text{ft}^2}$$

$$= 2.98 \text{ ft-candles}$$

$$\boxed{\frac{\ell\text{m}}{\text{ft}^2} \times \frac{1 \text{ ft-candle}}{1\ \ell\text{m}/\text{ft}^2} = \frac{\cancel{\ell\text{m}}}{\cancel{\text{ft}^2}} \times 1 \text{ ft-candle} \times \frac{\cancel{\text{ft}^2}}{\cancel{\ell\text{m}}} = \text{ft-candle}}$$

In photography, photoelectric cells are used in light meters or exposure meters to measure illumination for taking photographs. The electricity produced is proportional to the illumination and is directly calibrated on the instrument scale. The units of measurement of such meters, however, are not standardized, and the scale may be arbitrarily selected.

PROBLEMS 19.5

1. $I = 48.0$ cd
$I =$ ______ ℓm
2. $I = 342$ cd
$I =$ ______ ℓm
3. $I = 765$ ℓm
$I =$ ______ cd
4. $I = 432$ ℓm
$I =$ ______ cd
5. $I = 75.0$ cd
$I =$ ______ ℓm
6. $I = 65\bar{0}$ ℓm
$I =$ ______ cd
7. $I = 90\bar{0}$ ℓm
$r = 7.00$ ft
$E = ?$
8. $I = 741$ ℓm
$r = 6.50$ m
$E = ?$
9. $I = 893$ ℓm
$r = 3.25$ ft
$E = ?$
10. $E = 4.32$ lux
$r = 9.00$ m
$I = ?$
11. $E = 10.5$ ft-candles
$r = 6.00$ ft
$I = ?$
12. Find the intensity of a light source that produces an illumination of 5.50 ft-candles at 9.85 ft from the source.
13. Find the intensity of a light source that produces an illumination of 2.39 lux at 4.50 m from the source.
14. Find the intensity of a light source that produces an illumination of 5.28 lux at 6.50 m from the source.

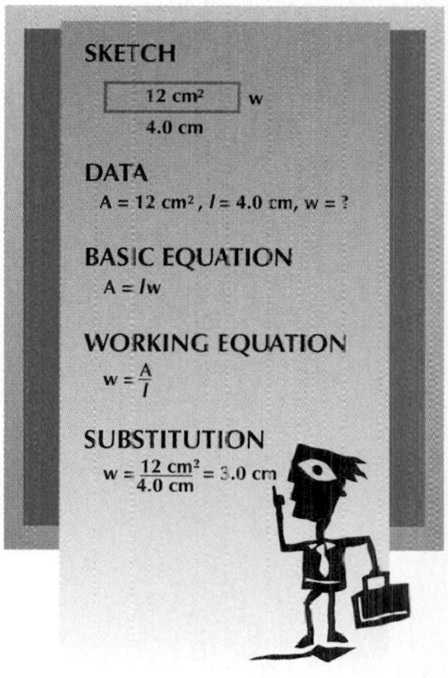

15. If an observer triples her distance from a light source:
    (a) Does the illumination at that point increase or decrease?
    (b) In what proportion does the illumination increase or decrease?
16. If the illuminated surface is slanted at an angle of 35.0°, what part of the full-front illumination is lost?
17. Find the illumination on a surface by three light sources, each with intensity 150 ℓm, located at distances of 2.00 m, 2.70 m, and 2.98 m from the surface, respectively.
18. Find the intensity of two identical light sources located 1.40 m and 1.96 m, respectively, from a point where the illumination is 3.54 $\ell m/m^2$.
19. Find the intensity of two identical light sources located 0.880 m and 1.12 m from a point where the illumination is 5.86 $\ell m/m^2$.
20. A desk is 3.35 m below an 1850-ℓm incandescent lamp. What is the illumination on the desktop?

Glossary

Frequency The number of complete vibrations or cycles per second of a wave. (p. 545)
Illumination The luminous intensity per unit area. (p. 550)
Laser A light source that produces a narrow beam with high intensity. An acronym for "light amplification by stimulated emission of radiation." (p. 547)
Light Radiant energy that can be seen by the human eye. (p. 540)
Light-Year The distance that light travels in one earth year: 9.45×10^{15} m or 5.87×10^{12} mi. (p. 544)
Luminous Intensity A measure of the brightness of a light source. (p. 549)
Particle Theory Theory that light consists of streams of particles. (p. 541)
Photoelectric Effect The emission of electrons by a surface when struck by electromagnetic radiation. (p. 542)
Photometry The study of the measurement of light. (p. 549)
Photons Wave packets of energy that carry light and other forms of electromagnetic radiation. (p. 542)
Planck's Constant A fundamental constant of quantum theory (6.626×10^{-34} J s). (p. 547)
Quantum Theory Theory initiated by Planck and Einstein that energy, including electromagnetic radiation, is radiated or absorbed in multiples of certain units of energy. (p. 542)
Reflection The turning back of all or part of a beam of light at a surface. (p. 540)
Refraction The bending of light as it passes through the boundary between two media, such as air and water. (p. 540)
Speed of Light The speed at which light and other forms of electromagnetic radiation travel. Equal to 3.00×10^8 m/s in a vacuum. (p. 543)
Wave Theory Theory that light consists of waves traveling out from light sources, like water waves traveling out from the point at which a stone is dropped into water. (p. 541)
Wavelength The distance between two successive corresponding points on a wave. (p. 545)

Formulas

19.2 $s = ct$

19.3 $c = \lambda f$

19.4 $E = hf$

19.5 $E = \dfrac{I}{4\pi r^2}$

Review Questions

1. Which of the following are examples of electromagnetic radiation?
   (a) Gamma rays (b) Sound waves
   (c) Radio waves (d) Water waves
   (e) Visible light
2. The particle theory of light explains
   (a) diffraction of light around a sharp edge. (b) refraction of light at a boundary.
   (c) the photoelectric effect. (d) none of the above.
3. A light-year equals
   (a) the time it takes light to travel from the sun to the earth.
   (b) the distance from the sun to the earth.
   (c) the distance to the nearest star other than the sun.
   (d) the distance light travels in one earth year.
4. Light behaves
   (a) as a massive particle. (b) always as a wave.
   (c) sometimes as a wave, sometimes as a particle. (d) as none of the above.
5. Does the wavelength of light depend on its frequency? Explain.
6. How does the energy of a photon of light depend on its frequency?
7. How does the intensity of illumination depend on the distance from a source radiating uniformly in all directions?
8. In your own words, explain how the speed of light has been measured.
9. Does light always travel at the same velocity? Explain.
10. What name is given to the entire range of waves that are similar to visible light?
11. Who proposed the particle theory of light?
12. Who developed the wave packet theory of light?
13. Who made the first estimate of the speed of light?
14. How was the first estimate of the speed of light made?
15. What are the units of luminous intensity?
16. In your own words, explain luminous intensity.

Review Problems

1. Find the distance (in metres) traveled by a radio wave in 21.5 h.
2. A radar wave that is bounced off an airplane returns to the radar receiver in 3.78×10^{-5} s. How far (in miles) is the airplane from the radar receiver?
3. How long does it take for a police radar beam to travel to a car and back if the car is 0.245 mi from the radar unit?
4. How long does it take for a pulse of laser light to return to a police speed detector after bouncing off a speeding car 0.274 mi away?
5. How long does it take for a radio signal to travel from the earth to a communications satellite 22,500 mi above the surface of the earth?
6. Find the wavelength of a radio wave from an AM station broadcasting at a frequency of 1230 kHz.
7. Find the frequency of a radio wave if its wavelength is 46.5 m.
8. Find the frequency of a light wave if its wavelength is 5.415×10^{-8} m.
9. What is the energy of a photon with frequency 1.45×10^{11} Hz?
10. What is the frequency of a photon with energy of 4.75×10^{-23} J?
11. What is the energy of a photon with frequency 8.25×10^{-15} Hz?
12. Find the intensity of the light source necessary to produce an illumination of 3.75 ft-candles at 6.75 ft from the source.
13. Find the intensity of the light source necessary to produce an illumination of 4.86 lux at 9.25 m from the source.

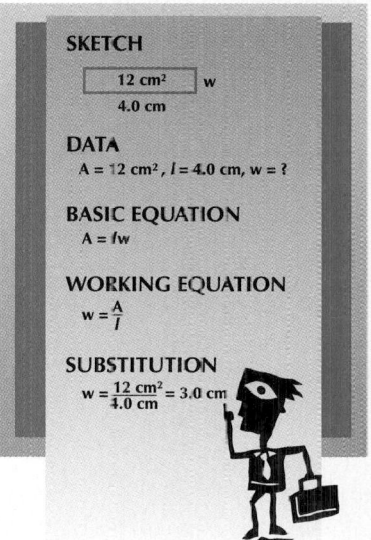

SKETCH

12 cm² w
4.0 cm

DATA

$A = 12\text{ cm}^2$, $l = 4.0$ cm, $w = ?$

BASIC EQUATION

$A = lw$

WORKING EQUATION

$w = \frac{A}{l}$

SUBSTITUTION

$w = \frac{12\text{ cm}^2}{4.0\text{ cm}} = 3.0\text{ cm}$

14. What is the intensity of the light source required to produce the illumination of Problem 13 if the distance from the light source is doubled?
15. What are the maximum and minimum transit times for light traveling from Jupiter to Mars? The orbital radii are 215 million kilometres for Mars and 725 million kilometres for Jupiter. Assume the planetary orbits are circular. Also make the (nonphysical) assumption that the sun is transparent to the transmission of light between the planets.
16. Find the intensity of two identical light sources located 0.454 m and 0.538 m, respectively, from a point where the illumination is 8.46 $\ell m/m^2$.
17. Find the illumination on a surface by three light sources, each with intensity 125 ℓm, located at 1.85 m, 1.92 m, and 2.43 m from the surface, respectively.

APPLIED CONCEPTS

1. The distance between New York City and London is 3470 mi. (a) If a radio wave from New York City is transmitted directly across the ocean, how long will it take to reach a receiver in London? (b) In fact, radio waves cannot be transmitted directly across such large distances on the earth. A signal is typically transmitted to a communications satellite located 2.20×10^4 mi above the surface of the earth. If the satellite is located midway between the two cities, how long will the radio wave take to reach London? (Ignore the effect of the curvature of the earth.)
2. (a) When the Apollo astronauts landed on the moon, it took the radio signal 1.28 s to reach Mission Control on the earth. How far away were the astronauts from the earth? (b) Since the sun is 1.50×10^{11} m from the earth, how much time does it take for light to travel from the sun to the earth? (c) Light from our next closest star, Alpha Centauri, takes 4.31 years to reach the earth. How far away from the earth is Alpha Centauri?
3. The range of electromagnetic wave frequencies on the FM radio band is 88.0 MHz to 108 MHz. (a) What is the range of wavelengths for the FM radio band? (b) What is the range of photon energy for the FM radio band? (c) Explain the relationship between frequency and photon energy.
4. The individual rods on rooftop antennas are designed to be one-quarter of a wavelength for each television frequency. What is the range of rod lengths needed for television Channels 2 through 6 if their frequencies are between 54.0 MHz and 88.0 MHz?
5. An illumination of 180 lux on student desks and other work areas is the standard that architects use when designing lighting systems for schools. If a ceiling is 2.50 m above a student work area, what intensity light source must be installed? (b) If the ceiling were twice its original height, what light intensity would be needed to meet the standard requirement?

CHAPTER 20

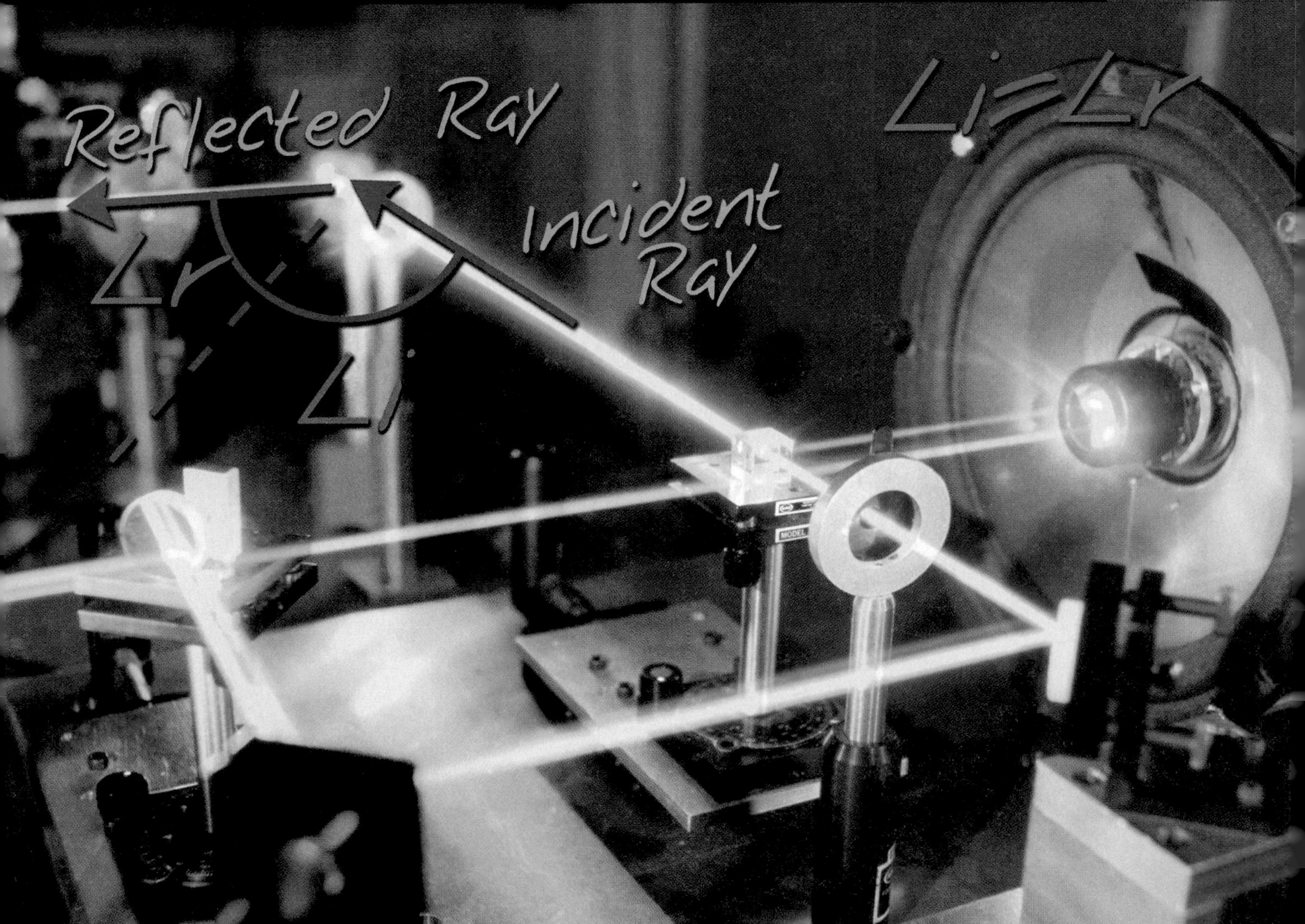

REFLECTION AND REFRACTION

The nature of light may still be somewhat of a mystery. However, its characteristics have been the subject of intensive study for hundreds of years. Light may be transmitted, reflected, or absorbed by a medium.

Anyone wearing glasses can appreciate the refraction of light as it bends upon passing from one medium to another. The index of refraction is a tool of the scientist to describe the ability of certain substances to bend light as it passes through them.

Our examination of the behavior of light begins with the study of images and reflection.

Objectives

The major goals of this chapter are to enable you to:

1. Describe the laws of reflection.
2. Locate and describe images formed by plane, convex, and concave mirrors.
3. Apply the mirror formula to image formation.
4. Describe the law of refraction.
5. Describe total internal reflection.
6. Locate and describe images formed by converging and diverging lenses.

20.1 Mirrors and Images

Although the nature of light is complex, we know very well how light behaves. Every day we experience and unconsciously use our knowledge of what light does and depend on the ways it works. **Reflection** is the turning or turning back of all or a part of a beam of light from a surface. Unlike sound, light does not require a medium (some kind of matter) to travel through and may be transmitted through empty space. When light does strike a medium, the light may be reflected, absorbed, transmitted, or undergo a combination of the three.

Mirrors show how light may be reflected. Any dark cloth shows how light may be absorbed. Window glass illustrates how light may be transmitted through a medium.

A medium can be classified according to how well light is transmitted through it:

1. ***Transparent:*** almost all light passes through
   Examples: window glass, clear water
2. ***Translucent:*** some but not all light passes through
   Examples: murky water, light fog, skylight panels for farm buildings, stained glass
3. ***Opaque:*** almost all light reflected or absorbed
   Examples: wood, metal, plaster

These classifications are relative because for any medium, some light is reflected from the surface and some passes into or through it.

In studying reflection we observe what happens when light is turned back from a surface. The beam of a flashlight directed at a mirror shows several things about reflection. First, upon striking the surface of the glass, some of the light is reflected in all directions. This is called *scattering*. If there were no scattering, no light would reach our eye and we would be unable to observe the beam at all. However, only a very small part of the beam of light is scattered. Rough or uneven surfaces produce more scattering than do smooth ones. This scattering of light by uneven surfaces is called **diffusion.** Diffused lighting has many applications at home and in industry where bright glare is not desirable.

Nearly complete reflection (with very little scattering) is called **regular** (or specular) **reflection.** Regular reflection occurs when parallel or nearly parallel rays of light (such as sunlight and spotlight beams) remain parallel after being reflected from a surface (Fig. 20.1). Note that the incoming rays are referred to as *incident rays*.

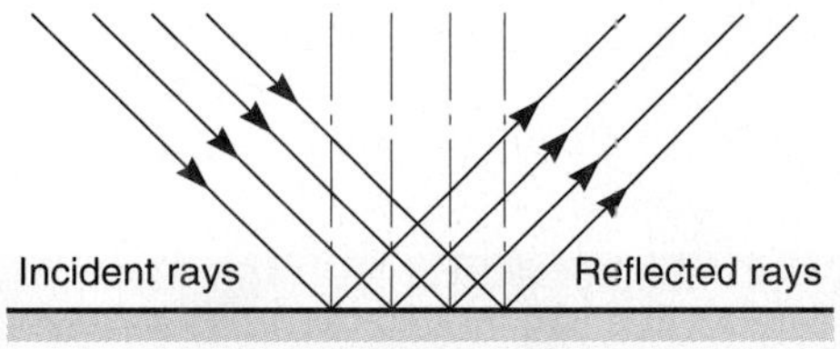

Figure 20.1 Regular reflection (Reflected rays are parallel.)

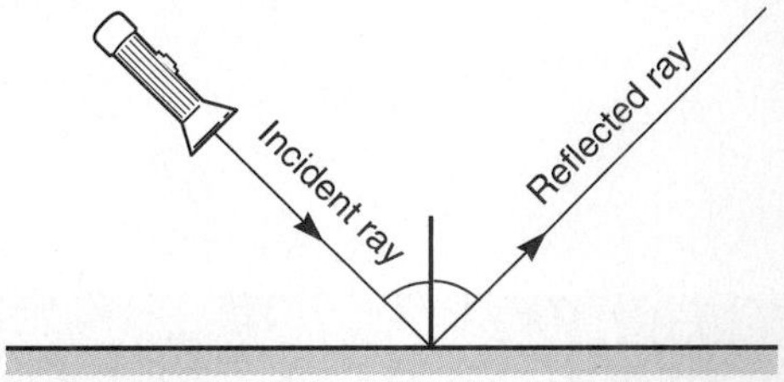

Figure 20.2 On a regular surface, the reflected rays leave at the same angle as the incident rays.

A flashlight beam on a mirror in a darkened room also shows something else about light striking a regular reflecting surface: The reflected rays of light leave the surface at the same angle at which the incident (incoming) rays strike the surface (Fig. 20.2). Expressed another way, the angles measured from the normal (the perpendicular) to the reflecting surface are equal. These angles are shown in Fig. 20.3.

The same principle applies to curved surfaces (Fig. 20.4). This behavior of light rays is defined by the following law:

FIRST LAW OF REFLECTION

The angle of incidence, *i*, is equal to the angle of reflection, *r*; that is,

$$\angle i = \angle r$$

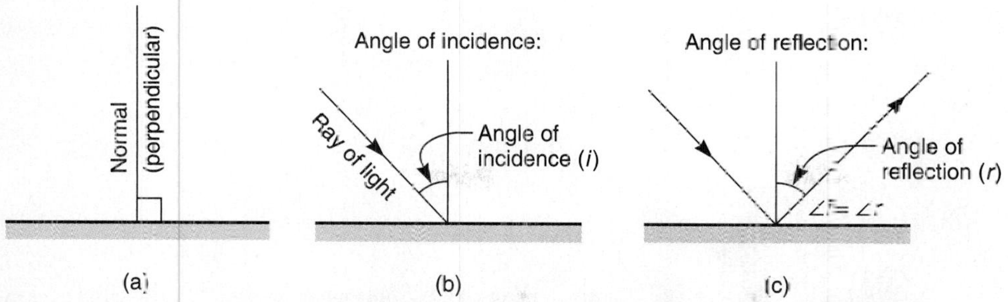

Figure 20.3 The angle of incidence is equal to the angle of reflection.

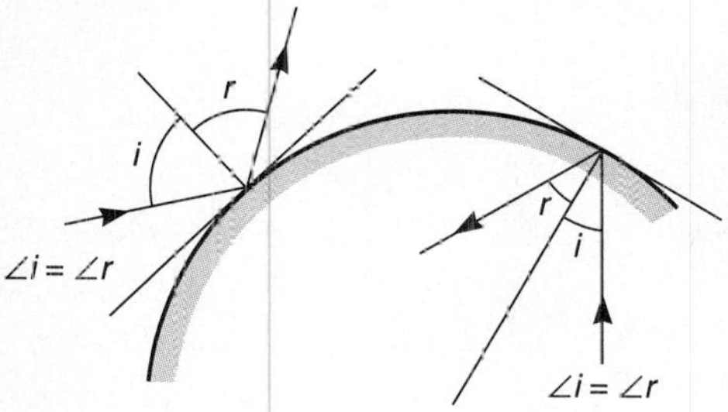

Figure 20.4 Likewise, for curved surfaces, $\angle i = \angle r$

Further observation of the light beam readily shows a second law:

SECOND LAW OF REFLECTION

The incident ray, the reflected ray, and the normal (perpendicular) to the surface all lie in the same plane.

These laws of reflection apply not only to light, but to all kinds of waves.

We look next at how images are formed by three widely used kinds of mirrors: plane, concave, and convex. **Plane mirrors** are flat. **Concave mirrors** are curved away from the observer [like the inside of a bowl; Fig. 20.5(a)] and **convex mirrors** are curved toward the observer [like ball-shaped Christmas tree ornaments; Fig. 20.5(b)].

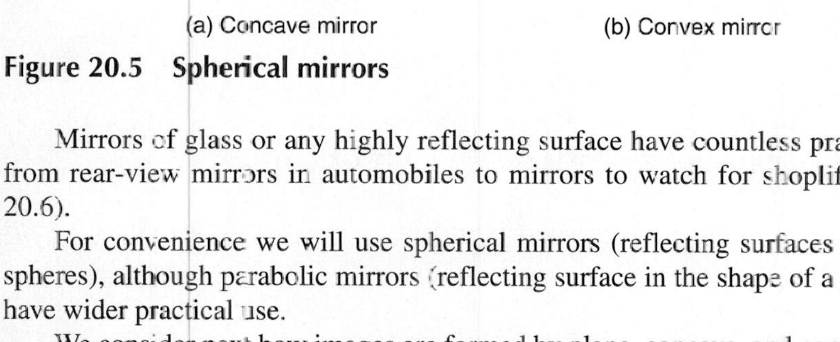

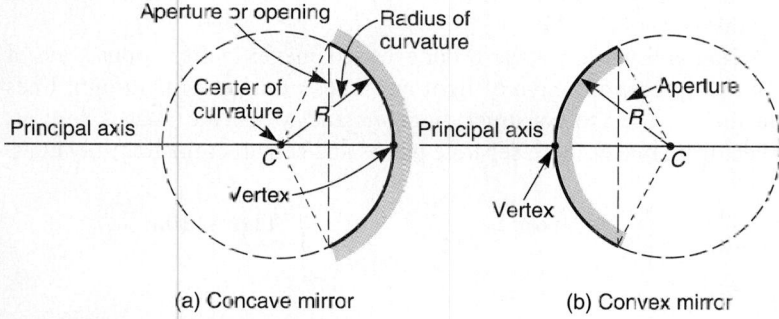

Figure 20.5 Spherical mirrors

Mirrors of glass or any highly reflecting surface have countless practical applications, from rear-view mirrors in automobiles to mirrors to watch for shoplifting in stores (Fig. 20.6).

For convenience we will use spherical mirrors (reflecting surfaces that are sections of spheres), although parabolic mirrors (reflecting surface in the shape of a parabola; Fig. 20.7) have wider practical use.

We consider next how images are formed by plane, concave, and convex mirrors. Images formed by mirrors may be **real images** (images formed by rays of light) or **virtual images** (images that only appear to the eye to be formed by rays of light).

Real images made by a single mirror are always inverted (upside down) and may be larger than, smaller than, or the same size as the object. They can be shown on a screen. Virtual

Figure 20.6 Convex security mirrors are used to watch for shoplifting because they reflect images from over wide areas.

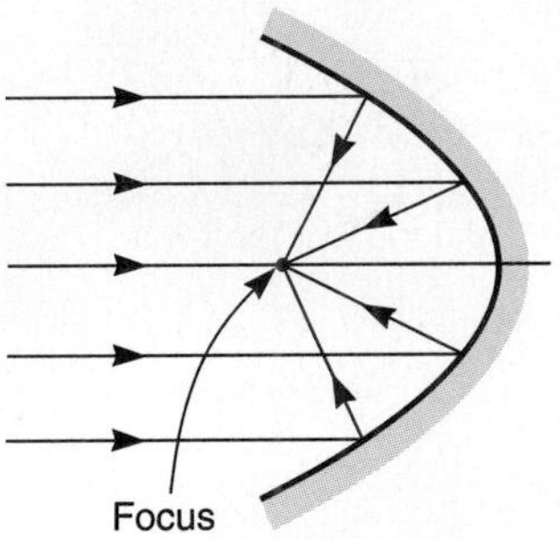

Figure 20.7 Parabolic mirror

images are always erect and may be larger than, smaller than, or the same size as the object. They cannot be shown on a screen.

20.2 Images Formed by Plane Mirrors

Plane mirror images are always erect and virtual and appear as far behind the mirror as the distance the object is in front of the mirror. Note that plane mirrors also reverse right and left, so the right hand held in front of a plane mirror appears in the mirror to be the left hand.

Look at your image in a mirror. Then look at something on the mirror's surface like a speck of dirt. Notice that you have to adjust your eyes and refocus from looking at the image of your face to looking at the mirror surface. It should be clear that the image is not on the surface of the mirror but behind it.

We use light-ray diagrams to illustrate how our eyes see images in the various kinds of mirrors. We do this by representing rays of light and lines of sight with straight lines (Fig. 20.8). This method is used to construct diagrams and locate the images formed. Simply view the object from two or more separate places and construct the light-ray lines.

Figure 20.8

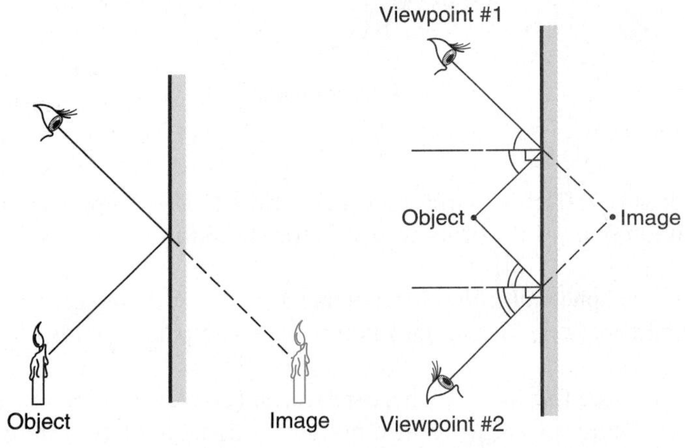

(a) Plane mirror image

(b) Location of an image in a plane mirror

20.3 Images Formed by Concave Mirrors

Find a shiny tablespoon and look at your image in it (Fig. 20.9). Now turn it over and look again. The images are very different; one is erect and the other, inverted (upside down). We use ray diagrams to show why this happens. As we shall see, the kind of image produced depends on the location of the object with respect to the mirror.

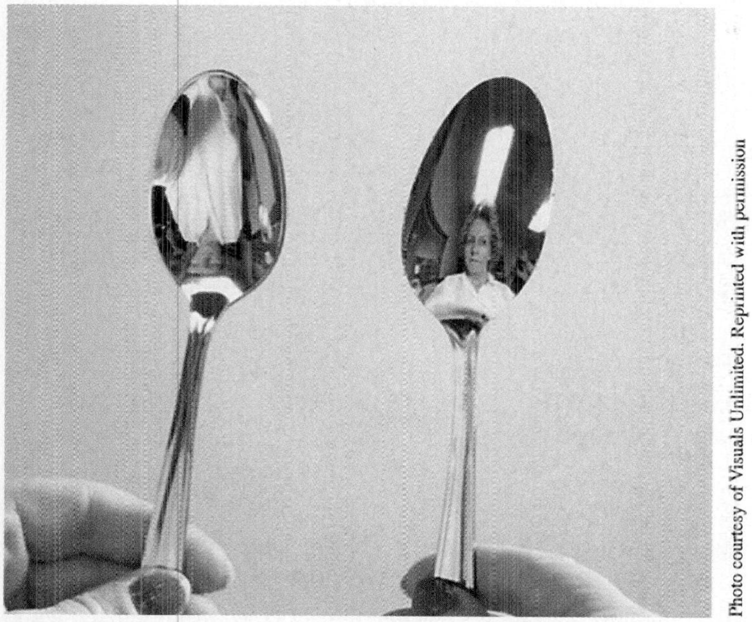

Figure 20.9 Reflected images as seen on opposite sides of a large spoon

We need to define some terms before we discuss how images are formed. Figure 20.10(a) shows a spherical mirror with the key terms identified. The *center of curvature, C*, is the center of the sphere that forms a part of the spherical mirror. The *vertex, V*, is the center of the mirror (sometimes called its optical center). The *principal axis* is the line *CV* drawn through the center of curvature and the vertex. The *principal focus, F*, is the point on the principal axis through which all rays parallel to the principal axis converge in a concave mirror as shown in Fig. 20.10(b) or from which they diverge in a convex mirror. Note that any ray through the center of curvature is along a radius of a spherical mirror and is reflected straight back because any such ray is perpendicular to the surface of the mirror at the point the ray strikes the mirror. The **focal length** is the distance between the principal

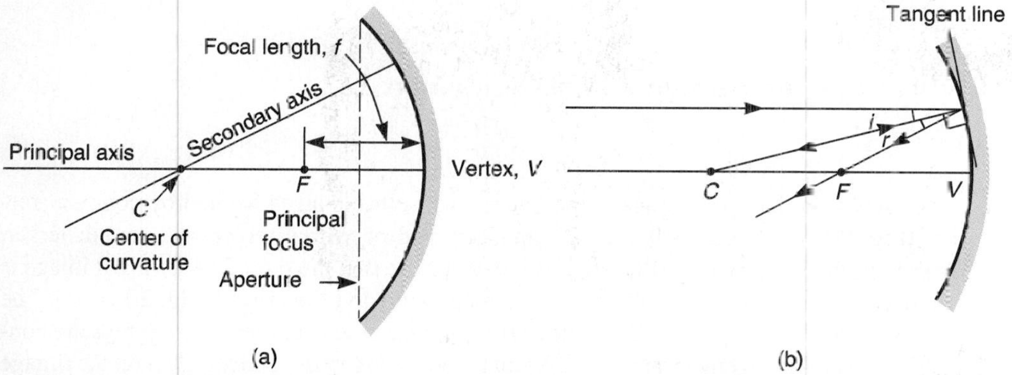

Figure 20.10 Spherical mirror

focus of a mirror (or lens) and its vertex. The *aperture* is that very small portion of the sphere that forms the actual mirror. For mirrors with small apertures, the focal length, f, is one-half the radius of curvature, R. That is, $f = \frac{R}{2}$.

If the object is placed at the focal point, no image will be formed because the rays of light will be reflected parallel to the principal axis (Fig. 20.11). The location of the reflected ray may be found by using the laws of reflection (Fig. 20.12). The angle of incidence is equal to the angle of reflection.

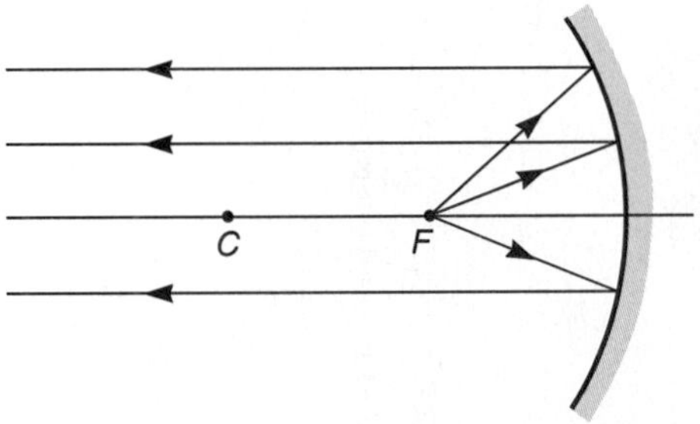

Figure 20.11 No image is formed if the object is located at the focal point.

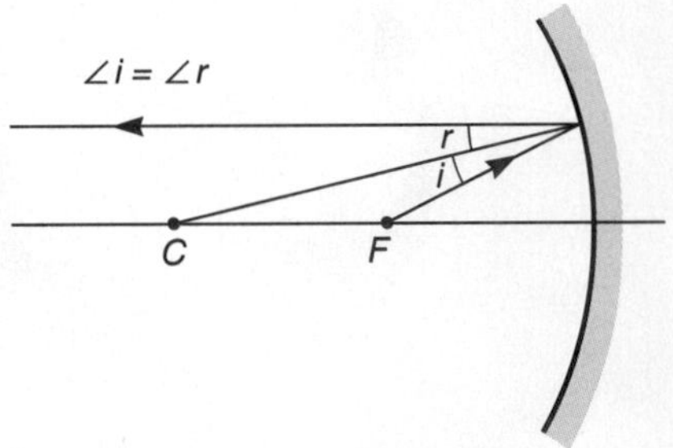

Figure 20.12

Now consider the more common case, in which the object is beyond the center of curvature [for example, looking into a tablespoon as in Fig. 20.9, whose mirror diagram is shown in Fig. 20.13(a)]. Note that again we use the fact that a ray parallel to the principal axis is reflected through the focal point and a ray through the focal point is reflected parallel to the principal axis. Then, where the two rays intersect, a point on the image is formed. (In this case, it is the flame of the candle. Note that the candle base image lies on the principal axis because the object base is also on that line.) The same method is used for the case where the candle base extends below the principal axis [Fig. 20.13(b)].

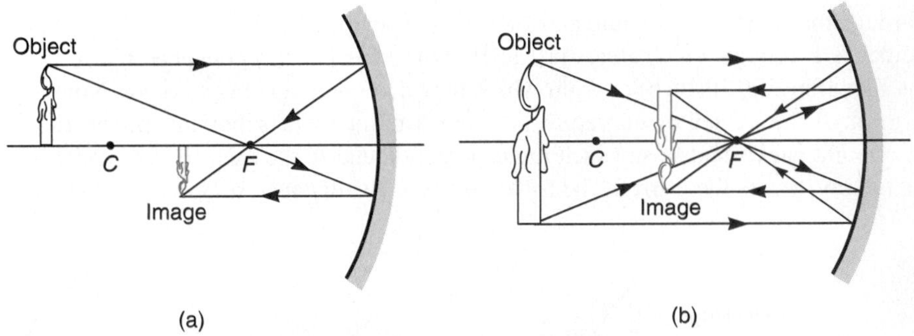

Figure 20.13 Images formed in spherical mirrors

Now apply these principles to the diagrams of other images formed by concave mirrors (Fig. 20.14). Decide whether the image is real or virtual; erect or inverted; larger, smaller, or the same size; and where it is located. Note that the only time a virtual image is produced is when the object is between the focal point and the mirror [Fig. 20.14(f)]. The construction of the diagram is the same as the other cases except that the light rays are converging (coming together) and must be extended behind the mirror (where the image appears to be) forming the virtual image.

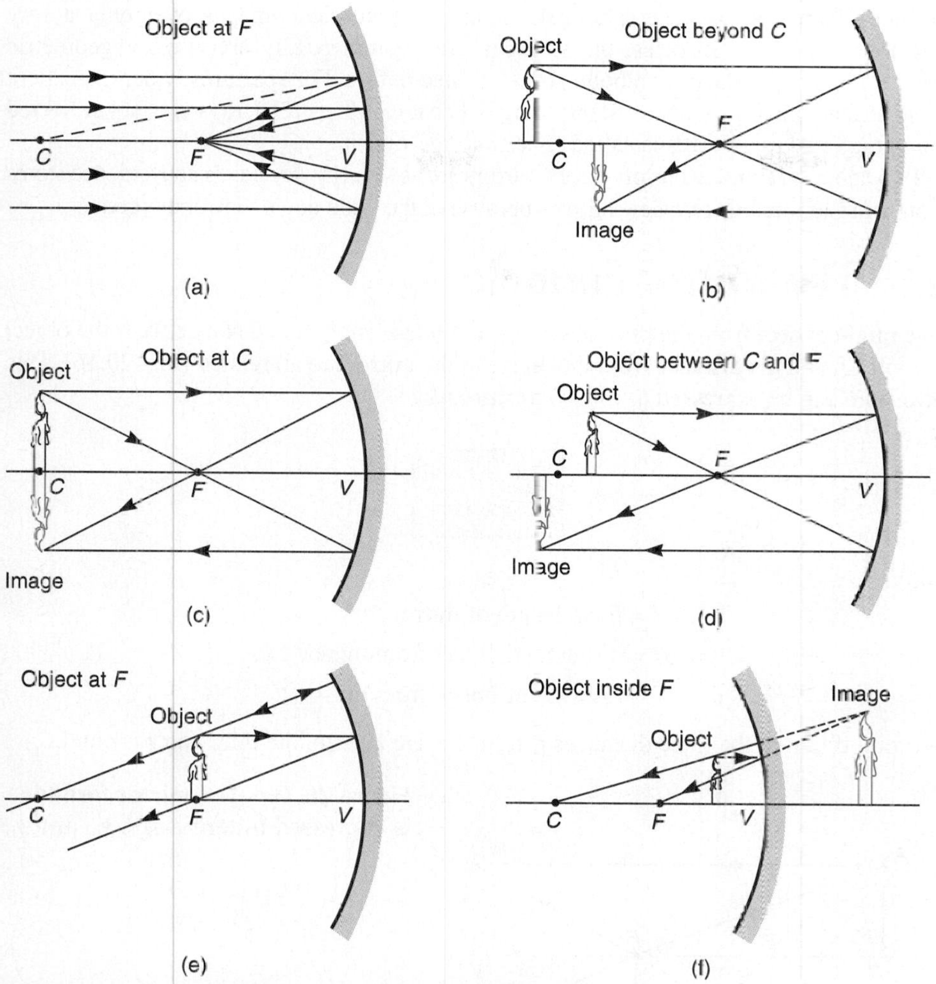

Figure 20.14 Formation of images in concave mirrors

20.4 Images Formed by Convex Mirrors

By looking into the back side of our tablespoon in Fig. 20.9, we see an erect, virtual, smaller image. Use the mirror diagram shown in Fig. 20.15 to see how such an image is formed. Curved surface mirrors are used in some telescopes, spotlights, and automobile

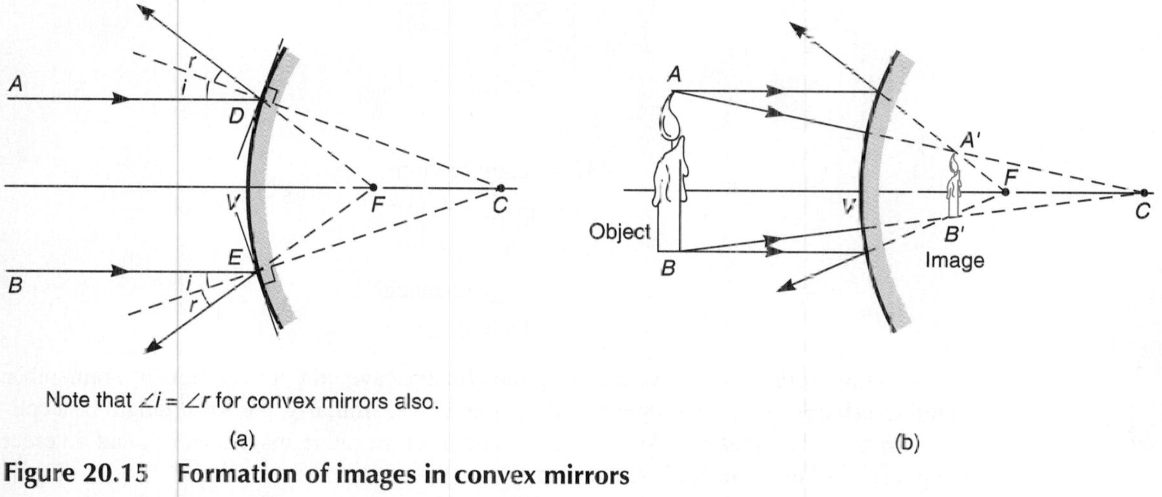

Figure 20.15 Formation of images in convex mirrors

headlights. However, because spherical mirrors produce clear images over only a very small portion of their surfaces, the surfaces used commercially are another geometric shape, that of a parabola, or parabolic, as mentioned before. For apertures wider than about 10°, spherical mirrors produce fuzzy images because all parallel rays are not reflected through the focal point. This is called *spherical aberration.*

Two common applications of convex mirrors are as security devices in convenience stores and on some automobile rearview mirrors because of the wide-angle view they produce.

20.5 The Mirror Formula

As we might expect from the previous cases, the focal length, the distance from the object to the mirror, and the distance from the image to the mirror are all related (Fig. 20.16). This relationship can be expressed as the *mirror formula:*

$$\frac{1}{f} = \frac{1}{s_o} + \frac{1}{s_i}$$

where

f = focal length of mirror
s_o = distance of object from mirror
s_i = distance of image from mirror

Therefore, if two of the three distances f, s_o, and s_i are known, the third can be found.

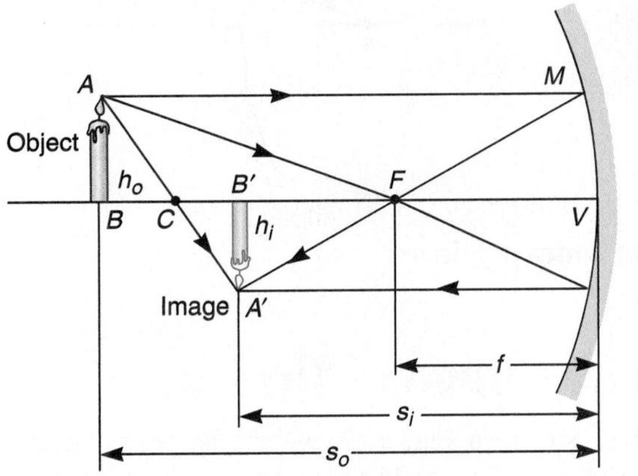

Figure 20.16 The mirror formula is expressed in terms of *f*, s_o, and s_i.

A second formula shows the magnification of the mirror and how the height of the object and the height of the image depend on the object distance and the image distance:

$$M = \frac{h_i}{h_o} = \frac{-s_i}{s_o}$$

where

M = magnification
h_i = image height
h_o = object height
s_i = image distance
s_o = object distance

In using *both* of the preceding formulas for concave and convex mirrors, remember that the distance to a virtual image is always negative; similarly, the focal length of a convex mirror is also negative. An inverted image has a negative magnification and an erect image has a positive magnification.

EXAMPLE

An object 10.0 cm in front of a convex mirror forms an image 5.00 cm behind the mirror. What is the focal length of the mirror?

Sketch:

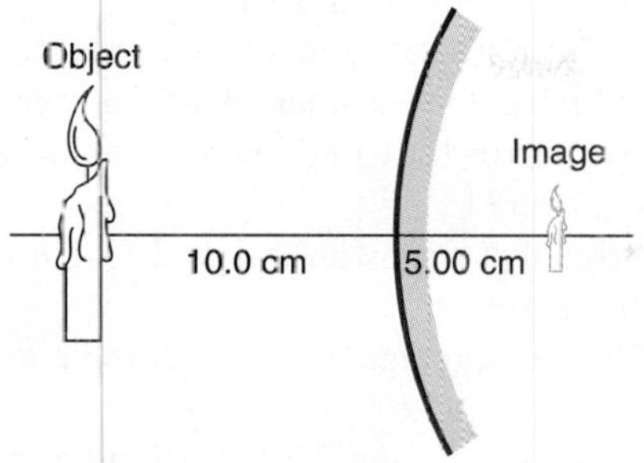

Data:

$s_o = 10.0$ cm

$s_i = -5.00$ cm ***Note:*** The image is virtual (appears behind the mirror) so s_i is given a (−) sign to show this. [Won't f also be (−)?]

$f = ?$

Basic Equation:

$$\frac{1}{f} = \frac{1}{s_o} + \frac{1}{s_i}$$

Working Equation: Same

Substitution:

$$\frac{1}{f} = \frac{1}{10.0 \text{ cm}} + \frac{1}{-5.00 \text{ cm}} = \frac{1}{10.0 \text{ cm}} - \frac{1}{5.00 \text{ cm}}$$

$$f = -10.0 \text{ cm}$$

The key entry sequence on a scientific calculator for this calculation is

10 [x^{-1}] [−] 5 [x^{-1}] [=] [x^{-1}] [=]

[-10]

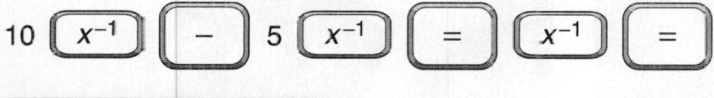

Remember that f and s_i may be negative only when forming virtual images and/or using convex mirrors.

PROBLEMS 20.5

Use the formulas $\frac{1}{f} = \frac{1}{s_o} + \frac{1}{s_i}$ and $M = \frac{h_i}{h_o} = \frac{-s_i}{s_o}$ for Problems 1–8.

1. Given $s_o = 1.65$ cm and $s_i = 6.00$ cm, find f.
2. Given $f = 15.0$ cm and $s_i = 3.00$ cm, find s_o.
3. Given $s_i = 14.5$ cm and $f = 10.0$ cm, find s_o.
4. Given $s_i = -10.0$ cm and $f = -5.00$ cm, find s_o.
5. Given $s_o = 7.35$ cm and $s_i = 17.0$ cm, find f.
6. Given $h_i = 2.75$ cm, $h_o = 4.50$ cm, and $s_i = 5.00$ cm, find s_o.
7. Given $h_o = 12.0$ cm, $s_i = 13.0$ cm, and $s_o = 25.0$ cm, find h_i.
8. Given $h_i = 3.50$ cm, $h_o = 2.50$ cm, and $s_i = 15.5$ cm, find s_o.

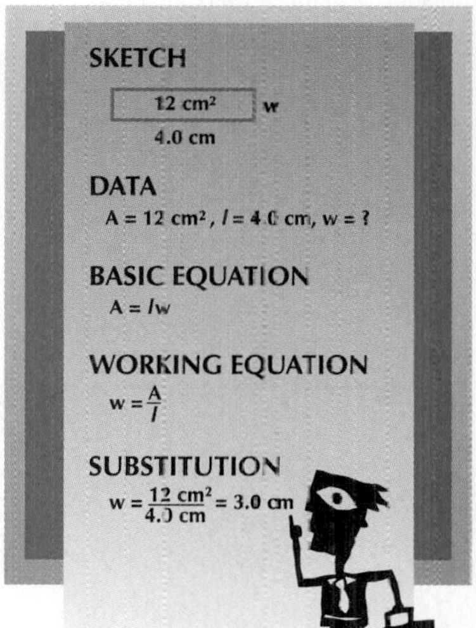

9. If an object is 2.50 m tall and 8.60 m from a large mirror with an image formed 3.75 m from the mirror, find the height of the image.
10. An object 30.0 cm tall is located 10.5 cm from a concave mirror with focal length 16.0 cm. (a) Where is the image located? (b) How high is it?
11. An object and its image in a concave mirror are the same height, yet inverted, when the object is 20.0 cm from the mirror. What is the focal length of the mirror?
12. An object 12.6 cm in front of a convex mirror forms an image 6.00 cm behind the mirror. What is the focal length of the mirror?
13. Find the focal length of a convex mirror that forms an image 3.55 cm behind the mirror of an object 24.5 cm in front of the mirror.
14. Find the focal length of a mirror that forms an image 5.66 cm behind a mirror of an object 34.4 cm in front of the mirror.
15. Find the focal length of a mirror that forms an image 2.30 m behind a mirror of an object 6.50 m in front of the mirror.
16. An image of a statue appears to be 11.5 cm behind a convex mirror with focal length 13.5 cm. Find the distance from the statue to the mirror.
17. (a) What is the height of a figurine 7.33 cm in front of a concave mirror that produces an image −2.75 cm high? The image appears to be 5.03 cm in front of the mirror. (b) Find the focal length of the mirror. (c) What distance would an image appear to be from the mirror with double the focal length?

20.6 The Law of Refraction

Does light travel in a straight line? Most people's first answer would be "Yes." But does it always? See Fig. 20.17. Why does a straw appear to bend at the surface when placed in water as in Fig. 20.17(b)?

(a) (b)

Figure 20.17 Refraction is the bending of light as it passes from one medium to another.

TRY THIS ACTIVITY

Bent Pencil

Place a pencil in a glass of water so that it rests against the side of the glass. Observe the glass from a variety of angles. Why does the pencil appear to be broken or bent at some angles and straight at other angles?

The answers to our questions may be found in the study of another property of light—refraction. **Refraction** is the bending of light as it passes at an angle from one medium to another of different optical density. **Optical density** is a property of a transparent material that is a measure of the speed of light through the given material. For example, water is optically denser than air and the speed of light in water is less than the speed of light in air. This change of speed of light when it passes from one medium to another produces refraction. The wave shown in Fig. 20.18 illustrates how this occurs. Note that when passing from one medium to another perpendicular to the surface (called the *interface*) the wave is not bent, although the speed of the wave is slowed.

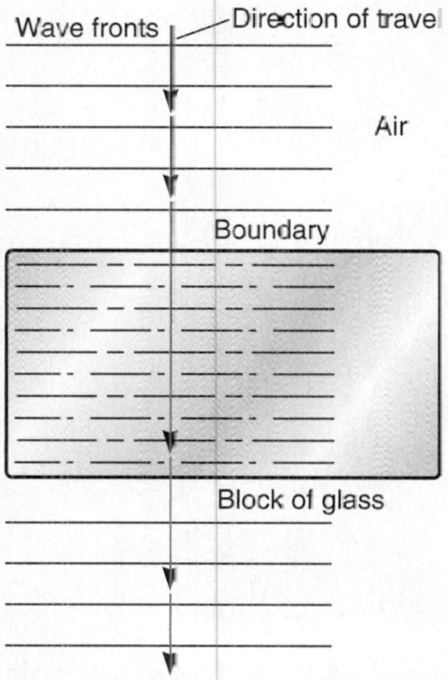

Figure 20.18 The speed of light is different in different mediums.

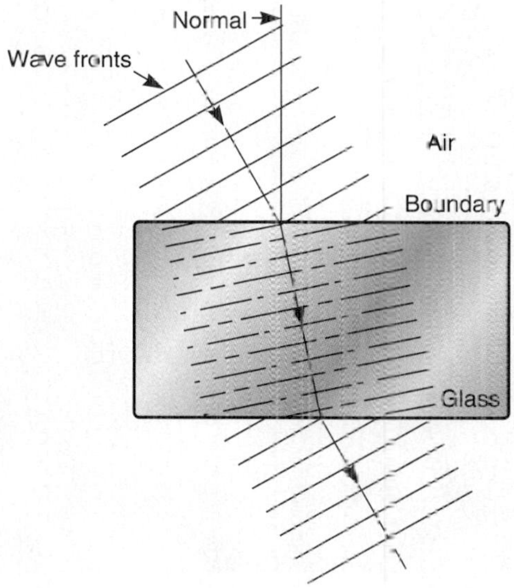

Figure 20.19 The wave bends when all parts of the wave don't strike the glass at the same time. It also bends when leaving the glass.

When the wave passes through at an angle, the entire wave front does not all strike the surface at the same time. The first part of the wave to strike the glass is slowed before the part striking later—thus the bending of the wave (Fig. 20.19). Draw your own diagram to show whether a fish in a pond, as viewed from the bank, is actually nearer the surface or the bottom of the pond than it appears to be. (see Fig. 19.2)

Another way to understand refraction is to compare it to a child's wagon rolling along a sloping sidewalk. The front wheel rolls off the edge and goes into the grass. The interaction of the wheel with the rougher surface, the grass, causes the wagon to change direction and to be pulled toward the grass. The wheel that hits the grass slows down first while the opposite wheel is still rolling on the sidewalk. The wagon pivots, and its path is bent toward the grass.

A mirage is not a figment in a person's imagination. It is actually formed by light and can be photographed (Fig. 20.20). Light travels faster through the very hot air near the ground on a hot day, and an image appears to the observer to shimmer as if it had been reflected from the surface of water. It is not reflected, however, but refracted. The twinkling of stars is also very real, and occurs as light passes through unstable layers in the atmosphere.

As we might expect, since the speed of light increases when a light wave leaves a denser medium to enter the air, bending also occurs. In this case, however, instead of the light bending toward the perpendicular or normal, the light is bent away from the normal when passing from the denser medium to the less dense one (Fig. 20.19). This illustrates the following law:

LAW OF REFRACTION

When a beam of light passes at an angle from a medium of lower optical density to a denser medium, the light is bent *toward* the normal. When a beam of light passes at an angle from a medium of greater optical density to one less dense, the light is bent *away from* the normal.

Figure 20.20 A mirage is actually formed by light and can be photographed.

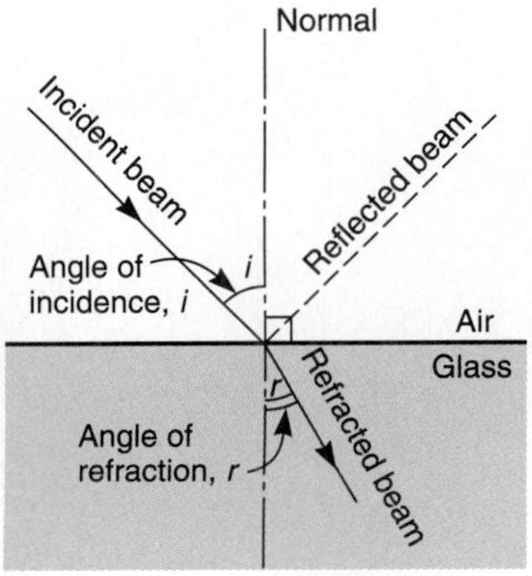

Figure 20.21 The angles of incidence and refraction are measured from the normal.

Note: The angles of incidence and refraction are measured from the *normal* as in Fig. 20.21.

Willebord Snell found a formula for a measure of the optical density of a material that can pass light called the **index of refraction.** This index is a constant for a particular material and is independent of the angle at which the light strikes. **Snell's law** may be expressed in two ways:

Willebord Snell (1580–1626), *mathematician, was born in The Netherlands. He discovered the law of refraction known as Snell's law and extensively developed the use of triangulation in surveying.*

1. The index of refraction, n, equals the sine of the angle of incidence divided by the sine of the angle of refraction, where the incident ray travels through a vacuum or air:

$$n = \frac{\sin i}{\sin r}$$

2. The index of refraction, n, for a given substance is the ratio of the speed of light in a vacuum (nearly the same as in air) to the speed of light in the substance:

$$n = \frac{\text{speed of light in vacuum}}{\text{speed of light in substance}}$$

EXAMPLE 1

The angle of incidence of light passing from air to water is 61.0°. The angle of refraction is 41.0°. What is the index of refraction of the water?

Data:

$$i = 61.0°$$
$$r = 41.0°$$
$$n = ?$$

Basic Equation:

$$n = \frac{\sin i}{\sin r}$$

Working Equation: Same

Substitution:

$$n = \frac{\sin 61.0°}{\sin 41.0°}$$

$$= 1.33$$

EXAMPLE 2

The index of refraction of water is 1.33. What is the speed of light in water?

Data:

$$n = 1.33$$

$$c = 3.00 \times 10^8 \text{ m/s}$$

$$v_{\text{water}} = ?$$

Basic Equation:

$$n = \frac{\text{speed of light in vacuum}}{\text{speed of light in substance}}$$

Working Equation:

$$\text{speed of light in water} = \frac{\text{speed of light in vacuum}}{n}$$

Substitution:

$$\text{speed of light in water} = \frac{3.00 \times 10^8 \text{ m/s}}{1.33}$$

$$= 2.26 \times 10^8 \text{ m/s}$$

The index of refraction for some common substances is given in Table 20.1.

Table 20.1 Indices of Refraction for Various Substances

Substance	Index of Refraction
Air, dry (STP)	1.00029
Alcohol, ethyl	1.360
Benzene	1.501
Carbon dioxide (STP)	1.00045
Carbon disulfide	1.625
Carbon tetrachloride	1.459
Diamond	2.417
Glass, crown flint	1.575
Lucite	1.50
Quartz, fused	1.45845
Water, distilled	1.333
Water vapor (STP)	1.00025
Zircon	1.92

20.7 Total Internal Reflection

If the angle of refraction is 90° or greater, a beam of light does not leave the medium but is reflected inside it (Fig. 20.22). This is called **total internal reflection.** Also note the total internal reflection of the blue light beam in Fig. 20.17. Total internal reflection occurs when the angle of incidence is greater than the critical angle. The **critical angle** is the smallest angle of incidence at which all light striking the surface is totally internally reflected. It may be expressed by the formula

$$\sin i_c = \frac{1}{n}$$

where

i_c = critical angle of incidence

n = index of refraction of denser medium

Note: The incident ray travels through a vacuum or air.

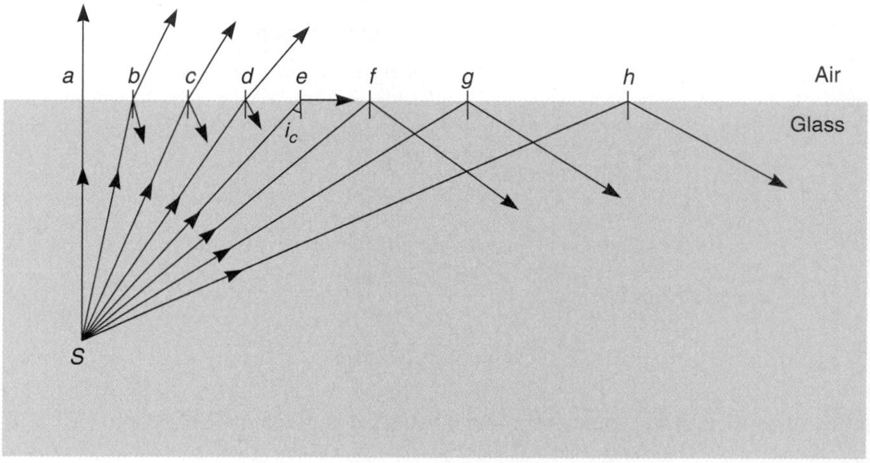

Figure 20.22 A point light source *S* is shown in glass. For all angles of incidence less than the critical angle i_c, the light ray is partially reflected but most refracted and leaves the glass as shown at *a*, *b*, *c*, and *d*. When the angle of incidence equals the critical angle i_c, the refracted ray points along the glass–air surface as shown at *e*. For all angles of incidence greater than the critical angle i_c, total internal reflection of the light ray occurs as shown at *f*, *g*, and *h*.

TRY THIS ACTIVITY

Water Fiber Optics

A 2-L plastic soda pop bottle and a pen laser are needed for this activity. Make a small, pencil-sized hole in the lower wall of an empty plastic soda pop bottle. Placing a finger over the hole, fill the bottle with water. Shine the laser pen through the bottle so the laser light strikes the hole where your finger is located. Remove your finger from the hole. Have someone stand in front of the hole and observe what is happening to the laser light as the water and the light emerge from the hole. Explain, using internal reflection principles, why the laser light behaves as it does when it emerges from the bottle.

EXAMPLE

What is the critical angle of incidence for water that has an index of refraction of 1.33?

Data:

$$n = 1.33$$
$$i_c = ?$$

Basic Equation:

$$\sin i_c = \frac{1}{n}$$

Working Equation: Same

Substitution:

$$\sin i_c = \frac{1}{1.33} = 0.752$$
$$i_c = 49°$$

Where there is total internal reflection, no light enters the air; it is totally reflected within the glass [see Figs. 20.23(a) and 20.26]. The property of having a very small critical angle gives a diamond its brilliance due to multiple internal reflections occurring before the light passes out through the top. An example of the practical application of this principle is fiber optics (Fig. 20.23). Light may be transferred inside flexible glass or plastic fibers, which are transparent but keep nearly all the light inside because the light is reflected along the inside surface of the fibers.

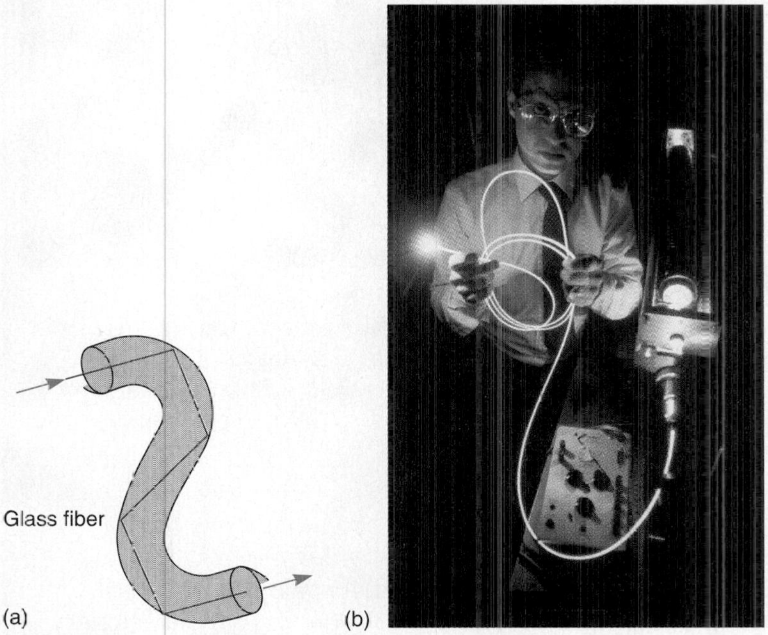

Figure 20.23 **(a) Light travels inside an optical or glass fiber like a light pipe in which the light is always incident at an angle greater than the critical angle. Thus, no light escapes the optical fiber by refraction. Fiber optics is essential to light-wave communications systems. (b) Total internal reflection within the tiny fibers of this light pipe makes it possible to transmit light along complex paths with minimal loss.**

PHYSICS CONNECTIONS

Fiber Optic Cables

Most transmission of information travels as electric impulses through electric and telephone lines and fiber optic cables. Electric signals travel relatively slowly, cause wires to heat up, and need transformers to boost the voltage of the signal traveling over long distances. Electric signals and wires are being replaced with light signals traveling through flexible, low-cost strands of glass. Because light travels through glass optical fibers, there is no resistance to weaken the light signal, and the signal travels at the speed of light, which is much faster than the speed of conventional electric signals. Such advances in fiber optics communications are revolutionizing the way we communicate.

Light traveling in the same medium travels in a straight line, whereas fiber optic cables can transmit a signal while twisting and turning, because of total internal reflection. The angle at which the light strikes the cladding of the fiber is always greater than the critical angle of the cladding and the core. The low critical angle allows the light to continually reflect and travel great distances without needing to be reamplified. In order for the cable to maintain a low critical angle, the glass must contain no imperfections or bubbles that would cause the light to be directed out or backward through the cable (see Fig. 20.24).

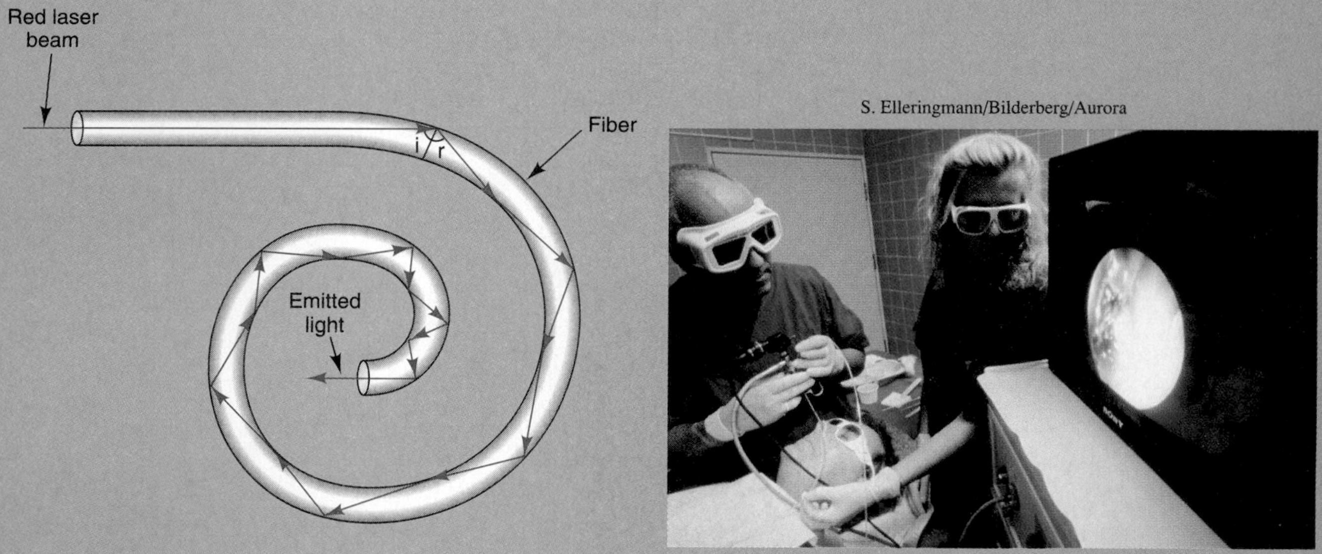

S. Elleringmann/Bilderberg/Aurora

(a) (b)

Figure 20.24 (a) The red laser light entering the fiber optic cable is totally internally reflected, which results in the light emerging at the end of the cable. (b) An endoscope is a bundle of fiber optic cables used in many minimally invasive surgeries. Here an endoscope is used in the removal of nose adenoids with a laser therapy procedure.

Fiber optic cables are used in telecommunications, computer networks, and medicine. A few strands of glass fiber can carry thousands of separate digital telephone conversations by slightly altering the frequency of the light for each phone conversation. A digital signal transmitted at one frequency cannot be confused by a signal carried at another frequency. Many computer networks and internal components in computers use fiber optic cables to carry data. By eliminating electric wiring, the fiber optic cable helps to reduce the temperature inside computers and servers. Finally, physicians use fiber optic bundles to perform minimally invasive procedures. A tool called an endoscope, composed of a bundle of fiber optic cables, transmits light into a patient's body while another bundle of fibers on the endoscope functions as a digital camera. The camera picks up the image and sends it back through the fiber optic cable to a monitor in the procedure room.

20.8 Types of Lenses

Many technical applications, ranging from apparatus to test the nature of liquids to microscopes to eyeglasses, use the principles of refraction. We now consider the use of refraction in applications using lenses. Lenses may be converging or diverging. **Converging lenses** bend the light passing through them to some point beyond the lens. Converging lenses are always thicker in the center than on the edges [Fig. 20.25(a)]. **Diverging lenses** bend the light passing through them so as to spread the light. Diverging lenses are thicker on the edges than at the center [Fig. 20.25(b)].

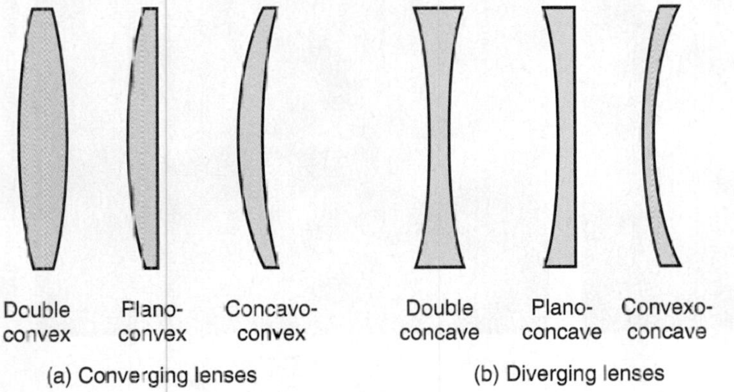

Figure 20.25

Understanding how light is bent in lenses may be made easier by observing the bending of light passing through a prism (Fig. 20.26). Recall and apply the law of refraction, noting that light is bent toward the normal when passing into the glass and away from the normal when passing from the glass back to the air.

Figure 20.26 Bending of light passing through a prism. Note that some point light sources are refracted up through the prism and some are totally internally reflected inside the prism before being refracted through the prism to the right.

20.9 Images Formed by Converging Lenses

As with mirrors, we use light-ray diagrams to show how light can be bent with lenses. Every lens has a focal length—that distance from the lens center to the point where parallel beams directed through the lens come together if converging or *appear* to come together if diverging (Fig. 20.27).

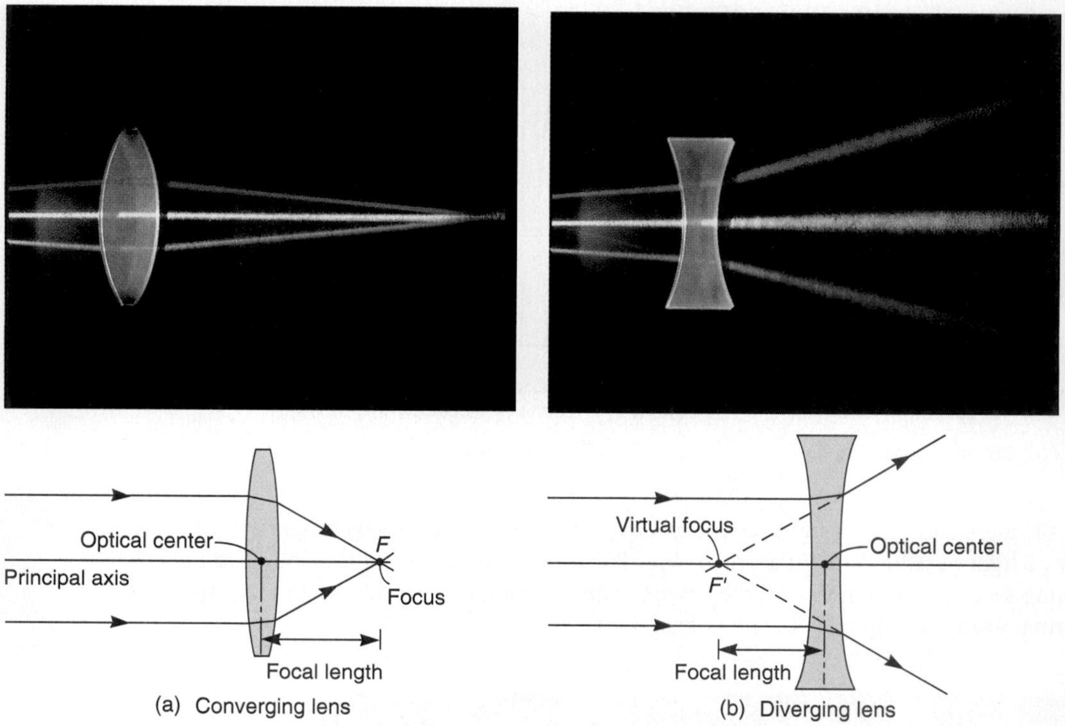

Figure 20.27

The location of the focus depends upon the curvature of the lens and the index of refraction of the glass or material of which the lens is made. Rays of light passing through the optical center of the lens are refracted so little that we may consider them as going straight through (Fig. 20.28).

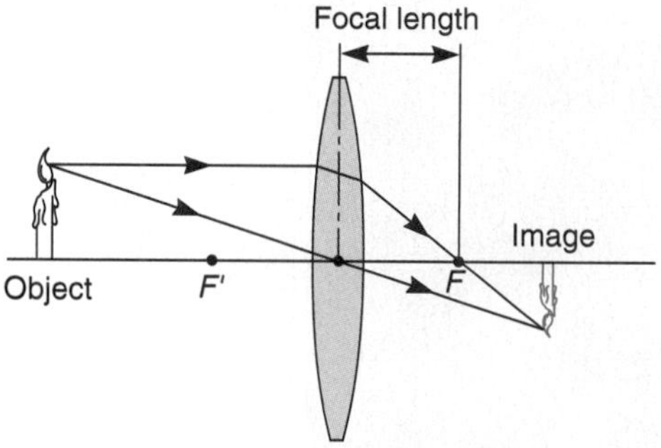

Figure 20.28 Lens diagram for the formation of an image using a convex lens.

Now apply these principles to the diagrams shown in Fig. 20.29 of other images formed by converging lenses, depending on the object location, and decide whether the image is real or virtual or no image formed at all; and whether the image is erect or inverted; larger, smaller, or the same size; and where it is located.

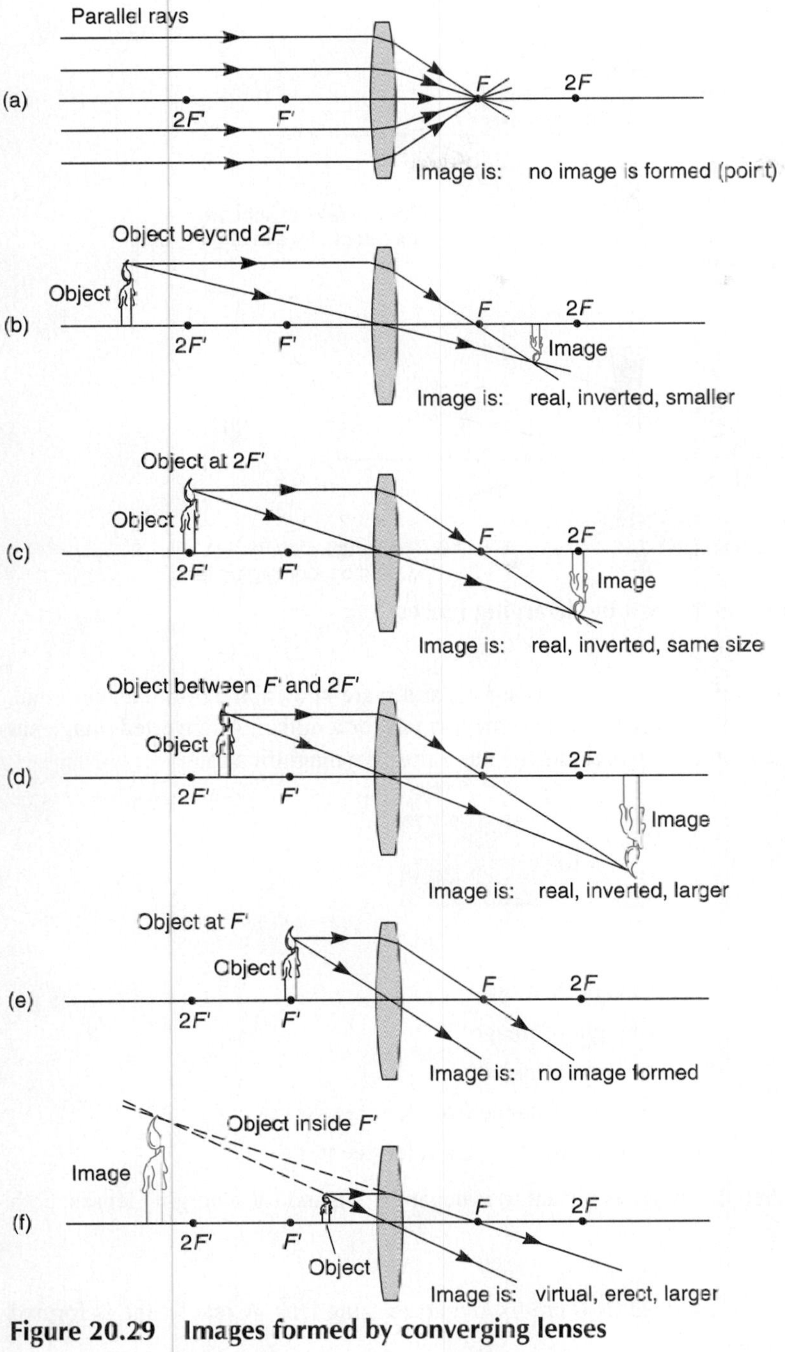

Figure 20.29 Images formed by converging lenses

20.10 Images Formed by Diverging Lenses

Virtual images are the only images produced by diverging lenses (Fig. 20.30). The same formulas that apply to mirrors also apply to converging and diverging lenses.

$$\frac{1}{f} = \frac{1}{s_o} + \frac{1}{s_i}$$

where

f = focal length

s_o = object distance from lens center

s_i = image distance from lens center

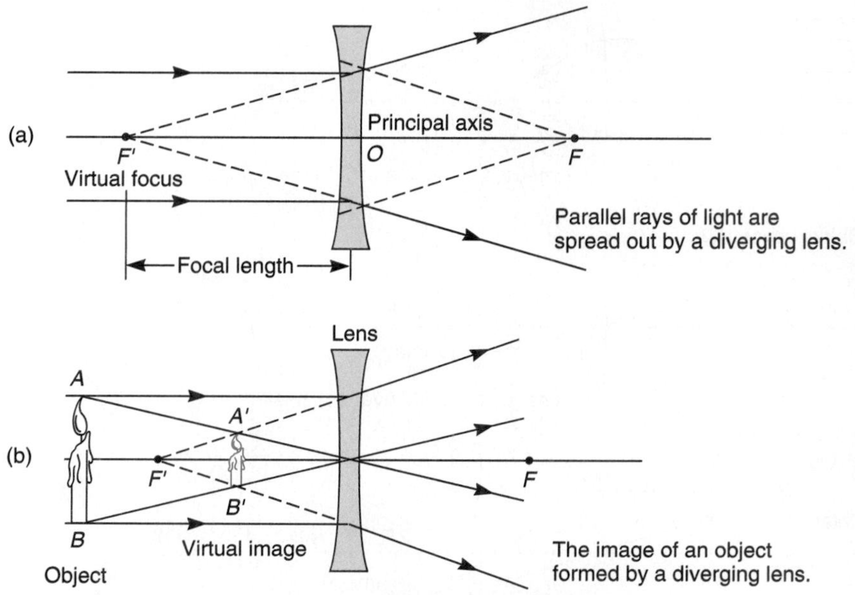

Figure 20.30 Images formed by diverging lenses

Therefore, if two of the three distances, f, s_o, and s_i are known, the third can be found. Also, the magnification of a lens is in the same form as for a mirror. An inverted image has a negative magnification and an erect image has a positive magnification:

$$M = \frac{h_i}{h_o} = \frac{-s_i}{s_o}$$

where

M = magnification
h_i = height of image
h_o = height of object
s_i = image distance from lens center
s_o = object distance from lens center

Remember, when the image is virtual, s_i is negative (−), and for diverging lenses, both s_i and f are negative.

EXAMPLE

An object 3.00 cm tall is placed 24.0 cm from a converging lens. A real image is formed 8.00 cm from the lens.

(a) What is the focal length of the lens?
(b) What is the size of the image?

(a) Data:

$s_o = 24.0$ cm
$s_i = 8.00$ cm
$f = ?$

Basic Equation:

$$\frac{1}{f} = \frac{1}{s_o} + \frac{1}{s_i}$$

Working Equation: Same

Substitution:

$$\frac{1}{f} = \frac{1}{24.0\text{ cm}} + \frac{1}{8.00\text{ cm}}$$

$$f = 6.00\text{ cm}$$

The key entry sequence on a scientific calculator for this calculation is

24 [x^{-1}] [+] 8 [x^{-1}] [=] [x^{-1}] [=]

[6]

(b) Data:

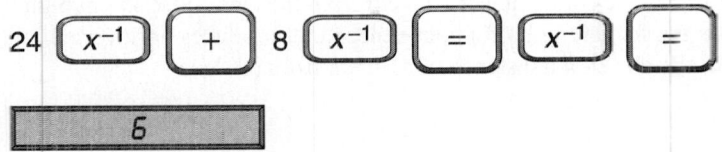

$$s_o = 24.0\text{ cm}$$
$$s_i = 8.00\text{ cm}$$
$$h_o = 3.00\text{ cm}$$
$$h_i = ?$$

Basic Equation:

$$\frac{h_i}{h_o} = \frac{-s_i}{s_o}$$

Working Equation:

$$h_i = \frac{-s_i h_o}{s_o}$$

Substitution:

$$h_i = \frac{-(8.00\text{ cm})(3.00\text{ cm})}{24.0\text{ cm}}$$
$$= -1.00\text{ cm}$$

PROBLEMS 20.10

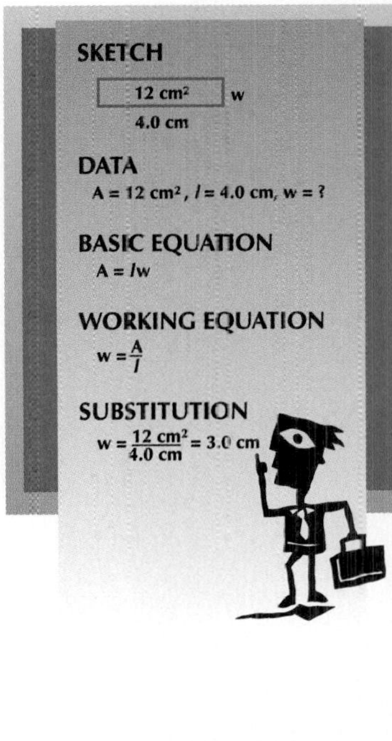

1. Find the index of refraction of a medium for which the angle of incidence of a light beam is 31.5° and angle of refraction is 25.6°.
2. If the index of refraction of a medium is 2.40 and the angle of incidence is 14.6°, what is the angle of refraction?
3. If the index of refraction of a liquid is 1.50, find the speed of light in that liquid.
4. The angle of incidence of light passing from air to a liquid is 38.0°. The angle of refraction is 24.5°. What is the index of refraction of the liquid?
5. If the critical angle of a liquid is 42.4°, find the index of refraction for that liquid.
6. If the index of refraction of a substance is 2.45, find its critical angle of incidence.
7. A converging lens has a focal length of 15.0 cm. If it is placed 48.0 cm from an object, how far from the lens will the image be formed?
8. An object 2.50 cm tall is placed 20.0 cm from a converging lens. A real image is formed 9.00 cm from the lens. (a) What is the focal length of the lens? (b) What is the size of the image?
9. The focal length of a lens is 5.00 cm. How far from the lens must the object be to produce an image 1.50 cm from the lens?
10. If the distance from the lens in your eye to the retina is 19.0 mm, what is the focal length of the lens when reading a sign 40.0 cm from the lens?

11. An object 5.00 cm tall is placed 15.0 cm from a converging lens, and a real image is formed 7.50 cm from the lens. (a) What is the focal length of the lens? (b) What is the size of the image?
12. An object 4.50 cm tall is placed 18.0 cm from a converging lens with a focal length of 26.0 cm. (a) What is the location of the image? (b) What is its size?
13. What are the size and location of an image produced by a converging lens with a focal length of 19.5 cm of an object 5.76 cm from the lens and 1.45 cm high?
14. What are the size and location of an image produced by a convex lens with a focal length of 14.5 cm of an object 10.5 cm from the lens and 2.35 cm high?
15. What is the focal length of a convex lens that produces an inverted image twice as large as the object at a distance of 13.3 cm from the lens?

Glossary

Concave Mirror A mirror with a surface that curves away from an observer. (p. 559)

Converging Lens A lens that bends the light passing through it to some point beyond the lens. Converging lenses are thicker in the center. (p. 573)

Convex Mirror A mirror with a surface that curves inward toward an observer. (p. 559)

Critical Angle The smallest angle of incidence at which all light striking a surface is totally internally reflected. (p. 570)

Diffusion Scattering of light by an uneven surface. (p. 558)

Diverging Lens A lens that bends the light passing through it so as to spread the light. Diverging lenses are thicker at the edges than at the center. (p. 573)

First Law of Reflection The angle of incidence equals the angle of reflection. (p. 558)

Focal Length The distance between the principal focus of a mirror or lens and its vertex. (p. 561)

Index of Refraction A measure of the optical density of a material. Equal to the ratio of the speed of light in vacuum to the speed of light in the material. (p. 568)

Law of Refraction When a beam of light passes at an angle from a medium of lower optical density to a denser medium, the light is bent toward the normal. When a beam passes from a medium of greater optical density to one less dense, the light is bent away from the normal. (p. 568)

Opaque Absorbing or reflecting almost all light. (p. 558)

Optical Density A property of a transparent material that is a measure of the speed of light through the given material. (p. 567)

Plane Mirror A mirror with a flat surface. (p. 559)

Real Image An image formed by rays of light. (p. 559)

Reflection The turning or turning back of all or part of a beam of light at a surface. (p. 558)

Refraction The bending of light as it passes at an angle from one medium to another of different optical density. (p. 567)

Regular Reflection Reflection of light with very little scattering. (p. 558)

Second Law of Reflection The incident ray, the reflected ray, and the normal (perpendicular) to the reflecting surface all lie in the same plane. (p. 559)

Snell's Law The index of refraction equals the sine of the angle of incidence divided by the sine of the angle of refraction. (p. 568)

Total Internal Reflection A condition such that light striking a surface does not pass through the surface but is completely reflected inside it. (p. 570)

Translucent Allowing some but not all light to pass through. (p. 558)

Transparent Allowing almost all light to pass through so that objects or images can be seen clearly. (p. 558)

Virtual Image An image that only appears to the eye to be formed by rays of light. (p. 559)

Formulas

20.5 $$\frac{1}{f} = \frac{1}{s_o} + \frac{1}{s_i}$$

$$M = \frac{h_i}{h_o} = \frac{-s_i}{s_o}$$

20.6 $$n = \frac{\sin i}{\sin r}$$

$$n = \frac{\text{speed of light in vacuum}}{\text{speed of light in substance}}$$

20.7 $$\sin i_c = \frac{1}{n}$$

20.10 $$\frac{1}{f} = \frac{1}{s_o} + \frac{1}{s_i}$$

$$M = \frac{h_i}{h_o} = \frac{-s_i}{s_o}$$

Review Questions

1. Stained glass is an example of
   (a) a transparent material. (b) a translucent material.
   (c) an opaque material. (d) none of the above.
2. A virtual image may be
   (a) larger than the object. (b) smaller than the object.
   (c) erect. (d) all of the above.
   (e) none of the above.
3. A real image may be
   (a) erect. (b) shown on a screen.
   (c) formed by a plane mirror. (d) none of the above.
4. Explain the difference between diffusion and regular reflection.
5. In your own words, explain the first law of reflection.
6. In your own words, explain the second law of reflection.
7. Describe the type of images formed by plane mirrors.
8. Explain the difference between real and virtual images.
9. Explain the difference between a concave and a convex mirror.
10. Explain the effect of spherical aberration.
11. For a mirror of given focal length, how does the image distance change if the object distance is decreased?
12. For a given object distance from a mirror, how does the image distance change if the focal length is increased?
13. The index of refraction depends on
    (a) the focal length. (b) the speed of light.
    (c) the image distance. (d) none of the above.
14. Snell's law involves
    (a) the lens equation. (b) the index of refraction.
    (c) the focal length. (d) none of the above.
15. Explain the difference between converging and diverging lenses.
16. Give several examples of total internal reflection.
17. In your own words, explain the law of refraction.
18. How does the speed of light in a high-index-of-refraction material compare to the speed of light in a vacuum?
19. In your own words, explain why light waves are refracted at a boundary between two materials.

20. What types of images are formed by diverging lenses?
21. What types of images are formed by converging lenses?
22. How do water waves affect the escape of light from below the surface of the water?
23. Explain why a fish under water appears to be at a different depth below the surface than it actually is. Does it appear deeper or shallower?
24. Does light always travel in a straight line? Explain.
25. Explain how total internal reflection allows light in a glass fiber to be guided along the fiber.
26. Under what conditions will a converging lens form a virtual image?
27. Under what conditions will a converging lens form a real image that is the same size as the object?
28. Under what conditions will a diverging lens form a virtual image that is smaller than the object?

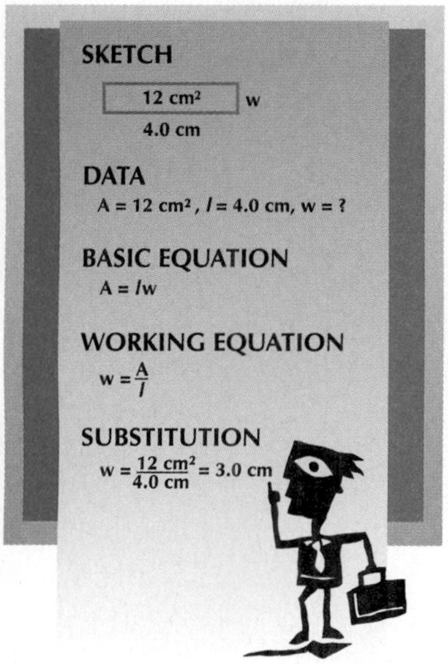

Review Problems

1. Using $\frac{1}{f} = \frac{1}{s_o} + \frac{1}{s_i}$, $s_o = 3.50$ cm, and $s_i = 7.25$ cm, find f.
2. Using $\frac{1}{f} = \frac{1}{s_o} + \frac{1}{s_i}$, $s_o = 8.50$ cm, and $f = 25.0$ cm, find s_i.
3. Using $M = \frac{h_i}{h_o} = \frac{-s_i}{s_o}$, $h_o = 6.50$ cm, $s_i = 7.50$ cm, and $s_o = 14.0$ cm, find h_i.
4. If an object is 3.75 m tall and 7.35 m from a large mirror with an image formed 4.35 m from the mirror, what is the height of the image?
5. An object 43.0 cm tall is located 23.4 cm from a concave mirror with focal length 21.4 cm. (a) Where is the image located? (b) How high is the image?
6. An object and its image in a concave mirror are the same height, but inverted, when the object is 45.3 cm from the mirror. What is the focal length of the mirror?
7. The angle of incidence of light passing from air to a liquid is 41.0°. The angle of refraction is 29.0°. Find the index of refraction of the liquid.
8. If the index of refraction of a liquid is 1.44, find the speed of light in that liquid.
9. If the critical angle of a liquid is 45.6°, find the index of refraction for that liquid.
10. If the index of refraction of a substance is 1.50, find its critical angle of incidence.
11. A converging lens has a focal length of 12.0 cm. If it is placed 36.0 cm from an object, how far from the lens will the image be formed?
12. An object 4.50 cm tall is placed 20.0 cm from a converging lens. A real image is formed 12.0 cm from the lens. (a) What is the focal length of the lens? (b) What is the size of the image?
13. The focal length of a lens is 4.00 cm. How far from the lens must the object be to produce an image 7.20 cm from the lens?
14. What is the focal length of a convex lens that produces an image three times as large as the object at a distance of 25.0 cm from the lens?
15. What is the focal length of a mirror that forms an image 3.44 m behind a convex mirror of an object 5.33 m in front of the mirror?
16. What are the size and location of an image produced by a convex lens with a focal length of 21.0 cm of an object 11.5 cm from the lens and 3.25 cm high?
17. What is the speed of light passing through a diamond? See Table 20.1.
18. Find the critical angle of incidence for Lucite. See Table 20.1.
19. Find the focal length of a concave mirror with an object 39.3 cm in front of it that projects an image 17.8 cm in front of the mirror.
20. (a) Find the height and location of an image produced by a concave mirror with focal length 8.70 cm and an object that is 13.2 cm tall and 19.3 cm from the mirror. (b) Find the height of the image produced by the mirror if the object is twice as far from the mirror.

APPLIED CONCEPTS

1. Tamera uses a concave mirror when applying makeup. (a) The mirror has a radius of curvature of 38.0 cm. What is the focal length of the mirror? (b) Tamera's face is located 12.5 cm from the mirror. Where will her image appear? (c) Will the image be upright or inverted? (d) How long will her face appear to be in the mirror if her face is actually 25.0 cm long?
2. A convex security mirror has a radius of curvature of −1.50 m. (a) A shoplifter picks up a 0.255-m-high purse 11.5 m from the mirror. How far behind the mirror is the image of the purse? (b) Is the image real or virtual? (c) How high is the image of the purse? (d) Are convex security mirrors better for viewing detailed images or large fields of view?
3. A fish tank made of crown glass is full of fresh water. (a) What is the angle of refraction when the light travels from the air and enters the crown glass at an incident angle of 30.0°? (b) Determine the angle of refraction when the light leaves the crown glass and enters the water. (c) If the aquarium is full of saltwater, will the angle of refraction be greater or less than the angle of refraction for fresh water?
4. Diamonds are cut to take advantage of internal reflections of light. The angles of a diamond are cut so that all of the light that enters it reflects out the top. The light or brilliance that emerges from the top of a diamond is one of the qualities that determines its monetary value. (a) What is the critical angle for light traveling from a diamond to air? (b) Would diamond jewelers want to create angles smaller or larger than the critical angle? (c) Zircon is an inexpensive substitute for diamond. Zircon's index of refraction is 1.92. What is its critical angle? (d) If a zircon has the same cut as a diamond, which will have a greater brilliance by internally reflecting more light to the top of the stone?
5. A photographer uses a 60.0-mm lens. (a) How far away should the lens be from the film if the object is located 9.20 m from the lens? (b) What will be the height of the image if the object is 2.00 m tall?

CHAPTER 21

Colors of the visible Spectrum

COLOR

Understanding why a beam of sunlight forms a spectrum when directed through a glass prism, why the sky is blue, why sunsets are red, why clouds are white, why the ocean is blue, why we see rainbows, how colors of paint are mixed, how Polaroid sunglasses work, and how the LCD screen in a calculator works are basic questions most people commonly ask. Our examination of these basic questions forms the basis of the development and discussion of basic color properties.

Objectives

The major goals of this chapter are to enable you to:

1. Describe how the colors of the visible spectrum are formed through dispersion of light.
2. Describe color as a property of light and how it is related to its frequency or its wavelength.
3. Describe the result of mixing colors.
4. Describe diffraction as a property of light.
5. Describe interference as a property of light.
6. Describe polarization of light and some of its applications.

21.1 The Color of Light

Figure 21.1 A narrow beam of sunlight directed into a clear glass prism in a dark room is dispersed into the visible spectrum.

Let us observe a narrow beam of sunlight that is directed into and passes through a glass prism in a dark room as in Fig. 21.1. Note the band of colors with one color shade gradually blending into another. This band of colors is called the **visible spectrum.** This spreading of white light into the full spectrum is called **dispersion.** Dispersion was described by Newton, who observed six colors: red, orange, yellow, green, blue, and violet. Sunlight is an example of white light. The *color* of the light is related to its wavelength or its frequency. Red light has lower frequency and longer wavelength, whereas at the opposite end of the spectrum violet light has higher frequency and shorter wavelength. *Polychromatic light* is light consisting of several colors. *Monochromatic light* consists of only one color.

As you can see from the dispersion of light, the refraction of the red light is less than that of the others and the refraction of the violet light is the greatest (Fig. 21.2). The shorter wavelengths are refracted more than the longer wavelengths. Thus, **color** is a property of the light that reaches our eyes and is determined by its wavelength or its frequency. Light waves with wavelengths outside the visible spectrum as shown in Fig. 19.5 are not visible to the human eye. Light waves within the visible spectrum have wavelengths in the range of 4.0×10^{-7} m (for violet) to 7.5×10^{-7} m (for red). Some photographic films are sensitive to ultraviolet (UV) light and others, to infrared (IR) light.

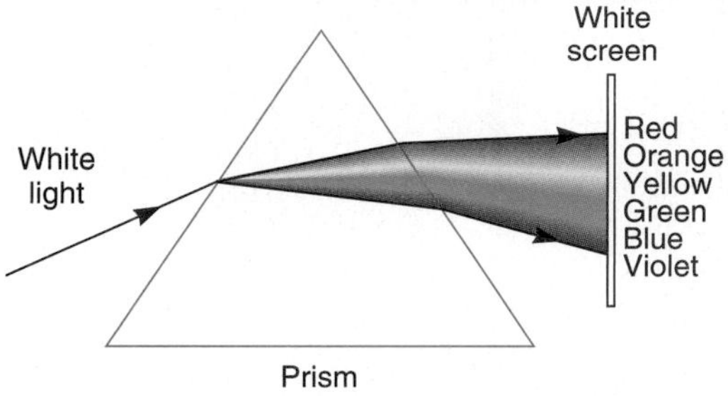

Figure 21.2 White light is dispersed by a prism into the visible spectrum on a white screen.

Most of the objects we see reflect rather than emit light. A shirt is called red if it reflects only red light and absorbs all other colors [Fig. 21.3(a)]. A shirt is called blue if it reflects only blue light and absorbs all other colors [Fig. 21.3(b)]. The color of an opaque object depends on the color of the light it reflects.

A white object reflects all the colors (light rays of various wavelengths) it receives. A black object absorbs all the colors that fall upon it. A shirt that is blue in sunlight appears

(a) (b)

Figure 21.3 (a) A red shirt reflects only red light and absorbs light of all other colors. (b) A blue shirt reflects only blue light and absorbs light of all other colors.

black in red light because there is no blue light present for it to reflect, and it absorbs all the other colors. Similarly, this is why a red shirt appears black in blue light.

Regular window glass transmits all colors and is often called colorless. Red glass absorbs all colors except red, which it transmits. Blue glass absorbs all colors except blue, which it transmits (Fig. 21.4). The color of a transparent object depends on the color of the light it transmits.

Figure 21.4 Red glass transmits only red light and absorbs light of all other colors. Blue glass transmits only blue light and absorbs light of all other colors.

Mixing Colors of Light

Just as polychromatic light, such as sunlight (sometimes called white light), can be dispersed into its separate colors, the separate colors can be combined to form polychromatic light. White light can also be produced by projecting overlapping red, green, and blue light onto a white screen as shown in Fig. 21.5. For this reason, red, green, and blue are called the **primary colors**. This is an additive mixture of the three colors of light.

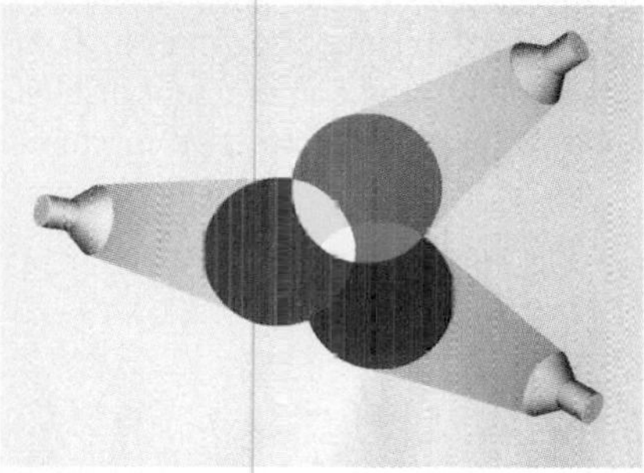

Figure 21.5 When the three primary colors overlap (light mixed by addition), white light is produced.

Note in Fig. 21.5 the colors cyan, magenta, and yellow are produced when only two of the primary colors overlap, that is,

$$\text{blue} + \text{green} = \text{cyan}$$
$$\text{red} + \text{blue} = \text{magenta}$$
$$\text{red} + \text{green} = \text{yellow}$$

The color cyan is often called the opposite of red, magenta is the opposite of green, and yellow is the opposite of blue. Thus,

$$\text{cyan} + \text{red} = \text{white} = (\text{blue} + \text{green}) + \text{red}$$
$$\text{magenta} + \text{green} = \text{white} = (\text{red} + \text{blue}) + \text{green}$$
$$\text{yellow} + \text{blue} = \text{white} = (\text{red} + \text{green}) + \text{blue}$$

When any two colors are combined to form white, they are called **complementary colors.** Thus, cyan and red, magenta and green, and yellow and blue are complementary colors.

Colors of light are commonly mixed on theatrical stages to produce many different color effects (Fig. 21.6).

Figure 21.6 Bright spotlights at this night club show multiple colors using overlapping colored lights.

Mixing Pigments

Mixing pigments in paints and dyes is completely different from mixing lighting. Artists and painters know that mixing red, green, and blue paint results in a murky dark brown—not white! Pigments are actually small particles that absorb, rather than reflect, given colors. When paint pigments are mixed, each one subtracts, or absorbs, certain colors from white light with the resulting reflected color depending on the light waves *not* absorbed. The **primary pigments** are the complements of the three primary colors, namely, cyan (the complement of red), magenta (the complement of green), and yellow (the complement of blue) (see Fig. 21.7). Comparing additive mixtures and the corresponding subtractive mixtures, we have the following:

Additive Mixture	*Corresponding Subtractive Mixture*
cyan + red = white	red = white − cyan
magenta + green = white	green = white − magenta
yellow + blue = white	blue = white − yellow

That is, pigments that produce red absorb its complement, cyan, and reflect red. In other words, taking cyan away from white produces red. Pigments that produce green absorb its

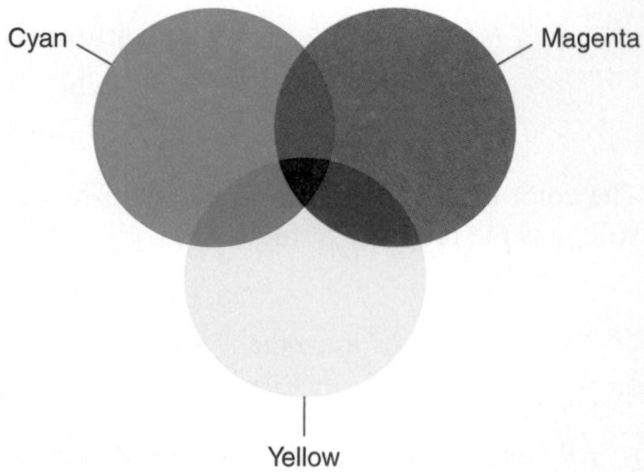

Figure 21.7 Primary pigments. The results of the subtractive combination of cyan, magenta, and yellow filters are shown. Note that the combination of any two primary pigments gives the complement of the third by subtraction, namely, red, green, and blue. The subtractive combination of all three primary pigments transmits no light.

complement, magenta, and reflect green. Or, taking magenta away from white produces green. Pigments that produce blue absorb its complement, yellow, and reflect blue. Or, taking yellow away from white produces blue.

When the three primary pigments are mixed in appropriate proportions, a large range of colors can be produced.

Color Printing

All color photographs printed in books, magazines, and newspapers use only four colors of ink: cyan, magenta, and yellow for the colors and black for the shadow areas and definition. Four negatives are made using different filters so that each negative has the given amount of its color corresponding to the original color photograph. A red filter is used for making the cyan-printer negative, a green filter is used for making the magenta-printer negative, and a blue filter is used for making the yellow-printer negative. Each filter makes the color to be printed by its corresponding plate photograph as black and the other two colors photograph as white. The negative for the black plate is made by partial exposures through each of the three color filters so that black and shadows will print where all three colors are present.

A metal printing plate is made from each of these four negatives. Each metal printing plate is treated to hold the printer's ink in areas that have been exposed to light. Each of the four plates is printed in the same position, one after the other, with great precision on white paper. (Recall seeing a color photograph in a newspaper or magazine where the plates or the paper was not within the proper precision and the colors were out of synchronization.) The lightness and the darkness of each color are controlled by the size of the very small printing dots. Use a magnifying glass to examine the very small printing dots in any color photograph in this text, a magazine, or a newspaper. Note that not only do the dots provide the basic colors, but also the overlapping dots of these basic colors give the appearance of many other colors. Figure 21.8 illustrates how only four colors produce color photographs and illustrations with many vibrant colors.

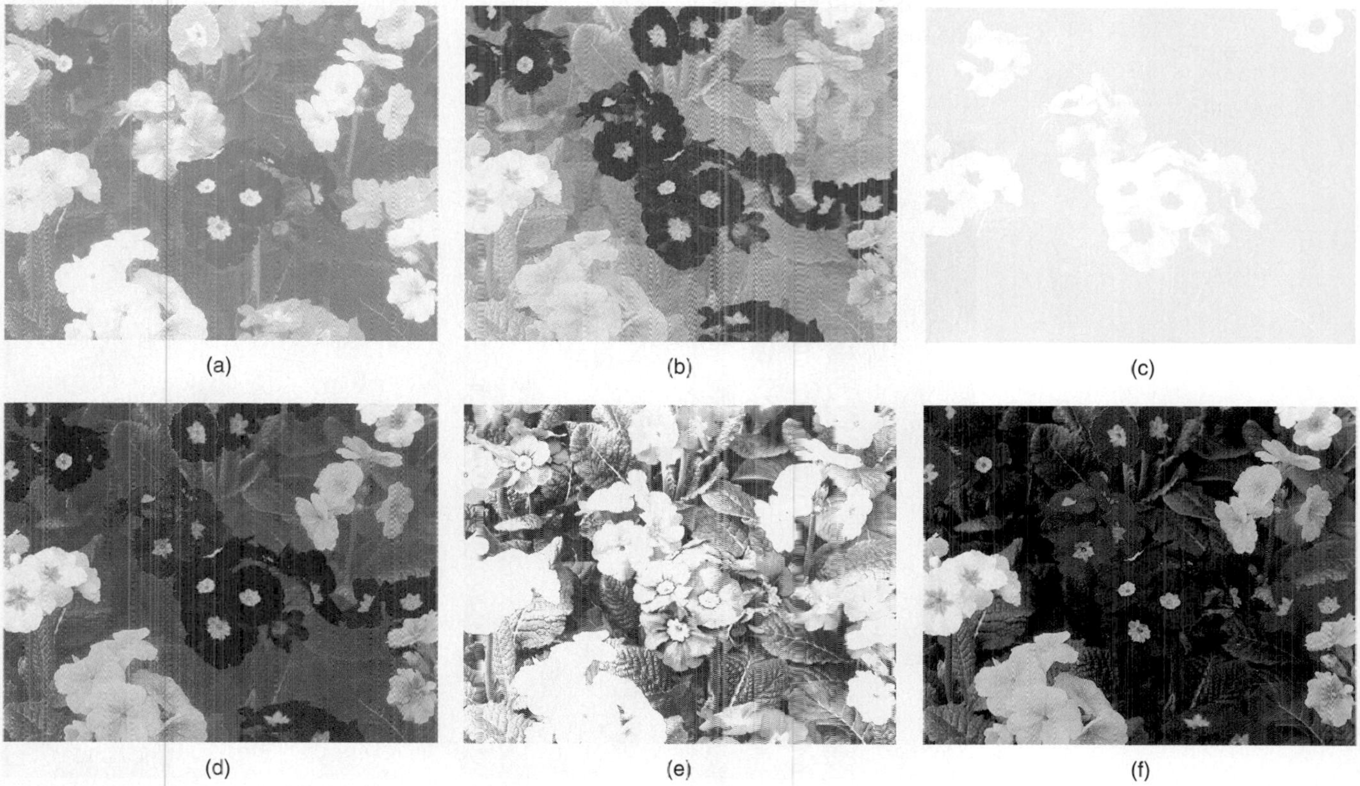

(a) (b) (c) (d) (e) (f)

Figure 21.8 Four-color printing process. The colors (a) cyan, (b) magenta, and (c) yellow printed separately. (d) Only these three colors combined. (e) The fourth (black) color. (f) The combined, finished result. Note the importance of the black color for the background, shadows, and definition.

Inkjet printers deposit simultaneous combinations of cyan, magenta, yellow, and black inks on white paper, which results in a color photograph or illustration using the same four-color principle.

Color digital cameras use filters and record as pixels the three primary colors red, blue, and green of different intensities to create the variety of colors. A pixel is a tiny unit or dot that makes up the image on a television screen, computer monitor, or similar display. The greater the number of pixels per inch, the greater is the resolution, that is, the better the clarity of the detail that can be distinguished in the image on the screen.

Why the Sky Is Blue

When sunlight passes through the atmosphere, the light is *scattered* in all directions by air molecules and other small particles in the atmosphere. The amount of scattering depends on the wavelength of the light. Air molecules and particles much smaller than the wavelengths of the various colors of light are less of an obstruction to long wavelengths than to short wavelengths. Thus, the longer wavelengths of red, orange, yellow, and green light are scattered much less than the shorter wavelengths of blue and violet light. Since our eyes are not very sensitive to violet light, the scattered blue light in the lower atmosphere is the dominate color (Fig. 21.9).

A deep blue sky is most likely on a clear, low-humidity day. When the atmosphere contains water molecules, dust, and other particles larger than the oxygen and nitrogen molecules, the longer wavelength light (red, orange, yellow) is also significantly scattered, which turns the sky into a gray-white haze. This haze is caused by very small particles that scatter and absorb light before it reaches our eyes. As the number of particles increases, more light is absorbed and scattered, which results in less clarity, color, and visual range. Factories and car and truck engines emit pollution in the form of many billions of particles per second. Although these particles are very small, other particles adhere to them and scatter the longer wavelength light, which produces gray-white or brown hazes (Fig. 21.10). A rainstorm often cleanses the lower atmosphere of the polluted particles so that we can enjoy the freshened air and see a clear blue sky.

Figure 21.9 The sky is blue because sunlight is scattered as it passes through the atmosphere. We see the most scattered, shorter wavelengths of blue light, which makes the sky appear blue.

(a)

(b)

Figure 21.10 (a) Small particles in the air due to pollution scatter longer wavelength sunlight that produces a gray-white haze. Here a smog alert day in the Los Angeles Basin is triggered as the downtown freeways generate hazardous air. (b) Heavy air pollution particles actually absorb and block out more of the sunlight to produce a brown haze. Here clouds of pollution belch smoke from factory chimneys in Copsa Mica, Romania, one of the worst-polluted cities in the world when this photo was taken. The smoke and exhaust from high sulfur coal block out the generally polluted sky.

Why Sunsets Are Red

At sunset (and sunrise), sunlight travels through significantly more of the atmosphere than at other times of the day (Fig. 21.11). This results in (1) even more scattering of the shorter wavelengths of violet, blue, and green, which are taken out before reaching us, and (2) some orange but primarily red light—the light of longest wavelength and least scattered—most readily passing through the atmosphere to reach our eyes for a red sunset (Fig. 21.12).

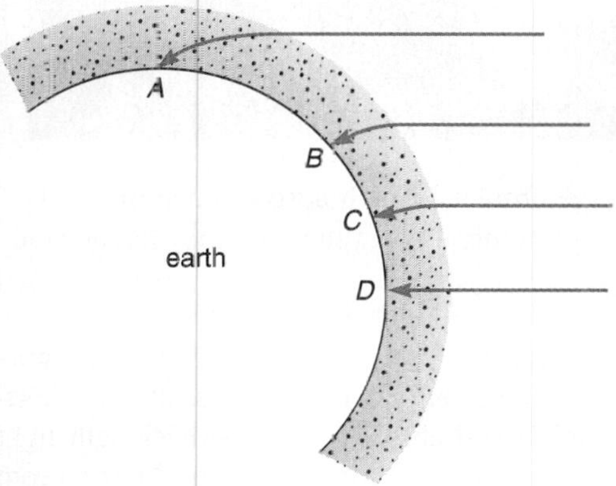

Figure 21.11 A ray of sunlight must travel through more of the atmosphere to a given point on the earth's surface at sunset than at any other time during the day. At point *A*, the shorter wavelengths of violet, blue, and some green have been scattered so that the light that reaches us tends to be the longer wavelengths of some orange and mostly red. At point *D*, a ray of sunlight travels most directly through the least amount of atmosphere with the least amount of scattering.

Figure 21.12 A vivid red sunset over the blue Pacific Ocean.

Why Clouds and Snow Are White

Clouds and snow consist of many different-sized water droplets. Small droplets tend to scatter the short-wavelength (blue) light, whereas the medium droplets tend to scatter the medium-wavelength (green) light, and the large droplets tend to scatter the long-wavelength (red) light. The overall result is that all of the colors are scattered and combined so that we see a general white reflection (Fig. 21.13).

Figure 21.13 Clouds consist of all sizes of water droplets. Small droplets scatter the short-wavelength (blue) light, the medium-size droplets scatter the medium-wavelength (green) light, and the large droplets scatter the long-wavelength (red) light. The overall result is that all of the colors are scattered and we see a general white reflection.

Dark clouds are mostly composed of large droplets, which absorb much of the light passing through them. As the size of these droplets increases even more, they fall as rain.

Why the Ocean Is Blue

The ocean and most bodies of water appear blue because (1) the water surface acts like a mirror and reflects the blue color of the sky and (2) the water tends to reflect and scatter the short-wavelength light (yellow, green, blue, and violet) and absorb the long-wavelength light (orange and red) as well as infrared, which helps increase its temperature (Fig. 21.14). From Table 13.2, we see that the thermal conductivity of water is relatively low, which is why, due

Figure 21.14 Ocean water reflects the blue color of the sky, reflects and scatters the short-wavelength (blue and green) light, and absorbs the long-wavelength (orange and red) light. This results in blue to blue-green water.

to the infrared absorption of heat, the water near the surface in large bodies of water is much warmer in the summer due to the infrared absorption and much colder in the winter than the deeper water.

The spray above water waves breaking up is white because, like clouds, it contains many small water droplets, which scatter all colors of sunlight. Bodies of water that take on other colors usually contain other particles, such as mud, minerals, or pollution.

Why We See Rainbows

A **rainbow** is a spectrum of light formed when sunlight strikes raindrops, refracts into them, reflects within them, and then refracts out of them. Figure 21.15 shows the path of a ray of sunlight in passing through a single raindrop. As the ray enters at point *A*, it is refracted and dispersed with the red end of the spectrum to point *C* and the violet end of the spectrum to point *D*, where the spectrum is then reflected to the raindrop surface near *B*, where the spectrum is again refracted upon leaving the water drop. For a person to see a rainbow, (1) the person must be between the sun and the raindrops and (2) the angle between the sun to the raindrops and back to the person's eyes must be between 40° and 42°. A full rainbow is shown in Fig. 21.16.

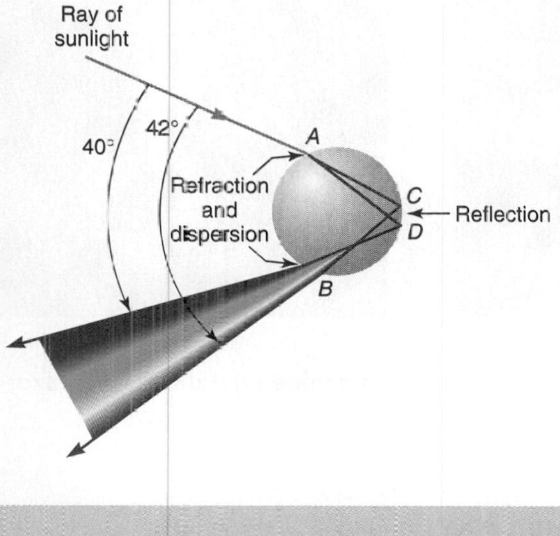

Figure 21.15 The path of a ray of sunlight being refracted and dispersed, reflected, and then refracted and dispersed again as it passes through one raindrop of a rainbow. The angle between the ray entering the raindrop and the ray to the person's eyes must be 42° for the red color and 40° for violet as shown.

Figure 21.16 A full rainbow

21.2 Diffraction of Light

As we saw in Section 15.1, **diffraction** is a property of a wave that describes its ability to bend around obstacles in its path. We used water waves to illustrate this property and noted that diffraction is commonly observed only when the obstacle or opening is nearly the same size as the wavelength. The diffraction of sound waves at doorways and other openings makes it possible to "hear around corners."

When a narrow beam of light passes by a sharp edge, around a fine wire, or through a narrow slit or a pinhole, the beam tends to spread or flare out and a distinct shadow is formed (Fig. 21.17). This distinct shadow is called a *diffraction pattern*. For best diffraction effects, monochromatic light or light of a single wavelength from a point source should

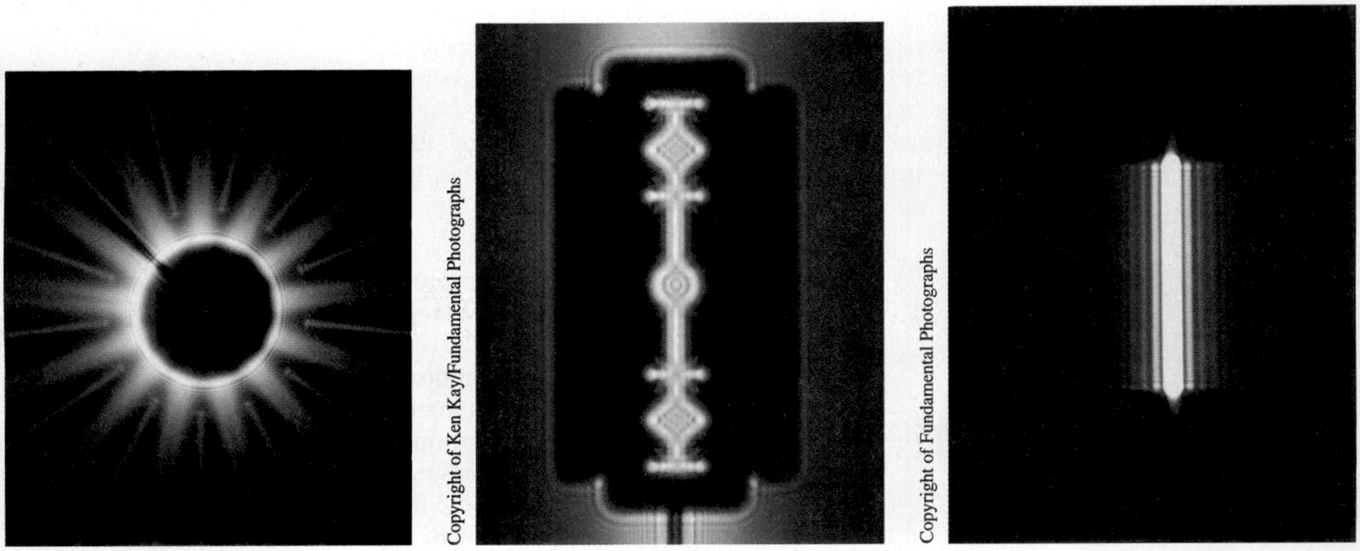

(a) (b) (c)

Figure 21.17 Diffraction patterns resulting from a point source precisely centered behind (a) a dime, (b) a razor blade, and (c) a single slit be used. We often do not notice diffraction patterns because our most common sources of light are not monochromatic or point sources, so that the light coming from different points of the source or light from multiple sources washes out any diffraction pattern.

Huygens' principle (Section 19.1) also explains what happens when a wave front hits an obstacle and is partially obstructed, such as by a sharp edge, around a fine wire, or through a narrow opening. It shows that waves bend, or diffract, behind the obstacle. Since diffraction only occurs for waves and not for particles, it verifies the wave nature of light.

21.3 Interference

We saw the results of the interference of water waves in a ripple tank in Fig. 15.9 in Section 15.1. We saw in Section 15.3 that when sound waves interfere with each other, beats are produced. Now let's study the interference of light waves. In 1801, Thomas Young first performed his famous *double-slit experiment*. Light from a single source (Young used the sun) falls on a screen that contains two closely spaced narrow slits, S_1 and S_2, as shown in a schematic diagram of his experiment in Fig. 21.18. If light consisted of particles, two bright lines would be expected to appear on a viewing screen placed behind the screen with the two slits. However, Young found not two, but a series of bright lines. Young explained his result as a *wave-interference* phenomenon.

Thomas Young (1773–1829),

English physicist and physician, is often acknowledged for his famous double-slit experiment and his work on light and interference, which established the wave theory of light. He also contributed to better understanding in the areas of work and energy, the elastic properties of materials, fluids, hemodynamics (the study of the forces involved in the circulation of blood), color, and egyptology, in which he provided the key to the deciphering of Egyptian hieroglyphics.

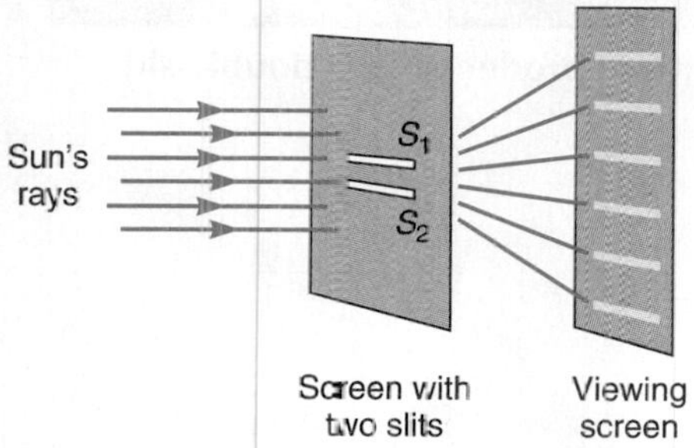

Figure 21.18 Basic schematic of Young's double-slit experiment. Young found not two, but a series of bright lines behind the screen with the two slits.

Young's double-slit experiment was conducted using sunlight as shown in Fig. 21.19. The narrow slits, S_1 and S_2, are about 1 mm apart on a black piece of paper. A series of dark and bright narrow bands is seen on the viewing screen similar to the one in Fig. 21.20. We have used red laser light to produce a dramatic series of red and dark bands to illustrate the constructive and destructive interference patterns.

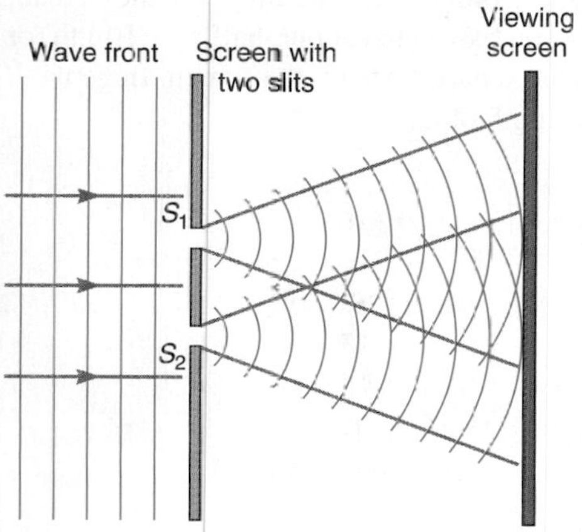

Figure 21.19 The light wave fronts in Young's experiment

As we saw in Section 15.1, when two waves of a similar type pass through the same medium, a new wave is created by the *superposition of waves*. This new wave is the algebraic sum of the separate displacements of the individual waves as in Fig. 15.6. *Constructive interference* occurs when the waves add together to form a larger displacement [more intense light (red bands) in Fig. 21.20]. *Destructive interference* occurs when the waves add together to form a smaller displacement [less intense light (dark bands) in Fig. 21.20]. The

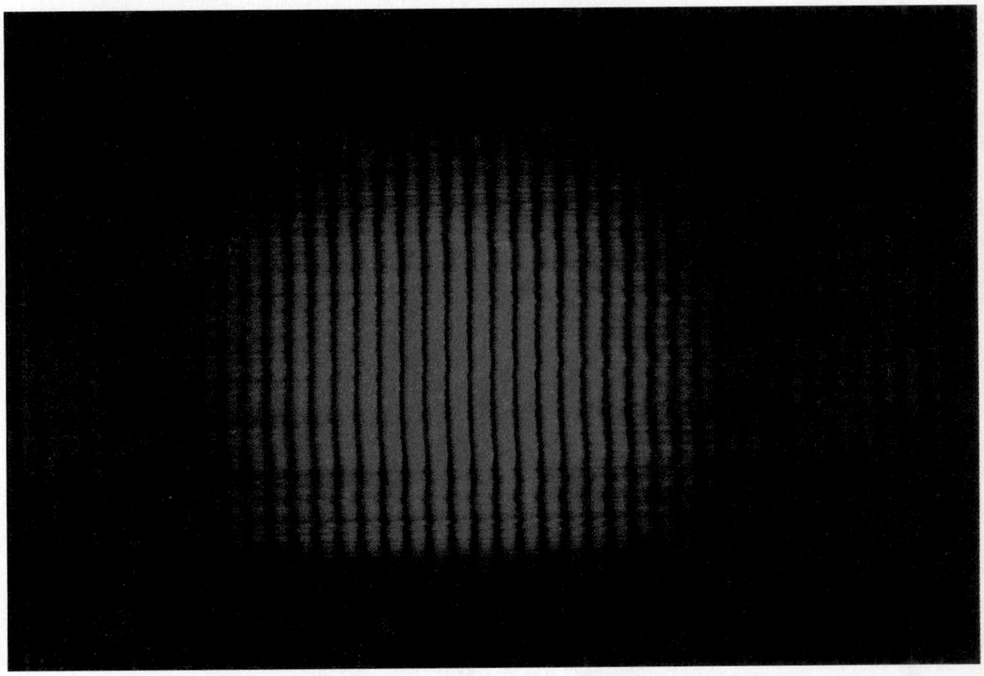

Figure 21.20 Photograph of an interference pattern produced by a double-slit experiment with red laser light.

series of bright and dark bands in Fig. 21.20 results from the different wave paths. Bright areas occur when the waves from both slits arrive in phase, whereas dark areas occur when the overlapping waves arrive out of phase (see Fig. 21.21). If the waves from the two slits travel the same distance, they are in phase and produce the brightest color at the center of the screen. Constructive interference also occurs when one wave travels an extra distance that is a whole number multiple of the wavelength and produces bright lines on the screen. Destructive interference occurs when one wave travels a distance of one-half wavelength (or 3/2, 5/2, 7/2, etc.) more than the other and produces dark lines on the screen. In general, constructive interference occurs for a given wavelength of light when

$$d \sin \theta = n\lambda$$

where

d = distance between slits
θ = angle the light rays make with the horizontal
n = 1, 2, 3, ... (called the *order* of the interference fringes)
λ = given wavelength of light

Destructive interference occurs for a given wavelength of light when

$$d \sin \theta = (n + \tfrac{1}{2}\lambda)$$

The first-order fringes ($n = 1$) occur on either side of the central bright spot; the second-order fringes occur next; and so forth. The intensity is greatest for $n = 1$ and decreases for higher orders.

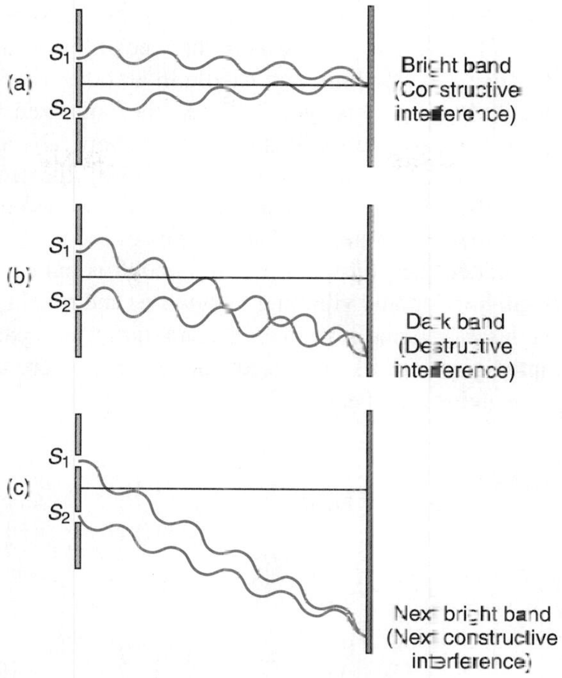

Figure 21.21 The bright and dark bands of light on the viewing screen result from the constructive and destructive interference, respectively, of the light waves after they pass through the double slits.

21.4 Polarization of Light

Light waves are usually emitted in all directions with many orientations. Some are vertical, some are horizontal, and some are at other angles. An ordinary incandescent light bulb and the sun emit light waves in multiple orientations. **Polarized light** has its waves restricted to a single plane that is perpendicular to the direction of the wave motion. Ordinary light may become polarized using a Polaroid filter (Polaroid materials were invented by Edward Land in 1929). Figure 21.22 shows a simple model to illustrate polarized light. Figure 21.22(a) shows a vibrating rope attached to a wall through two vertically aligned pieces of picket fence. Note that only a vertical transverse wave can pass through the vertical slats of the first fence and then continue *through* the vertical slats of the second. In Fig. 21.22(b)

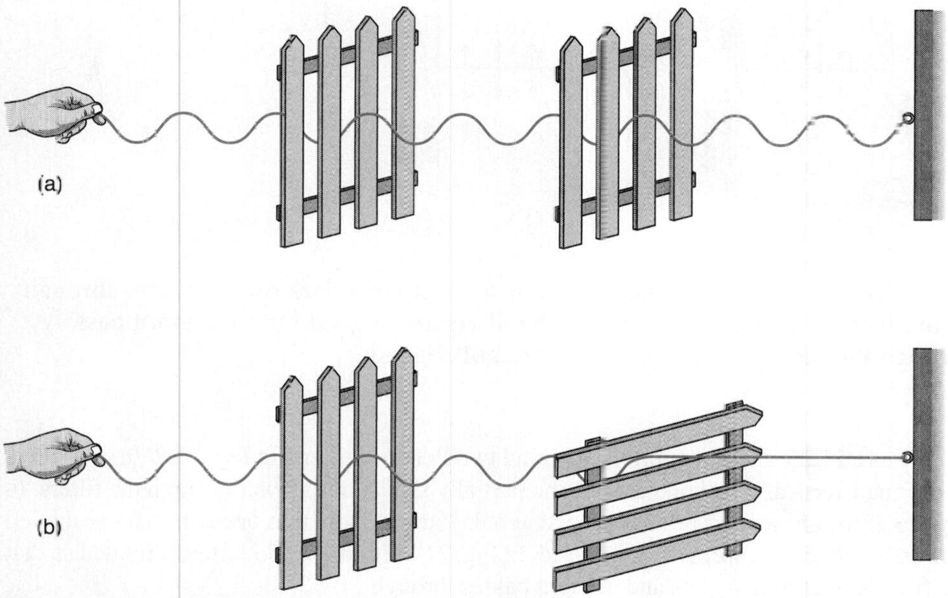

Figure 21.22 A uniform transverse wave generated by a rope (a) passes through uniformly when the filters are aligned but (b) cannot pass through the second filter when they are not aligned.

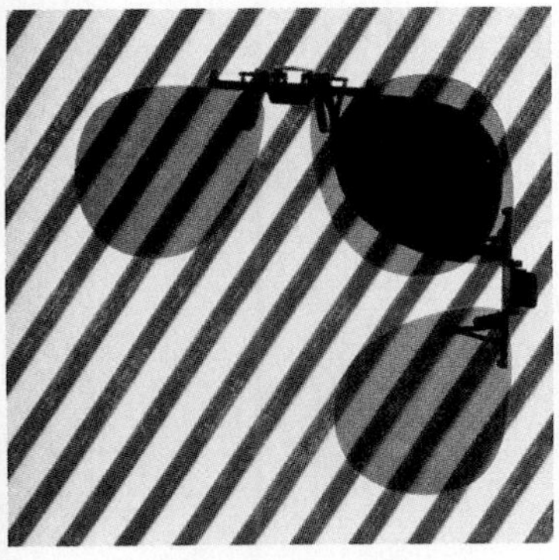

Figure 21.24 When Polaroid sunglasses overlap at right angles, no light gets through.

note that the vertical transverse wave that passes through the vertical slats of the first fence is *stopped* by the horizontal slats of the second fence and cannot pass through.

Similarly, if you look at a candle in an otherwise dark room through two vertically aligned Polaroid filters, you will see the polarized light from the candle as in Fig. 21.23(a). The two filters will allow only the vertically aligned light waves to pass through from the candle to your eyes. In Fig. 21.23(b), the second filter is horizontally aligned. Note that the vertically aligned light waves now pass through only the first filter and not through the second filter, so that you cannot see the candle. In this arrangement, the first filter is called the *polarizer* and the second is called the *analyzer*. Polaroid sunglasses block out horizontally vibrating light waves and roughly 50% of the light. When two pairs of sunglasses overlap at right angles, no light waves pass through (Fig. 21.24). Whenever unwanted glare occurs, polarized sunglasses are very useful, especially for driving, sailing, or walking on a beach.

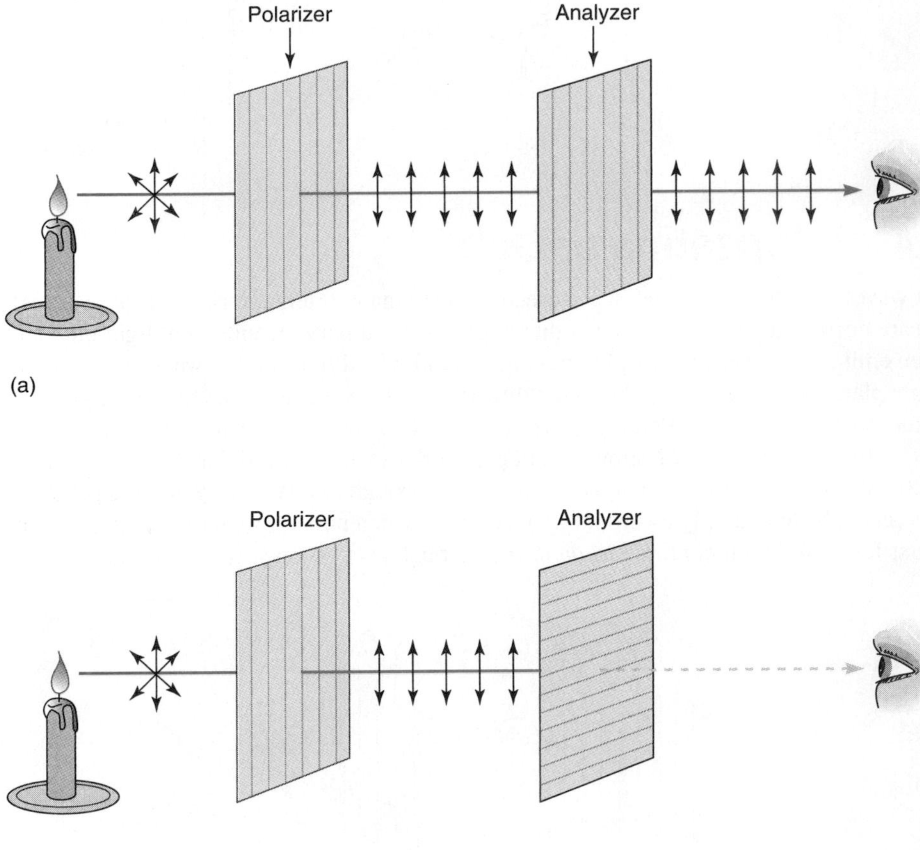

Figure 21.23 Light waves generated by a candle in a dark room (a) pass through both filters as polarized light when the filters are aligned but (b) cannot pass through the second filter when they are not aligned.

Figure 21.25 shows a red flower behind two Polaroid filters. In Fig. 21.25(a) the filters are aligned vertically and one can see the partially shaded red flower through the filters. In Fig. 21.25(b) the second filter is rotated at a 45° angle. Note that one can still see the red flower, but the red flower is more shaded. In Fig. 21.25(c) the second filter is rotated so that the filters are at right angles and no light passes through.

Much of the light that is reflected from nonmetallic surfaces is polarized. Good examples are the glare off smooth water or the glass of a windshield. The reflected light waves from the sun tend to be the horizontal-component waves parallel to the reflecting

(c)

Figure 21.25 When a red flower is behind two overlapping Polaroid filters that (a) are aligned, one can easily see the red flower; (b) are 45° out of alignment, one can see the partially shaded red flower; and (c) are at right angles, one cannot see the red flower.

surface as shown in Fig. 21.26, whereas the vertical-component waves tend to be refracted through the surface into the water or glass. Because most reflected nonmetallic surfaces are horizontal, Polaroid sunglasses are aligned so that the polarized slits are vertical. Figure 21.27(a) shows a photograph of an automobile with glare from its windows.

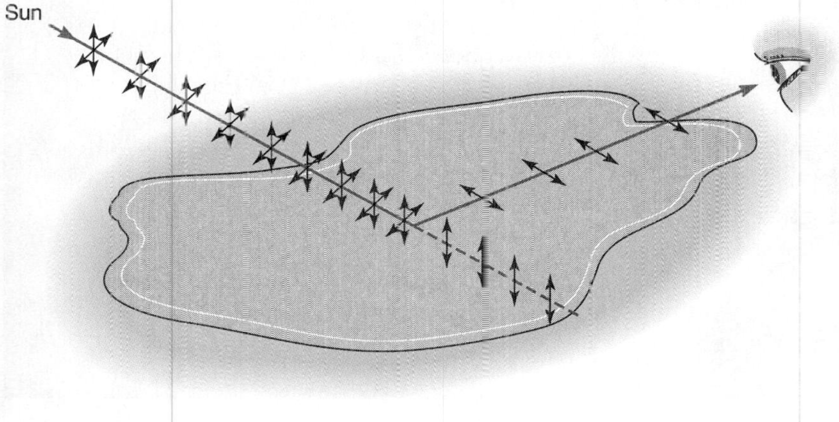

Figure 21.26 Much of the sunlight reflected from water is polarized. The horizontal-component waves reflect parallel to the water's surface, whereas the vertical-component waves tend to be refracted through the surface into the water.

Figure 21.27(b) shows a photograph of the same automobile through a polarized filter; note the absence of glare, so one can see into the automobile.

Figure 21.27 (a) An automobile with glare from its windows. (b) The same automobile seen through a polarized filter; note there is no glare, so one can see into the automobile.

Electronic devices that display information using a liquid crystal display (LCD) use polarized filters to form the black segments that form the numbers and letters (see Fig. 21.28).

Figure 21.28 The segments that form letters and numbers appearing in an LCD are formed using rotating polarized filters.

Courtesy of Research In Motion (RIM)

PHYSICS CONNECTIONS

Polarized Calculators

Polarized filters are often found in sunglasses, but are also used in devices such as stress analyzers, flat-screen computer displays, digital clocks, and calculators. Polarized filters play a part in displaying numbers on calculator screens.

There are three main components in calculators with liquid crystal display screens. A reflective mirror is placed at the rear of the display to reflect the light coming into the display back out to the user. A liquid crystal is located in front of the mirror. Liquid crystals have the ability to gradually twist the orientation of the light 90° from its original orientation. When a small voltage is applied to the liquid crystal, it untwists and maintains the original orientation of the light. The third component of the display is a pair of polarized filters. The polarized filters are oriented perpendicular to one another and are placed on each side of the liquid crystal. The inner polarized filter is aligned horizontally and the outer filter is aligned vertically (Fig. 21.29).

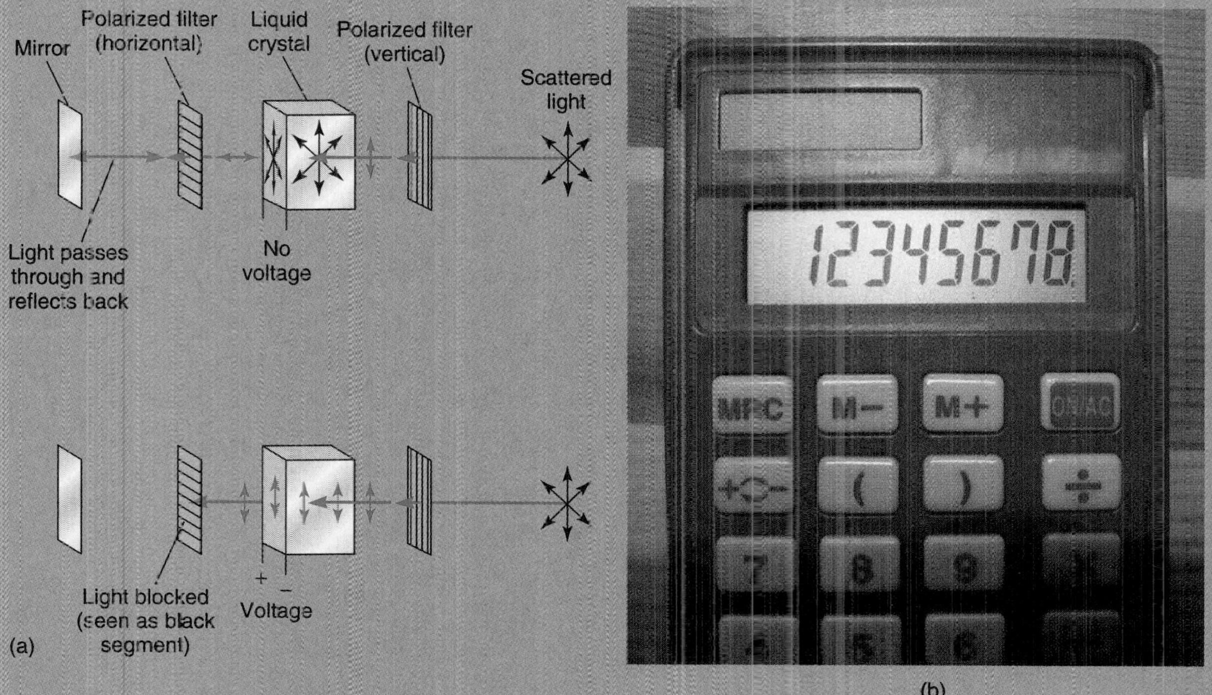

Figure 21.29 (a) The path that light takes as it enters the liquid crystal display of a calculator. (b) A standard liquid crystal display found on many calculators.

When light enters the display and passes through the first filter, only the vertically polarized light is permitted to pass through the filter and reach the liquid crystal. If there is no voltage placed across the crystal, the light is forced to alter its orientation by 90° as it passes through the liquid crystal. Since the light's orientation is now horizontal, the light passes through the second polarized filter, reflects off the mirror, passes through the second filter again, moves through the liquid crystal, changes its orientation, passes through the initial filter, and is directed to the user. The light emitted is seen as the gray background of the LCD display.

If the calculator is programmed to apply a voltage across the liquid crystal, the crystal will untwist and prevent the light from passing through the second filter and reflecting off the mirror. A black segment would appear to the calculator user because the horizontally polarized filter absorbs the vertically oriented light. Therefore, when a calculator displays a number, voltage must be placed across the liquid crystal to create that black segment.

Glossary

Color A property of the light that reaches our eyes and is determined by its wavelength or its frequency. (p. 584)

Complementary Colors Two colors that, when combined, form white; for example, cyan and red, magenta and green, and yellow and blue. (p. 585)

Diffraction A property of a wave that describes its ability to bend around obstacles in its path. (p. 592)

Dispersion The spreading of white light into the full spectrum. (p. 584)

Polarized Light Light waves restricted to a single plane that is perpendicular to the direction of the wave motion. (p. 595)

Primary Colors The colors red, green, and blue; an additive mixture of the three colors of light resulting in white. (p. 585)

Primary Pigments The complements of the three primary colors; namely, cyan (the complement of red), magenta (the complement of green), and yellow (the complement of blue). (p. 586)

Rainbow A spectrum of light formed when sunlight strikes raindrops, refracts into them, reflects within them, and then refracts out of them. (p. 591)

Visible Spectrum The colors resulting from the dispersion of white light through a glass prism: red, orange, yellow, green, blue, and violet. (p. 584)

Review Questions

1. Name the colors of the visible spectrum.
2. What property of light determines its color?
3. Name the light waves whose wavelengths are (a) slightly longer than visible light and (b) slightly shorter than visible light.
4. What is the apparent color of a green dress in a closed room with only a red light? Explain.
5. What would the U.S. flag look like through a piece of (a) red glass and (b) blue glass?
6. What are the primary colors?
7. What is a complementary color? Name the complement of each primary color.
8. How are the primary pigments related to the primary colors?
9. What are the four colors used to print color photographs in books, magazines, and newspapers?
10. Why is the sky blue?
11. Why is a sunset red?
12. Why are clouds white?
13. Why is the ocean blue?
14. True or false: For you to see a rainbow, the raindrops must be between the sun and you.
15. Name the property of a wave that describes its ability to bend around obstacles in its path.
16. When a narrow beam of light passes by a sharp edge, around a fine wire, or through a narrow slit or a pinhole, the beam tends to spread or flare out. The distinct shadow formed is called a ______.
17. Name the person who first performed the famous double-slit experiment.
18. When waves add together to form a larger displacement or more intense light, this is called _____.
19. When waves add together to form a smaller displacement or less intense light, this is called _____.
20. How do Polaroid sunglasses reduce the glare of bright sunlight?

CHAPTER 22

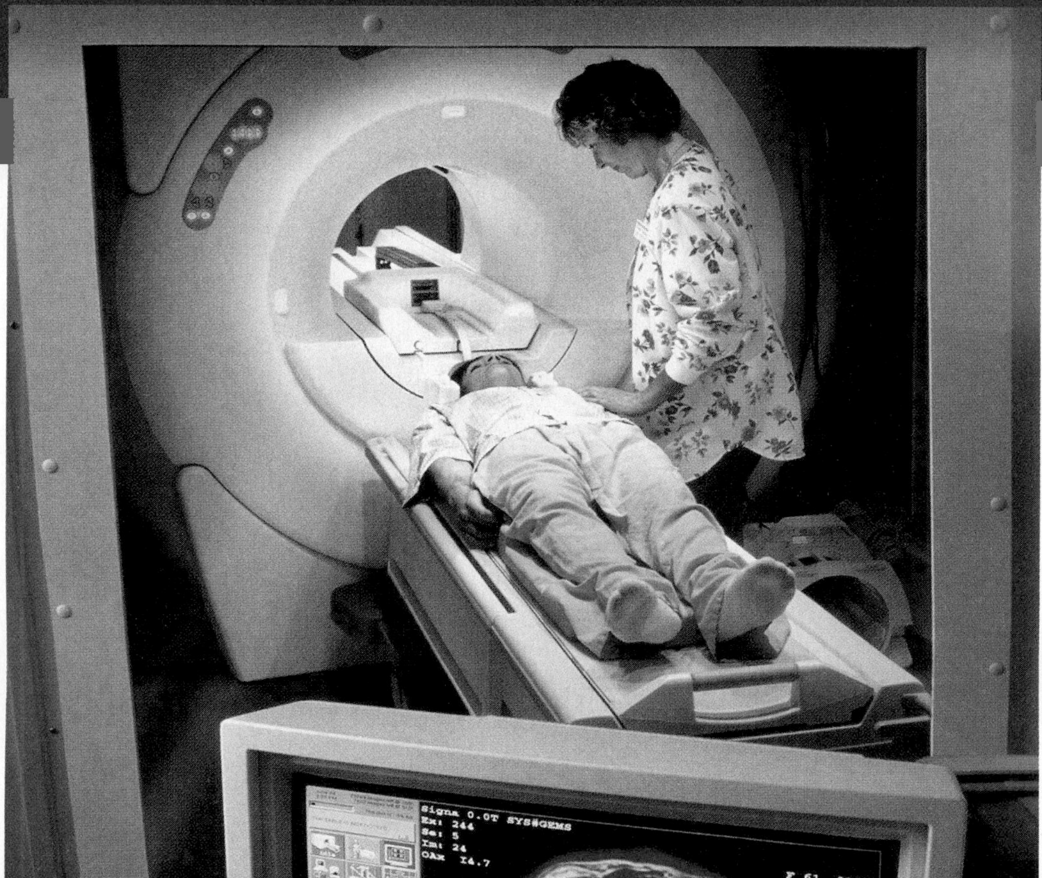

Photo courtesy of Corbis Corporation. Reprinted with permission

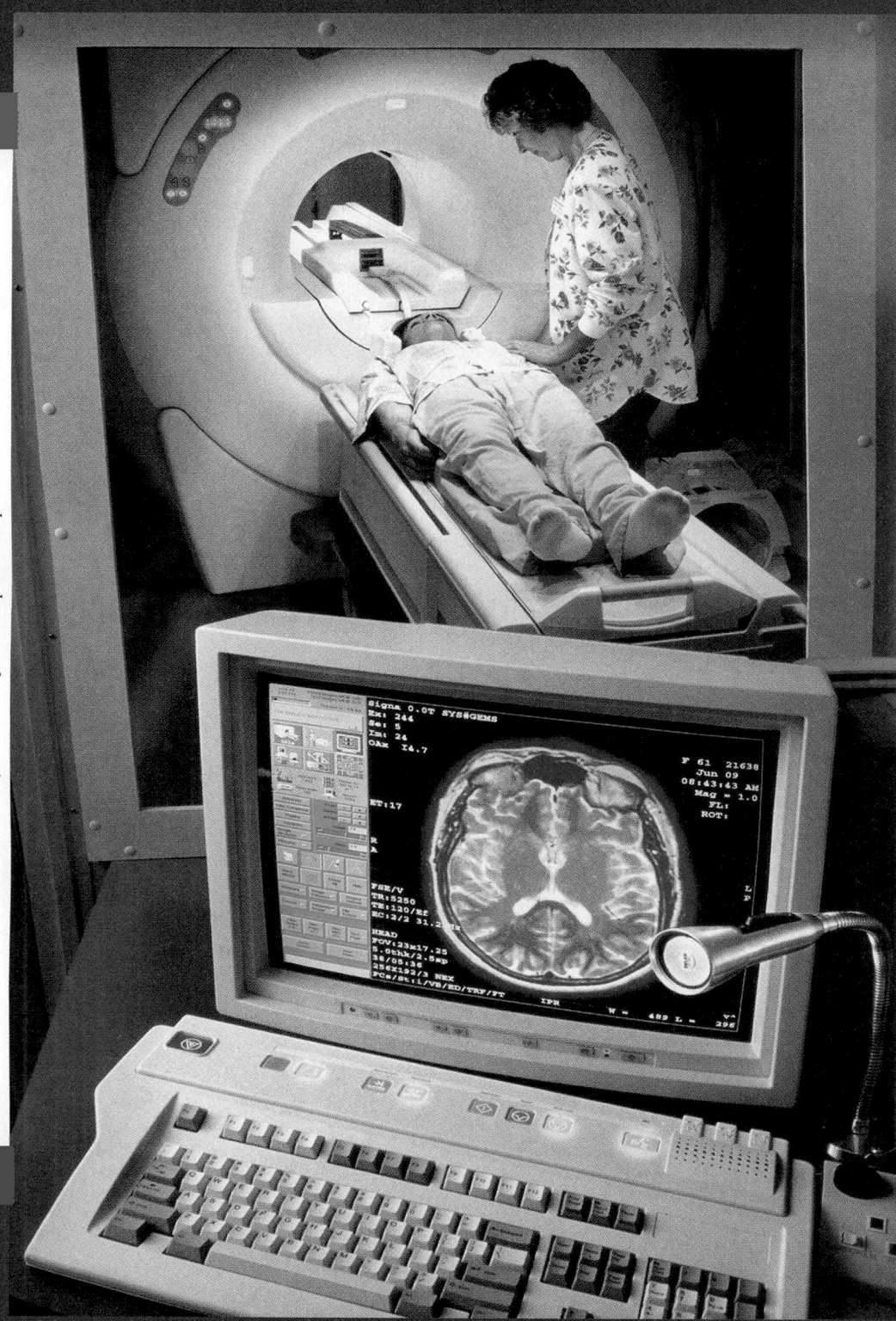

SURVEY OF MODERN PHYSICS

To this point we have studied classical or Newtonian physics. Physics is also concerned with the building blocks of all matter—atoms—and the subatomic particles that make up atoms and the forces between them.

Ancient Greek philosophers thought that all matter could be reduced to four basic elements: air, earth, fire, and water. Modern physics probes deeply into the nature of matter.

Quantum theory, the atom, radioactivity, nuclear fission, and fusion are all topics being studied by physicists today. While an in-depth study of these topics is beyond the scope of this text, some familiarity with the concepts is essential to those in all technical fields.

Objectives

The major goals of this chapter are to enable you to:

1. Describe the basis of quantum theory.
2. Describe the development of the current model of the atom.
3. Describe the structure and properties of the atomic nucleus.
4. Analyze problems of radioactive decay.
5. Describe nuclear fission and fusion.
6. Describe principles of detection and measurement of radioactivity.

22.1 Quantum Theory

James C. Maxwell (1831–1879),

physicist, was born in Scotland. His early work led to the mathematical development of the theory of electricity and magnetism. His greatest work was his theory of electromagnetic radiation and the basic equations of electromagnetism, which established him as the leading theoretical physicist of the nineteenth century.

In the 1860s, **James Clerk Maxwell** developed a theory relating magnetic fields and electric fields in the transmission of energy across empty space, which became known as the electromagnetic wave theory. Maxwell's theory led to a complete description of electricity and magnetism and was the basis for the development of radio, television, and countless electronic devices. It was also the starting point in the study of the nature of the atom.

Later experiments by Heinrich Hertz confirmed Maxwell's predictions and confirmed that light was a form of wave—almost. Only two problems remained: the wave theory of light could not explain why hot objects change color when heated and why ultraviolet light could discharge electrically charged metal plates. Solving these two problems led to the development of the quantum theory. Its confirmation forever changed our view of the physical world.

Particles Behave Like Waves and Waves Behave Like Particles

Incandescent objects (like a glowing light bulb filament) emit light because of the vibrations of charged particles inside their atoms. The spectrum of color produced by the light depends on the heat of the object and the frequency range of the light. Max Planck explained this with the then-revolutionary idea that particles can have only certain energies, which are whole-number multiples of a constant now known as Planck's constant. Planck proposed that atoms could only emit radiation when their vibration energy changed in multiples of this constant.

Louis Victor de Broglie (1892–1987),

physicist, was born in France. In 1923, while a graduate student at the University of Paris, he theorized that since light could be seen to behave under some conditions as particles and other times as waves, atomic particles have both particle and wave properties. This led to his pioneering work on the wave nature of the electron.

In addition, certain metals emit electrons when they are exposed to light. This is called the *photoelectric effect* and could not be explained by simple wave theory. The light produced an effect like a stream of particles. Einstein explained this effect by suggesting that light consists of packets of energy, called *photons,* that act like particles. He thought that although photons have no mass, they travel at the speed of light, have energy related to their frequency, and have momentum. Later experiments confirmed his theory.

Later, **Louis Victor de Broglie** in 1923 suggested that particles have wave properties. This revolutionary idea was largely ignored until Einstein supported it; de Broglie won a Nobel Prize for it in 1929. Two wave properties exhibited by particles are diffraction and interference. These can easily be reproduced in the laboratory.

Most physicists today believe that the particle and wave aspects of light are complementary. That is, they must be viewed together to have a complete picture of the true nature of light. This complementary nature in exhibiting properties of both waves and particles is the essence of the quantum theory.

22.2 The Atom

In Section 11.1 we saw that an **atom** is the smallest particle of an element that can exist in a stable or independent state. The three most important particles of the atom are (1) the **proton**, a particle with a positive charge, (2) the **electron**, a particle with a negative charge, and (3) the **neutron**, a particle that does not have an electric charge. The **nucleus** is the center part of an atom, made up of protons and neutrons, while the electrons surround the nucleus. Historically, how was this current concept of the atom developed?

J. J. Thomson (1856–1940),

physicist, was born in England. His experiments in 1897 led to the discovery of the electron, which he found to be 2000 times smaller in mass than the lightest known atomic particle, the hydrogen ion. He received the Nobel Prize in Physics in 1906.

By 1900, most scientists agreed atoms exist. In 1890, **J. J. Thomson** discovered small, negatively charged particles, which he called electrons. Between 1909 and 1911, Ernest Rutherford discovered that atoms have a relatively more massive, positively charged center, or nucleus. Another 30 years was required for agreement to be reached on how these parts of the atom were arranged.

The Rutherford Model

Thomson originally thought that electrons were embedded in a spherical volume of positive charge like raisins in a muffin. However, Rutherford, with Hans Geiger and Ernst Marsden, was able to show that the atom was mostly space with most of the mass concentrated in a relatively small region he called the nucleus. He thought electrons orbited the nucleus the way planets orbit the sun, which is the most popular oversimplification of atomic structure still found today. The flaw in Rutherford's model is that if electrons were like planets, they would quickly lose energy and collapse into the nucleus, which does not happen.

Scientists sought a model of the atom that could explain the mystery of atomic spectra. An emission spectrum is a spectrum of electromagnetic radiation emitted by a luminous source. Many substances emit a characteristic color when heated. For example, applying a high voltage to gas atoms, as in a neon light, causes them to emit their characteristic color. The light can be studied with a diffraction grating or prism. A diffraction grating is a glass or polished metal surface containing many closely spaced, very fine parallel lines in the surface, which separate the colors of light by interference and produces a light spectrum. In contrast to an emission spectrum, as a gas cools, the opposite process occurs and an absorption spectrum is produced. An absorption spectrum is a spectrum of electromagnetic radiation absorbed by matter when radiation of all frequencies is passed through it. Based on the above concepts, atomic spectra were used to determine the structure of the atom.

Niels Bohr developed the next recognized model of the atom, using Einstein's idea that since emission spectra contain only specific wavelengths, an electron can emit or absorb only specific amounts of energy; that is, the energy is quantized. Bohr knew his model was not complete and continued to search for a better model. The Bohr model did, however, provide an explanation for some of the chemical properties of elements. The foundation of much of our knowledge of chemical reactions and bonding is based on the idea that each atom has a unique electron arrangement.

In 1926, **Erwin Schroedinger** used de Broglie's idea of matter behaving like waves to create a quantum model of the atom based on waves. **Werner Heisenberg** and others developed the theory into a complete description of the atom. The theory does not provide a simple picture like the planetary model. The wave–particle nature of matter means it is impossible to know both the momentum and the position of an electron at the same time. The quantum model of the atom predicts only the probability that an electron is at a specific location. This region in which there is a high probability of finding an electron is called an electron cloud. Though difficult to visualize, quantum mechanics has been very successful in predicting many details of the structure of atoms.

Niels Bohr (1885–1962),

Danish physicist, developed an early model of atomic structure in which electrons travel around the nucleus in given stable orbits determined by quantum conditions.

Erwin Schroedinger (1887–1961),

physicist, was born in Vienna. His study of the wave behavior of matter within quantum mechanics led to the understanding of matter at the subatomic level. He also made significant contributions to molecular biology.

Werner Heisenberg (1901–1976),

theoretical physicist, was born in Germany. He developed a method of expressing quantum mechanics mathematically and formulated his revolutionary uncertainty principle—increasing the accuracy of measurement of one observable quantity increases the uncertainty with which a related quantity may be known.

22.3 Atomic Structure and Atomic Spectra

Niels Bohr suggested the energy of an electron is quantized, that is, restricted only to certain fixed values called *energy levels*. Each of these energy levels was shown by Bohr to be given by the equation

$$E = -\frac{kZ^2}{n^2}$$

where

E = energy of electron
k = a constant (2.179×10^{-18} J or 13.595 eV)
Z = atomic number (the number of positive charges in the nucleus)
n = integer that characterizes the energy level, ($n = 1, 2, 3, 4, \ldots$), also called the quantum number.

An *electron volt* (eV) is the energy acquired by an electron as a result of moving through a potential difference of 1 V. One eV is equal to 1.602×10^{-19} J. So, one MeV (one million electron volts) is equal to 1.602×10^{-13} J. The distance from an electron in an atom to the nucleus increases as the integer n increases. (The formula is $r_n = n^2 r_1$ for the hydrogen atom.) In the lowest-energy level or lowest orbit, its energy is minimum. This lowest-energy level ($n = 1$) is called the stable or **ground state** of the electron. The higher-energy levels ($n = 2, 3, 4, \ldots$) are called **excited states** of the electron.

The **Bohr model** was an early model of atomic structure in which electrons travel around the nucleus in a number of discrete stable energy levels determined by quantum conditions. Of the several early proposed models of atomic structure, Bohr's was the first successful model for the atomic structure of the hydrogen atom (Fig. 22.1). The Bohr model is clearly limited in scope, but it was the forerunner to the modern quantum theory.

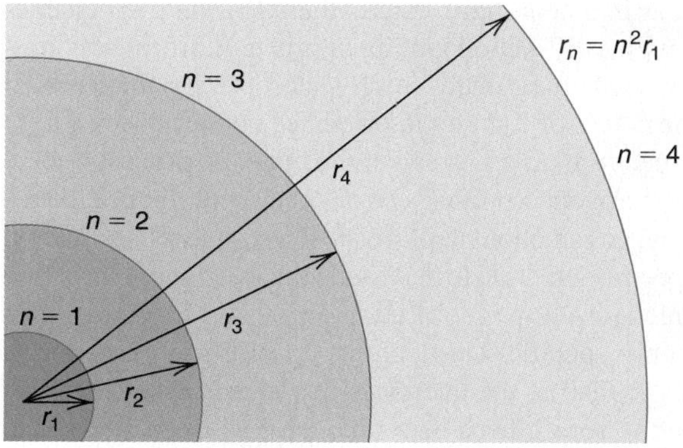

Figure 22.1 The Bohr model of the hydrogen atom

EXAMPLE 1

Find the energy of an electron in the $n = 2$ energy level in a hydrogen atom.

Data:

$$n = 2$$
$$Z = 1$$
$$E = ?$$

Basic Equation:

$$E = -\frac{kZ^2}{n^2}$$

Working Equation: Same

Substitution:

$$E = -\frac{(13.595 \text{ eV})(1)^2}{2^2}$$
$$= -3.3988 \text{ eV}$$

All objects emit radiation with intensity that is proportional to the fourth power of the Kelvin temperature (T_K^4). At normal temperatures, we are not aware of this electromagnetic radiation because of its low intensity. At higher temperatures, sufficient infrared radiation exists so that we feel the heat if we are close to the object. At still higher temperatures (approximately 1000 K), an object, such as molten steel, an electric toaster element, or an electric stove burner, glows. The filament of a light bulb glows at temperatures above 2000 K. This radiation is due to the rapid vibrations of the atoms and molecules.

Rarefied (low density and low pressure) gases can also be excited to emit light by intense heating in a tube. The radiation from excited gases emits light of only certain wavelengths. When the light is analyzed through the slit of a spectroscope, a line spectrum is observed instead of a continuous color spectrum.

Neon lights work on the principle that certain colors of light are given off by different atoms when the electrons in the atoms undergo a transition from the ground state to an excited state due to an electric current passing through the neon tube containing the gas at low pressure. As the electrons return to their ground state, they give off their excess energy in the form of electromagnetic radiation. A neon tube containing hydrogen gives off the color blue. Other atoms give off other colors, leading to the rich colors available through this type of light tube.

Fluorescence is a property of certain substances in which radiation is absorbed at one frequency and then reemitted at a lower frequency. When radiation is absorbed by a particle (an atom or a molecule), the energy of the electrons is increased to a higher energy level. This results in the particle being in an abnormal, excited state. Some absorbed energy is lost by collisions with other particles so that the particle cannot maintain its energy level in its excited state. The electrons give up their remaining extra energy and the particle returns to its ground state in a series of two or more steps and emits radiation in the form of visible light. The image on a television screen is the result of fluorescence due to electron bombardment of the screen.

A *fluorescent lamp* is a source of artificial light that consists of a long, sealed glass tube with a small amount of mercury and an electrode at each end. The inside of the tube is coated with a mixture of fluorescent powders. An electric current through the tube vaporizes the mercury and causes it to emit ultraviolet radiation. That radiation is absorbed by the fluorescent coating, which then emits visible light. See Fig. 22.2.

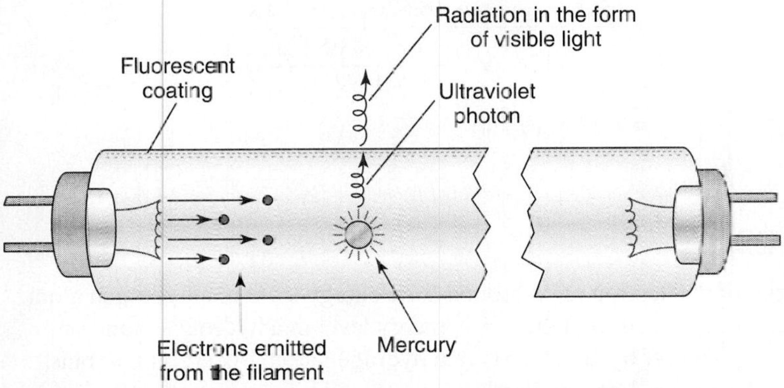

Figure 22.2 Fluorescent lamp. An electric current through the glass tube vaporizes the mercury and causes it to emit ultraviolet light. That radiation is absorbed by the fluorescent coating, which then emits visible light.

The wavelength of a photon emitted during the transition between energy levels can be determined using concepts developed in earlier chapters: $E = hf$ gives the energy of a photon and $c = \lambda f$ relates the wavelength and frequency of light.

EXAMPLE 2

What is the wavelength of a photon given off by a hydrogen atom undergoing a transition from the $n = 2$ to the $n = 1$ energy level?

Data: For $n = 2$ energy level

$$E = -3.3988 \text{ eV} \qquad \text{(from Example 1)}$$

For $n = 1$ energy level

$$n = 1$$
$$Z = 1$$

Basic Equations:

$$E = -\frac{kZ^2}{n^2}$$
$$E = hf$$
$$c = \lambda f$$

Working Equations:

$$E_1 = -\frac{kZ^2}{n^2} \qquad E_{\text{transition}} = E_2 - E_1$$

$$\lambda = \frac{c}{f} = \frac{c}{E/h} = \frac{ch}{E}$$

Substitutions:

$$E_1 = -\frac{(13.595 \text{ eV})(1)^2}{1^2}$$
$$= -13.595 \text{ eV}$$

Therefore,

$$E_{\text{transition}} = -3.3988 \text{ eV} - (-13.595 \text{ eV})$$
$$= 10.196 \text{ eV}$$

So,

$$\lambda = \frac{(3.00 \times 10^8 \text{ m/s})(6.626 \times 10^{-34} \text{ J s})}{10.196 \text{ eV} \times \dfrac{1.602 \times 10^{-19} \text{ J}}{1 \text{ eV}}}$$
$$= 1.22 \times 10^{-7} \text{ m}$$

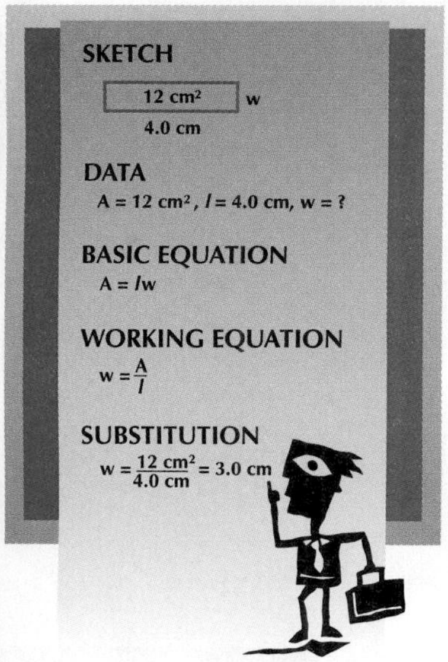

PROBLEMS 22.3

1. Find the energy of an electron in its ground-state energy level in a hydrogen atom.
2. Find the energy of an electron in the $n = 3$ energy level in a hydrogen atom.
3. Find the energy given off by an electron in a hydrogen atom undergoing a transition from the $n = 3$ to the $n = 1$ energy level.
4. Find the wavelength of the photon given off in the transition in Problem 3.
5. Find the wavelength of a photon given off in a hydrogen-filled neon tube undergoing a transition from the $n = 3$ to the $n = 2$ energy level.

22.4 Quantum Mechanics and Atomic Properties

The agreement between the Bohr model and the energy levels of the hydrogen atom does not account for the properties of the more complex atoms, that is, those with more than one electron. The quantum model of the atom better fits the more complex atoms. **Quantum mechanics** is a theory that unifies the wave–particle dual nature of electromagnetic radiation. The quantum model is based on the theory that matter, just like light, behaves sometimes as a particle and sometimes as a wave. The wave theory of the atom can be used in a simple way to understand the Bohr model structure of an atom. In the Bohr model, the electron can have only specific discrete energy levels; other energy levels are not possible. When an electron in an atom behaves as a wave, its distance from the nucleus or its energy level must consist of an integral number of wavelengths; otherwise, destructive interference occurs (Fig. 22.3).

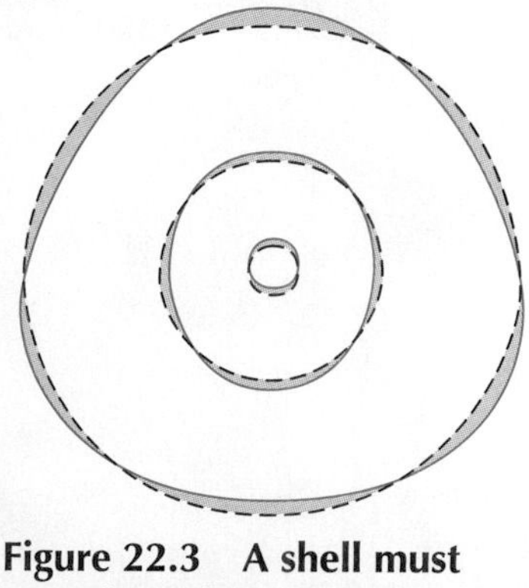

Figure 22.3 A shell must contain an integral number of wavelengths of the electron.

When an electron behaves as a wave, its exact location cannot be determined; instead, the probability can be calculated, using wave equations, of where it is likely to be found. This probability describes for any position in space how likely the electron is to be found at that point at any given time (Fig. 22.4). A thin spherical shell region of space (or electron cloud) surrounding the atomic nucleus in which an electron is likely to be found is its energy level. In the ground state of the hydrogen atom, the electron spends most of its time in a spherical shell region, or energy level, centered about 0.5 Å (angstrom) from the nucleus as shown in Fig. 22.5. ***Note:*** 1 Å = 1×10^{-10} m.

The shell regions, or energy levels, are characterized by a number, n, which is called the *principal quantum number*, where $n = 1, 2, 3, 4, \ldots$. Just as in the Bohr model, the shells extend farther from the nucleus as n increases. As the shells extend farther from the nucleus, there are different shell shapes that become possible for each principal quantum number. Each of these shapes is described by a *secondary quantum number* or letter (s, p, d, f, ...) as shown in Fig. 22.6.

Each possible shape represents an energy level, or shell, which can contain up to two electrons. In the ground state (lowest energy state) of an atom, the electrons fill up the lowest energy levels. The types of energy levels, or shells, for each principal quantum number are shown in Table 22.1. Table 22.2 summarizes the number of electrons in each shell for the atomic elements with up to 21 protons in the nucleus and therefore up to 21 electrons in energy levels. Note in Table 22.2 that the $n = 4$ shell starts filling up (see potassium) before the $n = 3$ shell is completely filled. This is because the 4s shell has lower energy than the 3d shells.

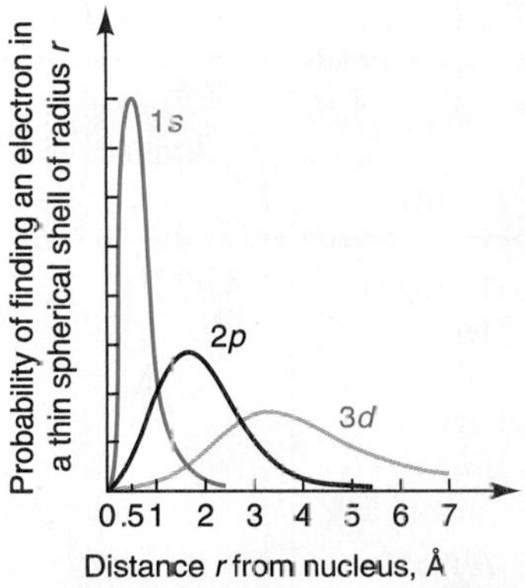

Figure 22.4 The probability of finding the electron at a distance *r* from the nucleus for the three lowest shells of a hydrogen atom

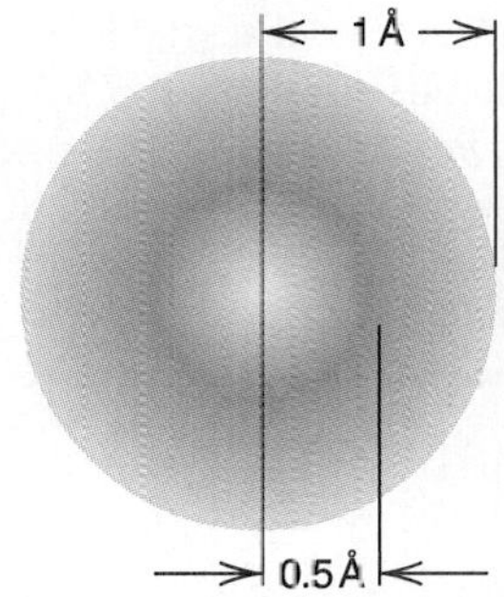

Figure 22.5 Electron density for an electron in the ground state of a hydrogen atom

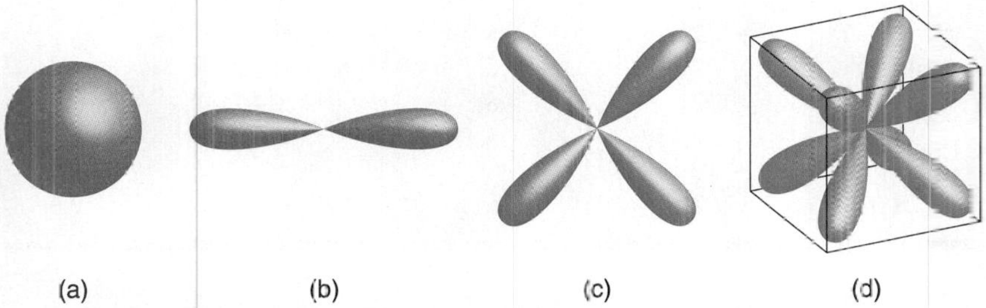

Figure 22.6 Shapes for (a) an *s* subshell, (b) a *p* subshell, (c) a *d* subshell, and (d) an *f* subshell

Table 22.1 Maximum Number of Electrons in Atomic Shells or Energy Levels

n Value	Types of Energy Levels	Maximum Number of Electrons
1	1s	2
2	2s, 2p	8
3	3s, 3p, 3d	18
4	4s, 4p, 4d, 4f	32
5	5s, 5p, 5d, 5f, 5g	50

The theory of electron shells provides a basis for understanding the properties of different atoms, based on how many electrons are in the outer shell and how much energy it takes to add or remove an electron from that shell during chemical reactions between atoms. Each atom in all matter attempts to have its outer shell complete and does this by attempting to borrow, lend, or share electrons with other atoms surrounding it.

An atom is classified as a metal if it tends to lend electrons. Copper is an example of a metal. If it tends to borrow electrons, it is called a nonmetal. Atoms are classified as inert when they have complete outer shells and therefore tend not to borrow or share electrons. Helium and neon are examples of inert atoms. They both have completely filled outer shells (see Table 22.2). The fewer electrons an atom tends to borrow, lend, or share, the more reactive it tends to be in chemical reactions.

Table 22.2 Table of the First 21 Elements

Element	Atomic No.	Number of Protons	Number of Electrons	Electrons in $n = 1$ Shell	Electrons in $n = 2$ Shell	Electrons in $n = 3$ Shell	Electrons in $n = 4$ Shell
Hydrogen	1	1	1	1			
Helium	2	2	2	2			
Lithium	3	3	3	2	1		
Beryllium	4	4	4	2	2		
Boron	5	5	5	2	3		
Carbon	6	6	6	2	4		
Nitrogen	7	7	7	2	5		
Oxygen	8	8	8	2	6		
Fluorine	9	9	9	2	7		
Neon	10	10	10	2	8		
Sodium	11	11	11	2	8	1	
Magnesium	12	12	12	2	8	2	
Aluminum	13	13	13	2	8	3	
Silicon	14	14	14	2	8	4	
Phosphorus	15	15	15	2	8	5	
Sulfur	16	16	16	2	8	6	
Chlorine	17	17	17	2	8	7	
Argon	18	18	18	2	8	8	
Potassium	19	19	19	2	8	8	1
Calcium	20	20	20	2	8	8	2
Scandium	21	21	21	2	8	9	2

PROBLEMS 22.4

1. What atom has the $n = 2$ energy level filled?
2. What atom has the $n = 1$ energy level filled?
3. What atom is one electron short of having the $n = 2$ energy level filled?
4. What atom has just one electron in the $n = 3$ energy level?

22.5 The Nucleus—Structure and Properties

Henri Becquerel discovered radioactivity in 1896 when he was working with compounds containing uranium and found that his photographic plates were partially exposed when these uranium compounds were anywhere near his plates. The plates were exposed by penetrating rays that went through the plate coverings. This phenomenon was called radiation. Many scientists then began to study radiation, and one of the first results was an understanding of the composition of the atomic nucleus.

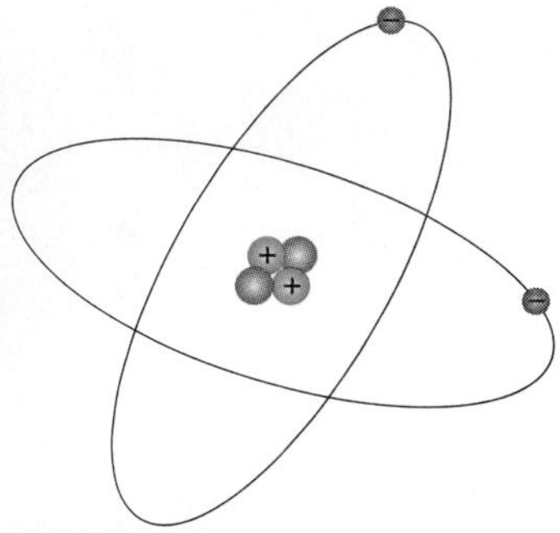

Figure 22.7 The nucleus is composed of neutrons and protons. Electrons surround the nucleus.

The nucleus of an atom contains over 99.9% of the mass of an atom, yet it occupies less than one-trillionth of 1% of the volume of an atom. The fundamental particles that make up the nucleus of an atom are called *nucleons*. There are two types of nucleons: *protons,* which carry a positive charge, and *neutrons,* which are electrically neutral (Fig. 22.7). Protons have a mass of

$$m_p = 1.6726 \times 10^{-27} \text{ kg}$$

while neutrons have a mass of

$$m_n = 1.6749 \times 10^{-27} \text{ kg}$$

Both are much more massive than the third fundamental particle of an atom, the negatively charged electron, which has a mass of

$$m_e = 9.1094 \times 10^{-31} \text{ kg}$$

Negatively charged electrons are attracted to the nucleus by the positively charged protons (Fig. 22.8). The *size* of the positive charge of a proton is the same as the *size* of the negative charge of an electron. An electrically neutral atom contains an equal number of protons and electrons.

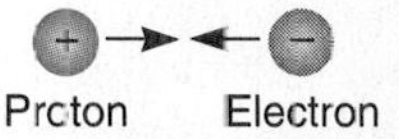

Figure 22.8 The attractive strong force operates at short distances.

The simplest atom, hydrogen, contains one proton in its nucleus. Other atoms contain two or more protons together with some number of neutrons in their nuclei. The role of neutrons in the nucleus of these more complex atoms is to bind the positively charged protons together and prevent the nucleus from flying apart under the repulsive electric force between the protons. In addition to the repulsive **Coulomb force** (electric force) between protons, there is a nuclear force, referred to as the **strong force**, which is a very short-range attractive force among all nucleons (neutrons and protons) independent of their charge. The neutrons provide additional strong force to overcome Coulomb repulsion. Two neutrons in a nucleus with two protons provide the additional attractive strong force that holds the nucleus together. For heavier atoms it is necessary to have more neutrons than protons to hold the nucleus together (Fig. 22.9). For example, there are 82 protons and 126 neutrons in an atom of common lead. Atoms heavier than lead cannot be completely stabilized even by the addition of extra neutrons.

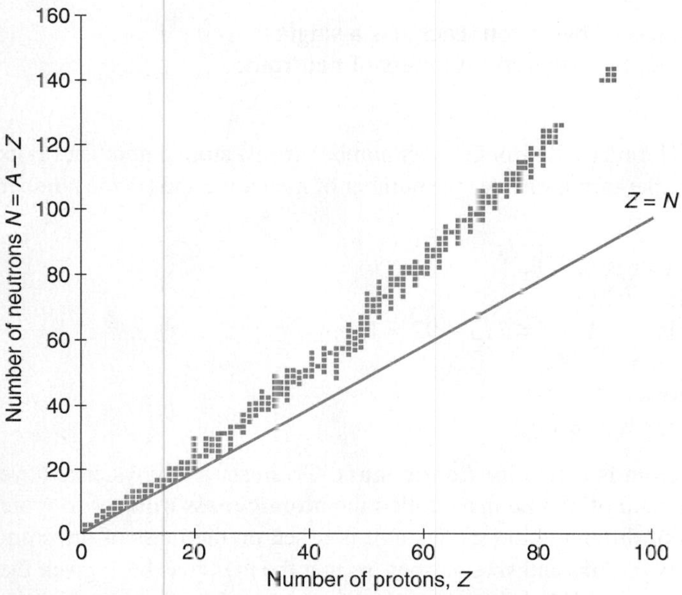

Figure 22.9 Number of neutrons versus number of protons for stable nuclei (dots)

The number of protons in a nucleus is called its **atomic number**, *Z*. The total number of nucleons (protons and neutrons) in a nucleus is called the **atomic mass number**, *A*. A **nuclide** is a specific type of atom characterized by its nuclear properties, such as its number of neutrons and protons and the energy state of its nucleus. The **neutron number**, *N*, is the number of neutrons in an atomic nucleus of a particular nuclide. Also, $N = A - Z$, the difference between the atomic mass number and the atomic number. *N* can be different for nuclei with the same atomic number. Each type of nuclide is specified using a symbol of the form

$$^{A}_{Z}X$$

where

X = chemical symbol for the given element

A = atomic mass number

Z = atomic number

For example, $^{14}_{6}C$ refers to a carbon nucleus containing 6 protons with an atomic mass number of 14. This atom thus contains 8 neutrons. Other carbon atoms exist that contain 5, 6, 7, 9, or 10 neutrons. All carbon atoms contain 6 protons. (Otherwise, they wouldn't be carbon!) Nuclei that contain the same number of protons but a different number of neutrons are called **isotopes**. See Fig. 22.10 for an illustration of the isotopes of hydrogen.

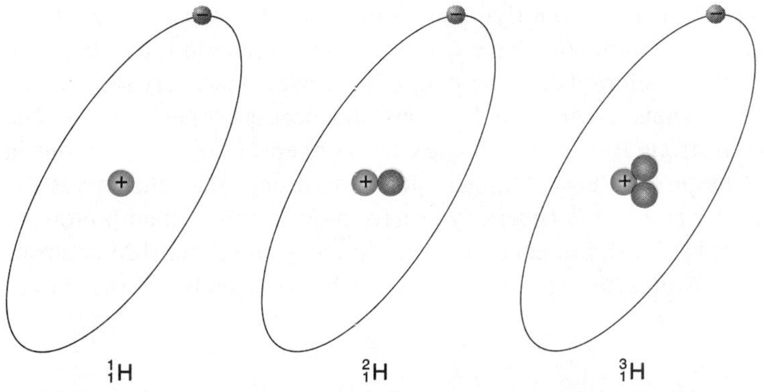

Figure 22.10 Three isotopes of hydrogen. Each has a single electron and a single proton, but different numbers of neutrons.

EXAMPLE 1

For the uranium isotope $^{238}_{92}U$, find (a) its atomic mass number, (b) its atomic number, (c) its neutron number, (d) the number of protons, (e) the number of nucleons, and (f) the number of neutrons.

(a) The atomic mass number is A, 238.
(b) The atomic number is Z, 92.
(c) The neutron number is $N = A - Z = 238 - 92 = 146$.
(d) The number of protons is $Z = 92$.
(e) The number of nucleons is $A = 238$.
(f) The number of neutrons is $N = 146$.

The mass of a single atom is called its **atomic mass**. Chemists and physicists have developed a unit of the measure of atomic mass called the **atomic mass unit** (u) to more easily compare the masses of different atoms. This unit is based on the mass of the common carbon atom with its six protons and six neutrons, so that the mass has been given the exact value of 12 u. Thus, 1 u is 1/12 the mass of the common carbon atom and so the approximate mass of a single proton or a single neutron is 1 u. The mass of an atom in atomic mass units is simply the sum of its protons and its neutrons. More precisely, the mass of a proton is 1.007276 u, the mass of a neutron is 1.008665 u, the mass of an electron is 0.00054858 u, and the mass of a neutral hydrogen atom is 1.007825 u.

Periodic Table

Dmitri Mendeleev (1834–1907), *chemist, was born in Russia. He developed the periodic table of the elements and predicted the existence of several elements that were subsequently discovered.*

In 1889, years before the atomic theory of the atom was developed, **Dmitri Mendeleev** proposed a table containing the elements arranged in a periodic manner that showed various trends in the properties of atoms. This **Periodic Table** (see Table 21 of Appendix C) contains all of the atomic elements arranged according to their atomic numbers, which can be used to predict their chemical properties. The existence of eight columns in his table and the trends in the properties of the atoms were later understood in terms of the atomic structure described in the last section.

The horizontal rows of the Periodic Table are called periods or rows. There are seven periods, each of which starts with an atom having one electron in its outer shell and ends with an atom with a complete outer shell structure (an inert element). The first three rows consist of 2, 8, and 8 elements. Rows 4 and 5 are longer rows consisting of 18 elements, and row 6 has 32 elements. Most row 7 elements are radioactive and do not occur in nature.

The number of elements in each row can be understood in terms of the atomic theory described earlier and the number of electrons in each energy level, or shell (see Table 22.2).

Metals are found on the left side of the table, with the most active metals in the lower left corner. Nonmetals are found on the right side. The noble or inert gases are on the far right. The acid-forming properties increase toward the right. Base-forming properties increase toward the left. Other properties, including the atomic weight and the atomic size, are given in a common form in the Periodic Table.

PROBLEMS 22.5

For each given isotope, find (a) its atomic mass number, (b) its atomic number, (c) its neutron number, (d) the number of protons, (e) the number of nucleons, and (f) the number of neutrons.

1. ${}^{12}_{6}C$ 2. ${}^{13}_{7}N$ 3. ${}^{48}_{22}Ti$ 4. ${}^{141}_{58}Ba$

From the Periodic Table, find (a) the atomic number and (b) the atomic mass for each element.

5. Na (sodium) 6. Fe (iron) 7. Pb (lead) 8. Cl (chlorine)
9. The nucleus of a certain element contains 6 protons and 8 neutrons. Write its chemical symbol.
10. The nucleus of a certain element contains 92 protons and 142 neutrons. Write its chemical symbol.

Use the Periodic Table to determine the following:

11. Which of the elements Br and Ca would be expected to be more acidic?
12. Which of the elements Br and Ca would be expected to be a stronger base?
13. Which of the elements Cl, Ar, I, Ca, and K is inert?
14. Which of the elements C, N, O, and Fe is a metal?

22.6 Nuclear Mass and Binding Energy

When the total mass of a nucleus is compared with the sum of the individual masses of the protons and neutrons when unbound, or broken apart, the nuclear mass is found to be smaller. How can this be? Early in the twentieth century, Einstein stated as a physical law that mass and energy are equivalent forms of "matter." According to this **mass–energy equivalence**, mass can be converted into energy as follows:

$$E = \Delta mc^2$$

where

$$E = \text{energy}$$
$$\Delta m = \text{change in mass}$$
$$c = \text{speed of light, } 3.00 \times 10^8 \text{ m/s}$$

When the total mass in a reaction is decreased by some amount, energy is given off in the form of radiation or kinetic energy. The difference in mass, converted using the equation above, is called the **binding energy of the nucleus**. This energy represents the total energy required to break a nucleus into its separate nucleons. For instance, if the mass of a ${}^{4}_{2}He$ nucleus were exactly equal to the mass of two protons and two neutrons, the nucleus would break apart without any additional energy. To be stable, the mass of the nucleus must be less than the mass of its component nucleons. You may think of the binding energy as the energy a nucleus lacks relative to the total mass of its separate nucleon components. The binding energies of nuclei are approximately 10^6 times greater than the binding energies of the electrons in atoms and therefore are much more significant. ***Note:*** In this chapter we often use more than three significant digits because of the precision of atomic and nuclear measurements.

EXAMPLE 1

Find the binding energy of a ^4_2He nucleus. The mass of the neutral ^4_2He atom is 6.6463×10^{-27} kg.

Data: The mass of two individual neutrons, two individual protons, and two individual electrons is

$$\begin{aligned} m &= 2m_p + 2m_n + 2m_e \\ &= 2(1.6726 \times 10^{-27}\text{ kg}) + 2(1.6749 \times 10^{-27}\text{ kg}) + 2(9.1094 \times 10^{-31}\text{ kg}) \\ &= 6.6968 \times 10^{-27}\text{ kg} \end{aligned}$$

Since the mass of neutral ^4_2He atom $= 6.6463 \times 10^{-27}$ kg,

$$\begin{aligned} \Delta m &= 6.6968 \times 10^{-27}\text{ kg} - 6.6463 \times 10^{-27}\text{ kg} \\ &= 0.0505 \times 10^{-27}\text{ kg} = 5.05 \times 10^{-29}\text{ kg} \end{aligned}$$

Basic Equation:

$$E = \Delta mc^2$$

Working Equation: Same

Substitution:

$$\begin{aligned} E &= (5.05 \times 10^{-29}\text{ kg})(3.00 \times 10^8\text{ m/s})^2 \\ &= 4.55 \times 10^{-12}\text{ kg m}^2/\text{s}^2 \qquad (1\text{ J} = 1\text{ kg m}^2/\text{s}^2) \\ &= 4.55 \times 10^{-12}\text{ J} \end{aligned}$$

The conversion factor relating atomic mass units and metric mass is

$$1\text{ u} = 1.6605 \times 10^{-27}\text{ kg}$$

The mass of atomic and nuclear particles is also expressed in million electron volts (MeV). Recall that an electron volt is the energy that one electron would gain in passing through an electric potential difference of 1 V. One MeV is given by

$$1\text{ MeV} = 1.602 \times 10^{-13}\text{ J}$$

Thus, the binding energy of a ^4_2He nucleus from Example 1 in MeV is

$$4.55 \times 10^{-12}\text{ J} \times \frac{1\text{ MeV}}{1.602 \times 10^{-13}\text{ J}} = 28.4\text{ MeV}$$

EXAMPLE 2

Express the MeV unit in terms of its equivalent mass in kilograms and atomic mass units using the equation $E = \Delta mc^2$.

Data:

$$\begin{aligned} E &= 1\text{ MeV} = 1.602 \times 10^{-13}\text{ J} \\ c &= 2.998 \times 10^8\text{ m/s} \\ \Delta m &= ? \end{aligned}$$

(***Note:*** we need to use 4 significant digits here.)

Basic Equation:

$$E = \Delta mc^2$$

Working Equation:

$$\Delta m = \frac{E}{c^2}$$

Substitution:

$$\begin{aligned} \Delta m &= \frac{1.602 \times 10^{-13}\text{ J}}{(2.998 \times 10^8\text{ m/s})^2} \times \frac{1\text{ kg m}^2/\text{s}^2}{1\text{ J}} \\ &= 1.782 \times 10^{-30}\text{ kg} \end{aligned}$$

Next, convert kilograms to atomic mass units:

$$1.782 \times 10^{-30}\text{ kg} \times \frac{1\text{ u}}{1.6605 \times 10^{-27}\text{ kg}} = 1.073 \times 10^{-3}\text{ u}$$

Thus, 1 MeV (1.602×10^{-13}J) has an equivalent mass of 1.782×10^{-30} kg or 1.073×10^{-3} u.

The masses of some atomic particles are listed in the following table.

	Mass	
Particle	**kg**	**u**
Electron	9.1094×10^{-31}	0.00054858
Proton	1.67262×10^{-27}	1.007276
Neutron	1.67493×10^{-27}	1.008665
$^1_1$H atom	1.67353×10^{-27}	1.007825

The **average binding energy per nucleon** in a nucleus is the total binding energy of the nucleus divided by the total number of nucleons, *A*. Fig. 22.11 gives the average binding energy per nucleon in MeV for stable nuclei. For $^4_2$He the average binding energy is its total binding energy in MeV, which we found following Example 1, divided by its total number of nucleons; that is, 28.4 MeV/4 = 7.1 MeV. Binding energies are largest for the number of nucleons, *A*, between about 30 and 60. This allows nuclei below 30 and above 60 to undergo reactions or decay, which produce substantial amounts of energy.

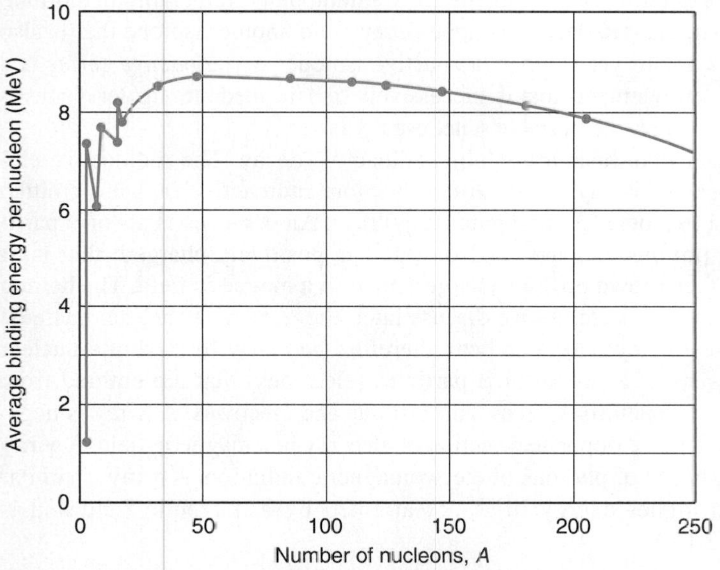

Figure 22.11 Average binding energy per nucleon for a given mass number *A* for stable nuclei

EXAMPLE 3

Estimate the average binding energy for $^{238}_{92}$U from Fig. 22.11.

Note that $^{238}_{92}$U has 238 nucleons as indicated by its atomic mass number, the upper number. Place a ruler or straight edge vertically at 238 on the horizontal axis. Note its point of intersection with the graph. Then, read the number on the vertical axis, 7.6 MeV, corresponding to this point of intersection. Thus, the average binding energy for $^{238}_{92}$U is approximately 7.6 MeV.

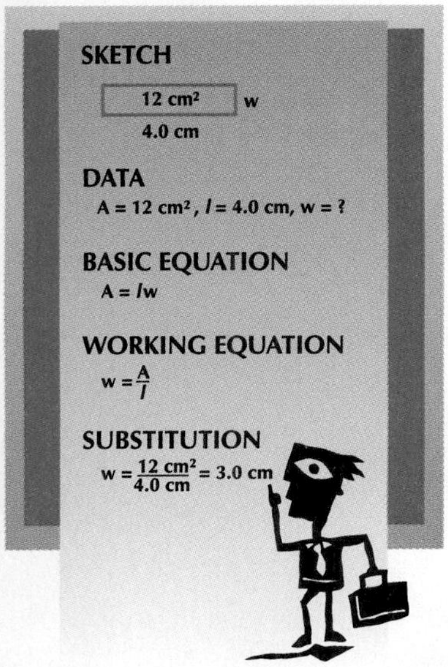
SKETCH
12 cm²
w
4.0 cm
DATA
A = 12 cm², l = 4.0 cm, w = ?
BASIC EQUATION
A = lw
WORKING EQUATION
w = A/l
SUBSTITUTION
w = 12 cm²/4.0 cm = 3.0 cm

PROBLEMS 22.6

1. Find the mass in kilograms of the ${}^{232}_{92}\text{U}$ atom if its mass in atomic mass units is 232.037131 u.
2. Find the mass in kilograms of the ${}^{228}_{90}\text{Th}$ atom if its mass in atomic mass units is 228.028716 u.
3. Find the binding energy of a ${}^{14}_{7}\text{N}$ nucleus in MeV if the mass of the neutral N atom is 14.00307 u.
4. Find the binding energy of a ${}^{232}_{92}\text{U}$ nucleus in MeV if the mass of the neutral U atom is 232.037131 u.
5. Find the binding energy of a ${}^{228}_{90}\text{Th}$ nucleus in MeV if the mass of the neutral Th atom is 228.028716 u.

Estimate the average binding energy per nucleon for each of the following nuclei from Fig. 22.11.

6. ${}^{48}_{22}\text{Ti}$
7. ${}^{141}_{58}\text{Ba}$
8. ${}^{12}_{6}\text{C}$

22.7 Radioactive Decay

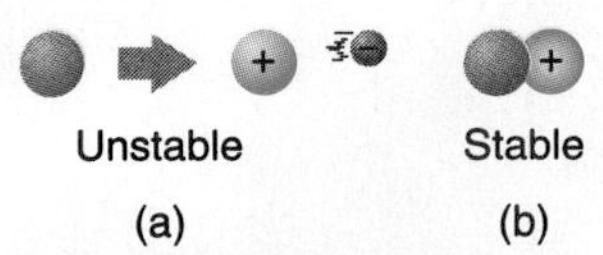
Unstable
Stable
(a)
(b)

Figure 22.12 A neutron by itself is unstable, but it is stable with a proton. The lone neutron decays into a proton and an electron.

The size of an atomic nucleus is limited by the fact that neutrons are unstable. On average, an isolated neutron will decay into a proton and an electron in about 12 min. In the presence of protons, the neutron is more stable (Fig. 22.12). In many atomic nuclei, neutrons are stable for many billions of years. For heavy atoms, the large number of neutrons required to hold the protons together leads to a situation where not enough protons exist to prevent one or more of the neutrons from decaying into a proton and an electron. Such unstable nuclei are called *radioactive*. **Radioactive decay** occurs when an unstable nucleus of an atom is transformed into a new element through the spontaneous disintegration of its atomic nuclei. Examples of radioactive elements are uranium, plutonium, radium, thorium, and their products. Often, one radioactive isotope decays into another isotope that is also radioactive, which decays into yet another radioactive isotope. A *radioactive series* is a series of isotopes of various elements that is successively transformed into lighter elements before the series of elements reaches a stable state, usually lead.

All elements heavier than bismuth exhibit radioactive decay. The radioactive elements exhibit three types of decay in which one or more radioactive rays are emitted from the nucleus: alpha (α), beta (β), and gamma (γ) rays. An **α ray** consists of α particles, each having two protons and two neutrons, and is positively charged; that is, it curves in the direction that known positive charges curve in a magnetic field. The helium nucleus is identical to an α particle. As we discuss later, the α particle (helium nucleus) is the most stable nuclear combination and can therefore be easily formed in a nuclear reaction. A **β ray** consists of a stream of β particles (electrons) that are emitted from neutrons in a nucleus as the neutrons decay into protons and electrons. A β ray is negatively charged and curves in the opposite direction of an α ray in a magnetic field. A **γ ray** has no mass and is composed of photons of electromagnetic radiation. A γ ray is similar to light but has much higher energy; it is not affected by a magnetic field and is uncharged. (See Fig. 22.13.)

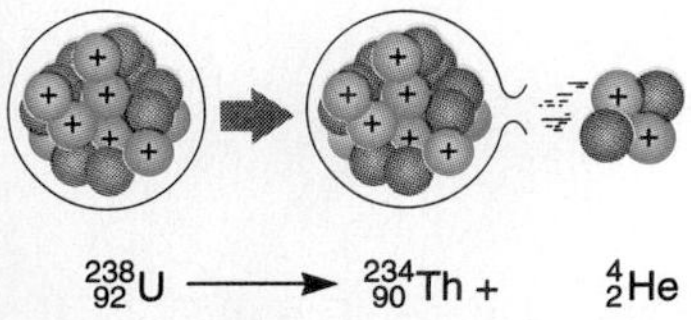
${}^{238}_{92}\text{U} \longrightarrow {}^{234}_{90}\text{Th} + {}^{4}_{2}\text{He}$

Figure 22.14 Uranium 238 is unstable and decays by giving off an α particle.

Alpha Decay

A nucleus that emits an α particle becomes different because it has lost two protons and two neutrons (Fig. 22.14).When radium 226 (${}^{226}_{88}\text{Ra}$) gives off an α particle, it becomes a nucleus with $A = 226 - 4 = 222$ and $Z = 88 - 2 = 86$. This nucleus is that of the element radon (${}^{222}_{86}\text{Rn}$). This nuclear reaction is written in the form

$${}^{226}_{88}\text{Ra} \longrightarrow {}^{222}_{86}\text{Rn} + {}^{4}_{2}\text{He}$$

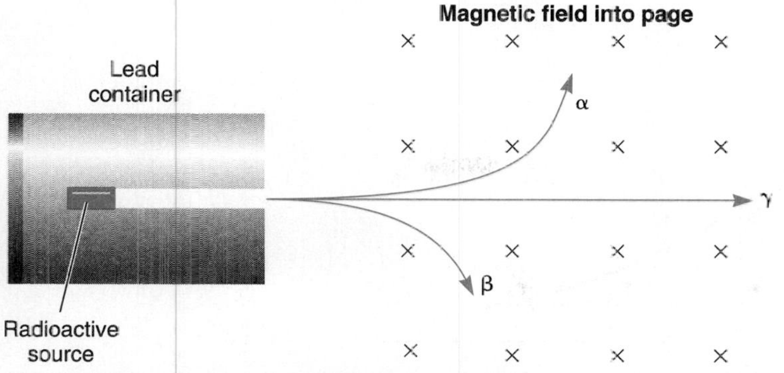

Figure 22.13 Alpha and beta rays curve in opposite directions in a magnetic field; gamma rays are undeflected.

Alpha decay occurs because there are not enough neutrons in the nucleus to keep it stable. The total energy released in radioactive decay is called the **disintegration energy**, Q, and is given by

$$Q = (M_p - M_d - m_\alpha)c^2$$

where

Q = disintegration energy
M_p = mass of the parent nucleus
M_d = mass of the daughter nucleus
m_α = mass of the α particle
c = speed of light

The emission of α particles changes the *parent* nucleus into a different nucleus, called a *daughter* nucleus. This disintegration energy is in the form of kinetic energy of (1) the α particle as it moves away from the nucleus and (2) the recoiling nucleus that moves away in the opposite direction.

Beta Decay

A nucleus can also decay by the emission of a β particle (an electron). How can a nucleus made up of only protons and neutrons emit an electron? One of the neutrons in the nucleus can break up into a proton and an electron and give off the electron in a manner similar to the decay of free neutrons. In β decay, a nucleus thus changes its charge, and the atom will have to pick up an additional electron in its charge clouds to remain neutral. An element involved in this process is thereby changed into a different element with one more electron. An example of β decay is the changing of carbon into nitrogen (Fig. 22.15):

$$^{14}_{6}\text{C} \longrightarrow {}^{14}_{7}\text{N} + e^-$$

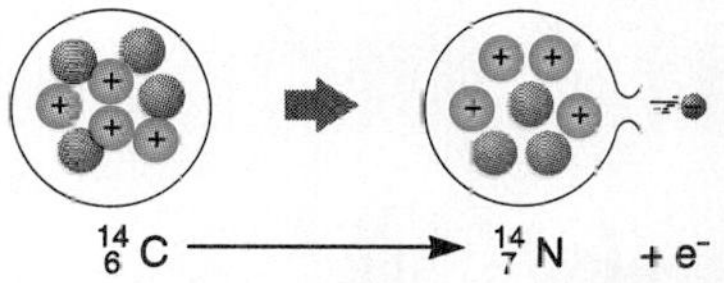

Figure 22.15 Carbon 14 can decay into a nitrogen 14 atom by giving off a β particle (an electron).

The radioactivity of an isotope is commonly measured by the decay rate or the half-life (Fig. 22.16). The **decay rate** is the probability per unit time that a decay of radioactive isotopes will occur. The decay of radioactive isotopes is a completely random process. That is, it cannot be predicted exactly when a given nucleus will decay. For a large number of nuclei in a sample, the **half-life** is the length of time required for one-half of the original amount of the radioactive atoms in the sample to decay (Fig. 22.16). This means that after the first half-life, one-half of the original amount remains. After the second half-life, one-half of one-half, or one-fourth, of the original amount remains. After the third half-life, one-half of one-fourth, or one-eighth, of the original amount remains. This process continues.

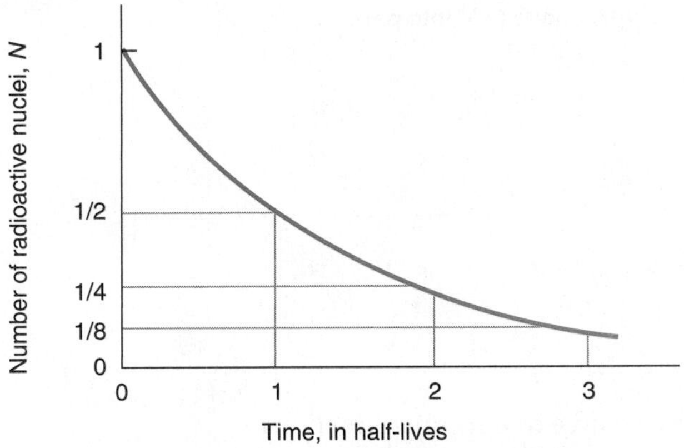

Figure 22.16 An exponential decrease of radioactive nuclei is observed.

EXAMPLE 1

Draw a graph through four half-lives of the radioactive decay of 80.0 g of radium 226, which has a half-life of 1620 years (See Fig. 22.17).

Note that the amount of radium is halved as we move down the vertical axis in Fig. 22.17, whereas the number of years doubles as we move across the horizontal axis. That is, one-half of the original amount (80.0 g) of radium 226 decays in the first 1620 years (leaving 40.0 g). During the second half-life, one-half of the remaining 40.0 g decays in the next 1620 years (leaving 20.0 g). During the third half-life, one-half of the remaining 20.0 g decays in the next 1620 years (leaving 10.0 g). During the fourth half-life, one-half of the remaining 10.0 g decays in the next 1620 years (leaving 5.0 g).

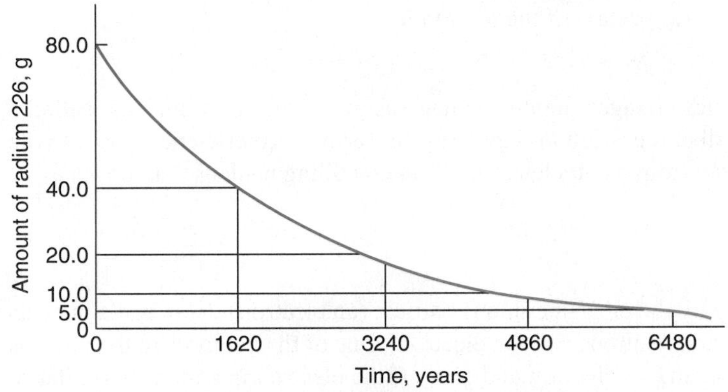

Figure 22.17 Radioactive decay of 80.0 g of radium 226 through four half-lives

Radioactive decay can be expressed by the equation

$$N = N_0 e^{-\lambda t}$$

where

N = number of remaining radioactive atoms

N_0 = original number of radioactive atoms

e = the natural number (e = 2.718 approximately)

λ = decay constant

t = elapsed time

(Note that calculations involving the natural number e can easily be performed on calculators with an e^x key.) The half-life $T_{1/2}$ is related to the decay constant by the equation

$$T_{1/2} = \frac{0.693}{\lambda}$$

where

$$T_{1/2} = \text{half-life}$$
$$\lambda = \text{decay constant}$$

EXAMPLE 2

Find the half-life of ${}^{13}_{7}N$ if its decay constant is 1.16×10^{-3} decay/s.

Data:

$$\lambda = 1.16 \times 10^{-3}/\text{s}$$
$$T_{1/2} = ?$$

Basic Equation:

$$T_{1/2} = \frac{0.693}{\lambda}$$

Working Equation: Same

Substitution:

$$T_{1/2} = \frac{0.693}{1.16 \times 10^{-3}/\text{s}}$$
$$= 597 \text{ s}$$

That is, one-half of the original amount will decay in 597 s.

EXAMPLE 3

Find the remaining amount of radioactive radium after 2450 years if the original amount was 3.54×10^{23} atoms. The half-life of radium is 1620 years.

Data:

$$N_0 = 3.54 \times 10^{23}$$
$$\lambda = \frac{0.693}{T_{1/2}} = \frac{0.693}{1620 \text{ yr}} = 4.28 \times 10^{-4}/\text{yr}$$
$$t = 2450 \text{ yr}$$
$$N = ?$$

Basic Equation:

$$N = N_0 e^{-\lambda t}$$

Working Equation: Same

Substitution:

$$N = (3.54 \times 10^{23})e^{-(4.28 \times 10^{-4}/\text{yr})(2450 \text{ yr})}$$
$$= (3.54 \times 10^{23})e^{-1.0486}$$
$$= 1.24 \times 10^{23} \text{ atoms}$$

Note: $\frac{1.24 \times 10^{23}}{3.54 \times 10^{23}} = 35.0\%$ of the radioactive atoms remain after 2450 years.

Half-lives of atoms range from a small fraction of a second to billions of years. For short times, radioactive half-lives can be measured by waiting until one-half of the atoms decay. For longer times, the rate at which a sample decays can be measured by using one of several types of radiation detectors.

Radiocarbon Dating

Radiocarbon dating is a method used to obtain age estimates of organic materials. The method is used in archaeology, geology, and other branches of science to find the age of wood, charcoal, bone and antler, peat and organic-bearing sediments, and other organic materials. Radioactive carbon 14 is produced when nitrogen 14 is bombarded by cosmic rays in the atmosphere, drifts down to earth, and is absorbed from the air by plants. Animals

PHYSICS CONNECTIONS

Food Irradiation

The widespread threat of food-borne diseases and contamination of food from bacteria and insects as well as food's relatively short shelf-life have resulted in significant economic losses and health problems. In the 1950s, the U.S. Army began irradiating food to prolong its shelf-life for soldiers. In the 1960s, the U.S. Food and Drug Administration (FDA) approved the irradiation of wheat flour to limit the growth of mold. Since then, the FDA has tested and approved the irradiation of fruits, vegetables, spices, pork, chicken, and red meat to increase shelf-life and reduce or eliminate insects, parasites, and bacteria in food (Fig. 22.18).

(a)

(b)

Figure 22.18 (a) This symbol must be placed on all food that is exposed to food irradiation. (b) Food that has not been exposed and that has been exposed to irradiation, respectively.

Most food is irradiated using γ radiation from cobalt 60 and cesium 137 sources. The γ radiation produced from the radioactive sources breaks up the DNA molecules in the microbes, insects, and bacteria. The exposure to the radiation either kills the organisms or prevents them from reproducing. Insects are quickly killed upon breakup of their DNA, whereas bacteria may take longer to destroy. If fruits and vegetables are exposed to radiation, their DNA is destroyed, which delays unwanted sprouting and/or ripening.

Upon seeing the irradiated food symbol, many people will not purchase the item. When food has been irradiated, it means that the food was exposed to radiation to destroy unwanted bacteria, insects, and microbes, but the food itself does not become radioactive. Irradiating food to increase shelf-life reduces the cost of rushed preparation and shipping costs. Such benefits would be realized not only by individuals in the United States, but also by individuals in third-world countries, where a great deal of food is prevented from being imported because it spoils before arriving at the ports. A few studies have indicated that irradiation reduces some essential vitamins and other nutrient levels that were present before the irradiation process. However, other studies have shown that such reductions are also found when food is canned or cooked.

Irradiation, although relatively new for the majority of our population, has been used for years by the military and on NASA space missions. A crop can be preserved for a significant amount of time as a result of irradiation.

eat the plants and take the carbon into their bodies. Humans take the carbon into their bodies by eating both plants and animals. When a living organism dies, it stops absorbing the carbon; the carbon already in the organism begins to disintegrate. Scientists then can measure how much carbon 14 has disintegrated and how much remains. Carbon 14 decays at a slow, steady rate (half-life 5730 years) and reverts to nitrogen 14. The method involves counting the number of β radiations per minute per gram of material. Modern carbon 14 emits approximately 15 β radiations per gram of material; carbon material that is 5730 years old will emit only half that amount. Radiocarbon dating of organic materials older than 50,000 years is not reliable because the amount of carbon 14 remaining is so small that the fossil cannot be dated reliably. The accuracy of radiocarbon dating for some materials less than 50,000 years old has been verified through comparison with ancient historical records.

PROBLEMS 22.7

1. Find the half-life of a radioactive sample if its decay constant is 1.72×10^4 decays/s.
2. Find the half-life of a radioactive sample if its decay constant is 8.25×10^{-6} decay/s.
3. Find the remaining quantity of $^{124}_{55}$Cs from an original sample of 50.0 g after 4.00 min. Its half-life is 30.8 s.
4. Find the remaining quantity of radon 222 from an original sample of 75.0 g after 10.0 days. Its half-life is 3.82 days.
5. Find the percent of a sample of $^{124}_{55}$Cs that will decay in the next 10.0 s. Its half-life is 30.8 s.
6. Find the percent of a sample of $^{238}_{92}$U that will decay in the next 975 years. Its half-life is 4.47×10^9 years.
7. Find the remaining quantity of uranium 238 atoms from an original sample of 5.50×10^{20} atoms after 2.45 billion years. Its half-life is 4.50 billion years.
8. Find the remaining quantity of $^{14}_{6}$C from an original sample of 3.75×10^{21} atoms after $10\bar{0}0$ years. Its half-life is 5730 years.
9. Find the percent of a $^{14}_{6}$C sample that will decay in the next $30\bar{0}0$ years. Its half-life is 5730 years.
10. Find the percent of a radioactive sample of half-life 2.35 s that will decay in the next second.

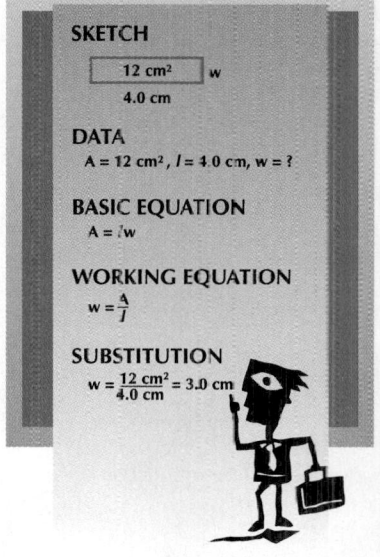

SKETCH

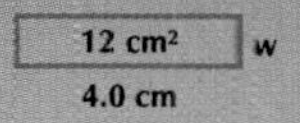

DATA

$A = 12\text{ cm}^2$, $l = 4.0$ cm, $w = ?$

BASIC EQUATION

$A = lw$

WORKING EQUATION

$w = \frac{A}{l}$

SUBSTITUTION

$w = \frac{12\text{ cm}^2}{4.0\text{ cm}} = 3.0\text{ cm}$

22.8 Nuclear Reactions—Fission and Fusion

Nuclear reactions take place when a nucleus is struck by a neutron, a γ ray, or another nucleus, causing an interaction. Many nuclear reactions have been viewed since Ernest Rutherford's 1919 observation of a reaction in which nitrogen, when bombarded by α particles, emitted a proton and was transformed into an isotope of oxygen. One of the most significant findings occurred during an experiment using neutron bombardment of existing elements in an attempt to produce new elements. Scientists, while attempting to form heavier elements using neutron bombardment of uranium, the heaviest known element at that time, made an amazing discovery, which has had a great impact on the world. This discovery led to the development of nuclear bombs and nuclear reactors used to produce heat energy for electric power, propulsion of nuclear submarines, various industrial processes, and other applications. They found that the bombardment of uranium with neutrons produces lighter nuclei, each nearly half the size of uranium, along with a number of neutrons (Fig. 22.19). The reaction can be written

$$n + {}^{235}_{92}\text{U} \longrightarrow {}^{236}_{92}\text{U} \longrightarrow {}^{141}_{58}\text{Ba} + {}^{92}_{36}\text{Kr} + 3n$$

This nuclear reaction, in which an atomic nucleus splits into fragments with the release of energy, is known as **nuclear fission**. A substantial amount of energy is released in this process because the mass of the uranium nucleus plus the mass of the bombarding neutron is larger than the combined mass of the fission fragments produced by the reaction. The total energy release per fission is approximately 200 MeV. On the nuclear scale, this is an extremely large amount of energy per atom!

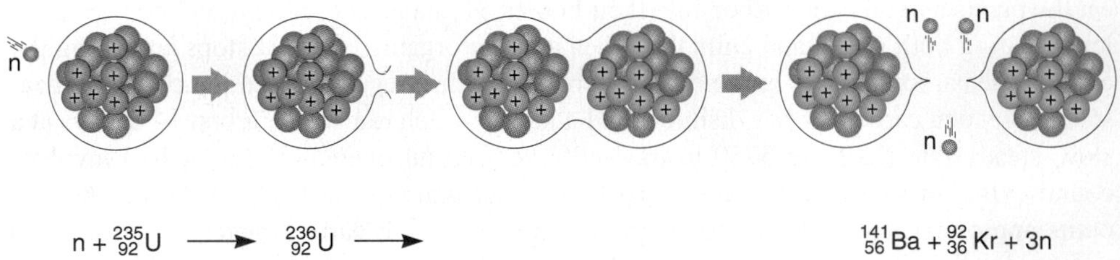

Figure 22.19 Fission of $^{235}_{92}U$ after neutron capture

The neutrons released in each nuclear fission reaction can be used to create further reactions. The process is called a **chain reaction**. **Enrico Fermi** and his associates succeeded in producing the first self-sustaining chain reaction at the University of Chicago in 1942. Chain reactions are used in nuclear reactors to produce electric power, to produce intense neutron sources for research and medical use, and in atomic bombs. **Robert J. Oppenheimer** led the research that resulted in the construction of the first atomic bomb. Many of the end products of fission reactions are radioactive for long periods of time. These radioactive wastes must be disposed of with extreme care.

Enrico Fermi (1901–1954),

physicist, was born in Italy. He helped explain the theoretical behavior of atomic particles, developed the theory of beta decay, and produced the first controlled nuclear chain reaction at the University of Chicago in 1942.

Another process for releasing energy is **nuclear fusion**, a nuclear reaction in which light nuclei interact to form heavier nuclei with the release of energy. Nuclear fusion can occur because the total mass of the reaction products can be less than that of the initial nuclei and particles. Very high temperatures are required for nuclear fusion to occur. An example of a sequence of fusion reactions is

$$
\begin{array}{lll}
{}^{1}_{1}\mathrm{H} + {}^{1}_{1}\mathrm{H} \longrightarrow {}^{2}_{1}\mathrm{H} + e^{+} & & (0.42\ \mathrm{MeV}) \\
{}^{1}_{1}\mathrm{H} + {}^{2}_{1}\mathrm{H} \longrightarrow {}^{3}_{2}\mathrm{He} + \gamma & & (5.49\ \mathrm{MeV}) \\
{}^{3}_{2}\mathrm{He} + {}^{3}_{2}\mathrm{He} \longrightarrow {}^{4}_{2}\mathrm{He} + {}^{1}_{1}\mathrm{H} + {}^{1}_{1}\mathrm{H} & & (12.86\ \mathrm{MeV})
\end{array}
$$

Robert J. Oppenheimer (1904–1967),

physicist, was born in New York City. He led the scientific research that resulted in the construction of the first atomic bomb.

This reaction sequence is constantly going on in many stars, including our sun. Fusion is the source of energy in stars. The sun's energy is believed to be produced largely by this sequence of reactions, which starts with protons and produces helium nuclei (α particles) with the release of substantial energy. Nuclear fusion has been used in hydrogen bombs. Fusion is also a subject of considerable research as a "clean" energy source because the end products of the reaction are not radioactive.

22.9 Detection and Measurement of Radiation

Since we cannot see, touch, or feel nuclear radiation, it is necessary to have instruments available to detect its presence. The most common detector of radiation is the Geiger counter (Fig. 22.20), named after **Hans Wilhelm Geiger**. This instrument contains a cylindrical metal tube filled with gas. A wire runs along the center of the metal tube. A high voltage (approximately 1000 V) is maintained across the wire and cylinder. When a

Hans Wilhelm Geiger (1882–1945),

physicist, was born in Germany. He performed experiments on beta-ray radioactivity and devised an instrument (the Geiger counter) to measure it.

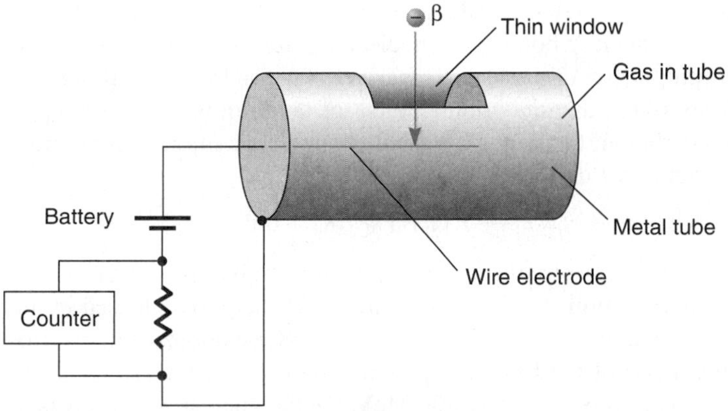

Figure 22.20 Diagram of a Geiger counter

charged particle enters the tube and ionizes some of the gas, the voltage on the tube causes the gas to "break down" and conduct an electric pulse between the wire and the cylinder. This electric pulse is detected by electronics that count the number of pulses. A loudspeaker is often hooked up to the electronics, allowing a "click" to be heard each time a charged particle enters the tube.

Since radiation can cause severe damage to living organisms and materials, it is necessary to be able to measure the amount, or *dose*, of radiation. There are two common units of *source activity:* the **curie** (Ci), which is defined as

$$1 \text{ Ci} = 3.70 \times 10^{10} \text{ disintegrations/s}$$

and the **becquerel** (Bq), which is the SI unit and defined as

$$1 \text{ Bq} = 1 \text{ disintegration/s}$$

Marie Curie (1867–1934), *Polish physicist, and husband,* **Pierre Curie (1859–1906),**

French physicist, were pioneers in the study of magnetism and radioactivity and discovered radium. The Curies and Becquerel shared the 1903 Nobel Prize in Physics for the discovery of radioactivity and Marie Curie was given the Nobel Prize for Chemistry in 1911 for isolating polonium and radium in 1910.

The **source activity** is the strength of a source of radiation that can be specified at a given time according to the equation

$$\boxed{A = \lambda N = \lambda N_0 e^{-\lambda t} = A_0 e^{-\lambda t}}$$

where

A = source activity
N = remaining quantity of radioactive atoms
N_0 = original quantity
λ = decay constant
t = time
A_0 = original source activity

Henri Becquerel (1852–1908),

physicist, was born in France. His work led to the isolation of radium and the beginnings of modern physics.

EXAMPLE

Find the source activity of a $^{222}_{88}\text{Ra}$ (radium) source 6.54 days after it was originally certified to have an activity of 0.356 Ci. Its half-life is 3.82 days.

Data:

$$t = 6.54 \text{ days}$$
$$A_0 = 0.356 \text{ Ci}$$
$$T_{1/2} = 3.82 \text{ days}$$
$$\lambda = \frac{0.693}{T_{1/2}} = \frac{0.693}{3.82 \text{ days}} = 0.181/\text{day}$$
$$A = ?$$

Basic Equation:

$$A = A_0 e^{-\lambda t}$$

Working Equation: Same

Substitution:

$$\begin{aligned} A &= (0.356 \text{ Ci})e^{-(0.181/\text{day})(6.54 \text{ day})} \\ &= (0.356 \text{ Ci})e^{-1.184} \\ &= 0.109 \text{ Ci} \end{aligned}$$

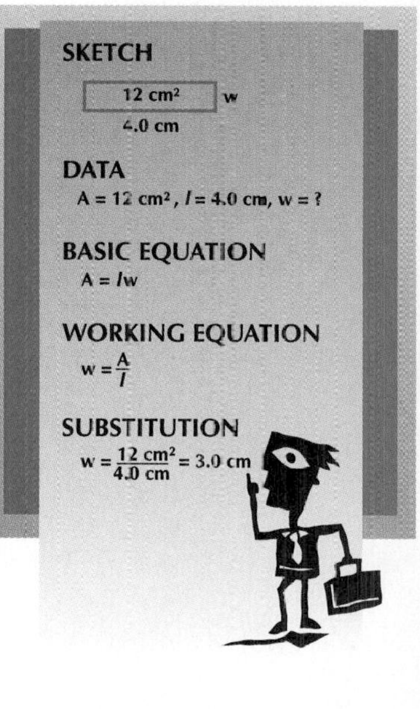

PROBLEMS 22.9

1. Find the source activity of a 1.24-Ci sample of $^{13}_{7}\text{N}$ (nitrogen) 20.0 min after certification. Its half-life is 10.0 min.
2. Find the source activity of a 2.64-Ci sample of $^{14}_{6}\text{C}$ (carbon) $40\bar{0}0$ years after certification. Its half-life is 5370 yr.

3. Find the source activity of a 0.476-Ci sample of $^{24}_{11}Na$ (sodium) 36.5 h after certification. Its half-life is 14.95 h.
4. Find the source activity of a 3.98-Ci sample of $^{11}_{6}C$ (carbon) 10.3 h after certification. Its half-life is 20.4 min.
5. Find the source activity of a 10.0-Ci sample of $^{226}_{88}Ra$ (radium) 125 yr after certification. Its half-life is 1590 yr.
6. Find the source activity of a 4.00-Ci sample of $^{131}_{53}I$ (iodine) 6.00 h after certification. Its half-life is 8.06 days.
7. Find the source activity of a 75.0-μCi sample of $^{214}_{84}Po$ (polonium) 5.00 μs after certification. Its half-life is 1.50×10^{-4} s.

22.10 Radiation Penetrating Power

The three types of radiation rays (α, β, and γ) can be stopped by matter. Some are stopped more easily than others (Fig. 22.21). Alpha rays, which are massive and carry a double positive charge, are stopped quite easily during "collisions" with atoms in the material. These collisions occur when the positively charged α particle comes close enough to an atom to feel a Coulomb force from some of the electrons or protons. The α particle gives up its energy to the material through these "collisions." A few pieces of paper or a few centimetres of air are sufficient to stop an α particle. As soon as an α particle slows enough to pick up two electrons while passing through matter, it becomes a harmless helium atom.

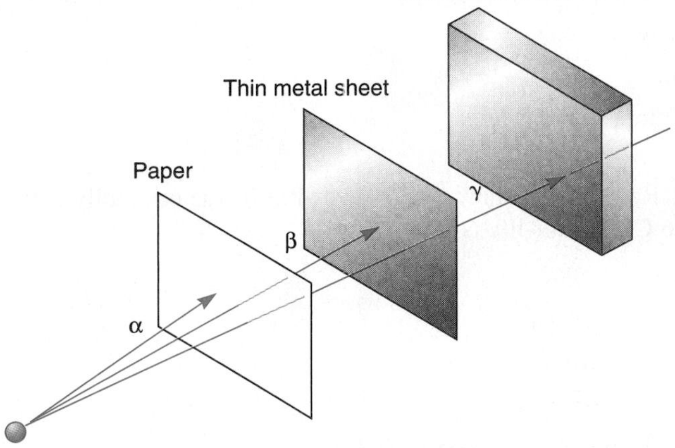

Figure 22.21 Penetrating power of different forms of nuclear radiation

Beta rays, which are electrons, can easily be stopped by a thin sheet of metal, such as aluminum. They lose their energy due to many collisions with electrons in atoms in the material they are passing through. They become electrons in the material, indistinguishable from others.

Gamma rays have the greatest penetrating power because they have no charge and therefore cannot be electrically attracted or repelled by electrons or protons. The γ ray, which is an energetic photon, is slowed only through a direct hit of an electron or a nucleus. A dense material, such as a heavy metal, is the best absorber of γ radiation.

As radiation passes through matter and loses its energy, it can do much damage. Many materials become weakened and brittle upon exposure to substantial levels of radiation. This major engineering problem had to be solved in order to build safe nuclear reactors.

Radiation damage to living organisms can be very severe. As radiation is absorbed in material, atoms become ionized as electrons are captured or emitted. If these atoms are basic in bonding molecules together, the molecule may break apart. The functioning of the living cell may be altered substantially if a large number of key molecules are damaged. If the damage occurs to the DNA molecule, which controls the growth and replication of the cell, the cell may be severely damaged and die. Obviously, if a large number of cells die, the organism itself will become sick (radiation sickness) or die. If the damaged cell replicates itself rapidly with faulty DNA, cancer may result. To prevent this type of damage during normal medical X-ray procedures, it is common for patients to wear lead aprons around tissue not being imaged by the X ray. This is also why X-ray technicians step behind a lead shield when X rays are being taken.

Radiation can be useful for medical diagnostics (X rays and radioactive tracers) and radiation therapy. In radiation therapy, cancer cells can be destroyed by the localized application to a tumor.

Glossary

Alpha Ray A ray consisting of alpha particles, each having two protons and two neutrons, and positively charged, that is, it curves in the direction that known positive charges curve in a magnetic field. (p. 616)

Atomic Mass The mass of a single atom, usually expressed in atomic mass units, u. (p. 612)

Atomic Mass Number The total number of nucleons (protons and neutrons) in a nucleus, A. (p. 611)

Atomic Mass Unit (u) A unit of measure of atomic mass based on the mass of the common carbon atom, which has six protons and six neutrons, so the mass of the carbon 12 atom has been given the exact value of 12 u. Thus, the approximate mass of a single proton or a single neutron is 1 u, and the mass of an atom in atomic mass units is simply the sum of the number of its protons and its neutrons. (p. 612)

Atomic Number The number of protons in a nucleus, Z. (p. 611)

Average Binding Energy per Nucleon The total binding energy of the nucleus divided by the total number of nucleons. (p. 615)

Becquerel (Bq) Unit of source activity; 1 Bq = 1 disintegration/s. (p. 623)

Beta Ray A ray consisting of a stream of beta particles (electrons), which are emitted from neutrons in a nucleus as they decay into protons and electrons. This ray is negatively charged and curves in the opposite direction of an alpha ray in a magnetic field. (p. 616)

Binding Energy of the Nucleus The total energy required to break a nucleus apart into separate nucleons. (p. 613)

Bohr Model An early model of atomic structure in which electrons travel around the nucleus in a number of discrete stable energy levels determined by quantum conditions. (p. 606)

Chain Reaction The process of using the neutrons released in each nuclear fission reaction to create further reactions. (p. 622)

Coulomb Force An electric, repulsive force between protons. (p. 611)

Curie (Ci) Unit of source activity; 1 Ci = 3.70×10^{10} disintegrations/s. (p. 623)

Decay Rate The probability per unit time that a decay of radioactive isotopes will occur. (p. 617)

Disintegration Energy The total energy released in radioactive decay in the form of kinetic energy. (p. 617)

Electron A fundamental particle of an atom; negatively charged. (p. 604)

Excited State A high-energy level for an electron in an atom. (p. 605)

Fluorescence A property of certain substances in which radiation is absorbed at one frequency and then reemitted at a lower frequency. (p. 607)

Gamma Ray A ray composed of photons of electromagnetic radiation, which have no mass. (p. 616)

Ground State The lowest energy level for an electron in an atom. (p. 605)

Half-Life The length of time required for one-half of the original amount of the radioactive atoms in a sample to decay. (p. 617)

Isotopes Nuclei that contain the same number of protons but a different number of neutrons. (p. 612)

Mass–Energy Equivalence A physical law stating that mass and energy are equivalent forms of matter; $E = \Delta mc^2$; stated by Einstein. (p. 613)

Neutron A fundamental particle in the nucleus of an atom; neutrally charged. (p. 604)

Neutron Number The number of neutrons in an atomic nucleus of a particular nuclide, N. Also, $N = A - Z$, the difference between the atomic mass number and the atomic number. N can be different for nuclei with the same atomic number. (p. 611)

Nuclear Fission A nuclear reaction in which an atomic nucleus splits into fragments with the release of energy. (p. 621)

Nuclear Fusion A nuclear reaction in which light nuclei interact to form heavier nuclei with the release of energy. (p. 622)

Nuclide A specific type of atom characterized by its nuclear properties, such as the number of neutrons and protons and the energy state of its nucleus. (p. 611)

Periodic Table A table that contains all of the atomic elements arranged according to their atomic numbers and which can be used to predict their chemical properties. (p. 612)

Proton A fundamental particle in the nucleus of the atom; positively charged. (p. 604)

Quantum Mechanics A theory that unifies the wave–particle dual nature of electromagnetic radiation. The quantum model is based on the idea that matter, just like light, behaves sometimes as a particle and sometimes as a wave. (p. 608)

Radioactive Decay A type of nuclear decay that occurs when an unstable atom is transformed into a new element through the spontaneous disintegration of its nucleus. (p. 616)

Radiocarbon Dating A method used to obtain age estimates of organic materials using carbon 14 decay. (p. 620)

Source Activity The strength of a source of radiation that can be specified at a given time. (p. 623)

Strong Force An attractive force among all nucleons (neutrons and protons) independent of their charge. (p. 611)

Formulas

22.3 $$E = -\frac{kZ^2}{n^2}$$

22.6 $$E = \Delta mc^2$$

22.7 $$Q = (M_p - M_d - m_\alpha)c^2$$

$$N = N_0 e^{-\lambda t}$$

$$T_{1/2} = \frac{0.693}{\lambda}$$

22.9 $$A = \lambda N = \lambda N_0 e^{-\lambda t} = A_0 e^{-\lambda t}$$

Review Questions

1. Which of the following are nuclear particles?
   (a) neutrons (b) protons
   (c) nucleons (d) atoms
   (e) all of the above
2. Einstein's equivalence principle relates to
   (a) weight and time. (b) space and gravity.
   (c) mass and energy. (d) all of the above.
   (e) none of the above.
3. The amount of radioactive material remaining after a period of time is related to the
   (a) atomic mass. (b) pressure.
   (c) half-life. (d) volume.
   (e) none of the above.
4. Explain the difference between the ground state and the excited states of electrons in atoms.
5. Describe the Bohr model of the atom.
6. Describe the similarities of protons and neutrons. Describe the differences.

7. Describe the differences between the electric force and the strong force.
8. If the strong force suddenly became much weaker in a nucleus while the electric force remained unchanged, what might happen to the nucleus?
9. Explain the principle of mass–energy equivalence in your own words.
10. What is the difference among the following atoms: ${}^{11}_{6}C$, ${}^{12}_{6}C$, and ${}^{13}_{6}C$? What are the similarities?
11. Explain the term *electron volt* in your own words.
12. Describe the importance of the neutron in atomic nuclei.
13. Describe an α ray in your own words.
14. Describe a γ ray in your own words.
15. Describe a β ray in your own words.
16. What important discovery was made by Enrico Fermi?
17. Explain a self-sustaining chain reaction.
18. Describe nuclear fusion.
19. Describe nuclear fission.
20. What fraction of a radioactive sample has not decayed after four half-lives have elapsed?
21. What damage can be caused to living organisms by radiation?
22. What medical uses does radiation have?

Review Problems

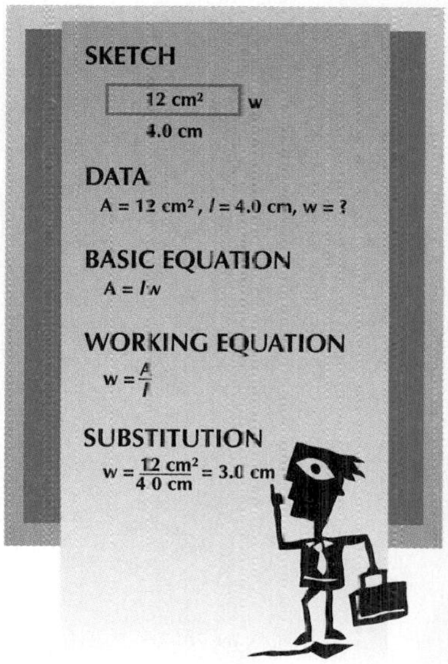
SKETCH
12 cm²
w
4.0 cm
DATA
A = 12 cm², l = 4.0 cm, w = ?
BASIC EQUATION
A = lw
WORKING EQUATION
w = A/l
SUBSTITUTION
w = 12 cm² / 4.0 cm = 3.0 cm

1. Find the energy of the electron in the $n = 5$ energy level of a hydrogen atom.
2. Find the energy of the photon given off when the electron in Problem 1 transitions down to the $n = 3$ energy level.
3. List four inert atoms from the Periodic Table.
4. List four metals from the Periodic Table.
5. Find the mass in kilograms of the ${}^{15}_{8}O$ atom. Its mass in atomic mass units is 15.003065 u.
6. Find the mass in kilograms of the ${}^{19}_{9}F$ atom. Its mass in atomic mass units is 18.998404 u.
7. Find the mass in kilograms of the ${}^{166}_{68}Er$ atom. Its mass in atomic mass units is 165.930292 u.
8. Find the binding energy in MeV of a ${}^{86}_{38}Sr$ atom. The mass of the neutral Sr atom is 85.909266 u.
9. Estimate the average binding energy for ${}^{102}_{44}Ru$ from Fig. 22.11
10. Estimate the average binding energy for ${}^{153}_{63}Eu$ from Fig. 22.11.
11. Estimate the average binding energy for ${}^{187}_{75}Re$ from Fig. 22.11.
12. Find the remaining quantity of osmium 191 atoms from an original sample of 8.25×10^{13} atoms after 54 days. Its half-life is 15.4 days.
13. Find the remaining quantity of iodine 131 atoms from an original sample of 8.33×10^{18} atoms after 34.4 days. Its half-life is 8.04 days.
14. Find the percent of a strontium 88 sample that will decay in the next 2.40 yr. Its half-life is 29.1 yr.
15. Find the percent of an osmium 191 sample that will decay in the next 2.00 h. Its half-life is 15.4 days.
16. Find the source activity of a 2.43-Ci sample of osmium 191 43.3 days after certification. Its half-life is 15.4 days.
17. Find the source activity of a 3.79-Ci sample of nitrogen 13 43.0 min after certification. Its half-life is 10.0 min.
18. Find the source activity of a 9.41-Ci sample of carbon 11 95.4 min after certification. Its half-life is 20.4 min.
19. Find the source activity of a 6.75-Ci sample of ${}^{238}_{92}U$ (uranium) one billion years after certification. Its half-life is 4.5×10^{9} yr.
20. Find the source activity of a 50.0-μCi sample of ${}^{214}_{82}Pb$ (lead) 1.00 h after certification. Its half-life is 26.8 min.

APPLIED CONCEPTS

1. To produce the blue neon light used in many storefront signs, an electron from a hydrogen atom must undergo a transition from one energy level to another. (a) What is the transition energy (eV) if the electron changes its energy level from $n = 4$ to $n = 2$? (b) What is the frequency for the photon that is released? (c) What is the photon's velocity? (d) What is the wavelength for the released photon?
2. A photon is released from a hydrogen atom when an electron moves from a higher energy level to a lower energy level. At what level does an electron begin its journey to the level $n = 1$ if the emitted photon has a wavelength of 1.22×10^{-7} m?
3. The binding energy for a ${}^{4}_{2}He$ nucleus is 28.40 MeV. (a) What is the binding energy (in J)? (b) How many protons, neutrons, and electrons are in the ${}^{4}_{2}He$ atom? (c) Find the total mass of the various subatomic particles in the ${}^{4}_{2}He$ atom. (d) What is the mass of the neutral ${}^{4}_{2}He$ atom?
4. A source of radon 222 has 2.45×10^{23} atoms remaining after 10.0 days. (a) If its half-life is 3.82 days, what is its decay rate? (b) Find the original number of atoms before the decay.
5. Find the source activity of a 0.875-Ci sample of ${}^{24}_{11}Na$ (sodium) 32.4 h before certification if its half-life is 14.95 h.

CHAPTER 23

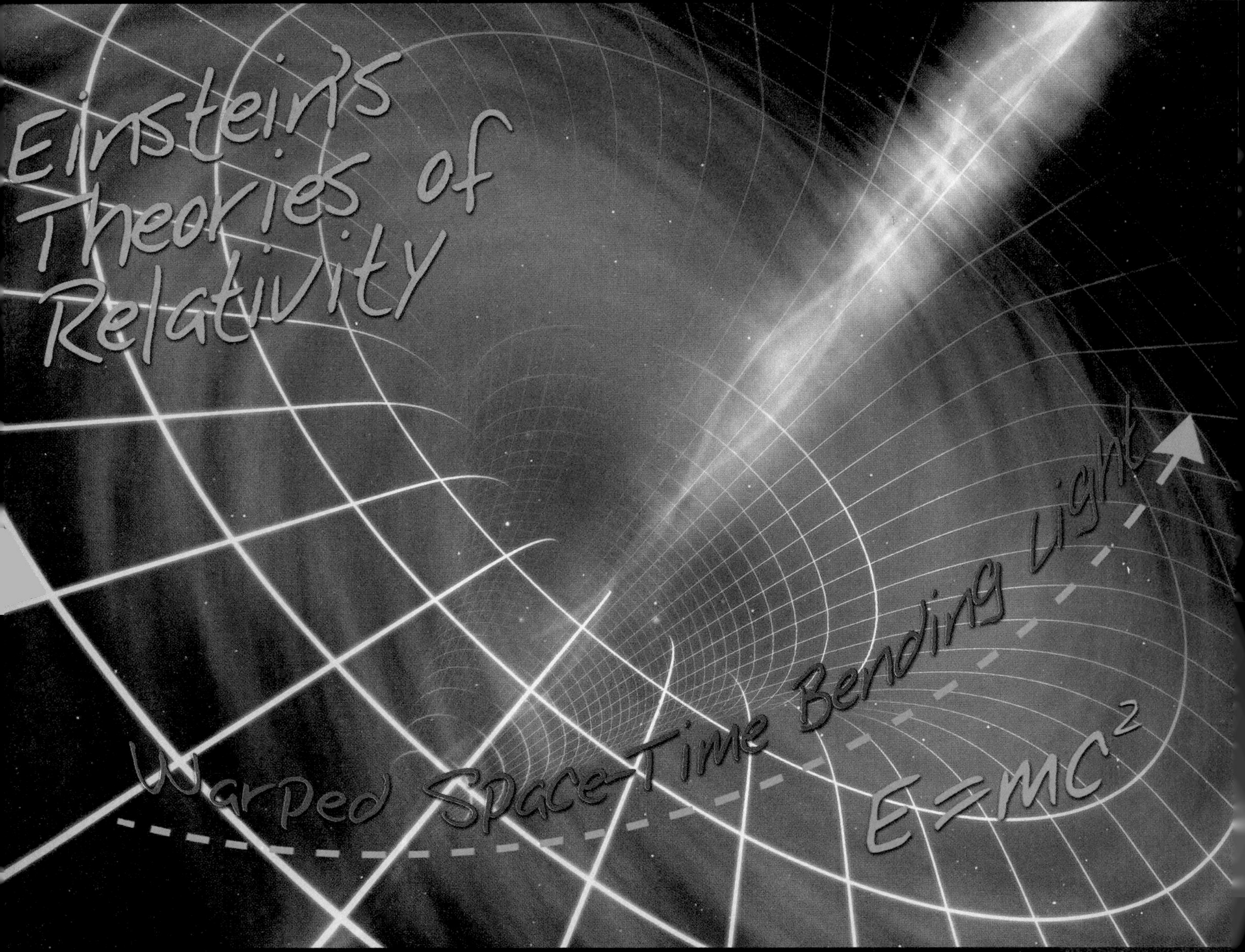

Einstein's Theories of Relativity
Warped Space-Time Bending Light
E = MC²

SPECIAL AND GENERAL RELATIVITY

At the age of 16, Albert Einstein dreamed of what it would be like to ride on a beam of light. His enthusiasm for science, specifically the topics of light, space, and time, led him to develop several theories that would make him the most famous scientist of all time. In this chapter we will explore the life of Einstein, the warping of space-time, and the formula that changed the world, $E = mc^2$.

Objectives

The major goals of this chapter are to enable you to:

1. Discuss the theories, successes, and milestones that made Albert Einstein an influential scientist.
2. Explain that the speed of light is constant, but the rate at which time passes can vary according to motion.
3. Find the amount of energy contained in a particular amount of mass.
4. Describe the theory of a four-dimensional space-time universe.
5. Express Einstein's theory that acceleration and gravity are equivalent.

Photo courtesy of Corbis Corporation/Betteman. Reprinted with permission

Figure 23.1 Albert Einstein (right) as a boy with his sister Maja.

Photo courtesy of the National Archives and Records Administration

Figure 23.2 Albert Einstein at work in his study at Princeton University in 1943.

23.1 Albert Einstein

Born in Ulm, Germany, in 1879, Albert Einstein spent most of his younger years (Fig. 23.1) reading and engaging in science experiments, while rebelling against the traditional regimentation found in many German schools at the time. He was given a compass at the age of five by his father and was intrigued by how the compass needle would move even though nothing was physically touching it. The physical mystery surrounding the compass fascinated Einstein and set him on the course of revolutionizing science. However, his path was not easy and included being expelled, dropping out of school, and failing academic entrance exams. Einstein was eventually able to enroll in the Zurich Polytechnic Institute, where he was largely left to explore the wonders of science on his own.

After graduation, Einstein was unable to find a full-time job as an assistant in a laboratory. Instead, he obtained a job at the Swiss Patent Office. During this period he did some of his greatest work, including his 1905 paper about the **Special Theory of Relativity,** which is based on the following two basic assumptions: (1) *the laws of physics are the same in both moving and nonmoving frames of reference and* (2) *the speed of light is constant no matter what the speed of the observer or the source of light.* The Special Theory boosted Einstein's physics career and caused many universities to take notice of the previously unknown patent clerk.

Ten years later, Einstein produced another paper, which extended the special theory to accelerated frames of reference. The **General Theory of Relativity** is based on the assumption that *gravity and acceleration are equivalent and that light has mass and its path can be warped by gravity.* In 1921, Einstein won the Nobel Prize in Physics and began to focus on a *theory that would link what came to be understood as the five fundamental forces: gravitational, electric, magnetic, strong, and weak forces.* The work that Einstein began in the search for such a theory (now called a **Grand Unified Theory**) is continued today by physicists around the world.

During the turbulent years leading up to World War II, Germany was not a safe place for the world's most famous Jewish scientist. In the 1930s, Einstein decided to leave Germany and live in the United States. He relocated to Princeton, New Jersey, where he continued to lecture and work on his unified theory. Although his famous formula $E = mc^2$ was the scientific principle behind the atomic bomb, and he had urged the development of such a weapon because he feared that Germany might develop one, after the war he opposed its use.

Albert Einstein died in Princeton in 1955. Throughout his life, he addressed many major scientific matters and was also a strong moral voice on social and political issues. Einstein showed great compassion for humankind and was known throughout the world as the genius professor who attempted, as Einstein said, "to read the mind of God" (Fig. 23.2).

23.2 Special Theory of Relativity

Let's say you are reading this textbook while sitting on the ground. Are you moving? You would probably say no because when you look around, nothing seems to be changing its position. However, you suddenly hear a noise and see a man driving a car down the street. You see that the car is changing position, but how do you know that the car is moving and you are not?

The driver of the car looks out the window and sees you moving toward and then away from him after passing you. The driver can adopt the perspective that you are moving while he is at rest. In fact, he can see everything outside the car as moving and himself as stationary. This is a simplistic example of **relative motion**, which is the concept that motion can be described differently depending upon the observer's perspective. In order for a motion to be fully described, it must be described in reference to something or someone's perspective. Both the driver and you see the same relative motion in two different ways. You see the car moving toward you, while the driver can argue that you are moving toward him. The difference can be expressed through the notion of *relativity:* the same event being observed differently depending on the perspective of the observer.

PHYSICS CONNECTIONS

Albert Einstein: "The Person of the Century"

"Why is it that nobody understands me, yet everybody likes me?" This quote from Einstein, taken from an interview with *The New York Times* in 1944, is reflective of Einstein's fame and genius. Although Albert Einstein is recognized around the world as a genius, very few people understand what his genius meant for the scientific world. Despite his superior intellect, he was not able to understand why so many people, who did not understand his scientific work, adored him.

Einstein became world famous after two British scientists experimentally proved Einstein's theory that the path of light could be warped due to gravity. This occurred in 1919, just after the end of World War I, when the world needed something to rally around and get excited about. His revolutionary scientific discovery, his "crazy professor" appearance, and his outspoken views on life, human nature, and politics carried his fame to nonscientists throughout the world.

Today, almost a half-century after his death, Albert Einstein's name is still synonymous with "genius." Einstein's unique quirkiness, funny facial expressions, and thoughtful quotes can now be seen on coffee mugs, T-shirts, posters, books, and magazines. In addition, numerous fictional and documentary films have been made about his life and scientific work. *Time Magazine* recognized Albert Einstein's impact on the world by awarding him the title, "Person of the Century."

As a result of Einstein's searching thoughts on motion and light, he concluded that the laws of physics must be the same for all nonaccelerating objects. This is called Einstein's **First Postulate of Special Relativity:** *the laws of physics are the same for both moving and nonmoving frames of reference.* If an individual in a moving train drops a ball while the train is moving at a constant velocity, it will appear to that person that the ball accelerates directly down toward the floor of the train. A ball also will fall directly toward the ground if a person standing on the side of the tracks drops it. In other words, Einstein believed that the laws of physics are the same for objects at rest as they are for objects moving at a constant velocity.

Einstein also proposed the **Second Postulate of Special Relativity:** *the speed of light is constant regardless of the speed of the observer or the light source.*

$$c = \text{speed of light} = 3.00 \times 10^8 \text{ m/s} = 1.86 \times 10^5 \text{ mi/s}$$

Einstein created a "thought experiment" to show the implications of the postulate that the speed of light is constant. In his thought experiment, two telephone poles are struck simultaneously by lightning at the instant a train moving close to the speed of light is passing the midpoint of the two poles. From the perspective of a person on the side of the tracks, the lightning strikes both poles at exactly the same time [Fig. 23.3(a)]. A person on the train sees the lightning first strike pole B in front of the train and then strike pole A behind the train an instant later [Fig. 23.3(b)] because, as the train moves closer to pole B, the light from pole B takes less time than the light from pole A to reach the person on the train [Fig. 23.3(c)]. Einstein's thought experiment showed, in theory, that two people can see the same event happen at different points in time. The distance and time may vary but the speed of light will always be constant. By changing your state of motion, as was the case with the individual on the train relative to the person on the side of the tracks, the temporary relationship of events can be altered.

$E = mc^2$

Albert Einstein's development of the equation $E = mc^2$ became one of the most important scientific achievements of the 20th century. This equation defines the equivalence between energy and mass.

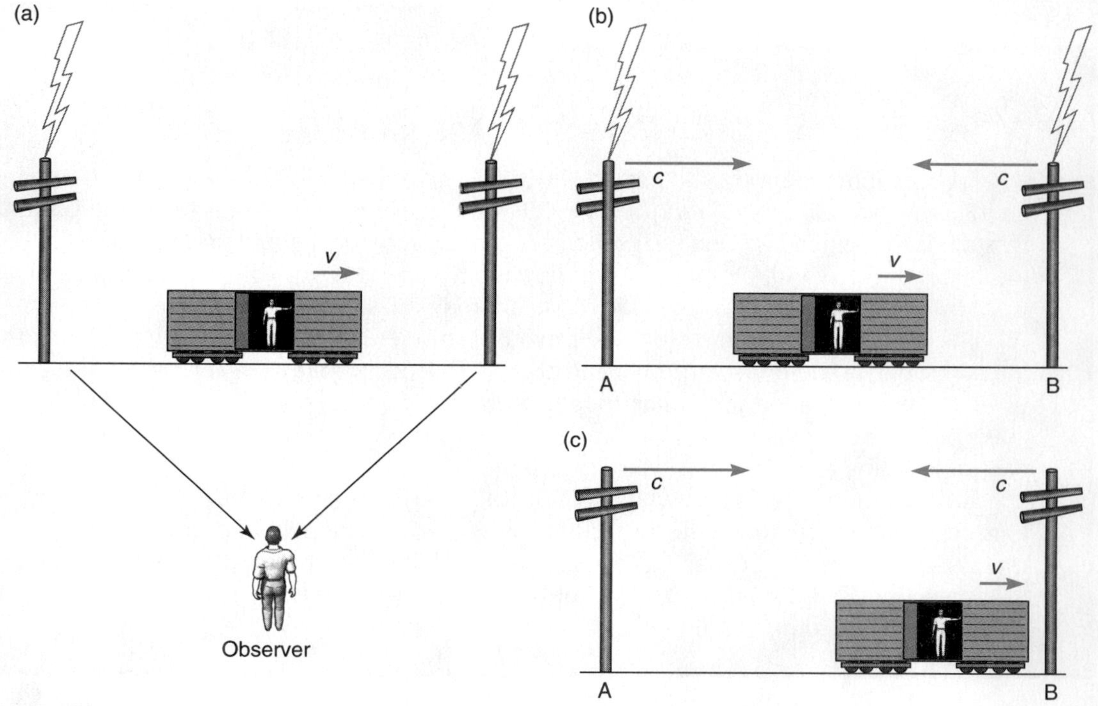

Figure 23.3 (a) Light travels the same distance from each telephone pole to the observer on the ground, who sees the two lightning strikes at the same time. (b) A person on the train sees the lightning first strike pole B in front of the train and then strike pole A behind the train an instant later. (c) As the train moves closer to pole B, the light from pole B first reaches the person on the train sooner than that from pole A because the light from pole B takes less time to reach this observer.

$$E = mc^2$$

where

E = energy
m = mass
$c = 3.00 \times 10^8$ m/s $= 1.86 \times 10^5$ mi/s (the speed of light)

A very little mass is equivalent to a tremendous amount of energy. However, converting mass to 100% energy is virtually impossible. If you take your pencil eraser and convert it into energy by burning it, some of the eraser's mass would be converted into energy in the form of heat, light, and sound. Yet the majority of the eraser would remain as burnt ash, which still has mass. Nuclear reactions, as discussed in Chapter 22, are the best methods scientists have for converting mass to pure energy.

EXAMPLE 1

How much energy is contained in 1.00 g of matter?

Data:

$$m = 1.00 \text{ g} = 1.00 \times 10^{-3} \text{ kg}$$
$$c = 3.00 \times 10^8 \text{ m/s}$$
$$E = ?$$

Basic Equation:

$$E = mc^2$$

Working Equation: Same
Substitution:

$$E = (1.00 \times 10^{-3} \text{ kg})(3.00 \times 10^8 \text{ m/s})^2$$
$$= 9.00 \times 10^{13} \text{ J}$$

This very large amount of energy, produced from only 1.00 g of mass, can light a 60-W light bulb for 47,600 yr!

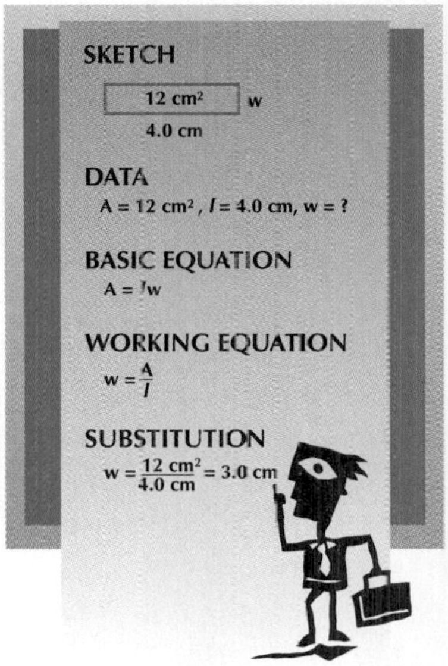

PROBLEMS 23.2

1. If the tip of a pencil has a mass of 2.30 g, how much energy can be produced if all the mass is converted to energy?
2. If a textbook has a mass of 1.30 kg, how much energy can be released if the textbook is converted to energy?
3. How much mass is needed to create $60\overline{0}$ J of energy?
4. How much mass is needed to create 67.0 J of energy?

23.3 Space-Time

Scientists and mathematicians use coordinates to locate objects in our physical world. A pilot defines her location by specifying her longitude, latitude, and altitude. Similarly, mathematicians and scientists use coordinates in relation to the x-, y-, and z-axes to define a position in three dimensions (Fig. 23.4).

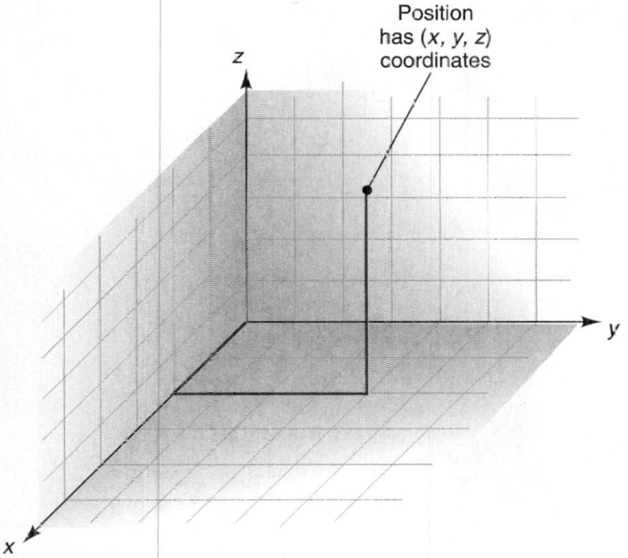

Figure 23.4 Spatial location can be determined on a three-axis coordinate system.

Albert Einstein agreed that objects exist in three spatial dimensions, but felt that three spatial dimensions alone could not accurately locate an object at any particular instant. He felt that a time dimension needed to be taken into consideration when defining the dimensions of the physical universe. His argument was that anything could be located in a particular spatial location, but it could also change its position over time. **Space-time** allows an object's position in the universe to be defined using the three spatial dimensions (x, y, z) and one time dimension.

Space-time plays an important role in Einstein's Theory of General Relativity. Einstein's theory describes the universe as a place where massive objects such as the sun, planets, and stars, warp the four dimensions of space-time. The warping of space-time influences the motion of an object traveling through the universe (Fig. 23.5). Black holes warp space to such an extent that mass and light cannot escape.

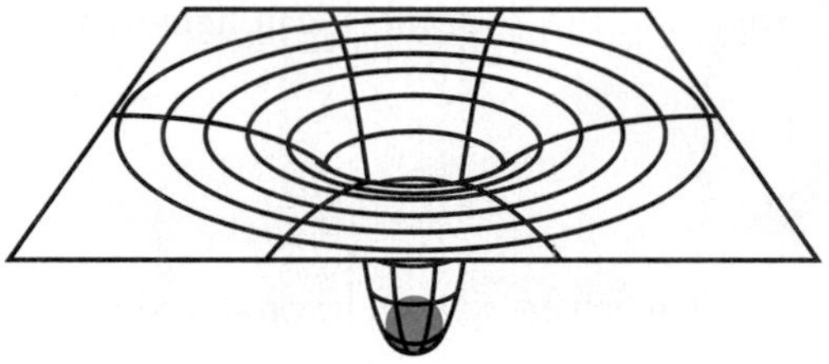

Figure 23.5 Any massive object will warp the space around it, causing objects of mass to deviate from their initial path.

23.4 The General Theory of Relativity

Although the Special Theory of Relativity focuses on the relationship between motion and time, Einstein's **General Theory of Relativity** states that gravity is equivalent to acceleration and that the path of light is warped by gravity.

Einstein designed several thought experiments to illustrate General Relativity. In one thought experiment, he pictured a person in a spaceship accelerating at 9.80 m/s^2. In such a situation, the spaceship applies a force on the astronaut equal to the force of gravity on the earth. As a result, Einstein concluded, gravity and acceleration are equivalent. Einstein continued in his thought experiment. If a beam of light comes through a window as the spaceship accelerates, the light will strike the opposite wall just a bit lower than the height at which it entered. As a result of the accelerating spaceship, the light appears to bend in the accelerating/gravitational field. Finally, since gravity only alters the path of objects with mass, Einstein concluded that light must have mass (Fig. 23.6).

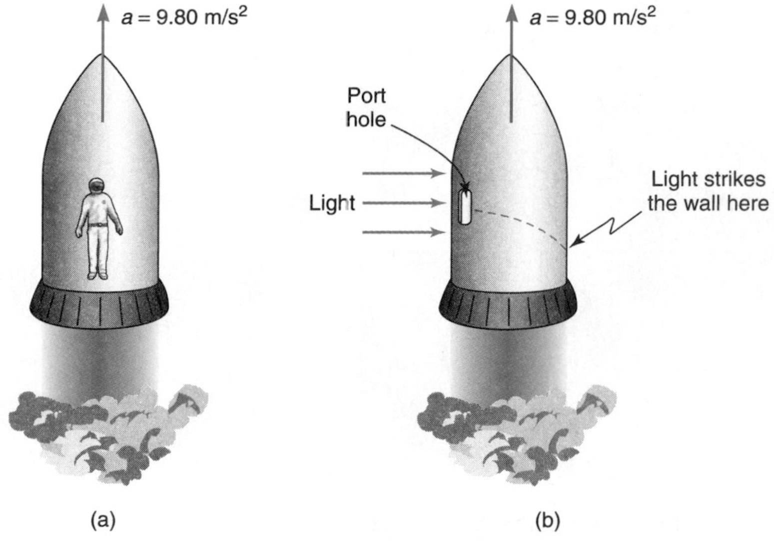

Figure 23.6 (a) If the space ship accelerates at 9.80 m/s^2, the force on the person inside will feel the same as the pull of gravity on earth, showing that acceleration is the same as gravity. (b) Although this depiction is exaggerated, the light would strike the wall lower than where it entered, showing that gravity bends light.

In 1919, British scientists traveled to South America to observe a solar eclipse and experimentally prove Einstein's Theory of Relativity. On the night before the solar eclipse, the scientists observed the exact positions of several stars. The next day, when the moon eclipsed the sun, the scientists observed the same stars in the darkened sky. However, the stars appeared to have slightly changed their positions in the sky. The light from the stars curved around the warp in space-time created by the massive sun and moon. The apparent shift in the stars' positions experimentally proved Einstein's Theory of General Relativity and made him a world-renowned celebrity (Fig. 23.7).

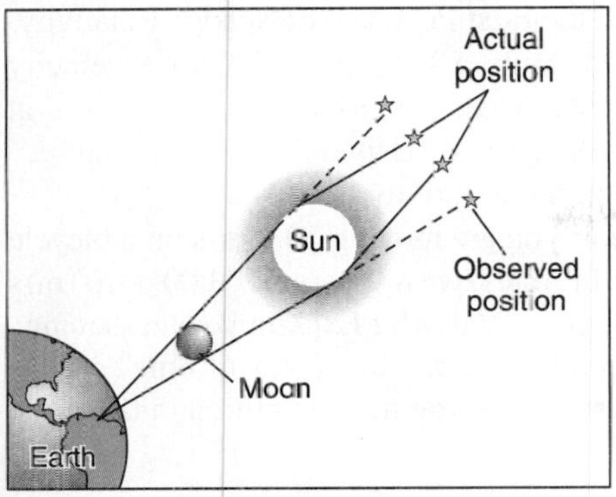

Figure 23.7 Massive objects warp space, causing light from distant stars to bend, altering their observed position in the sky.

Glossary

$E = mc^2$ The equation that illustrates and defines the equivalence between energy and mass. (p. 633)

First Postulate of Special Relativity The laws of physics are the same for both moving and nonmoving frames of reference. (p. 633)

General Theory of Relativity Extends the special theory of relativity to accelerated frames of reference with the assumption that gravity and acceleration are equivalent and that light has mass and its path can be warped by gravity. (p. 632)

Grand Unified Theory A theory linking the five fundamental forces: gravitational, electric, magnetic, strong, and weak forces. (p. 632)

Relative Motion The concept that motion can be described differently depending upon the observer's perspective. (p. 632)

Second Postulate of Special Relativity The speed of light is constant regardless of the speed of the observer or of the light source. (p. 633)

Space-Time An object's position in the universe can be pinpointed using three spatial dimensions and one time dimension. (p. 635)

Special Theory of Relativity The laws of physics are the same in moving and nonmoving frames of reference and the speed of light is constant no matter what the speed of the observer or the light source. (p. 632)

Formula

23.2 $E = mc^2$

Review Questions

1. What field or fields of physics intrigued Einstein at a very young age?
   (a) Momentum
   (b) Thermodynamics
   (c) Electricity and magnetism
   (d) Sound and waves
2. Which of the following did Albert Einstein not complete?
   (a) General relativity
   (b) Special relativity
   (c) Warping of space-time
   (d) Grand unified theory
3. Describe Einstein's First Postulate of Special Relativity.

4. Explain a situation where you experienced the First Postulate of Special Relativity.
5. If you are riding a bike at 10.0 m/s and throw a ball forward with a velocity of 4.00 m/s, will the ball travel 4.00 m/s relative to the ground, or will it travel at 10.0 m/s plus another 4.00 m/s relative to the ground? Explain your reasoning.
6. Describe Einstein's Second Postulate of Special Relativity.
7. According to Einstein's Second Postulate, if you are moving 10.0 m/s on a bicycle with a headlight on, will the light from the bicycle have a velocity of 3.00×10^8 m/s or will it have a velocity of 3.00×10^8 m/s plus 10.0 m/s? Explain your reasoning.
8. What does traveling close to the speed of light do to the dimension of time?
9. While you are sitting and reading this question, are you traveling through the space dimensions or the time dimension?
10. Explain what $E = mc^2$ represents.
11. What are the four dimensions of our space-time universe? Why is time an important dimension?
12. What does Einstein's General Theory of Relativity state about the relationship between gravity and acceleration?
13. What happens to light and other electromagnetic waves around massive objects such as the sun?
14. Explain how the solar eclipse of 1919 proved Einstein's theory that light has mass.

Review Problems

1. A train is moving at a speed of 65.0 mi/h. The ticket collector is walking 2.00 mi/h toward the front of the train. How fast is the ticket collector moving from the point of view of a person on the train?
2. How fast is the ticket collector in Problem 1 moving relative to an observer on the side of the tracks?
3. The ticket collector in Problem 1 turns around and walks toward the rear of the train at 2.00 mi/h. How fast is he walking relative to an observer on the side of the tracks?
4. Convert the mass of one electron ($m = 9.10 \times 10^{-31}$ kg) to energy.
5. Convert the mass of one proton ($m = 1.67 \times 10^{-27}$ kg) to energy.
6. A particular task requires 9.80 J of energy. Using $E = mc^2$, determine how much mass is needed to accomplish this task.

APPENDIX A

MATHEMATICS REVIEW

A.1 Signed Numbers

Signed numbers have many applications in the study of physics. The rules for working with signed numbers follow.

Adding Signed Numbers

To add two positive numbers, add their absolute values.* A positive sign may or may not be placed before the result. It is usually omitted.

EXAMPLE 1

Add:

(a) $\begin{array}{r} +4 \\ +7 \\ \hline +11 \end{array}$ or 11 (b) $(+3) + (+5) = +8$ or 8

To add two negative numbers, add their absolute values and place a negative sign before the result.

EXAMPLE 2

Add:

(a) $\begin{array}{r} -2 \\ -5 \\ \hline -7 \end{array}$ (b) $(-6) + (-7) = -13$ (c) $(-8) + (-4) = -12$

To add a negative number and a positive number, find the difference of their absolute values. Place the sign of the number having the larger absolute value before the result.

EXAMPLE 3

Add:

(a) $\begin{array}{r} +4 \\ -6 \\ \hline -2 \end{array}$ (b) $\begin{array}{r} -2 \\ +8 \\ \hline +6 \end{array}$ (c) $\begin{array}{r} -8 \\ +3 \\ \hline -5 \end{array}$ (d) $\begin{array}{r} +9 \\ -4 \\ \hline +5 \end{array}$

(e) $(+7) + (-2) = +5$ (f) $(-9) + (-6) = -3$

(g) $(-3) + (+10) = +7$ (h) $(+4) + (-12) = -8$

*The absolute value of a number is its nonnegative value. For example, the absolute value of −6 is 6; the absolute value of +10 is 10; and the absolute value of 0 is 0.

To add three or more signed numbers:

1. Add the positive numbers.
2. Add the negative numbers.
3. Add the sums from steps 1 and 2 according to the rules for addition of signed numbers.

EXAMPLE 4

Add: (−2) + 4 + (−6) + 10 + (−7).

$$\text{Step 1: } \begin{array}{r} +4 \\ \underline{+10} \\ +14 \end{array} \qquad \text{Step 2: } \begin{array}{r} -2 \\ -6 \\ \underline{-7} \\ -15 \end{array} \qquad \text{Step 3: } \begin{array}{r} -15 \\ \underline{+14} \\ -1 \end{array}$$

Therefore, (−2) + 4 + (−6) + 10 + (−7) = −1.

Subtracting Signed Numbers

To subtract two signed numbers, change the sign of the *number being subtracted* and *add* according to the rules for addition.

EXAMPLE 5

Subtract:

(a) Subtract: $\begin{array}{r} +3 \\ \underline{+7} \\ -4 \end{array}$ ↔ Add: $\begin{array}{r} +3 \\ \underline{-7} \\ -4 \end{array}$ To subtract, change the sign of the number being subtracted, +7, and add.

(b) Subtract: $\begin{array}{r} -9 \\ \underline{-6} \\ -3 \end{array}$ ↔ Add: $\begin{array}{r} -9 \\ \underline{+6} \\ -3 \end{array}$ To subtract, change the sign of the number being subtracted, −6, and add.

(c) Subtract: $\begin{array}{r} +8 \\ \underline{-4} \\ +12 \end{array}$ ↔ Add: $\begin{array}{r} +8 \\ \underline{+4} \\ +12 \end{array}$

(d) Subtract: $\begin{array}{r} -6 \\ \underline{+8} \\ -14 \end{array}$ ↔ Add: $\begin{array}{r} -6 \\ \underline{-8} \\ -14 \end{array}$

(e) (+6) − (+8) = (+6) + (−8) = −2 To subtract, change the sign of the number being subtracted, +8, and add.

(f) (−3) − (−5) = (−3) + (+5) = +2
(g) (+10) − (−3) = (+10) + (+3) = +13
(h) (−5) − (+2) = (−5) + (−2) = −7

When more than two signed numbers are involved in subtraction, change the sign of *each* number being subtracted and add the resulting signed numbers.

EXAMPLE 6

Subtract: (−2) − (+4) − (−1) − (−3) − (+5)
= (−2) + (−4) + (+1) + (+3) + (−5).

Step 1:	+1	Step 2:	−2	Step 3:	+4
	+3		−4		−11
	+4		−5		−7
			−11		

Therefore, (−2) − (+4) − (−1) − (−3) − (+5) = −7.

When combinations of addition and subtraction of signed numbers occur in the same problem, change *only* the sign of each number being subtracted. Then add the resulting signed numbers.

EXAMPLE 7

Find the result:

(−2) − (−4) − (−3) − (+6) + (+1) − (+2) + (−7) − (−5)
= (−2) − (−4) + (+3) + (−6) + (+1) + (−2) + (−7) + (+5)

Step 1:	+3	Step 2:	−2	Step 3:	+9
	+1		−4		−21
	+5		−6		−12
	+9		−2		
			−7		
			−21		

Therefore, (−2) + (−4) − (−3) − (+6) + (+1) − (+2) + (−7) − (−5) = −12

Multiplying Signed Numbers

To multiply two signed numbers:

1. If the signs of the numbers are both positive or both negative, find the product of their absolute values. This product is always positive.
2. If the signs of the numbers are unlike, find the product of their absolute values and place a negative sign before the result.

EXAMPLE 8

Multiply:

(a)	+3	(b)	−5	(c)	−6	(d)	+2
	+4		−8		+7		−3
	+12		+40		−42		−6

(e) (+3)(+5) = +15 (f) (−7)(−8) = +56
(g) (−1)(+6) = −6 (h) (+4)(−2) = −8

To multiply more than two signed numbers, first multiply the absolute values of the numbers. If there is an odd number of negative factors, place a negative sign before the result. If there is an even number of negative factors, the product is positive. ***Note***: An *even* number is a number divisible by 2.

EXAMPLE 9

Multiply:

(a) (+5)(−6)(+2)(−1) = +60
(b) (−3)(−3)(+4)(−5) = −180

Dividing Signed Numbers

The rules for dividing signed numbers are similar to those for multiplying signed numbers.

To divide two signed numbers:

1. If the signs of the numbers are both positive or both negative, divide their absolute values. This quotient is always positive.
2. If the two numbers have different signs, divide their absolute values and place a negative sign before the quotient.

Note: Division by 0 is undefined.

EXAMPLE 10

Divide:

(a) $\frac{+10}{+2} = +5$ (b) $\frac{-18}{-3} = +6$ (c) $\frac{+20}{-4} = -5$ (d) $\frac{-24}{+2} = -12$

PROBLEMS A.1

Perform the indicated operations.

1. $(-5) + (-6)$
2. $(+1) + (-10)$
3. $(-3) + (+8)$
4. $(+5) + (+7)$
5. $(-5) + (+3)$
6. $0 + (-3)$
7. $(-7) - (-3)$
8. $(+2) - (-9)$
9. $(-4) - (+2)$
10. $(+4) - (+7)$
11. $0 - (+3)$
12. $0 - (-2)$
13. $(-9)(-2)$
14. $(+4)(+6)$
15. $(-7)(+3)$
16. $(+5)(-8)$
17. $(+6)(0)$
18. $(0)(-4)$
19. $\frac{+36}{+12}$
20. $\frac{-9}{-3}$
21. $\frac{+16}{-2}$
22. $\frac{-15}{+3}$
23. $\frac{0}{+6}$
24. $\frac{4}{0}$
25. $(+2) + (-1) + (+10)$
26. $(-7) + (+2) + (+9) + (-8)$
27. $(-9) + (-3) + (+3) + (-8) + (+4)$
28. $(+8) + (-2) + (-6) + (+7) + (-6) + (+9)$
29. $(-4) - (+5) - (-4)$
30. $(+3) - (-5) - (-6) - (+5)$
31. $(-7) - (-4) - (+6) - (+4) - (-5)$
32. $(-8) - (+7) - (+3) - (-7) - (-8) - (-2)$
33. $(+5) + (-2) - (+7)$
34. $(-3) - (-8) - (+3) + (-9)$
35. $(-2) - (+1) - (-10) + (+12) + (-9)$
36. $(-1) - (-11) + (+2) - (-10) + (+8)$
37. $(+3)(-5)(+3)$
38. $(-1)(+2)(+2)(-1)$
39. $(+2)(-4)(-6)(-3)(+2)$
40. $(-1)(+3)(-2)(-4)(+5)(-1)$

A.2 Powers of 10

The ability to work quickly and accurately with powers of 10 is important in scientific and technical fields.

When multiplying two powers of 10, add the exponents. That is,

$$10^a \times 10^b = 10^{a+b}$$

EXAMPLE 1

Multiply:

(a) $(10^6)(10^3) = 10^{6+3} = 10^9$
(b) $(10^4)(10^2) = 10^{4+2} = 10^6$
(c) $(10^1)(10^{-3}) = 10^{1+(-3)} = 10^{-2}$
(d) $(10^{-2})(10^{-5}) = 10^{[-2+(-5)]} = 10^{-7}$

When dividing two powers of 10, subtract the exponents as follows:

$$10^a \div 10^b = 10^{a-b}$$

EXAMPLE 2

Divide:

(a) $\dfrac{10^7}{10^4} = 10^{7-4} = 10^3$
(b) $\dfrac{10^3}{10^5} = 10^{3-5} = 10^{-2}$
(c) $\dfrac{10^{-2}}{10^{+3}} = 10^{(-2)-(+3)} = 10^{-5}$
(d) $\dfrac{10^4}{10^{-2}} = 10^{4-(-2)} = 10^6$

To raise a power of 10 to a power, multiply the exponents as follows:

$$(10^a)^b = 10^{ab}$$

EXAMPLE 3

Find each power:

(a) $(10^2)^3 = 10^{(2)(3)} = 10^6$
(b) $(10^{-3})^2 = 10^{(-3)(2)} = 10^{-6}$
(c) $(10^4)^{-5} = 10^{(4)(-5)} = 10^{-20}$
(d) $(10^{-3})^{-4} = 10^{(-3)(-4)} = 10^{12}$

Next, we will show that $10^0 = 1$. To do this, we need to use the substitution principle, which states that

$$\text{if} \quad a = b \quad \text{and} \quad a = c \quad \text{then} \quad b = c$$

First,

$$\frac{10^n}{10^n} = 10^{n-n} \qquad \text{To divide powers, subtract the exponents.}$$
$$= 10^0$$

Second,

$$\frac{10^n}{10^n} = 1 \qquad \text{Any number other than zero divided by itself equals 1.}$$

That is, since

$$\frac{10^n}{10^n} = 10^0 \quad \text{and} \quad \frac{10^n}{10^n} = 1$$

then $10^0 = 1$.

We also will use the fact that $\dfrac{1}{10^a} = 10^{-a}$. To show this, we write

$$\frac{1}{10^a} = \frac{10^0}{10^a} \qquad (1 = 10^0)$$
$$= 10^{0-a} \qquad \text{To divide powers, subtract the exponents.}$$
$$= 10^{-a}$$

We also need to show that $\frac{1}{10^{-a}} = 10^a$. We write

$$\frac{1}{10^{-a}} = \frac{10^0}{10^{-a}}$$
$$= 10^{0-(-a)}$$
$$= 10^a$$

In summary,

$$10^0 = 1 \qquad \frac{1}{10^a} = 10^{-a} \qquad \frac{1}{10^{-a}} = 10^a$$

Actually, any number (except zero) raised to the zero power equals 1.

PROBLEMS A.2

Do as indicated. Express the results using positive exponents.

1. $(10^5)(10^3)$
2. $10^6 \div 10^2$
3. $(10^2)^4$
4. $(10^{-2})(10^{-3})$
5. $\frac{10^3}{10^6}$
6. $(10^{-3})^3$
7. $10^5 \div 10^{-2}$
8. $(10^{-2})^{-3}$
9. $(10^4)(10^{-1})$
10. $\frac{10^0}{10^{-4}}$
11. $(10^0)(10^{-4})$
12. $\frac{10^{-4}}{10^{-3}}$
13. $(10^0)^{-2}$
14. 10^{-3}
15. $\frac{1}{10^{-5}}$
16. $\frac{(10^4)(10^{-2})}{(10^6)(10^3)}$
17. $\frac{(10^{-2})(10^{-3})}{(10^3)^2}$
18. $\frac{(10^2)^4}{(10^{-3})^2}$
19. $\left(\frac{1}{10^3}\right)^2$
20. $\left(\frac{10^2}{10^{-3}}\right)^2$
21. $\left(\frac{10 \cdot 10^2}{10^{-1}}\right)^2$
22. $\left(\frac{1}{10^{-3}}\right)^2$
23. $\frac{(10^4)(10^{-2})}{10^{-8}}$
24. $\frac{(10^4)(10^6)}{(10^0)(10^{-2})(10^3)}$

A.3 Scientific Notation

Scientists and technicians often need to use very large or very small numbers. For example, the thickness of an oil film on water is about 0.0000001 m. **Scientific notation** is a useful method of expressing such very small (or very large) numbers. Expressed this way, the thickness of the film is 1×10^{-7} m or 10^{-7} m. For example:

$$0.1 = 1 \times 10^{-1} \quad \text{or} \quad 10^{-1}$$
$$10{,}000 = 1 \times 10^4 \quad \text{or} \quad 10^4$$
$$0.001 = 1 \times 10^{-3} \quad \text{or} \quad 10^{-3}$$

A number in scientific notation is written as a product of a number between 1 and 10 and a power of 10. General form: $M \times 10^n$, where

M = a number between 1 and 10
n = the exponent or power of 10

EXAMPLE 1

The following numbers are written in scientific notation:

(a) $325 = 3.25 \times 10^2$
(b) $100{,}000 = 1 \times 10^5$ or 10^5

To write any decimal number in scientific notation:

1. Place a decimal point after the first nonzero digit reading from left to right.
2. Place a caret (^) at the position of the *original* decimal point.
3. If the newly added decimal point is to the *left* of the caret, the exponent of 10 is the number of places from the caret to the decimal point.
   Example: $83{,}662 = 8.3662_{\wedge} \times 10^{4}$ (4 places)
4. If the newly added decimal point is to the *right* of the caret, the exponent of 10 is the negative of the number of places from the caret to the decimal point.
   Example: $0.00683 = {}_{\wedge}006.83 \times 10^{-3}$ (3 places)
5. If the decimal point and the caret coincide, the exponent of 10 is zero.
   Example: $5.12 = 5.12 \times 10^0$

A number greater than 10 is expressed in scientific notation as a product of a decimal between 1 and 10 and a *positive* power of 10.

EXAMPLE 2

Write each number greater than 10 in scientific notation.

(a) $2580 = 2.58 \times 10^3$
(b) $54{,}600 = 5.46 \times 10^4$
(c) $42{,}000{,}000 = 4.2 \times 10^7$
(d) $715.8 = 7.158 \times 10^2$
(e) $34.775 = 3.4775 \times 10^1$

A number between 0 and 1 is expressed in scientific notation as a product of a decimal between 1 and 10 and a *negative* power of 10.

EXAMPLE 3

Write each positive number less than 1 in scientific notation.

(a) $0.0815 = 8.15 \times 10^{-2}$
(b) $0.00065 = 6.5 \times 10^{-4}$
(c) $0.73 = 7.3 \times 10^{-1}$
(d) $0.0000008 = 8 \times 10^{-7}$

A number between 1 and 10 is expressed in scientific notation as a product of a decimal between 1 and 10 and the *zero* power of 10.

EXAMPLE 4

Write each number between 1 and 10 in scientific notation.

(a) $7.33 = 7.33 \times 10^0$
(b) $1.06 = 1.06 \times 10^0$

To change a number from scientific notation to decimal form:

1. Multiply the decimal part by the power of 10 by moving the decimal point *to the right* the same number of decimal places as indicated by the power of 10 if it is *positive*.
2. Multiply the decimal part by the power of 10 by moving the decimal point *to the left* the same number of decimal places as indicated by the power of 10 if it is *negative*.
3. Supply zeros as needed.

EXAMPLE 5

Write 7.62×10^2 in decimal form.

$7.62 \times 10^2 = 762$ Move the decimal point two places to the right.

EXAMPLE 6

Write 6.15×10^{-4} in decimal form.

$6.15 \times 10^{-4} = 0.000615$ Move the decimal point four places to the left and insert three zeros.

EXAMPLE 7

Write each number in decimal form.

(a) $3.75 \times 10^2 = 375$
(b) $1.09 \times 10^5 = 109{,}000$
(c) $2.88 \times 10^{-2} = 0.0288$
(d) $9.4 \times 10^{-6} = 0.0000094$
(e) $6.7 \times 10^0 = 6.7$

Since calculators used in science and technology accept numbers entered in scientific notation and give some results in scientific notation, it is essential that you fully understand this topic before going to the next section. See Appendix B, Section B.1, for using a calculator with numbers in scientific notation.

PROBLEMS A.3

Write each number in scientific notation.

1. 326
2. 798
3. 2650
4. 14,500
5. 826.4
6. 24.97
7. 0.00413
8. 0.00053
9. 6.43
10. 482,300
11. 0.000065
12. 0.00224
13. 540,000
14. 1,400,000
15. 0.0000075
16. 0.0000009
17. 0.00000005
18. 3,500,000,000
19. 732,000,000,000,000,000
20. 0.00000000000000618

Write each number in decimal form.

21. 8.62×10^4
22. 8.67×10^2
23. 6.31×10^{-4}
24. 5.41×10^3
25. 7.68×10^{-1}
26. 9.94×10^1
27. 7.77×10^8
28. 4.19×10^{-6}
29. 6.93×10^1
30. 3.78×10^{-2}
31. 9.61×10^4
32. 7.33×10^3
33. 1.4×10^0
34. 9.6×10^{-5}
35. 8.4×10^{-6}

36. 9×10^8 37. 7×10^{11} 38. 4.05×10^0

39. 7.2×10^{-7} 40. 8×10^{-9} 41. 4.5×10^{12}

42. 1.5×10^{11} 43. 5.5×10^{-11} 44. 8.72×10^{-10}

A.4 Solving Linear Equations

An equation is a mathematical sentence stating that two quantities are equal. To solve an equation means to find the number or numbers that can replace the variable in the equation to make the equation a true statement. The value we find that makes the equation a true statement is called the *root* of the equation. When the root of an equation is found, we say we have *solved* the equation.

> If $a = b$, then $a + c = b + c$ or $a - c = b - c$. (If the same quantity is added to or subtracted from both sides of an equation, the resulting equation is equivalent to the original equation.)

To solve an equation using this rule, think first of undoing what has been done to the variable.

EXAMPLE 1

Solve: $x - 5 = -9$.

$$x - 5 = -9$$

$$x - 5 + 5 = -9 + 5 \quad \text{Undo the subtraction by adding 5 to both sides.}$$

$$x = -4$$

EXAMPLE 2

Solve: $x + 4 = 29$.

$$x + 4 = 29$$

$$x + 4 - 4 = 29 - 4 \quad \text{Undo the addition by subtracting 4 from both sides.}$$

$$x = 25$$

> If $a = b$, then $ac = bc$ or $a/c = b/c$ with $c \neq 0$. (If both sides of an equation are multiplied or divided by the same nonzero quantity, the resulting equation is equivalent to the original equation.)

EXAMPLE 3

Solve: $3x = 18$.

$$3x = 18$$

$$\frac{3x}{3} = \frac{18}{3} \quad \text{Undo the multiplication by dividing both sides by 3.}$$

$$x = 6$$

EXAMPLE 4

Solve: $x/4 = 9$.

$$\frac{x}{4} = 9$$

$$4\left(\frac{x}{4}\right) = 4 \cdot 9$$ Undo the division by multiplying both sides by 4.

$$x = 36$$

When more than one operation is indicated on the variable in an equation, undo the additions and subtractions first, then undo the multiplications and divisions.

EXAMPLE 5

Solve: $3x + 5 = 17$.

$$3x + 5 = 17$$

$$3x + 5 - 5 = 17 - 5$$ Subtract 5 from both sides.

$$3x = 12$$

$$\frac{3x}{3} = \frac{12}{3}$$ Divide both sides by 3.

$$x = 4$$

EXAMPLE 6

Solve: $2x - 7 = 10$.

$$2x - 7 = 10$$

$$2x - 7 + 7 = 10 + 7$$ Add 7 to both sides.

$$2x = 17$$

$$\frac{2x}{2} = \frac{17}{2}$$ Divide both sides by 2.

$$x = \frac{17}{2} = 8.5$$

EXAMPLE 7

Solve: $(x/5) - 10 = 22$.

$$\frac{x}{5} - 10 = 22$$

$$\frac{x}{5} - 10 + 10 = 22 + 10$$ Add 10 to both sides.

$$\frac{x}{5} = 32$$

$$5\left(\frac{x}{5}\right) = 5(32)$$ Multiply both sides by 5.

$$x = 160$$

To solve an equation with variables on both sides:

1. Add or subtract either variable term from both sides of the equation.
2. Add or subtract from both sides of the equation the constant term that now appears on the same side of the equation with the variable. Then solve.

EXAMPLE 8

Solve: $3x + 6 = 7x - 2$.

$$3x + 6 = 7x - 2$$

$$3x + 6 - 3x = 7x - 2 - 3x$$ Subtract $3x$ from both sides.

$$6 = 4x - 2$$

$$6 + 2 = 4x - 2 + 2 \qquad \text{Add 2 to both sides.}$$

$$8 = 4x$$

$$\frac{8}{4} = \frac{4x}{4} \qquad \text{Divide both sides by 4.}$$

$$2 = x$$

EXAMPLE 9

Solve: $4x - 2 = -5x + 10$.

$$4x - 2 = -5x + 10$$

$$4x - 2 + 5x = -5x + 10 + 5x \qquad \text{Add } 5x \text{ to both sides.}$$

$$9x - 2 = 10$$

$$9x - 2 + 2 = 10 + 2 \qquad \text{Add 2 to both sides.}$$

$$9x = 12$$

$$\frac{9x}{9} = \frac{12}{9} \qquad \text{Divide both sides by 9.}$$

$$x = \frac{4}{3}$$

To solve equations containing parentheses, first remove the parentheses and then proceed as before. The rules for removing parentheses follow:

1. If the parentheses are preceded by a plus (+) sign, they may be removed without changing any signs.

   Examples:

   $$2 + (3 - 5) = 2 + 3 - 5$$

   $$3 + (x + 4) = 3 + x + 4$$

   $$5x + (-6x + 9) = 5x - 6x + 9$$

2. If the parentheses are preceded by a minus (−) sign, the parentheses may be removed if *all* the signs of the numbers (or letters) within the parentheses are changed.

   Examples:

   $$2 - (3 - 5) = 2 - 3 + 5$$

   $$5 - (x - 7) = 5 - x + 7$$

   $$7x - (-4x - 11) = 7x + 4x + 11$$

3. If the parentheses are preceded by a number, the parentheses may be removed if each of the terms inside the parentheses is multiplied by that (signed) number.

   Examples:

   $$2(x + 4) = 2x + 8$$

   $$-3(x - 5) = -3x + 15$$

   $$2 - 4(3x - 5) = 2 - 12x + 20$$

EXAMPLE 10

Solve: $3(x - 4) = 15$.

$$3(x - 4) = 15$$

$$3x - 12 = 15 \qquad \text{Remove parentheses.}$$

$$3x - 12 + 12 = 15 - 12 \qquad \text{Add 12 to both sides.}$$

$$3x = 27$$

$$\frac{3x}{3} = \frac{27}{3} \qquad \text{Divide both sides by 3.}$$

$$x = 9$$

EXAMPLE 11

Solve: $2x - (3x + 15) = 4x - 1$.

$$2x - (3x + 15) = 4x - 1$$
$$2x - 3x - 15 = 4x - 1 \qquad \text{Remove parentheses.}$$
$$-x - 15 = 4x - 1 \qquad \text{Combine like terms.}$$
$$-x - 15 + x = 4x - 1 + x \qquad \text{Add } x \text{ to both sides.}$$
$$-15 = 5x - 1$$
$$-15 + 1 = 5x - 1 + 1 \qquad \text{Add 1 to both sides.}$$
$$-14 = 5x$$
$$\frac{-14}{5} = \frac{5x}{5} \qquad \text{Divide both sides by 5.}$$
$$-2.8 = x$$

PROBLEMS A.4

Solve each equation.

1. $3x = 4$
2. $\frac{y}{2} = 10$
3. $x - 5 = 12$
4. $x + 1 = 9$
5. $2x + 10 = 10$
6. $4x = 28$
7. $2x - 2 = 33$
8. $4 = \frac{x}{10}$
9. $172 - 43x = 43$
10. $9x + 7 = 4$
11. $6y - 24 = 0$
12. $3y + 15 = 75$
13. $15 = \frac{105}{y}$
14. $6x = x - 15$
15. $2 = \frac{50}{2y}$
16. $9y = 67.5$
17. $8x - 4 = 36$
18. $10 = \frac{136}{4x}$
19. $2x + 22 = 75$
20. $9x + 10 = x - 26$
21. $4x + 9 = 7x - 18$
22. $2x - 4 = 3x + 7$
23. $-2x + 5 = 3x - 10$
24. $5x + 3 = 2x - 18$
25. $3x + 5 = 5x - 11$
26. $-5x + 12 = 12x - 5$
27. $13x + 2 = 20x - 5$
28. $5x + 3 = -9x - 39$
29. $-4x + 2 = -10x - 20$
30. $9x + 3 = 6x + 8$
31. $3x + (2x - 7) = 8$
32. $11 - (x + 12) = 100$
33. $7x - (13 - 2x) = 5$
34. $20(7x - 2) = 240$
35. $-3x + 5(x - 6) = 12$
36. $3(x + 117) = 201$
37. $5(2x - 1) = 8(x + 3)$
38. $3(x + 4) = 8 - 3(x - 2)$
39. $-2(3x - 2) = 3x - 2(5x + 1)$
40. $\frac{x}{5} - 2\left(\frac{2x}{5} + 1\right) = 28$

A.5 Solving Quadratic Equations

A quadratic equation in one variable is one in which the largest exponent of the variable is 2. The most general quadratic equation in variable x is written as

$$ax^2 + bx + c = 0 \qquad (\text{where } a \neq 0)$$

EXAMPLE 1

Solve: $x^2 = 16$.

To solve a quadratic equation of this type, take the square root of both sides of the equation.

$$x^2 = 16$$
$$x = \pm 4 \qquad \text{Take the square root of both sides.}$$

In general, solve equations of the form $ax^2 = b$, where $a \neq 0$, as follows:

$$ax^2 = b$$

$$x^2 = \frac{b}{a} \qquad \text{Divide both sides by } a.$$

$$x = \pm\sqrt{\frac{b}{a}} \qquad \text{Take the square root of both sides.}$$

EXAMPLE 2

Solve: $2x^2 - 18 = 0$.

$$2x^2 - 18 = 0$$

$$2x^2 = 18 \qquad \text{Add 18 to both sides.}$$

$$x^2 = 9 \qquad \text{Divide both sides by 2.}$$

$$x = \pm 3 \qquad \text{Take the square root of both sides.}$$

EXAMPLE 3

Solve: $5y^2 = 100$.

$$5y^2 = 100$$

$$y^2 = 20 \qquad \text{Divide both sides by 5.}$$

$$y = \pm\sqrt{20} \qquad \text{Take the square root of both sides.}$$

$$y = \pm 4.47$$

The solutions of the general quadratic equation

$$ax^2 + bx + c = 0 \qquad (\text{where } a \neq 0)$$

are given by the formula (called the *quadratic formula*)

$$x = \frac{-b \pm \sqrt{b^2 - 4ac}}{2a}$$

where

a = coefficient of the x^2 term
b = coefficient of the x term
c = constant term

The symbol ($\pm$) is used to combine two expressions or equations into one. For example, $a \pm 2$ means $a + 2$ or $a - 2$. Similarly,

$$x = \frac{-b \pm \sqrt{b^2 - 4ac}}{2a}$$

means

$$x = \frac{-b + \sqrt{b^2 - 4ac}}{2a} \quad \text{or} \quad x = \frac{-b - \sqrt{b^2 - 4ac}}{2a}$$

EXAMPLE 4

In the equation $4x^2 - 3x - 7 = 0$, identify a, b, and c.

$$a = 4, \quad b = -3, \quad \text{and} \quad c = -7$$

EXAMPLE 5

Solve $x^2 + 2x - 8 = 0$ using the quadratic formula.

First, $a = 1$, $b = 2$, and $c = -8$. Then

$$x = \frac{-b \pm \sqrt{b^2 - 4ac}}{2a}$$

$$x = \frac{-2 \pm \sqrt{(2)^2 - 4(1)(-8)}}{2(1)}$$

$$= \frac{-2 \pm \sqrt{4 - (-32)}}{2}$$

$$= \frac{-2 \pm \sqrt{36}}{2}$$

$$= \frac{-2 \pm 6}{2}$$

$$= \frac{-2 + 6}{2} \text{ or } \frac{-2 - 6}{2}$$

$$= \frac{4}{2} \text{ or } \frac{-8}{2}$$

$$= 2 \text{ or } -4$$ The solutions are 2 and −4.

If the number under the radical sign is not a perfect square, find the square root of the number by using a calculator and proceed as before.

EXAMPLE 6

Solve $4x^2 - 7x = 32$ using the quadratic formula.

Before identifying a, b, and c, the equation must be set equal to zero. That is,

$$4x^2 - 7x - 32 = 0$$

First, $a = 4$, $b = -7$, and $c = -32$. Then

$$x = \frac{-b \pm \sqrt{b^2 - 4ac}}{2a}$$

$$= \frac{-(-7) \pm \sqrt{(-7)^2 - 4(4)(-32)}}{2(4)}$$

$$= \frac{7 \pm \sqrt{49 - (-512)}}{8}$$

$$= \frac{7 \pm \sqrt{561}}{8}$$

$$= \frac{7 \pm 23.7}{8} \qquad (\sqrt{561} = 23.7)$$

$$= \frac{7 + 23.7}{8} \text{ or } \frac{7 - 23.7}{8}$$

$$= 3.84 \text{ or } -2.09$$

The approximate solutions are 3.84 and −2.09.

PROBLEMS A.5

Solve each equation.

1. $x^2 = 36$
2. $y^2 = 100$
3. $2x^2 = 98$
4. $5x^2 = 0.05$
5. $3x^2 - 27 = 0$
6. $2y^2 - 15 = 17$
7. $10x^2 + 4.9 = 11.3$
8. $2(32)(48 - 15) = v^2 - 27^2$
9. $2(107) = 9.8t^2$
10. $65 = \pi r^2$
11. $2.50 = \pi r^2$
12. $24^2 = a^2 + 16^2$

Find the values of a, b, and c, in each quadratic equation.

13. $3x^2 + x - 5 = 0$
14. $-2x^2 + 7x + 4 = 0$
15. $6x^2 + 8x + 2 = 0$
16. $5x^2 - 2x - 15 = 0$
17. $9x^2 + 6x = 4$
18. $6x^2 = x + 9$
19. $5x^2 + 6x = 0$
20. $7x^2 - 45 = 0$
21. $9x^2 = 64$
22. $16x^2 = 49$

Solve each quadratic equation using the quadratic formula.

23. $x^2 - 10x + 21 = 0$
24. $2x^2 + 13x + 15 = 0$
25. $6x^2 - 7x = 20$
26. $15x^2 = 4x + 4$
27. $6x^2 - 2x = 19$
28. $4x^2 = 28x - 49$
29. $18x^2 - 15x = 26$
30. $48x^2 + 9 = 50x$
31. $16.5x^2 + 8.3x - 14.7 = 0$
32. $125x^2 - 167x + 36 = 0$

A.6 Formulas

A **formula** is an equation, usually expressed in letters (called *variables* in algebra) and numbers. A **variable** is a symbol, usually a letter, used to represent some unknown number or quantity.

EXAMPLE 1

The formula $s = vt$ states that the distance traveled, s, equals the product of the velocity, v, and the time, t.

EXAMPLE 2

The formula $I = \frac{Q}{t}$ states that the current, I, equals the quotient of the charge, Q, and the time, t.

To solve a formula for a given letter means to express the given letter or variable in terms of all the remaining letters. That is, by using the equation-solving principles, rewrite the formula so that the given letter appears on one side of the equation by itself and all the other letters appear on the other side.

EXAMPLE 3

Solve $s = vt$ for v.

$$s = vt$$

$$\frac{s}{t} = \frac{vt}{t} \quad \text{Divide both sides by } t.$$

$$\frac{s}{t} = v$$

EXAMPLE 4

Solve $I = Q/t$

(a) for Q. (b) for t.

(a)

$$I = \frac{Q}{t}$$

$$(I)t = \left(\frac{Q}{t}\right)t \qquad \text{Multiply both sides by } t.$$

$$It = Q$$

(b) Starting with $It = Q$, we obtain

$$\frac{It}{I} = \frac{Q}{I} \qquad \text{Divide both sides by } I.$$

$$t = \frac{Q}{I}$$

EXAMPLE 5

Solve $V = E - Ir$ for r.

Method 1:

$$V = E - Ir$$

$$V - E = E - Ir - E \qquad \text{Subtract } E \text{ from both sides.}$$

$$V - E = -Ir$$

$$\frac{V - E}{-I} = \frac{-Ir}{-I} \qquad \text{Divide both sides by } -I.$$

$$\frac{V - E}{-I} = r$$

Method 2:

$$V = E - Ir$$

$$V + Ir = E - Ir + Ir \qquad \text{Add } Ir \text{ to both sides.}$$

$$V + Ir = E$$

$$V + Ir - V = E - V \qquad \text{Subtract } V \text{ from both sides.}$$

$$Ir = E - V$$

$$\frac{Ir}{I} = \frac{E - V}{I} \qquad \text{Divide both sides by } I.$$

$$r = \frac{E - V}{I}$$

Note that the two results are equivalent. Take the first result and multiply both numerator and denominator by -1. That is,

$$\frac{V - E}{-I} = \left(\frac{V - E}{-I}\right)\left(\frac{-1}{-1}\right) = \frac{-V + E}{I} = \frac{E - V}{I}$$

which is the same as the second result.

We often use the same quantity in more than one way in a formula. For example, we may wish to use a certain measurement of a quantity, such as velocity, at a given time, say at $t = 0$ s, then use the velocity at a later time, say at $t = 6$ s. To write these desired values of the velocity is rather awkward. We simplify this written statement by using *subscripts* (small letters or numbers printed a half space below the printed line and to the right of the variable) to shorten what we must write.

For the example given, v at time $t = 0$ s will be written as v_i (initial velocity); v at time $t = 6$ s will be written as v_f (final velocity). Mathematically, v_i and v_f are two different quantities, which in most cases are unequal. The sum of v_i and v_f is written as $v_i + v_f$. The product of v_i and v_f is written as $v_i\,v_f$. The subscript notation is used only to distinguish the general quantity, v, velocity, from the measure of that quantity at certain specified times.

EXAMPLE 6

Solve the formula $x = x_i + v_i t + \frac{1}{2}at^2$ for v_i.

Method 1:

$$x = x_i + v_i t + \tfrac{1}{2}at^2$$

$$x - v_i t = x_i + v_i t + \tfrac{1}{2}at^2 - v_i t \qquad \text{Subtract } v_i t \text{ from both sides.}$$

$$x - v_i t = x_i + \tfrac{1}{2}at^2$$

$$x - v_i t - x = x_i + \tfrac{1}{2}at^2 - x \qquad \text{Subtract } x \text{ from both sides.}$$

$$-v_i t = x_i + \tfrac{1}{2}at^2 - x$$

$$\frac{-v_i t}{-t} = \frac{x_i + \frac{1}{2}at^2 - x}{-t} \qquad \text{Divide both sides by } -t.$$

$$v_i = \frac{x_i + \frac{1}{2}at^2 - x}{-t}$$

Method 2:

$$x = x_i + v_i t + \tfrac{1}{2}at^2$$

$$x - x_i - \tfrac{1}{2}at^2 = x_i + v_i t + \tfrac{1}{2}at^2 - x_i - \tfrac{1}{2}at^2 \qquad \text{Subtract } x_i \text{ and } \tfrac{1}{2}at^2 \text{ from both sides.}$$

$$x - x_i - \tfrac{1}{2}at^2 = v_i t$$

$$\frac{x - x_i - \frac{1}{2}at^2}{t} = \frac{v_i t}{t} \qquad \text{Divide both sides by } t.$$

$$\frac{x - x_i - \frac{1}{2}at^2}{t} = v_i$$

EXAMPLE 7

Solve the formula $v_{avg} = \frac{1}{2}(v_f + v_i)$ for v_f (avg is used here as a subscript meaning average).

$$v_{avg} = \tfrac{1}{2}(v_f + v_i)$$

$$2v_{avg} = v_f + v_i \qquad \text{Multiply both sides by 2.}$$

$$2v_{avg} - v_i = v_f \qquad \text{Subtract } v_i \text{ from both sides.}$$

EXAMPLE 8

Solve $A = \dfrac{\pi d^2}{4}$ for d, where d is a diameter.

$$A = \frac{\pi d^2}{4}$$

$$4A = \left(\frac{\pi d^2}{4}\right)(4) \qquad \text{Multiply both sides by 4.}$$

$$4A = \pi d^2$$

$$\frac{4A}{\pi} = \frac{\pi d^2}{\pi} \qquad \text{Divide both sides by } \pi.$$

$$\frac{4A}{\pi} = d^2$$

$$\pm\sqrt{\frac{4A}{\pi}} = d \qquad \text{Take the square root of both sides.}$$

In this case, a negative diameter has no physical meaning, so the result is

$$d = \sqrt{\frac{4A}{\pi}}$$

PROBLEMS A.6

Solve each formula for the quantity given.

1. $v = \frac{s}{t}$ for s
2. $a = \frac{v}{t}$ for v
3. $w = mg$ for m
4. $F = ma$ for a
5. $E = IR$ for R
6. $V = lwh$ for w
7. $\text{PE} = mgh$ for g
8. $\text{PE} = mgh$ for h
9. $v^2 = 2gh$ for h
10. $X_L = 2\pi fL$ for f
11. $P = \frac{W}{t}$ for W
12. $p = \frac{F}{A}$ for F
13. $P = \frac{W}{t}$ for t
14. $p = \frac{F}{A}$ for A
15. $\text{KE} = \frac{1}{2}mv^2$ for m
16. $\text{KE} = \frac{1}{2}mv^2$ for v^2
17. $W = Fs$ for s
18. $v_f = v_i + at$ for a
19. $V = E - Ir$ for I
20. $v_2 = v_1 + at$ for t
21. $R = \frac{\pi}{2P}$ for P
22. $R = \frac{kL}{d^2}$ for L
23. $F = \frac{9}{5}C + 32$ for C
24. $C = \frac{5}{9}(F - 32)$ for F
25. $X_C = \frac{1}{2\pi fC}$ for f
26. $R = \frac{\rho L}{A}$ for L
27. $R_T = R_1 + R_2 + R_3 + R_4$ for R_3
28. $Q_1 = P(Q_2 - Q_1)$ for Q_2
29. $\frac{I_S}{I_P} = \frac{N_P}{N_S}$ for I_P
30. $\frac{V_P}{V_S} = \frac{N_P}{N_S}$ for N_S
31. $v_{\text{avg}} = \frac{1}{2}(v_f + v_i)$ for v_i
32. $2a(s - s_i) = v^2 - v_i^2$ for a
33. $2a(s - s_i) = v^2 - v_i^2$ for s
34. $Ft = m(V_2 - V_1)$ for V_1
35. $Q = \frac{I^2Rt}{J}$ for R
36. $x = x_i + v_i t + \frac{1}{2}at^2$ for x_i
37. $A = \pi r^2$ for r, where r is a radius
38. $V = \pi r^2 h$ for r, where r is a radius
39. $R = \frac{kL}{d^2}$ for d, where d is a diameter
40. $V = \frac{1}{3}\pi r^2 h$ for r, where r is a radius
41. $Q = \frac{I^2Rt}{J}$ for I
42. $F = \frac{mv^2}{r}$ for v

A.7 Substituting Data Into Formulas

An important part of problem solving is substituting the given data into the appropriate formula to find the value of the unknown quantity. Basically, there are two ways of substituting data into formulas to solve for the unknown quantity:

1. Solve the formula for the unknown quantity and then make the substitution of the data.
2. Substitute the data into the formula first and then solve for the unknown quantity.

When using a calculator, the first way is more useful. We will be using this way most of the time in this text.

EXAMPLE 1

Given the formula $A = bh$, $A = 120\text{ m}^2$, and $b = 15$ m, find h.

First, solve for h:

$$A = bh$$

$$\frac{A}{b} = \frac{bh}{b} \quad \text{Divide both sides by } b.$$

$$\frac{A}{b} = h$$

Then substitute the data:

$$h = \frac{A}{b} = \frac{120\text{ m}^2}{15\text{ m}} = 8.0\text{ m}$$

(Remember to follow the rules of measurement discussed in Chapter 1 We use them consistently throughout.)

EXAMPLE 2

Given the formula $P = 2a + 2b$, $P = 824$ cm, and $a = 292$ cm, find b.

First, solve for b:

$$P = 2a + 2b$$

$$P - 2a = 2a + 2b - 2a \quad \text{Subtract } 2a \text{ from both sides.}$$

$$P - 2a = 2b$$

$$\frac{P - 2a}{2} = \frac{2b}{2} \quad \text{Divide both sides by 2.}$$

$$\frac{P - 2a}{2} = b \quad \left(\text{or } b = \frac{P}{2} - a\right)$$

Then substitute the data:

$$b = \frac{P - 2a}{2} = \frac{824\text{ cm} - 2(292\text{ cm})}{2}$$

$$= \frac{824\text{ cm} - 584\text{ cm}}{2}$$

$$= \frac{24\bar{0}\text{ cm}}{2} = 12\bar{0}\text{ cm}$$

EXAMPLE 3

Given the formula $A = \left(\frac{a + b}{2}\right)h$, $A = 15\bar{0}\text{ m}^2$, $b = 18.0$ m, and $h = 10.0$ m, find a.

First, solve for a:

$$A = \left(\frac{a + b}{2}\right)h$$

$$2A = \left[\left(\frac{a+b}{2}\right)h\right](2) \quad \text{Multiply both sides by 2.}$$

$$2A = (a+b)h$$

$$2A = ah + bh \quad \text{Remove the parentheses.}$$

$$2A - bh = ah + bh - bh \quad \text{Subtract } bh \text{ from both sides.}$$

$$2A - bh = ah$$

$$\frac{2A - bh}{h} = \frac{ah}{h} \quad \text{Divide both sides by } h.$$

$$\frac{2A - bh}{h} = a \quad \left(\text{or } a = \frac{2A}{h} - b\right)$$

Then substitute the data:

$$a = \frac{2A - bh}{h}$$

$$= \frac{2(15\bar{0}\text{ m}^2) - (18.0\text{ m})(10.0\text{ m})}{10.0\text{ m}}$$

$$= \frac{30\bar{0}\text{ m}^2 - 18\bar{0}\text{ m}^2}{10.0\text{ m}}$$

$$= \frac{12\bar{0}\text{ m}^2}{10.0\text{ m}} = 12.0\text{ m}$$

EXAMPLE 4

Given the formula $V = \frac{1}{3}\pi r^2 h$, $V = 64{,}400\text{ mm}^3$, and $h = 48.0$ mm, find r, where r is a radius.

First, solve for r:

$$V = \tfrac{1}{3}\pi r^2 h$$

$$3V = \left(\tfrac{1}{3}\pi r^2 h\right)(3) \quad \text{Multiply both sides by 3.}$$

$$3V = \pi r^2 h$$

$$\frac{3V}{\pi h} = \frac{\pi r^2 h}{\pi h} \quad \text{Divide both sides by } \pi h.$$

$$\frac{3V}{\pi h} = r^2$$

$$\pm\sqrt{\frac{3V}{\pi h}} = r \quad \text{Take the square root of both sides.}$$

In this case, a negative radius has no physical meaning, so the result is

$$r = \sqrt{\frac{3V}{\pi h}}$$

Then substitute the data:

$$r = \sqrt{\frac{3(64{,}400\text{ mm}^3)}{\pi(48.0\text{ mm})}}$$

$$= 35.8\text{ mm}$$

PROBLEMS A.7

For each formula, (a) solve for the indicated letter and then (b) substitute the given data to find the value of the indicated letter. Follow the rules of calculations with measurements. ***Note:*** In Problems 14 and 16, r is a radius, and in Problem 15, b is the length of the side of a square.

Formula	*Data*	*Find*
1. $A = bh$	$b = 14.5$ cm, $h = 11.2$ cm	A
2. $V = lwh$	$l = 16.7$ m, $w = 10.5$ m, $h = 25.2$ m	V
3. $A = bh$	$A = 34.5\ \text{cm}^2$, $h = 4.60$ cm	b
4. $P = 4b$	$P = 42\bar{0}$ in.	b
5. $P = a + b + c$	$P = 48.5$ cm, $a = 18.2$ cm, $b = 24.3$ cm	c
6. $C = \pi d$	$C = 495$ ft	d
7. $C = 2\pi r$	$C = 68.5$ yd	r
8. $A = \frac{1}{2}bh$	$A = 468\ \text{m}^2$, $b = 36.0$ m	h
9. $P = 2(a + b)$	$P = 88.7$ km, $a = 11.2$ km	b
10. $V = \pi r^2 h$	$r = 61.0$ m, $h = 125.3$ m	V
11. $V = \pi r^2 h$	$V = 368\ \text{m}^3$, $r = 4.38$ m	h
12. $A = 2\pi rh$	$A = 51\bar{0}\ \text{cm}^2$, $r = 14.0$ cm	h
13. $V = Bh$	$V = 2185\ \text{m}^3$, $h = 14.2$ m	B
14. $A = \pi r^2$	$A = 463.5\ \text{m}^2$	r
15. $A = b^2$	$A = 465\ \text{in}^2$	b
16. $V = \frac{1}{3}\pi r^2 h$	$V = 2680\ \text{m}^3$, $h = 14.7$ m	r
17. $C = 2\pi r$	$r = 19.36$ m	C
18. $V = \frac{4}{3}\pi r^3$	$r = 25.65$ m	V
19. $V = \frac{1}{3}Bh$	$V = 19{,}850\ \text{ft}^3$, $h = 486.5$ ft	B
20. $A = \left(\frac{a+b}{2}\right)h$	$A = 205.2\ \text{m}^2$, $a = 16.50$ m, $b = 19.50$ m	h

A.8 Right-Triangle Trigonometry

A **right triangle** is a triangle with one right angle (90°), two acute angles (less than 90°), two legs, and a hypotenuse (the side opposite the right angle) (Fig. A.1).

When it is necessary to label a triangle, the vertices are often labeled using capital letters and the sides opposite the vertices are labeled using the corresponding lowercase letters (Fig. A.2).

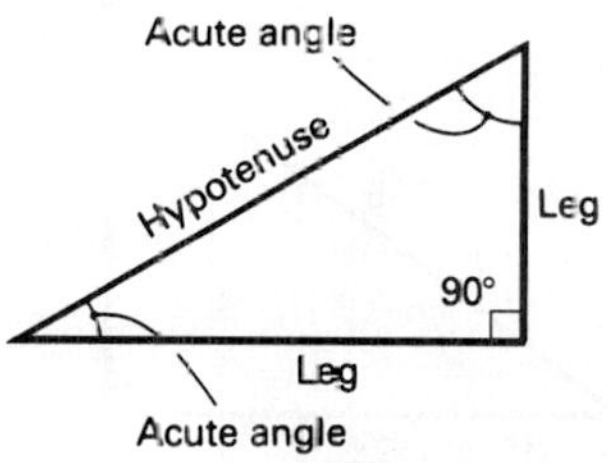

Figure A.1 Parts of a right triangle

The side opposite angle A is a.
The side adjacent to angle A is b.
The side opposite angle B is b.
The side adjacent to angle B is a.
The side opposite angle C is c and is called the *hypotenuse*.

If we consider a certain acute angle of a right triangle, the two legs can be identified as the side opposite or the side adjacent to the acute angle.

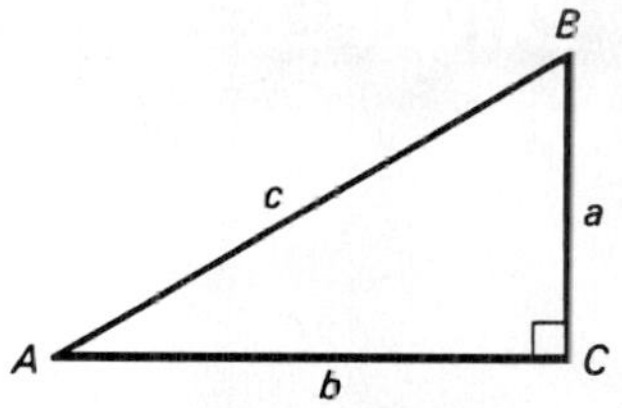

Figure A.2 Labeling a right triangle

The side opposite angle A is the same as the side adjacent to angle B.
The side adjacent to angle A is the same as the side opposite angle B.
The side opposite angle B is the same as the side adjacent to angle A.
The side adjacent to angle B is the same as the side opposite angle A.

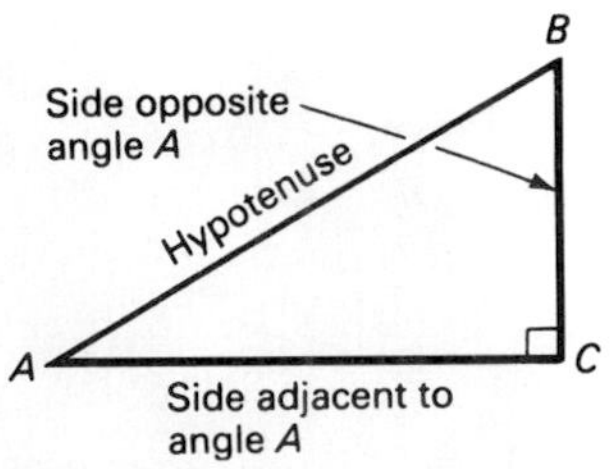

Figure A.3

A **ratio** is a comparison of two quantities by division. In a right triangle (Fig. A.3), there are three very important ratios:

$\dfrac{\text{side opposite angle } A}{\text{hypotenuse}}$ is called **sine A** (abbreviated sin *A*)

$\dfrac{\text{side adjacent to angle } A}{\text{hypotenuse}}$ is called **cosine A** (abbreviated cos *A*)

$\dfrac{\text{side opposite angle } A}{\text{side adjacent to angle } A}$ is called **tangent A** (abbreviated tan *A*)

$$\sin A = \frac{\text{side opposite angle } A}{\text{hypotenuse}}$$
$$\cos A = \frac{\text{side adjacent to angle } A}{\text{hypotenuse}}$$
$$\tan A = \frac{\text{side opposite angle } A}{\text{side adjacent to angle } A}$$

The ratios are defined similarly for angle *B*:

$$\sin B = \frac{\text{side opposite angle } B}{\text{hypotenuse}}$$
$$\cos B = \frac{\text{side adjacent to angle } B}{\text{hypotenuse}}$$
$$\tan B = \frac{\text{side opposite angle } B}{\text{side adjacent to angle } B}$$

EXAMPLE 1

Find the three trigonometric ratios of angle *A* in Fig. A.4.

$$\sin A = \frac{\text{side opposite angle } A}{\text{hypotenuse}} = \frac{3}{5} = 0.60$$
$$\cos A = \frac{\text{side adjacent to angle } A}{\text{hypotenuse}} = \frac{4}{5} = 0.80$$
$$\tan A = \frac{\text{side opposite angle } A}{\text{side adjacent to angle } A} = \frac{3}{4} = 0.75$$

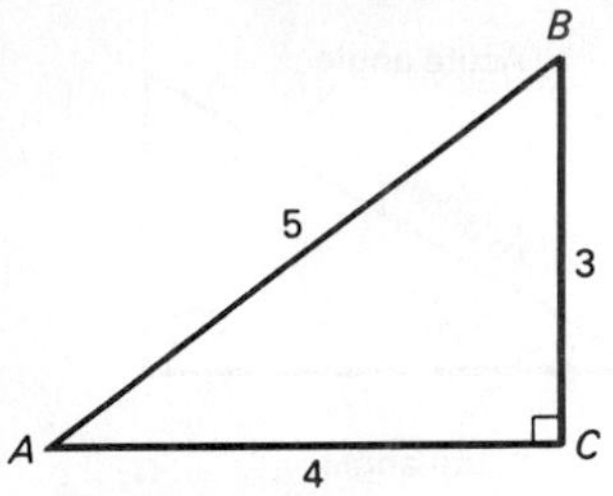

Figure A.4

EXAMPLE 2

Find the three trigonometric ratios of angle *B* in Fig. A.4.

$$\sin B = \frac{\text{side opposite angle } B}{\text{hypotenuse}} = \frac{4}{5} = 0.80$$
$$\cos B = \frac{\text{side adjacent to angle } B}{\text{hypotenuse}} = \frac{3}{5} = 0.60$$
$$\tan B = \frac{\text{side opposite angle } B}{\text{side adjacent to angle } B} = \frac{4}{3} = 1.33$$

Note: Every acute angle has three trigonometric ratios associated with it.

In this book we assume that you will be using a calculator. When calculations involve a trigonometric ratio, we will use the following generally accepted practice for significant digits:

Angle Expressed to Nearest:	Length of Side Contains:
1°	Two significant digits
0.1°	Three significant digits
0.01°	Four significant digits

A useful and time-saving fact about right triangles is that *the sum of the two acute angles of any right triangle is always* 90°. That is,

$$A + B = 90°$$

Why is this true? We know that the sum of the three interior angles of any triangle is 180°. A right triangle must contain a right angle, whose measure is 90°. This leaves 90° to be divided between the two acute angles. Therefore, if one acute angle is known, the other acute angle may be found by subtracting the known angle from 90°. That is,

$$A = 90° - B$$
$$B = 90° - A$$

EXAMPLE 3

Find angle B and side a in the right triangle in Fig. A.5.

To find angle B, we use

$$B = 90° - A = 90° - 30.0° = 60.0°$$

To find side a, we use a trigonometric ratio. Note that we are looking for the *side opposite* angle A and that the *hypotenuse* is given. The trigonometric ratio having these two quantities is the sine. We write

$$\sin A = \frac{\text{side opposite angle } A}{\text{hypotenuse}}$$

$$\sin 30.0° = \frac{a}{20.0 \text{ m}}$$

$$(\sin 30.0°)(20.0 \text{ m}) = \left(\frac{a}{20.0 \text{ m}}\right)(20.0 \text{ m}) \quad \text{Multiply both sides by 20.0 m.}$$

$$10.0 \text{ m} = a$$

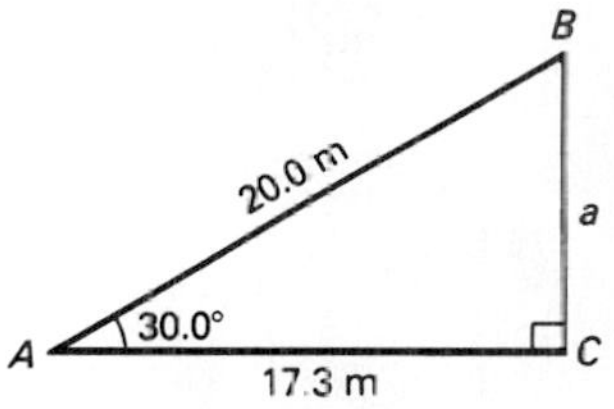

Figure A.5

EXAMPLE 4

Find angle A, angle B, and side a in the right triangle in Fig. A.6.

First, find angle A. The *side adjacent* to angle A and the *hypotenuse* are given. Therefore, we use cos A to find angle A because cos A uses these two quantities:

$$\cos A = \frac{\text{side adjacent to } A}{\text{hypotenuse}}$$

$$\cos A = \frac{13 \text{ ft}}{19 \text{ ft}} = 0.684$$

Using a calculator as in Section B.3 of Appendix B, we find that $A = 47°$.

To find angle B, we use

$$B = 90° - A = 90° - 47° = 43°$$

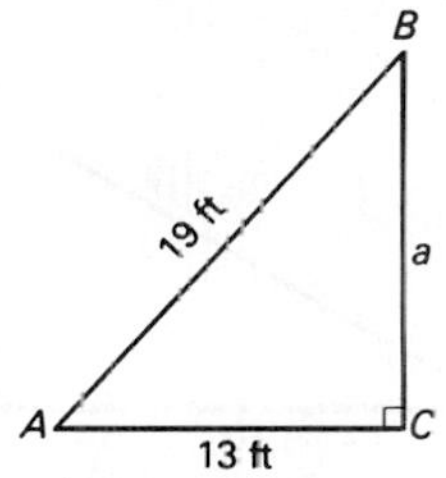

Figure A.6

To find side a, we use $\sin A$ because the *hypotenuse* is given and side a is the *side opposite* angle A:

$$\sin A = \frac{\text{side opposite angle } A}{\text{hypotenuse}}$$

$$\sin 47° = \frac{a}{19 \text{ ft}}$$

$$(\sin 47°)(19 \text{ ft}) = \left(\frac{a}{19 \text{ ft}}\right)(19 \text{ ft}) \qquad \text{Multiply both sides by 19 ft.}$$

$$14 \text{ ft} = a$$

EXAMPLE 5

Find angle A, angle B, and the hypotenuse in the right triangle in Fig. A.7.

To find angle A, use $\tan A$:

$$\tan A = \frac{\text{side opposite angle } A}{\text{side adjacent to angle } A}$$

$$\tan A = \frac{12.00 \text{ km}}{19.00 \text{ km}} = 0.6316$$

$$A = 32.28°$$

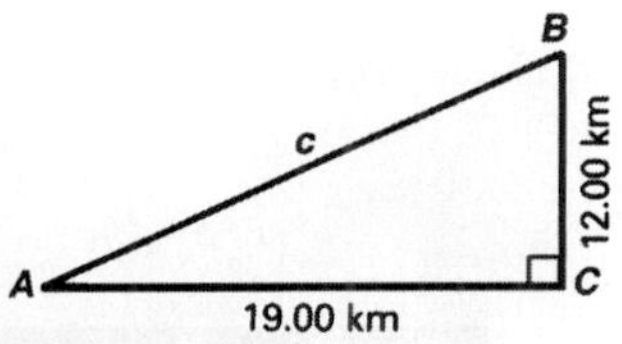

Figure A.7

To find angle B, write

$$B = 90° - A = 90° - 32.28° = 57.72°$$

To find the hypotenuse, use $\sin A$:

$$\sin A = \frac{\text{side opposite angle } A}{\text{hypotenuse}}$$

$$\sin 32.28° = \frac{12.00 \text{ km}}{c}$$

$$(\sin 32.28°)(c) = \left(\frac{12.00 \text{ km}}{c}\right)(c) \qquad \text{Multiply both sides by } c.$$

$$(\sin 32.28°)(c) = 12.00 \text{ km}$$

$$\frac{c(\sin 32.28°)}{\sin 32.28°} = \frac{12.00 \text{ km}}{\sin 32.28°} \qquad \text{Divide both sides by } \sin 32.28°.$$

$$c = \frac{12.00 \text{ km}}{\sin 32.28°}$$

$$= 22.47 \text{ km}$$

When the two legs of a right triangle are given, the hypotenuse can be found without using trigonometric ratios. From geometry, *the sum of the squares of the legs of a right triangle is equal to the square of the hypotenuse* (**Pythagorean theorem;** see Fig. A.8):

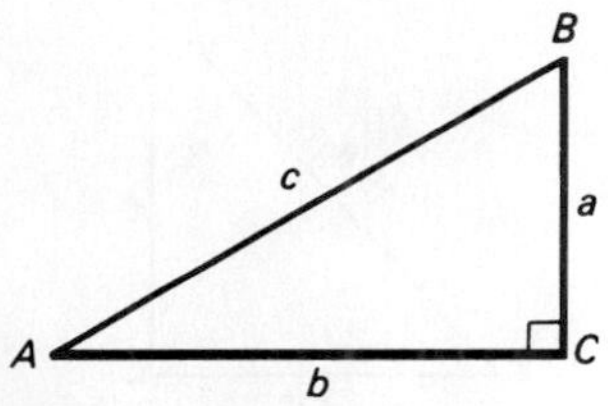

Figure A.8

$$a^2 + b^2 = c^2$$

or, taking the square root of each side of the equation,

$$\boxed{c = \sqrt{a^2 + b^2}}$$

Also, if one leg and the hypotenuse are given, the other leg can be found by using

$$a = \sqrt{c^2 - b^2}$$
$$b = \sqrt{c^2 - a^2}$$

EXAMPLE 6

Find the hypotenuse of the right triangle in Fig. A.9.

$$\begin{aligned} c &= \sqrt{a^2 + b^2} \\ c &= \sqrt{(13.0\text{ m})^2 + (11.0\text{ m})^2} \\ &= \sqrt{169\text{ m}^2 + 121\text{ m}^2} \\ &= \sqrt{29\bar{0}\text{ m}^2} \\ &= 17.0\text{ m} \end{aligned}$$

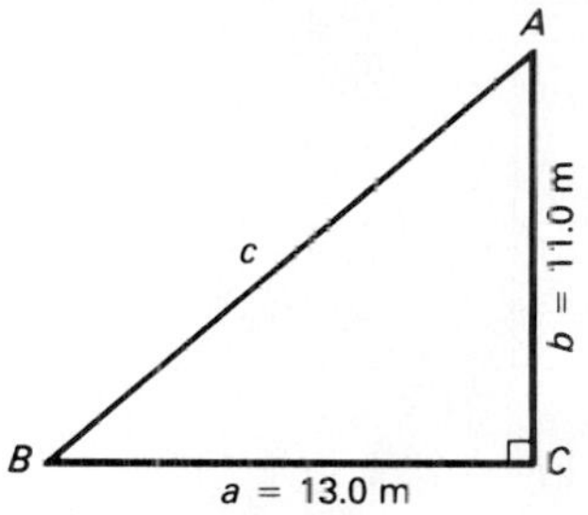

Figure A.9

EXAMPLE 7

Find side b in the right triangle in Fig. A.10.

$$\begin{aligned} b &= \sqrt{c^2 - a^2} \\ b &= \sqrt{(12.2\text{ km})^2 - (7.30\text{ km})^2} \\ &= 9.77\text{ km} \end{aligned}$$

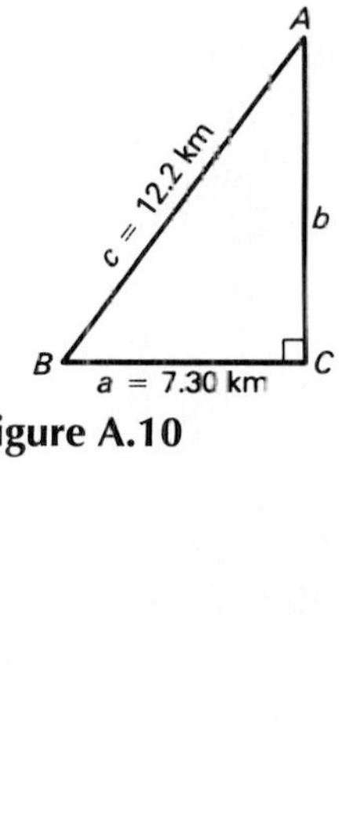

Figure A.10

PROBLEMS A.8

Use right triangle ABC in Fig. A.11 to fill in each blank.

1. The side opposite angle A is ______.
2. The side opposite angle B is ______.
3. The hypotenuse is ______.
4. The side adjacent to angle A is ______.
5. The side adjacent to angle B is ______.
6. The angle opposite side a is ______.
7. The angle opposite side b is ______.
8. The angle opposite side c is ______.
9. The angle adjacent to side a is ______.
10. The angle adjacent to side b is ______.

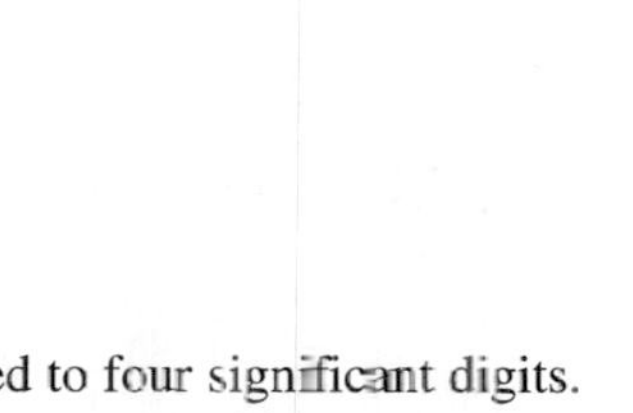

Figure A.11

Use a calculator to find each trigonometric ratio rounded to four significant digits.

11. sin 71°	12. cos 40°	13. tan 61°
14. tan 41.2°	15. cos 11.5°	16. sin 79.4°
17. cos 49.63°	18. tan 53.45°	19. tan 17.04°
20. cos 34°	21. sin 27.5°	22. cos 58.72°

Find each angle rounded to the nearest whole degree.

23. $\sin A = 0.2678$ 24. $\cos B = 0.1046$ 25. $\tan A = 0.9237$

26. $\sin B = 0.9253$ 27. $\cos B = 0.6742$ 28. $\tan A = 1.351$

Find each angle rounded to the nearest tenth of a degree.

29. $\sin B = 0.5963$ 30. $\cos A = 0.9406$ 31. $\tan B = 1.053$

32. $\sin A = 0.9083$ 33. $\cos A = 0.8660$ 34. $\tan B = 0.9433$

Find each angle rounded to the nearest hundredth of a degree.

35. $\sin A = 0.3792$ 36. $\cos B = 0.06341$ 37. $\tan B = 0.3010$

38. $\sin A = 0.4540$ 39. $\cos B = 0.8141$ 40. $\tan A = 2.369$

Solve each triangle (find the missing angles and sides) using trigonometric ratios.

41.

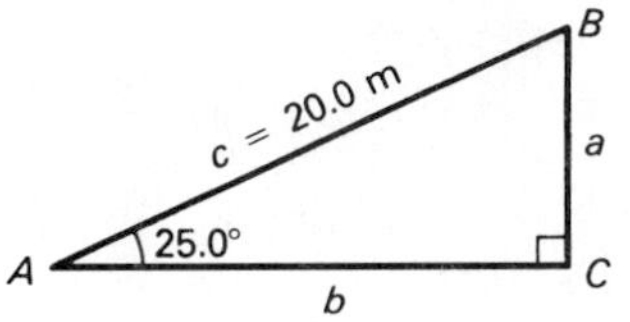

42.

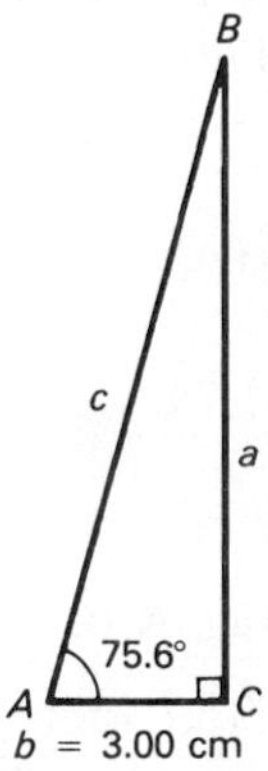

43.

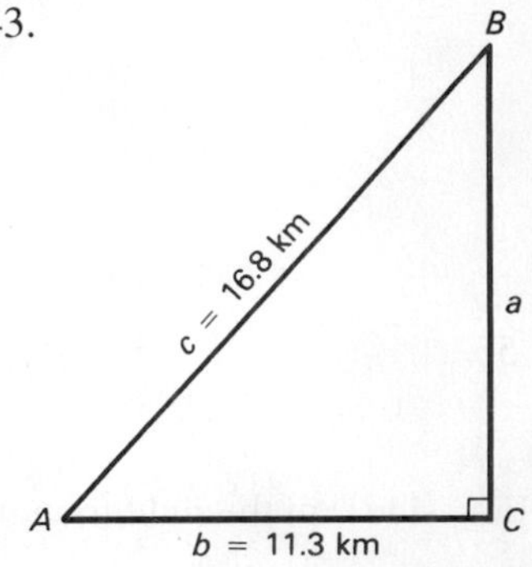

44.

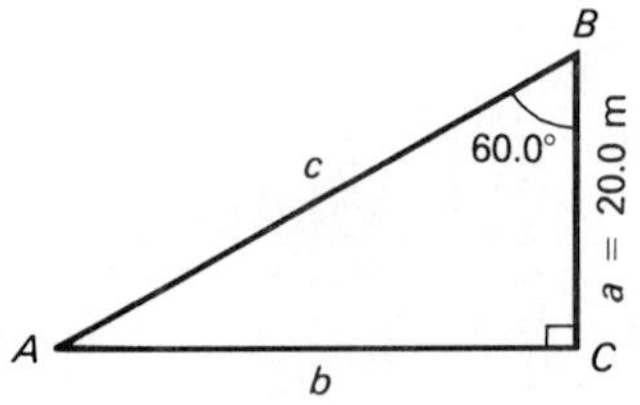

45.

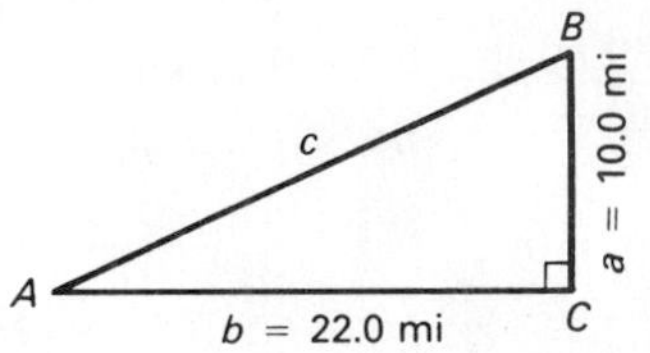

46.

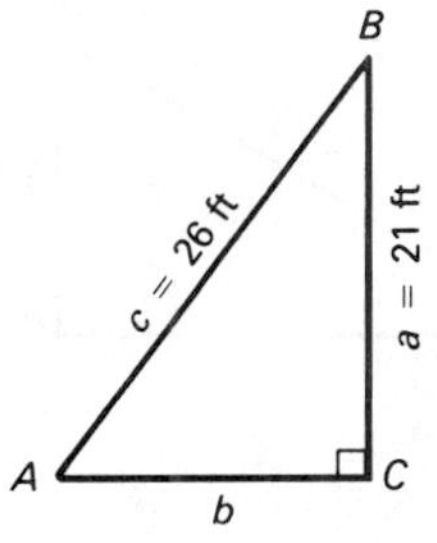

47.

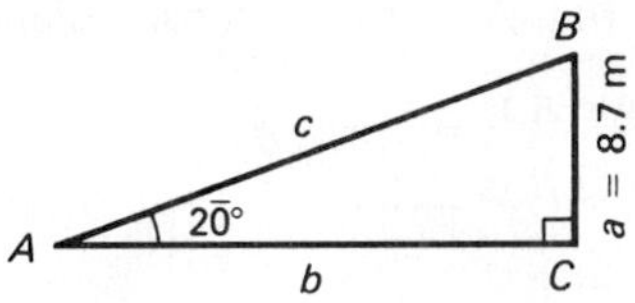

48.

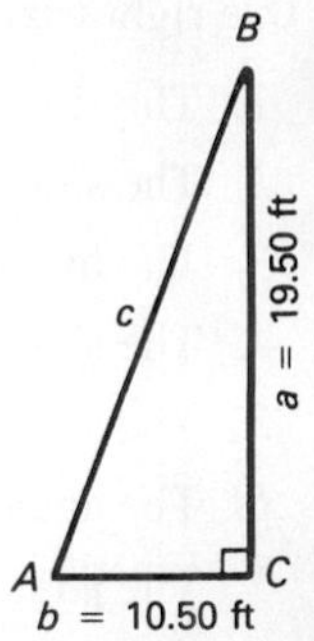

49.

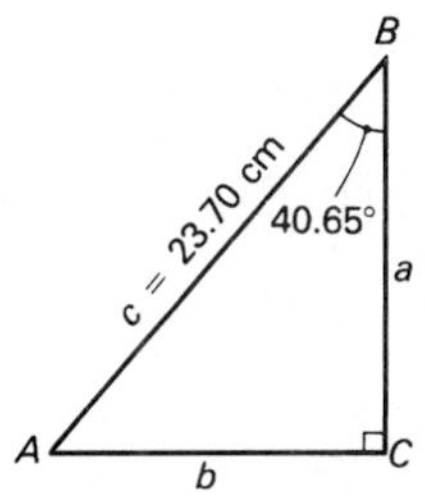

50.

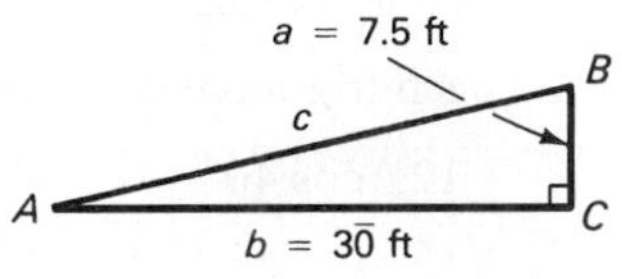

Find the missing side in each right triangle using the Pythagorean theorem.

51.

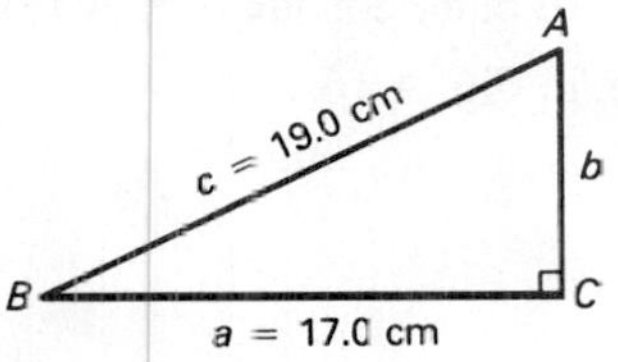

52.

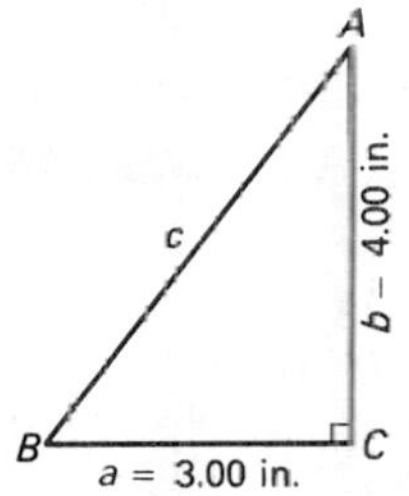

53.

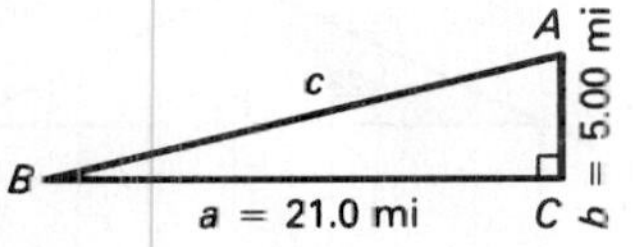

54.

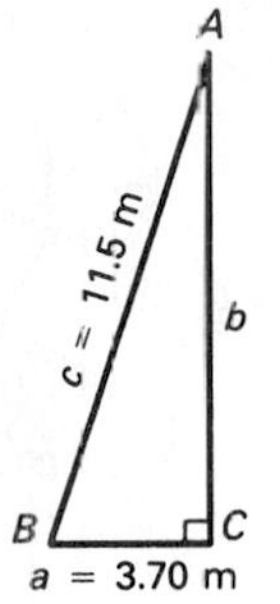

55.

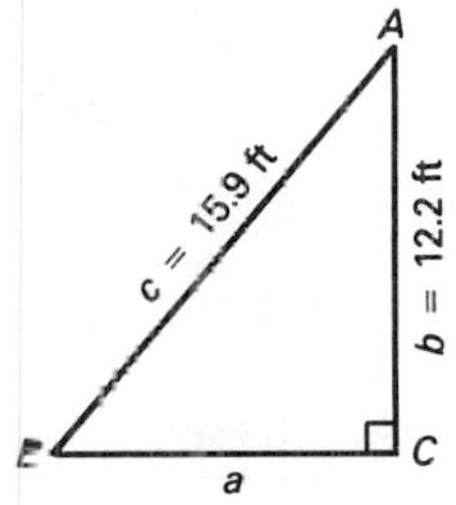

56.

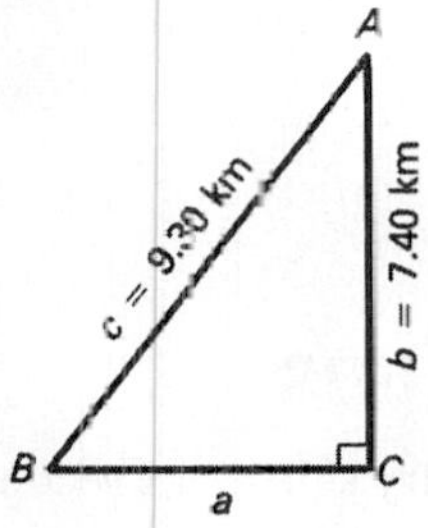

57.

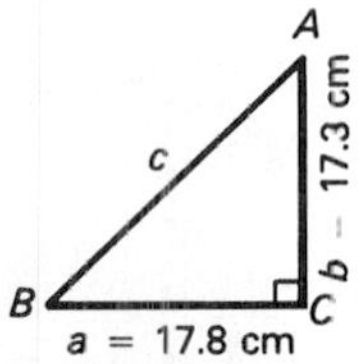

58.

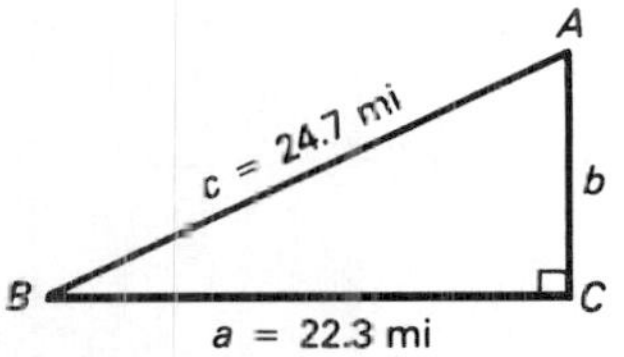

59.

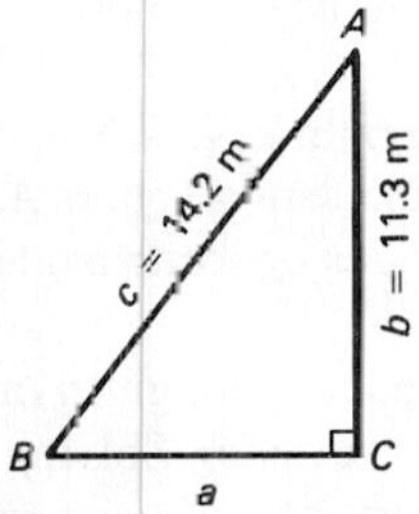

60.

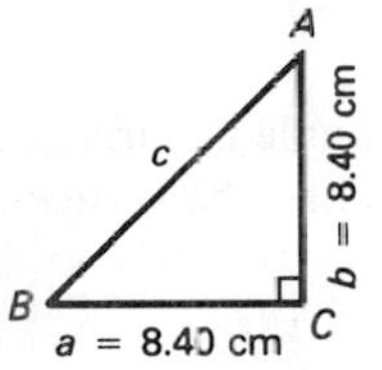

61. A round taper is shown in Fig. A.12.
   (a) Find $\angle BAC$. (b) Find the length BC.
   (c) Find the diameter of end x.

62. The distance between two parallel flat surfaces, a, of a hexagonal nut is $\frac{3}{4}$ in. Find the distance between any two farthest corners, b (Fig. A.13).

63. Find distances c and d between the holes of the plate shown in Fig. A.14.

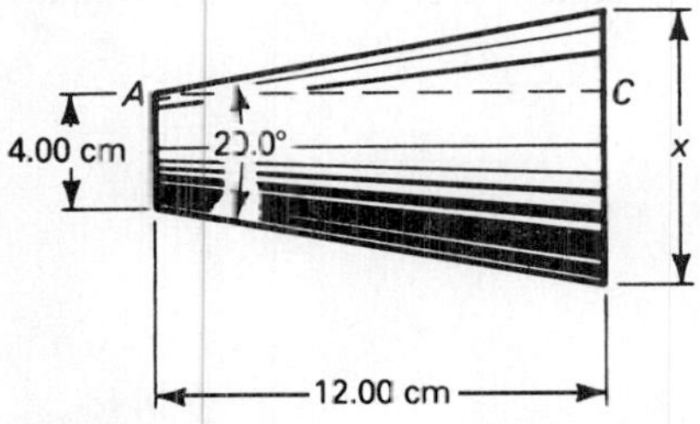

Figure A.12

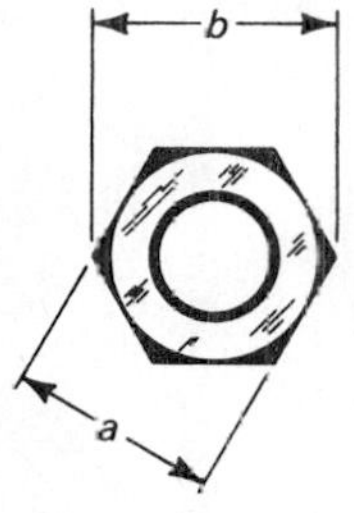

Figure A.13

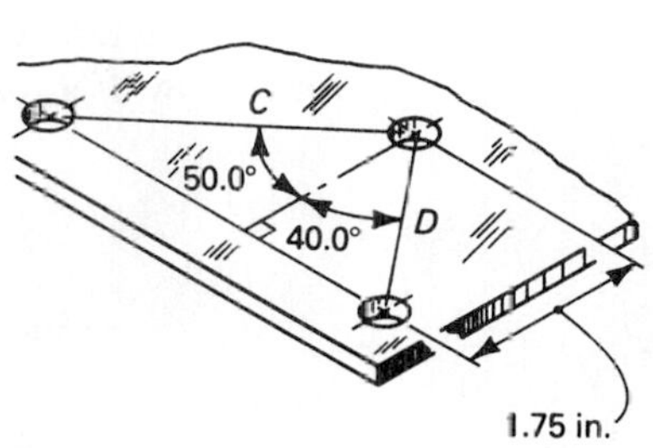

Figure A.14

64. A piece of electric conduit cuts across a corner of a room $24\bar{0}$ cm from the corner. It meets the adjoining wall $35\bar{0}$ cm from the corner. Find length *AB* of the conduit (Fig. A.15).
65. Find the distances between holes on the plate shown in Fig. A.16.
66. Find length *x* in Fig. A.17. (*Note*: *AB* = *BC*.)

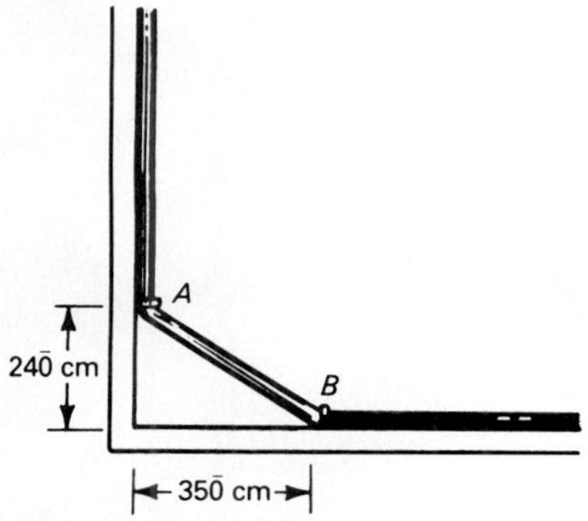

Figure A.15

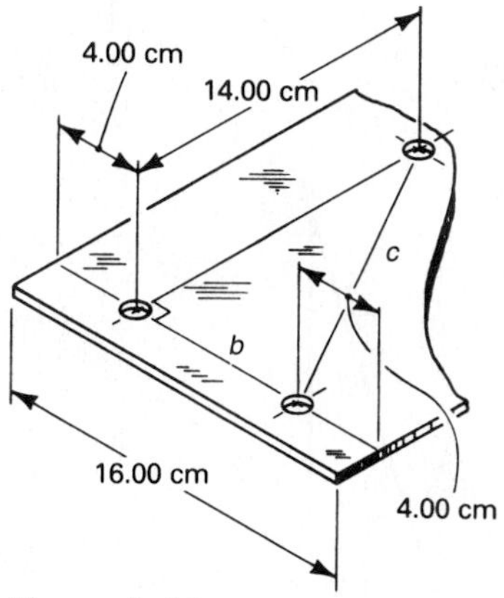

Figure A.16

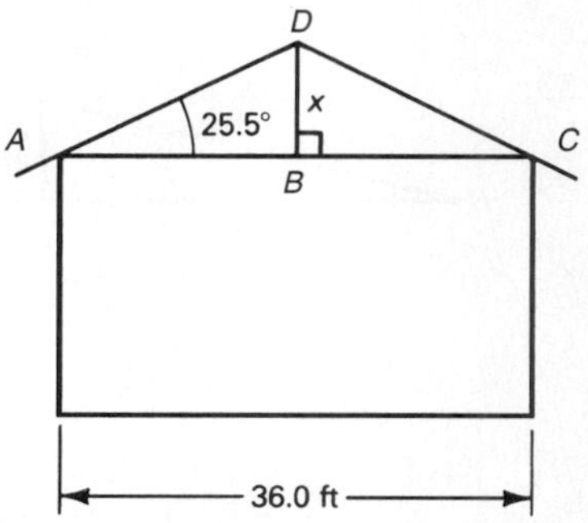

Figure A.17

A.9 Law of Sines and Law of Cosines

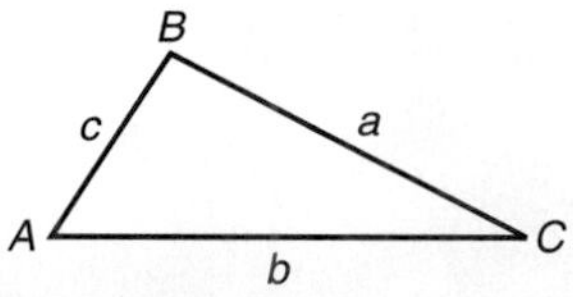

Figure A.18 Oblique, or general, triangle ABC

For those classes and individuals with general triangle trigonometry skills, this section is written as a review. This text is written so that no more than right triangle trigonometry needs to be used, but those with the prerequisite skills may prefer to use the more advanced trigonometry techniques.

An **oblique,** or **general,** triangle is a triangle that contains no right angles. We shall use the standard notation of labeling the vertices of a triangle by the capital letters *A, B,* and *C* and using the small letters *a, b,* and *c* as the labels for the sides opposite angles *A, B,* and *C,* respectively (see Fig. A.18).

Solving a triangle means finding all those sides and angles that are not given or known. To solve a triangle, we need three parts (including at least one side). Solving any oblique triangle falls into one of four cases where the following parts of a triangle are known:

1. Two sides and an angle opposite one of them (SSA).
2. Two angles and a side opposite one of them (AAS).
3. Two sides and the included angle (SAS).
4. Three sides (SSS).

One law that we use to solve triangles is called the **law of sines.** In words, for any triangle the ratio of any side to the sine of the opposite angle is a constant. The formula for the law of sines is as follows:

LAW OF SINES

$$\frac{a}{\sin A} = \frac{b}{\sin B} = \frac{c}{\sin C}$$

In order to use the law of sines, we must know either of the following:

1. Two sides and an angle opposite one of them (SSA).
2. Two angles and a side opposite one of them (AAS). *Note:* Knowing two angles and any side is sufficient because knowing two angles, we can easily find the third.

You must select the proportion that contains three parts that are known and the unknown part.

When calculations with measurements involve a trigonometric function, we shall use the following rule for significant digits:

Angle Expressed to Nearest	Length of Side Contains
1°	Two significant digits
0.1°	Three significant digits
0.01°	Four significant digits

The following relationship is very helpful as a check when solving a general triangle: The longest side of any triangle is opposite the largest angle and the shortest side is opposite the smallest angle.

EXAMPLE 1

If $A = 65.0°$, $a = 20.0$ m, and $b = 15.0$ m, solve the triangle.

First, draw a triangle as in Fig. A.19 and find angle B by using the law of sines:

$$\frac{a}{\sin A} = \frac{b}{\sin B}$$

$$\frac{20.0 \text{ m}}{\sin 65.0°} = \frac{15.0 \text{ m}}{\sin B}$$

$$\sin B = \frac{(15.0 \text{ m})(\sin 65.0°)}{20.0 \text{ m}} = 0.6797$$

$$B = 42.8°$$

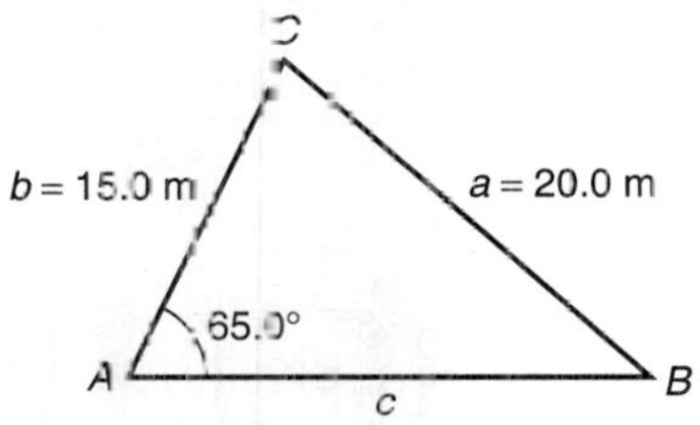

Figure A.19

This angle may be found using a calculator as follows:

15 [×] [sin] 65 [)] [÷] 20 [=] [sin⁻¹] [ANS] [=]

42.82261353

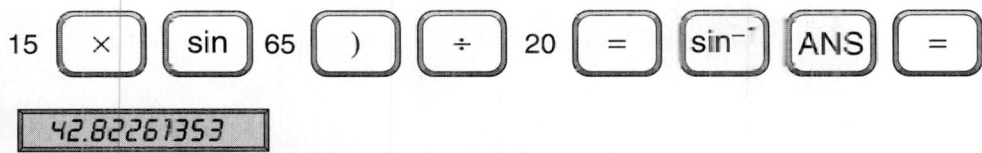

So $B = 42.8°$ rounded to the nearest tenth of a degree.

To find C, use the fact that the sum of the angles of any triangle is 180°. Therefore,

$$C = 180° - 65.0° - 42.8°$$
$$= 72.2°$$

Finally, find c using the law of sines.

$$\frac{a}{\sin A} = \frac{c}{\sin C}$$

$$\frac{20.0 \text{ m}}{\sin 65.0°} = \frac{c}{\sin 72.2°}$$

$$c = \frac{(20.0 \text{ m})(\sin 72.2°)}{\sin 65.0°} = 21.0 \text{ m} \qquad \text{(Rounded to three significant digits)}$$

This side may be found using a calculator as follows:

20 [×] [sin] 72.2 [)] [÷] [sin] 65 [=]

21.01117096

That is, side $c = 21.0$ m rounded to three significant digits.
The solution is $B = 42.8°$, $C = 72.2°$, and $c = 21.0$ m.

EXAMPLE 2

If $C = 25°$, $c = 59$ ft, and $B = 108°$, solve the triangle.

First, draw a triangle as in Fig. A.20 and find b:

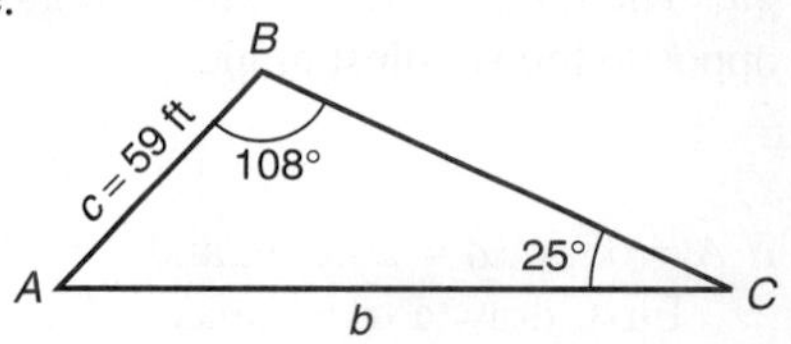

Figure A.20

$$\frac{c}{\sin C} = \frac{b}{\sin B}$$

$$\frac{59 \text{ ft}}{\sin 25°} = \frac{b}{\sin 108°}$$

$$b = \frac{(59 \text{ ft})(\sin 108°)}{\sin 25°} = 130 \text{ ft} \qquad \text{(Rounded to two significant digits)}$$

59 [×] [sin] 108 [)] [÷] [sin] 25 [=]

132.7730946

$$A = 180° - 25° - 108° = 47°$$

Find a:

$$\frac{a}{\sin A} = \frac{c}{\sin C}$$

$$\frac{a}{\sin 47°} = \frac{59 \text{ ft}}{\sin 25°}$$

$$a = \frac{(59 \text{ ft})(\sin 47°)}{\sin 25°} = 1\bar{0}0 \text{ ft}$$

The solution is $A = 47°$, $a = 1\bar{0}0$ ft, and $b = 130$ ft.

The Ambiguous Case

The solution of a triangle when two sides and an angle opposite one of the sides (SSA) are given requires special care. There may be one, two, or no triangles formed from the given data. By construction and discussion, let's study the possibilities.

EXAMPLE 3

Construct a triangle given that $A = 35°$, $b = 10$, and $a = 7$.

As you can see from Fig. A.21, two triangles that satisfy the given information can be drawn: triangles ACB and ACB'. Note that in one triangle angle B is acute and in the other triangle angle B is obtuse.

Figure A.21

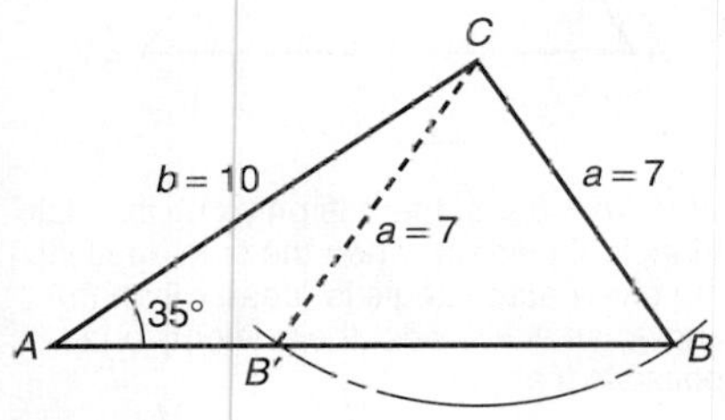

EXAMPLE 4

Construct a triangle given that $A = 45°$, $b = 10$, and $a = 5$.

As you can see from Fig. A.22, no triangle can be drawn that satisfies the given information. Side a is simply not long enough to reach the side opposite angle C.

Figure A.22

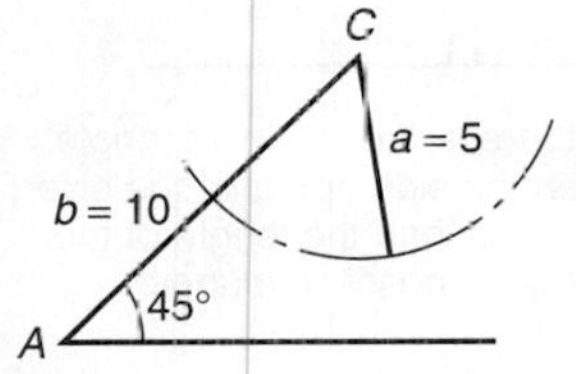

EXAMPLE 5

Construct a triangle given that $A = 60°$, $b = 6$, and $a = 10$.

As you can see from Fig. A.23, only one triangle that satisfies the given information can be drawn. Side a is too long for two solutions.

Figure A.23

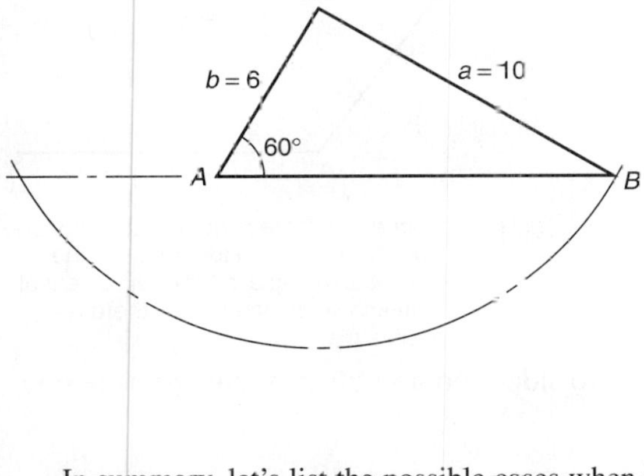

In summary, let's list the possible cases when two sides and an angle opposite one of the sides are given. Assume that acute angle A and adjacent side b are given. As a result of $h = b \sin A$, h is also determined. Depending on the length of the opposite side, a, we have the four cases shown in Fig. A.24.

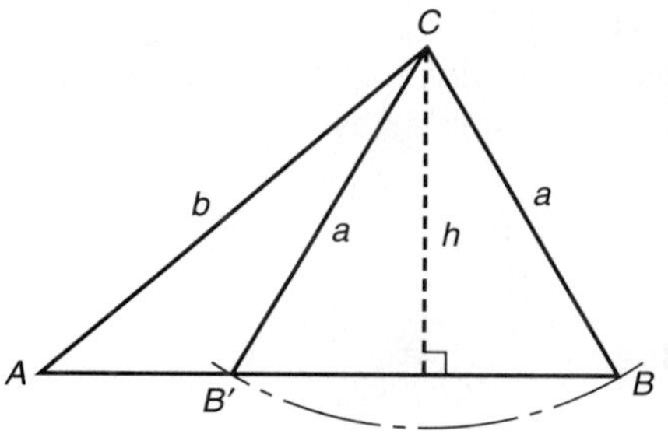

(a) When $h < a < b$, there are two possible triangles. In words, when the side opposite the given *acute* angle is less than the known adjacent side but greater than the altitude, there are two possible triangles.

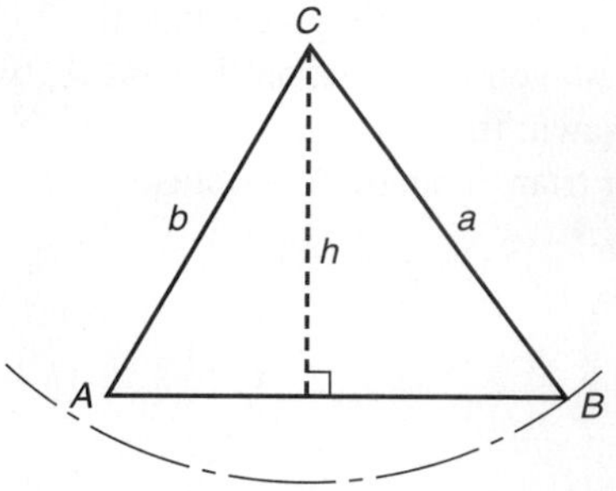

(b) When $h < b < a$, there is only one possible triangle. In words, when the side opposite the given *acute* angle is greater than the known adjacent side, there is only one possible triangle.

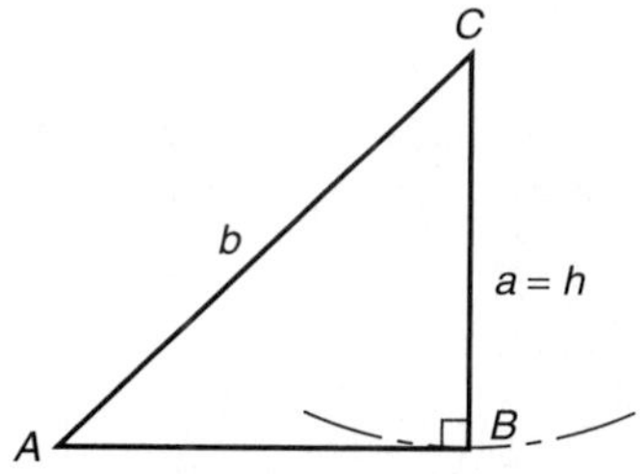

(c) When $a = h$, there is one possible (right) triangle. In words, when the side opposite the given *acute* angle equals the length of the altitude, there is only one possible (right) triangle.

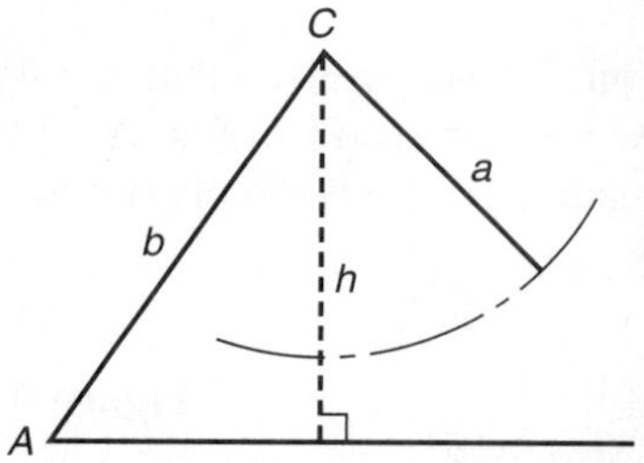

(d) When $a < h$, there is no possible triangle. In words, when the side opposite the given *acute* angle is less than the length of the altitude, there is no possible triangle.

Figure A.24 Possible triangles when two sides and an acute angle opposite one of the sides are given

If angle A is *obtuse*, we have two possible cases (Fig. A.25).

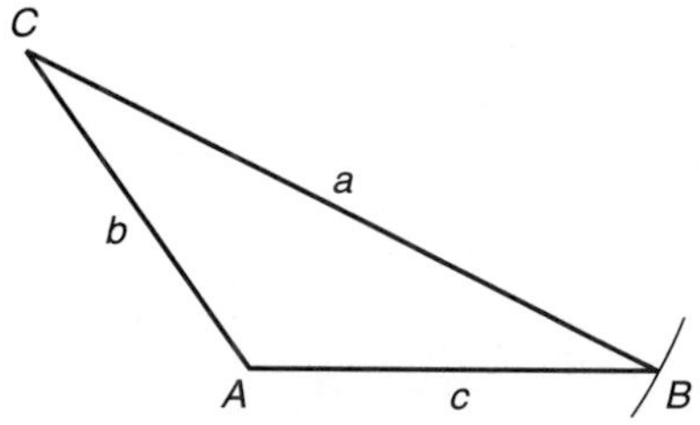

(a) When $a > b$, there is one possible triangle. In words, when the side opposite the given *obtuse* angle is greater than the known adjacent side, there is only one possible triangle.

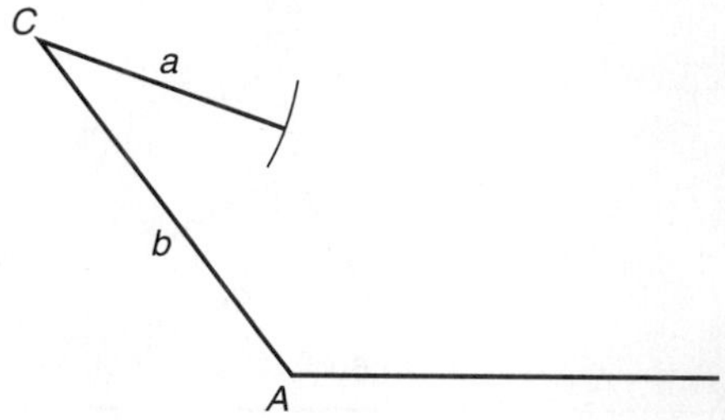

(b) When $a \leq b$, there is no possible triangle. In words, when the side opposite the given *obtuse* angle is less than or equal to the known adjacent side, there is no possible triangle.

Figure A.25 Possible triangles when two sides and an obtuse angle opposite one of the sides are given

Note: If the given parts are not angle A, side opposite a, and side adjacent b as in our preceding discussions, then you must substitute the given angle and sides accordingly. This is why it is so important to understand the general word description corresponding to each case.

EXAMPLE 6

If $A = 26°$, $a = 25$ cm, and $b = 41$ cm, solve the triangle.

First, find h:

$$h = b \sin A = (41 \text{ cm})(\sin 26°) = 18 \text{ cm}$$

Since $h < a < b$, there are two solutions. First, let's find B in triangle ACB in Fig. A.26:

$$\frac{a}{\sin A} = \frac{b}{\sin B}$$

$$\frac{25 \text{ cm}}{\sin 26°} = \frac{41 \text{ cm}}{\sin B}$$

$$\sin B = \frac{(41 \text{ cm})(\sin 26°)}{25 \text{ cm}} = 0.7189$$

$$B = 46°$$

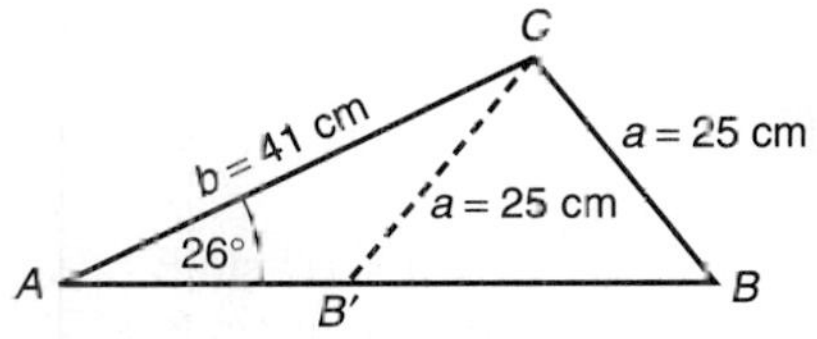

Figure A.26

41 [×] [sin] 26 [)] [÷] 25 [=] [sin⁻¹] [ANS] [=]

45.96610103

$$C = 180° - 26° - 46° = 108°$$

Find c:

$$\frac{c}{\sin C} = \frac{a}{\sin A}$$

$$\frac{c}{\sin 108°} = \frac{25 \text{ cm}}{\sin 26°}$$

$$c = \frac{(25 \text{ cm})(\sin 108°)}{\sin 26°} = 54 \text{ cm}$$

Therefore, the first solution is $B = 46°$, $C = 108°$, and $c = 54$ cm.

The second solution occurs when B is obtuse, as in triangle ACB'. That is, find the obtuse angle whose sine is 0.7189. Write

$$B' = 180° - 46° = 134°$$

Then $C = 180° - 26° - 134° = 2\bar{0}°$.

For c,

$$\frac{c}{\sin C} = \frac{a}{\sin A}$$

$$\frac{c}{\sin 2\bar{0}°} = \frac{25 \text{ cm}}{\sin 26°}$$

$$c = \frac{(25 \text{ cm})(\sin 2\bar{0}°)}{\sin 26°} = 2\bar{0} \text{ cm}$$

The second solution is $B' = 134°$, $C = 2\bar{0}°$, and $c = 2\bar{0}$ cm.

EXAMPLE 7

If $A = 62.0°$, $a = 415$ m, and $b = 855$ m, solve the triangle.

First, find h:

$$h = b \sin A$$
$$h = (855 \text{ m})(\sin 62.0°)$$
$$= 755 \text{ m}$$

Since $a < h$, there is no possible solution. What would happen if you applied the law of sines anyway? You would obtain

$$\frac{a}{\sin A} = \frac{b}{\sin B}$$

$$\frac{415 \text{ m}}{\sin 62.0°} = \frac{855 \text{ m}}{\sin B}$$

$$\sin B = \frac{(855 \text{ m})(\sin 62.0°)}{415 \text{ m}} = 1.819 \qquad \text{(Tilt!)}$$

Note: sin B = 1.819 is impossible because $-1 \leq \sin B \leq 1$.

In summary:

1. ***Given two angles and one side (AAS):*** There is only one possible triangle.
2. ***Given two sides and an angle opposite one of them (SSA):*** There are three possibilities. If the side opposite the given angle is:
   (a) greater than the known adjacent side, there is only one possible triangle.
   (b) less than the known adjacent side but greater than the altitude, there are two possible triangles.
   (c) less than the altitude, there is no possible triangle.

Since solving a general triangle requires several operations, errors are often introduced. The following points may be helpful in avoiding some of these errors:

1. Always choose a given value over a calculated value when doing calculations.
2. Always check your results to see that the largest angle is opposite the largest side and the smallest angle is opposite the smallest side.
3. Avoid finding the largest angle by the law of sines whenever possible because it is often not clear whether the resulting angle is acute or obtuse.

Law of Cosines

When the law of sines cannot be used, we use the **law of cosines.** In words, the square of any side of a triangle is equal to the sum of the squares of the other two sides minus twice the product of these two sides and the cosine of their included angle (see Fig. A.27). By formula, the law is stated as follows.

LAW OF COSINES

$$a^2 = b^2 + c^2 - 2bc \cos A$$
$$b^2 = a^2 + c^2 - 2ac \cos B$$
$$c^2 = a^2 + b^2 - 2ab \cos C$$

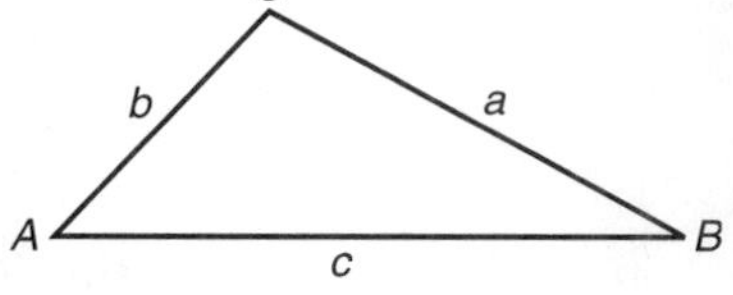

Figure A.27

There are two cases when the law of sines does not apply and we use the law of cosines to solve triangles:

1. Two sides and the included angle are known (SAS).
2. All three sides are known (SSS).

Do you see that when the law of cosines is used, there is no possibility of an ambiguous case? If not, draw a few triangles for each of these two cases (SAS and SSS) to convince yourself intuitively.

EXAMPLE 8

If $a = 112$ m, $b = 135$ m, and $C = 104.3°$, solve the triangle.

First, draw a triangle as in Fig. A.28 and find c by using the law of cosines:

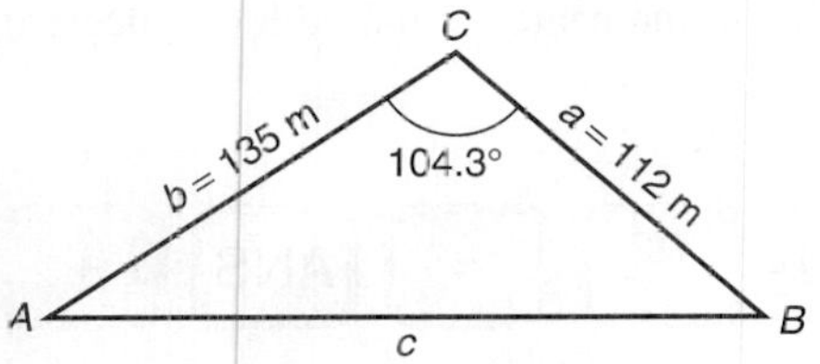

Figure A.28

$$c^2 = a^2 + b^2 - 2ab\cos C$$
$$c^2 = (112\text{ m})^2 + (135\text{ m})^2 - 2(112\text{ m})(135\text{ m})(\cos 104.3°)$$
$$c = 196\text{ m}$$

This side may be found using a calculator as follows:

112 [x^2] [+] 135 [x^2] [−] 2 [×] 112 [×] 135 [×]

[cos] 104.3 [=] [√] [ANS] [=]

195.5460308

So $c = 196$ m rounded to three significant digits.

To find A, use the law of sines since it requires less computation:

$$\frac{a}{\sin A} = \frac{c}{\sin C}$$
$$\frac{112\text{ m}}{\sin A} = \frac{196\text{ m}}{\sin 104.3°}$$
$$\sin A = \frac{(112\text{ m})(\sin 104.3°)}{196\text{ m}} = 0.5537$$
$$A = 33.6°$$
$$B = 180° - 104.3° - 33.6° = 42.1°$$

The solution is $A = 33.6°$, $B = 42.1°$, and $c = 196$ m.

EXAMPLE 9

If $a = 375.0$ ft, $b = 282.0$ ft, and $c = 114.0$ ft, solve the triangle.

First, draw a triangle as in Fig. A.29 and find A by using the law of cosines:

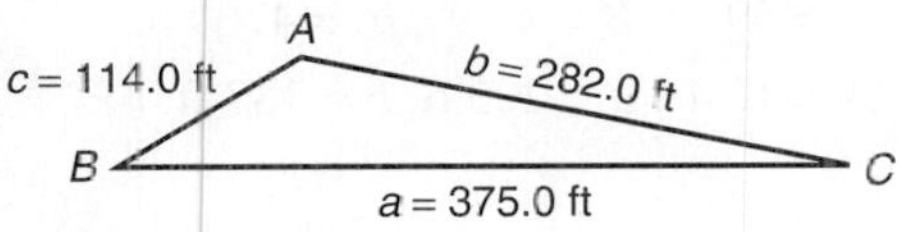

Figure A.29

$$a^2 = b^2 + c^2 - 2bc\cos A$$
$$(375.0\text{ ft})^2 = (282.0\text{ ft})^2 + (114.0\text{ ft})^2 - 2(282.0\text{ ft})(114.0\text{ ft})\cos A$$

$$\cos A = \frac{(375.0\text{ ft})^2 - (282.0\text{ ft})^2 - (114.0\text{ ft})^2}{-2(282.0\text{ ft})(114.0\text{ ft})}$$

$$\cos A = -0.7482$$

$$A = 138.43° \qquad \text{(Rounded to the nearest hundredth of a degree)}$$

375 [x^2] [−] 282 [x^2] [−] 114 [x^2] [=] [ANS] [÷]

(−) 2 [÷] 282 [÷] 114 [=] [$\cos^{-1}$] [ANS] [=]

138.432994

Next, to find B, use the law of sines:

$$\frac{a}{\sin A} = \frac{b}{\sin B}$$

$$\frac{375.0\text{ ft}}{\sin 138.43°} = \frac{282.0\text{ ft}}{\sin B}$$

$$\sin B = \frac{(282.0\text{ ft})(\sin 138.43°)}{375.0\text{ ft}} = 0.4990$$

$$B = 29.93°$$

$$C = 180° - 138.43° - 29.93° = 11.64°$$

The solution is $A = 138.43°$, $B = 29.93°$, and $C = 11.64°$.

PROBLEMS A.9

Solve each triangle using the labels as shown in Fig. A.30*.

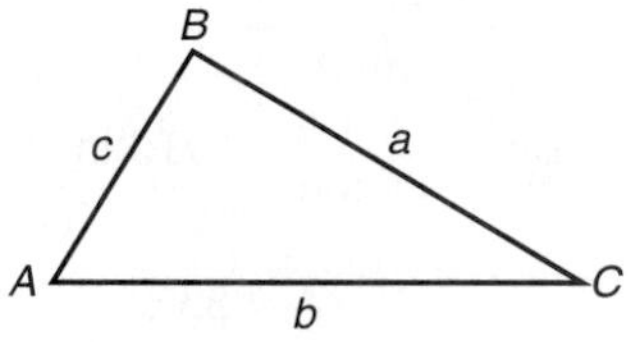

Figure A.30

Express the lengths of sides to three significant digits and the angles to the nearest tenth of a degree.

1. $A = 69.0°$, $a = 25.0$ m, $b = 16.5$ m
2. $C = 57.5°$, $c = 166$ mi, $b = 151$ mi
3. $B = 61.4°$, $b = 124$ cm, $c = 112$ cm
4. $A = 19.5°$, $a = 487$ km, $c = 365$ km
5. $B = 75.3°$, $A = 57.1°$, $b = 257$ ft
6. $C = 59.6°$, $B = 43.9°$, $b = 4760$ m
7. $A = 115.0°$, $a = 5870$ m, $b = 4850$ m
8. $A = 16.4°$, $a = 205$ ft, $b = 187$ ft

*Because of differences in rounding, your answers may differ slightly from the answers in the text if you choose to solve for the parts of a triangle in an order different from that chosen by the authors.

Express the lengths of sides to four significant digits and the angles to the nearest hundredth of a degree.

9. $C = 72.58°$, $b = 28.63$ cm, $c = 42.19$ cm
10. $A = 58.95°$, $a = 3874$ m, $c = 2644$ m
11. $B = 28.76°$, $C = 19.30°$, $c = 39{,}750$ mi
12. $A = 35.09°$, $B = 48.64°$, $a = 8.362$ km

Express the lengths of sides to two significant digits and the angles to the nearest degree.

13. $A = 25°$, $a = 5\overline{0}$ cm, $b = 4\overline{0}$ cm
14. $B = 42°$, $b = 5.3$ km, $c = 4.6$ km
15. $C = 8°$, $c = 16$ m, $a = 12$ m
16. $A = 105°$, $a = 460$ mi, $c = 380$ mi

For each general triangle, (a) determine the number of solutions and (b) solve the triangle, if possible. Express the lengths of sides to three significant digits and the angles to the nearest tenth of a degree.

17. $A = 37.0°$, $a = 21.5$ cm, $b = 16.4$ cm
18. $B = 55.0°$, $b = 182$ m, $c = 203$ m
19. $C = 26.5°$, $c = 42.7$ km, $a = 47.2$ km
20. $B = 40.4°$, $b = 81.4$ m, $c = 144$ m
21. $A = 71.5°$, $a = 3.45$ m, $c = 3.50$ m
22. $C = 17.2°$, $c = 2.20$ m, $b = 2.00$ m
23. $B = 105.0°$, $b = 16.5$ mi, $a = 12.0$ mi
24. $A = 98.8°$, $a = 707$ ft, $b = 585$ ft

Express the lengths of sides to two significant digits and the angles to the nearest degree.

25. $C = 18°$, $c = 24$ mi, $a = 45$ mi
26. $B = 36°$, $b = 75$ cm $a = 95$ cm
27. $C = 6\overline{0}°$, $c = 150$ m, $b = 180$ m
28. $A = 3\overline{0}°$, $a = 4800$ ft, $c = 3600$ ft
29. $B = 8°$, $b = 450$ m, $c = 850$ m
30. $B = 45°$, $c = 2.5$ m, $b = 3.2$ m

Express the lengths of sides to four significant digits and the angles to the nearest hundredth of a degree.

31. $B = 41.50°$, $b = 14.25$ km, $a = 18.50$ km
32. $A = 15.75°$, $a = 642.5$ m, $c = 592.7$ m
33. $C = 63.85°$, $c = 29.50$ cm, $b = 38.75$ cm
34. $B = 50.00°$, $b = 41{,}250$ km, $c = 45{,}650$ km
35. $C = 8.75°$, $c = 89.30$ m, $a = 61.93$ m
36. $A = 31.50°$, $a = 375.0$ mm, $b = 405.5$ mm

Express the lengths of sides to three significant digits and the angles to the nearest tenth of a degree.

37. $A = 60.0°$, $b = 19.5$ m, $c = 25.0$ m
38. $B = 19.5°$, $a = 21.5$ ft, $c = 12.5$ ft
39. $C = 109.0°$, $a = 14\overline{0}$ km, $b = 215$ km
40. $A = 94.7°$, $c = 875$ yd, $b = 185$ yd
41. $a = 19.2$ m, $b = 21.3$ m, $c = 27.2$ m
42. $a = 125$ km, $b = 195$ km, $c = 145$ km
43. $a = 4.25$ ft, $b = 7.75$ ft, $c = 5.50$ ft
44. $a = 3590$ m, $b = 7950$ m, $c = 4650$ m

Express the lengths of sides to two significant digits and the angles to the nearest degree.

45. $A = 45°$, $b = 51$ m, $c = 39$ m
46. $B = 6\overline{0}°$, $a = 160$ cm, $c = 230$ cm
47. $a = 7\overline{0}00$ m, $b = 5600$ m, $c = 4800$ m
48. $a = 5.8$ cm, $b = 5.8$ cm, $c = 9.6$ cm
49. $C = 135°$, $a = 36$ ft, $b = 48$ ft
50. $A = 5°$, $b = 19$ m, $c = 25$ m

Express the lengths of sides to four significant digits and the angles to the nearest hundredth of a degree.

51. $B = 19.25°$, $a = 4815$ m, $c = 1925$ m
52. $C = 75.00°$, $a = 37{,}550$ mi, $b = 45{,}250$ mi
53. $C = 108.75°$, $a = 405.0$ mm, $b = 325.0$ mm
54. $A = 111.05°$, $b = 1976$ ft, $c = 325\bar{0}$ ft
55. $a = 207.5$ km, $b = 105.6$ km, $c = 141.5$ km
56. $a = 19.45$ m, $b = 36.50$ m, $c = 25.60$ m

APPENDIX B

SCIENTIFIC CALCULATOR

There are several kinds and brands of calculators. Some are very simple to use; others do more difficult calculations. We demonstrate various operations on common basic scientific calculators that use algebraic logic, which follows the steps commonly used in mathematics. Yours may differ. If so, consult your manual.

To demonstrate how to use a calculator, we show what buttons are pushed and the order in which they are pushed. We assume that you know how to add, subtract, multiply, and divide on the calculator.

B.1 Scientific Notation

Numbers expressed in scientific notation can be entered into many calculators. The results may then also be given in scientific notation.

EXAMPLE 1

Multiply $(6.5 \times 10^{8})(1.4 \times 10^{-15})$ and write the result in scientific notation.

6.5 [EE] 8 [×] 1.4 [EE] [(−)] 15 [=]

[9.1 x 10^-7]

The result is 9.1×10^{-7}.

EXAMPLE 2

Divide $\dfrac{3.24 \times 10^{-5}}{7.2 \times 10^{-12}}$ and write the result in scientific notation.

3.24 [EE] [(−)] 5 [÷] 7.2 [EE] [(−)] 12 [=]

[4.5 x 10^6]

The result is 4.5×10^{6}.

EXAMPLE 3

Find the value of $\dfrac{(-6.3 \times 10^{4})(-5.07 \times 10^{-9})(8.11 \times 10^{-6})}{(5.63 \times 10^{12})(-1.84 \times 10^{7})}$ and write the result in scientific notation rounded to three significant digits.

(−) 6.3 EE 4 × (−) 5.07 EE (−) 9 × 8.11 EE (−) 6 ÷ 5.63 EE 12 ÷ (−) 1.84 EE 7 =

-2.500593772 10-29

The result rounded to three significant digits is -2.50×10^{-29}.

B.2 Squares and Square Roots

EXAMPLE 1

Find the value of $(46.8)^2$.

46.8 x^2 =

2190.24

The result is 2190.24.

EXAMPLE 2

Find the value of $(6.3 \times 10^{-18})^2$.

6.3 EE (−) 18 x^2 =

3.969 x 10-35

The result is 3.969×10^{-35}.

EXAMPLE 3

Find the value of $\sqrt{158.65}$ and round to four significant digits.

√ 158.65 =

12.59563416

The result rounded to four significant digits is 12.60.

EXAMPLE 4

Find the value of $\sqrt{6.95 \times 10^{-15}}$ and round to three significant digits.

√ 6.95 EE (−) 15 =

8.336666 x 10-8

The result rounded to three significant digits is 8.34×10^{-8}.

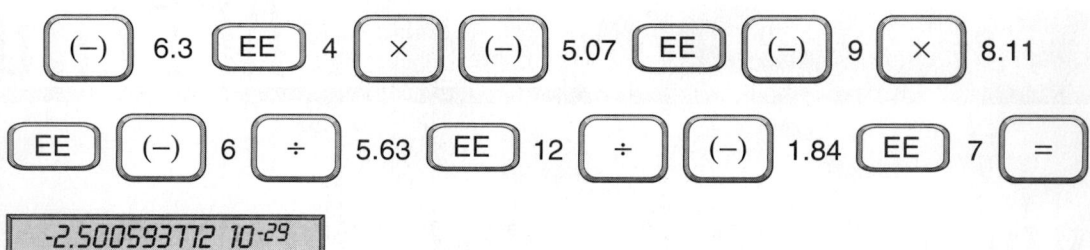

EXAMPLE 5

Find the value of $\sqrt{15.7^2 + 27.6^2}$ and round to three significant digits.

[$\sqrt{\ }$] 15.7 [x^2] [+] 27.6 [x^2] [=]

`31.75295262`

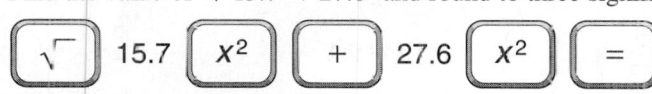

The result is 31.8 rounded to three significant digits.

Note: You may need to use parentheses with some calculators.

EXAMPLE 6

Find the value of $\frac{14}{\sqrt{5}} - \sqrt{\frac{15}{8}}$ and round the result to three significant digits.

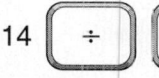

14 [÷] [$\sqrt{\ }$] 5 [ )]* [−] [$\sqrt{\ }$] 15 [÷] 8 [ )]* [=]

`4.891583543`

The result is 4.89 rounded to three significant digits.

EXAMPLE 7

Find the value of $\sqrt{\left(\frac{16}{1.3}\right)^2 + \left[\frac{1}{2\pi(60)(6 \times 10^{-5})}\right]^2}$ rounded to three significant digits.

[$\sqrt{\ }$] [( ] 16 [÷] 1.3 [ )] [x^2] [+] [( ] [( ] 2
[×] [π] [×] 60 [×] 6 [EE] [(−)] 5 [ )] [x^{-1}]
[ )] [x^2] [ )] [=]

`45.89092973`

Note: You will need to insert parentheses to clarify the order of operations. The result is 45.9 rounded to three significant digits. You may need to supply other parentheses.

B.3 Trigonometric Operations

Calculators must have sine, cosine, and tangent buttons for use in this book.

EXAMPLE 1

Find sin 26° rounded to four significant digits.

[sin] 26 [=]

`0.4383711467`

That is, sin 26° = 0.4384 rounded to four significant digits.

**Note:* You may need to insert a right parenthesis to clarify the order of operations. The square root key sometimes includes the left parenthesis. If not, you may need to supply it.

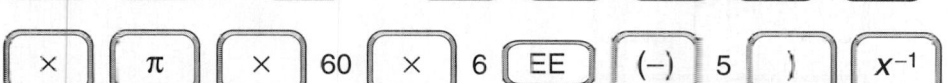

EXAMPLE 2

Find cos 36.75° rounded to four significant digits.

[cos] 36.75 [=]

0.801253813

That is, cos 36.75° = 0.8013 rounded to four significant digits.

EXAMPLE 3

Find tan 70.6° rounded to four significant digits.

[tan] 70.6 [=]

2.839653913

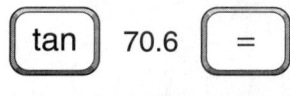

That is, tan 70.6° = 2.840 rounded to four significant digits.

Since we use right triangles almost exclusively, we show how to find the angle of a right triangle when the value of a given trigonometric ratio is known. That is, we will find angle A when $0° \leq A \leq 90°$.

EXAMPLE 4

Given sin A = 0.4321, find angle A to the nearest tenth of a degree.

[$\sin^{-1}$] .4321 [=]

25.60090542

Note: Make certain your calculator is in the degree mode.
Thus, A = 25.6° to the nearest tenth of a degree.

EXAMPLE 5

Given cos B = 0.6046, find angle B to the nearest tenth of a degree.

[$\cos^{-1}$] .6046 [=]

52.79993633

Thus, B = 52.8° to the nearest tenth of a degree.

EXAMPLE 6

Given tan A = 2.584, find angle A to the nearest tenth of a degree.

[$\tan^{-1}$] 2.584 [=]

68.8437168

Thus, A = 68.8° to the nearest tenth of a degree.

Trigonometric functions often occur in expressions that must be evaluated.

EXAMPLE 7

Given $a = (\tan 54°)(25.6\text{ m})$, find a rounded to three significant digits.

[tan] 54 [)] [×] 25.6 [=]

[35.23537716]

Thus, $a = 35.2$ m rounded to three significant digits.

EXAMPLE 8

Given $b = \dfrac{452\text{ m}}{\cos 37.5°}$, find b rounded to three significant digits.

452 [÷] [cos] 37.5 [=]

[569.7335311]

Thus, $b = 5\overline{7}0$ m rounded to three significant digits.

B.4 Finding a Power

To raise a number to a power, use the [^] button as follows.

EXAMPLE 1

Find the value of 4^5.

4 [^] or [y^x] 5 [=]

[1024]

That is, $4^5 = 1024$.

EXAMPLE 2

Find the value of 1.5^{-4} rounded to three significant digits.

1.5 [^] or [y^x] [(−)] 4 [=]

[0.197530864]

That is, $1.5^{-4} = 0.198$ rounded to three significant digits.

PROBLEMS B.4

Do as indicated and round each result to three significant digits.

1. $(6.43 \times 10^3)(5.16 \times 10^{10})$
2. $(4.16 \times 10^{-5})(3.45 \times 10^{-7})$
3. $(1.456 \times 10^{12})(-4.69 \times 10^{-18})$
4. $(-5.93 \times 10^9)(7.055 \times 10^{-12})$

5. $(7.45 \times 10^8) \div (8.92 \times 10^{18})$

6. $(1.38 \times 10^{-6}) \div (4.324 \times 10^6)$

7. $\dfrac{-6.19 \times 10^{12}}{7.755 \times 10^{-8}}$

8. $\dfrac{1.685 \times 10^{10}}{1.42 \times 10^{24}}$

9. $\dfrac{(5.26 \times 10^{-8})(8.45 \times 10^6)}{(-6.142 \times 10^9)(1.056 \times 10^{-12})}$

10. $\dfrac{(-2.35 \times 10^{-9})(1.25 \times 10^{11})(4.65 \times 10^{17})}{(8.75 \times 10^{23})(-5.95 \times 10^{-6})}$

11. $(68.4)^2$

12. $(3180)^2$

13. $\sqrt{46{,}500}$

14. $\sqrt{0.000634}$

15. $(1.45 \times 10^5)^2$

16. $(1.095 \times 10^{-18})^2$

17. $\sqrt{4.63 \times 10^{18}}$

18. $\sqrt{9.49 \times 10^{-15}}$

19. $\sqrt{(4.68)^2 + (9.63)^2}$

20. $\sqrt{(18.4)^2 - (6.5)^2}$

21. $18\sqrt{3} + \left(\dfrac{28.1}{19}\right)^2$

22. $\dfrac{8}{\sqrt{2}} + \sqrt{\dfrac{58}{14.5}}$

23. $25^2 - \sqrt{\dfrac{29.8}{0.0256}}$

24. $\dfrac{18.3}{6\sqrt{5}} - \left(\dfrac{225}{147}\right)^2$

25. $(12.6^2 + 21.5^2)^2 + (34.2^2 - 26.4^2)^2$

26. $\sqrt{21.4^2 + 18.7^2} + \sqrt{31.5^2 - 16.3^2}$

27. $\dfrac{91.4 - 48.6}{91.4 - 15.9}$

28. $\dfrac{14.7 + 9.6}{45.7 + 68.2}$

29. $\sqrt{\left(\dfrac{80.5}{25.6}\right)^2 + \left[\dfrac{1}{2\pi(60)(1.5 \times 10^{-7})}\right]^2}$

30. $\sqrt{\left(\dfrac{175}{36.5}\right)^2 + \left[\dfrac{1}{2\pi(60)(8.5 \times 10^{-10})}\right]^2}$

31. $\dfrac{(17.2)(11.6) + (8)(17.6) - (6)(16)}{(5)(15) + (8.5)(15) + (10)(26.5)}$

32. $\dfrac{(18.8)(5.5) + (7.75)(16.5) - (9.25)(13.85)}{(6.25)(12.5) + (4.75)(16.5) + (11.5)(14.1)}$

33. $\sin 13^\circ$

34. $\cos 22^\circ$

35. $\tan 52.3^\circ$

36. $\tan 31.25^\circ$

37. $\cos 59.36^\circ$

38. $\sin 84.55^\circ$

39. $\sin 48^\circ$

40. $\cos 48^\circ$

41. $\tan 75^\circ$

42. $\sin 8^\circ$

43. $\sin 8.7^\circ$

44. $\cos 35^\circ$

Find each angle rounded to the nearest tenth of a degree.

45. $\sin A = 0.6527$

46. $\cos B = 0.2577$

47. $\tan A = 0.4568$

48. $\sin B = 0.4658$

49. $\cos A = 0.5563$

50. $\tan B = 1.496$

51. $\sin B = 0.1465$

52. $\cos A = 0.4968$

53. $\tan B = 1.987$

54. $\sin A = 0.2965$

55. $\cos B = 0.3974$

56. $\tan A = 0.8885$

Find each angle to the nearest tenth of a degree between 0° and 90° and each side to three significant digits.

57. $b = (\sin 58.2^\circ)(296\text{ m})$

58. $a = (\cos 25.2^\circ)(54.5\text{ m})$

59. $c = \dfrac{37.5\text{ m}}{\cos 65.2^\circ}$

60. $b = \dfrac{59.7\text{ m}}{\tan 41.2^\circ}$

61. $\tan A = \dfrac{512\text{ km}}{376\text{ km}}$

62. $\cos B = \dfrac{75.2\text{ m}}{89.5\text{ m}}$

63. $a = (\cos 19.5^\circ)(15.7\text{ cm})$

64. $c = \dfrac{235\text{ km}}{\sin 65.2^\circ}$

65. $b = \dfrac{36.7\text{ m}}{\tan 59.2^\circ}$

66. $a = (\tan 5.7^\circ)(135\text{ m})$

Find the value of each power and round each result to three significant digits.

67. 12^4

68. 1.8^3

69. 0.46^5

70. 9^{-3}

71. 14^{-5}

72. 0.65^{-4}

APPENDIX C

TABLES

Table 1 U.S. Weights and Measures

Units of Length	Units of Volume	Units of Weight
Standard unit—inch (in. or ″) 12 inches = 1 foot (ft or ′) 3 feet = 1 yard (yd) $5\frac{1}{2}$ yards or $16\frac{1}{2}$ feet = 1 rod (rd) 5280 feet = 1 mile (mi)	*Liquid* 16 ounces (fl oz) = 1 pint (pt) 2 pints = 1 quart (qt) 4 quarts = 1 gallon (gal) *Dry* 2 pints (pt) = 1 quart (qt) 8 quarts = 1 peck (pk) 4 pecks = 1 bushel (bu)	Standard unit—pound (lb) 16 ounces (oz) = 1 pound 2000 pounds = 1 ton (T)

Table 2 Conversion Table for Length

	cm	m	km	in.	ft	mi
1 centimetre =	1	10^{-2}	10^{-5}	0.394	3.28×10^{-2}	5.21×10^{-6}
1 metre =	100	1	10^{-3}	39.4	3.28	5.21×10^{-4}
1 kilometre =	10^5	1000	1	3.94×10^4	3280	0.621
1 inch =	2.54	2.54×10^{-2}	2.54×10^{-5}	1	8.33×10^{-2}	1.58×10^{-5}
1 foot =	30.5	0.305	3.05×10^{-4}	12	1	1.89×10^{-4}
1 mile =	1.61×10^5	1610	1.61	6.34×10^4	5280	1

Table 3 Conversion Table for Area

Metric	U.S.
$1\ m^2 = 10{,}000\ cm^2$ $= 1{,}000{,}000\ mm^2$ $1\ cm^2 = 100\ mm^2$ $= 0.0001\ m^2$ $1\ km^2 = 1{,}000{,}000\ m^2$	$1\ ft^2 = 144\ in^2$ $1\ yd^2 = 9\ ft^2$ $1\ rd^2 = 30.25\ yd^2$ $1\ acre = 160\ rd^2$ $= 4840\ yd^2$ $= 43{,}560\ ft^2$ $1\ mi^2 = 640$ acres

	m^2	cm^2	ft^2	in^2
1 square metre =	1	10^4	10.8	1550
1 square centimetre =	10^{-4}	1	1.08×10^{-3}	0.155
1 square foot =	9.29×10^{-2}	929	1	144
1 square inch =	6.45×10^{-4}	6.45	6.94×10^{-3}	1

1 circular mil = $5.07 \times 10^{-6}\ cm^2 = 7.85 \times 10^{-7}\ in^2$

1 hectare = $10{,}000\ m^2$ = 2.47 acres

Table 4 Conversion Table for Volume

Metric	U.S.
1 m^3 = 10^6 cm^3 1 cm^3 = 10^{-6} m^3 = 10^3 mm^3	1 ft^3 = 1728 in^3 1 yd^3 = 27 ft^3

	m^3	cm^3	L	ft^3	in^3
1 m^3 =	1	10^6	1000	35.3	6.10×10^4
1 cm^3 =	10^{-6}	1	1.00×10^{-3}	3.53×10^{-5}	6.10×10^{-2}
1 litre =	1.00×10^{-3}	1000	1	3.53×10^{-2}	61.0
1 ft^3 =	2.83×10^{-2}	2.83×10^4	28.3	1	1728
1 in^3 =	1.64×10^{-5}	16.4	1.64×10^{-2}	5.79×10^{-4}	1

1 U.S. fluid gallon = 4 U.S. fluid quarts = 8 U.S. pints = 128 U.S. fluid ounces = 231 in^3 = 0.134 ft^3
1 L = 1000 cm^3 = 1.06 qt   1 fl oz = 29.5 cm^3 1 ft^3 = 7.47 gal = 28.3 L

Table 5 Conversion Table for Mass

	g	kg	slug	oz	lb	ton
1 gram =	1	0.001	6.85×10^{-5}	3.53×10^{-2}	2.21×10^{-3}	1.10×10^{-6}
1 kilogram =	1000	1	6.85×10^{-2}	35.3	2.21	1.10×10^{-3}
1 slug =	1.46×10^4	14.6	1	515	32.2	1.61×10^{-2}
1 ounce =	28.4	2.84×10^{-2}	1.94×10^{-3}	1	6.25×10^{-2}	3.13×10^{-5}
1 pound =	454	0.454	3.11×10^{-2}	16	1	5.00×10^{-4}
1 ton =	9.07×10^5	907	62.2	3.2×10^4	2000	1

1 metric ton = 1000 kg = 2205 lb

Quantities in the shaded areas are not mass units. When we write, for example, 1 kg"="2.21 lb, this means that a kilogram is a mass that weighs 2.21 pounds under standard conditions of gravity ($g = 9.80 \text{ m/s}^2 = 32.2 \text{ ft/s}^2$).

Table 6 Conversion Table for Density

	$slug/ft^3$	kg/m^3	g/cm^3	lb/ft^3	lb/in^3
1 slug per ft^3 =	1	515	0.515	32.2	1.86×10^{-2}
1 kilogram per m^3 =	1.94×10^{-3}	1	0.001	6.24×10^{-2}	3.61×10^{-5}
1 gram per cm^3 =	1.94	1000	1	62.4	3.61×10^{-2}
1 pound per ft^3 =	3.11×10^{-2}	16.0	1.60×10^{-2}	1	5.79×10^{-4}
1 pound per in^3 =	53.7	2.77×10^4	27.7	1728	1

Quantities in the shaded areas are weight densities and, as such, are dimensionally different from mass densities.

Note that

$$D_w = D_m g$$

where

D_w = weight density
D_m = mass density
$g = 9.80 \text{ m/s}^2 = 32.2 \text{ ft/s}^2$

Table 7 Conversion Table for Time

	yr	day	h	min	s
1 year =	1	365	8.77×10^3	5.26×10^5	3.16×10^7
1 day =	2.74×10^{-3}	1	24	1440	8.64×10^4
1 hour =	1.14×10^{-4}	4.17×10^{-2}	1	60	3600
1 minute =	1.90×10^{-6}	6.94×10^{-4}	1.67×10^{-2}	1	60
1 second =	3.17×10^{-8}	1.16×10^{-5}	2.78×10^{-4}	1.67×10^{-2}	1

Table 8 Conversion Table for Speed

	ft/s	km/h	m/s	mi/h	cm/s
1 foot per second =	1	1.10	0.305	0.682	30.5
1 kilometre per hour =	0.911	1	0.278	0.621	27.8
1 metre per second =	3.28	3.60	1	2.24	100
1 mile per hour =	1.47	1.61	0.447	1	44.7
1 centimetre per second =	3.28×10^{-2}	3.60×10^{-2}	0.01	2.24×10^{-2}	1

1 mi/min = 88.0 ft/s = 60.0 mi/h

Table 9 Conversion Table for Force

	N	lb	oz
1 newton =	1	0.225	3.60
1 pound =	4.45	1	16
1 ounce =	0.278	0.0625	1

Table 10 Conversion Table for Power

	Btu/h	ft lb/s	hp	kW	W
1 British thermal unit per hour =	1	0.216	3.93×10^{-4}	2.93×10^{-4}	0.293
1 foot pound per second =	4.63	1	1.82×10^{-3}	1.36×10^{-3}	1.36
1 horsepower =	2550	550	1	0.746	746
1 kilowatt =	3410	738	1.34	1	1000
1 watt =	3.41	0.738	1.34×10^{-3}	0.001	1

Table 11 Conversion Table for Pressure

	atm	Inches of Water	mm Hg	N/m² (Pa)	lb/in²	lb/ft²
1 atmosphere =	1	407	76$\bar{0}$	1.01×10^5	14.7	2120
1 inch of water[a] at 4°C =	2.46×10^{-3}	1	1.87	249	3.61×10^{-2}	5.20
1 millimetre of mercury[a] at 0°C =	1.32×10^{-3}	0.535	1	133	1.93×10^{-2}	2.79
1 newton per metre² (pascal) =	9.87×10^{-6}	4.02×10^{-3}	7.50×10^{-3}	1	1.45×10^{-4}	2.09×10^{-2}
1 pound per in² =	6.81×10^{-2}	27.7	51.7	6.90×10^3	1	144
1 pound per ft² =	4.73×10^{-4}	0.192	0.359	47.9	6.94×10^{-3}	1

[a]Where the acceleration of gravity has the standard value, 9.80 m/s² = 32.2 ft/s².

Table 12 Mass and Weight Density

Substance	Mass Density (kg/m³)	Weight Density (lb/ft³)
Solids		
Copper	8,890	555
Iron	7,800	490
Lead	11,300	708
Aluminum	2,700	169
Ice	917	57
Wood, white pine	420	26
Concrete	2,300	140
Cork	240	15
Liquids		
Water	1,0$\bar{0}$0[a]	62.4
Seawater	1,025	64.0
Oil	870	54.2
Mercury	13,600	846
Alcohol	790	49.4
Gasoline	680	42.0
	At 0°C and 1 atm Pressure	**At 32°F and 1 atm Pressure**
Gases[b]		
Air	1.29	0.081
Carbon dioxide	1.96	0.123
Carbon monoxide	1.25	0.078
Helium	0.178	0.011
Hydrogen	0.0899	0.0056
Oxygen	1.43	0.089
Nitrogen	1.25	0.078
Ammonia	0.760	0.047
Propane	2.02	0.126

[a]Metric weight density of water = 98$\bar{0}$0 N/m³.
[b]The density of a gas is found by pumping the gas into a container, measuring its volume and mass or weight, and then using the appropriate density formula.

Table 13 Specific Gravity of Certain Liquids[a]

Liquid	Specific Gravity
Benzene	0.90
Ethyl alcohol	0.79
Gasoline	0.68
Kerosene	0.82
Mercury	13.5
Seawater	1.025
Sulfuric acid	1.84
Turpentine	0.87
Water	1.000

[a]At room temperature (20°C or 68°F).

Table 14 Conversion Table for Energy, Work, and Heat

	Btu	ft lb	J	cal	kWh
1 British thermal unit =	1	778	1060	252	2.93×10^{-4}
1 foot pound =	1.29×10^{-3}	1	1.36	0.324	3.77×10^{-7}
1 joule =	9.48×10^{-4}	0.738	1	0.239	2.78×10^{-7}
1 calorie =	3.97×10^{-3}	3.09	4.19	1	1.16×10^{-6}
1 kilowatt-hour =	3410	2.66×10^{6}	3.60×10^{6}	8.60×10^{5}	1

Table 15 Heat Constants

			Specific Heat		Heat of Fusion		Vaporization	
	Melting Point (°C)	Boiling Point (°C)	cal/g °C or kcal/kg °C or Btu/lb °F	J/kg °C	cal/g or kcal/kg	J/kg	cal/g or kcal/kg	J/kg
Alcohol, ethyl	−117	78.5	0.58	2400	24.9	1.04×10^{5}	204	8.54×10^{5}
Aluminum	660	2057	0.22	920	76.8	3.21×10^{5}		
Brass	840		0.092	390				
Copper	1083	2330	0.092	390	49.0	2.05×10^{5}		
Glass			0.21	880				
Ice	0		0.51	2100	$8\overline{0}$	3.35×10^{5}		
Iron (steel)	1540	3000	0.115	481	7.89	3.30×10^{4}		
Lead	327	1620	0.031	130	5.86	2.45×10^{4}		
Mercury	−38.9	357	0.033	140	2.82	1.18×10^{4}	65.0	2.72×10^{5}
Silver	961	1950	0.056	230	26.0	1.09×10^{5}		
Steam			0.48	$2\overline{0}00$				
Water (liquid)	0	$10\overline{0}$	1.00	4190	$8\overline{0}$	3.35×10^{5}	$54\overline{0}$	2.26×10^{6}
Zinc	419	907	0.092	390	23.0	9.63×10^{4}		

Table 16 Coefficient of Linear Expansion

Material	α (metric)	α (U.S.)
Aluminum	2.3×10^{-5}/C°	1.3×10^{-5}/F°
Brass	1.9×10^{-5}/C°	1.0×10^{-5}/F°
Concrete	1.1×10^{-5}/C°	6.0×10^{-6}/F°
Copper	1.7×10^{-5}/C°	9.5×10^{-6}/F°
Glass	9.0×10^{-6}/C°	5.1×10^{-6}/F°
Pyrex	3.0×10^{-6}/C°	1.7×10^{-6}/F°
Steel	1.3×10^{-5}/C°	6.5×10^{-6}/F°
Zinc	2.6×10^{-5}/C°	1.5×10^{-5}/F°

Table 17 Coefficient of Volume Expansion

Liquid	β (metric)	β (U.S.)
Acetone	1.49×10^{-3}/C°	8.28×10^{-4}/F°
Alcohol, ethyl	1.12×10^{-3}/C°	6.62×10^{-4}/F°
Carbon tetrachloride	1.24×10^{-3}/C°	6.89×10^{-4}/F°
Mercury	1.8×10^{-4}/C°	1.0×10^{-4}/F°
Petroleum	9.6×10^{-4}/C°	5.33×10^{-4}/F°
Turpentine	9.7×10^{-4}/C°	5.39×10^{-4}/F°
Water	2.1×10^{-4}/C°	1.17×10^{-4}/F°

Table 18 Charge

Charge on one electron = 1.60×10^{-19} coulomb

1 coulomb = 6.25×10^{18} electrons of charge

1 ampere-hour = 3600 C

Table 19 Relationships of Metric SI Base and Derived Units

This chart shows graphically how the 17 SI-derived units with special names are derived from the base and supplementary units. It was provided by the National Institute of Standards and Technology.

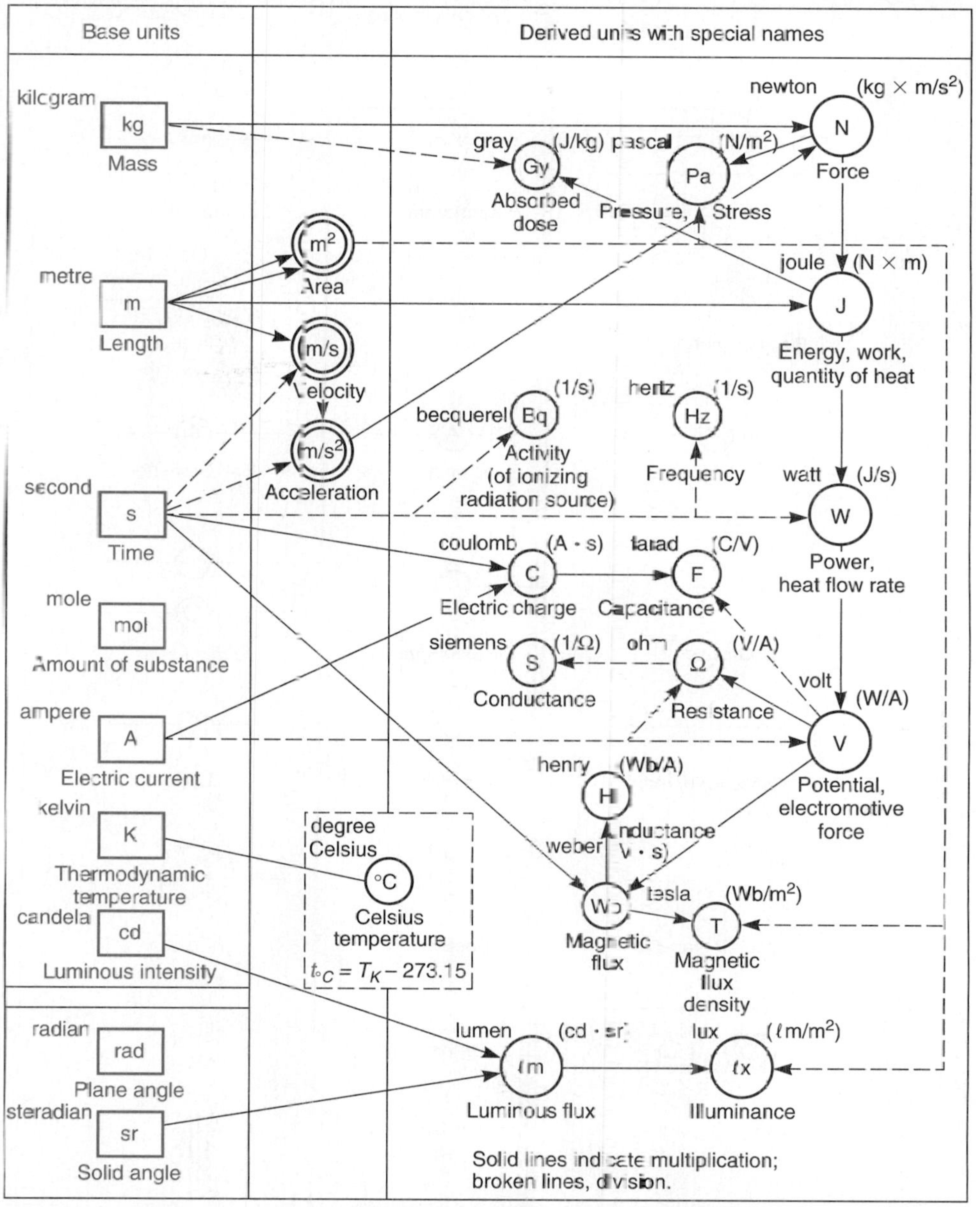

Table 20 Electric Symbols

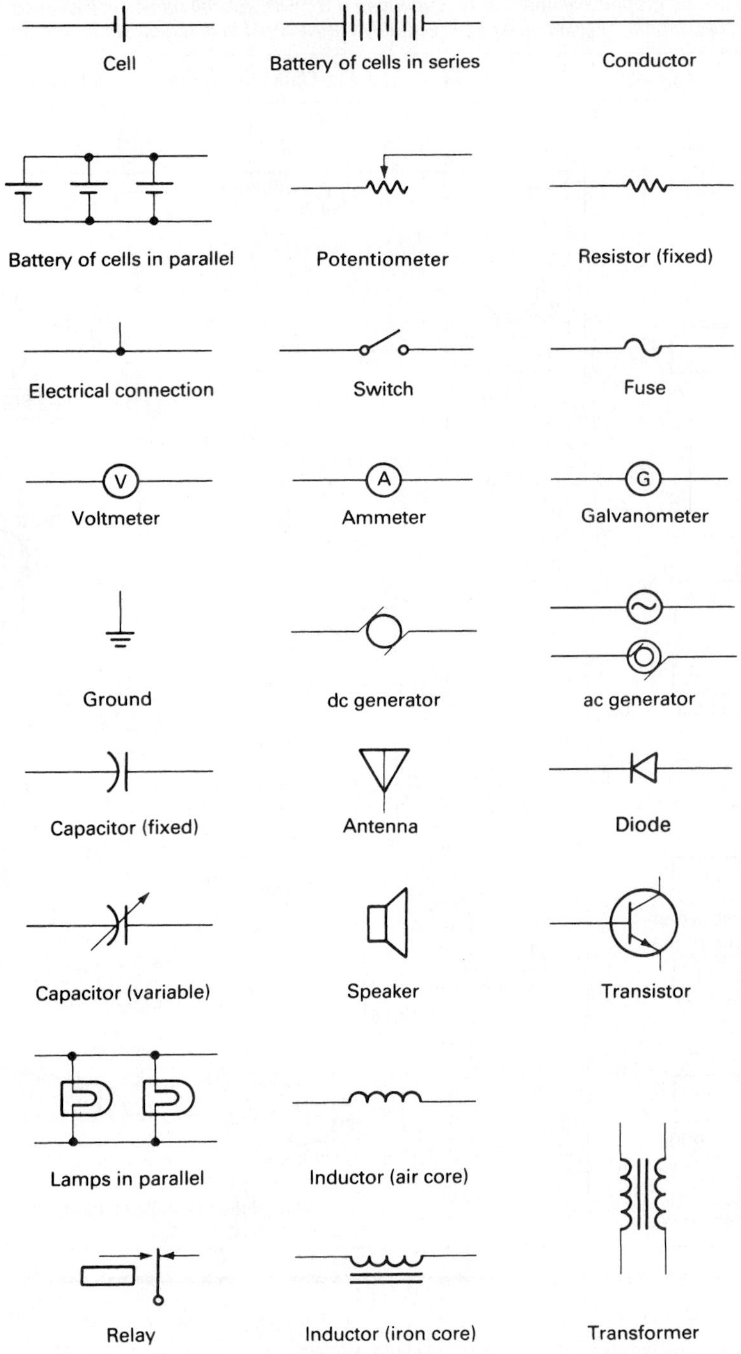
Cell
Battery of cells in series
Conductor
Battery of cells in parallel
Potentiometer
Resistor (fixed)
Electrical connection
Switch
Fuse
V
Voltmeter
A
Ammeter
G
Galvanometer
Ground
dc generator
ac generator
Capacitor (fixed)
Antenna
Diode
Capacitor (variable)
Speaker
Transistor
Lamps in parallel
Inductor (air core)
Relay
Inductor (iron core)
Transformer

Table 21 Periodic Table

Symbol — Cl 17 — Atomic number
Atomic mass* — 35.453
$3p^5$ — Electron configuration

Group I	*Group II*	*Transition Elements*										*Group III*	*Group IV*	*Group V*	*Group VI*	*Group VII*	*Group 0*
H 1 1.0079 $1s^1$																	He 2 4.00260 $1s^2$
Li 3 6.94 $2s^1$	Be 4 9.01218 $2s^2$											B 5 10.81 $2p^1$	C 6 12.011 $2p^2$	N 7 14.0067 $2p^3$	O 8 15.9994 $2p^4$	F 9 18.9984 $2p^5$	Ne 10 20.18 $2p^6$
Na 11 22.9898 $3s^1$	Mg 12 24.305 $3s^2$											Al 13 26.9815 $3p^1$	Si 14 28.0855 $3p^2$	P 15 30.974 $3p^3$	S 16 32.06 $3p^4$	Cl 17 35.453 $3p^5$	Ar 18 39.948 $3p^6$
K 19 39.0983 $4s^1$	Ca 20 40.08 $4s^2$	Sc 21 44.9559 $3d^14s^2$	Ti 22 47.9 $3d^24s^2$	V 23 50.9415 $3d^34s^2$	Cr 24 51.996 $3d^54s^1$	Mn 25 54.938 $3d^54s^2$	Fe 26 55.847 $3d^64s^2$	Co 27 58.9332 $3d^74s^2$	Ni 28 58.7 $3d^84s^2$	Cu 29 63.546 $3d^{10}4s^1$	Zn 30 65.39 $3d^{10}4s^2$	Ga 31 69.73 $4p^1$	Ge 32 72.6 $4p^2$	As 33 74.9216 $4p^3$	Se 34 78.96 $4p^4$	Br 35 79.904 $4p^5$	Kr 36 83.80 $4p^6$
Rb 37 85.47 $5s^1$	Sr 38 87.62 $5s^2$	Y 39 88.9059 $4d^15s^2$	Zr 40 91.22 $4d^25s^2$	Nb 41 92.9064 $4d^45s^1$	Mo 42 95.94 $4d^55s^1$	Tc 43 (98) $4d^55s^2$	Ru 44 101.07 $4d^75s^1$	Rh 45 102.906 $4d^85s^1$	Pd 46 106.4 $4d^{10}5s^0$	Ag 47 107.868 $4d^{10}5s^1$	Cd 48 112.41 $4d^{10}5s^2$	In 49 114.82 $5p^1$	Sn 50 118.7 $5p^2$	Sb 51 121.75 $5p^3$	Te 52 127.60 $5p^4$	I 53 126.90 $5p^5$	Xe 54 131.3 $5p^6$
Cs 55 132.905 $6s^1$	Ba 56 137.33 $6s^2$	57-71†	Hf 72 178.49 $5d^26s^2$	Ta 73 180.95 $5d^36s^2$	W 74 183.85 $5d^46s^2$	Re 75 186.207 $5d^56s^2$	Os 76 190.2 $5d^66s^2$	Ir 77 192.22 $5d^76s^2$	Pt 78 195.08 $5d^96s^1$	Au 79 196.97 $5d^{10}6s^1$	Hg 80 200.59 $5d^{10}6s^2$	Tl 81 204.38 $6p^1$	Pb 82 207.2 $6p^2$	Bi 83 208.980 $6p^3$	Po 84 (209) $6p^4$	At 85 (210) $6p^5$	Rn 86 (222) $6p^6$
Fr 87 (223) $7s^1$	Ra 88 226.025 $7s^2$	89-103‡	Rf 104 (261) $6d^27s^2$	Ha 105 (262) $6d^37s^2$	106 (263)	107 (262)	108 (265)	109 (266)									

†*Lanthanide series*

La 57 138.906 $5d^16s^2$	Ce 58 140.12 $5d^14f^16s^2$	Pr 59 140.908 $4f^36s^2$	Nd 60 144.24 $4f^46s^2$	Pm 61 (145) $4f^56s^2$	Sm 62 150.4 $4f^66s^2$	Eu 63 151.96 $4f^76s^2$	Gd 64 157.25 $5d^14f^76s^2$	Tb 65 158.925 $5d^14f^86s^2$	Dy 66 162.50 $4f^{10}6s^2$	Ho 67 164.930 $4f^{11}6s^2$	Er 68 167.26 $4f^{12}6s^2$	Tm 69 168.934 $4f^{13}6s^2$	Yb 70 173.04 $4f^{14}6s^2$	Lu 71 174.967 $5d^14f^{14}6s^2$

‡*Actinide series*

Ac 89 (227) $6d^17s^2$	Th 90 232.038 $6d^27s^2$	Pa 91 231.036 $5f^26d^17s^2$	U 92 238.029 $5f^36d^17s^2$	Np 93 237.048 $5f^46d^17s^2$	Pu 94 (244) $5f^66d^07s^2$	Am 95 (243) $5f^76d^07s^2$	Cm 96 (247) $5f^76d^17s^2$	Bk 97 (247) $5f^96d^07s^2$	Cf 98 (251) $5f^{10}6d^07s^2$	Es 99 (252) $5f^{11}6d^07s^2$	Fm 100 (257) $5f^{12}6d^07s^2$	Md 101 (258) $5f^{13}6d^07s^2$	No 102 (259) $6d^07s^2$	Lr 103 (260) $6d^17s^2$

*Atomic mass values averaged over isotopes in percentages they occur on the earth's surface. For many unstable elements, mass number of the most stable known isotope is given in parentheses.

Table 22 The Greek Alphabet

Capital	Lowercase	Name
Α	α	alpha
Β	β	beta
Γ	γ	gamma
Δ	δ	delta
Ε	ϵ	epsilon
Ζ	ζ	zeta
Η	η	eta
Θ	θ	theta
Ι	ι	iota
Κ	κ	kappa
Λ	λ	lambda
Μ	μ	mu
Ν	ν	nu
Ξ	ξ	xi
Ο	ο	omicron
Π	π	pi
Ρ	ρ	rho
Σ	σ	sigma
Τ	τ	tau
Υ	υ	upsilon
Φ	ϕ	phi
Χ	χ	chi
Ψ	ψ	psi
Ω	ω	omega

APPENDIX D

Glossary

Absolute Pressure The actual air pressure given by the gauge reading plus the normal atmospheric pressure.

Absolute Zero The lowest possible temperature.

Acceleration Change in velocity per unit time.

Acceleration Due to Gravity The acceleration of a freely falling object. On the earth's surface the acceleration due to gravity is 9.80 m/s^2 (metric) or 32.2 ft/s^2 (U.S.)

Accuracy The number of digits, called significant digits, in a measurement, which indicates the number of units that we are reasonably sure of having counted. The greater the number of significant digits, the better is the accuracy.

Actual Power A measure of the actual power available to be converted into other forms of energy.

Adhesion The force of attraction between different or unlike molecules.

Alpha Ray A ray consisting of alpha particles, each having two protons and two neutrons, and positively charged, that is, it curves in the direction that known positive charges curve in a magnetic field.

Alternating Current A current that flows in one direction in a conductor, changes direction, and then flows in the other direction.

Ammeter An instrument that measures the current flowing in a circuit.

Ampère's Rule To find the direction of a magnetic field near a current and a straight wire, hold the wire in your right hand with your thumb extended in the direction of the current. Your fingers circle the wire in the direction of the flux lines.

Amplification The process of increasing the strength of an electronic signal.

Amplitude The maximum displacement of any part of a wave or a vibration from its equilibrium, or rest, position.

Angular Acceleration The rate of change of angular velocity (change in angular velocity/time).

Angular Displacement The angle through which any point on a rotating body moves.

Angular Momentum For a rotating body about a fixed axis, the angular momentum is the product of the moment of inertia and the angular velocity of the body.

Angular Velocity The rate of angular displacement (angular displacement/time).

Apparent Power The product of the effective values of alternating current and voltage.

Approximate Number A number that has been determined by some measurement or estimation process.

Archimedes' Principle Any object placed in a fluid apparently loses weight equal to the weight of the displaced fluid.

Area The number of square units contained in a figure.

Armature The rotating coil or electromagnet in a generator.

Astronomy The branch of science that studies everything that takes place outside of the earth's atmosphere.

Atmospheric Pressure The pressure caused by the weight of the air in the atmosphere.

Atom The smallest particle of an element that can exist in a stable or independent state.

Atomic Mass The mass of a single atom, usually expressed in atomic mass units, u.

Atomic Mass Number The total number of nucleons (protons and neutrons) in a nucleus, A.

Atomic Mass Unit (u) A unit of measure of atomic mass based on the mass of the common carbon atom, which has six protons and six neutrons, so the mass of the carbon 12 atom has been given the exact value of 12 u. Thus, the approximate mass of a single proton or a single neutron is 1 u, and the mass of an atom in atomic mass units is simply the sum of the number of its protons and its neutrons.

Atomic Number The number of protons in a nucleus, Z.

Average Binding Energy per Nucleon The total binding energy of the nucleus divided by the total number of nucleons.

Becquerel (Bq) Unit of source activity; 1 Bq = 1 disintegration/s.

Bending Consists of both tension and compression stresses. It occurs when a force is placed on a beam causing it to sag.

Bernoulli's Principle For the horizontal flow of a fluid through a tube, the sum of the pressure and energy of motion (kinetic energy) per unit volume of the fluid is constant.

Beta Ray A ray consisting of a stream of beta particles (electrons), which are emitted from neutrons in a nucleus as they decay into protons and electrons. This ray is negatively charged and curves in the opposite direction of an alpha ray in a magnetic field.

Binding Energy of the Nucleus The total energy required to break a nucleus apart into separate nucleons.

Biology The branch of science that studies living organisms.

Bohr Model An early model of atomic structure in which electrons travel around the nucleus in a number of discrete stable energy levels determined by quantum conditions.

Boyle's Law If the temperature of a gas is constant, the volume is inversely proportional to the absolute pressure, $V/V' = P'/P$.

Brinell Method Common industrial method used to measure the hardness of a metal.

Btu (British thermal unit) The amount of heat (energy) necessary to raise the temperature of 1 lb of water 1°F.

Buoyant Force The upward force exerted on a submerged or partially submerged object.

Calorie The amount of heat necessary to raise the temperature of 1 g of water 1°C.

Capacitance The ratio of the charge on either plate of a capacitor to the potential difference between the plates.

Capacitive Reactance A measure of the opposition to ac current flow by a capacitor.

Capacitor A circuit component consisting of two parallel plates separated by a thin insulator used to build up and store charge.

Capillary Action The behavior of liquids that causes the liquid level in small tubes to be different than in larger tubes. This behavior is due both to adhesion of the liquid molecules with the tube and to the surface tension of the liquid.

Celsius Scale The metric temperature scale on which ice melts at 0° and water boils at 100°.

Center of Gravity The point of any body at which all of its weight can be considered to be concentrated.

Centripetal Force The force acting on a body in circular motion that causes it to move in a circular path. This force is exerted toward the center of the circle.

Chain Reaction The process of using the neutrons released in each nuclear fission reaction to create further reactions.

Change of Phase (sometimes called *change of state*) A change in a substance from one form of matter (solid, liquid, or gas) to another.

Charles' Law If the pressure on a gas is constant, the volume is directly proportional to its Kelvin or Rankine temperature, $V/T = V'/T'$.

Chemistry The branch of science that studies the composition, structure, properties, and reactions of matter.

Coefficient of Friction The ratio between the frictional force and the normal force of an object. The number represents how rough or smooth two surfaces are when moving across one another.

Coefficient of Linear Expansion A constant that indicates the amount by which a solid expands or contracts when its temperature is changed 1 degree.

Cohesion The force of attraction between like molecules that holds the closely packed molecules of a solid together.
Color A property of the light that reaches our eyes and is determined by its wavelength or its frequency.
Commutator A device in an ac generator that produces a direct current. Composed of a split ring that replaces the slip rings in an ac generator and produces a direct current in the circuit connected to the split ring of the generator.
Compass A small magnetic needle that is free to rotate on a bearing.
Complementary Colors Two colors that, when combined, form white; for example, cyan and red, magenta and green, and yellow and blue.
Component Vector When two or more vectors are added, each of the vectors is called a component of the resultant, or sum, vector.
Compound A substance containing two or more elements.
Compound Machine A combination of simple machines. Its total mechanical advantage is the product of the mechanical advantage of each simple machine.
Compression A stress caused by two forces acting directly toward each other. This stress tends to cause objects to become shorter and thicker.
Concave Mirror A mirror with a surface that curves away from an observer.
Concurrent Forces Two or more forces applied to, or acting at, the same point.
Condensation The change of phase from gas or vapor to a liquid.
Conduction A form of heat transfer from a warmer part of a substance to a cooler part as a result of molecular collisions, which cause the slower-moving molecules to move faster. A transfer of charge from one place to another.
Conductor A material through which an electron charge is readily transferred.
Constructive Interference The superposition of waves to form a larger disturbance (wave) in a medium. Occurs when two crests or troughs of superimposed waves meet.
Convection A form of heat transfer by the movement of warm molecules from one region of a gas or a liquid to another.
Converging Lens A lens that bends the light passing through it to some point beyond the lens. Converging lenses are thicker in the center.
Conversion Factor An expression used to convert from one set of units to another. Often expressed as a fraction whose numerator and denominator are equal to each other although in different units.
Convex Mirror A mirror with a surface that curves inward toward an observer.
Coulomb Force An electric, repulsive force between protons.
Coulomb's Law The force between two point charges is directly proportional to the product of their magnitudes and inversely proportional to the square of the distance between them.
Critical Angle The smallest angle of incidence at which all light striking a surface is totally internally reflected.
Curie (Ci) Unit of source activity; 1 Ci = 3.70×10^{10} disintegrations/s.
Current The flow of charge that passes through a conductor.
Curvilinear Motion Motion along a curved path.
Decay Rate The probability per unit time that a decay of radioactive isotopes will occur.
Deceleration An acceleration that indicates an object is slowing down.
Degree An angular unit of measure. Defined as 1/360 of one complete revolution.
Destructive Interference The superposition of waves to form a smaller disturbance (wave) in a medium.
Dew Point The temperature at which air becomes saturated with water vapor and condensation occurs.
Diffraction The property of a wave that describes its ability to bend around obstacles in its path.
Diffusion The process by which molecules of a gas mix with the molecules of a solid, a liquid, or another gas. Scattering of light by an uneven surface.

Diode A device that allows current to flow through it in only one direction.
Direct Current Current that flows in one direction.
Disintegration Energy The total energy released in radioactive decay in the form of kinetic energy.
Dispersion The spreading of white light into the full spectrum.
Displacement The net change in position of an object, or the direct distance and direction it moves; a vector.
Displacement The distance of an object is simple harmonic motion from its equilibrium or rest, position.
Diverging Lens A lens that bends the light passing through it so as to spread the light. Diverging lenses are thicker at the edges than at the center.
Doppler Effect The variation of the frequency heard when a source of sound and the ear are moving relative to each other.
Dry Cell A voltage-generating cell that consists of a chemical paste and two electrodes of unlike materials, one of which reacts chemically with the electrolyte.
Ductility A property of a metal that enables it to be drawn through a die to produce a wire.
Effective Value The number of amperes of alternating current that produce the same amount of heat in a resistance as an equal number of amperes of a steady direct current.
Efficiency The ratio of the work output to the work input of a machine.
Effort The force applied to a machine.
Effort Arm The distance from the effort force to the fulcrum of a lever.
Elastic Collision A collision in which two objects return to their original shape without being permanently deformed.
Elastic Limit The point beyond which a deformed object cannot return to its original shape.
Elasticity A measure of a deformed object's ability to return to its original size and shape once the deforming force is removed.
Electric Circuit A conducting loop in which electrons carrying electric energy may be transferred from a suitable source to do useful work and returned to the source.
Electric Field An electric field exists where an electric force acts on a charge brought into the area.
Electrolyte An acid solution that produces large numbers of free electrons at the negative pole of a cell.
Electromagnet A combination of a solenoid and a magnetic material, such as iron, in the core of the solenoid. When a current is passed through the solenoid, the magnetic fields of the atoms in the magnetic material line up to produce a strong magnetic field.
Electromagnetic Spectrum The entire range of electromagnetic waves classified according to frequency.
Electromagnetic Wave A transverse wave resulting from a periodic disturbance in an electromagnetic field having an electric component and a magnetic component, each being perpendicular to the other and both perpendicular to the direction of travel.
Electron A fundamental particle of an atom; negatively charged.
Element A substance that cannot be separated into simpler substances.
emf The potential difference across a source.
Energy The ability to do work. There are many forms of energy, such as mechanical, electrical, thermal, fluid, chemical, atomic, and sound. Work delivered to an electric component or appliance (power × time).
Equilibrant Force The force that produces equilibrium.
Equilibrium An object is in equilibrium when the net force acting on it is zero. A body that is in equilibrium is either at rest or moving at a constant velocity.
Equivalent Resistance The single resistance that can replace a series and/or parallel combination of resistances in a circuit and provide the same current flow and voltage drop.
Evaporation The process by which high-energy molecules of a liquid continually leave its surface.

Exact Number A number that has been determined as a result of counting or by some definition, such as 1 h = 60 min.
Excited State A high-energy level for an electron in an atom.
Expansion Property of a gas in which the rapid random movement of its molecules causes the gas to completely occupy the volume of its container.
Experimental Physicist A physicist who performs experiments to develop and confirm physical theories.
$E = mc^2$ The equation that illustrates and defines the equivalence between mass and energy forms of matter.
Fahrenheit Scale The U.S. temperature scale on which ice melts at 32° and water boils at 212°.
First Condition of Equilibrium The sum of all parallel forces on a body in equilibrium must be zero.
First Law of Reflection The angle of incidence equals the angle of reflection.
First Postulate of Special Relativity The laws of physics are the same for both moving and nonmoving frames of reference.
Fixed Pulley A pulley that is fastened to a fixed object.
Flow Rate The volume of fluid flowing past a given point in a pipe per unit time.
Fluid A substance that takes the shape of its container. Either a liquid or a gas.
Fluorescence A property of certain substances in which radiation is absorbed at one frequency and then reemitted at a lower frequency.
Flux Lines Lines indicating the direction of the magnetic field near a magnetic pole.
Focal Length The distance between the principal focus of a mirror or lens and its vertex.
Force A push or a pull that tends to change the motion of an object or prevent an object from changing motion. Force is a vector quantity with both magnitude and direction.
Freezing The change of phase from liquid to solid. Also called *solidification*.
Frequency The number of complete vibrations or cycles per second of a wave.
Friction A force that resists the relative motion of two objects in contact caused by the irregularities of two surfaces sliding or rolling across each other.
Fulcrum A pivot about which a lever is free to turn.
Fusion The change of phase from solid to liquid. Also called *melting*.
Gamma Ray A ray composed of photons of electromagnetic radiation, which have no mass.
Gas A substance that takes the shape of its container and has the same volume as its container.
Gauge Pressure The amount of air pressure excluding the normal atmospheric pressure.
Gear Train A series of gears that transfers rotational motion from one gear to another.
General Theory of Relativity Extends the special theory of relativity to accelerated frames of reference with the assumption that gravity and acceleration are equivalent and that light has mass and its path can be warped by gravity.
Generator An apparatus consisting of a coil of wire rotating in a magnetic field. A current is induced in the coil, converting mechanical energy into electric energy.
Geology The branch of science that studies the origin, history, and structure of the earth.
Grand Unified Theory A theory linking the five fundamental forces: gravitational, electric, magnetic, strong, and weak forces.
Gravitational Field The area around a massive body in which an object experiences a gravitational force. The more massive and closer an object is to that body, the stronger is the gravitational field.
Gravitational Potential Energy The energy determined by the position of an object relative to a particular reference level.
Ground State The lowest energy level for an electron in an atom.
Half-Life The length of time required for one-half of the original amount of the radioactive atoms in a sample to decay.
Hardness A measure of the internal resistance of the molecules of a solid being forced farther apart or closer together.

Heat A form of internal kinetic and potential energy contained in an object associated with the motion of its atoms or molecules and which may be transferred from an object at a higher temperature to one at a lower temperature.

Heat of Fusion The heat required to melt 1 g or 1 kg or 1 lb of a liquid.

Heat of Vaporization The amount of heat required to vaporize 1 g or 1 kg or 1 lb of a liquid.

Heat Pump A pump containing a vapor (refrigerant) that is easily condensed to a liquid when under pressure. Produces heat during compression and cooling during vaporization.

Hooke's Law A principle of elasticity in solids: The ratio of the force applied to an object to its change in length (resulting in its being stretched or compressed by the applied force) is constant as long as the elastic limit has not been exceeded.

Hydraulic Principle (Pascal's Principle) The pressure applied to a confined liquid is transmitted without measurable loss throughout the entire liquid to all inner surfaces of the container.

Hydrometer A sealed glass tube weighted at one end so that it floats vertically in a liquid.

Hydrostatic Pressure The pressure a liquid at rest exerts on a submerged object.

Hypothesis A scientifically based prediction that needs testing to verify its validity.

Illumination The luminous intensity per unit area.

Impedance A measure of the total opposition to current flow in an ac circuit resulting from the effect of both the resistance and the inductive reactance on the circuit.

Impulse The product of the force exerted and the time interval during which the force acts on the object. Impulse equals the change in momentum of an object in response to the exerted force.

Impulse–Momentum Theorem If the mass of an object is constant, then a change in its velocity results in a change of its momentum. That is, $F\Delta t = \Delta p = m\Delta v = mv_f - mv_i$.

Inclined Plane A plane surface set at an angle from the horizontal used to raise objects that are too heavy to lift vertically.

Index of Refraction A measure of the optical density of a material. Equal to the ratio of the speed of light in vacuum to the speed of light in the material.

Induced Current A current produced in a circuit by the emf produced by motion of the circuit through the flux lines of a magnetic field.

Induced Magnetism Magnetism produced in a magnetic material such as iron when the material is placed in a magnetic field, such as that produced in the core of a current-carrying solenoid.

Inductance A measure of the tendency of a coil of wire to resist a change in the current because the magnetism produced by one part of the coil acts to oppose the change of current in other parts of the coil.

Induction A method of charging one object by bringing a charged object near to, but not touching, it.

Induction Motor An ac motor with an electromagnetic current induced by the moving magnetic field of the ac current.

Inductive Reactance A measure of the opposition to ac current flow in an inductor.

Inductor A circuit component, such as a coil, in which an induced emf opposes any current change in the circuit.

Inelastic Collision A collision in which two objects couple together.

Inertia The property of a body that causes it to remain at rest if it is at rest or to continue moving with a constant velocity unless an unbalanced force acts upon it.

Instantaneous Current The current at any instant of time.

Instantaneous Voltage The voltage at any instant of time.

Insulator A substance that does not allow electric current to flow readily through it.

Intensity The energy transferred by sound per unit time through unit area.

Interference The effect of two intersecting waves resulting in a loss of displacement in certain areas and an increase in displacement in others.

Internal Potential Energy The energy determined by the nature or condition of a substance.

Internal Resistance The resistance within a cell that opposes movement of the electrons.

Isotopes Nuclei that contain the same number of protons but a different number of neutrons.

Kelvin Scale The metric absolute temperature scale on which absolute zero is 0 K and the units are the same as on the Celsius scale.

Kilocalorie The amount of heat necessary to raise the temperature of 1 kg of water 1°C.

Kilogram The basic metric unit of mass.

Kinetic Energy The energy due to the mass and the velocity of a moving object.

Laser A light source that produces a narrow beam with high intensity. An acronym for "light amplification by stimulated emission of radiation."

Lateral Surface Area The area of all the lateral (side) faces of a geometric solid.

Law The highest level of certainty for an explanation of physical occurrences. A law is often accompanied by a formula.

Law of Acceleration The total force acting on a body is equal to the mass of the body times its acceleration (Newton's second law).

Law of Action and Reaction For every force applied by object *A* to object *B* (action), there is an equal but opposite force exerted by object *B* on object *A* (reaction) (Newton's third law).

Law of Conservation of Angular Momentum The angular momentum of a system remains unchanged unless an outside torque acts on it.

Law of Conservation of Mechanical Energy The sum of the kinetic energy and the potential energy in a system is constant if no resistant forces do work.

Law of Conservation of Momentum When no outside forces are acting on a system of moving objects, the total momentum of the system remains constant.

Law of Inertia A body that is in motion continues in motion with the same velocity (at constant speed and in a straight line) and a body at rest continues at rest unless an unbalanced (outside) force acts upon it (Newton's first law).

Law of Refraction When a beam of light passes at an angle from a medium of lower optical density to a denser medium, the light is bent toward the normal. When a beam passes from a medium of greater optical density to one less dense, the light is bent away from the normal.

Law of Simple Machines Resistance force × resistance distance = effort force × effort distance.

Lever A rigid bar free to turn on a pivot called a fulcrum.

Light Radiant energy that can be seen by the human eye.

Light-Year The distance that light travels in one earth year: 9.45×10^{15} m or 5.87×10^{12} mi.

Liquid A substance that takes the shape of its container and has a definite volume.

Load The object in a circuit that converts electric energy into other forms of energy or work.

Longitudinal Wave A disturbance in a medium in which the motion of the particles is along the direction of the wave travel.

Loudness The strength of the sensation of sound to an observer.

Luminous Intensity A measure of the brightness of a light source.

Machine An object or system that is used to transfer energy from one place to another and allows work to be done that could not otherwise be done or could not be done as easily.

Magnetic Property of metals or other materials that can attract iron or steel.

Magnetic Field A field of force near a magnetic pole or a current that can be detected using a magnet.

Malleability A property of a metal that enables it to be hammered and rolled into a sheet.

Mass A measure of the inertia of a body. A measure of the quantity of material making up an object.

Mass Density The mass per unit volume of a substance.

Mass–Energy Equivalence A physical law stating that mass and energy are equivalent forms of matter; $E = \Delta mc^2$; stated by Einstein.

Matter Anything that occupies space and has mass.

Mechanical Advantage The ratio of the resistance force to the effort force.

Mechanical Equivalent of Heat The relationship between heat and mechanical work.

Melting The change of phase from solid to liquid. Also called *fusion*.

Meniscus The crescent-shaped surface of a liquid column in a tube.

Method of Mixtures When two substances at different temperatures are "mixed together," heat flows from the warmer body to the cooler body until they reach the same temperature. Part of the heat lost by the warmer body is transferred to the cooler body and to surrounding objects. If the two substances are well insulated from surrounding objects, the heat lost by the warmer body is equal to the heat gained by the cooler body.

Metre The basic metric unit of length.

Molecule The smallest particle of a substance that exists in a stable and independent state.

Moment of Inertia Rotational inertia; the property of a rotating body that causes it to continue to turn until a torque causes it to change its rotational motion.

Momentum A measure of the amount of inertia and motion an object has or the difficulty in bringing a moving object to rest. Momentum equals the mass times the velocity of an object.

Motion A change of position.

Motor A device that is composed of an armature and a stator. When a current is passed through the armature, the armature rotates in the magnetic field of the stator and converts electric energy to mechanical energy.

Movable Pulley A pulley that is fastened to the object to be moved.

Multimeter An instrument used to measure current flow, voltage drop, and resistance.

Natural Frequency The frequency at which an object vibrates when struck by another object, such as a rubber hammer.

Neutron A fundamental particle in the nucleus of an atom; neutrally charged

Neutron Number The number of neutrons in an atomic nucleus of a particular nuclide, N. Also, $N = A - Z$, the difference between the atomic mass number and the atomic number. N can be different for nuclei with the same atomic number.

Newton's Law of Universal Gravitation All objects that have mass are attracted to one another by a gravitational force.

Normal Force Force perpendicular to the contact surface.

Nuclear Fission A nuclear reaction in which an atomic nucleus splits into fragments with the release of energy.

Nuclear Fusion A nuclear reaction in which light nuclei interact to form heavier nuclei with the release of energy.

Nucleus The center part of an atom made up of protons and neutrons.

Nuclide A specific type of atom characterized by its nuclear properties, such as the number of neutrons and protons and the energy state of its nucleus.

Number Plane A plane determined by the horizontal line called the x-axis and a vertical line called the y-axis intersecting at right angles. These two lines divide the number plane into four quadrants. The x-axis contains positive numbers to the right of the origin and negative numbers to the left of the origin. The y-axis contains positive numbers above the origin and negative numbers below the origin.

Ohm's Law When a voltage is applied across a resistance in an electric circuit, the current equals the voltage drop across the resistance divided by the resistance, $I = V/R$.

Ohmmeter An instrument that measures the resistance of a circuit component.

Opaque Absorbing or reflecting almost all light.

Optical Density A property of a transparent material that is a measure of the speed of light through the given material.

Orbit The path taken by an object during its revolution around another object, such as the path of the moon or a satellite about the earth or of a planet about the sun.

Parallel Circuit An electric circuit with more than one path for the current to flow. The current is divided among the branches of the circuit.

Particle Theory Theory that light consists of streams of particles.

Pendulum An object suspended so that it swings freely back and forth about a pivot.

Period The time required for a single wave to pass a given point or the time required for one complete vibration of an object in simple harmonic motion.

Periodic Table A table that contains all of the atomic elements arranged according to their atomic numbers and which can be used to predict their chemical properties.

Phase Angle The angle between the resistance and impedance vectors in a circuit.

Photoelectric Effect The emission of electrons by a surface when struck by electromagnetic radiation.

Photometry The study of the measurement of light.

Photons Wave packets of energy that carry light and other forms of electromagnetic radiation.

Physicist A person who is an expert in or who studies physics.

Physics The branch of science that describes the motion and energy of all matter throughout the universe.

Pitch The distance a screw advances in one revolution of the screw. Also the distance between two successive threads. The effect of the frequency of sound waves on the ear.

Planck's Constant A fundamental constant of quantum theory (6.626×10^{-34} J s).

Plane Mirror A mirror with a flat surface.

Platform Balance An instrument consisting of two platforms connected by a horizontal rod that balances on a knife edge. The pull of gravity on objects placed on the two platforms is compared.

Polarized Light Light waves restricted to a single plane that is perpendicular to the direction of the wave motion.

Potential Energy The stored energy of a body due to its internal characteristics or its position.

Power The rate of doing work (work divided by time). Energy per unit time consumed in a circuit.

Power Factor The ratio of the actual power to the apparent power.

Precision Refers to the smallest unit with which a measurement is made, that is, the position of the last significant digit.

Pressure The force applied per unit area.

Primary Cell A cell that cannot be recharged.

Primary Coil The coil of a transformer that carries an alternating current and induces a current in the secondary coil.

Primary Colors The colors red, green, and blue; an additive mixture of the three colors of light resulting in white.

Primary Pigments The complements of the three primary colors; namely, cyan (the complement of red), magenta (the complement of green), and yellow (the complement of blue).

Principle A rule or fundamental assumption that has been proven in the laboratory.

Problem-Solving Method An orderly procedure that aids in understanding and solving problems.

Projectile A propelled object that travels through the air but has no capacity to propel itself.

Projectile Motion The motion of a projectile as it travels through the air influenced only by its original velocity and gravitational acceleration.

Propagation Velocity The velocity of energy transfer of a wave, given by the distance traveled by the wave in one period divided by the period.

Proton A fundamental particle in the nucleus of the atom; positively charged.

Pulley A grooved wheel that turns readily on an axle and is supported in a frame.

Pulse Nonrepeated disturbance that carries energy through a medium or through space.

Quantum Mechanics A theory that unifies the wave–particle dual nature of electromagnetic radiation. The quantum model is based on the idea that matter, just like light, behaves sometimes as a particle and sometimes as a wave.

Quantum Theory Theory initiated by Planck and Einstein that energy, including electromagnetic radiation, is radiated or absorbed in multiples of certain units of energy.

Radian An angular unit of measurement. Defined as that angle with its vertex at the center of a circle whose sides cut off an arc on the circle equal to its radius. Equal to approximately 57.3°.

Radiation A form of heat transfer through energy being radiated or transmitted in the forms of rays, waves, or particles.

Radioactive Decay A type of nuclear decay that occurs when an unstable atom is transformed into a new element through the spontaneous disintegration of its nucleus.

Radiocarbon Dating A method used to obtain age estimates of organic materials using carbon 14 decay.

Rainbow A spectrum of light formed when sunlight strikes raindrops, refracts into them, reflects within them, and then refracts out of them.

Range The horizontal distance that a projectile will travel before striking the ground.

Rankine Scale The U.S. absolute temperature scale on which absolute zero is 0° R and the degree units are the same as on the Fahrenheit scale.

Real Image An image formed by rays of light.

Recharging The passing of an electric current through a secondary cell to restore the original chemicals.

Rectification The process of changing ac to dc.

Rectifier A device that changes ac to dc.

Rectilinear Motion Motion in a straight line.

Reflection The turning or turning back of all or part of a beam of light at a surface.

Refraction The bending of light as it passes at an angle from one medium to another of different optical density.

Regular Reflection Reflection of light with very little scattering.

Relative Humidity Ratio of the actual amount of vapor in the atmosphere to the amount of vapor required to reach 100% of saturation at the existing temperature.

Relative Motion The concept that motion can be described differently depending upon the observer's perspective.

Resistance The force overcome by a machine. The opposition to current flow.

Resistance Arm The distance from the resistance force to the fulcrum of a lever.

Resistivity The resistance per unit length of a material with uniform cross section.

Resonance A sympathetic vibration of an object caused by the transfer of energy from another object vibrating at the natural frequency of vibration of the first object. A condition in a circuit when the inductive reactance equals the capacitive reactance and they nullify each other. The current that flows in the circuit is then at its maximum value.

Resultant Force The sum of the forces applied at the same point, the single force that has the same effect as the two or more forces acting together.

Resultant Vector The sum of two or more vectors.

Revolution A unit of measurement in rotational motion. One complete rotation of a body.

Rotational Motion Spinning motion of a body.

Rotor The rotating coil in a generator.

Scalar A physical quantity that can be completely described by a number (called its magnitude) and a unit.

Science A system of knowledge that is concerned with establishing accurate conclusions about the behavior of everything in the universe.

Scientific Method An orderly procedure used by scientists in collecting, organizing, and analyzing new information which refutes or supports a scientific hypothesis.

Screw An inclined plane wrapped around a cylinder.

Second The basic unit of time.

Second Condition of Equilibrium The sum of the clockwise torques on a body in equilibrium must be equal to the sum of the counterclockwise torques about any point.

Second Law of Reflection The incident ray, the reflected ray, and the normal (perpendicular) to the reflecting surface all lie in the same plane.

Second Postulate of Special Relativity The speed of light is constant regardless of the speed of the observer or of the light source.

Secondary Cell A rechargeable type of cell.

Secondary Coil The coil of a transformer in which a current is induced by the current in the primary coil.

Semiconductors A small number of materials that fall between conductors and insulators in their ability to conduct electric current.

Series Circuit An electric circuit with only one path for the current to flow. The current in a series circuit is the same throughout.

Shearing A stress caused by two forces applied in parallel, opposite directions.

SI (Système International d'Unités) The international modern metric system of units of measurement.

Significant Digits The number of digits in a measurement, which indicates the number of units we are reasonably sure of having counted.

Simple Harmonic Motion A type of linear motion of an object in which the acceleration is directly proportional to its displacement from its equilibrium position and the motion is always directed to the equilibrium position.

Simple Machine Any one of six mechanical devices in which an applied force results in useful work. The six simple machines are the lever, the wheel and axle, the pulley, the inclined plane, the screw, and the wedge.

Snell's Law The index of refraction equals the sine of the angle of incidence divided by the sine of the angle of refraction.

Solenoid A coil of tightly wrapped wire. Commonly used to create a strong magnetic field by passing current through the wire.

Solid A substance that has a definite shape and a definite volume.

Solidification The change of phase from liquid to solid. Also called *freezing*.

Sound Those waves transmitted through a medium with frequencies capable of being detected by the human ear.

Source The object that supplies electric energy for the flow of electric charge (electrons) in a circuit.

Source Activity The strength of a source of radiation that can be specified at a given time.

Space-Time An object's position in the universe can be pinpointed using three spatial dimensions and one time dimension.

Special Theory of Relativity The laws of physics are the same in moving and nonmoving frames of reference and the speed of light is constant no matter what the speed of the observer or the light source.

Specific Gravity The ratio of the density of any material to the density of water.

Specific Heat The amount of heat necessary to change the temperature of 1 kg of a substance 1°C in the metric system or 1 lb of a substance 1°F in the U.S. system.

Speed The distance traveled per unit of time. A scalar described by a number and a unit.

Speed of Light The speed at which light and other forms of electromagnetic radiation travel: 3.00×10^8 m/s in a vacuum.

Speed of Sound The speed at which sound waves travel in a medium: 331 m/s in dry air at 1 atm pressure and 0°C.
Spring Balance An instrument containing a spring, which stretches in proportion to the force applied to it, and a pointer attached to the spring with a calibrated scale read directly in given units.
Standard Position A vector is in standard position when its initial point is at the origin of the number plane. The vector is expressed in terms of its length and its angle, measured counterclockwise from the positive x-axis to the vector.
Standard Temperature and Pressure (STP) A commonly used reference in gas laws. Standard temperature is the freezing point of water. Standard pressure is equivalent to atmospheric pressure.
Standards of Measure A set of units of measurement for length, weight, and other quantities defined in such a way as to be useful to a large number of people.
Standing Waves A special case of superposition of two waves when no energy propagation occurs along the wave. The two waves of equal amplitude and wavelength do not appear to be traveling.
Statics The study of objects that are in equilibrium.
Stator The field magnets in a generator.
Step-Down Transformer A transformer used to lower voltage; it has more turns in the primary coil.
Step-Up Transformer A transformer used to increase voltage; it has more turns in the secondary coil.
Strain The deformation of an object due to an applied force.
Streamline Flow The smooth flow of a fluid through a tube.
Stress The ratio of an outside applied distorting force to the area over which the force acts.
Strong Force An attractive force among all nucleons (neutrons and protons) independent of their charge.
Superconductor A material that continuously conducts electric current without resistance when cooled to typically very low temperatures, often near absolute zero.
Superposition of Waves The algebraic sum of the separate displacements of two or more individual waves passing through a medium.
Surface Tension The ability of the surface of a liquid to act like a thin, flexible film.
Synchronous Motor An ac motor whose speed of rotation is constant and is directly proportional to the frequency of its ac power supply
Technology The field that uses scientific knowledge to develop material products or processes that satisfy human needs and desires.
Temperature A measure of the hotness or coldness of an object.
Tensile Strength A measure of a solid's resistance to being pulled apart.
Tension A stress caused by two forces acting directly opposite each other. This stress tends to cause objects to become longer and thinner.
Terminal Speed The speed attained by a freely falling body when the air resistance equals its weight and no further acceleration occurs.
Theoretical Physicist A physicist who predominantly uses previous theories and mathematical models to form new theories in physics.
Theory A scientifically accepted principle that attempts to explain natural occurrences.
Thermal Conductivity The ability of a material to transfer heat by conduction.
Torque The tendency to produce change in rotational motion. Equal to the applied force times the length of the torque arm.
Torsion A stress related to a twisting motion. This type of stress severely compromises the strength of most materials.
Total Internal Reflection A condition such that light striking a surface does not pass through the surface but is completely reflected inside it.
Total Surface Area The total area of all the surfaces of a geometric solid; that is, the lateral surface area plus the area of the bases.

Transformer A device composed of two coils (primary and secondary) and a magnetic core. Used to step up or step down a voltage.
Translucent Allowing some but not all light to pass through.
Transparent Allowing almost all light to pass through so that objects or images can be seen clearly.
Transverse Wave A disturbance in a medium in which the motion of the particles is perpendicular to the direction of the wave motion.
Turbulent Flow The erratic, unpredictable flow of a fluid resulting from excessive speed of the flow or sudden changes in direction or size of the tube.
Universal Motor A motor that can be run on either ac or dc power.
Vaporization The change of phase from liquid to a gas or vapor.
Vector A physical quantity that requires both magnitude (size) and direction to be completely described.
Velocity The rate of motion in a particular direction. The time rate of change of an object's displacement. Velocity is a vector that gives the direction of travel and the distance traveled per unit of time.
Virtual Image An image that only appears to the eye to be formed by rays of light.
Viscosity The internal friction of a fluid caused by molecular attraction, which makes it resist a tendency to flow.
Visible Spectrum The colors resulting from the dispersion of white light through a glass prism: red, orange, yellow, green, blue, and violet.
Volatility A measure of a liquid's ability to vaporize. The more volatile the liquid, the greater is its rate of evaporation.
Voltage Drop The potential difference across a load in a circuit.
Voltmeter An instrument that measures the difference in potential (voltage drop) between two points in a circuit.
Volume The number of cubic units contained in a figure.
Wave A disturbance that moves through a medium or through space.
Wavelength The distance between two successive corresponding points on a wave.
Wave Theory Theory that light consists of waves traveling out from light sources, like water waves traveling out from the point at which a stone is dropped into water.
Wedge An inclined plane in which the plane is moved instead of the resistance.
Weight A measure of the gravitational force or pull exerted on an object by the earth or by another large body.
Weight Density The weight per unit volume of a substance.
Wheel-and-Axle A large wheel attached to an axle so that both turn together.
Work The product of the force in the direction of motion and the displacement.
***x*-component** The horizontal component of a vector that lies along the x-axis.
***y*-component** The vertical component of a vector that lies along the y-axis.

APPENDIX E

ANSWERS TO ODD-NUMBERED PROBLEMS AND TO CHAPTER REVIEW QUESTIONS AND PROBLEMS

Chapter 0 Review Questions Page 10

1. d **2.** b **3.** c **4.** b **5.** d

6. Archimedes conducted and documented his physical theories.

7. Science is a system of knowledge while technology uses that knowledge to develop material products or processes that satisfy human needs and desires.

8. Behavior of light—fiber optics; electricity—light bulb.

9. The scientific method is used to discover facts about the natural world. The problem-solving method uses the scientific method to create something useful.

10. Knowledgeable of the physical world. It allows one to understand and answer new questions about everyday occurrences. It is also important in many technical fields.

Chapter 1

1.2 Page 17

1. kilo **3.** hecto **5.** milli **7.** mega **9.** h **11.** m **13.** M **15.** c
17. 135 mm **19.** 28 kL **21.** 49 cg **23.** 75 hm **25.** 24 metres
27. 59 grams **29.** 27 millimetres **31.** 45 dekametres **33.** 26 megametres
35. metre **37.** litre and cubic metre **39.** second

1.3 Pages 21–22

1. 1 metre **3.** 1 kilometre **5.** 1 kilometre **7.** cm **9.** m **11.** mm
13. km **15.** mm **17.** m **19.** km **21.** cm **23.** km **25.** cm **27.** km
29. mm **31.** cm **33.** 1000 **35.** 100 **37.** 0.1 **39.** 1000 **41.** 100
43. 0.001 **45.** 10 **47.** 0.25 km **49.** 178,000 m **51.** 8 3 m **53.** 3750 mm
55. 4,000,000 μm **57.** (a) 9 ft (b) 3 yd **59.** (a) 11,000 yd (b) 33,000 ft
61. 412.16 km **63.** 2.80 in. **65.** 6.1 m **67.** 9

1.4 Pages 29–31

1. 40 cm^2 **3.** 39 in^2 **5.** 22 in^2 **7.** 72 in^3 **9.** 40 cm^3 **11.** 1 litre
13. 1 cubic centimetre **15.** 1 square kilometre **17.** L **19.** m^2 **21.** m^3
23. ha **25.** mL **27.** m^3 **29.** L **31.** mL **33.** L **35.** m^2 **37.** L
39. ha **41.** m^3 **43.** m^2 **45.** L **47.** 1000 **49.** 0.1 **51.** 0.01 **53.** 10
55. 1 **57.** 1,000,000 **59.** 0.001 **61.** 10,000 **63.** 100 **65.** 0.01
67. 100 **69.** 7.5 L **71.** 1600 mL **73.** 275,000 mm^3 **75.** 4×10^9 mm^3
77. 275 mL **79.** 1000 L **81.** 7500 cm^3 **83.** 50 cm^2 **85.** 50,000 cm^2
87. 400 km^2 **89.** 45 ft^2 **91.** 13,935 cm^2 **93.** 0.75 ft^2 **95.** 155,816 in^2

97. 513 ft^3 **99.** 30.1 yd^3 **101.** 13,824 in^3 **103.** 623.2 cm^3 **105.** 36 cm^2
107. 76 cm^2 **109.** 40 mL

1.5 Pages 35–36

1. 1 gram **3.** 1 kilogram **5.** 1 kilogram **7.** kg **9.** kg **11.** metric ton **13.** g **15.** mg **17.** kg **19.** g **21.** kg **23.** g **25.** g **27.** kg **29.** kg **31.** metric ton **33.** kg **35.** 1000 **37.** 100 **39.** 0.1 **41.** 1000 **43.** 100 **45.** 0.001 **47.** 1,000,000 **49.** 575,000 mg **51.** 0.65 g **53.** 5 g **55.** 30,000,000 mg **57.** 2.5 kg **59.** 0.4 mg **61.** 750 g **63.** 15,575 N **65.** 890 N **67.** 450 lb **69.** 7.5 lb **71.** 36 oz **73.** 418.3 N **75.** second, s **77.** newton, N **79.** 1 millisecond **81.** 1 ms **83.** 45 ns **85.** 3.45×10^{-4} s **87.** 15,915 s **89.** 4×10^9 ns

1.6 Page 38

1. 3 **3.** 4 **5.** 2 **7.** 3 **9.** 5 **11.** 3 **13.** 5 **15.** 4 **17.** 3 **19.** 2 **21.** 1 **23.** 4 **25.** 2 **27.** 3 **29.** 4

1.7 Page 40

1. 1 ft **3.** 1 m **5.** 0.0001 in. **7.** 10 km **9.** 0.01 m **11.** 0.00001 in. **13.** 0.01 m **15.** 1 kg **17.** 0.0001 in. **19.** 1000 N **21.** 1 N **23.** 0.01 m^2 **25.** 100 kg **27.** 0.000001 kg or 1×10^{-6} kg **29.** 10,000 kg or 1×10^4 kg
31. (a) 15.7 in. (b) 0.018 in. **33.** (a) 16.01 cm (b) 0.734 cm
35. (a) 0.0350 s (b) 0.00040 s **37.** (a) 27,0$\bar{0}$0 L (b) 4.75 L
39. (a) All have one significant digit. (b) 50 N **41.** (a) 0.05 in. (b) 16.4 in.
43. (a) 0.65 m (b) 27.5 m **45.** (a) 0.00005 g (b) 0.75 g
47. (a) 3 N (b) 45,000 N **49.** (a) 20 kg (b) 40$\bar{0}$,000 kg

1.8 Pages 44–45

1. 14,200 ft **3.** 83.3 cm **5.** 7$\bar{0}$,000 N **7.** 802 m or 80,200 cm **9.** 18 s **11.** 500 kg **13.** 41.0 g **15.** 3200 km **17.** 900,000 kg **19.** 0.40 m or 4$\bar{0}$ cm **21.** 4900 m^2 **23.** 1,4$\bar{0}$0,000 km^2 or 1.40×10^6 km^2 **25.** 737.7 m^2 **27.** 5560 cm^3 **29.** 2.91×10^7 in^3 **31.** 3$\bar{0}$ ft **33.** 3.06 cm **35.** 75 km/h **37.** 1$\bar{0}$00 mi/h **39.** 1100 ft lb/s **41.** 370 mi/h **43.** 43.2 m **45.** 4530 kg m/s^2 **47.** 10,300 m^3 **49.** 6100 m^2 **51.** 28,800 ft

1.9 Pages 50–51

1. 25,900 cm^3 **3.** 284 cm^3 **5.** 102.1 cm^2 **7.** 10,100 ft^3 **9.** 864 ft^3 **11.** 12.0 cm^2 **13.** 1.58 cm^2 **15.** 36.0 m **17.** 9.39 m **19.** 137 m **21.** 65.5 ft **23.** 6$\bar{0}$ panels **25.** 24.1 m^3 **27.** 4.44 yd^3 **29.** 266 in^3

Chapter 1 Review Questions Pages 52–53

1. c **2.** b **3.** b **4.** c **5.** a
6. (1) Pieces made separately may not fit together.
(2) Workers could not communicate directions to each other.
(3) Workers could not tell each other how much material to buy.
7. It is based on the decimal system.
8. (1) The distance to the moon.
(2) The thickness of aluminum foil.
9. Yes
10. The surface that would be seen by cutting a geometric solid with a thin plate.
11. Yes **12.** Hectare **13.** Litre
14. (1) Medicines (2) Perfumes (3) Wine

15. Mass measures the quantity of matter. Weight measures the gravitational pull on an object. **16.** Newton (N) **17.** Millionth
18. Because nearly all measurements are approximate numbers rather than exact numbers.
19. No **20.** Yes
21. Most mistakes are made in problem solving by missing needed information or misinterpreting the information given.
22. Making a sketch helps to visualize what is happening in the problem.
23. The basic equation
24. The working equation is found by solving the basic equation for the unknown quantity.
25. Carrying the units through a problem shows whether the answer is the kind expected.
26. Making an estimate of the correct answer shows whether the solution is reasonable.

Chapter 1 Review Problems Pages 53–54

1. k **2.** m **3.** μ **4.** M **5.** 45 mg **6.** 138 cm **7.** 1 L
8. 1 kg **9.** 1 m^3 **10.** 0.25 **11.** 0.85 **12.** 5400 **13.** 550,000
14. 25,000 **15.** 75,000 **16.** 27,500 **17.** 0.035 **18.** 150,000 **19.** 500
20. 68.2 **21.** 11.0 **22.** 98.4 **23.** 968 **24.** 216 **25.** 212 **26.** 71.2
27. 4 h 20 min **28.** 3 **29.** 4 **30.** 2 **31.** 3 **32.** 0.1 ft **33.** 0.0001 s
34. 1000 mi **35.** 100,000 N **36.** (a) 12.00 m (b) 0.008 m (c) 0.150 m (d) 2600 m **37.** (a) 18,050 L (b) 0.75 L (c) 0.75 L (d) 18,050 L **38.** 0.125 s
39. 63,000 N **40.** 1,800,000 cm^3 **41.** 150 m^2 **42.** 9.73 kg m/s^2
43. 9.90 m^2 **44.** $7\bar{0}0$ cm^3 **45.** 12.0 cm **46.** 42.3 mm **47.** 606 cm^2
48. 6.0 cm **49.** 6.27 m **50.** 14.4 m **51.** 430 cm^3 **52.** 4580 m^2

Chapter 1 Applied Concepts Page 55

1. 2.62¢/yd^2 **2.** 2.13 ft^3/s **3.** 2.91 **4.** 16 **5.** (a) 6.73 m^3 (b) Not safe; the spool's mass is 52,200 kg.

Chapter 2

2.1 Pages 63–64

1. 2.0 **3.** 2.8 **5.** 6.3

7.
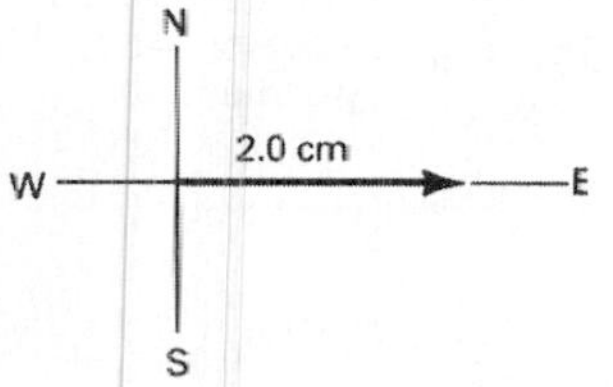

9.
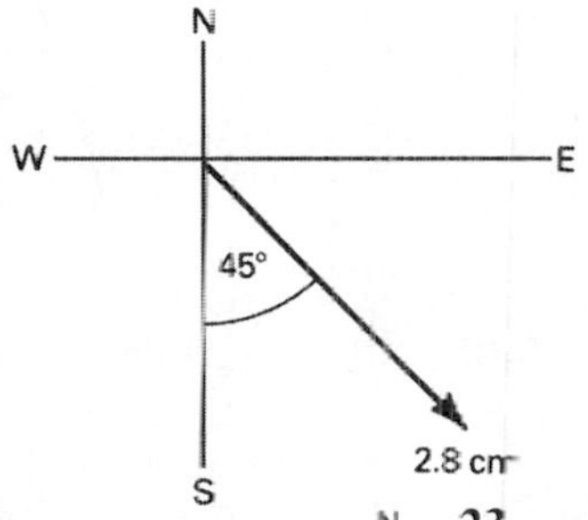

11.
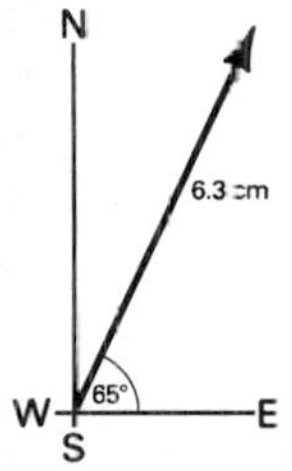

13. $1\frac{1}{4}$ **15.** $2\frac{5}{8}$ **17.** $\frac{15}{16}$

19.
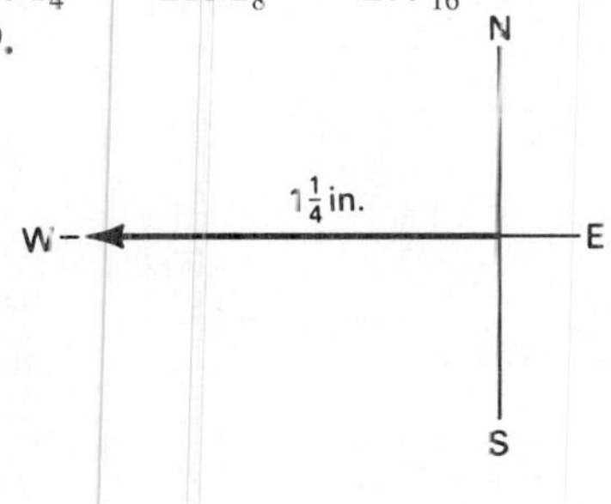

21.
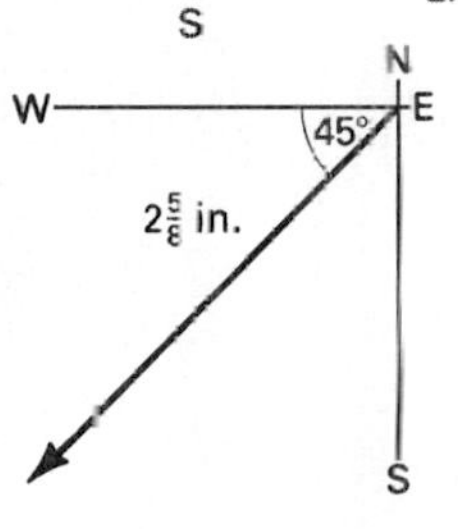

23
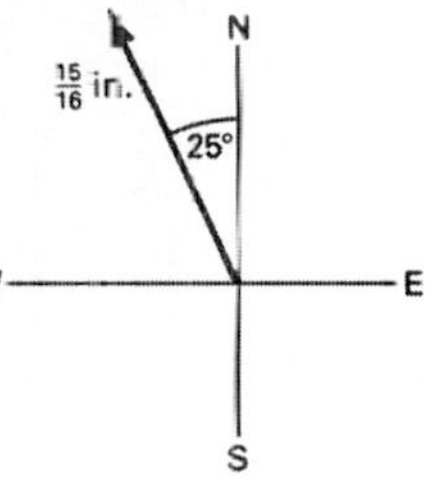

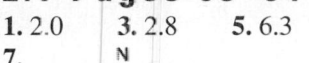

25. 61 km at 55° north of east
27. 1300 mi at 1° west of south
29. 36 km at 5° east of north
31. 38 km at 25° north of west
33. 1500 mi at 71° north of east
35. 120 km at 72° south of east
37. 47 mi at 49° north of east

2.2 Pages 70–72

1.

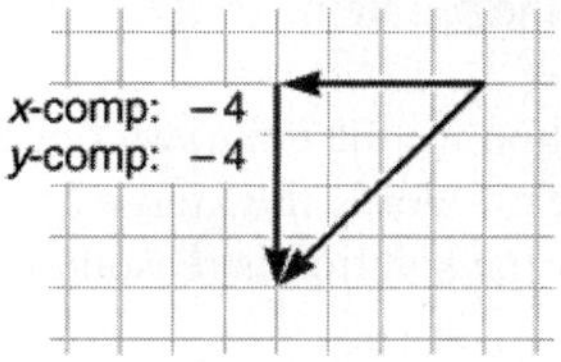

3.

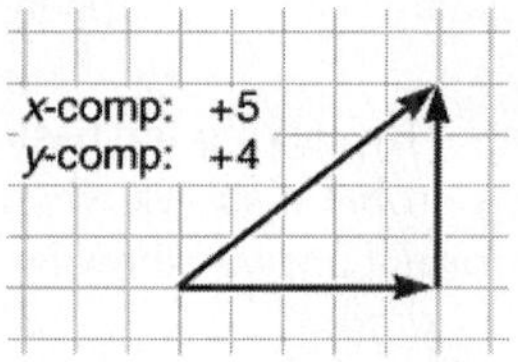

5.

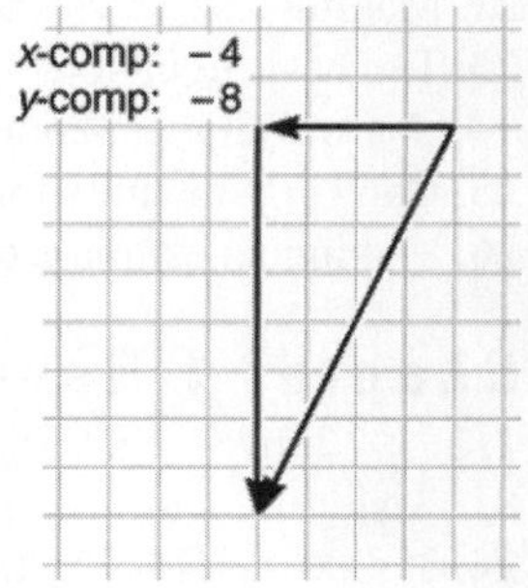

7.

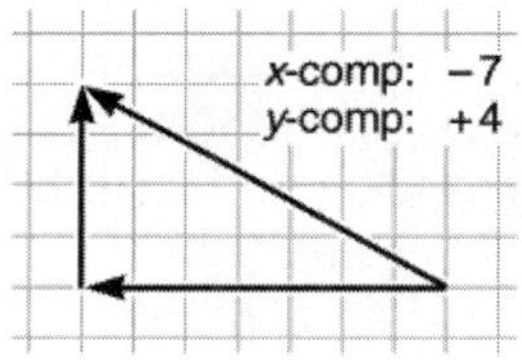

9.

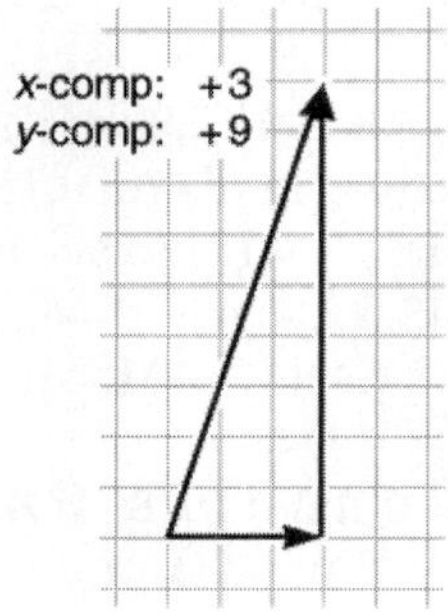

11.

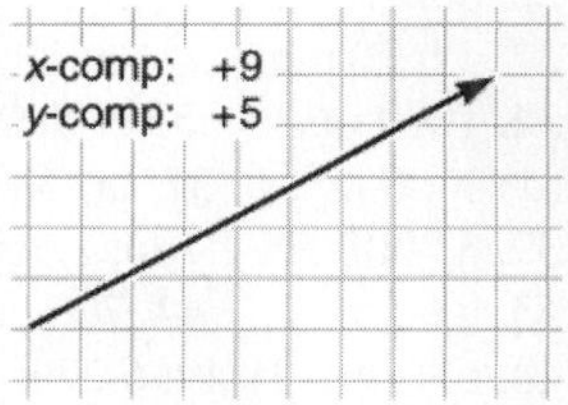

13.

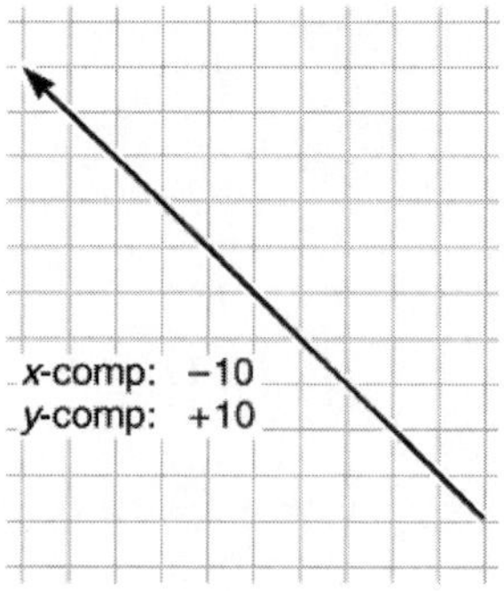

15.

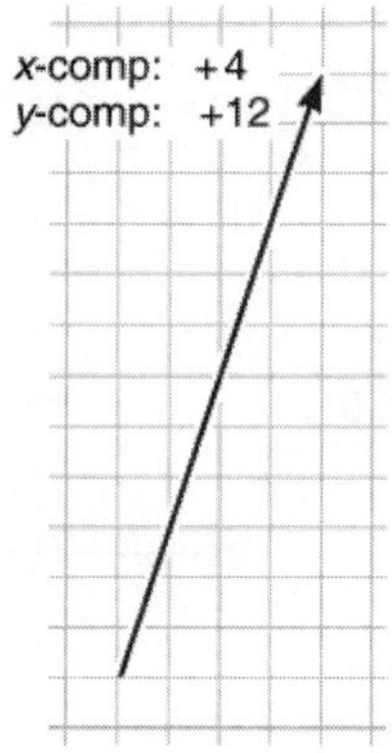

17.

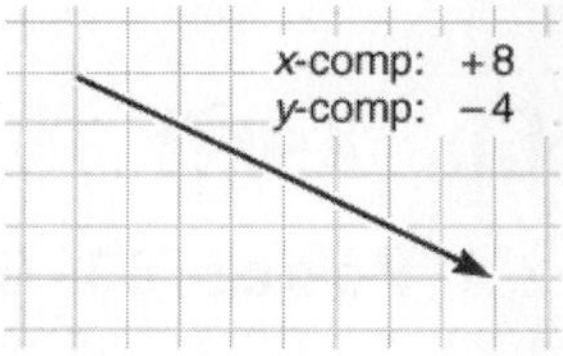

19.

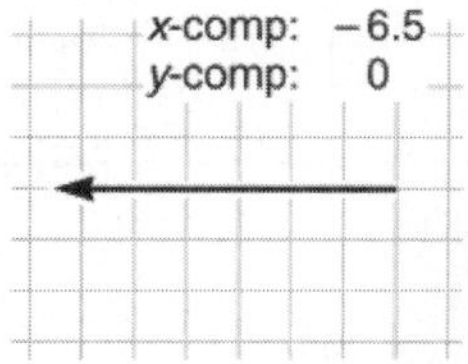

21.

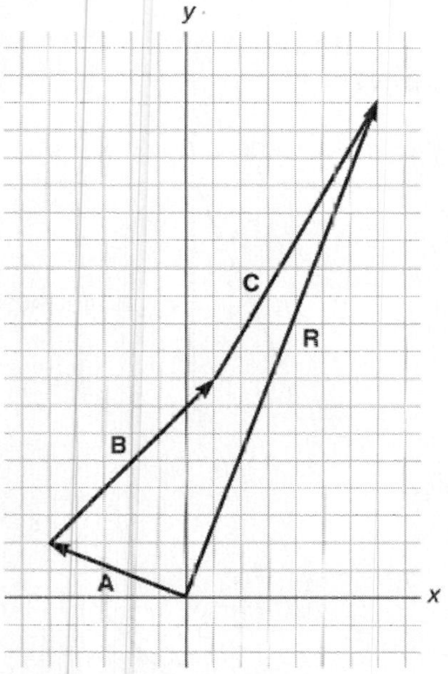

A + B + C = R

A + B + C = R

Vector	*x-component*	*y-component*
A	−5	2
B	6	6
C	6	10
R	7	18

23.

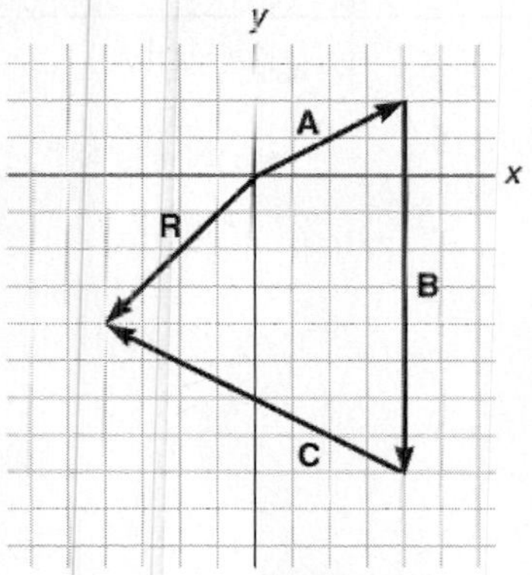

A + B + C = R

A + B + C = R

Vector	*x-component*	*y-component*
A	4	2
B	0	−10
C	−8	4
R	−4	−4

25.

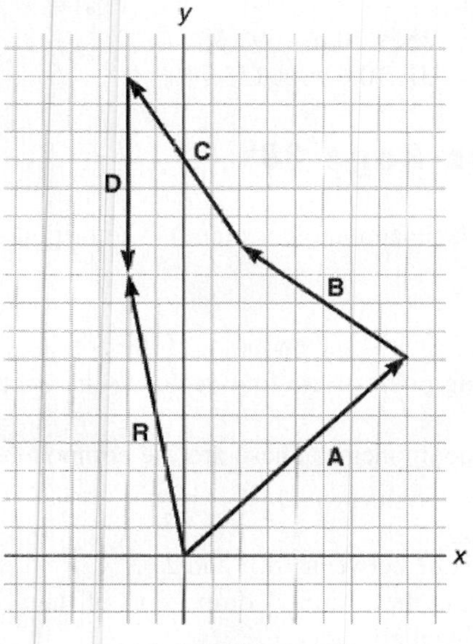

A + B + C + D = R

A + B + C − D = R

Vector	*x-component*	*y-component*
A	8	7
B	−6	4
C	−4	6
D	0	−7
R	−2	10

2.3 Pages 78–79

1.

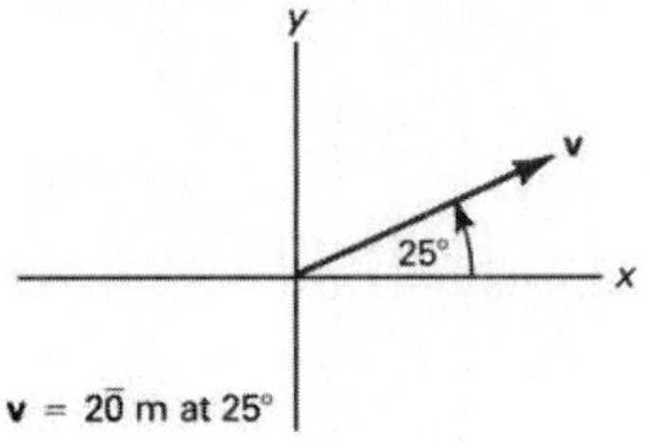

3.

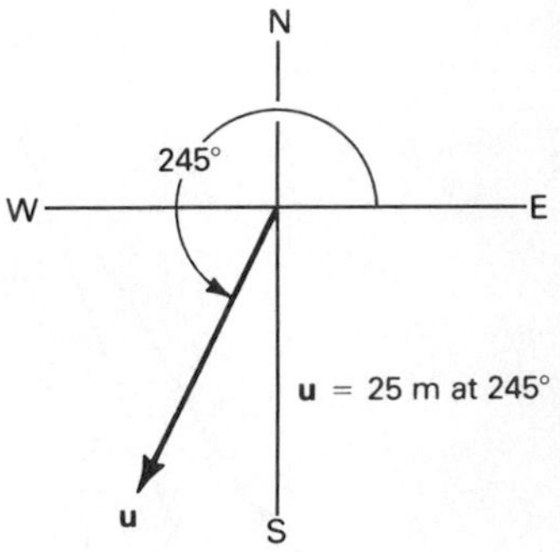

5.

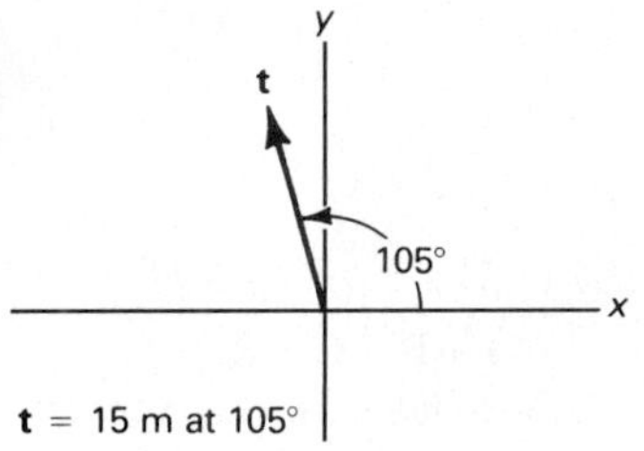

7.

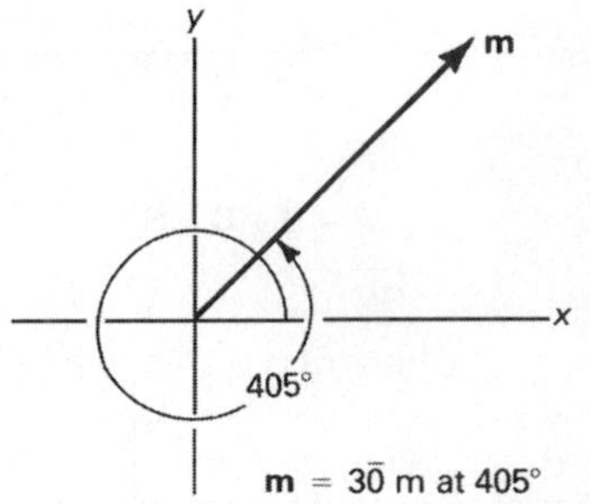

	x-component	*y-component*
9.	9.96 m	8.97 m
11.	−18.2 km	−45.1 km
13.	97.4 km	−14.4 km
15.	38.2 m	7.09 m
17.	6.17 km	−7.35 km
19.	−5.88 m	28.9 m

21. 10.0 m at 36.9° **23.** 26.2 mi at 315.0° **25.** 9.70 m at 98.3°
27. 53.3 m at 291.5° **29.** 22.3 mi at 48.8° **31.** 10.6 m at 155.7°

Chapter 2 Review Questions Page 80

1. d **2.** a **3.** a

4. Two intersecting lines (axes) determine a plane (the number plane). The origin is the point where the axes intersect.

5. Yes **6.** Yes

7. Graph each vector placing the initial point of each at the endpoint of the previous one; the resultant is the vector from the beginning point of the first vector to the endpoint of the last vector.

8. Add the *x*-components and then add the *y*-components. These are the components of the resultant.

9. No **10.** Counterclockwise

11. In the third quadrant, the angle measure must be between 180° and 270°.

12. Complete the right triangle with the legs being the *x*- and *y*-components of the vector. Then find the lengths of *x* and *y* and determine their signs.

13. Complete the right triangle with the legs being the *x*- and *y*-axes, respectively.

Chapter 2 Review Problems Pages 80–81

1. 3.00 km north **2.** 3.00 km south **3.** 284 km at 39.3° north of west
4. 318 km at 53.4° north of east **5.** 14.3 **6.** 11.2 **7.** 8.25 **8.** 5.00
9. $\mathbf{R}_x = 6.00$; $\mathbf{R}_y = -2.00$ **10.** $\mathbf{R}_x = 9.00$; $\mathbf{R}_y = 9.00$ **11.** $\mathbf{R}_x = -6.00$; $\mathbf{R}_y = 2.00$
12. $\mathbf{R}_x = -9.00$; $\mathbf{R}_y = -9.00$

13.

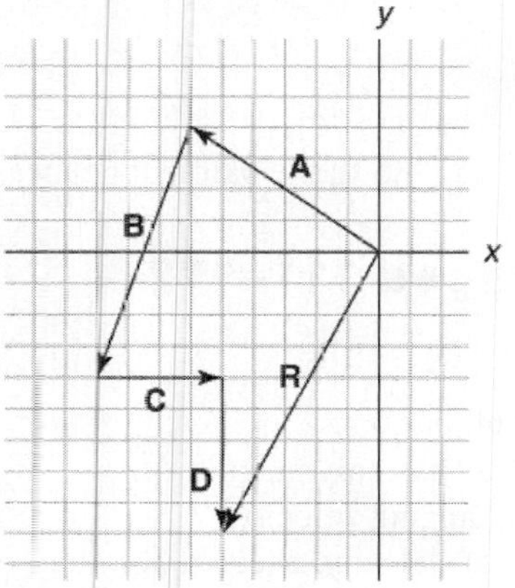

$\mathbf{A} + \mathbf{B} + \mathbf{C} + \mathbf{D} = \mathbf{R}$

Vector	*x-component*	*y-component*
A	−6	4
B	−3	−8
C	4	0
D	0	−5
R	−5	−9

14.

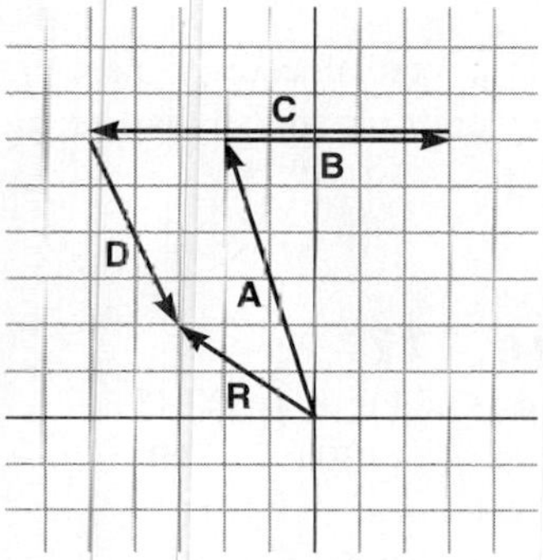

$\mathbf{A} + \mathbf{B} + \mathbf{C} + \mathbf{D} = \mathbf{R}$

Vector	*x-component*	*y-component*
A	−2	6
B	5	0
C	−8	0
D	2	−4
R	−3	2

15. $\mathbf{R}_x = 11.3$ cm; $\mathbf{R}_y = 6.50$ cm **16.** $\mathbf{R}_x = 5.00$ cm; $\mathbf{R}_y = 8.66$ cm
17. $\mathbf{R}_x = 17.3$ cm; $\mathbf{R}_y = 10.0$ cm **18.** $\mathbf{R}_x = -4.50$ cm; $\mathbf{R}_y = -7.79$ cm
19. $\mathbf{R}_x = 6.89$ cm; $\mathbf{R}_y = 5.79$ cm **20.** $\mathbf{R}_x = 10.3$ cm; $\mathbf{R}_y = -14.7$ cm

Chapter 2 Applied Concepts Page 82

1. 417 ft; 184 ft **2.** 1090 m **3.** 13.6 m **4.** 2.80 mi **5.** 556 km

Chapter 3

3.1 Page 92

1. 50 mi/h **3.** 21.6 m/s **5.** 68.3 ft/s **7.** 320 km
9. (a) 81 ft/s (b) 25 m/s (c) 89 km/h **11.** 80 km/h, east
13. 125 mi/h, south **15.** 61.1 km/h at 30° south of east
17. (a) 53.7 mi/h (b) 25.5 mi/h, west **19.** 370 km/h at 90.0°
21. 239 km/h at 190.8° **23.** 128 mi/h at 239.4° **25.** 227 km/h at 162.4°

3.2 Pages 96–97

1. 15 m/s^2 **3.** 10 ft/s^2 **5.** 3.2 m/s^2 **7.** 6.25 m/s^2 **9.** 0.205 m/s^2
11. 57.5 mi/h **13.** 720 km/h **15.** 16.7 s
17. 4.00 m/s^2 **19.** −8.33 m/s^2 **21.** −16.7 m/s^2

3.3 Pages 104–105

1. 5.05 m/s **3.** 127.8 m **5.** 2.13 ft/s^2 **7.** 9.00 mi/h **9.** 55$\overline{0}$ ft **11.** 1.1 m/s^2 **13.** –16.4 m/s^2 **15.** –1.6 m/s^2 **17.** (a) 23.5 m/s (b) 28.2 m **19.** (a) 3190 m (b) 25.5 s (c) 51.0 s **21.** (a) 1.02 s (b) 5.10 m (c) 24.3 m/s (d) 3.50 s **23.** (a) 44.0 ft/s (b) 4.21 s (c) 135 ft/s **25.** 2.06 s **27.** 116 km/h **29.** 9.05 s

3.4 Page 110

1. 62.5 m **3.** 118 m **5.** 124 m **7.** 381 m **9.** 288 ft; the ball will bounce.

Chapter 3 Review Questions Pages 111–112

1. c **2.** c **3.** a **4.** b **5.** d

6. Gravity accelerates all objects at the same rate.

7. Horizontal motion is independent of the vertical pull of gravity.

8. Velocity has magnitude and direction, whereas speed has only magnitude.

9. No; anything speeding up or slowing down has a changing velocity.

10. Vectors are necessary to determine the direction of velocity, acceleration, and other vector quantities. Examples: navigating airplanes, ships, and spacecraft.

11. (a) An automobile speeding up or slowing down; (b) anything being dropped to the ground; (c) a bullet being fired.

12. Acceleration is change in velocity; deceleration is a special case of acceleration where the object is slowing; acceleration need not be uniform and so may be subject to averaging.

13. 9.80 m/s^2 (metric) and 32.2 ft/s^2 (U.S.)

Chapter 3 Review Problems Page 112

1. 25.5 mi **2.** 2.73 h **3.** 226 km/h at 107.8° **4.** 33.4 km/h at 337.1° **5.** 4.00 ft/s^2 **6.** 6.9 s **7.** 15.0 km/h **8.** 3.90 m/s **9.** 3$\overline{0}$ h **10.** 13 m/s **11.** 4.00 m/s **12.** 2.6 m/s^2 **13.** 2.52 m/s^2 **14.** 60.4 m/s **15.** (a) 205 m/s (b) 20.9 s (c) 41.8 s **16.** (a) 37.0 m/s (b) 64.6 m **17.** 8.55 m **18.** Since $s_x = 52.8$ ft, the arrow will miss the bull's eye.

Chapter 3 Applied Concepts Page 113

1. (a) 3.76 m/s (b) 2.01 m/s **2.** 16.6 mi/h, 18.8° downriver **3.** 633 ft **4.** 4.71 m/s **5.** 25.2 m/s

Chapter 4

4.2 Pages 120–121

1. 30.0 N **3.** 744 lb **5.** 252 lb **7.** 23$\overline{0}$ N **9.** 40.0 m/s^2 **11.** 11.7 m/s^2 **13.** 0.518 m/s^2 **15.** 1.39 ft/s^2 **17.** 5250 N **19.** 1320 lb **21.** 5740 N **23.** 6.00 kg **25.** 12.8 m/s^2 **27.** (a) 7.37 kg (b) It is halved (9.50 m/s^2). (c) It is doubled (38.0 m/s^2). **29.** (a) 9880 N in the opposite direction of the truck. (b) Double the force would be required to stop the truck in half the distance.

4.3 Pages 124–125

1. 380 N **3.** 1100 N **5.** 0.080 **7.** 25,000 lb **9.** 1600 N **11.** 0.350 **13.** (a) –3680 N (the – sign indicates the frictional force acts in opposite direction to that of the truck) (b) –4.91 m/s^2 (c) 91.6 m

4.4 Page 127

1. 3.0 N, right **3.** 15.0 N; left **5.** 4 N; left **7.** 4.00 ft/s^2 **9.** 0.509 m/s^2

4.5 Page 129

1. 294 N **3.** 322 lb **5.** 1.73 kg **7.** 1220 kg **9.** 6.8×10^{11} kg
11. 11,300 N **13.** 85.4 slugs **15.** (a) 13,200 N (b) $22\bar{0}0$ N
17. (a) 65.0 kg (b) 106 N **19.** Answers vary. **21.** Answers vary.
23. (a) 3.57 slugs (b) 303 lb **25.** Answers vary.

Chapter 4 Review Questions Pages 130–131

1. d **2.** b **3.** b **4.** c **5.** a
6. (a) A bridge being supported over a river (b) Isometric exercises (c) A magnet on a refrigerator
7. (a) No (b) No
8. A car hit from behind is forced into the car ahead of it.
9. A body in motion continues in motion and a body at rest continues at rest unless an outside force acts on it.
10. Acceleration is change of velocity.
11. Only if they have the same mass **12.** Yes; 3.00 lb = 13.35 N
13. More difficult; everything would slide.
14. Weight is a measure of gravitational pull. The moon, having less mass than the earth, exerts less gravitational pull.
15. For every force applied by object *A* to *B*, there is an equal force applied by *B* to *A* in the opposite direction.
16. For every action, there is an equal and opposite reaction.

Chapter 4 Review Problems Page 131

1. 3.00 m/s^2 **2.** 75.0 kg **3.** 19 ft/s^2 **4.** 19,000 N **5.** 640 N
6. 0.17 **7.** 3800 N **8.** 11 N **9.** 127 N **10.** 12 lb

Chapter 4 Applied Concepts Page 132

1. (a) 1.55 m/s^2 (b) 39.4 m/s (c) 68.2 m/s **2.** (a) 12.1 m/s^2 (b) 1.36×10^5 m or 136 km **3.** (a) $12\bar{0}$ lb (b) 137 lb (c) $12\bar{0}$ lb (d) 99.6 lb **4.** (a) 714 N (b) −2.94 m/s^2 (c) 13.7 s **5.** -1.23×10^4 N

Chapter 5

5.1 Pages 143–144

1. 80.0 kg m/s **3.** 765 slug ft/s **5.** 9.5×10^8 kg m/s **7.** 6.39×10^8 kg m/s
9. (a) 12,600 slug ft/s (b) 158 ft/s (c) $58\bar{0}0$ lb; 2580 lb
11. (a) 55,200 kg m/s (b) $17\bar{0}$ km/h **13.** 7.67 m/s **15.** (a) 14.0 m/s (b) 1750 kg m/s **17.** (a) 0.00287 s (b) $12{,}\bar{0}00$ N (c) 34.4 kg m/s (d) 34.5 kg m/s
19. 52.8 m

5.2 Pages 146–147

1. 8.58 m/s, right **3.** 1.75 m/s, right **5.** 1.71 m/s, north **7.** 3.75 m/s, right
9. 13.1 m/s **11.** (a) 3.95 m/s (b) 10.4 m/s

Chapter 5 Review Questions Pages 147–148

1. b **2.** d

3. The slow-moving truck has a large mass and a small velocity, whereas the rifle bullet has a small mass and a large velocity; in both cases, the product of mass and velocity is large.

4. They are the same.

5. The longer the bat (applied force) is on the ball, the greater is the impulse.

6. Total momentum in a system remains constant.

7. Momentum of the escaping gas molecules is equal to the momentum of the rocket.

8. Elastic **9.** Inelastic **10.** They are equal.

Chapter 5 Review Problems Page 148

1. 1.23×10^5 slug ft/s **2.** 0.204 m/s **3.** $15{,}\bar{0}00$ N s **4.** 8.5 kg m/s **5.** 0.556 m/s **6.** (a) 12 m/s (b) 1800 kg m/s **7.** (a) 4.62×10^{-4} s (b) 106,000 N (c) 49.0 kg m/s (d) 48.8 kg m/s **8.** (a) 70.4 m (b) 4.74 s **9.** 0.521 m/s, right **10.** 1.50 m/s, east **11.** 8.12 m/s, right

Chapter 5 Applied Concepts Page 149

1. (a) 8.46 slug ft/s (b) The outgoing velocity is less, thereby reducing the change in momentum and the impulse. **2.** (a) $p_{\text{adult}} = -1680$ N s; $p_{\text{child}} = 842$ N s (b) $F_{\text{adult}} =$ 2980 N; $F_{\text{child}} = 3240$ N **3.** (a) 3690 N (b) 1010 N (c) Bungee cords increase the time, thereby decreasing the force of the impulse. **4.** (a) 6.68 ft/s (b) It is easier to step out of a heavier canoe. The canoe has a greater inertia and does not move backward with as large a velocity. **5.** (a) -26.5 m/s $= -95.5$ km/h (the negative sign represents west) (b) Yes, the Jeep was speeding.

Chapter 6

6.1 Pages 155–157

1. $3\bar{0}$ N (right) **3.** (a) $40\bar{0}$ N (b) $5\bar{0}$ N **5.** 1640 N at 55.6° **7.** 2730 lb at 140.4° **9.** 4620 N at 123.2° **11.** 190 N at 126.3° from $\mathbf{F}_1$

6.2 Pages 164–165

1. $10\bar{0}$ N **3.** $26\bar{0}$ N **5.** $69\bar{0}0$ N **7.** $57\bar{0}$ N **9.** Yes
11. $\mathbf{F}_1 = 70.7$ N; $\mathbf{F}_2 = 70.7$ N **13.** $\mathbf{F}_1 = 577$ lb; $\mathbf{F}_2 = 289$ lb
15. $\mathbf{F}_1 = 433$ lb; $\mathbf{F}_2 = 50\bar{0}$ lb **17.** $\mathbf{T}_1 = \mathbf{T}_2 = 731$ lb **19.** $\mathbf{T}_1 = 565$ lb; $\mathbf{T}_2 = 613$ lb
21. $\mathbf{C} = 30\bar{0}0$ lb; $\mathbf{T} = 26\bar{0}0$ lb **23.** 5540 N **25.** $\mathbf{T} = 2330$ lb; $\mathbf{C} = 3690$ lb

6.3 Pages 168–169

1. 96.0 lb ft **3.** 2.00 m **5.** 187 N **7.** 1.60×10^3 N **9.** 0.357 m
11. 159 lb **13.** $40\bar{0}$ N **15.** (a) 25 lb (b) It is halved. **17.** 133 N

6.4 Pages 173–174

1. $10\bar{0}$ lb **3.** $50\bar{0}$ N **5.** $45\bar{0}$ N **7.** $40\bar{0}$ N **9.** 551 N; 331 N
11. 2.77×10^5 N; 1.07×10^4 N **13.** 2.67 m **15.** 872 N; 948 N

6.5 Pages 177–178

1. 22.6 **3.** $\mathbf{F}_1 = 99$ lb; $\mathbf{F}_2 = 88.0$ lb **5.** 1.52×10^4 N; 9.85×10^3 N
7. 9.90×10^4 N; 8.82×10^4 N **9.** 117 lb; 51 lb **11.** 8.99×10^4 N; 1.01×10^5 N
13. 197 N; 118 N **15.** 2390 N; 943 N **17.** 1020 N; 775 N
19. 1525 N (up) 4.54 m from *A*

Chapter 6 Review Questions Page 180

1. b **2.** b **3.** a **4.** c **5.** c **6.** a **7.** b **8.** b
9. No (e.g., bridge) **10.** They are equal.
11. Equilibrium is the condition of a body where the net force acting on it is zero.
12. Toward the center of the earth. **13.** They are in equilibrium.
14. A diagram showing how forces act on a body
15. No, only when the pedals are parallel to the ground.
16. Even if the vector sum of opposing forces is zero, they must also be positioned so there is no rotation in the system.
17. Choose a point through which a force acts to eliminate a variable.
18. (a) Stacking bricks (b) Riding a bicycle (c) Lifting any object (d) Hitting a baseball (e) Leaning into the wind
19. No; only if the object is of uniform composition and shape
20. The support closer to the bricks

Chapter 6 Review Problems Pages 180–183

1. 569 N (left) **2.** (a) $50\bar{0}$ lb (b) $5\bar{0}$ lb **3.** 3700 N at 156°
4. 8450 lb at 334.5° **5.** 94,600 N at 80.3° **6.** 1610 N at 114.2° from $\mathbf{F}_1$
7. 470 N **8.** 6.0 tons **9.** –525 **10.** 1080 N at 33.7°
11. $\mathbf{F}_1 = 645$ N; $\mathbf{F}_2 = 1520$ N **12.** $\mathbf{F}_1 = 5240$ lb; $\mathbf{F}_2 = 2210$ lb
13. $\mathbf{T} = 1790$ N; $\mathbf{C} = 1370$ N **14.** $\mathbf{T}_1 = 348$ lb; $\mathbf{T}_2 = 426$ lb; $\mathbf{T}_3 = 475$ lb
15. $\mathbf{T}_1 = 2900$ N; $\mathbf{T}_2 = 1900$ N; $\mathbf{T}_3 = 2200$ N
16. $\mathbf{T}_1 = 6250$ N; $\mathbf{T}_2 = 6250$ N; $\mathbf{C} = 6250$ N **17.** 26.5 N m **18.** 27.0 lb
19. 880 N **20.** 160 lb **21.** 100.0 kg **22.** 51.0 cm **23.** $56\bar{0}0$ kg
24. End closest to truck, 62,300 N; other end, 47,700 N **25.** 343 N; 483 N
26. 2580 N

Chapter 6 Applied Concepts Page 183

1. (a) $23\bar{0}0$ N m (b) 1150 N m (c) Continually push perpendicular to the door
2. (a) 18.6° (b) 58.3 lb (c) 79.5 lb (d) The angle between the ropes in increased so that Sean and Greg not only pull up but also horizontally.
3. (a) 2.28 N m more torque. (b) 18.0 N less force **4.** (a) $\mathbf{T} = 45.5$ N m, which will not support the torque. (b) Reduce the angle between the wall and the bracket.
5. (a) $\mathbf{F}_{fulcrum} = 2210$ N; $\mathbf{F}_{bracket} = -1420$ N (negative sign means that the bracket pulls down) (b) Between the bracket and the fulcrum

Chapter 7

7.1 Pages 190–191

1. 34.3 N m **3.** 917 kJ **5.** 24.4° **7.** 16,200 J **9.** 4410 J **11.** 34,000 ft lb
13. 24.1 kJ **15.** 163,000 ft lb **17.** 6.2 MJ **19.** 28,900 J **21.** (a) 903 J
(b) −903 J

7.2 Pages 196–197

1. 18.9 N m/s or 18.9 W **3.** 0.533 s **5.** 12.4 W **7.** (a) 1.58 hp (b) 219 N
9. 59.4 s **11.** 1.49 kW **13.** 6.17 kW **15.** (a) 6.9 kW (b) 9.2 hp (c) 9.3 kW
17. Four times **19.** (a) 110 passengers (b) 22 kW **21.** 1.15 kW

7.3 Pages 200–201

1. 2460 N m **3.** 217 J **5.** (a) 80.7 ft/s (b) 3.09×10^6 ft lb **7.** 14.2 ft/s
9. (a) 294 kJ (b) 392 kJ **11.** 342 m/s **13.** 220 kJ **15.** (a) 2650 kW

(b) 3550 hp (c) 1.59×10^8 J or 159 MJ **17.** 4 **19.** (a) 1960 J (b) 1960 J (c) 2160 N (d) 1.31×10^5 N **21.** (a) 3150 J (b) 3150 J (c) The increase in gravitational potential energy is equal to the work done in raising the painter. The work done by the painter is the source of the increase in energy.

7.4 Pages 205–207

1. 7.00 m/s **3.** 72.7 m/s **5.** 87.3 ft **7.** 49.5 m/s **9.** $19\bar{0}$ ft/s
11.

t	s	v	KE	h	PE	Total
0.000	0.00	0.00	0	300.0	11,760	11,760
1.000	4.90	9.80	190	295.1	11,570	11,760
2.000	19.6	19.60	770	280.4	10,990	11,760
3.000	44.1	29.40	1,730	255.9	10,030	11,760
4.000	78.4	39.20	3,070	221.6	8690	11,760
5.000	122.5	49.00	4,800	177.5	6960	11,760
6.000	176.4	58.80	6,910	123.6	4850	11,760
7.000	240.1	68.60	9,410	59.9	2350	11,760
7.800	300.0	76.68	11,760	0.00	0.00	11,760

Chapter 7 Review Questions Pages 207–208

1. c **2.** a **3.** a **4.** d **5.** c **6.** No **7.** No **8.** $J = \left(\frac{\text{kg m}}{\text{s}^2}\right)(\text{m})$
9. No **10.** No **11.** Yes
12. By measuring the force applied, the distance traveled, and the time taken
13. It possesses the ability to do work (e.g., turn a wheel) in falling to the lower level.
14. (a) Elevator counterweights (b) Roller coasters
15. Yes; $KE = \frac{1}{2}mv^2$ **16.** At its lowest point **17.** At its highest point
18. No **19.** Yes **20.** The bolt has accelerated to a higher velocity.

Chapter 7 Review Problems Page 208

1. 3.6×10^6 J **2.** 0 **3.** 10.2 m **4.** 9.80×10^5 J **5.** (a) 27.2 W (b) 0.0365 hp **6.** 0.102 m **7.** 2.00 ft **8.** 1.41 m/s **9.** 1.40 m/s **10.** 303,000 ft lb **11.** 2350 J **12.** $3\bar{0}00$ J **13.** 1670 J **14.** (a) $3\bar{0}00$ J (b) 1880 J **15.** 31.3 m/s

Chapter 7 Applied Concepts Page 209

1. (a) 4.49 W (b) The mass of the water, the distance the substance is carried, and the time to move the water **2.** (a) 2.09×10^5 J (b) A, 34.3 m/s; B, 28.0 m/s; C, 24.3 m/s; D, 19.8 m/s (c) The higher the elevation, the less is the velocity
3. (a) 4.80×10^7 J (same as the jet's original kinetic energy) (b) 4.17×10^5 N (c) The force needed to stop the jet is less because the time is more.
4. (a) 5.27×10^9 W (this is enough energy to power all of Paraguay) (b) 1.05×10^{10} W (c) The higher the dam, the more potential energy changes to kinetic energy.
5. (a) 1.56×10^5 J (b) 1.56×10^5 J (c) 3.57×10^5 N (d) When the wrecking ball is at its lowest point, all of its energy has been converted to kinetic energy, and thus it strikes the wall with the greatest velocity.

Chapter 8

8.1 Pages 217–219

1. (a) 40.8 rad (b) 2340° **3.** (a) 12.5 rev (b) 4500°

5. 154 rpm **7.** 8.38 rad/s **9.** 354 rev/s **11.** 6450 rpm
13. 14.2 rad/s **15.** $66\bar{0}$ rpm **17.** (a) 6.67 rev/s (b) $40\bar{0}$ rpm (c) 41.9 rad/s
19. (a) 0.0571 s (b) 87.5 rev **21.** 8.40 rad **23.** 0.131 m **25.** 1.33 m/s
27. (a) 68.6 rad/s (b) 12,300 rad (c) 157 cm (d) 3080 m (e) 17.2 m/s
29. (a) 50.5 rad/s (b) 1520 rad (c) 502 m (d) 502 m **31.** 188 in./s or 15.7 ft/s
33. (a) 1680 km/h (b) 1450 km/h (c) 1190 km/h (d) 838 km/h **35.** 1.33 rad/s^2
37. (a) –3.60 rad/s^2 (b) 432 m/s (c) 18.3 rev

8.3 Pages 223–224

1. 4350 N **3.** 79.4 slugs **5.** 5.53 m/s **7.** 1.92 m **9.** 3.60×10^3 lb
11. 28.8 m **13.** 5420 N **15.** 19 N **17.** $20\bar{0}0$ N

8.4 Page 227

1. 7260 **3.** 18.7 **5.** 1860 N m **7.** 343 hp **9.** 4.95/s
11. 4600 W **13.** (a) 299 hp (b) 599 hp **15.** 2.65 N m **17.** 37 kJ/min
19. 6.67/s **21.** 2800 W **23.** 7.17 N m

8.6 Pages 235–237

1. 52 rpm **3.** 24 teeth **5.** 207 rpm **7.** 42.5 rpm **9.** 75 teeth
11. 63 teeth **13.** 144 rpm **15.** 60 teeth **17.** 9 teeth
19. Counterclockwise **21.** Clockwise **23.** Clockwise **25.** Clockwise
27. Counterclockwise **29.** 1050 rpm **31.** 576 rpm **33.** 1480 rpm
35. 40 teeth **37.** 20 teeth **39.** Gear *B* is reversed in all problems.

8.7 Pages 239–240

1. 2250 **3.** $36\bar{0}0$ **5.** 147 **7.** 62.5 rpm **9.** 22.5 in. **11.** Clockwise
13. Counterclockwise **15.** Counterclockwise

Chapter 8 Review Questions Page 241

1. d **2.** a **3.** b
4. Curvilinear motion is motion along a curved path; rotational motion occurs when the body itself is spinning.
5. Radians and revolutions
6. A radian is an angle with its vertex at the center of a circle whose sides cut off an arc on the circle equal to its radius.
7. Angular displacement is the angle through which any point on a rotating body moves. It can be measured in radians or revolutions.
8. Linear velocity = angular velocity × radius
9. They are alike except for the substitution of θ for s, ω for v, and α for a.
10. Law of conservation of angular momentum
11. Yes
12. No; it tends to cause a body to continue in a straight line (tangent to the curve).
13. Number of teeth of driver times number of revolutions of driver = number of teeth of driven gear times number of revolutions of driven gear
14. An idler changes direction of rotation of the driver gear.
15. Opposite
16. Since the gears are connected, they both rotate together.
17. Diameters are used in similar equations rather than teeth.
18. The small pulley
19. The belt is one continuous piece of material.

Chapter 8 Review Problems Page 242

1. (a) 26π or 81.7 rad (b) 4680° **2.** 1.7 rad/s **3.** 0.885 m/s
4. 7.50 N **5.** 2.75 m/s **6.** 5.67 ft lb/s **7.** 6.00 rad/s
8. 6.00 rpm **9.** 20 teeth **10.** 57.5 rpm **11.** Yes **12.** 105 rpm
13. 2.00 cm **14.** Counterclockwise **15.** $72\bar{0}$ rpm **16.** 32 teeth

Chapter 8 Applied Concepts Page 243

1. (a) 1.78 rad/s = 17.0 rpm (b) 2.51 rad/s = 24.0 rpm (c) No, the mass of the person is unrelated to the rotational speed. **2.** (a) $24\bar{0}$ Nm (b) 405 kg m^2 (c) Increase the radius of the waterwheel or make the waterwheel a ring and not a solid disk. More mass farther away from the axis increases the rotational inertia. **3.** (a) B, 8.00 *g*'s; D, 2.00 *g*'s (b) B is not safe. The B loop should have a larger radius to reduce the centripetal force. D is safe. **4.** (a) 86.8 ft/s = 59.2 mi/h (b) 123 ft/s = 83.9 mi/h (c) As the radius increases, the centripetal force is reduced and the maximum velocity increases. **5.** (a) 2.26×10^4 lb ft/s = 41.1 hp (b) 43.2 lb ft (c) The lower the gear, the lower is the speed and the greater is the torque. More torque is required to move up steep hills.

Chapter 9

9.2 Pages 252–253

1. 14.8 **3.** 36.3 **5.** 52.4 **7.** 2.39 **9.** 48.8 **11.** 1.55 **13.** 2.27
15. (a) $11\bar{0}0$ lb (b) 4.00 **17.** (a) 1.50 ft (b) 4.00

9.3 Pages 254–255

1. 14.0 **3.** 271 **5.** 524 **7.** 48.8 **9.** 20.4 **11.** (a) 438 lb (b) 6.00
13. (a) 2.99 cm (b) 3.34

9.4 Pages 258–259

1. 1 **3.** 3 **5.** 6 **7.** 2
15. 2
17. (a) 388 N (b) 82.0 m
19. No

9.

11.

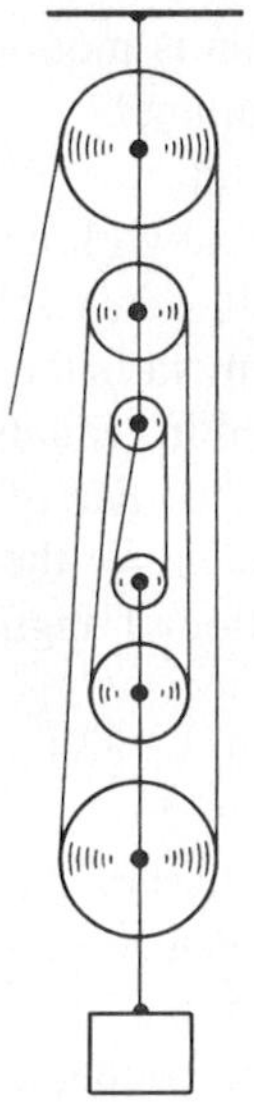

13.

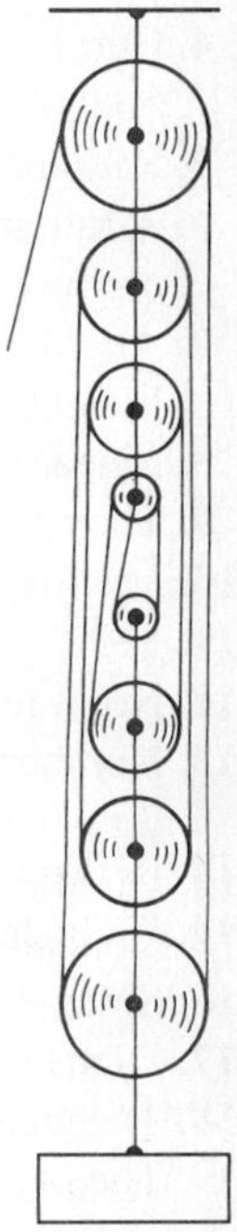

21. Mechanical Advantage (MA)

Number of Pulleys	1	2	3	4	5	6	7	8
Fixed	1	1	2	2	3	3	4	4
Movable	0	1	1	2	2	3	3	4
Fixed		0	1	1	2	2	3	3
Movable		1	1	2	2	3	3	4

9.5 Pages 261–262

1. 12.8 **3.** 21.2 **5.** 36.3 **7.** 4.62 **9.** 5.61 **11.** (a) 4.00 (b) 182 N (c) No **13.** (a) 1.33 m (b) 4.33

9.6 Page 264

1. 2.30 **3.** 14.2 **5.** 2.28 **7.** 29.3 **9.** 35.2 **11.** (a) 4.34 lb (b) 754 N **13.** (a) 37.7 (b) 566 lb (c) 188 lb

9.8 Pages 266–267

1. 15.0 **3.** 40.0 **5.** 75.0 lb **7.** 125 N **9.** 1.44×10^5 N

Chapter 9 Review Questions Pages 269–270

1. d **2.** a **3.** b **4.** d **5.** b **6.** b
7. (a) Bicycle (b) Auto transmission (c) High-speed drill **8.** Resistance force
9. Effort force × effort distance = resistance force × resistance distance
10. Mechanical advantage **11.** Efficiency **12.** No **13.** The fulcrum
14. $\text{MA} = \dfrac{\text{Effort arm length}}{\text{Resistance arm length}}$
15. First class **16.** $F_R \times s_R = F_E \times s_E$
17. The opposite end of the resistance force with the effort force in between
18. Resistance force × resistance radius = effort force × effort radius
19. No; it depends on the radii.
20. A fixed pulley does not move. It is suspended by its center axle. A movable pulley is free to move and is suspended by the strand around the groove.
21. No **22.** $\text{MA} = \dfrac{\text{Length of plane}}{\text{Height of plane}}$
23. The distance a screw advances into the wood in one revolution
24. It is greater because the handle of the jackscrew can be longer than the radius of the screwdriver.

Chapter 9 Review Problems Pages 270–271

1. 2.00 **2.** 73.0% **3.** 3.82 **4.** (a) 540 N (b) 360 N **5.** (a) 6.54 cm (b) 2.60 **6.** 12 **7.** 675 N **8.** (a) 292 N (b) 3.00 **9.** 2.50 ft **10.** 16,400 N **11.** 1.08 mm **12.** 151 **13.** (a) 0.112 (b) 0.106 (c) 16.4 N **14.** (a) The MA is doubled. (b) 1.57 cm **15.** 1.74 **16.** 81.4 cm **17.** 5 **18.** 4 **19.** (a) 54.0 (b) 4800 N **20.** 6.0

Chapter 9 Applied Concepts Pages 271–272

1. (a) 2.62×10^{31} m (b) $\text{MA}_{\text{lever}} = 6.82 \times 10^{22}$ This was clearly a theoretical statement by Archimedes. **2.** (a) 0.120 (b) 2.54 m (c) 4.58 m/s (d) Increases the speed of the resistance end. The tip of the fishing pole needs to be moving at a high speed to cast the lure. **3.** (a) 7.16 (b) 3.27 lb (c) The MA would increase.

4. (a) 0.118 (b) 6.42 lb (c) 10.4 in. (d) An MA less than one is advantageous for speed. The bike must be in a high gear and moving quickly. **5.** (a) 3.29 (b) 3.53 (c) Increase the length of the lever arm and/or the radius of the wheel.

Chapter 10

10.1 Pages 278–279

1. (a) 826 N (b) 0.497 N **3.** 1.23×10^{22} N **5.** 5.70×10^{16} N **7.** 3.37×10^{5} N **9.** 3.61×10^{-51} N

10.3 Page 283

1. 1020 m/s **3.** 4.79×10^{4} m/s **5.** 9630 m/s **7.** 7.59×10^{6} s or 0.241 yr **9.** 9.32×10^{8} s or 29.6 yr

Chapter 10 Review Questions Page 284

1. d **2.** b **3.** a **4.** c **5.** a

6. The earth's mass gives it a greater gravitational force.

7. The increased radius would cause you to weigh one fourth of your original weight.

8. The increased mass would cause you to weigh twice your original weight.

9. The satellite's horizontal velocity enables it to continually miss the earth as it falls.

10. With no horizontal velocity, the apple struck the ground. The moon's horizontal velocity enables it to maintain its orbit.

11. The force at the perigee would be greater than that at the apogee.

12. No, the mass in the period equation is the mass of the object being orbited.

Chapter 10 Review Problems Page 285

1. 9.60×10^{-11} N **2.** 2.82×10^{-7} N **3.** 636 N; 143 lb **4.** 1610 N; 362 lb **5.** 49.2 N; 11.1 lb **6.** 8.37×10^{5} s or 9.69 days **7.** 6.70×10^{6} s or 77.5 days **8.** 1.90×10^{7} s or $22\bar{0}$ days **9.** 832 N **10.** 745 N

Chapter 10 Applied Concepts Page 285

1. (a) 3.70 m/s^2 (b) 315 N; 833 N (c) The astronaut's muscles would weaken and have a difficult time supporting the returning astronaut on the earth. **2.** (a) 1.57×10^{7} m (b) 2.46 earth radii **3.** (a) $r = 4.22 \times 10^{7}$ m from the center of the earth; $r = 3.58 \times 10^{7}$ m from the surface of the earth (b) 3.07×10^{3} m/s $= 1.11 \times 10^{4}$ km/h **4.** (a)1.61×10^{3} m/s $= 5.80 \times 10^{3}$ km/h (b) 7370 s or 2.05 h (c) Increasing the altitude would decrease the velocity and increase the period. **5.** (a) 565 N (b) The astronauts only feel weightless because they are in continuous free fall.

Chapter 11

11.2 Pages 301–303

1. 2.50 m × 0.80 m: 71 kPa; 2.50 m × 0.45 m: 130 kPa; 0.80 m × 0.45 m: 390 kPa **3.** (a) 46.2 in. (b) 20.3 lb **5.** $20\bar{0}$ N/m. **7.** (a) 0.0250 cm (b) 1.56×10^{6} N

9. 3.57×10^5 lb **11.** 5.97×10^7 Pa or 59.7 MPa
13. (a)

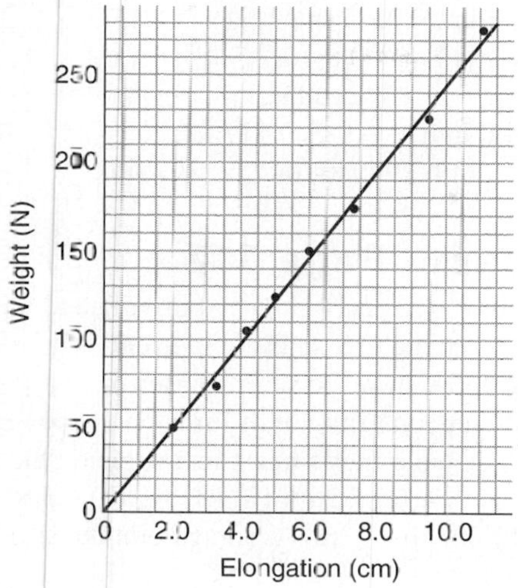

(b) 180 N (c) 9.0 cm (d) 25 N/cm **15.** 16.8 cm; 12.6 cm **17.** 489 N
The breaking point is dependent on the diameter of the wire and not on its length.
19. 0.827 J

11.5 Page 313

1. 2870 kg/m^3 **3.** 1750 lb **5.** 5600 cm^3 **7.** 1210 lb/ft^3 **9.** 2710 kg/m^3
11. 684 kg/m^3 **13.** 3.5 m^3 **15.** 58.8 lb/ft^3 **17.** 1.49 m^3 **19.** 2820 kg/m^3
21. (a) 1000 L (b) 1500 L (c) 73.5 L **23.** 0.917 **25.** 7.8
27. 0.68 **29.** Floats **31.** Floats **33.** 5.8×10^{15} **35.** 2330 kg/m^3

Chapter 11 Review Questions Pages 315–316

1. a, c, and e **2.** All **3.** d **4.** d **5.** d **6.** b **7.** d **8.** e
9. Mass density refers to the mass per unit volume; weight density refers to weight per unit volume.
10. Yes; no.
11. Capillary action refers to the effect of surface tension of liquids that causes the level of liquid in a small-diameter tube to be higher or lower than that of the liquid in a large-diameter tube due to the adhesive force between the liquid and the tube.
12. Adhesion refers to the attractive force between different molecules; cohesion refers to the attractive force between similar molecules.
13. The surface tension of water allows the base of many insects' legs to be supported by the surface of water, allowing them to "walk" on the surface of a pond.
14. 1800 **15.** Pressure
16. Stress is directly proportional to strain as long as the elastic limit is not exceeded.
17. kPa
18. The specific gravity of an object can be found by dividing the density of the object by the density of water. The density can be found by determining the mass of the object and dividing by its volume.
19. Mass density **20.** Viscosity **21.** Elastic limit **22.** Solid, liquid, gas
23. An atom consists of one nucleus and its surrounding electrons; a molecule consists of two or more atoms.
24. A proton has a positive charge; a neutron has a neutral charge.
25. Tension, compression, shear, bending, and twisting
26. The hydrometer measures the density of the battery electrolyte. The density is related to the amount of sulfuric acid in the electrolyte and therefore the charge on the battery. Temperature does affect the measurement.

Chapter 11 Review Problems Page 316

1. 165 N **2.** 24.6 cm **3.** 1.63×10^6 lb/in.; 0.0317 in.
4. 311 MPa **5.** 2020 lb/ft^3 **6.** 1.3 N **7.** 8950 kg/m^3
8. 83.4 lb/ft^3 **9.** 55.8 cm^3 **10.** 21.9 m^3 **11.** 2160 kg
12. 59,300 lb **13.** 18.9 lb/ft^3 **14.** 54.8 lb/ft^3 **15.** 0.117 ft^3
16. 0.694 **17.** 77.8 lb **18.** 9.26 kg **19.** 2960 kg/m^3 **20.** Sink

Chapter 11 Applied Concepts Page 317

1. (a) 3.08×10^5 N/m^2 (b) 1.27×10^5 N/m^2 (c) High-speed driving and the high-pressure tire together increase the heat energy in the tire. This could eventually break down the tire, resulting in a blowout. **2.** (a) A shearing strain causes the warping of the lines. (b) An automobile exerts a tremendous amount of force on the pavement to begin its acceleration. On the open road, the vehicle exerts much less force on the pavement to continue its motion. **3.** 3.72 ft^3 **4.** (a) 7.26×10^4 kg (b) 6.3×10^4 kg (c) 4.9×10^4 kg **5.** (a) 178 lb/in. (b) 201 lb (c) At some point the spring will not be able to compress any farther.

Chapter 12

12.1 Pages 325–326

1. 21.7 lb/in^2 **3.** 16.5 lb/ft^3 **5.** 55.2 lb/in^2 **7.** 245 kPa **9.** 757 kg/m^3
11. 57.8 lb/in^2 **13.** 233 lb/in^2 **15.** 2.22×10^8 N **17.** 49$\bar{0}$ kPa
19. 2870 kPa

12.2 Pages 329–330

1. 60.0 lb **3.** 22.0 **5.** 24.0 **7.** 48 N **9.** 23,000 N
11. (a) 146 N (b) 11.6 N/cm^2 (c) 36.0
13. (a) 48$\bar{0}$0 N (b) 9.55 kPa (c) 9.55 kPa (d) 64.0
(e) The lift will exert twice the force on the large piston; the MA and the pressure will double.
(f) The lift will exert four times the force on the large piston; the MA and the pressure will increase by a factor of 4.
15. Increased by a factor of 4. **17.** 2610 N

12.3 Pages 332–333

1. 118 lb/in^2 **3.** 293 kPa **5.** (a) 303.96 kPa (b) 202.64 kPa (c) 88.2 lb/in^2
(d) 506.6 kPa (e) 33.77 kPa (f) 25.33 kPa **7.** 586 kPa **9.** 30.3 lb/in^2
11. 306 kPa **13.** 1174 kPa **15.** 401 kPa

12.4 Pages 335–336

1. 13.0 lb **3.** 3.9 N **5.** 75.5 lb **7.** 3.20 N **9.** 105 lb **11.** 431 N
13. (a) 3.06×10^4 ft^3 (b) 795 tons **15.** 7.65 m^3 **17.** 440 N

12.5 Pages 340–341

1. 358 L/min **3.** (a) 9.6 cm (b) 0.34 m/s **5.** (a) 34.1 ft/min (b) 136 ft/min
7. 16.7 cm

Chapter 12 Review Questions Pages 341–342

1. c **2.** b **3.** b **4.** c **5.** a **6.** kPa
7. Pressure is the force applied per unit area.
8. $F_s = \frac{1}{2}AhD_w$ or one-half the force exerted by the water on the bottom of the tank.

9. External pressure applied to a fluid is transmitted to all inner surfaces of the liquid's container.
10. A ship floats because it is lighter than an equal volume of displaced water.
11. The spinning baseball creates a different air velocity on one side than on the other. This creates a higher pressure on one side, which causes the ball to "curve."
12. The top side of the wing is curved more than the bottom, so the velocity of air rushing past the top must be larger than that going past the bottom. This creates a low-pressure area at the top of the wing and creates lift.
13. In streamline flow, all particles of the fluid passing a given point follow the same path. In turbulent flow, the particles passing a given point may follow different paths from that point.
14. A balloon filled with a gas with lower density than air, such as helium, rises.
15. Absolute pressure is the actual air pressure given by the gauge reading plus normal atmospheric pressure. Gauge pressure is the pressure measured relative to atmospheric pressure. Gauge pressure is used in an automobile tire gauge.
16. The pressures are identical. The forces are different.
17. This force depends on the horizontal surface area and the average pressure on that surface. The average pressure depends on the density and the height of the liquid.
18. The pressure at the bottom is greater than at the top.
19. Smaller. The total force exerted on the brake pads must be larger than the force applied to the brake pedal.
20. Yes, but only if the fluid to be drawn through the straw is in an airtight sealed container.

Chapter 12 Review Problems Pages 342–343

1. 31.8 kPa **2.** 455 ft **3.** 30.3 m **4.** $93\bar{0}0$ lb **5.** 196 kPa
6. 9.60×10^7 N **7.** 7.01×10^8 N **8.** 2.80×10^4 N **9.** 2610 kPa
10. 192 lb/in^2 **11.** 9410 kPa **12.** (a) 77.2 lb (b) 8.42 **13.** 35.5 N
14. (a) 103 N (b) 111 N/cm^2 (c) 42.2 **15.** 303 kPa **16.** 554 kPa
17. 299 lb/in^2 **18.** 3.3 N **19.** 6.30 N **20.** 2.4 N **21.** $26{,}\bar{0}00$ ft^3; 654 tons
22. 188 L/min **23.** 3610 L/min; 90,300 L **24.** (a) 49.3 N/m^3 (b) Alcohol

Chapter 12 Applied Concepts Page 343

1. (a) 372 m^2 (b) 2.76×10^6 N (c) The bands should be spaced closer together because of the increased pressure at the bottom of the tank. **2.** (a) 44.6 lb/in^2 (b) 429 N on each piston or brake drum **3.** (a) 0.703 m^3 (b) 16,000 N (c) 7060 N (d) 8900 N **4.** (a) 50.5 mi/h (b) The space between the buildings acts as a Venturi tube by limiting the area that the air can pass through. **5.** (a) 201 in./s = 16.8 ft/s (b) Limit the number of holes or make the holes smaller. Limiting the area for the water to come out causes it to travel faster.

Chapter 13

13.1 Page 349

1. 25°C **3.** 125°C **5.** 293°F **7.** −12.2°C **9.** 203°F **11.** 298 K
13. 133°C **15.** 223 K **17.** 5727°C **19.** −38.9°C **21.** 29°C **23.** −321°F
25. $12\bar{0}0$ **27.** 941 **29.** 1175 **31.** $59\bar{0}$

13.2 Page 351

1. 23 cal **3.** 1.21×10^6 ft lb **5.** 3.21×10^6 J or 3.21 MJ **7.** 4450 Btu
9. 2.62 MJ **11.** 42.2 MJ **13.** (a) 2.77×10^5 J/s (b) 277 kW

13.3 Pages 356–357

1. 0.021 ft^2 °F h/Btu **3.** 0.45 ft^2 °F h/Btu **5.** 1.7 ft^2 °F h/Btu
7. 8.1×10^6 Btu **9.** 11 J **11.** 1400 Btu **13.** 14 MJ **15.** 6800 J

13.4 Pages 359–360

1. 173 Btu **3.** $38\bar{0}0$ Btu **5.** 2.1×10^5 J or 210 kJ **7.** 280 cal
9. 3.01×10^4 J or 30.1 kJ **11.** 64.0 kcal **13.** 0.259 kg **15.** 29,000 kcal
17. 1.89×10^8 J or 189 MJ **19.** 8.07×10^7 J or 80.7 MJ **21.** 252°C
23. 5770 J **25.** (a) 10.1°C (b) 22°C (c) Water is the better coolant because its temperature increase is only about half that of alcohol in absorbing the same amount of heat.

13.5 Pages 362–363

1. 428°F **3.** 0.051 cal/g °C **5.** 95°F **7.** 81°C **9.** 0.104 cal/g °C
11. 286°C **13.** 0.259 kg

13.6 Pages 367–368

1. 0.30 ft **3.** 0.10 m **5.** 200.10 m **7.** 0.54 ft **9.** 0.752 in.
11. 1.8 cm^2 **13.** 88.97 cm^2 **15.** 60.10 cm^3 **17.** 4.9 mm
19. 1.08×10^{-5}/°C **21.** 15 cm^3 **23.** 0.3397 m^3

13.7 Page 370

1. 653 L **3.** 12.2 m^3 **5.** 3754 ft^3 **7.** 0.58 cm^3 **9.** $215 **11.** 23°C

13.8 Pages 379–380

1. 1100 cal **3.** 10,700 Btu **5.** 26,000 cal **7.** 6.70×10^6 J or 6.70 MJ
9. 3.39×10^6 J or 3.39 MJ **11.** 6070 Btu **13.** 11,650 Btu **15.** $5\bar{0}00$ kcal
17. 3.12×10^6 J or 3.12 MJ **19.** 467,000 cal **21.** 3090 kcal

Chapter 13 Review Questions Pages 382–383

1. a, b, and d **2.** b and d **3.** a, b, and d **4.** a, b, and d **5.** d **6.** c
7. The total heat lost by warm objects is gained by cold objects.
8. 778 ft lb of work is equivalent to 1 Btu.
9. The Rankine scale. **10.** The Kelvin scale.
11. Each Celsius degree is 1.8 times the Fahrenheit degree. The freezing point of water is 0°C and 32°F. The boiling point of water is 100°C and 212°F.
12. Heat is an amount of energy. Temperature is a measure of the average kinetic energy of atoms and molecules in matter.
13. Burning fuel in a furnace. Burning fuel in an engine. Conversion of heat into electricity in a steam-driven generator.
14. Heat generated in a solid by a drill bit. Heat generated by rubbing two objects together. Heat generated in automobile brakes. All are due to friction.
15. Light clothing should be worn. It absorbs less heat.
16. It increases. The metal block increases in size, as does the hole.
17. (a) 4 degrees Celsius. (b) Most other liquids have their highest density at the temperature at which the change of state to a solid occurs. (c) Water below 4 degrees Celsius is less dense than water at 4 degrees Celsius and rises. Freezing therefore occurs at the top of a body of water.

18. Ten kilograms of ice at 0°C. The difference is the heat of fusion, which must be added to the ice to turn it to water.
19. The large amount of heat released when steam changes to water can cause severe burns.
20. Because water expands as it solidifies.
21. Conduction is the transfer of kinetic energy from atoms or molecules in a warmer location through nearby atoms or molecules to atoms or molecules in a colder section of the material. Convection is the transfer of kinetic energy from one region to another via the motion of warmer atoms or molecules from one region to another. Radiation is the transfer of energy through the emission and absorption of electromagnetic radiation.
22. Coolants boil at a higher temperature at high pressure. Therefore the engine coolant can be at a higher temperature and transfer more heat from the engine to the atmosphere because of the greater temperature difference.
23. Heat is extracted from the outside air in winter to vaporize the refrigerant. The heat is given up by the condensing fluid to warm the inside air. In summer, heat is extracted from the inside air to vaporize the refrigerant. The condensing fluid then gives up this heat to the outside air.

Chapter 13 Review Problems Page 383

1. 71°C **2.** 297 K **3.** 9230°F **4.** 335°C **5.** 10.3 cal
6. 1.56 kcal **7.** 3.38×10^5 ft lb **8.** 1.4×10^5 Btu **9.** 2.4×10^4 J or 24 kJ
10. 4.18×10^4 Btu **11.** 1.5×10^4 kcal **12.** 1.22×10^4 kcal **13.** 1300°F
14. 0.353 cal/g °C **15.** 85°C **16.** 12.524 m **17.** 7.51 mm; 44.3 mm^2
18. 0.0295 cm^3 **19.** 2.57 L **20.** 123 ft^3 **21.** 11,600 kcal
22. 12$\bar{0}$0 Btu **23.** 3440 kcal **24.** 4.80×10^5 J

Chapter 13 Applied Concepts Page 384

1. 1.25×10^4 J **2.** 4.52×10^5 kcal **3.** (a) 8.4×10^6 kcal (b) 8.9×10^6 kcal (c) 1.8×10^5 kcal **4.** (a) Steel will contract less. (b) 0.0161 ft = 0.193 in.
5. (a) 2.43 gal (b) $0.14

Chapter 14

14.1 Pages 389–390

1. 288 K **3.** 44°C **5.** 532°R **7.** 9$\bar{0}$°F **9.** 258 cm^3 **11.** −16°F
13. 38$\bar{0}$ m^3 **15.** 1430 L **17.** −59°C **19.** 30.5 L **21.** 107 K
23. −111°C **25.** 154 L

14.2 Pages 392–393

1. 265 cm^3 **3.** 75.0 kPa **5.** 2.08 kg/m^3 **7.** 0.180 lb/ft^3 **9.** 90.4 kPa
11. 3.34 kg/m^3 **13.** 801 kPa **15.** 19.0 ft^3 **17.** 162 psi
19. (a) 37.5 in^3 (b) 25.0 in^3 (c) 15$\bar{0}$ in^3 **21.** 4.37 kg/m^3
23. 1.16 L **25.** 0.352 kg/m^3 **27.** (a) Lower (b) 12.4 kPa

14.3 Page 396

1. 1270 in^3 **3.** 506 m^3 **5.** −39°C **7.** 143°C **9.** 22$\bar{0}$0 psi **11.** 399 kPa
13. (a) 7$\bar{0}$0 kPa (b) 891°C (c) 3.59 m^3 (d) 0.563 m^3 **15.** 12.3 L

Chapter 14 Review Questions Page 397

1. c **2.** a **3.** c **4.** c **5.** b **6.** b
7. Standard pressure is 101.32 kPa or 14.7 lb/in^2 and standard temperature is 0°C or 32°F.

8. A temperature increase tends to cause a volume increase. A pressure increase tends to cause a volume decrease. If both temperature and pressure are increased, the volume change is given by the combined Charles' and Boyle's laws.
9. The temperature will increase.
10. Heating a gas increases the kinetic energy of the gas molecules. This causes an increase in pressure and volume.
11. When a gas is compressed, work must be done on it to decrease its volume. This work is transferred into increased kinetic energy of the molecules of the gas, causing a temperature increase.
12. When the volume is increased, the number of gas molecules striking the surface of the container per unit area decreases. Thus the pressure exerted by the gas molecules on the container decreases.

Chapter 14 Review Problems Pages 397–398

1. 14.9 ft^3 **2.** 4.40 m^3 **3.** 131°F **4.** 30.9°C **5.** 34.1°C **6.** 138 kPa **7.** 23.9 ft^3 **8.** 3.68 kg/m^3 **9.** 478 in^3 **10.** 19.9°C **11** (a) 2180 psi (b) 18°F **12.** 304 kPa **13.** 135 kPa **14.** 822 L **15.** 3.07 kg/m^3

Chapter 14 Applied Concepts Page 398

1. (a) The temperature of the helium begins to decrease, and the volume of the balloon decreases. (b) 1.66 ft^3 (c) The volume would increase to 1.97 ft^3. (d) Overinflate the balloon in winter; underinflate the balloon in summer. **2.** (a) 39 lb/in^2 (b) 26 lb/in^2 **3.** (a) 1.88×10^{-4} m^3 (b) 138 kPa (c) 148 kPa **4.** (a) 6.68 m^3 (b) 1980 m^3 **5.** (a) $98\bar{0}$ L (b) No, the balloon should not be fully inflated. If it were, the balloon would probably burst on the way up.

Chapter 15

15.2 Page 411

1. 2.00 ms **3.** $8\bar{0}0$ m/s **5.** 1.93 m/s **7.** 0.813 m **9.** 12.0 Hz **11.** 2.00×10^8 Hz **13.** 5.21×10^{-3} m **15.** 2.06×10^{10} m/s

15.4 Pages 417–418

1. 337 m/s **3.** 317 m/s **5.** 5.00 s **7.** 2640 m **9.** 622 Hz **11.** 507 Hz **13.** 653 Hz **15.** 0.325 s **17.** (a) 695 Hz (b) 522 Hz

15.6 Page 423

1. 55.9 cm **3.** 2.24 s **5.** 4.13 in. **7.** 1.36 s **9.** It is $\sqrt{2}$ times the original period. **11.** 1.74 s

Chapter 15 Review Questions Pages 424–425

1. a, b, and c **2.** c **3.** a **4.** b **5.** b **6.** a **7.** c and d
8. Interference is the result of two or more waves traveling through the same region at the same time. Diffraction is the bending of a single wave passing near an obstacle with an opening nearly the same size as the wavelength.
9. The addition of two or more waves to form a larger wave is constructive interference. The formation of a smaller wave is destructive interference.
10. Sound would not be heard if some obstacle came between you and the stereo speakers and there were no nearby reflecting surfaces.
11. Waves passing through a break in a seawall.
12. It increases.

13. A wave is a periodic disturbance. A pulse is a one-time disturbance.
14. A sharp explosion creates a sound pulse.
15. The blue and violet light from the sun is diffracted by small molecules and dust particles in the air. The red light passes through without diffraction.
16. The blue light from the sun is diffracted by small molecules and dust particles in the atmosphere. This diffracted blue light gives the sky a blue color.
17. The speed of sound increases. The higher kinetic energy of air molecules at high temperature leads to a higher velocity of sound through the air.
18. A seismograph detects slight vibrations of the earth's surface by detecting the relative motion of the earth's surface and a massive object.
19. The speed of sound is higher in water than in air. The higher density of water provides a higher speed of sound.
20. Motion toward an oncoming sound wave increases the frequency at which the maximum-pressure regions in the sound wave strike an observer, therefore producing a higher frequency sound. Motion away from an oncoming sound wave decreases the frequency at which the maximum-pressure regions strike an observer, therefore producing a lower frequency sound.
21. Sympathetic vibrations occur when an object vibrates at its natural resonance frequency in response to the vibration of a nearby object at some other frequency. Forced vibration occurs when an object vibrates at the same frequency as another nearby vibrating object.
22. Resonance occurs when an object vibrates at its natural resonance frequency in response to the vibration at the same frequency of a nearby object.
23. The light from these stars is shifted toward the red as a result of their motion away from the earth.
24. Amplitude is maximum displacement.
25. Period is the time required for one full vibration. Frequency is the number of complete vibrations per unit of time.
26. No; the period of a pendulum is independent of its mass.

Chapter 15 Review Problems Pages 425–426

1. 2.82×10^{-6} s or 2.82 μs **2.** 3.13 Hz **3.** (a) 6.67×10^{14} Hz (b) 1.50×10^{-15} s
4. 5.87 m/s **5.** 1.24 m **6.** 65 Hz **7.** 4.58 m **8.** 383 m/s **9.** $31\bar{0}$ m/s
10. 0.967 s **11.** 29 s **12.** 602 Hz **13.** (a) 5360 Hz (b) $46\bar{0}0$ Hz
14. 368 Hz **15.** 5.59×10^{14} Hz **16.** 1.35 s **17.** 4.80 in.

Chapter 15 Applied Concepts Pages 426–427

1. (a) 0.248 m (b) No (c) 0.0411 m
2. (a) One wavelength (b) 560 ft/s (c) 1400 ft
3. (a) AM: 182 to 545 m; FM: 2.78 to 3.41 m (b) FM wavelengths are closer to the sizes of openings of tunnels and underpasses. As stated in the text, "Wave diffraction is commonly observed only when the opening is nearly the same size as the wavelength."
4. (a) 1.16×10^5 W (b) 2.50 W/m^2 (c) Twenty-five percent of the initial intensity
5. (a) 29 m/s (b) 706 Hz

Chapter 16

16.3 Page 435

1. 922 N **3.** 1.3 cm **5.** 1.5×10^{-8} C **7.** (a) −0.93 N (b) +0.707 N
(c) −2.08 N

16.4 Page 438

1. 1.50×10^4 N/C **3.** 0.500 N/C **5.** 1.78×10^{-3} C **7.** 0.676 N

16.5 Pages 443–444

1. 1.07 Ω **3.** 0.0165 Ω/ft **5.** 3.95 Ω **7.** 0.684 Ω **9.** 0.0131 cm^2

16.6 Pages 446–447

1. 4.79 A **3.** 22$\bar{0}$ V **5.** 153 Ω **7.** 17.6 Ω **9.** 1.26 A **11.** (a) 0.067 A (b) 3$\bar{0}$ V (c) 0.13 A **13.** 18.8 Ω **15.** 1.85 A

16.7 Pages 450–451

1. 13.50 Ω **3.** 60.0 Ω **5.** 0.750 A **7.** 378 V **9.** 23.0 Ω **11.** 10.8 Ω, 14.2 Ω, 19.0 Ω **13.** 2.78 Ω, 40.0 V, 3.40 Ω

16.8 Page 457

1. (a) 4.28 Ω (b) 11.7 A (c) 4.55 A (d) 7.14 A **3.** (a) 10$\bar{0}$ Ω (b) 1.67 A (c) 0.250 A

16.9 Pages 461–462

1. (a) R_2, R_3 (b) 3.00 Ω **3.** 8.89 A **5.** (a) 2.23 A (b) 6.68 A **7.** 21.24 Ω **9.** 45 V **11.** 5.41 A **13.** 10.0 Ω **15.** 17.3 Ω **17.** 30.0 V **19.** 6.93 V **21.** 13.3 Ω **23.** 60.2 V **25.** 0.370 A

16.10 Pages 463–465

1.

	V	I	R
Batt.	12.0 V	9.00 A	1.33 Ω
R_1	12.0 V	6.00 A	2.00 Ω
R_2	12.0 V	3.00 A	4.00 Ω

3.

	V	I	R
Batt.	36.0 V	6.00 A	6.00 Ω
R_1	36.0 V	2.00 A	18.0 Ω
R_2	36.0 V	3.00 A	12.0 Ω
R_3	36.0 V	1.00 A	36.0 Ω

5.

	V	I	R
Batt.	50.0 V	5.00 A	10.0 Ω
R_1	25.0 V	2.00 A	12.5 Ω
R_2	25.0 V	2.00 A	12.5 Ω
R_3	10.0 V	3.00 A	3.33 Ω
R_4	40.0 V	3.00 A	13.3 Ω

7.

	V	I	R
Batt.	36.0 V	9.00 A	4.00 Ω
R_1	12.0 V	6.00 A	2.00 Ω
R_2	12.0 V	3.00 A	4.00 Ω
R_3	24.0 V	6.00 A	4.00 Ω
R_4	24.0 V	3.00 A	8.00 Ω

9.

	V	I	R
Batt.	80.0 V	12.0 A	6.67 Ω
R_1	18.0 V	4.00 A	4.50 Ω
R_2	18.0 V	2.00 A	9.00 Ω
R_3	18.0 V	6.00 A	3.00 Ω
R_4	48.0 V	12.0 A	4.00 Ω
R_5	8.00 V	4.00 A	2.00 Ω
R_6	8.00 V	8.00 A	1.00 Ω
R_7	6.00 V	12.0 A	0.500 Ω

11.

	V	I	R
Batt.	65.0 V	5.00 A	13.0 Ω
R_1	10.0 V	0.500 A	20.0 Ω
R_2	10.0 V	1.00 A	10.0 Ω
R_3	10.0 V	2.50 A	4.00 Ω
R_4	10.0 V	1.00 A	10.0 Ω
R_5	25.0 V	5.00 A	5.00 Ω
R_6	30.0 V	5.00 A	6.00 Ω

16.12 Pages 469–470

1. 1.49 V **3.** 1.33 A **5.** (a) 3.75 A (b) 9.00 V (c) 0.0200 Ω **7.** 0.160 A
9. 0.120 A

16.13 Pages 473–474

1. 957 W **3.** 0.682 A **5.** 6.82 A **7.** $0.56 **9.** Yes **11.** $0.026
13. $0.046 **15.** (a) Microwave oven, two fluorescent bulbs, two light bulbs (b) Projection TV, personal computer, two lightbulbs.
17. 2.36 W **19.** 60.0 W **21.** $90\bar{0}$ J **23.** (a) $24\bar{0}$ W (b) 50.4 kWh (c) $5.54
25. (a) 9.58 mA (b) 1.10 W (c) $0.07

Chapter 16 Review Questions Pages 476–478

1. b **2.** c **3.** c **4.** b **5.** e **6.** a and e **7.** a **8.** c **9.** b **10.** c
11. c **12.** b and c

13. Materials can become charged when a charged object is brought nearby, inducing a polarization (separation) of charge on the material. If one side of the material is touched by another object, the charge at one side of the material can then be "drained" off, leaving the charge at the other side of the material.

14. Protons, electrons, and neutrons **15.** Protons and neutrons

16. Electrons are located in charge clouds surrounding the nucleus.

17. Positive and negative. Protons carry a positive charge; electrons carry a negative charge.

18. A charged object is brought into contact with the electroscope, thereby providing some of that charge to the electroscope.

19. A charged object is brought near the conducting ball on an electroscope, causing a polarization (separation) of charge on the electroscope. The conducting ball is touched by a "ground," allowing one type of charge to leave the electroscope and go to ground, leaving the other charge behind.

20. Coulomb's law states that the force between two charges is directly proportional to the product of the magnitude of the charges and inversely proportional to the square of the separation between the two charges.

21. An electric field at a point represents the magnitude and direction of the force that would be exerted on a single unit of charge if placed at that point.

22. Lightning is the discharge of built-up static charge on a portion of a cloud.

23. Current **24.** (a) Ampere (b) Volt (c) Ohm

25. It decreases the resistance by a factor of 4.

26. The voltage drop across a segment of a circuit equals the product of the current through that segment and the resistance of that segment of the circuit.

27. Current flows sequentially through each portion of a series circuit. The current is divided among different segments of a parallel circuit.

28. The equivalent resistance in a series circuit is the sum of the resistances in the circuit. In a parallel circuit, the equivalent resistance is given by the reciprocal of the sum of the reciprocals of all resistances.

29. The highest range

30. Water flow is split and flows in parallel through different segments of a water distribution system. In a similar manner, current is divided and flows in parallel through resistors connected in parallel.

31. The current decreases by a factor of 2. **32.** The current increases by a factor of 2.

33. The resistance increases by a factor of 2.

34. Electrical charges move from regions of higher potential to regions of lower potential. Chemical reactions in batteries raise charges to higher potential energy. These charges can flow through a circuit and do work on circuit elements (create heat, light, or motion).

35. Chemical energy is transferred into electrical potential energy in the dry cell. Charges flow through the two lamps, giving up their energy in the form of light and heat.
36. Primary cells are not rechargeable, whereas secondary cells are.
37. An electric current flows in the reverse direction through a secondary cell, causing the normal chemical reaction to proceed in reverse, thus restoring charge in the battery.
38. The electrolyte causes a chemical reaction at the plates and conducts current between the plates.
39. An electrolyte causes a chemical reaction at the plates that releases energy to force electrical charge to move through the battery and the outside circuit.
40. It decreases the voltage through the outside circuit. **41.** Watt
42. Power is given by the product of the voltage and the current.
43. We pay for our energy use. Energy consumed is the total work done. Power is the instantaneous use of electricity; energy is the power multiplied by the duration of time the power is used.
44. Power is the square of the voltage divided by the resistance.
45. The power increases by a factor of 2. **46.** The power increases by a factor of 4.
47. The power decreases by a factor of 4. **48.** The cost increases by a factor of 2.

Chapter 16 Review Problems Pages 478–480

1. 8.10×10^4 N **2.** 9.13×10^{-6} C **3.** 0.0671 m **4.** 2.00×10^5 N/C
5. 3.33×10^6 C **6.** 1.06 N **7.** 1.17 Ω **8.** 3.47 Ω **9.** 1.11 Ω
10. 145 ft **11.** 0.0105 cm^2 **12.** 7.47 A **13.** 25.1 Ω **14.** 0.491 A
15. 19.60 Ω **16.** 0.612 A **17.** 48.5 V **18.** 16.2 Ω **19.** 3.5 Ω
20. 2.45 Ω **21.** 44.9 A **22.** 19.5 A **23.** 25.5 A **24.** 5.00 Ω
25. 10.0 A **26.** 5.95 A; 4.05 A **27.** 4.42 Ω **28.** 17.1 A **29.** 58.5 V
30. 2.83 A **31.** 12.3 V
32.

	V	*I*	*R*
Batt.	35.0 V	4.70 A	7.45 Ω
R_1	5.00 V	2.75 A	1.82 Ω
R_2	5.00 V	1.95 A	2.56 Ω
R_3	13.2 V	4.70 A	2.80 Ω
R_4	7.50 V	0.97 A	7.73 Ω
R_5	7.50 V	3.73 A	2.01 Ω
R_6	9.30 V	4.70 A	1.98 Ω

33. 1.43 V **34.** 12.2 V **35.** (a) 3.95 A (b) 6.00 V (c) 0.0165 Ω **36.** 0.528 A
37. 6.67 Ω **38.** 39.3 W **39.** 1.36 A **40.** $1.01 **41.** 240 h **42.** 0.91 A

Chapter 16 Applied Concepts Page 480

1. (a) -8.23×10^{-8} N (b) 3.63×10^{-47} N (c) We would need an extremely large amount of charge on our bodies to feel that attraction. The earth has so much mass that we are noticeably attracted to the earth via gravity.
2. (a) 3.06×10^{-4} N (b) 3.06×10^{-4} N (c) 8.38×10^{-9} C (d) As the distance between the particles decreases, the force increases, which results in a stronger electric field in the sawdust. **3.** (a) 2.20×10^{-4} A (b) 1.10 A
4. (a) $I_{microwave} = 8.33$ A; $I_{bulb} = 0.333$ A; $I_{computer} = 4.58$ A (b) $R_{microwave} = 14.4\ \Omega$; $R_{bulb} = 36\bar{0}\ \Omega$; $R_{computer} = 26.2\ \Omega$ **5.** 3.32×10^{-4} m

Chapter 17

17.2 Page 491

1. 1.20×10^{-5} T **3.** 57.5 A **5.** 1.71×10^{-6} T **7.** 0.0196 T **9.** 0.249 A
11. 0.0126 T

Chapter 17 Review Questions Page 502

1. d **2.** b **3.** b **4.** Tesla

5. A tightly wound solenoid with many turns per unit length carrying a large current produces a strong magnetic field. A magnetic core such as iron significantly increases the magnetic field.

6. Use Ampère's rule to find the direction of the flux line from any single turn in the solenoid. The magnetic field direction of the solenoid is in this direction.

7. The current-carrying coil produces a magnetic field that causes the magnetic domains in the magnetic material to align in the direction of the coil's field. This produces a stronger induced field in the core.

8. A moving magnet induces a current in the generator's coil.

9. The commutator is a split ring that allows the current produced by a generator always to flow in the same direction.

10. An induced magnetic field in the motor's electromagnet is repelled by the permanent magnet, causing the rotor to spin. The commutator in a dc motor allows the current through the electromagnet to change polarity, causing the rotor to continue spinning.

11. A synchronous motor rotates at a frequency that depends on the number of coils and the frequency of the ac power source. The motor has a number of poles along the stator, which cause the rotor to spin at a fixed frequency.

12. A universal motor can operate on either dc or ac. The induction motor can operate only on ac.

13. The stator is a static magnet. The armature is an electromagnet that is free to rotate.

14. An electromagnet produces a strong magnetic field when a current is run through a solenoid. This magnetic field in turn induces a stronger magnetic field in a magnetic core.

15. The magnetic field increases by a factor of 2.

16. The field does not change as long as the length of the solenoid is much greater than the original diameter.

17. The magnetic field increases by a factor of 4.

18. The flux lines can be found by placing a small compass or magnetic filings near the magnet.

19. A spinning armature in the field of a stator crosses the magnetic flux lines of the stator, thereby inducing a current to flow in the armature coil. As the armature rotates and is reversed in the field, the direction of the current in the coil is reversed.

Chapter 17 Review Problems Page 502

1. 1.08×10^{-6} T **2.** 4.90×10^{-6} T **3.** 41.6 A **4.** 0.0253 T **5.** 12.4 A
6. 15.4°

Chapter 17 Applied Concepts Page 503

1. (a) 1.98×10^{-5} T (b) 3.17 m **2.** (a) 7.77×10^{-6} T (b) The wire's magnetic field is 14.9% of the strength of the earth's magnetic field. **3.** The magnetic field of the inner cable cancels out the opposite magnetic field orientation of the outer braid.
4. (a) For both solenoids, N is on top and S is on the bottom. (b) "b" The electron is coming out of the page. **5.** (a) 0.525 Ω (b) 8.57 A (c) 98.0 loops (d) 0.0896 m (e) 1.05×10^{-3} T

Chapter 18

18.1 Pages 508–509

1. 47.1 V **3.** 117 V **5.** 9.19 A **7.** 83.2 V **9.** 5.66 A **11.** 70.3°
13. 1590 V **15.** 117 V **17.** 12.0 A **19.** 12.0 A **21.** 95.4 V
23. (a) 24.0 V (b) 0.120 A (c) $20\bar{0}$ Ω **25.** (a) 0.495 A (b) 242 Ω **27.** $20\bar{0}0$ W

18.2 Pages 516–517

1. 21.9 Ω **3.** $35\bar{0}0$ W or 3.50 kW **5.** 1320 W or 1.32 kW **7.** 588 W
9. 2750 W or 2.75 kW **11.** 10.0 turns **13.** 58.5 V **15.** $30\bar{0}$ turns
17. 6.00 A **19.** (a) 0.0587 A or 58.7 mA (b) $88\bar{0}$ W **21.** 155 V
23. (a) $12\bar{0}$ V (b) 0.600 A

18.3 Pages 519–520

1. 1.13 Ω **3.** $44\bar{0}0$ Ω **5.** 40.1 kΩ **7.** 0.594 A **9.** 0.796 A **11.** 0.116 A

18.4 Pages 522–523

1. (a) 215 Ω (b) 21.4° (c) 0.209 A **3.** (a) $33\bar{0}0$ Ω (b) 72.3° (c) 4.55 mA
5. (a) 302 Ω (b) 7.2° (c) 19.9 mA

18.5 Page 525

1. 7.96 Ω **3.** 2650 Ω or 2.65 kΩ **5.** 0.796 Ω **7.** 58.9 Ω **9.** 1.61 Ω

18.6 Page 527

1. (a) 1880 Ω (b) 57.9° (c) 53.2 mA **3.** (a) $48\bar{0}0$ Ω (b) 0.0241° (c) 3.13 mA
5. (a) 3.18 Ω (b) 89.8° (c) 4.72 mA

18.7 Page 529

1. 42.4 Ω; 0.118 A **3.** 1180 Ω; 12.7 mA **5.** $44\bar{0}$ Ω; 13.6 mA **7.** 206 Ω; 24.3 mA **9.** 529 Ω; 47.3 mA

18.8 Page 531

1. 79.6 kHz **3.** 38.0 kHz **5.** 3.86 kHz **7.** 4.20 kHz **9.** 5.63 kHz

18.10 Page 533

1. 10,700 kW **3.** 133,000 kVA **5.** 10,100 kW **7.** 19,300 kW
9. 407,000 kVA **11.** 0.880

Chapter 18 Review Questions Pages 535–536

1. d **2.** c **3.** a and e **4.** a, b, and c
5. The maximum current is the maximum instantaneous current. The effective value of an alternating current is the number of amperes that produce the same amount of heat in a resistance as an equal number of amperes of a steady direct current.
6. The maximum voltage is the maximum instantaneous voltage. The instantaneous voltage is the voltage at any instant.
7. Power is the product of the effective values of the voltage and the current.
8. Power is the square of the effective value of the voltage divided by the resistance.
9. The output voltage doubles. **10.** Henry
11. Inductive reactance allows the analysis of circuits containing inductors.
12. The inductive reactance is directly proportional to the frequency.

13. The current lags the voltage in an inductive circuit.
14. Energy is stored in the form of potential energy associated with a sheet of positive charge on one side of the capacitor and a sheet of negative charge on the other side.
15. The current leads the voltage in a capacitive circuit.
16. The frequency and reactance of a capacitor are inversely proportional.
17. Resonance occurs when the inductive reactance equals the capacitive reactance. The current is then given by its maximum value.
18. A diode allows current to flow in one direction but not in the reverse direction.
19. Amplification produces an increase in the value of a voltage or current in a circuit. Rectification produces a current or voltage in only one direction.
20. No; the phase angle depends on the frequency.

Chapter 18 Review Problems Pages 536–537

1. 110 V **2.** 120 V **3.** 68 V **4.** 95.4 V **5.** 8.95 A **6.** 727 W **7.** 495 W **8.** 1310 W **9.** (a) 1340 turns (b) 0.255 A (c) 523 W **10.** (a) 3.92 A (b) 38.2° (c) 71.0 V **11.** (a) 2.29 A (b) 18.9° (c) 109 V **12.** (a) 2.20 A (b) 67.1° (c) 42.9 V (d) 102 V **13.** (a) 1.52 A (b) 32.8 V (c) 72.6° **14.** 194 Ω; 0.567 A **15.** 126 Ω; 0.873 A **16.** 30,200 Hz, 633 Ω, 0.174 A **17.** 3.46×10^5 Hz, 565 Ω, 0.195 A **18.** 3.41×10^6 kVA

Chapter 18 Applied Concepts Page 537

1. (a) 130̄0 W (b) 16.6 A (c) $i = 0$ A because the current is reversing at this instant.
2. (a) 23.0 turns (b) $P_p = 114$ W; $P_s = 114$ W (c) The power is conserved in a transformer. **3.** (a) 386 A (b) 14.2 A (c) $P_{L\,220V} = 37{,}200$ W; $P_{L\,6000V} = 50.4$ W (d) The 60̄00-V power line is better because it only loses 50.4 W of power.
4. (a) 0.332 H (b) 0.524 A (c) 34.3 W **5.** (a) 1.19×10^{-13} F (b) 9.04×10^{-14} F (c) Tuning to a high frequency lowers the capacitance because the plates are smaller.

Chapter 19

19.2 Pages 544–545

1. 1.50×10^9 m **3.** 0.108 s **5.** 50̄0 s **7.** 16.1 ms **9.** 7.57×10^{-7} s **11.** 25̄0 s; 12̄00 s **13.** (a) 1940 s (b) 2890 s **15.** 3.84×10^8 m

19.3 Pages 546–547

1. 6.59×10^{12} Hz **3.** 3.09×10^{-4} m **5.** 6.58 m **7.** 4.55×10^{-5} Hz **9.** 214 m **11.** 3.51 MHz **13.** 7.5×10^{14} Hz **15.** 5.2×10^{14} Hz

19.4 Page 549

1. 5.93×10^{-23} J **3.** 3.01×10^{-25} J **5.** 83.0 MHz **7.** 5.51×10^{10} Hz **9.** 2.7×10^{-19} J **11.** 3.4×10^{-19} J **13.** 1410 kHz

19.5 Pages 551–552

1. 603 ℓm **3.** 60.9 cd **5.** 942 ℓm **7.** 1.46 ft-cd **9.** 6.73 ft-cd **11.** 4750 ℓm **13.** 608 ℓm **15.** (a) Decrease (b) $\frac{1}{9}$ **17.** 5.96 lux **19.** 35.3 ℓm

Chapter 19 Review Questions Page 553

1. a, c, and e **2.** c **3.** d **4.** c
5. Yes; the wavelength varies inversely with the frequency.
6. The energy is directly proportional to the frequency.

7. The intensity falls off as 1 divided by the square of the distance.
8. The speed of light is measured by determining the time light takes to travel a measured distance and using the relationship $v = s/t$.
9. It always travels at the same velocity in a vacuum. It has a different velocity (slower) in any medium.
10. Electromagnetic radiation 11. Max Planck
12. Albert Einstein and Max Planck 13. Olaus Roemer
14. By measuring the time difference for the start of the eclipse of the moons of Jupiter as viewed from different parts of earth's orbit 15. Candela and lumen
16. Luminous intensity is the brightness of a light source.

Chapter 19 Review Problems Pages 553–554

1. 2.32×10^{13} m **2.** 3.52 mi **3.** 2.63×10^{-6} s or 2.63 μs **4.** 1.47×10^{-6}s or 1.47 μs **5.** 0.121 s **6.** 244 m **7.** 6.45×10^{6} Hz or 6.45 MHz **8.** 5.54×10^{15} Hz **9.** 9.61×10^{-23} J **10.** 7.17×10^{10} Hz **11.** 5.47×10^{-18} J **12.** 2150 ℓm **13.** 5230 ℓm **14.** 20,900 ℓm **15.** 3130 s; $17\bar{0}0$ s **16.** 12.8 ℓm **17.** 7.29 lux

Chapter 19 Applied Concepts Page 554

1. (a) 0.0187 s (b) 0.238 s **2.** (a) 3.84×10^{5} km or 239,000 mi (b) $50\bar{0}$ s or 8.33 min (c) 4.08×10^{16} m or 2.53×10^{13} mi **3.** (a) 3.41 to 2.78 m (b) 5.83×10^{-26} to 7.16×10^{-26} J (c) The higher the frequency, the greater is the energy.
4. Channel 2: $\frac{1}{4}\lambda = 1.39$ m to Channel 6: $\frac{1}{4}\lambda = 0.852$ m **5.** (a) 14,100 ℓm (b) 56,500 ℓm (4 times the original)

Chapter 20

20.5 Pages 565–566

1. 1.29 cm **3.** 32.2 cm **5.** 5.13 cm **7.** −6.24 cm **9.** −1.09 m **11.** 10.0 cm **13.** −4.15 cm **15.** −3.56 m **17.** (a) 4.01 cm (b) 2.98 cm (c) 31.9 cm

20.10 Pages 577–578

1. 1.21 **3.** 2.00×10^{8} m/s **5.** 1.48 **7.** 21.8 cm **9.** −2.14 cm **11.** (a) 5.00 cm (b) −2.50 cm **13.** $s_i = -8.17$ cm, $h_i = 2.06$ cm **15.** 4.43 cm

Chapter 20 Review Questions Pages 579–580

1. a 2. d 3. a and b
4. Parallel light rays reflected off a rough surface may scatter in many different directions (diffusion), whereas parallel light rays reflect off a smooth surface as parallel rays.
5. At a smooth surface, the incident angle of a light ray is the same as the angle of the reflected ray.
6. At a smooth surface, the normal to the surface and both the incident and reflected rays lie in the same plane.
7. They are virtual, erect, and lie as far behind the mirror as the object is in front of the mirror.
8. A real image can be shown on a screen. A virtual image cannot.
9. Concave mirrors curve away from the observer; convex mirrors curve out toward the observer.
10. For large apertures, not all parallel rays are reflected through the focal point. This produces a fuzzy or aberrant image.

11. The image distance is increased. **12.** The image distance is increased.
13. b **14.** b
15. Converging lenses convert parallel rays into converging rays. Diverging lenses convert parallel rays into diverging rays.
16. Light propagating in an optical fiber, light reflected into a swimming pool from an underwater light
17. Light passing at an angle through an interface from a medium of low optical density to a medium of higher optical density is bent toward the normal to the interface.
18. The speed of light is lower in the high-index material.
19. A wave passing at an angle from one medium to another with a different wave velocity will be bent toward or away from the normal depending on whether the speed is higher or lower in the new material, respectively.
20. Virtual **21.** Real or virtual
22. They roughen the surface and scatter the light.
23. It appears shallower because of the refraction of light at the surface.
24. Light travels in a straight line unless it is reflected or refracted.
25. Light traveling at an angle greater than the critical angle is reflected from one side of the fiber to another as it travels down the length of the fiber.
26. When the object is closer to the lens than the focal length.
27. When the object is located a distance from the lens equal to the focal length.
28. When the object is located outside the focal point.

Chapter 20 Review Problems Page 580

1. 2.36 cm **2.** −12.9 cm **3.** −3.48 cm **4.** −2.22 m **5.** (a) $25\bar{0}$ cm from the mirror (b) −459 cm **6.** 22.7 cm **7.** 1.35 **8.** 2.08×10^8 m/s **9.** 1.40 **10.** 41.8° **11.** 18.0 cm **12.** (a) −7.50 cm (b) 2.70 cm **13.** 9.00 cm **14.** 6.25 cm **15.** −9.70 m **16.** −25.4 cm, 7.18 cm high **17.** 1.24×10^8 m/s **18.** 41.8° **19.** 12.3 cm **20.** (a) $s_i = 15.8$ cm; $h_i = -10.8$ cm (b) −3.83 cm

Chapter 20 Applied Concepts Page 581

1. (a) 19.0 cm (b) $s_i = -36.5$ cm (the minus sign represents the opposite side of the mirror) (c) Upright (d) 73.0 cm **2.** (a) −0.704 m (b) Virtual (c) 0.0156 m (d) Better for increasing the field of view **3.** (a) 18.5° (b) 22.[illegible]° (c) The angle will be smaller because light refracts toward the normal line when entering a denser medium. **4.** (a) 24.44° (b) Larger than 24.44° (c) 31.4° (d) Diamond; a zircon will allow light to escape when angles are less than 31.4°. **5.** (a) 60.4 mm (b) −[illegible]3.1 mm

Chapter 21 Review Questions Page 60[illegible]

1. Red, orange, yellow, green, blue, and violet
2. The color of light is determined by its wavelength or its frequency
3. (a) Infrared (b) Ultraviolet
4. Black. The cloth in the dress absorbs all colors except green and reflects only green. The red light is absorbed and no light is reflected.
5. (a) Red stars on a black field, rest of the flag would be red (b) Blue field with no stars, rest of the flag would have blue and black stripes
6. Red, green, and blue
7. Two colors that combine to form white are called complementary colors. The complement of red is cyan, the complement of green is magenta, and the complement of blue is yellow.
8. The primary pigments are the complements of the three primary colors.
9. Cyan, magenta, and yellow for the colors and black for the shadow areas and definition
10. The sky is blue because sunlight is scattered as it passes through the atmosphere. The amount of scattering depends on the wavelength of the light. The longer wavelengths

of red, orange, and yellow are scattered much less than the shorter wavelengths of blue and violet. Because our eyes are not very sensitive to violet light, the scattered blue light in the atmosphere dominates our vision so that we see a blue sky.

11. At sunset, sunlight travels through significantly more of the atmosphere than at other times of the day. This results in even more scattering of the shorter wavelengths of violet, blue, and green, which are taken out before reaching us, with some orange but primarily red, the light of longest wavelength and least scattered, most readily passing through the atmosphere to reach our eyes for a red sunset.

12. Clouds consist of many different-sized water droplets. The small droplets tend to scatter the short-wavelength (blue) light, the medium droplets tend to scatter the medium-wavelength (green) light, and the large droplets tend to scatter the long-wavelength (red) light. The overall result is that all of the colors are scattered and combined so that we see a general white reflection.

13. The ocean is blue because the water surface acts like a mirror and reflects the blue color of the sky and the water tends to reflect and scatter the short-wavelength yellow, green, blue, and violet light and absorb the long-wavelength orange and red light.

14. False

15. Diffraction

16. Diffraction pattern

17. Thomas Young

18. Constructive interference

19. Destructive interference

20. Polarized sunglasses restrict the light waves passing through to a single plane rather than normal sunlight, which is emitted in all directions with many orientations.

Chapter 22

22.3 Page 608

1. −13.595 eV **3.** 12.084 eV **5.** 6.57×10^{-7} m

22.4 Page 610

1. Neon **3.** Fluorine

22.5 Page 613

1. (a) 12 (b) 6 (c) 6 (d) 6 (e) 12 (f) 6 **3.** (a) 48 (b) 22 (c) 26 (d) 22 (e) 48 (f) 26 **5.** (a) 11 (b) 22.9898 u **7.** (a) 82 (b) 207.2 u **9.** ^{14}C
11. Bromine **13.** Argon

22.6 Page 616

1. 3.8530×10^{-25} kg **3.** 6680 MeV **5.** 1750 MeV **7.** 8.5 MeV

22.7 Page 621

1. 4.03×10^{-5} s **3.** 0.226 g **5.** 79.9% **7.** 3.77×10^{20} atoms **9.** 30.4%

22.9 Pages 623–624

1. 0.310 Ci **3.** 0.0875 Ci **5.** 9.47 Ci **7.** 73.3 μCi

Chapter 22 Review Questions Pages 626–627

1. a, b, and c **2.** c **3.** c

4. The ground state is the lowest energy level for the electron in the atom. The excited states are the higher energy levels, which are unstable. An electron in an excited state will in time decay through lower-energy-level excited states to the ground state. The ground state has the lowest energy level.

5. In the Bohr theory of the atom, the energy levels of the electrons are restricted to certain values; that is, the energy is quantized. The energy levels in a hydrogen atom are given by the equation $E = -kZ^2/n^2$.
6. Protons and neutrons are both nucleons, which exert the attractive strong force on other nucleons. They have similar masses, although the neutron is slightly more massive. The proton is a stable particle as an individual nucleon. The neutron, however, is unstable by itself. The proton has a positive charge, equal in magnitude to that of the electron. The neutron is uncharged.
7. The electric force can be either attractive or repulsive and is exerted between charged particles. The strong force is always attractive and is exerted between nucleons, whether charged or not. The strong force is a very short-range force. The electric force is exerted over larger distances.
8. The nucleus would expand in size due to the repulsive electric force between protons. The nucleus might even break apart.
9. Mass and energy are equivalent forms. Mass can be changed into energy under the proper conditions and vice versa.
10. All three atoms have six protons in the nucleus and six electrons in the orbital shells. They all exhibit the chemical properties of carbon. They each have a different number of neutrons in the nucleus and therefore have a different mass.
11. An electron volt is the energy that is gained or lost by an electron in passing through a potential difference of 1 volt.
12. The neutron is the "glue" that binds the positively charged protons together in the nucleus. Without neutrons, the positively charged protons would be repelled by their similar electric charge.
13. An α ray is composed of particles that have a double positive charge and are composed of two protons and two neutrons. The α particles are identical to the nucleus of a helium atom.
14. A γ ray is composed of very energetic photons of uncharged electromagnetic radiation.
15. A β ray is composed of negatively charged particles (electrons).
16. Enrico Fermi discovered that the neutron bombardment of uranium can result in the formation of lighter nuclei that are approximately one-half the mass of uranium.
17. A self-sustaining chain reaction is a nuclear reaction in which a sufficient number of neutrons are produced to cause subsequent nuclear reactions to continue at a fixed rate.
18. In nuclear fusion reactions, nuclei bombard each other, producing heavier nuclei. The original nuclei "fuse" together.
19. In nuclear fission reactions, nuclei are bombarded by nuclear particles such as neutrons or α particles, causing the original nuclei to split apart.
20. 6.25%
21. Molecules in plant or animal cells can be damaged as a result of changes in the chemicals caused by nuclear reactions produced by the radiation. Some of these damaged molecules may be in the genetic material, which might cause cells produced by the division of this cell to be defective.
22. Radiation is used for diagnostic purposes by the injection or ingestion of radioactive tracer materials that may allow the identification of cancerous or defective cells. It can also be used to treat cancer by bombardment of the cancer cells, therefore destroying the cancerous material.

Chapter 22 Review Problems Page 627

1. −0.5438 eV **2.** 0.967 eV **3.** He, Ne, Ar, Kr, Xe, Rn
4. Cu, Ag, Au, Fe, Ti, Cr, Mn, etc. **5.** 2.4913×10^{-26} kg
6. 3.1547×10^{-26} kg **7.** 2.7553×10^{-25} kg **8.** 749 MeV
9. 8.7 MeV **10.** 8.3 MeV **11.** 8.1 MeV **12.** 7.26×10^{12} atoms

13. 4.29×10^{17} atoms **14.** 5.6% **15.** 0.37% **16.** 0.346 Ci
17. 0.193 Ci **18.** 0.367 Ci **19.** 5.8 Ci **20.** 10.6 μCi

Chapter 22 Applied Concepts Page 628

1. (a) 2.5491 eV (b) 6.163×10^{14} Hz (c) 3.00×10^{8} m/s (d) 4.87×10^{-7} m
2. $n = 2$ **3.** (a) 4.550×10^{-12} J (b) two protons, two neutrons, two electrons (c) 6.6968×10^{-27} kg (d) 6.64×10^{-27} kg
4. (a) 0.181/day (b) 1.50×10^{24} atoms **5.** 0.393 Ci

Chapter 23

23.2 Page 635

1. 2.07×10^{14} J **3.** 6.67×10^{-15} kg

Chapter 23 Review Questions Pages 637–638

1. c **2.** d
3. Physics is the same for moving and nonmoving objects.
4. If you were in a moving car and flipped a coin, the coin would flip into the air and fall back into your hand the same as would happen if you flipped a coin while standing on the side of the road.
5. It would travel at 14 m/s because the ball was already traveling at 10 m/s while attached to the bike.
6. The speed of light is not relative, but is constant.
7. The light would only travel at 3.00×10^{8} m/s.
8. Time passes more slowly than for someone not traveling close to the speed of light.
9. Relative to the ground (considering I am not walking and reading), just the time dimension.
10. That energy and mass are equivalent, just in different forms. Mass has energy and energy has mass.
11. Spatially we exist in three dimensions (x, y, z), yet we also move through the dimension of time. Space and time are needed to locate a particular occurrence in the universe.
12. Gravity and acceleration are the same.
13. Light can be warped around massive objects.
14. The light curved around the warping of space-time that was created by the mass of the sun and the moon.

Chapter 23 Review Problems Page 638

1. 2.00 mi/h **2.** 67.0 mi/h **3.** 63.0 mi/h **4.** 8.19×10^{-14} J **5.** 1.50×10^{-10} J
6. 1.09×10^{-16} kg

Appendix A

A.1 Page 642

1. −11 **3.** 5 **5.** −2 **7.** −4 **9.** −6 **11.** −3 **13.** 18 **15.** −21
17. 0 **19.** 3 **21.** −8 **23.** 0 **25.** 11 **27.** −13 **29.** −5 **31.** −8
33. −4 **35.** 10 **37.** −45 **39.** −288

A.2 Page 644

1. 10^8 **3.** 10^8 **5.** $\frac{1}{10^3}$ **7.** 10^7 **9.** 10^3 **11.** $\frac{1}{10^4}$ **13.** 1 **15.** 10^5
17. $\frac{1}{10^{11}}$ **19.** $\frac{1}{10^6}$ **21.** 10^8 **23.** 10^{10}

A.3 Pages 646–647

1. 3.26×10^2 **3.** 2.65×10^3 **5.** 8.264×10^2 **7.** 4.13×10^{-3} **9.** 6.43×10^0
11. 6.5×10^{-5} **13.** 5.4×10^5 **15.** 7.5×10^{-6} **17.** 5×10^{-8} **19.** 7.32×10^{17}
21. 86,200 **23.** 0.000631 **25.** 0.768 **27.** 777,000,000 **29.** 69.3
31. 96,100 **33.** 1.4 **35.** 0.0000084 **37.** 700,000,000,000 **39.** 0.00000072
41. 4,500,000,000,000 **43.** 0.000000000055

A.4 Page 650

1. $\frac{4}{3}$ **3.** 17 **5.** 0 **7.** 17.5 **9.** 3 **11.** 4 **13.** 7 **15.** 12.5 **17.** 5
19. 26.5 **21.** 9 **23.** 3 **25.** 8 **27.** 1 **29.** $-\frac{11}{3}$ or $-3\frac{2}{3}$ **31.** 3
33. 2 **35.** 21 **37.** $\frac{29}{2}$ or $14\frac{1}{2}$ **39.** -6

A.5 Page 653

1. ± 6 **3.** ± 7 **5.** ± 3 **7.** ± 0.8 **9.** ± 4.67 **11.** ± 0.892
13. $a = 3$; $b = 1$; $c = -5$ **15.** $a = 6$; $b = 8$; $c = 2$ **17.** $a = 9$; $b = 6$; $c = -4$
19. $a = 5$; $b = 6$; $c = 0$ **21.** $a = 9$; $b = 0$; $c = -64$ **23.** 3; 7
25. $\frac{4}{3}$; $-\frac{5}{2}$ **27.** 1.95; -1.62 **29.** 1.69; -0.855 **31.** 0.725; -1.23

A.6 Page 656

1. $s = vt$ **3.** $m = \dfrac{w}{g}$ **5.** $R = \dfrac{E}{I}$ **7.** $g = \dfrac{PE}{mh}$ **9.** $h = \dfrac{v^2}{2g}$ **11.** $W = Pt$

13. $t = \dfrac{W}{P}$ **15.** $m = \dfrac{2(KE)}{v^2}$ **17.** $s = \dfrac{W}{F}$ **19.** $I = \dfrac{V - E}{-r}$ or $I = \dfrac{E - V}{r}$

21. $P = \dfrac{\pi}{2R}$ **23.** $C = \dfrac{5F - 160}{9}$ or $C = \dfrac{5}{9}(F - 32)$ **25.** $f = \dfrac{1}{2\pi CX_C}$

27. $R_3 = R_T - R_1 - R_2 - R_4$ **29.** $I_P = \dfrac{I_S N_S}{N_P}$ **31.** $v_i = 2v_{avg} - v_f$

33. $s = \dfrac{v_f^2 - v_i^2 + 2as_i}{2a}$ **35.** $R = \dfrac{QJ}{I^2 t}$ **37.** $r = \sqrt{\dfrac{A}{\pi}}$ **39.** $d = \sqrt{\dfrac{kL}{R}}$

41. $I = \pm\sqrt{\dfrac{QJ}{Rt}}$

A.7 Page 659

1. (a) $A = bh$ (b) 162 cm^2 **3.** (a) $b = \dfrac{A}{h}$ (b) 7.50 cm **5.** (a) $c = P - a - b$
(b) 6.0 cm **7.** (a) $r = \dfrac{C}{2\pi}$ (b) 10.9 yd **9.** (a) $b = \dfrac{P - 2a}{2}$ or $b = \dfrac{P}{2} - a$
(b) 33.2 km **11.** (a) $h = \dfrac{V}{\pi r^2}$ (b) 6.11 m **13.** (a) $B = \dfrac{V}{h}$ (b) 154 m^2
15. (a) $b = \sqrt{A}$ (b) 21.6 in. **17.** (a) $C = 2\pi r$ (b) 121.6 m **19.** (a) $B = \dfrac{3V}{h}$
(b) 122.4 ft^2

A.8 Pages 663–666

1. *a* **3.** *c* **5.** *a* **7.** *B* **9.** *B* **11.** 0.9455 **13.** 1.8[illegible]4 **15.** 0.9799
17. 0.6477 **19.** 0.3065 **21.** 0.4617 **23.** 16° **25.** 43° **27.** 48°
29. 36.6° **31.** 46.5° **33.** 30.0° **35.** 22.28° **37.** 16.7[illegible]° **39.** 35.50°
41. $B = 65.0°$; $a = 8.45$ m; $b = 18.1$ m **43.** $A = 47.7°$; $B = 42.3°$; $a = 12.4$ km
45. $A = 24.4°$; $B = 65.6°$; $c = 24.2$ mi **47.** $B = 70°$; $b = 24$ m; $c = 25$ m
49. $A = 49.35°$; $a = 17.98$ cm; $b = 15.44$ cm **51.** $b = 8.49$ cm **53.** $c = 21.6$ mi

55. $a = 10.2$ ft **57.** $c = 24.8$ cm **59.** $a = 8.60$ m **61.** (a) 10.0° (b) 2.12 cm (c) 8.24 cm **63.** $C = 2.72$ in.; $D = 2.28$ in. **65.** $b = 8.00$ cm; $c = 16.1$ cm

A.9 Pages 674–676

1. $B = 38.0°$, $C = 73.0°$, $c = 25.6$ m **3.** $C = 52.5°$, $A = 66.1°$, $a = 129$ cm **5.** $C = 47.6°$, $a = 223$ ft, $c = 196$ ft **7.** $B = 48.5°$, $C = 16.5°$, $c = 1840$ m **9.** $A = 67.07°$, $B = 40.35°$, $a = 40.72$ cm **11.** $A = 131.94°$, $a = 89{,}460$ mi, $b = 57{,}870$ mi **13.** $B = 2\bar{0}°$, $C = 135°$, $c = 84$ cm **15.** $A = 6°$, $B = 166°$, $b = 28$ m **17.** $B = 27.3°$, $C = 115.7°$, $c = 32.2$ cm **19.** $A = 29.6°$, $B = 123.9°$, $b = 79.4$ km; or $A = 150.4°$, $B = 3.1°$, $b = 5.18$ km **21.** $B = 34.3°$, $C = 74.2°$, $b = 2.05$ m; or $B = 2.7°$, $C = 105.8°$, $b = 0.171$ m **23.** $A = 44.6°$, $C = 30.4°$, $c = 8.64$ mi **25.** $A = 35°$, $B = 127°$, $b = 62$ mi; or $A = 145°$, $B = 17°$, $b = 23$ mi **27.** No triangle
29. $A = 157°$, $C = 15°$, $a = 1300$ m; or $A = 7°$, $C = 165°$, $a = 390$ m
31. $A = 59.34°$, $C = 79.16°$, $c = 21.12$ km; or $A = 120.66°$, $C = 17.84°$, $c = 6.588$ km
33. No triangle **35.** $A = 6.06°$, $B = 165.19°$, $b = 150.1$ m
37. $B = 47.8°$, $C = 72.2°$, $a = 22.8$ m **39.** $A = 27.0°$, $B = 44.0°$, $c = 292$ km
41. $A = 44.6°$, $B = 51.2°$, $C = 84.2°$ **43.** $A = 32.1°$, $B = 104.6°$, $C = 43.3°$
45. $B = 85°$, $C = 5\bar{0}°$, $a = 36$ m **47.** $A = 84°$, $B = 53°$, $C = 43°$
49. $A = 19°$, $B = 26°$, $c = 78$ ft **51.** $A = 148.80°$, $C = 11.95°$, $b = 3064$ m
53. $A = 40.11°$, $B = 31.14°$, $c = 595.2$ mm **55.** $A = 113.43°$, $B = 27.84°$, $C = 38.73°$

Appendix B

B.4 Pages 681–682

1. 3.32×10^{19} **3.** -6.83×10^{-6} **5.** 8.35×10^{-11} **7.** -7.98×10^{19}
9. −68.5 **11.** 4680 **13.** 216 **15.** 2.10×10^{10} **17.** 2.15×10^{9} **19.** 10.7
21. 33.4 **23.** 591 **25.** 609,000 **27.** 0.567 **29.** 1.77×10^{4} **31.** 0.523
33. 0.225 **35.** 1.29 **37.** 0.510 **39.** 0.743 **41.** 3.73 **43.** 0.151
45. 40.7° **47.** 24.6° **49.** 56.2° **51.** 8.4° **53.** 63.3° **55.** 66.6°
57. 252 m **59.** 89.4 m **61.** 53.7° **63.** 14.8 cm **65.** 21.9 m **67.** 20,700
69. 0.0206 **71.** 1.86×10^{-6}

INDEX

Absolute pressure, 332, 341
Absolute temperature, 348
Absolute zero, 350, 380
Acceleration:
  angular, 216, 240
  defined, 93, 111
  and distance, 100
  equations for, 97
  of gravity, 100, 111
  units of, 94
  and velocity, 93
Accuracy, 36, 52
Acoustics, 8
Action and reaction, 129
Active noise cancellation, 6
Actual power, 532, 534
Adhesion, 290, 313
Air pressure, 330
Alpha particle, 616
Alpha rays, 616, 625
Alternating current, 439, 474, 506, 534
  effective value, 508, 534
  generators, 494, 532
  power, 509
Ammeter, 462, 474
Ampere, Andre, 486
Ampere, unit of current, 441
Ampere's rule, 486, 501
Amplification, 532, 534
Amplitude, 402, 423
Angular acceleration, 216, 240
Angular displacement, 213, 240
Angular measurement, 213
Angular momentum, 220, 240
Angular motion, 213
Angular velocity, 213, 240
Anode, 531
Antilock brakes, 124
Apparent power, 532, 534
Approximate number, 36, 52
Archimedes, 5
Archimedes' principle, 333, 341
Armature, 495, 501
Astronomy 4, 10

Astrophysics, 8
Atom, 288, 313
Atomic:
  mass, 612, 625
  mass number, 611, 625
  mass units, 612, 625
  number, 611, 625
  physics, 8
  spectra, 605
  structure, 605
Atmospheric pressure, 330, 341
Auto transformer, 514
Average binding energy per nucleon, 615, 625

Bardeen, John, 532
Barometric pressure, 331
Battery storage, 466
Beats, 414
Becquerel, 623, 625
Becquerel, Henri, 623
Bell, Alexander Graham, 414
Belts and pulleys, 255
Bending, 296, 313
Bernoulli, Daniel, 338
Bernoulli's principle, 338, 341
Beta ray, 616, 625
Binding energy, nuclear, 613, 625
Biology, 4, 10
Bohr model, 606, 625
Bohr, Niels, 605
Boiling, 347
Boyle, Robert, 390
Boyle's law, 390, 396
Brinell, Johan, 290
Brinell method, 290, 314
British thermal unit (Btu), 350, 380
Buoyant force, 333, 341

Calculations with measurements, 40
Calorie, 350, 380
Candela, 549
Capacitance, 523, 534
Capacitive reactance, 524, 534

Capacitor, 523, 534
Capillary action, 305, 314
Cartesian coordinate system, 65
Cathode, 531
Cell:
  dry, 466
  series and parallel connections, 467
  storage, 466
Celsius, 347, 380
Celsius, Anders, 347
Center of gravity, 174, 178
Centripetal force, 222, 240
Chain, 14
Chain reaction, 622, 625
Change of phase, 370, 380
Charge, electric, 430
  on electron, 434
Charles, Jacques, 388
Charles' law, 388, 396
Chemistry, 4, 10
Circuits:
  capacitive, 523
  electric, 439
  integrated, 531
  series and parallel, 447, 451
  simple, 439
Coefficient of expansion, 365, 380
Coefficient of friction, 122, 130
Cohesion, 290, 314
Collisions, 144
Color, 584, 599
Commutator, 496, 501
Compass, 484, 501
Complementary colors, 585, 599
Components of vector, 65, 79
Compound, 288, 314
Compound machine, 265, 268
Compression, 159, 178, 295, 314
Concave mirrors, 559, 578
Concurrent forces, 152, 178
Condensation, 372, 380
Conduction:
  of electricity, 430, 474
  of heat, 352, 380

Conductors, electrical, 441, 474
Conservation of energy, law of, 202
Conservation of momentum, 141, 147
Convection, 353, 380
Converging lenses, 573, 578
Conversion factor, 18, 19, 52
Convex mirror, 559, 578
Copper losses, 514
Coulomb, Charles, 434
Coulomb, definition of, 434
Coulomb force, 611, 625
Coulomb's law, 434, 474
Critical angle, 570, 578
Cryogenics, 7
Cubic measurement, 25
Curie, 623, 625
Curie, Marie and Pierre, 623
Current:
- alternating (*see* Alternating current)
- defined, 441, 474
- direct, 439
- effective, 508
- unit of, 441

Current, electric, 441
Curvilinear motion, 212, 240

deBroglie, Louis, 604
Decay:
- alpha, 616
- beta, 617
- radioactive, 616
- rate, 617, 625

Deceleration, 95, 111
Degree:
- unit of angular measure, 212, 240
- unit of temperature, 347

Density, 306
Derived units, 16
Descartes, Rene, 65
Dew point, 373, 380
Diffraction, 407, 423, 592, 599
Diffusion, 306, 314, 558, 578
Diodes, 531, 534
Direct current, 439, 474
Disintegration energy, 617, 625
Dispersion, 584, 599
Displacement, 58, 79
Diverging lenses, 573, 578
Doppler, Christian, 415
Doppler effect, 415, 423
Dry cell, 466, 474
Ductility, 291, 314

$E = mc^2$, 633, 637
Eddy currents, 514
Edison effect, 531
Edison, Thomas, 6, 532
Effective value, 508, 534
Efficiency, 248, 268
Effort, 246, 268
Effort arm, 248, 268
Einstein, Albert, 632
Elastic collisions, 145, 147
Elastic limit, 292, 314
Elasticity, 292, 314
Electric:
- circuit, 439, 474
- field, 436, 474
- instruments, 462
- power, 470

Electric current (*see* Current)
Electric, charges, 430
Electricity, theory of, 434
Electrolyte, 466, 475
Electromagnetic:
- radiation, 544
- spectrum, 410, 423
- theory, 542
- waves, 409, 423

Electromagnetism, 8, 492, 501
Electromotive force (emf), 442, 475
- induced, 493

Electron, 288, 314, 430, 475, 604, 625
Electrostatic attraction, 432
Electrostatics, 430
Element, 288, 314
Emf, 442, 475
Energy:
- conservation of, 202
- defined, 197, 207
- electric, 472, 475
- heat, 350, 381
- kinetic, 197, 198
- potential, 197
- units of, 197

Equilibrant force, 158, 179
Equilibrium:
- of concurrent forces, 157, 179
- conditions of, 159
- of parallel forces, 159

Evaporation, 378, 380
Exact number, 36, 52
Excited state, 605, 625
Expansion:
- area, 366
- coefficient of, 365
- linear, 364
- of gases, 306, 314
- of liquids, 368
- of solids, 363
- of volume, 366
- of water, 368

Experimental physicist, 5, 9

Fahrenheit, Gabriel, 348
Fahrenheit scale, 348, 381
Farad, 524
Faraday, Michael, 524
Fermi, Enrico, 622
Field:
- electric, 436
- gravitational, 279, 284
- magnetic, 484, 489

First postulate of special relativity, 633, 637
Fission, nuclear, 622
Fixed pulley, 255, 268
Flow rate, 336, 341
Fluid dynamics, 8
Fluids, 306, 314
Fluorescence, 607, 625
Flux lines, 484, 501
Focal length, 561, 578
Foot-candle, 551
Foot-pound, 166
Force:
- buoyant, 333
- centripetal, 222
- concurrent, 152, 178
- definition of, 115, 116, 130
- electromotive, 442, 475
- of friction, 121, 130
- of gravity, 100
- vector addition of, 152

Formula, 653
Franklin, Benjamin, 430, 431
Freezing, 370, 381
Frequency, 403, 423, 545, 552
- natural, 418, 423

Friction, 121, 130
Fulcrum of lever, 248, 268
Fusion, 370, 381
Fusion, nuclear, 622, 626

g, acceleration of gravity, 100
Galilei, Galileo, 543
Galvani, Luigi, 466
Gamma ray, 616, 625
Gases, 289, 306
- Boyle's law, 390
- Charles' law, 388
- defined, 289, 314
- density of, 306

expansion, 306
pressure exerted by, 390
Gauge pressure, 332, 341
Gear train, 230, 240
Gears, 229
Geiger, Hans, 622
General theory of relativity, 632, 637
Generator: 494, 501
ac, 494
dc, 496
Geology, 4, 10
Geophysics, 8
Global positioning systems, 64
Gram, 32
Grand unified theory, 632, 637
Graphical addition of vectors, 60
Gravitational field, 279, 284
Gravitational potential energy, 197, 207
Gravity:
acceleration of, 100
center of, 174, 178
force of, 100
Ground state, 505, 625
Gyroscope, 6

Half-life, 617, 625
Hardness, 290, 314
Harmonic motion, simple, 420
Heat:
as energy, 350, 381
of fusion, 372, 381
mechanical equivalent, 350, 381
pump, 378, 381
specific, 358
units of, 350
of vaporization, 373, 381
Heat transfer:
conduction, 352, 380
convection, 353, 380
radiation, 353, 380
Hectare, 24
Heisenberg, Werner, 605
Henry, Joseph, 517
Henry, unit of inductance, 517
Hertz, Heinrich, 403
High energy physics, 8
Hooke, Robert, 298
Hooke's law, 298, 314
Horsepower, defined, 192
Humidity:
absolute, 395
relative, 372, 395
Huygens, Christiaan, 541
Huygens' principle, 541
Hydraulic principle, 326, 341
Hydrometer, 312, 314
Hydrostatic pressure, 320, 341
Hypothesis, 5, 10
Hysteresis, 514

Illumination, 550, 552
Images:
lens, 574, 575
mirror, 560, 561
Impedance, 520, 534
Impulse, 137, 147
Impulse-momentum theorem, 142, 147
Incident ray, 558
Inclined plane, 259, 268
Index of refraction, 568, 578
Induced current, 493, 501
Induced magnetism, 492, 501
Inductance, 517, 534
Induction, electromagnetic, 431, 474
Induction motor, 498, 501
Inductive reactance, 518, 534
Inductor, 517, 534
Inelastic collisions, 145, 147
Inertia, 117, 130
moment of, 219, 240
Instantaneous:
current, 507, 534
voltage, 507, 534
Insulator, 441, 475
Integrated circuit, 531
Intensity:
of illumination, 549, 550
sound, 414, 423
Interference, 407, 423
constructive, 405, 423
destructive, 405, 423
Interferometer, 543
Internal potential energy, 197, 207
Internal resistance, 466, 474
Isotope, 612, 625

Joule, James P., 187
Joule, unit of work, 187

Kelvin, Lord, 348
Kelvin scale of absolute temperature, 348, 381
Kilocalorie, 350, 381
Kilogram, 32, 52
Kilowatt, 192
Kilowatt hour, 472
Kinetic energy, 197, 207

Laser, 547, 552
Law, 9, 10
Lead storage cell, 466
Length, measurement and units of, 18
Lenses, 573
Levers, 248, 268
Light, 540, 552
polarized, 595, 599
speed of, 543, 552
Light-years, 544, 552
Liquid crystal displays, 6
Liquids, 289, 314
Litre, 26
Load, 441, 475
Lodestone, 484
Longitudinal wave, 402, 423
Loudness, 414, 423
Lumen, 549
Luminous intensity, 549, 552
Lux, 550

Machines:
compound, 265
efficiency of, 248
mechanical advantage of, 247, 260, 262
simple, 246, 268
Magnetic domain, 492
Magnetic field, 484, 501
around current-carrying wire, 486
losses, 514
Magnetic levitation, 6, 498
Magnetism, 484, 501
Malleability, 292, 314
Marconi, Guglielmo, 542
Mass:
defined, 32, 52, 118, 130
density, 306, 314
and force and acceleration, 118
and weight, 128
Mass-energy equivalence, 613, 625
Mathematical physics, 8
Matter, defined, 288, 314
Maxwell, James C., 604
Mechanical advantage, defined, 247, 268
mechanics, 7
Melting, 370, 381
Mendeleev, Dmitri, 612
Meniscus, 306, 314
Method of mixtures, 360, 381
Metre, 18, 52
Metric-English conversions, 20
area, 22
length, 20
volume, 25
MeV, 614

Michaelson, Albert, 543
Milligram, 32
Mirror formula, 564
Mirrors, 558
Mixtures, method of, 360, 381
Molecule, 288, 314
Molecular physics, 8
Moment of inertia, 219, 240
Momentum, 136, 147
Motion:
 accelerated, 93
 definition of, 85, 86, 111
 rectilinear, 212, 240
 rotational, 212
Motor principle, 497
Motors, 498, 501
Mouton, Gabriel, 16
Movable pulleys, 255, 268
Multimeter, 462, 475

Natural frequency, 418, 423
Neutron, 288, 314, 430, 475, 604, 626
Neutron number, 611, 626
Newton, defined, 32
Newton, Isaac, 116, 117
Newton's laws, 116
Nonconcurent forces, 169
Normal force, 122, 130
Nuclear fission, 621, 626
Nuclear fusion, 622, 626
Nuclear mass, 613
Nuclear physics, 8
Nucleons, 610
Nucleus, 289, 314
Nuclides, 611, 626
Number plane, 65, 79

Oersted, Hans Christian, 486
Ohm, Georg Simon, 444
Ohm, unit of electrical resistance, 444
Ohmmeter, 463, 475
Ohm's law, 444, 475
Opaque, 558, 578
Oppenheimer, Robert K., 622
Optical density, 567, 578
Optics, 8
Orbit, 280, 284
Origin, 65

Parallel connection, of cells, 469
 circuit, 451, 475
 of resistances, 451
Parallel forces, 169
Particle theory, 541, 552
Pascal, Blaise, 293
Pascal's principle, 326, 341
Pendulum, 421, 423
Period, 403, 423
Periodic Table, 612, 626
Phase angle, 521, 534
Phase relations in ac circuits, 521
Photoelectric effect, 542, 552
Photometry, 549, 552
Photon, 542, 547
Physicist, 4, 10
Physics, 4, 10
Pitch, 262, 268, 414, 423
Planck, Max, 542
Planck's Constant, 547, 552
Plane mirror, 559, 578
Plasma physics, 7
Platform balance, 33, 52
Polarized light, 595, 599
Potential difference, 442
Potential energy, 197, 207
Power:
 in ac circuits, 509
 defined, 191, 207
 electric, 470, 475
 factor, 532, 534
 rotational, 224
 units of, 192
Precision, 38, 52
Pressure:
 absolute, 332
 air, 330
 and density, 391
 defined, 320, 341
 gauge, 332
 liquid, 326
 transmission in liquids, 326
 units of, 320
 vapor, 394
Primary cells, 466, 475
Primary coil, 512, 534
Primary colors, 585, 600
Primary pigments, 586, 600
Principal focus, 561
Principle, 8
Principle of Equivalence, 613
Problem-solving method, 9, 45, 52
Projectile, 106, 111
Projectile motion, 106, 111
Propagation velocity, 404, 423
Proton, 288, 314, 430, 474, 604, 626
Pulleys and pulley systems, 255, 268
Pulse, wave, 402, 424

Quantum mechanics, 608, 626
Quantum number, 609
Quantum physics, 8
Quantum theory, 542, 552, 608

Radian, 212, 240
Radiation, 353, 381
 damage, 624
 detection, 622
 half-life, 617
 measurement, 622
 penetrating power, 624
Radioactive decay, 616, 626
Radiocarbon dating, 620, 626
Radioactive half-life, 617
Rainbow, 591, 600
Range, 108, 111
Rankine scale of temperature, 348, 381
Rankine, William, 348
RCL circuits, 527
Reactance, 518, 524
Reaction, 129
Reactive kVA, 532
Real images, 559, 578
Recharging, 466, 475
Rectification, 531, 534
Rectifier, 510, 534
Rectilinear motion, 212, 240
Reflection, 540, 552, 558, 578
 first law, 558, 578
 second law, 559, 578
Refraction, 540, 552, 567, 578
Refrigerant, 379
Relative humidity, 372, 381
Relative motion, 632, 637
Resistance arm, 248, 268
Resistance, electric, 442, 475
Resistance, equivalent, 448, 475
Resistance, machines, 256, 268
Resistivity, electrical, 443, 475
Resolution of vectors, 65
Resonance:
 electrical, 529, 534
 sound, 418, 424
Resultant force, 152, 179
Resultant vector, 65, 79
Revolution, 212, 240
Robotics, 5
Rod, 14
Roemer, Ole, 543

Rotational motion, 212, 240
Rotor, 495, 501
Rounding numbers, 25
Rutherford, Ernest, 289
R value, 355

Scalar, 58, 79
Scattering, 409, 558
Schroedinger, Erwin, 605
Science, 5, 10
Scientific method, 9, 10
Scientific notation, 644
Screw, 262, 268
Second, 33, 52
Second condition of equilibrium, 171, 179
Second postulate of special relativity, 633, 637
Secondary cell, 466, 475
Secondary coil, 512, 534
Seismograph, 412
Semiconductor, 441, 475
Series circuit, 447, 475
Shearing, 295, 314
Significant digits, 36, 37, 52
SI metric units, 16, 52
Simple circuit, 439
Simple harmonic motion, 420, 424
Simple machines, 246, 247, 268
Snell, Willebord, 568
Snell's Law, 568, 578
Solenoid, 490, 501
Solidification, 370, 381
Solid state physics, 8
Solids, 289, 314
Sound, 412, 424
Source, 440, 475
Source activity, 623, 626
Space-Time, 635, 637
Special theory of relativity, 632, 637
Specific gravity, 311, 314
Specific heat, defined, 358, 381
Spectrum, electromagnetic, 410
  visible, 584, 600
Speed, defined, 86, 111
  of light, 409, 424, 543, 552
  of sound, 412, 424
Spherical aberration, 564
Spring balance, 32, 52
Standard position of vectors, 73, 79
Standard temperature and pressure, 394, 396
Standards of measure, 14, 52
Standing waves, 406, 424

Static electricity, 430
Statics, 157, 179
Statistical mechanics, 8
Stator, 495, 501
Step-down transformer, 514, 534
Step-up transformer, 513, 534
Storage cell, 466
Strain, 296, 314
Streamline flow, 336, 341
Stress, 293, 314
Strong force, 611, 626
Superconductor, 441, 475
Superposition of waves, 405, 424
Surface area, 28
  lateral, 28, 52
  total, 28, 52
Surface tension, 304, 314
Sympathetic vibration, 418
Synchronous motor, 498, 501

Technology, 5, 10
Temperature, 346, 381
Tensile strength, 290, 314
Tension, 159, 179, 295, 314
Terminal speed, 101, 111
Tesla, 486
Tesla, Nikola, 487
Theoretical physicist, 5, 10
Theory, 8, 10
Thermal conductivity, 354, 381
Thermodynamics, 7
Thermometers, 346
Thompson, J. J., 604
Time, measurement of, 33
Torque, defined, 166, 179
Torsion, 296, 314
Total internal reflection, 570, 578
Transformers, 512, 534
  step-down, 514, 534
  step-up, 513, 534
Transistor, 532
Translucent, 558, 578
Transparent, 558, 578
Transverse waves, 402, 424
Turbulent flow, 336, 341

Units, conversion of, 19
Universal gravitation, 276, 284
Universal motor, 498, 501

Van de Graaff, Robert, 433
Vapor pressure, 394
Vaporization, 372, 381
Variable, 653

Vectors:
  components of, 65, 79
  defined, 58, 79
  graphical addition, 60
  magnitude, 58
  resultant, 65, 79
  standard position, 73, 79
  x-component, 65, 79
  y-component, 65, 79
Velocity:
  angular, 213
  defined, 86, 111
  units of, 86
Venturi, Giovanni Battista, 338
Vibration:
  forced, 418
  sympathetic, 418
Virtual images, 559, 578
Viscosity, 304, 314
Visible spectrum, 584, 600
Volatility, 378, 381
Volt, 442
Volta, Allessandro, 442
Voltage drop, 442, 475
Voltmeter, 462, 475
Volume, 25, 52, 129
von Braun, Werner, 141

Watt, James, 192
Watt, unit of power, 192
Wavelength, 545, 552
Waves, 402, 424
  electromagnetic, 409
  longitudinal, 402
  sound, 412
  standing, 406, 424
  theory, 541, 552
  transverse, 402, 424
Wedge, 264, 268
Weight:
  defined, 32, 52, 127, 130
  density, 306, 314
  measurement, 32
Wheel and axle, 253, 268
Work:
  defined, 186, 207
  units of, 186

Z, impedance, 520
Zero, absolute, 350, 380

Intermediate Accounting

TENTH EDITION

J. DAVID SPICELAND
University of Memphis

MARK W. NELSON
Cornell University

WAYNE B. THOMAS
University of Oklahoma

Mc Graw Hill Education

Dedicated to:

David's wife Charlene, two daughters Denise and Jessica, and three sons Mike, Michael, and David

Mark's wife Cathy, and daughters Liz and Clara

Wayne's wife Julee, daughter Olivia, and three sons Jake, Eli, and Luke

INTERMEDIATE ACCOUNTING, TENTH EDITION

Published by McGraw-Hill Education, 2 Penn Plaza, New York, NY 10121.

This book is printed on acid-free paper.

1 2 3 4 5 6 7 8 9 LWI 21 20 19 18

ISBN 978-1-260-31017-7
MHID 1-260-31017-5

Executive Portfolio Manager: *Rebecca Olson*
Product Developers: *Christina Sanders* and *Danielle McLimore*
Marketing Manager: *Zachary Rudin*
Content Project Managers: *Pat Frederickson* and *Angela Norris*
Buyer: *Laura Fuller*
Design: *Matt Diamond*
Content Licensing Specialist: *Jacob Sullivan*
Cover Image: ©*dibrova/Shutterstock*
Compositor: *SPi Global*

All credits appearing on page or at the end of the book are considered to be an extension of the copyright page. Additional icons include the following: Data Analytics graph diagram, ©Iasagnaforone/Getty Images; vector buttons, ©KristinaVelickovic/Getty Images; IFRS globe, ©geopaul/Getty Images; compass, ©pictafolio/Getty Images; Ethical Dilemma scales, ©alexsl/Getty Images.

Library of Congress Cataloging-in-Publication Data

Cataloging-in-Publication Data has been requested from the Library of Congress.

About the Authors

DAVID SPICELAND

David Spiceland is Professor Emeritus in the School of Accountancy where he taught financial accounting at the undergraduate, master's, and doctoral levels for 36 years. He received his BS degree in finance from the University of Tennessee, his MBA from Southern Illinois University, and his PhD in accounting from the University of Arkansas.

Professor Spiceland has published articles in a variety of academic and professional journals including *The Accounting Review, Accounting and Business Research, Journal of Financial Research, Advances in Quantitative Analysis of Finance and Accounting,* and most accounting education journals: *Issues in Accounting Education, Journal of Accounting Education, Advances in Accounting Education, The Accounting Educators' Journal, Accounting Education, The Journal of Asynchronous Learning Networks,* and *Journal of Business Education.*, and is an author of McGraw-Hill's *Financial Accounting* with Wayne Thomas and Don Herrmann. Professor Spiceland has received university and college awards and recognition for his teaching, research, and technological innovations in the classroom.

MARK NELSON

Mark Nelson is the Anne and Elmer Lindseth Dean and Professor of Accounting at Cornell University's Samuel Curtis Johnson Graduate School of Management. He received his BBA degree from Iowa State University and his MA and PhD degrees from The Ohio State University. Professor Nelson has won ten teaching awards, including an inaugural Cook Prize from the American Accounting Association.

Professor Nelson's research focuses on decision making in financial accounting and auditing. His research has been published in the *Accounting Review;* the *Journal of Accounting Research; Contemporary Accounting Research; Accounting, Organizations and Society;* and several other journals. He has received the American Accounting Association's Notable Contribution to Accounting Literature Award, as well as the AAA's Wildman Medal for work judged to make a significant contribution to practice.

Professor Nelson served three terms as an area editor of *The Accounting Review* and is a member of the editorial boards of several journals. He also served for four years on the FASB's Financial Accounting Standards Advisory Council.

WAYNE THOMAS

Wayne Thomas is the W.K. Newton Chair and George Lynn Cross Research Professor of Accounting at the University of Oklahoma's Price College of Business. He received his BS degree from Southwestern Oklahoma State University and his MS and PhD from Oklahoma State University. He has received teaching awards at the university, college, and departmental levels, and has received the Outstanding Educator Award from the Oklahoma Society of CPAs. He is an author of McGraw-Hill's *Financial Accounting* with David Spiceland and Don Herrmann.

His research focuses on various financial reporting issues and has been published in *The Accounting Review, Journal of Accounting Research, Journal of Accounting and Economics, Contemporary Accounting Research, Review of Accounting Studies, Accounting Organizations and Society,* and others. He has served as an editor for *The Accounting Review* and has won the American Accounting Association's Competitive Manuscript Award and Outstanding International Accounting Dissertation.

Professor Thomas enjoys various activities such as tennis, basketball, golf, and crossword puzzles, and most of all, he enjoys spending time with his wife and kids.

Intermediate Accounting Tenth Edition:

Welcome to the new standard in intermediate accounting! Instructors recognize the "Spiceland advantage" in content that's intensive and thorough, as well as in writing that's fluid and precise—together, these combine to form a resource that's rigorous yet readable. By blending a comprehensive approach, clear conversational tone, current updates on key standards, and the market-leading technological innovations of Connect®, the Spiceland team delivers an unrivaled experience. As a result of Spiceland's rigorous yet readable learning system, students develop a deeper and more complete understanding of intermediate accounting topics.

"The textbook is readable and easy to follow since the authors present basic concepts and then cover advanced issues. Conceptually-oriented and dependable as the authors are timely in updating new accounting standards."

—Hong Pak, California State Polytechnic University, Pomona

The *Intermediate Accounting* learning system is built around four key attributes: current, comprehensive, clear, and Connect.

Current: Few disciplines see the rapid changes that accounting experiences. The Spiceland team is committed to keeping instructors' courses up to date. The tenth edition fully integrates the latest FASB updates, including:

- ASU No. 2018-02—Income Statement—Reporting Comprehensive Income (Topic 220)—Reclassification of Certain Tax Effects from Accumulated Other Comprehensive Income
- ASU No. 2016–013, Financial Instruments—Credit Losses (Topic 326) on "Current Expected Credit Loss" (CECL) model for accounting for credit losses, as well as current GAAP requirements for recognizing impairments of investments
- Comprehensive revision of Chapter 16, Accounting for Income Taxes, improving pedagogy as well as covering effects of the Tax Cuts and Jobs Act of 2017
- FASB ASC 842-10-15-42A: Leases-Overall-Scope and Scope Exceptions-Lessor
- FASB Accounting Standards Update, Compensation–Retirement Benefits (Topic 715): Improving the Presentation of Net Periodic Pension Cost and Net Periodic Postretirement Benefit Cost, FASB: December, 2016. SEC Amendment to Rule 15c6-1, Ex-Dividend Date, September 2017
- FASB Accounting Standards Update No. 2017-12, Derivatives and Hedging (Topic 815): Targeted Improvements to Accounting for Hedging Activities
- ASU No. 2017-04, Intangibles—Goodwill and Other (Topic 350): Simplifying the Test for Goodwill Impairment
- ASU No. 2015-05—Intangibles—Goodwill and Other—Internal-Use Software (Subtopic 350-40): Customer's Accounting for Fees Paid in a Cloud Computing Arrangement

The Spiceland team also ensures that Intermediate Accounting stays current with the latest pedagogy and digital tools. The authors have incorporated new Data Analytics cases featuring Tableau at the end of each chapter. These cases are auto-gradable in Connect and help students develop in-demand skills in analyzing and interpreting data, and effectively communicating findings.

Data Analytics connect

The phrase *scientia est potentia* is a Latin aphorism meaning "knowledge is power!" In a business sense, this might be paraphrased as "Information is money!" Better information . . . better business decisions! This is the keystone of data analytics.

Data analytics is the process of examining data sets in order to draw conclusions about the information they contain. Data analytics is widely used in business to enable organizations to make better-informed business decisions. Increasingly, this is accomplished with the aid of specialized data visualization software such as Tableau.

The New Standard

Current events regularly focus public attention on the key role of accounting in providing information useful to financial decision makers. The CPA exam, too, has changed to emphasize the professional skills needed to critically evaluate accounting method alternatives. *Intermediate Accounting* provides a **decision makers' perspective,** highlighting the professional judgment and critical thinking skills required of accountants in today's business environment. New in the 10th edition, many of these cases have been translated to an auto-graded format in Connect.

"The use of real-world examples throughout the text helps bring accounting to life for students. For example, the use of Amazon Prime is excellent, illustrating the complexity of current business with an engaging example."

— Jennifer Winchel, University of Virginia

Comprehensive: The Spiceland team ensures comprehensive coverage and quality throughout the learning system by building content and assets with a unified methodology that meets rigorous standards. Students are challenged through diverse examples and carefully crafted problem sets which promote in-depth understanding and drive development of critical-thinking skills.

The author team is committed to providing a learning experience that fully prepares students for the future by solidifying core comprehension and enabling confident application of key concepts. Students can feel confident that the conceptual underpinnings and practical skills conveyed in the tenth edition will prepare them for a wide range of real world scenarios.

Clear: Reviewers, instructors, and students have all hailed *Intermediate Accounting's* ability to explain both simple and complex topics in language that is coherent and approachable. Difficult topics are structured to provide a solid conceptual foundation and unifying framework that is built upon with thorough coverage of more advanced topics. As examples, see chapters 6 (Revenue Recognition), 15 (Leases) and 16 (Income Taxes). The author team's highly acclaimed conversational writing style establishes a friendly dialogue—establishing the impression of a conversation with students, as opposed to lecturing at them.

This tone remains consistent throughout the learning system, as authors Spiceland, Nelson, and Thomas write not only the primary content, but also every major supplement: instructor's resource manual, solutions manual, and test bank. All end-of-chapter material, too, is written by the author team and tested in their classrooms. *Intermediate Accounting* is written to be the most complete, coherent, and student-oriented resource on the market.

Connect: Today's accounting students expect to learn in multiple modalities. As a result, the tenth edition of Spiceland's learning system features the following: Connect, SmartBook's adaptive learning and reading experience, **NEW** Concept Overview Videos, Guided Examples, **NEW** Excel® simulations, and General Ledger problems.

Quality assessment continues to be a focus of Connect, with over **2,500 questions** available for assignment, including more than 1,125 algorithmic questions.

McGraw-Hill Education is continually updating and improving our digital resources. To that end, our partnership with Roger CPA, provides multiple choice practice questions directly within our Connect banks, as well as assignable links to the Roger CPA site for complementary access to selected simulations.

Spiceland's Financial Accounting Series

Intermediate Accounting forms a complete learning system when paired with *Financial Accounting* by authors David Spiceland, Wayne Thomas, and Don Herrmann. Now in its fifth edition, *Financial Accounting* uses the same proven approach that has made *Intermediate Accounting* a success—a conversational writing style with real-world focus and author-prepared supplements, combined with Connect's market leading technology solutions and assessment.

What Keeps SPICELAND Users Coming Back?

Financial Reporting Cases

Each chapter opens with a Financial Reporting Case that places the student in the role of the decision maker, engaging the student in an interesting situation related to the accounting issues to come. Then, the cases pose questions for the student in the role of decision maker. Marginal notations throughout the chapter point out locations where each question is addressed. The case questions are answered at the end of the chapter.

Financial Reporting Case Solution

©goodluz/123RF

1. **What purpose do adjusting entries serve?** *(p. 64)* Adjusting entries help ensure that all revenues are recognized in the period goods or services are transferred to customers, regardless of when cash is received. In this instance, for example, $13,000 cash has been received for services that haven't yet been performed. Also, adjusting entries enable a company to recognize all expenses incurred during a period, regardless of when cash is paid. Without depreciation, the friends' cost of using the equipment is not taken into account. Conversely, without adjustment, the cost of rent is overstated by $3,000 paid in advance for part of next year's rent.

With adjustments, we get an accrual income statement that provides a more complete measure of a company's operating performance and a better measure for predicting future operating cash f...

Where We're Headed

These boxes describe the potential financial reporting effects of many of the FASB's proposed projects that have not yet been adopted, as well as joint proposed projects with the IASB. Where We're Headed boxes allow instructors to deal with ongoing projects to the extent they desire.

Where We're Headed

As part of its ongoing disclosure framework project, the FASB has proposed using the U.S. Supreme Court's description of materiality. Under that definition, which comes from court cases and interpretations, qualitative and quantitative disclosures are material if there is a substantial likelihood that omitting a disclosure would have been viewed by a reasonable user as having significantly altered the total mix of information made available in making a decision. This change is controversial because it could affect the amount of information that companies disclose. The FASB has committed to seek more input before moving forward with this proposal.

Additional Consideration Boxes

These are "on the spot" considerations of important, but incidental or infrequent, aspects of the primary topics to which they relate.

Additional Consideration

Solving for the unknown factor in either of these examples could just as easily be done using the future value tables. The number of years is the value of *n* that will provide a present value of $10,000 when $16,000 is the future amount and the interest rate is 10%.

$16,000 (future value) = $10,000 (present value) × ?*

*Future value of $1: n = ?, i = 10%

Rearranging algebraically, the future value table factor is 1.6.

$16,000 (future value) ÷ $10,000 (present value) = 1.6*

*Future value of $1: n = ?, i = 10%

When you consult the future value table, Table 1, you search the 10% column (i = 10%) for this value and find 1.61051 in row five. So it would take approximately five years to accumulate $16,000 in the situation described.

Decision Makers' Perspective

These sections appear throughout the text to illustrate how accounting information is put to work in today's firms. With the CPA exam placing greater focus on application of skills in realistic work settings, these discussions help your students gain an edge that will remain with them as they enter the workplace.

Decision Makers' Perspective

Cash often is called a *nonearning* asset because it earns little or no interest. For this reason, managers invest idle cash in either cash equivalents or short-term investments, both of which provide a larger return than a checking account. Management's goal is to hold the minimum amount of cash necessary to conduct normal business operations, meet its obligations, and take advantage of opportunities. Too much cash reduces profits through lost returns, while too little cash increases risk. This trade-off between risk and return is an ongoing choice made by management (internal decision makers). Whether the choice made is appropriate is an ongoing assessment made by investors and creditors (external decision makers).

A company must have cash available for the compensating balances we discussed in the previous section as well as for planned disbursements related to normal operating, investing, and financing cash flows. However, because cash ...

Companies hold cash to pay for planned and unplanned transactions

In talking with so many intermediate accounting faculty, we heard more than how to improve the book—there was much, much more that both users and nonusers insisted we not change. Here are some of the features that have made Spiceland such a phenomenal success.

> "Our students need to think about topics critically and broadly. Spiceland is superior in terms of presenting big picture cases and scenarios that help students understand how to think beyond simple journal entries and worksheet problems."
>
> **—James Brushwood, Colorado State University**

Decision Makers' Perspective

©ImageGap/Getty Images

Apply your critical-thinking ability to the knowledge you've gained. These cases will provide you an opportunity to develop your research, analysis, judgment, and communication skills. You also will work with other students, integrate what you've learned, apply it in real-world situations, and consider its global and ethical ramifications. This practice will broaden your knowledge and further develop your decision-making abilities.

Judgment Case 4–1
Earnings quality
● LO4–2, LO4–3

The financial community in the United States has become increasingly concerned with the quality of reported company earnings.

Required:

1. Define the term *earnings quality.*

Decision Makers' Perspective Cases

Designed to further develop students' decision-making abilities, each chapter includes a robust set of powerful and effective cases, asking students to analyze, evaluate, and communicate findings, further building their critical thinking skills. Many of these cases are now auto-gradable in Connect.

Ethical Dilemma

You recently have been employed by a large retail chain that sells sporting goods. One of your tasks is to help prepare periodic financial statements for external distribution. The chain's largest creditor, National Savings & Loan, requires quarterly financial statements, and you are currently working on the statements for the three-month period ending June 30, 2021.

During the months of May and June, the company spent $1,200,000 on a hefty radio and TV advertising campaign. The $1,200,000 included the costs of producing the commercials as well as the radio and TV time purchased to air the commercials. All of the costs were charged to advertising expense. The company's chief financial officer (CFO) has asked you to prepare a June 30 adjusting entry to remove the costs from advertising expense and to set up an asset called *prepaid advertising* that will be expensed in July. The CFO explained that "This advertising campaign has led to significant sales in May and June and I think it

Ethical Dilemmas

Because ethical ramifications of business decisions impact so many individuals as well as the core of our economy, Ethical Dilemmas are incorporated within the context of accounting issues as they are discussed. These features lend themselves very well to impromptu class discussions and debates, and are complemented by Ethics Cases found in the Decision Makers' Perspective Case section at the end of each chapter.

P 15–25
Operating lease; uneven lease payments
● LO15–4, LO15–7
★

On January 1, 2021, Harlon Consulting entered into a three-year lease for new office space agreeing to lease payments of $5,000 in 2021, $6,000 in 2022, and $7,000 in 2023. Payments are due on December 31 of each year with the first payment being made on December 31, 2021. Harlon is aware that the lessor used a 5% interest rate when calculating lease payments.

Required:

1. Prepare the appropriate entries for Harlon Consulting on January 1, 2021, to record the lease.
2. Prepare all appropriate entries for Harlon Consulting on December 31, 2021, related to the lease.
3. Prepare all appropriate entries for Harlon Consulting on December 31, 2022, related to the lease.
4. Prepare all appropriate entries for Harlon Consulting on December 31, 2023, related to the lease.

Star Problems ★

In each chapter, particularly rigorous problems, designated by a ★, require students to combine multiple concepts or require significant use of judgment.

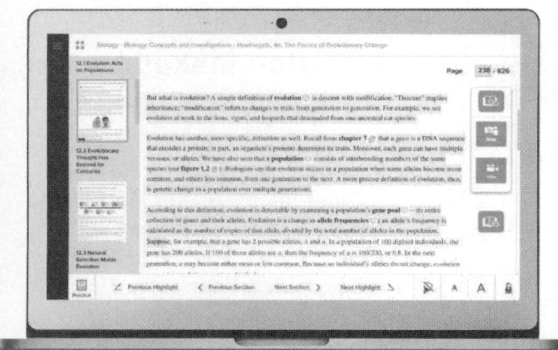

For Students

Effective, efficient studying.

Connect helps you be more productive with your study time and get better grades using tools like SmartBook, which highlights key concepts and creates a personalized study plan. Connect sets you up for success, so you walk into class with confidence and walk out with better grades.

©Shutterstock/wavebreakmedia

"I really liked this app—it made it easy to study when you don't have your text-book in front of you."

- Jordan Cunningham,
Eastern Washington University

Study anytime, anywhere.

Download the free ReadAnywhere app and access your online eBook when it's convenient, even if you're offline. And since the app automatically syncs with your eBook in Connect, all of your notes are available every time you open it. Find out more at **www.mheducation.com/readanywhere**

No surprises.

The Connect Calendar and Reports tools keep you on track with the work you need to get done and your assignment scores. Life gets busy; Connect tools help you keep learning through it all.

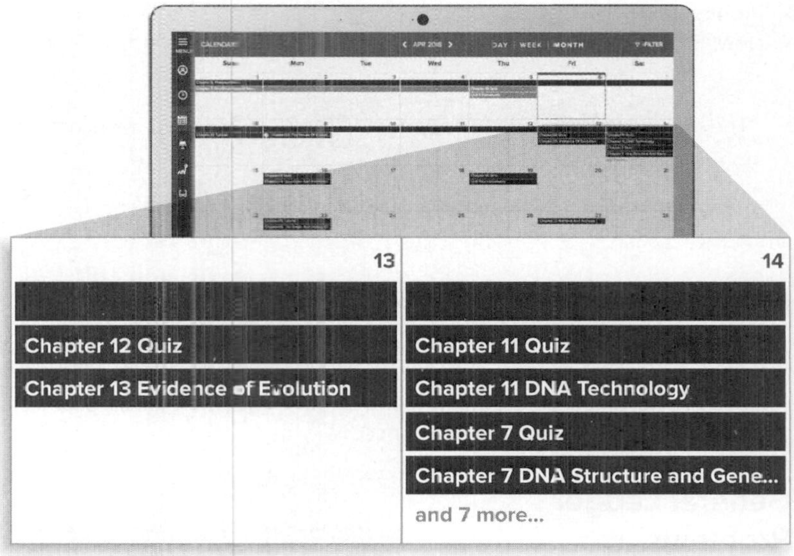

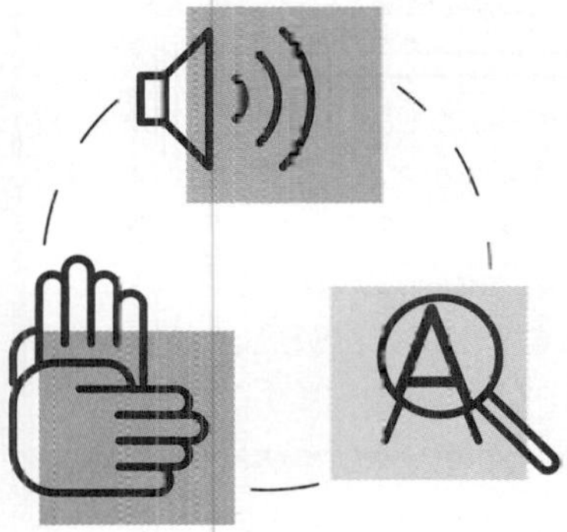

Learning for everyone.

McGraw-Hill works directly with Accessibility Services Departments and faculty to meet the learning needs of all students. Please contact your Accessibility Services office and ask them to email accessibility@mheducation.com, or visit **www.mheducation.com/accessibility** for more information.

THE NEW STANDARD:

Online Assignments

Connect helps students learn more efficiently by providing feedback and practice material when they need it, where they need it. Connect grades homework automatically and gives immediate feedback on any questions students may have missed. The extensive assignable, gradable end-of-chapter content includes a general journal application that looks and feels like what one would find in a general ledger software package. For this edition, numerous questions have been redesigned to test students' knowledge more fully. New to this edition, many Decision Makers' Perspective Cases have been incorporated as auto-gradable in Connect.

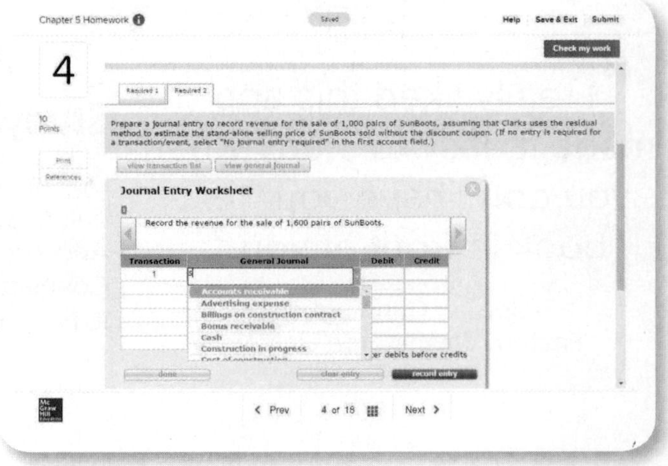

End-of-chapter questions in Connect include:

- Brief Exercises
- Exercises
- Problems
- New! Select Decision Makers' Perspective Cases, Target and Air France Cases, and Data Analytics Cases

"The digital materials are an important learning resource for my students, and the constant updates between new editions are VERY useful."

—Pamela Trafford, University of Massachusetts, Amherst

NEW! Data Analytics Cases

Data analytics is an enormously in-demand skill among employers. Students who can interpret data and effectively communicate their findings to help business makes better-informed decisions are in high-demand. New Data Analysis Cases featuring Tableau are now incorporated at the end of most chapters. These cases can easily be assigned in Connect and are auto-gradable for the instructor's convenience.

General Ledger Problems

General Ledger Problems allow students to see how transactions flow through the various financial statements. Students can audit their mistakes by easily linking back to their original journal entries. Many General Ledger Problems include an analysis tab that allows students to demonstrate their critical thinking skills and a deeper understanding of accounting concepts.

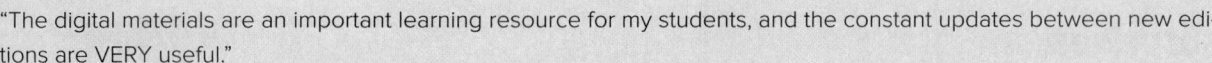

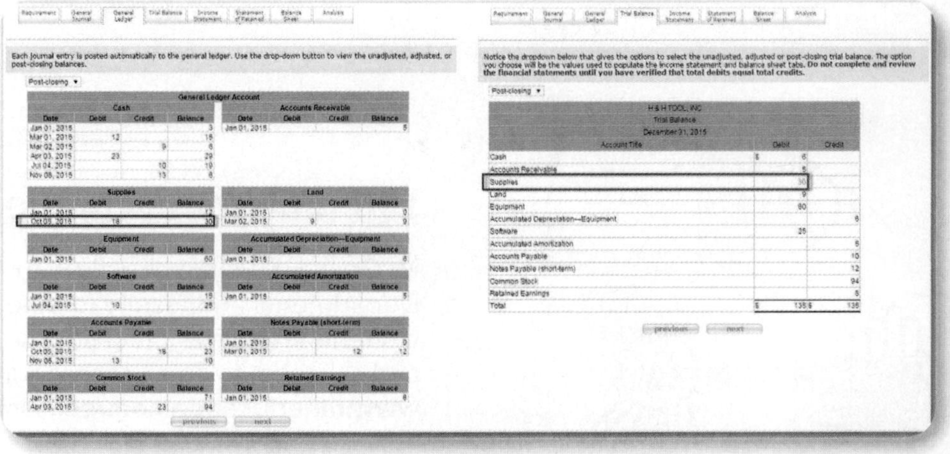

POWERFUL ONLINE TOOLS & ASSESSMENTS

Concept Overview Videos

Concept Overview Videos provide engaging narratives of key topics in an assignable and interactive online format. These videos follow the structure of the text and are available with all learning objectives within each chapter of *Intermediate Accounting*. The Concept Overview Videos provide additional explanation of material in the text, allowing students to learn at their own pace – and test their knowledge with assignable questions.

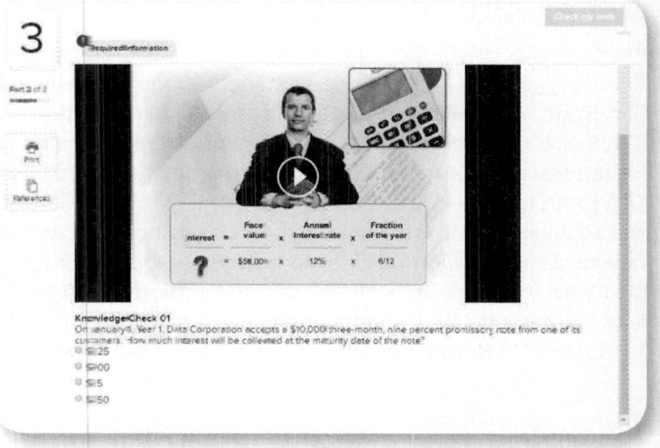

Excel Simulations

Simulated Excel Questions, assignable within Connect, allow students to practice their Excel skills—such as basic formulas and formatting—within the content of financial accounting. These questions feature animated, narrated Help and Show Me tutorials (when enabled), as well as automatic feedback and grading for both students and professors

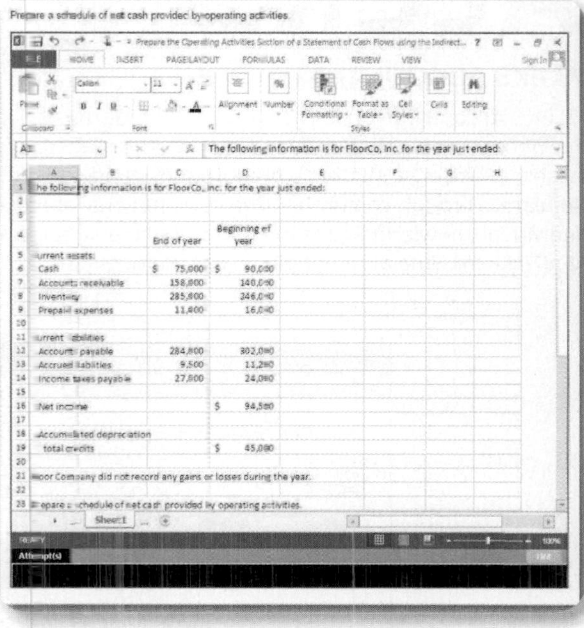

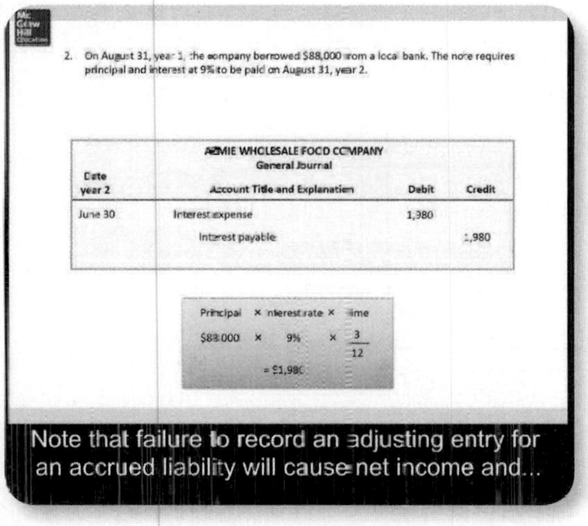

Guided Example/Hint Videos

The **Guided Examples** in Connect provide a narrated, animated, step-by-step walk-through of select exercises similar to those assigned, These short videos are presented to students as hints and provide reinforcement when students need it most. Instructors have the option of turning them on or off.

ROGER CPA Review

CPA SIMULATIONS

McGraw-Hill Education has partnered with Roger CPA Review, a global leader in CPA Exam preparation, to provide students a smooth transition from the accounting classroom to successful completion of the CPA Exam. While many aspiring accountants wait until they have completed their academic studies to begin preparing for the CPA Exam, research shows that those who become familiar with exam content earlier in the process have a stronger chance of successfully passing the CPA Exam. Accordingly, students using these McGraw-Hill materials will have access to sample CPA Exam Multiple-Choice questions and Task-based Simulations from Roger CPA Review, with expert-written explanations and solutions. All questions are either directly from the AICPA or are modeled on AICPA questions that appear in the exam. Task-based Simulations are delivered via the Roger CPA Review platform, which mirrors the look, feel and functionality of the actual exam. McGraw-Hill Education and Roger CPA Review are dedicated to supporting every accounting student along their journey, ultimately helping them achieve career success in the accounting profession. For more information about the full Roger CPA Review program, exam requirements and exam content, visit www.rogercpareview.com.

ALEKS®

ALEKS ACCOUNTING CYCLE

ALEKS Accounting Cycle is a web-based program that provides targeted coverage of prerequisite and introductory material necessary for student success in Intermediate Accounting. ALEKS uses artificial intelligence and adaptive questioning to assess precisely a student's preparedness and deliver personalized instruction on the exact topics the student is **most ready to learn.** Through comprehensive explanations, practice, and immediate feedback, ALEKS enables students to quickly fill individual knowledge gaps in order to build a strong foundation of critical accounting skills. Better prepared students save you valuable time at the beginning of your course!

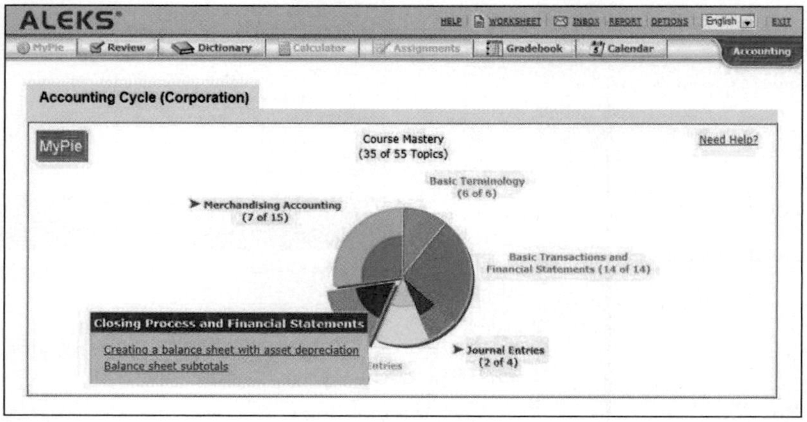

Use ALEKS Accounting Cycle as a pre-course assignment or during the first weeks of the term to see improved student confidence and performance, as well as fewer drops.

ALEKS Accounting Cycle Features:

- **Artificial Intelligence:** Targets Gaps in Prerequisite Knowledge
- **Individualized Learning and Assessment:** Ensure Student Preparedness
- **Open-Response Environment:** Avoids Multiple-Choice and Ensures Mastery
- **Dynamic, Automated Reports:** Easily Identify Struggling Students

For more information, please visit: www.aleks.com/highered/business.

Read ALEKS Success Stories: www.aleks.com/highered/business/success_stories.

The New Standard:

INSTRUCTOR LIBRARY

The Connect Instructor Library is a repository of additional resources to improve student engagement in and out of class. You can select and use any asset that enhances your lecture. The Connect Instructor Library includes:

In-Class Presentation Tools

PowerPoints

Three types of PowerPoint decks are available, each responding to a different instructional need:

- **Lecture PowerPoints with Concept Checks** allow instructors to intersperse short exercises that students can solve individually or do in groups before an answer is "revealed"
- **Lecture PowerPoints without Concept Checks**: No questions included, mirror presentation from book with key illustrations and notes
- **Accessible PowerPoints**: Allow slide content to be read by a screen reader and provide alternative text descriptions for any image files. Accessible PowerPoints are also designed with high-contrast color palettes and use texture instead of color whenever possible to denote different aspects of imagery.

Please note: All of our illustrations in the PowerPoint decks are fully editable so that you can change numbers, content, and easily customize our figures to your teaching. Teaching Tip: Use your students' names for company names in the Illustrations or Concept Checks!

Exercise Presentations

PowerPoint slides created from the Guided Example/Hint videos, these allow you to walk through a version of one of the book exercises in class without giving away the answers.

Digital Image Library

High-resolution images of all illustrations from the text.

Teaching Resources

New! Instructor's Edition

We know Intermediate Accounting is a daunting course not only for students, but also for instructors! This all-new digital guide to the Spiceland Intermediate learning program contains resources to help instructors think through effective course planning, including:

- Example syllabi
- Guides to Connect setup
- Course planning materials to help identify resources that align with your course goals
- Tips from the authors on key illustrations, teaching points, and end-of-chapter content they highlight and assign
- Bloom's rubrics for end-of-chapter content

Instructor's Manual

Specific to each chapter, contains learning outcomes, a full lecture outline, and suggestions for in-class activities including real world scenarios, group research activities, IFRS activities, and professional skills development activities. Assignment charts are also provided with topics and estimated completion times.

Solutions Manual

Created by the authors, includes solutions to end-of-chapter content.

Updates

Stop here for all the most recent updates from FASB. Our authors work tirelessly to keep you current — for instance, within a month of the 2017 Tax Reform bill, our authors had posted an updated PowerPoint deck, video walkthrough, and teaching tip material on how to address the new updates. We have your back!

Test Bank

Connect Test Bank

Multiple-choice, true/false and worksheet questions are all available to help assess students throughout all levels from understanding to evaluation. Create an effective test in 3 easy steps:

Flexible Instructor Resources

1. Tailor: Use Connect's robust filters to quickly identify the content you want based on difficulty, learning objective, Bloom's, AICPA and AACSB standards, and accessibility level.
2. Secure: Once you've identified the questions you want, Connect's policy setting help you create a secure test through features like question pooling, time limits, scrambling, feedback constraint, and Tegrity proctoring.
3. Report: Connect's rich reporting features allow you to quickly compile thorough reports that will wow your assessment committees. Teaching tip: Add your department learning objectives as tags in Connect for reporting completely aligned with your institution's objectives.

TestGen TestGen is a complete, state-of-the-art test generator and editing application software that allows instructors to quickly and easily select test items from McGraw Hill's TestGen testbank content and to organize, edit and customize the questions and answers to rapidly generate paper tests. Questions can include stylized text, symbols, graphics, and equations that are inserted directly into questions using built-in mathematical templates. TestGen's random generator provides the option to display different text or calculated number values each time questions are used. With both quick-and-simple test creation and flexible and robust editing tools, TestGen is a test generator system for today's educators.

MCGRAW-HILL EDUCATION CUSTOMER EXPERIENCE GROUP CONTACT INFORMATION

At McGraw-Hill Education, we understand that getting the most from new technology can be challenging. That's why our services don't stop after you purchase our products. You can contact our Product Specialists 24 hours a day to get product training online. Or you can search the knowledge bank of Frequently Asked Questions on our support website. For Customer Support, call **800-331-5094,** or visit www.mhhe.com/support. One of our Technical Support Analysts will be able to assist you in a timely fashion.

ASSURANCE OF LEARNING

Many educational institutions today are focused on the notion of *assurance of learning,* an important element of some accreditation standards. *Intermediate Accounting* is designed specifically to support your assurance of learning initiatives with a simple, yet powerful solution.

Each test bank question for *Intermediate Accounting* maps to a specific chapter learning objective listed in the text. You can use Connect to easily query for learning outcomes/objectives that directly relate to the learning objectives for your course. You can then use the reporting features of Connect to aggregate student results in a similar fashion, making the collection and presentation of assurance of learning data simple and easy.

AACSB STATEMENT

McGraw-Hill Education is a proud corporate member of AACSB International. Understanding the importance and value of AACSB accreditation, *Intermediate Accounting* recognizes the curricula guidelines detailed in the AACSB standards for business accreditation by connecting selected questions in the test bank to the eight general knowledge and skill guidelines in the AACSB standards.

The statements contained in *Intermediate Accounting* are provided only as a guide for the users of this textbook. The AACSB leaves content coverage and assessment within the purview of individual schools, the mission of the school, and the faculty. While *Intermediate Accounting* and the teaching package make no claim of any specific AACSB qualification or evaluation, within the Test Bank to accompany *Intermediate Accounting* we have labeled selected questions according to the eight general knowledge and skill areas.

What's New in the Tenth Edition?

Spiceland is the new global standard for providing students the most accessible, comprehensive, and current Intermediate Accounting learning system. We take seriously the confidence the marketplace has accorded our text. Each revision carefully considers how the print and digital content work together to coordinate improvements in content and industry-leading technology to provide the most robust learning solution. The Spiceland team implements only those changes that constitute real improvements as identified through extensive research with users. The result is a learning system that enhances our reputation for providing the best preparation for passing the CPA exam and successful accounting careers.

Improvements in this edition include the following:

- **NEW! Updated content to reflect the latest GAAP and Accounting Standards Updates including:**
  - **Income taxes**
  - **Leases**
  - **Financial instruments**
  - **Revenue recognition**
- **NEW!** Most **Decision Makers' Perspective cases** are now *auto-gradable in Connect.*
- **NEW! Data Analytics Cases** providing students the opportunity to experience the power and efficacy of data analytics in the context of each chapter's topics, using Tableau as a tool, that are *auto-gradable in Connect.*
- Enhanced partnership with **Roger CPA Review,** with new CPA Exam Review multiple-choice questions that are *auto-gradable in Connect* and access to CPA Exam Review simulations.
- Updated and revised **real-world** illustrations, assignments, and discussions.
- Revised Continuing Cases featuring **Target Corporation** financial statements prepared using U.S. GAAP, now *auto-gradable in Connect.* A comprehensive version of the case is available in Appendix B.
- Revised Continuing Cases featuring **Air France–KLM** financial statements prepared using IFRS, now *auto-gradable in Connect.* A comprehensive version of the case is available in Appendix C.
- Incorporated the latest technology, including:
  - **NEW! Connect interface for students,** along with **Connect Insight for students**
  - an updated **SmartBook**
  - **General Ledger Problems** that auto-post from journal entries to T-accounts to trial balances (*auto-gradable in Connect*)
  - **Excel Simulations** that allow students to practice their Excel skills within the content of financial accounting with animated, narrated Help and Show Me tutorials (*auto-gradable in Connect*)
  - **NEW! Concept Overview Videos** that provide engaging narratives of key topics in an assignable and interactive online format (*assignable in Connect*)
  - **Guided Examples/Hint Videos** in Connect that provide a narrated, animated, step-by-step walk-through of select exercises that provide reinforcement when students need it most (can be turned on or off by instructors)
  - **NEW! Instructor's Edition** to help instructors to more easily design the course and organize chapter resources

Chapter 1

ENVIRONMENT AND THEORETICAL STRUCTURE OF FINANCIAL ACCOUNTING

- Provided more focused discussion of the convergence process.
- Revised discussion of FASB standard setting and ongoing projects (disclosure framework, materiality).
- Added or updated problems and cases relevant to The Gap.

Chapter 2

REVIEW OF THE ACCOUNTING PROCESS

- Revised and reorganized the presentation of the accounting processing cycle.
- Added the use of a Dividends account.
- Eliminated the use of Income Summary in the Closing Process.
- Added a Data Analytics case based on topics in the chapter, using Tableau as a tool (auto-gradable in Connect).

Chapter 3

THE BALANCE SHEET AND FINANCIAL DISCLOSURES

- Changed opening balance sheet to Nike.
- Clarified distinction between book value, market value, and fair value.
- Updated terminology in chapter and end-of-chapter material to reflect ASU 2016-01 for investments.
- Updated definition of long-term liability to include FASB's proposed definition.

- Discussed and provided an example of the new format for the auditor's report.
- Revised discussion of executive compensation to reflect the more prominent role of restricted stock compared to stock options.
- Added a Data Analytics case based on topics in the chapter, using Tableau as a tool (auto-gradable in Connect).

Chapter 4

THE INCOME STATEMENT, COMPREHENSIVE INCOME, AND THE STATEMENT OF CASH FLOWS

- Updated all illustrations and end-of-chapter assignments involving income taxes to a new overall rate of 25%.
- Revised definition and discussion of other comprehensive income.
- Added a Data Analytics case based on topics in the chapter, using Tableau as a tool (auto-gradable in Connect).

Chapter 5

TIME VALUE OF MONEY CONCEPTS

- Revised the Financial Reporting Case
- Added **LO5–5** Explain the role of present value techniques in the valuation of notes in the Preview of Accounting Applications of Present Value Techniques—Single Cash Amount section.
- Eliminated the Expected Cash Flow section in the Preview of Accounting Applications of Present Value Techniques—Single Cash Amount section.
- Added a Valuation of Long Term Notes section in the Preview of Accounting Applications of Present Value Techniques—Single Cash Amount section.
- Added a Valuation of Long Term Notes section in the Preview of Accounting Applications of Present Value Techniques—Annuities section

Chapter 6

REVENUE RECOGNITION

- Reordered chapters 5 and 6 to have time value of money (now 5) precede coverage of revenue recognition (now 6) and have revenue recognition immediately precede coverage of receivables (7).
- Expanded coverage of time value of money considerations, including numerical examples of significant financing components for prepayments and receivables, along with new end of chapter material in brief exercises and exercises.
- Added Microsoft real-world example to illustrate materiality of new treatment of licenses.
- Added new Trueblood cases.
- Added or updated problems and cases relevant to Expedia, Priceline, and Alphabet.
- Added a Data Analytics case based on topics in the chapter, using Tableau as a tool (auto-gradable in Connect).

Chapter 7

CASH AND RECEIVABLES

- In chapter, as well as Appendix 7B, enhanced coverage of ASU 2016–13's CECL model for accounting for credit losses.
- Modified coverage of IFRS to focus on IFRS No. 9.
- Enhanced end of chapter material with respect to credit losses and accounts receivable.
- Revised coverage on noninterest-bearing notes receivable.
- Added or updated problems and cases relevant to General Mills, Microsoft, Amdahl, Nike, Avon Products, Cisco, Sanofi-Aventis, Tyson Foods, and Pilgrim's Pride Corp.
- Added a Data Analytics case based on topics in the chapter, using Tableau as a tool (auto-gradable in Connect).

Chapter 8

INVENTORIES: MEASUREMENT

- Clarified discussion of consignment arrangements and related costs.
- Added details about calculation of cost of goods sold and ending inventory for specific identification.
- Clarified discussion of calculating of cost of goods sold and ending inventory using perpetual average cost.
- Clarified the use of LIFO calculations under periodic versus perpetual system in practice.
- Added discussion of the impact of technology and the use of the perpetual inventory system
- Modified the Concept Review Exercise on inventory cost flow to include LIFO reserve.
- Revised Illustration 8-5 to reflect entries under a perpetual inventory system.
- Revised discussion of purchase discounts under the gross versus net method
- Added BE 8-10 for LIFO reserve.
- Modified E 8-19 and P 8-1 to include LIFO reserve adjustment from perpetual FIFO to periodic LIFO.
- Modified E 8-16 to compare FIFO and LIFO when costs are increasing and when costs are decreasing.
- Added a Data Analytics case based on topics in the chapter, using Tableau as a tool (auto-gradable in Connect).

Chapter 9

INVENTORIES: ADDITIONAL ISSUES

- Modified Concept Review Exercise for lower of cost or net realizable value to include unit values.
- Clarified treatment of employee discounts in the conventional retail method and revised Illustration 9-13.
- Revised discussion of Dollar-Value LIFO Retail.
- Added a Data Analytics case based on topics in the chapter, using Tableau as a tool (auto-gradable in Connect).

Chapter 10

PROPERTY, PLANT, AND EQUIPMENT AND INTANGIBLE ASSETS: ACQUISITION

- Revised discussion of nonmonetary exchanges to include four steps, and added summary Illustration 10-16.
- Revised Illustration 10-14A to clarify recording of a nonmonetary exchange.
- Moved discussion and illustration of amortization of software development costs to chapter 11.
- Added discussion of accounting for cloud computing arrangements and related implementation costs.
- Added Brief Exercise 10-18 on accounting for software development costs for internal purposes.
- Added Brief Exercise 10-19 on accounting for software development costs in cloud computing arrangements.
- Added a Data Analytics case based on topics in the chapter, using Tableau as a tool (auto-gradable in Connect).

Chapter 11

PROPERTY, PLANT, AND EQUIPMENT AND INTANGIBLE ASSETS: UTILIZATION AND DISPOSITION

- Moved discussion and illustration of amortization of software development costs from chapter 10.
- Modified Illustration 11–4B to simplify and clarify calculation of partial year depreciation.
- Added discussion of amortization of software development costs for internal purposes and in cloud computing arrangements.
- Updated discussion of impairment for goodwill based on ASU No. 2017-04.
- Added Exercise 11-13 on reporting assets held for sale.

- Updated discussion of MACRS depreciation for changes enacted by the Tax Cuts and Jobs Act.
- Added a Data Analytics case based on topics in the chapter, using Tableau as a tool (auto-gradable in Connect).

Chapter 12

INVESTMENTS

- In chapter, as well as Appendix 12B, enhanced coverage of ASU 2016–13's CECL model for accounting for credit losses.
- Revised wording of account titles to provide more streamlined and cohesive presentation of accounting for HTM, TS, AFS, equity, and equity method investments.
- Provided new Decision Makers' Perspective cases for Intel, FCA and Merck.
- Added a Data Analytics case based on topics in the chapter, using Tableau as a tool (auto-gradable in Connect).

Chapter 13

CURRENT LIABILITIES AND CONTINGENCIES

- Updated General Mills example used in Illustration 13–1 and throughout the chapter.
- Updated contingent liability examples.
- Added a Data Analytics case based on topics in the chapter, using Tableau as a tool (auto-gradable in Connect).

Chapter 14

BONDS AND LONG-TERM NOTES

- Updated Real World Financials in Bonds section.
- Revised the journal entry in Illustration 14-A2 and added additional explanation.
- Replaced illustration of zero coupon securities in Ill. 14-7 with a newer real world example, Coca Cola.
- Added a Data Analytics case based on topics in the chapter, using Tableau as a tool (auto-gradable in Connect).

Chapter 15

LEASES

- Added discussion to the operating leases section *related* to our *recording* both interest and amortization even though, for an operating lease, the *lessee* will *report* a single lease expense rather than the separate interest and amortization as with a finance lease.
- Added a *Why Lease?* section to Part A and added related EOC and TB questions.
- Added an *Is it a Lease?* section to Part C and added related EOC and TB questions.
- Added a paragraph at end of Nonlease Payments section in Part C to describe new ASU on simplification option for lessors.
- Added a Data Analytics case based on topics in the chapter, using Tableau as a tool (auto-gradable in Connect).

Chapter 16

ACCOUNTING FOR INCOME TAXES

- Reordered early coverage to provide more conceptual basis before walking through treatment of individual temporary differences.
- Revised illustrations to highlight the four-step process for calculating tax expense, using color-coded steps in all examples and solutions to end-of-chapter material.
- Revised coverage to walk through each combination of deferred tax assets and liabilities and revenue- and expense-related temporary differences.
- Modified all examples and end of chapter material to reflect new tax rates.
- Modified coverage of net operating loss carrybacks and carryforwards to reflect new tax act.
- Modified coverage of non-temporary differences to reflect new tax act, including Additional Consideration covering earning repatriation.
- Updated Real World "Shoe Carnival" case covering linkage between tax expense journal entry and changes in deferred tax assets, liabilities, and the valuation allowance.
- Added a Data Analytics case based on topics in the chapter, using Tableau as a tool (auto-gradable in Connect).

Chapter 17

PENSIONS AND OTHER POSTRETIREMENT BENEFIT PLANS

- Added Additional Consideration box to note the trend toward the "spot rate" method of determining the interest rate.
- Added Additional Consideration box to indicate that some companies are voluntarily choosing to recognize pension gains and losses immediately rather than amortizing them.
- Most Decision Makers' Perspective cases are now auto-gradable in Connect.
- Added a Data Analytics case based on topics in the chapter, using Tableau as a tool (auto-gradable in Connect).

Chapter 18

SHAREHOLDERS' EQUITY

- Revised discussion and assignment material to reflect the SEC's revision of the ex-dividend date from two business days before the date of record to one.
- Replaced Alcoa Case with Nike Case.
- Added a Data Analytics case based on topics in the chapter, using Tableau as a tool (auto-gradable in Connect).

Chapter 19

SHARE-BASED COMPENSATION AND EARNINGS PER SHARE

- Modified all illustrations and end of chapter material to reflect new tax rates.
- Added a Data Analytics case based on topics in the chapter, using Tableau as a tool (auto-gradable in Connect).

Chapter 20

ACCOUNTING CHANGES AND ERROR CORRECTIONS

- Revised discussion of approaches to account for accounting changes to include the modified retrospective approach.
- Modified all illustrations and end of chapter material to reflect new tax rates.

Chapter 21

STATEMENT OF CASH FLOWS REVISITED

- Revised a CVS Caremark Corp illustration of presenting cash flows from operating activities by the direct method.
- Added an enhanced Additional Consideration box on reporting bad debt expense in the SCF.
- Added a real world illustration of presenting cash flows from operating activities by the indirect method.

- Modified all illustrations and end of chapter material to reflect new tax rates.
- Revised a Research Case related to FedEx's investing and financing activities.
- Added a Real World Case on Staples reporting of its SCF.
- Added a Data Analytics case based on topics in the chapter, using Tableau as a tool (auto-gradable in Connect).

Appendix A
DERIVATIVES

- Extensively revised all illustrations, discussions, and assignment material to reflect changes emanating from the new FASB Accounting Standards Update No. 2017-12, Derivatives and Hedging (Topic 815): Targeted Improvements to Accounting for Hedging Activities.
- Added new discussion, illustration and assignment material for a cash flow hedge, interest rate swap.
- Added new discussion, illustration and assignment material for an option contract,
- Added new discussion, illustration and assignment material for a nonfinancial forward contract
- Revised a Real World Case related to the Chicago Mercantile Exchange.
- Revised a Johnson & Johnson Real World Case on hedging transactions.

Acknowledgments

Intermediate Accounting is the work not just of its talented authors, but of the more than 750 faculty reviewers who shared their insights, experience, and insights with us. Our reviewers helped us to build *Intermediate Accounting* into the very best learning system available. A blend of Spiceland users and nonusers, these reviewers explained how they use texts and technology in their teaching, and many answered detailed questions about every one of Spiceland's 21 chapters. The work of improving *Intermediate Accounting* is ongoing—even now, we're scheduling new symposia and reviewers' conferences to collect even more opinions from faculty.

We would like to acknowledge and highlight the Special Reviewer role that Ilene Leopold Persoff of Long Island University (LIU Post) took on the tenth edition. Utilizing her accounting and reviewing expertise, Ilene verified the accuracy of the manuscript and promoted our efforts toward quality and consistency. Her deep subject-matter knowledge, keen eye for detail, and professional excellence in all aspects were instrumental in ensuring a current, comprehensive, and clear edition. Her contributions are deeply appreciated.

We are especially grateful for the contributions of Charlene Parnell Spiceland of Simmons University in developing the Data Analyitics Case sequence that is a key enhancement to the tenth edition of this textbook. Utilizing her expertise in Tableau, she ensured that these cases provide students the opportunity to glimpse the power and efficacy of data analytics. The cases encourage students to use Tableau to analyze data sets to make business decisions in the context of chapter topics as well as to appreciate the most effective and efficient ways to communicate their findings.

In addition, we want to recognize the valuable input of all those who helped guide our developmental decisions for the tenth edition.

H. Kyle Anderson, *Clemson University*
Matthew Anderson, *Michigan State University*
Avinash Arya, *William Paterson University*
Patricia Ball, *Massasoit Community College*
Michael Baker, *Southern New Hampshire University*
Brian Bratten, *University of Kentucky, Lexington*
Maureen Breen, *Drexel University*
Stephen Brown, *University of Maryland*
Esther Bunn, *Stephen F. Austin State University*
John Capka, *Cuyahoga Community College*
Kam Chan, *Pace University*
Chiaho Chang, *Montclair State University*
Carolyn Christesen, *SUNY Westchester Community College*
Simona Citron, *Northwestern University Southern New Hampshire University*
Mariah Dar, *John Tyler Community College*
Amy Donnelly, *Clemson University*
Jeffrey T. Doyle, *Utah State University*
Dov Fischer, *Brooklyn College*
Corinne Frad, *Eastern Iowa Community College*
Shari Fowler, *Indiana University East*
Dana Garner, *Virginia Tech*
Paul Goodchild, *University of Central Missouri*
Pamela J. Graybeal, *University of Central Florida Orlando*
Rick Johnston, *University of Alabama at Birmingham*
Beth Kane, *Columbia College of Missouri*
Gordon Klein, *University of California Los Angeles*
Melissa Larson, *Brigham Young University*
Ming Lu, *Santa Monica College*
Stephen McCarthy, *Kean University*
Dawn McKinley, *Harper College*
Jeanette Milius, *Nebraska Wesleyan University*
Anita Morgan, *Indiana University*
Ahmad (Sia) Nassiripour, *William Paterson University*
Anne Mary Nash-Haruna, *Bellevue College*
Sewon O, *Texas Southern University*
Hong S. Pak, *California State Polytechnic University, Pomona*
Debra Petrizzo-Wilkins, *Franklin University*
Kris Portz, *St. Cloud State University*
Linda Quick, *East Carolina University*
David Ravetch, *University of California Los Angeles*
Cecile Roberti, *Community College of Rhode Island*
Mark Ross, *Western Kentucky University*
Sheldon Smith, *Utah Valley University*
Larry G. Stephens, *Austin Community College District*
Vicki M Stewart, *Texas A & M University Commerce*
Peter Theuri, *Northern Kentucky University*
Geoffrey Tickell, *Indiana University of Pennsylvania*
Dick Williams, *Missouri State University*
Jennifer Winchel, *University of Virginia*
Lori Zaher, *Bucks County Community College*

We Are Grateful

We would like to acknowledge and thank the following individuals for their contributions in developing, reviewing and shaping the extensive ancillary package: Jeannie Folk, *College of DuPage;* Julie Hankins, User Euphoria; Patrick Lee, University of Connecticut; Beth Kobylarz of Accuracy Counts; Mark McCarthy, *East Carolina University;* Barbara Muller, *Arizona State University;* Patricia Plumb; Helen Roybark, *Radford University;* Jacob Shortt, *Virginia Tech*; Pamela Simmons, Cuyahoga Community College; Kevin Smith, *Utah Valley University;* Emily Vera, *University of Colorado–Denver;* Marcia Watson, University of North Carolina Charlotte; Jennifer Winchel, *University of Virginia* and Teri Zuccaro, *Clarke University,* who contributed new content and accuracy checks of Connect and LearnSmart. We greatly appreciate everyone's hard work on these products!

We are most grateful for the talented assistance and support from the many people at McGraw-Hill Education. We would particularly like to thank Tim Vertovec, managing director; Natalie King, marketing director; Rebecca Olson, executive brand manager; Christina Sanders and Danielle McLimore, senior product developers; Zach Rudin, marketing manager; Kevin Moran, director of digital content; Xin Lin, digital product analyst; Daryl Horrocks, program manager; Pat Frederickson, lead content project manager; Angela Norris, senior content project manager; Laura Fuller, buyer; Matt Diamond, senior designer; and Jacob Sullivan, content licensing specialists.

Finally, we extend our thanks to Roger CPA Review for their assistance developing simulations for inclusion in the end-of-chapter material, as well as Target and Air France–KLM for allowing us to use their Annual Reports throughout the text. We also acknowledge permission from the AICPA to adapt material from the Uniform CPA Examination, and the IMA for permission to adapt material from the CMA Examination.

David Spiceland   Mark Nelson   Wayne Thomas

Contents in Brief

The Role of Accounting as an Information System

SECTION 1

1. Environment and Theoretical Structure of Financial Accounting 2
2. Review of the Accounting Process 46
3. The Balance Sheet and Financial Disclosures 110
4. The Income Statement, Comprehensive Income, and the Statement of Cash Flows 166
5. Time Value of Money Concepts 238
6. Revenue Recognition 276

Assets

SECTION 2

7. Cash and Receivables 338
8. Inventories: Measurement 402
9. Inventories: Additional Issues 458
10. Property, Plant, and Equipment and Intangible Assets: Acquisition 512
11. Property, Plant, and Equipment and Intangible Assets: Utilization and Disposition 570
12. Investments 644

Liabilities and Shareholders' Equity

SECTION 3

13. Current Liabilities and Contingencies 716
14. Bonds and Long-Term Notes 772
15. Leases 834
16. Accounting for Income Taxes 908
17. Pensions and Other Postretirement Benefits 976
18. Shareholders' Equity 1046

Additional Financial Reporting Issues

SECTION 4

19. Share-Based Compensation and Earnings Per Share 1104
20. Accounting Changes and Error Corrections 1170
21. The Statement of Cash Flows Revisited 1214

Appendix A: Derivatives A-2

Appendix B: GAAP Comprehensive Case B-0

Appendix C: IFRS Comprehensive Case C-0

Glossary G-1

Index I-1

Present and Future Value Tables P-1

Contents

1 The Role of Accounting as an Information System

1 CHAPTER Environment and Theoretical Structure of Financial Accounting 2

Part A: Financial Accounting Environment 3

The Economic Environment and Financial Reporting 5
- The Investment-Credit Decision—A Cash Flow Perspective 6
- Cash versus Accrual Accounting 7

The Development of Financial Accounting and Reporting Standards 8
- Historical Perspective and Standards 9
- The Standard-Setting Process 12

Encouraging High-Quality Financial Reporting 14
- The Role of the Auditor 14
- Financial Reporting Reform 15

A Move Away from Rules-Based Standards? 16
- Ethics in Accounting 17

Part B: The Conceptual Framework 18

Objective of Financial Reporting 20

Qualitative Characteristics of Financial Reporting Information 20
- Fundamental Qualitative Characteristics 20
- Enhancing Qualitative Characteristics 22
- Key Constraint: Cost Effectiveness 23

Elements of Financial Statements 23

Underlying Assumptions 23
- Economic Entity Assumption 23
- Going Concern Assumption 24
- Periodicity Assumption 25
- Monetary Unit Assumption 25

Recognition, Measurement, and Disclosure Concepts 25
- Recognition 26
- Measurement 27
- Disclosure 30

Evolving GAAP 31

2 CHAPTER Review of the Accounting Process 46

The Basic Model 48
- The Accounting Equation 48
- Account Relationships 50

The Accounting Processing Cycle 51
- Brief Overview of Accounting Processing Cycle 52
- Illustration of Accounting Processing Cycle 55
- Journal, Ledger, and Trial Balance 55

Concept Review Exercise: Journal Entries for External Transactions 61

Adjusting Entries 63
- Prepayments 64
- Accruals 67
- Estimates 69

Concept Review Exercise: Adjusting Entries 71

Preparing the Financial Statements 72
- The Income Statement and the Statement of Comprehensive Income 72
- The Balance Sheet 73
- The Statement of Cash Flows 75
- The Statement of Shareholders' Equity 76

The Closing Process 76

Concept Review Exercise: Financial Statement Preparation and Closing 78

Conversion from Cash Basis to Accrual Basis 80

APPENDIX 2A: Using a Worksheet 84

APPENDIX 2B: Reversing Entries 85

APPENDIX 2C: Subsidiary Ledgers and Special Journals 87

3 CHAPTER The Balance Sheet and Financial Disclosures 110

Part A: The Balance Sheet 112
Usefulness 112
Limitations 113
Classification of Elements 113
Assets 114
Liabilities 118
Shareholders' Equity 119
Concept Review Exercise: Balance Sheet Classification 121
Part B: Annual Report Disclosures 122
Disclosure Notes 122
Summary of Significant Accounting Policies 123
Subsequent Events 123
Noteworthy Events and Transactions 124
Management's Discussion and Analysis 125
Management's Responsibilities 125
Compensation of Directors and Top Executives 126
Auditor's Report 128
Part C: Risk Analysis 130
Using Financial Statement Information 130
Liquidity Ratios 131
Solvency Ratios 133
APPENDIX 3: Reporting Segment Information 138

4 CHAPTER The Income Statement, Comprehensive Income, and the Statement of Cash Flows 166

Part A: The Income Statement and Comprehensive Income 168
Income from Continuing Operations 168
Revenues, Expenses, Gains, and Losses 168
Operating Income versus Nonoperating Income 170
Income Tax Expense 170
Income Statement Formats 171
Earnings Quality 173
Income Smoothing and Classification Shifting 173
Operating Income and Earnings Quality 173
Nonoperating Income and Earnings Quality 175
Non-GAAP Earnings 176
Discontinued Operations 177
What Constitutes a Discontinued Operation? 177
Reporting Discontinued Operations 178
Accounting Changes 182
Change in Accounting Principle 182
Change in Depreciation, Amortization, or Depletion Method 183
Change in Accounting Estimate 183
Correction of Accounting Errors 183
Prior Period Adjustments 183
Earnings per Share 184
Comprehensive Income 185
Flexibility in Reporting 185
Accumulated Other Comprehensive Income 187
Concept Review Exercise: Income Statement Presentation; Comprehensive Income 188
Part B: The Statement of Cash Flows 189
Usefulness of the Statement of Cash Flows 190
Classifying Cash Flows 190
Operating Activities 190
Investing Activities 193
Financing Activities 194
Noncash Investing and Financing Activities 195
Concept Review Exercise: Statement of Cash Flows 196
Part C: Profitability Analysis 197
Activity Ratios 197
Profitability Ratios 199
Profitability Analysis—An Illustration 201
APPENDIX 4: Interim Reporting 204

5 CHAPTER Time Value of Money Concepts 238

Part A: Basic Concepts 239
Time Value of Money 239
Simple versus Compound Interest 240
Future Value of a Single Amount 241
Present Value of a Single Amount 242
Solving for Other Values When FV and PV are Known 243
Concept Review Exercise: Valuing A Single Cash Flow Amount 245
Preview of Accounting Applications of Present Value Techniques—Single Cash Amount 246
Part B: Basic Annuities 248
Future Value of an Annuity 248
Future Value of an Ordinary Annuity 248
Future Value of an Annuity Due 250
Present Value of an Annuity 250
Present Value of an Ordinary Annuity 250
Present Value of an Annuity Due 252
Present Value of a Deferred Annuity 253
Financial Calculators and Excel 255
Solving for Unknown Values in Present Value Situations 255
Concept Review Exercise: Annuities 257
Preview of Accounting Applications of Present Value Techniques—Annuities 259
Valuation of Long-Term Bonds 259
Valuation of Long-Term Leases 260
Valuation of Installment Notes 260
Valuation of Pension Obligations 261
Summary of Time Value of Money Concepts 262

6 CHAPTER Revenue Recognition 276

Part A Introduction to Revenue Recognition 278

Recognizing Revenue at a Single Point in Time 280

Recognizing Revenue over a Period of Time 281
- Criteria for Recognizing Revenue over Time 281
- Determining Progress toward Completion 282

Recognizing Revenue for Contracts that Contain Multiple Performance Obligations 283
- Step 2: Identify the Performance Obligation(s) 283
- Step 3: Determine the Transaction Price 284
- Step 4: Allocate the Transaction Price to Each Performance Obligation 284
- Step 5: Recognize Revenue When (Or As) Each Performance Obligation Is Satisfied 284

Concept Review Exercise: Revenue Recognition for Contracts with Multiple Performance Obligations 286

Part B: Special Issues in Revenue Recognition 287
- Special Issues for Step 1: Identify the Contract 287
- Special Issues for Step 2: Identify the Performance Obligation(s) 289
- Special Issues for Step 3: Determine the Transaction Price 290
- Special Issues for Step 4: Allocate the Transaction Price to the Performance Obligations 297
- Special Issues for Step 5: Recognize Revenue When (Or As) Each Performance Obligation Is Satisfied 298
- Disclosures 302

Part C: Accounting for Long-Term Contracts 304
- Accounting for a Profitable Long-Term Contract 305
- A Comparison of Revenue Recognized Over the Term of the Contract and at the Completion of Contract 310
- Long-Term Contract Losses 311

Concept Review Exercise: Long-Term Construction Contracts 314

2 Assets

7 CHAPTER Cash and Receivables 338

Part A: Cash and Cash Equivalents 339

Internal Control 340
- Internal Control Procedures—Cash Receipts 341
- Internal Control Procedures—Cash Disbursements 341

Restricted Cash and Compensating Balances 341

Part B: Current Receivables 343

Accounts Receivable 344
- Initial Valuation of Accounts Receivable 344
- Subsequent Valuation of Accounts Receivable 350

Concept Review Exercise: Uncollectible Accounts Receivable 355

Notes Receivable 356
- Short-Term Interest-Bearing Notes 357
- Short-Term Noninterest-Bearing Notes 357
- Long-Term Notes Receivable 359
- Subsequent Valuation of Notes Receivable 360

Financing with Receivables 362
- Secured Borrowing 362
- Sale of Receivables 363
- Transfers of Notes Receivable 365
- Deciding Whether to Account for a Transfer as a Sale or a Secured Borrowing 366
- Disclosures 367

Concept Review Exercise: Financing with Receivables 369

APPENDIX 7A: Cash Controls 373

APPENDIX 7B: Accounting for a Troubled Debt Restructuring 377

8 CHAPTER Inventories: Measurement 402

Part A: Recording and Measuring Inventory 403

Types of Inventory 404
- Merchandising Inventory 404
- Manufacturing Inventories 404

Types of Inventory Systems 405
- Perpetual Inventory System 405
- Periodic Inventory System 407
- A Comparison of the Perpetual and Periodic Inventory Systems 408

What is Included in Inventory? 408
- Physical Units Included in Inventory 408
- Transactions Affecting Net Purchases 409

Inventory Cost Flow Assumptions 412
- Specific Identification 414
- Average Cost 414
- First-In, First-Out (FIFO) 416
- Last-In, First-Out (LIFO) 417
- Comparison of Cost Flow Methods 419
- Factors Influencing Method Choice 420
- LIFO Reserves and LIFO Liquidation 422

Concept Review Exercise: Inventory Cost Flow Methods 424

Part B: Methods of Simplifying LIFO 429

LIFO Inventory Pools 429

Dollar-Value LIFO 430
Cost Indexes 431
The DVL Inventory Estimation Technique 431
Concept Review Exercise: Dollar-Value Lifo 432
Advantages of DVL 433

9 CHAPTER Inventories: Additional Issues 458

Part A: Subsequent Measurement of Inventory 459
Lower of Cost or Net Realizable Value (LCNRV) 460
Applying Lower of Cost or Net Realizable Value 461
Adjusting Cost to Net Realizable Value 462
Concept Review Exercise: Lower of Cost or Net Realizable Value 463
Lower of Cost or Market (LCM) 465
Part B: Inventory Estimation Techniques 466
The Gross Profit Method 467
A Word of Caution 467
The Retail Inventory Method 469
Retail Terminology 470
Cost Flow Methods 471
Other Issues Pertaining to the Retail Method 473
Concept Review Exercise: Retail Inventory Method 475
Part C: Dollar-Value LIFO Retail 477
Concept Review Exercise: Dollar-Value LIFO Retail Method 479
Part D: Change in Inventory Method and Inventory Errors 480
Change in Inventory Method 480
Most Inventory Changes 480
Change to the LIFO Method 481
Inventory Errors 482
When the Inventory Error Is Discovered the Following Year 483
When the Inventory Error Is Discovered Two Years Later 484
Concept Review Exercise: Inventory Errors 484
Earnings Quality 485
APPENDIX 9: Purchase Commitments 487

10 CHAPTER Property, Plant, and Equipment and Intangible Assets: Acquisition 512

Part A: Valuation at Acquisition 514
Types of Assets 514
Costs to be Capitalized 516
Property, Plant, and Equipment 516
Intangible Assets 521
Lump-Sum Purchases 525
Part B: Noncash Acquisitions 526
Deferred Payments 526
Issuance of Equity Securities 528
Donated Assets 528
Exchanges 531
Fair Value Not Determinable 532
Lack of Commercial Substance 532
Concept Review Exercise: Exchanges 534
Part C: Self-Constructed Assets and Research and Development 535
Self-Constructed Assets 535
Overhead Allocation 535
Interest Capitalization 536
Research and Development (R&D) 541
Determining R&D Costs 541
Software Development Costs 544
R&D Performed for Others 545
R&D Purchased in Business Acquisitions 546
Start-Up Costs 546
APPENDIX 10: Oil And Gas Accounting 548

11 CHAPTER Property, Plant, and Equipment and Intangible Assets: Utilization and Disposition 570

Part A: Depreciation, Depletion, and Amortization 571
Cost Allocation—an Overview 571
Measuring Cost Allocation 572
Service Life 573
Allocation Base 573
Allocation Method 574
Depreciation 574
Time-Based Depreciation Methods 574
Activity-Based Depreciation Methods 577
Concept Review Exercise: Depreciation Methods 580
Partial Period Depreciation 581
Dispositions 583
Group and Composite Depreciation Methods 585
Depletion of Natural Resources 589
Amortization of Intangible Assets 591
Intangible Assets Subject to Amortization 591
Intangible Assets Not Subject to Amortization 593
Concept Review Exercise: Depletion and Amortization 595
Part B: Additional Issues 596
Change in Estimates 596
Change in Depreciation, Amortization, or Depletion Method 597
Error Correction 598

Impairment of Value 600
- Assets Held and Used 600
- Assets Held for Sale 608
- Impairment Losses and Earnings Quality 609

Concept Review Exercise: Impairment 610

Part C: Subsequent Expenditures 611

Expenditures Subsequent to Acquisition 611
- Repairs and Maintenance 611
- Additions 612
- Improvements 612
- Rearrangements 613
- Costs of Defending Intangible Rights 613

APPENDIX 11A: Comparison with MACRS (Tax Depreciation) 615

APPENDIX 11B: Retirement and Replacement Methods of Depreciation 616

12 CHAPTER Investments 644

Part A: Accounting for Debt Investments 646

Example of a Debt Investment 646
- Recording the Purchase of a Debt Investment 647
- Recording Interest Revenue 647

Three Classifications of Debt Investments 649

Debt Investments to Be Held to Maturity (HTM) 650
- Unrealized Holding Gains and Losses Are Not Recognized For HTM Investments 650
- Sale of HTM Investments 650
- Impairment of HTM Investments 651
- Financial Statement Presentation 651

Debt Investments Classified as Trading Securities 652
- Adjust Trading Security Investments to Fair Value (2021) 653
- Sale of Trading Security Investments 654
- Financial Statement Presentation 656

Debt Investments Classified as Available-for-Sale Securities 656
- Comprehensive Income 656
- Rationale for AFS Treatment of Unrealized Holding Gains and Losses 657
- Adjust AFS Investments to Fair Value (2021) 657
- Sale of AFS Investments 658
- Impairment of AFS Investments 660
- Financial Statement Presentation 661
- Comparison of HTM, TS, and AFS Approaches 662
- Transfers between Reporting Categories 663
- Fair Value Option 664

Concept Review Exercise: Debt Investment Securities 665
- Financial Statement Presentation and Disclosure 668

Part B: Accounting for Equity Investments 669

When the Investor Does not have Significant Influence: Fair Value through Net Income 669
- Purchase Investments 670
- Recognize Investment Revenue 670
- Adjust Equity Investments to Fair Value (2021) 670
- Sell the Equity Investment 671
- Adjust Remaining Equity Investments to Fair Value (2022) 672
- Financial Statement Presentation 673

When the Investor Has Significant Influence: The Equity Method 674
- Control and Significant Influence 674
- Purchase of Investment 676
- Recording Investment Revenue 676
- Receiving Dividends 676

Further Adjustments 676
- Adjustments for Additional Depreciation 677
- No Adjustments for Land or Goodwill 678
- Adjustments for Other Assets and Liabilities 678
- Reporting the Investment 678
- When the Investment is Acquired in Mid-Year 679
- When the Investee Reports a Net Loss 679
- Impairment of Equity Method Investments 679
- What If Conditions Change? 680
- Fair Value Option 682

Concept Review Exercise: The Equity Method 683

APPENDIX 12A: Other Investments (Special Purpose Funds, Investments in Life Insurance Policies) 687

APPENDIX 12B: Impairment of Debt Investments 688

3 Liabilities and Shareholders' Equity

13 CHAPTER Current Liabilities and Contingencies 716

Part A: Current Liabilities 718

Characteristics of Liabilities 718

What is a Current Liability? 719

Open Accounts and Notes 719
- Accounts Payable and Trade Notes Payable 719
- Short-Term Notes Payable 720
- Commercial Paper 723

Accrued Liabilities 723
- Accrued Interest Payable 724
- Salaries, Commissions, and Bonuses 724

Liabilities from Advance Collections 726
- Deposits and Advances from Customers 726
- Gift Cards 728
- Collections for Third Parties 728

A Closer Look at the Current and Noncurrent Classification 729
- Current Maturities of Long-Term Debt 729
- Obligations Callable by the Creditor 729
- When Short-Term Obligations Are Expected to Be Refinanced 730

Concept Review Exercise: Current Liabilities 731

Part B: Contingencies 732

Loss Contingencies 732
- Product Warranties and Guarantees 735
- Litigation Claims 738
- Subsequent Events 739
- Unasserted Claims and Assessments 740

Gain Contingencies 742

Concept Review Exercise: Contingencies 743

APPENDIX 13: Payroll-Related Liabilities 747

14 CHAPTER Bonds and Long-Term Notes 772

The Nature of Long-Term Debt 773

Part A: Bonds 774

The Bond Indenture 774

Recording Bonds at Issuance 775
- Determining the Selling Price 775
- Bonds Issued at a Discount 777

Determining Interest—Effective Interest Method 778
- Zero-Coupon Bonds 780
- Bonds Issued at a Premium 780
- When Financial Statements are Prepared between Interest Dates 782
- The Straight-Line Method—A Practical Expediency 783

Concept Review Exercise: Issuing Bonds and Recording Interest 784

Debt Issue Costs 785

Part B: Long-Term Notes 786

Note Issued for Cash 786

Note Exchanged for Assets or Services 786

Installment Notes 789

Concept Review Exercise: Note with an Unrealistic Interest Rate 790

Financial Statement Disclosures 791

Part C: Debt Retired Early, Convertible Into Stock, or Providing an Option to Buy Stock 795

Early Extinguishment of Debt 795

Convertible Bonds 795
- When the Conversion Option Is Exercised 797
- Induced Conversion 798

Bonds with Detachable Warrants 798

Concept Review Exercise: Debt Disclosures and Early Extinguishment of Debt 800

Part D: Option to Report Liabilities at Fair Value 801
- Determining Fair Value 801
- Reporting Changes in Fair Value 802
- Mix and Match 803

Appendix 14A: Bonds Issued Between Interest Dates 805

APPENDIX 14B: Troubled Debt Restructuring 807

15 CHAPTER Leases 834

Part A: Accounting by the Lessor and Lessee 835
- Why Lease? 835
- Lease Classification 836
- Finance Leases and Installment Notes Compared 839

Finance/Sales-Type Leases 841
- Recording Interest Expense/Interest Revenue 843
- Recording Amortization of the Right-of-Use Asset 843

Concept Review Exercise: Finance Lease/Sales-Type Lease: No Selling Profit 844

Sales-Type Leases with Selling Profit 846

Operating Leases 847
- Recording lease Expense/Lease Revenue 848
- Reporting Lease Expense and Lease Revenue 850
- Discount Rate 853

Concept Review Exercise: Operating Lease 853

Short-Term Leases—A ShortCut Method 855

Part B: Uncertainty in Lease Transactions 855
- What if the Lease Term is Uncertain? 856
- What if the Lease Payments are Uncertain? 857
- What if Lease Terms are Modified? 858
- Residual Value 858
- Purchase Option 865
- Summary of the Lease Uncertainties 868
- Remeasurement of the Lease Liability 868

Part C: Other Lease Accounting Issues and Reporting Requirements 869

Is It a Lease? 869

Nonlease Components of Lease Payments 871

Concept Review Exercise: Various Lease Accounting Issues: Finance/Sales-Type Lease 874

Statement of Cash Flow Impact 876
- Operating Leases 876
- Finance Leases—Lessee 876
- Sales-Type Leases—Lessor 876

Lease Disclosures 877
Qualitative Disclosures 877
Quantitative Disclosures 877
APPENDIX 15: Sale-Leaseback Arrangements 879

16 CHAPTER
Accounting for Income Taxes 908

Part A: Temporary Differences 909
Conceptual Underpinnings 910
The 4-Step Process 911
Types of Temporary Differences 911
Deferred Tax Liabilities 912
Expense-Related Deferred Tax Liabilities 912
Balance Sheet and Income Statement Perspectives 915
Revenue-Related Deferred Tax Liabilities 916
Deferred Tax Assets 919
Expense-Related Deferred Tax Assets 919
Revenue-Related Deferred Tax Assets 922
Valuation Allowance 925
Disclosures Linking Tax Expense with Changes in Deferred Tax Assets and Liabilities 927
Part B: Permanent Differences 929
Concept Review Exercise: Temporary and Permanent Differences 932
Part C: Other Tax Accounting Issues 933
Tax Rate Considerations 933
When Enacted Tax Rates Differ Between Years 933
Changes in Enacted Tax Laws or Rates 933
Multiple Temporary Differences 935
Net Operating Losses 937
Net Operating Loss Carryforward 937
Financial Statement Presentation 941
Balance Sheet Classification 941
Disclosure Notes 941
Part D: Coping with Uncertainty in Income Taxes 942
Intraperiod Tax Allocation 946
Concept Review Exercise: Multiple Differences and Net Operating Loss 948

17 CHAPTER
Pensions and Other Postretirement Benefits 976

Part A: The Nature of Pension Plans 978
Defined Contribution Pension Plans 980
Defined Benefit Pension Plans 981
Pension Expense—An Overview 982
Part B: The Pension Obligation and Plan Assets 983
The Pension Obligation 983
Accumulated Benefit Obligation 984
Vested Benefit Obligation 984
Projected Benefit Obligation 984
Illustration Expanded to Consider the Entire Employee Pool 989
Pension Plan Assets 990
Reporting the Funded Status of the Pension Plan 992
Part C: Determining Pension Expense 992
The Relationship between Pension Expense and Changes in the PBO and Plan Assets 992
Components of Pension Expense 992
Income Smoothing 996
Part D: Reporting Issues 998
Recording Gains and Losses 998
Recording the Pension Expense 1000
Reporting Pension Expense in the Income Statement 1000
Recording the Funding of Plan Assets 1001
Comprehensive Income 1003
Income Tax Considerations 1004
Putting the Pieces Together 1004
Settlement or Curtailment of Pension Plans 1006
Concept Review Exercise: Pension Plans 1006
Part E: Postretirement Benefits Other Than Pensions 1009
What Is a Postretirement Benefit Plan? 1010
Postretirement Health Benefits and Pension Benefits Compared 1010
Determining the Net Cost of Benefits 1010
Postretirement Benefit Obligation 1011
Measuring the Obligation 1011
Attribution 1012
Accounting for Postretirement Benefit Plans Other Than Pensions 1012
A Comprehensive Illustration 1014
Concept Review Exercise: Other Postretirement Benefits 1016
APPENDIX 17: Service Method of Allocating Prior Service Cost 1019

18 CHAPTER
Shareholders' Equity 1046

Part A: The Nature of Shareholders' Equity 1047
Financial Reporting Overview 1048
Paid-in Capital 1048
Retained Earnings 1049
Treasury Stock 1050
Accumulated Other Comprehensive Income 1050
Reporting Shareholders' Equity 1051

The Corporate Organization 1051
Limited Liability 1053
Ease of Raising Capital 1053
Disadvantages 1054
Types of Corporations 1054
Hybrid Organizations 1054
The Model Business Corporation Act 1055
Part B: Paid-in Capital 1055
Fundamental Share Rights 1055
Distinguishing Classes of Shares 1056
Typical Rights of Preferred Shares 1056
Is It Equity or Is It Debt? 1057
The Concept of Par Value 1057
Accounting for the Issuance of Shares 1058
Par value Shares Issued For Cash 1058
No-Par Shares Issued for Cash 1059
Shares Issued for Noncash Consideration 1059
More Than One Security Issued for a Single Price 1059
Share Issue Costs 1060
Concept Review Exercise: Expansion of Corporate Capital 1060
Share Repurchases 1062
Shares Formally Retired or Viewed as Treasury Stock 1063
Accounting for Retired Shares 1063
Accounting for Treasury Stock 1065
Resale of Shares 1065
Concept Review Exercise: Treasury Stock 1067
Part C: Retained Earnings 1068
Characteristics of Retained Earnings 1068
Dividends 1068
Liquidating Dividend 1069
Retained Earnings Restrictions 1069
Cash Dividends 1069
Property Dividends 1070
Stock Dividends and Splits 1070
Stock Dividends 1070
Stock Splits 1072
Stock Splits Effected in the Form of Stock Dividends (Large Stock Dividends) 1073
Concept Review Exercise: Changes in Retained Earnings 1076
APPENDIX 18: Quasi Reorganizations 1079

4 Additional Financial Reporting Issues

19 CHAPTER Share-Based Compensation and Earnings Per Share 1104

Part A: Share-Based Compensation 1105
Restricted Stock Plans 1106
Restricted Stock Awards 1106
Restricted Stock Units 1106
Stock Option Plans 1108
Expense—The Great Debate 1108
Recognizing the Fair Value of Options 1109
When Options are Exercised 1112
When Unexercised Options Expire 1112
Plans with Graded Vesting 1114
Plans with Performance or Market Conditions 1115
Employee Share Purchase Plans 1117
Concept Review Exercise: Share-Based Compensation Plans 1118
Part B: Earnings Per Share 1119
Basic Earnings Per Share 1119
Issuance of New Shares 1120
Stock Dividends and Stock Splits 1120
Reacquired Shares 1122
Earnings Available to Common Shareholders 1122
Diluted Earnings Per Share 1123
Potential Common Shares 1123
Options, Rights, and Warrants 1123
Convertible Securities 1126
Antidilutive Securities 1128
Options, Warrants, Rights 1128
Convertible Securities 1130
Order of Entry for Multiple Convertible Securities 1130
Concept Review Exercise: Basic and Diluted EPS 1131
Additional EPS Issues 1133
Components of the "Proceeds" in the Treasury Stock Method 1133
Restricted Stock Awards in EPS Calculations 1133
Contingently Issuable Shares 1134
Summary of the Effect of Potential Common Shares on Earnings Per Share 1135
Actual Conversions 1136
Financial Statement Presentation of Earnings Per Share Data 1136
Concept Review Exercise: Additional Eps Issues 1139
APPENDIX 19A: Option-Pricing Theory 1142
APPENDIX 19B: Stock Appreciation Rights 1143

20 CHAPTER Accounting Changes and Error Corrections 1170

Part A: Accounting Changes 1171

Change in Accounting Principle 1173

The Retrospective Approach: Most Changes in Accounting Principle 1174

The Modified Retrospective Approach 1178

The Prospective Approach 1178

Change in Accounting Estimate 1180

Changing Depreciation, Amortization, and Depletion Methods 1181

Change in Reporting Entity 1182

Errors 1183

Concept Review Exercise: Accounting Changes 1184

Part B: Correction of Accounting Errors 1185

Prior Period Adjustments 1186

Error Correction Illustrated 1187

Error Discovered in the Same Reporting Period That It Occurred 1187

Error Affecting Previous Financial Statements, but Not Net Income 1187

Error Affecting a Prior Year's Net Income 1188

Concept Review Exercise: Correction of Errors 1192

21 CHAPTER The Statement of Cash Flows Revisited 1214

Part A: The Content and Value of the Statement of Cash Flows 1215

Cash Inflows and Outflows 1215

Structure of the Statement of Cash Flows 1217

Cash, Cash Equivalents, and Restricted Cash 1219

Primary Elements of the Statement of Cash Flows 1219

Preparation of the Statement of Cash Flows 1225

Part B: Preparing the SCF: Direct Method of Reporting Cash Flows from Operating Activities 1228

Using a Spreadsheet 1228

Income Statement Accounts 1230

Balance Sheet Accounts 1238

Concept Review Exercise: Comprehensive Review 1247

Part C: Preparing the SCF: Indirect Method of Reporting Cash Flows from Operating Activities 1250

Getting There through the Back Door 1250

Components of Net Income that do not Increase or Decrease Cash 1251

Components of Net Income That Do Increase or Decrease Cash 1251

Comparison with the Direct Method 1252

Reconciliation of Net Income to Cash Flows from Operating Activities 1254

APPENDIX 21A: Spreadsheet for the Indirect Method 1258

APPENDIX 21B: The T-Account Method of Preparing the Statement of Cash Flows 1262

Appendix A: Derivatives A-2

Appendix B: GAAP Comprehensive Case B-0

Appendix C: IFRS Comprehensive Case C-0

Glossary G-1

Index I-1

Present and Future Value Tables P-1

CHAPTER

1 Environment and Theoretical Structure of Financial Accounting

OVERVIEW

The primary function of financial accounting is to provide useful financial information to users who are external to the business enterprise, particularly investors and creditors. These users make critical resource allocation decisions that affect the global economy. The primary means of conveying financial information to external users is through financial statements and related notes.

In this chapter you explore such important topics as the reason why financial accounting is useful, the process by which accounting standards are produced, and the conceptual framework that underlies financial accounting. The perspective you gain in this chapter serves as a foundation for more detailed study of financial accounting.

LEARNING OBJECTIVES

After studying this chapter, you should be able to:

- **LO1–1** Describe the function and primary focus of financial accounting. *(p. 3)*
- **LO1–2** Explain the difference between cash and accrual accounting. *(p. 7)*
- **LO1–3** Define generally accepted accounting principles (GAAP) and discuss the historical development of accounting standards, including convergence between U.S. and international standards. *(p. 8)*
- **LO1–4** Explain why establishing accounting standards is characterized as a political process. *(p. 12)*
- **LO1–5** Explain factors that encourage high-quality financial reporting. *(p. 14)*
- **LO1–6** Explain the purpose of the conceptual framework. *(p. 18)*
- **LO1–7** Identify the objective and qualitative characteristics of financial reporting information and the elements of financial statements. *(p. 20)*
- **LO1–8** Describe the four basic assumptions underlying GAAP. *(p. 23)*
- **LO1–9** Describe the recognition, measurement, and disclosure concepts that guide accounting practice. *(p. 25)*
- **LO1–10** Contrast a revenue/expense approach and an asset/liability approach to accounting standard setting. *(p. 31)*
- **LO1–11** Discuss the primary differences between U.S. GAAP and IFRS with respect to the development of accounting standards and the conceptual framework underlying accounting standards. *(p. 11, 14 and 19)*

©Rawpixel.com/Shutterstock

FINANCIAL REPORTING CASE

Misguided Marketing Major

During a class break in your investments class, a marketing major tells the following story to you and some friends:

The chief financial officer (CFO) of a large company is interviewing three candidates for the top accounting position with his firm. He asks each the same question:

CFO:	What is two plus two?
First candidate:	Four.
CFO:	What is two plus two?
Second candidate:	Four.
CFO:	What is two plus two?
Third candidate:	What would you like it to be?
CFO:	You're hired.

After you take some good-natured ribbing from the non-accounting majors, your friend says, "Seriously, though, there must be ways the accounting profession prevents that kind of behavior. Aren't there some laws or rules, or something? Is accounting based on some sort of theory, or is it just arbitrary?"

QUESTIONS

By the time you finish this chapter, you should be able to respond appropriately to the questions posed in this case. Compare your response to the solution provided at the end of the chapter.

1. What should you tell your friend about the presence of accounting standards in the United States and the rest of the world? Who has the authority for standard setting? Who has the responsibility? (p. 8)
2. What is the economic and political environment in which standard setting occurs? (p. 12)
3. What is the relationship among management, auditors, investors, and creditors that tends to preclude the "What would you like it to be?" attitude? (p. 14)
4. In general, what is the conceptual framework that underlies accounting principles, and how does it encourage high-quality financial reporting? (p. 18)

PART A

Financial Accounting Environment

LO1–1

In 1902, George Dayton took ownership of the Dayton Dry Goods Company, the fourth largest department store in Minneapolis, Minnesota. Successive generations of Daytons were innovative managers, flying in inventory to prevent shortages (1920), committing to giving 5 percent of profits back to the community (1946), and creating the nation's first enclosed shopping mall (1956). In 1962, George's grandchildren transformed Dayton's, by then a regional department store chain, into the **Target Corporation**, promising "a quality store with quality merchandise at discount prices."[1] Today Target has grown to be the second largest general merchandise retailer in America, with over 1,800 stores, almost 350,000 employees, and **www.target.com** reaching the online market. However, Target still stands by its "Expect More, Pay Less" motto, and still donates 5 percent of profits back to the community (giving more than $4 million per week).

[1] "Target Through the Years" at https://corporate.target.com/about/history/Target-through-the-years.

Many factors contributed to Target's success. The Daytons were visionary in their move into the upscale discount retail market. The company's commitment to quality products, customer service, and community support also played an important role. But the ability to raise money from investors and lenders at various times also was critical to Target's evolution. Target used proceeds from its 1967 initial public stock offering to expand nationally. Creditors (lenders) also supplied needed capital at various times. In fact, without access to capital, the Target Corporation we know today likely would not exist.

Investors and creditors use many different kinds of information before supplying capital to businesses like Target. They use the information to predict the future risk and potential return of their prospective investments or loans.[2] For example, information about the enterprise's products and its management is key to this assessment. Investors and creditors also rely on various kinds of accounting information.

Think of accounting as a special "language" that companies like Target use to communicate financial information to help people inside and outside of the business to make decisions. The Pathways Commission of the American Accounting Association developed an illustration to help visualize this important role of accounting.[3] As shown in Illustration 1–1, accounting provides useful information about economic activity to help produce good decisions and foster a prosperous society. Economic activity is complex, and decisions have real consequences, so critical thinking and many judgments are needed to produce the most useful accounting information possible.

The primary focus of *financial accounting* is on the information needs of investors and creditors.

This book focuses on **financial accounting**, which is chiefly concerned with providing financial information to various *external* users.[4] The chart in Illustration 1–2 lists a number of groups that provide financial information as well as several external user groups. For these groups, the primary focus of financial accounting is on the financial information

Illustration 1–1

Pathways Commission visualization: "THIS is accounting!"

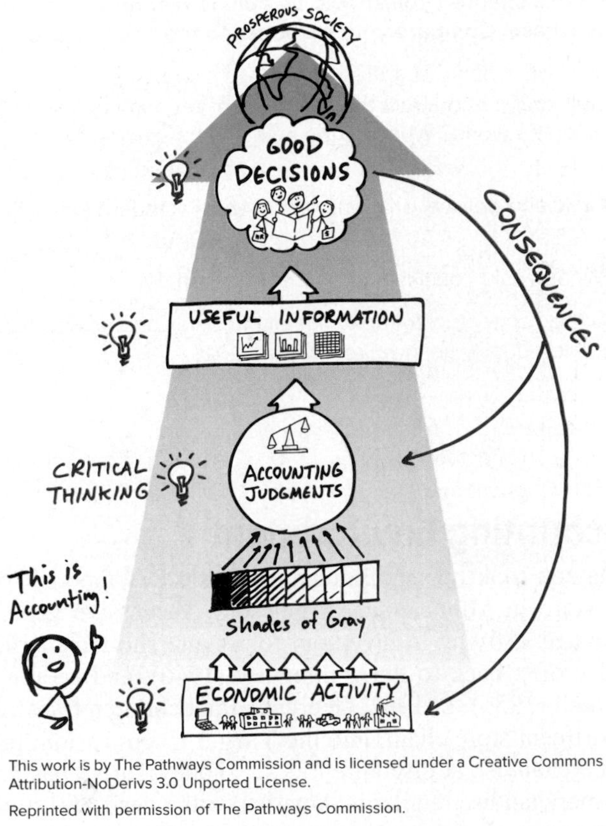

[2]Risk refers to the variability of possible outcomes from an investment. Return is the amount received over and above the investment.

[3]Reprinted with permission from the American Accounting Association. This work is by The Pathways Commission and is licensed under a Creative Commons Attribution-NoDerivs 3.0 Unported License.

[4]In contrast, *managerial* accounting deals with the concepts and methods used to provide information to an organization's *internal users*, that is, its managers. You study managerial accounting elsewhere in your curriculum.

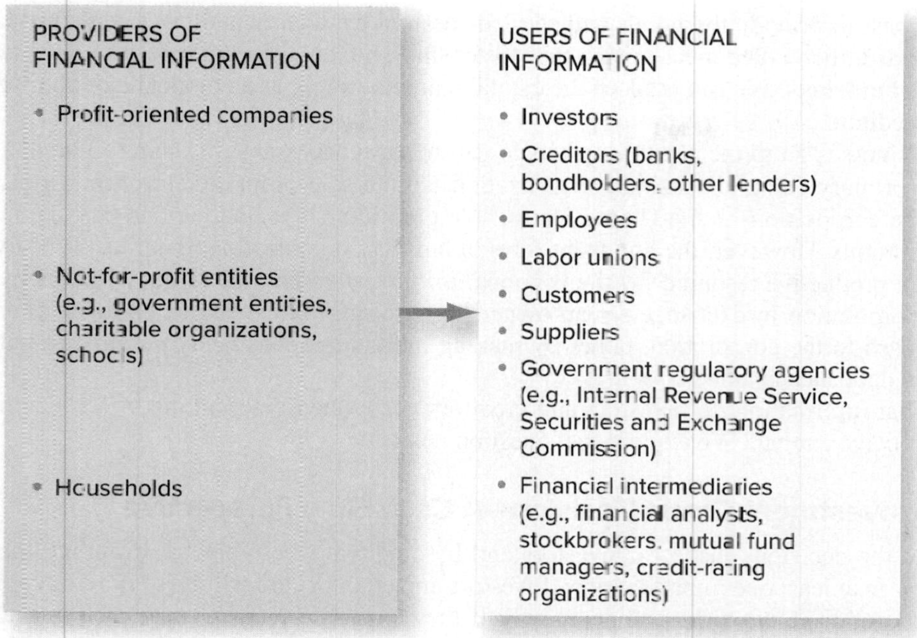

PROVIDERS OF FINANCIAL INFORMATION
• Profit-oriented companies
• Not-for-profit entities (e.g., government entities, charitable organizations, schools)
• Households
USERS OF FINANCIAL INFORMATION
• Investors
• Creditors (banks, bondholders, other lenders)
• Employees
• Labor unions
• Customers
• Suppliers
• Government regulatory agencies (e.g., Internal Revenue Service, Securities and Exchange Commission)
• Financial intermediaries (e.g., financial analysts, stockbrokers, mutual fund managers, credit-rating organizations)

Illustration 1–2

Financial Information Providers and External User Groups

provided by *profit-oriented companies to their present and potential investors and creditors.* One external user group, often referred to as *financial intermediaries,* includes financial analysts, stockbrokers, mutual fund managers, and credit rating organizations. These users provide advice to investors and creditors and/or make investment-credit decisions on their behalf.

The primary means of conveying financial information to investors, creditors, and other external users is through financial statements and related disclosure notes. The financial statements most frequently provided are (1) the balance sheet, also called the statement of financial position, (2) the income statement, also called the statement of operations, (3) the statement of cash flows, and (4) the statement of shareholders' equity. Also, companies must either provide a statement of other comprehensive income immediately following the income statement or present a combined statement of comprehensive income that includes the information normally contained in both the income statement and the statement of other comprehensive income.[5] As you progress through this book, you will review and expand your knowledge of the information in these financial statements, the way the elements in these statements are measured, and the concepts underlying these measurements and related disclosures. We use the term **financial reporting** to refer to the process of providing this information to external users. Keep in mind, though, that external users receive important financial information in a variety of other formats as well, including news releases and management forecasts, prospectuses, and reports filed with regulatory agencies.

Target's financial statements for the fiscal year ended January 30, 2016, and related disclosure notes are provided in Connect. You also can access these statements and notes under the Investor Relations link at the company's website (**Target.com**). A Target case is included among the Real-Word Cases that accompany each chapter, so you can see how each chapter's topics relate to a single familiar company.

The Economic Environment and Financial Reporting

In the United States, we have a highly developed free-enterprise economy with the majority of productive resources privately owned rather than government owned. For the economy to operate efficiently, these resources should be allocated to private enterprises that will use

[5]FASB ASC 220-45: Comprehensive Income—Other Presentation Matters (originally "Presentation of Comprehensive Income," *Accounting Standards Update No. 2011-05* (Norwalk, CT: FASB, June 2011)).

them best to provide the goods and services desired by society and not to enterprises that will waste them. The mechanisms that foster this efficient allocation of resources are the **capital markets**. We can think of the capital markets simply as a composite of all investors and creditors.

The *capital markets* provide a mechanism to help our economy allocate resources efficiently.

Businesses go to the capital markets to get the cash necessary for them to function. The three primary forms of business organization are the sole proprietorship, the partnership, and the corporation. In the United States, sole proprietorships and partnerships outnumber corporations. However, the dominant form of business organization, in terms of the ownership of productive resources, is the **corporation**. Investors provide resources, usually cash, to a corporation in exchange for an ownership interest, that is, shares of stock. Creditors lend cash to the corporation, either by making individual loans or by purchasing publicly traded debt such as bonds.

Corporations acquire capital from investors in exchange for ownership interest and from creditors by borrowing.

What information do investors and creditors need when determining which companies will receive capital? We explore that question next.

The Investment-Credit Decision—A Cash Flow Perspective

While the decisions made by investors and by creditors are somewhat different, they are similar in at least one important way. Investors and creditors are willing to provide capital to a corporation (buy stocks or bonds) only if they expect to receive more cash in return at some time in the future. A corporation's shareholders will receive cash from their investment through the ultimate sale of the ownership shares of stock. In addition, many corporations distribute cash to their shareholders in the form of periodic dividends. For example, if an investor provides a company with \$10,000 cash by purchasing stock at the end of 2020, receives \$400 in dividends from the company during 2021, and sells the ownership interest (shares) at the end of 2021 for \$10,600, the investment would have generated a **rate of return** of 10% for 2021, calculated as follows:

The expected rate of return and the uncertainty, or risk, of that return are key variables in the investment decision.

$$\frac{\$400 \text{ dividends} + \$600 \text{ share price appreciation}}{\$10{,}000 \text{ initial investment}} = 10\%$$

All else equal, investors and creditors would like to invest in stocks or bonds that provide the highest expected rate of return. However, there are many variables to consider before making an investment decision. For example, the *uncertainty,* or *risk,* of that expected return also is important. To illustrate, consider the following two investment options:

1. Invest \$10,000 in a savings account insured by the U.S. government that will generate a 5% rate of return.
2. Invest \$10,000 in a profit-oriented company.

A company will be able to provide a positive return to investors and creditors only if it can generate a profit from selling its products or services.

While the rate of return from option 1 is known with virtual certainty, the return from option 2 is uncertain. The amount and timing of the cash to be received in the future from option 2 are unknown. The company in option 2 will be able to provide investors with a return only if it can generate a profit. That is, it must be able to use the resources provided by investors and creditors to generate cash receipts from selling a product or service that exceed the cash disbursements necessary to provide that product or service. Therefore, potential investors require information about the company that will help them estimate the potential for future profits, as well as the return they can expect on their investment and the risk that is associated with it. If the potential return is high enough, investors will prefer to invest in the profit-oriented company, even if that return has more risk associated with it.

The objective of financial accounting is to provide investors and creditors with useful information for decision making.

In summary, the primary objective of financial accounting is to provide investors and creditors with information that will help them make investment and credit decisions. That information should help investors and creditors evaluate the *amounts, timing,* and *uncertainty* of the enterprise's future cash receipts and disbursements. The better this information is, the more efficient will be investor and creditor resource allocation decisions. But financial accounting doesn't only benefit companies and their investors and creditors. By providing key information to capital market participants, financial accounting plays a vital role that helps direct society's resources to the companies that will utilize those resources most effectively.

Cash versus Accrual Accounting

● LO1–2

Even though predicting future cash flows is the primary goal of many users of financial reporting, the model best able to achieve that goal is **accrual accounting**. A competing model is **cash-basis accounting**. Each model produces a periodic measure of performance that could be used by investors and creditors for predicting future cash flows.

CASH-BASIS ACCOUNTING Cash-basis accounting produces a measure called **net operating cash flow**. This measure is the difference between cash receipts and cash payments from transactions related to providing goods and services to customers during a reporting period.

Net operating cash flow is the difference between cash receipts and cash disbursements from providing goods and services.

Over the life of a company, net operating cash flow definitely is the measure of concern. However, over short periods of time, operating cash flows may not be indicative of the company's long-run cash-generating ability. Sometimes a company pays or receives cash in one period that relates to performance in multiple periods. For example, in one period a company receives cash that relates to prior period sales, or makes advance payments for costs related to future periods.

To see this more clearly, consider Carter Company's net operating cash flows during its first three years of operations, shown in Illustration 1–3. Carter's operations for these three years included the following:

1. Credit sales to customers were $100,000 each year ($300,000 total), while cash collections were $50,000, $125,000, and $125,000. Carter's customers owe Carter nothing at the end of year 3.
2. At the beginning of year 1, Carter prepaid $60,000 for three years' rent ($20,000 per year).
3. Employee salaries of $50,000 were paid in full each year.
4. Utilities cost was $10,000 each year, but $5,000 of the cost in year 1 was not paid until year 2.
5. In total, Carter generated positive net operating cash flow of **$60,000**.

Is the three-year pattern of net operating cash flows indicative of the company's year-by-year performance? No. Sales to customers and costs of operating the company (rent, salaries, and utilities) occurred evenly over the three years, but net operating cash flows occurred at an uneven rate. Net operating cash flows varied each year because Carter (a) didn't collect cash from customers in the same pattern that sales occurred and (b) didn't pay for rent and utilities in the same years in which those resources were actually consumed. This illustration also shows why operating cash flows may not predict the company's long-run cash-generating ability. Net operating cash flow in year 1 (negative **$65,000**)[6] is not an accurate predictor of Carter's future cash-generating ability in year 2 (positive $60,000) or year 3 (positive $65,000).

Over short periods of time, operating cash flow may not be an accurate predictor of future operating cash flows.

Illustration 1–3

Cash-Basis Accounting

	Year 1	Year 2	Year 3	Total
Sales (on credit)	$100,000	$100,000	$100,000	$300,000
Net Operating Cash Flows				
Cash receipts from customers	$ 50,000	$125,000	$125,000	$300,000
Cash disbursements:				
Prepayment of three years' rent	(60,000)	–0–	–0–	(60,000)
Salaries to employees	(50,000)	(50,000)	(50,000)	(150,000)
Utilities	(5,000)	(15,000)	(10,000)	(30,000)
Net operating cash flow	**$(65,000)**	$ 60,000	$ 65,000	**$ 60,000**

[6] If cash flow from operating the company is negative, the company can continue to operate by using cash obtained from investors or creditors to make up the difference.

Illustration 1–4
Accrual Accounting

CARTER COMPANY
Income Statements

	Year 1	Year 2	Year 3	Total
Revenues	$100,000	$100,000	$100,000	$300,000
Expenses:				
Rent	20,000	20,000	20,000	60,000
Salaries	50,000	50,000	50,000	150,000
Utilities	10,000	10,000	10,000	30,000
Total expenses	80,000	80,000	80,000	240,000
Net Income	**$ 20,000**	$ 20,000	$ 20,000	**$ 60,000**

***Net income** is the difference between **revenues** and **expenses**.*

ACCRUAL ACCOUNTING If we measure Carter's activities by the accrual accounting model, we get a more accurate prediction of future operating cash flows and a more reasonable portrayal of the periodic operating performance of the company. The accrual accounting model doesn't focus only on cash flows. Instead, it also reflects other resources provided and consumed by operations during a period. The accrual accounting model's measure of resources provided by business operations is called *revenues,* and the measure of resources sacrificed to produce revenues is called *expenses.* The difference between revenues and expenses is **net income**, or net loss if expenses are greater than revenues.[7]

Illustration 1–4 shows how we would measure revenues and expenses in this very simple situation.

Net income is considered a better indicator of future operating cash flows than is current net operating cash flow.

Revenue for year 1 is the $100,000 sales. Given that sales eventually are collected in cash, the year 1 revenue of $100,000 is a better measure of the inflow of resources from company operations than is the $50,000 cash collected from customers. Also, net income of **$20,000** for year 1 appears to be a reasonable predictor of the company's cash-generating ability, as total net operating cash flow for the three-year period is a positive **$60,000**. Comparing the three-year pattern of net operating cash flows in Illustration 1–3 to the three-year pattern of net income in Illustration 1–4, the net income pattern is more representative of Carter Company's steady operating performance over the three-year period.[8]

While this example is somewhat simplistic, it allows us to see the motivation for using the accrual accounting model. Accrual income attempts to measure the resource inflows and outflows generated by operations during the reporting period, which may not correspond to cash inflows and outflows. Does this mean that information about cash flows from operating activities is not useful? No. Indeed, one of the basic financial statements—the statement of cash flows—reports information about cash flows from operating, investing, and financing activities, and provides important information to investors and creditors.[9] Focusing on accrual accounting as well as cash flows provides a more complete view of a company and its operations.

The Development of Financial Accounting and Reporting Standards

● LO1–3

FINANCIAL Reporting Case

Q1, p. 3

Accrual accounting is the financial reporting model used by the majority of profit-oriented companies and by many not-for-profit companies. The fact that companies use the same model is important to investors and creditors, allowing them to *compare* financial information among companies. To facilitate these comparisons, financial accounting employs a body of standards known as **generally accepted accounting principles**, often abbreviated as **GAAP** (and pronounced *gap*). GAAP is a dynamic set of both broad and specific guidelines that companies should follow when measuring and reporting the information in their

[7]Net income also includes gains and losses, which are discussed later in the chapter.

[8]Empirical evidence that accrual accounting provides a better measure of short-term performance than cash flows is provided by Patricia Dechow, "Accounting Earnings and Cash Flows as Measures of Firm Performance: The Role of Accrual Accounting," *Journal of Accounting and Economics* 18 (1994), pp. 3–42.

[9]The statement of cash flows is discussed in detail in Chapters 4 and 21.

financial statements and related notes. The more important concepts underlying GAAP are discussed in a subsequent section of this chapter and revisited throughout this book in the context of particular accounting topics.

Historical Perspective and Standards

Pressures on the accounting profession to establish uniform accounting standards began after the stock market crash of 1929. Some felt that insufficient and misleading financial statement information led to inflated stock prices and that this contributed to the stock market crash and the subsequent depression.

The 1933 Securities Act and the 1934 Securities Exchange Act were designed to restore investor confidence. The 1933 Act sets forth accounting and disclosure requirements for initial offerings of securities (stocks and bonds). The 1934 Act applies to secondary market transactions and mandates reporting requirements for companies whose securities are publicly traded on either organized stock exchanges or in over-the-counter markets.[10]

The 1934 Act also created the **Securities and Exchange Commission (SEC)**. Congress gave the SEC the authority to set accounting and reporting standards for companies whose securities are publicly traded. However, the SEC, a government appointed body, has *delegated* the task of setting accounting standards to the private sector. It is important to understand that the power still lies with the SEC. If the SEC does not agree with a particular standard issued by the private sector, it can force a change in the standard. In fact, it has done so in the past.[11]

The Securities and Exchange Commission (SEC) has the authority to set accounting standards for companies, but it relies on the private sector to do so.

EARLY U.S. STANDARD SETTING The first private sector body to assume the task of setting accounting standards was the **Committee on Accounting Procedure (CAP)**. The CAP was a committee of the **American Institute of Accountants (AIA)**. The AIA was renamed the **American Institute of Certified Public Accountants (AICPA)** in 1957, which is the national professional organization for certified professional public accountants.

From 1938 to 1959, the CAP issued 51 *Accounting Research Bulletins (ARBs)* which dealt with specific accounting and reporting problems. No theoretical framework for financial accounting was established. This piecemeal approach of dealing with individual issues without a framework led to criticism.

In 1959 the **Accounting Principles Board (APB)** replaced the CAP. The APB operated from 1959 through 1973 and issued 31 *Accounting Principles Board Opinions (APBOs),* various *Interpretations,* and four *Statements.* The *Opinions* also dealt with specific accounting and reporting problems. Many *ARBs* and *APBOs* still represent authoritative GAAP.

The APB suffered from a variety of problems. It was never able to establish a conceptual framework for financial accounting and reporting that was broadly accepted. Also, members served on the APB on a voluntary, part-time basis, so the APB was not able to act quickly enough to keep up with financial reporting issues as they developed. Perhaps the most important flaw of the APB was a perceived lack of independence. Because the APB was composed almost entirely of certified public accountants and supported by the AICPA, critics charged that the clients of the represented public accounting firms exerted self-interested pressure on the board and inappropriately influenced decisions. A related complaint was that other interest groups lacked an ability to provide input to the standard-setting process.

THE FASB Criticism of the APB led to the creation in 1973 of the **Financial Accounting Standards Board (FASB)** and its supporting structure. There are seven full-time members of the FASB. FASB members represent various constituencies concerned with accounting standards, and have included representatives from the auditing profession, profit-oriented companies, accounting educators, financial analysts, and government. The FASB is supported by its parent organization, the **Financial Accounting Foundation (FAF)**, which is

The *FASB* was established to set U.S. accounting standards.

[10]Reporting requirements for SEC registrants include Form 10-K, the annual report form, and Form 10-Q, the report that must be filed for the first three quarters of each fiscal year.

[11]The SEC issues *Financial Reporting Releases (FRRs),* which regulate what information companies must report to it. The SEC staff also issues *Staff Accounting Bulletins* that provide the SEC's interpretation of standards previously issued by the private sector. To learn more about the SEC, consult its Internet site at **www.sec.gov**.

responsible for selecting the members of the FASB and its Financial Accounting Standards Advisory Council (FASAC), ensuring adequate funding of FASB activities and exercising general oversight of the FASB's activities.[12,13]

In 1984, the FASB's **Emerging Issues Task Force (EITF)** was formed to improve financial reporting by resolving narrowly defined financial accounting issues within the framework of existing GAAP. The EITF primarily addresses implementation issues, thereby speeding up the standard-setting process and allowing the FASB to focus on pervasive long-term problems. EITF rulings are ratified by the FASB and are considered part of GAAP.

Illustration 1–5 summarizes this discussion on accounting standards. The graphic shows the hierarchy of accounting standard setting in order of authority.

CODIFICATION Present-day GAAP includes a huge amount of guidance. The FASB has developed a **conceptual framework** (discussed in Part B of this chapter) that is not authoritative GAAP but provides an underlying structure for the development of accounting standards. The FASB also has issued many accounting standards, currently called *Accounting Standards Updates* (*ASUs*) and previously called *Statements of Financial Accounting Standards* (*SFASs*), as well as numerous FASB *Interpretations, Staff Positions, Technical Bulletins,* and *EITF Issue Consensuses.* The SEC also has issued various important pronouncements. Determining the appropriate accounting treatment for a particular event or transaction might require an accountant to research several of these sources.

The *FASB Accounting Standards Codification* is the only source of authoritative U.S. GAAP, other than rules and interpretive releases of the SEC.

To simplify the task of researching an accounting topic, in 2009 the FASB implemented its *FASB Accounting Standards Codification.* The Codification integrates and topically organizes all relevant accounting pronouncements comprising GAAP in a searchable, online database. It represents the single source of authoritative nongovernmental U.S. GAAP, and also includes portions of SEC accounting guidance that are relevant to financial reports filed with the SEC. When the FASB issues a new ASU, it becomes authoritative when it is entered into the Codification. The Codification is organized into nine main topics and approximately 90 subtopics. The main topics and related numbering system are presented in Illustration 1–6.[14] The Codification can be located at **www.fasb.org**.

Illustration 1–5
Accounting Standard Setting

HIERARCHY OF STANDARD-SETTING AUTHORITY

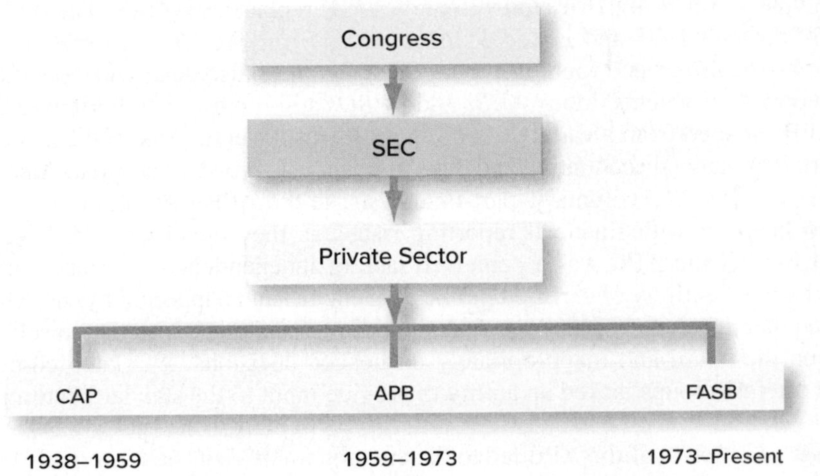

[12]The FAF's primary sources of funding are fees assessed against issuers of securities under the *Public Company Accounting Reform and Investor Protection Act of 2002,* commonly referred to as the *Sarbanes-Oxley Act.* The FAF is governed by trustees, the majority of whom are appointed from the membership of eight sponsoring organizations. These organizations represent important constituencies involved with the financial reporting process. For example, one of the founding organizations is the CFA Institute which represents financial information users, and another is the Financial Executives International which represents financial information preparers. The FAF also raises funds to support the activities of the Governmental Accounting Standards Board (GASB).

[13]The major responsibility of the FASAC is to advise the FASB on the priorities of its projects, including the suitability of new projects that might be added to its agenda. FASAC includes approximately 35 representatives from auditing firms, private companies, various user groups, and academia.

[14]FASB ASC 105–10: Generally Accepted Accounting Principles—Overall (previously "The FASB Accounting Standards Codification® and the Hierarchy of Generally Accepted Accounting Principles—a replacement of FASB Statement No. 162," *Statement of Financial Accounting Standards No. 168* (Norwalk, Conn.: FASB: 2009)).

Illustration 1–6
FASB Accounting Standards Codification Topics

FASB Accounting Standards Codification Topics

Topic	Numbered
General Principles	100–199
Presentation	200–299
Assets	300–399
Liabilities	400–499
Equity	500–599
Revenues	600–699
Expenses	700–799
Broad Transactions	800–899
Industry	900–999

Additional Consideration

Accounting standards and the standard-setting process discussed above relate to profit-oriented organizations and nongovernmental not-for-profit entities. In 1984, the **Governmental Accounting Standards Board (GASB)** was created to develop accounting standards for governmental units such as states and cities. The FAF oversees and funds the GASB, and the Governmental Accounting Standards Advisory Council (GASAC) provides input to it.

Throughout this book, we use the Accounting Standards Codification System (ASC) in footnotes when referencing generally accepted accounting principles (FASB ASC followed by the appropriate number). Each footnote also includes a reference to the original accounting standard that is codified in the ASC.

● LO1–11

INTERNATIONAL STANDARD SETTING Most industrialized countries have organizations responsible for determining accounting and reporting standards. In some countries, the United Kingdom, for instance, the responsible organization is a private sector body similar to the FASB in the United States. In other countries, the organization is a governmental body. Historically, these different organizations often produced different accounting standards, which complicated accounting by multinational companies, reduced comparability between companies using different standards, and potentially made it harder for companies to raise capital in international markets.

*The **International Accounting Standards Board (IASB)** is dedicated to developing a single set of global accounting standards.*

In response to these problems, the **International Accounting Standards Committee (IASC)** was formed in 1973 to develop global accounting standards. The IASC reorganized itself in 2001 and created a new standard-setting body called the **International Accounting Standards Board (IASB)**. The IASB's main objective is to develop a single set of high-quality, understandable, and enforceable global accounting standards to help participants in the world's capital markets and other users make economic decisions.[15]

As shown in Illustration 1–7, the way international standard setting is structured is similar in many respects to the way standard setting is structured in the United States.

International Financial Reporting Standards are issued by the IASB.

The IASC issued 41 International Accounting Standards (IASs), and the IASB endorsed these standards when it was formed in 2001. Since then, the IASB has revised many IASs and has issued new standards of its own, called **International Financial Reporting Standards (IFRS)**. More and more countries are basing their national accounting standards on IFRS. By 2018, more than 120 jurisdictions, including Hong Kong, Egypt, Canada, Australia, and the countries in the European Union (EU), require or permit the use of IFRS or a local variant of IFRS.[16]

---

[15] **www.ifrs.org**.

[16] See **http://www.ifrs.com/ifrs_faqs.html**.

Illustration 1–7
Comparison of Organizations of U.S. and International Standard Setters

	U.S. GAAP	IFRS
Regulatory oversight provided by:	Securities Exchange Commission (SEC)	Monitoring Board
Foundation providing oversight, appointing members, raising funds:	Financial Accounting Foundation (FAF): 14–18 trustees	IFRS Foundation: 22 trustees
Standard-setting board:	Financial Accounting Standards Board (FASB): 7 members	International Accounting Standards Board (IASB): 14 members
Advisory council providing input on agenda and projects:	Financial Accounting Standards Advisory Council (FASAC): 30–40 members	IFRS Advisory Council: approx. 50 members
Group to deal with emerging issues:	Emerging Issues Task Force (EITF): approx. 15 members	IFRS Interpretations Committee: 14 members

EFFORTS TO CONVERGE U.S. AND INTERNATIONAL STANDARDS Should the United States also adopt IFRS? Many argue that a single set of global standards would improve comparability of financial reporting and facilitate access to capital. In 2007, the SEC signaled its view that IFRS are of high quality by eliminating a requirement for foreign companies that issue stock in the United States to include in their financial statements a reconciliation of IFRS to U.S. GAAP. However, others argue that U.S. standards should remain customized to fit the stringent legal and regulatory requirements of the U.S. business environment. There also is concern that differences in implementation and enforcement from country to country make accounting under IFRS appear more uniform and comparable than actually is the case. Another argument is that competition between alternative standard-setting regimes is healthy and can lead to improved standards.[17]

The FASB and IASB have worked for many years to converge to one global set of accounting standards. Much progress has occurred, and in this book you will learn about already-converged standards that deal with such topics as revenue recognition, earnings per share, share-based compensation, nonmonetary exchanges, inventory costs, and the calculation of fair value. **Where We're Headed** boxes throughout the book describe additional projects that are ongoing.

However, at the time this book is being written, recent events suggest that full convergence will not be achieved in the foreseeable future. For example, as discussed further in Chapter 12, the FASB and IASB eventually concluded that full convergence was not possible with respect to accounting for financial instruments. While it appears likely that the FASB and IASB will continue to work together to converge where possible, some differences between IFRS and U.S. GAAP will remain.

Nonetheless, you should be aware of important differences that exist between U.S. GAAP and IFRS. Therefore, **International Financial Reporting Standards** boxes are included throughout the book to highlight circumstances in which IFRS differs from U.S. GAAP. Throughout this book, and also in the end-of-chapter questions, exercises, problems, and cases, IFRS-related material is marked with the globe icon that you see beside this paragraph. And, similar to the **Target** case, an **Air France–KLM (AF)** case is included among the Real-World Cases that accompany each chapter, so you can see how each chapter's IFRS material relates to a single, familiar company.

● LO1–4

The Standard-Setting Process

FINANCIAL Reporting Case

Q2, p. 3

DUE PROCESS When developing accounting standards, a standard setter must understand the nuances of the economic transactions the standards address and the views of key constituents concerning how accounting would best capture that economic reality. Therefore, the FASB undertakes a series of elaborate information-gathering steps before issuing

[17]For a comprehensive analysis of the pros and cons of U.S. adoption of IFRS, see L. Hail, C. Leuz and P. Wysocki, "Global Accounting Convergence and the Potential Adoption of IFRS in the US (Part 1): An Analysis of Economic and Policy Factors," *Accounting Horizons 24* (No 3.), September 2010, pp. 355–394, and ". . . (Part 2): Political Factors and Future Scenarios for U.S. Accounting Standards," *Accounting Horizons* 24 (No. 4), December 2010, pp. 567–588.

Step	Explanation
1.	The Board identifies financial reporting issues based on requests/recommendations from stakeholders or through other means.
2.	The Board decides whether to add a project to the technical agenda based on a staff-prepared analysis of the issues.
3.	The Board deliberates at one or more public meetings the various issues identified and analyzed by the staff.
4.	The Board issues an Exposure Draft. (In some projects, a Discussion Paper may be issued to obtain input at an early stage that is used to develop an Exposure Draft.)
5.	The Board holds a public roundtable meeting on the Exposure Draft, if necessary.
6.	The staff analyzes comment letters, public roundtable discussion, and any other information. The Board redeliberates the proposed provisions at public meetings.
7.	The Board issues an Accounting Standards Update describing amendments to the Accounting Standards Codification.

Illustration 1–8

The FASB's Standard-Setting Process

The FASB undertakes a series of information-gathering steps before issuing an Accounting Standards Update.

an Accounting Standards Update. These steps include open hearings, deliberations, and requests for written comments from interested parties. Illustration 1–8 outlines the FASB's standard-setting process.[18]

These steps help the FASB acquire information to determine the preferred method of accounting. However, as a practical matter, this information gathering also exposes the FASB to much political pressure by various interest groups who want an accounting treatment that serves their economic best interest. As you will see later in this chapter, the FASB's concepts statements indicate that standards should present information in a neutral manner, rather than being designed to favor particular economic consequences, but sometimes politics intrudes on the standard-setting process.

POLITICS IN STANDARD SETTING A change in accounting standards can result in a substantial redistribution of wealth within our economy. Therefore, it is no surprise that the FASB has had to deal with intense political pressure over controversial accounting standards and sometimes has changed standards in response to that pressure.

One example of the effect of politics on standard setting occurred in the mid-1990s with respect to accounting for employee stock options. The accounting standards in place at that time typically did not recognize compensation expense if a company paid their employees with stock options rather than cash. Yet, the company was sacrificing something of value to compensate its employees. Therefore, the FASB proposed that companies recognize compensation expense in an amount equal to the fair value of the options, with some of the expense recognized in each of the periods in which the employee earned the options. Numerous companies (particularly in California's Silicon Valley, where high-tech companies had been compensating employees with stock options to a great extent) applied intense political pressure against this proposal, and eventually the FASB backed down and required only disclosure of options-related compensation expense in the notes to the financial statements. Nearly a decade later, this contentious issue resurfaced in a more amenable political climate, and the FASB issued a standard requiring expense recognition as originally proposed. This issue is discussed at greater length in Chapter 19.

Another example of the political process at work in standard setting is the controversy surrounding the implementation of the fair value accounting standard issued in 2007. Many financial assets and liabilities are reported at fair value in the balance sheet, and many types of fair value changes are included in net income. Some have argued that fair values were estimated in a manner that exacerbated the financial crisis of 2008–2009 by forcing financial institutions to take larger than necessary write-downs of financial assets in the illiquid markets that existed at that time. As discussed further in Chapter 12, pressure from lobbyists and politicians influenced the FASB to revise its guidance on recognizing investment losses in these situations, and ongoing pressure remains to reduce the extent to which fair value changes are included in the determination of net income.

[18] www.FASB.org.

International Financial Reporting Standards

● LO1–11

Politics in International Standard Setting. Political pressures on the IASB's standard-setting process are severe. Politicians from countries that use IFRS lobby for the standards they prefer. The European Union (EU) is a particularly important adopter of IFRS and utilizes a formal evaluation process for determining whether an IFRS standard will be endorsed for use in EU countries. Economic consequences for EU member nations are an important consideration in that process.

For example, in 2008 the EU successfully pressured the IASB to suspend its due process and immediately allow reclassification of investments so that EU banks could avoid recognizing huge losses during a financial crisis.[19] Commenting on standards setting at that time, Charlie McCreevy, European Commissioner for Internal Markets and Service, stated that "Accounting is now far too important to be left to . . . accountants!"[20]

Additional Consideration

Private Company Council (PCC). Are the complex, comprehensive standards that are necessary to reflect the activities of a huge multinational conglomerate like **General Electric** also appropriate for a private company that, say, just needs to provide financial statements to its bank to get a loan? Private companies might be able to avoid much of that complexity. They don't sell securities like stocks and bonds to the general public, and they usually can identify the information needs of the specific users who rely on their financial statements and provide direct access to management to answer questions. Private companies typically also have a smaller accounting staff than do public companies. For those reasons, private companies have long sought a version of GAAP that is less costly to apply and better meets the information needs of the users of their financial statements.

In 2012, the Financial Accounting Foundation responded to this concern by establishing the Private Company Council (PCC). The ten-member PCC determines whether changes to existing GAAP are necessary to meet the needs of users of private company financial statements but a proposed exception or modification for private companies must be endorsed by the FASB before being issued as an Accounting Standards Update and added to the Codification. The PCC also advises the FASB about its current projects that affect private companies.

● LO1–5

Encouraging High-Quality Financial Reporting

FINANCIAL Reporting Case

Q3, p. 3

Numerous factors affect the quality of financial reporting. In this section, we discuss the role of the auditor, recent reforms in financial reporting, and the debate about whether accounting standards should emphasize rules or underlying principles.

The Role of the Auditor

***Auditors* express an opinion on the compliance of financial statements with GAAP.**

It is the responsibility of management to apply GAAP appropriately. Another group, **auditors**, examine (audit) financial statements to express a professional, independent opinion about whether the statements fairly present the company's financial position, its results of operations, and its cash flows in compliance with GAAP. Audits add credibility to the financial statements, increasing the confidence of those who rely on the information. Auditors, therefore, play an important role in the capital markets.

[19]Sarah Deans and Dane Mott, "Lowering Standards," **www.morganmarkets.com**, 10/14/2008.

[20]Charlie McCreevy, Keynote Address, "Financial Reporting in a Changing World" Conference, Brussels, 5/7/2009.

Most companies receive what's called an unmodified audit report. For example, **Apple Inc.**'s 2017 audit report by **Ernst & Young LLP** states, "In our opinion, the financial statements referred to above present fairly, in all material respects, the consolidated financial position of Apple Inc. at September 30, 2017 and September 24, 2016, and the consolidated results of its operations and its cash flows for each of the three years in the period ended September 30, 2017, in conformity with U.S. generally accepted accounting principles." This is known as a clean opinion. Had there been any material departures from GAAP or other problems that caused the auditors to question the fairness of the statements, the report would have been modified to inform readers. Normally, companies correct any material misstatements that auditors identify in the course of an audit, so companies usually receive clean opinions. The audit report for public companies also provides the auditors' opinion on the effectiveness of the company's internal control over financial reporting.

In most states, only individuals licensed as **certified public accountants (CPAs)** can represent that the financial statements have been audited in accordance with generally accepted auditing standards. Requirements to be licensed as a CPA vary from state to state, but all states specify education, testing, and experience requirements. The testing requirement is to pass the Uniform CPA Examination.

***Certified public accountants (CPAs)* are licensed by states to provide audit services.**

Financial Reporting Reform

The dramatic collapse of **Enron** in 2001 and the dismantling of the international public accounting firm of **Arthur Andersen** in 2002 severely shook U.S. capital markets. The credibility of the accounting profession itself as well as of corporate America was called into question. Public outrage over accounting scandals at high-profile companies like **WorldCom**, **Xerox**, **Merck**, **Adelphia Communications**, and others increased the pressure on lawmakers to pass measures that would restore credibility and investor confidence in the financial reporting process.

Driven by these pressures, Congress acted swiftly and passed the *Public Company Accounting Reform and Investor Protection Act of 2002*, commonly referred to as the *Sarbanes-Oxley Act*, or *SOX*, for the two congressmen who sponsored the bill. SOX applies to public securities-issuing entities. It provides for the regulation of auditors and the types of services they furnish to clients, increases accountability of corporate executives, addresses conflicts of interest for securities analysts, and provides for stiff criminal penalties for violators. Illustration 1–9 outlines key provisions of the Act.

Section 404 is perhaps the most controversial provision of SOX. It requires that company management document internal controls and report on their adequacy. Auditors also must express an opinion on whether the company has maintained effective control over financial reporting.

No one argues the importance of adequate internal controls, but many argued that the benefits of Section 404 did not justify the costs of complying with it. Research provides evidence that 404 reports affect investors' risk assessments and companies' stock prices, indicating these reports are seen as useful by investors.[21] Unfortunately, it is not possible to quantify the more important benefit of potentially avoiding business failures like Enron by focusing attention on the implementation and maintenance of adequate internal controls.

The costs of 404 compliance initially were quite steep. For example, one survey of Fortune 1,000 companies estimated that large companies spent, on average, approximately $8.5 million and $4.8 million (including internal costs and auditor fees) during the first two years of the act to comply with 404 reporting requirements.[22] As expected, the costs dropped significantly in the second year, and continued to drop as the efficiency of internal control audits increased. Many companies now perceive that the benefits of these internal control reports exceed their costs.[23]

We revisit Section 404 in Chapter 7 in the context of an introduction to internal controls.

[21]Hollis Ashbaugh Skaife, Daniel W. Collins, William R. Kinney, Jr., and Ryan LaFond, "The Effect of SOX Internal Control Deficiencies on Firm Risk and Cost of Equity," *Journal of Accounting Research* 47 (2009), pp. 1–43.

[22]"Sarbanes-Oxley 404 Costs and Implementation Issues: Spring 2006 Survey Update," CRA International (April 17, 2006).

[23]Protiviti, Inc., *2011 Sarbanes-Oxley Compliance Survey* (June, 2011).

Illustration 1–9

Public Company Accounting Reform and Investor Protection Act of 2002 (Sarbanes-Oxley)

Key Provisions of the Sarbanes-Oxley Act:

- **Oversight board.** The five-member (two accountants) Public Company Accounting Oversight Board has the authority to establish standards dealing with auditing, quality control, ethics, independence, and other activities relating to the preparation of audit reports, or can choose to delegate these responsibilities to the AICPA. Prior to the act, the AICPA set auditing standards. The SEC has oversight and enforcement authority.
- **Corporate executive accountability.** Corporate executives must personally certify the financial statements and company disclosures with severe financial penalties and the possibility of imprisonment for fraudulent misstatement.
- **Nonaudit services.** The law makes it unlawful for the auditors of public companies to perform a variety of nonaudit services for audit clients. Prohibited services include bookkeeping, internal audit outsourcing, appraisal or valuation services, and various other consulting services. Other nonaudit services, including tax services, require preapproval by the audit committee of the company being audited.
- **Retention of work papers.** Auditors of public companies must retain all audit or review work papers for seven years or face the threat of a prison term for willful violations.
- **Auditor rotation.** Lead audit partners are required to rotate every five years. Mandatory rotation of audit firms came under consideration.
- **Conflicts of interest.** Audit firms are not allowed to audit public companies whose chief executives worked for the audit firm and participated in that company's audit during the preceding year.
- **Hiring of auditor.** Audit firms are hired by the audit committee of the board of directors of the company, not by company management.
- **Internal control.** Section 404 of the act requires that company management document and assess the effectiveness of all internal control processes that could affect financial reporting. The PCAOB's *Auditing Standard No. 2* (since replaced by *AS 2201*) requires that the company auditors express an opinion on whether the company has maintained effective internal control over financial reporting.

A Move Away from Rules-Based Standards?

A *principles-based,* or *objectives-oriented,* approach to standard-setting stresses professional judgment, as opposed to following a list of rules.

The accounting scandals at Enron and other companies involved managers using elaborately structured transactions to try to circumvent specific rules in accounting standards. One consequence of those scandals was a rekindled debate over **principles-based**, or more recently termed **objectives-oriented**, versus **rules-based accounting standards**. In fact, a provision of the Sarbanes-Oxley Act required the SEC to study the issue and provide a report to Congress on its findings. That report, issued in July 2003, recommended that accounting standards be developed using an objectives-oriented approach.[24]

An objectives-oriented approach to standard setting emphasizes using professional judgment, as opposed to following a list of rules, when choosing how to account for a transaction. Proponents of an objectives-oriented approach argue that a focus on professional judgment means that there are few rules to sidestep and we are more likely to arrive at an appropriate accounting treatment. Detractors, on the other hand, argue that the absence of detailed rules opens the door to even more abuse, because management can use the latitude provided by objectives to justify their preferred accounting approach. Even in the absence of intentional misuse, reliance on professional judgment might result in different interpretations for similar transactions, raising concerns about comparability. Also, detailed rules help auditors withstand pressure from clients who want a more favorable accounting treatment and help companies ensure that they are complying with GAAP and avoid litigation or SEC

[24]"Study Pursuant to Section 108 (d) of the Sarbanes-Oxley Act of 2002 on the Adoption by the United States Financial Reporting System of a Principles-Based Accounting System," Securities and Exchange Commission (July 2003).

inquiry. For these reasons, it's challenging to avoid providing detailed rules in the U.S. reporting environment.

Regardless of whether accounting standards are based more on rules or on objectives, prior research highlights that there is some potential for abuse, either by structuring transactions around precise rules or opportunistically interpreting underlying principles.[25] The key is whether management is dedicated to high-quality financial reporting. It appears that poor ethical values on the part of management are at the heart of accounting abuses and scandals, so we now turn to a discussion of ethics in the accounting profession.

Ethics in Accounting

***Ethics* deals with the ability to distinguish right from wrong.**

Ethics is a term that refers to a code or moral system that provides criteria for evaluating right and wrong. An ethical dilemma is a situation in which an individual or group is faced with a decision that tests this code. Many of these dilemmas are simple to recognize and resolve. For example, have you ever been tempted to call your professor and ask for an extension on the due date of an assignment by claiming a pretended illness? Temptation like this will test your personal ethics.

Accountants, like others in the business world, are faced with many ethical dilemmas, some of which are complex and difficult to resolve. For instance, the capital markets' focus on near-term profits may tempt a company's management to bend or even break accounting rules to inflate reported net income. In these situations, technical competence is not enough to resolve the dilemma.

ETHICS AND PROFESSIONALISM One characteristic that distinguishes a profession from other occupations is the acceptance by its members of a responsibility for the interests of those it serves. Ethical behavior is expected of those engaged in a profession. That expectation often is articulated in a code of ethics. For example, law and medicine are professions that have their own codes of professional ethics. These codes provide guidance and rules to members in the performance of their professional responsibilities.

Public accounting has achieved widespread recognition as a profession. The AICPA, the national organization of certified public accountants, has its own Code of Professional Conduct that prescribes the ethical conduct members should strive to achieve. Similarly, the **Institute of Management Accountants (IMA)**—the primary national organization of accountants working in industry and government—has its own code of ethics, as does the **Institute of Internal Auditors**—the national organization of accountants providing internal auditing services for their own organizations.

ANALYTICAL MODEL FOR ETHICAL DECISIONS Ethical codes are informative and helpful, but the motivation to behave ethically must come from within oneself and not just from the fear of penalties for violating professional codes. Presented below is a sequence of steps that provide a framework for analyzing ethical issues. These steps can help you apply your own sense of right and wrong to ethical dilemmas:[26]

Step 1. Determine the facts of the situation. This involves determining the who, what, where, when, and how.

Step 2. Identify the ethical issue and the stakeholders. Stakeholders may include shareholders, creditors, management, employees, and the community.

---

[25] Mark W. Nelson, John A. Elliott, and Robin L. Tarpley, "Evidence From Auditors About Manager' and Auditors Earnings Management Decisions," *The Accounting Review* 77 (2002), pp. 175–202.

[26] Adapted from Harold Q. Langenderfer and Joanne W. Rockness, "Integrating Ethics into the Accounting Curriculum: Issues, Problems, and Solutions," *Issues in Accounting Education* (Spring 1989). These steps are consistent with those provided by the American Accounting Association's Advisory Committee on Professionalism and Ethics in their publication *Ethics in the Accounting Curriculum: Cases and Readings, 1990.*

Ethical Dilemma

You recently have been employed by a large retail chain that sells sporting goods. One of your tasks is to help prepare periodic financial statements for external distribution. The chain's largest creditor, National Savings & Loan, requires quarterly financial statements, and you are currently working on the statements for the three-month period ending June 30, 2021.

During the months of May and June, the company spent $1,200,000 on a hefty radio and TV advertising campaign. The $1,200,000 included the costs of producing the commercials as well as the radio and TV time purchased to air the commercials. All of the costs were charged to advertising expense. The company's chief financial officer (CFO) has asked you to prepare a June 30 adjusting entry to remove the costs from advertising expense and to set up an asset called *prepaid advertising* that will be expensed in July. The CFO explained that "This advertising campaign has led to significant sales in May and June and I think it will continue to bring in customers through the month of July. By recording the ad costs as an asset, we can match the cost of the advertising with the additional July sales. Besides, if we expense the advertising in May and June, we will show an operating loss on our income statement for the quarter. The bank requires that we continue to show quarterly profits in order to maintain our loan in good standing."

Step 3. Identify the values related to the situation. For example, in some situations confidentiality may be an important value that might conflict with the right to know.

Step 4. Specify the alternative courses of action.

Step 5. Evaluate the courses of action specified in step 4 in terms of their consistency with the values identified in step 3. This step may or may not lead to a suggested course of action.

Step 6. Identify the consequences of each possible course of action. If step 5 does not provide a course of action, assess the consequences of each possible course of action for all of the stakeholders involved.

Step 7. Make your decision and take any indicated action.

Ethical dilemmas are presented throughout this book. The analytical steps outlined above provide a framework you can use to evaluate these situations.

PART B

The Conceptual Framework

● LO1–6

FINANCIAL Reporting Case

Q4, p. 3

Sturdy buildings are built on sound foundations. The U.S. Constitution is the foundation for the laws of our land. The **conceptual framework** has been described as an "Accounting Constitution" because it provides the underlying foundation for U.S. accounting standards. The conceptual framework provides structure and direction to financial accounting and reporting but does not directly prescribe GAAP. It is a coherent system of interrelated objectives and fundamentals that is intended to lead to consistent standards and that prescribes the nature, function, and limits of financial accounting and reporting. The fundamentals are the underlying concepts of accounting that guide the selection of events to be accounted for, the measurement of those events, and the means of summarizing and communicating them to interested parties.[27]

The *conceptual framework* does not prescribe GAAP. It provides an underlying foundation for accounting standards.

The FASB disseminates this framework through *Statements of Financial Accounting Concepts (SFACs)*. *SFAC 8* discusses the objective of financial reporting and the qualitative characteristics of useful financial information. *SFAC 7* describes how cash flows and present values are used when making accounting measurements, *SFAC 6* defines the accounts and

[27]"Conceptual Framework for Financial Accounting and Reporting: Elements of Financial Statements and Their Measurement," *Discussion Memorandum* (Stamford, Conn.: FASB, 1976), p. 2.

Illustration 1–10

The Conceptual Framework

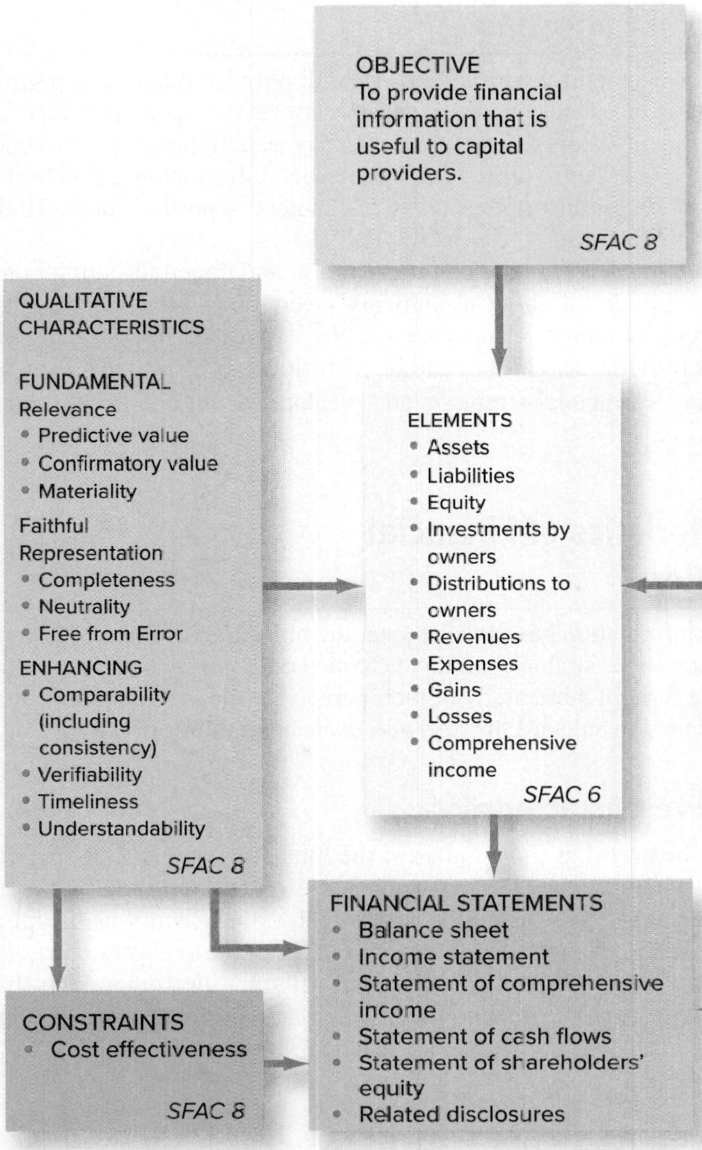

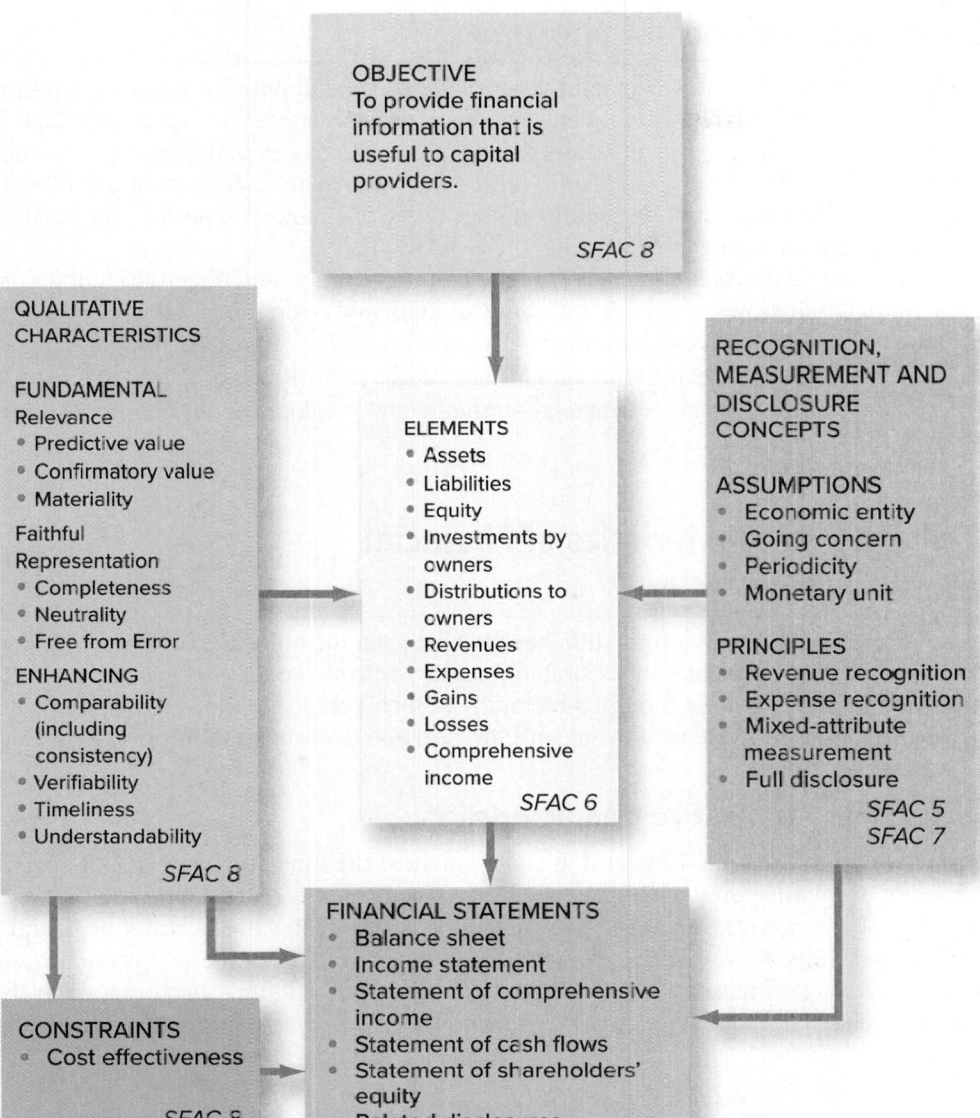

accrual accounting concepts that appear in financial statements, and *SFAC 5* discusses recognition and measurement concepts. Earlier *SFAC*s either have been superseded or involve nonbusiness organizations that aren't considered in this book.

In the remainder of this section we discuss the components of the conceptual framework, as depicted in Illustration 1–10.

International Financial Reporting Standards

Role of the conceptual framework. The conceptual frameworks in U.S. GAAP and IFRS are very similar and are converging even more with ongoing efforts by the FASB and IASB. However, in U.S. GAAP, the conceptual framework primarily provides guidance to standard setters to help them develop high-quality standards. In IFRS, the conceptual framework guides standard setting, but in addition it provides a basis for practitioners to make accounting judgments when another IFRS standard does not apply. Also, IFRS emphasizes the overarching concept of the financial statements providing a "fair presentation" of the company. U.S. GAAP does not include a similar requirement, but U.S. auditing standards require this consideration.

● LO1–11

Objective of Financial Reporting

LO1–7

As indicated in Part A of this chapter, the objective of general purpose financial reporting is to provide financial information about companies that is useful to capital providers in making decisions. For example, investors decide whether to buy, sell, or hold equity or debt securities, and creditors decide whether to provide or settle loans. Information that is useful to capital providers may also be useful to other users of financial reporting information, such as regulators or taxing authorities.

Investors and creditors are interested in the amount, timing, and uncertainty of a company's future cash flows. Information about a company's economic resources (assets) and claims against resources (liabilities) also is useful. Not only does this information about resources and claims provide insight into future cash flows, it also helps decision makers identify the company's financial strengths and weaknesses and assess liquidity and solvency.

Qualitative Characteristics of Financial Reporting Information

What characteristics should information have to best meet the objective of financial reporting? Illustration 1–11 indicates the desirable qualitative characteristics of financial reporting information, presented in the form of a hierarchy of their perceived importance. Notice that these characteristics are intended to enhance the **decision usefulness** of information.

Decision usefulness requires that information possess the qualities of *relevance* and *faithful representation*.

Fundamental Qualitative Characteristics

For financial information to be useful, it should possess the fundamental decision-specific qualities of **relevance** and **faithful representation**. Both are critical. Information is of little value if it's not relevant. And even if information is relevant, it is not as useful if it doesn't faithfully represent the economic phenomenon it purports to represent. Let's look closer at each of these two qualitative characteristics, including the components that make those characteristics desirable. We also consider other characteristics that enhance usefulness.

Illustration 1–11

Hierarchy of Qualitative Characteristics of Financial Information

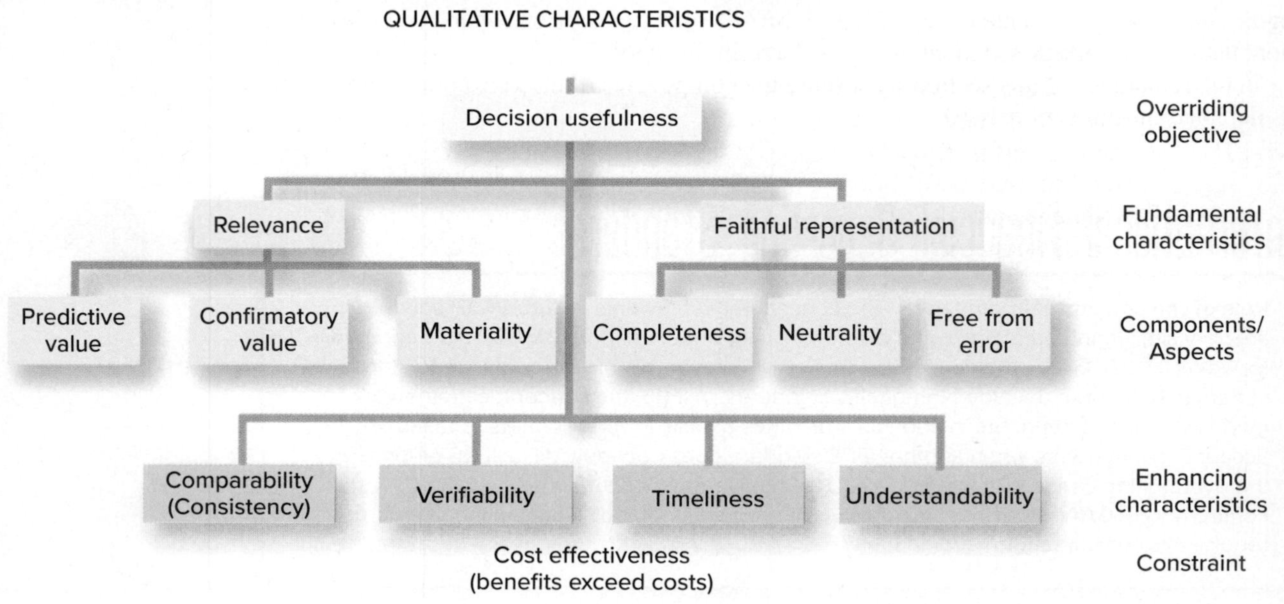

RELEVANCE Obviously, to make a difference in the decision process, information must be relevant to the decision. Relevance in the context of financial reporting means that information must possess **predictive value** and/or **confirmatory value**, typically both. For example, current-period net income has predictive value if it helps users predict a company's future cash flows, and it has confirmatory value if it helps investors confirm or change their prior assessments regarding a company's cash-flow generating ability. Predictive and confirmatory value are central to the concept of "earnings quality," the ability of reported earnings (income) to predict a company's future earnings.

Relevance requires that information have predictive and confirmatory value.

Financial information is **material** if it is probable that including or correcting the information would affect a user's judgment decisions. Materiality is an aspect of relevance that depends on a company's particular situation and is based on the nature or magnitude of the item that is being reported. Recall that **Apple Inc.**'s audit report discussed earlier only expressed an opinion about material items. If information is immaterial, it's not relevant.

Information is *material* if it could affect users' judgment.

The threshold for materiality often depends on the *relative* dollar amount of the transaction. For example, $10,000 in total anticipated bad debts for a multibillion dollar company would not be considered material. This same $10,000 amount, however, might easily be material for a neighborhood pizza parlor. Because of the context-specific nature of materiality, the FASB has been reluctant to establish any quantitative materiality guidelines. The threshold for materiality has been left to the subjective judgment of the company preparing the financial statements and its auditors.

Materiality often relates to the nature of the item as well. It has to be evaluated in light of surrounding circumstances, and it depends on qualitative as well as quantitative considerations. For example, an illegal payment of a $10,000 bribe to an official of a foreign government to secure a valuable contract probably would be considered material qualitatively even if the amount is small relative to the size of the company. Similarly, a small dollar amount that changes a net loss to a net income for the reporting period could be viewed as material to financial statement users for qualitative reasons.

Professional judgment determines what amount is material in each situation.

FAITHFUL REPRESENTATION **Faithful representation** exists when there is agreement between a measure or description and the phenomenon it purports to represent. For example, the term *inventory* in the balance sheet of a retail company is understood by external users to represent items that are intended for sale in the ordinary course of business. If inventory includes, say, accounts receivable, it lacks faithful representation.

Faithful representation means agreement between a measure and a real-world phenomenon that the measure is supposed to represent.

To break it down further, faithful representation requires that information be *complete, neutral,* and *free from error.* A depiction of an economic phenomenon is **complete** if it includes all the information necessary for faithful representation of the economic phenomenon that it purports to represent. Omitting a portion of that information can cause it to be false or misleading.

A depiction is *complete* if it includes all information necessary for faithful representation.

A financial accounting standard, and the standard-setting process, is **neutral** if it is free from bias. You learned earlier that changes in accounting standards can lead to adverse economic consequences for certain companies and that political pressure is sometimes brought to bear on the standard-setting process in hopes of achieving particular outcomes. Accounting standards should be established with the goal of providing high-quality information, and should not try to achieve particular social outcomes or favor particular groups or companies. The FASB faces a difficult task in maintaining neutrality in the face of economic consequences and resulting political pressures.

Neutral implies freedom from bias.

Information is *free from error* if it contains no errors or omissions.

Representational faithfulness also is enhanced if information is **free from error**, meaning that there are no errors or omissions in the description of the amount or the process used to report the amount. Uncertainty is a fact of life when we measure many items of financial information included in financial statements. Estimates are common, and some inaccuracy is likely. An estimate is represented faithfully if it is described clearly and accurately as being an estimate, and financial statement users are given enough information to understand the potential for inaccuracy that exists.

Conservatism requires greater verification for good news than for bad news.

Many accountants have recommended that we deal with the potential for error by employing **conservatism**. Conservatism means that accountants require greater verification before recognizing good news than bad news. The result is that losses are reflected in net income more quickly than are gains, and net assets tend to be biased downwards.

SFAC 8 explicitly rejects conservatism as a desirable characteristic of accounting information, stating that conservatism undermines representational faithfulness by being inconsistent with neutrality. Nevertheless, some accounting practices appear to be generated by a desire to be conservative. For example, companies are required to recognize losses for declines in the value of inventory, buildings, and equipment, but aren't allowed to recognize gains for increases in those values. One justification for these practices is that investors and creditors who lose money on their investments are less likely to sue the company if bad news has been exaggerated and good news underestimated. Another justification is that conservative accounting can trigger debt covenants that allow creditors to protect themselves from bad management. So, despite the lack of support for conservatism in the conceptual framework, it is likely to persist as an important consideration in accounting practice and in the application of some accounting standards.

Enhancing Qualitative Characteristics

Illustration 1–11 identifies four *enhancing* qualitative characteristics: *comparability (including consistency), verifiability, timeliness,* and *understandability.*

Information is *comparable* if similar items are treated the same way and different items are treated differently.

Comparability helps users see similarities and differences between events and conditions. We already have discussed the importance of investors and creditors being able to compare information *among companies* to make their resource allocation decisions. Closely related to comparability is the notion that **consistency** of accounting practices over time permits valid comparisons *among different reporting periods.* The predictive and confirmatory value of information is enhanced if users can compare the performance of a company over time.[28] Companies typically include as Note 1 to the financial statements a summary of significant accounting policies and provide full disclosure of any changes in those policies to alert users to the potential for diminished consistency.

Information is *consistent* if it is measured and reported the same way in each time period.

Information is *verifiable* if different measurers would reach consensus about whether it is representationally faithful.

Verifiability implies that different knowledgeable and independent measurers would reach consensus regarding whether information is a faithful representation of what it is intended to depict. Direct verification involves observing the item being depicted. For example, the historical cost of a parcel of land to be reported in a company's balance sheet usually is highly verifiable. The cost can be traced to an exchange transaction, the purchase of the land. On the other hand, the fair value of that land is much more difficult to verify. Appraisers could differ in their assessment of fair value. Verification of their estimates would be indirect, involving examination of their valuation models and assessments of the reasonableness of model inputs. The term *objectivity* often is linked to verifiability. The historical cost of the land is objective and easy to verify, but the land's fair value is subjective, and may be influenced by the measurer's past experience and biases. A measurement that is subjective is more difficult to verify, which may make users doubt its representational faithfulness.

Information is *timely* if it is available to users before a decision is made.

Timeliness also is important for information to be useful. Information is timely when it's available to users early enough to allow them to use it in their decision process. To enhance

[28]Companies occasionally do change their accounting practices, which makes it difficult for users to make comparisons among different reporting periods. Chapter 4 and Chapter 20 describe the disclosures that a company makes in this situation to restore consistency among periods.

timeliness, the SEC requires its registrants to submit financial statement information on a quarterly as well as on an annual basis for each fiscal year.

Understandability means that users must be able to comprehend the information within the context of the decision being made. This is a user-specific quality because users will differ in their ability to comprehend any set of information. The overriding objective of financial reporting is to provide comprehensible information to those who have a *reasonable understanding* of business and economic activities and are diligent in studying the information.

Information is *understandable* if users can comprehend it.

Key Constraint: Cost Effectiveness

Most of us learn early in life that we can't get everything we want. The latest electronic gadget may have all the qualitative characteristics that current technology can provide, but limited resources may lead us to buy a model with fewer bells and whistles. **Cost effectiveness** constrains the accounting choices we make. The benefits of endowing financial information with all the qualitative characteristics we've discussed must exceed the costs of doing so.

Information is *cost effective* if the benefit of increased *decision usefulness* exceeds the costs of providing that information.

The costs of providing financial information include those of gathering, processing, and disseminating information. There also are costs to users when interpreting information. In addition, costs include possible adverse economic consequences of implementing accounting standards. For example, consider the requirement that companies having more than one operating segment must disclose certain disaggregated financial information.[29] In addition to the costs of information gathering, processing, and communicating that information, many companies feel that this reporting requirement imposes what could be called *competitive disadvantage costs*. These companies are concerned that their competitors will gain some advantage from having access to the disaggregated data.

The costs of providing financial information include any possible adverse economic consequences of accounting standards.

The perceived benefit from this or any accounting standard is increased *decision usefulness* of the information provided, which, ideally, improves the resource allocation process. It is inherently impossible to quantify this benefit. The elaborate information-gathering process undertaken by the FASB in setting accounting standards is an attempt to assess both costs and benefits of a proposed accounting standard, even if in a subjective, nonquantifiable manner.

Elements of Financial Statements

SFAC 6 defines 10 elements of financial statements. These elements are "the building blocks with which financial statements are constructed—the classes of items that financial statements comprise."[30] They focus directly on items related to reporting financial position and measuring performance. The *accrual accounting* model is embodied in the element definitions. For now, we list and define the elements in Illustration 1–12. You will learn much more about these elements in subsequent chapters.

The 10 elements of financial statements defined in *SFAC 6* describe financial position and periodic performance.

Underlying Assumptions

● LO1–8

Though not emphasized in the FASB's concepts statements, four basic assumptions underlie GAAP: (1) the economic entity assumption, (2) the going concern assumption, (3) the periodicity assumption, and (4) the monetary unit assumption. These assumptions identify the entity that is being reported on, the assumption that the entity will continue to exist, and the frequency and denomination in which reports occur.

Illustration 1–13 summarizes the four assumptions underlying GAAP.

Economic Entity Assumption

The **economic entity assumption** presumes that all economic events can be identified with a particular *economic entity*. Investors desire information about an economic entity that

The *economic entity assumption* presumes that economic events can be identified with a particular economic entity.

[29]FASB ASC 280: Segment Reporting (previously "Disclosures about Segments of an Enterprise and Related Information," *Statement of Financial Accounting Standards No. 131* (Norwalk, Conn.: FASB, 1997)).

[30]"Elements of Financial Statements," *Statement of Financial Accounting Concepts No. 6* (Stamford, Conn.: FASB, 1985), par. 5.

Illustration 1–12

Elements of Financial Statements

	Elements of Financial Statements
Assets	Probable future economic benefits obtained or controlled by a particular entity as a result of past transactions or events.
Liabilities	Probable future sacrifices of economic benefits arising from present obligations of a particular entity to transfer assets or provide services to other entities in the future as a result of past transactions or events.
Equity (or net assets)	Called shareholders' equity or stockholders' equity for a corporation, it is the residual interest in the assets of an entity that remains after deducting its liabilities.
Investments by owners	Increases in equity of a particular business enterprise resulting from transfers to it from other entities of something of value to obtain or increase ownership interests in it.
Distributions to owners	Decreases in equity of a particular enterprise resulting from transfers to owners.
Comprehensive income	The change in equity of a business enterprise during a period from transactions and other events and circumstances from nonowner sources. It includes all changes in equity during a period except those resulting from investments by owners and distributions to owners.
Revenues	Inflows or other enhancements of assets of an entity or settlements of its liabilities during a period from delivering or producing goods, rendering services, or other activities that constitute the entity's ongoing major or central operations.
Expenses	Outflows or other using up of assets or incurrences of liabilities during a period from delivering or producing goods, rendering services, or other activities that constitute the entity's ongoing major or central operations.
Gains	Increases in equity from peripheral or incidental transactions of an entity.
Losses	Represent decreases in equity arising from peripheral or incidental transactions of an entity.

Illustration 1–13

Summary of Assumptions Underlying GAAP

Assumptions	Description
Economic entity	All economic events can be identified with a particular economic entity.
Going concern	In the absence of information to the contrary, it is anticipated that a business entity will continue to operate indefinitely.
Periodicity	The life of a company can be divided into artificial time periods to provide timely information to external users.
Monetary unit	In the United States, financial statement elements should be measured in terms of the U.S. dollar.

corresponds to their ownership interest. For example, if you were considering buying some ownership stock in **Alphabet** (the parent company of **Google**), you would want information on the various operating units that constitute Alphabet. You would need information not only about its United States operations but also about its European and other international operations. The financial information for the various companies (subsidiaries) in which Alphabet owns a controlling interest (greater than 50% ownership of voting stock) should be combined with that of Alphabet (the parent) to provide a complete picture. The parent and its subsidiaries are separate *legal* entities but one *accounting* entity.

Another key aspect of this assumption is the distinction between the economic activities of owners and those of the company. For example, the economic activities of a sole proprietorship, Uncle Jim's Restaurant, should be separated from the activities of its owner, Uncle Jim. Uncle Jim's personal residence, for instance, is not an asset of the business.

Going Concern Assumption

Another necessary assumption is that, in the absence of information to the contrary, we anticipate that a business entity will continue to operate indefinitely. Accountants realize that the **going concern assumption** does not always hold since there certainly are many business failures. However, this assumption is critical to many broad and specific accounting principles. For example, the assumption provides justification for measuring many assets based on their historical costs. If it were known that an enterprise would cease operations in the near future, assets and liabilities would be measured at their current liquidation values. Similarly, when we depreciate a building over an estimated life of 40 years, we assume the business will operate that long.

The *going concern* assumption presumes that a business will operate indefinitely.

Periodicity Assumption

The **periodicity assumption** relates to the qualitative characteristic of *timeliness*. External users need *periodic* information to make decisions. This need for periodic information requires that the economic life of a company (presumed to be indefinite) be divided into artificial time periods for financial reporting. Corporations whose securities are publicly traded are required to provide financial information to the SEC on a quarterly and annual basis.[31] Financial statements often are prepared on a monthly basis for banks and others that might need more timely information.

The *periodicity assumption* allows the life of a company to be divided into artificial time periods to provide timely information.

For many companies, the annual time period (the fiscal year) is the calendar year. However, other companies have chosen a fiscal year that does not correspond to the calendar year. The accounting profession and the SEC advocate that companies adopt a fiscal year that corresponds to their natural business year. A natural business year is the 12-month period that ends when the business activities of a company reach their lowest point in the annual cycle. For example, many retailers, such as **Walmart**, have adopted a fiscal year ending around January 31. Business activity in January generally is quite slow following the very busy Christmas period. The **Campbell Soup Company**'s fiscal year ends in July; **Clorox**'s in June; and **Monsanto**'s in August.

Monetary Unit Assumption

The **monetary unit assumption** requires that financial statement elements be measured in nominal units of money, without any adjustment for changes in purchasing power. In the United States, the U.S. dollar is the monetary unit used in financial statements. In the European Union (EU), the euro is the monetary unit. Other countries use other currencies as their monetary units.

The *monetary unit assumption* states that financial statement elements should be measured in a particular monetary unit (in the United States, the U.S. dollar).

One problem with use of a monetary unit like the dollar or the euro is that it is presumed to be stable over time. That is, the value of the dollar, in terms of its ability to purchase certain goods and services, is assumed to be constant over time. This assumption obviously does not strictly hold. The U.S. economy has experienced periods of rapidly changing prices. To the extent that prices are unstable, and machines, trucks, and buildings were purchased at different times, the monetary unit used to measure them is not the same. The effect of changing prices on financial information generally is discussed elsewhere in your accounting curriculum, often in an advanced accounting course.

Recognition, Measurement, and Disclosure Concepts

● LO1–9

Now that we have identified the various elements and underlying assumptions of the financial statements, we discuss *when* the elements should be recognized (recorded) and how they should be *measured* and *disclosed*. For example, an asset was previously defined as a probable future economic benefit obtained or controlled by a company as a result of past transactions or events. But *when* should the asset be recorded, at *what* amount, and what other important information about the asset should be provided in the financial statements? *SFAC 5* addresses these issues. **Recognition** refers to the process of admitting information into the financial statements. **Measurement** is the process of associating numerical amounts with the elements. **Disclosure** refers to including pertinent information in the financial statements and accompanying notes.

[31]The report that must be filed for the first three quarters of each fiscal year is Form 10-Q and the annual report is Form 10-K.

Recognition

Recognition criteria determine whether items should be included in the financial statements.

GENERAL RECOGNITION CRITERIA According to *SFAC 5,* an item should be recognized in the basic financial statements when it meets the following four criteria, subject to a cost effectiveness constraint and materiality threshold:

1. *Definition.* The item meets the definition of an element of financial statements.
2. *Measurability.* The item has a relevant attribute measurable with sufficient reliability.
3. *Relevance.* The information about it is capable of making a difference in user decisions.
4. *Reliability.* The information is representationally faithful, verifiable, and neutral.[32]

SFAC 5 provides further guidance with respect to revenue and expense recognition, and you will learn about more specific guidelines throughout this book.

Revenues are inflows of assets or settlements of liabilities as a result of providing goods or services to customers.

REVENUE RECOGNITION **Revenues** are inflows of assets or settlements of liabilities resulting from providing a product or service to a customer. An income statement should report the results of these activities only for the time period specified in the financial statements. Therefore, the *timing* of revenue recognition is a key element of earnings measurement. Not adhering to revenue recognition criteria could result in overstating revenue and hence net income in one reporting period and, consequently, understating revenue and net income in another period.

Until recently, revenue recognition was guided by the *realization principle,* which requires that two criteria be satisfied before revenue can be recognized:

1. The earnings process is judged to be complete or virtually complete.
2. There is reasonable certainty as to the collectibility of the asset to be received (usually cash).

As discussed further in Chapter 6, *ASU No. 2014-09* changed how we determine the timing and measurement of revenue.[33] That standard requires that companies recognize revenue when goods or services are transferred to customers for the amount the company expects to be entitled to receive in exchange for those goods or services. Revenue is recognized at a point in time or over a period of time, depending on when goods or services are transferred to customers. For example, revenue for the sale of most goods is recognized upon delivery, but revenue for services like renting apartments or lending money is recognized over time as those services are provided. No revenue is recognized if it isn't probable that the seller will collect the amounts it's entitled to receive. While that standard doesn't rely on the realization principle, you can see that aspects of the realization principle remain—we still focus on the seller fulfilling its obligations to its customers, and before revenue can be recognized we still require a relatively high likelihood that the seller will be paid.

Notice that these criteria help implement the accrual accounting model. Revenue is recognized when the seller transfers goods or services to a customer, which isn't necessarily at the same time the seller is paid by the customer.

The timing of revenue recognition also affects the timing of asset recognition. When revenue is recognized by crediting a revenue account, the corresponding debit typically increases some asset, usually cash or an account receivable.

Expenses are outflows or other using up of assets or incurrences of liabilities from providing goods or services.

EXPENSE RECOGNITION **Expenses** are outflows or other using up of assets or incurrences of liabilities from providing goods or services. When are expenses recognized? In practice, expense recognition often matches revenues and expenses that arise from the same

---

[32] "Recognition and Measurement in Financial Statements," *Statement of Financial Accounting Concepts No. 5* (Stamford, Conn.: FASB, 1984), par. 63. *SFAC 8* has replaced reliability with faithful representation as the second primary qualitative characteristic of financial information.

[33] "Revenue from Contracts with Customers (Topic 606)," *Accounting Standards Update 2014–09* (Norwalk, Conn: FASB, 2014).

transactions or other events.[34] There is a cause-and-effect relationship between revenue and expense recognition implicit in this approach. The net result is a measure—net income—that identifies the amount of profit or loss for the period provided by operations.

Although these concepts are straightforward, their implementation can be difficult, because many expenses are not incurred *directly* to produce a particular amount of revenue. Instead, the association between revenue and many expenses is indirect. Therefore, expense recognition is implemented by one of four different approaches, depending on the nature of the specific expense:[35]

- **Based on an exact cause-and-effect relationship.** This approach is appropriate for *cost of goods sold,* as one example. There is a definite cause-and-effect relationship between **PetSmart**'s revenue from selling dog food and its costs to purchase that dog food from suppliers. Commissions paid to salespersons for obtaining revenues also is an example of an expense recognized based on this approach.
- **By associating an expense with the revenues recognized in a specific time period.** Many expenses can be related only to periods of time during which revenue is earned. For example, the monthly salary paid to an office worker is not directly related to any specific revenue event. Instead, the employee provides benefits to the company for that one month that *indirectly* relate to the revenue recognized in that same period.
- **By a systematic and rational allocation to specific time periods.** Some costs are incurred to acquire assets that provide benefits to the company for more than one reporting period, so we recognize expenses over those time periods. For example, straight-line depreciation is a "systematical and rational" way to allocate the cost of equipment to the periods in which that equipment is used to produce revenue.
- **In the period incurred, without regard to related revenues.** Sometimes costs are incurred, but it is impossible to determine in which period or periods, if any, related revenues will occur. For example, let's say **Google** spends $1 million for a series of television commercials. It's difficult to determine when, how much, or even whether additional revenues occur as a result of that particular series of ads. As a result, we recognize advertising expenditures as expenses in the period incurred.

The timing of expense recognition also affects the timing of asset and liability recognition and de-recognition. When we debit an expense, the corresponding credit usually either decreases an asset (for example, decreasing cash because it was used to pay an employee's salary) or increases a liability (for example, increasing salaries payable to accrue wages that will be paid at a later date).

Measurement

If an amount is to be recognized, it also must be measured. As indicated in *SFAC 5,* GAAP currently employs a "mixed attribute" measurement model. If you look at a balance sheet, for instance, you might see land measured at historical cost, inventory at net realizable value, a liability at the present value of future cash payments, and an investment at fair value. The attribute chosen to measure a particular item should be the one that maximizes the combination of relevance and representational faithfulness. *SFAC 5* lists five measurement attributes employed in GAAP:

1. Historical cost
2. Net realizable value
3. Current cost

[34]The term *matching principle* is sometimes used to refer to the practice of first recognizing revenue and then recognizing all expenses that were incurred to generate that revenue. However, the conceptual framework does not include that term. Rather, *SFACs 5* and *6* discuss matching as a result of recognizing expenses and revenues that arise from the same underlying transactions or events. Standard setters are reluctant to apply matching more broadly, because they are concerned that doing so could result in inappropriately recognizing as assets some amounts that do not provide "probable future economic benefits," and therefore don't meet the definition of an asset. We discuss this topic more in the "Evolving GAAP" section at the end of this chapter.

[35]"Elements of Financial Statements—a replacement of FASB Concepts Statement No. 3 (incorporating an amendment of FASB Concepts Statement No. 2)," *Statement of Financial Accounting Concepts No. 6* (Norwalk, Conn. FASB, 1985).

4. Present (or discounted) value of future cash flows
5. Fair value

These different measurement attributes often indicate the same amount, particularly when the amount is initially recognized. However, sometimes they differ in important ways.

Historical cost **bases measurements on the amount given or received in the original exchange transaction.**

HISTORICAL COST We often measure assets and liabilities based on their *original transaction value,* that is, their **historical cost**. Some accountants refer to this practice as applying the *historical cost principle.* For an asset, historical cost equals the value of what is given in exchange (usually cash) for the asset at its initial acquisition. For liabilities, it is the current cash equivalent received in exchange for assuming the liability. Historical cost for long-lived, revenue-producing assets such as equipment typically is adjusted subsequent to its initial measurement by recognizing depreciation or amortization.

Why base measurement on historical costs? First, historical cost provides important cash flow information as it represents the cash or cash equivalent paid for an asset or received in exchange for the assumption of a liability. Second, because historical cost valuation is the result of an exchange transaction between two independent parties, the agreed-upon exchange value is objective and highly verifiable.

Net realizable value **bases measurements on the net amount of cash into which the asset or liability will be converted in the ordinary course of business.**

NET REALIZABLE VALUE Some assets are measured at their **net realizable value**, which is defined by the FASB as the estimated selling price in the ordinary course of business, less reasonably predictable costs of completion, disposal, and transportation. Intuitively, net realizable value is the net amount of cash into which an asset or liability is expected to be converted in the ordinary course of business. For example, if inventory could be sold for $10,000, and the company would incur additional costs of $2,000 to complete it and transport it to a customer, the inventory's net realizable value is $8,000. Departures from historical cost measurement such as this provide useful information to aid in the prediction of future cash flows.

Current cost **is the cost that would be incurred to purchase or reproduce an asset.**

CURRENT COST Companies sometimes report **current costs**, particularly if they operate in inflationary economies. The current cost of an asset is the cost that would be incurred to purchase or reproduce the asset.

Present value **bases measurement on future cash flows discounted for the time value of money.**

PRESENT VALUE Because of its importance to many accounting measurements, **present value** is the focus of a FASB concept statement, *SFAC 7,* which provides a framework for using future cash flows as the basis for accounting measurement and also indicates that the objective in valuing an asset or liability using present value is to approximate its fair value.[36] We explore the topic of present value in depth in Chapter 5 and the application of present value in accounting measurement in subsequent chapters.

Fair value **bases measurements on the price that would be received to sell assets or transfer liabilities in an orderly market transaction.**

FAIR VALUE We measure many financial assets and liabilities at **fair value** (called *current market value* originally in *SFAC 5*). Also, we use fair values when determining whether the value of nonfinancial assets like property, plant, and equipment and intangible assets has been impaired. Given the complexity and growing importance of this measurement attribute, we discuss it in some detail.

Fair value is defined as the price that would be received to sell assets or paid to transfer a liability in an orderly transaction between market participants at the measurement date. A key aspect of this definition is its focus on the perspective of *market participants.* For instance, if a company buys a competitor's patent, not intending to use it but merely to keep the competitor from using it, the company still will have to assign a value to the asset because a market participant would find value in using the patent.

The FASB has provided a framework for measuring fair value whenever fair value is called for in applying generally accepted accounting principles.[37] The IASB uses the same

[36]"Using Cash Flow Information and Present Value in Accounting Measurements," *Statement of Financial Accounting Concepts No. 7* (Norwalk, Conn.: FASB, 2000).

[37]FASB ASC 820: Fair Value Measurements and Disclosures (previously "Fair Value Measurements," *Statement of Financial Accounting Standards No. 157* (Norwalk, Conn.: FASB, 2006)).

framework.[38] In the framework, three types of valuation techniques can be used to measure fair value. *Market approaches* base valuation on market information. For example, the value of a share of a company's stock that's not traded actively could be estimated by multiplying the earnings of that company by the P/E (price of shares/earnings) multiples of similar companies. *Income approaches* estimate fair value by first estimating future amounts (for example, earnings or cash flows) and then mathematically converting those amounts to a single present value. You will see how to apply such techniques in Chapter 5 when we discuss time value of money concepts. *Cost approaches* determine value by estimating the amount that would be required to buy or construct an asset of similar quality and condition. A firm can use one or more of these valuation approaches, depending on availability of information, and should try to use them consistently unless changes in circumstances require a change in approach.

Fair value can be measured using:
1. Market approaches.
2. Income approaches.
3. Cost approaches.

To increase consistency and comparability in applying this definition, the framework provides a "hierarchy" that prioritizes the inputs companies should use when determining fair value. The priority is based on three broad preference levels. The higher the level (Level 1 is the highest), the more preferable the input. The framework encourages companies to strive to obtain the highest level input available for each situation. Illustration 1–14 describes the type of inputs and provides an example for each level.

Companies also must provide detailed disclosures about their use of fair value measurements. The disclosures include a description of the inputs used to measure fair value. For recurring fair value measurements that rely on significant *unobservable* inputs (within Level 3 of the fair value hierarchy), companies should disclose the effect of the measurements on earnings (or changes in net assets) for the period.

Illustration 1–14

Fair Value Hierarchy

Fair Value Hierarchy

Level	Inputs	Example
1 Most Desirable	Quoted market prices in active markets for identical assets or liabilities.	In Chapter 12 you will learn that certain investments in marketable securities are reported at their *fair values*. Fair value in this case would be measured using the quoted market price from the NYSE, NASDAQ, or other exchange on which the security is traded.
2	Inputs other than quoted prices that are *observable* for the asset or liability. These inputs include quoted prices for *similar* assets or liabilities in active or inactive markets and inputs that are derived principally from or corroborated by observable related market data.	In Chapter 10 we discuss how companies sometimes acquire assets with consideration other than cash. In any noncash transaction, each element of the transaction is recorded at its *fair value*. If one of the assets in the exchange is a building, for instance, then quoted market prices for similar buildings recently sold could be used to value the building or, if there were no similar buildings recently exchanged from which to obtain a comparable market price, valuation could be based on the price per square foot derived from observable market data.
3 Least Desirable	*Unobservable* inputs that reflect the entity's own assumptions about the assumptions market participants would use in pricing the asset or liability developed based on the best information available in the circumstances.	Asset retirement obligations (AROs), discussed in Chapter 10, are measured at *fair value*. Neither Level 1 nor Level 2 inputs would be possible in most ARO valuation situations. Fair value would be estimated using Level 3 inputs to include the present value of expected cash flows estimated using the entity's own data if there is no information indicating that market participants would use different assumptions.

[38]"Fair Value Measurement," *International Financial Reporting Standard No. 13* (London, UK: IASCF, 2011).

The use of the fair value measurement attribute is increasing, both under U.S. GAAP and IFRS. This trend, though, is controversial. Proponents of fair value cite its relevance and are convinced that historical cost information may not be useful for many types of decisions. Opponents of fair value counter that estimates of fair value may lack representational faithfulness, particularly when based on inputs from Level 3 in the fair value hierarchy, and that managers might be tempted to exploit the unverifiability of such inputs to manipulate earnings. They argue that accounting should emphasize verifiability by recognizing only those gains and other increases in fair value that actually have been realized in transactions or are virtually certain to exist.

The *fair value option* lets companies choose whether to value some financial assets and liabilities at fair value.

FAIR VALUE OPTION Usually the measurement attribute we use for a particular financial statement item is not subject to choice. However, GAAP allows a **fair value option** in some circumstances which permits companies to choose whether to report specified *financial* assets and liabilities at fair value.[39] For example, in Chapter 14 you will learn that a company normally would report bonds payable at historical cost (adjusted for unamortized premium or discount), but the fair value option allows that company to choose instead to report the bonds payable at fair value. If a company chooses the fair value option, future changes in fair value are reported as gains and losses in the income statement.

Why allow the fair value option for financial assets and liabilities and not for, say, buildings or land? Financial assets and liabilities are cash and other assets and liabilities that convert directly into known amounts of cash. These include investments in stocks and bonds of other entities, notes receivable and payable, bonds payable, and derivative securities.[40] Some of these financial assets and liabilities currently are *required* under GAAP to be reported at fair value, and others are not, leading to some potential inconsistencies in how similar or related items are treated. The fair value option provides companies a way to reduce volatility in reported earnings without having to comply with complex hedge accounting standards. It also helps in the convergence with international accounting standards we discussed earlier in the chapter as the IASB also has adopted a fair value option for financial instruments.

Disclosure

Remember, the purpose of accounting is to provide information that is useful to decision makers. So, naturally, if there is accounting information not included in the primary financial statements that would benefit users, that information should be provided too. The **full-disclosure principle** means that the financial reports should include any information that could affect the decisions made by external users. Of course, the benefits of that information should exceed the costs of providing the information. Such information is disclosed in a variety of ways, including:

The *full-disclosure principle* requires that any information useful to decision makers be provided in the financial statements, subject to the cost effectiveness constraint.

1. **Parenthetical comments** or **modifying comments** placed on the face of the financial statements.
2. **Disclosure notes** conveying additional insights about company operations, accounting principles, contractual agreements, and pending litigation.
3. **Supplemental schedules and tables** that report more detailed information than is shown in the primary financial statements.

We discuss and illustrate disclosure requirements as they relate to specific financial statement elements in later chapters as those elements are discussed.

Illustration 1–15 provides an overview of key recognition, measurement and disclosure concepts.

[39] FASB ASC 825–10–25–1: Financial Instruments—Overall—Recognition—Fair Value Option.

[40] The fair value option does not apply to certain specified financial instruments, including pension obligations and assets or liabilities arising from leases.

Illustration 1–15
Summary of Recognition, Measurement, and Disclosure Concepts

Concept	Description
Recognition	General criteria: 1. Meets the definition of an element 2. Has a measurement attribute 3. Is relevant 4. Is reliable (representationally faithful) Examples of recognition timing: 1. Revenues 2. Expenses
Measurement	Mixed attribute model in which the attribute used to measure an item is chosen to maximize relevance and representational faithfulness. These attributes include: 1. Historical cost 2. Net realizable value 3. Current cost 4. Present (or discounted) value of future cash flows 5. Fair value
Disclosure	Financial reports should include all information that could affect the decisions made by external users. Examples of disclosures: 1. Parenthetical amounts 2. Notes to the financial statements 3. Supplemental schedules and tables

Where We're Headed

"Disclosure overload" is a frequent complaint by companies and investors alike. The notes to the financial statements can be very useful, but they are costly for companies to prepare and difficult for many users to sift through and understand. In response to that concern, the FASB has been developing a framework intended to make disclosures more effective and less redundant. In August of 2018 the FASB issued an addition to Concepts Statement No. 8, titled *Chapter 8: Notes to Financial Statements*, which suggests a series of questions that the FASB and its staff should consider when determining what notes should be required by new standards.[41] A separate part of the project will develop further guidance to help companies apply judgment when meeting disclosure requirements.

Evolving GAAP

● LO1–10

U.S. and international GAAP have been evolving over time from an emphasis on revenues and expenses to an emphasis on assets and liabilities. Of course, you know from introductory accounting that the balance sheet and income statement are intertwined and must reconcile with each other. For example, the revenues reported in the income statement depict inflows of assets whose balances at a particular point in time are reported in the balance sheet. But which comes first, identifying revenues and expenses, or identifying assets and liabilities? That emphasis can affect accounting standards in important ways. To help you understand the changes taking place, we start by discussing the revenue/expense approach and then discuss the asset/liability approach.

[41] "Conceptual Framework for Financial Reporting—Chapter 8, *Notes to Financial Statements*," *Statement of Financial Accounting Concepts No. 8* (Norwalk, Conn.: FASB, 2018).

With the *revenue/expense approach*, recognition and measurement of revenues and expenses are emphasized.

Under the **revenue/expense approach**, we emphasize principles for recognizing revenues and expenses, with some assets and liabilities recognized as necessary to make the balance sheet reconcile with the income statement. For example, when accounting for sales revenue our focus would be on whether a good or service has been delivered, and if we determine that to be the case, we would record an asset (usually cash or accounts receivable) that is associated with that revenue.[42] We also would identify the expenses associated with delivering those goods and services, and then would adjust assets and liabilities accordingly.

With the *asset/liability approach*, recognition and measurement of assets and liabilities drives revenue and expense recognition.

Under the **asset/liability approach**, on the other hand, we first recognize and measure the assets and liabilities that exist at a balance sheet date and, secondly, recognize and measure the revenues, expenses, gains and losses needed to account for the changes in these assets and liabilities from the previous measurement date. Proponents of this approach point out that, since revenues and expenses are defined in terms of inflows and outflows of assets and liabilities, the fundamental concepts underlying accounting are assets and liabilities. Therefore, we should try to recognize and measure assets and liabilities appropriately, and as a result will also capture their inflows and outflows in a manner that provides relevant and representationally faithful information about revenues and expenses.

For example, when accounting for a sales transaction, our focus would be on whether a potential accounts receivable meets the definition of an asset (a probable future economic benefit). We would consider such factors as whether the receivable is supported by an enforceable contract and whether the seller has performed its obligations enough to be able to expect receipt of cash flows. The key would be determining if the seller has an asset and then recognizing whatever amount of revenue is implied by the inflow of that asset. Also, we would not attempt to match expenses to revenues. Rather, we would determine those net assets that had decreased as part of operations during the period, and recognize those decreases as expenses.

In subsequent chapters you will see that recent standards involving accounting for revenue, investments, and income taxes follow this asset/liability approach. These changes are controversial. It may seem like it shouldn't matter whether standard setters use the revenue/expense or asset/liability approach, given that both approaches affect both the income statement and balance sheet, and it is true that these approaches often will result in the same accounting outcomes. However, the particular approach used by a standard setter can affect recognition and measurement in important ways. In particular, the asset/liability approach encourages us to focus on accurately measuring assets and liabilities. It perhaps is not surprising, then, that a focus on assets and liabilities has led standard setters to lean more and more toward fair value measurement. The future changes to the conceptual framework discussed in the following Where We're Headed box are likely to continue this emphasis on the asset/liability approach.

Where We're Headed

Since 2010, the FASB and IASB worked separately on their conceptual frameworks. The FASB has worked on relatively narrow projects, such as the definition of materiality and a more general disclosure framework. The IASB, in contrast, has moved forward with a more comprehensive overhaul of its conceptual framework. The IASB's 2015 exposure draft of its concepts statement addressed a range of issues, including the objective and qualitative characteristics of financial reporting, the definitions of key financial-statement elements, when to recognize and de-recognize assets and liabilities, when to use various measurement approaches, the distinction between net income and other comprehensive income, and what constitutes a reporting entity. The IASB anticipates issuing a concepts statement sometime in 2018.

[42]Some assets and liabilities aren't related to revenue or expense. For example, issuance of shares of stock increases cash as well as shareholders' equity. The treatment of these sorts of transactions is not affected by whether GAAP emphasizes revenues and expenses or assets and liabilities.

Financial Reporting Case Solution

©Rawpixel.com/Shutterstock

1. **What should you tell your friend about the presence of accounting standards in the United States and the rest of the world? Who has the authority for standard setting? Who has the responsibility?** *(p. 8)* In the United States we have a set of standards known as generally accepted accounting principles (GAAP). GAAP is a dynamic set of both broad and specific guidelines that companies should follow when measuring and reporting the information in their financial statements and related notes. The Securities and Exchange Commission has the authority to set accounting standards for companies whose securities are publicly traded but it relies on the private sector to accomplish that task. At present, the Financial Accounting Standards Board is the private sector body responsible for standard setting.
2. **What is the economic and political environment in which standard setting occurs?** *(p. 12)* The setting of accounting and reporting standards often has been characterized as a *political process.* Standards, particularly changes in standards, can have significant differential effects on companies, investors and creditors, and other interest groups. A change in an accounting standard or the introduction of a new standard can result in a substantial redistribution of wealth within our economy. The FASB's due process is designed to obtain information from all interested parties to help determine the appropriate accounting approach, but standards are supposed to be neutral with respect to the interests of various parties. Nonetheless, both the FASB and IASB sometimes come under political pressure that sways the results of the standard-setting process.
3. **What is the relationship among management, auditors, investors, and creditors that tends to preclude the "What would you like it to be?" attitude?** *(p. 14)* It is the responsibility of management to apply accounting standards when communicating with investors and creditors through financial statements. Auditors serve as independent intermediaries to help ensure that the management-prepared statements are presented fairly in accordance with GAAP. In providing this assurance, the auditor precludes the "What would you like it to be?" attitude.
4. **In general, what is the conceptual framework that underlies accounting principles, and how does it encourage high-quality financial reporting?** *(p. 18)* The conceptual framework is a coherent system of interrelated objectives and fundamentals that improves financial reporting by encouraging consistent standards and by prescribing the nature, function, and limits of financial accounting and reporting. The fundamentals are the underlying concepts of accounting, concepts that guide the selection of events to be accounted for, the measurement of those events, and the means of summarizing and communicating them to interested parties. ●

The Bottom Line

- **LO1–1** Financial accounting is concerned with providing relevant financial information to various external users. However, the primary focus is on the financial information provided by profit-oriented companies to their present and potential investors and creditors. *(p. 3)*
- **LO1–2** Cash-basis accounting provides a measure of periodic performance called *net operating cash flow,* which is the difference between cash receipts and cash disbursements from transactions related to providing goods and services to customers. Accrual accounting provides a measure of performance called *net income,* which is the difference between revenues and expenses. Periodic net income is considered a better indicator of future operating cash flows than is current net operating cash flows. *(p. 7)*
- **LO1–3** Generally accepted accounting principles (GAAP) comprise a dynamic set of both broad and specific guidelines that companies follow when measuring and reporting the information in their financial statements and related notes. The Securities and Exchange Commission (SEC) has the authority to set accounting standards in the United States. However, the SEC has always delegated the task to a private sector body, at this time the Financial Accounting Standards Board (FASB). The International Accounting

Standards Board (IASB) sets global accounting standards and works with national accounting standard setters to achieve convergence in accounting standards around the world. *(p. 8)*

- **LO1–4** Accounting standards can have significant differential effects on companies, investors, creditors, and other interest groups. Various interested parties sometimes lobby standard setters for their preferred outcomes. For this reason, the setting of accounting standards often has been characterized as a political process. *(p. 12)*
- **LO1–5** Factors encouraging high-quality financial reporting include conceptually based financial accounting standards, external auditors, financial reporting reforms (such as the Sarbanes-Oxley Act), ethical management, and professional accounting organizations that prescribe ethical conduct and license practitioners. *(p. 14)*
- **LO1–6** The FASB's conceptual framework is a set of cohesive objectives and fundamental concepts on which financial accounting and reporting standards can be based. *(p. 18)*
- **LO1–7** The objective of financial reporting is to provide useful financial information to capital providers. The primary decision-specific qualities that make financial information useful are relevance and faithful representation. To be relevant, information must possess predictive value and/or confirmatory value, and all material information should be included. Completeness, neutrality, and freedom from error enhance faithful representation. The 10 elements of financial statements are assets, liabilities, equity, investments by owners, distributions to owners, revenues, expenses, gains, losses, and comprehensive income. *(p. 20)*
- **LO1–8** The four basic assumptions underlying GAAP are (1) the economic entity assumption, (2) the going concern assumption, (3) the periodicity assumption, and (4) the monetary unit assumption. *(p. 23)*
- **LO1–9** Recognition determines whether an item is reflected in the financial statements, and measurement determines the amount of the item. Measurement involves choice of a monetary unit and choice of a measurement attribute. In the United States, the monetary unit is the dollar. Various measurement attributes are used in GAAP, including historical cost, net realizable value, current cost, present value, and fair value. *(p. 25)*
- **LO1–10** A revenue/expense approach to financial reporting emphasizes recognition and measurement of revenues and expenses, while an asset/liability approach emphasizes recognition and measurement of assets and liabilities. *(p. 31)*
- **LO1–11** IFRS and U.S. GAAP are similar in the organizations that support standard setting and in the presence of ongoing political pressures on the standard-setting process. U.S. GAAP and IFRS also have similar conceptual frameworks, although the role of the conceptual framework in IFRS is to provide guidance to preparers as well as to standard setters, while the role of the conceptual framework in U.S. GAAP is more to provide guidance to standard setters. *(pp. 11, 14* and *19)* ●

Questions For Review of Key Topics

Q 1–1 What is the function and primary focus of financial accounting?

Q 1–2 What is meant by the phrase *efficient allocation of resources?* What mechanism fosters the efficient allocation of resources in the United States?

Q 1–3 Identify two important variables to be considered when making an investment decision.

Q 1–4 What must a company do in the long run to be able to provide a return to investors and creditors?

Q 1–5 What is the primary objective of financial accounting?

Q 1–6 Define net operating cash flows. Briefly explain why periodic net operating cash flows may not be a good indicator of future operating cash flows.

Q 1–7 What is meant by GAAP? Why should all companies follow GAAP in reporting to external users?

Q 1–8 Explain the roles of the SEC and the FASB in the setting of accounting standards.

Q 1–9 Explain the role of the auditor in the financial reporting process.

Q 1–10 List three key provisions of the Sarbanes-Oxley Act of 2002. Order your list from most important to least important in terms of the likely long-term impact on the accounting profession and financial reporting.

Q 1–11 Explain what is meant by *adverse economic consequences* of new or changed accounting standards.

Q 1–12 Why does the FASB undertake a series of elaborate information-gathering steps before issuing a substantive accounting standard?

Q 1–13 What is the purpose of the FASB's conceptual framework?

Q 1–14 Discuss the terms *relevance* and *faithful representation* as they relate to financial accounting information.

Q 1–15 What are the components of relevant information? What are the components of faithful representation?

Q 1–16 Explain what is meant by: The benefits of accounting information must exceed the costs.

Q 1–17 What is meant by the term *materiality* in financial reporting?

Q 1–18 Briefly define the financial accounting elements: (1) assets, (2) liabilities, (3) equity, (4) investments by owners, (5) distributions to owners, (6) revenues, (7) expenses, (8) gains, (9) losses, and (10) comprehensive income.

Q 1–19 What are the four basic assumptions underlying GAAP?

Q 1–20 What is the going concern assumption?

Q 1–21 Explain the periodicity assumption.

Q 1–22 What are four key accounting practices that often are referred to as principles in current GAAP?

Q 1–23 What are two advantages to basing the valuation of assets and liabilities on their historical cost?

Q 1–24 Describe how revenue recognition relates to transferring goods or services.

Q 1–25 What are the four different approaches to implementing expense recognition? Give an example of an expense that is recognized under each approach.

Q 1–26 In addition to the financial statement elements arrayed in the basic financial statements, what are some other ways to disclose financial information to external users?

Q 1–27 Briefly describe the inputs that companies should use when determining fair value. Organize your answer according to preference levels, from highest to lowest priority.

Q 1–28 What measurement attributes are commonly used in financial reporting?

Q 1–29 Distinguish between the revenue/expense and the asset/liability approaches to setting financial reporting standards.

IFRS **Q 1–30** What are the functions of the conceptual framework under IFRS?

IFRS **Q 1–31** What is the standard-setting body responsible for determining IFRS? How does it obtain its funding?

IFRS **Q 1–32** In its Final Staff Report (issued in 2012), what type of convergence between U.S. GAAP and IFRS did the SEC staff argue was not feasible? What reasons did the SEC staff give for that conclusion?

Brief Exercises

connect

BE 1–1
Accrual accounting
● LO1–2

Cash flows during the first year of operations for the Harman-Kardon Consulting Company were as follows: Cash collected from customers, $340,000; Cash paid for rent, $40,000; Cash paid to employees for services rendered during the year, $120,000; Cash paid for utilities, $50,000.

In addition, you determine that customers owed the company $60,000 at the end of the year and no bad debts were anticipated. Also, the company owed the gas and electric company $2,000 at year-end, and the rent payment was for a two-year period. Calculate accrual net income for the year.

BE 1–2
Financial statement elements
● LO1–7

For each of the following items, identify the appropriate financial statement element or elements: (1) probable future sacrifices of economic benefits; (2) probable future economic benefits owned by the company; (3) inflows of assets from ongoing, major activities; (4) decrease in equity from peripheral or incidental transactions.

BE 1–3
Basic assumptions and principles
● LO1–7 through LO1–9

Listed below are several statements that relate to financial accounting and reporting. Identify the accounting concept that applies to each statement.

1. **SiriusXM Radio Inc.** files its annual and quarterly financial statements with the SEC.
2. The president of **Applebee's International, Inc.**, travels on the corporate jet for business purposes only and does not use the jet for personal use.
3. Jackson Manufacturing does not recognize revenue for unshipped merchandise even though the merchandise has been manufactured according to customer specifications.
4. Lady Jane Cosmetics depreciates the cost of equipment over their useful lives.

BE 1–4
Basic assumptions and principles
● LO1–7 through LO1–9

Identify the accounting concept that was violated in each of the following situations.

1. Astro Turf Company recognizes an expense, cost of goods sold, in the period the product is manufactured.
2. McCloud Drug Company owns a patent that it purchased three years ago for $2 million. The controller recently revalued the patent to its approximate market value of $8 million.
3. Philips Company pays the monthly mortgage on the home of its president Larry Crosswhite and charges the expenditure to miscellaneous expense.

BE 1–5
Basic assumptions and principles
● LO1–7 through LO1–9

For each of the following situations, (1) indicate whether you agree or disagree with the financial reporting practice employed and (2) state the accounting concept that is applied (if you agree) or violated (if you disagree).

1. Winderl Corporation did not disclose that it was the defendant in a material lawsuit because the trial was still in progress.
2. Alliant Semiconductor Corporation files quarterly and annual financial statements with the SEC.
3. Reliant Pharmaceutical paid rent on its office building for the next two years and charged the entire expenditure to rent expense.
4. Rockville Engineering records revenue only after products have been shipped, even though customers pay Rockville 50% of the sales price in advance.

BE 1–6
IFRS
● LO1–11
IFRS

Indicate the organization related to IFRS that performs each of the following functions:

1. Obtains funding for the IFRS standard-setting process.
2. Determines IFRS.
3. Oversees the IFRS Foundation.
4. Provides input about the standard-setting agenda.
5. Provides implementation guidance about relatively narrow issues.

Exercises

connect

E 1–1
Accrual accounting
● LO1–2

Listed below are several transactions that took place during the first two years of operations for the law firm of Pete, Pete, and Roy.

	Year 1	Year 2
Amounts billed to clients for services rendered	$170,000	$220,000
Cash collected from clients	160,000	190,000
Cash disbursements		
Salaries paid to employees for services rendered during the year	90,000	100,000
Utilities	30,000	40,000
Purchase of insurance policy	60,000	–0–

In addition, you learn that the firm incurred utility costs of $35,000 in year 1, that there were no liabilities at the end of year 2, no anticipated bad debts on receivables, and that the insurance policy covers a three-year period.

Required:

1. Calculate the net operating cash flow for years 1 and 2.
2. Prepare an income statement for each year similar to Illustration 1–4 according to the accrual accounting model.
3. Determine the amount of receivables from clients that the firm would show in its year 1 and year 2 balance sheets prepared according to the accrual accounting model.

E 1–2
Accrual accounting
● LO1–2

Listed below are several transactions that took place during the second and third years of operations for the RPG Company.

	Year 2	Year 3
Amounts billed to customers for services rendered	$350,000	$450,000
Cash collected from credit customers	260,000	400,000
Cash disbursements:		
Payment of rent	80,000	–0–
Salaries paid to employees for services rendered during the year	140,000	160,000
Utilities	30,000	40,000
Advertising	15,000	35,000

In addition, you learn that the company incurred advertising costs of $25,000 in year 2, owed the advertising agency $5,000 at the end of year 1, and there were no liabilities at the end of year 3. Also, there were no anticipated bad debts on receivables, and the rent payment was for a two-year period, year 2 and year 3.

Required:

1. Calculate accrual net income for both years.
2. Determine the amount due the advertising agency that would be shown as a liability on RPG's balance sheet at the end of year 2.

E 1–3
FASB codification research
● LO1–3

Access the *FASB Accounting Standards Codification* at the FASB website (**www.fasb.org**).

Required:

1. Identify the Codification topic number that provides guidance on fair value measurements.
2. What is the specific seven-digit Codification citation (XXX-XX-XX) that lists the disclosures required in the notes to the financial statements for each major category of assets and liabilities measured at fair value?

E 1–4
FASB codification research
● LO1–3

Access the *FASB Accounting Standards Codification* at the FASB website (**www.fasb.org**). Determine the specific citation for each of the following items:

1. The topic number for business combinations
2. The topic number for related party disclosures
3. The specific seven-digit Codification citation (XXX-XX-XX) for the initial measurement of internal-use software
4. The specific seven-digit Codification citation (XXX-XX-XX) for the subsequent measurement of asset retirement obligations
5. The specific seven-digit Codification citation (XXX-XX-XX) for the recognition of stock compensation

E 1–5
Participants in establishing GAAP
● LO1–3

Three groups that participate in the process of establishing GAAP are users, preparers, and auditors. These groups are represented by various organizations. For each organization listed below, indicate which of these groups it primarily represents.

1. Securities and Exchange Commission
2. Financial Executives International
3. American Institute of Certified Public Accountants
4. Institute of Management Accountants
5. Association of Investment Management and Research

E 1–6
Financial statement elements
● LO1–7

For each of the items listed below, identify the appropriate financial statement element or elements.

1. Obligation to transfer cash or other resources as a result of a past transaction
2. Dividends paid by a corporation to its shareholders
3. Inflow of an asset from providing a good or service
4. The financial position of a company
5. Increase in equity during a period from nonowner transactions
6. Increase in equity from peripheral or incidental transaction
7. Sale of an asset used in the operations of a business for less than the asset's book value
8. The owners' residual interest in the assets of a company
9. An item owned by the company representing probable future benefits
10. Revenues plus gains less expenses and losses
11. An owner's contribution of cash to a corporation in exchange for ownership shares of stock
12. Outflow of an asset related to the production of revenue

E 1–7
Concepts; terminology; conceptual framework
● LO1–7

Listed below are several terms and phrases associated with the FASB's conceptual framework. Pair each item from List A (by letter) with the item from List B that is most appropriately associated with it.

	List A	List B
______	1. Predictive value	a. Decreases in equity resulting from transfers to owners.
______	2. Relevance	b. Requires consideration of the costs and value of information.
______	3. Timeliness	c. Important for making interfirm comparisons.
______	4. Distribution to owners	d. Applying the same accounting practices over time.
______	5. Confirmatory value	e. Users understand the information in the context of the decision being made.
______	6. Understandability	f. Agreement between a measure and the phenomenon it purports to represent.
______	7. Gain	g. Information is available prior to the decision.
______	8. Faithful representation	h. Pertinent to the decision at hand.
______	9. Comprehensive income	i. Implies consensus among different measurers.
______	10. Materiality	j. Information confirms expectations.
______	11. Comparability	k. The change in equity from nonowner transactions.
______	12. Neutrality	l. The process of admitting information into financial statements.
______	13. Recognition	m. The absence of bias.
______	14. Consistency	n. Increases in equity from peripheral or incidental transactions of an entity.
______	15. Cost effectiveness	o. Information is useful in predicting the future.
______	16. Verifiability	p. Concerns the relative size of an item and its effect on decisions.

E 1–8
Qualitative characteristics
● LO1–7

The conceptual framework indicates the desired fundamental and enhancing qualitative characteristics of accounting information. Several constraints impede achieving these desired characteristics. Answer each of the following questions related to these characteristics and constraints.

1. Which component would allow a large company to record the purchase of a $120 printer as an expense rather than capitalizing the printer as an asset?
2. Donald Kirk, former chairman of the FASB, once noted that ". . . there must be public confidence that the standard-setting system is credible, that selection of board members is based on merit and not the influence of special interests . . ." Which characteristic is implicit in Mr. Kirk's statement?
3. Allied Appliances, Inc., changed its revenue recognition policies. Which characteristic is jeopardized by this change?
4. National Bancorp, a publicly traded company, files quarterly and annual financial statements with the SEC. Which characteristic is relevant to the timing of these periodic filings?
5. In general, relevant information possesses which qualities?
6. When there is agreement between a measure or description and the phenomenon it purports to represent, information possesses which characteristic?
7. Jeff Brown is evaluating two companies for future investment potential. Jeff's task is made easier because both companies use the same accounting methods when preparing their financial statements. Which characteristic does the information Jeff will be using possess?
8. A company should disclose information only if the perceived benefits of the disclosure exceed the costs of providing the information. Which constraint does this statement describe?

E 1–9
Basic assumptions, principles, and constraints
● LO1–7 through LO1–9

Listed below are several terms and phrases associated with the accounting concepts. Pair each item from List A (by letter) with the item from List B that is most appropriately associated with it.

List A	List B
______ 1. Expense recognition	a. The enterprise is separate from its owners and other entities.
______ 2. Periodicity assumption	b. A common denominator is the dollar.
______ 3. Historical cost principle	c. The entity will continue indefinitely.
______ 4. Materiality	d. Record expenses in the period the related revenue is recognized.
______ 5. Revenue recognition	e. The original transaction value upon acquisition.
______ 6. Going concern assumption	f. All information that could affect decisions should be reported.
______ 7. Monetary unit assumption	g. The life of an enterprise can be divided into artificial time periods.
______ 8. Economic entity assumption	h. Criteria usually satisfied for products at point of sale.
______ 9. Full-disclosure principle	i. Concerns the relative size of an item and its effect on decisions.

E 1–10
Basic assumptions and principles
● LO1–7 through LO1–9

Listed below are several statements that relate to financial accounting and reporting. Identify the accounting concept that applies to each statement.

1. Jim Marley is the sole owner of Marley's Appliances. Jim borrowed $100,000 to buy a new home to be used as his personal residence. This liability was not recorded in the records of Marley's Appliances.
2. **Apple Inc.** distributes an annual report to its shareholders.
3. **Hewlett-Packard Corporation** depreciates machinery and equipment over their useful lives.
4. Crosby Company lists land on its balance sheet at $120,000, its original purchase price, even though the land has a current fair value of $200,000.
5. **Honeywell International Inc.** records revenue when products are delivered to customers, even though the cash has not yet been received.
6. Liquidation values are not normally reported in financial statements even though many companies do go out of business.
7. **IBM Corporation,** a multibillion dollar company, purchased some small tools at a cost of $800. Even though the tools will be used for a number of years, the company recorded the purchase as an expense.

E 1–11
Basic assumptions and principles
● LO1–8, LO1–9

Identify the accounting concept that was violated in each of the following situations.

1. Pastel Paint Company purchased land two years ago at a price of $250,000. Because the value of the land has appreciated to $400,000, the company has valued the land at $400,000 in its most recent balance sheet.
2. Atwell Corporation has not prepared financial statements for external users for over three years.
3. The Klingon Company sells farm machinery. Revenue from a large order of machinery from a new buyer was recorded the day the order was received.
4. Don Smith is the sole owner of a company called Hardware City. The company recently paid a $150 utility bill for Smith's personal residence and recorded a $150 expense.

5. Golden Book Company purchased a large printing machine for $1,000,000 (a material amount) and recorded the purchase as an expense.
6. Ace Appliance Company is involved in a major lawsuit involving injuries sustained by some of its employees in the manufacturing plant. The company is being sued for $2,000,000, a material amount, and is not insured. The suit was not disclosed in the most recent financial statements because no settlement had been reached.

E 1–12
Basic assumptions and principles
● LO1–7 through LO1–9

For each of the following situations, indicate whether you agree or disagree with the financial reporting practice employed and state the accounting concept that is applied (if you agree) or violated (if you disagree).

1. Wagner Corporation adjusted the valuation of all assets and liabilities to reflect changes in the purchasing power of the dollar.
2. Spooner Oil Company changed its method of accounting for oil and gas exploration costs from successful efforts to full cost. No mention of the change was included in the financial statements. The change had a material effect on Spooner's financial statements.
3. Cypress Manufacturing Company purchased machinery having a five-year life. The cost of the machinery is being expensed over the life of the machinery.
4. Rudeen Corporation purchased equipment for $180,000 at a liquidation sale of a competitor. Because the equipment was worth $230,000, Rudeen valued the equipment in its subsequent balance sheet at $230,000.
5. Davis Bicycle Company received a large order for the sale of 1,000 bicycles at $100 each. The customer paid Davis the entire amount of $100,000 on March 15. However, Davis did not record any revenue until April 17, the date the bicycles were delivered to the customer.
6. Gigantic Corporation purchased two small calculators at a cost of $32.00. The cost of the calculators was expensed even though they had a three-year estimated useful life.
7. Esquire Company provides financial statements to external users every three years.

E 1–13
Basic assumptions and principles
● LO1–7 through LO1–9

For each of the following situations, state whether you agree or disagree with the financial reporting practice employed, and briefly explain the reason for your answer.

1. The controller of the Dumars Corporation increased the carrying value of land from its original cost of $2 million to its recently appraised value of $3.5 million.
2. The president of Vosburgh Industries asked the company controller to charge miscellaneous expense for the purchase of an automobile to be used solely for personal use.
3. At the end of its 2020 fiscal year, Dower, Inc., received an order from a customer for $45,350. The merchandise will ship early in 2021. Because the sale was made to a long-time customer, the controller recorded the sale in 2020.
4. At the beginning of its 2020 fiscal year, Rossi Imports paid $48,000 for a two-year lease on warehouse space. Rossi recorded the expenditure as an asset to be expensed equally over the two-year period of the lease.
5. The Reliable Tire Company included a note in its financial statements that described a pending lawsuit against the company.
6. The Hughes Corporation, a company whose securities are publicly traded, prepares monthly, quarterly, and annual financial statements for internal use but disseminates to external users only the annual financial statements.

E 1–14
Basic assumptions and principles
● LO1–7 through LO1–9

Listed below are accounting concepts discussed in this chapter.

a. Economic entity assumption
b. Going concern assumption
c. Periodicity assumption
d. Monetary unit assumption
e. Historical cost principle
f. Conservatism
g. Expense recognition
h. Full-disclosure principle
i. Cost effectiveness
j. Materiality

Identify by letter the accounting concept that relates to each statement or phrase below.

______ 1. Inflation causes a violation of this assumption.
______ 2. Information that could affect decision making should be reported.
______ 3. Recognizing expenses in the period they were incurred to produce revenue.

_____ 4. The basis for measurement of many assets and liabilities.
_____ 5. Relates to the qualitative characteristic of timeliness.
_____ 6. All economic events can be identified with a particular entity.
_____ 7. The benefits of providing accounting information should exceed the cost of doing so.
_____ 8. A consequence is that GAAP need not be followed in all situations.
_____ 9. Not a qualitative characteristic, but a practical justification for some accounting choices.
_____ 10. Assumes the entity will continue indefinitely.

E 1–15
Multiple choice; concept statements, basic assumptions, principles
● LO1–6 through LO1–9

Determine the response that best completes the following statements or questions.

1. The primary objective of financial reporting is to provide information
   a. About a firm's management team.
   b. Useful to capital providers.
   c. Concerning the changes in financial position resulting from the income-producing efforts of the entity.
   d. About a firm's financing and investing activities.
2. *Statements of Financial Accounting Concepts* issued by the FASB
   a. Represent GAAP.
   b. Have been superseded by *SFAS*s.
   c. Are subject to approval of the SEC.
   d. Identify the conceptual framework within which accounting standards are developed.
3. In general, revenue is recognized when
   a. The sales price has been collected.
   b. A purchase order has been received.
   c. A good or service has been delivered to a customer.
   d. A contract has been signed.
4. In depreciating the cost of an asset, accountants are most concerned with
   a. Conservatism.
   b. Recognizing revenue in the appropriate period.
   c. Full disclosure.
   d. Recognizing expense in the appropriate period.
5. The primary objective of expense recognition is to
   a. Provide full disclosure.
   b. Record expenses in the period that related revenues are recognized.
   c. Provide timely information to decision makers.
   d. Promote comparability between financial statements of different periods.
6. The separate entity assumption states that, in the absence of contrary evidence, all entities will survive indefinitely.
   a. True
   b. False

Decision Makers' Perspective

©ImageGap/Getty Images

Apply your critical-thinking ability to the knowledge you've gained. These cases will provide you an opportunity to develop your research, analysis, judgment, and communication skills. You will also work with other students, integrate what you've learned, apply it in real-world situations, and consider its global and ethical ramifications. This practice will broaden your knowledge and further develop your decision-making abilities.

Judgment Case 1–1
The development of accounting standards
● LO1–3

In 1934, Congress created the Securities and Exchange Commission (SEC).

Required:

1. Does the SEC have more authority or less authority than the FASB with respect to standard setting? Explain the relationship between the SEC and the FASB.
2. Can you think of any reasons why the SEC relies on private sector bodies to set accounting standards, rather than undertaking the task itself?

Research Case 1–2
Accessing SEC information through the Internet
● LO1–3

The purpose of this case is to introduce you to the information available on the website of the Securities and Exchange Commission (SEC) and its EDGAR database.

Required:
Access the SEC home page on the Internet. The web address is **www.sec.gov**. Answer the following questions:
1. Choose the subaddress "About the SEC." What are the two basic objectives of the 1933 Securities Act?
2. Return to the SEC home page and access EDGAR. Describe the contents of the database.

Research Case 1–3
Accessing FASB information through the Internet
● LO1–4

The purpose of this case is to introduce you to the information available on the website of the Financial Accounting Standards Board (FASB).

Required:
Access the FASB home page on the Internet. The web address is **www.fasb.org**. Answer the following questions:
1. The FASB has how many board members?
2. How many board members of the FASB have a background that is primarily drawn from accounting education?
3. Describe the mission of the FASB.
4. How are topics added to the FASB's technical agenda?

Research Case 1–4
Accessing IASB information through the Internet
● LO1–3

The purpose of this case is to introduce you to the information available on the website of the International Accounting Standards Board (IASB).

Required:
Access the IASB home page on the Internet. The web address is **www.iasb.org**. Answer the following questions.
1. The IASB has how many board members?
2. In what city is the IASB located?
3. Describe the mission of the IASB.

Communication Case 1–5
Relevance and representational faithfulness
● LO1–7

Some theorists contend that companies that create pollution should report the social cost of that pollution in income statements. They argue that such companies are indirectly subsidized as the cost of pollution is borne by society while only production costs (and perhaps minimal pollution fines) are shown in the income statement. Thus, the product sells for less than would be necessary if all costs were included.

Assume that the FASB is considering a standard to include the social costs of pollution in the income statement. The process would require considering both relevance and faithful representation of the information produced by the new standard. Your instructor will divide the class into two to six groups depending on the size of the class. The mission of your group is to explain how the concepts of relevance and faithful representation relate to this issue.

Required:
Each group member should consider the question independently and draft a tentative answer prior to the class session for which the case is assigned.

In class, each group will meet for 10 to 15 minutes in different areas of the classroom. During that meeting, group members will take turns sharing their suggestions for the purpose of arriving at a single group treatment.

After the allotted time, a spokesperson for each group (selected during the group meetings) will share the group's solution with the class. The goal of the class is to incorporate the views of each group into a consensus answer to the question.

Communication Case 1–6
Accounting standard setting
● LO1–4

One of your friends is a financial analyst for a major stock brokerage firm. Recently she indicated to you that she had read an article in a weekly business magazine that alluded to the political process of establishing accounting standards. She had always assumed that accounting standards were established by determining the approach that conceptually best reflected the economics of a transaction.

Required:
Write a one to two-page article for a business journal explaining what is meant by the political process for establishing accounting standards. Be sure to include in your article a discussion of the need for the FASB to balance accounting considerations and economic consequences.

Ethics Case 1–7
The auditors' responsibility
● LO1–4

Auditors often earn considerable fees from a company for examining (auditing) its financial statements. In addition, it's not uncommon for auditors to earn additional fees from the company by providing consulting, tax, and other advisory services.

Required:
1. Which party has primary responsibility—auditors or company executives—for properly applying accounting standards when communicating with investors and creditors through financial statements?
2. Are auditors considered employees of the company?

3. Does the fact that clients compensate auditors for providing audits and other consulting services have the potential to jeopardize an auditor's independence?
4. What pressures on a typical audit engagement might affect an auditor's independence?

Judgment Case 1–8
Qualitative characteristics
● LO1–7

A friend asks you about whether GAAP requires companies to disclose forecasts of financial variables to external users. She thinks that information could be very useful to investors.

Required:

1. What are the two primary qualitative characteristics of accounting information?
2. Does GAAP routinely require companies to disclose forecasts of financial variables to external users? Indicate yes or no and explain how your answer relates to the qualitative characteristics of accounting information.

Judgment Case 1–9
Cost effectiveness
● LO1–7

Assume that the FASB is considering revising an important accounting standard.

Required:

1. What constraint applies to the FASB's consideration of whether to require companies to provide new information?
2. In what Concepts Statement is that constraint discussed?
3. What are some of the possible costs that could result from a revision of an accounting standard?
4. What does the FASB do in order to assess possible benefits and costs of a proposed revision of an accounting standard?

Judgment Case 1–10
Revenue recognition
● LO1–9

A new client, the Wolf Company, asks your advice concerning the point in time that the company should recognize revenue from the rental of its office buildings under generally accepted accounting principles. Renters usually pay rent on a quarterly basis at the beginning of the quarter. The owners contend that the critical event that motivates revenue recognition should be the date the cash is received from renters. After all, the money is in hand and is very seldom returned.

Required:

Do you agree or disagree with the position of the owners of Wolf Company? State whether you agree or disagree, and support your answer by relating it to accrual accounting under GAAP.

Real World Case 1–11
Elements; disclosures; Gap Inc.
● LO1–7, LO1–9

Real World Financials

Access the financial statements for the year ended January 28, 2017, for **Gap Inc.** by downloading them from www.gapinc.com, and use them to answer the following questions.

Required:

1. What amounts did Gap Inc. report for the following items for the fiscal year ended January 28, 2017?
   a. Total net revenues
   b. Total operating expenses
   c. Net income (earnings)
   d. Total assets
   e. Total stockholders' equity
2. How many shares of common stock had been issued by Gap Inc. as of January 28, 2017?
3. Does Gap Inc. report more than one year of data in its financial statements? Explain why or why not.

Case 1–12
Convergence
● LO1–11
IFRS

Consider the question of whether the United States should converge accounting standards with IFRS.

Required:

1. Make a list of arguments that favor convergence.
2. Make a list of arguments that favor nonconvergence.
3. Indicate your own conclusion regarding whether the United States should converge with IFRS, and indicate the primary considerations that determined your conclusion.

Data Analytics

connect

The phrase *scientia est potentia* is a Latin aphorism meaning "knowledge is power!" In a business sense, this might be paraphrased as "Information is money!" Better information . . . better business decisions! This is the keystone of data analytics.

Data analytics is the process of examining data sets in order to draw conclusions about the information they contain. Data analytics is widely used in business to enable organizations to make better-informed business decisions. Increasingly, this is accomplished with the aid of specialized data visualization software such as Tableau.

In each chapter of this textbook, you will have the opportunity to experience the power and efficacy of data analytics in the context of that chapter's topics, using Tableau as a tool.

To prepare for this case—and those in subsequent chapters—using Tableau, you will need to download Tableau to your computer. Tableau provides free instructor and student licenses as well as free videos and support for utilizing and learning the software. You will receive further information on setting up Tableau from your instructor. Once you are set up with Tableau, watch the three "Getting Started" Tableau videos. Future cases will build off what you learn in these videos, and additional videos will be suggested. All short training videos can be found at **www.tableau.com/learn/training#getting-started**.

Data Analytics Case
Net Income and Cash Flows
● LO1–2

This case introduces you to data analysis in a very basic way by allowing you to quickly extract, visualize, and compare financial data over a span of ten years for two (hypothetical) publicly traded companies: Discount Goods and Big Store. For this case, assume you are an analyst conducting introductory research into the relative merits of investing in these companies. For this initial look, you want to visualize the pattern of net income and of net operating cash flows over the past ten years for the two companies.

Required:

Use Tableau to create charts depicting (a) the trend of net income and (b) the trend of net operating cash flows over the period 2012–2021 for each of the two companies. Based upon what you find, answer the following questions:

1. One of the companies reported a huge loss on its income statement related to a foreign expatriation of properties in one of the years. Which company, which year, and what was the amount of net income or net loss that resulted for that year?
2. What was the amount of net operating cash flow in the year of the loss for the company reporting the loss?
3. Which company reported the largest percentage change in net operating cash flows between 2012 and 2021, and what was the percentage change for that company (rounded to nearest whole percentage point)?
4. Which company reported the largest percentage change in net income between 2012 and 2021, and what was the percentage change for that company (rounded to nearest whole percentage point)?

Resources:

You have available to you an extensive data set that includes detailed financial data for Discount Goods and Big Store and for 2012–2021. The data set is in the form of an Excel file available to download from Connect, or under Student Resources within the Library tab. There are four Excel files available. The one for use in this chapter is named "Discount_Goods_Big_StoreCase1.xlsx." Download this file and save it to the computer on which you will be using Tableau.

After you view the training videos, follow these steps to create the charts you'll use for this case:

- Open Tableau and connect to the Excel spreadsheet you downloaded. You should see the data in the "canvas" area of Tableau.
- Click on the "Sheet 1" tab at the bottom of the canvas, to the right of the Data Source at the bottom of the screen.
- Move "Account name" into the Filters shelf. In the Filter dialog box that opens, if all accounts are selected, un-check all accounts by selecting the "None" button.
- Select only the account "Net cash flows from operating activities." Click OK. Note: the accounts are listed alphabetically.
- Drag "Store" and "Measure Names" into the "Columns" row at the top of the canvas.
- Drag "Measure Values" into "Rows".
- Remove "Number of Records" from "Measure Values" by clicking the pull-down menu arrow from the Measures Values dimension. Select "Edit Filter", unselect "Number of Records", and click "OK."
- On the "Show Me" menu in the upper right corner of the toolbar, select the side-by-side bars if they do not already appear in that chart format. Add labels to the bars by clicking on "Label" in the "Marks" card and clicking the box "Show mark label." In the same box, format the labels to Times New Roman, bold, black, 10-point by selecting those options using "Font".
- Change the title of the sheet to be "Operating Cash Flows Discount Goods vs. Big Store" by right-clicking "Sheet 1" and selecting "Edit Title." Format the store names to Times New Roman, bold, black, and 15-point font. (We'll create the Net income/(loss) sheet next.)
- Change the colors of the bars by dragging "Measure Names" onto "Color" in the "Marks" card.
- Format the years by right-clicking on one year, selecting format in the Format Measures Name window (on the left)–Default Font dropdown box, select Times New Roman, bold, black, and 10-point options.
- To save time completing the next page, right-click the "Sheet 1" tab at the bottom of the canvas and select "Duplicate."
- Click on the "Sheet 2" tab. You should now see a duplicate of the "Sheet 1" canvas. Click on "Account name" on "Filter" in the "Marks" card. When the filter dialog box opens, un-check all accounts by selecting the "None" button. Note: The filter dialog box may open to the right of the canvas.
- Select only the account "Net income/(loss)." Change the title to "Net Income / (Loss): Discount Goods vs. Big Store" by editing the title as instructed above.
- Once complete, save the file as "DA1_Your initials.twbx."

Continuing Cases

Target Case

● LO1–9

Target Corporation prepares its financial statements according to U.S. GAAP. Target's financial statements and disclosure notes for the year ended February 3, 2018, are available in Connect. This material also is available under the Investor Relations link at the company's website (**www.target.com**).

Required:

1. What amounts did Target report for the following items for the year ended February 3, 2018?
   a. Total revenues
   b. Income from current operations
   c. Net income or net loss
   d. Total assets
   e. Total equity
2. What was Target's basic earnings per share for the year ended February 3, 2018?
3. What is Target's fiscal year-end? Why do you think Target chose that year-end?
4. Regarding Target's audit report:
   a. Who is Target's auditor?
   b. Did Target receive a "clean" (unmodified) audit opinion?

Air France–KLM Case

● LO1–11

IFRS

Air France–KLM (AF), a Franco-Dutch company, prepares its financial statements according to International Financial Reporting Standards. AF's financial statements and disclosure notes for the year ended December 31, 2017, are available in Connect. This material is also available under the Finance link at the company's website (**www.airfranceklm.com**).

Required:

1. What amounts did AF report for the following items for the year ended December 31, 2017?
   a. Total revenues
   b. Income from current operations
   c. Net income or net loss (AF equity holders)
   d. Total assets
   e. Total equity
2. What was AF's basic earnings or loss per share for the year ended December 31, 2017?

CPA Exam Questions and Simulations

ROGER CPA Review

Sample CPA Exam questions from Roger CPA Review are available in Connect as support for the topics in this chapter. These multiple-choice questions and task-based simulations include expert-written explanations and solutions, and provide a starting point for students to become familiar with the content and functionality of the actual CPA Exam.

CHAPTER

2

Review of the Accounting Process

OVERVIEW

Chapter 1 explained that the primary means of conveying financial information to investors, creditors, and other external users is through financial statements and related notes. The purpose of this chapter is to review the fundamental accounting process used to produce the financial statements. This review establishes a framework for the study of the concepts covered in intermediate accounting.

Actual accounting systems differ significantly from company to company. This chapter focuses on the many features that tend to be common to any accounting system.

LEARNING OBJECTIVES

After studying this chapter, you should be able to:

- **LO2–1** Understand routine economic events—transactions—and determine their effects on a company's financial position and on specific accounts. *(p. 48)*
- **LO2–2** Describe the steps in the accounting processing cycle. *(p. 51)*
- **LO2–3** Analyze and record transactions using journal entries. *(p. 55)*
- **LO2–4** Post the effects of journal entries to general ledger accounts and prepare an unadjusted trial balance. *(p. 60)*
- **LO2–5** Identify and describe the different types of adjusting journal entries. *(p. 63)*
- **LO2–6** Record adjusting journal entries in general journal format, post entries, and prepare an adjusted trial balance. *(p. 64)*
- **LO2–7** Describe the four basic financial statements. *(p. 72)*
- **LO2–8** Explain the closing process. *(p. 76)*
- **LO2–9** Convert from cash-basis net income to accrual-basis net income. *(p. 80)*

©goodluz/123RF

FINANCIAL REPORTING CASE

Engineering Profits

After graduating from college last year, two of your engineering-major friends started an Internet consulting practice. They began operations on July 1 and felt they did quite well during their first year. Now they would like to borrow $20,000 from a local bank to buy new computing equipment and office furniture. To support their loan application, the friends presented the bank with the following income statement for their first year of operations ending June 30:

Consulting revenue		$ 96,000
Operating expenses:		
Salaries	$32,000	
Rent	9,000	
Supplies	4,800	
Utilities	3,000	
Advertising	1,200	(50,000)
Net income		$ 46,000

The bank officer noticed that there was no depreciation expense in the income statement and has asked your friends to revise the statement after making year-end adjustments. After agreeing to help, you discover the following information:

a. The friends paid $80,000 for equipment when they began operations. They think the equipment will be useful for five years.

b. They pay $500 a month to rent office space. In January, they paid a full year's rent in advance. This is included in the $9,000 rent expense.

c. Included in consulting revenue is $13,000 they received from a customer in June as a deposit for work to be performed in August.

QUESTIONS

By the time you finish this chapter, you should be able to respond appropriately to the questions posed in this case. Compare your response to the solution provided at the end of the chapter.

1. What purpose do adjusting entries serve? *(p. 64)*

2. What year-end adjustments are needed to revise the income statement? Did your friends do as well their first year as they thought? *(p. 64)*

A solid foundation is vital to a sound understanding of intermediate accounting. So, we review the fundamental accounting process here to serve as a framework for the new concepts you will learn in this course.

Chapter 1 introduced the theoretical structure of financial accounting and the environment within which it operates. The primary function of financial accounting—to provide financial information to external users that possesses the fundamental decision-specific qualities of relevance and faithful representation—is accomplished by periodically disseminating financial statements and related notes. In this chapter we review the process used to identify, analyze, record, summarize, and then report the economic events affecting a company's financial position.

Keep in mind as you study this chapter that the accounting information systems businesses actually use are quite different from company to company. Larger companies generally use more complex systems than smaller companies use. The types of economic events affecting companies also cause differences in systems. We focus on the many features that tend to be common to all accounting systems.

Computers are used to process accounting information. In this chapter we provide an overview of the basic model that underlies computer software programs.

It's important to understand that this chapter and its appendices are not intended to describe actual accounting systems. In most business enterprises, the sheer volume of data that must be processed precludes a manual accounting system. Fortunately, the computer provides a solution. *We describe and illustrate a manual accounting information system to provide an overview of the basic model that underlies the computer software programs actually used to process accounting information.*

Electronic data processing is fast, accurate, and affordable. Many large and medium-sized companies own or rent their own mainframe computers and company-specific data processing systems. Smaller companies can take advantage of technology with relatively inexpensive desktop and laptop computers and generalized data software packages such as QuickBooks and Peachtree Accounting Software. Enterprise Resource Planning (ERP) systems are now being installed in companies of all sizes. The objective of ERP is to create a customized software program that integrates all departments and functions across a company onto a single computer system that can serve the information needs of those different departments, including the accounting department.

LO2–1

The Basic Model

Economic events cause changes in the financial position of the company.

The first objective of any accounting system is to identify the **economic events** that can be expressed in financial terms by the system.[1] An economic event for accounting purposes is any event that *directly* affects the financial position of the company. Recall from Chapter 1 that financial position comprises assets, liabilities, and owners' equity. Broad and specific accounting principles determine which events should be recorded, when the events should be recorded, and the dollar amount at which they should be measured.

External events involve an exchange between the company and another entity.

Economic events can be classified as either external events or internal events. **External events** involve an exchange between the company and a separate economic entity. Examples are transactions involving purchasing merchandise inventory for cash, borrowing cash from a bank, and paying salaries to employees. In each instance, the company receives something (merchandise, cash, and services) in exchange for something else (cash, assumption of a liability, or both).

Internal events do not involve an exchange transaction but do affect the company's financial position.

On the other hand, **internal events** directly affect the financial position of the company but don't involve an exchange transaction with another entity. Examples are the depreciation of equipment and the use of supplies. As we will see later in the chapter, these events must be recorded to properly reflect a company's financial position and results of operations in accordance with the accrual accounting model.

The Accounting Equation

The **accounting equation** underlies the process used to capture the effect of economic events.

$$\text{Assets} = \text{Liabilities} + \text{Owners' Equity}$$

This general expression portrays the equality between the total economic resources of an entity (its assets)—shown on the left side of the equation—and the total claims against the entity (liabilities and equity)—shown on the right side. In other words, the resources of an enterprise are provided by creditors and owners. Look at that equation again, it's the key to all the discussion that follows.

[1]There are many economic events that affect a company *indirectly* and are not recorded. For example, when the Federal Reserve changes its discount rate, it is an important economic event that can affect the company in many ways, but it is not recorded by the company.

As discussed in Chapter 1, owners of a corporation are its shareholders, so owners' equity for a corporation is referred to as *shareholders'* equity. Shareholders' equity for a corporation arises primarily from two sources: (1) amounts *invested* by shareholders in the corporation and (2) amounts *earned* by the corporation (on behalf of its shareholders). These are reported as (1) **paid-in capital** and (2) **retained earnings**. Retained earnings equals net income less distributions to shareholders (primarily dividends) since the inception of the corporation. Illustration 2–1 shows the basic accounting equation for a corporation with shareholders' equity expanded to highlight its composition. We use the corporate format throughout the remainder of the chapter.

Owners' equity, for a corporation called shareholders' equity, is classified by source as either paid-in capital or retained earnings.

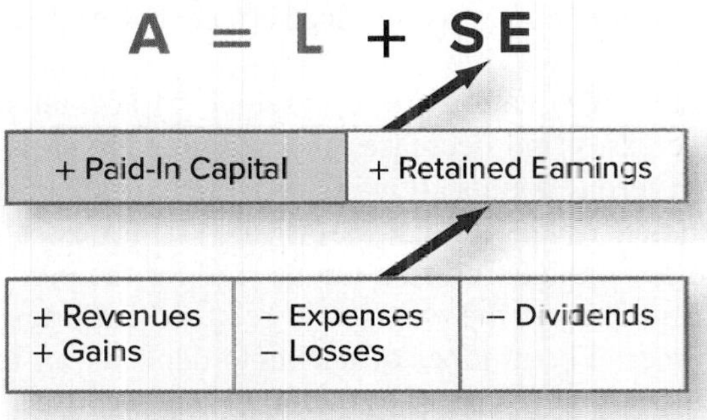

Illustration 2–1

Accounting Equation for a Corporation

The equation also implies that each economic event affecting this equation will have a dual effect because resources always must equal claims. For illustration, consider the events (we refer to these throughout the text as **transactions**) in Illustration 2–2. As we analyze each transaction to determine its effect on the equation, we also look at its effect on specific

Each event, or *transaction*, has a dual effect on the accounting equation.

Illustration 2–2

Transaction analysis

1. **An attorney invested $50,000 to open a law office.**
   An investment by the owner causes both assets and shareholders' (owners') equity to increase.

Assets	=	Liabilities	+	Shareholders' Equity
+$50,000 (cash)				+$50,000 (investment by owner)

2. **$40,000 was borrowed from a bank and a note payable was signed.**
   This transaction causes assets and liabilities to increase. A bank loan increases cash and creates an obligation to repay it.

Assets	=	Liabilities	+	Shareholders' Equity
+$40,000 (cash)		+$40,000 (note payable)		

3. **Supplies costing $3,000 were purchased on account.**
   Buying supplies on credit also increases both assets and liabilities.

Assets	=	Liabilities	+	Shareholders' Equity
+$3,000 (supplies)		+$3,000 (accounts payable)		

Transactions 4, 5, and 6 are revenue and expense transactions. Revenues and expenses (and gains and losses) are events that cause shareholders' equity to change. Revenues and gains describe inflows of assets, causing shareholders' equity to increase. Expenses and losses describe outflows of assets (or increases in liabilities), causing shareholders' equity to decrease.

4. **Services were performed on account for $10,000.**

Assets	=	Liabilities	+	Shareholders' Equity
+$10,000 (accounts receivable)				+$10,000 (revenue)

5. **Salaries of $5,000 were paid to employees.**

Assets	=	Liabilities	+	Shareholders' Equity
–$5,000 (cash)				–$5,000 (expense)

6. **$500 of supplies were used.**

Assets	=	Liabilities	+	Shareholders' Equity
–$500 (supplies)				–$500 (expense)

7. **$1,000 was paid on account to the supplies vendor.**
   This transaction causes assets and liabilities to decrease.

Assets	=	Liabilities	+	Shareholders' Equity
–$1,000 (cash)		–$1,000 (accounts payable)		

financial elements, which we call **accounts**, that will be classified and reported to decision makers. To visualize these effects, we show the transactions in a columnar fashion within the equation layout.

Now let's see how these accounts are related.

Account Relationships

The *double-entry system* is used to process transactions.

All transactions could be recorded in columnar fashion as increases or decreases to elements of the accounting equation. However, even for a very small company with few transactions, this would become cumbersome. So, most companies use a process called the **double-entry system**. The term *double-entry* refers to the dual effect that each transaction has on the accounting equation.

A *general ledger* is a collection of storage areas, called *accounts*, used to keep track of increases, decreases, and balances in financial position elements.

Elements of the accounting equation are represented by accounts which are contained in a **general ledger**. Increases and decreases in each element of a company's financial position are recorded in these accounts allowing us to keep track of their balances. A separate account is maintained for individual assets and liabilities, retained earnings, and paid-in capital. Also, to accumulate information needed for the income statement, we use separate accounts to keep track of the changes in retained earnings caused by revenues, expenses, gains, and losses. The number of accounts depends on the complexity of the company's operations.

An account includes the account title, an account number to aid the processing task, and columns or fields for increases, decreases, the cumulative balance, and the date. For instructional purposes we use **T-accounts** instead of formal ledger accounts. A T-account has space at the top for the account title and two sides for recording increases and decreases.

In the double-entry system, *debit* means *left* side of an account and *credit* means *right* side of an account.

Account Title	

For centuries, accountants have effectively used a system of **debits** and **credits** to increase and decrease account balances in the ledger. Debits merely represent the *left* side of the account and credits the *right* side, as shown below.

Account Title	
Debit side	Credit side

Asset *increases* are entered on the *debit* side of accounts and *decreases* are entered on the *credit* side. Liability and equity account *increases* are *credits* and *decreases* are *debits*.

Whether a debit or a credit represents an increase or a decrease depends on the type of account. Accounts on the left side of the accounting equation (assets) are *increased* (+) by *debit* entries and *decreased* (−) by *credit* entries. Accounts on the right side of the accounting equation (liabilities and shareholders' equity) are *increased* (+) by *credit* entries and *decreased* (−) by *debit* entries. This arbitrary, but effective, procedure ensures that for each transaction the net impact on the left sides of accounts always equals the net impact on the right sides of accounts.

For example, consider the bank loan in our earlier illustration. An asset, cash, increased by **$40,000**. Increases in assets are *debits.* Liabilities also increased by **$40,000**. Increases in liabilities are *credits.*

Assets		=	**Liabilities**		+	**Shareholders' Equity**
Cash			**Note Payable**			
Debit	Credit		Debit	Credit		
+ 40,000				40,000 +		

The debits equal the credits in every transaction (dual effect), so both before and after a transaction the accounting equation is in balance.

Prior exposure to the terms debit and credit probably comes from your experience with a bank account. For example, when a bank debits your checking account for service charges, it decreases your account balance. When you make a deposit, the bank credits your account, increasing your account balance. You must remember that from the bank's perspective, your bank account balance is a liability—it represents the amount that the bank owes you. Therefore, when the bank debits your account, it is decreasing its liability. When the bank credits your account, its liability increases.

Illustration 2–3 demonstrates the relationship among the accounting equation, debits and credits, and the increases and decreases in financial position elements.

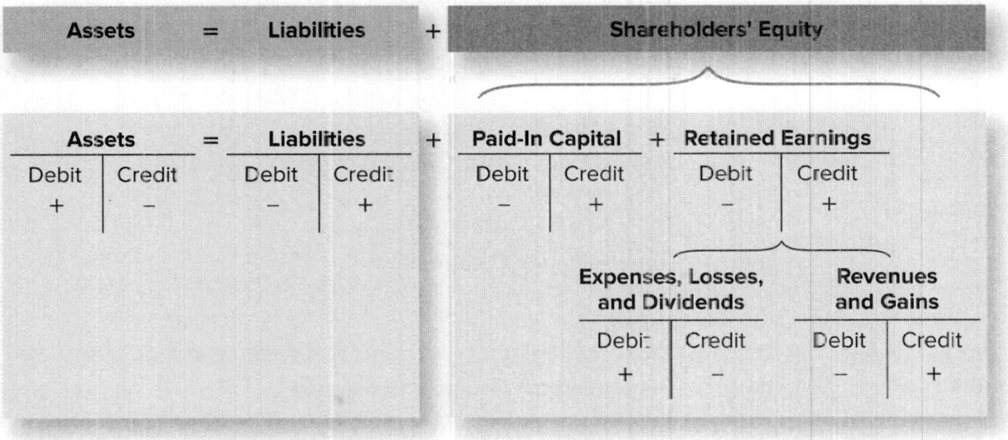

Illustration 2–3

Accounting Equation, Debits and Credits, Increases and Decreases

Notice that increases and decreases in retained earnings are recorded *indirectly*. For example, an expense represents a decrease in retained earnings, which requires a debit. That debit, however, is recorded in an appropriate expense account rather than in retained earnings itself. This allows the company to maintain a separate record of expenses incurred during an accounting period. The debit to retained earnings for the expense is recorded in a closing entry (reviewed later) at the end of the period, only after the expense total is reflected in the income statement. Similarly, an increase in retained earnings due to revenue is recorded indirectly with a credit to a revenue account, which is later reflected as a credit to retained earnings.

The general ledger accounts serve as control accounts. Subsidiary accounts associated with a particular general ledger control account are maintained in separate subsidiary ledgers. For example, a subsidiary ledger for accounts receivable contains individual account receivable accounts for each of the company's credit customers and the total of all subsidiary accounts would equal the amount in the control account. Subsidiary ledgers are discussed in more detail in Appendix 2C.

Each general ledger account can be classified as either *permanent* or *temporary*. **Permanent accounts** represent assets, liabilities, and shareholders' equity at a point in time. **Temporary accounts** represent changes in the retained earnings component of shareholders' equity for a corporation caused by revenue, expense, gain, loss, and dividend transactions. It would be cumbersome to record each revenue/expense, gain/loss, and dividend transaction directly into the retained earnings account. The different types of events affecting retained earnings should be kept separate to facilitate the preparation of the financial statements. The balances in these temporary accounts are periodically, usually once a year, closed (zeroed out), and the net effect is recorded in the permanent retained earnings account. The temporary accounts need to be zeroed out to measure income on an annual basis. This closing process is discussed in a later section of this chapter.

Permanent accounts **represent the basic financial position elements of the accounting equation.**

Temporary accounts **keep track of the changes in the retained earnings component of shareholders' equity.**

The Accounting Processing Cycle

Now that we've reviewed the basics of the double-entry system, let's look closer at the process used to identify, analyze, record, and summarize transactions and prepare financial statements. This section deals only with *external transactions,* those that involve an exchange transaction with another entity and that change financial position when the event occurs. Internal transactions are discussed in a later section.

● LO2–2

The 10 steps in the accounting processing cycle are listed in Illustration 2–4. Steps 1–4 occur during the accounting period while steps 5–8 are applied at the end of the accounting period. Steps 9 and 10 are needed only at the end of the year.

Illustration 2–4

The Accounting Processing Cycle

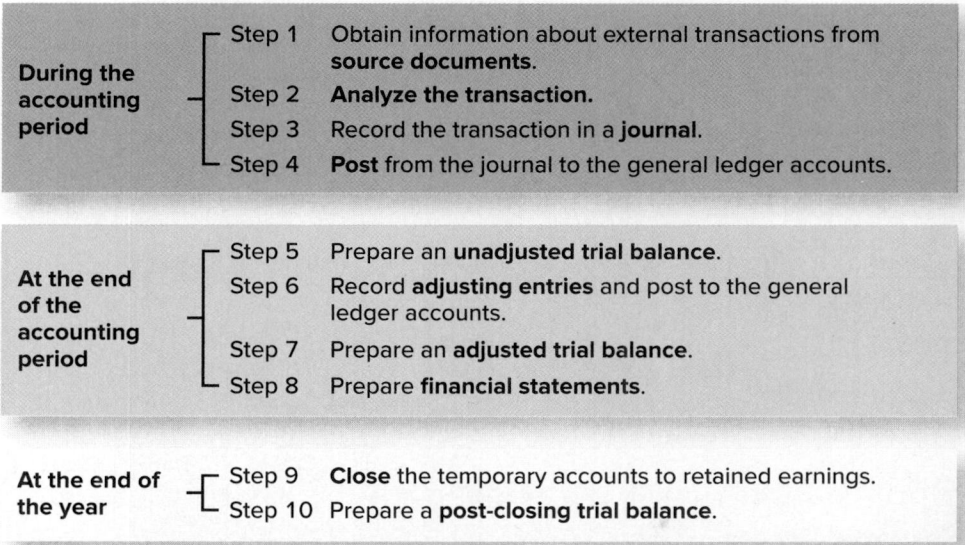

The Steps of the Accounting Processing Cycle

During the accounting period	Step 1	Obtain information about external transactions from **source documents**.
	Step 2	**Analyze the transaction.**
	Step 3	Record the transaction in a **journal**.
	Step 4	**Post** from the journal to the general ledger accounts.
At the end of the accounting period	Step 5	Prepare an **unadjusted trial balance**.
	Step 6	Record **adjusting entries** and post to the general ledger accounts.
	Step 7	Prepare an **adjusted trial balance**.
	Step 8	Prepare **financial statements**.
At the end of the year	Step 9	**Close** the temporary accounts to retained earnings.
	Step 10	Prepare a **post-closing trial balance**.

Brief Overview of Accounting Processing Cycle

Now we will take a closer look at each of these steps. First, let's see a brief overview of each step, and then we will explore each one in more detail.

STEP 1

Obtain information about transactions from *source documents.*

The first step in the process is to *identify* external transactions affecting the accounting equation. An accountant usually does not directly witness business transactions. A mechanism is needed to relay the essential information about each transaction to the accountant. **Source documents** such as sales invoices, bills from suppliers, and cash register tapes serve this need.

STEP 2

Analyze the transaction.

These source documents usually identify the date and nature of each transaction, the participating parties, and the monetary terms. For example, a sales invoice identifies the date of sale, the customer, the specific goods sold, the dollar amount of the sale, and the payment terms. With this information, the second step in the processing cycle, **transaction analysis**, can proceed. Transaction analysis is the process of reviewing the source documents to determine the dual effect on the accounting equation and the specific elements (which we'll be calling accounts) that will be used to classify the item of analysis.

Let's revisit the seven transactions described previously in Illustration 2–2 and expand that analysis to include their effects of specific account balances represented by T-accounts. We do this in Illustration 2–5. The item in each T-account is numbered to show the related transaction. We don't use dollar signs next to numbers in the accounting records (journal entries, journals, ledgers, trial balances)—only in the actual financial statements.

STEP 3

Record the transaction in a *journal.*

The third step in the process is to record the transaction in a **journal**. A journal provides a chronological record of all economic events affecting a firm. Each journal entry is expressed in terms of equal debits and credits to accounts affected by the transaction being recorded. Debits and credits represent increases or decreases to specific accounts, depending on the type of account, as explained earlier. For example, for credit sales, we record a debit to accounts receivable and a credit to sales revenue in a sales journal.

A sales journal is an example of a **special journal** used to record a repetitive type of transaction. In Appendix 2C we discuss the use of special journals in more depth. In this chapter and throughout the text, we use the **general journal** format to record all transactions.

Any type of transaction can be recorded in a general journal. It has a place for the date of the transaction, a place for account titles, account numbers, and supporting explanations, as well as a place for debit entries, and a place for credit entries. A simplified journal entry is used throughout the text that lists the account titles to be debited and credited and the dollar

Illustration 2–5 Transaction Analysis, the Accounting Equation, and Debits and Credits

Transaction	Transaction Analysis	Assets	=	Liabilities	+	Shareholders' Equity	Account Entry	
		Accounting Equation						
1. An attorney invested $50,000 to open a law office.	Assets (cash) and shareholders' equity each increased by $50,000.	**+50,000**				**+50,000**	**Cash** 1. 50,000 |	**Shareholders' Equity** | 50,000 1.
	Cumulative balances	50,000	=			50,000		
2. $40,000 was borrowed from a bank and a notes payable was signed.	Assets (cash) and liabilities (note payable) each increased by $40,000.	**+40,000**		**+40,000**			**Cash** 1. 50,000 | 2. 40,000 |	**Notes Payable** | 40,000 2.
	Cumulative balances	90,000	=	40,000	+	50,000		
3. Supplies costing $3,000 were purchased on account.	Assets (supplies) and liabilities (accounts payable) each increased by $3,000.	**+3,000**		**+3,000**			**Supplies** 3. 3,000 |	**Accounts Payable** | 3,000 3.
	Cumulative balances	93,000	=	43,000	+	50,000		
4. Services were performed on account for $10,000.	Assets (accounts receivable) and shareholders' equity (revenue) each increased by $10,000.	**+10,000**				**+10,000**	**Accounts Receivable** 4. 10,000 |	**Shareholders' Equity (Revenue)** | 10,000 4.
	Cumulative balances	103,000	=	43,000	+	60,000		
5. Salaries of $5,000 were paid to employees.	Assets (cash) decreased and shareholders' equity decreased (salaries expense increased) by $5,000.	**−5,000**				**−5,000**	**Cash** 1. 50,000 | 5,000 5. 2. 40,000 |	**Shareholders' Equity (Salaries Expense)** 5. 5,000 |
	Cumulative balances	98,000	=	43,000	+	55,000		
6. $500 of supplies were used.	Assets (supplies) decreased and shareholders' equity decreased (supplies expense increased) by $500.	**−500**				**−500**	**Supplies** 3. 3,000 | 500 6.	**Shareholders' Equity (Supplies Expense)** 6. 500 |
	Cumulative balances	97,500	=	43,000	+	54,500		
7. $1,000 was paid on account to the supplies vendor.	Assets (cash) and liabilities (accounts payable) each decreased by $1,000.	**−1,000**		**−1,000**			**Cash** 1. 50,000 | 5,000 5. 2. 40,000 | 1,000 7.	**Accounts Payable** 7. 1,000 | 3,000 3.
	Cumulative balances	96,500	=	42,000	+	54,500		

amounts. A common convention is to list the debited accounts first, indent the credited accounts, and use the first of two columns for the debit amounts and the second column for the credit amounts. An explanation is entered for each journal entry (for ease in this example the explanation is located below the entry). For example, the **journal entry** for the bank loan in Illustration 2–1, which requires a debit to cash and a credit to notes payable, is recorded as follows:

Cash	40,000	
Notes payable		40,000
To record the borrowing of cash and the signing of a note payable.		

STEP 4

Post from the journal to the general ledger accounts.

The fourth step is to periodically transfer or *post* the debit and credit information from the journal to individual ledger accounts. Recall that a ledger is simply a collection of all of the company's various accounts. Each account provides a summary of the effects of all events and transactions on that individual account. **Posting** involves transferring debits and credits recorded in individual journal entries to the specific accounts affected. As discussed earlier in the chapter, most accounting systems today are computerized. For these systems, the journal input information creates a stored journal and simultaneously posts each entry to the ledger accounts.

STEP 5

Prepare an unadjusted trial balance.

The fifth step is to prepare an unadjusted trial balance. The general ledger accounts provide the information needed. A trial balance is simply a list of general ledger accounts and their balances at a particular date. The unadjusted trial balance allows us to check for completeness, verify that the total of all debits is equal to the total of all credits, and detect some errors that may have occurred. We call it an *unadjusted* trial balance because, as we discuss later, we will prepare another trial balance after adjusting the accounts for items not yet reflected in the balances.

STEP 6

Record and post adjusting entries.

The sixth step is to record adjusting entries and post those to the general ledger accounts. Before we can prepare the financial statements, we need to bring our account balances up-to-date. We do this using adjusting entries at the end of the period. Adjusting entries are used to record changes in assets and liabilities (and their related revenues and expenses) that have occurred during the period but which we have not yet recorded.

STEP 7

Prepare an adjusted trial balance.

The seventh step is to prepare an adjusted trial balance. After posting the adjusting journal entries to the general ledger accounts, we prepare another trial balance, this time after all account balances are updated by the adjusting entries.

STEP 8

Prepare the financial statements.

The eigth step is to prepare the financial statements. In fact, the purpose of all the steps up to this point is so we have the information needed to do this. These statements:

- the income statement and statement of comprehensive income,
- the balance sheet,
- the statement of cash flows, and
- the statement of shareholders' equity

are the primary means of providing information to investors and other decision makers, which is the purpose of accounting. The adjusted trial balance supplies the necessary information to produce these statements.

STEP 9

Close the temporary accounts to retained earnings.

The ninth step is to record closing entries and post those to the general ledger accounts. Closing entries serve two purposes: (1) to transfer the balances of temporary accounts (revenues, expenses, gains, losses, and dividends) to the retained earnings account, and (2) to reduce the balances of these temporary accounts to zero to "wipe the slate clean" and ready them for measuring activity in the next period.

STEP 10

Prepare a post-closing trial balance.

The final step is to prepare a post-closing trial balance. After the temporary account balances (revenues, expenses, and dividends) are transferred to retained earnings and the closing entries are posted to the ledger accounts, we prepare another trial balance. This post-closing trial balance is comprised of final balances in permanent accounts that will be used to open (begin) the records in the next period.

Illustration of Accounting Processing Cycle

Now that we have looked at a brief overview of the ten steps of the accounting processing cycle, let's take a closer look at each one. We start by illustrating the first four steps in the processing cycle using the external transactions described in Illustration 2–6 that occurred during the month of July 2021, the first month of operations for Dress Right Clothing Corporation. The company operates a retail store that sells men's and women's clothing. Dress Right is organized as a corporation so owners' equity is classified by source as either paid-in capital or retained earnings.

Illustration 2–6

External Transactions for July 2021

July	1	Two individuals each invested $30,000 in the corporation. Each investor was issued 3,000 shares of common stock.
	1	Borrowed $40,000 from a local bank and signed two notes. The first note for $10,000 requires payment of principal and 10% interest in six months. The second note for $30,000 requires the payment of principal in two years. Interest at 10% is payable each year on July 1, 2022, and July 1, 2023.
	1	Paid $24,000 in advance for one year's rent on the store building.
	1	Purchased office equipment from eTronics for $12,000 cash.
	3	Purchased $60,000 of clothing inventory on account from the Birdwell Wholesale Clothing Company.
	6	Purchased $2,000 of supplies for cash.
	4–31	During the month, sold merchandise costing $20,000 for $35,000 cash.
	9	Sold clothing on account to Briarfield School for Girls for $3,500. The clothing cost $2,000.
	16	Subleased a portion of the building to a jewelry store. Received $1,000 in advance for the first two months' rent beginning on July 16.
	20	Paid Birdwell Wholesale Clothing $25,000 on account.
	20	Paid salaries to employees for the first half of the month, $5,000.
	25	Received $1,500 on account from Briarfield.
	30	The corporation paid its shareholders a cash dividend of $1,000.

The local bank requires that Dress Right furnish financial statements on a monthly basis. The transactions listed in the illustration are used to demonstrate the accounting processing cycle for the month of July, 2021.

Journal, Ledger, and Trial Balance

Now that we've overviewed the ten steps in the accounting processing cycle, we look closer at the process.

STEPS 1, 2, AND 3 Obtain information about transactions from *source documents.*
Analyze the transaction.
Record the transaction in a journal.

For each transaction, a source document provides the necessary information to complete steps two and three in the processing cycle: transaction analysis and recording the appropriate journal entry. Each transaction described in Illustration 2–6 is analyzed below, preceded by the necessary journal entry.

JOURNAL ENTRIES

● LO2–3

To record the issuance of common stock.

July 1

Cash	60,000	
Common stock		60,000

This first transaction is an investment by owners that increases an asset, cash, and also increases shareholders' equity. Increases in assets are recorded as debits and increases in shareholders' equity are recorded as credits. We use the paid-in capital account called common stock because stock was issued in exchange for cash paid in.[2]

To record the borrowing of cash and the signing of a note payable.

July 1		
Cash	40,000	
Notes payable		40,000

This transaction causes increases in both cash and the liability, notes payable. Increases in assets are debits and increases in liabilities are credits. The note requires payment of \$40,000 in principal and \$6,500[(\$10,000 × 10% × $\frac{6}{12}$ = \$500) + (\$30,000 × 10% × 2 years = \$6,000)] in interest. However, at this point we are concerned only with the external transaction that occurs when the cash is borrowed and the note is signed. Later we discuss how the interest is recorded.

To record the payment of one year's rent in advance.

July 1		
Prepaid rent	24,000	
Cash		24,000

This transaction increased an asset called prepaid rent, which is debited, and decreased the asset cash (a credit). Dress Right acquired the right to use the building for one full year. This is an asset because it represents a future benefit to the company. As we will see later, this asset expires over the one-year rental period.

To record the purchase of office equipment.

July 1		
Office equipment	12,000	
Cash		12,000

This transaction increases one asset, office equipment, and decreases another, cash.

To record the purchase of merchandise inventory.

July 3		
Inventory	60,000	
Accounts payable		60,000

This purchase of merchandise on account is recorded by a debit to inventory, an asset, and a credit to accounts payable, a liability. Increases in assets are debits. Increases in liabilities are credits.

The Dress Right Clothing Company uses the *perpetual inventory system* to keep track of its merchandise inventory. This system requires that the cost of merchandise purchased be recorded in inventory, an asset account. When inventory is sold, the inventory account is decreased by the cost of the item sold. An alternative method, the *periodic system,* is briefly discussed later in this chapter. We explore the topic of inventory in depth in Chapters 8 and 9.

To record the purchase of supplies.

July 6		
Supplies	2,000	
Cash		2,000

We record the acquisition of supplies as a debit to the asset account supplies (an increase) and as a credit to the asset cash (a decrease). Supplies are an asset because they represent future benefits.

[2]The different types of stock are discussed in Chapter 18.

July 4–31

Cash	35,000	
Sales revenue		35,000
Cost of goods sold (expense)	20,000	
Inventory		20,000

To record the month's cash sales and the cost of those sales.

During the month of July, cash sales to customers totaled $35,000. The company's assets (cash) increase by this amount as does shareholders' equity. This increase in equity is recorded by a credit to the temporary account sales revenue.

At the same time, an asset, inventory, decreases and retained earnings decreases. Recall that expenses are outflows or using up of assets from providing goods and services. Dress Right incurred an expense equal to the cost of the inventory sold. The temporary account cost of goods sold increases. However, this increase in an expense represents a *decrease* in shareholders' equity—retained earnings—and accordingly the account is debited. Both of these transactions are *summary* transactions. Normally each sale made during the month requires a separate and similar entry in a special journal, which we discuss in Appendix 2C.

July 9

Accounts receivable	3,500	
Sales revenue		3,500
Cost of goods sold	2,000	
Inventory		2,000

To record a credit sale and the cost of that sale.

This transaction is similar to the cash sale shown earlier. The only difference is that the asset acquired in exchange for merchandise is accounts receivable rather than cash.

Additional Consideration

Periodic Inventory System

The principal alternative to the perpetual inventory system is the periodic system. By this approach, we record the cost of merchandise purchased in a temporary account called *purchases*. When inventory is sold, the inventory account is not decreased and cost of goods sold is not recorded. Instead, we determine the balance in the inventory account only at the end of a reporting period (periodically), and then we can determine the cost of goods sold for the period.

For example, the purchase of $60,000 of merchandise on account by Dress Right Clothing is recorded as follows:

Purchases	60,000	
Accounts payable		60,000

No cost of goods sold entry is recorded when sales are made in the periodic system.

At the end of July, the amount of ending inventory is determined (either by means of a physical count of goods on hand or by estimation) to be $38,000 and cost of goods sold for the month is determined as follows:

Beginning inventory	$ –0–
Plus: Purchases	60,000
Less: Ending inventory	(38,000)
Cost of goods sold	$ 22,000

We record the cost of goods sold for the period and adjust the inventory account to the actual amount on hand (in this case from zero to $38,000) this way:

Cost of goods sold	22,000	
Inventory	38,000	
Purchases		60,000

We discuss inventory in depth in Chapters 8 and 9.

To record the receipt of rent in advance.

July 16		
Cash	1,000	
Deferred rent revenue (liability)		1,000

Cash increases by $1,000 so we debit the cash account. At this point, Dress Right does not recognize revenue yet even though cash has been received. Instead, revenue is recorded only after Dress Right has provided the jewelry store with the use of facilities; that is, as the rental period expires. On receipt of the cash, a liability called *deferred rent revenue* increases and is credited. This liability represents Dress Right's obligation to provide the use of facilities to the jewelry store.

To record the payment of accounts payable.

July 20		
Accounts payable	25,000	
Cash		25,000

This transaction decreases both an asset (cash) and a liability (accounts payable). A debit decreases the liability, and a credit decreases the asset.

To record the payment of salaries for the first half of the month.

July 20		
Salaries expense	5,000	
Cash		5,000

Employees were paid for services rendered during the first half of the month. The cash expenditure did not create an asset since no future benefits resulted. Cash decreases and is credited; shareholders' equity decreases and is debited. The reduction in shareholders' equity is recorded in the temporary account, salaries expense.

To record receipt of cash on account.

July 25		
Cash	1,500	
Accounts receivable		1,500

This transaction is an exchange of one asset, accounts receivable, for another asset, cash.

To record the payment of a cash dividend.

July 30		
Dividends	1,000	
Cash		1,000

The payment of a cash dividend is a distribution to owners that reduces both cash and retained earnings. Dividends is a temporary account that later is closed (transferred) to retained earnings along with the other temporary accounts (revenues and expenses) at the end of the fiscal year. We discuss and illustrate the closing process later in the chapter.

Additional Consideration

An alternative method of recording a cash dividend is to debit retained earnings at the time the dividend is paid rather than reducing retained earnings during the closing process. The entry to record the dividend using this approach is as follows:

Retained earnings	1,000	
Cash		1,000

As an expedient, we apply this approach when discussing dividends in Chapter 18.

Illustration 2–7 summarizes each of the transactions we just discussed as they would appear in a general journal. In addition to the date, account titles, and debit and credit columns, the journal also has a column titled Post Ref. (Posting Reference). In this column we enter the number assigned to the general ledger account that is being debited or credited. For purposes of this illustration, all asset accounts have been assigned numbers in the 100s, all liabilities are 200s, permanent shareholders' equity accounts are 300s, revenues are 400s, expenses are 500s and dividends is 600.

Illustration 2–7
The General Journal

General Journal				Page 1
Date 2021	**Account Title and Explanation**	**Post Ref.**	**Debit**	**Credit**
July 1	Cash	100	60,000	
	Common stock	300		60,000
	To record the issuance of common stock.			
1	Cash	100	40,000	
	Notes payable	220		40,000
	To record the borrowing of cash and the signing of notes payable.			
1	Prepaid rent	130	24,000	
	Cash	100		24,000
	To record the payment of one year's rent in advance.			
1	Office equipment	150	12,000	
	Cash	100		12,000
	To record the purchase of office equipment.			
3	Inventory	140	60,000	
	Accounts payable	210		60,000
	To record the purchase of merchandise inventory.			
6	Supplies	125	2,000	
	Cash	100		2,000
	To record the purchase of supplies.			
4–31	Cash	100	35,000	
	Sales revenue	400		35,000
	To record cash sales for the month.			
4–31	Cost of goods sold	500	20,000	
	Inventory	140		20,000
	To record the cost of cash sales.			
9	Accounts receivable	110	3,500	
	Sales revenue	400		3,500
	To record credit sale.			
9	Cost of goods sold	500	2,000	
	Inventory	140		2,000
	To record the cost of a credit sale.			
16	Cash	100	1,000	
	Deferred rent revenue	230		1,000
	To record the receipt of rent in advance.			
20	Accounts payable	210	25,000	
	Cash	100		25,000
	To record the payment of accounts payable.			
20	Salaries expense	510	5,000	
	Cash	100		5,000
	To record the payment of salaries for the first half of the month.			
25	Cash	100	1,500	
	Accounts receivable	110		1,500
	To record the receipt of cash on account.			
30	Dividends	600	1,000	
	Cash	100		1,000
	To record the payment of a cash dividend.			

The ledger accounts also contain a posting reference, usually the page number of the journal in which the journal entry was recorded. This allows for easy cross-referencing between the journal and the ledger.

● LO2–4 **LEDGER ACCOUNTS** Step 4 in the processing cycle is to transfer (post) the debit/credit information from the journal to the general ledger accounts. Illustration 2–8 contains the ledger accounts (in T-account form) for Dress Right *after* all the general journal transactions have been

Illustration 2–8

General Ledger Accounts

Balance Sheet Accounts

Cash 100

July 1 GJ1	60,000	24,000	July 1 GJ1
1 GJ1	40,000	12,000	1 GJ1
4-31 GJ1	35,000	2,000	6 GJ1
16 GJ1	1,000	25,000	20 GJ1
25 GJ1	1,500	5,000	20 GJ1
		1,000	30 GJ1
July 31 Bal.	**68,500**		

Prepaid Rent 130

July 1 GJ1	24,000		
July 31 Bal.	**24,000**		

Accounts Receivable 110

July 9 GJ1	3,500	1,500	July 25 GJ1
July 31 Bal.	**2,000**		

Inventory 140

July 3 GJ1	60,000	20,000	July 4–31
		2,000	9 GJ1
July 31 Bal.	**38,000**		

Supplies 125

July 6 GJ1	2,000		
July 31 Bal.	**2,000**		

Office equipment 150

July 1 GJ1	12,000		
July 31 Bal.	**12,000**		

Accounts Payable 210

July 20 GJ1	25,000	60,000	July 3 GJ1
		35,000	**July 31 Bal.**

Notes Payable 220

		40,000	July 1 GJ1
		40,000	**July 31 Bal.**

Deferred Rent Revenue 230

		1,000	July 16 GJ1
		1,000	**July 31 Bal.**

Common Stock 300

		60,000	July 1 GJ1
		60,000	**July 31 Bal.**

Retained Earnings 310

July 31 Bal.	**0**		

Income Statement Accounts

Sales Revenue 400

		35,000	July 4–31 GJ1
		3,500	9 GJ1
		38,500	**July 31 Bal.**

Cost of Goods Sold 500

July 4-31 GJ1	20,000		
9 GJ1	2,000		
July 31 Bal.	**22,000**		

Salaries Expense 510

July 20 GJ1	5,000		
July 31 Bal.	**5,000**		

Dividends

Dividends 600

July 30 GJ1	1,000		
July 31 Bal.	**1,000**		

posted. The reference GJ1 next to each of the posted amounts indicates that the source of the entry is page 1 of the general journal. An alternative is to number each of the entries in chronological order and reference them by number. Note that each account shows the balance that we call a *normal balance*—that is, the balance is the debit or credit side used to increase the account.

TRIAL BALANCE Before preparing financial statements and adjusting entries (internal transactions) at the end of an accounting period, we prepare an **unadjusted trial balance**—step 5. A trial balance is simply a list of the general ledger accounts along with their balances at a particular date, listed in the order that they appear in the ledger. The purpose of the trial balance is to allow us to check for completeness and to verify that the sum of the accounts with debit balances equals the sum of the accounts with credit balances. The fact that the debits and credits are equal, though, does not necessarily ensure that the equal balances are correct. The trial balance could contain offsetting errors. As we will see later in the chapter, this trial balance also helps with preparing adjusting entries.

STEP 5

Prepare an unadjusted trial balance.

The unadjusted trial balance at July 31, 2021, for Dress Right appears in Illustration 2–9. Notice that dividends has a debit balance of $1,000. This reflects the payment of the cash dividend to shareholders. We record increases and decreases in retained earnings from revenue, expense, gain and loss transactions *indirectly* in temporary accounts. Before the start of the next year, we transfer these increases and decreases to the retained earnings account.

Illustration 2–9

Unadjusted Trial Balance

DRESS RIGHT CLOTHING CORPORATION
Unadjusted Trial Balance
July 31, 2021

Account Title	Debits	Credits
Cash	68,500	
Accounts receivable	2,000	
Supplies	2,000	
Prepaid rent	24,000	
Inventory	38,000	
Office equipment	12,000	
Accounts payable		35,000
Notes payable		40,000
Deferred rent revenue		1,000
Common stock		60,000
Retained earnings		0
Sales revenue		38,500
Cost of goods sold	22,000	
Salaries expense	5,000	
Dividends	1,000	
Totals	174,500	174,500

At any time, the total of all debit balances should equal the total of all credit balances.

Concept Review Exercise

JOURNAL ENTRIES FOR EXTERNAL TRANSACTIONS

The Wyndham Wholesale Company began operations on August 1, 2021. The following transactions occur during the month of August.

a. Owners invest $50,000 cash in the corporation in exchange for 5,000 shares of common stock.
b. Equipment is purchased for $20,000 cash.
c. On the first day of August, $6,000 rent on a building is paid for the months of August and September.
d. Merchandise inventory costing $38,000 is purchased on account. The company uses the perpetual inventory system.
e. $30,000 is borrowed from a local bank, and a note payable is signed.

f. Credit sales for the month are $40,000. The cost of merchandise sold is $22,000.
g. $15,000 is collected on account from customers.
h. $20,000 is paid on account to suppliers of merchandise.
i. Salaries of $7,000 are paid to employees for August.
j. A bill for $2,000 is received from the local utility company for the month of August.
k. $20,000 cash is loaned to another company, evidenced by a note receivable.
l. The corporation pays its shareholders a cash dividend of $1,000.

Required:

1. Prepare a journal entry for each transaction.
2. Prepare an unadjusted trial balance as of August 31, 2021.

Solution:

1. Prepare a journal entry for each transaction.
   a. The issuance of common stock for cash increases both cash and shareholders' equity (common stock).

Cash	50,000	
Common stock		50,000

b. The purchase of equipment increases equipment and decreases cash.

Equipment	20,000	
Cash		20,000

c. The payment of rent in advance increases prepaid rent and decreases cash.

Prepaid rent	6,000	
Cash		6,000

d. The purchase of merchandise on account increases both inventory and accounts payable.

Inventory	38,000	
Accounts payable		38,000

e. Borrowing cash and signing a note increases both cash and notes payable.

Cash	30,000	
Notes payable		30,000

f. The sale of merchandise on account increases both accounts receivable and sales revenue. Also, cost of goods sold increases and inventory decreases.

Accounts receivable	40,000	
Sales revenue		40,000
Cost of goods sold	22,000	
Inventory		22,000

g. The collection of cash on account increases cash and decreases accounts receivable.

Cash	15,000	
Accounts receivable		15,000

h. The payment to suppliers on account decreases both accounts payable and cash.

Accounts payable	20,000	
Cash		20,000

i. The payment of salaries for the period increases salaries expense (decreases retained earnings) and decreases cash.

Salaries expense	7,000	
Cash		7,000

j. The receipt of a bill for services rendered increases both an expense (utilities expense) and accounts payable. The expense decreases retained earnings.

Utilities expense	2,000	
Accounts payable		2,000

k. The lending of cash to another entity and the signing of a note increases notes receivable and decreases cash.

Notes receivable	20,000	
Cash		20,000

l. Cash dividends paid to shareholders reduce both retained earnings and cash.

Dividends[3]	1,000	
Cash		1,000

2. Prepare an unadjusted trial balance as of August 31, 2021.

Account Title	Debits	Credits
Cash	21,000	
Accounts receivable	25,000	
Prepaid rent	6,000	
Inventory	16,000	
Notes receivable	20,000	
Equipment	20,000	
Accounts payable		20,000
Notes payable		30,000
Common stock		50,000
Retained earnings		0
Sales revenue		40,000
Cost of goods sold	22,000	
Salaries expense	7,000	
Utilities expense	2,000	
Dividends	1,000	
Totals	140,000	140,000

Adjusting Entries

LO2–5

STEP 6

Record adjusting entries and post to the ledger accounts.

Step 6 in the processing cycle is to record in the general journal and post to the ledger accounts the effect of *internal events* on the accounting equation. These transactions do not involve an exchange transaction with another entity and, therefore, are not initiated by a source document. They are recorded *at the end of any period when financial statements are prepared.* These transactions are commonly referred to as **adjusting entries**.

Even when all transactions and events are analyzed, corrected, journalized, and posted to appropriate ledger accounts, some account balances will require updating. Adjusting entries

[3]An alternative is to debit retained earnings at the time the dividend is distributed.

FINANCIAL Reporting Case

Q1, p. 47

are required to implement the *accrual accounting model.* More specifically, these entries help ensure that all revenues are recognized in the period goods or services are transferred to customers, regardless of when the cash is received. Also, they enable a company to recognize all expenses incurred during a period, regardless of when cash payment is made. As a result, a period's income statement provides a more complete measure of a company's operating performance and a better measure for predicting future operating cash flows. The balance sheet also provides a more complete assessment of assets and liabilities as sources of future cash receipts and disbursements. You might think of adjusting entries as a method of bringing the company's financial information up-to-date before preparing the financial statements.

FINANCIAL Reporting Case

Q2, p. 47

Adjusting entries are necessary for three situations:

1. **Prepayments**, sometimes referred to as *deferrals*
2. **Accruals**
3. **Estimates**

Prepayments

Prepayments are transactions in which the cash flow *precedes* expense or revenue recognition.

Prepayments occur when the cash flow *precedes* either expense or revenue recognition. For example, a company may buy supplies in one period but use them in a later period. The cash outflow creates an asset (supplies) which then must be expensed in a future period as the asset is used up. Similarly, a company may receive cash from a customer in one period but provide the customer with a good or service in a future period. For instance, magazine publishers usually receive cash in advance for magazine subscriptions. The cash inflow creates a liability (deferred revenue) that is recognized as revenue in a future period when the goods or services are transferred to customers.

Prepaid expenses represent assets recorded when a cash disbursement creates benefits beyond the current reporting period.

PREPAID EXPENSES **Prepaid expenses** are the costs of assets acquired in one period and expensed in a future period. Whenever cash is paid, and it is not to (1) satisfy a liability or (2) pay a dividend or return capital to owners, it must be determined whether or not the payment creates future benefits or whether the payment benefits only the current period. The purchase of buildings, equipment, or supplies or the payment of rent in advance are examples of payments that create future benefits and should be recorded as assets. The benefits provided by these assets expire in future periods and their cost is expensed in future periods as related revenues are recognized.

The adjusting entry required for a prepaid expense is a *debit to an expense* and a *credit to an asset.*

To illustrate this concept, assume that a company paid a radio station $2,000 in July for advertising. If that $2,000 were for advertising provided by the radio station during the month of July, the entire $2,000 would be expensed in the same period as the cash disbursement. If, however, the $2,000 was a payment for advertising to be provided in a future period, say the month of August, then the cash disbursement creates an asset called *prepaid advertising.* Then, an adjusting entry is required at the end of August to increase advertising expense (decrease shareholders' equity) and to decrease the asset, prepaid advertising, by $2,000. So, the adjusting entry for a prepaid expense is a *debit to an expense* and a *credit to an asset.*

● LO2–6

Supplies		
Beg. bal.	0	
	2,000	**800**
End bal.	1,200	

Supplies Expense		
Beg. bal.	0	
	800	
End. bal.	800	

The unadjusted trial balance can provide a starting point for determining which adjusting entries are required for a period, particularly for prepayments. Review the July 31, 2021, unadjusted trial balance for the Dress Right Clothing Corporation in Illustration 2–9 and try to anticipate the required adjusting entries for prepaid expenses.

The first asset that requires adjustment is supplies, $2,000 of which were purchased during July. This transaction created an asset as the supplies will be used in future periods. The company could either track the supplies used or simply count the supplies at the end of the period and determine the dollar amount of supplies remaining. Assume that Dress Right determines that at the end of July, $1,200 of supplies remain. The following adjusting journal entry is required.

To record the cost of supplies used during the month of July.

July 31		
Supplies expense	800	
Supplies		800

The next prepaid expense requiring adjustment is rent. Recall that at the beginning of July, the company paid $24,000 to its landlord representing one year's rent in advance. As it is reasonable to assume that the rent services provided each period are equal, the monthly rent is $2,000. At the end of July 2021, one month's prepaid rent has expired and must be recognized as expense.

July 31		
Rent expense ($24,000 ÷ 12)	2,000	
Prepaid rent		2,000

To record the cost of expired rent for the month of July.

After this entry is recorded and posted to the ledger accounts, the prepaid rent account will have a debit balance of $22,000, representing 11 remaining months at $2,000 per month, and the rent expense account will have a $2,000 debit balance.

Rent Expense

	Debit	Credit
Beg. bal.	0	
	2,000	
End bal.	2,000	

Prepaid Rent

	Debit	Credit
Beg. bal.	0	
	24,000	2,000
End. bal.	22,000	

The final prepayment involves the asset represented by office equipment that was purchased for $12,000. This asset has a long life but nevertheless will expire over time. For the previous two adjusting entries, it was fairly straightforward to determine the amount of the asset that expired during the period.

However, it is difficult, if not impossible, to determine how much of the benefits from using the office equipment expired during any particular period. Recall from Chapter 1 that one approach is to recognize an expense "by a systematic and rational allocation to specific time periods."

Assume that the office equipment has a useful life of five years (60 months) and will be worthless at the end of that period, and that we choose to allocate the cost equally over the period of use. The amount of monthly expense, called *depreciation expense,* is $200 ($12,000 ÷ 60 months = $200), and the following adjusting entry is recorded.

July 31		
Depreciation expense	200	
Accumulated depreciation		200

To record depreciation of office equipment for the month of July.

The entry reduces an asset, office equipment, by $200. However, the asset account is not reduced directly. Instead, the credit is to an account called *accumulated depreciation.* This is a contra account to office equipment. The normal balance in a contra asset account will be a credit, that is, "contra," or opposite, to the normal debit balance in an asset account. The purpose of the contra account is to keep the original cost of the asset intact while reducing it indirectly. In the balance sheet, office equipment is reported net of accumulated depreciation. When we have multiple depreciable assets, it's helpful to differentiate accumulated depreciation accounts, like accumulated depreciation—office equipment, accumulated depreciation—buildings, etc. This topic is covered in depth in Chapter 11.

After this entry is recorded and posted to the ledger accounts, the accumulated depreciation account will have a credit balance of $200 and the depreciation expense account will have a $200 debit balance. If a required adjusting entry for a prepaid expense is not recorded, net income, assets, and shareholders' equity (retained earnings) will be overstated.

DEFERRED REVENUES Deferred revenues are created when a company receives cash from a customer in one period for goods or services that are to be provided in a future period. The cash receipt, an external transaction, is recorded as a debit to cash and a credit to a liability. This liability reflects the company's obligation to provide goods or services in the future.

Deferred revenues represent liabilities recorded when cash is received from customers in advance of providing a good or service.

To illustrate a deferred revenue transaction, assume that during the month of June a magazine publisher received $24 in cash for a 24-month subscription to a monthly magazine. The subscription begins in July. On receipt of the cash, the publisher records a liability, deferred subscription revenue, of $24. Subsequently, revenue of $1 is recognized as each

The adjusting entry required when deferred revenues are recognized is a *debit to a liability* and a *credit to revenue.*

monthly magazine is published and mailed to the customer. An adjusting entry is required each month to increase shareholders' equity (revenue) to recognize the $1 in revenue and to decrease the liability. Assuming that the cash receipt entry included a credit to a liability, the adjusting entry for deferred revenues, therefore, is a *debit to a liability,* in this case deferred subscription revenue, and a *credit to revenue.*

Once again, the unadjusted trial balance provides information concerning deferred revenues. For Dress Right Clothing Corporation, the only deferred revenue in the trial balance is deferred rent revenue. Recall that the company subleased a portion of its building to a jewelry store for $500 per month. On July 16, the jewelry store paid Dress Right $1,000 in advance for the first two months' rent. The transaction was recorded as a debit to cash and a credit to deferred rent revenue.

At the end of July, how much of the $1,000 must be recognized? Approximately one-half of one month's rent service has been provided, or $250, requiring the following adjusting entry.

To record previously deferred rent revenue recognized during July.

July 31

Deferred rent revenue	250	
Rent revenue		250

Deferred Rent Revenue

	0 Beg. bal.
250	1,000
	750 End bal.

Rent Revenue

	0 Beg. bal.
	250
	250 End bal.

After this entry is recorded and posted to the ledger accounts, the deferred rent revenue account is reduced to a credit balance of $750 for the remaining one and one-half months' rent, and the rent revenue account will have a $250 credit balance. If this entry is not recorded, net income and shareholders' equity (retained earnings) will be understated, and liabilities will be overstated.

ALTERNATIVE APPROACH TO RECORD PREPAYMENTS The same end result can be achieved for prepayments by recording the external transaction directly into an expense or revenue account. In fact, many companies prefer this approach. For simplicity, bookkeeping instructions might require all cash payments for expenses to be debited to the appropriate expense accounts and all cash payments for revenues to be credited to the appropriate revenue accounts. In the adjusting entry, then, the *unexpired* prepaid expense (asset) or *deferred* revenue (liability) as of the end of the period are recorded.

For example, on July 1, 2021, Dress Right paid $24,000 in cash for one year's rent on its building. The entry included a debit to prepaid rent. The company could have debited rent expense instead of prepaid rent.

Rent Expense

Beg. bal. 0	
24,000	22,000
End bal. 2,000	

Alternative Approach

July 1

Rent expense	24,000	
Cash		24,000

The adjusting entry then records the amount of prepaid rent as of the end of July, $22,000, and reduces rent expense to $2,000, the cost of rent for the month of July.

Prepaid Rent

Beg. bal. 0	
22,000	
End bal. 22,000	

Alternative Approach

July 31

Prepaid rent	22,000	
Rent expense		22,000

The net effect of handling the transactions in this manner is the same as the previous treatment. Either way, the prepaid rent account will have a debit balance at the end of July of $22,000 to represent 11 months remaining prepaid rent, and the rent expense account will have a debit balance of $2,000 to represent the one month of rent used in the period of this income statement being prepared. What's important is that an adjusting entry is recorded to ensure the appropriate amounts are reflected in both the expense and asset *before financial statements are prepared.*

Similarly, the July 16 cash receipt from the jewelry store representing an advance for two months' rent initially could have been recorded by Dress Right as a credit to rent revenue instead of deferred rent revenue (a liability).

Alternative Approach
July 16

Cash	1,000	
Rent revenue		1,000

Rent Revenue

Debit	Credit
	0 Beg. bal.
750	1,000
	250 End bal.

If Dress Right records the entire $1,000 as rent revenue in this way, it would then use the adjusting entry to record the amount of deferred revenue as of the end of July, $750 to represent the one and one-half month of rent revenue remaining as collected in advance, and the rent revenue account will have a credit balance of $250, which is for the one-half month of July that has passed in the current period.

Alternative Approach
July 31

Rent revenue	750	
Deferred rent revenue		750

Deferred Rent Revenue

Debit	Credit
	0 Beg. bal.
	750
	750 End bal.

Accruals

Accruals occur when the cash flow comes *after* either expense or revenue recognition. For example, a company often uses the services of another entity in one period and pays for them in a subsequent period. An expense must be recognized in the period incurred and an accrued liability recorded. Also, goods and services often are provided to customers on credit. In such instances, a revenue is recognized in the period goods or services are transferred to customers and an asset, a receivable, is recorded.

Accruals involve transactions where the cash outflow or inflow takes place in a period subsequent to expense or revenue recognition.

Many accruals involve external transactions that automatically are recorded from a source document. For example, a sales invoice for a credit sale provides all the information necessary to record the debit to accounts receivable and the credit to sales revenue. However, there are some accruals that involve internal transactions and thus require adjusting entries. Because accruals involve recognition of expense or revenue before cash flow, the unadjusted trial balance will not be as helpful in identifying required adjusting entries as with prepayments.

ACCRUED LIABILITIES For **accrued liabilities**, we are concerned with expenses incurred but not yet paid. Dress Right Clothing Corporation requires two adjusting entries for accrued liabilities at July 31, 2021.

Accrued liabilities represent liabilities recorded when an expense has been incurred prior to cash payment.

The first entry is for employee salaries for the second half of July. Recall that on July 20 the company paid employees $5,000 for salaries for the first half of the month. Salaries for the second half of July probably will be paid in early August. Nevertheless, an expense is incurred in July for services rendered to the company by its employees. The accrual income statement for July must reflect these services for the entire month regardless of when the cash payment is made.

Therefore, an obligation exists at the end of July to pay the salaries earned by employees for the last half of that month. An adjusting entry is required to increase salaries expense (decrease shareholders' equity) and to increase liabilities for the salaries payable. The adjusting entry for an accrued liability always includes a *debit to an expense,* and a *credit to a liability.* Assuming that salaries for the second half of July are $5,500, the following adjusting entry is recorded.

The adjusting entry required to record an accrued liability is a debit to an expense and a credit to a liability.

July 31

Salaries expense	5,500	
Salaries payable		5,500

Salaries Payable

Debit	Credit
	0 Beg. bal.
	5,500
	5,500 End bal.

Salaries Expense

	Debit	Credit
Beg. bal.	0	
July 20	5,000	
	5,500	
End bal.	10,500	

After this entry is recorded and posted to the general ledger, the salaries expense account will have a debit balance of $10,500 ($5,000 + 5,500), and the salaries payable account will have a credit balance of $5,500.

The unadjusted trial balance does provide information about the second required accrued liability entry. In the trial balance we can see a balance in the notes payable account of $40,000. The company borrowed this amount on July 1, 2021, evidenced by two notes, each requiring the payment of 10% interest. Whenever the trial balance reveals interest-bearing debt, and interest is not paid on the last day of the period, an adjusting entry is required for the amount of interest that has built up (accrued) since the last payment date or the last date interest was accrued. In this case, we calculate interest as follows:

$$\text{Principal} \times \text{Interest rate} \times \text{Time} = \text{Interest}$$
$$\$40{,}000 \times 10\% \times 1/12 = \$333 \text{ (rounded)}$$

Interest rates always are stated as the annual rate. Therefore, the above calculation uses this annual rate multiplied by the principal amount multiplied by the amount of time outstanding, in this case one month or one-twelfth of a year.

To accrue interest expense for July on notes payable.

July 31

Interest expense	333	
Interest payable		333

After this entry is recorded and posted to the ledger accounts, the interest expense account will have a debit balance of $333, and the interest payable account will have a credit balance of $333. Failure to record a required adjusting entry for an accrued liability will cause net income and shareholders' equity (retained earnings) to be overstated and liabilities to be understated.[4]

It's important to note that we differentiate accrued liabilities like salaries payable and interest payable from accounts payable accounts. Accounts payable is the account we use for purchases of merchandise that will be resold in the ordinary course of business and for property, plant, and equipment that will be not paid for with notes payable or other types of financing.

Also, sometimes we use an account titled "accrued liabilities" instead of specific payable accounts. This is particularly useful when grouping several relatively small liabilities that aren't large enough to warrant separate reporting.

Accrued receivables **involve situations when the revenue is recognized in a period prior to the cash receipt.**

ACCRUED RECEIVABLES Accrued receivables involve the recognition of revenue for goods or services transferred to customers *before* cash is received. An example of an internal accrued revenue event is the recognition of interest earned on a loan to another entity. For example, assume that Dress Right loaned another corporation $30,000 at the beginning of August, evidenced by a note receivable. Terms of the note call for the payment of principal, $30,000, and interest at 8% in three months. An external transaction records the cash disbursement—a debit to note receivable and a credit to cash of $30,000.

The adjusting entry required to record an accrued revenue is a *debit to an asset,* a receivable, and a *credit to revenue.*

What adjusting entry would be required at the end of August? Dress Right needs to record the interest revenue earned but not yet received along with the corresponding receivable. Interest receivable increases and interest revenue (shareholders' equity) also increases. The adjusting entry for accrued receivables always includes a *debit to an asset,* a receivable, and a *credit to revenue.* In this case, at the end of August Dress Right recognizes $200 in interest revenue ($30,000 × 8% × 1/12) and makes the following adjusting entry. If this entry is not recorded, net income, assets, and shareholders' equity (retained earnings) will be understated.

To accrue interest revenue earned in August on a note receivable.

August 31

Interest receivable	200	
Interest revenue		200

There are no accrued revenue adjusting entries required for Dress Right at the end of July.

The required adjusting entries for prepayments and accruals are recapped with the aid of T-accounts in Illustration 2–10. In each case an expense or revenue is recognized in a period that differs from the period in which cash was paid or received. These adjusting

[4] Dress Right Clothing is a corporation. Corporations are income-tax-paying entities. Income taxes—federal, state, and local—are assessed on an annual basis and payments are made throughout the year. An additional adjusting entry would be required for Dress Right to accrue the amount of estimated income taxes payable that are applicable to the month of July. Accounting for income taxes is introduced in Chapter 4 and covered in depth in Chapter 16.

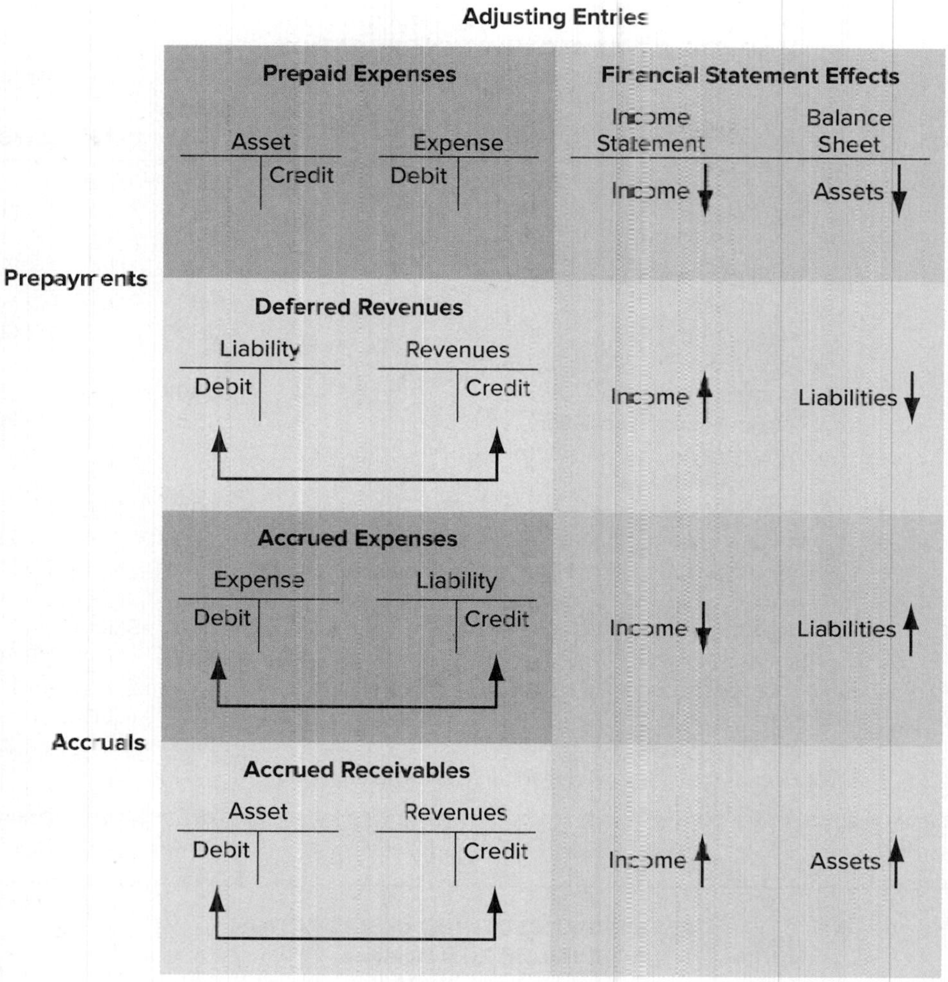

Illustration 2–10
Adjusting Entries

entries are necessary to properly measure operating performance and financial position on an accrual basis.

Estimates

A third classification of adjusting entries is **estimates**. Accountants often must make estimates of future events to comply with the accrual accounting model. For example, the calculation of depreciation expense requires an estimate of expected useful life of the asset being depreciated as well as its expected residual value. We discussed the adjusting entries for depreciation expense in the context of its being a prepayment, but it also could be thought of as an estimate.

Accountants often must make *estimates* in order to comply with the accrual accounting model.

One adjusting-entry situation involving an estimate that does not fit neatly into either the prepayment or accrual classification is bad debts. Accounting for bad debts requires a company to estimate the amount of accounts receivable that ultimately will prove to be uncollectible and to reduce accounts receivable by that estimated amount. This is neither a prepayment nor an accrual because it does not involve the payment of cash either before or after income is reduced. We explore accounts receivable and bad debts in depth in Chapter 7.

Illustration 2–11 recaps the July 31, 2021, adjusting entries for Dress Right Clothing Corporation as they would appear in a general journal. The journal entries are numbered (1) to (6) corresponding to the numbers used in the worksheet illustrated in Appendix 2A.

After the adjusting entries are posted to the general ledger accounts, the next step—step 7—in the processing cycle is to prepare an **adjusted trial balance**. The term adjusted refers to the fact that adjusting entries have now been posted to the accounts. Recall that the column titled Post. Ref. (Posting Reference) is the number assigned to the general ledger account that is being debited or credited. Illustration 2–12 shows the July 31, 2021, adjusted trial balance for Dress Right Clothing Corporation.

Illustration 2–11

The General Journal—Adjusting Entries

DRESS RIGHT CLOTHING CORPORATION
General Journal
Page 2

Date 2021		Account Title and Explanation	Post. Ref.	Debit	Credit
July (1)	31	Supplies expense	520	800	
		Supplies	125		800
		To record the cost of supplies used during the month of July.			
(2)	31	Rent expense	530	2,000	
		Prepaid rent	130		2,000
		To record the cost of expired rent for the month of July.			
(3)	31	Depreciation expense	540	200	
		Accumulated depreciation	155		200
		To record depreciation of office equipment for the month of July.			
(4)	31	Deferred rent revenue	230	250	
		Rent revenue	410		250
		To record previously deferred rent revenue recognized during July.			
(5)	31	Salaries expense	510	5,500	
		Salaries payable	230		5,500
		To record accrued salaries at the end of July.			
(6)	31	Interest expense	550	333	
		Interest payable	240		333
		To accrue interest expense for July on notes payable.			

Illustration 2–12

Adjusted Trial Balance

STEP 7

Prepare an *adjusted trial balance.*

DRESS RIGHT CLOTHING CORPORATION
Adjusted Trial Balance
July 31, 2021

Account Title	Debits	Credits
Cash	68,500	
Accounts receivable	2,000	
Supplies	1,200	
Prepaid rent	22,000	
Inventory	38,000	
Office equipment	12,000	
Accumulated depreciation		200
Accounts payable		35,000
Notes payable		40,000
Deferred rent revenue		750
Salaries payable		5,500
Interest payable		333
Common stock		60,000
Retained earnings		0
Sales revenue		38,500
Rent revenue		250
Cost of goods sold	22,000	
Salaries expense	10,500	
Supplies expense	800	
Rent expense	2,000	
Depreciation expense	200	
Interest expense	333	
Dividends	1,000	
Totals	180,533	180,533

Concept Review Exercise

ADJUSTING ENTRIES

The Wyndham Wholesale Company needs to prepare financial statements at the end of August 2021 for presentation to its bank. An unadjusted trial balance as of August 31, 2021, was presented in a previous Concept Review Exercise.

The following information also is available:

a. The notes payable account contains one note. The note requires the entire $30,000 in principal plus interest at 10% to be paid on July 31, 2022. The date of the note is August 1, 2021.

b. Depreciation on the equipment for the month of August is $500.

c. The notes receivable account contains one note. It is dated August 16, 2021. The note requires the entire $20,000 in principal plus interest at 12% to be repaid in four months (the loan was outstanding for one-half month during August).

d. The prepaid rent of $6,000 represents rent for the months of August and September.

Required:

1. Prepare any necessary adjusting entries at August 31, 2021
2. Prepare an adjusted trial balance as of August 31, 2021.
3. What is the total net effect on income (overstated or understated) if the adjusting entries are not recorded?

Solution:

1. Prepare any necessary adjusting entries at August 31, 2021
   a. An adjusting entry is required to accrue the interest expense on the note payable for the month of August. Accrued interest is calculated as follows:

$$\$30,000 \times 10\% \times 1/12 = \$250$$

Interest expense	250	
Interest payable		250

   b. Depreciation expense on the equipment must be recorded.

Depreciation expense	500	
Accumulated depreciation		500

   c. An adjusting entry is required for the one-half month of accrued interest revenue earned on the note receivable. Accrued interest is calculated as follows:

$$\$20,000 \times 12\% \times 1/12 \times 1/2 = \$100$$

Interest receivable	100	
Interest revenue		100

   d. An adjusting entry is required to recognize the amount of prepaid rent that expired during August.

Rent expense	3,000	
Prepaid rent		3,000

2. Prepare an adjusted trial balance as of August 31, 2021.

Account Title	Debits	Credits
Cash	21,000	
Accounts receivable	25,000	
Prepaid rent	3,000	
Inventory	16,000	
Interest receivable	100	
Notes receivable	20,000	
Equipment	20,000	
Accumulated depreciation		500
Accounts payable		20,000
Interest payable		250
Notes payable		30,000
Common stock		50,000
Retained earnings		0
Sales revenue		40,000
Interest revenue		100
Cost of goods sold	22,000	
Salaries expense	7,000	
Utilities expense	2,000	
Interest expense	250	
Depreciation expense	500	
Rent expense	3,000	
Dividends	1,000	
Totals	140,850	140,850

3. What is the effect on income (overstated or understated), if the adjusting entries are not recorded?

Adjusting Entry	Income overstated (understated)
Interest expense	$ 250
Depreciation expense	500
Interest revenue	(100)
Rent expense	3,000
Net effect, income overstated by	$3,650

We now turn our attention to the preparation of financial statements.

● LO2–7

Preparing the Financial Statements

STEP 8

Preparation of *financial statements.*

The purpose of each of the steps in the processing cycle to this point is to provide information for step 8—preparation of the **financial statements**. The adjusted trial balance contains the necessary information. After all, the financial statements are the primary means of communicating financial information to external parties.

The Income Statement and the Statement of Comprehensive Income

The *income statement* is a *change* statement that summarizes the profit-generating transactions that caused shareholders' equity (retained earnings) to change during the period.

The purpose of the **income statement** is to summarize the profit-generating activities of a company that occurred during a particular period of time. It is a *change* statement in that it reports the changes in shareholders' equity (retained earnings) that occurred during the period as a result of revenues, expenses, gains, and losses. Illustration 2–13 shows the income statement for Dress Right Clothing Corporation for the month of July 2021.

The income statement indicates a profit for the month of July of **$2,917**. During the month, the company was able to increase its net assets (equity) from activities related to selling its product. Dress Right is a corporation and subject to the payment of income tax on its profits. We ignore this required accrual here and address income taxes in a later chapter.

The components of the income statement usually are classified, that is, grouped according to common characteristics. A common classification scheme is to separate operating items from nonoperating items, as we do in Dress Right's income statement. Operating items

Illustration 2–13
Income Statement

DRESS RIGHT CLOTHING CORPORATION Income Statement For the Month of July 2021		
Sales revenue		$38,500
Cost of goods sold		22,000
Gross profit		16,500
Operating expenses:		
Salaries expense	$10,500	
Supplies expense	800	
Rent expense	2,000	
Depreciation expense	200	
Total operating expenses		13,500
Operating income		3,000
Other income (expense):		
Rent revenue	250	
Interest expense	(333)	(83)
Net income		$ 2,917

include revenues and expenses directly related to the principal revenue-generating activities of the company. For example, operating items for a manufacturing company include sales revenues from the sale of products and all expenses related to this activity. Companies that sell products like Dress Right often report a subtotal within operating income, sales less cost of goods sold, called *gross profit.* Nonoperating items include certain gains and losses and revenues and expenses from peripheral activities. For Dress Right Clothing, rent revenue and interest expense are nonoperating items because they do not relate to the principal revenue-generating activity of the company, selling clothes. In Chapter 4 we discuss the format and content of the income statement in more depth.

The *statement of comprehensive income* extends the income statement by reporting all changes in shareholders' equity during the period that were not a result of transactions with owners. A few types of gains and losses, called **other comprehensive income (OCI) or loss** items, are excluded from the determination of net income and the income statement, but are included in the broader concept of **comprehensive income**. Comprehensive income can be reported in one of two ways: (1) in a single, continuous statement of comprehensive income, or (2) in two separate, but consecutive statements.[5]

In the single statement approach, net income is a subtotal within the statement followed by these OCI items, culminating in a final total of comprehensive income. In the two statement approach, a company presents an income statement immediately followed by a statement of comprehensive income. The statement of comprehensive income begins with net income as the first component followed by OCI items to arrive at comprehensive income. Obviously, the approaches are quite similar: in the separate statement approach, we separate the continuous statement into two parts, but the content is the same.

Dress Right Clothing has no OCI items so the company presents only an income statement in Illustration 2–13. An entity that has no OCI items is not required to report OCI or comprehensive income. We discuss comprehensive income and the alternative approaches to its presentation in more depth in Chapter 4.

The Balance Sheet

The purpose of the **balance sheet** is to present the financial position of the company on a particular date. Unlike the income statement, which is a change statement reporting events that occurred *during a period of time*, the balance sheet is a statement that

The *balance sheet* presents an organized list of assets, liabilities, and equity at a particular point in time.

[5]FASB ASC 220–10–45–1: Comprehensive Income—Overall—Other Presentation Matters (*Accounting Standards Update No. 2011–05* (Norwalk, Conn.: FASB, June 2011)).

presents an organized list of assets, liabilities, and shareholders' equity *at a point in time.* To provide a quick overview, Illustration 2–14 shows the balance sheet for Dress Right at July 31, 2021.

Illustration 2–14
Balance Sheet

DRESS RIGHT CLOTHING CORPORATION
Balance Sheet
At July 31, 2021

Assets		
Current assets:		
Cash		$ 68,500
Accounts receivable		2,000
Supplies		1,200
Inventory		38,000
Prepaid rent		22,000
Total current assets		131,700
Property and equipment:		
Office equipment	$12,000	
Less: Accumulated depreciation	200	11,800
Total assets		$143,500
Liabilities and Shareholders' Equity		
Current liabilities:		
Accounts payable		$ 35,000
Salaries payable		5,500
Deferred rent revenue		750
Interest payable		333
Notes payable		10,000
Total current liabilities		51,583
Long-term liabilities:		
Notes payable		30,000
Shareholders' equity:		
Common stock (6,000 shares issued and outstanding)	$60,000	
Retained earnings	**1,917***	
Total shareholders' equity		61,917
Total liabilities and shareholders' equity		$143,500

*Beginning retained earnings + Net income − Dividends
$0 + 2,917 − 1,000 = **$1,917**

Balance sheet items usually are classified (grouped) according to common characteristics.

As we do in the income statement, we group the balance sheet elements into meaningful categories. For example, most balance sheets include the classifications of **current assets** and **current liabilities**. Current assets are those assets that are cash, will be converted into cash, or will be used up within one year from the balance sheet date (or operating cycle, if longer). Current liabilities are those liabilities that will be satisfied within one year from the balance sheet date (or operating cycle, if longer). For a manufacturing company, the operating cycle refers to the period of time necessary to convert cash to raw materials, raw materials to a finished product, the finished product to receivables, and then finally receivables back to cash. For most companies, this period is less than a year.

Examples of assets not classified as current include property and equipment and long-term receivables and investments. The only noncurrent asset that Dress Right has at July 31, 2021, is office equipment, which is classified under the property and equipment category.

All liabilities not classified as current are listed as long term. Dress Right's liabilities at July 31, 2021, include the $30,000 note payable due to be paid in 23 months. This liability is classified as long term.

Shareholders' equity lists the *paid-in capital* portion of equity—common stock—and *retained earnings*. Notice that the income statement we looked at in Illustration 2–13 ties in to the balance sheet through retained earnings. Specifically, the revenue, expense, gain, and loss transactions that make up net income in the income statement ($2,917) become the major components of retained earnings. Later in the chapter we discuss the closing process we use to transfer, or close, these *temporary* income statement accounts along with the temporary account, dividends, to the *permanent* retained earnings account.

During the month, retained earnings, which increased by the amount of net income, also decreased by the amount of the cash dividend paid to shareholders, $1,000. The net effect of these two changes is an increase in retained earnings from zero at the beginning of the period to $1,917 ($2,917 – $1,000) at the end of the period and also is reported in the statement of shareholders' equity in Illustration 2–16.

The Statement of Cash Flows

Similar to the income statement, the **statement of cash flows** also is a change statement. The purpose of the statement is to report the events that caused cash to change during the period. The statement classifies all transactions affecting cash into one of three categories: (1) **operating activities**, (2) **investing activities**, and (3) **financing activities**. Operating activities are inflows and outflows of cash related to transactions entering into the determination of net income. Investing activities involve the acquisition and sale of (1) long-term assets used in the business and (2) nonoperating investment assets. Financing activities involve cash inflows and outflows from transactions with creditors and owners.

The purpose of the *statement of cash flows* is to report the events that caused cash to change during the period.

The statement of cash flows for Dress Right for the month of July 2021 is shown in Illustration 2–15. As this is the first period of operations for Dress Right, the cash balance at the beginning of the period is zero. The net increase in cash of $68,500, therefore, equals the ending balance of cash disclosed in the balance sheet.

Illustration 2–15

Statement of Cash Flows

DRESS RIGHT CLOTHING CORPORATION
Statement of Cash Flows
For the Month of July 2021

Cash Flows from Operating Activities		
Cash inflows:		
From customers	$36,500	
From rent	1,000	
Cash outflows:		
For rent	(24,000)	
For supplies	(2,000)	
To suppliers of merchandise	(25,000)	
To employees	(5,000)	
Net cash flows from operating activities		$(18,500)
Cash Flows from Investing Activities		
Purchase of office equipment		(12,000)
Cash Flows from Financing Activities		
Issue of common stock	$60,000	
Increase in notes payable	40,000	
Payment of cash dividend	(1,000)	
Net cash flows from financing activities		99,000
Net increase in cash		$ 68,500

There are two generally accepted formats that can be used to report operating activities, the direct method and the indirect method. In Illustration 2–15 we use the direct method. These two methods are discussed and illustrated in subsequent chapters.

The Statement of Shareholders' Equity

The *statement of shareholders' equity* discloses the sources of changes in the permanent shareholders' equity accounts.

The final statement, the **statement of shareholders' equity**, also is a change statement. Its purpose is to disclose the sources of the changes in the various permanent shareholders' equity accounts that occurred during the period from investments by owners, distributions to owners, net income, and other comprehensive income. Illustration 2–16 shows the statement of shareholders' equity for Dress Right for the month of July 2021.[6]

Illustration 2–16

Statement of Shareholders' Equity

DRESS RIGHT CLOTHING CORPORATION
Statement of Shareholders' Equity
For the Month of July 2021

	Common Stock	Retained Earnings	Total Shareholders' Equity
Balance at July 1, 2021	$ –0–	$ –0–	$ –0–
Issue of common stock	60,000		60,000
Net income for July 2021		2,917	2,917
Less: Dividends		(1,000)	(1,000)
Balance at July 31, 2021	$ 60,000	$ 1,917	$ 61,917

The individual profit-generating transactions causing retained earnings to change are summarized in the income statement. Therefore, the statement of shareholders' equity only shows the net effect of these transactions on retained earnings, in this case an increase of **$2,917**. In addition, the company paid its shareholders a cash dividend that reduced retained earnings.

The Closing Process

STEP 9

Close the temporary accounts to retained earnings (at year-end only).

At the end of any interim reporting period, the accounting processing cycle is now complete. An interim reporting period is any period when financial statements are produced other than at the end of the fiscal year. However, at the end of the fiscal year, two final steps are necessary, closing the temporary accounts—step 9—and preparing a post-closing trial balance—step 10.

● LO2–8

The **closing process** serves a *dual purpose:* (1) the temporary accounts (revenues, expenses, gains and losses, and dividends) are reduced to *zero balances,* ready to measure activity in the upcoming accounting period, and (2) these temporary account balances are *closed (transferred) to retained earnings* to reflect the changes that have occurred in that account during the period.

To illustrate the closing process, assume that the fiscal year-end for Dress Right Clothing Corporation is July 31. Using the adjusted trial balance in Illustration 2–12, we can prepare the following closing entries.

To close the revenue accounts to retained earnings.

July 31		
Sales revenue	38,500	
Rent revenue	250	
Retained earnings		**38,750**

The first closing entry transfers the revenue account balances to retained earnings. Because revenue accounts have credit balances, we debit each one to bring their balances to zero. After this entry is posted to the accounts, both revenue accounts have a zero balance.

[6]Some companies choose to disclose the changes in the retained earnings component of shareholders' equity in a separate statement or in a combined statement of income and retained earnings.

To close the expense accounts to retained earnings.

July 31		
Retained earnings	**35,833**	
Cost of goods sold		22,000
Salaries expense		10,500
Supplies expense		800
Rent expense		2,000
Depreciation expense		200
Interest expense		333

The second closing entry transfers the expense account balances to retained earnings. Because expense accounts have debit balances, we credit them to bring their balances to zero. After this entry is posted to the accounts, the expense accounts have a zero balance. Also, because retained earnings in this example began the reporting period with a zero balance, after we add (credit) revenues to the account and subtract (debit) expenses, the retained earnings account has a credit balance equal to net income for the period, in this case $2,917.

Retained Earnings

Expenses	**35,833**	**38,750**	Revenues
		2,917	Net income

Additional Consideration

A company can choose to prepare closing entries by closing revenues and expenses to a temporary account called income summary. If so, the income summary account is then closed to retained earnings. This alternative set of closing entries provides the same effect of closing revenues and expenses directly to the retained earnings account as in our illustration. Most companies, though, have computerized accounting systems that perform the closing process automatically without the income summary account.

In the third entry we close the dividends account to retained earnings. Like expenses, dividends reduce retained earnings.

To close the dividends account to retained earnings.

July 31		
Retained earnings	1,000	
Dividends		1,000

STEP 10

Prepare a *post-closing trial balance* (at year-end only).

After this entry is posted to the accounts, the temporary accounts have zero balances and retained earnings has increased by the amount of the net income and decreased by dividends distributed from that net income. It's important to remember that the temporary accounts are closed only at year-end and not at the end of any interim period. Closing the temporary accounts during the year would make it difficult to prepare the annual income statement.

Additional Consideration

A previous additional consideration indicated that an alternative method of recording a cash dividend is to debit retained earnings when the dividends are declared. If this approach is used, this third closing entry to close the dividends account to retained earnings isn't needed.

The net result of a cash dividend is the same—a reduction in retained earnings and a reduction in cash.

After the closing entries are posted to the ledger accounts, a **post-closing trial balance** is prepared. The purpose of this trial balance is to verify that the closing entries were prepared and posted correctly and that the accounts are now ready for next year's transactions. Illustration 2–17 shows the post-closing trial balance for Dress Right at July 31, 2021, assuming a July 31 fiscal year-end.

Illustration 2–17
Post-Closing Trial Balance

DRESS RIGHT CLOTHING CORPORATION
Post-Closing Trial Balance
July 31, 2021

Account Title	Debits	Credits
Cash	68,500	
Accounts receivable	2,000	
Supplies	1,200	
Prepaid rent	22,000	
Inventory	38,000	
Office equipment	12,000	
Accumulated depreciation—office equipment		200
Accounts payable		35,000
Notes payable		40,000
Deferred rent revenue		750
Salaries payable		5,500
Interest payable		333
Common stock		60,000
Retained earnings		1,917
Totals	143,700	143,700

Concept Review Exercise

FINANCIAL STATEMENT PREPARATION AND CLOSING

Refer to the August 31, 2021, adjusted trial balance of the Wyndham Wholesale Company presented in the previous Concept Review Exercise.

Required:

1. Prepare an income statement and a statement of shareholders' equity for the month ended August 31, 2021, and a classified balance sheet as of August 31, 2021.
2. Assume that August 31 is the company's fiscal year-end. Prepare the necessary closing entries and a post-closing trial balance.

Solution:

1. Prepare an income statement and a statement of shareholders' equity for the month ended August 31, 2021, and a classified balance sheet as of August 31, 2021.

WYNDHAM WHOLESALE COMPANY
Income Statement
For the Month of August 2021

Sales revenue		$40,000
Cost of goods sold		22,000
Gross profit		18,000
Operating expenses:		
Salaries	$7,000	
Utilities	2,000	
Depreciation	500	
Rent	3,000	
Total operating expenses		12,500
Operating income		5,500
Other income (expense):		
Interest revenue	100	
Interest expense	(250)	(150)
Net income		$ 5,350

WYNDHAM WHOLESALE COMPANY
Statement of Shareholders' Equity
For the Month of August 2021

	Common Stock	Retained Earnings	Total Shareholders' Equity
Balance at August 1, 2021	$ –0–	$ –0–	$ –0–
Issue of common stock	50,000		50,000
Net income for August 2021		5,350	5,350
Less: Dividends		(1,000)	(1,000)
Balance at August 31, 2021	$50,000	$4,350	$54,350

WYNDHAM WHOLESALE COMPANY
Balance Sheet
At August 31, 2021

Assets

Current assets:		
Cash		$ 21,000
Accounts receivable		25,000
Inventory		16,000
Interest receivable		100
Notes receivable		20,000
Prepaid rent		3,000
Total current assets		85,100
Property and equipment:		
Equipment	$20,000	
Less: Accumulated depreciation	500	19,500
Total assets		$104,600

Liabilities and Shareholders' Equity

Current liabilities:		
Accounts payable		$ 20,000
Interest payable		250
Notes payable		30,000
Total current liabilities		50,250
Shareholders' equity:		
Common stock, 5,000 shares issued and outstanding	$50,000	
Retained earnings	4,350	
Total shareholders' equity		54,350
Total liabilities and shareholders' equity		$104,600

2. Assume that August 31 is the company's fiscal year-end. Prepare the necessary closing entries and a post-closing trial balance.

	Debit	Credit	
August 31			
Sales revenue	40,000		To close the revenue accounts to retained earnings.
Interest revenue	100		
Retained earnings		40,100	
August 31			
Retained earnings	34,750		To close the expense accounts to retained earnings.
Cost of goods sold		22,000	
Salaries expense		7,000	
Utilities expense		2,000	
Depreciation expense		500	
Rent expense		3,000	
Interest expense		250	
August 31			
Retained earnings	1,000		To close the dividends account to retained earnings.
Dividends		1,000	

Post-Closing Trial Balance

Account Title	Debits	Credits
Cash	21,000	
Accounts receivable	25,000	
Prepaid rent	3,000	
Inventory	16,000	
Interest receivable	100	
Notes receivable	20,000	
Equipment	20,000	
Accumulated depreciation—equipment		500
Accounts payable		20,000
Interest payable		250
Notes payable		30,000
Common stock		50,000
Retained earnings		4,350
Totals	105,100	105,100

Conversion from Cash Basis to Accrual Basis

● LO2–9 In Chapter 1, we discussed and illustrated the differences between cash and accrual accounting. Cash-basis accounting produces a measure called *net operating cash flow.* This measure is the difference between cash receipts and cash disbursements during a reporting period from transactions related to providing goods and services to customers. On the other hand, the accrual-accounting model measures an entity's accomplishments and resource sacrifices during the period, regardless of when cash is received or paid. At this point, you might wish to review the material in Chapter 1 to reinforce your understanding of the motivation for using the accrual-accounting model.

Adjusting entries, for the most part, are conversions from cash basis to accrual basis. Prepayments and accruals occur when cash flow precedes or follows expense or revenue recognition.

Accountants sometimes are called upon to convert cash-basis financial statements to accrual-basis financial statements, particularly for small businesses. You now have all of the tools you need to make this conversion. For example, if a company paid $20,000 cash for insurance during the fiscal year and you determine that there was $5,000 in prepaid insurance at the beginning of the year and $3,000 at the end of the year, then you can determine (accrual basis) *insurance expense* for the year. Prepaid insurance decreased by $2,000 during the year, so insurance expense must be $22,000 ($20,000 in cash paid *plus* the decrease in prepaid insurance). You can visualize as follows:

	Prepaid Insurance
Balance, beginning of year	$ 5,000
Plus: Cash paid	20,000
Less: Insurance expense	?
Balance, end of year	$ 3,000

Insurance expense of $22,000 completes the explanation of the change in the balance of prepaid insurance. Prepaid insurance of $3,000 is reported as an asset in an accrual-basis balance sheet.

Suppose a company paid $150,000 for salaries to employees during the year and you determine that there were $12,000 and $18,000 in salaries payable at the beginning and end of the year, respectively. What was salaries expense for the year?

Salaries Payable

Balance, beginning of year	$ 12,000
Plus: Salaries expense	?
Less: Cash paid	(150,000)
Balance, end of year	$ 18,000

Salaries payable increased by $6,000 during the year, so *salaries expense* must be $156,000 ($150,000 in cash paid *plus* the increase in salaries payable). Salaries payable of $18,000 is reported as a liability in an accrual-basis balance sheet.

Using T-accounts is a convenient approach for converting from cash to accrual accounting.

Salaries Payable

		12,000	Beg. balance
Cash paid	150,000		
		?	Salaries expense
		18,000	End. balance

Salaries Expense

?	

The debit to salaries expense and credit to salaries payable must have been $156,000 to balance the salaries payable account.

For another example using T-accounts, assume that the amount of cash collected from customers during the year was $220,000, and you know that accounts receivable at the beginning of the year was $45,000 and $33,000 at the end of the year. You can use T-accounts to determine that *sales revenue* for the year must have been $208,000, the necessary debit to accounts receivable and credit to sales revenue to balance the accounts receivable account.

Accounts Receivable

Beg. balance	45,000		
Credit sales	?		
		220,000	Cash collections
End. balance	33,000		

Sales Revenue

	?	Credit sales

Now suppose that, on occasion, customers pay in advance of receiving a product or service. Recall from our previous discussion of adjusting entries that this event creates a liability called deferred revenue. Assume the same facts in the previous example except you also determine that deferred revenues were $10,000 and $7,000 at the beginning and end of the year, respectively. A $3,000 decrease in deferred revenues means that the company recognized an additional $3,000 in sales revenue for which the cash had been collected in a previous year. So, *sales revenue* for the year must have been $211,000, the $208,000 determined in the previous example *plus* the $3,000 decrease in deferred revenue.

Illustration 2–18 provides another example of converting from cash-basis net income to accrual-basis net income.

Notice a pattern in the adjustments to cash-basis net income. When converting from cash-basis to accrual-basis income, we add increases and deduct decreases in assets. For example, an increase in accounts receivable means that the company recognized more revenue than cash collected, requiring the addition to cash-basis income. Conversely, we add decreases and deduct increases in accrued liabilities. For example, a decrease in interest payable means that the company incurred less interest expense than the cash interest it paid, requiring the addition to cash-basis income. These adjustments are summarized in Illustration 2–19.

Most companies must convert from an accrual basis to a cash basis when preparing the statement of cash flows.

Most companies keep their books on an accrual basis.[7] A more important conversion for these companies is from the accrual basis to the cash basis. This conversion, essential for the preparation of the statement of cash flows, is discussed and illustrated in Chapters 4

[7]Generally accepted accounting principles require the use of the accrual basis. Some small, nonpublic companies might use the cash basis in preparing their financial statements as an other basis of accounting.

Illustration 2–18

Cash to Accrual

The Krinard Cleaning Services Company maintains its records on the cash basis, with one exception. The company reports equipment as an asset and records depreciation expense on the equipment. During 2021, Krinard collected $165,000 from customers, paid $92,000 in operating expenses, and recorded $10,000 in depreciation expense, resulting in net income of **$63,000**. The owner has asked you to convert this $63,000 in net income to full accrual net income. You are able to determine the following information about accounts receivable, prepaid expenses, accrued liabilities, and deferred revenues:

	January 1, 2021	December 31, 2021
Accounts receivable	$16,000	$25,000
Prepaid expenses	7,000	4,000
Accrued liabilities (for operating expenses)	2,100	1,400
Deferred revenues	3,000	4,200

Accrual net income is $68,500, determined as follows:

Cash-basis net income		**$63,000**
Add:	Increase in accounts receivable	9,000
Deduct:	Decrease in prepaid expenses	(3,000)
Add:	Decrease in accrued liabilities	700
Deduct:	Increase in deferred revenues	(1,200)
Accrual-basis net income		$ 68,500

Illustration 2–19

Converting Cash-Basis to Accrual-Basis Income

Converting Cash-Basis Income to Accrual-Basis Income

	Increases	Decreases
Assets	Add	Deduct
Liabilities	Deduct	Add

and 21. The lessons learned here, though, will help you with that conversion. For example, if sales revenue for the period is $120,000 and beginning and ending accounts receivable are $20,000 and $24,000, respectively, how much cash did the company collect from its customers during the period? The answer is $116,000. An increase in accounts receivable of $4,000 means that the company collected $4,000 less from customers than accrual sales revenue, and cash-basis income is $4,000 less than accrual-basis income.

Financial Reporting Case Solution

©goodluz/123RF

1. **What purpose do adjusting entries serve?** *(p. 64)* Adjusting entries help ensure that all revenues are recognized in the period goods or services are transferred to customers, regardless of when cash is received. In this instance, for example, $13,000 cash has been received for services that haven't yet been performed. Also, adjusting entries enable a company to recognize all expenses incurred during a period, regardless of when cash is paid. Without depreciation, the friends' cost of using the equipment is not taken into account. Conversely, without adjustment, the cost of rent is overstated by $3,000 paid in advance for part of next year's rent.

With adjustments, we get an accrual income statement that provides a more complete measure of a company's operating performance and a better measure for predicting future operating cash flows. Similarly, the balance sheet provides a more complete assessment of assets and liabilities as sources of future cash receipts and disbursements.

2. What year-end adjustments are needed to revise the income statement? Did your friends do as well their first year as they thought? *(p. 64)* Three year-end adjusting entries are needed:

1. Depreciation expense ($80,000 ÷ 5 years)	16,000	
Accumulated depreciation		16,000
2. Prepaid rent [$500 × 6 months (July–Dec.)]	3,000	
Rent expense		3,000
3. Consulting revenue	13,000	
Deferred consulting revenue		13,000

No, your friends did not fare as well as their cash-based statement would have indicated. With appropriate adjustments, their net income is actually only $20,000:

Consulting revenue ($96,000 – $13,000)		$83,000
Operating expenses:		
Salaries expense	$32,000	
Rent expense ($9,000 – $3,000)	6,000	
Supplies expense	4,800	
Utilities expense	3,000	
Advertising expense	1,200	
Depreciation expense	16,000	63,000
Net income		$20,000

The Bottom Line

- **LO2–1** The accounting equation underlies the process used to capture the effect of economic events. The equation (Assets = Liabilities + Owners' Equity) implies an equality between the total economic resources of an entity (its assets) and the total claims against the entity (liabilities and equity). It also implies that each economic event affecting this equation will have a dual effect because resources always must equal claims. (*p. 48*)
- **LO2–2** The accounting processing cycle is the process used to identify, analyze, record, and summarize transactions and prepare financial statements. (*p. 51*)
- **LO2–3** After determining the dual effect of external events on the accounting equation, the transaction is recorded in a journal. A journal is a chronological list of transactions in debit/credit form. (*p. 55*)
- **LO2–4** The next step in the processing cycle is to periodically transfer, or *post,* the debit and credit information from the journal to individual general ledger accounts. A general ledger is simply a collection of all of the company's various accounts. Each account provides a summary of the effects of all events and transactions on that individual account. The process of entering items from the journal to the general ledger is called *posting.* An unadjusted trial balance is then prepared. (*p. 60*)
- **LO2–5** The next step in the processing cycle is to record the effect of *internal events* on the accounting equation. These transactions are commonly referred to as *adjusting entries.* Adjusting entries can be classified into three types: (1) prepayments, (2) accruals, and (3) estimates. Prepayments are transactions in which the cash flow *precedes* expense or revenue recognition. Accruals involve transactions where the cash outflow or inflow takes place in a period *subsequent* to expense or revenue recognition. Estimates for items such as future bad debts on receivables often are required to comply with the accrual accounting model. (*p. 63*)
- **LO2–6** Adjusting entries are recorded in the general journal and posted to the ledger accounts at the end of any period when financial statements must be prepared for external use. After these entries are posted to the general ledger accounts, an adjusted trial balance is prepared. (*p. 64*)
- **LO2–7** The adjusted trial balance is used to prepare the financial statements. The basic financial statements are: (1) the income statement, (2) the statement of comprehensive income, (3) the balance sheet, (4) the statement of cash flows, and (5) the statement of shareholders' equity. The purpose of the income statement is to summarize the profit-generating activities of the company that occurred during a particular period of time. A company also must report its other comprehensive income (OCI) or loss items either in a single, continuous statement or in a separate statement of comprehensive income. In the single-statement approach, net income is a subtotal within the statement followed by these OCI items, culminating in a

final total of comprehensive income. In the two-statement approach, a company presents an income statement immediately followed by a statement of comprehensive income. The statement of comprehensive income begins with net income as the first component followed by OCI items to arrive at comprehensive income. The balance sheet presents the financial position of the company on a particular date. The statement of cash flows discloses the events that caused cash to change during the reporting period. The statement of shareholders' equity discloses the sources of the changes in the various permanent shareholders' equity accounts that occurred during the period. (*p. 72*)

- **LO2–8** At the end of the fiscal year, a final step in the accounting processing cycle, closing, is required. The closing process serves a *dual purpose:* (1) the temporary accounts (revenues, expenses. and dividends) are reduced to *zero balances,* ready to measure activity in the upcoming accounting period, and (2) these temporary account balances are *closed (transferred) to retained earnings* to reflect the changes that have occurred in that account during the period. After the entries are posted to the general ledger accounts, a post-closing trial balance is prepared. (*p. 76*)
- **LO2–9** Cash-basis accounting produces a measure called *net operating cash flow.* This measure is the difference between cash receipts and cash disbursements during a reporting period from transactions related to providing goods and services to customers. On the other hand, the accrual-accounting model measures an entity's accomplishments and resource sacrifices during the period, regardless of when cash is received or paid. Accountants sometimes are called upon to convert cash-basis financial statements to accrual-basis financial statements, particularly for small businesses. (*p. 80*)

APPENDIX 2A Using a Worksheet

A *worksheet* can be used as a tool to facilitate the preparation of adjusting and closing entries and the financial statements.

A **worksheet** often is used to organize the accounting information needed to prepare adjusting and closing entries and the financial statements. It is an informal tool only and is not part of the accounting system. There are many different ways to design and use worksheets. We will illustrate a representative method using the financial information for the Dress Right Clothing Corporation presented in the chapter. Software such Excel facilitates the use of worksheets.

Illustration 2A–1 presents the completed worksheet. The worksheet is utilized in conjunction with Step 5 in the processing cycle, preparation of an unadjusted trial balance.

Illustration 2A–1 Worksheet, Dress Right Clothing Corporation, July 31, 2021

Source: Microsoft Excel 2007

Worksheet, Dress Right Clothing Corporation, July 31, 2021

	Unadjusted Trial Balance		Adjusting Entries		Adjusted Trial Balance		Income Statement		Balance Sheet	
Account Titles	Dr.	Cr.	Dr.	Cr.	Dr.	Cr.	Dr.	Cr.	Dr.	Cr.
Cash	68,500				68,500				68,500	
Accounts receivable	2,000				2,000				2,000	
Supplies	2,000			(1) 800	1,200				1,200	
Prepaid Rent	24,000			(2) 2,000	22,000				22,000	
Inventory	38,000				38,000				38,000	
Office equipment	12,000				12,000				12,000	
Accumulated depreciation		0		(3) 200		200				200
Accounts payable		35,000				35,000				35,000
Notes payable		40,000				40,000				40,000
Deferred rent revenue		1,000	(4) 250			750				750
Salaries payable		0		(5) 5,500		5,500				5,500
Interest payable		0		(6) 333		333				333
Common stock		60,000				60,000				60,000
Retained earnings		0				0		0		0
Sales revenue		38,500				38,500		38,500		
Rent revenue		0		(4) 250		250		250		
Cost of goods sold	22,000				22,000		22,000			
Salaries expense	5,000		(5) 5,500		10,500		10,500			
Supplies expense	0		(1) 800		800		800			
Rent expense	0		(2) 2,000		2,000		2,000			
Depreciation expense	0		(3) 200		200		200			
Interest expense	0		(6) 333		333		333			
Dividends	1,000				1,000				1,000	
Totals:	174,500	174,500	9,083	9,083	180,533	180,533				
Net income							2,917			2,917
Totals							38,750	38,750	144,700	144,700

Step 1. The account titles as they appear in the general ledger are entered in column A and the balances of these accounts are copied onto columns B and C, titled Unadjusted Trial Balance. The accounts are copied in the same order as they appear in the general ledger, which usually is assets, liabilities, shareholders' equity permanent accounts, revenues, and expenses. The debit and credit columns are totaled to make sure that they balance. This procedure is repeated for each set of columns in the worksheet to check for accuracy.

The first step is to enter account titles in column A and the unadjusted account balances in columns B and C.

Step 2. The end-of-period adjusting entries are determined and entered directly on the worksheet in columns D and E, titled Adjusting Entries. The adjusting entries for Dress Right Clothing Corporation were discussed in detail in the chapter and exhibited in general journal form in Illustration 2–11. You should refer back to this illustration and trace each of the entries to the worksheet. For worksheet purposes, the entries have been numbered from (1) to (6) for easy referencing.

The second step is to determine end-of-period adjusting entries and enter them in columns D and E.

For example, entry (1) records the cost of supplies used during the month of July with a debit to supplies expense and a credit to supplies for **$800**. A (1) is placed next to the **$800** in the debit column in the supplies expense row as well as next to the $800 in the credit column in the supplies row. This allows us to more easily reconstruct the entry for general journal purposes and locate errors if the debit and credit columns do not balance.

Step 3. The effects of the adjusting entries are added to or deducted from the account balances listed in the Unadjusted Trial Balance columns and copied across to columns F and G, titled Adjusted Trial Balance. For example, supplies had an unadjusted balance of **$2,000**. Adjusting entry (1) credited this account by $800, reducing the balance to **$1,200**.

The third step adds or deducts the effects of the adjusting entries on the account balances.

Step 4. The balances in the temporary retained earnings accounts, revenues and expenses, are transferred to columns H and I, titled Income Statement. The difference between the total debits and credits in these columns is equal to net income or net loss. In this case, because credits (revenues) exceed debits (expenses), a net income of $2,917 results. To balance the debits and credits in this set of columns, a **$2,917** debit entry is made in the line labeled Net income.

The fourth step is to transfer the temporary retained earnings account balances to columns H and I.

Step 5. The balances in the permanent accounts are transferred to columns J and K, titled Balance Sheet. To keep the debits and credits equal in the worksheet, a **$2,917** credit must be entered to offset the $2,917 debit entered in step 4 and labeled as net income. This credit represents the fact that when the temporary revenue and expense accounts are closed out to retained earnings, a $2,917 credit to retained earnings will result. The credit in column K, therefore, represents an increase in retained earnings for the period, that is, net income. The balance in the temporary account, dividends, is also transferred to column J, titled Balance Sheet. This debit represents the fact that when the dividends account is closed to retained earnings, the $2,917 balance in that account resulting from net income will be reduced by the $1,000 of net income distributed to shareholders as a dividend.

The fifth step is to transfer the balances in the permanent accounts to columns J and K.

After the worksheet is completed, the financial statements can be prepared directly from columns H–K. The financial statements for Dress Right Clothing Corporation are shown in Illustrations 2–13 through 2–16 . The accountant must remember to then record the adjusting entries and the closing entries in the general journal and post them to the general ledger accounts. An adjusted trial balance should then be prepared identical to the one in the worksheet, which is used to prepare the financial statements. At fiscal year-end, the income statement columns can then be used to prepare closing entries. After that, a post-closing trial balance is prepared to wipe the slate clean for the next reporting period. ●

APPENDIX 2B

Reversing Entries

Accountants sometimes use **reversing entries** at the beginning of a reporting period. These optional entries remove the effects of some of the adjusting entries recorded at the end of the previous reporting period for the sole purpose of simplifying journal entries recorded during the new period. If the accountant does use reversing entries, these entries are recorded in the general journal and posted to the general ledger accounts on the first day of the new period.

Reversing entries are used most often with accruals. For example, the following adjusting entry for accrued salaries was recorded at the end of July 2021 for the Dress Right Clothing Corporation in the chapter:

To record accrued salaries at the end of July.

July 31		
Salaries expense	5,500	
Salaries payable		5,500

If reversing entries are not used, when the salaries actually are paid in August, the accountant needs to remember to debit salaries payable and not salaries expense.

The account balances before and after salary payment can be seen below with the use of T-accounts.

Salaries Expense

	Debit	Credit
Bal. July 31	10,500	

Salaries Payable

	Debit	Credit	
		5,500	Bal. July 31
Cash payment	5,500		
		–0–	Balance

If the accountant for Dress Right employs reversing entries, the following entry is recorded on August 1, 2021:

To reverse accrued salaries expense recorded at the end of July.

August 1		
Salaries payable	5,500	
Salaries expense		5,500

This entry reduces the salaries payable account to zero and reduces the salary expense account by $5,500. When salaries actually are paid in August, the debit is to salaries expense, thus increasing the account by $5,500.

Salaries Expense

	Debit	Credit
Bal. July 31	10,500	
(Reversing entry)		5,500
Cash payment	5,500	
Balance	10,500	

Salaries Payable

	Debit	Credit	
		5,500	Bal. July 31
(Reversing entry)	5,500		
		–0–	Balance

We can see that balances in the accounts after cash payment is made are identical. The use of reversing entries for accruals, which is optional, simply allows cash payments or cash receipts to be entered directly into the temporary expense or revenue accounts without regard to the accruals recorded at the end of the previous period.

Reversing entries also can be used with prepayments and deferred revenues. For example, earlier in the chapter Dress Right Clothing Corporation used the following entry to record the purchase of supplies on July 6:

To record the purchase of supplies.

July 6		
Supplies	2,000	
Cash		2,000

If reversing entries are not used, an adjusting entry is needed at the end of July to record the amount of supplies consumed during the period. In the illustration, Dress Right recorded this adjusting entry at the end of July:

To record the cost of supplies used during the month of July.

July 31		
Supplies expense	800	
Supplies		800

T-accounts help us visualize the account balances before and after the adjusting entry.

Supplies		
(Cash payment)	2,000	
		800
Bal. July 31	1,200	

Supplies Expense		
(Adjusting entry)	800	
Bal. July 31	800	

If the accountant for Dress Right employs *reversing entries,* the purchase of supplies is recorded as follows:

To record the purchase of supplies.

July 6		
Supplies expense	2,000	
Cash		2,000

The adjusting entry then is used to establish the balance in the supplies account at $1,200 (amount of supplies still on hand at the end of the month) and reduce the supplies expense account from the amount purchased to the amount used.

To record the cost of supplies on hand at the end of July.

July 31		
Supplies (balance on hand)	1,200	
Supplies expense ($2,000 – $800)		1,200

T-accounts make the process easier to see before and after the adjusting entry.

Supplies		
(Adjusting entry)	1,200	
Bal. July 31	1,200	

Supplies Expense		
Cash payment	2,000	
(Adjusting entry)		1,200
Bal. July 31	800	

Notice that the ending balances in both accounts are the same as when reversing entries are not used. Up to this point, this approach is the alternate approach to recording prepayments discussed in a previous section of this chapter. The next step is an optional expediency.

On August 1, the following reversing entry can be recorded:

To reverse the July adjusting entry for supplies on hand.

August 1		
Supplies expense	1,200	
Supplies		1,200

This entry reduces the supplies account to zero and increases the supplies expense account to $2,000. Subsequent purchases would then be entered into the supplies expense account and future adjusting entries would record the amount of supplies still on hand at the end of the period. At the end of the fiscal year, the supplies expense account, along with all other temporary accounts, is closed to retained earnings.

Using reversing entries for prepayments, which is optional, simply allows cash payments to be entered directly into the temporary expense accounts without regard to whether only the current, or both the current and future periods, are benefitted by the expenditure. Adjustments are then recorded at the end of the period to reflect the amount of the unexpired benefit (asset). ●

APPENDIX 2C

Subsidiary Ledgers and Special Journals

Subsidiary Ledgers

Accounting systems employ a *subsidiary ledger,* which contains a group of subsidiary accounts associated with particular general ledger control accounts.

The general ledger contains what are referred to as *control accounts.* In addition to the general ledger, a **subsidiary ledger** contains a group of subsidiary accounts associated with a particular general ledger control account. For example, there will be a subsidiary ledger for accounts receivable that keeps track of the increases and decreases in the account receivable balance for each of the company's customers purchasing goods or services on credit. After all of the postings are made from the appropriate journals, the balance in the accounts receivable control account should equal the sum of the balances in the accounts receivable subsidiary ledger accounts. Subsidiary ledgers also are used for accounts payable, property and equipment, investments, and other accounts.

Special Journals

For most external transactions, *special journals* are used to capture the dual effect of the transaction in debit/credit form.

An actual accounting system employs many different types of journals. The purpose of each journal is to record, in chronological order, the dual effect of a transaction in debit/credit form. The chapter used the general journal format to record each transaction. However, even for small companies with relatively few transactions, the general journal is used to record only a few types of transactions.[8]

The majority of transactions are recorded in **special journals**. These journals capture the dual effect of *repetitive* types of transactions. For example, cash receipts are recorded in a **cash receipts journal**, cash disbursements in a **cash disbursements journal**, credit sales in a **sales journal**, and the purchase of merchandise on account in a **purchases journal**.

Special journals simplify the recording process in the following ways:

1. Journalizing the effects of a particular transaction is made more efficient through the use of specifically designed formats.
2. Individual transactions are not posted to the general ledger accounts but are accumulated in the special journals and a summary posting is made on a periodic basis.
3. The responsibility for recording journal entries for the repetitive types of transactions is placed on individuals who have specialized training in handling them.

The concepts of subsidiary ledgers and special journals are illustrated using the *sales journal* and the *cash receipts journal.*

Sales Journal

All credit sales are recorded in the *sales journal.*

The purpose of the **sales journal** is to record all credit sales. Cash sales are recorded in the cash receipts journal. Every entry in the sales journal has exactly the same effect on the accounts; the sales revenue account is credited and the accounts receivable control account is debited. Therefore, there is only one column needed to record the debit/credit effect of these transactions. Other columns are needed to capture information for updating the accounts receivable subsidiary ledger. Illustration 2C–1 presents the sales journal for Dress Right Clothing Corporation for the month of August 2021.

Illustration 2C–1

Sales Journal, Dress Right Clothing Corporation, August 2021

Page 1

Date	Accounts Receivable Subsidiary Account No.	Customer Name	Sales Invoice No.	Cr. Sales Revenue (400) Dr. Accounts Receivable (110)
2021				
Aug. 5	801	Leland High School	10-221	1,500
9	812	Mr. John Smith	10-222	200
18	813	Greystone School	10-223	825
22	803	Ms. Barbara Jones	10-224	120
29	805	Hart Middle School	10-225	650
				3,295

[8] For example, end-of-period adjusting entries would be recorded in the general journal.

During the month of August, the company made five credit sales, totaling $3,295. This amount is posted as a debit to the accounts receivable control account, account number 110, and a credit to the sales revenue account, account number 400. The T-accounts for accounts receivable and sales revenue appear below. The reference SJ1 refers to page 1 of the sales journal.

General Ledger

Accounts Receivable	110		Sales Revenue	400
July 31 Balance	2,000			
Aug. 31 SJ1	3,295		3,295	Aug. 31 SJ1

In a computerized accounting system, as each transaction is recorded in the sales journal, the subsidiary ledger accounts for the customer involved will automatically be updated. For example, the first credit sale of the month is to Leland High School for $1,500. The sales invoice number for this sale is 10-221 and the customer's subsidiary account number is 801. As this transaction is entered, the subsidiary account 801 for Leland High School is debited for $1,500

Accounts Receivable Subsidiary Ledger

Leland High School		801
August 5 SJ1	1,500	

As cash is collected from this customer the cash receipts journal records the transaction with a credit to the accounts receivable control account and a debit to cash. At the same time, the accounts receivable subsidiary ledger account number 801 also is credited. After the postings are made from the special journals, the balance in the accounts receivable control account should equal the sum of the balances in the accounts receivable subsidiary ledger accounts.

Cash Receipts Journal

The purpose of the **cash receipts journal** is to record all cash receipts, regardless of the source. Every transaction recorded in this journal produces a debit entry to the cash account with the credit to various other accounts. Illustration 2C–2 shows a cash receipts journal using transactions of the Dress Right Clothing Corporation for the month of August 2021.

All cash receipts are recorded in the *cash receipts journal*.

Illustration 2C–2

Cash Receipts Journal, Dress Right Clothing Corporation, August 2021

Page 1

Date	Explanation or Account Name	Dr. Cash (100)	Cr. Accounts Receivable (110)	Cr. Sales Revenue (400)	Cr. Other	Other Accounts
2021						
Aug. 7	Cash sale	500		500		
11	Borrowed cash	10,000			10,000	Notes payable (220)
17	Leland High School	750	750			
20	Cash sale	300		300		
25	Mr. John Smith	200	200			
		11,750	950	800	10,000	

Because every transaction results in a debit to the cash account, No. 100, a column is provided for that account. At the end of August, an $11,750 debit is posted to the general ledger cash account with the source labeled CR1, cash receipts journal, page 1.

Because cash and credit sales are common, separate columns are provided for these accounts. At the end of August, a **$950** credit is posted to the accounts receivable general ledger account, No. 110, and an **$800** credit is posted to the sales revenue account, No. 400. Two additional credit columns are provided for uncommon cash receipt transactions, one for the credit amount and one for the account being credited. We can see that in August, Dress Right borrowed **$10,000** requiring a credit to the notes payable account, No. 220.

In addition to the postings to the general ledger control accounts, each time an entry is recorded in the accounts receivable column, a credit is posted to the accounts receivable subsidiary ledger account for the customer making the payment. For example, on August 17, Leland High School paid $750 on account. The subsidiary ledger account for Leland High School is credited for **$750**.

Accounts Receivable Subsidiary Ledger

	Leland High School		801
August 5 SJ1	1,500		
		750	August 17 CR1

Questions For Review of Key Topics

Q 2–1 Explain the difference between external events and internal events. Give an example of each type of event.

Q 2–2 Each economic event or transaction will have a dual effect on financial position. Explain what is meant by this dual effect.

Q 2–3 What is the purpose of a journal? What is the purpose of a general ledger?

Q 2–4 Explain the difference between permanent accounts and temporary accounts. Why does an accounting system include both types of accounts?

Q 2–5 Describe how debits and credits affect assets, liabilities, and permanent owners' equity accounts.

Q 2–6 Describe how debits and credits affect temporary owners' equity accounts.

Q 2–7 What is the first step in the accounting processing cycle? What role do source documents fulfill in this step?

Q 2–8 Describe what is meant by transaction analysis.

Q 2–9 Describe what is meant by posting, the fourth step in the processing cycle.

Q 2–10 Describe the events that correspond to the following two journal entries:

1. Inventory	20,000	
Accounts payable		20,000
2. Accounts receivable	30,000	
Sales revenue		30,000
Cost of goods sold	18,000	
Inventory		18,000

Q 2–11 What is an unadjusted trial balance? An adjusted trial balance?

Q 2–12 Define adjusting entries and discuss their purpose.

Q 2–13 Define closing entries and their purpose.

Q 2–14 Define prepaid expenses and provide at least two examples.

Q 2–15 Deferred revenues represent liabilities recorded when cash is received from customers in advance of providing a good or service. What adjusting journal entry is required at the end of a period to recognize the amount of deferred revenues that were recognized during the period?

Q 2–16 Define accrued liabilities. What adjusting journal entry is required to record accrued liabilities?

Q 2–17 Describe the purpose of each of the five primary financial statements.

Q 2–18 [Based on Appendix A] What is the purpose of a worksheet? In a columnar worksheet similar to Illustration 2A–1, what would be the result of incorrectly transferring the balance in a liability account to column I, the credit column under Income Statement?

Q 2–19 [Based on Appendix B] Define reversing entries and discuss their purpose.

Q 2–20 [Based on Appendix C] What is the purpose of special journals? In what ways do they simplify the recording process?

Q 2–21 [Based on Appendix C] Explain the difference between the general ledger and a subsidiary ledger.

Brief Exercises

connect

BE 2–1
Transaction analysis
● LO2–1

The Marchetti Soup Company entered into the following transactions during the month of June: (1) purchased inventory on account for $165,000 (assume Marchetti uses a perpetual inventory system); (2) paid $40,000 in salaries to employees for work performed during the month; (3) sold merchandise that cost $120,000 to credit customers for $200,000; (4) collected $180,000 in cash from credit customers; and (5) paid suppliers of inventory $145,000. Analyze each transaction and show the effect of each on the accounting equation for a corporation.

BE 2–2
Journal entries
● LO2–3

Prepare journal entries for each of the transactions listed in BE 2–1.

BE 2–3
T-accounts
● LO2–4

Post the journal entries prepared in BE 2–2 to T-accounts. Assume that the opening balances in each of the accounts is zero except for cash, accounts receivable, and accounts payable that had opening balances of $65,000, $43,000, and $22,000, respectively.

BE 2–4
Journal entries
● LO2–3

Prepare journal entries for each of the following transactions for a company that has a fiscal year-end of December 31: (1) on October 1, $12,000 was paid for a one-year fire insurance policy; (2) on June 30 the company advanced its chief financial officer $10,000; principal and interest at 6% on the note are due in one year; and (3) equipment costing $60,000 was purchased at the beginning of the year for cash.

BE 2–5
Adjusting entries
● LO2–6

Prepare the necessary adjusting entries at December 31 for each of the items listed in BE 2–4. Depreciation on the equipment is $12,000 per year.

BE 2–6
Adjusting entries; income determination
● LO2–5, LO2–6

If the adjusting entries prepared in BE 2–5 were not recorded, would net income be higher or lower and by how much?

BE 2–7
Adjusting entries
● LO2–6

Prepare the necessary adjusting entries at its year-end of December 31, 2021, for the Jamesway Corporation for each of the following situations. No adjusting entries were recorded during the year.

1. On December 10, 2021, Jamesway received a $4,000 payment from a customer for services begun on that date and which were completed by December 31, 2021. Deferred service revenue was credited.
2. On December 1, 2021, the company paid a local radio station $2,000 for 40 radio ads that were to be aired, 20 per month, throughout December and January. Prepaid advertising was debited.
3. Employee salaries for the month of December totaling $16,000 will be paid on January 7, 2022.
4. On August 31, 2021, Jamesway borrowed $60,000 from a local bank. A note was signed with principal and 8% interest to be paid on August 31, 2022.

BE 2–8
Income determination
● LO2–5

If none of the adjusting journal entries prepared in BE 2–7 were recorded, would assets, liabilities, and shareholders' equity on the 12/31/2021 balance sheet be higher or lower and by how much?

BE 2–9
Adjusting entries
● LO2–6

Prepare the necessary adjusting entries for Johnstone Controls at the end of its December 31, 2021, fiscal year-end for each of the following situations. No adjusting entries were recorded during the year.

1. On March 31, 2021, the company lent $50,000 to another company. A note was signed with principal and interest at 6% payable on March 31, 2022.
2. On September 30, 2021, the company paid its landlord $12,000 representing rent for the period September 30, 2021, to September 30, 2022. Johnstone debited prepaid rent.
3. Supplies on hand at the end of 2020 totaled $3,000. Additional supplies costing $5,000 were purchased during 2021 and debited to the supplies account. At the end of 2021, supplies costing $4,200 remain on hand.
4. Vacation pay of $6,000 for the year that had been earned by employees was not paid or recorded. The company records vacation pay as salaries expense.

BE 2–10
Financial statements
● LO2–7

The following account balances were taken from the 2021 adjusted trial balance of the Bowler Corporation: sales revenue, $325,000; cost of goods sold, $168,000; salaries expense, $45,000; rent expense, $20,000; depreciation expense, $30,000; and miscellaneous expense, $12,000. Prepare an income statement for 2021.

BE 2–11
Financial statements
● LO2–7

The following account balances were taken from the 2021 post-closing trial balance of the Bowler Corporation: cash, $5,000; accounts receivable, $10,000; inventory, $16,000; equipment, $100,000; accumulated depreciation, $40,000; accounts payable, $20,000; salaries payable, $12,000; retained earnings, $9,000; and common stock, $50,000. Prepare a 12/31/2021 balance sheet.

BE 2–12
Closing entries
● LO2–8

The year-end adjusted trial balance of the Timmons Tool and Die Corporation included the following account balances: retained earnings, $220,000; dividends, $12,000; sales revenue, $850,000; cost of goods sold, $580,000; salaries expense, $180,000; rent expense, $40,000; and interest expense, $15,000. Prepare the necessary closing entries.

BE 2–13
Cash versus accrual accounting
● LO2–9

Newman Consulting Company maintains its records on a cash basis. During 2021 the following cash flows were recorded: cash received for services rendered to clients, $420,000; and cash paid for salaries, utilities, and advertising, $240,000, $35,000, and $12,000, respectively. You also determine that customers owed the company $52,000 and $60,000 at the beginning and end of the year, respectively, and that the company owed the utility company $6,000 and $4,000 at the beginning and end of the year, respectively. Determine accrual net income for the year.

Exercises

connect

E 2–1
Transaction analysis
● LO2–1

The following transactions occurred during March 2021 for the Wainwright Corporation. The company owns and operates a wholesale warehouse.

1. Issued 30,000 shares of common stock in exchange for $300,000 in cash.
2. Purchased equipment at a cost of $40,000. $10,000 cash was paid and a note payable to the seller was signed for the balance owed.
3. Purchased inventory on account at a cost of $90,000. The company uses the perpetual inventory system.
4. Credit sales for the month totaled $120,000. The cost of the goods sold was $70,000.
5. Paid $5,000 in rent on the warehouse building for the month of March.
6. Paid $6,000 to an insurance company for fire and liability insurance for a one-year period beginning April 1, 2021.
7. Paid $70,000 on account for the merchandise purchased in 3.
8. Collected $55,000 from customers on account.
9. Recorded depreciation expense of $1,000 for the month on the equipment.

Required:

Analyze each transaction and show the effect of each on the accounting equation for a corporation.

Example:

	Assets	= Liabilities + Paid-In Capital + Retained Earnings
1.	+300,000 (cash)	+ 300,000 (common stock)

E 2–2
Journal entries
● LO2–3

Prepare journal entries to record each of the transactions listed in E 2–1.

E 2–3
T-accounts and trial balance
● LO2–4

Post the journal entries prepared in E 2–2 to T-accounts. Assume that the opening balances in each of the accounts is zero. Prepare a trial balance from the ending account balances.

E 2–4
Journal entries
● LO2–3

The following transactions occurred during the month of June 2021 for the Stridewell Corporation. The company owns and operates a retail shoe store.

1. Issued 100,000 shares of common stock in exchange for $500,000 cash.
2. Purchased office equipment at a cost of $100,000. $40,000 was paid in cash and a note payable was signed for the balance owed.
3. Purchased inventory on account at a cost of $200,000. The company uses the perpetual inventory system.
4. Credit sales for the month totaled $280,000. The cost of the goods sold was $140,000.
5. Paid $6,000 in rent on the store building for the month of June.

6. Paid $3,000 to an insurance company for fire and liability insurance for a one-year period beginning June 1, 2021.
7. Paid $120,000 on account for the merchandise purchased in 3.
8. Collected $55,000 from customers on account.
9. Paid shareholders a cash dividend of $5,000.
10. Recorded depreciation expense of $2,000 for the month on the office equipment.
11. Recorded the amount of prepaid insurance that expired for the month.

Required:

Prepare journal entries to record each of the transactions and events listed above.

E 2–5
The accounting processing cycle
● LO2–2

Listed below are several terms and phrases associated with the accounting processing cycle. Pair each item from List A (by letter) with the item from List B that is most appropriately associated with it.

List A	List B
______ 1. Source documents	a. Record of the dual effect of a transaction in debit/credit form.
______ 2. Transaction analysis	b. Internal events recorded at the end of a reporting period.
______ 3. Journal	c. Primary means of disseminating information to external decision makers.
______ 4. Posting	d. To zero out the temporary accounts.
______ 5. Unadjusted trial balance	e. Determine the dual effect on the accounting equation.
______ 6. Adjusting entries	f. List of accounts and their balances before recording adjusting entries.
______ 7. Adjusted trial balance	g. List of accounts and their balances after recording closing entries.
______ 8. Financial statements	h. List of accounts and their balances after recording adjusting entries.
______ 9. Closing entries	i. A means of organizing information: not part of the formal accounting system.
______ 10. Post-closing trial balance	j. Transferring balances from the journal to the ledger.
______ 11. Worksheet	k. Used to identify and process external transactions.

E 2–6
Debits and credits
● LO2–1

Indicate whether a *debit* will increase (I) or decrease (D) each of the following accounts listed in items 1 through 15.

Increase (I) or Decrease (D)	Account
1. ______	Inventory
2. ______	Depreciation expense
3. ______	Accounts payable
4. ______	Prepaid rent
5. ______	Sales revenue
6. ______	Common stock
7. ______	Salaries payable
8. ______	Cost of goods sold
9. ______	Utilities expense
10. ______	Equipment
11. ______	Accounts receivable
12. ______	Utilities payable
13. ______	Rent expense
14. ______	Interest expense
15. ______	Interest revenue

E 2–7
Journal entries; debits and credits
● LO2–3

Some of the ledger accounts for the Sanderson Hardware Company are numbered and listed below. For each of the October 2021 transactions numbered 1 through 10 below, indicate by account number which accounts should be debited and which should be credited when preparing journal entries. The company uses the perpetual inventory system.

(1) Accounts payable	(2) Equipment	(3) Inventory
(4) Accounts receivable	(5) Cash	(6) Supplies
(7) Supplies expense	(8) Prepaid rent	(9) Sales revenue
(10) Retained earnings	(11) Notes payable	(12) Common stock
(13) Deferred sales revenue	(14) Rent expense	(15) Salaries payable
(16) Cost of goods sold	(17) Salaries expense	(18) Interest expense
(19) Dividends		

	Account(s) Debited	Account(s) Credited
Example: Purchased inventory for cash	3	5

1. Paid a cash dividend.
2. Paid rent for the next three months.
3. Sold goods to customers on account.
4. Purchased inventory on account.
5. Purchased supplies for cash.
6. Issued common stock in exchange for cash.
7. Collected cash from customers for goods sold in 3.
8. Borrowed cash from a bank and signed a note.
9. At the end of October, recorded the amount of supplies that had been used during the month.
10. Received cash for advance payment from customer.

E 2–8
Adjusting entries
● LO2–6

Prepare the necessary adjusting entries at December 31, 2021, for the Falwell Company for each of the following situations. Assume that no financial statements were prepared during the year and no adjusting entries were recorded.

1. A three-year fire insurance policy was purchased on July 1, 2021, for $12,000. The company debited insurance expense for the entire amount.
2. Depreciation on equipment totaled $15,000 for the year.
3. Employee salaries of $18,000 for the month of December will be paid in early January 2022.
4. On November 1, 2021, the company borrowed $200,000 from a bank. The note requires principal and interest at 12% to be paid on April 30, 2022.
5. On December 1, 2021, the company received $3,000 in cash from another company that is renting office space in Falwell's building. The payment, representing rent for December, January, and February was credited to deferred rent revenue.
6. In the previous transaction, suppose the company credited rent revenue rather than deferred rent revenue for $3,000 on December 1, 2021. What would be the appropriate adjusting entry at December 31, 2021?

E 2–9
Adjusting entries
● LO2–6

Prepare the necessary adjusting entries at December 31, 2021, for the Microchip Company for each of the following situations. Assume that no financial statements were prepared during the year and no adjusting entries were recorded.

1. On October 1, 2021, Microchip lent $90,000 to another company. A note was signed with principal and 8% interest to be paid on September 30, 2022.
2. On November 1, 2021, the company paid its landlord $6,000 representing rent for the months of November through January. Prepaid rent was debited.
3. On August 1, 2021, collected $12,000 in advance rent from another company that is renting a portion of Microchip's factory. The $12,000 represents one year's rent and the entire amount was credited to deferred rent revenue.
4. Depreciation on office equipment is $4,500 for the year.
5. Vacation pay for the year that had been earned by employees but not paid to them or recorded is $8,000. The company records vacation pay as salaries expense.
6. Microchip began the year with $2,000 in its asset account, supplies. During the year, $6,500 in supplies were purchased and debited to supplies. At year-end, supplies costing $3,250 remain on hand.

E 2–10
Adjusting entries; solving for unknowns
● LO2–5, LO2–6

The Eldorado Corporation's controller prepares adjusting entries only at the end of the reporting year. The following adjusting entries were prepared on December 31, 2021:

	Debit	Credit
Interest expense	7,200	
Interest payable		7,200
Rent expense	35,000	
Prepaid rent		35,000
Interest receivable	500	
Interest revenue		500

Additional information:

1. The company borrowed $120,000 on March 31, 2021. Principal and interest are due on March 31, 2022. This note is the company's only interest-bearing debt.
2. Rent for the year on the company's office space is $60,000. The rent is paid in advance.
3. On October 31, 2021, Eldorado lent money to a customer. The customer signed a note with principal and interest at 6% due in one year.

Required:

Determine the following:

1. What is the interest rate on the company's note payable?
2. The 2021 rent payment was made at the beginning of which month?
3. How much did Eldorado lend its customer on October 31?

E 2–11
Adjusting entries; fiscal year
● LO2–6

The Mazzanti Wholesale Food Company's fiscal year-end is June 30. The company issues quarterly financial statements requiring the company to prepare adjusting entries at the end of each quarter. Assuming all quarterly adjusting entries were properly recorded, prepare the necessary year-end adjusting entries at the end of June 30, 2021, for the following situations.

1. On December 1, 2020, the company paid its annual fire insurance premium of $6,000 for the year beginning December 1 and debited prepaid insurance.
2. On August 31, 2020, the company borrowed $80,000 from a local bank. The note requires principal and interest at 8% to be paid on August 31, 2021.
3. Mazzanti owns a warehouse that it rents to another company. On January 1, 2021, Mazzanti collected $24,000 representing rent for the 2021 calendar year and credited deferred rent revenue.
4. Depreciation on the office building is $20,000 for the fiscal year.
5. Employee salaries for the month of June 2021 of $15,000 will be paid on July 20, 2021.

E 2–12
Financial statements and closing entries
● LO2–7, LO2–8

The December 31, 2021, adjusted trial balance for the Blueboy Cheese Corporation is presented below.

Account Title	Debits	Credits
Cash	21,000	
Accounts receivable	300,000	
Prepaid rent	10,000	
Inventory	50,000	
Office equipment	600,000	
Accumulated depreciation		250,000
Accounts payable		60,000
Notes payable (due in six months)		60,000
Salaries payable		8,000
Interest payable		2,000
Common stock		400,000
Retained earnings		100,000
Sales revenue		800,000
Cost of goods sold	480,000	
Salaries expense	120,000	
Rent expense	30,000	
Depreciation expense	60,000	
Interest expense	4,000	
Advertising expense	5,000	
Totals	1,680,000	1,680,000

Required:

1. Prepare an income statement for the year ended December 31, 2021, and a classified balance sheet as of December 31, 2021.
2. Prepare the necessary closing entries at December 31, 2021.

E 2–13
Closing entries
● LO2–8

American Chip Corporation's reporting year-end is December 31. The following is a partial adjusted trial balance as of December 31, 2021.

Account Title	Debits	Credits
Retained earnings		80,000
Sales revenue		750,000
Interest revenue		3,000
Cost of goods sold	420,000	
Salaries expense	100,000	
Rent expense	15,000	
Depreciation expense	30,000	
Interest expense	5,000	
Insurance expense	6,000	

Required:

Prepare the necessary closing entries at December 31, 2021.

E 2–14
Closing entries
● LO2–8

Presented below is income statement information of the Schefter Corporation for the year ended December 31, 2021.

Sales revenue	$492,000	Cost of goods sold	284,000
Salaries expense	80,000	Insurance expense	12,000
Interest revenue	6,000	Interest expense	4,000
Advertising expense	10,000	Income tax expense	30,000
Gain on sale of investments	8,000	Depreciation expense	20,000

Required:
Prepare the necessary closing entries at December 31, 2021.

E 2–15
Cash versus accrual accounting; adjusting entries
● LO2–5, LO2–6, LO2–9

The Righter Shoe Store Company prepares monthly financial statements for its bank. The November 30 and December 31, 2021, trial balances contained the following account information:

	Nov. 30		Dec. 31	
	Dr.	Cr.	Dr.	Cr.
Supplies	1,500		3,000	
Prepaid insurance	6,000		4,500	
Salaries payable		10,000		15,000
Deferred rent revenue		2,000		1,000

The following information also is known:

a. The December income statement reported $2,000 in supplies expense.
b. No insurance payments were made in December.
c. $10,000 was paid to employees during December for salaries.
d. On November 1, 2021, a tenant paid Righter $3,000 in advance rent for the period November through January. Deferred rent revenue was credited.

Required:
1. What was the cost of supplies purchased during December?
2. What was the adjusting entry recorded at the end of December for prepaid insurance?
3. What was the adjusting entry recorded at the end of December for accrued salaries?
4. What was the amount of rent revenue recognized in December? What adjusting entry was recorded at the end of December for deferred rent revenue?

E 2–16
External transactions and adjusting entries
● LO2–3, LO2–6

The following transactions occurred during 2021 for the Beehive Honey Corporation:

Feb. 1	Borrowed $12,000 from a bank and signed a note. Principal and interest at 10% will be paid on January 31, 2022.
Apr. 1	Paid $3,600 to an insurance company for a two-year fire insurance policy.
July 17	Purchased supplies costing $2,800 on account. The company records supplies purchased in an asset account. At the year-end on December 31, 2021, supplies costing $1,250 remained on hand.
Nov. 1	A customer borrowed $6,000 and signed a note requiring the customer to pay principal and 8% interest on April 30, 2022.

Required:
1. Record each transaction in general journal form. Omit explanations.
2. Prepare any necessary adjusting entries at the year-end on December 31, 2021. No adjusting entries were recorded during the year for any item.

E 2–17
Accrual accounting income determination
● LO2–5, LO2–9

During the course of your examination of the financial statements of the Hales Corporation for the year ended December 31, 2021, you discover the following:

a. An insurance policy covering three years was purchased on January 1, 2021, for $6,000. The entire amount was debited to insurance expense and no adjusting entry was recorded for this item.
b. During 2021, the company received a $1,000 cash advance from a customer for merchandise to be manufactured and shipped in 2022. The $1,000 was credited to sales revenue. No entry was recorded for the cost of merchandise.
c. There were no supplies listed in the balance sheet under assets. However, you discover that supplies costing $750 were on hand at December 31.
d. Hales borrowed $20,000 from a local bank on October 1, 2021. Principal and interest at 12% will be paid on September 30, 2022. No accrual was recorded for interest.
e. Net income reported in the 2021 income statement is $30,000 before reflecting any of the above items.

Required:
Determine the proper amount of net income for 2021.

E 2–18
Cash versus accrual accounting
● LO2–9

Stanley and Jones Lawn Service Company (S&J) maintains its books on a cash basis. However, the company recently borrowed $100,000 from a local bank, and the bank requires S&J to provide annual financial statements prepared on an accrual basis. During 2021, the following cash flows were recorded:

Cash collected for: Services to customers		$320,000
Cash paid for:		
Salaries	$180,000	
Supplies	25,000	
Rent	12,000	
Insurance	6,000	
Miscellaneous	20,000	243,000
Net operating cash flow		$ 77,000

You are able to determine the following information about accounts receivable, prepaid expenses, and accrued liabilities:

	January 1, 2021	December 31, 2021
Accounts receivable	$32,000	$27,000
Prepaid insurance	–0–	2,000
Supplies	1,000	1,500
Accrued liabilities (for miscellaneous expenses)	2,400	3,400

In addition, you learn that the bank loan was dated September 30, 2021, with principal and interest at 6% due in one year. Depreciation on the company's equipment is $10,000 for the year.

Required:
Prepare an accrual basis income statement for 2021. (Ignore income taxes.)

E 2–19
Cash versus accrual accounting
● LO2–9

Haskins and Jones, Attorneys-at-Law, maintains its books on a cash basis. During 2021, the law firm collected $545,000 for services rendered to its clients and paid out $412,000 in expenses. You are able to determine the following information about accounts receivable, prepaid expenses, deferred service revenue, and accrued liabilities:

	January 1, 2021	December 31, 2021
Accounts receivable	$62,000	$55,000
Prepaid insurance	4,500	6,000
Prepaid rent	9,200	8,200
Deferred service revenue	9,200	11,000
Accrued liabilities (for various expenses)	12,200	15,600

In addition, 2021 depreciation expense on office equipment is $22,000.

Required:
Determine accrual basis net income for 2021.

E 2–20
Worksheet
● Appendix 2A

The December 31, 2021, unadjusted trial balance for the Wolkstein Drug Company is presented below. December 31 is the company's year-end reporting date.

Account Title	Debits	Credits
Cash	20,000	
Accounts receivable	35,000	
Prepaid rent	5,000	
Inventory	50,000	
Equipment	100,000	
Accumulated depreciation		30,000
Accounts payable		25,000
Salaries payable		–0–
Common stock		100,000
Retained earnings		29,000
Sales revenue		323,000
Cost of goods sold	180,000	
Salaries expense	71,000	
Rent expense	30,000	
Depreciation expense	–0–	
Utilities expense	12,000	
Advertising expense	4,000	
Totals	507,000	507,000

The following year-end adjusting entries are required:

a. Depreciation expense for the year on the equipment is $10,000.

b. Salaries at year-end should be accrued in the amount of $4,000.

Required:

1. Prepare and complete a worksheet similar to Illustration 2A–1.
2. Prepare an income statement for 2021 and a balance sheet as of December 31, 2021.

E 2–21
Reversing entries
● Appendix 2B

The employees of Xitrex, Inc., are paid each Friday. The company's fiscal year-end is June 30, which falls on a Wednesday for the current year. Salaries are earned evenly throughout the five-day work week, and $10,000 will be paid on Friday, July 2.

Required:

1. Prepare an adjusting entry to record the accrued salaries as of June 30, a reversing entry on July 1, and an entry to record the payment of salaries on July 2.
2. Prepare journal entries to record the accrued salaries as of June 30 and the payment of salaries on July 2 assuming a reversing entry is not recorded.

E 2–22
Reversing entries
● Appendix 2B

Refer to E 2–9 and respond to the following requirements.

Required:

1. If Microchip's accountant employed reversing entries for accruals, which adjusting entries would she likely reverse at the beginning of the following year?
2. Prepare the adjusting entries at the end of 2021 for the adjustments you identified in requirement 1.
3. Prepare the appropriate reversing entries at the beginning of 2022.

E 2–23
Special journals
● Appendix 2C

The White Company's accounting system consists of a general journal (GJ), a cash receipts journal (CR), a cash disbursements journal (CD), a sales journal (SJ), and a purchases journal (PJ). For each of the following, indicate which journal should be used to record the transaction.

Transaction	Journal
1. Purchased merchandise on account.	______
2. Collected an account receivable.	______
3. Borrowed $20,000 and signed a note.	______
4. Recorded depreciation expense.	______
5. Purchased equipment for cash.	______
6. Sold merchandise for cash (the sale only, not the cost of the merchandise).	______
7. Sold merchandise on credit (the sale only, not the cost of the merchandise).	______
8. Recorded accrued salaries payable.	______
9. Paid employee salaries.	______
10. Sold equipment for cash.	______
11. Sold equipment on credit.	______
12. Paid a cash dividend to shareholders.	______
13. Issued common stock in exchange for cash.	______
14. Paid accounts payable.	______

E 2–24
Special journals
● Appendix 2C

The accounting system of K and M Manufacturing consists of a general journal (GJ), a cash receipts journal (CR), a cash disbursements journal (CD), a sales journal (SJ), and a purchases journal (PJ). For each of the following, indicate which journal should be used to record the transaction.

Transaction	Journal
1. Paid interest on a loan.	______
2. Recorded depreciation expense.	______
3. Purchased office equipment for cash.	______
4. Purchased merchandise on account.	______
5. Sold merchandise on credit (the sale only, not the cost of the merchandise).	______
6. Sold merchandise for cash (the sale only, not the cost of the merchandise).	______
7. Paid rent.	______
8. Recorded accrued interest payable.	______
9. Paid advertising bill.	______
10. Sold a factory building in exchange for a note receivable.	______
11. Collected cash from customers on account.	______
12. Paid employee salaries.	______
13. Collected interest on the note receivable.	______

Problems

connect

P 2–1
Accounting cycle through unadjusted trial balance
● LO2–3, LO2–4

Halogen Laminated Products Company began business on January 1, 2021. During January, the following transactions occurred:

Jan.	1	Issued common stock in exchange for $100,000 cash.
	2	Purchased inventory on account for $35,000 (the perpetual inventory system is used).
	4	Paid an insurance company $2,400 for a one-year insurance policy. Prepaid insurance was debited for the entire amount.
	10	Sold merchandise on account for $12,000. The cost of the merchandise was $7,000.
	15	Borrowed $30,000 from a local bank and signed a note. Principal and interest at 10% is to be repaid in six months.
	20	Paid employees $6,000 salaries for the first half of the month.
	22	Sold merchandise for $10,000 cash. The cost of the merchandise was $6,000.
	24	Paid $15,000 to suppliers for the merchandise purchased on January 2.
	26	Collected $6,000 on account from customers.
	28	Paid $1,000 to the local utility company for January gas and electricity.
	30	Paid $4,000 rent for the building. $2,000 was for January rent, and $2,000 for February rent. Prepaid rent and rent expense were debited for their appropriate amounts.

Required:

1. Prepare general journal entries to record each transaction. Omit explanations.
2. Post the entries to T-accounts.
3. Prepare an unadjusted trial balance as of January 30, 2021.

P 2–2
Accounting cycle through unadjusted trial balance
● LO2–3, LO2–4

The following is the post-closing trial balance for the Whitlow Manufacturing Corporation as of December 31, 2020.

Account Title	Debits	Credits
Cash	5,000	
Accounts receivable	2,000	
Inventory	5,000	
Equipment	11,000	
Accumulated depreciation		3,500
Accounts payable		3,000
Accrued liabilities		–0–
Common stock		10,000
Retained earnings		6,500
Sales revenue		–0–
Cost of goods sold	–0–	
Salaries expense	–0–	
Rent expense	–0–	
Advertising expense	–0–	
Totals	23,000	23,000

The following transactions occurred during January 2021:

Jan.	1	Sold merchandise for cash, $3,500. The cost of the merchandise was $2,000. The company uses the perpetual inventory system.
	2	Purchased equipment on account for $5,500 from the Strong Company.
	4	Received a $150 invoice from the local newspaper requesting payment for an advertisement that Whitlow placed in the paper on January 2.
	8	Sold merchandise on account for $5,000. The cost of the merchandise was $2,800.
	10	Purchased merchandise on account for $9,500.
	13	Purchased equipment for cash, $800.
	16	Paid the entire amount due to the Strong Company.
	18	Received $4,000 from customers on account.
	20	Paid $800 to the owner of the building for January's rent.
	30	Paid employees $3,000 for salaries for the month of January.
	31	Paid a cash dividend of $1,000 to shareholders.

Required:

1. Set up T-accounts and enter the beginning balances as of January 1, 2021.
2. Prepare general journal entries to record each transaction. Omit explanations.
3. Post the entries to T-accounts.
4. Prepare an unadjusted trial balance as of January 31, 2021.

P 2–3
Adjusting entries
● LO2–6

Pastina Company sells various types of pasta to grocery chains as private label brands. The company's reporting year-end is December 31. The unadjusted trial balance as of December 31, 2021, appears below.

Account Title	Debits	Credits
Cash	30,000	
Accounts receivable	40,000	
Supplies	1,500	
Inventory	60,000	
Notes receivable	20,000	
Interest receivable	–0–	
Prepaid rent	2,000	
Prepaid insurance	6,000	
Office equipment	80,000	
Accumulated depreciation		30,000
Accounts payable		31,000
Salaries payable		–0–
Notes payable		50,000
Interest payable		–0–
Deferred sales revenue		2,000
Common stock		60,000
Retained earnings		28,500
Dividends	4,000	
Sales revenue		146,000
Interest revenue		–0–
Cost of goods sold	70,000	
Salaries expense	18,900	
Rent expense	11,000	
Depreciation expense	–0–	
Interest expense	–0–	
Supplies expense	1,100	
Insurance expense	–0–	
Advertising expense	3,000	
Totals	347,500	347,500

Information necessary to prepare the year-end adjusting entries appears below.

1. Depreciation on the office equipment for the year is $10,000.
2. Employee salaries are paid twice a month, on the 22nd for salaries earned from the 1st through the 15th, and on the 7th of the following month for salaries earned from the 16th through the end of the month. Salaries earned from December 16 through December 31, 2021, were $1,500.
3. On October 1, 2021, Pastina borrowed $50,000 from a local bank and signed a note. The note requires interest to be paid annually on September 30 at 12%. The principal is due in 10 years.
4. On March 1, 2021, the company lent a supplier $20,000, and a note was signed requiring principal and interest at 8% to be paid on February 28, 2022.
5. On April 1, 2021, the company paid an insurance company $6,000 for a two-year fire insurance policy. The entire $6,000 was debited to prepaid insurance.
6. $800 of supplies remained on hand at December 31, 2021.
7. A customer paid Pastina $2,000 in December for 1,500 pounds of spaghetti to be delivered in January 2022. Pastina credited deferred sales revenue.
8. On December 1, 2021, $2,000 rent was paid to the owner of the building. The payment represented rent for December 2021 and January 2022 at $1,000 per month. The entire amount was debited to prepaid rent.

Required:
Prepare the necessary December 31, 2021, adjusting journal entries.

P 2–4
Accounting cycle; adjusting entries through post-closing trial balance
● LO2–4, LO2–6 through LO2–8

Refer to P 2–3 and complete the following steps:

1. Enter the unadjusted balances from the trial balance into T-accounts.
2. Post the adjusting entries prepared in P 2–3 to the accounts.
3. Prepare an adjusted trial balance.
4. Prepare an income statement and a statement of shareholders' equity for the year ended December 31, 2021, and a classified balance sheet as of December 31, 2021. Assume that no common stock was issued during the year and that $4,000 in cash dividends were paid to shareholders during the year.
5. Prepare closing entries and post to the accounts.
6. Prepare a post-closing trial balance.

P 2–5
Adjusting entries
● LO2–6

Howarth Company's reporting year-end is December 31. Below are the unadjusted and adjusted trial balances for December 31, 2021.

	Unadjusted		Adjusted	
Account Title	**Debits**	**Credits**	**Debits**	**Credits**
Cash	50,000		50,000	
Accounts receivable	35,000		35,000	
Prepaid rent	2,000		1,200	
Supplies	1,500		800	
Inventory	60,000		60,000	
Notes receivable	30,000		30,000	
Interest receivable	–0–		1,500	
Office equipment	45,000		45,000	
Accumulated depreciation		15,000		21,500
Accounts payable		34,000		34,000
Salaries payable		–0–		6,200
Notes payable		50,000		50,000
Interest payable		–0–		2,500
Deferred rent revenue		2,000		–0–
Common stock		46,000		46,000
Retained earnings		20,000		20,000
Sales revenue		244,000		244,000
Rent revenue		4,000		6,000
Interest revenue		–0–		1,500
Cost of goods sold	126,000		126,000	
Salaries and wages expense	45,000		51,200	
Rent expense	11,000		11,800	
Depreciation expense	–0–		6,500	
Supplies expense	1,100		1,800	
Interest expense	5,400		7,900	
Advertising expense	3,000		3,000	
Totals	415,000	415,000	431,700	431,700

Required:
Prepare the adjusting journal entries that were recorded at December 31, 2021.

P 2–6
Accounting cycle
● LO2–3 through LO2–8

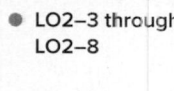

The general ledger of the Karlin Company, a consulting company, at January 1, 2021, contained the following account balances:

Account Title	**Debits**	**Credits**
Cash	30,000	
Accounts receivable	15,000	
Equipment	20,000	
Accumulated depreciation		6,000
Salaries payable		9,000
Common stock		40,500
Retained earnings		9,500
Total	65,000	65,000

The following is a summary of the transactions for the year:

a. Service revenue, $100,000, of which $30,000 was on account and the balance was received in cash.
b. Collected on accounts receivable, $27,300.
c. Issued shares of common stock in exchange for $10,000 in cash.
d. Paid salaries, $50,000 (of which $9,000 was for salaries payable at the end of the prior year).
e. Paid miscellaneous expense for various items, $24,000.
f. Purchased equipment for $15,000 in cash.
g. Paid $2,500 in cash dividends to shareholders.

Required:
1. Set up the necessary T-accounts and enter the beginning balances from the trial balance.
2. Prepare a general journal entry for each of the summary transactions listed above.

3. Post the journal entries to the accounts.
4. Prepare an unadjusted trial balance.
5. Prepare and post adjusting journal entries. Accrued salaries at year-end amounted to $1,000. Depreciation for the year on the equipment is $2,000.
6. Prepare an adjusted trial balance.
7. Prepare an income statement for 2021 and a balance sheet as of December 31, 2021.
8. Prepare and post-closing entries.
9. Prepare a post-closing trial balance.

P 2–7
Adjusting entries and income effects
● LO2–5, LO2–6

The information necessary for preparing the December 31, 2021 year-end adjusting entries for Vito's Pizza Parlor appears below.

a. On July 1, 2021, purchased $10,000 of **IBM Corporation** bonds at face value. The bonds pay interest twice a year on January 1 and July 1. The annual interest rate is 12%.
b. Vito's depreciable equipment has a cost of $30,000, a five-year life, and no salvage value. The equipment was purchased in 2019. The straight-line depreciation method is used.
c. On November 1, 2021, the bar area was leased to Jack Donaldson for one year. Vito's received $6,000 representing the first six months' rent and credited deferred rent revenue.
d. On April 1, 2021, the company paid $2,400 for a two-year fire and liability insurance policy and debited insurance expense.
e. On October 1, 2021, the company borrowed $20,000 from a local bank and signed a note. Principal and interest at 12% will be paid on September 30, 2022.
f. At year-end, there is a $1,800 debit balance in the supplies (asset) account. Only $700 of supplies remain on hand.

Required:

1. Prepare the necessary adjusting journal entries at December 31, 2021.
2. Determine the amount by which net income would be misstated if Vito's failed to record these adjusting entries. (Ignore income tax expense.)

P 2–8
Adjusting entries
● LO2–6

Excalibur Corporation sells video games for personal computers. The unadjusted trial balance as of December 31, 2021, appears below. December 31 is the company's reporting year-end. The company uses the perpetual inventory system.

Account Title	Debits	Credits
Cash	23,300	
Accounts receivable	32,500	
Supplies	–0–	
Prepaid rent	–0–	
Inventory	65,000	
Office equipment	75,000	
Accumulated depreciation		10,000
Accounts payable		26,100
Salaries payable		3,000
Notes payable		30,000
Common stock		80,000
Retained earnings		22,050
Dividends	6,000	
Sales revenue		180,000
Cost of goods sold	95,000	
Interest expense	–0–	
Salaries expense	32,350	
Rent expense	14,000	
Supplies expense	2,000	
Utilities expense	6,000	
Totals	351,150	351,150

Information necessary to prepare the year-end adjusting entries appears below.

1. The office equipment was purchased in 2019 and is being depreciated using the straight-line method over a ten-year useful life with no salvage value.
2. Accrued salaries at year-end should be $4,500.

3. The company borrowed $30,000 on September 1, 2021. The principal is due to be repaid in 10 years. Interest is payable twice a year on each August 31 and February 28 at an annual rate of 10%.
4. The company debits supplies expense when supplies are purchased. Supplies on hand at year-end cost $500.
5. Prepaid rent at year-end should be $1,000.

Required:

Prepare the necessary December 31, 2021, adjusting entries.

P 2–9
Accounting cycle; unadjusted trial balance through closing
● LO2–4, LO2–6, LO2–8

The unadjusted trial balance as of December 31, 2021, for the Bagley Consulting Company appears below. December 31 is the company's reporting year-end.

Account Title	Debits	Credits
Cash	8,000	
Accounts receivable	9,000	
Prepaid insurance	3,000	
Land	200,000	
Buildings	50,000	
Accumulated depreciation—buildings		20,000
Office equipment	100,000	
Accumulated depreciation—office equipment		40,000
Accounts payable		35,050
Salaries payable		–0–
Deferred rent revenue		–0–
Common stock		200,000
Retained earnings		56,450
Service revenue		90,000
Interest revenue		3,000
Rent revenue		7,500
Salaries expense	37,000	
Depreciation expense	–0–	
Insurance expense	–0–	
Utilities expense	30,000	
Maintenance expense	15,000	
Totals	452,000	452,000

Required:

1. Enter the account balances in T-accounts.
2. From the trial balance and information given, prepare adjusting entries and post to the accounts.
   a. The buildings have an estimated useful life of 50 years with no salvage value. The company uses the straight-line depreciation method.
   b. The office equipment is depreciated at 10 percent of original cost per year.
   c. Prepaid insurance expired during the year, $1,000.
   d. Accrued salaries at year-end, $1,500.
   e. Deferred rent revenue at year-end should be $1,200.
3. Prepare an adjusted trial balance.
4. Prepare closing entries
5. Prepare a post-closing trial balance.

P 2–10
Accrual accounting; financial statements
● LO2–5, LO2–7, LO2–9

McGuire Corporation began operations in 2021. The company purchases computer equipment from manufacturers and then sells to retail stores. During 2021, the bookkeeper used a check register to record all cash receipts and cash disbursements. No other journals were used. The following is a recap of the cash receipts and disbursements made during the year.

Cash receipts:	
Issue of common stock	$ 50,000
Collections from customers	320,000
Borrowed from local bank on April 1, note signed requiring principal and interest at 12% to be paid on March 31, 2022	40,000
Total cash receipts	$410,000

(continued)

(concluded)

Cash disbursements:	
Purchase of merchandise	$220,000
Payment of salaries	80,000
Purchase of office equipment	30,000
Payment of rent on building	14,000
Miscellaneous expense	10,000
Total cash disbursements	$354,000

You are called in to prepare financial statements at December 31, 2021. The following additional information was provided to you:

1. Customers owed the company $22,000 at year-end.
2. At year-end, $30,000 was still due to suppliers of merchandise purchased on credit.
3. At year-end, merchandise inventory costing $50,000 still remained on hand.
4. Salaries owed to employees at year-end amounted to $5,000.
5. On December 1, $3,000 in rent was paid to the owner of the building used by McGuire. This represented rent for the months of December through February.
6. The office equipment, which has a 10-year life and no salvage value, was purchased on January 1, 2021. Straight-line depreciation is used.

Required:

Prepare an income statement for 2021 and a balance sheet as of December 31, 2021.

P 2–11
Cash versus accrual accounting
● LO2–9

Selected balance sheet information for the Wolf Company at November 30, and December 31, 2021, is presented below. The company uses the perpetual inventory system and all sales to customers are made on credit.

	Nov. 30		Dec. 31	
	Debits	**Credits**	**Debits**	**Credits**
Accounts receivable	10,000		3,000	
Prepaid insurance	5,000		7,500	
Inventory	7,000		6,000	
Accounts payable		12,000		15,000
Salaries payable		5,000		3,000

The following cash flow information also is available:

a. Cash collected from credit customers, $80,000
b. Cash paid for insurance, $5,000
c. Cash paid to suppliers of inventory, $60,000 (the entire accounts payable amounts relate to inventory purchases)
d. Cash paid to employees for salaries, $10,000

Required:

1. Determine the following for the month of December:
   a. Sales revenue
   b. Cost of goods sold
   c. Insurance expense
   d. Salaries expense
2. Prepare summary journal entries to record the month's sales and cost of those sales.

P 2–12
Cash versus accrual accounting
● LO2–9

Zambrano Wholesale Corporation maintains its records on a cash basis. At the end of each year the company's accountant obtains the necessary information to prepare accrual basis financial statements. The following cash flows occurred during the year ended December 31, 2021:

Cash receipts:	
From customers	$675,000
Interest on note	4,000
Loan from a local bank	100,000
Total cash receipts	$779,000
Cash disbursements:	
Purchase of merchandise	$390,000
Annual insurance payment	6,000
Payment of salaries	210,000
Dividends paid to shareholders	10,000
Annual rent payment	24,000
Total cash disbursements	$640,000

Selected balance sheet information:

	12/31/2020	12/31/2021
Cash	$ 25,000	$ 164,000
Accounts receivable	62,000	92,000
Inventory	80,000	62,000
Prepaid insurance	2,500	?
Prepaid rent	11,000	?
Interest receivable	3,000	?
Notes receivable	50,000	50,000
Equipment	100,000	100,000
Accumulated depreciation	(40,000)	(50,000)
Accounts payable (for merchandise)	110,000	122,000
Salaries payable	20,000	24,000
Notes payable	–0–	100,000
Interest payable	–0–	?

Additional information:

1. On March 31, 2020, Zambrano lent a customer $50,000. Interest at 8% is payable annually on each March 31. Principal is due in 2024.
2. The annual insurance payment is paid in advance on April 30. The policy period begins on May 1.
3. On October 31, 2021, Zambrano borrowed $100,000 from a local bank and signed a note promising repayment. Principal and interest at 6% are due on October 31, 2022.
4. Annual rent on the company's facilities is paid in advance on June 30. The rental period begins on July 1.

Required:

1. Prepare an accrual basis income statement for 2021 (ignore income taxes).
2. Determine the following balance sheet amounts on December 31, 2021:
   a. Prepaid insurance
   b. Prepaid rent
   c. Interest receivable
   d. Interest payable

P 2–13
Worksheet
● Appendix 2A

Using the information from P 2–8, prepare and complete a worksheet similar to Illustration 2A–1. Use the information in the worksheet to prepare an income statement and a statement of shareholders' equity for 2021 and a balance sheet as of December 31, 2021. Cash dividends paid to shareholders during the year amounted to $6,000. Also prepare the necessary closing entries assuming that adjusting entries have been correctly posted to the accounts.

Decision Makers' Perspective

©ImageGap/Getty Images

Apply your critical-thinking ability to the knowledge you've gained. These cases will provide you an opportunity to develop your research, analysis, judgment, and communication skills. You also will work with other students, integrate what you've learned, apply it in real-world situations, and consider its global and ethical ramifications. This practice will broaden your knowledge and further develop your decision-making abilities.

Discussion Case 2–1
Cash versus accrual accounting; adjusting entries; Chapters 1 and 2
● LO2–5, LO2–9

You have recently been hired by Davis & Company, a small public accounting firm. One of the firm's partners, Alice Davis, has asked you to deal with a disgruntled client, Mr. Sean Pitt, owner of the city's largest hardware store. Mr. Pitt is applying to a local bank for a substantial loan to remodel his store. The bank requires accrual-based financial statements, but Mr. Pitt has always kept the company's records on a cash basis. He doesn't see the purpose of accrual-based statements. His most recent outburst went something like this: "After all, I collect cash from customers, pay my bills in cash, and I am going to pay the bank loan with cash. And, I already show my building and equipment as assets and depreciate them. I just don't understand the problem."

Required:

Draft a memo to Mr. Pitt providing an explanation of why the bank requests accrual-based financial statements for loan requests such as his. Include in the memo.

- An explanation of the difference between a cash basis and an accrual basis measure of performance.
- Why, in most cases, that accrual-basis net income provides a better measure of performance than net operating cash flow.
- The purpose of adjusting entries as they relate to the difference between cash and accrual accounting.

Judgment Case 2–2
Cash versus accrual accounting
● LO2–9

Refer to Case 2–1 above. Mr. Pitt has relented and agrees to provide you with the information necessary to convert his cash basis financial statements to accrual basis statements. He provides you with the following transaction information for the fiscal year ending December 31, 2021:

1. A new comprehensive insurance policy requires an annual payment of $12,000 for the upcoming year. Coverage began on September 1, 2021, at which time the first payment was made.
2. Mr. Pitt allows customers to pay using a credit card. At the end of the current year, various credit card companies owed Mr. Pitt $6,500. At the end of last year, customer credit card charges outstanding were $5,000.
3. Employees are paid once a month, on the 10th of the month following the work period. Cash disbursements to employees were $8,200 and $7,200 for January 10, 2022, and January 10, 2021, respectively.
4. Utility bills outstanding totaled $1,200 at the end of 2021 and $900 at the end of 2020.
5. A physical count of inventory is always taken at the end of the fiscal year. The merchandise on hand at the end of 2021 cost $35,000. At the end of 2020, inventory on hand cost $32,000.
6. At the end of 2020, Mr. Pitt did not have any bills outstanding to suppliers of merchandise. However, at the end of 2021, he owed suppliers $4,000.

Required:

1. Mr. Pitt's 2021 cash basis net income (after one adjustment for depreciation expense) is $26,000. Determine net income applying the accrual accounting model.
2. Explain the effect on Mr. Pitt's balance sheet of converting from cash to accrual. That is, would assets, liabilities, and owner's equity be higher or lower and by what amounts?

Communication Case 2–3
Adjusting entries
● LO2–5

"I don't understand," complained Chris, who responded to your bulletin board posting for tutoring in introductory accounting. The complaint was in response to your statements that recording adjusting entries is a critical step in the accounting processing cycle, and the two major classifications of adjusting entries are prepayments and accruals.

Required:

Respond to Chris.

1. When do prepayments occur? Accruals?
2. Describe the appropriate adjusting entry for prepaid expenses and for deferred revenues. What is the effect on net income, assets, liabilities, and shareholders' equity of not recording a required adjusting entry for prepayments?
3. Describe the required adjusting entry for accrued liabilities and for accrued receivables. What is the effect on net income, assets, liabilities, and shareholders' equity of not recording a required adjusting entry for accruals?

Data Analytics

connect

Data analytics is the process of examining data sets in order to draw conclusions about the information they contain. Data analytics is widely used in business to enable organizations to make better-informed business decisions. Increasingly, this is accomplished with the aid of specialized software such as Tableau. In each chapter of our textbook, you will have the opportunity to glimpse the power and efficacy of data analytics in the context of that chapter's topics, using Tableau as a tool.

To prepare for this case—and those in subsequent chapters—using Tableau, you will need to download Tableau to your computer. Tableau provides free instructor and student licenses as well as free videos and support for utilizing and learning the software. You will receive further information on setting up Tableau from your instructor. Once you are set up with Tableau, watch the first few training videos on the Tableau website. If you completed the first Data Analytics Case, you already have watched the three "Getting Started" videos. Now to prepare for this case you should view the "Ways to Filter," "Using the Filter Shelf," and "The Formatting Pane" videos listed under "Visual Analytics." Future cases will build off of your knowledge gained from these, and additional videos will be suggested. All short training videos can be found at **www.tableau.com/learn/training#getting-started**.

Data Analytics Case
Composition of the Balance Sheet
● LO2–7

In the Chapter 1 Data Analytics Case, you applied Tableau to examine a data set and create bar charts to compare ten-year trends in operating cash flows and net income for our two (hypothetical) publicly traded companies: Discount Goods and Big Store. In this case you continue in your role as an analyst conducting introductory research into the relative merits of investing in one or both of these companies, this time comparing the proportional composition of the assets, liabilities, and stockholders' equity of the two companies. You want to create pie charts to visualize the relative size of the components of the balance sheets for each of the two companies to garner some initial insights.

Required:

Total assets are equal to Total liabilities plus Total shareholders' equity. This equation is based on the fact that a company's assets come from these two sources: (1) amounts borrowed and (2) amounts provided by owners ([a] amounts *invested* by shareholders in the corporation, reported as paid-in capital and [b] amounts *earned* by the corporation on behalf of its shareholders, reported as retained earnings). The relative proportion of assets provided by liabilities can be an indication of the riskiness of an investment, making these relationships valuable when analyzing a company's financial position. Use Tableau to create pie charts depicting the relative size of the assets, liabilities, and stockholders' equity for each of the two companies in 2021. Based upon what you find, answer the following questions:

1. For Discount Goods, do liabilities or shareholders' equity provide the greater proportion of the company's total assets?
2. For Big Store, do liabilities or shareholders' equity provide the greater proportion of the company's total assets?
3. Which of the two companies has the highest ratio of current liabilities to total liabilities? What is the ratio for that company (rounded to nearest whole percentage point)?
4. Which of the two companies has the highest ratio of current assets to current liabilities? What is the ratio for that company (rounded to nearest whole percentage point)?

Resources:

You have available to you an extensive data set that includes detailed financial data for Discount Goods and Big Store and for 2012 - 2021. The data set is in the form of an Excel file available to download from Connect, or under Student Resources within the Library tab. There are four Excel files available. The one for use in this chapter is named "Discount_Goods_Big_StoreCase1.xlsx." Download this file and save it to the computer on which you will be using Tableau.

For this case, you will need to review the raw data in the Excel file for both companies before beginning. Review the questions above to determine which account classifications, like "Total current assets," to include in your pie charts. These will be the items you include in what's called "Account name." Notice that the two companies have somewhat different names for their accounts, but the major groups are the same.

After you view the training videos, follow these steps to create the charts you'll use for this case:

- Open Tableau and connect to the Excel spreadsheet you downloaded. Starting on the "Sheet 1" tab, at the bottom of the canvas to the right of the data source at the bottom of the screen, double click on "2021" under "Measures" to add it to the canvas.
- Click the pull-down menu arrow on "Store" under "Dimensions" and click "Show Filter." The filter box will appear on the right side. Un-click the "(All)" button and click only Discount Goods.
- Move "Account name" into the Filter shelf. In the Filter dialog box that opens, click all Discount Goods' account classifications you've identified as needed for the case. You can determine this information by checking the raw data as instructed above. Hint: You should have selected a total of 5 account classifications.
- Drag "Account name" to "Color" in the "Marks" card to add more colors to the chart. On the "Show me" tab in the upper right corner of the toolbar, select the pie chart.
- Enlarge the pie chart by holding down the shift-control-and-clicking the "B" key several times until your pie chart fills most of the screen (at least 25 times).
- Add "Labels" to the pie sections by clicking on "Label" in the "Marks" card and clicking the box "Show mark label."
- Format the label amounts to Times New Roman, bold, black, 12-point font by selecting the label, right-clicking, and clicking on "Format."
- Change the title of the sheet to be "Discount Goods 2021 Balance Sheet" by right-clicking and selecting "Edit title." Format the title to Times New Roman, bold, black and 15-point font.
- To save time completing the next page, right-click the "Sheet 1" tab at the bottom of the canvas and select "Duplicate."
- Click on the "Sheet 2" tab. You should now see a duplicate of the "Sheet 1" canvas. In the filter box on the right-hand side, click only the Big Store button.
- Change the title of the sheet to be "Big Store 2021 Balance Sheet" by right-clicking and selecting "Edit title." Format the title to Times New Roman, bold, black and 15-point font.
- Once complete, save the file as "*DA2_Your initials*.twbx."

Continuing Cases

Target Case

● LO2–4, LO2–8

Target Corporation prepares its financial statements according to U.S. GAAP. Target's financial statements and disclosure notes for the year ended February 3, 2018, are available in Connect. This material also is available under the Investor Relations link at the company's website (**www.target.com**).

Required:

1. Refer to Target's balance sheet for the years ended February 3, 2018, and January 28, 2017. Based on the amounts reported for accumulated depreciation, and assuming no depreciable assets were sold during the year, prepare an adjusting entry to record Target's depreciation for the year.

2. Refer to Target's statement of cash flows for the year ended February 3, 2018. Assuming your answer to requirement 1 includes all depreciation expense recognized during the year, how much amortization expense was recognized during the year?
3. Note 13 provides information on Target's current assets. Assume all prepaid expenses are for prepaid insurance and that insurance expense comprises $50 million of the $14,665 million of selling, general, and administrative expenses reported in the income statement for the year ended February 3, 2018. How much cash did Target pay for insurance coverage during the year? Prepare the adjusting entry Target would make to record all insurance expense for the year. What would be the effect on the income statement and balance sheet if Target didn't record an adjusting entry for prepaid expenses?

Air France–KLM Case
● LO2–4

Air France-KLM (AF), a Franco-Dutch company, prepares its financial statements according to International Financial Reporting Standards. AF's financial statements and disclosure notes for the year ended December 31, 2017, are provided in Connect. This material also is available under the Finance link at the company's website (**www.airfranceklm.com**).

Required:

1. Refer to AF's balance sheet and compare it to the balance sheet presentation in Illustration 2–14. What differences do you see in the format of the two balance sheets?
2. What differences do you see in the terminology used in the two balance sheets?

CPA Exam Questions and Simulations

ROGER CPA Review

Sample CPA Exam questions from Roger CPA Review are available in Connect as support for the topics in this chapter. These multiple-choice questions and task-based aimulations include expert-written explanations and solutions, and provide a starting point for students to become familiar with the content and functionality of the actual CPA Exam.

CHAPTER

3

The Balance Sheet and Financial Disclosures

OVERVIEW

Chapter 1 stressed the importance of the financial statements to investors and creditors, and Chapter 2 reviewed the preparation of those financial statements. In this chapter, we'll take a closer look at the balance sheet, along with accompanying disclosures. The balance sheet provides relevant information useful in helping investors and creditors not only to predict future cash flows, but also to make the related assessments of liquidity and long-term solvency. In Chapter 4 we'll continue our conversation about the financial statements with discussion of the income statement, statement of comprehensive income, and statement of cash flows.

LEARNING OBJECTIVES

After studying this chapter, you should be able to:

- **LO3–1** Describe the purpose of the balance sheet and understand its usefulness and limitations. (*p. 112*)
- **LO3–2** Identify and describe the various asset classifications. (*p. 114*)
- **LO3–3** Identify and describe the various liability and shareholders' equity classifications. (*p. 118*)
- **LO3–4** Explain the purpose of financial statement disclosures. (*p. 122*)
- **LO3–5** Describe disclosures related to management's discussion and analysis, responsibilities, and compensation. (*p. 125*)
- **LO3–6** Explain the purpose of an audit and describe the content of the audit report. (*p. 128*)
- **LO3–7** Describe the techniques used by financial analysts to transform financial information into forms more useful for analysis. (*p. 130*)
- **LO3–8** Identify and calculate the common liquidity and solvency ratios used to assess risk. (*p. 131*)
- **LO3–9** Discuss the primary differences between U.S. GAAP and IFRS with respect to the balance sheet, financial disclosures, and segment reporting. (*pp. 120 and 139*)

©TY Lim/Shutterstock

FINANCIAL REPORTING CASE

What's It Worth?

"I can't believe it. Why don't you accountants prepare financial statements that are relevant?" Your friend Jerry is a finance major and is constantly badgering you about what he perceives to be a lack of relevance of financial statements prepared according to generally accepted accounting principles. "For example, take a look at this balance sheet for **Nike** that I just downloaded off the Internet. The equity (or book value) of the company according to the 2017 balance sheet was about $12.4 billion. But if you multiply the number of outstanding shares by the stock price per share at the same point in time, the company's market value was nearly $88 billion. I thought equity was supposed to represent the value of the company, but those two numbers aren't close." You decide to look at the company's balance sheet and try to set Jerry straight.

QUESTIONS

By the time you finish this chapter, you should be able to respond appropriately to the questions posed in this case. Compare your response to the solution provided at the end of the chapter.

1. Respond to Jerry's criticism that shareholders' equity does not represent the market value of the company. What information does the balance sheet provide? *(p. 113)*
2. The usefulness of the balance sheet is enhanced by classifying assets and liabilities according to common characteristics. What are the classifications used in the balance sheet and what elements do those categories include? *(p. 113)*

The SEC's EDGAR system improves the efficiency with which company information is collected and disseminated.

Most companies provide their financial statements and accompanying disclosures on their website in an "Investor Relations" link, but this information also can be found at the SEC's website (**www.sec.gov**) through its electronic filing system known as EDGAR (Electronic Data Gathering, Analysis, and Retrieval system). Companies are required to file their financial information in a timely manner using the EDGAR system. As stated by the SEC, "Its (EDGAR system's) primary purpose is to increase the efficiency and fairness of the securities market for the benefit of investors, corporations, and the economy by accelerating the receipt, acceptance, dissemination, and analysis of time-sensitive corporate information filed with the agency."

The first part of this chapter begins our discussion of the financial statements by providing an overview of the balance sheet. The balance sheet reports a company's assets, liabilities, and shareholders' equity. You can see an example of **Nike**'s balance sheet on the next page.

In the second part of this chapter, we discuss additional disclosures that companies are required to provide beyond the basic financial statements. These disclosures are critical to understanding the financial statements and to evaluating a firm's performance and financial health.

In the third part of this chapter, we discuss how information in the financial statements can be used by decision makers to assess business risk (liquidity and long-term solvency).

Real World Financials

NIKE, INC.
Consolidated Balance Sheet
May 31, 2017
($ in millions)

Assets		
Current assets:		
Cash and equivalents	$3,808	
Short-term investments	2,371	
Accounts receivable (net)	3,677	
Inventories	5,055	
Prepaid expenses and other current assets	1,150	
Total current assets		$16,061
Property, plant and equipment (net)	3,989	
Identifiable intangible assets (net)	283	
Goodwill	139	
Deferred income taxes and other assets	2,787	
Total long-term assets		7,198
Total assets		$23,259
Liabilities and Shareholders' Equity		
Current liabilities:		
Current portion of long-term debt	$ 6	
Notes payable	325	
Accounts payable	2,048	
Accrued liabilities	3,011	
Income taxes payable	84	
Total current liabilities		$ 5,474
Long-term debt	3,471	
Deferred income taxes and other liabilities	1,907	
Total long-term liabilities		5,378
Total liabilities		$10,852
Shareholders' equity:		
Common stock	3	
Additional paid-in capital	8,638	
Retained earnings	3,979	
Accumulated other comprehensive loss	(213)	
Total shareholders' equity		12,407
Total liabilities and shareholders' equity		$23,259

PART A

The Balance Sheet

• LO3–1

The **balance sheet**, sometimes referred to as the **statement of financial position**, presents an organized list of assets, liabilities, and equity *at a point in time*. It is a freeze frame or snapshot of a company's financial position at the end of a particular day marking the end of an accounting period.

Usefulness

The balance sheet provides information useful for assessing future cash flows, liquidity, and long-term solvency.

The balance sheet provides a list of assets and liabilities that are classified (grouped) according to common characteristics. These classifications, which we explore in the next section, along with related disclosure notes, help the balance sheet to provide useful information about liquidity and long-term solvency. **Liquidity** most often refers to the ability of a company to convert its assets to cash to pay its *current* liabilities. **Long-term solvency** refers to an assessment of whether a company will be able to pay all its liabilities, which includes *long-term* liabilities as well. Other things being equal, the risk that a company will not be able to pay its debt increases as its liabilities, relative to equity, increases.

Solvency also provides information about *financial flexibility*—the ability of a company to alter cash flows in order to take advantage of unexpected investment opportunities and needs.

For example, the higher the percentage of a company's liabilities to its equity, the more difficult it typically will be to borrow additional funds either to take advantage of a promising investment opportunity or to meet obligations. In general, the less financial flexibility, the more risk there is that an enterprise will fail. In Part C of this chapter, we introduce some common ratios used to assess liquidity and long-term solvency.[1]

Limitations

Despite its usefulness, the balance sheet has limitations. One important limitation is that a company's **book value**, its reported assets minus liabilities as shown in the balance sheet, usually *will not directly measure the company's market value*. Market value represents the price at which something could be sold in a given market. In the case of a public corporation, market value is represented by the trading price of a share of the corporation's stock. We can get an idea of the corporation's overall market value by multiplying the share price times the number of shares outstanding.

FINANCIAL Reporting Case

Q1, p. 111

Assets minus liabilities, measured according to GAAP, is not likely to be representative of the market value of the entity.

The two primary reasons that a company's book value in the balance sheet does not equal its market value are:

1. Many assets, like land and buildings, are measured at their historical costs rather than amounts for which the assets could be sold (often referred to as the assets' fair values). For example, suppose a company owns land and the amount for which the land could be sold increases. The increase in the land's fair value is not reported in the balance sheet, so it has no effect on the company's book value. However, to the extent the increase in the land's fair value is known by investors, that increase will be reflected in the company's overall market value.
2. Many aspects of a company may represent valuable resources (such as trained employees, experienced management team, loyal customer relationships, and product knowledge). These items, however, are not recorded as assets in the balance sheet and therefore have zero book value. Investors understand the ability of these resources to generate future profits and therefore these resources will be reflected in a company's overall market value.

As an example of the second item, in 2017, **Facebook** spent nearly $8 billion on research and development to design new products and improve existing services. While these costs likely improve the company's ability to generate future profits (and therefore increase the market value of the company to shareholders), these items are not reported in the balance sheet as an asset. They typically are expensed in the year incurred. As a result of this and other items, Facebook's ratio of market value to book value at the end of 2017 was 7.2. In comparison, the average ratio of market value to book value for companies in the S&P 500 index was about 3.0.

Another limitation of the balance sheet is that many items in the balance sheet are heavily reliant on *estimates and judgments rather than determinable amounts*. For example, companies estimate (a) the amount of receivables they will be able to actually collect, (b) the amount of warranty costs they will eventually incur for products already sold, (c) the residual values and useful lives of their long-term assets, and (d) amounts used to calculate employee pension obligations. Each of these estimates affects amounts reported in the balance sheet.

In summary, even though the balance sheet *does not directly measure* the market value of the entity, it provides valuable information that can be used to *help judge* market value.

Classification of Elements

The usefulness of the balance sheet is enhanced when assets and liabilities are grouped according to common characteristics. *The broad distinction made in the balance sheet is the current versus long-term (noncurrent) classification of both assets and liabilities.* The remainder of Part A provides an overview of the balance sheet. We discuss each of the three primary elements of the balance sheet (assets, liabilities, and shareholders' equity) in the

FINANCIAL Reporting Case

Q2, p. 111

The key classification of assets and liabilities in the balance sheet is the current versus long-term distinction.

[1]Another way the balance sheet is useful is in combination with income statement items. We explore some of these relationships in Chapter 4.

order they are reported in the statement as well as the classifications typically made within the elements. The balance sheet elements were defined in Chapter 1 as follows:

> **Assets** are probable future economic benefits obtained or controlled by a particular entity as a result of past transactions or events. Simply, these are the economic resources of a company.
>
> **Liabilities** are probable future sacrifices of economic benefits arising from present obligations of a particular entity to transfer assets or provide services to other entities in the future as a result of past transactions or events. Simply, these are the obligations of a company.
>
> **Equity** (or net assets), called **shareholders' equity** or **stockholders' equity** for a corporation, is the residual interest in the assets of an entity that remains after deducting liabilities. Stated another way, equity equals total assets minus total liabilities.

Illustration 3–1 shows the relationship among assets, liabilities, and shareholders' equity, often referred to as the **accounting equation**. Included in the illustration are the subclassifications of each element. We will discuss each of these subclassifications next.

Illustration 3–1
Classification of Elements within a Balance Sheet

Assets	=	Liabilities	+	Shareholders' Equity
1. Current assets		1. Current liabilities		1. Paid-in capital
2. Long-term assets		2. Long-term liabilities		2. Retained earnings

Assets

• LO3–2

CURRENT ASSETS Current assets include cash and other assets that are reasonably expected to be converted to cash or consumed within one year from the balance sheet date, or within the normal operating cycle of the business if that's longer than one year. The **operating cycle** for a typical merchandising or manufacturing company refers to the period of time from the initial outlay of cash for the purchase of inventory until the time the company collects cash from a customer from the sale of inventory.

Current assets include cash and all other assets expected to become cash or be consumed within one year from the balance sheet date (or *operating cycle*, if longer).

For a merchandising company, the initial purchase of inventory often is for a finished good, although some preparation may be necessary to get the inventory ready for sale (such as packaging or distribution). *For a manufacturing company,* the initial outlay of cash often involves the purchase of raw materials, which are then converted into a finished product through the manufacturing process. The concept of an operating cycle is shown in Illustration 3–2.

Illustration 3–2
Operating Cycle of a Typical Merchandising or Manufacturing Company

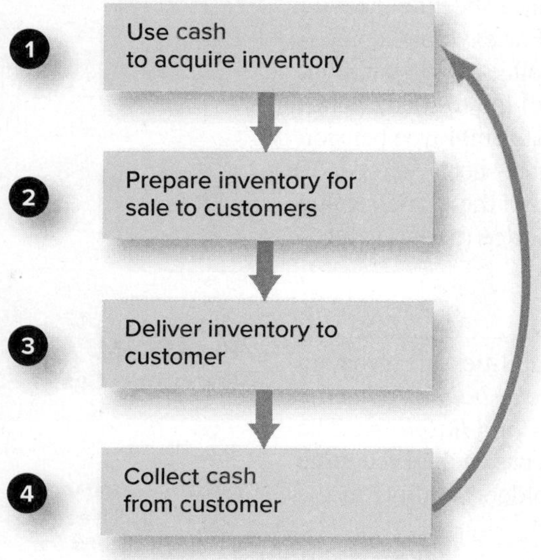
1
Use cash to acquire inventory
2
Prepare inventory for sale to customers
3
Deliver inventory to customer
4
Collect cash from customer

In some businesses, such as shipbuilding or distilleries, the operating cycle extends far beyond one year. For example, if it takes two years to build an oil-carrying supertanker, then the shipbuilder will classify as current those assets that will be converted to cash or consumed within two years from the balance sheet date. But for most businesses, the operating cycle will be shorter than one year. In these situations, the one-year convention is used to classify both assets and liabilities. Where a company has no clearly defined operating cycle, the one-year convention is used.

Illustration 3–3 presents the current assets section of **Nike**'s balance sheet for the years ended May 31, 2017, and May 31, 2016. In keeping with common practice, the individual current assets are listed in the order of their liquidity (the ability to convert the asset to cash).

Cash and Cash Equivalents. The most liquid asset, cash, is listed first. Cash includes cash on hand and in banks that is available for use in the operations of the business and such items as bank drafts, cashier's checks, and money orders. **Cash equivalents** are defined as short-term investments that have a maturity date no longer than three months *from the date of purchase.* Examples include certain negotiable items such as

($ in millions)	May 31, 2017	May 31, 2016
Current assets:		
Cash and equivalents	$ 3,808	$ 3,138
Short-term investments	2,371	2,319
Accounts receivable (net)	3,677	3,241
Inventory	5,055	4,838
Prepaid expenses and other current assets	1,150	1,489
Total current assets	$16,061	$15,025

Source: Nike, Inc.

Illustration 3–3

Current Assets—Nike, Inc.

Real World Financials

commercial paper, money market funds, and U.S. treasury bills. These items are listed as cash equivalents because they are highly liquid investments that can be quickly converted into cash with little risk of loss. **Nike**'s policy for classifying items as cash equivalents is stated in its disclosure note on summary of significant accounting policies, as shown in Illustration 3–4.

Cash and equivalents represent cash and short-term, highly liquid investments, including commercial paper, U.S. Treasury, U.S. Agency, money market funds, time deposits, and corporate debt securities with maturities of 90 days or less at the date of purchase.

Source: Nike, Inc.

Illustration 3–4

Disclosure of Cash Equivalents—Nike, Inc.

Real World Financials

Cash that is restricted for a special purpose and not available for current operations should not be included in the primary balance of cash and cash equivalents. These restrictions could include future plans to repay debt, purchase equipment, or make investments. Restricted cash is classified as a current asset if it is expected to be used within one year from the balance sheet date. Otherwise, restricted cash is classified as a long-term asset.

Short-Term Investments. Investments not classified as cash equivalents that the company has the ability and intent to sell within one year (or operating cycle, if longer) are reported as **short-term investments**. These investments include items such as equity investments in the common stock of other corporations, as well as debt investments in commercial paper and U.S. Treasury securities. We discuss accounting for equity and debt investments in Chapter 12.

Accounts Receivable. **Accounts receivable** result from the sale of goods or services on account (discussed in Chapter 7). Accounts receivable often are referred to as *trade receivables* because they arise in the course of a company's normal trade. *Nontrade receivables* result from loans or advances by the company to individuals and other entities. When receivables are supported by a formal agreement or note that specifies payment terms they are called **notes receivable**.

Accounts receivable usually are due in 30 to 60 days, depending on the terms offered to customers and are, therefore, classified as current assets. Any receivable, regardless of the source, not expected to be collected within one year (or operating cycle, if longer) is classified as a long-term investment. In addition, receivables are typically reported at the net amount expected to be collected. The net amount is calculated as total receivables less an allowance for the estimate of uncollectible accounts. **Nike** reported net accounts receivable of $3,677 million in the 2017 balance sheet. In a disclosure note, the company indicated that total accounts receivable were $3,704 million, indicating an allowance for uncollectible accounts of $27 million.

Inventory. **Inventory** for a wholesale or retail company consists of finished goods for sale to customers. For example, you buy finished goods such as shoes and athletic wear from **Nike**, potato chips at **Costco**, school supplies at **Staples**, and a new shirt at **Gap**. However, the inventory of a manufacturer will include not only finished goods but also goods in the

***Inventory* consists of assets that a retail or wholesale company acquires for resale or goods that manufacturers produce for sale.**

course of production (work in process) and goods to be consumed directly or indirectly in production (raw materials). Manufacturers typically report all three types of inventory either directly in the balance sheet or in a disclosure note. Illustration 3–5 demonstrates how **Intel Corp.**, a semiconductor chip manufacturer, discloses its components of inventory.

Illustration 3–5

Inventory Disclosure—Intel Corp.

Real World Financials

($ in millions)	December 30, 2017	December 31, 2016
Raw materials	$1,098	$ 695
Work in process	3,893	3,190
Finished goods	1,992	1,668
Total inventory	$6,983	$5,553

Source: Intel Corp.

Inventory is reported as a current asset because it normally is sold within the operating cycle.

Prepaid Expenses. Recall from Chapter 2 that a **prepaid expense** arises when a company incurs a cost of acquiring an asset in one period that won't be expensed until a future period. Examples are supplies, prepaid rent, and prepaid insurance. These assets are not converted to cash, like receivables collected and inventory sold, but they instead are consumed in the future. Supplies are used, prepaid rent expires over the rental period, and prepaid insurance expires over the period of insurance coverage.

Whether a prepaid expense is current or noncurrent depends on the period in which the item is consumed. For example, if rent on an office building were prepaid for one year, then the prepayment is classified as a current asset. However, if rent were prepaid for a period extending beyond the coming year, that portion of the prepayment is classified as a long-term asset.[2] Nike includes prepaid expenses with other current assets. Other current assets could include assets such as nontrade receivables that, because their amounts are not material, do not warrant separate disclosure.

LONG-TERM ASSETS When assets are expected to be converted to cash or consumed in more than one year (or operating cycle, if longer), they are reported as *long-term* (or *noncurrent*) *assets*. Typical classifications of long-term assets are as follows:

1. Investments
2. Property, plant, and equipment
3. Intangible assets
4. Other long-term assets

Next, we'll discuss each of these categories of long-term assets.

Investments are assets not used directly in operations.

Investments. Most companies occasionally acquire assets that are not used directly in the operations of the business. These assets include investments in debt and equity securities of other corporations, land held for speculation, long-term receivables, and cash set aside for special purposes (such as for future plant expansion). These assets are classified as long-term because management does not intend to convert the assets into cash in the next year (or operating cycle, if longer).

Tangible, long-lived assets used in the operations of the business are classified as *property, plant, and equipment.*

Property, Plant, and Equipment. Virtually all companies own assets classified as **property, plant, and equipment**. The common characteristics these assets share are that they are *tangible, long-lived,* and *used in the operations of the business.* Property, plant, and equipment often are the primary revenue-generating assets of the business.

[2]Companies often include prepayments for benefits extending beyond one year in the other assets category. When amounts are not material, some companies simply report all prepayments as current assets.

Property, plant, and equipment include land, buildings, equipment, machinery, furniture, and vehicles, as well as natural resources, such as mineral mines, timber tracts, and oil wells. These various assets often are reported in the balance sheet by showing their original cost on one line and their accumulated depreciation (or depletion for natural resources) to date on the next line. The difference between original cost and accumulated depreciation is the net amount of property, plant, and equipment. Sometimes, companies combine the original cost and accumulated depreciation for all property, plant, and equipment and report a single net amount in the balance sheet. In this case, a separate disclosure note provides details of the original cost and accumulated depreciation for each major asset category included in property, plant, and equipment. Land is included as part of property, plant, and equipment, but it has an unlimited useful life and therefore is not depreciated.

Illustration 3–6 shows the long-term asset section of **Nike**'s balance sheets, including its property, plant, and equipment, net of accumulated depreciation.

Illustration 3–6

Long-term Assets—Nike, Inc.

Real World Financials

($ in millions)	May 31, 2017	May 31, 2016
Long-term assets:		
Property, plant and equipment (net)	$3,989	$3,520
Identifiable intangible assets (net)	283	281
Goodwill	139	131
Deferred income taxes and other assets	2,787	2,422
Total long-term assets	$7,198	$6,354

Source: Nike, Inc.

Intangible Assets. Some assets used in the operations of a business have no physical substance. These assets are appropriately called **intangible assets**. Many intangible assets grant an exclusive right to a company to provide a product or service. This right can be a valuable resource in generating future revenues. Patents, copyrights, franchises, and trademarks are examples.

***Intangible assets* generally represent exclusive rights that a company can use to generate future revenues.**

Many intangible assets are reported in the balance sheet at their purchase price less accumulated *amortization*. The calculation is similar to how we report property, plant, and equipment at their purchase price less accumulated *depreciation*. Companies most often combine the purchase price and accumulated amortization to report a single net amount for intangible assets. A disclosure note is used to detail amounts for individual types of intangible assets. In 2017, **Nike** reported identifiable intangible assets, net of their amortization, of $283 million. These consist primarily of acquired trade names and trademarks.

Another common type of intangible asset, also reported by Nike, is *goodwill*. Goodwill isn't associated with any specific identifiable right, but instead arises when one company acquires another company. The amount reported for goodwill equals the acquisition price above the fair value of the identifiable net assets acquired. We'll discuss goodwill and other intangible assets in more detail in Chapter 10.

Not all intangible assets are purchased; some are developed internally. For example, instead of purchasing a patent granting the exclusive right to manufacture a certain drug, a pharmaceutical company may spend significant amounts in research and development to discover the drug and obtain a patent on its own. For internally developed intangibles, none of the research and development costs incurred in developing the intangible asset are included in reported cost. Instead, research and development costs are expensed as incurred. **Pfizer**, one of the world's largest pharmaceutical companies, typically spends $7 to $8 billion each year on researching and developing new medicines, vaccines, medical devices, and other health care products. These costs may help to generate future profits, but none of them were reported as an asset. Similarly, Nike spends large amounts on advertising and marketing to create brand recognition and help maintain its long-term competitiveness, but these costs are expensed in the period incurred (rather than reported as an intangible asset).

Other Long-Term Assets. This category of long-term assets (reported by most companies) represents a catchall classification of long-term assets that were not reported separately in one of the other long-term classifications. This amount most often includes long-term prepaid expenses, called *deferred charges.* For instance, **Nike** includes promotional expenditures related to long-term endorsement contracts and long-term advertising in this category. Nike also includes deferred charges related to income taxes.

This category might also include any long-term investments that are not material in amount and that were not reported separately in the long-term investments category discussed earlier. In the disclosure notes to its financial statements, Nike revealed that it did have a small amount of long-term investments as well as long-term receivables that were combined with deferred charges and reported in this category (instead of being reported separately in a long-term investments category).

Asset classification is affected by management intent.

A key to understanding which category an asset is reported is *management intent.* For example, in which category will land be reported? It depends on management intent. Management may intend to use land for long-term operating purposes (property, plant, and equipment), hold it for future resale (investment), or sell it in its ordinary course of business (inventory for a real estate company).

Liabilities

• LO3–3

Liabilities represent obligations to other entities. The information value of reporting these amounts is enhanced by classifying them as current liabilities and long-term liabilities. Illustration 3–7 shows the liability section of Nike's balance sheets.

Illustration 3–7

Liabilities—Nike, Inc.

Real World Financials

($ in millions)	May 31, 2017	May 31, 2016
Current liabilities:		
Current portion of long-term debt	$ 6	$ 44
Notes payable	325	1
Accounts payable	2,048	2,191
Accrued liabilities	3,011	3,037
Income taxes payable	84	85
Total current liabilities	5,474	5,358
Long-term debt	3,471	1,993
Deferred income taxes and other liabilities	1,907	1,770
Total liabilities	$10,852	$9,121

Source: Nike, Inc.

Current liabilities are expected to be satisfied within one year from the balance sheet date (or operating cycle, if longer).

CURRENT LIABILITIES Current liabilities are those obligations that are expected to be satisfied through the use of current assets or the creation of other current liabilities. So, this classification includes all liabilities that are expected to be satisfied within one year from the balance sheet date (or within the operating cycle if that's longer than one year). As of May 31, 2017, Nike had current liabilities of $5,474 million that it planned to pay in the next 12 months.

The most common current liabilities are accounts payable, notes payable (short-term borrowings), deferred revenues, accrued liabilities, and the currently maturing portion of long-term debt.

Accounts payable are obligations to suppliers of merchandise or of services purchased on account, with payment usually due in 30 to 60 days.

Notes payable are written promises to pay cash at some future date (I.O.U.s). Unlike accounts payable, notes usually require the payment of explicit interest in addition to the original obligation amount.

Deferred revenues, sometimes called *unearned revenues,* represent cash received from a customer for goods or services to be provided in a future period. For example, a company records deferred revenue when it sells gift cards. Revenue is not recorded until those gift cards are redeemed by the customer for merchandise or expire without being used.

Accrued liabilities represent obligations created when expenses have been incurred but amounts owed will not be paid until a subsequent reporting period. For example, a company might owe salaries at the end of the fiscal year to be paid some time in the following year. In this case, the company would report *salaries payable* as an accrued liability in the current year's balance sheet (as well as the related salaries expense in the income statement). Other common examples of accrued liabilities include interest payable, taxes payable, utilities payable, and legal fees payable. In most financial statements, these items are reported in the balance sheet for a single amount as total accrued liabilities, and the individual account balances may be found listed in the disclosure notes.

Current maturities of long-term debt refer to the portion of long-term notes, loans, mortgages, and bonds payable that is payable within the next year (or operating cycle, if longer).[3] For example, a $1,000,000 note payable requiring $100,000 in principal payments to be made in each of the next 10 years is classified as a $100,000 current liability and a $900,000 long-term liability. Nike classifies the current portion of its long-term debt as a current liability.

An exception for the current liability classification is a liability that management intends to refinance on a long-term basis. For example, if management intends to refinance a six-month note payable by substituting a two-year note payable and has the ability to do so, then the six-month note payable would not be classified as current even though it's due within the coming year. This exception and issues related to current liabilities are discussed in more detail in Chapter 13.

LONG-TERM LIABILITIES Long-term liabilities are obligations that are (a) due to be settled or (b) have a contractual right by the borrowing company to be settled in more than one year (or operating cycle, if longer) after the balance sheet date. Examples are long-term notes, bonds, pension obligations, and lease obligations.

Noncurrent, or long-term liabilities, usually are those payable beyond one year from the balance sheet date (or operating cycle, if longer).

But simply classifying a liability as long-term doesn't provide complete information to external users. For instance, long-term could mean anything from 2 to 20, 30, or 40 years. Payment terms, interest rates, and other details needed to assess the impact of these obligations on future cash flows and long-term solvency are reported in a disclosure note. Nike reports long-term liabilities related to corporate bonds and promissory notes.

Shareholders' Equity

Owners' equity is simply total assets minus total liabilities. For that reason, it's sometimes referred to as the *net assets* or *book value* of a company. Because owners of a corporation are its shareholders, owners' equity for a corporation typically is referred to as *shareholders' equity* or *stockholders' equity*. Here's a simple way to think of equity. If someone buys a house for $200,000 by making an initial $50,000 payment and borrowing the remaining $150,000, then the house's owner has an asset of $200,000, a liability of $150,000, and equity of $50,000.

Shareholders' equity for a corporation arises primarily from

1. Paid-in capital.
2. Retained earnings.

Paid-in capital is the amount that shareholders have invested in the company. It most often arises when the company issues stock. As shown in Illustration 3–8, the shareholders' equity section of **Nike**'s balance sheets reports the full amount of paid-in capital in two accounts—common stock and additional paid-in capital. Information about the number of shares the company has authorized and how many shares have been issued and are outstanding also must be disclosed either directly in the balance sheet or in a note.

Illustration 3–8

Shareholders' Equity—Nike, Inc.

Real World Financials

($ in millions)	May 31, 2017	May 31, 2016
Shareholders' equity:		
Common stock (1,643 and 1,682 shares outstanding)	$ 3	$ 3
Additional paid-in capital	8,638	7,786
Retained earnings	3,979	4,151
Accumulated other comprehensive (loss) income	(213)	318
Total shareholders' equity	$12,407	$12,258

Source: Nike, Inc.

[3] Payment can be with current assets or the creation of other current liabilities.

Retained earnings represents the accumulated net income reported by a company since its inception minus all dividends distributed to shareholders. In other words, it's the accumulated lifetime profits a company has earned for its shareholders but has not yet distributed to those shareholders. The fact that a company does not distribute all of its profits each year as dividends is not necessarily a bad thing from the shareholders' perspective. Instead of paying additional cash dividends, Nike's management can put those undistributed profits to productive use, such as buying additional inventory or equipment or paying liabilities as they come due. Nike has been quite profitable over its lifetime (since 1964) and has distributed some of those profits as dividends to investors and retained the remaining portion of $3,979 million (that's almost $4 billion). Some other companies have not been as profitable. Companies that report net losses could end up with a negative balance in the retained earnings account. When this occurs, the retained earnings account often is referred to as the *accumulated deficit* account.

Nike also reports a *third component of stockholders' equity*—**accumulated other comprehensive income (AOCI)**. Other comprehensive income refers to changes in stockholders' equity other than transactions with owners and other than transactions that affect net income. We accumulate items of other comprehensive income in the *accumulated other comprehensive (loss) income* account, just like we accumulate each year's net income (that hasn't been distributed as dividends) in the *retained earnings* account. The AOCI account is discussed in detail in Chapter 4.

While accumulated other comprehensive income typically is not a large portion of total shareholders' equity, companies are required to report this amount separately. We will discuss specific examples of items included in other comprehensive income in Chapters 4, 12, and 18.

We also will discuss other transactions affecting equity, such as a company's purchase of its own stock that is not retired (*treasury stock*) in Chapter 18. Purchased shares are essentially the same as shares that never were issued at all. A company may decide to resell those shares in the future, but until then, they are reported as negative (or contra) shareholders' equity. Nike does not report any treasury stock but many companies do.

International Financial Reporting Standards

• LO3–9

Balance Sheet Presentation. There are more similarities than differences in balance sheets prepared according to U.S. GAAP and those prepared applying IFRS. Some of the differences are

- International standards specify a minimum list of items to be presented in the balance sheet. U.S. GAAP has no minimum requirements.
- *IAS No. 1, revised,*[4] changed the title of the balance sheet to *statement of financial position,* although companies are not required to use that title. Some U.S. companies use the statement of financial position title as well.
- Under U.S. GAAP, we present current assets and liabilities before noncurrent assets and liabilities. IAS No. 1 doesn't prescribe the format of the balance sheet, but balance sheets prepared using IFRS often report noncurrent items first. A recent survey of large companies that prepare their financial statements according to IFRS reports that 73% of the surveyed companies list noncurrent items first.[5] For example, **H&M**, a Swedish-based clothing company, reported assets, liabilities, and shareholders' equity in its balance sheet in the following order:

	(SEK in millions)
Noncurrrent assets (including property, plant, and equipment)	47,916
Current assets	50,663
Total assets	98,579
Shareholders' equity	61,236
Long-term liabilities	5,638
Current liabilities	31,705
Total liabilities and equity	98,579

[4] "Financial Statement Presentation," *International Accounting Standard No. 1* (IASCF), as amended effective January 1, 2018.
[5] *IFRS Accounting Trends and Techniques—2011* (New York, AICPA, 2011), p.133.

Concept Review Exercise

BALANCE SHEET CLASSIFICATION

The following is a post-closing trial balance for the Sepia Paint Corporation at December 31, 2021, the end of the company's fiscal year:

Account Title	Debits	Credits
Cash	$ 80,000	
Accounts receivable	200,000	
Allowance for uncollectible accounts		$ 20,000
Inventory	300,000	
Prepaid expenses	30,000	
Notes receivable (due in one month)	60,000	
Investment in equity securities	50,000	
Land	120,000	
Buildings	550,000	
Machinery	500,000	
Accumulated depreciation—buildings		200,000
Accumulated depreciation—machinery		250,000
Patent (net of amortization)	50,000	
Accounts payable		170,000
Salaries payable		40,000
Interest payable		10,000
Notes payable		100,000
Bonds payable (due in 10 years)		500,000
Common stock (no par)		400,000
Retained earnings		250,000
Totals	$1,940,000	$1,940,000

The company intends to hold the $50,000 investment in equity securities of other corporations for at least three years. The $100,000 note payable is an installment loan. $10,000 of the principal, plus interest, is due on each July 1 for the next 10 years. At the end of the year, 100,000 shares of common stock were issued and outstanding. The company has 500,000 shares of common stock authorized.

Required:

Prepare a classified balance sheet for the Sepia Paint Corporation at December 31, 2021.

Solution:

SEPIA PAINT CORPORATION
Balance Sheet
At December 31, 2021
Assets

Current assets:		
Cash		$ 80,000
Accounts receivable	$ 200,000	
Less: Allowance for uncollectible accounts	(20,000)	180,000
Notes receivable		60,000
Inventory		300,000
Prepaid expenses		30,000
Total current assets		650,000
Investments:		
Investment in equity securities		50,000
Property, plant, and equipment:		
Land	120,000	
Buildings	550,000	
Machinery	500,000	
	1,170,000	
Less: Accumulated depreciation—buildings	(200,000)	
Less: Accumulated depreciation—machinery	(250,000)	
Net property, plant, and equipment		720,000
Intangible assets:		
Patent (net)		50,000
Total assets		$1,470,000

(continued)

(concluded)

Liabilities and Shareholders' Equity		
Current liabilities:		
Accounts payable		$ 170,000
Salaries payable		40,000
Interest payable		10,000
Current maturities of long-term debt		10,000
Total current liabilities		230,000
Long-term liabilities:		
Notes payable	$ 90,000	
Bonds payable	500,000	
Total long-term liabilities		590,000
Total liabilities		820,000
Shareholders' equity:		
Common stock (no par, 500,000 shares authorized, 100,000 shares issued and outstanding)	400,000	
Retained earnings	250,000	
Total shareholders' equity		650,000
Total liabilities and shareholders' equity		$1,470,000

The usefulness of the balance sheet, as well as the other financial statements, is significantly enhanced by financial statement disclosures. We now turn our attention to these disclosures.

PART B

Annual Report Disclosures

At the end of each fiscal year, companies with public securities are required to provide shareholders with an *annual report.* The annual report includes financial statements such as the balance sheet. Financial statements, though, are only part of the information provided in the annual report. Critical to understanding the financial statements and to evaluating a company's performance and financial health are additional disclosures included as part of the financial statements and also as part of the annual reporting requirements to the SEC.

The amount of information provided by annual report disclosures can be significant. For example, **Nike**'s 2017 annual report filed with the SEC (known as the Form 10-K) included five pages of financial statements followed by 19 pages of related disclosure notes. There were another 50 pages of disclosures related to business conditions, risk factors, legal proceedings, the company's stock performance, management's discussion and analysis, and internal control procedures. We will discuss some of the annual report disclosures in the next section.

Disclosure Notes

LO3–4

The full-disclosure principle requires that financial statements provide all material relevant information concerning the reporting entity.

Some financial statement disclosures are provided by including additional information, often parenthetically, on the face of the statement. Common examples of disclosures included on the face of the balance sheet are the allowance for uncollectible accounts and information about common stock. Other disclosures include supporting discussion, calculations, and schedules in notes following the financial statements. These notes are the most common means of providing additional disclosure. For instance, the fair values of financial instruments and "off-balance-sheet" risk associated with financial instruments are disclosed in notes. Information providing details of many financial statement items is provided using disclosure notes. Some examples include

- Pension plans.
- Leases.
- Long-term debt.
- Investments.

- Income taxes
- Property, plant, and equipment.
- Employee benefit plans.

Disclosure notes must include certain specific notes such as a summary of significant accounting policies, descriptions of subsequent events, and related third-party transactions, but many notes are fashioned to suit the disclosure needs of the particular reporting enterprise. Actually, any explanation that contributes to investors' and creditors' understanding of the results of operations, financial position, and cash flows of the company should be included. Let's take a look at just a few disclosure notes.

Summary of Significant Accounting Policies

There are many areas where management chooses from among equally acceptable alternative accounting methods. For example, management chooses whether to use accelerated or straight-line depreciation, whether to use FIFO, LIFO, or average cost to measure inventories, and whether to measure certain financial investments at fair value or cost. The company also defines which securities it considers to be cash equivalents and its policies regarding the timing of recognizing revenues and the estimated useful lives of its depreciable assets. Typically, the first disclosure note consists of a summary of significant accounting policies that discloses the choices the company makes.[6] Illustration 3–9 shows a portion of a typical summary note from a recent annual report of the **Starbucks Corporation**.

The *summary of significant accounting policies* conveys valuable information about the company's choices from among various alternative accounting methods.

Illustration 3–9 Summary of Significant Accounting Policies—Starbucks Corporation

Real World Financials

Note 1: Summary of Accounting Policies (in part)

Principles of Consolidation
The consolidated financial statements reflect the financial position and operating results of Starbucks, including wholly owned subsidiaries and investees that we control.

Cash and Cash Equivalents
We consider all highly liquid instruments with a maturity of three months or less at the time of purchase to be cash equivalents.

Inventories
Inventories are stated at the lower of cost (primarily moving average cost) or market. We record inventory reserves for obsolete and slow-moving inventory and for estimated shrinkage between physical inventory counts.

Property, Plant, and Equipment
Property, plant and equipment are carried at cost less accumulated depreciation. Cost includes all direct costs necessary to acquire and prepare assets for use, including internal labor and overhead in some cases. Depreciation of property, plant, and equipment, which includes assets under capital leases, is provided on the straight-line method over estimated useful lives, generally ranging from 2 to 15 years for equipment and 30 to 40 years for buildings. Leasehold improvements are amortized over the shorter of their estimated useful lives or the related lease life, generally 10 years.

Revenue Recognition
Consolidated revenues are presented net of intercompany eliminations for wholly-owned subsidiaries and investees controlled by us and for product sales to and royalty and other fees from licensees accounted for under the equity method. Additionally, consolidated revenues are recognized net of any discounts, returns, allowances, and sales incentives, including coupon redemptions and rebates.

Source: Starbucks

Studying this note is an essential step in analyzing financial statements. Obviously, knowing which methods were used to derive certain accounting numbers is critical to assessing the adequacy of those amounts.

Subsequent Events

When an event that has a material effect on the company's financial position occurs after the fiscal year-end but before the financial statements are issued or "available to be issued," the

[6]FASB ASC 235–10–50: Notes to Financial Statements—Overall—Disclosure (previously "Disclosure of Accounting Policies," *Accounting Principles Board Opinion No. 22* (New York: AICPA, 1972)).

A *subsequent event* is a significant development that occurs after a company's fiscal year-end but before the financial statements are issued or available to be issued.

event is described in a **subsequent event** disclosure note.[7] Examples include the issuance of debt or equity securities, a business combination or the sale of a business, the sale of assets, an event that sheds light on the outcome of a loss contingency, or any other event having a material effect on operations. Illustration 3–10 illustrates an event that **Target Corporation** disclosed in its annual report for the year ending January 31, 2015, but before the release of those financial statements. We cover subsequent events in more depth in Chapter 13.

Illustration 3–10

Subsequent Event—Target Corp.

Real World Financials

29. Subsequent Events

In March 2015, we announced a headquarters workforce reduction. As a result, we expect to record approximately $100 million of severance and other benefits-related charges within SG&A in the first quarter of 2015, the vast majority of which are expected to require cash expenditures.

Source: Target Corp.

Noteworthy Events and Transactions

Some transactions and events occur only occasionally but, when they do occur, they are potentially important to evaluating a company's financial statements. In this category are related-party transactions, errors and fraud, and illegal acts. The most frequent of these is related-party transactions.

Sometimes a company will engage in transactions with owners, management, families of owners or management, affiliated companies, and other parties that can significantly influence or be influenced by the company. The potential problem with **related-party transactions** is that their economic substance may differ from their legal form. For instance, borrowing or lending money at an interest rate that differs significantly from the market interest rate is an example of a transaction that could result from a related-party involvement. As a result of the potential for misrepresentation, financial statement users are particularly interested in more details about these transactions.

The economic substance of *related-party* transactions should be disclosed, including dollar amounts involved.

When related-party transactions occur, companies must disclose the nature of the relationship, provide a description of the transactions, and report the dollar amounts of transactions and any amounts due from or to related parties.[8] Illustration 3–11 shows a disclosure

Illustration 3–11

Related-Party Transactions Disclosure—Champions Oncology, Inc.

Real World Financials

Note 4. Related-Party Transactions

Related-party transactions include transactions between the Company and its shareholders, management, or affiliates. The following transactions were in the normal course of operations and were measured and recorded at the exchange amount, which is the amount of consideration established and agreed to by the parties.

Consulting Services

For both years ended April 30, 2017 and 2016, the Company paid a member of its Board of Directors $72,000 for consulting services unrelated to his duties as board member. During the years ended April 30, 2017 and 2016, the Company paid a board member's company $0 and $8,800, respectively, for consulting services unrelated to his duties as a board member. During the year ended April 30, 2017, the Company paid a board member $48,214 and granted 45,000 options that vest annually over a three-year period and have a fair value of $94,192 for consulting services unrelated to his duties as a board member. All of the amounts paid to these related parties have been recognized in expense in the period the services were performed.

Source: Champions Oncology, Inc.

[7]Financial statements are viewed as issued if they have been widely distributed to financial statement users in a format consistent with GAAP. Some entities (for example, private companies) don't widely distribute their financial statements to users. For those entities, the key date for subsequent events is not the date of issuance but rather the date upon which the financial statements are "available to be issued," which occurs when the financial statements are complete, in a format consistent with GAAP, and have obtained the necessary approvals for issuance. Companies must disclose the date through which subsequent events have been evaluated. FASB ASC 855: Subsequent Events (previously "Subsequent Events," *Statement of Financial Accounting Standards No. 165* (Stamford, Conn.: FASB, 2009)).

[8]FASB ASC 850–10–50: Related Party Disclosures—Overall—Disclosure (previously "Related Party Disclosures," *Statement of Financial Accounting Standards No. 57* (Stamford, Conn.: FASB, 1982)).

note from a recent annual report of **Champions Oncology, Inc.** The note describes payments for consulting services to a member of its Board of Directors.

Less frequent events are errors and **fraud**. The distinction between these two terms is that *errors are unintentional* while *fraud is intentional* misappropriation of assets or incorrect financial reporting.[9] Errors and fraud may require disclosure (e.g., of assets lost through either errors or fraud). Obviously, the existence of fraud involving management might cause a user to approach financial analysis from an entirely different and more cautious viewpoint.

Closely related to fraud are **illegal acts** such as bribes, kickbacks, illegal contributions to political candidates, and other violations of the law. Accounting for illegal practices has been influenced by the Foreign Corrupt Practices Act passed by Congress in 1977. The Act is intended to discourage illegal business practices through tighter controls and also encourage better disclosure of those practices when encountered. The nature of such disclosures should be influenced by the materiality of the impact of illegal acts on amounts disclosed in the financial statements.[10] However, the SEC issued guidance expressing its view that exclusive reliance on quantitative benchmarks to assess materiality in preparing financial statements is inappropriate.[11] A number of other factors, including whether the item in question involves an unlawful transaction, should also be considered when determining materiality.

As you might expect, any disclosures of related-party transactions, fraud, and illegal acts can be quite sensitive. Although auditors must be considerate of the privacy of the parties involved, that consideration cannot be subordinate to users' needs for full disclosure.

We've discussed only a few of the disclosure notes most frequently included as an integral part of the financial statements. Other common disclosures include details concerning earnings per share calculations, income taxes, property and equipment, contingencies, long-term debt, leases, pensions, stock compensation, changes in accounting methods, fair values of financial instruments, and exposure to market risk and credit risk. We discuss and illustrate these in later chapters in the context of related financial statement elements.

Disclosure notes for some financial statement elements are required. Others are provided when required by specific situations in the interest of full disclosure.

Management's Discussion and Analysis

• LO3–5

In addition to the financial statements and accompanying disclosure notes, each annual report of a public company requires a fairly lengthy discussion and analysis provided by the company's management. In this section, which precedes the financial statements and the auditor's report, management provides its views on significant events, trends, and uncertainties pertaining to the company's (a) operations, (b) liquidity, and (c) capital resources. Although the **management's discussion and analysis (MD&A)** section may embody management's biased perspective, it can offer an informed insight that might not be available elsewhere. As an example, Illustration 3–12 contains a portion of **GameStop Corporation**'s MD&A regarding liquidity and capital resources that is in its annual report.

Management's discussion and analysis provides a biased but informed perspective of a company's (a) operations, (b) liquidity, and (c) capital resources.

Management's Responsibilities

Management prepares and is responsible for the financial statements and other information in the annual report. To enhance the awareness of the users of financial statements concerning the relative roles of management and the auditor, annual reports of public companies include a management's responsibilities section that asserts the responsibility of management for the information contained in the annual report as well as an assessment of the company's internal control procedures.

Management acknowledges responsibility and certifies accuracy of financial statements.

[9]"Consideration of Fraud in a Financial Statement Audit," *AICPA Professional Standards AU 240* (New York: AICPA, 2012).

[10]"Consideration of Laws and Regulations in an Audit of Financial Statements," *AICPA Professional Standards AU 250* (New York: AICPA, 2012).

[11]FASB ASC 250–10–S99–1, SAB Topic 1.M: Assessing Materiality (originally "Materiality," *Staff Accounting Bulletin No. 99* (Washington, D.C.: SEC, August 1999)).

Illustration 3–12

Management's Discussion and Analysis—GameStop Corp.

Real World Financials

Management's Discussion and Analysis of Financial Condition and Results of Operations
(In part: Liquidity and Capital Resources)

Overview

Based on our current operating plans, we believe that available cash balances, cash generated from our operating activities and funds available under our $400.0 million asset-based revolving credit facility (the "Revolver") together will provide sufficient liquidity to fund our operations, store openings and remodeling activities and corporate capital allocation programs, including acquisitions, share repurchases and the payment of dividends declared by the Board of Directors, for at least the next 12 months.

Cash Flow

During fiscal 2016, cash provided by operations was $537.1 million, compared to cash provided by operations of $656.8 million in fiscal 2015. The decrease in cash provided by operations of $119.7 million from fiscal 2015 to fiscal 2016 was primarily due to the decrease in net income combined with a decrease in cash provided by changes in operating assets and liabilities of $99.6 million, due primarily to the timing of payments for income taxes and accounts payable.

Acquisitions and Capital Expenditures

We opened or acquired 583 Technology Brands stores and opened 73 Video Game Brands stores in fiscal 2016. We expect to open approximately 35 stand-alone collectibles stores and approximately 65 stores in our Technology Brands segment in fiscal 2017.

In fiscal 2016, in connection with the continued expansion of our Technology Brands segment, Spring Mobile completed the acquisition of 507 AT&T authorized retailer stores for an aggregate of $440.3 million in cash (net of cash acquired), which includes working capital adjustments, and future contingent consideration, which we estimate will range from $40.0 million to $50.0 million. The majority of the store acquisitions were funded by our 2021 Senior Notes. The contingent consideration will be paid in two installments. The first installment of $20 million is contingent on the relocation of certain stores and is due the latter of August 2017 or when relocations are completed. The second installment, which we expect to range from $20.0 million to $30.0 million, is contingent on sales performance of certain stores and is due in March 2018.

Source: GameStop Corp.

Illustration 3–13 contains Management's Report on Internal Control over Financial Reporting for **Nike**, included with the company's financial statements for the year ended May 31, 2017. Recall from our discussion of financial reporting reform in Chapter 1, that the *Sarbanes-Oxley Act of 2002* requires corporate executives to personally certify the financial statements. Submission of false statements carries a penalty of up to 20 years in jail. Mark G. Parker, Nike's president and chief executive officer, and Donald W. Blair, the chief financial officer, signed the required certifications.

Compensation of Directors and Top Executives

The compensation large U.S. corporations pay their top executives is an issue of considerable public debate and controversy. Shareholders, employees, politicians, and the public in general sometimes question the huge pay packages received by company officials at the same time that more and more rank-and-file employees are being laid off as a result of company cutbacks. Contributing to the debate is the realization that the compensation gap between executives and lower-level employees is much wider in the United States than in most other industrialized countries.

A substantial portion of executive pay often is in the form of stock options or restricted stock awards. Executive stock options give their holders the right to buy the company's

Management's Responsibility for Financial Statements (in part)
Management of NIKE, Inc. is responsible for the information and representations contained in this report. The financial statements have been prepared in conformity with accounting principles generally accepted in the United States of America ("U.S. GAAP") and include certain amounts based on our best estimates and judgments. Other financial information in this report is consistent with these financial statements.

The Audit Committee is responsible for the appointment of the independent registered public accounting firm and reviews, with the independent registered public accounting firm, management and the internal audit staff, the scope and the results of the annual audit, the effectiveness of the accounting control system and other matters relating to the financial affairs of NIKE as the Audit Committee deems appropriate. The independent registered public accounting firm and the internal auditors have full access to the Audit Committee, with and without the presence of management, to discuss any appropriate matters.

Management's Report on Internal Control over Financial Reporting (in part)
Management is responsible for establishing and maintaining adequate internal control over financial reporting, as such term is defined in Rule 13(a)-15(f) and Rule 15(d)-15(f) of the Securities Exchange Act of 1934, as amended. Internal control over financial reporting is a process designed to provide reasonable assurance regarding the reliability of financial reporting and the preparation of the financial statements for external purposes in accordance with generally accepted accounting principles in the United States of America.

While "reasonable assurance" is a high level of assurance, it does not mean absolute assurance. Because of its inherent limitations, internal control over financial reporting may not prevent or detect every misstatement and instance of fraud. Controls are susceptible to manipulation, especially in instances of fraud caused by the collusion of two or more people, including our senior management.

Under the supervision and with the participation of our Chief Executive Officer and Chief Financial Officer, our management conducted an evaluation of the effectiveness of our internal control over financial reporting based upon the framework in Internal Control — Integrated Framework (2013) issued by the Committee of Sponsoring Organizations of the Treadway Commission (COSO). Based on the results of our evaluation, our management concluded that our internal control over financial reporting was effective as of May 31, 2017.

Mark G. Parker
President & Chief Executive Officer

Donald W. Blair
Chief Financial Officer

Source: Nike, Inc.

Illustration 3–13

Management's Responsibilities and Certification—Nike, Inc.

Real World Financials

stock at a set price, regardless of how high the stock price rises. Restricted stock is a unit of stock given to an employee, but that unit of stock is not fully transferable until certain conditions are met (such as length of employment or attainment of performance goals). In recent years, restricted stock as a form of compensation has become more popular than stock options. Both forms of stock compensation are discussed in depth in Chapter 19.

To help shareholders and others sort out the content of executive pay packages and better understand the commitments of the company in this regard, SEC requirements provide for disclosures on compensation to directors and executives. A **proxy statement** is provided each year and includes compensation information for directors and top executives. The statement also invites shareholders to the annual meeting to elect board members and to vote on issues before the shareholders or to vote by proxy. Illustration 3–14 shows a portion of **Best Buy**'s summary compensation table included in a recent proxy statement.

The proxy statement contains disclosures on compensation to directors and executives.

Illustration 3–14 Summary Compensation Table—Best Buy Co., Inc. Real World Financials

Summary Compensation Table (in part)

Name and Principal Position	Salary	Stock Awards	Option Awards	Short-Term Incentive Plan Payout	All Other Compensation	Total
Hubert July Chairman and Chief Executive Officer	$1,175,000	$7,689,879	$1,800,076	$2,878,750	$494,275	$14,037,980
Corie S. Barry Chief Financial Officer	713,462	1,689,495		1,184,167	14,893	3,602,017
Shari L. Ballard President, Multi-Channel Retail and Operations	800,000	1,930,865		1,470,000	62,737	4,263,602
R. Michael Mohan Chief Merchandising and Marketing Officer	833,654	2,895,073		1,531,251	55,284	5,315,262
Keith Nelsen General Counsel and Secretary	650,000	1,592,960		796,250	68,761	3,107,971

Source: Best Buy Co., Inc.

Auditor's Report

LO3–6

Auditors examine financial statements and the internal control procedures designed to support the content of those statements. Their role is to attest to the fairness of the financial statements based on that examination. The auditor's attest function for public business entities results in an opinion stated in the **auditor's report**.

There are four basic types of auditor's reports, as follows:

1. Unqualified
2. Unqualified with an explanatory or emphasis paragraph
3. Qualified
4. Adverse or disclaimer

An auditor issues an *unqualified* (or "clean") opinion when the auditor has undertaken professional care to ensure that the financial statements are presented in conformity with generally accepted accounting principles (GAAP). Professional care would include sufficient planning of the audit, understanding the company's internal control procedures, and gathering evidence to attest to the accuracy of the amounts reported in the financial statements.

The unqualified auditor's report prepared by **Ernst & Young (EY)** for the financial statements of **Facebook** is shown in Illustration 3–15.

In most cases, including the report for Facebook, the auditors will be satisfied that the financial statements "present fairly" the financial position, results of operations, and cash flows and are "in conformity with U.S. generally accepted accounting principles." These situations prompt an unqualified opinion. Notice that the report also references the auditor's opinion on the effectiveness of the company's internal control over financial reporting.[12]

[12]The auditors' reports of public companies must be in compliance with the specifications of the PCAOB as specified in AS 2201: An Audit of Internal Control over Financial Reporting That Is Integrated with An Audit of Financial Statements.

Illustration 3–15

Auditor's Report—Facebook, Inc.

Real World Financials

Report of Independent Registered Public Accounting Firm (in part)

To the Stockholders and Board of Directors of Facebook, Inc.

Opinion on the Financial Statements

We have audited the accompanying consolidated balance sheets of Facebook, Inc. (the Company) as of December 31, 2017 and 2016, and the related consolidated statements of income, comprehensive income, stockholders' equity and cash flows for each of the three years in the period ended December 31, 2017, and the related notes (collectively referred to as the "financial statements"). In our opinion, the financial statements present fairly, in all material respects, the consolidated financial position of the Company at December 31, 2017 and 2016, and the consolidated results of its operations and its cash flows for each of the three years in the period ended December 31, 2017, in conformity with U.S. generally accepted accounting principles.

We also have audited, in accordance with the standards of the Public Company Accounting Oversight Board (United States) (PCAOB), the Company's internal control over financial reporting as of December 31, 2017, based on criteria established in Internal Control—Integrated Framework issued by the Committee of Sponsoring Organizations of the Treadway Commission (2013 framework) and our report dated February 1, 2018 expressed an unqualified opinion thereon.

Basis for Opinion

These financial statements are the responsibility of the Company's management. Our responsibility is to express an opinion on the Company's financial statements based on our audits. We are a public accounting firm registered with the PCAOB and are required to be independent with respect to the Company in accordance with the U.S. federal securities laws and the applicable rules and regulations of the Securities and Exchange Commission and the PCAOB.

We conducted our audits in accordance with the standards of the PCAOB. Those standards require that we plan and perform the audit to obtain reasonable assurance about whether the financial statements are free of material misstatement, whether due to error or fraud. Our audits included performing procedures to assess the risks of material misstatement of the financial statements, whether due to error or fraud, and performing procedures that respond to those risks. Such procedures included examining, on a test basis, evidence regarding the amounts and disclosures in the financial statements. Our audits also included evaluating the accounting principles used and significant estimates made by management, as well as evaluating the overall presentation of the financial statements. We believe that our audits provide a reasonable basis for our opinion.

/s/ Ernst & Young LLP

We have served as the Company's auditor since 2007.
San Francisco, California
February 1, 2018

Source: Facebook

Sometimes circumstances cause the auditor to issue an opinion that is *unqualified with an explanatory paragraph.* In these circumstances, the auditor believes the financial statements are in conformity with GAAP (unqualified), but the auditor feels that other important information needs to be emphasized to financial statement users. Most notably, these situations include the following:

- *Lack of consistency* due to a change in accounting principle such that comparability is affected even though the auditor concurs with the desirability of the change.
- *Going concern* when the auditor determines there is significant doubt as to whether the company will be able to pay its debts as they come due. Indicators of a going concern include significant operating losses, loss of a major customer, or legal proceedings that might jeopardize the company's ability to continue operations.[13]
- *Material misstatement* in previously issued financial statements has been corrected.

[13]"The Auditor's Consideration of an Entity's Ability to Continue as a Going Concern (Redrafted)," *Statement on Auditing Standards No. 126, AICPA Professional Standards AU 570* (New York: AICPA, 2012). Management is also required to assess the company's ability to continue as a going concern and provide certain disclosures when substantial doubt exists (ASC 205-40).

An audit opinion may also include a paragraph on *emphasis of a matter*. This discussion would include items such as significant transactions with related parties, important events subsequent to the balance sheet date, or uncertainty relating to the future outcome of significant litigation or regulatory actions.

An example of a going concern paragraph in the 2016 auditor's report of **Timberline Resources** is shown in Illustration 3–16.

Illustration 3–16

Going Concern Paragraph—Timberline Resources Corp.

Real World Financials

> The accompanying consolidated financial statements have been prepared assuming that the Company will continue as a going concern. As discussed in Note 2 to the financial statements, the Company has accumulated losses since inception and has negative working capital. These factors raise substantial doubt about its ability to continue as a going concern. Management's plans in regard to these matters are also described in Note 2. The financial statements do not include any adjustments that might result from the outcome of this uncertainty.
>
> Source: Timberline Resources Corp.

The auditor's report calls attention to problems that might exist in the financial statements.

Some audits result in the need to issue other than an unqualified opinion due to exceptions such as (a) nonconformity with generally accepted accounting principles, (b) inadequate disclosures, and (c) a limitation or restriction of the scope of the examination. In these situations the auditor will issue one of the following:

- *Qualified opinion* when either the audit process has been limited (scope limitation) or there has been a departure from GAAP, but neither is of sufficient seriousness to invalidate the financial statements as a whole.
- *Adverse opinion* when the auditor has specific knowledge that financial statements or disclosures are seriously misstated or misleading. Adverse opinions are rare because auditors usually are able to persuade management to rectify problems to avoid this undesirable report.
- *Disclaimer* when the auditor is not able to gather sufficient information that financial statements are in conformity with GAAP.

Obviously, the auditor's report is most informative when any of these deviations from the standard unqualified opinion are present. These departures from the norm should raise a red flag to a financial analyst and prompt additional search for information.

PART C Risk Analysis

Using Financial Statement Information

• LO3–7

Investors and others use information that companies provide in corporate financial reports to make decisions. Although the financial reports focus primarily on the past performance and the present financial condition of the reporting company, users are most interested in information about the future. Investors want to know a company's **default risk**. This is the risk that a company won't be able to pay its obligations when they come due. Another aspect of risk is **operational risk**, which relates more to how adept a company is at withstanding various events and circumstances that might impair its ability to earn profits.

Trying to gain a glimpse of the future from past and present data entails using various tools and techniques to formulate predictions. This is the goal of financial statement analysis. Common methods for analyzing financial statements include the following:

1. **Comparative financial statements**. Financial statements that are accompanied by the corresponding financial statement of the preceding year, and often the previous two years.
2. **Horizontal analysis**. Each item in a financial statement is expressed as a percentage of that same item in the financial statements of another year (base amount). For example,

comparing inventory this year to inventory last year would provide the percentage change in inventory.

3. **Vertical analysis.** Each item in the financial statements is expressed as a percentage of an appropriate corresponding total, or base amount, but within the same year. For example, cash, receivables, and inventory in the current year can be restated as a percentage of total assets in the current year.
4. **Ratio analysis.** Financial statement items are converted to ratios for evaluating the performance and risk of a company.

As an example, we presented the current asset section of Nike's comparative balance sheets in Illustration 3–3. From this, we could easily calculate the current year's percentage change in cash (horizontal analysis) or cash as a percentage of total current assets (vertical analysis). Understanding Nike's cash position, either relative to the previous year or relative to other current assets, provides some information for assessing the cash available to pay debt or to respond to other operating concerns.

However, the most common way to analyze financial statements is ratio analysis. Next, we'll analyze Nike's risk by using ratio analysis to investigate **liquidity** and **long-term solvency**. In Chapter 4, we introduce ratios related to profitability analysis. You will also employ ratios in Decision Makers' Perspective sections of many of the chapters in this text. Analysis cases that benefit from ratio analysis are included in many of these chapters as well.

We use ratios every day. Batting averages indicate how well our favorite baseball players are performing. We evaluate basketball players by field goal percentage and rebounds per game. Speedometers measure the speed of our cars in terms of miles per hour. We compare grocery costs on the basis of price per pound or ounce. In each of these cases, the ratio is more meaningful than a single number by itself. Do 45 hits indicate satisfactory performance? It depends on the number of at-bats. Is $2 a good price for cheese? It depends on how many ounces the $2 buys. Ratios make these measurements meaningful.

Likewise, we can use ratios to help evaluate a firm's performance and financial position. Is net income of $4 million a cause for shareholders to celebrate? Probably not if shareholders' equity is $10 billion. But if shareholders equity is $10 million, that's a return on equity of 40% (net income divided by shareholders' equity)! Although ratios provide more meaningful information than absolute numbers alone, the ratios are most useful when analyzed relative to some standard of comparison. That standard of comparison may be previous performance of the same company, the performance of a competitor company, or an industry average for the particular ratio. Such comparisons are useful to investors and analysts in evaluating the company's performance, to management for planning and control purposes, and to auditors in assessing the reasonableness of amounts in financial statements.

Evaluating information in ratio form allows analysts to control for size differences over time and among firms.

Liquidity Ratios

LO3–8

Liquidity most often refers to the ability of a company to convert its assets to cash to pay its *current* obligations. By examining a company's liquidity, we can obtain a general idea of the firm's ability to pay its short-term debts as they come due.

Current assets usually are thought of as the most liquid of a company's assets, because these assets are easier to convert to cash than are long-term assets. Obviously, though, some current assets are more liquid than others, so it's important also to evaluate the specific makeup of current assets.

Two common measures of liquidity are (1) the current ratio and (2) the acid-test ratio (or quick ratio) calculated as follows:

$$\text{Current ratio} = \frac{\text{Current assets}}{\text{Current liabilities}}$$

$$\text{Acid-test (or quick ratio)} = \frac{\text{Quick assets}}{\text{Current liabilities}}$$

CURRENT RATIO Implicit in the definition of a current liability is the relationship between current assets and current liabilities. The difference between current assets and current liabilities is called **working capital**. By comparing a company's obligations that will

***Working capital*, the difference between current assets and current liabilities, is a popular measure of a company's ability to satisfy its short-term obligations.**

shortly become due with the company's cash and other assets that, by definition, will shortly be converted to cash or used to generate cash, the analysis offers some indication of an ability to pay current debts. This ratio is particularly useful to lenders considering whether to offer short-term credit. **Nike**'s working capital (in millions) at the end of its May 31, 2017, fiscal year is $10,587. This amount is computed as Nike's current assets of $16,061 (Illustration 3–3) minus its current liabilities of $5,474 (Illustration 3–7).

$$\text{Working capital} = \$16{,}061 - \$5{,}474 = \$10{,}587$$

The **current ratio** equals current assets divided by current liabilities. Nike's current ratio indicates the company has $2.93 of current assets for each $1 of current liabilities.

$$\text{Current ratio} = \frac{\$16{,}061}{\$5{,}474} = 2.93$$

Care should be taken, however, in assessing liquidity based solely on the current ratio. Liabilities usually are paid with cash, not other types of current assets. A company could have difficulty paying its liabilities even with a current ratio significantly greater than 1.0. For example, if a significant portion of current assets consisted of inventories, and inventories usually are not converted to cash for several months, there could be a problem in paying accounts payable due in 30 days. On the other hand, a current ratio of less than 1.0 doesn't necessarily mean the company will have difficulty meeting its current obligations. A line of credit, for instance, which the company can use to borrow funds, provides financial flexibility. That also must be considered in assessing liquidity.

The *acid-test ratio* provides a more stringent indication of a company's ability to pay its current obligations.

ACID-TEST RATIO (OR QUICK RATIO). Some analysts like to modify the current ratio to consider only current assets that are readily available to pay current liabilities. One such variation in common use is the **acid-test ratio**. This ratio excludes inventories, prepaid items, and restricted cash from current assets before dividing by current liabilities. The numerator, then, consists of (unrestricted) cash, short-term investments, and accounts receivable, which are referred to as the "quick assets." By eliminating current assets that are less readily convertible into cash, the acid-test ratio provides a more rigorous indication of liquidity than does the current ratio.

Illustration 3–3 shows that **Nike**'s quick assets (in millions) total $9,856 ($3,808 + $2,371 + $3,677). The acid-test ratio can be computed as follows:

$$\text{Acid-test ratio} = \frac{\$9{,}856}{\$5{,}474} = 1.80$$

Are these liquidity ratios adequate? It's generally difficult to say without some point of comparison. As indicated previously, common standards for such comparisons are industry averages for similar ratios or ratios of the same company in prior years. Industry averages for the above two ratios are as follows:

Industry Average

Current ratio = 1.43

Acid-test ratio = 1.15

Nike's ratios are higher than the industry average, so Nike's liquidity appears to be in good shape. What if the ratios were lower? Would that indicate a liquidity problem? Not necessarily, but it would raise a red flag that calls for caution in analyzing other areas. Remember that each ratio is but one piece of the entire puzzle. For instance, profitability is perhaps the best indication of liquidity in the long run. We discuss ratios that measure profitability in Chapter 4.

Liquidity ratios should be assessed in the context of both profitability and efficiency of managing assets.

Also, management may be very efficient in managing current assets so that, let's say, receivables are collected faster than normal or inventory is sold faster than normal, making those assets more liquid than they otherwise would be. Higher turnover ratios, relative to those of a competitor or the industry, generally indicate a more liquid position for a given level of the current ratio. We discuss these turnover ratios in Chapter 4.

Ethical Dilemma

The Raintree Cosmetic Company has several loans outstanding with a local bank. The debt agreements all contain a covenant stipulating that Raintree must maintain a current ratio of at least 0.9. Jackson Phillips, company controller, estimates that the 2021 year-end current assets and current liabilities will be $2,100,000 and $2,400,000, respectively. These estimates provide a current ratio of only 0.875. Violation of the debt agreement will increase Raintree's borrowing costs as the loans are renegotiated at higher rates.

Jackson proposes to the company president that Raintree purchase inventory of $600,000 on credit before year-end. This will cause both current assets and current liabilities to increase by the same amount, but the current ratio will increase to 0.9. The extra $600,000 in inventory will be used over the later part of 2022. However, the purchase will cause warehousing costs and financing costs to increase.

Jackson is concerned about the ethics of his proposal. What do you think?

Solvency Ratios

Investors and creditors, particularly long-term creditors, are vitally interested in long-term solvency, a company's ability to pay its long-term debts. Two common solvency ratios are (1) the debt to equity ratio and (2) the times interest earned ratio:

$$\text{Debt to equity ratio} = \frac{\text{Total liabilities}}{\text{Shareholders' equity}}$$

$$\text{Times interest earned ratio} = \frac{\text{Net income} + \text{Interest expense} + \text{Income taxes}}{\text{Interest expense}}$$

The *debt to equity ratio* indicates the extent of reliance on creditors, rather than owners, in providing resources.

DEBT TO EQUITY RATIO The **debt to equity ratio** compares resources provided by creditors with resources provided by owners. It is calculated by dividing total liabilities (current and long-term) by total shareholders' equity (including retained earnings). The mixture of liabilities and shareholders' equity in a company refers to its **capital structure**.[14]

Other things being equal, the higher the ratio, the higher the company's risk. The higher the ratio, the greater the creditor claims on assets, so the higher the likelihood an individual creditor would not be paid in full if the company is unable to meet its obligations.

Nike's liabilities (in millions) are $10,852 (Illustration 3–7), and stockholders' equity is $12,407 (Illustration 3–8). The debt to equity ratio can be computed as follows:

$$\text{Debt to equity ratio} = \frac{\$10,852}{\$12,407} = 0.87$$

As with all ratios, the debt to equity ratio is more meaningful if compared to some standard such as an industry average or a competitor. For example, an industry average debt to equity ratio of 0.95 would indicate that Nike has a lower portion of liabilities in its capital structure than does the average firm in its industry. Does this mean that Nike's default risk is lower? Other things equal—yes. Is that good? Not necessarily. As discussed in the next section, it may be that debt is being underutilized by Nike. More debt might increase the potential for return to shareholders, but with higher debt comes higher risk. This is a fundamental trade-off faced by virtually all firms when trying to settle on the optimal amount of debt versus equity in its capital structure.

The makeup of liabilities also is important. For example, liabilities could include deferred revenues. Recall that deferred revenues are liabilities recorded when cash is received from customers in advance of providing a good or service. Companies satisfy these liabilities not by paying cash, but by providing the good or service to their customers.

[14]A commonly used variation of the debt to equity ratio is found by dividing total liabilities by *total assets*, rather than by shareholders' equity only. Of course, in this configuration the ratio measures precisely the same attribute of the firm's capital structure but can be interpreted as the percentage of a company's total assets provided by funds from creditors, rather than by owners.

The *times interest earned ratio* indicates the margin of safety provided to creditors.

TIMES INTEREST EARNED RATIO A ratio that is commonly used in conjunction with the debt to equity ratio is the **times interest earned ratio**. This ratio is calculated as income before subtracting interest expense and income taxes, divided by interest expense. To remain solvent or to take on more debt if needed, a company needs to have funds available in the current year to pay interest charges. The ability of a company to "cover" its interest charges commonly is measured by the extent to which income exceeds interest charges in the current period.

If income is many times greater than interest expense, creditors' interests are more protected than if income just barely covers this expense. For this purpose, income should be the amount available to pay interest, which is income before subtracting interest and income taxes, calculated by adding back to net income the interest and income taxes that were deducted.

As an example, **Nike** reports the following:

	($ in millions)
Net income	$4,240
Interest expense	+ 59
Income taxes	+ 646
Income before interest and taxes	$4,945

The times interest earned ratio can be computed as follows:

$$\text{Times interest earned ratio} = \frac{\$4{,}945}{\$59} = 83.81$$

The ratio of 83.81 indicates a considerable margin of safety for creditors. Income could decrease many times and the company would still be able to meet its interest payment obligations.[15] Nike is a highly profitable company with little interest-bearing debt. In comparison, the average times interest earned ratio for its industry is approximately 36.4 times.

Relationship Between Risk and Profitability. Now that we've assessed a company's liquidity and solvency, let's look at why a company might want to borrow money. While there are default risks associated with borrowing, a company can use those borrowed funds to provide greater returns to its shareholders. This is referred to as favorable **financial leverage** and is a very common (but risky) business activity.

To see how financial leverage works, consider a newly formed corporation attempting to determine the appropriate mix of debt and equity. The initial capitalization goal is $50 million. The capitalization mix alternatives have been narrowed to two: (1) $10 million in debt and $40 million in equity and (2) $30 million in debt and $20 million in equity.

Also assume that regardless of the capitalization mix chosen, the corporation will be able to generate a 16% annual return, *before payment of interest and income taxes,* on the $50 million in assets acquired. In other words, income before interest and taxes will be $8 million (16% × $50 million). If the interest rate on debt is 8% and the income tax rate is 25%, comparative net income for the first year of operations for the two capitalization alternatives can be calculated as follows:

	Alternative 1 **Debt = $10 million** **Equity = $40 million**	**Alternative 2** **Debt = $30 million** **Equity = $20 million**
Income before interest and income taxes	$ 8,000,000	$ 8,000,000
Less: Interest expense	(800,000)*	(2,400,000)†
Income before income taxes	7,200,000	5,600,000
Less: Income tax expense (25%)	(1,800,000)	(1,400,000)
Net income	**$5,400,000**	**$4,200,000**

*8% × $10,000,000
†8% × $30,000,000

[15]Of course, interest is paid with cash, not with "income." The times interest earned ratio often is calculated by using cash flow from operations before subtracting either interest payments or tax payments as the numerator and interest payments as the denominator.

Would shareholders be in favor of alternative 1? Probably not. Although alternative 1 provides a higher net income, the return on the shareholders' equity (net income divided by shareholders' equity) is higher for alternative 2. Under alternative 1, shareholders had to invest $40 million to earn $5.4 million. Under alternative 2, shareholders had to invest only $20 million to earn $4.2 million.

Favorable financial leverage means earning a return on borrowed funds that exceeds the cost of borrowing the funds.

	Alternative 1	Alternative 2
Return on equity[16] =	$5,400,000 / $40,000,000	$4,200,000 / $20,000,000
=	13.5%	21.0%

Alternative 2 generated a higher return for each dollar invested by shareholders. This is because the company leveraged its $20 million equity investment with additional debt. *Any time the cost of the additional debt (8%) is less than the return on assets invested (16%), the return to shareholders is higher with borrowing.* This is the essence of favorable financial leverage.

Be aware, though, leverage is risky and not always favorable; the cost of borrowing the funds might exceed the returns they provide. If the return on assets invested turned out to be less than expected, the additional debt could result in a lower return on equity for alternative 2. If, for example, the return on assets invested (before interest and income taxes) had been 6% of $50,000,000 (or $3,000,000), rather than 16%, alternative 1 would have provided the better return on equity:

	Alternative 1 Debt = $10 million Equity = $40 million	Alternative 2 Debt = $30 million Equity = $20 million
Income before interest and income taxes	$ 3,000,000	$ 3,000,000
Less: Interest expense	(800,000)*	(2,400,000)†
Income before income taxes	2,200,000	600,000
Less: Income tax expense (25%)	(550,000)	(150,000)
Net income	**$1,650,000**	**$ 450,000**

*8% × $10,000,000
†8% × $30,000,000

	Alternative 1	Alternative 2
Return on equity[17] =	$1,650,000 / $40,000,000	$450,000 / $20,000,000
=	4.125%	2.250%

If the return on assets are too low and the company has become too leveraged, it faces the risk of not being able to make its interest and debt payments. So, shareholders typically are faced with a trade-off between the risk that high debt denotes and the potential for a higher return from having the higher debt. Liquidity and solvency ratios can help with that decision.

[16]If return is calculated on *average* shareholders' equity, we're technically assuming that all income is paid to shareholders in cash dividends, so that beginning, ending, and average shareholders' equity are the same. If we assume *no* dividends are paid, rates of return would be

	Alternative 1	Alternative 2
Return on equity =	$5,400,000 / ($40,000,000 + $45,400,000)/2	$4,200,000 / ($20,000,000 + $24,200,000)/2
=	12.65%	19.00%

In any case, our conclusions are the same.

[17]If we assume *no* dividends are paid, rates of return would be

	Alternative 1	Alternative 2
Return on equity =	$1,650,000 / ($40,000,000 + $41,650,000)/2	$450,000 / ($20,000,000 + $20,450,000)/2
=	4.04%	2.23%

In any case, our conclusions are the same.

Financial Reporting Case Solution

©TY Lim/Shutterstock

1. **Respond to Jerry's criticism that shareholders' equity does not represent the market value of the company. What information does the balance sheet provide?** *(p. 113)* Jerry is correct. The financial statements are supposed to help investors and creditors value a company. However, the balance sheet is not intended to portray the market value of the company. The assets of a company minus its liabilities as shown in the balance sheet (shareholders' equity) usually will not equal the company's market value for several reasons. For example, many assets are measured at their historical costs rather than their fair values. Also, many company resources, including its trained employees, its experienced management team, and its reputation are not recorded as assets at all. The balance sheet must be used in conjunction with other financial statements, disclosure notes, and other publicly available information.

   The balance sheet does, however, provide valuable information that can be used by investors and creditors to help determine market value. After all, it is the balance sheet that describes many of the resources a company has available for generating future cash flows. The balance sheet also provides important information about liquidity and long-term solvency.
2. **The usefulness of the balance sheet is enhanced by classifying assets and liabilities according to common characteristics. What are the classifications used in the balance sheet and what elements do those categories include?** *(p. 113)* Nike's balance sheet contains the following classifications:

Assets:

- *Current assets* include cash and several other assets that are reasonably expected to be converted to cash or consumed within the coming year or within the normal operating cycle of the business if that's longer than one year.
- *Property, plant and equipment* are the tangible long-lived assets used in the operations of the business. This category includes land, buildings, equipment, machinery, and furniture, as well as natural resources.
- *Intangible assets* are assets that represent exclusive rights to something such as a product, a process, or a name. Patents, copyrights, and franchises are examples.
- *Goodwill* is a unique intangible asset in that its cost can't be directly associated with any specifically identifiable right and is not separable from the company as a whole. It represents the unique value of the company as a whole over and above all identifiable tangible and intangible assets.
- *Deferred income taxes* result from temporary differences between taxable income and accounting income.
- *Other long-term assets* is a "catchall" classification of long-term (noncurrent) assets and could include long-term prepaid expenses and any long-term asset not included in one of the other categories.

Liabilities:

- *Current liabilities* are those obligations that are expected to be satisfied through the use of current assets or the creation of other current liabilities. Usually, this means liabilities that are expected to be paid within one year from the balance sheet date (or within the operating cycle if that's longer than one year).
- *Long-term liabilities* are obligations that will *not* be satisfied in the next year or operating cycle, whichever is longer. Examples are long-term notes, bonds, pension obligations, and lease obligations.

Shareholders' equity:

- *Common stock* and *additional paid-in capital* collectively equal the amounts invested by shareholders in the corporation.

- *Retained earnings* represents the accumulated net income or net loss reported since inception of the corporation less dividends distributed to shareholders. If this amount is negative, it is called *accumulated deficit*.
- *Accumulated other comprehensive income/loss* is the cumulative amount of other comprehensive income/loss items. This topic is addressed in subsequent chapters. ●

The Bottom Line

- **LO3–1** The balance sheet is a position statement that presents an organized list of assets, liabilities, and equity at a point in time. The statement does not portray the market value of the entity. However, the information in the statement can be useful in assessing market value, as well as in providing important information about liquidity and long-term solvency. (*p. 112*)
- **LO3–2** Current assets include cash and other assets that are reasonably expected to be converted to cash or consumed within one year from the balance sheet date, or within the normal operating cycle of the business if that's longer than one year. All other assets are classified as various types of noncurrent assets. In addition to cash and cash equivalents, current assets include short-term investments, accounts receivable, inventory, and prepaid expenses. Long-term asset classifications include investments; property, plant, and equipment; intangible assets; and other assets. (*p. 114*)
- **LO3–3** Current liabilities are those obligations that are expected to be satisfied through the use of current assets or the creation of other current liabilities. All other liabilities are classified as long term. Current liabilities include notes and accounts payable, deferred revenues, accrued liabilities, and the current maturities of long-term debt. Long-term liabilities include long-term notes, loans, mortgages, bonds, pension and lease obligations, as well as deferred income taxes. Shareholders' equity for a corporation arises primarily from two sources: (1) paid-in capital—amounts invested by shareholders in the corporation, and (2) retained earnings—accumulated net income reported by a company since its inception minus all dividends distributed to shareholders. (*p. 118*)
- **LO3–4** Financial statement disclosures are used to convey additional information about the account balances in the basic financial statements as well as to provide supplemental information. This information is disclosed, often parenthetically in the basic financial statements, or in disclosure notes that often include supporting schedules. (*p. 122*)
- **LO3–5** Annual reports of public companies will include management's discussion and analysis of key aspects of the company's business. The purpose of this disclosure is to provide external parties with management's insight into certain transactions, events, and circumstances that affect the enterprise, including their financial impact. (*p. 125*)
- **LO3–6** The purpose of an audit is to provide a professional, independent opinion as to whether or not the financial statements are prepared in conformity with generally accepted accounting principles. The standard audit report of a public company identifies the financial statements and their dates, issues an opinion on whether those financial statements are presented fairly, explains the basis for that opinion, and states that the financial statements are the responsibility of management. Explanatory language and emphasis of matters might be included. (*p. 128*)
- **LO3–7** Financial analysts use various techniques to transform financial information into forms more useful for analysis. Horizontal analysis provides a useful way of analyzing year-to-year changes, while vertical analysis allows individual items to be expressed as a percentage of some base amount within the same year. Ratio analysis allows analysts to control for size differences over time and among firms while investigating important relationships among financial variables. (*p. 130*)
- **LO3–8** The balance sheet provides information that can be useful in assessing risk. A key element of risk analysis is investigating a company's ability to pay its obligations when they come due. Liquidity ratios and solvency ratios provide information about a company's ability to pay its obligations. (*p. 131*)
- **LO3–9** There are more similarities than differences in balance sheets and financial disclosures prepared according to U.S. GAAP and those prepared applying IFRS. Balance sheet presentation is one important difference. Under U.S. GAAP, we present current assets and liabilities before long-term assets and liabilities. IFRS doesn't prescribe the format of the balance sheet, but balance sheets prepared using IFRS often report long-term items first. Reportable segment disclosures also are similar. However, IFRS requires an additional disclosure, the amount of segment liabilities (Appendix 3). (*pp. 120* and *139*) ●

APPENDIX 3 Reporting Segment Information

Many companies operate in several business segments as a strategy to achieve growth and to reduce operating risk through diversification.

Financial analysis of diversified companies is especially difficult. Consider, for example, a company that operates in several distinct business segments including computer peripherals, home health care systems, textiles, and consumer food products. The results of these distinctly different activities will be aggregated into a single set of financial statements, making difficult an informed projection of future performance. It may well be that the five-year outlook differs greatly among the areas of the economy represented by the different segments. To make matters worse for an analyst, the integrated financial statements do not reveal the relative investments in each of the business segments nor the success the company has had within each area. Given the fact that so many companies these days have chosen to balance their operating risks through diversification, aggregated financial statements pose a widespread problem for analysts, lending and credit officers, and other financial forecasters.

Reporting by Operating Segment

Segment reporting facilitates the financial statement analysis of diversified companies.

To address the problem, the accounting profession requires public business entities to provide supplemental information concerning individual operating segments. The supplemental disaggregated data do not include complete financial statements for each reportable segment, only certain specified items.

WHAT IS A REPORTABLE OPERATING SEGMENT? According to U.S. GAAP guidelines, a *management approach* is used in determining which segments of a company are reportable. This approach is based on the way that management organizes the segments within the company for making operating decisions and assessing performance. The segments are, therefore, evident from the structure of the company's internal organization.

More formally, the following characteristics define an **operating segment**[18] as a component of a public business entity:

- That engages in business activities from which it may recognize revenues and incur expenses (including revenues and expenses relating to transactions with other components of the same entity).
- Whose operating results are regularly reviewed by the entity's chief operating decision maker to make decisions about resources to be allocated to the segment and assess its performance.
- For which discrete financial information is available.

The FASB hopes that this approach provides insights into the risk and opportunities management sees in the various areas of company operations. Also, reporting information based on a company's internal organization should reduce the incremental cost to companies of providing the data. In addition, there are quantitative thresholds for the definition of an operating segment to limit the number of reportable segments. Only segments of material size (10% or more of total company revenues, assets, or net income) must be disclosed. However, a company must account for at least 75% of consolidated revenue through segment disclosures.

WHAT AMOUNTS ARE REPORTED BY AN OPERATING SEGMENT? For areas determined to be reportable operating segments, the following disclosures are required:

a. General information about the operating segment.
b. Information about reported segment profit or loss, including certain revenues and expenses included in reported segment profit or loss, segment assets, and the basis of measurement.
c. Reconciliations of the totals of segment revenues, reported profit or loss, assets, and other significant items to corresponding entity amounts.
d. Interim period information.[19]

[18] FASB ASC 280-10-50-1: Segment Reporting—Overall—Disclosure (previously "Disclosures about Segments of an Enterprise and Related Information," *Statement of Financial Accounting Standards No. 131* (Norwalk, Conn.: FASB, 1997), par. 10).

[19] FASB ASC 280-10-50-20 through 26 and 280-10-50-32: Segment Reporting—Overall—Disclosure (previously "Disclosures about Segments of an Enterprise and Related Information," *Statement of Financial Accounting Standards No. 131* (Norwalk, Conn.: FASB, 1997), par. 25).

Illustration 3A–1 shows the business segment information reported by **Abbott Laboratories**, in its 2017 annual report.

Illustration 3A–1
Business Segment Information Disclosure—Abbott Laboratories, Inc.

Real World Financials

Business Segment Information
($ in millions)

Segments	Net Sales	Operating Earnings	Total Assets	Depr. and Amort.	Capital Expenditures
Established Pharmaceuticals	$ 4,287	$ 848	$ 2,728	$ 90	$ 181
Nutritionals	6,925	1,589	3,160	164	147
Diagnostics	5,616	1,468	4,226	300	374
Cardiovascular	8,911	2,720	5,074	298	206
Other	1,651	—	—	194	227
Total	**$27,390**	$6,625	$15,188	$1,046	$1,135

Source: Abbott Laboratories, Inc.

International Financial Reporting Standards

Segment Reporting. U.S. GAAP requires companies to report information about reported segment profit or loss, including certain revenues and expenses included in reported segment profit or loss, segment assets, and the basis of measurement. The international standard on segment reporting, *IFRS No. 8*,[20] requires that companies also disclose total *liabilities* of its reportable segments.

• LO3–9

REPORTING BY GEOGRAPHIC AREA In today's global economy it is sometimes difficult to distinguish domestic and foreign companies. Most large U.S. firms conduct significant operations in other countries in addition to having substantial export sales from this country. Differing political and economic environments from country to country means risks and associated rewards sometimes vary greatly among the various operations of a single company. For instance, manufacturing facilities in a South American country embroiled in political unrest pose different risks from having a plant in Vermont, or even Canada. Without disaggregated financial information, these differences cause problems for analysts.

U.S. GAAP requires a public business entity to report certain geographic information unless it is impracticable to do so. This information includes the following:

a. Revenues from external customers (1) attributed to the entity's country of domicile and (2) attributed to all foreign countries in total from which the entity derives revenues, and

b. Long-lived assets other than financial instruments, long-term customer relationships of a financial institution, mortgage and other servicing rights, deferred policy acquisition costs, and deferred tax assets (1) located in the entity's country of domicile and (2) located in all foreign countries in total in which the entity holds material assets.[21]

Abbott Laboratories reports its geographic sales by separate countries in 2017, as shown in Illustration 3A–2. Notice that both the business segment (Illustration 3A–1) and geographic information disclosures include a reconciliation to company totals. In both illustrations, net sales of both the business segments and the geographic areas are reconciled to the company's total net sales of **$27,390 million**.

[20] "Operating Segments," *International Financial Reporting Standard No. 8* (IASCF), as amended effective January 1, 2018.
[21] FASB ASC 280–10–50–41: Segment Reporting—Overall—Disclosure (previously "Disclosures about Segments of an Enterprise and Related Information," *Statement of Financial Accounting Standards No. 131* (Norwalk, Conn.: FASB, 1997), par. 38).

Illustration 3A–2

Geographic Area Sales Disclosure—Abbott Laboratories, Inc.

Real World Financials

Geographic Areas
($ in millions)

	Net Sales		
U.S.	$9,673	France	628
China	2,146	Brazil	541
Germany	1,366	Italy	507
Japan	1,255	United Kingdom	498
India	1,237	Columbia	494
Netherlands	929	Canada	443
Switzerland	841	Vietnam	427
Russia	664	All Others	5,741
		Total	**$27,390**

Source: Abbott Laboratories, Inc.

Revenues from major customers must be disclosed.

INFORMATION ABOUT MAJOR CUSTOMERS Financial analysts are extremely interested in information concerning the extent to which a company's prosperity depends on one or more major customers. For this reason, if 10% or more of the revenue of an entity is derived from transactions with a single customer, the entity must disclose that fact, the total amount of revenue from each such customer, and the identity of the operating segment or segments reporting the revenue. The identity of the major customer or customers need not be disclosed, although companies routinely provide that information. In its 2017 annual report, Abbott Laboratories did not report any major customer information.

Lockheed Martin Corporation reported a major customer in its 2017 segment disclosures. The company reported that the U.S. government accounts for 69% of its revenues. Many companies in the defense industry derive substantial portions of their revenues from contracts with the Defense Department. When cutbacks occur in national defense or in specific defense systems, the impact on a company's operations can be considerable. ●

Questions For Review of Key Topics

Q 3–1 Describe the purpose of the balance sheet.

Q 3–2 Explain why the balance sheet does not portray the market value of the entity.

Q 3–3 Define current assets and list the typical asset categories included in this classification.

Q 3–4 Define current liabilities and list the typical liability categories included in this classification.

Q 3–5 Describe what is meant by an operating cycle for a typical manufacturing company.

Q 3–6 Explain the difference(s) between investments in equity securities classified as current assets versus those classified as long-term (noncurrent) assets.

Q 3–7 Describe the common characteristics of assets classified as property, plant, and equipment and identify some assets included in this classification.

Q 3–8 Distinguish between property, plant, and equipment and intangible assets.

Q 3–9 Explain how each of the following liabilities would be classified in the balance sheet:

- A note payable of $100,000 due in five years
- A note payable of $100,000 payable in annual installments of $20,000 each, with the first installment due next year

Q 3–10 Define the terms *paid-in capital* and *retained earnings.*

Q 3–11 Disclosure notes are an integral part of the information provided in financial statements. In what ways are the notes critical to understanding the financial statements and to evaluating the firm's performance and financial health?

Q 3–12 A summary of the company's significant accounting policies is a required disclosure. Why is this disclosure important to external financial statement users?

Q 3–13 Define a subsequent event.

Q 3–14 Every annual report of a public company includes an extensive discussion and analysis provided by the company's management. Specifically, which aspects of the company must this discussion address? Isn't management's perspective too biased to be of use to investors and creditors?

Q 3–15 What is a proxy statement? What information does it provide?

Q 3–16 The auditor's report provides the analyst with an independent and professional opinion about the fairness of the representations in the financial statements. What are the four main types of opinion an auditor of a public company might issue? Describe each.

Q 3–17 Define the terms *working capital, current ratio,* and *acid-test ratio* (or *quick ratio*).

Q 3–18 Show the calculation of the following solvency ratios: (1) the debt to equity ratio, and (2) the times interest earned ratio.

IFRS Q 3–19 Where can we find authoritative guidance for balance sheet presentation under IFRS?

IFRS Q 3–20 Describe at least two differences between U.S. GAAP and IFRS in balance sheet presentation.

Q 3–21 (Based on Appendix 3) Segment reporting facilitates the financial statement analysis of diversified companies. What determines whether an operating segment is a reportable segment for this purpose?

Q 3–22 (Based on Appendix 3) For segment reporting purposes, what amounts are reported by each operating segment?

IFRS Q 3–23 (Based on Appendix 3) Describe any differences in segment disclosure requirements between U.S. GAAP and IFRS.

Brief Exercises

connect

BE 3–1
Current versus long-term classification
● LO3–2, LO3–3

Indicate whether each of the following assets and liabilities typically should be classified as current or long-term: (a) accounts receivable; (b) prepaid rent for the next six months; (c) notes receivable due in two years; (d) notes payable due in 90 days; (e) notes payable due in five years; and (f) patent.

BE 3–2
Balance sheet classification
● LO3–2, LO3–3

The trial balance for K and J Nursery, Inc., listed the following account balances at December 31, 2021, the end of its fiscal year: cash, $ 6,000; accounts receivable, $11,000; inventory, $25,000; equipment (net), $80,000; accounts payable, $14,000; salaries payable, $9,000; interest payable, $1,000; notes payable (due in 18 months), $30,000; common stock, $50,000. Calculate total current assets and total current liabilities that would appear in the company's year-end balance sheet.

BE 3–3
Balance sheet classification
● LO3–2, LO3–3

Refer to the situation described in BE 3–2. Determine the year-end balance in retained earnings for K and J Nursery, Inc.

BE 3–4
Balance sheet classification
● LO3–2, LO3–3

Refer to the situation described in BE 3–2. Prepare a classified balance sheet for K and J Nursery, Inc. The equipment originally cost $140,000.

BE 3–5
Balance sheet classification
● LO3–2, LO3–3

The following is a December 31, 2021, post-closing trial balance for Culver City Lighting, Inc. Prepare a classified balance sheet for the company.

Account Title	Debits	Credits
Cash	$ 55,000	
Accounts receivable	39,000	
Inventory	45,000	
Prepaid insurance	15,000	
Equipment	100,000	
Accumulated depreciation		$ 34,000
Patent (net)	40,000	
Accounts payable		12,000
Interest payable		2,000
Notes payable (due in 10 years)		100,000
Common stock		70,000
Retained earnings		76,000
Totals	$294,000	$294,000

BE 3–6
Balance sheet classification
● LO3–2, LO3–3

You have been asked to review the December 31, 2021, balance sheet for Champion Cleaning. After completing your review, you list the following three items for discussion with your superior:

1. An investment of $30,000 is included in current assets. Management has indicated that it has no intention of liquidating the investment in 2022.
2. A $100,000 note payable is listed as a long-term liability, but you have determined that the note is due in 10 equal annual installments with the first installment due on March 31, 2022.
3. Deferred revenue of $60,000 is included as a current liability even though only two-thirds will be recognized as revenue in 2022, and the other one-third in 2023.

Determine the appropriate classification of each of these items.

BE 3–7
Balance sheet preparation; missing elements
● LO3–2, LO3–3

Use the following information from the balance sheet of Raineer Plumbing to determine the missing amounts.

Cash and cash equivalents	$40,000	Accounts payable	$32,000
Retained earnings	?	Accounts receivable	120,000
Inventory	?	Notes payable (due in 2 years)	50,000
Common stock	100,000	Property, plant and equipment (net)	?
Total current assets	235,000	Total assets	400,000

BE 3–8
Financial statement disclosures
● LO3–4

For each of the following note disclosures, indicate whether the disclosure would likely appear in (A) the summary of significant accounts policies or (B) a separate note: (1) depreciation method; (2) contingency information; (3) significant issuance of common stock after the fiscal year-end; (4) cash equivalent designation; (5) long-term debt information; and (6) inventory costing method.

BE 3–9
Calculating ratios
● LO3–8

Refer to the trial balance information in BE 3–5. Calculate the (a) current ratio, (b) acid-test ratio, and (c) debt to equity ratio.

BE 3–10
Effect of decisions on ratios
● LO3–8

At the end of 2021, Barker Corporation's preliminary trial balance indicated a current ratio of 1.2. Management is contemplating paying some of its accounts payable balance before the end of the fiscal year. Determine whether the effect this transaction would increase or decrease the current ratio. Would your answer be the same if the preliminary trial balance indicated a current ratio of 0.8?

BE 3–11
Calculating ratios; solving for unknowns
● LO3–8

The current asset section of Stibbe Pharmaceutical Company's balance sheet included cash of $20,000 and accounts receivable of $40,000. The only other current asset is inventory. The company's current ratio is 2.0 and its acid-test ratio is 1.5. Determine the ending balance in inventory and total current liabilities.

Exercises

connect

E 3–1
Balance sheet; missing elements
● LO3–2, LO3–3, LO3–8

The following December 31, 2021, fiscal year-end account balance information is available for the Stonebridge Corporation:

Cash and cash equivalents	$ 5,000
Accounts receivable (net)	20,000
Inventory	60,000
Property, plant, and equipment (net)	120,000
Accounts payable	44,000
Salaries payable	15,000
Paid-in capital	100,000

The only asset not listed is short-term investments. The only liabilities not listed are $30,000 notes payable due in two years and related accrued interest of $1,000 due in four months. The current ratio at year-end is 1.5:1.

Required:

Determine the following at December 31, 2021:

1. Total current assets
2. Short-term investments
3. Retained earnings

E 3–2
Balance sheet classification
● LO3–2, LO3–3

The following are the typical classifications used in a balance sheet:

a. Current assets
b. Investments
c. Property, plant, and equipment
d. Intangible assets
e. Other assets
f. Current liabilities
g. Long-term liabilities
h. Paid-in capital
i. Retained earnings

Required:
For each of the following balance sheet items, use the letters above to indicate the appropriate classification category. If the item is a contra account, place a minus sign before the chosen letter.

1. ______ Equipment
2. ______ Accounts payable
3. ______ Allowance for uncollectible accounts
4. ______ Land (held for investment)
5. ______ Notes payable (due in 5 years)
6. ______ Deferred revenue (for the next 12 months)
7. ______ Notes payable (due in 6 months)
8. ______ Accumulated amount of net income less dividends
9. ______ Investment in XYZ Corp. (long-term)
10. ______ Inventory
11. ______ Patent
12. ______ Land (used in operations)
13. ______ Accrued liabilities (due in 6 months)
14. ______ Prepaid rent (for the next 9 months)
15. ______ Common stock
16. ______ Building (used in operations)
17. ______ Cash
18. ______ Income taxes payable

E 3–3
Balance sheet classification
● LO3–2, LO3–3

The following are the typical classifications used in a balance sheet:

a. Current assets
b. Investments
c. Property, plant, and equipment
d. Intangible assets
e. Other assets
f. Current liabilities
g. Long-term liabilities
h. Paid-in capital
i. Retained earnings

Required:
For each of the following 2021 balance sheet items, use the letters above to indicate the appropriate classification category. If the item is a contra account, place a minus sign before the chosen letter.

1. ______ Interest payable (due in 3 months)
2. ______ Franchise
3. ______ Accumulated depreciation
4. ______ Prepaid insurance (for 2022)
5. ______ Bonds payable (due in 10 years)
6. ______ Current maturities of long-term debt
7. ______ Notes payable (due in 3 months)
8. ______ Long-term receivables
9. ______ Restricted cash (will be used to retire bonds in 10 years)
10. ______ Supplies
11. ______ Machinery
12. ______ Land (used in operations)
13. ______ Deferred revenue (for 2022)
14. ______ Copyrights
15. ______ Common stock
16. ______ Land (held for speculation)
17. ______ Cash equivalents
18. ______ Salaries payable

E 3–4
Balance sheet preparation
● LO3–2, LO3–3

The following is a December 31, 2021, post-closing trial balance for the Jackson Corporation.

Account Title	Debits	Credits
Cash	$ 40,000	
Accounts receivable	34,000	
Inventory	75,000	
Prepaid rent (for the next 8 months)	16,000	
Investment in equity securities (short term)	10,000	
Machinery	145,000	
Accumulated depreciation		$ 11,000
Patent (net)	83,000	
Accounts payable		8,000
Salaries payable		4,000
Income taxes payable		32,000
Bonds payable (due in 10 years)		200,000
Common stock		100,000
Retained earnings		48,000
Totals	$403,000	$403,000

Required:
Prepare a classified balance sheet for Jackson Corporation at December 31, 2021, by properly classifying each of the accounts.

E 3–5
Balance sheet preparation
● LO3–2, LO3–3

The following are the ending balances of accounts at December 31, 2021, for the Valley Pump Corporation.

Account Title	Debits	Credits
Cash	$ 25,000	
Accounts receivable	56,000	
Inventory	81,000	
Interest payable		$ 10,000
Investment in equity securities	44,000	
Land	120,000	
Buildings	300,000	
Accumulated depreciation—buildings		100,000
Equipment	75,000	
Accumulated depreciation—equipment		25,000
Copyright (net)	12,000	
Prepaid expenses (next 12 months)	32,000	
Accounts payable		65,000
Deferred revenue (next 12 months)		20,000
Notes payable		250,000
Allowance for uncollectible accounts		5,000
Common stock		200,000
Retained earnings		70,000
Totals	$745,000	$745,000

Additional Information:
1. The $120,000 balance in the land account consists of $100,000 for the cost of land where the plant and office buildings are located. The remaining $20,000 represents the cost of land being held for speculation.
2. The $44,000 balance in the investment in equity securities account represents an investment in the common stock of another corporation. Valley intends to sell one-half of the stock within the next year.
3. The notes payable account consists of a $100,000 note due in six months and a $150,000 note due in three annual installments of $50,000 each, with the first payment due in August of 2022.

Required:
Prepare a classified balance sheet for the Valley Pump Corporation at December 31, 2021. Use the additional information to help determine appropriate classifications and account balances.

E 3–6
Balance sheet; Current versus long-term classification
● LO3–2, LO3–3

Presented next are the ending balances of accounts for the Kansas Instruments Corporation at December 31, 2021.

Account Title	Debits	Credits
Cash	$ 20,000	
Accounts receivable	130,000	
Raw materials	24,000	
Notes receivable	100,000	
Interest receivable	3,000	
Interest payable		$ 5,000
Investment in debt securities	32,000	
Land	50,000	
Buildings	1,300,000	
Accumulated depreciation—buildings		620,000
Work in process	42,000	
Finished goods	89,000	
Equipment	300,000	
Accumulated depreciation—equipment		130,000
Patent (net)	120,000	
Prepaid rent (for the next two years)	60,000	
Deferred revenue (next 12 months)		36,000
Accounts payable		180,000
Notes payable		400,000
Restricted cash (for payment of notes payable)	80,000	
Allowance for uncollectible accounts		13,000
Sales revenue		800,000
Cost of goods sold	450,000	
Rent expense	28,000	

Additional Information:

1. The notes receivable, along with any accrued interest, are due on November 22, 2022.
2. The notes payable are due in 2025. Interest is payable annually.
3. The investment in debt securities consist of treasury bills, all of which mature next year.
4. Deferred revenue will be recognized as revenue equally over the next two years.

Required:

Determine the company's working capital (current assets minus current liabilities) at December 31, 2021.

E 3–7
Balance sheet preparation; errors
● LO3–2, LO3–3

The following balance sheet for the Los Gatos Corporation was prepared by a recently hired accountant. In reviewing the statement you notice several errors.

LOS GATOS CORPORATION
Balance Sheet
At December 31, 2021

Assets	
Cash	$ 40,000
Accounts receivable	80,000
Inventory	55,000
Machinery (net)	120,000
Franchise (net)	30,000
Total assets	$325,000
Liabilities and Shareholders' Equity	
Accounts payable	$ 50,000
Allowance for uncollectible accounts	5,000
Notes payable	55,000
Bonds payable	110,000
Shareholders' equity	105,000
Total liabilities and shareholders' equity	$325,000

Additional Information:

1. Cash includes a $20,000 restricted amount to be used for repayment of the bonds payable in 2025.
2. The cost of the machinery is $190,000.
3. Accounts receivable includes a $20,000 note receivable from a customer due in 2024.
4. The notes payable balance includes accrued interest of $5,000. Principal and interest are both due on February 1, 2022.
5. The company began operations in 2016. Net income less dividends since inception of the company totals $35,000.
6. 50,000 shares of no par common stock were issued in 2016. 100,000 shares are authorized.

Required:

Prepare a corrected, classified balance sheet. Use the additional information to help determine appropriate classifications and account balances. The cost of machinery and its accumulated depreciation are shown separately.

E 3–8
Balance sheet; current versus long-term classification
● LO3–2, LO3–3

Cone Corporation is in the process of preparing its December 31, 2021, balance sheet. There are some questions as to the proper classification of the following items:

a. $50,000 in cash restricted in a savings account to pay bonds payable. The bonds mature in 2025.
b. Prepaid rent of $24,000, covering the period January 1, 2022, through December 31, 2023.
c. Notes payable of $200,000. The notes are payable in annual installments of $20,000 each, with the first installment payable on March 1, 2022.
d. Accrued interest payable of $12,000 related to the notes payable.
e. Investment in equity securities of other corporations, $80,000. Cone intends to sell one-half of the securities in 2022.

Required:

Prepare the asset and liability sections of a classified balance sheet to show how each of the above items should be reported.

E 3–9
Balance sheet preparation
● LO3–2, LO3–3

The following is the balance sheet of Korver Supply Company at December 31, 2020 (prior year).

KORVER SUPPLY COMPANY
Balance Sheet
At December 31, 2020

Assets	
Cash	$120,000
Accounts receivable	300,000
Inventory	200,000
Furniture and fixtures (net)	150,000
Total assets	$770,000
Liabilities and Shareholders' Equity	
Accounts payable (for merchandise)	$190,000
Notes payable	200,000
Interest payable	6,000
Common stock	100,000
Retained earnings	274,000
Total liabilities and shareholders' equity	$770,000

Transactions during 2021 (current year) were as follows:

1. Sales to customers on account	$800,000
2. Cash collected from customers	780,000
3. Purchase of merchandise on account	550,000
4. Cash payment to suppliers	560,000
5. Cost of merchandise sold	500,000
6. Cash paid for operating expenses	160,000
7. Cash paid for interest on notes	12,000

Additional Information:

The notes payable are dated June 30, 2020, and are due on June 30, 2022. Interest at 6% is payable annually on June 30. Depreciation on the furniture and fixtures for 2021 is $20,000. The furniture and fixtures originally cost $300,000.

Required:

Prepare a classified balance sheet at December 31, 2021, by updating ending balances from 2020 for transactions during 2021 and the additional information. The cost of furniture and fixtures and their accumulated depreciation are shown separately.

E 3–10
Financial statement disclosures
● LO3–4

The following are typical disclosures that would appear in the notes accompanying financial statements. For each of the items listed, indicate where the disclosure would likely appear—either in (A) the significant accounting policies note or (B) a separate note.

1. Inventory costing method	A
2. Information on related party transactions	____
3. Composition of property, plant, and equipment	____
4. Depreciation method	____
5. Subsequent event information	____
6. Measurement basis for certain financial instruments	____
7. Important merger occurring after year-end	____
8. Composition of receivables	____

E 3–11
Disclosure notes
● LO3–4

Hallergan Company produces car and truck batteries that it sells primarily to auto manufacturers. Dorothy Hawkins, the company's controller, is preparing the financial statements for the year ended December 31, 2021. Hawkins asks for your advice concerning the following information that has not yet been included in the statements. The statements will be issued on February 28, 2022.

1. Hallergan leases its facilities from the brother of the chief executive officer.
2. On January 8, 2022, Hallergan entered into an agreement to sell a tract of land that it had been holding as an investment. The sale, which resulted in a material gain, was completed on February 2, 2022.
3. Hallergan uses the straight-line method to determine depreciation on all of the company's depreciable assets.
4. On February 8, 2022, Hallergan completed negotiations with its bank for a $10,000,000 line of credit.
5. Hallergan uses the first-in, first-out (FIFO) method to value inventory.

Required:

For each of the above items, discuss any additional disclosures that Hawkins should include in Hallergan's financial statements.

E 3–12
Financial statement disclosures
● LO3–4

Parkman Sporting Goods is preparing its annual report for its 2021 fiscal year. The company's controller has asked for your help in determining how best to disclose information about the following items:

1. A related-party transaction.
2. Depreciation method.
3. Allowance for uncollectible accounts.
4. Composition of investments.
5. Composition of long-term debt.
6. Inventory costing method.
7. Number of shares of common stock authorized, issued, and outstanding.
8. Employee benefit plans.

Required:
Indicate whether the above items should be disclosed (A) in the summary of significant accounting policies note, (B) in a separate disclosure note, or (C) on the face of the balance sheet.

E 3–13
FASB codification research
● LO3–4
CODE

Access the *FASB Accounting Standards Codification* at the FASB website (**www.fasb.org**). Determine each of the following:

1. The topic number (Topic XXX) that provides guidance on information contained in the notes to the financial statements.
2. The specific seven-digit Codification citation (XXX-XX-XX) that requires a company to identify and describe in the notes to the financial statements the accounting principles and methods used to prepare the financial statements.
3. Describe the disclosure requirements.

E 3–14
FASB codification research
● LO3–2, LO3–4
CODE

Access the *FASB Accounting Standards Codification* at the FASB website (**www.fasb.org**). Determine the specific eight-digit Codification citation (XXX-XX-XX-X) for each of the following:

1. The balance sheet classification for a note payable due in six months that was used to purchase a building.
2. The assets that may be excluded from current assets.
3. Whether a note receivable from a related party would be included in the balance sheet with notes receivable or accounts receivable from customers.
4. The items that are nonrecognized subsequent events that require a disclosure in the notes to the financial statements.

E 3–15
Concepts; terminology
● LO3–2 through LO3–4, LO3–6

Listed below are several terms and phrases associated with the balance sheet and financial disclosures. Pair each item from List A (by letter) with the item from List B that is most appropriately associated with it.

List A

_____ 1. Balance sheet
_____ 2. Liquidity
_____ 3. Current assets
_____ 4. Operating cycle
_____ 5. Current liabilities
_____ 6. Cash equivalent
_____ 7. Intangible asset
_____ 8. Working capital
_____ 9. Accrued liabilities
_____ 10. Summary of significant accounting policies
_____ 11. Subsequent events
_____ 12. Unqualified opinion
_____ 13. Qualified opinion

List B

a. Will be satisfied through the use of current assets.
b. Items expected to be converted to cash or consumed within one year or the operating cycle, whichever is longer.
c. The statements are presented fairly in conformity with GAAP.
d. An organized array of assets, liabilities, and equity.
e. Important to a user in comparing financial information across companies.
f. Scope limitation or a departure from GAAP.
g. Recorded when an expense is incurred but not yet paid.
h. Refers to the ability of a company to convert its assets to cash to pay its current obligations.
i. Occurs after the fiscal year-end but before the statements are issued.
j. Period of time from payment of cash to collection of cash.
k. One-month U.S. Treasury bill.
l. Current assets minus current liabilities.
m. Lacks physical substance.

E 3–16
Calculating ratios
● LO3–8

The 2021 balance sheet for Hallbrook Industries, Inc., is shown below.

HALLBROOK INDUSTRIES, INC.
Balance Sheet
December 31, 2021
($ in thousands)

Assets	
Cash	$ 200
Short-term investments	150
Accounts receivable	200
Inventory	350
Property, plant, and equipment (net)	1,000
Total assets	$1,900
Liabilities and Shareholders' Equity	
Current liabilities	$ 400
Long-term liabilities	350
Paid-in capital	750
Retained earnings	400
Total liabilities and shareholders' equity	$1,900

The company's 2021 income statement reported the following amounts ($ in thousands):

Net sales	$4,600
Interest expense	40
Income tax expense	100
Net income	160

Required:
Determine the following ratios for 2021:
1. Current ratio
2. Acid-test ratio
3. Debt to equity ratio
4. Times interest earned ratio

E 3–17
Calculating ratios; Best Buy
● LO3–8

Real World Financials

Best Buy Co, Inc., is a leading retailer specializing in consumer electronics. A condensed income statement and balance sheet for the fiscal year ended January 28, 2017, are shown next.

Best Buy Co., Inc.
Balance Sheet
At January 28, 2017
($ in millions)

Assets	
Current assets:	
Cash and cash equivalents	$ 2,240
Short-term investments	1,681
Accounts receivable (net)	1,347
Inventory	4,864
Other current assets	384
Total current assets	10,516
Long-term assets	3,340
Total assets	$13,856
Liabilities and Shareholders' Equity	
Current liabilities:	
Accounts payable	$ 4,984
Other current liabilities	2,138
Total current liabilities	7,122
Long-term liabilities	2,025
Shareholders' equity	4,709
Total liabilities and shareholders' equity	$13,856

Best Buy Co., Inc.
Income Statement
For the Year Ended January 28, 2017
($ in millions)

Revenues	$39,403
Costs and expenses	37,549
Operating income	1,854
Other income (expense)*	(38)
Income before income taxes	1,816
Income tax expense	609
Net income	$ 1,207

*Includes $72 of interest expense.

Liquidity and solvency ratios for the industry are as follows:

	Industry Average
Current ratio	1.43
Acid-test ratio	1.15
Debt to equity	0.68
Times interest earned	8.25 times

Required:

1. Determine the following ratios for Best Buy for its fiscal year ended January 28, 2017.
   a. Current ratio
   b. Acid-test ratio
   c. Debt to equity ratio
   d. Times interest earned ratio
2. Using the ratios from requirement 1, assess Best Buy's liquidity and solvency relative to its industry.

E 3–18
Calculating ratios; solve for unknowns
● LO3–8

The current asset section of the Excalibur Tire Company's balance sheet consists of cash, marketable securities, accounts receivable, and inventory. The December 31, 2021, balance sheet revealed the following:

Inventory	$ 840,000
Total assets	$2,800,000
Current ratio	2.25
Acid-test ratio	1.2
Debt to equity ratio	1.8

Required:

Determine the following 2021 balance sheet items:

1. Current assets
2. Shareholders' equity
3. Long-term assets
4. Long-term liabilities

E 3–19
Calculating ratios; solve for unknowns
● LO3–8

The current asset section of Guardian Consultant's balance sheet consists of cash, accounts receivable, and prepaid expenses. The 2021 balance sheet reported the following: cash, $1,300,000; prepaid expenses, $360,000; long-term assets, $2,400,000; and shareholders' equity, $2,500,000. The current ratio at the end of the year was 2.0 and the debt to equity ratio was 1.4.

Required:

Determine the following 2021 amounts and ratios:

1. Current liabilities
2. Long-term liabilities
3. Accounts receivable
4. The acid-test ratio

E 3–20
Effect of management decisions on ratios
● LO3–8

Most decisions made by management impact the ratios analysts use to evaluate performance. Indicate (by letter) whether each of the actions listed below will immediately increase (I), decrease (D), or have no effect (N) on the ratios shown. Assume each ratio is less than 1.0 before the action is taken.

Action	Current Ratio	Acid-Test Ratio	Debt to Equity Ratio
1. Issuance of long-term bonds	_____	_____	_____
2. Issuance of short-term notes	_____	_____	_____
3. Payment of accounts payable	_____	_____	_____
4. Purchase of inventory on account	_____	_____	_____
5. Purchase of inventory for cash	_____	_____	_____
6. Purchase of equipment with a 4-year note	_____	_____	_____
7. Repayment of long-term notes payable	_____	_____	_____
8. Issuance of common stock	_____	_____	_____
9. Payment for advertising expense	_____	_____	_____
10. Purchase of short-term investment for cash	_____	_____	_____
11. Reclassification of long-term notes payable to current notes payable	_____	_____	_____

E 3–21
Segment reporting
● Appendix 3

The Canton Corporation operates in four distinct business segments. The segments, along with 2021 information on revenues, assets, and net income, are listed below ($ in thousands):

Segment	Revenues	Assets	Net Income
Pharmaceuticals	$2,000	$1,000	$200
Plastics	3,000	1,500	270
Farm equipment	2,500	1,250	320
Electronics	500	250	40
Total company	$8,000	$4,000	$830

Required:

1. For which segments must Canton report supplementary information according to U.S. GAAP?
2. What amounts must be reported for the segments you identified in requirement 1?

E 3–22
Segment reporting
● Appendix 3
LO3–9

IFRS

Refer to E 3–21.

Required:

How might your answers differ if Canton Corporation prepares its segment disclosure according to International Financial Reporting Standards?

Problems

connect

P 3–1
Balance sheet preparation
● LO3–2, LO3–3

Presented below is a list of balance sheet accounts.

Accounts payable	Cash
Accounts receivable	Common stock
Accumulated depreciation—buildings	Copyright (net)
Accumulated depreciation—equipment	Equipment
Allowance for uncollectible accounts	Interest receivable (due in three months)
Restricted cash (to be used in 10 years)	Inventory
Bonds payable (due in 10 years)	Land (in use)
Buildings	Long-term investments
Notes payable (due in 6 months)	Interest payable (current)
Notes receivable (due in 2 years)	Retained earnings
Patent (net)	Short-term investments
Additional paid-in capital	Income taxes payable
Prepaid expenses	Salaries payable

Required:

Prepare a classified balance sheet ignoring monetary amounts. Include headings for each classification, as well as titles for each classification's subtotal. An example of a classified balance sheet can be found in the Concept Review Exercise at the end of Part A of this chapter.

P 3–2
Balance sheet preparation; missing elements
● LO3–2, LO3–3

The data listed below are taken from a balance sheet of Trident Corporation at December 31, 2021. Some amounts, indicated by question marks, have been intentionally omitted.

	($ in thousands)
Cash and cash equivalents	$ 239,186
Short-term investments	353,700
Accounts receivable	504,944
Inventory	?
Prepaid expenses (current)	83,259
Total current assets	1,594,927
Long-term receivables	110,800
Equipment (net)	?
Total assets	?
Notes payable (current)	31,116
Accounts payable	?
Accrued liabilities	421,772
Other current liabilities	181,604
Total current liabilities	693,564
Long-term debt	?
Total liabilities	956,140
Common stock	370,627
Retained earnings	1,000,000

Required:
1. Determine the missing amounts.
2. Prepare Trident's classified balance sheet. Include headings for each classification, as well as titles for each classification's subtotal. An example of a classified balance sheet can be found in the Concept Review Exercise at the end of Part A of this chapter.

P 3–3
Balance sheet preparation
● LO3–2, LO3–3

The following is a December 31, 2021, post-closing trial balance for Almway Corporation.

Account Title	Debits	Credits
Cash	$ 45,000	
Investment in equity securities	110,000	
Accounts receivable	60,000	
Inventory	200,000	
Prepaid insurance (for the next 9 months)	9,000	
Land	90,000	
Buildings	420,000	
Accumulated depreciation—buildings		$ 100,000
Equipment	110,000	
Accumulated depreciation—equipment		60,000
Patent (net)	10,000	
Accounts payable		75,000
Notes payable		130,000
Interest payable		20,000
Bonds payable		240,000
Common stock		300,000
Retained earnings		129,000
Totals	$1,054,000	$1,054,000

Additional Information:
1. The investment in equity securities account includes an investment in common stock of another corporation of $30,000 which management intends to hold for at least three years. The balance of these investments is intended to be sold in the coming year.
2. The land account includes land which cost $25,000 that the company has not used and is currently listed for sale.
3. The cash account includes $15,000 restricted in a fund to pay bonds payable that mature in 2024 and $23,000 restricted in a three-month Treasury bill.
4. The notes payable account consists of the following:
   a. a $30,000 note due in six months.
   b. a $50,000 note due in six years.
   c. a $50,000 note due in five annual installments of $10,000 each, with the next installment due February 15, 2022.

5. The $60,000 balance in accounts receivable is net of an allowance for uncollectible accounts of $8,000.
6. The common stock account represents 100,000 shares of no par value common stock issued and outstanding. The corporation has 500,000 shares authorized.

Required:

Prepare a classified balance sheet for the Almway Corporation at December 31, 2021. Include headings for each classification, as well as titles for each classification's subtotal. An example of a classified balance sheet can be found in the Concept Review Exercise at the end of Part A of this chapter.

P 3–4
Balance sheet preparation
● LO3–2, LO3–3

The following is the ending balances of accounts at December 31, 2021, for the Weismuller Publishing Company.

Account Title	Debits	Credits
Cash	$ 65,000	
Accounts receivable	160,000	
Inventory	285,000	
Prepaid expenses	148,000	
Equipment	320,000	
Accumulated depreciation		$ 110,000
Investments	140,000	
Accounts payable		60,000
Interest payable		20,000
Deferred revenue		80,000
Income taxes payable		30,000
Notes payable		200,000
Allowance for uncollectible accounts		16,000
Common stock		400,000
Retained earnings		202,000
Totals	$1,118,000	$1,118,000

Additional Information:

1. Prepaid expenses include $120,000 paid on December 31, 2021, for a two-year lease on the building that houses both the administrative offices and the manufacturing facility.
2. Investments include $30,000 in Treasury bills purchased on November 30, 2021. The bills mature on January 30, 2022. The remaining $110,000 is an investment in equity securities that the company intends to sell in the next year.
3. Deferred revenue represents customer prepayments for magazine subscriptions. Subscriptions are for periods of one year or less.
4. The notes payable account consists of the following:
   a. a $40,000 note due in six months.
   b. a $100,000 note due in six years.
   c. a $60,000 note due in three annual installments of $20,000 each, with the next installment due August 31, 2022.
5. The common stock account represents 400,000 shares of no par value common stock issued and outstanding. The corporation has 800,000 shares authorized.

Required:

Prepare a classified balanced sheet for the Weismuller Publishing Company at December 31, 2021. Include headings for each classification, as well as titles for each classification's subtotal. An example of a classified balance sheet can be found in the Concept Review Exercise at the end of Part A of this chapter.

P 3–5
Balance sheet preparation
● LO3–2, LO3–3

The following is the ending balances of accounts at June 30, 2021, for Excell Company.

Account Title	Debits	Credits
Cash	$ 83,000	
Short-term investments	65,000	
Accounts receivable (net)	280,000	
Prepaid expenses (for the next 12 months)	32,000	
Land	75,000	
Buildings	320,000	
Accumulated depreciation—buildings		$ 160,000
Equipment	265,000	
Accumulated depreciation—equipment		120,000
Accounts payable		173,000

(continued)

(concluded)

Account Title	Debits	Credits
Accrued liabilities		45,000
Notes payable		100,000
Mortgage payable		250,000
Common stock		100,000
Retained earnings		172,000
Totals	$1,120,000	$1,120,000

Additional Information:

1. The short-term investments account includes $18,000 in U.S. treasury bills purchased in May. The bills mature in July, 2021.
2. The accounts receivable account consists of the following:

a. Amounts owed by customers	$225,000
b. Allowance for uncollectible accounts—trade customers	(15,000)
c. Nontrade notes receivable (due in three years)	65,000
d. Interest receivable on notes (due in four months)	5,000
Total	$280,000

3. The notes payable account consists of two notes of $50,000 each. One note is due on September 30, 2021, and the other is due on November 30, 2022.
4. The mortgage payable is a loan payable to the bank in *semiannual* installments of $5,000 each plus interest. The next payment is due on October 31, 2021. Interest has been properly accrued and is included in accrued expenses.
5. Five hundred thousand shares of no par common stock are authorized, of which 200,000 shares have been issued and are outstanding.
6. The land account includes $50,000 representing the cost of the land on which the company's office building resides. The remaining $25,000 is the cost of land that the company is holding for investment purposes.

Required:

Prepare a classified balance sheet for the Excell Company at June 30, 2021. Include headings for each classification, as well as titles for each classification's subtotal. An example of a classified balance sheet can be found in the Concept Review Exercise at the end of Part A of this chapter.

P 3–6
Balance sheet preparation; disclosures
● LO3–2 through LO3–4

The following is the ending balances of accounts at December 31, 2021, for the Vosburgh Electronics Corporation.

Account Title	Debits	Credits
Cash	$ 67,000	
Short-term investments	182,000	
Accounts receivable	123,000	
Long-term investments	35,000	
Inventory	215,000	
Receivables from employees	40,000	
Prepaid expenses (for 2022)	16,000	
Land	280,000	
Building	1,550,000	
Equipment	637,000	
Patent (net)	152,000	
Franchise (net)	40,000	
Notes receivable	250,000	
Interest receivable	12,000	
Accumulated depreciation—building		$ 620,000
Accumulated depreciation—equipment		210,000
Accounts payable		189,000
Dividends payable (payable on 1/16/2022)		10,000
Interest payable		16,000
Income taxes payable		40,000
Deferred revenue		60,000
Notes payable		300,000
Allowance for uncollectible accounts		8,000
Common stock		2,000,000
Retained earnings		146,000
Totals	$3,599,000	$3,599,000

Additional Information:

1. The common stock represents 1 million shares of no par stock authorized, 500,000 shares issued and outstanding.
2. The receivables from employees are due on June 30, 2022.
3. The notes receivable are due in installments of $50,000, payable on each September 30. Interest is payable annually.
4. Short-term investments consist of securities that the company plans to sell in 2022 and $50,000 in treasury bills purchased on December 15 of the current year that mature on February 15, 2022. Long-term investments consist of securities that the company does not plan to sell in the next year.
5. Deferred revenue represents payments from customer for extended service contracts. Eighty percent of these contracts expire in 2022, the remainder in 2023.
6. Notes payable consists of two notes, one for $100,000 due on January 15, 2023, and another for $200,000 due on June 30, 2024.

Required:

1. Prepare a classified balance sheet for Vosburgh at December 31, 2021. Include headings for each classification, as well as titles for each classification's subtotal. An example of a classified balance sheet can be found in the Concept Review Exercise at the end of Part A of this chapter.
2. Identify the items that would require additional disclosure, either on the face of the balance sheet or in a disclosure note.

P 3–7
Balance sheet preparation; errors
● LO3–2, LO3–3

The following balance sheet for the Hubbard Corporation was prepared by the company:

HUBBARD CORPORATION
Balance Sheet
At December 31, 2021
Assets

Buildings	$ 750,000
Land	250,000
Cash	60,000
Accounts receivable (net)	120,000
Inventory	240,000
Machinery	280,000
Patent (net)	100,000
Investment in equity securities	60,000
Total assets	$1,860,000

Liabilities and Shareholders' Equity

Accounts payable	$ 215,000
Accumulated depreciation	255,000
Notes payable	500,000
Appreciation of inventory	80,000
Common stock (authorized and issued 100,000 shares of no par stock)	430,000
Retained earnings	380,000
Total liabilities and shareholders' equity	$1,860,000

Additional Information:

1. The buildings, land, and machinery are all stated at cost except for a parcel of land that the company is holding for future sale. The land originally cost $50,000 but, due to a significant increase in market value, is listed at $120,000. The increase in the land account was credited to retained earnings.
2. The investment in equity securities account consists of stocks of other corporations and are recorded at cost, $20,000 of which will be sold in the coming year. The remainder will be held indefinitely.
3. Notes payable are all long term. However, a $100,000 note requires an installment payment of $25,000 due in the coming year.
4. Inventory is recorded at current resale value. The original cost of the inventory is $160,000.

Required:

Prepare a corrected classified balance sheet for the Hubbard Corporation at December 31, 2021. Include headings for each classification, as well as titles for each classification's subtotal. An example of a classified balance sheet can be found in the Concept Review Exercise at the end of Part A of this chapter.

P 3–8
Balance sheet; errors; missing amounts
● LO3–2, LO3–3

The following incomplete balance sheet for the Sanderson Manufacturing Company was prepared by the company's controller. As accounting manager for Sanderson, you are attempting to reconstruct and revise the balance sheet.

Sanderson Manufacturing Company
Balance Sheet
At December 31, 2021
($ in thousands)

Assets

Current assets:		
Cash		$ 1,250
Accounts receivable		3,500
Allowance for uncollectible accounts		(400)
Finished goods inventory		6,000
Prepaid expenses		1,200
Total current assets		11,550
Long-term assets:		
Investments		3,000
Raw materials and work in process inventory		2,250
Equipment		15,000
Accumulated depreciation		(4,200)
Patent (net)		?
Total assets		$?

Liabilities and Shareholders' Equity

Current liabilities:		
Accounts payable		$ 5,200
Notes payable		4,000
Interest payable (on notes)		100
Deferred revenue		3,000
Total current liabilities		12,300
Long-term liabilities:		
Bonds payable		5,500
Interest payable (on bonds)		200
Shareholders' equity:		
Common stock	$?	
Retained earnings	?	?
Total liabilities and shareholders' equity		?

Additional Information ($ in thousands):

1. Certain records that included the account balances for the patent and shareholders' equity items were lost. However, the controller told you that a complete, preliminary balance sheet prepared before the records were lost showed a debt to equity ratio of 1.2. That is, total liabilities are 120% of total shareholders' equity. Retained earnings at the beginning of the year was $4,000. Net income for 2021 was $1,560 and $560 in cash dividends were declared and paid to shareholders.
2. Management intends to sell the investments in the next six months.
3. Interest on both the notes and the bonds is payable annually.
4. The notes payable are due in annual installments of $1,000 each.
5. Deferred revenue will be recognized as revenue equally over the next two fiscal years.
6. The common stock represents 400,000 shares of no par stock authorized, 250,000 shares issued and outstanding.

Required:
Prepare a complete, corrected, classified balance sheet. Include headings for each classification, as well as titles for each classification's subtotal. An example of a classified balance sheet can be found in the Concept Review Exercise at the end of Part A of this chapter.

P 3–9
Balance sheet preparation
● LO3–2, LO3–3

Presented below is the balance sheet for HHD, Inc., at December 31, 2021.

Current assets	$ 600,000	Current liabilities	$ 400,000
Investments	500,000	Long-term liabilities	1,100,000
Property, plant, and equipment	2,000,000	Shareholders' equity	1,800,000
Intangible assets	200,000		
Total assets	$3,300,000	Total liabilities and shareholders' equity	$3,300,000

The captions shown in the summarized statement above include the following:

a. Current assets: cash, $150,000; accounts receivable (net), $200,000; inventory, $225,000; and prepaid insurance, $25,000.

b. Investments: investment in equity securities, short term, $90,000, and long term, $160,000; and restricted cash, long term, $250,000.

c. Property, plant, and equipment: buildings, $1,500,000 less accumulated depreciation, $600,000; equipment, $500,000 less accumulated depreciation, $200,000; and land, $800,000.

d. Intangible assets net of amortization: patent, $110,000; and copyright, $90,000.

e. Current liabilities: accounts payable, $100,000; notes payable, short term, $150,000, and long term, $90,000; and income taxes payable, $60,000.

f. Long-term liabilities: bonds payable due 2023.

g. Shareholders' equity: common stock, $1,000,000; retained earnings, $800,000. Five hundred thousand shares of no par common stock are authorized, of which 200,000 shares were issued and are outstanding.

Required:

Prepare a corrected classified balance sheet for HHD, Inc., at December 31, 2021. Include headings for each classification, as well as titles for each classification's subtotal. An example of a classified balance sheet can be found in the Concept Review Exercise at the end of Part A of this chapter.

P 3–10
Balance sheet preparation
● LO3–2, LO3–3

Melody Lane Music Company was started by John Ross early in 2021. Initial capital was acquired by issuing shares of common stock to various investors and by obtaining a bank loan. The company operates a retail store that sells records, tapes, and compact discs. Business was so good during the first year of operations that John is considering opening a second store on the other side of town. The funds necessary for expansion will come from a new bank loan. In order to approve the loan, the bank requires financial statements.

John asks for your help in preparing the balance sheet and presents you with the following information for the year ending December 31, 2021:

a. Cash receipts consisted of the following:

From customers	$360,000
From issue of common stock	100,000
From bank loan	100,000

b. Cash disbursements were as follows:

Purchase of inventory	$300,000
Rent	15,000
Salaries	30,000
Utilities	5,000
Insurance	3,000
Purchase of equipment	40,000

c. The bank loan was made on March 31, 2021. A note was signed requiring payment of interest and principal on March 31, 2022. The interest rate is 12%.

d. The equipment was purchased on January 3, 2021, and has an estimated useful life of 10 years with no anticipated salvage value. Depreciation per year is $4,000.

e. Inventory on hand at the end of the year cost $100,000.

f. Amounts owed at December 31, 2021, were as follows:

To suppliers of inventory	$20,000
To the utility company	1,000

g. Rent on the store building is $1,000 per month. On December 1, 2021, four months' rent was paid in advance.

h. Net income for the year was $76,000. Assume that the company is not subject to federal, state, or local income tax.

i. One hundred thousand shares of no par common stock are authorized, of which 20,000 shares were issued and are outstanding.

Required:

Prepare a balance sheet at December 31, 2021. Include headings for each classification, as well as titles for each classification's subtotal. An example of a classified balance sheet can be found in the Concept Review Exercise at the end of Part A of this chapter.

Decision Makers' Perspective

©ImageGap/Getty Images

Apply your critical-thinking ability to the knowledge you've gained. These cases will provide you an opportunity to develop your research, analysis, judgment, and communication skills. You also will work with other students, integrate what you've learned, apply it in real-world situations, and consider its global and ethical ramifications. This practice will broaden your knowledge and further develop your decision-making abilities.

Communication Case 3–1
Current versus long-term classification
● LO3–2

A first-year accounting student is confused by a statement made in a recent class. Her instructor stated that the assets listed in the balance sheet of the **IBM Corporation** include computers that are classified as current assets as well as computers that are classified as long-term (noncurrent) assets. In addition, the instructor stated that investments in equity securities of other corporations could be classified in the balance sheet as either current or long-term assets.

Required:

Explain to the student the distinction between current and long-term assets pertaining to the IBM computers and the investments in equity securities.

Analysis Case 3–2
Current versus long-term classification
● LO3–2, LO3–3

The usefulness of the balance sheet is enhanced when assets and liabilities are grouped according to common characteristics. The broad distinction made in the balance sheet is the current versus long-term classification of both assets and liabilities.

Required:

1. Discuss the factors that determine whether an asset or liability should be classified as current or long term in a balance sheet.
2. Identify six items that under different circumstances could be classified as either current or long term. Indicate the factors that would determine the correct classification.

Communication Case 3–3
FASB codification research; inventory or property, plant, and equipment
● LO3–2

CODE

The Red Hen Company produces, processes, and sells fresh eggs. The company is in the process of preparing financial statements at the end of its first year of operations and has asked for your help in determining the appropriate treatment of the cost of its egg-laying flock. The estimated life of a laying hen is approximately two years, after which they are sold to soup companies.

The controller considers the company's operating cycle to be two years and wants to present the cost of the egg-producing flock as inventory in the current asset section of the balance sheet. He feels that the hens are "goods awaiting sale." The chief financial officer does not agree with this treatment. He thinks that the cost of the flock should be classified as property, plant, and equipment because the hens are used in the production of product—the eggs.

The focus of this case is the balance sheet presentation of the cost of the egg-producing flock. Your instructor will divide the class into two to six groups depending on the size of the class. The mission of your group is to reach consensus on the appropriate presentation.

Required:

1. Each group member should deliberate the situation independently and draft a tentative argument prior to the class session for which the case is assigned.
2. In class, each group will meet for 10 to 15 minutes in different areas of the classroom. During that meeting, group members will take turns sharing their suggestions for the purpose of arriving at a single group treatment.
3. After the allotted time, a spokesperson for each group (selected during the group meetings) will share the group's solution with the class. The goal of the class is to incorporate the views of each group into a consensus approach to the situation.

IFRS Case 3–4
Balance sheet presentation; Vodafone Group, Plc.
● LO3–2, LO3–3, LO3–9

IFRS

Real World Financials

Vodafone Group, Plc., a U.K. company, is the largest mobile telecommunications network company in the world. The company prepares its financial statements in accordance with International Financial Reporting Standards. Below are partial company balance sheets (statements of financial position) included in a recent annual report:

Vodafone Group, Plc.
Consolidated Statements of Financial Position
At March 31

	2017	2016
	£m	£m
Long-term assets:		
Goodwill	26,808	23,238
Other intangible assets	19,412	31,326
Property, plant, and equipment	30,204	35,515

(continued)

(concluded)

	2017	2016
	£m	£m
Investments in associates and joint ventures	3,138	479
Other investments	3,459	4,631
Deferred tax assets	24,300	28,306
Post employment benefits	57	224
Trade and other receivables	4,569	5,793
	111,947	133,512
Current assets:		
Inventory	576	716
Taxation recoverable	150	1,402
Trade and other receivables	9,861	11,561
Other investments	6,120	5,337
Cash and cash equivalents	8,835	12,922
Assets held for sale	17,195	3,657
	42,737	35,595
Total assets	154,684	169,107
Equity (details provided in complete statements)	73,719	85,136
Long-term liabilities:		
Long-term borrowings	34,523	37,089
Deferred tax liabilities	535	564
Post employment benefits	651	565
Provisions	1,130	1,619
Trade and other payables	1,737	1,899
	38,576	41,736
Current liabilities:		
Short-term borrowings	12,051	20,260
Taxation liabilities	661	683
Provisions	1,049	958
Trade and other payables	16,834	19,896
Liabilities held for sale	11,794	438
	42,389	42,235
Total equity and liabilities	154,684	169,107

Required:

1. Describe the differences between Vodafone's balance sheets and a typical U.S. company balance sheet.
2. What type of liabilities do you think are included in the *provisions* category in Vodafone's balance sheets?

Judgment Case 3–5
Balance sheet; errors
● LO3–2 through LO3–4

You recently joined the internal auditing department of Marcus Clothing Corporation. As one of your first assignments, you are examining a balance sheet prepared by a staff accountant.

MARCUS CLOTHING CORPORATION
Balance Sheet
At December 31, 2021
Assets

Current assets:		
Cash		$ 137,000
Accounts receivable (net)		80,000
Notes receivable		53,000
Inventory		240,000
Investments		66,000
Total current assets		576,000
Other assets:		
Land	$ 200,000	
Equipment (net)	320,000	
Prepaid expenses (for the next 12 months)	27,000	
Patent (net)	22,000	
Total other assets		569,000
Total assets		$1,145,000

(continued)

(concluded)

Liabilities and Shareholders' Equity

Current liabilities:		
Accounts payable		$ 125,000
Salaries payable		32,000
Total current liabilities		157,000
Long-term liabilities:		
Notes payable	$ 100,000	
Bonds payable (due in 5 years)	300,000	
Interest payable	20,000	
Total long-term liabilities		420,000
Shareholders' equity:		
Common stock	500,000	
Retained earnings	68,000	
Total shareholders' equity		568,000
Total liabilities and shareholders' equity		$1,145,000

In the course of your examination you uncover the following information pertaining to the balance sheet:

1. The company rents its facilities. The land that appears in the statement is being held for future sale.
2. The notes receivable account contains one note that is due in 2023. The balance of $53,000 includes $3,000 of accrued interest. The next interest payment is due in July 2022.
3. The notes payable account contains one note that is due in installments of $20,000 per year. All interest is payable annually.
4. The company's investments consist of marketable equity securities of other corporations. Management does not intend to liquidate any investments in the coming year.

Required:

Identify and explain the deficiencies in the statement prepared by the company's accountant. Include in your answer items that require additional disclosure, either on the face of the statement or in a note.

Judgment Case 3–6
Financial disclosures
● LO3–4

You recently joined the auditing staff of Best, Best, and Krug, CPAs. You have been assigned to the audit of Clearview, Inc., and have been asked by the audit senior to examine the balance sheet prepared by Clearview's accountant.

CLEARVIEW, INC.
Balance Sheet
At December 31, 2021
($ in millions)
Assets

Current assets:		
Cash		$ 10.5
Accounts receivable		112.1
Inventory		220.6
Prepaid expenses		5.5
Total current assets		348.7
Investments		22.0
Property, plant, and equipment (net)		486.9
Total assets		$857.6
Liabilities and Shareholders' Equity		
Current liabilities:		
Accounts payable		$ 83.5
Accrued taxes and interest		25.5
Current maturities of long-term debt		20.0
Total current liabilities		129.0
Long-term liabilities:		420.0
Total liabilities		549.0
Shareholders' equity:		
Common stock	$100.0	
Retained earnings	208.6	
Total shareholders' equity		308.6
Total liabilities and shareholders' equity		$857.6

Required:

Identify the items in the statement that most likely would require further disclosure either on the face of the statement or in a note. Further identify those items that would require disclosure in the significant accounting policies note.

Real World Case 3–7

Balance sheet and significant accounting policies disclosure; Walmart

● LO3–2 through LO3–4, LO3–8

Real World Financials

The balance sheet and disclosure of significant accounting policies taken from the 2017 annual report of **Walmart Stores Inc.** appear below. Use this information to answer the following questions:

1. Does Walmart separately report current assets versus long-term assets, and current liabilities versus long-term liabilities (yes/no)?
2. What amounts did Walmart report for the following items for 2017:
   a. Total assets
   b. Current assets
   c. Current liabilities
   d. Total equity
   e. Retained earnings
   f. Inventory
3. What is Walmart's largest current asset? What is its largest current liability?
4. Compute Walmart's current ratio for 2017.
5. Identify the following items from the summary of significant accounting policies:
   a. Does the company have any securities classified as cash equivalents (yes/no)?
   b. What cost method does the company use for its U.S. inventory?
   c. When does the company recognize revenue from service transactions?

WAL-MART STORES, INC.
Consolidated Balance Sheets
($ in millions except per share data)

	As of January 31,	
	2017	**2016**
Assets		
Current assets:		
Cash and cash equivalents	$ 6,867	$ 8,705
Receivables (net)	5,835	5,624
Inventories	43,046	44,469
Prepaid expenses and other	1,941	1,441
Total current assets	57,689	60,239
Property and equipment:		
Property and equipment	179,492	176,958
Less accumulated depreciation	(71,782)	(66,787)
Property and equipment (net)	107,710	110,171
Property under capital leases:		
Property under capital lease and financing obligations	11,637	11,096
Less accumulated amortization	(5,169)	(4,751)
Property under capital leases and financing obligations (net)	6,468	6,345
Goodwill	17,037	16,695
Other assets and deferred charges	9,921	6,131
Total assets	$198,825	$199,581
Liabilities, Redeemable Noncontrolling Interest and Equity		
Current liabilities:		
Short-term borrowings	$ 1,099	$ 2,708
Accounts payable	41,433	38,487
Accrued liabilities	20,654	19,607
Accrued income taxes	921	521
Long-term debt due within one year	2,256	2,745
Capital lease and financing obligations due within one year	565	551
Total current liabilities	66,928	64,619
Long-term debt	36,015	38,214
Long-term capital lease and financing obligations	6,003	5,816
Deferred income taxes and other	9,344	7,321
Commitments and contingencies Equity:		
Common stock	305	317
Capital in excess of par value	2,371	1,805
Retained earnings	89,354	90,021
Accumulated other comprehensive income (loss)	(14,232)	(11,597)
Total Walmart shareholders' equity	77,798	80,546
Nonredeemable noncontrolling interest	2,737	3,065
Total equity	80,535	83,611
Total liabilities, redeemable noncontrolling interest and equity	$198,825	$199,581

Source: Wal-Mart

NOTES TO CONSOLIDATED FINANCIAL STATEMENTS
WAL-MART STORES, INC.

1 Summary of Significant Accounting Policies (in part)

Cash and Cash Equivalents

The Company considers investments with a maturity when purchased of three months or less to be cash equivalents.

Inventories

The Company values inventories at the lower of cost or market as determined primarily by the retail inventory method of accounting, using the last-in, first-out ("LIFO") method for substantially all of the Walmart U.S. segment's inventories. The inventory at the Walmart International segment is valued primarily by the retail inventory method of accounting, using the first-in, first-out ("FIFO") method. At January 31, 2017 and January 31, 2016, the Company's inventories valued at LIFO approximated those inventories as if they were valued at FIFO.

Revenue Recognition

The Company recognizes sales revenue, net of sales taxes and estimated sales returns, at the time it sells merchandise to the customer. Digital retail sales include shipping revenue and are recorded upon delivery to the customer. Customer purchases of shopping cards, to be utilized in our stores or on our e-commerce websites, are not recognized as revenue until the card is redeemed and the customer purchases merchandise using the shopping card. The Company recognizes revenue from service transactions at the time the service is performed. Generally, revenue from services is classified as a component of net sales in the Company's Consolidated Statements of Income.

Source: Wal-Mart

Judgment Case 3–8
Post fiscal year-end events
● LO3–4

The fiscal year-end for the Northwest Distribution Corporation is December 31. The company's 2021 financial statements were issued on March 15, 2022. The following events occurred between December 31, 2021, and March 15, 2022.

1. On January 22, 2022, the company negotiated a major merger with Blandon Industries. The merger will be completed by the middle of 2022.
2. On February 3, 2022, Northwest negotiated a $10 million long-term note with the Credit Bank of Ohio. The amount of the note is material.
3. On February 25, 2022, a flood destroyed one of the company's manufacturing plants causing $600,000 of uninsured damage.

Required:

Determine whether each of these events should be disclosed with the 2021 financial statements of Northwest Distribution Corporation.

Research Case 3–9
FASB codification; locate and extract relevant information and cite authoritative support for a financial reporting issue; related-party disclosures; Enron Corporation
● LO3–4
Real World Financials

Enron Corporation was a darling in the energy-provider arena, and in January 2001 its stock price rose above $100 per share. A collapse of investor confidence in 2001 and revelations of accounting fraud led to one of the largest bankruptcies in U.S. history. By the end of the year, Enron's stock price had plummeted to less than $1 per share. Investigations and lawsuits followed. One problem area concerned transactions with related parties that were not adequately disclosed in the company's financial statements. Critics stated that the lack of information about these transactions made it difficult for analysts following Enron to identify problems the company was experiencing.

Access the *FASB Accounting Standards Codification* at the FASB website (**www.fasb.org**). Determine each of the following:

1. The specific eight-digit Codification citation (XXX-XX-XX-X) that outlines the required information on related-party disclosures that must be included in the notes to the financial statements?
2. Describe the disclosures required for related-party transactions.
3. Use EDGAR (**www.gov.sec**) or another method to locate the December 31, 2000, financial statements of Enron. Search for the related-party disclosure. Briefly describe the relationship central to the various transactions described.
4. Why is it important that companies disclose related-party transactions? Use the Enron disclosure of the sale of dark fiber inventory in your answer.

Real World Case 3–10
Disclosures; proxy statement; Coca-Cola
● LO3–4, LO3–5
Real World Financials

EDGAR, the Electronic Data Gathering, Analysis, and Retrieval system, performs automated collection, validation, indexing, and forwarding of submissions by companies and others who are required by law to file forms with the SEC. All publicly traded domestic companies use EDGAR to make the majority of their filings. (Some foreign companies file voluntarily.) Form 10-K, which includes the annual report, is required to be filed on EDGAR. The SEC makes this information available on the Internet.

CODE

Required:

1. Access EDGAR on the Internet. The web address is **www.sec.gov**.
2. Search for **The Coca-Cola Company**. Access the 10-K for the year ended December 31, 2016. Search or scroll to find the disclosure notes and audit report.
3. Answer the following questions:
   a. Describe the subsequent events disclosed by the company.
   b. Which firm is the company's auditor? What type of audit opinion did the auditor render?
4. Access the proxy statement filed with the SEC on March 9, 2017 (the proxy statement designation is Def 14A), locate the executive officers summary compensation table and answer the following questions:
   a. What is the principal position of Muhtar Kent?
   b. What was the salary paid to Mr. Kent during the year ended December 31, 2016?

Judgment Case 3–11
Debt versus equity
● LO3–7

A common problem facing any business entity is the debt versus equity decision. When funds are required to obtain assets, should debt or equity financing be used? This decision also is faced when a company is initially formed. What will be the mix of debt versus equity in the initial capital structure? The characteristics of debt are very different from those of equity as are the financial implications of using one method of financing as opposed to the other.

Cherokee Plastics Corporation is formed by a group of investors to manufacture household plastic products. Their initial capitalization goal is $50,000,000. That is, the incorporators have decided to raise $50,000,000 to acquire the initial assets of the company. They have narrowed down the financing mix alternatives to two:

1. All equity financing
2. $20,000,000 in debt financing and $30,000,000 in equity financing

No matter which financing alternative is chosen, the corporation expects to be able to generate a 10% annual return, before payment of interest and income taxes, on the $50,000,000 in assets acquired. The interest rate on debt would be 8%. The effective income tax rate will be approximately 25%.

Alternative 2 will require specified interest and principal payments to be made to the creditors at specific dates. The interest portion of these payments (interest expense) will reduce the taxable income of the corporation and hence the amount of income tax the corporation will pay. The all-equity alternative requires no specified payments to be made to suppliers of capital. The corporation is not legally liable to make distributions to its owners. If the board of directors does decide to make a distribution, it is not an expense of the corporation and does not reduce taxable income and hence the taxes the corporation pays.

Required:

1. Prepare abbreviated income statements that compare first-year profitability for each of the two alternatives.
2. Which alternative would be expected to achieve the highest first-year profits? Why?
3. Which alternative would provide the highest rate of return on equity? Why?
4. Which alternative is considered to be riskier, all else equal?

Analysis Case 3–12
Obtain and critically evaluate an actual annual report
● LO3–4, LO3–6 through LO3–8

Real World Financials

Financial reports are the primary means by which corporations report their performance and financial condition. Financial statements are one component of the annual report mailed to their shareholders and to interested others.

Required:

Obtain an annual report from a corporation with which you are familiar. Using techniques you learned in this chapter and any analysis you consider useful, respond to the following questions:

1. Do the firm's auditors provide a clean opinion on the financial statements?
2. Has the company made changes in any accounting methods it uses?
3. Have there been any subsequent events, errors and fraud, illegal acts, or related-party transactions that have a material effect on the company's financial position?
4. What are two trends in the company's operations or capital resources that management considers significant to the company's future?
5. Is the company engaged in more than one significant line of business? If so, compare the relative profitability of the different segments.
6. How stable are the company's operations?
7. Has the company's situation deteriorated or improved with respect to liquidity, solvency, asset management, and profitability?

Note: You can obtain a copy of an annual report from a local company, from a friend who is a shareholder, from the investor relations department of the corporation, from a friendly stockbroker, or from EDGAR (Electronic Data Gathering, Analysis, and Retrieval) on the Internet (**www.sec.gov**).

Analysis Case 3–13
Obtain and compare annual reports from companies in the same industry
● LO3–4, LO3–7, LO3–8
Real World Financials

Insight concerning the performance and financial condition of a company often comes from evaluating its financial data in comparison with other firms in the same industry.

Required:

Obtain annual reports from three corporations in the same primary industry. Using techniques you learned in this chapter and any analysis you consider useful, respond to the following questions:

1. Are there differences in accounting methods that should be taken into account when making comparisons?
2. How do earnings trends compare in terms of both the direction and stability of income?
3. Which of the three firms had the greatest earnings relative to resources available?
4. Which corporation has made most effective use of financial leverage?
5. Of the three firms, which seems riskiest in terms of its ability to pay short-term obligations? Long-term obligations?

Note: You can obtain copies of annual reports from friends who are shareholders, from the investor relations department of the corporations, from a friendly stockbroker, or from EDGAR (Electronic Data Gathering, Analysis, and Retrieval) on the Internet (**www.sec.gov**).

Analysis Case 3–14
Balance sheet information
● LO3–2 through LO3–4
Real World Financials

Target Corporation prepares its financial statements according to U.S. GAAP. Target's financial statements and disclosure notes for the year ended February 3, 2018, are available in the Connect library. This material also is available under the Investor Relations link at the company's website (**www.target.com**).

Required:

1. Does the company separately report current assets and long-term assets, as well as current liabilities and long-term liabilities?
2. Are any investments shown as a current asset? Why?
3. In which liability account would the company report the balance of its gift card liability?
4. What method does the company use to depreciate its property and equipment?
5. What purpose do the disclosure notes serve?

Analysis Case 3–15
Segment reporting concepts
● Appendix 3, LO3–9
IFRS

Levens Co. operates in several distinct business segments. The company does not have any reportable foreign operations or major customers.

Required:

1. What is the purpose of operating segment disclosures?
2. Define an operating segment.
3. List the amounts to be reported by operating segment.
4. How would your answer to requirement 3 differ if Levens Co. prepares its segment disclosure according to International Financial Reporting Standards?

Ethics Case 3–16
Segment reporting
● Appendix 3

You are in your third year as an accountant with McCarver-Lynn Industries, a multidivisional company involved in the manufacturing, marketing, and sales of surgical prosthetic devices. After the fiscal year-end, you are working with the controller of the firm to prepare geographic area disclosures. Yesterday you presented her with the following summary information:

	($ in millions)					
	Domestic	Libya	Egypt	France	Cayman Islands	Total
Revenues	$ 845	$222	$265	$343	$2,311	$3,986
Operating income	145	76	88	21	642	972
Assets	1,005	301	290	38	285	1,919

Upon returning to your office after lunch, you find the following memo:

Nice work. Let's combine the data this way:

	($ in millions)			
	Domestic	Africa	Europe and Other Foreign	Total
Revenues	$ 845	$487	$2,654	$3,986
Capital expenditures	145	164	663	972
Assets	1,005	591	323	1,919

Because of political instability in North Africa, let's not disclose specific countries. In addition, we restructured most of our French sales and some of our U.S. sales to occur through our offices in the Cayman Islands. This allows us to avoid paying higher taxes in those countries. The Cayman Islands has a 0% corporate income tax rate. We don't want to highlight our ability to shift profits to avoid taxes.

Required:

Data Analytics Case 3–17

Do you perceive an ethical dilemma? What would be the likely impact of following the controller's suggestions? Who would benefit? Who would be injured?

Data Analytics

connect

Data analytics is the process of examining data sets in order to draw conclusions about the information they contain. To prepare for this case, you will need to download Tableau to your computer. If you have not completed any of the prior data analytics cases, follow the instructions listed in the Chapter 1 Data Analytics case to get set up. You will need to watch the videos referred to in both the Chapter 1 and Chapter 2 Data Analytics cases. To prepare for this video, you should watch "Getting Started with Calculations" and "Calculation Syntax" videos listed under "Calculations." Future cases will build off of your knowledge gained from these, and additional videos will be suggested. All short training videos can be found here: **www.tableau.com/learn/training#getting-started**.

Data Analytics Case
Liquidity
● LO3–7

In the Chapter 1 and Chapter 2 Data Analytics Cases, you applied Tableau to examine a data set and create charts to compare our two (hypothetical) publicly traded companies: Discount Goods and Big Store. In this case, you continue in your role as an analyst conducting introductory research into the relative merits of investing in one or both of these companies. This time you assess the companies' liquidity: i.e., their ability to pay short-term debts as they come due. You do so by comparing liquidity ratios in a line chart.

Required:

Use Tableau to calculate the liquidity ratios for each of the two companies in each year from 2012 to 2021. Based upon what you find, answer the following questions:

1. Analyzing the liquidity ratios over the ten-year period, is Discount Goods current ratio (a) generally increasing, (b) roughly the same, or (c) generally decreasing from year to year?
2. Analyzing the liquidity ratios over the ten-year period, is Big Store's current ratio (a) generally increasing, (b) roughly the same, or (c) generally decreasing from year to year?
3. What is the current ratio in 2021 for Discount Goods? For Big Store?
4. What is the acid-test (or quick) ratio in 2021 for Discount Goods? For Big Store?
5. Which company has more favorable liquidity ratios in 2021?

Note: As you assess the liquidity aspect of the companies, keep in mind that liquidity ratios should be evaluated in the context of both profitability and efficiency of managing assets.

Resources:

You have available to you an extensive data set that includes detailed financial data for Discount Goods and Big Store and for 2012 - 2021. The data set is in the form of an Excel file available to download from Connect, or under Student Resources within the Library tab. There are four Excel files available. The one for use in this chapter is named "Discount_Goods_Big_Store_Financials.xlsx" Download this file and save it to the computer on which you will be using Tableau. Open Tableau and connect to the Excel spreadsheet you downloaded.

For this case, you will create several calculations to produce a line chart of liquidity ratios to allow you to compare and contrast the two companies. You also will need to change data from *continuous* to *discrete*. Continuous data are data that is identified as a range of numeric values. A bar chart uses continuous data. "Measures" are usually identified by Tableau as continuous data and produce a range of data when dragged to the row shelf. If you want only one number, you must change the data to be discrete data. The *pill* that represents the data on the shelves will be blue if it is discrete and green if it is continuous.

After you view the training videos, follow these steps to create the charts you'll use for this case:

- Open Tableau and connect to the Excel spreadsheet you downloaded.
- Click on the "Sheet 1" tab at the bottom of the canvas, to the right of the Data Source at the bottom of the screen. Drag "Company" under "Dimensions" to the Columns shelf.
- Drag the "Year" under "Dimensions" to the Columns shelf. If the "Year" pill is green, Tableau has identified it as continuous data. Select the drop-down menu box on the "Year" pill and click "Discrete" instead of "Continuous." Note the pill is now blue.
- Drag "Total current assets" under "Measures" to the Rows shelf. Change to "Discrete" data following the same instructions above. Note: You must drag accounts into the Rows shelf if you are going to use them in calculated fields and need an *aggregated* sum. If this is not done first, your answer will not match the solutions.

- Drag "Total current liabilities" under Measures to the Rows shelf. Follow the same instructions above for "Total current assets."
- Create a calculated field by clicking the "Analysis" tab from the Toolbar at the top of the screen and selecting "Create Calculated field." A calculation box will pop up. Name the calculation "Quick Assets." Drag "Cash and cash equivalents" under "Measures" into the calculation box. Add a plus sign and drag "Receivables, net" into the calculation box. Make sure the box says that the calculation is valid and click OK. The new field "Quick Assets" will now appear under "Measures."
- Create another calculated field named "Current Ratio" by dragging "Total current assets" into the calculation box, typing a division sign, then dragging "Total current liabilities" beside it. Make sure the box says that the calculation is valid and click OK. The new field "Current Ratio" will now appear under Measures. Drag "Current Ratio" to the Rows shelf. Click on the dropdown from the "Current Ratio" pill, click "Format", set the formatting options. From the 'Font' dropdown select Times New Roman, 10-point font, Bold, Black. From the 'Numbers' dropdown select 'Number (custom) "3" decimal places.
- Create a calculated field "Quick Ratio" following the same process. Drag the newly created "Quick Assets" under "Measures" into the calculation box. Type a division sign, then drag Total current liabilities" under "Measures" beside it. Make sure the box says that the calculation is valid and click OK. Drag "Quick Ratio" to the Rows shelf. Click on the dropdown from the "Quick Ratio" pill, set Formats as noted above.
- Drag "Sum: Total current assets" and "Sum: Total current liabilities" from the "Rows" shelf back under "Measures." This will remove them from the canvas.
- In the "Marks" card, select the drop-down menu for presentation type and select "Line" (it defaults to "Automatic"). In the "Label" box, select "Show mark labels." Format to Times New Roman, bold, and 10-point font.
- Change the title of the sheet to be "Liquidity" by right-clicking and selecting "Edit title." Format the title to Times New Roman, bold, black and 20-point font. Change the title of "Sheet 1" to match the sheet title by right-clicking, selecting "Rename" and typing in the new title.
- Once complete, save the file as "DA3_Your initials.twbx."

Continuing Cases

Target Case
- LO3–2, LO3–3, LO3–8

Target Corporation prepares its financial statements according to U.S. GAAP. Target's financial statements and disclosure notes for the year ended February 3, 2018, are available in Connect. This material is also available under the Investor Relations link at the company's website (**www.target.com**).

Required:

1. By what name does Target label its balance sheet?
2. What amounts did Target report for the following items on February 3, 2018?
   a. Current assets
   b. Long-term assets
   c. Total assets
   d. Current liabilities
   e. Long-term liabilities
   f. Total liabilities
   g. Total shareholders' equity
3. What was Target's largest current asset? What was its largest current liability?
4. Compute Target's current ratio and debt to equity ratio in 2018?
5. Assuming Target's industry had an average current ratio of 1.0 and an average debt to equity ratio of 2.5, comment on Target's liquidity and long-term solvency.

Air France–KLM Case
- LO3–9

IFRS

Air France-KLM (AF), a Franco-Dutch company, prepares its financial statements according to International Financial Reporting Standards. AF's financial statements and disclosure notes for the year ended December 31, 2017, are provided in Connect. This material is also available under the Finance link at the company's website (**www.airfranceklm.com**).

Required:

Describe the apparent differences in the order of presentation of the components of the balance sheet between IFRS as applied by Air France–KLM (AF) and a typical balance sheet prepared in accordance with U.S. GAAP.

CPA Exam Questions and Simulations

ROGER CPA Review

Sample CPA Exam questions from Roger CPA Review are available in Connect as support for the topics in this chapter. These multiple-choice questions and task-based simulations include expert-written explanations and solutions, and provide a starting point for students to become familiar with the content and functionality of the actual CPA Exam.

CHAPTER

4 The Income Statement, Comprehensive Income, and the Statement of Cash Flows

OVERVIEW

This chapter has three purposes: (1) to consider important issues dealing with the content, presentation, and disclosure of net income and other components of comprehensive income; (2) to provide an *overview* of the statement of cash flows, which is covered in depth in Chapter 21; and (3) to examine common ratios used in profitability analysis.

The income statement summarizes the profit-generating activities that occurred during a particular reporting period. Comprehensive income includes net income as well as other gains and losses that are not part of net income.

The statement of cash flows provides information about the cash receipts and cash disbursements of an enterprise's operating, investing, and financing activities that occurred during the period.

Profitability ratios measure how well a company manages its operations and utilizes resources to generate a profit. Profitability is a key metric in understanding the company's ability to generate cash in the future.

LEARNING OBJECTIVES

After studying this chapter, you should be able to:

- **LO4–1** Discuss the importance of income from continuing operations and describe its components. (*p. 168*)
- **LO4–2** Describe earnings quality and how it is impacted by management practices to alter reported earnings. (*p. 173*)
- **LO4–3** Discuss the components of operating and nonoperating income and their relationship to earnings quality. (*p. 173*)
- **LO4–4** Define what constitutes discontinued operations and describe the appropriate income statement presentation for these transactions. (*p. 177*)
- **LO4–5** Discuss additional reporting issues related to accounting changes, error corrections, and earnings per share (EPS). (*p. 182*)
- **LO4–6** Explain the difference between net income and comprehensive income and how we report components of the difference. (*p. 185*)
- **LO4–7** Describe the purpose of the statement of cash flows. (*p. 189*)
- **LO4–8** Identify and describe the various classifications of cash flows presented in a statement of cash flows. (*p. 190*)
- **LO4–9** Discuss the primary differences between U.S. GAAP and IFRS with respect to the income statement, statement of comprehensive income, and statement of cash flows. (*pp. 172, 186, and 195*)
- **LO4–10** Identify and calculate the common ratios used to assess profitability. (*p. 197*)

©GaudiLab/Shutterstock

FINANCIAL REPORTING CASE

Abbott Laboratories

Your friend, Becky Morgan, just received a generous gift from her grandfather. Accompanying a warm letter were 200 shares of stock of **Abbott Laboratories,** a global health care company, along with the most recent annual financial statements of the company. Becky knows that you are an accounting major and pleads with you to explain some items in the company's income statement. "I remember studying the income statement in my introductory accounting course," says Becky, "but I am still confused. What is this item *discontinued operations?* I also read in the annual report the company has *restructuring costs?* These don't sound good. Are they something I should worry about? We studied earnings per share briefly, but what does *earnings per common share—diluted* mean?" You agree to try to help.

ABBOTT LABORATORIES AND SUBSIDIARY COMPANIES
Statements of Earnings For the Year Ended December 31
($ in millions, except per share data)

	2016	2015
Net sales	$ 20,853	$20,405
Costs and expenses:		
Cost of products sold	9,024	8,747
Amortization of intangible assets	550	601
Research and development expenses	1,422	1,405
Selling, general and administrative	6,672	6,785
Operating income	3,185	2,867
Other income (deductions)	(1,776)	316
Income from continuing operations before income taxes	1,413	3,183
Provision for taxes on income	350	577
Income from continuing operations	1,063	2,606
Income from discontinued operations, net of tax	337	1,817
Net income	$ 1,400	$ 4,423
Earnings per common share—basic:		
Continuing operations	$ 0.71	$ 1.73
Discontinued operations	0.23	1.21
Net income	$ 0.94	$ 2.94
Earnings per common share—diluted:		
Continuing operations	$ 0.71	$ 1.72
Discontinued operations	0.23	1.20
Net income	$ 0.94	$ 2.92

Source: Abbott Laboratories, Inc.

By the time you finish this chapter, you should be able to respond appropriately to the questions posed in this case. Compare your response to the solution provided at the end of the chapter.

QUESTIONS

1. How would you explain restructuring costs to Becky? Are restructuring costs something Becky should worry about?
2. Explain to Becky what is meant by discontinued operations and describe to her how that item is reported in an income statement.
3. Describe to Becky the difference between basic and diluted earnings per share

In Chapter 1, we discussed the critical role of financial accounting information in allocating resources within our economy. Ideally, resources should be allocated to private enterprises that will (1) provide the goods and services our society desires and (2) at the same time provide a fair rate of return to those who supply the resources. A company will be able to achieve these goals only if it can generate enough cash to stay in business. A company's ability to generate cash relates to its ability to sell products and services for amounts greater than the costs of providing those products and services (that is, generate a profit).

Two financial statements that are critical for understanding the company's ability to earn profits and generate cash in the future are as follows:

1. **Income statement** (also called *statement of operations* or *statement of earnings*).
2. **Statement of cash flows.**

The **income statement** reports a company's profit during a particular reporting period. Profit equals revenues and gains minus expenses and losses. A few types of gains and losses are excluded from the income statement but are included in the broader concept of **comprehensive income**. We refer to these other gains and losses as **other comprehensive income (OCI)**.

The **statement of cash flows** provides information about the cash receipts and cash payments of a company during a particular reporting period. The difference between cash receipts and cash payments represents the change in cash for the period. To help investors and creditors better understand the sources and uses of cash during the period, the statement of cash flows distinguishes among operating, investing, and financing activities.

PART A

The Income Statement and Comprehensive Income

Before we discuss the specific components of an income statement in much depth, let's take a quick look at the general makeup of the statement. Illustration 4–1 offers an income statement for McAllister's Manufacturing, a hypothetical company, that you can refer to as we proceed through the chapter. At this point, our objective is only to gain a general perspective on the items reported and classifications contained in corporate income statements. In addition, each income statement should include in the heading the name of the company, the title of the statement, and the date or time period. McAllister's income statement is for the year ended December 31. This means that amounts in the income statement are the result of transactions from January 1 to December 31 of that year. In reality, many companies have reporting periods (often referred to as *fiscal years*) that end in months other than December, as you'll see demonstrated later in this chapter. Illustration 4–1 shows comparative income statements for two consecutive years.

Let's first look closer at the components of net income. At the end of Part A, we'll see how net income fits within the concept of comprehensive income and how comprehensive income is reported.

Income from Continuing Operations

Revenues, Expenses, Gains, and Losses

LO4–1

Unlike the balance sheet, which is a position statement *at a point in time,* the income statement measures activity *over a period of time.* The income statement reports the revenues, expenses, gains, and losses that have occurred during the reporting period. For example, if a company reports revenues of $100 million in its income statement for the year ended December 31, 2021, this means that the company had revenue transactions from January 1, 2021, to December 31, 2021, equal to $100 million.

Revenues are inflows of resources resulting from providing goods or services to customers. For merchandising companies like **Walmart**, the main source of revenue is sales revenue derived from selling merchandise. Service companies such as **FedEx** and **State Farm Insurance** generate revenue by providing services.

Illustration 4–1

Income Statement

McAllister's Manufacturing
Income Statement
($ in millions, except per share data)

		Year Ended December 31	
		2021	**2020**
Income from Continuing Operations	Sales revenue	$1,450.6	$1,380.0
	Cost of goods sold	832.6	800.4
	Gross profit	618.0	579.6
	Operating expenses:		
	Selling expense	123.5	110.5
	General and administrative expense	147.8	139.1
	Research and development expense	55.0	65.0
	Restructuring costs	125.0	—
	Total operating expenses	451.3	314.6
	Operating income	166.7	265.0
	Other income (expense):		
	Interest revenue	12.4	11.1
	Interest expense	(25.9)	(24.8)
	Gain on sale of investments	18.0	19.0
	Income from continuing operations before income taxes	171.2	270.3
	Income tax expense	59.9	94.6
	Income from continuing operations	111.3	175.7
Discontinued Operations	Discontinued operations:		
	Loss from operations of discontinued component (including gain on disposal in 2021 of $47)	(7.6)	(45.7)
	Income tax benefit	2.0	13.0
	Loss on discontinued operations	(5.6)	(32.7)
	Net income	$ 105.7	$ 143.0
Earnings per Share	Earnings per common share—basic:		
	Income from continuing operations	$ 2.14	$ 3.38
	Discontinued operations	(0.11)	(0.63)
	Net income	$ 2.03	$ 2.75
	Earnings per common share—diluted:		
	Income from continuing operations	$ 2.06	$ 3.25
	Discontinued operations	(0.10)	(0.61)
	Net income	$ 1.96	$ 2.64

Expenses are outflows of resources incurred while generating revenue. They represent the costs of providing goods and services. When recognizing expenses, we attempt to establish a causal relationship between revenues and expenses. If causality can be determined, expenses are reported in the same period that the related revenue is recognized. If a causal relationship cannot be established, we relate the expense to a particular period, allocate it over several periods, or expense it as incurred.

Gains and **losses** are increases or decreases in equity from peripheral or incidental transactions of an entity. In general, these gains and losses do not reflect normal operating activities of the company, but they nevertheless represent transactions that affect a company's financial position. For example, gains and losses can arise when a company sells investments or property, plant, and equipment for an amount that differs from their recorded amount. Losses can occur on inventory due to obsolescence, on assets for impaired values, and on litigation claims. We will discuss many types of gains and losses throughout this book.

Income from continuing operations includes the revenues, expenses, gains and losses from operations that are more likely to continue into the future.

To understand a company's ability to generate cash in the future, investors and creditors assess which components of net income are likely to continue into the future. **Income from continuing operations** includes revenues, expenses (including income taxes), gains, and losses arising from operations that are more likely to continue. In contrast, income from *discontinued operations* will not continue into the future.[1]

The three major components of income from continuing operations include the following:

1. Operating income
2. Nonoperating income
3. Income tax expense

Operating Income versus Nonoperating Income

Operating income includes revenues, expenses, gains, and losses directly related to the *primary revenue-generating activities* of the company. For a manufacturing company, operating income includes sales revenue from selling the products it manufactures minus cost of goods sold and operating expenses related to its primary activities.[2] As shown in Illustration 4–1, operating income is often presented as gross profit (sales revenue minus cost of goods sold) minus operating expenses. In practice, many companies include *net sales* on the first line of the income statement to represent total sales revenue less sales discounts, returns, and allowances, which we discuss in depth in chapters 6 and 7. For a service company, operating income would include service revenue minus operating expenses (and no cost of goods sold).

Nonoperating income includes revenues, expenses, gains, and losses related to *peripheral or incidental activities* of the company. For example, a manufacturer would include interest revenue, gains and losses from selling investments, and interest expense in nonoperating income. These items are not directly related to the primary revenue-generating activities of a manufacturing company. On the other hand, a financial institution like a bank would consider those items to be a part of operating income because they relate to the primary revenue-generating activities for that type of business. As shown in Illustration 4–1, nonoperating items often are included in the income statement under the heading *Other income (expense).*

Income Tax Expense

Income taxes are levied on taxpayers in proportion to the amount of taxable income that is reported to taxing authorities. Like individuals, corporations are income-tax-paying entities.[3] Because of the importance and size of income tax expense (sometimes called *provision for income taxes*), it always is reported in a separate line in corporate income statements. In addition, as we'll discuss in more detail later in this chapter, companies are required to report income tax expense associated with operations that are *continuing* separately from income tax expense associated with operations that are being *discontinued.* Separately reporting income tax expense in this way is known as *intrapreriod tax allocation.*

Income tax expense is reported in a separate line in the income statement.

When tax rules and GAAP differ regarding the timing of revenue or expense recognition, the actual payment of taxes may occur in a period different from when income tax expense is reported in the income statement. We discuss this and other issues related to accounting for income taxes in Chapter 16. At this point, simply consider income tax expense to be a percentage of income before taxes.

Illustration 4–2 presents the 2017 income statement for **The Home Depot, Inc.**, a retail company specializing in home improvement products. Notice that Home Depot distinguishes between operating income, nonoperating income, and income tax expense. Operating income includes revenues and expenses from primary business activities related to merchandise sales. After operating income is determined, nonoperating items are added or subtracted. For a retail company like Home Depot, activities related to interest and investments are not primary operations and therefore are listed as nonoperating items. Finally, income tax expense is subtracted to arrive at net income.

Now let's consider the format used to report the components of net income.

---

[1]Discontinued operations are addressed in a subsequent section.

[2]In certain situations, operating income might also include gains and losses from selling equipment and other assets used in the manufacturing process. FASB ASC 360-10-45-5: Property, plant, and equipment—Overall—Other Presentation Matters (previously "Accounting for the Impairment of Long-Lived Assets and for Long-Lived Assets to Be Disposed Of," *Statement of Financial Accounting Standards No. 144* (Norwalk, Conn.: FASB, 2001)).

[3]Partnerships are not tax-paying entities. Their taxable income or loss is included in the taxable income of the individual partners.

Illustration 4–2

Income Statement–The Home Depot, Inc.

Real World Financials

Consolidated Statement of Earnings
($ in millions)

	Year ended January 29, 2017
Net sales	$94,595
Cost of goods sold	62,282
Gross profit	32,313
Selling, general, and administrative expenses	17,132
Depreciation and amortization	1,754
Operating income	13,427
Interest and investment income	36
Interest expense	(972)
Income before income taxes	12,491
Income tax expense	4,534
Net income	$ 7,957

Source: Home Depot, Inc.

Income Statement Formats

No specific standards dictate how income from continuing operations must be displayed, so companies have considerable latitude in how they present the components of income from continuing operations. This flexibility has resulted in a variety of income statement presentations. However, we can identify two general approaches, the single-step and the multiple-step formats, that might be considered the two extremes, with the income statements of most companies falling somewhere in between.

The **single-step** format first lists all the revenues and gains included in income from continuing operations. Then, expenses and losses are grouped, subtotaled, and subtracted—in a single step—from revenues and gains to derive income from continuing operations. In a departure from that, though, companies usually report income tax expense in a separate line in the statement. In a single-step income statement, operating and nonoperating items are not separately classified. Illustration 4–3 shows an example of a single-step income statement for a hypothetical manufacturing company, Maxwell Gear Corporation.

A *single-step* income statement format groups all revenues and gains together and all expenses and losses together.

Illustration 4–3

Single-Step Income Statement

MAXWELL GEAR CORPORATION
Income Statement
For the Year Ended December 31, 2021

Revenues and gains:		
Sales revenue	$573,522	
Interest revenue	5,500	
Gain on sale of investments	26,400	
Total revenues and gains		$605,422
Expenses and losses:		
Cost of goods sold	302,371	
Selling expense	47,341	
General and administrative expense	24,388	
Research and development expense	16,300	
Interest expense	14,522	
Total expenses and losses		405,422
Income before income taxes		200,000
Income tax expense		50,000
Net income		$150,000

A *multiple-step* income statement format includes a number of intermediate subtotals before arriving at income from continuing operations.

The **multiple-step** format reports a series of intermediate subtotals such as gross profit, operating income, and income before taxes. Most of the real-world income statements are in this format. Illustration 4–4 presents a multiple-step income statement for the Maxwell Gear Corporation.

Illustration 4–4

Multiple-Step Income Statement

MAXWELL GEAR CORPORATION
Income Statement
For the Year Ended December 31, 2021

Sales revenue		$573,522
Cost of goods sold		302,371
Gross profit		271,151
Operating expenses:		
Selling expense	$47,341	
General and administrative expense	24,888	
Research and development expense	16,300	
Total operating expenses		88,529
Operating income		182,622
Other income (expense):		
Interest revenue	5,500	
Gain on sale of investments	26,400	
Interest expense	(14,522)	
Total other income, net		17,378
Income before income taxes		200,000
Income tax expense		50,000
Net income		$150,000

A primary advantage of the multiple-step format is that, by separately classifying operating and nonoperating items, it provides information that might be useful in analyzing trends. Similarly, the classification of expenses by function also provides useful information. For example, reporting *gross profit* for merchandising companies highlights the important relationship between sales revenue and cost of goods sold. *Operating income* provides a measure of profitability for core (or normal) operations, a key performance measure for predicting the future profit-generating ability of the company. *Income before taxes* could be useful for comparing the performance of companies in different tax jurisdictions or comparing corporations (tax-paying entities) with sole proprietorships or partnerships (typically non-tax-paying entities).

It is important to note that the difference between the single-step and multiple-step income statement is one of presentation. The bottom line, *net income,* is the same regardless of the format used. Most companies use the multiple-step format. We use the multiple-step format for illustration purposes throughout the remainder of this chapter.

International Financial Reporting Standards

Income Statement Presentation. There are more similarities than differences between income statements prepared according to U.S. GAAP and those prepared applying international standards. Some of the differences are as follows:

● LO4–9

- International standards require certain minimum information to be reported on the face of the income statement. U.S. GAAP has no minimum requirements.
- International standards allow expenses to be classified either by function (e.g., cost of goods sold, general and administrative, etc.), or by natural description (e.g., salaries, rent, etc.). SEC regulations require that expenses be classified by function.
- In the United States, the "bottom line" of the income statement usually is called either *net income* or *net loss.* The descriptive term for the bottom line of the income statement prepared according to international standards is either *profit* or *loss.*

Earnings Quality

Investors, creditors, and financial analysts are concerned with more than just the bottom line of the income statement—net income. The presentation of the components of net income and the related supplemental disclosures provide clues to the user of the statement in an assessment of *earnings quality*. Earnings quality is used as a framework for more in-depth discussions of operating and nonoperating income.

● LO4–2

One meaning of **earnings quality** is the ability of reported earnings (income) to predict a company's future earnings. The relevance of any historical-based financial statement hinges on its predictive value. To enhance predictive value, analysts try to separate a company's *temporary earnings* from its *permanent earnings*. Temporary earnings arise from transactions or events that are not likely to occur again in the foreseeable future or that are likely to have a different impact on earnings in the future. In contrast, permanent earnings arise from operations that are expected to generate similar profits in the future. Analysts begin their assessment of permanent earnings with income before discontinued operations, that is, income from continuing operations. Later in the chapter we address discontinued operations that, because of their nature, are required to be reported separately at the bottom of the income statement.

Earnings quality refers to the ability of reported earnings (income) to predict a company's future earnings.

It would be a mistake, though, to assume that all items included in income from continuing operations reflect permanent earnings. Some income items that fall under this category may be temporary. In a sense, the label *continuing* may be misleading.

Income Smoothing and Classification Shifting

An often-debated contention is that, within GAAP, managers have the power to change reported income by altering assumptions and estimates. And these alternatives are not always in the direction of higher income. Survey evidence suggests that managers often alter income upwards in one year but downward in other years.[4] For example, in a year when income is high, managers may create reserves by overestimating certain expenses (such as future bad debts or warranties). These reserves reduce reported income in the current year. Then, in later years, they can use those reserves by underestimating expenses, which will increase reported income. By shifting income in this manner, managers effectively smooth the pattern in reported income over time, portraying a steadier income stream to investors, creditors, and other financial statement users.[5]

Management's *income smoothing* behavior is controversial. While some believe that a smoother income pattern helps investors and creditors to better predict future performance, others believe that managers are doing this to hide the true risk (volatility) of operations. By hiding this underlying volatility through manipulation of the income pattern over time, managers may be "fooling" investors and creditors into believing that the company's operations are lower-risk than they really are.

Income smoothing may help investors to predict future performance but it could also hide underlying risk.

Another way that managers affect reported income is through *classification shifting* in the income statement.[6] The most common example of this involves misclassifying operating expenses as nonoperating expenses. By shifting operating expenses to a nonoperating expense classification (often referred to as "special charges" or "special items"), managers report fewer operating expenses and therefore higher operating income. This type of manipulation creates the appearance of stronger performance for core operations. While bottom-line net income remains unaffected, investors and creditors may believe the core business is stronger than it really is.

Classification shifting inflates core performance.

Operating Income and Earnings Quality

● LO4–3

Should all items of revenue and expense included in operating income be considered indicative of a company's permanent earnings? No, not necessarily. Sometimes a company will

[4]See, for example, I. Dichev, J. Graham, H. Campbell, and S. Rajgopal, "Earnings Quality: Evidence from the Field," *Journal of Accounting and Economics* 56 (2013), pp. 1–33.

[5]J. Graham, H. Campbell, and S. Rajgopal, "The Economic Implications of Corporate Financial Reporting," *Journal of Accounting and Economics* 40 (2005), pp. 3–73.

[6] S. McVay, "Earnings Management Using Classification Shifting: An Examination of Core Earnings and Special Items," *The Accounting Review* 81 (2006), pp. 501–31.

have an unusual or infrequent event. Even though these events may be unlikely to occur again in the near future, we report them as part of operating income because they are so closely related to the company's core business.[7]

What kind of items might be included in this category? Look closely at the partial income statements of **The Hershey Company**, the largest producer of chocolate in North America, presented in Illustration 4–5. Which items appear unusual? Certainly not net sales, cost of sales, or selling, marketing and administrative expenses. But what about "Goodwill and other intangible asset impairment charges" and "Business realignment charges"? Let's consider both.

Illustration 4–5

Partial Income Statement—The Hershey Company

Real World Financials

Income Statements (in part)
($ in thousands)

	Year Ended	
	December 31, 2016	**December 31, 2015**
Net sales	$7,440,181	$7,386,626
Cost of sales	4,282,290	4,003,951
Selling, marketing and administrative expenses	1,915,378	1,969,308
Goodwill and other intangible asset impairment charges	4,204	280,802
Business realignment charges	32,526	94,806
Operating profit	$1,205,783	$1,037,759

Source: The Hershey Company

FINANCIAL Reporting Case

Q1, p. 167

***Restructuring* costs include costs associated with shutdown or relocation of facilities or downsizing of operations.**

GAAP requires that restructuring costs be recognized only in the period incurred.

Should restructuring costs be considered part of a company's permanent earnings stream?

RESTRUCTURING COSTS It's not unusual for a company to reorganize its operations to attain greater efficiency. When this happens, the company often incurs significant **restructuring costs** (sometimes referred to as *reorganization costs* or *realignment costs*). Restructuring costs are associated with management's plans to materially change the scope of business operations or the manner in which they are conducted.[8] For example, facility closings and related employee layoffs translate into costs incurred for severance pay and relocation costs. **The Hershey Company** had restructuring costs related to reorganization of operations in Brazil to enhance distribution of the company's products, as well as employee severance costs related to eliminating several positions as part of the company's Productivity Initiative.

Restructuring costs are recognized in the period the exit or disposal cost obligation actually is incurred. Suppose, as part of a restructuring plan, employees to be terminated are offered various benefits but only if they complete a certain period of work for the company. In that case, a liability for termination benefits, and corresponding expense, should be accrued in the required period(s) of work. On the other hand, if future work by the employee is not required to receive the termination benefits, the liability and corresponding expense for benefits are recognized at the time the company communicates the arrangement to employees. Similarly, costs associated with closing facilities and relocating employees are recognized when goods or services associated with those activities are received.

Because it usually takes considerable time to sell or terminate a line of business, or to close a location or facility, many restructuring costs represent long-term liabilities. GAAP requires initial measurement of these liabilities to be at fair value, which often is determined as the present value of future estimated cash outflows. Companies also are required to provide many disclosures in the notes, including the years over which the restructuring is expected to take place.

Now that we understand the nature of restructuring costs, we can address the important question: Should financial statement users attempting to forecast future earnings consider these costs to be part of a company's permanent earnings stream, or are they unlikely to occur again? There is no easy answer. For example, Hershey has reported some amount of restructuring costs in each year from 2005–2016. However, the amount reported in 2016

[7]GAAP previously required events that were both unusual and infrequent to be classified as "extraordinary items" and reported net of tax after income from continuing operations. As part of the FASB's Simplification Initiative, Accounting Standards Update No. 2015-01 eliminated the extraordinary classification. International accounting standards also do not recognize the extraordinary classification.

[8]FASB ASC 420-10-20: Exit or Disposal Cost Obligations—Overall—Glossary (previously "Accounting for Costs Associated with Exit or Disposal Activities," *Statement of Financial Accounting Standards No. 146* (Norwalk, Conn.: FASB, 2002)).

was substantially less than the amount in 2015. Will the company incur these costs again in the near future? Probably. A recent survey reports that of the 500 companies surveyed, 40% included restructuring costs in their income statements.[9] The inference: A financial statement user must interpret restructuring charges in light of a company's past history and financial statement note disclosures that outline the plan and the period over which it will take place. In general, the more frequently these sorts of unusual charges occur, the more appropriate it is that financial statement users include them in their estimation of the company's permanent earnings stream.[10]

OTHER UNUSUAL ITEMS Two other expenses in Hershey's income statements that warrant additional scrutiny are *goodwill impairments* and *asset impairments.* Any long-lived asset, whether tangible or intangible, should have its balance reduced if there has been a significant impairment of value. We explore property, plant, and equipment and intangible assets in Chapters 10 and 11. After discussing this topic in more depth in those chapters, we revisit the concept of earnings quality as it relates to asset impairment.

These aren't the only components of operating expenses that call into question this issue of earnings quality. For example, in Chapter 9 we discuss the write-down of inventory that can occur with obsolete or damaged inventory. Other possibilities include losses from natural disasters such as earthquakes and floods and gains and losses from litigation settlements. Earnings quality also is influenced by the way a company records income from investments (Chapter 12) and accounts for its pension plans (Chapter 17).

Unusual items included in operating income require investigation to determine their permanent or temporary nature.

Earnings quality is affected by revenue issues as well. As an example, suppose that toward the end of its fiscal year, a company loses a major customer that can't be replaced. That would mean the current year's revenue number includes a component that will not occur again next year. Of course, in addition to its effect on revenues, losing the customer would have implications for certain related expenses and net income.

Another issue affecting earnings quality is the intentional misstatement of revenue. In 2016, the Securities and Exchange Commission (SEC) charged **Lime Energy** with reporting nonexistent revenues. The allegations include executives faking "pipeline" revenue that didn't really exist. Executives were alleged to have recognized revenue on contract proposals that had been presented to customers, but those customers hadn't yet agreed to or signed the contract. We explore additional issues of revenue recognition in Chapter 6. Now, though, let's discuss earnings quality issues related to *nonoperating* items.

Nonoperating Income and Earnings Quality

Most of the components of earnings in an income statement relate directly to the ordinary, continuing operations of the company. Some, though such as interest or the gains and losses on the sale of investments relate only tangentially to normal operations. We refer to these as nonoperating items. How should these items be interpreted in terms of their relationship to future earnings? Are these expenses likely to occur again next year? Investors need to understand that some of these items may recur, such as interest expense, while others are less likely to recur, such as gains and losses on investments.

Gains and losses from the sale of investments typically relate only tangentially to normal operations.

Home Depot's partial income statement is shown in Illustration 4–6. There are two nonoperating amounts reported after operating income. The first one is "Interest and investment income" that primarily includes gains on the sale of investments in another company's stock. Because Home Depot's primary business includes selling home improvement products, sales of investments are not considered normal operations. Therefore, Home Depot reports these amounts as nonoperating items. Home Depot disclosed in the notes of its 2016 report that gains on the sale of investments were $144 million. Once these investments were sold, their gains were not expected to continue into future profitability, so investors would not consider these gains to be a permanent component of profit. In 2017, we see that there were no gains on the sale of investments, consistent with this source of income not continuing. The remaining relatively minor amount in this nonoperating category in 2017 comes from interest income.

[9] *U.S. GAAP Financial Statements—Best Practices in Presentation and Disclosure*—2013 (New York: AICPA, 2013).
[10] Arthur Levitt, Jr. "The Numbers Game," *The CPA Journal*, December 1998, p. 16.

Illustration 4–6
Income Statements (in part)—The Home Depot, Inc.

Real World Financials

Income Statements (in part)
($ in millions)

	Year Ended	
	January 29, 2017	**January 31, 2016**
Operating income	$13,427	$11,774
Interest and investment income	36	166
Interest expense	(972)	(919)
Income before taxes	$12,491	$11,021

Source: Home Depot, Inc.

Another large nonoperating item reported by most companies is interest expense. In its 2017 report, Home Depot reported $972 million in interest expense. The company also reported long-term debt of nearly $22 billion in the balance sheet. Because this long-term debt will not be repaid for several years, the company will have to pay interest for several years. Therefore, interest expense represents a type of nonoperating item that is expected by investors to be a more permanent component of future profitability.

Non-GAAP Earnings

Many companies voluntarily provide *non-GAAP earnings*—management's assessment of permanent earnings.

Companies are required to report earnings based on Generally Accepted Accounting Principles (GAAP). This number includes *all revenues and expenses.* Most companies, however, also voluntarily provide **non-GAAP earnings** when they announce annual or quarterly earnings. Non-GAAP earnings *exclude certain expenses* and sometimes certain revenues. Common expenses excluded are restructuring costs, acquisition costs, write-downs of impaired assets, and stock-based compensation. Supposedly, non-GAAP earnings are management's view of "permanent earnings," in the sense of being a better long-run measure of its company's performance.

Nearly all major companies report non-GAAP earnings. For example, **Yelp** regularly reports non-GAAP earnings, defined as net income excluding stock-based compensation expense, amortization of intangibles, gains (losses) on the disposal of business units, restructuring and integration costs, and certain tax effects. For its year ended 2016, the effect of these exclusions was to increase reported performance from a GAAP loss of nearly $5 million to a non-GAAP profit of $60 million. **Microsoft** reported GAAP earnings in 2017 of $21.2 billion. At the same time, the company announced that its non-GAAP earnings for the year were $25.9 billion (22% higher than GAAP earnings). The company stated that the difference in earnings numbers related primarily to management adjusting for Windows 10 revenue deferrals.

Non-GAAP earnings are controversial.

Non-GAAP earnings are controversial because determining which expenses to exclude is at the discretion of management. By removing certain expenses from reported GAAP earnings, management has the potential to report misleadingly higher profits. In 2015, companies in the S&P 500 reported non-GAAP earnings that were 33% *higher* than GAAP earnings.[11] Some companies that reported non-GAAP profits actually had GAAP losses. The issue is: Do non-GAAP earnings represent management's true belief of core, long-term performance (so excluding certain temporary expenses is helpful to investors), or do non-GAAP earnings represent management's attempt to mislead investors into believing the company is more profitable than it actually is (and therefore harming investors)? Many are concerned that the latter is more likely.

The Sarbanes-Oxley Act requires reconciliation between non-GAAP earnings and earnings determined according to GAAP.

The Sarbanes-Oxley Act addressed non-GAAP earnings in its Section 401. One of the act's important provisions requires that if non-GAAP earnings are included in any periodic or other report filed with the SEC or in any public disclosure or press release, the company also must provide a reconciliation with earnings determined according to GAAP.[12]

We now turn our attention to discontinued operations, an item that is not part of a company's permanent earnings and, appropriately, is excluded from continuing operations.

[11] "S&P 500 Earnings: Far Worse Than Advertised," *The Wall Street Journal* (February 24, 2016).

[12] The Congress of the United States of America, *The Sarbanes-Oxley Act of 2002,* Section 401 (b) (2), Washington, D.C., 2004.

Discontinued Operations

LO4–4

Companies sometimes decide to sell or dispose of a component of their business. The operations of that business component are known as **discontinued operations**. For example, as we discussed in the introduction, **Abbott Laboratories** decided to sell its businesses related to developed markets branded generic pharmaceuticals and animal health. Obviously, profits from these discontinued operations *will not continue*. Because discontinued operations represent a material[13] component of the company, the results from discontinued operations are reported separately in the income statement to allow financial statement users to more clearly understand results from continuing operations.

For example, suppose a company has total pretax profits for the year of $1,200. This includes $1,000 from operations that are continuing, and $200 from operations that are discontinued by the end of the year. Assuming a 25% tax rate, the presentation of discontinued operations is mandated as follows:[14]

Income from continuing operations before income taxes	$1,000
Income tax expense (25%)	250
Income from continuing operations	**750**
Income from discontinued operations ($200 net of $50 tax expense)	**150**
Net income	$ 900

FINANCIAL Reporting Case

Q2, p. 167

The objective of this format is to inform financial statements users of which components of net income are continuing. We do this by separately reporting income from continuing operations (**$750**) and income from discontinued operations (**$150**). All else the same, investors should not expect next year's net income to be $900, because only $750 of profits from this year are part of continuing operations.

Income from discontinued operations (and its tax effect) are reported separately.

Separate reporting includes taxes as well. The income tax expense associated with continuing operations is reported separately from the income tax of discontinued operations. Also, in the case that there is a loss from discontinued operations, there would be an *income tax benefit* (instead of income tax expense); losses from discontinued operations are tax deductible and would reduce overall taxes owed, thereby providing a benefit. The process of associating income tax effects with the income statement components that create those effects is referred to as *intraperiod tax allocation*, something we discuss in depth in Chapter 16.

What are some examples of discontinued operations? In recent years, **General Electric** sold its financial services businesses. **Campbell Soup Company** sold its European simple meals business to **Soppa Investments**. **Google** sold its Motorola Mobility smartphone subsidiary to **Lenovo. Procter & Gamble** sold its batteries division. These are all examples of discontinued operations.

What Constitutes a Discontinued Operation?

Discontinued operations are reported when

1. A *component of an entity* or group of components has been sold or disposed of, or is considered held for sale,
2. If the disposal represents a *strategic shift* that has, or will have, a major effect on a company's operations and financial results.[15]

For the first item, a *component of an entity* includes activities and cash flows that can be clearly distinguished, operationally and for financial reporting purposes, from the rest of the

[13]We discussed the concept of materiality in Chapter 1.

[14]The presentation of discontinued operations is the same for single-step and multiple-step income statement formats. The single-step versus multiple-step distinction applies to items included in income from continuing operations.

[15]A discontinued operation is also defined as business or nonprofit activity that is considered held for sale *when acquired* (FASB ASC 205-20-15-2). A business is a set of activities and assets that is managed for purposes of providing economic benefits to the company. A nonprofit activity is similar to a business but is not intended to provide goods and services to customers at a profit.

company. A component could include an operating segment, a reporting unit, a subsidiary, or an asset group.[16]

For the second item, whether the disposal represents a *strategic shift* requires the judgment of company management. Examples of possible strategic shifts include the disposal of operations in a major geographical area, a major line of business, a major equity method investment,[17] or other major parts of the company.

As part of the continuing process to converge U.S. GAAP and international standards, the FASB and IASB have developed a common definition and a common set of disclosures for discontinued operations.[18]

Reporting Discontinued Operations

By definition, the income or loss stream from a discontinued operation no longer will continue. A financial statement user is more interested in the results of a company's operations that will continue. It is informative, then, for companies to separate the effects of the discontinued operations from the results of operations that will continue. For this reason, the revenues, expenses, gains, losses, and income tax related to a *discontinued* operation must be removed from *continuing* operations and reported separately *for all years presented.*

For example, even though Abbott Laboratories did not sell its generic pharmaceuticals and animal health businesses until 2015, it's important for comparative purposes to separate the effects for any prior years presented. This allows an apples-to-apples comparison of income from *continuing* operations. So, in its 2015 three-year comparative income statements, the 2014 and 2013 income statements reclassified income from generic pharmaceuticals and animal health businesses to income from discontinued operations. In addition, there was a disclosure note to inform readers that prior years were reclassified.[19] Furthermore, you can see in the introduction to this chapter how the results of those discontinued operations didn't continue into 2016. The amount reported in 2016 reflected the remaining income tax benefit of a prior discontinued operation in 2013 (**AbbVie**).

Sometimes a discontinued component actually has been sold by the end of a reporting period. Often, though, the disposal transaction has not yet been completed as of the end of the reporting period. We consider these two possibilities next.

WHEN THE COMPONENT HAS BEEN SOLD When the discontinued component is sold before the end of the reporting period, the reported income effects of a discontinued operation will include two elements.

1. Income or loss from operations (revenues, expenses, gains, and losses) of the component from the beginning of the reporting period *to the disposal date*
2. Gain or loss on disposal of the component's assets

The first element would consist primarily of income from daily operations of this discontinued component of the company. This would include typical revenues from sales to customers and ordinary expenses such as cost of goods sold, salaries, rent, and insurance. The second element includes gains and losses on the sale of assets, such as selling a building or office equipment of this discontinued component.

These two elements can be combined or reported separately, net of their tax effects. If combined, the gain or loss component must be indicated. In our illustrations to follow, we combine the income effects. Illustration 4–7 describes a situation in which the discontinued component is sold before the end of the reporting period.

Notice that an *income tax benefit* occurs because a *loss* reduces taxable income, saving the company $500,000 in taxes.

On the other hand, suppose Duluth's discontinued division had a pretax loss from operations of only $1,000,000 (and still had a gain on disposal of $3,000,000). In this case,

[16]FASB ASC 205-20-20 : Presentation of Financial Statements—Discontinued Operations—Glossary.

[17]Equity method investments are discussed in Chapter 12.

[18]"Noncurrent Assets Held for Sale and Discontinued Operations," *International Financial Reporting Standard No. 5* (IASCF), as amended effective January 1, 2016.

[19]The presentation of discontinued operations in comparative income statements enhances the qualitative characteristics of comparability and consistency. We discussed the concepts of comparability and consistency in Chapter 1.

Illustration 4–7

Discontinued Operations Sold—Loss

In October 2021, management of Duluth Holding Company decided to sell one of its divisions that qualifies as a separate component according to generally accepted accounting principles. On December 18, 2021, **the division was sold**. Consider the following facts related to the division:

1. From January 1 through disposal, the division had a pretax loss from operations of $5,000,000.
2. The assets of the division had a net selling price of $15,000,000 and book value of $12,000,000.

Duluth's income statement for 2021, beginning with after-tax income from continuing operations of $20,000,000, would be reported as follows (assuming a 25% tax rate):

Income from continuing operations		$20,000,000
Discontinued operations:		
Loss from operations of discontinued component (including gain on disposal of $3,000,000*)	**$(2,000,000)†**	
Income tax benefit	**500,000‡**	
Loss on discontinued operations		**(1,500,000)**
Net income		$18,500,000

*Net selling price of $15 million less book value of $12 million
†Loss from operations of $5 million less gain on disposal of $3 million
‡$2,000,000 × 25%

the combined amount of $2,000,000 represents *income* from operations of the discontinued component, and the company would have an additional *income tax expense* of $500,000, as demonstrated below.

Income from continuing operations		$20,000,000
Discontinued operations:		
Income from operations of discontinued component (including gain on disposal of $3,000,000*)	**$2,000,000†**	
Income tax expense	**(500,000)‡**	
Income on discontinued operations		1,500,000
Net income		$21,500,000

*Net selling price of $15 million less book value of $12 million
†Loss from operations of $1 million plus gain on disposal of $3 million
‡$2,000,000 × 25%

Additional Consideration

For reporting discontinued operations in the income statement, some companies separate the income/loss from operations and the gain/loss on disposal. For example, in Illustration 4–7, Duluth Holding Company could have reported the $5,000,000 loss from operations separately from the $3,000,000 gain on disposal, with each shown net of their tax effects (25%).

Income from continuing operations		$ 20,000,000
Discontinued operations:		
Loss from operations (net of tax benefit)	**$(3,750,000)†**	
Gain on disposal (net of tax expense)	**2,250,000‡**	
Loss on discontinued operations		**(1,500,000)**
Net income		$ 18,500,000

†$5,000,000 − ($5,000,000 × 25%)
‡$15,000,000 − $12,000,000 = $3,000,000; $3,000,000 − ($3,000,000 × 25%) = $2,250,000

(continued)

(concluded)

The amount of the loss on discontinued operations of **$1,500,000** is the same with either presentation. In practice, most companies report on the face of the income statement a single net amount for discontinued operations, with a note disclosure providing details of the calculation.

Income from continuing operations	$20,000,000
Loss on discontinued operations (net of tax)	**(1,500,000)**
Net income	$18,500,000

If a component to be discontinued has not yet been sold, its income effects, including any impairment loss, usually still are reported separately as discontinued operations.

WHEN THE COMPONENT IS CONSIDERED HELD FOR SALE What if a company has decided to discontinue a component but, when the reporting period ends, the component has not yet been sold? If the situation indicates that the component is likely to be sold within a year, the component is considered "held for sale."[20] In that case, the income effects of the discontinued operation still are reported, but the two components of the reported amount are modified as follows:

1. Income or loss from operations (revenues, expenses, gains and losses) of the component from the beginning of the reporting period *to the end of the reporting period.*
2. An impairment loss if the book value (sometimes called carrying value or carrying amount) of the assets of the component is more than fair value minus cost to sell.

The two income elements can be combined or reported separately, net of their tax effects. In addition, if the amounts are combined and there is an impairment loss, the loss must be disclosed, either parenthetically on the face of the statement or in a disclosure note. Consider the example in Illustration 4–8.

Illustration 4–8

Discontinued Operations Held for Sale—Impairment Loss

In October 2021, management of Duluth Holding Company decided to sell one of its divisions that qualifies as a separate component according to generally accepted accounting principles. On December 31, 2021, the end of the company's fiscal year, **the division had not yet been sold.** Consider the following facts related to the division:

1. For the year, the division reported a pretax loss from operations of $5,000,000.
2. On December 31, assets of the division had a book value of $12,000,000 and a fair value, minus anticipated cost to sell, of $9,000,000.

Duluth's income statement for 2021, beginning with after-tax income from continuing operations of $20,000,000, would be reported as follows (assuming a 25% tax rate):

Income from continuing operations		$20,000,000
Discontinued operations:		
Loss from operations of discontinued component (including impairment loss of $3,000,000*)	**$(8,000,000)'**	
Income tax benefit	2,000,000'	
Loss on discontinued operations		**(6,000,000)**
Net income		$14,000,000

*Book value of $12 million less fair value net of cost to sell of $9 million
'Loss from operations of $5 million plus impairment loss of $3 million
'$8,000,000 × 25%

Also, the net-of-tax income or loss from operations of the component being discontinued is reported separately from continuing operations for any prior year that is presented for comparison purposes along with the 2021 income statement. Then, in the year of actual

[20]Six criteria are used to determine whether the component is likely to be sold and therefore considered "held for sale." You can find these criteria in FASB ASC 360-10-45-9: Property, Plant, and Equipment—Overall—Other Presentation Matters—Long-Lived Assets Classified as Held for Sale (previously "Accounting for the Impairment or Disposal of Long-Lived Assets," *Statement of Financial Accounting Standards No. 144* (Norwalk, Conn.: FASB, 2001), par. 30).

disposal, the discontinued operations section of the income statement will include the final gain or loss on the sale of the discontinued segment's assets. The gain or loss is determined relative to the revised book values of the assets after the impairment write-down.

Important information about discontinued operations, whether sold or held for sale, is reported in a disclosure note. The note provides additional details about the discontinued component, including its identity, its major classes of assets and liabilities, the major revenues and expenses constituting pretax income or loss from operations, the reason for the discontinuance, and the expected manner of disposition if held for sale.[21]

In Illustration 4–8, if the fair value of the division's assets minus cost to sell exceeded the book value of $12,000,000, there is no impairment loss and the income effects of the discontinued operation would include only the loss from operations of $5,000,000, less the income tax benefit.

The balance sheet is affected, too. The assets and liabilities of the component considered held for sale are reported at the lower of their book value or fair value minus cost to sell. And, because it's not in use, an asset classified as held for sale is no longer reported as part of property, plant, and equipment or intangible assets and is not depreciated or amortized.[22]

For example, in 2016, **Abbott Laboratories** decided to sell its businesses related to vision care. The 2016 year-end balance sheet reported **$513** million in "Current assets held for disposition," and **$245** million in "Current liabilities held for disposition." Information about the discontinued operations was included in the disclosure note shown in Illustration 4–9.

Illustration 4–9

Discontinued Operations Disclosure—Abbott Laboratories.

Real World Financials

NOTE 3—DISCONTINUED OPERATIONS (in part)

The following is a summary of the assets and liabilities held for disposition

($ in millions)	December 31, 2016
Trade receivables, net	$ 222
Total inventories	240
Prepaid expenses and other current assets	51
Current assets held for disposition	**$513**
Trade accounts payable	$ 71
Salaries, wages, commissions and other accrued liabilities	174
Current liabilities held for disposition	**$245**

Source: Abbott Laboratories.

Notice that the assets and liabilities held for sale are classified as *current* because the company expects to complete the transfer of these assets and liabilities in the next fiscal year.

INTERIM REPORTING Remember that companies whose ownership shares are publicly traded in the United States must file quarterly reports with the Securities and Exchange Commission. If a component of an entity is considered held for sale at the end of a quarter, the income effects of the discontinued component must be separately reported in the quarterly income statement. These effects would include the income or loss from operations for the quarter as well as an impairment loss if the component's assets have a book value more than fair value minus cost to sell. If the assets are impaired and written down, any gain or loss on disposal in a subsequent quarter is determined relative to the new, written-down book value.

[21]For a complete list of disclosure requirements, see FASB ASC 205-20-50: Presentation of Financial Statements—Discontinued Operations—Disclosure.

[22]The assets and liabilities held for sale are not offset and presented as a single net amount, but instead are listed separately (ASC 205-20-45-10).

Accounting Changes

LO4–5

Accounting changes fall into one of three categories: (1) a change in an accounting principle, (2) a change in estimate, or (3) a change in reporting entity. The correction of an error is another adjustment that is accounted for in the same way as certain accounting changes. A brief overview of a change in accounting principle, a change in estimate, and correction of errors is provided here. We cover accounting changes, including changes in reporting entities, and accounting errors in detail in subsequent chapters, principally in Chapter 20.

Change in Accounting Principle

A change in accounting principle refers to a change from one acceptable accounting method to another. There are many situations that allow alternative treatments for similar transactions. Common examples of these situations include the choice among FIFO, LIFO, and average cost for the measurement of inventory and among alternative revenue recognition methods. New accounting standard updates issued by the FASB also may require companies to change their accounting methods.

MANDATED CHANGES IN ACCOUNTING PRINCIPLES Sometimes the FASB requires a change in accounting principle. These changes in accounting principles potentially hamper the ability of external users to compare financial information among reporting periods because information lacks consistency. The board considers factors such as this, as well as the cost and complexity of adopting new standards, and chooses among various approaches to require implementation by companies.

1. **Retrospective approach.** The new standard is applied to all periods presented in the financial statements. That is, we restate prior period financial statements as if the new accounting method had been used in those prior periods. We revise the balance of each account affected to make those statements appear as if the newly adopted accounting method had been applied all along.
2. **Modified retrospective approach.** The new standard is applied to the adoption period only. Prior period financial statements are not restated. The cumulative effect of the change on prior periods' net income is shown as an adjustment to the beginning balance of retained earnings in the adoption period.
3. **Prospective approach.** This approach requires neither a modification of prior period financial statements nor an adjustment to account balances. Instead, the change is simply implemented in the current period and all future periods.

VOLUNTARY CHANGES IN ACCOUNTING PRINCIPLES Occasionally, without being required by the FASB, a company will change from one generally accepted accounting principle to another. For example, a company may decide to change its inventory method from LIFO to FIFO. When this occurs, inventory and cost of goods sold are measured in one reporting period using LIFO, but then are measured using FIFO in a subsequent period. Inventory and cost of goods sold, and hence net income, for the two periods are not comparable. To improve comparability and consistency, GAAP typically requires that voluntary accounting changes be accounted for retrospectively.[23,24]

We will see these aspects of accounting for the change in accounting principle demonstrated in Chapter 9 in the context of our discussion of inventory methods. We'll also discuss changes in accounting principles in depth in Chapter 20.

[23]FASB ASC 250-10-45-5: Accounting Changes and Error Corrections—Overall—Other Presentation Matters (previously "Accounting Changes and Error Corrections—a replacement of APB Opinion No. 20 and FASB Statement No. 3," *Statement of Financial Accounting Standard No. 154* (Norwalk, Conn.: FASB, 2005)).

[24]Sometimes a lack of information makes it impracticable to report a change retrospectively so the new method is simply applied prospectively, that is, we simply use the new method from now on. Also, if a new standard specifically requires prospective accounting, that requirement is followed.

Change in Depreciation, Amortization, or Depletion Method

A change in depreciation, amortization, or depletion method is considered to be a change in accounting estimate that is achieved by a change in accounting principle. We account for this change prospectively, almost exactly as we would any other change in estimate. One difference is that most changes in estimate don't require a company to justify the change. However, this change in estimate is a result of changing an accounting principle and therefore requires a clear justification as to why the new method is preferable. Chapter 11 provides an illustration of a change in depreciation method.

Changes in depreciation, amortization, or depletion methods are accounted for the same way as a change in an accounting estimate.

Change in Accounting Estimate

Estimates are a necessary aspect of accounting. A few of the more common accounting estimates are the amount of future bad debts on existing accounts receivable, the useful life and residual value of a depreciable asset, and future warranty expenses.

A *change in accounting estimate* is reflected in the financial statements of the current period and future periods.

Because estimates require the prediction of future events, it's not unusual for them to turn out to be wrong. When an estimate is modified as new information comes to light, accounting for the change in estimate is quite straightforward. We do not revise prior years' financial statements to reflect the new estimate. Instead, we merely incorporate the new estimate in any related accounting determinations from that point on; that is, we account for a change in accounting estimate prospectively.[25] If the effect of the change is material, a disclosure note is needed to describe the change and its effect on both net income and earnings per share. Chapters 11 and 20 provide illustrations of changes in accounting estimates.

Correction of Accounting Errors

Errors occur when transactions are either recorded incorrectly or not recorded at all. We briefly discuss the correction of errors here as an overview and in later chapters in the context of the effect of errors on specific chapter topics. In addition, Chapter 20 provides comprehensive coverage of the correction of errors.

Accountants employ various control mechanisms to ensure that transactions are accounted for correctly. In spite of this, errors occur. When errors do occur, they can affect any one or several of the financial statement elements on any of the financial statements a company prepares. In fact, many kinds of errors simultaneously affect more than one financial statement. When errors are discovered, they should be corrected

Most errors are discovered in the same year that they are made. These errors are simple to correct. The original erroneous journal entry is reversed and the appropriate entry is recorded. If an error is discovered in a year subsequent to the year the error is made, the accounting treatment depends on whether or not the error is material with respect to its effect on the financial statements. In practice, the vast majority of errors are not material and are, therefore, simply corrected in the year discovered. However, material errors that are discovered in subsequent periods require a prior period adjustment.

Prior Period Adjustments

Assume that after its financial statements are published and distributed to shareholders, Roush Distribution Company discovers a material error in the statements. What does it do? Roush must make a **prior period adjustment**.[26] Roush would record a journal entry that adjusts any balance sheet accounts to their appropriate levels and would account for the income effects of the error by increasing or decreasing the beginning retained earnings balance in a statement of shareholders' equity. Remember, net income in prior periods was closed to retained earnings so, by adjusting retained earnings, the prior period adjustment accounts for the error's effect on prior periods' net income.

[25] If the original estimate had been based on erroneous information or calculations or had not been made in good faith, the revision of that estimate would constitute the correction of an error (discussed in the next section).

[26] FASB ASC 250-10-45-23: Accounting Changes and Error Corrections-Overall-Other Presentation Matters (previously "Prior Period Adjustments," *Statement of Financial Accounting Standards No. 16* (Norwalk, Conn.: FASB, 1977)).

Simply reporting a corrected retained earnings amount might cause a misunderstanding for someone familiar with the previously reported amount. Explicitly reporting a prior period adjustment in the statement of shareholders' equity (or statement of retained earnings if that's presented instead) highlights the adjustment and avoids this confusion.

In addition to reporting the prior period adjustment to retained earnings, previous years' financial statements that are incorrect as a result of the error are retrospectively restated to reflect the correction. Also, a disclosure note communicates the impact of the error on prior periods' net income.

Earnings per Share

We've discussed that the income statement reports a company's net income for the period. Net income is reported in total dollars (total dollars of revenues minus total dollars of expenses) and represents the total profits that the company has generated for *all shareholders* during the period. However, for individual decision making, investors want to know how much profit has been generated for *each shareholder.* To know this, we calculate **earnings per share (EPS)** to relate the amount of net income a company generates to the number of common shares outstanding.

EPS provides a convenient way for investors to link the company's profitability to the value of an individual share of ownership. The ratio of stock price per share to earnings per share (the PE ratio) is one of the most widely used financial metrics in the investment world. EPS also makes it easier to compare the performance of the company over time or with other companies. Larger companies may naturally have larger dollar amounts of net income, but they do not always generate more profit for each shareholder.

All corporations whose common stock is publicly traded must disclose EPS.

FINANCIAL Reporting Case

Q3, p. 167

U.S. GAAP requires that public companies report two specific calculations of EPS: (1) basic EPS and (2) diluted EPS. **Basic EPS** equals total net income (less any dividends to preferred shareholders) divided by the weighted-average number of common shares outstanding. Dividends to preferred shareholders are subtracted from net income in the numerator because those dividends are distributions of the company not available to common shareholders. The denominator is the weighted-average number of common shares outstanding, rather than the number of shares outstanding at the beginning or end of the period, because the goal is to relate performance for the period to the shares that were in place throughout that period. The number of common shares may change over the year from additional issuances or company buybacks, so a weighted average better reflects the number of shares outstanding for the period. The resulting EPS provides a measure of net income generated for each share of common stock during the period.

For example, suppose the Fetzer Corporation reported net income of $600,000 for its fiscal year ended December 31, 2021. Preferred stock dividends of $75,000 were declared during the year. Fetzer had 1,000,000 shares of common stock outstanding at the beginning of the year and issued an additional 1,000,000 shares on March 31, 2021. Basic EPS of $0.30 per share for 2021 is computed as follows:

$$\frac{\$600{,}000 - \$75{,}000}{\underset{\text{Shares at Jan. 1}}{1{,}000{,}000} + \underset{\text{New shares}}{1{,}000{,}000\ (9/12)}} = \frac{\$525{,}000}{1{,}750{,}000} = \$0.30$$

Diluted EPS incorporates the dilutive effect of all *potential* common shares in the calculation of EPS. Dilution refers to the reduction in EPS that occurs as the number of common shares outstanding increases. Companies may have certain securities outstanding that could be converted into common shares, or they could have stock options outstanding that create additional common shares if the options were exercised. Because these items could cause the number of shares in the denominator to increase, they potentially decrease EPS. We devote a substantial portion of Chapter 19 to understanding these two measures of EPS. Here, we provide only an overview.

When the income statement includes discontinued operations, we report per-share amounts for both income (loss) from continuing operations and for net income (loss), as well as for the discontinued operations. We see this demonstrated for **Abbott Laboratories** in Illustration 4–10.

Illustration 4–10

EPS Disclosures—Abbott Laboratories.

Real World Financials

Abbott Laboratories
Statements of Earnings For the Year Ended December 31 (in part)

($ in millions, except per share amounts)	2016	2015
Income from continuing operations	$1,063	$2,606
Income from discontinued operations, net of tax	337	1,817
Net income	$1,400	$4,423
Earnings per common share—basic:		
Continuing operations	$ 0.71	$ 1.73
Discontinued operations	0.23	1.21
Net income	$ 0.94	$ 2.94
Earnings per common share—diluted:		
Continuing operations	$ 0.71	$ 1.72
Discontinued operations	0.23	1.20
Net income	$ 0.94	$ 2.92

Source: Abbott Laboratories.

Comprehensive Income

● LO4–6

Changes in shareholders' equity arise from two sources—transactions with owners (shareholders) and transactions with nonowners. Transactions with owners include events such as increasing equity by issuing stock to shareholders or decreasing equity by purchasing stock from shareholders and paying dividends. All other changes in equity during the period arise from nonowner sources and represent **comprehensive income**.

Comprehensive income is the total change in equity for a reporting period other than from transactions with owners.

The transactions and events that lead to changes in equity from nonowner sources are recorded as revenues, expenses, gains, and losses for the period. You are already familiar with many of these items; they are reported in the income statement and are used to calculate net income. However, there are a few gains and losses that are not reported in the income statement but nevertheless affect equity. These other gains and losses are reported as **other comprehensive income (OCI)**. Therefore, the broader concept of comprehensive income includes net income plus *other* changes in shareholders' equity from nonowner sources:

Comprehensive income = Net income + Other comprehensive income

Accounting professionals have engaged in an ongoing debate concerning whether certain gains and losses should be included as components of net income or as part of other comprehensive income. For example, in Chapter 12 you will learn that investments in certain securities are reported in the balance sheet at their fair values. The gains and losses from adjusting investments to their fair values are considered to be unrealized because the securities have not been sold. For many debt and equity securities, these unrealized gains and losses are included in net income. However, for certain investments in debt securities (known as *available-for-sale debt securities*), unrealized gains and losses are reported as other comprehensive income in shareholders' equity.

Currently, the FASB has established *no conceptual basis* for determining which gains and losses qualify for net income versus other comprehensive income. To help avoid confusion about which amounts are included in net income versus other comprehensive income, companies are required to provide a reconciliation from net income to comprehensive income.[27] The reconciliation simply extends net income from the income statement to include other comprehensive income items, reported net of tax, as shown in Illustration 4–11.

Reporting comprehensive income can be accomplished with a single, continuous statement or in two separate, but consecutive statements.

Flexibility in Reporting

The information in the income statement and other comprehensive income items shown in Illustration 4–11 can be presented either (1) in a single, continuous statement of

[27] FASB ASC 220-10-45-1A and 1B: Comprehensive Income-Overall-Other Presentation Matters (previously "Reporting Comprehensive Income," *Statement of Financial Accounting Standards No. 130* (Norwalk, Conn.: FASB, 1997)).

Illustration 4–11
Comprehensive Income

Net income		$xxx
Other comprehensive income, net of tax:		
Gain (loss) on debt securities*	$ x	
Gain (loss) on projected benefit obligation†	(x)	
Gain (loss) on derivatives‡	(x)	
Foreign currency translation adjustment§	x	
Total other comprehensive income (loss)		xx
Comprehensive income		$xxx

*Changes in the fair value of investments in debt securities (described in Chapter 12)
†Gain or loss due to revising assumptions of the employee pension plan (described in Chapter 17).
‡When a derivative designated as a cash flow hedge is adjusted to fair value, the gain or loss is deferred as a component of comprehensive income and included in earnings later, at the same time as earnings are affected by the hedged transaction (described in the Derivatives Appendix to the text).
§The amount could be an addition to or reduction in shareholders' equity. (This item is discussed elsewhere in your accounting curriculum.)

comprehensive income or (2) in two separate, but consecutive statements, an income statement and a statement of comprehensive income. Each component of other comprehensive income can be displayed net of tax, as in Illustration 4–11, or alternatively, before tax with one amount shown for the aggregate income tax expense (or benefit).[28]

AstroNova, Inc., a manufacturer of a broad range of specialty technology products, chose to report a separate statement of comprehensive income, as shown in Illustration 4–12.

Illustration 4–12
Comprehensive Income Presented as a Separate Statement—AstroNova, Inc.

Real World Financials

ASTRONOVA, INC.
Consolidated Statements of Comprehensive Income
For the Years Ended January 31

($ in thousands)	**2017**	**2016**
Net income	**$4,228**	$4,525
Other comprehensive income, net of taxes:		
Foreign currency translation adjustments	(65)	(269)
Unrealized gain (loss) on securities available for sale	(16)	(7)
Other comprehensive income	**(81)**	(276)
Comprehensive income	$ 4,147	$4,249

Source: AstroNova, Inc.

International Financial Reporting Standards

Comprehensive Income. Both U.S. GAAP and IFRS allow companies to report comprehensive income in either a single statement of comprehensive income or in two separate statements.

● LO4–9

Other comprehensive income items are similar under the two sets of standards. However, an additional OCI item, *changes in revaluation surplus,* is possible under IFRS. In Chapter 11 you will learn that *IAS No. 16*[29] permits companies to value property, plant, and equipment at (1) cost less accumulated depreciation or (2) fair value (revaluation). *IAS No. 38*[30] provides a similar option for the valuation of intangible assets. U.S. GAAP prohibits revaluation.

If the revaluation option is chosen and fair value is higher than book value, the difference, changes in revaluation surplus, is reported as *other comprehensive income* and then accumulates in a revaluation surplus account in equity.

[28]GAAP does not require the reporting of comprehensive earnings per share.
[29]"Property, Plant and Equipment," *International Accounting Standard No. 16* (IASCF), as amended effective January 1, 2018.
[30]"Intangible Assets," *International Accounting Standard No. 38* (IASCF), as amended effective January 1, 2018.

Accumulated Other Comprehensive Income

In addition to reporting OCI that occurs in the current reporting period, we must also report these amounts on a cumulative basis in the balance sheet. This is consistent with the way we report net income for the period in the income statement and also report accumulated net income (that hasn't been distributed as dividends) in the balance sheet as retained earnings. Similarly, we report OCI for the period in the statement of comprehensive income and also report **accumulated other comprehensive income (AOCI)** in the balance sheet as a component of shareholders' equity. This is demonstrated in Illustration 4–13 for AstroNova, Inc.

The cumulative total of OCI (or comprehensive loss) is reported as accumulated other comprehensive income (AOCI), an additional component of shareholders' equity that is displayed separately.

Illustration 4–13

Shareholders' Equity—AstroNova, Inc.

Real World Financials

ASTRONOVA, INC.
Consolidated Balance Sheets (in part)
As of January 31

($ in thousands)	2017	2016
Shareholders' equity:		
Common stock	492	483
Additional paid-in capital	47,524	45,675
Retained earnings	44,358	42,212
Treasury stock, at cost	(20,781)	(20,022)
Accumulated other comprehensive income	**(1,056)**	**(975)**
Total shareholders' equity	$70,537	$67,373

Supplementing information in Illustration 4–13 with numbers reported in Illustration 4–12 along with dividends declared by AstroNova, we can reconcile the changes in both retained earnings and AOCI:

($ in thousands)	Retained Earnings	Accumulated Other Comprehensive Income
Balance, 1/31/2016	$42,212	$ (975)
Add: Net income	4,228	
Deduct: Dividends	(2,082)	
Other comprehensive income		(81)
Balance, 1/31/2017	$44,358	$(1,056)

To further understand the relationship between net income and other comprehensive income, consider the following example. Philips Corporation began the year with retained earnings of $700 million and accumulated other comprehensive income of $30 million. Let's also assume that net income for the year, before considering the gain discussed below, is $100 million, of which $40 million was distributed to shareholders as dividends. Now assume that Philips had a $10 million net-of-tax gain that was also reported in one of two ways:

1. As a gain in net income, or
2. As a gain in other comprehensive income.

Under the first alternative, the gain will be included in shareholders' equity through retained earnings.

($ in millions)	Retained Earnings	Accumulated Other Comprehensive Income
Beginning Balance	$700	$30
Net income ($100 + 10)	110	
Dividends	(40)	
Other comprehensive income		–0–
Ending Balance	$770	$30

$800

Under the second alternative, the net-of-tax gain of $10 million is reported as a component of *other comprehensive income (loss)*. The gain will be included in shareholders' equity through *accumulated other comprehensive income (loss)*, rather than retained earnings, as demonstrated below. The total of retained earnings and accumulated other comprehensive income is $800 million either way.

($ in millions)	Retained Earnings		Accumulated Other Comprehensive Income
Beginning Balance	$700		$30
Net income	100		
Dividends	(40)		
Other comprehensive income			10
Ending Balance	$760		$40
		$800	

Net income and comprehensive income are identical for an enterprise that has no other comprehensive income items. When this occurs for all years presented, a statement of comprehensive income is not required. Components of other comprehensive income are described in subsequent chapters.

Concept Review Exercise

INCOME STATEMENT PRESENTATION; COMPREHENSIVE INCOME

The Barrington Company sells clothing and furniture. On September 30, 2021, the company decided to sell the entire furniture business for $40 million. The sale was completed on December 15, 2021. Income statement information for 2021 is provided below for the two components of the company.

	($ in millions)	
	Clothing Component	Furniture Component
Sales revenue	$650	$425
Cost of goods sold	200	225
Gross profit	450	200
Operating expenses	226	210
Operating income	224	(10)
Other income (loss)*	16	(30)
Income (loss) before income taxes	240	(40)
Income tax expense (benefit)†	60	(10)
Net income (loss)	$180	$ (30)

*For the furniture component, the entire Other income (loss) amount represents the loss on sale of assets of the component for $40 million when their book value was $70 million.
†A 25% tax rate applies to all items of income or loss.

In addition, in 2021 the company had a pretax net unrealized gain on debt securities of $6 million and a positive foreign currency translation adjustment of $2 million.

Required:

1. Prepare a single, continuous 2021 statement of comprehensive income for the Barrington Company including EPS disclosures. There were 100 million shares of common stock outstanding throughout 2021. The company had no potentially dilutive securities outstanding or stock options that could cause additional common shares. Use the multiple-step approach for the income statement portion of the statement.

2. Prepare a separate 2021 statement of comprehensive income.

Solution:

1. Prepare a single, continuous 2021 statement of comprehensive income.

BARRINGTON COMPANY
Statement of Comprehensive Income
For the Year Ended December 31, 2021
($ in millions, except per share amounts)

Sales revenue		$ 650
Cost of goods sold		200
Gross profit		450
Operating expenses		226
Operating income		224
Other income		16
Income from continuing operations before income taxes		240
Income tax expense		60
Income from continuing operations		180
Discontinued operations:		
Loss from operations of discontinued furniture component (including loss on disposal of $30)	$(40)	
Income tax benefit	10	
Loss on discontinued operations		(30)
Net income		150
Other comprehensive income, net of tax:		
Gain on debt securities	4.5	
Foreign currency translation adjustment	1.5	
Total other comprehensive income		6
Comprehensive income		$ 156
Earnings per share:		
Income from continuing operations		$1.80
Discontinued operations		(0.30)
Net income		$1.50

2. Prepare a separate 2021 statement of comprehensive income.

BARRINGTON COMPANY
Statement of Comprehensive Income
For the Year Ended December 31, 2021
($ in millions)

Net income		$150
Other comprehensive income, net of tax:		
Gain on debt securities	$4.5	
Foreign currency translation adjustment	1.5	
Total other comprehensive income		6
Comprehensive income		$156

Now that we have discussed the presentation and content of the income statement, we turn our attention to the statement of cash flows.

The Statement of Cash Flows

PART B

When a balance sheet and an income statement are presented, a **statement of cash flows (SCF)** is required for each income statement period.[31] The purpose of the SCF is to provide information about the cash receipts and cash disbursements of an enterprise. Similar to the income statement, it is a *change* statement, summarizing the transactions that affected cash during the period. The term *cash* in the statement of cash flows refers to the total of cash,

● LO4–7

A statement of cash flows is presented for each period for which an income statement is provided.

[31]FASB ASC 230-10-45: Statement of Cash Flows-Overall-Other Presentation Matters (previously "Statement of Cash Flows," *Statement of Financial Accounting Standards No. 95* (Norwalk, Conn.: FASB, 1987)).

cash equivalents, and restricted cash. Cash equivalents, discussed in Chapter 3, include highly liquid (easily converted to cash) investments such as Treasury bills. Chapter 21 is devoted exclusively to the SCF. A brief overview is provided here.

Usefulness of the Statement of Cash Flows

We discussed the difference between cash and accrual accounting in Chapter 1. It was pointed out and illustrated that over short periods of time, operating cash flows may not be indicative of the company's long-run cash-generating ability, and that accrual-based net income provides a more accurate prediction of future operating cash flows. Nevertheless, information about cash flows from operating activities, when combined with information about cash flows from other activities, can provide information helpful in assessing future profitability, liquidity, and long-term solvency. After all, a company must pay its debts with cash, not with income.

Of particular importance is the amount of cash generated from operating activities. In the long run, a company must be able to generate positive cash flow from activities related to selling its product or service. These activities must provide the necessary cash to pay debts, provide dividends to shareholders, and provide for future growth.

Classifying Cash Flows

● LO4–8

A list of cash flows is more meaningful to investors and creditors if they can determine the type of transaction that gave rise to each cash flow. Toward this end, the statement of cash flows classifies all transactions affecting cash into one of three categories: (1) operating activities, (2) investing activities, and (3) financing activities.

Operating Activities

Operating activities **are inflows and outflows of cash related to the transactions entering into the determination of net operating income.**

The inflows and outflows of cash that result from activities reported in the income statement are classified as cash flows from **operating activities**. In other words, this classification of cash flows includes the elements of net income reported on a cash basis rather than an accrual basis.[32]

Cash inflows include cash received from the following:

1. Customers from the sale of goods or services.
2. Interest and dividends from investments.

These amounts may differ from sales revenue and investment income reported in the income statement. For example, sales revenue measured on the accrual basis reflects revenue recognized during the period, not necessarily the cash actually collected. Revenue will not equal cash collected from customers if receivables from customers or deferred revenue changed during the period.

Cash outflows include cash paid for the following:

1. The purchase of inventory.
2. Salaries, wages, and other operating expenses.
3. Interest on debt.
4. Income taxes.

Likewise, these amounts may differ from the corresponding accrual expenses reported in the income statement. Expenses are reported when incurred, not necessarily when cash is actually paid for those expenses. Also, some revenues and expenses, like depreciation expense, don't affect cash at all and aren't included as cash outflows from operating activities.

The difference between the inflows and outflows is called *net cash flows from operating activities.* This is equivalent to net income if the income statement had been prepared on a cash basis rather than an accrual basis.

[32]Cash flows related to gains and losses from the sale of assets shown in the income statement are reported as investing activities in the SCF.

DIRECT AND INDIRECT METHODS OF REPORTING Two generally accepted formats can be used to report operating activities, the direct method and the indirect method. Under the **direct method**, the cash effect of each operating activity is reported directly in the statement. For example, *cash received from customers* is reported as the cash effect of sales activities. Income statement transactions that have no cash flow effect, such as depreciation, are simply not reported.

By the *direct method*, the cash effect of each operating activity is reported directly in the SCF.

By the **indirect method**, on the other hand, we arrive at net cash flow from operating activities indirectly by starting with reported net income and working backwards to convert that amount to a cash basis. Two types of adjustments to net income are needed.

By the *indirect method*, cash flow from operating activities is derived indirectly by starting with reported net income and adding or subtracting items to convert that amount to a cash basis.

1. Components of net income that do not affect operating cash are reversed. That means that noncash revenues and gains are subtracted, while noncash expenses and losses are added. For example, depreciation expense does not reduce cash but it is subtracted in the income statement. To reverse this, then, we add back depreciation expense to net income to arrive at the amount that we would have had if depreciation had not been subtracted in the first place.
2. Net income is adjusted for changes in operating assets and liabilities during the period. These changes represent amounts that are included as components of net income but that are not cash flows. For instance, suppose accounts receivable increase during the period because revenue from credit sales is greater than cash collected from customers. This increase in accounts receivable would then be subtracted from net income to arrive at *cash flow from operating activities.* In the indirect method, positive adjustments to net income are made for decreases in related assets and increases in related liabilities, while negative adjustments are made for increases in those assets and decreases in those liabilities.

To contrast the direct and indirect methods further, consider the income statement and balance sheet for Arlington Lawn Care (ALC) in Illustration 4–14. We'll use these to construct the operating activities section of the statement of cash flows.

DIRECT METHOD Let's begin with the direct method of presentation. We illustrated this method previously in Chapter 2. In that chapter, specific cash transactions were provided and we simply included them in the appropriate cash flow category in the SCF. Here, we start with account balances, so the direct method requires a bit more reasoning.

From the income statement, we see that ALC's net income has four components. Three of those—service revenue, general and administrative expense, and income tax expense—affect cash flows, but not by the accrual amounts reported in the income statement. One component—depreciation—reduces net income but not cash; it's simply an allocation over time of a prior year's expenditure for a depreciable asset. So, to report these operating activities on a cash basis, rather than an accrual basis, we take the three items that affect cash and adjust the amounts to reflect cash inflow rather than revenue recognized, and cash outflows rather than expenses incurred. Let's start with service revenue.

Service revenue is $100,000, but ALC did not collect that much cash from its customers. We know that because accounts receivable increased from $0 to $12,000. ALC must have collected to date only **$88,000**.

Accounts receivable

	Debit	Credit	
Beg. bal.	0		
Revenue	100		
		88	Cash
End. bal.	12		

Similarly, general and administrative expense of $32,000 was incurred, but $7,000 of that hasn't yet been paid. We know that because accrued liabilities increased by $7,000. Also, prepaid insurance increased by $4,000 so ALC must have paid $4,000 more cash for insurance coverage than the amount that expired and was reported as insurance expense. That means cash paid thus far for general and administrative expense was only $29,000 ($32,000 less the $7,000 increase in accrued liabilities plus the $4,000 increase in prepaid insurance). The other expense, income tax, was $15,000, but the increase in income taxes payable indicates that $5,000 hasn't yet been paid. This means the amount paid must have been only $10,000.

We can report ALC's cash flows from operating activities using the direct method as shown in Illustration 4–14A.

INDIRECT METHOD To report operating cash flows using the indirect method, we take a different approach. We start with ALC's net income but realize that the $45,000 includes both cash and noncash components. We need to adjust net income, then, to eliminate the

Illustration 4–14
Contrasting the Direct and Indirect Methods of Presenting Cash Flows from Operating Activities

Net income is $45,000, but cash flow from these same activities is not necessarily the same

Arlington Lawn Care (ALC) began operations at the beginning of 2021. ALC's 2021 income statement and its year-end balance sheet are shown below ($ in thousands).

ARLINGTON LAWN CARE
Income Statement
For the Year Ended December 31, 2021

Service revenue		$100
Operating expenses:		
General and administrative expense	$32*	
Depreciation expense	8	
Total operating expenses		40
Income before income taxes		60
Income tax expense		15
Net income		$ 45

*Includes $6 in insurance expense

Changes in assets and liabilities can indicate that cash inflows are different from revenues and cash outflows are different from expenses.

ARLINGTON LAWN CARE
Balance Sheet
At December 31, 2021

Assets		**Liabilities and Shareholders' Equity**	
Current assets:		Current liabilities:	
Cash	$ 54	Accrued liabilities**	$ 7
Accounts receivable	12	Income taxes payable	5
Prepaid insurance	4	Total current liabilities	12
Total current assets	70	Shareholders' equity:	
Equipment	40	Common stock	50
Less: Accumulated depreciation	(8)	Retained earnings	40†
Total assets	$102	Total liabilities and shareholders' equity	$102

**For various items of general and administrative expense
†Net income of $45 less $5 in cash dividends paid

Illustration 4–14A
Direct Method of Presenting Cash Flows from Operating Activities

By the direct method, we report the components of net income on a cash basis.

ARLINGTON LAWN CARE
Statement of Cash Flows
For the Year Ended December 31, 2021

	($ in thousands)
Cash Flows from Operating Activities	
Cash received from customers*	$ 88
Cash paid for general and administrative expense**	(29)
Cash paid for income taxes***	(10)
Net cash flows from operating activities	$ 49

*Service revenue of $100 thousand, less increase of $12 thousand in accounts receivable.
**General and administrative expense of $32 thousand, less increase of $7 thousand in accrued liabilities, plus increase of $4 thousand in prepaid insurance.
***Income tax expense of $15 thousand, less increase of $5 thousand in income taxes payable.

Depreciation expense does not reduce cash, but is subtracted in the income statement. So, we add back depreciation expense to net income to eliminate it.

noncash effects so that we're left with only the cash flows. We start by eliminating the only noncash component of net income in our illustration—depreciation expense. As shown in Illustration 4-14B, depreciation of $8,000 was subtracted in the income statement, so to eliminate its negative effect on net income, we simply add it back.

That leaves us with three components of net income that do affect cash but not necessarily by the amounts reported—service revenue, general and administrative expense, and income tax expense. For those, we need to make adjustments to net income to cause it to reflect cash flows rather than accrual amounts. For instance, we saw earlier that even though

$100,000 in service revenue is reflected in net income, only $88,000 cash was received from customers. That means we need to include an adjustment to reduce net income by $12,000, the increase in accounts receivable.

In a similar manner, we include adjustments for the changes in prepaid insurance, accrued liabilities, and income taxes payable to adjust net income to reflect cash payments rather than expenses incurred. For prepaid insurance, because insurance expense in the income statement was less than cash paid for insurance, we need to subtract the difference, which equals the increase in prepaid insurance. If this asset had decreased, we would have added, rather than subtracted, the change. For accrued liabilities and income taxes payable, because the related expense in the income statement was more than cash paid for those expenses, we need to add back the differences. If these liabilities had decreased, we would have subtracted, rather than added, the changes.

We make adjustments for changes in assets and liabilities that indicate that components of net income are not the same as cash flows.

Cash flows from operating activities using the indirect method are shown in Illustration 4–14B.

Illustration 4–14B
Indirect Method of Presenting Cash Flows from Operating Activities

ARLINGTON LAWN CARE
Statement of Cash Flows
For the Year Ended December 31, 2021

		($ in thousands)
Cash Flows from Operating Activities		
Net income		$45
Adjustments for noncash effects:		
Depreciation expense	$ 8	
Changes in operating assets and liabilities:		
Increase in accounts receivable	(12)	
Increase in prepaid insurance	(4)	
Increase in accrued liabilities	7	
Increase in income taxes payable	5	4
Net cash flows from operating activities		$49

By the indirect method, we start with net income and work backwards to convert that amount to a cash basis.

Both the direct and the indirect methods produce the same net cash flows from operating activities ($49,000 in our illustration); they are merely alternative approaches to reporting the cash flows. The FASB, in promulgating GAAP for the statement of cash flows, stated its preference for the direct method. However, nearly all U.S. companies use the indirect method.

The choice of presentation method for cash flow from operating activities has no effect on how investing activities and financing activities are reported. We now look at how cash flows are classified into those two categories.

Investing Activities

Cash flows from **investing activities** include inflows and outflows of cash related to the acquisition and disposition of long-lived assets used in the operations of the business (such as property, plant, and equipment) and investment assets (except those classified as cash equivalents and trading securities). The purchase and sale of inventory are not considered investing activities. Inventory is purchased for the purpose of being sold as part of the company's operations, so the purchase and sale are included with operating activities rather than investing activities.

Investing activities involve the acquisition and sale of (1) long-term assets used in the business and (2) nonoperating investment assets.

Cash outflows from investing activities include cash paid for the following:

1. The purchase of long-lived assets used in the business.
2. The purchase of investment securities like stocks and bonds of other entities (other than those classified as cash equivalents and trading securities).
3. Loans to other entities.

Later, when the assets are disposed of, cash inflow from the sale of the assets (or collection of loans and notes) also is reported as cash flows from investing activities.

As a result, cash inflows from these transactions are considered investing activities as follows:

1. The sale of long-lived assets used in the business.
2. The sale of investment securities (other than cash equivalents and trading securities).
3. The collection of a nontrade receivable (excluding the collection of interest, which is an operating activity).

Net cash flows from investing activities equal the difference between the inflows and outflows. The only investing activity indicated in Illustration 4–14 is ALC's investment of $40,000 cash for equipment. We know $40,000 was paid to buy equipment because that balance sheet account increased from $0 to $40,000.

Financing Activities

Financing activities involve cash inflows and outflows from transactions with creditors (excluding trade payables) and owners.

Financing activities relate to the external financing of the company. Cash inflows occur when cash is borrowed from creditors or invested by owners. Cash outflows occur when cash is paid back to creditors or distributed to owners. The payment of interest to a creditor, however, is classified as an operating activity.

Financing cash inflows include cash received from the following:

1. Owners when shares are sold to them.
2. Creditors when cash is borrowed through notes, loans, mortgages, and bonds.

Financing cash outflows include cash paid to the following:

1. Owners in the form of dividends or other distributions.
2. Owners for the reacquisition of shares previously sold.
3. Creditors as repayment of the principal amounts of debt (excluding trade payables that relate to operating activities).

Net cash flows from financing activities equal the difference between the inflows and outflows. The only financing activities indicated in Illustration 4–14 are ALC's receipt of $50,000 cash from issuing common stock and the payment of $5,000 in cash dividends. Because common stock increased from $0 to $50,000, we know that common stock was issued for that amount. For dividends, we know the balance of retained earnings increases by the amount of net income minus dividends. This means that the increase in retained earnings of $40,000 represents net income of $45,000 minus dividends of $5,000.

The 2021 statement of cash flows for ALC, beginning with net cash flows from operating activities, is shown in Illustration 4–15.

Illustration 4–15

Statement of Cash Flows (beginning with net cash flows from operating activities)

ARLINGTON LAWN CARE
Statement of Cash Flows (in part)
For the Year Ended December 31, 2021

		($ in thousands)
Net cash flows from operating activities (from Illustration 4–14A or 4–14B)		$49
Cash flows from investing activities:		
Purchase of equipment		(40)
Cash flows from financing activities:		
Issuance of common stock	$50	
Dividends paid to shareholders	(5)	
Net cash flows from financing activities		45
Net increase in cash		54
Cash balance, January 1		0
Cash balance, December 31		$54

International Financial Reporting Standards

● LO4–9

Classification of Cash Flows. Like U.S. GAAP, international standards also require a statement of cash flows. Consistent with U.S. GAAP, cash flows are classified as operating, investing, or financing. However, the U.S. standard designates cash outflows for interest payments and cash inflows from interest and dividends received as operating cash flows. Dividends paid to shareholders are classified as financing cash flows.

IAS No. 7,[33] on the other hand, allows more flexibility. Companies can report interest and dividends paid as either operating or financing cash flows and interest and dividends received as either operating or investing cash flows. Interest and dividend payments usually are reported as financing activities. Interest and dividends received normally are classified as investing activities.

Typical Classification of Cash Flows from Interest and Dividends

U.S. GAAP	IFRS
Operating Activities	*Operating Activities*
Dividends received	
Interest received	
Interest paid	
Investing Activities	*Investing Activities*
	Dividends received
	Interest received
Financing Activities	*Financing Activities*
Dividends paid	Dividends paid
	Interest paid

Siemens AG, a German company, prepares its financial statements according to IFRS. In its statement of cash flows for 2017, the company reported interest and dividends received as operating cash flows, as would a U.S. company. However, Siemens classified **interest paid** as a financing cash flow.

SIEMENS AG
Statement of Cash Flows (partial)
For the Year Ended September 30, 2017

Cash flows from financing activities:	(€ in millions)
Transactions with owners	191
Issuance of long-term debt	6,958
Repayment of long-term debt	(4,868)
Change in short-term debt and other financing activities	260
Interest paid	(1,000)
Dividends paid	(3,101)
Cash flows from financing activities	(1,560)

Noncash Investing and Financing Activities

As we just discussed, the statement of cash flows provides useful information about the investing and financing activities in which a company is engaged. Even though these primarily result in cash inflows and cash outflows, there may be significant investing and financing activities occurring during the period that do not involve cash flows at all. In order to provide complete information about these activities, any significant noncash investing and financing activities (that is, noncash exchanges) are reported either on the face of the SCF or in a disclosure note. An example of a significant noncash investing and financing activity is the acquisition of equipment (an investing activity) by simultaneously issuing either a long-term note payable or equity securities (a financing activity) to the seller of the equipment.

[33]"Statement of Cash Flows," *International Accounting Standard No. 7* (IASCF), as amended effective January 1, 2016.

Concept Review Exercise

STATEMENT OF CASH FLOWS

Dublin Enterprises, Inc. (DEI), owns a chain of retail electronics stores located in shopping malls. The following are the company's 2021 income statement and comparative balance sheets ($ in millions):

Income Statement
For the Year Ended December 31, 2021

Sales revenue		$2,100
Cost of goods sold		1,400
Gross profit		700
Operating expenses:		
Selling and administrative expense	$ 355	
Depreciation expense	85	
Total operating expenses		440
Income before income taxes		260
Income tax expense		65
Net income		$ 195

Comparative Balance Sheets	12/31/2021	12/31/2020
Assets:		
Cash	$ 300	$ 220
Accounts receivable	227	240
Inventory	160	120
Property, plant, and equipment	960	800
Less: Accumulated depreciation	(405)	(320)
Total assets	$1,242	$1,060
Liabilities and shareholders' equity:		
Accounts payable	$ 145	$ 130
Accrued liabilities (for selling and admin. expense)	147	170
Income taxes payable	82	50
Long-term debt	–0–	100
Common stock	463	400
Retained earnings	405	210
Total liabilities and shareholders' equity	$1,242	$1,060

Required:

1. Prepare DEI's 2021 statement of cash flows using the direct method.
2. Prepare the cash flows from operating activities section of DEI's 2021 statement of cash flows using the indirect method.

Solution:

1. Prepare DEI's 2021 statement of cash flows using the direct method.

DUBLIN ENTERPRISES, INC.
Statement of Cash Flows
For the Year Ended December 31, 2021
($ in millions)

Cash Flows from Operating Activities		
Cash received from customers*	$2,113	
Cash paid for inventory**	(1,425)	
Cash paid for selling and administrative expense†	(378)	
Cash paid for income taxes‡	(33)	
Net cash flows from operating activities		$277
Cash Flows from Investing Activities		
Purchase of property, plant, and equipment		(160)
Cash Flows from Financing Activities		
Issuance of common stock	63	
Payment on long-term debt	(100)	

(continued)

(concluded)

Net cash flows from financing activities	(37)
Net increase in cash	80
Cash, January 1	220
Cash, December 31	$300

*Sales revenue of $2,100 million, plus $13 million decrease in accounts receivable.
**Cost of goods sold of $1,400 million, plus $40 million increase in inventory, less $15 million increase in accounts payable.
†Selling and administrative expense of $355 million, plus $23 million decrease in accrued liabilities.
‡Income tax expense of $65 million, less $32 million increase in income taxes payable.

2. Prepare the cash flows from operating activities section of DEI's 2021 statement of cash flows using the indirect method.

DUBLIN ENTERPRISES, INC.
Statement of Cash Flows
For the Year Ended December 31, 2021
($ in millions)

Cash Flows from Operating Activities		
Net Income	$195	
Adjustments for noncash effects:		
Depreciation expense	85	
Changes in operating assets and liabilities:		
Decrease in accounts receivable	13	
Increase in inventory	(40)	
Increase in accounts payable	15	
Increase in income taxes payable	32	
Decrease in accrued liabilities	(23)	
Net cash flows from operating activities		$277

Profitability Analysis

PART C

Chapter 3 provided an overview of financial statement analysis and introduced some of the common ratios used in risk analysis to investigate a company's liquidity and long-term solvency. We now look at ratios related to profitability analysis.

• LO4–10

Activity Ratios

One key to profitability is how well a company manages and utilizes its assets. Some ratios are designed to evaluate a company's effectiveness in managing assets. Of particular interest are the activity, or turnover ratios, of certain assets. The greater the number of times an asset turns over, the fewer assets that are required to maintain a given level of activity (revenue). Therefore, high turnover ratios usually are preferred.

Activity ratios measure a company's efficiency in managing its assets.

Although, in concept, the activity or turnover can be measured for any asset, activity ratios are most frequently calculated for total assets, accounts receivable, and inventory. These ratios are calculated as follows:

$$\text{Asset turnover ratio} = \frac{\text{Net sales}}{\text{Average total assets}}$$

$$\text{Receivables turnover ratio} = \frac{\text{Net sales}}{\text{Average accounts receivable (net)}}$$

$$\text{Inventory turnover ratio} = \frac{\text{Cost of goods sold}}{\text{Average inventory}}$$

ASSET TURNOVER A broad measure of asset efficiency is the **asset turnover ratio**. The ratio is computed by dividing a company's net sales by the average total assets available for use during a period. The denominator, average assets, is determined by adding beginning and ending total assets and dividing by two. The asset turnover ratio provides an indication of how efficiently a company utilizes all of its assets to generate revenue.

The *asset turnover ratio* measures a company's efficiency in using assets to generate revenue.

The *receivables turnover ratio* offers an indication of how quickly a company is able to collect its accounts receivable.

RECEIVABLES TURNOVER The **receivables turnover ratio** is calculated by dividing a period's net credit sales by the average net accounts receivable. Because income statements seldom distinguish between cash sales and credit sales, this ratio usually is computed using total net sales as the numerator. The denominator, average accounts receivable, is determined by adding beginning and ending net accounts receivable (gross accounts receivable less allowance for uncollectible accounts) and dividing by two.[34]

The receivables turnover ratio provides an indication of a company's efficiency in collecting cash from customers. The ratio shows the number of times during a period that the average accounts receivable balance is collected. The higher the ratio, the shorter the average time between sales and cash collection.

A convenient extension is the **average collection period**. This measure is computed by dividing 365 days by the receivables turnover ratio. The result is an approximation of the number of days the average accounts receivable balance is outstanding.

$$\text{Average collection period} = \frac{365}{\text{Receivables turnover ratio}}$$

The *average collection period* indicates the average age of accounts receivable.

Monitoring the receivables turnover ratio (and average collection period) over time can provide useful information about a company's future prospects. For example, a decline in the receivables turnover ratio (an increase in the average collection period) could be an indication that sales are declining because of customer dissatisfaction with the company's products. Another possible explanation is that the company has changed its credit policy and is granting extended credit terms in order to maintain customers. Either explanation could signal a future increase in bad debts. Ratio analysis does not explain what is wrong. It does provide information that highlights areas for further investigation. We cover additional details of the receivables turnover ratio and average collection period in Chapter 7.

The *inventory turnover ratio* measures a company's efficiency in managing its investment in inventory.

INVENTORY TURNOVER An important activity measure for a merchandising company (a retail, wholesale, or manufacturing company) is the **inventory turnover ratio**. The ratio shows the number of times the average inventory balance is sold during a reporting period. It indicates how quickly inventory is sold. The more frequently a business is able to sell, or turn over, its inventory, the lower its investment in inventory must be for a given level of sales. The ratio is computed by dividing the period's cost of goods sold by the average inventory balance. The denominator, average inventory, is determined by adding beginning and ending inventory and dividing by two.[35]

A relatively high ratio, say compared to a competitor, usually is desirable. A high ratio indicates comparative strength, perhaps caused by a company's superior sales force or maybe a successful advertising campaign. However, it might also be caused by a relatively low inventory level, which could mean either very efficient inventory management or stockouts and lost sales in the future.

On the other hand, a relatively low ratio, or a decrease in the ratio over time, usually is perceived to be unfavorable. Too much capital may be tied up in inventory. A relatively low ratio may result from overstocking, the presence of obsolete items, or poor marketing and sales efforts.

Similar to the receivables turnover, we can divide the inventory turnover ratio into 365 days to compute the **average days in inventory**. This measure indicates the number of days it normally takes to sell inventory.

$$\text{Average days in inventory} = \frac{365}{\text{Inventory turnover ratio}}$$

---

[34]Although net accounts receivable typically is used in practice for the denominator of receivables turnover, some prefer to use gross accounts receivable. Why? As the allowance for bad debts increases, net accounts receivable decreases, so if net accounts receivable is in the denominator, more bad debts have the effect of decreasing the denominator and therefore increasing receivables turnover. All else equal, an analyst would rather see receivables turnover improve because of more sales or less gross receivables, and not because of an increase in the allowance for bad debts.

[35] Notice the consistency in the measure used for the numerator and denominator of the two turnover ratios. For the receivables turnover ratio, both numerator and denominator are based on sales dollars, whereas they are both based on cost for the inventory turnover ratio.

Profitability Ratios

A fundamental element of an analyst's task is to develop an understanding of a firm's profitability. Profitability ratios attempt to measure a company's ability to earn an adequate return relative to sales or resources devoted to operations. Resources devoted to operations can be defined as total assets or only those assets provided by owners, depending on the evaluation objective.

Profitability ratios assist in evaluating various aspects of a company's profit-making activities.

Three common profitability measures are (1) the profit margin on sales, (2) the return on assets, and (3) the return on equity. These ratios are calculated as follows:

$$\text{Profit margin on sales} = \frac{\text{Net income}}{\text{Net sales}}$$

$$\text{Return on assets} = \frac{\text{Net income}}{\text{Average total assets}}$$

$$\text{Return on equity} = \frac{\text{Net income}}{\text{Average shareholders' equity}}$$

Notice that for all of the profitability ratios, our numerator is net income. Recall our discussion earlier in this chapter on earnings quality. The relevance of any historical-based financial statement hinges on its predictive value. To enhance predictive value, analysts often adjust net income in these ratios to separate a company's *temporary earnings* from its *permanent earnings*. Analysts begin their assessment of permanent earnings with income from continuing operations. Then, adjustments are made for any unusual, one-time gains or losses included in income from continuing operations. It is this adjusted number that they use as the numerator in these ratios.

When calculating profitability ratios, analysts often adjust net income for any temporary income effects.

PROFIT MARGIN ON SALES The **profit margin on sales** is simply net income divided by net sales. The ratio measures an important dimension of a company's profitability. It indicates the portion of each dollar of revenue that is available after all expenses have been covered. It offers a measure of the company's ability to withstand either higher expenses or lower revenues.

The *profit margin on sales* measures the amount of net income achieved per sales dollar.

What is considered to be a desirable profit margin is highly sensitive to the nature of the business activity For instance, you would expect a specialty shop to have a higher profit margin than, say, **Walmart**. A low profit margin can be compensated for by a high asset turnover rate, and vice versa, which brings us to considering the trade-offs inherent in generating return on assets.

RETURN ON ASSETS The **return on assets (ROA)** ratio expresses income as a percentage of the average total assets available to generate that income. Because total assets are partially financed with debt and partially by equity funds, this is an inclusive way of measuring earning power that ignores specific sources of financing.

A company's return on assets is related to both profit margin and asset turnover. Specifically, profitability can be achieved by either a high profit margin, high asset turnover, or a combination of the two. In fact, the return on assets can be calculated by multiplying the profit margin and the asset turnover.

Profit margin and asset turnover combine to yield *return on assets*, which measures the return generated by a company's assets.

$$\textbf{Return on assets} = \textbf{Profit margin} \times \textbf{Asset turnover}$$

$$\frac{\text{Net income}}{\text{Average total assets}} = \frac{\text{Net income}}{\text{Net sales}} \times \frac{\text{Net sales}}{\text{Average total assets}}$$

Industry standards are particularly important when evaluating asset turnover and profit margin. Some industries are characterized by low turnover but typically make up for it with higher profit margins. Others have low profit margins but compensate with high turnover. Grocery stores typically have relatively low profit margins but relatively high asset turnover. In comparison, a manufacturer of specialized equipment will have a higher profit margin but a lower asset turnover ratio.

Additional Consideration

The return on assets ratio often is computed as follows:

$$\text{Return on assets} = \frac{\text{Net income} + \text{Interest expense (1} - \text{Tax rate)}}{\text{Average total assets}}$$

The reason for adding back interest expense (net of tax) is that interest represents a return to suppliers of debt capital and should not be deducted in the computation of net income when computing the return on assets. In other words, the numerator is the total amount of income available to both debt and equity capital.

Return on equity measures the return to suppliers of equity capital.

The DuPont framework shows that return on equity depends on profitability, activity, and financial leverage.

RETURN ON SHAREHOLDERS' EQUITY Shareholders are concerned with how well management uses their equity to generate a profit. A closely watched measure that captures this concern is **return on equity (ROE)**, calculated as net income divided by average shareholders' equity.

In addition to monitoring return on equity, investors want to understand how that return can be improved. The **DuPont framework** provides a convenient basis for analysis that breaks return on equity into three key components:[36]

- **Profitability,** measured by the profit margin (Net income ÷ Sales). As discussed already, a higher profit margin indicates that a company generates more profit from each dollar of sales.
- **Activity,** measured by asset turnover (Sales ÷ Average total assets). As discussed already, higher asset turnover indicates that a company uses its assets efficiently to generate more sales from each dollar of assets.
- **Financial Leverage,** measured by the equity multiplier (Average total assets ÷ Average total equity). A high equity multiplier indicates that relatively more of the company's assets have been financed with debt; that is, the company is more leveraged. As discussed in Chapter 3, leverage can provide additional return to the company's equity holders.

In equation form, the DuPont framework looks like this:

$$\textbf{Return on equity} = \textbf{Profit margin} \times \textbf{Asset turnover} \times \textbf{Equity multiplier}$$

$$\frac{\text{Net income}}{\text{Avg. total equity}} = \frac{\text{Net income}}{\text{Net sales}} \times \frac{\text{Net sales}}{\text{Avg. total assets}} \times \frac{\text{Avg. total assets}}{\text{Avg. total equity}}$$

Notice that net sales and average total assets appear in the numerator of one ratio and the denominator of another, so they cancel to yield net income ÷ average total equity, or ROE.

We have already seen that ROA is determined by profit margin and asset turnover, so another way to compute ROE is by multiplying ROA by the equity multiplier:

$$\textbf{Return on equity} = \textbf{Return on assets} \times \textbf{Equity multiplier}$$

$$\frac{\text{Net income}}{\text{Avg. total equity}} = \frac{\text{Net income}}{\text{Avg. total assets}} \times \frac{\text{Avg. total assets}}{\text{Avg. total equity}}$$

We can see from this equation that an equity multiplier of greater than 1 will produce a return on equity that is higher than the return on assets. However, as with all ratio analysis, there are trade-offs. If leverage is too high, creditors become concerned about the potential for default on the company's debt and require higher interest rates. Because interest is recognized as an expense, net income is reduced, so at some point the benefits of a higher equity multiplier are offset by a lower profit margin. Part of the challenge of managing a

[36]DuPont analysis is so named because the basic model was developed by F. Donaldson Brown, an electrical engineer who worked for DuPont in the early part of the twentieth century.

Additional Consideration

Sometimes when return on equity is calculated, shareholders' equity is viewed more narrowly to include only common shareholders. In that case, preferred stock is excluded from the denominator, and preferred dividends are deducted from net income in the numerator. The resulting rate of return on equity focuses on profits generated on resources provided by common shareholders.

company is to identify the combination of profitability, activity, and leverage that produces the highest return for equity holders.

Illustration 4–16 provides a recap of the ratios we have discussed.

Illustration 4–16

Summary of Profitability Analysis Ratios

Activity ratios		
Asset turnover	=	$\frac{\text{Net sales}}{\text{Average total assets}}$
Receivables turnover	=	$\frac{\text{Net sales}}{\text{Average accounts receivable (net)}}$
Average collection period	=	$\frac{365}{\text{Receivables turnover ratio}}$
Inventory turnover	=	$\frac{\text{Cost of goods sold}}{\text{Average inventory}}$
Average days in inventory	=	$\frac{365}{\text{Inventory turnover ratio}}$
Profitability ratios		
Profit margin on sales	=	$\frac{\text{Net income}}{\text{Net sales}}$
Return on assets	=	$\frac{\text{Net income}}{\text{Average total assets}}$
Return on equity	=	$\frac{\text{Net income}}{\text{Average shareholders' equity}}$
Leverage ratio		
Equity multiplier	=	$\frac{\text{Average total assets}}{\text{Average total equity}}$

Profitability Analysis—An Illustration

To illustrate the application of the DuPont framework and the computation of the activity and profitability ratios, we analyze the 2017 financial statements of two well-known retailers, **Costco Wholesale Corporation** and **Walmart Stores Inc.**[37] The operations of these two companies are similar in their focus on operating large general merchandising and food discount stores. Illustration 4–17 presents selected financial statement information for the two companies (all numbers are in millions of dollars).

In absolute dollars, it appears that Walmart is far more profitable than Costco. As shown at the bottom of Illustration 4–17, Walmart's 2017 net income was $14,293 million, compared to Costco's $2,679 million. But that's not the whole story. Even though both are very large companies, Walmart is more than five times the size of Costco in terms of total assets, so how can they be compared? Focusing on financial ratios helps adjust for size differences, and the DuPont framework helps identify the determinants of profitability from the perspective of shareholders.

Illustration 4–18 includes the DuPont analysis for Walmart and Costco, as well as some additional activity ratios we've discussed. The first item to notice is that Walmart's profit

[37] Walmart's financial statements are for the fiscal year ended January 31, 2017. Walmart refers to this as its 2017 fiscal year. Costco's financial statements are for the fiscal year ended August 30, 2017.

Illustration 4–17 Selected Financial Information for Costco Wholesale Corporation and Walmart Stores Inc. **Real World Financials**

	Costco			Walmart		
	2017		**2016**	**2017**		**2016**
($ in millions)						
Balance Sheet—2017						
Accounts receivable (net)	$ 1,432		$ 1,252	$ 5,835		$ 5,624
Inventories	9,834		8,969	43,046		44,469
Total assets	36,347		33,163	198,825		199,581
Total liabilities	25,268		20,831	118,290		115,970
Total shareholders' equity	10,778		12,079	80,535		83,611
Average for 2017:						
Accounts receivable (net)		$ 1,342.0			$ 5,729.5	
Inventories		9,401.5			43,757.5	
Total assets		34,755.0			199,203.0	
Total shareholders' equity		11,428.5			82,073.0	
Income Statement—2017						
Net sales		126,172			481,317	
Cost of goods sold		111,882			361,256	
Net Income		2,679			14,293	

Source: Costco; Walmart

Illustration 4–18 DuPont Framework and Activity Ratios—Costco Wholesale Corporation and Walmart Stores Inc. **Real World Financials**

DuPont analysis	**Costco**	**Walmart**	**Industry Average***
Profit margin on sales	$= \frac{\$2,679}{\$126,172} = 2.12\%$	$\frac{\$14,293}{\$481,317} = 2.97\%$	4.38%
×	×	×	
Asset turnover	$= \frac{\$126,172}{\$34,755} = 3.63$	$\frac{\$481,317}{\$199,203} = 2.42$	1.82
=	=	=	
Return on assets	$= \frac{\$2,679}{\$34,755} = 7.71\%$	$\frac{\$14,293}{\$199,203} = 7.18\%$	9.10%
×	×	×	
Equity Multiplier	$= \frac{\$34,755}{\$11,428.5} = 3.04$	$\frac{\$199,203}{\$82,073} = 2.43$	2.08
=	=	=	
Return on equity	$= \frac{\$2,679}{\$11,428.5} = 23.44\%$	$\frac{\$14,293}{\$82,073} = 17.41\%$	18.95%
Other activity ratios			
Receivables turnover	$= \frac{\$126,172}{\$1,342} = 94.02$	$\frac{\$481,317}{\$5,729.5} = 84.01$	80.24
Average collection period	$= \frac{365}{94.02} = 3.88$ days	$\frac{365}{84.01} = 4.34$ days	4.55 days
Inventory turnover	$= \frac{\$111,882}{\$9,401.5} = 11.90$	$\frac{\$361,256}{\$43,757.5} = 8.26$	6.96
Average days in inventory	$= \frac{365}{11.90} = 30.67$ days	$\frac{365}{8.26} = 44.19$ days	52.44 days

*www.reuters.com.

Source: www.reuters.com

margin is higher than Costco's (2.97% for Walmart compared to 2.12% for Costco). This means that Walmart generates more profit for each $1 of sales. However, by looking at the asset turnover ratio, we see that Costco is better able to use its assets to generate sales. Costco's asset turnover ratio of 3.63 is noticeably higher than Walmart's ratio of only 2.42. Then, we see from the equity multiplier of each company that Costco's strategy is to use more debt (or conversely, less equity) to finance the purchase of its assets. This strategy has paid off for its shareholders. By using financial leverage to purchase more productive assets, Costco is able to generate a higher return on equity (ROE) for its shareholders (23.44% for Costco versus only 17.41% for Walmart). The DuPont analysis helps us understand how this occurs.

Other activity ratios further reveal Costco's greater operating efficiency. Two components of asset turnover are inventory turnover and receivables turnover. Based on the inventory turnover ratio, we see that inventory sells much more quickly at Costco. Inventory takes only 31 days on average before being sold at Costco, compared with 44 days at Walmart. Costco also turns over its accounts receivable slightly faster than Walmart does. While accounts receivable are relatively small for both companies, the difference in the receivables turnover could provide investors with another signal of Costco management's greater overall operating efficiency.

The essential point of our discussion here, and in Part C of Chapter 3, is that raw accounting numbers alone mean little to decision makers. The numbers gain value when viewed in relation to other numbers. Similarly, the financial ratios formed by those relationships provide even greater perspective when compared with similar ratios of other companies, or with averages for several companies in the same industry. Accounting information is useful in making decisions. Financial analysis that includes comparisons of financial ratios enhances the value of that information.

Financial Reporting Case Solution

©GaudiLab/Shutterstock

1. **How would you explain restructuring costs to Becky? Are restructuring costs something Becky should worry about?** *(p. 174)* Restructuring costs include employee severance and termination benefits plus other costs associated with the shutdown or relocation of facilities or downsizing of operations. Restructuring costs are not necessarily bad. In fact, the objective is to make operations more efficient. The costs are incurred now in hopes of better earnings later.
2. **Explain to Becky what is meant by discontinued operations and describe to her how that item is reported in an income statement.** *(p. 177)* Separate reporting as a discontinued operation is required when the disposal of a component represents a strategic shift that has, or will have, a major effect on a company's operations and financial results. The net-of-tax effect of discontinued operations is separately reported below income from continuing operations. If the component has been disposed of by the end of the reporting period, the income effects include: (1) income or loss from operations of the discontinued component from the beginning of the reporting period through the disposal date and (2) gain or loss on disposal of the component's assets. If the component has not been disposed of by the end of the reporting period, the income effects include: (1) income or loss from operations of the discontinued component from the beginning of the reporting period through the end of the reporting period, and (2) an impairment loss if the fair value minus cost to sell of the component's assets is less than their book value.
3. **Describe to Becky the difference between basic and diluted earnings per share.** *(p 184)* Basic earnings per share is computed by dividing net income available to common shareholders (net income less any preferred stock dividends) by the weighted-average number of common shares outstanding for the period. Diluted earnings per share reflects the potential dilution that could occur for companies that have certain securities outstanding that are convertible into common shares or stock options that could create additional common shares if the options were exercised. These items could cause earnings per share to decrease (become diluted). Because of the complexity of the calculation and the importance of earnings per share to investors, the text devotes a substantial portion of Chapter 19 to this topic. ●

The Bottom Line

- **LO4–1** The components of income from continuing operations are revenues, expenses (including income taxes), gains, and losses, excluding those related to discontinued operations. Companies often distinguish between operating and nonoperating income within continuing operations. (*p. 168*)
- **LO4–2** The term *earnings quality* refers to the ability of reported earnings (income) to predict a company's future earnings. The relevance of any historical-based financial statement hinges on its predictive value. To enhance predictive value, analysts try to separate a company's *temporary earnings* from its *permanent earnings*. Many believe that manipulating income reduces earnings quality because it can mask permanent earnings. Two major methods used by managers to manipulate earnings are (1) income shifting and (2) income statement classification. (*p. 173*)
- **LO4–3** Analysts begin their assessment of permanent earnings with income from continuing operations. It would be a mistake to assume income from continuing operations reflects permanent earnings entirely. In other words, there may be temporary earnings effects included in both operating and nonoperating income. (*p. 173*)
- **LO4–4** A discontinued operation refers to the disposal or planned disposal of a component of the entity. The net-of-tax effect of discontinued operations is separately reported below income from continuing operations. (*p. 177*)
- **LO4–5** Accounting changes include changes in principle, changes in estimate, or changes in reporting entity. Their effects on the current period and prior period financial statements are reported using various approaches—retrospective, modified retrospective, and prospective. Error corrections are made by restating prior period financial statements. Any correction to retained earnings is made with an adjustment to the beginning balance in the current period. Earnings per share (EPS) is the amount of income achieved during a period expressed per share of common stock outstanding. EPS must be disclosed for income from continuing operations and for discontinued operations. (*p. 182*)
- **LO4–6** The FASB's Concept Statement 6 defines the term *comprehensive income* as the change in equity from nonowner transactions. The calculation of net income, however, excludes certain transactions that are included in comprehensive income. To convey the relationship between the two measures, companies must report both net income and comprehensive income and reconcile the difference between the two. The presentation can be (1) in a single, continuous statement of comprehensive income, or (2) in two separate, but consecutive statements—an income statement and a statement of comprehensive income. (*p. 185*)
- **LO4–7** When a company provides a balance sheet and income statement, a statement of cash flows also is provided. The purpose of the statement of cash flows is to provide information about the cash receipts and cash disbursements that occurred during the period. (*p. 189*)
- **LO4–8** To enhance the usefulness of the information, the statement of cash flows classifies all transactions affecting cash into one of three categories: (1) operating activities, (2) investing activities, or (3) financing activities. (*p. 190*)
- **LO4–9** There are more similarities than differences between income statements and statements of cash flows prepared according to U.S. GAAP and those prepared applying international standards. In a statement of cash flows, some differences are possible in the classifications of interest and divided revenue, interest expense, and dividends paid. (*pp. 172, 186, and 195*)
- **LO4–10** Activity and profitability ratios provide information about a company's profitability. Activity ratios include the receivables turnover ratio, the inventory turnover ratio, and the asset turnover ratio. Profitability ratios include the profit margin on sales, the return on assets, and the return on equity. DuPont analysis explains return on equity as determined by profit margin, asset turnover, and the extent to which assets are financed with equity versus debt. (*p. 197*)

APPENDIX 4 Interim Reporting

Interim reports **are issued for periods of less than a year, typically as quarterly financial statements.**

Financial statements covering periods of less than a year are called *interim reports.* Companies registered with the SEC, which includes most public companies, must submit quarterly reports, and you will see excerpts from these reports throughout this book.[38] Though there is no requirement to do so, most also provide quarterly reports to their shareholders

[38] Quarterly reports are filed with the SEC on Form 10-Q. Annual reports to the SEC are on Form 10-K.

and typically include abbreviated, unaudited interim reports as supplemental information within their annual reports. For instance, Illustration 4A–1 shows the quarterly information disclosed in the annual report of **The Home Depot, Inc.**, for the fiscal year ended January 29, 2017. Amounts for each of the four quarters sum to the reported amount for the full year. Compare these numbers to the annual income statement in Illustration 4–2.

Illustration 4A–1

Interim Data in Annual Report—The Home Depot, Inc.

Real World Financials

NOTE 16 – SUMMARY OF QUARTERLY RESULTS OF OPERATIONS (UNAUDITED)

Year Ended January 29, 2017	1st Quarter	2nd Quarter	3rd Quarter	4th Quarter	Full Year
	($ in millions, except per share data)				
Net sales	$22,762	$26,472	$23,154	$22,207	$ 94,595
Gross profit	7,791	8,927	8,042	7,553	32,313
Net income	1,803	2,441	1,969	1,744	7,957
Basic	$ 1.45	$ 1.98	$ 1.61	$ 1.45	$ 6.47*
Diluted	$ 1.44	$ 1.97	$ 1.60	$ 1.44	$ 6.45

*Rounding error of $0.02
Source: Home Depot, Inc.

For accounting information to be useful to decision makers, it must be available on a timely basis. One of the objectives of interim reporting is to enhance the timeliness of financial information. In addition, quarterly reports provide investors and creditors with additional insight on the seasonality of business operations that might otherwise get lost in annual reports. Why are sales and net income higher in the 2nd and 3rd quarters (compared to the 1st and 4th quarters) for Home Depot? Most home improvement projects occur in the warmer months. These months occur in the 2nd and 3rd quarters, so it is expected that sales would be higher in these quarters. Because the company sells home products for a profit, profitability is also higher in quarters with greater sales.

However, the downside to interim reporting is that the amounts often are less reliable. With a shorter reporting period, questions associated with estimation and allocation are magnified. For example, certain expenses often benefit an entire year's operations and yet are incurred primarily within a single interim period. Similarly, should smaller companies use lower tax rates in the earlier quarters and higher rates in later quarters as higher tax brackets are reached? Another result of shorter reporting periods is the intensified effect of unusual events such as material gains and losses. A second quarter casualty loss, for instance, that would reduce annual profits by 10% might reduce second quarter profits by 40% or more. Is it more realistic to allocate such a loss over the entire year? These and similar questions tend to hinge on the way we view an interim period in relation to the fiscal year. More specifically, should each interim period be viewed as a *discrete* reporting period or as an *integral part* of the annual period?

The fundamental debate regarding interim reporting centers on the choice between the *discrete* and *integral part* approaches.

Reporting Revenues and Expenses

Existing practice and current reporting requirements for interim reporting generally follow the viewpoint that interim reports are an integral part of annual statements, although the discrete approach is applied to some items. Most revenues and expenses are recognized using the same accounting principles applicable to annual reporting. Some modifications are necessary to help cause interim statements to relate better to annual statements. This is most evident in the way costs and expenses are recognized. Most are recognized in interim periods as incurred. But when an expenditure clearly benefits more than just the period in which it is incurred, the expense should be allocated among the periods benefited on an allocation basis consistent with the company's annual allocation procedures. For example, annual repair expenses, property tax expense, and advertising expenses incurred in the first quarter that clearly benefit later quarters are assigned to each quarter through the use of accruals and deferrals. Costs and expenses subject to year-end adjustments, such as depreciation expense, are estimated and allocated to interim periods in a systematic way. Similarly, income tax expense at each interim date should be based on estimates of the effective tax

With only a few exceptions, the same accounting principles applicable to annual reporting are used for interim reporting.

rate for the whole year. This would mean, for example, that if the estimated effective rate has changed since the previous interim period(s), the tax expense in the current period would be determined as the new rate times the cumulative pretax income to date, less the total tax expense reported in previous interim periods.

Reporting Unusual Items

Discontinued operations and unusual items are reported entirely within the interim period in which they occur.

On the other hand, major events such as discontinued operations should be reported separately in the interim period in which they occur. That is, these amounts should not be allocated among individual quarters within the fiscal year. The same is true for items that are unusual. Treatment of these items is more consistent with the discrete view than the integral part view.

International Financial Reporting Standards

● LO4–9

Interim Reporting. *IAS No. 34* requires that a company apply the same accounting policies in its interim financial statements as it applies in its annual financial statements. Therefore, IFRS takes much more of a discrete-period approach than does U.S. GAAP. For example, costs for repairs, property taxes, and advertising that do not meet the definition of an asset at the end of an interim period are expensed entirely in the period in which they occur under IFRS, but are accrued or deferred and then charged to each of the periods they benefit under U.S. GAAP. This difference would tend to make interim period income more volatile under IFRS than under U.S. GAAP. However, as in U.S. GAAP, income taxes are accounted for based on an estimate of the tax rate expected to apply for the entire year.[39]

Earnings Per Share

Quarterly EPS calculations follow the same procedures as annual calculations.

A second item that is treated in a manner consistent with the discrete view is earnings per share. EPS calculations for interim reports follow the same procedures as annual calculations that you will study in Chapter 19. The calculations are based on conditions actually existing during the particular interim period rather than on conditions estimated to exist at the end of the fiscal year.

Reporting Accounting Changes

Accounting changes made in an interim period are reported by retrospectively applying the changes to prior financial statements.

Recall that we account for a change in accounting principle retrospectively, meaning we recast prior years' financial statements when we report those statements again in comparative form. In other words, we make those statements appear as if the newly adopted accounting method had been used in those prior years. It's the same with interim reporting. We retrospectively report a change made during an interim period in similar fashion. Then in financial reports of subsequent interim periods of the same fiscal year, we disclose how that change affected (a) income from continuing operations, (b) net income, and (c) related per share amounts for the postchange interim period.

Minimum Disclosures

Complete financial statements are not required for interim period reporting, but certain minimum disclosures are required as follows:[40]

- Sales, income taxes, and net income.
- Earnings per share.
- Seasonal revenues, costs, and expenses.
- Significant changes in estimates for income taxes.
- Discontinued operations and unusual items.

[39]Interim Financial Reporting," *International Accounting Standard No. 34* (IASCF), as amended effective January 1, 2016, par. 28–30.

[40]FASB ASC 270-10-50: Interim Reporting—Overall—Disclosure (previously "Interim Financial Reporting," *Accounting Principles Board Opinion No 28* (New York: AICPA, 1973)).

- Contingencies.
- Changes in accounting principles or estimates.
- Information about fair value of financial instruments and the methods and assumptions used to estimate fair values.
- Significant changes in financial position.

When fourth quarter results are not separately reported, material fourth quarter events, including year-end adjustments, should be reported in disclosure notes to annual statements.

Questions For Review of Key Topics

Q 4–1 The income statement is a change statement. Explain what is meant by this.

Q 4–2 What transactions are included in income from continuing operations? Briefly explain why it is important to segregate income from continuing operations from other transactions affecting net income.

Q 4–3 Distinguish between operating and nonoperating income in relation to the income statement.

Q 4–4 Briefly explain the difference between the single-step and multiple-step income statement formats.

Q 4–5 Explain what is meant by the term *earnings quality*.

Q 4–6 What are restructuring costs and where are they reported in the income statement?

Q 4–7 Define intraperiod tax allocation. Why is the process necessary?

Q 4–8 How are discontinued operations reported in the income statement?

Q 4–9 What is meant by a change in accounting principle? Describe the possible accounting treatments for a mandated change in accounting principle.

Q 4–10 Accountants very often are required to make estimates, and very often those estimates prove incorrect. In what period(s) is the effect of a change in an accounting estimate reported?

Q 4–11 The correction of a material error discovered in a year subsequent to the year the error was made is considered a prior period adjustment. Briefly describe the accounting treatment for prior period adjustments.

Q 4–12 Define earnings per share (EPS). For which income statement items must EPS be disclosed?

Q 4–13 Define comprehensive income. What are the two ways companies can present comprehensive income?

Q 4–14 Describe the purpose of the statement of cash flows.

Q 4–15 Identify and briefly describe the three categories of cash flows reported in the statement of cash flows.

Q 4–16 Explain what is meant by noncash investing and financing activities pertaining to the statement of cash flows. Give an example of one of these activities.

Q 4–17 Distinguish between the direct method and the indirect method for reporting the results of operating activities in the statement of cash flows.

IFRS **Q 4–18** Describe the potential statement of cash flows classification differences between U.S. GAAP and IFRS.

Q 4–19 Show the calculation of the following activity ratios: (1) the receivables turnover ratio, (2) the inventory turnover ratio, and (3) the asset turnover ratio. What information about a company do these ratios offer?

Q 4–20 Show the calculation of the following profitability ratios: (1) the profit margin on sales, (2) the return on assets, and (3) the return on equity. What information about a company do these ratios offer?

Q 4–21 Show the DuPont framework's calculation of the three components of return on equity. What information about a company do these ratios offer?

Q 4–22 Interim reports are issued for periods of less than a year, typically as quarterly financial statements. Should these interim periods be viewed as separate periods or integral parts of the annual period?

IFRS **Q 4–23** [Based on Appendix 4] What is the primary difference between interim reports under IFRS and U.S. GAAP?

Brief Exercises

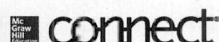

connect

BE 4–1
Single-step income statement
● LO4–1

The adjusted trial balance of Pacific Scientific Corporation on December 31, 2021, the end of the company's fiscal year, contained the following income statement items ($ in millions): sales revenue, $2,106; cost of goods sold, $1,240; selling expense, $126; general and administrative expense, $105; interest expense, $40; and gain on sale of investments, $45. Income tax expense has not yet been recorded. The income tax rate is 25%.

Using the account balances, prepare a single-step income statement for 2021. An example of a single-step income statement can be found in Illustration 4–3 of this chapter.

BE 4–2
Multiple-step income statement
● LO4–1, LO4–3

Refer to the situation described in BE 4–1. If the company's accountant prepared a multiple-step income statement, what amount would appear in that statement for (a) operating income and (b) nonoperating income?

BE 4–3
Multiple-step income statement
● LO4–1, LO4–3

Refer to the situation described in BE 4–1. Using the account balances, prepare a multiple-step income statement for 2021. An example of a multiple-step income statement can be found in Illustration 4–4 of this chapter.

BE 4–4
Multiple-step income statement
● LO4–1, LO4–3

The following is a partial year-end adjusted trial balance.

Account Title	Debits	Credits
Sales revenue		$300,000
Loss on sale of investments	$ 22,000	
Interest revenue		4,000
Cost of goods sold	160,000	
General and administrative expense	40,000	
Restructuring costs	50,000	
Selling expense	25,000	
Income tax expense	?	

Income tax expense has not yet been recorded. The income tax rate is 25%. Determine the following: (a) operating income (loss), (b) income (loss) before income taxes, and (c) net income (loss).

BE 4–5
Income from continuing operations
● LO4–3, LO4–5

The following are partial income statement account balances taken from the December 31, 2021, year-end trial balance of White and Sons, Inc.: restructuring costs, $300,000; interest revenue, $40,000; before-tax loss on discontinued operations, $400,000; and loss on sale of investments, $50,000. Income tax expense has not yet been recorded. The income tax rate is 25%.

Prepare the lower portion of the 2021 income statement beginning with $800,000 income from continuing operations before income taxes. Include appropriate EPS disclosures. The company had 100,000 shares of common stock outstanding throughout the year.

BE 4–6
Discontinued operations
● LO4–4

On December 31, 2021, the end of the fiscal year, Revolutionary Industries completed the sale of its robotics business for $9 million. The robotics business segment qualifies as a component of the entity according to GAAP. The book value of the assets of the segment was $7 million. The income from operations of the segment during 2021 was $4 million. Pretax income from continuing operations for the year totaled $12 million. The income tax rate is 25%. Prepare the lower portion of the 2021 income statement beginning with income from continuing operations before income taxes. Ignore EPS disclosures.

BE 4–7
Discontinued operations
● LO4–4

On December 31, 2021, the end of the fiscal year, California Microtech Corporation completed the sale of its semiconductor business for $10 million. The semiconductor business segment qualifies as a component of the entity according to GAAP. The book value of the assets of the segment was $8 million. The loss from operations of the segment during 2021 was $3.6 million. Pretax income from continuing operations for the year totaled $5.8 million. The income tax rate is 25%. Prepare the lower portion of the 2021 income statement beginning with income from continuing operations before income taxes. Ignore EPS disclosures.

BE 4–8
Discontinued operations
● LO4–4

Refer to the situation described in BE 4–7. Assume that the semiconductor segment was not sold during 2021 but was held for sale at year-end. The estimated fair value of the segment's assets, less costs to sell, on December 31 was $10 million. Prepare the lower portion of the 2021 income statement beginning with income from continuing operations before income taxes. Ignore EPS disclosures.

BE 4–9
Discontinued operations
● LO4–4

Refer to the situation described in BE 4–8. Assume instead that the estimated fair value of the segment's assets, less costs to sell, on December 31 was $7 million rather than $10 million. Prepare the lower portion of the 2021 income statement beginning with income from continuing operations before income taxes. Ignore EPS disclosures.

BE 4–10
Comprehensive income
● LO4–6

O'Reilly Beverage Company reported net income of $650,000 for 2021. In addition, the company deferred a $60,000 pretax loss on derivatives and had pretax net unrealized gains on debt securities of $40,000. Prepare a separate statement of comprehensive income for 2021. The company's income tax rate is 25%.

BE 4–11
Statement of cash flows; direct method
● LO4–8

The following are summary cash transactions that occurred during the year for Hilliard Healthcare Co. (HHC):

Cash received from:	
Customers	$660,000
Interest on notes receivable	12,000
Collection of notes receivable	100,000
Sale of land	40,000
Issuance of common stock	200,000
Cash paid for:	
Interest on notes payable	18,000
Purchase of equipment	120,000
Operating expenses	440,000
Dividends to shareholders	30,000

Prepare the cash flows from operating activities section of HHC's statement of cash flows using the direct method.

BE 4–12
Statement of cash flows; investing and financing activities
● LO4–8

Refer to the situation described in BE 4–11. Prepare the cash flows from investing and financing activities sections of HHC's statement of cash flows.

BE 4–13
Statement of cash flows; indirect method
● LO4–8

Net income of Mansfield Company was $45,000. The accounting records reveal depreciation expense of $80,000 as well as increases in prepaid rent, salaries payable, and income taxes payable of $60,000, $15,000, and $12,000, respectively. Prepare the cash flows from operating activities section of Mansfield's statement of cash flows using the indirect method.

BE 4–14
IFRS; Statement of cash flows
● LO4–8, LO4–9
IFRS

Refer to the situation described in BE 4–11 and BE 4–12. How might your solution to those brief exercises differ if Hilliard Healthcare Co. prepares its statement of cash flows according to International Financial Reporting Standards?

BE 4–15
Receivables and inventory turnover ratios
● LO4–10

Universal Calendar Company began the year with accounts receivable (net) and inventory balances of $100,000 and $80,000, respectively. Year-end balances for these accounts were $120,000 and $60,000, respectively. Sales for the year of $600,000 generated a gross profit of $200,000. Calculate the receivables and inventory turnover ratios for the year.

BE 4–16
Profitability ratios
● LO4–10

The 2021 income statement for Anderson TV and Appliance reported net sales of $420,000 and net income of $65,000. Average total assets for 2021 was $800,000. Shareholders' equity at the beginning of the year was $500,000, and $20,000 was paid to shareholders as dividends. There were no other shareholders' equity transactions that occurred during the year. Calculate the profit margin on sales, return on assets, and return on equity for 2021.

BE 4–17
Profitability ratios
● LO4–10

Refer to the facts described in BE 4–16. Show the DuPont framework's calculation of the three components of the 2021 return on equity for Anderson TV and Appliance.

BE 4–18
Inventory turnover ratio
● LO4–10

During 2021, Rogue Corporation reported net sales of $600,000. Inventory at both the beginning and end of the year totaled $75,000. The inventory turnover ratio for the year was 6.0. What amount of gross profit did the company report in its 2021 income statement?

Exercises

E 4–1
Operating versus nonoperating Income
● LO4–1

Pandora Corporation operates several factories in the Midwest that manufacture consumer electronics. The December 31, 2021, year-end trial balance contained the following income statement items:

Account Title	Debits	Credits
Sales revenue		$12,500,000
Interest revenue		50,000
Loss on sale of investments	$ 100,000	
Cost of goods sold	6,200,000	
Selling expense	620,000	
General and administrative expense	1,520,000	
Interest expense	40,000	
Research and development expense	1,200,000	
Income tax expense	900,000	

Required:
Calculate the company's operating income for the year.

E 4–2
Income statement format; single step and multiple step
● LO4–1, LO4–5

The following is a partial trial balance for the Green Star Corporation as of December 31, 2021:

Account Title	Debits	Credits
Sales revenue		$1,300,000
Interest revenue		30,000
Gain on sale of investments		50,000
Cost of goods sold	$720,000	
Selling expense	160,000	
General and administrative expense	75,000	
Interest expense	40,000	
Income tax expense	130,000	

There were 100,000 shares of common stock outstanding throughout 2021.

Required:
1. Prepare a single-step income statement for 2021, including EPS disclosures, by inserting the amounts above into the appropriate section.
2. Prepare a multiple-step income statement for 2021, including EPS disclosures, by inserting the amounts above into the appropriate section.

E 4–3
Income statement format; single step and multiple step
● LO4–1, LO4–3, LO4–5

The following is a partial trial balance for General Lighting Corporation as of December 31, 2021:

Account Title	Debits	Credits
Sales revenue		$2,350,000
Interest revenue		80,000
Loss on sale of investments	$ 22,500	
Cost of goods sold	1,200,300	
Loss on inventory write-down (obsolescence)	200,000	
Selling expense	300,000	
General and administrative expense	150,000	
Interest expense	90,000	

There were 160,000 shares of common stock outstanding throughout 2021. Income tax expense has not yet been recorded. The income tax rate is 25%.

Required:
1. Prepare a single-step income statement for 2021, including EPS disclosures. Be sure to include appropriate headings and subtotal titles. An example of a single-step income statement can be found in Illustration 4–3 of this chapter.

2. Prepare a multiple-step income statement for 2021, including EPS disclosures. Be sure to include appropriate headings and subtotal titles. An example of a multiple-step income statement can be found in Illustration 4–4 of this chapter.

E 4–4
Multiple-step continuous statement of comprehensive income
● LO4–1, LO4–5, LO4–6

The trial balance for Lindor Corporation, a manufacturing company, for the year ended December 31, 2021, included the following accounts:

Account Title	Debits	Credits
Sales revenue		$2,300,000
Cost of goods sold	$1,400,000	
Selling and administrative expense	420,000	
Interest expense	40,000	
Gain on debt securities		80,000

The gain on debt securities is unrealized and classified as other comprehensive income. The trial balance does not include the accrual for income taxes. Lindor's income tax rate is 25%. There were 1,000,000 shares of common stock outstanding throughout 2021.

Required:
Prepare a single, continuous multiple-step statement of comprehensive income for 2021, including appropriate EPS disclosures.

E 4–5
Income statement presentation
● LO4–1, LO4–5

The following *incorrect* income statement was prepared by the accountant of the Axel Corporation:

AXEL CORPORATION
Income Statement
For the Year Ended December 31, 2021

Revenues and gains:		
Sales revenue		$592,000
Interest revenue		32,000
Gain on sale of investments		86,000
Total revenues and gains		710,000
Expenses and losses:		
Cost of goods sold	$325,000	
Selling expense	67,000	
Administrative expense	87,000	
Interest expense	16,000	
Restructuring costs	55,000	
Income tax expense	40,000	
Total expenses and losses		590,000
Net Income		$120,000
Earnings per share		$ 1.20

Required:
Prepare a multiple-step income statement for 2021 applying generally accepted accounting principles. The income tax rate is 25%. Be sure to include appropriate headings and subtotal titles. An example of a multiple-step income statement can be found in Illustration 4–4 of this chapters.

E 4–6
Discontinued operations
● LO4–4, LO4–5

Chance Company had two operating divisions, one manufacturing farm equipment and the other office supplies. Both divisions are considered separate components as defined by generally accepted accounting principles. The farm equipment component had been unprofitable, and on September 1, 2021, the company adopted a plan to sell the assets of the division. The actual sale was completed on December 15, 2021, at a price of $600,000. The book value of the division's assets was $1,000,000, resulting in a before-tax loss of $400,000 on the sale.

The division incurred a before-tax operating loss from operations of $120,000 from the beginning of the year through December 15. The income tax rate is 25%. Chance's after-tax income from its continuing operations is $550,000.

Required:
Prepare an income statement for 2021 beginning with income from continuing operations. Include appropriate EPS disclosures assuming that 100,000 shares of common stock were outstanding throughout the year.

E 4–7
Income statement presentation; discontinued operations; restructuring costs
● LO4–1, LO4–3, LO4–4

Esquire Comic Book Company had income before tax of $1,000,000 in 2021 *before* considering the following material items:

1. Esquire sold one of its operating divisions, which qualified as a separate component according to generally accepted accounting principles. The before-tax loss on disposal was $340,000. The division generated before-tax income from operations from the beginning of the year through disposal of $500,000.
2. The company incurred restructuring costs of $80,000 during the year.

Required:
Prepare a 2021 income statement for Esquire beginning with income from continuing operations. Assume an income tax rate of 25%. Ignore EPS disclosures.

E 4–8
Discontinued operations; disposal in subsequent year
● LO4–4

Kandon Enterprises, Inc., has two operating divisions; one manufactures machinery and the other breeds and sells horses. Both divisions are considered separate components as defined by generally accepted accounting principles. The horse division has been unprofitable, and, on November 15, 2021, Kandon adopted a formal plan to sell the division. The sale was completed on April 30, 2022. At December 31, 2021, the component was considered held for sale.

On December 31, 2021, the company's fiscal year-end, the book value of the assets of the horse division was $240,000. On that date, the fair value of the assets, less costs to sell, was $200,000. The before-tax loss from operations of the division for the year was $140,000. The company's effective tax rate is 25%. The after-tax income from continuing operations for 2021 was $400,000.

Required:
1. Prepare a partial income statement for 2021 beginning with income from continuing operations. Ignore EPS disclosures.
2. Repeat requirement 1 assuming that the estimated net fair value of the horse division's assets was $400,000, instead of $200,000.

E 4–9
Discontinued operations; disposal in subsequent year; solving for unknown
● LO4–4

On September 17, 2021, Ziltech, Inc., entered into an agreement to sell one of its divisions that qualifies as a component of the entity according to generally accepted accounting principles. By December 31, 2021, the company's fiscal year-end, the division had not yet been sold, but was considered held for sale. The net fair value (fair value minus costs to sell) of the division's assets at the end of the year was $11 million. The pretax income from operations of the division during 2021 was $4 million. Pretax income from continuing operations for the year totaled $14 million. The income tax rate is 25%. Ziltech reported net income for the year of $7.2 million.

Required:
Determine the book value of the division's assets on December 31, 2021.

E 4–10
Earnings per share
● LO4–5

The Esposito Import Company had 1 million shares of common stock outstanding during 2021. Its income statement reported the following items: income from continuing operations, $5 million; loss from discontinued operations, $1.6 million. All of these amounts are net of tax.

Required:
Prepare the 2021 EPS presentation for the Esposito Import Company.

E 4–11
Comprehensive income
● LO4–6

The Massoud Consulting Group reported net income of $1,354,000 for its fiscal year ended December 31, 2021. In addition, during the year the company experienced a positive foreign currency translation adjustment of $240,000 and an unrealized loss on debt securities of $80,000. The company's effective tax rate on all items affecting comprehensive income is 25%. Each component of other comprehensive income is displayed net of tax.

Required:
Prepare a separate statement of comprehensive income for 2021.

E 4–12
Statement of cash flows; classifications
● LO4–8

The statement of cash flows classifies all cash inflows and outflows into one of the three categories shown below and lettered from a through c. In addition, certain transactions that do not involve cash are reported in the statement as noncash investing and financing activities, labeled d.

a. Operating activities
b. Investing activities
c. Financing activities
d. Noncash investing and financing activities

Required:
For each of the following transactions, use the letters above to indicate the appropriate classification category.

1._____ Purchase of equipment for cash.
2._____ Payment of employee salaries.
3._____ Collection of cash from customers.
4._____ Cash proceeds from notes payable.
5._____ Purchase of common stock of another corporation for cash.
6._____ Issuance of common stock for cash.
7._____ Sale of equipment for cash.
8._____ Payment of interest on notes payable.
9._____ Issuance of bonds payable in exchange for land and building.
10._____ Payment of cash dividends to shareholders.
11._____ Payment of principal on notes payable

E 4–13
Statement of cash flows preparation; direct method
● LO4–8

The following summary transactions occurred during 2021 for Bluebonnet Bakers:

Cash Received from:	
Collections from customers	$380,000
Interest on notes receivable	6,000
Collection of notes receivable	50,000
Sale of investments	30,000
Issuance of notes payable	100,000
Cash Paid for:	
Purchase of inventory	160,000
Interest on notes payable	5,000
Purchase of equipment	85,000
Salaries to employees	90,000
Payment of notes payable	25,000
Dividends to shareholders	20,000

The balance of cash and cash equivalents at the beginning of 2021 was $17,000.

Required:
Prepare a statement of cash flows for 2021 for Bluebonnet Bakers. Use the direct method for reporting operating activities.

E 4–14
IFRS; statement of cash flows
● LO4–8, LO4–9
IFRS

Refer to the situation described in E 4–13.

Required:
Prepare the statement of cash flows assuming that Bluebonnet prepares its financial statements according to International Financial Reporting Standards. Where IFRS allows flexibility, use the classification used most often in IFRS financial statements.

E 4–15
Indirect method; reconciliation of net income to net cash flows from operating activities
● LO4–8

The accounting records of Hampton Company provided the data below ($ in thousands).

Net income	$17,300
Depreciation expense	7,800
Increase in accounts receivable	4,000
Decrease in inventory	5,500
Decrease in prepaid insurance	1,200
Decrease in salaries payable	2,700
Increase in interest payable	800

Required:
Prepare a reconciliation of net income to net cash flows from operating activities.

E 4–16
Statement of cash flows; directly from transactions
● LO4–8

The following transactions occurred during March 2021 for the Wainwright Corporation. The company owns and operates a wholesale warehouse. [These are the same transactions analyzed in Exercise 2–1, when we determined their effect on elements of the accounting equation.]

1. Issued 30,000 shares of common stock in exchange for $300,000 in cash.

2. Purchased equipment at a cost of $40,000. $10,000 cash was paid and a note payable to the seller was signed for the balance owed.
3. Purchased inventory on account at a cost of $90,000. The company uses the perpetual inventory system.
4. Credit sales for the month totaled $120,000. The cost of the goods sold was $70,000.
5. Paid $5,000 in rent on the warehouse building for the month of March.
6. Paid $6,000 to an insurance company for fire and liability insurance for a one-year period beginning April 1, 2021.
7. Paid $70,000 on account for the merchandise purchased in 3.
8. Collected $55,000 from customers on account.
9. Recorded depreciation expense of $1,000 for the month on the equipment.

Required:

1. Analyze each transaction and classify each as a financing, investing, and/or operating activity (a transaction can represent more than one type of activity). In doing so, also indicate the cash effect of each, if any. If there is no cash effect, simply place a check mark (√) in the appropriate column(s).

   Example:

Operating	Investing	Financing
1.		$300,000

2. Prepare a statement of cash flows, using the direct method to present cash flows from operating activities. Assume the cash balance at the beginning of the month was $40,000.

E 4–17
Statement of cash flows; indirect method
● LO4–8

Cemptex Corporation prepares its statement of cash flows using the indirect method to report operating activities. Net income for the 2021 fiscal year was $624,000. Depreciation and amortization expense of $87,000 was included with operating expenses in the income statement. The following information describes the changes in current assets and liabilities other than cash:

Decrease in accounts receivable	$22,000
Increase in inventory	9,200
Increase prepaid expenses	8,500
Increase in salaries payable	10,000
Decrease in income taxes payable	14,000

Required:
Prepare the operating activities section of the 2021 statement of cash flows.

E 4–18
Statement of cash flows; indirect method
● LO4–8

Chew Corporation prepares its statement of cash flows using the indirect method of reporting operating activities. Net income for the 2021 fiscal year was $1,250,000. Depreciation expense of $140,000 was included with operating expenses in the income statement. The following information describes the changes in current assets and liabilities other than cash:

Increase in accounts receivable	$152,000
Decrease in inventory	108,000
Decrease prepaid expenses	62,000
Decrease in salaries payable	30,000
Increase in income taxes payable	44,000

Required:
Calculate cash flows from operating activities for 2021.

E 4–19
IFRS; statement of cash flows
● LO4–8, LO4–9
IFRS

The statement of cash flows for the year ended December 31, 2021, for Bronco Metals is presented below.

BRONCO METALS
Statement of Cash Flows
For the Year Ended December 31, 2021

Cash flows from operating activities:	
Collections from customers	$353,000
Interest on notes receivable	4,000
Dividends received from investments	2,400
Purchase of inventory	(186,000)
Payment of operating expenses	(67,000)

(continued)

(concluded)

Payment of interest on notes payable	(8,000)	
Net cash flows from operating activities		$ 98,400
Cash flows from investing activities:		
Collection of notes receivable	100,000	
Purchase of equipment	(154,000)	
Net cash flows from investing activities		(54,000)
Cash flows from financing activities:		
Proceeds from issuance of common stock	200,000	
Dividends paid to shareholders	(40,000)	
Net cash flows from financing activities		160,000
Net increase in cash		204,400
Cash and cash equivalents, January 1		28,600
Cash and cash equivalents, December 31		$233,000

Required:
Prepare the statement of cash flows assuming that Bronco prepares its financial statements according to International Financial Reporting Standards. Where IFRS allows flexibility, use the classification used most often in IFRS financial statements.

E 4–20
Statement of cash flows; indirect method
● LO4–8

Presented below is the 2021 income statement and comparative balance sheet information for Tiger Enterprises.

TIGER ENTERPRISES
Income Statement
For the Year Ended December 31, 2021

($ in thousands)		
Sales revenue		$7,000
Operating expenses:		
Cost of goods sold	$3,360	
Depreciation expense	240	
Insurance expense	100	
General and administrative expense	1,800	
Total operating expenses		5,500
Income before income taxes		1,500
Income tax expense		(600)
Net income		$ 900

Balance Sheet Information ($ in thousands)	Dec. 31, 2021	Dec. 31, 2020
Assets:		
Cash	$ 300	$ 200
Accounts receivable	750	830
Inventory	640	600
Prepaid insurance	50	20
Equipment	2,100	1,800
Less: Accumulated depreciation	(840)	(600)
Total assets	$3,000	$2,850
Liabilities and Shareholders' Equity:		
Accounts payable	$ 300	$ 360
Accrued liabilities (for general & admin. expense)	300	400
Income taxes payable	200	150
Notes payable (due 12/31/2022)	800	600
Common stock	900	800
Retained earnings	500	540
Total liabilities and shareholders' equity	$3,000	$2,850

Required:
Prepare Tiger's statement of cash flows, using the indirect method to present cash flows from operating activities. (*Hint:* You will have to calculate dividend payments).

E 4–21
Statement of cash flows; direct method
● LO4–8

Refer to the situation described in E 4–20.

Required:
Prepare the cash flows from operating activities section of Tiger's 2021 statement of cash flows using the direct method. Assume that all purchases and sales of inventory are on account, and that there are no anticipated bad

debts for accounts receivable. (*Hint:* Use the Concept Review Exercise at the end of Part B as a guide to help determine your answers).

E 4–22
FASB codification research
● LO4–5

Access the *FASB Accounting Standards Codification* at the FASB website (**www.fasb.org**). Determine each of the following:

1. The topic number (Topic XXX) that provides the accounting for earnings per share.
2. The specific eight-digit Codification citation (XXX-XX-XX-X) that describes the additional information for earnings per share that must be included in the notes to the financial statements.
3. The specific eight-digit Codification citation (XXX-XX-XX-X) that requires disclosure of transactions affecting the number of common shares outstanding that occur after the most recent reporting period but before the financial statements are issued.

E 4–23
FASB codification research
● LO4–5, LO4–6, LO4–8

Access the *FASB Accounting Standards Codification* at the FASB website (**www.fasb.org**). Determine the specific eight-digit Codification citation (XXX-XX-XX-X) for each of the following:

1. The calculation of the weighted average number of shares for basic earnings per share purposes.
2. The alternative formats permissible for reporting comprehensive income.
3. The classifications of cash flows required in the statement of cash flows.

E 4–24
Concepts; terminology
● LO4–1, LO4–2, LO4–3, LO4–4, LO4–5, LO4–6, LO4–7, LO4–8

Listed below are several terms and phrases associated with income statement presentation and the statement of cash flows. Pair each item from List A (by letter) with the item from List B that is most appropriately associated with it.

List A	List B
_____ 1. Intraperiod tax allocation	a. An other comprehensive income item.
_____ 2. Comprehensive income	b. Starts with net income and works backwards to convert to cash.
_____ 3. Unrealized gain on debt securities	c. Reports the cash effects of each operating activity directly on the statement.
_____ 4. Operating income	d. Correction of a material error of a prior period.
_____ 5. A discontinued operation	e. Related to the external financing of the company.
_____ 6. Earnings per share	f. Associates tax with income statement item.
_____ 7. Prior period adjustment	g. Total nonowner change in equity.
_____ 8. Financing activities	h. Related to the transactions entering into the determination of net income.
_____ 9. Operating activities (SCF)	i. Related to the acquisition and disposition of long-term assets.
_____ 10. Investing activities	j. Required disclosure for publicly traded corporation.
_____ 11. Direct method	k. A component of an entity.
_____ 12. Indirect method	l. Directly related to principal revenue-generating activities.

E 4–25
Inventory turnover; calculation and evaluation
● LO4–10

The following is a portion of the condensed income statement for Rowan, Inc., a manufacturer of plastic containers:

Net sales		$2,460,000
Less: Cost of goods sold:		
Inventory, January 1	$ 630,000	
Net purchases	1,900,000	
Inventory, December 31	(690,000)	1,840,000
Gross profit		$ 620,000

Required:

1. Determine Rowan's inventory turnover.
2. What information does this ratio provide?

E 4–26
Evaluating efficiency of asset management
● LO4–10

The 2021 income statement of Anderson Medical Supply Company reported net sales of $8 million, cost of goods sold of $4.8 million, and net income of $800,000. The following table shows the company's comparative balance sheets for 2021 and 2020:

	($ in thousands)	
	2021	2020
Assets		
Cash	$ 300	$ 380
Accounts receivable	700	500
Inventory	900	700
Property, plant, and equipment (net)	2,400	2,120
Total assets	$4,300	$3,700
Liabilities and Shareholders' Equity		
Current liabilities	$ 960	$ 830
Bonds payable	1,200	1,200
Common stock	1,000	1,000
Retained earnings	1,140	670
Total liabilities and shareholders' equity	$4,300	$3,700

Some industry averages for Anderson's line of business are

Inventory turnover	5 times
Average collection period	25 days
Asset turnover	1.8 times

Required:

1. Determine the following ratios for 2021:
   a. Inventory turnover
   b. Receivables turnover
   c. Average collection period
   d. Asset turnover
2. Assess Anderson's asset management relative to its industry.

E 4–27
Profitability ratios
● LO4–10

The following condensed information was reported by Peabody Toys, Inc., for 2021 and 2020:

	($ in thousands)	
	2021	2020
Income statement information		
Net sales	$5,200	$4,200
Net income	180	124
Balance sheet information		
Current assets	$ 800	$ 750
Property, plant, and equipment (net)	1,100	950
Total assets	$1,900	$1,700
Current liabilities	$ 600	$ 450
Long-term liabilities	750	750
Common stock	400	400
Retained earnings	150	100
Liabilities and shareholders' equity	$1,900	$1,700

Required:

1. Determine the following ratios for 2021:
   a. Profit margin on sales
   b. Return on assets
   c. Return on equity
2. Determine the amount of dividends paid to shareholders during 2021.

E 4–28
DuPont analysis
● LO4–10

This exercise is based on the Peabody Toys, Inc., data from E 4–27.

Required:

1. Determine the following components of the DuPont framework for 2021:
   a. Profit margin on sales

b. Asset turnover
c. Equity multiplier
d. Return on equity

2. Write an equation that relates these components in calculating ROE. Use the Peabody Toys data to show that the equation is correct.

E 4–29
Interim financial statements; income tax expense
● Appendix 4

Joplin Laminating Corporation reported income before income taxes during the first three quarters, and management's estimates of the annual effective tax rate at the end of each quarter as shown below:

	Quarter		
	First	**Second**	**Third**
Income before income taxes	$50,000	$40,000	$100,000
Estimated annual effective tax rate	22%	25%	24%

Required:
Determine the income tax expense to be reported in the income statement in each of the three quarterly reports.

E 4–30
Interim reporting; recognizing expenses
● Appendix 4

Security-Rand Corporation determines executive incentive compensation at the end of its fiscal year. At the end of the first quarter, management estimated that the amount will be $300 million. Depreciation expense for the year is expected to be $60 million. Also during the quarter, the company realized a gain of $23 million from selling two of its manufacturing plants.

Required:
What amounts for these items should be reported in the first quarter's income statement?

E 4–31
Interim financial statements; reporting expenses
● Appendix 4

Shields Company is preparing its interim report for the second quarter ending June 30. The following payments were made during the first two quarters:

Required:

Expenditure	Date	Amount
Annual advertising	January	$800,000
Property tax for the fiscal year	February	350,000
Annual equipment repairs	March	260,000
One-time research and development fee to consultant	May	96,000

For each expenditure, indicate the amount that would be reported in the quarterly income statements for the periods ending March 31, June 30, September 30, and December 31.

E 4–32
Interim financial statements
● Appendix 4
IFRS

Assume the same facts as in E 4–31, but that Shields Company reports under IFRS. For each expenditure, indicate the amount that would be reported in the quarterly income statements for the periods ending March 31, June 30, September 30, and December 31.

Problems

connect

P 4–1
Comparative income statements; multiple-step format
● LO4–1, LO4–3, LO4–4, LO4–5

Selected information about income statement accounts for the Reed Company is presented below (the company's fiscal year ends on December 31).

	2021	2020
Sales revenue	$4,400,000	$3,500,000
Cost of goods sold	2,860,000	2,000,000
Administrative expense	800,000	675,000
Selling expense	360,000	302,000
Interest revenue	150,000	140,000
Interest expense	200,000	200,000
Loss on sale of assets of discontinued component	48,000	—

On July 1, 2021, the company adopted a plan to discontinue a division that qualifies as a component of an entity as defined by GAAP. The assets of the component were sold on September 30, 2021, for $48,000 less than their book value. Results of operations for the component (*included* in the above account balances) were as follows:

	1/1/2021–9/30/2021	2020
Sales revenue	$400,000	$500,000
Cost of goods sold	(290,000)	(320,000)
Administrative expense	(50,000)	(40,000)
Selling expense	(20,000)	(20,000)
Operating income before taxes	$ 40,000	$120,000

In addition to the account balances above, several events occurred during 2021 that have *not* yet been reflected in the above accounts:

1. A fire caused $50,000 in uninsured damages to the main office building. The fire was considered to be an unusual event.
2. Inventory that had cost $40,000 had become obsolete because a competitor introduced a better product. The inventory was written down to its scrap value of $5,000.
3. Income taxes have not yet been recorded.

Required:

Prepare a multiple-step income statement for the Reed Company for 2021 showing 2020 information in comparative format, including income taxes computed at 25% and EPS disclosures assuming 300,000 shares of outstanding common stock.

P 4–2
Discontinued operations
● LO4–4

The following condensed income statements of the Jackson Holding Company are presented for the two years ended December 31, 2021 and 2020:

	2021	2020
Sales revenue	$15,000,000	$9,600,000
Cost of goods sold	9,200,000	6,000,000
Gross profit	5,800,000	3,600,000
Operating expenses	3,200,000	2,600,000
Operating income	2,600,000	1,000,000
Gain on sale of division	600,000	—
	3,200,000	1,000,000
Income tax expense	1,280,000	400,000
Net income	$ 1,920,000	$ 600,000

On October 15, 2021, Jackson entered into a tentative agreement to sell the assets of one of its divisions. The division qualifies as a component of an entity as defined by GAAP. The division was sold on December 31, 2021, for $5,000,000. Book value of the division's assets was $4,400,000. The division's contribution to Jackson's operating income before-tax for each year was as follows:

2021	$400,000
2020	$300,000

Assume an income tax rate of 25%.

Required:

1. Prepare revised income statements according to generally accepted accounting principles, beginning with income from continuing operations before income taxes. Ignore EPS disclosures.
2. Assume that by December 31, 2021, the division had not yet been sold but was considered held for sale. The fair value of the division's assets on December 31 was $5,000,000. What would be the amount presented for discontinued operations?
3. Assume that by December 31, 2021, the division had not yet been sold but was considered held for sale. The fair value of the division's assets on December 31 was $3,900,000. What would be the amount presented for discontinued operations?

P 4–3
Income statement presentation; discontinued operations; accounting error
● LO4–4, LO4–5

For the year ending December 31, 2021, Olivo Corporation had income from continuing operations before taxes of $1,200,000 before considering the following transactions and events. All of the items described below are before taxes and the amounts should be considered material.

1. In November 2021, Olivo sold its PizzaPasta restaurant chain that qualified as a component of an entity. The company had adopted a plan to sell the chain in May 2021. The income from operations of the chain from January 1, 2021, through November was $160,000 and the loss on sale of the chain's assets was $300,000.
2. In 2021, Olivo sold one of its six factories for $1,200,000. At the time of the sale, the factory had a book value of $1,100,000. The factory was not considered a component of the entity.
3. In 2019, Olivo's accountant omitted the annual adjustment for patent amortization expense of $120,000. The error was not discovered until December 2021.

Required:
Prepare Olivo's income statement, beginning with income from continuing operations before taxes, for the year ended December 31, 2021. Assume an income tax rate of 25%. Ignore EPS disclosures.

P 4–4
Restructuring costs; discontinued operations; accounting error
● LO4–3, LO4–4, LO4–5

The preliminary 2021 income statement of Alexian Systems, Inc., is presented below:

ALEXIAN SYSTEMS, INC.
Income Statement
For the Year Ended December 31, 2021
($ in millions, except earnings per share)

Revenues and gains:	
Sales revenue	$ 425
Interest revenue	4
Other income	126
Total revenues and gains	555
Expenses:	
Cost of goods sold	245
Selling and administrative expense	154
Income tax expense	39
Total expenses	438
Net Income	$ 117
Earnings per share	$5.85

Additional Information:
1. Selling and administrative expense includes $26 million in restructuring costs.
2. Included in other income is $120 million in income from a discontinued operation. This consists of $90 million in operating income and a $30 million gain on disposal. The remaining $6 million is from the gain on sale of investments.
3. Cost of goods sold was increased by $40 million to correct an error in the calculation of 2020's ending inventory. The amount is material.

Required:
For each of the three additional facts listed in the additional information, discuss the appropriate presentation of the item described. Do not prepare a revised statement.

P 4–5
Income statement presentation; restructuring costs; discontinued operations; accounting error
● LO4–1, LO4–3, LO4–4, LO4–5

[This is a variation of the previous problem focusing on income statement presentation.]

Required:
Refer to the information presented in P 4–4. Prepare a revised income statement for 2021 reflecting the additional facts. Use a multiple-step format similar to Illustration 4–4 of this chapter to prepare income from continuing operations, and then add to this the discontinued operations portion of the income statement. Assume that an income tax rate of 25% applies to all income statement items, and that 20 million shares of common stock were outstanding throughout the year.

P 4–6
Income statement presentation; discontinued operations; EPS
● LO4–1, LO4–3, LO4–4, LO4–5

Rembrandt Paint Company had the following income statement items for the year ended December 31, 2021 ($ in thousands):

Sales revenue	$18,000
Interest revenue	100
Interest expense	300
Cost of goods sold	10,500
Selling and administrative expense	2,500
Restructuring costs	800

In addition, during the year, the company completed the disposal of its plastics business and incurred a loss from operations of $1.6 million and a gain on disposal of the component's assets of $2 million. There were 500,000 shares of common stock outstanding throughout 2021. Income tax expense has not yet been recorded. The income tax rate is 25% on all items of income (loss)

Required:

Prepare a multiple-step income statement for 2021, including EPS disclosures. Use a multiple-step format similar to Illustration 4–4 of this chapter to prepare income from continuing operations, and then add to this the discontinued operations portion of the income statement.

P 4–7
Income statement presentation; statement of comprehensive income; unusual items
● LO4–1, LO4–3 through LO4–6

The following income statement items appeared on the adjusted trial balance of Schembri Manufacturing Corporation for the year ended December 31, 2021 ($ in thousands): sales revenue, $15,300; cost of goods sold, $6,200; selling expenses, $1,300; general and administrative expenses, $800; interest revenue, $40; interest expense, $180. Income taxes have not yet been recorded. The company's income tax rate is 25% on all items of income or loss. These revenue and expense items appear in the company's income statement every year. The company's controller, however, has asked for your help in determining the appropriate treatment of the following nonrecurring transactions that also occurred during 2021 ($ in thousands). All transactions are material in amount.

1. Investments were sold during the year at a loss of $220. Schembri also had an unrealized gain of $320 for the year on investments in debt securities that qualify as components of comprehensive income.
2. One of the company's factories was closed during the year. Restructuring costs incurred were $1,200.
3. During the year, Schembri completed the sale of one of its operating divisions that qualifies as a component of the entity according to GAAP. The division had incurred a loss from operations of $560 in 2021 prior to the sale, and its assets were sold at a gain of $1,400.
4. In 2021, the company's accountant discovered that depreciation expense in 2020 for the office building was understated by $200.
5. Negative foreign currency translation adjustment for the year totaled $240.

Required:

1. Prepare Schembri's single, continuous multiple-step statement of comprehensive income for 2021, including earnings per share disclosures. There were 1,000,000 shares of common stock outstanding at the beginning of the year and an additional 400,000 shares were issued on July 1, 2021. Use a multiple-step format similar to the one in the Concept Review Exercise at the end of Part A of this chapter.
2. Prepare a separate statement of comprehensive income for 2021.

P 4–8
Multiple-step statement of income and comprehensive income
● LO4–1, LO4–3, LO4–5, LO4–6

Duke Company's records show the following account balances at December 31, 2021:

Sales revenue	$15,000,000
Cost of goods sold	9,000,000
General and administrative expense	1,000,000
Selling expense	500,000
Interest expense	700,000

Income tax expense has not yet been determined. The following events also occurred during 2021. All transactions are material in amount.

1. $300,000 in restructuring costs were incurred in connection with plant closings.
2. Inventory costing $400,000 was written off as obsolete. Material losses of this type are considered to be unusual.
3. It was discovered that depreciation expense for 2020 was understated by $50,000 due to a mathematical error.
4. The company experienced a negative foreign currency translation adjustment of $200,000 and had an unrealized gain on debt securities of $180,000.

Required:

Prepare a single, continuous multiple-step statement of comprehensive income for 2021. The company's effective tax rate on all items affecting comprehensive income is 25%. Each component of other comprehensive income should be displayed net of tax. Ignore EPS disclosures. Use a multiple-step format similar to the one in the Concept Review Exercise at the end of Part A of this chapter (excluding discontinued operations shown there).

P 4–9
Statement of cash flows
● LO4–8

The Diversified Portfolio Corporation provides investment advice to customers. A condensed income statement for the year ended December 31, 2021, appears below:

Service revenue	$900,000
Operating expenses	700,000
Income before income taxes	200,000
Income tax expense	50,000
Net income	$150,000

The following balance sheet information also is available:

	12/31/2021	12/31/2020
Cash	$305,000	$ 70,000
Accounts receivable	120,000	100,000
Accrued liabilities (for operating expenses)	70,000	60,000
Income taxes payable	10,000	15,000

In addition, the following transactions took place during the year:

1. Common stock was issued for $100,000 in cash.
2. Long-term investments were sold for $50,000 in cash. The original cost of the investments also was $50,000.
3. $80,000 in cash dividends was paid to shareholders.
4. The company has no outstanding debt, other than those payables listed above.
5. Operating expenses include $30,000 in depreciation expense.

Required:

1. Prepare a statement of cash flows for 2021 for the Diversified Portfolio Corporation. Use the direct method for reporting operating activities. Use a format similar to the one in the Concept Review Exercise at the end of Part B of this chapter.
2. Prepare the cash flows from operating activities section of Diversified's 2021 statement of cash flows using the indirect method. Use a format similar to the one in the Concept Review Exercise at the end of Part B of this chapter.

P 4–10
Integration of financial statements; Chapters 3 and 4
● LO4–8

The chief accountant for Grandview Corporation provides you with the company's 2021 statement of cash flows and income statement. The accountant has asked for your help with some missing figures in the company's comparative balance sheets. These financial statements are shown next ($ in millions).

GRANDVIEW CORPORATION
Statement of Cash Flows
For the Year Ended December 31, 2021

Cash Flows from Operating Activities:		
Collections from customers	$71	
Payment to suppliers	(30)	
Payment of general & administrative expense	(18)	
Payment of income taxes	(9)	
Net cash flows from operating activities		$14
Cash Flows from Investing Activities:		
Sale of investments		65
Cash Flows from Financing Activities:		
Issuance of common stock	10	
Payment of dividends	(3)	
Net cash flows from financing activities		7
Net increase in cash		$86

GRANDVIEW CORPORATION
Income Statement
For the Year Ended December 31, 2021

Sales revenue		$80
Cost of goods sold		32
Gross profit		48
Operating expenses:		
General and administrative expense	$18	
Depreciation expense	10	
Total operating expenses		28
Operating income		20
Other income:		
Gain on sale of investments		15
Income before income taxes		35
Income tax expense		7
Net income		$28

(continued)

(concluded)

GRANDVIEW CORPORATION
Balance Sheets
At December 31

	2021	2020
Assets:		
Cash	$ 145	$?
Accounts receivable	?	84
Investments	—	50
Inventory	60	?
Property, plant & equipment	150	150
Less: Accumulated depreciation	(65)	?
Total assets	?	?
Liabilities and Shareholders' Equity:		
Accounts payable	$ 40	$ 30
Accrued liabilities (for selling & admin. expense)	9	9
Income taxes payable	22	?
Common stock	240	230
Retained earnings	?	47
Total liabilities and shareholders' equity	?	?

Required:

1. Calculate the missing amounts.
2. Prepare the operating activities section of Grandview's 2021 statement of cash flows using the indirect method. Use a format similar to the one in the Concept Review Exercise at the end of Part B of this chapter.

P 4–11
Statement of cash flows; indirect method
● LO4–8

Presented below are the 2021 income statement and comparative balance sheets for Santana Industries.

SANTANA INDUSTRIES
Income Statement
For the Year Ended December 31, 2021
($ in thousands)

Sales revenue	$14,250	
Service revenue	3,400	
Total revenue		$17,650
Operating expenses:		
Cost of goods sold	7,200	
Selling expense	2,400	
General and administrative expense	1,500	
Total operating expenses		11,100
Operating income		6,550
Interest expense		150
Income before income taxes		6,400
Income tax expense		1,600
Net income		$ 4,800

Balance Sheet Information ($ in thousands)	**Dec. 31, 2021**	**Dec. 31, 2020**
Assets:		
Cash	$ 8,300	$ 2,200
Accounts receivable	2,500	2,200
Inventory	4,000	3,000
Prepaid rent	150	300
Equipment	14,500	12,000
Less: Accumulated depreciation	(5,100)	(4,500)
Total assets	$24,350	$15,200
Liabilities and shareholders' equity:		
Accounts payable	$ 1,400	$ 1,100
Interest payable	100	0
Deferred revenue	800	600
Income taxes payable	550	800
Notes payable (due 12/31/2023)	5,000	0
Common stock	10,000	10,000
Retained earnings	6,500	2,700
Total liabilities and shareholders' equity	$24,350	$15,200

Additional information for the 2021 fiscal year ($ in thousands):

1. Cash dividends of $1,000 were declared and paid.
2. Equipment costing $4,000 was purchased with cash.
3. Equipment with a book value of $500 (cost of $1,500 less accumulated depreciation of $1,000) was sold for $500.
4. Depreciation of $1,600 is included in operating expenses.

Required:

Prepare Santana Industries' 2021 statement of cash flows, using the indirect method to present cash flows from operating activities. Use a format similar to the one in the Concept Review Exercise at the end of Part B of this chapter.

P 4–12
Calculating activity and profitability ratios
● LO4–10

Financial statements for Askew Industries for 2021 are shown below (in thousands):

2021 Income Statement

Net sales	$9,000
Cost of goods sold	(6,300)
Gross profit	2,700
Operating expenses	(2,100)
Interest expense	(200)
Income tax expense	(100)
Net income	$ 300

Comparative Balance Sheets

	Dec. 31 2021	2020
Assets		
Cash	$ 600	$ 500
Accounts receivable	600	400
Inventory	800	600
Property, plant, and equipment (net)	2,000	2,100
	$4,000	$3,600
Liabilities and Shareholders' Equity		
Current liabilities	$1,100	$ 850
Bonds payable	1,400	1,400
Common stock	600	600
Retained earnings	900	750
	$4,000	$3,600

Required:

Calculate the following ratios for 2021.

1. Inventory turnover ratio
2. Average days in inventory
3. Receivables turnover ratio
4. Average collection period
5. Asset turnover ratio
6. Profit margin on sales
7. Return on assets
8. Return on equity
9. Equity multiplier
10. Return on equity (using the DuPont framework)

P 4–13
Use of ratios to compare two companies in the same industry
● LO4–10

Presented below are condensed financial statements adapted from those of two actual companies competing in the pharmaceutical industry—**Johnson and Johnson (J&J)** and **Pfizer, Inc.** ($ in millions, except per share amounts).

Required:

Evaluate and compare the two companies by responding to the following questions. ***Note:*** Because two-year comparative statements are not provided, you should use year-end balances in place of average balances as appropriate.

1. Which of the two companies appears more efficient in collecting its accounts receivable and managing its inventory?
2. Which of the two firms had greater earnings relative to resources available?
3. Have the two companies achieved their respective rates of return on assets with similar combinations of profit margin and turnover?
4. From the perspective of a common shareholder, which of the two firms provided a greater rate of return?
5. From the perspective of a common shareholder, which of the two firms appears to be using leverage more effectively to provide a return to shareholders above the rate of return on assets?

Balance Sheets
($ in millions, except per share data)

	J&J	Pfizer
Assets:		
Cash	$ 5,377	$ 1,520
Short-term investments	4,146	10,432
Accounts receivable (net)	6,574	8,775
Inventory	3,588	5,837
Other current assets	3,310	3,177
Current assets	22,995	29,741
Property, plant, and equipment (net)	9,846	18,287
Intangibles and other assets	15,422	68,747
Total assets	$48,263	$116,775
Liabilities and Shareholders' Equity:		
Accounts payable	$ 4,966	$2,601
Short-term notes	1,139	8,818
Other current liabilities	7,343	12,238
Current liabilities	13,448	23,657
Long-term debt	2,955	5,755
Other long-term liabilities	4,991	21,986
Total liabilities	21,394	51,398
Common stock (par and additional paid-in capital)	3,120	67,050
Retained earnings	30,503	29,382
Accumulated other comprehensive income (loss)	(590)	195
Less: Treasury stock and other equity adjustments	(6,164)	(31,250)
Total shareholders' equity	26,869	65,377
Total liabilities and shareholders' equity	$48,263	$116,775
Income Statements		
Net sales	$41,862	$45,188
Cost of goods sold	12,176	9,832
Gross profit	29,686	35,356
Operating expenses	19,763	28,486
Other (income) expense—net	(385)	3,610
Income before taxes	10,308	3,260
Income tax expense	3,111	1,621
Net income	$ 7,197	$ 1,639*
Basic net income per share	$ 2.42	$ 0.22

*This is before income from discontinued operations.

P 4–14
Creating a balance sheet from ratios; Chapters 3 and 4
● LO4–10

Cadux Candy Company's income statement for the year ended December 31, 2021, reported interest expense of $2 million and income tax expense of $12 million. Current assets listed in its balance sheet include cash, accounts receivable, and inventory. Property, plant, and equipment is the company's only noncurrent asset. Financial ratios for 2021 are listed below. Profitability and turnover ratios with balance sheet items in the denominator were calculated using year-end balances rather than averages.

Debt to equity ratio	1.0
Current ratio	2.0
Acid-test ratio	1.0
Times interest earned ratio	17 times
Return on assets	10%
Return on equity	20%

(continued)

(concluded)

Profit margin on sales	5%
Gross profit margin (gross profit divided by net sales)	40%
Inventory turnover	8 times
Receivables turnover	20 times

Required:
Prepare a December 31, 2021, balance sheet for the Cadux Candy Company.

P 4–15
Compare two companies in the same industry; Chapters 3 and 4
● LO4–10

Presented below are condensed financial statements adapted from those of two actual companies competing as the primary players in a specialty area of the food manufacturing and distribution industry ($ in millions, except per share amounts).

	BalanceSheets	
	Metropolitan	**Republic**
Assets:		
Cash	$ 179.3	$ 37.1
Accounts receivable (net)	422.7	325.0
Short-term investments	—	4.7
Inventory	466.4	635.2
Prepaid expenses and other current assets	134.6	476.7
Current assets	1,203.0	1,478.7
Property, plant, and equipment (net)	2,608.2	2,064.6
Intangibles and other assets	210.3	464.7
Total assets	$4,021.5	$4,008.0
Liabilities and Shareholders' Equity		
Accounts payable	$ 467.9	$ 691.2
Short-term notes	227.1	557.4
Accruals and other current liabilities	585.2	538.5
Current liabilities	1,280.2	1,787.1
Long-term debt	535.6	542.3
Deferred tax liability	384.6	610.7
Other long-term liabilities	104.0	95.1
Total liabilities	2,304.4	3,035.2
Common stock (par and additional paid-in capital)	144.9	335.0
Retained earnings	2,476.9	1,601.9
Less: Treasury stock	(904.7)	(964.1)
Total liabilities and shareholders' equity	$4,021.5	$4,008.0
Income Statements		
Net sales	$5,698.0	$7,768.2
Cost of goods sold	(2,909.0)	(4,481.7)
Gross profit	2,789.0	3,286.5
Operating expenses	(1,743.7)	(2,539.2)
Interest expense	(56.8)	(46.6)
Income before taxes	988.5	700.7
Income tax expense	(394.7)	(276.1)
Net income	$ 593.8	$ 424.6
Net income per share	$ 2.40	$ 6.50

Required:
Evaluate and compare the two companies by responding to the following questions.

Note: Because comparative statements are not provided you should use year-end balances in place of average balances as appropriate.

1. Which of the two firms had greater earnings relative to resources available?
2. Have the two companies achieved their respective rates of return on assets with similar combinations of profit margin and turnover?
3. From the perspective of a common shareholder, which of the two firms provided a greater rate of return?
4. Which company is most highly leveraged and which has made most effective use of financial leverage?
5. Of the two companies, which appears riskier in terms of its ability to pay short-term obligations?
6. How efficiently are current assets managed?

7. From the perspective of a creditor, which company offers the most comfortable margin of safety in terms of its ability to pay fixed interest charges?

P 4–16
Interim financial reporting
● Appendix 4

Branson Electronics Company is a small, publicly traded company preparing its first quarter interim report to be mailed to shareholders. The following information for the quarter has been compiled:

Sales revenue		$180,000
Cost of goods sold		35,000
Operating expenses:		
Fixed	$59,000	
Variable	43,000	107,000

Fixed operating expenses include payments of $50,000 to an advertising firm to promote Branson through various media throughout the year. The income tax rate for Branson's level of operations in the first quarter is 20%, but management estimates the effective rate for the entire year will be 25%.

Required:

Prepare the income statement to be included in Branson's first quarter interim report.

Decision Makers' Perspective

©ImageGap/Getty Images

Apply your critical-thinking ability to the knowledge you've gained. These cases will provide you an opportunity to develop your research, analysis, judgment, and communication skills. You also will work with other students, integrate what you've learned, apply it in real-world situations, and consider its global and ethical ramifications. This practice will broaden your knowledge and further develop your decision-making abilities.

Judgment Case 4–1
Earnings quality
● LO4–2, LO4–3

The financial community in the United States has become increasingly concerned with the quality of reported company earnings.

Required:

1. Define the term *earnings quality*.
2. Explain the distinction between permanent and temporary earnings as it relates to the concept of earnings quality.
3. How do earnings management practices affect the quality of earnings?
4. Assume that a manufacturing company's annual income statement included a large gain from the sale of investment securities. What factors would you consider in determining whether or not this gain should be included in an assessment of the company's permanent earnings?

Judgment Case 4–2
Restructuring costs
● LO4–3

The appearance of restructuring costs in corporate income statements increased significantly in the 1980s and 1990s and continues to be relevant today.

Required:

1. What types of costs are included in restructuring costs?
2. When are restructuring costs recognized?
3. How would you classify restructuring costs in a multi-step income statement?
4. What factors would you consider in determining whether or not restructuring costs should be included in an assessment of a company's permanent earnings?

Judgment Case 4–3
Earnings management
● LO4–2, LO4–3

Companies often are under pressure to meet or beat Wall Street earnings projections in order to increase stock prices and also to increase the value of stock options. Some resort to earnings management practices to artificially create desired results.

Required:

Is *earnings management* always intended to produce higher income? Explain.

Real World Case 4–4
Earnings quality and non-GAAP earnings
● LO4–3

Companies often voluntarily provide non-GAAP earnings when they announce annual or quarterly earnings.

Required:

1. What is meant by the term *non-GAAP earnings* in this context?
2. How do non-GAAP earnings relate to the concept of earnings quality?

Research Case 4–5
FASB codification; locate and extract relevant information and cite authoritative support for a financial reporting issue; restructuring costs; exit or disposal cost obligations
● LO4–2, LO4–3

Access the *FASB Accounting Standards Codification* at the FASB website (**www.fasb.org**). Determine each of the following:

1. The topic number (Topic XXX) that addresses exit or disposal cost obligations.
2. The specific eight-digit Codification citation (XXX-XX-XX-X) that addresses the initial measurement of these obligations.
3. The amount for which these obligations and related costs are measured.
4. The specific eight-digit Codification citation (XXX-XX-XX-X) that describes the disclosure requirements in the notes to the financial statements for exit or disposal obligations.
5. List the required disclosures.

Judgment Case 4–6
Income statement presentation
● LO4–3, LO4–4, LO4–5

Each of the following situations occurred during 2021 for one of your audit clients:

1. An inventory write-down due to obsolescence.
2. Discovery that depreciation expenses were omitted by accident from 2020's income statement.
3. The useful lives of all machinery were changed from eight to five years.
4. The depreciation method used for all equipment was changed from the declining-balance to the straight-line method.
5. Restructuring costs were incurred.
6. The Stridewell Company, a manufacturer of shoes, sold all of its retail outlets. It will continue to manufacture and sell its shoes to other retailers. A loss was incurred in the disposition of the retail stores. The retail stores are considered a component of the entity.
7. The inventory costing method was changed from FIFO to average cost.

Required:

1. For each situation, identify the appropriate reporting treatment from the list below (consider each event to be material):
   a. As an unusual gain or loss
   b. As a prior period adjustment
   c. As a change in accounting principle
   d. As a discontinued operation
   e. As a change in accounting estimate
   f. As a change in accounting estimate achieved by a change in accounting principle
2. Indicate whether each situation would be included in the income statement in continuing operations (CO) or below continuing operations (BC), or if it would appear as an adjustment to retained earnings (RE). Use the format shown below to answer requirements 1 and 2.

Situation	Treatment (a–f)	Financial Statement Presentation (CO, BC, or RE)
1.		
2.		
3.		
4.		
5.		
6.		
7.		

Judgment Case 4–7
Income statement presentation
● LO4–3, LO4–4, LO4–5

The following events occurred during 2021 for various audit clients of your firm. Consider each event to be independent and the effect of each event to be material.

1. A manufacturing company recognized a loss on the sale of investments.

2. An automobile manufacturer sold all of the assets related to its financing component. The operations of the financing business is considered a component of the entity.
3. A company changed its depreciation method from the double-declining-balance method to the straight-line method.
4. Due to obsolescence, a company engaged in the manufacture of high-technology products incurred a loss on inventory write-down.
5. One of your clients discovered that 2020's depreciation expense was overstated. The error occurred because of a miscalculation of depreciation for the office building.
6. A cosmetics company decided to discontinue the manufacture of a line of women's lipstick. Other cosmetic lines will be continued. A loss was incurred on the sale of assets related to the lipstick product line. The operations of the discontinued line is not considered a component of the entity.

Required:

Determine whether each of the above events would be reported as income from continuing operations, income from discontinued operations, or not reported in the current year's income statement.

IFRS Case 4–8
Statement of cash flows; GlaxoSmithKline Plc.
● LO4–8, LO4–9
Real World Financials

GlaxoSmithKline Plc. (GSK) is a global pharmaceutical and consumer health-related products company located in the United Kingdom. The company prepares its financial statements in accordance with International Financial Reporting Standards. Below is a portion of the company's statements of cash flows included in recent financial statements:

GLAXOSMITHKLINE PLC.
Consolidated Cash Flow Statement
For the Year Ended 31 December

	£m
Cash flow from operating activities	
Profit after taxation for the year	5,628
Adjustments reconciling profit after tax to operating cash flows	2,871
Cash generated from operations	8,499
Taxation paid	(1,277)
Net cash inflow from operating activities	7,222
Cash flow from investing activities	
Purchase of property, plant and equipment	(1,188)
Proceeds from sale of property, plant and equipment	46
Purchase of intangible assets	(513)
Proceeds from sale of intangible assets	136
Purchase of equity investments	(133)
Proceeds from sale of equity investments	59
Purchase of businesses, net of cash acquired	(247)
Disposal of businesses	1,851
Investments in associates and joint ventures	(8)
Proceeds from disposal of subsidiary and interest in associate	429
Decrease in liquid investments	15
Interest received	59
Dividends from associates and joint ventures	18
Net cash inflow/(outflow) from investing activities	524
Cash flow from financing activities	
Proceeds from own shares for employee share options	–
Shares acquired by ESOP Trusts	(45)
Issue of share capital	585
Purchase of own shares for cancellation or to be held as Treasury shares	(1,504)
Purchase of non-controlling interests	(588)
Increase in long-term loans	1,913
Increase in short-term loans	–
Repayment of short-term loans	(1,872)
Net repayment of obligations under finance leases	(31)
Interest paid	(749)
Dividends paid to shareholders	(3,680)
Distributions to non-controlling interests	(238)
Other financing cash flows	(64)

(continued)

(concluded)

Net cash outflow from financing activities	(6,273)
Increase/(decrease) in cash and bank overdrafts	1,473
Cash and bank overdrafts at beginning of year	3,906
Exchange adjustments	(148)
Increase/(decrease) in cash and bank overdrafts	1,473
Cash and bank overdrafts at end of year	5,231
Cash and bank overdrafts at end of year comprise:	
Cash and cash equivalents	5,534
Overdrafts	(303)
	5,231

Required:

Identify the items in the above statement that would be reported differently if GlaxoSmithKline prepared its financial statements according to U.S. GAAP rather than IFRS.

Judgment Case 4–9
Income statement presentation; unusual items; comprehensive income
● LO4–3, LO4–4, LO4–5, LO4–6

Norse Manufacturing Inc. prepares an annual single, continuous statement of income and comprehensive income. The following situations occurred during the company's 2021 fiscal year:

1. Restructuring costs were incurred due to the closing of a factory.
2. Investments were sold, and a loss was recognized.
3. A positive foreign currency translation adjustment was recognized.
4. Interest expense was incurred.
5. A division was sold that qualifies as a separate component of the entity according to GAAP.
6. Obsolete inventory was written off.
7. The controller discovered an error in the calculation of 2020's patent amortization expense.

Required:

1. For each situation, identify the appropriate reporting treatment from the list below (consider each event to be material):
   a. As a component of operating income
   b. As a nonoperating income item (other income or expense)
   c. As a discontinued operation
   d. As an other comprehensive income item
   e. As an adjustment to retained earnings
2. Identify the situations that would be reported net-of-tax.

Judgment Case 4–10
Management incentives for change
● LO4–2

It has been suggested that not all accounting choices are made by management in the best interest of fair and consistent financial reporting.

Required:

What motivations can you think of for management's choice of accounting methods?

Research Case 4–11
Non-GAAP earnings
● LO4–3

Companies often voluntarily provide non-GAAP earnings when they announce annual or quarterly earnings. These numbers are controversial as they represent management's view of permanent earnings. The Sarbanes-Oxley Act (SOX), issued in 2002, requires that if non-GAAP earnings are included in any periodic or other report filed with the SEC or in any public disclosure or press release, the company also must provide a reconciliation with earnings determined according to GAAP.

Professors Entwistle, Feltham, and Mbagwu, in "Financial Reporting Regulation and the Reporting of Pro Forma Earnings," examine whether firms changed their reporting practice in response to the regulations included in SOX.

Required:

1. In your library or from some other source, locate the indicated article in *Accounting Horizons,* March 2006.
2. What sample of firms did the authors use in their examination?
3. What percent of firms reported non-GAAP earnings (referred to as *pro forma earnings* by the authors) in 2001? In 2003?
4. What percent of firms had non-GAAP earnings greater than GAAP earnings in 2001? In 2003?

5. What was the most frequently reported adjusting item in 2001? In 2003?
6. What are the authors' main conclusions of the impact of SOX on non-GAAP reporting?

Integrating Case 4–12
Balance sheet and income statement
Chapters 3 and 4
● LO4–3

Rice Corporation is negotiating a loan for expansion purposes and the bank requires financial statements. Before closing the accounting records for the year ended December 31, 2021, Rice's controller prepared the following financial statements:

RICE CORPORATION
Balance Sheet At December 31, 2021
($ in thousands)

Assets	
Cash	$ 275
Investments	78
Accounts receivable	487
Inventory	425
Allowance for uncollectible accounts	(50)
Property and equipment (net)	160
Total assets	$1,375
Liabilities and Shareholders' Equity	
Accounts payable and accrued liabilities	$ 420
Notes payable	200
Common stock	260
Retained earnings	495
Total liabilities and shareholders' equity	$1,375

RICE CORPORATION
Income Statement
For the Year Ended December 31, 2021
($ in thousands)

Sales revenue		$1,580
Expenses:		
Cost of goods sold	$755	
Selling and administrative expense	385	
Miscellaneous expense	129	
Income tax expense	100	
Total expenses		1,369
Net income		$ 211

Additional Information:

1. The company's common stock is traded on an organized stock exchange.
2. The investment portfolio consists of short-term investments valued at $57,000. The remaining investments will not be sold until the year 2023.
3. Notes payable consist of two notes: Note 1: $80,000 face value dated September 30, 2021. Principal and interest at 10% are due on September 30, 2022.

   Note 2: $120,000 face value dated April 30, 2021. Principal is due in two equal installments of $60,000 plus interest on the unpaid balance. The two payments are scheduled for April 30, 2022, and April 30, 2023.

   Interest on both loans has been correctly accrued and is included in accrued liabilities on the balance sheet and selling and administrative expense in the income statement.
4. Selling and administrative expense includes $90,000 representing costs incurred by the company in restructuring some of its operations. The amount is material.

Required:

Identify and explain the deficiencies in the presentation of the statements prepared by the company's controller. Do not prepare corrected statements. Include in your answer a list of items that require additional disclosure, either on the face of the statement or in a note.

Analysis Case 4–13
Income statement information
● LO4–1

Refer to the income statement of **The Home Depot, Inc.**, in Illustration 4–2 of this chapter.

Required:

1. Is this income statement presented in the single-step or multiple-step format?
2. What is the company's approximate income tax rate?
3. What is the percentage of net income relative to net sales?

Real World Case 4–14
Income statement information
● LO4–1, LO4–3, LO4–4

Real World Financials

EDGAR, the Electronic Data Gathering, Analysis, and Retrieval system, performs automated collection, validation, indexing, and forwarding of submissions by companies and others who are required by law to file forms with the U.S. Securities and Exchange Commission (SEC). All publicly traded domestic companies use EDGAR to make the majority of their filings. (Some foreign companies file voluntarily.) Form 10-K, which includes the annual report, is required to be filed on EDGAR. The SEC makes this information available on the Internet.

Required:

1. Access EDGAR on the Internet. The web address is **www.sec.gov.**
2. Search for a public company with which you are familiar. Access the most recent 10-K filing. Search or scroll to find the financial statements and related notes.
3. Answer the following questions related to the company's income statement:
   a. Does the company use the single-step or multiple-step format, or a variation?
   b. Does the income statement contain any income or loss on discontinued operations? If it does, describe the component of the company that was discontinued. (*Hint:* there should be a related disclosure note.)
   c. Describe the trend in net income over the years presented.
4. Repeat requirements 2 and 3 for two additional companies.

Real World Case 4–15
Income statement format; restructuring costs; earnings per share; comprehensive income; statement of cash flows; Ralph Lauren
● LO4–1, LO4–3, LO4–5, LO4–6, LO4–8

Real World Financials

Ralph Lauren Corporation is a global leader in the design, marketing, and distribution of premium lifestyle products, including men's, women's and children's apparel. Below are selected financial statements taken from a recent 10-K filing.

Required:

Use the information in the financial statements to answer the following questions.

1. Does the company use the single-step or multiple-step format to present its income statements?
2. Does the company report restructuring costs (yes/no)? What are restructuring costs?
3. Does the company report asset impairments (yes/no)? What are asset impairments?
4. Does the company chose to report comprehensive income in two consecutive statements or a combined statement?
5. What "other comprehensive items (OCI)" did the company report?
6. What method does the company use to report net cash provided by operating activities? What other method(s) could the company have used?
7. What is the largest cash outflow from investing activities?

RALPH LAUREN CORPORATION
CONSOLIDATED STATEMENTS OF INCOME

($ in millions, except per share data)	Fiscal Year Ended April 1, 2017
Net sales	$ 6,652.8
Cost of goods sold	(3,001.7)
Gross profit	3,651.1
Other costs and expenses:	
Selling, general, and administrative expenses	(3,149.4)
Amortization of intangible assets	(24.1)
Impairments of assets	(253.8)
Restructuring and other costs	(318.6)
Total other costs and expenses, net	(3,745.9)
Operating income	(94.8)
Foreign currency losses	(1.1)
Interest expense	(12.4)
Interest and other income, net	8.7
Equity in losses of equity-method investees	(5.3)
Income before provision for income taxes	(104.9)
Provision for income taxes	(5.6)
Net income	$ (99.3)
Net income per common share:	
Basic	$ (1.20)
Diluted	$ (1.20)

(continued)

(concluded)

RALPH LAUREN CORPORATION
CONSOLIDATED STATEMENTS OF COMPREHENSIVE INCOME

	Fiscal Year Ended April 1, 2017
Net income	$ (99.3)
Other comprehensive income, net of tax:	
Foreign currency translation adjustments	(48.6)
Net gains (losses) on cash flow hedges	26.6
Net gains (losses) on defined benefit plans	5.1
Other comprehensive income (loss)	(16.9)
Total comprehensive income	$(116.2)

RALPH LAUREN CORPORATION
CONSOLIDATED STATEMENTS OF CASH FLOWS

	Fiscal Year Ended April 1, 2017
Cash flows from operating activities:	
Net income	$ (99.3)
Adjustments to reconcile net income to net cash provided by operating activities:	
Depreciation and amortization expense	307.5
Deferred income tax expense (benefit)	(38.9)
Equity in losses of equity-method investees	5.2
Non-cash stock-based compensation expense	63.6
Non-cash impairment of assets	253.8
Non-cash inventory charges	197.9
Other non-cash charges, net	29.2
Excess tax benefits from stock-based compensation arrangements	(0.3)
Changes in operating assets and liabilities:	
Accounts receivable	54.1
Inventories	120.4
Prepaid expenses and other current assets	(27.8)
Accounts payable and accrued liabilities	112.9
Income tax receivables and payables	(34.0)
Deferred income	(20.7)
Other balance sheet changes, net	28.7
Net cash provided by operating activities	952.3
Cash flows from investing activities:	
Capital expenditures	(284.0)
Purchases of investments	(860.4)
Proceeds from sales and maturities of investments	942.4
Acquisitions and ventures, net of cash acquired	(6.1)
Change in restricted cash deposits	0.3
Net cash used in investing activities	(207.8)
Cash flows from financing activities:	
Proceeds from issuance of debt	3,735.2
Repayment of debt	(3,851.3)
Payments of capital lease obligations	(27.3)
Payments of dividends	(164.8)
Repurchases of common stock, including shares surrendered for tax withholdings	(215.2)
Proceeds from exercise of stock options	5.0
Excess tax benefits from stock-based compensation arrangements	0.3
Other financing activities	—
Net cash used in financing activities	(518.1)
Effect of exchange rate changes on cash and cash equivalents	(14.4)
Net increase (decrease) in cash and cash equivalents	212.0
Cash and cash equivalents at beginning of period	456.3
Cash and cash equivalents at end of period	$ 668.3

Analysis Case 4–16
Evaluating profitability and asset management; obtain and compare annual reports from companies in the same industry
● LO4–10

Performance and profitability of a company often are evaluated using the financial information provided by a firm's annual report in comparison with other firms in the same industry. Ratios are useful in this assessment.

Required:

Obtain annual reports from two corporations in the same primary industry. Using techniques you learned in this chapter and any analysis you consider useful, respond to the following questions:

1. How do earnings trends compare in terms of both the direction and stability of income?
2. Which of the two firms had greater earnings relative to resources available?
3. How efficiently are current assets managed?
4. Has each of the companies achieved its respective rate of return on assets with similar combinations of profit margin and turnover?
5. Are there differences in accounting methods that should be taken into account when making comparisons?

Note: You can obtain copies of annual reports from friends who are shareholders, the investor relations department of the corporations, from a friendly stockbroker, or from EDGAR (Electronic Data Gathering, Analysis, and Retrieval) on the Internet (**www.sec.gov**).

Analysis Case 4–17
Relationships among ratios; Chapters 3 and 4
● LO4–10

You are a part-time financial advisor. A client is considering an investment in common stock of a waste recycling firm. One motivation is a rumor the client heard that the company made huge investments in a new fuel creation process. Unable to confirm the rumor, your client asks you to determine whether the firm's assets had recently increased significantly.

Because the firm is small, information is sparse. Last quarter's interim report showed total assets of $324 million, approximately the same as last year's annual report. The only information more current than that is a press release last week in which the company's management reported "record net income for the year of $21 million, representing a 14.0% return on equity. Performance was enhanced by the Company's judicious use of financial leverage on a debt/equity ratio of 2 to 1."

Required:

Use the information available to calculate total assets, total liabilities, and total shareholders' equity.

Integrating Case 4–18
Using ratios to test reasonableness of data; Chapters 3 and 4
● LO4–10

You are a new staff accountant with a large regional CPA firm, participating in your first audit. You recall from your auditing class that CPAs often use ratios to test the reasonableness of accounting numbers provided by the client. Since ratios reflect the relationships among various account balances, if it is assumed that prior relationships still hold, prior years' ratios can be used to estimate what current balances should approximate. However, you never actually performed this kind of analysis until now. The CPA in charge of the audit of Covington Pike Corporation brings you the list of ratios shown below and tells you these reflect the relationships maintained by Covington Pike in recent years.

Profit margin on sales = 5%
Return on assets = 7.5%
Gross profit margin = 40%
Inventory turnover ratio = 6 times
Receivables turnover ratio = 25 times
Acid-test ratio = 0.9 to one
Current ratio = 2 to 1
Return on equity = 10%
Debt to equity ratio = 1/3
Times interest earned ratio = 12 times

Jotted in the margins are the following notes:

- Net income $15,000.
- Only one short-term note ($5,000); all other current liabilities are trade accounts.
- Property, plant, and equipment are the only noncurrent assets.
- Bonds payable are the only noncurrent liabilities.
- The effective interest rate on short-term notes and bonds is 8%.
- No investment securities.
- Cash balance totals $15,000.

Required:

You are requested to approximate the current year's balances in the form of a balance sheet and income statement, to the extent the information allows. Accompany those financial statements with the calculations you use to estimate each amount reported.

Data Analytics

Data Analytics Case
Profitability Ratios
● LO4-10

Data analytics is the process of examining data sets in order to draw conclusions about the information they contain. To prepare for this case, you will need to download Tableau to your computer. If you have not completed any of the prior data analytics cases, follow the instructions listed in the Chapter 1 Data Analytics case to get set up. You will need to watch the videos referred to in the Chapter 1, Chapter 2, and Chapter 3 Data Analytics cases. No additional videos are required for this case. Future cases will build off of your knowledge gained from these, and additional videos will be suggested. All short training videos can be found here: **www.tableau.com/learn/training#getting-started**.

In the Chapter 3 Data Analytics Case, you applied Tableau to examine a data set and create calculations to compare two companies' liquidity. In this case, you continue in your role as an analyst conducting introductory research into the relative merits of investing in one or both of these companies. This time assess the companies' return on assets and how that return is related to both profit margin and asset turnover.

Required:

Use Tableau to calculate (a) profit margin on sales, (b) asset turnover, and (c) return on assets for each of the two companies in each year from 2012 to 2021. Based upon what you find, answer the following questions:

1. What is the return on assets for Big Store (a) in 2012 and (b) in 2021?
2. What is the return on assets for Discount Goods (a) in 2012 and (b) in 2021?

To help determine why the relative profitability of the two companies has shifted over the ten-year period and to get a better company-to-company comparison, drag the second "pill": (**Year** Dimension) to the left of the "pill": (**Company** Dimension in the text chart).

3. The return on assets is a result of the profit margin and the asset turnover. Demonstrate this for Big Store in 2021 by showing that the profit margin times the asset turnover equals return on assets.
4. Analyzing the asset turnover ratios over the ten-year period, is Big Store's asset turnover (a) generally increasing, (b) roughly the same, or (c) generally decreasing from year to year?
5. Analyzing the asset turnover ratios over the ten-year period, is Discount Goods' turnover (a) generally increasing, (b) roughly the same, or (c) generally decreasing from year to year?
6. As of 2021, which company reports a more favorable return on assets and is this primarily attributable to its asset turnover or profit margin?

Note: As you assess the profitability aspect of the companies, keep in mind that profitability should be evaluated in the context of both risk and efficiency of managing assets.

Resources:

You have available to you an extensive data set that includes detailed financial data for Discount Goods and Big Store and for 2012-2021. The data set is in the form of an Excel file available to download from Connect, under Student Resources within the Library tab. There are four Excel files available. The one for use in this chapter is named "Discount_Goods_Big_Store_Financials.xlsx." Download this file and save it to the computer on which you will be using Tableau.

For this case, you will create several calculations to produce a text chart of profitability ratios to allow you to compare and contrast the two companies. You also will need to change data from continuous to discrete (see Chapter 3 Data Analytics case for more information on this).

After you view the training videos, follow these steps to create the charts you'll use for this case:

- Open Tableau and connect to the Excel spreadsheet you downloaded.
- Click on the "Sheet 1" tab at the bottom of the canvas, to the right of the Data Source at the bottom of the screen.
- Drag "Year" and "Company" under "Dimensions" to the Rows shelf. Change "Year" to discrete by right-clicking and selecting "Discrete."
- Drag the "Net income/(loss)," "Net sales," and "Average total assets" under "Measures" to the Rows shelf. Change each to *discrete*, following the process above.
- Create a calculated field by clicking the "Analysis" tab at the top of the screen and selecting "Create Calculated Field." A calculation box will pop up. Name the calculation "Return on Assets." In the calculation box, from the Rows shelf, drag "Net income/(loss)," type a division sign, then drag "Average total assets" beside it. Make sure the box says that the calculation is valid and click OK.
- Repeat the process by creating a calculated field named "Profit Margin on Sales" that consists of "Net income/(loss)," divided by "Net sales."
- Repeat the process one more time by creating a calculated field "Asset Turnover" that consists of "Net sales" divided by "Average total assets."
- Drag the newly created "Return on Assets" to add it to the Rows shelf. Format it to Discrete, Times New Roman, 10-point font, black, center alignment, and percentage.
- Drag the newly created "Profit Margin on Sales" and "Asset Turnover" to the Rows shelf. Format them both to Discrete, Times New Roman, 10-point font, black, center alignment, and percentage.

- Right click the "Net income / (loss)," "Net sales," and "Average total assets" on the Rows shelf and un-click "Show header." This will hide these items from view but still allow them to be used in the formulas.
- Change the title of the sheet to be "Profitability Ratios" by right-clicking and selecting "Edit title." Format the title to Times New Roman, bold, black and 15-point font. Change the title of "Sheet 1" to match the sheet title by right-clicking, selecting "Rename" and typing in the new title.
- Click on the New Worksheet tab on the lower left ("Sheet 2" should open). Drag "Year" to the Columns shelf and "Return on Assets," "Profit Margin on Sales" and "Asset Turnover" to the Rows shelf.
- Drag "Company" under "Dimensions" to "Color" on the "Marks" card. You should see two lines in the three graphs.
- Format the labels on the left of the sheet ("Return on Assets," "Profit Margin on Sales," and "Asset Turnover") to Times New Roman, 10-point font, bold, and black.
- Change the title of the sheet to be "Graph of Profitability Ratios" by right-clicking and selecting "Edit title." Format the title to Times New Roman, bold, black and 15-point font. Change the title of "Sheet 2" to match the sheet title by right-clicking, selecting "Rename" and typing in the new title.
- Once complete, save the file as "DA4_Your initials.twbx."

Continuing Cases

Target Case

● LO4–3, LO4–4, LO4–6, LO4–8

Target Corporation prepares its financial statements according to U.S. GAAP. Target's financial statements and disclosure notes for the year ended February 3, 2018, are available in Connect. This material is also available under the Investor Relations link at the company's website (**www.target.com**).

Required:

1. By what name does Target label its income statement?
2. What amounts did Target report for the following items for the year ended February 3, 2018?
   a. Sales
   b. Gross margin
   c. Earnings from continuing operations before income taxes
   d. Net earnings from continuing operations
   e. Net earnings
3. Does Target report any items as part of its comprehensive income? If so, what are they.
4. Does Target prepare the statement of cash flows using the direct method or the indirect method?
5. Which is higher, net earnings or operating cash flows? Which line item is the biggest reason for this difference? Explain why.
6. What is the largest investing cash flow and the largest financing cash flow reported by the company for the year ended February 3, 2018?

Air France–KLM Case

● LO4–9

IFRS

Air France–KLM (AF), a Franco-Dutch company, prepares its financial statements according to International Financial Reporting Standards. AF's financial statements and disclosure notes for the year ended December 31, 2017, are provided in Connect. This material is also available under the Finance link at the company's website (**www.airfranceklm.com**).

Required:

1. How does AF classify operating expenses in its income statement? How are these expenses typically classified in a U.S. company income statement?
2. How does AF classify interest paid, interest received, and dividends received in its statement of cash flows? What other alternatives, if any, does the company have for the classification of these items? How are these items classified under U.S. GAAP?

CPA Exam Questions and Simulations

ROGER CPA Review

Sample CPA Exam questions from Roger CPA Review are available in Connect as support for the topics in this chapter. These multiple-choice questions and task-based simulations include expert-written explanations and solutions, and provide a starting point for students to become familiar with the content and functionality of the actual CPA Exam.

CHAPTER

5 Time Value of Money Concepts

OVERVIEW

Time value of money concepts, specifically future value and present value, are essential in a variety of accounting situations. These concepts and the related computational procedures are the subjects of this chapter. Present values and future values of *single amounts* and present values and future values of *annuities* (series of equal periodic payments) are described separately but shown to be interrelated.

LEARNING OBJECTIVES

After studying this chapter, you should be able to:

- **LO5–1** Explain the difference between simple and compound interest. (*p. 240*)
- **LO5–2** Compute the future value of a single amount. (*p. 241*)
- **LO5–3** Compute the present value of a single amount. (*p. 242*)
- **LO5–4** Solve for either the interest rate or the number of compounding periods when present value and future value of a single amount are known. (*p. 243*)
- **LO5–5** Apply present value techniques in the valuation of notes. (*p. 246*)
- **LO5–6** Explain the difference between an ordinary annuity and an annuity due situation. (*p. 248*)
- **LO5–7** Compute the future value of both an ordinary annuity and an annuity due. (*p. 248*)
- **LO5–8** Compute the present value of an ordinary annuity, an annuity due, and a deferred annuity. (*p. 250*)
- **LO5–9** Solve for unknown values in annuity situations involving present value. (*p. 255*)
- **LO5–10** Briefly describe how the concept of the time value of money is incorporated into the valuation of bonds, long-term leases, installment notes, and pension obligations. (*p. 259*)

©Gabriel Petrescu/Shutterstock

FINANCIAL REPORTING CASE

The Winning Ticket

"I was just there to buy it for luck. It was just chance, a chance I had to take," she said. On August 23, 2017, Mavis Wanczyk, 53, of Chicopee, MA, blissfully discovered she had purchased the winning $758.7 million Powerball ticket, the largest jackpot in U.S. history. Imagine her surprise, though, when she was told that, having chosen a lump-sum payout rather than 30-year installment payments, she would get a check for "only" $336 million. Sure, federal tax was withheld at a 25% rate and Massachusetts added another 5%, but this difference was far more than the amount of the tax.

QUESTIONS

By the time you finish this chapter, you should be able to respond appropriately to the questions posed in this case. Compare your response to the solution provided at the end of the chapter.

1. Why was Mavis to receive $336 million rather than the $758.7 million lottery prize? (*p. 251*)
2. What interest (discount) rate did the state of Massachusetts use to calculate the $336 million lump-sum payment? (*p. 256*)
3. What are some of the accounting applications that incorporate the time value of money into valuation? (*p. 259*)

Basic Concepts

PART A

Time Value of Money

The key to solving the problem described in the financial reporting case is an understanding of the concept commonly referred to as the **time value of money**. This concept means that money invested today will grow to a larger dollar amount in the future. For example, $100 invested in a savings account at your local bank yielding 6% annually will grow to $106 in one year. The difference between the $100 invested now—the present value of the investment—and its $106 future value represents the time value of money.

The *time value of money* means that money can be invested today to earn interest and grow to a larger dollar amount in the future.

This concept has nothing to do with the worth or buying power of those dollars. Prices in our economy can change. If the inflation rate were higher than 6%, then the $106 you would have in the savings account actually would be worth less than the $100 you had a year earlier. The time value of money concept concerns only the growth in the dollar amounts of money, ignoring inflation.

The concepts you learn in this chapter are useful in solving business decisions such as determining the lottery award presented in the financial reporting case. More important, the concepts are necessary when valuing assets and liabilities for financial reporting purposes. Most accounting applications that incorporate the time value of money involve the concept of present value. The valuation of leases, bonds, pension obligations, and certain notes receivable and payable are a few prominent examples. It's important that you master the concepts and tools we review here because they are essential for the remainder of your accounting education.

Time value of money concepts are useful in valuing several assets and liabilities.

Simple versus Compound Interest

LO5–1

Interest is the amount of money paid or received in excess of the amount borrowed or lent.

Interest is the "rent" paid for the use of money for some period of time. In dollar terms, it is the amount of money paid or received in excess of the amount of money borrowed or lent. If you lend someone $100 today and "receive" $106 a year from now, your interest would be $6. Interest also can be expressed as a rate at which money will grow. In this case, that rate is 6%. It is this interest that gives money its time value.

Simple interest is computed by multiplying an initial investment times both the applicable interest rate and the period of time for which the money is used. For example, simple interest earned each year on a $1,000 investment paying 10% is $100 ($1,000 × 10%).

Compound interest includes interest not only on the initial investment but also on the accumulated interest in previous periods.

Compound interest occurs when money remains invested for multiple periods. It results in increasingly larger interest amounts for each period of the investment. The reason is that interest is then being generated not only on the initial investment amount but also on the accumulated interest earned in previous periods.

For example, Cindy Johnson invested $1,000 in a savings account paying 10% interest *compounded* annually. How much interest will she earn each year, and what will be her investment balance after three years?

Date	Interest (Interest rate × Outstanding balance = Interest)	Balance
Initial deposit		$1,000
End of year 1	10% × $1,000 = $100	$1,100
End of year 2	10% × $1,100 = $110	$1,210
End of year 3	10% × $1,210 = $121	**$1,331**

With compound interest at 10% annually, the $1,000 investment would grow to **$1,331** at the end of the three-year period. If Cindy withdrew the interest earned each year, she would earn only $100 in interest each year (the amount of simple interest). If the investment period had been 20 years, 20 calculations would be needed. However, calculators, Excel, and compound interest tables, like those in an appendix to this textbook, make these calculations easier.

Interest rates are typically stated as annual rates.

Most banks compound interest more frequently than once a year. Daily compounding is common for savings accounts. More rapid compounding has the effect of increasing the actual rate, which is called the **effective rate**, at which money grows per year. It is important to note that interest is typically stated as an annual rate regardless of the length of the compounding period involved. In situations when the compounding period is less than a year, the interest rate per compounding period is determined by dividing the annual rate by the number of periods. Assuming an annual rate of 12%:

Compounded	Interest Rate Per Compounding Period
Semiannually	12% ÷ 2 = 6%
Quarterly	12% ÷ 4 = 3%
Monthly	12% ÷ 12 = 1%

As an example, now let's assume Cindy Johnson invested $1,000 in a savings account paying 10% interest *compounded* twice a year. There are two six-month periods paying interest at 5% (the annual rate divided by two periods). How much interest will she earn the first year, and what will be her investment balance at the end of the year?

Date	Interest (Interest rate × Outstanding balance = Interest)	Balance
Initial deposit		$ 1,000.00
After six months	5% × $1,000 = $50.00	$ 1,050.00
End of year 1	5% × $1,050 = $52.50	**$1,102.50**

The *effective interest rate* is the rate at which money actually will grow during a full year.

The $1,000 would grow by $102.50, the interest earned, to **$1,102.50**, $2.50 more than if interest were compounded only once a year. The effective annual interest rate, often referred to as the annual *yield,* is 10.25% ($102.50 ÷ $1,000).

Future Value of a Single Amount

● LO5–2

In the first Cindy example, in which \$1,000 was invested for three years at 10% compounded annually, the **\$1,331** is referred to as the **future value (FV)**. A time diagram is a useful way to visualize this relationship, with 0 indicating the date of the initial investment.

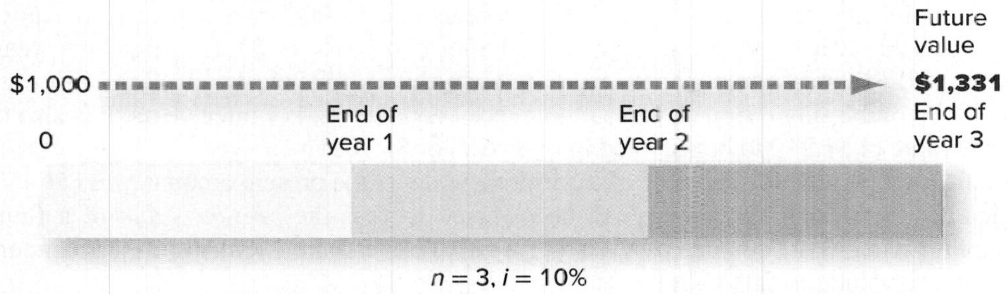

The future value after one year can be calculated as \$1,000 × 1.10 (1.00 + .10) = \$1,100. After three years, the future value is \$1,000 × 1.10 × 1.10 × 1.10 = \$1,331. In fact, the future value of any invested amount can be determined as follows:

$$FV = I(1 + i)^n$$

where: FV = Future value of the invested amount
I = Amount invested at the beginning of the period
i = Interest rate
n = Number of compounding periods

The *future value* of a single amount is the amount of money that a dollar will grow to at some point in the future.

The future value can be determined by using Table 1, Future Value of \$1, located at the end of this textbook. The table contains the future value of \$1 invested for various periods of time, n, and at various rates, i.

Use this table to determine the future value of any invested amount simply by multiplying that invested amount by the table value at the *intersection* of the column for the desired rate and the row for the number of compounding periods. Illustration 5–1 provides an excerpt from Table 1.

Illustration 5–1

Future Value of \$1 (excerpt from Table 1 located at the end of this book)

	Interest Rates (i)					
Periods (n)	**7%**	**8%**	**9%**	**10%**	**11%**	**12%**
1	1.07000	1.08000	1.09000	1.10000	1.11000	1.12000
2	1.14490	1.16640	1.18810	1.21000	1.23210	1.25440
3	1.22504	1.25971	1.29503	**1.33100**	1.36763	1.40493
4	1.31080	1.36049	1.41158	1.46410	1.51807	1.57352
5	1.40255	1.46933	1.53862	1.61051	1.68506	1.76234

The table shows various values of $(1 + i)^n$ for different combinations of i and n. From the table you can find the future value factor (table value) for three periods at 10% to be **1.331**. This means that \$1 invested at 10% compounded annually will grow to approximately \$1.33 in three years. So, the future value of \$1,000 invested for three years at 10% is **\$1,331**:

$$FV = I \times FV \text{ factor}$$
$$FV = \$1{,}000 \times \mathbf{1.331}^{*} = \$1{,}331$$

*Future value of \$1; $n = 3$, $i = 10\%$

The future value function in financial calculators or in an Excel spreadsheet calculates future values in the same way. Determining future values (and present values) electronically avoids the need for tables such as those in the appendix. But, we use the tables extensively in the book to make it easier for you to visualize how the amounts are actually determined. It's important to remember that the n in the future value formula refers to the number of compounding periods, not necessarily the number of years. For example, suppose you wanted to know the future value *two* years from today of \$1,000 invested at 12% with *quarterly* compounding. The number of periods is therefore eight, and the compounding rate is 3% (12% annual rate divided by four, the number of quarters in a year). The future value factor from Table 1 is 1.26677, so the future value is $1,266.77 ($1,000 × 1.26677).[1]

Present Value of a Single Amount

● LO5–3

The present value of a single amount is today's equivalent to a particular amount in the future.

The example used to illustrate future value reveals that $1,000 invested today is equivalent to $1,100 received after one year, $1,210 after two years, or $1,331 after three years, assuming 10% interest compounded annually. Thus, the **$1,000** investment (I) is the **present value (PV)** of the single sum of $1,331 to be received at the end of three years. It is also the present value of $1,210 to be received in two years or $1,100 in one year.

Remember that the future value of a present amount is the present amount *times* $(1 + i)^n$. Logically, then, that computation can be reversed to find the *present value* of a future amount to be the future amount *divided* by $(1 + i)^n$. We substitute PV for I (invested amount) in the future value formula above.

$$FV = PV(1 + i)^n$$

$$PV = \frac{FV}{(1 + i)^n}$$

In our example,

$$PV = \frac{\$1{,}331}{(1 + .10)^3} = \frac{\$1{,}331}{1.331} = \mathbf{\$1{,}000}$$

Of course, dividing by $(1 + i)^n$ is the same as multiplying by its reciprocal, $1/(1 + i)^n$.

$$PV = \$1{,}331 \times \frac{1}{(1 + .10)^3} = \$1{,}331 \times 0.75131 = \mathbf{\$1{,}000}$$

So, for instance, the future value amount $1,331 times the PV factor of 0.75131 gives us a present value of $1,000, and the present value amount of $1,000 times the FV factor of 1.331 gives us the future value of $1,331 three years from now.

As with future value, these computations are simplified by using calculators, Excel, or present value tables. Table 2, Present Value of $1, provides the solutions of $1/(1 + i)^n$ for various interest rates (i) and compounding periods (n). These amounts represent the present value of $1 to be received at the *end* of the different periods. The table can be used to find the present value of any single amount to be received in the future by *multiplying* that future amount by the value in the table that lies at the *intersection* of the column for the appropriate rate and the row for the number of compounding periods.[2] Illustration 5–2 provides an excerpt from Table 2.

Illustration 5–2

Present Value of $1 (excerpt from Table 2)

	Interest Rates (*i*)					
Periods (*n*)	**7%**	**8%**	**9%**	**10%**	**11%**	**12%**
1	0.93458	0.92593	0.91743	0.90909	0.90090	0.89286
2	0.87344	0.85734	0.84168	0.82645	0.81162	0.79719
3	0.81630	0.79383	0.77218	**0.75131**	0.73119	0.71178
4	0.76290	0.73503	0.70843	0.68301	0.65873	0.63552
5	0.71299	0.68058	0.64993	0.62092	0.59345	0.56743

We can see a couple of interesting patterns in this table. First, look at any of the *rows*. Notice that the present value factors get smaller and smaller as the interest rates increase.

[1]When interest is compounded more frequently than once a year, the effective annual interest rate, or yield, can be determined using the following equation:

$$\text{Yield} = (1 + \tfrac{i}{p})^p - 1$$

with i being the annual interest rate and p the number of compounding periods per year. In this example, the annual yield would be 12.55%, calculated as follows:

$$\text{Yield} = (1 + \tfrac{.12}{4})^4 - 1 = 1.1255 - 1 = .1255$$

Determining the yield is useful when comparing returns on investment instruments with different compounding period length.

[2]The factors in Table 2 are the reciprocals of those in Table 1. For example, the future value factor for 10%, three periods is 1.331, while the present value factor is .75131. $1 ÷ 1.331 = $.75131, and $1 ÷ **0.75131** = $1.331.

That demonstrates that the higher the time value of money, the lower is the present value of a future amount. That's logical. The higher the return you can get from putting to use money you have now (higher time value of money), the less desirable it is to wait to get the money (lower present value of the future amount).

Now, look at any of the *columns*. Notice that the further into the future the $1 is to be received, the less valuable it is now. That demonstrates that the longer you have to wait for your money, the more you give up in terms of return you could be getting if you could put the money to work now, and the lower is the present value of the future amount.

This is the essence of the concept of the time value of money. Given a choice between $1,000 now and $1,000 three years from now, you would choose to have the money now. If you have it now, you could put it to use. But the choice between, say, $740 now and $1,000 three years from now would depend on your time value of money. If your time value of money is 10%, you would choose the $1,000 in three years, because the $740 invested at 10% for three years would grow to only $984.94 [$740 × 1.331 (FV of $1, i = 10%, n = 3)]. On the other hand, if your time value of money is 11% or higher, you would prefer the $740 now.[3] Presumably, you would invest the $740 now and have it grow to $1,012.05 ($740 × 1.36763) in three years. So, if your time value of money is 11%, you should be indifferent with respect to receiving $740 now or $1,012.05 three years from now.

Using the present value table in Illustration 5–2, the present value of $1,000 to be received in three years assuming a time value of money of 10% is $751.31 [$1,000 × **0.75131** (PV of $1, i = 10% and n = 3)]. Because the present value of the future amount, $1,000, is higher than $740 we could have today, we again determine that with a time value of money of 10%, the $1,000 in three years is preferred to the $740 now.

In our earlier example, **$1,000** now is equivalent to **$1,331** in three years, assuming the time value of money is 10%. Graphically, the relation between the present value and the future value can be viewed this way:

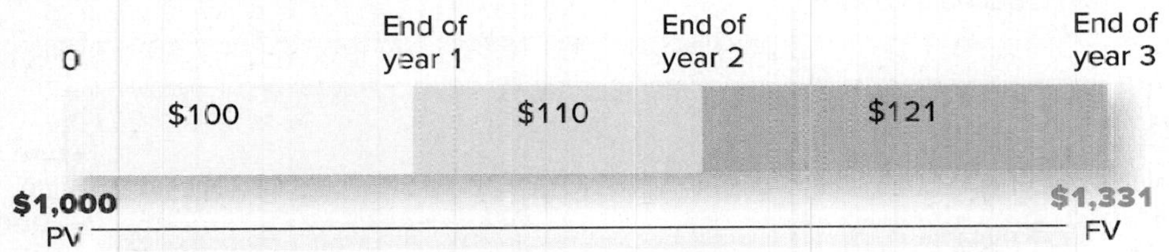

While the calculation of ***future value*** of a single sum invested today requires the *inclusion* of compound interest, ***present value*** problems require the *removal* of compound interest. The process of computing present value *removes* the $331 of interest earned over the three-year period from the future value of **$1,331**, just as the process of computing future value *adds* $331 of interest to the present value of ***$1,000*** to arrive at the future value of ***$1,331***.

The calculation of future value requires the addition of interest, while the calculation of present value requires the removal of interest.

As we demonstrate later in this chapter and in subsequent chapters, present value calculations are incorporated into accounting valuation much more frequently than future value.

Accountants use PV calculations much more frequently than FV.

Solving for Other Values When FV and PV are Known

● LO5–4

There are four variables in the process of adjusting single cash flow amounts for the time value of money: the present value (PV), the future value (FV), the number of compounding periods (n), and the interest rate (i). If you know any three of these, the fourth can be determined. Illustration 5–3 solves for an unknown interest rate and Illustration 5–4 determines an unknown number of periods.

DETERMINING THE UNKNOWN INTEREST RATE

Suppose a friend asks to borrow $500 today and promises to repay you $605 two years from now. What is the annual interest rate you would be agreeing to?

Illustration 5–3

Determining i When PV, FV, and n are Known

[3]The interest rate used to find the present value sometimes is called the *discount rate,* and finding the present value is sometimes called "discounting" a future amount to its present value.

The following time diagram illustrates the situation:

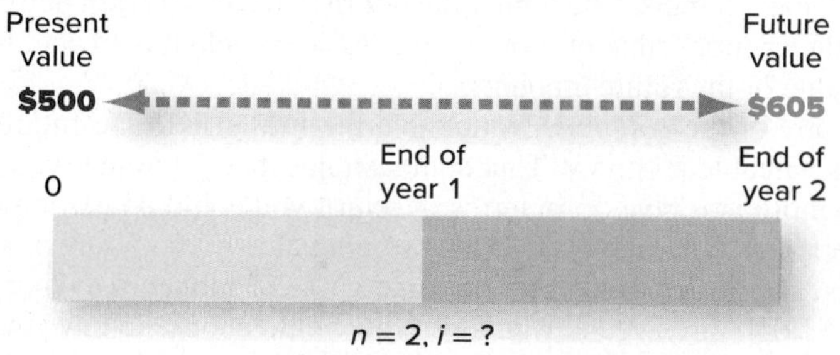

The interest rate is the rate that will provide a present value of **$500** when determining the present value of the **$605** to be received in two years:

$500 (present value) = **$605** (future value) × ?*

*Present value of $1: $n = 2$, $i = ?$

Rearranging algebraically, we find that the present value table factor is 0.82645.

The unknown variable is the interest rate.

$500 (present value) ÷ **$605** (future value) = 0.82645*

*Present value of $1: $n = 2$, $i = ?$

When you consult the present value table, Table 2, you search row two ($n = 2$) for this value and find it in the 10% column. So the effective interest rate is 10%. Notice that the computed factor value exactly equals the table factor value.[4]

DETERMINING THE UNKNOWN NUMBER OF PERIODS

Illustration 5–4

Determining *n* When PV, FV, and *i* are Known

> You want to invest $10,000 today to accumulate $16,000 for graduate school. If you can invest at an interest rate of 10% compounded annually, how many years will it take to accumulate the required amount?

The following time diagram illustrates the situation:

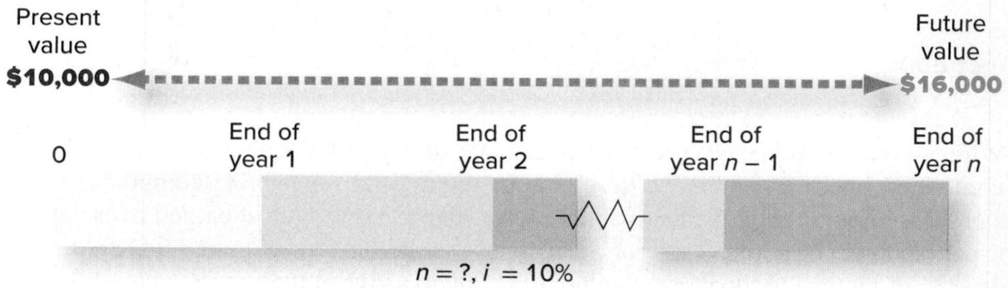

The years it will take is the value of *n* that will provide a present value of **$10,000** when finding the present value of **$16,000** at a rate of 10%:

The unknown variable is *n*, the number of periods.

$10,000 (present value) = **$16,000** (future value) × ?*

*Present value of $1; $n = ?$, $i = 10\%$

Rearranging algebraically, we find that the present value table factor is 0.625.

$10,000 (present value) ÷ **$16,000** (future value) = 0.625*

*Present value of $1: $n = ?$, $i = 10\%$

When you consult the present value table, Table 2, you search the 10% column ($i = 10\%$) for this value and find 0.62092 in row five. So it would take approximately five years to accumulate **$16,000** in the situation described.

[4] If the calculated factor lies between two table factors, interpolation is useful in finding the unknown value. For example, if the future value in our example is $600, instead of $605, the calculated PV factor is 0.83333 ($500 ÷ $600). This factor lies between the 9% factor of 0.84168 and the 10% factor of 0.82645. The total difference between these factors is 0.01523 (0.84168 – 0.82645). The difference between the calculated factor of 0.83333 and the 10% factor of 0.82645 is 0.00688. This is 45% of the difference between the 9% and 10% factors:

$$\frac{0.00688}{0.01523} = 0.45$$

Therefore, the interpolated interest rate is 9.55% (10 – 0.45).

Additional Consideration

Solving for the unknown factor in either of these examples could just as easily be done using the future value tables. The number of years is the value of n that will provide a present value of $10,000 when $16,000 is the future amount and the interest rate is 10%.

$16,000 (future value) = $10,000 (present value) × ?*

*Future value of $1: $n = ?, i = 10\%$

Rearranging algebraically, the future value table factor is 1.6.

$16,000 (future value) ÷ $10,000 (present value) = 1.6*

*Future value of $1: $n = ?, i = 10\%$

When you consult the future value table, Table 1, you search the 10% column ($i = 10\%$) for this value and find 1.61051 in row five. So it would take approximately five years to accumulate $16,000 in the situation described.

Concept Review Exercise

VALUING A SINGLE CASH FLOW AMOUNT

Using the appropriate table, answer each of the following independent questions.

1. What is the future value of $5,000 at the end of six periods at 8% compound interest?
2. What is the present value of $8,000 to be received eight periods from today assuming a compound interest rate of 12%?
3. What is the present value of $10,000 to be received two *years* from today assuming an annual interest rate of 24% and *monthly* compounding?
4. If an investment of $2,000 grew to $2,520 in three periods, what is the interest rate at which the investment grew? Solve using both present and future value tables.
5. Approximately how many years would it take for an investment of $5,250 to accumulate to $15,000, assuming interest is compounded at 10% annually? Solve using both present and future value tables.

Solution:

1. FV = $5,000 × 1.58687* = $7,934

   *Future value of $1: $n = 6, i = 8\%$ (from Table 1)

2. PV = $8,000 × 0.40388* = $3,231

   *Present value of $1: $n = 8, i = 12\%$ (from Table 2)

3. PV = $10,000 × 0.62172* = $6,217

   *Present value of $1: $n = 24, i = 2\%$ (from Table 2)

4. Using present value table,

   $$\frac{\$2,000}{\$2,520} = 0.7937^*$$

   *Present value of $1: $n = 3, i = ?$ (from Table 2, i approximately **8%**)

   Using future value table,

   $$\frac{\$2,520}{\$2,000} = 1.260^*$$

   *Future value of $1: $n = 3, i = ?$ (from Table 1, i approximately **8%**)

5. Using present value table,

   $$\frac{\$5,250}{\$15,000} = 0.35^*$$

   *Present value of $1: $n = ?, i = 10\%$ (from Table 2, n approximately **11 years**)

   Using future value table,

   $$\frac{\$15,000}{\$5,250} = 2.857^*$$

   *Future value of $1: $n = ?, i = 10\%$ (from Table 1, n approximately **11 years**)

Preview of Accounting Applications of Present Value Techniques—Single Cash Amount

Doug Smith switched off his television set immediately after watching the Super Bowl game and swore to himself that this would be the last year he would watch the game on his 10-year-old 24-inch TV. "Next year, a big screen 4K LED," he promised himself. Soon after, he saw an advertisement in the local newspaper from Slim Jim's TV and Appliance offering a Samsung 78-inch television on sale for $1,800. And the best part of the deal was that Doug could take delivery immediately but would not have to pay the $1,800 for one whole year! "In a year, I can easily save the $1,800," he thought.

In the above scenario, the seller, Slim Jim's TV and Appliance, records a sale when the TV is delivered to Doug. How should the company value its receivable and corresponding sales revenue? We provide a solution to this question at the end of this section. The following discussion will help you to understand that solution.

Most *monetary assets* and *monetary liabilities* are valued at the present value of future cash flows.

Many assets and most liabilities are monetary in nature. **Monetary assets** include money and claims to receive money, the amount of which is fixed or determinable. Examples include cash and most receivables. **Monetary liabilities** are obligations to pay amounts of cash, the amount of which is fixed or determinable. Most liabilities are monetary. For example, if you borrow money from a bank and sign a note payable, the amount of cash to be repaid to the bank is fixed. Monetary receivables and payables are valued based on the fixed amount of cash to be received or paid in the future taking into account the time value of money. In other words, we value most receivables and payables at the present value of future cash flows, reflecting an appropriate time value of money.[5]

● LO5–5

The example in Illustration 5–5 demonstrates the application of present value techniques in valuing a note receivable and note payable for which the time period is one year and the interest rate is known.

Illustration 5–5
Valuing a Note: One Payment, Explicit Interest

Explicit Interest
The Stridewell Wholesale Shoe Company manufactures athletic shoes for sale to retailers. The company recently sold a large order of shoes to Harmon Sporting Goods for $50,000. Stridewell agreed to accept a note in payment for the shoes requiring payment of $50,000 in one year plus interest at 10%.

How should Stridewell value the note receivable and corresponding sales revenue? How should Harmon value the note payable and corresponding inventory purchased? As long as the interest rate explicitly stated in the agreement properly reflects the time value of money, the answer is $50,000, the face value of the note. It's important to realize that this amount also equals the present value of future cash flows at 10%. Future cash flows equal **$55,000**, the $50,000 note itself plus $5,000 interest ($50,000 × 10%). Here's a time diagram:

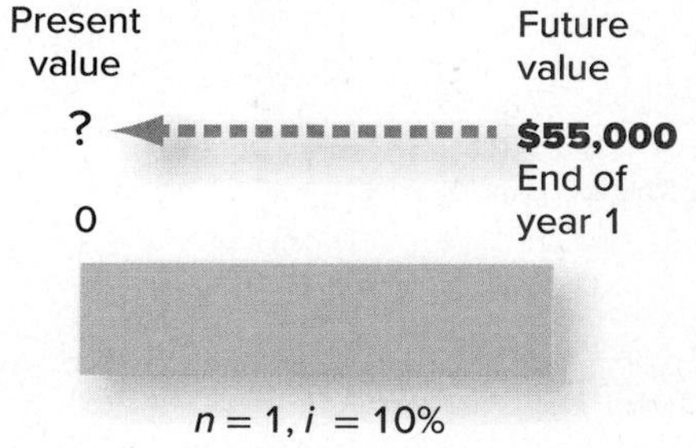

In equation form, we can solve for present value as follows:

$55,000 (future value) × 0.90909* = $50,000 (present value)

*Present value of $1: $n = 1$, $i = 10\%$

[5]FASB ASC 835–30: Interest—Imputation of Interest (previously "Interest on Receivables and Payables," *Accounting Principles Board Opinion No. 21* (New York: AICPA, 1971)). As you will learn in Chapter 7, we value normal trade accounts receivable and accounts payable at the amounts expected to be received or paid, not the present value of those amounts. The difference between the amounts expected to be received or paid and present values often is immaterial.

By calculating the present value of $55,000 to be received in one year, the interest of $5,000 is removed from the future value, resulting in the appropriate note receivable/sales revenue value of $50,000 for Stridewell and a $50,000 note payable/inventory value for Harmon.

While most notes, loans, and mortgages explicitly state an interest rate that will properly reflect the time value of money, there can be exceptions. Consider the example in Illustration 5–6 which uses present value techniques to value a note due in two years and for which the interest rate is not known.

No Explicit Interest

The Stridewell Wholesale Shoe Company recently sold a large order of shoes to Harmon Sporting Goods. Terms of the sale require Harmon to sign a noninterest-bearing note of $60,500 with payment due in two years.

Illustration 5–6

Valuing a Note: One Payment, No Explicit Interest

How should Stridewell and Harmon value the note receivable/payable and corresponding sales revenue/inventory? Even though the agreement states a noninterest-bearing note, the $60,500 does, in fact, include interest for the two-year period of the loan. We need to remove the interest portion of the $60,500 to determine the portion that represents the sales price of the shoes. We do this by calculating the present value. The following time diagram illustrates the situation assuming that a rate of 10% reflects the appropriate interest rate for a loan of this type:

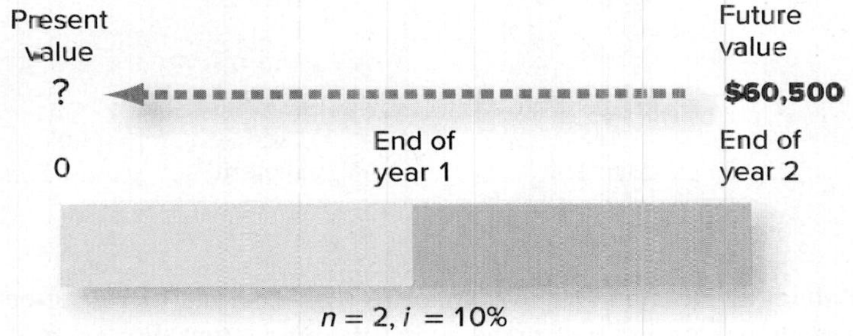

Again, using the present value of $1 table,

$60,500 (future value) × 0.82645* = $50,000 (present value)

*Present value of $1: $n = 2$, $i = 10\%$

Both the note receivable for Stridewell and the note payable for Harmon initially will be valued at $50,000. The difference of $10,500 (**$60,500** – $50,000) represents interest revenue/expense to be recognized over the life of the note. The appropriate journal entries are illustrated in later chapters.

Now can you answer the question posed in the scenario at the beginning of this section? Assuming that a rate of 10% reflects the appropriate interest rate in this situation, Slim Jim's TV and Appliance records a receivable and sales revenue of $1,636, which is the present value of the $1,800 to be received from Doug Smith one year from the date of sale.

$1,800 (future value) × 0.90909* = $1,636 (present value)

*Present value of $1: $n = 1$, $i = 10\%$ (from Table 2)

Additional Consideration

In Illustration 5–5, if Harmon Sporting Goods had prepaid Stridewell for delivery of the shoes in two years, rather than buying now and paying later, Harmon would be viewed as providing a two-year loan to Stridewell. Assuming that Harmon pays Stridewell $41,323, the present value of $50,000 for two-periods at 10%, Stridewell would record interest expense and Harmon would record interest revenue of $8,677 ($50,000 – $41,323) over the two-year period. When delivery occurs in two years, Stridewell records sales revenue of $50,000 and Harmon values the inventory acquired at $50,000.

PART B

Basic Annuities

● LO5–6

The previous examples involved the receipt or payment of a single future amount. Financial instruments frequently involve multiple receipts or payments of cash. If the same amount is to be received or paid each period, the series of cash flows is referred to as an annuity. A common annuity encountered in practice is a loan on which periodic interest is paid in equal amounts. For example, bonds typically pay interest semiannually in an amount determined by multiplying a stated rate by a fixed principal amount. Some loans and most leases are paid in equal installments during a specified period of time.

In an *ordinary annuity*, cash flows occur at the end of each period.

An agreement that creates an annuity can produce either an **ordinary annuity** or an **annuity due** (sometimes referred to as an annuity in advance) situation. The first cash flow (receipt or payment) of an ordinary annuity is made one compounding period *after* the date on which the agreement begins. The final cash flow takes place on the *last* day covered by the agreement. For example, an installment note payable dated December 31, 2021, might require the debtor to make three equal annual payments, with the first payment due on December 31, 2022, and the last one on December 31, 2024. The following time diagram illustrates an ordinary annuity:

Ordinary annuity

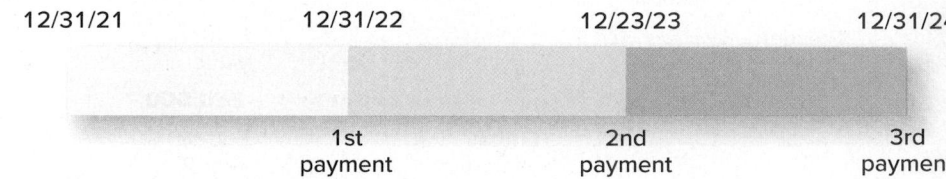

In an *annuity due*, cash flows occur at the *beginning* of each period.

The first payment of an annuity due is made on the *first* day of the agreement, and the last payment is made one period *before* the end of the agreement. For example, a three-year lease of a building that begins on December 31, 2021, and ends on December 31, 2024, may require the first year's lease payment in advance on December 31, 2021. The third and last payment would take place on December 31, 2023, the beginning of the third year of the lease. The following time diagram illustrates this situation:

Annuity due

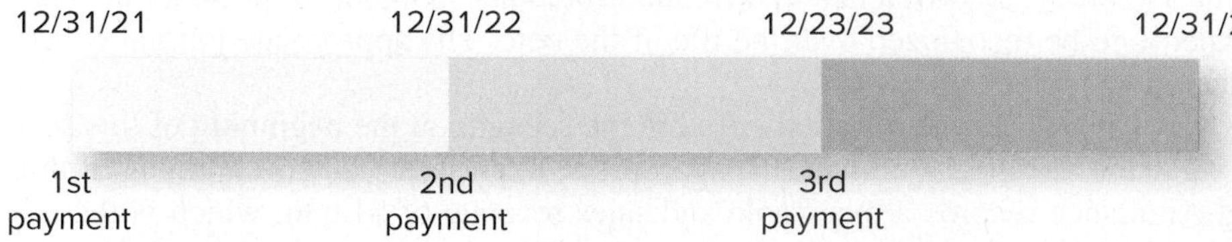

Future Value of an Annuity

Future Value of an Ordinary Annuity

● LO5–7

Let's first consider the future value of an ordinary annuity in Illustration 5–7.

Illustration 5–7
Future Value of an Ordinary Annuity

> Rita Grant wants to accumulate a sum of money to pay for graduate school. Rather than investing a single amount today that will grow to a future value, she decides to invest $10,000 a year over the next three years in a savings account paying 10% interest compounded annually. She decides to make the first payment to the bank one year from today.

The following time diagram illustrates this ordinary annuity situation. Time 0 is the start of the first period.

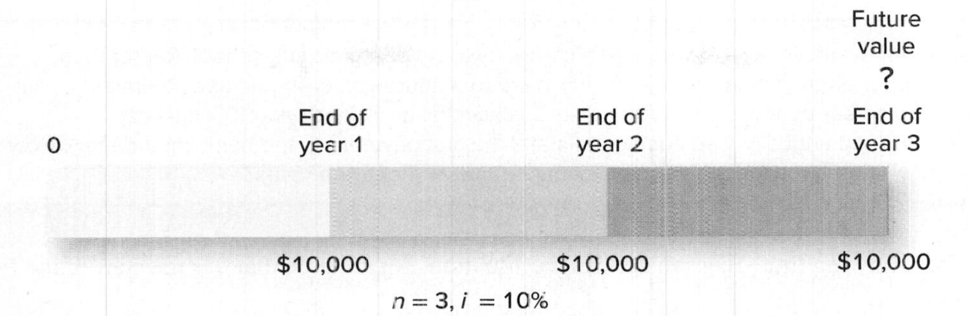

Using the FV of $1 factors from Table 1, we can calculate the future value of this annuity by calculating the future value of each of the individual payments as follows:

	Payment		FV of $1 $i = 10\%$		Future Value (at the end of year 3)	n
First payment	$10,000	×	1.21	=	$ 12,100	2
Second payment	10,000	×	1.10	=	11,000	1
Third payment	10,000	×	1.00	=	10,000	0
Total			**3.31**		**$33,100**	

From the time diagram, we can see that the first payment has two compounding periods to earn interest. The factor used, 1.21, is the FV of $1 invested for two periods at 10%. The second payment has one compounding period, and the last payment does not earn any interest because it is invested on the last day of the three-year annuity period. Therefore, the factor used is 1.00.

In an ordinary annuity, the last cash payment will not include any interest.

This illustration shows that it's possible to calculate the future value of the annuity by separately calculating the FV of each payment and then adding these amounts together. Fortunately, that's not necessary. Table 3, Future Value of an Ordinary Annuity of $1, simplifies the computation by summing the individual FV of $1 factors for various factors of n and i. Illustration 5–8 contains an excerpt from Table 3.

Illustration 5–8

Future Value of an Ordinary Annuity of $1 (excerpt from Table 3)

	Interest Rates (i)					
Periods (n)	**7%**	**8%**	**9%**	**10%**	**11%**	**12%**
1	1.0000	1.0000	1.0000	1.0000	1.0000	1.0000
2	2.0700	2.0800	2.0900	2.1000	2.1100	2.1200
3	3.2149	3.2464	3.2781	**3.3100**	3.3421	3.3744
4	4.4399	4.5061	4.5731	4.6410	4.7097	4.7793
5	5.7507	5.8666	5.9847	6.1051	6.2278	6.3528

The future value of $1 at the end of each of three periods invested at 10% is shown in Table 3 to be **$3.31**. We can simply multiply this factor by $10,000 to derive the FV of our ordinary annuity (FVA):

FVA = $10,000 (annuity amount) × **3.31*** = **$33,100**

*Future value of an ordinary annuity of $1: $n = 3$, $i = 10\%$

Future Value of an Annuity Due

Let's modify the previous illustration to create an annuity due in Illustration 5–9.

Illustration 5–9
Future Value of an Annuity Due

> Rita Grant wants to accumulate a sum of money to pay for graduate school. Rather than investing a single amount today that will grow to a future value, she decides to invest $10,000 a year over the next three years in a savings account paying 10% interest compounded annually. She decides to make the first payment to the bank immediately. How much will Rita have available in her account at the end of three years?

The following time diagram depicts the situation. Again, note that 0 is the start of the first period.

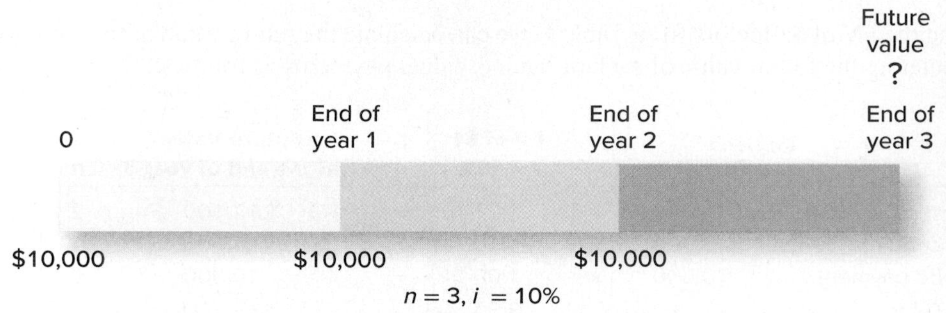

The future value can be found by separately calculating the FV of each of the three payments and then summing those individual future values:

In an annuity due, the last cash payment will include interest.

	Payment		FV of $1 $i = 10\%$		Future Value (at the end of year 3)	n
First payment	$10,000	×	1.331	=	$13,310	3
Second payment	10,000	×	1.210	=	12,100	2
Third payment	10,000	×	1.100	=	11,000	1
Total			**3.641**		**$36,410**	

And, again, this same future value can be found by using the future value of an annuity due (FVAD) factor from Table 5, Future Value of an Annuity Due of $1 as follows:

FVAD = $10,000 (annuity amount) × **3.641*** = **$36,410**

*Future value of an annuity due of $1: $n = 3$, $i = 10\%$

Of course, if *unequal* amounts are invested each year, we can't solve the problem by using the annuity tables. The future value of each payment would have to be calculated separately using the Future Value of $1 for each individual period and adding them together.

Present Value of an Annuity

Present Value of an Ordinary Annuity

● LO5–8 You will learn in later chapters that liabilities and receivables, with the exception of certain trade receivables and payables, are reported in financial statements at their present values. Most of these financial instruments specify equal periodic interest payments or installment payments (which represent a combination of interest and principal). As a result, the most common accounting applications of the time value of money involve determining the present value of annuities. As in the future value applications we discussed above, an annuity can be either an ordinary annuity or an annuity due. Let's look at an ordinary annuity first.

In Illustration 5–7, we determined that Rita Grant could accumulate $33,100 for graduate school by investing $10,000 at the end of each of three years at 10%. The $33,100 is the future value of the ordinary annuity described. Another alternative is to invest one single amount at the beginning of the three-year period. (See Illustration 5–10.) This single amount will equal the present value at the beginning of the three-year period of the $33,100 future value. It will also equal the present value of the $10,000 three-year annuity.

FINANCIAL Reporting Case

Q1, p. 239

Illustration 5–10

Present Value of an Ordinary Annuity

Rita Grant wants to accumulate a sum of money to pay for graduate school. She wants to invest a single amount today in a savings account earning 10% interest compounded annually that is equivalent to investing $10,000 at the end of each of the next three years.

The present value can be found by separately calculating the PV of each of the three payments and then summing those individual present values as follows:

	Payment		PV of $1 i = 10%		Present Value (at the beginning of year 1)	n
First payment	$10,000	×	0.90909	=	$ 9,091	1
Second payment	10,000	×	0.82645	=	8,264	2
Third payment	10,000	×	0.75131	=	7,513	3
Total			**2.48685**		**$24,868**	

A more efficient method of calculating present value is to use Table 4, Present Value of an Ordinary Annuity of $1. Illustration 5–11 contains an excerpt from Table 4.

Illustration 5–11

Present Value of an Ordinary Annuity of $1 (excerpt from Table 4)

	Interest Rates (i)					
Periods (n)	**7%**	**8%**	**9%**	**10%**	**11%**	**12%**
1	0.93458	0.92593	0.91743	0.90909	0.90090	0.89286
2	1.80802	1.78326	1.75911	1.73554	1.71252	1.69005
3	2.62432	2.57710	2.53129	**2.48685**	2.44371	2.40183
4	3.38721	3.31213	3.23972	3.16987	3.10245	3.03735
5	4.10020	3.99271	3.88965	3.79079	3.69590	3.60478

Using Table 4, we calculate the PV of the ordinary annuity (PVA) as follows:

PVA = $10,000 (annuity amount) × **2.48685*** = **$24,868**

*Present value of an ordinary annuity of $1: $n = 3$, $i = 10\%$

The relationship between the present value and the future value of the annuity can be depicted graphically as follows:

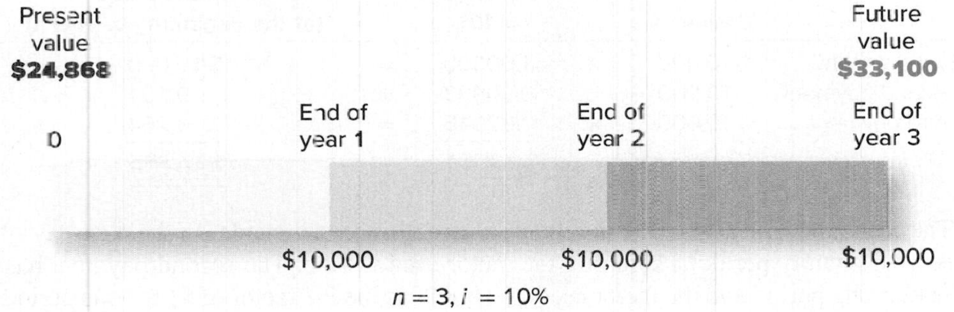

This can be interpreted in several ways.

1. $10,000 invested at 10% at the end of each of the next three years will accumulate to $33,100 at the end of the third year.
2. $24,868 invested at 10% now will grow to $33,100 after three years.
3. Someone whose time value of money is 10% would be willing to pay $24,868 now to receive $10,000 at the end of each of the next three years.
4. If your time value of money is 10%, you should be indifferent with respect to paying/receiving (a) $24,868 now, (b) $33,100 three years from now, or (c) $10,000 at the end of each of the next three years. In other words, at the end of the third year you would have the same $33,100, regardless of the option chosen to get there.

Additional Consideration

We also can verify that these are the present value and future value of the same annuity by calculating the present value of a single cash amount of $33,100 three years hence:

PV = $33,100 (future value) × 0.75131* = $24,868

*Present value of $1: $n = 3$, $i = 10\%$

Present Value of an Annuity Due

Illustration 5–12

Present Value of an Annuity Due

In the previous illustration, suppose that the three equal payments of $10,000 are to be made at the *beginning* of each of the three years. Recall from Illustration 5–9 that the future value of this annuity is $36,410. What is the present value?

The following time diagram depicts this situation:

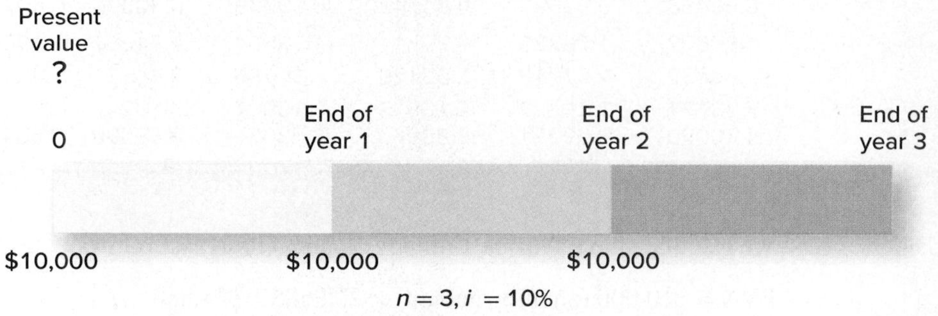

Once again, using individual PV factors of $1 from Table 2, the PV of the annuity due can be calculated as follows:

	Payment		PV of $1 $i = 10\%$		Present Value (at the beginning of year 1)	n
First payment	$10,000	×	1.00000	=	$ 10,000	0
Second payment	10,000	×	0.90909	=	9,091	1
Third payment	10,000	×	0.82645	=	8,264	2
Total			**2.73554**		**$27,355**	

In an annuity due, the first cash payment won't include interest.

The first payment does not contain any interest since it is made on the first day of the three-year annuity period. Therefore, the factor used is 1.00. The second payment has one compounding period and the factor used of 0.90909 is the PV factor of $1 for one period and

10%, and we need to remove two compounding periods of interest from the third payment. The factor used of 0.82645 is the PV factor of $1 for two periods and 10%.

The relationship between the present value and the future value of the annuity can be depicted graphically as follows:

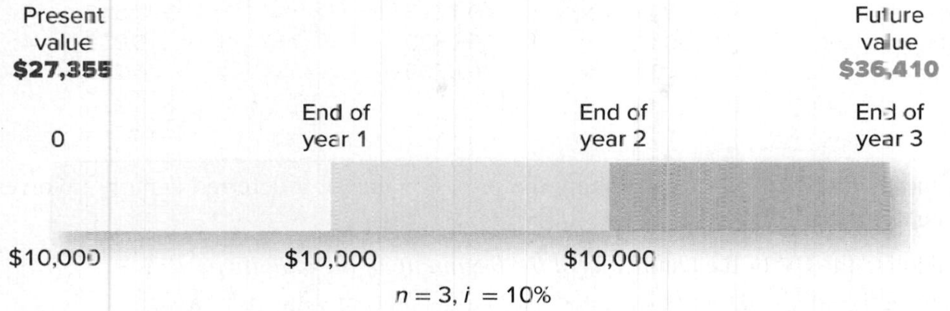

Using Table 6, Present Value of an Annuity Due, we can more efficiently calculate the PV of the annuity due (PVAD):

PVAD = $10,000 (annuity amount) × **2.73554*** = $27,355

*Present value of an annuity due of $1: $n = 3$, $i = 10\%$

To better understand the relationship between Tables 4 and 6, notice that the PVAD factor for three periods, 10%, from Table 6 is **2.73554**. This is simply the PVA factor for two periods, 10%, of 1.73554, plus 1.0. The addition of 1.0 reflects the fact that the first payment occurs immediately and thus includes no interest.

Of course, if payment amounts are *not the same* each year, we don't have an annuity and can't find a solution by using the annuity tables. In that case, we'd need to find the present value of each payment separately using the Present Value of $1 for each individual period and add them together.

Present Value of a Deferred Annuity

Accounting valuations often involve the present value of annuities in which the first cash flow is expected to occur more than one time period after the date of the agreement. As the inception of the annuity is deferred beyond a single period, this type of annuity is referred to as a **deferred annuity**.[6]

A *deferred annuity* exists when the first cash flow occurs more than one period after the date the agreement begins.

Illustration 5–13

Deferred Annuity

At January 1, 2021, you are considering acquiring an investment that will provide three equal payments of $10,000 each to be received at the end of three consecutive years. However, the first payment is not expected until *December 31, 2023*. The time value of money is 10%. How much would you be willing to pay for this investment?

The following time diagram depicts this situation:

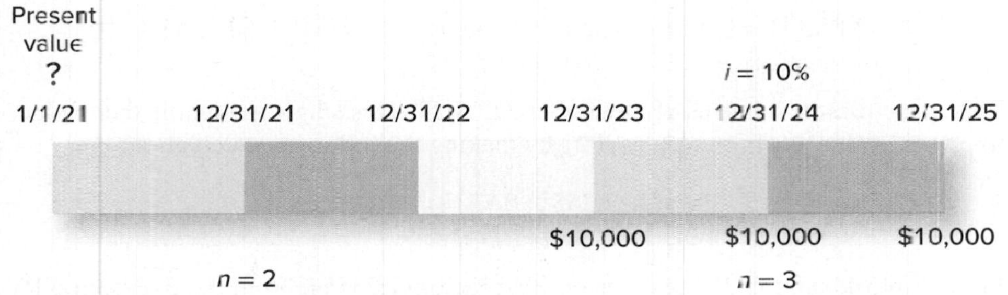

[6]The future value of a deferred annuity is the same as the future amount of an annuity not deferred. That's because there are no interest compounding periods prior to the beginning of the annuity period.

The present value of the deferred annuity can be calculated by summing the present values of the three individual cash flows, as of today.

	Payment		PV of \$1 $i = 10\%$		Present Value	n
First payment	\$10,000	×	0.75131	=	\$ 7,513	3
Second payment	10,000	×	0.68301	=	6,830	4
Third payment	10,000	×	0.62092	=	6,209	5
					\$20,552	

A more efficient way of calculating the present value of a deferred annuity involves a two-step process.

1. Calculate the PV of the annuity *as of the beginning of the annuity period.*
2. Reduce the single amount calculated in (1) to its present value *as of today.*

In this case, we compute the present value of the annuity as of December 31, 2022, by multiplying the annuity amount by the three-period ordinary annuity factor:

PVA = \$10,000 (annuity amount) × 2.48685* = **\$24,868**

*Present value of an ordinary annuity of \$1: $n = 3$, $i = 10\%$

This is the present value as of December 31, 2022. This single amount is then reduced to present value as of January 1, 2021, by making the following calculation:

PV = \$24,868 (future amount) × 0.82645* = **\$20,552**

*Present value of \$1: $n = 2$, $i = 10\%$

The following time diagram illustrates this two-step process:

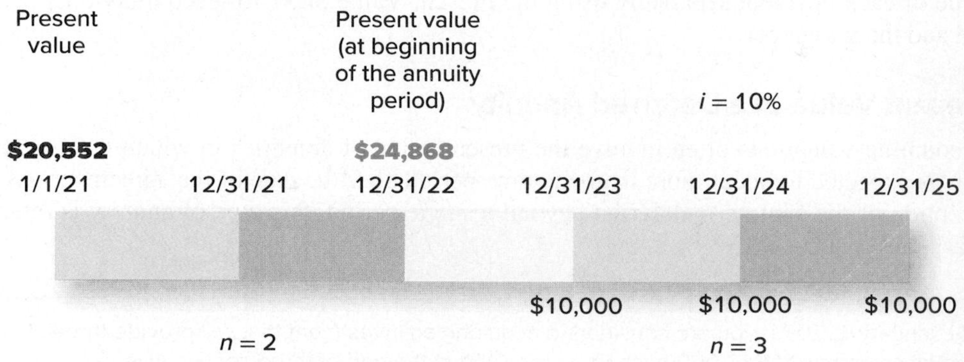

If you recall the concepts you learned in this chapter, you might think of other ways the present value of a deferred annuity can be determined. Among them:

1. Calculate the PV of an annuity due, rather than an ordinary annuity, and then reduce that amount **three** periods rather than two:

   PVAD = \$10,000 (annuity amount) × 2.73554* = \$27,355

   *Present value of an annuity due of \$1: $n = 3$, $i = 10\%$

   This is the present value as of December 31, 2023. This single amount is then reduced to present value as of January 1, 2021 by making the following calculation:

   PV = \$27,355 × 0.75131* = **\$20,552**

   *Present value of \$1: $n = 3$, $i = 10\%$

2. From Table 4, subtract the two-period PVA factor (1.73554) from the five-period PVA factor (3.79079) and multiply the difference (2.05525) by \$10,000 to get **\$20,552**.

Financial Calculators and Excel

As previously mentioned, financial calculators can be used to solve future and present value problems. For example, a Texas Instruments model BA-35 has the following pertinent keys:

N %I PV FV PMT CPT

These keys are defined as follows:

N = number of periods
%I = interest rate
PV = present value
FV = future value
PMT = annuity payments
CPT = compute button

To illustrate its use, assume that you need to determine the present value of a 10-period ordinary annuity of $200 using a 10% interest rate. You would enter N 10, %I 10, PMT −200, then press CPT and PV to obtain the answer of $1,229.

Many professionals choose to use spreadsheet software, such as Excel, to solve time value of money problems. These spreadsheets can be used in a variety of ways. A template can be created using the formulas shown in Illustration 5–22. An alternative is to use the software's built-in financial functions. For example, Excel has a function called PV that calculates the present value of an ordinary annuity. To use the function, you would select the pull-down menu for "Insert," click on "Function" and choose the category called "Financial." Scroll down to PV and double-click. You will then be asked to input the necessary variables—interest rate, the number of periods, and the payment amount.

In subsequent chapters we illustrate the use of both a calculator and Excel in addition to present value tables to solve present value calculations for selected examples and illustrations.

Using a calculator:
enter: N 10 I 10
PMT −200
Output: FV 1,229

Using Excel,
enter: FV(10,10,−200)
Output: 1,229

Solving for Unknown Values in Present Value Situations

● LO5–9

In present value problems involving annuities, there are four variables: (1) present value of an ordinary annuity (PVA) or present value of an annuity due (PVAD), (2) the amount of each annuity payment, (3) the number of periods, *n*, and (4) the interest rate, *i*. If you know any three of these, the fourth can be determined.

Illustration 5–14

Determining the Annuity Amount When Other Variables Are Known

Assume that you borrow $700 from a friend and intend to repay the amount in four equal annual installments beginning one year from today. Your friend wishes to be reimbursed for the time value of money at an 8% annual rate. What is the required annual payment that must be made (the annuity amount), to repay the loan in four years?

The following time diagram illustrates the situation:

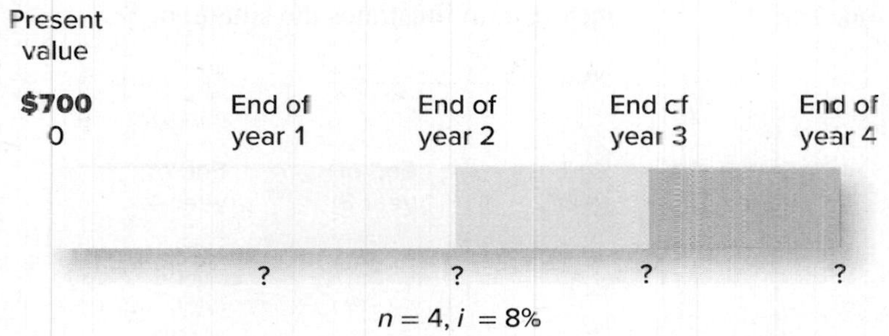

The required payment is the annuity amount that will provide a present value of $700 when using an interest rate of 8%.

$$\$700 \text{ (present value)} = 3.31213^{*} \times \text{annuity amount}$$

*Present value of an ordinary annuity of $1: $n = 4$, $i = 8\%$

Rearranging algebraically, we find that the annuity amount is $211.34.

$$\$700 \text{ (present value)} \div 3.31213^{*} = \$211.34 \text{ (annuity amount)}$$

*Present value of an ordinary annuity of $1: $n = 4$, $i = 8\%$

The unknown variable is the annuity amount.

You would have to make four annual payments of $211.34 to repay the loan. Total payments of $845.36 (4 × $211.34) would include $145.36 in interest ($845.36 – $700.00).

Illustration 5–15

Determining n When Other Variables Are Known

Assume that you borrow $700 from a friend and intend to repay the amount in equal installments of $100 per year over a period of years. The payments will be made at the end of each year beginning one year from now. Your friend wishes to be reimbursed for the time value of money at a 7% annual rate. How many years would it take before you repaid the loan?

Once again, this is an ordinary annuity situation because the first payment takes place one year from now. The following time diagram illustrates the situation:

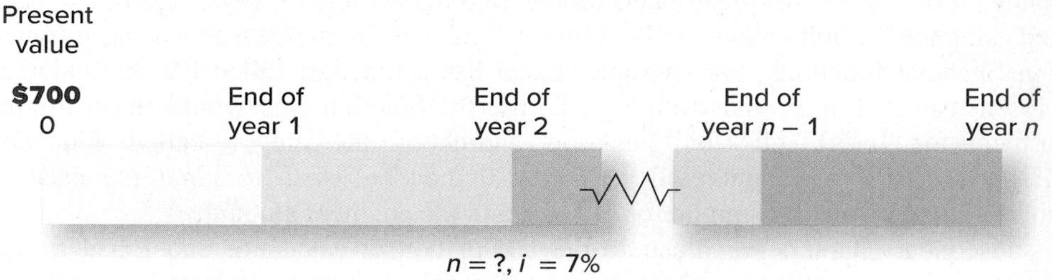

The number of years is the value of n that will provide a present value of $700 when finding the present value of $100 payments using an interest rate of 7%:

$$\$700 \text{ (present value)} = \$100 \text{ (annuity amount)} \times ?^{*}$$

*Present value of an ordinary annuity of $1: $n = ?$, $i = 7\%$

The unknown variable is the number of periods.

Rearranging algebraically, we find that the PVA table factor is 7.0.

$$\$700 \text{ (present value)} \div \$100 \text{ (annuity amount)} = 7.0^{*}$$

*Present value of an ordinary annuity of $1: $n = ?$, $i = 7\%$

When you consult the PVA table, Table 4, you search the 7% column ($i = 7\%$) for this value and find 7.02358 in row 10. So it would take approximately 10 years to repay the loan in the situation described.

Illustration 5–16

Determining i When Other Variables Are Known

Suppose that a friend asked to borrow $331 today (present value) and promised to repay you $100 (the annuity amount) at the end of each of the next four years. What is the annual interest rate implicit in this agreement?

FINANCIAL Reporting Case

Q2, p. 239

First of all, we are dealing with an ordinary annuity situation as the payments are at the end of each period. The following time diagram illustrates the situation:

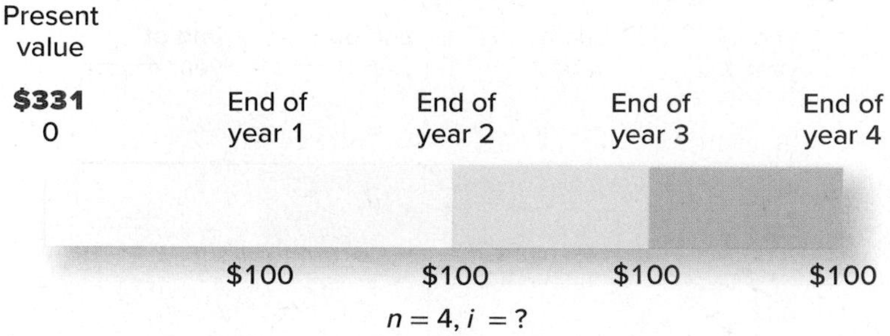

The interest rate is the rate that will provide a present value of $331 when finding the present value of the $100 four-year ordinary annuity.

$331 (present value) = $100 (annuity amount) × ?*

*Present value of an ordinary annuity of $1: $n = 4$, $i = ?$

The unknown variable is the interest rate.

Rearranging algebraically, we find that the PVA table factor is 3.31.

$331 (present value) ÷ $100 (annuity amount) = 3.31*

*Present value of an ordinary annuity of $1: $n = 4$, $i = ?$

When you consult the PVA table, Table 4, you search row four ($n = 4$) for this value and find it in the 8% column. So the effective interest rate is 8%.

Suppose that you borrowed $400 from a friend and promised to repay the loan by making three annual payments of $100 at the end of each of the next three years plus a final payment of $200 at the end of year four. What is the interest rate implicit in this agreement?

Illustration 5–17

Determining *i* When Other Variables Are Known—Unequal Cash Flows

The following time diagram illustrates the situation:

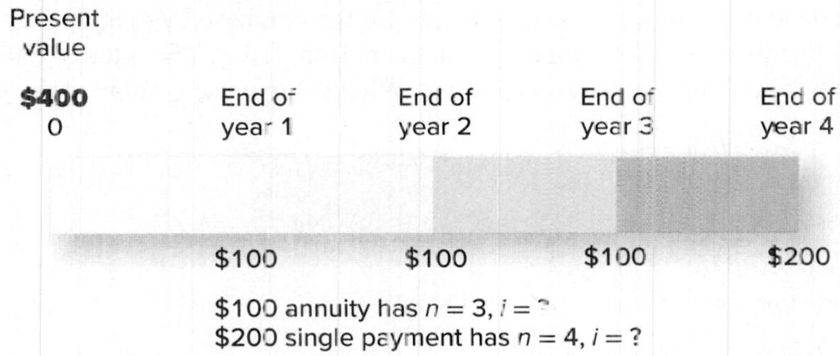

The interest rate is the rate that will provide a present value of $400 when finding the present value of the $100 three-year ordinary annuity plus the $200 to be received in four years:

$400 (present value) = $100 (annuity amount) × ?* + $200 (single payment) × ?†

*Present value of an ordinary annuity of $1: $n = 3$, $i = ?$

†Present value of $1: $n = 4$, $i = ?$

The unknown variable is the interest rate.

This equation involves two unknowns and is not as easily solved as the two previous examples. One way to solve the problem is to trial-and-error the answer. For example, if we assumed i to be 9%, the total PV of the payments would be calculated as follows:

PV = $100 (2.53129*) + $200 (0.70843†) = $395

*Present value of an ordinary annuity of $1: $n = 3$, $i = 9\%$

†Present value of $1: $n = 4$, $i = 9\%$

Because the present value computed is less than the $400 borrowed, using 9% removes too much interest. Recalculating PV with $i = 8\%$ results in a PV of $405. This indicates that the interest rate implicit in the agreement is between 8% and 9%.

Concept Review Exercise

ANNUITIES

Using the appropriate table, answer each of the following independent questions.

1. What is the future value of an annuity of $2,000 invested at the *end* of each of the next six periods at 8% interest?
2. What is the future value of an annuity of $2,000 invested at the *beginning* of each of the next six periods at 8% interest?

3. What is the present value of an annuity of \$6,000 to be received at the *end* of each of the next eight periods assuming an interest rate of 10%?
4. What is the present value of an annuity of \$6,000 to be received at the *beginning* of each of the next eight periods assuming an interest rate of 10%?
5. Jane bought a \$3,000 audio system and agreed to pay for the purchase in 10 equal annual installments of \$408 beginning one year from today. What is the interest rate implicit in this agreement?
6. Jane bought a \$3,000 audio system and agreed to pay for the purchase in 10 equal annual installments beginning one year from today. The interest rate is 12%. What is the amount of the annual installment?
7. Jane bought a \$3,000 audio system and agreed to pay for the purchase by making nine equal annual installments beginning one year from today plus a lump-sum payment of \$1,000 at the end of 10 periods. The interest rate is 10%. What is the required annual installment?
8. Jane bought an audio system and agreed to pay for the purchase by making four equal annual installments of \$800 beginning one year from today plus a lump-sum payment of \$1,000 at the end of five years. The interest rate is 12%. What was the cost of the audio system? (*Hint:* What is the present value of the cash payments?)
9. Jane bought an audio system and agreed to pay for the purchase by making five equal annual installments of \$1,100 beginning four years from today. The interest rate is 12%. What was the cost of the audio system? (*Hint:* What is the present value of the cash payments?)

Solution:

1. $\text{FVA} = \$2{,}000 \times 7.3359^{*} = \$14{,}672$

*Future value of an ordinary annuity of \$1: $n = 6$, $i = 8\%$ (from Table 3)

2. $\text{FVAD} = \$2{,}000 \times 7.9228^{*} = \$15{,}846$

*Future value of an annuity due of \$1: $n = 6$, $i = 8\%$ (from Table 5)

3. $\text{PVA} = \$6{,}000 \times 5.33493^{*} = \$32{,}010$

*Present value of ordinary annuity of \$1: $n = 8$, $i = 10\%$ (from Table 4)

4. $\text{PVAD} = \$6{,}000 \times 5.86842^{*} = \$35{,}211$

*Present value of an annuity due of \$1: $n = 8$, $i = 10\%$ (from Table 6)

5. $\dfrac{\$3{,}000}{\$408} = 7.35^{*}$

*Present value of an ordinary annuity of \$1: $n = 10$, $i = ?$ (from Table 4, i approximately 6%)

6. $\text{Each annuity payment} = \dfrac{\$3{,}000}{5.65022^{*}} = \$531$

*Present value of an ordinary annuity of \$1: $n = 10$, $i = 12\%$ (from Table 4)

7. $$\text{Each annuity payment} = \frac{\$3{,}000 - [\text{PV of } \$1{,}000\ (n = 10, i = 10\%)]}{5.75902^{*}}$$

$$\text{Each annuity payment} = \frac{\$3{,}000 - (\$1{,}000 \times .38554^{\dagger})}{5.75902^{*}}$$

$$\text{Each annuity payment} = \frac{\$2{,}614}{5.75902^{*}} = \$454$$

*Present value of an ordinary annuity of \$1: $n = 9$, $i = 10\%$ (from Table 4)
†Present value of \$1: $n = 10$, $i = 10\%$ (from Table 2)

8. $\text{PV} = \$800 \times 3.03735^{*} + \$1{,}000 \times 0.56743^{\dagger} = \$2{,}997$

*Present value of an ordinary annuity of \$1: $n = 4$, $i = 12\%$ (from Table 4)

†Present value of \$1: $n = 5$, $i = 12\%$ (from Table 2)

9. $\text{PVA} = \$1{,}100 \times 3.60478^{*} = \$3{,}965$

*Present value of an ordinary annuity of \$1: $n = 5$, $i = 12\%$ (from Table 4)

This is the present value three years from today (the beginning of the five-year ordinary annuity). This single amount is then reduced to present value as of today by making the following calculation:

$PV = \$3,965 \times 0.71178^{\dagger} = \$2,822$

†Present value of \$1: $n = 3$, $i = 12\%$, (from Table 2)

Preview of Accounting Applications of Present Value Techniques—Annuities

● LO5–10

The time value of money has many applications in accounting. Most of these applications involve the concept of present value. Because financial instruments typically specify equal periodic payments, these applications quite often involve annuity situations. For example, let's consider one accounting situation using both an ordinary annuity and the present value of a single amount (long-term bonds), two using an annuity due (long-term leases and installment notes), and a fourth using a deferred annuity (pension obligations).

FINANCIAL Reporting Case

Q3, p. 239

Valuation of Long-Term Bonds

You will learn in Chapter 14 that a long-term bond usually requires the issuing (borrowing) company to repay a specified amount at maturity and make periodic stated interest payments over the life of the bond. The *stated* interest payments are equal to the contractual stated rate multiplied by the face value of the bonds. At the date the bonds are issued (sold), the marketplace will determine the price of the bonds based on the *market* rate of interest for investments with similar characteristics. The market rate at date of issuance may not equal the bonds' stated rate in which case the price of the bonds (the amount the issuing company actually is borrowing) will not equal the bonds' face value. Bonds issued at more than face value are said to be issued at a premium, while bonds issued at less than face value are said to be issued at a discount. Consider the example in Illustration 5–18.

Illustration 5–18

Valuing a Long-Term Bond Liability

On June 30, 2021, Fumatsu Electric issued 10% stated rate bonds with a face amount of \$200 million. The bonds mature on June 30, 2041 (20 years). The market rate of interest for similar issues was 12%. Interest is paid semiannually (5%) on June 30 and December 31, beginning December 31, 2021. The interest payment is \$10 million (5% × \$200 million). What was the price of the bond issue? What amount of interest expense will Fumatsu record for the bonds in 2021?

To determine the price of the bonds, we calculate the present value of the 40-period annuity (40 semiannual interest payments of \$10 million) and the lump-sum (single amount) payment of \$200 million paid at maturity using the semiannual market rate of interest of 6%. In equation form,

$$PVA = \$10 \text{ million (annuity amount)} \times 15.04630^{*} = \$150,463,000$$
$$PV = \$200 \text{ million (lump-sum)} \times 0.09722^{\dagger} = \underline{19,444,000}$$
$$\text{Price of the bond issue} = \underline{\underline{\$169,907,000}}$$

*Present value of an ordinary annuity of \$1: $n = 40$, $i = 6\%$
†Present value of \$1: $n = 40$, $i = 6\%$

The bonds will sell for \$169,907,000, which represents a discount of \$30,093,000 (\$200,000,000 − \$169,907,000). The discount results from the difference between the semiannual stated rate of 5% and the market rate of 6%. Fumatsu records a \$169,907,000 increase in cash and a corresponding liability for bonds payable.

Interest expense for the first six months is determined by multiplying the carrying value (book value) of the bonds (\$169,907,000) by the semiannual effective rate (6%) as follows:

$$\$169,907,000 \times 6\% = \$10,194,420$$

The difference between interest expense ($10,194,420) and interest paid ($10,000,000) increases the carrying value of the bond liability. Interest for the second six months of the bonds' life is determined by multiplying the new carrying value by the 6% semiannual effective rate.[7]

We discuss the specific accounts used to record these transactions in Chapters 12 and 14.

Valuation of Long-Term Leases

Companies frequently acquire the use of assets by leasing rather than purchasing them. Leases usually require the payment of fixed amounts at regular intervals over the life of the lease. You will learn in Chapter 15 that certain leases are treated in a manner similar to an installment purchase by the lessee. In other words, the lessee records an asset and corresponding lease liability at the present value of the lease payments. Consider the example in Illustration 5–19.

Illustration 5–19

Valuing a Long-Term Lease Liability

> On January 1, 2021, the Stridewell Wholesale Shoe Company signed a 25-year lease agreement for an office building. Terms of the lease call for Stridewell to make annual lease payments of **$10,000** at the beginning of each year, with the first payment due on January 1, 2021. Assuming an interest rate of 10% properly reflects the time value of money in this situation, how should Stridewell value the asset acquired and the corresponding lease liability?

Leases require the recording of an asset and corresponding liability at the present value of future lease payments.

Once again, by computing the present value of the lease payments, we remove the portion of the payments that represents interest, leaving the portion that represents payment for the asset itself. Because the first payment is due immediately, as is common for leases, this is an annuity due situation. In equation form:

$$PVAD = \$10,000 \text{ (annuity amount)} \times 9.98474^{*} = \$99,847$$

*Present value of an annuity due of $1: $n = 25$, $i = 10\%$

Stridewell initially will value the leased asset and corresponding lease liability at $99,847.

Journal entry at the beginning of the lease.

Right-of-use asset	99,847	
Lease payable		99,847

The difference between this amount and total future cash payments of $250,000 ($10,000 × 25) represents the interest that is implicit in this agreement. That difference is recorded as interest over the life of the lease.

Valuation of Installment Notes

If you have purchased a car, or maybe a house, unless you paid cash, you signed a note promising to pay a portion of the purchase price over, say, five years for the car or 30 years for the house. Such notes usually call for payment in monthly installments rather than by a single amount at maturity. Corporations also often borrow using installment notes. Usually, installment payments are equal amounts each period so they constitute an annuity. Each payment includes both an amount that represents interest and an amount that represents a reduction of the outstanding balance (principal reduction). The periodic reduction of the balance is sufficient that, at maturity, the note is completely paid. If we know the amount of the periodic installment payments, we can calculate the amount at which to record the note in precisely the same way as we did in the previous illustration for a lease. But, suppose instead that we know the amount of the loan and want to calculate the amount of the installment payments. In that case, rather than multiplying the payment by the appropriate present value of an annuity factor to get the present value, we would invert the calculation and divide the amount of the loan by that present value factor. For instance, consider the example in Illustration 5–20.

[7]In Chapters 12 and 14, we refer to the process of determining interest as the effective interest rate times the loan balance as the *effective interest method.*

Illustration 5–20

Calculating Installment Note Payments

On January 1, 2021, the Stridewell Wholesale Shoe Company purchased a $35,000 machine, with a $5,000 down payment and a 5-year installment note for the remaining $30,000. Terms of the note call for Stridewell to make annual installment payments at the beginning, with the first payment due on January 1, 2021, and at each December 31 thereafter. Assuming an interest rate of 4%, what is the amount of the annual installment payments?

The annual installment payment amount that would pay a $30,000 loan over 5 years at a 4% interest rate is:

$$\$30,000 \div 4.62990^{*} = \$6,480$$

*present value of an annuity due of $1: $n = 5$, $i = 4\%$

Stridewell initially will record the machine at its $35,000 cost with credits to cash and notes payable.

Machine	35,000	
Cash		5,000
Notes payable		30,000

Stridewell will record the first installment payment at the date of purchase:

Notes payable	6,480	
Cash		6,480

No interest is needed for the first installment payment because no time had yet passed, so no interest had accrued. With the second payment, though, the payment represents part interest and part reduction of the loan:

Interest expense*	941	
Notes payable	5,539	
Cash		6,480

*Interest expense is 4% times the amount owed initially reduced by the first payment a year earlier, or 4% × ($30,000 – $6,480) = $941.

Valuation of Pension Obligations

Pension plans are important compensation vehicles used by many U.S. companies. These plans are essentially forms of deferred compensation as the pension benefits are paid to employees after they retire. You will learn in Chapter 17 that some pension plans create obligations during employees' service periods that must be paid during their retirement periods. These obligations are funded during the employment period. This means companies contribute cash to pension funds annually with the intention of accumulating sufficient funds to pay employees the retirement benefits they have earned. The amounts contributed are determined using estimates of retirement benefits. The actual amounts paid to employees during retirement depend on many factors including future compensation levels and length of life. Consider Illustration 5–21.

Illustration 5–21

Valuing a Pension Obligation

On January 1, 2021, the Stridewell Wholesale Shoe Company hired Terry Elliott. Terry is expected to work for 25 years before retirement on December 31, 2045. Annual retirement payments will be paid at the end of each year during his retirement period, expected to be 20 years. The first payment will be on December 31, 2046. During 2021 Terry earned an annual retirement benefit estimated to be **$2,000** per year. The company plans to contribute cash to a pension fund that will accumulate to an amount sufficient to pay Terry this benefit. Assuming that Stridewell anticipates earning 6% on all funds invested in the pension plan, how much would the company have to contribute at the end of 2021 to pay for pension benefits earned in 2021?

To determine the required contribution, we calculate the present value on December 31, 2021, of the deferred annuity of **$2,000** that begins on December 31, 2046, and is expected to end on December 31, 2065.

The following time diagram depicts this situation:

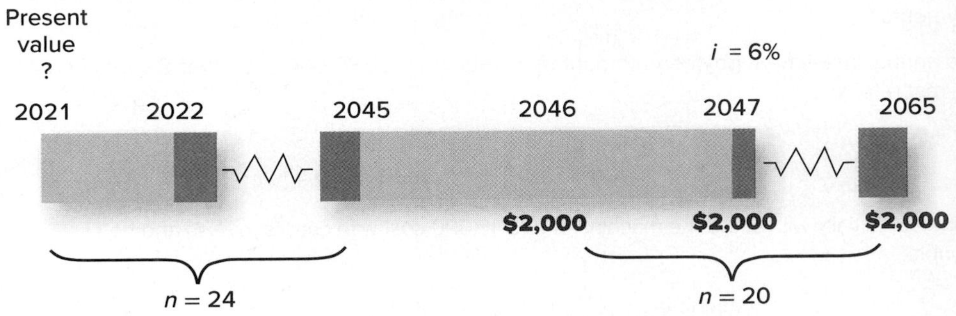

We can calculate the present value of the annuity using a two-step process. The first step computes the present value of the annuity as of December 31, 2045, by multiplying the annuity amount by the 20-period ordinary annuity factor.

PVA = **$2,000** (annuity amount) × 11.46992* = $22,940

*Present value of an ordinary annuity of $1: $n = 20$, $i = 6\%$

This is the present value as of December 31, 2045. This single amount is then reduced to present value as of December 31, 2021, by a second calculation.

PV = $22,940 (future amount) × 0.24698* = $5,666

*Present value of $1: $n = 24$, $i = 6\%$

Stridewell would have to contribute $5,666 at the end of 2021 to fund the estimated pension benefits earned by its employee in 2021. Viewed in reverse, $5,666 invested now at 6% will accumulate a fund balance of $22,940 at December 31, 2045. If the fund balance remains invested at 6%, $2,000 can be withdrawn each year for 20 years before the fund is depleted.

Among the other situations you'll encounter using present value techniques are valuing notes (Chapters 10 and 14) and other postretirement benefits (Chapter 17).

Summary of Time Value of Money Concepts

Illustration 5–22 summarizes the time value of money concepts discussed in this chapter.

Illustration 5–22

Summary of Time Value of Money Concepts

Concept	Summary	Formula	Table
Future value (FV) of $1	The amount of money that a dollar will grow to at some point in the future.	$FV = \$1(1 + i)^n$	1
Present value (PV) of $1	The amount of money today that is equivalent to a given amount to be received or paid in the future.	$PV = \frac{\$1}{(1 + i)^n}$	2
Future value of an ordinary annuity (FVA) of $1	The future value of a series of equal-sized cash flows with the first payment taking place at the end of the first compounding period.	$FVA = \frac{(1 + i)^n - 1}{i}$	3

(continued)

(concluded)

Concept	Summary	Formula	Table
Present value of an ordinary annuity (PVA) of $1	The present value of a series of equal-sized cash flows with the first payment taking place at the end of the first compounding period.	$PVA = \dfrac{1 - \dfrac{1}{(1+i)^n}}{i}$	4
Future value of an annuity due (FVAD) of $1	The future value of a series of equal-sized cash flows with the first payment taking place at the beginning of the annuity period.	$FVAD = \left[\dfrac{(1+i)^n - 1}{i}\right] \times (1+i)$	5
Present value of an annuity due (PVAD) of $1	The present value of a series of equal-sized cash flows with the first payment taking place at the beginning of the annuity period.	$PVAD = \left[\dfrac{1 - \dfrac{1}{(1+i)^n}}{i}\right] \times (1+i)$	6

Financial Reporting Case Solution

©Gabriel Petrescu/Shutterstock

1. **Why was Mavis to receive $336 million rather than the $758.7 million lottery prize?** *(p. 251)* Mavis chose to receive her lottery winnings in one lump-sum payment immediately rather than in 30 installment payments. The state calculates the present value of the payments, withholds the necessary federal and state income tax, and pays Mavis the remainder.

2. **What interest (discount) rate did the state of Massachusetts use to calculate the $336 million lump-sum payment** *(p. 256)* Assuming equal annual installment payments beginning immediately, the amount of the payments is determined by dividing $758.7 million by 30 periods:

$758.7 million ÷ 30 =	$25,290,000
Less: 30% income tax	(7,587,000)
Net-of-tax payment	$17,703,000

Because the first payment is made immediately, this is an annuity due situation. We must find the interest rate that provides a present value of $336 million. $336,000,000 ÷ $17,703,000 is 18.98, the present value factor that equates the payments and their present value. Searching for that factor in row 30 of Table 6, we find the factor closest to 18.98 in the 3.5% column. So, the discount rate used by the state was approximately 3.5%.

3. **What are some of the accounting applications that incorporate the time value of money into valuation?** *(p. 259)* Accounting applications that incorporate the time value of money techniques into valuation include the valuation of long-term notes receivable and various long-term liabilities that include bonds, notes, leases, pension obligations, and other postretirement benefits. We study these in detail in later chapters. ●

The Bottom Line

- **LO5–1** Compound interest includes interest not only on the initial investment but also on the accumulated interest in previous periods. (*p. 240*)
- **LO5–2** The future value of a single amount is the amount of money that a dollar will grow to at some point in the future. It is computed by *multiplying* the single amount by $(1 + i)^n$, where i is the interest rate and n the

number of compounding periods. The Future Value of $1 table allows for the calculation of future value for any single amount by providing the factors for various combinations of i and n. (*p. 241*)

- **LO5–3** The present value of a single amount is the amount of money today that is equivalent to a given amount to be received or paid in the future. It is computed by *dividing* the future amount by $(1 + i)^n$. The Present Value of $1 table simplifies the calculation of the present value of any future amount. (*p. 242*)
- **LO5–4** There are four variables in the process of adjusting single cash flow amounts for the time value of money: present value (PV), future value (FV), i and n. If we know any three of these, the fourth can be computed easily. (*p. 243*)
- **LO5–5** We value most notes receivable and notes payable at the present value of future cash flows they call for, reflecting an appropriate time value of money. (*p. 246*)
- **LO5–6** An annuity is a series of equal-sized cash flows occurring over equal intervals of time. An ordinary annuity exists when the cash flows occur at the end of each period. An annuity due exists when the cash flows occur at the beginning of each period. (*p. 248*)
- **LO5–7** The future value of an ordinary annuity (FVA) is the future value of a series of equal-sized cash flows with the first payment taking place at the end of the first compounding period. The last payment will not earn any interest since it is made at the end of the annuity period. The future value of an annuity due (FVAD) is the future value of a series of equal-sized cash flows with the first payment taking place at the beginning of the annuity period (the beginning of the first compounding period). (*p. 248*)
- **LO5–8** The present value of an ordinary annuity (PVA) is the present value of a series of equal-sized cash flows with the first payment taking place at the end of the first compounding period. The present value of an annuity due (PVAD) is the present value of a series of equal-sized cash flows with the first payment taking place at the beginning of the annuity period. The present value of a deferred annuity is the present value of a series of equal-sized cash flows with the first payment taking place more than one time period after the date of the agreement. (*p. 250*)
- **LO5–9** In present value problems involving annuities, there are four variables: PVA or PVAD, the annuity amount, the number of compounding periods (n) and the interest rate (i). If you know any three of these, you can determine the fourth. (*p. 255*)
- **LO5–10** Most accounting applications of the time value of money involve the present values of annuities. The initial valuation of long-term bonds is determined by calculating the present value of the periodic stated interest payments and the present value of the lump-sum payment made at maturity. Certain leases require the lessee to compute the present value of future lease payments to value the leased asset and corresponding lease obligation. Similarly, installment notes sometimes require us to calculate the present value of installment payments as the amount at which to record the note. Also, pension plans require the payment of deferred annuities to retirees. (*p. 259*)

Questions For Review of Key Topics

Q 5–1 Define interest.

Q 5–2 Explain compound interest.

Q 5–3 What would cause the annual interest rate to be different from the annual effective rate or yield?

Q 5–4 Identify the three items of information necessary to calculate the future value of a single amount.

Q 5–5 Define the present value of a single amount.

Q 5–6 Explain the difference between monetary and nonmonetary assets and liabilities.

Q 5–7 What is an annuity?

Q 5–8 Explain the difference between an ordinary annuity and an annuity due.

Q 5–9 Explain the relationship between Table 2, Present Value of $1, and Table 4, Present Value of an Ordinary Annuity of $1.

Q 5–10 Prepare a time diagram for the present value of a four-year ordinary annuity of $200. Assume an interest rate of 10% per year.

Q 5–11 Prepare a time diagram for the present value of a four-year annuity due of $200. Assume an interest rate of 10% per year.

Q 5–12 What is a deferred annuity?

Q 5–13 Assume that you borrowed $500 from a friend and promised to repay the loan in five equal annual installments beginning one year from today. Your friend wants to be reimbursed for the time value of money at an 8% annual rate. Explain how you would compute the required annual payment.

Q 5–14 Compute the required annual payment in Question 5–13.

Q 5–15 Explain how the time value of money concept is incorporated into the valuation of certain leases.

Brief Exercises

connect

BE 5–1
Simple versus compound interest
● LO5–1

Fran Smith has two investment opportunities. The interest rate for both investments is 8%. Interest on the first investment will compound annually while interest on the second will compound quarterly. Which investment opportunity should Fran choose? Why?

BE 5–2
Future value; single amount
● LO5–2

Bill O'Brien would like to take his wife, Mary, on a trip three years from now to Europe to celebrate their 40th anniversary. He has just received a $20,000 inheritance from an uncle and intends to invest it for the trip. Bill estimates the trip will cost $23,500 and he believes he can earn 5% interest, compounded annually, on his investment. Will he be able to pay for the trip with the accumulated investment amount?

BE 5–3
Future value; solving for unknown; single amount
● LO5–4

Refer to the situation described in BE 5–2. Assume that the trip will cost $26,600. What interest rate, compounded annually, must Bill earn to accumulate enough to pay for the trip?

BE 5–4
Present value; single amount
● LO5–3

John has an investment opportunity that promises to pay him $16,000 in four years. He could earn a 6% annual return investing his money elsewhere. What is the maximum amount he would be willing to invest in this opportunity?

BE 5–5
Present value; solving for unknown; single amount
● LO5–4

Refer to the situation described in BE 5–4. Suppose the opportunity requires John to invest $13,200 today. What is the interest rate John would earn on this investment?

BE 5–6
Future value; ordinary annuity
● LO5–7

Leslie McCormack is in the spring quarter of her freshman year of college. She and her friends already are planning a trip to Europe after graduation in a little over three years. Leslie would like to contribute to a savings account over the next three years in order to accumulate enough money to take the trip. Assuming an interest rate of 4%, compounded quarterly, how much will she accumulate in three years by depositing $500 at the *end* of each of the next 12 quarters, beginning three months from now?

BE 5–7
Future value; annuity due
● LO5–7

Refer to the situation described in BE 5–6. How much will Leslie accumulate in three years by depositing $500 at the *beginning* of each of the next 12 quarters?

BE 5–8
Present value; ordinary annuity; installment notes
● LO5–8, LO5–10

Canliss Mining Company borrowed money from a local bank. The note the company signed requires five annual installment payments of $10,000 beginning one year from today. The interest rate on the note is 7%. What amount did Canliss borrow?

BE 5–9
Present value; annuity due; installment notes
● LO5–8, LO5–10

Refer to the situation described in BE 5–8. What amount did Canliss borrow assuming that the first $10,000 payment was due immediately?

BE 5–10
Deferred annuity
● LO5–8, LO5–10

Refer to the situation described in BE 5–8. What amount did Canliss borrow assuming that the first of the five annual $10,000 payments was not due for three years?

BE 5–11
Solve for unknown; annuity
● LO5–9

Kingsley Toyota borrowed $100,000 from a local bank. The loan requires Kingsley to pay 10 equal annual installments beginning one year from today. Assuming an interest rate of 8%, what is the amount of each annual installment payment?

BE 5–12
Price of a bond
● LO5–10

On December 31, 2021, Interlink Communications issued 6% stated rate bonds with a face amount of $100 million. The bonds mature on December 31, 2051. Interest is payable annually on each December 31, beginning in 2022. Determine the price of the bonds on December 31, 2021, assuming that the market rate of interest for similar bonds was 7%.

BE 5–13
Lease payment
● LO5–10

On September 30, 2021, Ferguson Imports leased a warehouse. Terms of the lease require Ferguson to make 10 annual lease payments of $55,000 with the first payment due immediately. Accounting standards require the company to record a lease liability when recording this type of lease. Assuming an 8% interest rate, at what amount should Ferguson record the lease liability on September 30, 2021, before the first payment is made?

Exercises

connect

E 5–1
Future value; single amount
● LO5–2

Determine the future value of the following single amounts:

	Invested Amount	Interest Rate	No. of Periods
1.	$15,000	6%	12
2.	20,000	8	10
3.	30,000	12	20
4.	50,000	4	12

E 5–2
Future value; single amounts
● LO5–2

Determine the future value of $10,000 under each of the following sets of assumptions:

	Annual Rate	Period Invested	Interest Compounded
1.	10%	10 years	Semiannually
2.	12	5 years	Quarterly
3.	24	30 months	Monthly

E 5–3
Present value; single amounts
● LO5–3

Determine the present value of the following single amounts:

	Future Amount	Interest Rate	No. of Periods
1.	$20,000	7%	10
2.	14,000	8	12
3.	25,000	12	20
4.	40,000	10	8

E 5–4
Present value; multiple, unequal amounts
● LO5–3

Determine the combined present value as of December 31, 2021, of the following four payments to be received at the end of each of the designated years, assuming an annual interest rate of 8%.

Payment	Year Received
$5,000	2022
6,000	2023
8,000	2025
9,000	2027

E 5–5
Solving for unknowns; single amounts
● LO5–4

For each of the following situations involving single amounts, solve for the unknown (?). Assume that interest is compounded annually. (i = interest rate, and n = number of years)

	Present Value	Future Value	i	n
1.	?	$ 40,000	10%	5
2.	$36,289	65,000	?	10
3.	15,884	40,000	8	?
4.	46,651	100,000	?	8
5.	15,376	?	7	20

E 5–6
Noninterest-bearing note; single payment
● LO5–5

The Field Detergent Company sold merchandise to the Abel Company on June 30, 2021. Payment was made in the form of a noninterest-bearing note requiring Abel to pay $85,000 on June 30, 2023. Assume that a 10% interest rate properly reflects the time value of money in this situation.

Required:
Calculate the amount at which Field should record the note receivable and corresponding sales revenue on June 30, 2021.

E 5–7
Concepts; terminology
● LO5–1 through LO5–3, LO5–6

Listed below are several terms and phrases associated with concepts discussed in the chapter. Pair each item from List A with the item from List B (by letter) that is most appropriately associated with it.

List A	List B
______ 1. Interest	a. First cash flow occurs one period after agreement begins
______ 2. Monetary asset	b. The rate at which money will actually grow during a year
______ 3. Compound interest	c. First cash flow occurs on the first day of the agreement
______ 4. Simple interest	d. The amount of money that a dollar will grow to
______ 5. Annuity	e. Amount of money paid/received in excess of amount borrowed/lent
______ 6. Present value of a single amount	f. Obligation to pay a sum of cash, the amount of which is fixed
______ 7. Annuity due	g. Money can be invested today and grow to a larger amount
______ 8. Future value of a single amount	h. No fixed dollar amount attached
______ 9. Ordinary annuity	i. Computed by multiplying an invested amount by the interest rate
______ 10. Effective rate or yield	j. Interest calculated on invested amount plus accumulated interest
______ 11. Nonmonetary asset	k. A series of equal-sized cash flows
______ 12. Time value of money	l. Amount of money required today that is equivalent to a given future amount
______ 13. Monetary liability	m. Claim to receive a fixed amount of money

E 5–8
Future value; compound interest
● LO5–7

Wiseman Video plans to make four annual deposits of $2,000 each to a special building fund. The fund's assets will be invested in mortgage instruments expected to pay interest at 12% on the fund's balance. Using the appropriate time-value of money table, determine how much will be accumulated in the fund on December 31, 2024, under each of the following situations:

1. The first $2,000 deposit is made on December 31, 2021, and interest is compounded annually.
2. The first $2,000 deposit is made on December 31, 2020, and interest is compounded annually.
3. The first $2,000 deposit is made on December 31, 2020, and interest is compounded quarterly.
4. The first $2,000 deposit is made on December 31, 2020, interest is compounded annually, *and* interest earned is withdrawn at the end of each year.

E 5–9
Present value; annuities
● LO5–8

Using the appropriate present value table and assuming a 12% annual interest rate, determine the present value on December 31, 2021, of a five-period annual annuity of $5,000 under each of the following situations:

1. The first payment is received on December 31, 2022, and interest is compounded annually.
2. The first payment is received on December 31, 2021, and interest is compounded annually.
3. The first payment is received on December 31, 2022, and interest is compounded quarterly.

E 5–10
Future and present value
● LO5–3, LO5–7, LO5–8

Answer each of the following independent questions.

1. Alex Meir recently won a lottery and has the option of receiving one of the following three prizes: (1) $64,000 cash immediately, (2) $20,000 cash immediately and a six-period annuity of $8,000 beginning one year from today, or (3) a six-period annuity of $13,000 beginning one year from today. Assuming an interest rate of 6%, which option should Alex choose?
2. The Weimer Corporation wants to accumulate a sum of money to repay certain debts due on December 31, 2030. Weimer will make annual deposits of $100,000 into a special bank account at the end of each of 10 years beginning December 31, 2021. Assuming that the bank account pays 7% interest compounded annually, what will be the fund balance after the last payment is made on December 31, 2030?

E 5–11
Deferred annuities
● LO5–8

Lincoln Company purchased merchandise from Grandville Corp. on September 30, 2021. Payment was made in the form of a noninterest-bearing note requiring Lincoln to make six annual payments of $5,000 on each September 30, beginning on September 30, 2024.

Required:
Calculate the amount at which Lincoln should record the note payable and corresponding purchases on September 30, 2021, assuming that an interest rate of 10% properly reflects the time value of money in this situation.

E 5–12
Future value; solving for annuities and single amount
● LO5–4, LO5–9

John Rider wants to accumulate $100,000 to be used for his daughter's college education. He would like to have the amount available on December 31, 2026. Assume that the funds will accumulate in a certificate of deposit paying 8% interest compounded annually.

Required:
Answer each of the following independent questions.
1. If John were to deposit a single amount, how much would he have to invest on December 31, 2021?
2. If John were to make five equal deposits on each December 31, beginning on December 31, 2022, what is the required amount of each deposit?
3. If John were to make five equal deposits on each December 31, beginning on December 31, 2021, what is the required amount of each deposit?

E 5–13
Solving for unknowns; annuities
● LO5–9

For each of the following situations involving annuities, solve for the unknown (?). Assume that interest is compounded annually and that all annuity amounts are received at the *end* of each period. (i = interest rate, and n = number of years)

	Present Value	Annuity Amount	i	n
1.	?	$ 3,000	8%	5
2.	$242,980	75,000	?	4
3.	161,214	20,000	9	?
4.	500,000	80,518	?	8
5.	250,000	?	10	4

E 5–14
Solving for unknown annuity amount
● LO5–9

Sandy Kupchack just graduated from State University with a bachelor's degree in history. During her four years at the university, Sandy accumulated $12,000 in student loans. She asks for your help in determining the amount of the *quarterly* loan payment. She tells you that the loan must be paid back in five years and that the annual interest rate is 8%. Payments begin in three months.

Required:
Determine Sandy's quarterly loan payment.

E 5–15
Deferred annuities; solving for annuity amount
● LO5–8, LO5–9

On April 1, 2021, John Vaughn purchased appliances from the Acme Appliance Company for $1,200. In order to increase sales, Acme allows customers to pay in installments and will defer any payments for six months. John will make 18 equal monthly payments, beginning October 1, 2021. The annual interest rate implicit in this agreement is 24%.

Required:
Calculate the monthly payment necessary for John to pay for his purchases.

E 5–16
Price of a bond
● LO5–9, LO5–10

On September 30, 2021, the San Fillipo Corporation issued 8% stated rate bonds with a face amount of $300 million. The bonds mature on September 30, 2041 (20 years). The market rate of interest for similar bonds was 10%. Interest is paid semiannually on March 31 and September 30.

Required:
Determine the price of the bonds on September 30, 2021.

E 5–17
Price of a bond; interest expense
● LO5–9, LO5–10

On June 30, 2021, Singleton Computers issued 6% stated rate bonds with a face amount of $200 million. The bonds mature on June 30, 2036 (15 years). The market rate of interest for similar bond issues was 5% (2.5% semiannual rate). Interest is paid semiannually (3%) on June 30 and December 31, beginning on December 31, 2021.

Required:
1. Determine the price of the bonds on June 30, 2021.
2. Calculate the interest expense Singleton reports in 2021 for these bonds.

E 5–18
Solving for unknown annuity payment; installment notes
● LO5–9, LO5–10

Don James purchased a new automobile for $20,000. Don made a cash down payment of $5,000 and agreed to pay the remaining balance in 30 monthly installments, beginning one month from the date of purchase. Financing is available at a 24% *annual* interest rate.

Required:
Calculate the amount of the required monthly payment.

E 5–19
Solving for unknown interest rate; installment notes
● LO5–9, LO5–10

Lang Warehouses borrowed $100,000 from a bank and signed a note requiring 20 annual payments of $13,388 beginning one year from the date of the agreement.

Required:
Determine the interest rate implicit in this agreement.

E 5–20
Lease payments
● LO5–9, LO5–10

On June 30, 2021, Fly-By-Night Airlines leased a jumbo jet from **Boeing Corporation**. The terms of the lease require Fly-By-Night to make 20 annual payments of $400,000 on each June 30. Generally accepted accounting principles require this lease to be recorded as a liability for the present value of scheduled payments. Assume that a 7% interest rate properly reflects the time value of money in this situation.

Required:

1. At what amount should Fly-By-Night record the lease liability on June 30, 2021, assuming that the first payment will be made on June 30, 2022?
2. At what amount should Fly-By-Night record the lease liability on June 30, 2021, *before* any payments are made, assuming that the first payment will be made on June 30, 2021?

E 5–21
Lease payments; solve for unknown interest rate
● LO5–9, LO5–10

On March 31, 2021, Southwest Gas leased equipment from a supplier and agreed to pay $200,000 annually for 20 years beginning March 31, 2022. Generally accepted accounting principles require that a liability be recorded for this lease agreement for the present value of scheduled payments. Accordingly, at inception of the lease, Southwest recorded a $2,293,984 lease liability.

Required:

Determine the interest rate implicit in the lease agreement.

Problems

connect

P 5–1
Analysis of alternatives
● LO5–3, LO5–8

Esquire Company needs to acquire a molding machine to be used in its manufacturing process. Two types of machines that would be appropriate are presently on the market. The company has determined the following:

Machine A could be purchased for $48,000. It will last 10 years with annual maintenance costs of $1,000 per year. After 10 years the machine can be sold for $5,000.

Machine B could be purchased for $40,000. It also will last 10 years and will require maintenance costs of $4,000 in year three, $5,000 in year six, and $6,000 in year eight. After 10 years, the machine will have no salvage value.

Required:

Determine which machine Esquire should purchase. Assume an interest rate of 8% properly reflects the time value of money in this situation and that maintenance costs are paid at the end of each year. Ignore income tax considerations.

P 5–2
Present and future value
● LO5–7, LO5–8, LO5–10

Johnstone Company is facing several decisions regarding investing and financing activities. Address each decision independently.

1. On June 30, 2021, the Johnstone Company purchased equipment from Genovese Corp. Johnstone agreed to pay Genovese $10,000 on the purchase date and the balance in five annual installments of $8,000 on each June 30 beginning June 30, 2022. Assuming that an interest rate of 10% properly reflects the time value of money in this situation, at what amount should Johnstone value the equipment?
2. Johnstone needs to accumulate sufficient funds to pay a $400,000 debt that comes due on December 31, 2026. The company will accumulate the funds by making five equal annual deposits to an account paying 6% interest compounded annually. Determine the required annual deposit if the first deposit is made on December 31, 2021.
3. On January 1, 2021, Johnstone leased an office building. Terms of the lease require Johnstone to make 20 annual lease payments of $120,000 beginning on January 1, 2021. A 10% interest rate is implicit in the lease agreement. At what amount should Johnstone record the lease liability on January 1, 2021, *before* any lease payments are made?

P 5–3
Analysis of alternatives
● LO5–3, LO5–8

Harding Company is in the process of purchasing several large pieces of equipment from Danning Machine Corporation. Several financing alternatives have been offered by Danning:

1. Pay $1,000,000 in cash immediately.
2. Pay $420,000 immediately and the remainder in 10 annual installments of $80,000, with the first installment due in one year.
3. Make 10 annual installments of $135,000 with the first payment due immediately.
4. Make one lump-sum payment of $1,500,000 five years from date of purchase.

Required:

Determine the best alternative for Harding, assuming that Harding can borrow funds at an 8% interest rate.

P 5–4
Investment analysis; uneven cash flows
● LO5–3, LO5–8

John Wiggins is considering the purchase of a small restaurant. The purchase price listed by the seller is $800,000. John has used past financial information to estimate that the net cash flows (cash inflows less cash outflows) generated by the restaurant would be as follows:

Years	Amount
1–6	$80,000
7	70,000
8	60,000
9	50,000
10	40,000

If purchased, the restaurant would be held for 10 years and then sold for an estimated $700,000.

Required:
Assuming that John desires a 10% rate of return on this investment, should the restaurant be purchased? (Assume that all cash flows occur at the end of the year.)

P 5–5
Investment decision; varying rates
● LO5–3, LO5–8

John and Sally Claussen are considering the purchase of a hardware store from John Duggan. The Claussens anticipate that the store will generate cash flows of $70,000 per year for 20 years. At the end of 20 years, they intend to sell the store for an estimated $400,000. The Claussens will finance the investment with a variable rate mortgage. Interest rates will increase twice during the 20-year life of the mortgage. Accordingly, the Claussens' desired rate of return on this investment varies as follows:

Years 1–5	8%
Years 6–10	10%
Years 11–20	12%

Required:
What is the maximum amount the Claussens should pay John Duggan for the hardware store? (Assume that all cash flows occur at the end of the year.)

P 5–6
Solving for unknowns
● LO5–3, LO5–9

The following situations should be considered independently.

1. John Jamison wants to accumulate $60,000 for a down payment on a small business. He will invest $30,000 today in a bank account paying 8% interest compounded annually. Approximately how long will it take John to reach his goal?
2. The Jasmine Tea Company purchased merchandise from a supplier for $28,700. Payment was a noninterest-bearing note requiring Jasmine to make five annual payments of $7,000 beginning one year from the date of purchase. What is the interest rate implicit in this agreement?
3. Sam Robinson borrowed $10,000 from a friend and promised to pay the loan in 10 equal annual installments beginning one year from the date of the loan. Sam's friend would like to be reimbursed for the time value of money at a 9% annual rate. What is the annual payment Sam must make to pay back his friend?

P 5–7
Deferred annuities
● LO5–8

On January 1, 2021, the Montgomery Company agreed to purchase a building by making six payments. The first three are to be $25,000 each, and will be paid on December 31, 2021, 2022, and 2023. The last three are to be $40,000 each and will be paid on December 31, 2024, 2025, and 2026. Montgomery borrowed other money at a 10% annual rate.

Required:
1. At what amount should Montgomery record the note payable and corresponding cost of the building on January 1, 2021?
2. How much interest expense on this note will Montgomery recognize in 2021?

P 5–8
Deferred annuities
● LO5–8

John Roberts is 55 years old and has been asked to accept early retirement from his company. The company has offered John three alternative compensation packages to induce John to retire.

1. $180,000 cash payment to be paid immediately
2. A 20-year annuity of $16,000 beginning immediately
3. A 10-year annuity of $50,000 beginning at age 65

Required:
Which alternative should John choose assuming that he is able to invest funds at a 7% rate?

P 5–9
Noninterest-bearing note; annuity and lump-sum payment
● LO5–3, LO5–8

On January 1, 2021, The Barrett Company purchased merchandise from a supplier. Payment was a noninterest-bearing note requiring five annual payments of $20,000 on each December 31 beginning on December 31, 2021, and a lump-sum payment of $100,000 on December 31, 2025. A 10% interest rate properly reflects the time value of money in this situation.

Required:
Calculate the amount at which Barrett should record the note payable and corresponding merchandise purchased on January 1, 2021.

P 5–10
Solving for unknowns; installment notes
● LO5–9

Lowlife Company defaulted on a $250,000 loan that was due on December 31, 2021. The bank has agreed to allow Lowlife to repay the $250,000 by making a series of equal annual payments beginning on December 31, 2022.

Required:
Calculate the amount at which Barrett should record the note payable and corresponding merchandise purchased on January 1, 2021.

1. Calculate the required annual payment if the bank's interest rate is 10% and four payments are to be made.
2. Calculate the required annual payment if the bank's interest rate is 8% and five payments are to be made.
3. If the bank's interest rate is 10%, how many annual payments of $51,351 would be required to repay the debt?
4. If three payments of $104,087 are to be made, what interest rate is the bank charging Lowlife?

P 5–11
Solving for unknown lease payment
● LO5–9, LO5–10

Benning Manufacturing Company is negotiating with a customer for the lease of a large machine manufactured by Benning. The machine has a cash price of $800,000. Benning wants to be reimbursed for financing the machine at an 8% annual interest rate.

Required:

1. Determine the required lease payment if the lease agreement calls for 10 equal annual payments beginning immediately.
2. Determine the required lease payment if the first of 10 annual payments will be made one year from the date of the agreement.
3. Determine the required lease payment if the first of 10 annual payments will be made immediately and Benning will be able to sell the machine to another customer for $50,000 at the end of the 10-year lease.

P 5–12
Solving for unknown lease payment; compounding periods of varying length
● LO5–9, LO5–10

[This is a variation of P 5–11 focusing on compounding periods of varying length.]

Benning Manufacturing Company is negotiating with a customer for the lease of a large machine manufactured by Benning. The machine has a cash price of $800,000. Benning wants to be reimbursed for financing the machine at a 12% annual interest rate over the five-year lease term.

Required:

1. Determine the required lease payment if the lease agreement calls for 10 equal semiannual payments beginning six months from the date of the agreement.
2. Determine the required lease payment if the lease agreement calls for 20 equal quarterly payments beginning immediately.
3. Determine the required lease payment if the lease agreement calls for 60 equal monthly payments beginning one month from the date of the agreement. The present value of an ordinary annuity factor for $n = 60$ and $i = 1\%$ is 44.9550.

P 5–13
Lease vs. buy alternatives
● LO5–3, LO5–8, LO5–10

Kiddy Toy Corporation needs to acquire the use of a machine to be used in its manufacturing process. The machine needed is manufactured by Lollie Corp. The machine can be used for 10 years and then sold for $10,000 at the end of its useful life. Lollie has presented Kiddy with the following options:

1. *Buy machine.* The machine could be purchased for $160,000 in cash. All insurance costs, which approximate $5,000 per year, would be paid by Kiddy.
2. *Lease machine.* The machine could be leased for a 10-year period for an annual lease payment of $25,000 with the first payment due immediately. All insurance costs will be paid for by the Lollie Corp. and the machine will revert back to Lollie at the end of the 10-year period.

Required:
Assuming that a 12% interest rate properly reflects the time value of money in this situation and that all maintenance and insurance costs are paid at the end of each year, determine which option Kiddy should choose. Ignore income tax considerations.

P 5–14
Deferred annuities; pension obligation
● LO5–8, LO5–10

Three employees of the Horizon Distributing Company will receive annual pension payments from the company when they retire. The employees will receive their annual payments for as long as they live. Life expectancy for each employee is 15 years beyond retirement. Their names, the amount of their annual pension payments, and the date they will receive their first payment are shown below:

Employee	Annual Payment	Date of First Payment
Tinkers	$20,000	12/31/24
Evers	25,000	12/31/25
Chance	30,000	12/31/26

Required:

1. Compute the present value of the pension obligation to these three employees as of December 31, 2021. Assume an 11% interest rate.
2. The company wants to have enough cash invested at December 31, 2024, to provide for all three employees. To accumulate enough cash, they will make three equal annual contributions to a fund that will earn 11% interest compounded annually. The first contribution will be made on December 31, 2021. Compute the amount of this required annual contribution.

P 5–15
Bonds and leases; deferred annuities
● LO5–3, LO5–8, LO5–10

On the last day of its fiscal year ending December 31, 2021, the Sedgwick & Reams (S&R) Glass Company completed two financing arrangements. The funds provided by these initiatives will allow the company to expand its operations.

1. S&R issued 8% stated rate bonds with a face amount of $100 million. The bonds mature on December 31, 2041 (20 years). The market rate of interest for similar bond issues was 9% (4.5% semiannual rate). Interest is paid semiannually (4%) on June 30 and December 31, beginning on June 30, 2022.
2. The company leased two manufacturing facilities. Lease A requires 20 annual lease payments of $200,000 beginning on January 1, 2022. Lease B also is for 20 years, beginning January 1, 2022. Terms of the lease require 17 annual lease payments of $220,000 beginning on January 1, 2025. Generally accepted accounting principles require both leases to be recorded as liabilities for the present value of the scheduled payments. Assume that a 10% interest rate properly reflects the time value of money for the lease obligations.

Required:

What amounts will appear in S&R's December 31, 2021, balance sheet for the bonds and for the leases?

Decision Makers' Perspective

©ImageGap/Getty Images

Apply your critical-thinking ability to the knowledge you've gained. These cases will provide you an opportunity to develop your research, analysis, judgment, and communication skills. You also will work with other students, integrate what you've learned, apply it in real-world situations, and consider its global and ethical ramifications. This practice will broaden your knowledge and further develop your decision-making abilities.

Ethics Case 5–1
Rate of return
● LO5–1

The Damon Investment Company manages a mutual fund composed mostly of speculative stocks. You recently saw an ad claiming that investments in the funds have been earning a rate of return of 21%. This rate seemed quite high so you called a friend who works for one of Damon's competitors. The friend told you that the 21% return figure was determined by dividing the two-year appreciation on investments in the fund by the average investment. In other words, $100 invested in the fund two years ago would have grown to $121 ($21 ÷ $100 = 21%).

Required:

Discuss the ethics of the 21% return claim made by the Damon Investment Company.

Analysis Case 5–2
Bonus alternatives; present value analysis
● LO5–3, LO5–8

Sally Hamilton has performed well as the chief financial officer of the Maxtech Computer Company and has earned a bonus. She has a choice among the following three bonus plans:

1. A $50,000 cash bonus paid now.
2. A $10,000 annual cash bonus to be paid each year over the next six years, with the first $10,000 paid now.
3. A three-year $22,000 annual cash bonus with the first payment due three years from now.

Required:

Evaluate the three alternative bonus plans. Sally can earn a 6% annual return on her investments.

1. What is the present value of the first alternative?
2. What is the present value of the second alternative?
3. What is the present value of the third alternative?

Communication Case 5–3
Present value of annuities
● LO5–8

Harvey Alexander, an all-league professional football player, has just declared free agency. Two teams, the San Francisco 49ers and the Dallas Cowboys, have made Harvey the following offers to obtain his services:

49ers	$1 million signing bonus payable immediately and an annual salary of $1.5 million for the five-year term of the contract.
Cowboys	$2.5 million signing bonus payable immediately and an annual salary of $1 million for the five-year term of the contract.

With both contracts, the annual salary will be paid in one lump sum at the end of the football season.

Required:

You have been hired as a consultant to Harvey's agent, Phil Marks, to evaluate the two contracts. Write a short letter to Phil with your recommendation including the method you used to reach your conclusion. Assume that Harvey has no preference between the two teams and that the decision will be based entirely on monetary considerations. Also assume that Harvey can invest his money and earn an 8% annual return.

Communication Case 5–4
Present value of an annuity
● LO5–8

On a rainy afternoon two years ago, John Smiley left work early to attend a family birthday party. Eleven minutes later, a careening truck slammed into his SUV on the freeway causing John to spend two months in a coma. Now he can't hold a job or make everyday decisions and is in need of constant care. Last week, the 40-year-old Smiley won an out-of-court settlement from the truck driver's company. He was awarded payment for all medical costs and attorney fees, plus a lump-sum settlement of $2,330,716. At the time of the accident, John was president of his family's business and earned approximately $200,000 per year. He had anticipated working 25 more years before retirement.[8]

John's sister, Sara, an acquaintance of yours from college, has asked you to explain to her how the attorneys came up with the settlement amount. "They said it was based on his lost future income and a 7% rate of some kind," she explained. "But it was all 'legal-speak' to me."

Required:

Construct the text of an e-mail to Sara describing how the amount of the lump-sum settlement was determined. Include a calculation that might help Sara and John understand.

Judgment Case 5–5
Replacement decision
● LO5–3, LO5–8

Hughes Corporation is considering replacing a machine used in the manufacturing process with a new, more efficient model. The purchase price of the new machine is $150,000 and the old machine can be sold for $100,000. Output for the two machines is identical; they will both be used to produce the same amount of product for five years. However, the annual operating costs of the old machine are $18,000 compared to $10,000 for the new machine. Also, the new machine has a salvage value of $25,000, but the old machine will be worthless at the end of the five years.

You are deciding whether the company should sell the old machine and purchase the new model. You have determined that an 8% rate properly reflects the time value of money in this situation and that all operating costs are paid at the end of the year. For this initial comparison you ignore the effect of the decision on income taxes.

Required:

1. What is the incremental cash outflow required to acquire the new machine?
2. What is the present value of the benefits of acquiring the new machine?

Real World Case 5–6
Zero-coupon bonds; Johnson & Johnson
● LO5–3, LO5–10
Real World Financials

Johnson & Johnson is one of the world's largest manufacturers of health care products. The company's July 2, 2017, financial statements included the following information in the long-term debt disclosure note:

	($ in millions) July 2, 2017
Zero-coupon convertible subordinated debentures, due 2020	$69

The bonds were issued at the beginning of 2000. The disclosure note stated that the effective interest rate for these bonds is 3% annually. Some of the original convertible bonds have been converted into Johnson & Johnson shares of stock. The $69 million is the present value of the bonds not converted and thus reported in the financial statements. Each individual bond has a maturity value (face amount) of $1,000. The maturity value indicates the amount that Johnson & Johnson will pay bondholders at the beginning of 2020. Zero-coupon bonds pay no cash interest during the term to maturity. The company is "accreting" (gradually increasing) the issue price to maturity value using the bonds' effective interest rate computed on a semiannual basis.

[8]This case is based on actual events.

Required:

1. Determine to the nearest million dollars the maturity value of the zero-coupon bonds that Johnson & Johnson will pay bondholders at the beginning of 2020.
2. Determine to the nearest dollar the issue price at the beginning of 2000 of a single, $1,000 maturity-value bond.

Real World Case 5–7
Leases; Southwest Airlines
● LO5–3, LO5–10

Real World Financials

Southwest Airlines provides scheduled air transportation services in the United States. Like many airlines, Southwest leases many of its planes from **Boeing Company**. In its long-term debt disclosure note included in the financial statements for the year ended December 31, 2017, the company listed $885 million in lease obligations. The existing leases had an approximate ten-year remaining life and future lease payments average approximately $100 million per year.

Required:

1. Determine (to the nearest one-half percent) the effective interest rate the company used to determine the lease liability assuming that lease payments are made at the end of each fiscal year.
2. Repeat requirement 1 assuming that lease payments are made at the beginning of each fiscal year.

Continuing Cases

Target Case
● LO5–3, LO5–8 LO5–10

Target Corporation prepares its financial statements according to U.S. GAAP. Target's financial statements and disclosure notes for the year ended February 3, 2018, are available in Connect. This material is also available under the Investor Relations link at the company's website (**www.target.com**). Target leases most of its facilities.

Required:

1. Refer to disclosure note 22 following Target's financial statements. What is the amount reported for "capital" leases (shown as the present value of minimum lease payments)? What is the total of those lease payments? What accounts for the difference between the two amounts?
2. What is the total of the operating lease payments?
3. New lease accounting guidance (discussed in Chapter 15) will require companies to report a liability for operating leases at present value as well as for capital leases (now called finance leases). If Target had used the new lease accounting guidance in its fiscal 2017 financial statements, what would be the amount reported as a liability for operating leases? *Hint:* Assume the payments "after 2020" are to be paid evenly over a 16 years period and all payments are at the end of years indicated. Target indicates elsewhere in its financial statements that 6% is an appropriate discount rate for its leases.

Air France–KLM Case
● LO5–10

IFRS

Air France–KLM (AF), a Franco-Dutch company, prepares its financial statements according to International Financial Reporting Standards. AF's financial statements and disclosure notes for the year ended December 31, 2017, are available in Connect. This material is also available under the Finance link at the company's website (**www.airfranceklm.com**.) The presentation of financial statements often differs between U.S. GAAP and IFRS, particularly with respect to the balance sheet.

Using IFRS, companies discount their pension obligations using an interest rate approximating the average interest rate on high-quality corporate bonds. U.S. GAAP allows much more flexibility in the choice of a discount rate. By both sets of standards, the rate used is reported in the disclosure note related to pensions.

Required:

Refer to AF's Note 30.2 "Description of the actuarial assumptions and related sensitivities."

1. What are the average discount rates used to measure AF's (a) 10–15 year and (b) 15 year and more pension obligations in the "euro" geographic zone in 2017?
2. If the rate used had been 1% (100 basis points) higher, what change would have occurred in the pension obligation in 2017? What if the rate had been 1% lower?

CPA Exam Questions and Simulations

ROGER CPA Review

Sample CPA Exam questions from Roger CPA Review are available in Connect as support for the topics in this chapter. These multiple-choice questions and task-based simulations include expert-written explanations and solutions, and provide a starting point for students to become familiar with the content and functionality of the actual CPA Exam.

CHAPTER

6 Revenue Recognition

OVERVIEW

In Chapter 4, we discussed net income and its presentation in the income statement. In Chapter 6 we focus on revenue recognition, which determines when and how much revenue appears in the income statement. In Part A of this chapter, we discuss the general approach for recognizing revenue in three situations—at a point in time, over a period of time, and for contracts that include multiple parts that might require recognizing revenue at different times. In Part B, we see how to deal with special issues that affect the revenue recognition process. In Part C, we discuss how to account for revenue in long-term contracts.

LEARNING OBJECTIVES

After studying this chapter, you should be able to:

- **LO6–1** State the core revenue recognition principle and the five key steps in applying it. (*p. 278*)
- **LO6–2** Explain when it is appropriate to recognize revenue at a single point in time. (*p. 280*)
- **LO6–3** Explain when it is appropriate to recognize revenue over a period of time. (*p. 281*)
- **LO6–4** Allocate a contract's transaction price to multiple performance obligations. (*p. 283*)
- **LO6–5** Determine whether a contract exists, and whether some frequently encountered features of contracts qualify as performance obligations. (*p. 287*)
- **LO6–6** Understand how variable consideration and other aspects of contracts affect the calculation and allocation of the transaction price. (*p. 290*)
- **LO6–7** Determine the timing of revenue recognition with respect to licenses, franchises, and other common arrangements. (*p. 298*)
- **LO6–8** Understand the disclosures required for revenue recognition, accounts receivable, contract assets, and contract liabilities. (*p. 302*)
- **LO6–9** Demonstrate revenue recognition for long-term contracts, both at a point in time when the contract is completed and over a period of time according to the percentage completed. (*p. 304*)
- **LO6–10** Discuss the primary differences between U.S. GAAP and IFRS with respect to revenue recognition. (*p. 288, 291, 299*)

©Hero Images/Getty Images

FINANCIAL REPORTING CASE

Ask the Oracle

"Good news! I got the job," she said, closing the door behind her.

Your roommate, a software engineer, goes on to explain that she accepted a position at **Oracle Corporation**, a world leader in enterprise software, computer hardware, and cloud-computing services.

"The salary's good, too," she continued. "Plus, Mr. Watson, my supervisor, said I'll be getting a bonus tied to the amount of revenue my projects produce. So I started looking at Oracle's financial statements, but I can't even understand when they get to recognize revenue. Sometimes they recognize it all at once, sometimes over time, and sometimes they seem to break apart a sale and recognize revenue for different parts at different times. And sometimes they recognize revenue before they are even done with a project, according to the percentage they have completed so far. You're the accountant. What determines when Oracle gets to recognize revenue?"

QUESTIONS

By the time you finish this chapter, you should be able to respond appropriately to the questions posed in this case. Compare your response to the solution provided at the end of the chapter.

1. Under what circumstances do companies recognize revenue at a point in time? Over a period of time? (*p. 280*)
2. When do companies break apart a sale and treat its parts differently for purposes of recognizing revenue? (*p. 283*)
3. How do companies account for long-term contracts that qualify for revenue recognition over time? (*p. 305*)

What is **revenue**? According to the FASB's conceptual framework, "Revenues are inflows or other enhancements of assets of an entity or settlements of its liabilities (or a combination of both) from delivering or producing goods, rendering services, or other activities that constitute the entity's ongoing major or central operations."[1] In simpler terms, revenue is the inflow of cash or accounts receivable that a business receives when it provides goods or services to its customers.

For many companies, revenue is the single largest number reported in the financial statements. Its pivotal role in the picture painted by the financial statements makes measuring and reporting revenue one of the most critical aspects of financial reporting. It is important not only to determine *how much* revenue to recognize (record), but also *when* to recognize it. A one-year income statement should report a company's revenues for only that one-year period. Sometimes, though, it's difficult to determine how much revenue to recognize in a particular period. Also, you can imagine that a manager who is evaluated according to how much revenue she generates each period might be tempted to recognize more revenue than is appropriate. In fact, the SEC has cracked down on revenue-recognition abuses in the past, and its enforcement division continues to do so.[2]

Revenue recognition criteria help ensure that an income statement reflects the actual accomplishments of a company for the period.

Revenue recognition accounting standards help ensure that the appropriate amount of revenue appears in each period's income statement. That guidance has changed recently.

[1]Source: "Elements of Financial Statements," *Statement of Financial Concepts No. 6* (Stamford, Conn.: FASB, 1985, par. 78).
[2]For SEC guidance that provides examples of appropriate and inappropriate revenue recognition, see FASB ASC 605–10–S99: Revenue Recognition—Overall—SEC Materials (originally "Revenue Recognition in Financial Statements," *Staff Accounting Bulletin No. 101* (Washington, D.C.: SEC, December 1999) and *Staff Accounting Bulletin No. 104* (Washington, D.C.: SEC, December 2003)).

Revenue recognition previously was based on the "realization principle," which required that we recognize revenue when both the earnings process is virtually complete and there is reasonable certainty as to the collectibility of the assets to be received in exchange for goods and services. That approach to revenue recognition created some problems. Revenue recognition was poorly tied to the FASB's conceptual framework, which emphasizes recognizing assets and liabilities rather than the earnings process. Also, the focus on the earnings process led to similar transactions being treated differently in different industries. And, the realization principle was difficult to apply to complex arrangements that involved multiple goods or services.

To address these concerns, the FASB and IASB worked together to develop a new revenue recognition standard, which the FASB issued as *Accounting Standards Update (ASU) No. 2014–09,* "Revenue from Contracts with Customers," on May 28, 2014.[3] The ASU provides a unified approach that replaces more than 200 different pieces of specialized guidance that had developed over time in U.S. GAAP for revenue recognition under various industries and circumstances. Public companies reporting under U.S. GAAP must have adopted the ASU for periods beginning after December 15, 2017.[4] Companies had two options for adopting *ASU No 2014–09*: they could restate prior years presented in comparative financial statements to appear as if the company had always accounted for revenue under the ASU, or they could leave prior year financial statements unchanged and in the beginning of 2018 record the adjustments necessary to convert to the ASU.

Because you need to understand the revenue recognition requirements that will be in effect during your career, we focus on the revenue recognition approach described in *ASU No. 2014–09* (and clarified in subsequent guidance).

PART A Introduction to Revenue Recognition

LO6–1

Let's start with the core revenue recognition principle and the key steps we use to apply that principle. These are shown in Illustration 6–1.

Illustration 6–1 Core Revenue Recognition Principle and the Five Steps Used to Apply the Principle

Core Revenue Recognition Principle

Companies recognize revenue when goods or services are transferred to customers for the amount the company expects to be entitled to receive in exchange for those goods or services.*

Five Steps Used to Apply the Principle

Step 1	• Identify the contract with a customer.
Step 2	• Identify the performance obligation(s) in the contract.
Step 3	• Determine the transaction price.
Step 4	• Allocate the transaction price to each performance obligation.
Step 5	• Recognize revenue when (or as) each performance obligation is satisfied.

*FASB ASC 606-10-05-4: Revenue from Contracts with Customers–Overall–Overview and Background–General (previously "Revenue from Contracts with Customers (Topic 606)" Accounting Standards Update 2014–09 (Norwalk, Conn: FASB, 2014)).

[3]"Revenue from Contracts with Customers (Topic 606)" *Accounting Standards Update 2014–09* (Norwalk, Conn: FASB, 2014).

[4]The new revenue recognition standard was required for nonpublic companies for annual reporting periods beginning after December 15, 2018. All U.S. companies could adopt the ASU for periods starting after December 15, 2016, if they so chose. For companies issuing reports under IFRS, the IFRS version of this standard, IFRS 15, became effective for periods beginning January 1, 2018. Early adoption of IFRS 15 was permitted.

All revenue recognition starts with a contract between a seller and a customer. You may not have realized it, but you have been a party to several such contracts very recently. Maybe you bought a cup of **Starbucks** coffee or a breakfast biscuit at **McDonald's** this morning. Or maybe you bought this textbook through **Amazon** or had a checkup at your doctor's office. Even though these transactions weren't accompanied by written and signed agreements, they are considered contracts for purposes of revenue recognition. The key is that, implicitly or explicitly, you entered into an arrangement that specifies the legal rights and obligations of a seller and a customer.

Contracts between a seller and a customer contain one or more **performance obligations**, which are promises by the seller to transfer goods or services to a customer. The seller recognizes revenue when it satisfies a performance obligation by transferring the promised good or service. We consider transfer to have occurred when the customer has *control* of the good or service. *Control* means that the customer has direct influence over the use of the good or service and obtains its benefits.

Performance obligations **are promises to transfer goods or services to a customer.**

For many contracts, following this approach is very straightforward. In particular, if a contract includes only one performance obligation, we typically just have to decide when the seller delivers the good or provides the service to a customer, and then make sure that the seller recognizes revenue at that time.

Performance obligations are satisfied when the seller transfers *control* of goods or services to the customer.

As a simple example, assume **Macy's** sells a skirt to Susan for $75 that Macy's previously purchased from a wholesaler for $40. How would Macy's account for the sale to Susan?

1. **Identify the contract with a customer:** In this case, the contract may not be written, but it is clear—Macy's delivers the skirt to Susan, and Susan agrees to pay $75 to Macy's.
2. **Identify the performance obligation(s) in the contract:** Macy's has only a single performance obligation—to deliver the skirt.
3. **Determine the transaction price:** Macy's is entitled to receive $75 from Susan.
4. **Allocate the transaction price to each performance obligation:** With only one performance obligation, Macy's allocates the full transaction price of $75 to delivery of the skirt.
5. **Recognize revenue when (or as) each performance obligation is satisfied:** Macy's satisfies its performance obligation when it delivers the skirt to Susan, so Macy's records the following journal entries at that time:

Cash	75	
Sales revenue		75
Cost of goods sold*	40	
Inventory		40

*This second journal entry assumes that Macy's uses a "perpetual" inventory system, by which we record increases and decreases in inventory as they occur ("perpetually"). We reviewed this method briefly in Chapter 2 and explore it in more depth in Chapter 8.

Revenue recognition gets more complicated when a contract contains more than one performance obligation. For example, when **Verizon** signs up a new cell phone customer, the sales contract might require Verizon to provide (1) a smartphone, (2) related software, (3) a warranty on the phone, (4) ongoing network access, and (5) optional future upgrades. Verizon must determine which of these goods and services constitute performance obligations, allocate the transaction price to those performance obligations, and recognize revenue when (or as) each performance obligation is satisfied.

In Part A of this chapter, we apply the five steps for recognizing revenue to various types of contracts. First we'll focus on contracts that have only one performance obligation to deliver a good or service at a single point in time, like when Macy's sells a skirt to Susan. Then we'll consider situations in which one performance obligation to deliver goods and services is satisfied over time, like a landlord renting an apartment or a bank lending money. After that, we'll consider contracts that contain multiple performance obligations, like the Verizon example we just discussed. Illustration 6–2 summarizes some key considerations we will return to throughout this chapter.

Illustration 6–2 Key Considerations When Applying the Five Steps to Revenue Recognition

Five Steps to Recognizing Revenue	For Transactions Involving Single and Multiple Performance Obligations		
Step 1 Identify the contract	Legal rights of seller and customer established		
Step 2 Identify the performance obligation(s)	***Single*** performance obligation		***Multiple*** performance obligations
Step 3 Determine the transaction price	Amount seller is entitled to receive from customer		Amount seller is entitled to receive from customer
Step 4 Allocate the transaction price	No allocation required		Allocate a portion to each performance obligation
Step 5 Recognize revenue when (or as) each performance obligation is satisfied	At a point in time	Over a period of time	At whatever time is appropriate for each performance obligation

Recognizing Revenue at a Single Point in Time

● LO6–2

FINANCIAL Reporting Case

Q1, p. 277

First we consider a simple contract that includes only one performance obligation and is satisfied at a single point in time. The performance obligation is satisfied when control of the goods or services is transferred from the seller to the customer. Usually it's obvious that transfer occurs at the time of delivery. In our Macy's example above, for instance, the performance obligation is satisfied at the time of the sale when the skirt is transferred to Susan.

In other cases transfer of control can be harder to determine. Illustration 6–3 lists five key indicators we use to decide whether control has passed from the seller to the customer. Sellers should evaluate these indicators individually and in combination to decide whether control has been transferred and revenue can be recognized.

Illustration 6–3

Indicators that Control Has Been Transferred from the Seller to the Customer

The customer is more likely to control a good or service if the customer has:

- An obligation to pay the seller.
- Legal title to the asset.
- Physical possession of the asset.
- Assumed the risks and rewards of ownership.
- Accepted the asset.*

*These indicators apply to both goods and services. It may seem strange to talk about the customer accepting an asset with respect to a service, but think of a service as an asset that is consumed as the customer receives it.

In Illustration 6–4 we apply these indicators to TrueTech Industries, a company we will revisit throughout this chapter to illustrate revenue recognition.

Illustration 6–4

Recognizing Revenue at a Point in Time

TrueTech Industries sells the Tri-Box, a gaming console that allows users to play video games individually or in multiplayer environments over the Internet. A Tri-Box is only a gaming module and includes no other goods or services. When should TrueTech recognize revenue for the following sale of 1,000 Tri-Boxes to CompStores?

- **December 20, 2020: CompStores orders 1,000 Tri-Boxes at a price of $240 each, promising payment within 30 days after delivery.** TrueTech has received the order but hasn't fulfilled its performance obligation to deliver Tri-Boxes. In light of this and other indicators, TrueTech's judgment is that control has not been transferred and revenue should not be recognized.

(continued)

Illustration 6–4
(concluded)

- **January 1, 2021: TrueTech delivers 1,000 Tri-Boxes to CompStores, and title to the Tri-Boxes transfers to CompStores.** TrueTech has delivered the Tri-Boxes, and CompStores has accepted delivery, so CompStores has physical possession, legal title, the risks and rewards of ownership, and an obligation to pay TrueTech. TrueTech's performance obligation has been satisfied, so TrueTech can recognize revenue and a related account receivable of $240,000.*

Accounts receivable ($240 × 1,000)	240,000	
Sales revenue		240,000

- **January 25, 2021: TrueTech receives $240,000 from CompStores.** This transaction does not affect revenue. We recognize revenue when performance obligations are satisfied, not when cash is received. TrueTech simply records collection of the account receivable.

Cash	240,000	
Accounts receivable		240,000

*TrueTech also would debit cost of goods sold and credit inventory to recognize the cost of inventory sold.

Recognizing Revenue over a Period of Time

LO6–3

Services such as lending money, performing audits, and providing consulting advice are performed over a period of time. Some construction contracts require construction over months or even years. In these situations, should a company recognize revenue continuously over time as a product or service is being provided, or wait to recognize revenue at the single point in time when the company has finished providing the product or service? As we'll see next, in most situations like these, companies should recognize revenue over time as the service or product is being provided.

Criteria for Recognizing Revenue over Time

Let's assume once again that we have a contract with a customer that includes a single performance obligation and a known transaction price. As indicated in Illustration 6–5, we recognize revenue over time if *any* of three criteria is met.

Illustration 6–5
Criteria for Recognizing Revenue over Time

Revenue is recognized over time if either:

1. **The customer consumes the benefit of the seller's work as it is performed,** as when a company provides cleaning services to a customer for a period of time, or
2. **The customer controls the asset as it is created,** as when a contractor builds an extension onto a customer's existing building, or
3. **The seller is creating an asset that has no alternative use to the seller, and the seller has the legal right to receive payment for progress to date,** as when a company manufactures customized fighter jets for the U.S. Air Force.

We recognize revenue over time if any of three criteria is met.

If a performance obligation meets at least one of these criteria, we recognize revenue over time, in proportion to the amount of the performance obligation that has been satisfied. If, say, one-third of a service has been performed, then one-third of the performance obligation has been satisfied, so one-third of the revenue should be recognized. For example, **Gold's Gym** recognizes revenue from a two-year membership over the 24-month membership period, and **Six Flags Entertainment** recognizes revenue for season passes over the operating season. Often these arrangements involve receiving cash in advance of satisfying a performance obligation, which requires recognition of a liability, deferred revenue. For example, consider Illustration 6–6.

When a performance obligation is satisfied over time, revenue is recognized in proportion to the amount of the performance obligation that has been satisfied.

Most long-term construction contracts qualify for revenue recognition over time. For example, many long-term construction contracts are structured such that the customer owns the work-in-process (WIP) as it is constructed, which satisfies the second criterion in Illustration 6–5. Also, the third criterion is satisfied if the asset the seller is constructing has no alternate use to the

Most long-term construction contracts qualify for revenue recognition over time.

Illustration 6–6

Recognizing Revenue over a Period of Time

Deferred Revenue

	Dr.	Cr.
1/1		**60,000**
1/31	**5,000**	
2/28	5,000	
. . .	. . .	
12/31	5,000	
12/31		–0–

Service Revenue

	Dr.	Cr.
1/1		–0–
1/31		**5,000**
2/28		5,000
. . .		. . .
12/31		5,000
12/31		60,000

TrueTech Industries sells one-year subscriptions to the Tri-Net multiuser platform of Internet-based games. TrueTech sells 1,000 subscriptions for $60 each on January 1, 2021.

TrueTech has a single performance obligation—to provide a service to subscribers by allowing them access to the gaming platform for one year. Because Tri-Net users consume the benefits of access to that service over time, under the first criterion in Illustration 6–5 TrueTech recognizes revenue from the subscriptions over the one-year time period.

On January 1, 2021, TrueTech records the following journal entry:

Cash ($60 × 1,000)	60,000	
Deferred revenue		**60,000**

TrueTech recognizes no revenue on January 1. Rather, TrueTech recognizes a deferred revenue liability for $60,000 associated with receiving cash prior to satisfying its performance obligation to provide customers with access to the Tri-Net games for a year.

Tri-Net subscribers receive benefits each day they have access to the Tri-Net network, so TrueTech uses "proportion of time" as its measure of progress toward completion. At the end of each of the 12 months following the sale, TrueTech would record the following entry to recognize Tri-Net subscription revenue:

Deferred revenue ($60,000 ÷ 12)	**5,000**	
Service revenue		**5,000**

After 12 months TrueTech will have recognized the entire $60,000 of Tri-Net subscription revenue, and the deferred revenue liability will be reduced to zero.

seller and the contract stipulates that the seller is paid for performance. We discuss accounting for long-term construction contracts in more detail in Part C of this chapter.

If a performance obligation doesn't meet any of the three criteria for recognizing revenue over time, we recognize revenue at the point in time when the performance obligation has been completely satisfied, which usually occurs at the end of the contract.

Many services are so short term in nature that companies don't bother with recognizing revenue over time even if they qualify for doing so. For example, a house painting service provides benefit to its customer as work is being performed, so it meets the first criterion for revenue recognition over time. However, it likely is easier to simply recognize revenue for the painting service at the conclusion of a painting assignment. This departure from GAAP is immaterial given the short duration of the painting service and the lack of additional useful information that would be provided by more precise timing of revenue recognition.

Determining Progress toward Completion

Because progress toward completion is the basis for recognizing revenue over time, the seller needs to estimate that progress in a way that reflects when control of goods or services is transferred to the customer.

***Input* or *output* methods can be used to estimate progress toward completion when performance obligations are satisfied over time.**

Sellers sometimes use an *output-based* estimate of progress toward completion, measured as the proportion of the goods or services transferred to date. For our Tri-Net example in Illustration 6–6, output is measured by the passage of time, because the performance obligation being satisfied is to provide access to the Tri-Net gaming platform. Other times sellers use an *input-based* estimate of progress toward completion, measured as the proportion of effort expended thus far relative to the total effort expected to satisfy the performance obligation. For example, sellers often use the ratio of costs incurred to date compared to total costs estimated to complete the job.[5] In Part C of this chapter, we continue our discussion of output- and input-based measures of progress toward completion, and also consider how to deal with changes in estimates of progress toward completion.

[5]If for some reason the seller can't make a reasonable estimate of progress to completion using either input or output methods, the seller must wait to recognize revenue until the performance obligation has been completely satisfied. However, if the seller expects to be able to at least recover its costs from the customer, the seller can recognize an amount of revenue equal to the costs incurred until it can make a reasonable estimate of progress toward completion.

Recognizing Revenue for Contracts that Contain Multiple Performance Obligations

● LO6–4

Revenue recognition becomes more complicated when a contract contains multiple performance obligations. As an example, in Illustration 6–7 we combine the two TrueTech examples we already have discussed. In the first example (Illustration 6–4), TrueTech sold Tri-Box modules and recognized revenue at a single point in time (upon delivery). In the second example (Illustration 6–6), TrueTech sold one-year subscriptions to the Tri-Net platform and recognized revenue over time (one-twelfth each month over the year). Now, let's consider how TrueTech would recognize revenue if these two items were sold as a package deal for a single price.

FINANCIAL Reporting Case

Q2, p. 277

Illustration 6–7

Contract Containing Multiple Performance Obligations

TrueTech Industries manufactures the Tri-Box System, a multiplayer gaming system allowing players to compete with each other over the Internet.

- The Tri-Box System includes the physical Tri-Box module as well as a one-year subscription to the Tri-Net multiuser platform of Internet-based games and other applications.
- TrueTech sells individual one-year subscriptions to the Tri-Net platform for $60. Customers can access the Tri-Net using a Tri-Box as well as other gaming modules.
- TrueTech sells individual Tri-Box modules for $240. Customers can use a Tri-Box to access the Tri-Net as well as other multiuser gaming platforms.
- As a package deal, TrueTech sells the Tri-Box System (module plus subscription) for $250.

On January 1, 2021, TrueTech delivers 1,000 Tri-Box Systems to CompStores at a price of $250 per system. TrueTech receives $250,000 from CompStores on January 25, 2021.

We'll assume TrueTech has concluded that it has a contract with CompStores, so step 1 of revenue recognition is satisfied. We'll start with step 2.

Step 2: Identify the Performance Obligation(s)

Sellers account for a promise to provide a good or service as a performance obligation if the good or service is **distinct** from other goods or services in the contract. The idea is to separate contracts into parts that can be viewed on a stand-alone basis. That way the financial statements can better reflect the timing of the transfer of separate goods and services and the profit generated on each one. Goods or services that are not distinct are combined and treated as a single performance obligation.

Promises to provide goods and services are performance obligations when the goods and services are *distinct*.

A good or service is distinct if it is *both*

1. *Capable of being distinct.* The customer could use the good or service on its own or in combination with other goods or services it could obtain elsewhere, and
2. *Separately identifiable from other goods or services in the contract.* The promises to transfer goods and services are distinct in the *context of the contract*, because the seller is promising to provide goods and services *individually* as opposed to promising to provide a *combined* good or service for which the individual goods or services are inputs.

As an example of applying these criteria, think of going to a store like **The Home Depot** to purchase lumber, paint, and other building supplies for a home project. Each of those products is capable of being distinct, because you can buy it individually and use it however you desire. Each also is separately identifiable from other goods and services, because Home Depot's only performance obligation is to deliver the individual items. So, Home Depot can view its promise to deliver each of these items as a separate performance obligation.

Now think about signing an agreement with a construction contractor like **Toll Brothers** to build a house for you. Like Home Depot, the contractor is selling you lumber, paint, and other building supplies. However, while those items are capable of being distinct, they aren't separately identifiable in the context of the contract, because the contractor's performance obligation is to combine those inputs and deliver a completed building. Therefore, Toll Brothers views itself as having a single performance obligation. As we discuss further in Part C of this chapter, most long-term construction contracts are viewed as including a single performance obligation because the seller provides the service of combining goods and services into a combined output.

Construction contracts aren't the only ones that fail the "separately identifiable" criterion. Goods and services also aren't considered separately identifiable if they are highly interdependent, or if one significantly modifies or customizes another. For example, consider a company that offers online access to a clip-art library. The only way customers can access the clip art is by using the online access service. Even though a clip-art library and an online access service might be capable of being distinct outside the context of the contract, within the contract they are so intertwined that they are more appropriately thought of as a single performance obligation.[6]

In Illustration 6–8 we apply these criteria to identify the performance obligations for our TrueTech example.

Illustration 6–8

Determining Whether Goods or Services Are Distinct

Assume the same facts as in Illustration 6–7. Do the Tri-Box module and the Tri-Net subscription qualify as performance obligations in TrueTech's contract with CompStores?

Which of the goods and services promised in the contract are distinct? Both the Tri-Box module and the Tri-Net subscription can be used on their own by a customer, so they are capable of being distinct. The module and subscription are not highly interrelated and do not modify or customize each other, and the nature of TrueTech's promise is not to integrate the module and service into a combined unit, so they are separately identifiable in the context of the contract.

Conclusion: The module and subscription are distinct so the contract has two performance obligations: (1) delivery of a Tri-Box module and (2) fulfillment of a one-year Tri-Net subscription.

Step 3: Determine the Transaction Price

The *transaction price* is the amount the seller expects to be entitled to receive from the customer in exchange for providing goods or services.

The **transaction price** is the amount the seller expects to be entitled to receive from the customer in exchange for providing goods or services.[7] Determining the transaction price is simple if the customer pays a fixed amount immediately or soon after the sale. That's the case with our TrueTech example. The transaction price is $250,000, equal to $250 per system × 1,000 systems.

Step 4: Allocate the Transaction Price to Each Performance Obligation

We allocate the transaction price to performance obligations in proportion to their relative *stand-alone selling prices.*

If a contract includes more than one performance obligation, the seller allocates the transaction price to each one in proportion to the stand-alone selling prices of the goods or services underlying all the performance obligations in the contract. The **stand-alone selling price** is the amount at which the good or service is sold separately under similar circumstances. If a stand-alone selling price can't be directly observed, the seller should estimate it.[8]

Look at Illustration 6–9 to see how we allocate the transaction price to each of the performance obligations in our TrueTech example.

Step 5: Recognize Revenue When (Or As) Each Performance Obligation Is Satisfied

Revenue with respect to each performance obligation is recognized when (or as) that performance obligation is satisfied.

As we discussed earlier, performance obligations can be satisfied either at a point in time or over a period of time, and revenue with respect to a performance obligation is recognized when (or as) the performance obligation is satisfied. We determine the timing of revenue recognition for each performance obligation individually.

---

[6]Sellers also treat as a single performance obligation a series of distinct goods or services that are substantially the same and have the same pattern of transfer.

[7]Normally, sellers are immediately or eventually paid in cash, but sometimes sellers are paid with other assets like property. In that case, the seller measures the assets received at fair value at the start of the contract.

[8]A contractually stated "list price" doesn't necessarily represent a stand-alone selling price, because the seller might actually sell the good or service for a different amount. The seller has to reference actual stand-alone selling prices, or estimate those prices.

Illustration 6–9

Allocating the Transaction Price to Performance Obligations Based on Relative Selling Prices

Assume the same facts as in Illustration 6–7, specifically:

- The transaction price of one Tri-Box System is $250.
- The stand-alone price of a Tri-Box module is $240.
- The stand-alone price of a Tri-Net subscription is $60.

Because the stand-alone price of a Tri-Box module ($240) represents 80% of the sum of the stand-alone selling prices [$240 ÷ ($240 + $60)], and the stand-alone price of a Tri-Net subscription comprises 20% of the total [$60 ÷ ($240 + $60)], we allocate 80% of the transaction price to the Tri-Box module and 20% of the transaction price to the Tri-Net subscription, as follows:

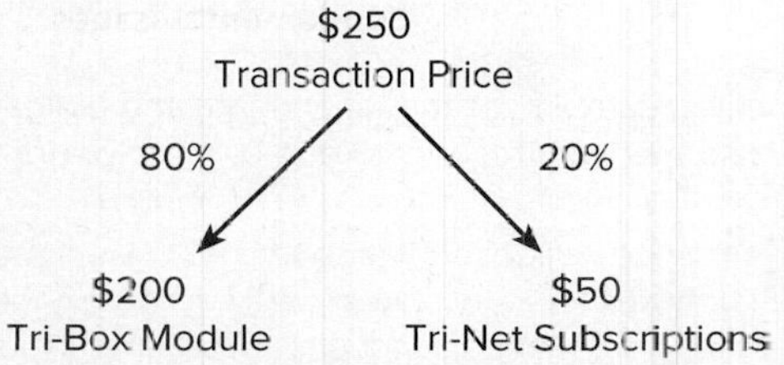

Additional Consideration

Discounts in Contracts with Multiple Performance Obligations. Note that Illustration 6–7 shows that Tri-Box systems are sold at a discount—TrueTech sells the system for a transaction price ($250) that's less than the $300 sum of the stand-alone selling prices of the Tri-Box module ($240) and the subscription to Tri-Net ($60). Because there is no evidence that the discount relates to only one of the performance obligations, it is spread between them in the allocation process. If TrueTech had clear evidence from sales of those goods and services that the discount related to only one of them, the entire discount would be allocated to that good or service.

Returning to our Tri-Box System example, the $200,000 of revenue associated with the Tri-Box modules is recognized when those modules are delivered to CompStores on January 1, but the $50,000 of revenue associated with the Tri-Net subscriptions is recognized over the one-year subscription term. The timing of revenue recognition for each performance obligation is shown in Illustration 6–10.

Illustration 6–10

Recognizing Revenue for Multiple Performance Obligations

Given the allocation of transaction price indicated in Illustration 6–9. TrueTech records the following journal entry at the time of the sale to CompStores (ignoring any entry to record the reduction in inventory and the corresponding cost of goods sold):

January 1, 2021:		
Accounts receivable	250,000	
Sales revenue ($250,000 × 80%)		200,000
Deferred revenue ($250,000 × 20%)		**50,000**

In each of the 12 months following the sale, TrueTech records the following entry to recognize Tri-Net subscription revenue:

Deferred revenue ($50,000 ÷ 12)	**4,167**	
Service revenue		4,167

After 12 months TrueTech will have recognized the entire $50,000 of Tri-Net subscription revenue, and the deferred revenue liability will have been reduced to zero.

Deferred Revenue

	Debit	Credit
1/1		**50,000**
1/31	**4,167**	
2/28	4,167	
. . .	. . .	
12/31	4,167	
12/31		–0–

Service Revenue

	Debit	Credit
1/1		–0–
1/31		4,167
2/28		4,167
. . .		. . .
12/31		4,167
12/31		50,000

It is interesting to compare the journal entries in Illustration 6–10 to those made in Illustrations 6-4 and 6-6. Because $200,000 of the $250,000 transaction price is allocated to the Tri-Box module, less revenue is recognized upon delivery than was the case when the module was sold separately for $240,000. Similarly, because $50,000 of the $250,000 transaction price is allocated to the Tri-Net subscription, less revenue is recognized each month for the subscription ($4,167 per month) than was the case when the subscription was sold separately for $60,000 ($5,000 per month).

Illustration 6–11 summarizes Part A's discussion of the fundamental issues related to recognizing revenue.

Illustration 6–11 Summary of Fundamental Issues Related to Recognizing Revenue

Revenue Recognition	Fundamental Issues	
Step 1 Identify the contract	A contract establishes the legal rights and obligations of the seller and customer with respect to one or more performance obligations.	
Step 2 Identify the performance obligation(s)	A performance obligation is a promise to transfer a good or service that is distinct, which is the case if the good or service is both (a) capable of being distinct and (b) separately identifiable.	
Step 3 Determine the transaction price	The transaction price is the amount the seller is entitled to receive from the customer.	
Step 4 Allocate the transaction price	The seller allocates the transaction price to performance obligations based on the relative stand-alone selling prices of the goods or services in each performance obligation.	
Step 5 Recognize revenue when (or as) each performance obligation is satisfied	The seller recognizes revenue **at a single point in time** when control passes to the customer, which is more likely if the customer has: • Obligation to pay the seller. • Legal title to the asset. • Possession of the asset. • Assumed the risks and rewards of ownership. • Accepted the asset.	The seller recognizes revenue **over a period of time** if: • Customer consumes benefit as work performed, • Customer controls asset as it's created, or • Seller is creating an asset that has no alternative use to the seller and the seller has right to receive payment for work completed.

Concept Review Exercise

REVENUE RECOGNITION FOR CONTRACTS WITH MULTIPLE PERFORMANCE OBLIGATIONS

Macrovision sells a variety of satellite TV packages. The popular $600 Basic Package includes a hardware component (consisting of a satellite dish and receiver) along with a twelve-month subscription to 130 TV channels. Macrovision sells the hardware component without a subscription for $180, and sells a twelve-month subscription to the same 130 channels without hardware for $540/year. Let's account for the sale of one Basic Package for $600 on January 1, 2021.

Required:

1. Identify the performance obligations in the Basic Package contract, and determine when revenue for each should be recognized.
2. For the single Basic Package sold on January 1, 2021, allocate the $600 transaction price to the performance obligations in the contract, and prepare a journal entry to record the sale (ignoring any entry to record the reduction in inventory and the corresponding cost of goods sold).
3. Prepare any journal entry necessary to record revenue related to the same contract on January 31, 2021.

Solution:

1. Identify the performance obligations in the Basic Package contract, and determine when revenue for each should be recognized.

The hardware component and the twelve-month subscription are *capable of being distinct* (they are sold separately) and are *separately identifiable* (the hardware and services are not highly intertwined so it makes sense to consider them separately). Therefore, the hardware component and the twelve-month subscription are distinct from each other and should be treated as separate performance obligations. Revenue for the hardware component should be recognized on January 1, 2021, because transfer of control of the hardware occurs when the hardware is delivered to the customer. Revenue for the subscription should be recognized over the next twelve months as the customer receives the benefit of having access to TV channels.

2. For the single Basic Package sold on January 1, 2021, allocate the $600 transaction price to the performance obligations in the contract, and prepare a journal entry to record the sale (ignoring any entry to record the reduction in inventory and the corresponding cost of goods sold).

Because the stand-alone price of the hardware component ($180) represents 25% of the total of all the stand-alone selling prices ($180 ÷ [$180 + $540]), and the stand-alone price of the twelve-month subscription comprises 75% of the total ($540 ÷ [$180 + $540]), we allocate 25% of the transaction price to the hardware component and 75% of the transaction price to the twelve-month subscription. The transaction price of $600 would be allocated as follows:

Hardware Component: $600 × 25% = $150.
Twelve-Month Subscription: $600 × 75% = $450.

The journal entry recorded on January 1, 2021, would be:

Cash	600	
Sales revenue (for delivery of hardware)		150
Deferred revenue (for subscription)		450

3. Prepare any journal entry necessary to record revenue for the same contract on January 31, 2021.

Deferred revenue ($450 ÷ 12)	37.50	
Service revenue		37.50

Special Issues in Revenue Recognition

PART B

LO6–5

Now that we've covered the basics, let's consider some important issues that occur in practice with respect to each of the five steps. We'll cover each step in turn.

Special Issues for Step 1: Identify the Contract

A **contract** is an agreement that creates legally enforceable rights and obligations. We normally think of a contract as being specified in a written document, but contracts can be oral rather than written. Contracts also can be *implicit* based on the typical business practices that a company follows. Remember from our example in Part A, just buying a skirt from Macy's implies a contract for purposes of recognizing revenue. The key is that all parties to the contract are committed to performing their obligations and enforcing their rights.[9]

A *contract* is an agreement that creates legally enforceable rights and obligations.

[9]Specifically, *ASC Topic 606* indicates that a contract exists for purposes of revenue recognition only if it (a) has commercial substance, affecting the risk, timing or amount of the seller's future cash flows, (b) has been approved by both the seller and the customer, indicating commitment to fulfilling their obligations, (c) specifies the seller's and customer's rights regarding the goods or services to be transferred, (d) specifies payment terms, and (e) is probable that the seller will collect substantially all of the amount it is entitled to receive. These criteria are very similar to requirements previously indicated by the SEC in the Staff Accounting Bulletins No. 101 and No. 104 mentioned earlier in this chapter.

A seller must believe collectibility is probable for a contract to exist for purposes of revenue recognition.

For a contract to exist for purposes of revenue recognition, the seller must believe it's probable that it will collect substantially all of the amount it's entitled to receive in exchange for the goods or services that it will provide to the customer. This collectibility threshold makes sure that revenue really reflects an inflow of net assets from the customer. However, even if a contract doesn't exist, the seller still can recognize an amount of revenue equal to any nonrefundable payments it has received, so long as the seller has transferred control of the goods or services and does not have any further obligations to transfer goods or services to the customer.

A contract does not exist if (a) neither the seller nor the customer has performed any obligations under the contract and (b) both the seller and the customer can terminate the contract without penalty. In other words, either the seller or the customer must have done something that has commercial substance for the seller to start accounting for revenue. Illustration 6–12 provides an example.

Illustration 6–12

Determining Whether a Contract Exists for Revenue Recognition Purposes

Recall from Illustration 6–7 that CompStores ordered 1,000 Tri-Box systems on December 20, 2020, at a price of $250 per unit. Assume that CompStores and TrueTech can cancel the order without penalty prior to delivery. TrueTech made delivery on January 1, 2021, and received $250,000 on January 25, 2021. When does TrueTech's arrangement with CompStores qualify as a contract for purposes of revenue recognition?

The arrangement qualifies as a contract on January 1, 2021. That's the date TrueTech makes delivery to CompStores. Prior to delivery, neither TrueTech nor CompStores had performed an obligation under the contract, and both parties could cancel the order without penalty, so the arrangement didn't qualify as a contract for purposes of revenue recognition.

International Financial Reporting Standards

LO6–10

ASU No. 2014–09 defines "probable" as "likely to occur." Similarly, *SFAC No. 6* defines "probable" to mean an amount can "reasonably be expected or believed on the basis of available evidence or logic but is neither certain nor proved," which implies a relatively high likelihood of occurrence. IFRS defines "probable" as a likelihood that is greater than 50%, which is lower than the definition in U.S. GAAP. Therefore, some contracts might not meet this threshold under U.S. GAAP that do meet it under IFRS.

Additional Consideration

Contract Modifications. A customer and seller might agree to modify a contract in some way. For instance, they might change the transaction price, change the performance obligations, or add another performance obligation. The way we account for a contract modification depends on the nature of the modification:

1. Sometimes a modification is really just a separate new contract. That happens when the modification adds another distinct good or service and requires the customer to pay an additional amount equal to the stand-alone selling price of the added good or service. In that case, we view the modification as a separate contract.
2. Other times a modification is to a contract for which the remaining goods and services are distinct from those already transferred, but the modification doesn't qualify as a separate contract. In that case, the seller acts as if the old contract has been terminated and a new contract has been created. The new contract includes whatever performance obligations remain after the modification, and its transaction price is equal to the amount that hasn't yet been recognized as revenue under the old contract plus or minus any change in price required by the modification. We allocate the revised transaction price to all performance obligations remaining in the contract based on their stand-alone selling prices at that time.
3. Finally, sometimes we modify a contract for which the remaining performance obligations are not distinct and therefore form a single performance obligation that is being satisfied over time. In that case, we need to update our assessment of progress toward completion and adjust revenue as appropriate to reflect progress to date, just like we treat other changes in estimates.

Special Issues for Step 2: Identify the Performance Obligation(s)

Previously we saw that promises to provide goods and services are treated as performance obligations when the goods and services are distinct. Now let's consider several aspects of contracts we often encounter and determine if they qualify as performance obligations. We discuss prepayments, warranties, and options.

PREPAYMENTS Some contracts require nonrefundable up-front fees for particular activities (for example, **LA Fitness** charges up-front registration fees for gym memberships). We don't consider such *prepayments* to be performance obligations because they aren't a promise to transfer a product or service to a customer. Instead, the up-front fee is an advance payment by the customer for future products or services and should be included in the transaction price, allocated to the various performance obligations in the contract, initially recorded as deferred revenue, and recognized as revenue when (or as) each performance obligation is satisfied.

A *prepayment* is not a performance obligation.

WARRANTIES Most products are sold with a warranty that obligates the seller to make repairs or replace products that later are found to be defective or unsatisfactory. These warranties are not sold separately, and either can be stated explicitly or be implicit based on normal business practice. We call these **quality-assurance warranties**. A quality-assurance warranty (sometimes called an "assurance-type warranty") is not a performance obligation. Rather, it is a cost of satisfying the performance obligation to deliver products of acceptable quality. The seller recognizes this cost in the period of sale as a warranty expense and related contingent liability. Because the exact amount of the cost usually is not known at the time of the sale, it must be estimated. For example, **Deere & Company**, which manufactures equipment used for construction, landscaping, and other purposes, reported a quality-assurance warranty liability of $1.007 billion at the end of its 2017 fiscal year.

A *quality-assurance warranty* obligates the seller to repair or replace defective products.

A quality-assurance warranty is not a performance obligation.

Extended warranties, on the other hand, are offered as an additional service that covers new problems arising after the customer takes control of the product. It's unusual these days to buy a phone, digital tablet, car, or almost any durable consumer product without being asked to buy an extended warranty. An extended warranty (sometimes called a "service-type warranty") provides protection beyond the manufacturer's quality-assurance warranty. Because an extended warranty usually is priced and sold separately from the product, it constitutes a performance obligation and can be viewed as a separate sales transaction. The price is recorded as a deferred revenue liability and then recognized as revenue over the extended warranty period. If an extended warranty is included along with the related product as part of a single contract, the extended warranty is treated as a separate performance obligation, allocated a portion of the transaction price, and that portion of the transaction price is recorded as deferred revenue. Deere & Company reported a liability for deferred extended warranty revenue of $461 million at the end of its 2017 fiscal year.

An *extended warranty* is an additional service that covers new problems arising after a customer takes control of a product.

An extended warranty is a separate performance obligation

How can you tell if a warranty should be treated as a quality-assurance warranty or an extended warranty? A warranty should be treated as an extended warranty if either (a) the customer has the option to purchase the warranty separately from the seller or (b) the warranty provides a service to the customer beyond only assuring that the seller delivered a product or service that was free from defects. The specifics of the warranty have to be considered when making this determination. For example, if the warranty period is very long, it's likely the warranty is covering more than just the quality of the product at the date of delivery, so it likely would represent an extended warranty.

We discuss accounting for warranties more in Chapter 13.

CUSTOMER OPTIONS FOR ADDITIONAL GOODS OR SERVICES In some contracts the seller grants to the customer an *option* to receive additional goods or services at no cost or at a discount. Examples include software upgrades, customer loyalty programs (frequent flier miles, credit card points), discounts on future goods or services, and contract renewal options. Options for additional goods or services are considered performance obligations if they provide a *material right* to the customer that the customer would not receive otherwise.[10]

[10]Be careful not to confuse these types of options with stock options, which are financial instruments that allow purchase of shares of stock at a specific price at a future date.

For example, if a shoe seller always discounts its products by 5% from list price, but advertises a special deal by which customers who purchase one pair of shoes receive a 20% discount off the next pair of shoes purchased at the same store, the extra discount of 15% (20% – 5%) is a material right, as it is a discount customers would not receive had they not bought the first pair of shoes. In that case, the seller has two performance obligations associated with sale of the first pair of shoes: (1) delivery of the shoes and (2) provision of the extra 15% discount on the next pair of shoes purchased at the same store.

An *option* for additional goods or services is a performance obligation if it confers a material right to the customer.

When a contract includes an option that provides a material right, the seller must allocate part of the contract's transaction price to the option. Just like for other performance obligations, that allocation process requires the seller to estimate the stand-alone selling price of the option, as well as taking into account the likelihood that the customer will actually exercise the option. The seller recognizes revenue associated with the option when the option is exercised or expires. Illustration 6–13 provides an example.

Illustration 6–13

Customer options for additional goods or services.

TrueTech offers a promotional coupon with every Tri-Box it sells for $240.

- The coupon gives the Tri-Box customer an opportunity to buy a headset that normally sells for $150 for only $90 (a 40% discount).
- The coupon must be redeemed within one year of the Tri-Box purchase.
- TrueTech estimates that 80% of customers will take advantage of the coupon.

How would TrueTech account for the cash sale of 100 Tri-Boxes sold under this promotion on January 1, 2021?

The coupon provides a material right to the customer, because it allows the customer to receive a discount of $150 × 40% = $60. Therefore, the coupon provides an option that constitutes a performance obligation, and TrueTech must allocate the $240 transaction price of the Tri-Box to two performance obligations: the Tri-Box and the right to acquire a headset at a discount upon presentation of the coupon.

Because TrueTech expects only 80% of the coupons to be used, it estimates the stand-alone selling price of the coupon to be the $48, computed as the value of the opportunity to receive a discount ($60) multiplied by the probability a customer will take advantage of that opportunity (80%).* The sum of the stand-alone selling prices of the performance obligations is $288, equal to the Tri-Box module ($240) plus the coupon ($48). The Tri-Box module ($240) represents five-sixths (or 83.33%) of the total ($240 ÷ $288), and the coupon comprises one-sixth (or 16.67%) of the total ($48 ÷ $288), so TrueTech allocates five-sixths of the $240 transaction price to the Tri-Box module and one-sixth to the coupon, as follows:

January 1, 2021:		
Cash	24,000	
Sales revenue ($240 × 5/6 × 100 units)		20,000
Deferred revenue—coupons ($240 × 1/6 × 100 units)		4,000

When the coupons are later redeemed or expire, TrueTech will debit deferred revenue—coupons and credit revenue.

*It may seem strange that we consider the likelihood that the customer will use the coupon when estimating the coupon's stand-alone selling price, but think about it from TrueTech's perspective. Each coupon saves a customer $60, but on average TrueTech will only have to provide discounts of $48, so $48 is its estimate of the average stand-alone value of its performance obligation.

Special Issues for Step 3: Determine the Transaction Price

● LO6–6

The *transaction price* is the amount the seller expects to be entitled to receive from the customer.

Until now we've assumed that contracts indicate a fixed **transaction price** that will be paid at or soon after delivery. However, in some contracts the transaction price is less clear. Specific situations affecting the transaction price are (a) variable consideration and the constraint on its recognition, (b) sales with a right of return (a particular type of variable consideration), (c) identifying whether the seller is acting as a principal or an agent, (d) the time value of money, and (e) payments by the seller to the customer. Let's consider these one at a time to see how recording revenue will be affected when there is variable consideration.

Variable consideration is the portion of a transaction price that depends on the outcome of future events.

VARIABLE CONSIDERATION Sometimes a transaction price is uncertain because some of the price depends on the outcome of future events. Contracts that include this **variable consideration** are commonplace in many industries, including construction (incentive

Additional Consideration

Shipping Costs. Amazon Prime includes "free" two-day shipping. You pay one price, but Amazon has two obligations: to provide a good and to ship that good. Are these considered separate performance obligations that require Amazon to allocate the transaction price for purposes of revenue recognition? It depends.

If shipping is provided *prior* to the seller transferring control of goods to the buyer (for example, if title passes to the customer upon delivery), then shipping is viewed as just another cost of doing business.[11] In that case, shipping is not treated as a separate performance obligation, and none of the transaction price would be allocated to it.

If shipping is provided *after* the customer has taken control of goods, then shipping could be viewed as a separate service and treated as a separate performance obligation. Note 1 of Amazon's 2017 10-K provides an example of this situation: "Retail sales to customers are made pursuant to a sales contract that provides for transfer of both title and risk of loss upon our delivery to the carrier or the customer." Amazon could view these arrangements as including two separate performance obligations: providing goods and then shipping those goods. However, allocating a small portion of the transaction price to "shipping revenue" might be more trouble than it's worth. Therefore, the FASB allows companies to choose whether to treat shipping provided after the customer takes control as either (a) just a cost of doing business or (b) a separate performance obligation (which requires sellers to allocate a portion of the transaction price to shipping and then to recognize shipping revenue when shipping is completed). The seller's policy must be disclosed clearly.[12] The IASB does not allow this choice, so some companies that apply IFRS might be forced to recognize "shipping revenue" when they would prefer not to go to the trouble of doing so.

payments), entertainment and media (royalties), health care (Medicare and Medicaid reimbursements), manufacturing (volume discounts and product returns), and telecommunications (rebates).

Estimating Variable Consideration. When an amount to be received depends on some uncertain future event, the seller still should include the uncertain amount in the transaction price by estimating it. A seller estimates variable consideration as either (a) the *expected value* (calculated as the sum of each possible amount multiplied by its probability), or (b) the *most likely amount*, depending on which estimation approach better predicts the amount that the seller will receive. If there are several possible outcomes, the expected value will be more appropriate. On the other hand, if only two outcomes are possible, the most likely amount might be the best indication of the amount the seller will likely receive. Illustration 6–14 provides an example.

Variable consideration is estimated as either the expected value or the most likely amount.

Illustration 6–14

Accounting for Variable Consideration

TrueTech enters into a contract with ProSport Gaming to add ProSport's online games to the Tri-Net network. ProSport offers popular games like Brawl of Bards, and wants those games offered on the Tri-Net so ProSport can sell gems, weapons, health potions, and other game features that allow players to advance more quickly in a game.

On January 1, 2021, ProSport pays TrueTech an up-front fixed fee of $300,000 for six months of featured access. ProSport also will pay TrueTech a bonus of $180,000 if Tri-Net users access ProSport games for at least 15,000 hours during the six-month period. TrueTech estimates a 75% chance that it will achieve the usage target and receive the $180,000 bonus.

TrueTech records the following entry for the receipt of the cash on January 1, 2021:

Cash	300,000	
Deferred revenue		300,000

Subsequent entries to recognize revenue depend on whether TrueTech estimates the transaction price as the expected value or the most likely amount.

(continued)

[11] In Chapter 8 we'll distinguish between situations in which title transfers before shipment and after shipment.

[12] FASB ASC 606-10-25-18B: Revenue from Contracts with Customers—Overall—Identifying Performance Obligations.

(concluded)

Bonus Receivable

1/1	–0–	
1/31	30,000	
2/28	30,000	
3/31	30,000	
4/30	30,000	
5/31	30,000	
6/30	30,000	180,000
6/30	–0–	

Service Revenue

1/1		–0–
1/31		80,000
2/28		80,000
3/31		80,000
4/30		80,000
5/31		80,000
6/30		80,000
6/30		480,000

Alternative 1: Expected Value

The expected value is calculated as a probability-weighted transaction price.

Possible Amounts	Probabilities	Expected Amounts
$480,000 ($300,000 fixed fee + 180,000 bonus)	× 75% =	$ 360,000
$300,000 ($300,000 fixed fee + 0 bonus)	× 25% =	75,000
Expected value of the contract price at inception		**$435,000**

Alternative 2: Most Likely Amount

Because there is a greater chance of qualifying for the bonus than of not qualifying for the bonus, a transaction price based on the most likely amount is $300,000 + $180,000, or **$480,000**.

Let's assume that TrueTech bases the estimate on the most likely amount, **$480,000**. In each successive month TrueTech recognizes one month's revenue based on a total transaction price of $480,000. Because it previously recorded $300,000 as deferred revenue, at the end of each month TrueTech reduces deferred revenue by one-sixth of the $300,000 as well as recognizing a bonus receivable for one-sixth of the $180,000 bonus it expects to receive.

Deferred revenue ($300,000 ÷ 6 months)	50,000	
Bonus receivable ($180,000 ÷ 6 months)	30,000	
Service revenue (**$480,000** ÷ 6 months)*		**80,000**

After six months, TrueTech's deferred revenue account has been reduced to a zero balance, and the bonus receivable account has a balance of $180,000 ($30,000 × 6). At that point, TrueTech knows if the usage of ProSport products had reached the bonus threshold and records one of the following two journal entries:

If TrueTech receives the bonus:			If TrueTech does not receive the bonus:		
Cash	180,000		Service revenue	180,000	
Bonus receivable		180,000	Bonus receivable		180,000

*If TrueTech instead used the expected value as its estimate of the transaction price, the journal entries would be the same except that the amount of revenue recognized each month would be $72,500 (**$435,000** ÷ 6 months). The reduction in the deferred revenue liability each month would still be $50,000, and the amount of bonus receivable accrued each month would be $22,500 ($135,000 ÷ 6 months).

The seller must reassess its estimate of the transaction price in each period to determine whether circumstances have changed. If the seller revises its estimate of the amount of variable consideration it will receive, it must revise any receivable it has recorded and reflect the adjustment in that period's revenue, as we see in Illustration 6–15.

Bonus Receivable

1/1	–0–	
1/31	30,000	
2/28	30,000	
3/31	30,000	
4/1		90,000
4/1	–0–	

Service Revenue

1/1		–0–
1/31		80,000
2/28		80,000
3/31		80,000
4/1	90,000	
4/30		50,000
5/31		50,000
6/30		50,000
6/30		300,000

Illustration 6–15 Accounting for Variable Consideration

Assume the same facts as in Illustration 6–14, but that, after three months, TrueTech concludes that, due to low usage of ProSport's games, the most likely outcome is that TrueTech will *not* receive the $180,000 bonus. TrueTech records the following entry at the start of April to reduce its bonus receivable to zero and reflect the adjustment in revenue:

Service revenue	**90,000**	
Bonus receivable (reducing the account to zero)		90,000

For the remainder of the contract, TrueTech only recognizes revenue in each month associated with the up-front fixed payment of $300,000.

Deferred revenue ($300,000 ÷ 6 months)	50,000	
Service revenue		50,000

Constraint on Recognizing Variable Consideration. Sometimes sellers lack sufficient information to make a good estimate of variable consideration. The concern is that a seller might overestimate variable consideration, recognize revenue based on a transaction price that is too high, and later have to reverse that revenue (and reduce net income) to correct the estimate. To guard against this, sellers only include an estimate of variable consideration in the transaction price to the extent it is "probable" that a significant reversal of revenue recognized to date will not occur when the uncertainty associated with the variable consideration is resolved in the future.

International Financial Reporting Standards

IFRS uses the term "highly probable" instead of "probable" in this case. Because IFRS defines "probable" to mean a likelihood greater than 50%, its use of "highly probable" is intended to convey the same likelihood as is conveyed by "probable" in U.S. GAAP.

● LO6–10

Applying this constraint requires judgment on the part of the seller, taking into account all information available. Indicators that a significant revenue reversal could occur include (a) poor evidence on which to base an estimate, (b) dependence of the estimate on factors outside the seller's control, (c) a history of the seller changing payment terms on similar contracts, (d) a broad range of outcomes that could occur, and (e) a long delay before uncertainty resolves.

Sellers are limited to recognizing variable consideration to the extent that it is probable that a significant revenue reversal will not occur in the future.

If a seller changes its opinion regarding whether a constraint on variable consideration is necessary, the seller should update the transaction price in the current reporting period, just as the seller would do for other changes in estimated variable consideration. Illustration 6–16 provides an example.

Illustration 6–16

Constraint on Recognizing Variable Consideration

Assume the same facts as in Illustration 6–14, but that initially TrueTech can't conclude that it is probable that a significant revenue reversal will not occur in the future. In that case, TrueTech is constrained from recognizing revenue associated with variable consideration. TrueTech includes only the up-front fixed payment of $300,000 in the transaction price, and recognizes revenue of $50,000 each month.

Deferred revenue ($300,000 ÷ 6 months)	50,000	
Service revenue		50,000

On April 1 after three months of the contract have passed, TrueTech concludes it can make an accurate enough bonus estimate for it to be probable that a significant revenue reversal will not occur. As in Illustration 6–14, TrueTech estimates a 75% likelihood it will receive the bonus and bases its estimate on the "most likely amount" of $180,000. Since on April 1 the contract is one-half finished (3 of the 6 months have passed), TrueTech records a bonus receivable and service revenue for $90,000 ($180,000 × 3/6), the cumulative amount that would have been recognized over the first three months of the contract if an estimate of variable consideration had been included in the transaction price to begin with:

Bonus receivable ($180,000 × 3/6)	**90,000**	
Service revenue		**90,000**

In the final three months of the contract, TrueTech recognizes the remaining revenue assuming a transaction price of $480,000, exactly as if it had included an estimate of variable consideration in the transaction price all along:

Deferred revenue ($300,000 ÷ 6 months)	50,000	
Bonus receivable ($180,000 ÷ 6 months)	30,000	
Service revenue ($480,000 ÷ 6 months)		80,000

Bonus Receivable

	Debit	Credit
1/1	–0–	
4/1	**90,000**	
4/30	30,000	
5/31	30,000	
6/30	30,000	
6/30	180,000	

Service Revenue

	Debit	Credit
1/1		–0–
1/31		50,000
2/28		50,000
3/31		50,000
4/1		**90,000**
4/30		80,000
5/31		80,000
6/30		80,000
6/30		480,000

RIGHT OF RETURN Retailers usually give customers the right to return merchandise if customers decide they don't want it, are not satisfied with it, or are unable to resell it. For example, video-game manufacturers like **Take-Two Interactive Software** often give customers a **right of return** for unsold products.

The right to return merchandise does not create a performance obligation for the seller. Instead, it represents a potential failure to satisfy the original performance obligation to provide goods that the customer wants to keep.

***A right of return* is not a performance obligation.**

Because the total amount of cash a seller is entitled to receive depends on the amount of returns, a right of return creates a situation involving variable consideration. The actual

***A right of return* is a form of variable consideration.**

consideration the seller receives will be sales revenue minus the portion of those sales returned by customers, which we call *sales returns.* The sales returns account is a "contra revenue" account that has the effect of reducing revenue. As a result, we report sales revenue net of the amount expected to be returned.

Sellers report net sales revenue in the income statement, equal to gross sales revenue less actual and estimated returns.

Based on past experience, a seller usually can estimate the returns that will result for a given volume of sales, so the seller reduces revenue by the estimated returns. For example, assume that TruTech sold 1,000 Tri-Boxes to CompStores for $240 each, for a total of $240,000 of revenue. If TruTech estimated that CompStores will return 5% of the Tri-Boxes purchased, TruTech would report net sales revenue of $228,000 in its income statement, equal to gross revenue of $240,000 less estimated returns of $12,000 ($240,000 × 5% estimated returns).

Sales revenue ($240 × 1,000 Tri-Boxes)	$240,000
Less: Estimated returns (5%)	(12,000)
Net sales revenue	$228,000

Conceptually, sellers should account for returns each time they make a sale. In practice, though, most companies find it impractical to do that. Instead, they simply record sales revenue at the time of sale, record sales returns during the period as customers return products and receive refunds, and account for additional estimated returns at the end of the accounting period. Importantly, though, that practical expedient takes us to the same amounts reported in the financial statements. The seller reports net sales revenue in the income statement, and also reports a refund liability in the balance sheet for any additional amounts it expects to refund to customers who make returns. We discuss these and other aspects of accounting for returns in more detail in Chapter 7.

What if the seller lacks sufficient information to be able to accurately estimate returns? In that case, the constraint on recognizing variable consideration we discussed earlier applies, and the seller should recognize revenue only to the extent it is probable that a significant revenue reversal will not occur later if the estimate of returns changes. In fact, the seller might postpone recognizing any revenue until the uncertainty about returns is resolved. Illustration 6–17 provides an example from **Intel Corporation's** 10-K dated February 16, 2018.

Illustration 6–17

Disclosure of Revenue Recognition Policy—Intel Corporation.

Real World Financials

Revenue Recognition

On sales made to distributors that allow for price protections or right of return until the distributor sells through the merchandise, we defer product revenue, and related costs of sales, due to sales price reductions and rapid technology obsolescence in our industry.

IS THE SELLER A PRINCIPAL OR AGENT? Sometimes more than one company is involved in providing goods or services to a customer. In those situations, we need to determine whether a company is acting as a **principal**, providing the good or service to the customer, or an **agent**, only arranging for another company to provide the good or service.

A *principal* controls goods or services and is responsible for providing them to the customer.

We view the seller as a principal if it obtains control of the goods or services before they are transferred to the customer. Various indicators help determine whether a seller obtains control. Control is indicated if the seller has primary responsibility for providing a product or service that the customer finds acceptable, if the seller has discretion in setting prices, and/or if the seller is vulnerable to risks associated with holding inventory or having inventory returned to it.

An *agent* doesn't control goods or services, but rather facilitates transfers between sellers and customers.

A principal's performance obligation is to provide goods and services. In contrast, an agent doesn't primarily provide goods or services, but acts as a facilitator that receives a commission for helping sellers provide goods and services to buyers. An agent's performance obligation is to facilitate a transaction between a principal and a customer.

Many examples of agents occur in business. One you're familiar with is a real estate agent. Real estate agents don't own the houses they sell, but rather charge a commission to help home owners transact with home buyers. Similarly, online auction houses like **eBay**, travel facilitators like **Expedia, Inc.** and **priceline.com**, and broad web-based retailers like **Amazon** act as agents for a variety of sellers. Complicating matters, these same companies also act as principals on some other arrangements, selling their own products and services directly to customers.

The distinction between a principal and an agent is important because it affects the amount of revenue that a company can record. If the company is a principal, it records revenue equal to the total sales price paid by customers as well as cost of goods sold equal to the cost of the item to the company. On the other hand, if the company is an agent, it records as revenue only the commission it receives on the transaction.

An agent only records its commission as revenue.

We see from Illustration 6–18 that whether the seller is a principal or an agent can have a significant effect on its revenue. This is particularly important for start-ups or growth-oriented companies that may be valued more for growth in revenue than for growth in net income.

Illustration 6–18

Comparison of Revenue Recognition by Principals and Agents

Mike buys a Tri-Box module from an online retailer for $290. Let's consider accounting for that sale by two retailers, PrinCo and AgenCo.

- PrinCo purchases Tri-Box modules directly from TrueTech for $240, has the modules shipped to its distribution center in Kansas, and then ships individual modules to buyers when a sale is made. PrinCo offers occasional price discounts according to its marketing strategy. Because PrinCo is responsible for fulfilling the contract, bears the risk of holding inventory, and has latitude in setting sales prices, the evidence suggests that PrinCo is a principal in this transaction.
- AgenCo serves as a web portal by which multiple game module manufacturers like TrueTech can offer their products for sale. The manufacturers ship directly to buyers when a sale is made. AgenCo receives a $50 commission on each sale that occurs via its web portal. Given that AgenCo is not primarily responsible for fulfilling the contract, bears no inventory risk, has no latitude in setting sales prices, and is paid on commission, the evidence suggests AgenCo is an agent in this transaction.

The first part of the income statement for each retailer is shown below. Notice that the same amount of gross profit, $50, is recognized by the principal and the agent. What differ are the amounts of revenue and expense that are recognized and reported.

A Principal Records Gross Revenue (PrinCo)		An Agent Records Net Revenue (AgenCo)	
Revenue	$290	Revenue	$50
Less: Cost of goods sold	(240)	Less: Cost of goods sold	0
Gross profit	$ 50	Gross profit	$50

THE TIME VALUE OF MONEY It's common for contracts to specify that payment occurs either before or after delivery. We recognize an account receivable when payment occurs *after* delivery. In that case, the seller can be viewed as making a loan to the customer between delivery and payment. We recognize deferred revenue if payment occurs *before* delivery. In that case, the customer can be viewed as making a loan to the seller by paying in advance.

As you learned in Chapter 5, there is an interest charge (a "time value of money") implicit in these arrangements. If the time value of money is a significant part of the contract, the seller should view the transaction price as consisting of two components:

If the time value of money is significant, a contract is viewed as including *delivery* and *financing* components.

1. A *delivery component*, equal to the cash price of the good or service, and
2. A *financing component*, which is interest considered paid to the customer (in the case of a prepayment) or to the seller (in the case of a receivable).

In Illustration 6–19 we see both the prepayment and receivable cases:

Illustration 6–19

Accounting for the Time Value of Money

On January 1, 2021, TrueTech enters into a contract with GameStop Stores to deliver four $240 Tri-Box modules that have a combined fair value of $960.

- ***Prepayment Case:*** GameStop pays TrueTech $873 on January 1, 2021, and TrueTech agrees to deliver the modules on December 31, 2021. Because GameStop pays in advance of delivery, we view TrueTech as borrowing money from GameStop, so TrueTech will incur interest expense.
- ***Receivable Case:*** TrueTech delivers the modules on January 1, 2021, and GameStop agrees to pay TrueTech $1,056 on December 31, 2021. Because TrueTech delivers the modules in advance of payment, we view TrueTech as lending money to GameStop, so TrueTech will earn interest revenue.

In both cases, TrueTech views the financing component to be significant, and the applicable interest rate is 10%. In the following table we compare TrueTech's accounting for the contract for the two cases (ignoring the entry for cost of goods sold):

Prepayment (collection *before* delivery)			**Receivable (collection *after* delivery)**		
January 1, 2021			January 1, 2021		
When collection occurs:			***When delivery occurs:***		
Cash	873		Notes receivable	1,056	
Deferred revenue		873*	Discount on notes receivable		96
			Sales revenue		**960****
December 31, 2021			December 31, 2021		
When subsequent delivery occurs:			***When subsequent collection occurs:***		
Interest expense ($837 × 10%)	87		Cash	1,056	
Deferred revenue	873		Discount on notes receivable	96	
Sales revenue		**960**	Interest revenue ($960 x 10%)		96
			Notes receivable		1,056

*$873 = $960 × 0.90909 (present value of $1, n = 1, I = 10%; from Table 2)
**$960 = $1,056 x 0.90909 (present value of $1, n = 1, I = 10%; from Table 2)

Notice in Illustration 6-19 that, even though in both cases the customer is buying a product with a cash price of $960, the customer pays less in the prepayment case ($873) than in the receivable case ($1,056). That's because the customer is *lending* money to the seller in the prepayment case and so is getting some compensation for the time value of money, but the customer is *borrowing* money from the seller in the receivable case and so is having to pay compensation for the time value of money.

When is a financing component considered *significant?* As with other applications of the materiality concept, that's a matter of professional judgment, but there are indicators that might suggest significance. If the customer would pay a substantially different amount if it paid cash at the time the good or service was delivered, the financing component likely is significant. Also, the financing component is more likely to be significant as the time between delivery and payment increases, or if the interest rate implicit in the contract is large. As a practical matter, sellers can assume the financing component is not significant if the period between delivery and payment is less than a year.

We discuss further how sellers make adjustments for the time value of money for accounts receivable in Chapter 7 and for prepayments in Chapter 13.

PAYMENTS BY THE SELLER TO THE CUSTOMER Usually it's the customer who pays the seller for goods or services. Occasionally, though, a *seller* also makes payments to a *customer*. For example, **Samsung** sells TVs, smartphones, tablets, and other products to **BestBuy**. However, Samsung also might pay BestBuy for dedicated space in BestBuy stores or to conduct special Samsung-focused advertising programs. The question is whether a payment by Samsung is a purchase of goods or services from BestBuy, or really just a refund of some of the price paid by BestBuy to purchase Samsung products.

The way we account for payments by a seller to a customer depends on the specifics of the arrangement. If the seller is purchasing distinct goods or services from the customer at the fair value of those goods or services, we account for that purchase as a separate transaction. If a seller pays more for distinct goods or services purchased from its customer than the fair value of those goods or services, those excess payments are viewed as a refund. They are subtracted from the amount the seller is entitled to receive when calculating the transaction price of the sale to the customer. In our Samsung example, if Samsung pays more for dedicated floor space at BestBuy than the fair value of that floor space, Samsung should treat that excess payment as a refund to BestBuy of part of the price paid by BestBuy for Samsung products.

Special Issues for Step 4: Allocate the Transaction Price to the Performance Obligations

We already discussed the need for the seller to allocate the transaction price to each performance obligation in a contract in proportion to the stand-alone selling prices of the goods or services. We also noted that when goods and services aren't normally sold separately, sellers must estimate those stand-alone selling prices. Various approaches are available to estimate stand-alone selling prices. Examples include the following:

1. **Adjusted market assessment approach:** The seller considers what it could sell the product or services for in the market in which it normally conducts business, perhaps referencing prices charged by competitors.
2. **Expected cost plus margin approach:** The seller estimates its costs of satisfying a performance obligation and then adds an appropriate profit margin.
3. **Residual approach:** The seller estimates an unknown (or highly uncertain) stand-alone selling price by subtracting the sum of the known or estimated stand-alone selling prices of other goods or services in the contract from the total transaction price of the contract. The residual approach is allowed only if the stand-alone selling price is highly uncertain, either because the seller hasn't previously sold the good or service and hasn't yet determined a price for it, or because the seller provides the same good or service to different customers at substantially different prices. Illustration 6–20 provides an example of the residual approach.

The *residual approach* is used to estimate a stand-alone selling price that is very uncertain.

Illustration 6–20
Allocating Transaction Price to Performance Obligations Using the Residual Approach

Assume the same facts as Illustration 6–7, with the Tri-Box System comprised of the Tri-Box module and the Tri-Net subscription. The normal selling price of each system is $250 and the standalone price of each Tri-Box is $240, but the stand-alone selling price of the one-year Tri-Net subscription is highly uncertain because TrueTech hasn't sold that service previously and hasn't established a price for it. Under the residual approach, the value of the subscription would be estimated as follows:

Total price of Tri-Box with Tri-Net subscription ($250 × 1,000)	$250,000
Stand-alone price of Tri-Box sold without subscription ($240 × 1,000)	(240,000)
Estimated stand-alone price of Tri-Net subscription	$ 10,000

Based on these relative stand-alone selling prices, if CompStores orders 1,000 Tri-Box Systems at the normal wholesale price of $250 each, TrueTech records the following journal entry (ignoring any entry to record the reduction in inventory and corresponding cost of goods sold):

Accounts receivable	250,000	
Sales revenue		240,000
Deferred revenue		10,000

TrueTech would convert the $10,000 of deferred revenue to revenue (debit deferred revenue; credit service revenue) over the one-year term of the Tri-Net subscription.

Additional Consideration

Allocating Variable Consideration. What if a contract that has variable consideration includes multiple performance obligations? Typically the seller would include the variable consideration in the transaction price that is allocated to each of those performance obligations according to their relative stand-alone selling prices. Also, changes in estimated variable consideration are allocated to performance obligations on the same basis. However, if the variable consideration relates only to one performance obligation, it is allocated to only that performance obligation.

Special Issues for Step 5: Recognize Revenue When (Or As) Each Performance Obligation Is Satisfied

● LO6–7

Previously, we discussed recognizing revenue at a point in time and over a period of time. Now let's look at a few commonplace arrangements that occur in practice that make it more difficult to determine when revenue should be recognized. In particular, we discuss licenses, franchises, bill-and-hold sales, consignment arrangements, and gift cards.

Licenses allow the customer to access the seller's intellectual property.

LICENSES Customers sometimes pay a licensing fee to access a company's intellectual property ("IP"). **Licenses** are common in the software, technology, media, and entertainment (including motion pictures and music) industries. The accounting question is when to recognize revenue for these arrangements, which the FASB recently addressed in *ASU 2016-10.*[13]

Licenses of functional IP transfer a *right of use,* so sellers typically recognize revenue at a point in time.

Functional intellectual property: Some licenses transfer a *right of use* to IP that has *significant stand-alone functionality,* meaning that it can perform a function or a task, or be played or aired over various types of media. The benefit the customer receives from the license isn't affected by the seller's ongoing activity. Examples of this *functional IP* include software, drug formulas, and media content like books, music, and movies. For licenses of functional IP, sellers typically recognize revenue at the *point in time* that the customer can start using the IP.

For example, once you download a Rihanna hit from **Apple**'s iTunes, you can enjoy listening to that song as often as you like, regardless of future actions by Rihanna or iTunes. You probably didn't realize it, but you had just purchased a license to use IP (and that license came with some restrictions, like not being able to copy and sell the music or broadcast it publicly). Apple recognized revenue at the point in time that the download occurred, because that is when you could start listening to the downloaded music.

Licenses of symbolic IP transfer a *right of access,* so sellers recognize revenue over time.

Symbolic intellectual property: Other licenses provide the customer with the *right of access* to the seller's IP with the understanding that the IP does not have significant stand-alone functionality. Rather, the seller will undertake ongoing activities during the license period that benefit the customer. The seller might make changes to the IP over the course of the license, or could perform marketing or other activities that affect the value of the license to the customer. Examples of this *symbolic IP* include trademarks, logos, brand names, and franchise rights. For licenses of symbolic IP, sellers recognize revenue *over time,* because that is when they satisfy their performance obligation.

For example, suppose the **NBA** sells a five-year license that allows **Adidas** to manufacture shirts and hats with NBA team logos. Adidas agrees to the license arrangement with the understanding that the NBA will continue to play games and promote the league during the license period. These activities affect the value of the shirts and hats and thus the license to Adidas. Therefore, the NBA satisfies its performance obligation under the license agreement through its ongoing activities and will recognize revenue over the *period of time* for which access is provided.

Additional Consideration

This distinction between functional and symbolic intellectual property can have very material consequences. For example, when **Microsoft** early-adopted the revenue recognition ASU on July 1, 2017, it determined that Windows 10 qualified as functional intellectual property. Therefore, Microsoft had to change to recognizing revenue upon delivery, rather than deferring revenue at the time of delivery and recognizing revenue over the life of the related device, ranging from 2 to 4 years, as it had been doing. Microsoft reported that the impact of adopting the new standard would be to increase its revenue by $6.6 billion in 2017 and $5.8 billion in 2016.

[13]FASB ASC 606-10-55-59-62: Revenue from Contracts with Customers—Overall— Implementation Guidance and Illustrations—Determining the Nature of the Entity's Promise.

Sometimes functional IP also requires revenue recognition over time. This happens when the seller is expected to change the functionality over the license period and the customer is required to use the updated version. For example, that's the case with some software products, like virus protection. In that case, even though the license involves functional IP, we view the license as transferring a right of access, so revenue must be recognized over the license period.

Finally, sometimes a license isn't considered to be a separate performance obligation because it's not *distinct* from other goods or services provided in the same transaction. For example, an online service might grant a license to customers to access content at a website. In that case, the license isn't distinct from the website content, because the purpose of the license is to access the content. As a result, the website access and license would be treated as a single performance obligation, and revenue would be recognized over time as customers are provided access to the website.

International Financial Reporting Standards

The IASB's licensing guidance doesn't rely on the functional/symbolic classifications, so it differs from U.S. GAAP by not requiring revenue recognition over time for all symbolic intellectual property. Instead, the IASB requires the seller to recognize revenue over time only if the seller's ongoing activities affect the benefits the customer obtains from the IP. This distinction usually doesn't matter, but there are exceptions. For example, given that the Brooklyn Dodgers played their last baseball game in 1957, it is unlikely that their ongoing activities affect the benefit of licensing the Brooklyn Dodgers logo. Therefore, under IFRS, the seller would recognize revenue for licensing that logo at the point in time the customer can use the logo. However, because U.S. GAAP would focus on the logo being symbolic IP, U.S. GAAP would require the seller to recognize revenue over the license period.

● LO6–10

Additional Consideration

Variable Consideration and Licenses. Previously you learned about sellers being able to recognize revenue associated with variable consideration. There's an exception if variable consideration is based on sales or usage of a license. Those amounts (often called "royalties") are only included in the transaction price when the sales or usage has actually occurred, such that they are known rather than needing to be estimated.

FRANCHISES Many retail outlets for fast food, restaurants, hotels, and auto rental agencies are operated as franchises. In franchise arrangements, the **franchisor**, such as **Subway**, grants to the **franchisee**, often an individual, a right to sell the franchisor's products and use its name for a specified period of time. The franchisor also typically provides initial start-up services (such as identifying locations, remodeling or constructing facilities, selling equipment, and providing training to the franchisee) as well as providing ongoing products and services (such as franchise-branded products and advertising and administrative services). So, a franchise involves a *license* to use the franchisor's intellectual property, but also involves *initial sales* of products and services as well as *ongoing sales* of products and services. The franchisor must evaluate each part of the franchise arrangement to identify the performance obligations. Illustration 6–21 gives an example.

In a *franchise* arrangement, a franchisor grants to the franchisee the right to sell the franchisor's products and use its name.

BILL-AND-HOLD ARRANGEMENTS A **bill-and-hold** arrangement exists when a customer purchases goods but requests that the seller retain physical possession of the goods until a later date. For example, a customer might buy equipment and ask the seller to store the equipment until an installation site has been prepared.

A *bill-and-hold* arrangement occurs when a customer purchases goods but requests that shipment occur at a later date.

Illustration 6–21
Franchise Arrangements

Assume that TrueTech starts selling TechStop franchises. TrueTech charges franchisees an initial fee in exchange for (a) the exclusive right to operate the only TechStop in a particular area for a five-year period, (b) the equipment necessary to distribute and repair TrueTech products, and (c) training services to be provided over a two-year period. Similar equipment and training can be purchased elsewhere. What are the performance obligations in this arrangement, and when would TrueTech recognize revenue for each of them?

1. The exclusive five-year right to operate the only TechStop in a particular area is distinct because it can be used with other goods or services (furnishings, equipment, products) that the customer could obtain elsewhere.
2. The equipment is distinct because similar equipment is sold separately.
3. The training is distinct because similar training could be acquired elsewhere.

TrueTech would allocate the initial franchise fee to three separate performance obligations based on their relative stand-alone prices: (1) the right to operate a TechStop, (2) equipment, and (3) training. TrueTech would recognize revenue for the right to operate a TechStop over the five-year license period because the TechStop name qualifies as symbolic IP (TrueTech's ongoing activities over the license period affect the value of the right to run a TechStop). TrueTech would recognize revenue for the equipment at the time the equipment is delivered to the franchisee. TrueTech would recognize revenue for the training over the two-year period that the training is provided, just as it would treat revenue from any ongoing services provided to franchisees.

For bill-and-hold arrangements, the key issue is that the customer doesn't have physical possession of the asset until the seller has delivered it. Remember, physical possession is one of the indicators that control may have been transferred as listed in Illustration 6–3. Bill-and-hold arrangements might arise normally in the course of business, but they also have been abused by some companies in the past. Managers at companies like **Sunbeam**, **NutraCea**, and **Nortel Networks** are alleged to have overstated revenue by falsely claiming that unsold inventory had been sold under a bill-and-hold arrangement.

Revenue recognition usually occurs at delivery for a *bill-and-hold* arrangement.

The physical possession indicator normally overshadows other control indicators in a bill-and-hold arrangement, so sellers usually conclude that control has not been transferred and revenue should not be recognized until actual delivery to the customer occurs. Consistent with SEC guidance, sellers can recognize revenue prior to delivery only if (a) they conclude that the customer controls the product, (b) there is a good reason for the bill-and-hold arrangement, and (c) the product is specifically identified as belonging to the customer and is ready for shipment.[14]

Additional **Consideration**

Vaccine Stockpiles. Drug manufacturers supply vaccines that the U.S. Federal Government holds in reserve in case an epidemic occurs. These could be viewed as bill-and-hold arrangements because delivery to the end user (the person who eventually is vaccinated) is delayed, perhaps for years. What if drug companies stopped manufacturing vaccines because they couldn't recognize revenue for such a long time? To avoid this dangerous possibility, the SEC clarified in 2017 that control of specified vaccines passes to a customer when the vaccine is transferred to a stockpile, allowing revenue recognition at that point in time.

[14]FASB ASC 605–10–S99: Revenue Recognition—Overall—SEC Materials (originally "Revenue Recognition in Financial Statements," *Staff Accounting Bulletin No. 101* (Washington, D.C.: SEC, 1999) and *Staff Accounting Bulletin No. 104* (Washington, D.C.: SEC, 2003)).

Ethical Dilemma

The Precision Parts Corporation manufactures automobile parts. The company has reported a profit every year since the company's inception in 1980. Management prides itself on this accomplishment and believes one important contributing factor is the company's incentive plan that rewards top management a bonus equal to a percentage of operating income if the operating income goal for the year is achieved. However, 2021 has been a tough year, and prospects for attaining the income goal for the year are bleak.

Tony Smith, the company's chief financial officer, has determined a way to increase December sales by an amount sufficient to boost operating income over the goal for the year and secure bonuses for all top management. A reputable customer ordered $120,000 of normally stocked parts to be shipped on January 5, 2022. Tony told the rest of top management "I know we can get that order ready by December 31. We can then just leave the order on the loading dock until shipment. I see nothing wrong with recognizing the sale in 2021, since the parts will have been manufactured and we do have a firm order from a reputable customer." The company's normal procedure is to ship goods f.o.b. destination and to recognize sales revenue when the customer receives the parts.

CONSIGNMENT ARRANGEMENTS Sometimes a company arranges for another company to sell its product under **consignment**. In these arrangements, the "consignor" physically transfers the goods to the other company (the consignee), but the consignor retains legal title. If a buyer is found, the consignee remits the selling price (less commission and approved expenses) to the consignor. If the consignee can't find a buyer within an agreed-upon time, the consignee returns the goods to the consignor.

Revenue recognition occurs upon sale to an end customer in a *consignment* arrangement.

When does control transfer from the consignor, allowing the consignor to recognize revenue? Referring to the indicators listed in Illustration 6–3, the consignor still has title and retains many of the risks and rewards of ownership for goods it has placed on consignment. Therefore, it's likely that the consignor would be judged to retain control after transfer to the consignee and would postpone recognizing revenue until sale to an end customer occurs. Illustration 6–22 provides an example of a consignment arrangement by **Boston Scientific Corporation** from its 10-K dated February 20, 2018. We discuss accounting for consignment arrangements further in Chapter 8.

Note A — Significant Accounting Policies: Revenue Recognition (in part)

Revenue is recognized upon passage of title and risk of loss to customers, unless we are required to provide additional services and provided we can form an estimate for sales returns. We recognize revenue from consignment arrangements based on product usage, or implant, which indicates that the sale is complete.

Source: Boston Scientific Corporation

Illustration 6–22

Disclosure of Revenue Recognition Policy for Consignment Arrangements—Boston Scientific Corporation

Real World Financials

GIFT CARDS Let's assume you received an iTunes **gift card** that allows you to download songs or audiobooks later. When your friend bought that gift card, **Apple** recorded a deferred revenue liability in anticipation of recording revenue when you used your gift card to get songs. But, what if you lose the card or fail to redeem it for some other reason? Sellers like Apple, **Target**, **Amazon**, and others will recognize revenue at the point when they have concluded based on past experience that there is only a "remote likelihood" that customers will use the cards.[15] We discuss accounting for gift card liabilities further in Chapter 13.

Sales of gift cards are recognized as deferred revenue.

[15]This is sometimes referred to as "breakage" of the gift card.

Disclosures

LO6–8

INCOME STATEMENT DISCLOSURE Of course, a seller reports revenue in its income statement. In addition, that seller is required to either include in its income statement or disclosure notes any bad debt expense and any interest revenue or interest expense associated with significant financing components of long-term contracts.

The seller recognizes *contract liabilities, contract assets,* and *accounts receivable* on separate lines of its balance sheet.

BALANCE SHEET DISCLOSURE A seller reports accounts receivable, "contract liabilities," and "contract assets" on separate lines of its balance sheet. We discuss each in turn.

The seller has a contract *liability* if it received payment prior to satisfying a performance obligation.

A seller has an *account receivable* if it has an unconditional right to receive payment after satisfying a performance obligation.

A seller has a *contract asset* if it has a conditional right to receive payment after satisfying a performance obligation.

If a customer pays the seller before the seller has satisfied a performance obligation, we saw earlier that the seller records deferred revenue. For example, we recorded deferred revenue in Illustration 6–6 when TrueTech received payment for Tri-Net subscriptions prior to providing that service. A **contract liability** is a label we give to deferred revenue (or unearned revenue) accounts.

On the other hand, if the seller satisfies a performance obligation *before* the customer has paid for it, the seller records either a contract asset or accounts receivable. The seller recognizes an **account receivable** if the seller has an unconditional right to receive payment, which is the case if only the passage of time is required before the payment is due. In other words, the seller has satisfied all of its performance obligations and is just waiting to be paid.

If, instead, the seller satisfies a performance obligation but payment depends on something other than the passage of time, the seller recognizes a **contract asset**. For example, construction companies sometimes complete a significant amount of work prior to when the construction contract indicates they can bill their clients for progress payments. As we will see in Part C of this chapter, a construction company in that situation reports a contract asset called "construction-in-progress in excess of billings" to reflect that the company will be able to bill its client in the future for the work that has been completed.

DISCLOSURE NOTES Several important aspects of revenue recognition must be disclosed in the notes to the financial statements. For example, sellers must separate their revenue into categories that help investors understand the nature, amount, timing, and uncertainty of revenue and cash flows. Categories might include product lines, geographic regions, types of customers, or types of contracts. Sellers also must disclose amounts included in revenue that were previously recognized as deferred revenue or that resulted from changes in transaction prices.

Companies provide detailed disclosures about revenues.

Sellers also must describe their outstanding performance obligations, discuss how performance obligations typically are satisfied, and describe important contractual provisions like payment terms and policies for refunds, returns, and warranties. They also must disclose any significant judgments used to estimate transaction prices, to allocate transaction prices to performance obligations, and to determine when performance obligations have been satisfied.

Sellers also must explain significant changes in contract assets and contract liabilities that occurred during the period.

The objective of these disclosures is to help users of financial statements understand the revenue and cash flows arising from contracts with customers. Of course, the downside of these disclosures is that sellers also are providing information to competitors, suppliers, and customers.

Illustration 6–23 provides a summary of both Parts A and B of this chapter to provide a comprehensive review of revenue recognition.

Illustration 6–23 Summary of Fundamental and Special Issues Related to Recognizing Revenue

Revenue Recognition	Fundamental Issues (Part A)	Special Issues (Part B)
Step 1 Identify the contract	A contract establishes the legal rights and obligations of the seller and customer with respect to one or more performance obligations.	A contract exists if it (a) has commercial substance, (b) has been approved by both the seller and the customer, (c) specifies the seller's and customer's rights and obligations, (d) specifies payment terms, and (e) is probable that the seller will collect the amounts it is entitled to receive. A contract does ***not*** exist if (a) neither the seller nor the customer has performed any obligations under the contract, and (b) both the seller and the customer can terminate the contract without penalty.
Step 2 Identify the performance obligation(s)	A performance obligation is a promise to transfer a good or service that is distinct, which is the case if the good or service is both (a) capable of being distinct and (b) separately identifiable.	The following ***do not*** qualify as performance obligations: • Quality assurance warranties • Customer prepayments The following ***do*** qualify as performance obligations: • Extended warranties • Customer options for additional goods and services that provide a material right
Step 3 Determine the transaction price	The transaction price is the amount the seller is entitled to receive from the customer.	The seller adjusts the transaction price for: • Variable consideration (estimated as either the expected value or the most likely amount). Constraint: Variable consideration is recognized only to the extent it is probable that a significant revenue reversal will not occur in the future. • Whether the seller is acting as a principal or agent • A significant financing component • Any payments by the seller to the customer
Step 4 Allocate the transaction price	The seller allocates the transaction price to performance obligations based on relative stand-alone selling prices of the goods or services in each performance obligation.	Various approaches are available to estimate stand-alone selling prices: • Adjusted market assessment approach • Expected cost plus margin approach • Residual approach
Step 5 Recognize revenue when (or as) each performance obligation is satisfied	The seller recognizes revenue **at a single point in time** when control passes to the customer, which is more likely if the customer has: • Obligation to pay seller. • Legal title to the asset. • Possession of the asset. • Assumed the risks and rewards of ownership. • Accepted the asset. The seller recognizes revenue **over a period of time** if: • Customer consumes benefit as work performed, • Customer controls asset as it's created, or • Seller is creating an asset that has no alternative use to the seller and the seller has right to receive payment for work completed.	The seller must determine the timing of revenue recognition for: • Licenses (if functional intellectual property, usually recognize revenue at beginning of license; if symbolic intellectual property, recognize revenue over license period). • Franchises (initial fees recognized when goods and services are transferred; continuing fees recognized over time). • Bill-and-hold arrangements (typically do not transfer control, so recognize upon delivery of goods to customer). • Consignment arrangements (do not transfer control, so recognize after sale to end customer occurs). • Gift cards (initially deferred and then recognized as redeemed or expire).

PART C

Accounting for Long-Term Contracts

LO6–9

A survey of reporting practices of 500 large public companies indicates that approximately one in every eight companies participates in long-term contracts.[16] These are not only construction companies. Illustration 6–24 lists just a sampling of companies that use long-term contracts, many of which you might recognize.

Illustration 6–24
Companies Engaged in Long-Term Contracts

Company	Type of Industry or Product
Oracle Corp.	Computer software, license and consulting fees
Lockheed Martin Corporation	Aircraft, missiles, and spacecraft
HP	Information technology
Northrop Grumman Corporation	Shipbuilding
Nortel Networks Inc.	Networking solutions and services to support the Internet
SBA Communications Corp.	Telecommunications
Layne Christensen Company	Water supply services and geotechnical construction
KB Home	Commercial and residential construction
Raytheon Company	Defense electronics
Halliburton	Construction, energy services

The five-step process for recognizing revenue described in Parts A and B of this chapter also applies to long-term contracts. However, steps 2 and 5 merit special attention.

Step 2, "Identify the performance obligation(s) in the contract," is important because long-term contracts typically include many products and services that could be viewed as separate performance obligations. For example, constructing a building requires the builder to deliver many different materials and other products (concrete, lumber, furnace, bathroom fixtures, carpeting) and to provide many different services (surveying, excavating, construction, fixture installation, painting, landscaping). These products and services are capable of being distinct, but they are not separately identifiable, because the seller's role is to combine those products and services for purposes of delivering a completed building to the customer. Therefore, it's the bundle of products and services that comprise a single performance obligation. Most long-term contracts should be viewed as including a single performance obligation.

Step 5, "Recognize revenue when (or as) each performance obligation is satisfied," is important because there can be a considerable difference for long-term contracts between recognizing revenue over time and recognizing revenue only when the contract has been completed. Imagine a builder who spends years constructing a skyscraper but only gets to recognize revenue at the end of the contract. Such delayed revenue recognition would do a poor job of informing financial statement users about the builder's economic activity. Fortunately, most long-term contracts qualify for revenue recognition over time. Often the customer owns the seller's work in process, such that the seller is creating an asset that the customer controls as it is completed. Also, often the seller is creating an asset that is customized for the customer, so the seller has no other use for the asset and has the right to be paid for progress even if the customer cancels the contract. In either of those cases, the seller recognizes revenue over time.

Long-term contracts are complex, and specialized accounting approaches have been developed to handle that complexity. For many years, long-term contracts that qualified for revenue recognition over time were accounted for using an approach called the *percentage-of-completion method,* which recognized revenue in each year of the contract according to the progress toward completion that occurred during that year. Long-term contracts that didn't qualify for revenue recognition over time were accounted for using an approach called the *completed contract method,* because all revenue was recognized at a single point in time—upon completion of the contract.

[16] *U.S. GAAP Financial Statements–Best Practices in Presentation and Disclosure–2013* (New York: AICPA, 2013).

ASU No. 2014–09 removed the terms "percentage-of-completion method" and "completed contract method" from the Accounting Standards Codification, and changed the criteria that determine whether revenue should be recognized over a period of time or at a point in time. However, the journal entries necessary to account for revenue over time are the same as those that were used under the percentage-of-completion method, and those used to account for revenue at a point in time are the same as those used under the completed contract method. We demonstrate those journal entries next.

Accounting for a Profitable Long-Term Contract

FINANCIAL Reporting Case

Q3, p. 277

Much of the accounting for long-term contracts is the same regardless of whether we recognize revenue over the contract period or upon completion of the contract. So, we start by discussing the similarities between the two approaches, and then the differences. You'll see that we recognize the same total amounts of revenue and profit over the life of the contract either way. Only the timing of recognition differs.

Illustration 6–25 provides information for a typical long-term construction contract that we'll use to consider accounting for long-term contracts.

Illustration 6–25

Example of Long-Term Construction Contract

At the beginning of 2021, the Harding Construction Company received a contract to build an office building for $5 million. Harding will construct the building according to specifications provided by the buyer, and the project is estimated to take three years to complete. According to the contract, Harding will bill the buyer in installments over the construction period according to a prearranged schedule. Information related to the contract is as follows:

	2021	2022	2023
Construction costs incurred during the year	$ 1,500,000	$1,000,000	$ 1,500,000
Construction costs incurred in prior years	–0–	1,500,000	2,500,000
Cumulative actual construction costs	1,500,000	2,500,000	4,100,000
Estimated costs to complete at end of year	2,250,000	1,500,000	–0–
Total estimated and actual construction costs	**$3,750,000**	**$4,000,000**	**$4,100,000**
Billings made during the year	$ 1,200,000	$ 2,000,000	$ 1,800,000
Cash collections during year	1,000,000	1,400,000	2,600,000

Construction costs include the labor, materials, and overhead costs directly related to the construction of the building. Notice how the total of estimated and actual construction costs changes from period to period. Cost revisions are typical in long-term contracts because costs are estimated over long periods of time.

ACCOUNTING FOR THE COST OF CONSTRUCTION AND ACCOUNTS RECEIVABLE Summary journal entries are shown in Illustration 6–25A for actual construction costs, billings, and cash receipts. These journal entries are not affected by the timing of revenue recognition.

The first journal entry shows Harding incurring various costs during the construction process and recording them in an asset account called **construction in progress** (or "CIP" for short). This asset account is equivalent to work-in-process inventory in a manufacturing company. This is logical because the construction project is essentially an inventory item in process for the contractor.

***Construction in progress* (CIP) is the contractor's work-in-process inventory.**

The second journal entry occurs when Harding bills its customer according to whatever schedule the contract permits. Notice that periodic billings are credited to **billings on construction contract**. This account is a contra account to the CIP asset. At the end of each period, the balances in these two accounts are compared. If the net amount is a debit, it is reported in the balance sheet as a contract asset, and might be named "Construction in progress in excess of billings" or something similar. Conversely, if the net amount is a credit, it is reported as a contract liability, and might be named "Billings in excess of construction in progress" or something similar.[17]

Accounting for costs, billings, and cash receipts does not depend on the timing of revenue recognition.

[17]If the company is engaged in more than one long-term contract, all contracts for which construction in progress exceeds billings are reported in the balance sheet as contract assets, and all contracts for which billings exceed construction in progress are reported as contract liabilities.

Illustration 6–25A Journal Entries—Costs, Billings, and Cash Collections

	2021		2022		2023	
Construction in progress (CIP)............	1,500,000		1,000,000		1,600,000	
Cash, materials, etc.		1,500,000		1,000,000		1,600,000
To record construction costs.						
Accounts receivable....................	1,200,000		2,000,000		1,800,000	
Billings on construction contract		1,200,000		2,000,000		1,800,000
To record progress billings.						
Cash..................................	1,000,000		1,400,000		2,600,000	
Accounts receivable..................		1,000,000		1,400,000		2,600,000
To record cash collections.						

To understand why we use the billings on construction contract account (or "billings" for short), consider a key difference between accounting for a long-term contract and accounting for a more normal sale of inventory to a customer. In the normal case, the seller debits an account receivable and credits revenue, and also debits cost of goods sold and credits inventory. Thus, the seller gives up its physical asset (inventory) and recognizes cost of goods sold at the same time it gets a financial asset (an account receivable) and recognizes revenue. First the physical asset is in the balance sheet, and then the financial asset, but the two are not in the balance sheet at the same time.

The *billings on construction contract* account prevents "double counting" assets by reducing CIP whenever an account receivable is recognized.

Now consider our Harding Construction example. Harding is creating a physical asset (CIP) in the same periods it recognizes a financial asset (first recognizing accounts receivable when the customer is billed and then recognizing cash when the receivable is collected). Having both the physical asset and the financial asset in the balance sheet at the same time constitutes double counting the same arrangement. The billings account solves this problem. Whenever an account receivable is recognized, the other side of the journal entry increases the billings account, which is contra to (and thus reduces) CIP. As a result, the financial asset (accounts receivable) increases and the physical asset (the net amount of CIP and billings) decreases, and no double counting occurs.

Remember, we recognize accounts receivable when the seller has an unconditional right to receive payment, which is the case if only the passage of time is required before the payment is due, and we recognize a contract asset when the seller's right to receive payment depends on something other than the passage of time. Consistent with those definitions, Harding will report an account receivable for amounts it has billed the client and not yet been paid, and will report a contract asset (computed as CIP – Billings) for the remaining amount of work completed, for which it eventually will be paid once it is able to bill the client.

REVENUE RECOGNITION—GENERAL APPROACH Now let's consider revenue recognition. The top portion of Illustration 6–25B shows the single journal entry to recognize

Illustration 6–25B Journal Entries—Revenue Recognition

	2021		2022		2023	
Recognizing Revenue upon Completion						
Construction in progress (CIP).............					900,000	
Cost of construction					4,100,000	
Revenue from long-term contracts........						5,000,000
To record gross profit.						
Recognizing Revenue Over Time According to Percentage of Completion						
Construction in progress (CIP).............	500,000		125,000		275,000	
Cost of construction	1,500,000		1,000,000		1,600,000	
Revenue from long-term contracts........		2,000,000		1,125,000		1,875,000
To record gross profit.						

revenue, cost of construction (think of this as cost of goods sold), and gross profit when recognizing revenue upon completion of the contract, while the bottom portion shows the journal entries that achieve this when recognizing revenue over the term of the contract. At this point focus on the structure of the journal entries (what is debited and credited). We'll discuss how to calculate the specific amounts later in the chapter.

It's important to understand two key aspects of Illustration 6–25B. First, the same amounts of revenue (the $5 million contract price), cost, and gross profit are recognized whether it's over the term of the contract or only upon completion. The only difference is timing. To check this, we can add together all of the revenue recognized for both methods over the three years, as follows:

Revenue Recognition

	Over Time	Upon Completion
Revenue recognized:		
2021	$2,000,000	$ –0–
2022	1,125,000	–0–
2023	1,875,000	5,000,000
Total revenue	$5,000,000	$5,000,000

The same total amount of revenue is recognized whether it's over the term of the contract or only upon completion, but the timing of recognition differs.

Second, notice that, regardless of the timing of revenue recognition, we add gross profit (the difference between revenue and cost) to the CIP asset. That seems odd—why add profit to what is essentially an inventory account? The key here is that, when Harding recognizes gross profit, Harding is acting like it has sold some portion of the asset to the customer, but Harding keeps the asset in Harding's own balance sheet (in the CIP account) until delivery to the customer. Putting recognized gross profit into the CIP account just updates that account to reflect the total value (cost + gross profit = sales price) of the customer's asset. But don't forget that the billings account is contra to the CIP account. Over the life of the construction project, Harding will bill the customer for the entire sales price of the asset. Therefore, at the end of the contract, the CIP account (containing total cost and gross profit) and the billings account (containing all amounts billed to the customer) will have equal balances that exactly offset to create a net value of zero.

CIP includes profits and losses on the contract that have been recognized to date.

Now let's discuss the timing of revenue recognition in more detail.

REVENUE RECOGNITION UPON THE COMPLETION OF THE CONTRACT If a contract doesn't qualify for revenue recognition over time, revenue is recognized at the point in time that control transfers from the seller to the customer, which typically occurs when the contract has been completed. At that time, the seller views itself as selling the asset and recognizes revenues and expenses associated with the sale. As shown in Illustration 6–25B and in the T-accounts below, completion occurs in 2023 for our Harding example. Prior to then, CIP includes only costs, showing a cumulative balance of $1,500,000 and $2,500,000 at the end of 2021 and 2022, respectively, and totaling $4,100,000 ($1,500,000 + $1,000,000 + $1,600,000) when the project is completed in 2023. Upon completion, Harding recognizes revenue of $5,000,000 and cost of construction (similar to cost of goods sold) of $4,100,000, because the asset is viewed as "sold" on that date. Harding includes the resulting **$900,000** gross profit in CIP, increasing its balance to the $5,000,000 total cost + gross profit for the project.

If a contract doesn't qualify for revenue recognition over time, revenue is recognized when the project is completed.

Recognizing Revenue upon Completion

	Construction in Progress (CIP)		Billings on Construction Contract		
2021 construction costs	1,500,000			1,200,000	2021 billings
End balance, 2021	1,500,000			1,200,000	End balance, 2021
2022 construction costs	1,000,000			2,000,000	2022 billings
End balance, 2022	2,500,000			3,200,000	End balance, 2022
2023 construction costs	1,600,000			1,800,000	2023 billings
Total gross profit	**900,000**				
Balance, before closing	5,000,000			5,000,000	Balance, before closing

If revenue is recognized upon completion of the contract, CIP is updated to include gross profit at that point in time.

If a contract qualifies for revenue recognition over time, revenue is recognized over time as the project is completed.

RECOGNIZING REVENUE OVER TIME ACCORDING TO PERCENTAGE OF COMPLETION If a contract qualifies for revenue recognition over time, revenue is recognized based on progress towards completion. How should progress to date be estimated?

As discussed in Part A of this chapter, one approach to estimating progress towards completion is to use output-based measures, like number of units produced or delivered, achievement of milestones, and surveys or appraisals of performance completed to date. A shortcoming of output measures is that they may provide a distorted view of actual progress to date.[18] For example, an output measure for highway construction might be finished miles of road, but that measure could be deceptive if not all miles of road require the same effort. A highway contract for the state of Arizona would likely pay the contractor more for miles of road blasted through the mountains than for miles paved across flat desert. Another shortcoming of some output measures is that the information they require, such as surveys or appraisals, might be costly to obtain.

Another way to estimate progress is to base it on the seller's *input,* measured as the proportion of effort expended thus far relative to the total effort expected to satisfy the performance obligation. Measures of effort include costs incurred, labor hours expended, machine hours used, or time lapsed. The most common approach to estimating progress toward completion is to use a "cost-to-cost ratio" that compares total cost incurred to date to the total estimated cost to complete the project.[19] When using that approach, sellers have to make sure to exclude from the ratio costs that don't reflect progress toward completion. For example, inefficiencies in production could lead to wasted materials, labor, or other resources. Those costs must be expensed as incurred, but not included in the cost-to-cost ratio.

Regardless of the specific approach used to estimate progress towards completion, we determine the amount of revenue recognized in each period using the following logic:

$$\begin{array}{c}\text{Revenue}\\\text{recognized this}\\\text{period}\end{array} = \underbrace{\left(\begin{array}{c}\text{Total}\\\text{estimated}\\\text{revenue}\end{array} \times \begin{array}{c}\text{Percentage}\\\text{completed}\\\text{to date}\end{array}\right)}_{\text{Cumulative revenue to be recognized to date}} - \begin{array}{c}\text{Revenue}\\\text{recognized in}\\\text{prior periods}\end{array}$$

When recognizing revenue over the term of the contract, CIP is updated each period to include gross profit.

Illustration 6–25C shows the calculation of revenue for each of the years for our Harding Construction Company example, with progress to date estimated using the cost-to-cost ratio. Notice that this approach automatically includes changes in estimated cost to complete the job, and therefore in estimated percentage completion, by first calculating the cumulative amount of revenue to be recognized to date and then subtracting revenue recognized in prior periods to determine revenue recognized in the current period. Refer to the following T-accounts to see that the gross profit recognized in each period is added to the CIP account.

Recognizing Revenue Over the Term of the Contract

	Construction in Progress (CIP)		Billings on Construction Contract		
2021 construction costs	1,500,000			1,200,000	2021 billings
2021 gross profit	**500,000**				
End balance, 2021	2,000,000			1,200,000	End balance, 2021
2022 construction costs	1,000,000			2,000,000	2022 billings
2022 gross profit	**125,000**				
End balance, 2022	3,125,000			3,200,000	End balance, 2022
2023 construction costs	1,600,000			1,800,000	2023 billings
2023 gross profit	**275,000**				
Balance, before closing	5,000,000			5,000,000	Balance, before closing

[18] Number of units produced or delivered is not an appropriate basis for measuring progress toward completion if these measures are distorted by the seller having material amounts of work-in-progress or finished-goods inventory at the end of the period.

[19] R. K. Larson and K. L. Brown, "Where Are We with Long-Term Contract Accounting?" *Accounting Horizons,* September 2004, pp. 207–219.

Illustration 6–25C

Allocation of Revenue to Each Period

	2021	2022	2023
Construction costs:			
Construction costs incurred during the year	$1,500,000	$1,000,000	$1,600,000
Construction costs incurred in prior years	–0–	1,500,000	2,500,000
Actual costs to date	1,500,000	2,500,000	4,100,000
Estimated remaining costs to complete	2,250,000	1,500,000	–0–
Total cost (estimated + actual)	$3,750,000	$4,000,000	$4,100,000
Contract price	$5,000,000	$5,000,000	$5,000,000
Multiplied by:	×	×	×
Percentage of completion Actual costs to date / Total cost (est. + actual)	($1,500,000 / $3,750,000 = 40%)	($2,500,000 / $4,000,000 = 62.5%)	($4,100,000 / $4,100,000 = 100%)
Equals:			
Cumulative revenue to be recognized to date	$2,000,000	$3,125,000	$5,000,000
Less:			
Revenue recognized in prior periods	–0–	(2,000,000)	(3,125,000)
Equals:			
Revenue recognized in the current period	**$2,000,000**	**$1,125,000**	**$1,875,000**
Journal entries to recognize revenue:			
Construction in progress (CIP)	500,000	125,000	275,000
Cost of construction	1,500,000	1,000,000	1,600,000
Revenue from long-term contracts	2,000,000	1,125,000	1,875,000

If a contract qualifies for revenue recognition over time, the income statement for each year will report the appropriate revenue and cost of construction amounts. For example, in 2021, the income statement will report revenue of $2,000,000 (40% of the $5,000,000 contract price) less $1,500,000 cost of construction, yielding gross profit of **$500,000**.[20] The table in Illustration 6–25D shows the revenue, cost of construction, and gross profit recognized in each of the three years of our example.

The income statement includes revenue, cost of construction, and gross profit.

Illustration 6–25D

Recognition of Revenue and Cost of Construction in Each Period

2021		
Revenue recognized ($5,000,000 × 40%)		$2,000,000
Cost of construction		(1,500,000)
Gross profit		$ 500,000
2022		
Revenue recognized to date ($5,000,000 × 62.5%)	$3,125,000	
Less: Revenue recognized in 2021	(2,000,000)	
Revenue recognized		$1,125,000
Cost of construction		(1,000,000)
Gross profit		$ 125,000
2023		
Revenue recognized to date ($5,000,000 × 100%)	$5,000,000	
Less: Revenue recognized in 2021 and 2022	(3,125,000)	
Revenue recognized		$1,875,000
Cost of construction		(1,600,000)
Gross profit		$ 275,000

COMPLETION OF THE CONTRACT After the job is finished, the only task remaining is for Harding to officially transfer title to the finished asset to the customer. At that time, Harding will prepare a journal entry that removes the contract from its balance sheet

We record the same journal entry to close out the billings and CIP accounts regardless of whether revenue is recognized over time or upon completion.

[20]In most cases, cost of construction also equals the construction costs incurred during the period. Cost of construction does not equal the construction costs incurred during the year when a loss is projected on the entire project. This situation is illustrated later in the chapter.

by debiting billings and crediting CIP for the entire value of the contract. As shown in Illustration 6–25E, the same journal entry is recorded to close out the billings on construction contract and CIP accounts whether revenue is recognized over the term of the contract or at the completion of the contract.

Illustration 6–25E
Journal Entry to Close Billings and CIP Accounts Upon Contract Completion

	2021	2022	2023	
Billings on construction contract			5,000,000	
Construction in progress (CIP)				5,000,000
To close accounts.				

A Comparison of Revenue Recognized Over the Term of the Contract and at the Completion of Contract

INCOME RECOGNITION Illustration 6–25B shows the journal entries that would determine the amount of revenue, cost, and therefore gross profit that would appear in the income statement when we recognize revenue over the term of the contract and at the completion of contract. Comparing the gross profit patterns produced by each method of revenue recognition demonstrates the essential difference between them:

Timing of revenue recognition does not affect the total amount of profit or loss recognized.

	Revenue Recognition	
	Over Time	**Upon Completion**
Gross profit recognized:		
2021	**$500,000**	$ –0–
2022	**125,000**	–0–
2023	**275,000**	900,000
Total gross profit	$ 900,000	$900,000

Whether revenue is recognized over time or upon completion does not affect the total gross profit of $900,000 recognized over the three-year contract, but the timing of gross profit recognition is affected. When the contract does not qualify for recognizing revenue over time, we defer all gross profit to 2022, when the project is completed. Obviously, recognizing revenue over the term of the contract provides a better measure of the company's economic activity and progress over the three-year term. As indicated previously, most long-term contracts qualify for revenue recognition over time.[21] Revenue is deferred until the completion of the contract only if the seller doesn't qualify for revenue recognition over time according to the criteria listed in Illustration 6–5.

BALANCE SHEET RECOGNITION The balance sheet presentation for the construction-related accounts for both methods is shown in Illustration 6–25F.

Illustration 6–25F
Balance Sheet Presentation

Balance Sheet (End of Year)	**2021**	**2022**
Projects for which Revenue Recognized Upon Completion:		
Current assets:		
Accounts receivable	$200,000	$800,000
Costs ($1,500,000) in excess of billings ($1,200,000)	300,000	
Current liabilities:		
Billings ($3,200,000) in excess of costs ($2,500,000)		$700,000
Projects for which Revenue Recognized Over Time:		
Current assets:		
Accounts receivable	$200,000	$800,000
Costs and profit ($2,000,000) in excess of billings ($1,200,000)	800,000	
Current liabilities:		
Billings ($3,200,000) in excess of costs and profit ($3,125,000)		$ 75,000

[21]For income tax purposes, revenue can be recognized at completion for home construction contracts and certain other real estate construction contracts. All other contracts must recognize revenue over time according to percentage of completion.

Billings on construction contracts are subtracted from CIP to determine balance sheet presentation.

In the balance sheet, the construction in progress (CIP) account is offset against the billings on construction contract account, with CIP > Billings viewed as a contract asset and labeled as "CIP in excess of Billings", and Billings > CIP viewed as a contract liability and labeled as "Billing in excess of CIP". Rather than referring to CIP companies sometimes refer to what the CIP account contains, as is done in Illustration 6–25F. When revenue is recognized over the term of the contract, CIP contains cost and gross profit; if revenue is recognized upon the completion of the contract, CIP typically contains only costs. Because a company may have some contracts that have a net asset position and others that have a net liability position, it is not unusual to see both contract assets and contract liabilities shown in a balance sheet at the same time.

CIP in excess of billings is treated as a contract asset rather than an accounts receivable because something other than the passage of time must occur for the company to be paid for that amount. Although Harding has incurred construction costs (and is recognizing gross profit over the term of the contract) for which it will be paid by the buyer, those amounts are not yet billable according to the construction contract. Once Harding has made progress sufficient to bill the customer it will debit accounts receivable and credit billings, which will increase the accounts receivable asset and reduce CIP in excess of billings (by increasing billings).

On the other hand, *Billings in excess of CIP* is treated as a contract liability. It reflects that Harding has billed its customer for more work than it actually has done. This is similar to the deferred revenue liability that is recorded when a customer pays for a product or service in advance. The advance is properly shown as a liability that represents the obligation to provide the good or service in the future.

Long-Term Contract Losses

The Harding Construction Company example above involves a situation in which an overall profit was anticipated at each stage of the contract. Unfortunately, losses sometimes occur on long-term contracts. As facts change, sellers must update their estimates and recognize losses if necessary to properly account for the amount of revenue that should have been recognized to date. How we treat losses in any one period depends on whether the contract is profitable overall.

PERIODIC LOSS OCCURS FOR PROFITABLE PROJECT When a project qualifies for revenue recognition over time, a loss sometimes must be recognized in at least one period along the way, even though the project as a whole is expected to be profitable. We determine the loss in precisely the same way we determine the profit in profitable years. For example, assume the same $5 million contract for Harding Construction Company described earlier in Illustration 6–25 but with the following cost information:

	2021	2022	2023
Construction costs incurred during the year	$1,500,000	$ 1,260,000	$1,840,000
Construction costs incurred in prior years	–0–	1,500,000	2,760,000
Cumulative construction costs	1,500,000	**2,760,000**	4,600,000
Estimated costs to complete at end of year	2,250,000	1,840,000	–0–
Total estimated and actual construction costs	$3,750,000	**$4,600,000**	$4,600,000

At the end of 2021, 40% of the project is complete ($1,500,000 ÷ $3,750,000). Revenue of $2,000,000 – cost of construction of $1,500,000 = gross profit of $500,000 is recognized in 2021, as previously determined.

At the end of 2022, though, the company now forecasts a total profit of $400,000 (equal to $5,000,000 – $4,600,000) on the project and, at that time, the project is estimated to be 60% complete (**$2,760,000** ÷ **$4,600,000**). Applying this percentage to the anticipated revenue of $5,000,000 results in revenue *to date* of $3,000,000. This implies that gross profit recognized to date should be $240,000 (equal to $3,000,000 – **$2,760,000**). But remember, gross profit of $500,000 was recognized the previous year.

We treat a situation like this as a *change in accounting estimate* because it results from a change in the estimation of costs to complete at the end of 2021. Total estimated costs to complete the project at the end of 2022 of $4,600,000 were much higher than the 2021 year-end estimate of $3,750,000. Recall from our discussion of changes in accounting estimates in Chapter 4 that we don't go back and restate the prior year's income statement. Instead, the 2022 income statement reports *a loss of $260,000* (computed as $500,000 – $240,000) so that the cumulative amount of gross profit recognized to date is $240,000. The loss consists of 2022 revenue of $1,000,000 (computed as $5,000,000 × 60% = $3,000,000 revenue to be recognized by end of 2022 less 2021 revenue of $2,000,000) less cost of construction of $1,260,000 (cost incurred in 2022). Just as gross profit gets debited to the CIP account, this gross loss gets credited to the CIP account, so that CIP includes the cumulative amount of gross profit or loss recognized to date. The following journal entry records the loss in 2022:

Recognized losses on long-term contracts reduce the CIP account.

Cost of construction	1,260,000	
Revenue from long-term contracts (below)		1,000,000
Construction in progress (CIP)		260,000

In 2023 the company recognizes $2,000,000 in revenue (equal to $5,000,000 less revenue of $3,000,000 recognized in 2021 and 2022) and $1,840,000 in cost of construction (cost incurred in 2023), yielding a gross profit of $160,000.

Revenue	$2,000,000
Less: Cost of construction	(1,840,000)
Gross profit	$ 160,000

Of course, if revenue is instead recognized upon the completion of the contract, rather than over time, no profit or loss is recorded in 2021 or 2022. Instead, revenue of $5,000,000, cost of construction of $4,600,000, and gross profit of $400,000 are recognized in 2023.

LOSS IS PROJECTED ON THE ENTIRE PROJECT If an overall loss is projected on the entire contract, the total loss must be recognized in the period in which that loss becomes evident, regardless of whether revenue is recognized over the term of the contract or only upon the completion of the contract. Again, consider the Harding Construction Company example but with the following cost information:

	2021	2022	2023
Construction costs incurred during the year	$1,500,000	$ 1,260,000	$2,440,000
Construction costs incurred in prior years	–0–	1,500,000	2,760,000
Cumulative construction costs	1,500,000	2,760,000	5,200,000
Estimated costs to complete at end of year	2,250,000	2,340,000	–0–
Total estimated and actual construction costs	$3,750,000	**$5,100,000**	$5,200,000

An estimated loss on a long-term contract is fully recognized in the first period the loss is anticipated, regardless of the whether revenue is recognized over time or upon completion.

At the end of 2022, revised costs indicate an estimated loss of $100,000 for the entire project (computed as contract revenue of $5,000,000 less estimated construction costs of **$5,100,000**). In this situation, the *total* anticipated loss must be recognized in 2022 regardless of whether revenue is recognized over the term of the contract or only upon the completion of the contract. If revenue is being recognized over the term of the contract, a gross profit of $500,000 was recognized in 2021, so a *$600,000 loss is recognized in 2022* to make the cumulative amount recognized to date total a $100,000 loss. Once again, this situation is treated as a change in accounting estimate, with no restatement of 2021 income. On the other hand, if revenue is being recognized only upon the completion of the contract, no gross profit is recognized in 2021, and the $100,000 loss for the project is recognized in 2022. This is accomplished by debiting a loss from long-term contracts and crediting CIP for $100,000.

Why recognize the estimated overall loss of $100,000 in 2022, rather than at the end of the contract? If the loss is not recognized in 2022, CIP would be valued at an amount greater than the company expects to realize from the contract. To avoid that problem, the loss reduces the CIP account to $2,660,000 (computed as $2,760,000 in costs to date less $100,000 estimated total loss). This amount combined with the estimated costs to complete of $2,340,000 equals the contract price of $5,000,000. Recognizing losses on long-term contracts in the period the losses become known is similar to measuring inventory at the lower of cost or net realizable value, a concept we will study in Chapter 9.

The pattern of gross profit (loss) over the contract term is summarized in the following table. Notice that in 2023 an additional unanticipated increase in costs of $100,000 causes a further loss of $100,000 to be recognized.

	Revenue Recognition	
	Over Time	Upon Completion
Gross profit (loss) recognized:		
2021	**$500,000**	$ -0-
2022	**(600,000)**	(100,000)
2023	**(100,000)**	(100,000)
Total project loss	$(200,000)	$(200,000)

The table in Illustration 6–25G shows the revenue and cost of construction recognized in each of the three years, assuming the contract qualifies for recognizing revenue over time. Revenue is recognized in the usual way by multiplying a percentage of completion by the total contract price. In situations where a loss is expected on the entire project cost of construction for the period will no longer be equal to cost incurred during the period. The easiest way to compute the cost of construction is to add the amount of the loss recognized to the amount of revenue recognized. For example, in 2022 revenue recognized of $706,000 is added to the loss of $600,000 to arrive at the cost of construction of $1,306,000.[22]

Illustration 6–25G

Allocation of Revenue and Cost of Construction to Each Period—Loss on Entire Project

2021		
Revenue recognized ($5,000,000 × 40%)		$2,000,000
Cost of construction		(1,500,000)
Gross profit		**$ 500,000**
2022		
Revenue recognized to date ($5,000,000 × 54.12%)*	$2,706,000	
Less: Revenue recognized in 2021	(2,000,000)	
Revenue recognized		$ 706,000
Cost of construction†		(1,306,000)
Loss		**$ (600,000)**
2023		
Revenue recognized to date ($5,000,000 × 100%)	$5,000,000	
Less: Revenue recognized in 2021 and 2022	(2,706,000)	
Revenue recognized		$2,294,000
Cost of construction†		(2,394,000)
Loss		**$ (100,000)**

*$2,760,000 ÷ $5,100,000 = 54.12%
†The difference between revenue and loss

The journal entries to record the losses in 2022 and 2023 are as follows:

2022		
Cost of construction	1,306,000	
Revenue from long-term contracts		706,000
Construction in progress (CIP)		600,000
2023		
Cost of construction	2,394,000	
Revenue from long-term contracts		2,294,000
Construction in progress (CIP)		100,000

[22]The cost of construction for 2022 also can be determined as follows:

Loss to date (100% recognized)		$ 100,000
Add:		
Remaining total project cost, not including the loss ($5,100,000 – 100,000)	$5,000,000	
Multiplied by the percentage of completion	× .5412*	2,706,000
Total		2,806,000
Less: Cost of construction recognized in 2021		(1,500,000)
Cost of construction recognized in 2022		$ 1,306,000

*$2,760,000 ÷ 5,100,000

Recognizing revenue over time in this case produces a large overstatement of income in 2021 and a large understatement in 2022 because of a change in the estimation of future costs. These estimate revisions happen occasionally when revenue is recognized over time.

When the contract does not qualify for recognizing revenue over time, no revenue or cost of construction is recognized until the contract is complete. In 2022, a loss on long-term contracts (an income statement account) of $100,000 is recognized. In 2023, the income statement will report revenue of $5,000,000 and cost of construction of $5,100,000, combining for an additional loss of $100,000. The journal entries to record the losses in 2022 and 2023 are as follows:

2022		
Loss on long-term contracts	100,000	
Construction in progress (CIP)		100,000
2023		
Cost of construction	5,100,000	
Revenue from long-term contracts		5,000,000
Construction in progress (CIP)		100,000

Concept Review Exercise

LONG-TERM CONSTRUCTION CONTRACTS

During 2021, the Samuelson Construction Company began construction on an office building for the City of Gernon. The contract price is $8,000,000 and the building will take approximately 18 months to complete. Completion is scheduled for early in 2023. The company's fiscal year ends on December 31.

The following is a year-by-year recap of construction costs incurred and the estimated costs to complete the project as of the end of each year. Progress billings and cash collections also are indicated.

	2021	2022	2023
Actual costs incurred during the year	$1,500,000	$4,500,000	$1,550,000
Actual costs incurred in prior years	–0–	1,500,000	6,000,000
Cumulative actual costs incurred to date	1,500,000	6,000,000	7,550,000
Estimated costs to complete at end of year	4,500,000	1,500,000	–0–
Total costs (actual + estimated)	$6,000,000	$7,500,000	$7,550,000
Billings made during the year	$1,400,000	$5,200,000	$1,400,000
Cash collections during year	1,000,000	4,000,000	3,000,000

Required:

1. Determine the amount of construction revenue, construction cost, and gross profit or loss to be recognized in each of the three years assuming (a) the contract qualifies for recognizing revenue over time and (b) the contract does *not* qualify for recognizing revenue over time.
2. Assuming the contract qualifies for recognizing revenue over time, prepare the necessary summary journal entries for each of the three years to account for construction costs, construction revenue, contract billings, and cash collections, and to close the construction accounts in 2023.
3. Assuming the contract qualifies for recognizing revenue over time, prepare a partial balance sheet for 2021 and 2022 that includes all construction-related accounts.

Solution:

1. Determine the amount of construction revenue, construction cost, and gross profit or loss to be recognized in each of the three years assuming (a) the contract qualifies for recognizing revenue over time and (b) the contract does *not* qualify for recognizing revenue over time.

The Contract Qualifies for Recognizing Revenue Over Time

	2021	2022	2023
Contract price	$ 8,000,000	$8,000,000	$8,000,000
Multiplied by % of completion*	25%	80%	100%
Cumulative revenue to be recognized to date	2,000,000	6,400,000	8,000,000
Less revenue recognized in prior years	–0–	(2,000,000)	(6,400,000)
Revenue recognized this year	2,000,000	4,400,000	1,600,000
Less actual costs incurred this year	(1,500,000)	(4,500,000)	1,550,000
Gross profit (loss) recognized this year	$ 500,000	$ (100,000)	$ 50,000

*Estimated percentage of completion:

2021	2022	2023
$\frac{1,500,000}{6,000,000} = 25\%$	$\frac{6,000,000}{7,500,000} = 80\%$	Project complete = 100%

The Contract Does *Not* Qualify for Recognizing Revenue Over Time

	2021	2022	2023
Revenue recognized	$–0–	$–0–	$8,000,000
Less cost of construction recognized	–0–	–0–	(7,550,000)
Gross profit recognized	$–0–	$–0–	$ 450,000

2. Assuming the contract qualifies for recognizing revenue over time, prepare the necessary summary journal entries for each of the three years to account for construction costs, construction revenue, contract billings, and cash collections, and to close the construction accounts in 2023.

	2021		2022		2023	
Construction in progress (CIP)	1,500,000		4,500,000		1,550,000	
Cash, materials, etc.		1,500,000		4,500,000		1,550,000
To record construction costs.						
Construction in progress (CIP)	500,000				50,000	
Cost of construction	1,500,000				1,550,000	
Revenue		2,000,000				1,600,000
To record revenue and gross profit.						
Cost of construction			4,500,000			
Revenue				4,400,000		
Construction in progress (CIP)				100,000		
To record revenue and gross loss.						
Accounts receivable	1,400,000		5,200,000		1,400,000	
Billings on construction contract		1,400,000		5,200,000		1,400,000
To record progress billings.						
Cash	1,000,000		4,000,000		3,000,000	
Accounts receivable		1,000,000		4,000,000		3,000,000
To record cash collections.						
Billings on construction contract					8,000,000	
Construction in progress (CIP)						8,000,000
To close accounts.						

3. Assuming the contract qualifies for recognizing revenue over time, prepare a partial balance sheet for 2021 and 2022 that includes all construction-related accounts.

Balance Sheet (End of Year)

	2021	2022
Current assets:		
Accounts receivable	$400,000	$1,600,000
Costs and profit ($2,000,000) in excess of billings ($1,400,000)	600,000	
Current liabilities:		
Billings ($6,600,000) in excess of costs and profit ($6,400,000)		$ 200,000

Financial Reporting Case Solution

©Hero Images/Getty Images

1. **Under what circumstances do companies recognize revenue at a point in time? Over a period of time?** *(p. 280)* A seller recognizes revenue when it satisfies a performance obligation, which happens when the seller transfers control of a good or service to the customer. Indicators that transfer of control has occurred include customer acceptance and physical possession of the good or service as well as the seller having a right to receive payment. Some performance obligations are satisfied at a point in time, when the seller has finished transferring the good or service to the customer. Other performance obligations are satisfied over time. For example, the customer might consume the benefit of the seller's work as it is performed, or the customer might control an asset as the seller creates it.
2. **When do companies break apart a sale and treat its parts differently for purposes of recognizing revenue?** *(p. 283)* Sellers must break apart a contract if the contract contains more than one performance obligation. Goods and services are viewed as separate performance obligations if they are both capable of being distinct (for example, if the goods and services could be sold separately) and if they are separately identifiable (which isn't the case if the point of the contract is to combine various goods and services into a completed product, as occurs with construction contracts). To account for contracts with multiple performance obligations, the seller allocates the contract's transaction price to the performance obligations according to their stand-alone selling prices and then recognizes revenue for each performance obligation when it is satisfied.
3. **How do companies account for long-term contracts that qualify for revenue recognition over time?** *(p. 305)* When contracts qualify for revenue recognition over time, the seller recognizes revenue each period as the contract is being fulfilled. The amount recognized is based on progress to date, which usually is estimated as the fraction of the project's cost incurred to date divided by total estimated costs. To calculate the total amount of revenue that should be recognized up to a given date, the estimated percentage of completion is multiplied by the contract price. To calculate the amount of revenue to be recognized in the current period, the amount of revenue recognized in prior periods is subtracted from the total amount of revenue that should be recognized as of the end of the current period. ●

The Bottom Line

- **LO6–1** Companies recognize revenue when goods or services are transferred to customers for the amount the company expects to be entitled to receive in exchange for those goods or services. That core principle is implemented by (1) identifying a contract with a customer, (2) identifying the performance obligations in the contract, (3) determining the transaction price of the contract, (4) allocating that price to the performance obligations, and (5) recognizing revenue when (or as) each performance obligation is satisfied. (*p. 278*)
- **LO6–2** Revenue should be recognized *at a single point in time* when control of a good or service is transferred to the customer on a specific date. Indicators that transfer has occurred and that revenue should be recognized include the seller having the right to receive payment, the customer having legal title and physical possession of the asset, the customer formally accepting the asset, and the customer assuming the risks and rewards of ownership. (*p. 280*)
- **LO6–3** Revenue should be recognized *over time* when a performance obligation is satisfied over time. That occurs if (1) the customer consumes the benefit of the seller's work as it is performed, (2) the customer controls the asset as the seller creates it, or (3) the asset has no alternative use to the seller and the seller can be paid for its progress even if the customer cancels the contract. (*p. 281*)
- **LO6–4** A contract's transaction price is allocated to its performance obligations. The allocation is based on the *stand-alone selling prices* of the goods and services underlying those performance obligations. The stand-alone selling price must be estimated if a good or service is not sold separately. (*p. 283*)

- **LO6–5** A contract exists when it has commercial substance and all parties to the contract are committed to performing the obligations and enforcing the rights that it specifies. Performance obligations are promises by the seller to transfer goods or services to a customer. A promise to transfer a good or service is a separate performance obligation if it is *distinct,* which is the case if it is both *capable of being distinct* (meaning that the customer could use the good or service on its own or in combination with other goods and services it could obtain elsewhere), and *it is separately identifiable* (meaning that the good or service is not highly interrelated with other goods and services in the contract, so it is distinct in the context of the contract). Prepayments, rights to return merchandise, and normal quality-assurance warranties do not qualify as performance obligations, because they don't transfer a good or service to the customer. On the other hand, extended warranties and customer options to receive goods or services in some preferred manner (for example, at a discount) qualify as performance obligations. (*p. 287*)
- **LO6–6** When a contract includes consideration that depends on the outcome of future events, sellers estimate that variable consideration and include it in the contract's transaction price. The seller's estimate is based either on the most likely outcome or the expected value of the outcome. However, a constraint applies—variable consideration only should be included in the transaction price to the extent it is probable that a significant revenue reversal will not occur. The estimate of variable consideration is updated each period to reflect changes in circumstances. A seller also needs to determine if it is a principal (and recognizes as revenue the amount received from the customer) or an agent (and recognizes its commission as revenue), consider time value of money, and consider the effect of any payments by the seller to the customer. Once the transaction price is estimated, we allocate it to performance obligations according to their stand-alone selling prices, which can be estimated using the adjusted market assessment approach, the expected cost plus margin approach, or the residual approach. (*p. 290*)
- **LO6–7** If the seller's activity over the license period is expected to affect the benefits the customer receives from the intellectual property being licensed, as with symbolic IP, the seller recognizes revenue over the license period. Otherwise, the seller recognizes revenue at the point in time that the customer obtains access to the seller's intellectual property. Franchises are an example of contracts that typically include licenses as well as other performance obligations. Revenue for bill-and-hold sales should be recognized when the seller transfers control of goods to the customer, even if the seller retains physical possession of the goods, and for consignment sales when goods are delivered to the end customer. Revenue for gift cards should be recognized when the gift card is redeemed, expires, or viewed as broken. (*p. 298*)
- **LO6–8** Much disclosure is required for revenue recognition. For example, a seller recognizes contract liabilities, contract assets, and accounts receivable on separate lines of its balance sheet. If the customer makes payment to the seller before the seller has satisfied performance obligations, the seller records a contract liability, such as deferred revenue. If the seller satisfies a performance obligation before the customer has paid for it, the seller records either a contract asset or an account receivable. The seller recognizes an account receivable if only the passage of time is required before the payment is due. If instead the seller's right to payment depends on something other than the passage of time, the seller recognizes a contract asset. (*p. 302*)
- **LO6–9** Long-term contracts usually qualify for revenue recognition over time. We recognize revenue over time by assigning a share of the project's revenues and costs to each reporting period over the life of the project according to the percentage of the project completed to date. If long-term contracts don't qualify for revenue recognition over time, we recognize revenues and expenses at the point in time when the project is complete. (*p. 304*)
- **LO6–10** IFRS interprets the word "probable" as indicating a lower probability threshold than does U.S. GAAP. As a consequence, a seller could conclude a contract meets the "probable that it will collect the amounts it's entitled to receive" threshold and recognize revenue under IFRS but not under U.S. GAAP. Also, IFRS doesn't tie the timing of revenue recognition for licenses to whether a license involves "functional" versus "symbolic" intellectual property, so IFRS might allow revenue recognition at a point in time for "symbolic IP" licenses that would require revenue recognition over time under U.S. GAAP. (*p. 288, 291, 299*)

Questions For Review of Key Topics

Q 6–1 What are the five key steps a company follows to apply the core revenue recognition principle?

Q 6–2 What indicators suggest that a performance obligation has been satisfied at a single point in time?

Q 6–3 What criteria determine whether a company can recognize revenue over time?

Q 6–4 We recognize service revenue either at one point in time or over a period of time. Explain the rationale for recognizing service revenue using these two approaches.

Q 6–5 What characteristics make a good or service a performance obligation?

Q 6–6 How does a seller allocate a transaction price to a contract's performance obligations?

Q 6–7 What must a contract include for the contract to exist for purposes of revenue recognition?

IFRS **Q 6–8** How might the definition of "probable" affect determining whether a contract exists under IFRS as compared to U.S. GAAP?

Q 6–9 When a contract includes an option to buy additional goods or services, when does that option give rise to a performance obligation?

Q 6–10 Is variable consideration included in the calculation of a contract's transaction price? If so, how is the amount of variable consideration estimated?

Q 6–11 How are sellers constrained from recognizing variable consideration, and under what circumstances does the constraint apply?

Q 6–12 Is a customer's right to return merchandise a performance obligation of the seller? How should sellers account for a right of return?

Q 6–13 What is the difference between a principal and an agent for determining the amount of revenue to recognize?

Q 6–14 Under what circumstances should sellers consider the time value of money when recognizing revenue?

Q 6–15 When should a seller view a payment to its customer as a refund of part of the price paid by the customer for the seller's products or services?

Q 6–16 What are three methods for estimating stand-alone selling prices of goods and services that normally are not sold separately?

Q 6–17 When is revenue recognized with respect to licenses?

Q 6–18 In a franchise arrangement, what are a franchisor's typical performance obligations?

Q 6–19 When does a company typically recognize revenue for a bill-and-hold sale?

IFRS **Q 6–20** How might a license for symbolic intellectual property be treated differently under IFRS as compared to U.S. GAAP?

Q 6–21 When does a consignor typically recognize revenue for a consignment sale?

Q 6–22 When does a company recognize revenue for a sale of a gift card?

Q 6–23 Must bad debt expense be reported on its own line on the income statement? If not, how should it be disclosed?

Q 6–24 Explain the difference between contract assets, contract liabilities, and accounts receivable.

Q 6–25 Explain how to account for revenue on a long-term contract over time as opposed to at a point in time. Under what circumstances should revenue be recognized at the point in time a contract is completed?

Q 6–26 Periodic billings to the customer for a long-term construction contract are recorded as billings on construction contract. How is this account reported in the balance sheet?

Q 6–27 When is an estimated loss on a long-term contract recognized, both for contracts that recognize revenue over time and those that recognize revenue at the point in time the contract is completed?

Brief Exercises

connect

BE 6–1
Revenue recognition at a point in time
● LO6–2

On July 1, 2021, Apache Company, a real estate developer, sold a parcel of land to a construction company for $3,000,000. The book value of the land on Apache's books was $1,200,000. Terms of the sale required a down payment of $150,000 and 19 annual payments of $150,000 plus interest at an appropriate interest rate due on each July 1 beginning in 2022. How much revenue will Apache recognize for the sale (ignoring interest), assuming that it recognizes revenue at the point in time at which it transfers the land to the construction company?

BE 6–2
Timing of revenue recognition
● LO6–3

Estate Construction is constructing a building for CyberB, an online retailing company. Under the construction agreement, if for any reason Estate can't complete construction, CyberB would own the partially completed building and could hire another construction company to complete the job. When should Estate recognize revenue: as the building is constructed, or after construction is completed?

BE 6–3
Timing of revenue recognition
● LO6–3

On May 1, 2021,Varga Tech Services signed a $6,000 consulting contract with Shaffer Holdings. The contract requires Varga to provide computer technology support services whenever requested over the period from May 1, 2021, to April 30, 2022, with Shaffer paying the entire $6,000 on May 1, 2021. How much revenue should Varga recognize in 2021?

BE 6–4
Allocating the transaction price
● LO6–4

Sarjit Systems sold software to a customer for $80,000. As part of the contract, Sarjit promises to provide "free" technical support over the next six months. Sarjit sells the same software without technical support for $70,000 and a stand-alone six-month technical support contract for $30,000, so these products would sell for $100,000 if sold separately. Prepare Sarjit's journal entry to record the sale of the software.

BE 6–5
Existence of a contract
● LO6–5

Tulane Tires wrote a contract for a $100,000 sale of tires to the new Garden District Tour Company. Tulane only anticipates a slightly greater than 50 percent chance that Garden will be able to pay the amounts that Tulane is entitled to receive under the contract. Upon delivery of the tires, assuming no payment has yet been made by Garden, how much revenue should Tulane recognize under U.S. GAAP?

BE 6–6
Existence of a contract; IFRS
● LO6–5, LO6–10
IFRS

Assume the same facts as in BE 6–5 but that Tulane Tires reports under IFRS. How much revenue should Tulane recognize under IFRS?

BE 6–7
Performance obligations; prepayments
● LO6–5

eLean is an online fitness community, offering access to workout routines, nutrition advice, and eLean coaches. Customers pay a $50 fee to become registered on the website, and then pay $5 per month for access to all eLean services. How many performance obligations exist in the implied contract when a customer registers for the services?

BE 6–8
Performance obligations; warranties
● LO6–5

Vroom Vacuums sells the Tornado vacuum cleaner. Each Tornado has a one-year warranty that covers any product defects. When customers purchase a Tornado, they also have the option to purchase an extended three-year warranty that covers any breakage or maintenance. The extended warranty sells for the same amount regardless of whether it is purchased at the same time as the Tornado or at some other time. How many performance obligations exist in the implied contract for the purchase of a vacuum cleaner?

BE 6–9
Performance obligations; warranties
● LO6–5

Assume the same facts as in BE 6–8 but that customers pay 20% less for the extended warranty if they buy it at the same time they buy a Tornado. How many performance obligations exist in the implied contract for the purchase of a vacuum cleaner?

BE 6–10
Performance obligations; options
● LO6–5

McAfee sells a subscription to its antivirus software along with a subscription renewal option that allows renewal at half the prevailing price for a new subscription. How many performance obligations exist in this contract?

BE 6–11
Performance obligations; construction
● LO6–5

Precision Equipment, Inc. specializes in designing and installing customized manufacturing equipment. On February 1, 2021, it signs a contract to design a fully automated wristwatch assembly line for $2 million, which will be settled in cash upon completion of construction. Precision Equipment will install the equipment on the client's property, furnish it with a customized software package that is integral to operations, and provide consulting services that integrate the equipment with Precision's other assembly lines. How many performance obligations exist in this contract?

BE 6–12
Performance obligations; construction
● LO6–5

On January 1, 2021, Lego Construction Company signed a contract to build a custom garage for a customer and received $10,000 in advance for the job. The new garage will be built on the customer's land. To complete this project, Lego must first build a concrete floor, construct wooden pillars and walls, and finally install a roof. Lego normally charges stand-alone prices of $3,000, $4,000, and $5,000, respectively, for each of these three smaller tasks if done separately. How many performance obligations exist in this contract?

BE 6–13
Performance obligations; right of return
● LO6–5, LO6–6

Aria Perfume, Inc., sold 3,210 boxes of white musk soap during January of 2021 at the price of $90 per box. The company offers a full refund to unsatisfied customers for any product returned within 30 days from the date of purchase. Based on historical experience, Aria expects that 3% of sales will be returned. How many performance obligations are there in each sale of a box of soap? How much revenue should Aria recognize in January?

BE 6–14
Variable consideration
● LO6–6

Leo Consulting enters into a contract with Highgate University to restructure Highgate's processes for purchasing goods from suppliers. The contract states that Leo will earn a fixed fee of $25,000 and earn an additional $10,000 if Highgate achieves $100,000 of cost savings. Leo estimates a 50% chance that Highgate will achieve $100,000 of cost savings. Assuming that Leo determines the transaction price as the expected value of expected consideration, what transaction price will Leo estimate for this contract?

BE 6–15
Variable consideration
● LO6–6

In January 2021, Continental Fund Services, Inc., enters into a one-year contract with a client to provide investment advisory services. The company will receive a management fee, prepaid at the beginning of the contract, that is calculated as 1% of the client's $150 million total assets being managed. In addition, the contract specifies that Continental will receive a performance bonus of 20% of any returns in excess of the return on the Dow Jones Industrial Average market index. Continental estimates that it will earn a $2 million performance bonus, but is very uncertain of that estimate, given that the bonus depends on a highly volatile stock market. On what transaction price should Continental base revenue recognition?

BE 6–16
Right of return
● LO6–6

Finerly Corporation sells cosmetics through a network of independent distributors. Finerly shipped cosmetics to its distributors and is considering whether it should record $300,000 of revenue upon shipment of a new line of cosmetics. Finerly expects the distributors to be able to sell the cosmetics, but is uncertain because it has little experience with selling cosmetics of this type. Finerly is committed to accepting the cosmetics back from the distributors if the cosmetics are not sold. How much revenue should Finerly recognize upon delivery to its distributors?

BE 6–17
Principal or agent
● LO6–6

Assume that **Amazon.com** sells the MacBook Pro, a computer produced by **Apple**, for a retail price of $1,500. Amazon arranges its operations such that customers receive products directly from Apple Stores rather than Amazon. Customers purchase from Amazon using credit cards, and Amazon forwards cash to Apple equal to the retail price minus a $150 commission that Amazon keeps. In this arrangement, how much revenue will Amazon recognize for the sale of one MacBook Pro?

BE 6–18
Time value of money
● LO6–6

On January 1, 2021, Wooten Technology Associates sold computer equipment to the Denison Company. Delivery was made on January 1, 2021, but payment for the equipment of $10,000 is not due until December 31, 2021. Assuming that Wooten views the time value of money to be a significant component of this transaction and that an 8% interest rate is applicable, how much sales revenue would Wooten recognize on January 1, 2021?

BE 6–19
Time value of money
● LO6–6

On January 1, 2021, Hodge Beanery received $8,000 from the Kennedy Company in exchange for a coffee roaster that it will deliver to Kennedy on December 31, 2021. Assuming that Hodge views the time value of money to be a significant component of this transaction, and that a 9% interest rate is applicable, how much deferred revenue would Hodge recognize on January 1, 2021?

BE 6–20
Payments by the seller to the customer
● LO6–6

Lewis Co. sold merchandise to AdCo for $60,000 and received $60,000 for that sale one month later. One week prior to receiving payment from AdCo, Lewis made a $10,000 payment to AdCo for advertising services that have a fair value of $7,500. After accounting for any necessary adjustments, how much revenue should Lewis Co. record for the merchandise sold to AdCo?

BE 6–21
Estimating stand-alone selling prices: adjusted market assessment approach
● LO6–6

O'Hara Associates sells golf clubs, and with each sale of a full set of clubs provides complementary club-fitting services. A full set of clubs with the fitting services sells for $1,500. Similar club-fitting services are offered by other vendors for $110, and O'Hara generally charges approximately 10% more than do other vendors for similar services. Estimate the stand-alone selling price of the club-fitting services using the adjusted market assessment approach.

BE 6–22
Estimating stand-alone selling prices: expected cost plus margin approach
● LO6–6

O'Hara Associates sells golf clubs, and with each sale of a full set of clubs provides complementary club-fitting services. A full set of clubs with the fitting services sells for $1,500. O'Hara estimates that it incurs $60 of staff compensation and other costs to provide the fitting services, and normally earns 30% over cost on similar services. Assuming that the golf clubs and the club-fitting services are separate performance obligations, estimate the stand-alone selling price of the club-fitting services using the expected cost plus margin approach.

BE 6–23
Estimating stand-alone selling prices; residual approach
● LO6–6

O'Hara Associates sells golf clubs, and with each sale of a full set of clubs provides complementary club-fitting services. A full set of clubs with the fitting services sells for $1,500. O'Hara sells the same clubs without the fitting service for $1,400. Assuming that the golf clubs and the club-fitting services are separate performance obligations, estimate the stand-alone selling price of the club fitting services using the residual approach.

BE 6–24
Timing of revenue recognition; licenses
● LO6–7

Saar Associates sells two licenses to Kim & Company on September 1, 2021. First, in exchange for $100,000, Saar provides Kim with a copy of its proprietary investment management software, which Saar does not anticipate updating and which Kim can use permanently. Second, in exchange for $90,000, Saar provides Kim with a three-year right to market Kim's financial advisory services under the name of Saar Associates, which Saar advertises on an ongoing basis. How much revenue will Saar recognize in 2021 under this arrangement?

BE 6–25
Timing of revenue recognition; licenses
● LO6–7
IFRS

Assume the same facts as in BE 6–24 except that the trade name "Saar Associates" is not well known in the marketplace and the owner provides no advertising or other benefits to a licensee of the Saar Associates trade name during the license period. How much revenue will Saar recognize in 2021 under this arrangement if Saar reports under U.S. GAAP?

BE 6–26
Timing of revenue recognition; licenses
● LO6–7, LO6–10
IFRS

Assume the same facts as in BE 6–25. How much revenue will Saar recognize in 2021 under this arrangement if Saar reports under IFRS?

BE 6–27
Timing of revenue recognition; franchises
● LO6–7

TopChop sells hairstyling franchises. TopChop receives $50,000 from a new franchisee for providing initial training, equipment, and furnishings that have a stand-alone selling price of $50,000. TopChop also receives $30,000 per year for use of the TopChop name and for ongoing consulting services (starting on the date the franchise is purchased). Carlos became a TopChop franchisee on July 1, 2021, and on August 1, 2021, had completed training and was open for business. How much revenue in 2021 will TopChop recognize for its arrangement with Carlos?

BE 6–28
Timing of revenue recognition; bill-and-hold
● LO6–7

Dowell Fishing Supply, Inc., sold $50,000 of Dowell Rods on December 15, 2021, to Bassadrome. Because of a shipping backlog, Dowell held the inventory in Dowell's warehouse until January 12, 2022 (having assured Bassadrome that it would deliver sooner if necessary). How much revenue should Dowell recognize in 2021 for the sale to Bassadrome?

BE 6–29
Timing of revenue recognition; consignment
● LO6–7

Kerianne paints landscapes, and in late 2021 placed four paintings with a retail price of $250 each in the Holmstrom Gallery. Kerianne's arrangement with Holmstrom is that Holmstrom will earn a 20% commission on paintings sold to gallery patrons. As of December 31, 2021, one painting had been sold by Holmstrom to gallery patrons. How much revenue with respect to these four paintings should Kerianne recognize in 2021?

BE 6–30
Timing of revenue recognition; gift card
● LO6–7

GoodBuy sells gift cards redeemable for GoodBuy products either in store or online. During 2021, GoodBuy sold $1,000,000 of gift cards, and $840,000 of the gift cards were redeemed for products. As of December 31, 2021, $30,000 of the remaining gift cards had passed the date at which GoodBuy concludes that the cards will never be redeemed. How much gift card revenue should GoodBuy recognize in 2021?

BE 6–31
Contract assets and contract liabilities
● LO6–8

Holt Industries received a $2,000 prepayment from the Ramirez Company for the sale of new office furniture. Holt will bill Ramirez an additional $3,000 upon delivery of the furniture to Ramirez. Upon receipt of the $2,000 prepayment, how much should Holt recognize for a contract asset, a contract liability, and accounts receivable?

BE 6–32
Contract assets and contract liabilities
● LO6–8, LO6–9

As of December 31, 2021, Cady Construction has one construction job for which the construction in progress (CIP) account has a balance of $20,000 and the billings on construction contract account has a balance of $14,000. Cady has another construction job for which the construction in progress account has a balance of $3,000 and the billings on construction contract account has a balance of $5,000. Indicate the amount of contract asset and/or contract liability that Cady would show in its December 31, 2021, balance sheet.

BE 6–33
Long-term contract; revenue recognition over time; profit recognition
● LO6–9

A construction company entered into a fixed-price contract to build an office building for $20 million. Construction costs incurred during the first year were $6 million and estimated costs to complete at the end of the year were $9 million. The company recognizes revenue over time according to percentage of completion. How much revenue and gross profit or loss will appear in the company's income statement in the first year of the contract?

BE 6–34
Long-term contract; revenue recognition over time; balance sheet
● LO6–9

Refer to the situation described in BE 6–33. Assume that, during the first year the company billed its customer $7 million, of which $5 million was collected before year-end. What would appear in the year-end balance sheet related to this contract?

BE 6–35
Long-term contract; revenue recognition upon completion
● LO6–9

Refer to the situation described in BE 6–33. Assume that the building was completed during the second year, and construction costs incurred during the second year were $10 million. How much revenue and gross profit or loss will the company recognize in the first and second year if it recognizes revenue upon contract completion?

BE 6–36
Long-term contract; revenue recognition; loss on entire project
● LO6–9

Franklin Construction entered into a fixed-price contract to build a freeway-connecting ramp for $30 million. Construction costs incurred in the first year were $16 million and estimated remaining costs to complete at the end of the year were $17 million. How much gross profit or loss will Franklin recognize in the first year if it recognizes revenue over time according to percentage of completion? What if instead Franklin recognizes revenue upon contract completion?

Exercises

connect

E 6–1
FASB codification research
● LO6–1, LO6–2, LO6–3

CODE

Access the *FASB's Accounting Standards Codification* at the FASB website (**www.fasb.org**).

Required:

Determine the specific nine-digit Codification citation (XXX-XX-XX-XX) for accounting for each of the following items:

1. What are the five key steps to applying the revenue recognition principle?
2. What are indicators that control has passed from the seller to the buyer, such that it is appropriate to recognize revenue at a point in time?
3. Under what circumstances can sellers recognize revenue over time?

E 6–2
Service revenue
● LO6–3

Ski West, Inc., operates a downhill ski area near Lake Tahoe, California. An all-day adult lift ticket can be purchased for $85. Adult customers also can purchase a season pass that entitles the pass holder to ski any day during the season, which typically runs from December 1 through April 30. Ski West expects its season pass holders to use their passes equally throughout the season. The company's fiscal year ends on December 31.

On November 6, 2021, Jake Lawson purchased a season pass for $450.

Required:

1. When should Ski West recognize revenue from the sale of its season passes?
2. Prepare the appropriate journal entries that Ski West would record on November 6 and December 31.
3. What will be included in the Ski West 2021 income statement and balance sheet related to the sale of the season pass to Jake Lawson?

E 6–3
Allocating transaction price
● LO6–4

Video Planet (VP) sells a big screen TV package consisting of a 60-inch plasma TV, a universal remote, and on-site installation by VP staff. The installation includes programming the remote to have the TV interface with other parts of the customer's home entertainment system. VP concludes that the TV, remote, and installation service are separate performance obligations. VP sells the 60-inch TV separately for $1,700, sells the remote separately for $100, and offers the installation service separately for $200. The entire package sells for $1,900.

Required:

How much revenue would be allocated to the TV, the remote, and the installation service?

E 6–4
FASB codification research
● LO6–4, LO6–5

CODE

Access the *FASB Standards Codification* at the FASB website (**www.fasb.org**).

Required:

Determine the specific nine-digit Codification citation (XXX-XX-XX-XX) for accounting for each of the following items:

1. On what basis is a contract's transaction price allocated to its performance obligations?
2. What are indicators that a promised good or service is separately identifiable from other goods and services promised in the contract?
3. Under what circumstances is an option viewed as a performance obligation?

E 6–5
Performance obligations
● LO6–2, LO6–4, LO6–5

On March 1, 2021, Gold Examiner receives $147,000 from a local bank and promises to deliver 100 units of certified 1-oz. gold bars on a future date. The contract states that ownership passes to the bank when Gold Examiner delivers the products to Brink's, a third-party carrier. In addition, Gold Examiner has agreed to provide a replacement shipment at no additional cost if the product is lost in transit. The stand-alone price of a gold bar is $1,440 per unit, and Gold Examiner estimates the stand-alone price of the replacement insurance service to be $60 per unit. Brink's picked up the gold bars from Gold Examiner on March 30, and delivery to the bank occurred on April 1.

Required:

1. How many performance obligations are in this contract?
2. Prepare the journal entry Gold Examiner would record on March 1.
3. Prepare the journal entry Gold Examiner would record on March 30.
4. Prepare the journal entry Gold Examiner would record on April 1.

E 6–6
Performance obligations; customer option for additional goods or services
● LO6–2, LO6–4, LO6–5

Clarks Inc., a shoe retailer, sells boots in different styles. In early November the company starts selling "SunBoots" to customers for $70 per pair. When a customer purchases a pair of SunBoots, Clarks also gives the customer a 30% discount coupon for any additional future purchases made in the next 30 days. Customers can't obtain the discount coupon otherwise. Clarks anticipates that approximately 20% of customers will utilize the coupon, and that on average those customers will purchase additional goods that normally sell for $100.

Required:

1. How many performance obligations are in a contract to buy a pair of SunBoots?
2. Prepare a journal entry to record revenue for the sale of 1,000 pairs of SunBoots, assuming that Clarks uses the residual method to estimate the stand-alone selling price of SunBoots sold without the discount coupon.

E 6–7
Performance obligations; customer option for additional goods or services; prepayment
● LO6–3, LO6–4, LO6–5

A New York City daily newspaper called "Manhattan Today" charges an annual subscription fee of $135. Customers prepay their subscriptions and receive 260 issues over the year. To attract more subscribers, the company offered new subscribers the ability to pay $130 for an annual subscription that also would include a coupon to receive a 40% discount on a one-hour ride through Central Park in a horse-drawn carriage. The list price of a carriage ride is $125 per hour. The company estimates that approximately 30% of the coupons will be redeemed.

Required:

1. How much revenue should Manhattan Today recognize upon receipt of the $130 subscription price?
2. How many performance obligations exist in this contract?
3. Prepare the journal entry to recognize sale of 10 new subscriptions, clearly identifying the revenue or deferred revenue associated with each performance obligation.

E 6–8
Performance obligations; customer option for additional goods or services
● LO6–4, LO6–5

On May 1, 2021, Meta Computer, Inc., enters into a contract to sell 5,000 units of Comfort Office Keyboard to one of its clients, Bionics, Inc., at a fixed price of $95,000, to be settled by a cash payment on May 1. Delivery is scheduled for June 1, 2021. As part of the contract, the seller offers a 25% discount coupon to Bionics for any purchases in the next six months. The seller will continue to offer a 5% discount on all sales during the same time period, which will be available to all customers. Based on experience, Meta Computer estimates a 50% probability that Bionics will redeem the 25% discount voucher, and that the coupon will be applied to $20,000 of purchases. The stand-alone selling price for the Comfort Office Keyboard is $19.60 per unit.

Required:

1. How many performance obligations are in this contract?
2. Prepare the journal entry that Meta would record on May 1, 2021.
3. Assume the same facts and circumstances as above, except that Meta gives a 5% discount option to Bionics instead of 25%. In this case, what journal entry would Meta record on May 1, 2021?

E 6–9
Variable consideration; estimation and constraint
● LO6–6

Thomas Consultants provided Bran Construction with assistance in implementing various cost-savings initiatives. Thomas's contract specifies that it will receive a flat fee of $50,000 and an additional $20,000 if Bran reaches a prespecified target amount of cost savings. Thomas estimates that there is a 20% chance that Bran will achieve the cost-savings target.

Required:

1. Assuming Thomas uses the expected value as its estimate of variable consideration, calculate the transaction price.
2. Assuming Thomas uses the most likely value as its estimate of variable consideration, calculate the transaction price.
3. Assume Thomas uses the expected value as its estimate of variable consideration, but is very uncertain of that estimate due to a lack of experience with similar consulting arrangements. Calculate the transaction price.

E 6–10
Variable consideration—most likely amount; change in estimate
● LO6–3, LO6–6

Rocky Guide Service provides guided 1–5 day hiking tours throughout the Rocky Mountains. Wilderness Tours hires Rocky to lead various tours that Wilderness sells. Rocky receives $1,000 per tour day, and shortly after the end of each month Rocky learns whether it will receive a $100 bonus per tour day it guided during the previous month if its service during that month received an average evaluation of "excellent" by Wilderness customers. The $1,000 per day and any bonus due are paid in one lump payment shortly after the end of each month.

- On July 1, based on prior experience, Rocky estimated there is a 30% chance it will earn the bonus for July tours. It guided a total of 10 days from July 1–July 15.
- On July 16, based on Rocky's view that it had provided excellent service during the first part of the month, Rocky revised its estimate to an 80% chance it would earn the bonus for July tours. Rocky also guided customers for 15 days from July 16–July 31.
- On August 5, Rocky learned it did not receive an average evaluation of "excellent" for its July tours, so it would not receive any bonus for July, and received all payment due for the July tours.

Rocky bases estimates of variable consideration on the most likely amount it expects to receive.

Required:

1. Prepare Rocky's July 15 journal entry to record revenue for tours given from July 1–July 15.
2. Prepare Rocky's July 31 journal entry to record revenue for tours given from July 16–July 31.
3. Prepare Rocky's August 5 journal entry to record any necessary adjustments to revenue and receipt of payment from Wilderness.

E 6–11
Variable consideration–expected value; change in estimate
● LO6–3, LO6–6

Assume the same facts as in E 6–10.

Required:

Complete the requirements of E 6–10 assuming that Rocky bases estimates of variable consideration on the expected value it expects to receive.

E 6–12
Time value of money for accounts receivable
● LO6–6

Arctic Cat sold Seneca Motor Sports a shipment of snowmobiles. The snowmobiles were delivered on January 1, 2021, and Arctic received a note from Seneca indicating that Seneca will pay Arctic $40,000 on a future date. Unless informed otherwise, assume that Arctic views the time value of money component of this arrangement to be significant and that the relevant interest rate is 8%.

Required:

1. Assume the note indicates that Seneca is to pay Arctic the $40,000 due on the note on December 31, 2021. Prepare the journal entry for Arctic to record the sale on January 1, 2021.
2. Assume the same facts as in requirement 1, and prepare the journal entry for Arctic to record collection of the payment on December 31, 2021.
3. Assume instead that Seneca is to pay Arctic the $40,000 due on the note on December 31, 2022. Prepare the journal entry for Arctic to record the sale on January 1, 2021.
4. Assume instead that Arctic does not view the time value of money component of this arrangement to be significant, and that the note indicates that Seneca is to pay Arctic the $40,000 due on the note on December 31, 2021. Prepare the journal entry for Arctic to record the sale on January 1, 2021.

E 6–13
Time value of money for deferred revenue
● LO6–6

[This is a variation of E 6–12 focusing on deferred revenue.]

Arctic Cat sold Seneca Motor Sports a shipment of snowmobiles that have a fair market value of $40,000. Seneca paid for the snowmobiles on January 1, 2021, with delivery to occur subsequently. Unless informed otherwise, assume that Arctic views the time value of money component of this arrangement to be significant, and that the relevant interest rate is 8%.

Required:

1. Assume that, on January 1, 2021, Seneca prepays Arctic for a December 31, 2021 delivery of the snowmobiles. Prepare the journal entry for Arctic to record collection on January 1, 2021, assuming Seneca prepays the present value of the snowmobiles.
2. Prepare the journal entry for Arctic to record delivery of the snowmobiles on December 31, 2021.
3. Assume instead that delivery is to occur on December 31, 2022. Prepare the journal entry for Arctic to record collection on January 1, 2021, assuming Seneca prepays the present value of the snowmobiles.
4. Assume instead that Arctic does not view the time value of money component of this arrangement to be significant. Also assume that, on January 1, 2021, Seneca prepays Arctic for a December 31, 2021 delivery of the snowmobiles, and that Seneca prepays the present value of the snowmobiles. Prepare the journal entry for Arctic to record collection on January 1, 2021.

E 6–14
Consideration payable to customer; collectibility of transaction price
● LO6–2, LO6–5, LO6–6

Furtastic manufactures imitation fur garments. On June 1, 2021, Furtastic made a sale to Willett's Department Store under terms that require Willett to pay $150,000 to Furtastic on June 30, 2021. In a separate transaction on June 15, 2021, Furtastic purchased brand advertising services from Willett for $12,000. The fair value of those advertising services is $5,000. Furtastic expects that 3% of all sales will prove uncollectible.

Required:

1. Prepare the journal entry to record Furtastic's sale on June 1, 2021.
2. Prepare the journal entry to record Furtastic's purchase of advertising services from Willett on June 15, 2021. Assume all of the advertising services are delivered on June 15, 2021.
3. Prepare the journal entry to record Furtastic's receipt of $150,000 from Willett on June 30, 2021.
4. How would Furtastic's expectation regarding uncollectible accounts affect its recognition of revenue from the sale to Willett's Department Store on June 1, 2021? Explain briefly.

E 6–15
Approaches for estimating stand-alone selling prices
● LO6–6

(This exercise is a variation of E 6–3.)

Video Planet (VP) sells a big screen TV package consisting of a 60-inch plasma TV, a universal remote, and on-site installation by VP staff. The installation includes programming the remote to have the TV interface with other parts of the customer's home entertainment system. VP concludes that the TV, remote, and installation service are separate performance obligations. VP sells the 60-inch TV separately for $1,750 and sells the remote separately for $100, and offers the entire package for $1,900. VP does not sell the installation service separately. VP is aware that other similar vendors charge $150 for the installation service. VP also estimates that it incurs approximately $100 of compensation and other costs for VP staff to provide the installation service. VP typically charges 40% above cost on similar sales.

Required:

1. Estimate the stand-alone selling price of the installation service using the adjusted market assessment approach.
2. Estimate the stand-alone selling price of the installation service using the expected cost plus margin approach.
3. Estimate the stand-alone selling price of the installation service using the residual approach.

E 6–16
FASB codification research
● LO6–6, LO6–7

Access the *FASB Accounting Standards Codification* at the FASB website (**www.fasb.org**).

Required:

Determine the specific nine-digit Codification citation (XXX-XX-XX-XX) for accounting for each of the following items:

1. What alternative approaches can be used to estimate variable consideration?
2. What alternative approaches can be used to estimate the stand-alone selling price of performance obligations that are not sold separately?
3. What determines the timing of revenue recognition with respect to licenses of symbolic intellectual property?
4. What indicators suggest that a seller is a principal rather than an agent?

E 6–17
Franchises; residual method
● LO6–6, LO6–7

Monitor Muffler sells franchise arrangements throughout the United States and Canada. Under a franchise agreement, Monitor receives $600,000 in exchange for satisfying the following separate performance obligations: (1) franchisees have a five-year right to operate as a Monitor Muffler retail establishment in an exclusive sales territory, (2) franchisees receive initial training and certification as a Monitor Mechanic, and (3) franchisees receive a Monitor Muffler building and necessary equipment. The stand-alone selling price of the initial training and certification is $15,000, and $450,000 for the building and equipment. Monitor estimates the stand-alone selling price of the five-year right to operate as a Monitor Muffler establishment using the residual approach.

Monitor received $75,000 on July 1, 2021, from Perkins and accepted a note receivable for the rest of the franchise price. Monitor will construct and equip Perkins's building and train and certify Perkins by September 1, and Perkins's five-year right to operate as a Monitor Muffler establishment will commence on September 1 as well.

Required:

1. What amount would Monitor calculate as the stand-alone selling price of the five-year right to operate as a Monitor Muffler retail establishment?
2. What journal entry would Monitor record on July 1, 2021, to reflect the sale of a franchise to Dan Perkins?
3. How much revenue would Monitor recognize in the year ended December 31, 2021, with respect to its franchise arrangement with Perkins? (Ignore any interest on the note receivable.)

E 6–18
FASB codification research
● LO6–8

Access the *FASB Accounting Standards Codification* at the FASB website (**www.fasb.org**).

Required:

Determine the specific nine-digit Codification citation (XXX-XX-XX-XX) for accounting for each of the following items:

1. What disclosures are required with respect to performance obligations that the seller is committed to satisfying but that are not yet satisfied?

2. What disclosures are required with respect to uncollectible accounts receivable, also called impairments of receivables?
3. What disclosures are required with respect to significant changes in contract assets and contract liabilities?

E 6–19
Long-term contract; revenue recognition over time and at a point in time
● LO6–9

Assume **Nortel Networks** contracted to provide a customer with Internet infrastructure for $2,000,000. The project began in 2021 and was completed in 2022. Data relating to the contract are summarized below:

	2021	2022
Costs incurred during the year	$ 300,000	$1,575,000
Estimated costs to complete as of 12/31	1,200,000	–0–
Billings during the year	380,000	1,620,000
Cash collections during the year	250,000	1,750,000

Required:
1. Compute the amount of revenue and gross profit or loss to be recognized in 2021 and 2022 assuming Nortel recognizes revenue over time according to percentage of completion.
2 Compute the amount of revenue and gross profit or loss to be recognized in 2021 and 2022 assuming this project does not qualify for revenue recognition over time.
3. Prepare a partial balance sheet to show how the information related to this contract would be presented at the end of 2021 assuming Nortel recognizes revenue over time according to percentage of completion.
4. Prepare a partial balance sheet to show how the information related to this contract would be presented at the end of 2021 assuming this project does not qualify for revenue recognition over time.

E 6–20
Long-term contract; revenue recognition over time vs. upon project completion
● LO6–9

On June 15, 2021, Sanderson Construction entered into a long-term construction contract to build a baseball stadium in Washington, D.C., for $220 million. The expected completion date is April 1, 2023, just in time for the 2023 baseball season. Costs incurred and estimated costs to complete at year-end for the life of the contract are as follows ($ in millions):

	2021	2022	2023
Costs incurred during the year	$ 40	$ 80	$ 50
Estimated costs to complete as of December 31	120	60	—

Required:
1. How much revenue and gross profit will Sanderson report in its 2021, 2022, and 2023 income statements related to this contract assuming Sanderson recognizes revenue over time according to percentage of completion?
2. How much revenue and gross profit will Sanderson report in its 2021, 2022, and 2023 income statements related to this contract assuming this project does not qualify for revenue recognition over time?
3. Suppose the estimated costs to complete at the end of 2022 are $80 million instead of $60 million. Determine the amount of revenue and gross profit or loss to be recognized in 2022 assuming Sanderson recognizes revenue over time according to percentage of completion.

E 6–21
Long-term contract; revenue recognition over time; loss projected on entire project
● LO6–9

On February 1, 2021, Arrow Construction Company entered into a three-year construction contract to build a bridge for a price of $8,000,000. During 2021, costs of $2,000,000 were incurred with estimated costs of $4,000,000 yet to be incurred. Billings of $2,500,000 were sent, and cash collected was $2,250,000.

In 2022, costs incurred were $2,500,000 with remaining costs estimated to be $3,600,000. 2022 billings were $2,750,000, and $2,475,000 cash was collected. The project was completed in 2023 after additional costs of $3,800,000 were incurred. The company's fiscal year-end is December 31. Arrow recognizes revenue over time according to percentage of completion.

Required:
1. Calculate the amount of revenue and gross profit or loss to be recognized in each of the three years.
2. Prepare journal entries for 2021 and 2022 to record the transactions described (credit "various accounts" for construction costs incurred).
3. Prepare a partial balance sheet to show the presentation of the project as of December 31, 2021 and 2022.

E 6–22
Long-term contract; revenue recognition upon project completion; loss projected on entire project
● LO6–8, LO6–9

[This is a variation of E 6–21 focusing on the revenue recognition upon project completion.]

On February 1, 2021, Arrow Construction Company entered into a three-year construction contract to build a bridge for a price of $8,000,000. During 2021, costs of $2,000,000 were incurred, with estimated costs of $4,000,000 yet to be incurred. Billings of $2,500,000 were sent, and cash collected was $2,250,000.

In 2022, costs incurred were $2,500,000 with remaining costs estimated to be $3,600,000. 2022 billings were $2,750,000, and $2,475,000 cash was collected. The project was completed in 2023 after additional costs of $3,800,000 were incurred. The company's fiscal year-end is December 31. This project does not qualify for revenue recognition over time.

Required:

1. Calculate the amount of gross profit or loss to be recognized in each of the three years.
2. Prepare journal entries for 2021 and 2022 to record the transactions described (credit "various accounts" for construction costs incurred).
3. Prepare a partial balance sheet to show the presentation of the project as of December 31, 2021 and 2022. Indicate whether any of the amounts shown are contract assets or contract liabilities.

E 6–23
Income (loss) recognition; Long-term contract; revenue recognition over time vs. upon project completion
● LO6–9

Brady Construction Company contracted to build an apartment complex for a price of $5,000,000. Construction began in 2021 and was completed in 2023. The following is a series of independent situations, numbered 1 through 6, involving differing costs for the project. All costs are stated in thousands of dollars.

	Costs Incurred During Year			Estimated Costs to Complete (As of the End of the Year)		
Situation	**2021**	**2022**	**2023**	**2021**	**2022**	**2023**
1	$1,500	$ 2,100	$ 900	$3,000	$ 900	—
2	1,500	900	2,400	3,000	2,400	—
3	1,500	2,100	1,600	3,000	1,500	—
4	500	3,000	1,000	3,500	875	—
5	500	3,000	1,300	3,500	1,500	—
6	500	3,000	1,800	4,600	1,700	—

Required:

Copy and complete the following table:

	Gross Profit (Loss) Recognized					
	Over Time			Upon Completion		
Situation	**2021**	**2022**	**2023**	**2021**	**2022**	**2023**
1						
2						
3						
4						
5						
6						

E 6–24
Long-term contract; revenue recognition over time; solve for unknowns
● LO6–9

In 2021, Long Construction Corporation began construction work under a three-year contract. The contract price is $1,600,000. Long recognizes revenue over time according to percentage of completion for financial reporting purposes. The financial statement presentation relating to this contract at December 31, 2021, is as follows:

Balance Sheet		
Accounts receivable (from construction progress billings)		$30,000
Construction in progress	$100,000	
Less: Billings on construction contract	(94,000)	
Cost and profit of uncompleted contracts in excess of billings		6,000
Income Statement		
Income (before tax) on the contract recognized in 2021		$20,000

Required:

1. What was the cost of construction actually incurred in 2021?
2. How much cash was collected in 2021 on this contract?
3. What was the estimated cost to complete as of the end of 2021?
4. What was the estimated percentage of completion used to calculate revenue in 2021? *(AICPA adapted)*

Problems

P 6–1
Upfront fees; performance obligations
● LO6–4, LO6–5

Fit & Slim (F&S) is a health club that offers members various gym services.

Required:

1. Assume F&S offers a deal whereby enrolling in a new membership for $700 provides a year of unlimited access to facilities and also entitles the member to receive a voucher redeemable for 25% off yoga classes for one year. The yoga classes are offered to gym members as well as to the general public. A new membership

normally sells for $720, and a one-year enrollment in yoga classes sells for an additional $500. F&S estimates that approximately 40% of the vouchers will be redeemed. F&S offers a 10% discount on all one-year enrollments in classes as part of its normal promotion strategy.

a. How many performance obligations are included in the new member deal?

b. How much of the contract price would be allocated to each performance obligation? Explain your answer.

c. Prepare the journal entry to recognize revenue for the sale of a new membership. Clearly identify revenue or deferred revenue associated with each performance obligation.

2. Assume F&S offers a "Fit 50" coupon book with 50 prepaid visits over the next year. F&S has learned that Fit 50 purchasers make an average of 40 visits before the coupon book expires. A customer purchases a Fit 50 book by paying $500 in advance, and for any additional visits over 50 during the year after the book is purchased, the customer can pay a $15 visitation fee per visit. F&S typically charges $15 to nonmembers who use the facilities for a single day.

a. How many separate performance obligations are included in the Fit 50 member deal? Explain your answer.

b. How much of the contract price would be allocated to each separate performance obligation? Explain your answer.

c. Prepare the journal entry to recognize revenue for the sale of a new Fit 50 book.

P 6–2
Performance obligations; warranties; option
● LO6–2, LO6–4, LO6–5

Creative Computing sells a tablet computer called the Protab. The $780 sales price of a Protab Package includes the following:

- One Protab computer.
- A 6-month limited warranty. This warranty guarantees that Creative will cover any costs that arise due to repairs or replacements associated with defective products for up to six months.
- A coupon to purchase a Creative Probook e-book reader for $200, a price that represents a 50% discount from the regular Probook price of $400. It is expected that 20% of the discount coupons will be utilized.
- A coupon to purchase a one-year extended warranty for $50. Customers can buy the extended warranty for $50 at other times as well. Creative estimates that 40% of customers will purchase an extended warranty.
- Creative does not sell the Protab without the limited warranty, option to purchase a Probook, and the option to purchase an extended warranty, but estimates that if it did so, a Protab alone would sell for $760.

Required:

1. How many performance obligations are included in a Protab Package? Explain your answer.
2. List the performance obligations in the Protab Package in the following table, and complete it to allocate the transaction price of 100,000 Protab Packages to the performance obligations in the contract.

Performance obligation:	Stand-alone selling price of the performance obligation:	Percentage of the sum of the stand-alone selling prices of the performance obligations (to two decimal places):	Allocation of total transaction price to the performance obligation:

3. Prepare a journal entry to record sales of 100,000 Protab Packages (ignore any sales of extended warranties).

P 6–3
Performance obligations; warranties; option
● LO6–2, LO6–4, LO6–5

Assume the same facts as in P 6–2, except that customers must pay $75 to purchase the extended warranty if they don't purchase it with the $50 coupon that was included in the Protab Package. Creative estimates that 40% of customers will use the $50 coupon to purchase an extended warranty. Complete the same requirements as in P 6–2.

P 6–4
Performance obligations; customer options for additional goods and services
● LO6–2, LO6–4, LO6–5

Supply Club, Inc., sells a variety of paper products, office supplies, and other products used by businesses and individual consumers. During July 2021 it started a loyalty program through which qualifying customers can accumulate points and redeem those points for discounts on future purchases. Redemption of a loyalty point reduces the price of one dollar of future purchases by 20% (equal to 20 cents). Customers do not earn additional loyalty points for purchases on which loyalty points are redeemed. Based on past experience, Supply Club estimates a 60% probability that any point issued will be redeemed for the discount. During July 2021, the company records $135,000 of revenue and awards 125,000 loyalty points. The aggregate stand-alone selling price of the purchased products is $135,000. Eighty percent of sales were cash sales, and the remainder were credit sales.

Required:

1. Prepare Supply Club's journal entry to record July sales.
2. During August, customers redeem loyalty points on $60,000 of merchandise. Seventy-five percent of those sales were for cash, and the remainder were credit sales. Prepare Supply Club's journal entry to record those sales.

P 6–5
Variable consideration
● LO6–3, LO6–6

On January 1, Revis Consulting entered into a contract to complete a cost reduction program for Green Financial over a six-month period. Revis will receive $20,000 from Green at the end of each month. If total cost savings reach a specific target, Revis will receive an additional $10,000 from Green at the end of the contract, but if total cost savings fall short, Revis will refund $10,000 to Green. Revis estimates an 80% chance that cost savings will reach the target and calculates the contract price based on the expected value of future payments to be received.

Required:

Prepare the following journal entries for Revis:

1. Prepare the journal entry on January 31 to record the collection of cash and recognition of the first month's revenue.
2. Assuming total cost savings exceed target, prepare the journal entry on June 30 to record receipt of the bonus.
3. Assuming total cost savings fall short of target, prepare the journal entry on June 30 to record payment of the penalty.

P 6–6
Variable consideration; change of estimate
● LO6–3, LO6–6

Since 1970, Super Rise, Inc., has provided maintenance services for elevators. On January 1, 2021, Super Rise obtains a contract to maintain an elevator in a 90-story building in New York City for 10 months and receives a fixed payment of $80,000. The contract specifies that Super Rise will receive an additional $40,000 at the end of the 10 months if there is no unexpected delay, stoppage, or accident during the year. Super Rise estimates variable consideration to be the most likely amount it will receive.

Required:

1. Assume that, because the building sees a constant flux of people throughout the day, Super Rise is allowed to access the elevators and related mechanical equipment only between 3 a.m. and 5 a.m. on any given day, which is insufficient to perform some of the more time-consuming repair work. As a result, Super Rise believes that unexpected delays are likely and that it will not earn the bonus. Prepare the journal entry Super Rise would record on January 1.
2. Assume instead that Super Rise knows at the inception of the contract that it will be given unlimited access to the elevators and related equipment each day, with the right to schedule repair sessions any time. When given these terms and conditions, Super Rise has never had any delays or accidents in the past. Prepare the journal entry Super Rise would record on January 31 to record one month of revenue.
3. Assume the same facts as requirement 1. In addition assume that, on May 31, Super Rise determines that it does not need to spend more than two hours on any given day to operate the elevator safely because the client's elevator is relatively new. Therefore, Super Rise believes that unexpected delays are very unlikely. Prepare the journal entry Super Rise would record on May 31 to recognize May revenue and any necessary revision in its estimated bonus receivable.

P 6–7
Variable consideration; constraint and change of estimate
● LO6–3, LO6–6

Assume the same facts as P 6–6.

Required:

1. Assume that Super Rise anticipates it will earn the performance bonus, but is highly uncertain about its estimate given unfamiliarity with the building and uncertainty about its access to the elevators and related equipment. Prepare the journal entry Super Rise would record on January 1.
2. Assume the same facts as requirement 1. In addition assume that, on May 31, Super Rise determines that it has sufficient experience with the company to make an accurate estimate of the likelihood that it will earn the performance bonus, and concludes that it is likely to earn the performance bonus. Prepare the journal entry Super Rise would record on May 31 to recognize May revenue and any necessary revision in its estimated bonus receivable.

P 6–8
Variable transaction price
● LO6–3, LO6–6

Velocity, a consulting firm, enters into a contract to help Burger Boy, a fast-food restaurant, design a marketing strategy to compete with **Burger King.** The contract spans eight months. Burger Boy promises to pay $60,000 at the beginning of each month. At the end of the contract, Velocity either will give Burger Boy a refund of $20,000 or will be entitled to an additional $20,000 bonus, depending on whether sales at Burger Boy at year-end have increased to a target level. At the inception of the contract, Velocity estimates an 80% chance that it will earn the $20,000 bonus and calculates the contract price based on the expected value of future payments to be received. After four months, circumstances change, and Velocity revises to 60% its estimate of the probability that it will earn the bonus. At the end of the contract, Velocity receives the additional consideration of $20,000.

Required:

1. Prepare the journal entry to record revenue each month for the first four months of the contract.
2. Prepare the journal entry that the Velocity Company would record after four months to recognize the change in estimate associated with the reduced likelihood that the $20,000 bonus will be received.
3. Prepare the journal entry to record the revenue each month for the second four months of the contract.
4. Prepare the journal entry after eight months to record receipt of the $20,000 cash bonus.

P 6–9
Variable transaction price
● LO6–3, LO6–6, LO6–7

CODE

Tran Technologies licenses its functional intellectual property to Lyon Industries. Terms of the arrangement require Lyon to pay Tran $500,000 on April 1, 2021, when Lyon first obtains access to Tran's intellectual property, and then in the future to pay Tran a royalty of 4% of future sales of products that utilize that intellectual property. Tran anticipates receiving sales-based royalties of $1,000,000 during 2021 and $1,500,000/year for the years 2022–2026. Assume Tran accounts for the Lyon license as a right of use, because Tran's actions subsequent to April 1, 2021, will affect the benefits that Lyon receives from access to Tran's intellectual property.

Required:

1. Access the *FASB Accounting Standards Codification* at the FASB website (**www.fasb.org**). Identify the specific nine-digit Codification citation (XXX-XX-XX-XX) for accounting for variable consideration arising from sales-based royalties on licenses of intellectual property, and consider the relevant GAAP. When can Tran recognize revenue from sales-based royalties associated with the Lyon license?
2. What journal entry would Tran record on April 1, 2021, when it receives the $500,000 payment from Lyon?
3. Assume on December 31, 2021, Tran receives $1,000,000 for all sales-based royalties from Lyon in 2021. What journal entry would Tran record on December 31, 2021, to recognize any revenue that should be recognized in 2021 with respect to the Lyon license that it has not already recognized?
4. Assume Tran accounts for the Lyon license as a five-year right to access Tran's symbolic intellectual property from April 1, 2021, through March 31, 2026. Tran expects that its ongoing marketing efforts will affect the value of the license to Lyon during the five-year license period. Repeat requirements 2 and 3.

P 6–10
Long-term contract; revenue recognition over time
● LO6–8, LO6–9

In 2021, the Westgate Construction Company entered into a contract to construct a road for Santa Clara County for $10,000,000. The road was completed in 2023. Information related to the contract is as follows:

	2021	2022	2023
Cost incurred during the year	$2,400,000	$3,600,000	$2,200,000
Estimated costs to complete as of year-end	5,600,000	2,000,000	–0–
Billings during the year	2,000,000	4,000,000	4,000,000
Cash collections during the year	1,800,000	3,600,000	4,600,000

Westgate recognizes revenue over time according to percentage of completion.

Required:

1. Calculate the amount of revenue and gross profit to be recognized in each of the three years.
2. Prepare all necessary journal entries for each of the years (credit "Cash, Materials, etc." for construction costs incurred).
3. Prepare a partial balance sheet for 2021 and 2022 showing any items related to the contract. Indicate whether any of the amounts shown are contract assets or contract liabilities.
4. Calculate the amount of revenue and gross profit to be recognized in each of the three years assuming the following costs incurred and costs to complete information:

	2021	2022	2023
Costs incurred during the year	$2,400,000	$3,800,000	$3,200,000
Estimated costs to complete as of year-end	5,600,000	3,100,000	–0–

5. Calculate the amount of revenue and gross profit to be recognized in each of the three years assuming the following costs incurred and costs to complete information:

	2021	2022	2023
Costs incurred during the year	$2,400,000	$3,800,000	$3,900,000
Estimated costs to complete as of year-end	5,600,000	4,100,000	–0–

P 6–11
Long-term contract; revenue recognition upon completion
● LO6–9

[This is a variation of P 6–10 modified to focus on revenue recognition upon project completion.]

Required:

Complete the requirements of P 6–10 assuming that Westgate Construction's contract with Santa Clara County does *not* qualify for revenue recognition over time.

P 6–12
Long-term contract; revenue recognized over time; loss projected on entire project
● LO5–9

Curtiss Construction Company, Inc., entered into a fixed-price contract with Axelrod Associates on July 1, 2021, to construct a four-story office building. At that time, Curtiss estimated that it would take between two and three years to complete the project. The total contract price for construction of the building is $4,000,000. Curtiss concludes that the contract does not qualify for revenue recognition over time. The building was completed on December 31, 2023. Estimated percentage of completion, accumulated contract costs incurred, estimated costs to complete the contract, and *accumulated* billings to Axelrod under the contract were as follows:

	At 12-31-2021	At 12-31-2022	At 12-31-2023
Percentage of completion	10%	60%	100%
Costs incurred to date	$ 350,000	$2,500,000	$4,250,000
Estimated costs to complete	3,150,000	1,700,000	–0–
Billings to Axelrod, to date	720,000	2,170,000	3,600,000

Required:
1. For each of the three years, prepare a schedule to compute total gross profit or loss to be recognized as a result of this contract.
2. Assuming Curtiss recognizes revenue over time according to percentage of completion, compute gross profit or loss to be recognized in each of the three years.
3. Assuming Curtiss recognizes revenue over time according to percentage of completion, compute the amount to be shown in the balance sheet at the end of 2021 and 2022 as either cost in excess of billings or billings in excess of costs.

(AICPA adapted)

P 6–13
Long-term contract; revenue recognition over time vs. upon project completion
● LO6–9

Citation Builders, Inc., builds office buildings and single-family homes. The office buildings are constructed under contract with reputable buyers. The homes are constructed in developments ranging from 10–20 homes and are typically sold during construction or soon after. To secure the home upon completion, buyers must pay a deposit of 10% of the price of the home with the remaining balance due upon completion of the house and transfer of title. Failure to pay the full amount results in forfeiture of the down payment. Occasionally, homes remain unsold for as long as three months after construction. In these situations, sales price reductions are used to promote the sale.

During 2021, Citation began construction of an office building for Altamont Corporation. The total contract price is $20 million. Costs incurred, estimated costs to complete at year-end, billings, and cash collections for the life of the contract are as follows

	2021	2022	2023
Costs incurred during the year	$ 4,000,000	$ 9,500,000	$4,500,000
Estimated costs to complete as of year-end	12,000,000	4,500,000	—
Billings during the year	2,000,000	10,000,000	8,000,000
Cash collections during the year	1,800,000	8,600,000	9,600,000

Also during 2021, Citation began a development consisting of 12 identical homes. Citation estimated that each home will sell for $600,000, but individual sales prices are negotiated with buyers. Deposits were received for eight of the homes, three of which were completed during 2021 and paid for in full for $600,000 each by the buyers. The completed homes cost $450,000 each to construct. The construction costs incurred during 2021 for the nine uncompleted homes totaled $2,700,000.

Required:
1. Briefly explain the difference between recognizing revenue over time and upon project completion when accounting for long-term construction contracts.
2. Answer the following questions assuming that Citation concludes it does not qualify for revenue recognition over time for its office building contracts:
   a. How much revenue related to this contract will Citation report in its 2021 and 2022 income statements?
   b. What is the amount of gross profit or loss to be recognized for the Altamont contract during 2021 and 2022?
   c. What will Citation report in its December 31, 2021, balance sheet related to this contract? (Ignore cash.)
3. Answer requirements 2a through 2c assuming that Citation recognizes revenue over time according to percentage of completion for its office building contracts.

4. Assume the same information for 2021 and 2022, but that as of year-end 2022 the estimated cost to complete the office building is $9,000,000. Citation recognizes revenue over time according to percentage of completion for its office building contracts.
   a. How much revenue related to this contract will Citation report in the 2022 income statement?
   b. What is the amount of gross profit or loss to be recognized for the Altamont contract during 2022?
   c. What will Citation report in its 2022 balance sheet related to this contract? (Ignore cash.)
5. When should Citation recognize revenue for the sale of its single-family homes?
6. What will Citation report in its 2021 income statement and 2021 balance sheet related to the single-family home business (ignore cash in the balance sheet)?

Decision Makers' Perspective

©ImageGap/Getty Images

Apply your critical-thinking ability to the knowledge you've gained. These cases will provide you an opportunity to develop your research, analysis, judgment, and communication skills. You also will work with other students, integrate what you've learned, apply it in real-world situations, and consider its global and ethical ramifications. This practice will broaden your knowledge and further develop your decision-making abilities.

Research Case 6–1
Earnings management with respect to revenues
● LO6–1

An article published in *Accounting Horizons* describes various techniques that companies use to manage their earnings.

Required:

In your library, on the Internet, or from some other source, locate the article "How Are Earnings Managed? Evidence from Auditors" in *Accounting Horizons,* 2003 (Supplement), and answer the following questions:

1. What are the four most common revenue-recognition abuses identified by auditors in that article? From the examples provided in the article, briefly explain each abuse.
2. What is the revenue-recognition abuse identified in the article related to the percentage-of-completion method?
3. What effect did these revenue-recognition abuses tend to have on net income in the year they occurred? Indicate "increase" or "decrease" and the percentage of the time they had that effect.
4. Did auditors tend to require their clients to make adjustments that reduced the revenue-recognition abuses they detected? Indicate "yes" or "no" and the percentage of the time auditors required adjustment.

Judgment Case 6–2
Satisfaction of performance obligations
● LO6–2

Assume **McDonald's** enters into a contract to sell Billy Bear dolls for Toys4U Stores. Based on the contract, McDonald's displays the dolls in selected stores. Toys4U is not paid until the dolls have been sold by McDonald's, and unsold dolls are returned to Toys4U.

Required:

Has Toys4U satisfied its performance obligation when it delivers the dolls to McDonald's? Explain your answer.

Judgment Case 6–3
Satisfaction of performance obligations
● LO6–2

Cutler Education Corporation developed a software product to help children under age 12 learn mathematics. The software contains two separate parts: Basic Level (Level I) and Intermediate Level (Level II). Parents purchase each level separately and are eligible to purchase the access code for Level II only if their children pass the Level I exam.

Kerry purchases the Level I software at a price of $50 for his son, Tom, on December 1. Suppose Tom passed the Level I test on December 10, and Kerry immediately purchased the access code for Level II for an additional $30. Cutler provided Kerry with the access code to Level II on December 20.

Required:

Indicate the date upon which Cutler would recognize revenue for the sale of Level I and Level II software.

Ethics Case 6–4
Revenue recognition
● LO6–2

Horizon Corporation manufactures personal computers. The company began operations in 2016 and reported profits for the years 2016 through 2019. Due primarily to increased competition and price slashing in the industry, 2020's income statement reported a loss of $20 million. Just before the end of the 2021 fiscal year, a memo from the company's chief financial officer to Jim Fielding, the company controller, included the following comments:

If we don't do something about the large amount of unsold computers already manufactured, our auditors will require us to write them off. The resulting loss for 2021 will cause a violation of our debt covenants and force the company into bankruptcy. I suggest that you ship half of our inventory to J.B. Sales, Inc., in Oklahoma City. I know the company's president and he will accept the merchandise and acknowledge the shipment as a purchase. We can record the sale in 2021 which will boost profits to an acceptable level. Then J.B. Sales will simply return the merchandise in 2022 after the financial statements have been issued.

Required:

Discuss the ethical dilemma faced by Jim Fielding.

Judgment Case 6–5
Satisfying performance obligations
● LO6–2, LO6–3

Consider each of the following scenarios separately:

Scenario 1: Crown Construction Company entered into a contract with Star Hotel for building a highly sophisticated, customized conference room to be completed for a fixed price of $400,000. Nonrefundable progress payments are made on a monthly basis for work completed during the month. Legal title to the conference room equipment is held by Crown until the end of the construction project, but if the contract is terminated before the conference room is finished, Star retains the partially completed job and must pay for any work completed to date.

Scenario 2: Regent Company entered into a contract with Star Hotel for constructing and installing a standard designed gym for a fixed price of $400,000. Nonrefundable progress payments are made on a monthly basis for work completed during the month. Legal title to the gym passes to Star upon completion of the building process. If Star cancels the contract before the gym construction is completed, Regent removes all the installed equipment and Star must compensate Regent for any loss of profit on sale of the gym to another customer.

Scenario 3: On January 1, the CostDriver Company, a consulting firm, entered into a three-month contract with Coco Seafood Restaurant to analyze its cost structure in order to find a way to reduce operating costs and increase profits. CostDriver promises to share findings with the restaurant every two weeks and to provide the restaurant with a final analytical report at the end of the contract. This service is customized to Coco, and CostDriver would need to start from scratch if it provided a similar service to another client. Coco promises to pay $5,000 per month. If Coco chooses to terminate the contract, it is entitled to receive a report detailing analyses to that stage.

Scenario 4: Assume International Tower (Phase II) is developing luxury residential real estate and begins to market individual apartments during their construction. The Tower entered into a contract with Edwards for the sale of a specific apartment. Edwards pays a deposit that is refundable only if the Tower fails to deliver the completed apartment in accordance with the contract. The remainder of the purchase price is paid on completion of the contract when Edwards obtains possession of the apartment.

Required:

For each of the scenarios, determine whether the seller should recognize revenue (a) over time or (b) when the product or service is completed. Explain your answer.

Trueblood Accounting Case 6–6
Applying the five-step revenue recognition process

The following Trueblood case is recommended for use with this chapter. The case provides an excellent opportunity for class discussion, group projects, and writing assignments. The case, along with Professor's Discussion Material, can be obtained from the Deloitte Foundation at its website **www.deloitte.com/us/truebloodcases**.

Case 17.7: *Mesmerizing Marketers*

This case concerns application of ASC 606 for purposes of recognizing revenue from contracts with customers.

Judgment Case 6–7
Performance obligation; licensing
● LO6–5, LO6–7

Assume that **Pfizer,** a large research-based pharmaceutical company, enters into a contract with a start-up biotechnology company called HealthPro and promises to

1. Grant HealthPro the exclusive rights to use Pfizer's Technology A for the life of its patent. The license gives HealthPro the exclusive right to market, distribute, and manufacture Drug B as developed using Technology A. Pfizer views the patent as functional intellectual property.
2. Assign four full-time equivalent employees to perform research and development services for HealthPro in a specially designated Pfizer lab facility. The primary objective of these services is to receive regulatory approval to market and distribute Drug B using Technology A.

HealthPro is required to use Pfizer's lab to perform the research and development services necessary to develop Drug B using Technology A, because the expertise related to Technology A is proprietary to Pfizer and not available elsewhere.

Required:

Are the license and R&D development services separate performance obligations? Indicate "yes" or "no," and explain your reasoning.

Communication Case 6–8
Performance obligations; loyalty program
● LO6–5

Jerry's Ice Cream Parlor is considering a marketing plan to increase sales of ice cream cones. The plan will give customers a free ice cream cone if they buy 10 ice cream cones at regular prices. Customers will be issued a card that will be punched each time an ice cream cone is purchased. After 10 punches, the card can be turned in for a free cone.

Jerry Donovan, the company's owner, is not sure how the new plan will affect accounting procedures. He realizes that the company will be incurring costs each time a free ice cream cone is awarded, but there will be no corresponding revenue or cash inflow.

The focus of this case is on how to account for revenue if the new plan is adopted. Your instructor will divide the class into two to six groups depending on the size of the class. The mission of your group is to reach consensus on the appropriate accounting treatment for the new plan. That treatment should describe when revenue is recognized and how it will be calculated.

Required:

1. Each group member should deliberate the situation independently and draft a tentative argument prior to the class session for which the case is assigned.
2. In class, each group will meet for 10–15 minutes in different areas of the classroom. During that meeting, group members will take turns sharing their suggestions for the purpose of arriving at a single group treatment.
3. After the allotted time, a spokesperson for each group (selected during the group meetings) will share the group's solution with the class. The goal of the class is to incorporate the views of each group into a consensus approach to the situation.

Judgment Case 6–9
Principal or agent
● LO6–6

AuctionCo.com sells used products collected from different suppliers. Assume a customer purchases a used bicycle through AuctionCo.com for $300. AuctionCo.com agrees to pay the supplier $200 for the bicycle. The bicycle will be shipped to the customer by the original bicycle owner.

Required:

1. Assume AuctionCo.com takes control of this used bicycle before the sale and pays $200 to the supplier. Under this assumption, how much revenue would AuctionCo.com recognize at the time of the sale to the customer?
2. Assume AuctionCo.com never takes control of this used bicycle before the sale. Instead, the bicycle is shipped directly to the customer by the original bicycle owner, and then AuctionCo.com pays $200 to the supplier. Under this assumption, how much revenue would AuctionCo.com recognize at the time of the sale to the customer?
3. Assume AuctionCo.com promises to pay $200 to the supplier regardless of whether the bicycle is sold, but the bicycle will continue to be shipped directly from the supplier to the customer. Under this assumption, how much revenue would AuctionCo.com recognize at the time of the sale to the customer?

Real World Case 6–10
Principal agent considerations
● LO6–6

EDGAR, the Electronic Data Gathering, Analysis, and Retrieval system, performs automated collection, validation, indexing, and forwarding of submissions by companies and others who are required by law to file forms with the U.S. Securities and Exchange Commission (SEC). All publicly traded domestic companies use EDGAR to make the majority of their filings. (Some foreign companies file voluntarily.) Form 10-K, which includes the annual report, is required to be filed on EDGAR. The SEC makes this information available on the Internet.

Required:

1. Access EDGAR on the Internet. The web address is **www.sec.gov**.
2. Search for the most recent 10-K's of **Expedia, Inc.,** and **Booking Holdings Inc.** (which includes Priceline). Search or scroll to find the revenue recognition note in the financial statements.
3. For each of the following types of revenue, indicate whether the amount shown in the income statement is "net" or "gross" (the terms used with respect to revenue recognition in the chapter), and briefly explain your answer.
   a. Expedia's "merchant hotel model" revenues
   b. Priceline's "'Name Your Own Price' services"
   c. Priceline's "Merchant Retail Services"
4. Consider your responses to 3a through 3c. Does it look like there is the potential for noncomparability when readers consider Expedia and Priceline? Indicate "yes" or "no," and briefly explain your answer.

Trueblood Accounting Case 6–11
Principal or agent

The following Trueblood case is recommended for use with this chapter. The case provides an excellent opportunity for class discussion, group projects, and writing assignments. The case, along with Professor's Discussion Material, can be obtained from the Deloitte Foundation at its website **www.deloitte.com/us/truebloodcases**.

Case 17.5: ***Stanley and Sons***
This case concerns the presentation of revenue on a gross or net basis under U.S. GAAP and IFRS.

Research Case 6–12
FASB codification; locate and extract relevant information and authoritative support for a financial reporting issue; reporting revenue as a principal or as an agent
● LO6–6

The birth of the Internet in the 1990s led to the creation of a new industry of online retailers such as **Amazon, Overstock.com**, and **PCM, Inc.** Many of these companies often act as intermediaries between the manufacturer and the customer without ever taking possession of the merchandise sold. Revenue recognition for this type of transaction has been controversial.

Assume that **Overstock.com** sold you a product for $200 that cost $150. The company's profit on the transaction clearly is $50. Should Overstock recognize $200 in revenue and $150 in cost of goods sold (the gross method), or should it recognize only the $50 in gross profit (the net method) as commission revenue?

Required:

1. Access the *FASB Accounting Standards Codification* at the FASB website (**www.fasb.org**). What is the specific nine-digit Codification citation (XXX-XX-XX-XX) that indicates what an entity assesses to determine whether the nature of its promise is to act as a principal or agent?
2. What indicators does the Codification list that suggest an entity is a principal? Determine the specific nine-digit Codification citation (XXX-XX-XX-XX).

3. Using EDGAR (**www.sec.gov**), access **Alphabet, Inc.**'s 2017 10-K. Locate the disclosure note that discusses the company's revenue recognition policy with respect to ads placed on Goggle Network Members' properties.
4. Do you agree with Alphabet's reasoning with respect to choosing whether it reports revenue gross versus net with respect to these advertising services? Indicate "yes" or "no," and explain.

Real World Case 6–13
Chainsaw Al; revenue recognition and earnings management
● LO6–7

In May 2001, the Securities and Exchange Commission sued the former top executives at Sunbeam, charging the group with financial reporting fraud that allegedly cost investors billions in losses. **Sunbeam Corporation** is a recognized designer, manufacturer, and marketer of household and leisure products, including Coleman, Eastpak, First Alert, Grillmaster, Mixmaster, Mr. Coffee, Oster, Powermate, and Campingaz. In the mid-1990s, Sunbeam needed help: its profits had declined by over 80 percent, and in 1996, its stock price was down over 50 percent from its high. To the rescue: Albert Dunlap, also known as "Chainsaw Al" based on his reputation as a ruthless executive known for his ability to restructure and turn around troubled companies, largely by eliminating jobs.

The strategy appeared to work. In 1997, Sunbeam's revenues had risen by 18 percent. However, in April 1998, the brokerage firm of **Paine Webber** downgraded Sunbeam's stock recommendation. Why the downgrade? Paine Webber had noticed unusually high accounts receivable, massive increases in sales of electric blankets in the third quarter 1997, which usually sell best in the fourth quarter, as well as unusually high sales of barbeque grills for the fourth quarter. Soon after, Sunbeam announced a first quarter loss of $44.6 million, and Sunbeam's stock price fell 25 percent.

It eventually came to light that Dunlap and Sunbeam had been using a "bill-and-hold" strategy with retail buyers. This involved selling products at large discounts to retailers before they normally would buy and then holding the products in third-party warehouses, with delivery at a later date.

Many felt Sunbeam had deceived shareholders by artificially inflating earnings and the company's stock price. A class-action lawsuit followed, alleging that Sunbeam and Dunlap violated federal securities laws, suggesting the motivation to inflate the earnings and stock price was to allow Sunbeam to complete hundreds of millions of dollars of debt financing in order to complete some ongoing mergers. Shareholders alleged damages when Sunbeam's subsequent earnings decline caused a huge drop in the stock price.

Required:

1. How might Sunbeam's 1997 "bill-and-hold" strategy have contributed to artificially high earnings in 1997?
2. How would the strategy have led to the unusually high accounts receivable Paine Webber noticed?
3. How might Sunbeam's 1997 "bill-and-hold" strategy have contributed to a 1998 earnings decline?
4. How does earnings management of this type affect earnings quality?

Judgment Case 6–14
Revenue recognition; long-term construction contracts
● LO6–9

Two accounting students were discussing the timing of revenue recognition for long-term construction contracts. The discussion focused on which method was most like the typical revenue recognition method of recognizing revenue at the point of product delivery. Bill argued that recognizing revenue upon project completion was preferable because it was analogous to recognizing revenue at the point of delivery. John disagreed and supported recognizing revenue over time, stating that it was analogous to accruing revenue as a performance obligation was satisfied. John also pointed out that an advantage of recognizing revenue over time is that it provides information sooner to users.

Required:

Which argument do you support—Bill's or John's? Why?

Communication Case 6–15
Long-term contract revenue recognition over time vs. upon project completion
● LO6–9

Willingham Construction is in the business of building high-priced, custom, single-family homes. The company, headquartered in Anaheim, California, operates throughout the Southern California area. The construction period for the average home built by Willingham is six months, although some homes have taken as long as nine months.

You have just been hired by Willingham as the assistant controller and one of your first tasks is to evaluate the company's revenue recognition policy. The company presently recognizes revenue upon completion for all of its projects and management is now considering whether revenue recognition over time is appropriate.

Required:

Write a 1- to 2-page memo to Virginia Reynolds, company controller, describing the differences between the effects of recognizing revenue over time and upon project completion on the income statement and balance sheet. Indicate any criteria specifying when revenue should be recognized. Be sure to include references to GAAP as they pertain to the choice of method. Do not address the differential effects on income taxes nor the effect on the financial statements of switching between methods.

Data Analytics

connect

Data analytics is the process of examining data sets in order to draw conclusions about the information they contain. If you haven't completed any of the prior data analytics cases, follow the instructions listed in the Chapter 1 Data Analytics case to get set up. You will need to watch the videos referred to in the Chapters 1–3 Data Analytics cases. No additional videos are required for this case. All short training videos can be found here: **https://www.tableau.com/learn/training#getting-started**.

Data Analytics Case
Sales Returns
● LO6–6

In prior chapters, you applied Tableau to examine a data set and create calculations to compare two companies' financial information. In this case, you continue in your role as an analyst conducting introductory research into the relative merits of investing in one or both of these companies. You will assess the companies' ability to satisfy their performance obligations with respect to product sales by examining their sales returns.

Required:

Use Tableau to create a "combination chart" comparing sales and sales returns for each of the two companies in each year from 2012 to 2021 as described in the Resources section below. Once the chart is created, move your cursor to hover above various data points in the chart. Notice that an information box appears to reveal the pertinent sales and return measures for that company in that year.

1. Which company exhibited a more favorable sales returns percentage in *2012,* Big Store or Discount Goods? What percent of that company's customer purchases were returned?
2. During the period 2013-2016, did Big Store's sales return percentage become (a) more favorable or (b) less favorable?
3. During the period 2013-2016, did Discount Goods' sales return percentage become (a) more favorable or (b) less favorable?
4. Which company exhibited a more favorable sales returns percentage in *2021,* Big Store or Discount Goods? What percent of that company's customer purchases were returned?

Resources:

You have available to you an extensive data set that includes detailed financial data for 2012-2021 for both Discount Goods and Big Store. The data set is in the form of four Excel files available to download from Connect, or under Student Resources within the Library tab. Download the file "Discount_Goods_Big_Store_Financials.xlsx" to the computer, save it, and open it in Tableau.

For this case, you will create two calculations to produce a right of return sales analysis chart to allow you to compare and contrast the two companies' returns management performance.

After you view the training videos and review instructions in Chapters 1-3, follow these steps to create the chart you will use for this case:

- Open Tableau and connect to the Excel spreadsheet you downloaded. Click on the "Sheet 1" tab at the bottom of the canvas, to the right of the Data Source at the bottom of the screen.
- Drag "Year" to the Column shelf and "Company" to the Rows shelf. Change Year to discrete data type by selecting the drop-down menu box on the "Year" pill box and clicking "Discrete" instead of "Continuous."
- Drag "Sales revenue" and "Less sales returns" under "Measures" to the Rows shelf. Change each to discrete. Format each to Times New Roman, 10-point font, center alignment, bold, and currency (custom), 0 decimal places by selecting "Format" from the pill drop-down menu and making selections on the menu to the left. Select blue font for "Sales revenue" and red font for "Less sales returns."
- Create a calculated field by clicking the "Analysis" tab in the toolbar at the top of the screen and clicking "Create Calculated Field." A calculation window will pop up. Name the calculation "Net of returns." In the Calculation Editor window, drag "Sales revenue" from the Rows shelf to the window, type a plus sign, then drag "Less sales returns" beside it. Make sure the window says that the calculation is valid and click OK. Drag the newly created "Net of Returns" to the Rows shelf. Change to Discrete data type. Format to Times New Roman, 10-point font, bold, purple, center alignment, and currency (custom), 0 decimal places following the process above.
- Repeat the process one more time by creating a calculated field "Sales return %" that consists of typing a negative sign and then dragging "Less sales returns" divided by "Sales revenue" from the Rows shelf to the calculation window. Make sure the window says that the calculation is valid and click OK. Change to Discrete data type. Format to Times New Roman, 10-point font, bold, green, center alignment, and percentage.
- Drag the "Sales return %" to be right next to "Company" on the Rows shelf. Format to Times New Roman, 10-point font, bold, green, center alignment, and percentage.
- From the Measures shelf, drag "Sales return %" to "Color" on the "Marks" card. The graph of the "Sales return %" will appear under Year.
- Drag company to "Color" on the "Marks" card. You will now see the graph divide the companies by color.
- Click the arrow on "Sales return %" in the Rows shelf and select "sort." If the graph does not sort "descending" (largest to smallest) then select descending from the Toolbar at the top.
- Format all the labels on the sheet to Times New Roman, 12-point font, and black.
- Change the title of the sheet to be "Right of Return – Variable Consideration" by right-clicking and selecting "Edit title." Format the title to Times New Roman, bold, green and 15-point font. Change the title of "Sheet 1" to match the sheet title by right-clicking, selecting "Rename" and typing in the new title.
- Once complete, save the file as "*DA6_Your initials*.twbx."

Continuing Cases

Target Case
- LO6–2, LO6–6, LO6–7

Target Corporation prepares its financial statements according to U.S. GAAP. Target's financial statements and disclosure notes for the year ended February 3, 2018, are available in Connect. This material also is available under the Investor Relations link at the company's website (**www.target.com**).

Required:

1. On what line of Target's income statement is revenue reported? What was the amount of revenue Target reported for the fiscal year ended February 3, 2018?
2. Disclosure Note 2 indicates that Target generally records revenue in retail stores at the point of sale. Does that suggest that Target generally records revenue at a point in time or over a period of time? Explain.
3. Disclosure Note 2 indicates that customers ("guests") can return some merchandise within 90 days of purchase and can return other merchandise within a year of purchase. How is Target's revenue and net income affected by returns, given that it does not know at the time a sale is made which items will be returned?
4. Disclosure Note 2 indicates that "Commissions earned on sales generated by leased departments are included within sales and were $44 million . . . in 2017." Do you think it likely that Target is accounting for those sales as a principal or an agent? Explain.
5. Disclosure Note 2 discusses Target's accounting for gift card sales. Does Target recognize revenue when it sells a gift card to a customer? If not, when does it recognize revenue? Explain.
6. Disclosure Note 4 discussed how Target accounts for consideration received from vendors, which they call "vendor income." Does that consideration produce revenue for Target? Does that consideration produce revenue for Target's vendors? Explain.

Air France–KLM Case
- LO6–2, LO6–4, LO6–5

Air France–KLM (AF), a Franco-Dutch company, prepares its financial statements according to International Financial Reporting Standards. AF's financial statements and disclosure notes for the year ended December 31, 2017, are available in Connect. This material is also available under the Finance link at the company's website (**www.airfranceklm.com**).

Required:

1. In note 4.6, AF indicates that "Sales related to air transportation are recognized when the transportation service is provided," so passenger and freight tickets "are consequently recorded as 'Deferred revenue upon issuance date'."
   a. Examine AF's balance sheet. What is the total amount of deferred revenue on ticket sales as of December 31, 2017?
   b. When transportation services are provided with respect to the deferred revenue on ticket sales, what journal entry would AF make to reduce deferred revenue?
   c. Does AF's treatment of deferred revenue under IFRS appear consistent with how these transactions would be handled under U.S. GAAP? Explain.
2. AF has a frequent flyer program, "Flying Blue," which allows members to acquire "miles" as they fly on AF or partner airlines that are redeemable for free flights or other benefits.
   a. How does AF account for these miles?
   b. Does AF report any liability associated with these miles as of December 31, 2017?
   c. Although AF's 2017 annual report was issued prior to the effective date of *ASU No. 2014-09*, consider whether the manner in which AF accounts for its frequent flier program appears consistent with the revenue recognition guidelines included in the ASU.

CPA Exam Questions and Simulations

ROGER CPA Review

Sample CPA Exam questions from Roger CPA Review are available in Connect as support for the topics in this chapter. These multiple-choice questions and task-based simulations include expert-written explanations and solutions, and provide a starting point for students to become familiar with the content and functionality of the actual CPA Exam.

CHAPTER

7 Cash and Receivables

OVERVIEW

We begin our study of assets by looking at cash and receivables—the two assets typically listed first in a balance sheet. For cash, the key issues are internal control and classification in the balance sheet. For receivables, the key issues are valuation and the related income statement effects of transactions involving accounts receivable and notes receivable.

LEARNING OBJECTIVES

After studying this chapter, you should be able to:

- **LO7–1** Define what is meant by internal control and describe some key elements of an internal control system for cash receipts and disbursements. *(p. 340)*
- **LO7–2** Explain the possible restrictions on cash and their implications for classification in the balance sheet. *(p. 341)*
- **LO7–3** Distinguish between the gross and net methods of accounting for cash discounts. *(p. 344)*
- **LO7–4** Describe the accounting treatment for merchandise returns. *(p. 346)*
- **LO7–5** Describe the accounting treatment of anticipated uncollectible accounts receivable. *(p. 350)*
- **LO7–6** Describe how to estimate the allowance for uncollectible accounts and introduce the CECL model. *(p. 352)*
- **LO7–7** Describe the accounting treatment of notes receivable. *(p. 356)*
- **LO7–8** Differentiate between the use of receivables in financing arrangements accounted for as a secured borrowing and those accounted for as a sale. *(p. 362)*
- **LO7–9** Describe the variables that influence a company's investment in receivables and calculate the key ratios used by analysts to monitor that investment. *(p. 370)*
- **LO7–10** Discuss the primary differences between U.S. GAAP and IFRS with respect to cash and receivables. *(pp. 342, 361, and 369)*

©PhotosIndia.com/Glow Images

FINANCIAL REPORTING CASE

Bad Debt Trouble

Your roommate, Karen Buckley, was searching for some information about her future employer, **Community Health Systems**. Karen, a nursing major, noticed an article online titled "CHS Missed Big in Q4." "This doesn't look good," Karen said. "This article says my new employer's provision for bad debts is $200 million. Does that mean those patients haven't paid? What is CHS doing for cash—I want my paycheck to clear!" You look over the article and start to explain. "First of all, the term *provision* just means expense. The company uses what is called the allowance method to account for bad debts. It looks like CHS recorded more in expense and increased the allowance for uncollectible accounts. That doesn't necessarily mean they have a cash flow problem." Karen was not happy with your answer. "So, it isn't that the patients haven't paid, it's that they aren't going to pay? How does CHS know what they are going to do? And here is another article saying that CHS is securitizing its receivables to get more cash. Does that mean they are making the receivables more secure?" "Okay," you offer, "let's start at the beginning."

QUESTIONS

By the time you finish this chapter, you should be able to respond appropriately to the questions posed in this case. Compare your response to the solution provided at the end of the chapter.

1. Explain the allowance method of accounting for bad debts. (*p. 350*)
2. What approaches might CHS have used to arrive at the $200 million bad debt provision? (*p. 351*)
3. Are there any alternatives to the allowance method? (*p. 354*)
4. What does it mean for CHS to securitize its receivables? (*p. 364*)

In the earlier chapters of this text, we studied the underlying measurement and reporting concepts for the basic financial statements presented to external decision makers. Now we turn our attention to the elements of those financial statements. Specifically, we further explore the elements of the balance sheet, and also consider the income statement effects of transactions involving these elements. We first address assets, then liabilities, and finally shareholders' equity. This chapter focuses on the current assets cash and cash equivalents and receivables.

PART A

Cash and Cash Equivalents

Cash includes currency and coins, balances in checking accounts, and items acceptable for deposit in these accounts, such as checks and money orders received from customers. These forms of cash represent amounts readily available to pay off debt or to use in operations, without any legal or contractual restriction.

Managers typically invest temporarily idle cash to earn interest on those funds rather than keep an unnecessarily large checking account. These amounts are essentially equivalent to cash because they can quickly become available for use as cash. So, short-term, highly liquid investments that can be readily converted to cash with little risk of loss are viewed as cash equivalents. For financial reporting, we make no distinction between cash in the form of currency or bank account balances and amounts held in cash-equivalent investments.

A company's policy concerning which short-term, highly liquid investments it classifies as *cash equivalents* should be described in a disclosure note.

Cash equivalents include money market funds, treasury bills, and commercial paper. To be classified as cash equivalents, these investments must have a maturity date no longer than three months *from the date of purchase.* Companies are permitted flexibility in designating cash equivalents and must establish individual policies regarding which short-term, highly liquid investments are classified as cash equivalents. A company's policy should be consistent with the usual motivation for acquiring these investments. The policy should be disclosed in the notes to the financial statements.

Illustration 7–1 shows a note from the 2017 annual report of **Walgreens Boots Alliance, Inc.**, which operates the second largest drugstore chain in the United States.

Illustration 7–1

Disclosure of Cash Equivalents—Walgreens Boots Alliance, Inc.

Real World Financials

Note 2: Summary of Major Accounting Policies (in part) Cash and Cash Equivalents

Cash and cash equivalents include cash on hand and all highly liquid investments with an original maturity of three months or less. Credit and debit card receivables from banks, which generally settle within two to seven business days, of $98 million and $114 million were included in cash and cash equivalents at August 31, 2017 and 2016, respectively.

Source: Walgreens Boots Alliance, Inc.

Credit and debit card receivables often are included in cash equivalents.

The measurement and reporting of cash and cash equivalents are largely straightforward because cash generally presents no measurement problems. It is the standard medium of exchange and the basis for measuring assets and liabilities. Cash and cash equivalents usually are combined and reported as a single amount in the balance sheet. However, cash that is not available for use in current operations because it is restricted for a special purpose usually is classified in one of the noncurrent asset categories. Restricted cash is discussed later in this chapter.

All assets must be safeguarded against possible misuse. However, cash is the most liquid asset and the asset most easily stolen. As a result, a system of internal control of cash is a key accounting issue.

Internal Control

LO7–1

The success of any business enterprise depends on an effective system of **internal control**. Internal control refers to a company's plan to (a) encourage adherence to company policies and procedures, (b) promote operational efficiency, (c) minimize errors and theft, and (d) enhance the reliability and accuracy of accounting data. From a financial accounting perspective, the focus is on controls intended to improve the accuracy and reliability of accounting information and to safeguard the company's assets.

The Sarbanes-Oxley Act requires a company to document and assess its internal controls. Auditors express an opinion on management's assessment.

Recall from our discussion in Chapter 1 that Section 404 of the *Sarbanes-Oxley Act of 2002* requires that companies document their internal controls and assess their adequacy. The Public Company Accounting Oversight Board AS 2201 further requires the auditor to express its own opinion on whether the company has maintained effective internal control over financial reporting.

Many companies have incurred significant costs in an effort to comply with the requirements of Section 404.[1] A framework for designing an internal control system is provided by the *Committee of Sponsoring Organizations (COSO)* of the Treadway Commission.[2] Formed in 1985, the organization is dedicated to improving the quality of financial reporting through, among other things, effective internal controls.

COSO defines internal control as a process, undertaken by an entity's board of directors, management, and other personnel, designed to provide reasonable assurance regarding the achievement of objectives in the following categories:

- Effectiveness and efficiency of operations.
- Reliability of financial reporting.
- Compliance with applicable laws and regulations.[3]

[1]PCAOB AS 2201 emphasizes audit efficiency with a focused, risk-based testing approach that is intended to reduce the total costs of 404 compliance.

[2]The sponsoring organizations include the AICPA, the Financial Executives International, the Institute of Internal Auditors, the American Accounting Association, and the Institute of Management Accountants.

[3]**www.coso.org**.

Internal Control Procedures—Cash Receipts

As cash is the most liquid of all assets, a well-designed and functioning system of internal control must surround all cash transactions. **Separation of duties** is critical. Individuals that have physical responsibility for assets should not also have access to accounting records. So, employees who handle cash should not be involved in or have access to accounting records nor be involved in the reconciliation of cash book balances to bank balances.

Employees involved in recordkeeping should not also have physical access to the assets.

Consider the cash receipts process. Most nonretail businesses receive payment for goods by checks received through the mail. An approach to internal control over cash receipts that utilizes separation of duties might include the following steps:

1. Employee A opens the mail each day and prepares a multicopy listing of all checks including the amount and payor's name.
2. Employee B takes the checks, along with one copy of the listing, to the person responsible for depositing the checks in the company's bank account.
3. A second copy of the check listing is sent to the accounting department, where Employee C enters receipts into the accounting records.

Good internal control helps ensure accuracy as well as safeguard against theft. The bank-generated deposit slip can be compared with the check listing to verify that the amounts received were also deposited. And, because the person opening the mail is not the person who maintains the accounting records, it's impossible for one person to steal checks and alter accounting records to cover up their theft.

Internal Control Procedures—Cash Disbursements

Proper controls for cash disbursements should be designed to prevent any unauthorized payments and ensure that disbursements are recorded in the proper accounts. Important elements of a cash disbursement control system include the following:

1. All disbursements, other than very small disbursements from petty cash, should be made by check. This provides a permanent record of all disbursements.
2. All expenditures should be *authorized* before a check is prepared. For example, a vendor invoice for the purchase of inventory should be compared with the purchase order and receiving report to ensure the accuracy of quantity, price, part numbers, and so on. This process should include verification of the proper ledger accounts to be debited.
3. Checks should be signed only by authorized individuals.

Once again, separation of duties is important. Responsibilities for check signing, check writing, check mailing, cash disbursement documentation, and recordkeeping should be separated whenever possible. That way, a single person can't write checks to himself and disguise that theft as a payment to an approved vendor.

An important part of any system of internal control of cash is the periodic reconciliation of book balances and bank balances to the correct balance. In addition, a petty cash system is employed by many business enterprises. We cover these two topics in Appendix 7A.

Restricted Cash and Compensating Balances

● LO7–2

We discussed the classification of assets and liabilities in Chapter 3. You should recall that only cash available for current operations or to satisfy current liabilities is classified as a current asset. Cash that is restricted in some way and not available for current use usually is reported as a noncurrent asset such as *investments* or *other assets*.

Restrictions on cash can be informal, arising from management's intent to use a certain amount of cash for a specific purpose. For example, a company may set aside funds for future plant expansion. This cash, if material, should be classified as investments or other assets. Sometimes restrictions are contractually imposed. Debt instruments, for instance, frequently require the borrower to set aside funds (often referred to as a sinking fund) for the future payment of a debt. In these instances, the restricted cash is classified as noncurrent investments or other assets if the debt is classified as noncurrent. On the other hand, if the liability is current, the restricted cash also is classified as current. Disclosure notes should describe any material restrictions of cash and indicate the amounts and line items in which restricted cash appears in the balance sheet. Also, on the statement of cash flows, restricted cash and cash equivalents should be included with cash and cash equivalents when reconciling the beginning-of-period and end-of-period cash balances.

Banks frequently require cash restrictions in connection with loans or loan commitments (lines of credit). Typically, the borrower is asked to maintain a specified balance in a low interest or noninterest-bearing account at the bank (creditor). The required balance usually is some percentage of the committed amount (say 2% to 5%). These are known as **compensating balances** because they compensate the bank for granting the loan or extending the line of credit.

The effect of a *compensating balance* is a higher effective interest rate on the debt.

A compensating balance results in the borrower's paying an effective interest rate higher than the stated rate on the debt. For example, suppose that a company borrows $10,000,000 from a bank at an interest rate of 12%. If the bank requires a compensating balance of $2,000,000 to be held in a noninterest-bearing checking account, the company really is borrowing only $8,000,000 (the loan less the compensating balance). This means an effective interest rate of 15% ($1,200,000 interest divided by $8,000,000 cash available for use).

A material compensating balance must be disclosed regardless of the classification of the cash.

The classification and disclosure of a compensating balance depends on the nature of the restriction and the classification of the related debt.[4] If the restriction is legally binding, the cash is classified as either current or noncurrent (investments or other assets) depending on the classification of the related debt. In either case, note disclosure is appropriate.

If the compensating balance arrangement is informal with no contractual agreement that restricts the use of cash, the compensating balance can be reported as part of cash and cash equivalents, with note disclosure of the arrangement.

Illustration 7–2 provides an example of a note disclosure from **Walgreens**' annual report.

Illustration 7–2

Disclosure of Restricted Cash—Walgreens Boots Alliance, Inc.

Real World Financials

Note 2: Summary of Major Accounting Policies (in part) Restricted Cash

The Company is required to maintain cash deposits with certain banks which consist of deposits restricted under contractual agency agreements and cash restricted by law and other obligations. As of August 31, 2017 and 2016, the amount of such restricted cash was $202 million and $185 million, respectively, and is reported in other current assets on the Consolidated Balance Sheets.

Source: Walgreens Boots Alliance, Inc.

International Financial Reporting Standards

● LO7–10

Cash and Cash Equivalents. In general, cash and cash equivalents are treated similarly under IFRS and U.S. GAAP. One difference relates to bank overdrafts, which occur when withdrawals from a bank account exceed the available balance. U.S. GAAP requires that overdrafts typically be treated as liabilities. In contrast, *IAS No. 7* allows bank overdrafts to be offset against other cash accounts when overdrafts are payable on demand and fluctuate between positive and negative amounts as part of the normal cash management program that a company uses to minimize its cash balance.[5] For example, LaDonia Company has two cash accounts with the following balances as of December 31, 2021:

National Bank	$300,000
Central Bank	(15,000)

Under U.S. GAAP, LaDonia's 12/31/21 balance sheet would report a cash asset of $300,000 and an overdraft current liability of $15,000. Under IFRS, LaDonia would report a cash asset of $285,000.

[4] FASB ASC 210–10–S99–2: SAB Topic 6.H–Balance Sheet—Overall—SEC Materials, *Accounting Series Release 148*.

[5] "Statement of Cash Flows," *International Accounting Standard No. 7* (IASCF), as amended effective January 1, 2018, par. 8.

Decision Makers' Perspective

Cash often is called a *nonearning* asset because it earns little or no interest. For this reason, managers invest idle cash in either cash equivalents or short-term investments, both of which provide a larger return than a checking account. Management's goal is to hold the minimum amount of cash necessary to conduct normal business operations, meet its obligations, and take advantage of opportunities. Too much cash reduces profits through lost returns, while too little cash increases risk. This trade-off between risk and return is an ongoing choice made by management (internal decision makers). Whether the choice made is appropriate is an ongoing assessment made by investors and creditors (external decision makers).

A company must have cash available for the compensating balances we discussed in the previous section as well as for planned disbursements related to normal operating, investing, and financing cash flows. However, because cash inflows and outflows can vary from planned amounts, a company needs an additional cash cushion as a precaution against unexpected events. The size of the cushion depends on the company's ability to convert cash equivalents and short-term investments into cash quickly, along with its short-term borrowing capacity.

Companies hold cash to pay for planned and unplanned transactions and to satisfy compensating balance requirements.

Liquidity is a measure of a company's cash position and overall ability to obtain cash in the normal course of business to pay liabilities as they come due. A company is assumed to be liquid if it has sufficient cash or is capable of converting its other assets to cash in a relatively short period of time so that current needs can be met. Frequently, liquidity is measured with respect to the ability to pay currently maturing debt. The current ratio is one of the most common ways of measuring liquidity and is calculated by dividing current assets by current liabilities. By comparing liabilities that must be satisfied in the near term with assets that either are cash or will be converted to cash in the near term we have a base measure of a company's liquidity. We can refine the measure by adjusting for the implicit assumption of the current ratio that all current assets are equally liquid. In the acid-test or quick ratio, the numerator consists of "quick assets," which include only cash and cash equivalents, short-term investments, and accounts receivable. By eliminating inventories and prepaid expenses, the current assets that are less readily convertible into cash, we get a more precise indication of a company's short-term liquidity than with the current ratio. We discussed and illustrated these liquidity ratios in Chapter 3.

We should evaluate the adequacy of any ratio in the context of the industry in which the company operates and other specific circumstances. Bear in mind, though, that industry averages are only one indication of acceptability and any ratio is but one indication of liquidity. Profitability, for instance, is perhaps the best long-run indication of liquidity. And a company may be very efficient in managing its current assets so that, say, receivables are more liquid than they otherwise would be. The receivables turnover ratio we discuss in Part B of this chapter offers a measure of management's efficiency in this regard.

There are many techniques that a company can use to manage cash balances. A discussion of these techniques is beyond the scope of this text. However, it is sufficient here to understand that management must make important decisions related to cash that have a direct impact on a company's profitability and risk. Because the lack of prudent cash management can lead to the failure of an otherwise sound company, it is essential that managers, as well as outside investors and creditors, maintain close vigil over this facet of a company's health. ●

A manager should actively monitor the company's cash position.

Current Receivables

PART B

Receivables represent a company's claims to the future collection of cash, other assets, or services. Receivables resulting from the sale of goods or services on account (also called *credit sales*) are called **accounts receivable** and often are referred to as *trade receivables. Nontrade receivables* are those other than trade receivables and include tax refund claims, interest receivable, and advances to employees. When a receivable, trade or nontrade, is accompanied by a formal promissory note, it is referred to as a *note receivable*. We consider notes receivable after first discussing accounts receivable.

An account receivable and an account payable reflect opposite sides of the same transaction.

As you study receivables, realize that one company's claim to the future collection of cash corresponds to another company's (or individual's) obligation to pay cash. For example, one company's account receivable will be the mirror image of another company's account payable. Chapter 13 addresses accounts payable and other current liabilities.

Accounts Receivable

Typically, revenue and related accounts receivable are recognized at the point of delivery of the product or service.

Accounts receivable are created when sellers recognize revenue associated with a credit sale. Recall from Chapter 6 that revenue is recognized when a seller satisfies a performance obligation. For most products or services, the performance obligation is satisfied at the point of delivery of the product or service, so revenue and the related receivable are recognized at that time.

Accounts receivable are current assets because their normal collection period falls within one year (or the company's operating cycle, if longer).

Most businesses provide credit to their customers, either because it's not practical to require immediate cash payment or to encourage customers to purchase the company's product or service. Accounts receivable are *informal* credit arrangements supported by an invoice and normally are due in 30 to 60 days after the sale. They almost always are classified as current assets because their normal collection period falls within one year (or the company's operating cycle, if longer).

Initial Valuation of Accounts Receivable

You learned in Chapter 6 that sellers recognize an amount of revenue equal to the amount they are entitled to receive in exchange for satisfying a performance obligation. Sellers allocate the transaction price to the various performance obligations in a contract and then recognize revenue (and the corresponding receivable for credit sales) when performance obligations are satisfied. Clearly, revenue recognition and accounts receivable recognition are closely related. That means that some of the complexities that affect revenue recognition also affect accounts receivable.

Accounts receivable typically are not shown at present value.

One potential complexity relates to time value of money. Because credit sales allow a customer to get goods or services now but pay for them in the future, you can view a credit sale as providing a loan in addition to whatever goods or services are included in a contract. However, as you learned in Chapter 6, sellers can ignore this "financing component" when it is not significant, which typically is the case when receivables are due in less than one year. Therefore, sellers usually record relatively short-term accounts receivable at the entire amount the seller expects to receive, rather than at the present value of that amount.[6] For long-term receivables, the financing component is more significant and sellers have to account for it, as you will see when we cover notes receivable later in this section.

Another type of complexity relates to variable consideration. As you learned in Chapter 6, contracts can include some aspect of variable consideration that must be estimated when determining the transaction price, and therefore the amount of the receivable. In particular, contracts can allow cash discounts as well as sales returns and allowances. Let's discuss each of those complexities in turn.

Trade discounts allow a customer to pay an amount that is below the list price.

DISCOUNTS There are two types of discounts that companies commonly offer, trade discounts and cash discounts. Companies frequently offer **trade discounts** to customers, usually a percentage reduction from the list price. For example, a manufacturer might list a machine part at $2,500 but sell it to an important customer at a 10% discount. That discount of $250 is reflected by recording the sale at the agreed-upon price of $2,250.

Trade discounts are not variable consideration.

● LO7–3

Trade discounts are not variable consideration, because the amount the company is entitled to receive from the customer doesn't depend on the outcome of a *future* event. Trade discounts are simply a way to specify the current transaction price and thus affect reported revenue at that time. Many companies use trade discounts to offer reduced prices to certain customers (such as college students at the nearby pizza restaurant or senior citizens at the

[6]FASB ASC 606-10-32-15: Revenue from Contracts with Customers—Overall—Measurement—The Existence of a Significant Financing Component (previously "Revenue from Contracts with Customers (Topic 606)" *Accounting Standards Update 2014-09* (Norwalk, Conn: FASB, 2014)).

local ballgame), to give quantity discounts to large customers, or perhaps to disguise real prices from competitors.

Sales discounts, often called *cash discounts,* represent reductions in the amount to be received by the company from a customer that makes payment within a specified period of time. A sales discount provides an incentive for quick payment. The amount of a sales discount and the time period within which it's available usually are conveyed by terms like 2/10, n/30 (meaning a 2% discount if paid within 10 days, otherwise full payment within 30 days).

Sales discounts reduce the amount to be received from the customer if payment occurs within a specified short period of time.

Sales discounts are variable consideration, because the amount to be received by the company depends on a future event—whether cash is received from the customer within the discount period. However, it is usually difficult for a seller to estimate the amount of discount that will be taken with every sale. Therefore, sellers use two methods in practice that simplify the process of recording sales discounts: the *gross method* and the *net method.* To see how these methods work, consider the example in Illustration 7–3.

Sales discounts are variable consideration.

Illustration 7–3

Sales Discounts Using the Gross Method versus Net Method

The Hawthorne Manufacturing Company offers credit customers a 2% sales discount if the sales price is paid within 10 days. Any amounts not paid within 10 days are due in 30 days. These payment terms are stated as 2/10, n/30. The following events occurred:

1. On October 5, 2021, Hawthorne sold merchandise at a price of $20 000.
2. The customer paid $13,720 ($14,000 less the 2% cash discount) on October 14.
3. The customer paid the remaining balance of $6,000 on November 4.

The appropriate journal entries to record the sale and cash collection, comparing the gross and net methods are as follows:

Gross Method			Net Method		
October 5, 2021			**October 5, 2021**		
Accounts receivable..........	20,000		Accounts receivable..........	19,600	
Sales revenue..............		20,000	Sales revenue..............		19,600
October 14, 2021			**October 14, 2021**		
Cash..................................	13,720		Cash..................................	13,720	
Sales discounts...............	280		Accounts receivable..........		13,720
Accounts receivable....		14,000			
November 4, 2021			**November 4, 2021**		
Cash..................................	6,000		Cash..................................	6,000	
Accounts receivable....		6,000	Accounts receivable..........		5,880
			Sales discounts forfeited....		120

Using the **gross method**, we initially record the revenue and related receivable at the full **$20,000** price. Using the **net method**, we record revenue and the related accounts receivable at the agreed-upon price *less* the 2% discount offered, yielding **$19,600** of revenue at the time of sale. Subsequent accounting under both methods depends on whether payment occurs within the discount period.

Payment within the discount period. On October 14, Hawthorne receives payment for $14,000 of the $20,000 merchandise sold. Because payment is within the 10-day discount period, the company allows a discount of $280 ($14,000 × 2%) and receives only $13,720 ($14,000 − $280). Under the gross method, we record the discount as a debit to an account called *sales discounts.* This is a contra account to sales revenue and is deducted from sales revenue to derive the net sales revenue reported in the income statement. Under the net method, we simply debit cash and credit accounts receivable for $13,720, because the discount was recorded at the time of the sale.

The *gross method* assumes customers won't take sales discounts and then reduces revenue for any discounts taken.

Payment *not* within the discount period. On November 4, Hawthorne receives final payment for the remaining $6,000 of merchandise sold. This payment is after the 10-day discount period, and the customer forfeits the possible discount of **$120** ($6,000 × 2%).

The *net method* assumes customers will take sales discounts and then increases revenue for discounts forfeited.

Under the gross method, we simply record the collection on account for $6,000, the gross amount of receivable originally recorded. Under the net method, we reduce the receivable by $5,880 and record the $120 discount not taken by the customer as a credit to an account called *sales discounts forfeited.* This account is added to sales revenue to calculate net sales revenue.[7]

With both methods, net sales revenue ends up being reduced by only those discounts that are actually taken. Therefore, both methods get us to the same place from the perspective of net sales revenue and total net income recognized, as shown in the table below.

Revenue comparison of the gross method and the net method.

	Gross Method	Net Method
Sales revenue	$20,000	$19,600
Less: Sales discounts	(280)	–0–
Add: Discounts forfeited	0	120
Net sales revenue	$ 19,720	$ 19,720

The net method is more correct conceptually.

Which method is correct? Remember from Chapter 6 that sales revenue and the corresponding accounts receivable should be stated at the amount of consideration the seller expects to be entitled to receive. The net method typically better reflects that amount, because the discount is a savings that prudent customers are unwilling to forgo. To appreciate the size of that savings, consider a 2/10, n/30 discount offer on a $100 sale. In order to save $2 (equal to $100 × 2%), the customer must pay $98 twenty days earlier than otherwise due, effectively "investing" $98 to "earn" $2. That equals a rate of return of 2.04% ($2/$98) for a 20-day period. To convert this 20-day rate to an annual rate, we multiply by 365/20:

$$2.04\% \times 365/20 = 37.23\% \text{ effective rate}$$

Wouldn't you like to earn a sure return of over 37%? You can see why customers try to take the discount if at all possible. That's why recording the receivable net of the discount more accurately reflects the amount the seller expects to be entitled to receive. Still, even though the net method is more correct conceptually, both methods are used in practice, because many sellers prefer to record receivables at gross and then make downward adjustments at the time the receivable is collected. The dollar value of the difference between methods usually is viewed as immaterial.

LO7–4

SALES RETURNS Customers frequently are given the right to return merchandise they purchase. When practical, a customer might be given a special price reduction as an incentive to keep the merchandise they want to return.[8] We use the term **sales returns** to refer to these returns and other adjustments. Sales returns are common in industries such as food products, publishing, and retailing.

Sales returns are variable consideration.

As we discussed in Chapter 6, sales returns are a form of variable consideration. Because products might be returned or prices adjusted, there is uncertainty as to the final amount the seller will be entitled to receive (the transaction price). Recognizing returns only at the time they happen might cause revenue and profit to be overstated in the period the sale is made and understated in the return period. For example, assume merchandise is sold to a customer for $10,000 in December 2021, the last month in the selling company's fiscal year, and that the merchandise cost $6,000. The company would recognize gross profit of $4,000 in 2021 ($10,000 – $6,000). If all of the merchandise is returned in 2022, after financial statements for 2021 are issued, gross profit will be overstated in 2021 and understated in 2022 by $4,000. The 2021 balance sheet also is affected, with a refund liability understated by S10,000 and inventory understated by $6,000.

[7]Conceptually, sales discounts forfeited is similar to interest revenue, since it is extra revenue received because a receivable is outstanding for a longer period of time. In fact, prior to the effective date of *ASU 2014-09* (which specified how variable consideration should be treated in revenue recognition), sales discounts forfeited were often disclosed as interest revenue or other revenue in the income statement.

[8]Price reductions sometimes are referred to as *sales allowances* and are distinguished from situations when the products actually are returned for a refund or credit (sales returns).

To avoid overstating revenue and net assets in 2021, the seller should reduce revenue for all returns of current-period sales, including those that occur in the current period as well as those the seller estimates will occur eventually. We accomplish this by debiting a contra revenue account, **sales returns**, which reduces sales revenue indirectly. This way, the income statement will report net sales revenue (sales revenue minus actual and estimated sales returns).

Sellers should reduce revenue for estimated future sales returns.

A contra revenue account used to reduce revenue for actual and estimated sales returns.

Sales revenue	
Less: Sales returns	
Net sales revenue	

When should estimated returns be recognized? Technically, the seller should estimate returns each time a sale is made and adjust the transaction price accordingly at that time. However, as we noted in Chapter 6, it is impractical for most sellers to estimate returns every time they make a sale. For that reason, sellers typically account for any returns during the reporting period when the returns actually occur, and then use an adjusting entry at the end of the reporting period to recognize additional future returns expected to occur as a result of the current period's sales transactions. In that adjusting entry, the seller debits sales returns and credits a **refund liability** for the amount the seller estimates will be refunded to customers who make returns. We see that approach in Illustration 7–4.

A *refund liability* reflects the amount the seller estimates will be refunded to customers who make returns.

As shown in Illustration 7–4, Hawthorne initially sells merchandise to customers for $2,000,000. The company records the cash collected and sales revenue. At the same time, the company also records cost of goods sold and a decrease in inventory of $1,200,000.

Illustration 7–4

Accounting for Sales Returns

During 2021, its first year of operations, the Hawthorne Manufacturing Company sold merchandise for $2,000,000 cash. This merchandise cost Hawthorne $1,200,000 (60% of the selling price). Industry experience indicates that 10% of all sales will be returned, which equals $200,000 ($2,000,000 × 10%) in this case. Customers returned $130,000 of sales during 2021. Hawthorne uses a perpetual inventory system.

Sales of $2,000,000 occurred in 2021, with cost of goods sold of $1,200,000.	Cash	2,000,000	
	Sales revenue		2,000,000
	Cost of goods sold	1,200,000	
	Inventory		1,200,000
Sales returns of $130,000 occurred during 2021. The cost of returned inventory is $78,000 ($130,000 × 60%).	Sales returns	130,000	
	Cash		130,000
	Inventory	78,000	
	Cost of goods sold		78,000
At the end of 2021, an additional $70,000 of sales returns are expected. The cost of the inventory expected to be returned is $42,000 ($70,000 × 60%).	Sales returns	70,000	
	Refund liability		**70,000**
	Inventory—est. returns	42,000	
	Cost of goods sold		42,000
Sales returns of $70,000 occurred during 2022. The cost of returned inventory is $42,000.	**Refund liability**	**70,000**	
	Cash		70,000
	Inventory	42,000	
	Inventory—est. returns		42,000

Actual Returns. During the period, as actual returns occur, the company needs to reduce the sales revenue previously recognized. This is done by debiting sales returns, a contra revenue account. The credit to cash represents the amount of the refund. If the sales initially had been made on credit, Hawthorne would have debited accounts receivable rather than cash at the time of sale. Then, if the receivable was still outstanding at the time the return occurred, Hawthorne would credit accounts receivable to show that the receivable is no longer outstanding.

Hawthorne also must adjust inventory to account for the returned merchandise. In a perpetual inventory system we record increases (debits) and decreases (credits) in the inventory account as events occur. So, when actual returns occur, Hawthorne increases inventory to include the $78,000 cost of returned items. The return also means that the previous amount for cost of goods sold should be reduced.

Estimated Returns. At the end of the period, Hawthorne estimates that another $70,000 of sales will be returned. One common way of making this estimate would be to base it on the difference between an estimate of the total amount of returns of current year sales ($200,000) and the amount already returned during the year ($130,000). To account for this estimate of $70,000 of future returns, we reduce revenue in the current period by debiting sales returns and recognizing a refund liability. The refund liability represents an estimate of additional cash that will be refunded when customers return products in the future. Even if the initial sale was for credit and cash had not yet been received, Hawthorne still would recognize a refund liability for estimated future returns.[9] The $42,000 of inventory *expected* to be returned in 2022 is included as an asset, *Inventory—estimated returns,* which reflects Hawthorne's right to any inventory that is returned. That asset is shown in Hawthorne's 2021 balance sheet, separate from Hawthorne's normal inventory.

Be sure to notice that this accounting approach reports Hawthorne's 2021 income net of *actual* and *estimated* returns. Hawthorne's 2021 income statement reports net sales revenue of $1,800,000, which is gross sales revenue of $2,000,000 reduced by actual and estimated returns of $200,000 ($130,000 + $70,000). Hawthorne's 2021 cost of goods sold equals $1,080,000, which is $1,200,000 reduced by the cost of actual and estimated returns of $120,000 ($78,000 + $42,000).[10]

Sometimes a customer will return damaged or defective merchandise. This possibility is included in a company's estimate of returns. As we will see in Chapter 9, the amount recorded in inventory for damaged or defective merchandise reflects a value less than its original cost.

Changes in estimated returns are recorded in whatever period the estimate changes.

What happens if Hawthorne's estimate of future returns is wrong, such that returns end up being more or less than $70,000? Hawthorne won't revise prior years' financial statements to reflect the new estimate. Instead, as you learned in Chapter 6, Hawthorne must adjust the accounts to reflect the change in estimated returns, with any effect on income recognized in the period in which the adjustment is made. For example, suppose in our illustration that only $60,000 of 2021 sales were returned in 2022, instead of the $70,000 that Hawthorne anticipated, and that Hawthorne estimates that no more returns will occur. In that case, the refund liability still has a pre-adjustment balance of **$10,000** remaining from 2021 sales. Likewise, the inventory—estimated returns account still has a pre-adjustment balance of **$6,000** (60% × $10,000) remaining from 2021 sales.

---

[9]ASC 606-10-32-10 requires that sellers recognize a refund liability even if the seller has not yet been paid and has a receivable outstanding. That may seem odd, as the seller is showing a liability to refund cash it has not yet received. Depending on the specifics of the sales contract, some sellers may conclude that offsetting the refund liability against the receivable is appropriate, such that accounts receivable are shown net of refund liabilities in the balance sheet.

[10]As indicated in Chapter 6, companies technically should estimate all variable consideration, including sales returns, at the time a sale is made. The financial statement outcomes of that approach are the same as those shown in Illustration 7–4, but the journal entries that get us there are somewhat different. To adapt our example to account for estimated returns at the time of sale, look at the journal entries that Hawthorne makes at the end of 2021 to account for estimated future returns. Hawthorne would record the same journal entries at the time of sale, but for the full amount of estimated sales returns ($200,000) and estimated inventory to be returned ($200,000 × 60% = $120,000). After recording the estimated returns at the time of the sale, all *actual* returns would be accounted for as shown in our example for actual returns occurring in 2022, debiting the refund liability and crediting cash as individual returns occur.

Refund liability

1/1/2022		70,000
2022 returns	60,000	
Pre-adjustment balance		**10,000**
Change in estimate	**10,000**	
12/31/2022		-0-

Inventory—estimated returns

1/1/2022	42,000	
2022 returns		36,000
Pre-adjustment balance	**6,000**	
Change in estimate		6,000
12/31/2022	-0-	

In 2022, Hawthorne would record its change in estimated returns of 2021 sales as follows:

Refund liability	10,000	
Sales returns ($70,000 – $60,000)		10,000
Cost of goods sold (60% × $10,000)	6,000	
Inventory—estimated returns		6,000

These adjustments remove the remaining 2021 balance from the refund liability and inventory—estimated returns. They also increase net sales revenue (by reducing sales returns) and cost of goods sold in 2022, the period in which the change in estimate occurs.

How do companies estimate returns? They rely on past history, but they also consider any changes that might affect future experience. For example, changes in customer base, payment terms offered to customers, and overall economic conditions might suggest that future returns will differ from past returns. The task of estimating returns is made easier for many large retail companies whose fiscal year-end is the end of January. Since retail companies generate a large portion of their annual sales during the Christmas season, most returns from these sales already have been accounted for by the end of January. In fact, that's an important motivation for those companies to choose to end their fiscal years at the end of January.

Experience guides firms when estimating returns.

AVX Corporation is a leading manufacturer of electronic components. Illustration 7–5, drawn from AVX's 2017 annual report, describes the company's approach to estimating returns.

Revenue Recognition and Accounts Receivable (in part): Returns

Sales revenue and cost of sales reported in the statement of operations are reduced to reflect estimated returns. We record an estimated sales allowance for returns at the time of sale based on historical trends, current pricing and volume information, other market specific information, and input from sales, marketing, and other key management personnel. The amount accrued reflects the return of value of the customer's inventory. These procedures require the exercise of significant judgments. We believe that these procedures enable us to make reliable estimates of future returns. Our actual results have historically approximated our estimates. When the product is returned and verified, the customer is given credit against their accounts receivable.

Source: AVX Corporation

Illustration 7–5

Disclosure of Sales Returns Policy—AVX Corporation

Real World Financials

In some industries, returns typically are small and infrequent. Companies in these industries usually don't bother to estimate returns and simply record returns in the period they occur, because the effect on income measurement and asset valuation is immaterial. Also, companies sometimes lack sufficient information to make a good estimate of returns. As you learned in Chapter 6, companies must recognize only the amount of revenue that is probable to not require reversal in the future. That way, sales are less likely to be overstated in the period of the transfer of goods or services. Difficulty estimating returns requires a larger estimate of returns (and therefore a smaller amount of net sales revenue recognized) than would be the case if a more precise estimate was possible.

Subsequent Valuation of Accounts Receivable

● LO7–5

You learned in Chapter 6 that revenue and a corresponding receivable are recognized for the amount the seller is *entitled* to receive for satisfying a performance obligation. However, being entitled to receive payment doesn't necessarily mean that the seller *will* be paid. In fact, **credit losses** (bad debts) are an inherent cost of granting credit. How should we account for the fact that not every account receivable is likely to be collected?

DIRECT WRITE-OFF METHOD (NOT GAAP) A simple approach to recognizing bad debts that is ***not*** allowed by GAAP is to wait until a particular account is deemed uncollectible and write it off at that time. This approach is called the direct write-off method. For example, if a customer goes bankrupt and it becomes clear that a $15,000 account receivable will not be collected, the company would reduce net income for the amount of the bad debt, and eliminate the receivable from total assets, because the account is no longer expected to be collected:

Bad debt expense	15,000	
Accounts receivable		15,000

The two shortcomings of the direct write-off method are that it:

1. Overstates the balance in accounts receivable in the periods prior to the write-off, because it fails to anticipate that some accounts receivable will prove uncollectible.
2. Distorts net income by postponing recognition of any bad debt expense until the period in which the customer actually fails to pay, even though some bad debt expense was predictable before that time.

FINANCIAL Reporting Case

Q1, p. 339

These conceptual problems are why the direct write-off method isn't allowed for financial reporting purposes unless the amount of bad debt is not material. However, the direct write-off method is required for income tax purposes for most companies.

The carrying value of accounts receivable is reduced by the allowance for uncollectible accounts.

ALLOWANCE METHOD (GAAP) GAAP requires use of the **allowance method** whenever the amount of bad debts is material. Under the allowance method, companies use a contra-asset account, the **allowance for uncollectible accounts**, to reduce the carrying value of accounts receivable to the amount of cash they expect to collect.[11] Both the carrying value and the amount of the allowance typically are shown on the face of the balance sheet. For example, Illustration 7–6 shows how **Johnson & Johnson**, the large pharmaceutical company, reported accounts receivable in its comparative balance sheets for 2017 and 2016.

Illustration 7–6

Disclosure of Accounts Receivable—Johnson & Johnson

Real World Financials

	($ in millions)	
	2017	**2016**
Accounts receivable trade, less allowances for doubtful accounts $291 (2016, $252)	13,490	11,699

Source: Johnson & Johnson

Johnson & Johnson's balance sheet communicates that, as of year-end 2017, it had net accounts receivable of $13,490 and an allowance for doubtful accounts of $291, which implies a gross accounts receivable of $13,490 + $291 = $13,781.

The allowance method recognizes bad debt expense when the allowance is adjusted for estimated bad debts, not when specific accounts are written off.

Under the allowance method, bad debt expense is ***not*** recognized when specific accounts are written off. Rather, bad debt expense is recognized earlier, when accounts are ***estimated*** to be uncollectible and the allowance is created. Later, when a specific account receivable is deemed *actually* uncollectible, both the allowance and the specific account receivable are reduced to write off the receivable.

[11]You may see this carrying value referred to as "net realizable value". However, as you will learn in Chapter 9, that term has a very specific definition in GAAP that should not be misused.

An example will clarify how the allowance method works. Assume the Hawthorne Manufacturing Company started operations in 2021. It had sales of $1,200,000 and collections of $895,000, leaving a balance of **$305,000** in accounts receivable as of December 31, 2021.

Accounts Receivable

1/1/2021	-0-	
Sales	1,200,000	
Collections		895,000
12/31/2021	**305,000**	

Recognizing allowance for uncollectible accounts (end of first year). Hawthorne's analysis indicates it expects to collect $280,000 of its accounts receivable, so it must establish an allowance for uncollectible accounts of $25,000 (**$305,000** – $280,000). (Later we'll talk about how Hawthorne would estimate the allowance. For now let's assume that number.)

Hawthorne establishes the necessary allowance with the following journal entry:

Allowance for Uncollectible Accounts

1/1/2021		-0-
Bad debt expense		25,000
12/31/2021		25,000

Bad debt expense	25,000	
Allowance for uncollectible accounts		**25,000**

The balances in accounts receivable (**$305,000**) and the contra asset allowance for uncollectible accounts (**$25,000**) offset to carry receivables at a net amount of $280,000 on the balance sheet.

Of course, at this point Hawthorne doesn't know which particular accounts receivable will prove uncollectible. (If it could predict that perfectly, it wouldn't have made those sales to begin with!) Hawthorne only can record an **estimate** of the amount of uncollectible accounts for its outstanding receivables. That estimate both reduces net income (through bad debt expense) and reduces the carrying value of Hawthorne's accounts receivable.

When accounts are deemed uncollectible. On April 24, 2022, Hawthorne concludes that it will not collect a $15,000 account receivable from a customer that recently has gone bankrupt. Hawthorne would make the following journal entry:

Allowance for uncollectible accounts	**15,000**	
Accounts receivable		15,000

Accounts Receivable

1/1/2022	305,000	
Specific write-off		15,000
	290,000	

Allowance for Uncollectible Accounts

1/1/2022		25,000
Specific write-off	**15,000**	
		10,000

Note that the journal entry didn't affect net income, and that accounts receivable still are carried at a net amount of $280,000 (**$290,000** – $10,000) on the balance sheet.

The write-off of an account receivable reduces both gross receivables and the allowance, thus having no effect on net income or financial position.

When previously written-off accounts are reinstated. Occasionally, a receivable that has been written off is later reinstated because the company gets new information and now believes the receivable will be collected in part or in full. When this happens, the entry to write off the account is reversed. For example, assume Hawthorne learns on May 30 that the financial situation of its customer has improved and it will collect $1,200 of the receivable that previously was written off. Hawthorne reinstates that amount of the receivable with the following journal entry:

FINANCIAL Reporting Case

Q2, p. 339

Accounts receivable	1,200	
Allowance for uncollectible accounts		**1,200**

Accounts Receivable		
	290,000	
Reinstatement	1,200	
	291,200	

Allowance for Uncollectible Accounts		
		10,000
Reinstatement		**1,200**
		11,200

Reinstating a previously written-off accounts receivable increases both gross receivables and the allowance, thus having no effect on net income or financial position.

Note that, once again, the journal entry didn't affect net income and that accounts receivable still are carried at a net amount of $280,000 (**$291,200** – $11,200) on the balance sheet. Just as the carrying value of accounts receivable was not affected by writing off a specific receivable, it also is not affected by reversing the write-off of a specific receivable.

Once the receivable is reinstated, collection is recorded the same way it would be if the receivable had never been written off, debiting cash and crediting accounts receivable.

ESTIMATING THE ALLOWANCE FOR UNCOLLECTIBLE ACCOUNTS Basing bad debt expense on the appropriate carrying value of accounts receivable is very much a **balance sheet approach**. The company estimates what the ending balance of the allowance for uncollectible accounts should be to have the carrying value of accounts receivable reflect the net amount expected to be collected. The company then records whatever amount of bad debt expense that's necessary to adjust the allowance to that desired balance.

● LO7–6

Companies estimate the necessary balance in the allowance for uncollectible accounts using the **CECL ("Current Expected Credit Loss") model.**[12] That model does not specify a particular method for estimating bad debts ("credit losses"). Rather, it allows a company to apply any method that reasonably captures its expectation of credit losses, so that the resulting carrying value of net accounts receivable reflects the cash the company expects to collect. That estimate should consider all receivables and be based on all relevant information, including historical experience, current conditions, and reasonable and supportable forecasts. Estimation could be done by analyzing each customer account, by applying an estimate of the percentage of bad debts to the entire outstanding receivable balance, or by applying different percentages to accounts receivable balances depending on the length of time outstanding. This latter approach employs an accounts receivable aging schedule and is very common in practice. For example, Illustration 7–7 shows the aging schedule for Hawthorne's December 31, 2021 accounts receivable balance of $305,000.

Illustration 7–7

Accounts Receivable Aging Schedule

Customer	Accounts Receivable 12/31/2021	0–60 Days	61–90 Days	91–120 Days	Over 120 Days
Axel Manufacturing Co.	$ 20,000	$ 14,000	$ 6,000		
Banner Corporation	33,000		20,000	$10,000	$ 3,000
Dando Company	60,000	50,000	10,000		
...	...	...	...	...	...
...	...	...	...	...	...
Xicon Company	18,000	10,000	4,000	3,000	1,000
Totals	$305,000	$220,000	$50,000	$25,000	$10,000

Summary

Age Group	Amount	Estimated Percent Uncollectible	Estimated Allowance
0–60 days	$220,000	5%	$11,000
61–90 days	50,000	10%	5,000
91–120 days	25,000	20%	5,000
Over 120 days	10,000	40%	4,000
Allowance for uncollectible accounts			**$25,000**

[12]"Financial Instruments—Credit Losses (Topic 326)" *Accounting Standards Update 2016-13* (Norwalk, Conn: FASB, 2016).

The schedule classifies the year-end receivable balances according to their length of time outstanding. Typically, the longer an account has been outstanding, the more likely it will prove uncollectible. Therefore, the schedule applies higher estimated default percentages to groups of older receivables. Because the schedule calculates a necessary balance in the allowance of $25,000, Hawthorn would adjust the balance of the allowance to that amount, as shown previously in this section.

Recognizing allowance for uncollectible accounts (end of second year). At the end of the second year of operations (and each year thereafter), Hawthorne must once again estimate what the carrying value of its accounts receivable should be, and adjust the allowance as necessary to produce that carrying value. Let's suppose that at the end of 2022 Hawthorne has a gross accounts receivable balance of **$400,000** but believes it only will collect $360,000. That implies that Hawthorne needs an allowance of $40,000 (**$400,000 – $360,000**). Unlike the first year, the balance of the allowance for uncollectible accounts in later years may not be zero at the time Hawthorne makes the adjustment. In our example, the pre-adjustment balance is $11,200. What adjustment is needed to update the allowance account from $11,200 to the new estimate of $40,000? The adjusting entry is a "plug" of **$28,800**.

Allowance for Uncollectible Accounts

	Debit	Credit
(Pre-adjustment balance)		11,200
Bad debt expense		28,800
(Post-adjustment balance)		40,000

Bad debt expense	28,800	
Allowance for uncollectible accounts		28,800

Be sure to understand that the amount of bad debt expense recognized in this journal entry is purely a plug that is determined by the difference between the pre-adjustment balance and the necessary post-adjustment balance in the allowance. For example, if Hawthorne had written off so many accounts receivable during 2022 that the adjustment account had a pre-adjustment balance that was a *debit* of $2,000, Hawthorne would have to *credit* the allowance for **$42,000** (and debit bad debt expense for the same amount) to reach the necessary $40,000 post-adjustment balance.

Allowance for Uncollectible Accounts

	Debit	Credit	
Pre-adjustment balance	2,000		
		42,000	Bad debt expense
		40,000	Post-Adjustment Balance

On the other hand, if the pre-adjustment balance in the allowance was a *credit* of $11,200 and the quality of Hawthorne's receivables improved so much that Hawthorne believed the post-adjustment balance in the allowance should be only $5,000, Hawthorne would need to *debit* the allowance for **$6,200**.

Allowance for Uncollectible Accounts

	Debit	Credit	
		11,200	Pre-adjustment balance
Bad debt expense	6,200		
		5,000	Post-adjustment balance

A debit to the allowance requires a credit to bad debt expense. While crediting an expense looks weird, think of it as driven by a change in an estimate. The allowance was increased by too much in a prior period, so Hawthorne has to reduce it this period. As with changes in other estimates, we don't restate prior year financial statements, but instead show the effect of the change in estimate in current-period net income.

FINANCIAL Reporting Case

Q3, p. 339

Income statement approach. An alternative to the balance sheet approach is to estimate bad debt expense directly as a percentage of each period's net credit sales. This percentage usually is determined by reviewing the company's recent history of the relationship between credit sales and actual bad debts. For a relatively new company, this percentage may be estimated by referring to other sources such as industry averages. For example, if Hawthorne had sales of $1,200,000 in 2021, and estimated that 2% of those sales would prove uncollectible, it would debit bad debt expense and credit the allowance for uncollectible accounts for $24,000 (= $1,200,000 × 2%).

It's important to notice that this **income statement approach** focuses on the current year's credit sales, so the effect on the balance sheet—the allowance for uncollectible accounts and hence net accounts receivable—is an incidental result of estimating the expense. An income statement approach should be used only if it doesn't provide a distorted estimate of the net amount of cash that is expected to be collected from accounts receivable.

Combined approaches. Some companies use a combination of approaches when estimating bad debts. For example, Hawthorne might decide it's more convenient to estimate bad debts on a quarterly basis using the income statement approach and then refine the estimate by employing the balance sheet approach at the end of the year based on an aging of receivables. In our example, Hawthorne recognized $24,000 in the allowance and the bad debt expense accounts during the year under the income statement approach. If at the end of the year Hawthorne's aging schedule indicated that a total of $25,000 was needed in the allowance, Hawthorne would recognize an additional **$1,000** of allowance and bad debt expense to adjust the total to the necessary balance.

Allowance for Uncollectible Accounts

	24,000
	1,000
	25,000

Bad debt expense	1,000	
Allowance for uncollectible accounts		1,000

Ethical Dilemma

The management of the Auto Parts Division of the Santana Corporation receives a bonus if the division's income achieves a specific target. For 2021 the target will be achieved by a wide margin. Mary Beth Williams, the controller of the division, has been asked by Philip Stanton, the head of the division's management team, to try to reduce this year's income and "bank" some of the profits for future years. Mary Beth suggests that the division's bad debt expense as a percentage of the gross accounts receivable balance for 2021 be increased from 3% to 5%. She believes that 3% is the more accurate estimate but knows that both the corporation's internal auditors as well as the external auditors allow some flexibility when estimates are involved. Does Mary Beth's proposal present an ethical dilemma?

Illustration 7–8 summarizes the key issues involving measuring and reporting accounts receivable.

Illustration 7–8

Measuring and Reporting Accounts Receivable

Recognition	Depends on revenue recognition; for most credit sales, revenue and the related receivables are recognized at the point of delivery.
Initial valuation	Initially recorded at the amount of consideration the seller is entitled to receive. Affected by cash discounts and variable consideration such as sales discounts and sales returns.
Subsequent valuation	Initial valuation reduced by allowance for uncollectible accounts, so accounts receivable are shown at the amount of cash the seller expects to receive.
Classification	Almost always classified as a current asset.

Additional Consideration

Guidance in this area has changed. Companies are required to use the **CECL ("Current Expected Credit Loss") model** starting in 2020, and companies can choose to use that approach starting in 2019.[13] We emphasize the CECL model because it will be in effect during your career.

GAAP in place prior to the CECL model accounted for bad debts as loss contingencies, because the amount of loss depends on some future event, like a customer's failure to pay. As discussed further in Chapter 13, an allowance and corresponding expense is only recognized for a contingent loss if the loss is both probable to occur and can be reasonably estimated, based on events that have occurred to date. The CECL model was developed after the global financial crisis of 2008/2009 because of a concern that this preexisting approach could understate bad debt expense. The "probable" threshold for identifying bad debts means that sellers don't record an estimate of bad debts that they view as only possible. Also, basing estimates only on events that already have occurred could ignore forecasts of future economic conditions that suggest impending problems. Given these differences, it is likely that bad debts accruals will be larger under the CECL model than under preexisting GAAP.

Concept Review Exercise

UNCOLLECTIBLE ACCOUNTS RECEIVABLE

The Crowe Company offers 30 days of credit to its customers. At the end of the year, bad debts expense is estimated and the allowance for uncollectible accounts is adjusted based on an aging of accounts receivable. The company began 2022 with the following balances in its accounts:

Accounts receivable	$350,000
Allowance for uncollectible accounts	(30,000)
Net accounts receivable	$320,000

During 2022, sales on credit were $1,300,000, cash collections from customers were $1,253,000, and actual write-offs of accounts were $25,000.

Crowe Company Accounts Receivable Aging
December 31, 2022

		Summary
Age Group	**Amount**	**Estimated Percent Uncollectible**
0–60 days	$250,000	6%
61–90 days	80,000	15%
91–120 days	32,000	25%
Over 121 days	10,000	50%
Total	$372,000	

Required:

1. Determine the balances in accounts receivable and allowance for uncollectible accounts at the end of 2022.
2. Determine bad debt expense for 2022.
3. Prepare journal entries to write off receivables and to recognize bad debt expense for 2022.

[13] "Financial Instruments—Credit Losses (Topic 326)" *Accounting Standards Update 2016-13* (Norwalk, Conn.: FASB, 2016).

Solution:

1. Determine the balances in accounts receivable and allowance for uncollectible accounts at the end of 2022.

Accounts Receivable

	Debit	Credit	
12/31/2021	350,000		
Sales	1,300,000		
		1,253,000	Collections
		25,000	Write-offs
12/31/2022	**372,000**		

Crowe Company Accounts Receivable Aging December 31, 2022

Summary

Age Group	Amount	Estimated Percent Uncollectible	Estimated Allowance
0–60 days	$250,000	6%	$15,000
61–90 days	80,000	15%	12,000
91–120 days	32,000	25%	8,000
Over 121 days	10,000	50%	5,000
Total	**$372,000**		**$40,000**

Allowance for Uncollectible Accounts

	Debit	Credit	
		30,000	12/31/2021
Write-offs	25,000		
		35,000	Bad debt expense
		40,000	12/31/2022

As shown in the T-account, the ending balance in accounts receivable is **$372,000**. As shown in the calculation within the aging of accounts receivable, the ending balance in the allowance for uncollectible accounts is **$40,000.**

2. Determine bad debt expense for 2022.

   As shown in the T-account above, the plug necessary to reach the required ending balance in the allowance for uncollectible accounts is **$35,000,** so that is bad debt expense for 2022.

3. Prepare journal entries to write off receivables and to recognize bad debt expense for 2022.

	Debit	Credit
Allowance for uncollectible accounts	25,000	
Accounts receivable		25,000
Write-off of accounts receivable as they are determined uncollectible.		
Bad debt expense	35,000	
Allowance for uncollectible accounts		**35,000**
Year-end adjusting entry for bad debts.		

Notes Receivable

LO7–7

Notes receivable are formal credit arrangements between a creditor (lender) and a debtor (borrower) that specify payment terms. Notes arise from loans made to people and companies who need cash, including stockholders and employees, as well as from the sale of merchandise, other assets, or services under relatively long-term payment arrangements. Notes receivable are classified as either current or noncurrent depending on the expected collection date(s).

When the term of a note is less than a year, it is reported as a short-term note. We start by discussing short-term notes, and then discuss long-term notes.

Short-Term Interest-Bearing Notes

The typical **interest-bearing note receivable** involves collection of a specified face amount, also called principal, at a specified maturity date or dates. In addition, interest is received at a stated percentage of the face amount. Interest on notes is calculated as

Face amount × Annual rate × Fraction of the annual period

For an example, consider Illustration 7–9.

Illustration 7–9
Interest-Bearing Notes Receivable

The Stridewell Wholesale Shoe Company manufactures athletic shoes that it sells to retailers. On May 1, 2021, the company sold shoes to Harmon Sporting Goods. Stridewell agreed to accept a $700,000, 6-month, 12% note in payment for the shoes. Interest is receivable at maturity. Assume that an interest rate of 12% is appropriate for a note of this type. Stridewell would account for the note as follows:*

May 1, 2021		
Notes receivable	700,000	
Sales revenue		700,000
To record the sale of goods in exchange for a note receivable.		
November 1, 2021		
Cash ($700,000 + $42,000)	742,000	
Interest revenue ($700,000 × 12% × 6/12)		42,000
Notes receivable		700,000
To record the collection of the note at maturity.		

*To focus on recording the note we intentionally omit the entry required for the cost of the goods sold if the perpetual inventory system is used.

If the sale in the illustration instead occurred on August 1, 2021, and the company's fiscal year-end is December 31, a year-end adjusting entry accrues interest earned.

December 31, 2021		
Interest receivable	35,000	
Interest revenue ($700,000 × 12% × 5/12)		35,000

The February 1 collection is then recorded as follows:

February 1, 2022		
Cash ($700,000 + [$700,000 × 12% × 6/12])	742,000	
Interest revenue ($700,000 × 12% × 1/12)		7,000
Interest receivable (accrued at December 31, 2021)		35,000
Notes receivable		700,000

Short-Term Noninterest-Bearing Notes

Sometimes a receivable assumes the form of a so-called **noninterest-bearing note**. The name is a misnomer, though. Noninterest-bearing notes actually do bear interest, but the interest is deducted (or *discounted*) from the face amount to determine the cash proceeds made available to the borrower at the outset. For example, what if Stridewell accepted a six-month, $700,000 noninterest-bearing note in exchange for delivering goods that have a cash sales price of $658,000?

You learned in Chapter 6 that, as a practical matter, a seller can assume the time value of money is not significant if the period between delivery and payment is less than a year. If Stridewell views the financing component of the note to be insignificant, it would ignore the interest rate and simply record sales revenue and an account receivable for $700,000. However, Stridewell could view the financing component of the contract to be significant. In that case, **$42,000** (= $700,000 − $658,000) interest would be discounted at the outset rather than explicitly stated, and sales revenue of $658,000 recognized upon delivery. Stridewell still receives total revenue of $700,000, but that total revenue is recognized as sales revenue of $658,000 and interest revenue of $42,000. Assuming a May 1, 2021 sale, the transaction is recorded as follows:[14]

May 1, 2021		
Notes receivable (face amount)	700,000	
Discount on notes receivable (to balance)		**42,000**
Sales revenue		658,000
November 1, 2021		
Discount on notes receivable	**42,000**	
Interest revenue		42,000
Cash	700,000	
Notes receivable (face amount)		700,000

The discount on notes receivable is a contra account to the notes receivable account. That is, the notes receivable would be reported in the balance sheet net (less) any remaining discount. The discount represents future interest revenue that will be recognized as it is earned over time. We can calculate the effective interest rate as follows.

	$ 42,000	Interest for 6 months
÷	$658,000	Sales price
=	6.383%	Rate for 6 months
×	2*	To annualize the rate
=	12.766%	Effective interest rate

*Two 6-month periods

Using a calculator, enter: N 1 I .06383 PMT − 700000
Output: PV 658,000

Using Excel, enter:
= PV (.06383,1,700000)
Output: 658,000

This note should be valued at the present value of future cash receipts. The present value of $700,000 to be received in six months using an effective interest rate of 6.383% is $658,000 (= $700,000 ÷ 1.06383 = $658,000). The use of present value techniques for valuation purposes was introduced in Chapter 5, and we'll use these techniques extensively in subsequent chapters to value various long-term notes receivable, investments, and liabilities.

In the illustration, if the sale occurs on August 1, the December 31, 2021, adjusting entry and the entry to record the cash collection on February 1, 2022, are recorded as follows:

December 31, 2021		
Discount on notes receivable	35,000	
Interest revenue ($42,000 × 5/12)*		35,000
February 1, 2022		
Discount on notes receivable	7,000	
Interest revenue ($42,000 × 1/12)		7,000
Cash	700,000	
Notes receivable (face amount)		700,000

*We also can calculate interest revenue by multiplying the net notes receivable balance by the effective interest rate ($658,000 × 12.766% × 5/12 = $35,000).

[14] The entries shown assume the note is recorded by the gross method. By the net method, the interest component is netted against the face amount of the note as follows:

May 1, 2016		
Notes receivable	658,000	
Sales revenue		658,000
November 1, 2016		
Cash	700,000	
Notes receivable		658,000
Interest revenue ($700,000 × 12% × 6/12)		42,000

In the December 31, 2021 balance sheet, the note receivable is shown at $693,000: the face amount ($700,000) less remaining discount ($7,000).

Long-Term Notes Receivable

We account for long-term notes receivable the same way we account for short-term notes receivable, but the time value of money has a larger effect. Therefore, if long-term notes are received in exchange for goods and services, the financing component of the transaction typically is viewed as significant for revenue recognition purposes. To provide an example, Illustration 7–10 modifies the facts in Illustration 7–9 to have Stridewell agree on January 1, 2021, to accept a two-year noninterest-bearing note.

Illustration 7–10

Long-Term Noninterest-Bearing Notes Receivable

The Stridewell Wholesale Shoe Company manufactures athletic shoes that it sells to retailers. On January 1, 2021 the company sold shoes to Harmon Sporting Goods. Stridewell agreed to accept a $700,000, two-year note in payment for the shoes. Assuming a 12% effective interest rate, Stridewell would account for the note as follows:*

January 1, 2021		
Notes receivable	700,000	
Discount on notes receivable		141,964
Sales revenue†		**558,036**
To record the sale of goods in exchange for a two-year note receivable.		
December 31, 2021		
Discount on notes receivable	66,964	
Interest revenue (**$558,036** × 12%)		66,964
To record interest revenue in 2021.		
December 31, 2022		
Cash	700,000	
Discount on notes receivable	75,000	
Interest revenue [(**$558,036** + **$66,964**) × 12%]		75,000
Notes receivable		700,000
To record interest revenue in 2022 and collection of the note.		

*To focus on recording the note we intentionally omit the entry required for the cost of the goods sold if the perpetual inventory system is used.

†$700,000 × Present value of $1; $n = 2$, $i = 12\%$

Using a calculator, enter: N 2 I .12 PMT – 700000
Output: PV 558,036

Using Excel, enter: = PV(.12,2,700000)
Output: 558,036

It's useful to consider a couple of aspects of Illustration 7–10. First, notice that the sales revenue in 2021 is not recorded for the full $700,000. Instead, sales revenue is calculated as (**$558,036**), the *present value* of the amount ($700,000) to be received in two years. The net amount of the note receivable equals its present value ($558,036), which is the notes receivable of $700,000 less the discount on notes receivable of $141,964. Recall from Chapter 6 that sellers need to recognize the financing component of a contract when the time value of money is significant. That is what Stridewell is doing. It earns sales revenue of $558,036 from selling shoes, and interest revenue totaling $141,964 (= $66,964 + $75,000) for financing the transaction.

Second, interest revenue each period is calculated based on the *net* notes receivable as of the beginning of that period. As a consequence, interest revenue differs between periods because the net notes receivable increases over time. In our example, Stridewell's 2021 interest revenue of **$66,964** is calculated based on the initial net notes receivable balance ($558,036 × 12%). We reduce the discount on notes receivable for 2021 interest revenue, which increases the net notes receivable balance to $625,000 (= **$558,036** + **$66,964**). So, Stridewell's 2022 interest revenue of $75,000 is calculated based on that higher net notes receivable balance ($625,000 × 12%), reducing the discount on notes receivable to zero by the end of 2022. At that point, the book value of the note, sometimes called the carrying value, carrying amount, or amortized cost basis, is $700,000, and that is the amount of cash that is collected. The effective interest rate stays the same over time, but interest revenue increases as that rate is multiplied by a receivable balance that increases over time. This is an application of the **effective interest method**, by which we calculate interest revenue or interest expense by multiplying the outstanding balance of a long-term receivable or

In the *effective interest method*, interest is determined by multiplying the outstanding balance by the effective interest rate.

liability by the effective interest rate. As you will see in later chapters, that method is used for all long-term receivables and liabilities.

We discuss long-term notes receivable in greater detail in Chapter 14 at the same time we discuss accounting for long-term notes payable. That discussion provides more examples of accounting for long-term noninterest-bearing notes, as well as long-term interest-bearing notes, and also considers accounting for notes receivable that are collected in equal installment payments.

Marriott International accepts both interest-bearing and noninterest-bearing notes from developers and franchisees of new hotels. The 2016 disclosure note shown in Illustration 7–11 describes the company's accounting policy for these notes.

Illustration 7–11

Disclosure of Notes Receivable—Marriott International

Real World Financials

Summary of Significant Accounting Policies (in part)

We may make senior, mezzanine, and other loans to owners of hotels that we operate or franchise, generally to facilitate the development of a hotel and sometimes to facilitate brand programs or initiatives. We expect the owners to repay the loans in accordance with the loan agreements, or earlier as the hotels mature and capital markets permit.

Note 9: Notes Receivable (in part)

Notes Receivable Principal Payments (net of reserves and unamortized discounts) and Interest Rates ($ in millions)	**Amount**
Balance at year-end 2016	$248
Range of stated interest rates at year-end 2016	0 to 18%

When a noninterest-bearing note is received solely in exchange for cash, the amount of cash exchanged is the basis for valuing the note.

NOTES RECEIVED SOLELY FOR CASH If a note with an unrealistic interest rate—even a noninterest-bearing note—is received *solely* in exchange for cash, the cash paid to the issuer is considered to be the note's present value.[15] Even if this means recording interest at a ridiculously low or zero rate, the amount of cash exchanged is the basis for valuing the note. If the noninterest-bearing note in the previous example had been received solely in exchange for $700,000 cash, the transaction would be recorded as follows:

Notes receivable (face amount)	700,000	
Cash (given)		700,000

Subsequent Valuation of Notes Receivable

Similar to accounts receivable, companies are required to estimate bad debts (also called credit losses) on notes receivable, and use an allowance account to reduce the receivable to the appropriate carrying value. The process of recording bad debt expense is the same as with accounts receivable.

Under the CECL model, companies consider all relevant information when assessing credit losses, including reasonable and supportable forecasts about the future. Companies can use a variety of approaches to estimate uncollectible notes receivable, including the aging method illustrated for accounts receivable. They often use discounted cash flow techniques, estimating the amount of necessary allowance by comparing the balance in notes receivable to the present value of the cash flows expected to be received, discounted at the interest rate that was effective when each note was initially recognized. An example of that approach is included in Appendix 7B.

One of the more difficult measurement problems facing banks and other lending institutions is the estimation of bad debts on their long-term notes (loans). It's difficult to predict uncollectible accounts for these arrangements. As an example, **Wells Fargo & Company,** a large bank holding company, reported the following in the asset section of its December 31, 2016, balance sheet:

[15]This assumes that no other present or future considerations are included in the agreement. For example, a noninterest-bearing note might be given to a vendor in exchange for cash *and* a promise to provide future inventories at prices lower than anticipated market prices. The issuer values the note at the present value of cash payments using a realistic interest rate, and the difference between present value and cash payments is recognized as interest revenue over the life of the note. This difference also increases future inventory purchases to realistic market prices.

($ in millions)	December 31, 2017	December 31, 2016
Loans	$956,770	$967,604
Allowance for loan losses	(11,004)	(11,419)
Net loans	$945,766	$956,185

A disclosure note, reproduced in Illustration 7–12, describes Wells Fargo's loan loss policy.

Illustration 7–12
Disclosure of Allowance for Loan Losses—Wells Fargo & Company

Real World Financials

Allowance for Credit Losses (ACL) (in part)

The allowance for credit losses is management's estimate of credit losses inherent in the loan portfolio, including unfunded credit commitments, at the balance sheet date. We have an established process to determine the appropriateness of the allowance for credit losses that assesses the losses inherent in our portfolio and related unfunded credit commitments. We develop and document our allowance methodology at the portfolio segment level—commercial loan portfolio and consumer loan portfolio. While we attribute portions of the allowance to our respective commercial and consumer portfolio segments, the entire allowance is available to absorb credit losses inherent in the total loan portfolio and unfunded credit commitments.

Source: Wells Fargo & Company

Sometimes a company grants a concession to a borrower because the borrower faces financial difficulties. In the case of such a *troubled debt restructuring,* the creditor modifies the terms of the receivable and no longer expects to collect all amounts that were due to it originally. Accounting for troubled debt restructurings is discussed in Appendix 7B.

GAAP requires that companies disclose the fair value of their notes receivable in the disclosure notes (they don't have to disclose the fair value of accounts receivable when the book value of the receivables approximates fair value).[16] Also, companies can choose to carry receivables at fair value in their balance sheets, with changes in fair value recognized as gains or losses in the income statements.[17] This "fair value option" is discussed in Chapter 12.

International Financial Reporting Standards

● LO7–10

Accounts and Notes Receivable. *IFRS No. 9* governs treatment of accounts and notes receivable. Its requirements are similar to U.S. GAAP with respect to trade and cash discounts, sales returns, recognition of interest on notes receivable, and use of an allowance for uncollectible accounts (which typically is called a "provision for bad debts" under IFRS).[18] A few key differences remain. IFRS and U.S. GAAP both allow a "fair value option" for accounting for receivables, but the IFRS standard restricts the circumstances in which that option is allowed (we discuss this more in Chapter 12). Also, U.S. GAAP allows "available for sale" accounting for investments in debt securities, but *IFRS No. 9* does not allow that approach (we also discuss that approach further in Chapter 12). Also, U.S. GAAP requires more disaggregation of accounts and notes receivable in the balance sheet or notes to the financial statements. For example, companies need to separately disclose accounts receivable from customers, from related parties, and from others. IFRS recommends but does not require separate disclosure.

A final important difference relates to estimating bad debts. IFRS uses the ECL ("Expected Credit Loss") model. For most receivables, the ECL model reports a "12-month ECL," which bases expected credit losses only on defaults that could occur within the next twelve months. Only if a receivable's credit quality has deteriorated significantly does the creditor instead report the "lifetime ECL," which also includes credit losses expected to occur from defaults after twelve months, as is done for all receivables under the CECL model used in U.S. GAAP. As a result of this lack of convergence, it is likely that accruals for credit losses under IFRS will be lower, and occur later, than under U.S. GAAP.

[16]FASB ASC 825–10–50–10: Financial Instruments—Overall—Disclosure—Fair Value of Financial Instruments.
[17]FASB ASC 825–10–25: Financial Instruments—Overall—Recognition Fair Value Option.
[18]Source: "Financial Instruments," *International Financial Reporting Standard No. 9* (IASCF), as amended effective January 1, 2018.

Financing with Receivables

● LO7–8

Financial institutions have developed a wide variety of ways for companies to use their receivables to obtain immediate cash. Companies can find this attractive because it shortens their operating cycles by providing cash immediately rather than having to wait until credit customers pay the amounts due. Also, many companies avoid the difficulties of servicing (billing and collecting) receivables by having financial institutions take on that role. Of course, financial institutions require compensation for providing these services, usually interest and/or a finance charge.

The various approaches used to finance with receivables differ with respect to which rights and risks are retained by the *transferor* (the company who was the original holder of the receivables) and which are passed on to the *transferee* (the new holder, the financial institution). Despite this diversity, any of these approaches can be described as either

1. A *secured borrowing.* Under this approach, the transferor (borrower) simply acts like it borrowed money from the transferee (lender), with the receivables remaining in the transferor's balance sheet and serving as collateral for the loan. On the other side of the transaction, the transferee recognizes a note receivable.
2. *A sale of receivables.* Under this approach, the transferor (seller) "derecognizes" (removes) the receivables from its balance sheet, acting like it sold them to the transferee (buyer). On the other side of the transaction, the transferee recognizes the receivables as assets in its balance sheet and measures them at their fair value.

As you will see in the examples that follow, the transferor (borrower) debits cash regardless of whether the transaction is treated as a secured borrowing or a sale of receivables. What differs is whether the borrower credits a liability (for a secured borrowing) or credits the receivable asset (for a sale of receivables). Let's discuss each of these approaches in more detail as they apply to accounts receivable and notes receivable. Then we'll discuss the circumstances under which GAAP requires each approach.

Secured Borrowing

When companies *pledge* accounts receivable as collateral for debt, a disclosure note describes the arrangement.

Sometimes companies **pledge** accounts receivable as collateral for a loan. Assets that serve as collateral are forfeited if the borrower defaults on the loan. No particular receivables are associated with the loan. Rather, the entire receivables balance serves as collateral. The responsibility for collection of the receivables remains solely with the company. No special accounting treatment is needed for pledged receivables, but the arrangement should be described in a disclosure note. For example, Illustration 7–13 shows a portion of the long-term debt disclosure note included in the 2016 annual report of **Virco Mfg. Corporation,** a manufacturer of office furniture.

Illustration 7–13

Disclosure of Receivables Used as Collateral—Virco Mfg. Corporation

Real World Financials

Liquidity and Capital Resources (in part)

The Revolving Credit Facility is an asset-based line of credit that is subject to a borrowing base limitation and generally provides for advances of up to 85% of eligible accounts receivable . . .

Source: Virco Mfg. Corporation

Alternatively, financing arrangements can require that companies **assign** particular receivables to serve as collateral for loans. You already may be familiar with the concept of assigning an asset as collateral if you or someone you know has a mortgage on a home. The bank or other financial institution holding the mortgage will require that, if the homeowner defaults on the mortgage payments, the home be sold and the proceeds used to pay off the mortgage debt. Similarly, in the case of an assignment of receivables, nonpayment of a debt will require the proceeds from collecting the assigned receivables to go directly toward repayment of the debt.

In these arrangements, the lender typically lends an amount of money that is less than the amount of receivables assigned by the borrower. The difference provides some protection for the lender to allow for possible uncollectible accounts. Also, the lender (sometimes called an *assignee*) usually charges the borrower (sometimes called an *assignor*) an upfront finance charge in addition to stated interest on the loan. The receivables might be collected either by the lender or the borrower, depending on the details of the arrangement. Illustration 7–14 provides an example.

Illustration 7–14

Assignment of Accounts Receivable

On December 1, 2021, the Santa Teresa Glass Company borrowed $500,000 from Finance Bank and signed a promissory note. Interest at 12% is payable monthly. The company assigned **$620,000** of its receivables as collateral for the loan. Finance Bank charges a finance fee equal to 1.5% of the accounts receivable assigned.

Santa Teresa Glass records the borrowing as follows:

Cash (difference)	490,700	
Finance charge expense* (1.5% × $620,000)	9,300	
Liability—financing arrangement		500,000

*In theory this fee should be allocated over the entire period of the loan rather than recorded as an expense in the initial period. However, amounts usually are small and the loan period usually is short. For expediency, then, we expense the entire fee immediately.

Santa Teresa will continue to collect the receivables, and will record any discounts, sales returns, and bad debt write-offs, but will remit the cash to Finance Bank, usually on a monthly basis. If $400,000 of the receivables assigned are collected in December, Santa Teresa Glass records the following entries:

Cash	400,000	
Accounts receivable		400,000
Interest expense ($500,000 × 12% × 1/12)	5,000	
Liability—financing arrangement	400,000	
Cash		405,000

Accounts Receivable

620,000	
	400,000
220,000	

Liability—Financing Arrangement

	500,000
400,000	
	100,000

In Santa Teresa's financial statements, the arrangement is described in a disclosure note.

Sale of Receivables

Accounts and notes receivable, like any other assets, can be sold at a gain or a loss. The basic accounting treatment for the sale of receivables is similar to accounting for the sale of other assets. The seller (transferor) (a) removes from the accounts the receivables (and any allowance for bad debts associated with them), (b) recognizes at fair value any assets acquired or liabilities assumed by the seller in the transaction, and (c) records the difference as a gain or loss.

The sale of accounts receivable is a popular method of financing. A technique once used by companies in a few industries or with poor credit ratings the sale of receivables is now a common occurrence for many different types of companies. For example, **General Motors, Deere & Co.,** and **Bank of America** all sell receivables. The two most common types of selling arrangements are **factoring** and **securitization**. We'll now discuss each type.

Two popular arrangements used for the sale of receivables are *factoring* and *securitization*.

In a **factoring** arrangement, the company sells its accounts receivable to a financial institution. The financial institution typically buys receivables for cash, handles the billing and collection of the receivables, and charges a fee for this service. Actually, credit cards like **VISA** and **Mastercard** are forms of factoring arrangements. The seller relinquishes all rights to the future cash receipts in exchange for cash from the buyer (the *factor*).

As an example, Illustration 7–15 shows an excerpt from the website of **BusinessCash.Com,** a financial institution that offers factoring as one of its services.

Illustration 7–15

Advertisement of Factoring BusinessCash.Com

Real World Financials

Accounts Receivable Factoring

0.79-1.99% 30-Day Rates! **Funding in 24–48 hours!**

Accounts Receivable Financing is the sale of a company's receivables at a discount in exchange for immediate cash. The percentage of working capital a company can receive upfront ranges from 70%–92%. AR financing can be a perfect solution for many industries, including to textiles, to wine distributors, and to start-ups.

Notice that the factor, BusinessCash.com, advances only between 70%–92% of the factored receivables. The remaining balance is retained as security until all of the receivables are collected and then remitted to the transferor, net of the factor's fee. The interest rate charged by this factor might seem low, but realize it is a 30-day rate. Multiply it by 12, and you will see a range of annual interest rates from 9.5% to almost 24%. The specific rate charged depends on, among other things, the quality of the receivables and the length of time before payment is required.

FINANCIAL Reporting Case

Q4, p. 339

Another popular arrangement used to sell receivables is **securitization**. In a typical accounts receivable securitization, the company creates a "special purpose entity" (SPE), usually a trust or a subsidiary. The SPE buys a pool of trade receivables, credit card receivables, or loans from the company, and then sells related securities, typically debt such as bonds or commercial paper, that are backed (collateralized) by the receivables. Securitizing receivables using an SPE can provide significant economic advantages, allowing companies to reach a large pool of investors and to obtain more favorable financing terms.[19]

As an example of a securitization, Illustration 7–16 shows a portion of the disclosure note included in the 2017 annual report of **Flextronics International Limited,** a worldwide leader in design, manufacturing, and logistics services, describing the securitization of its trade accounts receivables.

Illustration 7–16

Description of Securitization Program—Flextronics International Limited

Real World Financials

Note 10: Trade Receivables Securitization

The Company continuously sells designated pools of trade receivables . . . to affiliated special purpose entities, each of which in turn sells 100% of the receivables to unaffiliated financial institutions . . . . The company services, administers and collects the receivables on behalf of the special purpose entities and receives a servicing fee of 0.1% to 0.5% of serviced receivables per annum.

Source: Flextronics International Limited

The specifics of sale accounting vary depending on the particular arrangement between the seller and buyer (transferee).[20] One key feature is whether the receivables are transferred **without recourse** or **with recourse**.

The buyer assumes the risk of uncollectibility when accounts receivable are sold *without recourse*.

SALE WITHOUT RECOURSE If a factoring arrangement is made **without recourse**, the buyer can't ask the seller for more money if the receivables prove to be uncollectible. Therefore, the buyer assumes the risk of bad debts. Illustration 7–17 provides an example of receivables factored without recourse.

Note that in Illustration 7–17 the fair value (**$50,000**) of the last 10% of the receivables to be collected is less than 10% of the total book value of the receivables (10% × $600,000 = $60,000). That's common, because the last receivables to be collected are likely to be reduced by sales returns and allowances, and therefore have a lower fair value.

The seller retains the risk of uncollectibility when accounts receivable are sold *with recourse*.

SALE WITH RECOURSE When a company sells accounts receivable **with recourse**, the seller retains all of the risk of bad debts. In effect, the seller guarantees that the buyer will be paid even if some receivables prove to be uncollectible. To compensate the seller for retaining the risk of bad debts, the buyer usually charges a lower factoring fee when receivables are sold with recourse.

In Illustration 7–17, even if the receivables were sold with recourse, Santa Teresa Glass still could account for the transfer as a sale so long as the conditions for sale treatment are met. The only difference is the additional requirement that Santa Teresa record the estimated

[19] Although SPEs usually aren't viewed as separate entities for accounting purposes, they typically are separate entities for legal purposes. As a consequence, an SPE typically is viewed as "bankruptcy remote," meaning that the transferor's creditors can't access the receivables if the transferor goes bankrupt. This increases the safety of the SPE's assets and typically allows it to obtain more favorable financing terms than could the transferor.

[20] FASB ASC 860: Transfers and Servicing.

Illustration 7–17

Accounts Receivable Factored without Recourse

In December 2021, the Santa Teresa Glass Company entered into the following factoring arrangement with Factor Bank:

- Santa Teresa transfered accounts receivable that had a book value of $600,000.
- The transfer was made without recourse.
- Factor immediately remitted to Santa Teresa cash equal to 90% of the factored amount (90% × $600,000 = $540,000).
- Factor retains the remaining 10% to cover its factoring fee (equal to 4% of the total factored amount; 4% × $600,000 = $24,000) and to provide a cushion against potential sales returns and allowances.
- After Factor has collected cash equal to the amount advanced to Santa Teresa plus the factoring fee, Factor will remit the excess to Santa Teresa. Therefore, Santa Teresa has a "beneficial interest" in the transferred receivables equal to the fair value of the last 10% of the receivables to be collected (which management estimates to equal **$50,000**, less the 4% factoring fee.*

Santa Teresa Glass records the transfer as follows:

Cash (90% × $600,000)	540,000	
Loss on sale of receivables (to balance)	34,000	
Receivable from factor (**$50,000** − $24,000 fee)	26,000	
Accounts receivable (book value sold)		600,000

*Illustration 7–17 depicts an arrangement in which the factor's fee is paid out of the 10% of receivables retained by the factor. Alternatively, a factoring arrangement could be structured to have the factor's fee withheld from the cash advanced to the company at the start of the arrangement. In that case, in Illustration 7–17 the journal entry recorded by Santa Teresa would be

Cash ([90% × $600,000] − $24,000 fee)	516,000	
Loss on sale of receivables (to balance)	34,000	
Receivable from factor	50,000	
Accounts receivable (book value sold)		600,000

fair value of its recourse obligation as a liability. The recourse obligation is the estimated amount that Santa Teresa will have to pay Factor Bank as a reimbursement for uncollectible receivables. Illustration 7–18 provides an example of receivables factored with recourse.

Illustration 7–18

Accounts Receivable Factored with Recourse

Assume the same facts as in Illustration 7–17, except that Santa Teresa sold the receivables to Factor Bank *with recourse* and estimates the fair value of the recourse obligation to be $5,000. Santa Teresa records the transfer as follows:

Cash (90% × $600,000)	540,000	
Loss on sale of receivables (to balance)	39,000	
Receivable from factor ($50,000 − $24,000 fee)	26,000	
Recourse liability		5,000
Accounts receivable (book value sold)		600,000

When comparing Illustration 7–17 and 7–18, notice that the estimated recourse liability of **$5,000** increases the loss on sale by $5,000. If the factor eventually collects all of the receivables, Santa Teresa eliminates the recourse liability and recognizes a gain.

Transfers of Notes Receivable

We handle transfers of notes receivable in the same manner as transfers of accounts receivable. A note receivable can be used to obtain immediate cash from a financial institution either by pledging the note as collateral for a loan or by selling the note. Notes also can be securitized.

The transfer of a note receivable to a financial institution is called *discounting*.

The transfer of a note to a financial institution is referred to as **discounting**. The financial institution accepts the note and gives the seller cash equal to the maturity value of the note reduced by a discount. The discount is computed by applying a discount rate to the maturity value and represents the financing fee the financial institution charges for the transaction. Illustration 7–19 provides an example of the calculation of the proceeds received by the transferor.

Illustration 7–19

Discounting a Note Receivable

On December 31, 2021, the Stridewell Wholesale Shoe Company sold land in exchange for a nine-month, 10% note. The note requires the payment of $200,000 plus interest on September 30, 2022. The company's fiscal year-end is December 31. The 10% rate properly reflects the time value of money for this type of note. On March 31, 2022, Stridewell discounted the note at the Bank of the East. The bank's discount rate is 12%.

Because the note had been outstanding for three months before it was discounted at the bank, Stridewell first records the interest that has accrued prior to being discounted:

STEP 1: Accrue interest earned on the note receivable prior to its being discounted.

March 31, 2022		
Interest receivable	5,000	
Interest revenue ($200,000 × 10% × 3/12)		5,000

Next, the value of the note if held to maturity is calculated. Then the discount for the time remaining to maturity is deducted to determine the cash proceeds from discounting the note:

STEP 2: Add interest to maturity to calculate maturity value.

STEP 3: Deduct discount to calculate cash proceeds.

$ 200,000	Face amount
15,000	Interest to maturity ($200,000 × 10% × 9/12)
215,000	Maturity value
(12,900)	Discount ($215,000 × 12% × 6/12)
$202,100	Cash proceeds

Similar to accounts receivable, Stridewell potentially could account for the transfer as a sale or a secured borrowing. For example, Illustration 7–20 shows the appropriate journal entry to account for the transfer as a sale without recourse.

Illustration 7–20

Discounted Note Treated as a Sale

Cash (proceeds determined above)	**202,100**	
Loss on sale of notes receivable (to balance)	2,900	
Notes receivable (face amount)		200,000
Interest receivable (accrued interest determined above)		5,000

Deciding Whether to Account for a Transfer as a Sale or a Secured Borrowing

Transferors usually prefer to use the sale approach rather than the secured borrowing approach to account for the transfer of a receivable because the sale approach makes the transferor seem less leveraged, more liquid, and perhaps more profitable than does the secured borrowing approach. Illustration 7–21 explains why by describing particular effects on key accounting metrics.

Illustration 7–21

Why Do Transferors of Receivables Generally Want to Account for the Transfer as a Sale?

	Transfer of Receivables Accounted for as		
Does the Accounting Approach	**Sale**	**Secured Borrowing**	**Why Sales Approach is Preferred by the Transferor**
Derecognize A/R, reducing assets?	Yes	No	Sale approach produces lower total assets and higher return on assets (ROA).
Recognize liability for cash received?	No	Yes	Sale approach produces lower liabilities and less leverage (debt/equity).
Where is cash received shown in the statement of cash flows?	May be in operating or financing sections	Always in financing section	Sale approach can produce higher cash flow from operations at time of transfer.
Recognize gain on transfer?	More likely	Less likely	Sale approach can produce higher income at time of transfer.

So when is a company allowed to account for the transfer of receivables as a sale? The most critical element is the extent to which the company (the transferor) *surrenders control over the assets transferred.* For some arrangements, surrender of control is clear (e.g., when a receivable is sold without recourse and without any other involvement by the transferor). However, for other arrangements this distinction is not obvious. Indeed, some companies appear to structure transactions in ways that qualify for sale treatment but retain enough involvement to have control. This led the FASB to provide guidelines designed to constrain inappropriate use of the sale approach. Specifically, the transferor (defined to include the company, its consolidated affiliates, and people acting on behalf of the company) is determined to have surrendered control over the receivables if and only if all of the following conditions are met:[21]

a. The transferred assets have been isolated from the transferor—beyond the reach of the transferor and its creditors.
b. Each transferee has the right to pledge or exchange the assets it received.
c. The transferor does not maintain *effective control* over the transferred assets, for example, by structuring the transfer such that the assets are likely to end up returned to the transferor.

If the transferor is deemed to have surrendered control over the transferred receivables, the arrangement is accounted for as a sale; otherwise as a secured borrowing.

If *all* of these conditions are met, the transferor accounts for the transfer as a sale. If *any* of the conditions are not met, the transferor treats the transaction as a secured borrowing.

It is not surprising that some companies have aggressively tried to circumvent these conditions by creating elaborate transactions to qualify for sale treatment. The most famous recent case was Lehman Brothers' use of "Repo 105" transactions, discussed in Illustration 7–22.

Illustration 7–22

Repo 105 Transactions—Lehman Brothers

Real World Financials

Lehman Brothers' bankruptcy in 2008 was the largest ever to occur in the United States. One factor that likely contributed to investor losses was Lehman's use of "Repo 105" transactions that concealed how overburdened with liabilities the company had become. Here is how a Repo 105 transaction worked. Near the end of each quarter, Lehman would transfer financial assets like receivables to a bank or other financial institution in exchange for cash, and would account for that transfer as a sale of the financial assets. Lehman would use the cash obtained from the transfer to pay down liabilities, so the net effect of the transaction was to reduce assets, reduce liabilities, and therefore make Lehman appear less leveraged and less risky. Lehman also agreed to repurchase ("repo") the assets in the next quarter for an amount of cash that exceeded the amount it initially received.

In substance, this transaction is a loan, since Lehman ended up retaining the financial assets and paying amounts equivalent to principal and interest. However, Lehman argued that the assets were beyond its *effective control,* because the cash it received for transferring the assets was insufficient to enable Lehman to repurchase those assets (the "105" in "Repo 105" refers to the assets being worth at least 105% of the cash Lehman was getting for them). Although Lehman's interpretation was supported by the GAAP in effect at the time, these transactions were very poorly disclosed, and when they eventually came to light, the financial markets and investing public reacted very negatively. In response to the Lehman debacle, the FASB has taken steps to close the loophole that allowed Repo 105 transactions to be accounted for as sales.*

*FASB ASC 860–10–55: Transfers and Servicing—Overall—Implementation Guidance and Illustrations.

Source: Lehman Brothers

Illustration 7–23 summarizes the decision process that is used to determine whether a transfer of a receivable is accounted for as a secured borrowing or a sale.

Disclosures

Much disclosure is required when the transferor has continuing involvement in the transferred assets but accounts for the transfer as a sale. Why? Those are the circumstances under

[21] FASB ASC 860–10–40: Transfers and Servicing–Overall–Derecognition (previously "Accounting for Transfers and Servicing of Financial Assets and Extinguishments of Liabilities," *Statement of Financial Accounting Standards No. 140* (Norwalk, Conn.: FASB, 2000), as amended by *SFAS No. 166*).

Illustration 7–23 Accounting for the Financing of Receivables

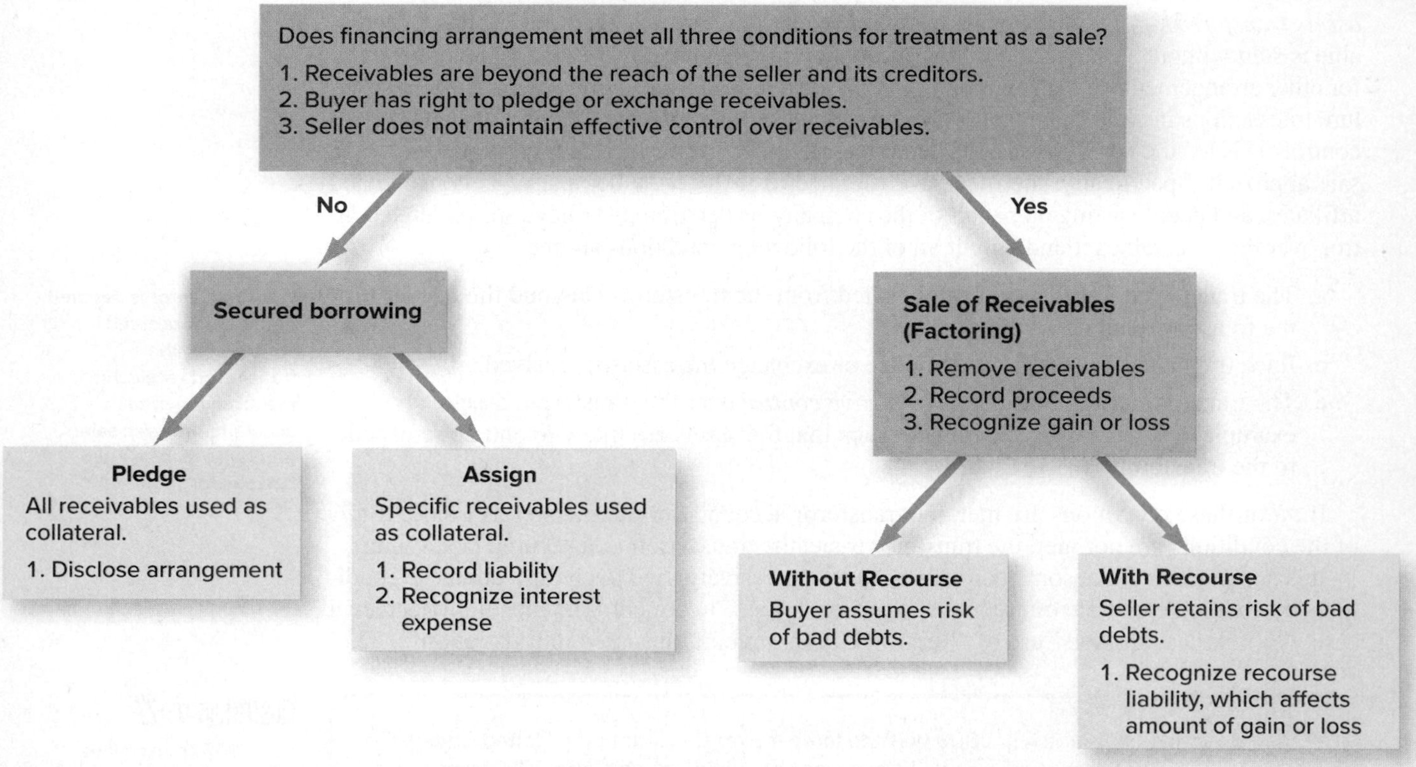

which it's most likely that the transferor may still bear significant risk associated with the arrangement, so those are the arrangements that analysts often view instead as a secured borrowing. As a result, transferors must provide enough information about the transfer to allow financial statement users to fully understand (a) the transfer, (b) any continuing involvement with the transferred assets, and (c) any ongoing risks to the transferor.

The company also has to provide information about the quality of the transferred assets. For example, for transferred receivables, the company needs to disclose the amount of receivables that are past due and any credit losses occurring during the period. Among the other information the company must disclose are these:

- How fair values were estimated when recording the transaction.
- Any cash flows occurring between the transferor and the transferee.
- How any continuing involvement in the transferred assets will be accounted for on an ongoing basis.[22]

Additional Consideration

Participating Interests. What if, rather than transferring all of a particular receivable, a company transfers only part of it? For example, what if a company transfers the right to receive future interest payments on a note, but retains the right to receive the loan principal? U.S. GAAP requires that a partial transfer be treated as a secured borrowing unless the amount transferred qualifies as a "participating interest" as well as meeting the "surrender of control" requirements described above. Participating interests are defined as having a proportionate ownership interest in the receivable and sharing proportionally in the cash flows of the receivable. Many common securitization arrangements do not qualify as participating interests, so this change in GAAP makes it harder for partial transfers to qualify for the sale approach.

[22]FASB ASC 860–10–40: Transfers and Servicing—Overall—Derecognition (previously "Accounting for Transfers of Financial Assets, an amendment of FASB Statement No. 140," *Statement of Financial Accounting Standards No. 166* (Norwalk, Conn.: FASB, 2009), par. 17).

International Financial Reporting Standards

Transfers of Receivables. *IFRS No. 9* and *FASB ASC 860* cover financing with receivables under IFRS and U.S. GAAP, respectively.[23,24] The international and U.S. guidance often lead to similar accounting treatments. Both seek to determine whether an arrangement should be treated as a secured borrowing or a sale, and, having concluded which approach is appropriate, both account for the approaches in a similar fashion.

● LO7–10

Where IFRS and U.S. GAAP most differ is in the conceptual basis for their choice of accounting approaches and in the decision process they require to determine which approach to use. As you have seen in this chapter, U.S. GAAP focuses on whether control of assets has shifted from the transferor to the transferee. In contrast, IFRS requires a more complex decision process. The company has to have transferred the rights to receive the cash flows from the receivable and then considers whether the company has transferred "substantially all of the risks and rewards of ownership," as well as whether the company has transferred control. Under IFRS,

1. If the company *transfers* substantially all of the risks and rewards of ownership, the transfer is treated as a sale.
2. If the company *retains* substantially all of the risks and rewards of ownership, the transfer is treated as a secured borrowing.
3. If neither conditions 1 or 2 hold, the company accounts for the transaction as a sale if it has transferred control, and as a secured borrowing if it has retained control.

Whether risks and rewards have been transferred is evaluated by comparing how variability in the amounts and timing of the cash flows of the transferred asset affect the company before and after the transfer.

This is a broad overview of the IFRS guidance. Application of the detailed rules is complex, and, depending on the specifics of an arrangement, a company could have different accounting under IFRS and U.S. GAAP.

Source: International Financial Reporting Standards

Concept Review Exercise

FINANCING WITH RECEIVABLES

The Hollywood Lumber Company obtains financing from the Midwest Finance Company by factoring (or discounting) its receivables. During June 2021, the company factored $1,000,000 of accounts receivable to Midwest. The transfer was made *without* recourse. The factor, Midwest Finance, remits 80% of the factored receivables and retains 20%. When the receivables are collected by Midwest, the retained amount, less a 3% fee (3% of the total factored amount), will be remitted to Hollywood Lumber. Hollywood estimates that the fair value of the amount retained by Midwest is $180,000.

In addition, on June 30, 2021, Hollywood discounted a note receivable without recourse. The note, which originated on March 31, 2021, requires the payment of $150,000 *plus* interest at 8% on March 31, 2022. Midwest's discount rate is 10%. The company's fiscal year-end is December 31.

Required:

Prepare journal entries for Hollywood Lumber for the factoring of accounts receivable and the note receivable discounted on June 30. Assume that the required criteria are met and the transfers are accounted for as sales.

---

[23] "Financial Instruments," *International Financial Reporting Standard No. 9* (IASCF), as amended effective January 1, 2018.

[24] FASB ASC 860: Transfers and Servicing.

Solution:

The Factoring of Receivables

Cash ($1,000,000 × 80%)	800,000	
Loss on sale of receivables (to balance)	50,000	
Receivable from factor ($180,000 − $30,000 fee)	150,000	
Accounts receivable (balance sold)		1,000,000

The Note Receivable Discounted

Interest receivable	3,000	
Interest revenue ($150,000 × 8% × 3/12)		3,000
Cash (proceeds determined below)	**149,850**	
Loss on sale of notes receivable (difference)	3,150	
Notes receivable (face amount)		150,000
Interest receivable (accrued interest determined above)		3,000

$150,000	Face amount
12,000	Interest to maturity ($150,000 × 8%)
162,000	Maturity value
(12,150)	Discount ($162,000 × 10% × 9/12)
$149,850	Cash proceeds

Decision Makers' Perspective

● LO7–9

RECEIVABLES MANAGEMENT A company's investment in receivables is influenced by several variables, including the level of sales, the nature of the product or service sold, and credit and collection policies. These variables are, of course, related. For example, a change in credit policies could affect sales. In fact, more liberal credit policies—allowing customers a longer time to pay or offering cash discounts for early payment—often are initiated with the specific objective of increasing sales volume.

Management must evaluate the costs and benefits of any change in credit and collection policies.

Management's choice of credit and collection policies often involves trade-offs. For example, offering cash discounts may increase sales volume, accelerate customer payment, and reduce bad debts. These benefits are not without cost. The cash discounts reduce the amount of cash collected from customers who take advantage of the discounts. Extending payment terms also may increase sales volume. However, this creates an increase in the required investment in receivables and may increase bad debts.

The ability to use receivables as a method of financing also offers management alternatives. Assigning, factoring, and discounting receivables are alternative methods of financing operations that must be evaluated relative to other financing methods such as lines of credit or other types of short-term borrowing.

Investors, creditors, and financial analysts can gain important insights by monitoring a company's investment in receivables. Chapter 4 introduced the receivables turnover ratio and the related average collection period, ratios designed to monitor receivables. Recall that these ratios are calculated as follows:

$$\text{Receivables turnover ratio} = \frac{\text{Net sales}}{\text{Average accounts receivable (net)}}$$

$$\text{Average collection period} = \frac{\text{365 days}}{\text{Receivables turnover ratio}}$$

The turnover ratio shows the number of times during a period that the average accounts receivable balance is collected, and the average collection period is an approximation of the number of days the average accounts receivable balance is outstanding.

As a company's sales grow, receivables also will increase. If the percentage increase in receivables is greater than the percentage increase in sales, the receivables turnover ratio will decline (the average collection period will increase). This could indicate customer dissatisfaction with the product or that the company has extended too generous payment terms in order to attract new customers, which, in turn, could increase sales returns and bad debts.

These ratios also can be used to compare the relative effectiveness of companies in managing the investment in receivables. Of course, it would be meaningless to compare the receivables turnover ratio of a computer products company such as **IBM** with that of, say, a food products company like **Hershey**. A company selling high-priced, low-volume products like mainframe computers generally will grant customers longer payment terms than a company selling lower-priced, higher-volume food products. Illustration 7–24 lists the 2017 receivables turnover ratio for some well-known companies. The differences are as expected, given the nature of the companies' products and operations. In particular, companies designing expensive products for medical and business applications turn over their receivables less frequently than do consumer-goods manufacturers and wholesalers.

Company	2017 Receivables Turnover Ratio
Medtronic (medical technology)	5.33
Autodesk (design software)	3.67
General Mills (wholesale consumer foods)	11.19

Illustration 7–24
Receivables Turnover Ratios

To illustrate receivables analysis in more detail, let's compute the 2017 receivables turnover ratio and the average collection period for two companies in the software industry, **Symantec Corp.** and **CA, Inc.**

($ in millions)

	Symantec Corp.			CA, Inc.		
	2017		2016	2017		2016
Accounts receivable (net)	$649		$556	$764		$625
Two-year averages		$ 603			$ 695	
Net sales—2017		**$4,019**			**$4,036**	

	Symantec Corp.	CA, Inc.	Industry Average
Receivables Turnover	$= \frac{\$4{,}019}{\$603} = 6.67$ times	$\frac{\$4{,}036}{\$695} = 5.81$ times	8.38 times
Average Collection Period	$= \frac{365}{6.67} = 54.72$ days	$\frac{365}{5.81} = 62.82$ days	43.56 days

On average, Symantec collects its receivables 8 days sooner than does CA, but 11 days slower than the industry average. A major portion of Symantec's sales of products like Norton antivirus software are made directly to consumers online who pay immediately with credit cards, significantly accelerating payment. CA, on the other hand, sells primarily to businesses, which take longer to pay.

EARNINGS QUALITY Recall our discussion in Chapter 4 concerning earnings quality. We learned that managers have the ability, to a limited degree, to manipulate reported income and that many observers believe this practice diminishes earnings quality because it can mask "permanent" earnings. Former SEC Chairman Arthur Levitt listed discretionary accruals, which he called "Miscellaneous Cookie Jar Reserves," as one of the most popular methods companies use to manipulate income.

Bad debt expense is one of a variety of discretionary accruals that provide management with the opportunity to manipulate income.

Sometimes financial statement users can examine accounts receivable, the allowance for bad debts, and other accounts to detect low earnings quality. For example, in an analysis that eventually led to an important SEC fraud case, **PaineWebber Inc.** downgraded its stock recommendation for **Sunbeam, Inc.**, after noticing unusually high accounts receivable and unexpected increases in sales of certain products. Also, Sunbeam's allowance for uncollectible accounts had shown large increases in prior periods. It eventually came to light that Sunbeam had been manipulating its income by using a "bill and hold" strategy with retail buyers. This involved selling products at large discounts to retailers before they normally would buy and then holding the products in third-party warehouses, with delivery at a later date. More generally, research indicates that companies sometimes manage earnings by building up large allowances for bad debts in good periods and then reducing those allowances to increase earnings in bad periods.[25]

Another area for accounting-quality concern is the sale method used to account for transfers of receivables. Research suggests that some companies manage earnings by distorting the fair value estimates that are made as part of recording securitizations.[26] Also, some firms classify cash flows associated with selling their accounts receivable in the operating section of the statement of cash flows, such that changes in the extent to which accounts receivable are sold can be used to manipulate cash flow from operations. In fact, evidence suggests that sophisticated investors and bond-rating agencies undo sales accounting to treat transfers of receivables as secured borrowings before assessing the riskiness of a company's debt.[27] Still, Wall Street is very good at identifying clever ways to structure transactions around accounting standards, so it is important to be vigilant regarding the accounting for these transactions. ●

Financial Reporting Case Solution

©PhotosIndia.com/Glow Images

1. **Explain the allowance method of accounting for bad debts.** *(p. 350)* The allowance method is used to report the carrying value of accounts receivable at the net amount expected to be collected. In an adjusting entry, we record bad debt expense and reduce accounts receivable indirectly by crediting a contra account to accounts receivable for an estimate of the amount that eventually will prove uncollectible.
2. **What approaches might CHS have used to arrive at the $200 million bad debt provision?** *(p. 351)* CHS would determine the balance in the allowance for uncollectible accounts that is necessary to show net accounts receivable at the appropriate carrying value and then would record whatever adjustment and corresponding bad debt expense resulted in that balance. It likely would use an aging of accounts receivable to calculate the appropriate balance. It might also use a percentage of net sales to estimate bad debt expense, particularly on an interim basis, but that approach would only be allowed if it produced a similar carrying value as the balance sheet approach.
3. **Are there any alternatives to the allowance method?** *(p. 354)* An alternative to the allowance method is the direct write-off method. Using this method, adjusting entries are not recorded and any bad debt that does arise simply is written off as bad debt expense. The direct write-off method is not permitted by GAAP except in limited circumstances.
4. **What does it mean for CHS to securitize its receivables?** *(p. 364)* Securitization is a method used to sell accounts receivable. In a typical securitization, a company creates a "special purpose entity" that buys receivables from the company and issues securities that are collateralized by the receivables. Securitization allows CHS to obtain cash for its receivables sooner than it could otherwise, and to reach a large pool of investors at favorable financing rates. ●

[25] S. B. Jackson and X. Liu, "The Allowance for Uncollectible Accounts, Conservatism, and Earnings Management," *Journal of Accounting Research* 48, No. 3 (2010). pp. 565–601.

[26] P. M. Dechow, L. A. Myers, and C. Shakespeare, "Fair Value Accounting and Gains from Asset Securitizations: A Convenient Earnings Management Tool with Compensation Side-Benefits," *Journal of Accounting and Economics* 49, No. 2 (2010), pp. 2–25.

[27] Other recent research provides evidence that investors treat securitizations as loans rather than asset sales, suggesting that they are unconvinced that a sale has truly taken place. For example, see W. R. Landsman, K. Peasnell, and C. Shakespeare, "Are Asset Securitizations Sales or Loans?" *The Accounting Review* 83, No. 5 (2008), pp. 1251–72.

The Bottom Line

- LO7–1 Internal control refers to the plan designed to encourage adherence to company policies and procedures; promote operational efficiency; minimize errors, thefts, or fraud; and maximize the reliability and accuracy of accounting data. Key elements of an internal control system for cash receipts and disbursements include separation of record keeping from control of cash duties and the periodic preparation of a bank reconciliation. *(p. 340)*
- LO7–2 Cash can be informally restricted by management for a particular purpose. Restrictions also can be contractually imposed. If restricted cash is available for current operations or to pay current liabilities, it's classified as a current asset; otherwise, it's classified as investments and funds or other assets. *(p. 341)*
- LO7–3 The gross method of accounting for cash discounts considers a discount not taken as part of sales revenue. The net method considers a discount not taken as sales discount forfeited. *(p. 344)*
- LO7–4 When merchandise returns are anticipated, a refund liability should be recorded, and sales revenue should be reduced by anticipated sales returns. *(p. 346)*
- LO7–5 At the end of each period, the company estimates the necessary balance in the allowance for uncollectible accounts and then records whatever adjustment to the allowance and corresponding bad debt expense is necessary to reach that balance. *(p. 350)*
- LO7–6 The balance sheet approach determines bad debt expense by estimating the appropriate carrying value of accounts receivable to be reported in the balance sheet and then adjusting the allowance for uncollectible accounts as necessary to reach that carrying value. The income statement approach estimates bad debt expense based on the notion that a certain percentage of each period's credit sales will prove to be uncollectible. *(p. 352)*
- LO7–7 Notes receivable are formal credit arrangements between a creditor (lender) and a debtor (borrower). The typical note receivable requires the payment of a specified face amount, also called principal, at a specified maturity date or dates. In addition, interest is paid at a stated percentage of the face amount. Interest on notes is calculated by multiplying the face amount by the annual rate by the fraction of the annual period. *(p. 356)*
- LO7–8 A wide variety of methods exists for companies to use their receivables to obtain immediate cash. These methods can be described as either a secured borrowing or a sale of receivables. If three conditions indicating surrender of control are met, the transferor accounts for the transfer of receivables as a sale; otherwise as a secured borrowing. *(p. 362)*
- LO7–9 A company's investment in receivables is influenced by several related variables, to include the level of sales, the nature of the product or service, and credit and collection policies. Investors, creditors, and financial analysts can gain important insights by monitoring a company's investment in receivables. The receivables turnover and average collection period ratios are designed to monitor receivables. *(p. 370)*
- LO7–10 Accounting for cash and accounts receivable are similar under U.S. GAAP and IFRS. Other than some differences in terminology and balance sheet classifications, the most important differences involve accounting for transfers of receivables. Both IFRS and U.S. GAAP seek to distinguish between determining whether a sales treatment or secured borrowing treatment is appropriate, but they use different conceptual frameworks to guide that choice. U.S. GAAP focuses on whether control of the receivables is transferred, while IFRS uses a more complex decision process that also considers whether substantially all of the risks and rewards of ownership have been transferred. *(p. 342, 361, and 369)* ●

Cash Controls

APPENDIX 7A

Bank Reconciliation

One of the most important tools used in the control of cash is the **bank reconciliation**. Since all cash receipts are deposited into the bank account and cash disbursements are made by check, the bank account provides a separate record of cash. It's desirable to periodically compare the bank balance with the balance in the company's own records and reconcile any differences.

You probably know from your own personal experience that the ending balance in your checking account reported on your monthly bank statement rarely equals the balance you have recorded in your checkbook. Differences arise from two types of items: timing differences and errors.

Differences between the cash book and bank balance occur due to differences in the timing of recognition of certain transactions and errors.

Timing differences occur when the company and the bank record transactions at different times. At any point in time the company may have adjusted the cash balance for items of which the bank is not yet aware. Likewise, the bank may have adjusted its record of that balance by items of which the company is not yet aware. For example, checks written and cash deposits are not all processed by the bank in the same month that they are recorded by the company. Also, the bank may adjust the company's account for items such as service charges that the company is not aware of until the bank statement is received.

Errors can be made either by the company or the bank. For example, a check might be written for $210 but recorded on the company's books as a $120 disbursement; a deposit of $500 might be processed incorrectly by the bank as a $50 deposit. In addition to serving as a safeguard of cash, the bank reconciliation also uncovers errors such as these and helps ensure that the proper cash balance is reported in the balance sheet.

STEP 1: Adjust the bank balance to the corrected cash balance.

Bank reconciliations include adjustments to the balance per bank for timing differences involving transactions already reflected in the company's accounting records that have not yet been processed by the bank. These adjustments usually include *checks outstanding* and *deposits outstanding* (also called *deposits in transit*). In addition, the balance per bank would be adjusted for any bank errors discovered. These adjustments produce an adjusted bank balance that represents the corrected cash balance.

STEP 2: Adjust the book balance to the corrected cash balance.

The balance per books is similarly adjusted for timing differences involving transactions already reflected by the bank of which the company is unaware until the bank statement is received. These would include service charges, charges for NSF (nonsufficient funds) checks, and collections made by the bank on the company's behalf. In addition, the balance per books is adjusted for any company errors discovered, resulting in an adjusted book balance that will also represent the corrected cash balance. *Each of these adjustments requires a journal entry to correct the book balance.* Only adjustments to the book balance require journal entries. Illustration 7A–1 recaps these reconciling items.

Illustration 7A–1

Bank Reconciliation—Reconciling Items

Balance per Bank
+ Deposits outstanding
− Checks outstanding
± Errors
Corrected balance

Balance per Book
+ Collections by bank
− Service charges
− NSF checks
± Errors
Corrected balance

The two corrected balances must equal.

Step 1: Adjustments to Bank Balance

1. *Add deposits outstanding.* These represent cash amounts received by the company and debited to cash that have not been deposited in the bank by the bank statement cutoff date and cash receipts deposited in the bank near the end of the period that are not recorded by the bank until after the cutoff date.
2. *Deduct checks outstanding.* These represent checks written and recorded by the company as credits to cash that have not yet been processed by the bank before the cutoff date.
3. *Bank errors.* These will either be increases or decreases depending on the nature of the error.

Step 2: Adjustments to Book Balance

1. *Add collections made by the bank* on the company's behalf and other increases in cash that the company is unaware of until the bank statement is received.
2. *Deduct service and other charges* made by the bank that the company is unaware of until the bank statement is received.
3. *Deduct NSF (nonsufficient funds) checks.* These are checks previously deposited for which the payors do not have sufficient funds in their accounts to cover the amount of the checks. The checks are returned to the company whose responsibility it is to seek payment from payors.
4. *Company errors.* These will either be increases or decreases depending on the nature of the error.

To demonstrate the bank reconciliation process, consider Illustration 7A–2.

Illustration 7A–2
Bank Reconciliation

The Hawthorne Manufacturing Company maintains a general checking account at the First Pacific Bank. First Pacific provides a bank statement and canceled checks once a month. The cutoff date is the last day of the month. The bank statement for the month of May is summarized as follows:

Balance, May 1, 2021	$ 32,120
Deposits	82,140
Checks processed	(78,433)
Service charges	(80)
NSF checks	(2,187)
Note payment collected by bank (includes $120 interest)	1,120
Balance, May 31, 2021	**$34,680**

The company's general ledger cash account has a balance of **$35,276** at the end of May. A review of the company records and the bank statement reveals the following:

1. Cash receipts not yet deposited totaled $2,965.
2. A deposit of $1,020 was made on May 31 that was not credited to the company's account until June.
3. All checks written in April have been processed by the bank. Checks written in May that had not been processed by the bank total $5,536.
4. A check written for $1,790 was incorrectly recorded by the company as a $790 disbursement. The check was for payment to a supplier of raw materials.

The bank reconciliation prepared by the company appears as follows:

Step 1: Bank Balance to Corrected Balance	
Balance per bank statement	**$34,680**
Add: Deposits outstanding	3,985*
Deduct: Checks outstanding	(5,536)
Corrected cash balance	$ 33,129
Step 2: Book Balance to Corrected Balance	
Balance per books	**$35,276**
Add: Note collected by bank	1,120
Deduct:	
Service charges	(80)
NSF checks	(2,187)
Error—understatement of check	(1,000)
Corrected cash balance	$ 33,129

*$2,965 + 1,020 = $3,985

The next step is to prepare adjusting journal entries to reflect each of the adjustments to the balance per books. These represent amounts the company was not previously aware of until receipt of the bank statement. No adjusting entries are needed for the adjustments to the balance per bank because the company has already recorded these items. However, the bank needs to be notified of any errors discovered.

To record the receipt of principal and interest on note collected directly by the bank

Cash	1,120	
Notes receivable		1,000
Interest revenue		120

(continued)

(concluded)

To record credits to cash revealed by the bank reconciliation		
Miscellaneous expense (bank service charges)	80	
Accounts receivable (NSF checks)	2,187	
Accounts payable (error in check to supplier)	1,000	
Cash		3,267

After these entries are posted, the general ledger cash account will equal the corrected balance of $33,129.

Petty Cash

Most companies keep a small amount of cash on hand to pay for low-cost items such as postage, office supplies, delivery charges, and entertainment expenses. It would be inconvenient, time-consuming, and costly to process a check each time these small payments are made. A petty cash fund provides a more efficient way to handle these payments.

The petty cash fund always should have cash and receipts that together equal the amount of the fund.

A petty cash fund is established by transferring a specified amount of cash from the company's general checking account to an employee designated as the petty cash custodian. The amount of the fund should approximate the expenditures made from the fund during a relatively short period of time (say a week or a month). The custodian disburses cash from the fund when the appropriate documentation is presented, such as a receipt for the purchase of office supplies. At any point in time, the custodian should be in possession of cash and appropriate receipts that sum to the amount of the fund. The receipts serve as the basis for recording appropriate expenses each time the fund is replenished. Consider the example in Illustration 7A–3.

Illustration 7A–3

Petty Cash Fund

On May 1, 2021, the Hawthorne Manufacturing Company established a $200 petty cash fund. John Ringo is designated as the petty cash custodian. The fund will be replenished at the end of each month. On May 1, 2021, a check is written for $200 made out to John Ringo, petty cash custodian. During the month of May, John paid bills totaling $160 summarized as follows:

Postage	$ 40
Office supplies	35
Delivery charges	55
Entertainment	30
Total	**$160**

In journal entry form, the transaction to establish the fund would be recorded as follows:

May 1, 2021		
Petty Cash	200	
Cash (checking account)		200

A petty cash fund is established by writing a check to the custodian.

No entries are recorded at the time the actual expenditures are made from the fund. The expenditures are recorded when reimbursement is requested at the end of the month. At that time, a check is written to John Ringo, petty cash custodian, for the total of the fund receipts, **$160** in this case. John cashes the check and replenishes the fund to $200. In journal entry form, replenishing the fund would be recorded as follows:

May 31, 2021		
Postage expense	40	
Office supplies expense	35	
Delivery expense	55	
Entertainment expense	30	
Cash (checking account)		**160**

The petty cash account is not debited when replenishing the fund. If, however, the size of the fund is increased at time of replenishment, the account is debited for the increase. Similarly, petty cash would be credited if the size of the fund is decreased.

The appropriate expense accounts are debited when the petty cash fund is reimbursed.

To maintain the control objective of separation of duties, the petty cash custodian should not be involved in the process of writing or approving checks, nor in recordkeeping. In addition, management should arrange for surprise counts of the fund.

APPENDIX 7B

Accounting for a Troubled Debt Restructuring[28]

Sometimes a creditor changes the original terms of a debt agreement in response to the debtor's financial difficulties. The creditor makes concessions to the debtor that make it easier for the debtor to pay, with the goal of maximizing the amount of cash that the creditor can collect. In that case, the new arrangement is referred to as a **troubled debt restructuring**. Because identifying an arrangement as a troubled debt restructuring requires recognizing any loss associated with the arrangement, creditors might be reluctant to conclude that a troubled debt restructuring has occurred, so the FASB provides guidance to help ensure that all troubled debt restructurings are properly identified.[29]

A *troubled debt restructuring* occurs when the creditor makes concessions to the debtor in response to the debtor's financial difficulties.

WHEN THE RECEIVABLE IS CONTINUED, BUT WITH MODIFIED TERMS In a troubled debt restructuring, it's likely that the creditor allows the receivable to continue but with the terms of the debt agreement modified to make it easier for the debtor to comply. The creditor might agree to reduce or delay the scheduled interest payments. Or, it may agree to reduce or delay the maturity amount. Often a troubled debt restructuring will call for some combination of these concessions.

The CECL approach used to account for credit losses also applies to accounting for troubled debt restructurings. The creditor uses its normal methodology to calculate credit losses. Often that methodology involves calculating discounted cash flows, so that is what we illustrate in Illustration 7B–1.

Note that the interest rate used in the illustration is the original rate (10%) included in the original credit agreement. That is because a troubled debt restructuring does not result in a new loan, but rather is part of the creditor's ongoing efforts to collect the original loan.

As First Prudent receives the reduced interest payments and principal, it will record receipt of cash and reduce the allowance for uncollectible accounts for the difference between the amount of interest it was supposed to receive originally and the amount it receives under the troubled debt restructuring:

December 31, 2021		
Cash (per agreement)	2,000,000	
Allowance for uncollectible accounts (to balance)	413,233	
Interest revenue (($30,000,000 – $5,867,670) × 10%)		2,413,233

Reducing the allowance increases the carrying value of the note receivable, so under the effective interest method the interest revenue associated with the receivable is higher in the next period:

December 31, 2022		
Cash (per agreement)	2,000,000	
Allowance for uncollectible accounts (to balance)	454,556	
Interest revenue (($30,000,000 – ($5,867,670 – 413,233)) × 10%)		2,454,556

[28]When a receivable has been securitized, or when a company has elected to account for a receivable under the fair value option, the receivable is viewed as an investment and different GAAP applies, as described in Appendix 12B.

[29]These terms have been clarified (ASC 310–40–15: Receivables—Troubled Debt Restructurings by Creditors—Scope and Scope Exceptions). A debtor is viewed as experiencing *financial difficulties* if it is probable that the debtor will default on any of its liabilities unless the creditor restructures the debt. A *concession* has occurred if, as a result of the restructuring, the creditor does not expect to collect all amounts due, including accrued interest. A concession also can occur if the creditor restructures the terms of the debt in a way that provides the debtor with funds at a better rate of interest than the debtor could receive if the debtor tried to obtain new debt with similar terms (for example, a similar payment schedule, collateral, and guarantees) as the restructured debt. But, not all changes are concessions. For example, a restructuring that results in an insignificant delay of payment is not a concession.

Illustration 7B–1

Troubled Debt Restructuring with Modified Terms

The date is January 1, 2021. Brillard Properties owes First Prudent Bank $30 million under a 10% note with two years remaining to maturity. Due to Brillard's financial difficulties, the previous year's interest ($3 million) was not paid. First Prudent and Brillard renegotiated Brillard's debt as follows:

- forgive the interest accrued from last year,
- reduce the two remaining interest payments from $3 million each to $2 million each, and
- reduce the face amount from $30 million to $25 million

How would First Prudent account for this troubled debt restructuring?

Analysis

The discounted present value of the cash flows prior to the loss is the same as the receivable's book value.

The discounted present value of the cash flows after the loss is less than book value.

The difference is a credit loss, debited to bad debt expense.

Previous Value			
Accrued interest (10% × $30,000,000)		$ 3,000,000	
Principal		30,000,000	
Book value of the receivable			$33,000,000
New Value (based on estimated cash flows to be received)			
Present value of accrued interest	=	$ 0	
Present value of future interest	$ 2 million × 1.73554* =	3,471,080	
Present value of estimated principal	$25 million × 0.82645† =	20,661,250	
Present value of the receivable			(24,132,330)
Loss			**$ 8,867,670**

*Present value of an ordinary annuity of $1: $n = 2, i = 10\%$
†Present value of $1: $n = 2, i = 10\%$

Journal Entry on January 1, 2021

Bad debt expense (to balance)#	8,867,670	
Interest receivable		3,000,000
Allowance for uncollectible accounts@ ($30,000,000 – $24,132,330)		5,867,670

#Rather than debiting bad debt expense, First Prudent might debit Loss on troubled debt restructuring or some similar title.
@This account might also be called Allowance for loan losses, Allowance for credit losses, or some similar title.

When the note matures, First Prudent records receipt of cash and reduces the note receivable and related allowance for uncollectible accounts to zero.

WHEN THE RECEIVABLE IS SETTLED OUTRIGHT Sometimes a receivable in a troubled debt restructuring is actually settled at the time of the restructuring by the debtor making a payment of cash, some other noncash assets, or even shares of the debtor's stock. In that case, the creditor records a loss for the difference between the carrying amount of the receivable and the fair value of the asset(s) or equity securities received. Illustration 7B–2 provides an example.

Illustration 7B–2

Debt Settled at the Time of a Restructuring

First Prudent Bank is owed $30 million by Brillard Properties under a 10% note with two years remaining to maturity. Due to Brillard's financial difficulties, the previous year's interest ($3 million) was not received. The bank agrees to settle the receivable (and not receive the accrued interest receivable) in exchange for property having a fair value of $20 million. Assume that $2 million of bad debt expense has previously been accrued for this loan, resulting in a balance of $2 million in the allowance for uncollectible accounts. First Prudent, the creditor, would record the following journal entry:

	($ in millions)	
Land (fair value)	20	
Allowance for uncollectible accounts@	2	
Bad debt expense (to balance)*	11	
Interest receivable (10% × $30 million)		3
Notes receivable (account balance)		30

@This account might also be called Allowance for loan losses, Allowance for credit losses, or some similar title.
*Rather than debiting bad debt expense, First Prudent might debit Loss on troubled debt restructuring or some similar title.

You may be wondering whether recognizing a loss on a troubled debt restructuring leads to a double counting of bad debt expense. After all, the receivable first was included in a group of receivables for which bad debt expense was estimated, and then was singled out for recognition of a loss. Creditors avoid this problem by separately estimating bad debt expense for the restructured troubled debt receivable and for the rest of the receivables. That way there is no double counting.

In this appendix we have focused on creditors' accounting for troubled debt restructurings. We discuss those topics from the standpoint of the debtor in Chapter 14, Appendix B.

Questions For Review of Key Topics

Q 7–1 Define cash equivalents.

Q 7–2 Explain the primary functions of internal controls procedures in the accounting area. What is meant by separation of duties?

Q 7–3 What are the responsibilities of management described in Section 404 of the Sarbanes-Oxley Act? What are the responsibilities of the company's auditor?

Q 7–4 Is restricted cash included in the reconciliation of cash balances on a statement of cash flows? Explain.

Q 7–5 Define a compensating balance. How are compensating balances reported in financial statements?

IFRS Q 7–6 Do U.S. GAAP and IFRS differ in how bank overdrafts are treated? Explain.

Q 7–7 Explain the difference between a trade discount and a cash discount.

Q 7–8 Distinguish between the gross and net methods of accounting for cash discounts.

Q 7–9 Briefly explain the accounting treatment for sales returns.

Q 7–10 Explain the typical way companies account for uncollectible accounts receivable (bad debts). When is it permissible to record bad debt expense only at the time when receivables actually prove uncollectible?

Q 7–11 Define the CECL model for accounts receivable. On what does it base the estimate of the allowance for uncollectible accounts?

IFRS Q 7–12 Define the ECL model for accounts receivable. How does it differ from the CECL model?

Q 7–13 Briefly explain the difference between the income statement approach and the balance sheet approach to estimating bad debts.

IFRS Q 7–14 If a company has accounts receivable from ordinary customers and from related parties, can they combine those receivables in their financial statements under U.S. GAAP? Under IFRS?

Q 7–15 Is any special accounting treatment required for the assigning of accounts receivable in general as collateral for debt?

Q 7–16 Explain any possible differences between accounting for an account receivable factored with recourse compared with one factored without recourse.

IFRS Q 7–17 Do U.S. GAAP and IFRS differ in the criteria they use to determine whether a transfer of receivables is treated as a sale? Explain.

Q 7–18 What is meant by the discounting of a note receivable? Describe the four-step process used to account for discounted notes.

Q 7–19 What are the key variables that influence a company's investment in receivables? Describe the two ratios used by financial analysts to monitor a company's investment in receivables.

Q 7–20 (Based on Appendix 7A) In a two-step bank reconciliation, identify the items that might be necessary to adjust the bank balance to the corrected cash balance. Identify the items that might be necessary to adjust the book balance to the corrected cash balance.

Q 7–21 (Based on Appendix 7A) How is a petty cash fund established? How is the fund replenished?

Q 7–22 (Based on Appendix 7B) Marshall Companies, Inc., holds a note receivable from a former subsidiary. Due to financial difficulties, the former subsidiary has been unable to pay the previous year's interest on the note. Marshall agreed to restructure the debt by both delaying and reducing remaining cash payments. The concessions result in a credit loss on the creditor's investment in the receivable. How is the credit loss on the troubled debt restructuring calculated?

Brief Exercises

BE 7–1
Internal control
● LO7–1

Janice Dodds opens the mail for the Ajax Plumbing Company. She lists all customer checks on a spreadsheet that includes the name of the customer and the check amount. The checks, along with the spreadsheet, are then sent to Jim Seymour in the accounting department who records the checks and deposits them daily in the company's checking account. How could the company improve its internal control procedure for the handling of its cash receipts?

BE 7–2
Bank overdrafts
● LO7–2, LO7–10
IFRS

Cutler Company has a cash account with a balance of $250,000 with Wright Bank and a cash account with an overdraft of $5,000 at Lowe Bank. What would the current assets section of Cutler's balance sheet include for "cash" under IFRS? Under U.S. GAAP?

BE 7–3
Cash and cash equivalents
● LO7–2

The following items appeared on the year-end trial balance of Consolidated Freight Corporation: cash in a checking account, U.S. Treasury bills that mature in six months, undeposited customer checks, cash in a savings account, and currency and coins. Which of these items would be included in the company's balance sheet as cash and cash equivalents?

BE 7–4
Cash discounts; gross method
● LO7–3

On December 28, 2021, Tristar Communications sold 10 units of its new satellite uplink system to various customers for $25,000 each. The terms of each sale were 1/10, n/30. Tristar uses the gross method to account for sales discounts. In what year will income before tax be affected by discounts, assuming that all customers paid the net-of-discount amount on January 6, 2022? By how much?

BE 7–5
Cash discounts; net method
● LO7–3

Refer to the situation described in BE 7–4. Answer the questions assuming that Tristar uses the net method to account for sales discounts.

BE 7–6
Sales returns
● LO7–4

During 2021, its first year of operations, Hollis Industries recorded sales of $10,600,000 and experienced returns of $720,000. Cost of goods sold totaled $6,360,000 (60% of sales). The company estimates that 8% of all sales will be returned. Prepare the year-end adjusting journal entries to account for anticipated sales returns under the assumption that all sales are made for cash (no accounts receivable are outstanding).

BE 7–7
Sales returns
● LO7–4

Refer to the situation described in BE 7–6. Prepare the year-end adjusting journal entries to account for anticipated sales returns, assuming that all sales are made on credit and all accounts receivable are outstanding.

BE 7–8
Accounts receivable classification
● LO7–5, LO7–10
IFRS

Singletary Associates has accounts receivable due from normal credit customers and also has an account receivable due from a director of the company. Singletary would like to combine both of those receivables on one line in the current assets section of their balance sheet and in the disclosure notes. Is that permissible under U.S. GAAP? Under IFRS? Explain.

BE 7–9
Uncollectible accounts; establishing an allowance
● LO7–5, LO7–6

At the end of the first year of operations, Gaur Manufacturing had gross accounts receivable of $300,000. Gaur's management estimates that 6% of the accounts will prove uncollectible. What journal entry should Gaur record to establish an allowance for uncollectible accounts?

BE 7–10
Uncollectible accounts; adjusting the allowance, credit balance
● LO7–5, LO7–6

At the end of the year, Breyer Associates had a *credit* balance in its allowance for uncollectible accounts of $12,000 before adjustment. The balance in Breyer's gross accounts receivable is $600,000. Breyer's management estimates that 10% of its accounts receivable balance will not be collected. What journal entry should Breyer record to adjust its allowance for uncollectible accounts?

BE 7–11
Uncollectible accounts; adjusting the allowance, debit balance
● LO7–5, LO7–6

At the end of the year, Syer Associates had a *debit* balance in its allowance for uncollectible accounts of $12,000 before adjustment. The balance in Syer's gross accounts receivable is $600,000. Syer's management estimates that 10% of its accounts receivable balance will not be collected. What journal entry should Syer record to adjust its allowance for uncollectible accounts?

BE 7–12
Calculate uncollectible accounts using the aging method
● LO7–5, LO7–6

Crell Computers categorizes its accounts receivable into four age groups for purposes of estimating its allowance for uncollectible accounts.

1. Accounts not yet due = $60,000; estimated uncollectible = 5%.
2. Accounts 1–30 days past due = $15,000; estimated uncollectible = 10%.
3. Accounts 31–60 days past due = $10,000; estimated uncollectible = 20%.
4. Accounts more than 60 days past due = $5,000; estimated uncollectible = 30%.

What should be the balance in Crell's allowance for uncollectible accounts?

BE 7–13
Uncollectible accounts; income statement approach
● LO7–5, LO7–6

The following information relates to a company's accounts receivable: gross accounts receivable balance at the beginning of the year, $300,000; allowance for uncollectible accounts at the beginning of the year, $25,000 (credit balance); credit sales during the year, $1,500,000; accounts receivable written off during the year, $16,000; cash collections from customers, $1,450,000. Assuming the company estimates bad debts at an amount equal to 2% of credit sales, calculate (1) bad debt expense for the year and (2) the year-end balance in the allowance for uncollectible accounts.

BE 7–14
Uncollectible accounts; balance sheet approach
● LO7–5, LO7–6

Refer to the situation described in BE 7–13. Answer the two questions assuming the company estimates that future bad debts will equal 10% of the year-end balance in accounts receivable.

BE 7–15
Uncollectible accounts; solving for unknown
● LO7–5, LO7–6

A company's year-end balance in accounts receivable is $2,000,000. The allowance for uncollectible accounts had a beginning-of-year credit balance of $30,000. An aging of accounts receivable at the end of the year indicates a required allowance of $38,000. If bad debt expense for the year was $40,000, what was the amount of bad debts written off during the year?

BE 7–16
Uncollectible accounts; solving for unknown
● LO7–5, LO7–6

Refer to the situation described in BE 7–15. If credit sales for the year were $8,200,000 and $7,950,000 was collected from credit customers, what was the beginning-of-year balance in accounts receivable?

BE 7–17
Note receivable
● LO7–7

On December 1, 2021, Davenport Company sold merchandise to a customer for $20,000. In payment for the merchandise, the customer signed a 6% note requiring the payment of interest and principal on March 1, 2022. How much interest revenue will the company recognize during 2021? In 2022?

BE 7–18
Long-term notes receivable
● LO7–4

On April 19, 2021, Millipede Machinery sold a tractor to Thomas Hartwood, accepting a note promising payment of $120,000 in five years. The applicable effective interest rate is 7%. What amount of sales revenue would Millipede recognize on April 19, 2021, for the Hartwood transaction?

BE 7–19
Factoring of accounts receivable
● LO7–8

Logitech Corporation transferred $100,000 of accounts receivable to a local bank. The transfer was made without recourse. The local bank remits 85% of the factored amount to Logitech and retains the remaining 15%. When the bank collects the receivables, it will remit to Logitech the retained amount less a fee equal to 3% of the total amount factored. Logitech estimates a fair value of its 15% interest in the receivables of $11,000 (not including the 3% fee). What is the effect of this transaction on the company's assets, liabilities, and income before income taxes?

BE 7–20
Factoring of accounts receivable
● LO7–8

Refer to the situation described in BE 7–19. Assuming that the sale criteria are not met, describe how Logitech would account for the transfer.

BE 7–21
Transfers of accounts receivable
● LO7–8, LO7–10
IFRS

Huling Associates plans to transfer $300,000 of accounts receivable to Mitchell Inc. in exchange for cash. Huling has structured the arrangement so that it retains substantially all the risks and rewards of ownership but shifts control over the receivables to Mitchell. Assuming all other criteria are met for recognizing the transfer as a sale, how would Huling account for this transaction under IFRS? Under U.S. GAAP?

BE 7–22
Discounting a note
● LO7–8

On March 31, Dower Publishing discounted a $30,000 note at a local bank. The note was dated February 28 and required the payment of the principal amount and interest at 6% on May 31. The bank's discount rate is 8%. How much cash will Dower receive from the bank on March 31?

BE 7–23
Receivables turnover
● LO7–8

Camden Hardware's credit sales for the year were $320,000. Accounts receivable at the beginning and end of the year were $50,000 and $70,000, respectively. Calculate the accounts receivable turnover ratio and the average collection period for the year.

BE 7–24
Bank reconciliation
● Appendix 7A

Marin Company's general ledger indicates a cash balance of $22,340 as of September 30, 2021. Early in October Marin received a bank statement indicating that during September Marin had an NSF check of $1,500 returned to a customer and incurred service charges of $45. Marin also learned it had incorrectly recorded a check received from a customer on September 15 as $500 when in fact the check was for $550. Calculate Marin's correct September 30, 2021, cash balance.

BE 7–25
Bank reconciliation
● Appendix 7A

Shan Enterprises received a bank statement listing its May 31, 2021, bank balance as $47,582. Shan determined that as of May 31 it had cash receipts of $2,500 that were not yet deposited and checks outstanding of $7,224. Calculate Shan's correct May 31, 2021, cash balance.

BE 7–26
Troubled debt restructuring
● Appendix 7B

Thaler Inc. holds a $1 million receivable ($800,000 principal, $200,000 accrued interest) from Einhorn Industries, and agrees to settle the receivable outright for $900,000 given Einhorn's difficult financial situation. How much gain or loss should Thaler recognize on this troubled debt restructuring?

Exercises

connect

E 7–1
Cash and cash equivalents; restricted cash
● LO7–2

The controller of the Red Wing Corporation is in the process of preparing the company's 2021 financial statements. She is trying to determine the correct balance of cash and cash equivalents to be reported as a current asset in the balance sheet. The following items are being considered:

a. Balances in the company's accounts at the First National Bank; checking $13,500, savings $22,100.
b. Undeposited customer checks of $5,200.
c. Currency and coins on hand of $580.
d. Savings account at the East Bay Bank with a balance of $400,000. This account is being used to accumulate cash for future plant expansion (in 2023).
e. $20,000 in a checking account at the East Bay Bank. The balance in the account represents a 20% compensating balance for a $100,000 loan with the bank. Red Wing may not withdraw the funds until the loan is due in 2024.
f. U.S. Treasury bills; 2-month maturity bills totaling $15,000, and 7-month bills totaling $20,000.

Required:

1. Determine the correct balance of cash and cash equivalents to be reported in the current asset section of the 2021 balance sheet.
2. For each of the items not included in your answer to requirement 1, explain the correct classification of the item.

E 7–2
Cash and cash equivalents
● LO7–2

Delta Automotive Corporation has the following assets listed in its 12/31/2021 trial balance:

Cash in bank—checking account	$22,500
U.S. Treasury bills (mature in 60 days)*	5,000
Cash on hand (currency and coins)	1,350
U.S. Treasury bills (mature in six months)*	10,000
Undeposited customer checks	1,840

*Purchased on 11/30/2021

Required:

1. Determine the correct balance of cash and cash equivalents to be reported in the current asset section of the 2021 balance sheet.
2. For each of the items not included in your answer to requirement 1, explain the correct classification of the item.

E 7–3
FASB codification research
● LO7–2, LO7–6, LO7–7

CODE

Access the *FASB Accounting Standards Codification* at the FASB website (**www.fasb.org**).

Required:

Indicate the specific seven-digit Codification citation (XXX-XX-XX) for each of the following items:

1. Accounts receivables from related parties should be shown separately from trade receivables
2. The definition of cash equivalents (Hint: see Statement of Cash Flows.)
3. The requirement to value notes exchanged for cash at the cash proceeds
4. The two conditions that must be met to accrue a loss on an accounts receivable

E 7–4
Bank overdrafts
● LO7–2, LO7–10

IFRS

Parker Inc. has the following cash balances:

First Bank	$150,000
Second Bank	(10,000)
Third Bank	25,000
Fourth Bank	(5,000)

Required:

1. Prepare the current assets and current liabilities section of Parker's 2021 balance sheet, assuming Parker reports under U.S. GAAP.
2. Prepare the current assets and current liabilities section of Parker's 2021 balance sheet, assuming Parker reports under IFRS.

E 7–5
Trade and cash discounts; the gross method and the net method compared
● LO7–3

Tracy Company, a manufacturer of air conditioners, sold 100 units to Thomas Company on November 17, 2021. The units have a list price of $600 each, but Thomas was given a 30% trade discount. The terms of the sale were 2/10, n/30.

Required:

1. Prepare the journal entries to record the sale on November 17 (ignore cost of goods) and collection on November 26, 2021, assuming that the gross method of accounting for cash discounts is used.
2. Prepare the journal entries to record the sale on November 17 (ignore cost of goods) and collection on December 15, 2021, assuming that the gross method of accounting for cash discounts is used.
3. Repeat requirements 1 and 2 assuming that the net method of accounting for cash discounts is used.

E 7–6
Cash discounts; the gross method
● LO7–3

Harwell Company manufactures automobile tires. On July 15, 2021, the company sold 1,000 tires to the Nixon Car Company for $50 each. The terms of the sale were 2/10, n/30. Harwell uses the gross method of accounting for cash discounts.

Required:

1. Prepare the journal entries to record the sale on July 15 (ignore cost of goods) and collection on July 23, 2021.
2. Prepare the journal entries to record the sale on July 15 (ignore cost of goods) and collection on August 15, 2021.

E 7–7
Cash discounts; the net method
● LO7–3

[This is a variation of E 7–6 modified to focus on the net method of accounting for cash discounts.]

Harwell Company manufactures automobile tires. On July 15, 2021, the company sold 1,000 tires to the Nixon Car Company for $50 each. The terms of the sale were 2/10, n/30. Harwell uses the net method of accounting for cash discounts.

Required:

1. Prepare the journal entries to record the sale on July 15 (ignore cost of goods) and payment on July 23, 2021.
2. Prepare the journal entries to record the sale on July 15 (ignore cost of goods) and payment on August 15, 2021.

E 7–8
Sales returns
● LO7–4

Halifax Manufacturing allows its customers to return merchandise for any reason up to 90 days after delivery and receive a credit to their accounts. All of Halifax's sales are for credit (no cash is collected at the time of sale). The company began 2021 with a refund liability of $300,000. During 2021, Halifax sold merchandise on account for $11,500,000. Halifax's merchandise costs it 65% of merchandise selling price. Also during the year, customers returned $450,000 in sales for credit, with $250,000 of those being returns of merchandise sold prior to 2021, and the rest being merchandise sold during 2021. Sales returns, estimated to be 4% of sales, are recorded as an adjusting entry at the end of the year.

Required:

1. Prepare entries to (a) record actual returns in 2021 of merchandise that was sold ***prior to*** 2021; (b) record actual returns in 2021 of merchandise that was sold ***during*** 2021; and (c) adjust the refund liability to its appropriate balance at year end.
2. What is the amount of the year-end refund liability after the adjusting entry is recorded?

E 7–9
FASB codification research
● LO7–5

The *FASB Accounting Standards Codification* represents the single source of authoritative U.S. generally accepted accounting principles.

Required:

1. Obtain the relevant authoritative literature on accounting for accounts receivable using the FASB's Codification Research System at the FASB website (**www.fasb.org**). What is the specific eight-digit Codification citation (XXX-XX-XX-X) that describes the information about loans and trade receivables that is to be disclosed in the summary of significant accounting policies?
2. List the disclosure requirements.

E 7–10
Establish an allowance and account for bad debts
● LO7–5, LO7–6

Pincus Associates uses the allowance method to account for bad debts. During 2021, its first year of operations, Pincus provided a total of $250,000 of services on account. In 2021, the company wrote off uncollectible accounts of $10,000. By the end of 2021, cash collections on accounts receivable totaled $210,000. Pincus estimates that 20% of the accounts receivable balance at 12/31/2021 will prove uncollectible.

Required:

1. What journal entry did Pincus record to write off uncollectible accounts during 2021?
2. What journal entry should Pincus record to recognize bad debt expense for 2021?

E 7–11
Uncollectible accounts; allowance method vs. direct write-off method
● LO7–5, LO7–6

Johnson Company calculates its allowance for uncollectible accounts as 10% of its ending balance in gross accounts receivable. The allowance for uncollectible accounts had a credit balance of $30,000 at the beginning of 2021. No previously written-off accounts receivable were reinstated during 2021. At 12/31/2021, gross accounts receivable totaled $500,000, and prior to recording the adjusting entry to recognize bad debts expense for 2021, the allowance for uncollectible accounts had a debit balance of 55,000.

Required:

1. What was the balance in gross accounts receivable as of 12/31/2020?
2. What journal entry should Johnson record to recognize bad debt expense for 2021?
3. Assume Johnson made no other adjustment of the allowance for uncollectible accounts during 2021. Determine the amount of accounts receivable written off during 2021.
4. If Johnson instead used the direct write-off method, what would bad debt expense be for 2021?

E 7–12
Uncollectible accounts; allowance method estimating bad debts as percentage of net sales vs. direct write-off method
● LO7–5, LO7–6

Ervin Company uses the allowance method to account for uncollectible accounts receivable. Bad debt expense is established as a percentage of credit sales. For 2021, net credit sales totaled $4,500,000, and the estimated bad debt percentage is 1.5%. No previously written-off accounts receivable were reinstated during 2021. The allowance for uncollectible accounts had a credit balance of $42,000 at the beginning of 2021 and $40,000, after adjusting entries, at the end of 2021.

Required:

1. What is bad debt expense for 2021 as a percent of net credit sales?
2. Assume Ervin makes no other adjustment of bad debt expense during 2021. Determine the amount of accounts receivable written off during 2021.
3. If the company uses the direct write-off method, what would bad debt expense be for 2021?

E 7–13
Calculate uncollectible accounts using the aging method; record adjustment
● LO7–5, LO7–6

Dhaliwal Digital categorizes its accounts receivable into three age groups for purposes of estimating its allowance for uncollectible accounts.

1. Accounts not yet due = $180,000; estimated uncollectible = 10%.
2. Accounts 1–45 days past due = $25,000; estimated uncollectible = 20%.
3. Accounts more than 45 days past due = $10,000; estimated uncollectible = 30%.

Before recording any adjustments, Dhaliwal has a debit balance of $45,000 in its allowance for uncollectible accounts.

Required:

1. Estimate the appropriate 12/31/2021 balance for Dhaliwal's allowance for uncollectible accounts.
2. What journal entry should Dhaliwal record to adjust its allowance for uncollectible accounts?

E 7–14
Calculate uncollectible accounts using the aging method; record adjustments
● LO7–5, LO7–6

Zuo Software categorizes its accounts receivable into four age groups for purposes of estimating its allowance for uncollectible accounts.

1. Accounts not yet due = $400,000; estimated uncollectible = 8%.
2. Accounts 1–30 days past due = $50,000; estimated uncollectible = 15%.
3. Accounts 31–90 days past due = $40,000; estimated uncollectible = 30%
4. Accounts more than 90 days past due = $30,000; estimated uncollectible = 50%.

At 12/31/2021, before recording any adjustments, Zuo has a credit balance of $22,000 in its allowance for uncollectible accounts.

Required:

1. Estimate the appropriate 12/31/2021 balance for Zuo's allowance for uncollectible accounts.
2. What journal entry should Zuo record to adjust its allowance for uncollectible accounts?
3. Calculate Zuo's 12/31/2021 net accounts receivable balance.

E 7–15
Uncollectible accounts; allowance method; balance sheet approach
● LO7–5, LO7–6

Colorado Rocky Cookie Company offers credit terms to its customers. At the end of 2021, accounts receivable totaled $625,000. The allowance method is used to account for uncollectible accounts. The allowance for uncollectible accounts had a credit balance of $32,000 at the beginning of 2021 and $21,000 in receivables were written off during the year as uncollectible. Also, $1,200 in cash was received in December from a customer whose account previously had been written off. The company estimates bad debts by applying a percentage of 10% to accounts receivable at the end of the year.

Required:

1. Prepare journal entries to record the write-off of receivables, the collection of $1,200 for previously written off receivables, and the year-end adjusting entry for bad debt expense.
2. How would accounts receivable be shown in the 2021 year-end balance sheet?

E 7–16
Uncollectible accounts allowance method and direct write-off method compared; solving for unknown
● LO7–6

Castle Company provides estimates for its uncollectible accounts. The allowance for uncollectible accounts had a credit balance of $17,280 at the beginning of 2021 and a $22,410 credit balance at the end of 2021 (after adjusting entries). If the direct write-off method had been used to account for uncollectible accounts (bad debt expense equals actual write-offs), the income statement for 2021 would have included bad debt expense of $17,100 and revenue of $2,200 from the collection of previously written off bad debts.

Required:

Determine bad debt expense for 2021 according to the allowance method.

E 7–17
Uncollectible accounts; allowance method; solving for unknowns; General Mills
● LO7–5, LO7–6
Real World Financials

General Mills reported the following information in its 2017 financial statements ($ in millions)

	2017	2016
Balance Sheet:		
Accounts receivable, net	$ 1,430.1	$1,360.8
Income statement:		
Sales revenue	$15,619.8	

A note disclosed that the allowance for uncollectible accounts had a balance of $24.3 million and $29.6 million at the end of 2017 and 2016, respectively. Bad debt expense for 2017 was $16.6 million.

Required:

Determine the amount of cash collected from customers during 2017.

E 7–18
Notes receivable
● LO7–7

On June 30, 2021, the Esquire Company sold some merchandise to a customer for $30,000. In payment, Esquire agreed to accept a 6% note requiring the payment of interest and principal on March 31, 2022. The 6% rate is appropriate in this situation.

Required:

1. Prepare journal entries to record the sale of merchandise (omit any entry that might be required for the cost of the goods sold), the December 31, 2021 interest accrual, and the March 31, 2022 collection.
2. If the December 31 adjusting entry for the interest accrual is not prepared, by how much will income before income taxes be over- or understated in 2021 and 2022?

E 7–19
Noninterest-bearing notes receivable
● LO7–7

[This is a variation of E 7–18 modified to focus on a noninterest-bearing note.]

On June 30, 2021, the Esquire Company sold merchandise to a customer and accepted a noninterest-bearing note in exchange. The note requires payment of $30,000 on March 21, 2022. The fair value of the merchandise exchanged is $28,200. Esquire views the financing component of this contract as significant.

Required:

1. Prepare journal entries to record the sale of merchandise (omit any entry that might be required for the cost of the goods sold), any December 31, 2021 interest accrual, and the March 31, 2022 collection.
2. What is the *effective* interest rate on the note?

E 7–20
Long-term notes receivable
● LO7–7

On January 1, 2021, Wright Transport sold four school buses to the Elmira School District. In exchange for the buses, Wright received a note requiring payment of $515,000 by Elmira on December 31, 2023. The effective interest rate is 8%.

Required:

1. How much sales revenue would Wright recognize on January 1, 2021, for this transaction?
2. Prepare journal entries to record the sale of merchandise on January 1, 2021 (omit any entry that might be required for the cost of the goods sold), the December 31, 2021, interest accrual, the December 31, 2022, interest accrual, and receipt of payment of the note on December 31, 2023.

E 7–21
Interest-bearing notes receivable; solving for unknown rate
● LO7–7

On January 1, 2021, the Apex Company exchanged some shares of common stock it had been holding as an investment for a note receivable. The note principal plus interest is due on January 1, 2022. The 2021 income statement reported $2,200 in interest revenue from this note and a $6,000 gain on sale of investment in stock. The stock's book value was $16,000. The company's fiscal year ends on December 31.

Required:

1. What is the note's effective interest rate?
2. Reconstruct the journal entries to record the sale of the stock on January 1, 2021, and the adjusting entry to record interest revenue at the end of 2021. The company records adjusting entries only at year-end.

E 7–22
Assigning of specific accounts receivable
● LO7–8

On June 30, 2021, the High Five Surfboard Company had outstanding accounts receivable of $600,000. On July 1, 2021, the company borrowed $450,000 from the Equitable Finance Corporation and signed a promissory note. Interest at 10% is payable monthly. The company assigned specific receivables totaling $600,000 as collateral for the loan. Equitable Finance charges a finance fee equal to 1.8% of the accounts receivable assigned.

Required:
Prepare the journal entry to record the borrowing on the books of High Five Surfboard.

E 7–23
Factoring of accounts receivable without recourse
● LO7–8

Mountain High Ice Cream Company transferred $60,000 of accounts receivable to the Prudential Bank. The transfer was made *without recourse.* Prudential remits 90% of the factored amount to Mountain High and retains 10%. When the bank collects the receivables, it will remit to Mountain High the retained amount (which Mountain estimates has a fair value of $5,000) less a 2% fee (2% of the total factored amount).

Required:
Prepare the journal entry to record the transfer on the books of Mountain High assuming that the sale criteria are met.

E 7–24
Factoring of accounts receivable with recourse
● LO7–8

[This is a variation of E 7–23 modified to focus on factoring with recourse.]

Mountain High Ice Cream Company transferred $60,000 of accounts receivable to the Prudential Bank. The transfer was made *with recourse.* Prudential remits 90% of the factored amount to Mountain High and retains 10% to cover sales returns and allowances. When the bank collects the receivables, it will remit to Mountain High the retained amount (which Mountain estimates has a fair value of $5,000). Mountain High anticipates a $3,000 recourse obligation. The bank charges a 2% fee (2% of $60,000), and requires that amount to be paid at the start of the factoring arrangement.

Required:
Prepare the journal entry to record the transfer on the books of Mountain High assuming that the sale criteria are met.

E 7–25
Factoring of accounts receivable with recourse under IFRS
● LO7–8, LO7–10
IFRS

[This is a variation of E 7–24 modified to focus on factoring with recourse under IFRS.]

Mountain High Ice Cream Company reports under IFRS. Mountain High transferred $60,000 of accounts receivable to the Prudential Bank. The transfer was made *with recourse.* Prudential remits 90% of the factored amount to Mountain High and retains 10% to cover sales returns and allowances. When the bank collects the receivables, it will remit to Mountain High the retained amount (which Mountain estimates has a fair value of $5,000). Mountain High anticipates a $3,000 recourse obligation. The bank charges a 2% fee (2% of $60,000), and requires that amount to be paid at the start of the factoring arrangement. Mountain High has transferred control over the receivables, but determines that it still retains substantially all risks and rewards associated with them.

Required:
Prepare the journal entry to record the transfer on the books of Mountain High, considering whether the sale criteria under IFRS have been met.

E 7–26
Discounting a note receivable
● LO7–8

Selkirk Company obtained a $15,000 note receivable from a customer on January 1, 2021. The note, along with interest at 10%, is due on July 1, 2021. On February 28, 2020, Selkirk discounted the note at Unionville Bank. The bank's discount rate is 12%.

Required:
Prepare the journal entries required on February 28, 2021, to accrue interest and to record the discounting (round all calculations to the nearest dollar) for Selkirk. Assume that the discounting is accounted for as a sale.

E 7–27
Concepts; terminology
● LO7–1 through LO7–8

Listed below are several terms and phrases associated with cash and receivables. Pair each item from List A (by letter) with the item from List B that is most appropriately associated with it.

List A	List B
_____ 1. Internal control	a. Restriction on cash.
_____ 2. Trade discount	b. Cash discount not taken is sales revenue.
_____ 3. Cash equivalents	c. Includes separation of duties.
_____ 4. Allowance for uncollectibles	d. Bad debt expense a % of credit sales.
_____ 5. Cash discount	e. Recognizes bad debts as they occur.
_____ 6. Balance sheet approach	f. Sale of receivables to a financial institution.
_____ 7. Income statement approach	g. Include highly liquid investments.
_____ 8. Net method	h. Estimate of bad debts.
_____ 9. Compensating balance	i. Reduction in amount paid by credit customer.
_____ 10. Discounting	j. Reduction below list price.
_____ 11. Gross method	k. Cash discount not taken is sales discount forfeited.
_____ 12. Direct write-off method	l. Bad debt expense determined by estimating amount of accounts receivable expected to be received.
_____ 13. Factoring	m. Sale of note receivable to a financial institution.

E 7–28
Receivables; transaction analysis
● LO7–3, LO7–5 through LO7–8

Weldon Corporation's fiscal year ends December 31. The following is a list of transactions involving receivables that occurred during 2021:

Mar. 17	Accounts receivable of $1,700 were written off as uncollectible. The company uses the allowance method.
30	Loaned an officer of the company $20,000 and received a note requiring principal and interest at 7% to be paid on March 30, 2022.
May 30	Discounted the $20,000 note at a local bank. The bank's discount rate is 8%. The note was discounted without recourse and the sale criteria are met.
June 30	Sold merchandise to the Blankenship Company for $12,000. Terms of the sale are 2/10, n/30. Weldon uses the gross method to account for cash discounts.
July 8	The Blankenship Company paid its account in full.
Aug. 31	Sold stock in a nonpublic company with a book value of $5,000 and accepted a $6,000 noninterest-bearing note with a discount rate of 8%. The $6,000 payment is due on February 28, 2022. The stock has no ready market value.
Dec. 31	Weldon estimates that the allowance for uncollectible accounts should have a balance in it at year-end equal to 2% of the gross accounts receivable balance of $700,000. The allowance had a balance of $12,000 at the start of 2021.

Required:

1. Prepare journal entries for each of the above transactions (round all calculations to the nearest dollar).
2. Prepare any additional year-end adjusting entries indicated.

E 7–29
Ratio analysis; Microsoft
● LO7–9
Real World Financials

Microsoft Corporation reported the following information in its financial statements for three successive quarters ($ in millions):

	Three Months Ended		
	9/30/2017 (Q1)	6/30/2017 (Q4)	3/31/2017 (Q3)
Balance sheets:			
Accounts receivable, net	$14,561	$19,792	$12,882
Income statements:			
Sales revenue	$24,538	$23,317	$22,090

Required:

Compute the receivables turnover ratio and the average collection period for Q1 and Q4. Assume that each quarter consists of 91 days.

E 7–30
Ratio analysis
● LO7–9

The current asset section of the Moorcroft Outboard Motor Company's balance sheet reported the following amounts:

	12/31/2021	12/31/2020
Accounts receivable, net	$400,000	$300,000

The average collection period for 2021 is 50 days.

Required:

Determine net sales for 2021.

E 7–31
Petty cash
● Appendix 7A

Loucks Company established a $200 petty cash fund on October 2, 2021. The fund is replenished at the end of each month. At the end of October 2021, the fund contained $37 in cash and the following receipts:

Office supplies	$76
Advertising	48
Postage	20
Miscellaneous	19

Required:
Prepare the necessary general journal entries to establish the petty cash fund on October 2 and to replenish the fund on October 31.

E 7–32
Petty cash
● Appendix 7A

The petty cash fund of Ricco's Automotive contained the following items at the end of September 2021:

Currency and coins		$ 58
Receipts for the following expenditures:		
Delivery charges	$16	
Printer paper	11	
Paper clips and rubber bands	8	35
Lent money to an employee		25
Postage		32
Total		$150

The petty cash fund was established at the beginning of September with a transfer of $150 from cash to the petty cash account.

Required:
Prepare the journal entry to replenish the fund at the end of September.

E 7–33
Bank reconciliation
● Appendix 7A

Jansen Company's general ledger showed a checking account balance of $23,820 at the end of May 2021. The May 31 cash receipts of $2,340, included in the general ledger balance, were placed in the night depository at the bank on May 31 and were processed by the bank on June 1. The bank statement dated May 31, 2021, showed bank service charges of $38. All checks written by the company had been processed by the bank by May 31 and were listed on the bank statement except for checks totaling $1,890.

Required:
Prepare a bank reconciliation as of May 31, 2021. [*Hint:* You will need to compute the balance that would appear on the bank statement.]

E 7–34
Bank reconciliation and adjusting entries
● Appendix 7A

Harrison Company maintains a checking account at the First National City Bank. The bank provides a bank statement along with canceled checks on the last day of each month. The July 2021 bank statement included the following information:

Balance, July 1, 2021	$ 55,678
Deposits	179,500
Checks processed	(192,610)
Service charges	(30)
NSF checks	(1,200)
Monthly payment on note, deducted directly by bank from account (includes $320 in interest)	(3,320)
Balance, July 31, 2021	$ 38,018

The company's general ledger account had a balance of $38,918 at the end of July. Deposits outstanding totaled $6,300 and all checks written by the company were processed by the bank except for those totaling $8,420. In addition, a $2,000 July deposit from a credit customer was recorded as a $200 debit to cash and credit to accounts receivable, and a check correctly recorded by the company as a $30 disbursement was incorrectly processed by the bank as a $300 disbursement.

Required:
1. Prepare a bank reconciliation for the month of July.
2. Prepare the necessary journal entries at the end of July to adjust the general ledger cash account.

E 7–35
Troubled debt restructuring
● Appendix 7B

At January 1, 2021, Clayton Hoists Inc. owed Third BancCorp $12 million, under a 10% note due December 31, 2022. Interest was paid last on December 31, 2019. Clayton was experiencing severe financial difficulties and asked Third BancCorp to modify the terms of the debt agreement. After negotiation Third BancCorp agreed to do the following:

- Forgive the interest accrued for the year just ended.
- Reduce the remaining two years' interest payments to $1 million each.
- Reduce the principal amount to $11 million.

Required:

Prepare the journal entries by Third BancCorp necessitated by the restructuring of the debt at

1. January 1, 2021.
2. December 31, 2021.
3. December 31, 2022.

E 7–36
General Ledger Exercise
● LO7–5, LO7–6, LO7–7, LO7–9

On January 1, 2021, the general ledger of 3D Family Fireworks includes the following account balances:

Accounts	Debit	Credit
Cash	$ 23,900	
Accounts Receivable	13,600	
Allowance for Uncollectible Accounts		$ 1,400
Supplies	2,500	
Notes Receivable (6%, due in 2 years)	20,000	
Land	77,000	
Accounts Payable		7,200
Common Stock		96,000
Retained Earnings		32,400
Totals	$137,000	$137,000

During January 2021, the following transactions occur:

1. January 2	Provide services to customers for cash, $35,100.
2. January 6	Provide services to customers on account, $72,400.
3. January 15	Write off accounts receivable as uncollectible, $1,000.
4. January 20	Pay cash for salaries, $31,400.
5. January 22	Receive cash on accounts receivable, $70,000.
6. January 25	Pay cash on accounts payable, $5,500.
7. January 30	Pay cash for utilities during January, $13,700.

The following information is available on January 31, 2021.

1. At the end of January, $5,000 of accounts receivable are past due, and the company estimates that 20% of these accounts will not be collected. Of the remaining accounts receivable, the company estimates that 5% will not be collected. The note receivable of $20,000 is considered fully collectible and therefore is not included in the estimate of uncollectible accounts.
2. Supplies at the end of January total $700.
3. Accrued interest revenue on notes receivable for January. Interest is expected to be received each December 31.
4. Unpaid salaries at the end of January are $33,500.

Required:

1. Record each of the transactions listed above in the 'General Journal' tab (these are shown as items 1–7). Review the 'General Ledger' and the 'Trial Balance' tabs to see the effect of the transactions on the account balances.
2. Record adjusting entries on January 31. in the 'General Journal' tab (these are shown as items 8-11).
3. Review the adjusted 'Trial Balance' as of January 31, 2021, in the 'Trial Balance' tab.
4. Prepare an income statement for the period ended January 31, 2021, in the 'Income Statement' tab).
5. Prepare a classified balance sheet as of January 31, 2021, in the 'Balance Sheet' tab.
6. Record closing entries in the 'General Journal' tab (these are shown as items 12 and 13).
7. Using the information from the requirements above, complete the 'Analysis' tab.

Problems

connect

P 7–1
Uncollectible accounts; allowance method; income statement and balance sheet approach
● LO7–5, LO7–6

Swathmore Clothing Corporation grants its customers 30 days' credit. The company uses the allowance method for its uncollectible accounts receivable. During the year, a monthly bad debt accrual is made by multiplying 3% times the amount of credit sales for the month. At the fiscal year-end of December 31, an aging of accounts receivable schedule is prepared and the allowance for uncollectible accounts is adjusted accordingly.

At the end of 2020, accounts receivable were $574,000 and the allowance account had a credit balance of $54,000. Accounts receivable activity for 2021 was as follows:

Beginning balance	$ 574,000
Credit sales	2,620,000
Collections	(2,483,000)
Write-offs	(68,000)
Ending balance	$ 643,000

The company's controller prepared the following aging summary of year-end accounts receivable:

	Summary	
Age Group	Amount	Percent Uncollectible
0–60 days	$430,000	4%
61–90 days	98,000	15
91–120 days	60,000	25
Over 120 days	55,000	40
Total	$643,000	

Required:
1. Prepare a summary journal entry to record the monthly bad debt accrual and the write-offs during the year.
2. Prepare the necessary year-end adjusting entry for bad debt expense.
3. What is total bad debt expense for 2021? How would accounts receivable appear in the 2021 balance sheet?

P 7–2
Uncollectible accounts; EMC Corporation
● LO7–5
Real World Financials

EMC Corporation manufactures large-scale, high-performance computer systems. In a recent annual report, the balance sheet included the following information ($ in millions):

	2015	2014
Current assets:		
Receivables, less allowances of $90 in 2015 and $72 in 2014	$3,977	$4,413

In addition, the income statement reported sales revenue of $24,704 ($ in millions) for the current year. All sales are made on a credit basis. The statement of cash flows indicates that cash collected from customers during the current year was $25,737 ($ in millions). There were no recoveries of accounts receivable previously written off.

Required:
1. Compute the following ($ in millions):
   a. The amount of bad debts written off by EMC during 2015.
   b. The amount of bad debt expense that EMC included in its income statement for 2015.
   c. The approximate percentage that EMC used to estimate bad debts for 2015, assuming that it used the income statement approach
2. Suppose that EMC had used the direct write-off method to account for bad debts. Compute the following ($ in millions):
   a. The accounts receivable information that would be included in the 2015 year-end balance sheet.
   b. The amount of bad debt expense that EMC would include in its 2015 income statement.

P 7–3
Bad debts; Nike, Inc.
● LO7–5
Real World Financials

Nike, Inc., is a leading manufacturer of sports apparel, shoes, and equipment. The company's 2017 financial statements contain the following information ($ in millions):

	2017	2016
Balance sheets:		
Accounts receivable, net	$ 3,677	$ 3,241
Income statements:		
Sales revenue	$34,350	$32,376

A note disclosed that the allowance for uncollectible accounts had a balance of $19 million and $43 million at the end of 2017 and 2016, respectively. Bad debt expense for 2017 was $40 million. Assume that all sales are made on a credit basis.

Required:

1. What is the amount of gross (total) accounts receivable due from customers at the end of 2017 and 2016?
2. What is the amount of bad debt write-offs during 2017?
3. Analyze changes in the gross accounts receivable account to calculate the amount of cash received from customers during 2017.
4. Analyze changes in net accounts receivable to calculate the amount of cash received from customers during 2017.

P 7–4
Uncollectible accounts
● LO7–5, LO7–6

Raintree Cosmetic Company sells its products to customers on a credit basis. An adjusting entry for bad debt expense is recorded only at December 31, the company's fiscal year-end. The 2020 balance sheet disclosed the following:

Current assets:	
Receivables, net of allowance for uncollectible accounts of $30,000	**$432,000**

During 2021, credit sales were $1,750,000, cash collections from customers $1,830,000, and $35,000 in accounts receivable were written off. In addition, $3,000 was collected from a customer whose account was written off in 2020. An aging of accounts receivable at December 31, 2021, reveals the following:

Age Group	Percentage of Year-End Receivables in Group	Percent Uncollectible
0–60 days	65%	4%
61–90 days	20	15
91–120 days	10	25
Over 120 days	5	40

Required:

1. Prepare summary journal entries to account for the 2021 write-offs and the collection of the receivable previously written off.
2. Prepare the year-end adjusting entry for bad debts according to each of the following situations:
   a. Bad debt expense is estimated to be 3% of credit sales for the year.
   b. Bad debt expense is estimated by adjusting the allowance for uncollectible accounts to the balance that reduces the carrying value of accounts receivable to the amount of cash expected to be collected. The allowance for uncollectible accounts is estimated to be 10% of the year-end balance in accounts receivable.
   c. Bad debt expense is estimated by adjusting the allowance for uncollectible accounts to the balance that reduces the carrying value of accounts receivable to the amount of cash expected to be collected. The allowance for uncollectible accounts is determined by an aging of accounts receivable.
3. For situations (a)–(c) in requirement 2 above, what would be the net amount of accounts receivable reported in the 2021 balance sheet?

P 7–5
Receivables; bad debts and returns; Avon Products, Inc.
● LO7–4, LO7–5
Real World Financials

Avon Products, Inc., located in New York City, is one of the world's largest producers of beauty and related products. The company's consolidated balance sheets for the 2016 and 2015 fiscal years included the following ($ in thousands):

	2016	2015
Current assets:		
Receivables, less allowances of $131,100 in 2016 and $86,700 in 2015	$458,900	$440,000

A disclosure note accompanying the financial statements reported the following ($ in thousands):

	Year Ended	
	2016	**2015**
Calculation of account receivables, net:		
Receivables	$ 590,000	$ 526,700
Less: allowance for doubtful accounts	(122,900)	(77,600)
Less: reserve for product returns	(8,200)	(9,100)
Trade accounts receivable, net	$ 458,900	$ 440,000

Assume that the company reported bad debt expense in 2016 of $191,000 and had products returned for credit totaling $187,000 (sales price). Net sales for 2016 were $5,578,800 ($ in thousands).

Required:

1. What is the amount of accounts receivable due from customers at the end of 2016 and 2015?
2. What amount of accounts receivable did Avon write off during 2016?
3. What is the amount of Avon's gross sales for the 2016 fiscal year?
4. Assuming that all sales are made on a credit basis, what is the amount of cash Avon collected from customers during the 2016 fiscal year?

P 7–6
Notes receivable; solving for unknowns
● LO7–7

Cypress Oil Company's December 31, 2021, balance sheet listed $645,000 of notes receivable and $16,000 of interest receivable included in current assets. The following notes make up the notes receivable balance:

Note 1	Dated 8/31/2021, principal of $300,000 and interest at 10% due on 2/28/2022.
Note 2	Dated 6/30/2021, principal of $150,000 and interest due 3/31/2022.
Note 3	$200,000 face value noninterest-bearing note dated 9/30/2021, due 3/31/2022. Note was issued in exchange for merchandise.

The company records adjusting entries only at year-end. There were no other notes receivable outstanding during 2021.

Required:

1. Determine the rate used to discount the noninterest-bearing note.
2. Determine the explicit interest rate on Note 2.
3. What is the amount of interest revenue that appears in the company's 2021 income statement related to these notes?

P 7–7
Factoring versus assigning of accounts receivable
● LO7–8

Lonergan Company occasionally uses its accounts receivable to obtain immediate cash. At the end of June 2021, the company had accounts receivable of $780,000. Lonergan needs approximately $500,000 to capitalize on a unique investment opportunity. On July 1, 2021, a local bank offers Lonergan the following two alternatives:

a. Borrow $500,000, sign a note payable, and assign the entire receivable balance as collateral. At the end of each month, a remittance will be made to the bank that equals the amount of receivables collected plus 12% interest on the unpaid balance of the note at the beginning of the period.
b. Transfer $550,000 of specific receivables to the bank without recourse. The bank will charge a 2% factoring fee on the amount of receivables transferred. The bank will collect the receivables directly from customers. The sale criteria are met.

Required:

1. Prepare the journal entries that would be recorded on July 1 for each of the alternatives.
2. Assuming that 80% of all June 30 receivables are collected during July, prepare the necessary journal entries to record the collection and the remittance to the bank.
3. For each alternative, explain any required note disclosures that would be included in the July 31, 2021, financial statements.

P 7–8
Factoring of accounts receivable; without recourse
● LO7–8

Samson Wholesale Beverage Company regularly factors its accounts receivable with the Milpitas Finance Company. On April 30, 2021, the company transferred $800,000 of accounts receivable to Milpitas. The transfer was made without recourse. Milpitas remits 90% of the factored amount and retains 10%. When Milpitas collects the receivables, it remits to Samson the retained amount less a 4% fee (4% of the total factored amount). Samson estimates the fair value of the last 10% of its receivables to be $60,000.

Required:

Prepare the journal entry for Samson Wholesale Beverage for the transfer of accounts receivable on April 30, assuming the sale criteria are met.

P 7–9
Cash and accounts receivable under IFRS
● LO7–2, LO7–5, LO7–8, LO7–10
IFRS

The following facts apply to Walken Company during December 2021:

a. Walken began December with an accounts receivable balance (net of bad debts) of €25,000.
b. Walken had credit sales of €85,000.
c. Walken had cash collections of €30,000.
d. Walken factored €20,000 of net accounts receivable with Reliable Factor Company, transferring all risks and rewards associated with the receivable, and otherwise meeting all criteria necessary to qualify for treating the transfer of receivables as a sale.

e. Walken factored €15,000 of net accounts receivable with Dependable Factor Company, retaining all risks and rewards associated with the receivable, and otherwise meeting all criteria necessary to qualify for treating the transfer of receivables as a sale.

f. Walken did not recognize any additional bad debts expense, and had no write-offs of bad debts during the month.

g. At December 31, 2021, Walken had a balance of €40,000 of cash at M&V Bank and an overdraft of (€5,000) at First National Bank. (That cash balance includes any effects on cash of the other transactions described in this problem.)

Required:

Prepare the cash and accounts receivable lines of the current assets section of Walken's balance sheet, as of December 31, 2021.

P 7–10
Miscellaneous receivable transactions
● LO7–3, LO7–4, LO7–7, LO7–8

Evergreen Company sells lawn and garden products to wholesalers. The company's fiscal year-end is December 31. During 2021, the following transactions related to receivables occurred:

Feb. 28	Sold merchandise to Lennox, Inc., for $10,000 and accepted a 10%, 7-month note. 10% is an appropriate rate for this type of note.
Mar. 31	Sold merchandise to Maddox Co. that had a fair value of $7,200, and accepted a noninterest-bearing note for which $8,000 payment is due on March 31, 2022.
Apr. 3	Sold merchandise to Carr Co. for $7,000 with terms 2/10, n/30. Evergreen uses the gross method to account for cash discounts.
11	Collected the entire amount due from Carr Co.
17	A customer returned merchandise costing $3,200. Evergreen reduced the customer's receivable balance by $5,000, the sales price of the merchandise. Sales returns are recorded by the company as they occur.
30	Transferred receivables of $50,000 to a factor without recourse. The factor charged Evergreen a 1% finance charge on the receivables transferred. The sale criteria are met.
June 30	Discounted the Lennox, Inc., note at the bank. The bank's discount rate is 12%. The note was discounted without recourse.
Sep. 30	Lennox, Inc., paid the note amount plus interest to the bank.

Required:

1. Prepare the necessary journal entries for Evergreen for each of the above dates. For transactions involving the sale of merchandise, ignore the entry for the cost of goods sold (round all calculations to the nearest dollar).
2. Prepare any necessary adjusting entries at December 31, 2021. Adjusting entries are only recorded at year-end (round all calculations to the nearest dollar).
3. Prepare a schedule showing the effect of the journal entries in requirements 1 and 2 on 2021 income before taxes.

P 7–11
Discounting a note receivable
● LO7–7

Descriptors are provided below for six situations involving notes receivable being discounted at a bank. In each case, the maturity date of the note is December 31, 2021, and the principal and interest are due at maturity. For each, determine the proceeds received from the bank on discounting the note.

Note	Note Face Value	Date of Note	Interest Rate	Date Discounted	Discount Rate
1	$50,000	3/31/2021	8%	6/30/2021	10%
2	50,000	3/31/2021	8	9/30/2021	10
3	50,000	3/31/2021	8	9/30/2021	12
4	80,000	6/30/2021	6	10/31/2021	10
5	80,000	6/30/2021	6	10/31/2021	12
6	80,000	6/30/2021	6	11/30/2021	10

P 7–12
Accounts and notes receivable; discounting a note receivable; receivables turnover ratio
● LO7–5, LO7–6, LO7–7, LO7–8, LO7–9

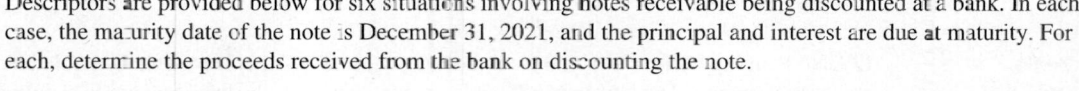

Chamberlain Enterprises Inc. reported the following receivables in its December 31, 2021, year-end balance sheet:

Current assets:	
Accounts receivable, net of $24,000 in allowance for uncollectible accounts	$218,000
Interest receivable	6,800
Notes receivable	260,000

Additional Information:

1. The notes receivable account consists of two notes, a $60,000 note and a $200,000 note. The $60,000 note is dated October 31, 2021, with principal and interest payable on October 31, 2022. The $200,000 note is dated June 30, 2021, with principal and 6% interest payable on June 30, 2022.

2. During 2022, sales revenue totaled $1,340,000, $1,280,000 cash was collected from customers, and $22,000 in accounts receivable were written off. All sales are made on a credit basis. Bad debt expense is recorded at year-end by adjusting the allowance account to an amount equal to 10% of year-end accounts receivable.
3. On March 31, 2022, the $200,000 note receivable was discounted at the Bank of Commerce. The bank's discount rate is 8%. Chamberlain accounts for the discounting as a sale.

Required:

1. In addition to sales revenue, what revenue and expense amounts related to receivables will appear in Chamberlain's 2022 income statement?
2. What amounts will appear in the 2022 year-end balance sheet for accounts receivable?
3. Calculate the receivables turnover ratio for 2022.

P 7–13
Bank reconciliation and adjusting entries; cash and cash equivalents
● Appendix 7A

The bank statement for the checking account of Management Systems Inc. (MSI) showed a December 31, 2021, balance of $14,632.12. Information that might be useful in preparing a bank reconciliation is as follows:

a. Outstanding checks were $1,320.25.
b. The December 31, 2021, cash receipts of $575 were not deposited in the bank until January 2, 2022.
c. One check written in payment of rent for $246 was correctly recorded by the bank but was recorded by MSI as a $264 disbursement.
d. In accordance with prior authorization, the bank withdrew $450 directly from the checking account as payment on a note payable. The interest portion of that payment was $350. MSI has made no entry to record the automatic payment.
e. Bank service charges of $14 were listed on the bank statement.
f. A deposit of $875 was recorded by the bank on December 13, but it did not belong to MSI. The deposit should have been made to the checking account of MIS, Inc.
g. The bank statement included a charge of $85 for an NSF check. The check was returned with the bank statement and the company will seek payment from the customer.
h. MSI maintains a $200 petty cash fund that was appropriately reimbursed at the end of December.
i. According to instructions from MSI on December 30, the bank withdrew $10,000 from the account and purchased U.S. Treasury bills for MSI. MSI recorded the transaction in its books on December 31 when it received notice from the bank. Half of the Treasury bills mature in two months and the other half in six months.

Required:

1. Prepare a bank reconciliation for the MSI checking account at December 31, 2021. You will have to compute the balance per books.
2. Prepare any necessary adjusting journal entries indicated.
3. What amount would MSI report as cash and cash equivalents in the current asset section of the December 31, 2021, balance sheet?

P 7–14
Bank reconciliation and adjusting entries
● Appendix 7A

El Gato Painting Company maintains a checking account at American Bank. Bank statements are prepared at the end of each month. The November 30, 2021, reconciliation of the bank balance is as follows:

Balance per bank, November 30		$3,231
Add: Deposits outstanding		1,200
Less: Checks outstanding		
#363	$123	
#365	201	
#380	56	
#381	86	
#382	340	(806)
Adjusted balance per bank, November 30		$3,625

The company's general ledger checking account showed the following for December:

Balance, December 1	$ 3,625
Receipts	42,650
Disbursements	(41,853)
Balance, December 31	$ 4,422

The December bank statement contained the following information:

Balance, December 1	$ 3,231
Deposits	43,000
Checks processed	(41,918)
Service charges	(22)
NSF checks	(440)
Balance, December 31	$ 3,851

The checks that were processed by the bank in December include all of the outstanding checks at the end of November except for check #365. In addition, there are some December checks that had not been processed by the bank by the end of the month. Also, you discover that check #411 for $320 was correctly recorded by the bank but was incorrectly recorded on the books as a $230 disbursement for advertising expense Included in the bank's deposits is a $1,300 deposit incorrectly credited to the company's account. The deposit should have been posted to the credit of the Los Gatos Company. The NSF checks have not been redeposited and the company will seek payment from the customers involved.

Required:

1. Prepare a bank reconciliation for the El Gato checking account at December 31, 2021.
2. Prepare any necessary adjusting journal entries indicated.

P 7–15
Troubled debt restructuring
● Appendix 7B

Rothschild Chair Company, Inc., was indebted to First Lincoln Bank under a $20 million, 10% unsecured note. The note was signed January 1, 2011, and was due December 31, 2024. Annual interest was last paid on December 31, 2019. At January 1, 2021, Rothschild Chair Company was experiencing severe financial difficulties and negotiated a restructuring of the terms of the debt agreement.

Required:

Prepare all journal entries by First Lincoln Bank to record the restructuring and any remaining transactions, for current and future years, relating to the debt under each of the independent circumstances below:

1. First Lincoln Bank agreed to settle the debt in exchange for land having a fair value of $16 million but carried on Rothschild Chair Company's books at $13 million.
2. First Lincoln Bank agreed to (a) forgive the interest accrued from last year, (b) reduce the remaining four interest payments to $1 million each, and (c) reduce the principal to $15 million.

Decision Makers' Perspective

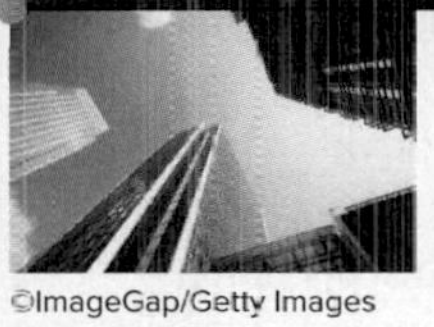

©ImageGap/Getty Images

Apply your critical-thinking ability to the knowledge you've gained. These cases will provide you an opportunity to develop your research, analysis, judgment, and communication skills. You also will work with other students, integrate what you've learned, apply it in real-world situations, and consider its global and ethical ramifications. This practice will broaden your knowledge and further develop your decision-making abilities.

Communication Case 7–1
Uncollectible accounts
● LO7–5

You have been hired as a consultant by a parts manufacturing firm to provide advice as to the proper accounting methods the company should use in some key areas. In the area of receivables, the company president does not understand your recommendation to use the allowance method for uncollectible accounts. She stated, "Financial statements should be based on objective data rather than the guesswork required for the allowance method. Besides, since my uncollectibles are fairly constant from period to period, with significant variations occurring infrequently, the direct write-off method is just as good as the allowance method."

Required:

Draft a one-page response in the form of a memo to the president in support of your recommendation for the company to use the allowance method.

Judgment Case 7–2
Accounts receivable
● LO7–3, LO7–7, LO7–8

Hogan Company uses the net method of accounting for sales discounts. Hogan offers trade discounts to various groups of buyers.

On August 1, 2021, Hogan factored some accounts receivable on a without recourse basis. Hogan incurred a finance charge.

Hogan also has some notes receivable bearing an appropriate rate of interest. The principal and total interest are due at maturity. The notes were received on October 1, 2021, and mature on September 30, 2022. Hogan's operating cycle is less than one year.

Required:

1. a. Using the net method, do sales discounts affect the amount recorded as sales revenue and accounts receivable at the time of sale? What is the rationale for the amount recorded as sales under the net method?
   b. Using the net method, is there an effect on Hogan's sales revenues and net income when customers do not take the sales discounts?
2. Do trade discounts affect the amount recorded as sales revenue and accounts receivable? Why?
3. Should Hogan decrease accounts receivable to account for the receivables factored on August 1, 2021? Why?
4. Would Hogan classify the interest-bearing notes receivable as current or non-current in its December 31, 2021, balance sheet? Why?

(AICPA adapted)

Real World Case 7–3
Sales returns; Green Mountain Coffee Roasters
● LO7–4

Real World Financials

The following is an excerpt from Sam Antar, "Is Green Mountain Coffee Roasters Shuffling the Beans to Beat Earnings Expectations?" (*Phil's Stock World delivered by Newstex,* May 9, 2011.)

> On May 3, 2011, Green Mountain Coffee Roasters (NASDAQ: GMCR) beat analysts' earnings estimates by $0.10 per share for the thirteen-week period ended March 26, 2011. The next day, the stock price had risen to $11.91 per share to close at $75.98 per share, a staggering 18.5% increase over the previous day's closing stock price. CNBC Senior Stocks Commentator Herb Greenberg raised questions about the quality of Green Mountain Coffees earnings because its provision for sales returns dropped $22 million in the thirteen-week period. He wanted to know if there was a certain adjustment to reserves ("a reversal") that helped Green Mountain Coffee beat analysts' earnings estimates. . . .
>
> During the thirteen-week period ended March 26, 2011, it was calculated that Green Mountain Coffee had a negative $22.259 million provision for sales returns. In its latest 10-Q report, Green Mountain Coffee disclosed that its provision for sales returns was $5.262 million for the twenty-six week period ending March 26, 2011, but the company did not disclose amounts for the thirteen-week period ended March 26, 2011. In its previous 10-Q report for the thirteen-week period ended December 25, 2010, Green Mountain Coffee disclosed that its provision for sales returns was $27.521 million. Therefore, the provision for sales returns for the thirteen-week period ended March 26, 2011 was a negative $22.259 million ($5.262 million minus $27.521 million).

Required:

1. Access EDGAR on the Internet. The web address is **www.sec.gov**.
2. Search for **Green Mountain Coffee Roasters, Inc.**'s 10-K for the fiscal year ended September 25, 2010 (filed December 9, 2010). (*Note:* the company now is named **Keurig Green Mountain, Inc.**) Answer the following questions related to the company's 2010 accounting for sales returns:
   a. Note: in 2010, companies could reduce the carrying value of accounts receivable for estimated returns, rather than recognizing a refund liability. What type of an account (for example, asset, contra liability) is Sales Returns Reserve? Explain.
   b. Prepare a T-account for fiscal 2010's sales returns account. Include entries for the beginning and ending balance, acquisitions, amounts charged to cost and expense, and deductions.
   c. Prepare journal entries for amounts charged to cost and expense and for deductions. Provide a brief explanation of what each of those journal entries represents.
   d. For any of the amounts included in your journal entries that appear in Green Mountain's operations section of its statement of cash flows on page F-8, indicate whether the amount appears as an increase or decrease to cash flows and explain why it appears as an increase or decrease.
3. Now consider the information provided by Antar in the excerpt at the beginning of this case.
   a. Prepare a T-account for the first quarter of fiscal 2011's sales returns reserve. Assume amounts associated with acquisitions and deductions are zero, such that the only entry affecting the account during the first quarter of fiscal 2011 is to record amounts charged or recovered from cost and expense. Compute the ending balance of the account.
   b. Prepare a T-account for the second quarter of fiscal 2011's sales returns reserve. Assume amounts associated with acquisitions and deductions are zero, such that the only entry affecting the account during the first quarter of fiscal 2011 is to record amounts charged or recovered from cost and expense. Compute the ending balance of the account.
   c. Assume that actual returns were zero during the second quarter of fiscal 2011. Prepare a journal entry to record amounts charged or recovered from cost and expense during the second quarter of fiscal 2011. How would that journal entry affect 2011 net income?
   d. Speculate as to what might have caused the activity in Green Mountain's sales returns account during the second quarter of fiscal 2011. Consider how this result could occur unintentionally, or why it might occur intentionally as a way to manage earnings.

Ethics Case 7–4
Uncollectible accounts
● LO7–5

You have recently been hired as the assistant controller for Stanton Industries, a large, publicly held manufacturing company. Your immediate superior is the controller who, in turn, is responsible to the vice president of finance.

The controller has assigned you the task of preparing the year-end adjusting entries. In the receivables area, you have prepared an aging of accounts receivable and have applied historical percentages to the balances of each of the age categories. The analysis indicates that an appropriate balance for the allowance for uncollectible accounts is $180,000. The existing balance in the allowance account prior to any adjusting entry is a $20,000 credit balance.

After showing your analysis to the controller, he tells you to change the aging category of a large account from over 120 days to current status and to prepare a new invoice to the customer with a revised date that agrees with the new aging category. This will change the required allowance for uncollectible accounts from $180,000 to $135,000. Tactfully, you ask the controller for an explanation for the change and he tells you "We need the extra income; the bottom line is too low."

Required:

1. What is the effect on income before taxes of the change requested by the controller?
2. Discuss the ethical dilemma you face. Consider your options and responsibilities along with the possible consequences of any action you might take.

Judgment Case 7–5
Internal control
● LO7–1

For each of the following independent situations, indicate whether there is an apparent internal control weakness, and, if one exists, suggest alternative procedures to eliminate the weakness.

1. John Smith is the petty cash custodian. John approves all requests for payment out of the $200 fund, which is replenished at the end of each month. At the end of each month, John submits to his supervisor a list of all accounts and amounts to be charged, along with supporting documentation. Once the supervisor indicates approval, a check is written to replenish the fund for the total amount. John's supervisor performs surprise counts of the fund to ensure that the cash and/or receipts equal $200 at all times.
2. All of the company's cash disbursements are made by check. Each check must be supported by an approved voucher, which is in turn supported by the appropriate invoice and, for purchases, a receiving document. The vouchers are approved by Dean Leiser, the chief accountant, after reviewing the supporting documentation. Betty Hanson prepares the checks for Leiser's signature. Leiser also maintains the company's check register (the cash disbursements journal) and reconciles the bank account at the end of each month.
3. Fran Jones opens the company's mail and makes a listing of all checks and cash received from customers. A copy of the list is sent to Jerry McDonald who maintains the general ledger accounts. Fran prepares and makes the daily deposit at the bank. Fran also maintains the subsidiary ledger for accounts receivable, which is used to generate monthly statements to customers.

Real World Case 7–6
Receivables; bad debts; Cisco Systems, Inc.
● LO7–5
Real World Financials

EDGAR, the Electronic Data Gathering, Analysis, and Retrieval system, performs automated collection, validation, indexing, and forwarding of submissions by companies and others who are required by law to file forms with the U.S. Securities and Exchange Commission (SEC). All publicly traded domestic companies use EDGAR to make the majority of their filings. (Some foreign companies file voluntarily). Form 10-K or 10-KSB, which include the annual report, is required to be filed on EDGAR. The SEC makes this information available on the Internet.

Required:

1. Access EDGAR on the Internet. The web address is **www.sec.gov**.
2. Search for **Cisco Systems, Inc.** Access the 10-K filing for the fiscal year ended July 31, 2017. Search or scroll to find the financial statements.
3. Answer the following questions related to the company's accounts receivable and bad debts:
   a. What is the amount of gross trade accounts receivable at the end of the year?
   b. What is the amount of bad debt expense for the year? (*Hint:* check the statement of cash flows).
   c. Determine the amount of actual bad debt write-offs made during the year. Assume that all bad debts relate only to trade accounts receivable.
   d. Using only information from the balance sheets, income statements, and your answer to requirement 3(c), determine the amount of cash collected from customers during the year. Assume that all sales are made on a credit basis, that the company provides no allowances for sales returns, that no previously written-off receivables were collected, and that all sales relate to trade accounts receivable.

Integrating Case 7–7
Change in estimate of bad debts
● LO7–5

McLaughlin Corporation uses the allowance method to account for bad debts. At the end of the company's fiscal year, accounts receivable are analyzed and the allowance for uncollectible accounts is adjusted. At the end of 2021, the company reported the following amounts:

Accounts receivable	$10,850,000
Less: Allowance for uncollectible accounts	(450,000)
Accounts receivable, net	$10,400,000

In 2022, it was determined that $1,825,000 of year-end 2021 receivables had to be written off as uncollectible. This was due in part to the fact that Hughes Corporation, a long-standing customer that had always paid its bills, unexpectedly declared bankruptcy in 2022. Hughes owed McLaughlin $1,400,000. At the end of 2021, none of the Hughes receivable was considered uncollectible.

Required:

Should McLaughlin's underestimation of bad debts be treated as an error correction (requiring retroactive restatement) or a change in estimate (and accounted for prospectively)? Describe the appropriate accounting treatment and required disclosures in the financial statements issued for the 2021 fiscal year.

Real World Case 7–8
Financing with receivables; Sanofi-Aventis
• LO7–5, LO7–8, LO7–10
IFRS

Search on the Internet for the 2016 annual report for **Sanofi-Aventis**. Find the accounts receivable disclosure note.

Required:

1. Has Sanofi-Aventis factored or securitized accounts receivable? How do you know?
2. Assume that Sanofi-Aventis decided to increase the extent to which it securitizes its accounts receivable, changing to a policy of securitizing accounts receivable immediately upon making a sale and treating the securitization as a sale of accounts receivable. Indicate the likely effect of that change in policy on
   a. Accounts receivable in the period of the change (reduce, increase, or no change from prior period)
   b. Cash flow from operations in the period of the change (reduce, increase, or no change from prior period)
   c. Accounts receivable in subsequent periods (stabilize at higher, lower or same level as in periods prior to the change)
   d. Cash flow from operations in subsequent periods (stabilize at higher, lower or same level as in periods prior to the change)
3. Given your answers to requirement 2, could a company change the extent to which it factors or securitizes receivables to create one-time changes in its cash flow? Answer yes or no, and explain.

Research Case 7–9
Locate and extract relevant information and authoritative support for a financial reporting issue; financing with receivables
• LO7–8
CODE

You are spending the summer working for a local wholesale furniture company, Samson Furniture, Inc. The company is considering a proposal from a local financial institution, Old Reliant Financial, to factor Samson's receivables. The company controller is unfamiliar with the prevailing GAAP that deals with accounting for the transfer of financial assets and has asked you to do some research. The controller wants to make sure the arrangement with the financial institution is structured in such a way as to allow the factoring to be accounted for as a sale.

Old Reliant has offered to factor all of the company's receivables on a "without recourse" basis. Old Reliant will remit to Samson 90% of the factored amount, collect the receivables from Samson's customers, and retain the remaining 10% until all of the receivables have been collected. When Old Reliant collects all of the receivables, it will remit to Samson the retained amount, less a 4% fee (4% of the total factored amount).

Required:

1. Access the relevant authoritative literature on accounting for the transfer of financial assets using the *FASB Accounting Standards Codification*. You might gain access at the FASB website (**www.fasb.org**), from your school library, or some other source. What conditions must be met for a transfer of receivables to be accounted for as a sale (or in accounting terms, "derecognized")? What is the specific seven-digit Codification citation (XXX-XX-XX) that Samson would rely on in applying that accounting treatment?
2. Assuming that the conditions for treatment as a sale are met, prepare Samson's journal entry to record the factoring of $400,000 of receivables. Assume that the fair value of the last 10% of Samson's receivables is equal to $25,000.
3. An agreement that both entitles and obligates the transferor, Samson, to repurchase or redeem transferred assets from the transferee, Old Reliant, maintains the transferor's effective control over those assets and the transfer is accounted for as a secured borrowing, not a sale, if and only if what conditions are met?

Analysis Case 7–10
Compare receivables management using ratios; Tyson Foods Inc. and Pilgrim's Pride Corp.
• LO7–9

Real World Financials

The table below contains selected financial information included in the 2016 financial statements of **Tyson Foods Inc.** and **Pilgrim's Pride Corp**.

	($ in millions)			
	Tyson		**Pilgrim's Pride**	
	2016	**2015**	**2016**	**2015**
Balance sheets:				
Accounts receivable, net	$ 1,542	$ 1,620	$ 317	$ 349
Income statements:				
Net sales	$36,881	$41,373	$7,931	$8,180

Required:

1. Calculate the 2016 receivables turnover ratio and average collection period for both companies. Evaluate the management of each company's investment in receivables
2. Obtain annual reports from three corporations in the same primary industry and compare the management of each company's investment in receivables.

Note: You can obtain copies of annual reports from your library, from the investor relations department of the corporations, or from EDGAR (Electronic Data Gathering, Analysis, and Retrieval) on the Internet (**www.sec.gov**).

Data Analytics

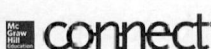

connect

Data analytics is the process of examining data sets in order to draw conclusions about the information they contain. If you have not completed any of the prior data analytics cases, follow the instructions listed in the Chapter 1 Data Analytics case to get set up. You will need to watch the videos referred to in the Chapters 1 - 3 Data Analytics cases. No additional videos are required for this case. All short training videos can be found here: **https://www.tableau.com/learn/training#getting-started**.

Data Analytics Case
Receivables Management
● LO7–9

In prior chapters, you applied Tableau to examine a data set and create calculations to compare two companies' financial information. In this case, you continue in your role as an analyst conducting introductory research into the relative merits of investing in one or both of these companies. This time you assess the companies' receivables management and how effective the companies are in managing the investment in receivables.

Required:

Use Tableau to calculate (a) receivables turnover ratios, and (b) average collection periods for each of the two companies in each year from 2012 to 2021. Based upon what you find, answer the following questions:

1. What is the receivables turnover ratio for Big Store (a) in 2012 and (b) in 2021?
2. What is the receivables turnover ratio for Discount Goods (a) in 2012 and (b) in 2021?
3. Which company exhibited more favorable receivables management, as measured by average collection period, in *2020*, Big Store or Discount Goods? What was that company's average collection period, and is it longer or shorter than the industry average in 2020?

   The average collection period for the retail industry sector is 4.0 in 2020 and 5.1 in 2021.
4. Which company exhibited more favorable receivables management, as measured by average collection period, in *2021*, Big Store or Discount Goods? What was that company's average collection period, and is it longer or shorter than the industry average in 2021?

Resources:

You have available to you an extensive data set that includes detailed financial data for 2012-2021 for both Discount Goods and Big Store. The data set is in the form of four Excel files available to download from Connect, or under Student Resources within the Library tab. Download the file "Discount_Goods_Big_Store_Financials.xlsx" to the computer, save it, and open it in Tableau.

For this case, you will create several calculations to produce a text and line chart of the receivables analysis to allow you to compare and contrast the two companies' receivables management.

After you view the training videos and review instructions in Chapters 1-3, follow these steps to create the charts you'll use for this case:

- Click on the "Sheet 1" tab at the bottom of the canvas, to the right of the Data Source at the bottom of the screen.
- Drag "Year" and "Company" under "Dimensions" to the Rows shelf. Change "Year" to discrete data type by right-clicking and selecting "Discrete."
- Drag the "Net sales" and "Average accounts receivable net" under Measures to the Rows shelf. Change each to discrete. Format each to Times New Roman, 10-point font, black, center alignment, and Currency (custom), 0 decimal places by right-clicking each and selecting "Format." Format selections will open to the left of the canvas.
- Create a calculated field by clicking the "Analysis" tab from the Toolbar at the top of the screen and selecting "Create Calculated field." A calculation box will pop up. Name the calculation "Receivables Turnover." In the Calculation Editor window, from the Rows shelf, drag "Net sales" to the window, type a division sign, then drag "Average accounts receivable net" beside it. Make sure the window says that the calculation is valid and click OK.

- Repeat the process one more time by creating a calculated field "Average Collection Period" that consists of the number 365 divided by the newly created "Receivables Turnover."
- Drag the "Receivables Turnover" and "Average Collection Period" to the Rows shelf. Change both to discrete data type. Format both to Times New Roman, 10-point font, black, center alignment, and Number (Custom), 2 decimal places.
- Change the title of the sheet to be "Receivables Management" by right-clicking and selecting "Edit title." Format the title to Times New Roman, bold, black and 15-point font. Change the title of "Sheet 1" to match the sheet title by right-clicking, selecting "Rename" and typing in the new title.
- Click on the "Sheet 2" tab and drag "Year" to the Columns shelf and "Receivables Turnover" and "Average Collection Period" to the Rows shelf.
- Drag "Company" under "Dimensions" to the Color in the Marks card. You should now see two colored lines in the two graphs.
- Format all the labels on the sheet ("Receivables Turnover," "Average Collection Period," and "Year") to Times New Roman, 10-point font, and black.
- Change the title of the sheet to be "Graph of Receivables Management" by right-clicking and selecting "Edit title." Format the title to Times New Roman, bold, black and 15-point font. Change the title of "Sheet 2" to match the sheet title by right-clicking, selecting "Rename" and typing in the new title.
- Once complete, save the file as "*DA7_Your initials*.twbx."

Continuing Cases

Target Case
- LO7–2, LO7–5

Target Corporation prepares its financial statements according to U.S. GAAP. Target's financial statements and disclosure notes for the year ended February 3, 2018, are available in Connect. This material also is available under the Investor Relations link at the company's website (**www.target.com**).

Required:

1. What is Target's policy for designating investments as cash equivalents?
2. What is Target's balance of cash equivalents for the fiscal year ended February 3, 2018?
3. What is Target's policy with respect to accounting for merchandise returns?
4. Does Target have accounts receivable? Speculate as to why it has the balance that it has. (*Hint:* See Disclosure Notes 9, 11, and 13).

Air France–KLM Case
- LO7–8

IFRS

Air France–KLM (AF), a Franco-Dutch company, prepares its financial statements according to International Financial Reporting Standards. AF's financial statements and disclosure notes for the year ended December 31, 2017, are available in Connect. This material is also available under the Finance link at the company's website (**www.airfranceklm.com**).

Required:

1. In note 4.11, AF describes how it values trade receivables. How does the approach used by AF compare to U.S. GAAP?
2. In note 25, AF reconciles the beginning and ending balances of its valuation allowances for trade accounts receivable. Prepare a T-account for the valuation allowance and include entries for the beginning and ending balances as well as any reconciling items that affected the account during 2017.
3. Examine note 27. Does AF have any bank overdrafts? If so, are the overdrafts shown in the balance sheet the same way they would be shown under U.S. GAAP?

CPA Exam Questions and Simulations

ROGER CPA Review

Sample CPA Exam questions from Roger CPA Review are available in Connect as support for the topics in this chapter. These multiple-choice questions and task-based simulations include expert-written explanations and solutions, and provide a starting point for students to become familiar with the content and functionality of the actual CPA Exam.

CHAPTER

8 Inventories: Measurement

OVERVIEW

The next two chapters continue our study of assets by investigating the measurement and reporting issues involving inventories and the related expense—cost of goods sold. Inventory refers to the assets a company (1) intends to sell in the normal course of business, (2) has in production for future sale, or (3) uses currently in the production of goods to be sold.

LEARNING OBJECTIVES

After studying this chapter, you should be able to:

- **LO8–1** Explain the types of inventory and the differences between a perpetual inventory system and a periodic inventory system. *(p. 404)*
- **LO8–2** Explain which physical units of goods should be included in inventory. *(p. 408)*
- **LO8–3** Account for transactions that affect net purchases and prepare a cost of goods sold schedule. *(p. 409)*
- **LO8–4** Differentiate between the specific identification, FIFO, LIFO, and average cost methods used to determine the cost of ending inventory and cost of goods sold. *(p. 412)*
- **LO8–5** Discuss the factors affecting a company's choice of inventory method. *(p. 420)*
- **LO8–6** Understand supplemental disclosures of LIFO reserves and the effect of LIFO liquidations on net income. *(p. 422)*
- **LO8–7** Calculate the key ratios used by analysts to monitor a company's investment in inventories. *(p. 427)*
- **LO8–8** Determine ending inventory using the dollar-value LIFO inventory method. *(p. 431)*
- **LO8–9** Discuss the primary difference between U.S. GAAP and IFRS with respect to determining the cost of inventory. *(p. 420)*

© Blend Images/Dave and Les Jacobs/Getty Images

FINANCIAL REPORTING CASE

Inventory Measurement at Kroger Company

As you were reading the annual report of **Kroger Company**, one of the world's largest grocery retailers, you notice the company accounts for nearly all of its grocery inventory based on a LIFO method (last-in, first-out). This means the company reports its most recent inventory purchases as sold *first*. However, you understand that most grocery inventory consists of perishable food items, meaning that the company in reality almost certainly sells its most recent purchases *last*. Otherwise, there would be considerable inventory spoilage. You decide to look a little deeper and notice the following discussion in the company's annual report:

Inventories (in part)

Inventories are stated at the lower of cost (principally on a last-in, first-out "LIFO" basis) or market. In total, approximately 93% of inventories in 2017 and 89% of inventories in 2016 were valued using the LIFO method. Replacement cost was higher than the carrying amount by $1,248 at February 3, 2018, and $1,291 at January 28, 2017.

($ in millions)	2018	2017
FIFO Inventory	$ 7,781	$ 7,852
LIFO reserve	(1,248)	(1,291)
Reported inventories	$ 6,533	$ 6,561

After seeing this information, you are further confused because you don't understand why Kroger would choose to report inventory for more than $1 billion below its replacement cost. Doesn't that make the company's inventory look less valuable and therefore the company less profitable?

You do some more research and find that other retail companies, as well as companies in industries such as automobiles, consumer products, energy, manufacturing, and mining, also use the LIFO method. Each of these companies most likely sells its actual inventory on a first-in, first-out (FIFO) basis, so why are they assuming the opposite?

QUESTIONS

By the time you finish this chapter, you should be able to respond appropriately to the questions posed in this case. Compare your response to the solution provided at the end of the chapter.

1. How is the LIFO method used to calculate inventories? Is this permissible according to GAAP? (*p. 419*)
2. What is the purpose of disclosing the difference between the reported LIFO inventory amounts and replacement cost, assuming that replacement cost is equivalent to a FIFO basis? (*p. 422*)
3. Are you correct that, by using LIFO, Kroger reports lower inventory and lower profits? Why would Kroger do that? (*p. 427*)

Recording and Measuring Inventory

PART A

Inventory refers to the assets a company (1) intends to sell in the normal course of business, (2) has in production for future sale (work in process), or (3) uses currently in the production of goods to be sold (raw materials). The computers produced by **Hewlett-Packard (HP)** that are intended for sale to customers are inventory, as are partially completed components,

Inventories consist of assets that a retail or wholesale company acquires for resale or goods that manufacturers produce for sale.

the computer chips, and memory modules that will go into computers produced later. The computers *used* by HP's employees to maintain its accounting system and other company operations, however, are not available for sale to customers and therefore are classified and accounted for as equipment. Similarly, the stocks and bonds a securities dealer holds for sale are inventory, whereas HP would classify the securities it holds as investments.

Cost of goods sold is the expense related to inventory. The inventory amount in the balance sheet represents the cost of the inventory still on hand (not yet sold) at the end of the period, while cost of goods sold in the income statement represents the cost of the inventory sold during the period.

Inventory usually is one of the most valuable assets listed in the balance sheet for manufacturing, wholesale, and retail companies (enterprises that produce revenue by selling goods). Similarly, cost of goods sold typically is the largest expense in the income statement of these companies. For example, a recent balance sheet for **Best Buy** reported inventories of $5.2 billion, which represented 53% of current assets. The company's income statement reported cost of goods sold of $32 billion, representing 80% of operating expenses.

As we'll see in this and the next chapter, it's usually difficult to measure inventory and cost of goods sold at the exact cost of the actual physical quantities on hand and sold. Fortunately, accountants can use one of several techniques to approximate the desired result and satisfy our measurement objectives.

Types of Inventory

Merchandising Inventory

LO8–1

Wholesale and retail companies purchase goods that are primarily in finished form. These companies are intermediaries in the process of moving goods from the manufacturer to the end-user. They often are referred to as merchandising companies and their inventory as merchandise inventory. *The cost of merchandise inventory includes the purchase price plus any other costs necessary to get the goods in condition and location for sale.* We discuss the concept of condition and location and the types of costs that typically constitute inventory later in this chapter.

Manufacturing Inventories

Inventory for a manufacturing company consists of raw materials, work in process, and finished goods.

Unlike merchandising companies, manufacturing companies actually produce the goods they sell to wholesalers, retailers, other manufacturers, or consumers. Inventory for a manufacturer consists of (1) raw materials, (2) work in process, and (3) finished goods.

Raw materials represent the cost of components purchased from suppliers that will become part of the finished product. For example, Hewlett-Packard's raw materials inventory includes semiconductors, circuit boards, plastic, and glass that go into the production of personal computers.

Work-in-process inventory refers to the products that are not yet complete in the manufacturing process. The cost of work in process includes the cost of raw materials used in production, the cost of labor that can be directly traced to the goods in process, and an allocated portion of other manufacturing costs, called *manufacturing overhead.* Overhead costs include electricity and other utility costs to operate the manufacturing facility, depreciation of manufacturing equipment, and many other manufacturing costs that cannot be directly linked to the production of specific goods.

Finished goods are goods that have been completed in the manufacturing process but have not yet been sold. They have reached their final stage and now await sale to a customer. Their cost includes the cost of all raw materials and work in process used in production.

Manufacturing companies generally disclose, either in a note or directly in the balance sheet, the dollar amount of each inventory category. For example, **Intel**, one of the world's largest semiconductor chip manufacturers, reports inventory as shown in Illustration 8–1.

	($ in millions)	
	December 30, 2017	**December 31, 2016**
Raw materials	$1,098	$ 695
Work in process	3,893	3,190
Finished goods	1,992	1,668
Total Inventories	$6,983	$5,553

Source: Intel Corporation

Illustration 8–1
Inventories Disclosure—Intel Corporation

Real World Financials

The inventory accounts and the cost flows for a typical manufacturing company are shown using T-accounts in Illustration 8–2. The costs of raw materials used, direct labor applied, and manufacturing overhead applied flow into work in process and then to finished goods. When the goods are sold, the cost of those goods flows to cost of goods sold.

Illustration 8–2
Inventory Components and Cost Flow for a Manufacturing Company

The costs of inventory units follow their physical movement from one stage of activity to another.

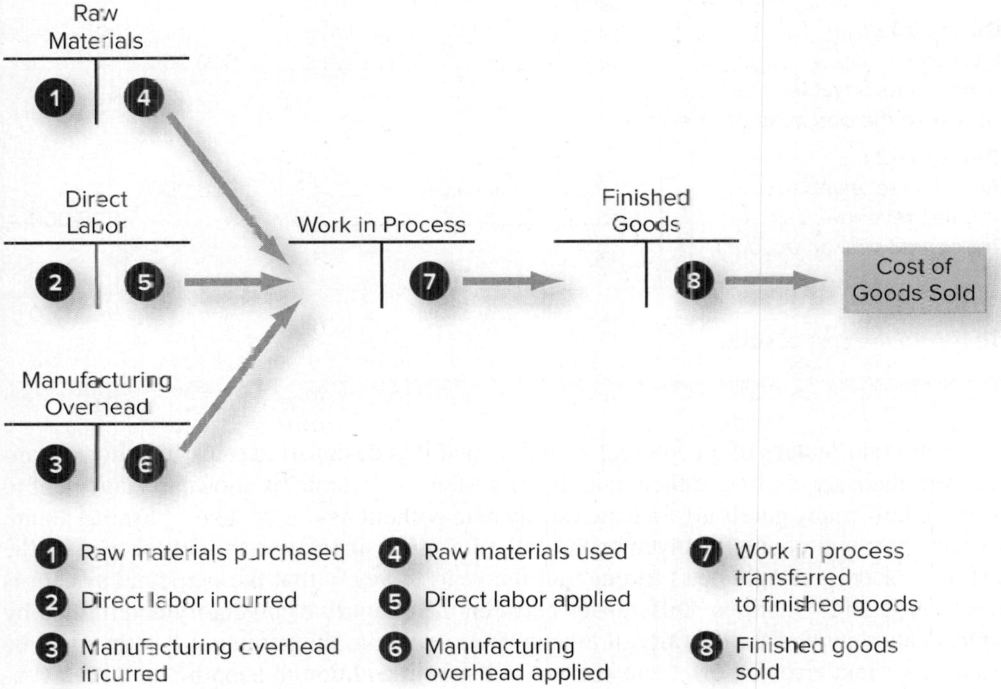

We focus in this text primarily on merchandising companies (wholesalers and retailers). Still, most of the accounting principles and procedures discussed here also apply to manufacturing companies. The unique problems involved with accumulating the direct costs of raw materials and labor and with allocating manufacturing overhead are addressed in managerial and cost accounting textbooks.

Types of Inventory Systems

Perpetual Inventory System

Two accounting systems are used to record transactions involving inventory: the **perpetual inventory system** and the **periodic inventory system**. The perpetual system was introduced in Chapter 2. The system is aptly termed perpetual because the *inventory* account is continually adjusted for each change in inventory, whether it's caused by a purchase, a sale, or a return of inventory (a *purchase return* for the buyer, a *sales return* for the seller).[1] In addition, the *cost of goods sold* account is adjusted each time goods are sold or are

[1] We discussed accounting for sales returns in Chapter 7.

returned by a customer. The perpetual inventory system is applied to the Lothridge Wholesale Beverage Company for which inventory information is provided in Illustration 8–3. This hypothetical company also will be used in the next several illustrations.

Illustration 8–3

Perpetual Inventory System

The Lothridge Wholesale Beverage Company purchases soft drinks from producers and then sells them to retailers. The company begins 2021 with inventory of $120,000 on hand. The following information relates to inventory transactions during 2021:

1. Additional soft drink inventory is purchased on account at a cost of $600,000.
2. Sales for the year, all on account, totaled $820,000.
3. The cost of the soft drink inventory sold is $540,000.

Lothridge uses the **perpetual inventory system** to keep track of both inventory quantities and inventory costs. The system indicates that the cost of inventory on hand at the end of the year is $180,000.

The following summary journal entries record the inventory transactions for the Lothridge Company:

During 2021		
Inventory	600,000	
Accounts payable		600,000
To record the purchase of inventory.		
During 2021		
Accounts receivable	820,000	
Sales revenue		820,000
To record sales on account.		
Cost of goods sold	540,000	
Inventory		540,000
To record the cost of sales.		

An important feature of a perpetual system is that it is designed to track inventory quantities from their acquisition to their sale. If the system is accurate, it allows management to determine how many goods are on hand on any date without having to take a physical count. However, physical counts of inventory usually are made anyway, either at the end of the fiscal year or on a sample basis throughout the year, to verify that the perpetual system is correctly tracking quantities. Differences between the quantity of inventory determined by the physical count and the quantity of inventory according to the perpetual system could be caused by system errors, theft, breakage, or spoilage. In addition to keeping up with inventory purchases, a perpetual system also directly determines how many items are sold during a period.

A *perpetual inventory system* continuously tracks both inventory quantities and inventory costs.

When a company uses a perpetual inventory system to record inventory and cost of goods sold transactions, merchandise cost data also is included in the system. That way, when merchandise is purchased/sold, the system records not only the addition/reduction in inventory *quantity* but also the addition/reduction in inventory *cost.*

Nearly all major companies use a perpetual inventory system to maintain a record of inventory transactions.

Technology and inventory accounting. You're familiar with the barcode scanning mechanisms used at grocery stores and other checkout counters. Scanners, such as these, are linked to a company's accounting records to allow for continuous tracking of inventory. Each time inventory is purchased or sold, employees scan barcodes attached to merchandise. This provides management with real-time information about inventory levels and cost. In addition to barcode scanning, companies have developed techniques to further automate their management and recording of inventory. For example, some companies use inventory software linked to radio frequency identification (RFID) tags, allowing inventory to be recorded automatically as it moves in and out of a company. Companies are developing "smart" systems to use with "big data" to monitor customer trends and automatically initiate orders in real time as those trends change. These types of technological advances help to reduce the burden of physical inventory counts and manual record keeping, and make the use of a perpetual inventory system easier and more efficient. *That's why nearly all major companies use a perpetual inventory system, and that's why we focus on the use of the perpetual system in this chapter.* Nevertheless, even with the advances in technology to improve the automation of inventory accounting, many companies still must physically count the inventory on hand at the end of the period to verify the accuracy of the accounting records and to account for inventory damaged or stolen.

Periodic Inventory System

A **periodic inventory system** does not maintain a continual record of inventory quantity or cost during the period. Instead, the quantity of inventory is determined only at the end of the period based on a physical count. The balance of the inventory account is updated for purchases and sales that occurred during the year using a year-end adjusting entry. A periodic system also does not maintain a record of the cost of goods sold with each sale. The balance is established in the year-end adjusting entry for an amount that equals beginning inventory plus net purchases during the year (total purchases plus freight-in less purchase returns and discounts) minus ending inventory:

A *periodic inventory system* adjusts inventory and records cost of goods sold only at the end of each reporting period.

Beginning inventory + Net purchases − Ending inventory = Cost of goods sold

Cost of goods sold equation

Implied in the calculation of cost of goods sold is that any beginning inventory or net purchases that are not included in ending inventory must have been sold. This may not be the case if inventory items were either damaged or stolen. If damaged and stolen inventory are identified, they may be removed from beginning inventory or purchases before calculating cost of goods sold and then classified as a separate expense item.

Illustration 8–4 looks at the periodic system using the Lothridge Wholesale Beverage Company example.

Illustration 8–4

Periodic Inventory System

The Lothridge Wholesale Beverage Company purchases soft drinks from producers and then sells them to retailers. The company begins 2021 with inventory of $120,000 on hand. The following information relates to inventory transactions during 2021:

1. Additional soft drink inventory is purchased on account at a cost of $600,000.
2. Sales for the year, all on account, totaled $820,000.

Lothridge uses the **periodic inventory system.** After a physical count of inventory at the end of the year, the cost of ending inventory is determined to be $180,000.

The following summary journal entries record the inventory transactions for the Lothridge Company:

During 2021		
Purchases	600,000	
Accounts payable		600,000
To record the purchase of inventory.		
During 2021		
Accounts receivable	820,000	
Sales revenue		820,000
To record sales on account.		

No entry is recorded for the cost of inventory sold.

Because cost of goods sold isn't determined continually by the periodic system, it must be determined indirectly after a physical inventory count at the end of the year. Cost of goods sold for 2021 is determined as follows:

Beginning inventory	$ 120,000
Plus: Purchases	600,000
Cost of goods available for sale	720,000
Less: Ending inventory (per physical count)	(180,000)
Cost of goods sold	**$540,000**

We then make the following entry to update the balance in the inventory account and combine the components of cost of goods sold into a single expense account:

December 31, 2021		
Cost of goods sold	**540,000**	
Inventory (ending)	180,000	
Inventory (beginning)		120,000
Purchases		600,000

To adjust inventory, close the purchases account, and record cost of goods sold.

This entry adjusts the inventory account to the correct period-end amount, closes the temporary purchases account, and records cost of goods sold. Now let's compare the two inventory accounting systems.

A Comparison of the Perpetual and Periodic Inventory Systems

Beginning inventory plus net purchases during the period is the *cost of goods available for sale.* The main difference between a perpetual and a periodic system is that the periodic system allocates cost of goods available for sale between ending inventory and cost of goods sold *only at the end of the period* (periodically). In contrast, the perpetual system updates the balance of inventory and cost of goods sold *each time goods are sold* (perpetually).

The impact on the financial statements of choosing one system over the other generally is not significant. The choice between the two approaches usually is motivated by management control considerations as well as the comparative costs of implementation. Perpetual systems can provide more information about the dollar amounts of inventory levels on a continuous basis. They also facilitate the preparation of interim (for example, quarterly) financial statements by providing fairly accurate information without the necessity of a physical count of inventory. Periodic systems require a physical count of inventory, complicating preparation of interim financial statements unless an inventory estimation technique is used.[2]

A perpetual system provides more timely information but generally is more costly.

On the other hand, a perpetual system may be more expensive to implement than a periodic system. This is particularly true for inventories consisting of large numbers of low-cost items. Perpetual systems are more workable with inventories of high-cost items such as construction equipment or automobiles. However, with the help of computers and electronic sales devices such as cash register systems with barcode scanners or RFID tags, the perpetual inventory system now is available to many small businesses that previously could not afford them and is economically feasible for a broader range of inventory items than before.

What is Included in Inventory?

Physical units Included in Inventory

LO8–2

Regardless of the system used, the measurement of inventory and cost of goods sold starts with determining the physical units of goods. Typically, determining the physical units that should be included in inventory is a simple matter because it consists of items in the possession of the company. However, in some situations the identification of items that should be included in inventory is more difficult. Consider, for example, goods in transit, goods on consignment, and sales returns.

GOODS IN TRANSIT At the end of a reporting period, it's important to ensure a proper inventory cutoff. This means ownership must be determined for goods that are in transit between the company and its customers as well as between the company and its suppliers.

For example, in December 2021, the Lothridge Wholesale Beverage Company sold goods to the Jabbar Company.

1. The goods were shipped from Lothridge on December 29, 2021, but
2. The goods didn't arrive at Jabbar until January 3, 2022.

[2]In Chapter 9 we discuss inventory estimation techniques that avoid the necessity of a physical count to determine ending inventory and cost of goods sold.

If both companies have December 31 fiscal year ends, whose inventory is it on December 31, 2021? In other words, which company will report these goods in ending inventory in the balance sheet as of December 31, 2021? The answer depends on who owns the goods on December 31. Ownership depends on the terms of the agreement between the two companies.

If the goods are shipped **f.o.b. (free on board) shipping point**, then legal title to the goods passes from the seller to the buyer at the *point of shipment* (when the seller hands over the goods to the delivery company, such as **FedEx**). In this case, the buyer is responsible for shipping costs and transit insurance. Lothridge records the sale of inventory on December 29, 2021, and Jabbar records the purchase of inventory on that same day. Jabbar will include these goods in its 2021 ending inventory even though the company is not in physical possession of the goods on the last day of the fiscal year.

Inventory shipped *f.o.b. shipping point* is included in the purchaser's inventory as soon as the merchandise is shipped.

On the other hand, if the goods are shipped **f.o.b. destination**, then legal title to the goods does not pass from the seller to the buyer until the goods arrive at their *destination* (the customer's location). In this case, the seller is responsible for shipping costs and transit insurance. In our example, if the goods are shipped f.o.b. destination, Lothridge waits to record the sale until January 3, 2022, and Jabbar records the purchase of inventory on that same day. Lothridge includes the goods in its 2021 ending inventory even though the inventory has already been shipped as of the last day of the financial statement reporting period.

Inventory shipped *f.o.b. destination* is included in the purchaser's inventory only after it reaches the purchaser's destination.

GOODS ON CONSIGNMENT Sometimes a company arranges for another company to sell its product under **consignment**. The goods are physically transferred to the other company (the consignee), but the transferor (consignor) retains legal title. If the consignee can't find a buyer, the goods are returned to the consignor. If a buyer is found, the consignee remits the selling price (less commission and approved expenses) to the consignor.

Goods held on *consignment* are included in the inventory of the consignor until sold by the consignee.

For example, suppose Pratt Clothing (consignor) ships merchandise to Regal Outlets (consignee). The arrangement specifies that Regal will attempt to sell the merchandise, and in return, Pratt will pay to Regal a 10% sales commission on any merchandise sold. Any inventory not sold within six months will be returned to Pratt. In this arrangement, Regal obtains physical possession of the inventory and has responsibility to sell to customers, but Pratt retains legal title to the inventory and risk of ownership and therefore keeps this inventory in its own records until the merchandise is sold to a customer. We discussed this issue in Chapter 6. The sale is not complete (revenue is not recognized) until an eventual sale to a third party occurs. In addition, any costs paid by Pratt to ship the inventory to Regal would be included in Pratt's cost of inventory at the time the cost was incurred. However, other costs incurred by Pratt, such as advertising or sales commissions to Regal, would be expensed as incurred and not included in the cost of inventory.

SALES RETURNS Recall from our discussions in Chapters 6 and 7 that when the right of return exists, a seller must be able to estimate those returns before revenue can be recognized. The adjusting entry for estimated sales returns includes a debit to sales returns and a credit to refund liability. At the same time, cost of goods sold is reduced and an estimate of inventory to be returned is made (see Illustration 7–4). As a result, a company includes in ending inventory the cost of merchandise sold that it anticipates will be returned.

Now that we've considered which goods are part of inventory, let's examine the types of costs associated with those inventory units.

Transactions Affecting Net Purchases

As mentioned earlier, the cost of inventory includes all necessary expenditures to acquire the inventory and bring it to its desired *condition* and *location* for sale or for use in the manufacturing process. Obviously, the cost includes the purchase price of the goods. But usually the cost of acquiring inventory also includes freight charges on incoming goods borne by the buyer; insurance costs incurred by the buyer while the goods are in transit (if shipped f.o.b. shipping point); and the costs of unloading, unpacking, and preparing merchandise

● LO8–3

Expenditures necessary to bring inventory to its *condition* and *location* for sale or use are included in its cost.

inventory for sale or raw materials inventory for use.[3] The costs included in inventory are called **product costs**. They are associated with products and *expensed as cost of goods sold only when the related products are sold.*[4]

The cost of *freight-in* paid by the purchaser generally is part of the cost of inventory.

FREIGHT-IN ON PURCHASES Freight-in on purchases is commonly included in the cost of inventory. These costs clearly are necessary to get the inventory in location for sale or use and can generally be associated with particular goods. Freight costs are added to the *inventory* account in a perpetual system.

In a periodic system, freight costs generally are recorded in a temporary account called **freight-in** or **transportation-in**, which is added to total purchases in determining net purchases for inclusion in cost of goods sold. (See also the cost of goods sold schedule in Illustration 8–6 later in this section.) From a control perspective, by recording freight-in as a separate item, management can more easily track its freight costs. The same perspectives pertain to purchase returns and purchase discounts, which are discussed in the next sections.

Shipping charges on outgoing goods are reported either as part of cost of goods sold or as an operating expense, usually among selling expenses.

Shipping charges on outgoing goods (freight-out) are not included in the cost of inventory. They are reported in the income statement either as part of cost of goods sold or as an operating expense, usually among selling expenses. If a company adopts a policy of not including shipping charges in cost of goods sold, both the amounts incurred during the period as well as the income statement classification of the expense must be disclosed.[5]

PURCHASE RETURNS In Chapter 7 we discussed merchandise returns from the perspective of the selling company. At the time a customer returns merchandise, the seller records a sales return (a contra revenue account). We now address returns from the buyer's point of view. When merchandise is returned, the buyer records a **purchase return**. In a perpetual inventory system, the purchase return is recorded directly as a reduction to the *inventory* account. In a periodic system, we use a *purchase returns* account to temporarily accumulate these amounts. Recall our earlier discussion of the cost of goods sold calculation (Beginning inventory + *Net purchases* − Ending inventory = Cost of goods sold). Purchase returns are subtracted from total purchases to calculate net purchases. (See also the cost of goods sold schedule in Illustration 8–6 later in this section.)

Purchase discounts represent reductions in the amount to be paid if remittance is made within a designated period of time.

PURCHASE DISCOUNTS Cash discounts were discussed from the seller's perspective in Chapter 7. Here, we discuss them from the buyer's perspective. Discounts offer incentives to the buyer to make quick payment. The amount of the discount and the time period within which it's available are conveyed by terms like 2/10, n/30 (meaning a 2% discount if paid within 10 days, otherwise full payment within 30 days). As with the seller, the purchaser can record these **purchase discounts** using either the **gross method** or the **net method**.

Consider Illustration 8–5. On October 5, Lothridge purchases $20,000 of inventory and is offered a discount of 2% for any amount paid within 10 days. The full invoice is due within 30 days. Under the *gross method,* we record the purchase for the full (or gross) amount of the inventory's cost. Under the *net method,* we record the purchase of inventory for its $20,000 cost minus the possible discount of $400 ($20,000 × 2%), resulting in a net purchase amount of $19,600.

Payment within the discount period. On October 14, Lothridge decides to make payment for $14,000 of the $20,000 purchase. Because payment is within the 10-day discount period, the company receives a discount of $280 ($14,000 × 2%) and must pay only $13,720

[3]Generally accepted accounting principles require that abnormal amounts of certain costs be recognized as current period expenses rather than being included in the cost of inventory, specifically idle facility costs, freight, handling costs, and waste materials (spoilage). FASB ASC 330–10–30: Inventory—Overall—Initial Measurement.

[4]For practical reasons, though, some of these expenditures often are not included in inventory cost and are treated as **period costs.** They often are immaterial or it is impractical to associate the expenditures with particular units of inventory (for example, unloading and unpacking costs). Period costs are not associated with products and are expensed in the *period* incurred.

[5]FASB ASC 605–45–50–2: Revenue Recognition—Principal Agent Considerations—Disclosure—Shipping and Handling Fees and Costs.

Illustration 8–5

Purchase Discounts Using the Gross Method versus Net Method—Perpetual System

Consider the following transactions for the Lothridge Wholesale Beverage Company that uses a perpetual inventory system:

1. October 5—Purchased merchandise with a cost of $20,000. The payment terms are stated as 2/10, n/30.
2. October 14—Paid $13,720 ($14,000 of the invoice less a 2% cash discount on that amount).
3. November 4—Paid the remaining $6,000 balance of the invoice.

Gross Method			Net Method		
October 5			**October 5**		
Inventory*	20,000		Inventory*	19,600	
Accounts payable		20,000	Accounts payable		19,600
October 14			**October 14**		
Accounts payable	14,000		Accounts payable	13,720	
Inventory**		280	Cash		13,720
Cash		13,720			
November 4			**November 4**		
Accounts payable	6,000		Accounts payable	5,880	
Cash		6,000	Purchase discounts lost	120	
			Cash		6,000

*The purchases account is used in a periodic inventory system.
**The purchase discounts account is used in a periodic inventory system.

($14,000 − $280). The discount effectively reduces the cost of inventory. Under the gross method, we record the discount by decreasing the inventory account. Under the net method, the discount was recorded at the time of the purchase, so no further reduction to the inventory account is needed at the time of payment.

Payment *not* within the discount period. On November 4, Lothridge makes final payment for the remaining $6,000 of inventory. This payment is after the 10-day discount period. Lothridge loses the possible discount of $120 ($6,000 × 2%). Under the gross method, we simply record the payment on account. Under the net method, we record the purchase discount lost. Most companies that use the net method expect to pay within the discount period. In practice, purchase discounts lost are generally immaterial.[6]

RECORDING TRANSACTIONS IN A PERPETUAL AND PERIODIC SYSTEM Illustration 8–6 compares the perpetual and periodic inventory systems, using the gross method. A schedule to demonstrate the calculation of cost of goods sold is provided at the end.

Illustration 8–6

Inventory Transactions Using the Gross Method—Perpetual and Periodic Systems

The Lothridge Wholesale Beverage Company purchases soft drinks from producers and then sells them to retailers. The company began the year with merchandise inventory of $120,000 on hand. During the year, additional inventory transactions include:

- Purchases of merchandise on account totaled $620,000, with terms 2/10, n/30.
- Freight charges paid by Lothridge were $16,000.
- Merchandise with a cost of $20,000 was returned to suppliers for credit.
- All purchases on account were paid within the discount period.
- Sales on account totaled $830,000. The cost of soft drinks sold was $550,000.
- Inventory remaining on hand at the end of the year totaled $174,000.

The above transactions are recorded in summary form according to both the perpetual and periodic inventory systems using the gross method:

(continued)

[6]There is variation in practice of how the purchase discounts lost account is classified. Many companies report this account as part of other expenses, similar to the treatment of interest expense. Other companies include the discount lost as part of goods sold, either at the time of payment or once those goods are sold.

(concluded)

($ in thousands)

Perpetual System			Periodic System		
Purchases					
Inventory	620		Purchases	620	
Accounts payable		620	Accounts payable		620
Freight					
Inventory	16		Freight-in	16	
Cash		16	Cash		16
Returns					
Accounts payable	20		Accounts payable	20	
Inventory		20	Purchase returns		20
Discounts					
Accounts Payable	600		Accounts Payable	600	
Inventory ($600 × 2%)		12	Purchase discounts ($600 × 2%)		12
Cash		588	Cash		588
Sales					
Accounts receivable	830		Accounts receivable	830	
Sales revenue		830	Sales revenue		830
Cost of goods sold	550		No entry		
Inventory		550			
End of period					
No entry			Cost of goods sold (below)	**550**	
			Inventory (ending)	174	
			Purchase returns	20	
			Purchase discounts	12	
			Inventory (beginning)		120
			Purchases		620
			Freight-in		16

Supporting Schedule

Cost of goods sold:		
Beginning Inventory		$120
Net purchases:		
Purchases	$620	
Plus: Freight-in	16	
Less: Returns	(20)	
Less: Discounts	(12)	604
Cost of good available		724
Less: Ending inventory		(174)
Cost of goods sold		**$550**

Inventory Cost Flow Assumptions

Regardless of whether the perpetual or periodic system is used, it's necessary to assign dollar amounts to the physical quantities of goods sold and goods remaining in ending inventory. Unless each item of inventory is specifically identified and traced through the system, assigning dollars is accomplished by making an assumption regarding how units of goods (and their associated costs) flow through the system. We examine the common cost flow assumptions next. In previous illustrations, dollar amounts of the cost of goods sold and the cost of ending inventory were assumed known. However, if various portions of inventory are acquired at different costs, we need a way to decide which units were sold and which remain in inventory. Illustration 8–7 will help explain.

● LO8–4

Illustration 8–7
Inventory Information

Goods available for sale include beginning inventory plus purchases.

The Browning Company has the following inventory information for 2021:

Beginning Inventory and Purchases During 2021

Date	Units	Unit Cost*	Total Cost
Jan. 1 (Beginning Inventory)	4,000	$5.50	$22,000
Purchases:			
Jan. 17	1,000	6.00	6,000
Mar. 22	3,000	7.00	21,000
Oct. 15	3,000	7.50	22,500
Goods available for sale	**11,000**		**$71,500**

Sales

Date of Sale	Units
Jan. 10	2,000
Apr. 15	1,500
Nov. 20	3,000
Total sales	6,500

*Includes purchase price and cost of freight.

Browning began the year with 4,000 units and purchased another 7,000 units, so there were **11,000** units available for sale. Of this amount, 6,500 units were sold. This means 4,500 units remain in ending inventory. This allocation of units is depicted in Illustration 8–7A.

Illustration 8–7A
Allocation of Units Available for Sale

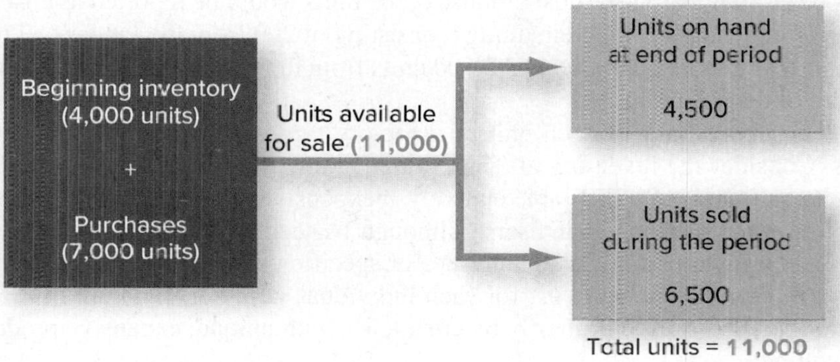

But what is the *cost* of the 6,500 units sold? What is the *cost* of the 4,500 units remaining on hand at the end of the year? To answer this, let's first consider that the total cost of the 11,000 goods available for sale (beginning inventory plus purchases during the year) is **$71,500**. This total amount will be allocated to ending inventory and cost of goods sold. The allocation decision is depicted in Illustration 8–7B.

Illustration 8–7B
Allocation of Cost of Goods Available for Sale

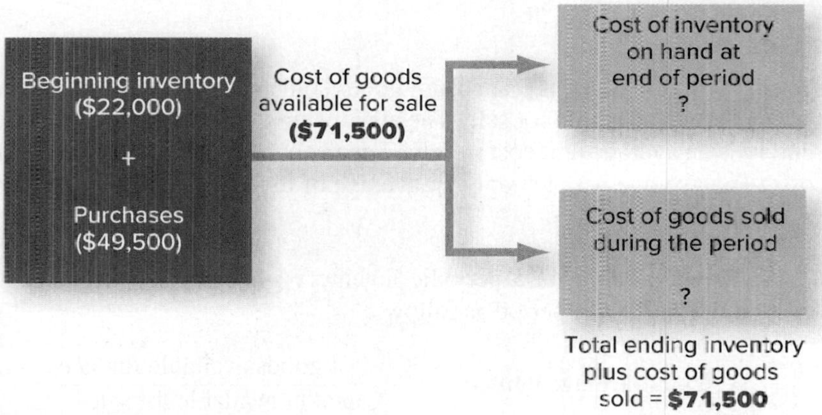

The allocation of the total cost of goods available for sale to ending inventory and cost of goods sold is also depicted in the schedule below. We'll use this schedule in the following sections.

Beginning inventory (4,000 units @ $5.50)	$ 22,000
Plus: Purchases (7,000 units @ various prices)	49,500
Cost of goods available for sale (11,000 units)	**$71,500**
Less: Ending inventory (4,500 units @ ?)	?
Cost of goods sold (6,500 units @ ?)	?

Let's turn our attention now to the various inventory costing methods that can be used to achieve the allocation between ending inventory and cost of goods sold.

Specific Identification

The specific identification method records *actual* units sold.

It's sometimes possible for each unit sold during the period or each unit on hand at the end of the period to be matched with its actual cost. Actual costs can be determined by reference to the invoice representing the purchase of the item. The **specific identification method** is used frequently by companies selling unique, expensive products with low sales volume which makes it relatively easy and economically feasible to associate each item with its actual cost. For example, automobiles have unique serial numbers that can be used to match a specific auto with the invoice identifying the actual purchase price.

In our example for Browning Company, we might have been able to track each of the 6,500 units sold. Suppose the *actual* units sold include 4,000 units of beginning inventory, 800 units of the January 17 purchase, 1,400 units from the March 22 purchase, and 300 units of the October 15 purchase. The cost of those 6,500 units would be reported as cost of goods sold. The cost of the 4,500 units remaining (consisting of 200 from the January 17 purchase, 1,600 from the March 22 purchase, and 2,700 units from the October 15 purchase) would be reported as ending inventory.

However, keeping track of each unit of inventory typically is not practicable for most companies. Consider the inventory at **The Home Depot, Inc.** or **Macy's**: large stores and numerous items, many of which are relatively inexpensive. Specific identification would be very difficult for such merchandisers. Although bar codes and RFID tags now make it possible to instantly track purchases and sales of specific types of inventory, it may be too costly to know the specific unit cost for each individual sale. For that reason, the specific identification method is used primarily by companies with unique, expensive products with low sales volume.

FIFO, LIFO, and weighted-average *assume* which units are sold.

Most companies instead use one of the three inventory cost flow assumptions—average cost, FIFO, or LIFO. Note the use of the word *assumptions.* FIFO, LIFO, and weighted-average cost *assume* a particular pattern of inventory cost flows. However, the *actual* flow of inventory does not need to match the *assumed* cost flow in order for the company to use a particular method. That's okay. Companies are allowed to report inventory costs by *assuming* which units of inventory are sold and not sold, even if this does not match the *actual* flow. This is another example of using estimates in financial accounting.

Average Cost

The *average cost method* assumes that items sold and items in ending inventory come from a mixture of all the goods available for sale.

The **average cost method** assumes that cost of goods sold and ending inventory consist of a mixture of all the goods available for sale. The average unit cost applied to goods sold or to ending inventory is an average unit cost *weighted* by the number of units acquired at the various unit costs. The average is not simply an average of the various unit costs of purchases during the period.

In a periodic inventory system, the average cost is computed only at the end of the period.

PERIODIC AVERAGE COST In a periodic inventory system, this weighted average is calculated only at the end of the period as follows:

$$\text{Weighted-average unit cost} = \frac{\text{Cost of goods available for sale}}{\text{Quantity available for sale}}$$

The calculation of average cost is demonstrated in Illustration 8–7C using data from Illustration 8–7.

Illustration 8–7C
Average Cost—Periodic Inventory System

Beginning inventory (4,000 units @ $5.50)	$22,000
Plus: Purchases (7,000 units @ various prices)	49,500
Cost of goods available for sale (11,000 units)	71,500
Less: Ending inventory* (determined below)	(29,250)
Cost of goods sold (6,500 units)	**$42,250**

***Cost of Ending Inventory:**

Weighted-average unit cost = $\frac{\$71,500}{11,000 \text{ units}}$ = $6.50

4,500 units × $6.50 = **$29,250**

Cost of goods sold also could be determined directly by multiplying the weighted-average unit cost of $6.50 by the number of units sold ($6.50 × 6,500 units = **$42,250**).

PERPETUAL AVERAGE COST The weighted-average unit cost in a perpetual inventory system becomes a moving-average unit cost. *A new weighted-average unit cost is calculated each time additional units are purchased.* The new average is determined after each purchase by (1) summing the cost of the previous inventory balance and the cost of the new purchase, and (2) dividing this new total cost (cost of goods available for sale) by the number of units on hand (the inventory units that are available for sale). This average is then used to cost any units sold before the next purchase is made. The moving-average concept is applied in Illustration 8–7D.

In a perpetual inventory system, the average cost method is applied by computing a moving-average unit cost each time additional inventory is purchased.

Illustration 8–7D Average Cost—Perpetual Inventory System

	Date	Purchased	Sold	Balance
	Beginning inventory	4,000 @ $5.50 = $22,000		4,000 @ $5.50 = $22,000
	Jan. 10		2,000 @ $5.50 = $ 11,000	2,000 @ $5.50 = $11,000
	Jan. 17	1,000 @ $6.00 = $6,000		$11,000 + $6,000 = $17,000
Average cost per unit	→	$\left[\frac{\$17,000}{3,000 \text{ units}}\right]$ = **$5.667**/unit		2,000 + 1,000 = 3,000 units
	Mar. 22	3,000 @ $7.00 = $21,000		$17,000 + $21,000 = $38,000
Average cost per unit	→	$\left[\frac{\$38,000}{6,000 \text{ units}}\right]$ = **$6.333**/unit		3,000 + 3,000 = 6,000 units
	Apr. 15		1,500 @ **$6.333** = $9,500	4,500 @ **$6.333** = $28,500
	Oct. 15	3,000 @ $7.50 = $22,500		$28,500 + $22,500 = $51,000
Average cost per unit	→	$\left[\frac{\$51,000}{7,500 \text{ units}}\right]$ = $6.80/unit		4,500 + 3,000 = 7,500 units
	Nov. 20		3,000 @ $6.80 = $20,400	4,500 @ $6.80 = **$30,600**
		Total cost of goods sold	= **$40,900**	

On January 17 the new average of **$5.667** (rounded) is calculated by dividing the $17,000 cost of goods available ($11,000 from beginning inventory + $6,000 purchased on January 17) by the 3,000 units available (2,000 units from beginning inventory + 1,000 units acquired on January 17). The average is updated to **$6.333** (rounded) with the March 22 purchase. The 1,500 units sold on April 15 are then costed at the average cost of **$6.333**.

The average is updated once again to $6.80 with the October 15 purchase, and 3,000 units are costed at that unit cost on November 20.

Cost of goods sold for the year equals the assumed cost of inventory sold for all sales, which equals **$40,900** in this example. Ending inventory equals the number of units in ending inventory times the most recently computed average inventory cost. That amount is **$30,600.** Periodic average cost and perpetual average cost generally produce different allocations to cost of goods sold and ending inventory.

First-In, First-Out (FIFO)

Ending inventory applying FIFO consists of the most recently acquired items.

The **first-in, first-out (FIFO) method** assumes that the first units purchased are the first ones sold. Beginning inventory is sold first, followed by purchases during the period in the chronological order of their acquisition. Using the example in Illustration 8–7, 6,500 units were sold during 2021. Applying FIFO, these would be the 4,000 units in beginning inventory, the 1,000 units purchased on January 17, and 1,500 of the 3,000 units purchased on March 22.

Ending inventory consists of the remaining units assumed not to be sold. In this case, the 4,500 units in ending inventory consist of 1,500 of the 3,000 units purchased on March 22 and all of the 3,000 units purchased on October 15. Graphically, the flow is as follows:

Units Available		
Beg. inv.	4,000	6,500 units sold
Jan. 17	1,000	
Mar. 22	1,500	
Mar. 22	1,500	4,500 units in ending inventory
Oct. 15	3,000	
Total	11,000	

PERIODIC FIFO Recall that we determine physical quantities on hand in a periodic inventory system by taking a physical count. Costing the 4,500 units in ending inventory this way automatically gives us the cost of goods sold as well. Using the numbers from our illustration, we determine cost of goods sold to be **$38,500** by subtracting the $33,000 ending inventory from $71,500 cost of goods available for sale as shown in Illustration 8–7E.

Illustration 8–7E

FIFO—Periodic Inventory System

Beginning inventory (4,000 units @ $5.50)	$22,000
Plus: Purchases (7,000 units @ various prices)	49,500
Cost of goods available for sale (11,000 units)	71,500
Less: Ending inventory* (determined below)	(33,000)
Cost of goods sold (6,500 units)	**$38,500**

***Cost of Ending Inventory:**

Date of Purchase	Units	Unit Cost	Total Cost
Mar. 22	1,500	$7.00	$ 10,500
Oct. 15	3,000	7.50	22,500
Total ending inventory	4,500		$33,000

Of course, the 6,500 units sold could be costed directly as follows:

Date of Purchase	Units	Unit Cost	Total Cost
Jan. 1 (Beginning inventory)	4,000	$5.50	$22,000
Jan. 17	1,000	6.00	6,000
Mar. 22	1,500	7.00	10,500
Total goods sold	6,500		**$38,500**

PERPETUAL FIFO *The ending inventory and cost of goods sold will have the same amounts in a perpetual inventory system as in a periodic inventory system when FIFO is used.* This is because the same units and costs are first in and first out whether cost of goods sold is determined as each sale is made or at the end of the period as a residual amount. The application of FIFO in a perpetual system is shown in Illustration 8–7F.

Perpetual FIFO results in cost of goods sold and ending inventory amounts that are the same as those obtained using periodic FIFO.

Illustration 8–7F

FIFO—Perpetual Inventory System

Date	Purchased	Sold	Balance	
Beginning inventory	4,000 @ $5.50 = $22,000		4,000 @ $5.50 = $22,000	
Jan. 10		2,000 @ $5.50 = $ 11,000	2,000 @ $5.50 = $11,000	
Jan. 17	1,000 @ $6.00 = $ 6,000		2,000 @ $5.50 1,000 @ $6.00	$17,000
Mar. 22	3,000 @ $7.00 = $21,000		2,000 @ $5.50 1,000 @ $6.00 3,000 @ $7.00	$38,000
Apr. 15		1,500 @ $5.50 = $ 8,250	500 @ $5.50 1,000 @ $6.00 3,000 @ $7.00	$29,750
Oct. 15	3,000 @ $7.50 = $22,500		500 @ $5.50 1,000 @ $6.00 3,000 @ $7.00 3,000 @ $7.50	$52,250
Nov. 20		500 @ $5.50 + 1,000 @ $6.00 + 1,500 @ $7.00 = $ 19,250	1,500 @ $7.00 3,000 @ $7.50	$33,000
	Total cost of goods sold	= $38,500		

Last-In, First-Out (LIFO)

The **last-in, first-out (LIFO) method** assumes that the last units purchased are the first ones sold. This is opposite the assumption of FIFO. In reality, virtually no company actually sells its inventory on a LIFO basis, but companies are allowed to assume LIFO for reporting purposes.

PERIODIC LIFO Let's use the information in Illustration 8–7 to determine LIFO amounts under a periodic inventory system. The 6,500 units sold for the year are assumed to consist of the 6,500 units most recently purchased *from the end of the year.* This includes the 3,000 units purchased on October 15, the 3,000 units purchased on March 22, and 500 of the 1,000 units purchased on January 17. Ending inventory, then, consists of the remaining units assumed not sold; in this case, 500 of the 1,000 units purchased on January 17 and all of the 4,000 units from beginning inventory. Graphically, the flow is as follows:

Ending inventory applying LIFO consists of the items acquired first.

Units Available		
Beg. inv.	4,000	4,500 units in ending inventory
Jan. 17	500	
Jan. 17	500	6,500 units sold
Mar. 22	3,000	
Oct. 15	3,000	
Total	11,000	

As shown in Illustration 8–7G, the cost of the 4,500 units assumed in ending inventory is $25,000. We subtract this amount from the cost of goods available for sale to calculate cost of goods sold of **$46,500**.

Illustration 8–7G

LIFO—Periodic Inventory System

Beginning inventory (4,000 units @ $5.50)	$22,000
Plus: Purchases (7,000 units @ various prices)	49,500
Cost of goods available for sale (11,000 units)	71,500
Less: Ending inventory* (determined below)	(25,000)
Cost of goods sold (6,500 units)	**$46,500**

***Cost of Ending Inventory:**

Date of Purchase	Units	Unit Cost	Total Cost
Jan. 1 (Beginning inventory)	4,000	$5.50	$22,000
Jan. 17	500	6.00	3,000
Total ending inventory	4,500		**$25,000**

Cost of goods sold could be calculated directly as follows:

Date of Purchase	Units	Unit Cost	Total Cost
Jan. 17	500	$6.00	$ 3,000
Mar. 22	3,000	7.00	21,000
Oct. 15	3,000	7.50	22,500
Total goods sold	6,500		**$46,500**

PERPETUAL LIFO In contrast to the periodic system, a perpetual system using LIFO assumes that units are sold based on the last units purchased *at the time of the sale*. However, for reasons we discuss later, *companies rarely, if ever, actually use a perpetual inventory system when reporting on a LIFO basis.* In practice, companies using LIFO report inventory and cost of goods sold based on calculations similar to the periodic system presented previously in Illustration 8–7G. We demonstrate the perpetual system here to emphasize the differences between periodic and perpetual amounts.

The application of LIFO in a perpetual system is shown in Illustration 8–7H. The 2,000 units sold on January 10 are assumed to include the units most recently purchased—beginning inventory. However, as we saw previously when using the periodic system, those 2,000 units sold on January 10 were assumed to come from the last purchase of the year—the November 20 purchase.

Additional purchases of 1,000 units on January 17 and 3,000 units on March 22 add to inventory. Then the sale of 1,500 units on April 15 is assumed to include the units most recently purchased—those purchased on March 22. This process continues with each purchase and sale throughout the year.

Illustration 8–7H

LIFO—Perpetual Inventory System

Date	Purchased	Sold	Balance	
Beginning inventory	4,000 @ $5.50 = $22,000		4,000 @ $5.50 =	$22,000
Jan. 10		2,000 @ $5.50 = $ 11,000	2,000 @ $5.50 =	$11,000
Jan. 17	1,000 @ $6.00 = $6,000		2,000 @ $5.50 1,000 @ $6.00	$17,000
Mar. 22	3,000 @ $7.00 = $21,000		2,000 @ $5.50 1,000 @ $6.00 3,000 @ $7.00	$38,000

(continued)

(concluded)

Date	Purchased	Sold	Balance	
Apr. 15		1,500 @ $7.00 = $ 10,500	2,000 @ $5.50 1,000 @ $6.00 1,500 @ $7.00	$27,500
Oct. 15	3,000 @ $7.50 = $22,500		2,000 @ $5.50 1,000 @ $6.00 1,500 @ $7.00 3,000 @ $7.50	$50,000
Nov. 20		3,000 @ $7.50 = $ 22,500	2,000 @ $5.50 1,000 @ $6.00 1,500 @ $7.00	$27,500
	Total cost of goods sold	= $44,000		

Notice that the total cost of goods available for sale is allocated **$44,000** to cost of goods sold and **$27,500** to ending inventory (the balance after the last transaction), which is different from the periodic LIFO result of **$46,500** and **$25,000**. *If inventory costs are rising throughout the year, periodic LIFO will generally result in lower cost of ending inventory and higher cost of goods sold than when applying perpetual LIFO.*

Perpetual LIFO generally results in cost of goods sold and inventory amounts that are different from those obtained by applying periodic LIFO.

Comparison of Cost Flow Methods

The three cost flow methods are compared below assuming a periodic inventory system.

	Average	FIFO	LIFO
Cost of goods sold	$42,250	$38,500	$46,500
Ending inventory	29,250	33,000	25,000
Goods available for sale	$71,500	$71,500	$71,500

Notice that the average cost method in this example produces amounts that fall between the FIFO and LIFO amounts for both cost of goods sold and ending inventory. This will usually be the case. Whether it will be FIFO or LIFO that produces the highest or lowest value of cost of goods sold depends on the pattern of the actual unit cost changes during the period.

During periods of rising costs, as in our example, FIFO results in a lower cost of goods sold because the lower costs of the earliest purchases are assumed sold. LIFO cost of goods sold will include the more recent higher cost purchases. On the other hand, FIFO ending inventory includes the most recent higher cost purchases, which results in a higher ending inventory. LIFO ending inventory includes the lower costs of the earliest purchases. Conversely, if costs are declining, then FIFO will result in a higher cost of goods sold and lower ending inventory.[7]

If unit costs are increasing, LIFO will result in a higher cost of goods sold and lower ending inventory.

All three methods are permissible according to generally accepted accounting principles and are frequently used in practice. Also, a company need not use the same method for its entire inventory. For example, **International Paper Company** uses LIFO for its raw materials and finished pulp and paper products, and both the FIFO and average cost methods for other inventories. Because of the importance of inventories and the possible differential effects of different methods on the financial statements, a company must identify in a disclosure note the method(s) it uses. The chapter's opening case included an example of this disclosure for **Kroger**, and you will encounter additional examples later in the chapter.

FINANCIAL Reporting Case

Q1, p. 403

A company must disclose the inventory method(s) it uses.

Illustration 8–8 shows the results of a survey of inventory methods used by 500 large public companies.[8] FIFO is the most popular method, but both LIFO and average cost are used

[7]The differences between the various methods also hold when a perpetual inventory system is used.

[8]*U.S. GAAP Financial Statements—Best Practices in Presentation and Disclosure—2013* (New York, New York: AICPA, 2013).

by many companies. Notice that the column total for the number of companies is greater than 500, indicating that many companies included in this sample do use multiple methods.

Illustration 8–8

Inventory Cost Flow Methods Used in Practice

Real World Financials

	# of Companies	% of Companies
FIFO	312	47%
LIFO	163	24
Average	133	20
Other* and not disclosed	61	9
Total	669	100%

*"Other" includes the specific identification method and miscellaneous less popular methods.

International Financial Reporting Standards

● LO8–9

Real World Financials

Inventory Cost Flow Assumptions. *IAS No. 2* does not permit the use of LIFO.[9] Because of this restriction, many U.S. multinational companies use LIFO only for their domestic inventories and FIFO or average cost for their foreign subsidiaries. A disclosure note included in a recent annual report of **General Mills** provides an example:

Inventories (in part)

All inventories in the United States other than grain are valued at the lower of cost, using the last-in, first-out (LIFO) method, or market. Inventories outside of the United States generally are valued at the lower of cost, using the first-in, first-out (FIFO) method, or net realizable value.

This difference could prove to be a significant impediment to U.S. convergence to international standards. Unless the U.S. Congress repeals the LIFO conformity rule, convergence would cause many corporations to lose a valuable tax shelter, the use of LIFO for tax purposes. If these companies were immediately taxed on the difference between LIFO inventories and inventories valued using another method, it would cost companies billions of dollars. Some industries would be particularly hard hit. Most oil companies and auto manufacturers, for instance, use LIFO. The government estimates that the repeal of the LIFO method would increase federal tax revenues by $76 billion over a ten-year period.[10] The companies affected most certainly will lobby heavily to retain the use of LIFO for tax purposes.

Factors Influencing Method Choice

● LO8–5

What factors motivate companies to choose one method over another? What factors have caused the increased popularity of LIFO? Choosing among alternative accounting methods is a complex issue. Often such choices are not made in isolation but in such a way that the combination of inventory cost flow assumptions, depreciation methods, pension assumptions, and other choices meet a particular objective. Also, many believe managers sometimes make these choices to maximize their own personal benefits rather than those of the company or its external constituents. But regardless of the motive, the impact on reported numbers is an important consideration in each choice of method. The inventory choice determines (a) how closely reported costs reflect the actual physical flow of inventory, (b) the timing of reported income and income tax expense, and (c) how well costs are matched with associated revenues.

PHYSICAL FLOW If a company wanted to choose a method that most closely approximates specific identification, then the actual physical flow of inventory in and out of the company would motivate the choice of method.

[9] "Inventories," *International Accounting Standard No. 2* (IASCF), as amended effective January 1, 2018.

[10] Andrew Lundeen, "Proposed Tax Changes in President Obama's Fiscal Year 2016 Budget," (taxfoundation.org).

For example, companies often attempt to sell the oldest goods in inventory first for some of their products. This certainly is the case with perishable goods such as many grocery items. The FIFO method best mirrors the physical flow in these situations. The average cost method might be used for liquids such as chemicals where items sold are taken from a mixture of inventory acquired at different times and different prices. There are very few inventories that actually flow in a LIFO manner. It is important for you to understand that there is no requirement that companies choose an inventory method that approximates actual physical flow and few companies make the choice on this basis. In fact, as we discuss next, the effect of inventory method on income and income taxes is the primary motivation that influences method choice.

A company is not required to choose an inventory method that approximates actual physical flow.

INCOME TAXES AND NET INCOME If the unit cost of inventory changes during a period, the inventory method chosen can have a significant effect on the amount of income reported by the company to external parties and also on the amount of income taxes paid to the Internal Revenue Service (IRS) and state and local taxing authorities. Over the entire life of a company, cost of goods sold for all years will equal actual costs of items sold regardless of the inventory method used. However, as we have discussed, different inventory methods can produce significantly different results in each particular year.

When inventory costs rise and inventory quantities are not decreasing, LIFO produces a higher cost of goods sold and therefore lower net income than the other methods. The company's income tax returns will report a lower taxable income using LIFO and lower taxes will be paid currently. Taxes are not reduced permanently, only deferred. The reduced amount will be paid to the taxing authorities when either the unit cost of inventory or the quantity of inventory subsequently declines. However, we know from our discussion of the time value of money that it is advantageous to save a dollar today even if it must be paid back in the future. In the past, high inflation (increasing prices) periods motivated many companies to switch to LIFO in order to gain this tax benefit. This is also why companies using LIFO calculate cost of goods sold using a periodic system instead of a perpetual system. As inventory costs rise throughout the year, companies are able to reduce taxes by a greater amount by assuming the last purchase of the year (highest cost units) are the first ones sold.

Many companies choose LIFO in order to reduce income taxes in periods when prices are rising.

A corporation's taxable income comprises revenues, expenses (including cost of goods sold), gains, and losses measured according to the regulations of the appropriate taxing authority. Income before tax as reported in the income statement does not always equal taxable income. In some cases, differences are caused by the use of different measurement methods.[11] However, IRS regulations, which determine federal taxable income, require that if a company uses LIFO to measure taxable income, the company also must use LIFO for external financial reporting. This is known as the **LIFO conformity rule** with respect to inventory methods.

If a company uses LIFO to measure its taxable income, IRS regulations require that LIFO also be used to measure income reported to investors and creditors (the *LIFO conformity rule*).

Because of the LIFO conformity rule, to obtain the tax advantages of using LIFO in periods of rising prices, lower net income is reported to shareholders, creditors, and other external parties. The income tax motivation for using LIFO may be offset by a desire to report higher net income. Reported net income could have an effect on a corporation's share price,[12] on bonuses paid to management, or on debt agreements with lenders. For example, research has indicated that the managers of companies with bonus plans tied to income measures are more likely to choose accounting methods that maximize their bonuses (often those that increase net income).[13]

---

[11]For example, a corporation can take advantage of incentives offered by Congress by deducting more depreciation in the early years of an asset's life in its federal income tax return than it reports in its income statement.

[12]The concept of capital market efficiency has been debated for many years. In an efficient capital market, the market is not fooled by differences in accounting method choice that do not translate into real cash flow differences. The only apparent cash flow difference caused by different inventory methods is the amount of income taxes paid currently. In an efficient market, we would expect the share price of a company that switched its method to LIFO and saved tax dollars to increase even though it reported lower net income than if LIFO had not been adopted. Research on this issue is mixed. For example, see William E. Ricks, "Market's Response to the 1974 LIFO Adoptions," *Journal of Accounting Research* (Autumn 1982); and Robert Moren Brown, "Short-Range Market Reaction to Changes to LIFO Using Preliminary Earnings Announcement Dates," *Journal of Accounting Research* (Spring 1980).

[13]For example, see P. M. Healy, "The Effect of Bonus Schemes on Accounting Decisions," *Journal of Accounting and Economics* (April 1985); and D. Dhaliwal, G. Salamon, and E. Smith, "The Effect of Owner Versus Management Control on the Choice of Accounting Methods," *Journal of Accounting and Economics* (July 1982).

FINANCIAL Reporting Case

Q2, p. 403

The LIFO conformity rule permits LIFO users to report non-LIFO inventory valuations in a disclosure note, but not on the face of the income statement. For example, Illustration 8–9 shows the notes provided in a recent annual report of **Dollar General Corporation**, a large variety store chain, disclosing its use of LIFO to value its inventories.

Illustration 8–9

Inventories Disclosures—Dollar General Corporation

Real World Financials

> **Inventories**
> Inventories are stated at the lower of cost or market with cost determined using the retail last-in, first-out ("LIFO") method as this method results in a better matching of costs and revenues.
> The excess of current cost over LIFO cost was approximately $78.5 million and $80.7 million at February 2, 2018, and February 3, 2017, respectively. Current cost is determined using the RIM on a first-in, first-out basis.
>
> Source: Dollar General Corporation

LIFO Reserves and LIFO Liquidation

● LO8–6

LIFO RESERVES Many companies maintain their internal records throughout the reporting period using FIFO or average cost but use LIFO for external reporting and income tax purposes at the end of the reporting period. There are many reasons that companies choose not to maintain internal records on a LIFO basis, including: (1) the high record-keeping costs for LIFO, (2) contractual agreements such as bonus or profit sharing plans that calculate net income with a method other than LIFO, and (3) using FIFO or average cost information for pricing decisions.

Generally, the conversion to LIFO from the internal records occurs at the end of the reporting period without actually entering the adjustment into the company's records. Some companies, though, enter the conversion adjustment—the difference between the internal method and LIFO—directly into the records as a "contra account" to inventory. This contra account is called the **LIFO reserve** or the *LIFO allowance.*

For illustration, let's say that the Doubletree Corporation uses perpetual FIFO throughout the year to maintain internal records but at the end of the year adjusts these amounts to LIFO for financial reporting purposes. Assume the company began 2021 with a balance of $475,000 in its LIFO reserve account. This balance means that at the beginning of the year, the inventory balance reported under LIFO was $475,000 lower than it would be under FIFO. By the end of 2021, assume the difference between LIFO and FIFO inventory balances increased to $535,000. The LIFO reserve needs to be adjusted to reflect the increase in the reserve during the year. The LIFO reserve (a contra asset) has a normal credit balance, so the entry to increase its balance is recorded as follows:

Cost of goods sold	60,000	
LIFO reserve ($535,000 − $475,000)		60,000

An increase in the LIFO reserve reduces reported profits.

The entry to increase the LIFO reserve also involves an increase to cost of goods sold. The debit to cost of goods sold increases total expenses and therefore lowers reported profitability.

A decrease in the LIFO reserve increases reported profits.

If the difference between inventory valued internally using FIFO and inventory valued using LIFO had *decreased* over the year, the LIFO reserve would need to be decreased with a debit. The corresponding credit would be to cost of goods sold. In this situation, when cost of goods sold is decreased, profits will increase. This situation demonstrates that even though reported inventory is lower under LIFO than FIFO, LIFO profits might be larger than FIFO profits when the LIFO reserve decreases during the year.

As an example of the disclosure of the LIFO reserve, Illustration 8–10 provides a disclosure note of **McKesson Corporation**, an American company distributing pharmaceuticals at a retail sale level and providing health information technology, medical supplies, and care management tools. The note shows the company's inventories valued at FIFO (the internal method), less the LIFO reserve, to arrive at the LIFO amount reported in the company's balance sheet.

Illustration 8–10

Inventories Disclosure—McKesson Corporation

Real World Financials

Inventories (in part)

($ in millions)	2017	2016
Inventories (at FIFO)	$16,283	$16,347
Less: LIFO reserve	(1,005)	(1,012)
Net inventories (at LIFO)	$15,278	$15,335

Source: McKesson Corporation

Under LIFO, inventory is reported in the 2017 balance sheet at a lower amount, which indicates the need for a LIFO reserve of $1,005 million. The LIFO reserve at the beginning of 2017 was $1,012 million, so its balance needs to be decreased by $7 million ($1,012 – $1,005). The decrease is recorded with a debit to the LIFO reserve. At the same time, we decrease (or credit) cost of goods sold, thereby increasing reported profit.

LIFO LIQUIDATIONS Earlier in the text, we demonstrated the importance of matching revenues and expenses in creating an income statement that is useful in predicting future cash flows. Under LIFO, sales reflect the most recent selling prices, and cost of goods sold includes the costs of the most recent purchases.

Proponents of LIFO argue that it results in a better match of revenues and expenses.

For the same reason, though, inventory costs in the balance sheet with LIFO generally are out of date because they reflect old purchase transactions. It is not uncommon for a company's LIFO inventory balance to be based on unit costs actually incurred several years earlier.

This distortion sometimes carries over to the income statement as well. When inventory quantities decline during a period, then these out-of-date inventory layers are liquidated and cost of goods sold will partially match noncurrent costs with current selling prices. This occurrence is known as a **LIFO liquidation**. If costs have been increasing (decreasing), LIFO liquidations produce higher (lower) net income than would have resulted if the liquidated inventory were included in cost of goods sold at current costs. The paper profits (losses) caused by including out-of-date, low (high) costs in cost of goods sold is referred to as the effect on income of liquidations of LIFO inventory.

To illustrate this problem, consider the example in Illustration 8–11.

Illustration 8–11

LIFO Liquidation

National Distributors, Inc., uses the LIFO inventory method. The company began the year with inventory of 20,000 units that cost $16 per unit. During the year, 30,000 units were purchased for $20 each. The cost of goods available for sale is determined as:

Beginning inventory	**20,000** units @ $16 per unit =	$320,000
Purchases during the year	**30,000** units @ $20 per unit =	600,000
Goods available for sale	50,000 units	$920,000

By the end of the year, 45,000 units were sold. Under the LIFO assumption, cost of goods sold is determined as:

From beginning inventory	**15,000** units @ $16 per unit =	$240,000
From purchases	**30,000** units @ $20 per unit =	600,000
Cost of goods sold	45,000 units	$840,000

Ending inventory equals $80,000 (= **5,000** units of beginning inventory @ $16 per unit).

Under the LIFO assumption, the 45,000 units sold include the **30,000** units purchased during the year and **15,000** units of beginning inventory. If the company had purchased at least 45,000 units, no liquidation in beginning inventory would have occurred. Then cost of goods sold would have been $900,000 (45,000 units × $20 per unit) instead of $840,000. The difference between these two cost of goods sold figures is $60,000 ($900,000 – $840,000). This is the before-tax income effect of the LIFO liquidation. We also can determine the $60,000 before-tax LIFO liquidation profit by multiplying the **15,000** beginning units liquidated by the difference between the $20 *current cost* per unit and the $16 *acquisition cost* per unit we included in cost of goods sold (15,000 units × [$20 − $16] = $60,000).

The after-tax LIFO liquidation effect (assuming a 25% income tax rate) is an increase to net income of $45,000 [$60,000 × (1 − 0.25)]. The lower the costs of the units liquidated, the more severe the effect on cost of goods sold and net income. National Distributors must disclose that the LIFO liquidation increased net income by $45,000 in 2021, assuming that this effect is considered material.

Illustration 8–12 shows the disclosure note included with recent financial statements of **Genuine Parts Company**, a distributor of automotive replacement parts, industrial replacement parts, office products, and electrical parts.

Illustration 8–12

LIFO Liquidation Disclosure—Genuine Parts Company

Real World Financials

Summary of Significant Accounting Policies (in part)

Merchandise Inventories

During 2017 and 2016, reductions in industrial parts inventories resulted in liquidations of LIFO inventory layers. The effect of the LIFO liquidations in 2017 and 2016 reduced cost of goods sold by approximately $2,000,000 and $6,000,000, respectively. There were no LIFO liquidations in 2015.

Source: Genuine Parts Company

We've discussed several factors that influence companies in their choice of inventory method. A company could be influenced by the actual physical flow of its inventory, by the effect of inventory method on reported net income and the amount of income taxes payable currently, or by a desire to provide a better match of expenses with revenues. You've seen that the direction of the change in unit costs determines the effect of using different methods on net income and income taxes. While the United States has experienced persistent inflation for many years (increases in the general price-level), the prices of many goods and services have experienced periods of declining prices (for example, personal computers).

Concept Review Exercise

INVENTORY COST FLOW METHODS

Inventory information for the Rogers Company for the first six months of the year is as follows:

Date	Transaction
Jan. 1	Beginning inventory of 10 million units at a cost of $5 each.
Feb. 15	Purchased, on account, 5 million units at a cost of $6 each.
Mar. 20	Sold, on account, 12 million units at a selling price of $12 each.
Apr. 30	Purchased, on account, 5 million units at a cost of $7 each.

On June 30, units on hand were 8 million.

Required:

1. Record the inventory purchases and sales during the year, assuming the company uses a perpetual FIFO system to maintain internal records of inventory transactions.
2. Calculate the balance of ending inventory using perpetual FIFO.

3. Assume the company uses periodic LIFO for financial reporting purposes.
   a. Calculate the amounts to be reported for cost of goods sold and ending inventory.
   b. Record the adjusting entry for the LIFO reserve on June 30, assuming the balance of the reserve at the beginning of the year was $9 million.
4. Calculate the amounts to be reported for cost of goods sold and ending inventory assuming the company instead uses
   a. A perpetual average cost system.
   b. A periodic average cost system.

Solution:

1. Record the inventory purchases and sales during the year, assuming the company uses a perpetual FIFO system to maintain internal records of inventory transactions.

	($ in millions)	
February 15		
Inventory (5 million × $6)	30	
Accounts payable		30
To record the purchase of inventory.		
March 20		
Accounts receivable (12 million × $12)	144	
Sales revenue		144
To record sales on account.		
Cost of goods sold (determined below)	62	
Inventory		62
To record cost of goods sold.		
April 30		
Inventory (5 million × $7)	35	
Accounts payable		35
To record the purchase of inventory.		

Cost of goods sold assuming FIFO:

Units Sold	Cost of Units Sold	Total Cost
10 million (from Beg. inv.)	$5	$50
2 million (from Feb. 15 purchase)	6	12
12 million		**$62**

2. Calculate the balance of ending inventory using perpetual FIFO.

Units in Ending Inventory	Cost of Units	Total Cost
3 million (from Feb. 15 purchase)	$6	$18
5 million (from Apr. 30 purchase)	7	35
8 million		**$53**

3. Assume the company uses periodic LIFO for financial reporting purposes.
   a. Calculate the amounts to be reported for cost of goods sold and ending inventory.
   Cost of goods sold:

Units Sold	Cost of Units Sold	Total Cost
5 million (from Apr. 30 purchase)	$7	$35
5 million (from Feb. 15 purchase)	6	30
2 million (from Beg. inv.)	5	10
12 million		**$75**

Cost of ending inventory:

Units in Ending Inventory	Cost of Units	Total Cost
8 million (from Beg. inv.)	$5	**$40**

b. Record the adjusting entry for the LIFO reserve on June 30, assuming the balance of the reserve at the beginning of the year was $9 million.

June 30	($ in millions)	
Cost of goods sold	4	
LIFO reserve (determined below)		4
To record the LIFO reserve adjustment.		

Ending inventory under FIFO (above)	$ 53	
Less: Ending inventory under LIFO (above)	(40)	
Ending LIFO reserve		$13
Less: Beginning LIFO reserve		(9)
LIFO reserve adjustment needed		$ 4

4. Calculate the amounts to be reported for cost of goods sold and ending inventory assuming the company instead uses:

a. A perpetual average cost system.

Cost of goods sold:

Date	Purchased	Sold	Balance
Beg. inv.	10 million @ $5 = $50		$50
Feb. 15	5 million @ $6 = $30		$50 + $30 = $80
	$\frac{\$80}{15 \text{ million units}} = \$5.33/\text{unit}$		
Mar. 20		12 million @ $5.33 = **$64**	

Cost of ending inventory:

Units in Ending Inventory	Cost of Units	Total Cost
3 million (from Beg. inv. and Feb. 15 purchase)	$5.33	$16
5 million (from Apr. 30 purchase)	7.00	35
8 million		**$51**

b. A periodic average cost system.

Date	Units	Unit Cost	Total Cost
Beg. inv.	10 million	$5	$ 50
Feb. 15	5 million	6	30
April 30	5 million	7	35
Goods available for sale	20 million		$ 115

$$\text{Weighted-average unit cost} = \frac{\$115}{20 \text{ million units}} = \$5.75$$

Cost of good sold: 12 million × $5.75 (above) = **$69**

Cost of ending inventory: 8 million × $5.75 (above) = **$46**

Decision Makers' Perspective— Inventory management

● LO8–7

Managers closely monitor inventory levels to (1) ensure that the inventories needed to sustain operations are available and to (2) hold the cost of ordering and carrying inventories to the lowest possible level.[14] Unfortunately, these objectives often conflict with one another. Companies must maintain sufficient quantities of inventory to meet customer demand. However, maintaining inventory is costly. Fortunately, a variety of tools are available, including computerized inventory control systems and the outsourcing of inventory component production, to help balance these conflicting objectives.[15]

A just-in-time (JIT) system is another valuable technique that many companies have adopted to assist them with inventory management. JIT is a system used by a manufacturer to coordinate production with suppliers so that raw materials or components arrive just as they are needed in the production process. **Harley-Davidson** is a company known for its custom-ordered motorcycles, and the company's JIT inventory system in an important part of the company's success. This system enables Harley-Davidson to maintain relatively low inventory balances. At the same time, the company's efficient production techniques, along with its excellent relationships with suppliers ensuring prompt delivery of components, enables it to quickly meet customer demand. For the year ended December 31, 2017, Harley-Davidson reported sales of motorcycles and related parts of $4,915.0 million and cost of goods sold of only $3,261.7 million. That's a gross profit of $1,653.3 (or 33.6% of sales).

A company should maintain sufficient inventory quantities to meet customer demand while at the same time minimizing inventory ordering and carrying costs.

As we discussed in Chapter 4, one factor financial analysts use to evaluate management's success is how well the company utilizes its assets. This evaluation, which often is based on the calculation of certain ratios, is influenced by the company's inventory method choice. The different inventory methods affect reported amounts, requiring analysts to make adjustments when comparing companies that use different methods. For companies that use LIFO, supplemental disclosures help to convert inventory amounts to those that would have been reported using FIFO, allowing a better comparison to a company that reports using FIFO.

FINANCIAL Reporting Case

Q3, p. 403

For example, **Harley-Davidson** uses the LIFO method. Additional inventory information from the company's recent financial statements is provided below.

Real World Financials

	For the Year Ended	
($ in millions)	December 31, 2017	December 31, 2016
Balance sheets:		
Inventories	$ 538.2	$ 499.9
Income statements:		
Net sales	$ 4,915.0	$ 5,271.4
Cost of goods sold	3,261.7	3,419.7

Suppose an analyst wanted to compare Harley-Davidson with a competitor that used all FIFO, or that used both LIFO and FIFO but with different percentages of LIFO and FIFO. To compare apples with apples, we can convert Harley-Davidson's inventories and cost of goods sold (and the competitor's if necessary) to a 100% FIFO basis before comparing the two companies. The information necessary for this conversion is provided as a supplemental disclosure by-Harley Davidson:

Supplemental LIFO disclosures can be used to convert LIFO inventories and cost of goods sold amounts.

	2017	2016
Inventories (LIFO)	$538.2	$499.9
Add: conversion to FIFO	52.4	48.3
Inventories (100% FIFO)	$590.6	$548.2

[14]The cost of carrying inventory includes the possible loss from the write-down of obsolete inventory. We discuss inventory write-downs in Chapter 9. There are analytical models available to determine the appropriate amount of inventory a company should maintain. A discussion of these models is beyond the scope of this text.

[15]Eugene Brigham and Joel Houston, *Fundamentals of Financial Management*, 12th ed. Florence, Kentucky: South-Western, 2010).

If Harley-Davidson had used FIFO instead of LIFO, ending inventory in 2017 would have been $52.4 million higher. The large difference in reported inventory can have a material effect on financial ratios (as demonstrated below).

The use of FIFO versus LIFO also affects the calculation of cost of goods sold and therefore gross profit and net income. For Harley-Davidson, the *difference* between FIFO and LIFO inventory increased by $4.1 million during 2017 (from $48.3 to $52.4 million). This means the company would have recorded an increase of $4.1 million in the LIFO reserve in 2017, along with an increase in cost of goods sold (see earlier discussion of the LIFO reserve adjustment). The increase in cost of goods sold indicates that reported profits under LIFO were lower than if FIFO had been used. If the LIFO reserve had decreased during the year, profits would have been higher under LIFO than FIFO.

One useful profitability indicator that involves cost of goods sold is the **gross profit ratio**. The ratio is computed as follows:

$$\text{Gross profit ratio} = \frac{\text{Gross profit}}{\text{Net sales}} \text{ or } \frac{\text{Net sales} - \text{Cost of goods sold}}{\text{Net sales}}$$

The *gross profit ratio* provides a measure of how much of each sales dollar is available to pay for expenses other than cost of goods.

This ratio provides a measure of how much of each sales dollar is available to pay for expenses other than cost of goods. For example, inventory that costs $100 and sells for $125 provides a gross profit of $25 ($125 − $100) and a gross profit ratio of 20% ($25 ÷ $125). This means that each $1 of sales generates $0.20 that can be used to help pay for expenses other than cost of goods sold or remain as profit. If that same inventory can be sold for $150, gross profit increases to $50 and the gross profit ratio increases to 33% ($50 ÷ $150).

As we discussed previously, the 2017 gross profit for Harley-Davidson's motorcycle sales is $1,653.3 million (equal to sales of $4,915.0 million minus cost of goods sold of $3,261.7 million). That's a gross profit ratio of 33.6%. The industry average is 20.0%. Harley-Davidson is able to sell its products at significantly higher markups. If the LIFO reserve adjustment that increased cost of goods sold in 2017 had not been made, cost of goods sold on a FIFO basis would have been $4.1 million lower and gross profit would have been even higher than reported for the year.

Monitoring the gross profit ratio over time can provide valuable insights. For example, a declining ratio might indicate that the company is unable to offset rising costs with corresponding increases in selling price, or perhaps that sales prices are declining without a commensurate reduction in costs. In either case, the decline in the ratio has important implications for future profitability.

In Chapter 4 we were introduced to an important ratio, the **inventory turnover ratio**, which is designed to evaluate a company's effectiveness in managing its investment in inventory. The ratio shows the number of times the average inventory balance is sold during a reporting period. The more frequently a business is able to sell or turn over its inventory, the lower its investment in inventory must be for a given level of sales. Monitoring the inventory turnover ratio over time can highlight potential problems. A declining ratio generally is unfavorable and could be caused by the presence of obsolete or slow-moving products, or poor marketing and sales efforts.

Recall that the ratio is computed as follows:

$$\text{Inventory turnover ratio} = \frac{\text{Cost of goods sold}}{\text{Average inventory}}$$

If the analysis is prepared for the fiscal year reporting period, we can divide the inventory turnover ratio into 365 days to calculate the **average days in inventory**, which indicates the average number of days it normally takes the company to sell its inventory. In 2017, Harley-Davidson's inventory turnover ratio using reported LIFO amounts is 6.28 {$3,261.7 ÷ [($538.2 + $499.9) ÷ 2]} and the average days in inventory is 58.1 days (365 ÷ 6.28).

Alternatively, if we convert amounts to those under FIFO, we see that Harley-Davidson's inventory turnover ratio would have been only 5.72 {$3,257.6 ÷ [($590.6 + $548.2) ÷ 2]} and the average days in inventory would have been 63.8 days (365 ÷ 5.72). The difference in inventory turnover ratios between LIFO and FIFO is caused primarily by lower average inventory under LIFO and demonstrates the noticeable effect that the choice of inventory method can have on financial ratios.

EARNINGS QUALITY. Changes in the ratios we discussed above often provide information about the quality of a company's current period earnings. For example, a slowing turnover ratio combined with higher than normal inventory levels may indicate the potential for decreased production, obsolete inventory, or a need to decrease prices to sell inventory (which will then decrease gross profit ratios and net income). This proposition was tested in an important academic research study. Professors Lev and Thiagarajan empirically demonstrated the importance of a set of 12 fundamental variables in valuing companies' common stock. The set of variables included inventory (change in inventory minus change in sales). The inventory variable was found to be a significant indicator of returns on investments in common stock, particularly during high and medium inflation years.[16]

The choice of which inventory method to use also affects earnings quality, particularly in times of rapidly changing prices. Earlier in this chapter we discussed the effect of a LIFO liquidation on company profits. A LIFO liquidation profit (or loss) reduces the quality of current period earnings. Fortunately for analysts, companies must disclose these profits or losses, if material. In addition, LIFO cost of goods sold determined using a periodic inventory system is more susceptible to manipulation than is FIFO. Year-end purchases can have a dramatic effect on LIFO cost of goods sold in rapid cost-change environments. Recall again our discussion in Chapter 4 concerning earnings quality. Many believe that manipulating income reduces earnings quality because it can mask permanent earnings. Inventory write-downs and changes in inventory method are two additional inventory-related techniques a company could use to manipulate earnings. We discuss these issues in the next chapter. ●

Methods of Simplifying LIFO

PART B

The LIFO method described and illustrated to this point is called *unit LIFO*[17] because the last-in, first-out concept is applied to individual units of inventory. One problem with unit LIFO is that it can be very costly to implement. It requires records of each unit of inventory. The costs of maintaining these records can be significant, particularly when a company has numerous individual units of inventory and when unit costs change often during a period.

The recordkeeping costs of unit LIFO can be significant.

In the previous section, a second disadvantage of unit LIFO was identified—the possibility that LIFO layers will be liquidated if the quantity of a particular inventory unit declines below its beginning balance. Even if a company's total inventory quantity is stable or increasing, if the quantity of any particular inventory unit declines, unit LIFO will liquidate all or a portion of a LIFO layer of inventory. When inventory quantity declines in a period of rising costs, noncurrent lower costs will be included in cost of goods sold and matched with current selling prices, resulting in LIFO liquidation profit.

Another disadvantage of unit LIFO is the possibility of LIFO liquidation.

This part of the chapter discusses techniques that can be used to significantly reduce the record-keeping costs of LIFO and to minimize the probability of LIFO inventory layers being liquidated. Specifically, we discuss the use of inventory pools and the dollar-value LIFO method.

LIFO Inventory Pools

The objectives of using **LIFO inventory pools** are to simplify record-keeping by grouping inventory units into pools based on physical similarities of the individual units and to reduce the risk of LIFO layer liquidation. For example, a glass company might group its various grades of window glass into a single window pool. Other pools might be auto glass and sliding-door glass. A lumber company might pool its inventory into hardwood, framing lumber, paneling, and so on.

A pool consists of inventory units grouped according to natural physical similarities.

This allows a company to account for a few inventory pools rather than every specific type of inventory separately. Within pools, all purchases during a period are considered to

The average cost for all of the pool purchases during the period is applied to the current year's LIFO layer.

[16]B. Lev and S. R. Thiagarajan, "Fundamental Information Analysis," *Journal of Accounting Research* (Autumn 1993). The main conclusion of the study was that fundamental variables, not just earnings, are useful in firm valuation, particularly when examined in the context of macroeconomic conditions such as inflation.

[17]Unit LIFO sometimes is called *specific goods LIFO*.

have been made at the same time and at the same cost. Individual unit costs are converted to an average cost for the pool. If the quantity of ending inventory for the pool increases, then ending inventory will consist of the beginning inventory plus a single layer added during the period at the average acquisition cost for that pool.

Here's an example. Let's say Diamond Lumber Company has a rough-cut lumber inventory pool that includes three types: pine, oak, and maple. The beginning inventory consisted of the following:

	Quantity (Board Feet)	Cost (Per Foot)	Total Cost
Pine	16,000	$2.20	$35,200
Oak	10,000	3.00	30,000
Maple	14,000	2.40	33,600
	40,000		$98,800

The average cost for this pool is $2.47 per board foot ($98,800 ÷ 40,000 board feet). Now assume that during the next reporting period Diamond purchased 50,000 board feet of lumber as follows:

	Quantity (Board Feet)	Cost (Per Foot)	Total Cost
Pine	20,000	$2.25	$ 45,000
Oak	14,000	3.00	42,000
Maple	16,000	2.50	40,000
	50,000		$127,000

The average cost for this pool is **$2.54** per board foot ($127,000 ÷ 50,000 board feet). Assuming that Diamond sold 46,000 board feet during this period, the quantity of inventory for the pool increased by **4,000** board feet (50,000 purchased less 46,000 sold). Ending inventory includes the beginning 40,000 feet of inventory and a new LIFO layer consisting of the **4,000** board feet purchased this period. We would add this LIFO layer at the average cost of purchases made during the period, **$2.54**. The ending inventory of $108,960 now consists of two layers:

	Quantity (Board Feet)	Cost (Per Foot)	Total Cost
Beginning inventory	40,000	$2.47	$ 98,800
LIFO layer added	**4,000**	**2.54**	10,160
Ending inventory	44,000		$108,960

Despite the advantages of LIFO inventory pools, it's easy to imagine situations in which its benefits are not achieved. Suppose, for instance, that a company discontinues a certain product included in one of its pools. The old costs that existed in prior layers of inventory would be recognized as cost of goods sold and produce LIFO liquidation profit. Even if the product is replaced with another product, the replacement may not be similar enough to be included in the same inventory pool. In fact, the process itself of having to periodically redefine pools as changes in product mix occur can be expensive and time-consuming. Next we discuss the dollar-value LIFO approach which helps overcome these problems.

Dollar-Value LIFO

A DVL pool is made up of items that are likely to face the same cost change pressures.

Many companies that report inventory using LIFO actually use a method called **dollar-value LIFO (DVL)**. DVL extends the concept of inventory pools by allowing a company to combine a large variety of goods into one pool. Pools are not based on physical units. Instead, an inventory pool is viewed as comprising layers of dollar value from different periods. Specifically, a pool should consist of those goods that are likely to be subject to the same cost change pressures.

● LO8–8

Cost Indexes

In either the unit LIFO approach or the pooled LIFO approach, we determine whether a new LIFO layer was added by comparing the ending quantity with the beginning quantity. The focus is on *units* of inventory. Under DVL, we determine whether a new LIFO layer was added by comparing the ending dollar amount with the beginning dollar amount. The focus is on inventory *value,* not units. However, if the price level has changed, we need a way to determine whether an observed increase is a real increase (an increase in the quantity of inventory) or one caused by an increase in prices. *So before we compare the beginning and ending inventory amounts, we need to deflate inventory dollar amounts by any increase in prices so that both the beginning and ending amounts are measured in terms of the same price level.* We accomplish this by using cost indexes. A cost index for a particular layer year is determined as follows:

$$\text{Cost index in layer year} = \frac{\text{Cost in layer year}}{\text{Cost in base year}}$$

The base year is the year in which the DVL method is adopted and the layer year is any subsequent year in which an inventory layer is created. The cost index for the base year is set at 1.00. Subsequent years' indexes reflect cost changes relative to the base year. For example, if a "basket" of inventory items cost $120 at the end of the current year, and $100 at the end of the base year, the cost index for the current year would be: $120 ÷ $100 = 120%, or 1.20. This index simply tells us that costs in the layer year are 120% of what they were in the base year (i.e., costs increased by 20%).

The cost index for the base year (the year DVL is initially adopted) is set at 1.00.

There are several techniques that can be used to determine an index for a DVL pool. An external index like the Consumer Price Index (CPI) or the Producer Price Index (PPI) can be used. For example, assume that a company adopted the DVL method on January 1, 2021, when the CPI was 200. This amount is set equivalent to 1.00, the base year index. Then, the index in the layer year, say the end of 2021, would be determined relative to 200. So, if the CPI is 210 at the end of 2021, the 2021 index for DVL purposes would be 1.05 (210 ÷ 200).

However, in most cases these indexes would not properly reflect cost changes for any individual DVL pool. Instead, most companies use an internally generated index. These indexes can be calculated using one of several techniques such as the *double-extension method* or the *link-chain method.* A discussion of these methods is beyond the scope of this text. In our examples and illustrations, we assume cost indexes are given.

The DVL Inventory Estimation Technique

To see the calculation of ending inventory using DVL, consider the following example. Assume that Hanes Company adopted DVL on January 1, 2021, when the inventory cost was $400,000. On December 31, 2021, Hanes determines the year-end cost of inventory is $462,000. This amount is obtained by taking the physical units of inventory on hand at the end of the year and multiplying by year-end costs. It's not necessary for Hanes to track the item-by-item cost of purchases during the year.

Assuming a cost index for 2021 of 1.05 (105%), we'll use three steps to calculate the amount to report for ending inventory using dollar-value LIFO.

STEP 1: Convert ending inventory to base year cost. Notice inventory increased from $400,000 at the beginning of the year to $462,000 at the end of the year. Does this $62,000 increase in inventory represent an increase in the *quantity* and/or an increase in the *cost* of inventory? To determine this, the first step is to convert the ending inventory from year-end costs to base year costs. We do this by dividing ending inventory by the year's cost index.

$$\text{Ending inventory at } \textit{base year} \text{ cost} = \frac{\$462{,}000}{1.05} = \$440{,}000$$

STEP 2: Identify the layers of ending inventory created each year. The 2021 ending inventory deflated to base year cost is $440,000. From this, we can determine that the $62,000 increase in the cost of total inventory for the year consists of a $40,000 increase due to quantity (new inventory layer added in 2021) plus $22,000 due to an increase in the cost index of inventory (= $440,000 × 5%).

The calculation above is important for identifying the two layers of inventory *quantity* for applying the LIFO concept: (1) beginning inventory layer of $400,000 and (2) $40,000 layer added in 2021. These are the costs as if each layer was acquired at base year prices. The increase in inventory due to rising costs ($22,000) does not represent more units purchased.

$400,000	(beginning inventory cost in 2021; beginning layer)
+ 40,000	(increase in *quantity* in 2021; new layer)
+ 22,000	(increase in *cost* = $440,000 × 5%)
$462,000	(ending inventory cost in 2021)

STEP 3: Restate each layer using the cost index in the year acquired. Once the layers are identified, each is restated to prices existing when the layers were acquired. This is done by multiplying each layer by the cost index for the year it was acquired. The $400,000 layer was acquired when prices were 1.00, and the $40,000 layer was acquired when prices were 1.05. All layers are added, and ending inventory under DVL would be reported at **$442,000.**[18]

Date	Ending Inventory at Base Year Cost	×	Cost Index	=	Ending Inventory at DVL Cost
1/1/2021	$400,000		1.00		$400,000
2021 layer	40,000		1.05		42,000
Totals	$440,000				**$442,000**

In cases where the quantity of inventory *decreases,* no layer would be added in the current year. In our example, if inventory at the end of 2021 had a base year cost of only $380,000 rather than $440,000, then no layer would have been added in 2021. Instead, the existing layer at the beginning of the year is reduced by $20,000. If more than one layer existed at the beginning of the year, then layers are reduced in LIFO order. The remaining layers are multiplied by the cost index that existed in the year those layers were added. Then, any future increases in inventory quantity (in 2022 or after) would add new layers beyond the $380,000 at the cost index relating to the year of the additional layer.

The identification of inventory layers is demonstrated further in the Concept Review Exercise below. In years when there is an increase in inventory at base year cost (2022 and 2024), a new inventory layer is added. In years when there is a decrease in inventory at base year cost (2023), we use LIFO and assume the inventory sold was from the last layer added.

Concept Review Exercise

DOLLAR-VALUE LIFO

On January 1, 2021, the Johnson Company adopted the dollar-value LIFO method. The inventory value on this date was $500,000. Inventory data for 2021 through 2024 are as follows:

Date	Ending Inventory at Year-End Costs	Cost Index
12/31/2021	$556,500	1.05
12/31/2022	596,200	1.10
12/31/2023	615,250	1.15
12/31/2024	720,000	1.25

[18] It is important to note that the costs of the year's layer are only an approximation of actual acquisition cost. DVL assumes that all inventory quantities added during a particular year were acquired at a single cost.

Required:

Calculate Johnson's ending inventory for the years 2021 through 2024.

Solution:

JOHNSON COMPANY

Date	Ending Inventory at Year-End Cost	Step 1: Ending Inventory at Base Year Cost	Step 2: Inventory Layers at Base Year Cost	Step 3: Inventory Layers Converted to Acquisition Year Cost	Ending Inventory at DVL Cost
1/1/2021	$500,000 (base year)	$\frac{\$500,000}{1.00} = \$500,000$	$500,000 (base)	$500,000 × 1.00 = $500,000	$500,000
12/31/2021	556,500	$\frac{\$556,500}{1.05} = \$530,000$	$500,000 (base)	$500,000 × 1.00 = $500,000	
			30,000 (2021)	30,000 × 1.05 = 31,500	531,500
12/31/2022	596,200	$\frac{\$596,200}{1.10} = \$542,000$	$500,000 (base)	$500,000 × 1.00 = $500,000	
			30,000 (2021)	30,000 × 1.05 = 31,500	
			12,000 (2022)	12,000 × 1.10 = 13,200	544,700
12/31/2023	615,250	$\frac{\$615,250}{1.15} = \$535,000$*	$500,000 (base)	$500,000 × 1.00 = $500,000	
			30,000 (2021)	30,000 × 1.05 = 31,500	
			5,000 (2022)	5,000 × 1.10 = 5,500	537,000
12/31/2024	720,000	$\frac{\$720,000}{1.25} = \$576,000$	$500,000 (base)	$500,000 × 1.00 = $500,000	
			30,000 (2021)	30,000 × 1.05 = 31,500	
			5,000 (2022)	5,000 × 1.10 = 5,500	
			41,000 (2024)	41,000 × 1.25 = 51,250	588,250

*Since inventory declined during 2023 (from $542,000 to $535,000 at base year costs), no new layer is added. Instead the most recently acquired layer, 2022, is reduced by $7,000 (from $12,000 to $5,000).

Advantages of DVL

The DVL method has important advantages. First, it simplifies the record-keeping procedures compared to unit LIFO because no information is needed about unit flows. Second, it minimizes the probability of the liquidation of LIFO inventory layers, even more so than the use of pools alone, through the aggregation of many types of inventory into larger pools. In addition, the method can be used by firms that do not replace units sold with new units of the same kind. For firms whose products are subject to annual model changes, for example, the items in one year's inventory are not the same as those of the prior year. Under pooled LIFO, however, the new replacement items must be substantially identical to previous models to be included in the same pool. Under DVL, no distinction is drawn between the old and new merchandise on the basis of their physical characteristics, so a much broader range of goods can be included in the pool. That is, the acquisition of the new items is viewed as replacement of the dollar value of the old items. Because the old layers are maintained, this approach retains the benefits of LIFO by matching the most recent acquisition cost of goods with sales measured at current selling prices.

Financial Reporting Case Solution

©Blend Images/Dave and Les Jacobs/Getty Images

1. **How is the LIFO method used to calculate inventories? Is this permissible according to GAAP?** *(p. 419)* The LIFO method uses the assumption that the most recent inventory purchased is sold first. Yes, this method is permissible according to generally accepted accounting principles. A company need not use the same method that represents its actual inventory flow.

2. **What is the purpose of disclosing the difference between the reported LIFO inventory amounts and replacement cost, assuming that replacement cost is equivalent to a FIFO basis?** *(p. 422)* The LIFO conformity rule requires that if a company uses LIFO to measure taxable income, it also must use LIFO for external financial reporting. The choice of using LIFO instead of FIFO for financial reporting, however, will affect amounts reported for inventory and cost of goods sold. This difference makes it harder to compare companies using LIFO to companies using FIFO. To make this comparison easier, companies using LIFO can disclose the amount of inventory they would have reported under FIFO. Kroger's disclosure note offers this additional information.
3. **Are you correct that, by using LIFO, Kroger reports lower inventory and lower profits? Why would Kroger do that?** *(p. 427)* Because Kroger uses LIFO instead of FIFO, reported inventory in the balance sheet is lower. For the year ended February 3, 2018, Kroger reported LIFO inventory of only $6,533 million, compared to $7,781 million that it would have reported under FIFO. This is a difference of $1,248 million. In the previous year, the difference between LIFO and FIFO was $1,291 million. This means the LIFO effect (or LIFO reserve) has decreased by $43 million (from $1,291 million to $1,248 million). A decrease in the LIFO reserve is recorded as a reduction of cost of goods sold, increasing profits. However, if the LIFO reserve had increased in the current year, cost of goods sold would be increased and profits would have been lower. ●

The Bottom Line

- **LO8–1** Inventory for a manufacturing company includes raw materials, work in process, and finished goods. Inventory for a merchandising company includes goods primarily in finished form ready for sale. In a perpetual inventory system, inventory is continually adjusted for each change in inventory. Cost of goods sold is adjusted each time goods are sold or returned by a customer. A periodic inventory system adjusts inventory and records cost of goods sold only at the end of a reporting period. (*p. 404*)
- **LO8–2** Generally, determining the physical quantity that should be included in inventory is a simple matter because it consists of items in the possession of the company. However, at the end of a reporting period it's important to determine the ownership of goods that are in transit between the company and its customers as well as between the company and its suppliers. Also, goods on consignment should be included in inventory of the consignor even though the company doesn't have physical possession of the goods. In addition, a company anticipating sales returns includes in inventory the cost of merchandise it estimates will be returned. (*p. 408*)
- **LO8–3** The cost of inventory includes all expenditures necessary to acquire the inventory and bring it to its desired condition and location for sale or use. Generally, these expenditures include the purchase price of the goods reduced by any returns and purchase discounts, plus freight-in charges. (*p. 409*)
- **LO8–4** Once costs are determined, the cost of goods available for sale must be allocated between cost of goods sold and ending inventory. Unless each item is specifically identified and traced through the system, the allocation requires an assumption regarding the flow of costs. First-in, first-out (FIFO) assumes that units sold are the first units acquired. Last-in, first-out (LIFO) assumes that the units sold are the most recent units purchased. The average cost method assumes that cost of goods sold and ending inventory consist of a mixture of all the goods available for sale. (*p. 412*)
- **LO8–5** A company's choice of inventory method will be influenced by (a) how closely cost flow reflects the actual physical flow of its inventory, (b) the timing of income tax expenses, and (c) how costs are matched with revenues. (*p. 420*)
- **LO8–6** The LIFO conformity rule requires that if a company uses LIFO to measure taxable income, it also must use LIFO for external financial reporting. LIFO users often provide a disclosure note describing the effect on inventories of using another method for inventory valuation rather than LIFO. If a company uses LIFO and inventory quantities decline during a period, then out-of-date inventory layers are liquidated and the cost of goods sold will partially match noncurrent costs with current selling prices. If costs have been increasing (decreasing), LIFO liquidations produce higher (lower) net income than would have resulted if the liquidated inventory were included in cost of goods sold at current costs. The paper profits (losses)

caused by including out-of-date, low (high) costs in cost of goods sold is referred to as the effect on income of liquidations of LIFO inventory. (*p. 422*)

- **LO8–7** Investors, creditors, and financial analysts can gain important insights by monitoring a company's investment in inventories. The gross profit ratio, inventory turnover ratio, and average days in inventory are designed to monitor inventories. (*p. 427*)
- **LO8–8** The dollar-value LIFO method converts ending inventory at year-end cost to base year cost using a cost index. After identifying the layers in ending inventory with the years they were created, each year's base year cost measurement is converted to layer year cost measurement using the layer year's cost index. The layers are then summed to obtain total ending inventory at cost. (*p. 431*)
- **LO8–9** The primary difference between U.S. GAAP and IFRS with respect to determining the cost of inventory is that IFRS does not allow the use of the LIFO method to value inventory. (*p. 420*)

Questions For Review of Key Topics

Q 8–1 Describe the three types of inventory of a manufacturing company.

Q 8–2 What is the main difference between a perpetual inventory system and a periodic inventory system? Which system is used more often by major companies?

Q 8–3 The Cloud Company employs a perpetual inventory system and the McKenzie Corporation uses a periodic system. Describe the differences between the two systems in accounting for the following events: (1) purchase of merchandise, (2) sale of merchandise, (3) return of merchandise to supplier, and (4) payment of freight charge on merchandise purchased. Indicate which inventory-related accounts would be debited or credited for each event.

Q 8–4 The Bockner Company shipped merchandise to Laetner Corporation on December 28, 2021. Laetner received the shipment on January 3, 2022. December 31 is the fiscal year-end for both companies. The merchandise was shipped f.o.b. shipping point. Explain the difference in the accounting treatment of the merchandise if the shipment had instead been designated f.o.b. destination.

Q 8–5 What is a consignment arrangement? Explain the accounting treatment of goods held on consignment.

Q 8–6 Distinguish between the gross and net methods of accounting for purchase discounts.

Q 8–7 The Esquire Company employs a periodic inventory system. Indicate the effect (increase or decrease) of the following items on cost of goods sold:

1. Beginning inventory
2. Purchases
3. Ending inventory
4. Purchase returns
5. Freight-in

Q 8–8 Identify four inventory costing methods for assigning cost to ending inventory and cost of goods sold and briefly explain the difference in the methods.

Q 8–9 It's common in the electronics industry for unit costs of raw materials inventories to decline over time. In this environment, explain the difference between LIFO and FIFO in terms of the effect on cost of goods sold and ending inventory. Assume that inventory quantities remain the same for the period.

Q 8–10 Explain why proponents of LIFO argue that it provides a better match of revenue and expenses. In what situation would it not provide a better match?

Q 8–11 Explain what is meant by the Internal Revenue Service conformity rule with respect to the inventory method choice.

Q 8–12 Describe the ratios used by financial analysts to monitor a company's investment in inventories.

Q 8–13 What is a LIFO inventory pool? How is the cost of ending inventory determined when pools are used?

Q 8–14 Identify two advantages of dollar-value LIFO compared with unit LIFO.

Q 8–15 The Austin Company uses the dollar-value LIFO inventory method with internally developed price indexes. Assume that ending inventory at year-end cost has been determined. Outline the remaining steps used in the dollar-value LIFO computations.

IFRS Q 8–16 Identify any differences between U.S. GAAP and International Financial Reporting Standards in the methods allowed to value inventory.

Brief Exercises

BE 8–1
Determining ending inventory
● LO8–1

A company began its fiscal year with inventory of $186,000. Purchases and cost of goods sold for the year were $945,000 and $982,000, respectively. What was the amount of ending inventory?

BE 8–2
Perpetual system; journal entries
● LO8–1

Litton Industries uses a perpetual inventory system. The company began its fiscal year with inventory of $267,000. Purchases of merchandise on account during the year totaled $845,000. Merchandise costing $902,000 was sold on account for $1,420,000. Prepare the journal entries to record these transactions.

BE 8–3
Goods in transit
● LO8–2

Kelly Corporation shipped goods to a customer f.o.b. destination on December 29, 2021. The goods arrived at the customer's location in January. In addition, one of Kelly's major suppliers shipped goods to Kelly f.o.b. shipping point on December 30. The merchandise arrived at Kelly's location in January. Which shipments should be included in Kelly's December 31 inventory?

BE 8–4
Purchase discounts; gross method
● LO8–3

On December 28, 2021, Videotech Corporation (VTC) purchased 10 units of a new satellite uplink system from Tristar Communications for $25,000 each. The terms of each sale were 1/10, n/30. VTC uses the gross method to account for purchase discounts and a perpetual inventory system. VTC paid the net-of-discount amount on January 6, 2022. Prepare the journal entries on December 28 and January 6 to record the purchase and payment.

BE 8–5
Purchase discounts; net method
● LO8–3

Refer to the situation described in BE 8–4. Prepare the necessary journal entries assuming that VTC uses the net method to account for purchase discounts.

BE 8–6
Inventory cost flow methods; perpetual system
● LO8–4

Samuelson and Messenger (SAM) began 2021 with 200 units of its one product. These units were purchased near the end of 2020 for $25 each. During the month of January, 100 units were purchased on January 8 for $28 each and another 200 units were purchased on January 19 for $30 each. Sales of 125 units and 100 units were made on January 10 and January 25, respectively. There were 275 units on hand at the end of the month. SAM uses a *perpetual* inventory system. Calculate ending inventory and cost of goods sold for January using (1) FIFO and (2) average cost.

BE 8–7
Inventory cost flow methods; periodic system
● LO8–4

Refer to the situation described in BE 8–6. SAM uses a *periodic* inventory system. Calculate ending inventory and cost of goods sold for January using (1) FIFO and (2) average cost.

BE 8–8
LIFO method
● LO8–4

Esquire Inc. uses the LIFO method to report its inventory. Inventory at January 1, 2021, was $500,000 (20,000 units at $25 each). During 2021, 80,000 units were purchased, all at the same price of $30 per unit. 85,000 units were sold during 2021. Calculate the December 31, 2021, ending inventory and cost of goods sold for 2021 based on a periodic inventory system.

BE 8–9
LIFO method
● LO8–4

AAA Hardware uses the LIFO method to report its inventory. Inventory at the beginning of the year consisted of 10,000 units of the company's one product. These units cost $15 each. During the year, 60,000 units were purchased at a cost of $18 each and 64,000 units were sold. Near the end of the fiscal year, management is considering the purchase of an additional 5,000 units at $18. What would be the effect of this purchase on income before income taxes? Would your answer be the same if the company used FIFO instead of LIFO?

BE 8–10
LIFO liquidation
● LO8–6

Refer to the situation described in BE 8–8. Assuming an income tax rate of 25%, what is LIFO liquidation profit or loss that the company would report in a disclosure note accompanying its financial statements?

BE 8–11
LIFO Reserve
● LO8–6

King Supply maintains its internal inventory records using perpetual FIFO, but for financial reporting purposes, reports ending inventory and cost of goods sold using periodic LIFO. At the beginning of the year, the company had a balance of $60,000 in its LIFO reserve account. By the end of the year, internal records reveal that FIFO ending inventory is $75,000 greater than LIFO ending inventory. Record the year-end adjusting entry for the LIFO reserve.

BE 8–12
Supplemental LIFO reserve disclosures; Walgreens
● LO8–6
Real World Financials

Walgreens Boots Alliance, Inc., reported inventories of $8,899 million and $8,956 million in its August 31, 2017, and August 31, 2016, balance sheets, respectively. Cost of goods sold for the year ended August 31, 2017, was $89,052 million. The company uses primarily the LIFO inventory method. A disclosure note reported that if FIFO had been used instead of LIFO, inventory would have been higher by $3,000 million and $2,800 million at the end of the August 31, 2017, and August 31, 2016, periods, respectively. Calculate cost of goods sold for the year ended August 31, 2017, assuming Walgreens used FIFO instead of LIFO.

BE 8–13
Ratio analysis
● LO8–7

Selected financial statement data for Schmitzer Inc. is shown below:

	2021	2020
Balancesheet:		
Inventories	$60,000	$48,000
Ratios:		
Gross profit ratio for 2021	40%	
Inventory turnover ratio for 2021	5	

What was the amount of net sales for 2021?

BE 8–14
Dollar-value LIFO
● LO8–8

At the beginning of 2021, a company adopts the dollar-value LIFO inventory method for its one inventory pool. The pool's value on that date was $1,400,000. The 2021 ending inventory valued at year-end costs was $1,664,000 and the year-end cost index was 1.04. Calculate the inventory value at the end of 2021 using the dollar-value LIFO method.

Exercises

connect

E 8–1
Perpetual inventory system; journal entries
● LO8–1

John's Specialty Store uses a perpetual inventory system. The following are some inventory transactions for the month of May:

1. John's purchased merchandise on account for $5,000. Freight charges of $300 were paid in cash.
2. John's returned some of the merchandise purchased in (1). The cost of the merchandise was $600 and John's account was credited by the supplier.
3. Merchandise costing $2,800 was sold for $5,200 in cash.

Required:
Prepare the necessary journal entries to record these transactions.

E 8–2
Periodic inventory system; journal entries
● LO8–1

[This is a variation of E 8–1 modified to focus on the periodic inventory system.]

John's Specialty Store uses a periodic inventory system. The following are some inventory transactions for the month of May:

1. John's purchased merchandise on account for $5,000. Freight charges of $300 were paid in cash.
2. John's returned some of the merchandise purchased in (1). The cost of the merchandise was $600 and John's account was credited by the supplier.
3. Merchandise costing $2,800 was sold for $5,200 in cash.

Required:
Prepare the necessary journal entries to record these transactions.

E 8–3
Determining cost of goods sold
● LO8–1

The June 30, 2021, year-end trial balance for Askew Company contained the following information:

Account	Debit	Credit
Inventory, 7/1/2020	32,000	
Sales revenue		380,000
Sales returns	12,000	
Purchases	240,000	
Purchase discounts		6,000
Purchase returns		10,000
Freight-in	17,000	

In addition, you determine that the June 30, 2021, inventory balance is $40,000.

Required:
Calculate the cost of goods sold for the Askew Company for the year ending June 30, 2021.

E 8–4
Perpetual and periodic inventory systems compared
● LO8–1

The following information is available for the Johnson Corporation:

Beginning inventory	$ 25,000
Inventory purchases (on account)	155,000
Freight charges on purchases (paid in cash)	10,000
Inventory returned to suppliers (for credit)	12,000
Ending inventory	30,000
Sales (on account)	250,000
Cost of inventory sold	148,000

Required:
Applying both a perpetual and a periodic inventory system, prepare the journal entries that summarize the transactions that created these balances. Include all end-of-period adjusting entries indicated.

E 8–5
Inventory transactions; missing data
● LO8–1

The Playa Company has the following information in its records. Certain data have been intentionally omitted ($ in thousands).

	2021	2022	2023
Beginning inventory	?	?	225
Cost of goods sold	627	621	?
Ending inventory	?	225	216
Cost of goods available for sale	876	?	800
Purchases (gross)	630	?	585
Purchase discounts	18	15	?
Purchase returns	24	30	14
Freight-in	13	32	16

Required:
Determine the missing numbers. Show computations where appropriate.

E 8–6
Goods in transit
● LO8–2

The Kwok Company's inventory balance on December 31, 2021, was $165,000 (based on a 12/31/2021 physical count) *before* considering the following transactions:

1. Goods shipped to Kwok f.o.b. destination on December 20, 2021, were received on January 4, 2022. The invoice cost was $30,000.
2. Goods shipped to Kwok f.o.b. shipping point on December 28, 2021, were received on January 5, 2022. The invoice cost was $17,000.
3. Goods shipped from Kwok to a customer f.o.b. destination on December 27, 2021, were received by the customer on January 3, 2022. The sales price was $40,000 and the merchandise cost $22,000.
4. Goods shipped from Kwok to a customer f.o.b. destination on December 26, 2021, were received by the customer on December 30, 2021. The sales price was $20,000 and the merchandise cost $13,000.
5. Goods shipped from Kwok to a customer f.o.b. shipping point on December 28, 2021, were received by the customer on January 4, 2022. The sales price was $25,000 and the merchandise cost $12,000.

Required:
Determine the correct inventory amount to be reported in Kwok's 2021 balance sheet.

E 8–7
Goods in transit; consignment
● LO8–2

The December 31, 2021, year-end inventory balance of the Raymond Corporation is $210,000. You have been asked to review the following transactions to determine if they have been correctly recorded.

1. Goods shipped to Raymond f.o.b. destination on December 26, 2021, were received on January 2, 2022. The invoice cost of $30,000 *is* included in the preliminary inventory balance.
2. At year-end, Raymond held $14,000 of merchandise on consignment from the Harrison Company. This merchandise *is* included in the preliminary inventory balance.
3. On December 29, merchandise costing $6,000 was shipped to a customer f.o.b. shipping point and arrived at the customer's location on January 3, 2022. The merchandise is *not* included in the preliminary inventory balance.
4. At year-end, Raymond had merchandise costing $15,000 on consignment with the Joclyn Corporation. The merchandise is *not* included in the preliminary inventory balance.

Required:
Determine the correct inventory amount to be reported in Raymond's 2021 balance sheet.

E 8–8
Physical quantities and costs included in inventory
● LO8–2

The Phoenix Corporation's fiscal year ends on December 31. Phoenix determines inventory quantity by a physical count of inventory on hand at the close of business on December 31. The company's controller has asked for your help in deciding if the following items should be included in the year-end inventory count.

1. Merchandise held on consignment for Trout Creek Clothing.
2. Goods shipped f.o.b. destination on December 28 that arrived at the customer's location on January 4.
3. Goods purchased from a vendor shipped f.o.b. shipping point on December 26 that arrived on January 3.
4. Goods shipped f.o.b. shipping point on December 28 that arrived at the customer's location on January 5.
5. Phoenix had merchandise on consignment at Lisa's Markets, Inc.
6. Goods purchased from a vendor shipped f.o.b. destination on December 27 that arrived on January 3.
7. Freight charges on goods purchased in 3.

Required:
Determine if each of the items above should be included or excluded from the company's year-end inventory.

E 8–9
Purchase discounts; the gross method
● LO8–3

On July 15, 2021, the Nixon Car Company purchased 1,000 tires from the Harwell Company for $50 each. The terms of the sale were 2/10, n/30. Nixon uses a perpetual inventory system and the *gross* method of accounting for purchase discounts.

Required:
1. Prepare the journal entries to record the (a) purchase on July 15 and (b) payment on July 23, 2021.
2. Prepare the journal entry for the payment, assuming instead that it was made on August 15, 2021.
3. If Nixon instead uses a periodic inventory system, explain any changes to the journal entries created in requirements 1 and 2.

E 8–10
Purchase discounts; the net method
● LO8–3

[This is a variation of E 8-9 modified to focus on the net method of accounting for purchase discounts.]

On July 15, 2021, the Nixon Car Company purchased 1,000 tires from the Harwell Company for $50 each. The terms of the sale were 2/10, n/30. Nixon uses a perpetual inventory system and the *net* method of accounting for purchase discounts.

Required:
1. Prepare the journal entries to record the (a) purchase on July 15 and (b) payment on July 23, 2021.
2. Prepare the journal entry for the payment, assuming instead that it was made on August 15, 2021.
3. If Nixon instead uses a periodic inventory system, explain any changes to the journal entries created in requirements 1 and 2.

E 8–11
Trade and purchase discounts; the gross method and the net method compared
● LO8–3

Tracy Company, a manufacturer of air conditioners, sold 100 units to Thomas Company on November 17, 2021. The units have a list price of $500 each, but Thomas was given a 30% trade discount. The terms of the sale were 2/10, n/30. Thomas uses a perpetual inventory system.

Required:
1. Prepare the journal entries to record the (a) purchase by Thomas on November 17 and (b) payment on November 26, 2021. Thomas uses the gross method of accounting for purchase discounts.
2. Prepare the journal entry for the payment, assuming instead that it was made on December 15, 2021.
3. Repeat requirements 1 and 2 using the net method of accounting for purchase discounts.

E 8–12
FASB codification research
● LO8–2, LO8–3

Access the *FASB Accounting Standards Codification* at the FASB website (**www.fasb.org**). Determine each of the following:

1. The specific eight-digit Codification citation (XXX-XX-XX-X) that describes the meaning of cost as it applies to the initial measurement of inventory.
2. The specific nine-digit Codification citation (XXX-XXX-XX-X) that describes the circumstances when it is appropriate to initially measure agricultural inventory at fair value.
3. The specific eight-digit Codification citation (XXX-XX-XX-X) that describes the major objective of accounting for inventory.
4. The specific eight-digit Codification citation (XXX-XX-XX-X) that describes the abnormal freight charges included in the cost of inventory.

E 8–13
Inventory cost flow methods; periodic system
● LO8–1, LO8–4

Altira Corporation provides the following information related to its merchandise inventory during the month of August 2021:

Aug. 1	Inventory on hand—2,000 units; cost $5.30 each.
8	Purchased 8,000 units for $5.50 each.
14	Sold 6,000 units for $12.00 each.
18	Purchased 6,000 units for $5.60 each.
25	Sold 7,000 units for $11.00 each.
28	Purchased 4,000 units for $5.80 each.
31	Inventory on hand—7,000 units.

Required:
Using calculations based on a periodic inventory system, determine the inventory balance Altira would report in its August 31, 2021, balance sheet and the cost of goods sold it would report in its August 2021 income statement using each of the following cost flow methods:
1. First-in, first-out (FIFO)
2. Last-in, first-out (LIFO)
3. Average cost

E 8–14
Inventory cost flow methods; perpetual system
● LO8–1, LO8–4

[This is a variation of E 8–13 modified to focus on the perpetual inventory system and alternative cost flow methods.]

Altira Corporation provides the following information related to its merchandise inventory during the month of August 2021:

Aug. 1	Inventory on hand—2,000 units; cost $5.30 each.
8	Purchased 8,000 units for $5.50 each.
14	Sold 6,000 units for $12.00 each.
18	Purchased 6,000 units for $5.60 each.
25	Sold 7,000 units for $11.00 each.
28	Purchased 4,000 units for $5.80 each.
31	Inventory on hand—7,000 units.

Required:
Using calculations based on a perpetual inventory system, determine the inventory balance Altira would report in its August 31, 2021, balance sheet and the cost of goods sold it would report in its August 2021 income statement using each of the following cost flow methods:
1. First-in, first-out (FIFO)
2. Average cost

E 8–15
LIFO; perpetual system
● LO8–1, LO8–4

[This is a variation of E 8–14 modified to focus on LIFO in a perpetual inventory system.]

Altira Corporation provides the following information related to its merchandise inventory during the month of August 2021:

Aug. 1	Inventory on hand—2,000 units; cost $5.30 each.
8	Purchased 8,000 units for $5.50 each.
14	Sold 6,000 units for $12.00 each.
18	Purchased 6,000 units for $5.60 each.
25	Sold 7,000 units for $11.00 each.
28	Purchased 4,000 units for $5.80 each.
31	Inventory on hand—7,000 units.

Required:
Using calculations based on a perpetual inventory system, determine the inventory balance Altira would report in its August 31, 2021, balance sheet and the cost of goods sold it would report in its August 2021 income statement using last-in, first-out (LIFO).

E 8–16
Comparison of FIFO and LIFO; periodic system
● LO8–1, LO8–4

Alta Ski Company's inventory records contained the following information regarding its latest ski model. The company uses a periodic inventory system.

Beginning inventory, January 1, 2021	600 units @ $80 each
Purchases:	
January 15	1,000 units @ $95 each
January 21	800 units @ $100 each
Sales:	
January 5	400 units @ $120 each
January 22	800 units @ $130 each
January 29	400 units @ $135 each
Ending inventory, January 31, 2021	800 units

Required:

1. Which method, FIFO or LIFO, will result in the highest cost of goods sold figure for January 2021? Why? Which method will result in the highest ending inventory balance? Why?
2. Compute cost of goods sold for January and the ending inventory using both the FIFO and LIFO methods.
3. Now assume that inventory costs were *declining* during January. The inventory purchased on January 15 had a unit cost of $70, and the inventory purchased on January 21 had a unit cost of $65. All other information is the same. Repeat requirements 1 and 2.

E 8–17
Average cost method; periodic and perpetual systems
● LO8–1, LO8–4

The following information is taken from the inventory records of the CNB Company for the month of September:

Beginning inventory, 9/1/2021	5,000 units @ $10.00
Purchases:	
9/7	3,000 units @ $10.40
9/25	8,000 units @ $10.75
Sales:	
9/10	4,000 units
9/29	5,000 units
7,000 units were on hand at the end of September.	

Required:

1. Assuming that CNB uses a periodic inventory system and employs the average cost method, determine cost of goods sold for September and September's ending inventory.
2. Repeat requirement 1 assuming that the company uses a perpetual inventory system.

E 8–18
FIFO, LIFO, and average cost methods
● LO8–1, LO8–4

Causwell Company began 2021 with 10,000 units of inventory on hand. The cost of each unit was $5.00. During 2021, an additional 30,000 units were purchased at a single unit cost, and 20,000 units remained on hand at the end of 2021 (20,000 units therefore were sold during 2021). Causwell uses a periodic inventory system. Cost of goods sold for 2021, applying the average cost method, is $115,000. The company is interested in determining what cost of goods sold would have been if the FIFO or LIFO methods were used.

Required:

1. Determine the cost of goods sold for 2021 using the FIFO method. [*Hint:* Determine the cost per unit of 2021 purchases.]
2. Determine the cost of goods sold for 2021 using the LIFO method.

E 8–19
Perpetual FIFO adjusted to periodic LIFO; LIFO reserve
● LO8–1, LO8–4, LO8–6

To more efficiently manage its inventory, Treynor Corporation maintains its internal inventory records using first-in, first-out (FIFO) under a perpetual inventory system. The following information relates to its merchandise inventory during the year:

Jan. 1	Inventory on hand—20,000 units; cost $12.20 each.
Feb. 12	Purchased 70,000 units for $12.50 each.
Apr. 30	Sold 50,000 units for $20.00 each.
Jul. 22	Purchased 50,000 units for $12.80 each.
Sep. 9	Sold 70,000 units for $20.00 each.
Nov. 17	Purchased 40,000 units for $13.20 each.
Dec. 31	Inventory on hand—60,000 units.

Required:

1. Determine the amount Treynor would calculate internally for ending inventory and cost of goods sold using first-in, first-out (FIFO) under a perpetual inventory system.
2. Determine the amount Treynor would report externally for ending inventory and cost of goods sold using last-in, first-out (LIFO) under a periodic inventory system.
3. Determine the amount Treynor would report for its LIFO reserve at the end of the year.
4. Record the year-end adjusting entry for the LIFO reserve, assuming the balance at the beginning of the year was $10,000.

E 8–20
Perpetual average cost adjusted to periodic LIFO; LIFO reserve
● LO8–1, LO8–4, LO8–6

Tipton Processing maintains its internal inventory records using average cost under a perpetual inventory system. The following information relates to its inventory during the year:

Jan. 1	Inventory on hand—80,000 units; cost $4.25 each.
Feb. 14	Purchased 120,000 units for $4.50 each.
Mar. 5	Sold 150,000 units for $14.00 each.
Aug. 27	Purchased 50,000 units for $4.80 each.
Sep. 12	Sold 60,000 units for $14.00 each.
Dec. 31	Inventory on hand—40,000 units.

Required:

1. Determine the amount Tipton would calculate internally for ending inventory and cost of goods sold using average cost under a perpetual inventory system.
2. Determine the amount Tipton would report externally for ending inventory and cost of goods sold using last-in, first-out (LIFO) under a periodic inventory system.
3. Determine the amount Tipton would report for its LIFO reserve at the end of the year.
4. Record the year-end adjusting entry for the LIFO reserve, assuming the balance at the beginning of the year was $8,000.

E 8–21
Supplemental LIFO disclosures; LIFO reserve; AmerisourceBergen
● LO8–6
Real World Financials

AmerisourceBergen is an American drug wholesale company. The company uses the LIFO inventory method for external reporting but maintains its internal records using FIFO. The following disclosure note was included in a recent quarterly report:

5. Inventories (in part)
Inventories are comprised of the following ($ in millions):

	September 30, 2017	September 30, 2016
Inventories (under FIFO)	$12,928	$12,349
Less: LIFO reserve	(1,467)	(1,625)
Inventories (under LIFO)	$11,461	$10,724

The company's income statements reported cost of goods sold of $148,598 million for the quarter ended September 30, 2017.

Required:

1. Assume that AmerisourceBergen adjusts the LIFO reserve at the end of its quarter. Prepare the September 30, 2017, adjusting entry to record the cost of goods sold adjustment.
2. If AmerisourceBergen had used FIFO to report its inventories, what would cost of goods sold have been for the quarter ended September 30, 2017?

E 8–22
LIFO liquidation
● LO8–1, LO8–4, LO8–6

The Reuschel Company began 2021 with inventory of 10,000 units at a cost of $7 per unit. During 2021, 50,000 units were purchased for $8.50 each. Sales for the year totaled 54,000 units leaving 6,000 units on hand at the end of 2021. Reuschel uses a periodic inventory system and the LIFO inventory cost method.

Required:

1. Calculate cost of goods sold for 2021.
2. From a financial reporting perspective, what problem is created by the use of LIFO in this situation? Describe the disclosure required to report the effects of this problem.

E 8–23
LIFO liquidation
● LO8–4, LO8–6

The Churchill Corporation uses a periodic inventory system and the LIFO inventory cost method for its one product. Beginning inventory of 20,000 units consisted of the following, listed in chronological order of acquisition:

12,000 units at a cost of $8.00 per unit = $96,000
8,000 units at a cost of $9.00 per unit = 72,000

During 2021, inventory quantity declined by 10,000 units. All units purchased during 2021 cost $12.00 per unit.

Required:

Calculate the before-tax LIFO liquidation profit or loss that the company would report in a disclosure note, assuming the amount determined is material.

E 8–24
FASB codification research
● LO8–6

Access the *FASB Accounting Standards Codification* at the FASB website (**www.fasb.org**). Determine each of the following:

1. The specific nine-digit Codification citation (XXX-XX-XXX-X) that describes the disclosure requirements that must be made by publicly traded companies for a LIFO liquidation. (Hint: Look in the SEC Materials section.)
2. Describe the disclosure requirements.

E 8–25
Ratio analysis; Home Depot and Lowe's
● LO8–7
Real World Financials

The table below contains selected information from recent financial statements of **The Home Depot, Inc.**, and **Lowe's Companies, Inc.**, two companies in the home improvement retail industry ($ in millions):

	Home Depot		Lowe's	
	1/28/2018	1/29/2017	2/2/2018	2/3/2017
Net sales	$100,904	$94,595	$68,619	$65,017
Cost of goods sold	66,548	62,282	45,210	42,553
Year-end inventory	12,748	12,549	11,393	10,458
Industry averages:				
Gross profit ratio	27.25%			
Inventory turnover ratio	3.63 times			
Average days in inventory	101 days			

Required:
Calculate the gross profit ratio, the inventory turnover ratio, and the average days in inventory for the two companies for their fiscal years ending in 2018. Compare your calculations for the two companies, taking into account the industry averages.

E 8–26
Dollar-value LIFO
● LO8–8

On January 1, 2021, the Haskins Company adopted the dollar-value LIFO method for its one inventory pool. The pool's value on this date was $660,000. The 2021 and 2022 ending inventory valued at year-end costs were $690,000 and $760,000, respectively. The appropriate cost indexes are 1.04 for 2021 and 1.08 for 2022.

Required:
Calculate the inventory value at the end of 2021 and 2022 using the dollar-value LIFO method.

E 8–27
Dollar-value LIFO
● LO8–8

Mercury Company has only one inventory pool. On December 31, 2021, Mercury adopted the dollar-value LIFO inventory method. The inventory on that date using the dollar-value LIFO method was $200,000. Inventory data are as follows:

Year	Ending Inventory at Year-End Costs	Ending Inventory at Base Year Costs
2022	$231,000	$220,000
2023	299,000	260,000
2024	300,000	250,000

Required:
Compute the inventory at December 31, 2022, 2023, and 2024, using the dollar-value LIFO method.

(AICPA adapted)

E 8–28
Dollar-value LIFO
● LO8–8

Carswell Electronics adopted the dollar-value LIFO method on January 1, 2021, when the inventory value of its one inventory pool was $720,000. The company decided to use an external index, the Consumer Price Index (CPI), to adjust for changes in the cost level. On January 1, 2021, the CPI was 240. On December 31, 2021, inventory valued at year-end cost was $880,000 and the CPI was 264.

Required:
Calculate the inventory value at the end of 2021 using the dollar-value LIFO method.

E 8–29
Concepts; terminology
● LO8–1 through LO8–5

Listed below are several terms and phrases associated with inventory measurement. Pair each item from List A with the item from List B (by letter) that is most appropriately associated with it.

List A

_____ 1. Perpetual inventory system
_____ 2. Periodic inventory system
_____ 3. F.o.b. shipping point
_____ 4. Gross method
_____ 5. Net method
_____ 6. Cost index
_____ 7. F.o.b. destination
_____ 8. FIFO
_____ 9. LIFO
_____ 10. Consignment
_____ 11. Average cost
_____ 12. IRS conformity rule

List B

a. Legal title passes when goods are delivered to common carrier.
b. Goods are transferred to another company but title remains with transferor.
c. Purchases are recorded for the full cost of inventory.
d. If LIFO is used for taxes, it must be used for financial reporting.
e. Assumes items sold are those acquired first.
f. Assumes items sold are those acquired last.
g. Purchase discounts are recorded for the full cost of inventory less any possible discount.
h. Used to convert ending inventory at year-end cost to base year cost.
i. Continuously records changes in inventory.
j. Assumes items sold come from a mixture of goods acquired during the period.
k. Legal title passes when goods arrive at location.
l. Adjusts inventory at the end of the period.

E 8–30
General ledger exercise; inventory transactions
● LO8–1 through LO8–8

On January 1, 2021, Displays Incorporated had the following account balances:

Account Title	Debits	Credits
Cash	$ 22,000	
Accounts receivable	19,000	
Supplies	25,000	
Inventory	60,000	
Land	227,000	
Accounts payable		$ 18,000
Notes payable (5%, due next year)		20,000
Common stock		186,000
Retained earnings		129,000
Totals	$353,000	$353,000

From January 1 to December 31, the following summary transactions occurred:

a. Purchased inventory on account for $330,000.

b. Sold inventory on account for $570,000. The cost of the inventory sold was $310,000.

c. Received $540,000 from customers on accounts receivable.

d. Paid freight on inventory received, $24,000.

e. Paid $320,000 to inventory suppliers on accounts payable of $325,000. The difference reflects purchase discounts of $5,000.

f. Paid rent for the current year, $42,000. The payment was recorded to Rent Expense.

g. Paid salaries for the current year, $150,000. The payment was recorded to Salaries Expense.

Required:

1. Record each of the transactions listed above in the 'General Journal' tab (these are shown as items 1-8) assuming a perpetual inventory system. Review the 'General Ledger' and the 'Trial Balance' tabs to see the effect of the transactions on the account balances.
2. Record adjusting entries on December 31 in the 'General Journal' tab (these are shown as items 9-11).
   a. Supplies on hand at the end of the year are $8,000.
   b. Accrued interest expense on notes payable for the year.
   c. Accrued income taxes at the end of December are $18,000.
3. Review the adjusted 'Trial Balance' as of December 31, 2018, in the 'Trial Balance' tab.
4. Prepare a multiple-step income statement for the period ended December 31, 2021, in the 'Income Statement' tab.
5. Prepare a classified balance sheet as of December 31, 2021, in the 'Balance Sheet' tab.
6. Record the closing entries in the 'General Journal' tab (these are shown as items 12-13).
7. Using the information from the requirements above, complete the 'Analysis' tab.
   a. Suppose Displays Incorporated decided to maintain its internal records using FIFO but to use LIFO for external reporting. Assuming the ending balance of inventory under LIFO would have been $85,000, calculate the LIFO reserve.
   b. Assume Displays Incorporated $60,000 beginning balance of inventory comes from the base year with a cost index of 1.00. The cost index at the end of 2021 of 1.10. Calculate the amount the company would report for inventory using dollar-value LIFO.
   c. Indicate whether each of the following amounts below would be *higher* or *lower* when reporting inventory using LIFO (or dollar-value LIFO) instead of FIFO in periods of rising inventory costs and stable inventory quantities: Inventory turnover ratio, average days in inventory, gross profit ratio.

E 8–31
General ledger exercise; receivable and inventory transactions
● LO8–1, LO8–3, LO8–5

On January 1, 2021, the general ledger of Tripley Company included the following account balances:

Account Title	Debits	Credits
Cash	$ 70,000	
Accounts receivable	40,000	
Allowance for uncollectible accounts		$ 5,000
Inventory	30,000	
Building	70,000	

Account Title	Debits	Credits
Accumulated depreciation		10,000
Land	200,000	
Accounts payable		20,000
Notes payable (8%, due in 3 years)		36,000
Common stock		100,000
Retained earnings		239,000
Totals	$410,000	$410,000

The $30,000 beginning balance of inventory consists of 300 units, each costing $100. During January 2021, the company had the following transactions:

Jan. 2	Lent $20,000 to an employee by accepting a 6% note due in six months.
5	Purchased 3,500 units of inventory on account for $385,000 ($110 each) with terms 1/10, n/30.
8	Returned 100 defective units of inventory purchased on January 5.
15	Sold 3,300 units of inventory on account for $429,000 ($130 each) with terms 2/10, n/30.
17	Customers returned 200 units sold on January 15. These units were initially purchased by the company on January 5. The units are placed in inventory to be sold in the future.
20	Received cash from customers on accounts receivable. This amount includes $36,000 from 2020 plus amount receivable on sale of 2,700 units sold on January 15.
21	Wrote off remaining accounts receivable from 2020.
24	Paid on accounts payable. The amount includes the amount owed at the beginning of the period plus the amount owed from purchase of 3,100 units on January 5.
28	Paid cash for salaries during January, $28,000.
29	Paid cash for utilities during January, $10,000.
30	Paid dividends, $3,000.

Required:

1. Record each of the transactions listed above in the 'General Journal' tab (these are shown as items 1-10) assuming a perpetual FIFO inventory system. Purchases and sales of inventory are recorded using the gross method for cash discounts. Review the 'General Ledger' and the 'Trial Balance' tabs to see the effect of the transactions on the account balances.
2. Record adjusting entries on January 31 in the 'General Journal' tab (these are shown as items 11-14):
   a. Of the remaining accounts receivable, the company estimates that 10% will not be collected.
   b. Accrued interest revenue on notes receivable for January.
   c. Accrued interest expense on notes payable for January.
   d. Accrued income taxes at the end of January for $5,000.
   e. Depreciation on the building, $2,000
3. Review the adjusted 'Trial Balance' as of January 31, 2021, in the 'Trial Balance' tab.
4. Prepare a multiple-step income statement for the period ended January 31, 2021, in the 'Income Statement' tab.
5. Prepare a classified balance sheet as of January 31, 2021, in the 'Balance Sheet' tab.
6. Record closing entries in the 'General Journal' tab (these are shown as items 15-16).
7. Using the information from the requirements above, complete the 'Analysis' tab.
   a. Calculate the inventory turnover ratio for the month of January. If the industry average of the inventory turnover ratio for the month of January is 4.5 times, is the company selling its inventory *more* or *less* quickly than other companies in the same industry?
   b. Calculate the gross profit ratio for the month of January. If the industry average gross profit ratio is 33%, is the company *more* or *less* profitable per dollar of sales than other companies in the same industry?
   c. Used together, what might the inventory turnover ratio and gross profit ratio suggest about the company's business strategy? Is the company's strategy to sell a *higher volume of less expensive* items or does the company appear to be selling a *lower volume of more expensive* items?

Problems

connect

P 8–1
Various inventory transactions; journal entries
● LO8–1 through LO8–3

James Company began the month of October with inventory of $15,000. The following inventory transactions occurred during the month:

a. The company purchased merchandise on account for $22,000 on October 12. Terms of the purchase were 2/10, n/30. James uses the net method to record purchases. The merchandise was shipped f.o.b. shipping point and freight charges of $500 were paid in cash.
b. On October 31, James paid for the merchandise purchased on October 12.
c. During October merchandise costing $18,000 was sold on account for $28,000.
d. It was determined that inventory on hand at the end of October cost $19,060.

Required:

1. Assuming that the James Company uses a perpetual inventory system, prepare journal entries for the above transactions.
2. Assuming that the James Company uses a periodic inventory system, prepare journal entries for the above transactions including the adjusting entry at the end of October to record cost of goods sold. James considers purchase discounts lost as part of interest expense.

P 8–2
Items to be included in inventory
● LO8–2

The following inventory transactions took place near December 31, 2021, the end of the Rasul Company's fiscal year-end:

1. On December 27, 2021, merchandise costing $2,000 was shipped to the Myers Company on consignment. The shipment arrived at Myers's location on December 29, but none of the merchandise was sold by the end of the year. The merchandise was *not* included in the 2021 ending inventory.
2. On January 5, 2022, merchandise costing $8,000 was received from a supplier and recorded as a purchase on that date and *not* included in the 2021 ending inventory. The invoice revealed that the shipment was made f.o.b. shipping point on December 28, 2021.
3. On December 29, 2021, the company shipped merchandise costing $12,000 to a customer f.o.b. destination. The goods, which arrived at the customer's location on January 4, 2022, were *not* included in Rasul's 2021 ending inventory. The sale was recorded in 2021.
4. Merchandise costing $4,000 was received on December 28, 2021, on consignment from the Aborn Company. A purchase was *not* recorded and the merchandise was *not* included in 2021 ending inventory.
5. Merchandise costing $6,000 was received and recorded as a purchase on January 8, 2022. The invoice revealed that the merchandise was shipped from the supplier on December 28, 2021, f.o.b. destination. The merchandise was *not* included in 2021 ending inventory.

Required:

State whether Rasul correctly accounted for each of the above transactions. Give the reason for your answer.

P 8–3
Costs included in inventory
● LO8–2, LO8–3

Reagan Corporation is a wholesale distributor of truck replacement parts. Initial amounts taken from Reagan's records are as follows:

Inventory at December 31 (based on a physical count of goods in Reagan's warehouse on December 31)		$1,250,000
Accounts payable at December 31:		
Vendor	**Terms**	**Amount**
Baker Company	2%, 10 days, net 30	$ 265,000
Charlie Company	Net 30	210,000
Dolly Company	Net 30	300,000
Eagler Company	Net 30	225,000
Full Company	Net 30	—
Greg Company	Net 30	—
Accounts payable, December 31		$1,000,000
Sales for the year		$9,000,000

Additional Information:

1. Parts held by Reagan on consignment from Charlie, amounting to $155,000, were included in the physical count of goods in Reagan's warehouse and in accounts payable at December 31.
2. Parts totaling $22,000, which were purchased from Full and paid for in December, were sold in the last week of the year and *appropriately* recorded as sales of $28,000. The parts were included in the physical count of goods in Reagan's warehouse on December 31 because the parts were on the loading dock waiting to be picked up by customers.

3. Parts in transit on December 31 to customers, shipped f.o.b. shipping point on December 28, amounted to $34,000. The customers received the parts on January 5 of the following year. Sales of $40,000 to the customers for the parts were recorded by Reagan on January 2.
4. Retailers were holding goods on consignment from Reagan, which had a cost of $210,000 and a retail value of $250,000.
5. Goods were in transit from Greg to Reagan on December 31. The cost of the goods was $25,000, and they were shipped f.o.b. shipping point on December 29.
6. A freight bill in the amount of $2,000 specifically relating to merchandise purchased in December, all of which was still in the inventory at December 31, was received on January 3. The freight bill was not included in either the inventory or in accounts payable at December 31.
7. All the purchases from Baker occurred during the last seven days of the year. These items have been recorded in accounts payable and accounted for in the physical inventory at cost before discount. Reagan's policy is to pay invoices in time to take advantage of all discounts, adjust inventory accordingly, and record accounts payable net of discounts.

Required:

Prepare a schedule of adjustments to the initial amounts using the format shown below. Show the effect, if any, of each of the transactions separately and if the transactions would have no effect on the amount shown, state *none*.

	Inventory	Accounts Payable	Sales
Initial amounts	$1,250,000	$ 1,000,000	$9,000,000
Adjustments—increase (decrease):			
1.			
2.			
3.			
4.			
5.			
6.			
7.			
Total adjustments			
Adjusted amounts	$	$	$

(*AICPA adapted*)

P 8–4
Various inventory transactions; determining inventory and cost of goods; LIFO reserve
● LO8–1 through LO8–4, LO8–6

Johnson Corporation began the year with inventory of 10,000 units of its only product. The units cost $8 each. The company uses a perpetual inventory system and the FIFO cost method. The following transactions occurred during the year:

a. Purchased 50,000 additional units at a cost of $10 per unit. Terms of the purchases were 2/10, n/30, and 100% of the purchases were paid for within the 10-day discount period. The company uses the gross method to record purchase discounts. The merchandise was purchased f.o.b. shipping point and freight charges of $0.50 per unit were paid by Johnson.
b. 1,000 units purchased during the year were returned to suppliers for credit. Johnson was also given credit for the freight charges of $0.50 per unit it had paid on the original purchase. The units were defective and were returned two days after they were received.
c. Sales for the year totaled 45,000 units at $18 per unit.
d. On December 28, Johnson purchased 5,000 additional units at $10 each. The goods were shipped f.o.b. destination and arrived at Johnson's warehouse on January 4 of the following year.
e. 14,000 units were on hand at the end of the year.

Required:

1. Determine ending inventory and cost of goods sold at the end of the year.
2. Assuming that operating expenses other than those indicated in the above transactions amounted to $150,000, determine income before income taxes for the year.
3. For financial reporting purposes, the company uses LIFO (amounts based on a periodic inventory system). Record the year-end adjusting entry for the LIFO reserve, assuming the balance in the LIFO reserve at the beginning of the year is $15,000.
4. Determine the amount the company would report as income before taxes for the year under LIFO. Operating expenses other than those indicated in the above transactions amounted to $150,000.

P 8–5
Various inventory costing methods
● LO8–1, LO8–4

Ferris Company began January with 6,000 units of its principal product. The cost of each unit is $8. Merchandise transactions for the month of January are as follows:

		Purchases	
Date of Purchase	**Units**	**Unit Cost***	**Total Cost**
Jan. 10	5,000	$ 9	$ 45,000
Jan. 18	6,000	10	60,000
Total purchases	11,000		$ 105,000

*Includes purchase price and cost of freight.

Sales	
Date of Sale	**Units**
Jan. 5	3,000
Jan. 12	2,000
Jan. 20	4,000
Total sales	9,000

8,000 units were on hand at the end of the month.

Required:

Calculate January's ending inventory and cost of goods sold for the month using each of the following alternatives:

1. FIFO, periodic system
2. LIFO, periodic system
3. FIFO, perpetual system
4. Average cost, periodic system
5. Average cost, perpetual system

P 8–6
Various inventory costing methods; gross profit ratio
● LO8–1, LO8–4, LO8–7

Topanga Group began operations early in 2021. Inventory purchase information for the quarter ended March 31, 2021, for Topanga's only product is provided below. The unit costs include the cost of freight. The company uses a periodic inventory system to report inventory and cost of goods sold.

Date of Purchase	Units	Unit Cost	Total Cost
Jan. 7	5,000	$4.00	$ 20,000
Feb. 16	12,000	4.50	54,000
March 22	17,000	5.00	85,000
Total purchases	34,000		$159,000

Sales for the quarter, all at $7.00 per unit, totaled 20,000 units leaving 14,000 units on hand at the end of the quarter.

Required:

1. Calculate Topanga's gross profit ratio for the first quarter using:
   a. FIFO
   b. LIFO
   c. Average cost
2. Comment on the relative effect of each of the three inventory methods on the gross profit ratio.

P 8–7
Various inventory costing methods
● LO8–1, LO8–4

Carlson Auto Dealers Inc. sells a handmade automobile as its only product. Each automobile is identical; however, they can be distinguished by their unique ID number. At the beginning of 2021, Carlson had three cars in inventory, as follows:

Car ID	Cost
203	$60,000
207	60,000
210	63,000

During 2021, each of the three autos sold for $90,000. Additional purchases (listed in chronological order) and sales for the year were as follows:

Car ID	Cost	Selling Price
211	$63,000	$ 90,000
212	63,000	93,000
213	64,500	not sold
214	66,000	96,000
215	69,000	100,500
216	70,500	not sold
217	72,000	105,000
218	72,300	106,500
219	75,000	not sold

Required:

1. Calculate 2021 ending inventory and cost of goods sold assuming the company uses the specific identification inventory method.
2. Calculate ending inventory and cost of goods sold assuming FIFO and a periodic inventory system.
3. Calculate ending inventory and cost of goods sold assuming LIFO and a periodic inventory system.
4. Calculate ending inventory and cost of goods sold assuming the average cost method and a periodic inventory system.

P 8–8
Supplemental LIFO disclosures; Caterpillar
● LO8–4, LO8–6
Real World Financials

Caterpillar, Inc., is one of the world's largest manufacturers of construction, mining, and forestry machinery. The following disclosure note is included in the company's 2017 financial statements:

> **D. Inventories ($ in millions)**
> Inventories are stated at the lower of cost or net realizable value. Cost is principally determined using the last-in, first-out (LIFO) method. The value of inventories on the LIFO basis represented about 65 percent and 60 percent of total inventories at December 31, 2017 and 2016. If the FIFO (first-in, first-out) method had been in use, inventories would have been $1,934 million and $2,139 million higher than reported at December 31, 2017 and 2016, respectively.

Required:

1. The company reported LIFO cost of goods sold of $31,049 million. Calculate the amount that would be reported for cost of goods sold had Caterpillar used the FIFO inventory method for all of its inventory during 2017.
2. How does the amount in requirement 1 affect income before taxes?
3. Why might the information contained in the disclosure note be useful to a financial analyst?

P 8–9
LIFO liquidation
● LO8–4, LO8–6

Taylor Corporation reports inventory and cost of goods sold based on calculations from a LIFO periodic inventory system. The company's records under this system reveal the following inventory layers at the beginning of 2021 (listed in chronological order of acquisition):

10,000 units @ $15	$150,000
15,000 units @ $20	300,000
Beginning inventory	$450,000

During 2021, 30,000 units were purchased for $25 per unit. Due to unexpected demand for the company's product, 2021 sales totaled 40,000 units at various prices, leaving 15,000 units in ending inventory.

Required:

1. Calculate the amount to report for cost of goods sold for 2021.
2. Determine the amount of LIFO liquidation profit that the company must report in a disclosure note to its 2021 financial statements. Assume an income tax rate of 25%.
3. If the company decided to purchase an additional 10,000 units at $25 per unit at the end of the year, how much income tax currently payable would be saved?

P 8–10
LIFO liquidation
● LO8–4, LO8–6

Cansela Corporation reports inventory and cost of goods sold based on calculations from a LIFO periodic inventory system. The company began 2021 with inventory of 4,500 units of its only product. The beginning inventory balance of $64,000 consisted of the following layers:

2,000 units at $12 per unit	=	$24,000
2,500 units at $16 per unit	=	40,000
Beginning inventory		$64,000

During the three years 2021–2023 the cost of inventory remained constant at $18 per unit. Unit purchases and sales during these years were as follows:

	Purchases	Sales
2021	10,000	11,000
2022	13,000	14,500
2023	12,000	13,000

Required:
1. Calculate cost of goods sold for 2021, 2022, and 2023.
2. Disregarding income tax, determine the LIFO liquidation profit or loss, if any, for each of the three years.
3. Prepare the company's LIFO liquidation disclosure note that would be included in the 2023 financial statements to report the effects of any liquidation on cost of goods sold and net income. Assume any liquidation effects are material and that Cansela's effective income tax rate is 25%. Cansela's 2023 financial statements include income statements for two prior years for comparative purposes.

P 8–11
Inventory cost flow methods: LIFO liquidation; ratios
● LO8–4, LO8–6, LO8–7

Cast Iron Grills, Inc., manufactures premium gas barbecue grills. The company reports inventory and cost of goods sold based on calculations from a LIFO periodic inventory system. Cast Iron's December 31, 2021, fiscal year-end inventory consisted of the following (listed in chronological order of acquisition):

Units	Unit Cost
5,000	$700
4,000	800
6,000	900

The replacement cost of the grills throughout 2022 was $1,000. Cast Iron sold 27,000 grills during 2022. The company's selling price is set at 200% of the current replacement cost.

Required:
1. Compute the gross profit (sales minus cost of goods sold) and the gross profit ratio for 2022 assuming that Cast Iron purchased 28,000 units during the year.
2. Repeat requirement 1 assuming that Cast Iron purchased only 15,000 units.
3. Why does the number of units purchased affect your answers to the above requirements?
4. Repeat requirements 1 and 2 assuming that Cast Iron reports the FIFO inventory cost method rather than the LIFO method.
5. Why does the number of units purchased have no effect on your answers to requirements 1 and 2 when the FIFO method is used for reporting purposes?

P 8–12
Integrating problem; inventories and accounts receivable; Chapters 7 and 8
● LO8–4, LO8–6, LO8–7

Inverness Steel Corporation is a producer of flat-rolled carbon, stainless and electrical steels, and tubular products. The company's income statement for the 2021 fiscal year reported the following information ($ in millions):

Sales	$6,255
Cost of goods sold	5,190

The company's balance sheets for 2021 and 2020 included the following information ($ in millions):

	2021	2020
Current assets:		
Accounts receivable, net	$703	$583
Inventories	880	808

The statement of cash flows reported bad debt expense for 2021 of $8 million. The summary of significant accounting policies included the following notes ($ in millions):

> **Accounts Receivable (in part)**
>
> The allowance for uncollectible accounts was $10 and $7 at December 31, 2021 and 2020, respectively. All sales are on credit.

> **Inventories**
>
> Inventories are valued at the lower of cost or market. The cost of the majority of inventories is measured using the last-in, first-out (LIFO) method. Other inventories are measured principally at average cost and consist mostly of foreign inventories and certain raw materials. If the entire inventory had been valued on an average cost basis, inventory would have been higher by $480 and $350 at the end of 2021 and 2020, respectively.
>
> During 2021, 2020, and 2019, liquidation of LIFO layers generated income of $6, $7, and $25, respectively.

Required:

Using the information provided:

1. Determine the amount of accounts receivable Inverness wrote off during 2021.
2. Calculate the amount of cash collected from customers during 2021.
3. Calculate what cost of goods sold would have been for 2021 if the company had used average cost to value its entire inventory.
4. Calculate the following ratios for 2021:
   a. Receivables turnover ratio
   b. Inventory turnover ratio
   c. Gross profit ratio
5. Explain briefly what caused the income generated by the liquidation of LIFO layers. Assuming an income tax rate of 25%, what was the effect of the liquidation of LIFO layers on cost of goods sold in 2021?

P 8–13
Dollar-value LIFO
● LO8–8

On January 1, 2021, the Taylor Company adopted the dollar-value LIFO method. The inventory value for its one inventory pool on this date was $400,000. Inventory data for 2021 through 2023 are as follows:

Date	Ending Inventory at Year-End Costs	Cost Index
12/31/2021	$441,000	1.05
12/31/2022	487,200	1.12
12/31/2023	510,000	1.20

Required:

Calculate Taylor's ending inventory for 2021, 2022, and 2023.

P 8–14
Dollar-value LIFO
● LO8–8

Kingston Company uses the dollar-value LIFO method of computing inventory. An external price index is used to convert ending inventory to base year. The company began operations on January 1, 2021, with an inventory of $150,000. Year-end inventories at year-end costs and cost indexes for its one inventory pool were as follows:

Year Ended December 31	Ending Inventory at Year-End Costs	Cost Index (Relative to Base Year)
2021	$200,000	1.08
2022	245,700	1.17
2023	235,980	1.14
2024	228,800	1.10

Required:

Calculate inventory amounts at the end of each year.

P 8–15
Dollar-value LIFO
● LO8–8

On January 1, 2021, Avondale Lumber adopted the dollar-value LIFO inventory method. The inventory value for its one inventory pool on this date was $260,000. An internally generated cost index is used to convert ending inventory to base year. Year-end inventories at year-end costs and cost indexes for its one inventory pool were as follows:

Year Ended December 31	Inventory Year-End Costs	Cost Index (Relative to Base Year)
2021	$340,000	1.02
2022	350,000	1.06
2023	400,000	1.07
2024	430,000	1.10

Required:
Calculate inventory amounts at the end of each year.

P 8–16
Dollar-value LIFO; solving for unknowns
● LO8–8

At the beginning of 2021, Quentin and Kopps (Q&K) adopted the dollar-value LIFO (DVL) inventory method. On that date the value of its one inventory pool was $84,000. The company uses an internally generated cost index to convert ending inventory to base year. Inventory data for 2021 through 2024 are as follows:

Year Ended December 31	Ending Inventory at Year-End Costs	Ending Inventory at Base Year Costs	Cost Index	Ending Inventory at DVL cost
2021	$100,800	$ 96,000	1.05	?
2022	136,800	?	1.14	?
2023	150,000	125,000	?	?
2024	?	?	1.25	$133,710

Required:
Determine the missing amounts.

Decision Makers' Perspective

©ImageGap/Getty Images

Apply your critical-thinking ability to the knowledge you've gained. These cases will provide you an opportunity to develop your research, analysis, judgment, and communication skills. You also will work with other students, integrate what you've learned, apply it in real-world situations, and consider its global and ethical ramifications. This practice will broaden your knowledge and further develop your decision-making abilities.

Judgment Case 8–1
Riding the Merry-Go-Round
● LO8–7

Real World Financials

Merry-Go-Round Enterprises, the clothing retailer for dedicated followers of young men's and women's fashion, was looking natty as a company. It was March 1993, and the Joppa, Maryland-based outfit had just announced the acquisition of Chess King, a rival clothing chain, a move that would give it the biggest share of the young men's clothing market. Merry-Go-Round told brokerage firm analysts that the purchase would add $13 million, or 15 cents a share, to profits for the year. So some Wall Street analysts raised their earnings estimates for Merry-Go-Round. The company's stock rose $2.25, or 15%, to $17 on the day of the Chess King news. Merry-Go-Round was hot—$100 of its stock in January 1988 was worth $804 five years later. In 1993, the chain owned 1,460 stores in 44 states, mostly under the Cignal, Chess King, and Merry-Go-Round names.

Merry-Go-Round's annual report for the fiscal year ended January 30, 1993, reported a 15% sales growth, to $877.5 million from $761.2 million. A portion of the company's balance sheet is reproduced below:

	Jan. 30, 1993	Feb. 1, 1992
Assets		
Cash and cash equivalents	$40,115,000	$29,781,000
Marketable securities	—	9,703
Receivables	6,466,000	6,195
Merchandise inventories	82,197,000	59,971,000

But Merry-Go-Round spun out. The company lost $544,000 in the first six months of 1993, compared with earnings of $13.5 million in the first half of 1992. In the fall of 1992, Leonard "Boogie" Weinglass, Merry-Go-Round's flamboyant founder and chairman who had started the company in 1968, boarded up his Merry-Go-Ranch in Aspen, Colorado, and returned to management after a 12-year hiatus. But the pony-tailed, shirtsleeved entrepreneur—the inspiration for the character Boogie in the movie *Diner*—couldn't save his company from bankruptcy. In January 1994, the company filed for Chapter 11 protection in Baltimore. Shares crumbled below $3.

Required:

In retrospect, can you identify any advance warning at the date of the financial statements of the company's impending bankruptcy?

[Adapted from Jonathan Burton, "Due Diligence," *Worth*, June 1994, pp. 89–96.]

Real World Case 8–2
Physical quantities and costs included in inventory; Dick's Sporting Goods
● LO8–2
Real World Financials

Determining the physical quantity that should be included in inventory normally is a simple matter because that amount consists of items in the possession of the company. The cost of inventory includes all necessary expenditures to acquire the inventory and bring it to its desired *condition* and *location* for sale or for use in the manufacturing process.

Required:

1. Identify and describe the situations in which physical quantity included in inventory is more difficult than simply determining items in the possession of the company.
2. In addition to the direct acquisition costs such as the price paid and transportation costs to obtain inventory, what other expenditures might be necessary to bring the inventory to its desired condition and location?
3. Access EDGAR on the Internet. The web address is **www.sec.gov**. Search for **Dick's Sporting Goods**, a leading omni-channel sporting goods retailer. Access the 10-K filing for the most recent fiscal year. Search or scroll to find the disclosure notes (footnotes). What types of costs does the company include in its inventory?

Judgment Case 8–3
The specific identification inventory method; inventoriable costs
● LO8–3, LO8–4

Happlia Co. imports household appliances. Each model has many variations and each unit has an identification number. Happlia pays all costs for getting the goods from the port to its central warehouse in Des Moines. After repackaging, the goods are consigned to retailers. A retailer makes a sale, simultaneously buys the appliance from Happlia, and pays the balance due within one week.

To alleviate the overstocking of refrigerators at a Minneapolis retailer, some were reshipped to a Kansas City retailer where they were still held in inventory at December 31, 2021. Happlia paid the costs of this reshipment. Happlia uses the specific identification inventory costing method.

Required:

1. In regard to the specific identification inventory costing method:
   a. Describe its key elements.
   b. Discuss why it is appropriate for Happlia to use this method.
2. a. What general criteria should Happlia use to determine inventory carrying amounts at December 31, 2021?
   b. Give four examples of costs included in these inventory carrying amounts.
3. What costs should be reported in Happlia's 2021 income statement?

(AICPA adapted)

Communication Case 8–4
LIFO versus FIFO
● LO8–4, LO8–5

You have just been hired as a consultant to Tangier Industries, a newly formed company. The company president, John Meeks, is seeking your advice as to the appropriate inventory method Tangier should use to value its inventory and cost of goods sold. Mr. Meeks has narrowed the choice to LIFO and FIFO. He has heard that LIFO might be better for tax purposes, but FIFO has certain advantages for financial reporting to investors and creditors. You have been told that the company will be profitable in its first year and for the foreseeable future.

Required:

Prepare a report for the president describing the factors that should be considered by Tangier in choosing between LIFO and FIFO.

Communication Case 8–5
LIFO versus FIFO
● LO8–4, LO8–5

An accounting intern for a local CPA firm was reviewing the financial statements of a client in the electronics industry. The intern noticed that the client used the FIFO method of determining ending inventory and cost of goods sold. When she asked a colleague why the firm used FIFO instead of LIFO, she was told that the client used FIFO to minimize its income tax liability. This response puzzled the intern because she thought that LIFO would minimize income tax liability.

Required:

What would you tell the intern to resolve the confusion?

Judgment Case 8–6
Goods in transit
● LO8–2

At the end of 2021, the Biggie Company performed its annual physical inventory count. John Lawrence, the manager in charge of the physical count, was told that an additional $22,000 in inventory that had been sold and was in transit to the customer should be included in the ending inventory balance. John was of the opinion that the merchandise shipped should be excluded from the ending inventory since Biggie was not in physical possession of the merchandise.

Required:

Discuss the situation and indicate why John's opinion might be incorrect.

Ethics Case 8–7
Profit manipulation
● LO8–4

In 2020, the Moncrief Company purchased from Jim Lester the right to be the sole distributor in the western states of a product called Zelenex. In payment, Moncrief agreed to pay Lester 20% of the gross profit recognized from the sale of Zelenex in 2021.

Moncrief reports inventory using the periodic LIFO assumption. Late in 2021, the following information is available concerning the inventory of Zelenex:

Beginning inventory, 1/1/2021 (10,000 units @ $30)	$ 300,000
Purchases (40,000 units @ $30)	1,200,000
Sales (35,000 units @ $60)	2,100,000

By the end of the year, the purchase price of Zelenex had risen to $40 per unit. On December 28, 2021, three days before year-end, Moncrief is in a position to purchase 20,000 additional units of Zelenex at the $40 per unit price. Due to the increase in purchase price, Moncrief will increase the selling price in 2022 to $80 per unit. Inventory on hand before the purchase, 15,000 units, is sufficient to meet the next six months' sales and the company does not anticipate any significant changes in purchase price during 2022.

Required:

1. Determine the effect of the purchase of the additional 20,000 units on the 2021 gross profit from the sale of Zelenex and the payment due to Jim Lester.
2. Discuss the ethical dilemma Moncrief faces in determining whether or not the additional units should be purchased.

Real World Case 8–8
Effects of inventory valuation methods; supplemental LIFO disclosures; Wolverine World Wide, Inc.
● LO8–4, LO8–6

Real World Financials

Income statement and balance sheet information abstracted from a recent annual report of **Wolverine World Wide, Inc.**, appears below:

Balance Sheets
($ in millions)

	December 30, 2017	December 31, 2016
Current assets:		
Inventories	$276.7	$348.7

Income Statements
($ in millions)

	For the Year Ended	
	December 30, 2017	December 31, 2016
Net sales	$2,350.0	$2,494.6
Cost of goods sold	1,426.6	1,526.4
Gross profit	$ 923.4	$ 968.2

The significant accounting policies note disclosure contained the following:

Inventories

The Company used the LIFO method to value inventories of $53.2 million at December 30, 2017 and $66.2 million at December 31, 2016. During fiscal 2017, a reduction in inventory quantities resulted in a liquidation of applicable LIFO inventory quantities carried at lower costs in prior years. This LIFO liquidation decreased cost of goods sold by $6.0 million. If the FIFO method had been used, inventories would have been $16.4 million and $22.4 million higher than reported at December 30, 2017 and December 31, 2016, respectively.

Required:

1. Why is Wolverine disclosing the FIFO cost of its LIFO inventory?
2. Calculate what beginning inventory and ending inventory would have been for the year ended December 31, 2017, if Wolverine had used FIFO for all of its inventories.
3. Calculate what cost of goods sold and gross profit would have been for the year ended December 31, 2017, if Wolverine had used FIFO for all of its inventories.
4. In 2017, Wolverine reported a LIFO liquidation. Why do companies separately disclose LIFO liquidations due to declines in the quantity of ending inventory?

Real World Case 8–9
Effects of inventory valuation methods; Whole Foods Market
● LO8–4, LO8–5, LO8–7

Real World Financials

EDGAR, the Electronic Data Gathering, Analysis, and Retrieval system, performs automated collection, validation, indexing, and forwarding of submission by companies and others who are required by law to file forms with the U.S. Securities and Exchange Commission (SEC). All publicly traded domestic companies use EDGAR to make the majority of their filings. (Some foreign companies file voluntarily.) Form 10-K, which includes the annual report, is required to be filed on EDGAR. The SEC makes this information available on the Internet.

Required:

1. Access EDGAR on the Internet. The web address is **www.sec.gov**.
2. Search for **Whole Foods Market, Inc**. Access the 10-K filing for the year ended September 24, 2017. Search or scroll to find the financial statements and related notes.
3. Answer the following questions related to the company's inventories:
   a. What method(s) does the company use to value its inventories?
   b. Calculate what cost of sales would have been for the year if the company had used FIFO to value its inventories.
   c. Calculate inventory turnover for the year using the reported numbers.

Communication Case 8–10
Dollar-value LIFO method
● LO8–8

Maxi Corporation uses the unit LIFO inventory method. The costs of the company's products have been steadily rising since the company began operations in 2008 and cost increases are expected to continue. The chief financial officer of the company would like to continue using LIFO because of its tax advantages. However, the controller, Sally Hamel, would like to reduce the record-keeping costs of LIFO that have steadily increased over the years as new products have been added to the company's product line. Sally suggested the use of the dollar-value LIFO method. The chief financial officer has asked Sally to describe the dollar-value LIFO procedure.

Required:

Describe the dollar-value LIFO procedure.

Research Case 8–11
FASB codification; locate and extract relevant information and authoritative support for a financial reporting issue; product financing arrangement
● LO8–2, LO8–3

You were recently hired to work in the controller's office of the Balboa Lumber Company. Your boss, Alfred Eagleton, took you to lunch during your first week and asked a favor. "Things have been a little slow lately, and we need to borrow a little cash to tide us over. Our inventory has been building up and the CFO wants to pledge the inventory as collateral for a short-term loan. But I have a better idea." Mr. Eagleton went on to describe his plan. "On July 1, 2021, the first day of the company's third quarter, we will sell $100,000 of inventory to the Harbaugh Corporation for $160,000. Harbaugh will pay us immediately and then we will agree to repurchase the merchandise in two months for $164,000. The $4,000 is Harbaugh's fee for holding the inventory and for providing financing. I already checked with Harbaugh's controller and he has agreed to the arrangement. Not only will we obtain the financing we need, but the third quarter's before-tax profits will be increased by $56,000, the gross profit on the sale less the $4,000 fee. Go research the issue and make sure we would not be violating any specific accounting standards related to product financing arrangements."

Required:

1. Access the *FASB Accounting Standards Codification* at the FASB website (**www.fasb.org**). Determine the specific eight-digit Codification citation (XXX-XX-XX-X) that provides guidance for determining whether an arrangement involving the sale of inventory is "in substance" a financing arrangement?
2. Determine the specific eight-digit Codification citation (XXX-XX-XX-X) that addresses the recognition of a product financing arrangement?
3. Determine the appropriate treatment of product financing arrangements like the one proposed by Mr. Eagleton.
4. Prepare the journal entry for Balboa Lumber to record the "sale" of the inventory and subsequent repurchase.

Analysis Case 8–12
Compare inventory management using ratios; Kohl's and Dillards
● LO8–7

Real World Financials

The table below contains selected financial information included in the 2017 financial statements of **Kohl's Corporation** and **Dillards, Inc.**, two companies in the department store industry.

($ in millions)	Kohl's Corp.		Dillards, Inc.	
	2017	**2016**	**2017**	**2016**
Balance sheet:				
Inventories	$ 3,542	$3,795	$1,464	$1,406
Income statement—2017:				
Net sales	$19,095		$6,261	
Cost of goods sold	12,176		4,200	

Required:

1. Calculate the 2017 gross profit ratio, inventory turnover ratio, and average days in inventory for both companies. Evaluate the management of each company's investment in inventory.
2. Obtain annual reports from three corporations in an industry other than department stores and compare the management of each company's investment in inventory.

Note: You can obtain copies of annual reports from your library, from friends who are shareholders, from the investor relations department of the corporations, from a friendly stockbroker, or from EDGAR (Electronic Data Gathering, Analysis, and Retrieval) on the Internet (**www.gov.sec**).

Data Analytics

Mc Graw Hill Education connect

Data analytics is the process of examining data sets in order to draw conclusions about the information they contain. If you have not completed any of the prior data analytics cases, follow the instructions listed in the Chapter 1 Data Analytics case to get set up. You will need to watch the videos referred to in the Chapters 1 - 3 Data Analytics cases. No additional videos are required for this case. All short training videos can be found here: **https://www.tableau.com/learn/training#getting-started**.

Data Analytics Case
Gross Profit and Gross Profit Ratio
● LO8–7

In prior chapters, you used Tableau to examine a data set and create calculations to compare two companies' financial information. In this case you are seeking to gain insight into the relative profitability of a company in various regions of the country and various states in which it sells. Your focus is on assessing the company's ability to generate gross profit on the sale of its inventory. Gross profit represents a company's ability to cover expenses other than cost of goods sold.

Required:
Use Tableau to calculate (a) gross profit and (b) gross profit ratios for the company in each of the five regions of the country designated in the company's data base and each of the states in which the company sells products. Based upon what you find, answer the following questions:

1. Which of the five regions' operations has the highest gross profit?
2. Which of the five regions' operations has the lowest gross profit?

To help determine the relative profitability of each state's sales, get a better state-by-state comparison by analyzing the "map" chart. Once the chart is created, move your cursor to hover above various circles within states in the chart. Notice that an information box appears to reveal the pertinent gross profit and gross profit ratios for that state.

3. Which state's operations provides the highest gross profit ratio?
4. Which state's operations provides the lowest gross profit ratio?

Resources:
You have available to you an extensive data set that includes detailed financial data for the company's inventory sales. The data set is in the form of an Excel file available to download from Connect, or under Student Resources within the Library tab. There are four Excel files available. The one for use in this case is named "Inventory.xlsx." Download this file to the computer, save it, and open it in Tableau.

For this case, you will create several calculations to produce a bar chart and a map of the gross profit to allow you to compare and contrast a company's inventory measurement.

After you view the training videos and review instructions in Chapters 1-3, follow these steps to create the charts you'll use for this case:

- Click on the "Sheet 1" tab, at the bottom of the canvas, to the right of the Data Source at the bottom of the screen. Drag "Region" under Dimensions to the Columns shelf.
- Drag "Status" under Dimensions to the Filter's card. When the Filter window opens, click the "S" so only the sold items will be viewed. Click OK.
- Click on the "Analysis" tab from the Toolbar at the top of the screen and select "Create Calculated Field." A calculation box will pop up. Name the calculation "Total Sales." Drag "Number" to the Calculation Editor window, type an asterisk sign for multiplication, and then drag "Selling Price" beside it. Make sure the window says that the calculation is valid and click OK.
- Repeat the process by creating a calculated field for "Cost of Goods Sold" that consists of "Number" times "Cost."
- Repeat the process by creating a calculated field for "Gross Profit" that consists of "Total Sales" minus "Cost of Goods Sold."
- Drag "Total Sales," "Cost of Goods Sold," and "Gross Profit" to the Rows shelf.
- Click the "Show Me" box on right side of the toolbar and select the "side-by-side bars."

- Show the labels by clicking on the Label box on the "Marks" card and checking "Show mark labels." Format the labels to Times New Roman, bold, black and 9-point font.
- Change the title of the sheet to be "Total Sales, COGS, and Gross Profit by Region" by right-clicking and selecting "Edit title." Format the title to Times New Roman, bold, black and 15-point font. Change the title of "Sheet 1" to match the sheet title by right-clicking, selecting "Rename" and typing in the new title.
- Format all the labels on the sheet to Times New Roman, 10-point font, and black.
- Click on the New Worksheet tab on the lower left ("Sheet 2" should open).
- Drag "State" from Dimensions to the Columns shelf.
- Drag "Gross Profit" and "Total Sales" to the Rows shelf.
- Create a calculated field "Gross Profit Ratio by State" by dragging "Sum(Gross Profit)" from the Rows shelf and then dividing by "Sum(Total Sales)" from the Rows shelf to calculate an "aggregated" measure for each state.
- Remove "Sum(Total Sales)" from the Rows shelf by dragging it back over to Dimensions.
- In the "Show Me" box on right side of the toolbar, select the "symbol map" and you should see circles appear in states with data. Drag "Gross Profit Ratio by State" from "Measures" to the bottom of the "Marks" card. Click on the circle symbol beside the "Gross Profit Ratio by State" and select "Color." The dots should now vary in size and color based on the size of the gross profit ratio and gross profit. Enlarge the dots by clicking "Size" on the "Marks" card and adjusting the slide. You can also change the color variation by clicking "Color" on the "Marks" card and editing the color.
- Change the title of the sheet to be "Gross Profit Ratio by State" by right-clicking and selecting "Edit title." Format the title to Times New Roman, bold, black and 15-point font. Change the title of "Sheet 2" to match the sheet title by right-clicking, selecting "Rename" and typing in the new title.
- Once complete, save the file as "*DA8_Your initials*.twbx."

Continuing Cases

Target Case

● LO8–1, LO8–4, LO8–7

Target Corporation prepares its financial statements according to U.S. GAAP. Target's financial statements and disclosure notes for the year ended February 3, 2018, are available in Connect. This material is also available under the Investor Relations link at the company's website (**www.target.com**).

Required:

1. Does Target use average cost, FIFO, or LIFO as its inventory cost flow assumption?
2. In addition to the purchase price, what additional expenditures does the company include in the initial cost of merchandise?
3. Calculate the gross profit ratio and the inventory turnover ratio for the fiscal year ended February 3, 2018. Compare Target's ratios with the industry averages of 24.5% and 7.1 times. Determine whether Target's ratios indicate the company is more/less profitable and sells its inventory more/less frequently compared to the industry average.

Air France–KLM Case

● LO8–9

IFRS

Air France–KLM (AF), a Franco-Dutch company, prepares its financial statements according to International Financial Reporting Standards. AF's financial statements and disclosure notes for the year ended December 31, 2017, are available in Connect. This material is also available under the Finance link at the company's website (**www.airfranceklm.com**).

Required:

What method does the company use to value its inventory? What other alternatives are available under IFRS? Under U.S. GAAP?

CPA Exam Questions and Simulations

ROGER CPA Review

Sample CPA Exam questions from Roger CPA Review are available in Connect as support for the topics in this chapter. These multiple-choice questions and task-based simulations include expert-written explanations and solutions, and provide a starting point for students to become familiar with the content and functionality of the actual CPA Exam.

CHAPTER

9 Inventories: Additional Issues

OVERVIEW

We covered most of the principal measurement and reporting issues involving the asset inventory and the corresponding expense cost of goods sold in the previous chapter. In this chapter, we complete our discussion of inventory measurement by explaining how inventories are measured at the end of the period. In addition, we investigate inventory estimation techniques, methods of simplifying LIFO, changes in inventory method, and inventory errors.

LEARNING OBJECTIVES

After studying this chapter, you should be able to:

- **LO9–1** Understand and apply rules for measurement of inventory at the end of the reporting period. *(p. 459)*
- **LO9–2** Estimate ending inventory and cost of goods sold using the gross profit method. *(p. 467)*
- **LO9–3** Estimate ending inventory and cost of goods sold using the retail inventory method, applying the various cost flow methods. *(p. 469)*
- **LO9–4** Explain how the retail inventory method can be made to approximate the lower of cost or market rule. *(p. 472)*
- **LO9–5** Determine ending inventory using the dollar-value LIFO retail inventory method. *(p. 477)*
- **LO9–6** Explain the appropriate accounting treatment required when a change in inventory method is made. *(p. 480)*
- **LO9–7** Explain the appropriate accounting treatment required when an inventory error is discovered. *(p. 482)*
- **LO9–8** Discuss the primary differences between U.S. GAAP and IFRS with respect to the lower of cost or net realizable value rule for valuing inventory. *(p. 463)*

©QualityHD/Shutterstock.com

FINANCIAL REPORTING CASE

Does It Count?

Today you drove over to **Dollar General** to pick up a few items. You recall from class yesterday that your accounting professor had discussed inventory measurement issues and the different methods (FIFO, LIFO, and average) used by companies to determine ending inventory and cost of goods sold. As of March 2, 2018, Dollar General had 14,609 store locations. You can't imagine actually counting the inventory in all of the stores around the country. You consider that there must be some way Dollar General can avoid counting all of that inventory every time they want to produce financial statements. When you get home, you check their financial statements on the Internet to see what kind of inventory method they use. You find the following in the summary of significant accounting policies included in Dollar General's most recent financial statements:

Merchandise Inventories:

Inventories are stated at the lower of cost or market ("LCM") with cost determined using the retail last-in, first-out ("LIFO") method as this method results in a better matching of costs and revenues. Under the Company's retail inventory method ("RIM"), the calculation of gross profit and the resulting valuation of inventories at cost are computed by applying a calculated cost-to-retail inventory ratio to the retail value of sales at a department level. The use of the RIM will result in valuing inventories at LCM if markdowns are currently taken as a reduction of the retail value of inventories. Costs directly associated with warehousing and distribution are capitalized into inventory.

The excess of current cost over LIFO cost was approximately $78.5 million and $80.7 million at February 2, 2018, and February 3, 2017, respectively. Current cost is determined using the RIM on a first-in, first-out basis. Under the LIFO inventory method, the impacts of rising or falling market price changes increase or decrease cost of sales (the LIFO provision or benefit). The Company recorded a LIFO provision (benefit) of $(2.2) million in 2017, $(12.2) million in 2016, and $(2.3) million in 2015, which is included in cost of goods sold in the consolidated statements of income.

QUESTIONS

By the time you finish this chapter, you should be able to respond appropriately to the questions posed in this case. Compare your response to the solution provided at the end of the chapter.

1. How does Dollar General avoid counting all its inventory every time it produces financial statements? *(p. 477)*
2. What is the index formulation used for? *(p. 477)*

Subsequent Measurement of Inventory

PART A

● LO9–1

As you would expect, companies hope to sell their inventory for more than its cost. Yet, this doesn't always happen. Sometimes circumstances arise after (or *subsequent* to) the purchase or production of inventory that indicate the company will have to sell its inventory for less than its cost. This might happen because of inventory damage, physical deterioration, obsolescence, changes in price levels, or any situation that lessens demand for the inventory. Consider, for example, the value of unsold electronics inventory when the next generation comes out, or the leftover clothing inventory at the end of the selling season. Usually, the only way these items can be sold is at deeply discounted prices (well-below their purchase cost).

GAAP requires that companies evaluate their unsold inventory at the end of each reporting period (for reasons mentioned above). When the expected benefit of unsold inventory is estimated to have fallen below its cost, companies must depart from the cost basis of reporting ending inventory; an adjusting entry is needed to reduce the reported amount of inventory and to reduce net income for the period. This end-of-period adjusting entry is known as an *inventory write-down*.

An inventory write-down has the effect of reducing inventory and reducing net income.

The two measurement approaches for recording an inventory write-down are listed in Illustration 9–1. The approach chosen by a company depends on which inventory costing method the company uses.[1] For companies that use a cost method other than LIFO or the retail inventory method, we report inventory at the *lower of cost or net realizable value.* For companies that use LIFO or the retail inventory method (discussed later in this chapter), we report inventory at the *lower of cost or market.*

Illustration 9–1

Subsequent Inventory Measurement

Measurement Approach	For Companies that Use	Financial Statement Effects of Inventory Write-Downs
1. Lower of cost or net realizable values (LCNRV)	FIFO, average cost, or any other method besides LIFO or the retail inventory method	a. Reduce reported inventory b. Reduce net income
2. Lower of cost or market (LCM)	LIFO or retail inventory method	a. Reduce reported inventory b. Reduce net income

For both measurement approaches, the financial statement effects of an inventory write-down are the same: (a) reduce reported inventory and (b) reduce net income. Both approaches have the same conceptual purpose of reporting inventory conservatively at the lower of two amounts. The difference between the two approaches is the measurement of the *amount* of the inventory write-down. We discuss those measurements next.[2]

Lower of Cost or Net Realizable Value (LCNRV)

Net realizable value (NRV) is the estimated selling price reduced by any costs of completion, disposal, and transportation.

Companies that use FIFO, average cost, or any other method besides LIFO or the retail inventory method report inventory using the **lower of cost or net realizable value** (LCNRV) approach. To do this, a company compares the inventory's cost to the inventory's **net realizable value** (NRV). NRV is the estimated selling price of the inventory in the ordinary course of business reduced by reasonably predictable costs of completion, disposal, and transportation (such as sales commissions and shipping costs).

Another way to think about NRV is that it's the *net* amount a company expects to *realize* (or collect in cash) from the sale of the inventory. Companies often estimate the "costs to sell" by applying a predetermined percentage to the selling price. For example, if the selling price of Product A is $10 per unit, and the company estimates that sales commissions and shipping costs average approximately 10% of selling price, NRV would be $9 [$10 – ($10 × 10%)].

After comparing cost and NRV, a company reports inventory at the lower of the two amounts.

When NRV is lower than cost, an inventory write-down is recorded.

1. If NRV is lower than cost, we need an adjusting entry to reduce inventory from its already recorded purchase cost to the lower NRV. NRV then becomes the new carrying value of inventory reported in the balance sheet.
2. If cost is lower than NRV, no adjusting entry is needed. Inventory already is recorded at cost at the time it was purchased, and this cost is lower than total NRV at the end of the period.

The LCNRV approach avoids reporting inventory at an amount greater than the cash it can provide to the company. Reporting inventories this way causes income to be reduced in the period the value of inventory declines below its cost rather than in the period in which the goods ultimately are sold.[3]

[1]FASB ASC 330-10-35-1A through 35-1C: Inventory—Overall—Subsequent Measurement.

[2]Interestingly, the two approaches to subsequent measurement of inventory arose out of *Accounting Standards Update (ASU) No. 2015–11,* which was part of the FASB's Simplification Initiative aimed at *reducing* reporting complexity. However, *ASU 2015–11 increases* the number of subsequent measurement approaches. Prior to *ASU 2015 – 11,* all companies reported inventory using a single approach—the lower of cost or market. Some may argue that the introduction of a second approach increases reporting complexity. Some companies use multiple inventory methods and may need to use both approaches, and financial statement users must be aware of different methods when comparing companies.

[3]In other words, if the inventory was not written down to NRV (and instead remained at cost) at the end of the current period, the full cost amount would become cost of goods sold and thus reduce net income when the inventory is sold in a subsequent period.

Applying Lower of Cost or Net Realizable Value

The LCNRV rule can be applied to individual inventory items, major inventory categories, or the entire inventory.

For financial reporting purposes, LCNRV can be applied (a) to individual inventory items, (b) to major categories of inventory, or (c) to the entire inventory.[4] Illustration 9–2 demonstrates the LCNRV approach with each of the three possible applications.

Illustration 9–2
Lower of Cost or Net Realizable Value—Application at Different Levels of Aggregation

The Collins Company has five inventory items on hand at the end of the year. The year-end cost (determined by applying the FIFO cost method) and net realizable value (current selling prices less costs of completion, disposal, and transportation) for each of the items are presented below.

Items A and B are a collection of similar items, and items C, D, and E are another collection of similar items. Each collection can be considered a category of inventory.

			Lower of Cost or NRV		
Item	**Cost**	**NRV**	**By Individual Items**	**By Category**	**By Total Inventory**
A	$ 50,000	$ 85,000	$ 50,000		
B	100,000	90,000	90,000		
Total A and B	$150,000	$175,000		$ 150,000	
C	$ 80,000	$ 75,000	75,000		
D	90,000	85,000	85,000		
E	95,000	96,000	95,000		
Total C, D, and E	$265,000	$256,000		256,000	
Total	$415,000	$431,000	**$395,000**	**$406,000**	**$415,000**

If we first consider LCNRV *by individual items,* we see that cost is lower than NRV for items A and E (see "By Individual Items" column). These inventory items were initially recorded at cost at the time of purchase. So, under the LCNRV approach, they are currently recorded at their proper amounts. No adjustment is needed. However, for items B, C, and D, we see that NRV is below cost. Therefore, we need to adjust the carrying value of these items downward from cost to NRV to report them at NRV in the balance sheet. After determining the lower of cost or NRV for each individual item, the total amount to report for inventory is **$395,000**.

Now let's see what happens if we apply the LCNRV approach *by inventory category* (see "By Category" column). The first category of inventory (items A and B) has a combined cost of $150,000 and a combined NRV of $175,000, so we report this second inventory group at its cost of $150,000. For the second category (items C, D, and E), the combined NRV ($256,000) is lower than the combined cost ($265,000), so we report this second inventory group at its NRV of $256,000. We report inventory at **$406,000** ($150,000 + $256,000).

Finally, if we apply the LCNRV approach *by total inventory,* we see that inventory's total cost ($415,000) is lower than its total NRV ($431,000). Inventory would be reported at its cost of **$415,000**.

As shown in this illustration, applying the LCNRV rule to groups of inventory items will cause a higher inventory valuation than if applied on an item-by-item basis. The reason is that group application permits increases in the net realizable value of some items to offset decreases in others. Each approach is acceptable but should be applied consistently from one period to another.

[4]FASB ASC 330–10–35–8: Inventory—Overall—Subsequent Measurement. In addition, for income tax purposes, the rule must be applied on an individual item basis.

Ethical Dilemma

The Hartley Company, owned and operated by Bill Hartley, manufactures and sells high-end ergonomic office chairs and custom bookshelves. The company has reported profits in the majority of years since the company's inception in 1975 and is projecting a profit in 2021 of $65,000, down from $96,000 in 2020.

Near the end of 2021, the company is in the process of applying for a bank loan. The loan proceeds will be used to replace manufacturing equipment to modernize the manufacturing operation. In preparing the financial statements for the year, the chief accountant, Don Davis, mentioned to Bill Hartley that net realizable value (NRV) of the bookshelf inventory is below its cost by $40,000 and should be written down in 2021. However, no write-down is necessary for office chairs because their NRV is $50,000 above cost.

Bill is worried that the write-down would lower 2021 net income to a level that might cause the bank to refuse the loan. Without the loan, it would be difficult for the company to compete. This could decrease future business, and employees might have to be laid off. Bill suggests to Don that the company combine the office chairs and bookshelves into a single inventory category (office furniture) for reporting purposes, so that the combined NRV is above the combined cost. In this case, no inventory write-down would be needed. The company has not previously combined these inventory items and has no stated policy on the matter. Don is contemplating his responsibilities in this situation.

Adjusting Cost to Net Realizable Value

When NRV is below cost, companies are required to write down inventory to the lower NRV. These write-downs usually are included as part of cost of goods sold because they are a natural consequence of holding inventory and therefore part of the inventory's normal cost. However, when a write-down is substantial and unusual, the write-down should be recorded in a separate loss account instead. That loss must be expressly disclosed. This could be accomplished with a disclosure note alone or also by reporting the loss as a separate line in the income statement, usually among operating expenses.

Referring back to Illustration 9–2, assume that we report inventory using the LCNRV approach by individual items. The recorded cost of inventory ($415,000) needs to be written down to its NRV (**$395,000**) at the end of the period. The amount of the reduction is $20,000. The period-end adjusting entry for the typical occurrence of decline to net realizable value would be the following:

Inventory

415,000	
	20,000
395,000	

Cost of goods sold*	20,000	
Inventory		20,000

*In situations when a write-down is substantial and unusual, the write-down should be recorded in a loss on inventory write-down account.

After the adjustment from original cost to net realizable value, the reduced inventory amount becomes the new cost basis for subsequent reporting. If the inventory value later increases prior to its sale, we do not write it back up.[5]

[5]The SEC, in *Staff Accounting Bulletin No. 100,* "Restructuring and Impairment Charges" (Washington, D.C.: SEC, November, 1999), paragraph B.B., (FASB ASC 330–10–S35–1: SAB Topic 5.BB), reaffirmed the provisions of GAAP literature on this issue. For interim reporting purposes, however, recoveries of losses on the same inventory in subsequent interim periods of the same fiscal year through market price recoveries should be recognized as gains in the later interim period, not to exceed the previously recognized losses.

International Financial Reporting Standards

Lower of cost or net realizable value. You just learned that in the United States some companies report inventory at the *lower of cost or net realizable value.* This is the same approach used under IFRS. However, there are some differences between U.S. GAAP and IFRS in the application of lower of cost or net realizable value.

First, *IAS No. 2* specifies that if circumstances indicate that an inventory write-down is no longer appropriate, it must be reversed.[6] Reversals are not permitted under U.S. GAAP.

Second, under U.S. GAAP, the lower of cost or net realizable value rule can be applied to individual items, inventory categories, or the entire inventory. Under the international standard, the assessment usually is applied to individual items, although using inventory categories is allowed under certain circumstances.

Siemens AG, a German electronics and electrical engineering company, prepares its financial statements according to IFRS. The following disclosure note illustrates the valuation of inventory at the lower of cost or net realizable value.

Inventories (in part)
Inventory is valued at the lower of acquisition or production cost and net realizable value, cost being generally determined on the basis of an average or first-in, first-out method.

● LO9–8

Concept Review Exercise

LOWER OF COST OR NET REALIZABLE VALUE

The Strand Company sells four products that can be grouped into two major categories and employs the FIFO cost method. Information needed to apply the lower of cost or NRV (LCNRV) rule on December 31, 2021 (end of the period), for each of the four products is presented below. Sales commissions and transportation costs average 10% of selling price. Inventory write-downs are a normal occurrence for Strand.

Product	Quantity	Unit Cost	Unit Selling Price	Total Cost	Total Selling Price
101	4,000	$30	$40	$120,000	$160,000
102	5,000	35	36	175,000	180,000
201	3,200	50	50	160,000	160,000
202	15,000	3	4	45,000	60,000

Products 101 and 102 are in category A, and products 201 and 202 are in category B.

Required:

Determine the reported amount of ending inventory and record any necessary year-end adjusting entry to write down inventory, applying the LCNRV approach to the following:

1. Individual items
2. Major categories
3. Total inventory

Solution:

Determine the reported amount of ending inventory and record any necessary year-end adjusting entry to write down inventory, applying the LCNRV approach.

[6] "Inventories," *International Accounting Standard No. 2* (IASCF), as amended effective January 1, 2018.

			Lower of Cost or NRV		
Product	**Cost**	**NRV***	**By Individual Products**	**By Category**	**By Total Inventory**
101	$120,000	$144,000	$120,000		
102	175,000	162,000	162,000		
Total 101 + 102	$295,000	$306,000		$295,000	
201	$160,000	$144,000	144,000		
202	45,000	54,000	45,000		
Total 201 + 202	$205,000	$198,000		198,000	
Total	$500,000	$504,000	$471,000	$493,000	$500,000

*NRV = Selling price less costs to sell. For product 101, $160,000 – ($160,000 × 10%) = $ 144, 000.

The NRV for both the individual product and category applications are lower than cost so inventory write-downs are needed. On the other hand, cost is lower than NRV at the total inventory level, so no adjustment would be needed.

1. Individual items

Reported ending inventory = $471,000 (NRV)

December 31, 2021		
Cost of goods sold†	29,000*	
Inventory		29,000
*$500,000 (recorded cost) – $471,000 (NRV)		

2. Major categories

Reported ending inventory = $493,000 (NRV)

December 31, 2021		
Cost of goods sold†	7,000*	
Inventory		7,000
*$500,000 (recorded cost) – $493,000 (NRV)		

†For the two entries above, we record the adjustment to the cost of goods sold account because the inventory write-down is considered usual. If the write-down had been substantial and unusual, we record the adjustment to a loss on inventory write-down account.

3. Total inventory

Reported ending inventory = $500,000 (Cost)

Because inventory already is recorded at cost as of December 31, 2021, and because total cost is lower than total NRV ($504,000), no year-end adjustment is needed.

Additional **Consideration**

Critics of reporting inventory lower than its cost contend that this causes losses to be recognized that haven't actually occurred. Others maintain that it introduces needless inconsistency in order to be conservative, because decreases in value are recognized as they occur, but not increases. As you learned in Chapter 1, conservatism is not part of the conceptual framework. So, why not record increases as well? Recall the revenue recognition guidance we discussed in Chapter 6. Recognizing increases in the value of inventory prior to sale would, in most cases, result in premature revenue recognition, because the seller would be acting as if it had satisfied a performance obligation (selling inventory) before that actually occurred. For example, let's say that merchandise costing $100 now has a net realizable value of $150. Recognizing a gain for the increase in value would increase pretax income by $50. This is equivalent to recognizing revenue of $150, cost of goods sold of $100, and gross profit of $50. The effect is to increase pretax income in a period prior to sale of the product. That's not allowed.

Lower of Cost or Market (LCM)

Companies that use LIFO or the retail inventory method report inventory using the **lower of cost or market** (LCM) approach. You might interpret the term *market* to mean the amount that could be realized if the inventory were sold. This would be similar to the concept of *net realizable value* discussed above. However, market is defined differently.

Market is current replacement cost, but not above the ceiling or below the floor.

Market is the inventory's current replacement cost (by purchase or reproduction) except that:

1. Market should not be greater than the net realizable value (this forms a "ceiling" on market), and
2. Market should not be less than net realizable value reduced by an allowance for an approximately normal profit margin (this forms a "floor" on market).

In effect, we have a ceiling and a floor between which market (that is, replacement cost) must fall. If replacement cost is between the ceiling and the floor, it represents market; if replacement cost is above the ceiling or below the floor, the ceiling or the floor becomes market. The designated market amount is compared with cost, and the lower of the two is used to value inventory.

To see an example, let's look at Illustration 9–3. In the top part of the illustration, we calculate the ceiling and the floor. In the bottom part, we calculate LCM.

Illustration 9–3

Lower of Cost or Market (LCM)

The Collins Company has five inventory items on hand at the end of the year. The year-end selling prices, and estimated costs of completion, disposal, and transportation (selling costs) for each of the items are given below. The normal profit for each of the products is 20% of selling price. These amounts are used to calculate the ceiling and floor as follows:

Item	Selling Price	Estimated Selling Costs	NRV [Ceiling]*	Normal Profit Margin (20% of Selling Price)	NRV – NPM [Floor]†
A	$100,000	$15,000	$85,000	$20,000	$65,000
B	120,000	30,000	90,000	24,000	66,000
C	90,000	15,000	75,000	18,000	57,000
D	100,000	15,000	85,000	20,000	65,000
E	110,000	14,000	96,000	22,000	74,000

Additional information related to year-end inventory cost (determined by applying the LIFO cost method) and replacement cost are given in the first two columns. Determination of LCM is a two-step process: (1) calculate the market amount using replacement cost, subject to a ceiling and floor, and (2) select the lower of cost or market.

		Market				
	(1)	(2)	(3)	(4)	(5)	
Item	Cost	Replacement Cost	NRV [Ceiling]	NRV – NPM [Floor]	Market [Middle of (2), (3), (4)]	LCM [Lower of (1) or (5)]
A	$ 50,000	$55,000	$85,000	$65,000	**$65,000**	$ 50,000
B	100,000	97,000	90,000	66,000	**90,000**	90,000
C	80,000	70,000	75,000	57,000	**70,000**	70,000
D	90,000	95,000	85,000	65,000	**85,000**	85,000
E	95,000	92,000	96,000	74,000	**92,000**	92,000
Total	**$415,000**					**$387,000**

*NRV = Estimated selling price less estimated selling costs. For item A, $100,000 – $15,000 = $85,000.
†NRV – NPM = NRV less a normal profit margin. For item A, $85,000 – ($100,000 selling price × 20%) = $65,000.

Notice the market amount for each inventory item is simply the middle value among replacement cost, NRV (ceiling), and NRV – NPM (floor). When replacement cost is

a. *Below the floor* (item A), we select the floor as the market.
b. *Above the ceiling* (items B and D), we select the ceiling as the market.
c. *Between the ceiling and the floor* (items C and E), we select replacement cost as the market.

We then compare each item's designated market amount to its cost and choose the lower of the two. After doing this, ending inventory under LCM is **$387,000**. This means that the recorded cost of $415,000 needs to be reduced by $28,000 ($415,000 – $387,000). We record that adjustment to cost of goods sold if the decrease to market is considered normal for this company.

Inventory

Debit	Credit
415,000	
	28,000
387,000	

	Debit	Credit
Cost of goods sold*	28,000	
Inventory		28,000

*In situations when a write-down is substantial and unusual, the write-down should be recorded in a loss on inventory write-down account.

The example in Illustration 9–3 calculates LCM on an individual items basis. The LCM method also can be applied to major categories of inventory or to the entire inventory, just like we saw in Illustration 9–2 under the LCNRV approach. As shown in Illustration 9–3A, inventory using the LCM approach applied *by category* would be reported as **$397,000**. If we calculate LCM *by total inventory,* the amount to report for ending inventory would be **$402,000**.

Illustration 9–3A

Lower of Cost or Market—Application at Different Levels of Aggregation

From the information in Illustration 9–3, also assume items A and B are a collection of similar items, and items C, D, and E are another collection of similar items. Each collection can be considered a category of inventory.

			Lower of Cost or Market		
Item	**Cost**	**Market**	**By Individual Items**	**By Category**	**By Total Inventory**
A	$ 50,000	$ 65,000	$ 50,000		
B	100,000	90,000	90,000		
Total A and B	$150,000	$155,000		$ 150,000	
C	$ 80,000	$ 70,000	70,000		
D	90,000	85,000	85,000		
E	95,000	92,000	92,000		
Total C, D, and E	$265,000	$247,000		247,000	
Total	$415,000	$402,000	**$387,000**	**$397,000**	**$402,000**

PART B

Inventory Estimation Techniques

For some companies or in certain situations, it becomes difficult or impossible to physically count each unit of inventory. For example, consider a large retail company that has locations over the entire nation and sells many different items. Trying to track the cost of each item would be nearly impossible. In these situations, companies have developed methods for estimating inventory. We'll study those methods next.

The Gross Profit Method

The **gross profit method**, also known as the **gross margin method**, is useful in situations where estimates of inventory are desirable. The technique is valuable in a variety of situations, including the following: ● LO9–2

1. In determining the cost of inventory that has been lost, destroyed, or stolen.
2. In estimating inventory and cost of goods sold for interim reports, avoiding the expense of a physical inventory count.
3. In auditors' testing of the overall reasonableness of inventory amounts reported by clients.
4. In budgeting and forecasting.

The technique relies on a relationship you learned in the previous chapter—ending inventory and cost of goods sold always equal the cost of goods available for sale. Even when inventory is unknown, we can estimate it because accounting records usually indicate the cost of goods available for sale (beginning inventory plus net purchases), and the cost of goods sold can be estimated from available information. So by subtracting the cost of goods sold estimate from the cost of goods available for sale, we obtain an estimate of ending inventory. Let's compare that with the way inventory and cost of goods sold normally are determined.

Usually, in a periodic inventory system, ending inventory is known from a physical count and cost of goods sold is *derived* as follows:

Beginning inventory	(from the accounting records)
Plus: Net purchases	(from the accounting records)
Goods available for sale	
Less: Ending inventory	(from a physical count)
Cost of goods sold	

However, when using the gross profit method, the ending inventory is *not* known. Instead, the amount of sales is known—from which we can estimate the cost of goods sold—and ending inventory is the amount calculated.

Beginning inventory	(from the accounting records)
Plus: Net purchases	(from the accounting records)
Goods available for sale	
Less: Cost of goods sold	(estimated)
Ending inventory	(estimated)

So, a first step in estimating inventory is to estimate cost of goods sold. This estimate relies on the historical relationship among (a) net sales, (b) cost of goods sold, and (c) gross profit. Gross profit, you will recall, is simply net sales minus cost of goods sold. So, if we know what net sales are, and if we know what percentage of net sales the gross profit is, we can fairly accurately estimate cost of goods sold. Companies often sell products that have similar gross profit ratios. As a result, accounting records usually provide the information necessary to estimate the cost of ending inventory, even when a physical count is impractical.

Suppose a company began 2021 with inventory of $600,000, and on March 17 a warehouse fire destroyed the entire inventory. Company records indicate net purchases of $1,500,000 and net sales of **$2,000,000** prior to the fire. The gross profit ratio in each of the previous three years has been very close to 40%. Illustration 9–4 shows how the company can estimate the cost of the inventory destroyed for its insurance claim.

A Word of Caution

The gross profit method provides only an estimate. The key to obtaining good estimates is the reliability of the gross profit ratio. The ratio usually is estimated from relationships between sales and cost of goods sold. However, the current relationship may differ from the past. In that case, all available information should be used to make necessary adjustments. For example, the company may have made changes in the markup percentage of some of its

The key to obtaining good estimates is the reliability of the gross profit ratio.

Illustration 9–4

Gross Profit Method

Beginning inventory (from records)		$ 600,000
Plus: Net purchases (from records)		1,500,000
Goods available for sale		2,100,000
Less: Cost of goods sold:		
Net sales	**$2,000,000**	
Less: Estimated gross profit of 40%	(800,000)	
Estimated cost of goods sold*		(1,200,000)
Estimated ending inventory		$ 900,000

*Alternatively, cost of goods sold can be calculated as $2,000,000 × (1 − 0.40) = $1,200,000.

products. Very often different products have different markups. In these situations, a blanket ratio should not be applied across the board. The accuracy of the estimate can be improved by grouping inventory into pools of products that have similar gross profit relationships rather than using one gross profit ratio for the entire inventory.

The company's cost flow assumption should be implicitly considered when estimating the gross profit ratio. For example, if LIFO is used and the relationship between cost and selling price has changed for recent acquisitions, this would suggest a ratio different from one where the average cost method was used.

Another difficulty with the gross profit method is that it does not explicitly consider possible theft or spoilage of inventory. The method assumes that if the inventory was not sold, then it must be on hand at the end of the period. Suspected theft or spoilage would require an adjustment to estimates obtained using the gross profit method.

The *gross profit method* is not acceptable for the preparation of annual financial statements.

Because of these deficiencies, the gross profit method is not allowed under generally accepted accounting principles for annual financial statements. The method can be used for interim reports.

Additional Consideration

The gross profit ratio is, by definition, a percentage of sales. Sometimes, though, the gross profit is stated as a percentage of cost instead. In that case, it is referred to as the **markup on cost**. For instance, a 66% markup on cost is equivalent to a gross profit ratio of 40%. Here's why:

A gross profit ratio of 40% can be formulated as follows:

$$\text{Sales} = \text{Cost} + \text{Gross profit}$$
$$100\% = 60\% + 40\%$$

Now, expressing gross profit as a percentage of cost we get the following:

$$\text{Gross profit \%} \div \text{Cost\%} = \text{Gross profit as a \% of cost}$$
$$40\% \div 60\% = 66\tfrac{2}{3}\%$$

Conversely, gross profit as a percentage of cost can be converted to gross profit as a percentage of sales (the gross profit ratio) as follows:

$$\text{Gross profit as a \% of sales} = \frac{\text{Gross profit as a \% of cost}}{1 + \text{Gross profit as a \% of cost}}$$

$$\frac{66\tfrac{2}{3}\%}{1 + 66\tfrac{2}{3}\%} = 40\%$$

Be careful to note which way the percentage is being stated. If stated as a markup on cost, it can be converted to the gross profit ratio, and the gross profit method can be applied the usual way.

The Retail Inventory Method

● LO9–3

The retail inventory method is used to estimate ending inventory and cost of goods sold.

As the name implies, the retail inventory method is used by many retail companies such as **Target**, **Walmart**, **Sears Holding Corporation**, **J.C. Penney**, and **Macy's**. Certain retailers like auto dealers and jewelry stores, whose inventory consists of few, high-priced items, can economically use the specific identification inventory method. However, high-volume retailers selling many different items at low unit prices find the retail inventory method ideal. Its principal benefit is that a physical count of inventory is not required to estimate ending inventory and cost of goods sold.[7]

In its simplest form, the retail inventory method first estimates the amount of ending inventory (at retail) by subtracting sales (at retail) from goods available for sale (at retail). *Retail* amounts refer to current selling prices. Ending inventory (at retail) is then multiplied by the current **cost-to-retail percentage** to estimate ending inventory (at cost). The cost-to-retail percentage is found by dividing goods available for sale at *cost* by goods available for sale at *current selling price*.

Illustration 9–5 provides an example of the retail inventory method. It shows how Home Improvement Stores, Inc., can use the relation between its inventory's cost and retail to estimate ending inventory and cost of goods sold for the month of June.

Illustration 9–5

Retail Inventory Method—Estimating ending inventory and cost of goods sold

Home Improvement Stores, Inc., uses a periodic inventory system and the retail inventory method to estimate ending inventory and cost of goods sold. The following data are available from the company's records for the month of June:

	Cost	Retail
Beginning inventory	$ 60,000	$ 100,000
Plus: Net purchases	287,200	460,000
Goods available for sale	347,200	560,000
Cost-to-retail percentage: $\frac{\$347,200}{\$560,000} = 62\%$		
Less: Net sales		(400,000)
Estimated ending inventory at retail		**$160,000**
Estimated ending inventory at cost ($160,000 × 62%)	**(99,200)**	
Estimated cost of goods sold	**$248,000***	

*Goods available for sale (at cost) minus ending inventory (at cost).

Home Improvement Stores first estimates ending inventory at retail (**$160,000**) by subtracting net sales from goods available for sale (at retail). Ending inventory at retail is multiplied by the cost-to-retail percentage (62%), which is found by dividing goods available for sale at *cost* by goods available for sale at *current selling price*. This multiplication leads to an estimate of ending inventory of **$99,200**. This amount is subtracted from the cost of goods available for sale to estimate cost of goods sold of **$248,000**.

The retail inventory method tends to provide a more accurate estimate than the gross profit method because it's based on the current relation between cost and selling prices rather than the historical gross profit ratio. This is one reason the retail inventory method is allowed for financial reporting purposes.

Another advantage of the retail inventory method is that different cost flow methods can be explicitly incorporated into the estimation technique. In other words, we can modify the application of the method to estimate ending inventory and cost of goods sold using FIFO, LIFO, or average cost.

As shown in Illustration 9–6, **American Eagle Outfitters** uses the retail inventory method with average cost to value its inventory.

[7]The retail inventory method is acceptable for external financial reporting if the results of applying the method are sufficiently close to what would have been achieved using a more rigorous determination of the cost of ending inventory. Also, it's allowed by the Internal Revenue Service as a method that can be used to determine cost of goods sold for income tax purposes.

Illustration 9–6
Inventory Method Disclosure—American Eagle Outfitters
Real World Financials

Summary of Significant Accounting Policies (in part)
Merchandise Inventory
Merchandise inventory is valued at the lower of average cost or market, utilizing the retail method.
Source: American Eagle Outfitters

Later in the chapter we illustrate average cost and LIFO with the retail inventory method. We do not illustrate the FIFO method with the retail inventory method because it is used infrequently in practice.

Like the gross profit method, the retail inventory method also can be used to estimate the cost of inventory lost, stolen, or destroyed; for testing the overall reasonableness of physical counts; in budgeting and forecasting as well as in generating information for interim financial statements. Even though the retail method provides fairly accurate estimates, a physical count of inventory usually is performed at least once a year to verify accuracy and detect spoilage, theft, and other irregularities.[8]

Retail Terminology

Changes in the selling prices must be included in the determination of ending inventory at retail.

Our example above is simplified in that we implicitly assumed that the selling prices of beginning inventory and of merchandise purchased did not change from date of acquisition to the end of the period. This frequently is an unrealistic assumption. After the initial markup of inventory but before it has been sold, companies sometimes increase the selling price further (additional markup) or reduce the selling price (markdown). For applying the retail inventory method, we need to track the movement in the selling price until it is sold. The terms in Illustration 9–7 are associated with changing retail prices of merchandise inventory.

Illustration 9–7
Terminology Used in Applying the Retail Method

Initial markup	Original amount of markup from cost to selling price.
Additional markup	Increase in selling price subsequent to initial markup.
Markup cancellation	Elimination of an additional markup.
Markdown	Reduction in selling price below the original selling price.
Markdown cancellation	Elimination of a markdown.

To illustrate, assume that a product purchased for $6 is initially listed with a selling price of $10 (that is, there is a $4 initial markup). If the selling price is subsequently increased to $12, the additional markup is $2. If the selling price is then subsequently decreased to **$10.50**, the markup cancellation is $1.50. A markup cancellation reduces an additional markup but not below the original selling price. We refer to the net effect of additional markups and markup cancellations ($2.00 – $1.50 = $0.50) as the **net markup**. Illustration 9–8A depicts these events.

Now, let's say the selling price of the product, purchased for $6 and initially marked up to $10, is reduced to $7. The markdown is $3. If the selling price is later increased to **$8**, the markdown cancellation is $1. A markdown cancellation reduces a markdown but not above the original selling price. We refer to the net effect of markdowns and markdown cancellations ($3 – $1 = $2) as the **net markdown**. Illustration 9–8B depicts this possibility.

Net markups and *net markdowns* are included in the retail column to determine ending inventory at retail.

When applying the retail inventory method, *net markups and net markdowns must be included in the determination of ending inventory at retail.* We now continue our illustration of the retail inventory method, but expand it to incorporate markups and markdowns as well as to approximate cost by each of the alternative inventory cost flow methods.

[8]The retail inventory method also is allowable under IFRS. "*Inventories,*" *International Accounting Standard No. 2 (IASCF),* as amended effective January 1, 2018, par. 22.

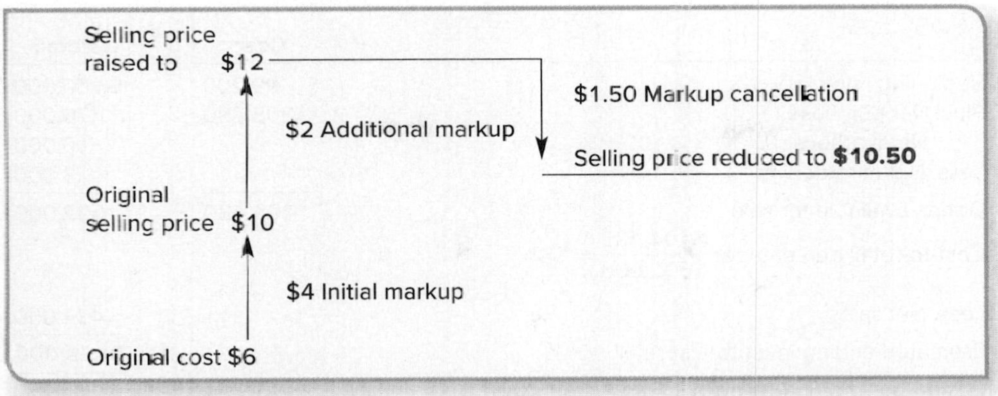

Illustration 9–8A

Retail Inventory Method Terminology

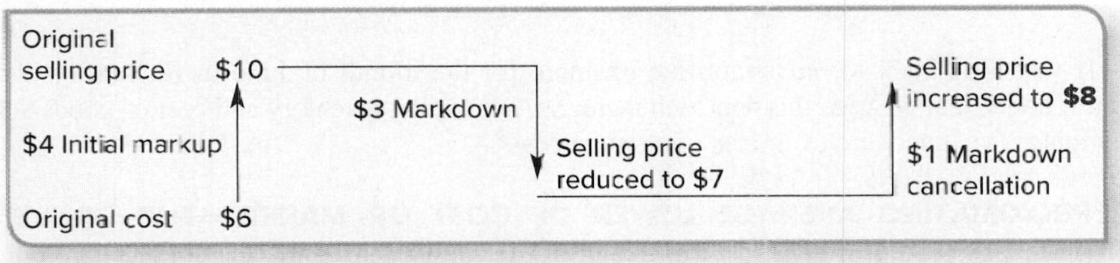

Illustration 9–8B

Retail Inventory Method Terminology

Cost Flow Methods

Let's continue our example of Home Improvement Stores, Inc., from Illustration 9–5. We'll use the data in Illustration 9–9 to see how the retail inventory method can be used to approximate different cost flow assumptions.

Illustration 9–9

The Retail Inventory Method—Various Cost Flow Methods

Home Improvement Stores, Inc., uses a periodic inventory system and the retail inventory method to estimate ending inventory and cost of goods sold. The following data are available from the company's records for the month of July:

	Cost	Retail
Beginning inventory	$ 99,200	$ 160,000
Net purchases	305,280*	470,000**
Net markups		10,000
Net markdowns		8,000
Net sales		434,000†

*Purchases at cost less returns, plus freight-in.
**Original selling price of purchased goods less returns at retail.
†Gross sales less returns.

APPROXIMATING AVERAGE COST Recall that the average cost method assumes that cost of goods sold and ending inventory each consist of a *mixture* of all the goods available for sale. So when we use the retail method to approximate average cost, the cost-to-retail percentage should be based on the weighted averages of the costs and retail amounts for *all* goods available for sale. This is achieved by calculating the cost-to-retail percentage by dividing the total cost of goods available for sale by total goods available for sale at retail. When this average percentage is applied to ending inventory at retail, we get an estimate of ending inventory at average cost.

To approximate average cost, the cost-to-retail percentage is determined for *all* goods available for sale.

We demonstrate the retail inventory method approximating average costs for July in Illustration 9–10. Notice that both markups and markdowns are included in the determination of goods available for sale at retail.

Illustration 9–10

Retail Inventory Method—Average Cost

	Cost	Retail
Beginning inventory	$ 99,200	$ 160,000
Plus: Net purchases	305,280	470,000
Net markups		10,000
Less: Net markdowns		(8,000)
Goods available for sale	404,480	632,000
Cost-to-retail percentage: $\frac{\$404,480}{\$632,000} = 64\%$		
Less: Net sales		(434,000)
Estimated ending inventory at retail		**$198,000**
Estimated ending inventory at cost ($198,000 × 64%)	**(126,720)**	
Estimated cost of goods sold	**$ 277,760**	

If you look back to our simplified example for the month of June in Illustration 9–5, you'll notice that we used this approach there. So, our ending inventory and cost-of-goods-sold estimates for June were estimates of average cost.[9]

● LO9–4

APPROXIMATING AVERAGE LOWER OF COST OR MARKET—THE CONVENTIONAL RETAIL METHOD Recall from our discussion earlier in the chapter that companies using the retail inventory method report inventory in the balance sheet at the lower of cost or market. Fortunately, we can apply the retail inventory method in such a way that the lower of average cost or market is approximated. This method often is referred to as the **conventional retail method**. We apply the method by *excluding markdowns from the calculation of the cost-to-retail percentage.* Markdowns still are subtracted in the retail column but only after the percentage is calculated. To approximate lower of average cost or market, the retail method is modified as shown in Illustration 9–11.

To approximate the lower of average cost or market, markdowns are not included in the calculation of the cost-to-retail percentage.

Illustration 9–11

Retail Inventory Method—Conventional

	Cost	Retail
Beginning inventory	$ 99,200	$ 160,000
Plus: Net purchases	305,280	470,000
Net markups		10,000
Goods available for sale (excluding net markdowns)	404,480	640,000
Cost-to-retail percentage: $\frac{\$404,480}{\$640,000} = 63.2\%$		
Less: Net markdowns		(8,000)
Goods available for sale (including net markdowns)		632,000
Less: Net sales		(434,000)
Estimated ending inventory at retail		**$198,000**
Estimated ending inventory at cost ($198,000 × 63.2%)	**(125,136)**	
Estimated cost of goods sold	**$ 279,344**	

Notice that by not subtracting net markdowns from the denominator, the cost-to-retail percentage is lower than it was previously (63.2% versus 64%). This always will be the case when markdowns exist. As a result, the cost approximation of ending inventory always will be less when markdowns exist.

The logic for using this approximation is that a markdown is evidence of a reduction in the utility of inventory.

To understand why this lower amount approximates the lower of average cost or market, we need to realize that markdowns usually occur when obsolescence, spoilage, overstocking, price declines, or competition has lessened the utility of the merchandise. To recognize this decline in utility in the period it occurs, we exclude net markdowns from the calculation

[9]We also implicitly assumed no net markups or markdowns in Illustration 9–5.

of the cost-to-retail percentage. It should be emphasized that this approach provides only an *approximation* of what ending inventory might be as opposed to applying the lower of cost or market rule in the more exact way described earlier in the chapter.

Also notice that the ending inventory at retail is the same using both approaches (**$198,000**). This will be the case regardless of the cost flow method used because in all approaches this amount reflects the ending inventory at current retail prices.

The conventional retail variation generally is not used in combination with LIFO. This does not mean that a company using LIFO ignores the lower of cost or market rule. Any obsolete or slow-moving inventory that has not been marked down by year-end can be written down to market after the estimation of inventory using the retail method. This usually is not a significant problem. If prices are rising, LIFO ending inventory includes old lower-priced items whose costs are likely to be lower than current market. The conventional retail variation could be applied to the FIFO method.

THE LIFO RETAIL METHOD The last-in, first-out (LIFO) method assumes that units sold are those most recently acquired. When there's a net increase in inventory quantity during a period, the use of LIFO results in ending inventory that includes the beginning inventory as well as one or more additional layers added during the period. When there's a net decrease in inventory quantity, LIFO layer(s) are liquidated. In applying LIFO to the retail method in the simplest way, we assume that the retail prices of goods remained stable during the period. This assumption, which is relaxed later in the chapter, allows us to look at the beginning and ending inventory in dollars to determine if inventory quantity has increased or decreased.

We'll use the numbers from our previous example to illustrate using the retail method to approximate LIFO so we can compare the results with those of the conventional retail method. Recall that beginning inventory at retail is $160,000 and ending inventory at retail is $198,000. If we assume stable retail prices, inventory quantity must have increased during the year. This means ending inventory includes the beginning inventory layer of $160,000 ($99,200 at cost) as well as some additional merchandise purchased during the period. To estimate total ending inventory at LIFO cost, we also need to determine the inventory layer added during the period. When using the LIFO retail method, we assume no more than one inventory layer is added per period if inventory increases.[10] Each layer will carry its own cost-to-retail percentage.

If inventory at retail increases during the year, a new layer is added.

Illustration 9–12 shows how Home Improvement Stores would estimate total ending inventory and cost of goods sold for the period using the LIFO retail method. The beginning inventory layer carries a cost-to-retail percentage of 62% ($99,200 ÷ $160,000). The layer of inventory added during the period is $38,000 at retail, which is determined by subtracting beginning inventory at retail from ending inventory at retail ($198,000 – $160,000). This layer will be converted to cost by multiplying it by its own cost-to-retail percentage reflecting the *current* period's ratio of cost to retail amounts, in this case 64.68%.

The next period's (August's) beginning inventory will include the two distinct layers (June and July), each of which carries its own unique cost-to-retail percentage. Notice in the illustration that both net markups and net markdowns are included in the calculation of the current period's cost-to-retail percentage.

Other Issues Pertaining to the Retail Method

To focus on the key elements of the retail method, we've so far ignored some of the details of the retail process. Fundamental elements such as returns and allowances, discounts, freight, spoilage, and shortages can complicate the retail method.

For calculating the cost-to-retail percentage, we use the cost of net purchases in the cost amount (numerator). Recall that net purchases is found by adding freight-in to purchases and subtracting both purchase returns and purchase discounts. This means that freight-in adds to the cost amount, while purchase returns and purchase discounts reduce the cost amount. For the retail amount (denominator), we calculate net purchases as total purchases (at retail) minus purchase returns (at retail). We subtract purchase returns for the simple reason that these items are no longer available for sale. We do not include freight-in and

[10]Of course, any number of layers at different costs can actually be added through the years. When using the regular LIFO method, rather than LIFO retail, we would keep track of each of those layers.

Illustration 9–12

LIFO Retail Method

		Cost	Retail
Beginning inventory		$ 99,200	**$160,000**
Plus: Net purchases		305,280	470,000
Net markups			10,000
Less: Net markdowns			(8,000)
Goods available for sale (excluding beginning inventory)		305,280	472,000
Goods available for sale (including beginning inventory)		404,480	632,000
Beginning inventory cost-to-retail percentage:	$\frac{\$99,200}{\$160,000} = $ **62.00%**		
July cost-to-retail percentage:	$\frac{\$305,280}{\$472,000} = $ **64.68%**		
Less: Net sales			(434,000)
Estimated ending inventory at retail			**$198,000**
Estimated ending inventory at cost (calculated below)		**(123,778)**	
Estimated cost of goods sold		**$280,702**	

Calculation of ending inventory at cost:

	Retail	Cost
Beginning inventory	**$160,000 × 62.00%** =	$ 99,200
Current period's layer	38,000 × **64.68%** =	24,578
Estimated ending Inventory	**$198,000**	**$123,778**

Each layer has its own cost-to-retail percentage.

purchase discounts in the calculation of net purchases for the retail amount because these are not items to be sold and therefore have no corresponding retail amounts.

Net sales, for purposes of applying the retail inventory method, include *sales returns* but exclude *sales discounts* and *employee discounts.*

After calculating the cost-to-retail percentage, we deduct net sales from goods available for sale (at retail). Net sales, for purposes of applying the retail inventory method, include *sales returns* but exclude *sales discounts* and *employee discounts.*

- Sales returns are subtracted from total sales because they represent inventory that has not been sold and therefore should not be included in the sales (or retail) amount.
- Sales discounts are not subtracted in calculating net sales because they do not represent an adjustment in selling prices but instead represent a financial incentive for customers to pay early. If sales initially were recorded net of sales discounts, those discounts are added back in calculating net sales.
- Employee discounts are not included in calculating net sales. The reason is that the cost-to-retail percentage did not include employee discounts, so the retail amount of ending inventory also should not be affected by employee discounts in estimating its cost. If sales to employees initially were recorded net of the employee-discounts, those discounts are added back in calculating net sales.

We also need to consider spoilage, breakage, and theft. So far we've assumed that by subtracting goods sold from goods available for sale, we find ending inventory. It's possible, though, that some of the goods available for sale were lost to such shortages and therefore do not remain in ending inventory.

Normal shortages are deducted in the retail column *after* the calculation of the cost-to-retail percentage.

Abnormal shortages are deducted in both the cost and retail columns *before* the calculation of the cost-to-retail percentage.

To take these shortages into account when using the retail method, we deduct the retail value of inventory lost due to spoilage, breakage, or theft in the retail column. These losses are expected for most retail ventures so they are referred to as *normal shortages* (spoilage, breakage, etc.), and are deducted from goods available for sale (at retail). Because these losses are anticipated, they are included implicitly in the determination of selling prices. Including normal spoilage in the calculation of the percentage would distort the normal relationship between cost and retail. In contrast, *abnormal shortages* are deducted from both the cost and retail amounts used to calculate the cost-to-retail percentage. These losses are not anticipated and are not included in the determination of selling prices.

We recap the treatment of special elements in the application of the retail method in Illustration 9–13 and illustrate the use of some of them in the concept review exercise that follows.

Illustration 9–13 Recap of Other Retail Method Elements

Element	Treatment
For calculating the cost-to-retail percentage:	
Freight-in	*Added* to the cost amount.
Purchase returns	*Deducted* from both the cost and retail amounts.
Purchase discounts taken (if gross method used to record purchases)	*Deducted* from the cost amount.
Abnormal shortages (spoilage, breakage, theft)	*Deducted* from both the cost and retail amounts.
After calculating the cost-to-retail percentage:	
Net sales (Total sales less sales returns)*	*Deducted* from goods available for sale (at retail).
Normal shortages (spoilage, breakage, theft)	*Deducted* from goods available for sale (at retail).

*The calculation of net sales for the retail inventory method does not include sales discounts or employee discounts.

Concept Review Exercise

RETAIL INVENTORY METHOD

The Henderson Company uses the retail inventory method to estimate ending inventory and cost of goods sold. The following data are available in Henderson's accounting records:

	Cost	Retail
Beginning inventory	$ 8,000	$12,000
Purchases	68,000	98,000
Freight-in	3,200	
Purchase returns	3,000	4,200
Net markups		6,000
Net markdowns		2,400
Normal spoilage		1,800
Sales		97,000
Sales returns		5,000
Employee discounts		2,300

Sales of $97,000 include a reduction for employee discounts of $2,300.

Required:

1. Estimate Henderson's ending inventory and cost of goods sold for the year using the average cost retail method.
2. Estimate Henderson's ending inventory and cost of goods sold for the year using the conventional retail method.
3. Estimate Henderson's ending inventory and cost of goods sold for the year using the LIFO retail method.

Solution:

1. Estimate Henderson's ending inventory and cost of goods sold for the year using the average cost retail method.

	Cost	Retail
Beginning inventory	$ 8,000	$ 12,000
Plus: Purchases	68,000	98,000
Freight-in	3,200	
Less: Purchase returns	(3,000)	(4,200)
Plus: Net markups		6,000
Less: Net markdowns		(2,400)
Goods available for sale	76,200	109,400

Cost-to-retail percentage: $\frac{\$76,200}{\$109,400} = 69.65\%$

(continued)

		Cost	Retail
Less: Normal spoilage			(1,800)
Less: Net sales			
Sales	$97,000		
Employee discounts	2,300		
Sales returns	(5,000)		(94,300)
Estimated ending inventory at retail			$ 13,300
Estimated ending inventory at cost (69.65% × $13,300)		(9,263)	
Estimated cost of goods sold		$66,937	

2. Estimate Henderson's ending inventory and cost of goods sold for the year using the conventional retail method.

		Cost	Retail
Beginning inventory		$ 8,000	$ 12,000
Plus: Purchases		68,000	98,000
Freight-in		3,200	
Less: Purchase returns		(3,000)	(4,200)
Plus: Net markups			6,000
			111,800

Cost-to-retail percentage: $\frac{\$76,200}{\$111,800} = 68.16\%$

		Cost	Retail
Less: Net markdowns			(2,400)
Goods available for sale		76,200	109,400
Less: Normal spoilage			(1,800)
Less: Net sales:			
Sales	$97,000		
Employee discounts	2,300		
Sales returns	(5,000)		(94,300)
Estimated ending inventory at retail			$ 13,300
Estimated ending inventory at cost (68.16% × $13,300)		(9,065)	
Estimated cost of goods sold		$67,135	

3. Estimate Henderson's ending inventory and cost of goods sold for the year using the LIFO retail method.

		Cost	Retail
Beginning inventory		$ 8,000	$ 12,000
Plus: Purchases		68,000	98,000
Freight-in		3,200	
Less: Purchase returns		(3,000)	(4,200)
Plus: Net markups			6,000
Less: Net markdowns			(2,400)
Goods available for sale (excluding beginning inventory)		68,200	97,400
Goods available for sale (including beginning inventory)		76,200	109,400

Cost-to-retail percentage: $\frac{\$68,200}{\$97,400} = 70.02\%$

		Cost	Retail
Less: Normal spoilage			(1,800)
Less: Net sales:			
Sales	$97,000		
Employee discounts	2,300		
Sales returns	(5,000)		(94,300)
Estimated ending inventory at retail			$ 13,300
Estimated ending inventory at cost (see below)		(8,910)	
Estimated cost of goods sold		$67,290	

	Retail	Cost
Beginning inventory	$12,000 × 66.67%* =	$8,000
Current period's layer	1,300 × 70.02% =	910
Estimated Ending Inventory	$13,300	$8,910

*$8,000 ÷ $12,000 = 66.67%

PART C

Dollar-Value LIFO Retail

● LO9–5

In our earlier discussion of the LIFO retail method, we assumed that the retail prices of the inventory remained stable during the period. If you recall from Illustration 9–12, we compared the ending inventory (at retail) with the beginning inventory (at retail) to see if inventory had increased. If the dollar amount of ending inventory exceeded the beginning amount, we assumed a new LIFO layer had been added. But this isn't necessarily true. It may be that the dollar amount of ending inventory exceeded the beginning amount simply because retail prices increased, without an actual change in the quantity of goods. So, to see if there's been a "real" increase in quantity, we need a way to eliminate the effect of any price changes before we compare the ending inventory with the beginning inventory. Fortunately, we can accomplish this by combining two methods we've already discussed—the LIFO retail method (Part B of this chapter) and dollar-value LIFO (previous chapter). The combination is called the **dollar-value LIFO retail method.**

FINANCIAL Reporting Case

Q1, p. 459

Allow for retail prices to change during the period.

Using the retail method to approximate LIFO is referred to as the *dollar-value LIFO retail method.*

To illustrate, we return to the Home Improvement Stores situation (Illustration 9–12) in which we applied LIFO retail. We keep the same inventory data, but change the illustration from the month of July to the fiscal year 2021. This allows us to build into Illustration 9–14 a significant change in retail prices over the year of 10% (an increase in the retail price index from 1.00 to 1.10). We follow the LIFO retail procedure up to the point of comparing the ending inventory with the beginning inventory. However, because prices have risen, the apparent increase in inventory is only partly due to an additional layer of inventory and partly due to the increase in retail prices. The real increase is found by deflating the ending inventory amount to beginning of the year prices before comparing beginning and ending amounts. We did this with the dollar-value LIFO technique discussed in the previous chapter.[11]

FINANCIAL Reporting Case

Q2, p. 459

Illustration 9–14

The Dollar-Value LIFO Retail Method

	Cost	Retail
Beginning inventory	$ 99,200	$160,000
Plus: Net purchases	305,280	470,000
Net markups		10,000
Less: Net markdowns		(8,000)
Goods available for sale (excluding beginning inventory)	305,280	472,000
Goods available for sale (including beginning inventory)	404,480	632,000
Base layer cost-to-retail percentage: $\frac{\$99,200}{\$160,000} = 62\%$		
2021 layer cost-to-retail percentage: $\frac{\$305,280}{\$472,000} = 64.68\%$		
Less: Net sales		(434,000)
Ending inventory at current year retail prices		$198,000
Estimated ending inventory at cost (calculated below)	**(113,430)**	
Estimated cost of goods sold	**$ 291,050**	

Calculation of ending inventory at cost:

Ending Inventory at Year-End Retail Prices	Step 1 Ending Inventory at Base Year Retail Prices	Step 2 Inventory Layers at Base Year Retail Prices	Step 3 Inventory Layers Converted to LIFO Cost
$198,000 (assumed) →	$\frac{\$198,000}{1.10}$ = $180,000 →	$180,000	
		160,000 (base) × 1.00 × 0.62	= $ 99,200
		20,000 (2021) × 1.10 × 0.6468 =	14,230
Total ending inventory at dollar-value LIFO retail cost			$113,430

[11]The index used here is analogous to the cost index used in regular DVL except that it reflects the change in retail prices rather than in acquisition costs.

Each layer year carries its unique retail price index and its unique cost-to-retail percentage.

In this illustration, the ending inventory (at retail) of **$198,000** is restated to base year prices ($180,000). Comparing this restated amount to beginning inventory layer of **$160,000** reveals that a layer of **$20,000** has been added in 2021. Multiplying each layer by its retail price index and by its cost-to-retail percentage converts it from retail to cost. The two layers are added to derive ending inventory at dollar-value LIFO retail cost.

When additional layers are added in subsequent years, their LIFO amounts are determined the same way. For illustration, let's assume ending inventory in 2022 is **$226,200** at current retail prices and the price level has risen to **1.16**. Also assume that the cost-to-retail percentage for 2022 net purchases is 63%. In Illustration 9–14A, the ending inventory converted to base year retail prices equals $195,000 (step 1). This amount represents the base year layer of $160,000 plus the layer of $20,000 added in 2021 plus a new layer of $15,000 added in 2022 (step 2). Each of these layers is multiplied by its retail price index and by its cost-to-retail percentage to calculate ending inventory of **$124,392** (step 3).

Illustration 9–14A

The Dollar-Value LIFO Retail Inventory Method—Layer Added in Subsequent Year

Ending Inventory at Year-End Retail Prices	Step 1 Ending Inventory at Base Year Retail Prices	Step 2 Inventory Layers at Base Year Retail Prices	Step 3 Inventory Layers Converted to LIFO Cost
$226,200 (assumed) →	$226,200 / **1.16** = $195,000 →	$195,000	
		160,000 (base) × 1.00 × 0.62 =	$ 99,200
		20,000 (2021) × 1.10 × 0.6468 =	14,230
		15,000 (2022) × 1.16 × 0.63 =	10,962
Total ending inventory at dollar-value LIFO retail cost			**$124,392**

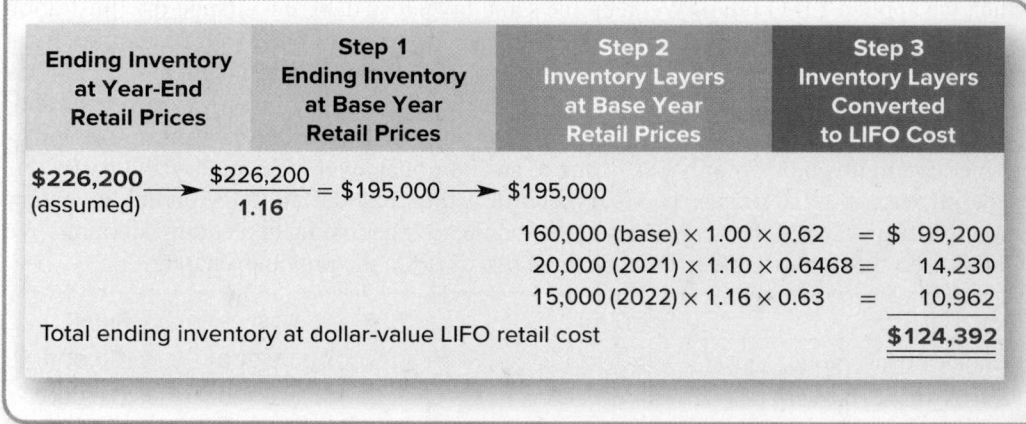

Now, let's assume that ending inventory in 2022 is **$204,160** at current retail prices (instead of $226,200 as in the previous example) and the price level has risen to **1.16**. Also assume that the cost-to-retail percentage for 2022 net purchases is 63%. As shown in step 1 of Illustration 9–14B, we convert ending inventory to a base year price of $176,000 ($204,160 ÷ 1.16).

When inventory decreases, no new layer is added and a previous layer is (partially) removed.

Next, we determine the layers of inventory (step 2). To do that, let's first recall from Illustration 9–14 that inventory layers at the end of 2021 consist of $160,000 from the base year and $20,000 added in 2021. Those layers totaled $180,000. In 2022, total inventory at base year prices has *declined* to $176,000. The decline means that no layer was added in 2022, and $4,000 of the $20,000 layer in 2021 has been sold, reducing the previous 2021 layer to a remainder of $16,000. Each layer is multiplied by its retail price index and by its cost-to-retail percentage to calculate ending inventory of **$110,584** (step 3).

Illustration 9–14B

The Dollar-Value LIFO Retail Inventory Method—No Layer Added in Subsequent Year

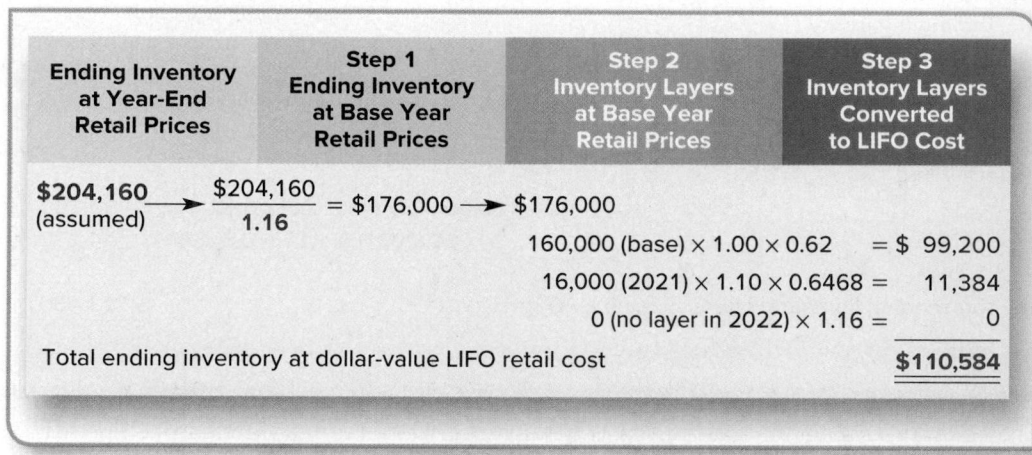

Ending Inventory at Year-End Retail Prices	Step 1 Ending Inventory at Base Year Retail Prices	Step 2 Inventory Layers at Base Year Retail Prices	Step 3 Inventory Layers Converted to LIFO Cost
$204,160 (assumed) →	$204,160 / **1.16** = $176,000 →	$176,000	
		160,000 (base) × 1.00 × 0.62 =	$ 99,200
		16,000 (2021) × 1.10 × 0.6468 =	11,384
		0 (no layer in 2022) × 1.16 =	0
Total ending inventory at dollar-value LIFO retail cost			**$110,584**

As we mentioned earlier in this section, many high-volume retailers selling many different items use the retail method. **Costco Wholesale Corporation**, for example, uses the dollar-value LIFO variation of the retail method. Illustration 9–15 shows the inventory disclosure note included in the company's recent financial statements.

Merchandise Inventories (in part)

Merchandise inventories are stated at the lower of cost or market. U.S. merchandise inventories are valued by the cost method of accounting, using the last-in, first-out (LIFO) basis. The Company believes the LIFO method more fairly presents the results of operations by more closely matching current costs with current revenues. The Company records an adjustment each quarter, if necessary, for the projected annual effect of inflation or deflation, and these estimates are adjusted to actual results determined at year-end, after actual inflation or deflation rates and inventory levels for the year have been determined.

Source: Costco Wholesale Corporation

Illustration 9–15

Disclosure of Inventory Method—Costco Wholesale Corporation

Real World Financials

Concept Review Exercise

DOLLAR-VALUE LIFO RETAIL METHOD

On January 1, 2021, the Nicholson Department Store adopted the dollar-value LIFO retail inventory method. Inventory transactions at both cost and retail and cost indexes for 2021 and 2022 are as follows:

	2021		2022	
	Cost	Retail	Cost	Retail
Beginning inventory	$16,000	$24,000		
Net purchases	42,000	58,500	$45,000	$58,700
Net markups		3,000		2,400
Net markdowns		1,500		1,100
Net sales		56,000		57,000
Price index:				
January 1, 2021	1.00			
December 31, 2021	1.08			
December 31, 2022	1.15			

Required:

Estimate the 2021 and 2022 ending inventory and cost of goods sold using the dollar-value LIFO retail inventory method.

Solution:

	2021		2022	
	Cost	Retail	Cost	Retail
Beginning inventory	$ 16,000	$ 24,000	$ 17,456	$ 28,000
Plus: Net purchases	42,000	58,500	45,000	58,700
Net markups		3,000		2,400
Less: Net markdowns		(1,500)		(1,100)
Goods available for sale (excluding beg. inv.)	42,000	60,000	45,000	60,000
Goods available for sale (including beg. inv.)	58,000	84,000	62,456	88,000
Base layer				
Cost-to-retail percentage: $\frac{\$16,000}{\$24,000} = 66.67\%$				
2021				
Cost-to-retail percentage: $\frac{\$42,000}{\$60,000} = 70.00\%$				
2022				
Cost-to-retail percentage: $\frac{\$45,000}{\$60,000} = 75.00\%$				
Less: Net sales		(56,000)		(57,000)
Estimated ending inv. at current year retail prices		$ 28,000		$ 31,000
Less: Estimated ending inventory at cost (below)	(17,456)		(18,345)	
Estimated cost of goods sold	$ 40,544		$ 44,111	

2021

Ending Inventory at Year-End Retail Prices	Step 1 Ending Inventory at Base Year Retail Prices	Step 2 Inventory Layers at Base Year Retail Prices	Step 3 Inventory Layers Converted to Cost
$28,000 (above)	$\frac{\$28,000}{1.08} = \$25,926$	\$24,000 (base) × 1.00 × 66.67% =	$16,000
		1,926 (2021) × 1.08 × 70.00% =	1,456
Total ending inventory at dollar-value LIFO retail cost			$17,456

2022

Ending Inventory at Year-End Retail Prices	Step 1 Ending Inventory at Base Year Retail Prices	Step 2 Inventory Layers at Base Year Retail Prices	Step 3 Inventory Layers Converted to Cost
$31,000 (above)	$\frac{\$31,000}{1.15} = \$26,957$	\$24,000 (base) × 1.00 × 66.67% =	$16,000
		1,926 (2021) × 1.08 × 70.00% =	1,456
		1,031 (2022) × 1.15 × 75.00% =	889
Total ending inventory at dollar-value LIFO retail cost			$18,345

PART D

Change in Inventory Method and Inventory Errors

Change in Inventory Method

LO9–6

Accounting principles should be applied consistently from period to period to allow for comparability of operating results. However, changes within a company as well as changes in the external economic environment may require a company to change an accounting method. As we mentioned in Chapter 8, in the past, high inflation periods motivated many companies to switch to LIFO for the tax benefit.

Specific accounting treatment and disclosures are prescribed for companies that change accounting principles. Chapter 4 introduced the subject of accounting changes and Chapter 20 provides in-depth coverage of the topic. Here we provide an overview of how changes in inventory methods are reported.

Most Inventory Changes

Changes in inventory methods, other than a change to LIFO, are accounted for retrospectively.

Recall from our brief discussion in Chapter 4 that most voluntary changes in accounting principles are reported retrospectively. This means reporting all previous periods' financial statements as if the new method had been used in all prior periods. Changes in inventory methods, other than a change to LIFO, are treated this way. We discuss the *change to LIFO* exception in the next section. In Chapter 4, we briefly discussed the steps a company undertakes to account for a change in accounting principle. We demonstrate those steps in Illustration 9–16.

Illustration 9–16
Change in Inventory Method

Autogeek, Inc., a wholesale distributor of auto parts, began business in 2018. Inventory reported in the 2020 year-end balance sheet, determined using the average cost method, was **$123,000**. In 2021, the company decided to change its inventory method to FIFO. If the company had used the FIFO method in 2020, ending inventory would have been **$146,000**. What steps should Autogeek take to report this change?

Step 1: Revise comparative financial statements.

The first step is to revise prior years' financial statements. That is, for each year reported in the comparative statements, Autogeek makes those statements appear as if the newly adopted inventory method, FIFO, had been applied all along. In its balance sheets, assuming that the company presents balance sheets for two years for comparative purposes, the company would report 2021 inventory by its newly adopted method, FIFO, and also would revise the amounts it reported last year for its 2020 inventory. In its 2021 and prior year income statements, cost of goods sold would also reflect the new method.

In its statements of shareholders' equity, Autogeek would report retained earnings each year as if it had used FIFO all along. And, for the earliest year reported, the company would revise beginning retained earnings that year to reflect the cumulative income effect of the difference in inventory methods for all prior years. You will see this step illustrated in Chapter 20 after you have studied the statement of shareholders' equity in more depth.

Step 2: The affected accounts are adjusted.

Autogeek also would record a journal entry to adjust the book balances from their current amounts to what those balances would have been using FIFO. Because differences in cost of goods sold and income are reflected in retained earnings, as are the income tax effects, the journal entry updates inventory, retained earnings, and the appropriate income tax account. We ignore the income tax effects here but include those effects in an illustration in Chapter 20. The journal entry below, *ignoring income taxes,* increases the 2021 beginning inventory to the FIFO basis amount of $146,000 and increases retained earnings by the same amount, because that's what the increase in prior years' income would have been had FIFO been used.

Inventory (**$146,000 − $123,000**)	23,000	
Retained earnings		23,000

Step 3: A disclosure note provides additional information.

Autogeek must provide in a disclosure note clear explanation of why the change to FIFO is preferable. The note also would indicate the effects of the change on (a) income from continuing operations, (b) net income, (c) each line-item affected, (d) earnings per share, and (e) the cumulative effect of the change on retained earnings or other components of equity as of the beginning of the earliest period presented.

We see in Illustration 9–17 an example of such a note in a recent annual report of **CVS Health Corporation** when it changed its inventory method for retail/LTC inventories to the average cost method.

Illustration 9–17

Disclosure of Change in Inventory Method—CVS Health Corporation

Real World Financials

Inventory (in part)

Effective January 1, 2015, the Company changed its methods of accounting for "front store" inventories in the Retail/LTC Segment. Prior to 2015, the Company valued front store inventories at the lower of cost or market on a first-in, first-out ("FIFO") basis in retail stores using the retail inventory method and in distribution centers using the FIFO cost method. Effective January 1, 2015, all front store inventories in the Retail/LTC Segment have been valued at the lower of cost or market using the weighted average cost method.

These changes were made primarily to provide the Company with better information to manage its retail front store operations and to bring all of the Company's inventories to a common inventory valuation methodology. The Company believes the weighted average cost method is preferable to the retail inventory method and the FIFO cost method because it results in greater precision in the determination of cost of revenues and inventories at the stock keeping unit ("SKU") level and results in a consistent inventory valuation method for all of the Company's inventories as all of the Company's remaining inventories, which consist of prescription drugs, were already being valued using the weighted average cost method.

The Company recorded the cumulative effect of these changes in accounting principle as of January 1, 2015. The effect of these changes in accounting principle as of January 1, 2015, was a decrease in inventories of $7 million, an increase in current deferred income tax assets of $3 million and a decrease in retained earnings of $4 million.

Source: CVS Health Corporation

Change to the LIFO Method

Accounting records usually are inadequate for a company changing to *LIFO* to report the change retrospectively.

When a company changes *to the LIFO inventory method* from any other method, it usually is impossible to calculate the income effect on prior years. To do so would require assumptions as to when specific LIFO inventory layers were created in years prior to the change.

As a result, a company changing to LIFO usually does not report the change retrospectively. Instead, the LIFO method simply is used from that point on. The base year inventory for all future LIFO determinations is the beginning inventory in the year the LIFO method is adopted.[12]

A disclosure note is needed to explain (a) the nature of and justification for the change, (b) the effect of the change on current year's income and earnings per share, and (c) why retrospective application was impracticable. When **Seneca Foods Corporation** adopted the LIFO inventory method, it reported the change in the note shown in Illustration 9–18.

Illustration 9–18

Change in Inventory Method Disclosure—Seneca Foods Corporation

Real World Financials

10. Inventories (in part)

The Company decided to change its inventory valuation method from the FIFO method to the LIFO method. In the high inflation environment that the Company is experiencing, the Company believes that the LIFO inventory method is preferable over the FIFO method because it better compares the cost of current production to current revenue. Selling prices are established to reflect current market activity, which recognizes the increasing costs. Under FIFO, revenue and costs are not aligned. Under LIFO, the current cost of sales is matched to the current revenue.

The Company determined that retrospective application of LIFO for periods prior to the current fiscal year was impracticable because the period-specific information necessary to analyze inventories, including inventories acquired as part of the prior fiscal year's Signature acquisition, were not readily available and could not be precisely determined at the appropriate level of detail, including the commodity, size and item code information necessary to perform the detailed calculations required to retrospectively compute the internal LIFO indices applicable to prior fiscal years. The effect of this change was to reduce net earnings by $37,917,000 and $18,307,000 in the current and prior fiscal year, respectively, below that which would have been reported using the Company's previous inventory method. The reduction in earnings per share was $3.12 ($3.09 diluted) and $1.50 per share ($1.49 diluted) in the current and prior fiscal year, respectively.

Source: Seneca Foods Corporation

As we discussed in Chapter 8, an important motivation for using LIFO in periods of rising costs is that it produces higher cost of goods sold and lowers net income and income taxes. Notice in the Seneca Foods disclosure note that the switch to LIFO did cause a decrease in net income and therefore income taxes in the year of the switch indicating an environment of increasing costs.

Additional Consideration

When changing from one generally accepted accounting principle to another, a company must justify that the change results in financial information that more properly portrays operating results and financial position. For income tax purposes, a company generally must obtain consent from the Internal Revenue Service before changing an accounting method. A special form also must be filed with the IRS when a company intends to adopt the LIFO inventory method. When a company changes from LIFO for tax purposes, it can't change back to LIFO until five tax returns have been filed using the non-LIFO method.

Inventory Errors

● LO9–7

Accounting errors must be corrected when they are discovered. In Chapter 4, we briefly discussed the correction of accounting errors, and Chapter 20 provides in-depth coverage. Here we provide an overview of the accounting treatment and disclosures in the context of inventory errors. Inventory errors include the over- or understatement of ending inventory

[12]A change to LIFO is handled the same way for income tax purposes.

due to a mistake in physical count or a mistake in pricing inventory quantities. Also, errors include the over- or understatement of purchases which could be caused by the cutoff errors described in Chapter 8.

If an inventory error is discovered in the same accounting period it occurred, the original erroneous entry should simply be reversed and the appropriate entry recorded. This situation presents no particular reporting problem.

If a *material* inventory error is discovered in an accounting period subsequent to the period in which the error was made, any previous years' financial statements that were incorrect as a result of the error are retrospectively restated to reflect the correction.[13] And, of course, any account balances that are incorrect as a result of the error are corrected by journal entry. If, due to an error affecting net income, retained earnings is one of the incorrect accounts, the correction is reported as a prior period adjustment to the beginning balance on the statement of shareholders' equity.[14] In addition, a disclosure note is needed to describe the nature of the error and the impact of its correction on net income, each line-item affected, and earnings per share.

For material errors, previous years' financial statements are retrospectively restated.

Incorrect balances are corrected.

When analyzing inventory errors, it's helpful to visualize the way cost of goods sold, net income, and retained earnings are determined (see Illustration 9–19). Beginning inventory and net purchases are *added* in the calculation of cost of goods sold. If either of these is overstated (understated), then cost of goods sold would be overstated (understated). On the other hand, ending inventory is *deducted* in the calculation of cost of goods sold, so if ending inventory is overstated (understated), then cost of goods sold is understated (overstated). Of course, errors that affect income also will affect income taxes. In the illustration that follows, we ignore the tax effects of the errors and focus on the errors themselves rather than their tax aspects.

A correction of retained earnings is reported as a prior period adjustment.

A disclosure note describes the nature and the impact of the error.

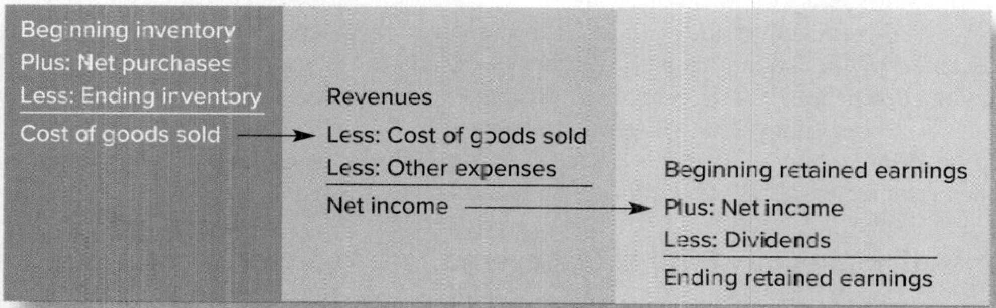

Illustration 9–19
Visualizing the Effect of Inventory Errors

Illustration 9-20 provides a numerical example of how an inventory error affects reported amounts in both the current year and the following year.

When the Inventory Error Is Discovered the Following Year

Now, let's assume the error in Illustration 9-20 is discovered in 2021. The 2020 financial statements that were incorrect as a result of the error are retrospectively restated to reflect the correct inventory amount, cost of goods sold, net income, and retained earnings when those statements are reported again for comparative purposes in the 2021 annual report. The following journal entry, *ignoring income taxes,* is needed in 2021 to correct the error in the company's records.

Previous years' financial statements are retrospectively restated.

Retained earnings	800,000	
Inventory		800,000

A journal entry corrects any incorrect account balance.

[13] If the effect of the error is not material, it is simply corrected in the year of discovery.

[14] The prior period adjustment is applied to beginning retained earnings for the year following the error, or for the earliest year being reported in the comparative financial statements when the error occurs prior to the earliest year presented. The retained earnings balances in years after the first year also are adjusted to what those balances would be if the error had not occurred, but a company may choose not to explicitly report those adjustments as separate line items.

Illustration 9–20

Inventory Error Correction

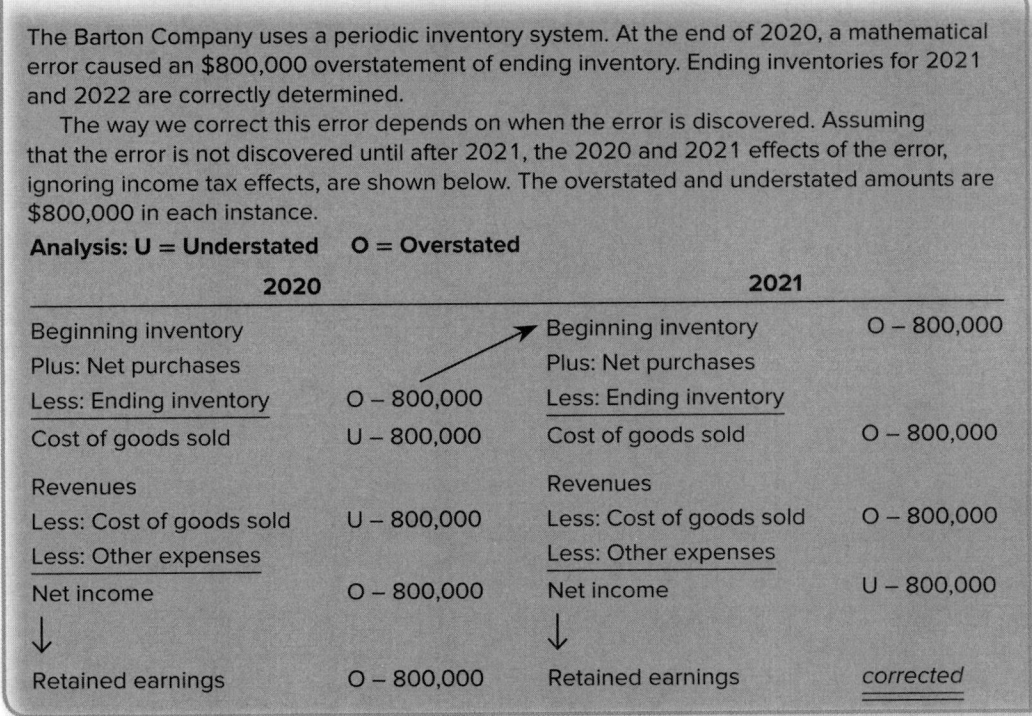

The Barton Company uses a periodic inventory system. At the end of 2020, a mathematical error caused an $800,000 overstatement of ending inventory. Ending inventories for 2021 and 2022 are correctly determined.

The way we correct this error depends on when the error is discovered. Assuming that the error is not discovered until after 2021, the 2020 and 2021 effects of the error, ignoring income tax effects, are shown below. The overstated and understated amounts are $800,000 in each instance.

Analysis: U = Understated O = Overstated

2020		2021	
Beginning inventory		Beginning inventory	O – 800,000
Plus: Net purchases		Plus: Net purchases	
Less: Ending inventory	O – 800,000	Less: Ending inventory	
Cost of goods sold	U – 800,000	Cost of goods sold	O – 800,000
Revenues		Revenues	
Less: Cost of goods sold	U – 800,000	Less: Cost of goods sold	O – 800,000
Less: Other expenses		Less: Other expenses	
Net income	O – 800,000	Net income	U – 800,000
↓		↓	
Retained earnings	O – 800,000	Retained earnings	*corrected*

When retained earnings requires correction, a *prior period adjustment* is made on the statement of shareholders' equity.

We debit retained earnings in 2021 to correct its balance. Net income in 2020 was overstated by the amount of the inventory error, and this caused retained earnings to be overstated. The correction to retained earnings is reported as a *prior period adjustment* to the 2021 beginning retained earnings balance in Barton's statement of shareholders' equity. Prior period adjustments do not flow through the income statement but directly adjust retained earnings. This adjustment is illustrated in Chapter 20.

When the Inventory Error Is Discovered Two Years Later

If the error in 2020 isn't discovered until 2022, the 2021 financial statements also are retrospectively restated to reflect the correct cost of goods sold and net income even though no correcting entry would be needed at that point. Inventory and retained earnings in 2022 would not require adjustment. The error has self-corrected and no prior period adjustment is needed.

A disclosure note describes the nature of the error and the impact of the correction on income.

Also, a disclosure note in Barton's annual report should describe the nature of the error and the impact of its correction on each year's net income (overstated by $800,000 in 2020; understated by $800,000 in 2021), the line items affected, and earnings per share.

Concept Review Exercise

INVENTORY ERRORS

In 2021, the controller of the Fleischman Wholesale Beverage Company discovered the following material errors related to the 2019 and 2020 financial statements:

a. Inventory at the end of 2019 was understated by $50,000.
b. Late in 2020, a $3,000 purchase was incorrectly recorded as a $33,000 purchase. The invoice has not yet been paid.
c. Inventory at the end of 2020 was overstated by $20,000.

The company uses a periodic inventory system.

Required:

1. Assuming that the errors were discovered after the 2020 financial statements were issued, analyze the effect of the errors on 2019 and 2020 cost of goods sold, net income, and retained earnings. Ignore income taxes.
2. Prepare a journal entry in 2021 to correct the errors.

Solution:

1.

Analysis: U = Understated O = Overstated

2019		2020	
Beginning inventory		Beginning inventory	U – 50,000
Plus: Net purchases		Plus: Net purchases	O – 30,000
Less: Ending inventory	U – 50,000	Less: Ending inventory	O – 20,000
Cost of goods sold	O – 50,000	Cost of goods sold	U – 40,000
Revenues		Revenues	
Less: Cost of goods sold	O – 50,000	Less: Cost of goods sold	U – 40,000
Less: Other expenses		Less: Other expenses	
Net income	U – 50,000	Net income	O – 40,000
↓		↓	
Retained earnings	U – 50,000	Retained earnings	U – 10,000

2. Prepare a journal entry in 2021 to correct the errors.

Accounts payable	30,000	
Inventory		20,000
Retained earnings		10,000

Earnings Quality

A change in the accounting method a company uses to value inventory is one way managers can artificially manipulate income. However, this method of income manipulation is transparent. As we learned in a previous section, the effect on income of switching from one inventory method to another must be disclosed. That disclosure restores comparability between periods and enhances earnings quality.

On the other hand, inventory write-downs are included in the broader category of "big bath" accounting techniques some companies use to manipulate earnings. By overstating the write-down, profits are increased in future periods as the inventory is used or sold. When the demand for many high-technology products decreased significantly in late 2000 and early 2001, several companies, including **Sycamore Networks**, **Cisco Systems**, **Lucent Technologies**, and **JDS Uniphase**, recorded large inventory write-offs, some in the billions of dollars. Certainly, these write-offs reflected the existing economic environment. However, some analysts questioned the size of some of the write-offs. For example, William Schaff, an investment officer at Bay Isle Financial, noted that Cisco's $2 billion write-off was approximately equal to the balance of inventory on hand at the end of the previous quarter and about equal to the cost of goods actually sold during the quarter.

Inventory write-downs often are cited as a method used to shift income between periods.

A financial analyst must carefully consider the effect of any significant asset write-down on the assessment of a company's permanent earnings.

Financial Reporting Case Solution

©QualityHD/Shutterstock.com

1. **How does Dollar General avoid counting all its inventory every time it produces financial statements?** *(p. 477)* Dollar General uses the dollar-value LIFO retail inventory method. The retail inventory estimation technique avoids the counting of ending inventory by keeping track of goods available for sale not only at cost but also at retail prices. Each period's sales, at sales prices, are deducted from the retail amount of goods available for sale to arrive at ending inventory at retail. This amount is then converted to cost using a cost-to-retail percentage.
2. **What is the index formulation used for?** *(p. 477)* The dollar-value LIFO retail method uses a price index to first convert ending inventory at retail to base year prices. Yearly LIFO layers are then determined, and each layer is converted to that year's current year retail prices using the year's price index and then to cost using the layer's cost-to-retail percentage. ●

The Bottom Line

- **LO9–1** Companies that use FIFO, average cost, or any other method besides LIFO or the retail inventory method report inventory at the lower of cost or net realizable value (NRV). Net realizable value is selling price less costs to sell. Companies that use LIFO or the retail inventory method report inventory at the lower of cost or market. Market equals replacement cost, except that market should not (a) be greater than NRV (ceiling) or (b) be less than NRV minus an approximately normal profit margin (floor). (*p. 459*)
- **LO9–2** The gross profit method estimates cost of goods sold which is then subtracted from cost of goods available for sale to estimate ending inventory. The estimate of cost of goods sold is determined by subtracting an estimate of gross profit from net sales. The estimate of gross profit is determined by multiplying the historical gross profit ratio times net sales. (*p. 467*)
- **LO9–3** The retail inventory method determines the amount of ending inventory at retail by subtracting sales for the period from goods available for sale at retail. Ending inventory at retail is then converted to *cost* by multiplying it by the cost-to-retail percentage, which is based on a current relationship between cost and selling price. (*p. 469*)
- **LO9–4** By the conventional retail method, we estimate average cost at lower of cost or market. Average cost is estimated by including beginning inventory in the calculation of the cost-to-retail percentage. The lower of average cost or market is estimated by excluding markdowns from the calculation. Markdowns are subtracted in the retail column after the percentage is calculated. (*p. 472*)
- **LO9–5** By the LIFO retail method, ending inventory includes the beginning inventory plus the current year's layer. To determine layers, we compare ending inventory at retail to beginning inventory at retail and assume that no more than one inventory layer is added if inventory increases. Each layer carries its own cost-to-retail percentage which is used to convert each layer from retail to cost. The dollar-value LIFO retail inventory method combines the LIFO retail method and the dollar-value LIFO method (Chapter 8) to estimate LIFO from retail prices when the price level has changed. (*p. 477*)
- **LO9–6** Most changes in inventory methods are reported retrospectively. This means revising all previous periods' financial statements to appear as if the newly adopted inventory method had been applied all along. An exception is a change to the LIFO method. In this case, it usually is impossible to calculate the income effect on prior years. To do so would require assumptions as to when specific LIFO inventory layers were created in years prior to the change. As a result, a company changing to LIFO usually does not report the change retrospectively. Instead, the LIFO method simply is used from that point on. (*p. 480*)
- **LO9–7** If a material inventory error is discovered in an accounting period subsequent to the period in which the error is made, previous years' financial statements that were incorrect as a result of the error are retrospectively restated to reflect the correction. Account balances are corrected by journal entry. A correction of retained earnings is reported as a prior period adjustment to the beginning balance in the statement of shareholders' equity. In addition, a disclosure note is needed to describe the nature of the error and the impact of its correction on continuing operations, net income, and earnings per share. (*p. 482*)

● LO9–8 *IAS No. 2* specifies that if circumstances reveal that an inventory write-down is no longer appropriate, it must be reversed. Reversals are not permitted under U.S. GAAP. Under U.S. GAAP, the lower of cost or net realizable value rule can be applied to individual items, inventory categories, or the entire inventory. Using the international standard, the assessment usually is applied to individual items, although using inventory categories is allowed under certain circumstances. (*p. 463*) ●

Purchase Commitments

APPENDIX 9

Purchase commitments are contracts that obligate a company to purchase a specified amount of merchandise or raw materials at specified prices on or before specified dates. Companies enter into these agreements to make sure they will be able to obtain important inventory as well as to protect against increases in purchase price. However, if the purchase price decreases before the agreement is exercised, the commitment has the disadvantage of requiring the company to purchase inventory at a higher than market price. If this happens, a loss on the purchase commitment is recorded.

Purchase commitments protect the buyer against price increases and provide a supply of product.

Because purchase commitments create the possibility of this kind of loss, the loss occurs when the market price falls below the commitment price rather than when the inventory eventually is sold. This means recording the loss when the product is purchased or, if the commitment is still outstanding, at the end of the reporting period. In other words, purchases are recorded at market price when that price is lower than the contract price, and a loss is recognized for the difference. Also, losses are recognized for any purchase commitments outstanding at the end of a reporting period when market price is less than contract price. This is best understood by the example in Illustration 9A–1.

Purchases made pursuant to a purchase commitment are recorded at the lower of contract price or market price on the date the contract is executed.

In July 2021, the Lassiter Company signed two purchase commitments. The first requires Lassiter to purchase inventory for $500,000 by November 15, 2021. The second requires Lassiter to purchase inventory for $600,000 by February 15, 2022. Lassiter's fiscal year-end is December 31. The company uses a perpetual inventory system.

Illustration 9A–1
Purchase Commitments

Contract Period within Fiscal Year

The contract period for the first commitment is contained within a single fiscal year. Lassiter would record the purchase at the contract price if the market price of inventory at date of acquisition is *equal to or greater than* the contract price of $500,000.[15]

Inventory (contract price)	500,000	
Cash (or accounts payable)		500,000

Purchase inventory at the contract price.

If the market price of inventory at acquisition is *less* than the contract price, inventory is recorded at the market price and a loss is recognized.[16] For example, if the market price is $425,000 at the time of acquisition, Lassiter must still pay $500,000 (contract price) and would record the following entry:

Inventory (market price)	425,000	
Loss on purchase commitment	75,000	
Cash (or accounts payable)		500,000

If market price is less than the contract price at acquisition, a loss is recorded.

[15]In each of the following situations, if a periodic inventory system is used, purchases is debited instead of inventory.

[16]Recall from the chapter that one method of recording losses from inventory write-downs is to report the loss as a line item in the income statement.

The objective of this treatment is to associate the loss with the period in which the price declines rather than with the period in which the company eventually sells the inventory. This is consistent with recording inventory at the lower of cost or market, as you studied in the chapter.

Contract Period Extends beyond Fiscal Year

Now let's consider Lassiter's second purchase commitment that is outstanding at the end of the fiscal year 2021 (that is, the purchases have not yet been made). If the market price of inventory at the end of the year is *equal to or greater than* the contract price of $600,000, no entry is recorded. However, if the market price at year-end is *less* than the contract price, a loss must be recognized. The objective is to associate the loss with the period in which the price declines rather than with the period in which the company eventually sells the inventory. Let's say the year-end market price of the inventory for Lassiter's second purchase commitment is $540,000. The following adjusting entry is recorded:

If the market price at year-end is less than the contract price, a loss is recorded for the difference.

December 31, 2021		
Estimated loss on purchase commitment ($600,000 − $540,000)	60,000	
Estimated liability on purchase commitment		60,000

A liability is credited for estimated losses on purchase commitments.

At this point, the loss is an *estimated* loss. The actual loss, if any, will not be known until the inventory actually is purchased. The best estimate of the market price on date of purchase is the current market price, in this case $540,000. Because no inventory has been acquired, we can't credit inventory for the loss. Instead, a liability is credited because, in a sense, the loss represents an obligation to purchase inventory above market price.

The entry to record the actual purchase on or before February 15, 2022, will vary depending on the market price of the inventory at date of purchase. If the market price is unchanged or has increased from the year-end price, the following entry is made:

If market price on purchase date has not declined from year-end price, the purchase is recorded at the year-end market price.

Inventory (accounting cost)	540,000	
Estimated liability on purchase commitment	60,000	
Cash (or accounts payable)		600,000

Even if the market price of the inventory increases, there is no recovery of the $60,000 loss recognized in 2021. Similar to the method of recording inventory at the lower of cost or market, the reduced inventory value, in this case the reduced value of purchases, is considered to be the new cost and any recovery of value is ignored.

If the market price declines even further from year-end levels, an additional loss is recognized. For example, if the market price of the inventory covered by the commitment declines to $510,000, the following entry is recorded:

If market price declines further from year-end, an additional loss is recorded at acquisition.

Inventory (market price)	510,000	
Loss on purchase commitment ($540,000 − $510,000)	30,000	
Estimated liability on purchase commitment	60,000	
Cash (or accounts payable)		600,000

The total loss on this purchase commitment of $90,000 is thus allocated between 2021 and 2022 according to when the decline in value of the inventory covered by the commitment occurred.

If there are material amounts of purchase commitments outstanding at the end of a reporting period, the contract details are disclosed in a note. This disclosure is required even if no loss estimate has been recorded. ●

Questions For Review of Key Topics

Q 9–1 Explain the (a) lower of cost or net realizable value (LCNRV) approach and the (b) lower of cost or market (LCM) approach to valuing inventory.

Q 9–2 What are the various levels of aggregation to which the LCNRV and LCM approaches can be applied?

Q 9–3 Describe the typical approach for recording inventory write-downs.

Q 9–4 Explain the gross profit method of estimating ending inventory.

Q 9–5 The Rider Company uses the gross profit method to estimate ending inventory and cost of goods sold. The cost percentage is determined based on historical data. What factors could cause the estimate of ending inventory to be overstated?

Q 9–6 Explain the retail inventory method of estimating ending inventory.

Q 9–7 Both the gross profit method and the retail inventory method provide a way to estimate ending inventory. What is the main difference between the two estimation techniques?

Q 9–8 Define each of the following retail terms: initial markup, additional markup, markup cancellation, markdown, markdown cancellation.

Q 9–9 Explain how to estimate the average cost of inventory when using the retail inventory method.

Q 9–10 What is the conventional retail method?

Q 9–11 Explain the LIFO retail inventory method.

Q 9–12 Discuss the treatment of freight-in, purchase returns, purchase discounts, normal spoilage, sales returns, sales discounts, and employee discounts in the application of the retail inventory method.

Q 9–13 Explain the difference between the retail inventory method using LIFO and the dollar-value LIFO retail method.

Q 9–14 Describe the accounting treatment for a change in inventory method other than to LIFO.

Q 9–15 When a company changes its inventory method to LIFO, an exception is made for the way accounting changes usually are reported. Explain the difference in the accounting treatment of a change *to* the LIFO inventory method from other inventory method changes.

Q 9–16 Explain the accounting treatment of material inventory errors discovered in an accounting period subsequent to the period in which the error is made.

Q 9–17 It is discovered in 2021 that ending inventory in 2019 was understated. What is the effect of the understatement on the following:

2019:	Cost of goods sold
	Net income
	Ending retained earnings
2020:	Net purchases
	Cost of goods sold
	Net income
	Ending retained earnings

IFRS **Q 9–18** Identify any differences between U.S. GAAP and IFRS when applying the lower of cost or net realizable value rule to inventory valuation.

Q 9–19 (Based on Appendix 9) Define purchase commitments. What is the advantage(s) of these agreements to buyers?

Q 9–20 (Based on Appendix 9) Explain how purchase commitments are recorded for the lower of contract price or market price.

Brief Exercises

connect

BE 9–1
Lower of cost or net realizable value
● LO9–1

Ross Electronics has one product in its ending inventory. Per unit data consist of the following: cost, $20; selling price, $30; selling costs, $4. What unit value should Ross use when applying the lower of cost or net realizable value rule to ending inventory?

BE 9–2
Lower of cost or net realizable value
● LO9–1

SLR Corporation has 1,000 units of each of its two products in its year-end inventory. Per unit data for each of the products are as follows:

	Product 1	Product 2
Cost	$50	$34
Selling price	70	36
Costs to sell	6	4

Determine the carrying value of SLR's inventory assuming that the lower of cost or net realizable value (LCNRV) rule is applied to individual products. What is the before-tax income effect of the LCNRV adjustment?

BE 9–3
Lower of cost or market
● LO9–1

[This is a variation of BE 9–1, modified to focus on the lower of cost or market.] Ross Electronics has one product in its ending inventory. Per unit data consist of the following: cost, $20; replacement cost, $18; selling price, $30; selling costs, $4. The normal profit is 30% of selling price. What unit value should Ross use when applying the lower of cost or market (LCM) rule to ending inventory?

BE 9–4
Lower of cost or market
● LO9–1

[This is a variation of BE 9–2, modified to focus on the lower of cost or market.] SLR Corporation has 1,000 units of each of its two products in its year-end inventory. Per unit data for each of the products are as follows:

	Product 1	Product 2
Cost	$50	$34
Replacement cost	48	26
Selling price	70	36
Selling costs	6	4
Normal profit	10	8

Determine the carrying value of SLR's inventory assuming that the lower of cost or market (LCM) rule is applied to individual products. What is the before-tax income effect of the LCM adjustment?

BE 9–5
Gross profit method
● LO9–2

On February 26, a hurricane destroyed the entire inventory stored in a warehouse owned by the Rockford Corporation. The following information is available from the records of the company's periodic inventory system: beginning inventory, $220,000; purchases and net sales from the beginning of the year through February 26, $400,000 and $600,000, respectively; gross profit ratio, 30%. Estimate the cost of the inventory destroyed by the hurricane using the gross profit method.

BE 9–6
Gross profit method; solving for unknown
● LO9–2

Adams Corporation estimates that it lost $75,000 in inventory from a recent flood. The following information is available from the records of the company's periodic inventory system: beginning inventory, $150,000; purchases and net sales from the beginning of the year through the date of the flood, $450,000 and $700,000, respectively. What is the company's gross profit ratio?

BE 9–7
Retail inventory method; average cost
● LO9–3

Kiddie World uses a periodic inventory system and the retail inventory method to estimate ending inventory and cost of goods sold. The following data are available for the quarter ending September 30, 2021:

	Cost	Retail
Beginning inventory	$300,000	$ 450,000
Net purchases	861,000	1,210,000
Freight-in	22,000	
Net markups		48,000
Net markdowns		18,000
Net sales		1,200,000

Estimate ending inventory and cost of goods sold (average cost) using the information provided.

BE 9–8
Retail inventory method; LIFO
● LO9–3

Refer to the situation described in BE 9–7. Estimate ending inventory and cost of goods sold (LIFO) using the information provided.

BE 9–9
Conventional retail method
● LO9–4

Refer to the situation described in BE 9–7. Estimate ending inventory and cost of goods sold using the conventional method and the information provided.

BE 9–10
Conventional retail method
● LO9–4

Roberson Corporation uses a periodic inventory system and the retail inventory method. Accounting records provided the following information for the 2021 fiscal year:

	Cost	Retail
Beginning inventory	$220,000	$ 400,000
Net purchases	640,000	1,180,000
Freight-in	17,800	
Net markups		16,000
Net markdowns		6,000
Normal spoilage		3,000
Sales		1,300,000

The company records sales to employees net of discounts. These discounts totaled $15,000 for the year. Estimate ending inventory and cost of goods sold using the conventional method and the information provided.

BE 9–11
Dollar-value LIFO retail
● LO9–5

On January 1, 2021, Sanderson Variety Store adopted the dollar-value LIFO retail inventory method. Accounting records provided the following information:

	Cost	Retail
Beginning inventory	$ 40,800	$ 68,000
Net purchases	155,440	270,000
Net markups		6,000
Net markdowns		8,000
Net sales		250,000
Retail price index, end of year		1.02

Estimate ending inventory using the dollar-value LIFO retail method and the information provided.

BE 9–12
Dollar-value LIFO retail
● LO9–5

This exercise is a continuation of BE 9–11. During 2022, purchases at cost and retail were $168,000 and $301,000, respectively. Net markups, net markdowns, and net sales for the year were $3,000, $4,000, and $280,000, respectively. The retail price index at the end of 2022 was 1.06. Estimate ending inventory in 2022 using the dollar-value LIFO retail method and the information provided.

BE 9–13
Change in inventory costing methods
● LO9–6

In 2021, Hopyard Lumber changed its inventory method from LIFO to FIFO. Inventory at the end of 2020 of $127,000 would have been $145,000 if FIFO had been used. Inventory at the end of 2021 is $162,000 using the new FIFO method but would have been $151,000 if the company had continued to use LIFO. Describe the steps Hopyard should take to report this change. What is the effect of the change on 2021 cost of goods sold?

BE 9–14
Change in inventory costing methods
● LO9–6

In 2021, Wade Window and Glass changed its inventory method from FIFO to LIFO. Inventory at the end of 2020 is $150,000. Describe the steps Wade Window and Glass should take to report this change

BE 9–15
Inventory error
● LO9–7

In 2021, Winslow International, Inc.'s controller discovered that ending inventories for 2019 and 2020 were overstated by $200,000 and $500,000, respectively. Determine the effect of the errors on retained earnings at January 1, 2021. (Ignore income taxes.)

BE 9–16
Inventory error
● LO9–7

Refer to the situation described in BE 9–15. What steps would be taken to report the error in the 2021 financial statements?

Exercises

connect

E 9–1
Lower of cost or net realizable value
● LO9–1

Herman Company has three products in its ending inventory. Specific per unit data at the end of the year for each of the products are as follows:

	Product 1	Product 2	Product 3
Cost	$20	$ 90	$50
Selling price	40	120	70
Costs to sell	6	40	10

Required:
What unit values should Herman use for each of its products when applying the lower of cost or net realizable value (LCNRV) rule to ending inventory?

E 9–2
Lower of cost or net realizable value
● LO9–1

The inventory of Royal Decking consisted of five products. Information about the December 31, 2021, inventory is as follows:

	Per Unit	
Product	**Cost**	**Selling Price**
A	$ 40	$ 60
B	80	100
C	40	80
D	100	130
E	20	30

Costs to sell consist of a sales commission equal to 10% of selling price and shipping costs equal to 5% of cost.

Required:
What unit value should Royal Decking use for each of its products when applying the lower of cost or net realizable value (LCNRV) rule to units of ending inventory?

E 9–3
Lower of cost or net realizable value
● LO9–1

Tatum Company has four products in its inventory. Information about the December 31, 2021, inventory is as follows:

Product	Total Cost	Total Net Realizable Value
101	$120,000	$100,000
102	90,000	110,000
103	60,000	50,000
104	30,000	50,000

Required:
1. Determine the carrying value of inventory at December 31, 2021, assuming the lower of cost or net realizable value (LCNRV) rule is applied to individual products.
2. Assuming that inventory write-downs are common for Tatum Company, record any necessary year-end adjusting entry.

E 9–4
Lower of cost or market
● LO9–1

[This is a variation of E 9-1, modified to focus on the lower of cost or market.] Herman Company has three products in its ending inventory. Specific per unit data at the end of the year for each of the products are as follows:

	Product 1	Product 2	Product 3
Cost	$20	$ 90	$50
Replacement cost	18	85	40
Selling price	40	120	70
Selling costs	6	40	10
Normal profit	5	30	12

Required:
What unit values should Herman use for each of its products when applying the lower of cost or market (LCM) rule to ending inventory?

E 9–5
Lower of cost or market
● LO9–1

[This is a variation of E 9-2, modified to focus on the lower of cost or market.] The inventory of Royal Decking consisted of five products. Information about the December 31, 2021, inventory is as follows:

	Per Unit		
Product	**Cost**	**Replacement Cost**	**Selling Price**
A	$ 40	$35	$ 60
B	80	70	100
C	40	55	80
D	100	70	130
E	20	28	30

Selling costs consist of a sales commission equal to 10% of selling price and shipping costs equal to 5% of cost. The normal profit is 30% of selling price.

Required:
What unit value should Royal Decking use for each of its products when applying the lower of cost or market (LCM) rule to units of ending inventory?

E 9–6
Lower of cost or market
● LO9–1

[This is a variation of E 9-3, modified to focus on the lower of cost or market.] Tatum Company has four products in its inventory. Information about the December 31, 2021, inventory is as follows:

Product	Total Cost	Total Replacement Cost	Total Net Realizable Value
101	$120,000	$100,000	$100,000
102	90,000	85,000	110,000
103	60,000	40,000	50,000
104	30,000	28,000	50,000

The normal profit is 25% of *total cost.*

Required:

1. Determine the carrying value of inventory at December 31, 2021, assuming the lower of cost or market (LCM) rule is applied to individual products.
2. Assuming that inventory write-downs are common for Tatum Company, record any necessary year-end adjusting entry.

E 9–7
FASB codification research
● LO9–3, LO9–6, LO9–7

Access the *FASB Accounting Standards Codification* at the FASB website (**www.fasb.org**). Determine each of the following:

1. The specific seven-digit Codification citation (XXX-XX-XX) that contains discussion of the measurement of ending inventory using the lower of cost or net realizable value (LCNRV) rule and the lower of cost or market (LCM) rule.
2. The specific eight-digit Codification citation (XXX-XX-XX-X) that describes the measurement of the ceiling using the lower of cost or market (LCM) rule.
3. The specific eight-digit Codification citation (XXX-XX-XX-X) that describes the measurement of the floor using the lower of cost or market (LCM) rule.

E 9–8
Gross profit method
● LO9–2

On September 22, 2021, a flood destroyed the entire merchandise inventory on hand in a warehouse owned by the Rocklin Sporting Goods Company. The following information is available from the records of the company's periodic inventory system:

Inventory, January 1, 2021	$140,000
Net purchases, January 1 through September 22	370,000
Net sales, January 1 through September 22	550,000
Gross profit ratio	25%

Required:

Estimate the cost of inventory destroyed in the flood using the gross profit method.

E 9–9
Gross profit method
● LO9–2

On November 21, 2021, a fire at Hodge Company's warehouse caused severe damage to its entire inventory of Product Tex. Hodge estimates that all usable damaged goods can be sold for $12,000. The following information was available from the records of Hodge's periodic inventory system:

Inventory, November 1	$100,000
Net purchases from November 1, to the date of the fire	140,000
Net sales from November 1, to the date of the fire	220,000

Based on recent history, Hodge's gross profit ratio on Product Tex is 35% of net sales.

Required:

Calculate the estimated loss on the inventory from the fire, using the gross profit method.

(*AICPA adapted*)

E 9–10
Gross profit method
● LO9–2

A fire destroyed a warehouse of the Goren Group, Inc., on May 4, 2021. Accounting records on that date indicated the following:

Merchandise inventory, January 1, 2021	$1,900,000
Purchases to date	5,800,000
Freight-in	400,000
Sales to date	8,200,000

The gross profit ratio has averaged 20% of sales for the past four years.

Required:

Use the gross profit method to estimate the cost of the inventory destroyed in the fire.

E 9–11
Gross profit method
● LO9–2

Royal Gorge Company uses the gross profit method to estimate ending inventory and cost of goods sold when preparing monthly financial statements required by its bank. Inventory on hand at the end of October was $58,500. The following information for the month of November was available from company records:

Purchases	$110,000
Freight-in	3,000
Sales	180,000
Sales returns	5,000
Purchases returns	4,000

In addition, the controller is aware of $8,000 of inventory that was stolen during November from one of the company's warehouses.

Required:

1. Calculate the estimated inventory at the end of November, assuming a gross profit ratio of 40%.
2. Calculate the estimated inventory at the end of November, assuming a markup on cost of 60%.

E 9–12
Gross profit method; solving for unknown cost percentage
● LO9–2

National Distributing Company uses a periodic inventory system to track its merchandise inventory and the gross profit method to estimate ending inventory and cost of goods sold for interim periods. Net purchases for the month of August were $31,000. The July 31 and August 31, 2021, financial statements contained the following information:

Income Statements
For the Months Ending

	August 31, 2021	July 31, 2021
Net sales	$50,000	$40,000

Balance Sheets at

	August 31, 2021	July 31, 2021
Assets:		
Merchandise inventory	$28,000	$27,000

Required:

Determine the company's cost percentage.

E 9–13
Retail inventory method; average cost
● LO9–3

San Lorenzo General Store uses a periodic inventory system and the retail inventory method to estimate ending inventory and cost of goods sold. The following data are available for the month of October 2021:

	Cost	Retail
Beginning inventory	$35,000	$50,000
Net purchases	19,120	31,600
Net markups		1,200
Net markdowns		800
Net sales		32,000

Required:

Estimate the average cost of ending inventory and cost of goods sold for October using the information provided.

E 9–14
Conventional retail method
● LO9–3

Campbell Corporation uses the retail method to value its inventory. The following information is available for the year 2021:

	Cost	Retail
Merchandise inventory, January 1, 2021	$190,000	$280,000
Purchases	600,000	840,000
Freight-in	8,000	
Net markups		20,000
Net markdowns		4,000
Net sales		800,000

Required:

Determine the December 31, 2021, inventory by applying the conventional retail method using the information provided.

E 9–15
Retail inventory method; LIFO
● LO9–3

Crosby Company owns a chain of hardware stores throughout the state. The company uses a periodic inventory system and the retail inventory method to estimate ending inventory and cost of goods sold. The following data are available for the three months ending March 31, 2021:

	Cost	Retail
Beginning inventory	$160,000	$280,000
Net purchases	607,760	840,000
Net markups		20,000
Net markdowns		4,000
Net sales		800,000

Required:
Estimate the LIFO cost of ending inventory and cost of goods sold for the three months ending March 31, 2021, using the information provided. Assume stable retail prices during the period.

E 9–16
Conventional retail method; normal spoilage
● LO9–4

Almaden Valley Variety Store uses the retail inventory method to estimate ending inventory and cost of goods sold. Data for 2021 are as follows:

	Cost	Retail
Beginning inventory	$ 12,000	$ 20,000
Purchases	102,600	165,000
Freight-in	3,480	
Purchase returns	4,000	7,000
Net markups		6,000
Net markdowns		3,000
Normal spoilage		4,200
Net sales		152,000

Required:
Estimate the ending inventory and cost of goods sold for 2021, applying the conventional retail method using the information provided.

E 9–17
Conventional and average cost retail methods; employee discounts
● LO9–3, LO9–4

LeMay Department Store uses the retail inventory method to estimate ending inventory for its monthly financial statements. The following data pertain to one of its largest departments for the month of March 2021:

	Cost	Retail
Beginning inventory	$ 40,000	$ 60,000
Purchases	207,000	400,000
Freight-in	14,488	
Purchase returns	4,000	6,000
Net markups		5,800
Net markdowns		3,500
Normal breakage		6,000
Net sales		280,000
Employee discounts		1,800

Sales are recorded net of employee discounts.

Required:
1. Compute estimated ending inventory and cost of goods sold for March applying the conventional retail method.
2. Recompute the cost-to-retail percentage using the average cost method.

E 9–18
Retail inventory method; solving for unknowns
● LO9–3

Adams Corporation uses a periodic inventory system and the retail inventory method to estimate ending inventory and cost of goods sold. The following data are available for the month of September 2021:

	Cost	Retail
Beginning inventory	$21,000	$35,000
Net purchases	10,500	?
Net markups		4,000
Net markdowns		1,000
Net sales		?

The company used the average cost flow method and estimated inventory at the end of September to be $17,437.50. If the company had used the LIFO cost flow method, the cost-to-retail percentage would have been 50%.

Required:
Compute net purchases at retail and net sales for the month of September using the information provided.

E 9–19
Dollar-value LIFO retail
● LO9–5

On January 1, 2021, the Brunswick Hat Company adopted the dollar-value LIFO retail method. The following data are available for 2021:

	Cost	Retail
Beginning inventory	$ 71,280	$ 132,000
Net purchases	112,500	255,000
Net markups		6,000
Net markdowns		11,000
Net sales		232,000
Retail price index, 12/31/2021		1.04

Required:
Calculate the estimated ending inventory and cost of goods sold for 2021 using the information provided.

E 9–20
Dollar-value LIFO retail
● LO9–5

Canova Corporation adopted the dollar-value LIFO retail method on January 1, 2021. On that date, the cost of the inventory on hand was $15,000 and its retail value was $18,750. Information for 2021 and 2022 is as follows:

Date	Ending Inventory at Retail	Retail Price Index	Cost-to-Retail Percentage
12/31/2021	$25,000	1.25	82%
12/31/2022	28,600	1.30	85

Required:
1. What is the cost-to-retail percentage for the inventory on hand at 1/1/2021?
2. Calculate the inventory value at the end of 2021 and 2022 using the dollar-value LIFO retail method.

E 9–21
Dollar-value LIFO retail
● LO9–5

Lance-Hefner Specialty Shoppes decided to use the dollar-value LIFO retail method to value its inventory. Accounting records provide the following information:

	Cost	Retail
Merchandise inventory, January 1, 2021	$160,000	$250,000
Net purchases	350,200	510,000
Net markups		7,000
Net markdowns		2,000
Net sales		380,000

Related retail price indexes are as follows:

January 1, 2021	1.00
December 31, 2021	1.10

Required:
Determine ending inventory and cost of goods sold using the information provided.

E 9–22
Dollar-value LIFO retail; solving for unknowns
● LO9–5

Bosco Company adopted the dollar-value LIFO retail method at the beginning of 2021. Information for 2021 and 2022 is as follows, with certain data intentionally omitted:

	Inventory			
Date	Cost	Retail	Retail Price Index	Cost-to-Retail Percentage
Inventory, 1/1/2021	$21,000	$28,000	1.00	?
Inventory, 12/31/2021	22,792	33,600	1.12	?
2022 net purchases	60,000	88,400		
2022 net sales		80,000		
Inventory, 12/31/2022	?	?	1.20	

Required:
Determine the missing data.

E 9–23
Change in inventory costing methods
● LO9–6

In 2021, CPS Company changed its method of valuing inventory from the FIFO method to the average cost method. At December 31, 2020, CPS's inventories were $32 million (FIFO). CPS's records indicated that the inventories would have totaled $23.8 million at December 31, 2020, if determined on an average cost basis.

Required:
1. Prepare the journal entry to record the adjustment. (Ignore income taxes.)
2. Briefly describe other steps CPS should take to report the change.

E 9–24
Change in inventory costing methods
● LO9–6

Goddard Company has used the FIFO method of inventory valuation since it began operations in 2018. Goddard decided to change to the average cost method for determining inventory costs at the beginning of 2021. The following schedule shows year-end inventory balances under the FIFO and average cost methods:

Year	FIFO	Average Cost
2018	$45,000	$54,000
2019	78,000	71,000
2020	83,000	78,000

Required:
1. Ignoring income taxes, prepare the 2021 journal entry to adjust the accounts to reflect the average cost method.
2. How much higher or lower would cost of goods sold be in the 2020 revised income statement?

E 9–25
Error correction; inventory error
● LO9–7

During 2021, WMC Corporation discovered that its ending inventories reported in its financial statements were misstated by the following material amounts:

2019	understated by	$120,000
2020	overstated by	150,000

WMC uses a periodic inventory system and the FIFO cost method.

Required:
1. Determine the effect of these errors on retained earnings at January 1, 2021, before any adjustments. Explain your answer. (Ignore income taxes.)
2. Prepare a journal entry to correct the errors.
3. What other step(s) would be taken in connection with the correction of the errors?

E 9–26
Inventory errors
● LO9–7

For each of the following inventory errors occurring in 2021, determine the effect of the error on 2021's cost of goods sold, net income, and retained earnings. Assume that the error is not discovered until 2022 and that a periodic inventory system is used. Ignore income taxes.

U = understated O = overstated NE = no effect

	Cost of Goods Sold	Net Income	Retained Earnings
1. Overstatement of ending inventory	U	O	O
2. Overstatement of purchases			
3. Understatement of beginning inventory			
4. Freight-in charges are understated			
5. Understatement of ending inventory			
6. Understatement of purchases			
7. Overstatement of beginning inventory			
8. Understatement of purchases plus understatement of ending inventory by the same amount			

E 9–27
Inventory error
● LO9–7

In 2021, the internal auditors of Development Technologies, Inc., discovered that a $4 million purchase of merchandise in 2021 was recorded in 2020 instead. The physical inventory count at the end of 2020 was correct.

Required:
Prepare the journal entry needed in 2021 to correct the error. Also, briefly describe any other measures Development Technologies would take in connection with correcting the error. (Ignore income taxes.)

E 9–28
Inventory errors
● LO9–7

In 2021, the controller of Sytec Corporation discovered that $42,000 of inventory purchases were incorrectly charged to advertising expense in 2020. In addition, the 2020 year-end inventory count failed to include $30,000 of company merchandise held on consignment by Erin Brothers. Sytec uses a periodic inventory system. Other than the omission of the merchandise on consignment, the year-end inventory count was correct. The amounts of the errors are deemed to be material.

Required:

1. Determine the effect of the errors on retained earnings at January 1, 2021. Explain your answer. (Ignore income taxes.)
2. Prepare a journal entry to correct the errors.
3. What other step(s) would be taken in connection with the correction of the errors?

E 9–29
Concepts; terminology
● LO9–1 through LO9–7

Listed below are several terms and phrases associated with inventory measurement. Pair each item from List A with the item from List B (by letter) that is most appropriately associated with it.

List A	List B
____ 1. Gross profit ratio	a. Reduction in selling price below the original selling price
____ 2. Cost-to-retail percentage	b. Beginning inventory is not included in the calculation of the cost-to-retail percentage
____ 3. Additional markup	c. Deducted in the retail column after the calculation of the cost-to-retail percentage
____ 4. Markdown	d. Requires base year retail to be converted to layer year retail and then to cost
____ 5. Net markup	e. Gross profit divided by net sales
____ 6. Retail method, FIFO and LIFO	f. Material inventory error discovered in a subsequent year
____ 7. Conventional retail method	g. Must be added to sales if sales are recorded net of discounts
____ 8. Change from LIFO	h. Deducted in the retail column to arrive at goods available for sale at retail
____ 9. Dollar-value LIFO retail	i. Divide cost of goods available for sale by goods available at retail
____ 10. Normal spoilage	j. Average cost, lower of cost or market
____ 11. Requires retrospective restatement	k. Added to the retail column to arrive at goods available for sale
____ 12. Employee discounts	l. Increase in selling price subsequent to initial markup
____ 13. Net markdowns	m. Selling price less estimated selling costs
____ 14. Net realizable value	n. Accounting change requiring retrospective treatment

E 9–30
Purchase commitments
● Appendix

On October 6, 2021, the Elgin Corporation signed a purchase commitment to purchase inventory for $60,000 on or before March 31, 2022. The company's fiscal year-end is December 31. The contract was exercised on March 21, 2022, and the inventory was purchased for cash at the contract price. On the purchase date of March 21, the market price of the inventory was $54,000. The market price of the inventory on December 31, 2021, was $56,000. The company uses a perpetual inventory system.

Required:

1. Prepare the necessary adjusting journal entry (if any is required) on December 31, 2021.
2. Prepare the journal entry to record the purchase on March 21, 2022.

E 9–31
Purchase commitments
● Appendix

In March 2021, the Phillips Tool Company signed two purchase commitments. The first commitment requires Phillips to purchase inventory for $100,000 by June 15, 2021. The second commitment requires the company to purchase inventory for $150,000 by August 20, 2021. The company's fiscal year-end is June 30. Phillips uses a periodic inventory system.

The first commitment is exercised on June 15, 2021, when the market price of the inventory purchased was $85,000. The second commitment was exercised on August 20, 2021, when the market price of the inventory purchased was $120,000.

Required:

Prepare the journal entries required on June 15, June 30, and August 20, 2021, to account for the two purchase commitments. Assume that the market price of the inventory related to the outstanding purchase commitment was $140,000 at June 30.

E 9–32
General ledger exercise; inventory transactions
● Chapter 8 and LO9–1

On January 1, 2021, the general ledger of Big Blast Fireworks included the following account balances:

Account Title	Debits	Credits
Cash	$ 21,900	
Accounts receivable	36,500	
Allowance for uncollectible accounts		$ 3,100
Inventory	30,000	
Land	61,600	
Accounts payable		32,400
Notes payable (8%, due in 3 years)		30,000
Common stock		56,000
Retained earnings		28,500
Totals	$150,000	$150,000

The $30,000 beginning balance of inventory consists of 300 units, each costing $100. During January 2021, Big Blast Fireworks had the following inventory transactions:

Jan. 3	Purchased 1,200 units for $126,000 on account ($105 each).
8	Purchased 1,300 units for $143,000 on account ($110 each).
12	Purchased 1,400 units for $161,000 on account ($115 each).
15	Returned 100 of the units purchased on January 12 because of defects.
19	Sold 4,000 units on account for $600,000. The cost of the units sold is determined using a FIFO perpetual inventory system.
22	Received $580,000 from customers on accounts receivable.
24	Paid $410,000 to inventory suppliers on accounts payable.
27	Wrote off accounts receivable as uncollectible, $2,500.
31	Paid cash for salaries during January, $128,000.

Required:

1. Record each of the transactions listed above in the 'General Journal' tab (these are shown as items 1-13) assuming a perpetual FIFO inventory system. Review the 'General Ledger' and the 'Trial Balance' tabs to see the effect of the transactions on the account balances.
2. Record adjusting entries on January 31 in the 'General Journal' tab (these are shown as items 14-18):
   a. At the end of January, the company estimates that the remaining units of inventory are expected to sell in February for only $100 each.
   b. At the end of January, $4,000 of accounts receivable are past due, and the company estimates that 40% of these accounts will not be collected. Of the remaining accounts receivable, the company estimates that 4% will not be collected.
   c. Accrued interest expense on notes payable for January. Interest is expected to be paid each December 31.
   d. Accrued income taxes at the end of January are $12,300.
3. Review the adjusted 'Trial Balance' as of January 31, 2021, in the 'Trial Balance' tab.
4. Prepare a multiple-step income statement for the period ended January 31, 2021, in the 'Income Statement' tab.
5. Prepare a classified balance sheet as of January 31, 2021, in the 'Balance Sheet' tab.
6. Record closing entries in the 'General Journal' tab (these are shown as items 19-21).
7. Using the information from the requirements above, complete the 'Analysis' tab.
   a. Calculate the inventory turnover ratio for the month of January. If the industry average of the inventory turnover ratio for the month of January is 18.5 times, is the company managing its inventory *more* or *less* efficiently than other companies in the same industry?
   b. Calculate the gross profit ratio for the month of January. If the industry average gross profit ratio is 33%, is the company *more* or *less* profitable per dollar of sales than other companies in the same industry?
   c. Used together, what might the inventory turnover ratio and gross profit ratio suggest about Big Blast Fireworks' business strategy? Is the company's strategy to sell a *higher volume of less expensive* items or does the company appear to be selling a *lower volume of more expensive* items?

Problems

connect

P 9–1
Lower of cost or net realizable value; by product and by total inventory
● LO9–1

Decker Company has five products in its inventory. Information about the December 31, 2021, inventory follows.

Product	Quantity	Unit Cost	Unit Selling Price
A	1,000	$10	$16
B	800	15	18
C	600	3	8
D	200	7	6
E	600	14	13

The cost to sell for each product consists of a 15 percent sales commission.

Required:

1. Determine the carrying value of inventory at December 31, 2021, assuming the lower of cost or net realizable value (LCNRV) rule is applied to individual products.

2. Determine the carrying value of inventory at December 31, 2021, assuming the LCNRV rule is applied to the entire inventory.
3. Assuming inventory write-downs are common for Decker, record any necessary year-end adjusting entry based on the amount calculated in requirement 2.

P 9–2
Lower of cost or net realizable value; by product, category, and total inventory
● LO9–1

Almaden Hardware Store sells two product categories, tools and paint products. Information pertaining to its 2021 year-end inventory is as follows:

Inventory, by Product Category	Quantity	Per Unit Cost	Net Realizable Value
Tools:			
Hammers	100	$ 5.00	$5.50
Saws	200	10.00	9.00
Screwdrivers	300	2.00	2.60
Paint products:			
1-gallon cans	500	6.00	5.00
Paint brushes	100	4.00	4.50

Required:

1. Determine the carrying value of inventory at year-end, assuming the lower of cost or net realizable value (LCNRV) rule is applied to (a) individual products, (b) product categories, and (c) total inventory.
2. Assuming inventory write-downs are common for Almaden, record any necessary year-end adjusting entry for each of the LCNRV applications in requirement 1.

P 9–3
Lower of cost or market; by product and by total inventory
● LO9–1

Forester Company has five products in its inventory. Information about the December 31, 2021, inventory follows.

Product	Quantity	Unit Cost	Unit Replacement Cost	Unit Selling Price
A	1,000	$10	$12	$16
B	800	15	11	18
C	600	3	2	8
D	200	7	4	6
E	600	14	12	13

The cost to sell for each product consists of a 15 percent sales commission. The normal profit for each product is 40 percent of the selling price.

Required:

1. Determine the carrying value of inventory at December 31, 2021, assuming the lower of cost or market (LCM) rule is applied to individual products.
2. Determine the carrying value of inventory at December 31, 2021, assuming the LCM rule is applied to the entire inventory.
3. Assuming inventory write-downs are common for Forester, record any necessary year-end adjusting entry based on the amount calculated in requirement 2.

P 9–4
Lower of cost or market; by product, category, and total inventory
● LO9–1

Home Stop sells two product categories, furniture and accessories. Information pertaining to its 2021 year-end inventory is as follows:

Inventory, by Product Category	Quantity	Per Unit Cost	Market
Furniture:			
Chairs	50	$25	$31
Desks	10	73	58
Tables	20	84	92
Accessories:			
Rugs	40	60	48
Lamps	30	22	18

Required:

1. Determine the carrying value of inventory at year-end, assuming the lower of cost or market (LCM) rule is applied to (a) individual products, (b) product categories, and (c) total inventory.
2. Assuming inventory write-downs are common for Home Stop, record any necessary year-end adjusting entry for each of the LCM applications in requirement 1.

P 9–5
Gross profit method
● LO9–2

Smith Distributors, Inc., supplies ice cream shops with various toppings for making sundaes. On November 17, 2021, a fire resulted in the loss of all of the toppings stored in one section of the warehouse. The company must provide its insurance company with an estimate of the amount of inventory lost. The following information is available from the company's accounting records:

	Fruit Toppings	Marshmallow Toppings	Chocolate Toppings
Inventory, January 1, 2021	$ 20,000	$ 7,000	$ 3,000
Net purchases through Nov. 17	150,000	36,000	12,000
Net sales through Nov. 17	200,000	55,000	20,000
Historical gross profit ratio	20%	30%	35%

Required:
1. Calculate the estimated cost of each of the toppings lost in the fire.
2. What factors could cause the estimates to be over- or understated?

P 9–6
Retail inventory method; average cost and conventional
● LO9–3, LO9–4

Sparrow Company uses the retail inventory method to estimate ending inventory and cost of goods sold. Data for 2021 are as follows:

	Cost	Retail
Beginning inventory	$ 90,000	$180,000
Purchases	355,000	580,000
Freight-in	9,000	
Purchase returns	7,000	11,000
Net markups		16,000
Net markdowns		12,000
Normal spoilage		3,000
Abnormal spoilage	4,800	8,000
Sales		540,000
Sales returns		10,000

The company records sales net of employee discounts. Employee discounts for 2021 totaled $4,000.

Required:
Estimate Sparrow's ending inventory and cost of goods sold for the year using the retail inventory method and the following applications:
1. Average cost
2. Conventional

P 9–7
Retail inventory method; conventional and LIFO
● LO9–3, LO9–4

Alquist Company uses the retail method to estimate its ending inventory. Selected information about its year 2021 operations is as follows:

a. January 1, 2021, beginning inventory had a cost of $100,000 and a retail value of $150,000.
b. Purchases during 2021 cost $1,387,500 with an original retail value of $2,000,000.
c. Freight costs were $10,000 for incoming merchandise.
d. Net additional markups were $300,000 and net markdowns were $150,000.
e. Based on prior experience, shrinkage due to shoplifting was estimated to be $15,000 of retail value.
f. Merchandise is sold to employees at a 20% of selling price discount. Employee sales are recorded in a separate account at the net selling price. The balance in this account at the end of 2021 is $250,000.
g. Sales to customers totaled $1,750,000 for the year.

Required:
1. Estimate ending inventory and cost of goods sold using the conventional retail method.
2. Estimate ending inventory and cost of goods sold using the LIFO retail method. (Assume stable prices.)

P 9–8
Retail inventory method; conventional
● LO9–4

Grand Department Store, Inc., uses the retail inventory method to estimate ending inventory for its monthly financial statements. The following data pertain to a single department for the month of October 2021:

Inventory, October 1, 2021:	
At cost	$ 20,000
At retail	30,000
Purchases (exclusive of freight and returns):	
At cost	100,151
At retail	146,495
Freight-in	5,100

(continued)

Purchase returns:	
At cost	2,100
At retail	2,800
Additional markups	2,500
Markup cancellations	265
Markdowns (net)	800
Normal spoilage and breakage	4,500
Sales	140,000
Sales returns	4,270

Required:

1. Using the conventional retail method, prepare a schedule computing estimated lower of cost or market (LCM) inventory for October 31, 2021.
2. A department store using the conventional retail inventory method estimates the cost of its ending inventory as $29,000. An accurate physical count reveals only $22,000 of inventory at lower of cost or market. List the factors that may have caused the difference between computed inventory and the physical count.

(AICPA adapted)

P 9–9
Retail method—average cost and conventional
● LO9–3, LO9–4

Smith-Kline Company maintains inventory records at selling prices as well as at cost. For 2021, the records indicate the following data:

	($ in thousands)	
	Cost	**Retail**
Beginning inventory	$ 80	$ 125
Purchases	671	1,006
Freight-in on purchases	30	
Purchase returns	1	2
Net markups		4
Net markdowns		8
Net sales		916

Required:

Use the retail method to approximate cost of ending inventory in each of the following ways:

1. Average cost
2. Conventional

P 9–10
Dollar-value LIFO retail method
● LO9–5

[This is a variation of P 9–9, modified to focus on the dollar-value LIFO retail method.] Smith-Kline Company maintains inventory records at selling prices as well as at cost. For 2021, the records indicate the following data:

	($ in thousands)	
	Cost	**Retail**
Beginning inventory	$ 80	$ 125
Purchases	671	1,006
Freight-in on purchases	30	
Purchase returns	1	2
Net markups		4
Net markdowns		8
Net sales		916

Required:

Assuming the price level increased from 1.00 at January 1 to 1.10 at December 31, 2021, use the dollar-value LIFO retail method to approximate cost of ending inventory and cost of goods sold.

P 9–11
Dollar-value LIFO retail
● LO9–5

On January 1, 2021, HGC Camera Store adopted the dollar-value LIFO retail inventory method. Inventory transactions at both cost and retail, and cost indexes for 2021 and 2022 are as follows:

	2021		**2022**	
	Cost	**Retail**	**Cost**	**Retail**
Beginning inventory	$28,000	$ 40,000		
Net purchases	85,000	108,000	$90,000	$114,000
Freight-in	2,000		2,500	
Net markups		10,000		8,000
Net markdowns		2,000		2,200
Net sales to customers		100,000		104,000
Sales to employees (net of 20% discount)		2,400		4,000
Price Index:				
January 1, 2021				1.00
December 31, 2021				1.06
December 31, 2022				1.10

Required:

Estimate the 2021 and 2022 ending inventory and cost of goods sold using the dollar-value LIFO retail inventory method.

P 9–12
Retail inventory method; various applications
● LO9–3 through LO9–5

Raleigh Department Store uses the conventional retail method for the year ended December 31, 2019. Available information follows:

a. The inventory at January 1, 2019, had a retail value of $45,000 and a cost of $27,500 based on the conventional retail method.

b. Transactions during 2019 were as follows:

	Cost	Retail
Gross purchases	$282,000	$490,000
Purchase returns	6,500	10,000
Purchase discounts	5,000	
Sales		492,000
Sales returns		5,000
Employee discounts		3,000
Freight-in	26,500	
Net markups		25,000
Net markdowns		10,000

Sales to employees are recorded net of discounts.

c. The retail value of the December 31, 2020, inventory was $56,100, the cost-to-retail percentage for 2020 under the LIFO retail method was 62%, and the appropriate price index was 102% of the January 1, 2020, price level.

d. The retail value of the December 31, 2021, inventory was $48,300, the cost-to-retail percentage for 2021 under the LIFO retail method was 61%, and the appropriate price index was 105% of the January 1, 2020, price level.

Required:

1. Estimate ending inventory for 2019 using the conventional retail method.
2. Estimate ending inventory for 2019 assuming Raleigh Department Store used the LIFO retail method.
3. Assume Raleigh Department Store adopts the dollar-value LIFO retail method on January 1, 2020. Estimate ending inventory for 2020 and 2021.

(AICPA adapted)

P 9–13
Retail inventory method; various applications
● LO9–3 through LO9–5

On January 1, 2021, Pet Friendly Stores adopted the retail inventory method. Inventory transactions at both cost and retail, and cost indexes for 2021 and 2022 are as follows:

	2021		2022	
	Cost	Retail	Cost	Retail
Beginning inventory	$ 90,000	$150,000		
Purchases	478,000	730,000	$511,000	$760,000
Purchase returns	2,500	3,500	2,200	4,000
Freight-in	6,960		8,000	
Net markups		8,500		10,000
Net markdowns		4,000		6,000
Net sales to customers		650,000		680,000
Sales to employees (net of 30% discount)		14,000		17,500
Normal spoilage		5,000		6,600
Price Index				
January 1, 2021	1.00			
December 31, 2021	1.03			
December 31, 2022	1.06			

Required:

1. Estimate the 2021 and 2022 ending inventory and cost of goods sold using the dollar-value LIFO retail method.
2. Estimate the 2021 ending inventory and cost of goods sold using the average cost retail method.
3. Estimate the 2021 ending inventory and cost of goods sold using the conventional retail method.

P 9–14
Change in methods
● LO9–6

Rockwell Corporation uses a periodic inventory system and has used the FIFO cost method since inception of the company in 1979. In 2021, the company decided to switch to the average cost method. Data for 2021 are as follows:

Beginning inventory, FIFO (5,000 units @ $30)		$150,000
Purchases:		
5,000 units @ $36	$180,000	
5,000 units @ $40	200,000	380,000
Cost of goods available for sale		$530,000
Sales for 2021 (8,000 units @ $70)		$560,000

Additional Information:

a. The company's effective income tax rate is 40% for all years.

b. If the company had used the average cost method prior to 2021, ending inventory for 2020 would have been $130,000.

c. 7,000 units remained in inventory at the end of 2021.

Required:

1. Ignoring income taxes, prepare the 2021 journal entry to adjust the accounts to reflect the average cost method.
2. What is the effect of the change in methods on 2021 net income?

P 9–15
Inventory errors
● LO9–7

You have been hired as the new controller for the Ralston Company. Shortly after joining the company in 2021, you discover the following errors related to the 2019 and 2020 financial statements:

a. Inventory at December 31, 2019, was understated by $6,000.

b. Inventory at December 31, 2020, was overstated by $9,000.

c. On December 31, 2020, inventory was purchased for $3,000. The company did not record the purchase until the inventory was paid for early in 2021. At that time, the purchase was recorded by a debit to purchases and a credit to cash.

The company uses a periodic inventory system.

Required:

1. Assuming that the errors were discovered after the 2020 financial statements were issued, analyze the effect of the errors on 2020 and 2019 cost of goods sold, net income, and retained earnings. (Ignore income taxes.)
2. Prepare a journal entry to correct the errors.
3. What other step(s) would be taken in connection with the errors?

P 9–16
Inventory errors
● LO9–7

The December 31, 2021, inventory of Tog Company, based on a physical count, was determined to be $450,000. Included in that count was a shipment of goods received from a supplier at the end of the month that cost $50,000. The purchase was recorded and paid for in 2022. Another supplier shipment costing $20,000 was correctly recorded as a purchase in 2021. However, the merchandise, shipped FOB shipping point, was not received until 2022 and was incorrectly omitted from the physical count. A third purchase, shipped from a supplier FOB shipping point on December 28, 2021, did not arrive until January 3, 2022. The merchandise, which cost $80,000, was not included in the physical count and the purchase has not yet been recorded.

The company uses a periodic inventory system.

Required:

1. Determine the correct December 31, 2021, inventory balance and, assuming that the errors were discovered after the 2021 financial statements were issued, analyze the effect of the errors on 2021 cost of goods sold, net income, and retained earnings. (Ignore income taxes.)
2. Prepare a journal entry to correct the errors.

P 9–17
Integrating problem; Chapters 8 and 9; inventory errors
● LO9–7

Capwell Corporation uses a periodic inventory system. The company's ending inventory on December 31, 2021, its fiscal-year end, based on a physical count, was determined to be $326,000. Capwell's unadjusted trial balance also showed the following account balances: Purchases, $620,000; Accounts payable; $210,000; Accounts receivable, $225,000; Sales revenue, $840,000.

The internal audit department discovered the following items:

1. Goods valued at $32,000 held on consignment from Dix Company were included in the physical count but not recorded as a purchase.
2. Purchases from Xavier Corporation were incorrectly recorded at $41,000 instead of the correct amount of $14,000. The correct amount was included in the ending inventory.

3. Goods that cost $25,000 were shipped from a vendor on December 28, 2021, terms FOB destination. The merchandise arrived on January 3, 2022. The purchase and related accounts payable were recorded in 2021.
4. One inventory item was incorrectly included in ending inventory as 100 units, instead of the correct amount of 1,000 units. This item cost $40 per unit.
5. The 2020 balance sheet reported inventory of $352,000. The internal auditors discovered that a mathematical error caused this inventory to be understated by $62,000. This amount is considered to be material. Comparative financial statements will be issued.
6. Goods shipped to a customer FOB destination on December 25, 2021, were received by the customer on January 4, 2022. The sales price was $40,000 and the merchandise cost $22,000. The sale and corresponding accounts receivable were recorded in 2021.
7. Goods shipped from a vendor FOB shipping point on December 27, 2021, were received on January 3, 2022. The merchandise cost $18,000. The purchase was not recorded until 2022.

Required:
1. Determine the correct amounts for 2021 purchases, accounts payable, sales revenue, and accounts receivable.
2. Calculate ending inventory and cost of goods sold for 2021.
3. Describe the steps Capwell would undertake to correct the error in the 2020 ending inventory. What was the effect of the error on 2020 before-tax income?

P 9–18
Purchase commitments
● Appendix

In November 2021, the Brunswick Company signed two purchase commitments. The first commitment requires Brunswick to purchase 10,000 units of inventory at $10 per unit by December 15, 2021. The second commitment requires the company to purchase 20,000 units of inventory at $11 per unit by March 15, 2022. Brunswick's fiscal year-end is December 31. The company uses a periodic inventory system. Both contracts were exercised on their expiration date.

Required:
1. Prepare the journal entry to record the December 15 purchase for cash assuming the following alternative unit market prices on that date:
   a. $10.50
   b. $ 9.50
2. Prepare any necessary adjusting entry at December 31, 2021, for the second purchase commitment assuming the following alternative unit market prices on that date:
   a. $12.50
   b. $10.30
3. Assuming that the unit market price on December 31, 2021, was $10.30, prepare the journal entry to record the purchase on March 15, 2022, assuming the following alternative unit market prices on that date:
   a. $11.50
   b. $10.00

Decision Maker's Perspective

©ImageGap/Getty Images

Apply your critical-thinking ability to the knowledge you've gained. These cases will provide you an opportunity to develop your research, analysis, judgment, and communication skills. You also will work with other students, integrate what you've learned, apply it in real-world situations, and consider its global and ethical ramifications. This practice will broaden your knowledge and further develop your decision-making abilities.

Judgment Case 9–1
Inventoriable costs; lower of cost or market; retail inventory method
● LO9–1, LO9–3, LO9–4

Hudson Company, which is both a wholesaler and a retailer, purchases its inventories from various suppliers. Additional facts for Hudson's wholesale operations are as follows:

a. Hudson incurs substantial warehousing costs.
b. Hudson values inventory at the lower of cost or market. Market is below cost of the inventories.

Additional facts for Hudson's retail operations are as follows:

a. Hudson determines the estimated cost of its ending inventories held for sale at retail using the conventional retail inventory method, which approximates lower of average cost or market.
b. Hudson incurs substantial freight-in costs.
c. Hudson has net markups and net markdowns.

Required:
1. Conceptually, in which account should Hudson report the warehousing costs related to its wholesale inventories? Why?

2. a. Inventory valued at the lower of cost or market is an example of which principle in accounting?
   b. At which amount should Hudson's wholesale inventories be reported in the balance sheet?
3. In the calculation of the cost-to-retail percentage used to determine the estimated cost of its ending retail inventories, how should Hudson treat:
   a. Freight-in costs?
   b. Net markups?
   c. Net markdowns?
4. How does Hudson's treatment of net markdowns affect the cost-to-retail percentage?

(AICPA adapted)

Communication Case 9–2
Lower of cost or net realizable value
● LO9–1

The lower of cost or net realizable value (LCNRV) approach to valuing inventory is a departure from the accounting principle of reporting assets at their historical costs. There are those who believe that inventory, as well as other assets, should be valued at NRV, regardless of whether NRV is above or below cost.

The focus of this case is the justification for the LCNRV rule for valuing inventories. Your instructor will divide the class into two to six groups depending on the size of the class. The mission of your group is to defend the LCNRV approach against the alternatives of valuing inventory at either historical cost or NRV.

Required:

1. Each group member should consider the situation independently and draft a tentative argument prior to the class session for which the case is assigned.
2. In class, each group will meet for 10 to 15 minutes in different areas of the classroom. During that meeting, group members will take turns sharing their suggestions for the purpose of arriving at a single group argument.
3. After the allotted time, a spokesperson for each group (selected during the group meetings) will share the group's solution with the class. The goal of the class is to incorporate the views of each group into a consensus approach to the situation.

Integrating Case 9–3
FIFO and lower of cost or net realizable value
● LO9–1

York Co. sells one product, which it purchases from various suppliers. York's trial balance at December 31, 2021, included the following accounts:

Sales (33,000 units @ $16)	$528,000
Sales discounts	7,500
Purchases	368,900
Purchase discounts	18,000
Freight-in	5,000
Freight-out	11,000

York Co.'s inventory purchases during 2021 were as follows:

	Units	Cost per Unit	Total Cost
Beginning inventory	7,000	$7.70	$ 53,900
Purchases, quarter ended March 31	13,000	7.50	97,500
Purchases, quarter ended June 30	15,000	7.90	118,500
Purchases, quarter ended September 30	12,000	8.25	99,000
Purchases, quarter ended December 31	8,000	8.20	65,600
	55,000		$434,500

Additional Information:

a. York's accounting policy is to report inventory in its financial statements at the lower of cost or net realizable value, applied to total inventory. Cost is determined under the first-in, first-out (FIFO) method.
b. York has determined that, at December 31, 2021, the net realizable value was $8.00 per unit.

Required:

1. Prepare York's schedule of cost of goods sold, with a supporting schedule of ending inventory. York includes inventory write-down losses in cost of goods sold.
2. Determine whether inventory should be reported at cost or net realizable value.

(AICPA adapted)

Judgment Case 9–4
The dollar-value LIFO method; the retail inventory method
● LO9–3, LO9–4

Huddell Company, which is both a wholesaler and retailer, purchases merchandise from various suppliers. The dollar-value LIFO method is used for the wholesale inventories.

Huddell determines the estimated cost of its retail ending inventories using the conventional retail inventory method, which approximates lower of average cost or market.

Required:

1. a. What are the advantages of using the dollar-value LIFO method as opposed to the traditional LIFO method?

   b. How does the application of the dollar-value LIFO method differ from the application of the traditional LIFO method?

2. a. In the calculation of the cost-to-retail percentage used to determine the estimated cost of its ending inventories, how should Huddell use
   - Net markups?
   - Net markdowns?

   b. Why does Huddell's retail inventory method approximate lower of average cost or market?

(AICPA adapted)

Communication Case 9–5
Retail inventory method
● LO9–3, LO9–4

The Brenly Paint Company, your client, manufactures paint. The company's president, Mr. Brenly, decided to open a retail store to sell paint as well as wallpaper and other items that would be purchased from other suppliers. He has asked you for information about the retail method of estimating inventories at the retail store.

Required:

Prepare a report to the president explaining the retail method of estimating inventories.

Analysis Case 9–6
Change in inventory method
● LO9–6

Generally accepted accounting principles should be applied consistently from period to period. However, changes within a company, as well as changes in the external economic environment, may force a company to change an accounting method. The specific reporting requirements when a company changes from one generally accepted inventory method to another depend on the methods involved.

Required:

Explain the accounting treatment for a change in inventory method (a) not involving LIFO, (b) from the LIFO method, and (c) to the LIFO method. Explain the logic underlying those treatments. Also, describe how disclosure requirements are designed to address the departure from consistency and comparability of changes in accounting principle.

Real World Case 9–7
Change in inventory method; Abercrombie & Fitch Co.
● LO9–6
Real World Financials

Abercrombie & Fitch Co. (A&F) is a specialty retail company operating over 1,000 stores globally. The following disclosure note was included in recent financial statements:

4. Change in Accounting Principle

The Company elected to change its method of accounting for inventory from the retail method to the weighted average cost method effective February 2, 2013. In accordance with generally accepted accounting principles, all periods have been retroactively adjusted to reflect the period-specific effects of the change to the weighted average cost method. The Company believes that accounting under the weighted average cost method is preferable as it better aligns with the Company's focus on realized selling margin and improves the comparability of the Company's financial results with those of its competitors. Additionally, it will improve the matching of cost of goods sold with the related net sales and reflect the acquisition cost of inventory outstanding at each balance sheet date. The cumulative adjustment as of January 30, 2010, was an increase in its inventory of $73.6 million and an increase in retained earnings of $47.3 million.

Required:

1. What approach should A&F take to account for its change in inventory accounting (retrospective, modified retrospective, or prospective)?
2. Will A&F need to restate financial statements prior to the year of the accounting change (yes/no)?
3. Which enhancing qualitative characteristic of accounting best describes the reason for A&F's approach to accounting for its change in inventory method?

Real World Case 9–8
Various inventory issues; Chapters 8 and 9; Fred's Inc.
● LO9–1, LO9–5, LO9–6

Real World Financials

Fred's Inc. operates general merchandise retail discount stores and full-service pharmacies in the Southeastern United States. Access the company's 10-K for the fiscal year ended January 28, 2017. You can find the 10-K by using EDGAR at **www.sec.gov** (Hint: do not use the apostrophe in the company name when doing an Edgar search.) Answer the following questions.

Required:

1. What inventory methods does Fred's use to value its inventory?
2. Which price index does the company use in applying the retail inventory method?
3. A company that uses LIFO is allowed to provide supplemental disclosures reporting the effect of using another inventory method rather than LIFO. Using the supplemental LIFO disclosures provided by Fred's, determine the pretax income effect of using LIFO versus another method for the current fiscal year.
4. Calculate the company's inventory turnover ratio for the fiscal year ended January 28, 2017.
5. Assume that in the next fiscal year the company decides to switch to the average cost method. What approach would Fred's take to account for this change (retrospective, modified retrospective, or prospective)?

Communication Case 9–9
Change in inventory method; disclosure note
● LO9–6

Mayfair Department Stores, Inc., operates over 30 retail stores in the Pacific Northwest. Prior to 2021, the company used the FIFO method to value its inventory. In 2021, Mayfair decided to switch to the dollar-value LIFO retail inventory method. One of your responsibilities as assistant controller is to prepare the disclosure note describing the change in method that will be included in the company's 2021 financial statements. Kenneth Meier, the controller, provided the following information:

a. Internally developed retail price indexes are used to adjust for the effects of changing prices.
b. If the change had not been made, cost of goods sold for the year would have been $22 million lower. The company's income tax rate is 40% and there were 100 million shares of common stock outstanding during 2021.
c. The cumulative effect of the change on prior years' income is not determinable.
d. The reasons for the change were (a) to provide a more consistent matching of merchandise costs with sales revenue, and (b) the new method provides a more comparable basis of accounting with competitors that also use the LIFO method.

Required:

1. Prepare for Kenneth Meier the disclosure note that will be included in the 2021 financial statements.
2. Explain why the "cumulative effect of the change on prior years' income is not determinable."

Judgment Case 9–10
Inventory errors
● LO9–7

Some inventory errors are said to be self-correcting in that the error has the opposite financial statement effect in the period following the error, thereby correcting the original account balance errors.

Required:

Despite this self-correcting feature, discuss why these errors should not be ignored and describe the steps required to account for the error correction.

Ethics Case 9–11
Overstatement of ending inventory
● LO9–7

Danville Bottlers is a wholesale beverage company. Danville uses the FIFO inventory method to determine the cost of its ending inventory. Ending inventory quantities are determined by a physical count. For the accounting year-end December 31, 2021, ending inventory was originally determined to be $3,265,000. However, on January 17, 2022, John Howard, the company's controller, discovered an error in the ending inventory count. He determined that the correct ending inventory amount should be $2,600,000.

Danville is a privately owned corporation with significant financing provided by a local bank. The bank requires annual audited financial statements as a condition of the loan. By January 17, the auditors had completed their review of the financial statements which are scheduled to be issued on January 25. They did not discover the inventory error.

John's first reaction was to communicate his finding to the auditors and to revise the financial statements before they are issued. However, he knows that his and his fellow workers' profit-sharing plans are based on annual pretax earnings, and he is uncertain what effect the error correction would have on pretax earnings.

Required:

1. What is the effect of the inventory error on pretax earnings?
2. How would correcting the error affect employee bonuses?
3. If the error is not corrected in the current year and is discovered by the auditors during the following year's audit, how will it be reported in the company's financial statements?
4. Discuss the ethical dilemma John Howard faces.

Analysis Case 9–12
Purchase commitments
● Appendix

The management of the Esquire Oil Company believes that the wholesale price of heating oil that they sell to homeowners will increase again as the result of increased political problems in the Middle East. The company is currently paying $0.80 a gallon. If they are willing to enter an agreement in November 2021 to purchase a million gallons of heating oil during the winter of 2022, their supplier will guarantee the price at $0.80 per gallon. However, if the winter is a mild one, Esquire would not be able to sell a million gallons unless they reduced their retail price and thereby increase the risk of a loss for the year. On the other hand, if the wholesale price did increase substantially, they would be in a favorable position with respect to their competitors. The company's fiscal year-end is December 31.

Required:
Discuss the accounting issues related to the purchase commitment that Esquire is considering.

Data Analytics

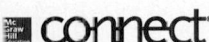

connect

Data analytics is the process of examining data sets in order to draw conclusions about the information they contain. If you have not completed any of the prior data analytics cases, follow the instructions listed in the Chapter 1 Data Analytics case to get set up. You will need to watch the videos referred to in the Chapters 1 - 3 Data Analytics cases. No additional videos are required for this case. All short training videos can be found here: **https://www.tableau.com/learn/training#getting-started**.

Data Analytics Case
Inventory Management
● LO9–3

In this case you are seeking to gain insight into the relative demand for the company's products in various regions of the country and various states in which it sells. Your focus is on assessing the company's inventory management relative to regional and local demand.

Required:
Use Tableau to calculate (a) additional markup percentage and (b) markdown percentage ratios for the company in each of the five regions of the country designated in the company's data base and in each of the states in which the company sells products. Calculate the ratios as the additional markup or markdown amounts as a percentage of sales. (A more precise measure that would produce similar results would replace sales in the ratio denominator with the original selling price of both sales and unsold inventory.) Based upon what you find, answer the following questions:

1. Which of the five regions' operations has the *most* favorable inventory management relative to demand as measured by the *additional markup* percentage ratio and what is that ratio (to the nearest one-tenth of a percent)?
2. Which of the five regions' operations has the *least* favorable inventory management relative to demand as measured by *markdown* percentage ratio and what is that ratio (to the nearest one-tenth of a percent)?

To view the state inventory management results more clearly, highlight the column "Additional Markup %" in the "Inventory Management by State" column and sort from largest to smallest. You can answer problem 4 below by following the same procedure for the "Markdown %" column.

3. Which state's operations has the *most* favorable inventory management relative to demand as measured by *additional markup* percentage ratio and what is that ratio (to the nearest one-tenth of a percent)?
4. Which state's operations has the *least* favorable inventory management relative to demand as measured by *markdown* percentage ratio and what is that ratio (to the nearest one-tenth of a percent)?

Resources:
You have available to you an extensive data set that includes detailed financial data for the company's inventory sales. The data set is in the form of four Excel files available to download from Connect, or under Student Resources within the Library tab. Download the file "Inventory.xlsx" to the computer, save it, and open it in Tableau.

For this case, you will create several calculations to produce a bar chart and a text chart to allow you to compare and contrast a company's inventory management.

After you view the training videos and review instructions in Chapters 1-3, follow these steps to create the charts you'll use for this case:

- Click on the "Sheet 1" tab at the bottom of the canvas, to the right of the Data Source at the bottom of the screen. Drag "Region" under Dimensions to the Columns shelf.
- Drag "Status" under Dimensions to the Filters card. When the Filter window opens, click the "S" so only the sold items will be viewed. Click ok.
- Under the Analysis tab, select Create Calculated Field. Create a measure named "Total Sales" by dragging "Number" to the Calculation Editor window, typing an asterisk sign for multiplication, and then dragging "Selling Price" beside it. Make sure the window says that the calculation is valid and click OK.
- Repeat the process by creating a calculated field "Total Markups" that consists of "Number" times "Net Markups." Make sure the window says that the calculation is valid and click OK.
- Repeat the process by creating a calculated field "Total Markdowns" that consists of "Number" times "Net Markdowns." Make sure the window says that the calculation is valid and click OK.
- Drag "Total Sales," "Total Markups," and "Total Markdowns" to the Rows shelf.
- Create a calculated field named "Additional Markup %" by dragging "Total Markups" from the Rows shelf into the Calculation Editor window, typing the division sign, and dragging "Total Sales" from the Rows shelf into the window next to it. Make sure the window says that the calculation is valid and click OK.

- Repeat the process one more time by creating a calculated field "Markdown %" by dragging "Total Markdowns" from the Rows shelf into the Calculation Editor window, typing the division sign, and dragging "Total Sales" from the Rows shelf into the window next to it.
- Drag "Total Sales," "Total Markups" and "Total Markdowns" from the Rows shelf back under Dimensions so they no longer show on the canvas.
- Drag "Additional Markup %" and "Markdown %," to the Rows shelf.
- Click the "Show Me" box on right side of the toolbar and select the "side-by-side bars."
- Show the labels by clicking on Label in the "Marks" card and checking "Show mark labels."
- Change the title of the sheet to be "Inventory Management by Region" by right-clicking and selecting "Edit title." Format the title to Times New Roman, bold, black and 15-point font. Change the title of "Sheet 1" to match the sheet title by right-clicking, selecting "Rename" and typing in the new title.
- Format all the labels on the sheet to Times New Roman, 10-point font, and black. You can also change the color variation by clicking "Color" on the "Marks" card and editing the color.
- Click on the New Worksheet tab on the lower right ("Sheet 2" should open) and drag "State" to the Rows shelf. Select the drop-down arrow on "State" to format the state names to Times New Roman, 10-point font, bold and blue (the format screen will open to the left).
- Double click "Additional Markup %" and "Markdown %." Right click on the drop-down menu of one of them and select "Format." Format to Times New Roman, bold, black, 10-point font, and center alignment (center alignment will be in the drop-down box beside alignment).
- Right click on one of the numbers and select "Format." Click on "Fields" in the upper-right corner of the Format window, select "Measure values," and format the numbers to "Percentage with two decimal places." Note that you can click on the right-side of each of the headings ("Additional Markup %" and "Markdown %") to sort from highest to lowest or lowest to highest. This may help you answer some of the questions.
- Change the title of the sheet to be "Inventory Management by State" by right-clicking and selecting "Edit title." Format the title to Times New Roman, bold, black and 15-point font. Change the title of "Sheet 2" to match the sheet title by right-clicking, selecting "Rename" and typing in the new title.
- Once complete, save the file as "*DA9_Your initials*.twbx."

Continuing Cases

Target Case
- LO9–3, LO9–4, LO9–5

Target Corporation prepares its financial statements according to U.S. GAAP. Target's financial statements and disclosure notes for the year ended February 3, 2018, are available in Connect. This material is also available under the Investor Relations link at the company's website (**www.target.com**).

Required:
1. What retail indices (internally measured or externally measured) does Target use to measure the LIFO provision?
2. Does Target adjust the retail value of inventory for permanent markups or permanent markdowns to effectively report inventory at the lower of cost or market?
3. Target has agreements with certain vendors whereby Target does not purchase or pay for merchandise until the merchandise is ultimately sold to a customer. Are sales and cost of sales of this inventory included in Target's income statement? Is unsold inventory at the end of the year included as part of ending inventory in the balance sheet?

Air France–KLM Case
- LO9–8

IFRS

Air France–KLM (AF), a Franco-Dutch company, prepares its financial statements according to International Financial Reporting Standards. AF's financial statements and disclosure notes for the year ended December 31, 2017, are available in Connect. This material is also available under the Finance link at the company's website (**www.airfranceklm.com**).

Required:
AF's inventories are valued at the lower of cost or net realizable value. Does this approach differ from U.S. GAAP?

CPA Exam Questions and Simulations

ROGER CPA Review

Sample CPA Exam questions from Roger CPA Review are available in Connect as support for the topics in this chapter. These multiple-choice questions and task-based simulations include expert-written explanations and solutions, and provide a starting point for students to become familiar with the content and functionality of the actual CPA Exam.

CHAPTER

10 Property, Plant, and Equipment and Intangible Assets: Acquisition

OVERVIEW

This chapter and the one that follows address the measurement and reporting issues involving property, plant, and equipment and intangible assets. These are the long-lived assets used in the production of goods and services. This chapter covers the valuation at date of acquisition. In Chapter 11, we discuss the allocation of the cost of property, plant, and equipment and intangible assets to the periods benefited by their use, the treatment of expenditures made over the life of these assets to maintain and improve them, impairment, and disposition.

LEARNING OBJECTIVES

After studying this chapter, you should be able to:

- **LO10–1** Identify the various costs included in the initial cost of property, plant, and equipment, natural resources, and intangible assets. (*p. 516*)
- **LO10–2** Determine the initial cost of individual property, plant, and equipment and intangible assets acquired as a group for a lump-sum purchase price. (*p. 525*)
- **LO10–3** Determine the initial cost of property, plant, and equipment and intangible assets acquired in exchange for a deferred payment contract. (*p. 526*)
- **LO10–4** Determine the initial cost of property, plant, and equipment and intangible assets acquired in exchange for equity securities or through donation. (*p. 528*)
- **LO10–5** Calculate the fixed-asset turnover ratio used by analysts to measure how effectively managers use property, plant, and equipment. (*p. 530*)
- **LO10–6** Determine the initial cost of property, plant, and equipment and intangible assets acquired in exchange for other nonmonetary assets. (*p. 531*)
- **LO10–7** Identify the items included in the cost of a self-constructed asset and determine the amount of capitalized interest. (*p. 535*)
- **LO10–8** Explain the difference in the accounting treatment of costs incurred to purchase intangible assets versus the costs incurred to internally develop intangible assets. (*p. 541*)
- **LO10–9** Discuss the primary differences between U.S. GAAP and IFRS with respect to the acquisition of property, plant, and equipment and intangible assets. (*p. 529, 543*)

© dealphotographer/123RF

FINANCIAL REPORTING CASE

A Disney Adventure

"Now I'm really confused," confessed Stan, your study partner, staring blankly at **The Walt Disney Company** balance sheet that your professor handed out last week. "I thought that interest is always expensed in the income statement. Now I see that Disney is capitalizing interest. I'm not even sure what *capitalize* means! And what about this other account called *goodwill?* What's that all about?" You respond, "We talked about these topics in our accounting class today. Let's take a look at the Disney financial statements and the disclosure note on capitalized interest and I'll try to explain it all to you."

Borrowings (in part):

The Company capitalizes interest on assets constructed for its parks and resorts and on certain film and television productions. In fiscal years 2017, 2016 and 2015, total interest capitalized was $87 million, $139 million, and $110 million, respectively.

QUESTIONS

By the time you finish this chapter, you should be able to respond appropriately to the questions posed in this case. Compare your response to the solution provided at the end of the chapter.

1. Describe to Stan what it means to capitalize an expenditure. What is the general rule for determining which costs are capitalized when property, plant, and equipment or an intangible asset is acquired? (*p. 516*)
2. Which costs might be included in the initial cost of equipment? (*p. 516*)
3. What is goodwill and how is it measured? (*p. 523*)
4. In what situations is interest capitalized rather than expensed? (*p. 536*)
5. What is the three-step process used to determine the amount of interest capitalized? (*p. 537*)

General Motors Corporation has significant investments in the production facilities it uses to manufacture the automobiles it sells. On the other hand, the principal revenue-producing assets of **Microsoft Corporation** are the copyrights on its computer software that permit it the exclusive rights to earn profits from those products. Timber reserves provide major revenues to **International Paper Company**. From a reporting perspective, we classify GM's production facilities as property, plant, and equipment; Microsoft's copyrights as intangible assets; and International Paper's timber reserves as natural resources.[1] Together, these three noncurrent assets constitute the *long-lived, revenue-producing assets* of a company. Unlike manufacturers, many service firms and merchandising companies rely primarily on people or investments in inventories rather than on property, plant, and equipment and intangible assets to generate revenues. Even nonmanufacturing firms, though, typically have at least modest investments in buildings and equipment.

The measurement and reporting issues pertaining to this group of assets include valuation at date of acquisition, disposition, the treatment of expenditures made over the life of these assets to maintain and improve them, the allocation of cost to reporting periods that benefit from their use, and impairment. We focus on initial valuation in this chapter. In Chapter 11, we examine subsequent expenditures, cost allocation, impairment, and disposition.

[1]These are sometimes called *plant assets* or *fixed assets.*

PART A

Valuation at Acquisition

Types of Assets

For financial reporting purposes, long-lived, revenue-producing assets typically are classified in two categories.

1. **Property, plant, and equipment.** Assets in this category include land, land improvements, buildings, machinery used in manufacturing, computers and other office equipment, vehicles, furniture, and fixtures. **Natural resources** such as oil and gas deposits, timber tracts, and mineral deposits also are included.
2. **Intangible assets.** Unlike property, plant, and equipment and natural resources, intangible assets lack physical substance, and the extent and timing of their future benefits typically are uncertain. They include items such as patents, copyrights, trademarks, franchises, and goodwill.

Of course, every company maintains its own unique mix of these assets. The way these assets are classified and combined for reporting purposes also varies from company to company. As an example, a recent balance sheet of **Semtech Corporation**, a leading supplier of semiconductor products, reported net property, plant, and equipment of **$124,586** thousand and **$108,910** thousand at the end of fiscal years 2018 and 2017, respectively. A disclosure note, shown in Illustration 10–1, provides the details.

Illustration 10–1

Property, Plant, and Equipment—Semtech Corporation

Real World Financials

Note 7. Property, Plant, and Equipment (in part):
The following is a summary of property and equipment, at cost less accumulated depreciation:

($ in thousands)	January 28, 2018	January 29, 2017
Property	$ 11,314	$ 11,314
Buildings	32,244	30,294
Leasehold improvements	10,050	9,566
Machinery and equipment	171,731	150,276
Enterprise resource planning systems	32,673	28,166
Furniture and office equipment	39,027	35,025
Construction in progress	7,151	5,505
Property, plant, and equipment, gross	304,190	270,146
Less: accumulated depreciation and amortization	(179,604)	(161,236)
Property, plant, and equipment, net	**$124,586**	**$108,910**

Source: Semtech Corporation

In practice, some companies report intangibles as part of property, plant, and equipment, and others show intangibles as a separate balance sheet category. For example, **The Walt Disney Company** reported goodwill of $31,426 million and other intangible assets of **$6,995** million and **$6,949** million at the end of fiscal years 2017 and 2016, respectively, as shown in Illustration 10–2.

In this chapter, we'll first discuss how to record the acquisition cost of several types of property, plant, and equipment, and then we'll discuss the acquisition cost of intangible assets. You should find it helpful to study the overview of typical acquisition costs for each type of asset in Illustration 10–3. In Part B of this chapter, we examine how to record the acquisition of these assets when no cash is involved. Self-constructed assets and special issues related to research and development costs are addressed in Part C of the chapter. In the next chapter, we'll see how asset acquisition costs are expensed over time in a process called *depreciation* for plant and equipment, *depletion* for natural resources, and *amortization* for intangibles.

($ in millions)	2017	2016
Character/franchise intangibles and copyrights	$ 5,829	$ 5,829
Other amortizable intangible assets	1,154	893
Accumulated amortization	(1,828)	(1,635)
Net amortizable intangible assets	5,155	5,087
FCC licenses	602	624
Trademarks	1,218	1,218
Other indefinite lived intangible assets	20	20
	$6,995	$6,949

Source: The Walt Disney Company

Illustration 10–2
Other Intangible Assets—The Walt Disney Company

Real World Financials

Illustration 10–3 Property, Plant, and Equipment and Intangible Assets and Their Acquisition Costs

Asset	Description	Typical Acquisition Costs
Property, plant, and equipment	Productive assets that derive their value from long-term use in operations rather than from resale.	All expenditures necessary to get the asset in condition and location for its intended use.
Equipment	Broad term that includes machinery, computers and other office equipment, vehicles, furniture, and fixtures.	Purchase price (less discounts), taxes, transportation, installation, testing, trial runs, and reconditioning.
Land	Real property used in operations (land held for speculative investment or future use is reported as investments or other assets).	Purchase price, attorney's fees, title, recording fees, commissions, back taxes, mortgages, liens, clearing, filling, draining, and removing old buildings.
Land improvements	Enhancements to property such as parking lots, driveways, private roads, fences, landscaping, and sprinkler systems.	Separately identifiable costs.
Buildings	Structures that include warehouses, plant facilities, and office buildings.	Purchase price, attorney's fees, commissions, and reconditioning.
Natural resources	Productive assets that are physically consumed in operations such as timber, mineral deposits, and oil and gas reserves.	Acquisition, exploration, development, and restoration costs.
Intangible Assets	Productive assets that lack physical substance and have long-term but typically uncertain benefits.	All expenditures necessary to get the asset in condition and location for its intended use.
Patents	Exclusive 20-year right to manufacture a product or use a process.	Purchase price, legal fees, filing fees, not including internal R&D.
Copyrights	Exclusive right to benefit from a creative work such as a song, film, painting, photograph, or book.	Purchase price, legal fees, filing fees, not including internal R&D.
Trademarks (tradenames)	Exclusive right to display a word, a slogan, a symbol, or an emblem that distinctively identifies a company, product, or a service.	Purchase price, legal fees, filing fees, not including internal R&D.
Franchises	A contractual arrangement under which a franchisor grants the franchisee the exclusive right to use the franchisor's trademark or tradename and certain product rights.	Franchise fee plus any legal fees.
Goodwill	The unique value of the company as a whole over and above all identifiable assets.	Excess of the fair value of the consideration given for a company over the fair value of the identifiable net assets acquired.

Costs to be Capitalized

● LO10–1

Property, plant, and equipment and intangible assets can be acquired through purchase, exchange, lease, donation, self-construction, or a business combination. We address acquisitions through leasing in Chapter 15.

The initial cost of property, plant, and equipment and intangible assets includes the purchase price and all expenditures necessary to bring the asset to its desired condition and location for use.

The initial valuation of property, plant, and equipment and intangible assets usually is quite simple. We know from prior study that assets are valued on the basis of their original costs. In Chapter 8 we introduced the concept of condition and location in determining the cost of inventory. For example, if Thompson Company purchased inventory for \$40,000 and incurred \$1,000 in freight costs to have the inventory shipped to its location, the initial cost of the inventory is \$41,000. This concept applies to the valuation of property, plant, and equipment and intangible assets as well. The initial cost of these assets includes the purchase price and all expenditures necessary to bring the asset to its desired condition and location for use. We discuss these additional expenditures in the next section.

FINANCIAL Reporting Case

Q1, p. 3

Our objective in identifying the costs of an asset is to distinguish the expenditures that produce future benefits from those that produce benefits only in the current period. The costs in the second group are recorded as expenses, but those in the first group are *capitalized;* that is, they are recorded initially as an asset and then expensed in future periods.[2] For example, the cost of a major improvement to a delivery truck that extends its useful life generally would be capitalized. On the other hand, the cost of an engine tune-up for the delivery truck simply allows the truck to continue its productive activity but does not increase future benefits. These maintenance costs would be expensed. Subsequent expenditures for these assets are discussed in Chapter 11.

The distinction is not trivial. This point was unmistakably emphasized in the summer of 2002 when **WorldCom, Inc.**, disclosed that it had improperly capitalized nearly \$4 billion in expenditures related to the company's telecom network. This massive fraud resulted in one of the largest financial statement restatements in history and triggered the collapse of the once powerful corporation. Capitalizing rather than expensing these expenditures caused a substantial understatement of expenses and overstatement of reported income for 2001 and the first quarter of 2002. If the deception had not been discovered, not only would income for 2001 and 2002 have been overstated but also income for many years into the future would have been understated as the fraudulent capitalized assets were depreciated. Of course, the balance sheet also would have overstated the assets and equity of the company.

Property, Plant, and Equipment

FINANCIAL Reporting Case

Q2, p. 3

COST OF EQUIPMENT Equipment is a broad term that encompasses machinery used in manufacturing, computers and other office equipment, vehicles, furniture, and fixtures. The cost of equipment includes the purchase price plus any sales tax (less any discounts received from the seller), transportation costs paid by the buyer to transport the asset to the location in which it will be used, expenditures for installation, testing, legal fees to establish title, and any other costs of bringing the asset to its condition and location for use. To the extent that these costs can be identified and measured, they should be included in the asset's initial valuation rather than expensed currently.

Although most costs can be identified easily, others are more difficult. For example, the costs of training personnel to operate machinery could be considered a cost necessary to make the asset ready for use. However, because it is difficult to measure the amount of training costs associated with specific assets, these costs usually are expensed. Consider Illustration 10–4.

[2]Exceptions are land and certain intangible assets that have indefinite useful lives. Costs to acquire these assets also produce future benefits and therefore are capitalized, but unlike other property, plant, and equipment and intangible assets, their costs are not systematically expensed in future periods as depreciation or amortization.

Illustration 10–4

Initial Cost of Equipment

Central Machine Tools purchased industrial equipment to be used in its manufacturing process. The purchase price was $62,000. Central paid a freight company $1,000 to transport the equipment to its plant location plus $300 shipping insurance. In addition, the equipment had to be installed and mounted on a special platform built specifically for the equipment at a cost of $1,200. After installation, several trial runs were made to ensure proper operation. The cost of these trials including wasted materials was $600. At what amount should Central capitalize the equipment?

Purchase price	$62,000
Freight and handling	1,000
Insurance during shipping	300
Special platform	1,200
Trial runs	600
Capitalized cost of equipment	$65,100

Each of the expenditures described was necessary to bring the equipment to its condition and location for use and should be capitalized. These costs will be expensed in the future periods in which the equipment is used.

COST OF LAND The cost of land also should include expenditures needed to get the land ready for its intended use. These include the purchase price plus closing costs such as fees for the attorney, real estate agent commissions, title and title search, and recording. If the property is subject to back taxes, liens, mortgages, or other obligations, these amounts are included also. In addition, any expenditures such as clearing, filling, draining, and even removing (razing) old buildings that are needed to prepare the land for its intended use are part of the land's cost. Proceeds from the sale of salvaged materials from old buildings torn down after purchase reduce the cost of land. Illustration 10–5 provides an example.

Illustration 10–5

Initial Cost of Land

The Byers Structural Metal Company purchased a six-acre tract of land and an existing building for $500,000. The company plans to remove the old building and construct a new office building on the site. In addition to the purchase price, the company made the following expenditures at closing of the purchase:

Title insurance	$ 3,000
Commissions	16,000
Property taxes	6,000*

*The $6,000 in property taxes included $4,000 of delinquent taxes paid by Byers on behalf of the seller as well as $2,000 attributable to the portion of the current fiscal year that remains *after the purchase date*.

In addition, shortly after closing, Byers paid a contractor $10,000 to tear down the old building and remove it from the site. An additional $5,000 was paid to grade the land. What should be the capitalized cost of the land?

Purchase price of land (and old building)	$500,000
Title insurance	3,000
Commissions	16,000
Delinquent property taxes	4,000
Cost of removing old building	10,000
Cost of grading	5,000
Capitalized cost of land	$538,000

Property taxes of $2,000 were not included. These relate only to the current operating period and should be expensed separately as property tax expense. Other costs were necessary to acquire the land and are capitalized.

LAND IMPROVEMENTS It's important to distinguish between the cost of land and the cost of **land improvements** because land has an indefinite life and land improvements usually have useful lives that are estimable. Examples of land improvements include the costs of establishing parking lots, driveways, and private roads and the costs of fences and lawn and garden sprinkler systems. Costs of these assets are separately identified and capitalized. We depreciate their cost over periods benefited by their use.

COST OF BUILDINGS The cost of acquiring a building usually includes realtor commissions and legal fees in addition to the purchase price. Quite often a building must be refurbished, remodeled, or otherwise modified to suit the needs of the new owner. These reconditioning costs are part of the building's acquisition cost. When a building is constructed rather than purchased, unique accounting issues are raised. We discuss these in Part C in the "Self-Constructed Assets" section.

COST OF NATURAL RESOURCES **Natural resources** that provide long-term benefits are reported as property, plant, and equipment. These include timber tracts, mineral deposits, and oil and gas deposits. They can be distinguished from other assets by the fact that their benefits are derived from their physical consumption. For example, mineral deposits are physically diminishing as the minerals are extracted from the ground and either sold or used in the production process.[3] On the contrary, equipment, land, and buildings produce benefits for a company through their *use* in the production of goods and services. Unlike those of natural resources, their physical characteristics usually remain unchanged during their useful lives.

Sometimes a company buys natural resources from another company. In that case, initial valuation is simply the purchase price plus any other costs necessary to bring the asset to condition and location for use. More frequently, though, the company will develop these assets. In this situation, the initial valuation can include (a) acquisition costs, (b) exploration costs, (c) development costs, and (d) restoration costs.

Acquisition costs are the amounts paid to acquire the rights to explore for undiscovered natural resources or to extract proven natural resources. **Exploration costs** are expenditures such as drilling a well, or excavating a mine, or any other costs of searching for natural resources. **Development costs** are incurred after the resource has been discovered but before production begins. They include a variety of costs such as expenditures for tunnels, wells, and shafts. **Restoration costs** include costs to restore land or other property to its original condition after extraction of the natural resource ends. Because restoration expenditures occur later—after production begins—they initially represent an obligation incurred in conjunction with an asset retirement. Restoration costs are one example of *asset retirement obligations,* the topic of the next subsection.

The cost of a natural resource includes the *acquisition costs* for the use of land, the *exploration* and *development costs* incurred before production begins, and *restoration costs* incurred during or at the end of extraction.

On the other hand, the costs of heavy equipment and other assets a company uses during drilling or excavation usually are not considered part of the cost of the natural resource itself. Instead, they are considered depreciable plant and equipment. However, if an asset used in the development of a natural resource cannot be moved and has no alternative use, its depreciable life is limited by the useful life of the natural resource.

ASSET RETIREMENT OBLIGATIONS Sometimes a company incurs obligations associated with the disposition of property, plant, and equipment and natural resources, often as a result of acquiring those assets. For example, an oil and gas exploration company might be required to restore land to its original condition after extraction is completed. Accounting guidelines were needed because there was considerable diversity in the ways companies accounted for these obligations. Some companies recognized these **asset retirement obligations (AROs)** gradually over the life of the asset while others did not recognize the obligations until the asset was retired or sold.

An asset retirement obligation (ARO) is measured at fair value and is recognized as a liability and corresponding increase in asset valuation.

Generally accepted accounting principles require that an existing legal obligation associated with the retirement of a tangible, long-lived asset be recognized as a liability and

[3]Because of this characteristic, natural resources sometimes are called *wasting assets.*

measured at fair value, if value can be reasonably estimated. When the liability is credited, the offsetting debit is to the related asset.[4] These retirement obligations could arise in connection with several types of assets. We introduce the topic here because it often arises with natural resources. Let's consider some of the provisions of the standard that addresses these obligations.

Scope. AROs arise only from *legal* obligations associated with the retirement of a tangible long-lived asset that result from the acquisition, construction, or development and (or) normal operation of a long-lived asset.

Recognition. A retirement obligation might arise at the inception of an asset's life or during its operating life. For instance, an offshore oil-and-gas production facility typically incurs its removal obligation when it begins operating. On the other hand, a landfill or a mining operation might incur a reclamation obligation gradually over the life of the asset as space is consumed with waste or as the mine is excavated.

Measurement. A company recognizes the fair value of an ARO in the period it's incurred. The amount of the liability increases the valuation of the related asset. Usually, the fair value is estimated by calculating the present value of estimated future cash outflows.

Present value calculations. Traditionally, the way uncertainty has been considered in present value calculations has been by discounting the "best estimate" of future cash flows applying a discount rate that has been adjusted to reflect the uncertainty or risk of those cash flows. That's not the approach we take here. Instead, we follow the approach described in the FASB's *Concept Statement No. 7,* which is to adjust the cash flows, not the discount rate, for the uncertainty or risk of those cash flows.[5] This **expected cash flow approach** incorporates specific probabilities of cash flows into the analysis. We use a discount rate equal to the *credit-adjusted risk free rate.* The higher a company's credit risk, the higher will be the discount rate. All other uncertainties or risks are incorporated into the cash flow probabilities. Illustration 10–6 demonstrates the approach in connection with the acquisition of a natural resource.

Illustration 10–6
Cost of Natural Resources

The Jackson Mining Company paid $1,000,000 for the right to explore for a coal deposit on 500 acres of land in Pennsylvania. Costs of exploring for the coal deposit totaled $800,000 and intangible development costs incurred in digging and erecting the mine shaft were $500,000. In addition, Jackson purchased new excavation equipment for the project at a cost of $600,000. After the coal is removed from the site, the equipment will be sold.

Jackson is required by its contract to restore the land to a condition suitable for recreational use after it extracts the coal. The company has provided the following three cash flow possibilities (A, B, and C) for the restoration costs to be paid in three years, after extraction is completed:

	Cash Outflow	Probability
A	$500,000	30%
B	600,000	50%
C	700,000	20%

(Continued)

[4]FASB ASC 410–20–25: Asset Retirement and Environmental Obligations—Asset Retirement Obligations—Recognition.
[5]"Using Cash Flow Information and Present Value in Accounting Measurements," *Statement of Financial Accounting Concepts No. 7* (Norwalk, Conn.: FASB, 2000).

Illustration 10–6
(*Concluded*)

The company's credit-adjusted risk free interest rate is 8%. What should be the capitalized cost of the coal deposit?

Purchase of rights to explore	$1,000,000
Exploration costs	800,000
Development costs	500,000
Restoration costs	**468,360***
Capitalized cost of the coal deposit	$2,768,360

*Present value of expected cash outflow for restoration costs (asset retirement obligation):

$500,000 × 30% = $150,000
600,000 × 50% = 300,000
700,000 × 20% = 140,000
$590,000 × 0.79383 = **$468,360**
(0.79383 is the present value of $1, $n = 3$, $i = 8\%$)

Journal Entries

Coal mine (determined above)	2,768,360	
Cash ($1,000,000 + $800,000 + $500,000)		2,300,000
Asset retirement liability (determined above)		**468,360**
Excavation equipment	600,000	
Cash (cost)		600,000

As we discuss in Chapter 11, the cost of the coal mine is allocated to future periods as *depletion* using a depletion rate based on the estimated amount of coal discovered. The $600,000 cost of the excavation equipment, less any anticipated residual value, is allocated to future periods as *depreciation.*

Additionally, the difference between the asset retirement liability of **$468,360** and the probability-weighted expected cash outflow of **$590,000** is recognized as **accretion expense**, an additional expense that accrues as an operating expense, over the three-year excavation period. This process increases the liability to $590,000 by the end of the excavation period.

Year	Accretion Expense	Increase in Balance	Asset Retirement Obligation
			468,360
1	8% (468,360) = 37,469	37,469	505,829
2	8% (505,829) = 40,466	40,466	546,295
3	8% (546,295) = 43,705*	43,705	**590,000**

*Rounded

The journal entry to record accretion expense for the first year is as follows:

Accretion expense	37,469	
Asset retirement liability		37,469

If the actual restoration costs are more (less) than the $590,000, we recognize a loss (gain) on retirement of the obligation for the difference. For example, if the actual restoration costs were $625,000, we would record the transaction as follows:

Asset retirement liability	590,000	
Loss ($625,000 – $590,000)	35,000	
Cash		625,000

Asset retirement obligations could result from the acquisition of many different types of tangible assets, not just natural resources.

SM Energy Company is engaged in the exploration, development, acquisition, and production of natural gas and crude oil. For the year ended December 31, 2017, SM reported $114.4 million in asset retirement obligations in its balance sheet. A disclosure note included in a recent annual report shown in Illustration 10–7 describes the company's policy and provides a summary of disclosure requirements.

Note 1—Asset Retirement Obligations

The Company recognizes an estimated liability for future costs associated with the abandonment of its oil and gas properties, including facilities requiring decommissioning. A liability for the fair value of an asset retirement obligation and corresponding increase to the carrying value of the related long-lived asset are recorded at the time a well is drilled or acquired and a facility is constructed. The increase in carrying value is included in proved oil and gas properties in the accompanying balance sheets. The Company depletes the amount added to proved oil and gas property costs and recognizes expense in connection with the accretion of the discounted liability over the remaining estimated economic lives of the respective oil and gas properties. Cash paid to settle asset retirement obligations is included in the operating section of the Company's accompanying statements of cash flows.

The Company's estimated asset retirement obligation liability is based on historical experience in plugging and abandoning wells, estimated economic lives, estimated plugging and abandonment cost, and federal and state regulatory requirements. The liability is discounted using the credit-adjusted risk-free rate estimated at the time the liability is incurred or revised. The credit-adjusted risk-free rates used to discount the Company's plugging and abandonment liabilities range from 5.5 percent to 12 percent. In periods subsequent to initial measurement of the liability, the Company must recognize period-to-period changes in the liability resulting from the passage of time, revisions to either the amount of the original estimate of undiscounted cash flows or changes in inflation factors or the Company's credit-adjusted risk-free rate as market conditions warrant. Please refer to Note 9 – Asset Retirement Obligations for a reconciliation of the Company's total asset retirement obligation liability as of December 31, 2017, and 2016.

Source: SM Energy Company

Illustration 10–7

Disclosure of Asset Retirement Obligations—SM Energy Company

Real World Financials

It is important to understand that asset retirement obligations could result from the acquisition of many different types of tangible assets, not just natural resources. For example, **Dow Chemical Company** reported a $96 million asset retirement liability in its 2016 balance sheet related to anticipated demolition and remediation activities at its manufacturing sites in the United States, Canada, Brazil, China, Argentina, Australia, and Europe; and capping activities at landfill sites in the United States, Canada, Brazil, and Italy.

Sometimes, after exploration or development, it becomes apparent that continuing the project is economically infeasible. If that happens, any costs incurred are expensed rather than capitalized. An exception is in the oil and gas industry, where we have two generally accepted accounting alternatives for accounting for projects that prove unsuccessful. We discuss these alternatives in Appendix 10.

Intangible Assets

Intangible assets are assets, other than financial assets, that lack physical substance. They include such items as patents, copyrights, trademarks, franchises, and goodwill. Despite their lack of physical substance, these assets can be extremely valuable resources for a company. For example, **Interbrand Sampson**, the world's leading branding consulting company, recently estimated the value of the **Coca-Cola** trademark to be $70 billion.[6] In general, intangible assets refer to the ownership of exclusive rights that provide benefits to the owner in the production of goods and services.

Intangible assets generally represent exclusive rights that provide benefits to the owner.

The issues involved in accounting for intangible assets are similar to those of property, plant, and equipment. One key difference, though, is that the future benefits that we attribute to intangible assets usually are much less certain than those attributed to tangible assets. For example, will the new toy for which a company acquires a patent be accepted by the market? If so, will it be a blockbuster like Silly Bandz or Rubik's Cube, or will it be only a moderate success? Will it have lasting appeal like Barbie dolls, or will it be a short-term fad? In short, it's often very difficult to anticipate the timing, and even the existence, of future benefits attributable to many intangible assets. In fact, this uncertainty is a discriminating

[6]This $70 billion represents an estimate of the fair value to the company at the time the estimate was made, not the historical cost valuation that appears in the balance sheet of Coca-Cola.

characteristic of intangible assets that perhaps better distinguishes them from tangible assets than their lack of physical substance. After all, other assets, too, do not exist physically but are not considered intangible assets. Accounts receivable and prepaid expenses, for example, have no physical substance and yet are reported among tangible assets.

Companies acquire intangible assets in two ways:

1. They *purchase* intangible assets like patents, copyrights, trademarks, or franchise rights from other companies.
2. They *develop* intangible assets internally, for instance, by developing a new product or process and obtaining a protective patent.

Purchased intangible assets are valued at their original cost plus other necessary costs.

The reporting rules for intangible assets vary depending on whether the company purchased the asset or developed it internally. Reporting purchased intangibles is similar to reporting purchased property, plant, and equipment. We record purchased intangible assets at their original cost plus all other costs, such as legal fees, necessary to get the asset ready for use. For example, if a company purchases a patent from another entity, it might pay legal fees and filing fees in addition to the purchase price. We value intangible assets acquired in exchange for stock, or for other nonmonetary assets, or with deferred payment contracts exactly as we do property, plant, and equipment.

The cost of an intangible asset is amortized over its useful life unless it has an indefinite useful life.[7] Also, just like property, plant, and equipment, intangibles are subject to asset impairment rules. We discuss amortization and impairment in Chapter 11.

Internally-developed intangible assets typically are expensed.

Reporting intangible assets that are developed internally is quite different. Rather than reporting these in the balance sheet as assets, we typically expense them in the income statement in the period we incur those internal development costs. For example, the research and development (R&D) costs incurred in developing a new product or process are not recorded as an intangible asset in the balance sheet. Instead, they are expensed directly in the income statement as incurred. The reason we expense all R&D costs is the difficulty in determining the portion of R&D that benefits future periods. Conceptually, we should record as an intangible asset the portion that benefits future periods. Due to the difficulties in arriving at this estimate, current U.S. accounting rules require firms to expense all R&D costs as incurred.

Let's look in detail at some specific types of intangible assets.

PATENTS A **patent** is an exclusive right to manufacture a product or to use a process. The U.S. Patent and Trademark Office grants this right for a period of 20 years. In essence, the holder of a patent has a monopoly on the use, manufacture, or sale of the product or process. If a patent is purchased from an inventor or another individual or company, the amount paid is its initial valuation. The cost might also include such other costs as legal and filing fees to secure the patent. Holders of patents often need to defend a patent in court against infringement. Any attorney fees and other costs of successfully defending a patent are added to the patent account.

In contrast, when a firm engages in its own research activities to develop a new product or process, it expenses those costs as it incurs them. For example, major pharmaceutical companies like **Amgen** and **Gilead Sciences** spend over a billion dollars each year developing new drugs. Most of these research and development costs are recorded as operating expenses in the income statement. An exception to this rule is legal fees. The firm will record in the Patent asset account the legal and filing fees to secure a patent, even if it developed the patented item or process internally. We discuss research and development in more detail in a later section.

COPYRIGHTS A **copyright** is an exclusive right of protection given by the U.S. Copyright Office to the creator of a published work such as a song, film, painting, photograph, book, or computer software. Copyrights give the creator/owner the exclusive right to reproduce and sell the artistic or published work for the life of the creator plus 70 years.

[7]FASB ASC 350–30–35–1: Intangibles—Goodwill and Other—General Intangibles Other Than Goodwill—Subsequent Measurement, and FASB ASC 350–20–35–1: Intangibles—Goodwill and Other—Goodwill—Subsequent Measurement (previously "Goodwill and Other Intangible Assets," *Statement of Financial Accounting Standards No. 142* (Norwalk, Conn.: FASB, 2001)).

A copyright also allows the holder to pursue legal action against anyone who attempts to infringe the copyright. Accounting for the costs of copyrights is virtually identical to that of patents.

TRADEMARKS A **trademark**, also called **tradename**, is a word, slogan, or symbol that distinctively identifies a company, product, or service. The firm can register its trademark with the U.S. Patent and Trademark Office to protect it from use by others for a period of 10 years. The registration can be renewed for an indefinite number of 10-year periods, so a trademark is an example of an intangible asset whose useful life can be indefinite.

Trademarks or tradenames often are considered to have indefinite useful lives.

Trademarks or tradenames often are acquired through a business combination. As an example, in 2002, **Hewlett-Packard Company (HP)** acquired all of the outstanding stock of **Compaq Computer Corporation** for $24 billion. Of that amount, $1.4 billion was assigned to the Compaq tradename. HP stated in a disclosure note that this ". . . intangible asset will not be amortized because it has an indefinite remaining useful life based on many factors and considerations, including the length of time that the Compaq name has been in use, the Compaq brand awareness and market position and the plans for continued use of the Compaq brand within a portion of HP's overall product portfolio."

Trademarks or tradenames can be very valuable. The estimated value of $70 billion for the Coca-Cola trademark mentioned previously is a good example. Note that the cost of the trademark reported in the balance sheet is far less than the estimate of its worth to the company. The Coca-Cola Company's balance sheet at December 31, 2017, disclosed all trademarks at a cost of only $7 billion.

FRANCHISES A **franchise** is a contractual arrangement under which the franchisor grants the franchisee the exclusive right to use the franchisor's trademark or tradename, as well as possibly product and formula rights, to operate a business within a geographical area and usually for a specified period of time. Many popular retail businesses such as fast-food outlets, automobile dealerships, and motels are franchises. For example, the last time you ordered a hamburger at McDonald's, you were probably dealing with a franchise.

Franchise operations are among the most common ways of doing business.

The owner of that McDonald's outlet paid **McDonald's Corporation** a fee in exchange for the exclusive right to use the McDonald's name and to sell its products within a specified geographical area. In addition, many franchisors provide other benefits to the franchisee, such as participating in the construction of the retail outlet, training of employees, and national advertising.

Payments to the franchisor usually include an initial payment plus periodic payments over the life of the franchise agreement. The franchisee capitalizes as an intangible asset the initial franchise fee plus any legal costs associated with the contract agreement. The franchise asset is then amortized over the life of the franchise agreement. The periodic payments usually relate to services provided by the franchisor on a continuing basis and are expensed as incurred.

Most purchased intangibles are *specifically identifiable*. That is, cost can be directly associated with a specific intangible right. An exception is goodwill, which we discuss next.

GOODWILL Goodwill is a unique intangible asset in that its cost can't be directly associated with any specific identifiable right and it is not separable from the company itself. It represents the unique value of a company as a whole over and above its identifiable tangible and intangible assets. Goodwill can emerge from a company's clientele and reputation, its trained employees and management team, its favorable business location, and any other unique features of the company that can't be associated with a specific asset.

Goodwill can only be purchased through the acquisition of another company.

FINANCIAL Reporting Case

Q3, p. 3

Because goodwill can't be separated from a company, it's not possible for a buyer to acquire it without also acquiring the whole company or a portion of it. Goodwill will appear as an asset in a balance sheet only when it was purchased in connection with the acquisition of control over another company. In that case, the capitalized cost of goodwill equals the fair value of the consideration given in exchange for the company (the acquisition price) less the fair value of the identifiable net assets acquired. The fair value of the identifiable net assets equals the fair value of all identifiable tangible and intangible assets less the fair value of any liabilities of the selling company assumed by the buyer. Goodwill is a residual asset; it's the amount left after other assets are identified and valued. Consider Illustration 10–8.

Goodwill is the excess of the fair value of the consideration given over the fair value of the identifiable net assets acquired.

Illustration 10–8
Goodwill

The Smithson Corporation acquired all of the outstanding common stock of the Rider Corporation in exchange for $180 million cash.* Smithson assumed all of Rider's long-term liabilities, which have a fair value of $120 million at the date of acquisition. The fair values of all identifiable assets of Rider are as follows ($ in millions):

Receivables	$ 50
Inventory	70
Property, plant, and equipment	90
Patent	40
Fair value of identifiable assets acquired	$250

The cost of the goodwill resulting from the acquisition is $50 million:

Fair value of consideration given (cash)		$180
Less: Fair value of identifiable net assets acquired		
Fair value of identifiable assets acquired	$250	
Less: Fair value of liabilities assumed	(120)	(130)
Goodwill		**$ 50**

The following journal entry captures the effect of the acquisition on Smithson's assets and liabilities:

Receivables (fair value)	50	
Inventory (fair value)	70	
Property, plant, and equipment (fair value)	90	
Patent (fair value)	40	
Goodwill (difference)	**50**	
Liabilities (fair value)		120
Cash (acquisition price)		180

*Determining the amount an acquirer is willing to pay for a company in excess of the identifiable net assets is a question of determining the value of a company as a whole. This question is addressed in most introductory and advanced finance textbooks.

Of course, a company can develop its own goodwill through advertising, training employees, and other efforts. In fact, most do. However, a company must expense all such costs incurred in the internal generation of goodwill. By not capitalizing these items, accountants realize that this results in an understatement of real assets because many of these expenditures do result in significant future benefits. Also, it's difficult to compare two companies when one has acquired goodwill and the other has not. But imagine how difficult it would be to associate these expenditures with any objective measure of goodwill. In essence, we have a situation where the characteristic of faithful representation overshadows relevance.

Goodwill, along with other intangible assets with indefinite useful lives, is not amortized.

Just like for other intangible assets that have indefinite useful lives, *we do not amortize goodwill.* This makes it imperative that companies make every effort to identify specific intangibles other than goodwill that they acquire in a business combination since goodwill is the amount left after other assets are identified.

Additional Consideration

It's possible for the consideration given to be *less than* the fair value of the identifiable net assets. This "bargain purchase" situation could result from an acquisition involving a "forced sale" in which the seller is acting under duress. The FASB previously required this bargain purchase to be allocated as a pro rata reduction of the amounts that otherwise would have been assigned to particular assets acquired. The allocation of a negative amount to assets then resulted in recording assets acquired at less than their fair values. However, current GAAP makes it mandatory that assets and liabilities acquired in a business combination be valued at their fair values.[8] Any bargain purchase is reported as a gain in the year of the acquisition.

[8]FASB ASC 805: Business Combinations.

In keeping with that goal, GAAP provides guidelines for determining which intangibles should be separately recognized and valued. Specifically, an intangible should be recognized as an asset apart from goodwill if it arises from contractual or other legal rights or is capable of being separated from the acquired entity. Possibilities are patents, trademarks, copyrights, and franchise agreements, and such items as customer lists, license agreements, order backlogs, employment contracts, and noncompetition agreements.[9] In past years, some of these intangibles, if present in a business combination, often were included in the cost of goodwill.[10]

In a business combination, an intangible asset must be recognized as an asset apart from goodwill if it arises from contractual or other legal rights or is separable.

Additional Consideration

Contract Acquisition Costs. Chapter 6 introduced you to the new guidance on revenue recognition. Under the new standard, sellers of goods and services are required to capitalize, as an intangible asset, the incremental costs of obtaining and fulfilling a long-term (longer than one year) contract. A sales commission is an example of a contract acquisition cost that could be capitalized, rather than expensed, under this new guidance.

If capitalized, the cost of the resulting intangible asset is amortized on a systematic basis that is consistent with the pattern of transfer of the goods or services to which the asset relates. The intangible asset also is evaluated for impairment using the same approach used for other intangible assets. We discuss amortization and impairment testing of intangible assets in Chapter 11.

Lump-Sum Purchases

● LO10–2

It's not unusual for a group of assets to be acquired for a single sum. If these assets are indistinguishable, for example 10 identical delivery trucks purchased for a lump-sum price of $150,000, valuation is obvious. Each of the trucks would be valued at $15,000 ($150,000 ÷ 10). However, if the lump-sum purchase involves different assets, it's necessary to allocate the lump-sum acquisition price among the separate items. The assets acquired may have different characteristics and different useful lives. For example, the acquisition of a factory may include assets that are significantly different such as land, building, and equipment.

The allocation is made in proportion to the individual assets' relative fair values. This process is best explained by an example in Illustration 10–9.

The relative fair value percentages are multiplied by the lump-sum purchase price to determine the initial valuation of each of the separate assets. Notice that the lump-sum purchase includes inventories. The procedure used here to allocate the purchase price in a lump-sum acquisition pertains to any type of asset mix, not just to property, plant, and equipment and intangible assets.

Ethical Dilemma

Grandma's Cookie Company purchased a factory building. The company controller, Don Nelson, is in the process of allocating the lump-sum purchase price between land and building. Don suggests to the company's chief financial officer, Judith Prince, that they fudge a little by allocating a disproportionately higher share of the price to land. Don reasons that this will reduce depreciation expense, boost income, increase their profit-sharing bonus, and hopefully, increase the price of the company's stock. Judith has some reservations about this because the higher reported income will also cause income taxes to be higher than they would be if a correct allocation of the purchase price is made.

What are the ethical issues? What stakeholders' interests are in conflict?

[9] Ibid.

[10] An assembled workforce is an example of an intangible asset that is not recognized as a separate asset. A workforce does not represent a contractual or legal right, nor is it separable from the company as a whole.

Illustration 10–9
Lump-Sum Purchase

The Smyrna Hand & Edge Tools Company purchased an existing factory for a single sum of **$2,000,000.** The price included title to the land, the factory building, the manufacturing equipment in the building, a patent on a process the equipment uses, and inventories of raw materials. An independent appraisal estimated the fair values of the assets (if purchased separately) as follows:

	Fair Values	
Land	$ 330,000	15%
Building	550,000	25%
Equipment	660,000	30%
Patent	440,000	20%
Inventories	220,000	10%
Total	$2,200,000	100%

Because the $2,200,000 fair value of the assets is greater than the **$2,000,000** purchase price, the purchase price is allocated to the separate assets as follows:

The total purchase price is allocated in proportion to the relative fair values of the assets acquired.

Land	(15% × $2,000,000)	300,000	
Building	(25% × $2,000,000)	500,000	
Equipment	(30% × $2,000,000)	600,000	
Patent	(20% × $2,000,000)	400,000	
Inventories	(10% × $2,000,000)	200,000	
Cash			**2,000,000**

PART B

Noncash Acquisitions

Companies sometimes acquire assets without paying cash at the time of the purchase. In Part B, we examine four situations where this occurs.

1. Deferred payments (notes payable)
2. Issuance of equity securities
3. Donated assets
4. Exchanges of nonmonetary assets for other assets

Assets acquired in noncash transactions are valued at the fair value of the assets given or the fair value of the assets received, whichever is more clearly evident.

The controlling principle in each of these situations is that in any noncash transaction, the asset acquired is recorded at its fair value. The first indicator of fair value is the fair value of the assets, debt, or equity securities given. Sometimes the fair value of the assets received is used when their fair value is more clearly evident than the fair value of the assets given.

Deferred Payments

LO10–3

A company can acquire an asset by giving the seller a promise to pay cash in the future and thus creating a liability, usually a note payable. The initial valuation of the asset is, again, quite simple as long as the note payable explicitly requires the payment of interest at a realistic interest rate. For example, suppose a machine is acquired for $15,000 and the buyer signs a note requiring the payment of $15,000 sometime in the future *plus* interest in the meantime at a realistic interest rate. The machine would be valued at $15,000 and the transaction recorded as follows:

Machine	15,000	
Notes payable		15,000

We know from our discussion of the time value of money in Chapter 5 that most liabilities are valued at the present value of future cash payments, reflecting an appropriate time value of money. As long as the note payable explicitly contains a realistic interest rate, the present value will equal the face value of the note, $15,000 in our previous example. This also should be equal to the fair value of the machine purchased. On the other hand, when an interest rate is not specified or is unrealistic, determining the cost of the asset is less straightforward. In that case, the accountant should look beyond the form of the transaction and record its substance. Consider Illustration 10–10.

On January 2, 2021, the Midwestern Steam Gas Corporation purchased industrial equipment. In payment, Midwestern signed a noninterest-bearing note requiring $50,000 to be paid on December 31, 2022 (two years later). If Midwestern had borrowed cash to buy the equipment, the bank would have required an interest rate of 10%.

Illustration 10–10

Asset Acquired with Debt—Present Value of Note Indicative of Fair Value

Some portion of the payment(s) required by a noninterest-bearing note in reality is interest.

On the surface, it might appear that Midwestern is paying $50,000 for the equipment, the eventual cash payment. However, when you recognize that the agreement specifies no interest even though the payment won't be made for two years, it becomes obvious that a portion of the $50,000 payment is not actually payment for the equipment, but instead is interest on the note. At what amount should Midwestern value the equipment and the related note payable?

Noncash transactions are recorded at the fair value of the items exchanged.

The answer is fair value, as it is for any noncash transaction. This might be the fair value of the equipment or the fair value of the note. Let's say, in this situation, that the equipment is custom-built, so its cash price is unavailable. But Midwestern can determine the fair value of the note payable by computing the present value of the cash payments at the appropriate interest rate of 10%. The amount actually paid for the equipment, then, is the present value of the cash flows called for by the loan agreement, discounted at the market rate—10% in this case.

$$PV = \$50{,}000\ (0.82645^{*}) = \$41{,}323$$

*Present value of $1: $n = 2$, $i = 10\%$ (from Table 2).

So the equipment should be recorded at its *real* cost, $41,323, as follows:[11]

Equipment (determined above)	41,323	
Discount on notes payable (difference)	8,677	
Notes payable (face amount)		50,000

The economic essence of a transaction should prevail over its outward appearance.

Notice that the note also is recorded at $41,323, its present value, but this is accomplished by using a contra account, called *discount on notes payable,* for the difference between the face amount of the note ($50,000) and its present value ($41,323). The difference of $8,677 is the portion of the eventual $50,000 payment that represents interest and is recognized as interest expense over the life of the note.

Assuming that Midwestern's fiscal year-end is December 31 and that adjusting entries are recorded only at the end of each year, the company would record the following entries at the end of 2021 and 2022 to accrue interest and the payment of the note:

December 31, 2021		
Interest expense ($41,323 × 10%)	4,132	
Discount on notes payable		4,132
December 31, 2022		
Interest expense [($41,323 + 4,132)* × 10%]	4,545	
Discount on notes payable		4,545
Notes payable (face amount)	50,000	
Cash		50,000

*The 2021 unpaid interest increases the amount owed by $4,132.

Discount on notes payable

8,677	Jan. 1, 2021
Dec. 31, 2021	4,132
Dec. 31, 2022	4,545
0	Dec. 31, 2022

Interest expense

4,132	Dec. 31, 2021
4,545	Dec. 31, 2022

[11]The entry shown assumes the note is recorded using the gross method. By the net method, a discount account is not used and the note is simply recorded at present value.

Equipment	41,323	
Notes payable		41,323

Sometimes, the fair value of an asset acquired in a noncash transaction is readily available from price lists, previous purchases, or otherwise. In that case, this fair value may be more clearly evident than the fair value of the note, and the fair value of the asset would serve as the best evidence of the exchange value of the transaction. As an example, let's consider Illustration 10–11.

Illustration 10–11

Noninterest-Bearing Note—Fair Value of Asset Is Known

On January 2, 2021, Dennison, Inc., purchased a machine and signed a noninterest-bearing note in payment. The note requires the company to pay $100,000 on December 31, 2023 (three years later). Dennison is not sure what interest rate appropriately reflects the time value of money. However, price lists indicate the machine could have been purchased for cash at a price of $79,383.

Dennison records both the asset and liability at $79,383 on January 2 as shown:

Machine (cash price)	79,383	
Discount on notes payable (difference)	20,617	
Notes payable (face amount)		100,000

In this situation, we infer the present value of the note from the fair value of the asset. Again, the difference between the note's $79,383 present value and the cash payment of $100,000 represents interest. We can determine the interest rate that is implicit in the agreement as follows:

$$\$79{,}383 \text{ (present value)} = \$100{,}000 \text{ (face amount)} \times \text{PV factor}$$

$$\$79{,}383 \div \$100{,}000 = 0.79383$$

*Present value of $1: $n = 3$, $i = ?$ (from Table 2, $i = 8\%$).

We refer to the **8%** rate as the *implicit rate of interest.* Dennison records interest each year at 8% in the same manner as demonstrated in Illustration 10–10 and discussed in greater depth in Chapter 14.

We now turn our attention to the acquisition of assets acquired in exchange for equity securities and through donation.

Issuance of Equity Securities

● LO10–4

The most common situation in which equity securities are issued for property, plant, and equipment and intangible assets occurs when small companies incorporate and the owner or owners contribute assets to the new corporation in exchange for ownership securities, usually common stock. Because the common shares are not publicly traded, it's difficult to determine their fair value. In that case, the fair value of the assets received by the corporation is probably the better indicator of the transaction's exchange value. In other situations, particularly those involving corporations whose stock is actively traded, the market value of the shares is the best indication of fair value. Consider Illustration 10–12.

Illustration 10–12

Asset Acquired by Issuing Equity Securities

Assets acquired by issuing common stock are valued at the fair value of the securities or the fair value of the assets, whichever is more clearly evident.

On March 31, 2021, the Elcorn Company issued 10,000 shares of its no-par common stock in exchange for land. On the date of the transaction, the fair value of the common stock, evidenced by its market price, was $20 per share. The journal entry to record this transaction is shown below:

Land	200,000	
Common stock (10,000 shares × $20)		200,000

If the fair value of the common stock had not been reliably determinable, the value of the land as determined through an independent appraisal would be used as the cost of the land and the value of the common stock.

Donated Assets

***Donated assets* are recorded at their fair values.**

On occasion, companies acquire assets through donation. The donation usually is an enticement to do something that benefits the donor. For example, the developer of an industrial park might pay some of the costs of building a manufacturing facility to entice a company to locate in its park. Companies record assets donated by unrelated parties at their fair values based on either an available market price or an appraisal value. This should not be considered a departure from historical cost valuation. Instead, it is equivalent to the donor contributing cash to the company and the company using the cash to acquire the asset.

Revenue is credited for the amount paid by an unrelated party.

As the recipient records the asset at its fair value, what account receives the offsetting credit? Over the years, there has been disagreement over this question. Should the recipient increase its paid-in capital—the part of shareholders' equity representing investments in the firm? Or, should the donated asset be considered revenue? GAAP requires that donated assets be recorded as *revenue*.[12] Recall that revenues generally are inflows of assets from delivering or producing goods, rendering services, or from other activities that constitute the entity's ongoing major or central operations. The rationale is that the company receiving the donation is performing a service for the donor in exchange for the asset donated.

Corporations occasionally receive incentives from governmental units. A local governmental unit might provide land or pay all or some of the cost of a new office building or manufacturing plant to entice a company to locate within its geographical boundaries. For example, the city of San Jose, California, paid a significant portion of the cost of a new office building for **IBM Corporation**. The new office building, located in downtown San Jose, brought jobs to a revitalized downtown area and increased revenues to the city. The City of San Jose did not receive an equity interest in IBM through its donation, but significantly benefited nevertheless. Illustration 10–13 provides an example.

Illustration 10–13
Asset Donation

Elcorn Enterprises decided to relocate its office headquarters to the city of Westmont. The city agreed to pay 20% of the $20 million cost of building the headquarters in order to entice Elcorn to relocate. The building was completed on May 3, 2021. Elcorn paid its portion of the cost of the building in cash. Elcorn records the transaction as follows:

Building	20,000,000	
Cash		16,000,000
Revenue—donation of asset (20% × $20 million)		4,000,000

International Financial Reporting Standards

● LO10–9

Government Grants. Both U.S. GAAP and IFRS require that companies value donated assets at their fair values. For government grants, though, the way that value is recorded is different under the two sets of standards. Unlike U.S. GAAP, donated assets are not recorded as revenue under IFRS. *IAS No. 20* requires that government grants be recognized in income over the periods necessary to match them on a systematic basis with the related costs that they are intended to compensate. So, for example, *IAS No. 20* allows two alternatives for grants related to assets.[13]

1. Deduct the amount of the grant in determining the initial cost of the asset.
2. Record the grant as a liability, deferred income, in the balance sheet and recognize it in the income statement systematically over the asset's useful life.

IFRS requires government grants to be recognized in income over the periods necessary to match them on a systematic basis with the related costs that they are intended to compensate.

In Illustration 10–13, if a company chose the first option, the building would be recorded at $16 million. If instead the company chose the second option, the building would be recorded at $20 million, but rather than recognizing $4 million in revenue as with U.S. GAAP, a $4 million credit to deferred income would be recorded and recognized as income over the life of the building.

Siemens, a global electronics and electrical engineering company based in Germany, prepares its financial statements according to IFRS, and sometimes receives government grants for the purchase or production of fixed assets. The following disclosure note included with recent financial statements indicates that Siemens uses the first option, deducting the amount of the grant from the initial cost of the asset.

Government Grants (in part)
Grants awarded for the purchase or the production of fixed assets (grants related to assets) are generally offset against the acquisition or production costs of the respective assets and reduce future depreciations accordingly.

[12] FASB ASC 958–505–15–2 and FASB ASC 958–605–25–2: Not-for-Profit Entities—Revenue Recognition—Scope and Scope Exceptions—Contributions Received.

[13] "Government Grants," *International Accounting Standard No. 20* (IASCF), as amended effective January 1, 2018.

Property, plant, and equipment and intangible assets also can be acquired in an exchange. Because an exchange transaction inherently involves a disposition of one asset as it is given up in exchange for another, we cover these transactions next under Exchanges.

Decision Makers' Perspective

The property, plant, and equipment and intangible asset acquisition decision is among the most significant decisions that management must make. A decision to acquire a new fleet of airplanes or to build or purchase a new office building or manufacturing plant could influence a company's performance for many years.

These decisions, often referred to as **capital budgeting** decisions, require management to forecast all future net cash flows (cash inflows minus cash outflows) generated by the asset(s). These cash flows are then used in a model to determine if the future cash flows are sufficient to warrant the capital expenditure. One such model, the net present value model, compares the present value of future net cash flows with the required initial acquisition cost of the asset(s). If the present value is higher than the acquisition cost, the asset is acquired. You have studied or will study capital budgeting in considerable depth in a financial management course. The introduction to the time value of money concept in Chapter 5 provided you with important tools necessary to evaluate capital budgeting decisions.

● LO10–5

A key to profitability is how well a company manages and utilizes its assets. Financial analysts often use activity, or turnover, ratios to evaluate a company's effectiveness in managing assets. This concept was illustrated with receivables and inventory in previous chapters. Property, plant, and equipment (PP&E) usually are a company's primary revenue-generating assets. Their efficient use is critical to generating a satisfactory return to owners. One ratio analysts often use to measure how effectively managers use PP&E is the **fixed-asset turnover ratio**. This ratio is calculated as follows:

The *fixed-asset turnover ratio* measures a company's effectiveness in managing property, plant, and equipment.

$$\text{Fixed-asset turnover ratio} = \frac{\text{Net sales}}{\text{Average fixed assets}}$$

The ratio indicates the level of sales generated by the company's investment in fixed assets. The denominator usually is the book value, sometimes called carrying value or carrying amount (cost less accumulated depreciation and depletion) of property, plant, and equipment.[14]

As with other turnover ratios, we can compare a company's fixed-asset turnover with that of its competitors, with an industry average, or with the same company's ratio over time. Let's compare the fixed-asset turnover ratios for **The Gap, Inc.**, and **Ross Stores, Inc.**, two companies in the retail apparel industry.

	($ in millions)			
	Gap		**Ross Stores**	
	2018	**2017**	**2018**	**2017**
Property, plant, and equipment (net)	$2,805	$2,616	$2,382	$2,328
Net sales—2018	$15,855		$14,135	

The 2018 fixed-asset turnover for Gap is 5.85 ($15,855 ÷ [($2,616 + 2,805) ÷ 2]) compared to the turnover for Ross Stores of 6.00 ($14,135 ÷ [($2,328 + 2,382) ÷ 2]). Ross Stores is able to generate $0.15 more in sales dollars than Gap for each dollar invested in fixed assets. ●

[14]If intangible assets are significant, their book value could be added to the denominator to produce a turnover that reflects all long-revenue-producing assets. The use of book value provides an approximation of the company's current investment in these assets.

Exchanges

Sometimes a company will acquire a new asset by giving up an existing asset. For example, a company might purchase a new delivery truck by trading-in its old delivery truck. This is referred to as a *nonmonetary asset exchange.*[15] If the values of those nonmonetary assets are not equal to one another, cash will be received or paid to equalize those values. When a company trades its old delievery truck for a new one, its likely that the fair value of the new delivery truck is higher, so the company trading-in the old delivery truck will pay cash to make up the difference.

An asset received in an exchange of nonmonetary assets generally is valued at fair value.

● LO10–6

The basic steps in recording nonmonetary asset exchanges follow:

Step 1: **Record the new asset at fair value.** Fair value is determined based on the fair value of the asset(s) given up or the fair value of the asset received. In a normal exchange, we expect those two fair values to equal.

Step 2: **Remove the book value of the nonmonetary asset given.** Book value equals the recorded cost of the old asset minus its accumulated depreciation.

Step 3: **Record any cash received or paid.** Cash is used to equalize the fair values of the nonmonetary assets in the exchange.

Step 4: **Record any gain or loss.** The gain or loss on the exchange equals the difference between fair value and book value of the asset given up in step 2. If fair value is greater, then a gain is recorded. If book value is greater, then a loss is recorded. The gain or loss also can be viewed as the "plug" that balances debits and credits in the journal entry.

To see an example, look at Illustration 10–14A.

Illustration 10–14A

Nonmonetary Asset Exchange—Gain

The Elcorn Company traded its old laser equipment for the newer air-cooled ion laser equipment manufactured by American Laser Corporation. The old equipment had a book value of $100,000 (cost of $500,000 less accumulated depreciation of $400,000) and a **fair value of $150,000.** Elcorn also paid American Laser **$430,000 in cash** as part of the exchange. The following journal entry records the transaction:

Equipment—new (fair value, determined below)	**580,000**	
Accumulated depreciation—old (account balance)	400,000	
Equipment—old (account balance)		500,000
Cash (amount paid)		430,000
Gain on exchange of assets (fair value − book value)		50,000

Assets given:	**Fair value**	**Book value**
Old equipment	$ 150,000	$100,000
Cash paid	430,000	430,000
Total	**$580,000**	$530,000

A gain is recognized when the fair value of an asset given is more than its book value.

Step 1: Record the new equipment for **$580,000**. This is the fair value of the assets given.
Step 2: Remove the book value of the old equipment with a credit to Equipment for $500,000 and a debit to Accumulated Depreciation for $400,000.
Step 3: Record the amount paid with a credit to Cash for $430,000.
Step 4: Record a gain of $50,000. The gain is computed as the fair value of the old equipment minus the book value of the old equipment ($150,000 – $100,000). Alternatively, the gain can be computed at the total fair value of the assets given minus their total book value ($580,000 – $530,000).

In Illustration 10–14A, the $150,000 fair value of the old asset was known. However, in a trade-in, quite often the fair value of the new asset is more clearly evident than the second-hand value of the asset traded in. For example, if it was known that the fair value of

[15] Monetary items are assets and liabilities whose *amounts are fixed*, by contract or otherwise, in terms of a specific number of dollars. Other assets are considered nonmonetary.

the new asset was $580,000, then the $150,000 fair value of the old asset could have been determined by subtracting the cash paid ($580,000 – $430,000).

Let's modify the illustration slightly by assuming that the fair value of the old equipment is $75,000 instead of $150,000. Illustration 10–14B shows the journal entry to record the transaction.

Illustration 10–14B

Nonmonetary Asset Exchange—Loss

The Elcorn Company traded its old laser equipment for the newer air-cooled ion laser equipment manufactured by American Laser Corporation. The old equipment had a book value of $100,000 (cost of $500,000 less accumulated depreciation of $400,000) and a **fair value of $75,000.** Elcorn also paid American Laser **$430,000 in cash** as part of the exchange. The following journal entry records the transaction:

Equipment—new (fair value determined below)	**505,000**	
Accumulated depreciation—old (account balance)	400,000	
Loss on exchange of assets (fair value – book value)	25,000	
Equipment—old (account balance)		500,000
Cash (amount paid)		430,000

Assets given:	Fair value	Book value
Old equipment	$ 75,000	$100,000
Cash paid	430,000	430,000
Total	**$505,000**	$530,000

The new equipment is recorded at **$505,000**, which is the fair value of the old equipment plus the cash paid ($75,000 + $430,000). Elcorn recognizes a loss of $25,000, the amount by which the fair value of the asset given ($75,000) is less than its book value ($100,000).

A loss is recognized when the fair value of an asset given is less than its book value.

Until 2005, the accounting treatment of nonmonetary asset exchanges depended on a number of factors including (1) whether the assets exchanged were similar or dissimilar, (2) whether a gain or loss was indicated in the exchange, and (3) whether cash was given or received. Then a new accounting standard[16] simplified accounting for exchanges by requiring the use of fair value except in rare situations in which the fair value can't be determined or the exchange lacks commercial substance.[17]

Let's discuss these two rare situations.

Fair Value Not Determinable

It would be unusual for a company to be unable to reasonably determine fair value of either asset in an exchange. Nevertheless, if the situation does occur, we modify step 1 above and use the *book value* of the assets given up to record the asset acquired. For example, if fair value had not been determinable in Illustration 10–14A, Elcorn would have recorded the exchange as follows:

If we can't determine the fair value of either asset in the exchange, the asset received is valued at the book value of the asset given.

Equipment—new (book value + cash: $100,000 + $430,000)	530,000	
Accumulated depreciation—old (account balance)	400,000	
Equipment—old (account balance)		500,000
Cash (amount paid)		430,000

The new equipment is recorded at the book value of the old equipment ($100,000) plus the cash given ($430,000). No gain or loss is recognized on an exchange when fair value is not determinable.

Lack of Commercial Substance

If we record an exchange at fair value, we recognize a gain or loss for the difference between the fair value and book value of the asset given up. To prevent a company from exchanging an asset whose fair value is greater than book value for the sole purpose of recognizing a gain, fair value can be used only in gain situations that have *commercial substance.*

[16]FASB ASC 845: Nonmonetary Transactions.

[17]There is a third situation which precludes the use of fair value in a nonmonetary exchange. The transaction is an exchange of inventories to facilitate sales to customers other than the parties to the exchange.

A nonmonetary exchange is considered to have commercial substance if future cash flows will change as a result of the exchange. Most exchanges are for business reasons and would not be transacted if there were no anticipated change in future cash flows. For example, newer models of equipment can increase production or improve manufacturing efficiency, causing an increase in revenue or a decrease in operating costs with a corresponding increase in future cash flows. The exchange of old laser equipment for the *newer* model in Illustration 10–14A is an example of an exchange transacted for business reasons.

Commercial substance occurs when future cash flows change as a result of the exchange.

GAIN SITUATION Suppose a company owned a tract of land that had a book value of $1 million and a fair value of $5 million. The only ways to recognize the $4 million difference as a gain are to either sell the land or to exchange the land for another nonmonetary asset for a legitimate business purpose. For example, if the land were exchanged for a different type of asset, say an office building, then future cash flows most likely will change. This exchange has commercial substance, and the $4 million gain can be recognized. On the other hand, if the land were exchanged for a tract of land that has the identical characteristics as the land given, then it is unlikely that future cash flows would change. This exchange lacks commercial substance, and the gain of $4 million cannot be recognized on the exchange. The new land is recorded at the book value of the old land plus any cash paid. Illustration 10–15 provides an example.

Illustration 10–15

Nonmonetary Asset Exchange—Exchange Lacks Commercial Substance

The Elcorn Company traded a tract of land to Sanchez Development for a similar tract of land. The old land had a book value of $2,500,000 and a fair value of $4,500,000. Elcorn also paid Sanchez $500,000 in cash. The following journal entry records the transaction, *assuming that the exchange lacks commercial substance:*

Land—new (book value + cash: $2,500,000 + $500,000)	**3,000,000**	
Land—old (account balance)		2,500,000
Cash (amount paid)		500,000

The new land is recorded at **$3,000,000**, the book value of the old land plus the cash given ($2,500,000 + $500,000). No gain is recognized because the exchange does not have commercial substance.

The FASB's intent in including the commercial substance requirement for the use of fair value was to avoid companies' trading property for no business reason other than to recognize the gain.

Additional Consideration

In Illustration 10–15, cash was *given*. What if cash was *received*? Suppose that $450,000 cash was received instead of the $500,000 given. The fair value of the old land was $4,500,000, which means that the fair value of the land received is $4,050,000 ($4,500,000 – $450,000). In that case, a portion of the transaction is considered monetary and we would recognize that portion of the $2,000,000 gain ($4,500,000 fair value – $2,500,000 book value). The amount of gain recognized is equal to the portion of cash received relative to the total received.[18]

$$\frac{\$450{,}000}{\$450{,}000 + 4{,}050{,}000} = 10\%$$

Elcorn would recognize a $200,000 gain ($2,000,000 × 10%) and would record the land received at **$2,250,000**, the book value of the land given plus the gain recognized less the cash received ($2,500,000 + $200,000 – $450,000). The following journal entry records the transaction:

Land—new ($2,500,000 + $200,000 – $450,000)	**2,250,000**	
Cash	450,000	
Land—old		2,500,000
Gain on exchange of assets		200,000

[18]If the amount of monetary consideration received is deemed significant, the transaction is considered to be monetary and the entire gain is recognized. In other words, the transaction is accounted for as if it had commercial substance. GAAP defines "significance" in this situation as 25% or more of the fair value of the exchange. FASB ASC 845–10–25–6: Nonmonetary transactions–Overall–Recognition.

When the fair value of the asset given is less than its book value, we always use fair value to record the exchange.

LOSS SITUATION In Illustration 10–15, what if the fair value of the land given was less than its book value? It's unlikely that a company would enter into this type of transaction unless there was a good reason. When a loss is indicated in a nonmonetary exchange, it's okay to record the loss and we use fair value to record the asset acquired.

Illustration 10–16 summarizes the treatment of gains and losses, as well as the amount to record for the asset received, for each type of exchange discussed above.

Illustration 10–16 Exchanges

	Type of Exchange		
	Fair value Determinable and Commercial Substance	**Fair Value Not Determinable**	**Lack of Commercial Substance**
Step 1: Record new asset	FV given + cash paid (or – cash received).†	BV given + cash paid (or – cash received).	If gain, BV given + cash paid (or – cash received) + gain recognized.* If loss, FV given + cash paid (or – cash received).
Step 2: Remove old asset	BV of cost minus accumulated depreciation.	Same	Same
Step 3: Record cash	Amount paid or received.	Same	Same
Step 4: Record gain or loss	FV given – BV given (Gain if FV > BV; Loss if FV < BV).	No gain or loss recognized.	Calculate gain to be recognized.* Recognize entire loss.

†FV = fair value; BV = book value. All references to FV and BV relate to the asset being given in the exchange.
*A gain situation occurs when FV of the assets given is greater than the BV of those assets. If no cash is received, no gain is recognized. If some cash is received, the amount of the gain recognized = (FV given – BV given) × (cash received ÷ total FV received). If cash constitutes at least 25% of the fair value received, the entire gain is recognized = FV given – BV given.

When fair value is determinable and commercial substance exists, we record the asset received at total fair value. Gains and losses are recorded for the difference between the fair value and book value of the asset given, depending on which is higher.

When fair value is not determinable, fair value cannot be used to record the asset being received, and therefore we don't record a gain or loss on the asset being given. The asset received is simply recorded for the book value of the asset given, adjusted for the effects of any cash exchanged.

The rules for exchanges that lack commercial substance reflect a bias toward preventing gains from being recognized unless some cash is received so that a portion of the transaction represents a monetary exchange. Losses in exchanges that lack commercial substance are fully recognized regardless of cash received or paid, just like losses when fair value is determinable.

Concept Review Exercise

EXCHANGES

The MD Corporation recently acquired new equipment to be used in its production process. In exchange, the company traded in an existing asset that had an original cost of $60,000 and accumulated depreciation on the date of the exchange of $45,000. The fair value of the old equipment is $17,000. In addition, MD paid $40,000 cash to the equipment manufacturer.

Required:

1. Prepare the journal entry MD would use to record the exchange transaction assuming that the transaction has commercial substance.
2. Prepare the journal entry MD would use to record the exchange transaction assuming that the transaction does *not* have commercial substance.
3. Prepare the journal entry MD would use to record the exchange transaction assuming that the fair value of the old equipment is only $10,000.

Solution:

1. Prepare the journal entry MD would use to record the exchange transaction assuming that the transaction has commercial substance.

Equipment—new ($17,000 + $40,000)	57,000	
Accumulated depreciation—old (account balance)	45,000	
Equipment—old (account balance)		60,000
Cash (amount paid)		40,000
Gain on exchange of assets ($17,000 FV – $15,000 BV)		2,000

2. Prepare the journal entry MD would use to record the exchange transaction assuming that the transaction does not have commercial substance.

Equipment—new ($15,000 + $40,000)	55,000	
Accumulated depreciation—old (account balance)	45,000	
Equipment—old (account balance)		60,000
Cash (amount paid)		40,000

3. Prepare the journal entry MD would use to record the exchange transaction assuming that the fair value of the old equipment is only $10,000.

Equipment—new ($10,000 + $40,000)	50,000	
Accumulated depreciation—old (account balance)	45,000	
Loss on exchange of assets ($10,000 FV – $15,000 BV)	5,000	
Equipment—old (account balance)		60,000
Cash (amount paid)		40,000

Self-Constructed Assets and Research and Development

PART C

Two types of expenditures relating to property, plant, and equipment and intangible assets whose accounting treatment has generated considerable controversy are interest costs pertaining to self-constructed assets and amounts spent for research and development. We now consider those expenditures and why those controversies have developed.

Self-Constructed Assets

● LO10–7

A company might decide to construct an asset for its own use rather than buy an existing one. For example, a retailer like **Nordstrom** might decide to build its own store rather than purchase an existing building. A manufacturing company like **Intel** could construct its own manufacturing facility. In fact, Nordstrom and Intel are just two of the many companies that self-construct assets. Other recognizable examples include **Walt Disney**, **Sears**, and **Caterpillar**. Quite often these companies act as the main contractor and then subcontract most of the actual construction work.

The critical accounting issue in these instances is identifying the cost of the self-constructed asset. The task is more difficult than for purchased assets because there is no external transaction to establish an exchange price. Actually, two difficulties arise in connection with assigning costs to self-constructed assets: (1) determining the amount of the company's indirect manufacturing costs (overhead) to be allocated to the construction and (2) deciding on the proper treatment of interest (actual or implicit) incurred during construction.

The cost of a self-constructed asset includes identifiable materials and labor and a portion of the company's manufacturing overhead costs.

Overhead Allocation

One difficulty of associating costs with self-constructed assets is the same difficulty encountered when determining cost of goods manufactured for sale. The costs of material and direct labor usually are easily identified with a particular construction project and are included in cost. However, the treatment of manufacturing overhead cost and its allocation between construction projects and normal production is a controversial issue.

Some accountants advocate the inclusion of only the *incremental* overhead costs in the total cost of construction. That is, the asset's cost would include only those additional costs that are incurred because of the decision to construct the asset. This would exclude such indirect costs as depreciation and the salaries of supervisors that would be incurred whether or not the construction project is undertaken. If, however, a new construction supervisor was hired specifically to work on the project, then that salary would be included in asset cost.

Others advocate assigning overhead on the same basis that is used for a regular manufacturing process. That is, all overhead costs are allocated both to production and to self-constructed assets based on the relative amount of a chosen cost driver (for example, labor hours) incurred. This is known as the *full-cost approach* and is the generally accepted method used to determine the cost of a self-constructed asset.

Interest Capitalization

FINANCIAL Reporting Case

Q4, p. 3

To reiterate, the cost of an asset includes all costs necessary to get the asset ready for its intended use. Unlike one purchased from another company, a self-constructed asset requires time to create it. During this construction period, the project must be financed in some way. This suggests the question as to whether interest costs during the construction period are one of the costs of acquiring the asset itself or simply costs of financing the asset.

On the one hand, we might point to interest charges to finance inventories during their period of manufacture or to finance the purchase of plant assets from others and argue that construction period interest charges are merely costs of financing the asset that should be expensed as incurred like all other interest costs.

On the other hand, we might argue that self-constructed assets are different in that during the construction period, they are not yet ready for their intended use for producing revenues. And, so, in keeping with the historical cost principle, all costs during this period, including interest, should be capitalized. Those costs are then allocated as depreciation during later periods when the assets are providing benefits.

Only assets that are constructed as discrete projects qualify for interest capitalization.

QUALIFYING ASSETS Generally accepted accounting principles are consistent with the second argument. Specifically, interest is capitalized during the construction period for (a) assets built for a company's own use as well as for (b) assets constructed *as discrete projects* for sale or lease (a ship or a real estate development, for example). This excludes from interest capitalization consideration inventories that are routinely manufactured in large quantities on a repetitive basis and assets that already are in use or are ready for their intended use.[19] Interest costs incurred during the productive life of the asset are expensed as incurred.

The interest capitalization period begins when construction begins and the first expenditure is made as long as interest costs are actually being incurred.

PERIOD OF CAPITALIZATION The capitalization period for a self-constructed asset starts with the first expenditure (materials, labor, or overhead) and ends either when the asset is substantially complete and ready for use or when interest costs no longer are being incurred. Interest costs incurred can pertain to borrowings other than those obtained specifically for the construction project. However, interest costs can't be imputed; actual interest costs must be incurred.

***Average accumulated expenditures* approximates the average debt necessary for construction.**

AVERAGE ACCUMULATED EXPENDITURES Because we consider interest to be a necessary cost of getting a self-constructed asset ready for use, the amount capitalized is only that portion of interest cost incurred during the construction period that *could have been avoided* if expenditures for the asset had not been made. In other words, if construction had not been undertaken, debt incurred for the project would not have been necessary and/or other interest-bearing debt could have been liquidated or employed elsewhere.

As a result, interest should be determined for only the construction expenditures *actually incurred* during the capitalization period. And unless all expenditures are made at the outset of the period, it's necessary to determine the *average* amount outstanding during the period. This is the amount of debt that would be required to finance the expenditures and thus the

[19]FASB ASC 835–20–25: Interest—Capitalization of Interest—Recognition.

amount on which interest would accrue. For instance, if a company accumulated $1,500,000 of construction expenditures fairly evenly throughout the construction period, the average expenditures would be

Total accumulated expenditures incurred evenly throughout the period	$1,500,000
	÷2
Average accumulated expenditures	$ 750,000

At the beginning of the period, no expenditures have accumulated, so no interest has accrued (on the equivalent amount of debt). But, by the end of the period interest is accruing on the total amount, $1,500,000. On average, then, interest accrues on half the total or $750,000.

If expenditures are not incurred evenly throughout the period, a simple average is insufficient. In that case, a *weighted average* is determined by time-weighting individual expenditures or groups of expenditures by the number of months from their incurrence to the end of the construction period.

Average accumulated expenditures is determined by time-weighting individual expenditures made during the construction period.

Let's use the expenditures and loan information in Illustration 10–17 to demonstrate the calculation of interest capitalization. This calculation will involve three steps: (1) determine the weighted-average accumulated expenditures, (2) calculate the amount of interest to be capitalized, and (3) compare calculated interest with actual interest incurred.

Illustration 10–17

Expenditures and Loan Information for Interest Capitalization

On January 1, 2021, the Mills Conveying Equipment Company began construction of a building to be used as its office headquarters. The building was completed on June 30, 2022. Expenditures on the project, mainly payments to subcontractors, were as follows:

January 1, 2021	$ 500,000
March 31, 2021	400,000
September 30, 2021	600,000
Accumulated expenditures at December 31, 2021 (before interest capitalization)	$1,500,000
January 31, 2022	600,000
April 30, 2022	300,000

On January 1, 2021, the company obtained a $1 million construction loan with an 8% interest rate. The loan was outstanding during the entire construction period. The company's other interest-bearing debt included two long-term notes of $2,000,000 and $4,000,000 with interest rates of 6% and 12%, respectively. Both notes were outstanding during the entire construction period.

The weighted-average accumulated expenditures by the end of 2021 are shown below:

FINANCIAL Reporting Case

Q5, p. 3

	Actual Expenditures		Portion of Year Outstanding		Time-Weighted Expenditures
January 1, 2021	$ 500,000	×	12/12	=	$ 500,000
March 31, 2021	400,000	×	9/12	=	300,000
September 30, 2021	600,000	×	3/12	=	150,000
	$1,500,000				**$950,000**

Notice that the weighted-average accumulated expenditures are less than the actual accumulated expenditures of $1,500,000. If Mills had borrowed exactly the amount necessary to finance the project, it would not have incurred interest on a loan of $1,500,000 for the whole year but only on an average loan of **$950,000.** The next step is to determine the interest to be capitalized for the weighted-average accumulated expenditures.

STEP 1: Determine the weighted-average accumulated expenditures.

INTEREST RATES In this situation, debt financing was obtained specifically for the construction project, and the amount borrowed is sufficient to cover the average accumulated expenditures. To determine the interest capitalized, then, we simply multiply the construction loan rate of 8% by the weighted-average accumulated expenditures.

STEP 2: Calculate the amount of interest to be capitalized.

Interest capitalized for 2021 = **$950,000** × 8% = **$76,000**

Notice that this is the same answer we would get by assuming separate 8% construction loans were made for each expenditure at the time each expenditure was made.

Actual Expenditures		Annual Rate		Portion of Year Outstanding		Calculated Interest
$500,000	×	8%	×	12/12	=	$ 40,000
400,000	×	8%	×	9/12	=	24,000
600,000	×	8%	×	3/12	=	12,000
Interest capitalized for 2021						**$76,000**

The interest of **$76,000** is added to the cost of the building, bringing accumulated expenditures at December 31, 2021, to $1,576,000 ($1,500,000 + **$76,000**). The remaining interest cost incurred but not capitalized is expensed.

It should be emphasized that interest capitalization does not require that funds actually be borrowed for this specific purpose, only that the company does have outstanding debt. The presumption is that even if the company doesn't borrow specifically for the project, funds from other borrowings must be diverted to finance the construction. Either way—directly or indirectly—interest costs are incurred. In our illustration, for instance, even without the construction loan, interest would be capitalized because other debt was outstanding. The capitalized interest would be the average accumulated expenditures multiplied by the weighted-average rate on these other loans. The weighted-average interest rate on all debt other than the construction loan would be 10%, calculated as follows:[20]

Loans		Annual Rate		Interest
$2,000,000	×	6%	=	$120,000
4,000,000	×	12%	=	480,000
$6,000,000				$600,000

$$\text{Weighted-average rate: } \frac{\$600{,}000}{\$6{,}000{,}000} = 10\%$$

This is a weighted average because total interest is $600,000 on total debt of $6,000,000. Therefore, in our illustration, without any specific construction loan, interest capitalized for 2021 would have been $95,000 ($950,000 × 10%).

Additional Consideration

The weighted-average rate isn't used for 2021 in our illustration because the specific construction loan is sufficient to cover the average accumulated expenditures. If the specific construction loan had been insufficient to cover the average accumulated expenditures, its 8% interest rate would be applied to the average accumulated expenditures up to the amount of the specific borrowing, and any remaining average accumulated expenditures in excess of specific borrowings would be multiplied by the weighted-average rate on all other outstanding interest-bearing debt. Suppose, for illustration, that the 8% construction loan had been only $500,000 rather than $1,000,000. We would calculate capitalized interest using both the specific rate and the weighted-average rate:

	Time-Weighted Expenditures		Annual Rate		Calculated Interest
Total	$950,000				
Specific borrowing	500,000	×	8%	=	$40,000
Excess	$450,000	×	10%	=	45,000
Capitalized interest					$85,000

In our illustration, it's necessary to use this approach in 2022.

[20] The same result can be obtained simply by multiplying the individual debt interest rates by the relative amount of debt at each rate. In this case, one-third of total debt is at 6% and two-thirds of the total debt is at 12% [(1/3 × 6%) + (2/3 × 12%) = 10%].

It's possible that the amount of interest calculated to be capitalized exceeds the amount of interest actually incurred. If that's the case, we limit the interest capitalized to the actual interest incurred. In our illustration, total interest cost incurred during 2021 far exceeds the **$76,000** of capitalized interest calculated, so it's not necessary to limit the capitalized amount.

Interest capitalized is limited to interest incurred.

STEP 3: Compare calculated interest with actual interest incurred.

Loans		Annual Rate		Actual Interest	Calculated Interest
$1,000,000	×	8%	=	$ 80,000	
2,000,000	×	6%	=	120,000	
4,000,000	×	12%	=	480,000	
				$680,000	**$76,000**

↑ Use lower amount

Continuing the example based on the information in Illustration 10–17, let's determine the amount of interest capitalized during 2022 for the building. The total accumulated expenditures by the end of the project follow:

Accumulated expenditures at the beginning of 2022 (including interest capitalization)	$1,576,000
January 31, 2022 expenditures	600,000
April 30, 2022 expenditures	300,000
Accumulated expenditures at June 30, 2022 (before 2022 interest capitalization)	$2,476,000

The weighted-average accumulated expenditures by the end of the project are as follows:

STEP 1: Determine the weighted-average accumulated expenditures.

	Actual Expenditures		Portion of Year by End of Project		Time-Weighted Expenditures
January 1, 2022	$1,576,000	×	6/6	=	$ 1,576,000
January 31, 2022	600,000	×	5/6	=	500,000
April 30, 2022	300,000	×	2/6	=	100,000
Weighted-average accumulated expenditures for 2022					**$2,176,000**

Notice that the 2022 expenditures are weighted relative to the construction period of six months because the project was finished on June 30, 2022. Interest capitalized for 2022 would be **$98,800**, calculated as follows:

STEP 2: Calculate the amount of interest to be capitalized.

	Time-Weighted Expenditures		Annual Rate		Portion of Year Outstanding		Calculated Interest
Total	$2,176,000						
Specific borrowing	1,000,000	×	8%	×	6/12	=	$ 40,000
Excess	$1,176,000	×	10%	×	6/12	=	58,800
Interest capitalized for 2022							**$98,800**

Multiplying by six-twelfths reflects the fact that the interest rates are annual rates (12-month rates) and the construction period is only 6 months. Next, we compare the calculated interest with the actual interest incurred.

STEP 3: Compare calculated interest with actual interest incurred.

Loans		Annual Rate		Portion of Year Oustanding		Actual Interest	Calculated Interest
$1,000,000	×	8%	×	6/12	=	$ 40,000	
2,000,000	×	6%	×	6/12	=	60,000	
4,000,000	×	12%	×	6/12	=	240,000	
						$340,000	**$98,800**

↑ Use lower amount

For the first six months of 2022, **$98,800** of interest would be capitalized, bringing the total capitalized cost of the building to $2,574,800 ($2,476,000 + $98,800), and $241,200 in interest would be expensed ($340,000 − $98,800).

Additional Consideration

To illustrate how the actual interest limitation might come into play, let's assume the nonspecific borrowings in our illustration were $200,000 and $400,000 (instead of $2,000,000 and $4,000,000). Our comparison would change as follows:

Loans		Annual Rate		Portion of Year Oustanding		Actual Interest	Calculated Interest
$1,000,000	×	8%	×	6/12	=	$40,000	
200,000	×	6%	×	6/12	=	6,000	
400,000	×	12%	×	6/12	=	24,000	
						$70,000	$98,800
						↑ Use lower amount	

Weighted-average method

The method of determining interest to capitalize that we've discussed is called the **specific interest method** because we use rates from specific construction loans to the extent of specific borrowings before using the average rate of other debt. Sometimes, though, it's difficult to associate specific borrowings with projects. In these situations, it's acceptable to just use the weighted-average rate on all interest-bearing debt, including all construction loans. This is known as the **weighted-average method**. In our illustration, for example, if the $1,000,000, 8% loan had not been specifically related to construction, we would calculate a single weighted-average rate as shown below.

Loans		Annual Rate		Interest
$1,000,000	×	8%	=	$ 80,000
2,000,000	×	6%	=	120,000
4,000,000	×	12%	=	480,000
$7,000,000				$680,000

$$\text{Weighted-average rate: } \frac{\$680{,}000}{\$7{,}000{,}000} = 9.7\%$$

If we were using the weighted-average method rather than the specific interest method, we would simply multiply this single rate times the average accumulated expenditures to determine capitalizable interest.

If material, the amount of interest capitalized during the period must be disclosed.

DISCLOSURE For an accounting period in which interest costs are capitalized, the total amount of interest costs capitalized, if the amount is material, should be disclosed. Illustration 10–18 shows an interest capitalization disclosure note that was included in a recent annual report of **Walmart Stores Inc.**, the world's largest retailer.

Illustration 10–18

Capitalized Interest Disclosure—Walmart Stores Inc.

Real World Financials

Property and equipment (in part)

Interest costs capitalized on construction projects were $36 million, $39 million, and $59 million in fiscal 2017, 2016, and 2015, respectively.

Source: Wal-Mart Stores, Inc.

Research and Development (R&D)

LO10–8

"Innovation" is a word commonly used in the marketing and production departments of most companies. Companies compete fiercely to dream up the next top-selling product or service. In 2015, **Volkswagen** spent more than $15 billion to research new technology for its hybrid vehicles and improve CO_2 emissions. **Microsoft** spent $11.4 billion on discovering new technology products. Of the 118,000 people employed by Miscrosoft, 39,000 are in product research and development. **Alphabet** (**Google**) spent $9.8 billion on futuristic ideas such as self-driving cars and computer eyewear. **Merck** invested $7.2 billion to test new drugs related to oncology, infectious diseases, vaccines, and diabetes.

R&D costs entail a high degree of uncertainty of future benefits and are difficult to match with future revenues.

Companies are willing to spend huge amounts on R&D because they believe the project will eventually provide benefits that exceed the current expenditures. Unfortunately, though, it's difficult to predict which individual research and development projects will ultimately provide benefits. Will Google's expenditures on self-driving cars ever generate enough revenue to cover its costs? Moreover, even for those projects that pan out, a direct relationship between research and development costs and specific future revenue is difficult to establish. In other words, even if R&D costs do lead to future benefits, it's difficult to objectively determine the size of the benefits and in which periods the costs should be expensed if they are capitalized. For these reasons, *the FASB takes a conservative approach and requires R&D costs to be expensed immediately*.[21]

R&D costs are expensed in the periods incurred.

The FASB's decision is controversial. Many companies would prefer to delay the recognition of these expenses until later years when presumably the expenditures bear fruit. In other words, many companies would prefer to record R&D expenditures initially as an asset, because at least some of the R&D expenditures will likely produce future benefits. The requirement to expense R&D immediately leads to assets being understated and expenses being overstated in the current period.

Determining R&D Costs

GAAP distinguishes research and development as follows:

- **Research** is planned search or critical investigation aimed at discovery of new knowledge with the hope that such knowledge will be useful in developing a new product or service or a new process or technique or in bringing about a significant improvement to an existing product or process.
- **Development** is the translation of research findings or other knowledge into a plan or design for a new product or process or for a significant improvement to an existing product or process whether intended for sale or use.[22]

R&D costs include salaries, wages, and other labor costs of personnel engaged in R&D activities, the costs of materials consumed, equipment, facilities, and intangibles used in R&D projects, the costs of services performed by others in connection with R&D activities, and a reasonable allocation of indirect costs related to those activities. General and administrative costs should not be included unless they are clearly related to the R&D activity.

R&D expense includes the depreciation and amortization of assets used in R&D activities.

If an asset is purchased specifically for a single R&D project, its cost is considered R&D and expensed immediately even though the asset's useful life extends beyond the current year. However, the cost of an asset that has an alternative future use beyond the current R&D project is *not* a current R&D expense. For example, a small building might be purchased to conduct R&D projects but will then be used for general storage once the projects are complete. The depreciation or amortization of these alternative-use assets is included as R&D expenses only in the periods the assets are used for R&D activities.

Costs incurred *before* the start of commercial production are expensed as R&D.

In general, R&D costs pertain to activities that occur prior to the start of commercial production. Commercial production most often refers to the point in time where the company has no other plans to materially change the product prior to beginning production for intended sales to customers.

[21]FASB ASC 730–10–25–1: Research and Development—Overall—Recognition.

[22]Ibid., section 730–10–20.

Costs incurred *after* commercial production begins would be either expensed or included in the cost of inventory.

Any costs incurred after the start of commercial production are not classified as R&D costs. These costs would be either expensed or treated as manufacturing overhead and included in the cost of inventory.

Illustration 10–19 captures this concept with a time line beginning with the start of an R&D project and ending with the ultimate sale of a developed product or the use of a developed process. The illustration also provides examples of activities typically included as R&D and examples of activities typically excluded from R&D.[23]

Illustration 10–19

Research and Development Expenditures

Examples of R&D Costs:	*Examples of Non-R&D Costs:*
• Laboratory research aimed at discovery of new knowledge	• Engineering follow-through in an early phase of commercial production
• Searching for applications of new research findings or other knowledge	• Quality control during commercial production including routine testing of products
• Design, construction, and testing of preproduction prototypes and models	• Routine ongoing efforts to refine, enrich, or otherwise improve on the qualities of an existing product
• Modification of the formulation or design of a product or process	• Adaptation of an existing capability to a particular requirement or customer's need as a part of a continuing commercial activity

Let's look at an example in Illustration 10–20.

Illustration 10–20

Research and Development Costs

The Askew Company made the following cash expenditures during 2021 related to the development of a new industrial plastic:

R&D salaries and wages	$ 10,000,000
R&D supplies consumed during 2021	3,000,000
Purchase of R&D equipment	5,000,000
Patent filing and legal costs	100,000
Payments to others for services in connection with R&D activities	1,200,000
Total	**$19,300,000**

The project resulted in a new product to be manufactured in 2022 (next year). A patent was filed with the U.S. Patent Office. Amortization of the patent's filing and legal costs will begin in 2022. The equipment purchased will be employed in other projects. Depreciation on the equipment for 2021 was $500,000.

Filing and legal costs for patents, copyrights, and other developed intangibles are capitalized and amortized in future periods.

The salaries and wages, supplies consumed, and payments to others for R&D services are expensed in 2021 as R&D. The equipment is capitalized and the 2021 depreciation is expensed as R&D. Even though the costs to develop the patented product are expensed, the filing and legal costs for the patent are capitalized and amortized in future periods, just as similar costs are capitalized for purchased intangibles. Amortization of the patent is discussed in Chapter 11.

[23]Ibid., section 730–10–55.

The various expenditures would be recorded as follows:

R&D expense ($10,000,000 + $3,000,000 + $1,200,000)	14,200,000	
Cash		14,200,000
To record R&D expenses.		
Equipment	5,000,000	
Cash		5,000,000
To record the purchase of equipment.		
R&D expense	500,000	
Accumulated depreciation—equipment		500,000
To record R&D depreciation.		
Patent	100,000	
Cash		100,000
To capitalize the patent filing and legal costs.		

Expenditures reconciliation:	
Recorded as R&D expense	$ 14,200,000
Capitalized as equipment*	5,000,000
Capitalized as patent	100,000
Total expenditures	**$19,300,000**

* $500,000 of the initial capitalized amount will be depreciated as R&D expense in the first year.

GAAP requires disclosure of total R&D expense incurred during the period.

GAAP requires that total R&D expense incurred must be disclosed either as a line item in the income statement or in a disclosure note. In our illustration, total R&D expense disclosed in 2021 would be $14,700,000 ($14,200,000 in expenditures and $500,000 in depreciation). Note that if Askew later sells this patent to another company for, say, $15 million, the buyer would capitalize the entire purchase price rather than only the filing and legal costs. Once again, the reason for the apparent inconsistency in accounting treatment of internally generated intangibles and externally purchased intangibles is the difficulty of associating costs and benefits.

International Financial Reporting Standards

IFRS requires companies to capitalize development expenditures that meet specified criteria.

● LO10–9

IAS No. 38 draws a distinction between research activities and development activities.[24] Research expenditures are *expensed* in the period incurred. However, development expenditures that meet specified criteria are *capitalized* as an intangible asset. Under both U.S. GAAP and IFRS, any direct costs to secure a patent, such as legal and filing fees, are capitalized.

Heineken, a company based in Amsterdam, prepares its financial statements according to IFRS. The following disclosure note describes the company's adherence to *IAS No. 38*.

Software, Research and Development and Other Intangible Assets (in part)

Expenditures on research activities, undertaken with the prospect of gaining new technical knowledge and understanding, are recognized in the income statement when incurred. Development activities involve a plan or design for the production of new or substantially improved products and processes. Development expenditures are capitalized only if development costs can be measured reliably, the product or process is technically and commercially feasible, future economic benefits are probable, and Heineken intends to, and has sufficient resources to, complete development and to use or sell the asset. The expenditures capitalized include the cost of materials, direct labor and overhead costs that are directly attributable to preparing the asset for its intended use, and capitalized borrowing costs.

Amortization of capitalized development costs begins when development is complete and the asset is available for use. Heineken disclosed that it amortizes its capitalized development costs using the straight-line method over an estimated three-year useful life.

Source: Heineken

[24] "Intangible Assets," *International Accounting Standard No. 38* (IASCF), as amended effective January 1, 2018.

We've just discussed that U.S. GAAP requires R&D costs to be expensed immediately. There are three types of costs related to R&D, however, that are capitalized (recorded as an asset). These exceptions are shown below:

1. Development costs for software that has reached the point of technological feasibility
2. R&D performed by the company for sale to others
3. R&D purchased in a business acquisition

Software Development Costs

The computer software industry has become a large and important U.S. business over the last two decades. Companies in this multibillion dollar industry include **Microsoft**, **Oracle**, **IBM**, **Adobe Systems**, and **Intuit**. A significant expenditure for these companies is the cost of developing software. In the early years of the software industry, some software companies were capitalizing software development costs and expensing them in future periods and others were expensing these costs in the period incurred.

Now GAAP establishes a timeline for purposes of accounting for software development costs. Any software costs incurred from initial development activity until **technological feasibility** of the software are treated like all other R&D costs (expensed as incurred). Technological feasibility refers to the point in time "when the enterprise has completed all planning, designing, coding, and testing activities that are necessary to establish that the product can be produced to meet its design specifications including functions, features, and technical performance requirements."[25]

Costs incurred after technological feasibility but before product release are capitalized.

Costs incurred after technological feasibility but before the software is available for general release to customers are capitalized as an intangible asset. These costs include items such as further coding and testing and the production of product masters. Any costs incurred after the software release date generally are expensed but not as part of R&D.

Illustration 10–21 shows the R&D time line introduced earlier in the chapter modified to include the point at which technological feasibility is established. Only the costs incurred between technological feasibility and the software release date are capitalized.

Illustration 10–21

Research and Development Expenditures—Computer Software

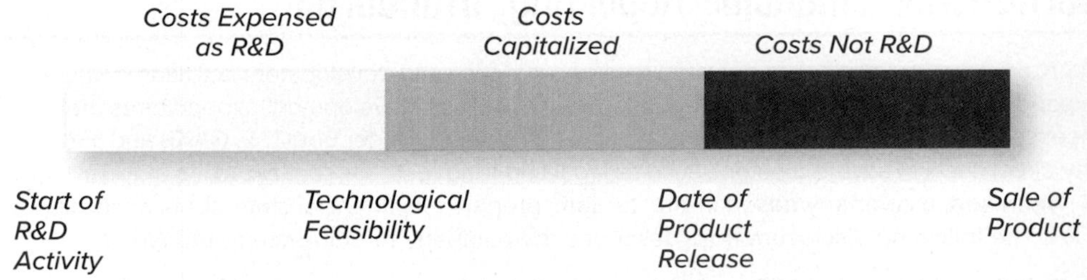

Illustration 10–22 shows the software disclosure included in a recent annual report of **CA, Inc**. The note provides a good summary of the accounting treatment of software development costs.

Why do generally accepted accounting principles allow this exception to the general rule of expensing all R&D? We could attribute it to the political process. Software is a very important industry to our economy and perhaps its lobbying efforts resulted in the standard allowing software companies to capitalize certain R&D costs.

We could also attribute the exception to the nature of the software business. Recall that R&D costs in general are expensed in the period incurred for two reasons: (1) they entail a high degree of uncertainty of future benefits, and (2) they are difficult to match with future benefits. With software, there is an important identifiable engineering milestone, technological feasibility. When this milestone is attained, the probability of the software product's

[25]FASB ASC 985–20–25–2: Software—Costs of Software to be Sold, Leased, or Marketed—Recognition.

Illustration 10–22
Software Disclosure—CA, Inc.

Real World Financials

Internally Developed Software Products

Internally developed software products, which are included in "Capitalized software and other intangible assets, net" in the Consolidated Balance Sheets, consist of capitalized costs associated with the development of computer software to be sold, leased or otherwise marketed. Software development costs associated with new products and significant enhancements to existing software products are expensed as incurred until technological feasibility, as defined in FASB ASC Topic 985-20, has been established. Costs incurred thereafter are capitalized until the product is made generally available. The stage during the Company's development process for a new product or new release at which technological feasibility requirements are established affects the amount of costs capitalized.

Source: CA, Inc.

success increases significantly. And because the useful life of software is fairly short (one to five years in most cases), it is much easier to determine the periods of increased revenues than for R&D projects in other industries. Compare this situation with, say, the development of a new drug. Even after the drug has been developed, it must go through extensive testing to meet approval by the Food and Drug Administration (FDA), which may never be attained. If attained, the useful life of the drug could be anywhere from a few months to many years.

The discussion above relates to costs incurred to develop or purchase computer software to be sold, leased, or otherwise marketed (that is, *for external purposes*).[26] We account for the costs incurred to develop computer software *to be used internally* in a similar manner. Costs incurred during the preliminary project stage are expensed as R&D. After the application development stage is reached (for example, at the coding stage or installation stage), we capitalize any further costs.[27] If, instead, we *purchase* computer software for internal use, those costs generally are capitalized.

Cloud computing arrangements. Cloud computing arrangements are becoming increasingly popular as a way for companies to utilize software without having to maintain the software and related hardware. These arrangements involve a company using software by accessing (via the Internet or dedicated line) a vendor's or a third party's hardware. The costs of cloud computing arrangements are treated as the purchase of intangible assets if both:

1. The customer has a contractual right to take possession of the software without significant penalty, and
2. The customer could run the software on its own or with an unrelated vendor.

If either of these criteria is not met, the arrangement is treated as a service contract, and costs are expensed as incurred.

Any implementation costs incurred during the application development phase (integration with the company's own software, coding, and configuration or customization) are capitalized. Interestingly, even if the arrangement itself does not qualify for capitalization as an intangible asset (because one of the two criteria above is not met), implementation costs still can be capitalized. Any costs related to preliminary planning or the post-implementation operation, however, would be expensed as incurred.

R&D Performed for Others

The principle requiring the immediate expensing of R&D does not apply to companies that perform R&D for other companies under contract. In these situations, the R&D costs are capitalized as inventory and carried forward into future years until the project is completed. Of course, justification is that the benefits of these expenditures are the contract revenues that will eventually be recognized. Revenue from these contracts can be recognized over time or at a point in time, depending on the specifics of the contract. We discussed these alternatives in Chapter 6.

[26]FASB ASC 985–20–25–1: Software—Costs of Software to be Sold, Leased, or Marketed—Recognition.
[27]FASB ASC 350–40–25: Intangibles—Goodwill and Other—Internal-Use Software—Recognition.

R&D Purchased in Business Acquisitions

It's not unusual for one company to buy another company in order to obtain technology that the acquired company has developed or is in the process of developing. Any time a company buys another, it values the tangible and intangible assets acquired at fair value. When part of the acquisition price involves technology, we distinguish between

1. Developed technology.
2. In-process research and development.

In business acquisitions, the fair value of developed technology is capitalized as an finite-life intangible asset.

To distinguish these two types of technology acquired, we borrow a criterion used in accounting for software development costs, and determine whether *technological feasibility* has been achieved. If it has, the technology is considered "developed," and we capitalize its fair value (record it as an asset) and amortize that amount over its useful life just like any other *finite-life* intangible asset.

In business acquisitions, the fair value of in-process research and development is capitalized as an indefinite-life intangible asset.

For in-process R&D (technology that has not reached the feasibility stage), GAAP also requires capitalization of its fair value. However, unlike developed technology, we view in-process R&D as an *indefinite-life* intangible asset.[28] As you will learn in Chapter 11, we don't amortize indefinite-life intangibles. Instead, we monitor these assets and test them for impairment when required by GAAP.

If the acquired R&D project is completed successfully, we switch to the way we account for developed technology and amortize the capitalized amount over the estimated period the product or process developed will provide benefits. If the project instead is abandoned, we expense the entire balance immediately.

R&D costs incurred after the acquisition to complete the project are expensed as incurred, consistent with the treatment of usual R&D expenditures. Similarly, the purchase of R&D from third parties not associated with a business combination is expensed as incurred, just as if the company had performed the R&D activity itself. An exception is when the purchased R&D has an alternative use beyond a specific project.

As an example of R&D associated with a business acquisition, in 2015 **AbbVie Inc.**, a global research-based biopharmaceutical company, acquired **Pharmacyclics Inc.** for approximately $20.8 billion. The fair values assigned included finite-life intangible assets of $11.4 billion, in-process R&D of $7.2 billion, and goodwill of $7.6 billion (as well as other identifiable assets of $1.7 billion and liabilities of $7.1 billion). As shown in Illustration 10-23, AbbVie classified projects acquired that have not yet received regulatory approval as in-process R&D.

Illustration 10–23

Acquired In-Process Research and Development—AbbVie, Inc.

Real World Financials

Acquired In-Process Research and Development

The initial costs of rights to IPR&D projects acquired in an asset acquisition are expensed as IPR&D in the consolidated statements of earnings unless the project has an alternative future use. These costs include initial payments incurred prior to regulatory approval in connection with research and development collaboration agreements that provide rights to develop, manufacture, market and/or sell pharmaceutical products. The fair value of IPR&D projects acquired in a business combination are capitalized and accounted for as indefinite-lived intangible assets until the underlying project receives regulatory approval, at which point the intangible asset will be accounted for as a definite-lived intangible asset, or discontinuation, at which point the intangible asset will be written off. Development costs incurred after the acquisition are expensed as incurred. Indefinite- and definite-lived assets are subject to impairment reviews as discussed previously.

Source: AbbVie, Inc.

Start-up costs are expensed in the period incurred.

Start-Up Costs

Whenever a company introduces a new product or service, or commences business in a new territory or with a new customer, it incurs **start-up costs**. Start-up costs also include **organization costs** related to organizing a new entity, such as legal fees and state filing fees to incorporate. As with R&D expenditures, companies are required to expense all the costs

[28]FASB ASC 805: Business Combinations.

related to a company's start-up and organization activities in the period incurred, rather than capitalize those costs as an asset.[29]

For the year ended December 31, 2017, **Chipotle Mexican Grill, Inc.**, opened 240 new restaurants. The company incurred a variety of one-time preopening costs for wages, benefits and travel for the training and opening teams, food, and other restaurant operating costs totaling $17 million. These costs were expensed immediately.

Financial Reporting Case Solution

©idealphotographer/123RF

1. **Describe to Stan what it means to capitalize an expenditure. What is the general rule for determining which costs are capitalized when property, plant, and equipment or an intangible asset is acquired?** *(p. 516)* To capitalize an expenditure simply means to record it as an asset. All expenditures other than payments to shareholders and debt repayments are either expensed as incurred or capitalized. In general, the choice is determined by whether the expenditure benefits more than just the current period. Exceptions to this general principle are discussed in the chapter. The initial cost of an asset includes all expenditures necessary to bring the asset to its desired condition and location for use.
2. **Which costs might be included in the initial cost of equipment?** *(p. 516)* In addition to the purchase price, the cost of equipment might include the cost of transportation, installation, testing, and legal fees to establish title.
3. **What is goodwill and how is it measured?** *(p. 523)* Goodwill represents the unique value of a company as a whole over and above its identifiable tangible and intangible assets. Because goodwill can't be separated from a company, it's not possible for a buyer to acquire it without also acquiring the whole company or a controlling portion of it. Goodwill will appear as an asset in a balance sheet only when it was purchased in connection with the acquisition of another company. In that case, the capitalized cost of goodwill equals the fair value of the consideration exchanged for the company less the fair value of the net assets acquired. Goodwill is a residual asset; it's the amount left after other assets are identified and valued. Just like for other intangible assets that have indefinite useful lives, we do not amortize goodwill.
4. **In what situations is interest capitalized rather than expensed?** *(p. 536)* Interest is capitalized only for assets constructed for a company's own use or for assets constructed as discrete products for sale or lease. For example, **Walt Disney** capitalizes interest on assets constructed for its theme parks, resorts, and other property, and on theatrical and television productions. During the construction period, interest is considered a cost necessary to get the asset ready for its intended use.
5. **What is the three-step process used to determine the amount of interest capitalized?** *(p. 537)* The first step is to determine the average accumulated expenditures for the period. The second step is to multiply the average accumulated expenditures by an appropriate interest rate or rates to determine the amount of interest capitalized. A final step compares the interest determined in step two with actual interest incurred. Interest capitalized is the lower of the two. ●

The Bottom Line

- **LO10–1** The initial cost of property, plant, and equipment and intangible assets acquired in an exchange transaction includes the purchase price and all expenditures necessary to bring the asset to its desired condition and location for use. The cost of a natural resource includes the acquisition costs for the use of land, the exploration and development costs incurred before production begins, and restoration costs incurred during or at the end of extraction. Purchased intangible assets are valued at their original cost to include the purchase price and legal and filing fees. (*p. 516*)

[29] FASB ASC 720–15–25–1: Other Expenses—Start-Up Costs—Recognition.

- **LO10–2** If a lump-sum purchase involves different assets, it is necessary to allocate the lump-sum acquisition price among the separate items according to some logical allocation method. A widely used allocation method is to divide the lump-sum purchase price according to the individual asset's relative fair values. *(p. 525)*
- **LO10–3** Assets acquired in exchange for deferred payment contracts are valued at their fair value or the present value of payments using a realistic interest rate. *(p. 526)*
- **LO10–4** Assets acquired through the issuance of equity securities are valued at the fair value of the securities if known; if not known, the fair value of the assets received is used. Donated assets are valued at their fair value. *(p. 528)*
- **LO10–5** A key to profitability is how well a company manages and utilizes its assets. Financial analysts often use activity, or turnover, ratios to evaluate a company's effectiveness in managing its assets. Property, plant, and equipment (PP&E) usually are a company's primary revenue-generating assets. Their efficient use is critical to generating a satisfactory return to owners. One ratio that analysts often use to measure how effectively managers use PP&E is the fixed-asset turnover ratio. This ratio is calculated by dividing net sales by average fixed assets. *(p. 530)*
- **LO10–6** The basic principle used for nonmonetary exchanges is to value the asset(s) received based on the fair value of the asset(s) given up. In certain situations, the valuation of the asset(s) received is based on the book value of the asset(s) given up. *(p. 531)*
- **LO10–7** The cost of a self-constructed asset includes identifiable materials and labor and a portion of the company's manufacturing overhead costs. In addition, GAAP provides for the capitalization of interest incurred during construction. The amount of interest capitalized is equal to the average accumulated expenditures for the period multiplied by the appropriate interest rates, not to exceed actual interest incurred. *(p. 535)*
- **LO10–8** Research and development costs incurred to internally develop an intangible asset are expensed in the period incurred. Filing and legal costs for developed intangibles are capitalized. *(p. 541)*
- **LO10–9** *IAS No. 20* requires that government grants be recognized in income over the periods necessary to match them on a systematic basis with the related costs that they are intended to compensate. Other than software development costs incurred after technological feasibility has been established, U.S. GAAP requires all research and development expenditures to be expensed in the period incurred. *IAS No. 38* draws a distinction between research activities and development activities. Research expenditures are expensed in the period incurred. However, development expenditures that meet specified criteria are capitalized as an intangible asset. *(pp. 529 and 543)*

APPENDIX 10 Oil And Gas Accounting

There are two generally accepted methods that companies can use to account for oil and gas exploration costs. The **successful efforts method** requires that exploration costs that are known *not* to have resulted in the discovery of oil or gas (sometimes referred to as *dry holes*) be included as expenses in the period the expenditures are made. The alternative, the **full-cost method**, allows costs incurred in searching for oil and gas within a large geographical area to be capitalized as assets and expensed in the future as oil and gas from the successful wells are removed from that area. Both of these methods are widely used. Illustration 10A–1 compares the two alternatives.

Illustration 10A–1
Oil and Gas Accounting

The Shannon Oil Company incurred $2,000,000 in exploration costs for each of 10 oil wells drilled in 2021 in west Texas. Eight of the 10 wells were dry holes.

The accounting treatment of the $20 million in total exploration costs will vary significantly depending on the accounting method used. The summary journal entries using each of the alternative methods are as follows:

Successful Efforts			Full Cost		
Oil Deposit	4,000,000		Oil Deposit	20,000,000	
Exploration Expense	16,000,000		Cash		20,000,000
Cash		20,000,000			

Using the full-cost method, Shannon would capitalize the entire $20 million which is expensed as oil from the two successful wells is depleted. On the other hand, using the successful efforts method, the cost of the unsuccessful wells is expensed in 2021, and only the $4 million cost related to the successful wells is capitalized and expensed in future periods as the oil is depleted.

Chapter 1 characterized the establishment of accounting and reporting standards as a political process. Standards, particularly changes in standards, can have significant differential effects on companies, investors and creditors, and other interest groups. The FASB must consider potential economic consequences of a change in an accounting standard or the introduction of a new standard. The history of oil and gas accounting provides a good example of this political process and the effect of possible adverse economic consequences on the standard-setting process.

In 1977 the FASB attempted to establish uniformity in the accounting treatment of oil and gas exploration costs. An accounting standard was issued requiring all companies to use the successful efforts method.[30]

This standard met with criticism from the oil and gas companies that were required to switch from full cost to successful efforts accounting. These companies felt that the switch would seriously depress their reported income over time. As a result, they argued, their ability to raise capital in the securities markets would be inhibited, which would result in a cutback of new exploration. The fear that the Standard would cause domestic companies to significantly reduce oil and gas exploration and thus increase our dependence on foreign oil was compelling to Congress, the SEC, and the U.S. Department of Energy.

Many feared that the requirement to switch to successful efforts accounting would cause a significant cutback in the exploration for new oil and gas in the United States.

Extensive pressure from Congress, the SEC, and affected companies forced the FASB to rescind the Standard. Presently, oil and gas companies can use either the successful efforts or full-cost method to account for oil and gas exploration costs. Of course, the method used must be disclosed. For example, Illustration 10A–2 shows how **Chevron Corp.** disclosed its use of the successful efforts method in its summary of significant accounting policies. ●

Properties, Plant, and Equipment (in part)

The successful efforts method is used for crude oil and natural gas exploration and production activities.

Source: Chevron Corp.

Illustration 10A–2

Oil and Gas Accounting Disclosure—Chevron Corp.

Real World Financials

Questions For Review of Key Topics

Q 10–1 Explain the difference between tangible and intangible long-lived, revenue-producing assets.

Q 10–2 What is included in the original cost of property, plant, and equipment and intangible assets acquired in an exchange transaction?

Q 10–3 Identify the costs associated with the initial valuation of a developed natural resource.

Q 10–4 Briefly summarize the accounting treatment for intangible assets, explaining the difference between purchased and internally developed intangible assets.

Q 10–5 What is goodwill and how is it measured?

Q 10–6 Explain the method generally used to allocate the cost of a lump-sum purchase to the individual assets acquired.

Q 10–7 When an asset is acquired and a note payable is assumed, explain how acquisition cost of the asset is determined when the interest rate for the note is less than the current market rate for similar notes.

Q 10–8 Explain how assets acquired in exchange for equity securities are valued.

Q 10–9 Explain how property, plant, and equipment and intangible assets acquired through donation are valued.

Q 10–10 What account is credited when a company receives donated assets? What is the rationale for this choice?

[30] The rescinded standard was "Financial Accounting and Reporting by Oil and Gas Producing Companies," *Statement of Financial Accounting Standards No. 19* (Stamford, Conn.: FASB, 1977). Authoritative guidance on this topic can now be found at FASB ASC 932: Extractive Activities—Oil and Gas.

Q 10–11 What is the basic principle for valuing property, plant, and equipment and intangible assets acquired in exchange for other nonmonetary assets?

Q 10–12 Identify the two exceptions to valuing property, plant, and equipment and intangible assets acquired in nonmonetary exchanges at the fair value of the asset(s) given up.

Q 10–13 In what situations is interest capitalized?

Q 10–14 Define average accumulated expenditures and explain how the amount is computed.

Q 10–15 Explain the difference between the specific interest method and the weighted-average method in determining the amount of interest to be capitalized.

Q 10–16 Define R&D according to U.S. GAAP.

Q 10–17 Explain the accounting treatment of equipment acquired for use in R&D projects.

Q 10–18 Explain the accounting treatment of costs incurred to develop computer software.

Q 10–19 Explain the difference in the accounting treatment of the cost of developed technology and the cost of in-process R&D in an acquisition.

IFRS Q 10–20 Identify any differences between U.S. GAAP and International Financial Reporting Standards in accounting for government grants received.

IFRS Q 10–21 Identify any differences between U.S. GAAP and International Financial Reporting Standards in the treatment of research and development expenditures.

IFRS Q 10–22 Identify any differences between U.S. GAAP and International Financial Reporting Standards in the treatment of software development costs.

Q 10–23 (Based on Appendix 10) Explain the difference between the successful efforts and the full-cost methods of accounting for oil and gas exploration costs.

Brief Exercises

connect

BE 10–1
Acquisition cost; equipment
● LO10–1

Beaverton Lumber purchased milling equipment for $35,000. In addition to the purchase price, Beaverton made the following expenditures: freight, $1,500; installation, $3,000; testing, $2,000; personal property tax on the equipment for the first year, $500. What is the initial cost of the equipment?

BE 10–2
Acquisition cost; land and building
● LO10–1

Fullerton Waste Management purchased land and a warehouse for $600,000. In addition to the purchase price, Fullerton made the following expenditures related to the acquisition: broker's commission, $30,000; title insurance, $3,000; miscellaneous closing costs, $6,000. The warehouse was immediately demolished at a cost of $18,000 in anticipation of the building of a new warehouse. Determine the amounts Fullerton should capitalize as the cost of the land and the building.

BE 10–3
Lump-sum acquisition
● LO10–2

Refer to the situation described in BE 10–2. Assume that Fullerton decides to use the warehouse rather than demolish it. An independent appraisal estimates the fair values of the land and warehouse at $420,000 and $280,000, respectively. Determine the amounts Fullerton should capitalize as the cost of the land and the building.

BE 10–4
Cost of a natural resource; asset retirement obligation
● LO10–1

Smithson Mining operates a silver mine in Nevada. Acquisition, exploration, and development costs totaled $5.6 million. After the silver is extracted in approximately five years, Smithson is obligated to restore the land to its original condition, including constructing a wildlife preserve. The company's controller has provided the following three cash flow possibilities for the restoration costs: (1) $500,000, 20% probability; (2) $550,000, 45% probability; and (3) $650,000, 35% probability. The company's credit-adjusted, risk-free rate of interest is 6%. What is the initial cost of the silver mine?

BE 10–5
Asset retirement obligation
● LO10–1

Refer to the situation described in BE 10–4. What is the book value of the asset retirement liability at the end of one year? Assuming that the actual restoration costs incurred after extraction is completed are $596,000, what amount of gain or loss will Smithson recognize on retirement of the liability?

BE 10–6
Goodwill
● LO10–1

Pro-tech Software acquired all of the outstanding stock of Reliable Software for $14 million. The book value of Reliable's net assets (assets minus liabilities) was $8.3 million. The fair values of Reliable's assets and liabilities equaled their book values with the exception of certain intangible assets whose fair values exceeded book values by $2.5 million. Calculate the amount paid for goodwill.

BE 10–7
Acquisition cost; noninterest-bearing note
● LO10–3

On June 30, 2021, Kimberly Farms purchased custom-made harvesting equipment from a local producer. In payment, Kimberly signed a noninterest-bearing note requiring the payment of $60,000 in two years. The fair value of the equipment is not known, but an 8% interest rate properly reflects the time value of money for this type of loan agreement. At what amount will Kimberly initially value the equipment? How much interest expense will Kimberly recognize in its income statement for this note for the year ended December 31, 2021?

BE 10–8
Acquisition cost; issuance of equity securities
● LO10–4

Shackelford Corporation acquired a patent from its founder, Jim Shackelford, in exchange for 50,000 shares of the company's no-par common stock. On the date of the exchange, the common stock had a fair value of $22 per share. Determine the cost of the patent.

BE 10–9
Fixed-asset turnover ratio
● LO10–5

Huebert Corporation and Winslow Corporation reported the following information:

	($ in millions)			
	Huebert		Winslow	
	2021	2020	2021	2020
Property, plant, and equipment (net)	$210	$220	$680	$650
Net sales—2021	$1,850		$5,120	

Calculate each companies fixed-asset turnover ratio and determine which company utilizes its fixed assets most efficiently to generate sales.

BE 10–10
Fixed-asset turnover ratio; solve for unknown
● LO10–5

The balance sheets of Pinewood Resorts reported net fixed assets of $740,000 and $940,000 at the end of 2020 and 2021, respectively. The fixed-asset turnover ratio for 2021 was 3.25. Calculate Pinewood's net sales for 2021.

BE 10–11
Nonmonetary exchange
● LO10–6

Calaveras Tire exchanged equipment for two pickup trucks. The book value and fair value of the equipment given up were $20,000 (original cost of $65,000 less accumulated depreciation of $45,000) and $17,000, respectively. Assume Calaveras paid $8,000 in cash and the exchange has commercial substance. (1) At what amount will Calaveras value the pickup trucks? (2) How much gain or loss will the company recognize on the exchange?

BE 10–12
Nonmonetary exchange
● LO10–6

Refer to the situation described in BE 10–11. Answer the questions assuming that the fair value of the equipment was $24,000, instead of $17,000.

BE 10–13
Nonmonetary exchange
● LO10–6

Refer to the situation described in BE 10–12. Answer the questions assuming that the exchange lacks commercial substance.

BE 10–14
Interest capitalization
● LO10–7

A company constructs a building for its own use. Construction began on January 1 and ended on December 30. The expenditures for construction were as follows: January 1, $500,000; March 31, $600,000; June 30, $400,000; October 30, $600,000. To help finance construction, the company arranged a 7% construction loan on January 1 for $700,000. The company's other borrowings, outstanding for the whole year, consisted of a $3 million loan and a $5 million note with interest rates of 8% and 6%, respectively. Assuming the company uses the *specific interest method,* calculate the amount of interest capitalized for the year.

BE 10–15
Interest capitalization
● LO10–7

Refer to the situation described in BE 10–14. Assuming the company uses the *weighted-average method,* calculate the amount of interest capitalized for the year.

BE 10–16
Research and development
● LO10–8

Maxtor Technology incurred the following costs during the year related to the creation of a new type of personal computer monitor:

Salaries	$220,000
Depreciation on R&D facilities and equipment	125,000
Utilities and other direct costs incurred for the R&D facilities	66,000
Patent filing and related legal costs	22,000
Payment to another company for performing a portion of the development work	120,000
Costs of adapting the new monitor for the specific needs of a customer	80,000

What amount should Maxtor report as research and development expense in its income statement?

BE 10–17
Software development costs; external purposes
● LO10–8

In February 2021, Culverson Company began developing a new software package to be sold to customers. The software allows people to enter health information and daily eating and exercise habits to track their health status. The project was completed in November 2021 at a cost of $800,000. Of this amount, $300,000 was spent before technological feasibility was established. Culverson expects a useful life of two years for the new product and total revenues of $1,500,000. Determine the amount that Culverson should capitalize as software development costs in 2021.

BE 10–18
Software development costs; internal purposes
● LO10–8

In March 2021, Price Company began developing a new software system to be used internally for managing its inventory. The software integrates customer orders with inventory on hand to automatically place orders for additional inventory when needed. The software then automatically records inventory as the shipment is received and read by scanners. The software development was completed in October 2021 at a cost of $150,000. Of this amount, $25,000 was spent before the application development stage was established. Price expects to use the software for at least five years. Determine the initial amount that Price should capitalize as software development costs.

BE 10–19
Software development costs; cloud computing arrangements
● LO10–8

Garrett Corporation began operations in 2021. To maintain its accounting records, Garrett entered into a two-year agreement with Accurite Company. The agreement specifies that Garrett will pay $35,000 to Accurite immediately, and in return, Accurite will make its accounting software accessible via the Internet to Garrett and maintain all infrastructure necessary to run the software and store records. At any time, Garrett Corporation can freely remove its records and run the software on its own hardware or that of another accounting services company. In addition to the cost of the agreement, Garrett incurred $5,000 early in the year devising a plan for its accounting software needs, $15,000 for customizing its own computers for integration with Accurite's software, and $10,000 after the software was implemented to train its employees. Determine the initial amount that Garrett should capitalize related to the software development costs.

BE 10–20
Research and development; various types
● LO10–8

Maltese Laboratories incurred the following research and developments costs related to its pharmaceutical business:

Internal projects (salaries, supplies, overhead for R&D facilities)	$620,000
Payment to acquire R&D from a third party related to a specific project	75,000
Costs of an R&D project to be sold under contract to Libo Pharmacy, a third party	82,000
In-process R&D associated with the acquisition of Curatics, an independent research company.	148,000

What amount should Maltese report as research and development expense in its income statement?

BE 10–21
Start-up costs
● LO10–8

In the current year, Big Burgers, Inc., expanded its fast-food operations by opening several new stores in Texas. The company incurred the following costs in the current year: market appraisal ($50,000), consulting fees ($72,000), advertising ($47,000), and traveling to train employees ($31,000). The company is willing to incur these costs because it foresees strong customer demand in Texas for the next several years. What amount should Big Burgers report as an expense in its income statement associated with these costs?

Exercises

connect

E 10–1
Acquisition costs; land and building
● LO10–1

On March 1, 2021, Beldon Corporation purchased land as a factory site for $60,000. An old building on the property was demolished, and construction began on a new building that was completed on December 15, 2021. Costs incurred during this period are listed below:

Demolition of old building	$ 4,000
Architect's fees (for new building)	12,000
Legal fees for title investigation of land	2,000
Property taxes on land (for period beginning March 1, 2021)	3,000
Construction costs	500,000
Interest on construction loan	5,000

Salvaged materials resulting from the demolition of the old building were sold for $2,000.

Required:

Determine the amounts that Beldon should capitalize as the cost of the land and the new building.

E 10–2
Acquisition cost; equipment
● LO10–1

Oaktree Company purchased new equipment and made the following expenditures:

Purchase price	$45,000
Sales tax	2,200
Freight charges for shipment of equipment	700
Insurance on the equipment for the first year	900
Installation of equipment	1,000

The equipment, including sales tax, was purchased on open account, with payment due in 30 days. The other expenditures listed above were paid in cash.

Required:

Prepare the necessary journal entries to record the above expenditures.

E 10–3
Acquisition costs; lump-sum acquisition
● LO10–1, LO10–2

Samtech Manufacturing purchased land and building for $4 million. In addition to the purchase price, Samtech made the following expenditures in connection with the purchase of the land and building:

Title insurance	$16,000
Legal fees for drawing the contract	5,000
Pro-rated property taxes for the period after acquisition	36,000
State transfer fees	4,000

An independent appraisal estimated the fair values of the land and building, if purchased separately, at $3.3 and $1.1 million, respectively. Shortly after acquisition, Samtech spent $82,000 to construct a parking lot and $40,000 for landscaping.

Required:

1. Determine the initial valuation of each asset Samtech acquired in these transactions.
2. Repeat requirement 1, assuming that immediately after acquisition, Samtech demolished the building. Demolition costs were $250,000 and the salvaged materials were sold for $6,000. In addition, Samtech spent $86,000 clearing and grading the land in preparation for the construction of a new building.

E 10–4
Cost of a natural resource; asset retirement obligation
● LO10–1

Jackpot Mining Company operates a copper mine in central Montana. The company paid $1,000,000 in 2021 for the mining site and spent an additional $600,000 to prepare the mine for extraction of the copper. After the copper is extracted in approximately four years, the company is required to restore the land to its original condition, including repaving of roads and replacing a greenbelt. The company has provided the following three cash flow possibilities for the restoration costs:

	Cash Outflow	Probability
1	$300,000	25%
2	400,000	40%
3	600,000	35%

To aid extraction, Jackpot purchased some new equipment on July 1, 2021, for $120,000. After the copper is removed from this mine, the equipment will be sold. The credit-adjusted, risk-free rate of interest is 10%.

Required:

1. Determine the cost of the copper mine.
2. Prepare the journal entries to record the acquisition costs of the mine and the purchase of equipment.

E 10–5
Intangibles
● LO10–1

In 2021, Bratten Fitness Company made the following cash purchases:

1. The exclusive right to manufacture and sell the X-Core workout equipment from Symmetry Corporation for $200,000. Symmetry created the unique design for the equipment. Bratten also paid an additional $10,000 in legal and filing fees to attorneys to complete the transaction.
2. An initial fee of $300,000 for a three-year agreement with Silver's Gym to use its name for a new facility in the local area. Silver's Gym has locations throughout the country. Bratten is required to pay an additional fee of $5,000 for each month it operates under the Silver's Gym name, with payments beginning in March 2021. Bratten also purchased $400,000 of exercise equipment to be placed in the new facility.
3. The exclusive right to sell *Healthy Choice,* a book authored by Kent Patterson, for $25,000. The book includes healthy recipes, recommendations for dietary supplements, and natural remedies. Bratten plans to display the book at the check-in counter at its new facility, as well as make it available online.

Required:
Prepare a summary journal entry to record expenditures related to initial acquisitions.

E 10–6
Goodwill
● LO10–1

On March 31, 2021, Wolfson Corporation acquired all of the outstanding common stock of Barney Corporation for $17,000,000 in cash. The book values and fair values of Barney's assets and liabilities were as follows:

	Book Value	Fair Value
Current assets	$ 6,000,000	$ 7,500,000
Property, plant, and equipment	11,000,000	14,000,000
Other assets	1,000,000	1,500,000
Current liabilities	4,000,000	4,000,000
Long-term liabilities	6,000,000	5,500,000

Required:
Calculate the amount paid for goodwill.

E 10–7
Goodwill
● LO10–1

Johnson Corporation acquired all of the outstanding common stock of Smith Corporation for $11,000,000 in cash. The book value of Smith's net assets (assets minus liabilities) was $7,800,000. The fair values of all of Smith's assets and liabilities were equal to their book values with the following exceptions:

	Book Value	Fair Value
Receivables	$1,300,000	$1,100,000
Property, plant, and equipment	8,000,000	9,400,000
Intangible assets	200,000	1,200,000

Required:
Calculate the amount paid for goodwill.

E 10–8
Lump-sum acquisition
● LO10–2

Pinewood Company purchased two buildings on four acres of land. The lump-sum purchase price was $900,000. According to independent appraisals, the fair values were $450,000 (building A) and $250,000 (building B) for the buildings and $300,000 for the land.

Required:
Determine the initial valuation of the buildings and the land.

E 10–9
Acquisition cost; noninterest-bearing note
● LO10–3

On January 1, 2021, Byner Company purchased a used tractor. Byner paid $5,000 down and signed a noninterest-bearing note requiring $25,000 to be paid on December 31, 2023. The fair value of the tractor is not determinable. An interest rate of 10% properly reflects the time value of money for this type of loan agreement. The company's fiscal year-end is December 31.

Required:
1. Prepare the journal entry to record the acquisition of the tractor. Round computations to the nearest dollar.
2. How much interest expense will the company include in its 2021 and 2022 income statements for this note?
3. What is the amount of the liability the company will report in its 2021 and 2022 balance sheets for this note?

E 10–10
Acquisition costs; noninterest-bearing note
● LO10–1, LO10–3

Teradene Corporation purchased land as a factory site and contracted with Maxtor Construction to construct a factory. Teradene made the following expenditures related to the acquisition of the land, building, and equipment for the factory:

Purchase price of the land	$1,200,000
Demolition and removal of old building	80,000
Clearing and grading the land before construction	150,000
Various closing costs in connection with acquiring the land	42,000
Architect's fee for the plans for the new building	50,000
Payments to Maxtor for building construction	3,250,000
Equipment purchased	860,000
Freight charges on equipment	32,000
Trees, plants, and other landscaping	45,000
Installation of a sprinkler system for the landscaping	5,000
Cost to build special platforms and install wiring for the equipment	12,000
Cost of trial runs to ensure proper installation of the equipment	7,000
Fire and theft insurance on the factory for the first year of use	24,000

In addition to the above expenditures, Teradene purchased four forklifts from **Caterpillar**. In payment, Teradene paid $16,000 cash and signed a noninterest-bearing note requiring the payment of $70,000 in one year. An interest rate of 7% properly reflects the time value of money for this type of loan.

Required:

Determine the initial valuation of each of the assets Teradene acquired in the above transactions.

E 10–11
IFRS; acquisition cost; issuance of equity securities and donation
● LO10–4, LO10–9
IFRS

On February 1, 2021, the Xilon Corporation issued 50,000 shares of its no-par common stock in exchange for five acres of land located in the city of Monrovia. On the date of the acquisition, Xilon's common stock had a fair value of $18 per share. An office building was constructed on the site by an independent contractor. The building was completed on November 2, 2021, at a cost of $6,000,000. Xilon paid $4,000,000 in cash and the remainder was paid by the city of Monrovia.

Required:

1. Prepare the journal entries to record the acquisition of the land and the building.
2. Assuming that Xilon prepares its financial statements according to International Financial Reporting Standards, explain the alternatives the company has for recording the acquisition of the office building.

E 10–12
IFRS; acquisition cost; acquisition by donation; government grant
● LO10–9
IFRS

Cranston LTD. prepares its financial statements according to International Financial Reporting Standards. In October 202[illegible], the company received a $2 million government grant. The grant represents 20% of the total cost of equipment that will be used to improve the roads in the local area. Cranston recorded the grant and the purchase of the equipment as follows:

Cash	2,000,000	
Revenue		2,000,000
Equipment	10,000,000	
Cash		10,000,000

Required:

1. Explain the alternative accounting treatments available to Cranston for accounting for this government grant.
2. Prepare any necessary correcting entries under each of the alternatives described in requirement 1.

E 10–13
Fixed-asset turnover ratio; Nvidia
● LO10–5
Real World Financials

Nvidia Corporation, a global technology company located in Santa Clara, California, reported the following information in its 2017 financial statements ($ in millions):

	2017	2016
Balance sheets		
Property, plant, and equipment (net)	$ 521	$466
Income statement		
Net sales for 2017	$6,910	

Required:

1. Calculate the company's 2017 fixed-asset turnover ratio.
2. How would you interpret this ratio?

E 10–14
Nonmonetary exchange
● LO10–6

Cedric Company recently traded in an older model of equipment for a new model. The old model's book value was $180,000 (original cost of $400,000 less $220,000 in accumulated depreciation) and its fair value was $200,000. Cedric paid $60,000 to complete the exchange which has commercial substance.

Required:

Prepare the journal entry to record the exchange.

E 10–15
Nonmonetary exchange
● LO10–6

[This is a variation of the previous exercise.]

Required:

Assume the same facts as in Exercise 10–14, except that the fair value of the old equipment is $170,000. Prepare the journal entry to record the exchange.

E 10–16
Nonmonetary exchange
● LO10–6

The Bronco Corporation exchanged land for equipment. The land had a book value of $120,000 and a fair value of $150,000. Bronco paid the owner of the equipment $10,000 to complete the exchange which has commercial substance.

Required:

1. What is the fair value of the equipment?
2. Prepare the journal entry to record the exchange.

E 10–17
Nonmonetary exchange
● LO10–6

[This is a variation of the previous exercise.]

Required:
Assume the same facts as in Exercise 10–16 except that Bronco *received* $10,000 from the owner of the equipment to complete the exchange.

1. What is the fair value of the equipment?
2. Prepare the journal entry to record the exchange.

E 10–18
Nonmonetary exchange
● LO10–6

The Tinsley Company exchanged land that it had been holding for future plant expansion for a more suitable parcel located farther from residential areas. Tinsley carried the land at its original cost of $30,000. According to an independent appraisal, the land currently is worth $72,000. Tinsley paid $14,000 in cash to complete the transaction.

Required:

1. What is the fair value of the new parcel of land received by Tinsley?
2. Prepare the journal entry to record the exchange assuming the exchange has commercial substance.
3. Prepare the journal entry to record the exchange assuming the exchange lacks commercial substance.
4. Prepare the journal entry to record the exchange except that Tinsley *received* $9,000 in the exchange, and the exchange lacks commercial substance.

E 10–19
Acquisition cost; multiple methods
● LO10–1, LO10–3, LO10–4, LO10–6

Connors Corporation acquired manufacturing equipment for use in its assembly line. Below are four *independent* situations relating to the acquisition of the equipment.

1. The equipment was purchased on account for $25,000. Credit terms were 2/10, n/30. Payment was made within the discount period and the company records the purchases of equipment net of discounts.
2. Connors gave the seller a noninterest-bearing note. The note required payment of $27,000 one year from date of purchase. The fair value of the equipment is not determinable. An interest rate of 10% properly reflects the time value of money in this situation.
3. Connors traded in old equipment that had a book value of $6,000 (original cost of $14,000 and accumulated depreciation of $8,000) and paid cash of $22,000. The old equipment had a fair value of $2,500 on the date of the exchange. The exchange has commercial substance.
4. Connors issued 1,000 shares of its no-par common stock in exchange for the equipment. The market value of the common stock was not determinable. The equipment could have been purchased for $24,000 in cash.

Required:
For each of the above situations, prepare the journal entry required to record the acquisition of the equipment. Round computations to the nearest dollar.

E 10–20
FASB codification research
● LO10–6

Access the *FASB Accounting Standards Codification* at the FASB website (www.fasb.org). Determine the specific eight-digit Codification citation (XXX-XX-XX-X) that describes each of the following items:

1. The basic principle for recording nonmonetary transactions at fair value.
2. Modifications of the basic principle for recording nonmonetary transactions when fair value is not determinable or the exchange lacks commercial substance.
3. The concept of commercial substance.
4. The required disclosures for nonmonetary transactions.

E 10–21
FASB codification research
● LO10–1, LO10–6, LO10–7, LO10–8

Access the *FASB Accounting Standards Codification* at the FASB website (**www.fasb.org**). Determine the specific eight-digit Codification citation (XXX-XX-XX-X) that describes each of the following items:

1. The disclosure requirements in the notes to the financial statements for depreciation on property, plant, and equipment.
2. The criteria for determining commercial substance in a nonmonetary exchange.
3. The disclosure requirements for interest capitalization.
4. The elements of costs to be included as R&D activities.

E 10–22
Interest capitalization
● LO10–7

On January 1, 2021, the Marjlee Company began construction of an office building to be used as its corporate headquarters. The building was completed early in 2022. Construction expenditures for 2021, which were incurred evenly throughout the year, totaled $6,000,000. Marjlee had the following debt obligations which were outstanding during all of 2021:

Construction loan, 10%	$1,500,000
Long-term note, 9%	2,000,000
Long-term note, 6%	4,000,000

Required:
Calculate the amount of interest capitalized in 2021 for the building using the specific interest method.

E 10–23
Interest capitalization
● LO10–7

On January 1, 2021, the Shagri Company began construction on a new manufacturing facility for its own use. The building was completed in 2022. The only interest-bearing debt the company had outstanding during 2021 was long-term bonds with a book value of $10,000,000 and an effective interest rate of 8%. Construction expenditures incurred during 2021 were as follows:

January 1	$500,000
March 1	600,000
July 31	480,000
September 30	600,000
December 31	300,000

Required:
Calculate the amount of interest capitalized for 2021.

E 10–24
Interest capitalization
● LO10–7

On January 1, 2021, the Highlands Company began construction on a new manufacturing facility for its own use. The building was completed in 2022. The company borrowed $1,500,000 at 8% on January 1 to help finance the construction. In addition to the construction loan, Highlands had the following debt outstanding throughout 2021:

$5,000,000, 12% bonds
$3,000,000, 8% long-term note

Construction expenditures incurred during 2021 were as follows:

January 1	$ 600,000
March 31	1,200,000
June 30	800,000
September 30	600,000
December 31	400,000

Required:
Calculate the amount of interest capitalized for 2021 using the specific interest method.

E 10–25
Interest capitalization; multiple periods
● LO10–7

Thornton Industries began construction of a warehouse on July 1, 2021. The project was completed on March 31, 2022. No new loans were required to fund construction. Thornton does have the following two interest-bearing liabilities that were outstanding throughout the construction period:

$2,000,000, 8% note
$8,000,000, 4% bonds

Construction expenditures incurred were as follows:

July 1, 2021	$400,000
September 30, 2021	600,000
November 30, 2021	600,000
January 30, 2022	540,000

The company's fiscal year-end is December 31.

Required:
Calculate the amount of interest capitalized for 2021 and 2022.

E 10–26
Research and development
● LO10–8

In 2021, Space Technology Company modified its model Z2 satellite to incorporate a new communication device. The company made the following expenditures:

Basic research to develop the technology	$2,000,000
Engineering design work	680,000
Development of a prototype device	300,000
Acquisition of equipment	60,000
Testing and modification of the prototype	200,000
Legal and other fees for patent application on the new communication system	40,000
Legal fees for successful defense of the new patent	20,000
Total	$3,300,000

The equipment will be used on this and other research projects. Depreciation on the equipment for 2021 is $10,000.

During your year-end review of the accounts related to intangibles, you discover that the company has capitalized all of the above as costs of the patent. Management contends that the device simply represents an improvement of the existing communication system of the satellite and, therefore, should be capitalized.

Required:

Prepare correcting entries that reflect the appropriate treatment of the expenditures.

E 10–27
Research and development
● LO10–8

Delaware Company incurred the following research and development costs during 2021:

Salaries and wages for lab research	$ 400,000
Materials used in R&D projects	200,000
Purchase of equipment	900,000
Fees paid to third parties for R&D projects	320,000
Patent filing and legal costs for a developed product	65,000
Salaries, wages, and supplies for R&D work performed for another company under a contract	350,000
Total	$2,235,000

The equipment has a seven-year life and will be used for a number of research projects. Depreciation for 2021 is $120,000.

Required:

Calculate the amount of research and development expense that Delaware should report in its 2021 income statement.

E 10–28
IFRS; research and development
● LO10–8, LO10–9

Janson Pharmaceuticals incurred the following costs in 2021 related to a new cancer drug:

Research for new formulas	$2,425,000
Development of a new formula	1,600,000
Legal and filing fees for a patent for the new formula	60,000
Total	$4,085,000

The development costs were incurred after technological and commercial feasibility was established and after the future economic benefits were deemed probable. The project was successfully completed and the new drug was patented before the end of the 2021 fiscal year.

Required:

1. Calculate the amount of research and development expense Janson should report in its 2021 income statement related to this project.
2. Repeat requirement 1 assuming that Janson prepares its financial statements according to International Financial Reporting Standards.

E 10–29
IFRS; research and development
● LO10–9
IFRS

NXS Semiconductor prepares its financial statements according to International Financial Reporting Standards. The company incurred the following expenditures during 2021 related to the development of a chip to be used in mobile devices:

Salaries and wages for basic research	$3,450,000
Materials used in basic research	330,000
Other costs incurred for basic research	1,220,000
Development costs	1,800,000
Legal and filing fees for a patent for the new technology	50,000

The development costs were incurred after NXS established technological and commercial feasibility and after NXS deemed the future economic benefits to be probable. The project was successfully completed, and the new chip was patented near the end of the 2021 fiscal year.

Required:

1. Which of the expenditures should NXS expense in its 2021 income statement?
2. Explain the accounting treatment of the remaining expenditures.

E 10–30
Concepts; terminology
● LO10–1, LO10–4, LO10–6, LO10–7

Listed below are several terms and phrases associated with property, plant, and equipment and intangible assets. Pair each item from List A with the item from List B (by letter) that is most appropriately associated with it.

List A	List B
____ 1. Property, plant, and equipment	a. Exclusive right to display a word, a symbol, or an emblem.
____ 2. Land improvements	b. Exclusive right to benefit from a creative work.
____ 3. Capitalize	c. Assets that represent rights.
____ 4. Average accumulated expenditures	d. Costs of establishing parking lots, driveways, and private roads.
____ 5. Revenue	e. Purchase price less fair value of net identifiable assets.
____ 6. Nonmonetary exchange	f. Assets such as land, buildings, and machinery.
____ 7. Natural resources	g. Approximation of average amount of debt if all construction funds were borrowed.
____ 8. Intangible assets	h. Account credited when assets are donated to a corporation.
____ 9. Copyright	i. Term meaning to record the cost as an asset.
____ 10. Trademark	j. Basic principle is to value assets acquired using fair value of assets given other than cash.
____ 11. Goodwill	k. Assets such as timber tracks and mineral deposits.

E 10–31
Software development costs
● LO10–8

Early in 2021, the Excalibur Company began developing a new software package to be marketed. The project was completed in December 2021 at a cost of $6 million. Of this amount, $4 million was spent before technological feasibility was established. Excalibur expects a useful life of five years for the new product with total revenues of $10 million. During 2022, revenue of $3 million was recognized.

Required:
Prepare a journal entry to record the 2021 development costs.

E 10–32
Software development costs
● LO10–8

On September 30, 2021, Athens Software began developing a software program to shield personal computers from malware and spyware. Technological feasibility was established on February 28, 2022, and the program was available for release on April 30, 2022. Development costs were incurred as follows:

September 30 through December 31, 2021	$2,200,000
January 1 through February 28, 2022	800,000
March 1 through April 30, 2022	400,000

Athens expects a useful life of four years for the software and total revenues of $5,000,000 during that time. During 2022, revenue of $1,000,000 was recognized.

Required:
Prepare a journal entry in each year to record development costs for 2021 and 2022.

E 10–33
Intangibles; start-up costs
● LO10–1, LO10–8

Freitas Corporation was organized early in 2021. The following expenditures were made during the first few months of the year:

Attorneys' fees in connection with the organization of the corporation	$ 12,000
State filing fees and other incorporation costs	3,000
Purchase of a patent	20,000
Legal and other fees for transfer of the patent	2,000
Purchase of equipment	30,000
Preopening salaries and employee training	40,000
Total	$107,000

Required:
Prepare a summary journal entry to record the $107,000 in cash expenditures.

E 10–34
Full-cost and successful efforts methods compared
● Appendix

The Manguino Oil Company incurred exploration costs in 2021 searching and drilling for oil as follows:

Well 101	$ 50,000
Well 102	60,000
Well 103	80,000
Wells 104–108	260,000
Total	$450,000

It was determined that Wells 104–108 were dry holes and were abandoned. Wells 101, 102, and 103 were determined to have sufficient oil reserves to be commercially successful.

Required:
1. Prepare a summary journal entry to record the indicated costs assuming that the company uses the full-cost method of accounting for exploration costs. All of the exploration costs were paid in cash.
2. Prepare a summary journal entry to record the indicated costs assuming that the company uses the successful efforts method of accounting for exploration costs. All of the exploration costs were paid in cash.

Problems

connect

P 10–1
Acquisition costs
● LO10–1 through LO10–4

Tristar Production Company began operations on September 1, 2021. Listed below are a number of transactions that occurred during its first four months of operations.

1. On September 1, the company acquired five acres of land with a building that will be used as a warehouse. Tristar paid $100,000 in cash for the property. According to appraisals, the land had a fair value of $75,000 and the building had a fair value of $45,000.
2. On September 1, Tristar signed a $40,000 noninterest-bearing note to purchase equipment. The $40,000 payment is due on September 1, 2022. Assume that 8% is a reasonable interest rate.
3. On September 15, a truck was donated to the corporation. Similar trucks were selling for $2,500.
4. On September 18, the company paid its lawyer $3,000 for organizing the corporation.
5. On October 10, Tristar purchased maintenance equipment for cash. The purchase price was $15,000 and $500 in freight charges also were paid.
6. On December 2, Tristar acquired various items of office equipment. The company was short of cash and could not pay the $5,500 normal cash price. The supplier agreed to accept 200 shares of the company's no-par common stock in exchange for the equipment. The fair value of the stock is not readily determinable.
7. On December 10, the company acquired a tract of land at a cost of $20,000. It paid $2,000 down and signed a 10% note with both principal and interest due in one year. Ten percent is an appropriate rate of interest for this note.

Required:

Prepare journal entries to record each of the above transactions.

P 10–2
Acquisition costs; land and building
● LO10–1, LO10–2, LO10–7

On January 1, 2021, the Blackstone Corporation purchased a tract of land (site number 11) with a building for $600,000. Additionally, Blackstone paid a real estate broker's commission of $36,000, legal fees of $6,000, and title insurance of $18,000. The closing statement indicated that the land value was $500,000 and the building value was $100,000. Shortly after acquisition, the building was razed at a cost of $75,000.

Blackstone entered into a $3,000,000 fixed-price contract with Barnett Builders, Inc., on March 1, 2021, for the construction of an office building on land site 11. The building was completed and occupied on September 30, 2022. Additional construction costs were incurred as follows:

Plans, specifications, and blueprints	$12,000
Architects' fees for design and supervision	95,000

To finance the construction cost, Blackstone borrowed $3,000,000 on March 1, 2021. The loan is payable in 10 annual installments of $300,000 plus interest at the rate of 14%. Blackstone's average amounts of accumulated building construction expenditures were as follows:

For the period March 1 to December 31, 2021	$ 900,000
For the period January 1 to September 30, 2022	2,300,000

Required:

1. Prepare a schedule that discloses the individual costs making up the balance in the land account in respect of land site 11 as of September 30, 2022.
2. Prepare a schedule that discloses the individual costs that should be capitalized in the office building account as of September 30, 2022.

Source: AICPA adapted

P 10–3
Acquisition costs
● LO10–1, LO10–4, LO10–6

The plant asset and accumulated depreciation accounts of Pell Corporation had the following balances at December 31, 2020:

	Plant Asset	Accumulated Depreciation
Land	$ 350,000	$ —
Land improvements	180,000	45,000
Building	1,500,000	350,000
Equipment	1,158,000	405,000
Automobiles	150,000	112,000

Transactions during 2021 were as follows:

a. On January 2, 2021, equipment were purchased at a total invoice cost of $260,000, which included a $5,500 charge for freight. Installation costs of $27,000 were incurred.

b. On March 31, 2021, a small storage building was donated to the company. The person donating the building originally purchased it three years ago for $25,000. The fair value of the building on the day of the donation was $17,000.

c. On May 1, 2021, expenditures of $50,000 were made to repave parking lots at Pell's plant location. The work was necessitated by damage caused by severe winter weather. The repair doesn't provide future benefits beyond those originally anticipated.

d. On November 1, 2021, Pell acquired a tract of land with an existing building in exchange for 10,000 shares of Pell's common stock that had a market price of $38 per share. Pell paid legal fees and title insurance totaling $23,000. Shortly after acquisition, the building was razed at a cost of $35,000 in anticipation of new building construction in 2022.

e. On December 31, 2021, Pell purchased a small storage building by giving $15,250 cash and an old automobile purchased for $18,000 in 2014. Depreciation on the old automobile recorded through December 31, 2021, totaled $13,500. The fair value of the old automobile was $3,750.

Required:
Prepare a schedule analyzing the changes in each of the plant assets during 2021, with detailed supporting computations.

(AICPA adapted)

P 10–4
Intangibles
● LO10–1, LO10–8

The Horstmeyer Corporation commenced operations early in 2021. A number of expenditures were made during 2021 that were debited to one account called *intangible asset*. A recap of the $144,000 balance in this account at the end of 2021 is as follows:

Date		Transaction	Amount
February	3	State incorporation fees and legal costs related to organizing the corporation	$ 7,000
March	1	Fire insurance premium for three-year period	6,000
March	15	Purchased a copyright	20,000
April	30	Research and development costs	40,000
June	15	Legal fees for filing a patent on a new product resulting from an R&D project	3,000
September	30	Legal fee for successful defense of patent developed above	12,000
October	13	Entered into a 10-year franchise agreement with franchisor	40,000
Various		Advertising costs	16,000
		Total	$144,000

Required:
Prepare the necessary journal entry to clear the intangible asset account and to set up accounts for separate intangible assets, other types of assets, and expenses indicated by the transactions.

P 10–5
Acquisition costs; journal entries
● LO10–1, LO10–3, LO10–6, LO10–8

Consider each of the transactions below. All of the expenditures were made in cash.

1. The Edison Company spent $12,000 during the year for experimental purposes in connection with the development of a new product.
2. In April, the Marshall Company lost a patent infringement suit and paid the plaintiff $7,500.
3. In March, the Cleanway Laundromat bought equipment. Cleanway paid $6,000 down and signed a noninterest-bearing note requiring the payment of $18,000 in nine months. The cash price for this equipment was $23,000.
4. On June 1, the Jamsen Corporation installed a sprinkler system throughout the building at a cost of $28,000.
5. The Mayer Company, plaintiff, paid $12,000 in legal fees in November, in connection with a successful infringement suit on its patent.
6. The Johnson Company traded its old equipment for new equipment. The new equipment has a fair value of $10,000. The old equipment had an original cost of $7,400 and a book value of $3,000 at the time of the trade. Johnson also paid cash of $8,000 as part of the trade. The exchange has commercial substance.

Required:
Prepare journal entries to record each of the above transactions.

P 10–6
Nonmonetary exchange
● LO10–6

Southern Company owns a building that it leases to others. The building's fair value is $1,400,000 and its book value is $800,000 (original cost of $2,000,000 less accumulated depreciation of $1,200,000). Southern exchanges this for a building owned by the Eastern Company. The building's book value on Eastern's books is $950,000 (original cost of $1,600,000 less accumulated depreciation of $650,000). Eastern also gives Southern $140,000 to complete the exchange. The exchange has commercial substance for both companies.

Required:
Prepare the journal entries to record the exchange on the books of both Southern and Eastern.

P 10–7
Nonmonetary exchange
● LO10–6

On September 3, 2021, the Robers Company exchanged equipment with Phifer Corporation. The facts of the exchange are as follows:

	Robers' Asset	Phifer's Asset
Original cost	$120,000	$140,000
Accumulated depreciation	55,000	63,000
Fair value	75,000	70,000

To equalize the exchange, Phifer paid Robers $5,000 in cash.

Required:
Record the exchange for both Robers and Phifer. The exchange has commercial substance for both companies.

P 10–8
Nonmonetary exchange
● LO10–6

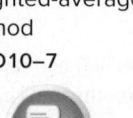

Case A. Kapono Farms exchanged an old tractor for a newer model. The old tractor had a book value of $12,000 (original cost of $28,000 less accumulated depreciation of $16,000) and a fair value of $9,000. Kapono paid $20,000 cash to complete the exchange. The exchange has commercial substance.

Required:
1. What is the amount of gain or loss that Kapono would recognize on the exchange? What is the initial value of the new tractor?
2. Repeat requirement 1 assuming that the fair value of the old tractor is $14,000 instead of $9,000.

Case B. Kapono Farms exchanged 100 acres of farmland for similar land. The farmland given had a book value of $500,000 and a fair value of $700,000. Kapono paid $50,000 cash to complete the exchange. The exchange has commercial substance.

Required:
1. What is the amount of gain or loss that Kapono would recognize on the exchange? What is the initial value of the new land?
2. Repeat requirement 1 assuming that the fair value of the farmland given is $400,000 instead of $700,000.
3. Repeat requirement 1 assuming that the exchange lacked commercial substance.

P 10–9
Interest capitalization; specific interest method
● LO10–7

On January 1, 2021, the Mason Manufacturing Company began construction of a building to be used as its office headquarters. The building was completed on September 30, 2022. Expenditures on the project were as follows:

January 1, 2021	$1,000,000
March 1, 2021	600,000
June 30, 2021	800,000
October 1, 2021	600,000
January 31, 2022	270,000
April 30, 2022	585,000
August 31, 2022	900,000

On January 1, 2021, the company obtained a $3 million construction loan with a 10% interest rate. The loan was outstanding all of 2021 and 2022. The company's other interest-bearing debt included two long-term notes of $4,000,000 and $6,000,000 with interest rates of 6% and 8%, respectively. Both notes were outstanding during all of 2021 and 2022. Interest is paid annually on all debt. The company's fiscal year-end is December 31.

Required:
1. Calculate the amount of interest that Mason should capitalize in 2021 and 2022 using the specific interest method.
2. What is the total cost of the building?
3. Calculate the amount of interest expense that will appear in the 2021 and 2022 income statements.

P 10–10
Interest capitalization; weighted-average method
● LO10–7

[This is a variation of the previous problem, modified to focus on the weighted-average interest method.]

Required:
Refer to the facts in Problem 10–9 but now assume the $3 million loan is not specifically tied to construction of the building. Answer the following questions:
1. Calculate the amount of interest that Mason should capitalize in 2021 and 2022 using the weighted-average method.
2. What is the total cost of the building?
3. Calculate the amount of interest expense that will appear in the 2021 and 2022 income statements.

P 10–11
Research and development
● LO10–8

In 2021, Starsearch Corporation began work on three research and development projects. One of the projects was completed and commercial production of the developed product began in December. The company's fiscal year-end is December 31. All of the following 2021 expenditures were included in the R&D expense account:

Salaries and wages for:	
Lab research	$ 300,000
Design and construction of preproduction prototype	160,000
Quality control during commercial production	20,000
Materials and supplies consumed for:	
Lab research	60,000
Construction of preproduction prototype	30,000
Purchase of equipment	600,000
Patent filing and legal fees for completed project	40,000
Payments to others for research	120,000
Total	$1,330,000

$200,000 of equipment was purchased solely for use in one of the projects. After the project is completed, the equipment will be abandoned. The remaining $400,000 in equipment will be used on future R&D projects. The useful life of equipment is five years. Assume that all of the equipment was acquired at the beginning of the year.

Required:

Prepare journal entries, reclassifying amounts in R&D expense, to reflect the appropriate treatment of the expenditures.

P 10–12
Acquisition costs; lump-sum acquisition; noninterest-bearing note; interest capitalization
● LO10–1, LO10–2, LO10–3, LO10–7

Early in its fiscal year ending December 31, 2021, San Antonio Outfitters finalized plans to expand operations. The first stage was completed on March 28 with the purchase of a tract of land on the outskirts of the city. The land and existing building were purchased by paying $200,000 immediately and signing a noninterest-bearing note requiring the company to pay $600,000 on March 28, 2023. An interest rate of 8% properly reflects the time value of money for this type of loan agreement. Title search, insurance, and other closing costs totaling $20,000 were paid at closing.

At the end of April, the old building was demolished at a cost of $70,000, and an additional $50,000 was paid to clear and grade the land. Construction of a new building began on May 1 and was completed on October 29. Construction expenditures were as follows:

May 1	$1,200,000
July 30	1,500,000
September 1	900,000
October 1	1,800,000

San Antonio borrowed $3,000,000 at 8% on May 1 to help finance construction. This loan, plus interest, will be paid in 2022. The company also had a $5,250,000, 8% long-term note payable outstanding throughout 2021.

In November, the company purchased 10 identical pieces of equipment and office furniture and fixtures for a lump-sum price of $600,000. The fair values of the equipment and the furniture and fixtures were $455,000 and $245,000, respectively. In December, San Antonio paid a contractor $285,000 for the construction of parking lots and for landscaping.

Required:

1. Determine the initial values of the various assets that San Antonio acquired or constructed during 2021. The company uses the specific interest method to determine the amount of interest capitalized on the building construction.
2. How much interest expense will San Antonio report in its 2021 income statement?

Decision Maker's Perspective

©ImageGap/Getty Images

Apply your critical-thinking ability to the knowledge you've gained. These cases will provide you an opportunity to develop your research, analysis, judgment, and communication skills. You also will work with other students, integrate what you've learned, apply it in real-world situations, and consider its global and ethical ramifications. This practice will broaden your knowledge and further develop your decision-making abilities.

Judgment Case 10–1
Acquisition costs
● LO10–1, LO10–3, LO10–6

A company may acquire property, plant, and equipment and intangible assets for cash, in exchange for a deferred payment contract, by exchanging other assets, or by a combination of these methods.

Required:

1. Identify six types of costs that should be capitalized as the cost of a parcel of land. For your answer, assume that the land has an existing building that is to be removed in the immediate future in order that a new building can be constructed on the site.
2. At what amount should a company record an asset acquired in exchange for a deferred payment contract?
3. In general, at what amount should assets received in exchange for other nonmonetary assets be valued? Specifically, at what amount should a company value new equipment acquired by exchanging older, similar equipment and paying cash?

(AICPA adapted)

Research Case 10–2
FASB codification; locate and extract relevant information and cite authoritative support for a financial reporting issue; restoration costs; asset retirement obligation
● LO10–1

Your client, Hazelton Mining, recently entered into an agreement to obtain the rights to operate a coal mine in West Virginia for $15 million. Hazelton incurred development costs of $6 million in preparing the mine for extraction, which began on July 1, 2021. The contract requires Hazelton to restore the land and surrounding area to its original condition after extraction is complete in three years.

The company controller, Alice Cushing, is not sure how to account for the restoration costs and has asked your advice. Alice is aware of an accounting standard addressing this issue, but is not sure of its provisions. She has narrowed down the possible cash outflows for the restoration costs to four possibilities.

Cash Outflow	Probability
$3 million	20%
4 million	30%
5 million	25%
6 million	25%

Alice also informs you that the company's credit-adjusted risk-free interest rate is 9%. Before responding to Alice, you need to research the issue.

Required:

1. Access the *FASB Accounting Standards Codification* at the FASB website (**www.fasb.org**). Determine the specific Codification citation for each of the following, based on the format indicated: (a) accounting for asset retirement obligations (XXX-XX), (b) recognition criteria related to asset retirement obligations (XXX-XX-XX), (c) the requirement to recognize the fair value of the asset retirement obligation as a liability (XXX-XX-XX-X), and (d) how to treat the capitalized cost of the asset retirement obligation for the related tangible long-lived asset (XXX-XX-XX-X).
2. Determine the capitalized cost of the coal mine.
3. Prepare a summary journal entry to record the acquisition costs of the mine.
4. How much accretion expense will the company record in its income statement for the 2021 fiscal year, related to this transaction? Determine the specific Codification citation for each of the following, based on the format indicated: (a) the calculation of accretion expense (XXX-XX-XX-X) and (b) the classification of accretion expense in the income statement (XXX-XX-XX-X).
5. Explain to Alice how Hazelton would account for the restoration if the restoration costs differed from the recorded liability in three years. By way of explanation, prepare the journal entry to record the payment of the retirement obligation in three years assuming that the actual restoration costs were $4.7 million.
6. Describe to Alice the necessary disclosure requirements for the obligation. What specific Codification citation contains these disclosure requirements (XXX-XX-XX-X)?

Judgment Case 10–3
Self-constructed assets
● LO10–7

Chilton Peripherals manufactures printers, scanners, and other computer peripheral equipment. In the past, the company purchased equipment used in manufacturing from an outside vendor. In March 2021, Chilton decided to design and build equipment to replace some obsolete equipment. A section of the manufacturing plant was set aside to develop and produce the equipment. Additional personnel were hired for the project. The equipment was completed and ready for use in September.

Required:

1. In general, what costs should be capitalized for a self-constructed asset?
2. Discuss two alternatives for the inclusion of overhead costs in the cost of the equipment constructed by Chilton. Which alternative is generally accepted for financial reporting purposes?
3. Under what circumstance(s) would interest be included in the cost of the equipment?

Judgment Case 10–4
Interest capitalization
● LO10–7

GAAP provides guidelines for the inclusion of interest in the initial cost of a self-constructed asset.

Required:

1. What assets qualify for interest capitalization? What assets do not qualify for interest capitalization?
2. Over what period should interest be capitalized?
3. Explain average accumulated expenditures.

4. Explain the two methods that could be used to determine the appropriate interest rate(s) to be used in capitalizing interest.
5. Describe the three steps used to determine the amount of interest capitalized during a reporting period.

Research Case 10–5
Goodwill
● LO10–1

Accounting for acquired goodwill has been a controversial issue for many years. In the United States, the amount of acquired goodwill is capitalized and not amortized. Globally, the treatment of goodwill varies significantly, with some countries not recognizing goodwill as an asset. Professors Johnson and Petrone, in "Is Goodwill an Asset?" discuss this issue.

Required:

1. In your library or from some other source, locate the indicated article in *Accounting Horizons*, September 1998.
2. Does goodwill meet the FASB's definition of an asset?
3. What are the key concerns of those that believe goodwill is not an asset?

Real World Case 10–6
Property, plant, and equipment; Norfolk Southern Corporation
● LO10–1

Real World Financials

Norfolk Southern Corporation, one of the nation's premier transportation companies, reported the following amounts in the asset section of its balance sheets for the years ended December 31, 2016 and 2015:

	($ in millions)	
	December 31, 2016	December 31, 2015
Property and equipment, net	$29,751	$28,992

In addition, information from the 2016 statement of cash flows and related notes reported the following items ($ in millions):

Depreciation	$1,030
Additions to property and equipment	1,887
Sales price of property and equipment	130

Required:

For what amount was the sales price above book value of property and equipment sold for the year ended December 31, 2016?

Judgment Case 10–7
Goodwill
● LO10–1

Athena Paper Corporation acquired for cash 100% of the outstanding common stock of Georgia, Inc., a supplier of wood pulp. The $4,500,000 amount paid was significantly higher than the book value of Georgia's net assets (assets less liabilities) of $2,800,000. The Athena controller recorded the difference of $1,700,000 as an asset, goodwill.

Required:

1. Discuss the meaning of the term goodwill
2. In what situation would the Athena controller be correct in her valuation of goodwill?

Judgment Case 10–8
Research and development
● LO10–8

Prior to 1974, accepted practice was for companies to either expense or capitalize R&D costs. In 1974, the FASB issued a Standard that requires all research and development costs to be charged to expense when incurred. This was a controversial standard, opposed by many companies who preferred delaying the recognition of these expenses until later years when presumably the expenditures bear fruit.

Several research studies have been conducted to determine if the Standard had any impact on the behavior of companies. One interesting finding was that, prior to 1974, companies that expensed R&D costs were significantly larger than those companies that capitalized R&D costs.

Required:

1. Explain the FASB's logic in deciding to require all companies to expense R&D costs in the period incurred.
2. Identify possible reasons to explain why, prior to 1974, companies that expensed R&D costs were significantly larger than those companies that capitalized R&D costs.

Judgment Case 10–9
Research and development
● LO10–8

Clonal, Inc., a biotechnology company, developed and patented a diagnostic product called Trouver. Clonal purchased some research equipment to be used for Trouver and subsequent research projects. Clonal defeated a legal challenge to its Trouver patent, and began production and marketing operations for the project.

Corporate headquarters' costs were allocated to Clonal's research division as a percentage of the division's salaries.

Required:

1. How should the equipment purchased for Trouver be reported in Clonal's income statement and balance sheet?
2. a. Describe the accounting treatment of research and development costs.
   b. What is the justification for the accounting treatment of research and development costs?
3. How should corporate headquarters' costs allocated to the research division be classified in Clonal's income statements? Why?
4. How should the legal expenses incurred in defending Trouver's patent be reported in Clonal's financial statements?

(AICPA adapted)

Communication Case 10–10
Research and development
● LO10–8

The focus of this case is the situation described in Case 10–9. What is the appropriate accounting for R&D costs? Do you believe that (1) capitalization is the correct treatment of R&D costs, (2) expensing is the correct treatment of R&D costs, or (3) that companies should be allowed to choose between expensing and capitalizing R&D costs?

Required:

1. Develop a list of arguments in support of your view prior to the class session for which the case is assigned. Do not be influenced by the method required by the FASB. Base your opinion on the conceptual merit of the options.
2. In class, your instructor will pair you (and everyone else) with a classmate who also has independently developed a position.
   a. You will be given three minutes to argue your view to your partner. Your partner likewise will be given three minutes to argue his or her view to you. During these three-minute presentations, the listening partner is not permitted to speak.
   b. Then after each person has had a turn attempting to convince his or her partner, the two partners will have a three-minute discussion in which they will decide which alternative is more convincing and arguments will be merged into a single view for each pair.
3. After the allotted time, a spokesperson for each of the three alternatives will be selected by the instructor. Each spokesperson will field arguments from the class as to the appropriate alternative. The class will then discuss the merits of the alternatives and attempt to reach a consensus view, though a consensus is not necessary.

Communication Case 10–11
Research and development
● LO10–8

Thomas Plastics is in the process of developing a revolutionary new plastic valve. A new division of the company was formed to develop, manufacture, and market this new product. As of year-end (December 31, 2021), the new product has not been manufactured for sale; however, prototype units were built and are in operation.

Throughout 2021, the new division incurred a variety of costs. These costs included expenses (including salaries of administrative personnel) and market research costs. In addition, approximately $500,000 in equipment (estimated useful life of 10 years) was purchased for use in developing and manufacturing the new valve. Approximately $200,000 of this equipment was built specifically for developing the design of the new product; the remaining $300,000 of the equipment was used to manufacture the preproduction prototypes and will be used to manufacture the new product once it is in commercial production.

The president of the company, Sally Rogers, has been told that research and development costs must be expensed as incurred, but she does not understand this treatment. She believes the research will lead to a profitable product and to increased future revenues. Also, she wonders how to account for the $500,000 of equipment purchased by the new division. "I thought I understood accounting," she growled. "Explain to me why expenditures that benefit our future revenues are expensed rather than capitalized!"

Required:

Write a one-to two-page report to Sally Rogers explaining the generally accepted accounting principles relevant to this issue. The report also should address the treatment of the equipment purchases.

(AICPA adapted)

Ethics Case 10–12
Research and development
● LO10–8

Mayer Biotechnical, Inc., develops, manufactures, and sells pharmaceuticals. Significant research and development (R&D) expenditures are made for the development of new drugs and the improvement of existing drugs. During 2021, $220 million was spent on R&D. Of this amount, $30 million was spent on the purchase of equipment to be used in a research project involving the development of a new antibiotic.

The controller, Alice Cooper, is considering capitalizing the equipment and depreciating it over the five-year useful life of the equipment at $6 million per year, even though the equipment likely will be used on only one project. The company president has asked Alice to make every effort to increase 2021 earnings because in 2022 the company will be seeking significant new financing from both debt and equity sources. "I guess we might use the equipment in other projects later," Alice wondered to herself.

Required:

1. Assuming that the equipment was purchased at the beginning of 2021, by how much would Alice's treatment of the equipment increase before tax earnings as opposed to expensing the equipment cost?
2. Discuss the ethical dilemma Alice faces in determining the treatment of the $30 million equipment purchase.

IFRS Case 10–13
Research and development; comparison of U.S. GAAP and IFRS; Siemens AG
● LO10–8, LO10–9

Siemens AG, a German company, is Europe's largest engineering and electronics company. The company prepares its financial statements according to IFRS.

Required:

1. Use the Internet to locate the most recent financial report for Siemens. The address is **www.siemens.com**. Locate the significant accounting policies disclosure note.
2. How does the company account for research and development expenditures? Does this policy differ from U.S. GAAP?

Analysis Case 10–14
Fixed-asset turnover ratio; Pier 1 Imports, Inc.
● LO10–5

Real World Financials

Pier 1 Imports, Inc., is a leading retailer of domestic merchandise and home furnishings. The company's 2017 fixed-asset turnover ratio, using the average book value of property, plant, and equipment (PP&E) as the denominator, was approximately 9.16. Additional information taken from the company's 2017 annual report is as follows:

	($ in thousands)
Book value of PP&E—beginning of 2017	$207,633
Net purchases of PP&E during 2017*	44,347
Depreciation of PP&E for 2017	60,504

*Net purchases include book value of equipment purchased minus book value of equipment sold.

Required:

1. How is the fixed-asset turnover ratio computed? How would you interpret Pier 1's ratio of 9.16?
2. Use the data to determine Pier 1's net sales for 2017.
3. Obtain annual reports from three corporations in the same primary industry as Pier 1 Imports, Inc. (Bed, Bath & Beyond and Williams-Sonoma, Inc., are two well-known companies in the same industry) and compare the management of each company's investment in property, plant, and equipment.

Note: You can obtain copies of annual reports from your library, from friends who are shareholders, from the investor relations department of the corporations, from a friendly stockbroker, or from EDGAR (Electronic Data Gathering, Analysis, and Retrieval) on the Internet (**www.sec.gov**).

Judgment Case 10–15
Computer software costs
● LO10–3

The Elegant Software Company recently completed the development and testing of a new software program that provides the ability to transfer data from among a variety of operating systems. The company believes this product will be quite successful and capitalized all of the costs of designing, developing, coding, and testing the software. These costs will be amortized over the expected useful life of the software on a straight-line basis.

Required:

1. Was Elegant correct in its treatment of the software development costs? Why?
2. Explain the appropriate method for determining the amount of periodic amortization for any capitalized software development costs.

Real World Case 10–16
Property, plant, and equipment; Home Depot
● LO10–1, LO10–7

Real World Financials

EDGAR, the Electronic Data Gathering, Analysis, and Retrieval system, performs automated collection, validation, indexing, and forwarding of submissions by companies and others who are required by law to file forms with the U.S. Securities and Exchange Commission (SEC). All publicly traded domestic companies use EDGAR to make the majority of their filings. (Some foreign companies file voluntarily.) Form 10-K, which includes the annual report, is required to be filed on EDGAR. The SEC makes this information available on the Internet.

Required:

1. Access EDGAR on the Internet. The web address is **www.sec.gov.**
2. Search for **Home Depot, Inc.** Access the 10-K filing for the most recent fiscal year. Search or scroll to find the financial statements and related notes.
3. Answer the following questions related to the company's property, plant, and equipment:
   a. Name the different types of assets the company lists in its balance sheet under property, plant, and equipment.
   b. How much cash was used for the acquisition of property, plant, and equipment during the year?
   c. What was the amount of interest capitalized during the year?
   d. Compute the fixed-asset turnover ratio for the fiscal year.

Data Analytics

connect

Data analytics is the process of examining data sets in order to draw conclusions about the information they contain. If you haven't completed any of the prior data analytics cases, follow the instructions listed in the Chapter 1 Data Analytics case to get set up. You will need to watch the videos referred to in the Chapters 1 - 3 Data Analytics cases. No additional videos are required for this case. All short training videos can be found here: **https://www.tableau.com/learn/training#getting-started.**

Data Analytics Case
Asset Turnover
● LO10–5

In the Chapter 3 Data Analytics Case, you applied Tableau to examine a data set and create calculations to compare two companies' profitability. For the case in this chapter, you continue in your role as an analyst conducting introductory research into the relative merits of investing in one or both of these companies. This time you assess the companies' fixed asset turnover ratios to determine which company utilizes its fixed assets most efficiently to generate sales. The fixed-asset turnover ratio is calculated as net sales divided by average fixed assets.

Required:

Use Tableau to calculate each company's fixed-asset turnover ratio in each year from 2012 to 2021. Based upon what you find, answer the following questions:

1. What is the fixed-asset turnover ratio for Big Store (a) in 2012 and (b) in 2021?
2. What is the fixed-asset turnover ratio for Discount Goods (a) in 2012 and (b) in 2021?
3. Comparing the two companies' fixed-asset turnover ratios over the ten-year period, which company exhibits the most favorable fixed asset turnover?
4. Comparing Big Store's fixed-asset turnover ratios over the ten-year period, is the company's turnover (a) generally increasing, (b) roughly the same, or (c) generally decreasing from year to year?

Resources:

You have available to you an extensive data set that includes detailed financial data for 2012–2021 for both Discount Goods and Big Store. The data set is in the form of four Excel files available to download from Connect, or under Student Resources within the Library tab. The file for use in this chapter is named "Discount_Goods_Big_Store_Financials.xlsx." Download this file and save it to the computer on which you will be using Tableau.

For this case, you will create a calculation to produce a bar chart of the fixed-asset turnover ratio that allows you to compare and contrast the two companies.

After you view the training videos, follow these steps to create the charts you'll use for this case:

- Open Tableau and connect to the Excel spreadsheet you downloaded.
- Click on the Sheet 1 tab, at the bottom of the canvas, to the right of the Data Source at the bottom of the screen. Drag "Company" and "Year" under Dimensions to the Rows shelf. Change "Year" to *discrete* by right-clicking and selecting "discrete."
- Drag the "Net sales" and "Average Fixed Assets" under Measures to the Rows shelf. Change each to *discrete.* Note: You must drag accounts into the Rows shelf if you are going to use them in calculated fields and need an *aggregated* sum. If this is not done first, your answer will not match the solutions.
- Under the Analysis tab, select Create Calculated Field. Create a measure named "Fixed Asset Turnover" by dragging "Net sales" from the Rows shelf to the Calculation Editor window, typing a division sign for division, and then dragging "Average Fixed Assets" from the Rows shelf beside it. Make sure the window says that the calculation is valid and click OK.
- Drag the newly created "Fixed Asset Turnover" to the Rows shelf.
- Right click the "Net sales" and "Average Fixed Assets" on the Rows shelf and unclick "Show header." This will hide these items from view but still allow them to be used in the formulas.
- Click on the "Show Me" tab in the upper right corner and select "side-by-side bars." Add labels to the bars by clicking on "Label" under the "Marks" card and clicking the box "Show mark label." Format the labels to Times New Roman, bold, black and 9-point font.
- Change the title of the sheet to be "Fixed Asset Turnover Ratio Bar Chart" by right-clicking and selecting "Edit title." Format the title to Times New Roman, bold, black and 15-point font. Change the title of "Sheet 1" to match the sheet title by right-clicking, selecting "Rename" and typing in the new title.
- Format the labels on the left of the sheet ("Fixed Asset Turnover") to Times New Roman, 9-point font, bold, and black.
- Format all other labels to be Times New Roman, bold, black and 10-point font.
- Once complete, save the file as "*DA10_Your initials*.twbx."

Continuing Cases

Target Case
● LO10–1, LO10–5

Target Corporation prepares its financial statements according to U.S. GAAP. Target's financial statements and disclosure notes for the year ended February 3, 2018, are available in Connect. This material is also available under the Investor Relations link at the company's website (**www.target.com**).

Required:

1. What amount ($ in millions) does Target report for net property and equipment for the year ended February 3, 2018? What is the largest category of property and equipment reported on the face of the balance sheet?
2. What amount ($ in millions) of cash was used in the fiscal year ended February 3, 2018, to purchase property and equipment? Is this an increase or decrease compared to the previous year?
3. Do you think a company like Target would have more research and development costs or more advertising costs? Explain.
4. What is Target's fixed-asset turnover ratio for the fiscal year ended February 3, 2018? What is the ratio intended to measure?
5. Does Target include any intangible assets in total assets (yes/no)? Hint: see Notes 15 and 16.

Air France–KLM Case

● LO10–9

IFRS

Air France–KLM (AF), a Franco-Dutch company, prepares its financial statements according to International Financial Reporting Standards. AF's financial statements and disclosure notes for the year ended December 31, 2017, are available in Connect. This material is also available under the Finance link at the company's website (www.airfranceklm.com).

Required:

1. What method does Air France–KLM use to amortize the cost of computer software development costs? How does this approach differ from U.S. GAAP?
2. AF does not report any research and development expenditures. If it did, its approach to accounting for research and development would be significantly different from U.S. GAAP. Describe the differences between IFRS and U.S. GAAP in accounting for research and development expenditures.
3. AF does not report the receipt of any governments grants. If it did, its approach to accounting for government grants would be significantly different from U.S. GAAP. Describe the differences between IFRS and U.S. GAAP in accounting for government grants. If AF received a grant for the purchase of assets, what alternative accounting treatments are available under IFRS?

CPA Exam Questions and Simulations

ROGER CPA Review

Sample CPA Exam questions from Roger CPA Review are available in Connect as support for the topics in this chapter. These multiple-choice questions and task-based simulations include expert-written explanations and solutions, and provide a starting point for students to become familiar with the content and functionality of the actual CPA Exam.

CHAPTER

11 Property, Plant, and Equipment and Intangible Assets: Utilization and Disposition

OVERVIEW

This chapter completes our discussion of accounting for property, plant, and equipment and intangible assets. We address the allocation of the cost of these assets to the periods benefited by their use.

The usefulness of most of these assets is consumed as the assets are applied to the production of goods or services. Cost allocation corresponding to this consumption of usefulness is known as *depreciation* for plant and equipment, *depletion* for natural resources, and *amortization* for intangibles.

We also consider other issues until final disposal such as impairment of these assets and the treatment of expenditures subsequent to acquisition.

LEARNING OBJECTIVES

After studying this chapter, you should be able to:

- **LO11–1** Explain the concept of cost allocation as it pertains to property, plant, and equipment and intangible assets. (*p. 571*)
- **LO11–2** Determine periodic depreciation using both time-based and activity-based methods and account for dispositions. (*p. 574*)
- **LO11–3** Calculate the periodic depletion of a natural resource. (*p. 589*)
- **LO11–4** Calculate the periodic amortization of an intangible asset. (*p. 591*)
- **LO11–5** Explain the appropriate accounting treatment required when a change is made in the service life or residual value of property, plant, and equipment and intangible assets. (*p. 596*)
- **LO11–6** Explain the appropriate accounting treatment required when a change in depreciation, amortization, or depletion method is made. (*p. 597*)
- **LO11–7** Explain the appropriate treatment required when an error in accounting for property, plant, and equipment and intangible assets is discovered. (*p. 598*)
- **LO11–8** Identify situations that involve a significant impairment of the value of property, plant, and equipment and intangible assets and describe the required accounting procedures. (*p. 600*)
- **LO11–9** Discuss the accounting treatment of repairs and maintenance, additions, improvements, and rearrangements to property, plant, and equipment and intangible assets. (*p. 593*)
- **LO11–10** Discuss the primary differences between U.S. GAAP and IFRS with respect to the utilization and impairment of property, plant, and equipment and intangible assets. (*pp. 580, 587, 590, 594, 603, 605, 607,* and *614*)

©Dragon Images/Shutterstock

FINANCIAL REPORTING CASE

What's in a Name?

"I don't understand this at all," your friend Penny Lane moaned. "Depreciation, depletion, amortization; what's the difference? Aren't they all the same thing?" Penny and you are part of a class team working on a case involving **Weyerhaeuser Company**, a large forest products company. Part of the project involves comparing reporting methods over a three-year period. "Look at these disclosure notes from last year's annual report. Besides mentioning those three terms, they also talk about asset impairment. How is that different?" Penny showed you the disclosure notes.

Property and Equipment and Timber and Timberlands (in part)

Depreciation is calculated using a straight-line method at rates based on estimated service lives. Logging roads are generally amortized—as timber is harvested—at rates based on the volume of timber estimated to be removed. We carry timber and timberlands at cost less depletion.

Depletion (in part)

To determine depletion rates, we divide the net carrying value by the related volume of timber estimated to be available over the growth cycle.

Impairment of Long-Lived Assets (in part)

We review long-lived assets—including certain identifiable intangibles—for impairment whenever events or changes in circumstances indicate that the carrying amount of the assets may not be recoverable.

QUESTIONS

By the time you finish this chapter, you should be able to respond appropriately to the questions posed in this case. Compare your response to the solution provided at the end of the chapter.

1. Is Penny correct? Do the terms *depreciation, depletion,* and *amortization* all mean the same thing? *(p. 572)*
2. Weyerhaeuser determines depletion based on the "volume of timber estimated to be available." Explain this approach. *(p. 577)*
3. Explain how asset impairment differs from depreciation, depletion, and amortization. How do companies measure impairment losses for property, plant, and equipment and intangible assets with finite useful lives? *(p. 600)*

Depreciation, Depletion, and Amortization

PART A

Cost Allocation—an Overview

Property, plant, and equipment and intangible assets are purchased with the expectation that they will provide future benefits. Specifically, they are acquired to be used as part of the revenue-generating operations, usually for several years. Logically, then, the cost of these acquisitions initially should be recorded as assets (as we saw in Chapter 10), and then these costs should be allocated to expense over the reporting periods benefited by their use. That is, their costs are reported with the revenues they help generate. ● LO11–1

Let's suppose that a company purchases a used truck for $8,200 to deliver products to customers. The company estimates that five years from the acquisition date the truck will be sold for $2,200. It is estimated, then, that **$6,000** ($8,200 – $2,200) of the truck's purchase cost will be used up (consumed) during a five-year useful life. The situation is portrayed in Illustration 11–1.

Illustration 11–1

Cost Allocation

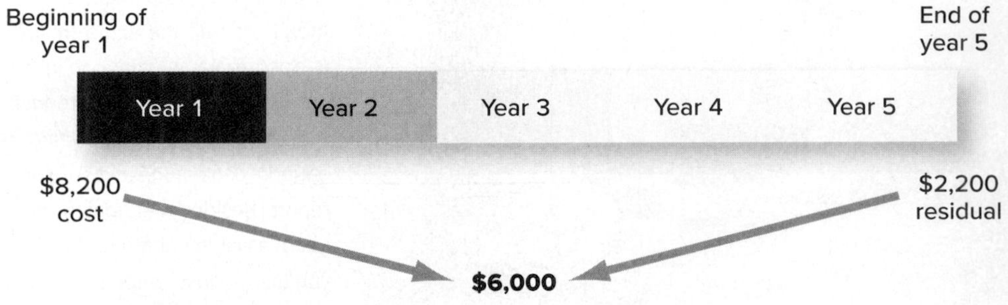

Because the truck will help to produce revenues over the next five years, an asset of $8,200 is recorded at the time of acquisition. Over the subsequent five years, **$6,000** of the truck's costs is expected to be consumed and, conceptually, should be allocated to expense in those years in direct proportion to the role the asset played in revenue production. However, very seldom is there a clear-cut relationship between the use of the asset and revenue production. In other words, we can't tell precisely the portion of the total benefits of the asset that was consumed in any particular period. As a consequence, we must resort to arbitrary allocation methods to approximate the portion of the asset's cost used each period. Contrast this situation with the $24,000 prepayment of one year's rent on an office building at $2,000 per month. In that case, we know precisely that the benefits of the asset (prepaid rent) are consumed at a rate of $2,000 per month. That's why we allocate $2,000 of prepaid rent to rent expense for each month that passes.

The process of allocating the cost of plant and equipment over the periods they are used to produce revenues is known as **depreciation**. The process of depreciation often is confused with measuring a decline in fair value of an asset. For example, let's say our delivery truck purchased for $8,200 can be sold for $5,000 at the end of one year but we intend to keep it for the full five-year estimated life. It has experienced a decline in value of $3,200 ($8,200 – $5,000). However, *depreciation is a process of cost allocation, not valuation.* We would not record depreciation expense of $3,200 for year one of the truck's life. Instead, we would distribute the cost of the asset, less any anticipated residual value, over the estimated useful life in a systematic and rational manner that attempts to associate revenues with the *use* of the asset, not the decline in its value. After all, the truck is purchased to be used in operations, not to be sold.

FINANCIAL Reporting Case

Q1, p. 571

Depreciation, depletion, and amortization are processes that allocate an asset's cost to periods of benefit.

For natural resources, we refer to cost allocation as **depletion**, and for intangible assets, we refer to it as **amortization**. While the terms *depreciation, depletion,* and *amortization* differ across types of assets, they conceptually refer to the same idea—the process of allocating an asset's cost over the periods it is used to produce revenues.

For assets used in the manufacture of a product, depreciation, depletion, or amortization is considered a product cost to be included as part of the cost of inventory. Eventually, when the product is sold, it becomes part of the cost of goods sold. For assets *not* used in production, primarily plant and equipment and certain intangibles used in the selling and administrative functions of the company, depreciation and amortization are reported as period expenses in the income statement. You might recognize this distinction between a product cost and a period cost. A product cost is reported as an expense (cost of goods sold) when the related product is sold; a period cost is reported as an expense in the reporting period in which it is incurred.

Measuring Cost Allocation

The process of cost allocation requires that three factors be established at the time the asset is put into use. These factors are

1. **Service life**—The estimated use that the company expects to receive from the asset.

2. **Allocation base**—The cost of the asset expected to be consumed during its service life.
3. **Allocation method**—The pattern in which the allocation base is expected to be consumed.

Let's consider these one at a time.

Service Life

The **service life**, or **useful life**, is the amount of use that the company expects to obtain from the asset before disposing of it. This use can be expressed in units of time or in units of activity. For example, the estimated service life of a delivery truck could be expressed in terms of years or in terms of the number of miles that the company expects the truck to be driven before disposition. We use the terms service life and useful life interchangeably throughout the chapter.

The *service life*, or *useful life*, can be expressed in units of time or in units of activity.

Physical life provides the upper bound for service life of tangible, long-lived assets. Physical life will vary according to the purpose for which the asset is acquired and the environment in which it is operated. For example, a diesel powered electric generator may last for many years if it is used only as an emergency backup or for only a few years if it is used regularly.

The service life of a tangible asset may be less than physical life for a variety of reasons. For example, the expected rate of technological change may shorten service life. If suppliers are expected to develop new technologies that are more efficient, the company may keep an asset for a period of time much shorter than physical life. Likewise, if the company sells its product in a market that frequently demands new products, the machinery and equipment used to produce products may be useful only for as long as its output can be sold. Similarly, a mineral deposit might be projected to contain 4 million tons of a mineral, but it may be economically feasible with existing extraction methods to mine only 2 million tons.

Expected obsolescence can shorten service life below physical life.

For intangible assets, legal or contractual life often is a limiting factor. For instance, a patent might be capable of providing enhanced profitability for 50 years, but the legal life of a patent is only 20 years.

Management intent also may shorten the period of an asset's usefulness below its physical, legal, or contractual life. For example, a company may have a policy of using its delivery trucks for a three-year period and then trading the trucks for new models.

Companies quite often disclose the range of service lives for different categories of assets. For example, Illustration 11–2 shows how **IBM Corporation** disclosed its service lives in a note accompanying recent financial statements.

Summary of Significant Accounting Policies (in part)
Depreciation and Amortization
The estimated useful lives of certain depreciable assets are as follows: buildings, 30 to 50 years; building equipment, 10 to 20 years; land improvements, 20 years; plant, laboratory, and office equipment, 2 to 20 years; and computer equipment, 1.5 to 5 years.

Source: International Business Machines Corporation

Illustration 11–2
Service Life Disclosure—International Business Machines Corporation

Real World Financials

Allocation Base

The **allocation base** is the amount of cost to be allocated over an asset's service life. The amount is the difference between the asset's capitalized cost at the date placed in service and the asset's estimated **residual value**. Residual value (sometimes called *salvage value*) is the amount the company expects to receive for the asset at the end of its service life, less any anticipated disposal costs.

Allocation base is the difference between the cost of the asset and its estimated *residual value*.

For plant and equipment, we commonly refer to the allocation base as the *depreciable base*. In our delivery truck example above, the depreciable base is $6,000 ($8,200 cost less $2,200 estimated residual value). We will allocate a portion of the $6,000 to each year of the truck's service life. For the depletion of natural resources, we refer to the allocation base as the *depletion base*. For amortization of an intangible asset, we refer to the allocation base as the *amortization base*.

In certain situations, residual value can be estimated by referring to a company's prior experience or to publicly available information concerning resale values of various types of assets. For example, if a company intends to trade its delivery truck in three years for the new model, approximations of the three-year residual value for that type of truck can be obtained from used truck values.

However, estimating residual value for many assets can be very difficult due to the uncertainty about the future. For this reason, along with the fact that residual values often are immaterial, many companies simply assume a residual value of zero. Companies usually do not disclose estimated residual values.

Allocation Method

The *allocation method* used should be systematic and rational and correspond to the pattern of asset use.

In determining how much cost to allocate to periods of an asset's use, a method should be selected that corresponds to the pattern of benefits received from the asset's use. Generally accepted accounting principles state that the chosen method should allocate the asset's cost "as equitably as possible to the periods during which services are obtained from [its] use." GAAP further specifies that the method should produce a cost allocation in a "systematic and rational manner."[1] The objective is to try to allocate cost to the period in an amount that is proportional to the amount of benefits generated by the asset during the period relative to the total benefits provided by the asset during its life.

In practice, there are two general approaches that attempt to obtain this systematic and rational allocation. The first approach allocates the cost base according to the *passage of time.* Methods following this approach are referred to as **time-based methods**. The second approach allocates an asset's cost base using a measure of the asset's *input* or *output.* This is an **activity-based method**. We compare these approaches first in the context of depreciation. Later we see that depletion of natural resources typically follows an activity-based approach, while the amortization of intangibles typically follows a time-based approach.

Depreciation

● LO11–2

To demonstrate and compare the most common depreciation methods, we refer to the situation described in Illustration 11–3.

Illustration 11–3
Depreciation Methods

At the beginning of Year 1, Hogan Manufacturing Company purchased a machine for $250,000 and made the following estimates at that time:

- Estimated residual value of $40,000.
- Estimated service life of 5 years.
- Estimated production of 140,000 units.

Actual production during the five years of the asset's life was as follows:

Year	Units Produced
1	20,000
2	32,000
3	44,000
4	28,000
5	26,000
Total	150,000

Time-Based Depreciation Methods

The *straight-line method* depreciates an equal amount of the depreciable base to each year of the asset's service life.

STRAIGHT-LINE METHOD By far the most easily understood and widely used technique for calculating depreciation is the **straight-line method**. In this approach, an equal amount of the depreciable base (or allocation base) is allocated to each year of the asset's service life. The depreciable base is simply divided by the number of years in the asset's life to

[1]Source: FASB ASC 360-10-35-4: Property, Plant, and Equipment-Overall-Subsequent Measurement (previously "Restatement and Revision of Accounting Research Bulletins," *Accounting Research Bulletin* No. 43 (New York: AICPA, 1953), Ch.9).

determine annual depreciation. Using the information given in Illustration 11–3, straight-line depreciation expense in each year is $42,000, calculated as follows:

$$\frac{\$250{,}000 - \$40{,}000}{5 \text{ years}} = \$42{,}000 \text{ per year}$$

The calculation of depreciation over the entire five-year life is demonstrated in detail in Illustration 11–3A. Notice the last three columns. **Depreciation expense** is the portion of the asset's cost that is allocated to an expense *in the current year.* **Accumulated depreciation** (a contra-asset account) represents the cumulative amount of the asset's cost that has been depreciated *in all prior years including the current year.* This amount represents the reduction in the asset's cost reported in the balance sheet. The asset is reported in the balance sheet at its **book value,** which is the asset's cost minus accumulated depreciation. Book value is sometimes called *carrying value* or carrying amount. The residual value ($40,000 in this example) does not affect the calculation of book value, but the residual value does set a limit on which book value cannot go below.

Illustration 11–3A Straight-Line Depreciation

Using the information given in Illustration 11–3:

Year	Depreciable Base ($250,000 – $40,000)	×	Depreciation Rate per Year	=	Depreciation Expense	Accumulated Depreciation	Book Value End of Year ($250,000 less Accum. Depreciation)
1	$210,000		⅕*		$ 42,000	$ 42,000	$208,000
2	210,000		⅕		42,000	84,000	166,000
3	210,000		⅕		42,000	126,000	124,000
4	210,000		⅕		42,000	168,000	82,000
5	210,000		⅕		42,000	210,000	40,000
Totals					$210,000		

*The rate equals 1 divided by the asset's 5-year estimated service life (⅕ = 20%).

The entry to record depreciation at the end of each year using the straight-line method would be:

Depreciation expense	42,000*	
Accumulated depreciation		42,000

*$42,000 = ($250,000 – $40,000) ÷ 5years.

Accelerated depreciation methods report higher depreciation in earlier years.

ACCELERATED METHODS Using the straight-line method implicitly assumes that the benefits derived from the use of the asset are the same each year. In some situations it might be more appropriate to assume that the asset will provide greater benefits in the early years of its life than in the later years. In these cases, a more appropriate matching of depreciation with revenues is achieved with a declining pattern of depreciation, with higher depreciation in the early years of the asset's life and lower depreciation in later years.

An accelerated depreciation method also would be appropriate when benefits derived from the asset are approximately equal over the asset's life, but repair and maintenance costs increase significantly in later years. The early years incur higher depreciation and lower repairs and maintenance expense, while the later years have lower depreciation and higher repairs and maintenance. One way to achieve such a declining pattern of depreciation is a declining-balance method, which we discuss next.

Declining-balance depreciation methods multiply beginning-of-year book value, not depreciable base, by an annual rate that is a multiple of the straight-line rate.

Declining-balance methods. One way to calculate higher depreciation in the early years of the asset's life is to multiply the asset's book value (cost less accumulated depreciation) at the beginning of the year by a constant percentage rate. Annual depreciation reduces book value each year. Therefore, multiplying a declining book value by a constant rate results in declining depreciation expense in each successive year.

Perhaps the most common declining-balance method is known as the **double-declining-balance (DDB) method**. Under this method, we multiply the asset's beginning-of-year book value by *double* (or 200%) the straight-line rate. For example, in our previous example using a five-year asset, the straight-line rate was 20% (= $^1/_5$). For the double-declining-balance method, we double that rate and use 40%. Various other multiples are used in practice, such as 125% or 150% of the straight-line rate.

Depreciation using the double-declining-balance method is calculated in Illustration 11–3B for the five years of the machine's life. Notice that book value at the beginning of the year, rather than the depreciable base, is used as the starting point. As book value decreases each year, so does depreciation. Further, notice that in year 4 we did not multiply $54,000 by 40%. If we had, annual depreciation would have been $21,600. This amount would have resulted in accumulated depreciation by the end of year 4 of $217,600 and book value of $32,400, which is below the asset's expected residual value of $40,000. Therefore, we instead use a plug amount that reduces book value to the expected residual value (book value beginning of year, $54,000, minus expected residual value, $40,000 = **$14,000**). This also means there is no depreciation in year 5 since book value has already been reduced to the expected residual value. Declining balance methods often allocate the asset's depreciable base over fewer years than the expected service life.

Illustration 11–3B Double-Declining-Balance Depreciation

Using the information given in Illustration 11–3:

Year	Book Value Beginning of Year	×	Depreciation Rate per Year	=	Depreciation Expense	Accumulated Depreciation	Book Value End of Year ($250,000 less Accum. Depreciation)
1	$250,000		40%*		$100,000	$100,000	$150,000
2	150,000		40%		60,000	160,000	90,000
3	90,000		40%		36,000	196,000	54,000
4	54,000				**14,000**†	210,000	40,000
5	40,000				0	210,000	40,000
Total					$210,000		

*Double the straight-line rate of 20%. The straight-line rate is 1 divided by the asset's 5-year estimated service life (⅕ = 20%).
†Amount necessary to reduce book value to residual value.

The *SYD method* multiplies depreciable base by a declining fraction.

Sum-of-the-years'-digits method. Another accelerated depreciation pattern can be achieved by multiplying the depreciable base by a declining fraction. One such method is the **sum-of-the-years'-digits (SYD) method**. This method is seldom used in practice. The denominator of the fraction remains constant and is the *sum of the digits* from one to *n*, where *n* is the number of years in the asset's service life. For example, if there are five years in the service life, the denominator is the sum of 1, 2, 3, 4, and 5, which equals 15.[2] The numerator decreases each year; it begins with the value of *n* in the first year and decreases by one each year until it equals one in the final year of the asset's estimated service life. The annual fractions for an asset with a five-year life are: $^5/_{15}$, $^4/_{15}$, $^3/_{15}$, $^2/_{15}$, and $^1/_{15}$. We calculate depreciation for the five years of the machine's life using the sum-of-the-years'-digits method in Illustration 11–3C.

Notice that total depreciation ($210,000) is the same for accelerated methods like DDB and SYD as it is for the straight-line method, as shown in Illustration 11–3A. The difference is the pattern in which this total cost is allocated to each year of the asset's service life.

[2]A formula useful when calculating the denominator is $n\ (n + 1)/2$.

Illustration 11–3C Sum-of-the-Years'-Digits Depreciation

Using the information given in Illustration 11–3:

Year	Depreciable Base ($250,000 – $40,000)	×	Depreciation Rate per Year	=	Depreciation Expense	Accumulated Depreciation	Book Value End of Year ($250,000 less Accum. Depreciation)
1	$210,000		5/15*		$ 70,000	$ 70,000	$180,000
2	210,000		4/15		56,000	126,000	124,000
3	210,000		3/15		42,000	168,000	82,000
4	210,000		2/15		28,000	196,000	54,000
5	210,000		1/15		14,000	210,000	40,000
Totals			15/15		$210,000		

$^{*}\frac{n(n+1)}{2} = \frac{5(5+1)}{2} = 15$

SWITCH FROM ACCELERATED TO STRAIGHT-LINE As a planned approach to depreciation, many companies have a formal policy to use accelerated depreciation for approximately the first half of an asset's service life and then switch to the straight-line method for the remaining life of the asset.

In Illustration 11–3B, the company would switch to the straight-line method in either year 3 or year 4. Assuming the switch is made at the beginning of year 4, and the book value at the beginning of that year is $54,000, an additional $14,000 ($54,000 – $40,000 in residual value) of depreciation must be recorded over the remaining life of the asset. Applying the straight-line concept, $7,000 ($14,000 divided by two remaining years) in depreciation is recorded in both year 4 and year 5.

It should be noted that this switch to straight-line is not a change in depreciation method. The switch is part of the company's planned depreciation approach. However, as you will learn later in the chapter, the accounting treatment is the same as a change in depreciation method.

Activity-Based Depreciation Methods

Activity-based depreciation methods estimate service life in terms of some measure of productivity.

The most logical way to allocate an asset's cost to periods of an asset's use is to measure the usefulness of the asset in terms of its productivity. For example, we could measure the service life of a machine in terms of its *output* (such as the estimated number of units it will produce) or in terms of its *input* (such as the number of hours it will operate). We have already mentioned that one way to measure the service life of a vehicle is to estimate the number of miles it will be driven. The most common activity-based method is called the **units-of-production method**.

FINANCIAL Reporting Case

Q2, p. 571

The *units-of-production method* computes a depreciation rate per measure of activity and then multiplies this rate by actual activity to determine periodic depreciation.

The measure of output used is the estimated number of units (pounds, items, barrels, etc.) to be produced by the machine. By the units-of-production method, we first compute the average depreciation rate per unit by dividing the depreciable base by the number of units expected to be produced. This per unit rate is then multiplied by the actual number of units produced each period. In our illustration, the depreciation rate per unit is $1.50, computed as follows:

$$\frac{\$250{,}000 - \$40{,}000}{140{,}000 \text{ units}} = \$1.50 \text{ per unit}$$

Each unit produced will require $1.50 of depreciation to be recorded. In other words, each unit produced is assigned $1.50 of the asset's cost.

Illustration 11–3D shows that depreciation each year is the actual units produced multiplied by the depreciation rate per unit. This means that the amount of depreciation each year varies proportionately with the number of units being produced, with one exception. Notice that the asset produced 26,000 units in year 5, causing total production over the life of the asset (150,000 units) to exceed its estimated production (140,000 units). In this case, we cannot record depreciation for the final 10,000 units produced. Depreciation in year five is limited to the amount that brings the book value of the asset down to its residual value (book value beginning of year, $64,000, minus expected residual value, $40,000, equals **$24,000**).

Illustration 11–3D

Units-of-Production Depreciation

Using the information given in Illustration 11–3:

Year	Units Produced	×	Depreciation Rate per Unit	=	Depreciation Expense	Accumulated Depreciation	Book Value End of Year ($250,000 less Accum. Depreciation)
1	20,000		$1.50*		$ 30,000	$ 30,000	$220,000
2	32,000		1.50		48,000	78,000	172,000
3	44,000		1.50		66,000	144,000	106,000
4	28,000		1.50		42,000	186,000	64,000
5	26,000				**24,000**†	210,000	40,000
Totals	150,000				$210,000		

*($250,000 – $40,000) / 140,000 units = $1.50 per unit.
†Amount necessary to reduce book value to residual value.

The machine may produce *fewer than* 140,000 units by the end of its useful life. For example, suppose production in year 5 had been only 6,000 units, bringing total production to 130,000 units, and management has no future plans to use the machine. We would record depreciation in year 5 for $9,000 (6,000 units × $1.50). If management then develops a formal plan to sell the machine, the machine is classified as "held for sale" (discussed in more detail below) and reported at the lower of its current book value or its fair value less any cost to sell. If management plans to retire the asset without selling it, a loss is recorded for the remaining book value.

Decision Makers' Perspective—Selecting A Depreciation Method

All methods provide the same total depreciation over an asset's life.

Illustration 11–3E compares periodic depreciation calculated using each of the alternatives we discussed and illustrated.

Illustration 11–3E

Comparison of Various Depreciation Methods

Year	Straight-Line	Double-Declining-Balance	Sum-of-the-Years'-Digits	Units of Production
1	$ 42,000	$100,000	$ 70,000	$ 30,000
2	42,000	60,000	56,000	48,000
3	42,000	36,000	42,000	66,000
4	42,000	14,000	28,000	42,000
5	42,000	0	14,000	24,000
Total	$210,000	$210,000	$210,000	$210,000

Activity-based methods are conceptually superior to time-based methods but often are impractical to apply in practice.

Conceptually, using an activity-based depreciation method provides a better matching of the asset's cost to the use of that asset to help produce revenues. Clearly, the productivity of an asset is more closely associated with the benefits provided by that asset than the mere passage of time. Also, these methods allow for patterns of depreciation to correspond with the patterns of asset use.

However, activity-based methods quite often are either infeasible or too costly to use. For example, buildings don't have an identifiable measure of productivity. Even for machinery, there may be an identifiable measure of productivity such as machine hours or units produced, but it frequently is more costly to determine each period than it is to simply measure the passage of time. For these reasons, most companies use time-based depreciation methods.

Illustration 11–4 shows the results of a recent survey of depreciation methods used by large public companies.[3]

Illustration 11–4
Use of Various Depreciation Methods

Real World Financials

Depreciation Method	Number of Companies
Straight-line	490
Declining-balance	9
Sum-of-the-years'-digits	2
Accelerated method—not specified	9
Units of production	12
Group/composite	17

Why do so many companies use the straight-line method as opposed to other time-based methods? Many companies perhaps consider the benefits derived from the majority of depreciable assets to be realized approximately evenly over these assets' useful lives. Certainly a contributing factor is that straight-line is the easiest method to understand and apply.

Another motivation is the positive effect on reported income. Straight-line depreciation produces a higher net income than accelerated methods in the early years of an asset's life. In Chapter 8, we pointed out that reported net income can affect bonuses paid to management or debt agreements with lenders.

A company does not have to use the same depreciation method for both financial reporting and income tax purposes.

Conflicting with the desire to report higher profits is the desire to reduce taxes by reducing taxable income. An accelerated method serves this objective by reducing taxable income more in the early years of an asset's life than straight-line. You probably recall a similar discussion from Chapter 8 in which the benefits of using the LIFO inventory method during periods of increasing costs were described. However, remember that the LIFO conformity rule requires companies using LIFO for income tax reporting to also use LIFO for financial reporting. *No such conformity rule exists for depreciation methods.* Income tax regulations allow firms to use different approaches to computing depreciation in their tax returns and in their financial statements. The method used for tax purposes is therefore not a constraint in the choice of depreciation methods for financial reporting. As a result, most companies use the straight-line method for financial reporting and an accelerated method for tax reporting (discussed in Appendix 11A). For example, Illustration 11–5 shows **Merck & Co.**'s depreciation policy as reported in a disclosure note accompanying recent financial statements.

Illustration 11–5
Depreciation Method Disclosure—Merck & Co.

Real World Financials

Summary of Accounting Policies (in part):
Depreciation
Depreciation is provided over the estimated useful lives of the assets, principally using the straight-line method. For tax purposes, accelerated methods are used.

Source: Merck & Co.

It is not unusual for a company to use different depreciation methods for different classes of assets. For example, Illustration 11–6 illustrates the **International Paper Company** depreciation policy disclosure contained in a note accompanying recent financial statements.

Illustration 11–6
Depreciation Method Disclosure—International Paper Company

Real World Financials

Summary of Accounting Policies (in part):
Plants, Properties, and Equipment
Plants, properties, and equipment are stated at cost, less accumulated depreciation. The units-of-production method of depreciation is used for major pulp and paper mills and the straight-line method is used for other plants and equipment.

Source: International Paper Company

[3] *U.S. GAAP Financial Statements—Best Practices in Presentation and Disclosure—2013* (New York: AICPA, 2013).

International Financial Reporting Standards

LO11–10

Depreciation. *IAS No. 16* requires that each component of an item of property, plant, and equipment must be depreciated separately if its cost is significant in relation to the total cost of the item.[4] In the United States, component depreciation is allowed but is not often used in practice.

Consider the following illustration:

Cavandish LTD. purchased a delivery truck for $62,000. The truck is expected to have a service life of six years and a residual value of $12,000. At the end of three years, the oversized tires, which have a cost of $6,000 (included in the $62,000 purchase price), will be replaced.

Under U.S. GAAP, the typical accounting treatment is to depreciate the $50,000 ($62,000 – $12,000) depreciable base of the truck over its six-year useful life. Using IFRS, the depreciable base of the truck is $44,000 ($62,000 – $12,000 – $6,000) and is depreciated over the truck's six-year useful life, and the $6,000 cost of the tires is depreciated separately over a three-year useful life.

U.S. GAAP and IFRS determine depreciable base in the same way, by subtracting estimated residual value from cost. However, IFRS requires a review of residual values at least annually.

Sanofi-Aventis, a French pharmaceutical company, prepares its financial statements using IFRS. In its property, plant, and equipment note, the company discloses its use of the component-based approach to accounting for depreciation.

Property, plant, and equipment (in part)

The component-based approach to accounting for property, plant, and equipment is applied. Under this approach, each component of an item of property, plant, and equipment with a cost which is significant in relation to the total cost of the item and which has a different useful life from the other components must be depreciated separately.

Depreciation Methods. *IAS No. 16* specifically mentions three depreciation methods: straight-line, units-of-production, and the diminishing balance method. The diminishing balance method is similar to the declining balance method sometimes used by U.S. companies. As in the United States, the straight-line method is used by most companies. A recent survey of large companies that prepare their financial statement according to IFRS reports 93% of the surveyed companies used the straight-line method.[5]

Concept Review Exercise

DEPRECIATION METHODS

The Sprague Company purchased a fabricating machine on January 1, 2021, at a net cost of $130,000. At the end of its four-year useful life, the company estimates that the machine will be worth $30,000. Sprague also estimates that the machine will run for 25,000 hours during its four-year life. The company's fiscal year ends on December 31.

Required:

Compute depreciation expense for 2021 through 2024 using each of the following methods:

1. Straight-line
2. Double-declining-balance
3. Sum-of-the-years'-digits
4. Units of production (using machine hours); actual production was as follows:

Year	Machine Hours
2021	6,000
2022	8,000
2023	5,000
2024	7,000

[4]"Property, Plant and Equipment," *International Accounting Standard No. 16* (IASCF), par. 42, as amended effective January 1, 2018.

[5]" *IFRS Accounting Trends and Techniques"* (New York, AICPA, 2011), p. 328.

Solution:

1. Straight-line:

$$\frac{\$130,000 - \$30,000}{4 \text{ years}} = \$25,000 \text{ per year}$$

2. Double-declining balance:

Year	Book Value Beginning of Year	×	Depreciation Rate per Year	=	Depreciation Expense	Book Value End of Year
2021	$130,000		50%*		$ 65,000	$65,000
2022	65,000		50%		32,500	32,500
2023	32,500				2,500†	30,000
2024	30,000				0	30,000
Total					$ 100,000	

*Double the straight-line rate of 25%. The straight-line rate is 1 divided by the asset's 4-year estimated service life (¼ = 25%).
†Amount necessary to reduce book value to residual value.

3. Sum-of-the-years'-digits:

Year	Depreciable Base	×	Depreciation Rate per Year	=	Depreciation Expense
2021	$ 100,000		4/10		$ 40,000
2022	100,000		3/10		30,000
2023	100,000		2/10		20,000
2024	100,000		1/10		10,000
Total					$100,000

4. Units of production (using machine hours):

Year	Machine Hours	×	Depreciation Rate per Hour	=	Depreciation Expense	Book Value End of Year
2021	6,000		$4*		$ 24,000	$106,000
2022	8,000		4		32,000	74,000
2023	5,000		4		20,000	54,000
2024	7,000				24,000†	30,000
Total					$ 100,000	

*($130,000 – $30,000)/25,000 hours = $4 per hour.
†Amount necessary to reduce book value to residual value.

Partial Period Depreciation

When acquisition and disposal of property, plant, and equipment occur at times other than the very beginning or very end of a company's fiscal year, a company theoretically must determine how much depreciation to record for the part of the year that each asset actually is used.

Let's repeat the Hogan Manufacturing Company illustration used earlier in Illustration 11–3 but modify it in Illustration 11–7 to assume that the asset was acquired *during* the company's fiscal year.

Illustration 11–7

Depreciation Methods—Partial Year

On April 1, 2021, Hogan Manufacturing Company purchased a machine for $250,000 and made the following estimates at that time:

- Estimated residual value of $40,000.
- Estimated service life of five years.
- Estimated production of 140,000 units.

(continued)

Illustration 11–7
(Concluded)

The company has a December 31 year-end. Actual production during the five years of the asset's life was as follows:

Year	Units Produced*
2021 (beginning April 1)	16,000
2022	30,000
2023	40,000
2024	32,000
2025	24,000
2026 (ending April 1)	8,000
Total	150,000

*The units produced in each year do not correspond to Years 1-5 in Illustration 11–3 because the asset was not purchased at the beginning of the first year. In both illustrations, total production occurs over five years for the same total amount of 150,000 units.

Depreciation per year of the asset's life calculated earlier in Illustration 11–3A, Illustration 11–3B, and Illustration 11–3D is summarized in Illustration 11–7A.

Illustration 11–7A
Yearly Depreciation

Year	Straight-Line	Double-Declining-Balance	Units-of-Production
1	$ 42,000	$ 100,000	$ 30,000
2	42,000	60,000	48,000
3	42,000	36,000	66,000
4	42,000	14,000	42,000
5	42,000	0	24,000
Total	**$210,000**	**$210,000**	**$210,000**

Illustration 11–7B shows how Hogan would depreciate the machinery by these three methods assuming an April 1 acquisition date.

Illustration 11–7B Partial-Year Depreciation

Year	Straight-Line		Double-Declining-Balance		Units-of-Production	
2021	$42,000 × ¾ =	$ 31,500	$250,000 × ⅖ × ¾ =	$ 75,000	16,000 units × $1.50^ =	$ 24,000
2022		42,000	$175,000 × ⅖ =	70,000*		45,000
2023		42,000	$105,000 × ⅖ =	42,000		60,000
2024		42,000		23,000†		48,000
2025		42,000		0		33,000†
2026	$42,000 × ¼ =	10,500		0		0
	Totals	**$210,000**		**$210,000**		**$210,000**

*Book value at the beginning of the year times double the straight-line rate: ($250,000 – $75,000) × ⅖ = $70,000.
^($250,000 – $40,000) / 140,000 units = $1.50 per unit. Depreciation each year equals actual units produced times $1.50.
†Amount necessary to reduce book value to residual value.

The first thing to notice is that even with partial-year depreciation, the total depreciation over the asset's total life ($210,000) is the same whether the asset is purchased at the beginning of a year or during the year. For straight-line, depreciation is ¾ of a full year's depreciation for the first year of the asset's life, because the asset was used only nine months, or ¾ of the year. The asset is depreciated for one full year in the following four years, and then the remaining ¼ of the asset's five-year useful life is depreciated in the final year.

For double-declining-balance, recall that depreciation equals the asset's book value at the beginning of the year times double the straight-line rate. In the first year, depreciation is ¾ of the full year's depreciation. After that, we calculate depreciation for a full year using the updated beginning-of-year book value. We continue each year until a final plug amount is needed to reduce book value to residual value. That final plug was reached in 2024 so no further depreciation is recorded over the remaining life of the asset (2025 and 2026).

For units-of-production, depreciation is not calculated based on time, so the portion of the year the asset is in operation does not matter. Instead, depreciation is calculated as the estimated depreciation rate per unit of output multiplied by the actual output for the year. Once the full allocation base has been depreciated, no further depreciation is recorded. For the example in Illustration 11–7B, the full allocation based was reached in 2025 (when actual production reached 140,000 units), so any further production does not affect depreciation.

Sometimes companies adopt a simplifying assumption, or convention, for computing partial year's depreciation and use it consistently. A common convention is to record one-half of a full year's depreciation in the year of acquisition and another half year in the year of disposal. This is known as the **half-year convention.**[6]

Dispositions

After using property, plant, and equipment, companies will sell or retire those assets. When selling property, plant, and equipment for monetary consideration (cash or a receivable), the seller recognizes a **gain** or **loss** for the difference between the consideration received and the book value of the asset sold.

A gain or loss is recognized for the difference between the consideration received and the asset's book value.

Selling price (consideration received)		$ xxx
Less: Book value of asset sold		
Original cost	$ xxx	
Accumulated depreciation	(xxx)	(xxx)
Gain/loss on sale of asset		$ xxx

We'll demonstrate this calculation next in Illustration 11–7C by modifying our earlier straight-line method example of Hogan Manufacturing Company in Illustration 11–3A. In Illustration 11–7C, Hogan sells a machine before the end of its service life and receives more cash than the asset's book value (cost minus accumulated depreciation) at the time of the sale. This causes a gain to be recognized.

Illustration 11–7C

Sale of Property, Plant, and Equipment

On January 1, 2021, Hogan Manufacturing Company purchased a machine for $250,000. The company expects the service life of the machine to be five years. The estimated residual value is $40,000. Hogan uses the straight-line depreciation method for machines recorded in the equipment account.

Suppose Hogan decides not to hold the machine for the expected five years but instead sells it on December 31, 2023 (three years later), for $140,000. We first need to update depreciation to the date of sale. Since depreciation for 2021 and 2022 has already been recorded in those years, we need to update depreciation only for the current year, 2023.

The entry to update depreciation for 2023:

Depreciation expense	42,000*	
Accumulated depreciation		**42,000**

*$42,000 = ($250,000 − $40,000) ÷ 5 years. See also Illustration 11–3A.

The balance of accumulated depreciation equals depreciation that has already been recorded in 2021 and 2022 ($42,000 + $42,000) plus the depreciation recorded above in 2023 ($42,000).

Accumulated Depreciation

Debit	Credit	Year
	42,000	2021
	42,000	2022
	42,000	2023
	126,000	

(continued)

[6]Another common method is the modified half-year convention. This method records a full year's depreciation when the asset is acquired in the first half of the year or sold in the second half. No depreciation is recorded if the asset is acquired in the second half of the year or sold in the first half. These half-year conventions are simple and, in most cases, will not result in material differences from a more precise calculation.

Illustration 11–7C
(Concluded)

We can now calculate the gain or loss on the sale as the difference between the selling price and the asset's book value. In this example, the amount of cash received is greater than the asset's book value, so a gain is recognized.

Selling price (cash received)		$ 140,000
Less: Book value of asset sold		
Original cost	$ 250,000	
Accumulated depreciation	**(126,000)**	(124,000)
Gain on sale of equipment (machine)		**$ 16,000**

Finally, the sale of the equipment requires that we do the following:

1. Record the cash received.
2. Remove the book value of the asset sold with a credit to the asset account and a debit to its accumulated depreciation.
3. Record the gain or loss.

The entry to record the sale on December 31, 2023, for $140,000:

Cash	140,000	
Accumulated depreciation (account balance)	**126,000**	
Equipment (account balance)		250,000
Gain on sale of equipment (selling price – book value)		**16,000**

The balances of the equipment account and the accumulated depreciation account for the machine will be $0 after this entry. The gain on the sale normally is reported in the income statement as a separate component of operating income.

Now, assume that Hogan sold the machine on December 31, 2023, for only $110,000. This amount is less than book value by **$14,000** and a loss on the sale would be recorded.

Selling price (cash received)		$ 110,000
Less: Book value of asset sold		
Original cost	$ 250,000	
Accumulated depreciation	**(126,000)**	(124,000)
Loss on sale of equipment (machine)		**$ (14,000)**

The entry to record the sale on December 31, 2023, for $110,000:

Cash	110,000	
Accumulated depreciation (account balance)	**126,000**	
Loss on sale of equipment (selling price – book value)	**14,000**	
Equipment (account balance)		250,000

Notice that the amounts of the equipment and accumulated depreciation removed from the books upon sale of the asset do not depend on whether a gain or loss is recorded; the asset's book value is written off completely. It's the amount of cash received relative to the asset's book value that determines the amount of the gain or loss.

Decision Makers' Perspective—Understanding gains and losses

It's tempting to think of a "gain" and "loss" on the sale of a depreciable asset as "good" and "bad" news. For example, we commonly use the term "gain" in everyday language to mean we sold something for more than we bought it. Gain could also be misinterpreted to mean the asset was sold for more than its fair value (we got a "good deal"). However, neither of these represents the meaning of a gain on the sale of assets. Refer back to our example in Illustration 11–7C. The sale of the machine resulted in a gain, but the machine was sold for *less than* its original cost, and there is no indication that Hogan sold the machine for more than its fair value.

A gain on the sale of a depreciable asset simply means the asset was sold for more than its book value. In other words, the asset being received and recorded (such as cash) is greater than the recorded book value of the asset being sold and written off. The net increase in the book value of total assets is an accounting gain (not an economic gain).

The same is true for losses. A loss signifies that the cash received is less than the book value of the asset being sold; there is a net decrease in the book value of total assets ●

ASSETS HELD FOR SALE Sometimes management plans to sell property, plant, and equipment or an intangible asset but that sale hasn't yet happened. In this case, the asset is classified as "held for sale" in the period in which all of the following criteria are met:[7]

Property, plant, and equipment or an intangible asset to be disposed of by sale is classified as held for sale and measured at the lower of the asset's book value or the asset's fair value less cost to sell.

1. Management commits to a plan to sell the asset.
2. The asset is available for immediate sale in its present condition.
3. An active plan to locate a buyer and sell the asset has been initiated.
4. The completed sale of the asset is probable and typically expected to occur within one year.
5. The asset is being offered for sale at a reasonable price relative to its current fair value.
6. Management's actions indicate the plan is unlikely to change significantly or be withdrawn.

An asset that is classified as held for sale is no longer depreciated or amortized. *An asset classified as held for sale is reported at the lower of its current book value or its fair value less any cost to sell.* If the fair value less cost to sell is below book value, we recognize a loss in the current period. If financial statements are again prepared prior to the sale, we reassess the asset's fair value less selling costs. If a further decline has occurred, we recognize another loss. If the fair value less selling costs has increased since the previous measurement, we recognize a gain, but limited to the cumulative amount of any previous losses.

RETIREMENTS Sometimes instead of selling a used asset, a company will retire (or abandon) the asset. Retirements are treated similarly to selling for monetary consideration. At the time of retirement, the asset account and the corresponding accumulated depreciation account are removed from the books and a loss equal to the remaining book value of the asset is recorded because there will be no monetary consideration received. When there is a formal plan to retire an asset, but before the actual retirement, there may be some revision in depreciation due to a change in the estimated service life or residual value.

Group and Composite Depreciation Methods

As you might imagine, depreciation records could become quite cumbersome and costly if a company has hundreds, or maybe thousands, of depreciable assets. However, the burden can be lessened if the company uses the group or composite method to depreciate assets collectively rather than individually. The two methods are the same except for the way the collection of assets is aggregated for depreciation. The **group depreciation method** defines the collection as depreciable assets that share similar service lives and other attributes. For example, group depreciation could be used for fleets of vehicles or collections of machinery. The **composite depreciation method** is used when assets are physically dissimilar but are aggregated anyway to gain the convenience of a collective depreciation calculation. For instance, composite depreciation can be used for all of the depreciable assets in one manufacturing plant, even though individual assets in the composite may have widely diverse service lives.

Group and composite depreciation methods aggregate assets to reduce the recordkeeping costs of determining periodic depreciation.

Both approaches are similar in that they involve applying a single straight-line rate based on the average service lives of the assets in the group or composite.[8] The process is demonstrated using Illustration 11–8.

[7]Source: FASB ASC 360–10–45–9: Property, Plant, and Equipment–Overall–Other Presentation Matters–Impairment or Disposal of Long-Lived Assets.

[8]A declining balance method could also be used with either the group or composite method by applying a multiple (e.g., 200%) to the straight-line group or composite rate.

Illustration 11–8
Group Depreciation

The Express Delivery Company began operations in 2021. It will depreciate its fleet of delivery vehicles using the group method. The cost of vehicles purchased early in 2021, along with residual values, estimated lives, and straight-line depreciation per year by type of vehicle, are as follows:

Asset	Cost	Residual Value	Depreciable Base	Estimated Life (years)	Depreciation per Year (straight line)
Vans	$150,000	$ 30,000	$120,000	6	$ 20,000
Trucks	120,000	16,000	104,000	5	20,800
Wagons	60,000	12,000	48,000	4	12,000
Totals	$330,000	$ 58,000	$272,000		**$52,800**

The *group depreciation* rate is determined by dividing the depreciation per year by the total cost. The group's *average service* life is calculated by dividing the depreciable base by the depreciation per year:

$$\text{Group depreciation rate} = \frac{\$52{,}800}{\$330{,}000} = 16\%$$

$$\text{Average service life} = \frac{\$272{,}000}{\$52{,}800} = 5.15 \text{ years (rounded)}$$

The depreciation rate is applied to the total cost of the group or composite for the period.

If there are no changes in the assets contained in the group, depreciation of **$52,800** per year ($330,000 × 16%) will be recorded for 5.15 years. This means the depreciation in the sixth year will be $7,920 (0.15 of a full year's depreciation = 15% × $52,800), which depreciates the cost of the group down to its estimated residual value. In other words, the group will be depreciated over the average service life of the assets in the group.

In practice, there very likely will be changes in the assets constituting the group as new assets are added and others are retired or sold. Additions are recorded by increasing the group asset account for the cost of the addition. Depreciation is determined by multiplying the group rate by the total cost of assets in the group for that period. Once the group or composite rate and the average service life are determined, they normally are continued despite the addition and disposition of individual assets. This implicitly assumes that the service lives of new assets approximate those of individual assets they replace.

No gain or loss is recorded when a group or composite asset is retired or sold.

Because depreciation records are not kept on an individual asset basis, dispositions are recorded under the assumption that the book value of the disposed item exactly equals any proceeds received and no gain or loss is recorded. For example, if a delivery truck in the above illustration that cost $15,000 is sold for $3,000 in the year 2024, the following journal entry is recorded:

Cash	3,000	
Accumulated depreciation (difference)	12,000	
Vehicles		15,000

Any actual gain or loss is included in the accumulated depreciation account. This practice generally will not distort income as the unrecorded gains tend to offset unrecorded losses.

The group and composite methods simplify the recordkeeping of depreciable assets. This simplification justifies any immaterial errors in income determination. Illustration 11–9 shows a disclosure note accompanying recent financial statements of the **El Paso Natural Gas Company (EPNG)** describing the use of the group depreciation method for its property that is regulated by federal statutes.

Illustration 11–9

Disclosure of Depreciation Method—El Paso Natural Gas Company

Real World Financials

Summary of Significant Accounting Policies (in part)
Property, Plant, and Equipment (in part)
We use the group method to depreciate property, plant, and equipment. Under this method, assets with similar lives and characteristics are grouped and depreciated as one asset. We apply the depreciation rate approved in our rate settlements to the total cost of the group until its net book value equals its salvage value. The majority of our property, plant, and equipment are on our El Paso Natural Gas Company (EPNG) system, which has depreciation rates ranging from one percent to 50 percent.
When we retire property, plant, and equipment, we charge accumulated depreciation and amortization for the original cost of the assets in addition to the cost to remove, sell, or dispose of the assets, less their salvage value. We do not recognize a gain or loss unless we sell an entire operating unit.

Source: El Paso Natural Gas Company

Additional group-based depreciation methods, the retirement and replacement methods, are discussed in Appendix 11B.

International Financial Reporting Standards

● LO11–10

Valuation of Property, Plant, and Equipment. As we've discussed, under U.S. GAAP a company reports property, plant, and equipment (PP&E) in the balance sheet at cost less accumulated depreciation (book value). *IAS No. 16* allows a company to report property, plant, and equipment at that amount or, alternatively, at its fair value (revaluation).[9] If a company chooses revaluation, all assets within a class of PP&E must be revalued on a regular basis. U.S. GAAP prohibits revaluation.

If the revaluation option is chosen, the way the company reports the difference between fair value and book value depends on which amount is higher:

- If fair value is higher than book value, the difference is reported as *other comprehensive income (OCI)*, which then accumulates in a "revaluation surplus" (sometimes called revaluation reserve) account in equity.
- If book value is higher than fair value, the difference is reported as an *expense in the income statement.* An exception is when a revaluation surplus account relating to the same asset has a balance from a previous *increase* in fair value, that balance is eliminated before debiting revaluation expense.

Consider the following illustration:

Candless Corporation prepares its financial statements according to IFRS. At the beginning of its 2021 fiscal year, the company purchased equipment for $100,000. The equipment is expected to have a five-year useful life with no residual value, so depreciation for 2021 is $20,000. At the end of the year, Candless chooses to revalue the equipment as permitted by *IAS No. 16*. Assuming that the fair value of the equipment at year-end is $84,000, Candless records depreciation and the revaluation using the following journal entries:

(a) Depreciation expense ($100,000 ÷ 5 years)	20,000	
Accumulated depreciation		20,000

After this entry, the book value of the equipment is $80,000; the fair value is $84,000. We use the ratio of the two amounts to adjust both the equipment and the accumulated depreciation accounts (and thus the book value) to fair value ($ in thousands):

December 31, 2021	Before Revaluation				After Revaluation
Equipment	$100	×	$84/80$	=	$105
Accumulated depreciation	20	×	$84/80$	=	21
Book value	$ 80	×	$84/80$	=	$ 84

(continued)

[9] "Property, Plant and Equipment," *International Accounting Standard No. 16* (IASCF), as amended effective January 1, 2018.

(concluded)

The entries to revalue the equipment and the accumulated depreciation accounts (and thus the book value) are as follows:

To record the revaluation of equipment to its fair value.

(b) Equipment ($105,000 – $100,000)	5,000	
Accumulated depreciation ($21,000 – $20,000)		1,000
Revaluation surplus—OCI ($84,000 – $80,000)		4,000

The new basis for the equipment is its fair value of $84,000 ($105,000 – $21,000), and the following years' depreciation is based on that amount. Thus, 2022 depreciation would be $84,000 divided by the four remaining years, or $21,000:[10]

(a) Depreciation expense ($84,000 ÷ 4 years)	21,000	
Accumulated depreciation		21,000

After this entry, the accumulated depreciation is $42,000 and the book value of the equipment is $63,000. Let's say the fair value now is $57,000. We use the ratio of the two amounts (fair value of $57,000 divided by book value of $63,000) to adjust both the equipment and the accumulated depreciation accounts (and thus the book value) to fair value ($ in thousands):

December 31, 2022	Before Revaluation				After Revaluation
Equipment	$105	×	57/63	=	$95
Accumulated depreciation	42	×	57/63	=	38
Book value	$ 63	×	57/63	=	$57

The entries to revalue the equipment and the accumulated depreciation accounts (and thus the book value) are as follows:

(b) Revaluation surplus—OCI ($57,000 – $63,000 = $6,000; limit: $4,000 balance)	4,000	
Revaluation expense (to balance)	2,000	
Accumulated depreciation ($38,000 – $42,000)	4,000	
Equipment ($95,000 – $105,000)		10,000

A decrease in fair value, as occurred in 2022, is expensed unless it reverses a revaluation surplus account relating to the same asset, as in this illustration. So, of the $6,000 decrease in value ($63,000 book value less $57,000 fair value), $4,000 is debited to the previously created revaluation surplus and the remaining $2,000 is recorded as revaluation expense in the income statement.

Investcorp, a provider and manager of alternative investment products headquartered in London, prepares its financial statements according to IFRS. The following disclosure note included in a recent annual report discusses the company's method of valuing its building and certain operating assets.

> **Premises and Equipment (in part)**
>
> The Bank carries its building on freehold land and certain operating assets at revalued amounts, being the fair value of the assets at the date of revaluation less any subsequent accumulated depreciation and subsequent accumulated impairment losses. Any revaluation surplus is credited to the asset revaluation reserve included in equity, except to the extent that it reverses a revaluation decrease of the same asset previously recognized in profit and loss, in which case the increase is recognized in profit or loss. A revaluation deficit is recognized directly in profit or loss, except that a deficit directly offsetting a previous surplus on the same asset is directly offset against the surplus in the asset revaluation reserve.

The revaluation alternative is used infrequently. A recent survey of large companies that prepare their financial statements according to IFRS reports that only 10 of the 160 surveyed companies used the revaluation alternative for at least one asset class.[11]

[10] *IAS No. 16* allows companies to choose between the method illustrated here and an alternative. The second method eliminates the entire accumulated depreciation account and adjusts the asset account (equipment in this illustration) to fair value. Using either method the revaluation surplus (or expense) would be the same.

[11] "*IFRS Accounting Trends and Techniques*" (New York, AICPA, 2011), p. 171.

Depletion of Natural Resources

Depletion of the cost of natural resources usually is determined using the units-of-production method.

Allocation of the cost of natural resources is called **depletion**. Because the usefulness of natural resources generally is directly related to the amount of the resources extracted, the activity-based units-of-production method is widely used to calculate periodic depletion. Service life is therefore the estimated amount of natural resource to be extracted (for example, tons of mineral or barrels of oil).

● LO11–3

The depletion base is cost less any anticipated residual value. Residual value could be significant if cost includes land that has a value after the natural resource has been extracted.

The example in Illustration 11–10 was first introduced in Chapter 10 in Illustration 10–6.

Illustration 11–10

Depletion of Natural Resources

In 2021, Jackson Mining Company has the following five costs related to 500 acres of land in Pennsylvania:

1. Payment for the right to explore for a coal deposit.	$ 1,000,000
2. Actual exploration costs for a coal deposit.	800,000
3. Intangible development costs in digging and constructing the mine shaft.	500,000
4. Expected cost to restore the land after extraction is completed.*	468,360
	$2,768,360
5. Purchase of excavation equipment for the project.	$ 600,000

* Determined using the expected cash flow approach.

The company's geologist estimates that 1,000,000 tons of coal will be extracted over a three-year period. After the coal is removed from the site, the excavation equipment will be sold for an anticipated residual value of $60,000. During 2021 300,000 tons were extracted.

The capitalized cost of the coal mine (natural resource), including the expected restoration costs, is **$2,768,360**. The $600,000 cost for equipment is capitalized separately. Since there is no residual value to the coal mine, the depletion base equals cost and the depletion rate per ton is calculated as follows:

$$\text{Depletion per ton} = \frac{\text{Depletion base}}{\text{Estimated extractable tons}}$$

$$\text{Depletion per ton} = \frac{\mathbf{\$2,768,360}}{1,000,000 \text{ tons}} = \$2.76836 \text{ per ton}$$

For each ton of coal extracted, $2.76836 in depletion is recorded. In 2021, the following journal entry records depletion for the 300,000 tons of coal actually extracted.

Depletion ($2.76836 × 300,000 tons)	830,508	
Coal mine		830,508

Notice that the credit is to the asset, coal mine, rather than to a contra account, accumulated depletion. Although this approach is traditional, the use of a contra account is acceptable.

Depletion is a product cost and is included in the cost of the inventory of coal, just as the depreciation on manufacturing equipment is included in inventory cost. The depletion is then included in cost of goods sold in the income statement when the coal is sold.

What about depreciation on the $600,000 cost of excavation equipment? If the equipment can be moved from the site and used on future projects, the equipment's depreciable base should be allocated over its useful life. If the asset is not movable, as in our illustration, then it should be depreciated over its useful life or the life of the natural resource, whichever is shorter.

The units-of-production method often is used to determine depreciation and amortization on assets used in the extraction of natural resources.

Quite often, companies use the units-of-production method to calculate depreciation and amortization on assets used in the extraction of natural resources. The activity base used is the same as that used to calculate depletion, the estimated recoverable natural resource. In our illustration, the depreciation rate would be $0.54 per ton, calculated as follows.

$$\text{Depreciation per ton} = \$0.54 \text{ per ton} = \frac{\$600{,}000 - \$60{,}000}{1{,}000{,}000 \text{ tons}}$$

In 2021, depreciation of $162,000 ($0.54 × 300,000 tons) is recorded and also included as part of the cost of the coal inventory.

The summary of significant accounting policies disclosure accompanying recent financial statements of **ConocoPhilips** shown in Illustration 11–11 provides a good summary of depletion, amortization, and depreciation for natural resource properties.

Illustration 11–11

Depletion Method Disclosure—ConocoPhilips

Real World Financials

Summary of Significant Accounting Policies (in part)

Depletion and Amortization—Leasehold costs of producing properties are depleted using the units-of-production method based on estimated proved oil and gas reserves. Amortization of intangible development costs is based on the units-of-production method using estimated proved developed oil and gas reserves.

Depreciation and Amortization—Depreciation and amortization of PP&E on producing hydrocarbon properties and certain pipeline assets (those which are expected to have a declining utilization pattern), are determined by the units-of-production method. Depreciation and amortization of all other PP&E are determined by either the individual-unit-straight-line method or the group-straight-line method (for those individual units that are highly integrated with other units).

Source: ConocoPhilips

Additional Consideration

Percentage Depletion

Depletion of cost less residual value required by GAAP should not be confused with percentage depletion (also called *statutory depletion*) allowable for income tax purposes for oil, gas, and most mineral natural resources. Under these tax provisions, a producer is allowed to deduct the greater of cost-based depletion or a fixed percentage of gross income as depletion expense. Over the life of the asset, depletion could exceed the asset's cost. The percentage allowed for percentage-based depletion varies according to the type of natural resource.

Because percentage depletion usually differs from cost depletion, a difference between taxable income and financial reporting income before tax results. Differences between taxable income and financial reporting income are discussed in Chapter 16.

International Financial Reporting Standards

● LO11–10

Biological Assets. Living animals and plants, including the trees in a timber tract or in a fruit orchard, are referred to as *biological assets.* Under U.S. GAAP, a timber tract is valued at cost less accumulated depletion and a fruit orchard at cost less accumulated depreciation. Under IFRS, biological assets are valued at their fair value less estimated costs to sell, with changes in fair value included in the calculation of net income.[12]

(continued)

[12]"Agriculture," *International Accounting Standard No. 41* (IASCF), as amended effective January 1, 2018.

(concluded)

Mondi Limited, an international paper and packing group headquartered in Johannesburg, South Africa, prepares its financial statements according to IFRS. The following disclosure note included in a recent annual report discusses the company's policy for valuing its forestry assets.

Owned Forestry Assets (in part)

Owned forestry assets are measured at fair value, calculated by applying the expected selling price, less costs to harvest and deliver, to the estimated volume of timber on hand at each reporting date.

Changes in fair value are recognized in the combined and consolidated income statement within other net operating expenses.

Amortization of Intangible Assets

Let's turn now to a third type of long-lived asset—intangible assets. As with other assets we have discussed, we allocate the cost of an intangible asset over its service or useful life. The allocation of intangible asset cost is called **amortization**. Below we distinguish those intangible assets with *finite* versus *indefinite* useful lives. Intangible assets with finite useful lives will be amortized, and those with indefinite useful lives will not be amortized.

● LO11–4

Intangible Assets Subject to Amortization

Most intangible assets have a finite useful life. This means their estimated useful life is limited in nature. We allocate the capitalized cost less any estimated residual value of an intangible asset to the periods in which the asset is expected to contribute to the company's revenue-generating activities. This requires that we determine the asset's useful life, its amortization base (cost less estimated residual value), and the appropriate allocation method, similar to our depreciating tangible assets.

The cost of an intangible asset with a *finite* useful life is *amortized.*

USEFUL LIFE Legal, regulatory, or contractual provisions often limit the useful life of an intangible asset. On the other hand, useful life might sometimes be less than the asset's legal or contractual life. For example, the useful life of a patent would be considerably less than its legal life of 20 years if obsolescence were expected to limit the longevity of a protected product.

RESIDUAL VALUE We discussed the cost of intangible assets in Chapter 10. The expected residual value of an intangible asset usually is zero. This might not be the case, though, if at the end of its useful life to the reporting entity the asset will benefit another entity. For example, if Quadra Corp. has a commitment from another company to purchase one of Quadra's patents at the end of its useful life at a determinable price, we use that price as the patent's residual value.

ALLOCATION METHOD The method of amortization should reflect the pattern of use of the asset in generating benefits. Most companies use the straight-line method to calculate amortization expense. **Intel Corporation** reported several intangible assets in a recent balance sheet. A note, shown in Illustration 11–12, disclosed the range of estimated useful lives.

Like depletion, amortization expense traditionally is credited to the asset account itself rather than to accumulated amortization. However, the use of a contra account is acceptable. Many companies choose to report intangible assets for their net amount on the face of the balance sheet and then report the amount of amortization in a disclosure note. Let's look at an example in Illustration 11–13A.

Similar to depreciation, amortization is either a product cost or a period cost depending on the use of the asset. For intangibles used in the manufacture of a product, amortization is a product cost and is included in the cost of inventory (and doesn't become an expense until the inventory is sold). For intangible assets not used in production, such as the franchise cost in our illustration, periodic amortization is expensed in the period incurred.

Illustration 11–12

Intangible Asset Useful Life Disclosure—Intel Corporation

Real World Financials

Summary of Significant Accounting Policies (in part)
Identified Intangible Assets (in part)
The estimated useful life ranges for identified intangible assets that are subject to amortization are as follows:

(in years)	Estimated Useful Life
Acquisition-related developed technology	3 – 9
Acquisition-related customer relationships	5 – 11
Acquisition-related brands	5 – 8
Licensed technology and patents	2 – 17

Source: Intel Corporation

Illustration 11–13A

Amortization of Intangibles—Franchise and Patent

Hollins Corporation began operations in 2021. Early in January, the company purchased the following two intangible assets:

1. A franchise from Ajax Industries for $200,000. The franchise agreement is for a period of 10 years.
2. A patent for $50,000. The remaining legal life of the patent is 13 years. However, due to expected technological obsolescence, the company estimates that the useful life of the patent is only 8 years.

Hollins uses the straight-line amortization method for all intangible assets. The company's fiscal year-end is December 31.

The journal entries to record a full year of amortization each year for these intangible assets are as follows:

Amortization expense ($200,000 ÷ 10 years)	20,000	
Franchise		20,000
To record amortization of franchise.		
Amortization expense ($50,000 ÷ 8 years)	6,250	
Patent		6,250
To record amortization of patent.		

Hollins decided to sell the patent on December 31, 2025 (five years after acquisition), for $21,000. Hollins had recorded annual amortization in each of the four prior years (2021-2024).

The journal entries to update amortization of the patent in 2025 and to sell the patent are as follows:

Amortization expense ($50,000 ÷ 8 years)	6,250	
Patent		6,250
To record amortization of patent in 2025.		
Cash	21,000	
Patent (account balance)*		18,750
Gain on sale of patent (selling price – book value)		2,250
To record sale of patent.		

*$50,000 – ($6,250 x 5 years)

In Chapter 10 we discussed that any software development costs incurred after the point of technological feasibility and before the product is available for sale are capitalized. These capitalized software development costs are amortized based on whichever of the following two methods produces a *greater* amount:

1. The ratio of current revenues to current and anticipated revenues (percentage of revenue method), or
2. The straight-line method based on the estimated useful life of the asset.

To see an example, let's look at Illustration 11–13B.

Illustration 11–13B

Amortization of Intangibles—Software Development Costs for External Purposes

The Astro Corporation develops computer software graphics programs for sale. A new development project started in 2020 and reached the point of technological feasibility on June 30, 2021. Costs in 2021 were as follows:

Prior to June 30, 2021	$1,200,000
From June 30 to December 31, 2021	800,000

The software was available for sale on January 1, 2022, and has the following related sales information:

Sales in 2022	$ 3,000,000
Estimated sales in 2023–2025	7,000,000
Total estimates sales over four years	$10,000,000

In 2021, Astro Corporation would expense the $1,200,000 in costs incurred prior to the establishment of technological feasibility and capitalize the $800,000 in costs incurred between technological feasibility and the product availability date. In 2022, amortization of the intangible asset, software development costs, is calculated as follows:

1. **Percentage-of-revenue method:**

$$\frac{\$3,000,000}{\$3,000,000 + \$7,000,000} = 30\% \times \$800,000 = \$240,000$$

2. **Straight-line method:**

$$\frac{1}{4} \text{ or } 25\% \times \$800,000 = \$200,000.$$

The percentage-of-revenue method is used because it produces the greater amortization, $240,000.

Amortization expense (above)	240,000	
Software development costs		240,000

To record amortization of software development costs in 2022.

The capitalized cost of software developed for internal purposes or as part of cloud computing arrangements is amortized over the software's expected useful life, generally using straight-line amortization.

International Financial Reporting Standards

Software Development Costs. The percentage we use to amortize computer software development costs under U.S. GAAP is the greater of (1) the ratio of current revenues to current and anticipated revenues or (2) the straight-line percentage over the useful life of the software. This approach is allowed under IFRS, but not required. Amortization under IFRS typically occurs over the useful life of the software, based on the pattern of benefits, with straight-line as the default.

● LO11–9

Intangible Assets Not Subject to Amortization

Intangible assets with an indefinite useful life are those with no foreseeable limit on the period of time over which the asset is expected to contribute to the cash flows of the entity.[13] In other words, there are no legal, contractual, or economic factors that are expected to limit their useful life to a company. Because of their indefinite lives, these intangible assets are not subject to periodic amortization.

The cost of an intangible asset with an *indefinite* useful life is *not* amortized.

For example, suppose Collins Corporation acquired a trademark in conjunction with the acquisition of a tire company. Collins plans to continue to produce the line of tires marketed under the acquired company's trademark. Recall from our discussion in Chapter 10 that

Trademarks or tradenames often are considered to have indefinite useful lives.

[13]FASB ASC 350–30–35–4: Intangibles–Goodwill and Other–General Intangibles Other than Goodwill–Subsequent Measurement.

trademarks have a legal life of 10 years, but the registration can be renewed for an indefinite number of 10-year periods. Therefore, the life of the purchased trademark is initially considered to be indefinite and the cost of the trademark is not amortized. However, if after several years management decides to phase out production of the tire line over the next three years, Collins would amortize the remaining book value over a three-year period.

In 2017, **Boeing Company** reported indefinite-lived intangible assets (other than goodwill) of $490 million. These indefinite-lived intangibles consist of brand and trade names acquired in business combinations. Illustration 11–14 provides another example in a disclosure made by **The Estee Lauder Companies Inc.**, in a recent annual report.

Illustration 11–14

Indefinite-Life Intangibles Disclosure—The Estee Lauder Companies Inc.

Real World Financials

Other Intangible Assets

Indefinite-lived intangible assets (e.g., trademarks) are not subject to amortization and are assessed at least annually for impairment during the fiscal fourth quarter, or more frequently if certain events or circumstances warrant.

Source: The Estee Lauder Companies Inc.

Goodwill is an intangible asset whose cost is *not* expensed through periodic amortization.

Goodwill is the most common intangible asset with an indefinite useful life. Recall that goodwill is measured as the difference between the purchase price of a company and the fair value of all of the identifiable net assets acquired (tangible and intangible assets minus the fair value of liabilities assumed). Does this mean that goodwill and other intangible assets with indefinite useful lives will remain in a company's balance sheet at their original capitalized values indefinitely? Not necessarily. Like other assets, intangibles are subject to the impairment rules we discuss in a subsequent section of this chapter.

International Financial Reporting Standards

Valuation of Intangible Assets. *IAS No. 38* allows a company to value an intangible asset subsequent to initial valuation at (1) cost less accumulated amortization or (2) fair value, if fair value can be determined by reference to an active market.[14] If revaluation is chosen, all assets within that class of intangibles must be revalued on a regular basis. Goodwill, however, cannot be revalued. U.S. GAAP prohibits revaluation of any intangible asset.

Notice that the revaluation option is possible only if fair value can be determined by reference to an active market, making the option relatively uncommon. However, the option possibly could be used for intangibles such as franchises and certain license agreements.

● LO11–10

If the revaluation option is chosen, the accounting treatment is similar to the way we applied the revaluation option for property, plant, and equipment earlier in this chapter. Recall that the way the company reports the difference between fair value and book value depends on which amount is higher. If fair value is higher than book value, the difference is reported as other comprehensive income (OCI) and then accumulates in a revaluation surplus account in equity. On the other hand, if book value is higher than fair value, the difference is expensed after reducing any existing revaluation surplus for that asset.

Consider the following illustration:

Amershan LTD. prepares its financial statements according to IFRS. At the beginning of its 2021 fiscal year, the company purchased a franchise for $500,000. The franchise has a 10-year contractual life and no residual value, so amortization in 2021 is $50,000. The company does not use an accumulated amortization account and credits the franchise account directly when amortization is recorded. At the end of the year, Amershan chooses to revalue the franchise as permitted by *IAS No. 38.* Assuming that the fair value of the

(continued)

[14] "Intangible Assets," *International Accounting Standard No. 38* (IASCF), as amended effective January 1, 2018.

(concluded)

franchise at year-end, determined by reference to an active market, is $600,000, Amershan records amortization and the revaluation using the following journal entries:

Amortization expense ($500,000 ÷ 10 years)	50,000	
Franchise		50,000
Franchise ($600,000 – $450,000)	150,000	
Revaluation surplus—OCI		50,000

To record the revaluation of franchise to its fair value.

With the second entry Amershan increases the book value of the franchise from $450,000 ($500,000 – $50,000) to its fair value of $600,000 and records a revaluation surplus for the difference. The new basis for the franchise is its fair value of $600,000, and the following years' amortization is based on that amount. Thus, 2022 amortization would be $600,000 divided by the nine remaining years, or $66,667.

Concept Review Exercise

DEPLETION AND AMORTIZATION

Part A:

On March 29, 2021, the Horizon Energy Corporation purchased the mineral rights to a coal deposit in New Mexico for $2 million. Development costs and the present value of estimated land restoration costs totaled an additional $3.4 million. The company removed 200,000 tons of coal during 2021 and estimated that an additional 1,600,000 tons would be removed over the next 15 months.

Required:

Compute depletion on the mine for 2021.

Solution:

Cost of Coal Mine:	($ in millions)
Purchase price of mineral rights	$2.0
Development and restoration costs	3.4
	$5.4

Depletion:

$$\text{Depletion per ton} = \frac{\$5.4 \text{ million}}{1.8 \text{ million tons*}} = \$3 \text{ per ton}$$

*200,000 + 1,600,000

$$2021 \text{ depletion} = \$3 \times 200{,}000 \text{ tons} = \$600{,}000$$

Part B:

On October 1, 2021, Advanced Micro Circuits, Inc., completed the purchase of Zotec Corporation for $200 million. Included in the allocation of the purchase price were the following identifiable intangible assets ($ in millions), along with the fair values and estimated useful lives:

Intangible Asset	Fair value	Useful Life (in years)
Patent	$10	5
Developed technology	50	4
Customer list	10	2

In addition, the fair value of acquired tangible assets was $100 million. Goodwill was valued at $30 million. Straight-line amortization is used for all purchased intangibles.

During 2021, Advanced finished work on a software development project. Development costs incurred after technological feasibility was achieved and before the product release date totaled $2 million. The software was available for release to the general public on September 29, 2021. At that time, the company estimates that the software will generate total revenue of $40 million over four years. For the final three months of 2021, revenue from the sale of the software was $4 million.

Required:

Compute amortization for purchased intangibles and software development costs for 2021.

Solution:

Amortization of Purchased Intangibles:

Patent	$10 million / 5 = $2 million × $^3/_{12}$ year = $0.5 million
Developed technology	$50 million / 4 = $12.5 million × $^3/_{12}$ year = $3.125 million
Customer list	$10 million / 2 = $5 million × $^3/_{12}$ year = $1.25 million
Goodwill	The cost of goodwill is not amortized.

Amortization of Software Development Costs:

(1) Percentage-of-revenue method:

$$\$2 \text{ million} \times \frac{\$4 \text{ million}}{\$40 \text{ million}} \text{ or } 10\% = \$200{,}000$$

(2) Straight-line:

$$\$2 \text{ million} \times \frac{3 \text{ months}}{48 \text{ months}} \text{ or } 6.25\% = \$125{,}000$$

Advanced will use the percentage-of-revenue method since it produces the greater amortization, $200,000.

PART B

Additional Issues

In this part of the chapter, we discuss the following issues related to cost allocation:

1. Change in estimates
2. Change in method
3. Error correction
4. Impairment of value

Change in Estimates

● LO11–5

The calculation of depreciation, depletion, or amortization requires estimates of both service life and residual value. It's inevitable that at least some estimates will prove incorrect. Chapter 4 briefly introduced the topic of changes in estimates along with coverage of changes in accounting principles and the correction of errors. Here and in subsequent sections of this chapter, we provide overviews of the accounting treatment and disclosures required for these changes and errors when they involve property, plant, and equipment and intangible assets.

A change in estimate should be reflected in the financial statements of the current period and future periods.

Changes in estimates are accounted for prospectively. When a company revises a previous estimate based on new information, prior financial statements are not restated. Instead, the company merely incorporates the new estimate in any related accounting determinations from then on. So, it usually will affect some aspects of both the balance sheet and the income statement in the current and future periods. Companies typically make these changes at the beginning of the year of the change, but they could be made at other times. A disclosure note should describe the effect of a change in estimate on net income and per share amounts for the current period.

Consider the example in Illustration 11–15.

The asset's book value is depreciated down to the anticipated residual value of $22,000 at the end of the revised eight-year service life. In addition, a note discloses the effect of the change in estimate on income, if material. The before-tax effect is an increase in income of $18,000 (depreciation of $42,000 if the change had not been made, less $24,000 depreciation after the change).

Illustration 11–15

Change in Accounting Estimate

On January 1, 2019, the Hogan Manufacturing Company purchased a machine for $250,000. At the time of purchase, the company estimated the following:

1. Service life of the machine to be five years.
2. Residual value to be $40,000.

On January 1, 2021 (two years later), the company revised its estimates as follows:

1. Service life from a total of **five to eight years.**
2. Residual value from **$40,000 to $22,000.**

The company's fiscal year-end is December 31 and the straight-line depreciation method is used for all depreciable assets.

Prior to the revision (2019 and 2020), depreciation was $42,000 per year [($250,000 – $40,000) ÷ 5 years] or $84,000 for the two years. The remaining book value at the beginning of 2021 is $166,000 ($250,000 – $84,000). Depreciation for 2021 and subsequent years is determined by allocating the remaining book value less the revised residual value equally over the remaining service life of six years (8 total years – 2 years completed). Depreciation for 2021 and subsequent years is recorded as follows:

Depreciation expense (below)	**24,000**	
Accumulated depreciation		24,000

	$250,000	Cost
$42,000		Previous annual depreciation ($210,000 ÷ 5 years)
× 2 years	(84,000)	Less: Depreciation to date (2019–2020)
	166,000	Book value as of January 1, 2021
	(22,000)	Less: Revised residual value
	144,000	Revised depreciable base
	÷ 6	Estimated remaining life (8 years – 2 years)
	$ 24,000	New annual depreciation

Verizon Communications Inc. recently revised its estimates of the service lives of certain property, plant, and equipment. Illustration 11–16 shows the note that disclosed the change.

Illustration 11–16

Change in Estimate Disclosure—Verizon Communications Inc.

Real World Financials

Plant and Depreciation (in part)

In connection with our ongoing review of the estimated useful lives of plant, property and equipment during 2016, we determined that the average useful lives of certain leasehold improvements would be increased from 5 to 7 years. This change resulted in a decrease to depreciation expense of $0.2 billion in 2016. We determined that changes were also necessary to the remaining estimated useful lives of certain assets as a result of technology upgrades, enhancements, and planned retirements. These changes resulted in an increase in depreciation expense of $0.3 billion, $0.4 billion and $0.6 billion in 2016, 2015 and 2014, respectively. While the timing and extent of current deployment plans are subject to ongoing analysis and modification, we believe the current estimates of useful lives are reasonable.

Source: Verizon Communications Inc.

Change in Depreciation, Amortization, or Depletion Method

Changes in depreciation, amortization, or depletion methods are accounted for the same way as a change in accounting estimate.

LO11–6

Generally accepted accounting principles allow a company to change from one depreciation method to another if the company can justify the change. For example, new information might become available to suggest that a different depreciation method would better represent the pattern of the asset's consumption relative to revenue production.

We account for these changes prospectively, exactly as we would any other change in estimate. One difference is that most changes in estimate do not require a company to justify the change. However, this change in estimate is a result of changing an accounting principle and therefore requires a clear justification as to why the new method is preferable. Consider the example in Illustration 11–17.

Illustration 11–17

Change in Depreciation Method

On January 1, 2019, the Hogan Manufacturing Company purchased a machine for $250,000. The company expects the service life of the machine to be five years and its anticipated residual value to be $30,000. The company's fiscal year-end is December 31 and the double-declining-balance (DDB) depreciation method is used. During 2021, the company switched from the DDB to the straight-line method. In 2021, the adjusting entry is

Depreciation expense (below)	**20,000**	
Accumulated depreciation		20,000

DDB depreciation:

2019	$ 100,000	($250,000 × 40%*)
2020	60,000	[($250,000 − $100,000) × 40%*]
Total	**$160,000**	

*Double the straight-line rate for 5 years [(⅕ = 20%) × 2 = 40%]

$ 250,000	Cost
(160,000)	Less: Depreciation to date, DDB (2019–2020)
90,000	Undepreciated cost as of January 1, 2021
(30,000)	Less: Residual value
60,000	Depreciable base
÷3 yrs.	Remaining life (5 years − 2 years)
$ 20,000	New annual depreciation, straight-line

A disclosure note reports the effect of the change on net income and earnings per share along with clear justification for changing depreciation methods.

Illustration 11–18 shows a disclosure note describing a recent change in depreciation method made by **Agrium Inc.**

Illustration 11–18

Change in Depreciation Method—Agrium Inc.

Real World Financials

Accounting standards and policy changes (in part)

We changed the method of depreciation from the straight-line basis to the units of production basis for our potash facility mining and milling assets beginning January 1, 2015, and our nitrogen and phosphate mining and plant assets beginning October 1, 2015. The change in method of depreciation reflects anticipated changes to our production schedule due to facility expansions, volatility in market conditions, and the frequency and duration of plant turnarounds. The current and expected reduction in depreciation expense is 2015 – $30 million and 2016 – $7 million.

Source: Agrium Inc.

Frequently, when a company changes depreciation method, the change will be effective only for assets placed in service after that date. Of course, that means depreciation schedules do not require revision because the change does not affect assets depreciated in prior periods. A disclosure note still is required to provide justification for the change and to report the effect of the change on the current year's income.

Error Correction

● LO11–7

Errors involving property, plant, and equipment and intangible assets include computational errors in the calculation of depreciation, depletion, or amortization and mistakes made in determining whether expenditures should be capitalized or expensed. These errors can affect many years. For example, let's say a major addition to equipment should be capitalized but incorrectly is expensed. Not only is income in the year of the error understated, but subsequent years' income is overstated because depreciation is omitted.

Recall from our discussion of inventory errors in Chapter 9 that if a material error is discovered in an accounting period subsequent to the period in which the error is made, any previous years' financial statements that were incorrect as a result of the error are retrospectively restated to reflect the correction. Any account balances that are incorrect as a result of the error are corrected by journal entry. If retained earnings is one of the incorrect accounts, the correction is reported as a *prior period adjustment* to the beginning balance in the statement of shareholders' equity.[15] In addition, a disclosure note is needed to describe the nature of the error and the impact of its correction on net income and earnings per share.

Here is a summary of the treatment of material errors occurring in a previous year:

- Previous years' financial statements are retrospectively restated.
- Account balances are corrected.
- If retained earnings requires correction, the correction is reported as a prior period adjustment.
- A note describes the nature of the error and the impact of the correction on income.

Consider Illustration 11–19.

Illustration 11–19

Error Correction

Sometimes, the analysis is easier if you re-create the entries actually recorded incorrectly and those that would have been recorded if the error hadn't occurred, and then compare them.

In 2021, the controller of the Hathaway Corporation discovered an error in recording $300,000 in legal fees to successfully defend a patent infringement suit in 2019. The $300,000 was charged to legal fee expense but should have been capitalized and amortized over the five-year remaining life of the patent. Straight-line amortization is used by Hathaway for all intangibles.

Correct (Should Have Been Recorded)			Incorrect (As Recorded)		
2019 Patent	300,000		Legal fee expense	300,000	
Cash		300,000	Cash		300,000
2019 Expense	60,000		Amortization entry omitted		
Patent		60,000			
2020 Expense	60,000		Amortization entry omitted		
Patent		60,000			

During the two-year period (2019 and 2020), amortization expense was *understated* by $120,000, but legal fee expenses were *overstated* by $300,000, so net income during the period was *understated* by $180,000 (ignoring income taxes). This means retained earnings by the end of 2020 is *understated* by $180,000. Patent is understated by $180,000. To correct each of these accounts in 2021, we need the following entry:

Patent	180,000	
Retained earnings		180,000

To correct incorrect accounts.

Because retained earnings is one of the accounts incorrect as a result of the error, a correction to that account of **$180,000** is reported as a prior period adjustment to the 2021 beginning retained earnings balance in Hathaway's comparative statements of shareholders' equity. Assuming that 2020 is included with 2021 in the comparative statements, a correction would be made to the 2020 beginning retained earnings balance as well. That prior period adjustment, though, would be for the pre-2020 difference: $300,000 – $60,000 = $240,000.

The 2019 and 2020 balance sheet and income statement that were incorrect as a result of the error are *retrospectively restated* to report the addition to the patent and to reflect the correct amount of amortization expense, assuming both statements are reported again for comparative purposes in the 2021 annual report.

[15] The prior period adjustment is applied to beginning retained earnings for the year following the error, or for the earliest year being reported in the comparative financial statements when the error occurs prior to the earliest year presented. The retained earnings balances in years after the first year also are adjusted to what those balances would be if the error had not occurred, but a company may choose not to explicitly report those adjustments as separate line items.

Also, a disclosure note accompanying Hathaway's 2021 financial statements should describe the nature of the error and the impact of its correction on each year's net income (understated by $240,000 in 2019 and overstated by $60,000 in 2020), and earnings per share.

Chapter 20 provides in-depth coverage of changes in estimates and methods, and of accounting errors. We cover the tax effect of these changes and errors in that chapter.

Impairment of Value

FINANCIAL Reporting Case

Q3, p. 571

LO11–8

Depreciation, depletion, and amortization reflect a gradual consumption of the benefits inherent in property, plant, and equipment and intangible assets. An implicit assumption in allocating the cost of an asset over its useful life is that there has been no significant reduction in the anticipated total benefits or service potential of the asset. Situations can arise, however, that cause a significant decline or impairment of those benefits or service potential. An extreme case would be the destruction of a plant asset—say a building destroyed by fire—before the asset is fully depreciated. The remaining book value of the asset in that case should be written off as a loss. Sometimes, though, the impairment of future value is more subtle.

The way we recognize and measure an impairment loss differs depending on whether the assets are classified as (1) held and used or (2) held for sale. Accounting is different, too, for assets with finite lives and those with indefinite lives. We consider those differences now.

Assets Held and Used

An increasingly common occurrence in practice is the partial write-down of property, plant, and equipment and intangible assets that remain in use. For example, in 2016, **Murphy Oil Corporation** recorded impairment charges of $95 million. The charges reflect the decline in asset values associated with lower oil and gas prices.

An asset held for use should be written down if there has been a significant impairment of value.

Conceptually, there is considerable merit for a policy requiring the write-down of an asset when there has been a significant decline in value. A write-down can provide important information about the future cash flows that a company can generate from using the asset. However, in practice, this process is very subjective. Even if it appears certain that significant impairment of value has occurred, it often is difficult to measure the amount of the required write-down.

For example, let's say a company purchased $2,000,000 of equipment to be used in the production of a new type of laser printer. Depreciation is determined using the straight-line method over a useful life of six years and the residual value is estimated at $200,000. At the beginning of year 3, the machine's book value has been depreciated to $1,400,000 [S2,000,000 – ($300,000 × 2)]. At that time, new technology is developed causing a significant reduction in the selling price of the new laser printer as well as a reduction in anticipated demand for the product. Management estimates that the equipment will be useful for only two more years and will have no significant residual value.

This situation is not simply a matter of a change in the estimates of useful life and residual value. Management must decide if the events occurring in year 3 warrant a write-down of the asset below $1,400,000. A write-down would be appropriate if the company decided that it would be unable to fully recover this amount through future use.

For assets to be held and used, different guidelines apply to (1) property, plant, and equipment and intangible assets with finite useful lives (subject to depreciation, depletion, or amortization) and (2) intangible assets with indefinite useful lives (not subject to amortization).

PROPERTY, PLANT, AND EQUIPMENT AND FINITE-LIFE INTANGIBLE ASSETS

Generally accepted accounting principles provide guidelines for when to recognize and how to measure impairment losses of long-lived tangible assets and intangible assets with finite useful lives.[16] For purposes of this recognition and measurement, assets are grouped at the

[16]FASB ASC 360–10–35–15 through 25: Property, Plant, and Equipment—Overall—Subsequent Measurement—Impairment or Disposal of Long-Lived Assets.

lowest level for which identifiable cash flows are largely independent of the cash flows of other assets. In simpler terms, assets should not be grouped for determining impairment unless those assets are dependent on one another. Determining which assets can or cannot be grouped requires considerable judgment by management.

When to Test for Impairment. It would be impractical to test all assets or asset groups for impairment at the end of every reporting period. GAAP requires investigation of possible impairment only *if events or changes in circumstances indicate that the book value of the asset or asset group may not be recoverable.* This might happen from the following:

Property, plant, and equipment and finite-life intangible assets are tested for impairment only when events or changes in circumstances indicate book value may not be recoverable.

a. A significant decrease in market price.
b. A significant adverse change in how the asset is being used or in its physical condition.
c. A significant adverse change in legal factors or in the business climate.
d. An accumulation of costs significantly higher than the amount originally expected for the acquisition or construction of an asset.
e. A current-period loss combined with a history of losses or a projection of continuing losses associated with the asset.
f. A realization that the asset will be disposed of significantly before the end of its estimated useful life.[17]

Determining whether an impairment loss has occurred and for how much to record the loss is a two-step process.

Step 1 **Recoverability Test.** An impairment occurs when the undiscounted sum of estimated future cash flows from an asset is less than the asset's book value.

Step 2 **Measurement.** If the recoverability test from step 1 indicates an impairment has occurred, an impairment loss is recorded for the amount by which the asset's fair value is less than its book value.

For step 2, fair value is the amount at which the asset could be bought or sold in a current transaction between willing parties. Quoted market prices could be used if they're available. If fair value is not determinable, it must be estimated.

If an impairment loss is recognized, the written-down book value becomes the new cost base for future cost allocation. Later recovery of an impairment loss is prohibited.

The process is best described by an example. Consider Illustration 11–20.

Illustration 11–20
Impairment Loss—Property, Plant, and Equipment

The Dakota Corporation operates several factories that manufacture medical equipment. Near the end of the company's 2021 fiscal year, a change in business climate related to a competitor's innovative products indicated to management that the $170 million book value (original cost of $300 million less accumulated depreciation of $130 million) of the assets of one of Dakota's factories may not be recoverable.

Management is able to identify cash flows from this factory and estimates that future cash flows over the remaining useful life of the factory will be $150 million. The fair value of the factory's assets is not readily available but is estimated to be $135 million.

Change in circumstances. A change in the business climate related to a competitor's innovative products requires Dakota to investigate for possible impairment.

Step 1. Recoverability Test. Because the undiscounted future cash flows of $150 million are less than book value of $170 million, an impairment loss is indicated.

(continued)

[17]FASB ASC 360–10–35–21: Property, Plant, and Equipment—Overall—Subsequent Measurement—Impairment or Disposal of Long-Lived Assets.

(concluded)

Step 2. Measurement of impairment loss. The impairment loss is **$35** million, determined as follows:

($ in millions)		
Fair value		$ 135
Less: Book value		
Original cost	$ 300	
Accumulated depreciation	(130)	(170)
Impairment loss		**$ (35)**

The entry to record the loss is ($ in millions):

Loss on impairment (above)	**35**	
Accumulated depreciation (reduce to zero)	130	
Factory assets (decrease to fair value; $300 – $135)		165

The loss normally is reported in the income statement as a separate component of operating expenses.

In the entry in Illustration 11–20, we reduce accumulated depreciation to zero and decrease the cost base of the assets by $165 million (from book value of $300 million to fair value of $135 million). The new carrying value of $135 million serves as the revised basis for subsequent depreciation over the remaining useful life of the assets, just as if the assets had been acquired on the impairment date for their fair values.

The present value of future cash flows often is used as a measure of fair value.

Because the fair value of the factory assets was not readily available to Dakota in Illustration 11–20, the $135 million had to be estimated. One method that can be used to estimate fair value is to compute the discounted present value of future cash flows expected from the asset. Keep in mind that we use *undiscounted* estimates of cash flows in step 1 to determine whether an impairment loss is indicated, but *discounted* estimates of cash flows in step 2 to determine the amount of the loss. In calculating present value, either a traditional approach or an expected cash flow approach can be used. The traditional approach is to incorporate risk and uncertainty into the discount rate. Recall from discussions in previous chapters that the expected cash flow approach incorporates risk and uncertainty instead into a determination of a probability-weighted cash flow expectation, and then discounts this expected cash flow using a risk-free interest rate. We discussed and illustrated the expected cash flow approach in previous chapters.

A disclosure note is needed to describe the impairment loss. The note should include a description of the impaired asset or asset group, the facts and circumstances leading to the impairment, the amount of the loss if not separately disclosed on the face of the income statement, and the method used to determine fair value.

Macy's, Inc., is a department store chain providing brand-name clothing, accessories, home furnishings, and housewares. Illustration 11–21 shows the company's disclosure notes describing recent impairment losses. The notes also provide a summary of the process used to identify and measure impairment losses for property, plant, and equipment and finite-life intangible assets.

Illustration 11–21

Asset Impairment Disclosure—Macy's, Inc.

Real World Financials

Property and Equipment (in part)

The carrying values of long-lived assets are periodically reviewed by the Company whenever events or changes in circumstances indicate that the carrying value may not be recoverable, such as historical operating losses or plans to close stores before the end of their previously estimated useful lives. Additionally, on an annual basis, the recoverability of the carrying values of individual stores are evaluated. A potential impairment has occurred if projected

(continued)

(concluded)

future undiscounted cash flows are less than the carrying value of the assets. The estimate of cash flows includes management's assumptions of cash inflows and outflows directly resulting from the use of those assets in operations. When a potential impairment has occurred, an impairment write-down is recorded if the carrying value of the long-lived asset exceeds its fair value.

Impairments, Store Closing and Other Costs (in part)

As a result of the Company's projected undiscounted future cash flows related to certain store locations and other assets being less than their carrying value, the Company recorded impairment charges, including properties that were the subject of announced store closings. The fair values of these assets were calculated based on the projected cash flows and an estimated risk-adjusted rate of return that would be used by market participants in valuing these assets or based on prices of similar assets. During 2016, long-lived assets held and used with a carrying value of $405 million were written down to their fair value of $147 million, resulting in asset impairment charges of $258 million.

Source: Macy's, Inc.

International Financial Reporting Standards

Impairment of Value: Property, Plant, and Equipment and Finite-Life Intangible Assets. Highlighted below are some important differences in accounting for impairment of value for property, plant, and equipment and finite-life intangible assets between U.S. GAAP and *IAS No. 36.*[18]

LO11–10

	U.S. GAAP	IFRS
When to Test	When events or changes in circumstances indicate that book value may not be recoverable.	Assets must be assessed for indicators of impairment at the end of each reporting period. Indicators of impairment are similar to U.S. GAAP.
Recoverability	An impairment loss is required when the undiscounted sum of estimated future cash flows from an asset is less than the asset's book value.	There is no equivalent recoverability test. An impairment loss is required when the "recoverable amount" (the higher of the asset's value-in-use and fair value less costs to sell) is less than the asset's book value. An asset's value-in-use is calculated as the present value of estimated future cash flows.
Measurement	The impairment loss is the amount by which fair value is less than book value.	The impairment loss is the amount by which the recoverable amount is less than book value.
Subsequent Reversal of Loss	Prohibited.	Required if the circumstances that caused the impairment are resolved.

Let's look at an illustration highlighting the important differences described above. The Jasmine Tea Company has a factory that has significantly decreased in value due to technological innovations in the industry. Below are data related to the factory's assets:

	($ in millions)
Book value	$18.5
Undiscounted sum of estimated future cash flows	19.0
Value-in-use (present value of future cash flows)	16.0
Fair value less cost to sell (determined by appraisal)	15.5

(continued)

[18]"Impairment of Assets," *International Accounting Standard No. 36* (IASCF), as amended effective January 1, 2018.

(concluded)

What amount of impairment loss should Jasmine Tea recognize, if any, under U.S. GAAP? Under IFRS?

U.S. GAAP There is no impairment loss. The sum of undiscounted estimated future cash flows exceeds the book value.

IFRS Jasmine should recognize an impairment loss of $2.5 million. Indicators of impairment are present and book value exceeds both value-in-use (present value of cash flows) and fair value less costs to sell. The recoverable amount is $16 million, the higher of value-in-use ($16 million) and fair value less costs to sell ($15.5 million). The impairment loss is the difference between book value of $18.5 million and the $16 million recoverable amount.

Nokia, a Finnish company, prepares its financial statements according to IFRS. The following disclosure note describes the company's impairment policy:

Assessment of the Recoverability of Long-Lived Assets, Intangible Assets, and Goodwill (in part)

The carrying value of identifiable intangible assets and long-lived assets is assessed if events or changes in circumstances indicate that such carrying value may not be recoverable. Factors that trigger an impairment review include, but are not limited to, underperformance relative to historical or projected future results, significant changes in the manner of the use of the acquired assets or the strategy for the overall business and significant negative industry or economic trends.

Nokia conducts its impairment testing by determining the recoverable amount for the asset. The recoverable amount of an asset is the higher of its fair value less costs to sell and its value-in-use. The recoverable amount is then compared to the asset's book value and an impairment loss is recognized if the recoverable amount is less than the book value. Impairment losses are recognized immediately in the income statement.

INDEFINITE-LIFE INTANGIBLE ASSETS Intangible assets with indefinite useful lives should be tested for impairment *annually and more frequently if events or changes in circumstances indicate that it is more likely than not that the asset is impaired.*

A company has the option of first undertaking a qualitative assessment. Companies selecting this option will evaluate relevant events and circumstances to determine whether it is "more likely than not" (a likelihood of more than 50 percent) that the fair value of the asset is less than its book value. Only if that's determined to be the case will the company move forward with measuring the impairment. A list of possible events and circumstances that a company should consider in this qualitative assessment is provided in ASC 350–20–35-3C.

If fair value is less than book value, an impairment loss is recognized for the difference.

The measurement of an impairment loss for indefinite-life intangible assets is a one-step process. If fair value is less than book value, an impairment loss is recognized for the difference. Notice that we omit the cash flow recoverability test with these assets. Because we anticipate cash flows to continue indefinitely, recoverability is not a good indicator of impairment.[19]

Similar to property, plant, and equipment and finite-life intangible assets, if an impairment loss is recognized, the written-down book value becomes the new cost base for future cost allocation. Recovery of the impairment loss is prohibited. Disclosure requirements also are similar.

[19]Until recently, the measurement of impairment for goodwill was a more complicated, two-step process. Step 1 was a recoverability test which involved comparing the fair value of the reporting unit to its book value. If step 1 indicated impairment, step 2 then required calculation of the goodwill's *implied fair value*. The implied fair value was computed as the fair value of the reporting unit minus the fair value of the reporting unit's net assets excluding goodwill. If the implied goodwill was less than the book value of goodwill, an impairment loss was recognized. Accounting Standards Update 2017-04 eliminated step 2 for measurement of impairment for goodwill. Now, the one-step process involves comparing only the fair value of the reporting unit to its book value, as described in the text.

International Financial Reporting Standards

LO11–10

Impairment of Value: Indefinite-Life Intangible Assets. Similar to U.S. GAAP, IFRS requires indefinite-life intangible assets to be tested for impairment at least annually. However, under U.S. GAAP, a company can choose first to provide only a qualitative assessment of the likelihood of impairment to determine if quantitative measurement is then necessary. Also, under U.S. GAAP, the impairment loss is measured as the difference between book value and fair value, while under IFRS the impairment loss is the difference between book value and the recoverable amount. The recoverable amount is the higher of the asset's value-in-use (present value of estimated future cash flows) and fair value less costs to sell.

IFRS requires the reversal of an impairment loss (other than for goodwill) if the circumstances that caused the impairment are resolved. Reversals are prohibited under U.S. GAAP.

Also, indefinite-life intangible assets may not be combined with other indefinite-life intangible assets for the required annual impairment test. Under U.S. GAAP, though, if certain criteria are met, indefinite-life intangible assets should be combined for the required annual impairment test.

Goodwill. Recall that goodwill is a unique intangible asset that is recorded only when one company acquires control of another company. Unlike other assets, its cost (a) can't be directly associated with any specific identifiable right and (b) is not separable from the company as a whole. Because of these unique characteristics, we discuss goodwill separately, although conceptually the measurement of impairment for goodwill is the same as for other indefinite-life intangible assets.

For all classifications of assets, we decide whether a write-down due to impairment is required by determining whether the value of an asset has fallen below its book value. However, in this comparison, the value of assets for property, plant, and equipment and finite-life intangible assets is considered to be value-in-use as measured by the sum of undiscounted cash flows expected from the asset. But due to its unique characteristics, the value of goodwill is not associated with any specific cash flows. By its very nature, goodwill is inseparable from a particular *reporting unit.* A reporting unit is an operating segment of a company or a component of an operating segment for which discrete financial information is available and segment management regularly reviews the operating results of that component. So, for purposes of impairment testing, we compare the value of the reporting unit itself with its book value. If the fair value of the reporting unit is less than its book value, an impairment loss is indicated.

A goodwill impairment loss is indicated when the fair value of the reporting unit is less than its book value.

The impairment loss recognized can't exceed the book value of goodwill. In other words, the impairment loss can't reduce the book value of goodwill below zero. Also, if goodwill is tested for impairment at the same time as other assets of the reporting unit, the other assets must be tested first, and any impairment losses and asset write-downs are recorded prior to testing goodwill. Subsequent reversal (recovery) of a previous goodwill impairment loss is not allowed. A goodwill impairment example is provided in Illustration 11–22.

The goodwill impairment loss can't exceed the book value of goodwill.

Illustration 11–22

Impairment Loss—Goodwill

In 2020, the Upjane Corporation acquired Pharmacopia Corporation for $500 million. Upjane recorded $100 million in goodwill related to this acquisition because the fair value of the identifiable net assets of Pharmacopia was $400 million. After the acquisition, Pharmacopia continues to operate as a separate company and is considered a reporting unit.

At the end of 2021, events and circumstances indicated that it is more likely than not that the fair value of Pharmacopia is less than its book value requiring Upjane to perform the goodwill impairment test. The book value of Pharmacopia's net assets at the end of 2021 is $440 million, including the $100 million in goodwill. On that date, the fair value of Pharmacopia is estimated to be $360 million.

(continued)

Illustration 11–22
(concluded)

Measurement of the impairment loss:		
Fair value of Pharmacopia		$360 million
Book value of the net assets of Pharmacopia		440 million
Impairment loss		**$ (80) million**
The entry to record the loss is ($ in millions):		
Loss on impairment of goodwill	**80**	
Goodwill		80

Recall that Upjane recorded $100 million in goodwill related to the initial acquisition of Pharmacopia. After the impairment loss of $80 million is recorded, the balance of goodwill is reduced to $20 million. If Pharmacopia's fair value was less than its book value by more than $100 million, then the goodwill impairment loss would have been limited to only $100 million to reduce the balance of goodwill to zero. The goodwill impairment loss normally is reported in the income statement as a separate component of operating expenses.

Source: Goodwill

Some examples of multibillion dollar goodwill impairment losses in recent years are shown in Illustration 11–23.

Illustration 11–23
Goodwill Impairment Losses

Real World Financials

Company	Goodwill Impairment Loss
General Motors	$27.1 billion
Hewlett-Packard	13.7 billion
Microsoft	11.3 billion
Yahoo	4.5 billion
Boston Scientific	4.4 billion
Community Health Systems	1.4 billion

Under Armour's disclosure of its goodwill impairment loss is shown in Illustration 11–24.

Illustration 11–24
Goodwill Impairment Disclosure—Under Armour, Inc.

Real World Financials

Goodwill Impairment Charge (in part)

Goodwill and indefinite lived intangible assets are not amortized and are required to be tested for impairment at least annually or sooner whenever events or changes in circumstances indicate that the assets may be impaired. In conducting an annual impairment test, the Company first reviews qualitative factors to determine whether it is more likely than not that the fair value of the reporting unit is less than its carrying amount. If factors indicate that is the case, the Company performs the goodwill impairment test. The Company compares the fair value of the reporting unit with its carrying amount.

During the third quarter of 2017, the Company made the strategic decision to not pursue certain planned future revenue streams in its Connected Fitness business in connection with the 2017 Restructuring Plan. The Company determined sufficient indication existed to trigger the performance of an interim impairment for the Company's Connected Fitness reporting unit resulting in goodwill impairment of $28.6 million, which represents all goodwill allocated to this reporting unit.

Where We're Headed

In January 2017, the Financial Accounting Standards Board (FASB) completed Phase 1 of its two-phase project on goodwill by issuing ASU 2017-04. This phase involved simplifying the measurement of goodwill impairment, as described in Illustration 11–22.

In Phase 2 of the project (which is not part of the ASU discussed above), the FASB will consider permitting or requiring companies to amortize goodwill and also consider making other changes in the impairment testing methodology.

Additional Consideration

Private Company GAAP—Accounting for Goodwill. The Private Company Council (PCC) sought feedback from private company stakeholders on the issue of goodwill accounting and found that most users of private company financial statements disregard goodwill and goodwill impairment losses. As a result, the PCC concluded that the cost and complexity of goodwill accounting outweigh the benefits for private companies.

In response to the PCC's conclusion, the FASB issued an Accounting Standards Update in 2014 that allows an accounting alternative for the subsequent measurement of goodwill for private companies that calls for goodwill to be amortized and also simplifies the goodwill impairment test.[20]

The main provisions of the alternative are

1. Amortizing goodwill on a straight-line basis over a maximum of 10 years.
2. Testing goodwill for impairment at either the company level or the reporting unit level.
3. Testing goodwill for impairment only when a triggering event occurs indicating that goodwill may be impaired.
4. The option of determining whether a quantitative impairment test is necessary when a triggering event occurs.
5. If a quantitative test is necessary, measuring the goodwill impairment loss as the amount by which fair value of the company (or reporting unit) is less than its book value, not to exceed the book value of goodwill.

The first provision is now being considered for public companies. We discussed this above in the previous "Where We're Headed" box.

International Financial Reporting Standards

● LO11–10

Impairment of Value—Goodwill. Highlighted below are some important differences in accounting for the impairment of goodwill between U.S. GAAP and *IAS No. 36*.

	U.S. GAAP	IFRS
Level of Testing	*Reporting unit*—a segment or a component of an operating segment for which discrete financial information is available.	*Cash-generating unit (CGU)*—the lowest level at which goodwill is monitored by management. A CGU can't be lower than a segment.
Measurement	Compare the fair value of the reporting unit with its book value. A loss is indicated if fair value is less than book value. Other assets must be tested first and any impairment loss and asset write-down is recorded prior to testing goodwill.	Compare the recoverable amount of the CGU to book value. If the recoverable amount is less, reduce goodwill first, then other assets. The recoverable amount is the higher of the asset's value-in-use and fair value less costs to sell. An asset's value-in-use is calculated as the present value of estimated future cash flows.

IAS No. 36 requires goodwill to be tested for impairment at least annually. However, under U.S. GAAP, a company can choose first to provide only a qualitative assessment of the likelihood of goodwill impairment to determine if quantitative measurement is then necessary. Both U.S. GAAP and *IAS No. 36* prohibit the reversal of goodwill impairment losses.

Let's look at an illustration highlighting these differences.

(continued)

[20]*Accounting Standards Update No. 2014–02,* "Intangibles—Goodwill and Other (Topic 350): Accounting for Goodwill," (Norwalk, Conn.: FASB, January 2014).

(concluded)

Canterbury LTD. has $38 million of goodwill in its balance sheet from the 2019 acquisition of Denton, Inc. At the end of 2021, Canterbury's management provided the following information for the year-end goodwill impairment test ($ in millions):

Fair value of Denton (determined by appraisal)	$132
Fair value of Denton's net assets (excluding goodwill)	120
Book value of Denton's net assets (including goodwill)	150
Value-in-use (present value of Denton's estimated future cash flows)	135

Assume that Denton is considered a reporting unit under U.S. GAAP and a cash-generating unit under IFRS, and that its fair value approximates fair value less costs to sell. What is the amount of goodwill impairment loss that Canterbury should recognize, if any, under U.S. GAAP? Under IFRS?

U.S. GAAP	Fair value of Denton (determined by appraisal)	$ 132
	Book value of Denton's net assets (including goodwill)	150
	Impairment loss	$ (18)
IFRS	The recoverable amount is $135 million, the higher of the $135 million value-in-use (present value of estimated future cash flows) and the $132 million fair value less costs to sell.	
	Recoverable amount	$ 135
	Book value of Denton's net assets (including goodwill)	(150)
	Impairment loss	$ (15)

Deutsche Bank is the largest bank in Germany and one of the largest financial institutions in Europe and the world. The company prepares its financial statements according to IFRS. The following disclosures describe the company's goodwill impairment policy as well as goodwill impairment loss:

Impairment of Goodwill (in part)

Goodwill is tested for impairment annually in the fourth quarter by comparing the recoverable amount of each goodwill-carrying cash-generating unit (CGU) with its carrying amount. In addition, in accordance with IAS 36, the Group tests goodwill whenever a triggering event is identified. The recoverable amount is the higher of a CGU's fair value less costs of disposal and its value in use.

Impairment charge during the period (in part)

The goodwill impairment test resulted in goodwill impairments totaling € 4,933 million, consisting of € 2,168 million, and € 2,765 million in the former CGUs, CB&S, and PBC, respectively. The impairment in CB&S was mainly driven by changes to the business mix in light of expected higher regulatory capital requirements, leading to a recoverable amount of approximately € 26.1 billion. The impairment in PBC was, in addition to the changed capital requirements, mainly driven by the disposal expectations regarding Hua Xia Bank Co. Ltd. and Postbank, which resulted in a recoverable amount of approximately € 12.3 billion for the CGU.

Assets Held for Sale

For assets held for sale, if fair value minus cost to sell is less than book value, an impairment loss is recognized for the difference.

We have been discussing the recognition and measurement for the impairment of value of assets to be held and used. We also test for impairment of assets held for sale. These are assets management has actively committed to immediately sell in their present condition and for which sale is probable.

An asset or group of assets classified as held for sale is measured at the lower of its book value, or fair value minus cost to sell. An impairment loss is recognized for any write-down to fair value minus cost to sell.[21] Except for including the cost to sell, notice the similarity

[21] If the asset is unsold at the end of a subsequent reporting period, a gain is recognized for any increase in fair value less cost to sell, but not in excess of the loss previously recognized.

to impairment of assets to be held and used. We don't depreciate or amortize these assets while classified as held for sale and we report them separately in the balance sheet. Recall from our discussion of discontinued operations in Chapter 4 that similar rules apply for a component of an entity that is classified as held for sale.

Illustration 11–25 summarizes the guidelines for the recognition and measurement of impairment losses.

Illustration 11–25
Summary of Asset Impairment Guidelines

Asset Classification	When to Test for Impairment	Impairment Test
Held and Used		
Property, plant, and equipment and finite-life intangible assets	When events or circumstances indicate book value may not be recoverable.	Step 1—An impairment loss is required only when book value is not recoverable (undiscounted sum of estimated future cash flows less than book value).
		Step 2—The impairment loss is the amount by which fair value is less than book value.
Indefinite-life intangible assets (other than goodwill)	At least annually, and more frequently if indicated. Option to first choose qualitative assessment to determine if quantitative measurement is then necessary.	If fair value is less than book value, an impairment loss is recognized for the difference.
Goodwill	At least annually, and more frequently if indicated. Option to first choose qualitative assessment to determine if quantitative measurement is then necessary.	If fair value of the reporting unit is less than its book value, a goodwill impairment loss is recognized for the difference. The impairment loss recognized can't exceed the book value of goodwill.
Held for Sale	At the time of classification as held for sale and thereafter.	If fair value minus cost to sell is less than book value, an impairment loss is recognized for the difference.

Impairment Losses and Earnings Quality

What do losses from the write-down of inventory and restructuring costs have in common? The presence of these items in a corporate income statement presents a challenge to an analyst trying to determine a company's permanent earnings—those likely to continue in the future. We discussed these issues in prior chapters.

We now can add asset impairment losses to the list of "big bath" accounting techniques some companies use to manipulate earnings. By writing off large amounts of assets, companies significantly reduce earnings in the year of the write-off but are able to increase future earnings by lowering future depreciation, depletion, or amortization. Here's how. We measure the impairment loss as the amount by which an asset's fair value is less than its book value. However, in most cases, fair value must be estimated, and the estimation process usually involves a forecast of future net cash flows the company expects to generate from the asset's use. If a company underestimates future net cash flows, fair value is understated. This has two effects: (1) current year's income is unrealistically low due to the impairment loss being overstated and (2) future income is unrealistically high because depreciation, depletion, and amortization are based on understated asset values.

An analyst must decide whether to consider asset impairment losses as temporary in nature or as a part of permanent earnings.

Concept Review Exercise

IMPAIRMENT

Part A:

Illumination Inc. owns a factory in Wisconsin that makes light bulbs. During 2021, due to increased competition from LED light bulb manufacturers, the company determined that an impairment test was appropriate. Management has prepared the following information for the assets of the factory ($ in millions):

Cost	$345
Accumulated depreciation	85
Estimated future undiscounted cash flows to be generated by the factory	230
Estimated fair value of the factory assets	170

Required:

1. Determine the amount of impairment loss Illumination should recognize, if any.
2. If a loss is indicated, prepare the journal entry to record the loss.
3. Repeat requirement 1 assuming that the estimated undiscounted future cash flows are $270 million instead of $230 million.

Solution:

1. Determine the amount of impairment loss Illumination should recognize, if any.

   Recoverability Test: Because the undiscounted future cash flows of $230 million are less than book value of $260 ($345 – $85) million, an impairment loss is indicated.

   Measurement: The impairment loss is $90 million, determined as follows:

	($ in millions)
Fair value	$170
Less: Book value	260
Impairment loss	$ (90)

2. If a loss is indicated, prepare the journal entry to record the loss.

	($ in millions)	
Loss on impairment (determined above)	90	
Accumulated depreciation (balance)	85	
Factory assets ($345 – $170)		175

3. Repeat requirement 1 assuming that the estimated undiscounted future cash flows are $270 million instead of $230 million.

   Because the undiscounted future cash flows of $270 million exceed book value of $260 million, the recoverability test indicates that there is no impairment, so no impairment loss is recorded.

Part B:

In 2019, Illumination Inc. acquired Zapo Lighting Company for $620 million, of which $80 million was allocated to goodwill. After the acquisition, Zapo continues to operate a separate company and is considered a reporting unit. At the end of 2021, management provided the following information for a required goodwill impairment test ($ in millions):

Fair value of Zapo Lighting	$540
Book value of Zapo's net assets (including goodwill)	600

Required:

Determine the amount of goodwill impairment loss that Illumination should recognize at the end of 2021, if any.

Solution:

Measurement of goodwill impairment loss:	($ in millions)
Fair value of Zapo	$540
Book value of Zapo's net assets (including goodwill)	600
Impairment loss	$ (60)

Subsequent Expenditures

PART C

Now that we have acquired and measured assets, we can address accounting issues incurred subsequent to their acquisition. This part of the chapter deals with the treatment of expenditures made over the life of these assets to maintain and/or improve them.

Expenditures Subsequent to Acquisition

Many long-lived assets require expenditures to repair, maintain, or improve them after their acquisition. These expenditures can present accounting problems if they are material. In general, a choice must be made between capitalizing the expenditures by either increasing the asset's book value or creating a new asset, or expensing them in the period in which they are incurred. Typically, we capitalize expenditures that are expected to produce benefits beyond the current fiscal year. In contrast, expenditures that simply maintain a given level of benefits are expensed in the period they are incurred.

● LO11–9

Expenditures related to assets can increase future benefits in the following ways:

1. An extension of the *useful life* of the asset.
2. An increase in the *operating efficiency* of the asset resulting in either an increase in the quantity of goods or services produced or a decrease in future operating costs.
3. An increase in the *quality* of the goods or services produced by the asset.

Expenditures that cause any of these results should be capitalized initially and then expensed in future periods through depreciation, depletion, or amortization. Of course, materiality is an important factor in the practical application of this approach.

For convenience, many companies set materiality thresholds for the capitalization of any expenditure. For example, a company might decide to expense all expenditures under $1,000 regardless of whether or not future benefits are increased. Judgment is required to determine the appropriate materiality threshold as well as the appropriate treatment of expenditures over $1,000. There often are practical problems in capitalizing these expenditures. For example, even if future benefits are increased by the expenditure, it may be difficult to determine how long the benefits will last. It's important for a company to establish a policy for treating these expenditures and apply it consistently.

Many companies do not capitalize any expenditure unless it exceeds a predetermined amount that is considered material.

We classify subsequent expenditures as (1) repairs and maintenance, (2) additions, (3) improvements, or (4) rearrangements.

Repairs and Maintenance

These expenditures are made to *maintain* a given level of benefits provided by the asset and do not *increase* future benefits. For example, the cost of an engine tune-up or the repair of an engine part for a delivery truck allows the truck to continue its productive activity. If the maintenance is not performed, the truck will not provide the benefits originally anticipated. In that sense, future benefits are provided; without the repair, the truck will no longer operate. The key, though, is that future benefits are not provided *beyond those originally anticipated.* Expenditures for these activities should be expensed in the period incurred.

Expenditures for *repairs and maintenance* generally are expensed when incurred.

Additional Consideration

If repairs and maintenance costs are seasonal, interim financial statements may be misstated. For example, suppose annual maintenance is performed on a company's fleet of delivery trucks. The annual income statement correctly includes one year's maintenance expense. However, for interim reporting purposes, if the entire expenditure is made in one quarter, should that quarter's income statement include as expense the entire cost of the annual maintenance? If these expenditures can be anticipated, they should be accrued evenly throughout the year by crediting an allowance account. The allowance account is then debited when the maintenance is performed.

Additions

As the term implies, **additions** involve adding a new major component to an existing asset and should be capitalized because future benefits are increased. For example, adding a refrigeration unit to a delivery truck increases the capability of the truck, thus increasing its future benefits. Other examples include the construction of a new wing on a building and the addition of a security system to an existing building.

The costs of *additions* usually are capitalized.

The capitalized cost includes all necessary expenditures to bring the addition to a condition and location for use. For a building addition, this might include the costs of tearing down and removing a wall of the existing building. The capitalized cost of additions is depreciated over the remaining useful life of the original asset or its own useful life, whichever is shorter.

Improvements

The costs of *improvements* usually are capitalized.

Expenditures classified as **improvements** involve the replacement of a major component of an asset. The replacement can be a new component with the same characteristics as the old component or a new component with enhanced operating capabilities. For example, an existing refrigeration unit in a delivery truck could be replaced with a new but similar unit or with a new and improved refrigeration unit. In either case, the cost of the improvement usually increases future benefits and should be capitalized by increasing the book value of the related asset (the delivery truck) and depreciated over the useful life of the improved asset. There are three methods used to record the cost of improvements.

1. *Substitution.* The improvement can be recorded as both (1) a disposition of the old component and (2) the acquisition of the new component. This approach is conceptually appealing but it is practical only if the original cost and accumulated depreciation of the old component can be separately identified.
2. *Capitalization of new cost.* Another way to record an improvement is to include the cost of the improvement (net of any consideration received from the disposition of the old component) as a debit to the related asset account, without removing the original cost and accumulated depreciation of the original component. This approach is acceptable only if the book value of the original component has been reduced to an immaterial amount through prior depreciation.
3. *Reduction of accumulated depreciation.* Another way to increase an asset's book value is to leave the asset account unaltered but decrease its related accumulated depreciation. The argument for this method is that many improvements extend the useful life of an asset and are equivalent to a partial recovery of previously recorded depreciation. This approach produces the same book value as the capitalization of cost to the asset account. However, cost and accumulated depreciation amounts will differ under the two methods.

The three methods are compared in Illustration 11–26.

Illustration 11–26

Improvements

The Palmer Corporation replaced the air conditioning system in one of its office buildings that it leases to tenants. The cost of the old air conditioning system, $200,000, is included in the cost of the building. However, the company has separately depreciated the air conditioning system. Depreciation recorded up to the date of replacement totaled $160,000. The old system was removed and the new system installed at a cost of $230,000, which was paid in cash. Parts from the old system were sold for $12,000.

Accounting for the improvement differs depending on the alternative chosen.

Alternative 1—Substitution

1. *Substitution* (a) Disposition of old component.

Cash	12,000	
Accumulated depreciation—buildings (remove old)	160,000	
Loss on disposal (difference)	28,000	
Buildings (remove old)		200,000

(continued)

Illustration 11–26
(concluded)

Buildings (add new)	230,000	
Cash		230,000

(b) Acquisition of new component.

Alternative 2—Capitalization of new cost

2. *Capitalization of new cost.*

Buildings	218,000	
Cash ($230,000 – $12,000)		218,000

Alternative 3—Reduction of accumulated depreciation

3. *Reduction of accumulated depreciation.*

Accumulated depreciation—buildings	218,000	
Cash ($230,000 – $12,000)		218,000

Rearrangements

Expenditures made to restructure an asset without addition, replacement, or improvement are termed **rearrangements**. The objective is to create a new capability for the asset and not necessarily to extend its useful life. Examples include the rearrangement of machinery on the production line to increase operational efficiency and the relocation of a company's operating plant or office building. If these expenditures are material and they clearly increase future benefits, they should be capitalized and expensed in the future periods benefited. If the expenditures are not material or if it's not certain that future benefits have increased, they should be expensed in the period incurred.

The costs of material *rearrangements* should be capitalized if they clearly increase future benefits.

Illustration 11–27 provides a summary of the accounting treatment for the various types of expenditures related to property, plant, and equipment.

Illustration 11–27
Expenditures Subsequent to Acquisition

Type of Expenditure	Definition	Usual Accounting Treatment
Repairs and maintenance	Expenditures to maintain a given level of benefits	Expense in the period incurred
Additions	The addition of a new major component to an existing asset	Capitalize and depreciate over the remaining useful life of the *original asset or its own useful* life, whichever is shorter
Improvements	The replacement of a major component	Capitalize and depreciate over the useful life of the improved asset
Rearrangements	Expenditures to restructure an asset without addition, replacement, or improvement	If expenditures are material and clearly increase future benefits, capitalize and depreciate over the future periods benefited

Costs of Defending Intangible Rights

Repairs, additions, improvements, and rearrangements generally relate to property, plant, and equipment. A possible significant expenditure incurred subsequent to the acquisition of intangible assets is the cost of defending the right that gives the intangible asset its value. If an intangible right is *successfully* defended, the litigation costs should be capitalized and amortized over the remaining useful life of the related intangible. This is the appropriate treatment of these expenditures even if the intangible asset was originally developed internally rather than purchased.

The costs incurred to *successfully* defend an intangible right should be capitalized.

If the defense of an intangible right is *unsuccessful,* then the intangible asset has no future value. In this case, the litigation costs provide no future benefit and should be expensed immediately. In addition, the book value of the intangible asset should be reduced to realizable value. For example, if a company is unsuccessful in defending a patent infringement suit, the patent's value may be eliminated and a loss recorded.

The costs incurred to *unsuccessfully* defend an intangible right should be expensed.

International Financial Reporting Standards

LO11–10

Costs of Defending Intangible Rights. Under U.S. GAAP, litigation costs to successfully defend an intangible right are capitalized and amortized over the remaining useful life of the related intangible. Under IFRS, these costs are expensed except in rare situations when an expenditure increases future benefits.[22]

Financial Reporting Case Solution

©Dragon Images/Shutterstock

1. **Is Penny correct? Do the terms *depreciation, depletion,* and *amortization* all mean the same thing?** *(p. 572)* Penny is correct. Each of these terms refers to the cost allocation of assets over their service lives. The term *depreciation* is used for plant and equipment, *depletion* for natural resources, and *amortization* for intangible assets.
2. **Weyerhaeuser determines depletion based on the "volume of timber estimated to be available." Explain this approach.** *(p. 577)* **Weyerhaeuser** is using the units-of-production method to determine depletion. The units-of-production method is an activity-based method that computes a depletion (or depreciation or amortization) rate per measure of activity and then multiplies this rate by actual activity to determine periodic cost allocation. The method is used by Weyerhaeuser to measure depletion of the cost of timber harvested and the amortization of logging roads. Logging roads are intangible assets because the company does not own the roads.
3. **Explain how asset impairment differs from depreciation, depletion, and amortization. How do companies measure impairment losses for property, plant, and equipment and intangible assets with finite useful lives?** *(p. 600)* Depreciation, depletion, and amortization reflect a gradual consumption of the benefits inherent in a long-lived asset. An implicit assumption in allocating the cost of an asset over its useful life is that there has been no significant reduction in the anticipated total benefits or service potential of the asset. Situations can arise, however, that cause a significant decline or *impairment* of those benefits or service potentials. Determining whether to record an impairment loss for an asset and actually recording the loss is a two-step process. The first step is a recoverability test—an impairment loss is required only when the undiscounted sum of estimated future cash flows from an asset is less than the asset's book value. The measurement of impairment loss—step 2—is the difference between the asset's book value and its fair value. If an impairment loss is recognized, the written-down book value becomes the new cost base for future cost allocation. ●

The Bottom Line

- LO11–1 The use of property, plant, and equipment and intangible assets represents a consumption of benefits, or service potentials, inherent in the assets. The cost of these inherent benefits or service potentials should be recognized as an expense over the periods they help to produce revenues. As there very seldom is a direct relationship between the use of assets and revenue production, accounting resorts to arbitrary methods to allocate these costs over the periods of their use. (*p. 571*)
- LO11–2 The allocation process for plant and equipment is called *depreciation.* Time-based depreciation methods estimate service life in years and then allocate depreciable base, cost less estimated residual value, using either a straight-line or accelerated pattern. Activity-based depreciation methods allocate the depreciable base by estimating service life according to some measure of productivity. When an item of property, plant, and equipment or an intangible asset is sold, a gain or loss is recognized for the difference between the consideration received and the asset's book value. (*p. 574*)

[22] "Intangible Assets," *International Accounting Standard No. 38* (IASCF), par. 20, as amended effective January 1, 2018.

- **LO11-3** The allocation process for natural resources is called *depletion*. The activity-based method called units-of-production usually is employed to determine periodic depletion. (*p. 589*)
- **LO11-4** The allocation process for intangible assets is called *amortization*. For an intangible asset with a finite useful life, the capitalized cost less any estimated residual value must be allocated to periods in which the asset is expected to contribute to the company's revenue-generating activities. An intangible asset that is determined to have an indefinite useful life is not subject to periodic amortization. Goodwill is perhaps the most typical intangible asset with an indefinite useful life. (*p. 591*)
- **LO11-5** A change in either the service life or residual value of property, plant, and equipment and intangible assets should be reflected in the financial statements of the current period and future periods by recalculating periodic depreciation, depletion, or amortization. (*p. 596*)
- **LO11-6** A change in depreciation, depletion, or amortization method is considered a change in accounting estimate that is achieved by a change in accounting principle. We account for these changes prospectively, exactly as we would any other change in estimate. One difference is that most changes in estimate do not require a company to justify the change. However, this change in estimate is a result of changing an accounting principle and therefore requires a clear justification as to why the new method is preferable. (*p. 597*)
- **LO11-7** A material error in accounting for property, plant, and equipment and intangible assets that is discovered in a year subsequent to the year of the error requires that previous years' financial statements that were incorrect as a result of the error be retrospectively restated to reflect the correction. Any account balances that are incorrect as a result of the error are corrected by journal entry. If retained earnings is one of the incorrect accounts, the correction is reported as a prior period adjustment to the beginning balance in the statement of shareholders' equity. In addition, a disclosure note is needed to describe the nature of the error and the impact of its correction on income. (*p. 598*)
- **LO11-8** Conceptually, there is considerable merit for a policy requiring the write-down of an asset when there has been a *significant* decline in value below book value. The write-down provides important information about the future cash flows to be generated from the use of the asset. However, in practice this policy is very subjective. GAAP [FASB ASC 360] establishes guidance for when to recognize and how to measure impairment losses of property, plant, and equipment and intangible assets that have finite useful lives. GAAP [FASB ASC 350] also provides guidance for the recognition and measurement of impairment for indefinite-life intangibles and goodwill. (*p. 600*)
- **LO11-9** Expenditures for repairs and maintenance generally are expensed when incurred. The costs of additions and improvements usually are capitalized. The costs of material rearrangements should be capitalized if they clearly increase future benefits. (*p. 593*)
- **LO11-10** Among the several differences between U.S. GAAP and IFRS with respect to the utilization and impairment of property, plant, and equipment and intangible assets pertains to reporting assets in the balance sheet. IFRS allows a company to value property, plant, and equipment (PP&E) and intangible assets subsequent to initial valuation at (1) cost less accumulated depreciation/amortization or (2) fair value (revaluation). U.S. GAAP prohibits revaluation. There also are differences in accounting for the impairment of property, plant, and equipment and intangible assets. (*pp. 580, 587, 590, 594, 603, 605, 607, and 614*)

Comparison with MACRS (Tax Depreciation)

APPENDIX 11A

Depreciation for financial reporting purposes is an attempt to distribute the cost of the asset, less any anticipated residual value, over the estimated useful life in a systematic and rational manner that attempts to match revenues with the use of the asset. Depreciation for income tax purposes is influenced by the revenue needs of government as well as the desire to influence economic behavior. For example, accelerated depreciation schedules currently allowed are intended to provide incentive for companies to expand and modernize their facilities thus stimulating economic growth.

The federal income tax code allows taxpayers to compute depreciation for their tax returns on assets acquired after 1986 using the **modified accelerated cost recovery** system (MACRS).[23] Key differences between the calculation of depreciation for financial reporting and the calculation using MACRS are

1. Estimated useful lives and residual values are not used in MACRS.
2. Firms can't choose among various accelerated methods under MACRS.
3. A half-year convention is used in determining the MACRS depreciation amounts.

Under MACRS, each asset generally is placed within a recovery period category. The six categories for personal property are 3, 5, 7, 10, 15, and 20 years. For example, the 5-year category includes automobiles, light trucks, and computers.

Depending on the category, fixed percentage rates are applied to the original cost of the asset. The rates for the 5-year asset category are as follows:

Year	Rate
1	20.00%
2	32.00
3	19.20
4	11.52
5	11.52
6	5.76
Total	100.00%

These rates are equivalent to applying the double-declining-balance (DDB) method with a switch to straight-line in the year straight-line yields an equal or higher deduction than DDB. In most cases, the half-year convention is used regardless of when the asset is placed in service.[24] The first-year rate of 20% for the five-year category is one-half of the DDB rate for an asset with a five-year life (2 × 20%). The sixth year rate of 5.76% is one-half of the straight-line rate established in year 4, the year straight-line depreciation exceeds DDB depreciation.

Companies have the option to use the straight-line method for the entire tax life of the asset, applying the half-year convention, rather than using MACRS depreciation schedules. Because of the differences discussed above, tax depreciation for a given year will likely be different from GAAP depreciation.

In December 2017, Congress passed the Tax Cuts and Jobs Act that substantially impacts a company's depreciation deduction for tax purposes. Under the new tax law, companies can choose to deduct 100% of the cost of many assets purchased between September 27, 2017, and December 31, 2022. This is called *bonus depreciation,* and for these assets, the MACRS depreciation schedule will not apply because the asset is fully depreciated in the year it is purchased. After 2022 and before 2027, the bonus deduction will still be allowed, but in smaller amounts each year. For qualified assets purchased in 2023, the first-year deduction drops to 80% of the asset's cost. The remaining 20% of the asset's cost would then be depreciated according to MACRS. The first-year deduction drops to 60% in 2024, 40% in 2025, and 20% in 2026. Any assets that do not qualify, or are not chosen, for bonus depreciation under the new tax law will be depreciated using MACRS. Beginning January 1, 2027, the first-year bonus depreciation is no longer available, and qualified assets will be depreciated fully using MACRS. ●

APPENDIX 11B Retirement and Replacement Methods of Depreciation

Retirement and replacement depreciation methods occasionally are used to depreciate relatively low-valued assets with short service lives. Under either approach, an aggregate asset account that represents a group of similar assets is increased at the time the initial collection is acquired.

[23]For assets acquired between 1981 and 1986, tax depreciation is calculated using the accelerated cost recovery system (ACRS), which is similar to MACRS. For assets acquired before 1981, tax depreciation can be calculated using any of the depreciation methods discussed in the chapter. Residual values are used in the calculation of depreciation for pre-1981 assets.

[24]In certain situations, mid-quarter and mid-month conventions are used.

Retirement Method

Using the **retirement depreciation method**, the asset account also is increased for the cost of subsequent expenditures. When an item is disposed of, the asset account is credited for its cost, and depreciation expense is recorded for the difference between cost and proceeds received, if any. No other entries are made for depreciation. As a consequence, one or more periods may pass without any expense recorded. For example, the following entry records the purchase of 100 handheld calculators at $50 acquisition cost each:

The retirement depreciation method records depreciation when assets are disposed of and measures depreciation as the difference between the proceeds received and cost.

Calculators (100 × $50)	5,000	
Cash		5,000
To record the acquisition of calculators.		

If 20 new calculators are acquired at $45 each, the asset account is increased.

Calculators (20 × $45)	900	
Cash		900
To record additional calculator acquisitions.		

Thirty calculators are disposed of (retired) by selling them secondhand to a bookkeeping firm for $5 each. The following entry reflects the retirement method:

Cash (30 × $5)	150	
Depreciation expense (difference)	1,350	
Calculators (30 × $50)		1,500
To record the sale/depreciation of calculators.		

Notice that the retirement system assumes a FIFO cost flow approach in determining the cost of assets, $50 each, that were disposed.

Replacement Method

By the **replacement depreciation method**, the initial acquisition of assets is recorded the same way as by the retirement method; that is, the aggregate cost is increased. However, depreciation expense is the amount paid for new or replacement assets. Any proceeds received from asset dispositions reduces depreciation expense. For our example, the acquisition of 20 new calculators at $45 each is recorded as depreciation as follows:

By the replacement method, depreciation is recorded when assets are replaced.

Depreciation expense (20 × $45)	900	
Cash		900
To record the replacement/depreciation of calculators.		

The sale of the old calculators is recorded as a reduction of depreciation:

Cash (30 × $5)	150	
Depreciation expense		150
To record the sale of calculators.		

The asset account balance remains the same throughout the life of the aggregate collection of assets.

Because these methods are likely to produce aggregate expense measurements that differ from individual calculations, retirement and replacement methods are acceptable only in situations where the distortion in depreciation expense does not have a material effect on income. These methods occasionally are encountered in regulated industries such as utilities. ●

Questions For Review of Key Topics

Q 11–1 Explain the similarities in and differences among depreciation, depletion, and amortization.

Q 11–2 Depreciation is a process of cost allocation, not valuation. Explain this statement.

Q 11–3 Identify and define the three characteristics of an asset that must be established to determine periodic depreciation, depletion, or amortization.

Q 11–4 Discuss the factors that influence the estimation of service life for a depreciable asset.

Q 11–5 What is meant by depreciable base? How is it determined?

Q 11–6 Briefly differentiate between activity-based and time-based allocation methods.

Q 11–7 Briefly differentiate between the straight-line depreciation method and accelerated depreciation methods.

Q 11–8 Why are time-based depreciation methods used more frequently than activity-based methods?

Q 11–9 What are some factors that could explain the predominant use of the straight-line depreciation method?

Q 11–10 When an item of property, plant, and equipment is disposed of, how is gain or loss on disposal computed?

Q 11–11 Briefly explain the differences and similarities between the group approach and composite approach to depreciating aggregate assets.

Q 11–12 Define depletion and compare it with depreciation.

Q 11–13 Compare and contrast amortization of intangible assets with depreciation and depletion.

Q 11–14 What are some of the simplifying conventions a company can use to calculate depreciation for partial years?

Q 11–15 Explain the accounting treatment required when a change is made to the estimated service life of equipment.

Q 11–16 Explain the accounting treatment and disclosures required when a change is made in depreciation method.

Q 11–17 Explain the steps required to correct an error in accounting for property, plant, and equipment and intangible assets that is discovered in a year subsequent to the year the error was made.

Q 11–18 Explain what is meant by the impairment of the value of property, plant, and equipment and intangible assets. How should these impairments be accounted for?

Q 11–19 Explain the differences in the accounting treatment of repairs and maintenance, additions, improvements, and rearrangements.

IFRS **Q 11–20** Identify any differences between U.S. GAAP and International Financial Reporting Standards in the subsequent valuation of property, plant, and equipment and intangible assets.

IFRS **Q 11–21** Briefly explain the difference between U.S. GAAP and IFRS in the *measurement* of an impairment loss for property, plant, and equipment and finite-life intangible assets.

IFRS **Q 11–22** Briefly explain the differences between U.S. GAAP and IFRS in the measurement of an impairment loss for goodwill.

IFRS **Q 11–23** Under U.S. GAAP, litigation costs to successfully defend an intangible right are capitalized and amortized over the remaining useful life of the related intangible. How are these costs typically accounted for under IFRS?

Brief Exercises

connect

BE 11–1
Cost allocation
● LO11–1

At the beginning of its fiscal year, Koeplin Corporation purchased equipment for $50,000. At the end of the year, the equipment had a fair value of $32,000. Koeplin's controller recorded depreciation of $18,000 for the year, the decline in the equipment's value. Is this the correct approach to measuring periodic depreciation (yes/no)? Discuss.

BE 11–2
Depreciation methods
● LO11–2

On January 1, 2021, Canseco Plumbing Fixtures purchased equipment for $30,000. Residual value at the end of an estimated four-year service life is expected to be $2,000. The company expects the equipment to operate for 10,000 hours. Calculate depreciation expense for 2021 and 2022 using each of the following depreciation methods: (1) straight line, (2) double-declining balance, and (3) units-of-production using hours operated. The equipment operated for 2,200 and 3,000 hours in 2021 and 2022, respectively.

BE 11–3
Depreciation methods; partial periods
● LO11–2

Refer to the situation described in BE 11–2. Assume the equipment was purchased on March 31, 2021, instead of January 1. Calculate depreciation expense for 2021 and 2022 using each of the following depreciation methods: (1) straight line, (2) double-declining balance, and (3) units-of-production using hours operated.

BE 11–4
Depreciation method; Sum-of-the-years'-digits
● LO11–2

Refer to the situation described in BE 11–2. Calculate depreciation expense for 2021 and 2022 using sum-of-the-years'-digits assuming the equipment was purchased on (1) January 1, 2021, and (2) March 31, 2021.

BE 11–5
Disposal of property, plant, and equipment
● LO11–2

Lawler Clothing sold manufacturing equipment for $16,000. Lawler originally purchased the equipment for $80,000, and depreciation through the date of sale totaled $71,000. What was the gain or loss on the sale of the equipment?

BE 11–6
Disposal of property, plant, and equipment
● LO11–2

Funseth Farms Inc. purchased a tractor in 2018 at a cost of $30,000. The tractor was sold for $3,000 in 2021. Depreciation recorded through the disposal date totaled $26,000. (1) Prepare the journal entry to record the sale. (2) Now assume the tractor was sold for $10,000; prepare the journal entry to record the sale.

BE 11–7
Disposal of property, plant, and equipment and intangible assets
● LO11–2

On July 15, 2021, Cottonwood Industries sold a patent and equipment to Roquemore Corporation for $750,000 and $325,000, respectively. On the date of the sale, the book value of the patent was $120,000, and the book value of the equipment was $400,000 (cost of $550,000 less accumulated depreciation of $150,000). Prepare separate journal entries to record (1) the sale of the patent and (2) the sale of the equipment.

BE 11–8
Assets held for sale
● LO11–2

On December 31, 2021, management of Jines Construction committed to a plan for selling an office building and its related equipment. Both are available for immediate sale. The building has a book value of $800,000 and a fair value of $900,000. The equipment has a book value of $240,000 and a fair value of $200,000. Calculate the amount that each asset will be reported at in the balance sheet and the amount of any gain or loss that will be reported in the income statement for the year ended December 31, 2021. Management expects to sell both assets in 2022.

BE 11–9
Group depreciation; disposal
● LO11–2

Mondale Winery depreciates its equipment using the group method. The cost of equipment purchased in 2021 totaled $425,000. The estimated residual value of the equipment was $40,000 and the group depreciation rate was determined to be 18%. (1) What is the annual depreciation for the group? (2) If equipment that cost $42,000 is sold in 2022 for $35,000, what amount of gain or loss will the company recognize for the sale?

BE 11–10
Depletion
● LO11–3

Fitzgerald Oil and Gas incurred costs of $8.25 million for the acquisition and development of a natural gas deposit. The company expects to extract 3 million cubic feet of natural gas during a four-year period. Natural gas extracted during years 1 and 2 were 700,000 and 800,000 cubic feet, respectively. What was the depletion for year 1 and for year 2?

BE 11–11
Amortization; Partial periods
● LO11–4

On June 28, Lexicon Corporation acquired 100% of the common stock of Gulf & Eastern. The purchase price allocation included the following items: $4 million, patent; $3 million, developed technology; $2 million, indefinite-life trademark; $5 million, goodwill. Lexicon's policy is to amortize intangible assets using the straight-line method, no residual value, and a five-year useful life. What is the total amount of expenses (ignoring taxes) that would appear in Lexicon's income statement for the year ended December 31 related to these items?

BE 11–12
Amortization; Software development costs
● LO11–4

[This is a continuation of Exercise 10–31 in Chapter 10 and includes amortization.]

Early in 2021, the Excalibur Company began developing a new software package to be marketed. The project was completed in December 2021 at a cost of $6 million. Of this amount, $4 million was spent before technological feasibility was established. Excalibur expects a useful life of five years for the new product with total revenues of $10 million. During 2022, revenue of $3 million was recognized.

Required:

(1) Prepare a journal entry to record the 2021 development costs. (2) Calculate the required amortization for 2022, and (3) determine the amount to report for the computer software costs in the December 31, 2022, balance sheet.

BE 11–13
Amortization; Software development costs
● LO11–4

[This is a continuation of Exercise 10–32 in Chapter 10 and includes amortization.]

On September 30, 2021, Athens Software began developing a software program to shield personal computers from malware and spyware. Technological feasibility was established on February 28, 2022, and the program was available for release on April 30, 2022. Development costs were incurred as follows:

September 30 through December 31, 2021	$2,200,000
January 1 through February 28, 2022	800,000
March 1 through April 30, 2022	300,000

Athens expects a useful life of four years for the software and total revenues of $5,000,000 during that time. During 2022, revenue of $1,000,000 was recognized.

Required:

(1) Prepare a journal entry to record the development costs in each year of 2021 and 2022. (2) Calculate the required amortization for 2022.

BE 11–14
Change in estimate; useful life of equipment
● LO11–5

At the beginning of 2019, Robotics Inc. acquired a manufacturing facility for $12 million. $9 million of the purchase price was allocated to the building. Depreciation for 2019 and 2020 was calculated using the straight-line method, a 25-year useful life, and a $1 million residual value. In 2021, the estimates of useful life and residual value were changed to 20 years and $500,000, respectively. What is depreciation on the building for 2021?

BE 11–15
Change in principle; change in depreciation method
● LO11–6

Refer to the situation described in BE 11–14. Assume that instead of changing the useful life and residual value, in 2021 the company switched to the double-declining-balance depreciation method. How should Robotics account for the change? What is depreciation on the building for 2021?

BE 11–16
Error correction
● LO11–7

Refer to the situation described in BE 11–14. Assume that 2019 depreciation was incorrectly recorded as $32,000. This error was discovered in 2021. (1) Record the journal entry needed in 2021 to correct the error. (2) What is depreciation on the building for 2021 assuming no change in estimate of useful life or residual value?

BE 11–17
Impairment; property, plant, and equipment
● LO11–8

Collison and Ryder Company (C&R) has been experiencing declining market conditions for its sportswear division. Management decided to test the assets of the division for possible impairment. The test revealed the following: book value of division's assets, $26.5 million; fair value of division's assets, $21 million; undiscounted sum of estimated future cash flows generated from the division's assets, $28 million. What amount of impairment loss should C&R recognize?

BE 11–18
Impairment; property, plant, and equipment
● LO11–8

Refer to the situation described in BE 11–17. Assume that the undiscounted sum of estimated future cash flows is $24 million instead of $28 million. What amount of impairment loss should C&R recognize?

BE 11–19
IFRS; impairment; property, plant, and equipment
● LO11–8, LO11–10
IFRS

Refer to the situation described in BE 11–17. Assume that the present value of the estimated future cash flows generated from the division's assets is $22 million and that their fair value approximates fair value less costs to sell. What amount of impairment loss should C&R recognize if the company prepares its financial statements according to IFRS?

BE 11–20
Impairment; goodwill
● LO11–8

WebHelper Inc. acquired 100% of the outstanding stock of Silicon Chips Corporation (SCC) for $45 million, of which $15 million was allocated to goodwill. At the end of the current fiscal year, an impairment test revealed the following: fair value of SCC, $40 million; book value of SCC's net assets (including goodwill), $42 million. What amount of impairment loss should WebHelper recognize?

BE 11–21
Impairment; goodwill
● LO11–8

Refer to the situation described in BE 11–20. Assume that the fair value of SCC is $44 million instead of $40 million. What amount of impairment loss should WebHelper recognize?

BE 11–22
IFRS; impairment; goodwill
● LO11–10
IFRS

Refer to the situation described in BE 11–20. Assume that SCC's fair value of $40 million approximates fair value less costs to sell and that the present value of SCC's estimated future cash flows is $41 million. If WebHelper prepares its financial statements according to IFRS and SCC is considered a cash-generating unit, what amount of impairment loss, if any, should WebHelper recognize?

BE 11–23
Subsequent expenditures
● LO11–9

Demmert Manufacturing incurred the following expenditures during the current fiscal year: annual maintenance on its equipment, $5,400; remodeling of offices, $22,000; rearrangement of the shipping and receiving area resulting in an increase in productivity, $35,000; addition of a security system to the manufacturing facility, $25,000. How should Demmert account for each of these expenditures?

Exercises

connect

E 11–1
Depreciation methods
● LO11–2

On January 1, 2021, the Excel Delivery Company purchased a delivery van for $33,000. At the end of its five-year service life, it is estimated that the van will be worth $3,000. During the five-year period, the company expects to drive the van 100,000 miles.

Required:

Calculate annual depreciation for the five-year life of the van using each of the following methods. Round all computations to the nearest dollar.

1. Straight line
2. Double-declining balance
3. Units of production using miles driven as a measure of output, and the following actual mileage:

Year	Miles
2021	22,000
2022	24,000
2023	15,000
2024	20,000
2025	21,000

E 11–2
Depreciation methods
● LO11–2

On January 1, 2021, the Allegheny Corporation purchased equipment for $115,000. The estimated service life of the equipment is 10 years and the estimated residual value is $5,000. The equipment is expected to produce 220,000 units during its life.

Required:

Calculate depreciation for 2021 and 2022 using each of the following methods. Round all computations to the nearest dollar.

1. Straight line
2. Double-declining balance
3. Units of production (units produced in 2021, 30,000; units produced in 2022, 25,000)

E 11–3
Depreciation methods; partial periods
● LO11–2

[This is a variation of Exercise 11–2 modified to focus on depreciation for partial years.]

On October 1, 2021, the Allegheny Corporation purchased equipment for $115,000. The estimated service life of the equipment is 10 years and the estimated residual value is $5,000. The equipment is expected to produce 220,000 units during its life.

Required:

Calculate depreciation for 2021 and 2022 using each of the following methods. Partial-year depreciation is calculated based on the number of months the asset is in service. Round all computations to the nearest dollar.

1. Straight line
2. Double-declining balance
3. Units of production (units produced in 2021, 10,000; units produced in 2022, 25,000)

E 11–4
Other depreciation methods
● LO11–2

[This is a variation of Exercise 11–2 modified to focus on other depreciation methods.]

On January 1, 2021, the Allegheny Corporation purchased equipment for $115,000. The estimated service life of the equipment is 10 years and the estimated residual value is $5,000. The equipment is expected to produce 220,000 units during its life.

Required:

Calculate depreciation for 2021 and 2022 using each of the following methods. Round all computations to the nearest dollar.

1. Sum-of-the-years'-digits
2. One hundred fifty percent declining balance
3. Assume instead the equipment was purchased on October 1, 2021. Calculate depreciation for 2021 and 2022 using each of the two methods. Partial-year depreciation is calculated based on the number of months the asset is in service. Round all computations to the nearest dollar.

E 11–5
Depreciation methods; asset addition; partial period
● LO11–2, LO11–9

Tristen Company purchased a five-story office building on January 1, 2019, at a cost of $5,000,000. The building has a residual value of $200,000 and a 30-year life. The straight-line depreciation method is used. On June 30, 2021, construction of a sixth floor was completed at a cost of $1,650,000.

Required:
Calculate the depreciation on the building and building addition for 2021 and 2022 assuming that the addition did not change the life or residual value of the building.

E 11–6
Depreciation methods; solving for unknowns
● LO11–2

For each of the following depreciable assets, determine the missing amount (?). Abbreviations for depreciation methods are SL for straight-line and DDB for double-declining-balance.

Asset	Cost	Residual Value	Service Life (Years)	Depreciation Method	Depreciation (Year 2)
A	?	$20,000	5	DDB	$24,000
B	$ 40,000	?	8	SL	4,500
C	65,000	5,000	?	SL	6,000
D	230,000	10,000	10	?	22,000
E	200,000	20,000	8	DDB	?

E 11–7
Depreciation methods; partial periods
● LO11–2

On March 31, 2021, Susquehanna Insurance purchased an office building for $12,000,000. Based on their relative fair values, one-third of the purchase price was allocated to the land and two-thirds to the building. Furniture and fixtures were purchased separately from office equipment on the same date for $1,200,000 and $700,000, respectively. The company uses the straight-line method to depreciate its buildings and the double-declining-balance method to depreciate all other depreciable assets. The estimated useful lives and residual values of these assets are as follows:

	Service Life	Residual Value
Building	30	10% of cost
Furniture and fixtures	10	10% of cost
Office equipment	5	$30,000

Required:
Calculate depreciation for 2021 and 2022.

E 11–8
IFRS; depreciation; partial periods
● LO11–2, LO11–10
IFRS

On June 30, 2021, Rosetta Granite purchased equipment for $120,000. The estimated useful life of the equipment is eight years and no residual value is anticipated. An important component of the equipment is a specialized high-speed drill that will need to be replaced in four years. The $20,000 cost of the drill is included in the $120,000 cost of the equipment. Rosetta uses the straight-line depreciation method for all equipment.

Required:
1. Calculate depreciation for 2021 and 2022 applying the typical U.S. GAAP treatment.
2. Repeat requirement 1 applying IFRS.

E 11–9
IFRS; revaluation of equipment; depreciation; partial periods
● LO11–10
IFRS

Dower Corporation prepares its financial statements according to IFRS. On March 31, 2021, the company purchased equipment for $240,000. The equipment is expected to have a six-year useful life with no residual value. Dower uses the straight-line depreciation method for all equipment. On December 31, 2021, the end of the company's fiscal year, Dower chooses to revalue the equipment to its fair value of $220,000.

Required:
1. Calculate depreciation for 2021.
2. Prepare the journal entry to record the revaluation of the equipment.
3. Calculate depreciation for 2022.
4. Repeat requirement 2 assuming that the fair value of the equipment at the end of 2021 is $195,000.

E 11–10
Disposal of property, plant, and equipment
● LO11–2

Mercury Inc. purchased equipment in 2019 at a cost of $400,000. The equipment was expected to produce 700,000 units over the next five years and have a residual value of $50,000. The equipment was sold for $210,000 part way through 2021. Actual production in each year was: 2019 = 100,000 units; 2020 = 160,000 units; 2021 = 80,000 units. Mercury uses units-of-production depreciation, and all depreciation has been recorded through the disposal date.

Required:
1. Calculate the gain or loss on the sale.
2. Prepare the journal entry to record the sale.
3. Assuming that the equipment was instead sold for $245,000, calculate the gain or loss on the sale.
4. Prepare the journal entry to record the sale in requirement 3.

E 11–11
Disposal of property, plant, and equipment; partial periods
● LO11–2

On July 1, 2016, Farm Fresh Industries purchased a specialized delivery truck for $126,000. At the time, Farm Fresh estimated the truck to have a useful life of eight years and a residual value of $30,000. On March 1, 2021, the truck was sold for $58,000. Farm Fresh uses the straight-line depreciation method for all of its plant and equipment. Partial-year depreciation is calculated based on the number of months the asset is in service.

Required:

1. Prepare the journal entry to update depreciation in 2021.
2. Prepare the journal entry to record the sale of the truck.
3. Assuming that the truck was instead sold for $80,000, prepare the journal entry to record the sale.

E 11–12
Depreciation methods disposal; partial periods
● LO11–2

Howarth Manufacturing Company purchased equipment on June 30, 2017, at a cost of $800,000. The residual value of the equipment was estimated to be $50,000 at the end of a five-year life. The equipment was sold on March 31, 2021, for $170,000. Howarth uses the straight-line depreciation method for all of its plant and equipment. Partial-year depreciation is calculated based on the number of months the asset is in service.

Required:

1. Prepare the journal entry to record the sale.
2. Assuming that Howarth had instead used the double-declining-balance method, prepare the journal entry to record the sale.

E 11–13
Assets held for sale; partial periods
● LO11–2

On March 31, 2021, management of Quality Appliances committed to a plan to sell equipment. The equipment was available for immediate sale, and an active plan to locate a buyer was initiated. The equipment had been purchased on January 1, 2019, for $260,000. The equipment had an estimated six-year service life and residual value of $20,000. The equipment was being depreciated using the straight-line method. Quality's fiscal year ends on December 31.

Required:

1. Calculate the equipment's book value as of March 31, 2021 (*Hint:* Depreciation for 2021 would include up to March 31).
2. By December 31, 2021, the equipment has not been sold, but management expects that it will be sold in 2022 for $150,000. For what amount is the equipment reported in the December 31, 2021, balance sheet?

E 11–14
Group depreciation
● LO11–2

Highsmith Rental Company purchased an apartment building early in 2021. There are 20 apartments in the building and each is furnished with major kitchen appliances. The company has decided to use the group depreciation method for the appliances. The following data are available:

Appliance	Cost	Residual Value	Service Life (in years)
Stoves	$15,000	$3,000	6
Refrigerators	10,000	1,000	5
Dishwashers	8,000	500	4

In 2024, three new refrigerators costing $2,700 were purchased for cash. In that same year, the old refrigerators, which originally cost $1,500, were sold for $200.

Required:

1. Calculate the group depreciation rate, group life, and depreciation for 2021.
2. Prepare the journal entries to record the purchase of the new refrigerators and the sale of the old refrigerators.

E 11–15
Double-declining-balance method; switch to straight line
● LO11–2, LO11–6

On January 2, 2021, the Jackson Company purchased equipment to be used in its manufacturing process. The equipment has an estimated life of eight years and an estimated residual value of $30,625. The expenditures made to acquire the asset were as follows:

Purchase price	$154,000
Freight charges	2,000
Installation charges	4,000

Jackson's policy is to use the double-declining-balance (DDB) method of depreciation in the early years of the equipment's life and then switch to straight line halfway through the equipment's life.

Required:

1. Calculate depreciation for each year of the asset's eight-year life.
2. Are changes in depreciation methods accounted for retrospectively or prospectively?

E 11–16
Depletion
● LO11–3

On April 17, 2021, the Loadstone Mining Company purchased the rights to a coal mine. The purchase price plus additional costs necessary to prepare the mine for extraction of the coal totaled $4,500,000. The company expects to extract 900,000 tons of coal during a four-year period. During 2021, 240,000 tons were extracted and sold immediately.

Required:
1. Calculate depletion for 2021.
2. Is depletion considered part of the product cost and included in the cost of inventory (yes/no)? Discuss.

E 11–17
Depreciation and depletion
● LO11–2, LO11–3

At the beginning of 2021, Terra Lumber Company purchased a timber tract from Boise Cantor for $3,200,000. After the timber is cleared, the land will have a residual value of $600,000. Roads to enable logging operations were constructed and completed on March 30, 2021. The cost of the roads, which have no residual value and no alternative use after the tract is cleared, was $240,000. During 2021, Terra logged 500,000 of the estimated five million board feet of timber.

Required:
Calculate the 2021 depletion of the timber tract and depreciation of the logging roads assuming the units-of-production method is used for both assets.

E 11–18
Cost of a natural resource; depletion and depreciation; Chapters 10 and 11
● LO11–2, LO11–3

[This exercise is a continuation of Exercise 10–4 in Chapter 10 focusing on depletion and depreciation.]

Jackpot Mining Company operates a copper mine in central Montana. The company paid $1,000,000 in 2021 for the mining site and spent an additional $600,000 to prepare the mine for extraction of the copper. After the copper is extracted in approximately four years, the company is required to restore the land to its original condition, including repaving of roads and replacing a greenbelt. The company has provided the following three cash flow possibilities for the restoration costs:

	Cash Outflow	Probability
1	$300,000	25%
2	400,000	40%
3	600,000	35%

To aid extraction, Jackpot purchased some new equipment on July 1, 2021, for $120,000. After the copper is removed from this mine, the equipment will be sold for an estimated residual amount of $20,000. There will be no residual value for the copper mine. The credit-adjusted risk-free rate of interest is 10%.

The company expects to extract 10 million pounds of copper from the mine. Actual production was 1.6 million pounds in 2021 and 3 million pounds in 2022.

Required:
1. Compute depletion and depreciation on the mine and mining equipment for 2021 and 2022. The units-of-production method is used to calculate depreciation.
2. Discuss the accounting treatment of the depletion and depreciation on the mine and mining equipment.

E 11–19
Amortization
● LO11–4, LO11–5

Janes Company provided the following information on intangible assets:

a. A patent was purchased from the Lou Company for $700,000 on January 1, 2019. Janes estimated the remaining useful life of the patent to be 10 years. The patent was carried on Lou's accounting records at a net book value of $350,000 when Lou sold it to Janes.
b. During 2021, a franchise was purchased from the Rink Company for $500,000. The contractual life of the franchise is 10 years and Janes records a full year of amortization in the year of purchase.
c. Janes incurred research and development costs in 2021 as follows:

Materials and supplies	$140,000
Personnel	180,000
Indirect costs	60,000
Total	$380,000

d. Effective January 1, 2021, based on new events that have occurred, Janes estimates that the remaining life of the patent purchased from Lou is only five more years.

Required:
1. Prepare the entries necessary for years 2019 through 2021 to reflect the above information.
2. Prepare a schedule showing the intangible asset section of Janes's December 31, 2021, balance sheet.

E 11–20
Patent amortization; patent defense
● LO11–4, LO11–9

On January 2, 2021, David Corporation purchased a patent for $500,000. The remaining legal life is 12 years, but the company estimated that the patent will be useful only for eight years. In January 2023, the company incurred legal fees of $45,000 in successfully defending a patent infringement suit. The successful defense did not change the company's estimate of useful life.

Required:
Prepare journal entries related to the patent for 2021, 2022, and 2023.

E 11–21
Change in estimate; useful life of patent
● LO11–4, LO11–5

Van Frank Telecommunications has a patent on a cellular transmission process. The company has amortized the patent on a straight-line basis since 2017, when it was acquired at a cost of $9 million at the beginning of that year. Due to rapid technological advances in the industry, management decided that the patent would benefit the company over a total of six years rather than the nine-year life being used to amortize its cost. The decision was made at the beginning of 2021.

Required:
Prepare the year-end journal entry for patent amortization in 2021. No amortization was recorded during the year.

E 11–22
IFRS; revaluation of patent; amortization
● LO11–10
IFRS

Saint John Corporation prepares its financial statements according to IFRS. On June 30, 2021, the company purchased a franchise for $1,200,000. The franchise is expected to have a 10-year useful life with no residual value. Saint John uses the straight-line amortization method for all intangible assets. On December 31, 2021, the end of the company's fiscal year, Saint John chooses to revalue the franchise. There is an active market for this particular franchise and its fair value on December 31, 2021, is $1,180,000.

Required:
1. Calculate amortization for 2021.
2. Prepare the journal entry to record the revaluation of the patent.
3. Calculate amortization for 2022.

E 11–23
Change in estimate; useful life and residual value of equipment
● LO11–2, LO11–5

Wardell Company purchased a minicomputer on January 1, 2019, at a cost of $40,000. The computer was depreciated using the straight-line method over an estimated five-year life with an estimated residual value of $4,000. On January 1, 2021, the estimate of useful life was changed to a total of 10 years, and the estimate of residual value was changed to $900.

Required:
1. Prepare the year-end journal entry for depreciation in 2021. No depreciation was recorded during the year.
2. Repeat requirement 1 assuming that the company uses the sum-of-the-years'-digits method instead of the straight-line method.

E 11–24
Change in principle; change in depreciation methods
● LO11–2, LO11–6

Alteran Corporation purchased office equipment for $1.5 million at the beginning of 2019. The equipment is being depreciated over a 10-year life using the double-declining-balance method. The residual value is expected to be $300,000. At the beginning of 2021 (two years later), Alteran decided to change to the straight-line depreciation method for this equipment.

Required:
Prepare the journal entry to record depreciation for 2021.

E 11–25
Change in principle; change in depreciation methods
● LO11–2, LO11–6

For financial reporting, Clinton Poultry Farms has used the declining-balance method of depreciation for conveyor equipment acquired at the beginning of 2018 for $2,560,000. Its useful life was estimated to be six years, with a $160,000 residual value. At the beginning of 2021, Clinton decides to change to the straight-line method. The effect of this change on depreciation for each year is as follows:

	($ in thousands)		
Year	**Straight Line**	**Declining Balance**	**Difference**
2018	$ 400	$ 853	$453
2019	400	569	169
2020	400	379	(21)
	$1,200	$1,801	$601

Required:
1. Briefly describe the way Clinton should report this accounting change in the 2019–2021 comparative financial statements.
2. Prepare any 2021 journal entry related to the change.

E 11–26
Error correction
● LO11–2, LO11–7

In 2021, internal auditors discovered that PKE Displays, Inc. had debited an expense account for the $350,000 cost of equipment purchased on January 1, 2018. The equipment's life was expected to be five years with no residual value. Straight-line depreciation is used by PKE.

Required:

1. Determine the cumulative effect of the error on net income over the three-year period from 2018 through 2020, and on retained earnings by the end of 2020.
2. Prepare the correcting entry assuming the error was discovered in 2021 before the adjusting and closing entries. (Ignore income taxes.)
3. Assume instead that the equipment was disposed of in 2022 and the original error was discovered in 2023 after the 2022 financial statements were issued. Prepare the correcting entry in 2023.

E 11–27
Impairment; property, plant, and equipment
● LO11–8

Chadwick Enterprises, Inc., operates several restaurants throughout the Midwest. Three of its restaurants located in the center of a large urban area have experienced declining profits due to declining population. The company's management has decided to test the assets of the restaurants for possible impairment. The relevant information for these assets is presented below.

Book value	$6.5 million
Estimated undiscounted sum of future cash flows	4.0 million
Fair value	3.5 million

Required:

1. Determine the amount of the impairment loss, if any.
2. Repeat requirement 1 assuming that the estimated undiscounted sum of future cash flows is $6.8 million and fair value is $5 million.

E 11–28
IFRS; impairment; property, plant, and equipment
● LO11–10
IFRS

Refer to the situation described in Exercise 11–27.

Required:

How might your solution differ if Chadwick Enterprises, Inc., prepares its financial statements according to International Financial Reporting Standards? Assume that the fair value amount given in the exercise equals both (a) the fair value less costs to sell and (b) the present value of estimated future cash flows.

E 11–29
IFRS; Impairment; property, plant, and equipment
● LO11–8, LO11–10
IFRS

Collinsworth LTD., a U.K. company, prepares its financial statements according to International Financial Reporting Standards. Late in its 2021 fiscal year, a significant adverse change in business climate indicated to management that the assets of its appliance division may be impaired. The following data relate to the division's assets:

	(£ in millions)
Book value	£220
Undiscounted sum of estimated future cash flows	210
Present value of future cash flows	150
Fair value less cost to sell (determined by appraisal)	145

Required:

1. What amount of impairment loss, if any, should Collinsworth recognize?
2. Assume that Collinsworth prepares its financial statements according to U.S. GAAP and that fair value less cost to sell approximates fair value. What amount of impairment loss, if any, should Collinsworth recognize?

E 11–30
Impairment; property, plant, and equipment
● LO11–8

General Optic Corporation operates a manufacturing plant in Arizona. Due to a significant decline in demand for the product manufactured at the Arizona site, an impairment test is deemed appropriate. Management has acquired the following information for the assets at the plant:

Cost	$32,500,000
Accumulated depreciation	14,200,000
General's estimate of the total cash flows to be generated by selling the products manufactured at its Arizona plant, not discounted to present value	15,000,000

The fair value of the Arizona plant is estimated to be $11,000,000.

Required:

1. Determine the amount of impairment loss, if any.
2. If a loss is indicated, prepare the entry to record the loss.
3. Repeat requirement 1 assuming that the estimated undiscounted sum of future cash flows is $12,000,000 instead of $15,000,000.

4. Repeat requirement 1 assuming that the estimated undiscounted sum of future cash flows is $19,000,000 instead of $15,000,000

E 11–31
Impairment; goodwill
● LO11–8

In 2019, Alliant Corporation acquired Centerpoint Inc. for $300 million, of which $50 million was allocated to goodwill. At the end of 2021, management has provided the following information for a required goodwill impairment test:

Fair value of Centerpoint Inc.	$220 million
Book value of Centerpoint's net assets (excluding goodwill)	200 million
Book value of Centerpoint's net assets (including goodwill)	250 million

Required:
1. Determine the amount of the impairment loss.
2. Repeat requirement 1 assuming that the fair value of Centerpoint is $270 million.

E 11–32
IFRS; impairment goodwill
● LO11–10
IFRS

Refer to the situation described in E 11–31, requirement 1. Alliant prepares its financial statements according to IFRS, and Centerpoint is considered a cash-generating unit. Assume that Centerpoint's fair value of $220 million approximates fair value less costs to sell and that the present value of Centerpoint's estimated future cash flows is $225 million.

Required:
Determine the amount of goodwill impairment loss Alliant should recognize.

E 11–33
Goodwill valuation and impairment; Chapters 10 and 11
● LO11–8

On May 28, 2021, Pesky Corporation acquired all of the outstanding common stock of Harman, Inc., for $420 million. The fair value of Harman's identifiable tangible and intangible assets totaled $512 million, and the fair value of liabilities assumed by Pesky was $150 million.

Pesky performed a goodwill impairment test at the end of its fiscal year ended December 31, 2021. Management has provided the following information:

Fair value of Harman, Inc.	$400 million
Fair value of Harman's net assets (excluding goodwill)	370 million
Book value of Harman's net assets (including goodwill)	410 million

Required:
1. Determine the amount of goodwill that resulted from the Harman acquisition.
2. Determine the amount of goodwill impairment loss that Pesky should recognize at the end of 2021, if any.
3. If an impairment loss is required, prepare the journal entry to record the loss.

E 11–34
FASB codification research
● LO11–8
CODE

Access the *FASB Accounting Standards Codification* at the FASB website (**www.fasb.org**). Determine each of the following:

Required:
1. The topic number (Topic XXX) that provides guidance on accounting for the impairment of long-lived assets.
2. The specific eight-digit Codification citation (XXX-XX-XX-X) that discusses the disclosures required in the notes to the financial statements for the impairment of long-lived assets classified as held and used.
3. Describe the disclosure requirements.

E 11–35
FASB codification research
● LO11–2, LO11–1, LO11–6, LO11–8
CODE

Access the *FASB Accounting Standards Codification* at the FASB website (**www.fasb.org**). Determine each of the following:
1. The specific eight-digit Codification citation (XXX-XX-XX-X) that discusses depreciation as a systematic and rational allocation of cost rather than a process of valuation.
2. The specific nine-digit Codification citation (XXX-XX-XX-XX) that involves the calculation of an impairment loss for property, plant, and equipment.
3. The specific nine-digit Codification citation (XXX-XX-XX-XX) that provides guidance on accounting for a change in depreciation method.
4. The specific eight-digit Codification citation (XXX-XX-XX-X) that indicates goodwill should not be amortized.

E 11–36
Subsequent expenditures
● LO11–9

Belltone Company made the following expenditures related to its 10-year-old manufacturing facility:
1. The heating system was replaced at a cost of $250,000. The cost of the old system was not known. The company accounts for improvements as reductions of accumulated depreciation.
2. A new wing was added at a cost of $750,000. The new wing substantially increases the productive capacity of the plant.

3. Annual building maintenance was performed at a cost of $14,000.
4. All of the equipment on the assembly line in the plant was rearranged at a cost of $50,000. The rearrangement clearly increases the productive capacity of the plant.

Required:

Prepare journal entries to record each of the above expenditures.

E 11–37
IFRS; amortization; cost to defend a patent
● LO11–4, LO11–9, LO11–10
IFRS

On September 30, 2019, Leeds LTD. acquired a patent in conjunction with the purchase of another company. The patent, valued at $6 million, was estimated to have a 10-year life and no residual value. Leeds uses the straight-line method of amortization for intangible assets. At the beginning of January 2021, Leeds successfully defended its patent against infringement. Litigation costs totaled $500,000.

Required:

1. Calculate amortization of the patent for 2019 and 2020.
2. Prepare the journal entry to record the 2021 litigation costs.
3. Calculate amortization for 2021.
4. Repeat requirements 1–3 assuming that Leeds prepares its financial statements according to IFRS.

E 11–38
Concepts; terminology
● LO11–1 through LO11–6, LO11–8

Listed below are several items and phrases associated with depreciation, depletion, and amortization. Pair each item from List A with the item from List B (by letter) that is most appropriately associated with it.

List A	List B
_____ 1. Depreciation	a. Cost allocation for natural resource
_____ 2. Service life	b. Accounted for prospectively
_____ 3. Depreciable base	c. When there has been a significant decline in value
_____ 4. Activity-based methods	d. The amount of use expected from plant and equipment and finite-life intangible assets
_____ 5. Time-based methods	e. Estimates service life in units of output
_____ 6. Double-declining balance	f. Cost less residual value
_____ 7. Group method	g. Cost allocation for plant and equipment
_____ 8. Composite method	h. Does not subtract residual value from cost
_____ 9. Depletion	i. Accounted for in the same way as a change in estimate
_____ 10. Amortization	j. Aggregates assets that are similar
_____ 11. Change in useful life	k. Aggregates assets that are physically unified
_____ 12. Change in depreciation method	l. Cost allocation for an intangible asset
_____ 13. Write-down of asset	m. Estimates service life in years

E 11–39
Retirement and replacement depreciation
● Appendix B

Cadillac Construction Company uses the retirement method to determine depreciation on its small tools. During 2019, the first year of the company's operations, tools were purchased at a cost of $8,000. In 2021, tools originally costing $2,000 were sold for $250 and replaced with new tools costing $2,500.

Required:

1. Prepare journal entries to record each of the above transactions.
2. Repeat requirement 1 assuming that the company uses the replacement depreciation method instead of the retirement method.

E 11–40
General ledger exercise; depreciation
● LO11–2

On January 1, 2021, the general ledger of TNT Fireworks included the following account balances:

Account Title	Debits	Credits
Cash	$ 58,700	
Accounts receivable	25,000	
Allowance for uncollectible accounts		$ 2,200
Inventory	36,300	
Notes receivable (5%, due in 2 years)	12,000	
Land	155,000	
Accounts payable		14,800
Common stock		220,000
Retained earnings		50,000
Totals	$287,000	$287,000

During January 2021, the following transactions occurred:

Jan. 1	Purchased equipment for $19,500. The company estimates a residual value of $1,500 and a five-year service life.
4	Paid cash on accounts payable $9,500.
8	Purchased additional inventory on account, $82,900.
15	Received cash on accounts receivable, $22,000.
19	Paid cash for salaries, $29,800
28	Paid cash for January utilities, $16,500.
30	Firework sales for January totaled $220,000. All of these sales were on account. The cost of the units sold was $115,000.

Required:

1. Record each of the transactions listed above in the 'General Journal' tab (these are shown as items 1–8) assuming a perpetual FIFO inventory system. Review the 'General Ledger' and the 'Trial Balance' tabs to see the effect of the transactions on the account balances.
2. Record adjusting entries on January 31 in the 'General Journal' tab (these are shown as items 9–13):
   a. Depreciation on the equipment for the month of January is calculated using the straight-line method.
   b. At the end of January, $3,000 of accounts receivable are past due, and the company estimates that 50% of these accounts will not be collected. Of the remaining accounts receivable, the company estimates that 3% will not be collected. The note receivable of $12,000 is considered fully collectible and therefore is not included in the estimate of uncollectible accounts.
   c. Accrued interest revenue on notes receivable for January.
   d. Unpaid salaries at the end of January are $32,600.
   e. Accrued income taxes at the end of January are $9,000.
3. Review the adjusted 'Trial Balance' as of January 31, 2021, in the 'Trial Balance' tab.
4. Prepare a multiple-step income statement for the period ended January 31, 2021, in the 'Income Statement' tab.
5. Prepare a classified balance sheet as of January 31, 2021, in the 'Balance Sheet' tab.
6. Record closing entries in the 'General Journal' tab (these are shown as items 14–15).
7. Using the information from the requirements above, complete the 'Analysis' tab.
   a. Calculate the return on assets ratio for the month of January. If the average return on assets for the industry in January is 2%, is the company *more* or *less* profitable than other companies in the same industry?
   b. Calculate the profit margin for the month of January. If the industry average profit margin is 4%, is the company *more* or *less* efficient at converting sales to profit than other companies in the same industry?
   c. Calculate the asset turnover ratio for the month of January. If the industry average asset turnover is 0.5 times per month, is the company *more* or *less* efficient at producing revenues with its assets than other companies in the same industry?

E 11–41

General ledger exercise; long-term asset transactions

● LO11–2, LO11–4, LO11–8, LO11–9

On January 1, 2021, the general ledger of Parts Unlimited included the following account balances:

Account Title	Debits	Credits
Cash	$162,400	
Accounts receivable	12,400	
Inventory	37,800	
Land	340,000	
Equipment	347,500	
Accumulated depreciation		$172,000
Accounts payable		14,800
Common stock		520,000
Retained earnings		193,300
Totals	$900,100	$900,100

From January 1 to December 31, the following summary transactions occurred:

a. Purchased inventory on account, $325,800.
b. Sold inventory on account, $567,200. The inventory cost $342,600.
c. Received cash from customers on account, $558,700.
d. Paid cash on account, $328,500.
e. Paid cash for salaries, $94,700, and for utilities, $52,700.

In addition, Parts Unlimited had the following transactions during the year:

April 1	Purchased equipment for $95,000 using a note payable, due in 12 months plus 8% interest. The company also paid cash of $3,200 for freight and $3,800 for installation and testing of the equipment. The equipment has an estimated residual value of $10,000 and a ten-year service life.
June 30	Purchased a patent for $40,000 from a third-party marketing company related to the packaging of the company's products. The patent has a 20-year useful life, after which it is expected to have no value.
October 1	Sold equipment for $30,200. The equipment cost $60,700 and had accumulated depreciation of $37,400 at the beginning of the year. Additional depreciation for 2021 up to the point of the sale is $8,500.
November 15	Several older pieces of equipment were improved by replacing major components at a cost of $54,100. These improvements are expected to enhance the equipment's operating capabilities. [Record this transaction using Alternative 2—capitalization of new cost.]

Required:

1. Record each of the transactions listed above in the 'General Journal' tab (these are shown as items 1–11) assuming a perpetual inventory system. Review the 'General Ledger' and the 'Trial Balance' tabs to see the effect of the transactions on the account balances.
2. Record adjusting entries on December 31 in the 'General Journal' tab (these are shown as items 12–17):
   a. Depreciation on the equipment purchased on April 1, 2021, calculated using the straight-line method.
   b. Depreciation on the remaining equipment, $21,500.
   c. Amortization of the patent purchased on June 30, 2021, using the straight-line method.
   d. Accrued interest payable on the note payable.
   e. Equipment with an original cost of $65,400 had the following related information at the end of the year: accumulated depreciation of $40,300, expected cash flows of $15,700, and a fair value of $10,800.
   f. Accrued income taxes at the end of the year are $12,600.
3. Review the adjusted 'Trial Balance' as of December 31, 2021, in the 'Trial Balance' tab.
4. Prepare a multiple-step income statement for the period ended December 31, 2021, in the 'Income Statement' tab.
5. Prepare a classified balance sheet as of December 31, 2021, in the 'Balance Sheet' tab.
6. Record closing entries in the 'General Journal' tab (these are shown as items 18–19).
7. Using the information from the requirements above, complete the 'Analysis' tab.
   a. Calculate the fixed asset turnover ratio for the year, using the total amount of property, plant, and equipment (net of accumulated deprecation). If the industry average fixed asset turnover is 0.75, is the company *more* or *less* efficient at generating sales with its fixed assets than other companies in the same industry?
   b. Suppose the equipment purchased on April 1, 2021, had been depreciated using the units of production method. At the time of purchase, expected output was 20,000 units, and actual production for 2021 was 2,000 units. Calculate the amount of depreciation expense that would have been recorded and determine the difference in net income and total assets for 2021 (ignoring tax effects).
   c. The transaction on June 30, 2021, shows the company purchased a patent for $40,000 from a third-party marketing company. Suppose the company instead spent $40,000 to internally develop the new packaging technology, which it then patented. Calculate the difference in net income and total assets for 2021 (ignoring tax effects).

Problems

connect

P 11–1
Depreciation methods; change in methods
● LO11–2, LO11–6

The fact that generally accepted accounting principles allow companies flexibility in choosing between certain allocation methods can make it difficult for a financial analyst to compare periodic performance from firm to firm.

Suppose you were a financial analyst trying to compare the performance of two companies. Company A uses the double-declining-balance depreciation method. Company B uses the straight-line method. You have the following information taken from the 12/31/2021 year-end financial statements for Company B:

Income Statement	
Depreciation expense	$ 10,000
Balance Sheet	
Assets:	
Plant and equipment, at cost	$200,000
Less: Accumulated depreciation	(40,000)
Net	$160,000

You also determine that all of the assets constituting the plant and equipment of Company B were acquired at the same time, and that all of the $200,000 represents depreciable assets. Also, all of the depreciable assets have the same useful life and residual values are zero.

Required:

1. In order to compare performance with Company A, estimate what B's depreciation expense would have been for 2021 if the double-declining-balance depreciation method had been used by Company B since acquisition of the depreciable assets.
2. If Company B decided to switch depreciation methods in 2021 from the straight line to the double-declining-balance method, prepare the 2021 journal entry to record depreciation for the year, assuming no journal entry for depreciation in 2021 has yet been recorded.

P 11–2
Comprehensive problem; Chapters 10 and 11
● LO11–2, LO11–4

At December 31, 2020, Cord Company's plant asset and accumulated depreciation and amortization accounts had balances as follows:

Category	Plant Asset	Accumulated Depreciation and Amortization
Land	$ 175,000	$ —
Buildings	1,500,000	328,900
Equipment	1,125,000	317,500
Automobiles and trucks	172,000	100,325
Leasehold improvements	216,000	108,000
Land improvements	—	—

Depreciation methods and useful lives:
Buildings—150% declining balance; 25 years.
Equipment—Straight line; 10 years.
Automobiles and trucks—200% declining balance; 5 years, all acquired after 2017.
Leasehold improvements—Straight line.
Land improvements—Straight line.

Depreciation is computed to the nearest month and residual values are immaterial. Transactions during 2021 and other information:

a. On January 6, 2021, a plant facility consisting of land and building was acquired from King Corp. in exchange for 25,000 shares of Cord's common stock. On this date, Cord's stock had a fair value of $50 a share. Current assessed values of land and building for property tax purposes are $187,500 and $562,500, respectively.
b. On March 25, 2021, new parking lots, streets, and sidewalks at the acquired plant facility were completed at a total cost of $192,000. These expenditures had an estimated useful life of 12 years.
c. The leasehold improvements were completed on December 31, 2017, and had an estimated useful life of eight years. The related lease, which would terminate on December 31, 2023, was renewable for an additional four-year term. On April 30, 2021, Cord exercised the renewal option.
d. On July 1, 2021, equipment was purchased at a total invoice cost of $325,000. Additional costs of $10,000 for delivery and $50,000 for installation were incurred.
e. On September 30, 2021, Cord purchased a new automobile for $12,500.
f. On September 30, 2021, a truck with a cost of $24,000 and a book value of $9,100 on date of sale was sold for $11,500. Depreciation for the nine months ended September 30, 2021, was $2,650.
g. On December 20, 2021, equipment with a cost of $17,000 and a book value of $2,975 at date of disposition was scrapped without cash recovery.

Required:

1. Prepare a schedule analyzing the changes in each of the plant asset accounts during 2021. This schedule should include columns for beginning balance, increase, decrease, and ending balance for each of the plant asset accounts. Do not analyze changes in accumulated depreciation and amortization.
2. For each asset category, prepare a schedule showing depreciation or amortization expense for the year ended December 31, 2021. Round computations to the nearest whole dollar.

(AICPA adapted)

P 11–3
Depreciation methods; partial periods Chapters 10 and 11
● LO11–2

[This problem is a continuation of Problem 10–3 in Chapter 10 focusing on depreciation.]

Required:

For each asset classification, prepare a schedule showing depreciation for the year ended December 31, 2021, using the following depreciation methods and useful lives:

Land improvements—Straight line; 15 years
Building—150% declining balance; 20 years
Equipment—Straight line; 10 years
Automobiles—Units-of-production; $0.50 per mile

Depreciation is computed to the nearest month and whole dollar amount, and no residual values are used. Automobiles were driven 38,000 miles in 2021.

(AICPA adapted)

P 11–4
Partial-year depreciation; asset addition; increase in useful life
● LO11–2, LO11–5, LO11–9

On April 1, 2019, the KB Toy Company purchased equipment to be used in its manufacturing process. The equipment cost $48,000, has an eight-year useful life, and has no residual value. The company uses the straight-line depreciation method for all manufacturing equipment.

On January 4, 2021, $12,350 was spent to repair the equipment and to add a feature that increased its operating efficiency. Of the total expenditure, $2,000 represented ordinary repairs and annual maintenance and $10,350 represented the cost of the new feature. In addition to increasing operating efficiency, the total useful life of the equipment was extended to 10 years.

Required:

Prepare journal entries for the following:

1. Depreciation for 2019 and 2020.
2. The 2021 expenditure.
3. Depreciation for 2021.

P 11–5
Property, plant, and equipment and intangible assets; comprehensive
● LO11–2

The Thompson Corporation, a manufacturer of steel products, began operations on October 1, 2019. The accounting department of Thompson has started the fixed-asset and depreciation schedule presented below. You have been asked to assist in completing this schedule. In addition to ascertaining that the data already on the schedule are correct, you have obtained the following information from the company's records and personnel:

a. Depreciation is computed from the first of the month of acquisition to the first of the month of disposition.
b. Land A and Building A were acquired from a predecessor corporation. Thompson paid $812,500 for the land and building together. At the time of acquisition, the land had a fair value of $72,000 and the building had a fair value of $828,000.
c. Land B was acquired on October 2, 2019, in exchange for 3,000 newly issued shares of Thompson's common stock. At the date of acquisition, the stock had a par value of $5 per share and a fair value of $25 per share. During October 2019, Thompson paid $10,400 to demolish an existing building on this land so it could construct a new building.
d. Construction of Building B on the newly acquired land began on October 1, 2020. By September 30, 2021, Thompson had paid $210,000 of the estimated total construction costs of $300,000. Estimated completion and occupancy are July 2022.
e. Certain equipment was donated to the corporation by the city. An independent appraisal of the equipment when donated placed the fair value at $16,000 and the residual value at $2,000.
f. Equipment A's total cost of $110,000 includes installation charges of $550 and normal repairs and maintenance of $11,000. Residual value is estimated at $5,500. Equipment A was sold on February 1, 2021.
g. On October 1, 2020, Equipment B was acquired with a down payment of $4,000 and the remaining payments to be made in 10 annual installments of $4,000 each beginning October 1, 2021. The prevailing interest rate was 8%.

THOMPSON CORPORATION
Fixed Asset and Depreciation Schedule
For Fiscal Years Ended September 30, 2020, and September 30, 2021

Assets	Acquisition Date	Cost	Residual	Depreciation Method	Estimated Life (in years)	Depreciation for Year Ended 9/30 2020	Depreciation for Year Ended 9/30 2021
Land A	10/1/2019	$(1)	N/A	N/A	N/A	N/A	N/A
Building A	10/1/2019	(2)	$47,500	SL	(3)	$14,000	$(4)
Land B	10/2/2019	(5)	N/A	N/A	N/A	N/A	N/A
Building B	Under construction	210,000 to date	—	SL	30	—	(6)
Donated Equipment	10/2/2019	(7)	2,000	DDB	10	(8)	(9)
Equipment A	10/2/2019	(10)	5,500	SYD	10	(11)	(12)
Equipment B	10/1/2020	(13)	—	SL	15	—	(14)

N/A = not applicable

Required:
Supply the correct amount for each numbered item on the schedule. For depreciation methods, SL indicates straight-line, DDB indicates double-declining-balance, and SYD indicates sum-of-the-years'-digits. Round each answer to the nearest dollar.

(AICPA adapted)

P 11–6
Depreciation methods; partial-year depreciation; sale of assets
● LO11–2

On March 31, 2021, the Herzog Company purchased a factory complete with vehicles and equipment. The allocation of the total purchase price of $1,000,000 to the various types of assets along with estimated useful lives and residual values are as follows:

Asset	Cost	Estimated Residual Value	Estimated Useful Life (in years)
Land	$ 100,000	N/A	N/A
Building	500,000	none	25
Equipment	240,000	10% of cost	8
Vehicles	160,000	$12,000	8
Total	$1,000,000		

On June 29, 2022, equipment included in the March 31, 2021, purchase that cost $100,000 was sold for $80,000. Herzog uses the straight-line depreciation method for buildings and equipment and the double-declining-balance method for vehicles. Partial-year depreciation is calculated based on the number of months an asset is in service.

Required:

1. Compute depreciation expense on the building, equipment, and vehicles for 2021.
2. Prepare the journal entries for 2022 to record (a) depreciation on the equipment sold on June 29, 2022, and (b) the sale of the equipment. Round to the nearest whole dollar amount.
3. Compute depreciation expense on the building, remaining equipment, and vehicles for 2022.

P 11–7
Depletion; change in estimate
● LO11–3, LO11–5

In 2021, the Marion Company purchased land containing a mineral mine for $1,600,000. Additional costs of $600,000 were incurred to develop the mine. Geologists estimated that 400,000 tons of ore would be extracted. After the ore is removed, the land will have a resale value of $100,000.

To aid in the extraction, Marion built various structures and small storage buildings on the site at a cost of $150,000. These structures have a useful life of 10 years. The structures cannot be moved after the ore has been removed and will be left at the site. In addition, new equipment costing $80,000 was purchased and installed at the site. Marion does not plan to move the equipment to another site, but estimates that it can be sold at auction for $4,000 after the mining project is completed.

In 2021, 50,000 tons of ore were extracted and sold. In 2022, the estimate of total tons of ore in the mine was revised from 400,000 to 487,500. During 2022, 80,000 tons were extracted, of which 60,000 tons were sold.

Required:

1. Compute depletion and depreciation of the mine and the mining facilities and equipment for 2021 and 2022. Marion uses the units-of-production method to determine depreciation on mining facilities and equipment.
2. Compute the book value of the mineral mine, structures, and equipment as of December 31, 2022.
3. Discuss the accounting treatment of the depletion and depreciation on the mine and mining facilities and equipment.

P 11–8
Amortization; partial period
● LO11–4

The following information concerns the intangible assets of Epstein Corporation:

a. On June 30, 2021, Epstein completed the acquisition of the Johnstone Corporation for $2,000,000 in cash. The fair value of the net identifiable assets of Johnstone was $1,700,000.

b. Included in the assets purchased from Johnstone was a patent that was valued at $80,000. The remaining legal life of the patent was 13 years, but Epstein believes that the patent will only be useful for another eight years.

c. Epstein acquired a franchise on October 1, 2021, by paying an initial franchise fee of $200,000. The contractual life of the franchise is 10 years.

Required:

1. Prepare year-end adjusting journal entries to record amortization expense on the intangibles at December 31, 2021.
2. Prepare the intangible asset section of the December 31, 2021, balance sheet.

P 11–9
Straight-line depreciation; disposal; partial period; change in estimate
● LO11–2, LO11–5

The property, plant, and equipment section of the Jasper Company's December 31, 2020, balance sheet contained the following:

Property, plant, and equipment:		
Land		$120,000
Building	$ 840,000	
Less: Accumulated depreciation	(200,000)	640,000
Equipment	180,000	
Less: Accumulated depreciation	?	?
Total property, plant, and equipment		?

The land and building were purchased at the beginning of 2016. Straight-line depreciation is used and a residual value of $40,000 for the building is anticipated.

The equipment is comprised of the following three machines:

Machine	Cost	Date Acquired	Residual Value	Life (in years)
101	$70,000	1/1/2018	$7,000	10
102	80,000	6/30/2019	8,000	8
103	30,000	9/1/2020	3,000	9

The straight-line method is used to determine depreciation on the equipment. On March 31, 2021, Machine 102 was sold for $52,500. Early in 2021, the useful life of machine 101 was revised to seven years in total, and the residual value was revised to zero.

Required:

1. Calculate the accumulated depreciation on the equipment at December 31, 2020.
2. Prepare the journal entry to record 2021 depreciation on machine 102 up to the date of sale.
3. Prepare a schedule to calculate the gain or loss on the sale of machine 102.
4. Prepare the journal entry for the sale of machine 102.
5. Prepare the 2021 year-end journal entries to record depreciation on the building and remaining equipment.

P 11–10
Accounting changes; three accounting situations
● LO11–2, LO11–5, LO11–6

Described below are three independent and unrelated situations involving accounting changes. Each change occurs during 2021 before any adjusting entries or closing entries are prepared.

a. On December 30, 2017, Rival Industries acquired its office building at a cost of $10,000,000. It has been depreciated on a straight-line basis assuming a useful life of 40 years and no residual value. Early in 2021, the estimate of useful life was revised to 28 years in total with no change in residual value.

b. At the beginning of 2017, the Hoffman Group purchased office equipment at a cost of $330,000. Its useful life was estimated to be 10 years with no residual value. The equipment has been depreciated by the straight-line method. On January 1, 2021, the company changed to the double-declining-balance method.

c. At the beginning of 2021, Jantzen Specialties, which uses the straight-line method, changed to the double-declining-balance method for newly acquired vehicles. The change decreased current year net income by $445,000.

Required:

For each situation:

1. Identify the type of change.
2. Prepare any journal entry necessary as a direct result of the change as well as any adjusting entry for 2021 related to the situation described. (Ignore income tax effects.)
3. Briefly describe any other steps that should be taken to appropriately report the situation.

P 11–11
Error correction change in depreciation method
● LO11–2, LO11–6, LO11–7

Collins Corporation purchased office equipment at the beginning of 2019 and capitalized a cost of $2,000,000. This cost figure included the following expenditures:

Purchase price	$1,850,000
Freight charges	30,000
Installation charges	20,000
Annual maintenance charge	100,000
Total	$2,000,000

The company estimated an eight-year useful life for the equipment. No residual value is anticipated. The double-declining-balance method was used to determine depreciation expense for 2019 and 2020.

In 2021, after the 2020 financial statements were issued, the company decided to switch to the straight-line depreciation method for this equipment. At that time, the company's controller discovered that the original cost of the equipment incorrectly included one year of annual maintenance charges for the equipment.

Required:

1. Ignoring income taxes, prepare the appropriate correcting entry for the equipment capitalization error discovered in 2021.
2. Ignoring income taxes, prepare any 2021 journal entry(s) related to the change in depreciation methods.

P 11–12
Depreciation and amortization; impairment
● LO11–2, LO11–4, LO11–8

At the beginning of 2019, Metatec Inc. acquired Ellison Technology Corporation for $600 million. In addition to cash, receivables, and inventory, the following assets and their fair values were also acquired:

Plant and equipment (depreciable assets)	$150 million
Patent	40 million
Goodwill	100 million

The plant and equipment are depreciated over a 10-year useful life on a straight-line basis. There is no estimated residual value. The patent is estimated to have a 5-year useful life, no residual value, and is amortized using the straight-line method.

At the end of 2021, a change in business climate indicated to management that the assets of Ellison might be impaired. The following amounts have been determined:

Plant and equipment:	
Undiscounted sum of future cash flows	$ 80 million
Fair value	60 million
Patent:	
Undiscounted sum of future cash flows	$ 20 million
Fair value	13 million
Goodwill:	
Fair value of Ellison Technology Corporation	$450 million
Book value of Ellison's net assets (excluding goodwill)	370 million
Book value of Ellison's net assets (including goodwill)	470 million*

*After first recording any impairment losses on plant and equipment and the patent.

Required:

1. Compute the book value of the plant and equipment and patent at the end of 2021.
2. When should the plant and equipment and the patent be tested for impairment?
3. When should goodwill be tested for impairment?
4. Determine the amount of any impairment loss to be recorded, if any, for the three assets.

P 11–13
Depreciation and depletion; change in useful life; asset retirement obligation; Chapters 10 and 11
● LO11–2, LO11–3, LO11–5

On May 1, 2021, Hecala Mining entered into an agreement with the state of New Mexico to obtain the rights to operate a mineral mine in New Mexico for $10 million. Additional costs and purchases included the following:

Development costs in preparing the mine	$3,200,000
Mining equipment	140,000
Construction of various structures on site	68,000

After the minerals are removed from the mine, the equipment will be sold for an estimated residual value of $10,000. The structures will be torn down.

Geologists estimate that 800,000 tons of ore can be extracted from the mine. After the ore is removed the land will revert back to the state of New Mexico.

The contract with the state requires Hecala to restore the land to its original condition after mining operations are completed in approximately four years. Management has provided the following possible outflows for the restoration costs:

Cash Outflow	Probability
$600,000	30%
700,000	30%
800,000	40%

Hecala's credit-adjusted risk-free interest rate is 8%. During 2021, Hecala extracted 120,000 tons of ore from the mine. The company's fiscal year ends on December 31.

Required:

1. Determine the amount at which Hecala will record the mine.
2. Calculate the depletion of the mine and the depreciation of the mining facilities and equipment for 2021, assuming that Hecala uses the units-of-production method for both depreciation and depletion. Round depletion and depreciation rates to four decimals.
3. How much accretion expense will the company record in its income statement for the 2021 fiscal year?
4. Are depletion of the mine and depreciation of the mining facilities and equipment reported as separate expenses in the income statement? Discuss the accounting treatment of these items in the income statement and balance sheet.
5. During 2022, Hecala changed its estimate of the total amount of ore originally in the mine from 800,000 to 1,000,000 tons. Briefly describe the accounting treatment the company will employ to account for the change *and* calculate the depletion of the mine and depreciation of the mining facilities and equipment for 2022 assuming Hecala extracted 150,000 tons of ore in 2022.

P 11–14
MACRS versus straight-line depreciation
● LO11–2, Appendix 11A

On April 1, 2023, Titan Corporation purchases office equipment for $50,000. For tax reporting, the company uses MACRS and classifies the equipment as 5-year personal property. In 2023, this type of equipment is eligible for 60% first-year bonus depreciation. For financial reporting, the company uses straight-line depreciation. Assume the equipment has no residual value.

Required:

1. Calculate annual depreciation for the five-year life of the equipment according to MACRS. The company uses the half-year convention for tax reporting purposes.
2. Calculate annual depreciation for the five-year life of the equipment according to straight-line depreciation. The company uses partial-year depreciation based on the number of months the asset is in service for financial reporting purposes.
3. In which year(s) is tax depreciation greater than financial reporting depreciation?

Decision Maker's Perspective

©ImageGap/Getty Images

Apply your critical-thinking ability to the knowledge you've gained. These cases will provide you an opportunity to develop your research, analysis, judgment, and communication skills. You also will work with other students, integrate what you've learned, apply it in real-world situations, and consider its global and ethical ramifications. This practice will broaden your knowledge and further develop your decision-making abilities.

Analysis Case 11–1
Depreciation, depletion, and amortization
● LO11–1

The terms depreciation, depletion, and amortization all refer to the process of allocating the cost of an asset to the periods the asset is used.

Required:

Discuss the differences between depreciation, depletion, and amortization as the terms are used in accounting for property, plant, and equipment and intangible assets.

Communication Case 11–2
Depreciation
● LO11–1

At a recent luncheon, you were seated next to Mr. Hopkins, the president of a local company that manufactures bicycle parts. He heard that you were a CPA and made the following comments to you:

> Why is it that I am forced to recognize depreciation expense in my company's income statement when I know that I could sell many of my assets for more than I paid for them? I thought that the purpose of the balance sheet was to reflect the value of my business and that the purpose of the income statement was to report the net change in value or wealth of a company. It just doesn't make sense to penalize my profits when there hasn't been any loss in value from using the assets.

At the conclusion of the luncheon, you promised to send him a short explanation of the rationale for current depreciation practices.

Required:

Prepare a letter to Mr. Hopkins. Explain the accounting concept of depreciation and include a brief example in your explanation showing that over the life of the asset the change in value approach to depreciation and the allocation of cost approach will result in the same total effect on income.

Judgment Case 11–3
Straight-line method; composite depreciation
● LO11–1, LO11–2

Portland Co uses the straight-line depreciation method for depreciable assets. All assets are depreciated individually except manufacturing equipment, which is depreciated by the composite method.

Required:

1. What factors should have influenced Portland's selection of the straight-line depreciation method?
2. a. What benefits should derive from using the composite method rather than the individual basis for manufacturing equipment?
   b. How should Portland have calculated the manufacturing equipment's annual depreciation in its first year of operation?

(AICPA adapted)

Judgment Case 11–4
Depreciation
● LO11–1, LO11–2

At the beginning of the year, Patrick Company acquired a computer to be used in its operations. The computer was delivered by the supplier, installed by Patrick, and placed into operation. The estimated useful life of the computer is five years, and its estimated residual value is significant.

Required:

1. a. What costs should Patrick capitalize for the computer?
   b. What is the objective of depreciation accounting?
2. What is the rationale for using accelerated depreciation methods?

(AICPA adapted)

Judgment Case 11–5
Capitalize or expense; materiality
● LO11–9

Redline Publishers, Inc. produces various manuals ranging from computer software instructional booklets to manuals explaining the installation and use of large pieces of industrial equipment. At the end of 2021, the company's balance sheet reported total assets of $62 million and total liabilities of $40 million. The income statement for 2021 reported net income of $1.1 million, which represents an approximate 3% increase from the prior year. The company's effective income tax rate is 30%.

Near the end of 2021, a variety of expenditures were made to overhaul the company's manufacturing equipment. None of these expenditures exceeded $750, the materiality threshold the company has set for the capitalization of any such expenditure. Even though the overhauls extended the service life of the equipment, the expenditures were expensed, not capitalized.

John Henderson, the company's controller, is worried about the treatment of the overhaul expenditures. Even though no individual expenditure exceeded the $750 materiality threshold, total expenditures were $70,000.

Required:

Should the overhaul expenditures be capitalized or expensed?

Communication Case 11–6
Capitalize or expense; materiality
● LO11–9

The focus of the case is the situation described in Judgment Case 11–5. Your instructor will divide the class into two to six groups depending on the size of the class. The mission of your group is to determine the treatment of the overhaul expenditures.

Required:

1. Each group member should deliberate the situation independently and draft a tentative argument prior to the class session for which the case is assigned.
2. In class, each group will meet for 10 to 15 minutes in different areas of the classroom. During the meeting, group members will take turns sharing their suggestions for the purpose of arriving at a single group treatment.
3. After the allotted time, a spokesperson for each group (selected during the group meetings) will share the group's solution with the class. The goal of the class is to incorporate the views of each group into a consensus approach to the situation.

Integrating Case 11–7
Errors; change in estimate; change in principle; inventory, patent, and equipment
● LO11–5 through LO11–7

Whaley Distributors is a wholesale distributor of electronic components. Financial statements for the year ended December 31, 2021, reported the following amounts and subtotals ($ in millions):

	Assets	Liabilities	Shareholders' Equity	Net Income	Expenses
2020	$640	$330	$310	$210	$150
2021	$820	$400	$420	$230	$175

In 2022 the following situations occurred or came to light:

a. Internal auditors discovered that ending inventories reported in the financial statements the two previous years were misstated due to faulty internal controls. The errors were in the following amounts:

2020 inventory	Overstated by $12 million
2021 inventory	Understated by $10 million

b. A patent costing $18 million at the beginning of 2020, expected to benefit operations for a total of six years, has not been amortized since acquired.

c. Whaley's conveyer equipment has been depreciated by the sum-of-the-years'-digits (SYD) method since constructed at the beginning of 2020 at a cost of $30 million. It has an expected useful life of five years and no expected residual value. At the beginning of 2022, Whaley decided to switch to straight-line depreciation.

Required:

For each situation:

1. Prepare any journal entry necessary as a direct result of the change or error correction as well as any adjusting entry for 2022 related to the situation described. (Ignore tax effects.)
2. Determine the amounts to be reported for each of the items shown above from the 2020 and 2021 financial statements when those amounts are reported again in the 2022, 2021, and 2020 comparative financial statements.

Judgment Case 11–8
Accounting changes
● LO11–5, LO11–6

There are various types of accounting changes, each of which is required to be reported differently.

Required:

1. What type of accounting change is a change from the double-declining-balance method of depreciation to the straight-line method for previously recorded assets as a result of new information related to production patterns? Under what circumstances does this type of accounting change occur?
2. What type of accounting change is a change in the expected service life of an asset arising because of more experience with the asset? Under what circumstances does this type of accounting change occur?

(AICPA adapted)

Research Case 11–9
FASB codification; locate and extract relevant information and cite authoritative support for a financial reporting issue; impairment of property, plant, and equipment and intangible assets
● LO11–8

CODE

The company controller, Barry Melrose, has asked for your help in interpreting the authoritative accounting literature that addresses the recognition and measurement of impairment losses for property, plant, and equipment and intangible assets. "We have a significant amount of goodwill on our books from last year's acquisition of Churchill Corporation. Also, I think we may have a problem with the assets of some of our factories out West. And one of our divisions is currently considering disposing of a large group of depreciable assets."

Your task as assistant controller is to research the issue. Access the *FASB Accounting Standards Codification* at the FASB website (**www.fasb.org**). Determine the specific nine-digit Codification citation (XXX-XX-XX-XX) that describes each of the following items:

Required:

1. The measurement of impairment losses for property, plant, and equipment.
2. When to test for impairment of property, plant, and equipment.
3. The new cost basis of impaired property, plant, and equipment and prohibiting later recovery of an impairment loss.
4. The recognition and measurement of impairment losses for intangible assets that are subject to amortization.
5. The requirement that intangible assets not subject to amortization be tested for impairment at least annually.

Ethics Case 11–10
Asset impairment
● LO11–8

At the beginning of 2019, the Healthy Life Food Company purchased equipment for $42 million to be used in the manufacture of a new line of gourmet frozen foods. The equipment was estimated to have a 10-year service life and no residual value. The straight-line depreciation method was used to measure depreciation for 2019 and 2020.

Late in 2021, it became apparent that sales of the new frozen food line were significantly below expectations. The company decided to continue production for two more years (2022 and 2023) and then discontinue the line. At that time, the equipment will be sold for minimal scrap values.

The controller, Heather Meyer, was asked by Harvey Dent, the company's chief executive officer (CEO), to determine the appropriate treatment of the change in service life of the equipment. Heather determined that there has been an impairment of value requiring an immediate write-down of the equipment of $12,900,000. The remaining book value would then be depreciated over the equipment's revised service life.

The CEO does not like Heather's conclusion because of the effect it would have on 2021 income. "Looks like a simple revision in service life from 10 years to 5 years to me," Dent concluded. "Let's go with it that way, Heather."

Required:

1. What is the difference in before-tax income between the CEO's and Heather's treatment of the situation?
2. Discuss Heather Meyer's ethical dilemma.

Judgment Case 11–11
Earnings management and accounting changes; impairment
● LO11–5, LO11–6, LO11–8

Companies often are under pressure to meet or beat Wall Street earnings projections in order to increase stock prices and also to increase the value of stock options. Some resort to earnings management practices to artificially create desired results.

Required:

1. How can a company manage earnings by changing its depreciation method? Is this an effective technique to manage earnings?
2. How can a company manage earnings by changing the estimated useful lives of depreciable assets? Is this an effective technique to manage earnings?
3. Using a fictitious example and numbers you make up, describe in your own words how asset impairment losses could be used to manage earnings. How might that benefit the company?

Trueblood Accounting Case 11–12
Accounting for impairment losses; property, plant, and equipment
● LO11–8

The following Trueblood case is recommended for use with this chapter. The case provides an excellent opportunity for class discussion, group projects, and writing assignments. The case, along with Professor's Discussion Material, can be obtained from the Deloitte Foundation at its website **www.deloitte.com/us/truebloodcases.**

Case 12–9: *Rough Waters Ahead*

This case concerns the impairment test for a cruise ship.

Judgment Case 11–13
Subsequent expenditures
● LO11–9

The Cummings Company charged various expenditures made during 2021 to an account called repairs and maintenance expense. You have been asked by your supervisor in the company's internal audit department to review the expenditures to determine if they were appropriately recorded. The amount of each of the transactions included in the account is considered material.

1. Engine tune-up and oil change on the company's 12 delivery trucks—$1,300.
2. Rearrangement of equipment on the main production line—$5,500. It is not evident that the rearrangement will increase operational efficiency.
3. Installation of aluminum siding on the manufacturing plant—$32,000.
4. Replacement of the old air conditioning system in the manufacturing plant with a new system—$120,000.
5. Replacement of broken parts on three machines—$1,500.
6. Annual painting of the manufacturing plant—$11,000.
7. Purchase of new forklift to move finished product to the loading dock—$6,000.
8. Patching leaks in the roof of the manufacturing plant—$6,500. The repair work did not extend the useful life of the roof.

Required:

For each of the transactions listed above, indicate whether the expenditure is appropriately charged to the repair and maintenance expense account, and if not, indicate the proper account to be charged.

Real World Case 11–14
Disposition and depreciation; Chapters 10 and 11; Amgen
● LO11–1

Real World Financials

Amgen, Inc. is an American multinational biopharmaceutical company headquartered in Thousand Oaks, California. Located in the Conejo Valley, Amgen is the world's largest independent biotechnology firm. Amgen reported the following in a disclosure note accompanying its 2016 financial statements ($ in millions):

	2016	2015
Property and equipment	$12,427	$12,172
Less: Accumulated depreciation	(7,466)	(7,265)
Property, plant, and equipment—Net	$ 4,961	$ 4,907

Also, the company disclosed that the total cost of property and equipment included $295 and $319 millions in land at the end of 2016 and 2015, respectively. In addition, the statement of cash flows for the year ended December 31, 2016, reported the following as cash flows from investing activities:

	($ in millions)
Purchases of property and equipment	$(738)
Proceeds from the sale of property and equipment	78

The statement of cash flows also reported 2016 depreciation and amortization of $2,105 million (depreciation of $619 and amortization of $1,486).

Required:

1. Assume that all property and equipment acquired during 2016 were purchased for cash. Determine the amount of gain or loss from sale of property and equipment that Amgen recognized during 2016.
2. Assume that Amgen uses the straight-line method to depreciate property and equipment (excluding land). What is the approximate average service life of depreciable assets?

Real World Case 11–15
Depreciation and depletion method; asset impairment; subsequent expenditures; Chevron
● LO11–2, LO11–3, LO11–8, LO11–9

Real World Financials

EDGAR, the Electronic Data Gathering, Analysis, and Retrieval system, performs automated collection, validation, indexing, and forwarding of submissions by companies and others who are required by law to file forms with the U.S. Securities and Exchange Commission (SEC). All publicly traded domestic companies use EDGAR to make the majority of their filings. (Some foreign companies file voluntarily.) Form 10-K, which includes the annual report, is required to be filed on EDGAR. The SEC makes this information available on the Internet.

Required:

1. Access EDGAR on the Internet. The web address is **www.sec.gov**.
2. Search for **Chevron Corporation**. Access the 10-K filing for most recent fiscal year. Search or scroll to find the financial statements and related notes.
3. Answer the following questions related to the company's property, plant, and equipment and intangible assets:
   a. Describe the company's depreciation and depletion policies.
   b. Describe the company's policy for subsequent expenditures made for plant and equipment.

IFRS Case 11–16
Subsequent valuation of property, plant, and equipment; comparison of U.S. GAAP and IFRS; GlaxoSmithKline
● LO11–10

Real World Financials

IFRS

GlaxoSmithKline is a global pharmaceutical and consumer health-related products company located in the United Kingdom. The company prepares its financial statements in accordance with International Financial Reporting Standards.

Required:

1. Use the Internet to locate GlaxoSmithKline's most recent annual report. The address is **www.gsk.com/en-gb/investors/**. Locate the significant accounting policies disclosure note.
2. How does the company value its property, plant, and equipment? Does the company have any other options under IFRS for valuing these assets? How do these options differ from U.S. GAAP?
3. What are the company's policies for possible reversals of impairment losses for goodwill and for other non-current assets? How do these policies differ from U.S. GAAP?

Data Analytics

connect

Data analytics is the process of examining data sets in order to draw conclusions about the information they contain. If you haven't completed any of the prior data analytics cases, follow the instructions listed in the Chapter 1 Data Analytics case to get set up. You will need to watch the videos referred to in the Chapters 1 - 3 Data Analytics cases. No additional videos are required for this case. All short training videos can be found here: **https://www.tableau.com/learn/training#getting-started**.

Data Analytics Case
Accumulated Depreciation to Fixed Assets

In the Chapter 10 Data Analytics Case, you used Tableau to examine a data set and create calculations to compare two companies' fixed-asset turnover. In this case you continue in your role as an analyst conducting introductory research into the relative merits of investing in one or both of these companies. This time you assess the companies' accumulated depreciation to fixed assets ratio to determine if new assets are being employed or utilized. The accumulated depreciation to fixed assets ratio = accumulated depreciation / fixed assets.

This ratio allows the investor or analyst to examine factors like the age of the assets and therefore their remaining useful life. A low ratio means that the assets have plenty of life left in them and should be able to be used for years to come. A high ratio means the opposite. Low metrics are not necessarily a good sign though. For example, a company with a very low ratio may be spending huge amounts of money replacing fixed assets too soon. So it's useful to track this metric over time to detect patterns. A continuing increase in the company's accumulated depreciation to fixed assets ratio can indicate that management is struggling to find the cash necessary to acquire new assets.

Required:

Use Tableau to calculate each company's annual accumulated depreciation to fixed assets ratios in each year from 2012 to 2021. Assume both companies use the same depreciation method over the entire period. Based upon what you find, answer the following questions:

1. Comparing the accumulated depreciation to fixed assets ratios over the first five years of the ten-year period, is Big Store's accumulated depreciation to fixed assets ratio (a) generally increasing, (b) roughly the same, or (c) generally decreasing from year to year?
2. Comparing the accumulated depreciation to fixed assets ratios over the second five years of the ten-year period, is Big Store's accumulated depreciation to fixed assets ratio (a) generally increasing, (b) roughly the same, or (c) generally decreasing from year to year?
3. Comparing the accumulated depreciation to fixed assets ratios over the first five years of the ten-year period, is Discount Goods' accumulated depreciation to fixed assets ratio (a) generally increasing, (b) roughly the same, or (c) generally decreasing from year to year?
4. Comparing the accumulated depreciation to fixed assets ratios over the second five years of the ten-year period, is Discount Goods' accumulated depreciation to fixed assets ratio (a) generally increasing, (b) roughly the same, or (c) generally decreasing from year to year?
5. Comparing the accumulated depreciation to fixed assets ratios over the entire ten-year period, which of the two companies appears to maintain fixed assets with longer remaining useful lives?

Resources:

You have available to you an extensive data set that includes detailed financial data for Discount Goods and Big Store and for 2012-2021. The data set is in the form of four Excel files available to download from Connect, or under Student Resources within the Library tab. The file for use in this chapter is named "Discount_Goods_Big_Store_Financials.xlsx." Download this file and save it to the computer on which you will be using Tableau.

For this case, you will create some calculations to produce a bar chart of the accumulated depreciation to fixed assets ratio to allow you to compare and contrast the two companies.

After you view the training videos, follow these steps to create the charts you'll use for this case:

- Open Tableau and connect to the Excel spreadsheet you downloaded.
- Click on the "Sheet 1" tab at the bottom of the canvas, to the right of the Data Source at the bottom of the screen. Drag "Company" and "Year" under Dimensions to the Rows shelf. Change "Year" to *discrete* by right-clicking and selecting "discrete."
- Drag the "Buildings and Improvements," "Computer hardware and software," "Construction in progress," "Fixtures and Equipment," "Land," and "Less accumulated depreciation" under Measures to the Rows shelf. Change each to *discrete.* Right click on the drop-down menu of each of the accounts and uncheck "Show Header" so they are not visible in the field.
- Create a calculated field named "AD to FA ratio" In the Calculation Editor window, from the Rows shelf, drag "Less accumulated depreciation" then type a multiplication sign (*) and a negative 1, followed by a division sign and an open parenthesis, then drag "Buildings and Improvements," type a plus sign, then "Computer hardware and software," plus sign, "Construction in progress," plus sign, "Fixtures and Equipment," plus sign, "Land," then a closed parenthesis. Make sure the window says that the calculation is valid and click OK.
- Drag the newly created "AD to FA ratio" to the Rows shelf.
- Click on the "Show Me" and select "side-by-side bars." Add labels to the bars by clicking on "Label" under the "Marks" card and clicking the box "Show mark labels." Format the labels to Times New Roman, bold, black and 9-point font.
- Change the title of the sheet to be "Accumulated Depreciation to Fixed Assets Ratio Bar Chart" by right-clicking and selecting "Edit title." Format the title to Times New Roman, bold, black and 15-point font. Change the title of "Sheet 1" to match the sheet title by right-clicking, selecting "Rename" and typing in the new title.
- Format the labels on the left of the sheet ("AD to FA ratio") to Times New Roman, 9-point font, bold, and black.
- Format all other labels to be Times New Roman, bold, black and 12-point font.
- Once complete, save the file as "*DA11_Your initials*.twbx."

Continuing Cases

Target Case
- LO11–2, LO11–8, LO11–9

Target Corporation prepares its financial statements according to U.S. GAAP. Target's financial statements and disclosure notes for the year ended February 3, 2018, are available in Connect. This material is also available under the Investor Relations link at the company's website (**www.target.com**).

Required:

1. Compare the property and equipment listed in the balance sheet with the list in Note 14. What are the estimated useful lives for recording depreciation? Is land listed in Note 14 (yes/no)?
2. In Note 14, which depreciation method does Target use for property and equipment for financial reporting? Which depreciation method is used for tax purposes? Why might these methods be chosen?
3. In Note 14, how does Target record repairs and maintenance expense?
4. In Note 14, does Target report any impairment of property and equipment for the year ended February 3, 2018? If so, what was the amount and what were the reasons for the impairments?
5. From Notes 15 and 16, were any impairments related to intangible assets recorded for the year ended February 3, 2018? If so, what was the amount and what were the reasons for the impairments?

Air France–KLM Case
- LO11–10

IFRS

Air France–KLM (AF), a Franco-Dutch company, prepares its financial statements according to International Financial Reporting Standards. AF's financial statements and disclosure notes for the year ended December 31, 2017, are available in Connect. This material is also available under the Finance link at the company's website (**www.airfranceklm.com**).

Required:

1. AF's property, plant, and equipment is reported at cost. The company has a policy of not revaluing property, plant, and equipment. Suppose AF decided to revalue its flight equipment on December 31, 2017, and that the fair value of the equipment on that date was €12,000 million. Prepare the journal entry to record the revaluation assuming that the journal entry to record annual depreciation had already been recorded. (*Hint:* you will need to locate the original cost and accumulated depreciation of the equipment at the end of the year in the appropriate disclosure note.)
2. Under U.S. GAAP, what alternatives do companies have to value their property, plant, and equipment?
3. AF calculates depreciation of plant and equipment on a straight-line basis, over the useful life of the asset. Describe any differences between IFRS and U.S. GAAP in the calculation of depreciation.
4. When does AF test for the possible impairment of fixed assets? How does this approach differ from U.S. GAAP?
5. Describe the approach AF uses to determine fixed asset impairment losses. (*Hint:* see Note 4.14) How does this approach differ from U.S. GAAP?
6. The following is included in AF's disclosure note 4.13: "Intangible assets are recorded at initial cost less accumulated amortization and any accumulated impairment losses." Assume that on December 31, 2017, AF decided to revalue its Other intangible assets (see Note 17) and that the fair value on that date was determined to be €500 million. Amortization expense for the year already has been recorded. Prepare the journal entry to record the revaluation.

CPA Exam Questions and Simulations

ROGER CPA Review

Sample CPA Exam questions from Roger CPA Review are available in Connect as support for the topics in this chapter. These multiple-choice questions and task-based simulations include expert-written explanations and solutions, and provide a starting point for students to become familiar with the content and functionality of the actual CPA Exam.

CHAPTER

12 Investments

OVERVIEW

In this chapter, you will learn about various approaches we use to account for investments that companies make in the debt and equity securities of other companies. An investing company always has the option to account for these investments at fair value, with changes in fair values reported in the income statement. However, depending on the nature of a *debt* investment, investors use accounting approaches that either ignore most fair value changes (*held-to-maturity* investments) or that include fair value changes only in other comprehensive income (*available-for-sale* investments) until the debt investment is sold. For an *equity* investment, when an investor owns enough stock to "significantly influence" an investee but does not control it, the investor uses the *equity method* of accounting, which ignores fair value changes but includes a portion of the investee's income in the investor's income. In appendices to this chapter, we discuss other types of investments as well as how to deal with an investment whose value has been impaired.

LEARNING OBJECTIVES

After studying this chapter, you should be able to:

- **LO12–1** Describe the key characteristics of a debt investment and demonstrate how to account for a purchase and for interest revenue. (*p. 646*)
- **LO12–2** Demonstrate how to identify and account for debt investments classified for reporting purposes as held-to-maturity. (*p. 650*)
- **LO12–3** Demonstrate how to identify and account for debt investments classified for reporting purposes as trading securities. (*p. 652*)
- **LO12–4** Demonstrate how to identify and account for debt investments classified for reporting purposes as available-for-sale securities. (*p. 656*)
- **LO12–5** Demonstrate how to identify and account for equity investments classified for reporting purposes as fair value through net income. (*p. 669*)
- **LO12–6** Demonstrate how to identify and account for equity investments accounted for under the equity method. (*p. 674*)
- **LO12–7** Explain the adjustments made in the equity method when the fair value of the net assets underlying an investment exceeds their book value at acquisition. (*p. 676*)
- **LO12–8** Explain how electing the fair value option affects accounting for investments. (*pp. 664 and 682*)
- **LO12–9** Discuss the primary differences between U.S. GAAP and IFRS with respect to investments. (*pp. 664–665, 673, 682, and 693*)

©8th.creator/Shutterstock

FINANCIAL REPORTING CASE

A Case of Coke

You are the one accounting major in your five-member group in your Business Policy class. A part of the case your group is working on is the analysis of the financial statements of **The Coca-Cola Company.**

The marketing major in the group is confused by the following disclosure note included in Coca-Cola's 2017 annual report:

NOTE 3: INVESTMENTS (in part)

Investments in debt securities that the Company has the positive intent and ability to hold to maturity are carried at amortized cost and classified as held-to-maturity. Investments in debt securities that are not classified as held-to-maturity are carried at fair value and classified as either trading or available-for-sale. Realized and unrealized gains and losses on trading securities and realized gains and losses on available-for-sale securities are included in net income. Unrealized gains and losses, net of deferred taxes, on available-for-sale securities are included in our consolidated balance sheets as a component of AOCI . . . .

"So they have held-to-maturity securities, trading securities, and available-for-sale securities. What's the difference? And they say that unrealized gains and losses on available-for-sale securities are reported as part of AOCI. What's that? I don't see these gains and losses in the income statement," he complained. "And what about equity method investments? On the balance sheet they have over $20 *billion* of investments accounted for under the equity method! They made more than $1 billion on those investments in 2017. Is that cash they can use?"

QUESTIONS

By the time you finish this chapter, you should be able to respond appropriately to the questions posed in this case. Compare your response to the solution provided at the end of the chapter.

1. Why are held-to-maturity securities treated differently from other investment securities? (*p. 650*)
2. Why are unrealized gains and losses on trading securities reported in the income statement? (*p. 652*)
3. Why are unrealized gains and losses on available-for-sale securities not reported in the income statement, but instead are reported in other comprehensive income, and then shown in accumulated other comprehensive income (AOCI) in the balance sheet? (*p. 653*)
4. Explain why Coke accounts for some of its investments by the equity method and what that means. (*p. 674*)

Corporations raise funds to finance their operations by selling equity securities (common and preferred stock) and debt securities (bonds and notes). These securities, also called financial instruments, are purchased as investments by individual investors, mutual funds, and also by other corporations. In later chapters we discuss equity and debt securities from the perspective of the issuing company. Our focus in this chapter is on the corporations that invest in debt and equity securities issued by other corporations as well as in debt securities issued by governmental units (bonds, Treasury bills, and Treasury bonds).

Most companies invest in financial instruments issued by other companies. For some investors, these investments represent ongoing affiliations with the companies whose securities are acquired. Examples of those sorts of investments include **Microsoft**'s acquisition of **LinkedIn** for over $25 billion and **Anheuser-Busch InBev**'s acquisition of **SABMiller** for over $100 billion. Some investments, though, are not made to obtain a favorable business relationship with another firm. Instead, companies seek only to earn a return from the dividends or interest the securities pay or from increases in the market prices of the securities—the same reasons that might motivate you to buy stocks, bonds, or other investment securities.

With such diversity in investment objectives, it's not surprising that there is diversity in the approaches used to account for investments. As you'll discover when reading this chapter, investments are accounted for in several different ways, depending on whether the investment is in a debt or equity security, the investor's purpose for holding the investment, and the nature of the investment relationship. In Part A, we discuss accounting for debt investments. In Part B, we discuss accounting for equity investments.

PART A

Accounting for Debt Investments

Example of a Debt Investment

LO12–1

Have you ever bought a CD (Certificate of Deposit) at a bank? If so, you've invested in a debt instrument. Let's say you buy a $500, 2-year, 4% CD. What's happened is that you are lending the bank $500 for 2 years, and the bank is promising to pay you 4% interest each year before returning your $500 after two years. The bank's debt (the $500 borrowed from you) is represented by a debt instrument (the CD), which specifies the maturity date (2 years), the principal ($500), and the annual rate of interest (4%).

Companies invest in debt too. Like a CD, a bond or other debt security has a specified date when it matures, and on that maturity date, the *principal* (also called the *face amount* or *maturity value*) is paid to investors. In the meantime, interest equal to some *stated interest rate* multiplied by the principal is paid to investors on specified interest dates (usually twice a year). Think of the principal and interest payments of the bond as a stream of cash flows that an investor will receive in the future in exchange for purchasing the bond today. The investor values that stream of future cash flows based on the prevailing *market interest rate* for debt of similar risk and maturity at the time the investor purchases the bond.

For an example of how an investor would determine how much to pay for a bond, see Illustration 12–1.

Illustration 12–1

Example of a Debt Investment: Bonds Purchased at a Discount

Because interest is paid semiannually, the present value calculations use:

a. one-half the stated rate (6%),
b. one-half the market rate (7%), and
c. 6 (= 3 × 2) semiannual periods.

On July 1, 2021, Masterwear Industries issued $700,000 of 12% bonds, dated July 1.

- Interest of $42,000 is payable semiannually on June 30 and December 31.
- The bonds mature in three years, on June 30, 2024.
- United Intergroup, Inc. purchased the entire bond issue on a date when the market interest rate for bonds of similar risk and maturity was 14%.*

Calculation of the Price of the Bonds		**Present Values**
Interest	$ 42,000 × 4.76654**	= $200,195
Principal (face amount)	$700,000 × 0.66634†	= 466,438
Present value (price) of the bonds		**$666,633**

*The numbers in this illustration are the same as those in Illustration 14–3 in Chapter 14 (except for some differences in dates between the two chapters). This helps us to better appreciate in Chapter 14 how Masterwear's accounting for its bond liability to United compares to United's accounting for its investment in Masterwear bonds. You can find further explanation of why we calculate the bond price this way in Chapter 14.
**Present value of an ordinary annuity of $1: $n = 6$, $i = 7\%$ (Table 4).
†Present value of $1: $n = 6$, $i = 7\%$ (Table 2).
Note: Present value tables are provided at the end of this textbook. If you need to review the concept of the time value of money, refer to the discussions in Chapter 5.

Illustration 12–1 shows that United will pay Masterwear **$666,633** on July 1, 2021. In return, United expects to receive from Masterwear $42,000 every six months for the next three years plus the principal of $700,000 when the bonds mature on June 30, 2024. Of course, if United sells the bonds to another investor, that investor will receive the remaining payments of interest and principal.

Notice that United paid $666,633 to purchase the $700,000 bonds. Why the difference? To determine the issue price of the bonds, it's important to understand how investors compare the bond's stated rate with the market rate. If the interest rate paid by the bond (the stated rate) is higher than the market rate, investors are willing to purchase the bond for more than its maturity value (so it's sold at a *premium*). If the bond's stated rate is lower than the market rate, then investors are willing to purchase the bond only at an amount less than its maturity value (so it's sold at a *discount*). Masterwear was offering its bonds for 12%, but investors could have obtained bonds of similar risk and maturity at a more favorable, higher rate of 14%. To attract investors, Masterwear had to sell its bonds at a discount.

The Masterwear bonds have the key characteristics of all debt investments. Over the three-year life of the bonds, United has to determine how it will account for four events:

1. Purchasing the debt investment.
2. Receiving interest every six months.
3. Holding the bonds during periods in which the bonds' fair value changes (and thus incurring *unrealized holding gains and losses,* since the bonds have not been sold).
4. Either selling the bonds before maturity or receiving the principal payment at their maturity date.

As we discuss below, companies classify debt investments as one of three types: held-to-maturity, trading, or available-for-sale. Accounting for the first two events—the purchase of a bond and the receipt of interest payments—is handled the same way regardless of how the debt investment is classified. We'll look at those events first. Then we'll look at the second two events, which are accounted for differently depending on how the debt investment is classified.

Recording the Purchase of a Debt Investment

When debt investments are purchased, they are recorded at cost—that is, the total amount paid for the investment, including any brokerage fees. Referring back to Illustration 12–1, we see that United paid **$666,633** to purchase Masterwear's $700,000 bonds. United would record the purchase as follows:

All investment securities are initially recorded at cost.

July 1, 2021

Investment in bonds (face amount)	700,000	
Discount on bond investment (difference)		33,367
Cash (price paid for the bonds)		**666,633**

Because United purchased the bonds for an amount that's less than their face amount, it credits *Discount on bond investment* for the difference. Discount on bond investment is a contra-asset to the investment account that serves to reduce the carrying value of the investment to its cost at the date of purchase. If United instead had purchased the bonds for an amount (say, $725,000) that is higher than the face amount of the bond ($700,000), it would instead debit *Premium on bond investment* (for $25,000) to record the investment at its cost at the date of purchase.

Recording Interest Revenue

United will record interest revenue using the **effective interest method**. Here's how it works.

Masterwear's bonds have a *stated rate* of 12%, payable semiannually. This means that every six months, United will receive exactly $42,000 in cash from Masterwear:

$700,000	×	(12% ÷ 2)	=	**$42,000**
Face amount		Stated rate		Interest received

[1]This is called the gross method. An alternative would be to record the cost ($666,633 or $725,000) directly in the investment account, called the net method.

However, recall that United purchased the $700,000 bonds at a discounted amount of $666,633. Why could United pay less than the $700,000 face value for the bonds (and why was Masterwear willing to sell the bonds for less than $700,000)? At the time United purchased the bonds, the *market rate* of interest for bonds of similar risk was 14%. So, United would only be willing to invest in these bonds if it could earn the 14% it could get elsewhere (not the 12% stated rate). To earn this higher rate, United needed to pay only $666,633. Paying the lower amount means United lowered its investment cost and effectively increased its rate of return to 14% (refer back to Illustration 12–1 to see the details of this calculation). While United will receive $42,000 (6%) in interest from Masterwear in the first six months, it will effectively earn interest revenue of $46,664 (7%) on its investment.

$666,633	×	(14% ÷ 2)	=	**$46,664**
Outstanding balance		Market rate		Interest revenue

Under the *effective interest method,* interest for a period equals the market rate of interest when the debt was purchased multiplied by the outstanding balance of the debt at the beginning of the period.

The amount by which interest revenue exceeds interest received ($46,664 − $42,000 = $4,664 in the first six months) represents a piece of the cost savings from purchasing the investment at a discount. This piece of the cost savings increases United's investment return from the rate the bond pays (12%) to the higher rate (14%) that investors could have earned on other similar bonds at the time they purchased the bond. In fact, this approach is called the *effective interest method* because interest revenue is based on the effective interest rate that the investment earns over its lifetime.

The journal entry to record the interest received for the first six months as investment revenue is

Discount on bond investment

	33,367
4,664	
	28,703

December 31, 2021		
Cash (stated rate × face amount)	**42,000**	
Discount on bond investment (difference)	4,664	
Interest revenue (market rate × outstanding balance)		**46,664**

This entry reduces the discount by $4,664 (from $33,367 to $28,703). Because the discount gets smaller, the "amortized cost" of the investment (equal to $700,000 less discount) gets larger by the same amount (from $666,633 to $671,297).

Illustration 12–2 demonstrates interest revenue being recorded at the effective rate over the life of this investment. As you can see, amortization of the discount gradually increases the amortized cost of the investment, until the investment reaches its principal amount of $700,000, which is the amount to be received when the debt matures.

Illustration 12–2

Amortization Schedule—Discount

If a bond is purchased at a discount, less cash is received each period than the effective interest earned by the investor, so the unpaid difference increases the outstanding balance of the investment.

Date	Cash Interest	Effective Interest (Interest Revenue)	Amortization of Discount	Discount Balance	Amortized Cost
	(6% × Face amount)	(7% × Outstanding balance)	(Difference)		(Face amount less discount)
7/1/2021				33,367	666,633
12/31/2021	42,000	0.07(666,633) = 46,664	4,664	28,703	671,297
6/30/2022	42,000	0.07(671,297) = 46,991	4,991	23,712	676,288
12/31/2022	42,000	0.07(676,288) = 47,340	5,340	18,372	681,628
6/30/2023	42,000	0.07(681,628) = 47,714	5,714	12,658	687,342
12/31/2023	42,000	0.07(687,342) = 48,114	6,114	6,544	693,456
6/30/2024	42,000	0.07(693,456) = 48,544*	6,544	0	700,000
	252,000	285,367	33,367		

*Rounded

If the bonds were instead purchased at a premium, a similar amortization schedule would *reduce* the premium over time until the bond's amortized cost reached the face amount of $700,000, which is the amount to be received when the debt matures.

Three Classifications of Debt Investments

The amortization schedule shown in Illustration 12–2 is based on the market interest rate (14%) that prevailed at the time United purchased the Masterwear bonds. United will use that amortization schedule for the life of the investment, and *won't change it* in response to changes in prevailing interest rates. However, the prevailing market rate may not always be 14%. The fair value of the bonds will change with changes in the prevailing market interest rate, because market participants will use the prevailing rate to compute the present value of the cash flows provided by the bond. United will incur **unrealized holding gains and losses** as a result of holding the bond during periods in which its fair value changes.

If the market rate of interest *rises* after a bond is purchased, the market will compute the present value of the cash flows provided by the bond using that higher discount rate, so the fair value of the bond falls. In that case, the person holding the bond suffers an *unrealized holding loss.* The fair value of the bond has decreased, but that loss hasn't been realized, because the investment has not been sold.

The fair value of a fixed-rate investment moves in the opposite direction of market interest rates.

Conversely, if the market rate of interest *falls* after a bond is purchased, the market will calculate the present value of the cash flows provided by the bond using that lower rate, so the fair value of the bond rises. In that case, the investor holding the bond enjoys an *unrealized holding gain.* The fair value of the bond has increased but that gain hasn't yet been realized, because the investment hasn't been sold.

Changes in fair value give rise to unrealized holding gains and losses.

Accounting for changes in fair value depends on the classification of the debt investment. As shown in Illustration 12–3, debt investments are classified in one of three categories: held-to-maturity (HTM) securities, trading securities (TS), or available-for-sale (AFS) securities.

Illustration 12–3
Accounting for Unrealized Holding Gains and Losses on Debt Investments

Reporting Approach	Treatment of Unrealized Holding Gains and Losses	Carried in Balance Sheet at
Held-to-maturity (HTM): used for debt for which the investor has the "positive intent and ability" to hold to maturity.	**Not recognized***	Amortized Cost
Trading (TS): used for debt that is held in an active trading account for immediate resale.	**Recognized in net income**, and therefore in retained earnings as part of shareholders' equity.	Fair Value
Available-for-sale (AFS): used for debt that does not qualify as held-to-maturity or trading.	**Recognized in other comprehensive income**, and therefore in accumulated other comprehensive income in shareholders' equity.*	Fair Value

*If the investor elects the *fair value option (FVO)*, this type of investment also can be accounted for using the same approach that's used for trading securities, with the investment reported at fair value and unrealized holding gains and losses included in net income. Also, as discussed in Appendix 12B, credit losses associated with HTM and AFS investments are recognized in net income, and other impairments of AFS investments are recognized in net income under some circumstances.

Illustration 12–4 provides a description from a recent annual report of how **General Motors** accounts for its investments in marketable securities.

Illustration 12–4
Disclosure about Investments—General Motors
Real World Financials

Note 2 (in part): Significant Accounting Policies

Marketable Securities We classify marketable securities as available-for-sale or trading. Various factors, including turnover of holdings and investment guidelines, are considered in determining the classification of securities. Available-for-sale securities are recorded at fair value with unrealized gains and losses recorded net of related income taxes in Accumulated other comprehensive loss until realized. Trading securities are recorded at fair value with changes in fair value recorded in Interest income and other non-operating income, net.

Source: General Motors Company

FINANCIAL Reporting Case

Q1, p. 645

Why treat unrealized holding gains and losses differently depending on the type of investment? As you know, the primary purpose of accounting is to provide information useful for making decisions. What's most relevant for that purpose is not necessarily the same for each investment a company might make. For example, a company might invest in corporate bonds to provide a steady return until the bonds mature, in which case day-to-day changes in fair value may not be viewed as very relevant, so the held-to-maturity approach is preferable. On the other hand, a company might invest in the same bonds because it plans to sell them at a profit in the near future, in which case the day-to-day changes in fair value could be viewed as very relevant, and the trading security or available-for-sale approach is preferable.

We'll discuss each reporting approach in turn, including how that approach accounts for unrealized holding gains and losses.

Debt Investments to Be Held to Maturity (HTM)

Unrealized Holding Gains and Losses Are Not Recognized For HTM Investments

Held to maturity (HTM) investments require the "positive intent and ability" to hold the investments to maturity.

LO12–2

Unrealized holding gains and losses are less important if sale before maturity isn't an alternative, because those gains and losses will never be realized by sale. For this reason, if an investor has the "positive intent and ability" to hold the securities to maturity, investments in debt securities typically are classified as **held-to-maturity (HTM)** and reported at their *amortized cost* in the balance sheet.[2] A debt security cannot be classified as held-to-maturity if the investor might sell it before maturity in response to changes in market prices or interest rates, to meet the investor's liquidity needs, or similar factors.

Held to maturity (HTM) investments are reported at amortized cost in the balance sheet.

To consider accounting for unrealized gains and losses, suppose that on December 31, 2021, the market interest rate for securities similar to the Masterwear bonds has fallen to 11%. An investor valuing the Masterwear bonds at that time would do so considering the current market interest rate (11%) because that's the rate of return available for similar bonds. Calculating the present value of the bonds using a lower discount rate results in a higher present value (price).

Let's say that checking market prices in *The Wall Street Journal* indicates that the fair value of the Masterwear bonds on December 31, 2021, is $714,943. As shown in Illustration 12–2, those same bonds have an amortized cost of $671,297. This means there is an unrealized gain of $43,646 for the difference. How will United account for this increase in the fair value of its debt investment? If United views the bonds as HTM investments, that change in fair value will be *ignored*. The investment simply will be shown in the balance sheet at amortized cost of $671,297. United will *disclose* the fair value of its HTM investments in a note to the financial statements, but will not recognize any fair value changes in the income statement or balance sheet.[3]

Sale of HTM Investments

Typically, held-to-maturity investments are—you guessed it—held to maturity. However, suppose that due to unforeseen circumstances the company decided to sell its debt investment for $725,000 on January 5, 2022.[4] United would record the sale as follows (for simplicity we ignore interest earned during the first five days of 2022):[5]

[2]FASB ASC 320–10–25–1: Investments–Debt Securities–Overall–Recognition.

[3]If United had chosen the fair value option for this investment, it would classify the investment as a trading security rather than as an HTM security. We'll illustrate the fair value option when we discuss trading securities.

[4]GAAP [FASB ASC 320–10–25–6: Investments—Debt Securities—Overall—Recognition, previously *Statement of Financial Accounting Standards No. 115*] lists major unforeseen events that could justify sale of an HTM investment. Sale for other reasons could call into question whether the company actually had the intent and ability to hold the investment to maturity. In that case, the company's HTM classification is viewed as "tainted," and the company can be required to reclassify *all* of its HTM investments as AFS investments and avoid using the HTM classification for two years. Similar provisions exist under IFRS for public companies.

[5]For purposes of this example, we ignore any interest accrued for the first five days of 2019. In practice, United would accrue five days of interest revenue and record five days of amortization of the discount, which would produce a revised amortized cost of the bonds that would be used to calculate any gain or loss on sale.

January 5, 2022		
Cash	725,000	
Discount on bond investment (account balance)	28,703	
Investment in bonds (account balance)		700,000
Gain on investments (NI) (to balance)		**53,703**

In other words, United would record this sale just like any other asset sale, with a realized gain or loss determined by comparing the cash received with the carrying value (in this case, the amortized cost) of the asset sold.

Impairment of HTM Investments

There is one important exception to the general rule that companies don't recognize unrealized gains and losses for HTM investments. You learned in Chapter 7 that companies are required to use the Current Expected Credit Loss (CECL) model to account for bad debts with respect to accounts receivable and notes receivable. Companies likewise are required to use the CECL model to account for credit losses on HTM investments. That requires companies to make an estimate of the amount of interest and principal payments they won't receive in the future. Companies account for that estimate by recognizing a credit loss in net income and reducing the carrying value of the HTM investment with an allowance for credit losses, just like they recognize bad debt expense and an allowance for uncollectible accounts for accounts receivable. In Appendix 12B we provide an illustration of recognizing credit losses for HTM investments.

Additional Consideration

Recall from Chapter 1 that GAAP identifies different ways a firm can determine fair value. If the Masterwear bonds are publicly traded, United can find the fair value by looking up the current market price (this way of obtaining fair value is consistent with "level 1" of the fair value hierarchy). On the other hand, if the bonds are not publicly traded, United can calculate the fair value by using the present value techniques shown in Illustration 12–1 (this way of obtaining fair value is consistent with "level 2" of the fair value hierarchy). With five interest periods remaining, and a current market rate of 11% (5.5% semiannually), the present value would be $714,943.

			Present Values
Interest	$ 42,000 × 4.27028*	=	$179,352
Principal	$700,000 × 0.76513†	=	535,591
Present value of the bonds			$714,943

*Present value of an ordinary annuity of $1: $n = 5$, $i = 5.5\%$. (Table 4)
†Present value of $1: $n = 5$, $i = 5.5\%$. (Table 2)

Using Excel, enter:
=PV(.055,5, − 42000, −700000)
Output: 714,946

Using a calculator:
enter: N 5 I 5.5 PMT −42000 FV − 700000
Output: = PV 714,946

Financial Statement Presentation

HTM securities appear in the financial statements as follows:

- **Income Statement and Statement of Comprehensive Income:** *Realized* gains and losses are shown in net income in the period in which securities are sold. *Unrealized* holding gains and losses are disclosed in the notes to financial statements. Investments in HTM securities do not affect other comprehensive income.
- **Balance Sheet:** Investments in HTM securities are reported at amortized cost, less any allowance for credit losses. Fair values of those investments are disclosed in the notes to financial statements.
- **Cash Flow Statement:** Cash flows from buying and selling HTM securities typically are classified as investing activities.

Assuming United sold its investment on January 5, 2022, United's 2021 and 2022 financial statements will include the amounts shown in Illustration 12–5.

Illustration 12–5

Reporting Held-to-Maturity Investments

(Ignoring income taxes)	2021	2022
Statement of Comprehensive Income		
Revenues	$ ◆	$ ◆
Expenses	◆	◆
Other income (expense):		
Interest revenue	46,664	–0–
Gain on investments	–0–	**53,703**
Net income	46,664	53,703
Other comprehensive income (OCI)	–0–	–0–
Comprehensive income (Net income + OCI)	$ 46,664	$ 53,703
Balance Sheet		
Assets:		
Investment in bonds (HTM)	671,297	–0–
Shareholders' equity:		
Retained Earnings	46,664	100,367*
Statement of Cash Flows (direct method)		
Operating Activities:		
Cash from interest received	42,000	–0–
Investing Activities:		
Purchase of HTM securities	(666,633)	–0–
Sale of HTM securities	–0–	725,000

*Net income of $46,664 (2021) + $53,703 (2022) = $100,367 accumulates in retained earnings by the end of 2022.

Only *realized* gains and losses are included in net income.

HTM securities are reported at amortized cost less any allowance for credit losses.

Cash flows from buying and selling HTM securities are classified as investing activities.

Debt Investments Classified as Trading Securities

***Trading securities* are actively managed in a trading account for the purpose of profiting from short-term price changes.**

● LO12–3

Some companies—primarily financial institutions—actively and frequently buy and sell securities, expecting to earn profits on short-term price fluctuations. Investments in debt acquired principally for the purpose of selling them in the near term are classified as **trading securities**. The holding period for trading securities generally is measured in hours and days rather than months or years. These investments typically are reported among the investor's current assets. Usually only banks and other financial operations invest in securities in the manner and for the purpose necessary to be categorized as trading securities.

Just like HTM investments, trading securities are recorded at cost when they are purchased, and any discount or premium is amortized to interest revenue over time as periodic interest payments are received. However, in subsequent periods, there are two important differences between trading securities and HTM investments.

1. Trading securities are written up or down to their fair value, or "marked to market," in the balance sheet. (HTM securities are kept at amortized cost.)
2. Corresponding unrealized holding gains and losses on trading securities are included in net income in the income statement. (HTM securities do not include unrealized holdings gains and losses in net income.)

FINANCIAL Reporting Case

Q2, p. 645

Be sure to notice that reporting trading securities at their fair value is a departure from amortized cost, which is the way many assets are reported in the balance sheet. Why the difference? For trading securities, fair value information is more relevant than for other assets intended primarily to be used in company operations, like buildings, land and equipment, or for debt investments intended to be held to maturity. Changes in fair values provide an indication of management's success in deciding when to acquire the investment, when to sell it, whether to invest in fixed-rate or variable-rate securities, and whether to invest in long-term or short-term securities. For that reason, it makes sense to report unrealized holding gains and losses on trading securities in net income during a period that fair values change, even though those gains and losses haven't yet been realized through the sale of the securities.

To see how we account for trading securities, let's return to our Masterwear bond example, but assume that those debt investments are held in an active trading portfolio. As of December 31, 2021, United has recorded the purchase of the bonds as well as receipt of the first semiannual interest payment, so the bonds have an amortized cost of $671,297 (refer back to Illustration 12–2).

Adjust Trading Security Investments to Fair Value (2021)

Unlike HTM securities, trading securities are carried at fair value in the balance sheet, so their carrying value must be adjusted to fair value by the end of every reporting period. In fact, many companies adjust trading securities to fair value at the end of every day. Rather than increasing or decreasing the investment account itself, we use a valuation allowance, *fair value adjustment*, to increase or decrease the carrying value of the investment. At the same time, we record an unrealized holding gain or loss that is included in net income in the period in which fair value changes (remember, the gain or loss is *unrealized* because the securities haven't been sold).

Trading securities are adjusted to their fair value in each reporting period.

Assuming the Masterwear bonds have a fair value of $714,943 as of December 31, 2021, the next table shows the calculation of the balance in the fair value adjustment account that is required on that date.

December 31, 2021

Security	Amortized Cost	Fair Value	Necessary Fair Value Adjustment Balance
Masterwear	$671,297	$714,943	**$43,646**

The bonds need to be reported at their fair value of $714,943. Because the bonds currently are recorded at their amortized cost of $671,297, the fair value adjustment account needs a debit balance of **$43,646**. United will recognize whatever unrealized holding gain or loss is necessary to move the fair value adjustment from its current balance of $0 (at purchase date) to $43,646 (on December 31, 2021). In this case, the calculation is simple:

	Fair Value Adjustment
Beginning balance on 7/1/2021	$ 0
± Adjustment needed to update fair value	?
Balance needed on 12/31/2021	**$43,646**

Fair Value Adjustment

Debit	Credit
0	
43,646	
43,646	

The journal entry to record the $43,646 change in United's fair value adjustment and the corresponding unrealized holding gain is:

December 31, 2021

	Debit	Credit
Fair value adjustment (calculated above)*	43,646	
Gain on investments (unrealized, NI)†		43,646

*Sometimes companies don't bother with a separate fair value adjustment account and simply adjust the investment account to fair value.

†We indicate "unrealized, NI" to highlight that, for trading securities, unrealized holding gains and losses are included in the income statement in the period in which they occur.

Each period United owns the Masterwear bonds, it will recognize whatever unrealized holding gain or loss is necessary to move the fair value adjustment to the value it needs to have at the end of the accounting period. Increases in the fair value adjustment produce gains on trading securities that increase net income; decreases produce losses that decrease net income.

Unrealized holding gains and losses for trading securities are included in net income in the period in which fair value changes.

Additional Consideration

Accounting for Portfolios

We have focused on accounting for an individual security, but United would use the same approach to account for a *portfolio* of trading securities. For example, assume that United has the following portfolio of trading securities as of December 31, 2021, as shown below:

Security	Amortized Cost	Fair Value	Necessary Fair Value Adjustment Balance
Miley Inc.	$ 800,000	$ 875,000	$ 75,000
Perry Corp.	950,000	790,000	(160,000)
Total	$1,750,000	$1,665,000	**$ (85,000)**

As of December 31, 2022, the portfolio's status is as follows:

Security	Amortized Cost	Fair Value	Necessary Fair Value Adjustment Balance
Miley Inc.	$ 600,000	$575,000	$ (25,000)
Swift Co.	450,000	325,000	(125,000)
Total	$1,050,000	$900,000	**$(150,000)**

Fair Value Adjustment

	85,000
	65,000
	150,000

On December 31, 2022, the balance of the fair value adjustment needs to change from a credit of **$85,000** to a credit of **$150,000,** requiring an additional credit of **$65,000** and recognition of a corresponding loss in net income.

To make that happen, United records the following journal entry:

Loss on investments (unrealized, NI)	65,000	
Fair value adjustment		65,000

Sale of Trading Security Investments

Now assume that United sells the bonds for $725,000 on January 5, 2022. To account for the sale, United needs to do two things. First, United needs to update the carrying value of the bonds to fair value and record in net income any unrealized holding gains and losses that occurred during 2022 up to the date of sale. Second, on the date of sale, United needs to record the receipt of cash and remove the amounts associated with the investment from the relevant balance sheet accounts. Let's record each of these entries. (As in our example for HTM investments, for simplicity we ignore interest earned during the first five days of 2022.)

For trading securities, unrealized holding gains and losses from fair value changes are recorded up to the date an investment is sold.

1. Adjust Trading Securities to Fair Value (2022). We first need to update the fair value adjustment and recognize any unrealized holding gains or losses that have occurred during the current reporting period prior to the date of sale. We already have accounted for fair value changes that occurred during 2021. Now we need to record the additional fair value changes and their related unrealized holding gains and losses that have occurred each day in 2022 up to the moment of sale. Companies might record these changes in fair value at the end of each day, but we use a single summary entry that captures those changes in fair value up to the moment of sale.

Remember that on December 31, 2021, the amortized cost of the Masterwear bonds was $671,297, and the fair value was $714,943. We recorded a fair value adjustment of $43,646 for this unrealized holding gain. Now, on January 5, 2022, the fair value has increased further to $725,000, so we need to update the fair value adjustment for the additional $10,057 increase in fair value.

January 5, 2022

Security	Amortized Cost	Fair Value	Necessary Fair Value Adjustment Balance
Masterwear	$671,297	$725,000	**$53,703**

	Fair Value Adjustment
Beginning balance on 12/31/2021	$43,646
± Adjustment needed to update fair value	?
Balance needed as of date of sale	**$53,703**

Fair Value Adjustment

43,646	
10,057	
53,703	

United needs to record an increase in the fair value adjustment and an additional unrealized holding gain of $10,057 that occurred during the first week of 2022. The journal entry is:

January 5, 2022		
Fair value adjustment	**10,057**	
Gain on investments (unrealized, NI) (to balance)		10,057

2. Record the Sale Transaction. After making the previous journal entry, the investment is carried at its fair value as of the date it is being sold, and already has included in net income any gain or loss arising from the difference between amortized cost and fair value as of the date of sale. All that remains is for United to record receipt of cash and remove the investment-related accounts from the balance sheet (again, for simplicity we ignore interest earned during the first five days of 2022).

When a trading security is sold, all of the gain or loss already has been included in net income, so no additional gain or loss is recognized.

January 5, 2022		
Cash	725,000	
Discount on bond investment (account balance)	28,703	
Investment in bonds (account balance)		700,000
Fair value adjustment (account balance)		**53,703**

As with the sale of the HTM investment, we record the receipt of $725,000 cash and remove all the balance sheet accounts associated with the investment. However, unlike the HTM investment our TS investment has an additional balance sheet account, the fair value adjustment, that needs to be removed when we record the sale. Another difference between the HTM and TS approach is that, because we carry trading securities at fair value as of the date of sale and already have included in net income the entire gain associated with changes in the fair value of the investment, there is no gain or loss to recognize on the date of sale.[6] However, over the life of the investments, United recognized the same amount of gain under the TS approach ($43,646 + 10,057 = $53,703) as it recognized upon sale under the HTM approach ($53,703). The only difference is timing, with trading securities recognizing unrealized holding gains and losses from fair value changes as they occur but the HTM approach recognizing gains or losses only when they are realized upon sale.

[6]As noted in FASB ASC 320-10-40: Investments—Debt Securities—Overall—Derecognition, for expediency companies may not update the fair value adjustment to the fair value as of the date of sale before recording the sale. In that case, the investment is carried at fair value as of the last balance sheet date, and this second entry would include a gain or loss based on the difference between the cash received and the carrying value of the investment. In our example, the fair value adjustment balance was $43,646 on December 31, 2021, and United would record the following sale entry on January 5, 2022:

Cash	725,000	
Discount on bond investment (account balance)	28,703	
Investment in bonds (account balance)		700,000
Fair value adjustment (account balance)		43,646
Gain on investments (NI) (to balance)		10,057

Financial Statement Presentation

Trading securities appear in the financial statements as follows:

- **Income Statement and Statement of Comprehensive Income:** For trading securities, gains and losses are included in the income statement in the periods in which fair value changes, *regardless of whether they are realized or unrealized.* Investments in trading securities do not affect other comprehensive income.
- **Balance Sheet:** Investments in trading securities are reported at fair value, typically as current assets.
- **Cash Flow Statement:** Cash flows from buying and selling trading securities typically are classified as operating activities, because the financial institutions that routinely hold trading securities consider them as part of their normal operations.

Assuming United sold its investment on January 5, 2022, United's 2021 and 2022 financial statements will include the amounts shown in Illustration 12–6.

Illustration 12–6
Reporting Trading Securities

For trading securities, fair value changes affect net income in the period in which they occur.

Trading securities are reported at fair value in the balance sheet.

Cash flows from buying and selling trading securities are classified as operating activities.

(Ignoring income taxes) **Statement of Comprehensive Income**	**2021**	**2022**
Revenues	$ ◆	$ ◆
Expenses	◆	◆
Other income (expense):		
Interest revenue	46,664	–0–
Gain on investments	**43,646**	**10,057**
Net income	90,310	10,057
Other comprehensive income (OCI)	–0–	–0–
Comprehensive income (Net income + OCI)	$ 90,310	$10,057
Balance Sheet		
Assets:		
Investments in bonds (TS)	714,943	–0–
Shareholders' equity:		
Retained Earnings	90,310	100,367*
Statement of Cash Flows (direct method)		
Operating Activities:		
Cash from interest received	42,000	–0–
Purchase of trading securities	(666,633)	–0–
Sale of trading securities	–0–	725,000

*Net income of $90,310 (2021) + $10,057 (2022) = $100,367 accumulates in retained earnings by the end of 2022.

● LO12–4

Debt Investments Classified as Available-for-Sale Securities

AFS investments aren't held for trading or designated as held to maturity.

AFS investments are reported at their fair values.

The HTM treatment assumes we hold the bonds for their entire life. The trading securities treatment assumes we are planning to sell the bonds in the very near future. Our third treatment, **available-for-sale (AFS) securities**, falls in the middle. We aren't planning to trade the debt investment actively, but the investment is available to sell if, for example, cash needs arise or the market is particularly favorable. In that case, the company classifies its debt investment as AFS. Like trading securities, we report AFS securities in the balance sheet at fair value. Unlike trading securities, though, unrealized holding gains and losses on AFS securities are *not* included in net income. Instead, they are reported in the statement of comprehensive income as other comprehensive income (OCI).

Comprehensive Income

Recall from Chapter 4 that comprehensive income is a more all-encompassing view of operations than net income. It includes not only net income but also all other changes in equity that

Additional Consideration

Don't Shoot the Messenger. Or, as written in *The Economist*, "Messenger, Shot: Accounting rules are under attack. Standard-setters should defend them. Politicians and banks should back off."[7] Using fair values that are hard to estimate is controversial. For example, during the financial crisis of 2008/2009 many financial-services companies had to recognize huge unrealized losses associated with their investments. Some blamed their losses on GAAP for requiring estimates of fair value that were driven by depressed current market prices, argued that those losses worsened the financial crisis, and lobbied for a move away from fair-value accounting. Others countered that these companies were using GAAP's requirement for fair value accounting as a "scapegoat" for their bad investment decisions. "Fair value accounting . . . does not create losses but rather reflects a firm's present condition," says Georgene Palacky, director of the CFA's financial reporting group."[8]

do not arise from transactions with owners.[9] Comprehensive income therefore includes net income as well as *other comprehensive income (OCI)*. You know that net income is closed to retained earnings at the end of each accounting period, and therefore accumulates in retained earnings over time in the shareholders' equity section of the balance sheet. Similarly, OCI is closed to *accumulated other comprehensive income (AOCI)* at the end of each accounting period, and therefore accumulates in AOCI in the shareholders' equity section of the balance sheet. OCI relates to AOCI the same way that net income relates to retained earnings.

Rationale for AFS Treatment of Unrealized Holding Gains and Losses

FINANCIAL Reporting Case

Q3, p. 645

Why use an approach for accounting for AFS securities that differs from that used for trading securities? Because AFS securities are likely to be held for multiple reporting periods, one could argue that there is sufficient time for unrealized holding gains in some periods to balance out with unrealized holding losses in other periods, so including unrealized holding gains and losses in income each period would confuse investors by making income appear more volatile than it really is over the long run.[10] But how can we show AFS investments at fair value in the balance sheet without recording in net income the unrealized gains and losses associated with changes in fair value? The solution is to show those unrealized gains and losses in OCI as they occur and then only include realized gains and losses in net income in the period in which an investment is actually sold.

Adjust AFS Investments to Fair Value (2021)

To see how we adjust AFS investments to their fair value, let's assume the Masterwear bond investment is classified as AFS. As of December 31, 2021, United has recorded the purchase of the bonds on July 1, 2021, as well as receipt of the first semiannual interest payment so the bonds have an amortized cost of $671,297 (refer back to Illustration 12–2). The fair value of the bonds on December 31, 2021, is $714,943. The next table shows the calculation of the balance in the fair value adjustment account that is required on that date.

December 31, 2021

Security	Amortized Cost	Fair Value	Necessary Fair Value Adjustment Balance
Masterwear	$671,297	$714,943	**$43,646**

[7]Transactions with owners primarily include dividends and the sale or purchase of shares of the company's stock.
[8]"Messenger, Shot," The Economist, April 8, 2009.
[9]Sarah Johnson, "The Fair Value Blame Game," CFO.com, March 19, 2009.
[10]Of course, one could counterargue that these unrealized holding gains and losses still are relevant, given that each period an investor has discretion over whether or not to continue holding the security or sell that security to realize a gain or loss. For that reason, many accountants would prefer that the FASB do away with the AFS approach and just treat these investments as trading securities.

United needs to adjust the balance of the fair value adjustment account from its current balance of $0 (at purchase date) to a debit balance of **$43,646** (on December 31, 2021).

Fair Value Adjustment

0	
43,646	
43,646	

	Fair Value Adjustment
Beginning balance on 7/1/2021	$ 0
± Adjustment needed to update fair value	?
Balance needed as of 12/31/2021	**$43,646**

The journal entry to record the $43,646 change in United's fair value adjustment and the corresponding unrealized holding gain is

For AFS securities, unrealized holding gains and losses from fair value changes are not included in net income, but instead are reported as OCI.

December 31, 2021		
Fair value adjustment*	43,646	
Gain on investments (unrealized, OCI)†		43,646

*Sometimes companies don't bother with a separate fair value adjustment account and simply adjust the investment account to fair value.
†We indicate "Unrealized, OCI" to highlight that, for available-for-sale securities, unrealized holding gains and losses are included in other comprehensive income (OCI) in the period in which they occur.

Notice that the amount of unrealized holding gain is the same for these AFS securities as it was for the trading securities in the previous section. What differs is that the unrealized holding gain is included in OCI for AFS securities instead of net income as it is for trading securities. At the end of the reporting period, the unrealized holding gain is closed to a shareholders' equity account for both approaches. What differs is that net income gets closed to retained earnings, and OCI gets closed to Accumulated Other Comprehensive Income (AOCI). As with trading securities, the fair value adjustment will be adjusted up or down each period, either for individual securities or for a portfolio of securities, and a corresponding unrealized holding gain or loss reported in OCI. Net income normally is not affected by AFS investments until the period an AFS investment is sold, as we'll discuss next.

Sale of AFS Investments

Let's once again assume that United sells its Masterwear bonds for $725,000 on January 5, 2022 (as with our HTM and TS examples, for simplicity we ignore any interest earned during 2022). For AFS securities, United needs to record three journal entries.[11]

For AFS investments, unrealized holding gains and losses from fair value changes are recorded up to the date an investment is sold.

1. **Adjust AFS Investments to Fair Value (2022).** As with trading securities, we first need to update the fair value adjustment and recognize any unrealized holding gains or losses that have occurred during the current reporting period prior to the date of sale. Remember that on December 31, 2021, the amortized cost of the Masterwear bonds was $671,297, and the fair value was $714,943. We recorded a fair value adjustment of $43,646 for this unrealized holding gain. Now, on January 5, 2022, the fair value has increased further to $725,000, so we need to update the fair value adjustment for the additional $10,057 increase in fair value.

January 5, 2022

Security	Amortized Cost	Fair Value	Necessary Fair Value Adjustment Balance
Masterwear	$671,297	$725,000	**$53,703**

[11]This description of accounting for sales of AFS investments is adapted from the example of journal entries and financial statement presentation shown by FASB ASC 220-10-55 (para. 24–27): Comprehensive Income—Overall—Implementation Guidance and Illustrations—Case B: Available-for-Sale Debt Securities. The three journal entries we describe could be combined into one or two entries in practice.

	Fair Value Adjustment
Balance as of 12/31/2021	$43,646
± Adjustment needed to update fair value	?
Balance needed as of date of sale	$53,703

Fair Value Adjustment

43,646	
10,057	
53,703	

United needs to record an increase in the fair value adjustment and an additional unrealized holding gain of $10,057 that occurred during the first week of 2022. The journal entry to record the gain is

January 5, 2022		
Fair value adjustment	10,057	
Gain on investments (unrealized, OCI)		10,057

At this point, the investment is carried in the balance sheet at its fair value as of the date it is being sold, and all unrealized gains and losses associated with the investment have been included in OCI. Because OCI gets closed to AOCI, the unrealized gains and losses accumulate in AOCI, which acts as a sort of "holding tank" in the shareholders' equity section of the balance sheet. Unrealized holding gains in some years offset unrealized losses in other years as they accumulate in the tank.

2. **Reverse Previous Fair Value Adjustments.** United has been recording changes in fair value over the life of the investment. If United now sells that investment, the effects of those fair value changes must be reversed. United reverses previous unrealized holding gains included in OCI by debiting a reclassification adjustment to OCI for the same amount. Similarly, the account balance of the fair value adjustment is eliminated.

When an AFS investment is sold, accumulated unrealized gains and losses are removed from AOCI using a reclassification entry.

January 5, 2022		
Reclassification adjustment (OCI)	53,703	
Fair value adjustment (account balance)		53,703

Fair Value Adjustment

43,646	
10,057	53,703
0	

After this journal entry is recorded, the fair value adjustment account has a zero balance. Also, after the reclassification adjustment is closed to AOCI, all of the unrealized gains that are associated with the investment have been removed from the AOCI holding tank in shareholders' equity.[12] It's as if no accounting for unrealized gains and losses had ever taken place.[13] That's important, because in the next entry United recognizes in net income a gain or loss on sale. If United didn't use the reclassification entry to back out the unrealized gains and losses from AOCI, it would end up having double counted them in comprehensive income and shareholders' equity after it records the sale in the next entry.

AOCI (after closing)

	43,646
53,703	10,057
	0

3. **Record the Sale Transaction.** Now that all of the unrealized holding gains and losses and the fair value adjustment have been cleared away, the final step is to "plug" for the realized gain or loss (a gain in this case).

January 5, 2022		
Cash	725,000	
Discount on bond investment (account balance)	28,703	
Investment in bonds (account balance)		700,000
Gain on investments (NI) (to balance)		53,703

[12]As noted in FASB ASC 320-10-40: Investments—Debt Securities—Overall—Derecognition, for expediency companies might not update the fair value adjustment to fair value before recording the sale. In that case, this second entry would be based on the fair value adjustment as of the last balance sheet date. In our example, the fair value adjustment balance would be $43,646, so that is the amount by which it would be reduced and that is the amount of reclassification adjustment that would be recorded in OCI.

[13]Companies may choose to make this journal entry at the end of the reporting period when they update their fair value adjustment. The order doesn't matter, so long as the journal entry is recorded in the correct reporting period.

When an AFS investment is sold, realized gains and losses are included in net income.

This entry is identical to the entry United made to record sale of the investment under the HTM approach. As with the HTM investments, no gain or loss is recognized in net income over the life of the investment. Instead, the entire gain or loss is recognized in net income at the time of the sale. The difference between HTM and AFS is that unrealized gains and losses are recognized in OCI prior to sale. That requires an investor to remove the amounts that are included in OCI and recognize them in net income at the time of sale. In United's case, the second and third entries essentially reclassify a $53,703 gain from OCI to net income (and thus to retained earnings). That's why the process is called reclassification.

Additional Consideration

More About Reclassification

Look back through the three entries that are shown for recording a sale of an AFS investment. Think about what has happened. First, United recorded unrealized holding gains in OCI each period as fair value changed over time, and at the end of each period it closed OCI to AOCI in shareholders' equity. Then, once the investment was sold, United backed those unrealized gains out of OCI (and thus out of AOCI) and included them in net income (which is closed to retained earnings in shareholders' equity). From the perspective of shareholders' equity, the unrealized gains were first accumulated in AOCI and then were reclassified from AOCI to retained earnings in the period of sale.

We can rearrange the second and third journal entries to highlight this reclassification process:

Reclassification adjustment (OCI)	**53,703**	
Gain on investments (NI) (to balance)		**53,703**
Cash	725,000	
Discount on bond investment (account balance)	28,703	
Investment in bonds (account balance)		700,000
Fair value adjustment (account balance)		53,703

Once the investment is sold, all the balance sheet accounts are removed and the unrealized holding gains and losses that have been accumulating in AOCI are transferred out of AOCI (via the reclassification entry) and end up in retained earnings.

It may seem odd that we bother to put all of the unrealized holding gains and losses into OCI and then take them out again. However, that approach makes it very clear how we account for AFS investment over time. We can see that unrealized gains and losses are included in OCI and accumulated in AOCI while an investment is held, and then upon sale are backed out of OCI (and AOCI) and included in net income (and retained earnings).

Impairment of AFS Investments

As with HTM investments, companies are required to account for impairments of AFS investments, but the accounting is somewhat more complex. If fair value is less than amortized cost (such that the fair value adjustment has a credit balance), some impairment exists. In that case, accounting for the impairment depends on management's belief about whether it will sell the investment. If management either intends to sell the investment or believes it is more likely than not that it will have to sell it before fair value recovers, the AFS investment is written down to fair value and the impairment loss recognized in net income. If, on the other hand, management does *not* intend to sell the investment and does not believe it is more likely than not it will have to sell the investment before fair value recovers, management is required to estimate and recognize credit losses and reduce the carrying value of the AFS investment with an allowance for credit losses, just as we do for HTM investments. Any remaining impairment is accounted for normally as an unrealized holding loss in other comprehensive income. In Appendix 12B, we provide an illustration of recognizing impairments of AFS investments.

Financial Statement Presentation

AFS securities appear in the financial statements as follows:

- **Income Statement and Statement of Comprehensive Income:** Gains and losses are shown in OCI in the periods in which changes in fair value occur. Those amounts are reclassified out of OCI and recognized in net income in the periods in which securities are sold.
- **Balance Sheet:** Investments in AFS securities are reported at fair value. *Unrealized* holding gains and losses become part of AOCI in shareholders' equity, and are reclassified out of AOCI in the periods in which securities are sold.
- **Cash Flow Statement:** Cash flows from buying and selling AFS securities typically are classified as investing activities.

Assuming United sold its investment on January 5, 2022, United's 2021 and 2022 financial statements will include the amounts shown in Illustration 12–7.

(Ignoring income taxes) **Statement of Comprehensive Income**	**2021**	**2022**
Revenues	$ •	$ ♦
Expenses	•	♦
Other income (expense):		
Interest revenue	46,664	–0–
Gain on investments	–0–	53,703
Net income	46,664	53,703
Other comprehensive income (loss) items (OCI):*		
Gain on investments (unrealized)	43,646	10,057
Reclassification adjustment for net gains and losses included in net income	–0–	(53,703)
Total OCI	43,646	(43,646)
Comprehensive income (Net income + OCI)	$ 90,310	$ 10,057
Balance Sheet		
Assets:		
Investment in bonds (AFS)	$714,943	$ –0–
Shareholders' equity:		
Accumulated other comprehensive income (AOCI)	43,646	–0–
Retained Earnings	46,664	100,367†
Statement of Cash Flows (direct method)		
Operating activities:		
Cash from interest received	42,000	–0–
Investing activities:		
Purchase of available-for-sale securities	(666,633)	–0–
Sale of available-for-sale securities	–0–	725,000

*As we discuss in more detail in Chapter 18, the statement of comprehensive income can be presented as a continuation of the income statement as shown here, or in a separate statement that immediately follows the income statement.
†Net income of $46,664 (2021) + $53,703 (2022) = $100,367 accumulates in retained earnings by the end of 2022.

Illustration 12–7
Reporting Available-for-Sale Securities

Only *realized* gains and losses are included in net income.

Other comprehensive income includes *unrealized* holding gains and losses *that occur during the reporting period.*

AFS securities are reported at fair value.

AOCI (in shareholders' equity) includes net unrealized holding gains or losses *accumulated over the current and prior periods.*

Cash flows from buying and selling AFS securities usually are classified as investing activities.

Individual securities available for sale are classified as either current or noncurrent assets, depending on how long they're likely to be held. An example from the 2017 annual report of **Cisco Systems** is shown in Illustration 12–8.

Illustration 12–8

Investments in Securities Available-for-Sale—Cisco Systems

Real World Financials

Item 1A: Risk Factors (in part)

We maintain an investment portfolio of various holdings, types, and maturities. These securities are generally classified as available-for-sale and, consequently, are recorded on our Consolidated Balance Sheets at fair value with unrealized gains or losses reported as a component of accumulated other comprehensive income (loss), net of tax.

Note 8: Investments (in part)

The following tables summarize the Company's available-for-sale investments ($ in millions):

July 29, 2017	**Amortized Cost**	**Gross Unrealized Gains**	**Gross Unrealized Losses**	**Fair Value**
Fixed income securities:				
U.S. Government securities	$19,880	$ 3	$ (60)	$19,823
U.S. Government agency securities	2,057	–	(5)	2,052
Non-U.S. Gov't and agency	389	–	(1)	388
Corporate debt securities	31,626	202	(93)	31,735
U.S. agency mortgage-backed	2,037	3	(17)	2,023
Commercial paper	996	–	–	996
Certificates of deposit	60	–	–	60
Total fixed income securities	57,045	208	(176)	57,077

Source: Cisco Systems

Comparison of HTM, TS, and AFS Approaches

Illustration 12–9 compares accounting for the Masterwear bonds under the three different approaches (using account title abbreviations).

Illustration 12–9 Comparison of HTM, TS, and AFS Approaches

	Held-to-Maturity (HTM)			**Trading (TS)**			**Available-for-Sale (AFS)**		
Purchase bonds at a discount	Investments	700,000		Same as HTM			Same as HTM		
	Discount		33,367						
	Cash		666,633						
Record interest revenue	Cash	42,000		Same as HTM			Same as HTM		
	Discount	4,664							
	Interest rev.		46,664						
Adjust to fair value, 2021	No entry			FV adjustment	43,646		FV adjustment	43,646	
				Gain (unrealized, NI)		**43,646**	Gain (unrealized, OCI)		43,646
Sell bonds in 2022									
1. Adjust to fair value, 2022	No entry			FV adjustment	10,057		FV adjustment	10,057	
				Gain (unrealized, NI)		**10,057**	Gain (unrealized, OCI)		10,057
2. Reclassify unrealized holding gains/ losses	No entry			No entry			Reclassification (OCI)	53,703	
							FV adjustment		53,703
3. Record sale of bonds	Cash	725,000		Cash	725,000		Cash	725,000	
	Discount	28,703		Discount	28,703		Discount	28,703	
	Investments		700,000	Investments		700,000	Investments		700,000
	Gain (NI)		**53,703**	FV Adjustment		53,703	**Gain (NI)**		**53,703**
	Note: Total gain of **$53,703** is recognized in net income in the period of the sale.			Note: Total gain of **$53,703** is recognized in net income over the periods in which the investment is held (**43,646** + **10,057**).			Note: Reclassification backs out unrealized gains from OCI (and therefore from AOCI), so total gain of **$53,703** is recognized in net income in the period of the sale.		

This side-by-side comparison highlights several aspects of these accounting approaches:

- To record the purchase of an investment and the receipt of interest revenue, we use identical entries in all three approaches.
- To record changes in fair value, the entries we use for TS and AFS securities have the same effect on the investment (via the fair value adjustment valuation allowance) and the same eventual effect on total shareholders' equity. What differs is whether the unrealized holding gain or loss is recognized in net income and then in retained earnings (TS) or recognized in OCI and then in AOCI (AFS). No fair value adjustment is reported for HTM securities.
- Regardless of approach, the cash flows are the same, and the same total amount of gain or loss is recognized in the income statement (TS: $43,646 in 2021 + $10,057 in 2022 = $53,703 total; AFS and HTM: $53,703 in 2022). The question is not *how much* total net income is recognized, but *when* the amounts are recognized in net income.

Additional Consideration

Available-for-Sale Investments and Income Taxes. When comparing accounting for TS and AFS securities, we saw that total shareholders' equity ends up being the same amount, regardless of whether unrealized gains and losses are included in net income and closed to retained earnings in shareholders' equity (for TS) or included only in OCI and shown in AOCI in shareholders' equity (for AFS securities). But what about taxes? Tax expense affects net income, so retained earnings includes after-tax amounts. For AOCI to be equivalent to retained earnings, it also should include only after-tax amounts. Therefore, adjustments must be made to OCI and AOCI to account for tax effects. Typically these adjustments also give rise to deferred tax assets and liabilities, as unrealized holding gains and losses rarely affect the current period's taxes payable. Deferred tax assets and liabilities are discussed in Chapter 16.

International Financial Reporting Standards

Accounting for Debt Investments. *IFRS No. 9* governs treatment of debt and equity investments.[14] *IFRS No. 9* eliminates the HTM and AFS classifications, replaced by new classifications that are more restrictive. Specifically, under *IFRS No. 9*, investments in debt securities are classified either as amortized cost (accounted for like HTM investments in U.S. GAAP), fair value through other comprehensive income ("FVOCI," accounted for like AFS investments, except for different impairment recognition criteria) or fair value through profit or loss ("FVPL," accounted for like trading securities). Classification depends on two criteria: (1) whether the investment's contractual cash flows consist solely of payments of principal and interest (SPPI), and (2) whether the business purpose of the investment is to hold it for purposes of collecting contractual cash flows, sell the investment at a profit, or both. If the investment qualifies as SPPI and is held only to collect cash flows, it is classified as amortized cost. If it qualifies as SPPI and is held both to collect cash flows and potentially be sold, it is classified as FVOCI. Otherwise it is classified as FVPL.

One other difference between U.S. GAAP and IFRS is worth noting. U.S. GAAP allows specialized accounting (beyond the scope of this textbook) for particular industries like securities brokers/dealers, investment companies, and insurance companies. IFRS does not.

Transfers between Reporting Categories

A transfer of a security between reporting categories is accounted for at fair value and in accordance with *the new reporting classification*.

At each reporting date, the appropriateness of the classification of a debt investment is reassessed. For instance, if the investor no longer has the ability to hold certain securities to maturity and will now hold them for resale, those securities would be reclassified from HTM to AFS. When a security is reclassified between two reporting categories, the security

[14] "Financial Instruments," *International Financial Reporting Standard No. 9* (IASCF), as amended effective January 1, 2018.

is transferred at its fair value on the date of transfer. Any unrealized holding gain or loss at reclassification should be accounted for *in a manner consistent with the classification into which the security is being transferred.* A summary is provided in Illustration 12–10.

Illustration 12–10
Transfer between Investment Categories

Transfer from:	To:	Unrealized Gain or Loss from Transfer at Fair Value
Either HTM or AFS	Trading	Include in current net income the total unrealized gain or loss, as if it all occurred in the current period.
Trading	Either HTM or AFS	Include in current net income any unrealized gain or loss that occurred in the current period prior to the transfer. (Unrealized gains and losses that occurred in prior periods already were included in net income in those periods.)
Held-to-maturity	Available-for-sale	No current income effect. Report total unrealized gain or loss as a separate component of shareholders' equity (in AOCI).
Available-for-sale	Held-to-maturity	No current income effect. Don't write off any existing unrealized holding gain or loss in AOCI, but amortize it to net income over the remaining life of the security (fair value amount becomes the security's amortized cost basis).

Reclassifications are quite unusual, so when they occur, disclosure notes should describe the circumstances that resulted in the transfers. Other disclosure notes are described in a later section.

International Financial Reporting Standards

LO12–9

Transfers Between Investment Categories. *IAS No. 39* allows transfers of debt investments out of the FVPL category into AFS or HTM in "rare circumstances." Under *IFRS No. 9,* transfers of debt investments between the amortized cost, FVOCI, and FVPL categories occurs if and only if the company changes its business model with respect to the debt investment.

Fair Value Option

Choosing the *fair value option* for HTM and AFS investments means accounting for them like trading securities.

You may recall from Chapter 1 that GAAP allows a **fair value option** (FVO) that permits companies to elect to account for most financial assets and liabilities at fair value. Under the FVO, HTM and AFS investments are shown in the balance sheet at their fair values, and unrealized gains and losses are recognized in net income in the period in which they occur. That accounting approach should sound familiar—it's the same approach we use to
LO12–8 account for trading securities. However, unlike trading securities, purchases and sales of investments accounted for under the FVO are likely to be classified as investing activities in the statement of cash flows, because those investments are not held for sale in the near term and therefore are not operational in nature.

The company decides whether to elect the FVO on the date the company purchases the investment. The company can elect the FVO for some securities and not for identical others—it's entirely up to the company, but the company has to explain in the notes why it made a partial election. The election is *irrevocable.* So, for example, if a company elects the FVO and later believes that the fair value of an investment is likely to decline, it can't change the election and discontinue use of fair value accounting to avoid recognizing a loss.

Why allow the FVO? As described in Appendix A of this book, companies sometimes enter into *hedging arrangements* that are intended to reduce earnings volatility by offsetting changes in the fair value of assets with changes in the fair value of liabilities. Complex rules apply to many hedging arrangements. The FVO simplifies this process by allowing companies to choose whether to use fair value for most types of financial assets and liabilities. Thus, when a company enters into a hedging arrangement, it just has to make sure to elect the FVO for each asset and liability in the hedging arrangement, and fair value changes of those assets and liabilities will be included in earnings.

International Financial Reporting Standards

● LO12–9

Fair Value Option. International accounting standards are more restrictive than U.S. standards for determining when firms are allowed to elect the FVO. Under both *IAS No. 39* and *IFRS No. 9*, companies can elect the FVO only in specific circumstances. For example, a firm could elect the FVO for an asset or liability in order to avoid the "accounting mismatch" that occurs when some parts of a fair value risk-hedging arrangement are accounted for at fair value and others are not. Although U.S. GAAP indicates that the intent of the FVO is to address these sorts of circumstances, it does not require that those circumstances exist.

Concept Review Exercise

DEBT INVESTMENT SECURITIES

Diversified Services, Inc., offers a variety of business services, including financial services through its escrow division. Diversified's fiscal year ends on December 31. Diversified entered into the following investment activities during the last month of 2021 and the first week of 2022:

2021

Dec. 1	Purchased $30 million of 12% bonds of Vince-Gill Amusement Corporation and $24 million of 10% bonds of Eastern Waste Disposal Corporation, both at face value. The Vince-Gill bonds are to be held until they mature. The Eastern Waste bonds are to be held but might be sold if cash needs require it. Interest on each bond issue is payable semiannually on November 30 and May 31.
30	Purchased U.S. Treasury bonds for $5.8 million as trading securities hoping to earn profits on short-term differences in prices.
31	Recorded the necessary adjusting entries relating to the investments.

As of December 31, 2021, the fair value of the Vince-Gill bonds was $32 million, the fair value of the Eastern Waste bonds was $25 million, and the fair value of the Treasury bonds was $5.7 million.

2022

Jan. 7	Sold the Eastern Waste bonds for $22 million and sold the U.S Treasury bonds for $6 million.

Required:

1. Prepare the appropriate journal entry for each transaction or event and show the amounts that would be reported in the company's 2021 income statement relative to these investments. Record interest accruing in 2021, but ignore any interest accruing in 2022.
2. Determine the effects of the Eastern Waste investment on net income, other comprehensive income, and comprehensive income for 2021, 2022, and combined over both years. Ignore interest revenue.

1. Journal entries:

2021

Dec. 1 **Purchased $30 million of 12% bonds of Vince-Gill Amusement Corporation and $24 million of 10% bonds of Eastern Waste Disposal Corporation, both at face value. The Vince-Gill bonds are to be held until they mature, but the Eastern Waste bonds are to be held but might be sold if cash needs require it. Interest on each bond issue is payable semiannually on November 30 and May 31.**

	($ in millions)	
Investment in bonds (HTM, Vince-Gill)	30	
Investment in bonds (AFS, Eastern Waste)	24	
Cash		54

Dec. 30 **Purchased U.S. Treasury bonds for $5.8 million as trading securities, hoping to earn profits on short-term differences in prices.**

	($ in millions)	
Investment in bonds (TS, U.S. Treasury bonds)	5.8	
Cash		5.8

Dec. 31 **Recorded the necessary adjusting entries relating to the investments.**

Accrued Interest (one month)	($ in millions)	
Interest receivable (Vince-Gill) ($30 million × 12% × 1/12)	0.3	
Interest receivable (Eastern Waste) ($24 million × 10% × 1/12)	0.2	
Interest revenue		0.5
Fair Value Adjustments		
Loss on investments (unrealized, NI) (TS) ($5.7 – $5.8)	0.1	
Fair value adjustment (TS)		0.1
Fair value adjustment (AFS) ($25 million cost – $24 million fair value)	1	
Gain on investments (unrealized, OCI) (AFS)		1

Note: Securities held-to-maturity are not adjusted to fair value.

Reported in the 2021 Income Statement	($ in millions)
Interest revenue ($0.5 interest)	$ 0.5
Loss on trading securities	(0.1)

Note: The $1 million unrealized holding gain for the Eastern Waste bonds is not included in net income because it pertains to available-for-sale securities rather than trading securities, and so is reflected in OCI.

2022

Jan. 7 **Sold the Eastern Waste bonds for $22 million and sold the U.S Treasury bonds for $6 million.**

First consider the AFS investment (Eastern Waste). The fair value of the Eastern Waste bonds at the time of sale is $22 million. Those bonds were carried at a fair value of $25 million as of December 31, 2021, so have suffered an unrealized loss of $3 million during the first part of 2022.

	($ in millions)	
Loss on investments (unrealized, OCI) (AFS) ($25 – $22)	3	
Fair value adjustment (AFS)		3

Second, given that the fair value adjustment for the Eastern Waste bonds now has a $2 million credit balance, we need to remove that amount and record the corresponding reclassification entry for OCI:

FV Adjustment (AFS)

Debit	Credit	
1		2021 unrealized gain
	3	2022 unrealized loss
	2	Preadjustment balance
2		Reclassification entry
	0	

	($ in millions)	
Fair value adjustment (AFS) (account balance)	2	
Reclassification adjustment (OCI)		2

Third, we need to record the receipt of cash and the loss realized upon sale of the investment:

	($ in millions)	
Cash	22	
Loss on investments (NI) (AFS) (to balance)	2	
Investment in bonds (AFS) (Eastern Waste)		24

Next consider the investment in trading securities (U.S. Treasuries). The fair value of the U.S. Treasury bonds at the time of sale is $6 million. Those bonds were carried at a fair value of $5.7 million as of December 31, 2021, so have enjoyed an unrealized holding gain of **$0.3** million during the first part of 2022.

	($ in millions)	
Fair value adjustment (TS)	**0.3**	
Gain on investments (unrealized, NI) (TS)		0.3

FV Adjustment (TS)

Debit	Credit	
	0.1	2021 unrealized loss
0.3		2022 unrealized gain
0.2		Preadjustment balance
	0.2	Sale entry
0		

On the date of sale, we also need to record the receipt of cash and remove the accounts associated with the trading security from the balance sheet:

	($ in millions)	
Cash	6	
Fair value adjustment (TS) (account balance)		**0.2**
Investment in bonds (TS) (U.S. Treasury bonds)		5.8

2. Effects of the Eastern Waste investments on net income, other comprehensive income, and comprehensive income for 2021, 2022, and combined over both years. Ignore interest revenue.

	2021	**2022**	**Cumulatively**
Net Income	$0	($2)	($2)
OCI	$1	($3) + $2 = ($1)	$0
Comprehensive Income	$1	($3)	($2)

Financial Statement Presentation and Disclosure

Trading securities, held-to-maturity securities, and available-for-sale securities are either current or noncurrent depending on when they are expected to mature or to be sold. However, it's not necessary that a company report individual amounts for the three categories of investments—held-to-maturity, available-for-sale, or trading—on the face of the balance sheet as long as that information is presented in the disclosure notes.[15]

Investors should disclose the following in the disclosure notes for each year presented:

- Aggregate fair value.
- Gross realized and unrealized holding gains.
- Gross realized and unrealized holding losses.
- Change in net unrealized holding gains and losses.
- Amortized cost basis by major security type.

The notes also include disclosures designed to help financial statement users understand the quality of the inputs companies use when determining fair values and to identify parts of the financial statements that are affected by those fair value estimates. For example, the notes should include the level of the fair value hierarchy (levels 1, 2, or 3) in which all fair value measurements fall. For level 2 or 3 fair values, the notes to the financial statements must include a description of the valuation technique(s) and the inputs used in the fair value measurement process. For level 3 fair values, the notes must indicate the significant inputs used in the fair value measurement, the sensitivity of fair values to significant changes in those inputs, and information about the effect of fair value measurements on earnings, including a reconciliation of beginning and ending balances of the investment that identifies the following:

- Total gains or losses for the period (realized and unrealized), unrealized gains and losses associated with assets and liabilities still held at the reporting date, and where those amounts are included in earnings or shareholders' equity.
- Purchases, sales, issuances, and settlements.
- Transfers in and out of the level 3 category.

All of this disclosure is designed to provide financial statement users with information about those fair values that are most vulnerable to bias or error in the estimation process. For example, as shown in Illustration 12–11, note 13 of **HP Inc.**'s 2017 annual report includes the following discussion of fair values.

Illustration 12–11

Fair Value Disclosures of Investment Securities—HP Incorporated

Real World Financials

Fair Value (partial)

	As of October 31, 2017			
	Fair Value Measured Using			
($ in millions)	**Level 1**	**Level 2**	**Level 3**	**Total**
Assets:				
Cash equivalents and Investments:				
Time deposits	$ –	$1,159	$ –	$1,159
Money market funds	5,592	–	–	5,592
Equity securities in public companies	–	–	–	–
Foreign bonds	9	214	–	223
Other debt securities	–	–	26	26
Derivative Instruments:				
Interest rate contracts	–	–	–	–
Foreign exchange contracts	–	259	–	259
Other derivatives	–	1	–	1
Total assets	$5,601	$1,633	$26	$7,260

Source: HP Incorporated

[15]FASB ASC 320-10-45-13: Investments—Debt and Equity Securities—Overall—Other Presentation Matters (previously *Statement of Financial Accounting Standards No. 115,* "Accounting for Certain Investments in Debt and Equity Securities" (Norwalk, Conn.: FASB, 1993), par. 18).

We can see from these disclosures that HP has relatively few level 3 investments (which are those with the most subjectively estimated fair values). That fact should give financial statement users more faith in the reliability of HP's fair value estimates.

PART B

Accounting for Equity Investments

The critical events over the life of an investment in the equity of another company, such as shares of common stock, include the following:

1. Purchasing the equity security.
2. Receiving dividends (for some equity securities).
3. Holding the investment during periods in which the investment's fair value changes (and thus incurring *unrealized holding* gains and losses, since the security has not yet been sold).
4. Selling the investment (and thus incurring *realized* gains and losses, since the security has been sold and the gains or losses actually incurred).

Also, equity investors typically get to vote on key decisions made by the company, such as who will serve on the company's board of directors. Each share of common stock gets a vote, so if an equity investor owns enough shares, the investor has enough votes to influence the operations of the company whose shares it owns. Therefore, accounting for equity investments also considers the extent to which the investor can influence the activities of the investee. As shown in Illustration 12–12, we use three different approaches to account for equity investments.

Illustration 12–12

Reporting Categories for Equity Investments

Characteristics of the Equity Investment	Reporting Method Used by the Investor
The investor ***does not have significant influence*** over the operating and financial policies of the investee (typically owns less than 20% of voting stock):	**Fair value through net income**—similar to the trading-securities approach used for debt; investment reported at fair value (with unrealized holding gains and losses included in net income), unless fair value is not readily determinable*
The investor ***has significant influence*** over the operating and financial policies of the investee (typically owns between 20% and 50% of the voting stock):	**Equity method**—investment reported at cost adjusted for investor's share of subsequent earnings and dividends of the investee**
The investor ***controls*** the investee (typically owns more than 50% of voting stock):	**Consolidation**—the financial statements of the investor and investee are combined as if they are a single company

*Later in this chapter we discuss in an Additional Consideration box the alternative approach that is used if fair value is not readily determinable.
**If the investor elects the *fair value option*, this type of investment also can be accounted for using the fair-value-through-net-income approach.

We'll first discuss the *fair value through net income* approach that is used when the investor lacks significant influence over the investee. Then we'll cover the *equity method* that's used when the investor does have significant influence. We won't cover *consolidation accounting*, which is used when the investor controls the investee, as that topic is beyond the scope of this book. However, we will discuss consolidations briefly when we discuss the equity method, as the two approaches are related.

When the Investor Does Not Have Significant Influence: Fair Value through Net Income

LO12–5

If an investor owns less than 20% of the voting shares of an investee, the investor is typically presumed to *not* have significant influence over the investee. In other words, the investor is in the same position you would be in if you bought a share of the company's stock—the investor hopes to receive dividends and that the value of the shares appreciate over time, but can't really tell the company what to do.

Until very recently, those sorts of equity investments were accounted for as either trading securities or available-for-sale investments, just like we would treat debt investments that weren't intended to be held to maturity. However, starting in 2018, investors are not allowed to use the AFS approach for equity investments. Instead, all equity investments are accounted for like trading securities, using an approach commonly referred to as *fair value through net income.* That means that equity investments for which the investor lacks significant influence are reported the same way we report debt investments that are classified as trading securities (covered in Part A of this chapter). The equity investments are carried at fair value in the balance sheet, with unrealized holding gains and losses recognized in net income in whatever period they occur.[16] Illustration 12–13 provides a simple example of an equity investment that we'll use throughout this section. Let's walk through the key events in the life of United's Arjent investment to see how the fair value through net income approach works for an equity investment.

Illustration 12–13

Example of an Equity Investment Accounted for as Fair Value Through Net Income

The following events during 2021 and 2022 pertain to United Intergroup's investment in the common stock of Arjent, Inc.

July 1, 2021	**Purchase** Arjent, Inc., common stock for $1,500,000.
December 31, 2021	**Recognize investment revenue** for a $75,000 cash dividend received from Arjent.
December 31, 2021	**Record a fair value adjustment** to recognize a decline in the value of the Arjent stock investment to $1,450,000.
January 5, 2022	**Sell** the Arjent stock for $1,446,000.

Purchase Investments

All equity investments are recorded initially at cost.

The journal entry to record the purchase of an equity investment is simple, just exchanging one asset (cash) for another (investment), as follows:

July 1, 2021		
Investment in equity securities	1,500,000	
Cash		1,500,000

Recognize Investment Revenue

Dividends received for equity investments are included in income.

The journal entry to record the receipt of dividends related to the Arjent equity investment also is straightforward.

December 31, 2021		
Cash	75,000	
Dividend revenue		75,000

Adjust Equity Investments to Fair Value (2021)

Equity investments are adjusted to their fair value at each reporting date.

The carrying value of equity investments must be adjusted to fair value at the end of every reporting period. As with trading securities, we use a valuation allowance, *fair value adjustment,* to increase or decrease the carrying value of the investment, and we simultaneously record an unrealized holding gain or loss that is included in net income in the period in which fair value changes. The next table shows the calculation of the balance in the fair value adjustment that is required on December 31, 2021.

December 31, 2021

Security	Cost	Fair Value	Necessary Fair Value Adjustment Balance
Arjent	$1,500,000	$1,450,000	**$(50,000)**

[16]FASB ASC 321–10–35–1: Investments—Equity Securities—Overall—General.

United needs to move the fair value adjustment from a balance of $0 (at purchase date) to a credit balance of **$50,000**:

	Fair Value Adjustment
Beginning balance	$ 0
± Adjustment needed to update fair value	?
Balance needed on 12/31/2021	**$(50,000)**

Fair Value Adjustment	
0	50,000
	50,000

The journal entry to record United's unrealized holding loss of **$50,000** and the corresponding decrease in United's fair value adjustment account is

December 31, 2021		
Loss on investments (unrealized, NI)*	50,000	
Fair value adjustment†		50,000

*We indicate "unrealized, NI" to highlight that, for equity investments accounted for as fair value through net income, unrealized holding gains and losses are included in net income in the period in which they occur.
†Sometimes companies don't bother with a separate fair value adjustment account and simply adjust the investment account to fair value.

What if Arjent's investment instead had a fair value of, say, $1,550,000 as of December 31, 2021? In that case, United would have an unrealized holding *gain* of $50,000, which would require a *debit* to the fair value adjustment account to *increase* the carrying value of the equity to fair value.

Sell the Equity Investment

Now assume United sells the Arjent stock for $1,446,000 on January 5, 2022. According to the FASB, "because all changes in an equity security's fair value are reported in earnings as they occur, the sale of an equity security does not necessarily give rise to a gain or loss. Generally, a debit to cash . . . is recorded for the sales proceeds, and a credit is recorded to remove the security at its fair value (or sales price)."[17] Thus, as with trading securities, United will make two journal entries. United first records in net income any unrealized holding gains and losses that occurred during 2022 prior to the date of sale. Then, on the date of sale, United records the receipt of cash and removes the amounts associated with the investment from the relevant balance sheet accounts.

1. **Adjust Securities to Fair Value (2022).** We first need to update the fair value adjustment and recognize any unrealized holding gains or losses that have occurred during the current reporting period prior to the date of sale.

January 5, 2022

Security	Cost	Fair Value	Necessary Fair Value Adjustment Balance
Arjent	$1,500,000	$1,446,000	**$(54,000)**

United needs to move the fair value adjustment from the credit balance of $50,000 existing at the end of 2021 to a credit balance of **$54,000** as of January 5, 2022, which requires a credit of **$4,000** to this asset valuation account:

	Fair Value Adjustment
Beginning balance on 1/1/2022	$ (50,000)
± Adjustment needed to update fair value	?
Balance needed as of date of sale	**$(54,000)**

Fair Value Adjustment	
0	50,000
	4,000
	54,000

[17]FASB ASC 321-10-40: Investments—Equity Securities—Overall—Derecognition.

The journal entry to record the investment's decrease in fair value is:

January 5, 2022		
Loss on investments (unrealized, NI)	4,000	
Fair value adjustment		4,000

2. **Record the Sale.** After making the previous journal entry, the investment is carried at its fair value as of the date it is being sold, and all gains and losses associated with the investment have been included in net income over the time the investment was held. All that remains is for United to record receipt of cash and remove the investment-related accounts from the balance sheet:

January 5, 2022		
Cash	1,446,000	
Fair value adjustment (account balance)	54,000	
Investment in equity securities (account balance)		1,500,000

If all of the unrealized holding gains or losses have been included in net income up to the time of sale, no additional gain or loss is recognized.

Because United carries the investment at fair value as of the date of sale, and already included in net income the entire loss associated with the investment, there is no additional gain or loss to recognize on the date of sale.[18] Over the life of the investment, United has recognized a total loss of $54,000, but that loss is spread over the life of the investment, recognized as the fair value of the investment changes:

2021 loss recognized in NI:	$(50,000)
2022 loss recognized in NI:	(4,000)
Total over 2021 and 2022:	$(54,000)

Adjust Remaining Equity Investments to Fair Value (2022)

If United hadn't sold the Arjent investment during 2022, it would just keep recording whatever fair value adjustment would be necessary to carry the investment at fair value in the balance sheet. For example, if the Arjent investment had a fair value of $1,300,000 at December 31, 2022, United would face the following situation:

December 31, 2022

Security	Cost	Fair Value	Necessary Fair Value Adjustment Balance
Arjent	$1,500,000	$1,300,000	$(200,000)

United would need to credit the fair value adjustment account by another $150,000 to account for the large drop in fair value that occurred during 2022.

Fair Value Adjustment

	50,000
	150,000
	200,000

	Fair Value Adjustment
12/31/2021 balance	$ (50,000)
± Adjustment needed to update fair value	?
12/31/2022 balance	$(200,000)

[18] As noted in FASB ASC 321-10-40: Investments—Equity Securities—Overall—Derecognition, for expediency companies may not update the fair value adjustment to the fair value that exists on the date of sale before recording the sale. In that case, this second entry would include a realized gain or loss based on the difference between the cash received and the fair value of the investment recorded on the last balance sheet date. In our example, the fair value adjustment balance was $50,000 as of December 31, 2018, and United would record the following sale entry on January 5, 2022:

Cash	1,446,000	
Fair value adjustment (account balance)	50,000	
Loss on investments (NI)	4,000	
Investment in equity securities		1,500,000

The necessary journal entry is shown below:

December 31, 2022		
Loss on investments (unrealized, NI)	150,000	
Fair value adjustment (calculated above)		150,000

Additional Consideration

What if Fair Value is Not Readily Determinable? Equity investments are accounted for using the *fair value through net income* approach when the fair value of shares is *readily determinable* by referencing prices on a security exchange or over-the-counter market. But what if fair value isn't readily determinable? In that case, investors can elect to base measurement of fair value on an adjusted cost of the security. Specifically, the investor estimates fair value as cost, minus any impairments that have been recognized previously, and plus or minus adjustments indicated by changes in the prices of similar equity issued by the same investee. Every accounting period, the investor needs to reevaluate whether fair value still is not readily determinable, and also needs to assess whether the investment has been impaired. The investor considers various qualitative factors to assess whether an investment is impaired, such as whether the earnings, cash flows, or business prospects of the investee have deteriorated. If the investor concludes the investment is impaired, it recognizes in net income whatever amount of impairment loss is necessary to reduce the carrying value of the investment to fair value.

International Financial Reporting Standards

● LO12–9

Accounting for Equity Investments When the Investor Does Not Have Significant Influence. *IFRS No. 9* governs treatment of debt and equity investments.[19] Under *IFRS No. 9*, investments in equity securities are classified as either FVPL (fair value through profit or loss) or FVOCI (fair value through other comprehensive income). If the equity is held for trading, it must be classified as FVPL, but otherwise the company can irrevocably elect to classify it as FVOCI. FVOCI is similar to the AFS treatment used for debt investments in U.S. GAAP, but there is an important difference. Unlike AFS, realized gains and losses are not reclassified out of OCI and into net income when the investment is later sold. Rather, the accumulated unrealized gain or loss associated with a sold investment is just transferred from AOCI to retained earnings (both shareholders' equity accounts), without passing through the income statement.

Financial Statement Presentation

Equity investments for which the investor does not have significant influence are classified as either current (short-term) or noncurrent (long-term) in the balance sheet. Those that are held with an intent for short-term profit are normally treated as operating activities in the statement of cash flows, similar to debt investments that are classified as trading securities. Other current equity investments, and long-term equity investments, are classified as investing activities in the statement of cash flows. Notes to the financial statements should disclose the portion of unrealized holding gains and losses for the period that relate to any equity securities still held by the company at the end of the reporting period. Notes also should provide information about how the carrying value was calculated for equity investments for which fair value is not readily determinable.

[19] "Financial Instruments," *International Financial Reporting Standard No. 9* (IASCF), as amended effective January 1, 2018.

Additional Consideration

Transition from Prior GAAP. Our discussion of investments in equity securities is based on *ASU 2016-01,* which becomes effective for fiscal years beginning after December 15, 2017.[20] Prior to that ASU, equity investments could be treated as trading securities or AFS investments, similar to debt investments. When companies adopted *ASU 2016-01,* they made a cumulative-effect adjustment to the balance sheet as of the beginning of the period of adoption, adjusting equity AFS investments to fair value on that date and adjusting retained earnings to appear as if those investments had been accounted for at fair value through net income all along.

When the Investor Has Significant Influence: The Equity Method

● LO12–6

Control and Significant Influence

Usually an investor can *control* the investee if it owns more than 50% of the investee's voting shares.

Consolidated financial statements combine the individual elements of the parent and subsidiary statements.

Usually an investor can exercise *significant influence* over the investee when it owns at least 20% of the investee's voting shares.

The equity method is used when an investor can't control, but can significantly influence, the investee.

Under the *equity method,* the investor recognizes on its own income statement its proportionate share of the investee's income.

FINANCIAL Reporting Case

Q4, p. 645

If a company acquires more than 50% of the voting stock of another company, it's said to have **control**, because by voting those shares, the investor actually can control the company acquired. The investor is called the *parent;* the investee is called the *subsidiary.* Both companies continue to operate as separate legal entities, and the subsidiary reports separate financial statements. However, because of the controlling interest, the parent company reports **consolidated financial statements** which treat the parent and the subsidiary as if there were only one company. This entails an item-by-item combination of the parent and subsidiary statements (after first eliminating any amounts that are shared by the separate financial statements).[21] For instance, if the parent has $8 million cash and the subsidiary has $3 million cash, the consolidated balance sheet would report $11 million cash.

Even if effective control is absent, the investor still may be able to exercise **significant influence** over the operating and financial policies of the investee. This would be the case if the investor owns a large percentage of the outstanding shares relative to other shareholders. By voting those shares as a block, decisions often can be swayed in the direction the investor desires. It is presumed, in the absence of evidence to the contrary, that the investor exercises significant influence over the investee when it owns at least 20% of the investee's voting shares.[22]

When significant influence exists but the investor does not have effective control, the investment should be accounted for by the **equity method**. Under the equity method, the investment is initially recorded at cost. After that, the investment balance is

- Increased by the investor's percentage share of the investee's net income (or decreased by its share of a loss).
- Decreased by the investor's percentage share of the investee's dividends paid.
- Potentially adjusted for other items (discussed next).

The rationale for this approach is the presumption that the fortunes of the investor and investee are so intertwined that, as the investee prospers, the investor prospers proportionately. Stated differently, as the investee earns additional net assets (income), the investor's share of those net assets increases. When the investee pays out assets (dividends), the investor's share of the remaining net assets decreases.

[20]FASB ASC 321-10-35-1: Investments—Equity Securities—Overall—General (previously *Accounting Standards Update No. 2016-1,* "Recognition and Measurement of Financial Assets and Financial Liabilities" (Norwalk, Conn.: FASB, 2016)).

[21]This avoids double counting those amounts in the consolidated statements. For example, amounts owed by one company to the other are represented by accounts payable in one set of financial statements and accounts receivable in the other. These amounts are not included in the statements of the consolidated entity because a company can't "owe itself."

[22]Shareholders are the owners of the corporation. By voting their shares, it is they who determine the makeup of the board of directors—who, in turn, appoint officers—who, in turn, manage the company. Common stock usually is the class of shares that has voting privileges. However, a corporation can create classes of preferred shares that also have voting rights. This is discussed at greater length in Chapter 18.

Additional Consideration

It's possible that a company owns more than 20% of the voting shares but still cannot exercise significant influence over the investee. If, for instance, another company or a small group of shareholders owns 51% or more of the shares, they control the investee regardless of how other investors vote their shares. GAAP provides this and other examples of indications that an investor may be unable to exercise significant influence.

- The investee challenges the investor's ability to exercise significant influence (through litigation or complaints to regulators).
- The investor surrenders significant shareholder rights in a signed agreement.
- The investor is unable to acquire sufficient information about the investee to apply the equity method.
- The investor tries and fails to obtain representation on the board of directors of the investee.[23]

Conversely, it's also possible that a company owns less than 20% of the voting shares but is able to exercise significant influence over the investee. Ability to exercise significant influence with less than 20% ownership might be indicated, for example, by having an officer of the investor corporation on the board of directors of the investee corporation or by having, say, 18% of the voting shares while no other single investor owns more than 50%. In such cases the equity method would be appropriate.

To see how the equity method works, let's turn to Illustration 12–14. In that illustration, we assume that United Intergroup purchased **30%** of Arjent, Inc.'s common stock for $1,500,000 cash on January 2, 2021. Let's start by thinking about Arjent overall, and then we'll account for United's 30% investment.

Illustration 12–14

Information for Arjent, Inc., at the time United Intergroup purchased 30% for $1,500,000.

	Book Value on Arjent's Financial Statements	Fair Value at Time of United's Investment	
Total fair value of Arjent (1/2/2021)			$5,000,000*
Buildings**	$1,000,000	$2,000,000	
Land	500,000	1,000,000	
Other identifiable net assets†	600,000	600,000	
Identifiable net assets	$2,100,000		3,600,000
Goodwill			$1,400,000
Other information (12/31/2021):			
Arjent's 2021 net income:	$ 500,000		
Arjent's 2021 dividends declared and paid:	$ 250,000		

*$5,000,000 × 30% purchased = **$1,500,000 purchase price.**
**10-year remaining useful life, no salvage value.
†Other net assets = other assets − total liabilities.

As shown in Illustration 12–14, buying 30% of Arjent for $1,500,000 implies that the full (100%) *fair value* of Arjent is $5,000,000 (because $5,000,000 × 30% purchased = $1,500,000 purchase price). However, notice that the *book value* of Arjent's net assets is only $2,100,000. Why is there a difference between the fair value of Arjent and the book value of Arjent's identifiable net assets? Part of the difference represents identifiable assets (in this case, buildings and land) that have fair values greater than their book values. Arjent recorded those assets at historical cost and recognized depreciation of the buildings over time, so the book values of those assets don't reflect their fair values. The remaining difference is previously unrecognized goodwill (e.g., because of loyal customers, well-trained

[23]FASB ASC 323-10-15-10: Investments—Equity Method and Joint Ventures—Overall—Scope and Scope Exceptions.

workers, etc.) that GAAP doesn't capture as separate identifiable assets but nevertheless represents value for which United was willing to pay. We will see that, under the equity method, all of these amounts are shown in a single investment account, but we still need to track their individual information to account for them correctly.

Purchase of Investment

Recording United's purchase of 30% of Arjent is straightforward. The investment is recorded at cost:

Investment in equity affiliate	1,500,000	
Cash		1,500,000

Recording Investment Revenue

As the investee earns additional net assets, the investor's investment in those net assets increases.

Under the equity method, the investor includes in net income its proportionate share of the investee's net income. The reasoning is that, as the investee's net assets increase, the value of the investor's share of those net assets also increases, so the investor increases its investment by the amount of income recognized. United's entry would be as follows:

Investment in equity affiliate	150,000	
Investment revenue (30% × $500,000)		150,000

Of course, if Argent had recorded a net loss rather than net income, United would *reduce* its investment in Arjent and recognize a *loss* on investment for its share of the loss. You won't always see these amounts called "investment revenue" or "investment loss." Rather, United might call this line "equity in earnings (losses) of affiliate" or some other title that suggests it is using the equity method.

Receiving Dividends

As the investee distributes net assets as dividends, the investor does not recognize revenue. Rather, the investor's investment in the investee's net assets is reduced.

Because we recognize investment revenue when net income is recognized by the investee, it would be inappropriate to recognize revenue again when that income is distributed as dividends. That would be double counting. Instead, we view the dividend distribution as reducing the investee's net assets. The rationale is that the investee is returning assets to its investors in the form of a cash payment, so each investor's equity interest in the remaining net assets declines proportionately.

Cash	75,000	
Investment in equity affiliate (30% × $250,000)		75,000

Further Adjustments

● LO12–7

When the investor's expenditure to acquire an equity-method investment exceeds the book value of the underlying net assets acquired, additional adjustments to both the investment account and investment revenue might be needed. The purpose is to approximate the effects of consolidation, without actually consolidating financial statements. More specifically, both the investment account and investment revenue are adjusted for differences between net income reported by the investee and what that amount would have been if consolidation procedures had been followed. This process is often referred to as "amortizing the differential," because it mimics the process of expensing some of the difference between the price paid for the investment and the book value of the investment. Let's look closer at what that means.

Consolidated financial statements report (a) the acquired company's assets at their fair values on the date of acquisition rather than their book values on the investee's balance sheet, and (b) goodwill for the excess of the acquisition price over the fair value of the identifiable net assets acquired. This matters because increasing asset balances to their fair values can result in higher expenses in the future. If it's land or goodwill that's increased, there is no income effect because we don't depreciate or amortize those assets over time.[24] On the other hand, if buildings, equipment, or other depreciable assets are recorded at higher values, depreciation expense will be higher during their remaining useful lives. Likewise, if the recorded amount of inventory is increased, cost of goods sold will be higher when the inventory is sold. When expenses rise, income falls. It is this negative effect on income that the equity method seeks to imitate.

In our example, United needs to make adjustments for the fact that, at the time it purchased its investment in Arjent, the fair values of Arjent's identifiable assets and liabilities were higher than the book values of those assets and liabilities in Arjent's balance sheet (refer back to Illustration 12–14). Illustration 12–15 shows United's 30% proportionate share of the amounts shown in Illustration 12–14:

	Investee Net Assets		Net Assets Purchased		
Purchase price	$5,000,000	× 30% =	$1,500,000		
Fair value (identifiable)	3,600,000	× 30% =	1,080,000		
Difference	$1,400,000	× 30% =	$ 420,000	Goodwill	
Fair value (identifiable)	$3,600,000	× 30% =	$1,080,000		
Book value (identifiable)	2,100,000	× 30% =	630,000	Buildings	$300,000
Difference	$1,500,000	× 30% =	$ 450,000	Land	$150,000

Illustration 12–15

Explanation for Differences Between the Investment and the Book Value of Net Assets Acquired

Notice in Illustration 12–15 that United paid $1,500,000 for 30% of the identifiable net assets that, sold separately, would have a fair value of $1,080,000. The $420,000 difference between the price paid and the fair value of United's share of Arjent's identifiable net assets is attributable to goodwill. The 30% of identifiable net assets with a fair value of $1,080,000 have a book value on Arjent's balance sheet of only $630,000. The $450,000 difference is attributable to undervalued buildings ($300,000) and land ($150,000). Now let's see what adjustments, if any, United needs to make for these differences.

Adjustments for Additional Depreciation

When Arjent determines its net income, it bases depreciation expense on the book value of its buildings on its own balance sheet. United, however, needs to depreciate its share of the *fair value* of those buildings at the time it made its investment. To account for this higher amount of depreciation expense, United reduces investment revenue as if Arjent had included the additional expense in its earnings.

The investor adjusts its share of the investee's net income to reflect revenues and expenses associated with differences between the fair value and book value of the investee's assets and liabilities that existed at the time the investment was made.

As shown in Illustration 12–14, the book value of Argent's buildings is $1,000,000 and the fair value is $2,000,000, which creates a difference of $1,000,000. United will need to recognize its 30% share of additional depreciation expense for this difference, totaling $300,000 over the remaining life of the buildings. Assuming a 10-year life of the buildings and straight-line depreciation, United must recognize $30,000 of additional depreciation each year for ten years. Had Arjent recorded that additional depreciation in its income statement, United's portion of Arjent's net income would have been lower by $30,000 (ignoring taxes). So, to act as if Arjent had recorded the additional depreciation, United reduces investment revenue and reduces its investment in Arjent stock by $30,000.

Investment revenue	30,000	
Investment in equity affiliate [30% × ($2,000,000 − $1,000,000) ÷ 10 years]		30,000

[24] Goodwill is not amortized but instead is tested annually for impairment. If the asset's value is judged to be impaired, all or a portion of the recorded amount be charged against earnings (FASB ASC 350-20-35: Intangibles—Goodwill and Other—Goodwill—Subsequent Measurement). For review see Chapter 11.

No Adjustments for Land or Goodwill

United makes no adjustments for land or goodwill. Land is not an asset we depreciate. As a result, the difference between the fair value and book value of the land would not cause higher expenses, and we have no need to adjust investment revenue or the investment in Arjent stock for the land.

Recall from Chapter 11 that goodwill, unlike most other intangible assets, is not amortized. In that sense, goodwill resembles land. Thus, acquiring goodwill will not cause higher expenses, so we have no need to adjust investment revenue or the investment in Arjent stock for goodwill.

Adjustments for Other Assets and Liabilities

If the fair value of purchased inventory exceeds its book value, we usually assume the inventory is sold in the next year and reduce investment revenue in the next year by the entire difference.

Also, because in our example there is no difference between the book value and fair value of the remaining net assets, we don't need an adjustment for them either. However, that often will not be the case. For example, Arjent's inventory could have had a fair value that exceeded its book value at the time United purchased its Arjent investment. To recognize expense associated with that higher fair value, United would need to identify the period in which that inventory is sold (usually the next year) and, in that period, reduce its investment revenue and its investment in Arjent stock by its 30% share of the difference between the fair value and book value of the inventory. If, for instance, the $1,000,000 difference between fair value and book value had been attributable to inventory rather than buildings, and that inventory was sold by Arjent in the year following United's investment, United would reduce investment revenue by its 30% share of the difference ($300,000) in the year following the investment. More generally, an equity method investor needs to make these sorts of adjustments whenever there are revenues or expenses associated with an asset or liability that had a difference between book value and fair value at the time the investment was made.

Additional Consideration

Effect on Deferred Income Taxes. Investment revenue is recorded by the equity method when net income is recognized by the investee, but that revenue is not taxed until it's actually received as cash dividends. This creates a temporary difference between book income and taxable income. You will learn in Chapter 16 that the investor must report a deferred tax liability for the income tax that ultimately will be paid when the income eventually is received as dividends.

Reporting the Investment

The carrying amount of the investment is its initial cost plus the investor's equity in the undistributed earnings of the investee.

The fair value of the investment shares at the end of the reporting period is not reported when using the equity method. The investment account is reported at its original cost, increased by the investor's share of the investee's net income (adjusted for additional expenses like depreciation), and decreased by the portion of those earnings actually received as dividends.

The balance of United's 30% investment in Arjent at December 31, 2021, would be calculated as follows:

Investment in Equity Affiliate

Purchase price	1,500,000		
Share of net income	150,000		
		75,000	Dividends
		30,000	Depreciation adjustment
	1,545,000		

In the statement of cash flows, we report the purchase and sale of the investment as outflows and inflows of cash in the investing activities section, and the receipt of dividends is reported as an inflow of cash in the operating activities section.[25]

Additional Consideration

Much like consolidation, the equity method views the investor and investee collectively as a special type of single entity (as if the two companies were one company). However, the equity method doesn't require the investor to record separate financial statement items of the investee on an item-by-item basis as in consolidation. Instead, the investor reports its equity interest in the investee as a single investment account. Also, the adjustments that investors make when applying the equity method are designed to mimic what would happen if an investment were consolidated. For those reasons, the equity method sometimes is referred to as a "one-line consolidation," because it essentially collapses the consolidation approach into single lines in the balance sheet and income statement, while having the same effect on total income and shareholders' equity.

When the Investment is Acquired in Mid-Year

Obviously, we've simplified the illustration by assuming the investment was acquired at the beginning of 2021, entailing a full year's income, dividends, and adjustments to account for the income effects of any differences between book value and fair value on the date the investment was acquired. In the more likely event that an investment is acquired sometime after the beginning of the year, applying the equity method is easily modified to include the appropriate fraction of each of those amounts. For example, if United's purchase of 30% of Arjent had occurred on October 1 rather than January 2, we would simply record income, dividends, and adjustments for three months (3/12) of the year. This would result in the following entries to the investment account:

Investment in Equity Affiliate

Cost	1,500,000		
Share of net income (3/12 × $150,000)	37,500		
		7,500	Depreciation adjustment (3/12 × $30,000)
		18,750	Dividends (3/12 × $75,000)
	1,511,250		

Changes in the investment account the first year are adjusted for the fraction of the year the investor has owned the investment.

AT&T reported its 2016 investments in affiliated companies for which it exercised significant influence using the equity method as shown in Illustration 12–16.

When the Investee Reports a Net Loss

Our illustration assumed the investee reported net income. If the investee reports a *net loss* instead, the investment account would be *decreased* by the investor's share of the investee's net loss (adjusted for additional expenses).

Impairment of Equity Method Investments

A series of losses or other factors could indicate that an equity-method investment's fair value has declined to an amount below its current carrying value. If that decline is viewed

[25]Most companies prepare a statement of cash flows using the indirect method of reporting operating activities. In that case, the operating section begins with net income and adjustments are made to back out the effects of accrual accounting and calculate cash from operations. For companies with equity method investments, net income will include investment revenue and gains or losses associated with sold investments, but cash from operations should include only cash dividends. As an example, because United's 2018 net income includes $120,000 of investment revenue from Arjent ($150,000 portion of income – $30,000 depreciation adjustment), but United received only $75,000 of dividends from Arjent, an indirect method statement of cash flows would include an adjustment, often titled "undistributed earnings of investee," that reduces net income by $45,000 ($75,000 – $120,000) when determining cash from operations.

Illustration 12–16

Equity Method Investments in the Balance Sheet—AT&T

Real World Financials

	Dec 31, 2016	Dec 31, 2015
Total current assets	$ 38,369	$ 35,992
Property, plant, and equipment—Net	124,899	124,450
Goodwill	105,207	104,568
Licenses	94,176	93,093
Customer lists and relationships—Net	14,243	18,208
Other intangible assets—net	8,441	9,409
Investments in equity affiliates	**1,674**	**1,606**
Other assets	16,812	15,346
Total assets	$403,821	$402,672

Source: AT&T

Additional Consideration

It's possible that the investor's proportionate share of investee losses could exceed the carrying amount of the investment. If this happens, the investor should discontinue applying the equity method until the investor's share of subsequent investee earnings has equaled losses not recognized during the time the equity method was discontinued. This avoids reducing the investment account below zero.

as other than temporary, the investor should recognize an impairment loss in net income and reduce the carrying value of the investment to fair value in the balance sheet. The investor then continues with accounting under the equity method.[26]

What If Conditions Change?

A CHANGE FROM THE EQUITY METHOD TO ANOTHER METHOD When the investor's level of influence changes, it may be necessary to change from the equity method to another method. For example, when **Air-France/KLM**'s ownership interest in **WAM (Amadeus)** declined from 22% to 15%, it had to stop using the equity method to account for its investment.

When this situation happens, *no adjustment* is made to the remaining carrying amount of the investment. Instead, the equity method is simply discontinued and the new method applied from then on. The balance in the investment account when the equity method is discontinued would serve as the new cost basis for writing the investment up or down to fair value in the next set of financial statements.

A CHANGE FROM ANOTHER METHOD TO THE EQUITY METHOD Sometimes companies change from another method to the equity method. For example, when the **Mitsubishi UFJ Financial Group** converted its investment in the convertible preferred stock of **Morgan Stanley** into common stock, its ownership interest was large enough to qualify for accounting for the investment under the equity method. When a change *to* the equity method is appropriate, the previous method is discontinued and the balance in the investment account at the date of the change (including any unrealized holding gains or losses that occurred prior to the date the investment qualifies for the equity method) is used as the starting balance for applying the equity method. Any cost of acquiring additional shares is added to that balance, and going forward that balance is adjusted for the investor's portion of investee earnings and dividends. A disclosure note also should describe the change.[27]

[26]FASB ASC 323-10-35-31: Investments—Equity Method and Joint Ventures—Overall—Decrease in Investment Value.

[27]Retroactive restatement used to be required when a company changed to the equity method, but as of 2017 that is no longer necessary, per FASB ASC 323-10-35-33: Investments—Equity Method and Joint Ventures—Overall—Increase in Level of Ownership or Degree of Influence.

IF AN EQUITY METHOD INVESTMENT IS SOLD When an investment reported by the equity method is sold, we recognize a gain or loss if the selling price is more or less than the carrying amount (book value) of the investment. For example, let's continue our illustration and assume United sells its investment in Arjent on January 1, 2022, for $1,446,000. A journal entry would record a loss as follows:

Cash	1,446,000	
Loss on investments (NI) (to balance)	99,000	
Investment in equity affiliate (account balance)		1,545,000

When an equity method investment is sold, a gain or loss is recognized for the difference between its selling price and its carrying amount.

COMPARISON OF FAIR VALUE AND THE EQUITY METHOD Illustration 12–17 compares accounting for the Arjent investment at fair value through net income and under the equity method (using account title abbreviations):

Illustration 12–17 Comparison of Fair Value and Equity Methods

	Fair Value Through Net Income			Equity Method		
Purchase equity investment	Investment in equity securities	1,500,000		Investment in equity affiliate	1,500,000	
	Cash		1,500,000	Cash		1,500,000
Recognize proportionate share of investee's net income and any related adjustments	No entry			Investment in equity affiliate	150,000	
				Investment revenue		**150,000**
				Investment revenue	**30,000**	
				Investment in equity affiliate		30,000
Adjust to fair value, 2021	**Loss (unrealized, NI)**	**50,000**		No entry		
	FV adjustment		50,000			
Receive dividend	Cash	75,000		Cash	75,000	
	Dividend revenue		**75,000**	Investment in equity affiliate		75,000
Sell equity investment						
1. Adjust to fair value, 2022	**Loss (unrealized, NI)**	**4,000**		No entry		
	FV adjustment		4,000			
2. Record sale	Cash	1,446,000		Cash	1,446,000	
	FV adjustment	54,000		**Loss (NI) (to balance)**	**99,000**	
	Investment in equity securities		1,500,000	Investment in equity affiliate		1,545,000

This side-by-side comparison highlights several aspects of these accounting approaches.

- To record the purchase of an investment, we use the same basic entry for both approaches.
- The two approaches differ in whether we record investment revenue when dividends are received and whether we recognize unrealized holding gains and losses associated with changes in the fair value of the investment.
- The differences in how the two approaches account for unrealized holding gains and losses result in different carrying values for the investment at the time the investment is sold, and therefore result in different realized gains or losses when the investment is sold.
- Regardless of approach, the same cash flows occur, and the same total amount of net income is recognized over the life of the investment. In the case of Arjent,
  - **Fair value through net income:** A total of $21,000 of net income is recognized over the life of the investment, equal to **$75,000** of dividend revenue minus $54,000 (**$50,000** + **$4,000**) loss on the investment.

- **Equity method:** A total of $21,000 of net income is recognized over the life of the investment, equal to $150,000 of United's portion of Arjent's net income minus $30,000 depreciation adjustment and minus $99,000 loss realized on sale of investment.
- Thus, the question is not how much total net income is recognized, but *when* that net income is recognized.

Fair Value Option

If the fair value option is chosen for investments otherwise accounted for by the equity method, the amount that is reported at fair value is clearly indicated.

LO12–8

Companies can choose the fair value option (FVO) for "significant influence" investments that otherwise would be accounted for under the equity method. The company makes an irrevocable decision about whether to elect the FVO and can make that election for some investments and not for others. As shown in Illustration 12–17, the company carries the investment at fair value in the balance sheet and includes unrealized gains and losses in net income. These investments are shown on their own line in the balance sheet or are combined with equity method investments with the amount at fair value shown parenthetically. Also, all of the disclosures that are required when reporting fair values as well as some of those that would be required under the equity method still must be provided.[28]

Exactly how a company does the bookkeeping necessary to comply with these broad requirements is up to the company. One alternative is to account for the investment using the entries that would be used if the investor lacked significant influence and accounted for it at fair value through net income. A second alternative is to record all of the accounting entries during the period under the equity method, and then record a fair value adjustment at the end of the period. For example, imagine the following scenario:

- an investment is purchased for $100,
- the investor's share of the investee's net income is $10,
- the investor's share of the investee's dividends is $2,
- the investor's amortization of differential is $3,
- fair value increases to $200.

If the investor accounted for the investment as fair value through net income, it would show dividend income of $2 as well as an unrealized gain of $100 in the income statement, increasing pre-tax income by a total of $102. On the other hand, if the investor first accounted for the investment under the equity method, it would show equity method investment income of $7 (calculated as income of $10 minus amortization of differential of $3), which would produce a pre-adjustment investment carrying value of $105 (calculated as $100 initial cost plus $7 income minus $2 dividends). Therefore, to get to a fair value of $200, the investor would record a fair value adjustment of $95. The fair value adjustment of $95 combines with the $7 of equity method investment income to increase pre-tax income by a total of $102. You can see that, regardless of which alternative the investor uses, the same total fair value is reported in the balance sheet at the end of the period, and the same total amount is shown in the income statement.

International Financial Reporting Standards

LO12–9

Equity Method. Like U.S. GAAP, international accounting standards require the equity method for use with significant influence investees (which they call "associates"), but there are a few important differences. First, *IAS No. 28* governs application of the equity method and requires that the accounting policies of investees be adjusted to correspond to those of the investor when applying the equity method.[29] U.S. GAAP has no such requirement.

Second, IFRS does not provide the fair value option for most investments that qualify for the equity method. U.S. GAAP provides the fair value option for all investments that qualify for the equity method.

[28]FASB ASC 825-10-50-28: Financial Instruments—Overall—Disclosure—Fair Value Option.

[29]"Investments in Associates," *International Accounting Standard 28* (London, UK: IASCF, 2003), as amended, effective January 1, 2018.

Concept Review Exercise

THE EQUITY METHOD

Delta Apparatus bought **40%** of Clay Crating Corp.'s outstanding common shares on January 2, 2021, for $540 million. The carrying amount of Clay Crating's net assets (shareholders' equity) at the purchase date totaled $900 million. Book values and fair values were the same for all financial statement items except for inventory and buildings, for which fair values exceeded book values by $25 million and $225 million, respectively. All inventory on hand at the acquisition date was sold during 2021. The buildings have average remaining useful lives of 18 years. During 2021, Clay Crating reported net income of $220 million and paid an $80 million cash dividend.

Required:

1. Prepare the appropriate journal entries during 2021 for the investment.
2. Determine the amounts relating to the investment that Delta Apparatus should report in the 2021 financial statements
   a. as an investment in the balance sheet.
   b. as investment revenue in the income statement.
   c. as investing and/or operating activities in the statement of cash flows (direct method).

Solution

1. Prepare the appropriate journal entries during 2021 for the investment.

	($ in millions)	
Purchase		
Investment in equity affiliate	540	
Cash		540
Net income		
Investment in equity affiliate (**40%** × $220 million)	88	
Investment revenue		88
Dividends		
Cash (**40%** × $80 million)	32	
Investment in equity affiliate		32
Inventory		
Investment revenue (as if 2021 cost of goods sold is higher because beginning inventory was adjusted to fair value)	10	
Investment in equity affiliate (**40%** × $25 million)		10
Buildings		
Investment revenue [($225 million × **40%**) ÷ 18 years]	5	
Investment in equity affiliate		5

	Investee Net Assets		Net Assets Purchased
Fair value (identifiable)	$ 1,150*	× 40% =	$ 460
Book value (identifiable)	900	× 40% =	360
Difference	$ 250	× 40% =	$ 100

*($900 + $25 + $225)

2. Determine the amounts that Delta Apparatus should report in the 2021 financial statements.
   a. As an investment in the balance sheet:

Investment in equity affiliate
($ in millions)

	Debit	Credit	
Purchase price	540		
40% of Clay Crating net income	88		
		32	Dividends
		10	Cost of goods sold adjustment for inventory (all sold in 2021)
		5	Depreciation adjustment for buildings ($90,000 ÷ 18)
Balance	581		

b. As investment revenue in the income statement:

$88 million (share of income) − ($10 + $5) million (adjustments) = $73 million

c. In the statement of cash flows (direct method):

- Investing activities: $540 million cash outflow to purchase investment
- Operating activities: $32 million cash inflow from dividends received

Decision Makers' Perspective

The various approaches used to account for investments can have very different effects on an investor's income statement and balance sheet. Consequently, it's critical that both managers and external decision makers clearly understand those effects and make decisions accordingly.

To highlight key considerations, suppose that, on January 1, 2021, BigCo spent $5,000,000 to purchase 20% of TechStart, a small start-up company that is developing products that apply an exciting new technology. The purchase price included $500,000 for BigCo's share of the difference between the fair value and book value of TechStart's inventory, all of which was then sold in 2021. TechStart paid a small dividend of $100,000 in 2021, so BigCo received 20% of it, or $20,000. TechStart incurs and expenses large amounts of research and development costs as it develops new technology, so it had a net loss in 2021 of $1,000,000. Yet, the future income-generating potential of the products that TechStart is developing has made TechStart a hot stock, and the fair value of BigCo's 20% investment increased to $5,500,000 by the end of 2021. Illustration 12–18 shows how BigCo's investment would be accounted for under two alternative approaches.

Illustration 12–18

Comparison of Methods Used to Account for Investments

	Fair Value Through Net Income	Equity Method
Share of investee net income (loss)*	$ –0–	$ (700,000)
Dividend revenue**	20,000	–0–
Increase in investee's fair value†	500,000	–0–
Total 2021 effect on net income (loss)	$ 520,000	$ (700,000)
December 31, 2021 investment book value††	$5,500,000	$4,280,000

*Not recognized if investment accounted for as fair value through net income. Under the equity method, investment revenue (loss) is 20% × ($1,000,000 loss) + ($500,000) additional expense for fair value inventory adjustment = ($700,000).
**Not recognized for equity method investments. Instead, dividends reduce book value of the investment.
†Recognized in net income if accounted for at fair value through net income. Not recognized under the equity method.
††Equals fair value if accounted for at fair value through net income. Equals initial cost plus income (or minus loss) and minus dividends for equity method.

The way an investment is accounted for affects net income, investment book value, and the amount of gain or loss recognized when the investment is sold.

The accounting method does not affect cash flows, but it has a big effect on net income in current and future periods. Also, because the accounting method affects the book value of the investment, it affects gain or loss on sale of that investment. In our example, if BigCo sold its TechStart investment at the beginning of 2022 for $5,000,000, it would recognize a $720,000 gain on sale if the investment was accounted for under the equity method, but a $500,000 loss if it was accounted for at fair value through net income. All of these income effects are predictable, but only if a user understands the relevant accounting methods and the fact that those methods all end up recognizing the same amount of total gain or loss over the life of an investment. Nevertheless, sometimes even experienced analysts get confused.[30]

[30]For example, in 2000, analysts were accustomed to including **Intel**'s investment gains as ordinary income, because those amounts were not particularly large and could be viewed as part of Intel's business. However, in the 2nd quarter of 2000, Intel recorded a net $2.1 billion gain from selling securities in its available-for-sale portfolio. Analysts were surprised and confused, with some eliminating the gain from their earnings estimates but others including them ("Intel Says Net Jumped 79%; Analysts Upset," *The Wall Street Journal,* July 19, 2000, p, A3).

One strength of the equity method is that it prevents the income manipulation that would be possible if a company recognized income when it received dividends and could significantly influence an investee to pay dividends whenever the company needed an income boost. Remember, under the equity method dividends aren't income, but rather reduce the book value of the investment. Nevertheless, users still need to realize that managers may choose and apply methods in ways that make their company appear most attractive. For example, research suggests that investments sometimes are structured to avoid crossing the 20 to 25% threshold that typically requires using the equity method, presumably to avoid the negative effect on earnings that comes from having to recognize the investor's share of investee losses and other income adjustments.[31] Also, a company might smooth income by timing the sale of equity method investments to realize gains in otherwise poor periods and realize losses in otherwise good periods. While consistent with GAAP, mixing these sorts of one-time gains and losses with operating income could encourage users to think that operating income is less volatile than it really is.

Managers may structure equity investments to qualify for their preferred accounting approach.

A benefit of fair value accounting is that it prevents managers from timing the sale of investments to recognize gains or losses in particular accounting periods.

Regarding the fair value through net income approach, of particular concern is the potential for inaccurate fair value estimates. Even if management is trying to provide the most accurate fair value estimate possible, there is much potential for error, particularly when making estimates at level 3 of the fair value hierarchy. Also, a company conceivably could use the discretion inherent in fair value estimation to manage earnings with respect to trading securities or other investments for which they have elected the fair value option. Given this potential for error and bias, it's not surprising that investors are nervous about the accuracy of fair value estimates. To address these sorts of concerns, the FASB has required extensive note disclosure about the quality of inputs associated with estimates of fair value, but financial statement users need to know to look for those disclosures and still must understand that they cannot assess fully the accuracy of fair value estimates. ●

A concern with fair value accounting is that management has much discretion over fair values, and may not be able to estimate fair values accurately.

FINANCIAL INSTRUMENTS AND INVESTMENT DERIVATIVES A **financial instrument** is defined as one of the following:

1. Cash
2. Evidence of an *ownership interest* in an entity.[32]
3. A contract that (a) imposes on one entity an obligation to *deliver* cash (say accounts payable) or another financial instrument and (b) conveys to the second entity a right to *receive* cash (say accounts receivable) or another financial instrument.
4. A contract that (a) imposes on one entity an obligation to *exchange* financial instruments on potentially unfavorable terms (say the issuer of a stock option) and (b) conveys to a second entity a right to *exchange* other financial instruments on potentially favorable terms (say the holder of a stock option).[33]

A complex class of financial instruments exists in financial markets in response to the desire of firms to manage risks. In fact, these financial instruments would not exist in their own right, but have been created solely to hedge against risks created by other financial instruments or by transactions that have yet to occur but are anticipated. Financial futures, interest rate swaps, forward contracts, and options have become commonplace.[34] These financial instruments often are called **derivatives** because they "derive" their values or contractually required cash flows from some other security or index. For instance, an option to buy an asset in the future at a preset price has a value that is dependent on, or derived from, the value of the underlying asset. Their rapid acceptance as indispensable components of the corporate capital structure has left the accounting profession scrambling to keep pace.

Derivatives are financial instruments that "derive" their values from some other security or index.

[31]E. E. Comiskey and C. W. Mulford, "Investment Decisions and the Equity Accounting Standard," *The Accounting Review* 61, no. 3 (July 1986), pp. 519–525
[32]This category includes not just shares of stock, but also partnership agreements and stock options.
[33]FASB ASC Master Glossary: Financial Instrument.
[34]Interest rate futures were traded for the first time in 1975 on the Chicago Board of Trade. Interest rate swaps were invented in the early 1980s. They now comprise over 70% of derivatives in use.

The urgency to establish accounting standards for financial instruments has been accelerated by headline stories in the financial press reporting multimillion-dollar losses on exotic derivatives by **Enron Corporation**, **Procter & Gamble**, **Orange County** (California), **Piper Jaffrey**, and **Gibson Greetings**, to mention a few. The headlines have tended to focus attention on the misuse of these financial instruments rather than their legitimate use in managing risk.

The FASB's ongoing financial instruments project is expected to lead to a consistent framework for accounting for all financial instruments.

The FASB has been involved for many years in a project to provide a consistent framework for resolving financial instrument accounting issues, including those related to derivatives and other "off-balance-sheet" instruments. The financial instruments project has three separate but related parts: disclosure, recognition and measurement, and distinguishing between liabilities and equities. Unfortunately, the issues to be resolved are extremely complex and will likely require several more years to resolve. To help fill the disclosure gap in the meantime, the FASB has offered a series of temporary, "patchwork" solutions. These are primarily in the form of additional disclosures for financial instruments. More recently, the FASB has tackled the issues of recognition and measurement of derivatives. We discuss these requirements in Appendix A to this book.

Financial Reporting Case Solution

©8th.creator/Shutterstock

1. **How should you respond? Why are held-to-maturity securities treated differently from other investment securities?** *(p. 650)* You should explain that if an investor has the positive intent and ability to hold the securities to maturity, investments in debt securities are classified as held-to-maturity and reported at amortized cost in the balance sheet. Increases and decreases in fair value are not reported in the financial statements. The reasoning is that the changes are not as relevant to an investor who will hold a security to its maturity regardless of those changes. Changes in the fair value between the time a debt security is acquired and the day it matures to a prearranged maturity value aren't as important if sale before maturity isn't an alternative.
2. **Why are unrealized gains and losses on trading securities reported in the income statement?** *(p. 652)* Trading securities are acquired for the purpose of profiting from short-term market price changes, so gains and losses from holding these securities while prices change are often viewed as relevant performance measures that should be included in net income.
3. **Why are unrealized gains and losses on available-for-sale securities not reported in the income statement, but instead are reported in other comprehensive income, and then shown in accumulated other comprehensive income (AOCI) in the balance sheet?** *(p. 657)* Available-for-sale securities are not acquired for the purpose of profiting from short-term market price changes, so gains and losses from holding these securities while prices change are viewed as insufficiently relevant performance measures to be included in net income. Instead, those amounts are shown in other comprehensive income (OCI) and accumulated in an owners' equity account (AOCI). It's likely that holding gains in some periods will be offset by holding losses in other periods. When the investment is sold, the net amount of gain or loss is removed from AOCI and recognized in net income.
4. **Explain why Coke accounts for some of its investments by the equity method and what that means.** *(p. 674)* When an investor does not have "control," but still is able to exercise *significant influence* over the operating and financial policies of the investee, the investment should be accounted for by the equity method. Apparently Coke owns between 20% and 50% of the voting shares of some of the companies it invests in. By the equity method, Coke recognizes investment income in an amount equal to its percentage share of the net income earned by those companies, instead of the amount of that net income it receives as cash dividends. The rationale is that as the investee earns additional net assets, Coke's share of those net assets increases. ●

The Bottom Line

- **LO12–1** Key events in the life of a debt investment are purchase, recording interest revenue, incurring unrealized holding gains and losses due to fair value changes, and recording sale or maturity. (*p. 646*)
- **LO12–2** If an investor has the positive intent and ability to hold the securities to maturity, investments in debt securities are classified as HTM and reported at amortized cost in the balance sheet. These investments are recorded at cost, and holding gains or losses from fair value changes are ignored. (*p. 650*)
- **LO12–3** Investments in debt securities acquired principally for the purpose of selling them in the near term are classified as trading securities. They are reported at their fair values. Holding gains and losses for trading securities are included in earnings. (*p. 652*)
- **LO12–4** Investments in debt securities that don't fit the definitions of the other reporting categories are classified as available-for-sale. They are reported at their fair values. Holding gains and losses from retaining securities during periods of price change are not included in the determination of income for the period; they are reported as a separate component of other comprehensive income in shareholders' equity. (*p. 656*)
- **LO12–5** Investments in equity securities for which the investor lacks the ability to exercise significant influence over the investee are accounted for using a fair value through net income approach. They are reported at their fair values. Holding gains and losses are included in earnings. (*p. 669*)
- **LO12–6** The equity method requires the investor to recognize investment income equal to its percentage share (based on share ownership) of the net income earned by the investee, rather than the amount received as cash dividends. The investment account is adjusted for the investor's percentage share of net income or loss reported by the investee. When the investor actually receives dividends, the investment account is reduced accordingly. (*p. 674*)
- **LO12–7** When the fair value of identifiable net assets acquired exceeds the book value of the underlying net assets acquired in the purchase of an equity investment, both the investment account and investment revenue are adjusted for differences between net income reported by the investee and what that amount would have been if consolidation procedures had been followed. (*p. 676*)
- **LO12–8** The fair value option allows companies to account for most financial assets and liabilities in the same way they account for trading securities, with unrealized holding gains and losses included in net income and the investment carried at fair value in the balance sheet. (*pp. 664* and *682*)
- **LO12–9** U.S. GAAP and IFRS are similar in most respects concerning how they account for investments. IFRS is more restrictive in terms of the circumstances in which the fair-value option can be used. IFRS allows accounting similar to AFS treatment for equity investments. Finally, as discussed in Appendix 12B, IFRS uses a somewhat different impairment recognition approach for HTM and AFS investments. (*pp. 664-665, 673, 682*, and *693*)

Other Investments (Special Purpose Funds, Investments in Life Insurance Policies)

APPENDIX 12A

Special Purpose Funds

It's often convenient for companies to set aside money to be used for specific purposes. You learned about one such special purpose fund in Chapter 7 when we discussed petty cash funds. Recall that a petty cash fund is money set aside to conveniently make small expenditures using currency rather than having to follow the time-consuming, formal procedures normally used to process checks. Similar funds sometimes are used to pay interest, payroll, or other short-term needs. Like petty cash, these short-term special purpose funds are reported as current assets.

Some special purpose funds—like petty cash—are current assets.

Special purpose funds also are sometimes established to serve longer-term needs. It's common, for instance, to periodically set aside cash into a fund designated to repay bonds and other long-term debt. Such funds usually accumulate cash over the debt's term to maturity and are composed of the company's periodic contributions plus interest or dividends from investing the money in various return-generating investments. In fact, some

Special purpose funds that serve longer-term needs are reported as noncurrent assets.

debt contracts require the borrower to establish such a fund to repay the debt. In similar fashion, management might voluntarily choose to establish a fund to accumulate money to expand facilities, provide for unexpected losses, buy back shares of stock, or any other special purpose that might benefit from an accumulation of funds. Of course, funds that won't be used within the upcoming operating cycle are noncurrent assets. They typically are reported as part of investments. The same criteria for classifying securities into reporting categories that we discussed previously should be used to classify securities in which funds are invested. Any investment revenue from these funds is reported as such in the income statement.

Investments in Life Insurance Policies

Certain life insurance policies can be surrendered while the *insured is still alive* in exchange for its *cash surrender value.*

Companies frequently buy life insurance policies on the lives of their key officers. Under normal circumstances, the company pays the premium for the policy and, as beneficiary, receives the proceeds when the officer dies. Of course, the objective is to compensate the company for the untimely loss of a valuable resource in the event the officer dies. However, some types of life insurance policies can be surrendered while the insured is still alive in exchange for a determinable amount of money, called the **cash surrender value**. In effect, a portion of each premium payment is not used by the insurance company to pay for life insurance coverage, but instead is invested on behalf of the insured company in a fixed-income investment. Accordingly, the cash surrender value increases each year by the portion of premiums invested plus interest on the previous amount invested. This is simply a characteristic of whole life insurance, unlike term insurance that has lower premiums and provides death benefits only.

From an accounting standpoint, the periodic insurance premium should not be expensed in its entirety. Rather, part of each premium payment, the investment portion, is recorded as an asset. Illustration 12A–1 provides an example.

Illustration 12A–1
Cash Surrender Value

Several years ago, American Capital acquired a $1 million insurance policy on the life of its chief executive officer, naming American Capital as beneficiary. Annual premiums are $18,000, payable at the beginning of each year. In 2021, the cash surrender value of the policy increased according to the contract from $5,000 to $7,000. The CEO died at the end of 2021.

Part of the annual premium represents a build-up in the cash surrender value.

Insurance expense (difference)	16,000	
Cash surrender value of life insurance ($7,000 – $5,000)	2,000	
Cash (2021 premium)		18,000
To record insurance expense and the increase in the investment.		

The cash surrender value is considered to be a noncurrent investment and would be reported in the investments section of the balance sheet. Of course when the insured officer dies, the corporation receives the death benefit of the insurance policy, and the cash surrender value ceases to exist because canceling the policy no longer is an option. The corporation recognizes a gain for the amount of the death benefit less the cash surrender value.

When the death benefit is paid, the cash surrender value becomes null and void.

Cash (death benefit)	1,000,000	
Cash surrender value of life insurance (balance)		7,000
Gain on life insurance settlement (difference)		993,000
To record the proceeds at death.		

APPENDIX 12B Impairment of Debt Investments

We saw in Chapter 11 that intangible assets and property, plant, and equipment are subject to impairment losses that reduce earnings. Similarly, you learned in Chapter 7 that companies recognize impairment losses for financial assets. These are known as credit losses (or bad debts) and are determined using the CECL ("Current Expected Credit Loss") model. Based on historical experience, current conditions, and reasonable forecasts, companies estimate the amount of their accounts receivable they expect to actually receive. They use an allowance for uncollectible accounts to reduce the carrying value of their accounts receivable to that amount, and each period recognize the amount of bad debt expense needed to adjust that allowance to its appropriate balance.

Companies also must account for credit losses with respect to debt investments. They don't need to worry about recognizing credit losses for trading securities or investments for which a company has chosen the fair value option, because all changes in the fair values of those investments, including credit losses, always are recognized in net income. However, that's not the case for held-to-maturity (HTM) and available-for-sale (AFS) debt investments, because changes in the fair value of those investments are not always recognized in net income. Rather, declines in fair value typically are ignored for HTM investments and are recorded in OCI for AFS investments, and only included in net income when HTM and AFS investments are sold. Therefore, companies do need to account for credit losses with respect to HTM and AFS investments. We discuss each in turn.

Credit Losses for Held-to-Maturity Investments

Companies recognize credit losses for HTM investments the same way they recognize bad debts for a note receivable. They use a contra asset account, an allowance for credit losses, to reduce the carrying value of HTM investments to the net amount expected to be collected. Each period they record whatever credit loss expense (bad debt expense) or recovery of credit loss is necessary to adjust that allowance to its appropriate balance.

The CECL model allows companies to choose from various methods to estimate credit losses for HTM debt investments. A very common approach is for the investor to estimate the future cash flows it expects to receive and then discount those cash flows at the effective interest rate that existed when the debt investment was purchased. The investor then compares that discounted cash flow estimate to the balance (amortized cost) of the debt, and adjusts the allowance for credit losses to reduce the carrying value of the debt to that estimate.

In Illustration 12B–1, we modify the Masterwear Industries example from Illustration 12-1 to demonstrate accounting for credit losses for an HTM investment. Later, we do the same for an AFS investment.

Illustration 12B–1

Credit Losses for an HTM Investment

On July 1, 2021, United Industries purchased bonds with a face value of $700,000 from Masterwear Industries. The stated rate of interest for the bonds is 12%, so $42,000 of interest is receivable semiannually on June 30 and December 31. The bonds mature in three years, on June 30, 2024 The market interest rate for bonds of similar risk and maturity is 14%. United purchased the bonds for $666,633, reflecting a discount of $33,367.

United received its $42,000 interest payment on December 31, 2021, so it amortized $4,664 of discount, leaving the amortized cost of the bond investment at $671,297.* When preparing its 2021 financial statements, United considered whether credit losses had occurred with respect to the Masterwear investment. United concluded that it was likely to receive interest payments of only $30,000 each period and a return of principal at maturity of only $600,000. To calculate its credit loss, United discounted those cash flows at the 7% effective interest rate that existed when the bonds were purchased, and compared that discounted amount to the amortized cost of the bonds at 12/31/2021:

Amortized cost of the bonds at 12/31/2021:		$ 671,297
Discounted cash flow of the bond	**Present Values**	
(using *effective interest rate on date of bond purchase*):		
Interest	$ 30,000 × 4.10020** = $123,006	
Principal (face amount)	$600,000 × 0.71299† = 427,794	
Total discounted cash flow		(550,800)
Necessary balance in the allowance for credit losses:		**$ 120,497**

**Refer to Illustration 12–2 for the amortization table that applies to the Masterwear bonds.*
***Present value of an ordinary annuity of $1: n = 5, i = 7% (Table 4).*
†Present value of $1: n = 5, i = 7% (Table 2).

Note: Present value tables are provided at the end of this textbook. If you need to review the concept of the time value of money, refer to the discussions in Chapter 5.

(continued)

Illustration 12B–1
(concluded)

To record the **$120,497** credit loss, United makes the following journal entry:

Credit loss expense	120,497	
Allowance for credit losses		**120,497**

United's 2021 before-tax net income will be reduced by the **$120,497** credit loss, and the Masterwear investment will be reported in the balance sheet at a carrying value of $550,800. In future periods, United will re-estimate the discounted cash flows associated with the Masterwear bonds and adjust the allowance for credit losses up or down as necessary to state the carrying value of the investment at the appropriate amount.

Credit Losses for Available-for-Sale Investments

The FASB requires that investors calculate credit losses for AFS investments using the discounted cash flow approach shown in Illustration 12B–1. However, accounting for credit losses is complicated by the fact that AFS investments must be carried at fair value, with unrealized gains and losses shown in OCI. To help us consider this difference, Illustration 12B–2 modifies Illustration 12B–1 to show how the *fair value* of the Masterwear bonds might be calculated.

Illustration 12B–2
Credit and Non-Credit Losses of an AFS Investment

Assume the same facts as in Illustration 12B–1. Also assume that United believes that, given the troubles Masterwear Industries has been facing, a discount rate of 20% (10% every six months) is appropriate for valuing those bonds in the current market. Using a discounted cash flow approach to value the Masterwear bonds,* the fair value of those bonds and the related unrealized loss would be calculated as follows at 12/31/2021:

Amortized cost of the bonds at 12/31/2021:		$ 671,297
Discounted cash flow of the bond (using *prevailing market interest rate*):	**Present Values**	
Interest	$ 30,000 × 3.79079** = $113,724	
Principal (face amount)	$600,000 × 0.62092† = 372,552	
Total discounted cash flow (used to estimate fair value)		(486,276)
Unrealized loss		**185,021**
Credit loss (from Illustration 12B–1):		**(120,497)**
Non-credit loss		**$ 64,524**

United could use other valuation approaches to estimate fair value, but assuming a discounted cash flow approach highlights the difference between credit losses and other unrealized losses.
***Present value of an ordinary annuity of $1: n = 5, i = 10% (Table 4).*
†Present value of $1: n = 5, i = 10% (Table 2).

Note: Present value tables are provided at the end of this textbook. If you need to review the concept of the time value of money, refer to the discussions in Chapter 5.

We see from Illustration 12B–2 that the fair value estimate can differ from the value used to calculate credit losses. The credit-loss calculation requires that the discount rate be the effective interest rate as of the date the investment was *purchased.* However, the fair value calculation uses the *current* discount rate appropriate for the investment. Market conditions change over time, so it makes sense that these two discount rates could differ. In this case, because United believes market participants would use a relatively high discount rate when valuing the Masterwear investment, it assigns a relatively low fair value to that investment.

How should United account for these losses? The flowchart in Illustration 12B–3 provides guidance:

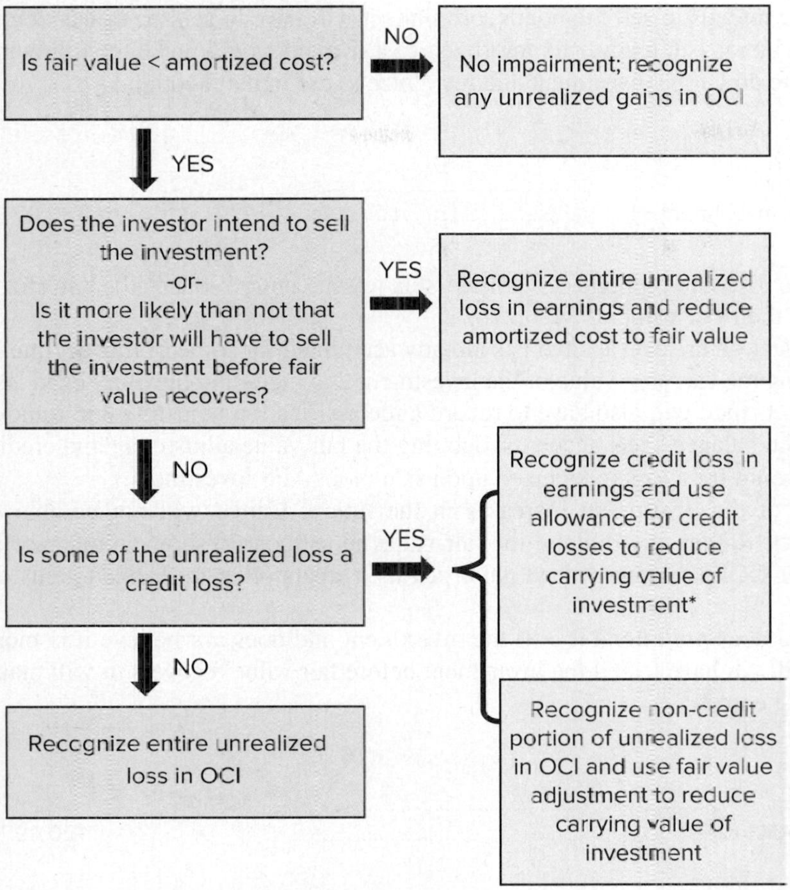
Is fair value < amortized cost?
NO
No impairment; recognize any unrealized gains in OCI
YES
Does the investor intend to sell the investment? -or- Is it more likely than not that the investor will have to sell the investment before fair value recovers?
YES
Recognize entire unrealized loss in earnings and reduce amortized cost to fair value
NO
Is some of the unrealized loss a credit loss?
YES
Recognize credit loss in earnings and use allowance for credit losses to reduce carrying value of investment*
Recognize non-credit portion of unrealized loss in OCI and use fair value adjustment to reduce carrying value of investment
NO
Recognize entire unrealized loss in OCI

*Note: The amount of credit loss is limited to the amount by which fair value is lower than amortized cost.

Illustration 12B–3

Decision Tree to Account for Impairment of AFS Investments

As shown in Illustration 12B–3, if an investment's fair value is not less than amortized cost, there is no impairment of the investment, and any unrealized gain should be shown in OCI.

If instead, fair value is less than amortized cost, there is some impairment of the investment. At this point, we need to consider what the investor intends to do with the investment. If the investor intends to sell the investment, or if the investor is more likely than not to have to sell the investment before the fair value of the investment can recover, we act as if the investor has realized that loss. The investor recognizes the entire loss in earnings and reduces the carrying value of the investment to its fair value. No recovery of that write-down is permitted if fair value recovers in the future. Rather, the new amortized cost will be compared to fair value and any unrealized gains or losses will be recorded in OCI as typically done for AFS investments.

If, on the other hand, the investor does not intend to sell the investment and does *not* believe it is more likely than not the investment will be sold before fair value recovers, the amount of the unrealized loss that is due to a credit loss must be identified. The credit loss portion is recognized in earnings, and an allowance for credit losses is established, just like we did for HTM investments.[34] If the credit loss decreases in the future, a reversal of credit loss can be included in earnings, just like for HTM investments. Any remaining non–credit-loss component is recognized as an unrealized loss in OCI, just as other unrealized losses are recognized for AFS investments.

Returning to Illustration 12B–2, let's consider how United should account for its AFS investment. We can see that the fair value of the Masterwear bonds is less than amortized cost, so it is clear that the investment is impaired. Let's consider two alternative cases.

[34]The amount of the credit loss is limited to the amount by which fair value is lower than amortized cost, because the company can always sell the investment at fair value to avoid additional credit losses.

Case 1: If United intends to sell the bonds, or thinks it will have to sell the bonds before the fair value of the bonds can recover to amortized cost, United would make the following journal entry to write down the investment and recognize a loss in net income:

December 31, 2021		
Loss on impairment (NI)	**185,021**	
Discount on bond investment		185,021

By increasing the discount on bond investment, this journal entry reduces the amortized cost of the bonds to their fair value of $486,276.

An added complication arises if United has already accounted for some of this decline in fair value by reducing the carrying value of the investment and including the unrealized loss in OCI. In that case, United will also have to record a reclassification adjustment to remove those effects from the balance sheet accounts, debiting the fair value adjustment and crediting OCI, just as it would for a loss recognized upon sale of an AFS investment.

If the fair value of the investment increases in the future, United will not be able to reverse that write-off. Rather, it will debit the fair value adjustment to show an increase in fair value and credit OCI as an unrealized gain, just as it treats other unrealized gains on AFS investments.

Case 2: If United does *not* intend to sell the investment and does *not* believe it is more likely than not that it will have to sell the investment before fair value recovers, it will make the following journal entries:

December 31, 2021		
Credit loss expense	**120,497**	
Allowance for credit losses		120,497

December 31, 2021		
Loss on AFS investment (unrealized, OCI)	**64,524**	
Fair value adjustment		64,524

With regard to the impairment, United is separately accounting for the credit loss and the non-credit loss. The credit loss reduces net income, and the non-credit loss reduces OCI. The allowance for credit losses and fair value adjustment are both contra accounts to the investment, and together reduce the carrying value of the bonds to their fair value of $486,276. Going forward, United will adjust the allowance for credit losses up or down as credit losses increase or are recovered, and will recognize any remaining (non-credit loss) unrealized gains or losses on its AFS investments as it would normally.

You can see that for both cases United carries its investment at fair value. They differ in how the impairment loss is reported. For case 1, the entire impairment loss reduces current-period earnings, but for case 2, only the credit loss reduces current-period earnings.

Companies are required to show AFS investment at fair value in the balance sheet, and to present parenthetically the amortized cost basis and the allowance for credit losses. Let's consider how each of our two cases would be presented in the balance sheet.

Case 1: In Case 1, United has reduced the amortized cost of its Masterwear investment to fair value. Therefore, United would show its Masterwear investment in the balance sheet as follows:

AFS Investment (amortized cost: $486,276)	$486,276

Case 2: In Case 2, United has recognized both an allowance for credit losses as well as a fair value adjustment. It must disclose amortized cost and the allowance as well as fair value. Therefore, United would show its Masterwear investment in the balance sheet as follows:

AFS Investment (amortized cost: $671,297, allowance for credit losses: $120,497)	$486,276

International Financial Reporting Standards

Accounting for Impairments. Under *IFRS No. 9,* companies recognize impairments for debt investments that are accounted for at amortized cost (rather than fair value through net income) or at fair value through other comprehensive income (FVOCI). The impairment is calculated using the ECL model discussed previously, and is measured either as the 12-month expected credit loss (if the credit risk on the investment has increased significantly) or the lifetime expected credit loss (if the credit risk on the investment has not increased significantly.) The entire impairment is recognized in earnings (there is no equivalent to recognizing in OCI any non-credit losses on debt investments), with an offsetting allowance reducing the carrying value of the investment to the appropriate amount. Companies can elect to always recognize lifetime credit losses, but only have to do so when a significant increase in credit loss has occurred. As in U.S. GAAP, IFRS allows recoveries of impairments to be recognized in earnings.

● LO12–9

Questions For Review of Key Topics

Q 12–1 All investments in *debt* securities are classified for reporting purposes in one of three categories, and can be accounted for differently depending on the classification. What are these three categories?

Q 12–2 When market rates of interest *rise* after a fixed-rate security is purchased, the value of the now-below-market, fixed-interest payments declines, so the market value of the investment falls. On the other hand, if market rates of interest *fall* after a fixed-rate security is purchased, the fixed-interest payments become relatively attractive, and the market value of the investment rises. Assuming these price changes are not viewed as giving rise to an other-than-temporary impairment, how are they reflected in the investment account for a security classified as held-to-maturity?

Q 12–3 Does GAAP distinguish between fair values that are readily determinable from current market prices versus those needing to be calculated based on the company's own assumptions? Explain how a user will know about the reliability of the inputs used to determine fair value.

Q 12–4 When a debt investment is acquired to be held for an unspecified period of time as opposed to being held to maturity, it is reported at the fair value of the investment securities on the reporting date. Why?

Q 12–5 Reporting an investment at its fair value means adjusting its carrying amount for changes in fair value after its acquisition (or since the last reporting date if it was held at that time). Such changes are called unrealized holding gains and losses because they haven't yet been realized through the sale of the security. If the security is classified as available-for-sale, how are unrealized holding gains and losses typically reported?

Q 12–6 What is "comprehensive income"? Its composition varies from company to company but may include which items related to available-for-sale investments that are not included in net income?

Q 12–7 Why are holding gains and losses treated differently for trading securities and securities available-for-sale?

Q 12–8 Western Die-Casting Company holds an investment in unsecured bonds of LGB Heating Equipment, Inc. When the investment was acquired, management's intention was to hold the bonds for resale. Now management has the positive intent and ability to hold the bonds to maturity. How should Western account for the reclassification of the investment?

Q 12–9 Is it necessary for an investor to report individual amounts for the three categories of investments—held-to-maturity, available-for-sale, or trading—in the financial statements? What information should be disclosed about these investments?

IFRS Q 12–10 Under *IFRS No. 9,* what reporting categories are used to account for debt investments?

IFRS Q 12–11 Under *IFRS No. 9,* which reporting categories are used to account for equity investments when the investor lacks the ability to significantly influence the operations of the investee?

Q 12–12 What is the effect of a company electing the fair value option with respect to a held-to-maturity investment or an available-for-sale investment?

IFRS Q 12–13 Do U.S. GAAP and IFRS differ in the amount of flexibility that companies have in electing the fair value option? Explain.

Q 12–14 Under what circumstances is the equity method used to account for an investment in stock?

Q 12–15 The equity method has been referred to as a *one-line consolidation.* What might prompt this description?

Q 12–16 In the application of the equity method, how should dividends from the investee be accounted for? Why?

Q 12–17 The fair value of depreciable assets of Penner Packaging Company exceeds their book value by $12 million. The assets' average remaining useful life is 10 years. They are being depreciated by the straight-line method. Finest Foods Industries buys 40% of Penner's common shares. When adjusting investment revenue and the investment by the equity method, how will the situation described affect those two accounts?

Q 12–18 Superior Company owns 40% of the outstanding stock of Bernard Company. During 2021, Bernard paid a $100,000 cash dividend on its common shares. What effect did this dividend have on Superior's 2021 financial statements?

Q 12–19 Sometimes an investor's level of influence changes, making it necessary to change from the equity method to another method. How should the investor account for this change in accounting method?

IFRS Q 12–20 How does IFRS differ from U.S. GAAP with respect to using the equity method?

Q 12–21 What is the effect of a company electing the fair value option with respect to an investment that otherwise would be accounted for using the equity method?

Q 12–22 Define a financial instrument. Provide three examples of current liabilities that represent financial instruments.

Q 12–23 Some financial instruments are called derivatives. Why?

Q 12–24 (Based on Appendix 12A) Northwest Carburetor Company established a fund in 2018 to accumulate money for a new plant scheduled for construction in 2021. How should this special purpose fund be reported in Northwest's balance sheet?

Q 12–25 (Based on Appendix 12A) Whole-life insurance policies typically can be surrendered while the insured is still alive in exchange for a determinable amount of money called the *cash surrender value.* When a company buys a life insurance policy on the life of a key officer to protect the company against the untimely loss of a valuable resource in the event the officer dies, how should the company account for the cash surrender value?

Q 12–26 (Based on Appendix 12B) When market rates of interest *rise* after a fixed-rate security is purchased, the value of the now-below-market, fixed-interest payments declines, so the market value of the investment falls. How would that drop in fair value be reflected in the investment account for a security classified as HTM? Would your answer change if the drop in fair value was due to worsened financial conditions at the investee?

Q 12–27 (Based on Appendix 12B) Answer Q12–26 but assume that the investment is classified as AFS.

IFRS Q 12–28 (Based on Appendix 12B) How does IFRS differ from current U.S. GAAP with respect to accounting for impairments?

Brief Exercises

connect

BE 12–1
Securities held-to-maturity; bond investment; effective interest
● LO12–1

Lance Brothers Enterprises acquired $720,000 of 3% bonds, dated July 1, on July 1, 2021, as a long-term investment. Management has the positive intent and ability to hold the bonds until maturity. The market interest rate (yield) was 4% for bonds of similar risk and maturity. Lance Brothers paid $600,000 for the investment in bonds and will receive interest semiannually on June 30 and December 31. Prepare the journal entries (a) to record Lance Brothers' investment in the bonds on July 1, 2021, and (b) to record interest on December 31, 2021, at the effective (market) rate.

BE 12–2
Trading securities
● LO12–3

S&L Financial buys and sells securities expecting to earn profits on short-term differences in price. On December 27, 2021, S&L purchased **Coca-Cola** bonds at par for $875,000 and sold the bonds on January 3, 2022, for $880,000. At December 31, the bonds had a fair value of $873,000. What pretax amounts did S&L include in its 2021 and 2022 net income as a result of this investment (ignoring interest)?

BE 12–3
Trading securities
● LO12–3

For the **Coca-Cola** bonds described in BE 12–2, prepare journal entries to record (a) any unrealized gains or losses occurring in 2021 and (b) the sale of the bonds in 2022.

BE 12–4
Available-for-sale securities
● LO12–4

S&L Financial buys and sells securities which it classifies as available-for-sale. On December 27, 2021, S&L purchased **Coca-Cola** bonds at par for $875,000 and sold the bonds on January 3, 2022, for $880,000. At December 31, the bonds had a fair value of $873,000, and S&L has the intent and ability to hold the investment until fair value recovers. What pretax amounts did S&L include in its 2021 and 2022 net income as a result of this investment?

BE 12–5
Available-for-sale securities
● LO12–4

For the **Coca-Cola** bonds described in BE 12–4, prepare journal entries to record (a) any unrealized gains or losses occurring in 2021 and (b) the sale of the bonds in 2022, including recognition of any unrealized gains in 2022 prior to sale and reclassification of amounts out of OCI.

BE 12–6
Fair value option; available-for-sale securities
● LO12–8

S&L Financial buys and sells securities that it typically classifies as available-for-sale. On December 27, 2021, S&L purchased **Coca-Cola** bonds at par for $875,000 and sold the bonds on January 3, 2022, for $880,000. At December 31, the bonds had a fair value of $873,000. When it purchased the Coca-Cola bonds, S&L Financial decided to elect the fair value option for this investment. What pretax amounts did S&L include in its 2021 and 2022 net income as a result of this investment (ignoring interest)?

BE 12–7
Available-for-sale securities
● LO12–4

For several years Fister Links Products has held **Microsoft** bonds, considered by the company to be securities available-for-sale. The bonds were acquired at a cost of $500,000. At the end of 2021, their fair value was $610,000 and their amortized cost was $510,000. At the end of 2022, their fair value was $600,000 and their amortized cost was $520,000. At what amount will the investment be reported in the December 31, 2022, balance sheet? What adjusting entry is required to accomplish this objective (ignore interest)?

BE 12–8
Debt investments under IFRS
● LO12–4, LO12–9
IFRS

Fowler Inc. purchased $75,000 of bonds on January 1, 2021. The bonds pay interest semiannually and mature in 20 years, at which time the $75,000 principal will be paid. The bonds do not pay any amounts other than interest and principal. Fowler's intention is to collect contractual cash flows and eventually sell the bonds within the next couple of years if the price is right. During 2021, the fair value of the bonds increased to $80,000. Fowler reports investments under *IFRS No. 9*. How much unrealized gain or loss will Fowler include in 2021 net income with respect to the bonds?

BE 12–9
Debt investments under IFRS
● LO12–2, LO12–9
IFRS

Assume the same facts as in BE 12–8, but that Fowler intends to hold the bonds until maturity. How much unrealized gain or loss would Fowler include in 2021 net income with respect to the bonds?

BE 12–10
Equity securities
● LO12–5

Adams Industries holds 40,000 shares of **FedEx** common stock, which is not a large enough ownership interest to allow Adams to exercise significant influence over FedEx. On December 31, 2021, and December 31, 2022, the market value of the stock is $95 and $100 per share, respectively. What is the appropriate reporting category for this investment and at what amount will it be reported in the 2022 balance sheet?

BE 12–11
Equity investments and dividends
● LO12–5

Turner Company owns 10% of the outstanding stock of ICA Company. During the current year, ICA paid a $5 million cash dividend on its common shares. What effect did this dividend have on Turner's 2021 financial statements? Explain the reasoning for this effect.

BE 12–12
Equity method and dividends
● LO12–6

Turner Company owns 40% of the outstanding stock of ICA Company. During the current year, ICA paid a $5 million cash dividend on its common shares. What effect did this dividend have on Turner's 2021 financial statements? Explain the reasoning for this effect.

BE 12–13
Equity method
● LO12–6 LO12–7

The fair value of Wallis, Inc.'s depreciable assets exceeds their book value by $50 million. The assets have an average remaining useful life of 15 years and are being depreciated by the straight-line method. Park Industries buys 30% of Wallis's common shares. When Park adjusts its investment revenue and the investment by the equity method, how will the situation described affect those two accounts?

BE 12–14
Equity method investments
● LO12–6, LO12–9
IFRS

Kim Company bought 30% of the shares of Phelps, Inc., at the start of 2021. Kim paid $10 million for the shares. Thirty percent of the book value of Phelps's net assets is $8 million, and the difference of $2 million is due to land that Phelps owns that has appreciated in value. During 2021, Phelps reported net income of $1 million and paid a cash dividend of $0.5 million. At what amount does Kim carry the Phelps investment on its balance sheet as of December 31, 2021?

BE 12–15
Change in principle; change to the equity method
● LO12–7

At the beginning of 2021, Pioneer Products' ownership interest in the common stock of LLB Co. increased to the point that it became appropriate to begin using the equity method of accounting for the investment. The balance in the investment account was $44 million at the time of the change but would have been $56 million if Pioneer had used the equity method since first investing in LLB. How should Pioneer report the change? Would your answer be the same if Pioneer is changing *from* the equity method rather than *to* the equity method?

BE 12–16
Fair value option; equity method investments
● LO12–8

Turner Company purchased 40% of the outstanding stock of ICA Company for $10,000,000 on January 2, 2021. Turner elects the fair value option to account for the investment. During 2021, ICA reports $750,000 of net income and on December 30 pays a dividend of $500,000. On December 31, 2021, the fair value of Turner's investment has increased to $11,500,000. What journal entries would Turner make to account for this investment during 2021, assuming Turner will account for the investment using the fair value through net income approach?

BE 12–17
HTM investments and impairment (Appendix 12B)
● LO12–2, LO12–8

LED Corporation owns $1,000,000 of Branch Pharmaceuticals bonds and classifies its investment as securities held to maturity. The market price of Branch's bonds fell by $450,000, due to concerns about one of the company's principal drugs. The concerns were justified when the FDA banned the drug. LED views $200,000 of the $450,000 loss as related to *credit* losses, and the other $250,000 as *noncredit* losses. LED thinks it is more likely than not that it will have to sell the investment before fair value recovers. What journal entries should LED record to account for any credit or noncredit losses in the current period? How should the decline affect net income and comprehensive income?

BE 12–18
AFS investments and impairment (Appendix 12B)
● LO12–4, LO12–8

LED Corporation owns $1,000,000 of Branch Pharmaceuticals bonds and classifies its investment as securities available-for-sale. The market price of Branch's bonds fell by $450,000, due to concerns about one of the company's principal drugs. The concerns were justified when the FDA banned the drug. $100,000 of that decline in value already had been included in OCI as a temporary unrealized loss in a prior period. LED views $200,000 of the $450,000 loss as related to *credit* losses, and the other $250,000 as *noncredit* losses. LED thinks it is more likely than not that it will have to sell the investment before fair value recovers. What journal entries should LED record to account for any credit or noncredit losses in the current period? How should the decline affect net income and comprehensive income?

BE 12–19
AFS investments and impairment (Appendix 12B)
● LO12–4, LO12–8

Assume the same facts as in BE 12–18, but that LED does not plan to sell the investment and does not think it is more likely than not that it will have to sell the investment before fair value recovers. What journal entries should LED record to account for the decline in market value in the current period? How should the decline affect net income and comprehensive income?

BE 12–20
Recovery of impairments under IFRS (Appendix 12B)
● LO12–3, LO12–8, LO12–9

Wickum Corporation reports under IFRS, and recognized a $500,000 impairment of an HTM debt investment in Right Corporation. Subsequently, the credit loss for Wickum's investment decreased by $300,000. How would Wickum account for that change?

Exercises

E 12–1
Securities held-to-maturity; bond investment; effective interest, discount
● LO12–1

Tanner-UNF Corporation acquired as a long-term investment $240 million of 6% bonds, dated July 1, on July 1, 2021. Company management has the positive intent and ability to hold the bonds until maturity. The market interest rate (yield) was 8% for bonds of similar risk and maturity. Tanner-UNF paid $200 million for the bonds. The company will receive interest semiannually on June 30 and December 31. As a result of changing market conditions, the fair value of the bonds at December 31, 2021, was $210 million.

Required:

1. Prepare the journal entry to record Tanner-UNF's investment in the bonds on July 1, 2021.
2. Prepare the journal entries by Tanner-UNF to record interest on December 31, 2021, at the effective (market) rate.
3. At what amount will Tanner-UNF report its investment in the December 31, 2021, balance sheet? Why?
4. Suppose Moody's bond rating agency downgraded the risk rating of the bonds motivating Tanner-UNF to sell the investment on January 2, 2022, for $190 million. Prepare the journal entry to record the sale.

E 12–2
Securities held-to-maturity; bond investment; effective interest, premium
● LO12–1

Mills Corporation acquired as a long-term investment $240 million of 6% bonds, dated July 1, on July 1, 2021. Company management has the positive intent and ability to hold the bonds until maturity. The market interest rate (yield) was 4% for bonds of similar risk and maturity. Mills paid $280 million for the bonds. The company will receive interest semiannually on June 30 and December 31. As a result of changing market conditions, the fair value of the bonds at December 31, 2021, was $270 million.

Required:

1. Prepare the journal entry to record Mills' investment in the bonds on July 1, 2021.
2. Prepare the journal entry by Mills to record interest on December 31, 2021, at the effective (market) rate.

3. At what amount will Mills report its investment in the December 31, 2021, balance sheet? Why?
4. Suppose Moody's bond rating agency upgraded the risk rating of the bonds, and Mills decided to sell the investment on January 2, 2022, for $290 million. Prepare the journal entry to record the sale.

E 12–3
Securities held-to-maturity
● LO12–1

FF&T Corporation is a confectionery wholesaler that frequently buys and sells securities to meet various investment objectives. The following selected transactions relate to FF&T's investment activities during the last two months of 2021. At November 1, FF&T held $48 million of 20-year, 10% bonds of Convenience, Inc., purchased May 1, 2021, at face value. Management has the positive intent and ability to hold the bonds until maturity. FF&T's fiscal year ends on December 31.

Nov. 1	Received semiannual interest of $2.4 million from the Convenience, Inc., bonds.
Dec. 1	Purchased 12% bonds of Facsimile Enterprises at their $30 million face value, to be held until they mature in 2024. Semiannual interest is payable May 31 and November 30.
31	Purchased U.S. Treasury bills to be held until they mature in two months for $8.9 million.
31	Recorded any necessary adjusting entry(s) relating to the investments.

The fair values of the investments at December 31 were:

Convenience bonds	$44.7 million
Facsimile Enterprises bonds	30.9 million
U.S. Treasury bills	8.9 million

Required:
Prepare the appropriate journal entry for each transaction or event.

E 12–4
FASB codification research
● LO12–2

Access the FASB Accounting Standards Codification at the FASB website (www.fasb.org).

Required:
1. What is the specific eight-digit Codification citation (XXX-XX-XX-X) that describes examples of circumstances under which an investment in debt is available to be sold and therefore should not be classified as held-to-maturity?
2. List the circumstances and conditions.

E 12–5
Trading securities
● LO12–1, LO12–3

[This is a variation of E 12–1 modified to focus on trading securities.]
Tanner-UNF Corporation acquired as an investment $240 million of 6% bonds, dated July 1, on July 1, 2021. Company management is holding the bonds in its trading portfolio. The market interest rate (yield) was 8% for bonds of similar risk and maturity. Tanner-UNF paid $200 million for the bonds. The company will receive interest semiannually on June 30 and December 31. As a result of changing market conditions, the fair value of the bonds at December 31, 2021, was $210 million.

Required:
1. Prepare the journal entry to record Tanner-UNF's investment in the bonds on July 1, 2021.
2. Prepare the journal entry by Tanner-UNF to record interest on December 31, 2021, at the effective (market) rate.
3. Prepare any additional journal entry necessary for Tanner-UNF to report its investment in the December 31, 2021, balance sheet.
4. Suppose Moody's bond rating agency downgraded the risk rating of the bonds motivating Tanner-UNF to sell the investment on January 2, 2022, for $190 million. Prepare the journal entries required on the date of sale.

E 12–6
Trading securities
● LO12–1, LO12–3

[This is a variation of E 12–2 modified to focus on trading securities.]
Mills Corporation acquired as an investment $240 million of 6% bonds, dated July 1, on July 1, 2021. Company management is holding the bonds in its trading portfolio. The market interest rate (yield) was 4% for bonds of similar risk and maturity. Mills paid $280 million for the bonds. The company will receive interest semiannually on June 30 and December 31. As a result of changing market conditions, the fair value of the bonds at December 31, 2021, was $270 million.

Required:
1. Prepare the journal entry to record Mills' investment in the bonds on July 1, 2021.
2. Prepare the journal entry by Mills to record interest on December 31, 2021, at the effective (market) rate.
3. Prepare the journal entry by Mills to record any fair value adjustment necessary for the year ended December 31, 2021.
4. Suppose Moody's bond rating agency upgraded the risk rating of the bonds, and Mills decided to sell the investment on January 2, 2022, for $290 million. Prepare the journal entries required on the date of sale..

E 12–7
Various transactions relating to trading securities
● LO12–1, LO12–3

Rantzow-Lear Company buys and sells debt securities expecting to earn profits on short-term differences in price, and holds these investments in its trading portfolio. The company's fiscal year ends on December 31. The following selected transactions relating to Rantzow-Lear's trading account occurred during December 2021 and the first week of 2022.

2021	
Dec. 17	Purchased 100 Grocers' Supply Corporation bonds at par for $350,000.
28	Received interest of $2,000 from the Grocers' Supply Corporation bonds.
31	Recorded any necessary adjusting entry relating to the Grocers' Supply Corporation bonds. The market price of the bond was $4,000 per bond.
2022	
Jan. 5	Sold the Grocers' Supply Corporation bonds for $395,000.

Required:

1. Prepare the appropriate journal entry for each transaction.
2. Indicate any amounts that Rantzow-Lear Company would report in its 2021 balance sheet and income statement as a result of this investment.

E 12–8
FASB codification research
● LO12–3, LO12–4, LO12–6, LO12–7

Access the *FASB's Codification Research System* at the FASB website **www.fasb.org.**

Required:

Indicate the specific Codification citation (XXX-XX-XX-X or XXX-XX-XX-XX) for accounting for each of the following items:

1. Unrealized holding gains for trading securities should be included in earnings.
2. Under the equity method, the investor accounts for its share of the earnings or losses of the investee in the periods they are reported by the investee in its financial statements.
3. Transfers of securities between categories are accounted for at fair value.
4. Disclosures for available-for-sale securities should include total losses for securities that have net losses included in accumulated other comprehensive income.

E 12–9
Securities available-for-sale; adjusting entries
● LO12–4

Loreal-American Corporation purchased several marketable securities during 2021. At December 31, 2021, the company had the investments in bonds listed below. None was held at the last reporting date, December 31, 2020, and all are considered securities available-for-sale.

	Cost	Fair Value	Unrealized Holding Gain (Loss)
Short term:			
Blair, Inc.	$ 480,000	$ 405,000	$ (75,000)
ANC Corporation	450,000	480,000	30,000
Totals	$ 930,000	$ 885,000	$ (45,000)
Long term:			
Drake Corporation	$ 480,000	$ 560,000	$ 80,000
Aaron Industries	720,000	660,000	(60,000)
Totals	$1,200,000	$1,220,000	$ 20,000

Required:

1. Prepare appropriate adjusting entry at December 31, 2021.
2. What amounts would be reported in the income statement at December 31, 2021, as a result of the adjusting entry?

E 12–10
Available-for-sale securities
● LO12–1, LO12–4

[This is a variation of E 12–1 modified to focus on available-for-sale securities.]
Tanner-UNF Corporation acquired as a long-term investment $240 million of 6% bonds, dated July 1, on July 1, 2021. Company management has classified the bonds as an available-for-sale investment. The market interest rate (yield) was 8% for bonds of similar risk and maturity. Tanner-UNF paid $200 million for the bonds. The company will receive interest semiannually on June 30 and December 31. As a result of changing market conditions, the fair value of the bonds at December 31, 2021, was $210 million.

Required:

1. Prepare the journal entry to record Tanner-UNF's investment in the bonds on July 1, 2021.
2. Prepare the journal entry by Tanner-UNF to record interest on December 31, 2021, at the effective (market) rate.

3. Prepare any additional journal entry necessary for Tanner-UNF to report its investment in the December 31, 2021, balance sheet.
4. Suppose Moody's bond rating agency downgraded the risk rating of the bonds motivating Tanner-UNF to sell the investment on January 2, 2022, for $190 million. Prepare the journal entries necessary to record the sale, including updating the fair-value adjustment, recording any reclassification adjustment, and recording the sale.

E 12–11
Available-for-sale securities
● LO12–1, LO12–4

[This is a variation of E 12–2 focusing on available-for-sale securities.]
Mills Corporation acquired as a long-term investment $240 million of 6% bonds, dated July 1, on July 1, 2021. Company management has classified the bonds as an available-for-sale investment. The market interest rate (yield) was 4% for bonds of similar risk and maturity. Mills paid $280 million for the bonds. The company will receive interest semiannually on June 30 and December 31. As a result of changing market conditions, the fair value of the bonds at December 31, 2021, was $270 million.

Required:
1. Prepare the journal entry to record Mills' investment in the bonds on July 1, 2021.
2. Prepare the journal entry by Mills to record interest on December 31, 2021, at the effective (market) rate.
3. At what amount will Mills report its investment in the December 31, 2021, balance sheet?
4. Suppose Moody's bond rating agency upgraded the risk rating of the bonds, and Mills decided to sell the investment on January 2, 2022, for $290 million. Prepare the journal entries required on the date of sale.

E 12–12
Available-for-sale securities
● LO12–1, LO12–4

Colah Company purchased $1 million of Jackson, Inc., 5% bonds at par on July 1, 2021, with interest paid semi-annually. Colah determined that it should account for the bonds as an available-for-sale investment. At December 31, 2021, the Jackson bonds had a fair value of $1.2 million. Colah sold the Jackson bonds on July 1, 2022 for $900,000.

Required:
1. Prepare Colah's journal entries to record
   a. The purchase of the Jackson bonds on July 1.
   b. Interest revenue for the last half of 2021.
   c. Any year-end 2021 adjusting entries.
   d. Interest revenue for the first half of 2022.
   e. Any entries necessary upon sale of the Jackson bonds on July 1, 2022, including updating the fair-value adjustment, recording any reclassification adjustment, and recording the sale.
2. Fill out the following table to show the effect of the Jackson bonds on Colah's net income, other comprehensive income, and comprehensive income for 2021, 2022, and cumulatively over 2021 and 2022.

	2021	2022	Total
Net Income			
OCI			
Comprehensive Income			

E 12–13
Classification of securities; adjusting entries
● LO12–4

On February 18, 2021, Union Corporation purchased 600 IBM bonds as a long-term investment at their face value for a total of $600,000. Union will hold the bonds indefinitely, and may sell them if their price increases sufficiently. On December 31, 2021, and December 31, 2022, the market value of the bonds was $580,000 and $610,000, respectively.

Required:
1. What is the appropriate reporting category for this investment? Why?
2. Prepare the adjusting entry for December 31, 2021.
3. Prepare the adjusting entry for December 31, 2022.

E 12–14
Various investment securities
● LO12–2, LO12–3, LO12–4

At December 31, 2021, Hull-Meyers Corp. had the following investments that were purchased during 2021, its first year of operations:

	Amortized Cost	Fair Value
Trading Securities:		
Security A	$ 900,000	$ 910,000
Security B	105,000	100,000
Totals	$1,005,000	$1,010,000

(continued)

(concluded)

Securities Available-for-Sale:		
Security C	$ 700,000	$ 780,000
Security D	900,000	915,000
Totals	$1,600,000	$1,695,000
Securities to Be Held-to-Maturity:		
Security E	$ 490,000	$ 500,000
Security F	615,000	610,000
Totals	$1,105,000	$1,110,000

No investments were sold during 2021. All securities except Security D and Security F are considered short-term investments. None of the fair value changes is considered permanent.

Required:

Determine the following amounts at December 31, 2021.

1. Investments reported as current assets.
2. Investments reported as noncurrent assets.
3. Unrealized gain (or loss) recognized in net income.
4. Unrealized gain (or loss) in accumulated other comprehensive income in shareholders' equity.

E 12–15
Equity investments; fair value through net income
● LO12–5

On March 31, 2021, Chow Brothers, Inc., bought 10% of KT Manufacturing's capital stock for $50 million. KT's net income for the year ended December 31, 2021, was $80 million. The fair value of the shares held by Chow was $35 million at December 31, 2021. KT did not declare or pay a dividend during 2021.

Required:

1. Prepare all appropriate journal entries related to the investment during 2021.
2. Assume that Chow sold the stock on January 20, 2022, for $30 million. Prepare the journal entry Chow would use to record the sale.

E 12–16
Equity investments; fair value through net income
● LO12–5

On January 2, 2021, Sanborn Tobacco Inc. bought 5% of Jackson Industry's capital stock for $90 million. Jackson Industry's net income for the year ended December 31, 2021, was $120 million. The fair value of the shares held by Sanborn was $98 million at December 31, 2021. During 2021, Jackson declared a dividend of $60 million.

Required:

1. Prepare all appropriate journal entries related to the investment during 2021.
2. Assume that Sanborn sold the stock on January 2, 2022 for $110 million. Prepare the journal entry Sanborn would use to record the sale.

E 12–17
Equity investments; fair value through net income
● LO12–5

The accounting records of Jamaican Importers, Inc., at January 1, 2021, included the following:

Assets:	
Investment in IBM common shares	$1,345,000
Less: Fair value adjustment	(145,000)
	$1,200,000

No changes occurred during 2021 in the investment portfolio.

Required:

Prepare appropriate adjusting entry(s) at December 31, 2021, assuming the fair value of the IBM common shares was

1. $1,175,000
2. $1,275,000
3. $1,375,000

E 12–18
Equity investments; fair value through net income
● LO12–5

The investments of Harlon Enterprises included the following cost and fair value amounts ($ in millions):

		Fair Value, Dec. 31	
Equity Investments	**Cost**	**2021**	**2022**
A Corporation shares	$ 20	$ 14	na
B Corporation shares	35	35	$ 37
C Corporation shares	15	na	14
D Industries shares	45	46	50
Totals	$115	$ 95	$101

Harlon accounts for its equity investment portfolio at fair value through net income. Harlon sold its holdings of A Corporation shares on June 1, 2022, for $15 million. On September 12, it purchased the C Corporation shares.

Required:

1. What is the effect of the sale of the A Corporation shares and the purchase of the C Corporation shares on Harlon's 2022 pretax earnings?
2. At what amount should Harlon's securities equity investment portfolio be reported in its 2022 balance sheet?

E 12–19
Investment securities and equity method investments compared
● LO12–5, LO12–6

As a long-term investment, Painters' Equipment Company purchased 20% of AMC Supplies Inc.'s 400,000 shares for $480,000 at the beginning of the fiscal year of both companies. On the purchase date, the fair value and book value of AMC's net assets were equal. During the year, AMC earned net income of $250,000 and distributed cash dividends of 25 cents per share. At year-end, the fair value of the shares is $505,000.

Required:

1. Assume no significant influence was acquired. Prepare the appropriate journal entries from the purchase through the end of the year.
2. Assume significant influence was acquired. Prepare the appropriate journal entries from the purchase through the end of the year.

E 12–20
Equity method; purchase; investee income; dividends
● LO12–6

As a long-term investment at the beginning of the 2021 fiscal year, Florists International purchased 30% of Nursery Supplies Inc.'s 8 million shares for $56 million. The fair value and book value of the shares were the same at that time. During the year, Nursery Supplies earned net income of $40 million and distributed cash dividends of $1.25 per share. At the end of the year, the fair value of the shares is $52 million.

Required:

Prepare the appropriate journal entries from the purchase through the end of the year.

E 12–21
Error corrections; equity method investment
● LO12–6, LO12–7

On December 12, 2021, an equity investment costing $80,000 was sold for $100,000. The investment was carried in the balance sheet at $75,000, and was accounted for under the equity method. An error was made in which the total of the sale proceeds was credited to the investment account.

Required:

1. Prepare the journal entry to correct the error assuming it is discovered before the books are adjusted or closed in 2021. (Ignore income taxes.)
2. Prepare the journal entry to correct the error assuming it is not discovered until early 2022. (Ignore income taxes.)

E 12–22
Equity method; adjustment for depreciation
● LO12–6, LO12–7

Fizer Pharmaceutical paid $68 million on January 2, 2021, for 4 million shares of Carne Cosmetics common stock. The investment represents a 25% interest in the net assets of Carne and gave Fizer the ability to exercise significant influence over Carne's operations. Fizer received dividends of $1 per share on December 21, 2021, and Carne reported net income of $40 million for the year ended December 31, 2021. The fair value of Carne's common stock at December 31, 2021, was $18.50 per share.

- The book value of Carne's net assets was $192 million.
- The fair value of Carne's depreciable assets exceeded their book value by $32 million. These assets had an average remaining useful life of eight years.
- The remainder of the excess of the cost of the investment over the book value of net assets purchased was attributable to goodwill.

Required:

Prepare all appropriate journal entries related to the investment during 2021.

E 12–23
Equity method
● LO12–6, LO12–7

On January 1, 2021, Cameron Inc. bought 20% of the outstanding common stock of Lake Construction Company for $300 million cash, giving Cameron the ability to exercise significant influence over Lake's operations. At the date of acquisition of the stock, Lake's net assets had a fair value of $900 million. Its book value was $800 million. The difference was attributable to the fair value of Lake's buildings and its land exceeding book value, each accounting for one-half of the difference. Lake's net income for the year ended December 31, 2021, was $150 million. During 2021, Lake declared and paid cash dividends of $30 million. The buildings have a remaining life of 10 years.

Required:

1. Prepare all appropriate journal entries related to the investment during 2021, assuming Cameron accounts for this investment by the equity method.
2. Determine the amounts to be reported by Cameron:
    a. As an investment in Cameron's 2021 balance sheet.

b. As net investment revenue in the income statement.
c. Among investing activities in the statement of cash flows.

E 12–24
Equity method, partial year
● LO12–6, LO12–7

On July 1, 2021, Gupta Corporation bought 25% of the outstanding common stock of VB Company for $100 million cash, giving Gupta the ability to exercise significant influence over VB's operations. At the date of acquisition of the stock, VB's net assets had a total fair value of $350 million and a book value of $220 million. Of the $130 million difference, $20 million was attributable to the appreciated value of inventory that was sold during the last half of 2021, $80 million was attributable to buildings that had a remaining depreciable life of 10 years, and $30 million related to equipment that had a remaining depreciable life of 5 years. Between July 1, 2021, and December 31, 2021, VB earned net income of $32 million and declared and paid cash dividends of $24 million.

Required:
1. Prepare all appropriate journal entries related to the investment during 2021, assuming Gupta accounts for this investment by the equity method.
2. Determine the amounts to be reported by Gupta:
   a. As an investment in Gupta's December 31, 2021, balance sheet.
   b. As net investment revenue or loss in Gupta's 2021 income statement.
   c. Among investing activities in Gupta's 2021 statement of cash flows.

E 12–25
Fair value option; held-to-maturity investments
● LO12–1, LO12–2, LO12–3, LO12–8

[This is a variation of E 12–1 modified to focus on the fair value option.]
Tanner-UNF Corporation acquired as a long-term investment $240 million of 6% bonds, dated July 1, on July 1, 2021. Company management has the positive intent and ability to hold the bonds until maturity, but when the bonds were acquired Tanner-UNF decided to elect the fair value option for accounting for its investment. The market interest rate (yield) was 8% for bonds of similar risk and maturity. Tanner-UNF paid $200 million for the bonds. The company will receive interest semiannually on June 30 and December 31. As a result of changing market conditions, the fair value of the bonds at December 31, 2021, was $210 million.

Required:
1. Would this investment be classified on Tanner-UNF's balance sheet as held-to-maturity securities, trading securities, available-for-sale securities, significant-influence investments, or other? Would it be reported at amortized cost or fair value?
2. Prepare the journal entry to record Tanner-UNF's investment in the bonds on July 1, 2021.
3. Prepare the journal entries by Tanner-UNF to record interest on December 31, 2021, at the effective (market) rate.
4. Prepare any journal entry necessary to recognize fair value changes as of December 31, 2021.
5. At what amount will Tanner-UNF report its investment in the December 31, 2021, balance sheet? Why?
6. Suppose Moody's bond rating agency downgraded the risk rating of the bonds motivating Tanner-UNF to sell the investment on January 2, 2022, for $190 million. Prepare the journal entries required on the date of sale.

E 12–26
Fair value option; available-for-sale investments
● LO12–2, LO12–3, LO12–8

[This is a variation of E 12–12 modified to focus on the fair value option.]
Colah Company purchased $1 million of Jackson, Inc., 5% bonds at their face amount on July 1, 2021, with interest paid semi-annually. The bonds mature in 20 years but Colah planned to keep them for less than 3 years, and classified them as available for sale investments. When the bonds were acquired Colah decided to elect the fair value option for accounting for its investment. At December 31, 2021, the Jackson bonds had a fair value of $1.2 million. Colah sold the Jackson bonds on July 1, 2022 for $900,000.

Required:
1. Prepare Colah's journal entries to record
   a. The purchase of the Jackson bonds on July 1.
   b. Interest revenue for the last half of 2021.
   c. Any year-end 2021 adjusting entries.
   d. Interest revenue for the first half of 2022.
   e. Any entry or entries necessary upon sale of the Jackson bonds on July 1, 2022.
2. Fill out the following table to show the effect of the Jackson bonds on Colah's net income, other comprehensive income, and comprehensive income for 2021, 2022, and cumulatively over 2021 and 2022.

	2021	2022	Total
Net Income			
OCI			
Comprehensive Income			

E 12–27
Fair value option; equity method investments
● LO12–5, LO12–8

[This is a variation of E 12–20 modified to focus on the fair value option.]
As a long-term investment at the beginning of the 2021 fiscal year, Florists International purchased 30% of Nursery Supplies Inc.'s 8 million shares of capital stock for $56 million. The fair value and book value of the shares were the same at that time. The company realizes that this investment typically would be accounted for under the equity method, but instead chooses to measure the investment at fair value. During the year, Nursery Supplies reported net income of $40 million and distributed cash dividends of $1.25 per share. At the end of the year, the fair value of the shares is $52 million.

Required:

1. Would this investment be classified on Florists' balance sheet as held-to-maturity securities, trading securities, available-for-sale securities, significant influence investments, or other? Explain.
2. Prepare all appropriate journal entries related to the investment during 2021, under the fair value option, and in a manner similar to what Florists would use for investments in equity securities for which it does not have significant influence.
3. Indicate the effect of this investment on 2021 income before taxes.

E 12–28
Life insurance policy (Appendix 12A)

Edible Chemicals Corporation owns a $4 million whole life insurance policy on the life of its CEO, naming Edible Chemicals as beneficiary. The annual premiums are $70,000 and are payable at the beginning of each year. The cash surrender value of the policy was $21,000 at the beginning of 2021.

Required:

1. Prepare the appropriate 2021 journal entry to record insurance expense and the increase in the investment assuming the cash surrender value of the policy increased according to the contract to $27,000.
2. The CEO died at the end of 2021. Prepare the appropriate journal entry.

E 12–29
Life insurance policy (Appendix 12A)

Below are two unrelated situations relating to life insurance.

Required:

Prepare the appropriate journal entry for each situation.

1. Ford Corporation owns a whole life insurance policy on the life of its president. Ford Corporation is the beneficiary. The insurance premium is $25,000. The cash surrender value increased during the year from $2,500 to $4,600.
2. Petroleum Corporation received a $250,000 life insurance settlement when its CEO died. At that time, the cash surrender value was $16,000.

E 12–30
Held-to-maturity securities; impairments (Appendix 12B)
● LO12–2, LO12–8

Bloom Corporation purchased $1,000,000 of Taylor Company 5% bonds, at their face amount, with the intent and ability to hold the bonds until they matured in 2025, so Bloom classifies its investment as HTM. Unfortunately, a combination of problems at Taylor Company and in the debt securities market caused the fair value of the Taylor investment to decline to $600,000 during 2021.

Required:

For each of the following scenarios, prepare the appropriate entry(s) at December 31, 2021, and indicate how the scenario will affect the 2021 income statement (ignoring income taxes).

1. Bloom now believes it is more likely than not that it will have to sell the Taylor bonds before the bonds have a chance to recover their fair value. Of the $400,000 decline in fair value, Bloom attributes $250,000 to credit losses, and $150,000 to noncredit losses.
2. Bloom does not plan to sell the Taylor bonds prior to maturity, and does not believe it is more likely than not that it will have to sell the Taylor bonds before the bonds have a chance to recover their fair value. Of the $400,000 decline in fair value, Bloom attributes $250,000 to credit losses, and $150,000 to noncredit losses.

E 12–31
Available-for-sale debt securities; impairments (Appendix 12B)
● LO12–4, LO12–8

Assume all of the same facts and scenarios as E 12–30, except that Bloom Corporation classifies its Taylor investment as AFS.

Required:

1. For each of the scenarios shown in E 12–30, prepare the appropriate entry(s) at December 31, 2021. Indicate how the scenario will affect the 2021 income statement, OCI, and comprehensive income.
2. Repeat requirement 1, but now assume that, at the end of 2020, Bloom had recorded an unrealized loss of $100,000 on the Taylor investment.

E 12–32
Accounting for impairments under IFRS (Appendix 12B)
● LO12–2, LO12–8, LO12–9
IFRS

Rell Corporation reports under IFRS No. 9. Rell has an investment in Tirish, Inc. bonds that Rell accounts for at amortized cost, given that the bonds pay only interest and principal and Rell's business purpose is to hold the bonds to maturity. Rell purchased the bonds for €10,000,000. As of December 31, 2021, Rell calculates €750,000 of credit losses expected for default events occurring during 2022 and €450,000 of credit losses expected for default events occurring after 2022.

Required:

1. Assume the Tirish bonds have not had a significant increase in credit risk. Prepare the journal entry to record any impairment loss as of December 31, 2021.

2. Assume the Tirish bonds have had a significant increase in credit risk. Prepare the journal entry to record any impairment loss as of December 31, 2021.
3. Assume the Tirish bonds have not had a significant increase in credit risk, and that as of December 31, 2022, Rell calculates €650,000 of credit losses expected for default events occurring during 2023 and €350,000 of credit losses expected for default events occurring after 2023. Prepare the journal entry Rell would make with respect to any impairment loss as of December 31, 2022.

Problems

connect

P 12–1
Securities held-to-maturity; bond investment; effective interest
● LO12–1, LO12–2

Fuzzy Monkey Technologies, Inc., purchased as a long-term investment $80 million of 8% bonds, dated January 1, on January 1, 2021. Management has the positive intent and ability to hold the bonds until maturity. For bonds of similar risk and maturity the market yield was 10%. The price paid for the bonds was $66 million. Interest is received semiannually on June 30 and December 31. Due to changing market conditions, the fair value of the bonds at December 31, 2021, was $70 million.

Required:

1. Prepare the journal entry to record Fuzzy Monkey's investment on January 1, 2021.
2. Prepare the journal entry by Fuzzy Monkey to record interest on June 30, 2021 (at the effective rate).
3. Prepare the journal entry by Fuzzy Monkey to record interest on December 31, 2021 (at the effective rate).
4. At what amount will Fuzzy Monkey report its investment in the December 31, 2021 balance sheet? Why?
5. How would Fuzzy Monkey's 2021 statement of cash flows be affected by this investment? (If more than one approach is possible, indicate the one that is most likely.)

P 12–2
Trading securities; bond investment; effective interest
● LO12–1, LO12–3

[This problem is a variation of P 12–1, modified to categorize the investment as trading securities.]

Fuzzy Monkey Technologies, Inc., purchased as a short-term investment $80 million of 8% bonds, dated January 1, on January 1, 2021. Management intends to include the investment in a short-term, active trading portfolio. For bonds of similar risk and maturity the market yield was 10%. The price paid for the bonds was $66 million. Interest is received semiannually on June 30 and December 31. Due to changing market conditions, the fair value of the bonds at December 31, 2021, was $70 million.

Required:

1. Prepare the journal entry to record Fuzzy Monkey's investment on January 1, 2021.
2. Prepare the journal entry by Fuzzy Monkey to record interest on June 30, 2021 (at the effective rate).
3. Prepare the journal entry by Fuzzy Monkey to record interest on December 31, 2021 (at the effective rate).
4. At what amount will Fuzzy Monkey report its investment in the December 31, 2021 balance sheet? Why? Prepare any entry necessary to achieve this reporting objective.
5. How would Fuzzy Monkey's 2021 statement of cash flows be affected by this investment? (If more than one approach is possible, indicate the one that is most likely.)

P 12–3
Securities available-for-sale; bond investment; effective interest
● LO 12–1, LO12–4

(Note: This problem is a variation of P 12–1, modified to categorize the investment as securities available-for-sale.)

Fuzzy Monkey Technologies, Inc., purchased as a long-term investment $80 million of 8% bonds, dated January 1, on January 1, 2021. Management intends to have the investment available for sale when circumstances warrant. For bonds of similar risk and maturity the market yield was 10%. The price paid for the bonds was $66 million. Interest is received semiannually on June 30 and December 31. Due to changing market conditions, the fair value of the bonds at December 31, 2021, was $70 million.

Required:

1. Prepare the journal entry to record Fuzzy Monkey's investment on January 1, 2021.
2. Prepare the journal entry by Fuzzy Monkey to record interest on June 30, 2021 (at the effective rate).
3. Prepare the journal entries by Fuzzy Monkey to record interest on December 31, 2021 (at the effective rate).
4. At what amount will Fuzzy Monkey report its investment in the December 31, 2021, balance sheet? Why? Prepare any entry necessary to achieve this reporting objective.
5. How would Fuzzy Monkey's 2021 statement of cash flows be affected by this investment? (If more than one approach is possible, indicate the one that is most likely.)

P 12–4
Fair value option; bond investment; effective interest
● LO12–1, LO12–2, LO12–3, LO12–4, LO12–8

[This problem is a variation of P 12–3, modified to cause the investment to be accounted for under the fair value option.]

Fuzzy Monkey Technologies, Inc., purchased as a long-term investment $80 million of 8% bonds, dated January 1, on January 1, 2021. Management intends to have the investment available for sale when circumstances warrant. When the company purchased the bonds, management elected to account for them under the fair value option. For bonds of similar risk and maturity the market yield was 10%. The price paid for the bonds was $66 million. Interest is received semiannually on June 30 and December 31. Due to changing market conditions, the fair value of the bonds at December 31, 2021, was $70 million.

Required:

1. Prepare the journal entry to record Fuzzy Monkey's investment on January 1, 2021.
2. Prepare the journal entry by Fuzzy Monkey to record interest on June 30, 2021 (at the effective rate).
3. Prepare the journal entries by Fuzzy Monkey to record interest on December 31, 2021 (at the effective rate).
4. At what amount will Fuzzy Monkey report its investment in the December 31, 2021, balance sheet? Why? Prepare any entry necessary to achieve this reporting objective.
5. How would Fuzzy Monkey's 2021 statement of cash flows be affected by this investment? (If more than one approach is possible, indicate the one that is most likely.)
6. How would your answers to requirements 1–5 differ if management had the intent and ability to hold the investments until maturity?

P 12–5
Various transactions related to trading securities
● LO12–1, LO12–3

The following selected transactions relate to investment activities of Ornamental Insulation Corporation during 2021. The company buys debt securities, intending to profit from short-term differences in price and maintaining them in an active trading portfolio. Ornamental's fiscal year ends on December 31. No investments were held by Ornamental on December 31, 2020.

Mar. 31	Acquired 8% Distribution Transformers Corporation bonds costing $400,000 at face value.
Sep. 1	Acquired $900,000 of American Instruments' 10% bonds at face value.
Sep. 30	Received semiannual interest payment on the Distribution Transformers bonds.
Oct. 2	Sold the Distribution Transformers bonds for $425,000.
Nov. 1	Purchased $1,400,000 of M&D Corporation 6% bonds at face value.
Dec. 31	Recorded any necessary adjusting entry(s) relating to the investments. The market prices of the investments are

American Instruments bonds	$ 850,000
M&D Corporation bonds	$1,460,000

(*Hint:* Interest must be accrued.)

Required:

1. Prepare the appropriate journal entry for each transaction or event during 2021, as well as any adjusting entries necessary at year end.
2. Indicate any amounts that Ornamental Insulation would report in its 2021 income statement, 2021 statement of comprehensive income, and 12/31/2021 balance sheet as a result of these investments. Include totals for net income, comprehensive income, and retained earnings as a result of these investments.

P 12–6
Various transactions related to securities available-for-sale
● LO12–1, LO12–4

(Note: This problem is a variation of P 12–5, modified to categorize the investments as securities available-for-sale.)

The following selected transactions relate to investment activities of Ornamental Insulation Corporation during 2021. The company buys debt securities, *not* intending to profit from short-term differences in price and *not* necessarily to hold debt securities to maturity, but to have them available for sale in years when circumstances warrant. Ornamental's fiscal year ends on December 31. No investments were held by Ornamental on December 31, 2020.

Mar. 31	Acquired 8% Distribution Transformers Corporation bonds costing $400,000 at face value.
Sep. 1	Acquired $900,000 of American Instruments' 10% bonds at face value.
Sep. 30	Received semiannual interest payment on the Distribution Transformers bonds.
Oct. 2	Sold the Distribution Transformers bonds for $425,000.
Nov. 1	Purchased $1,400,000 of M&D Corporation 6% bonds costing at face value.
Dec. 31	Recorded any necessary adjusting entry(s) relating to the investments. The market prices of the investments are:

American Instruments bonds	$ 850,000
M&D Corporation bonds	$1,460,000

(*Hint:* Interest must be accrued.)

Required:

1. Prepare the appropriate journal entry for each transaction or event during 2021, as well as any adjusting entries necessary at year-end. For any sales, prepare entries to update the fair-value adjustment, record any reclassification adjustment, and record the sale.
2. Indicate any amounts that Ornamental Insulation would report in its 2021 income statement, 2021 statement of comprehensive income, and 12/31/2021 balance sheet as a result of these investments. Include totals for net income, comprehensive income, and retained earnings as a result of these investments.

P 12–7
Various transactions related to equity investments: fair value through net income
● LO12–5

(Note: This problem is a variation of P 12–5, modified to consider equity investments.)

The following selected transactions relate to investment activities of Ornamental Insulation Corporation during 2021. The company buys equity securities as noncurrent investments. None of Ornamental's investments are large enough to exert significant influence on the investee. Ornamental's fiscal year ends on December 31. No investments were held by Ornamental on December 31, 2020.

Mar. 31	Acquired Distribution Transformers Corporation common stock for $400,000.	
Sep. 1	Acquired $900,000 of American Instruments' common stock.	
Sep. 30	Received a $16,000 dividend on the Distribution Transformers common stock.	
Oct. 2	Sold the Distribution Transformers common stock for $425,000.	
Nov. 1	Purchased $1,400,000 of M&D Corporation common stock.	
Dec. 31	Recorded any necessary adjusting entry(s) relating to the investments. The market prices of the investments are:	
	American Instruments common stock	$ 850,000
	M&D Corporation common stock	$1,460,000

Required:

1. Prepare the appropriate journal entry for each transaction or event during 2021, as well as any adjusting entry necessary at year-end.
2. Indicate any amounts that Ornamental Insulation would report in its 2021 income statement, 2021 statement of comprehensive income, and 12/31/2021 balance sheet as a result of these investments. Include totals for net income, comprehensive income, and retained earnings as a result of these investments.

P 12–8
Various transactions relating to trading securities and equity investments
● LO12–1, LO12–3, LO12–5

American Surety and Fidelity buys and sells securities expecting to earn profits on short-term differences in price. For the first 11 months of 2021, gains from selling trading securities totaled $8 million, losses from selling trading securities were $11 million, and the company had earned $5 million in interest revenue. The following selected transactions relate to American's investments in trading securities and equity securities during December 2021, and the first week of 2022. The company's fiscal year ends on December 31. No trading securities or equity investments were held by American on December 1, 2021.

2021	
Dec. 12	Purchased FF&G Corporation bonds for $12 million.
13	Purchased 2 million shares of Ferry Intercommunications common stock for $22 million. American does not have significant influence over Ferry's operations or policies.
15	Sold the FF&G Corporation bonds for $12.1 million.
22	Purchased U.S. Treasury bills for $56 million and Treasury bonds for $65 million.
23	Sold half the shares of Ferry Intercommunications common stock for $10 million.
26	Sold the U.S. Treasury bills for $57 million.
27	Sold the Treasury bonds for $63 million.
28	Received cash dividends of $200,000 from the Ferry Intercommunications common stock shares.
31	Recorded any necessary adjusting entry relating to the remaining investment. The market price of the Ferry Intercommunications stock was $10 per share.
2022	
Jan. 2	Sold the remaining Ferry Intercommunications common stock shares for $10.2 million.
5	Purchased Warehouse Designs Corporation bonds for $34 million.

Required:

1. Prepare the appropriate journal entry for each transaction or event during 2021 including any year-end adjusting entries.
2. Indicate any amounts that American would report in its 2021 balance sheet and income statement as a result of these investments.
3. Prepare the appropriate journal entry for each transaction or event during 2022.

P 12–9
Securities held-to-maturity; trading securities and equity investments
● LO12–1, LO12–2, LO12–3, LO12–5

Amalgamated General Corporation is a consulting firm that also offers financial services through its credit division. From time to time the company buys and sells securities. The following selected transactions relate to Amalgamated's investment activities during the last quarter of 2021 and the first month of 2022. The only securities held by Amalgamated at October 1, 2021 were $30 million of 10% bonds of Kansas Abstractors, Inc., purchased on May 1, 2021 at face value and held in Amalgamated's trading securities portfolio. The company's fiscal year ends on December 31.

2021	
Oct. 18	Purchased 2 million shares of Millwork Ventures Company common stock for $58 million. Millwork has a total of 30 million shares issued.
31	Received semiannual interest of $1.5 million from the Kansas Abstractors bonds.
Nov. 1	Purchased 10% bonds of Holistic Entertainment Enterprises at their $18 million face value, to be held until they mature in 2031. Semiannual interest is payable April 30 and October 31.
1	Sold the Kansas Abstractors bonds for $28 million because rising interest rates are expected to cause their fair value to continue to fall. No unrealized gains and losses had been recorded on these bonds previously.
Dec. 1	Purchased 12% bonds of Household Plastics Corporation at their $60 million face value, to be held until they mature in 2031. Semiannual interest is payable May 31 and November 30.
20	Purchased U. S. Treasury bonds for $5.6 million as trading securities, hoping to earn profits on short-term differences in prices.
21	Purchased 4 million shares of NXS Corporation's 50 million shares of common stock for $44 million, planning to hold these shares until market conditions encourage their sale.
23	Sold the Treasury bonds for $5.7 million.
29	Received cash dividends of $3 million from the Millwork Ventures Company shares of common stock.
31	Recorded any necessary adjusting entries relating to the investments. The market price of the Millwork Ventures Company common stock was $27.50 per share and $11.50 per share for the NXS Corporation common stock. The fair values of the bond investments were $58.7 million for Household Plastics Corporation and $16.7 million for Holistic Entertainment Enterprises.
2022	
Jan. 7	Sold the NXS Corporation common stock shares for $43 million.

Required:
Prepare the appropriate journal entry for each transaction or event. Use one summary entry on December 31 to adjust the portfolio of equity investments to fair value.

P 12–10
Investment securities and equity method investments compared
● LO12–6, LO12–7

On January 4, 2021, Runyan Bakery paid $324 million for 10 million shares of Lavery Labeling Company common stock. The investment represents a 30% interest in the net assets of Lavery and gave Runyan the ability to exercise significant influence over Lavery's operations. Runyan received dividends of $2.00 per share on December 15, 2021, and Lavery reported net income of $160 million for the year ended December 31, 2021. The market value of Lavery's common stock at December 31, 2021, was $31 per share. On the purchase date, the book value of Lavery's identifiable net assets was $800 million and:

a. The fair value of Lavery's depreciable assets, with an average remaining useful life of six years, exceeded their book value by $80 million.
b. The remainder of the excess of the cost of the investment over the book value of net assets purchased was attributable to goodwill.

Required:
1. Prepare all appropriate journal entries related to the investment during 2021, assuming Runyan accounts for this investment by the equity method.
2. Prepare the journal entries required by Runyan, assuming that the 10 million shares represent a 10% interest in the net assets of Lavery rather than a 30% interest.

P 12–11
Fair value option; equity method investments
● LO12–5, LO12–8

[This problem is a variation of P 12–10 focusing on the fair value option.]

On January 4, 2021, Runyan Bakery paid $324 million for 10 million shares of Lavery Labeling Company common stock. The investment represents a 30% interest in the net assets of Lavery and gave Runyan the ability to exercise significant influence over Lavery's operations. Runyan chose the fair value option to account for this investment. Runyan received dividends of $2.00 per share on December 15, 2021, and Lavery reported net income of $160 million for the year ended December 31, 2021. The market value of Lavery's common stock at December 31, 2021, was $31 per share. On the purchase date, the book value of Lavery's identifiable net assets was $800 million and:

a. The fair value of Lavery's depreciable assets, with an average remaining useful life of six years, exceeded their book value by $80 million.
b. The remainder of the excess of the cost of the investment over the book value of net assets purchased was attributable to goodwill.

Required:

Assuming Runyan accounts for this investment under the fair value option, prepare all appropriate journal entries in a manner similar to accounting for securities for which there is not significant influence.

P 12–12
Fair value option; equity method investments
● LO12–5, LO12–6, LO12–7, LO12–8

[This problem is an expanded version of P 12–11 that considers alternative ways in which a firm might apply the fair value option to account for significant-influence investments that would normally be accounted for under the equity method.]

Companies can choose the fair value option for investments that otherwise would be accounted for under the equity method. If the fair value option is chosen, the investment is shown at fair value in the balance sheet, and unrealized holding gains and losses are recognized in the income statement. However, exactly how a company complies with those broad requirements is up to the company. This problem requires you to consider alternative ways in which a company might apply the fair value option for investments that otherwise would be accounted for under the equity method.

On January 4, 2021, Runyan Bakery paid $324 million for 10 million shares of Lavery Labeling Company common stock. The investment represents a 30% interest in the net assets of Lavery and gave Runyan the ability to exercise significant influence over Lavery's operations. Runyan chose the fair value option to account for this investment. Runyan received dividends of $2.00 per share on December 15, 2021, and Lavery reported net income of $160 million for the year ended December 31, 2021. The market value of Lavery's common stock at December 31, 2021, was $31 per share. On the purchase date, the book value of Lavery's identifiable net assets was $800 million and:

a. The fair value of Lavery's depreciable assets, with an average remaining useful life of six years, exceeded their book value by $80 million.
b. The remainder of the excess of the cost of the investment over the book value of net assets purchased was attributable to goodwill.

Required:

1. Prepare all appropriate journal entries related to the investment during 2021, assuming Runyan accounts for this investment under the fair value option, and accounts for the Lavery investment in a manner similar to what it would use for securities for which there is not significant influence. Indicate the effect of these journal entries on 2021 net income, and indicate the amount at which the investment is carried in the December 31, 2021, balance sheet.
2. Prepare all appropriate journal entries related to the investment during 2021, assuming Runyan accounts for this investment under the fair value option, but uses equity method accounting to account for Lavery's income and dividends, and then records a fair value adjustment at the end of the year that allows it to comply with GAAP. Indicate the effect of these journal entries on 2021 net income, and indicate the amount at which the investment is carried in the December 31, 2021, balance sheet. (Note: You should end up with the same total 2021 income effect and same carrying value on the balance sheet for requirements 1 and 2.)

P 12–13
Equity method
● LO12–6, LO12–7

Northwest Paperboard Company, a paper and allied products manufacturer, was seeking to gain a foothold in Canada. Toward that end, the company bought 40% of the outstanding common shares of Vancouver Timber and Milling, Inc., on January 2, 2021, for $400 million.

At the date of purchase, the book value of Vancouver's net assets was $775 million. The book values and fair values for all balance sheet items were the same except for inventory and plant facilities. The fair value exceeded book value by $5 million for the inventory and by $20 million for the plant facilities.

The estimated useful life of the plant facilities is 16 years. All inventory acquired was sold during 2021.

Vancouver reported net income of $140 million for the year ended December 31, 2021. Vancouver paid a cash dividend of $30 million.

Required:

1. Prepare all appropriate journal entries related to the investment during 2021.
2. What amount should Northwest report as its income from its investment in Vancouver for the year ended December 31, 2021?
3. What amount should Northwest report in its balance sheet as its investment in Vancouver?
4. What should Northwest report in its statement of cash flows regarding its investment in Vancouver?

P 12–14
Equity method
● LO12–6, LO12–7

On January 2, 2021, Miller Properties paid $19 million for 1 million shares of Marlon Company's 6 million outstanding common shares. Miller's CEO became a member of Marlon's board of directors during the first quarter of 2021.

The carrying amount of Marlon's net assets was $66 million. Miller estimated the fair value of those net assets to be the same except for a patent valued at $24 million above cost. The remaining amortization period for the patent is 10 years.

Marlon reported earnings of $12 million and paid dividends of $6 million during 2021. On December 31, 2021, Marlon's common stock was trading on the NYSE at $18.50 per share.

Required:

1. When considering whether to account for its investment in Marlon under the equity method, what criteria should Miller's management apply?
2. Assume Miller accounts for its investment in Marlon using the equity method. Ignoring income taxes, determine the amounts related to the investment to be reported in its 2021
   a. Income statement.
   b. Balance sheet.
   c. Statement of cash flows.

P 12–15
Classifying investments
● LO12–2 through LO12–6

Indicate (by letter) the way each of the investments listed below most likely should be accounted for based on the information provided.

Item	Reporting Category
_____ 1. 35% of the nonvoting preferred stock of American Aircraft Company.	T. Trading securities
_____ 2. Treasury bills to be held to maturity.	M. Securities held-to-maturity
_____ 3. Two-year note receivable from affiliate.	A. Securities available-for-sale
_____ 4. Accounts receivable.	F. Fair value through net income
_____ 5. Treasury bond maturing in one week.	E. Equity method
_____ 6. Common stock held in an investment account for immediate resale.	C. Consolidation
_____ 7. Bonds acquired to profit from short-term differences in price.	N. None of these
_____ 8. 35% of the voting common stock of Computer Storage Devices Company.	
_____ 9. 90% of the voting common stock of Affiliated Peripherals, Inc.	
_____ 10. Corporate bonds of Primary Smelting Company to be sold if interest rates fall ½%.	
_____ 11. 25% of the voting common stock of Smith Foundries Corporation (51% family-owned by Smith family; fair value readily determinable).	
_____ 12. 17% of the voting common stock of Shipping Barrels Corporation (Investor's CEO on the board of directors of Shipping Barrels Corporation).	

P 12–16
Fair value option; held-to-maturity investments
● LO12–1, LO12–2, LO12–8

On January 1, 2021, Ithaca Corp. purchases Cortland Inc. bonds that have a face value of $150,000. The Cortland bonds have a stated interest rate of 6%. Interest is paid semiannually on June 30 and December 31, and the bonds mature in 10 years. For bonds of similar risk and maturity, the market yield on particular dates is as follows:

January 1, 2021	7.0%
June 30, 2021	8.0%
December 31, 2021	9.0%

Required:

1. Calculate the price Ithaca would have paid for the Cortland bonds on January 1, 2021 (ignoring brokerage fees), and prepare a journal entry to record the purchase.
2. Prepare all appropriate journal entries related to the bond investment during 2021, assuming Ithaca accounts for the bonds as a held-to-maturity investment. Ithaca calculates interest revenue at the effective interest rate as of the date it purchased the bonds.
3. Prepare all appropriate journal entries related to the bond investment during 2021, assuming that Ithaca chose the fair value option when the bonds were purchased, and that Ithaca determines fair value of the bonds semiannually. Ithaca calculates interest revenue at the effective interest rate as of the date it purchased the bonds.

P 12–17
Accounting for debt and equity investments
● LO12–1, LO12–4, LO12–5, LO12–9
IFRS

Feherty, Inc., accounts for its investments under *IFRS No. 9* and purchased the following investments during December 2021:

1. Fifty of Donald Company's $1,000 bonds. The bonds pay semiannual interest, return principal in eight years, and include no other cash flows or other features. Feherty plans to hold 10 of the bonds to collect contractual cash flows over the life of the investment and to hold 40, both to collect contractual cash flows but also to sell them if their price appreciates sufficiently. Subsequent to Feherty's purchase of the bonds, but prior to December 31, the fair value of the bonds increased to $1,040 per bond, and Feherty sold 10 of the 40 bonds. Feherty also sold 5 of the 10 bonds it had planned to hold to collect contractual cash flows over the life of the investment. The fair value of the bonds remained at $1,040 as of December 31, 2021.

2. $25,000 of Watson Company common stock. Feherty does not have the ability to significantly influence the operations of Watson. Feherty elected to account for this equity investment at fair value through OCI (FVOCI). Subsequent to Feherty's purchase of the stock, the fair value of the stock investment increased to $30,000 as of December 31, 2021.

Required:

1. Indicate how Feherty would account for its investments when it acquired the Donald bonds and Watson stock.
2. For each of the following categories of Feherty's investments, calculate the effect of realized and unrealized gains and losses on Feherty's net income, other comprehensive income, and comprehensive income for the year ended December 31, 2021:
   a. any Donald bonds accounted for at amortized cost that were purchased and held at year end,
   b. any Donald bonds accounted for at amortized cost that were purchased and sold,
   c. any Donald bonds accounted for at FVOCI that were purchased and held at year end,
   d. any Donald bonds accounted for at FVOCI that were purchased and sold, and
   e. the Watson stock. Ignore interest revenue and taxes.

P 12–18
Accounting for impairments (Appendix 12B)
● LO12–2, LO12–3, LO12–4, LO12–8

Stewart Enterprises has the following investments, all purchased prior to 2021:

1. Bee Company 5% bonds, purchased at face value, with an amortized cost of $4,000,000, and classified as held to maturity. At December 31, 2021, the Bee investment had a fair value of $3,500,000, and Stewart calculated that $240,000 of the fair value decline is a credit loss and $260,000 is a noncredit loss. At December 31, 2022, the Bee investment had a fair value of $3,700,000, and Stewart calculated that $140,000 of the difference between fair value and amortized cost was a credit loss and $160,000 was a noncredit loss.
2. Oliver Corporation 4% bonds, purchased at face value, with an amortized cost of $2,500,000, classified as a trading security. Because of unrealized losses prior to 2021, the Oliver bonds have a fair value adjustment account with a credit balance of $200,000, such that the carrying value of the Oliver investment is $2,300,000 prior to making any adjusting entries in 2021. At December 31, 2021, the Oliver investment had a fair value of $2,200,000, and Stewart calculated that $120,000 of the difference between amortized cost and fair value is a credit loss and $180,000 is a noncredit loss. At December 31, 2022, the Oliver investment had a fair value of $2,700,000.
3. Jones Inc. 6% bonds, purchased at face value, with an amortized cost of $3,500,000, and classified as an available-for-sale investment. Because of unrealized losses prior to 2021, the Jones bonds have a fair value adjustment account with a credit balance of $400,000, such that the carrying value of the Jones investment is $3,100,000 prior to making any adjusting entries in 2021. At December 31, 2021, the Jones investment had a fair value of $2,700,000, and Stewart calculated that $225,000 of the difference between amortized cost and fair value is a credit loss and $575,000 is a noncredit loss. At December 31, 2022, the Jones investment had a fair value of $2,875,000, and Stewart calculated that $125,000 of the difference between amortized cost and fair value is a credit loss and $500,000 is a noncredit loss.

Stewart does not intend to sell any of these investments and does not believe it is more likely than not that it will have to sell any of the bond investments before fair value recovers.

Required:

Prepare the appropriate adjusting journal entries to account for each investment for 2021 and 2022.

Decision Maker's Perspective

©ImageGap/Getty Images

Apply your critical-thinking ability to the knowledge you've gained. These cases will provide you an opportunity to develop your research, analysis, judgment, and communication skills. You also will work with other students, integrate what you've learned, apply it in real world situations, and consider its global and ethical ramifications. This practice will broaden your knowledge and further develop your decision-making abilities.

Real World Case 12–1
Intel's investments
● LO12–4

The following disclosure note appeared in the December 31, 2016, annual report of the **Intel Corporation.**

Note 5: Cash and Investments (partial)

Available-for-sale investments as of December 31, 2016, and December 26, 2015, were as follows:

($ in millions	December 31, 2016				December 26, 2015			
	Adjusted Cost	Gross Unrealized Gains	Gross Unrealized Losses	Fair Value	Adjusted Cost	Gross Unrealized Gains	Gross Unrealized Losses	Fair Value
Corporate debt	$ 3,847	$ 4	$(14)	$ 3,837	$ 4,169	$ 3	$(11)	$ 4,161
Financial institution instruments	6,098	5	(11)	6,092	11,140	1	(2)	11,139
Government debt	1,581	—	(8)	1,573	748	—	(1)	747
Marketable equity securities	2,818	3,363	(1)	6,180	3,254	2,706	—	5,960
Total available-for-sale investments	$14,344	$3,372	$(34)	$17,682	$19,311	$2,710	$(14)	$22,007

Intel also indicates the following: "During 2016, we sold available-for-sale investments for proceeds of $4.1 billion. . . . The gross realized gains on sales of available-for-sale investments were $530 million in 2016. Intel's Note 16 (Other Comprehensive Income) indicates unrealized holding gains of $1,170 million during 2016, and a reclassification adjustment of $530 for gains that had previously been included in OCI and recorded in the fair value adjustment but which were now being included in net income after being realized upon sale.

Required:

1. Draw a T-account that shows the change between the December 26, 2015, and December 31, 2016, balances for the fair value adjustment associated with Intel's AFS investments for 2016. By how much did the fair value adjustment change during 2016?
2. Prepare a journal entry that records any unrealized holding gains and losses that occurred during 2016. Ignore income taxes.
3. Prepare a journal entry that records any reclassification adjustment for available-for-sale investments sold during 2016. Ignore income taxes.
4. Using your journal entries from requirements 2 and 3, adjust your T-account from requirement 1. Have you accounted for the entire change in the fair value adjustment that occurred during 2016? Speculate as to the cause of any difference.

Real World Case 12–2
Reporting securities available-for-sale; obtain and critically evaluate an annual report
● LO12–4

All publicly traded domestic companies use EDGAR, the Electronic Data Gathering, Analysis, and Retrieval system, to make the majority of their filings with the SEC. You can access EDGAR at www.sec.gov.

Required:

1. Locate a recent annual report of a public company that includes a footnote that describes an investment in securities available-for-sale. You can use EDGAR at www.sec.gov.
2. Under what caption are the investments reported in the comparative balance sheets? Are they reported as current or noncurrent assets?
3. Are realized gains or losses reported in the comparative income statements?
4. Are unrealized gains or losses reported in the comparative statements and shareholders' equity?
5. Are accumulated unrealized gains or losses identifiable in the comparative balance sheets? If so, under what caption? Why are unrealized gains or losses reported here rather in the income statement?
6. Are cash flow effects of these investments reflected in the company's comparative statements of cash flows? If so, what information is provided by this disclosure?

International Case 12–3
Comparison of equity method between IFRS and U.S. GAAP
● LO12–5, LO12–6, LO12–7, LO12–9

The following are excerpts from the 2016 financial statements of **Renault**, a large French automobile manufacturer.

14 – INVESTMENT IN NISSAN

Renault and the Japanese automaker Nissan have chosen to develop a unique type of alliance between two distinct companies with common interests, uniting forces to achieve optimum performance. The Alliance is organized so as to preserve individual brand identities and respect each company's corporate culture.

Consequently:

- Renault is not assured of holding the majority of voting rights in Nissan's Shareholders' Meeting.
- The terms of the Renault-Nissan agreement do not entitle Renault to appoint the majority of Nissan directors, nor to hold the majority of voting rights at meetings of Nissan's Board of Directors; Renault cannot unilaterally appoint the President of Nissan; on December 31, 2016, Renault occupied two of the nine seats on Nissan's Board of Directors (unchanged since December 31, 2015).

- Renault-Nissan B.V., owned 50% by Renault and 50% by Nissan, is the Alliance's joint decision-making body for strategic issues concerning either group individually. Its decisions are applicable to both Renault and Nissan. This decision-making power was conferred on Renault-Nissan B.V. to generate synergies and bring both automakers worldwide economies of scale. This entity does not enable Renault to direct Nissan's financial and operating strategies, which are governed by Nissan's Board of Directors and cannot therefore be considered to represent contractual control by Renault over Nissan. The matters examined by Renault-Nissan B.V. since it was formed have remained strictly within this contractual framework, and are not an indication that Renault exercises control over Nissan.
- Renault can neither use nor influence the use of Nissan's assets in the same way as its own assets.
- Renault provides no guarantees in respect of Nissan's debt.

In view of this situation, Renault is considered to exercise significant influence over Nissan, and therefore uses the equity method to include its investment in Nissan in the consolidation.

Renault's Note D lists various restatements that Renault makes when accounting for its Nissan investment under the equity method. Some of those changes harmonize Nissan's accounting (under Japanese accounting standards). Others reflect adjustments to fair value of assets and liabilities applied by Renault at the time of acquisitions in 1999 and 2002.

Required:

1. Go to Deloitte's IAS Plus website and examine the summary of the IASB's *IAS No. 28* (**http://www.iasplus.com/standard/ias28.htm**), which governs application of the equity method. Focus on two areas: Identification of Associates and Applying the Equity Method of Accounting.
2. Evaluate Renault's decision to use the equity method to account for its investment in Nissan. Does Renault have insignificant influence, significant influence, or control?
3. Evaluate the fact that, when accounting for its investment in Nissan under the equity method, Renault makes adjustments that take into account the fair value of assets and liabilities at the time Renault invested in Nissan. Give an example of the sorts of adjustments that might be made. Are such adjustments consistent with IFRS? With U.S. GAAP? Explain.
4. Evaluate the fact that, when accounting for its investment in Nissan under the equity method, Renault makes adjustments for harmonization of accounting standards. Are such adjustments consistent with IFRS? With U.S. GAAP? Explain.

International Case 12–4
Comparison of equity method and proportionate consolidation under IFRS
● LO12–6, LO12–9
IFRS

Obtain the 2016 annual report of FCA Group (**www.fcagroup.com**), which manufactures Fiat-brand automobiles as well as other products.

Required:

Find FCA's discussion of "Basis of Consolidation" in the "Significant Accounting Policies" note that follows the financial statements. Is FCA accounting for its equity investments in a way that is consistent with U.S. GAAP in effect in 2018? Explain.

Research Case 12–5
Researching the way investments are reported; retrieving information from the Internet
● LO12–2, LO12–3, LO12–4, LO12–5, LO12–6

All publicly traded domestic companies use EDGAR, the Electronic Data Gathering, Analysis, and Retrieval system, to make the majority of their filings with the SEC. You can access EDGAR at **www.sec.gov**.

Required:

1. Search for a public company with which you are familiar. Access its most recent 10-K filing. Search or scroll to find financial statements and related notes.
2. Answer the following questions. (If the chosen company does not report investments in the securities of other companies, choose another company.)
   a. What is the amount and classification of any investment securities reported in the balance sheet? Are unrealized gains or losses reported in the shareholders' equity section?
   b. Are any investments reported by the equity method?
   c. What amounts from these investments are reported in the comparative income statements? Has that income increased or decreased over the years reported?
   d. Are any acquisitions or disposals of investments reported in the statement of cash flows?

Real World Case 12–6
Merck's investments
● LO12–4, LO12–5, LO12–6

Corporations frequently invest in securities issued by other corporations. Some investments are acquired to secure a favorable business relationship with another company. On the other hand, others are intended only to earn an investment return from the dividends or interest the securities pay or from increases in the market prices of the securities—the same motivations that might cause you to invest in stocks, bonds, or other securities. This diversity in investment objectives means no single accounting method is adequate to report every investment.

Merck & Co., Inc., invests in securities of other companies. Access Merck's 2016 10-K (which includes financial statements) using EDGAR at **www.sec.gov**. Note: Merck's 2016 financial statements were issued prior to the effective date of *ASU 2016-01,* so do not be surprised by the fact that Merck includes equity investments among its available-for-sale investments.

Required:

1. What is the amount and classification of any AFS investment securities reported in the balance sheet? In which current and noncurrent asset categories are investments reported by Merck? Is there an amount you can't find in the balance sheet but that you know must be there?
2. How are unrealized gains or losses on AFS investments reported, in net income or OCI? What about realized gains or losses on AFS investments?
3. Are any investments reported by the equity method?
4. What amounts from equity method investments are reported in the 2016 income statement?
5. Are cash flow effects of investments reflected in the company's statement of cash flows? If so, what information is provided by this disclosure?

Real World Case 12–7
Comprehensive income—Microsoft
● LO12–4

Microsoft's 2017 10-K includes the following information in Note 19—Accumulated Other Comprehensive Income relevant to its available-for-sale investments:

	Year Ended June 30,		
($ in millions)	**2017**	**2016**	**2015**
Investments			
Accumulated other comprehensive income balance, beginning of period	$2,941	$3,169	$3,531
Unrealized gains, net of tax effects of $267, $120 and $59	517	219	110
Reclassification adjustments for gains included in other income (expense), net	(2,513)	(688)	(728)
Tax expense included in provision for income taxes	880	241	256
Amounts reclassified from accumulated other comprehensive income	(1,633)	(447)	(472)
Net current period other comprehensive income (loss)	(1,116)	(228)	(362)
Accumulated other comprehensive income balance, end of period	$1,825	$2,941	$3,169

Required:

1. Prepare a journal entry to record unrealized gains for 2017. (*Hint:* $517 is net of tax effects, so you will need to add back tax effects to show the amount of pretax unrealized gain.)
2. Prepare a journal entry to record Microsoft's reclassification adjustment for 2017 (pretax).

Data Analytics

connect

Data analytics is the process of examining data sets in order to draw conclusions about the information they contain. If you haven't completed any of the prior data analytics cases, follow the instructions listed in the Chapter 1 Data Analytics case to get set up. You will need to watch the videos referred to in the Chapters 1 - 3 Data Analytics cases. No additional videos are required for this case. All short training videos can be found here: **https://www.tableau.com/learn/training#getting-started.**

Data Analytics Case
Investment in Equity Secureities
● LO12–5

In the Chapter 11 Data Analytics Case, you used Tableau to examine a data set and create calculations to compare the relative age of two companies' assets. In this case you continue in your role as an analyst conducting introductory research into the relative merits of investing in one or both of these companies. This time assess the extent to which the companies invest in equity securities on a long-term basis. To do so, you evaluate the percentage of noncurrent assets invested in equity securities. The investments in equity to assets formula is total equity investments/total noncurrent assets.

Required:

Use Tableau to calculate each company's annual percentage of noncurrent assets invested in equity securities in each year from 2018 to 2021. Based upon what you find, answer the following questions:

1. What is the percentage of noncurrent assets invested in equity securities for (a) Big Store and for (b) Discount Goods in 2021?
2. Comparing the percentage of noncurrent assets invested in equity securities ratios over the most recent four-year period, is Discount Good's equity investment (a) generally increasing, (b) roughly the same, or (c) generally decreasing from year to year?
3. In general, which company invests the higher amount in equity securities as a percentage of its noncurrent asset during the most recent four-year period?

Resources:

You have available to you an extensive data set that includes detailed financial data for 2012-2021 for both Discount Goods and Big Store. The data set is in the form of four Excel files available to download from Connect, or under Student Resources within the Library tab. The one for use in this chapter is named "Discount_Goods_Big_Store_Financials.xlsx." Download this file and save it to the computer on which you will be using Tableau.

For this case, you will create a calculation to produce a bar chart of the percentage of noncurrent assets invested in equity securities ratio to allow you to compare and contrast the two companies' investments.

After you view the training videos, follow these steps to create the charts you'll use for this case:

- Open Tableau and connect to the Excel spreadsheet you downloaded.
- Click on the "Sheet 1" tab at the bottom of the canvas, to the right of the Data Source at the bottom of the screen. Drag "Company" and "Year" under Dimensions to the Rows shelf. Change "Year" to *discrete* by right-clicking and selecting "discrete." From the same menu, select "Filter," unclick "All," and click only 2018, 2019, 2020, and 2021.
- Drag the "Investments in Equity Security" and "Total Noncurrent Assets" under Measures to the Rows shelf. Change each to *discrete.* Right click on the drop-down menu of each of the accounts and uncheck "Show Header" so they are not visible in the field.
- Under the Analysis tab, select Create Calculated Field. Create a measure named "Investments in Equity to Assets" by dragging "Investments in Equity Security" from the Rows shelf to the Calculation Editor window, typing a division sign for division, and then dragging "Total Noncurrent Assets" from the Rows shelf beside it. Make sure the window says that the calculation is valid and click OK.
- Drag the newly created "Investments in Equity to Assets" to the Rows shelf.
- Click on the "Show Me" and select "side-by-side bars." Add labels to the bars by clicking on "Label" under the "Marks" card and clicking the box "Show mark label." Format the labels to Times New Roman, bold, black and 9-point font.
- Change the title of the sheet to be "Fixed Asset Turnover Ratio Bar Chart" by right-clicking and selecting "Edit title." Format the title to Times New Roman, bold, black and 15-point font.
- Format the labels on the left of the sheet ("Fixed Asset Turnover") to Times New Roman, 9-point font, bold, and black. Change the title of "Sheet 1" to match the sheet title by right-clicking, selecting "Rename" and typing in the new title.
- Format all other labels to be Times New Roman, bold, black and 10-point font.
- Once complete, save the file as "*DA12_Your initials*.twbx."

Continuing Cases

Target Case

● LO12–4, LO12–6

Target Corporation prepares its financial statements according to U.S. GAAP. Target's financial statements and disclosure notes for the year ended February 3, 2018, are available in Connect. This material also is available under the Investor Relations link at the company's website (**www.target.com**). Target does not have investments in stock or bonds. However, **CVS Health Corp.**, which purchased Target's pharmacy and clinical business during 2015, does have some investments. Access CVS's 2017 10K (issued on February 14, 2018) at **investors.cvshealth.com** to answer the following questions.

Required:

1. CVS indicates in Note 1 that it has some short-term investments that consist of certificates of deposit (CDs).
   a. How has CVS classified those CDs for accounting purposes?
   b. Per CVS's balance sheet, what was the balance in CVS's short-term investments as of December 31, 2017, and December 31, 2016?
   c. Per CVS's statement of cash flows, what cash transactions affected short-term investments during 2017?
   d. Prepare a T-account that summarizes transactions affecting CVS's short-term investments during 2017. Speculate as to the explanation for any "plug" figure necessary to make the T-account balance.

2. Per Note 1, CVS has equity-method investments in SureScripts, LLC, and in Heartland Healthcare Services. CVS indicates that those investments are immaterial for the year ended December 31, 2017. Assuming that the Heartland investment is material,
   a. How would Heartland's earnings affect CVS's income statement?
   b. How would Heartland's earnings affect CVS's balance sheet?

Air France–KLM Case

● LO12–9

Air France–KLM (AF), a Franco-Dutch company, prepares its financial statements according to International Financial Reporting Standards. AF's financial statements and disclosure notes for the year ended December 31, 2017, are available in Connect. This material is also available under the Finance link at the company's website (**www.airfranceklm.com**).

Required:

1. Read Notes 23 and 35.4. Focusing on investments accounted for at fair value through profit and loss (FVTPL),
   a. As of December 31, 2017, what is the total balance of those investments in the balance sheet?
   b. How much of that balance is classified as current and how much as noncurrent?
   c. How much of the fair value of those investments is accounted for using level 1, level 2, and level 3 inputs of the fair value hierarchy? Given that information, assess the reliability (representational faithfulness) of this fair value estimate.
2. Complete requirement 1 again, but for investments accounted for as available for sale.
3. Given your answer to 1c and 2c, which type of investment has more reliable (representational faithful) fair value estimates: FVTPL investments or available for sale investments?
4. Read Notes 4.3 and 21.
   a. When AF can exercise significant influence over an investee, what accounting approach does it use to account for the investment? How does AF determine if it can exercise significant influence?
   b. If AF is involved in a joint venture, what accounting approach does it use to account for the investment?
   c. What is the carrying value of AF's equity-method investments in its December 31, 2017, balance sheet?
   d. How did AF's equity-method investments affect AF's 2017 net income from continuing operations?

CPA Exam Questions and Simulations

ROGER CPA Review

Sample CPA Exam questions from Roger CPA Review are available in Connect as support for the topics in this chapter. These multiple-choice questions and task-based simulations include expert-written explanations and solutions, and provide a starting point for students to become familiar with the content and functionality of the actual CPA Exam.

CHAPTER

13

Current Liabilities and Contingencies

OVERVIEW

Chapter 13 is the first of five chapters devoted to liabilities. In Part A of this chapter, we discuss liabilities that are classified appropriately as current. In Part B, we turn our attention to situations in which there is uncertainty as to whether an obligation really exists. These are designated as loss contingencies. Some loss contingencies are accrued as liabilities, but others only are disclosed in the notes.

LEARNING OBJECTIVES

After studying this chapter, you should be able to:

- **LO13–1** Define liabilities and distinguish between current and long-term liabilities. (*p. 718*)
- **LO13–2** Account for the issuance and payment of various forms of notes and record the interest on the notes. (*p. 720*)
- **LO13–3** Characterize accrued liabilities and liabilities from advance collection and describe when and how they should be recorded. (*p. 723*)
- **LO13–4** Determine when a liability can be classified as a noncurrent obligation. (*p. 729*)
- **LO13–5** Identify situations that constitute contingencies and the circumstances under which they should be accrued. (*p. 732*)
- **LO13–6** Demonstrate the appropriate accounting treatment for contingencies, including unasserted claims and assessments. (*p. 733*)
- **LO13–7** Discuss the primary differences between U.S. GAAP and IFRS with respect to current liabilities and contingencies. (*pp. 731, 741,* and *742*)

©Cultura/Image Source

FINANCIAL REPORTING CASE

Debbie's Dad

"My dad is confused," your friend Debbie Hirst proclaimed at the office one morning. "You see, we're competing against each other in that investment game I told you about, and one of his hot investments is Syntel Microsystems. When he got their annual report yesterday afternoon, he started analyzing it, you know, really studying it closely. Then he asked me about the current liability section of the balance sheet and related disclosure note":

SYNTEL MICROSYSTEMS, INC.
Balance Sheet
December 31, 2021 and 2020
($ in millions)

Current Liabilities	2021	2020
Accounts payable	$233.5	$241.6
Short-term borrowings (Note 3)	187.0	176.8
Accrued liabilities	65.3	117.2
Accrued loss contingency	76.9	—
Other current liabilities	34.6	45.2
Current portion of long-term debt	44.1	40.3
Total current liabilities	$641.4	$621.1

Note 3: Short-Term Borrowings (in part)
The components of short-term borrowings and their respective weighted average interest rates at the end of the period are as follows:

	2021		2020	
($ in millions)	Amount	Average Interest Rate	Amount	Average Interest Rate
Commercial paper	$ 34.0	5.2%	$ 27.1	5.3%
Bank loans	218.0	5.5	227.7	5.6
Amount reclassified to long-term liabilities	(65.0)	—	(78.0)	—
Total short-term borrowings	$187.0		$176.8	

The Company maintains bank credit lines sufficient to cover outstanding short-term borrowings. As of December 31, 2021, the Company had $200.0 million fee-paid lines available. At December 31, 2021 and 2020, the Company classified $65.0 million and $78.0 million, respectively, of commercial paper and bank notes as long-term debt. The Company has the intent and ability, through formal renewal agreements, to renew these obligations into future periods.

Note 6: Contingencies (in part)
Between 2019 and 2020, the Company manufactured cable leads that, the Company has learned, contribute to corrosion of linked components with which they are installed. At December 31, 2021, the Company accrued $132.0 million in anticipation of remediation and claims settlement deemed probable, of which $76.9 million is considered a current liability.

"So, what's the problem?" you asked.

"Well, he thinks I'm some sort of financial wizard because I'm in business."

"And because you tell him so all the time," you interrupted.

(continued)

(continued)

"Maybe so, but he's been told that current liabilities are riskier than long-term liabilities, and now he's focusing on that. He can't see why some long-term debt is reported here in the current section. And it also looks like some is reported the other way around; some current liabilities reported as long term. Plus, the contingency amount seems like it's not even a contractual liability. Then he wants to know what some of those terms mean. Lucky for me, I had to leave before I had to admit I didn't know the answers. You're the accountant; help me out."

QUESTIONS

By the time you finish this chapter, you should be able to respond appropriately to the questions posed in this case. Compare your response to the solution provided at the end of the chapter.

1. What are accrued liabilities? What is commercial paper? *(p. 723)*
2. Why did Syntel Microsystems include some long-term debt in the current liability section? *(p. 729)*
3. Did they also report some current amounts as long-term debt? Explain. *(p. 730)*
4. Must obligations be known contractual debts in order to be reported as liabilities? *(p. 733)*
5. Is it true that current liabilities are riskier than long-term liabilities? *(p. 745)*

PART A

Current Liabilities

Liabilities and owners' equity accounts represent specific sources of a company's assets.

Before a business can invest in an asset it first must acquire the money to pay for it. This can happen in either of two ways—funds can be provided by owners or the funds must be borrowed. You may recognize this as a description of the basic accounting equation: Assets = Liabilities + Owners' Equity. You studied assets in the chapters leading to this one and you will study owners' equity later. This chapter and the next four describe the various liabilities that constitute creditors' claims on a company's assets.

Characteristics of Liabilities

Most liabilities obligate the debtor to pay cash at specified times and result from legally enforceable agreements.

LO13–1

You already know what liabilities are. You encounter them every day. If you are paying for a car or a home with monthly payments, you have a personal liability. Similarly, when businesses issue notes and bonds, their creditors are the banks, individuals, and organizations that exchange cash for those securities. Each of these obligations represents the most common type of liability—one to be paid in cash and for which the amount and timing are specified by a legally enforceable contract.

Some liabilities are not contractual obligations and may not be payable in cash.

However, to be reported as a **liability**, an obligation need not be payable in cash. Instead, it may require the company to transfer other assets or to provide services. A liability doesn't have to be represented by a written agreement nor be legally enforceable. Even the amount and timing of repayment need not be precisely known.

From a financial reporting perspective, a liability has three essential characteristics. Liabilities

1. Are *probable, future* sacrifices of economic benefits.
2. Arise from *present* obligations (to transfer assets or provide services) to other entities.
3. Result from *past* transactions or events.[1]

Notice that the definition of a liability involves the past, the present, and the future. It is a present responsibility to sacrifice assets in the future because of a transaction or other event that happened in the past.

[1]"Elements of Financial Statements," *Statement of Financial Accounting Concepts No. 6* (Stamford, Conn.: FASB, 1985), par. 38.

What is a Current Liability?

In a classified balance sheet, we categorize liabilities as either **current liabilities** or long-term liabilities. We often characterize current liabilities as obligations payable within one year from the balance sheet date or within the firm's operating cycle, whichever is longer. This general definition usually applies. However, a more discriminating definition identifies current liabilities as those expected to be satisfied with *current assets* or by the creation of other *current liabilities*.[2]

Current liabilities are expected to require current assets and usually are payable within one year from the balance sheet date.

Classifying liabilities as either current or long term helps investors and creditors assess the risk that the liabilities will require expenditure of cash or other assets in the near term. Is the due date years in the future, permitting resources to be used for other purposes without risking default? Or, will payment require the use of current assets and reduce the amount of liquid funds available for other uses? If so, are sufficient liquid funds available to make necessary payments of liabilities in addition to meeting current operating needs, or must additional funds be obtained by raising capital? The answers to these questions can have significant implications. For example, a major factor contributing to the collapse of the financial giant **Bear Stearns** in 2008 was its reliance on short-term liabilities that it couldn't refinance when lenders, clients, and trading partners grew concerned about the quality of Bear's investments.[3]

Classifying liabilities as either current or long term helps investors and creditors assess the relative risk of a business's liabilities.

Conceptually, liabilities should be recorded at their present values. In other words, the amount recorded is the present value of all anticipated future cash payments resulting from the debt (specifically, principal and interest payments).[4] However, in practice, liabilities payable within one year from the balance sheet date ordinarily are recorded instead at their maturity amounts.[5] This inconsistency usually is inconsequential because the relatively short-term maturity of current liabilities makes the interest or time value component immaterial.

Current liabilities ordinarily are reported at their maturity amounts.

The most common obligations reported as current liabilities are accounts payable, notes payable, commercial paper, income tax liability, dividends payable, and accrued liabilities. Liabilities related to income taxes are the subject of Chapter 16. We discuss the others here.

Before we examine specific current liabilities, let's use the current liability section of the balance sheet of **General Mills, Inc.**, and related disclosure notes to provide perspective on some of the liabilities we discuss (Illustration 13–1). General Mills's presentation and supplemental note disclosures are fairly typical.

We'll refer back to portions of Illustration 13–1 as corresponding liabilities are described later in the chapter.

Open Accounts and Notes

Many businesses buy merchandise or supplies on credit. Most also find it desirable to borrow cash from time to time to finance their activities. In this section, we discuss the liabilities these borrowing activities create: trade accounts and trade notes, bank loans, and commercial paper.

Accounts Payable and Trade Notes Payable

Accounts payable are obligations to suppliers of merchandise or of services purchased on *open account*. Most trade credit is offered on open account. This means that the only formal credit instrument is the invoice. Because the time until payment usually is short (often 30, 45, or 60 days), these liabilities typically are noninterest-bearing and are reported at their face amounts. As shown in Illustration 13–1, General Mills's accounts payable in 2017 was $2,120 million. The key accounting considerations relating to accounts payable are determining their existence and ensuring that they are recorded in the appropriate accounting period. You studied these issues and learned how cash discounts are handled during your study of inventories in Chapter 8.

Buying merchandise on account in the ordinary course of business creates *accounts payable*.

[2] FASB ASC Master Glossary.

[3] William D. Cohan, *House of Cards: A Tale of Hubris and Wretched Excess on Wall Street* (New York: Doubleday, 2009).

[4] The concepts of the time value of money and the mechanics of present value calculations are covered in Chapter 5.

[5] In fact, those arising in connection with suppliers in the normal course of business and due within a year are specifically exempted from present value reporting by FASB ASC 835–30–15–3: Interest—Imputation of Interest—Scope and Scope Exceptions.

Illustration 13–1

Current Liabilities—General Mills

Real World Financials

GENERAL MILLS, INC.
Excerpt from Consolidated Balance Sheets
May 28, 2017 and May 29, 2016
Liabilities

	($ in millions)	
Current Liabilities	**2017**	**2016**
Accounts payable	$2,119.8	$2,046.5
Current portion of long-term debt	604.7	1,103.4
Notes payable	1,234.1	269.8
Other current liabilities	1,372.2	1,595.0
Total current liabilities	$5,330.8	$5,014.7

Note 8. Debt

Notes Payable The components of notes payable and their respective weighted-average interest rates at the end of the periods were as follows:

	2017		**2016**	
Dollars in Millions:	**Note Payable**	**Weighted Average Interest Rate**	**Note Payable**	**Weighted Average Interest Rate**
U.S. commercial paper	$ 954.7	1.1%	$ –	–%
Financial institutions	279.4	7.0	269.8	8.6
Total notes payable	$1,234.1	2.4%	$269.8	8.6%

To ensure availability of funds, we maintain bank credit lines sufficient to cover our outstanding notes payable. Commercial paper is a continuing source of short-term financing. We have commercial paper programs available to us in the United States and Europe. We also have uncommitted and asset-backed credit lines that support our foreign operations. The following table details the fee-paid committed and uncommitted credit lines we had available as of May 28, 2017:

Dollars in Billions:	**Facility Amount**	**Borrowed Amount**
Credit facility expiring:		
May 2022	$2.7	$ –
June 2019	0.2	0.1
Total committed credit facilities	2.9	0.1
Uncommitted credit facilities	0.5	0.1
Total committed and uncommitted credit facilities	$3.4	$0.2

Source: General Mills, Inc.

Trade notes payable differ from accounts payable in that they are formally recognized by a written promissory note. Often these are of a somewhat longer term than open accounts and bear interest.

Short-Term Notes Payable

LO13–2

The most common way for a corporation to obtain temporary financing is to arrange a short-term bank loan. When a company borrows cash from a bank and signs a promissory note (essentially an IOU), the firm's liability is reported as *notes payable* (sometimes *bank loans* or *short-term borrowings*). About two-thirds of bank loans are short term, but because many are routinely renewed, some tend to resemble long-term debt. In fact, in some cases we report them as long-term debt (as you'll see later in the chapter).

Very often, smaller firms are unable to tap into the major sources of long-term financing to the extent necessary to provide for their working capital needs. So they must rely heavily on short-term financing. Even large companies typically utilize short-term debt as a significant and indispensable component of their capital structure. One reason is that short-term funds usually offer lower interest rates than long-term debt. Perhaps most importantly, corporations desire flexibility. Managers want as many financing alternatives as possible.

CREDIT LINES Usually short-term bank loans are arranged under an existing **line of credit** with a bank or group of banks. A line of credit is an agreement to provide short-term financing, with amounts withdrawn by the borrower only when needed. Even though the loans are short-term, with amounts borrowed and repaid frequently, the agreement to provide a line of credit typically lasts several years. Lines of credit can be noncommitted or committed. A *noncommitted* line of credit is an informal agreement that permits a company to borrow up to a prearranged limit without having to follow formal loan procedures and paperwork. Banks sometimes require the company to maintain a compensating balance on deposit with the bank, say, 5% of the line of credit.[5] A *committed* line of credit is a more formal agreement that usually requires the company to pay a commitment fee to the bank to keep a credit line amount available to the company. A typical annual commitment fee is ¼% of the total committed funds, and may also require a compensating balance. A recent annual report of **IBM Corporation** illustrates noncommitted lines of credit (Illustration 13–2).

A *line of credit* allows a company to borrow cash without having to follow formal loan procedures and paperwork.

Note J. Borrowings (in part)

LINES OF CREDIT: In 2016, the company increased the size of its five-year Credit Agreement (the "Credit Agreement") to $10.25 billion and extended the term by one year to November 10, 2021. The total expense recorded by the company related to this global credit facility was $6.1 million in 2017, $5.5 million in 2016 and $5.3 million in 2015. The Credit Agreement permits the company and its Subsidiary Borrowers to borrow up to $10.25 billion on a revolving basis. Borrowings of the Subsidiary Borrowers will be unconditionally backed by the company. . . . As of December 31, 2017, there were no borrowings by the company, or its subsidiaries, under the Credit Facilities.

Source: BM Corporation

Illustration 13–2

Disclosure of Credit Lines—IBM Corporation

Real World Financials

General Mills's disclosure notes that we looked at in Illustration 13–1 indicate that the company has both noncommitted and committed lines of credit.

INTEREST When a company borrows money, it pays the lender **interest** in return for using the lender's money during the term of the loan. You might think of the interest as the "rent" paid for using money. Interest is stated in terms of a percentage rate to be applied to the face amount of the loan, which makes this an interest-bearing note. Because the stated rate typically is an annual rate, when calculating interest for a short-term note we must adjust for the fraction of the annual period the loan spans. Interest on notes is calculated as

Face amount × Annual rate × Time to maturity

This is demonstrated in Illustration 13–3.

On May 1, Affiliated Technologies, Inc., a consumer electronics firm, borrowed $700,000 cash from First BancCorp under a noncommitted short-term line of credit arrangement and issued a **six-month**, 12% promissory note. Interest was payable at maturity.

May 1		
Cash	700,000	
Notes payable		700,000
November 1		
Interest expense ($700,000 × 12% × 6⁄12)	42,000	
Notes payable	700,000	
Cash ($700,000 + $42,000)		742,000

Illustration 13–3

Interest-Bearing Notes Payable

[5]A compensating balance is a deposit kept by a company in a low-interest or noninterest-bearing account at the bank. The required deposit usually is some percentage of the committed amount or the amount used (say, 2% to 5%). The effect of the compensating balance is to increase the borrower's effective interest rate and the bank's effective rate of return.

Sometimes a note assumes the form of a so-called **noninterest-bearing note**. Obviously, though, nobody will lend money without interest. Noninterest-bearing loans actually do bear interest, but the interest is deducted (or discounted) from the face amount to determine the cash proceeds made available to the borrower at the outset.

You saw this in Chapter 7 when a company sold a product worth $658,000 and accepted a note receivable from the customer of $700,000 due in six months. While the note had no stated interest rate, implicit in the higher amount of the note is interest of $42,000 ($700,000 – $658,000). We can apply the same concept to our previous example in Illustration 13–3. Suppose Affiliated Technologies bought inventory worth $658,000 by issuing a $700,000 noninterest-bearing note due in six months. The difference of $42,000 is recorded as a discount on notes payable that will be recognized as interest expense over time.[7]

The proceeds of the note are reduced by the interest in a noninterest-bearing note.

May 1		
Inventory (difference)	658,000	
Discount on notes payable (to balance)	42,000	
Notes payable (face amount)		700,000
November 1		
Interest expense	42,000	
Discount on notes payable		42,000
Notes payable (face amount)	700,000	
Cash		700,000

Sometimes interest in such arrangements is described by referring to a discount rate that is applied to the face amount of the note. In this case, the six-month noninterest-bearing note would be described as "being discounted at issuance at a 12% discount rate," and the amount of interest would be calculated as the face amount times the discount rate times the fraction of the year the note is outstanding ($700,000 × 12% × 6/12 = $42,000). Notice, though, that the amount borrowed under this arrangement is only $658,000. This causes the *effective* interest rate to be higher than the 12% stated rate:

When interest is discounted from the face amount of a note, the effective interest rate is higher than the stated discount rate.

$$\frac{\$42{,}000 \text{ Interest for 6 months}}{\$658{,}000 \text{ Amount borrowed}} = 6.38\% \text{ Rate for 6 months}$$

To annualize:

$$6.38\% \times {}^{12}\!/_{6} = 12.76\% \text{ Effective interest rate}$$

You can refer back to Chapter 7 and verify that the accounting for a note receivable mirrors what we showed here for a note payable.

Inventory or accounts receivable often are pledged as security for short-term loans.

SECURED LOANS Sometimes short-term loans are *secured,* meaning a specified asset of the borrower is pledged as collateral or security for the loan. Although many kinds of assets can be pledged, the secured loans most frequently encountered in practice are secured by inventory or accounts receivable. For example, **Smithfield Foods, Inc.**, disclosed the secured notes described in Illustration 13–4.

When accounts receivable serve as collateral, we refer to the arrangement as *pledging* accounts receivable. Sometimes, the receivables actually are sold outright to a finance company as a means of short-term financing. This is called *factoring* receivables.[8]

[7]Be sure to understand that we are actually recording the note at 658,000, *not* 700,000, but are recording the interest portion separately in a contra-liability account, *discount on notes payable.* The entries shown reflect the gross method. By the net method, the interest component is netted against the face amount of the note as follows:

May 1		
Cash	658,000	
Notes payable		658,000
November 1		
Interest expense ($700,000 × 12% × 6/12)	42,000	
Notes payable	658,000	
Cash		700,000

[8] Both methods of accounts receivable financing are discussed in Chapter 7, "Cash and Receivables."

Note 7: Debt (in part)

Working Capital Facilities

The obligations under the Inventory Revolver Credit Agreement are guaranteed by substantially all of our subsidiary guarantors' personal property, including accounts receivable (other than those sold and financed pursuant to the Securitization Facility), inventory, cash and cash equivalents, deposit accounts, intercompany notes, intellectual property and certain capital stock and interests pledged by us and our subsidiary guarantors, and all proceeds thereof.

Source: Smithfield Foods, Inc.

Illustration 13–4

Disclosure of Notes Secured by Inventory—Smithfield Foods, Inc.

Real World Financials

Commercial Paper

Some large corporations obtain temporary financing by issuing **commercial paper**, often purchased by other companies as a short-term investment. Commercial paper refers to unsecured notes sold in minimum denominations of $25,000 with maturities ranging from 1 to 270 days (beyond 270 days the firm would be required to file a registration statement with the SEC). Interest often is discounted at the issuance of the note. Usually commercial paper is issued directly to the buyer (lender) and is backed by a line of credit with a bank. This allows the interest rate to be lower than in a bank loan. Commercial paper has become an increasingly popular way for large companies to raise funds, the total amount having expanded over fivefold in the last decade.

Large, highly rated firms sometimes sell *commercial paper* to borrow funds at a lower rate than through a bank loan.

Illustration 13–5 includes a disclosure note from the 2016 annual report of **Comcast Corporation** that describes Comcast's commercial paper program.

Note 10: (in part)

Commercial Paper Program

Our commercial paper programs provide a lower-cost source of borrowing to fund our short-term working capital requirements. The maximum borrowing capacity under the Comcast commercial paper program is $6.25 billion. We support this commercial paper program with unused capacity under the Comcast revolving credit facility. The maximum borrowing capacity under the NBCUniversal Enterprise commercial paper program is $1.35 billion. We support this commercial paper program with unused capacity under the NBCUniversal Enterprise revolving credit facility.

Source: Comcast Corporation

Illustration 13–5

Disclosure of Commercial Paper—Comcast Corporation

Real World Financials

The name *commercial paper* refers to the fact that a paper certificate traditionally is issued to the lender to signify the obligation, although there is a trend toward total computerization so that no paper is created. Since commercial paper is a form of notes payable, recording its issuance and payment is exactly the same as our earlier illustration.

In a statement of cash flows, the cash a company receives from using short-term notes to borrow funds as well as the cash it uses to repay the notes are reported among cash flows from financing activities. Most of the other liabilities we study in this chapter, such as accounts payable, interest payable, and bonuses payable, are integrally related to a company's primary operations and thus are part of operating activities. We discuss long-term notes in the next chapter.

FINANCIAL Reporting Case

Q1, p. 718

Accrued Liabilities

Accrued liabilities represent expenses already incurred but not yet paid (accrued expenses). These liabilities are recorded by adjusting entries at the end of the reporting period, prior to preparing financial statements. Common examples are salaries payable, income taxes payable, and interest payable. Although recorded in separate liability accounts, accrued liabilities usually are combined and reported under a single caption or perhaps two accrued liability captions in the balance sheet. General Mills includes accrued liabilities in other current liabilities (totaling $1,372.2 million as of May 28, 2017).

● LO13–3

Liabilities accrue for expenses that are incurred but not yet paid.

Accrued Interest Payable

Accrued interest payable arises in connection with notes payable (as well as other forms of debt). For example, to continue Illustration 13–3, let's assume the fiscal period for Affiliated Technologies ends on June 30, two months after the six-month note is issued. The issuance of the note, intervening adjusting entry, and note payment would be recorded as shown in Illustration 13–6.

Illustration 13–6

Note with Accrued Interest

Issuance of Note on May 1		
Cash	700,000	
Notes payable		700,000
Accrual of Interest on June 30		
Interest expense ($700,000 × 12% × 2/12)	14,000	
Interest payable		14,000
Note Payment on November 1		
Interest expense ($700,000 × 12% × 4/12)	28,000	
Interest payable (from June 30 accrual)	14,000	
Notes payable	700,000	
Cash ($700,000 + $42,000)		742,000

At June 30, two months' interest has accrued and is recorded to avoid misstating expenses and liabilities on the June 30 financial statements.

Salaries, Commissions, and Bonuses

Compensation for employee services can be in the form of hourly wages, salary, commissions, bonuses, stock compensation plans, and pensions.[9] Accrued liabilities arise in connection with compensation expense when employees have provided services but will be paid after the financial statement date. These accrued expenses/accrued liabilities are recorded by adjusting entries at the end of the reporting period, prior to preparing financial statements.

VACATIONS, SICK DAYS, AND OTHER PAID FUTURE ABSENCES Suppose a firm grants two weeks of paid vacation each year to employees. Some employees take their vacations during each period in which the vacations are earned and are compensated then. Some instead accumulate those days and use them in a future period. Should the firm recognize compensation expense during the period for only those employees who actually are paid that period for their absence? When you recall what you've learned about accrual accounting, you probably conclude otherwise.

An employer should accrue an expense and the related liability for employees' compensation for future absences (such as vacation pay) if the obligation meets *all* of the four conditions listed in Illustration 13–7.

Illustration 13–7

Conditions for Accrual of Paid Future Absences

1. The obligation is attributable to employees' services already performed.
2. The paid absence can be taken in a later year—the benefit vests (will be compensated even if employment is terminated) or the benefit can be accumulated over time.
3. Payment is probable.
4. The amount can be reasonably estimated.

If these conditions look familiar, it's because they are simply the characteristics of a liability we discussed earlier, adapted to relate to a potential obligation for future absences of employees. Also, these conditions are consistent with the requirement that we accrue loss contingencies only when the obligation is both (a) probable and (b) can be reasonably estimated, as discussed in Part B of this chapter.

[9]We discuss pensions in Chapter 17 and share-based compensation plans in Chapter 19.

The liability for paid absences usually is accrued at the existing wage rate rather than at a rate estimated to be in effect when absences occur.[10] So, if wage rates have risen, the difference between the accrual and the amount paid increases compensation expense that year. This situation is demonstrated in Illustration 13–8, in which vacation time carried over from 2021 is taken in 2022 and the actual amount paid to employees is $5,700,000.

Illustration 13–8

Paid Future Absences

Davidson-Getty Chemicals has 8,000 employees. Each employee earns two weeks of paid vacation per year. Vacation time not taken in the year earned can be carried over to subsequent years. During 2021, 2,500 employees took both weeks' vacation, but at the end of the year, 5,500 employees had vacation time carryovers as follows:

Employees	Vacation Weeks Earned But Not Taken	Total Carryover Weeks
2,500	0	0
2,000	1	2,000
3,500	2	7,000
8,000		**9,000**

During 2021, compensation averaged $600 a week per employee.

When Vacations Were Taken in 2021		
Salaries expense (2,500 × 2 weeks × $600) + (2,000 × 1 week × $600)	4,200,000	
Cash (or salaries payable)		4,200,000
December 31, 2021 (adjusting entry)		
Salaries expense (**9,000** carryover weeks × $600)	5,400,000	
Liability—compensated future absences		5,400,000

When the necessary conditions are met, compensated future absences are accrued in the year the compensation is earned.

Company policy and actual practice should be considered when deciding whether the rights to payment for absences have been earned by services already rendered. For example, scientists in a private laboratory are eligible for paid sabbaticals every seven years. Should a liability be accrued at the end of a scientist's sixth year? No—if sabbatical leave is granted only to perform research beneficial to the employer. Yes—if past practice indicates that sabbatical leave is intended to provide unrestricted compensated absence for past service and other conditions are met.

Customary practice should be considered when deciding whether an obligation exists.

When Year 2021 Vacations Are Taken in 2022		
Liability—compensated future absences (account balance)	5,400,000	
Salaries expense (difference)	300,000	
Cash (or salaries payable) (given)		5,700,000

Custom and practice also influence whether unused rights to paid absences expire or can be carried forward. Obviously, if rights vest (payable even if employment is terminated), they haven't expired. But holiday time, military leave, maternity leave, and jury time typically do not accumulate if unused, so a liability for those benefits usually is not accrued. On the other hand, if it's customary that a particular paid absence, say holiday time, can be carried forward—if employees work on holidays, in this case—a liability is accrued if it's probable that employees will be compensated in a future year.

Interestingly, sick pay quite often meets the conditions for accrual, but accrual is not mandatory because future absence depends on future illness, which usually is not a certainty. Similar to other forms of paid absences, the decision of whether to accrue nonvesting sick pay should be based on actual policy and practice. If company policy or custom is that employees are

Accrual of sick pay is not required, but may be appropriate in some circumstances.

[10]Actually, FASB ASC 710–10–25: Compensation—General—Overall—Recognition is silent on how the liability should be measured. In practice, most companies accrue at the current rate because it avoids estimates and usually produces a lower expense and liability. Then, later, they remeasure periodically at updated rates.

paid sick pay even when their absences are not due to illness, it's appropriate to record a liability for unused sick pay. For example, some companies routinely allow unused sick pay benefits to be accumulated and paid at retirement (or to beneficiaries if death comes before retirement). If each condition is met except that the company finds it impractical to reasonably estimate the amount of compensation for future absences, a disclosure note should describe the situation.

Bonuses sometimes take the place of permanent annual raises.

ANNUAL BONUSES Sometimes compensation packages include annual bonuses tied to performance objectives designed to provide incentive to executives. The most common performance measures are earnings per share, net income, and operating income, each being used by about a quarter of firms having bonus plans. Nonfinancial performance measures, such as customer satisfaction and product or service quality, also are used.[11] **Annual bonuses** also are popular, not just for executives, but for nonmanagerial personnel as well. Unfortunately for employees, bonuses often take the place of annual raises. This allows a company to increase employee pay without permanently locking in the increases in salaries. Bonuses are compensation expense of the period in which they are earned, so unpaid bonuses are accrued as a liability at year-end.

Liabilities from Advance Collections

Deposits and advances from customers, as well as collections for third parties, present situations where liabilities are created to either make payments or provide goods or services in the future.

Deposits and Advances from Customers

Collecting cash from a customer as a refundable deposit or as an advance payment for products or services creates a liability to return the deposit or to supply the products or services.[12]

REFUNDABLE DEPOSITS In some businesses it's typical to require customers to pay cash as a deposit that will be refunded when a specified event occurs. You probably have encountered such situations. When apartments are rented, security or damage deposits often are collected. Utility companies frequently collect deposits when service is begun. Similarly, deposits sometimes are required on returnable containers, to be refunded when the containers are returned. That situation is demonstrated in Illustration 13–9.

Illustration 13–9
Refundable Deposits

Rancor Chemical Company sells combustible chemicals in expensive, reusable containers. Customers are charged a deposit for each container delivered and receive a refund when the container is returned. Deposits collected on containers delivered during the year were $300,000. Deposits are forfeited if containers are not returned within one year. Ninety percent of the containers were returned within the allotted time. Deposits charged are twice the actual cost of containers. The inventory of containers remains on the company's books until deposits are forfeited.

When Deposits Are Collected		
Cash	300,000	
Liability—refundable deposits		300,000
When Containers Are Returned*		
Liability—refundable deposits	270,000	
Cash		270,000
When Deposits Are Forfeited*		
Liability—refundable deposits	30,000	
Revenue—sale of containers		30,000
Cost of goods sold	15,000	
Inventory of containers		15,000

*Of course, not all containers are returned at the same time, nor does the allotted return period expire at the same time for all containers not returned. These entries summarize the several individual returns and forfeitures.

[11] C. D. Ittner, D. F. Larcker, and M. V. Rajan, "The Choice of Performance Measures in Annual Bonus Contracts," *The Accounting Review*, Vol. 72, No. 2 (April 1997), pp. 231–255.

[12] *SFAC 6* specifically identifies customer advances and deposits as liabilities under the definition provided in that statement. "Elements of Financial Statements," *Statement of Financial Accounting Concepts No. 6* (Stamford, Conn.: FASB, 1985), par. 197.

ADVANCES FROM CUSTOMERS At times, businesses require advance payments from customers that will be applied to the purchase price when goods are delivered or services provided. Gift certificates, magazine subscriptions, layaway deposits, special order deposits, and airline tickets are examples. These customer advances, also called *deferred revenue* or *unearned revenue,* represent liabilities until the related product or service is provided. For instance, the **New York Times Company** reports deferred revenue from unexpired subscriptions of over $75 million in its 2017 balance sheet. Advances are demonstrated in Illustration 13–10.

A customer advance produces an obligation that is satisfied when the product or service is provided.

Illustration 13–10

Customer Advance

Tomorrow Publications collects magazine subscriptions from customers at the time subscriptions are sold. Subscription revenue is recognized over the term of the subscription. Tomorrow collected $20 million in subscription sales during its first year of operations. At December 31, the average subscription was one-fourth expired.

	($ in millions)	
When Advance Is Collected		
Cash	20	
Deferred subscription revenue		20
When Product Is Delivered		
Deferred subscription revenue	5	
Subscription revenue		5

This illustration highlights that deferred revenue gets reduced when the revenue associated with the advance has been recognized, either because the seller has delivered the goods or services as promised or because the buyer has forfeited the advance payment (for example, not redeeming a gift certificate before it expires). Less often, deferred revenue is reduced when an advance payment is returned by the seller to the buyer because the seller didn't deliver the goods or services as promised. This should sound familiar—recall from Chapter 5 that revenue recognition must be delayed until the seller has fulfilled its performance obligation. When revenue recognition is delayed, but some advanced payment is received, the seller reports a deferred revenue liability until delivery has occurred.

Additional Consideration

Accounting for Interest on Advances from Customers. Earlier in this chapter you saw that we calculate interest expense on "noninterest-bearing" notes. After all, there is a "time value of money," and nobody is willing to loan money for free. The same reasoning suggests that companies should calculate interest expense when a customer pays far in advance of delivery of a product or service. As indicated in Chapter 6, that's what GAAP for revenue recognition requires.[13]

For example, assume that on January 1, 2021, Lewis Manufacturing Co. enters into a contract to deliver equipment to Williams Surgical Supply, Inc. on December 31, 2022. Williams pays $907 (in thousands) to Lewis at the start of the contract (January 1, 2021). In this arrangement, Williams pays Lewis significantly in advance of delivery, so Williams can be viewed as loaning money to Lewis, and Lewis should recognize interest expense on the loan. Assuming an effective interest rate of 5%, Lewis would recognize interest expense on its deferred revenue liability as follows:

January 1, 2021
When prepayment occurs:

Cash	907*	
Deferred revenue		907

*$907 = $1,000 × 0.90703 (present value of $1, n = 2, I = 5%; from Table 2)

(continued)

[13]Source: FASB ASC 606-10-32-16: Revenue from Contracts with Customers-Overall-Measurement-The Existence of a Significant Financing Component in the Contract.

(concluded)

December 31, 2021

Accrual of year 1 interest expense:

Interest expense ($907 × 5%)	45	
Interest payable		45

December 31, 2022

When subsequent delivery occurs:

Interest expense (($907 + $45) × 5%)	48	
Interest payable	45	
Deferred revenue	907	
Sales revenue		1,000

Lewis recognizes revenue of $1,000 rather than $907 because that is the fair value it received from Williams, including the $907 received up front and $45 + $48 = $93 of interest that Williams is allowing Lewis to not pay. Put differently, the equipment must be worth $1,000 for Williams to be willing to accept it in exchange for giving up $907 cash and $93 of interest it is owed by Lewis.

As discussed in Chapter 6, companies can ignore the interest component of advance payments if it is not significant, which is assumed to be the case if the period between payment and delivery is less than one year.

Gift Cards

Gift cards or gift certificates are particularly common forms of advanced payments. As indicated in Chapter 6, when a company sells a gift card, it initially records the cash received as deferred revenue, and then recognizes revenue either when the gift card is redeemed or when the probability of redemption is viewed as remote (called *gift card breakage,* and based on expiration or the company's experience). The amounts involved can be significant. For example, **Best Buy**'s 2017 annual report lists a $472 million liability for unredeemed gift cards and $37 million of income from gift card breakage. Illustration 13–11 shows the journal entries used to account for gift cards.

Illustration 13–11

Accounting for Gift Cards

During May 2021, Great Buy, Inc., sold $2 million of gift cards. Also during May, $1.5 million of gift cards sold in May and in prior periods were redeemed by customers, and $1 million of gift cards sold in prior periods expired and unused. Great Buy would make the following journal entries:

To record sale of gift cards

Cash	2,000,000	
Deferred gift card revenue		2,000,000

To record redemption of gift cards (ignoring entries to inventory and cost of sales)

Deferred gift card revenue	1,500,000	
Revenue—gift cards		1,500,000

To record expiration of gift cards

Deferred gift card revenue	1,000,000	
Revenue—gift cards		1,000,000

Liabilities for deferred revenue are classified as current or long-term depending on when the obligation is expected to be satisfied.

Collections for Third Parties

Companies often make collections for third parties from customers or from employees and periodically remit these amounts to the appropriate governmental (or other) units. Amounts collected this way represent liabilities until remitted.

Additional Consideration

Escheatment laws specify the rights of a state to claim abandoned property. Unused gift cards are a type of abandoned (or unclaimed) property. While most states exempt gift cards from escheatment laws, some do not. In those states where gift cards are subject to escheatment laws, after a period of three to five years, retailers are required to turn over all or part of the unredeemed value of gift cards to the state. To see how this works, let's modify Illustration 13–11 and assume that Great Buy is subject to an escheatment law that requires payment of 100% of unredeemed gift cards to the state in which Great Buy is incorporated. The third journal entry of Illustration 13–11 would be modified as follows:

To record expiration of gift cards

Deferred gift card revenue (amount of expired gift cards)	1,000,000	
Cash (to balance)		1,000,000

An example is sales taxes. For illustration, assume a state sales tax rate of 4% and local sales tax rate of 3%. Adding the tax to a $100 sale creates a $7 liability until the tax is paid.

Cash (or accounts receivable)	107	
Sales revenue		100
Sales taxes payable [(4% + 3%) × $100]		7

Sales taxes collected from customers represent liabilities until remitted.

Payroll-related deductions such as withholding taxes, Social Security taxes, employee insurance, employee contributions to retirement plans, and union dues also create current liabilities until the amounts collected are paid to appropriate parties. These payroll-related liabilities are discussed further in the appendix to this chapter.

Amounts collected from employees in connection with payroll represent liabilities until remitted.

A Closer Look at the Current and Noncurrent Classification

● LO13–4

Given a choice, do you suppose management would prefer to report an obligation as a current liability or as a noncurrent liability? Other things being equal, most would choose the noncurrent classification. The reason is that in most settings outsiders (like banks, bondholders, and shareholders) consider debt that is payable currently to be riskier than debt that need not be paid for some time, because the current payable requires the company to be able to access the necessary cash relatively soon. Also, the long-term classification enables the company to report higher working capital (current assets minus current liabilities) and a higher current ratio (current assets/current liabilities). Working capital and the current ratio often are explicitly restricted in loan contracts. As you study this section, you should view the classification choice from this perspective. That is, a manager might not so much ask the question "What amount should I report as a current liability?" but rather "What amount can I exclude from classification as a current liability?"

Current Maturities of Long-Term Debt

FINANCIAL Reporting Case

Q2, p. 718

Long-term obligations (bonds, notes, lease liabilities, deferred tax liabilities) usually are reclassified and reported as current liabilities when they become payable within the upcoming year (or operating cycle, if longer than a year). For example, a 20-year bond issue is reported as a long-term liability for 19 years but normally is reported as a current liability on the balance sheet prepared during the 20th year of its term to maturity.[14] General Mills reported $604.7 million of its long-term debt as a current liability in 2017 (see Illustration 13–1).

Obligations Callable by the Creditor

The currently maturing portion of a long-term debt must be reported as a current liability.

The requirement to classify currently maturing debt as a current liability includes debt that is *callable* (in other words, due on demand) *by the creditor* in the upcoming year (or

[14]Debt to be refinanced is an exception we discuss later.

operating cycle, if longer), even if the debt is not expected to be called. The current liability classification also is intended to include situations in which the creditor has the right to demand payment because an *existing violation* of a provision of the debt agreement makes it callable (say, working capital has fallen below a minimum covenant specified by a debt agreement). This also includes situations in which debt is not yet callable but will be callable within the year if an existing violation is not corrected within a specified grace period (unless it's probable the violation will be corrected within the grace period or waived by the creditor).[15]

When Short-Term Obligations Are Expected to Be Refinanced

FINANCIAL Reporting Case

Q3, p. 718

Reconsider the 20-year bond issue we discussed earlier. Normally we would reclassify it as a current liability on the balance sheet prepared during its 20th year. But suppose a second 20-year bond issue is sold specifically to refund the first issue when it matures. Do we have a long-term liability for 19 years, then a current liability in year 20, and then another long-term liability in years 21 and beyond? Or, do we have a single 40-year, long-term liability? If we look beyond the outward form of the transactions, the substance of the events obviously supports a single, continuing, noncurrent obligation. The concept of substance over form influences the classification of obligations expected to be refinanced.

Short-term obligations can be reported as noncurrent liabilities if the company (a) *intends* to refinance on a long-term basis and (b) demonstrates the *ability* to do so by a refinancing agreement or by actual financing.

Short-term obligations (including the callable obligations we discussed in the previous section) that are expected to be refinanced on a long-term basis can be reported as noncurrent, rather than current, liabilities only if two conditions are met:

(1) The company must intend to refinance on a long-term basis, and

(2) the company must actually have demonstrated the ability to refinance on a long-term basis.

Ability to refinance on a long-term basis can be demonstrated by either an existing refinancing agreement or by actual financing prior to the issuance of the financial statements.[16] Illustration 13–12 provides an example.

Illustration 13–12

Short-Term Obligations that Are Expected to Be Refinanced on a Long-Term Basis

Brahm Bros. Ice Cream had $12 million of notes that mature in May 2022 and also had $4 million of bonds issued in 1992 that mature in February 2022. On December 31, 2021, the company's fiscal year-end, management intended to refinance both on a long-term basis.

On February 7, 2022, the company issued $4 million of 20-year bonds, applying the proceeds to repay the bond issue that matured that month. In early March, prior to the actual issuance of the 2021 financial statements, Brahm Bros. negotiated a line of credit with a commercial bank for up to $7 million any time during 2022. Any borrowings will mature two years from the date of borrowing. Interest is at the prime London interbank borrowing rate.*

	December 31, 2021 ($ in thousands)
Classification	
Current Liabilities	
Notes payable	$5,000
Long-Term Liabilities	
Notes payable	$7,000
Bonds payable	4,000

Management's ability to refinance the $4 million of bonds on a long-term basis was demonstrated by actual financing prior to the issuance of the financial statements. Management's ability to refinance $7 million of the $12 million of notes was demonstrated by a refinancing agreement. The remaining $5 million must be reported as a current liability.

*This is a widely available rate often used as a basis for establishing interest rates on lines of credit and often abbreviated as LIBOR.

[15]FASB ASC 470–10–45: Debt—Overall—Other Presentation Matters.

[16]FASB ASC 470–10–45–14: Debt—Overall—Other Presentation Matters.

If shares of stock had been issued to refinance the bonds in the illustration, the bonds still would be excluded from classification as a current liability. The specific form of the long-term refinancing (bonds, bank loans, equity securities) is irrelevant when determining the appropriate classification. Requiring companies to actually demonstrate the ability to refinance on a long-term basis in addition to merely intending to do so avoids intentional or unintentional understatements of current liabilities.

It's important to remember that several weeks usually pass between the end of a company's fiscal year and the date the financial statements for that year actually are issued. Events occurring during that period can be used to clarify the nature of financial statement elements at the reporting date. Here we consider refinancing agreements and actual securities transactions to support a company's ability to refinance on a long-term basis. Later in the chapter we use information that becomes available during this period to decide how loss contingencies are reported.

International Financial Reporting Standards

Classification of Liabilities to be Refinanced. Under U.S. GAAP, liabilities payable within the coming year are classified as long-term liabilities if refinancing is completed before the date of issuance of the financial statements. Under IFRS, refinancing must be completed before the balance sheet date.[17]

● LO13–7

Concept Review Exercise

CURRENT LIABILITIES

The following selected transactions relate to liabilities of Southern Communications, Inc., for portions of 2021 and 2022. Southern's fiscal year ends on December 31.

Required:

Prepare the appropriate journal entries for these transactions.

2021

July	1	Arranged an uncommitted short-term line of credit with First City Bank amounting to $25,000,000 at the bank's prime rate (11.5% in July). The company will pay no commitment fees for this arrangement.
Aug.	9	Received a $30,000 refundable deposit from a major customer for copper-lined mailing containers used to transport communications equipment.
Oct.	7	Received most of the mailing containers covered by the refundable deposit and a letter stating that the customer will retain containers represented by $2,000 of the deposit and will forfeit that amount. The cost of the forfeited containers was $1,500.
Nov.	1	Borrowed $7 million cash from First City Bank under the line of credit arranged in July and issued a nine-month promissory note. Interest at the prime rate of 12% was payable at maturity.
Dec.	31	Recorded appropriate adjusting entries for the liabilities described above.

2022

Feb.	12	Using the unused portion of the credit line as support, issued $9 million of commercial paper and issued a six-month promissory note. Interest was discounted at issuance at a 10% discount rate.
Aug.	1	Paid the 12% note at maturity.
	12	Paid the commercial paper at maturity.

2021

July 1

No entry is made for a line of credit until a loan actually is made. The existence and terms of the line would be described in a disclosure note.

August 9

Cash	30,000	
Liability—refundable deposits		30,000

[17]"Presentation of Financial Statements" *International Accounting Standard No. 1* (IASCF), as amended effective January 1, 2018.

October 7		
Liability—refundable deposits	30,000	
Cash		28,000
Revenue—sale of containers		2,000
Cost of goods sold	1,500	
Inventory of containers		1,500
November 1		
Cash	7,000,000	
Notes payable		7,000,000
December 31		
Interest expense ($7,000,000 × 12% × 2⁄12)	140,000	
Interest payable		140,000

2022

February 12		
Cash [$9,000,000 − ($9,000,000 × 10% × 6⁄12)]	8,550,000	
Discount on notes payable (difference)	450,000	
Notes payable		9,000,000
Note that the effective interest rate is [($9,000,000 × 10% × 6⁄12) ÷ $8,550,000] × 12⁄6 = $450,000 ÷ $8,550,000 × 2 = 10.53%.		
August 1		
Interest expense ($7,000,000 × 12% × 7⁄12)	490,000	
Interest payable (from adjusting entry)	140,000	
Notes payable (face amount)	7,000,000	
Cash ($7,000,000 + $630,000)		7,630,000
August 12		
Interest expense ($9,000,000 × 10% × 6⁄12)	450,000	
Discount on notes payable		450,000
Notes payable (face amount)	9,000,000	
Cash ($8,550,000 + $450,000)		9,000,000

PART B Contingencies

The feature that distinguishes the loss contingencies we discuss in this part of the chapter from the liabilities we discussed previously is uncertainty as to whether an obligation really exists. The circumstance giving rise to the contingency already has occurred, but there is uncertainty about whether a liability exists that will be resolved only when some future event occurs (or doesn't occur).

Loss Contingencies

● LO13–5

General Motors Company's 2017 financial statements indicate a variety of potential obligations, as shown in Illustration 13–13.

Illustration 13–13

Disclosure of Potential Contingent Losses—General Motors Company

Real World Financials

17. Commitments and Contingencies (in part)

Various other legal actions, governmental investigations, claims and proceedings are pending against us or our related companies or joint ventures, including matters arising out of alleged product defects; employment-related matters; governmental regulations relating to product and workplace safety, vehicle emissions, including CO2 and nitrogen oxide, fuel economy, and related governmental regulations; product warranties; financial services; dealer, supplier and other contractual relationships; government regulations relating to

(continued)

(concluded)

payments to foreign companies; government regulations relating to competition issues; tax-related matters not subject to the provision of ASC 740, Income Taxes (indirect tax-related matters); product design, manufacture and performance; consumer protection laws; and environmental protection laws, including laws regulating air emissions, water discharges, waste management and environmental remediation.

These "legal actions, governmental investigations, claims and proceedings" all relate to situations that already have occurred. However, it isn't clear that these situations have given rise to liabilities because GM doesn't know if it will have to make future payments. How likely is an unfavorable outcome, and how much will GM have to pay if an unfavorable outcome occurs?

A **loss contingency** is an existing, uncertain situation involving potential loss depending on whether some future event occurs. Whether a contingency is accrued and reported as a liability depends on (a) the likelihood that the confirming event will occur and (b) what can be determined about the amount of loss. For example, consider lawsuits filed against GM by customers who allege they suffered injuries as a result of faulty ignition switches in GM cars that were recalled in 2014. GM's note 17 discloses much information about such claims, including over 100 class-action lawsuits and several hundred personal-injury lawsuits filed against it, as of February 6, 2018. It may take several years for those lawsuits to be settled or litigated. When considering how to account for those lawsuits prior to their resolution, GM must assess the likelihood that it eventually will have to pay damages and, if so, what the amount of those damages will be.

A *loss contingency* arises when there is uncertainty about whether a past event will result in a future loss. The uncertainty will be resolved only when some future event occurs.

Note that we only account for a loss contingency when the event that gave rise to it occurred before the financial statement date. Otherwise, regardless of the likelihood of the eventual outcome, no liability existed at the statement date. Remember, one of the essential characteristics of a liability is that it results from past transactions or events. In our GM example, GM previously sold the cars to customers, so the event giving rise to potential litigation losses has occurred. The uncertainty relates not to that past event, but to the potential litigation losses that could occur in the future.

FINANCIAL Reporting Case

Q4, p. 718

Generally accepted accounting principles require that the likelihood that the future event(s) will confirm the incurrence of the liability be (somewhat arbitrarily) categorized as probable, reasonably possible, or remote:[18]

Probable	Confirming event is likely to occur.
Reasonably possible	The chance the confirming event will occur is more than remote but less than likely.
Remote	The chance the confirming event will occur is slight.

● LO13–6

Also key to reporting a contingent liability is its dollar amount. The amount of the potential loss is classified as either known, reasonably estimable, or not reasonably estimable.

A liability is accrued if it is both probable that the confirming event will occur and the amount can be at least reasonably estimated. A general depiction of the accrual of a loss contingency is

Loss (or expense)	x,xxx	
Liability		x,xxx

If one amount within a range of possible loss appears better than other amounts within the range, that amount is accrued. When no amount within the range appears more likely than others, the minimum amount should be recorded and the possible additional loss should be disclosed.[19]

[18]Because FASB ASC 740–10–25: Income Taxes—Overall—Recognition provides guidance on accounting for uncertainty in income taxes, FASB ASC 450–10: Contingencies—Loss Contingencies does not apply to income taxes. GAAP regarding uncertainty in income taxes changes the threshold for recognition of tax positions from the most probable amount to the amount that has a "more likely than not" chance of being sustained upon examination. We discuss accounting for uncertainty in income taxes in Chapter 16.
[19]FASB ASC 450–20–30: Contingencies—Loss Contingencies—Initial Measurement.

As an example, consider the following disclosure note from **Merck & Co., Inc.**'s 2017 financial statements (Illustration 13–14).

Illustration 13–14
Accrual of a Loss Contingency—Merck & Co., Inc.

Real World Financials

11. Contingencies and Environmental Liabilities (in part)

In management's opinion, the liabilities for all environmental matters that are probable and reasonably estimable have been accrued and totaled $82 million and $83 million at December 31, 2017 and 2016, respectively. These liabilities are undiscounted, do not consider potential recoveries from other parties and will be paid out over the periods of remediation for the applicable sites, which are expected to occur primarily over the next 15 years. Although it is not possible to predict with certainty the outcome of these matters, or the ultimate costs of remediation, management does not believe that any reasonably possible expenditures that may be incurred in excess of the liabilities accrued should exceed $63 million in the aggregate. Management also does not believe that these expenditures should result in a material adverse effect on the Company's financial position, results of operations, liquidity or capital resources for any year.

Source: Merck & Co., Inc.

Consistent with GAAP, Merck accrued and disclosed the $82 million loss that is probable and reasonably estimable, and then only disclosed the $63 million estimated range of reasonably possible losses above that amount.

It's important to note that some contingent losses don't involve liabilities at all. Rather, these contingencies, when resolved, cause a noncash asset to be impaired, so accruing the contingency means reducing the related asset rather than recording a liability:

Loss (or expense)	x,xxx	
Asset (or valuation account)		x,xxx

The most common loss contingency of this type is an uncollectible receivable. You have recorded these before without knowing you were accruing a loss contingency (*Debit:* bad debt expense; *Credit:* allowance for uncollectible accounts).[20]

A loss contingency is disclosed in notes to the financial statements if there is at least a reasonable possibility that the loss will occur.

Not all loss contingencies are accrued. If one or both criteria for accrual are not met, but there is at least a reasonable possibility that a loss will occur, a disclosure note should describe the contingency. It also should provide an estimate of the potential loss or range of loss, if possible. If an estimate cannot be made, a statement to that effect is needed.

As an example of only disclosing a contingent loss, consider a disclosure note accompanying a 2017 annual report of **United States Steel Corporation**. As shown in Illustration 13–15, U.S. Steel considered it to be impossible to estimate its potential environmental liability associated with previously sold properties and businesses, and so only disclosed that contingent loss.

Illustration 13–15
Disclosure of Loss Contingency—U. S. Steel

Real World Financials

Note 25: Contingencies and Commitments (in part)

Throughout its history, U.S. Steel has sold numerous properties and businesses and many of these sales included indemnifications and cost sharing agreements related to the assets that were divested. . . . The amount of potential environmental liability associated with these transactions and properties is not estimable due to the nature and extent of the unknown conditions related to the properties divested and deconsolidated.

Source: U.S. Steel

Illustration 13–16 highlights the appropriate accounting treatment for each possible combination of (a) the likelihood of an obligation being confirmed and (b) the determinability of its dollar amount.

[20]FASB ASC 310–10–35–7: Receivables—Overall—Subsequent Measurement—Losses from Uncollectible Receivables.

Likelihood	Dollar Amount of Potential Loss: Known	Reasonably Estimable	Not Reasonably Estimable
Probable	Liability accrued and disclosure note	Liability accrued and disclosure note	Disclosure note only
Reasonably possible	Disclosure note only	Disclosure note only	Disclosure note only
Remote	No disclosure required*	No disclosure required*	No disclosure required*

*Except for certain guarantees and other specified off-balance-sheet risk situations discussed in the next chapter.

Illustration 13–16
Accounting Treatment of Loss Contingencies

Product Warranties and Guarantees

MANUFACTURER'S QUALITY-ASSURANCE WARRANTY Satisfaction guaranteed! Your money back if not satisfied! If anything goes wrong in the first five years or 100,000 miles . . . ! Three-year guarantee! These and similar promises accompany most consumer goods. In Chapter 6 we called this sort of guarantee a *quality-assurance warranty.* When this type of warranty is included in a contract between a seller and a customer, it isn't a separate performance obligation for the seller. Rather, it is a guarantee by the seller that the customer will be satisfied with the goods or services that the seller provided.

Most consumer products are accompanied by a guarantee.

Why do sellers offer quality-assurance warranties? To boost sales. It follows, then, that any costs of making good on such guarantees should be estimated and recorded as expenses in the same accounting period the products are sold (matching expenses with revenues). Also, it is in the period of sale that the company becomes obligated to eventually make good on a guarantee, so it makes sense that it recognizes a liability in the period of sale. The challenge is that much of the cost of satisfying a guarantee usually occurs later, sometimes years later. So, this is a loss contingency. There may be a future sacrifice of economic benefits (cost of satisfying the guarantee) due to an existing circumstance (the guaranteed products have been sold) that depends on an uncertain future event (customer claim).

The costs of satisfying guarantees should be recorded as expenses in the same accounting period the products are sold.

The criteria for accruing a contingent loss (rather than only disclosing it) almost always are met for product warranties (or product guarantees). While we usually can't predict the liability associated with an individual sale, reasonably accurate estimates of the *total* liability for a period usually are possible, because prior experience makes it possible to predict how many warrantees or guarantees (on average) will need to be satisfied. Illustration 13–17 demonstrates accrual of the contingent liability for warranties in the reporting period in which the product under warranty is sold.

The contingent liability for quality-assurance warranties almost always is accrued.

Caldor Health, a supplier of in-home health care products, introduced a new therapeutic chair carrying a two-year warranty against defects. Estimates based on industry experience indicate warranty costs of 3% of sales during the first 12 months following the sale and 4% the next 12 months. During December 2021, its first month of availability, Caldor sold $2 million of chairs.

During December

Cash (and accounts receivable)	2,000,000	
Sales revenue		2,000,000

December 31, 2021 (adjusting entry)

Warranty expense [(3% + 4%) × $2,000,000]	140,000	
Warranty liability		140,000

When customer claims are made and costs are incurred to satisfy those claims, the liability is reduced (let's say $61,000 in 2022):

Warranty liability	61,000	
Cash (or salaries payable, parts and supplies, etc.)		61,000

Illustration 13–17
Product Warranty

Estimates of warranty costs cannot be expected to be precise. However, if the estimating method is monitored and revised when necessary, overestimates and underestimates should cancel each other over time. The estimated liability may be classified as current or as part current and part long-term, depending on when warranty claims are expected to be satisfied.

SFAC No. 7 **provides a framework for using future cash flows in accounting measurements.**

EXPECTED CASH FLOW APPROACH In Chapter 5, you learned of a framework for using future cash flows as the basis for measuring assets and liabilities, introduced by the FASB in 2000 with *Statement of Financial Accounting Concepts No. 7,* "Using Cash Flow Information and Present Value in Accounting Measurements."[21] The approach described in the Concept Statement offers a way to take into account *any uncertainty concerning the amounts and timing of the cash flows.* Although future cash flows in many instances are contractual and certain, the amounts and timing of cash flows are less certain in other situations, such as warranty obligations.

As demonstrated in Illustration 13–17, the traditional way of measuring a warranty obligation is to report the "best estimate" of future cash flows, ignoring the time value of money on the basis of immateriality. However, when the warranty obligation spans more than one year and we can associate probabilities with possible cash flow outcomes, the approach described by *SFAC No. 7* offers a more plausible estimate of the warranty obligation. This "expected cash flow approach" incorporates specific probabilities of cash flows into the analysis. Illustration 13–18 provides an example.

Illustration 13–18

Product Warranty

Caldor Health, a supplier of in-home health care products, introduced a new therapeutic chair carrying a two-year warranty against defects. During December of 2021, its first month of availability, Caldor sold $2 million of the chairs. Industry experience indicates the following probability distribution for the potential warranty costs:

Probabilities are associated with possible cash outcomes.

Warranty Costs	Probability
2022	
$50,000	20%
60,000	50%
70,000	30%
2023	
$70,000	20%
80,000	50%
90,000	30%

An arrangement with a service firm requires that costs for the two-year warranty period be settled at the end of 2022 and 2023. The risk-free rate of interest is 5%. Applying the expected cash flow approach, at the end of the 2021 fiscal year, Caldor would record a warranty liability (and expense) of $131,564, calculated as follows:

The probability-weighted cash outcomes provide the expected cash flows.

The present value of the expected cash flows is the estimated liability.

2022	$50,000 × 20% =	$10,000	
	60,000 × 50% =	30,000	
	70,000 × 30% =	21,000	
		$61,000	
		× 0.95238*	$ 58,095
2023	$70,000 × 20% =	$14,000	
	80,000 × 50% =	40,000	
	90,000 × 30% =	27,000	
		$81,000	
		× 0.90703†	73,469
			$131,564

*Present value of $1, $n = 1$, $i = 5\%$ (from Table 2)
†Present value of $1, $n = 2$, $i = 5\%$ (from Table 2)

December 31, 2021 (adjusting entry)

Warranty expense	131,564	
Warranty liability (calculated above)		131,564

[21]Source: Using Cash Flow Information and Present Value in Accounting Measurements," *Statement of Financial Accounting Concepts No. 7* (Norwalk, Conn.: FASB, 2000). Recall that Concept Statements do not directly prescribe GAAP, but instead provide structure and direction to financial accounting.

EXTENDED WARRANTY CONTRACTS It's difficult these days to buy a computer, a digital camera, a car, or almost any durable consumer product without being asked to buy an extended warranty agreement. An extended warranty provides warranty protection beyond the manufacturer's original warranty. As discussed in Chapter 6, because an extended warranty is priced and sold separately from the warranteed product, it constitutes a separate performance obligation. So, rather than only worrying about how to recognize the contingent liability associated with an extended warranty, we face another accounting question: When should the revenue from the sale of an extended warranty be recognized?

Revenue is recognized when performance obligations are satisfied, not necessarily when cash is received. Because an extended warranty provides coverage over a period of time, these arrangements typically qualify for revenue recognition over the period of coverage. However, cash typically is received up front, when the extended warranty is sold. Similar to other advanced payments for future products and services, revenue from extended warranty contracts is not recognized immediately, but instead is recorded as a deferred revenue liability at the time of sale and recognized as revenue over the contract period, typically on a straight-line basis. We demonstrate accounting for extended warranties in Illustration 13–19.

Illustration 13–19

Extended Warranty

Brand Name Appliances sells major appliances that carry a one-year manufacturer's warranty. Customers are offered the opportunity at the time of purchase to also buy a three-year extended warranty for an additional charge. On January 3, 2021, Brand Name sold a $60 extended warranty, covering years 2022, 2023, and 2024.

January 3, 2021		
Cash (or accounts receivable)	60	
Deferred revenue—extended warranties		60
December 31, 2022, 2023, 2024 (adjusting entries)		
Deferred revenue—extended warranties	20	
Revenue—extended warranties ($60 ÷ 3)		20

The costs incurred to satisfy customer claims under the extended warranties also will be recorded during the same three-year period. That way, net income in each year of the extended warranty will reflect both the warranty revenue recognized and the costs associated with that revenue.

Additional Consideration

Accounting for Rebates, Premiums, and Coupons. Sometimes a company promotes its products in ways that obligate the company to do something in the future. For example, at the time of a sale, customers obtain cash register receipts, bar codes on the product, or other proofs of purchase that later can be mailed to the manufacturer for *cash rebates*. Also, customers sometimes can show proof of purchase to obtain noncash items (called *premiums*), like toys, small appliances, or dishes. Companies may also mail out *coupons*, which entice customers to buy a company's products and services at a discounted price at some point in the future. How should we account for the future obligations created by these promotions?

Until very recently, all of these promotions were viewed as creating a contingent liability. After all, they obligate the seller to transfer some uncertain amount of cash, goods, or services at a future date. However, with recent changes in accounting for revenue recognition, the accounting in this area has changed. Now we view promises to provide rebates and premiums to be part of agreements between sellers and customers. This means we account for them using the revenue recognition process we learned about in Chapter 6.

Cash rebates are an obligation to return cash in the future, so they represent a reduction in the net amount paid by the customer. To record these, we estimate the amount of cash rebates that customers will take and reduce revenue by the amount of that estimate in the period revenue is recognized. We also recognize a liability for the estimated rebate, similar to how we recognize a refundable deposit.

(continued)

(concluded)

Premiums obligate a company to provide noncash items and are treated as separate performance obligations. We allocate a portion of the original sales price to the premiums (based on their relative stand-alone selling prices), record that amount as deferred revenue, and recognize that portion as revenue when premiums are delivered. No contingent liability is recognized because, as with extended warranties, we recognize the cost of premiums in the period in which we deliver them and recognize revenue.

Coupons that aren't offered as part of a sales contract aren't accounted for as part of revenue recognition. Instead, issuing coupons creates a contingent liability that is recognized in the period the coupons are issued. In practice, firms typically either (a) recognize the entire expense associated with estimated coupon redemptions in the period the coupons are issued, or (b) recognize no liability in the period the coupons are issued and instead record the expense when coupons are redeemed. The difference between these approaches typically is not viewed as material.

Litigation Claims

Pending litigation is not unusual. In fact, as you saw for **GM**, the majority of medium and large corporations annually report multiple loss contingencies due to litigation.

In practice, accrual of a loss from pending or ongoing litigation is rare. Can you guess why? Suppose you are chief financial officer of Feinz Foods. Feinz is the defendant in a $44 million class-action suit. The company's legal counsel informally advises you that the chance the company will win the lawsuit is quite doubtful. Counsel feels the company might lose $30 million. Now suppose you decide to accrue a $30 million loss in your financial statements. Later, in the courtroom, your disclosure that Feinz management feels it is probable that the company will lose $30 million would be welcome ammunition for the opposing legal counsel. Understanding this, most companies rely on the knowledge that in today's legal environment the outcome of litigation is highly uncertain, making likelihood predictions difficult.

Therefore, while companies may accrue estimated lawyer fees and other legal costs, they usually do not record a loss until after the ultimate settlement has been reached or negotiations for settlement are substantially completed. While companies should provide extensive disclosure of these contingent liabilities, they do not always do so in practice.[22] In 2010, the SEC began pressuring firms for more complete disclosure of litigation, including better descriptions of the range of losses that are reasonably possible to occur. As an example, Illustration 13–20 provides a few excerpts from the eight pages of litigation note appearing in **JP Morgan Chase & Company**'s 2016 annual report.

Illustration 13–20

Disclosure of Litigation Contingencies—JP Morgan Chase & Company

Real World Financials

Note 31: Litigation (in part)

As of December 31, 2016, the Firm and its subsidiaries and affiliates are defendants or putative defendants in numerous legal proceedings, including private, civil litigations and regulatory/government investigations. . . . The Firm believes the estimate of the aggregate range of reasonably possible losses, in excess of reserves established, for its legal proceedings is from $0 to approximately $3.0 billion at December 31, 2016. This estimated aggregate range of reasonably possible losses was based upon currently available information for those proceedings in which the Firm believes that an estimate of reasonably possible loss can be made. For certain matters, the Firm does not believe that such an estimate can be made, as of that date.

Source: JP Morgan Chase & Company

[22]For research indicating incomplete disclosure of litigation-related contingent liabilities, see R. Desir, K. Fanning, and R. Pfeiffer, "Are Revisions to SFAS No. 5 Needed?" *Accounting Horizons* 24(4), December 2010, pp. 525–546.

Even after a firm loses in court, it may not make an accrual. As you can see in Illustration 13–21, **Ameris Bancorp**, in its 2016 annual report, disclosed but did not accrue damages from a lawsuit it lost, because the company was appealing the verdict.

Illustration 13–21

Disclosure of a Lawsuit—Ameris Bancorp

Real World Financials

> **Note 19: Commitments and Contingent Liabilities (in part): Contingencies**
>
> As of December 31, 2016, the Company believes that it has valid bases in law and fact to overturn on appeal the verdict. As a result, the Company believes that the likelihood that the amount of the judgment will be affirmed is not probable, and, accordingly, that the amount of any loss cannot be reasonably estimated at this time. Because the Company believes that this potential loss is not probable or estimable, it has not recorded any reserves or contingencies related to this legal matter.
>
> Source: Ameris Bancorp

Subsequent Events

Several weeks usually pass between the end of a company's fiscal year and the date the financial statements for that year actually are issued or available to be issued.[23] As shown on the following time line, events occurring during this period can be used to clarify the nature of financial statement elements at the report date.

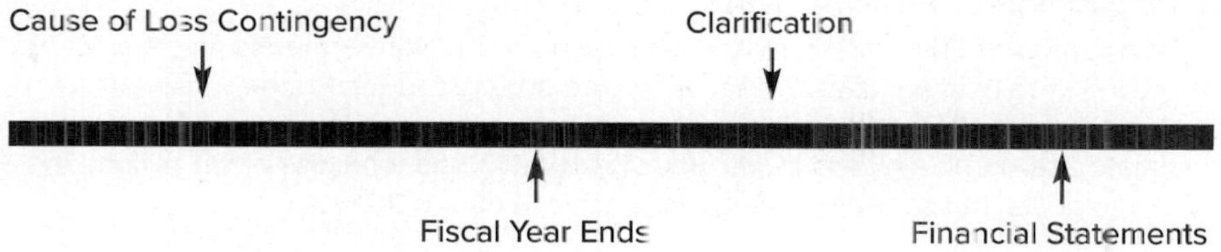

When the cause of a loss contingency occurs before the year-end, a clarifying event before financial statements are issued can be used to determine how the contingency is reported.

If information becomes available that sheds light on a claim that existed when the fiscal year ended, that information should be used in determining the probability of a loss contingency materializing and in estimating the amount of the loss.

The settlement of a lawsuit after **Starbucks**' September 29, 2013, fiscal year ended apparently influenced its accrual of a loss contingency (Illustration 13–22).

Illustration 13–22

Accrual of Litigation Contingencies—Starbucks

Real World Financials

> **Note 15: Commitments and Contingencies (in part)**
>
> ***Legal Proceedings***
>
> On December 6, 2010, Kraft commenced a federal court action against Starbucks, entitled Kraft Foods Global, Inc. v. Starbucks Corporation, in the U.S. District Court for the Southern District of New York. On November 12, 2013, the arbitrator ordered Starbucks to pay Kraft $2,227.5 million in damages plus prejudgment interest and attorney's fees. We have estimated prejudgment interest, which includes an accrual through the estimated payment date, and attorneys' fees to be approximately $556.6 million. As a result, we recorded a litigation charge of $2,784.1 million in our fiscal 2013 operating results.
>
> Source: Starbucks

For a loss contingency to be accrued, the cause of the lawsuit must have occurred before the accounting period ended. It's not necessary that the lawsuit actually was filed during that reporting period.

[23]Financial statements are viewed as *issued* if they have been widely distributed to financial statement users in a format consistent with GAAP. Some entities (for example, private companies) do not widely distribute their financial statements to users. For these entities, the key date for subsequent events is not the date of issuance but rather the date upon which the financial statements are *available to be issued,* which occurs when the financial statements are complete, in a format consistent with GAAP, and have obtained the necessary approvals for issuance. Entities must disclose the date through which subsequent events have been evaluated (FASB ASC 855: Subsequent Events.

Sometimes, the cause of a loss contingency occurs after the end of the year but before the financial statements are issued:

If an event giving rise to a contingency occurs after the year-end, a liability should not be accrued.

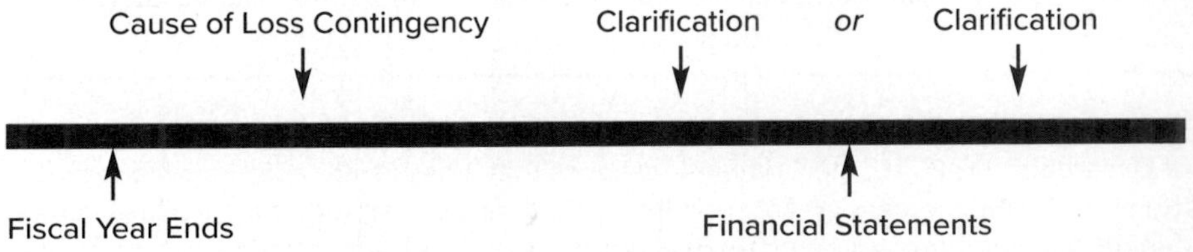

When a contingency comes into existence after the company's fiscal year-end, a liability cannot be accrued because it didn't exist at the end of the year. However, if the failure to disclose the possible loss would cause the financial statements to be misleading, the situation should be described in a disclosure note, including the effect of the possible loss on key accounting numbers affected.[24]

In fact, *any* event occurring after the fiscal year-end but before the financial statements are issued that has a material effect on the company's financial position must be disclosed in a subsequent events disclosure note. Examples are an issuance of debt or equity securities, a business combination, and discontinued operations.

A disclosure note from the **Coca-Cola Company**'s 2016 annual report is shown in Illustration 13–23 and describes events that occurred in the first quarter of 2017.

Illustration 13–23

Subsequent Events—Coca-Cola Company

Real World Financials

Note 21: Subsequent events (in part)

In the first quarter of 2017, additional bottling territories in North America met the criteria to be classified as held for sale. Therefore, we are required to record the related assets and liabilities at the lower of carrying value or fair value less any costs to sell based on the estimated selling prices, which will result in an estimated total loss of $9 million. The Company expects these territories to be refranchised during 2017.

Source: Coca-Cola Company

Additional **Consideration**

Contingent liabilities that a company takes on when it acquires another company may be treated differently from those that arise during the normal course of business. If the acquirer can determine the fair value of a contingent liability, the liability is measured at fair value. For example, the acquisition-date fair value of a warranty obligation often can be determined. If the acquirer cannot determine the fair value, then the acquirer uses the normal criteria that apply to contingent liabilities. That is, the contingent liability is accrued if (1) available information indicates that it is probable a liability has been incurred as of the acquisition date, and (2) the amount of the liability can be reasonably estimated.[25] Because some acquired contingencies could be accounted for at fair value, there could be a lack of comparability between contingencies that arise in the normal course of business and those that are obtained as part of an acquisition.

Unasserted Claims and Assessments

It must be probable that an unasserted claim or assessment will be asserted before considering whether and how to report the possible loss.

Sometimes companies are aware of a potential claim that has not yet been made. Such unasserted claims may require accrual or disclosure of a contingent liability. A two-step process is involved in deciding how the unasserted claim should be reported:

1. Is it probable that a claim will be asserted? If the answer to that question is "no," stop. No accrual or disclosure is necessary. If the answer is "yes," go on to step 2.
2. Treat the claim as if the claim has been asserted. That requires evaluating (a) the likelihood of an unfavorable outcome and (b) whether the dollar amount of loss can be estimated, just as we already have discussed for other loss contingencies for which a claim has already been asserted.

[24] FASB ASC 450–20–50v9: Contingencies—Loss Contingencies—Disclosure.

[25] FASB ASC 805-20-25: Business Combinations—Identifiable Assets and Liabilities, and Any Noncontrolling Interest—Recognition.

For example, suppose a trucking company frequently transports hazardous waste materials and is subject to environmental laws and regulations. Management has identified several sites at which it is or may be liable for remediation. For those sites for which no penalties have been asserted, management must assess the likelihood that a claim will be made, and if so, the likelihood that the company actually will be held liable. If management feels an assessment is probable, an estimated loss and contingent liability would be accrued only if an unfavorable outcome is probable and the amount can be reasonably estimated. However, a disclosure note alone would be appropriate if an unfavorable settlement is only reasonably possible or if the settlement is probable but cannot be reasonably estimated. No action is needed if chances of that outcome occurring are remote.

As described in a December 31, 2016, disclosure note (see Illustration 13–24), **Union Pacific** concluded that some unasserted claims met the criteria for accrual under this two-step decision process.

Illustration 13–24

Unasserted Claims—Union Pacific Corporation

Real World Financials

> **15. Commitments and Contingencies (in part)**
>
> **Asserted and Unasserted Claims—**Various claims and lawsuits are pending against us and certain of our subsidiaries. We cannot fully determine the effect of all asserted and unasserted claims on our consolidated results of operations, financial condition, or liquidity. To the extent possible, we have recorded a liability where asserted and unasserted claims are considered probable and where such claims can be reasonably estimated.
>
> Source: Union Pacific Corporation

Notice that the treatment of contingent liabilities is consistent with the accepted definition of liabilities as (a) probable, future sacrifices of economic benefits (b) that arise from present obligations to other entities and (c) that result from past transactions or events.[26] The inherent uncertainty involved with contingent liabilities means additional care is required to determine whether future sacrifices of economic benefits are probable and whether the amount of the sacrifices can be quantified.

International Financial Reporting Standards

● LO13–7

Loss Contingencies. Accounting for contingencies is part of a broader international standard, *IAS No. 37,* "Provisions, Contingent Liabilities and Contingent Assets." Overall, accounting for contingent losses under IFRS is quite similar to accounting under U.S. GAAP. A contingent loss is accrued if it's both probable and can be reasonably estimated, and disclosed if it's of at least a remote probability. However, there are some important differences:

- IFRS refers to accrued liabilities as "provisions," and refers to possible obligations that are not accrued as "contingent liabilities." The term "contingent liabilities" is used for all of these obligations in U.S. GAAP.
- IFRS requires disclosure (but not accrual) of two types of contingent liabilities: (1) possible obligations whose existence will be confirmed by some uncertain future events that the company does not control, and (2) a present obligation for which either it is not probable that a future outflow will occur or the amount of the future outflow cannot be measured with sufficient reliability. U.S. GAAP does not make this distinction but typically would require disclosure of the same contingencies.
- IFRS defines "probable" as "more likely than not" (greater than 50%), which is a lower threshold than typically associated with "probable" in U.S. GAAP.
- If a liability is accrued, IFRS measures the liability as the best estimate of the expenditure required to settle the present obligation. If there is a range of equally likely outcomes, IFRS would use the midpoint of the range, while U.S. GAAP requires use of the low end of the range.

(continued)

[26]"Elements of Financial Statements," *Statement of Financial Accounting Concepts No. 6* (Stamford, Conn.: FASB, 1985).

(concluded)

- If the effect of the time value of money is material, IFRS requires the liability to be stated at present value. U.S. GAAP allows using present values under some circumstances, but liabilities for loss contingencies like litigation typically are not discounted for time value of money.[27]
- IFRS recognizes provisions and contingencies for "onerous" contracts, defined as those in which the unavoidable costs of meeting the obligations exceed the expected benefits.[28] Under U.S. GAAP we generally don't disclose or recognize losses on such money-losing contracts, although there are some exceptions (for example, losses on long-term construction contracts are accrued, as are losses on contracts that have been terminated).

Here's a portion of a note from the 2016 financial statements of **Vodafone**, which reports under IFRS:

Note 17: Provisions (in part)

Provisions are recognised when the Group has a present obligation (legal or constructive) as a result of a past event, it is probable that the Group will be required to settle that obligation and a reliable estimate can be made of the amount of the obligation. Provisions are measured at the Directors' best estimate of the expenditure required to settle the obligation at the reporting date and are discounted to present value where the effect is material.

Where We're Headed

The controversy over accounting for contingencies and the mixture of approaches is likely to continue. The FASB recently removed from its agenda a project to reconsider the recognition and measurement of contingent losses, and has postponed further work on a project intended to enhance note disclosures of contingent losses. The IASB has a project ongoing, but progress on this project has been slow as the IASB has focused on more pressing projects.

Gain Contingencies

Gain contingencies are not accrued.

● LO13–7

A gain contingency is an uncertain situation that might result in a gain. For example, in a pending lawsuit, one side—the defendant—faces a loss contingency; the other side—the plaintiff—has a gain contingency. As we discussed earlier, loss contingencies are accrued when it's probable that an amount will be paid and the amount can reasonably be estimated. However, gain contingencies are not accrued. The nonparallel treatment of gain contingencies is an example of conservatism, following the reasoning that it's desirable to anticipate losses, but recognizing gains should await their realization.

Though gain contingencies are not recorded in the accounts, material ones are disclosed in notes to the financial statements. Care should be taken that the disclosure note not give "misleading implications as to the likelihood of realization."[29]

International Financial Reporting Standards

Gain Contingencies. Under U.S. GAAP, gain contingencies are never accrued. Under IFRS, gain contingencies are accrued if their future realization is "virtually certain" to occur. Both U.S. GAAP and IFRS *disclose* contingent gains when future realization is probable, but under IFRS "probable" is defined as "more likely than not" (greater than 50%), and so has a lower threshold than it does under U.S. GAAP.

[27]See, for example, FASB ASC 410–30–35–12: Asset Retirements and Environmental Obligations—Environmental Obligations—Subsequent Measurement.

[28]Paragraph 10 of "Provisions, Contingent Liabilities and Contingent Assets," *International Accounting Standard 37* (London, UK: IASB, 2008).

[29]Source: FASB ASC 450-30-50: Contingencies-Gain Contingencies-Disclosure.

Additional Consideration

Accounting for contingencies is controversial. Many accountants dislike the idea of only recognizing a liability for a contingent loss when it is probable, and then reporting the liability at the best estimate of the future expenditure. The obvious alternative is fair value, which does not incorporate probability into determining whether to *recognize* a loss but rather considers probability when *measuring* the amount of loss. Recent changes in GAAP have adopted fair value approaches for some types of events that are contingent losses (or appear to be close relatives). Consider the following examples:

- Illustration 13–18 uses a discounted expected cash flow approach to measure the contingent liability associated with a warranty. That approach approximates fair value.
- Recall from Chapter 10 that asset retirement obligations are recorded at fair value when an asset is acquired. The offsetting liability is similar to a contingent liability because it is an uncertain future amount arising from the past purchase of the asset but is recorded at fair value.
- Sometimes a company provides a guarantee that may require it to make payment to another party based on some future event. For example, a company might guarantee the debt of an affiliated company to make it easier for the affiliate to obtain financing. GAAP views this sort of guarantee as having two parts: (1) a certain "stand ready obligation" to meet the terms of the guarantee, and (2) the uncertain contingent obligation to make future payments depending on future events (for example, the affiliate defaulting on their debt). The "stand ready obligation" is recorded initially at fair value, while the contingent obligation is handled as an ordinary contingent loss.[30]

Concept Review Exercise

CONTINGENCIES

Hanover Industries manufactures and sells food products and food processing machinery. While preparing the December 31, 2021, financial statements for Hanover, the following information was discovered relating to contingencies and possible adjustments to liabilities. Hanover's 2021 financial statements were issued on April 1, 2022.

a. On November 12, 2021, a former employee filed a lawsuit against Hanover alleging age discrimination and asking for damages of $750,000. At December 31, 2021, Hanover's attorney indicated that the likelihood of losing the lawsuit was possible but not probable. On March 5, 2022, Hanover agreed to pay the former employee $125,000 in return for withdrawing the lawsuit.

b. Hanover believes there is a possibility a service provider may claim that it has been undercharged for outsourcing a processing service based on verbal indications of the company's interpretation of a negotiated rate. The service provider has not yet made a claim for additional fees as of April 2022, but Hanover thinks it will. Hanover's accountants and legal counsel believe the charges were appropriate but that if an assessment is made, there is a reasonable possibility that subsequent court action would result in an additional tax liability of $55,000.

c. Hanover grants a two-year warranty for each processing machine sold. Past experience indicates that the costs of satisfying warranties are approximately 2% of sales. During 2021, sales of processing machines totaled $21,300,000. 2021 expenditures for warranty repair costs were $178,000 related to 2021 sales and $220,000 related to 2020 sales. The January 1, 2021, balance of the warranty liability account was $250,000.

d. Hanover is the plaintiff in a $500,000 lawsuit filed in 2020 against Ansdale Farms for failing to deliver on contracts for produce. The suit is in final appeal. Legal counsel advises that it is probable that Hanover will prevail and will be awarded $300,000 (considered a material amount).

---

[30]Source: FASB ASC 460-10-25: Guarantees-Overall-Recobnition.

Required:

1. Determine the appropriate reporting for each situation. Briefly explain your reasoning.
2. Prepare any necessary journal entries and state whether a disclosure note is needed.

Solution:

a. This is a loss contingency. Hanover can use the information occurring after the end of the year in determining appropriate disclosure. The cause for the suit existed at the end of the year. Hanover should accrue the $125,000 loss because an agreement has been reached confirming the loss and the amount is known.

Loss—litigation	125,000	
Liability—litigation		125,000

A disclosure note also is appropriate.

b. At the time financial statements are issued, a claim is as yet unasserted. However, an assessment is probable. Thus, (a) the likelihood of an unfavorable outcome and (b) whether the dollar amount can be estimated are considered. No accrual is necessary because an unfavorable outcome is not probable. But because an unfavorable outcome is reasonably possible, a disclosure note is appropriate.

Note: If the likelihood of a claim being asserted is not probable, disclosure is not required even if an unfavorable outcome is thought to be probable in the event of an assessment and the amount is estimable.

c. The contingency for warranties should be accrued because it is probable that expenditures will be made and the amount can be estimated from past experience. When customer claims are made and costs are incurred to satisfy those claims the liability is reduced.

Warranty expense (2% × $21,300,000)	426,000	
Warranty liability		426,000
Warranty liability ($178,000 + $220,000)	398,000	
Cash (or salaries payable, parts and supplies, etc.)		398,000

The liability at December 31, 2021, would be reported as $278,000:

Warranty Liability
($ in thousands)

		250	Balance, Jan. 1
		426	2021 expense
2021 expenditures	398		
		278	Balance, Dec. 31

A disclosure note explaining the contingency also is appropriate.

d. This is a gain contingency. Gain contingencies cannot be accrued even if the gain is probable and reasonably estimable. The gain should be recognized only when realized. It can be disclosed, but care should be taken to avoid misleading language regarding the realizability of the gain.

Decision Makers' Perspective

Current liabilities impact a company's liquidity. Liquidity refers to a company's cash position and overall ability to obtain cash in the normal course of business. A company is said to be liquid if it has sufficient cash (or other assets convertible to cash in a relatively short time) to pay currently maturing debts. Because the lack of liquidity can cause the demise of an otherwise healthy company, it is critical that managers as well as outside investors and creditors maintain close scrutiny of this aspect of a company's well-being.

A risk analyst should be concerned with a company's ability to meet its short-term obligations.

Keeping track of the current ratio is one of the most common ways of doing this. The current ratio is intended as a measure of short-term solvency and is determined by dividing current assets by current liabilities.

When we compare liabilities that must be satisfied in the near term with assets that either are cash or will be converted to cash in the near term, we get a useful measure of a company's liquidity. A ratio of 1 to 1 or higher often is considered a rule-of-thumb standard, but like other ratios, acceptability should be evaluated in the context of the industry in which the company operates and other specific circumstances. Keep in mind, though, that industry averages are only one indication of adequacy and that the current ratio is but one indication of liquidity.

FINANCIAL Reporting Case

Q5, p. 718

We can adjust for the implicit assumption of the current ratio that all current assets are equally liquid. The acid-test, or quick, ratio is similar to the current ratio but is based on a more conservative measure of assets available to pay current liabilities. Specifically, the numerator, quick assets, includes only cash and cash equivalents, short-term investments, and accounts receivable. By eliminating current assets such as inventories and prepaid expenses that are less readily convertible into cash, the acid-test ratio provides a more rigorous indication of a company's short-term solvency than does the current ratio.

A manager should actively monitor a company's liquidity.

If either of these liquidity ratios is less than that of the industry as a whole, does that mean that liquidity is a problem? Not necessarily. It does, though, raise a red flag that suggests caution when assessing other areas. It's important to remember that each ratio is but one piece of the puzzle. For example, profitability is probably the best long-run indication of liquidity. Also, management may be very efficient in managing current assets so that some current assets—receivables or inventory—are converted to cash more quickly than they otherwise would be and are more readily available to satisfy liabilities. The turnover ratios discussed in earlier chapters help measure the efficiency of asset management in this regard.

A liquidity ratio is but one indication of a company's liquidity.

In fact, some companies view their accounts payable as a free loan from their suppliers, just as they view their accounts receivable as a free loan to their customers. These companies tend to pressure their customers to pay quickly, but try to obtain more extended terms with their suppliers. Although this would produce relatively low current assets and high current liabilities, and therefore a lower current ratio, it could be a very intelligent way to manage cash and decrease the overall amount of capital needed by the company to finance operations.

Some companies maintain relatively high current liabilities as part of a cash-management strategy.

Given the actual and perceived importance of a company's liquidity in the minds of analysts, it's not difficult to adopt a management perspective and imagine efforts to manipulate the ratios that measure liquidity. For instance, a company might use its economic muscle or persuasive powers to influence the timing of accounts payable recognition by asking suppliers to change their delivery schedules. Because accounts payable is included in the denominator in most measures of liquidity, such as the current ratio, the timing of their recognition could mean the difference between an unacceptable ratio and an acceptable one, or between violating a debt covenant and compliance with the terms of the debt agreement. For example, suppose a company with a current ratio of 1.25 (current assets of $5 million and current liabilities of $4 million) is in violation of a debt covenant requiring a minimum current ratio of 1.3. By delaying the delivery of $1 million of inventory, the ratio would be 1.33 (current assets of $4 million and current liability of $3 million).

Analysts should be alert for efforts to manipulate measures of liquidity.

It's important for creditors and analysts to be attentive for evidence of activities that would indicate timing strategies, such as unusual variations in accounts payable levels. You might notice that such timing strategies are similar to earnings management techniques we discussed previously—specifically, manipulating the timing of revenue and expense recognition in order to "smooth" income over time.

Finally, all financial statement users need to keep in mind the relation between the recognition of deferred revenue associated with advance payments and the later recognition of revenue. On the one hand, increases in deferred revenue can signal future revenue recognition because the deferred revenue likely will eventually be recognized as revenue. On the other hand, research suggests that some firms that report deferred revenue may manipulate the timing of revenue recognition to manage their earnings.[31]

Changes in deferred revenue provide information about future revenue, and may be used to manipulate it.

In the next chapter, we continue our discussion of liabilities. Our focus will shift from current liabilities to long-term liabilities in the form of bonds and long-term notes. ●

[31]Marcus Caylor, "Strategic Revenue Recognition to Achieve Earnings Benchmarks," *Journal of Accounting and Public Policy*, January–February 2010, pp. 82–95.

Financial Reporting Case Solution

©Cultura/Image Source

1. **What are accrued liabilities? What is commercial paper?** *(p. 723)* Accrued liabilities are reported for expenses already incurred but not yet paid (accrued expenses). These include salaries and wages payable, income taxes payable, and interest payable. Commercial paper is a form of notes payable sometimes used by large corporations to obtain temporary financing. It is sold to other companies as a short-term investment. It represents unsecured notes sold in minimum denominations of $25,000 with maturities ranging from 1 to 270 days. Typically, commercial paper is issued directly to the buyer (lender) and is backed by a line of credit with a bank.
2. **Why did Syntel Microsystems include some long-term debt in the current liability section?** *(p. 729)* Syntel Microsystems did include some long-term debt in the current liability section. The currently maturing portion of a long-term debt must be reported as a current liability. Amounts are reclassified and reported as current liabilities when they become payable within the upcoming year.
3. **Did they also report some current amounts as long-term debt? Explain.** *(p. 730)* Yes they did. It is permissible to report short-term obligations as noncurrent liabilities if the company (a) intends to refinance on a long-term basis and (b) demonstrates the ability to do so by a refinancing agreement or by actual financing. As the disclosure note explains, this is the case for a portion of Syntel's currently payable debt.
4. **Must obligations be known contractual debts in order to be reported as liabilities?** *(p. 733)* No. From an accounting perspective, it is not necessary that obligations be known, legally enforceable debts to be reported as liabilities. They must only be probable and the dollar amount reasonably estimable.
5. **Is it true that current liabilities are riskier than long-term liabilities?** *(p. 745)* Other things being equal, current liabilities generally are considered riskier than long-term liabilities. For that reason, management usually would rather report a debt as long term. Current debt, though, is not necessarily risky. The liquidity ratios we discussed in the chapter attempt to measure liquidity. Remember, any such measure must be assessed in the context of other factors: industry standards, profitability, turnover ratios, and risk management activities, to name a few. ●

The Bottom Line

- **LO13–1** Liabilities are present obligations to sacrifice assets in the future because of something that already has occurred. Current liabilities are expected to require current assets (or the creation of other current liabilities) and usually are payable within one year. *(p. 718)*
- **LO13–2** Short-term bank loans usually are arranged under an ongoing line of credit with a bank or group of banks. When interest is discounted from the face amount of a note (a type of noninterest-bearing note), the effective interest rate is higher than the stated discount rate. Large, highly rated firms sometimes sell commercial paper directly to the buyer (lender) to borrow funds at a lower rate than through a bank loan. *(p. 720)*
- **LO13–3** Accrued liabilities are recorded by adjusting entries for expenses already incurred, but for which cash has yet to be paid (accrued expenses). Familiar examples are advance collections, salaries payable, income taxes payable, and interest payable *(p. 723)*
- **LO13–4** Short-term obligations can be reported as noncurrent liabilities if the company (a) intends to refinance on a long-term basis and (b) demonstrates the ability to do so by actual financing or a formal agreement to do so. *(p. 729)*
- **LO13–5** A loss contingency is an existing, uncertain situation involving potential loss depending on whether some future event occurs. Whether a contingency is accrued and reported as a liability depends on (a) the likelihood that the confirming event will occur and (b) what can be determined about the amount of loss. It is accrued if it is both probable that the confirming event will occur and the amount can be at least reasonably estimated. *(p. 732)*
- **LO13–6** A clarifying event before financial statements are issued, but after the year-end, can be used to determine how a contingency is reported. An unasserted suit, claim, or assessment warrants accrual or disclosure if

it is probable it will be asserted. A gain contingency is a contingency that might result in a gain. A gain contingency is not recognized until it actually is realized. (*p. 733*)

LO13–7 IFRS and U.S. GAAP are relatively similar with respect to current liabilities and contingencies. Relatively minor differences relate to when financing must be in place for a liability expected to be refinanced to be classified as long term. Also, with respect to contingent losses, IFRS defines "probable" at a lower threshold, requires the accrual of the expected value of loss, and requires the use of present values when measuring amounts to be accrued. Contingent gains are not accrued under U.S. GAAP, but are accrued under IFRS when they are considered to be virtually certain to occur. (*p. 731, 741, and 742*)

Payroll-Related Liabilities

APPENDIX 13

All firms incur liabilities in connection with their payrolls. These arise primarily from legal requirements to withhold taxes from employees' paychecks and from payroll taxes on the firms themselves. Some payroll-related liabilities result from voluntary payroll deductions of amounts payable to third parties.

EMPLOYEES' WITHHOLDING TAXES Employers are required by law to withhold federal (and sometimes state) income taxes and Social Security taxes from employees' paychecks and remit these to the Internal Revenue Service. The amount withheld for federal income taxes is determined by a tax table furnished by the IRS and varies according to the amount earned and the number of exemptions claimed by the employee. Also, the Federal Insurance Contributions Act (FICA) requires employers to withhold a percentage of each employee's earnings up to a specified maximum. Both the percentage and the maximum are changed intermittently. As this text went to print, the deduction for Social Security was 6.2% of the first $128,400 an employee earns, and this ceiling amount changes each year for cost-of-living adjustments. Additionally, a deduction for Medicare tax is 1.45% with no limit on the base amount. The employer also must pay an equal (matching) amount on behalf of the employee. Individuals also must pay additional Medicare tax of 0.9% on wages in excess of $200,000 (there is no employer match against that additional Medicare tax). Self-employed persons must pay both the employer and employee portions (12.4% for Social Security and 2.9% for Medicare).

VOLUNTARY DEDUCTIONS Besides the required deductions for income taxes and Social Security taxes, employees often authorize their employers to deduct other amounts from their paychecks. These deductions might include union dues, contributions to savings or retirement plans, and insurance premiums. Amounts deducted this way represent liabilities until paid to the appropriate organizations.

EMPLOYERS' PAYROLL TAXES One payroll tax mentioned previously is the employer's matching amount of FICA taxes. The employer also must pay federal and state unemployment taxes on behalf of its employees. The Federal Unemployment Tax Act (FUTA) requires a tax of 6.0% of the first $7,000 earned by each employee. This amount is reduced by a 5.4% (maximum) credit for contributions to state unemployment programs, so the net federal rate often is 0.6%.[32] The most common state rate is 5.4%, but rates vary between states and also vary within states for different types of employers.

FRINGE BENEFITS In addition to salaries, withholding taxes, and payroll taxes, many companies provide employees a variety of fringe benefits. Most commonly, employers pay all or part of employees' insurance premiums and/or contributions to retirement income plans.

Representative payroll-related liabilities are presented in Illustration 13A–1. As you study the illustration, you should note the similarity among all payroll-related liabilities. Amounts withheld from paychecks—voluntarily or involuntarily—are liabilities until turned over to appropriate third parties. Payroll taxes and expenses for fringe benefits are incurred as a result of services performed by employees and also are liabilities until paid to appropriate third parties.

[32] All states presently have unemployment tax programs.

Illustration 13A–1
Payroll-Related Liabilities

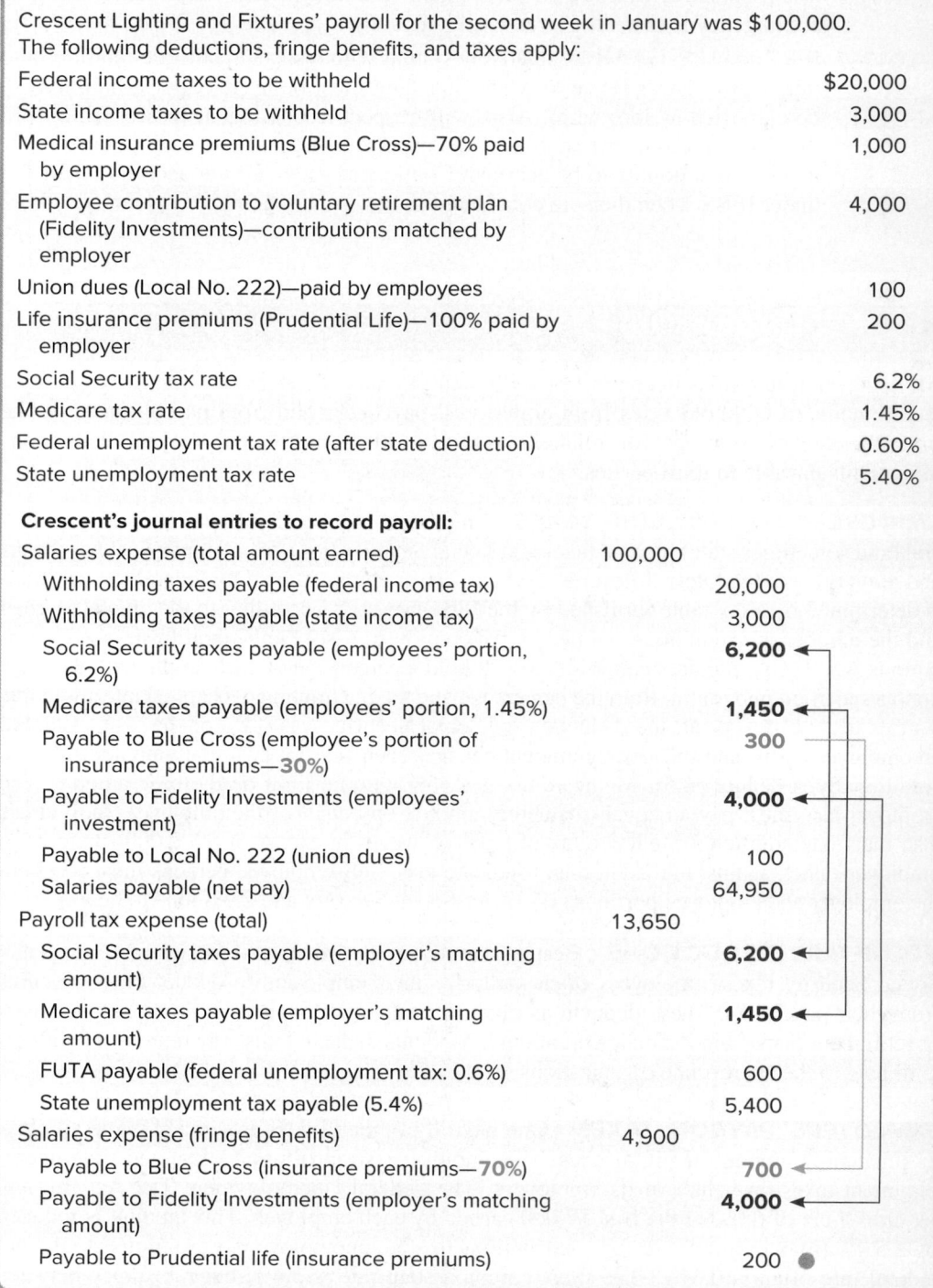

Crescent Lighting and Fixtures' payroll for the second week in January was $100,000. The following deductions, fringe benefits, and taxes apply:

Federal income taxes to be withheld	$20,000
State income taxes to be withheld	3,000
Medical insurance premiums (Blue Cross)—70% paid by employer	1,000
Employee contribution to voluntary retirement plan (Fidelity Investments)—contributions matched by employer	4,000
Union dues (Local No. 222)—paid by employees	100
Life insurance premiums (Prudential Life)—100% paid by employer	200
Social Security tax rate	6.2%
Medicare tax rate	1.45%
Federal unemployment tax rate (after state deduction)	0.60%
State unemployment tax rate	5.40%

Crescent's journal entries to record payroll:

Salaries expense (total amount earned)	100,000	
Withholding taxes payable (federal income tax)		20,000
Withholding taxes payable (state income tax)		3,000
Social Security taxes payable (employees' portion, 6.2%)		**6,200**
Medicare taxes payable (employees' portion, 1.45%)		**1,450**
Payable to Blue Cross (employee's portion of insurance premiums—**30%**)		**300**
Payable to Fidelity Investments (employees' investment)		**4,000**
Payable to Local No. 222 (union dues)		100
Salaries payable (net pay)		64,950
Payroll tax expense (total)	13,650	
Social Security taxes payable (employer's matching amount)		**6,200**
Medicare taxes payable (employer's matching amount)		**1,450**
FUTA payable (federal unemployment tax: 0.6%)		600
State unemployment tax payable (5.4%)		5,400
Salaries expense (fringe benefits)	4,900	
Payable to Blue Cross (insurance premiums—**70%**)		**700**
Payable to Fidelity Investments (employer's matching amount)		**4,000**
Payable to Prudential life (insurance premiums)		200

Amounts withheld from paychecks represent liabilities until remitted to third parties.

The employer's share of FICA and unemployment taxes constitute the employer's payroll tax expense.

Fringe benefits are part of salaries and wages expense and represent liabilities until remitted to third parties.

Questions For Review of Key Topics

Q 13–1 What are the essential characteristics of liabilities for purposes of financial reporting?

Q 13–2 What distinguishes current liabilities from long-term liabilities?

Q 13–3 Bronson Distributors owes a supplier $100,000 on open account. The amount is payable in three months. What is the theoretically correct way to measure the reportable amount for this liability? In practice, how will it likely be reported? Why?

Q 13–4 Bank loans often are arranged under existing lines of credit. What is a line of credit? How does a noncommitted line of credit differ from a committed line?

Q 13–5 Banks sometimes loan cash under noninterest-bearing notes. Is it true that banks lend money without interest?

Q 13–6 How does commercial paper differ from a bank loan? Why is the interest rate often less for commercial paper?

Q 13–7 Salaries of $5,000 have been earned by employees by the end of the period but will not be paid to employees until the following period. How should the expense and related liability be recorded? Why?

Q 13–8 Under what conditions should an employer accrue an expense and the related liability for employees' compensation for future absences? How do company custom and practice affect the accrual decision?

Q 13–9 How are refundable deposits and customer advances similar? How do they differ?

Q 13–10 How do companies account for gift cards?

Q 13–11 Amounts collected for third parties represent liabilities until remitted. Provide several examples of this kind of collection.

Q 13–12 When companies have debt that is not due to be paid for several years but that is callable (due on demand) by the creditor, do they classify the debt as current or as long-term?

Q 13–13 Long-term obligations usually are reclassified and reported as current liabilities when they become payable within the upcoming year (or operating cycle, if longer than a year). So, a 25-year bond issue is reported as a long-term liability for 24 years but normally is reported as a current liability on the balance sheet prepared during the 25th year of its term to maturity. Name a situation in which this would not be the case.

IFRS Q 13–14 How do IFRS and U.S. GAAP differ with respect to the classification of debt that is expected to be refinanced?

Q 13–15 Define a loss contingency. Provide three examples.

Q 13–16 List and briefly describe the three categories of likelihood that a future event(s) will confirm the incurrence of the liability for a loss contingency.

Q 13–17 Under what circumstances should a loss contingency be accrued?

IFRS Q 13–18 What is the difference between the use of the term *contingent liability* in U.S. GAAP and IFRS?

Q 13–19 Suppose the analysis of a loss contingency indicates that an obligation is not probable. What accounting treatment, if any, is warranted?

Q 13–20 Name two loss contingencies that almost always are accrued.

Q 13–21 Distinguish between the accounting treatment of a manufacturer's warranty and an extended warranty. Why the difference?

Q 13–22 At December 31, the end of the reporting period, the analysis of a loss contingency indicates that an obligation is only reasonably possible, though its dollar amount is readily estimable. During February, before the financial statements are issued, new information indicates the loss is probable. What accounting treatment is warranted?

Q 13–23 After the end of the reporting period, a contingency comes into existence. Under what circumstances, if any, should the contingency be reported in the financial statements for the period ended?

IFRS Q 13–24 How do U.S. GAAP and IFRS differ in their treatment of a range of equally likely losses?

IFRS Q 13–25 How do U.S. GAAP and IFRS differ in their use of present values when measuring contingent liabilities?

Q 13–26 Suppose the Environmental Protection Agency is in the process of investigating Ozone Ruination Limited for possible environmental damage but has not proposed a penalty as of December 31, 2020, the company's fiscal year-end. Describe the two-step process involved in deciding how this unasserted assessment should be reported.

Q 13–27 You are the plaintiff in a lawsuit. Your legal counsel advises that your eventual victory is inevitable. "You will be awarded $12 million," your attorney confidently asserts. Describe the appropriate accounting treatment.

IFRS Q 13–28 Answer Q 13–27, but assume that you report under IFRS.

Brief Exercises

connect

BE 13–1
Bank loan; accrued interest
● LO13–2

On October 1, Eder Fabrication borrowed $60 million and issued a nine-month, 12% promissory note. Interest was payable at maturity. Prepare the journal entry for the issuance of the note and the appropriate adjusting entry for the note at December 31, the end of the reporting period.

BE 13–2
Noninterest-bearing note; accrued interest
● LO13–2

On October 1, Eder Fabrication borrowed $60 million and issued a nine-month promissory note. Interest was discounted at issuance at a 12% discount rate. Prepare the journal entry for the issuance of the note and the appropriate adjusting entry for the note at December 31, the end of the reporting period.

BE 13–3
Determining accrued interest
● LO13–2

On July 1, Orcas Lab issued a $100,000, 12%, eight-month note. Interest is payable at maturity. What is the amount of interest expense that should be recorded in a year-end adjusting entry if the fiscal year-end is (a) December 31? (b) September 30?

BE 13–4
Commercial paper
● LO13–2

Branch Corporation issued $12 million of commercial paper on March 1 on a nine-month note. Interest was discounted at issuance at a 9% discount rate. Prepare the journal entry for the issuance of the commercial paper and its repayment at maturity.

BE 13–5
Noninterest-bearing note; effective interest rate
● LO13–2

Life.com issued $10 million of commercial paper on April 1 on a nine-month note. Interest was discounted at issuance at a 6% discount rate. What is the effective interest rate on the commercial paper?

BE 13–6
Advance collection
● LO13–3

On December 12, 2021, Pace Electronics received $24,000 from a customer toward a cash sale of $240,000 of diodes to be completed on January 16, 2022. What journal entries should Pace record on December 12 and January 16?

BE 13–7
Advance collection
● LO13–3

In Lizzie Shoes' experience, gift cards that have not been redeemed within 12 months are not likely to be redeemed. Lizzie Shoes sold gift cards for $18,000 during August 2021. $4,000 of cards were redeemed in September 2021, $3,000 in October, $2,500 in November, and $2,000 in December 2021. In 2022 an additional $1,000 of cards were redeemed in January and $500 in February. How much gift card revenue associated with the August 2021 gift card sales would Lizzie get to recognize in 2021 and 2022?

BE 13–8
Sales tax
● LO13–3

During December, Rainey Equipment made a $600,000 credit sale. The state sales tax rate is 6% and the local sales tax rate is 1.5%. Prepare the appropriate journal entry.

BE 13–9
Classifying debt
● LO13–4

Consider the following liabilities of Future Brands, Inc., at December 31, 2021, the company's fiscal year-end. Should they be reported as current liabilities or long-term liabilities?

1. $77 million of 8% notes are due on May 31, 2025. The notes are callable by the company's bank, beginning March 1, 2022.
2. $102 million of 8% notes are due on May 31, 2026. A debt covenant requires Future to maintain a current ratio (ratio of current assets to current liabilities) of at least 2 to 1. Future is in violation of this requirement but has obtained a waiver from the bank until May 2022, since both companies feel Future will correct the situation during the first half of 2022.

BE 13–10
Refinancing debt
● LO13–4

Coulson Company is in the process of refinancing some long-term debt. Its fiscal year ends on December 31, 2021, and its financial statements will be issued on March 15, 2022. Under current U.S. GAAP, how would the debt be classified if the refinancing is completed on December 15, 2021? What if instead it is completed on January 15, 2022?

BE 13–11
Refinancing debt
● LO13–4, LO13–7
IFRS

Fleener Company is in the process of refinancing some long-term debt. Its fiscal year ends on December 31, 2021, and its financial statements will be issued on March 15, 2022. Under current IFRS, how would the debt be classified if the refinancing is completed on December 15, 2021? What if instead it is completed on January 15, 2022?

BE 13–12
Warranties
● LO13–5, LO13–6

Right Medical introduced a new implant that carries a five-year warranty against manufacturer's defects. Based on industry experience with similar product introductions, warranty costs are expected to approximate 1% of sales. Sales were $15 million and actual warranty expenditures were $20,000 for the first year of selling the product. What amount (if any) should Right report as a liability at the end of the year?

BE 13–13
Product recall
● LO13–5, LO13–6

Consultants notified management of Goo Goo Baby Products that a crib toy poses a potential health hazard. Counsel indicated that a product recall is probable and is estimated to cost the company $5.5 million. How will this affect the company's income statement and balance sheet this period?

BE 13–14
Contingency
● LO13–5, LO13–6

Skill Hardware is the plaintiff in a $16 million lawsuit filed against a supplier. The litigation is in final appeal and legal counsel advises that it is virtually certain that Skill will win the lawsuit and be awarded $12 million. How should Skill account for this event?

BE 13–15
Contingency
● LO13–5, LO13–6

Bell International estimates that a $10 million loss will occur if a foreign government expropriates some company property. Expropriation is considered reasonably possible. How should Bell report the loss contingency?

BE 13–16
Contingencies
● LO13–5, LO13–6

Household Solutions manufactures kitchen storage products. During the year, the company became aware of potential costs due to (1) a recently filed lawsuit for patent infringement for which the probability of loss is remote and damages can be reasonably estimated, (2) another recently filed lawsuit for food contamination by the plastics used in Household Solutions' products for which a loss is probable but the amount of loss cannot be reasonably estimated, and (3) a new product warranty that is probable and can be reasonably estimated. Which, if any, of these costs should be accrued?

BE 13–17
Contingencies
● LO13–5, LO13–6, LO13–7
IFRS

Quandary Corporation has a major customer who is alleging a significant product defect. Quandary engineers and attorneys have analyzed the claim and have concluded that there is a 51% chance that the customer would be successful in court and that a successful claim would result in a range of damages from $10 million to $20 million, with each part of the range equally likely to occur. The damages would need to be paid soon enough that time-value-of-money considerations are not material. Would a liability be accrued under U.S. GAAP? Under IFRS? If a liability were accrued, what amount would be accrued under U.S. GAAP? Under IFRS?

BE 13–18
Unasserted assessment
● LO13–5, LO13–6

At March 13, 2022, the Securities Exchange Commission is in the process of investigating a possible securities law violation by Now Chemical. The SEC has not yet proposed a penalty assessment. Now's fiscal year ends on December 31, 2021, and its financial statements are published in March 2022. Management feels an assessment is *reasonably possible,* and if an assessment is made, an unfavorable settlement of $13 million is *probable.* What, if any, action should Now take for its financial statements?

Exercises

connect

E 13–1
Bank loan; accrued interest
● LO13–2

On November 1, 2021, Quantum Technology, a geothermal energy supplier, borrowed $16 million cash to fund a geological survey. The loan was made by Nevada BancCorp under a noncommitted short-term line of credit arrangement. Quantum issued a nine-month, 12% promissory note. Interest was payable at maturity. Quantum's fiscal period is the calendar year.

Required:

1. Prepare the journal entry for the issuance of the note by Quantum Technology.
2. Prepare the appropriate adjusting entry for the note by Quantum on December 31, 2021.
3. Prepare the journal entry for the payment of the note at maturity.

E 13–2
Determining accrued interest in various situations
● LO13–2

On July 1, 2021, Ross-Livermore Industries issued nine-month notes in the amount of $400 million. Interest is payable at maturity.

Required:

Determine the amount of interest expense that should be recorded in a year-end adjusting entry under each of the following independent assumptions:

	Interest Rate	Fiscal Year-End
1.	12%	December 31
2.	10%	September 30
3.	9%	October 31
4.	6%	January 31

E 13–3
Short-term notes
● LO13–2

The following selected transactions relate to liabilities of United Insulation Corporation. United's fiscal year ends on December 31.

Required:

Prepare the appropriate journal entries through the maturity of each liability.

2021	
Jan. 13	Negotiated a revolving credit agreement with Parish Bank that can be renewed annually upon bank approval. The amount available under the line of credit is $20 million at the bank's prime rate.
Feb. 1	Arranged a three-month bank loan of $5 million with Parish Bank under the line of credit agreement. Interest at the prime rate of 10% was payable at maturity.
May 1	Paid the 10% note at maturity.
Dec. 1	Supported by the credit line, issued $10 million of commercial paper on a nine-month note. Interest was discounted at issuance at a 9% discount rate.
31	Recorded any necessary adjusting entry(s).
2022	
Sept. 1	Paid the commercial paper at maturity.

E 13–4
Paid future absences
● LO13–3

JWS Transport Company's employees earn vacation time at the rate of 1 hour per 40-hour work period. The vacation pay vests immediately (that is, an employee is entitled to the pay even if employment terminates). During 2021, total salaries paid to employees equaled $404,000, including $4,000 for vacations actually taken in 2021 but not including vacations related to 2021 that will be taken in 2022. All vacations earned before 2021 were taken before January 1, 2021. No accrual entries have been made for the vacations. No overtime premium and no bonuses were paid during the period.

Required:
Prepare the appropriate adjusting entry for vacations earned but not taken in 2021.

E 13–5
Paid future absences
● LO13–3

On January 1, 2021, Poplar Fabricators Corporation agreed to grant its employees two weeks of vacation each year, with the stipulation that vacations earned each year can be taken the following year. For the year ended December 31, 2021, Poplar Fabricators' employees each earned an average of $900 per week. Seven hundred vacation weeks earned in 2021 were not taken during 2021.

Required:
1. Prepare the appropriate adjusting entry for vacations earned but not taken in 2021.
2. Suppose that, by the time vacations actually are taken in 2022, salary rates for employees have risen by an average of 5 percent from their 2021 level. Also, assume salaries earned in 2022 (including vacations earned and taken in 2022) were $31 million. Prepare a journal entry that summarizes 2022 salaries and the payment for 2021 vacations taken in 2022.

E 13–6
Gift cards; sales taxes
● LO13–3

Bavarian Bar and Grill opened for business in November 2021. During its first two months of operation, the restaurant sold gift cards in various amounts totaling $5,200, mostly as Christmas presents. They are redeemable for meals within two years of the purchase date, although experience within the industry indicates that 80% of gift cards are redeemed within one year. Gift cards totaling $1,300 were presented for redemption during 2021 for meals having a total price of $2,100. The sales tax rate on restaurant sales is 4%, assessed at the time meals (not gift cards) are purchased. Sales taxes will be remitted in January.

Required:
1. Prepare the appropriate journal entries (in summary form) for the gift cards and meals sold during 2021 (keeping in mind that, in actuality, each sale of a gift card or a meal would be recorded individually).
2. Determine the liability for gift cards to be reported on the December 31, 2021, balance sheet.
3. What is the appropriate classification (current or noncurrent) of the liabilities at December 31, 2021? Why?

E 13–7
Customer deposits
● LO13–3

Diversified Semiconductors sells perishable electronic components. Some must be shipped and stored in reusable protective containers. Customers pay a deposit for each container received. The deposit is equal to the container's cost. They receive a refund when the container is returned. During 2021, deposits collected on containers shipped were $850,000.

Deposits are forfeited if containers are not returned within 18 months. Containers held by customers at January 1, 2021, represented deposits of $530,000. In 2021, $790,000 was refunded and deposits forfeited were $35,000.

Required:
1. Prepare the appropriate journal entries for the deposits received, returned, and forfeited during 2021.
2. Determine the liability for refundable deposits to be reported on the December 31, 2021, balance sheet.

E 13–8
Various transactions involving advance collections
● LO13–3

The following selected transactions relate to liabilities of Interstate Farm Equipment Company for December 2021. Interstate's fiscal year ends on December 31.

Required:
Prepare the appropriate journal entries for these transactions.
1. On December 15, received $7,500 from Bradley Farms toward the sale by Interstate of a $98,000 tractor to be delivered to Bradley on January 6, 2022.
2. During December, received $25,500 of refundable deposits relating to containers used to transport equipment parts.
3. During December, credit sales totaled $800,000. The state sales tax rate is 5% and the local sales tax rate is 2%. (This is a summary journal entry for the many individual sales transactions for the period).

E 13–9
Gift cards
● LO13–3

CircuitTown commenced a gift card program in January 2021 and sold $10,000 of gift cards in January, $15,000 in February, and $16,000 in March 2021 before discontinuing further gift card sales. During 2021, gift card redemptions were $6,000 for the January gift cards sold, $4,500 for the February cards, and $4,000 for the March cards. CircuitTown considers gift cards to be "broken" (not redeemable) 10 months after sale.

Required:

1. How much revenue will CircuitTown recognize with respect to January gift card sales during 2021?
2. Prepare journal entries to record the sale of January gift cards, redemption of gift cards (ignore sales tax), and breakage (expiration) of gift cards.
3. How much revenue will CircuitTown recognize with respect to March gift card sales during 2021?
4. What liability for deferred revenue associated with gift card sales would CircuitTown show as of December 31, 2021?

E 13–10
FASB codification research
● LO13–3, LO13–4, LO13–5

CODE

Access the *FASB Accounting Standards Codification* at the FASB website (**www.fasb.org**).

Required:

Determine the specific eight- or nine-digit Codification citation (XXX-XX-XX-XX) that describes the following items:

1. If it is only reasonably possible that a contingent loss will occur the contingent loss should be disclosed.
2. Criteria allowing short-term liabilities expected to be refinanced to be classified as long-term liabilities.
3. Accounting for the revenue from separately priced extended warranty contracts.
4. The criteria to determine if an employer must accrue a liability for vacation pay.

E 13–11
Current–noncurrent classification of debt; Sprint Corporation
● LO13–1, LO13–4

An annual report of **Sprint Corporation** contained a rather lengthy narrative entitled "Review of Segmental Results of Operation." The narrative noted that short-term notes payable and commercial paper outstanding at the end of the year aggregated $756 million and that during the following year, "This entire balance will be replaced by the issuance of long-term debt or will continue to be refinanced under existing long-term credit facilities."

Required:

How did Sprint report the debt in its balance sheet? Why?

E 13–12
Current–noncurrent classification of debt; Sprint Corporation
● LO13–1, LO13–4, LO13–7

IFRS

Consider the information presented in E 13–11.

Required:

1. How would Sprint report the debt in its balance sheet if it reported under IFRS? Why?
2. Would your answer to requirement 1 change if Sprint obtained its long-term credit facility after the balance sheet date? Why?

E 13–13
Current–noncurrent classification of debt
● LO13–1, LO13–4

At December 31, 2021, Newman Engineering's liabilities include the following:

1. $10 million of 9% bonds were issued for $10 million on May 31, 1999. The bonds mature on May 31, 2029, but bondholders have the option of calling (demanding payment on) the bonds on May 31, 2022. However, the option to call is not expected to be exercised, given prevailing market conditions.
2. $14 million of 8% notes are due on May 31, 2022. A debt covenant requires Newman to maintain current assets at least equal to 175% of its current liabilities. On December 31, 2021, Newman is in violation of this covenant. Newman obtained a waiver from National City Bank until June 2022, having convinced the bank that the company's normal 2 to 1 ratio of current assets to current liabilities will be reestablished during the first half of 2022.
3. $7 million of 11% bonds were issued for $7 million on August 1, 1989. The bonds mature on July 31, 2022. Sufficient cash is expected to be available to retire the bonds at maturity.

Required:

What portion of each liability is reported as a current liability and as a noncurrent liability? Explain.

E 13–14
FASB codification research
● LO13–5

CODE

Access the *FASB Accounting Standards Codification* at the FASB website (**www.fasb.org**).

Required:

1. Obtain the relevant authoritative literature on recognition of contingent losses. What is the specific eight-digit Codification citation (XXX-XX-XX-X) that describes the guidelines for determining when an expense and liability should be accrued for a contingent loss?
2. List the guidelines.

E 13–15
Warranties
● LO13–5, LO13–6

Cupola Awning Corporation introduced a new line of commercial awnings in 2021 that carry a two-year warranty against manufacturer's defects. Based on their experience with previous product introductions, warranty costs are expected to approximate 3% of sales. Sales and actual warranty expenditures for the first year of selling the product were:

Sales	Actual Warranty Expenditures
$5,000,000	$37,500

Required:

1. Does this situation represent a loss contingency? Why or why not? How should Cupola account for it?
2. Prepare journal entries that summarize sales of the awnings (assume all credit sales) and any aspects of the warranty that should be recorded during 2021.
3. What amount should Cupola report as a liability at December 31, 2021?

E 13–16
Extended warranties
● LO13–5, LO13–6

Carnes Electronics sells consumer electronics that carry a 90-day manufacturer's warranty. At the time of purchase, customers are offered the opportunity to also buy a two-year extended warranty for an additional charge. During the year, Carnes received $412,000 for these extended warranties (approximately evenly throughout the year).

Required:

1. Does this situation represent a loss contingency? Why or why not? How should it be accounted for?
2. Prepare journal entries that summarize sales of the extended warranties and any aspects of the warranty that should be recorded during the year.

E 13–17
Contingency; product recall
● LO13–5, LO13–6

Sound Audio manufactures and sells audio equipment for automobiles. Engineers notified management in December 2021 of a circuit flaw in an amplifier that poses a potential fire hazard. An intense investigation indicated that a product recall is virtually certain, estimated to cost the company $2 million. The fiscal year ends on December 31.

Required:

1. Should this loss contingency be accrued, only disclosed, or neither? Explain.
2. What loss, if any, should Sound Audio report in its 2021 income statement?
3. What liability, if any, should Sound Audio report in its 2021 balance sheet?
4. Prepare any journal entry needed.

E 13–18
Impairment of accounts receivable
● LO13–5, LO13–6

The Manda Panda Company uses the allowance method to account for bad debts. At the beginning of 2021, the allowance account had a credit balance of $75,000. Credit sales for 2021 totaled $2,400,000 and the year-end accounts receivable balance was $490,000. During this year, $73,000 in receivables were determined to be uncollectible. Manda Panda anticipates that 3% of all credit sales will ultimately become uncollectible. The fiscal year ends on December 31.

Required:

1. Does this situation describe a loss contingency? Explain.
2. What is the bad debt expense that Manda Panda should report in its 2021 income statement?
3. Prepare the appropriate journal entry to record the contingency.
4. What is the net accounts receivable value Manda Panda should report in its 2021 balance sheet?

E 13–19
Unasserted assessment
● LO13–6

At April 1, 2022, the Food and Drug Administration is in the process of investigating allegations of false marketing claims by Hulkly Muscle Supplements. The FDA has not yet proposed a penalty assessment. Hulkly's fiscal year ends on December 31, 2021. The company's financial statements are issued in April 2022.

Required:

For each of the following scenarios, determine the appropriate way to report the situation. Explain your reasoning and prepare any necessary journal entry.

1. Management feels an assessment is *reasonably possible,* and if an assessment is made, an unfavorable settlement of $13 million is *reasonably possible.*
2. Management feels an assessment is *reasonably possible,* and if an assessment is made, an unfavorable settlement of $13 million is *probable.*
3. Management feels an assessment is *probable,* and if an assessment is made, an unfavorable settlement of $13 million is *reasonably possible.*
4. Management feels an assessment is *probable,* and if an assessment is made, an unfavorable settlement of $13 million is *probable.*

E 13–20
Various transactions involving contingencies
● LO13–5, LO13–6

The following selected transactions relate to contingencies of Classical Tool Makers, Inc., which began operations in July 2021. Classical's fiscal year ends on December 31. Financial statements are issued in April 2022.

Required:

Prepare the year-end entries for any amounts that should be recorded as a result of each of these contingencies and indicate whether a disclosure note is indicated.

1. Classical's products carry a one-year warranty against manufacturer's defects. Based on previous experience, warranty costs are expected to approximate 4% of sales. Sales were $2 million (all credit) for 2021. Actual warranty expenditures were $30,800 and were recorded as warranty expense when incurred.
2. Although no customer accounts have been shown to be uncollectible, Classical estimates that 2% of credit sales will eventually prove uncollectible.

3. In December 2021, the state of Tennessee filed suit against Classical, seeking penalties for violations of clean air laws. On January 23, 2022, Classical reached a settlement with state authorities to pay $1.5 million in penalties.
4. Classical is the plaintiff in a $4 million lawsuit filed against a supplier. The suit is in final appeal and attorneys advise that it is virtually certain that Classical will win the case and be awarded $2.5 million.
5. In November 2021, Classical became aware of a design flaw in an industrial saw that poses a potential electrical hazard. A product recall appears unavoidable. Such an action would likely cost the company $500,000.
6. Classical offered $25 cash rebates on a new model of jigsaw. Customers must mail in a proof-of-purchase seal from the package plus the cash register receipt to receive the rebate. Experience suggests that 60% of the rebates will be claimed. Ten thousand of the jigsaws were sold in 2021. Total rebates to customers in 2021 were $105,000 and were recorded as promotional expense when paid.

E 13–21
Various transactions involving contingencies
● LO13–5, LO13–6

The following selected circumstances relate to pending lawsuits for Erismus, Inc. Erismus's fiscal year ends on December 31. Financial statements are issued in March 2022. Erismus prepares its financial statements according to U.S. GAAP.

Required:
Indicate the amount of asset or liability that Erismus would record, and explain your answer.

1. Erismus is defending against a lawsuit. Erismus's management believes the company has a slightly worse than 50/50 chance of eventually prevailing in court, and that if it loses, the judgment will be $1,000,000.
2. Erismus is defending against a lawsuit. Erismus's management believes it is probable that the company will lose in court. If it loses, management believes that damages could fall anywhere in the range of $2,000,000 to $4,000,000, with any damage in that range equally likely.
3. Erismus is defending against a lawsuit. Erismus's management believes it is probable that the company will lose in court. If it loses, management believes that damages will eventually be $5,000,000, with a present value of $3,500,000.
4. Erismus is a plaintiff in a lawsuit. Erismus's management believes it is probable that the company eventually will prevail in court, and that if it prevails, the judgment will be $1,000,000.
5. Erismus is a plaintiff in a lawsuit. Erismus's management believes it is virtually certain that the company eventually will prevail in court, and that if it prevails, the judgment will be $500,000.

E 13–22
Various transactions involving contingencies; IFRS
● LO13–5, LO13–6, LO13–7
IFRS

[This exercise is a variation of E 13–21 focusing on reporting under IFRS]. Refer to the circumstances listed in E 13–21, but assume that Erismus prepares its financial statements according to International Financial Reporting Standards.

Required:
For each circumstance, indicate the amount of asset or liability that Erismus would record, and explain your answer.

E 13–23
Disclosures of liabilities
● LO13–1 though LO13–6

Indicate (by letter) the way each of the items listed below should be reported in a balance sheet at December 31, 2021.

Item	Reporting Method
_____ 1. Commercial paper	N. Not reported
_____ 2. Noncommitted line of credit	C. Current liability
_____ 3. Customer advances	L. Long-term liability
_____ 4. Estimated quality-assurance warranty cost	D. Disclosure note only
_____ 5. Accounts payable	A. Asset
_____ 6. Long-term bonds that will be callable by the creditor in the upcoming year unless an existing violation is not corrected (there is a reasonable possibility the violation will be corrected within the grace period)	
_____ 7. Note due March 3, 2022	
_____ 8. Interest accrued on note, December 31, 2021	
_____ 9. Short-term bank loan to be paid with proceeds of sale of common stock	
_____ 10. A determinable gain that is contingent on a future event that appears extremely likely to occur in three months	
_____ 11. Unasserted assessment of taxes owed on prior-year income that probably will be asserted, in which case there would probably be a loss in six months	
_____ 12. Unasserted assessment of taxes owed on prior-year income with a reasonable possibility of being asserted, in which case there would probably be a loss in 13 months	
_____ 13. A determinable loss from a past event that is contingent on a future event that appears extremely likely to occur in three months	
_____ 14. Note payable due April 4, 2024	
_____ 15. Long-term bonds callable by the creditor in the upcoming year that are not expected to be called	

E 13–24
Warranty expense; change in estimate
● LO13–5, LO13–6

Woodmier Lawn Products introduced a new line of commercial sprinklers in 2020 that carry a one-year warranty against manufacturer's defects. Because this was the first product for which the company offered a warranty, trade publications were consulted to determine the experience of others in the industry. Based on that experience, warranty costs were expected to approximate 2% of sales. Sales of the sprinklers in 2020 were $2.5 million. Accordingly, the following entries relating to the contingency for warranty costs were recorded during the first year of selling the product:

Accrued liability and expense		
Warranty expense (2% × $2,500,000)	50,000	
Warranty liability		50,000
Actual expenditures (summary entry)		
Warranty liability	23,000	
Cash		23,000

In late 2021, the company's claims experience was evaluated and it was determined that claims were far more than expected—3% of sales rather than 2%.

Required:

1. Assuming sales of the sprinklers in 2021 were $3.6 million and warranty expenditures in 2021 totaled $88,000, prepare any journal entries related to the warranty.
2. Assuming sales of the sprinklers were discontinued after 2020, prepare any journal entry(s) in 2021 related to the warranty.

E 13–25
Change in accounting estimate
● LO13–3

The Commonwealth of Virginia filed suit in October 2019 against Northern Timber Corporation, seeking civil penalties and injunctive relief for violations of environmental laws regulating forest conservation. When the 2020 financial statements were issued in 2021, Northern had not reached a settlement with state authorities, but legal counsel advised Northern Timber that it was probable the ultimate settlement would be $1,000,000 in penalties. The following entry was recorded:

Loss—litigation	1,000,000	
Liability—litigation		1,000,000

Late in 2021, a settlement was reached with state authorities to pay a total of $600,000 to cover the cost of violations.

Required:

1. Prepare any journal entries related to the change.
2. Briefly describe other steps Northern should take to report the change.

E 13–26
Contingency; DowDuPont Inc. disclosure
● LO13–5, LO13–6
Real World Financials

DowDuPont Inc. provides chemical, plastic, and agricultural products and services to various consumer markets. The following excerpt is taken from the disclosure notes of DowDuPont's 2017 annual report:

> At December 31, 2017, the Company had accrued obligations of $1,311 million for probable environmental remediation and restoration costs, including $219 million for the remediation of Superfund sites. These obligations are included in "Accrued and other current liabilities" and "Other noncurrent obligations" in the consolidated balance sheets. This is management's best estimate of the costs for remediation and restoration with respect to environmental matters for which the Company has accrued liabilities, although it is reasonably possible that the ultimate cost with respect to these particular matters could range up to approximately two times that amount.

Required:

1. Does the excerpt describe a loss contingency?
2. Under what conditions would DowDuPont accrue such a contingency?
3. What journal entry would DowDuPont use to record this amount of provision (loss)?

E 13–27
Payroll-related liabilities
● Appendix

Lee Financial Services pays employees monthly. Payroll information is listed below for January 2021, the first month of Lee's fiscal year. Assume that none of the employees exceeded any relevant base of pay, such that all benefit percentages apply to the entire $500,000 payroll.

Salaries	$500,000
Federal income taxes to be withheld	$100,000
Federal unemployment tax rate	0.60%
State unemployment tax rate (after SUTA deduction)	5.40%
Social Security tax rate	6.2%
Medicare tax rate	1.45%

Required:
Prepare the appropriate journal entries to record salaries expense and payroll tax expense for the January 2021 pay period.

Problems

connect

P 13–1
Bank loan; accrued interest
● LO13–2

Blanton Plastics, a household plastic product manufacturer, borrowed $14 million cash on October 1, 2021, to provide working capital for year-end production. Blanton issued a four-month, 12% promissory note to L&T Bank under a prearranged short-term line of credit. Interest on the note was payable at maturity. Each firm's fiscal period is the calendar year.

Required:
1. Prepare the journal entries to record (a) the issuance of the note by Blanton Plastics and (b) L&T Bank's receivable on October 1, 2021.
2. Prepare the journal entries by both firms to record all subsequent events related to the note through January 31, 2022.
3. Suppose the face amount of the note was adjusted to include interest (a noninterest-bearing note) and 12% is the bank's stated discount rate. (a) Prepare the journal entries to record the issuance of the noninterest-bearing note by Blanton Plastics on October 1, 2021, the adjusting entry at December 31, and payment of the note at maturity. (b) What would be the effective interest rate?

P 13–2
Various transactions involving liabilities
● LO13–2 through LO13–4

Camden Biotechnology began operations in September 2021. The following selected transactions relate to liabilities of the company for September 2021 through March 2022. Camden's fiscal year ends on December 31. Its financial statements are issued in April.

2021
a. On September 5, opened checking accounts at Second Commercial Bank and negotiated a short-term line of credit of up to $15,000,000 at the bank's prime rate (10.5% at the time). The company will pay no commitment fees.
b. On October 1, borrowed $12 million cash from Second Commercial Bank under the line of credit and issued a five-month promissory note. Interest at the prime rate of 10% was payable at maturity. Management planned to issue 10-year bonds in February to repay the note.
c. Received $2,600 of refundable deposits in December for reusable containers used to transport and store chemical-based products.
d. For the September–December period, sales on account totaled $4,100,000. The state sales tax rate is 3% and the local sales tax rate is 3%. (This is a summary journal entry for the many individual sales transactions for the period.)
e. Recorded the adjusting entry for accrued interest.

2022
f. In March, paid the entire amount of the note on its March 1 due date, using proceeds from a February issuance of $10 million of 10-year bonds at face value, along with other available cash.
g. The storage containers covered by refundable deposits are expected to be returned during the first nine months of the year. Half of the containers were returned in March 2022.

Required:
1. Prepare the appropriate journal entries for items a-g.
2. Prepare the current and long-term liability sections of the December 31, 2021, balance sheet. Trade accounts payable on that date were $252,000.

P 13–3
Current–noncurrent classification of debt
● LO13–1, LO13–4

The balance sheet at December 31, 2021, for Nevada Harvester Corporation includes the liabilities listed below:

a. 11% bonds with a face amount of $40 million were issued for $40 million on October 31, 2012. The bonds mature on October 31, 2032. Bondholders have the option of calling (demanding payment on) the bonds on October 31, 2022, at a redemption price of $40 million. Market conditions are such that the call is not expected to be exercised.
b. Management intended to refinance $6 million of its 10% notes that mature in May 2022. In early March, prior to the actual issuance of the 2021 financial statements, Nevada Harvester negotiated a line of credit with a commercial bank for up to $5 million any time during 2022. Any borrowings will mature two years from the date of borrowing.
c. Noncallable 12% bonds with a face amount of $20 million were issued for $20 million on September 30, 2002. The bonds mature on September 30, 2022. Sufficient cash is expected to be available to retire the bonds at maturity.
d. A $12 million 9% bank loan is payable on October 31, 2027. The bank has the right to demand payment after any fiscal year-end in which Nevada Harvester's ratio of current assets to current liabilities falls below a contractual minimum of 1.7 to 1 and remains so for six months. That ratio was 1.45 on December 31, 2021, due primarily to an intentional temporary decline in inventory levels. Normal inventory levels will be reestablished during the first quarter of 2022.

Required:

1. Determine the amount, for each item a through d, that can be reported as a current liability and as a noncurrent liability. Explain the reasoning behind your classifications.
2. Prepare the liability section of a classified balance sheet and any necessary note disclosure for Nevada Harvester at December 31, 2021. Accounts payable and accruals are $22 million.

P 13–4
Various liabilities
● LO13–1 through LO13–4

The unadjusted trial balance of the Manufacturing Equitable at December 31, 2021, the end of its fiscal year, included the following account balances. Manufacturing's 2021 financial statements were issued on April 1, 2022.

Accounts receivable	$ 92,500
Accounts payable	35,000
10% notes, payable to bank	600,000
Mortgage note payable	1,200,000

Other information:

a. The bank notes, issued August 1, 2021, are due on July 31, 2022, and pay interest at a rate of 10%, payable at maturity.
b. The mortgage note is due on March 1, 2022. Interest at 9% has been paid up to December 31 (assume 9% is a realistic rate). Manufacturing intended at December 31, 2021, to refinance the note on its due date with a new 10-year mortgage note. In fact, on March 1, Manufacturing paid $250,000 in cash on the principal balance and refinanced the remaining $950,000.
c. Included in the accounts receivable balance at December 31, 2021, were two subsidiary accounts that had been overpaid and had credit balances totaling $18,000. The accounts were of two major customers who were expected to order more merchandise from Manufacturing and apply the overpayments to those future purchases.
d. On November 1, 2021, Manufacturing rented a portion of its factory to a tenant for $30,000 per year, payable in advance. The payment for the 12 months ended October 31, 2022, was received as required and was credited to rent revenue.

Required:

1. Prepare any necessary adjusting journal entries at December 31, 2021, pertaining to each item of other information (a–d).
2. Prepare the current and long-term liability sections of the December 31, 2021, balance sheet.

P 13–5
Bonus compensation; algebra
● LO13–3

Sometimes compensation packages include bonuses designed to provide performance incentives to employees. The difficulty a bonus can cause accountants is not an accounting problem, but a math problem. The complication is that the bonus formula sometimes specifies that the calculation of the bonus is based in part on the bonus itself. This occurs any time the bonus is a percentage of income because expenses are components of income, and the bonus is an expense.

Regalia Fashions has an incentive compensation plan through which a division manager receives a bonus equal to 10% of the division's net income. Division income in 2021 before the bonus and income tax was $150,000. The tax rate is 30%.

Required:

1. Express the bonus formula as one or more algebraic equation(s).[33]
2. Using these formulas, calculate the amount of the bonus.
3. Prepare the adjusting entry to record the bonus compensation.
4. Bonus arrangements take many forms. Suppose the bonus specifies that the bonus is 10% of the division's income before tax, but after the bonus itself. Calculate the amount of the bonus.

P 13–6
Various contingencies
● LO13–5, LO13–6

Eastern Manufacturing is involved with several situations that possibly involve contingencies. Each is described below. Eastern's fiscal year ends December 31, and the 2021 financial statements are issued on March 15, 2022.

a. Eastern is involved in a lawsuit resulting from a dispute with a supplier. On February 3, 2022, judgment was rendered against Eastern in the amount of $107 million plus interest, a total of $122 million. Eastern plans to appeal the judgment and is unable to predict its outcome though it is not expected to have a material adverse effect on the company.
b. In November 2020, the State of Nevada filed suit against Eastern, seeking civil penalties and injunctive relief for violations of environmental laws regulating hazardous waste. On January 12, 2022, Eastern reached a settlement with state authorities. Based upon discussions with legal counsel, the Company feels it is probable that $140 million will be required to cover the cost of violations. Eastern believes that the ultimate settlement of this claim will not have a material adverse effect on the company.
c. Eastern is the plaintiff in a $200 million lawsuit filed against United Steel for damages due to lost profits from rejected contracts and for unpaid receivables. The case is in final appeal and legal counsel advises that it is probable that Eastern will prevail and be awarded $100 million.
d. At March 15, 2022, Eastern knows a competitor has threatened litigation due to patent infringement. The competitor has not yet filed a lawsuit. Management believes a lawsuit is reasonably possible, and if a lawsuit is filed, management believes damages of up to $33 million are reasonably possible.

Required:

Determine the appropriate way to report each situation, and prepare any necessary journal entries and disclosure notes. Explain your reasoning.

P 13–7
Various liabilities
● LO13–4, LO13–5, LO13–6, LO13–7
IFRS

HolmesWatson (HW) is considering what the effect would be of reporting its liabilities under IFRS rather than U.S. GAAP. The following facts apply:

a. HW is defending against a lawsuit and believes it is virtually certain to lose in court. If it loses the lawsuit, management estimates it will need to pay a range of damages that falls between $5,000,000 and $10,000,000, with each amount in that range equally likely.
b. HW is defending against another lawsuit that is identical to item (a), but the relevant losses will only occur far into the future. The present values of the endpoints of the range are $3,000,000 and $8,000,000, with the timing of cash flow somewhat uncertain. HW considers these effects of the time value of money to be material.
c. HW is defending against another lawsuit for which management believes HW has a slightly worse than 50/50 chance of losing in court. If it loses the lawsuit, management estimates HW will need to pay a range of damages that falls between $3,000,000 and $9,000,000, with each amount in that range equally likely.
d. HW has $10,000,000 of short-term debt that it intends to refinance on a long-term basis. Soon after the balance sheet date, but before issuance of the financial statements, HW obtained the financing necessary to refinance the debt.

Required:

1. For each item, indicate how treatment of the amount would differ between U.S. GAAP and IFRS.
2. Consider the total effect of items a–d. If HW's goal is to show the lowest total liabilities, which set of standards, U.S. GAAP or IFRS, best helps it meet that goal?

P 13–8
Expected cash flow approach; product recall
● LO13–6

The Heinrich Tire Company recalled a tire in its subcompact line in December 2021. Costs associated with the recall were originally thought to approximate $50 million. Now, though, while management feels it is probable the company will incur substantial costs, all discussions indicate that $50 million is an excessive amount. Based on prior recalls in the industry, management has provided the following probability distribution for the potential loss:

Loss Amount	Probability
$40 million	20%
$30 million	50%
$20 million	30%

[33] Remember when you were studying algebra and you wondered if you would ever use it?

An arrangement with a consortium of distributors requires that all recall costs be settled at the end of 2022. The risk-free rate of interest is 5%.

Required:

1. By the traditional approach to measuring loss contingencies, what amount would Heinrich record at the end of 2021 for the loss and contingent liability?
2. For the remainder of this problem, apply the expected cash flow approach of *SFAC No. 7.* Estimate Heinrich's liability at the end of the 2021 fiscal year.
3. Prepare the journal entry to record the contingent liability (and loss).
4. Prepare the journal entry to accrue interest on the liability at the end of 2022.
5. Prepare the journal entry to pay the liability at the end of 2022, assuming the actual cost is $31 million. Heinrich records an additional loss if the actual costs are higher or a gain if the costs are lower.

P 13–9
Subsequent events
● LO13–6

Lincoln Chemicals became involved in investigations by the U.S. Environmental Protection Agency in regard to damages connected to waste disposal sites. Below are four possibilities regarding the timing of (A) the alleged damage caused by Lincoln, (B) an investigation by the EPA, (C) the EPA assessment of penalties, and (D) ultimate settlement. In each case, assume that Lincoln is unaware of any problem until an investigation is begun. Also assume that once the EPA investigation begins, it is probable that a damage assessment will ensue and that once an assessment is made by the EPA, it is reasonably possible that a determinable amount will be paid by Lincoln.

Required:

For each case, decide whether (1) a loss should be accrued in the financial statements with an explanatory note, (2) a disclosure note only should be provided, or (3) no disclosure is necessary.

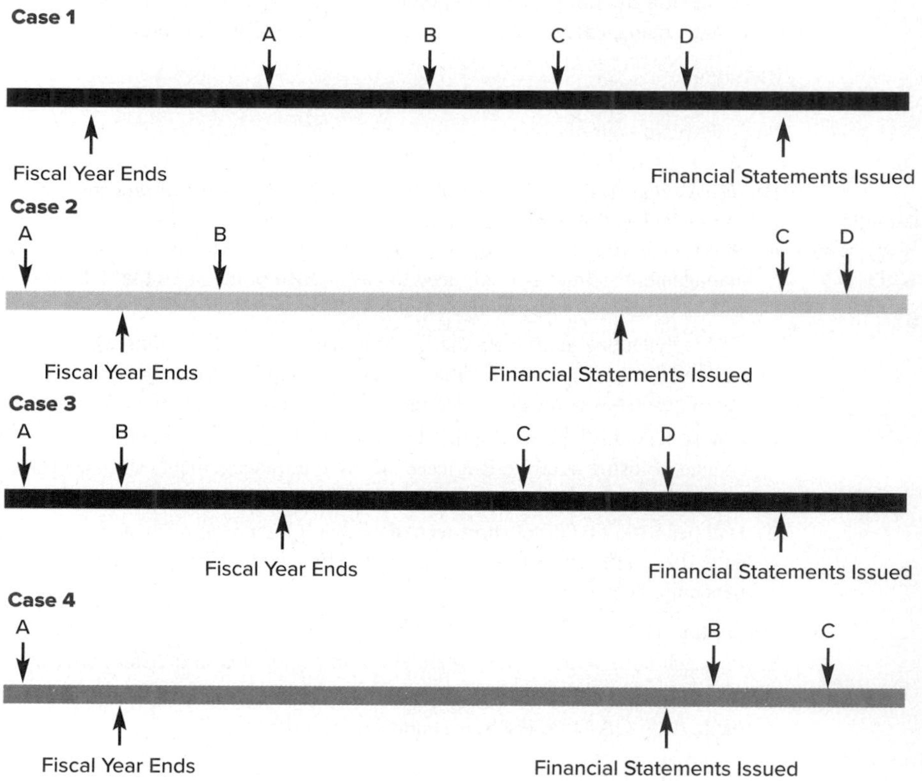

P 13–10
Subsequent events; classification of debt; loss contingency; financial statement effects
● LO13–4, LO13–5

Van Rushing Hunting Goods' fiscal year ends on December 31. At the end of the 2021 fiscal year, the company had notes payable of $12 million due on February 8, 2022. Rushing sold 2 million shares of its $0.25 par, common stock on February 3, 2022, for $9 million. The proceeds from that sale along with $3 million from the maturation of some 3-month CDs were used to pay the notes payable on February 8.

Through his attorney, one of Rushing's construction workers notified management on January 5, 2022, that he planned to sue the company for $1 million related to a work-site injury on December 20, 2021. As of December 31, 2021, management had been unaware of the injury, but reached an agreement on February 23, 2022, to settle the matter by paying the employee's medical bills of $75,000.

Rushing's financial statements were finalized on March 3, 2022.

Required:

1. What amount(s) if any, related to the situations described should Rushing report among current liabilities in its balance sheet at December 31, 2021? Why?
2. What amount(s) if any, related to the situations described should Rushing report among long-term liabilities in its balance sheet at December 31, 2021? Why?
3. How would your answers to requirements 1 and 2 differ if the settlement agreement had occurred on March 15, 2022, instead? Why?
4. How would your answers to requirements 1 and 2 differ if the work-site injury had occurred on January 3, 2022, instead? Why?

P 13–11
Concepts; terminology
● LO13–1 through LO13–4

Listed below are several terms and phrases associated with current liabilities. Pair each item from List A (by letter) with the item from List B that is most appropriately associated with it.

List A	List B
_____ 1. Face amount × Interest rate × Time	a. Informal agreement
_____ 2. Payable with current assets	b. Secured loan
_____ 3. Short-term debt to be refinanced with common stock	c. Refinancing prior to the issuance of the financial statements
_____ 4. Present value of interest plus present value of principal	d. Accounts payable
_____ 5. Noninterest-bearing	e. Accrued liabilities
_____ 6. Noncommitted line of credit	f. Commercial paper
_____ 7. Pledged accounts receivable	g. Current liabilities
_____ 8. Reclassification of debt	h. Long-term liability
_____ 9. Purchased by other corporations	i. Usual valuation of liabilities
_____ 10. Expenses not yet paid	j. Interest on debt
_____ 11. Liability until refunded	k. Customer advances
_____ 12. Liability until satisfy performance obligation	l. Customer deposits

P 13–12
Various liabilities; balance sheet classification; prepare liability section of balance sheet; write notes
● LO13–4, LO13–5

Transit Airlines provides regional jet service in the Mid-South. The following is information on liabilities of Transit at December 31, 2021. Transit's fiscal year ends on December 31. Its annual financial statements are issued in April.

1. Transit has outstanding 6.5% bonds with a face amount of $90 million. The bonds mature on July 31, 2027. Bondholders have the option of calling (demanding payment on) the bonds on July 31, 2022, at a redemption price of $90 million. Market conditions are such that the call option is not expected to be exercised.
2. A $30 million 8% bank loan is payable on October 31, 2024. The bank has the right to demand payment after any fiscal year-end in which Transit's ratio of current assets to current liabilities falls below a contractual minimum of 1.9 to 1 and remains so for six months. That ratio was 1.75 on December 31, 2021, due primarily to an intentional temporary decline in parts inventories. Normal inventory levels will be reestablished during the sixth week of 2022.
3. Transit management intended to refinance $45 million of 7% notes that mature in May 2022. In late February 2022, prior to the issuance of the 2021 financial statements, Transit negotiated a line of credit with a commercial bank for up to $40 million any time during 2022. Any borrowings will mature two years from the date of borrowing.
4. Transit is involved in a lawsuit resulting from a dispute with a food caterer. On February 13, 2022, judgment was rendered against Transit in the amount of $53 million plus interest, a total of $54 million. Transit plans to appeal the judgment and is unable to predict its outcome though it is not expected to have a material adverse effect on the company.

Required:

1. How should the 6.5% bonds be classified by Transit among liabilities in its balance sheet? Explain.
2. How should the 8% bank loan be classified by Transit among liabilities in its balance sheet? Explain.
3. How should the 7% notes be classified by Transit among liabilities in its balance sheet? Explain.
4. How should the lawsuit be reported by Transit? Explain.
5. Prepare the liability section of a classified balance sheet for Transit Airlines at December 31, 2021. Transit's accounts payable and accruals were $43 million.
6. Draft appropriate note disclosures for Transit's financial statements at December 31, 2021, for each of the five items described.

P 13–13
Payroll-related liabilities
● Appendix

Alamar Petroleum Company offers its employees the option of contributing retirement funds up to 5% of their salaries, with the contribution being matched by Alamar. The company also pays 80% of medical and life insurance premiums. Deductions relating to these plans and other payroll information for the first biweekly payroll period of February are listed as follows:

Wages and salaries	$2,000,000
Employee contribution to voluntary retirement plan	84,000
Medical insurance premiums	42,000
Life insurance premiums	9,000
Federal income taxes to be withheld	400,000
Local income taxes to be withheld	53,000
Payroll taxes:	
Federal unemployment tax rate	0.60%
State unemployment tax rate (after SUTA deduction)	5.40%
Social Security tax rate	6.20%
Medicare tax rate	1.45%

Required:
Prepare the appropriate journal entries to record salaries expense and payroll tax expense for the biweekly pay period. Assume that all employees' cumulative wages do not exceed the relevant wage bases for Social Security. Also assume that all employees' cumulative wages do exceed the relevant unemployment wage bases at the end of January.

Decision Maker's Perspective

©ImageGap/Getty Images

Apply your critical-thinking ability to the knowledge you've gained. These cases will provide you an opportunity to develop your research, analysis, judgment, and communication skills. You also will work with other students, integrate what you've learned, apply it in real-world situations, and consider its global and ethical ramifications. This practice will broaden your knowledge and further develop your decision-making abilities.

Research Case 13–1
Bank loan; accrued interest
● LO13–1, LO13–2

CODE

A fellow accountant has solicited your opinion regarding the classification of short-term obligations repaid prior to being replaced by a long-term security. Cheshire Foods, Inc., issued $5,000,000 of short-term commercial paper during 2020 to finance construction of a plant. At September 30, 2021, Cheshire's fiscal year-end, the company intends to refinance the commercial paper by issuing long-term bonds. However, because Cheshire temporarily has excess cash, in November 2021 it liquidates $2,000,000 of the commercial paper as the paper matures. In December 2021, the company completes a $10,000,000 long-term bond issue. Later during December, it issues its September 30, 2021, financial statements. The proceeds of the long-term bond issue are to be used to replenish $2,000,000 in working capital, to pay $3,000,000 of commercial paper as it matures in January 2022, and to pay $5,000,000 of construction costs expected to be incurred later that year to complete the plant.

You initially are hesitant because you don't recall encountering a situation in which short-term obligations were repaid prior to being replaced by a long-term security. However, you are encouraged by a vague memory that this general topic is covered by GAAP literature you came across when reading an Internet article.

Required:
Determine how the $5,000,000 of commercial paper should be classified by citing the specific nine-digit Codification citation (XXX-XX-XX-XX) in the FASB Codification. Before doing so, formulate your own opinion on the proper treatment.

Real World Case 13–2
Returnable containers
● LO13–1, LO13–3

Real World Financials

The **Zoo Doo Compost Company** processes a premium organic fertilizer made with the help of the animals at the Memphis Zoo. Zoo Doo is sold in a specially designed plastic pail that may be kept and used for household chores or returned to the seller. The fertilizer is sold for $12.50 per two-gallon pail (including the $1.76 cost of the pail). For each pail returned, Zoo Doo donates $1 to the Memphis Zoo and the pail is used again.[34]

Required:
The founder and president of this start-up firm has asked your opinion on how to account for the donations to be made when fertilizer pails are returned. (Ignore any tax implications.)

[34] Case based on Kay McCullen, "Take The Zoo Home With You!" *Head Lions,* July 1991; and a conversation with the Zoo Doo Compost Company president, Pierce Ledbetter.

Research Case 13–3
Relationship of liabilities to assets and owners' equity
● LO13–1

SFAC No. 6, "Elements of Financial Statements," states that "an entity's assets, liabilities, and equity (net assets) all pertain to the same set of probable future economic benefits." Explain this statement.

Judgment Case 13–4
Paid future absences
● LO13–3

Cates Computing Systems develops and markets commercial software for personal computers and workstations. Three situations involving compensation for possible future absences of Cates's employees are described below.

a. Cates compensates employees at their regular pay rate for time absent for military leave, maternity leave, and jury time. Employees are allowed predetermined absence periods for each type of absence.
b. Members of the new product development team are eligible for three months' paid sabbatical leave every four years. Five members of the team have just completed their fourth year of participation
c. Company policy permits employees four paid sick days each year. Unused sick days can accumulate and can be carried forward to future years.

Required:

1. What are the conditions that require accrual of an expense and related liability for employees' compensation for future absences?
2. For each of the three situations, indicate the circumstances under which accrual of an expense and related liability is warranted.

Ethics Case 13–5
Outdoors R Us
● LO13–1

Outdoors R Us owns several membership-based campground resorts throughout the Southwest. The company sells campground sites to new members, usually during a get-acquainted visit and tour. The campgrounds offer a wider array of on-site facilities than most. New members sign a multiyear contract, pay a down payment, and make monthly installment payments. Because no credit check is made and many memberships originate on a spur-of-the-moment basis, cancellations are not uncommon.

Business has been brisk during its first three years of operations, and since going public in 2008, the market value of its stock has tripled. The first sign of trouble came in 2021 when new sales dipped sharply.

One afternoon, two weeks before the end of the fiscal year, Diane Rice, CEO, and Gene Sun, controller, were having an active discussion in Sun's office.

Sun: I've thought more about our discussion yesterday. Maybe something can be done about profits.
Rice: I hope so. Our bonuses and stock value are riding on this period's performance.
Sun: We've been recording deferred revenues when new members sign up. Rather than recording liabilities at the time memberships are sold, I think we can justify reporting sales revenue for all memberships sold.
Rice: What will be the effect on profits?
Sun: I haven't run the numbers yet, but let's just say very favorable.

Required:

1. Why do you think liabilities had been recorded previously?
2. Is the proposal ethical?
3. Who would be affected if the proposal is implemented?

Trueblood Case 13–6
Contingencies
● LO13–5

The following Trueblood case is recommended for use with this chapter. The case provides an excellent opportunity for class discussion, group projects, and writing assignments. The case, along with Professor's Discussion Material, can be obtained from the Deloitte Foundation at its website: **www.deloitte.com/us/truebloodcases.**

Case 13–8: Accounting for a Loss Contingency for a Verdict Overturned on Appeal
This case gives students the opportunity to apply GAAP regarding accrual and disclosure of contingencies with respect to a complex lawsuit. As an addition to the case, students should cite the appropriate location of relevant GAAP in the FASB Codification.

Communication Case 13–7
Exceptions to the general classification guideline; group interaction
● LO13–4

Domestic Transfer and Storage is a large trucking company headquartered in the Midwest. Rapid expansion in recent years has been financed in large part by debt in a variety of forms. In preparing the financial statements for 2018, questions have arisen regarding the way certain of the liabilities are to be classified in the company's classified balance sheet.

A meeting of several members of the accounting area is scheduled for tomorrow, April 8, 2019. You are confident that that meeting will include the topic of debt classification. You want to appear knowledgeable at the meeting, but realizing it's been a few years since you have dealt with classification issues, you have sought out information you think relevant. Questionable liabilities at the company's fiscal year-end (January 31, 2019) include the following:

a. $15 million of 9% commercial paper is due on July 31, 2019. Management intends to refinance the paper on a long-term basis. In early April 2019, Domestic negotiated a credit agreement with a commercial bank for up to $12 million any time during the next three years, any borrowings from which will mature two years from the date of borrowing.
b. $17 million of 11% notes were issued on June 30, 2016. The notes are due on November 30, 2019. The company has investments of $20 million classified as "available for sale."
c. $25 million of 10% notes were due on February 28, 2019. On February 21, 2019, the company issued 30-year, 9.4% bonds in a private placement to institutional investors.
d. Recently, company management has considered reducing debt in favor of a greater proportion of equity financing. $20 million of 12% bonds mature on July 31, 2019. Discussions with underwriters, which began on January 4, 2019, resulted in a contractual arrangement on March 15 under which new common shares will be sold in July for approximately $20 million.

In order to make notes to yourself in preparation for the meeting concerning the classification of these items, you decide to discuss them with a colleague. Specifically, you want to know what portion of the debt can be excluded from classification as a current liability (that is, reported as a noncurrent liability) and why.

Required:

1. What is the appropriate classification of each liability? Develop a list of arguments in support of your view prior to the class session for which the case is assigned.
2. In class, your instructor will pair you (and everyone else) with a classmate (who also has independently developed a position). You will be given three minutes to argue your view to your partner. Your partner likewise will be given three minutes to argue his or her view to you. During these three-minute presentations, the listening partner is not permitted to speak.
3. Then after each person has had a turn attempting to convince his or her partner, the two partners will have a three-minute discussion to decide which classifications are more convincing. Arguments will be merged into a single view for each pair.
4. After the allotted time, a spokesperson for each of the four liabilities will be selected by the instructor. Each spokesperson will field arguments from the class as to the appropriate classification. The class then will discuss the merits of the classification and attempt to reach a consensus view, though a consensus is not necessary.

Communication Case 13–8
Various contingencies
● LO13–5, LO13–6

"I see an all-nighter coming on," Gayle grumbled. "Why did Mitch just now give us this assignment?" Your client, Western Manufacturing, is involved with several situations that possibly involve contingencies. The assignment Gayle refers to is to draft appropriate accounting treatment for each situation described below in time for tomorrow's meeting of the audit group. Western's fiscal year is the calendar year 2021, and the 2021 financial statements are issued on March 15, 2022.

1. During 2021, Western experienced labor disputes at three of its plants. Management hopes an agreement will soon be reached. However negotiations between the Company and the unions have not produced an acceptable settlement and, as a result, strikes are ongoing at these facilities since March 1, 2022. It is virtually certain that material costs will be incurred but the amount of possible costs cannot be reasonably ascertained.
2. In accordance with a 2019 contractual agreement with A. J. Conner Company, Western is entitled to $37 million for certain fees and expense reimbursements. These were written off as bad debts in 2020. A. J. Conner has filed for bankruptcy. The bankruptcy court on February 4, 2022, ordered A. J. Conner to pay $23 million immediately upon consummation of a proposed merger with Garner Holding Group.
3. Western warrants most products it sells against defects in materials and workmanship for a period of a year. Based on their experience with previous product introductions, warranty costs are expected to approximate 2% of sales. A warranty liability of $39 million was reported at December 31, 2020. Sales of warranted products during 2021 were $2,100 million and actual warranty expenditures were $40 million.
4. Western is involved in a suit filed in January 2022 by Crump Holdings seeking $88 million, as an adjustment to the purchase price in connection with the Company's sale of its textile business in 2021. The suit alleges that Western misstated the assets and liabilities used to calculate the purchase price for the textile division. Legal counsel advises that it is reasonably possible that Western could end up losing an indeterminable amount not expected to have a material adverse effect on the Company's financial position.

Required:

1. Determine the appropriate means of reporting each situation.
2. In a memo to the audit manager, Mitch Riley, explain your reasoning. Include any necessary journal entries and drafts of appropriate disclosure notes.

Judgment Case 13–9
Loss contingency and full disclosure
● LO13–5, LO13–6

In the March 2022 meeting of Valleck Corporation's board of directors, a question arose as to the way a possible obligation should be disclosed in the forthcoming financial statements for the year ended December 31. A veteran board member brought to the meeting a draft of a disclosure note that had been prepared by the controller's office for inclusion in the annual report. Here is the note:

> On May 9 2021, the United States Environmental Protection Agency (EPA) issued a Notice of Violation (NOV) to Valleck alleging violations of the Clean Air Act. Subsequently, in June 2021, the EPA commenced a civil action with respect to the foregoing violation seeking civil penalties of approximately $853,000. The EPA alleges that Valleck exceeded applicable volatile organic substance emission limits. The Company estimates that the cost to achieve compliance will be $190,000; in addition the Company expects to settle the EPA lawsuit for a civil penalty of $205,000 which will be paid in 2022.

"Where did we get the $205,000 figure?" he asked. On being informed that this is the amount negotiated last month by company attorneys with the EPA, the director inquires, "Aren't we supposed to report a liability for that in addition to the note?"

Required:
Explain whether Valleck should report a liability in addition to the note. Why or why not? For full disclosure, should anything be added to the disclosure note itself?

Communication Case 13–10
Change in loss contingency; write a memo
● LO13–5, LO13–6

Late in 2021, you and two other officers of Curbo Fabrications Corporation just returned from a meeting with officials of the City of Jackson. The meeting was unexpectedly favorable even though it culminated in a settlement with city authorities that required your company pay a total of $475,000 to cover the cost of violations of city construction codes. Jackson had filed suit in November 2019 against Curbo, seeking civil penalties and injunctive relief for violations of city construction codes regulating earthquake damage standards. Alleged violations involved several construction projects completed during the previous three years. When the financial statements were issued in 2020, Curbo had not reached a settlement with state authorities, but legal counsel had advised the Company that it was probable the ultimate settlement would be $750,000 in penalties. The following entry had been recorded:

Loss—litigation	750,000	
Liability—litigation		750,000

The final settlement, therefore, was a pleasant surprise. While returning from the meeting, your conversation turned to reporting the settlement in the 2021 financial statements. You drew the short straw and were selected to write a memo to Janet Zeno, the financial vice president, advising the proper course of action.

Required:
Write the memo. Include descriptions of any journal entries related to the change in amounts. Briefly describe other steps Curbo should take to report the settlement.

Research Case 13–11
Researching the way contingencies are reported; retrieving information from the Internet
● LO13–5, LO13–6

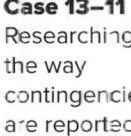

EDGAR (Electronic Data Gathering, Analysis and Retrieval system) performs automated collection, validation, indexing, acceptance, and forwarding of submissions by companies and others who are required by law to file forms with the U.S. Securities and Exchange Commission (SEC). All publicly traded domestic companies use EDGAR to make the majority of their filings. Form 10-K, which includes the annual report, is required to be filed on EDGAR. The SEC makes this information available on the Internet.

Required:
1. Access the *FASB Accounting Standards Codification* at the FASB website (**www.fasb.org**). Determine the specific eight-digit Codification citation (XXX-XX-XX-X) that provides guidance on accounting for contingent losses, and indicate the specific eight-digit Codification citation (XXX-XX-XX-X) citation that describes the guidelines for determining when an expense and liability associated with a contingent loss should be accrued versus only disclosed in the notes.
2. Access EDGAR on the Internet at: **www.sec.gov**.
3. Search for a public company with which you are familiar. Access its most recent 10-K filing. Search or scroll to find the financial statements and related notes.
4. Specifically, look for any contingency(s) reported in the disclosure notes. Identify the nature of the contingency(s) described and explain the reason(s) the loss or losses was or was not accrued.
5. Repeat requirements 2 and 3 for two additional companies.

Communication Case 13–12
Accounting changes
● LO13–5, LO13–6

Kevin Brantly is a new hire in the controller's office of Fleming Home Products. Two events occurred in late 2021 that the company had not previously encountered. The events appear to affect two of the company's liabilities, but there is some disagreement concerning whether they also affect financial statements of prior years. Each change occurred during 2021 before any adjusting entries or closing entries were prepared. The tax rate for Fleming is 40% in all years.

- Fleming Home Products introduced a new line of commercial awnings in 2020 that carry a one-year warranty against manufacturer's defects. Based on industry experience, warranty costs were expected to approximate 3% of sales. Sales of the awnings in 2020 were $3,500,000. Accordingly, warranty expense and a warranty liability of $105,000 were recorded in 2020. In late 2021, the company's claims experience was evaluated and it was determined that claims were far fewer than expected—2% of sales rather than 3%. Sales of the awnings in 2021 were $4,000,000 and warranty expenditures in 2021 totaled $91,000.
- In November 2019, the State of Minnesota filed suit against the company, seeking penalties for violations of clean air laws. When the financial statements were issued in 2020, Fleming had not reached a settlement with state authorities, but legal counsel advised Fleming that it was probable the company would have to pay $200,000 in penalties. Accordingly, the following entry was recorded:

Loss—litigation	200,000	
Liability—litigation		200,000

Late in 2021, a settlement was reached with state authorities to pay a total of $350,000 in penalties.

Required:
Kevin's supervisor, perhaps unsure of the answer, perhaps wanting to test Kevin's knowledge, e-mails the message, "Kevin, send me a memo on how we should handle our awning warranty and that clean air suit." Wanting to be accurate, Kevin consults his reference materials. What will he find? Prepare the memo requested.

Real World Case 13–13
Lawsuit settlement; Morgan Stanley
● LO13–5, LO13–6
Real World Financials

Morgan Stanley is a leading investment bank founded in 1935. The company's fiscal year ends December 31, 2013, and it filed its financial statements with the SEC on February 25, 2014. On February 5, 2014, *Bloomberg* reported that Morgan Stanley would make a $1.25 billion settlement with the Federal Housing Finance Agency:

> Morgan Stanley said yesterday it reached a $1.25 billion deal to end Federal Housing Finance Agency claims the bank sold faulty mortgage bonds to Fannie Mae (FNMA) and Freddie Mac (FMCC) before the firms' losses pushed them into U.S. conservatorship. . . . Morgan Stanley, which disclosed its settlement in a regulatory filing yesterday, was among 18 banks sued by the FHFA in 2011. Authorities sought to recoup some losses taxpayers covered when the government took control of the failing mortgage-finance companies in 2008.

Required:
1. From an accounting perspective, how should Morgan Stanley have treated the settlement?
2. Relying on the information provided by the news article, re-create the journal entry Morgan Stanley recorded for the settlement. Assume it had not made any prior liability accrual with respect to this litigation.
3. Suppose the settlement had occurred after February 25, 2014. How should Morgan Stanley have treated the settlement?

Ethics Case 13–14
Profits guaranteed
● LO13–5

This was Joel Craig's first visit to the controller's corner office since being recruited for the senior accountant position in May. Because he'd been directed to bring with him his preliminary report on year-end adjustments, Craig presumed he'd done something wrong in preparing the report. That he had not was Craig's first surprise. His second surprise was his boss's request to reconsider one of the estimated expenses.

S & G Fasteners was a new company, specializing in plastic industrial fasteners. All products carry a generous long-term warranty against manufacturer's defects. "Don't you think 4% of sales is a little high for our warranty expense estimate?" his boss wondered. "After all, we're new at this. We have little experience with product introductions. I just got off the phone with Blanchard (the company president). He thinks we'll have trouble renewing our credit line with the profits we're projecting. The pressure's on."

Required:
1. Should Craig follow his boss's suggestion?
2. Does revising the warranty estimate pose an ethical dilemma?
3. Who would be affected if the suggestion is followed?

IFRS Case 13–15
Current liabilities and contingencies; differences between U.S. GAAP and IFRS
● LO13–4, LO13–5, LO13–7

IFRS

As a second-year financial analyst for A.J. Straub Investments, you are performing an initial analysis on Fizer Pharmaceuticals. A difficulty you've encountered in making comparisons with its chief rival is that Fizer uses U.S. GAAP and the competing company uses International Financial Reporting Standards. Some areas of concern are the following:

1. Fizer has been designated as a potentially responsible party by the United States Environmental Protection Agency with respect to certain waste sites. These claims are in various stages of administrative or judicial proceedings and include demands for recovery of past governmental costs and for future investigations or remedial actions. Fizer accrues costs associated with environmental matters when they become probable and reasonably estimable. Counsel has advised that the likelihood of payments of about $70 million is slightly more than 50%. Accordingly, payment is judged reasonably possible and the contingency was disclosed in a note.
2. Fizer had $10 million of bonds issued in 1992 that mature in February 2022. On December 31, 2021, the company's fiscal year-end, management intended to refinance the bonds on a long-term basis. On February 7, 2022, Fizer issued $10 million of 20-year bonds, applying the proceeds to repay the bond issue that matured that month. The bonds were reported in Fizer's balance sheet as long-term debt.
3. Fizer reported in its 2021 financial statements a long-term contingency at its face amount rather than its present value even though the difference was considered material. The reason the cash flows were not discounted is that their timing is uncertain.

Required:
If Fizer used IFRS as does its competitor, how would the items described be reported differently?

Analysis Case 13–16
Analyzing financial statements; liquidity ratios
● LO13–1

IGF Foods Company is a large, primarily domestic, consumer foods company involved in the manufacture, distribution and sale of a variety of food products. Industry averages are derived from Troy's *The Almanac of Business and Industrial Financial Ratios*. Following are the 2021 and 2020 comparative balance sheets for IGF. (The financial data we use are from actual financial statements of a well-known corporation, but the company name used is fictitious and the numbers and dates have been modified slightly.)

IGF FOODS COMPANY
Comparative Balance Sheets Years Ended December 31, 2021 and 2020
($ in millions)

	2021	2020
Assets		
Current assets:		
Cash	$ 48	$ 142
Accounts receivable	347	320
Marketable securities	358	—
Inventories	914	874
Prepaid expenses	212	154
Total current assets	1,879	1,490
Property, plant, and equipment (net)	2,592	2,291
Intangibles (net)	800	843
Other assets	74	60
Total assets	$5,345	$4,684
Liabilities and Shareholders' Equity		
Current liabilities:		
Accounts payable	$ 254	$ 276
Accrued liabilities	493	496
Notes payable	518	115
Current portion of long-term debt	208	54
Total current liabilities	1,473	941
Long-term debt	534	728
Deferred income taxes	407	344
Total liabilities	2,414	2,013
Shareholders' equity:		
Common stock	180	180
Additional paid-in capital	21	63
Retained earnings	2,730	2,428
Total shareholders' equity	2,931	2,671
Total liabilities and shareholders' equity	$5,345	$4,684

Liquidity refers to a company's cash position and overall ability to obtain cash in the normal course of business. A company is said to be liquid if it has sufficient cash or is capable of converting its other assets to cash in a relatively short period of time so that currently maturing debts can be paid.

Required:

1. Calculate the current ratio for IGF for 2021. The average ratio for the stocks listed on the New York Stock Exchange in a comparable time period was 1.5. What information does your calculation provide an investor?
2. Calculate IGF's acid-test or quick ratio for 2021. The ratio for the stocks listed on the New York Stock Exchange in a comparable time period was .80. What does your calculation indicate about IGF's liquidity?

Trueblood Case 13–17
Accounting for litigation contingencies
● LO13–5, LO13–6, LO13–7

The following Trueblood case is recommended for use with this chapter. The case provides an excellent opportunity for class discussion, group projects, and writing assignments. The case, along with Professor's Discussion Material, can be obtained from the Deloitte Foundation at its website: **www.deloitte.com/us/truebloodcases**.

Case 14-07: eVade Pays Up

This case gives students an opportunity to consider the difference between a contractual or legal liability and a contingency.

Real World Case 13–18
Contingencies
● LO13–5
Real World Financials

The following is an excerpt from *USAToday.com:*

> Microsoft (MSFT) on Thursday extended the warranty on its Xbox 360 video game console and said it will take a charge of more than $1 billion to pay for "anticipated costs." Under the new warranty, Microsoft will pay for shipping and repairs for three years, worldwide, for consoles afflicted with what gamers call "the red ring of death." Previously, the warranty expired after a year for U.S. customers and two years for Europeans. The charge will be $1.05 billion to $1.15 billion for the quarter ended June 30. Microsoft reports its fourth-quarter results July 19.

Required:

1. Why must Microsoft report this charge of over $1 billion entirely in one quarter, the last quarter of the company's fiscal year?
2. When the announcement was made, analyst Richard Doherty stated that either a high number of Xbox 360s will fail or the company is being overly conservative in its warranty estimate. From an accounting standpoint, what will Microsoft do in the future if the estimate of future repairs is overly conservative (too high)?

Real World Case 13–19
Contingencies
● LO13–5
Real World Financials
IFRS

Reporting requirements for contingent liabilities under IFRS differ somewhat from those under U.S. GAAP.

Required:

For each of the following, access the online version of the indicated financial report and answer the question. Also compare reporting under IFRS to similar reporting under U.S. GAAP.

1. **AU Optronics** (Form 20-F, filed 3/21/2016): Where there is a continuous range of possible outcomes, with each point in the range as likely as any other, what amount is accrued as the estimate of the obligation?
2. **B Communications LTD** (Form 20-F, filed 4/19/2016): With respect to legal claims, at what probability level would B Communications accrue a liability for a possible litigation loss?

Trueblood Case 13–20
Subsequent Events
● LO13–5
CODE

The following Trueblood case is recommended for use with this chapter. The case provides an excellent opportunity for class discussion, group projects, and writing assignments. The case, along with Professor's Discussion Material, can be obtained from the Deloitte Foundation at its website: **www.deloitte.com/us/truebloodcases**.

Case 12-02: To Recognize or Not to Recognize, That is the Question

This case gives students the opportunity to apply GAAP regarding recognition of subsequent events. As an addition to the case, students should cite the appropriate location of relevant GAAP in the FASB Codification.

Data Analytics

connect

Data analytics is the process of examining data sets in order to draw conclusions about the information they contain. If you haven't completed any of the prior data analytics cases, follow the instructions listed in the Chapter 1 Data Analytics case to get set up. You will need to watch the videos referred to in the Chapters 1–3

Data Analytics cases. No additional videos are required for this case. All short training videos can be found here: **https://www.tableau.com/learn/training#getting-started**.

Data Analytics Case
Ability to Pay Short-Term Debt
● LO13–1

In the Chapter 3 Data Analytics Cases, you used Tableau to examine a data set and create charts to compare our two (hypothetical) publicly traded companies, Discount Goods and Big Store, as to their liquidity - their ability to pay short-term debts as they come due. You did so by comparing current ratios and acid-test ratios. In this case , you will expand your assessment of the companies' liquidity by also evaluating other aspects of liquidity as measured by their cash to current liabilities and current liabilities to net worth ratios.

The cash to current liabilities ratio (also known as the cash ratio) is measured as cash and cash equivalents / total current liabilities and tells us about the ability of a company to settle its current liabilities using only its cash and highly liquid investments.

The current liabilities to net worth ratio is measured as current liabilities / shareholders' equity and tells us about the relationship between capital contributed by current obligation creditors and capital contributed by owners. It indicates the ability of a company to safely meet the obligations of current creditors. The higher the ratio, the less security there is for current obligation creditors.

Required:

Use Tableau to create calculations for the current, acid-test, cash to current liabilities, and current liabilities to net worth ratios for each of the two companies in 2021. Based upon what you find, answer the following questions:

1. Other things being equal, which company appears to have the better liquidity position in terms of their ability to pay short-term debts as they come due as measured by the current ratio?
2. Which company appears to offer the better liquidity position in terms of their ability to pay short-term debts as they come due as measured by the acid-test or quick ratio?
3. Other things being equal, which company appears to have the better liquidity position in terms of ability of the company's current liabilities to be covered using its cash and cash equivalents?
4. Which company appears to offer the better security for its current obligation creditors as measured by the current liabilities to net worth ratio?

Note: As you assess the liquidity aspect of the companies, keep in mind that liquidity ratios should be evaluated in the context of both profitability and efficiency of managing assets and relative to average liquidity ratios of the industry.

Resources:

You have available to you an extensive data set that includes detailed financial data for Discount Goods and Big Store for 2012–2021. The data set is in the form of four Excel files available to download from Connect or under Student Resources within the Library tab.The one for use in this chapter is named "Discount_Goods_Big_Store_Financials.xlsx." Download this file and save it to the computer on which you will be using Tableau. For this case, you will create calculations to produce several liquidity ratios allowing you to compare and contrast the two companies.

After you view the training videos, follow these steps to create the charts you'll use for this case:

- Open Tableau and connect to the Excel spreadsheet you downloaded.
- Click on the "Sheet 1" tab at the bottom of the canvas, to the right of the Data Source at the bottom of the screen. Drag "Company" and "Year" under Dimensions to the Rows shelf. Change "Year" to *discrete* by right-clicking and selecting "Discrete." Select "Show Filter" and uncheck all the years except 2021. Click "Ok."
- Create the "Current Ratio" and "Quick Ratio" by following the steps in the Chapter 3 Data Analytics Case. After creating the two ratios, drag them to the Rows shelf. Change the ratios back to *continuous* by right-clicking and selecting "Continuous."
- Create a calculated field "Cash Ratio" by dragging "Cash and cash equivalents" under Measures into the Calculation Editor window. Type a division sign, then drag the "Total current liabilities" beside it. Make sure the window says that the calculation is valid and click OK. Drag the newly created "Cash Ratio" to the Rows shelf.
- Create a calculated field "Current Liabilities to Net Worth" by dragging "Total current liabilities" from the under Measures into the Calculation Editor window. Type a division sign, then drag the "Total shareholders' equity" beside it. Make sure the window says that the calculation is valid and click OK. Drag the newly created "Current Liabilities to Net Worth" to the Rows shelf.
- Click on the "Show Me" tab in the upper right corner and select "horizontal bars." Add labels to the bars by clicking on "Label" under the "Marks" card and clicking the box "Show mark label." Format the labels to Times New Roman, bold, black and 9-point font.
- Change the title of the sheet to be "Liquidity Ratios" by right-clicking and selecting "Edit title." Format the title to Times New Roman, bold, black and 15-point font. Format all other labels to be Times New Roman, bold, black and 12-point font. Change the title of "Sheet 1" to match the sheet title by right-clicking, selecting "Rename" and typing in the new title. Drag "Company" to the Color card in the "Marks" card to change the bars to two different colors.
- Once complete, save the file as *"DA13_Your initials.twbx."*

Continuing Cases

Target Case

- LO13–1, LO13–3, LO13–5

Target Corporation prepares its financial statements according to U.S. GAAP. Target's financial statements and disclosure notes for the year ended February 3, 2018, are available Connect. This material also is available under the Investor Relations link at the company's website (**www.target.com**).

Required:

1. Target's Consolidated Statement of Financial Position (its balance sheet) discloses its current assets and current liabilities.
   a. What are the three components of Target's current liabilities?
   b. Are current assets sufficient to cover current liabilities? What is the current ratio for the year ended February 3, 2018? How does the ratio compare with the prior year?
   c. Why might a company want to avoid having its current ratio be too low? Too high?
2. Disclosure Note 2 discusses Target's accounting for gift card sales. Disclosure Note 18 indicates the amount of gift card liability that is recognized in Target's balance sheet.
   a. By how much did Target's gift card liability change between February 3, 2018, and January 28, 2017?
   b. How would the following affect Target's gift card liability (indicate "increase," "decrease," or "no change" for each):
      i. Sale of a gift card
      ii. Redemption of a gift card (the holder using it to acquire goods or services)
      iii. Increase in breakage estimated for gift cards already sold
3. Disclosure Note 19 discusses Target's accounting for contingencies. What is Target's approach for accruing losses for litigation liabilities? Is their approach appropriate?

Air France–KLM Case

- LO13–7

IFRS

Air France–KLM (AF), a Franco-Dutch company, prepares its financial statements according to International Financial Reporting Standards. AF's financial statements and disclosure notes for the year ended December 31, 2017, are available in Connect. This material is also available under the Finance link at the company's website (**www.airfranceklm.com**).

Required:

1. Read Notes 4.6 and the Consolidated Balance Sheet. What do you think gave rise to deferred revenue on ticket sales of €2,889 as of the end of fiscal 2017? Would transactions of this type be handled similarly under U.S. GAAP?
2. Is the threshold for recognizing a provision under IFRS different than it is under U.S. GAAP? Explain.
3. Note 31 lists "other provisions."
   a. Do the beginning and ending balances of other provisions shown in Note 31 for fiscal 2017 tie to the balance sheet? By how much has the total amount of the AF's "other provisions" increased or decreased during fiscal 2017?
   b. Prepare journal entries for the following changes in the litigation provision that occurred during fiscal 2017, assuming any amounts recorded on the income statement are recorded as "provision expense," and any use of provision is paid for in cash. In each case, provide a brief explanation of the event your journal entry is capturing.
      i. New provision
      ii. Use of provision
   c. Is AF's treatment of its litigation provision under IFRS similar to how it would be treated under U.S. GAAP?
4. Note 31.2 lists a number of contingent liabilities. Are amounts for those items recognized as a liability on AF's balance sheet? Explain.

CPA Exam Questions and Simulations

ROGER CPA Review

Sample CPA Exam questions from Roger CPA Review are available in Connect as support for the topics in this chapter. These multiple-choice questions and task-based simulations include expert-written explanations and solutions, and provide a starting point for students to become familiar with the content and functionality of the actual CPA Exam.

CHAPTER

14 Bonds and Long-Term Notes

OVERVIEW

This chapter continues the presentation of liabilities. While the discussion focuses on the accounting treatment of long-term liabilities, the borrowers' side of the same transactions is presented as well. Long-term notes and bonds are discussed, as well as the extinguishment of debt and debt convertible into stock.

LEARNING OBJECTIVES

After studying this chapter, you should be able to:

- **LO14–1** Identify the underlying characteristics of debt instruments and describe the basic approach to accounting for debt. *(p. 773)*
- **LO14–2** Account for bonds issued at face value, at a discount, or at a premium, recording interest using the effective interest method or using the straight-line method. *(p. 775)*
- **LO14–3** Characterize the accounting treatment of notes, including installment notes, issued for cash or for noncash consideration. *(p. 786)*
- **LO14–4** Describe the disclosures appropriate to long-term debt in its various forms and calculate related financial ratios. *(p. 791)*
- **LO14–5** Record the early extinguishment of debt, its conversion into equity securities, and bond issues with warrants. *(p. 795)*
- **LO14–6** Understand the option to report liabilities at their fair values. *(p. 801)*
- **LO14–7** Discuss the primary differences between U.S. GAAP and IFRS with respect to accounting for bonds and long-term notes. *(pp. 797)*.

©monkeybusinessimages/Getty Images

FINANCIAL REPORTING CASE

Service Leader, Inc.

The mood is both upbeat and focused on this cool October morning. Executives and board members of Service Leader, Inc., are meeting with underwriters and attorneys to discuss the company's first bond offering in its 20-year history. You are attending in the capacity of company controller and two-year member of the board of directors. The closely held corporation has been financed entirely by equity, internally generated funds, and short-term bank borrowings.

Bank rates of interest, though, have risen recently and the company's unexpectedly rapid, but welcome, growth has prompted the need to look elsewhere for new financing. Under consideration are 15-year, 6.25% first mortgage bonds with a principal amount of $70 million. The bonds would be callable at 103 any time after June 30, 2023, and convertible into Service Leader common stock at the rate of 45 shares per $1,000 bond.

Other financing vehicles have been discussed over the last two months, including the sale of additional stock, nonconvertible bonds, and unsecured notes. This morning *The Wall Street Journal* indicated that market rates of interest for debt similar to the bonds under consideration are about 6.5%.

QUESTIONS

By the time you finish this chapter, you should be able to respond appropriately to the questions posed in this case. Compare your response to the solution provided at the end of the chapter.

1. What does it mean that the bonds are "first mortgage" bonds? What effect does that have on financing? *(p. 774)*
2. From Service Leader's perspective, why are the bonds callable? What does that mean? *(p. 775)*
3. How will it be possible to sell bonds paying investors 6.25% when other similar investments will provide the investors a return of 6.5%? *(p. 775)*
4. Would accounting differ if the debt were designated as notes rather than bonds? *(p. 786)*
5. Why might the company choose to make the bonds convertible into common stock? *(p. 795)*

The Nature of Long-Term Debt

● LO14–1

A company must raise funds to finance its operations and often the expansion of those operations. Presumably, at least some of the necessary funding can be provided by the company's own operations, though some funds must be provided by external sources. Ordinarily, external financing includes some combination of equity and debt funding. We explore debt financing first.

Liabilities signify *borrowers'* interests in a company's assets.

In the present chapter, we focus on debt in the form of bonds and notes. The following three chapters deal with liabilities also, namely those arising in connection with leases (Chapter 15), deferred income taxes (Chapter 16), and pensions and employee benefits (Chapter 17). Some employee benefits create equity rather than debt, which are discussed in Chapter 19. In Chapter 18, we examine shareholders' interests arising from external *equity* financing. In Chapter 21, we see that cash flows from both debt and equity financing are reported together in a statement of cash flows as "cash flows from financing activities."

As you read this chapter, you will find the focus to be on the liability side of the transactions we examine. Realize, though, that the mirror image of a liability is an asset (bonds payable/investment in bonds, note payable/note receivable, etc.). So as we discuss accounting for debts from the viewpoint of the issuers of the debt instruments (borrowers), we also will take the opportunity to see how the lender deals with the corresponding asset. Studying the two sides of the same transaction in tandem will emphasize their inherent similarities.

A note payable and a note receivable are two sides of the same coin.

Accounting for a liability is a relatively straightforward concept. This is not to say that all debt instruments are unchallenging, "plain vanilla" loan agreements. Quite the contrary, the financial community continually devises increasingly exotic ways to flavor financial instruments in the attempt to satisfy the diverse and evolving tastes of both debtors and creditors.

Periodic interest is the effective interest rate times the amount of the debt outstanding during the interest period.

Packaging aside, a liability requires the future payment of cash in specified (or estimated) amounts, at specified (or projected) dates. As time passes, interest accrues on debt. As a general rule, the periodic interest is the effective interest rate times the amount of the debt outstanding during the period. This same principle applies regardless of the specific form of the liability—note payable, bonds payable, lease liability, pension obligation, or other debt instruments. Also, as a general rule, long-term liabilities are reported at their present values. The present value of a liability is the present value of its related cash flows (principal and/or interest payments), discounted at the effective rate of interest at issuance.

We begin our study of long-term liabilities by examining accounting for bonds. We follow that section with a discussion of debt in the form of notes in Part B. It's important to note that, although particulars of the two forms of debt differ, the basic approach to accounting for each type is precisely the same. In Part C, we look at various ways bonds and notes are retired or converted into other securities. Finally, in Part D we discuss the option companies have to report liabilities at their fair values.

PART A

Bonds

A company can borrow cash from a bank or other financial institution by signing a promissory note. We discuss notes payable later in the chapter. Medium- and large-sized corporations often choose to borrow cash by issuing bonds to the public. In fact, the most common form of corporate debt is bonds. A bond issue, in effect, breaks down a large debt (large corporations often borrow hundreds of millions of dollars at a time) into manageable parts—usually $1,000 or $5,000 units. This avoids the necessity of finding a single lender who is both willing and able to loan a large amount of money at a reasonable interest rate. So rather than signing a $400 million note to borrow cash from a financial institution, a company may find it more economical to sell 400,000 $1,000 bonds to many lenders—theoretically up to 400,000 lenders.

A bond issue divides a large liability into many smaller liabilities.

Bonds obligate the issuing corporation to repay a stated amount (variously referred to as the *principal, par value, face amount,* or *maturity value*) at a specified *maturity date.* Maturities for bonds typically range from 10 to 40 years. In return for the use of the money borrowed, the company also agrees to pay *interest* to bondholders between the issue date and maturity. The periodic interest is a stated percentage of face amount (variously referred to as the *stated rate, coupon rate,* or *nominal rate*). Ordinarily, interest is paid semiannually on designated interest dates beginning six months after the day the bonds are "dated."

Corporations issuing bonds are obligated to repay a stated amount at a specified *maturity date* and to pay periodic *interest* between the issue date and maturity.

The Bond Indenture

The specific promises made to bondholders are described in a document called a **bond indenture**. Because it would be impractical for the corporation to enter into a direct agreement with each of the many bondholders, the bond indenture is held by a trustee, usually a commercial bank or other financial institution, appointed by the issuing firm to represent the rights of the bondholders. If the company fails to live up to the terms of the bond indenture, the trustee may bring legal action against the company on behalf of the bondholders.

A *bond indenture* describes the specific promises made to bondholders.

Most corporate bonds are debenture bonds. A **debenture bond** is backed only by the "full faith and credit" of the issuing corporation. No specific assets are pledged as security. Investors in debentures usually have the same standing as the firm's other general creditors.

FINANCIAL Reporting Case

Q1, p. 773

So in case of bankruptcy, debenture holders won't have priority over other general creditors. An exception is the **subordinated debenture**, which is not entitled to receive any liquidation payments until the claims of other specified debt issues are satisfied.

A **mortgage bond**, on the other hand, is backed by a lien on specified real estate owned by the issuer. Because a mortgage bond is considered less risky than debentures, it typically will command a lower interest rate. Alternatively, a bond might be "secured" by non-real estate assets in which case, if in the event the issuer does not pay the holder as promised, the holder has rights to take specified assets belonging to the issuer such as items of property, plant, and equipment.

Today most corporate bonds are registered bonds. Interest checks are mailed directly to the owner of the bond, whose name is registered with the issuing company. Years ago, it was typical for bonds to be structured as **coupon bonds** (sometimes called *bearer bonds*). The name of the owner of a coupon bond was not registered. Instead, to collect interest on a coupon bond the holder actually clipped an attached coupon and redeemed it in accordance with instructions in the indenture. A carryover effect of this practice is that we still sometimes see the term *coupon rate* in reference to the stated interest rate on bonds.

FINANCIAL Reporting Case

Q2, p. 773

Most corporate bonds are **callable** (or redeemable). The call feature allows the issuing company to buy back, or call, outstanding bonds from bondholders before their scheduled maturity date. This feature affords the company some protection against being stuck with relatively high-cost debt in the event interest rates fall during the period before maturity. The call price must be prespecified and often exceeds the bond's face amount (a call premium), sometimes declining as maturity is approached.

For example, a 2017 news release of **Mueller Industries** included this disclosure:

Real World Financials

The Debentures will be callable at the option of our company, in whole or in part, at any time or from time to time, subject to declining call premiums during the first five years.

Mandatory sinking fund redemptions retire a bond issue gradually over its term to maturity.

"No call" provisions usually prohibit calls during the first few years of a bond's life. Very often, calls are mandatory. That is, the corporation may be required to redeem the bonds on a prespecified, year-by-year basis. Bonds requiring such **sinking fund** redemptions often are labeled *sinking fund debentures*.

Serial bonds provide a more structured (and less popular) way to retire bonds on a piecemeal basis. Serial bonds are retired in installments during all or part of the life of the issue. Each bond has its own specified maturity date. So for a typical 30-year serial issue, 25 to 30 separate maturity dates might be assigned to specific portions of the bond issue.

Convertible bonds are retired as a consequence of bondholders choosing to convert them into shares of stock. We look closer at convertible bonds a little later in the chapter.

Recording Bonds at Issuance

● LO14–2

Bonds represent a liability to the corporation that issues the bonds and an asset to an investor who buys the bonds as an investment. Each side of the transaction is the mirror image of the other.[1] This is demonstrated in Illustration 14–1.

Most bonds these days are issued on the day they are dated (date printed in the indenture contract). On rare occasions, there may be a delay in issuing bonds that causes them to be issued between interest dates, in which case the interest that has accrued since the day they are dated is added to the price of the bonds. We discuss this infrequent event in Appendix 14A to this chapter.

Determining the Selling Price

FINANCIAL Reporting Case

Q3, p. 773

The price of a bond issue at any particular time is not necessarily equal to its face amount. The $700,000, 12% bond issue in the previous illustration, for example, may sell for more than face amount (at a **premium**) or less than face amount (at a **discount**), depending on

[1]You should recall from Chapter 12 that investments in bonds that are to be held to maturity by the investor are reported at amortized cost, which is the method described here. However, also remember that investments in debt securities *not* to be held to maturity are reported at the fair value of the securities held, as described in Chapter 12, with the interest determined by the effective interest method.

Illustration 14–1

Bonds Sold at Face Amount

On January 1, 2021, Masterwear Industries issued $700,000 of 12% bonds.

- Interest of $42,000 is payable semiannually on June 30 and December 31.
- The bonds mature in three years (an unrealistically short maturity to shorten the illustration).
- The entire bond issue was sold at the face amount.
- United Intergroup, Inc., the sole investor in the bonds, planned to hold the bonds to their maturity.

At Issuance (January 1)

Masterwear (Issuer)		
Cash	700,000	
Bonds payable (face amount)		700,000
United (Investor)		
Investment in bonds (face amount)	700,000	
Cash		700,000

how the 12% *stated* interest rate compares with the prevailing *market* or *effective rate* of interest (for securities of similar risk and maturity). For instance, if the 12% bonds are competing in a market in which similar bonds are providing a 14% return, the bonds could be sold only at a price less than $700,000. On the other hand, if the market rate is only 10%, the 12% stated rate would seem relatively attractive and the bonds would sell at a premium over face amount. The reason the stated rate often differs from the market rate, resulting in a *discount* or *premium,* is the inevitable delay between the date the terms of the issue are established and the date the issue comes to market.

Other things being equal, the lower the perceived riskiness of the corporation issuing bonds, the higher the price those bonds will command.

In addition to the characteristic terms of a bond agreement as specified in the indenture, the market rate for a specific bond issue is influenced by the creditworthiness of the company issuing the bonds. To evaluate the risk and quality of an individual bond issue, investors rely heavily on bond ratings provided by **Standard & Poor's Corporation** and by **Moody's Investors Service, Inc.** See the bond ratings in Illustration 14–2.

Illustration 14–2

Bond Ratings*

	S&P	Moody's
Investment Grades:		
Highest	AAA	Aaa
High	AA	Aa
Medium	A	A
Minimum investment grade	BBB	Baa
"Junk" Ratings:		
Speculative	BB	Ba
Very speculative	B	B
Default or near default	CCC	Caa
	CC	Ca
	C	C
	D	

*Adapted from **Moodys.com** and **SPRatings.com**.

A bond issue will be priced by the marketplace to yield the market rate of interest for securities of similar risk and maturity.

Forces of supply and demand cause a bond issue to be *priced to yield the market rate.* In other words, an investor paying that price will earn an effective rate of return on the investment equal to the market rate. The price is calculated as the present value of all the cash flows required of the bonds, where the discount rate used in the present value calculation is the market rate. Specifically, the price will be the *present value of the periodic cash interest payments (face amount × stated rate) plus the present value of the principal payable at maturity, both discounted at the market rate.*

Bonds priced at a discount are described in Illustration 14–3.

Illustration 14–3

Bonds Sold at a Discount

On January 1, 2021, Masterwear Industries issued $700,000 of 12% bonds, dated January 1.

- Interest of $42,000 is payable semiannually on June 30 and December 31.
- The bonds mature in three years.
- The market yield for bonds of similar risk and maturity is 14%.
- United Intergroup, Inc. purchased the entire bond issue, planning to hold the bonds until maturity.

Calculation of the Price of the Bonds

		Present Values
Interest	$ 42,000 × 4.76654* =	$200,195
Principal (face amount)	$700,000 × 0.66634† =	466,438
Present value (price) of the bonds		$666,633

*Present value of an ordinary annuity of $1: $n = 6$, $i = 7\%$ (Table 4).
†Present value of $1: $n = 6$, $i = 7\%$ (Table 2).
Because interest is paid semiannually, the present value calculations use: (a) one-half the stated rate (6%) to determine cash payments, (b) one-half the market rate (7%) as the discount rate, and (c) six (3 × 2) semiannual periods.
Note: Present value tables are provided at the end of this textbook. If you need to review the concept of the time value of money, refer to the discussions in Chapter 5.
Rounding: Because present value tables truncate decimal places, the solution may be slightly different if you use a calculator or Excel.

Using a calculator:
Enter: N 6 I 7
PMT – 42000
FV – 700000
Output: PV 666,634

Using Excel, enter:
=PV(.07,6,–42000, –700000
Output: 666,634

Bonds Issued at a Discount

The calculation is illustrated in Illustration 14–4.

Illustration 14–4

Cash Flows from a Bond Issue at a Discount

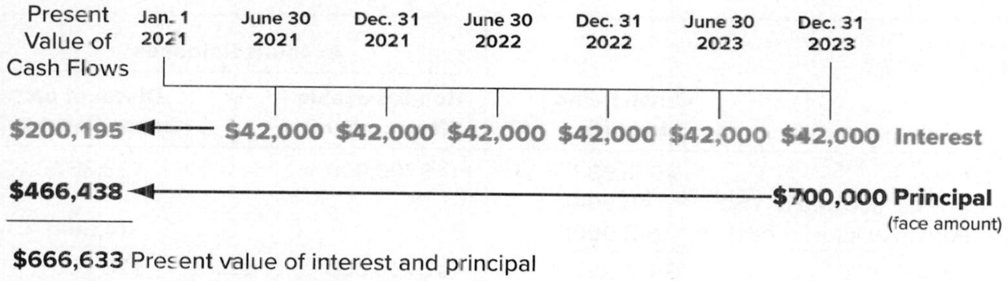

Although the cash flows total $952,000, the present value of those future cash flows as of January 1, 2021, is only $666,633. This is due to the time value of money. These bonds are issued at a discount because the present value of the cash flows is less than the face amount of the bonds.

Journal Entries at Issuance—Bonds Issued at a Discount

Masterwear (Issuer)		
Cash (price calculated above)	666,633	
Discount on bonds payable (difference)	33,367	
Bonds payable (face amount)		700,000
United (Investor)		
Investment in bonds (face amount)	700,000	
Discount on investment in bonds (difference)		33,367
Cash (price calculated above)		666,633

Note: A sometimes-used alternative way to record bonds is the "net method," shown next, in which the discount is included directly in the book value.

Masterwear (Issuer)		
Cash	666,633	
Bonds payable		666,633
United (Investor)		
Investment in bonds	666,633	
Cash		666,633

When bond prices are quoted in financial media, they typically are stated in terms of a percentage of face amounts. Thus, a price quote of 98 means one $1,000 bond will sell for $980; a bond priced at 101 will sell for $1,010.

In one of the largest U.S. corporate bond issuances in history, **CVS Health Corporation** issued $40,000,000,000 in bonds in 2018. CVS issued the 2.75% bonds to raise funds to pay for its acquisition of **Aetna Inc.** The bonds were priced at 95.97 to provide an annual yield of 3.75%.

Determining Interest—Effective Interest Method

The *effective interest* on debt is the market rate of interest multiplied by the outstanding balance of the debt.

Interest accrues on an outstanding debt at a constant percentage of the debt each period. Of course, under the concept of accrual accounting, the periodic effective interest is not affected by the time at which the cash interest actually is paid. Recording interest each period as the *effective market rate of interest multiplied by the outstanding balance of the debt* (during the interest period) is referred to as the **effective interest method**. Although giving this a label—the effective interest method—implies some specialized procedure, this simply is an application of the accrual concept, consistent with accruing all expenses as they are incurred.

Continuing our example, we determined that the amount of debt when the bonds are issued is $666,633. Since the effective interest rate is 14%, interest recorded (as expense to the issuer and revenue to the investor) for the first six-month interest period is $46,664:

$$\underset{\text{Outstanding balance}}{\$666{,}633} \times \underset{\text{Effective rate}}{(14\% \div 2)} = \underset{\text{Effective interest}}{\$46{,}664}$$

However, the bond indenture calls for semiannual interest payments of only $42,000—the *stated* rate (6%) times the *face amount* ($700,000). The remainder, $4,664, increases the liability and is reflected as a reduction in the discount (a contra-liability account). This is illustrated in Illustration 14–5.

Illustration 14–5
Change in Debt When Effective Interest Exceeds Cash Paid

The difference between the effective interest and the interest paid increases the existing liability.

			Account Balances		
	Outstanding Balance		**Bonds Payable (face amount)**		**Discount on Bonds Payable**
January 1	$666,633	=	$700,000	less	$33,367
Interest accrued at 7%	**46,664**				
Portion of interest paid	**(42,000)**				**(4,664)**
June 30	$671,297	=	$700,000	less	$28,703

Interest expense (issuer) and revenue (investor) are calculated on the outstanding debt balance at the effective (or market) rate. Interest *paid* is the amount specified in the bond indenture—the *stated* rate times the face amount. These amounts and the change in the outstanding debt are recorded as follows:

Journal Entries—The Interest Method

The effective interest is calculated each period as the market rate times the outstanding balance of the debt during the interest period.

At the First Interest Date (June 30)

Masterwear (Issuer)		
Interest expense (effective rate × outstanding balance)	46,664	
Discount on bonds payable (difference)		**4,664**
Cash (stated rate × face amount)		42,000
United (Investor)		
Cash (stated rate × face amount)	**42,000**	
Discount on investment in bonds (difference)	**4,664**	
Interest revenue (effective rate × outstanding balance)		**46,664**

In this example and others in the chapter we look at the investor's entries as well as the issuer's to see both sides of the same transactions.

Because the balance of the debt changes each period, the dollar amount of interest (balance × rate) also will change each period. To keep up with the changing amounts, it usually is convenient to prepare a schedule that reflects the changes in the debt over its term to maturity. An amortization schedule for the situation under discussion is shown in Illustration 14–6.

Illustration 14–6
Amortization Schedule—Discount

Date	Cash Interest	Effective Interest	Increase in Balance	Outstanding Balance
	(6% × Face amount)	(7% × Outstanding balance)	(Discount reduction)	
1/1/21				666,633
6/30/21	42,000	0.07 (666,633) = 46,664	**4,664**	671,297
12/31/21	42,000	0.07 (671,297) = 46,991	4,991	676,288
6/30/22	42,000	0.07 (676,288) = 47,340	5,340	681,628
12/31/22	42,000	0.07 (681,628) = 47,714	5,714	687,342
6/30/23	42,000	0.07 (687,342) = 48,114	6,114	693,456
12/31/23	42,000	0.07 (693,456) = 48,544*	6,544	700,000
	252,000	**285,367**	**33,367**	

*Rounded

Amounts for the journal entries each interest date are found in the second, third, and fourth columns of the schedule. The essential point to remember is that the effective interest method is a straightforward application of the accrual concept, whereby interest expense (or revenue) is accrued periodically at the effective rate. We record that amount of interest expense or revenue accrued even though the cash interest is a different amount.

Determining interest in this manner has a convenient side effect. It results in reporting the liability at the present value of future cash payments—the appropriate valuation method for any liability. This is obvious at issuance; we actually calculated the present value to be $666,633. What perhaps is not quite as obvious is that the outstanding amount of debt each subsequent period (shown in the right-hand column of the amortization schedule) is still the present value of the remaining cash flows, discounted at the original rate. The outstanding amount of the debt is its book value, sometimes called carrying value or carrying amount, which is the face amount minus the balance in the discount.

Before moving on, notice some key characteristics of the amortization schedule. As mentioned earlier, the unpaid interest each period ($4,664 the first period) adds to the balance. Since this happens each period, the balance continually increases, eventually becoming the face amount at maturity. Conveniently, that's the amount to be paid at maturity. Also, because the balance increases each period, so does the effective interest. That's because effective interest is the same percentage rate times a higher balance each period.

The balance continually increases, eventually becoming the face amount at maturity.

Now look at the column totals. The total interest expense (from the issuer's perspective) is equal to the sum of the total cash interest plus the total change in the balance (the discount). One way we might view this is to say the total interest paid ($285,367) is the $252,000 cash interest paid during the term to maturity plus the "extra" amount paid at maturity. That $33,367 amount is extra in the sense that, by selling the bonds, we borrow $666,633 but must repay $700,000 at maturity. That's why the effective interest on the bonds is 14% even though the cash interest is only 12% annually; the extra interest at maturity makes up the difference.

Additional Consideration

Although the reported amount each period is the present value of the bonds, at any date after issuance this amount is not necessarily equal to the market value (fair value) of the bonds. This is because the market rate of interest will not necessarily remain the same as the rate implicit in the original issue price (the effective rate). Of course, for negotiable financial instruments, the issue price is the market price at any given time. Differences between market values and present values based on the original rate are holding gains and losses. If we were to use the market rate to revalue bonds on each reporting date—that is, recalculate the present value using the market rate—the reported amount always would be the market value. As we will see later, companies have the option to report their liabilities at fair value.

Zero-Coupon Bonds

A zero-coupon bond pays no interest. Instead, it offers a return in the form of a "deep discount" from the face amount. For illustration, let's look at the zero-coupon bonds issued by **Coca Cola Enterprises.** One billion, nine hundred one million dollars face amount of the 25-year securities issued for two hundred forty-one million dollars. As the amortization schedule in Illustration 14–7 demonstrates, they were priced to yield 8.5%.[2]

Illustration 14–7

Zero-Coupon Securities—Coca Cola Enterprises

Real World Financials

($ in millions)	Cash Interest	Effective Interest		Increase in Balance	Outstanding Balance
	(0% × Face amount)	(8.5% × Outstanding debt)		(Discount reduction)	
					247
1995	0	0.085 (247) =	21	21	268
1996	0	0.085 (268) =	23	23	291
1997	0	0.085 (291) =	25	25	316
◆	◆	◆		◆	◆
◆	◆	◆		◆	◆
◆	◆	◆		◆	◆
2024	0	0.085 (1,614) =	137	137	1,751
2025	0	0.085 (1,751) =	149	149	1,900
			1,653	1,653	

Source: Coca Cola Enterprises.

Zero-coupon bonds provide us with a convenient opportunity to reinforce a key concept we just learned: that we accrue the interest expense (or revenue) each period at the effective rate regardless of how much cash interest actually is paid (zero in this case). An advantage of issuing zero-coupon bonds or notes is that the corporation can deduct for tax purposes the annual interest expense (see schedule) but has no related cash outflow until the bonds mature. However, the reverse is true for investors in "zeros." Investors receive no periodic cash interest, even though annual interest revenue is reportable for tax purposes. So those who invest in zero-coupon bonds usually have tax-deferred or tax-exempt status, such as pension funds, individual retirement accounts (IRAs), and charitable organizations. Zero-coupon bonds and notes have popularity but still constitute a relatively small proportion of corporate debt.

Bonds Issued at a Premium

In Illustration 14–3, Masterwear Industries sold the bonds at a price that would yield an effective rate higher than the stated rate. The result was a discount. On the other hand, if the 12% bonds had been issued when the market yield for bonds of similar risk and maturity was *lower* than the stated rate, say 10%, the issue would have been priced at a *premium.* Because the 12% rate would seem relatively attractive in a 10% market, the bonds would command an issue price of more than $700,000, calculated in Illustration 14–8.

Illustration 14–8

Bonds Sold at a Premium

On January 1, 2021, Masterwear Industries issued $700,000 of 12% bonds, dated January 1.

- Interest of $42,000 is payable semiannually on June 30 and December 31.
- The bonds mature in three years.
- The market yield for bonds of similar risk and maturity is 10%.
- United Intergroup, Inc. purchased the entire bond issue, planning to hold the bonds until maturity.

(continued)

[2]The present value of $1.901M, discounted at 8.5% for 25 years, is $247M. The bonds actually were issued in exchange for company warrants valued at $247M.

Calculation of the Price of the Bonds

	Present Values
Interest	$ 42,000 × 5.07569* = $213,179
Principal	$700,000 × 0.74622† = 522,354
Present value (price) of the bonds	$735,533

*Present value of an ordinary annuity of $1: $n = 6$, $i = 5\%$.
†Present value of $1: $n = 6$, $i = 5\%$.

Illustration 14–8

(concluded)

Because interest is paid *semiannually*, the present value calculations use

a. **one-half the stated rate (6%),**
b. **one-half the market rate (5%), and**
c. **6 (3 × 2) semiannual periods.**

Interest on bonds issued at a premium is determined in precisely the same manner as on bonds issued at a discount. Again, interest is the effective interest rate applied to the debt balance outstanding during each period (balance at the end of the previous interest period), and the cash paid is the stated rate times the face amount, as shown in Illustration 14–9. The difference between the two is the reduction (amortization) of the premium.

Journal Entries at Issuance—Bonds Sold at Premium

Masterwear (Issuer)		
Cash (price calculated above)	735,533	
Bonds payable (face amount)		700,000
Premium on bonds payable (difference)		35,533
United (Investor)		
Investment in bonds (face amount)	700,000	
Premium on investment in bonds (difference)	35,533	
Cash (price calculated above)		735,533

Illustration 14–9

Amortization Schedule—Premium

Date	Cash Interest	Effective Interest	Decrease in Balance	Outstanding Balance
	(6% × Face amount)	(5% × Outstanding balance)	(Premium reduction)	
1/1/21				735,533
6/30/21	42,000	0.05 (735,533) = 36,777	5,223	730,310
12/31/21	42,000	0.05 (730,310) = 36,516	5,484	724,826
6/30/22	42,000	0.05 (724,826) = 36,241	5,759	719,067
12/31/22	42,000	0.05 (719,067) = 35,953	6,047	713,020
6/30/23	42,000	0.05 (713,020) = 35,651	6,349	706,671
12/31/23	42,000	0.05 (706,671) = 35,329*	6,671	700,000
	252,000	**216,467**	**35,533**	

*Rounded

Since *more* cash is paid each period than the effective interest, the debt outstanding is reduced by the overpayment.

Notice that the debt declines each period. As the premium is reduced by amortization, the book value of the bonds declines toward face value. This is because the effective interest each period is less than the cash interest paid. Remember, this is precisely the opposite of when the bonds are at a discount, when the effective interest each period is more than the cash paid. As the discount is reduced by amortization, the book value of the bonds increases toward face value. This is illustrated in Illustration 14–10.

Whether bonds are issued at a premium or a discount, the outstanding balance becomes the face amount at maturity.

In practice, corporate bonds rarely are issued at a premium.[3] Because of the delay between the date the terms of the bonds are established and when the bonds are issued, it's difficult to set the stated rate equal to the ever-changing market rate. Knowing that, for marketing reasons, companies deliberately set the terms to more likely create a small discount rather than a premium at the issue date. Some investors are psychologically prone to prefer buying at a discount rather than a premium even if the yield is the same (the market rate).

[3]Only about 1% of corporate bonds are issued at a premium. D. Amiram, A. Kalay, and B Ozel, "The Bond Discount Puzzle," *Financial Accounting eJournal,* March 25, 2013.

Illustration 14–10

Premium and Discount Amortization Compared

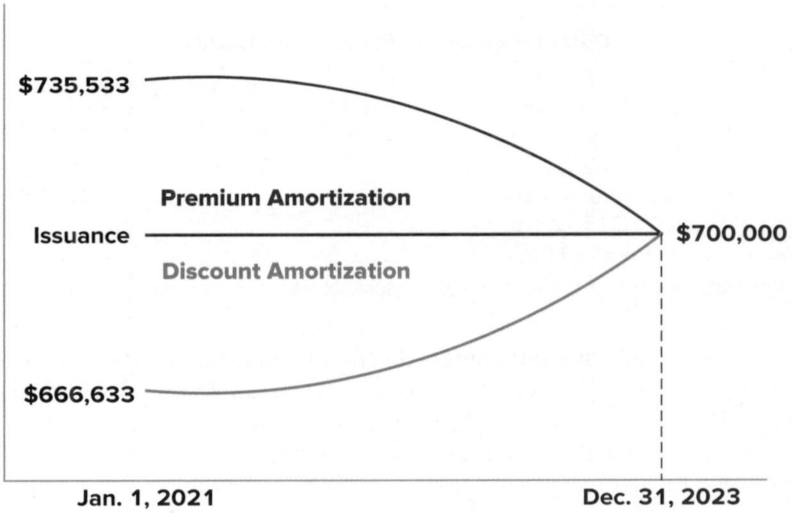

Additional Consideration

The preceding illustrations describe bonds sold at a discount and at a premium. The same concepts apply to bonds sold at face amount. But some of the procedures would be unnecessary. For instance, calculating the present value of the interest and the principal always will give us the face amount when the effective rate and the stated rate are the same:

Calculation of the Price of the Bonds

	Present Values
Interest	$ 42,000 × 4.91732* = $206,528
Principal	$700,000 × 0.70496† = 493,472
Present value (price) of the bonds	$700,000

*Present value of an ordinary annuity of $1: $n = 6$, $i = 6\%$.
†Present value of $1: $n = 6$, $i = 6\%$.

When Financial Statements are Prepared between Interest Dates

Any interest that has accrued since the last interest date must be recorded by an adjusting entry prior to preparing financial statements.

When an accounting period ends between interest dates, it is necessary to record interest that has accrued since the last interest date. As an example, refer again to Illustration 14–3. If the fiscal years of Masterwear and United end on October 31 and interest was last paid and recorded on June 30, four months' interest must be accrued in a year-end adjusting entry. Because interest is recorded for only a portion of a semiannual period, amounts recorded are simply the amounts shown in the amortization schedule (Illustration 14–6) times the appropriate fraction of the semiannual period (in this case 4/6).

Adjusting Entries—To Accrue Interest

To avoid understating interest in the financial statements, four months' interest is recorded at the end of the reporting period.

At October 31

Masterwear (Issuer)		
Interest expense (4/6 × 46,991)	31,327	
Discount on bonds payable (4/6 × 4,991)		3,327
Interest payable (4/6 × 42,000)		28,000
United (Investor)		
Interest receivable (4/6 × 42,000)	28,000	
Discount on investment in bonds (4/6 × 4,991)	3,327	
Interest revenue (4/6 × 46,991)		31,327

Two months later, when semiannual interest is paid next, the remainder of the interest is allocated to the first two months of the next fiscal year—November and December:

At the December 31 Interest Date

Masterwear (Issuer)		
Interest expense (2⁄6 × 46,991)	15,664	
Interest payable (from adjusting entry)	28,000	
Discount on bonds payable (2⁄6 × 4,991)		1,664
Cash (stated rate × face amount)		42,000
United (Investor)		
Cash (stated rate × face amount)	42,000	
Discount on investment in bonds (2⁄6 × 4,991)	1,664	
Interest receivable (from adjusting entry)		28,000
Interest revenue (2⁄6 × 46,991)		15,664

Of the six months' interest paid December 31, only the November and December interest is expensed in the new fiscal year.

The Straight-Line Method—A Practical Expediency

In some circumstances, the profession permits an exception to the conceptually appropriate method of determining interest for bond issues. A company is allowed to determine interest indirectly by allocating a discount or a premium equally to each period over the term to maturity—if doing so produces results that are not materially different from the usual (and preferable) interest method. The decision should be guided by whether the **straight-line method** would tend to mislead investors and creditors in the particular circumstance.

By the straight-line method, the discount in Illustration 14–3 and Illustration 14–6 would be allocated equally to the six semiannual periods (three years):

$33,367 ÷ 6 periods = **$5,561** per period

Journal Entries—Straight-Line Method

At Each of the Six Interest Dates

Masterwear (Issuer)		
Interest expense (to balance)	47,561	
Discount on bonds payable (discount ÷ 6 periods)		**5,561**
Cash (stated rate × face amount)		42,000
United (Investor)		
Cash (stated rate × face amount)	42,000	
Discount on bond investment (discount ÷ 6 periods)	**5,561**	
Interest revenue (to balance)		47,561

By the straight-line method, interest (expense and revenue) is a plug figure, resulting from calculating the amount of discount reduction.

Allocating the discount or premium equally over the life of the bonds by the straight-line method results in a constant dollar amount of interest each period. An amortization schedule, then, would serve little purpose. For example, if we prepared one for the straight-line method in this situation, it would provide the same amounts each period as shown in Illustration 14–11.

Illustration 14–11
Amortization Schedule—Straight-Line Method

	Cash Interest	Recorded Interest	Increase in Balance	Outstanding Balance
	(6% × Face amount)	(Cash + Discount reduction)	($33,367 ÷ 6)	
1/1/21				666,633
6/30/21	42,000	(42,000 + **5,561**) = 47,561	**5,561**	672,194
12/31/21	42,000	(42,000 + **5,561**) = 47,561	**5,561**	677,755
6/30/22	42,000	(42,000 + **5,561**) = 47,561	**5,561**	683,316
12/31/22	42,000	(42,000 + **5,561**) = 47,561	**5,561**	688,877
6/30/23	42,000	(42,000 + **5,561**) = 47,561	**5,561**	694,438
12/31/23	42,000	(42,000 + **5,561**) = 47,561	**5,561**	700,000*
	252,000	285,366	33,366	

*Rounded

By the straight-line method, the amount of the discount to be reduced periodically is calculated, and the recorded interest is the plug figure.

Determining interest by allocating the discount (or premium) on a straight-line basis is a practical expediency permitted in some situations by the materiality concept.

Remember, constant dollar amounts are not produced when the effective interest method is used. By that method, the dollar amounts of interest vary over the term to maturity because the percentage rate of interest remains constant but is applied to a changing debt balance.

Also, be sure to realize that the straight-line method is not an alternative method of determining interest in a conceptual sense. Instead, it is an application of the materiality concept, by which an appropriate application of GAAP (e.g., the effective interest method) can be bypassed for reasons of practical expediency in situations when doing so has no material effect on the results. Based on the frequency with which the straight-line method is used in practice, we can infer that managers very frequently conclude that its use has no material impact on investors' decisions.

Concept Review Exercise

ISSUING BONDS AND RECORDING INTEREST

On January 1, 2021, the Meade Group issued $8,000,000 of 11% bonds, dated January 1. Interest is payable semiannually on June 30 and December 31. The bonds mature in four years. The market yield for bonds of similar risk and maturity is 10%.

Required:

1. Determine the price these bonds issued to yield the 10% market rate and record their issuance by the Meade Group.
2. Prepare an amortization schedule that determines interest at the effective rate and record interest on the first interest date, June 30, 2021.

Solution

1. Determine the price these bonds sold for to yield the 10% market rate and record their issuance by the Meade Group.

Calculation of the Price of the Bonds

There are eight semiannual periods and one-half the market rate is 5%.		
Interest	$ 440,000 × 6.46321* =	$2,843,812
Principal	$ 8,000,000 × 0.67684† =	5,414,720
Present value (price) of the bonds		$8,258,532

*Present value of an ordinary annuity of $1: $n = 8$, $i = 5\%$.
†Present value of $1: $n = 8$, $i = 5\%$.

Journal Entries at Issuance

Cash (price calculated above)	8,258,532	
Bonds payable (face amount)		8,000,000
Premium on bonds payable (difference)		258,532

2. Prepare an amortization schedule that determines interest at the effective rate and record interest on the first interest date, June 30, 2021.

More cash is paid each period than the effective interest, so the debt outstanding is reduced by the "overpayment."

Amortization Schedule

Date	Cash Interest	Effective Interest		Decrease in Balance	Outstanding Balance
	(5.5% × Face amount)	(5% × Outstanding balance)		(Premium reduction)	
1/1/21					8,258,532
6/30/21	440,000	0.05 (8,258,532) =	412,927	27,073	8,231,459
12/31/21	440,000	0.05 (8,231,459) =	411,573	28,427	8,203,032
6/30/22	440,000	0.05 (8,203,032) =	410,152	29,848	8,173,184
12/31/22	440,000	0.05 (8,173,184) =	408,659	31,341	8,141,843
6/30/23	440,000	0.05 (8,141,843) =	407,092	32,908	8,108,935
12/31/23	440,000	0.05 (8,108,935) =	405,447	34,553	8,074,382
6/30/24	440.000	0.05 (8,074,382) =	403,719	36,281	8,038,101
12/31/24	440,000	0.05 (8,038,101) =	401,899*	38,101	8,000,000
	3,520,000		3,261,468	258,532	

*Rounded

Interest expense (5% x $8,258 532)	412,927	
Premium on bonds payable (difference)	27,073	
Cash (5.5% x $8,000,000)		440,000
To record interest for six months.		

Debt Issue Costs

Rather than sell bonds directly to the public, corporations usually sell an entire issue to an underwriter who then resells them to other security dealers and the public. By committing to purchase bonds at a set price, investment banks such as **JPMorgan Chase** and **Goldman Sachs** are said to underwrite any risks associated with a new issue. The underwriting fee is the spread between the price the underwriter pays and the resale price.

Alternatively, the issuing company may choose to sell the debt securities directly to a single investor (as we assumed in previous illustrations)—often a pension fund or an insurance company. This is referred to as *private placement.* Issue costs are less because privately placed securities are not subject to the costly and lengthy process of registering with the SEC that is required of public offerings. Underwriting fees also are avoided.[4]

Costs of issuing debt securities are called "debt issue costs" and are accounted for the same way as bond discount.

When issuing bonds or notes, the issuing company will incur costs, such as legal and accounting fees, printing costs, and registration and underwriting fees. These costs are recorded by combining them with any discount (or subtracting them from any premium) on the debt. The combined valuation account is reported in the balance sheet as a direct deduction from the liability and then amortized over the life of the debt.[5]

This approach has the appeal of reflecting the effect that debt issue costs have on the effective interest rate. **Debt issue costs** reduce the cash proceeds from the issuance of debt, so it makes sense that they should reflect a higher cost of borrowing. By deducting debt issue costs, we lower the carrying amount of the debt, which effectively increases the interest rate on that debt. In other words, with debt issue costs, the net amount borrowed is less, but interest payments are the same, so the effective rate of borrowing is higher.

For example, let's assume that in Illustration 14–11, we had debt issue costs of $14,000. The debt issue costs reduce the net cash the issuing company receives from the issuance of the bonds. The entry for the issuance would reflect the lower net cash proceeds and the higher valuation account for the discount and the debt issue costs:

Masterwear (Issuer)		
Cash ($666,633 price minus $14,000 issue costs)	652,633	
Discount and debt issue costs (difference)	47,367	
Bonds payable (face amount)		700,000

The price discount and the debt issue costs reduce the cash proceeds from the issuance of the bonds.

Interest expense is determined in the usual way; that is, the effective interest rate times the outstanding balance. The outstanding balance is the "net" liability, the face amount reduced by the discount and debt issue costs ($652,633 for the first interest period).[6] The debt issue costs of $14,000 have the effect of increasing the effective rate from 7% semiannually (as in Illustration 14–4) to 7.4389% semiannually (rounded). To see how the effective rate is determined, see the next Additional Consideration box. The first semiannual interest payment would be recorded as follows:

Interest expense ($652,633 x 7.4389%)	48,549	
Discount and debt issue costs (difference)		6,549
Cash ($700,000 x 6% stated rate)		42,000

The corporate bond is the basic long-term debt instrument for most large companies. But for many firms, the debt instrument often used is a *note.* We discuss notes next.

[4]Rule 144A of the Securities Act of 1933, as amended, allows for the private resale of unregistered securities to "qualified institutional buyers," which are generally large institutional investors with assets exceeding $100 million.

[5]Accounting Standards Update 2015-03: "Interest—Imputation of Interest (Subtopic 835-30) Simplifying the Presentation of Debt Issuance Costs," FASB, April 2015. This approach is consistent with the treatment of issue costs when equity securities are sold. You will see in Chapter 18 that the effect of share issue costs is to reduce the amount credited to stock accounts. It also is consistent with IFRS.

[6]If we have a premium instead of a discount, the valuation account is the premium reduced by the debt issue costs.

Additional Consideration

One important effect of the debt issue costs is that they reduce the amount of cash Masterwear is borrowing with the bond issuance (from $666,633 to $652,633). A lower (net) amount is borrowed at the same cost, making the *effective* interest rate higher than it would be without the debt issue costs. Previously, in Illustration 14–4 when we had no debt issue costs, the effective (market) rate was 14% (7% semiannually). That was the discount rate that caused the present value of the six interest payments and the maturity amount of $700,000 to be $666,633. Now, though, the effective rate is the one that would cause the present value of the six interest payments and the maturity amount of $700,000 to be $652,633. That annual rate is 14.8778% instead of 14%, and thus is a semiannual rate of 7.4389%, up from 7%. The following amortization schedule demonstrates this:

Cash Payments	Effective Interest	Increase or (decrease) in Balance	Outstanding Balance
6.0%	7.4389% × Balance		
			652,633
42,000	7.4389% × 652,633 = 48,549	6,549	659,182
42,000	7.4389% × 659,182 = 49,036	7,036	666,218
42,000	7.4389% × 666,218 = 49,559	7,559	673,777
42,000	7.4389% × 673,777 = 50,122	8,122	681,898
42,000	7.4389% × 681,898 = 50,726	8,726	690,624
42,000	7.4389% × 690,624 = 51,376	9,376	700,000

You should note that the balance in the last column of the schedule is the amount reported in the balance sheet for the bonds, reduced by both the discount and debt issue costs.

PART B

Long-Term Notes

● LO14–3

FINANCIAL Reporting Case

Q4, p. 773

When a company borrows cash from a bank and signs a promissory note, the firm's loan liability is reported as a *note payable.* Or a note might be issued in exchange for a noncash asset—perhaps to purchase equipment on credit. In concept, notes are accounted for in precisely the same way as bonds. In fact, we could properly substitute notes payable for bonds payable in each of our previous illustrations. For comparison, we continue to also present the lenders' entries (in blue) in the illustrations to follow.

As we discuss accounting for the borrower's notes *payable,* we also will consider the lender's perspective and look at notes *receivable* at the same time. By considering both sides of each transaction at the same time, we will see that the two sides are essentially mirror images of one another. This coverage of long-term notes complements our Chapter 7 discussion of *short-term* notes receivable.

Note Issued for Cash

The interest rate stated in a note is likely to be equal to the market rate because the rate usually is negotiated at the time of the loan. So discounts and premiums are less likely for notes than for bonds. Accounting for a note issued for cash is demonstrated in Illustration 14–12.

Note Exchanged for Assets or Services

Occasionally the *stated* interest rate is not indicative of the *market* rate at the time a note is negotiated. The value of the asset (cash or noncash) or service exchanged for the note establishes the market rate.[7] For example, let's assume Skill Graphics purchased a package-labeling machine from Hughes–Barker Corporation by issuing a 12%, $700,000, three-year note that requires interest to be paid semiannually. Let's also assume that the machine could have been purchased at a cash price of $666,633. You probably recognize this numerical situation as the one used earlier to illustrate bonds sold at a discount (Illustration 14–3).

[7]If the debt instrument is negotiable and a dependable exchange price is readily available, the market value of the debt may be better evidence of the value of the transaction than the value of a noncash asset, particularly if it has no established cash selling price. The value of the asset or the debt, whichever is considered more reliable, should be used to record the transaction.

Illustration 14–12

Note Issued for Cash

On January 1, 2021, Skill Graphics, Inc., a product-labeling and graphics firm, borrowed $700,000 cash from First BancCorp and issued a three-year $700,000 promissory note. Interest of $42,000 was payable semiannually on June 30 and December 31.

At Issuance		
Skill Graphics (Borrower)		
Cash	700,000	
Notes payable (face amount)		700,000
First BancCorp (Lender)		
Notes receivable (face amount)	700,000	
Cash		700,000
At Each of the Six Interest Dates		
Skill Graphics (Borrower)		
Interest expense	42,000	
Cash (stated rate × face amount)		42,000
First BancCorp (Lender)		
Cash (stated rate × face amount)	42,000	
Interest revenue		42,000
At Maturity		
Skill Graphics (Borrower)		
Notes payable	700,000	
Cash (face amount)		700,000
First BancCorp (Lender)		
Cash (face amount)	700,000	
Notes receivable		700,000

Reference to the earlier example will confirm that exchanging this $700,000 note for a machine with a cash price of $666,633 implies an annual market rate of interest of 14%. That is, 7% is one-half the discount rate that yields a present value of $666,633 for the note's cash flows (interest plus principal):

	Present Values
Interest	$ 42,000 × 4.76654* = $200,195
Principal	$700,000 × 0.66634† = 466,438
Present value of the note	$666,633

*Present value of an ordinary annuity of $1: $n = 6, i = 7\%$.
†Present value of $1: $n = 6, i = 7\%$.

This is referred to as the **implicit rate of interest**—the rate implicit in the agreement. It may be that the implicit rate is not apparent. Sometimes the value of the asset (or service) is not readily determinable, but the interest rate stated in the transaction is unrealistic relative to the rate that would be expected in a similar transaction under similar circumstances. Deciding what the appropriate rate should be is called *imputing* an interest rate.

For example, suppose the machine exchanged for the 12% note is custom-made for Skill Graphics so that no customary cash price is available with which to work backwards to find the implicit rate. In that case, the appropriate rate would have to be found externally. It might be determined, for instance, that a more realistic interest rate for a transaction of this type, at this time, would be 14%. Then it would be apparent that Skill Graphics actually paid less than $700,000 for the machine and that part of the face amount of the note in effect makes up for the lower than normal interest rate. You learned early in your study of accounting that the economic essence of a transaction should prevail over its outward appearance. In keeping with this basic precept, the accountant should look beyond the *form* of this transaction and record its *substance*. The amount actually paid for the machine is the present value of the cash flows called for by the loan agreement, discounted at the market rate—imputed in this case to be 14%. So both the asset acquired and the liability used to purchase it should be recorded at the real cost, $666,633.

A basic concept of accounting is *substance over form.*

Additional Consideration

For another example, let's assume the more realistic interest rate for a transaction of this type is, say, 16%. In that case we would calculate the real cost of the machine by finding the present value of both the interest and the principal, discounted at half the 16% rate (because interest is paid semiannually):

	Present Values
Interest	\$ 42,000 × 4.62288* = \$194,161
Principal	\$700,000 × 0.63017† = 441,119
Present value of the note	\$635,280

*Present value of an ordinary annuity of \$1: $n = 6, i = 8\%$.
†Present value of \$1: $n = 6, i = 8\%$.

Both the asset acquired and the liability used to purchase it would be recorded at \$635,280.

The accounting treatment is the same whether the amount is determined directly from the market value of the machine (and thus the note) or indirectly as the present value of the note (and thus the value of the asset):[8]

Journal Entries at Issuance—Note with Unrealistic Interest Rate

Skill Graphics (Buyer/Issuer)		
Machinery (cash price)	666,633	
Discount on notes payable (difference)	33,367	
Notes payable (face amount)		700,000
Hughes–Barker (Seller/Lender)		
Notes receivable (face amount)	700,000	
Discount on notes receivable (difference)		33,367
Sales revenue (cash price)		666,633

Note: The seller also would debit cost of goods sold and credit inventory for the amount spent to construct the machine.

Likewise, whether the effective interest rate is determined as the rate implicit in the agreement, given the asset's market value, or whether the effective rate is imputed as the appropriate interest rate if the asset's value is unknown, both parties to the transaction should record periodic interest (interest expense to the borrower, interest revenue to the lender) at the effective rate, rather than the stated rate.

Journal Entries—The Interest Method

The effective interest (expense to the issuer; revenue to the investor) is calculated each period as the effective rate times the amount of the debt outstanding during the interest period.

At the First Interest Date (June 30)

Skill Graphics (Borrower)		
Interest expense (effective rate × outstanding balance)	46,664	
Discount on notes payable (difference)		4,664
Cash (stated rate × face amount)		42,000
Hughes–Barker (Seller/Lender)		
Cash (stated rate × face amount)	42,000	
Discount on notes receivable (difference)	4,664	
Interest revenue (effective rate × outstanding balance)		46,664

[8]The method shown is the *gross method.* Alternatively, the *net method* can be used as follows:

Skill Graphics (Buyer/Issuer)		
Machinery (cas price)	666,633	
Notes payable		666,633
Hughes-Barker (Seller/Lender)		
Notes receivable	666,633	
Sales revenue (cash price)		666,633

Under the net method, the note is recorded at the face amount reduced by the discount on notes payable, which is the difference between face value and present value. As cash interest is paid, the note balance increases by the difference between the cash interest payment and the interest revenue (and interest expense). After the last payment, the note account balance will be equal to the face amount of the note.

The interest expense (interest revenue for the lender) varies as the balance of the note changes over time. See the amortization schedule in Illustration 14–13.[9] Be sure to notice that this amortization schedule is identical to the one in Illustration 14–6 for bonds issued at a discount.

Date	Cash Interest	Effective Interest	Increase in Balance	Outstanding Balance
	(6% × Face amount)	(7% × Outstanding balance)	(Discount reduction)	
1/1/21				666,633
6/30/21	42,000	0.07 (666,633) = 46,664	4,664	671,297
12/31/21	42,000	0.07 (671,297) = 46,991	4,991	676,288
6/30/22	42,000	0.07 (676,288) = 47,340	5,340	681,628
12/31/22	42,000	0.07 (681,628) = 47,714	5,714	687,342
6/30/23	42,000	0.07 (687,342) = 48,114	6,114	693,456
12/31/23	42,000	0.07 (693,456) = 48,544*	6,544	700,000
	252,000	285,367	33,367	

*Rounded

Illustration 14–13

Amortization Schedule—Note

Since less cash is paid each period than the effective interest, the unpaid difference (the discount reduction) increases the outstanding balance (book value) of the note.

Installment Notes

You may have recently purchased a car, or maybe a house. If so, unless you paid cash, you signed a note promising to pay a portion of the purchase price over, say, five years for the car, or 30 years for the house. Car and house notes usually call for payment in monthly installments rather than by a single amount at maturity. Corporations, too, often borrow using installment notes. Typically, installment payments are equal amounts each period. Each payment includes both an amount that represents interest and an amount that represents a reduction of the outstanding balance (principal reduction). The periodic reduction of the balance is sufficient that at maturity the note is completely paid. This amount is easily calculated by dividing the amount of the loan by the appropriate discount factor for the present value of an annuity. The installment payment amount that would pay the note described in the previous section is as follows:

$666,633	÷	4.76654	=	**$139,857**
Amount of loan		(from Table 4 $n = 6, i = 7.0\%$)		Installment payment

Consider Illustration 14–14.

Date	Cash Payment	Effective Interest	Decrease in Balance	Outstanding Balance
		(7% × Outstanding balance)		
1/1/21				666,633
6/30/21	**139,857**	0.07 (666,633) = 46,664	93,193	573,440
12/31/21	**139,857**	0.07 (573,440) = 40,141	99,716	473,724
6/30/22	**139,857**	0.07 (473,724) = 33,161	106,696	367,028
12/31/22	**139,857**	0.07 (367,028) = 25,692	114,165	252,863
6/30/23	**139,857**	0.07 (252,863) = 17,700	122,157	130,706
12/31/23	**139,857**	0.07 (130,706) = 9,151*	130,706	0
	839,142	172,509	666,633	

*Rounded

Illustration 14–14

Amortization Schedule—Installment Note

Each installment payment includes interest on the outstanding debt at the effective rate. The remainder of each payment reduces the outstanding balance.

[9]The creation of amortization schedules is simplified by an electronic spreadsheet such as Microsoft Excel.

For installment notes, the outstanding balance of the note does not eventually become its face amount as it does for notes with designated maturity amounts. Instead, at the maturity date the balance is zero. The procedure is the same as for a note whose principal is paid at maturity, but for an intallment note the periodic cash payments are larger and there is no lump-sum payment at maturity. We calculated the amount of the payments so that after covering the interest on the existing debt each period, the excess would exactly amortize the installment debt balance to zero at maturity (rather than to a designated maturity amount).

Consequently, the significance is lost of maintaining separate balances for the face amount (in a note account) and the discount (or premium). So an installment note typically is recorded at its net book value in a single note payable (or receivable) account:

Journal Entries at Issuance—Installment Note

Skill Graphics (Buyer/Issuer)		
Machinery	666,633	
Notes payable		666,633
Hughes–Barker (Seller/Lender)		
Notes receivable	666,633	
Sales revenue		666,633

At the First Interest Date (June 30)

Each payment includes both an amount that represents interest and an amount that represents a reduction of principal.

Skill Graphics (Borrower)		
Interest expense (effective rate × outstanding balance)	46,664	
Notes payable (difference)	93,193	
Cash (installment payment calculated above)		139,857
Hughes–Barker (Seller/Lender)		
Cash (installment payment calculated above)	139,857	
Notes receivable (difference)		93,193
Interest revenue (effective rate × outstanding balance)		46,664

Additional Consideration

You will learn in the next chapter that the liability associated with a lease is accounted for the same way as this installment note. In fact, if the asset described above had been leased rather than purchased, the cash payments would be designated lease payments rather than installment loan payments, and a virtually identical amortization schedule would apply.

The reason for the similarity is that we view a finance lease as being, in substance, equivalent to an installment purchase of the right to use an asset. Naturally, then, accounting treatment of the two essentially identical transactions should be consistent. Be sure to notice the parallel treatment as you study leases in the next chapter.

Concept Review Exercise

NOTE WITH AN UNREALISTIC INTEREST RATE

Cameron-Brown, Inc. constructed for Harmon Distributors a warehouse that was completed and ready for occupancy on January 2, 2021.

- Harmon paid for the warehouse by issuing a $900,000, four-year note that required 7% interest to be paid on December 31 of each year.
- The warehouse was custom-built for Harmon, so its cash price was not known.
- By comparison with similar transactions, it was determined that an appropriate interest rate was 10%.

Required:

1. Prepare the journal entry for Harmon's purchase of the warehouse on January 2, 2021.

2. Prepare (a) an amortization schedule for the four-year term of the note and (b) the journal entry for Harmon's first interest payment on December 31, 2021.
3. Suppose Harmon's note had been an installment note to be paid in four equal payments. What would be the amount of each installment if payable (a) at the end of each year, beginning December 31, 2021? or (b) at the beginning of each year, beginning on January 2, 2021?

Solution:

1. Prepare the journal entry for Harmon's purchase of the warehouse on January 2, 2021.

	Present Values
Interest	$ 63,000 × 3.16987* = $199,702
Principal	$900,000 × 0.68301† = 614,709
Present value of the note	$814,411

*Present value of an ordinary annuity of $1: $n = 4$, $i = 10\%$.
†Present value of $1: $n = 4$, $i = 10\%$.

Warehouse (price determined above)	814,411	
Discount on notes payable (difference)	85,589	
Notes payable (face amount)		900,000

2. Prepare (a) an amortization schedule for the four-year term of the note and (b) the journal entry for Harmon's first interest payment on December 31, 2021.

Dec. 31	Cash	Effective Interest	Increase in Balance	Outstanding Balance
	(7% × Face amount)	(10% × Outstanding balance)	(Discount reduction)	
				814,411
2021	63,000	0.10 (814,411) = 81,441	18,441	832,852
2022	63,000	0.10 (832,852) = 83,285	20,285	853,137
2023	63,000	0.10 (853,137) = 85,314	22,314	875,451
2024	63,000	0.10 (875,451) = 87,549*	24,549	900,000
	252,000	337,589	85,589	

*Rounded

Each period the unpaid interest increases the outstanding balance of the debt.

Interest expense (effective rate × outstanding balance)	81,441	
Discount on notes payable (difference)		18,441
Cash (stated rate × face amount)		63,000

The effective interest is the market rate times the amount of the debt outstanding during the year.

3. Suppose Harmon's note had been an installment note to be paid in four equal payments. What would be the amount of each installment if payable (a) at the end of each year, beginning December 31, 2021? or (b) at the beginning of each year, beginning on January 2, 2021?

Because money has a time value, installment payments delayed until the end of each period must be higher than if the payments are made at the beginning of each period.

a.	$814,411	÷	3.16987	=	$256,923
	Amount of loan		(from Table 4 $n = 4$, $i = 10\%$)		Installment payment
b.	$814,411	÷	3.48685	=	$233,566
	Amount of loan		(from Table 6 $n = 4$, $i = 10\%$)		Installment payment

Financial Statement Disclosures

● LO14–4

In the balance sheet, long-term debt (liability for the debtor; asset for the creditor) typically is reported as a single amount, net of any discount or increased by any premium, rather than at its face amount accompanied by a separate valuation account for the discount or premium. Any portion of the debt to be paid (received) during the upcoming year, or operating cycle if longer, should be reported as a current amount.

Note disclosure is required of the fair value of bonds, notes, and other financial instruments.

The fair value of financial instruments must be disclosed either in the body of the financial statements or in disclosure notes.[10] These fair values are available for bonds and other securities traded on market exchanges in the form of quoted market prices. On the other hand, financial instruments not traded on market exchanges require other evidence of market value. For example, the market value of a note payable might be approximated by the present value of principal and interest payments using a current discount rate commensurate with the risks involved.

The disclosure note for debt includes the nature of the company's liabilities, interest rates, maturity dates, call provisions, conversion options, restrictions imposed by creditors, and any assets pledged as collateral. For all long-term borrowings, disclosures also should include the aggregate amounts payable for each of the next five years. To comply, **Microsoft**'s annual report for the fiscal year ended June 30, 2017, included the disclosure shown in Illustration 14–15.

Illustration 14–15

Debt Disclosures—Microsoft Corporation

Real World Financials

Maturities of our long-term debt for each of the next five years and thereafter are as follows:

Year Ending June 30,	($ in millions)
2018	$ 2,190
2019	1,698
2020	1,180
2021	1,006
2022	932
Thereafter	3,100
Total	$10,106

Source: Microsoft Corporation

Borrowing is a financing activity; lending is an investing activity.

Paying or receiving interest is an operating activity.

In a statement of cash flows, issuing bonds or notes are reported as cash flows from financing activities by the issuer (borrower) and cash flows from investing activities by the investor (lender). Similarly, as the debt is repaid, the issuer (borrower) reports a financing activity while the investor (lender) reports an investing activity. However, because both interest expense and interest revenue are components of the income statement, both parties to the transaction report interest among operating activities. We discuss the cash flow reporting process in more depth in Chapter 21.

Decision Makers' Perspective

Business decisions involve risk. Failure to properly consider risk in those decisions is one of the most costly, yet one of the most common mistakes investors and creditors can make. Long-term debt is one of the first places decision makers should look when trying to get a handle on risk.

Generally speaking, debt increases risk.

In general, debt increases risk. As an owner, debt would place you in a subordinate position relative to creditors because the claims of creditors must be satisfied first in case of liquidation. In addition, debt requires payment, usually on specific dates. Failure to pay debt interest and principal on a timely basis may result in default and perhaps even bankruptcy. The debt to equity ratio, total liabilities divided by shareholders' equity, often is calculated to measure the degree of risk. Other things being equal, the higher the debt to equity ratio, the higher the risk. The type of risk this ratio measures is called *default risk* because it presumably indicates the likelihood a company will default on its obligations.

To evaluate a firm's risk, you might start by calculating its debt to equity ratio.

Debt also can be an advantage. It can be used to enhance the return to shareholders. This concept, known as leverage, was described and illustrated in Chapter 3. If a company earns a return on borrowed funds in excess of the cost of borrowing the funds, shareholders are provided with a total return greater than what could have been earned with equity funds

[10]FASB ASC 825–10–50–10: Financial Instruments—Overall—Disclosure.

alone. This desirable situation is called *favorable financial leverage*. Unfortunately, leverage is not always favorable. Sometimes the cost of borrowing the funds exceeds the returns they generate. This illustrates the typical risk-return trade-off faced by shareholders.

As a manager, you would try to create favorable financial leverage to earn a return on borrowed funds in excess of the cost of borrowing the funds.

Creditors demand interest payments as compensation for the use of their capital. Failure to pay interest as scheduled may cause several adverse consequences, including bankruptcy. Therefore, another way to measure a company's ability to pay its obligations is by comparing interest payments with income available to pay those charges. The times interest earned ratio does this by dividing income, before subtracting interest expense or income tax expense, by interest expense.

As an external analyst or a manager, you are concerned with a company's ability to repay debt.

Two points about this ratio are important. First, because interest is deductible for income tax purposes, income before interest and taxes is a better indication of a company's ability to pay interest than is income after interest and taxes (i.e., net income). Second, income before interest and taxes is a rough approximation for cash flow generated from operations. The primary concern of decision makers is, of course, the cash available to make interest payments. In fact, this ratio often is computed by dividing cash flow generated from operations by interest payments.

For illustration, let's compare the ratios for **Coca-Cola** and **PepsiCo.** Illustration 14–16 provides condensed financial statements adapted from 2017 annual reports of those companies.

Illustration 14–16

Condensed Financial Statements—Coca-Cola, PepsiCo

Real World Financials

Balance Sheets	($ in millions)	
	Coca-Cola	**PepsiCo**
Assets		
Current assets	$36,545	$31,027
Property, plant, and equipment (net)	8,203	17,240
Intangibles and other assets	43,148	31,537
Total assets	$87,896	$79,804
Liabilities and Shareholders' Equity		
Current liabilities	$27,194	$20,502
Long-term liabilities	41,725	48,321
Total liabilities	68,919	68,823
Shareholders' equity	18,977	10,981
Total liabilities and shareholders' equity	$87,896	$79,804
Income Statements		
Net sales	$35,410	$63,525
Cost of goods sold	(13,256)	(28,785)
Gross profit	22,154	34,740
Operating and other expenses	(14,423)	(23,987)
Interest expense	(841)	(1,151)
Income before taxes	6,890	9,602
Tax expense	(5,607)	(4,694)
Net income	$ 1,283	$ 4,908

Source: Coca-Cola, PepsiCo

The debt to equity ratio is much higher for PepsiCo:

The debt to equity ratio indicates the extent of trading on the equity, or financial leverage.

$$\text{Debt to equity ratio} = \frac{\text{Total liabilities}}{\text{Shareholders' equity}}$$

$$\text{Coca-Cola} = \frac{\$68,919}{\$18,977} = 3.6$$

$$\text{PepsiCo} = \frac{\$68,823}{\$10,981} = 6.3$$

Remember, that's not necessarily a positive or a negative. Let's look closer. When the return on equity is greater than the return on assets, management is using debt funds to enhance the earnings for shareholders. Both firms do this. We calculate return on assets as follows:

The rate of return on assets indicates profitability without regard to how resources are financed.

$$\text{Rate of return on assets}^{11} = \frac{\text{Net income}}{\text{Total assets}}$$

$$\text{Coca-Cola} = \frac{\$1,283}{\$87,896} = 1.5\%$$

$$\text{PepsiCo} = \frac{\$4,908}{\$79,804} = 6.2\%$$

The return on assets indicates a company's overall profitability, ignoring specific sources of financing. In this regard, PepsiCo's profitability is significantly higher than Coca-Cola's. That advantage continues when we compare the return to shareholders:

The return on equity indicates the effectiveness of employing resources provided by owners.

$$\text{Return on equity} = \frac{\text{Net income}}{\text{Shareholder's equity}}$$

$$\text{Coca-Cola} = \frac{\$1,283}{\$18,977} = 6.8\%$$

$$\text{PepsiCo} = \frac{\$4,908}{\$10,981} = 44.7\%$$

PepsiCo's return on assets is four times larger than Coca-Cola's, and its return on equity is six times larger. The reason is that higher leverage has been used by Pepsi to provide a relatively greater return to shareholders. PepsiCo increased its return to shareholders 7.2 times (44.7%/6.2%) the return on assets. Coca-Cola increased its return to shareholders 4.5 times (6.8%/1.5%) the return on assets. Interpret this with caution, though. PepsiCo's higher leverage means higher risk as well. In down times, PepsiCo's return to shareholders will suffer proportionally more than will Coca-Cola's.

From the perspective of a creditor, we might look at which company offers the most comfortable margin of safety in terms of its ability to pay fixed interest charges:

The times interest earned ratio indicates the margin of safety provided to creditors.

$$\text{Times interest earned ratio} = \frac{\text{Net income plus interest plus taxes}}{\text{Interest}}$$

$$\text{Coca-Cola} = \frac{\$1,283 + \$841 + \$5,607}{\$841} = 9.2 \text{ times}$$

$$\text{PepsiCo} = \frac{\$4,908 + \$1,151 + \$4,694}{\$1,151} = 9.3 \text{ times}$$

In this regard, Coca-Cola provides a much greater margin of safety. And Pepsi has more debt in its capital structure relative to Coca-Cola. However, Pepsi clearly is able to pay the cost of borrowing and provide an impressive return to its shareholders. Both firms, though, trade quite favorably on their leverage.

Decision makers should be alert to gains and losses that have nothing to do with a company's normal operating activities.

Liabilities also can have misleading effects on the income statement. Decision makers should look carefully at gains and losses produced by early extinguishment of debt. These have nothing to do with a company's normal operating activities. Unchecked, corporate management can be tempted to schedule debt buybacks to provide discretionary income in down years or even losses in up years to smooth income over time.

Outside analysts as well as managers should actively monitor risk management activities.

Alert investors and lenders also look outside the financial statements for risks associated with "off-balance-sheet" financing and other commitments that don't show up on the face of financial statements but nevertheless expose a company to risk. Relatedly, most companies attempt to actively manage the risk associated with these and other obligations. It is important for top management to understand and closely monitor risk management strategies. Some of the financial

[11] A more accurate way to calculate this ratio is: Rate of return on assets = Net income + Interest expense (1 − tax rate) ÷ Average total assets. The reason for adding back interest expense (net of tax) is that interest represents a return to suppliers of debt capital and should not be deducted in the computation of net income when computing the return on total assets. In other words, the numerator is the total amount of income available to both debt and equity capital.

losses that have grabbed headlines in recent years were permitted by a lack of oversight and scrutiny by senior management of companies involved. It is similarly important for investors and creditors to become informed about risks companies face and how well-equipped those companies are in managing that risk. The supplemental disclosures designed to communicate the degree of risk associated with the financial instruments we discuss in this chapter contribute to that understanding. We examine the significance of lease commitments in the next chapter. ●

Debt Retired Early, Convertible Into Stock, or Providing an Option to Buy Stock

PART C

Early Extinguishment of Debt

● LO14–5

As we saw in the previous section, debt paid in installments is systematically retired over the term to maturity so that at the designated maturity date the outstanding balance is zero. On the other hand, when a maturity amount is specified as in our earlier illustrations, any discount or premium has been systematically reduced to zero as of the maturity date and the debt is retired simply by paying the maturity amount. Sometimes, though, companies choose to retire debt before its scheduled maturity. In that case, a gain or a loss may result.

Earlier we noted that a call feature accompanies most bonds to protect the issuer against declining interest rates. Even when bonds are not callable, the issuing company can retire bonds early by purchasing them on the open market. Regardless of the method, when debt of any type is retired prior to its scheduled maturity date, the transaction is referred to as **early extinguishment of debt.**

To record the extinguishment, the account balances pertinent to the debt obviously must be removed from the books. Of course cash is credited for the amount paid—the call price or market price. The difference between the book value of the debt and the reacquisition price represents either a gain or a loss on the early extinguishment of debt. When the debt is retired for less than book value, the debtor is in a favorable position and records a gain. The opposite occurs for a loss. Let's continue an earlier example to illustrate the retirement of debt prior to its scheduled maturity (Illustration 14–17).

Any difference between the outstanding debt and the amount paid to retire that debt represents either a gain or a loss.

Illustration 14–17

Early Extinguishment of Debt

On January 1, 2022, Masterwear Industries called its $700,000, 12% bonds when their book value was $676,288.

- The indenture specified a call price of $685,000.
- The bonds were issued previously at a price to yield 14%.

Bonds payable (face amount)	700,000	
Loss on early extinguishment ($685,000 – $676,288)	8,712	
Discount on bonds payable ($700,000 – $676,288)		23,712
Cash (call price)		685,000

For instance, in its 2017 income statement, **Waste Management, Inc.** reported a $6 million loss on the early extinguishment of debt.

Convertible Bonds

FINANCIAL Reporting Case

Q5, p. 773

Sometimes corporations include a convertible feature as part of a bond offering. **Convertible bonds** can be converted into (that is, exchanged for) shares of stock at the option of the bondholder. Among the reasons for issuing convertible bonds rather than straight debt are (a) to sell the bonds at a higher price (which means a lower effective interest cost),[12] (b) to use as a medium of exchange in mergers and acquisitions, and (c) to enable smaller firms or debt-heavy companies to obtain access to the bond market. Sometimes convertible bonds

Convertible bonds can be exchanged for shares of stock at the option of the investor.

[12]Remember, there is an inverse relationship between bond prices and interest rates. When the price is higher, the rate (yield) is lower, and vice versa.

serve as an indirect way to issue stock when there is shareholder resistance to direct issuance of additional equity.

Convertible bonds have features of both debt and equity.

Central to each of these reasons for issuing convertible debt is that the conversion feature is attractive to investors. This hybrid security has features of both debt and equity. The owner has a fixed-income security that can become common stock if and when the firm's prosperity makes that feasible. This increases the investor's upside potential while limiting the downside risk. The conversion feature has monetary value. Just how valuable it is depends on both the conversion terms and market conditions. But from an accounting perspective the question raised is how to account for its value. To evaluate the question, consider Illustration 14–18.

Illustration 14–18
Convertible Bonds

On January 1, 2021, HTL Manufacturers issued $100 million of 8% convertible debentures due 2041 at 103 (103% of face value).

- The bonds are convertible at the option of the holder into no par common stock at a conversion ratio of 40 shares per $1,000 bond.
- HTL recently issued nonconvertible, 20-year, 8% debentures at 98.

Because of the inseparability of their debt and equity features, the entire issue price of convertible bonds is recorded as debt, as if they are nonconvertible bonds.

It would appear that the conversion feature is valued by the market at $5 million—the difference between the market value of the convertible bonds, $103 million, and the market value of the nonconvertible bonds, $98 million. Some accountants argue that we should record the value of the conversion option in a shareholders' equity account ($5 million in this case) and the debt value in the bond accounts ($100 million bonds payable less $2 million discount). However, counter to that intuitive argument, the currently accepted practice is to record the entire issue price as debt in precisely the same way as for nonconvertible bonds.[13] Treating the features as two inseparable parts of a single security avoids the practical difficulty of trying to measure the separate values of the debt and the conversion option.

Journal Entry at Issuance—Convertible Bonds

	($ in millions)	
Cash (103% × $100 million)	103	
Convertible bonds payable (face amount)		100
Premium on bonds payable (difference)		3

The value of the conversion feature is not separately recorded.

Since we make no provision for the separate value of the conversion option, all subsequent entries, including the periodic reduction of the premium, are exactly the same as if these were nonconvertible bonds. So, the illustrations and examples of bond accounting we discussed earlier would pertain equally to nonconvertible or convertible bonds.

Additional Consideration

While we normally don't separate the conversion option from the debt for convertible bonds, an exception is when the conversion option is deemed to be a "beneficial conversion feature." That's the case when the conversion option has a positive "intrinsic value" at the time the bonds are issued, meaning the fair value of the stock into which the bonds are convertible exceeds the face amount of the bonds.[14] Suppose, for instance, that HTL's stock in Illustration 14–18 has a fair value of $30 per share when the bonds are issued. This implies an intrinsic value of the conversion feature of **$5** per share:

Price per share to convert: $1,000 ÷ 40 shares = $25 per share	
Fair value of stock at issue date	$30
Conversion price for shares	25
Intrinsic value of beneficial conversion option per share	**$ 5**

(continued)

[13] FASB ASC 470–20–25: Debt—Debt with Conversion Options—Recognition.

[14] FASB ASC 470–20–25–5: Debt—Debt with Conversion and Other Options—Recognition—Beneficial Conversion Features.

(concluded)

Intrinsic value of beneficial conversion feature:

100,000 bonds × 40 shares = 4,000,000 shares × **$5** per share = **$20,000,000**

The intrinsic value of the conversion option is recorded separately:

	($ in millions)	
Cash (103% × $100 million)	103	
Discount on bonds payable (to balance)	17	
Convertible bonds payable (face amount)		100
Equity—conversion option (intrinsic value)		20

Where We're Headed

Accounting for convertible debt is likely to change. In 2017, the FASB added to its agenda a project on distinguishing liabilities from equity, including convertible debt, aiming to reduce complexity in financial statements and improve their understandability. A pre-Codification FASB Staff Position gives us another hint, perhaps, of the direction the project might take. For a limited subset of convertible securities not within the scope of current GAAP—those that could possibly be settled in cash rather than shares—companies must divide the proceeds from convertible securities into its two components and record the fair value of the debt as a liability and the conversion option in an equity account. Lending credence to that projection is the fact that international standards already require that convertible debt be divided into its liability and equity elements, so moving in that direction would be a step closer to convergence of U.S. GAAP and IFRS.

International Financial Reporting Standards

Convertible bonds. Under IFRS, unlike U.S. GAAP, convertible debt is divided into its liability and equity elements. In Illustration 14–18, if HTL prepared its financial statements in accordance with IFRS, it would record the convertible bonds as follows:

Journal Entry at Issuance—Convertible Bonds

LO14–7

	($ in millions)	
Cash (103% × $100 million)	103	
Convertible bonds payable (value of the debt only)		98*
Equity—conversion option (to balance)		5

*Notice that the discount is combined with the face amount of the bonds. This is the "net method". Using the "gross method" we would credit Convertible bonds payable $100 and debit Discount on bonds $2.

In essence, HTL is selling two securities—(1) bonds and (2) an option to convert to stock—for one package price. The bonds represent a liability; the option is shareholders' equity. Compound instruments such as this one are separated into their liability and equity components in accordance with *IAS No. 32*.[15] Because the bonds have a separate fair value of $98 million, we record that amount as the liability and the remaining $5 million (the difference between the fair value of the convertible bonds, $103 million, and the $98 million) as equity. If the fair value of the bonds cannot be determined from an active trading market, that value can be calculated as the present value of the bonds' cash flows, using the market rate of interest.

The components of compound financial instruments such as convertible bonds are valued and reported separately under IFRS.

When the Conversion Option Is Exercised

If and when the bondholder exercises his or her option to convert the bonds into shares of stock, the bonds are removed from the accounting records and the new shares issued are recorded at the same amount (in other words, at the book value of the bonds). To illustrate,

[15] "Financial Statements: Presentation," *International Accounting Standard (IAS) No. 32* (IASCF) as amended effective January 1, 2016.

assume that half the convertible bonds issued by HTL Manufacturers are converted at a time when the remaining unamortized premium is $2 million:

Journal Entry at Conversion

	($ in millions)	
Convertible bonds payable (½ the account balance)	50	
Premium on bonds payable (½ the account balance)	1	
Common stock (to balance)		51

Additional Consideration

The method just described is referred to as the *book value method,* since the new shares are recorded at the book value of the redeemed bonds. It is by far the most popular method in practice. Another acceptable approach, the *market value* method, records the new shares at the market value of the shares themselves or of the bonds, whichever is more determinable. Because the market value most likely will differ from the book value of the bonds, a gain or loss on conversion will result. Assume for illustration that the market value of HTL's stock is **$30** per share at the time of the conversion:

	($ in millions)	
Convertible bonds payable (½ the account balance)	50	
Premium on bonds payable (½ the account balance)	1	
Loss on conversion of bonds (to balance)	9	
Common stock [(50,000 bonds × 40 shares) × **$30**]		60

If the 50,000 convertible bonds were held by a single investor, that company would record the conversion as follows:

The 2 million shares issued are recorded at the $51 million book value of the bonds retired.

	($ in millions)	
Investment in common stock	51	
Investment in convertible bonds (account balance)		50
Premium on investment in bonds (account balance)		1

Induced Conversion

Investors often are reluctant to convert bonds to stock, even when share prices have risen significantly since the convertible bonds were purchased. This is because the market price of the convertible bonds will rise along with market prices of the stock. So companies sometimes try to induce conversion. The motivation might be to reduce debt and become a better risk to potential lenders or achieve a lower debt-to-equity ratio.

One way is through the call provision. As we noted earlier, most corporate bonds are callable by the issuing corporation. When the specified call price is less than the conversion value of the bonds (the market value of the shares), calling the convertible bonds provides bondholders with incentive to convert. Bondholders will choose the shares rather than the lower call price.

Any additional consideration provided to induce conversion of convertible debt is recorded as an expense of the period.

Occasionally, corporations may try to encourage voluntary conversion by offering an added inducement in the form of cash, stock warrants, or a more attractive conversion ratio. When additional consideration is provided to induce conversion, the fair value of that consideration is considered an expense incurred to bring about the conversion.[16]

Bonds with Detachable Warrants

Another (less common) way to sweeten a bond issue is to include **detachable stock purchase warrants** as part of the security issue. A stock warrant gives the investor an option to purchase a stated number of shares of common stock at a specified *option price,* often

[16]FASB ASC 470–20–40: Debt—Debt with Conversion Options—Derecognition.

within a given period of time. Like a conversion feature, warrants usually mean a lower interest rate and often enable a company to issue debt when borrowing would not be feasible otherwise.

However, unlike the conversion feature for convertible bonds, warrants can be separated from the bonds. This means they can be exercised independently or traded in the market separately from bonds, having their own market price. In essence, two different securities—the bonds and the warrants—are sold as a package for a single issue price. Accordingly, the issue price is allocated between the two different securities on the basis of their fair values. If the independent market value of only one of the two securities is reliably determinable, that value establishes the allocation. This is demonstrated in Illustration 14–19.

Illustration 14–19

Bonds with Detachable Warrants

On January 1, 2021, HTL Manufacturers issued $100 million of 8% debentures due 2025 at 103 (103% of face value).

- Accompanying each $1,000 bond were 20 warrants.
- Each warrant permitted the holder to buy one share of no par common stock at $25 per share.
- Shortly after issuance, the warrants were listed on the stock exchange at **$3** per warrant.

	($ in millions)	
Cash (103% × $100 million)	103	
Discount on bonds payable (difference)	3	
Bonds payable (face amount)		100
Equity—stock warrants* (**100,000** bonds × **20** warrants × **$3**)		**6**

*Reported as part of shareholders' equity rather than as a liability.

The issue price of bonds with detachable warrants is allocated between the two different securities on the basis of their fair values.

Additional Consideration

Market imperfections may cause the separate market values not to sum to the issue price of the package. In this event, allocation is achieved on the basis of the **relative market values** of the two securities. Let's say the bonds have a separate market price of $940 per bond (priced at 94):

Market Values	**Dollars** ($ in millions)	**Percent**
Bonds (100,000 bonds × $940)	$ 94	94%
Warrants (100,000 bonds × 20 warrants × $3)	6	6
Total	$100	100%

Proportion of Issue Price Allocated to Bonds:

$103 million × **94%** = $96,820,000

Proportion of Issue Price Allocated to Warrants:

$103 million × **6%** = $6,180,000

	($ in millions)	
Cash (103% × $100 million)	103.00	
Discount on bonds payable ($100 million – $96.82 million)	3.18	
Bonds payable (face amount)		100.00
Equity—stock warrants ($103 million × 6% = $6,180,000)		6.18

Notice that this is the same approach we used in Chapter 10 to allocate a single purchase price to two or more assets bought for that single price. We also will allocate the total selling price of two equity securities sold for a single issue in proportion to their relative market values in Chapter 18.

If one-half of the warrants (1 million) in Illustration 14–19 are exercised when the market value of HTL's common stock is $30 per share, 1 million shares would be issued for one warrant each plus the exercise price of $25 per share.

Journal Entry at Exercise of Detachable Warrants

	($ in millions)	
Cash (1,000,000 warrants × $25)	25	
Equity—stock warrants (1,000,000 warrants × $3)	3	
Common stock (to balance)		28

The $30 market value at the date of exercise is not used in valuing the additional shares issued. The new shares are recorded at the total of the previously measured values of both the warrants and the shares.

Concept Review Exercise

DEBT DISCLOSURES AND EARLY EXTINGUISHMENT OF DEBT

The disclosure notes to the 2021 financial statements of Olswanger Industries included the following:

Note 12: Bonds

On September 15, 2020, the Corporation sold bonds with an aggregate principal amount of $500,000,000 bearing a 14% interest rate. The bonds will mature on September 15, 2030, and are unsecured subordinated obligations of the Corporation. Interest is payable semiannually on March 15 and September 15. The Corporation may redeem the bonds at any time beginning September 15, 2020, as a whole or from time to time in part, through maturity, at specified redemption prices ranging from 112% of principal in declining percentages of principal amount through 2027 when the percentage is set at 100% of principal amount. The cost of issuing the bonds, totaling $11,000,000, and the discount of $5,000,000 are being amortized over the life of the bonds. Amortization of these items for the year ended December 31, 2021, was $1,212,000.

During the year ended December 31, 2021, the Corporation repurchased, in open market transactions, $200,000,000 in face amount of the bonds for $219,333,000. The unamortized cost of issuing these bonds and the unamortized discount, totaling $5,864,000 have been deducted in the current period.

From the information provided by Olswanger in Note 12, you should be able to recreate some of the journal entries the company recorded in connection with this bond issue.

Required:

1. Prepare the journal entry for the issuance of these bonds on September 15, 2020.
2. Prepare the journal entry for the repurchase of these bonds, assuming the date of repurchase was September 15, 2021. The cash paid to repurchase the bonds was $219,333,000.

Solution:

1. Prepare the journal entry for the issuance of these bonds on September 15, 2020.

	($ in thousands)	
Cash (to balance)	484,000	
Discount and debt issue costs (given in note)	16,000	
Bonds payable (face amount—given in note)		500,000

2. Prepare the journal entry for the repurchase of these bonds, assuming the date of repurchase was September 15, 2021.

	($ in thousands)	
Bonds payable (face amount repurchased)	200,000	
Loss on early extinguishment (to balance)	25,197	
Discount and debt issue costs (given in note)		5,864
Cash (given in requirement 2)		219,333

Option to Report Liabilities at Fair Value

PART D

● LO14–6
A company has the option to value financial assets and liabilities at fair value.

Companies are not required to, but have the option to, value some or all of their financial assets and liabilities at fair value.[17] In Chapter 12, we saw examples of the option applied to financial assets—specifically, companies reporting their investments in securities at fair value. Now, we see how liabilities, too, can be reported at fair value.

How does a liability's fair value change? Remember that there are two sides to every investment. For example, if a company has an investment in **General Motors**' bonds, that investment is an asset to the investor, and the same bonds are a liability to GM. So, the same market forces that influence the fair value of an investment in debt securities (interest rates, credit risk, etc.) influence the fair value of liabilities. For bank loans or other debts that aren't traded on a market exchange, the mix of factors will differ, but in any case, changes in the current market rate of interest often are a major contributor to changes in fair value.

Determining Fair Value

Changes in interest rates cause changes in the fair value of liabilities.

For demonstration, we revisit the Masterwear Industries bonds that sold at a discount in Illustration 14–3. Now, suppose it's six months later, the market rate of interest has fallen to 11%, and June 30 is the end of Masterwear's fiscal year. A decline in market interest rates means bond prices rise. Let's say that checking market prices in *The Wall Street Journal* indicates that the fair value of the Masterwear bonds on June 30, 2021, is **$714,943**. Referring to the amortization schedule, we see that on the same date, with five periods remaining to maturity, the present value of the bonds—their price—would have been $671,297 if the market rate still had been 14% (7% semiannually).

Additional Consideration

If the bonds are not traded on an open-market exchange, their fair value would not be readily observable. In that case, the next most preferable way to determine fair value would be to calculate the fair value as the present value of the remaining cash flows discounted at the current interest rate. If the rate is 11% (5.5% semiannually),[18] as we're assuming now, that present value would be $714,943:

	Present Values
Interest	\$ 42,000 × 4.27028* = \$179,352
Principal	\$700,000 × 0.76513† = 535,591
Present value of the bonds	**\$714,943**

*Present value of an ordinary annuity of \$1: $n = 5$, $i = 5.5\%$.
†Present value of \$1: $n = 5$, $i = 5.5\%$.

When the bonds were issued, Masterwear had a choice—report this liability (a) at its amortized initial measurement throughout the term to maturity or (b) at its current fair value on each reporting date. Had the company not elected the fair value option, on June 30 it would report the $671,297 we calculated earlier for the amortization schedule. On the other hand, if Masterwear had elected the fair value option, it would report the bonds at their current fair value, **$714,943.**

[17] FASB ASC 825–10–25–1: Financial Instruments—Overall—Recognition—Fair Value Option.

[18] The current market rate would consist of the general risk-free rate at the time, increased by a risk premium for the credit risk of the bonds.

Reporting Changes in Fair Value

Electing the FVO means reporting unrealized holding gains and losses in OCI to the extent the change is due to credit risk.

If a company chooses the option to report at fair value, a change in fair value will create a gain or loss. Any portion of that gain or loss that is a result of a change in the "credit risk" of the debt, rather than a change in general interest rates, is reported, not as part of net income, but instead as other comprehensive income (OCI). Credit risk is the risk that the investor in the bonds will not receive the promised interest and maturity amounts at the times they are due. Companies can assume that any change in fair value that exceeds the amount caused by a change in the general (risk-free) interest rate is the result of credit risk changes.

In our example, Masterwear would report the increase in fair value from $666,633 to **$714,943**, or $48,310. Note, though, that part of the change is due to the unpaid interest we discussed earlier. Here's a recap.

At June 30, 2021, the interest that accrued during the first six months was $46,664, but only $42,000 of that was paid in cash; so the book value increased by the $4,664 unpaid interest. We recorded the following entry:

Interest expense	46,664	
Discount on bonds payable		4,664
Cash		42,000

Amortizing the discount in this entry increased the book value of the liability by $4,664 to $671,297:

Bonds payable
Less: Discount
Book value

January 1 book value and fair value	$666,633
Increase from discount amortization	4,664
June 30 book value (amortized initial amount)	$671,297

FAIR VALUE RISES Comparing that book value with the fair value of the bonds on that date provides the amount needed to adjust the bonds to their fair value.

June 30 fair value	**$714,943**
June 30 book value (amortized initial amount)	671,297
Fair value adjustment needed	$ 43,646

Rather than increasing the bonds payable account itself, though, we instead adjust it *indirectly* with a credit to a valuation allowance (or contra) account. If general interest rates have not changed, we assume the change in fair value is due to the credit risk associated with the bonds and report the loss in OCI:

When the fair value option is elected, we report changes in fair value in OCI to the extent caused by changes in credit risk.

Loss on bonds payable (unrealized, OCI)	43,646	
Fair value adjustment (**$714,943** – $671,297)		43,646

The *credit* balance in the fair value adjustment *increases* the book value; the discount reduces it.

If any portion of the fair value change is due to a change in general interest rates, that portion would be reported in net income. For instance, if interest rate declines alone would have created a $20,000 increase in fair value, we would report $20,000 of the loss in in the income statement and $23,646 as OCI.

The new book value of the bonds is now the fair value:

Bonds payable	$700,000
Less: Discount ($33,367 – $4,664)	(28,703)
Amortization schedule value	$671,297
Plus: Fair value adjustment	43,646
Book value, June 30	**$714,943**

FAIR VALUE FALLS Suppose the fair value at June 30, 2021, had been $650,000 instead of **$714,943**. In that case, Masterwear would record a *reduction* in the liability from $671,297 to $650,000, or $21,297. Again assuming no change in interest rates, the entry would be as follows:

Fair value adjustment ($671,297 – **$650,000**)	21,297	
Gain on bonds payable (unrealized, OCI)		21,297

The new book value of the bonds is the fair value:

Bonds payable	$ 700,000
Less: Discount ($33,367 – $4,664)	(28,703)
Amortization schedule value	$ 671,297
Less: Fair value adjustment	(21,297)
Book value, June 30	**$650,000**

The *debit* balance in the fair value adjustment *reduces* the book value; as does the discount.

The outstanding balance in the last column of the amortization schedule at any date up to and including the balance at maturity will be the bonds payable less the discount (for instance $671,297 at June 30, 2021). But the amount we report in the balance sheet at any reporting date, the fair value, will be that amortized initial amount from the amortization schedule, plus or minus the fair value adjustment. That's the **$714,943** or the $650,000 in the two scenarios above.

To understand why we report credit risk-related fair value changes as OCI, it's useful to consider the consequence of reporting them in the income statement, which was the prescribed treatment prior to the FASB mandating their classification as OCI beginning in 2015. Here's one example. As a result of adjusting for fair value changes in its debt, **J.P. Morgan Chase & Co.** reported an impressive $1.9 billion gain in its 2011 third-quarter income statement. This was reason for the company's shareholders and creditors to celebrate, right? Not so fast. It's true that the company followed the proper procedure for recording a decline in the fair value of its debt, which was to record a gain in the income statement. As it turned out, though, the reason the fair value declined was that the company's financial situation had worsened, resulting in increased credit risk, and the increased credit risk caused the fair value of liabilities to decline. This, in fact, is the reason for most if not all of J.P. Morgan's huge gain. Does it make sense to you that worsening credit ratings should result in higher net income? Most observers, including the FASB, saw that as counterintuitive, and as a result, credit risk-related fair value changes are now reported as part of OCI rather as part of net income.

Mix and Match

Companies choose which financial instruments to report at fair value.

Remember from our discussions in prior chapters that if a company elects the fair value option, it's not necessary that the company elect the option to report all of its financial instruments at fair value or even all instruments of a particular type at fair value. They can "mix and match" on an instrument-by-instrument basis. So Masterwear, for instance, might choose to report these bonds at fair value but all its other liabilities at their amortized initial measurement. However, the company must make the election when the item originates, in this case when the bonds are issued, and is not allowed to switch methods once a method is chosen.

Here's an excerpt from the notes to the financial statements of **Citigroup Inc.** describing its fair value option election for some of its liabilities (omitting the portion of the note related to the fair value option election for financial assets):

25. FAIR VALUE ELECTIONS (In part)

The Company may elect to report most financial instruments and certain other items at fair value on an instrument-by-instrument basis. . . . The election is made upon the initial recognition of an eligible financial asset, financial liability or firm commitment or when certain specified reconsideration events occur. The fair value election may not be revoked once made. The changes in fair value are recorded in current earnings, other than DVA [Debit Valuation Adjustments], which from January 1, 2017, is reported in AOCI.

($ in millions)	Changes in fair value gains (losses) for the years ended December 31 2017	2016
Liabilities		
Interest-bearing deposits	$ (69)	$ (50)
Federal funds purchased and securities loaned or sold under agreements to repurchase selected portfolios of securities	223	25
Trading account liabilities	70	105
Short-term borrowings	(116)	(61)
Long-term debt	(1,491)	(935)
Total liabilities	$(1,383)	$(896)

Financial Reporting Case Solution

©monkeybusinessimages/Getty Images

1. **What does it mean that the bonds are "first mortgage" bonds? What effect does that have on financing?** *(p. 774)* A mortgage bond is backed by a lien on specified real estate owned by the issuer. This makes it less risky than unsecured debt, so Service Leader can expect to be able to sell the bonds at a higher price (lower interest rate).
2. **From Service Leader's perspective, why are the bonds callable? What does that mean?** *(p. 775)* The call feature gives Service Leader some protection against being stuck with relatively high-cost debt in case interest rates fall during the 15 years to maturity. Service Leader can buy back, or call, the bonds from bondholders before the 15-year maturity date, after June 30, 2020. The call price is prespecified at 103% of the face value—$1,030 per $1,000 bond.
3. **How will it be possible to sell bonds paying investors 6.25% when other, similar investments will provide the investors a return of 6.5%?** *(p. 775)* Service Leader will be able to sell its 6.25% bonds in a 6.5% market only by selling them at a discounted price, below face amount. Bonds are priced by the marketplace to yield the market rate of interest for securities of similar risk and maturity. The price will be the present value of all the periodic cash interest payments (face amount × stated rate) plus the present value of the principal payable at maturity, both discounted at the market rate.
4. **Would accounting differ if the debt were designated as notes rather than bonds?** *(p. 786)* No. Other things being equal, whether they're called bonds, notes, or some other form of debt, the same accounting principles apply. They will be recorded at present value and interest will be recorded at the market rate over the term to maturity.
5. **Why might the company choose to make the bonds convertible into common stock?** *(p. 795)* Convertible bonds can be converted at the option of the bondholders into shares of stock. Sometimes the motivation for issuing convertible bonds rather than straight debt is to use the bonds as a medium of exchange in mergers and acquisitions, as a way for smaller firms or debt-heavy companies to obtain access to the bond market, or as an indirect way to issue stock when there is shareholder resistance to direct issuance of additional equity. None of these seems pertinent to Service Leader. The most likely reason is to sell at a higher price. The conversion feature is attractive to investors. Investors have a fixed-income security that can become common stock if circumstances make that attractive. The investor has additional possibilities for higher returns, with downside risk limited by the underlying debt. ●

The Bottom Line

- **LO14–1** A liability requires the future payment of cash in specified amounts at specified dates. As time passes, interest accrues on debt at the effective interest rate times the amount of the debt outstanding during the period. This same principle applies regardless of the specific form of the liability. (*p. 773*)

- **LO14–2** Forces of supply and demand cause a bond to be priced to yield the market rate, calculated as the present value of all the cash flows required, where the discount rate is the market rate. Interest expense is calculated as the effective market rate of interest multiplied by the outstanding balance (during the interest period). A company is permitted to allocate a discount or a premium equally to each period over the term to maturity if doing so produces results that are not materially different from the interest method. (*p. 775*)
- **LO14–3** In concept, notes are accounted for in precisely the same way as bonds. When a note is issued with an unrealistic interest rate, the effective market rate is used both to determine the amount recorded in the transaction and to record periodic interest thereafter. (*p. 786*)
- **LO14–4** In the balance sheet, disclosure should include, for all long-term borrowings, the aggregate amounts maturing and sinking fund requirements (if any) for each of the next five years. Supplemental disclosures are needed for (a) off-balance-sheet credit or market risk, (b) concentrations of credit risk, and (c) the fair value of financial instruments. (*p. 791*)
- **LO14–5** A gain or loss on early extinguishment of debt should be recorded for the difference between the reacquisition price and the book value of the debt. Convertible bonds are accounted for as straight debt, but the value of the equity feature is recorded separately for bonds issued with detachable warrants. (*p. 795*)
- **LO14–6** Companies are not required to, but have the option to, value some or all of their liabilities at fair value. If the option is elected, an increase (or decrease) in fair value from one balance sheet to the next is reported as a loss (or gain) as other comprehensive income to the extent it's related to credit risk. Otherwise, it's reported in net income. It's a one-time election for each liability when the liability is created. (*p. 801*)
- **LO14–7** U.S. GAAP and IFRS are generally compatible with respect to accounting for bonds and long-term notes. Some differences exist in determining which securities are reported as debt, the way convertible securities are accounted for, and when the fair value option can be elected. (*pp. 797*).

APPENDIX 14A

Bonds Issued Between Interest Dates

In Part A of this chapter, we assumed that the bonds were issued on the day they were dated (date printed in the indenture contract). But suppose a weak market caused a delay in selling the bonds until two months after that date (four months before semiannual interest was to be paid). In that case, the buyer would be asked to pay the seller **accrued interest** for two months in addition to the price of the bonds.

When Bonds Are Issued at Face Amount between Interest Dates

For illustration, assume that in Illustration 14–1, Masterwear was unable to sell the bonds in the previous example until March 1—two months after they are dated. This variation is shown in Illustration 14A–1. United would pay the price of the bonds ($700,000) plus **$14,000** accrued interest:

$700,000	×	12%	×	2/12	=	**$14,000**
Face amount		Annual rate		Fraction of the annual period		Accrued interest

All bonds sell at their price plus any interest that has accrued since the last interest date.

Illustration 14A–1

Bonds Sold at Face Amount between Interest Dates

At Issuance (March 1)		
Masterwear (Issuer)		
Cash (price plus accrued interest)	714,000	
Bonds payable (face amount)		700,000
Interest payable (accrued interest determined above)		**14,000**
United (Investor)		
Investment in bonds (face amount)	700,000	
Interest receivable (accrued interest determined above)	**14,000**	
Cash (price plus accrued interest)		714,000

Since the investor will hold the bonds for only four months before receiving six months' interest, two months' accrued interest must be added to the price paid.

When Masterwear pays semiannual interest on June 30, a full six months' interest is paid. But having received two months' accrued interest in advance, Masterwear's net interest expense will be four months' interest, for the four months the bonds have been outstanding at that time. Likewise, when United receives six months' interest—after holding the bonds for only four months—United will net only the four months' interest to which it is entitled:

The issuer incurs interest expense, and the investor recognizes interest revenue, for only the four months the bonds are outstanding.

At the First Interest Date (June 30)

Masterwear (Issuer)		
Interest expense (6 mo. – 2 mo. = 4 mo.)	28,000	
Interest payable* (accrued interest determined above)	**14,000**	
Cash (stated rate × face amount)		42,000
United (Investor)		
Cash (stated rate × face amount)	**42,000**	
Interest receivable (accrued interest determined above)		**14,000**
Interest revenue (6 mo. – 2 mo. = 4 mo.)		**28,000**

*Some accountants prefer to credit interest expense, rather than interest payable, when the bonds are sold. When that is done, this entry would require a debit to interest expense and a credit to cash for $42,000. The interest expense account would then reflect the same *net* debit of four months' interest ($42,000 – **$14,000**).

Interest Expense

	2 months
6 months	
4 months	

Similarly, the investor could debit interest revenue, rather than interest receivable when buying the bonds.

When Bonds Are Issued at a Discount between Interest Dates

All bonds sell at their price plus any interest that has accrued since the last interest date.

Our objective is the same when the bonds are not issued at their face amount. For instance, in Illustration 14–3, the $700,000 of 12% bonds were issued on January 1 when the market rate was 14% and thus were priced at a discount, $666,633. If instead those bonds were issued March 1, two months after they are dated, the price will have increased. That's because the liability will have accrued two months' interest at the effective rate (7% × 2/6 × $666,633, or $15,555). $14,000 of that (6% × 2/6 × $700,000) represents stated interest that will be part of the first interest payment four months later, so the price of the bonds will have increased by the difference, $15,555 – $14,000, or $1,555. We can think of this as the time value of the bonds for 2 months. The cash flows remain the same (six semiannual interest payments plus the face amount at maturity), but because the payments now will be received two months sooner, the *present value* of those payments is greater. The investor would pay the January 1 value of the bonds ($666,633) plus the time value for 2 months ($1,555) plus $14,000 accrued interest (see Illustration 14A–2).

Illustration 14A–2

Bonds Sold at a Discount between Interest Dates

Masterwear (Issuer)		
Cash (price plus accrued interest)	682,188*	
Discount on bonds payable (difference)	31,812	
Bonds payable (face amount)		700,000
Interest payable (accrued interest determined above)		**14,000**
United (Investor)		
Investment in bonds (face amount)	**700,000**	
Interest receivable (accrued interest determined above)	**14,000**	
Discount on bond investment (difference)		**31,812**
Cash (price plus accrued interest)		**682,188***

*On March 1, the proceeds will be the price of the bonds based on the effective market interest rate on March 1, plus the $14,000 accrued interest. If the price of the bonds was $666,633 on January 1 as we assumed in Illustration 14–3, and the effective market interest rate remained 7% semiannually for the next two months, that price will have increased to $668,188:

$666,663	Balance, Jan. 1
15,555	7% × 2/6 × $666,633 2 mo. *effective* interest
$682,188	Balance, March 1
(14,000)	6% × 2/6 × $700,000 2 mo. *stated* interest
$668,188	Bond price, March 1
14,000	Accrued interest
$688,188	Proceeds: price plus accrued interest

By contract, a full six months' interest is paid on June 30 at the *stated* rate times the face amount. But having received two months' accrued interest in advance, Masterwear's *net* interest expense will be four months' interest, for the four months the bonds have been outstanding at that time. Interest is calculated on the outstanding debt balance at the effective (or market) rate. Interest for the first six months, then, would be \$668,188 × 14% × ½ = \$46,773. But because on June 30 only four months have passed since the date the bonds were issued, only four months' interest, 4/6 × 46,664, or \$31,182, is recorded as interest expense. Similarly, the reduction of the discount, which for a full six-month interest period would be \$46,773 – \$42,000, or \$46,773, will be only 4/6 of that amount. These amounts are recorded as follows:

Journal Entries—Interest between Interest Dates

At the First Interest Date (June 30)

Masterwear (Issuer)		
Interest expense (4/6 × 46,773)	31,182	
Interest payable (accrued interest determined above)	14,000	
Discount on bonds payable (4/6 × 46,773)		3,182
Cash (stated rate × face amount)		42,000
United (Investor)		
Cash (stated rate × face amount)	42,000	
Discount on bond investment (4/6 × 46,773)	3,182	
Interest receivable (accrued interest determined above)		14,000
Interest revenue (4/6 × 46,773)		31,182

Only four months' interest is recorded because the bonds were outstanding only four months.

Troubled Debt Restructuring

APPENDIX 14B

A respected real estate developer, Brillard Properties, was very successful developing and managing a number of properties in the southeastern United States. To finance these investments, the developer had borrowed hundreds of millions of dollars from several regional banks. For years, events occurred as planned. The investments prospered. Cash flow was high. Interest payments on the debt were timely and individual loans were repaid as they matured.

Almost suddenly, however, the real estate climate soured. Investments that had provided handsome profits now did not provide the cash flow necessary to service the debt. Bankers who had loaned substantial funds to Brillard now faced a dilemma. Because contractual interest payments were unpaid, the bankers had the legal right to demand payment, which would force the developer to liquidate all or a major part of the properties to raise the cash. Sound business practice? Not necessarily.

A sharp rise in debt restructurings accompanied the 2008/2009 economic crisis.

If creditors force liquidation, they then must share among themselves the cash raised from selling the properties—at forced sale prices. Believing the developer's financial difficulties were caused by temporary market forces, not by bad management, the bankers felt they could minimize their losses by *restructuring* the debt agreements, rather than by forcing liquidation.

When changing the original terms of a debt agreement is motivated by financial difficulties experienced by the debtor (borrower), the new arrangement is referred to as a **troubled debt restructuring**. By definition, a troubled debt restructuring involves some concessions on the part of the creditor (lender) that it wouldn't otherwise consider if not for the economic or legal reasons related to the debtor's financial difficulties. In the first half of 2009, the economic crisis prompted nearly 40 companies to negotiate such deals with their creditors, a huge increase over the handful that normally would do so during a six-month period.[19] A troubled debt restructuring may be achieved in either of two ways:

1. The debt may be *settled* at the time of the restructuring.
2. The debt may be *continued,* but with *modified terms.*

[19] David Henry, "The Time Bomb in Corporate Debt," *BusinessWeek,* July 15, 2009.

Debt Is Settled

In all areas of accounting, a noncash transaction is recorded at fair value.

In the situation described above, one choice the bankers had was to try to actually settle the debt outright at the time of the troubled debt restructuring. For instance, a bank holding a $30 million note from the developer might agree to accept a property valued at, let's say, $20 million as final settlement of the debt. In that case, the developer has a $10 million gain equal to the difference between the book value of the debt and the fair value of the property transferred. The debtor may need to adjust the book value of an asset to its fair value prior to recording its exchange for a debt. The developer in our example, for instance, would need to change the recorded amount for the property specified in the exchange agreement if it is carried at an amount other than its $20 million fair value. In such an instance, an ordinary gain or loss on disposition of assets should be recorded as shown in Illustration 14B–1.

Illustration 14B–1
Debt Settled

An asset is adjusted to fair value prior to recording its exchange for a debt.

First Prudent Bank agrees to settle Brillard's $30 million debt in exchange for property having a fair value of $20 million. The book value of the property on Brillard's books is $17 million:

	($ in millions)	
Land ($20 million minus $17 million)	3	
Gain on disposition of assets		3
Notes payable (book value)	30	
Gain on troubled debt restructuring		10
Land (fair value)		20

The payment to settle a debt in a troubled debt restructuring might be cash, or a noncash asset (as in the example here), or even shares of the debtor's stock. An example of shares of stock given in exchange for debt forgiveness is the celebrated reorganization of TWA in 1992 (since acquired by American Airlines), when creditors received a 55% stake in the company's common shares in return for forgiving about $1 billion of the airline's $1.5 billion debt. In any case, the debtor's gain is the difference between the book value of the debt and the fair value of the asset(s) or equity securities transferred.

Debt Is Continued, but with Modified Terms

The book value of a debt is the current balance of the primary debt plus any accrued (unpaid) interest.

We assumed in the previous example that First Prudent Bank agreed to accept property in full settlement of the debt. A more likely occurrence would be that the bank allows the debt to continue, but modifies the terms of the debt agreement to make it easier for the debtor to comply. The bank might agree to reduce or delay the scheduled *interest payments.* Or, it may agree to reduce or delay the *maturity amount.* Often a troubled debt restructuring will call for some combination of these concessions. Distressed by the 2008/2009 economic downturn, amusement park operator **Six Flags** persuaded a group of creditors to trim its debt by 5%, or $130 million, a restructuring designed to give Six Flags a financial cushion going into its busy 2009 summer season and an opportunity to recover its interest-paying ability.

Let's say the stated interest rate on the note in question is 10% and annual interest payments of $3 million (10% × $30 million) are payable in December of each of two remaining years to maturity. Also assume that the developer was unable to pay the $3 million interest payment for the year just ended. This means that the amount owed—the carrying amount (or book value) of the debt—is $33 million ($30 million plus one year's accrued interest).

Two quite different situations are created when the terms of a debt are modified, depending on whether the cash payments are reduced to the extent that interest is eliminated.

The way the debtor accounts for the restructuring depends on the extent of the reduction in cash payments called for by the restructured arrangement. More specifically, the accounting procedure depends on whether, under the new agreement, total cash payments (a) are *less than* the book value of the debt or (b) still *exceed* the book value of the debt.

WHEN TOTAL CASH PAYMENTS ARE LESS THAN THE BOOK VALUE OF THE DEBT By the original agreement, the debtor was to pay at maturity the $30 million loaned, plus enough periodic interest to provide a 10% effective rate of return. If the new agreement calls for less cash than the $33 million now owed, interest is presumed to have been eliminated.

As one of many possibilities, suppose the bank agrees to (1) forgive the interest accrued from last year, (2) reduce the two remaining interest payments from $3 million each to $2 million each, and (3) reduce the face amount from $30 million to $25 million. Clearly, the debtor will pay less by the new agreement than by the original one. In fact, if we add up the total payments called for by the new agreement, the total [($2 million × 2) plus $25 million] is less than the $33 million book value. Because the $29 million does not exceed the amount owed, the restructured debt agreement no longer provides interest on the debt. Actually, the new payments are $4 million short of covering the debt itself. So, after the debt restructuring, no interest expense is recorded. All subsequent cash payments are considered to be payment of the debt itself. Consider Illustration 14B–2.

Illustration 14B–2

Cash Payments Less than the Debt

Brillard Properties owes First Prudent Bank $30 million under a 10% note with two years remaining to maturity. Due to financial difficulties of the developer, the previous year's interest ($3 million) was not paid. First Prudent Bank agrees to:

1. Forgive the interest accrued from last year.
2. Reduce the remaining two interest payments to $2 million each.
3. Reduce the principal to $25 million.

Analysis:	Book value	$30 million + $3 million =	$33 million
	Future payments	($2 million × 2) + $25 million =	29 million
	Gain		$ 4 million

	($ in millions)	
Interest payable (10% × $30 million)	3	
Notes payable ($30 million – $29 million)	1	
Gain on debt restructuring		4

Book Value		
Before Restr.	**Adj.**	**After Restr.**
$30	(1)	$29
3	(3)	0
$33	(4)	$29

When the total future cash payments are less than the book value of the debt, the difference is recorded as a gain at the date of restructure. No interest should be recorded thereafter. That is, all subsequent cash payments result in reductions of principal.

After restructuring, no interest expense is recorded. All cash payments are considered to be payment of the note itself.

At Each of the Two Interest Dates	($ in millions)	
Notes payable	2	
Cash (revised "interest" amount)		2
At Maturity		
Notes payable	25	
Cash (revised principal amount)		25

The $25 million payment at maturity reduces the note to zero.

WHEN TOTAL CASH PAYMENTS EXCEED THE BOOK VALUE OF THE DEBT Let's modify the example in the previous section. Now suppose the bank agrees to delay the due date for all cash payments until maturity and accept $34,333,200 at that time in full settlement of the debt. Rather than just reducing the cash payments as in the previous illustration, the payments are delayed. It is not the nature of the change that creates the need to account differently for this situation, but the amount of the total cash payments under the agreement relative to the book value of the debt. This situation is demonstrated in Illustration 14B–3.

Illustration 14B–3

Cash Payments More than the Debt

Brillard Properties owes First Prudent Bank $30 million under a 10% note with two years remaining to maturity. Due to Brillard's financial difficulties, the previous year's interest ($3 million) was not paid. First Prudent Bank agrees to:

1. Delay the due date for all cash payments until maturity.
2. Accept $34,333,200 at that time in full settlement of the debt.

(continued)

Illustration 14B–3
(concluded)

The discount rate that equates the present value on the debt ($33 million) and its future value ($34,333,200) is the effective rate of interest.

Analysis:	Future payments		$34,333,200
	Book value	$30 million + $3 million =	33,000,000
	Interest		$ 1,333,200

Calculation of the New Effective Interest Rate

- $33,000,000 ÷ $34,333,200 = 0.9612, the Table 2 value for $n = 2, i = ?$
- In row 2 of Table 2, the number 0.9612 is in the **2%** column. So, this is the new effective interest rate.

As long as cash payments exceed the amount owed there will be interest—although at a lower effective rate.

Now the total payments called for by the new agreement, $34,333,200, exceed the $33 million book value. Because the payments exceed the amount owed, the restructured debt agreement still provides interest on the debt—but less than before the agreement was revised. No longer is the effective rate 10%. The accounting objective now is to determine what the new effective rate is and *record interest for the remaining term of the loan at that new, lower rate,* as shown in Illustration 14B–3.

Unpaid interest is accrued at the effective rate times the book value of the note.

Because the total future cash payments are not less than the book value of the debt, no reduction of the existing debt is necessary and no entry is required at the time of the debt restructuring. Even though no cash is paid until maturity under the restructured debt agreement, interest expense still is recorded annually—but at the new rate.

The book value of the debt is increased by the unpaid interest from the previous year.

The total of the accrued interest account plus the note account is equal to the amount scheduled to be paid at maturity.

At the End of the First Year		
Interest expense [**2%** × ($30,000,000 + $3,000,000)]	660,000	
Interest payable		660,000
At the End of the Second Year		
Interest expense [**2%** × ($30,000,000 + $3,660,000)]	673,200	
Interest payable		673,200
At Maturity (End of the Second Year)		
Notes payable	30,000,000	
Interest payable ($3,000,000 + $660,000 + $673,200)	4,333,200	
Cash (required by new agreement)		34,333,200

Additional Consideration

To keep up with the changing amounts, it may be convenient to prepare an amortization schedule for the debt.

Year	Cash Interest	Effective Interest	Increase in Balance	Outstanding Balance
		(2% × Outstanding balance)		
				33,000,000
1	0	0.02 (33,000,000) = 660,000	660,000	33,660,000
2	0	0.02 (33,660,000) = 673,200	673,200	34,333,200
	0	1,333,200	1,333,200	

An amortization schedule is particularly helpful if there are several years remaining to maturity.

In our example, the restructured debt agreement called for a single cash payment at maturity ($34,333,200). If more than one cash payment is required (as in the agreement in our earlier example), calculating the new effective rate is more difficult. The concept would remain straightforward: (1) determine the interest rate that provides a present value of all future cash payments that is equal to the current book value and (2) record the interest at that rate thereafter. Mechanically, though, the computation by hand would be cumbersome, requiring a time-consuming trial-and-error calculation. Since our primary interest is understanding the concepts involved, we will avoid the mathematical complexities of such a situation.

You also should be aware that when a restructuring involves modification of terms, accounting for a liability by the debtor, as described in this section, and accounting for a receivable by the creditor (essentially an impairment of a receivable), which was described in Chapter 7, are inconsistent. You may recall that when a creditor's investment in a receivable becomes impaired, due to a troubled debt restructuring or for any other reason, the receivable is remeasured based on the discounted present value of currently expected cash flows at the loan's *original* effective rate (regardless of the extent to which expected cash receipts have been reduced). For ease of comparison, the example in this chapter (Illustration 14B–3) describes the same situation as the example in Chapter 7 (Illustration 7B–2). There is no conceptual justification for the asymmetry between debtors' and creditors' accounting for troubled debt restructurings. ●

Questions For Review of Key Topics

Q 14–1 How is periodic interest determined for outstanding liabilities? For outstanding receivables? How does the approach compare from one form of debt instrument (say bonds payable) to another (say notes payable)?

Q 14–2 As a general rule, how should long-term liabilities be reported on the debtor's balance sheet?

Q 14–3 How are bonds and notes the same? How do they differ?

Q 14–4 What information is contained in a bond indenture? What purpose does it serve?

Q 14–5 On January 1, 2021, Brandon Electronics issued $85 million of 11.5% bonds, dated January 1. The market yield for bonds of maturity issued by similar firms in terms of riskiness is 12.25%. How can Brandon sell debt paying only 11.5% in a 12.25% market?

Q 14–6 How is the price determined for a bond (or bond issue)?

Q 14–7 A zero-coupon bond pays no interest. Explain.

Q 14–8 When bonds are issued at a premium the debt declines each period. Explain.

Q 14–9 Compare the two commonly used methods of determining interest on bonds.

Q 14–10 What are debt issue costs and how should they be reported?

Q 14–11 When a note's stated rate of interest is unrealistic relative to the market rate, the concept of substance over form should be employed. Explain.

Q 14–12 How does an installment note differ from a note for which the principal is paid as a single amount at maturity?

Q 14–13 Long-term debt can be reported either (a) as a single amount, net of any discount or increased by any premium or (b) at its face amount accompanied by a separate valuation account for the discount or premium. Any portion of the debt to be paid during the upcoming year, or operating cycle if longer, should be reported as a current amount. Regarding amounts to be paid in the future, what additional disclosures should be made in connection with long-term debt?

Q 14–14 Early extinguishment of debt often produces a gain or a loss. How is the gain or loss determined?

Q 14–15 Air Supply issued $6 million of 9%, 10-year convertible bonds at 101. The bonds are convertible into 24,000 shares of common stock. Bonds that are similar in all respects except that they are nonconvertible, currently are selling at 99 (that is, 99% of face amount). What amount should Air Supply record as equity and how much as a liability when the bonds are issued?

Q 14–16 Both convertible bonds and bonds issued with detachable warrants have features of both debt and equity. How does the accounting treatment differ for the two hybrid securities? Why is the accounting treatment different?

Q 14–17 At times, companies try to induce voluntary conversion by offering an added incentive—maybe cash, stock warrants, or a more favorable conversion ratio. How is such an inducement accounted for? How is it measured?

Q 14–18 Cordova Tools has bonds outstanding during a year in which the market rate of interest has risen. If Cordova has elected the fair value option for the bonds, will it report a gain or a loss on the bonds for the year? Explain.

IFRS Q 14–19 If a company prepares its financial statements according to International Financial Reporting Standards, how would it account for convertible bonds it issues for $12.5 million? What is the conceptual justification?

Q 14–20 (Based on Appendix 14A) Why will bonds always sell at their price plus any interest that has accrued since the last interest date?

Q 14–21 (Based on Appendix 14B) When the original terms of a debt agreement are changed because of financial difficulties experienced by the debtor (borrower), the new arrangement is referred to as a *troubled debt restructuring*. Such a restructuring can take a variety of forms. For accounting purposes, these possibilities are categorized. What are the accounting classifications of troubled debt restructurings?

Q 14–22 (Based on Appendix 14B) Pratt Industries owes First National Bank $5 million but, due to financial difficulties, is unable to comply with the original terms of the loan. The bank agrees to settle the debt in exchange for land having a fair value of $3 million. The book value of the property on Pratt's books is $2 million. For the reporting period in which the debt is settled, what amount(s) will Pratt report on its income statement in connection with the troubled debt restructuring?

Q 14–23 (Based on Appendix 14B) The way a debtor accounts for the restructuring depends on the extent of the reduction in cash payments called for by the restructured arrangement. Describe, in general, the accounting procedure for the two basic cases: when, under the new agreement, total cash payments (a) are less than the book value of the debt or (b) still exceed the book value of the debt.

Brief Exercises

connect

BE 14–1
Bond interest
● LO14–1

Holiday Brands issued $30 million of 6%, 30-year bonds for $27.5 million. What is the amount of interest that Holiday will pay semiannually to bondholders?

BE 14–2
Determining the price of bonds
● LO14–2

A company issued 5%, 20-year bonds with a face amount of $80 million. The market yield for bonds of similar risk and maturity is 6%. Interest is paid semiannually. At what price did the bonds sell?

BE 14–3
Determining the price of bonds
● LO14–2

A company issued 6%, 15-year bonds with a face amount of $75 million. The market yield for bonds of similar risk and maturity is 6%. Interest is paid semiannually. At what price did the bonds sell?

BE 14–4
Determining the price of bonds
● LO14–2

A company issued 5%, 20-year bonds with a face amount of $100 million. The market yield for bonds of similar risk and maturity is 4%. Interest is paid semiannually. At what price did the bonds sell?

BE 14–5
Effective interest on bonds
● LO14–2

On January 1, a company issued 7%, 15-year bonds with a face amount of $90 million for $82,218,585 to yield 8%. Interest is paid semiannually. What was interest expense at the effective interest rate on June 30, the first interest date?

BE 14–6
Effective interest on bonds
● LO14–2

On January 1, a company issued 3%, 20-year bonds with a face amount of $80 million for $69,033,776 to yield 4%. Interest is paid semiannually. What was the interest expense at the effective interest rate on the December 31 annual income statement?

BE 14–7
Straight-line interest on bonds
● LO14–2

On January 1, a company issued 3%, 20-year bonds with a face amount of $80 million for $69,033,776 to yield 4%. Interest is paid semiannually. What was the straight-line interest expense on the December 31 annual income statement?

BE 14–8
Investment in bonds
● LO14–2

On January 1, a company purchased 3%, 20-year corporate bonds for $69,033,776 as an investment. The bonds have a face amount of $80 million and are priced to yield 4%. Interest is paid semiannually. Prepare the journal entry to record revenue at the effective interest rate on December 31, the second interest payment date.

BE 14–9
Note issued for cash; borrower and lender
● LO14–3

On January 1, 2021, Nantucket Ferry borrowed $14,000,000 cash from BankOne and issued a four-year, $14,000,000, 6% note. Interest was payable annually on December 31. Prepare the journal entries for both firms to record interest at December 31, 2021.

BE 14–10
Note with unrealistic interest rate
● LO14–3

On January 1, Snipes Construction paid for earth-moving equipment by issuing a $300,000, 3-year note that specified 2% interest to be paid on December 31 of each year. The equipment's retail cash price was unknown, but it was determined that a reasonable interest rate was 5%. At what amount should Snipes record the equipment and the note? What journal entry should it record for the transaction?

BE 14–11
Installment note
● LO14–3

On January 1, a company borrowed cash by issuing a $300,000, 5%, installment note to be paid in three equal payments at the end of each year beginning December 31. What would be the amount of each installment? Prepare the journal entry for the second installment payment.

BE 14–12
Early extinguishment; effective interest
● LO14–5

A company retired $60 million of its 6% bonds at 102 ($61.2 million) before their scheduled maturity. At the time, the bonds had a remaining discount of $2 million. Prepare the journal entry to record the redemption of the bonds.

BE 14–13
Bonds with detachable warrants
● LO14–5

Hoffman Corporation issued $60 million of 5%, 20-year bonds at 102. Each of the 60,000 bonds was issued with 10 detachable stock warrants, each of which entitled the bondholder to purchase, for $20, one share of $1 par common stock. At the time of sale, the market value of the common stock was $25 per share and the market value of each warrant was $5. Prepare the journal entry to record the issuance of the bonds.

BE 14–14
Convertible bonds
● LO14–5

Hoffman Corporation issued $60 million of 5%, 20-year bonds at 102. Each of the 60,000 bonds was convertible into one share of $1 par common stock. Prepare the journal entry to record the issuance of the bonds.

BE 14–15
Reporting bonds at fair value
● LO14–6

AI Tool and Dye issued 8% bonds with a face amount of $160 million on January 1, 2021. The bonds sold for $150 million. For bonds of similar risk and maturity the market yield was 9%. Upon issuance, AI elected the option to report these bonds at their fair value. On June 30, 2021, the fair value of the bonds was $145 million as determined by their market value on the NASDAQ. Will AI report a gain or will it report a loss when adjusting the bonds to fair value? If the change in fair value is attributable to a change in the interest rate, did the rate increase or decrease? Will the gain or loss be reported in net income or as OCI?

Exercises

connect

E 14–1
Bond valuation
● LO14–2

Your investment department has researched possible investments in corporate debt securities. Among the available investments are the following $100 million bond issues, each dated January 1, 2021. Prices were determined by underwriters at different times during the last few weeks.

	Company	Bond Price	Stated Rate
1.	BB Corp.	$109 million	11%
2.	DD Corp.	$100 million	10%
3.	GG Corp.	$ 91 million	9%

Each of the bond issues matures on December 31, 2040, and pays interest semiannually on June 30 and December 31. For bonds of similar risk and maturity, the market yield at January 1, 2021, is 10%.

Required:

Other things being equal, which of the bond issues offers the most attractive investment opportunity if it can be purchased at the prices stated? The least attractive? Why?

E 14–2
Determine the price of bonds in various situations
● LO14–2

Determine the price of a $1 million bond issue under each of the following independent assumptions:

	Maturity	Interest Paid	Stated Rate	Effective (Market) Rate
1.	10 years	annually	10%	12%
2.	10 years	semiannually	10%	12%
3.	10 years	semiannually	12%	10%
4.	20 years	semiannually	12%	10%
5.	20 years	semiannually	12%	12%

E 14–3
Determine the price of bonds; issuance; effective interest
● LO14–2

The Bradford Company issued 10% bonds, dated January 1, with a face amount of $80 million on January 1, 2021. The bonds mature on December 31, 2030 (10 years). For bonds of similar risk and maturity, the market yield is 12%. Interest is paid semiannually on June 30 and December 31.

Required:

1. Determine the price of the bonds at January 1, 2021.
2. Prepare the journal entry to record their issuance by The Bradford Company on January 1, 2021.

3. Prepare the journal entry to record interest on June 30, 2021 (at the effective rate).
4. Prepare the journal entry to record interest on December 31, 2021 (at the effective rate).

E 14–4
Investor; effective interest
● LO14–2

(Note: This is a variation of E 14–3 modified to consider the investor's perspective.) The Bradford Company sold the entire bond issue described in the previous exercise to Saxton-Bose Corporation.

Required:

1. Prepare the journal entry to record the purchase of the bonds by Saxton-Bose on January 1, 2021.
2. Prepare the journal entry to record interest revenue on June 30, 2021 (at the effective rate).
3. Prepare the journal entry to record interest revenue on December 31, 2021 (at the effective rate).

E 14–5
Bonds; issuance; effective interest; financial statement effects
● LO14–2

Myriad Solutions, Inc. issued 10% bonds, dated January 1, with a face amount of $320 million on January 1, 2021, for $283,294,720. The bonds mature on December 31, 2030 (10 years). For bonds of similar risk and maturity the market yield is 12%. Interest is paid semiannually on June 30 and December 31.

Required:

1. What would be the net amount of the liability Myriad would report in its balance sheet at December 31, 2021?
2. What would be the amount related to the bonds that Myriad would report in its income statement for the year ended December 31, 2021?
3. What would be the amount(s) related to the bonds that Myriad would report in its statement of cash flows for the year ended December 31, 2021?

E 14–6
Bonds; issuance; effective interest
● LO14–2

The Gorman Group issued $900,000 of 13% bonds on June 30, 2021, for $967,707. The bonds were dated on June 30 and mature on June 30, 2041 (20 years). The market yield for bonds of similar risk and maturity is 12%. Interest is paid semiannually on December 31 and June 30.

Required:

1. Prepare the journal entry to record their issuance by The Gorman Group on June 30, 2021.
2. Prepare the journal entry to record interest on December 31, 2021 (at the effective rate).
3. Prepare the journal entry to record interest on June 30, 2022 (at the effective rate).

E 14–7
Determine the price of bonds; issuance; straight-line method
● LO14–2

Universal Foods issued 10% bonds, dated January 1, with a face amount of $150 million on January 1, 2021. The bonds mature on December 31, 2035 (15 years). The market rate of interest for similar issues was 12%. Interest is paid semiannually on June 30 and December 31. Universal uses the straight-line method.

Required:

1. Determine the price of the bonds at January 1, 2021.
2. Prepare the journal entry to record their issuance by Universal Foods on January 1, 2021.
3. Prepare the journal entry to record interest on June 30, 2021.
4. Prepare the journal entry to record interest on December 31, 2028.

E 14–8
Investor; straight-line method
● LO14–2

(Note: This is a variation of E 14–7 modified to consider the investor's perspective.) Universal Foods sold the entire bond issue described in the previous exercise to Wang Communications.

Required:

1. Prepare the journal entry to record the purchase of the bonds by Wang Communications on January 1, 2021.
2. Prepare the journal entry to record interest revenue on June 30, 2021.
3. Prepare the journal entry to record interest revenue on December 31, 2028.

E 14–9
Issuance of bonds; effective interest; amortization schedule; financial statement effects
● LO14–2

When Patey Pontoons issued 6% bonds on January 1, 2021, with a face amount of $600,000, the market yield for bonds of similar risk and maturity was 7%. The bonds mature December 31, 2024 (4 years). Interest is paid semiannually on June 30 and December 31.

Required:

1. Determine the price of the bonds at January 1, 2021.
2. Prepare the journal entry to record their issuance by Patey on January 1, 2021.
3. Prepare an amortization schedule that determines interest at the effective rate each period.
4. Prepare the journal entry to record interest on June 30, 2021.
5. What is the amount(s) related to the bonds that Patey will report in its balance sheet at December 31, 2021?
6. What is the amount(s) related to the bonds that Patey will report in its income statement for the year ended December 31, 2021? (Ignore income taxes.)
7. Prepare the appropriate journal entries at maturity on December 31, 2024.

E 14–10
Issuance of bonds; effective interest; amortization schedule
● LO14–2

National Orthopedics Co. issued 9% bonds, dated January 1, with a face amount of $500,000 on January 1, 2021. The bonds mature on December 31, 2024 (4 years). For bonds of similar risk and maturity the market yield was 10%. Interest is paid semiannually on June 30 and December 31.

Required:
1. Determine the price of the bonds at January 1, 2021.
2. Prepare the journal entry to record their issuance by National on January 1, 2021.
3. Prepare an amortization schedule that determines interest at the effective rate each period.
4. Prepare the journal entry to record interest on June 30, 2021.
5. Prepare the appropriate journal entries at maturity on December 31, 2024.

E 14–11
Bonds; effective interest; adjusting entry
● LO14–2

On February 1, 2021, Strauss-Lombardi issued 9% bonds, dated February 1, with a face amount of $800,000. The bonds sold for $731,364 and mature on January 31, 2041 (20 years). The market yield for bonds of similar risk and maturity was 10%. Interest is paid semiannually on July 31 and January 31. Strauss-Lombardi's fiscal year ends December 31.

Required:
1. Prepare the journal entry to record their issuance by Strauss-Lombardi on February 1, 2021.
2. Prepare the journal entry to record interest on July 31, 2021 (at the effective rate).
3. Prepare the adjusting entry to accrue interest on December 31, 2021.
4. Prepare the journal entry to record interest on January 31, 2022.

E 14–12
Bonds; straight-line method; adjusting entry
● LO14–2

On March 1, 2021, Stratford Lighting issued 14% bonds, dated March 1, with a face amount of $300,000. The bonds sold for $294,000 and mature on February 28, 2041 (20 years). Interest is paid semiannually on August 31 and February 28. Stratford uses the straight-line method and its fiscal year ends December 31.

Required:
1. Prepare the journal entry to record the issuance of the bonds by Stratford Lighting on March 1, 2021.
2. Prepare the journal entry to record interest on August 31, 2021.
3. Prepare the journal entry to accrue interest on December 31, 2021.
4. Prepare the journal entry to record interest on February 28, 2022.

E 14–13
Issuance of bonds; effective interest
● LO14–2

Federal Semiconductors issued 11% bonds, dated January 1, with a face amount of $800 million on January 1, 2021. The bonds sold for $739,814,813 and mature on December 31, 2040 (20 years). For bonds of similar risk and maturity the market yield was 12%. Interest is paid semiannually on June 30 and December 31.

Required:
1. Prepare the journal entry to record their issuance by Federal on January 1, 2021.
2. Prepare the journal entry to record interest on June 30, 2021 (at the effective rate).
3. Prepare the journal entry to record interest on December 31, 2021 (at the effective rate).
4. At what amount will Federal report the bonds among its liabilities in the December 31, 2021, balance sheet?

E 14–14
New debt issues; offerings announcements
● LO14–2

When companies offer new debt security issues, they publicize the offerings in the financial press and on Internet sites. Assume the following were among the debt offerings reported in December 2021:

New Securities Issues

Corporate

National Equipment Transfer Corporation—$200 million bonds via lead managers Second Tennessee Bank N.A. and Morgan, Dunavant & Co., according to a syndicate official. Terms: maturity, Dec. 15, 2030; coupon 7.46%; issue price, par; yield, 7.46%; noncallable; debt ratings: Ba-1 (Moody's Investors Service, Inc.), BBB+ (Standard & Poor's).

IgWig Inc.—$350 million of notes via lead manager Stanley Brothers, Inc., according to a syndicate official. Terms: maturity, Dec. 1, 2032; coupon, 6.46%; issue price, 99; yield, 6.56%; call date, NC; debt ratings: Baa-1 (Moody's Investors Service, Inc.), A (Standard & Poor's).

Required:
1. Prepare the appropriate journal entries to record the sale of both issues to underwriters. Ignore share issue costs and assume no accrued interest.
2. Prepare the appropriate journal entries to record the first semiannual interest payment for both issues.

E 14–15
Error correction; accrued interest on bonds
● LO14–2

At the end of 2020, Majors Furniture Company failed to accrue $61,000 of interest expense that accrued during the last five months of 2020 on bonds payable. The bonds mature in 2034. The discount on the bonds is amortized by the straight-line method. The following entry was recorded on February 1, 2021, when the semiannual interest was paid:

Interest expense	73,200	
Discount on bonds payable		1,200
Cash		72,000

Required:
Prepare any journal entry necessary to correct the errors as of February 2, 2021, when the errors were discovered. Also, prepare any adjusting entry at December 31, 2021, related to the situation described. (Ignore income taxes.)

E 14–16
Error in amortization schedule
● LO14–3

Wilkins Food Products, Inc., acquired a packaging machine from Lawrence Specialists Corporation. Lawrence completed construction of the machine on January 1, 2019. In payment for the machine Wilkins issued a three-year installment note to be paid in three equal payments at the end of each year. The payments include interest at the rate of 10%.

Lawrence made a conceptual error in preparing the amortization schedule, which Wilkins failed to discover until 2021. The error had caused Wilkins to understate interest expense by $45,000 in 2019 and $40,000 in 2020.

Required:
1. Determine which accounts are incorrect as a result of these errors at January 1, 2021, before any adjustments. Explain your answer. (Ignore income taxes.)
2. Prepare a journal entry to correct the error.
3. What other step(s) would be taken in connection with the error?

E 14–17
Note with unrealistic interest rate; borrower; amortization schedule
● LO14–3

Amber Mining and Milling, Inc., contracted with Truax Corporation to have constructed a custom-made lathe. The machine was completed and ready for use on January 1, 2021. Amber paid for the lathe by issuing a $600,000, three-year note that specified 4% interest, payable annually on December 31 of each year. The cash market price of the lathe was unknown. It was determined by comparison with similar transactions that 12% was a reasonable rate of interest.

Required:
1. Prepare the journal entry on January 1, 2021, for Amber Mining and Milling's purchase of the lathe.
2. Prepare an amortization schedule for the three-year term of the note.
3. Prepare the journal entries to record (a) interest for each of the three years and (b) payment of the note at maturity.

E 14–18
Note with unrealistic interest rate; lender; amortization schedule
● LO14–3

Refer to the situation described in E 14–17.

Required:
1. Prepare the journal entry on January 1, 2021, for Truax Corporation's sale of the lathe. Assume Truax spent $400,000 to construct the lathe.
2. Prepare an amortization schedule for the three-year term of the note.
3. Prepare the journal entries to record (a) interest for each of the three years and (b) payment of the note at maturity for Truax.

E 14–19
Installment note; lender; amortization schedule
● LO14–3

FinanceCo lent $8 million to Corbin Construction on January 1, 2021, to construct a playground. Corbin signed a three-year, 6% installment note to be paid in three equal payments at the end of each year.

Required:
1. Prepare the journal entry for FinanceCo's lending the funds on January 1, 2021.
2. Prepare an amortization schedule for the three-year term of the installment note.
3. Prepare the journal entry for the first installment payment on December 31, 2021.
4. Prepare the journal entry for the third installment payment on December 31, 2023.

E 14–20
Installment note; amortization schedule
● LO14–3

American Food Services, Inc., acquired a packaging machine from Barton and Barton Corporation. Barton and Barton completed construction of the machine on January 1, 2021. In payment for the $4 million machine, American Food Services issued a four-year installment note to be paid in four equal payments at the end of each year. The payments include interest at the rate of 10%.

Required:
1. Prepare the journal entry for American Food Services' purchase of the machine on January 1, 2021.
2. Prepare an amortization schedule for the four-year term of the installment note.
3. Prepare the journal entry for the first installment payment on December 31, 2021.
4. Prepare the journal entry for the third installment payment on December 31, 2023.

E 14–21
Installment note
● LO14–3

LCD Industries purchased a supply of electronic components from Entel Corporation on November 1, 2021. In payment for the $24 million purchase, LCD issued a 1-year installment note to be paid in equal monthly payments at the end of each month. The payments include interest at the rate of 12%.

Required:

1. Prepare the journal entry for LCD's purchase of the components on November 1, 2021.
2. Prepare the journal entry for the first installment payment on November 30, 2021.
3. What is the amount of interest expense that LCD will report in its income statement for the year ended December 31, 2021?

E 14–22
FASB codification research
● LO14–2, LO14–3, LO14–4

CODE

Access the *FASB Accounting Standards Codification* at the FASB website (**www.fasb.org**). Determine the specific eight- or nine-digit Codification citation (XXX-XX-XX-XX) for accounting for each of the following items:

1. Disclosure requirements for maturities of long-term debt.
2. How to estimate the value of a note when a note having no ready market and no interest rate is exchanged for a noncash asset without a readily available fair value.
3. When the straight-line method can be used as an alternative to the interest method of determining interest.

E 14–23
Early extinguishment
● LO14–5

The balance sheet of Indian River Electronics Corporation as of December 31, 2020, included 12.25% bonds having a face amount of $90 million. The bonds had been issued in 2013 and had a remaining discount of $3 million at December 31, 2020. On January 1, 2021, Indian River Electronics called the bonds before their scheduled maturity at the call price of 102.

Required:

Prepare the journal entry by Indian River Electronics to record the redemption of the bonds at January 1, 2021.

E 14–24
Convertible bonds; straight-line interest
● LO14–5

On January 1, 2021, Gless Textiles issued $12 million of 9%, 10-year convertible bonds at 101. The bonds pay interest on June 30 and December 31. Each $1,000 bond is convertible into 40 shares of Gless's no par common stock. Bonds that are similar in all respects, except that they are nonconvertible, currently are selling at 99 (that is, 99% of face amount). Century Services purchased 10% of the issue as an investment.

Required:

1. Prepare the journal entries for the issuance of the bonds by Gless and the purchase of the bond investment by Century.
2. Prepare the journal entries for the June 30, 2025, interest payment by both Gless and Century assuming both use the straight-line method.
3. On July 1 2026, when Gless's common stock had a market price of $33 per share, Century converted the bonds it held. Prepare the journal entries by both Gless and Century for the conversion of the bonds (book value method).

E 14–25
IFRS; convertible bonds
● LO14–5, LO14–7

Refer to the situation described in E 14–24.

Required:

How might your solution to requirement 1 for the issuer of the bonds differ if Gless Textiles prepares its financial statements according to International Financial Reporting Standards? Include any appropriate journal entry in your response.

E 14–26
Convertible bonds; induced conversion
● LO14–5

On January 1, 2021, Madison Products issued $40 million of 6%, 10-year convertible bonds at a net price of $40.8 million. Madison recently issued similar, but nonconvertible, bonds at 99 (that is, 99% of face amount). The bonds pay interest on June 30 and December 31. Each $1,000 bond is convertible into 30 shares of Madison's no par common stock. Madison records interest by the straight-line method.

On June 1, 2023, Madison notified bondholders of its intent to call the bonds at face value plus a 1% call premium on July 1, 2023. By June 30, all bondholders had chosen to convert their bonds into shares as of the interest payment date. On June 30, Madison paid the semiannual interest and issued the requisite number of shares for the bonds being converted.

Required:

1. Prepare the journal entry for the issuance of the bonds by Madison.
2. Prepare the journal entry for the June 30, 2021, interest payment.
3. Prepare the journal entries for the June 30, 2023, interest payment by Madison and the conversion of the bonds (book value method).

E 14–27
IFRS; convertible bonds
● LO14–5, LO14–7
IFRS

Refer to the situation described in E 14–26.

Required:

How might your solution for the issuer of the bonds differ if Madison Products prepares its financial statements according to International Financial Reporting Standards? Include any appropriate journal entries in your response.

E 14–28
Bonds with detachable warrants
● LO14–5

On August 1, 2021, Limbaugh Communications issued $30 million of 10% nonconvertible bonds at 104. The bonds are due on July 31, 2041. Each $1,000 bond was issued with 20 detachable stock warrants, each of which entitled the bondholder to purchase, for $60, one share of Limbaugh Communications' no par common stock. Interstate Containers purchased 20% of the bond issue. On August 1, 2021, the market value of the common stock was $58 per share and the market value of each warrant was $8.

In February 2032, when Limbaugh's common stock had a market price of $72 per share and the unamortized discount balance was $1 million, Interstate Containers exercised the warrants it held.

Required:

1. Prepare the journal entries on August 1, 2021, to record (a) the issuance of the bonds by Limbaugh and (b) the investment by Interstate.
2. Prepare the journal entries for both Limbaugh and Interstate in February 2032, to record the exercise of the warrants.

E 14–29
Reporting bonds at fair value
● LO14–6

(Note: This is a variation of E 14–13 modified to consider the fair value option for reporting liabilities.) Federal Semiconductors issued 11% bonds, dated January 1, with a face amount of $800 million on January 1, 2021. The bonds sold for $739,814,813 and mature on December 31, 2040 (20 years). For bonds of similar risk and maturity the market yield was 12%. Interest is paid semiannually on June 30 and December 31. Federal determines interest at the effective rate. Federal elected the option to report these bonds at their fair value. On December 31, 2021, the fair value of the bonds was $730 million as determined by their market value in the over-the-counter market.

Required:

1. Prepare the journal entry to adjust the bonds to their fair value for presentation in the December 31, 2021, balance sheet. Federal determined that none of the change in fair value was due to a decline in general interest rates.
2. Assume the fair value of the bonds on December 31, 2022, had risen to $736 million. Prepare the journal entry to adjust the bonds to their fair value for presentation in the December 31, 2022, balance sheet. Federal determined that one-half of the increase in fair value was due to a decline in general interest rates.

E 14–30
Reporting bonds at fair value
● LO14–6

On January 1, 2021, Rapid Airlines issued $200 million of its 8% bonds for $184 million. The bonds were priced to yield 10%. Interest is payable semiannually on June 30 and December 31. Rapid Airlines records interest at the effective rate and elected the option to report these bonds at their fair value. On December 31, 2021, the fair value of the bonds was $188 million as determined by their market value in the over-the-counter market. Rapid determined that $1,000,000 of the increase in fair value was due to a decline in general interest rates.

Required:

1. Prepare the journal entry to record interest on June 30, 2021 (the first interest payment).
2. Prepare the journal entry to record interest on December 31, 2021 (the second interest payment).
3. Prepare the journal entry to adjust the bonds to their fair value for presentation in the December 31, 2021, balance sheet.

E 14–31
Reporting bonds at fair value; calculate fair value
● LO14–6

On January 1, 2021, Essence Communications issued $800,000 of its 10-year, 8% bonds for $700,302. The bonds were priced to yield 10%. Interest is payable semiannually on June 30 and December 31. Essence Communications records interest at the effective rate and elected the option to report these bonds at their fair value. On December 31, 2021, the market interest rate for bonds of similar risk and maturity was 9%. The bonds are not traded on an active exchange. The decrease in the market interest rate was due to a 1% decrease in general (risk-free) interest rates.

Required:

1. Using the information provided, estimate the fair value of the bonds at December 31, 2021.
2. Prepare the journal entry to record interest on June 30, 2021 (the first interest payment).
3. Prepare the journal entry to record interest on December 31, 2021 (the second interest payment).
4. Prepare the journal entry to adjust the bonds to their fair value for presentation in the December 31, 2021, balance sheet.

E 14–32
Accrued interest
● Appendix A

On March 1, 2021, Brown-Ferring Corporation issued $100 million of 12% bonds, dated January 1, 2021, for $99 million (plus accrued interest). The bonds mature on December 31, 2040, and pay interest semiannually on June 30 and December 31. Brown-Ferring's fiscal period is the calendar year.

Required:

1. Determine the amount of accrued interest that was included in the proceeds received from the bond sale.
2. Prepare the journal entry for the issuance of the bonds by Brown-Ferring.

E 14–33
Troubled debt restructuring; debt settled
● Appendix B

At January 1, 2021, Transit Developments owed First City Bank Group $600,000, under an 11% note with three years remaining to maturity. Due to financial difficulties, Transit was unable to pay the previous year's interest.

First City Bank Group agreed to settle Transit's debt in exchange for land having a fair value of $450,000. Transit purchased the land in 2017 for $325,000.

Required:

Prepare the journal entry(s) to record the restructuring of the debt by Transit Developments.

E 14–34
Troubled debt restructuring; modification of terms
● Appendix B

At January 1, 2021, Brainard Industries, Inc., owed Second BancCorp $12 million under a 10% note due December 31, 2023. Interest was paid last on December 31, 2019. Brainard was experiencing severe financial difficulties and asked Second BancCorp to modify the terms of the debt agreement. After negotiation Second BancCorp agreed to:

a. Forgive the interest accrued for the year just ended.
b. Reduce the remaining two years' interest payments to $1 million each and delay the first payment until December 31, 2022.
c. Reduce the unpaid principal amount to $11 million.

Required:

Prepare the journal entries by Brainard Industries, Inc., necessitated by the restructuring of the debt at (1) January 1, 2021; (2) December 31, 2022; and (3) December 31, 2023.

E 14–35
Troubled debt restructuring; modification of terms; unknown effective rate
● Appendix B

At January 1, 2021, NCI Industries, Inc. was indebted to First Federal Bank under a $240,000, 10% unsecured note. The note was signed January 1, 2014, and was due December 31, 2022. Annual interest was last paid on December 31, 2019. NCI was experiencing severe financial difficulties and negotiated a restructuring of the terms of the debt agreement. First Federal agreed to reduce last year's interest and the remaining two years' interest payments to $11,555 each and delay all payments until December 31, 2022, the maturity date.

Required:

Prepare the journal entries by NCI Industries, Inc., necessitated by the restructuring of the debt at: (1) January 1, 2021; (2) December 31, 2021; and (3) December 31, 2022.

E 14–36
FASB codification research: legal fees in a troubled debt restructuring
● Appendix B

CODE

In negotiating and effecting a troubled debt restructuring, the creditor usually incurs various legal costs. The *FASB Accounting Standards Codification* represents the single source of authoritative U.S. generally accepted accounting principles.

Required:

1. Obtain the relevant authoritative literature on the accounting treatment of legal fees incurred by a creditor to effect a troubled debt restructuring using the *FASB Accounting Standards Codification* at the FASB website (**www.fasb.org**).
2. What is the specific eight-digit Codification citation (XXX-XX-XX-X) that describes the guidelines for reporting legal costs?
3. What is the appropriate accounting treatment?

E 14–37
General ledger exercise; bonds; installment note, early extinguishment
● LO14–2, LO14–3, LO14–5

GL

On January 1, 2021, the general ledger of Freedom Fireworks includes the following account balances:

Accounts	Debit	Credit
Cash	$101,200	
Accounts receivable	34,000	
Inventory	152,000	
Land	67,300	
Buildings	120,000	
Allowance for uncollectible accounts		$ 1,800
Accumulated depreciation		9,600
Accounts payable		17,700
Bonds payable		120,000
Discount on bonds payable	30,000	
Common stock		200,000
Retained earnings		155,400
Totals	$504,500	$504,500

During January 2021, the following transactions occurred:

January 1	Borrowed $100,000 from Captive Credit Corporation. The installment note bears interest at 7% annually and matures in 5 years. Payments of $1,980 are required at the end of each month for 60 months.
January 1	Called the bonds at the contractual call price of $100,000. The 6% bonds pay interest semiannually each June 30 and December 31.
January 4	Received $31,000 from customers on accounts receivable.
January 10	Paid cash on accounts payable, $11,000.
January 15	Paid cash for salaries, $28,900.
January 30	Firework sales for the month totalled $195,000. Sales included $65,000 for cash and $130,000 on account. The cost of the units sold was $112,500.
January 31	Paid the first monthly installment of $1,980 related to the $100,000 borrowed on January 1. (Round your interest calculation to the nearest dollar.)

The following information is available on January 31, 2021.

1. Depreciation on the building for the month of January is calculated using the straight-line method. At the time the building was purchased, the company estimated a service life of ten years and a residual value of $24,000.
2. At the end of January, $3,000 of accounts receivable are past due, and the company estimates that 50% of these accounts will not be collected. Of the remaining accounts receivable, the company estimates that 2% will not be collected. No accounts were written off as uncollectible in January.
3. Unpaid salaries at the end of January are $26,100.
4. Accrued income taxes at the end of January are $5,000.

Required:

1. Record each of the transactions listed above in the "General Journal" tab (these are shown as items 1–7) assuming a FIFO perpetual inventory system. The transaction on January 30 requires two entries: one to record sales revenue and one to record cost of goods sold. Review the "General Ledger" and the "Trial Balance" tabs to see the effect of the transactions on the account balances.
2. Record adjusting entries on January 31. in the "General Journal" tab (these are shown as items 8–11).
3. Review the adjusted "Trial Balance" as of January 31, 2021, in the "Trial Balance" tab.
4. Prepare an income statement for the period ended January 31, 2021, in the "Income Statement" tab).
5. Prepare a classified balance sheet as of January 31, 2021, in the "Balance Sheet" tab.
6. Record closing entries in the "General Journal" tab (these are shown as items 12 and 13).
7. Using the information from the requirements above, complete the "Analysis" tab.
   a. Calculate the debt to equity ratio. If the average debt to equity ratio for the industry is 1.0, is Freedom Fireworks more or less leveraged than other companies in the same industry?
   b. Calculate the times interest earned ratio. If the average times interest earned ratio for the industry is 20 times, is the company more or less able to meet interest payments than other companies in the same industry?

Problems

connect

P 14–1
Determining the price of bonds; discount and premium; issuer and investor
● LO14–2

On January 1, 2021, Instaform, Inc., issued 10% bonds with a face amount of $50 million, dated January 1. The bonds mature in 2040 (20 years). The market yield for bonds of similar risk and maturity is 12%. Interest is paid semiannually.

Required:

1. Determine the price of the bonds at January 1, 2021, and prepare the journal entry to record their issuance by Instaform.
2. Assume the market rate was 9%. Determine the price of the bonds at January 1, 2021, and prepare the journal entry to record their issuance by Instaform.
3. Assume Broadcourt Electronics purchased the entire issue in a private placement of the bonds. Using the data in requirement 2, prepare the journal entry to record the purchase by Broadcourt.

P 14–2
Effective interest; financial statement effects
● LO14–2

On January 1, 2021, Baddour, Inc., issued 10% bonds with a face amount of $160 million. The bonds were priced at $140 million to yield 12%. Interest is paid semiannually on June 30 and December 31. Baddour's fiscal year ends September 30.

Required:

1. What amount(s) related to the bonds would Baddour report in its balance sheet at September 30, 2021?
2. What amount(s) related to the bonds would Baddour report in its income statement for the year ended September 30, 2021?
3. What amount(s) related to the bonds would Baddour report in its statement of cash flows for the year ended September 30, 2021? In which section(s) should the amount(s) appear?

P 14–3
Straight-line and effective interest compared
● LO14–2

On January 1, 2021, Bradley Recreational Products issued $100,000, 9%, four-year bonds. Interest is paid semiannually on June 30 and December 31. The bonds were issued at $96,768 to yield an annual return of 10%.

Required:

1. Prepare an amortization schedule that determines interest at the effective interest rate.
2. Prepare an amortization schedule by the straight-line method.
3. Prepare the journal entries to record interest expense on June 30, 2023, by each of the two approaches.
4. Explain why the pattern of interest differs between the two methods.
5. Assuming the market rate is still 10%, what price would a second investor pay the first investor on June 30, 2023, for $10,000 of the bonds?

P 14–4
Bond amortization schedule
● LO14–2

On January 1, 2021, Tennessee Harvester Corporation issued debenture bonds that pay interest semiannually on June 30 and December 31. Portions of the bond amortization schedule appear below:

Payment	Cash Payment	Effective Interest	Increase in Balance	Outstanding Balance
				6,627,273
1	320,000	331,364	11,364	6,638,637
2	320,000	331,932	11,932	6,650,569
3	320,000	332,528	12,528	6,663,097
4	320,000	333,155	13,155	6,676,252
5	320,000	333,813	13,813	6,690,065
6	320,000	334,503	14,503	6,704,568
-	-	-	-	-
-	-	-	-	-
-	-	-	-	-
38	320,000	389,107	69,107	7,851,247
39	320,000	392,562	72,562	7,923,809
40	320,000	396,191	76,191	8,000,000

Required:

1. What is the face amount of the bonds?
2. What is the initial selling price of the bonds?
3. What is the term to maturity in years?
4. Interest is determined by what approach?
5. What is the stated annual interest rate?
6. What is the effective annual interest rate?
7. What is the total cash interest paid over the term to maturity?
8. What is the total effective interest expense recorded over the term to maturity?

P 14–5
Issuer and investor; effective interest; amortization schedule; adjusting entries
● LO14–2

On February 1, 2021, Cromley Motor Products issued 9% bonds, dated February 1, with a face amount of $80 million. The bonds mature on January 31, 2025 (4 years). The market yield for bonds of similar risk and maturity was 10%. Interest is paid semiannually on July 31 and January 31. Barnwell Industries acquired $80,000 of the bonds as a long-term investment. The fiscal years of both firms end December 31.

Required:

1. Determine the price of the bonds issued on February 1, 2021.
2. Prepare amortization schedules that indicate (a) Cromley's effective interest expense and (b) Barnwell's effective interest revenue for each interest period during the term to maturity.
3. Prepare the journal entries to record (a) the issuance of the bonds by Cromley and (b) Barnwell's investment on February 1, 2021.
4. Prepare the journal entries by both firms to record all subsequent events related to the bonds through January 31, 2023.

P 14–6
Issuer and investor; straight-line method; adjusting entries
● LO14–2, Appendix A

On April 1, 2021, Western Communications, Inc., issued 12% bonds, dated March 1, 2021, with face amount of $30 million. The bonds sold for $29.3 million and mature on February 28, 2024. Interest is paid semiannually on August 31 and February 28. Stillworth Corporation acquired $30,000 of the bonds as a long-term investment. The fiscal years of both firms end December 31, and both firms use the straight-line method.

Required:

1. Prepare the journal entries to record (a) issuance of the bonds by Western and (b) Stillworth's investment on April 1, 2021.
2. Prepare the journal entries by both firms to record all subsequent events related to the bonds through maturity.

P 14–7
Issuer and investor; effective interest
● LO14–2

McWherter Instruments sold $400 million of 8% bonds, dated January 1, on January 1, 2021. The bonds mature on December 31, 2040 (20 years). For bonds of similar risk and maturity, the market yield was 10%. Interest is paid semiannually on June 30 and December 31. Blanton Technologies, Inc., purchased $400,000 of the bonds as a long-term investment.

Required:

1. Determine the price of the bonds issued on January 1, 2021.
2. Prepare the journal entries to record (a) their issuance by McWherter and (b) Blanton's investment on January 1, 2021.
3. Prepare the journal entries by (a) McWherter and (b) Blanton to record interest on June 30, 2021 (at the effective rate).
4. Prepare the journal entries by (a) McWherter and (b) Blanton to record interest on December 31, 2021 (at the effective rate).

P 14–8
Bonds; effective interest; partial period interest; financial statement effects
● LO14–2

The fiscal year ends December 31 for Lake Hamilton Development. To provide funding for its Moonlight Bay project, LHD issued 5% bonds with a face amount of $500,000 on November 1, 2021. The bonds sold for $442,215, a price to yield the market rate of 6%. The bonds mature October 31, 2041 (20 years). Interest is paid semiannually on April 30 and October 31.

Required:

1. What amount of interest expense related to the bonds will LHD report in its income statement for the year ending December 31, 2021?
2. What amount(s) related to the bonds will LHD report in its balance sheet at December 31, 2021?
3. What amount of interest expense related to the bonds will LHD report in its income statement for the year ending December 31, 2022?
4. What amount(s) related to the bonds will LHD report in its balance sheet at December 31, 2022?

P 14–9
Zero-coupon bonds
● LO14–2

On January 1, 2021, Darnell Window and Pane issued $18 million of 10-year, zero-coupon bonds for $5,795,518.

Required:

1. Prepare the journal entry to record the bond issue.
2. Determine the effective rate of interest.
3. Prepare the journal entry to record annual interest expense at December 31, 2021.
4. Prepare the journal entry to record annual interest expense at December 31, 2022.
5. Prepare the journal entry to record the extinguishment at maturity.

P 14–10
Notes exchanged for assets; unknown effective rate
● LO14–3

At the beginning of the year, Lambert Motors issued the three notes described below. Interest is paid at year-end.

1. The company issued a two-year, 12%, $600,000 note in exchange for a tract of land. The current market rate of interest is 12%.
2. Lambert acquired some office equipment with a fair value of $94,643 by issuing a one-year, $100,000 note. The stated interest on the note is 6%. The current market rate of interest is 12%.
3. The company purchased a building by issuing a three-year installment note. The note is to be repaid in equal installments of $1 million per year beginning one year hence. The current market rate of interest is 12%.

Required:

Prepare the journal entries to record each of the three transactions and the interest expense at the end of the first year for each.

P 14–11
Note with unrealistic interest rate
● LO14–3

At January 1, 2021, Brant Cargo acquired equipment by issuing a five-year, $150,000 (payable at maturity), 4% note. The market rate of interest for notes of similar risk is 10%.

Required:

1. Prepare the journal entry for Brant Cargo to record the purchase of the equipment.
2. Prepare the journal entry for Brant Cargo to record the interest at December 31, 2021.
3. Prepare the journal entry for Brant Cargo to record the interest at December 31, 2022.

P 14–12
Noninterest-bearing installment note
● LO14–3

At the beginning of 2021, VHF Industries acquired a machine with a fair value of $6,074,700 by issuing a four-year, noninterest-bearing note in the face amount of $8 million. The note is payable in four annual installments of $2 million at the end of each year.

Required:

1. What is the effective rate of interest implicit in the agreement?
2. Prepare the journal entry to record the purchase of the machine.
3. Prepare the journal entry to record the first installment payment at December 31, 2021.

4. Prepare the journal entry to record the second installment payment at December 31, 2022.
5. Suppose the market value of the machine was unknown at the time of purchase, but the market rate of interest for notes of similar risk was 11%. Prepare the journal entry to record the purchase of the machine.

P 14–13
Note and installment note with unrealistic interest rate
● LO14–3

Braxton Technologies, Inc., constructed a conveyor for A&G Warehousers that was completed and ready for use on January 1, 2021. A&G paid for the conveyor by issuing a $100,000, four-year note that specified 5% interest to be paid on December 31 of each year, and the note is to be repaid at the end of four years. The conveyor was custom-built for A&G, so its cash price was unknown. By comparison with similar transactions it was determined that a reasonable interest rate was 10%.

Required:

1. Prepare the journal entry for A&G's purchase of the conveyor on January 1, 2021.
2. Prepare an amortization schedule for the four-year term of the note.
3. Prepare the journal entry for A&G's third interest payment on December 31, 2023.
4. If A&G's note had been an installment note to be paid in four equal payments at the end of each year beginning December 31, 2021, what would be the amount of each installment?
5. Prepare an amortization schedule for the four-year term of the installment note.
6. Prepare the journal entry for A&G's third installment payment on December 31, 2023.

P 14–14
Early extinguishment of debt
● LO14–5

Three years ago American Insulation Corporation issued 10%, $800,000, 10-year bonds for $770,000. American Insulation exercised its call privilege and retired the bonds for $790,000. The corporation uses the straight-line method to determine interest.

Required:

Prepare the journal entry to record the call of the bonds.

P 14–15
Early extinguishment; effective interest
● LO14–5

The long-term liability section of Twin Digital Corporation's balance sheet as of December 31, 2020, included 12% bonds having a face amount of $20 million and a remaining discount of $1 million. Disclosure notes indicate the bonds were issued to yield 14%.

Interest expense is recorded at the effective interest rate and paid on January 1 and July 1 of each year. On July 1, 2021, Twin Digital retired the bonds at 102 ($20.4 million) before their scheduled maturity.

Required:

1. Prepare the journal entry by Twin Digital to record the semiannual interest on July 1, 2021.
2. Prepare the journal entry by Twin Digital to record the redemption of the bonds on July 1, 2021.

P 14–16
Debt issue costs; issuance; expensing; early extinguishment; straight-line amortization
● LO14–2, LO14–5

Cupola Fan Corporation issued 10%, $400,000, 10-year bonds for $385,000 on June 30, 2021. Debt issue costs were $1,500. Interest is paid semiannually on December 31 and June 30. One year from the issue date (July 1, 2022), the corporation exercised its call privilege and retired the bonds for $395,000. The corporation uses the straight-line method both to determine interest expense and to amortize debt issue costs.

Required:

1. Prepare the journal entry to record the issuance of the bonds.
2. Prepare the journal entries to record the payment of interest and amortization of debt issue costs on December 31, 2021.
3. Prepare the journal entries to record the payment of interest and amortization of debt issue costs on June 30, 2022.
4. Prepare the journal entry to record the call of the bonds.

P 14–17
IFRS; transaction costs
● LO14–2, LO14–7
IFRS

Refer to the situation described in P 14–16.

Required:

How might your solution for the issuer of the bonds differ if Cupola prepares its financial statements according to International Financial Reporting Standards? Include any appropriate journal entries in your response.

P 14–18
Early extinguishment
● LO14–5

The long-term liability section of Eastern Post Corporation's balance sheet as of December 31, 2020, included 10% bonds having a face amount of $40 million and a remaining premium of $6 million. On January 1, 2021, Eastern Post retired some of the bonds before their scheduled maturity.

Required:

Prepare the journal entry by Eastern Post to record the redemption of the bonds under each of the independent circumstances below:

1. Eastern Post called half the bonds at the call price of 102 (102% of face amount).
2. Eastern Post repurchased $10 million of the bonds on the open market at their market price of $10.5 million.

P 14–19
Convertible bonds; induced conversion; bonds with detachable warrants
● LO14–5

Bradley-Link's December 31, 2021, balance sheet included the following items:

	($ in millions)
Long-Term Liabilities	
9.6% convertible bonds, callable at 101 beginning in 2022, due 2025 (net of unamortized discount of $2) [note 8]	$198
10.4% registered bonds callable at 104 beginning in 2031, due 2035 (net of unamortized discount of $1) [note 8]	49
Shareholders' Equity	
Equity—stock warrants	4

Note 8: Bonds (in part)
The 9.6% bonds were issued in 2008 at 97.5 to yield 10%. Interest is paid semiannually on June 30 and December 31. Each $1,000 bond is convertible into 40 shares of the Company's no par common stock.

The 10.4% bonds were issued in 2012 at 102 to yield 10%. Interest is paid semiannually on June 30 and December 31. Each $1,000 bond was issued with 40 detachable stock warrants, each of which entitles the holder to purchase one share of the Company's no par common stock for $25, beginning 2022.

On January 3, 2022, when Bradley-Link's common stock had a market price of $32 per share, Bradley-Link called the convertible bonds to force conversion. 90% were converted; the remainder were acquired at the call price. When the common stock price reached an all-time high of $37 in December of 2022, 40% of the warrants were exercised.

Required:
1. Prepare the journal entries that were recorded when each of the two bond issues was originally sold in 2005 and 2012.
2. Prepare the journal entry to record (book value method) the conversion of 90% of the convertible bonds in January 2022 and the retirement of the remainder.
3. Assume Bradley-Link induced conversion by offering $150 cash for each bond converted. Prepare the journal entry to record (book value method) the conversion of 90% of the convertible bonds in January 2022.
4. Assume Bradley-Link induced conversion by modifying the conversion ratio to exchange 45 shares for each bond rather than the 40 shares provided in the contract. Prepare the journal entry to record (book value method) the conversion of 90% of the convertible bonds in January 2022.
5. Prepare the journal entry to record the exercise of the warrants in December 2022.

P 14–20
Convertible bonds; zero coupon; potentially convertible into cash; FASB codification research
● LO14–5

The 2020 annual report of Mills General Corporation (MGC) included the following disclosure note:

Note 10: Borrowings (in part) Convertible Debt
On July 1, 2020, we issued $125 million of zero coupon convertible unsecured debt due on July 1, 2022 in a private placement offering, priced to yield 1.85%. Proceeds from the offering were $118.3115 million. Initially, each $1,000 principal amount of bonds was convertible into 30 shares of MGC common stock at a conversion price of $35 per share.

The bonds are convertible at any time. Upon conversion, we will pay cash up to the aggregate principal amount of the bonds and pay or deliver cash, shares of our common stock, or a combination of cash and shares of our common stock, at our election.

Because the convertible debt may be wholly or partially settled in cash, we are required to separately account for the liability and equity components of the bonds in a manner that reflects our nonconvertible debt borrowing rate when interest costs are recognized in subsequent periods. The net proceeds of $118.3 million were allocated between debt for $117.2 million and shareholders' equity for $1.1 million with the portion in shareholders' equity representing the fair value of the option to convert the debt.

Required:
1. Prepare the journal entry that was recorded when the bonds were issued on July 1, 2020.
2. What amount of interest expense, if any, did MGC record the fiscal year ended June 30, 2021?
3. Normally under U.S. GAAP, we record the entire issue price of convertible debt as a liability. However, MGC separately recorded the liability and equity components of the notes. Why?

4. Obtain the relevant authoritative literature on classification of debt expected to be financed using the FASB's Codification Research System. You might gain access from the FASB website (www.fasb.org), from your school library, or some other source. Determine the criteria for reporting debt potentially convertible into cash. What is the specific seven-digit Codification citation (XXX-XX-XX) that MGC would rely on in applying that accounting treatment?

P 14–21
Concepts; terminology
● LO14–1 through LO14–5

Listed below are several terms and phrases associated with long-term debt. Pair each item from List A with the item from List B (by letter) that is most appropriately associated with it.

	List A	List B
______	1. Effective rate times balance	a. Straight-line method
______	2. Promises made to bondholders	b. Discount
______	3. Present value of interest plus present value of principal	c. Liquidation payments after other claims satisfied
______	4. Call feature	d. Name of owner not registered
______	5. Debt issue costs	e. Premium
______	6. Market rate higher than stated rate	f. Checks are mailed directly
______	7. Coupon bonds	g. No specific assets pledged
______	8. Convertible bonds	h. Bond indenture
______	9. Market rate less than stated rate	i. Backed by a lien
______	10. Stated rate times face amount	j. Interest expense
______	11. Registered bonds	k. May become stock
______	12. Debenture bond	l. Legal, accounting, printing
______	13. Mortgage bond	m. Protection against falling rates
______	14. Materiality concept	n. Periodic cash payments
______	15. Subordinated debenture	o. Bond price

P 14–22
Determine bond price; record interest; report bonds at fair value
● LO14–6

On January 1, 2021, NFB Visual Aids issued $800,000 of its 20-year, 8% bonds. The bonds were priced to yield 10%. Interest is payable semiannually on June 30 and December 31. NFB Visual Aids records interest expense at the effective rate and elected the option to report these bonds at their fair value. On December 31, 2021, the fair value of the bonds was $668,000 as determined by their market value in the over-the-counter market. General (risk-free) interest rates did not change during 2021.

Required:

1. Determine the price of the bonds at January 1, 2021, and prepare the journal entry to record their issuance.
2. Prepare the journal entry to record interest on June 30, 2021 (the first interest payment).
3. Prepare the journal entry to record interest on December 31, 2021 (the second interest payment).
4. Prepare the journal entry to adjust the bonds to their fair value for presentation in the December 31, 2021, balance sheet.

P 14–23
Report bonds at fair value; quarterly reporting
● LO14–6

Appling Enterprises issued 8% bonds with a face amount of $400,000 on January 1, 2021. The bonds sold for $331,364 and mature in 2040 (20 years). For bonds of similar risk and maturity the market yield was 10%. Interest is paid semiannually on June 30 and December 31. Appling determines interest expense at the effective rate. Appling elected the option to report these bonds at their fair value. The fair values of the bonds at the end of each quarter during 2021 as determined by their market values in the over-the-counter market were the following:

March 31	$350,000
June 30	340,000
September 30	335,000
December 31	342,000

General (risk-free) interest rates did not change during 2021.

Required:

1. By how much will Appling's comprehensive income be increased or decreased by the bonds (ignoring taxes) in the March 31 *quarterly* financial statements?
2. By how much will Appling's comprehensive income be increased or decreased by the bonds (ignoring taxes) in the June 30 *quarterly* financial statements?
3. By how much will Appling's comprehensive income be increased or decreased by the bonds (ignoring taxes) in the September 30 *quarterly* financial statements?
4. By how much will Appling's comprehensive income be increased or decreased by the bonds (ignoring taxes) in the December 31 *annual* financial statements?

P 14–24
Investments in bonds; accrued interest; sale; straight-line interest
● LO14–2, Appendix A

The following transactions relate to bond investments of Livermore Laboratories. The company's fiscal year ends on December 31. Livermore uses the straight-line method to determine interest.

2021	
July 1	Purchased $16 million of Bracecourt Corporation 10% debentures, due in 20 years (June 30, 2041), for $15.7 million. Interest is payable on January 1 and July 1 of each year.
Oct. 1	Purchased $30 million of 12% Framm Pharmaceuticals debentures, due May 31, 2031, for $31,160,000 plus accrued interest. Interest is payable on June 1 and December 1 of each year.
Dec. 1	Received interest on the Framm bonds.
31	Accrued interest.
2022	
Jan. 1	Received interest on the Bracecourt bonds.
June 1	Received interest on the Framm bonds.
July 1	Received interest on the Bracecourt bonds.
Sept. 1	Sold $15 million of the Framm bonds at 101 plus accrued interest.
Dec. 1	Received interest on the remaining Framm bonds.
31	Accrued interest.
2023	
Jan. 1	Received interest on the Bracecourt bonds.
Feb. 28	Sold the remainder of the Framm bonds at 102 plus accrued interest.
Dec. 31	Accrued interest.

Required:
1. Prepare the appropriate journal entries for these long-term bond investments.
2. By how much will Livermore Labs' earnings increase in each of the three years as a result of these investments? (Ignore income taxes.)

P 14–25
Accrued interest; effective interest; financial statement effects
● Appendix A

On March 1, 2021, Baddour, Inc., issued 10% bonds, dated January 1, with a face amount of $160 million. The bonds were priced at $140 million (plus accrued interest) to yield 12%. Interest is paid semiannually on June 30 and December 31. Baddour's fiscal year ends September 30.

Required:
1. What would be the amount(s) related to the bonds Baddour would report in its balance sheet at September 30, 2021?
2. What would be the amount(s) related to the bonds that Baddour would report in its income statement for the year ended September 30, 2021?
3. What would be the amount(s) related to the bonds that Baddour would report in its statement of cash flows for the year ended September 30, 2021?

P 14–26
Troubled debt restructuring
● Appendix B

At January 1, 2021, Rothschild Chair Company, Inc., was indebted to First Lincoln Bank under a $20 million, 10% unsecured note. The note was signed January 1, 2018, and was due December 31, 2024. Annual interest was last paid on December 31, 2019. Rothschild Chair Company was experiencing severe financial difficulties and negotiated a restructuring of the terms of the debt agreement.

Required:
Prepare all journal entries by Rothschild Chair Company, Inc., to record the restructuring and any remaining transactions relating to the debt under each of the independent circumstances below:
1. First Lincoln Bank agreed to settle the debt in exchange for land having a fair value of $16 million but carried on Rothschild Chair Company's books at $13 million.
2. First Lincoln Bank agreed to (a) forgive the interest accrued from last year, (b) reduce the remaining four interest payments to $1 million each, and (c) reduce the principal to $15 million.
3. First Lincoln Bank agreed to defer all payments (including accrued interest) until the maturity date and accept $27,775,000 at that time in settlement of the debt.

Decision Maker's Perspective

©ImageGap/Getty Images

Apply your critical-thinking ability to the knowledge you've gained. These cases will provide you an opportunity to develop your research, analysis, judgment, and communication skills. You also will work with other students, integrate what you've learned, apply it in real-world situations, and consider its global and ethical ramifications. This practice will broaden your knowledge and further develop your decision-making abilities.

Real World Case 14–1
Zero-coupon debt; HP Inc.
● LO14–2

Real World Financials

HP Inc. (formerly Hewlett-Packard Company) issued zero-coupon notes at the end of its 1997 fiscal year that mature at the end of its 2017 fiscal year. One billion, eight hundred million dollars face amount of 20-year debt sold for $968 million, a price to yield 3.149%. In fiscal 2002, HP repurchased $257 million in face value of the notes for a purchase price of $127 million, resulting in a gain on the early extinguishment of debt.

Required:

1. What journal entry did HP Inc. use to record the sale in 1997?
2. Using Excel, prepare an amortization schedule for the notes. Assume interest is calculated annually and use numbers expressed in millions of dollars; that is, the face amount is $1,800.
3. What was the effect on HP's earnings in 1998?
4. From the amortization schedule, determine the book value of the debt at the end of 2002.
5. What journal entry did HP Inc. use to record the early extinguishment of debt in 2002, assuming the purchase was made at the end of the year?
6. If none of the notes is repaid prior to maturity, what entry would HP use to record their repayment at the end of 2017?

Source: Hewlett-Packard Company

Analysis Case 14–2
Issuance of bonds
● LO14–2

The following appeared in the October 15, 2021, issue of the *Financial Smarts Journal:*

This announcement is not an offer of securities for sale or an offer to buy securities.
New Issue October 15, 2021

$750,000,000
CRAFT FOODS, INC.
7.75% Debentures Due October 1, 2031
Price 99.57%
plus accrued interest if any from date of issuance

Copies of the prospectus and the related prospectus supplement may be obtained from such of the undersigned as may legally offer these securities under applicable securities laws.

Keegan Morgan & Co. Inc.
Coldwell Bros. & Co.
Robert Stacks & Co.
Sherwin-William & Co.

Required:

1. Based on the information provided in the announcement, indicate whether the market rate of interest is higher or lower than 7.75% when the Craft Foods bonds were issued.
2. If debt issue costs were $75,000 and the bonds were issued on an interest payment date, what entry did Craft use to record the sale?

Judgment Case 14–3
Noninterest-bearing debt
● LO14–3

While reading a recent issue of *Health & Fitness,* a trade journal, Brandon Wilde noticed an ad for equipment he had been seeking for use in his business. The ad offered oxygen therapy equipment under the following terms:

Model BL 44582
$204,000 zero interest loan
Quarterly payments of $17,000 for only 3 years

The ad captured Wilde's attention, in part, because he recently had been concerned that the interest charges incurred by his business were getting out of line. The price, though, was somewhat higher than prices for this model he had seen elsewhere.

Required:

You are asked to advise Mr. Wilde on the purchase he is considering.

1. Assume the market rate of interest at the time for this type of transaction is 8%. At what amount should he record the asset purchased and the liability used to purchase it?
2. Assume the market rate of interest is unknown but we know the cash price of the equipment is $185,430. What is the effective rate of interest?

Judgment Case 14–4
Noninterest-bearing note exchanged for cash and other privileges
● LO14–3

The Jaecke Group, Inc. manufactures various kinds of hydraulic pumps. In June 2021, the company signed a four-year purchase agreement with one of its main parts suppliers, Hydraulics, Inc. Over the four-year period, Jaecke has agreed to purchase 100,000 units of a key component used in the manufacture of its pumps. The agreement allows Jaecke to purchase the component at a price lower than the prevailing market price at the time of purchase. As part of the agreement, Jaecke will lend Hydraulics $200,000 to be repaid after four years with no stated interest (the prevailing market rate of interest for a loan of this type is 10%).

Jaecke's chief accountant has proposed recording the note receivable at $200,000. The parts inventory purchase from Hydraulics over the next four years would then be recorded at the actual prices paid.

Required:

You do not agree with the chief accountant's valuation of the note and his intention to value the parts inventory acquired over the four-year period of the agreement at actual prices paid.

1. What entry would you use to account for the initial transaction?
2. What entry would you use to account for the subsequent purchase of 25,000 units of the component for $650,000 when it's market value is 26.60 per unit?

Judgment Case 14–5
Analyzing financial statements; financial leverage; interest coverage
● LO14–1, LO14–4

AGF Foods Company is a large, primarily domestic, consumer foods company involved in the manufacture, distribution, and sale of a variety of food products. Industry averages are derived from Troy's *The Almanac of Business and Industrial Financial Ratios.* Following are the 2021 and 2020 comparative income statements and balance sheets for AGF. (The financial data we use are from actual financial statements of a well-known corporation, but the company name is fictitious and the numbers and dates have been modified slightly to disguise the company's identity.)

AGF FOODS COMPANY
Years Ended December 31, 2021 and 2020
($ in millions)

Comparative Income Statements	2021	2020
Net sales	$6,440	$5,800
Cost of goods sold	(3,667)	(3,389)
Gross profit	2,773	2,411
Operating expenses	(1,916)	(1,629)
Operating income	857	782
Interest expense	(54)	(53)
Income from operations before tax	803	729
Income taxes	(316)	(287)
Net income	$ 487	$ 442
Comparative Balance Sheets		
Assets		
Total current assets	$1,879	$1,490
Property, plant, and equipment (net)	2,592	2,291
Intangibles (net)	800	843
Other assets	74	60
Total assets	$5,345	$4,684
Liabilities and Shareholders' Equity		
Total current liabilities	$1,473	$ 941
Long-term debt	534	728
Deferred income taxes	407	344
Total liabilities	2,414	2,013
Shareholders' equity:		
Common stock	180	180
Additional paid-in capital	21	63
Retained earnings	2,730	2,428
Total shareholders' equity	2,931	2,671
Total liabilities and shareholders' equity	$5,345	$4,684

Long-term solvency refers to a company's ability to pay its long-term obligations. Financing ratios provide investors and creditors with an indication of this element of risk.

Required:

1. Calculate the debt to equity ratio for AGF for 2021.
2. The average ratio for the stocks listed on the New York Stock Exchange in a comparable time period was 1.0. Other things being equal, does AGF appear to have higher or lower default risk than others in its industry?
3. Is AGF experiencing favorable or unfavorable financial leverage?
4. Calculate AGF's times interest earned ratio for 2021.
5. The coverage for the stocks listed on the New York Stock Exchange in a comparable time period was 5.1. Other things being equal, does AGF appear to have higher or lower interest coverage than others in its industry?

Research Case 14–6
FASB codification research; researching the way long-term debt is reported; Macy's, Inc
● LO14–1 through LO14–4

Real World Financials

EDGAR, the Electronic Data Gathering, Analysis, and Retrieval system, performs automated collection, validation, indexing, acceptance and forwarding of submissions by companies and others who are required by law to file forms with the U.S. Securities and Exchange Commission (SEC). All publicly traded domestic companies use EDGAR to make the majority of their filings. (Some foreign companies do so voluntarily.) Form 10-K, including the annual report, is required to be filed on EDGAR. The SEC makes this information available on the Internet.

Required:

1. Access EDGAR on the Internet at **www.sec.gov** or from Investor Relations at the **Macy's, Inc. (www.macys.com)**. Search for Macy's. Access its 10-K filing for the year ended January 30, 2017. Search or scroll to find the financial statements and related notes. What is the total debt (including current liabilities and deferred taxes) reported in the balance sheet in the most recent two years?
2. Compare the total liabilities (including current liabilities and deferred taxes) with the shareholders' equity and calculate the debt to equity ratio for the most recent two years.
3. Does Macy's obtain more financing through notes, bonds, or commercial paper? Are required debt payments increasing or decreasing over the next three years?
4. Does Macy's classify any short-term debt as long-term?
5. **Note 6: Financing** includes the following statement: "On November 18, 2014, the Company issued $550 million aggregate principal amount of 4.5% senior notes due 2034. This debt was used to pay for the redemption of the $407 million of 7.875% senior notes due 2015 described above." Under some circumstances, Macy's could have reported the amounts due in 2015 as long-term debt at the end of the previous year even though these amounts were due within the coming year. Obtain the relevant authoritative literature on classification of debt expected to be financed using the *FASB Accounting Standards Codification*. You might gain access from the FASB website (**www.fasb.org**), from your school library, or some other source. Determine the criteria for reporting currently payable debt as long-term. What is the specific nine-digit Codification citation (XXX-XX-XX-XX) that Macy's would rely on in applying that accounting treatment?

Source: Macy's, Inc.

Analysis Case 14–7
Bonds; conversion; extinguishment
● LO14–5

On August 31, 2018, Chickasaw Industries issued $25 million of its 30-year, 6% convertible bonds dated August 31, priced to yield 5%. The bonds are convertible at the option of the investors into 1,500,000 shares of Chickasaw's common stock. Chickasaw records interest expense at the effective rate. On August 31, 2021, investors in Chickasaw's convertible bonds tendered 20% of the bonds for conversion into common stock that had a market value of $20 per share on the date of the conversion. On January 1, 2020, Chickasaw Industries issued $40 million of its 20-year, 7% bonds dated January 1 at a price to yield 8%. On December 31, 2021, the bonds were extinguished early through acquisition in the open market by Chickasaw for $40.5 million.

Required:

1. Using the book value method, does the conversion of the 6% convertible bonds into common stock result in a gain, a loss, or no gain or loss?
2. Using the market value method, by how much, does the conversion of the 6% convertible bonds into common stock result in a gain, a loss, or no gain or loss?
3. Were the 7% bonds issued at face value, at a discount, or at a premium?
4. In the second year of the term to maturity, will the amount of interest expense for the 7% bonds be higher than, lower than, or the same as in the first year?
5. Does the early extinguishment of the 7% bonds result in a gain, a loss, or no gain or loss?

Communication Case 14–8
Convertible securities and warrants; concepts
● LO14–5

It is not unusual to issue long-term debt in conjunction with an arrangement under which lenders receive an option to buy common stock during all or a portion of the time the debt is outstanding. Sometimes the vehicle is convertible bonds; sometimes warrants to buy stock accompany the bonds and are separable. Interstate Chemical is considering these options in conjunction with a planned debt issue.

"You mean we have to report $7 million more in liabilities if we go with convertible bonds? Makes no sense to me," your CFO said. "Both ways seem pretty much the same transaction. Explain it to me, will you?"

Required:

Write a memo. Include in your explanation each of the following:

1. The differences in accounting for proceeds from the issuance of convertible bonds and of debt instruments with separate warrants to purchase common stock.
2. The underlying rationale for the differences.
3. Arguments that could be presented for the alternative accounting treatment.

Communication Case 14–9
Is convertible debt a liability or is it shareholders' equity? Group Interaction
● LO14–5

Some financial instruments can be considered compound instruments in that they have features of both debt and shareholders' equity. The most common example encountered in practice is convertible debt—bonds or notes convertible by the investor into common stock. A topic of debate for several years has been whether:

- **View 1:** Issuers should account for an instrument with both liability and equity characteristics entirely as a liability or entirely as an equity instrument depending on which characteristic governs.
- **View 2:** Issuers should account for an instrument as consisting of a liability component and an equity component that should be accounted for separately.

In considering this question, you should disregard what you know about the current position of the FASB on the issue. Instead, focus on conceptual issues regarding the practicable and theoretically appropriate treatment, unconstrained by GAAP. Also, focus your deliberations on convertible bonds as the instrument with both liability and equity characteristics.

Required:

1. Which view do you favor? Develop a list of arguments in support of your view prior to the class session for which the case is assigned.
2. In class, your instructor will pair you (and everyone else) with a classmate (who also has independently developed an argument).
   a. You will be given three minutes to argue your view to your partner. Your partner likewise will be given three minutes to argue his or her view to you. During these three-minute presentations, the listening partner is not permitted to speak.
   b. After each person has had a turn attempting to convince his or her partner, the two partners will have a three-minute discussion in which they will decide which view is more convincing. Arguments will be merged into a single view for each pair.
3. After the allotted time, a spokesperson for each of the two views will be selected by the instructor. Each spokesperson will field arguments from the class in support of that view's position and list the arguments on the board. The class then will discuss the merits of the two lists of arguments and attempt to reach a consensus view, though a consensus is not necessary.

Communication Case 14–10
Note receivable exchanged for cash and other services
● LO14–3

The Pastel Paint Company recently loaned $300,000 to KIX 96, a local radio station. The radio station signed a noninterest-bearing note requiring the $300,000 to be repaid in three years. As part of the agreement, the radio station will provide Pastel with a specified amount of free radio advertising over the three-year term of the note.

The focus of this case is the valuation of the note receivable by Pastel Paint Company and the treatment of the "free" advertising provided by the radio station. Your instructor will divide the class into two to six groups depending on the size of the class. The mission of your group is to reach consensus on the appropriate note valuation and accounting treatment of the free advertising.

Required:

1. Each group member should deliberate the situation independently and draft a tentative argument prior to the class session for which the case is assigned.
2. In class, each group will meet for 10 to 15 minutes in different areas of the classroom. During that meeting, group members will take turns sharing their suggestions for the purpose of arriving at a single group treatment.
3. After the allotted time, a spokesperson for each group (selected during the group meetings) will share the group's solution with the class. The goal of the class is to incorporate the views of each group into a consensus approach to the situation.

Ethics Case 14–11
Debt for equity swaps; have your cake and eat it too
● LO14–5

The cloudy afternoon mirrored the mood of the conference of division managers. Claude Meyer, assistant to the controller for Hunt Manufacturing, wore one of the gloomy faces that were just emerging from the conference room. "Wow, I knew it was bad, but not that bad," Claude thought to himself. "I don't look forward to sharing those numbers with shareholders."

The numbers he discussed with himself were fourth-quarter losses which more than offset the profits of the first three quarters. Everyone had known for some time that poor sales forecasts and production delays had wreaked havoc on the bottom line, but most were caught off guard by the severity of damage.

Later that night he sat alone in his office, scanning and rescanning the preliminary financial statements on his computer monitor. Suddenly his mood brightened. "This may work," he said aloud, though no one could hear. Fifteen minutes later he congratulated himself, "Yes!"

The next day he eagerly explained his plan to Susan Barr, controller of Hunt for the last six years. The plan involved $300 million in convertible bonds issued three years earlier.

- *Meyer:* By swapping stock for the bonds, we can eliminate a substantial liability from the balance sheet, wipe out most of our interest expense, and reduce our loss. In fact, the book value of the bonds is significantly more than the market value of the stock we'd issue. I think we can produce a profit.
- *Barr:* But Claude, our bondholders are not inclined to convert the bonds.
- *Meyer:* Right. But, the bonds are callable. As of this year, we can call the bonds at a call premium of 1%. Given the choice of accepting that redemption price or converting to stock, they'll all convert. We won't have to pay a cent. And, since no cash will be paid, we won't pay taxes either.

Required:

Do you perceive an ethical dilemma? What would be the impact of following up on Claude's plan? Who would benefit? Who would be injured?

Data Analytics

connect

Data analytics is the process of examining data sets in order to draw conclusions about the information they contain. If you haven't completed any of the prior data analytics cases, follow the instructions listed in the Chapter 1 Data Analytics case to get set up. You will need to watch the videos referred to in the Chapters 1 - 3 Data Analytics cases. No additional videos are required for this case. All short training videos can be found here: **https://www.tableau.com/learn/training#getting-started**.

Data Analytics Case
Long-term Solvency
● LO14–4

In the Chapter 13 Data Analytics Cases, you used Tableau to examine a data set and create charts to compare our two (hypothetical) publicly traded companies: Discount Goods and Big Store as to their liquidity - their ability to pay short-term debts as they come due. In this case, you examine the companies' ability to pay their long-term obligations as measured by their debt to equity and times interest earned ratios.

The debt to equity ratio is measured as total liabilities / shareholder's equity and provides us an indication of the likelihood a company will default on its obligations. Other things being equal, the higher the debt to equity ratio, the higher the risk.

The times interest earned ratio compares interest payments with income available to pay those charges. It is calculated by adding interest plus taxes to net income and dividing by shareholder's equity. The higher the ratio, the greater the margin of safety provided to creditors.

Required:

Use Tableau to calculate and display the trends for the debt to equity and times interest earned ratios for each of the two companies in the period 2018-2021. the average debt to equity ratio and times interest earned ratio for companies in the General Retailers industry sector in a comparable time period are 1.92 and 10.6, respectively. Based upon what you find, answer the following questions:

1. Other things being equal, do both companies appear to have the ability to meet their obligations as measured by the debt to equity ratio?
2. Based solely on the times interest earned ratios, do you reach the same conclusion as in Requirement 1?
3. Is the margin of safety provided to creditors by Discount Goods improving or declining in recent years as measured by the average times interest earned ratio?

Resources:

You have available to you an extensive data set that includes detailed financial data for Discount Goods and Big Store for 2012–2021. The data set is in the form of four Excel files available to download from Connect or under Student Resources within the Library tab. The one for use in this chapter is named "Discount_Goods_Big_Store_Financials.xlsx." Download this file and save it to the computer on which you will be using Tableau.

For this case, you will create several calculations to produce several long-term solvency ratios to allow you to compare and contrast the two companies.

After you view the training videos, follow these steps to create the charts you'll use for this case:

- Open Tableau and connect to the Excel spreadsheet you downloaded.
- Click on the "Sheet 1" tab at the bottom of the canvas, to the right of the Data Source at the bottom of the screen. Drag "Company" and "Year" under Dimensions to the Rows shelf. Change "Year" to *discrete* by right-clicking and selecting "Discrete." Select "Show Filter" and uncheck all the years except 2018–2021. Click OK.
- Drag "Total liabilities" under Measures to the Rows shelf. Change to "Discrete" data following the same instructions.
- Drag "Total shareholders' equity" under Measures to the Rows shelf and change to discrete data.

- Create a calculated field by clicking the "Analysis" tab at the top of the screen and selecting "Create Calculated field." Name the calculation "Debt to Equity Ratio." In the Calculation Editor window drag "Total liabilities" from the Rows shelf, type a division sign (/), and then drag "Total shareholders' equity" from the Rows shelf to the right of the division sign. Make sure the window says that the calculation is valid and click OK.
- Drag the newly created "Debt to Equity Ratio" to the Rows shelf. Click on the "Show Me" and select "side-by-side bars." Add labels to the bars by clicking on "Label" under the "Marks" card and clicking the box "Show marks label." Format the labels to Times New Roman, bold, black and 9-point font. Edit the color on the "Marks" card if desired.
- Change the title of the sheet to be "Long-Term Solvency: Debt to Equity Ratio" by right-clicking and selecting "Edit title." Format the title to Times New Roman, bold, black and 15-point font. Change the title of "Sheet 1" to match the sheet title by right-clicking, selecting "Rename" and typing in the new title.
- Format all other labels to be Times New Roman, bold, black and 12-point font.
- On the Sheet 2 tab, follow the procedure above for the company and year.
- Drag "Net income / (loss)", "Interest expense, and "Provision for income taxes" under Measures into the Rows shelf. Change each to "Discrete" data. Create a calculated field "Times Interest Earned Ratio" following the same process above. In the Calculation Editor window, type an open parenthesis, then drag "Net income / (loss)" from the Rows shelf, type an addition sign (+), drag "Interest expense" from the Rows shelf, type an addition sign, and then drag "Provision for income taxes" from the Rows shelf and close the parenthesis. Next to the closed parenthesis, type a division sign (/) and drag "Interest expense" from the Rows shelf next to the right of the division sign. Make sure the window says that the calculation is valid and click OK.
- Right-click "Net income / loss," "Interest expense," and "Provision for income taxes" and uncheck "Show in Header" so they no longer show on the canvas.
- Drag the newly created "Times Interest Earned Ratio" into the Rows shelf. Click on the "Show Me" and select "side-by-side bars." Add labels to the bars by clicking on "Label" under the "Marks" card and clicking the box "Show mark labels." Format the labels to Times New Roman, bold, black and 10-point font. Edit the color on the Color marks card if desired.
- Change the title of the sheet to be "Long-Term Solvency: Times Interest Earned Ratio" by right-clicking and selecting "Edit title." Format the title to Times New Roman, bold, black and 15-point font. Change the title of "Sheet 2" to match the sheet title by right-clicking, selecting "Rename" and typing in the new title.
- Format all other labels to be Times New Roman, bold, black and 12-point font.
- Once complete, save the file as "*DA14_Your initials.twbx.*"

Continuing Cases

Target Case
- LO14–4

Target Corporation prepares its financial statements according to U.S. GAAP. Target's financial statements and disclosure notes for the year ended February 3, 2018, are available in Connect. This material is also available under the Investor Relations link at the company's website (**www.target.com**).

Long-term solvency refers to a company's ability to pay its long-term obligations. Financing ratios provide investors and creditors with an indication of this element of risk.

Required:

1. Calculate the debt to equity ratio for Target at February 3, 2018. The average ratio for companies in the Discount Retailers industry sector in a comparable time period was 2.0.
2. Calculate Target's times interest earned ratio for the year ended February 3, 2018. The coverage for companies in the Discount Retailers industry sector in a comparable time period was 6.9.

Air France–KLM Case
- LO14–7

IFRS

Air France–KLM (AF), a Franco-Dutch company, prepares its financial statements according to International Financial Reporting Standards. AF's financial statements and disclosure notes for the year ended December 31, 2017, are available in Connect. This material also is available under the Finance link at the company's website (**www.airfranceklm-finance.com**.)

Required:

1. Examine the long-term borrowings in AF's balance sheet and the related note (32.2.2). Note that AF has convertible bonds outstanding that it issued in 2013. Prepare the journal entry AF would use to record the issue of convertible bonds.
2. Prepare the journal entry AF would use to record the issue of the convertible bonds if AF used U.S. GAAP.
3. AF does not elect the fair value option (FVO) to report its financial liabilities. Examine Note 35.3 "Market value of financial instruments." If the company had elected the FVO for all of its debt measured at amortized cost, what would be the balance at December 31, 2017, in the fair value adjustment account?

CPA Exam Questions and Simulations

ROGER CPA Review

Sample CPA Exam questions from Roger CPA Review are available in Connect as support for the topics in this chapter. These multiple-choice questions and task-based simulations include expert-written explanations and solutions, and provide a starting point for students to become familiar with the content and functionality of the actual CPA Exam.

CHAPTER

15 Leases

OVERVIEW

In Chapter 14, we saw how companies account for their long-term debt. The focus of that discussion was *bonds* and *notes.* In this chapter, we continue our discussion of debt, but we now turn our attention to liabilities arising in connection with *leases.* A lease is a contract that gives a lessee (user) the right to control the use of an asset for a period of time. The lessee pays the lessor (owner) for that right, typically with a series of payments made over the lease term. The lessee initially accounts for this arrangement by recording a "right-of-use" asset and a lease liability at the beginning of the lease. Subsequent accounting depends on the nature of the lease contract. When the substance of the lease contract represents a temporary rental agreement between the lessee and lessor, we classify those as operating leases. However, in situations where the lease contract essentially represents the sale of an asset, the lessee has a finance lease, and the lessor has a sales-type lease. As we'll see in this chapter, the way we account for a lease contract depends on how the lease is classified.

LEARNING OBJECTIVES

After studying this chapter, you should be able to:

- **LO15–1** Explain why companies frequently choose to lease assets and describe the basis for each of the criteria used to classify leases. (*p. 835*)
- **LO15–2** Describe and demonstrate how the lessee accounts for a finance lease and the lessor accounts for a sales-type lease with no selling profit. (*p. 841*)
- **LO15–3** Describe and demonstrate how the lessor accounts for a sales-type lease with a selling profit. (*p. 846*)
- **LO15–4** Describe and demonstrate how the lessor and lessee account for all transactions associated with operating leases. (*p. 847*)
- **LO15–5** Explain when and how a lessee accounts for a lease by the shortcut method. (*p. xx*)
- **LO15–6** Explain the impact on lease accounting of uncertainties, including uncertain lease terms, variable lease payments, residual values, purchase options, and termination penalties. (*p. 855*)
- **LO15–7** Determine whether a contract contains a lease and explain the impact on lease accounting of other payments, including nonlease payments, initial direct costs, and leasehold improvements. (*p. 869*)
- **LO15–8** Describe the impact of leases on the statement of cash flows and disclosure requirements pertaining to leases. (*p. 876*)
- **LO15–9** Discuss the primary differences between U.S. GAAP and IFRS with respect to leases. (*p. 852, 855 and 859*)

©Pressmaster/Shutterstock

FINANCIAL REPORTING CASE

Lease or Buy?

"The days of off-balance-sheet financing are over," said Aaron Sanchez, controller for your new employer, Higher Graphics. "Now that we have to report liabilities for both finance leases and operating leases, maybe we shouldn't keep limiting our leases to short-term arrangements that would be considered operating leases. In fact, maybe we should start buying our equipment rather than leasing it."

Like most companies in your industry, HG leases much of the electronic equipment it uses to develop and test the software it produces. Eager to understand the implications of your new supervisor's concerns, you turn to the Internet to bolster your knowledge of the new lease accounting standards.

QUESTIONS

By the time you finish this chapter, you should be able to respond appropriately to the questions posed in this case. Compare your response to the solution provided at the end of the chapter.

1. Are there advantages of leasing its equipment that HG should consider? *(p. 835)*
2. Would the effect on HG's income statement be the same if HG used finance leases instead of operating leases? *(p. 846)*

PART A

Accounting by the Lessor and Lessee

If you ever have leased an apartment, you know that a lease is a contractual arrangement by which a **lessor** (owner) provides a **lessee** (user) the right to use an asset for a specified period of time. In return for this right, the lessee agrees to make periodic cash payments during the term of the lease. When a company reports a lease, the right to use the asset for a period of time is recorded as a "right-of-use" asset in the balance sheet, while the obligation to make payments over the lease period is recorded as a lease liability.

An apartment rental is a contractual arrangement constituting a *lease*.

You might be surprised to know that leases comprise more than $3.3 trillion of asset financing and have become the number one method of external financing by U.S. companies.[1] The airplane in which you last flew probably was leased, as was the gate from which it departed. Your cell phone service provider likely leases the space in which it operates. Many of the stores where you shop and the last hotel you stayed in were probably leased properties. So, why the popularity?

FINANCIAL Reporting Case

Q1 p. 835

Why Lease?

● LO15–1

1. Leasing reduces the upfront cash needed to use an asset.

The purchase of an asset can include several additional fees—loan origination fees, closing costs, brokerage fees, and certain taxes. You know that if you've ever bought a house or a car. Many leases, though, begin with the first lease payment including nothing more than the agreed-upon monthly amount. Relatedly, some companies, especially newer ones, might not have enough cash to pay the full purchase cost for an asset, but they likely have enough

Operational, financial, and tax incentives often make leasing an attractive alternative to purchasing.

[1] Jon Glover, "Balance Sheets to Swell $2.8 Trillion as Leasing Rules Tighten," *Bloomberg.com*, January 12, 2016.

cash to begin monthly payments. Also, companies that have high credit risk may not be able to obtain financing to purchase an asset. Leasing might be these companies' only option to acquire an asset.

2. Lease payments often are lower than installment payments.

Installment payments to buy an asset are based on the asset's full fair value. On the other hand, lease payments often are tied only to the portion of the asset's fair value expected to decline over the lease period. Because the lease period can be less than the asset's full life, the monthly payments associated with leasing often are lower.

3. Leasing offers flexibility and a lower cost when disposing of the asset.

Returning a leased asset at the end of the lease term requires little effort or cost. When you've rented a car for the week, you simply return the car to the rental company. When a company is finished with an office building, the company simply moves out. However, selling an asset that you've previously purchased usually isn't that easy. Some unique assets might not have an available market in which used items can be easily sold. Selling an asset also can require significant costs. For example, many realtors charge up to 6% to sell your home or office building.

4. Leasing might offer protection against the risk of declining asset values.

When a company *buys* an asset, the price it might eventually sell the asset for at the end of its productive life is uncertain. On the other hand, when a company *leases* an asset it avoids the risk of declining fair values (selling prices) but also misses out on any increase in fair value.

5. Leasing might offer tax advantages.

Sometimes leasing offers tax savings over outright purchases. For instance, a company with little or no taxable income—maybe a business just getting started, or one experiencing an economic downturn—will get little benefit from deducting depreciation on its tax return. Normally, depreciation reduces taxable income and thus reduces taxes, but if there's little or no taxable income to reduce, there's little or no tax savings. However, the company can benefit *indirectly* by leasing assets rather than buying. By allowing the *lessor* to retain ownership and thus benefit from depreciation deductions, the lessee often can negotiate lower lease payments. Lessees with sufficient taxable income to take advantage of the depreciation deductions, but still in lower tax brackets than lessors, also can achieve similar indirect tax benefits.

Lease Classification

A basic concept of accounting is substance over form.

The way we account for a lease depends on whether the lease more naturally resembles the purchase of an asset with debt financing or whether it's a rental agreement. In many cases, the true nature of the contract is obvious. For example, a 10-year noncancelable lease of a computer with a 10-year useful life, by which title passes to the lessee at the end of the lease term, obviously more nearly represents a purchase than a rental agreement. This type of lease is classified as a **finance lease** by the lessee, and a **sales-type lease** by the lessor.[2] In contrast, a 10-day rental of the same computer, with no passage of title to the lessee, obviously represents a rental and not a sale. These types of rental arrangements are referred to as **operating leases**. Illustration 15–1 compares the classification possibilities.

Illustration 15–1

Basic Lease Classifications

Lessee	Lessor
• Finance lease	• Sales-type lease ➡ without selling profit ➡ with selling profit
• Operating lease	• Operating lease*

*In a very specific situation described in Part B, a lease that doesn't qualify as a sales-type lease might be classified as a "direct financing" lease rather than an operating lease.

[2] An "Additional Consideration" in Part B describes a very specific situation in which a lease that doesn't qualify as a sales-type lease might be classified as a "direct financing" lease rather than an operating lease.

In some leasing arrangements, whether the company has engaged in a purchase versus rental agreement is not clear. What if the terms of the contract don't transfer title, and the lease term is for only seven years of the asset's 10-year life? What if contractual terms permit the lessee to obtain title under certain prearranged conditions? What if the present value of payments provided by the lease contract is nearly equal to the value of the asset under lease? These situations are less clear-cut.

Professional judgment is needed to differentiate between leases that in essence are installment purchases/sales and those that are not. It's important to note that from an accounting perspective *legal ownership is irrelevant* in the decision. As we do with other accounting decisions, we base our assessment on the *economic substance* of the transaction, even if that conflicts with its legal form. This reflects a basic concept of accounting referred to as "substance over form."

Accounting for leases attempts to see through the legal form of the agreements to determine their economic substance.

Although similar, a finance lease is not the same as the purchase of an asset. The rights a lease grants a lessee under any lease, whether classified as a finance lease or an operating lease, are different from the rights transferred in the outright purchase of an asset. As one example, a lessee can't sell the leased asset.

Nonetheless, a finance lease is economically similar to a purchase of the asset because the terms of a finance lease (a) normally allow the lessee to *direct the use* of the asset in a way that the lessee receives *substantially all of the remaining benefits* from the asset and (b) creates obligations for the lessee that are similar to those that financing the purchase of an asset would impose. An operating lease lacks those characteristics. Thus, the essential question when determining lease classification is whether a particular lease arrangement has the characteristics of a finance lease.

Determining lease classification based on judgment alone is likely to lead to inconsistencies in practice. The desire to encourage consistency in practice motivated the FASB to provide guidance for distinguishing between the two fundamental types of leases. As you study the classification criteria in the following paragraphs, keep in mind that some leases clearly fit the classifications we give them, but others fall in a gray area somewhere between the two extremes. For those, we end up classifying according to the best evidence available.

CLASSIFICATION CRITERIA We classify a lease transaction as a *finance lease* (*sales-type lease* from the lessor's perspective) if one or more of the five criteria listed in Illustration 15–2 is met. Otherwise, it is an *operating lease.*

1. The agreement specifies that ownership of the asset transfers to the lessee.
2. The agreement contains a purchase option that the lessee is reasonably certain to exercise.
3. The lease term is for the "major part" of the remaining economic life of the underlying asset.*
4. The present value of the total of the lease payments† equals or exceeds "substantially all" of the fair value of the underlying asset.
5. The underlying asset is of such a specialized nature that it is expected to have no alternative use to the lessor at the end of the lease term.

*If the lease begins at or near the end of the economic life of the asset, this criterion shouldn't be used for purposes of classifying the lease.
†This total includes any residual value of the asset at the end of the lease term that is guaranteed by the lessee. We discuss residual values, guaranteed and unguaranteed, later in the chapter.

Illustration 15–2
Criteria for Classification as a Finance Lease

Let's look closer at these criteria.

Criterion 1. Since our objective is to determine whether the lessor has, in substance, sold the asset to the lessee, the first criterion is self-evident. If legal title passes to the lessee during, or at the end of, the lease term, obviously ownership attributes are transferred.

Criterion 1: Transfer of ownership.

Criterion 2. A **purchase option** is a provision in the lease contract that gives the lessee the option to purchase the leased property at a specified price. Criterion 2 is met if the

Criterion 2: Purchase option the lessee is reasonably certain to exercise.

specified price is sufficiently lower than the expected fair value of the property when the option becomes exercisable that the exercise of the option by the lessee appears "reasonably certain" at the beginning of the lease.

For example, suppose instead of purchasing farmland for its fair value of $100,000, the farmland is leased for $20,000 per year for five years with an option at the end of the lease period to purchase it for an additional $25,000. If the farmland is expected to be worth, say, $45,000 in five years, the $25,000 would be considered to be a bargain price, so the lessee is reasonably certain to exercise the purchase option and buy the asset.

Although the lease accounting guidance (ASC Topic 842: "Leases") does not use the term, in this chapter we will refer to this as a bargain purchase option (BPO). Though not immediately obvious, the logic of the second criterion is similar to that of the first. That is, we again expect title to transfer, this time because the lessee is expected to exercise the option to purchase the asset. Applying criterion 2 in practice, though, often is more difficult because it's necessary to make a judgment *now* about whether a *future* fair value will be sufficiently higher than the exercise price so that exercise is "reasonably certain."

Criterion 3: Lease term is for the major part of economic life.

Criterion 3. The third criterion considers whether the asset is leased for the "major part" of its useful life. If that criterion is met, then most of the risks and rewards of ownership are deemed to have been transferred to the lessee, as would be the case if the asset had been purchased. The lease accounting guidance (ASC Topic 842: "Leases") suggests that one reasonable approach to assessing this criterion would be to conclude that 75% or more of the remaining economic life of the underlying asset constitutes a "major part" of the remaining economic life of that underlying asset. Note, though, that this is only one approach and not a precise indication.

Although the intent of this criterion is fairly straightforward, implementation sometimes is troublesome. First, the lease term may be uncertain. It may be renewable beyond its initial term. Or, the lease may be cancelable after a designated noncancelable period. If the lease includes these provisions, we need to consider whether a change in the lease term is reasonably certain to occur.

Additional Consideration

Periods covered by renewal options are not included in the lease term if a bargain purchase option (BPO) is present. This is because the lease term cannot extend beyond the date a purchase option is exercised. For example, assume a bargain purchase option allows a lessee to buy a leased delivery truck at the end of a noncancelable five-year lease term. Even if an option to renew the lease beyond that date is considered to be reasonably certain to be exercised in the event of not exercising the purchase option, that renewal period would not be included as part of the lease term. Remember, we presume the purchase option will be exercised after the initial five-year term, making the renewal option irrelevant.

A second implementation issue is estimating, at the beginning of the lease, the economic life of the leased property. This is the estimated remaining time the property is expected to be economically usable for its intended purpose, with normal maintenance and repairs. Estimates of the economic life of leased property are subject to the same uncertainty limitations of most estimates. This uncertainty presents the opportunity to arrive at estimates that cause this third criterion not to be met.

Finally, if the lease begins "at or near the end" of an asset's economic life, this third criterion does not apply. That's because most of the risks and rewards of ownership occur prior to that time.[3]

Criterion 4: Present value of payments is substantially all of fair value.

Criterion 4. The fourth criterion indicates that, if the **lease payments** have a total value that represents "substantially all" of the asset's fair value, it's logical to identify the lease

[3]ASC Topic 842: "Leases" specifies that one reasonable approach to determining whether this exception should be applied would be to conclude that a lease that commences in the final 25% of an asset's economic life is "at or near the end" of the underlying asset's economic life.

contract as equivalent to a sale. That is, we could say the lessee has paid enough to have purchased the asset. The lease accounting guidance suggests that one reasonable approach to assessing this criterion would be to conclude that payments with a present value of 90% or more of the fair value of the underlying asset represent "substantially all" of the fair value. (We review how to calculate present value later in the chapter.) In general, lease payments for the purpose of classifying a lease are payments the lessee is *required* to make in connection with the lease. We look closer at the makeup of lease payments later in the chapter when we discuss various uncertainties in lease transactions in more detail.

Criterion 5. The fifth criterion recognizes that, if the underlying asset is of such a specialized nature that it is expected to have *no alternative use* to the lessor at the end of the lease term, then only the lessee can derive the usual risks and rewards of ownership of the asset. Likewise, the lessor will achieve its desired return on investment only through the lease payments from that particular lease, indicating that the lessor intends a sale of ownership rights.

Criterion 5: Leased asset is so specialized that it cannot reasonably be repurposed.

Decision Makers' Perspective

Prior to recent lease accounting guidance, another reason to lease rather than buy was to obtain "off-balance-sheet financing." When funds are borrowed to purchase an asset, the liability has a detrimental effect on the company's debt to equity ratio and other quantifiable indicators of riskiness. Similarly, the purchased asset increases total assets and correspondingly lowers calculations of the rate of return on assets. To avoid looking more risky and less profitable, managers previously were able to keep assets and liabilities "off balance sheet" by leasing them rather than buying them. Under that preexisting GAAP, if the lessee could construct a leasing arrangement so that it qualified as an operating lease, neither the asset nor the liability had to be reported in the balance sheet.[4] That way, managers could avoid surpassing contractual limits on designated financial ratios (such as the debt to equity ratio).[5] Also, despite research that indicates otherwise, some managers might have thought they were fooling the financial markets into thinking their strategies were less risky and more profitable than actually was the case. However, GAAP now requires lessees to record assets and liabilities for all but short-term leases. In fact, the primary impetus for the new guidance was to curtail off-balance-sheet financing.

Traditionally, leasing was used as a means of off-balance-sheet financing.

Even without the incentive of off-balance-sheet financing, managers still are motivated to construct lease arrangements as an operating lease so they can get the favorable income statement effects those leases provide, as we see later in the chapter. Devising ways to tweak lease terms to avoid classification as finance leases is commonplace in practice. Analysts should be aware of this tactic when evaluating and comparing companies' reported profits. ●

Before we apply the classification criteria to identify a finance lease, it's useful to see how accounting for a finance lease is similar to the way we accounted for the acquisition of an asset with an installment note in the previous chapter.

Finance Leases and Installment Notes Compared

You learned in Chapter 14 how to account for an installment note. To a great extent, then, you already have learned how to account for a finance lease. Finance leases are agreements that we identify as being formulated outwardly as leases, but which in reality are essentially similar to installment purchases.

To illustrate, let's recall a situation we considered in the previous chapter. We assumed that Skill Graphics purchased a package-labeling machine from Hughes–Barker Corporation by issuing a three-year installment note that required six semiannual installment payments of $139,857 each. That arrangement provided for the purchase of the **$666,633** machine as well as interest at an annual rate of 14% (7% twice each year). Remember, too, that each installment payment consisted of part interest (7% times the outstanding balance) and part payment for the machine (the remainder of each payment).

[4]Instead, rent payments were simply reported as lease expense.

[5]It is common for debt agreements, particularly long-term ones, to include restrictions on the debtor as a way to provide some degree of protection to the creditor. Sometimes a minimum level is specified for current assets relative to current liabilities, debt as a ratio of equity, return on net assets, or many other financial ratios. Typically, the debt becomes due on demand when the debtor becomes in violation of such a debt covenant, often after a specified grace period.

Now let's suppose that Skill Graphics instead acquired control of the package-labeling machine from Hughes–Barker Corporation under a three-year *lease* that required six semi-annual lease payments of $139,857 each. The fundamental nature of the transaction remains the same regardless of whether it is negotiated as an installment purchase or as a lease. So, we account for this lease in fundamentally the same way as an installment purchase. To see this, let's first compare how we account for an installment note versus a finance lease at the beginning (or commencement) of the lease (see Illustration 15–3).

Illustration 15–3

Comparison of a Note and Finance Lease

In keeping with the basic accounting concept of substance over form, accounting for a finance lease parallels that for an installment purchase.

At Beginning (January 1)		
Installment Note		
Machinery	666,633	
Notes payable		**666,633**
Finance Lease		
Right-of-use asset	666,633	
Lease payable		**666,633**

Notice that we debit an asset account and credit a liability, regardless of whether the machinery is acquired through financing with a note payable or through a finance lease.

With each periodic payment, interest expense accrues at the effective rate times the outstanding balance, and the balance of the liability is reduced for the difference as shown in Illustration 15–3A.

Illustration 15–3A

Interest Compared for a Note and Finance Lease

Each payment includes both an amount that represents interest and an amount that represents a reduction of principal.

At the First Semiannual Payment Date (June 30) (7% × Outstanding balance)		
Installment Note		
Interest expense (7% × **$666,633**)	**46,664**	
Notes payable (difference)	93,193	
Cash (installment payment)		139,857
Finance Lease		
Interest expense (7% × **$666,633**)	**46,664**	
Lease payable (difference)	93,193	
Cash (lease payment)		139,857

Because the lease payable balance declines with each payment, the amount of interest expense decreases each period. An amortization schedule is convenient to track the changing amounts as shown in Illustration 15–3B.

Illustration 15–3B

Lease Amortization Schedule

Each lease payment includes interest on the outstanding balance at the effective rate. The remainder of each payment reduces the outstanding balance.

Date	Payments	Effective Interest	Decrease in Balance	Outstanding Balance
		(7% × Outstanding balance)	(Pmt. – Interest)	
				666,633
1	139,857	0.07 (**666,633**) = **46,664**	93,193	573,440
2	139,857	0.07 (573,440) = 40,141	99,716	473,724
3	139,857	0.07 (473,724) = 33,161	106,696	367,028
4	139,857	0.07 (367,028) = 25,692	114,165	252,863
5	139,857	0.07 (252,863) = 17,700	122,157	130,706
6	139,857	0.07 (130,706) = 9,151*	130,706	0
	839,142	172,509	**666,633**	

*Rounded

You should recognize this as the same amortization schedule we used in the previous chapter in connection with our installment note example. The reason for the similarity is that we view a finance lease as being, in substance, equivalent to an installment purchase. So naturally the accounting treatment of the two essentially identical transactions should be consistent.

In the remaining sections of Part A of this chapter, we'll apply the lease classification criteria to a basic lease situation and consider, in order, (a) finance leases to the lessee / sales-type leases to the lessor and (b) operating leases.

Finance/Sales-Type Leases

Let's look at an example that illustrates the application of the classification criteria listed in Illustration 15–2. The earlier example in Illustration 15–3B, comparing a finance lease to an installment purchase, assumed lease payments at the *end* of each period. A more typical leasing contract requires lease payments at the *beginning* of each period. This more realistic payment schedule is assumed in Illustration 15–4.

● LO15–2

Illustration 15–4

Finance Lease/Sales-Type Lease: No Selling Profit

On January 1, 2021, Sans Serif Publishers leased printing equipment from First LeaseCorp. First LeaseCorp purchased the equipment from CompuDec Corporation at a cost of $479,079.

- The lease agreement specifies six annual payments of $100,000 beginning January 1, 2021, the beginning of the lease, and at each December 31 from 2021 through 2025.
- The six-year lease term ending December 31, 2026, is equal to the estimated useful life of the equipment.
- First LeaseCorp routinely acquires electronic equipment for lease to other firms.
- The interest rate in these financing arrangements is 10%.

How should this lease be classified by the lessee (finance or operating)? By the lessor (sales-type or operating)? We apply the five classification criteria:

1. Does the agreement specify that ownership of the asset transfers to the lessee?	No
2. Does this agreement contain a purchase option reasonably certain to be exercised?	No
3. Is the lease term for the "major part" of the estimated economic life of the asset?	Yes (6-year lease term; 6-year life)
4. Is the present value of the lease payments equal to or greater than "substantially all" of the fair value of the asset?	Yes ($479,079 present value; $479,079 fair value) $100,000 × 4.79079* = $479,079
5. Is the asset of such a specialized nature that it is expected to have no alternative use to the lessor at the end of the lease term?	No (First LeaseCorp routinely leases this type of equipment)

Decision: Since at least one (two in this case) of the five classification criteria is met, this is a finance lease to the lessee and a sales-type lease to the lessor.

*Present value of an annuity due of $1: $n = 6$, $i = 10\%$. Recall from Chapter 5 that we refer to periodic payments at the beginning of each period as an *annuity due*.

If at least one of the five classification criteria is met, this is a finance/sales-type lease.

Let's first think about this contract from the perspective of Sans Serif, the lessee. Because *at least one of the five classification criteria is met,* Sans Serif would classify the lease as a finance lease. While Sans Serif does not have legal ownership of the printing equipment as with a typical purchase, Sans Serif did acquire the right to use the printing equipment. That's why Sans Serif records a "right-of-use asset." In addition, just as if it were an installment purchase, Sans Serif also would record a liability for the present value of the lease payments. The entry by Sans Serif to record the finance lease is shown in Illustration 15–4A.

On the flip side of the transaction, just as if it were an installment sale, First LeaseCorp records a lease receivable for the present value of the payments to be received and removes from its balance sheet (derecognizes) the asset being leased. Later, we'll look at situations in which the lease receivable (reflecting the "selling price") exceeds the asset's carrying value,

Illustration 15–4A

Finance Lease / Sales-Type Lease: No Selling Profit

Using Excel, enter:
=PV(.10,6, −100000, 1)
Output: 479079

Using a calculator, enter: BEG mode N 6
I 10
PMT −100000 FV
Output: PV 479079

An asset and liability are recorded by the lessee at the present value of the lease payments.

Notice that the lessor's entries are the flip side or mirror image of the lessee's entries.

On January 1, 2021, Sans Serif Publishers leased printing equipment from First LeaseCorp. First LeaseCorp purchased the equipment from CompuDec Corporation at a cost of $479,079.

- The lease agreement specifies six annual payments of $100,000 beginning January 1, 2021, the beginning of the lease, and at each December 31 from 2021 through 2025.
- The six-year lease term ending December 31, 2026, is equal to the estimated useful life of the equipment.
- First Lease Corp routinely acquires electronic equipment for lease to other firms.
- The interest rate in these financing contracts is 10%.

The price Sans Serif pays for the right to control the use of the equipment is the present value of the lease payments:

$$\$100{,}000 \times 4.79079^{*} = \$479{,}079$$

Lease Payments — Right-of-Use Asset

*Present value of an annuity due of $1: $n = 6$, $i = 10\%$.

Beginning of the Lease (January 1, 2021)

Sans Serif (Lessee)		
Right-of-use asset (present value of lease payments)	479,079	
Lease payable (present value of lease payments)		479,079
First LeaseCorp (Lessor)		
Lease receivable (present value of lease payments)	479,079	
Equipment (lessor's cost: carrying amount)		479,079

creating the need to also record a selling profit for the difference. Indeed, some lessors use leasing as an alternative way to sell their assets and have no intention of reusing or re-leasing assets leased under a sales-type lease.

In the journal entries in Illustrations 15–4A and 15–4B and throughout the chapter, we often look at the entries of the lessee and the lessor together. This way, we can be reminded that, usually, the entries for the lessor are essentially the mirror image of those for the lessee, the other side of the same coin.[6]

Illustration 15–4B

Journal Entries for the First and Second Lease Payment

Lease payable/receivable:

$479,079
(100,000) 1st payment
$379,079
(62,092) 2nd payment
$316,987

First Lease Payment (January 1, 2021)

Sans Serif (Lessee)		
Lease payable	100,000	
Cash (lease payment)		100,000
First LeaseCorp (Lessor)		
Cash (lease payment)	100,000	
Lease receivable		100,000

Second Lease Payment (December 31, 2021)

Sans Serif (Lessee)		
Interest expense [10% × ($479,079 – $100,000)]	37,908	
Lease payable (difference)	62,092	
Cash (lease payment)		100,000
First LeaseCorp (Lessor)		
Cash (lease payment)	100,000	
Lease receivable		62,092
Interest revenue [10% × ($479,079 – $100,000)]		37,908

Recording Interest Expense/Interest Revenue

As lease payments are made over the term of the lease, both the lessee and lessor record interest at the effective interest rate. In addition, the lessee, Sans Serif, will record amortization expense on its right-of-use asset over the term of the lease. Let's first consider interest.

[5]We see later in the chapter that sometimes the lessee and lessor have accounting that is not a mirror image, particularly when the lease is either a sales-type lease with a selling profit or an operating lease. However, even when the two sides are not mirror images, there are many similarities, so the comparison still is helpful.

As shown in Illustration 15–4B, the entire $100,000 first lease payment is applied to principal (lease payable/receivable) reduction.[7] That's because the payment occurred at the beginning of the lease, so no interest had yet accrued. Subsequent lease payments, though, include interest of 10% on the outstanding balance as well as a portion that reduces that outstanding balance. As of the second lease payment twelve months later, one year's interest of **$37,908** has accrued on the $379,079 ($479,079 – $100,000) balance outstanding during 2021. After recording that interest, **$62,092** of the $100,000 payment remains to reduce the outstanding balance to $316,987 ($379,079 – $62,092).

Interest accrues at the effective rate on the balance outstanding during the period.

Because the outstanding balance declines with each payment, the amount of interest decreases each period. An amortization schedule is convenient to track the changing amounts, as shown in Illustration 15–4C. That schedule shows how the lease balance and the effective interest change over the six-year lease term, using an effective interest rate of 10%. Each lease payment after the first one includes both an amount that represents interest and an amount that represents a reduction of the outstanding balance. The periodic reduction is sufficient that, *at the end of the lease term, the outstanding balance is zero.*

Both the lessee and lessor would use this same amortization schedule for recording interest. The lessee amortizes its lease payable and records interest *expense*. Similarly, the lessor amortizes its lease receivable and records interest *revenue*, reflecting the opposite side of the same transaction.

Illustration 15–4C

Lease Amortization Schedule

	Payments	Effective Interest	Decrease in Balance*	Outstanding Balance
		(10% × Outstanding balance)		
1/1/21				479,079
1/1/21	100,000		100,000	**379,079**
12/31/21	100,000	0.10 (**379,079**) = **37,908**	**62,092**	316,987
12/31/22	100,000	0.10 (316,987) = 31,699	68,301	248,686
12/31/23	100,000	0.10 (248,686) = 24,869	75,131	173,555
12/31/24	100,000	0.10 (173,555) = 17,355	82,645	90,910
12/31/25	100,000	0.10 (90,910) = 9,090*	90,910	0
	600,000	120,921†	479,079	

*Payment in first column minus the interest in the second column.
†Adjusted for rounding of other numbers in the schedule.

The first lease payment includes no interest.

The total of the cash payments ($600,000) provides for:
1. Payment for the equipment's use ($479,079).
2. Interest ($120,921) at an effective rate of 10%.

Now that we've seen how the lessee amortizes its *lease liability* to record interest and pay the liability, let's see how the lessee amortizes its *right-of-use asset* over the term of the lease.

Recording Amortization of the Right-of-Use Asset

Like other noncurrent assets, the lessee's right-of-use asset provides an economic benefit (the right to use a productive asset) over the period covered by the lease term. So, the lessee amortizes its right-of-use asset over the lease term. The amortization process usually is on a straight-line basis unless the lessee's pattern of using the asset is different.[8] That amortization results in an *expense* for the lessee. In Sans Serif's case, it amortizes its right-of-use asset over the six-year lease term.

[7] Another way to view this is to think of the first $100,000 as a down payment with the remaining $379,079 financed by 5 (i.e., 6 – 1) *year-end* lease payments.

[8] Output measures such as units produced or input measures such as hours used might provide a better indication of the reduction in some leased assets.

The lessee incurs an expense as it uses the asset.

December 31, 2021, and End of Next Five Years		
Amortization expense ($479,079 ÷ 6 years)	79,847	
Right-of-use asset		79,847

The amortization period is restricted to the lease term unless the lease provides for transfer of title or a BPO.

AMORTIZATION PERIOD The lessee normally should amortize a leased asset over the term of the lease. However, if ownership transfers, or exercise of a purchase option is reasonably certain (i.e., either of the first two classification criteria is met), the asset should be amortized over its useful life. Meeting either of those two criteria means the lessee will have the asset beyond the lease term, and thus the asset is amortized over the useful life of the asset *to the lessee* whether or not that useful life is limited by the term of the lease.

Concept Review Exercise

FINANCE LEASE / SALES-TYPE LEASE: NO SELLING PROFIT

United Cellular Systems leased a satellite transmission device from Pinnacle Leasing Services on January 1, 2021. Pinnacle paid $625,483 for the transmission device. Its fair value is $625,483.

Terms of the Lease Agreement and Related Information:	
Lease term	3 years (6 semiannual periods)
Semiannual lease payments	$120,000 at beginning of each period
Economic life of asset	3 years
Interest rate	12%

Required:

1. Prepare the appropriate entries for both United Cellular Systems and Pinnacle Leasing Services on January 1, 2021, the beginning of the lease.
2. Prepare an amortization schedule that shows the pattern of interest expense for United Cellular Systems and interest revenue for Pinnacle Leasing Services over the lease term.
3. Prepare the appropriate entries to record the second lease payment on July 1, 2021, and adjusting entries on December 31, 2021 (the end of both companies' fiscal years).

Solution:

1. Prepare the appropriate entries for both United Cellular Systems and Pinnacle Leasing Services on January 1, 2021, the beginning of the lease.

Calculation of the present value of lease payments.

Present value of periodic lease payments:

($120,000 × 5.21236*) = $625,483

* Present value of an annuity due of $: $n = 6$, $i = 6\%$.

January 1, 2021		
United Cellular Systems (Lessee)		
Right-of-use asset (calculated above)	625,483	
Lease payable (calculated above)		625,483
Lease payable	120,000	
Cash (lease payment)		120,000
Pinnacle Leasing Services (Lessor)		
Lease receivable (calculated above)	625,483	
Equipment (lessor's cost)		625,483
Cash (lease payment)	120,000	
Lease receivable		120,000

2. Prepare an amortization schedule that shows the pattern of interest expense for United Cellular Systems and interest revenue for Pinnacle Leasing Services over the lease term.

Date	Payments	Effective Interest	Decrease in Balance	Outstanding Balance
		(6% × Outstanding balance)		
1/1/21				625,483
1/1/21	120,000		120,000	505,483
7/1/21	120,000	0.06 (505,483) = 30,329	89,671	415,812
1/1/22	120,000	0.06 (415,812) = 24,949	95,051	320,761
7/1/22	120,000	0.06 (320,761) = 19,246	100,754	220,007
1/1/23	120,000	0.06 (220,007) = 13,200	106,800	113,207
7/1/23	120,000	0.06 (113,207) = 6,793*	113,207	0
	720,000	94,517	625,483	

*Adjusted for rounding of other numbers in the schedule.

3. Prepare the appropriate entries to record the second lease payment on July 1, 2021, and adjusting entries on December 31, 2021 (the end of both companies' fiscal years).

July 1, 2021		
United Cellular Systems (Lessee)		
Interest expense [6% × ($625,483 − $120,000)]	30,329	
Lease payable (difference)	89,671	
Cash (lease payment)		120,000
Pinnacle Leasing Services (Lessor)		
Cash (lease payment)	120,000	
Lease receivable (difference)		89,671
Interest revenue [6% × ($625,483 − $120,000)]		30,329
December 31, 2021		
United Cellular Systems (Lessee)		
Interest expense (6% × $415,812: from schedule)	24,949	
Interest payable		24,949
Amortization expense ($625,483 ÷ 3 years)	208,494	
Right-of-use asset		208,494
Pinnacle Leasing Services (Lessor)		
Interest receivable	24,949	
Interest revenue (6% × $415,812: from schedule)		24,949

In our illustrations above, the present value of the lease payments, or "selling price," is the same as the cost or carrying value of the asset "sold." For example, in Illustration 15–4A, we assumed that First LeaseCorp bought the equipment for $479,079 and then leased it for the same price. There was no profit on the "sale" itself. The only income derived by the lessor was interest revenue earned over the lease term. In effect, First LeaseCorp financed the acquisition of the equipment by Sans Serif Publishers.[9] That's usually the case when the lessor is a financial intermediary that provides financing by acquiring the asset and then leasing it to the lessee. In that case, the lessor records the "flip side" of the lessee's transaction. That is, the lessor records a receivable corresponding to the lessee's liability and removes the asset from its books as the lessee adds the asset to its books.

In some cases, though, the lessor is not a financial intermediary, but is a manufacturer or retailer using leases as a means of "selling" its products. In addition to interest revenue earned over the lease term, the lessor receives a selling profit on the "sale" of the asset. We look at this situation next.

[9]Under the GAAP in place prior to ASU 2016-02, this type of lease (no selling profit) was referred to as a *direct financing lease.* While that label still exists in the new lease guidance, it now refers to a situation that happens only when the lease includes a residual value and the residual value is *guaranteed by a third party* (not the lessee). We consider guaranteed residual values later in the chapter when we look at uncertainties in lease accounting.

Sales-Type Leases with Selling Profit

● LO15–3

FINANCIAL Reporting Case

Q2, p. 835

Selling profit exists when the fair value of the asset (usually the present value of the lease payments, or "selling price") exceeds the cost or carrying value of the asset sold. In addition to interest revenue earned over the lease term, the lessor recognizes a selling profit on the "sale" of the asset. Accounting is the same as for a sales-type lease without a selling profit except that profit is recognized at the beginning.[10]

As in the sale of any product, gross profit is the difference between sales revenue and cost of goods sold. **Dell Technologies Inc.**, "sells" some of its products using sales-type leases and disclosed its accounting policy in a recent annual report shown in Illustration 15–5.

Illustration 15–5

Sales-Type Leases—Dell Technologies Inc.

Real World Financials

Note 1 (in part)

The Company records revenue from the sale of equipment under sales-type leases as product revenue in an amount equal to the present value of minimum lease payments at the inception of the lease. Sales-type leases also produce financing income, which is included in net products revenue in the Consolidated Statements of Income (Loss) and is recognized at consistent rates of return over the lease term. Revenue from operating leases is recognized over the lease period.

Source: Dell Technologies Inc.

When there is a selling profit, all lessor entries, other than the entry at the beginning of the lease to include the selling profit, are precisely the same as the entries for a sales-type lease without a selling profit. On the other side of the transaction, accounting by the *lessee* is not affected by whether the *lessor* recognizes a profit or not. All lessee entries are exactly the same as in our previous illustration of a lessee's finance lease.

To illustrate accounting for sales-type leases with selling profit, let's modify our previous illustration. Assume all facts are the same except Sans Serif Publishers leased the equipment directly from CompuDec Corporation rather than through First LeaseCorp, the financing intermediary. Also assume that CompuDec's cost of the equipment was **$300,000**. If you recall that the lease payments (their present value) provide a selling price of $479,079, you see that CompuDec earns a gross profit on the sale of $479,079 – $300,000 = $179,079. This sales-type lease that provides a selling profit is demonstrated in Illustration 15–6.

Illustration 15–6

Finance Lease / Sales-Type Lease: With Selling Profit

Because the PV of lease payments > cost, the lessor has a profit that is recorded at the beginning of the lease.

On January 1, 2021, Sans Serif Publishers leased printing equipment from CompuDec Corporation.

- The lease agreement specifies six annual payments of $100,000 beginning January 1, 2021, the beginning of the lease, and at each December 31 from 2021 through 2025.
- The six-year lease term ending December 31, 2026, is equal to the estimated useful life of the equipment.
- CompuDec manufactured the equipment at a cost of **$300,000**.

Using an interest rate of 10% for financing this transaction, CompuDec calculates the present value of the lease payments to be received as **$479,079** ($100,000 × 4.79079*).

Sales revenue	**$479,079**
– COGS	**300,000**
Selling profit	**$179,079**

Beginning of the Lease (January 1, 2021)

CompuDec Corporation (Lessor)		
Lease receivable (present value of lease payments)	479,079	
Cost of goods sold (lessor's cost)	**300,000**	
Sales revenue (present value of lease payments)		**479,079**
Equipment (lessor's cost)		300,000

No interest has yet accrued when the first payment is made at the beginning of the lease.

First Lease Payment (January 1, 2021)

CompuDec Corporation (Lessor)		
Cash	100,000	
Lease receivable		100,000

*Present value of an annuity due of $1: $n = 6$, $i = 10\%$.

[10]It's possible that the asset's carrying value will exceed its fair value, in which case a selling loss should be recorded.

You should recognize the similarity between recording both the revenue and cost components of this "sale by lease" and recording the same components of other sales transactions. Remember, when a company sells a product on account, two entries are recorded: one to record the receivable and sales revenue, and another to record the cost of goods sold and corresponding reduction in inventory. Let's say you purchase a TV from **Best Buy** for $479 and pay for it with your Best Buy credit card. And, assume Best Buy paid the wholesale price of $300 to **Samsung** to acquire the TV. Here's the way Best Buy would record the sale to you:

Recording a sales-type lease is similar to recording a sale of merchandise on account:

A/R	(price)	
Sales rev		(price)
COGS	(cost)	
Inventory		(cost)

Accounts receivable (price)	479	
Sales revenue (price)		479
Cost of goods sold (cost)	300	
Inventory (cost)		300

Now compare those entries to the entry to record our sale by lease. We'll reorder a few lines in the entry to make that comparison easier:

Lease receivable (price)	479,079	
Sales revenue (price)		479,079
Cost of goods sold (cost)	300,000	
Equipment (cost)		300,000

In both cases, the seller is recording revenue and cost, and gross profit from the transaction will appear on the seller's income statement. Economically, leases are just one of the arrangements that sellers can use to help customers buy their products. Sales-type leases must be separately distinguished from nonlease sales in the financial statements.

Let's turn our attention now to situations in which at least one of the classification criteria is *not* met, so that the lease is classified as an operating lease.

Operating Leases

● LO15–4

FINANCIAL Reporting Case

In the finance/sales-type leases we've looked at, the lease is economically similar to the purchase of an asset, because the lessee has the ability to direct the use of the asset and obtain substantially all of its remaining benefits. If a lease doesn't meet any of the criteria for a finance/sales-type lease in Illustration 15–2, then it's considered to be more in the nature of a rental agreement for a period of time. We refer to this second type of arrangement as an **operating lease.**[11]

For the *lessor,* this distinction means *not* recording a lease receivable or derecognizing the leased asset (taking it off the balance sheet) for an operating lease as happens in a sales-type lease. It also means reporting *lease revenue* as an equal amount each period rather than a declining amount of interest revenue on a lease receivable as in a sales-type lease.

On the other hand, the *lessee* in an operating lease *does* report a right-of-use asset and a lease liability in its balance sheet exactly as in a finance lease. However, the lessee will calculate the amortization of the right-of-use asset a different way than it would for a finance lease (as we see demonstrated in the next illustration). Then, that amortization will be combined with the interest expense calculated on the lease liability, and the result will be an equal amount of *lease expense* reported in the lessee's income statement for each period of an operating lease.

So, the lessor and lessee both report straight-line amounts in their income statements for operating leases to reflect the fact that we consider an *operating lease* as a *straight-line rental* of the asset during the lease term.

To demonstrate accounting for operating leases, we need to change our example so none of the finance lease classification criteria in Illustration 15–2 is met. We'll do that by

[11]The term *operating lease* got its name long ago when a lessee routinely received from the lessor an operator along with leased equipment.

changing the lease term. In the previous illustrations, the lease term is equal to the expected useful life of the asset (six years). Now, in Illustration 15–7, the lease term is for only *four* years of the asset's six-year life, which is less than the "major part" of approximately 75%. Assuming none of the other criteria are met, the arrangement is classified as an operating lease. (See Illustration 15–7.)

Illustration 15–7
Operating Lease

On January 1, 2021, Sans Serif Publishers leased printing equipment from First LeaseCorp.

- First LeaseCorp purchased the equipment from CompuDec Corporation at a cost of $479,079.
- Sans Serif's borrowing rate for similar transactions is 10%.
- The lease agreement specifies **four** annual payments of $100,000 beginning January 1, 2021, the beginning of the lease, and at each December 31 from 2021 through 2023.
- The useful life of the equipment is estimated to be six years.

The price Sans Serif pays for the right to control the use of the equipment is the present value of the lease payments:

$100,000 × 3.48685* = **$348,685**
Lease Payments — Lessee's cost

*Present value of an annuity due of $1: $n = 4$, $i = 10\%$.

The lease doesn't meet any of the criteria listed in Illustration 15–2, so it is accounted for as an operating lease.

Beginning of the Lease (January 1, 2021)

Sans Serif (Lessee)

Right-of-use asset (present value of lease payments)	348,685	
Lease payable (present value of lease payments)		348,685

First LeaseCorp (Lessor)
(No entry to record a receivable or to derecognize asset)

By signing a lease, the lessee has an obligation to make payments (a liability), and in exchange for those payments, the lessee is receiving the right to use a specified asset (a right-of-use asset).

How do we account for an operating lease? The lessee still records an asset and a liability at the beginning of the lease. The reason is that Sans Serif, although not having effectively purchased the asset via the lease agreement, still has acquired the *right to use the asset* for four years of the asset's six-year useful life. Just as in the case of a finance lease, the right to use the leased asset can be a significant benefit, even if for a shorter period of time, and the promise to make the lease payments can be a significant obligation. So, just like our prior example in which the asset was leased for its entire life, Sans Serif still recognizes a right-of-use asset and lease liability. The only difference in the initial entry is that, because the lease includes fewer payments, a smaller present value is recorded at the beginning of the lease.

Although an operating lease liability is reported in the balance sheet of the lessee, it is designated as a "non-debt liability" in order to distinguish it from traditional liabilities, including finance lease liabilities. That separate classification is intended to prevent operating lease liabilities from causing companies to violate debt covenants, such as restrictions that the debt to equity ratio not exceed a preset limit.

The lessor does not record a lease receivable. Instead, the lessor views an operating lease as simply renting the asset to the lessee, and, as we'll see next, records rent revenue on a straight-line basis over the lease term.

Recording Lease Expense/Lease Revenue

For an operating lease, the *lessee* will report a single lease expense rather than the separate interest and amortization associated with a finance lease. However, it's convenient to think of this single lease expense as consisting of those two components (interest and amortization). Although ASC 842 doesn't specify this decomposition, the lessee must (a) calculate the interest component in order to determine the reduction of the lease payable over the lease term and (b) calculate the amortization component in order to determine the reduction in the balance of the right-of-use asset over the lease term. So, for convenience and to more easily compare accounting for operating leases with finance leases, in this chapter we determine and *record* these two "components," and then combine them to report a single lease expense amount.[12]

On the other hand, because the *lessor* in an operating lease does not record a lease receivable that would require determining a balance reduction (as happens in a sales-type lease), the lessor will simply record and report a single straight-line lease revenue. This lease revenue amount is the same as the lessee's lease expense.

INTEREST Sans Serif (the lessee) determines interest on the lease payable in an operating lease the same way as in a finance lease, as demonstrated in Illustration 15–7A.

Illustration 15–7A

Journal Entries for the First and Second Lease Payment

First Lease Payment (January 1, 2021)

Sans Serif (Lessee)		
Lease payable	100,000	
Cash (lease payment)		100,000
First LeaseCorp (Lessor)		
Cash (lease payment)	100,000	
Deferred lease revenue (lease revenue in 2021)		100,000

Second Lease Payment (December 31, 2021)

Sans Serif (Lessee)		
Interest expense [10% × (**$348,685** – **100,000**)]	**24,869**	
Lease payable (difference)	**75,131**	
Cash (lease payment)		100,000
First LeaseCorp (Lessor)		
Deferred lease revenue (Jan. 1 lease payment)	100,000	
Lease revenue		100,000
Cash (second lease payment)	100,000	
Deferred lease revenue (lease revenue in 2022)		100,000
Depreciation expense ($479,079 ÷ 6 years)	79,847	
Accumulated depreciation		79,847

Lease payable

$ 348,685
$(100,000) 1st payment
$ 248,685
(75,131) 2nd payment
$ 173,554

The lessor recognizes the lease revenue during the year after it is received.

The amortization schedule in Illustration 15–7B shows how the lease liability balance and the effective interest change over the four-year lease term.

Illustration 15–7B

Lease Amortization Schedule

	Payments	Effective Interest	Decrease in Balance	Outstanding Balance
		(10% × Outstanding balance)		
1/1/21				348,685
1/1/21	100,000		100,000	**248,685**
12/31/21	100,000	0.10 (**248,685**) = **24,869**	**75,131**	173,554
12/31/22	100,000	0.10 (173,554) = 17,355	82,645	90,909
12/31/23	100,000	0.10 (90,909) = 9,091	90,909	0
	400,000	**51,315**	**348,685**	

The first lease payment includes no interest.

The total of the cash payments ($400,000) provides for:

1. **Payment for the equipment's use ($348,685).**
2. **Interest ($51,315) at an effective rate of 10%.**

AMORTIZATION OF THE RIGHT-OF-USE ASSET In an operating lease we determine the amortization of the right-of-use asset differently from the way we amortize it in a finance lease. The disparity considers the nature of finance leases and operating leases.

[12]The FASB decided not to require companies to separately disclose interest and amortization components because, for many companies, it would be costly due to the fact that it would require new systems capabilities and processes.

In an operating lease, the lessee records interest the normal way and then "plugs" the right-of-use asset amortization at whatever amount is needed for interest plus amortization to equal the *straight-line* lease payment.

We can think of lease expenses in a *finance lease* as reflecting (a) the right to use the asset (amortization) plus (b) the financing of that right (interest). On the other hand, in an *operating lease,* lease expense is reported in a manner that is designed to mirror *straight-line rental* of the asset during the lease term. The way the lessee accomplishes that is to determine interest as in Illustration 15–7A and 15-7B and then to determine amortization of the right-of-use asset as the *amount needed to cause the total lease expense (interest plus amortization) to be an equal, straight-line amount over the lease term.* In other words, the lessee determines interest the normal way and then "plugs" the right-of-use asset amortization at whatever amount is needed for interest plus amortization to equal the straight-line lease payment. So, in Illustration 15–7C, Sans Serif, having determined interest to be **$24,869**, plugs an amortization amount of $75,131 to cause the total of the two components of the reported lease expense to be $100,000.

Interest	**$ 24,869**
Amortization	75,131
Lease expense	$100,000

Illustration 15–7C

Lessee Entries for the Second Lease Payment: Interest and Amortization

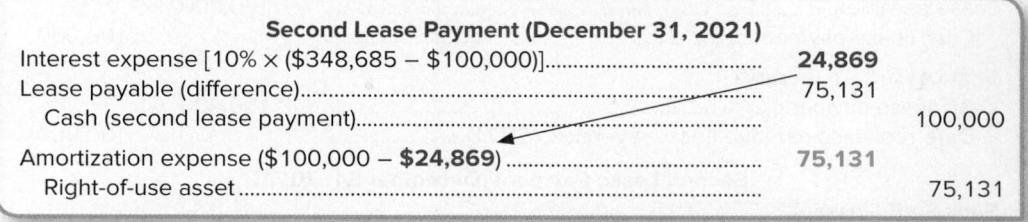

Second Lease Payment (December 31, 2021)		
Interest expense [10% × ($348,685 – $100,000)]	**24,869**	
Lease payable (difference)	75,131	
Cash (second lease payment)		100,000
Amortization expense ($100,000 – **$24,869**)	**75,131**	
Right-of-use asset		75,131

Total lease expense (amortization plus interest) for Sans Serif will equal $100,000 each year over the lease term.

The lessor recognizes the lease revenue during the year for which payment is received at the beginning of the year.

LESSOR On the flip side of the transaction, the lessor in an operating lease does not record a lease receivable at the beginning of the lease and does not remove from its balance sheet the asset being leased. With no receivable to accrue interest, First LeaseCorp simply records straight-line revenue as a single lease revenue amount equal to the $100,000 lease payments and equal to the $100,000 amount the lessee reports as lease expense. First LeaseCorp also depreciates the asset over the asset's useful life (six years in this example), just as it would any other asset it owns and keeps on its books.

Additional Consideration

Because ASC 842 doesn't require recording interest and amortization separately for an operating lease, the two entries in Illustration 15–7C can be combined and recorded as follows:

December 31, 2021		
Lease expense	100,000[a]	
Lease payable	75,131[b]	
Right-of-use asset		75,131[b]
Cash		100,000

[a]Includes the first year "interest component," $24,869 (= 10% × [$348,685 – $100,000]) and the "amortization component," $75,131 (= $100,000 – $24,869).
[b]$100,000 cash payment minus first year "interest component," $24,869 (= 10% × [$348,685 – $100,000])

In an operating lease, the lessee combines interest expense and amortization expense to report a single lease expense in the income statement.

Reporting Lease Expense and Lease Revenue

After recording these entries, the lessee will have two lease-related expenses—interest expense and amortization expense. However, the lessee combines these two accounts into a single *lease expense* and reports a single $100,000 amount each year in its income statement.

This is in keeping with the key objective of reporting a straight-line lease expense for an operating lease. Note that this is different than in a finance lease in which the lessee will report interest expense and amortization expense *separately* in the income statement.

First LeaseCorp, the lessor, has only a single lease revenue account in an operating lease and reports that straight-line amount, $100,000, each year in its income statement. Thus, in keeping with the presumption that an operating lease is considered to be more in the nature of a rental agreement, the lessee reports ***lease expense on a straight-line basis*** and the lessor reports ***lease revenue on a straight-line basis*** over the lease term. Illustration 15–7D shows this accounting for both the lessee and the lessor over the life of the lease.

Illustration 15–7D Operating Lease—Determining Lease Expense and Revenue

	Jan. 1, 2021		Dec. 31, 2021		Dec. 31, 2022		Dec. 31, 2023		Dec. 31, 2024	
Sans Serif (Lessee)										
Right-of-use asset	348,685									
Lease payable		348,685								
Interest expense	0		**24,869**		**17,355**		**9,091**		**0**	
Lease payable	100,000		75,131		82,645		90,909		0	
Cash		100,000		100,000		100,000		100,000		0
Amortization expense*			**75,131**		**82,645**		**90,909**		**100,000**	
Right-of-use asset				75,131		82,645		90,909		100,000
Total lease expenses each year =			**$100,000**		**$100,000**		**$100,000**		**$100,000**	
First LeaseCorp (Lessor)										
[No entry to record receivable or to derecognize asset]										
Cash	100,000		100,000		100,000		100,000			
Deferred revenue**		100,000		100,000		100,000		100,000		
Deferred revenue**			100,000		100,000		100,000		100,000	
Lease revenue				100,000		100,000		100,000		100,000
Depreciation expense†			79,847		79,847		79,847		79,847	
Accumulated depreciation				79,847		79,847		79,847		79,847

*Plug to cause the *total* lease expense to equal the straight-line amount: **$100,000 minus interest**.
**When the $100,000 is received at the beginning of the lease year, it is not yet earned and should be recorded as deferred lease revenue. When it has been earned by the end of the year, it becomes lease revenue.
†($479,079 ÷ 6). Depreciation expense also is recognized in 2025 and 2026, because the asset has a six-year useful life.

COMPARISON OF LESSEE'S EXPENSE RECOGNITION BETWEEN FINANCE AND OPERATING LEASES Now, let's turn to Illustration 15–8 to compare how the lessee reports its lease expense for a finance lease and an operating lease. Reflecting back to our illustrations for a finance lease, you should note that the lessee records more expense and the lessor records more revenue early in the life of a finance lease. This "front loading" of lease expense/revenue occurs due to the fact that interest is higher initially than it is in the later stages of a lease, while amortization expense for the lessee's right-of-use asset remains the same each period.[13] Operating leases avoid front loading. As you might imagine, lessees tend to prefer the operating lease classification. It defers expense recognition, making net income higher in the early years of the lease.

[13] Because the lessor, First LeaseCorp, removed the asset from its records at the beginning of the finance lease, it would not have depreciation (amortization) to record.

Illustration 15–8

Comparison of Lessee's Expense Recognition between Finance and Operating Leases

In an operating lease, it's the *total* lease expense, not the amortization component, that's a straight-line amount.

	Finance Lease Financing Approach			**Operating Lease Straight-Line Approach**		
	Interest Expense	**Amortization Expense**	**Total Expense**	**Interest Expense**	**Amortization Expense**	**Total Expense**
2021	**24,869**	87,171	112,040	**24,869**	**75,131**	100,000
2022	17,355	87,171	104,526	17,355	82,645	100,000
2023	9,091	87,171	96,262	9,091	90,909	100,000
2024	0	87,171	87,171	0	100,000	100,000
	51,315	348,685*	400,000*	51,315	348,685	400,000

*Adjusted for rounding of other numbers in the schedule.

International Financial Reporting Standards

● LO15–9

No Operating Leases for Lessees under IFRS. Under *IFRS No. 16,* all leases are accounted for as finance leases by the lessee (one-model approach). Only lessors apply the classification criteria to distinguish between finance and operating leases. Thus, even for leases that qualify under U.S. GAAP (two-model approach) as operating leases, the lessee amortizes the right-of-use asset on a straight-line basis rather than "plugging" that amount to cause the total of interest and amortization to be a straight-line amount. This means that those leases under IFRS will have a front-loaded expense profile (because interest expense is more at the beginning than at the end).

Decision Makers' Perspective—Financial Statement Impact

As indicated in the Decision Makers' Perspective earlier in the chapter, leasing can allow a firm to conserve assets, to avoid some risks of owning assets, and to obtain favorable tax benefits. These advantages are desirable. Accounting guidelines are designed to limit the ability of firms to obscure the realities of their financial position through off-balance-sheet financing or by avoiding violating terms of contracts that limit the amount of debt a company can have. Nevertheless, investors and creditors should be alert to the impact leases can have on a company's financial position and on its risk. Disclosure requirements include reporting finance lease liabilities and operating lease liabilities separately.

A noteworthy benefit to lessees of operating lease accounting is that expenses are recognized straight line, rather than front-loaded as in the case of finance leases for which interest expense and amortization are separately reported. In fact, the leasing industry lobbied the FASB extensively to make sure operating lease accounting was allowed. Lessees often went to extraordinary lengths to cause leases to qualify as operating leases under preexisting GAAP in order to avoid reporting lease liabilities in the balance sheet as well as to report straight-line lease expense. Now, following the new lease accounting guidance, many lessees continue to devise schemes to meet the criteria for operating leases so as to reap the benefits of straight-line lease expensing that those leases still convey.

From the perspective of an analyst, though, it's unclear why expenses associated with operating leases should be treated any differently than are expenses associated with finance leases. Consequently, even though a goal of ASC Topic 842: "Leases" was to reduce the need for investors to restate financial statements "as-if" leased assets and their related liabilities were on the balance sheet, it is likely that some sophisticated users now will restate financial statements to back out the straight-line recognition afforded operating leases and treat them like finance leases, similar to how they are treated under IFRS. ●

Discount Rate

An important factor in lease calculations that we've glossed over until now is the discount (interest) rate used in present value calculations. Because lease payments occur in future periods, we must consider the time value of money when evaluating their present value. The rate is important because it influences virtually every amount reported by both the lessor and the lessee in connection with the lease.

One discount rate is implicit in the lease agreement. This is the effective interest *rate of return the lease payments provide the lessor* under the lease. It is the desired rate of return the lessor has in mind when deciding the size of the lease payments. (Refer to our earlier calculations of the periodic lease payments.) If known by the lessee, the lessee also should make its calculations using the lessor's rate implicit in the lease agreement.

The lessee uses the interest rate implicit in the lease if known; otherwise the lessee uses its own incremental borrowing rate.

What if the lessee is unaware of the lessor's implicit rate? This is frequently the case in practice. This might happen, for example, if the leased asset has a relatively high residual value. As we will discuss later, if there's a residual value (guaranteed or not) it is an ingredient in the lessor's calculation of the lease payments. Sometimes it may be hard for the lessee to identify the residual value estimated by the lessor if the lessor chooses not to make it known.[14] However, as the lease term and risk of obsolescence increase, the residual value typically is less of a factor.

When the lessor's implicit rate is not known, the lessee should use its own incremental borrowing rate. This is the rate the lessee would expect to pay a bank if funds were borrowed to buy the asset.[15]

Concept Review Exercise

OPERATING LEASE

United Cellular Systems leased a satellite transmission device from Pinnacle Leasing Services on January 1, 2021. Pinnacle paid $800,000 for the transmission device. Its fair value is $800,000.

Terms of the Lease Agreement and Related Information:

Lease term	3 years (6 semiannual periods)
Semiannual lease payments	$120,000 at beginning of each period
Economic life of asset	5 years
Interest rate	12%

Required:

1. Prepare the appropriate entries for both United Cellular Systems and Pinnacle Leasing Services on January 1, 2021, the beginning of the lease.
2. Prepare an amortization schedule that shows the pattern of interest expense for United Cellular Systems and interest revenue for Pinnacle Leasing Services over the lease term.
3. Prepare the appropriate entries to *record* the second lease payment on July 1, 2021, and adjusting entries on December 31, 2021 (the end of both companies' fiscal years).
4. What will United Cellular Systems *report* in its income statement in connection with the lease for the year ended December 31, 2021?

Solution:

1. Prepare the appropriate entries for both United Cellular Systems and Pinnacle Leasing Services on January 1, 2021, the beginning of the lease.

[14] Disclosure requirements provide that the lessor must disclose the components of its investments in nonoperating leases, which would include any estimated residual values. But the disclosures are aggregate amounts, not amounts of individual leased assets.

[15] *Incremental borrowing rate* refers to the fact that lending institutions tend to view debt as being increasingly risky as the level of debt increases. Thus, additional (i.e. incremental) debt is likely to be loaned at a higher interest rate than existing debt, other things being equal.

Calculation of the present value of lease payments.

Present value of periodic lease payments:

$$(\$120{,}000 \times 5.21236^{*}) = \$625{,}483$$

*Present value of an annuity due of $1: $n = 6$, $i = 6\%$.

January 1, 2021

United Cellular Systems (Lessee)		
Right-of-use asset (calculated above)	625,483	
Lease payable (calculated above)		625,483
Lease payable	120,000	
Cash (lease payment)		120,000
Pinnacle Leasing Services (Lessor)		
[No entry to record a receivable or to remove asset]		
Cash (lease payment)	120,000	
Deferred lease revenue		120,000

2. Prepare an amortization schedule that shows the pattern of interest expense for United Cellular Systems and interest revenue for Pinnacle Leasing Services over the lease term.

Date	Payments	Effective Interest	Decrease in Balance	Outstanding Balance
		(6% × Outstanding balance)		
1/1/21				**625,483**
1/1/21	120,000		120,000	505,483
7/1/21	120,000	0.06 (505,483) = 30,329	89,671	415,812
1/1/22	120,000	0.06 (415,812) = 24,949	95,051	320,761
7/1/22	120,000	0.06 (320,761) = 19,246	100,754	220,007
1/1/23	120,000	0.06 (220,007) = 13,200	106,800	113,207
7/1/23	120,000	0.06 (113,207) = 6,793*	113,207	0
	720,000	94,517	**625,483**	

*Adjusted for rounding of other numbers in the schedule.

3. Prepare the appropriate entries to record the second lease payment on July 1, 2021, and adjusting entries on December 31, 2021 (the end of both companies' fiscal years).

July 1,2021

United Cellular Systems (Lessee)		
Interest expense [6% × ($625,483 – $120,000)]	30,329	
Lease payable (difference)	89,671	
Cash (lease payment)		120,000
Amortization expense ($120,000 – $30,329)	89,671	
Right-of-use asset		89,671
Pinnacle Leasing Services (Lessor)		
Deferred lease revenue (Jan. 1 lease payment)	120,000	
Lease revenue		120,000
Cash (second lease payment)	120,000	
Deferred lease revenue		120,000

December 31, 2021

United Cellular Systems (Lessee)		
Interest expense (6% × $415,812: from schedule)	24,949	
Interest payable		24,949
Amortization expense ($120,000 – $24,949)	95,051	
Right-of-use asset		95,051
Pinnacle Leasing Services (Lessor)		
Deferred lease revenue (July 1 lease payment)	120,000	
Lease revenue		120,000
Depreciation expense ($800,000 ÷ 5 years)	160,000	
Accumulated depreciation		160,000

4. What will United Cellular Systems report in its income statement in connection with the lease for the year ended December 31, 2021?

In an operating lease, the lessee combines interest expense and amortization expense to report a *single lease expense* in the income statement. In 2021, that will be a single lease expense of $240,000 (= $30,329 + $89,671 + $24,949 + $95,051).

Short-Term Leases—A ShortCut Method

LO15–5

It's not unusual to simplify accounting for situations in which doing so has no material effect on the financial statements. One such situation that permits a simpler application is a short-term lease. A lease is considered a "short-term lease" if it:

1. Has a lease term (including any options to renew or extend) of twelve months or less, *and*
2. Does not contain a purchase option that the lessee is reasonably certain to exercise, which would extend the term beyond twelve months.

In a short-term lease, the lessee can elect not to record a right-of-use asset and lease payable at the beginning of the lease term, but instead to simply record lease payments as expense as they occur.

The shortcut approach for short-term leases permits the lessee to choose *not* to record an asset and related liability associated with the lease at the beginning of the lease term. Instead, the lessee can simply record lease payments as rent expense over the lease term. Yes, this is the approach used by the lessor for lease revenue on the flip side of the transaction.

Let's look at an example that illustrates the relatively straightforward accounting for short-term leases. To do this, in Illustration 15–9 we modify Illustration 15–4A to assume the lease term is *twelve months.*

Illustration 15–9
Short-Term Lease

On January 1, 2021, Sans Serif Publishers leased printing equipment from First LeaseCorp. The lease agreement specifies twelve monthly payments of $8,333 beginning January 1, 2021, the beginning of the lease, and at the first day of each of the next eleven months. The useful life of the equipment is estimated to be six years.

Beginning of the Lease (January 1, 2021)

No entry to record a right-of-use asset and liability (lease term ≤ one year)

Monthly Lease Payments (January 1–December 1, 2021)

Lease expense	8,333	
Cash		8,333

International Financial Reporting Standards

LO15–9

Shortcut Method. As in the case for short-term leases in U.S. GAAP, IFRS allows the lessee to elect not to record a right-of-use asset and lease payable at the beginning of the lease term, but instead to simply record lease payments as an expense as they occur. U.S. GAAP, though, defines a short-term lease as a lease that has a lease term of 12 months or less and does not include a purchase option *that the lessee is reasonably certain to exercise.* IFRS precludes a lease from being considered short term if the lease includes a purchase option regardless of whether the lessee is reasonably certain to exercise it.

In addition, though, unlike U.S. GAAP, IFRS allows "small ticket leases" also to apply this shortcut method. Small ticket leases are defined as those having a value of $5,000 or less.

Uncertainty in Lease Transactions

PART B

LO15–6

Let's turn our attention now to several issues related to uncertainty in lease transactions that impact lessee and lessor accounting. Among those issues are variable lease terms, variable lease payments, residual values, purchase options, and termination penalties.

What if the Lease Term is Uncertain?

The lease term is the contractual lease term modified by any renewal or termination options that are reasonably certain to be exercised.

Sometimes the actual term of a lease is not obvious. Suppose, for instance, that the lease term is specified as four years, but it can be renewed at the option of the lessee for two additional years. Or maybe either party can terminate the lease after, say, three years. In such situations, we consider the lease term to be the contractual lease term adjusted for any periods covered by options to extend or terminate the lease for which there is a significant economic incentive to exercise the options. A company adjusts the lease term for an option only if it is "reasonably certain" that the lessee will exercise the option after considering the relevant economic factors.[16] In other words, if the benefits of exercising an option are sufficiently high that we think the lessee will exercise it, we adjust the contractual term by adding the additional period of renewal, or by subtracting the period that follows a termination option.

The lease term is reassessed only when a significant event or change in circumstances indicates a change in the economic incentive for extension or termination of the lease.

REASSESSMENT OF THE LEASE TERM Circumstances can change that require reassessment of how long the lease term will be. Reassessment requires a "triggering event" such that the lessee now has an economic incentive to exercise an option that extends or terminates the lease. For example, assume that a lessee had no significant economic incentive as of the beginning of a 6-year lease to exercise a 2-year extension option. However, by the end of the third year, the lessee has made significant improvements to the asset whose cost could be recovered only if it exercises the extension option, making extension of the lease "reasonably certain." In that case, the full term of the lease is now expected to be a total of eight years with five years remaining.

At the end of the third year, the lessee would remeasure the lease liability as the present value of the remaining five lease payments. Importantly, the discount rate for the new term is the incremental borrowing rate of the lessee using market interest rates *at the time of the reassessment,* rather than the rate used at the beginning of the lease. Assuming the present value of the remaining five lease payments totaled $400,000, while the current balance of the lease liability before remeasurement is $250,000, the lessee would make the following adjustment:

Reassessment of the Lease Term

Right-of-use asset	150,000	
Lease payable (increase in balance*)		150,000

*PV of *remaining* 5 *payments*, discounted at the *current* rate	$400,000
Liability balance after 3 years based on initial lease terms	($250,000)
Increase† in balance	$150,000

†In situations when the lease term is shortened (termination option), the journal entry would be reversed to *reduce* the account balances.

Also, when there is a change in the lease term, lessees are required to reassess the *classification* of a lease. In the current example, because, with the assumed renewal, the lease term is for the entire useful life of the asset, the lease classification might need to be changed. Suppose, for instance, that the lease had been classified initially as an operating lease. Now, with the increase in the lease term, it might meet the criteria for a finance lease rather than an operating lease as previously classified. If so, we would have had an operating lease for three years, and now have a five-year finance lease. That would mean amortizing the right-of-use asset as a straight-line allocation of the balance in that account at this point over the next five years, rather than "plugging" amortization as we would for an operating lease.

The discussion above pertains to a lessee. *A lessor is not permitted to reassess its initial determination of the lease term or discount rate.*

[16]Reasonably certain is a high threshold, essentially the same as "reasonably assured."

Decision Makers' Perspective—Considering the Economic Incentive for Exercising Options

What if we did not have the requirement to consider renewal options and the economic incentive for exercising them when we determine the lease term? Management might be tempted to structure leases with artificially short initial terms and numerous renewal options as a scheme to be able to use the shortcut method (which we discussed earlier) or to reduce significantly the amount of the lease liability to be reported (off-balance-sheet financing). ●

What if the Lease Payments are Uncertain?

Sometimes lease payments are scheduled to be increased (or decreased) at some future date during the lease term, depending on whether or not some specified event occurs. Usually the contingency is related to revenues, profitability, or asset usage above some designated level. For example in Illustration 15–10, a recent annual report of **The Gap** included the following note:

Rent Expense (in part)

Certain leases provide for contingent rents that are not measurable at inception. These contingent rents are primarily based on a percentage of sales that are in excess of a predetermined level and/or rent increase based on a change in the consumer price index or fair market value. These amounts are excluded from minimum rent and are included in the determination of rent expense when it is probable that the expense has been incurred and the amount can be reasonably estimated.

Illustration 15–10
Contingent Lease Payments—The Gap

Real World Financials

Why would a lease include a contingent payment provision? It's a way for lessees and lessors to share the risk associated with the asset's productivity. For example, let's assume a store owner, like at Gap, who pays for a premium mall location is doing so anticipating higher revenue. If the mall attracts a sufficiently higher number of shoppers, the lessee pays the lessor part of the resulting higher profits, but if not, the lessee makes only the normal lease payments. This arrangement also provides the lessor an incentive to attract shoppers to the mall, which is in the lessee's best interest. Because the amounts of future lease payments are uncertain and often avoidable, we don't consider them as part of the lease payments used to calculate the lessee's lease liability and the lessor's lease receivable. If and when lease payments increase, the change in the lease payments has no effect on balance sheet accounts and simply is reported as a separate lease expense (lessee) and lease revenue (lessor).[17]

Most variable lease payments are recognized when incurred rather than being estimated at lease commencement and included in the lessee's right-of-use asset and lease liability.

There are two exceptions to not including variable payments in the calculation of the lease liability recorded at the beginning of the lease. Let's take a look at both exceptions in order to understand when variable lease payments are included in the initial recording of the lease liability.

WHEN VARIABLE LEASE PAYMENTS ARE IN-SUBSTANCE FIXED PAYMENTS The first exception to not including variable payments is when apparent "variable" payments actually are in-substance fixed payments. We include these "fixed payments in disguise" as part of the lessee's lease payments. For example, assume a retail store's monthly lease payments will increase next year by the higher of $250 or 0.5% of monthly store revenue. In that case, we know that the lease payment will increase by at least $250, so those payments are deemed to be in-substance fixed payments and are included in the lease payments used in present value calculations.

If future lease payments are uncertain, we consider them as part of the lease payments only if they are "in-substance fixed payments" or if payments vary solely when an index or rate changes.

[17]The lease payment can also decrease. The decrease in the lease payments is reported as a reduction in lease expense (lessee) and a reduction in lease revenue (lessor).

WHEN VARIABLE LEASE PAYMENTS DEPEND ON AN INDEX OR RATE Another exception to not including variable payments when initially recording the lease is when the amount of the lease payments depends on an index or a rate, such as the Consumer Price Index or current interest rates. Even though lease payments will vary in the future, we use the initial lease payment amount, based on the current index (or rate), to discount to present value when determining the right-of-use asset, lease liability, and lease receivable. When lease payments do change in the future (because the index or rate changes), we don't remeasure the lease liability or leased asset at that time, but simply report the additional amount as a separate lease payment that produces expense for the lessee and revenue for the lessor.

Only if the lease asset and lease liability are later remeasured for another reason, will a change in payments based on the CPI or market interest rates affect the right-of-use asset and liability.

Only if and when the lessee remeasures the lease liability for reasons other than a change in the index or rate should the lessee adjust the right-of-use asset and lease liability for changes in the amount of the payments. This might happen, for example, because of a reassessment of the lease term (as described in the previous section) or a modification of the lease (as described in the next section). In that case, the leased asset and lease liability are recalculated by determining the present value of future lease payments using (a) the new lease payments as adjusted for changes in the index or rate and (b) the discount rate that applies as of the date of the reassessment.

The *lessor* never reassesses its lease receivable for variable lease payments.

What if Lease Terms are Modified?

If a modification grants the lessee an *additional right of use*, the original lease is terminated and a new lease is created based on the modified arrangement.

Sometimes the lessee and lessor will agree to modify the terms of a lease before the lease term ends. This creates two possibilities. First, the modification might grant the lessee an *additional right of use.* This would mean terminating the original lease and accounting for the modified arrangement as a new lease. Second, the modification might *alter the lessee's right to use* the asset rather than grant an additional right of use. This would mean adjusting, adding to, or deleting what has been recorded in order to conform to the new terms of the contract (say, a change in the lease term or lease payments) and perhaps reclassifying the lease from one type to another.

Modifying lease terms sometimes requires reclassifying operating leases to finance/sales-type leases, or vice versa.

As an example, assume a four-year operating lease of equipment with a useful life of six years (as we had in Illustration 15-7). Let's say that after two years, the lessee and lessor agree to extend the lease term by two years, and to alter the amount of the lease payments. The additional two years were not originally an option. In this case, the modification alters the lessee's right to use the equipment; it doesn't grant the lessee an additional right to use another asset. In addition, the modified lease term of two additional years (six years total) is now for a "major part" of the asset's six-year economic life, so classification changes from an operating lease to a finance/sales-type lease. Illustration 15–11 demonstrates how modifications to the contract would be recorded.

Sans Serif, the lessee, updates the balances of the right-of-use asset and lease liability for the increase in present value. In addition, because the lease is now classified as a finance lease, Sans Serif no longer will recognize straight-line lease expense of $100,000. Instead, at the end of each remaining year, Sans Serif will record interest expense and amortization of the right-of-use asset in the usual way for a finance lease.

First LeaseCorp, the lessor, records a lease receivable for the present value of the remaining lease payments and removes the equipment (and related accumulated depreciation) from its books. Because the sales revenue is higher than the carrying amount of the equipment, First LeaseCorp also recognizes selling profit for the difference: Sales revenue of $329,068 minus cost of goods sold of $319,386 equals $9,682. In addition, at the end of each remaining year, First LeaseCorp will record interest revenue at the now current effective rate of 9% for a sales-type lease (instead of straight-line rent revenue of $100,000 for an operating lease).

Residual Value

The residual value is an estimate of what a leased asset's commercial value will be at the end of the lease term.

The **residual value** of leased property is an estimate of what its commercial value will be at the end of the lease term. Typically, we will have a residual value in an *operating* lease because the lease term usually ends before the lease asset's value has been depleted. A residual value is less likely, but certainly not unusual, when the lease qualifies as a sales-type lease, because the lease term is for most, if not all, of the asset's life.

Illustration 15–11 Lease Modifications: Term, Payments, and Classification

On January 1, 2021, Sans Serif Publishers leased printing equipment from First LeaseCorp.

- First LeaseCorp purchased the equipment from CompuDec Corporation at a cost of $479,079.
- Sans Serif's borrowing rate for similar transactions is 10%.
- The lease agreement specifies four annual payments of $100,000 beginning January 1, 2021, the beginning of the lease, and at each December 31 from 2021 through 2023.
- The useful life of the equipment is estimated to be six years.
- The present value of those four payments at a discount rate of 10% is $348,685.
- On January 1, 2023 (after three payments*), the lessee and lessor agreed to (1) extend the lease term by two years and (2) increase the remaining three payments from $100,000 to $130,000 each.
- The market rate of interest at that time has decreased from 10% to 9%.

The table below summarizes the effect of four important contract modifications:

Contract	Original Lease	Modified Lease	Effect
1. Term	4 years	6 years	Increase in present value
2. Payments	$100,000	$130,000	Increase in present value
3. Discount rate	10%	9%	Increase in present value
PV on Jan. 1, 2023	$90,909**	**$329,068**†	**Lessee:** Increase PV balances by **$238,159**††
4. Classification	Operating	Finance/Sales-type	**Lessor:** Record sale of asset for **$329,068** (and remove asset's carrying amount)

Modification of the Lease (January 1, 2023)

Sans Serif (Lessee)			
Right-of-use asset	**238,159**		
Lease payable (increase in PV from $90,909 to $329,068)		**238,159**	
First LeaseCrop (Lessor)			
Lease receivable (PV of 3 remaining lease payments)	329,068		
Sales revenue (PV of 3 remaining lease payments)		**329,068**	profit 9,682
Cost of goods sold (cost minus accumulated depreciation)	**319,386**		
Accumulated depreciation [($479,079 ÷ 6 years) × 2 years]	159,693		
Equipment		479,079	

*The modification occurs two years into the lease, but there have been three payments since payments are at the beginning of each year.
**See Illustration 15–7B.
†Present value of remaining three lease payments = $130,000 × 2.53129. (Present value of an ordinary annuity of $1: $n = 3$, $i = 9\%$.)
††$329,068 − 90,909

International Financial Reporting Standards

Reassessment of the Right-of-Use Asset and Lease Liability. Under IFRS, a lessee will remeasure the variable lease payments that depend on an index or a rate not just when the lessee remeasures the right-of-use asset and lease liability for other reasons, but also whenever there is a change in the cash flows resulting from a change in the reference index or rate.

● LO15–9

Typically, the lessee promises to return the leased asset to the lessor at the end of a lease. An asset being returned will likely have some value.

GUARANTEE OF THE RESIDUAL VALUE Sometimes a lease agreement includes a guarantee by the lessee that the lessor will recover a specified residual value when custody of the asset reverts back to the lessor at the end of the lease term. This not only reduces the lessor's risk, but also provides an incentive for the lessee to exercise a higher degree of care in maintaining the leased asset to preserve the residual value. The lessee promises to return not only the property but possibly also sufficient cash to meet the guaranteed amount promised in the lease agreement.

A lessee sometimes will guarantee that the lessor will recover a specified residual value when custody of the asset reverts back to the lessor.

A residual value affects several aspects of lease accounting, including the size of the periodic lease payments, the classification of a lease, and the amounts recorded by both the lessee and lessor.

EFFECT OF A RESIDUAL VALUE ON THE SIZE OF LEASE PAYMENTS Suppose the printing equipment leased in our finance/sales-type lease (Illustration 15–4) was expected to be worth **$60,000** at the end of the six-year lease term. Should this influence the lessor's (First LeaseCorp) calculation of periodic rental payments? Yes! Here's why.

The leasing company purchased the equipment for $479,079. That's the amount the company needs to recover through the leasing contract along with interest revenue sufficient to achieve its business objectives. We are assuming that means a 10% return on investment. Where does that return come from? In some leases (our finance/sales-type lease in Illustration 15–4, for instance) the entire return comes from the lessee's lease payments, which must be enough over the lease term to pay the lessor its investment in the lease asset plus the desired amount of interest (10%).

In that situation the lessor calculated the lease payments as follows:

Amount to be recovered (present value)	$479,079
	÷4.79079*
Lease payments at the beginning of each of the next 6 years	$100,000

*Present value of an annuity due of $1: n = 6, i = 10%.

We calculated the necessary payments by dividing the $479,079 investment by the present value of an annuity due of $1, discounted at 10% for six years. That calculation by the lessor yielded $100,000 for each of the six payments.

However, if the lessor gets the asset back at the end of the lease, and the asset has commercial value at that time, the lessor has another source of return. The value of the asset itself, the residual value, will provide another source of recovery of the lessor's investment. That reduces the amount needed from periodic lessee payments for the lessor to generate its 10% return. To see how, in Illustration 15–12 let's assume a **$60,000** residual value at the end of the six-year lease term.

Illustration 15–12

Effect of Residual Value on the Calculation of Lease Payments

On January 1, 2021, Sans Serif Publishers leased printing equipment from First LeaseCorp.

- First LeaseCorp purchased the equipment from CompuDec Corporation at a cost of $479,079.
- Assume the lease includes six annual payments beginning January 1, 2021, and at each December 31 from 2021 through 2025.
- At the end of the six-year lease term ending December 31, 2026, the equipment is expected to have a **residual value of $60,000.**
- The estimated useful life of the equipment is **seven years.**

If the six lease payments are of an equal amount, what payment amount would provide First LeaseCorp with a return of 10%?

Amount to be recovered* (fair value)	$479,079
Less: Present value of the residual value (**$60,000** × 0.56447*)	(33,868)
Amount to be recovered through periodic rental payments	445,211
	÷ 4.79079**
Lease payments at the beginning of each of the six years	$ 92,931††

*The amount to be recovered would also include "initial direct costs" paid by the lessor. We discuss initial direct costs in Part C of this chapter.
*Present value of $1: $n = 6$, $i = 10\%$.
**Present value of an annuity due of $1: $n = 6$, $i = 10\%$.
††For another example, recall our four-year *operating* lease (Illustration 15–7) in which the lease payments are $100,000 each. Although not stated at the time, First LeaseCorp anticipated that the fair value of the equipment at the end of the lease term (residual value) would be **$190,911**, which is why the lease payments were $100,000:

Amount to be recovered (fair value)	$479,079
Less: Present value of the residual value (**$190,911** × 0.68301‡)	(130,394)
To be recovered through periodic lease payments (present value)	$348,685
	÷ 3.48685‡‡
Lease payments at the beginning of each of the four years	$100,000

‡ Present value of $1: $n = 4$, $i = 10\%$.
‡‡Present value of an annuity due of $1: $n = 4$, $i = 10\%$.

The present value of the residual value of a lease asset is called a **residual asset**. Since the property will revert back to the lessor, the lessee doesn't have the right of use of the entire value of the asset and thus doesn't view the residual asset as *its* asset. However, from the lessor's perspective, even if a residual value is not guaranteed, the lessor still expects to receive it. So, the lessor will view the residual asset as contributing to the amount needed to recover its $479,079 investment, causing the lessee's cash lease payments to be $92,931 rather than $100,000. *As the expected residual value increases, the size of the lease payments decreases.* The residual, whether guaranteed or unguaranteed, affects the size of the lease payments.

EFFECT OF A RESIDUAL VALUE ON AMOUNTS RECORDED BY THE LESSOR

Sales-type lease with residual value. The *lessee's accounting is unaffected* by the residual value other than its causing the payments to be lower. In the situation described in Illustration 15–12, Sans Serif, the lessee, would record a right-of-use asset and lease liability for the present value of the six lease payments ($445,211). Then, over the term of the lease, it would record interest expense and amortization in the usual way for this finance-type lease.

The *lessor's accounting is affected* by the residual value. In Illustration 15–12A we see how the lessor will record the lease introduced in Illustration 15–12.

Illustration 15–12A
Sales-Type Lease with Residual Value: Lessor

Beginning of the Lease (January 1, 2021)		
First LeaseCorp (Lessor)		
Lease receivable (PV of lease payments plus PV of **$60,000** residual value)*	479,079	
Equipment (lessor's cost: carrying amount)		479,079

*Alternatively, the lessor could record the receivable and the residual asset separately:

Lease receivable (PV of lease payments)	445,211	
Residual asset (PV of **$60,000** residual value)	33,868	
Equipment (lessor's cost: carrying amount)		479,079

The lessor's lease receivable includes the value of the asset expected at the end of the lease term.

Both (a) the present value of the lease payments and (b) the present value of the residual value combine to allow the lessor to recover its $479,079 investment and are recorded as the lease receivable at the beginning of the lease. The lessor includes the **$60,000** residual value in its amortization schedule, along with the lease payments as demonstrated in Illustration 15–12B.

Illustration 15–12B
Lessor's Amortization Schedule—with Residual Value

	Receipts	Effective Interest		Decrease in Balance	Outstanding Balance
		(10% × Outstanding balance)			
1/1/21					479,079
1/1/21	92,931			92,931	386,148
12/31/21	92,931	0.10 (386,148) =	38,615	54,316	331,832
12/31/22	92,931	0.10 (331,832) =	33,183	59,748	272,084
12/31/23	92,931	0.10 (272,084) =	27,208	65,723	206,361
12/31/24	92,931	0.10 (206,361) =	20,636	72,295	134,066
12/31/25	92,931	0.10 (134,066) =	13,407	79,524	54,542
12/31/26	**60,000**	0.10 (54,542) =	5,458*	54,542	0
	617,586		138,507	479,079	

*Adjusted for rounding of other numbers in the schedule.

The lessor expects to receive the residual value ($60,000) in the form of equipment at the end of the lease term.

The amortization schedule reveals several important points. Let's use the original information from Illustration 15-4A and the amortization schedule from Illustration 15-4C for comparison. In the original information First LeaseCorp entered into a lease for six cash payments that were calculated to be $100,000 with no residual value. Now, in our illustration of a lease with a residual value, the payments are calculated to be only $92,931, and there would also be a tangible asset returned to the lessor one year after the six cash payments. So,we now include in our calculation the **$60,000** residual value.

Despite the different composition of the amounts the lessor will receive, their present value ($479,079) is the same as when we assumed $100,000 periodic payments and no residual value. However, a greater amount of interest revenue will be recognized over the lease term: $138,507. (It was $120,921 before.) The higher interest reflects the fact that receipts are farther in the future, causing the outstanding lease balances (and interest on those balances) to be higher during the lease term. Also, note that a greater total amount of lease receipts is collected: $617,586. (It was $600,000 before.) This total is referred to as the lessor's *gross investment in the lease* and is included in the lessor's lease disclosure note. The present value of those same payments ($479,079) is referred to as the *net investment in the lease* and is the amount recorded as the lease receivable in the journal entry at the beginning of the lease.

The lessor's *gross investment in the lease* is the total of periodic lease payments and any residual value.

Remember, the final periodic cash payment on December 31, 2025, is at the beginning of the final year. Then, at the end of the year of the lease on December 31, 2026, the equipment is returned to the lessor and is reinstated on the lessor's books at its fair value, which we assume to be **$60,000**, the amount predicted when the lease began. This entry is demonstrated in Illustration 15-12C.

Illustration 15–12C

End of Lease Term—Actual Residual Value Equals the Estimated Amount: **Lessor**

End of the Lease (December 31, 2026)

First LeaseCorp (Lessor)		
Equipment (residual value)	**60,000**	
Lease receivable (account balance)		54,542
Interest revenue (10% × outstanding balance)		5,458

If the actual residual value at December 31, 2026, is less (or more) than **$60,000,** the lessor would record a loss (or gain). Notice, too, that the residual asset is returned to the lessor.

Additional Consideration

You might be wondering why we have a receivable balance of $54,542 in this last entry. It's the residual value of the asset, right? Yes, the original receivable balance represented both (a) the cash payments to be collected (now collected) and (b) the amount to be received in the form of a residual asset when the leased asset is returned to the lessor. But the original residual value was $60,000 and the amount we included in the receivable was the present value of that amount (the residual asset), which was $33,868 at the commencement of the lease. Now, though, that present value has grown to $54,542 at the end of 2025 (as we see in the amortization schedule and this final journal entry) and $60,000 at the end of 2026 (when the lessor gets the asset back at the end of the lease).

Sales-type lease with selling profit and residual value. When a sales-type lease that includes a residual value also has a selling profit, we need to take that into consideration when the lease is recorded initially. Recall that the lessor records both sales revenue and cost of goods sold to account for the selling profit. The sales revenue is the present value of only the periodic lease payments, not including the residual value. To visualize this, let's assume that our sales-type lease with a selling profit (Illustration 15–6) had included a residual value of **$60,000.** In that case, the initial lessor entry would be modified as shown in Illustration 15–13.

Illustration 15–13

Sales-Type Lease with Selling Profit and Residual Value: Lessor

On January 1, 2021, Sans Serif Publishers leased printing equipment from CompuDec Corporation.

- The lease agreement specifies six annual payments of $92,931 beginning January 1, 2021, the beginning of the lease, and at each December 31 from 2021 through 2025.
- At the end of the six-year lease term ending December 31, 2026 the equipment will be returned to the lessor and is expected to have a **residual value of $60,000.**
- The estimated useful life of the equipment is **seven years.**
- The interest rate in these financing arrangements is 10%.
- CompuDec manufactured the equipment at a cost of **$300,000.**

Beginning of the Lease (January 1, 2021)

CompuDec Corporation (Lessor)

Lease receivable (PV of lease payments* plus PV of **$60,000** residual value**)	479,079	
Cost of goods sold ($300,000 – **$33,868****)	266,132	
Sales revenue ($479,079 – **$33,868****)		445,211
Equipment (lessor's cost)		300,000

* $92,931 × 4.79079†† = $445,211 present value of lease payments
** **$60,000** × 0.56447† = $33,868 present value of residual value
† Present value of $1: $n = 6, i = 10\%$.
†† Present value of an annuity due of $1: $n = 6, i = 10\%$.

Sales revenue	$445,211
– COGS	266,132
Selling profit	$179,079

The lease provides the lessor with a selling profit if the PV of payments exceeds the asset's cost.

Sales revenue does not include the residual value because the revenue to be recovered from the lessee is lease payments only. The remainder of the lessor's investment is to be recovered—not from payment by the lessee, but by selling, releasing, or otherwise obtaining value from the asset when it reverts back to the lessor. Think of it this way: The portion of the asset sold is the portion not represented by the residual value. So, both the asset's cost of goods sold and its selling price are reduced by the present value of the portion *not sold.*

When a cash payment is predicted due to a lessee-guaranteed residual value. We saw in Illustration 15–4 that without a residual value contributing to the lessor recovering its investment, the amount of the six payments needed to recover the $479,079 investment would have been $100,000. We also saw in Illustration 15–12 that *with* a residual value guaranteed to the lessor at the end of the lease, the payments would be reduced to $92,931. Other than this reduction of lease payments, should the *lessee* (Sans Serif Publishers) be concerned with the residual value of the leased asset? The answer is maybe. It depends on whether the lessee has arranged to guarantee a specific value of the residual asset at the conclusion of the lease and, even then, how that value compares to the prediction of its actual value. We explore that possibility next.

As we discussed earlier, a lease agreement sometimes includes a guarantee by the lessee that the lessor will recover a specified residual value when custody of the asset reverts back to the lessor at the end of the lease term. The lessee promises to return not only the property but also sufficient cash to provide the lessor with a minimum combined value (residual value and cash). That doesn't necessarily mean, though, that the lessee will be required to make any cash payment. A cash payment would be expected as of the beginning of a lease only if the guaranteed amount exceeds the estimated residual value of the asset.

A cash payment that's predicted because of a lessee-guaranteed residual value is treated the same as another lease payment.

If a cash payment under a lessee-guaranteed residual value is predicted, the present value of that payment is added to the present value of the periodic lease payments that the lessee records as both a right-of-use asset and a lease liability. For instance, let's assume that in the situation described in Illustration 15–12 when the estimated residual value was $60,000, negotiations led to Sans Serif guaranteeing an $80,000 residual value. That means that if the property's value is less than $80,000 at the end of the six-year lease term, the lessee will pay cash for the difference. Since that value is expected to be $60,000, the expected *excess* guaranteed residual value is $80,000 – $60,000 = **$20,000.** The lessee views the expected excess guaranteed residual value as an additional cash flow to be paid to the lessor (addition to lease payments). The present value of the expected excess guaranteed residual is $11,289 ($20,000 × 0. 56447), and the lessee would add this amount to the present value of the periodic lease payments to record the right-of-use asset and lease liability at the beginning of the lease.[18]

Looking at it from the other side of the transaction, does the lessor also view the expected *excess* guaranteed residual value as an additional amount to be collected? Yes. In fact, the lessor expects to receive that payment ($20,000) *as well as the residual value itself* ($60,000). Those two amounts combine to equal the $80,000 guaranteed residual value that, if in a *sales-type* lease, the lessor includes in the lease receivable.[19]

Situations in which the lessee-guaranteed residual value exceeds the estimate of the actual residual value are rare in practice. It makes little economic sense for a lessee to agree to guarantee an amount greater than the estimated residual value, virtually ensuring an additional cash payment at the conclusion of the lease. The requirement to account for it in this way, though, serves as a deterrent to lessees and lessors who might be inclined to manipulate reported numbers by reducing lease payments while creating an excess *lessee-guaranteed* residual value to compensate for the reduced lease payments.

Additional Consideration

If a residual value is not guaranteed by the lessee or is guaranteed by the lessee but the guaranteed amount does not differ from the estimate of the actual fair value at the end of the lease term, it does not affect the lessee's calculation of the right-of-use asset or lease liability other than influencing the amount of each lease payment as we saw in Illustration 15–12.

EFFECT OF A RESIDUAL VALUE ON LEASE CLASSIFICATION We classify a lease as a finance/sales-type lease if, in substance, the lessor is transferring control of the asset to the lessee. Recall that one of the classification criteria is a comparison of the fair value of an asset with the present value of the payments coming from the lessee. Payments from the lessee are the periodic lease payments plus any portion of the residual value the lessee has guaranteed. That's the basis for classification criterion 4 (Illustration 15–2): If the present value of the lease payments, including any *lessee-guaranteed* residual value, constitutes "substantially all" of the fair value of the asset, it's a finance lease from the lessee's perspective and a sales-type lease from the lessor's perspective.

Additional Consideration

When the lessee does not control substantially all of the remaining benefits of the underlying asset, the lease usually is considered an operating lease.

Recall that by one of the lease classification criteria, we have a finance/sales-type lease if the sum of the present value of the lease payments and any residual value guaranteed by the lessee amounts to substantially all of the fair value of the underlying asset. When that's the case, the lessee, through its contractually obligated payments, guarantees recovery of substantially all the lessor's investment in the asset, so the lessee controls substantially all of the remaining benefits of the underlying asset. This control is consistent with the principle of a sale in ASC Topic 606: *Revenue from Contracts with Customers* for the lessor to recognize sales revenue on the lease.[20] On the other hand, when the lessee lacks that control, the lease is considered an operating lease.

An exception is when the lease qualifies as a **direct financing lease**. We have a direct financing lease only when a third party (typically an insurance company) is paid to guarantee all or part of the residual value. When that happens, it's possible that the combined present

(continued)

[18]Present value of $1: $n = 6$, $i = 10\%$.

[19]Remember, in an *operating* lease, the lessor doesn't record a receivable nor remove the asset from its books.

[20]As explained in more detail in Chapter 6, "control" requires that the customer has direct influence over the use of the good or service and obtains its benefits.

(concluded)

value of (a) the *lessee's* periodic payments, (b) the *lessee*-guaranteed residual value, and (c) the *third-party* guaranteed residual value will constitute substantially all of the asset's fair value, but that the lessee's obligated payments, (a) and (b), are insufficient to do so. That situation would indicate that the lessee *does not* control substantially all of the remaining benefits of the underlying asset, so from the lessee's perspective the lease is an operating lease.

However, from the lessor's perspective, the *total* of the obligated payments guarantees recovery of substantially all the lessor's investment in the asset (assuming collection is probable) indicating that it shouldn't be an operating lease, but the lessee's portion of the total is insufficient to indicate control, so it shouldn't be a sales-type lease either.

A lease is considered a direct financing lease if:

1. It doesn't qualify as a sales-type lease.
2. The combined present value of (a) the *lessee's* periodic payments, (b) the *lessee*-guaranteed residual value, and (c) the *third-party* guaranteed residual value will constitute substantially all of the asset's fair value.
3. It's probable that the lessor will collect the lease payments.

When a lease is a direct financing lease rather than a sales-type lease, the lessor's primary involvement in the lease is providing financing of the asset in exchange for interest revenue. The lessor records the lease the same as it would a sales-type lease with no selling profit:

Lease receivable (present value of lease payments*)	xxx	
Asset (carrying value)		xxx

*Including the lessee's periodic payments, any lessee-guaranteed residual value, and the third-party guaranteed residual value.

Although rare, if a direct financing lease gives rise to selling profit, the lessor does not recognize the profit at the beginning of the lease. Instead, the lessor would credit deferred profit (considered a contra asset reduction in the receivable) in the entry above. The lessor then would recognize the profit over the lease term in such a manner that the total of that profit and the interest revenue on the remainder of the lease receivable is a constant periodic rate of return on the lease. This can be accomplished by not recording the "deferred profit" separately, but subtracting it from the lease receivable. Decreasing the receivable causes the implicit rate (the interest rate that causes the present value of the lease payments to equal the receivable) to be higher. Determining interest revenue at this higher rate accomplishes the purpose of increasing interest revenue (actually interest plus a portion of profit) each period by a portion of the deferred revenue.

†As explained in more detail in Chapter 6, "control" requires that the customer has direct influence over the use of the good or service and obtains its benefits.

When control is not transferred to the lessee, but the present value of lease payments constitutes substantially all of the asset's fair value due to a third-party guaranteed residual value and it's probable that the lessor will collect the lease payments, the lease is considered a direct financing lease.

In the unlikely event of a profit in a direct financing lease, the revenue (and profit) must be deferred.

Purchase Option

A **purchase option** is a provision of some lease contracts that gives the lessee the option to purchase the lease asset during, or at the end of, the lease term at a specified exercise price. If it is "reasonably certain" that the lessee will exercise the purchase option, the accounting for the lease is affected in three ways: (1) the lease is classified as a finance/sales-type lease, (2) both the lessee and the lessor consider the exercise price of the option to be an additional cash payment, and (3) we assume the lease term ends on the date that the option is expected to be exercised. Both the additional cash payment and the shortened lease term impact the calculation of the lessee's right-of-use asset and lease liability and the lessor's lease receivable. Another implication is that, since the lessee is predicted to own the asset after the lease term, the right-of-use asset recognized by the lessee should be amortized over the economic life of the asset, rather than over the lease term.

The exercise price of a purchase option is considered to be an additional cash payment if exercise of the option is "reasonably certain."

In practice, a purchase option whose exercise is reasonably certain is often referred to as a "bargain" purchase option (BPO), and we'll use that term in this book. BPOs are so named because the exercise price of the purchase option is a good deal (a bargain) to the lessee, making it reasonably certain that the lessee will exercise the option. To see the accounting for a BPO, let's again modify Illustration 15–4, this time to assume the lease agreement allows Sans Serif the option to purchase the leased equipment for $60,000 at the end of the lease term. At this time, the fair value is expected to be sufficiently high ($75,000) to indicate reasonable certainty that Sans Serif will exercise the option. This situation is assumed in Illustration 15–14.

Illustration 15–14
Bargain Purchase Option

On January 1, 2021, Sans Serif Publishers leased printing equipment from First LeaseCorp. First LeaseCorp purchased the equipment from CompuDec Corporation at a cost of $479,079. The lease agreement specifies six annual payments of $92,931 beginning January 1, 2021, the beginning of the lease, and at each December 31 from 2021 through 2025. On December 31, 2026, at the end of the six-year lease term, the equipment is expected to be worth $75,000, and Sans Serif has the option to purchase it for **$60,000** on that date. The residual value after seven years is zero.† First LeaseCorp routinely acquires electronic equipment for lease to other firms. The interest rate in these financing arrangements is 10%.

The lessor *subtracts* the PV of the exercise price to determine lease payments.

Lessor's calculation of lease payments:	
Amount to be recovered (fair value)	$479,079
Less: Present value of the exercise price (**$60,000** × 0.56447*)	(33,868)
Amount to be recovered through periodic rental payments	445,211
	÷ 4.79079**
Lease payments at the beginning of each of the six years:	$ 92,931

The lessee *adds* the PV of the exercise price to determine its asset and liability.

Lessee's calculation of PV of lease payments:	
Present value of periodic lease payments ($92,931 × 4.79079**)	$445,211
Plus: Present value of the exercise price (**$60,000** × 0.56447*)	33,868
Present value of lease payments	$479,079
(Recorded as a leased asset and a lease liability)	

†Our discussion of the effect of a bargain purchase option (BPO) would be exactly the same if our illustration were of a sales-type lease with a selling profit (for instance, if the lessor's cost were $300,000) except that, of course, sales revenue and cost of goods sold would be recorded.
* Present value of $1: $n = 6$, $i = 10\%$.
**Present value of an annuity due of $1: $n = 6$, $i = 10\%$.

The lessor's calculation of periodic lease payments is precisely the same as when we had the $60,000 residual value in the previous illustration. In fact, the amortization schedule is exactly the same as in Illustration 15–12B when we had a residual value. This is because the exercise price is a component of the lease payments for both the lessor and lessee as was the residual value.

If the lessee obtains title, the lessor's computation of rental payments is unaffected by any residual value.

A question you might have at this point is: Why do we ignore the residual value now when we have a bargain purchase option? The reason is obvious when you recall an essential characteristic of such an option—it's expected to be exercised. So, when it is exercised, title to the leased asset passes to the lessee and, with title, any residual value also passes. When that happens, the residual value cannot be considered an additional lease payment to the lessor.

Recording the exercise of the option is similar to recording the periodic lease payments. That is, a portion of the payment covers interest for the year, and the remaining portion reduces the outstanding liability/receivable balance (to zero with this last payment), as shown in Illustration 15–14A.

Illustration 15–14A
Journal Entries—with Purchase Option

The cash payment expected when the BPO is exercised represents part interest and part principal just like the other cash payments.

Exercise of Purchase Option (December 31, 2026)

Sans Serif Publishers (Lessee)		
Interest expense (10% × $54,542*)	5,458	
Lease payable (difference)	54,542	
Cash (exercise price)		**60,000**
First LeaseCorp (Lessor)		
Cash (exercise price)	**60,000**	
Lease receivable (account balance)		**54,542**
Interest revenue (10% × outstanding balance)		**5,458**

*$54,542 is the balance of lease payable after all periodic lease payments have been made (see Illustration 15–12B).

Note also that amortization is affected by the BPO. As pointed out earlier, the lessee normally amortizes its right-of-use asset over the term of the lease. But if ownership transfers by contract or by the expected exercise of a purchase option, the lessee will have the asset beyond the lease term, and will amortize it over the longer useful life. This reflects the fact that the lessee anticipates using the leased equipment for its full useful life. In Illustration 15–14B, the lease term is six years but the equipment is expected to be useful for **seven years,** so amortization each year is $68,440 ($479,079 ÷ **7 years**).

Illustration 15–14B

Amortization Expense with Purchase Option

Each December 31 over the life of the asset

Sans Serif Publishers (Lessee)

Amortization expense ($479,079* ÷ **7 years**)	68,440	
Right-of-use asset		68,440

*The residual value is zero after the full seven-year useful life.

PURCHASE OPTION EXERCISABLE BEFORE THE END OF THE LEASE TERM In Illustration 15–14, we assumed that the purchase option was exercisable on December 31, 2026—the end of the lease term. Sometimes, though, the lease contract specifies that an option becomes exercisable before the designated lease term ends. If the option is reasonably certain to be exercised, *the lease term ends for accounting purposes when the option becomes exercisable.* Lease payments include only the periodic cash payments specified in the agreement that occur prior to the date a BPO becomes exercisable. We assume the option is exercised at that time and the lease ends.

When we have a BPO, the length of the lease term is limited to the time up to when the purchase option becomes exercisable.

TERMINATION PENALTIES Similar to a lease with a purchase option, if a lease contract includes a penalty payment if the lessee chooses to terminate the lease at a time specified in the contract, we consider the termination penalty to be an additional cash payment if the lessee is "reasonably certain" to terminate the lease. Again, if termination is predicted, we consider the lease term to be from the beginning of the lease to the expected termination date.

In Illustration 15–15, we see the factors that are included in the lease term and the lease payments used in accounting for leases.

Illustration 15–15

Composition of Lease Term and Payments

LEASE TERM INCLUDES:		**LEASE PAYMENTS INCLUDE:**	
Noncancelable period		Fixed payments	
PLUS:	Periods covered by renewal options if exercise is "reasonably certain"	*PLUS:*	Exercise price for purchase option if exercise is "reasonably certain"
MINUS:	Periods following date of purchase option if exercise is "reasonably certain"	*PLUS:*	Termination penalty for termination option if exercise is "reasonably certain"
PLUS:	Periods covered by renewal options if *under control of lessor*	*PLUS:*	Variable lease payments only if (a) deemed in-substance fixed payments or (b) based on an index or rate (with any changes in payments included only if and when the lessee remeasures the lease liability for another reason)
PLUS:	Periods following date of termination option if it's "reasonably certain" the option will ***not*** be exercised	*PLUS:*	Excess of guaranteed residual value over expected residual value

Summary of the Lease Uncertainties

Illustration 15–16 provides a summary of various uncertainties concerning lease arrangements that might influence the way lessees and lessors determine lease payments, the lessee's right-of-use asset and lease liability, and the lessor's lease receivable.

Illustration 15–16 Summary of Effects of Uncertainties

The lease term	Noncancelable period for which a lessee has the right to use an underlying asset, modified by any renewal or termination options that are "reasonably certain" to be exercised, or not exercised. Options whose exercise is under the control of the lessor are automatically included. Reassessed only when a significant occurrence indicates a change in the economic incentive.
Variable lease payments	Included only if payments are • in-substance fixed payments or • based on an index or rate, with changes due to the index or rate considered only if and when the lessee remeasures the lease liability for another reason
Unguaranteed residual value	Present value (called residual asset) • influences the size of lease payments • added to lease receivable • subtracted from sales revenue and COGS in sales-type lease
Lessee guaranteed residual value	Present value • considered when determining lease classification (criterion 4) • included in lease receivable • included in sales revenue
Excess of guaranteed residual value over expected residual value	Present value • influences the size of lease payments • added to right-of-use asset and lease liability
Purchase options	If exercise is "reasonably certain," • limits lease term • PV of exercise price added to right-of-use asset and lease liability • PV of exercise price added to lease receivable
Termination penalties	If exercise is "reasonably certain," • limits lease term • PV of penalty amount added to right-of-use asset and lease liability • PV of penalty amount added to lease receivable
Modification of a lease	• If a modification grants the lessee an *additional right of use,* the original lease is terminated and a new lease is created • Otherwise, it means adjusting, adding, or deleting accounts to conform with the new terms and perhaps reclassifying it from one type of lease to another

Remeasurement of the Lease Liability

When considering the various uncertainties surrounding leases in the previous sections, we've encountered several situations that cause us to remeasure a lease liability (and right-of-use asset). Illustration 15–17 provides us with a summary.

In each case, we compare the remeasured liability with its current balance to see the adjustment needed, calculating the new amount as the present value of the *remaining* lease payments:

Present value of *remaining* payments	$xxx
Liability balance (current)	xxx
Increase (or decrease) in balance	$xxx

Illustration 15–17 Situations Requiring Remeasurement of the Lease Liability

SITUATIONS REQUIRING REMEASUREMENT OF THE LEASE LIABILITY		Discount rate used in present value calculation updated from rate used at the beginning of the lease to rate current at remeasurement
1. There is a **change in the assessment** of	• the lease term*	Yes
	• whether exercise of a purchase or termination option is reasonably certain*	Yes
	• any cash payment because of a guaranteed residual value	No
	• whether a variable payment is now a fixed payment	No
2. There is a **modification** of the terms of the lease†	• not accounted for as a new lease	Yes

*We reassess the lease term or exercise of an option only when some new event, like a leasehold improvement, triggers a reassessment.
†Under prior GAAP, a modification was the only time the lease liability was remeasured.

Ethical Dilemma

"I know we had discussed that they're supposed to be worth $24,000 when our purchase option becomes exercisable," Ferris insisted. "That's why we agreed to the lease terms. But, Jenkins, you know how uncertain technology changes are. We can make a good case that they'll be worth $40,000 in three years."

The computers to which Ferris referred were leased to a customer this week. The lease meets none of the criteria for classification as a sales-type lease except that it contains a purchase option that is reasonably certain to be exercised if the computers could be purchased for $24,000 after three years.

"We could record revenue earlier that way," Jenkins agreed.

How could revenue be recorded earlier?

Do you perceive an ethical problem?

Other Lease Accounting Issues and Reporting Requirements

PART C

● LO15–7

Is It a Lease?

When does a contract meet the definition of a lease? Two key criteria must be met at the inception of the contract for an arrangement to constitute a lease for purposes of ASC 842:

1. There must be an *identified asset*. This means:
   - the asset must be property, plant, or equipment (not inventory, intangibles, or natural resources), and
   - the asset must be specified in the contract, either (a) *explicitly* with, say, a vehicle identification number or serial number or (b) *implicitly*, with enough information to recognize the physically distinct asset that is the subject of the lease.

2. The lessee must have the right to *control the use* of the identified asset, which requires that:
   - the customer can derive substantially all of the potential economic benefits from using the asset,
   - the customer can direct the use of the asset throughout the contract term, and
   - the lessor cannot have the right to substitute an alternative asset anytime during the period of use and possibly benefit economically from such a substitution.[21]

Often, judgment is needed to decide. For instance, some transportation contracts might explicitly or implicitly specify the exact truck, shipping vessel, or rail car that will be used to fulfill a service agreement. The contract might comprise a lease, depending on whether the customer can control the use of the transportation vehicle. These criteria are easier to understand if we look at an illustration (see Illustration 15–18).

Illustration 15–18

Does the Contract Contain a Lease?

CastCom Communications entered into a contract to supply Phast Franchising with next-generation BoostServers, model BS12, that promise to monitor Phast's IT infrastructure and boost data transfer rates among its various locations for a period of four years.

- As part of the contract, CastCom installed the BoostServers at Phast's corporate office and will be responsible for general maintenance and will replace the servers with an equivalent model in the event of a breakdown.
- Phast will provide the necessary interface with its IT infrastructure and will be responsible for decisions related to the configuration and use of the servers.

Does this contract meet the definition of a lease?

Our first criterion for an arrangement to constitute a lease is: Do we have an identified asset? It appears we do. The servers are equipment and are specifically identified (BoostServers, model BS12) in the contract. Even if the model number had not been provided, the servers could have been "implicitly" stated with enough information to recognize the asset that is the subject of the lease.

Now that we have determined that there is an identified asset, we check criterion 2: Does Phast have *the right to control the use of the asset*? For that to happen, the customer must "derive substantially all of the potential economic benefits from using the asset" *and* "direct the use of the asset throughout the contract term." In our illustration, Phast (a) has sole use of the servers throughout the four-year period, indicating that it has the right to obtain substantially all the economic benefits from using of the servers and (b) has the right to direct the use of the server because it is "responsible for decisions related to the configuration and use of the servers."

But this second criterion is a bit more specific. For Phast to *control the use* of the identified asset, CastCom cannot have the right to substitute alternative assets throughout the period of use and also benefit economically from that substitution (substantive substitution right). If so, Phast cannot *control the use* of the identified asset. That's because CastCom could substitute, let's say, a less costly asset whose use might not provide equivalent benefits to the customer who still makes the same payments. Having to replace an asset because of a malfunction, though, does not represent such a right because CastCom does not benefit economically from the substitution.

So, yes, both requirements are met: (1) There is an identified asset, and (2) the lessee has the right to control the use of the identified asset. We have a lease.

Now, let's modify the arrangement a bit (see Illustration 15–18A).

At first glance, we might say the first requirement is met: the *servers* constitute an identified asset. But let's look again. Apparently there is no type of identifier for the servers, such as a model number or serial number. Even that might be okay because the lease accounting

[21]referred to as a "substantive substitution right."

Illustration 15–18A

Does the Contract Contain a Lease?

CastCom Communications entered into a contract to supply network services to Phast Franchising promising to monitor its IT infrastructure and boost data transfer rates among its various locations.

- As part of the service, CastCom will install various servers at its discretion at Phast's corporate office through which it will proctor and maintain quality and data transfer rates.
- Phast cannot access or direct the use of the servers; CastCom is tasked with their operation and can replace or reconfigure them as needed to provide the contracted services.

Does this contract meet the definition of a lease?

guidance says that an asset needn't be explicitly stated in the contract (like with a model number); it can be "implicitly" stated (like just saying servers) as long as there's enough information to recognize the physically distinct servers that are the subject of the contract. In this contract, though, CastCom will install unspecified servers at its discretion to proctor and maintain unspecified quality and data transfer rates.

Even if we had an identified asset, we still wouldn't have a lease because criterion 2: *The lessee must have the right to control the use of the identified asset* isn't met either. The contract states that "Phast cannot access or direct the use of the servers."

If it's not a lease, we account for the arrangement as a service contract. We simply record each payment as an expense.

Some companies actually attempt to structure arrangements to intentionally fail the criteria that require lease accounting. That way they're accounted for as service contracts, which means avoiding the lease liability and right-of-use asset appearing on the balance sheet (off-balance-sheet financing). As we discussed earlier in the chapter, reporting a liability can have an unfavorable effect on the company's debt to equity ratio and other quantifiable indicators of riskiness. And, similarly, reporting the asset increases total assets and correspondingly lowers calculations of the rate of return on assets. So, managers often feel they can avoid looking more risky and less profitable if they can configure agreements in a way that causes it not to be considered a lease. In fact, many of the lease accounting guidelines we've looked at in this chapter are designed to keep managers from doing that.

Now that we've considered the type of contract that constitutes a lease, let's look at some other lease accounting issues.

Nonlease Components of Lease Payments

Service contracts, maintenance, hazard insurance, and property taxes are costs often associated with owning and operating an asset. Frequently, for convenience, a lease contract will specify that the lessor is to pay some or all of these costs, but in reality these additional costs are embedded in the periodic payments made by the lessee. The accounting question is whether to include these costs as separate components of the lease contract (to be expensed by the lessee) or, instead, as amounts to be capitalized as part of the right-of-use asset. The determining factor is whether the charge represents a *transfer of a good or service to the lessee* apart from the leased asset. If so, it qualifies as a "nonlease component" of the payment and is separated from the lease payments.

Nonlease components that are included in periodic lease payments to be paid by the lessor are, in effect, indirectly paid by the lessee—and expensed by the lessee.

For demonstration, let's modify Illustration 15–4A. Recall that the lease payments in that illustration were $100,000 each year. Now, let's assume the periodic lease payments were increased to $102,000 with the provision that the lessor (First LeaseCorp) pays an annual maintenance fee of $2,000 to a third-party maintenance firm. Illustration 15–19 demonstrates how the lessee and lessor would account for the maintenance portion of the lease payment.

Sans Serif receives two separate benefits in the lease contract—the right to use equipment as well as maintenance on that equipment. Thus, payments specified in the lease contract contain a lease component (use of equipment for $100,000) and a nonlease component (maintenance service of $2,000).

Illustration 15–19

Lease Payments That Include Nonlease Components Paid by the Lessor

On January 1, 2021, Sans Serif Publishers leased equipment from First LeaseCorp. First LeaseCorp purchased the equipment from CompuDec Corporation at a cost of $479,079.

- Six annual payments of $102,000 beginning January 1, 2021.
- Payments include **$2,000**, which First LeaseCorp will use to pay an annual maintenance fee.
- The interest rate in these financing arrangements is 10%.
- Finance lease to Sans Serif Publishers.
- Sales-type lease to First LeaseCorp.

First Payment (January 1, 2021)

Sans Serif Publishers (Lessee)		
Maintenance expense (2021 fee)	**2,000**	
Lease payable	100,000	
Cash (lease payment)		102,000
First LeaseCorp (Lessor)		
Cash (lease payment)	102,000	
Lease receivable		100,000
Maintenance fee payable*		**2,000**

*This assumes the **$2,000** maintenance fee hasn't yet been paid to the outside maintenance service.

At the beginning of the lease, Sans Serif would record a right-of-use asset and lease liability for the present value of the six $100,000 lease payments, $479,079 (see Illustration 15–4A). For the first payment of $102,000, **$2,000** is recorded as maintenance expense and the remaining $100,000 reduces the lease liability.

On the other hand, if the $2,000 had been a payment for *hazard insurance or property taxes,* that payment does not transfer to the lessee a separate good or service. Instead, payments for hazard insurance and property taxes are specifically identified in the lease accounting guidance as part of the lease payments rather than nonlease components. So, in that case, the right-of-use asset and lease liability (and the lessor's lease receivable) would be measured as the present value of the $102,000 lease payments, not $100,000.

As a practical expedient, the lessee is given the *option* to elect to *not* separate the nonlease components (like oil changes, snowplowing and other forms of maintenance) and, instead, account for the entire arrangement as a lease. For example, the lessee in our illustration could elect to treat the $2,000 maintenance cost, not as maintenance expense as shown, but instead as part of the lease payment like for insurance or taxes. If the amounts are material, the lessee likely will not elect this option because the effect is to increase the lease liability, something most lessees try to avoid.

The lessor, too, is permitted to elect to not separate the nonlease components and, instead, account for the entire arrangement as a lease. This practical expedient is allowed when:

1. the timing and pattern of transfer for the lease and nonlease components are identical (like when lease payments and payments for a maintenance contract both are equal straight-line amounts paid monthly), and
2. the lease would qualify as an operating lease if the lease payments alone (without the nonlease payments) are considered.[22]

If initial direct costs are incurred by the lessor: (a) recorded as a *selling expense* in a sales-type lease that *includes selling profit.* (b) deferred and expensed over the lease term in a *sales-type* lease with *no selling profit.* (c) deferred and expensed over the lease term in an *operating* lease.

INITIAL DIRECT COSTS The costs that (a) are associated directly with consummating a lease, (b) are essential to acquire the lease, and (c) would not have been incurred had the lease agreement not occurred are referred to as **initial direct costs**. They include legal fees, commissions, and preparing and processing lease documents. While legal and processing fees for executing the lease document are included, legal fees for negotiations and drafting documents are not initial direct costs and are expensed as incurred.

[22]FASB ASC 842-10-15-42A: Leases-Overall-Scope and Scope Exceptions-Lessor.

For the lessee, initial direct costs incurred are added to the right-of-use asset. For the lessor, accounting for initial direct costs incurred depends on the classification of the lease.

- In a **sales-type lease** ***that includes selling profit***, initial direct costs are expensed in the period of "sale"—that is, at the beginning of the lease. This treatment assumes that in a sales-type lease the primary reason for incurring these costs is to facilitate the sale of the leased asset.
- In a **sales-type lease** ***with no selling profit***, initial direct costs are deferred and expensed over the lease term. This can be accomplished by not recording the "prepaid expense" separately, but including it in the lease receivable. Increasing the receivable causes the implicit rate (the interest rate that causes the present value of the lease payments to equal the receivable) to be lower. Determining interest revenue at this lower rate accomplishes the purpose of reducing interest revenue each period by a portion of the prepaid expense.
- In an **operating lease**, initial direct costs are deferred and expensed over the lease term, generally on a straight-line basis.

ADVANCE PAYMENTS Often lease agreements call for **advance payments** to be made at the beginning of the lease. Those payments represent prepaid rent (lease costs). Such payments are included with other payments when the lessee determines the present value of lease payments to determine the right-of-use asset and lease liability and when the lessor calculates its lease receivable in a sales-type lease. Remember, though, the lessor doesn't record a lease receivable for an operating lease. So, in that case, these advance payments are recorded as deferred rent revenue and allocated (normally on a straight-line basis) to rent revenue over the lease term. The rent that is periodically reported in those cases consists of the periodic rent payments themselves plus an allocated portion of deferred rent revenue.

Advance payments represent prepaid rent.

LEASEHOLD IMPROVEMENTS Sometimes a lessee will make improvements to leased property that reverts back to the lessor at the end of the lease. If a lessee constructs a new building or makes modifications to existing structures, that cost represents an asset just like any other capital expenditure. Like other assets, its cost is allocated as amortization or depreciation expense over its useful life to the lessee, which will be the shorter of the lease term or the physical life of the asset. Theoretically, such assets can be recorded in accounts descriptive of their nature, such as buildings or plant. In practice, the traditional account title used is **Leasehold Improvements**.[23] In any case, the undepreciated cost usually is reported in the balance sheet under the caption *property, plant, and equipment.* Movable assets like office furniture and equipment that are not attached to the leased property are not considered leasehold improvements.

The cost of a *leasehold improvement* is depreciated over its useful life to the lessee.

As discussed previously in Part B of this chapter, the existence of leasehold improvements can affect the determination of the lease term. For instance, let's say a company leased a building for a noncancelable term of six years with a two-year renewal option. Before it takes possession of the building, the company pays for leasehold improvements that are expected to have significant value at the end of six years. Because that value can be realized only through continued occupancy of the leased building, at the beginning of the lease the company would determine that it is reasonably certain to exercise the renewal option because it would suffer a significant economic penalty if the leasehold improvements were abandoned at the end of the initial lease term. As a result, at the beginning of the lease, the company would conclude that the lease term is eight years. Similarly, if the leasehold improvement is made during the lease term, that additional cost might trigger a reconsideration of whether a purchase or renewal option is reasonably certain to be exercised, and thus a reassessment of the lease term.

[23]Also, traditionally, depreciation sometimes is labeled amortization when in connection with leased assets and leasehold improvements. This is of little consequence. Remember, both depreciation and amortization refer to the process of allocating an asset's cost over its useful life.

Hundreds of firms in 2005 corrected the way they accounted for leases.

Additional Consideration

During 2005, hundreds of companies, particularly in the retail and restaurant industries, underwent one of the most widespread accounting correction events ever. Corrections were required because of the way these companies, including **Pep Boys**, **Ann Taylor**, **Target**, and **Domino's Pizza**, had allocated the cost of leasehold improvements. Rather than expensing leasehold improvements properly over the lease terms, these firms for years had inappropriately expensed the cost over the longer estimated useful lives of the properties. Prompting the sweeping revisions was a Securities and Exchange Commission letter on February 7, 2005, urging companies to follow long-standing accounting standards in this area. A result of the improper practices was to defer expense, thereby accelerating earnings. For instance, **McDonald's Corp.** recorded a charge of $139 million to adjust for the difference.

Concept Review Exercise

VARIOUS LEASE ACCOUNTING ISSUES: FINANCE/SALES-TYPE LEASE

(This is an extension of the previous Concept Review Exercise.)

United Cellular Systems leased a satellite transmission device from Satellite Technology Corporation on January 1, 2021. Satellite Technology paid $500,000 for the transmission device. Its retail value is $653,681. The lease qualifies as a finance lease to the lessee and a sales-type lease with a selling profit to the lessor.

Terms of the Lease Agreement and Related Information:

Lease term	3 years (6 semiannual periods)
Semiannual lease payments	$123,000 at the beginning of each period
Economic life of asset	4 years
Interest rate	12%
Unguaranteed residual value	$40,000
Maintenance fees paid by lessor	$3,000/twice each year (included in lease payments)
Lessor's initial direct costs	$4,500
Contingent lease payments	Additional $4,000 if revenues exceed a specified base

Required:

1. Prepare an amortization schedule that describes the pattern of interest expense over the lease term for United Cellular Systems.
2. Prepare an amortization schedule that describes the pattern of interest revenue over the lease term for Satellite Technology.
3. Prepare the appropriate entries for both United Cellular Systems and Satellite Technology on January 1 and June 30, 2021. (Amortization is recorded only at year-end.)
4. Prepare the appropriate entries for both United Cellular Systems and Satellite Technology on December 31, 2023 (the end of the lease term), assuming the device is returned to the lessor and its actual residual value is $14,000 on that date.

Solution:

1. Prepare an amortization schedule that describes the pattern of interest expense over the lease term for United Cellular Systems.

Calculation of the Present Value of Lease Payments:

Present value of periodic lease payments excluding maintenance costs of $3,000:

($120,000 × 5.21236*) = $625,483

*Present value of an annuity due of $1: $n = 6$, $i = 6\%$.

Note: The residual value is excluded from lease payments for both the lessee and lessor but was influential in establishing the size of the lease payments.

Date	Payments	Effective Interest	Decrease in Balance	Outstanding Balance
		(6% × Outstanding balance)		
1/1/21				625,483
1/1/21	120,000		120,000	505,483
6/30/21	120,000	0.06 (505,483) = 30,329	89,671	415,812
1/1/22	120,000	0.06 (415,812) = 24,949	95,051	320,761
6/30/22	120,000	0.06 (320,761) = 19,246	100,754	220,007
1/1/23	120,000	0.06 (220,007) = 13,200	106,800	113,207
6/30/23	120,000	0.06 (113,207) = 6,793*	113,207	0
	720,000	94,517	625,483	

*Adjusted for rounding of other numbers in the schedule.

2. Prepare an amortization schedule that describes the pattern of interest revenue over the lease term for Satellite Technology.

Calculation of the Lessor's Net Investment:

Present value of periodic lease payments excluding maintenance costs of $3,000 ($120,000 × 5.21236*)	$625,483
Plus: Present value of the residual value ($40,000 × 0.70496†)	28,198
Lessor's net investment in lease	$653,681

*Present value of an annuity due of $1: $n = 6$, $i = 6\%$.
†Present value of $1: $n = 6$, $i = 6\%$.
Note: The residual value is excluded from lease payments, but is part of the lessor's gross and net invest ment in the lease.

Date	Payments	Effective Interest	Decrease in Balance	Outstanding Balance
		(6% × Outstanding balance)		
1/1/21				653,681
1/1/21	120,000		120,000	533,681
6/30/21	120,000	0.06 (533,681) = 32,021	87,979	445,702
1/1/22	120,000	0.06 (445,702) = 26,742	93,258	352,444
6/30/22	120,000	0.06 (352,444) = 21,147	98,853	253,591
1/1/23	120,000	0.06 (253,591) = 15,215	104,785	148,806
6/30/23	120,000	0.06 (148,806) = 8,928*	111,072	37,734
12/31/23	40,000	0.06 (37,734) = 2,266*	37,734	0
	760,000	106,319	653,681	

*Adjusted for rounding of other numbers in the schedule.

3. Prepare the appropriate entries for both United Cellular Systems and Satellite Technology on January 1 and June 30, 2021. (Amortization is recorded only at year-end.)

January 1, 2021

United Cellular Systems (Lessee)		
Right-of-asset (calculated above)	625,483	
Lease payable (calculated above)		625,483
Lease payable (payment less maintenance costs)	120,000	
Maintenance expense	3,000	
Cash (lease payment)		123,000
Satellite Technology (Lessor)		
Lease receivable (PV of lease payments)	653,681	
Cost of goods sold [$500,000 – ($40,000* × 0.70496)]	471,802	
Sales revenue (present value of lease payments)		625,483
Equipment (lessor's cost)		500,000
Selling expense	4,500	
Cash (initial direct costs†)		4,500
Cash (lease payment)	123,000	
Maintenance fee payable (or cash)		3,000
Lease receivable (payment less maintenance costs)		120,000

*This is the unguaranteed residual value.
†If incurred by the lessor, initial direct costs are recorded as a selling expense at the beginning of a sales-type lease.

June 30, 2021		
United Cellular Systems (Lessee)		
Interest expense [6% × ($625,483 – $120,000)]	30,329	
Lease payable (difference)	89,671	
Maintenance expense (annual fee)	3,000	
Cash (lease payment)		123,000
Satellite Technology (Lessor)		
Cash (lease payment)	123,000	
Maintenance fee payable (or cash)		3,000
Lease receivable (to balance)		87,979
Interest revenue [6% × ($653,681 – $120,000)]		32,021

4. Prepare the appropriate entries for both United Cellular Systems and Satellite Technology on December 31, 2023, (the end of the lease term) assuming the device is returned to the lessor and its actual residual value is $14,000 on that date.

December 31, 2023		
United Cellular Systems (Lessee)		
Amortization expense ($625,483 ÷ 3 years)	208,494	
Right-of-use asset		208,494
Satellite Technology (Lessor)		
Equipment (actual residual value)	14,000	
Loss on leased assets ($40,000 – $14,000)	26,000	
Lease receivable (account balance)		37,734
Interest revenue (6% × $37,734: from schedule)		2,266

● LO15–8

Statement of Cash Flow Impact

● LO15–8

Operating Leases

Both the lessee and lessor report cash payments for operating leases as operating activities.

Remember, lease payments for operating leases represent rent—expense to the lessee, revenue for the lessor. These amounts are included in net income, so both the lessee and lessor report cash payments for operating leases in a statement of cash flows as cash flows from operating activities.

Finance Leases—Lessee

The interest portion of a finance lease payment is a cash flow from operating activities and the principal portion is a cash flow from financing activities.

You've learned in this chapter that finance leases are agreements that we identify as being formulated outwardly as leases, but which are in reality installment purchases, so we account for them as such. Each lease payment (except the first if paid at the beginning) includes both an amount that represents interest and an amount that represents a reduction of principal. In a statement of cash flows, then, the lessee reports the interest portion as a cash outflow from operating activities and the principal portion as a cash outflow from financing activities.

At the beginning of the lease, the lessee reports the right-of-use asset and lease liability as a noncash investing/financing activity in the disclosure notes to the financial statements.

Sales-Type Leases—Lessor

Cash receipts from a sales-type lease are cash flows from operating activities.

In a sales-type lease we assume the lessor is actually selling its product. Consistent with reporting sales of products under installment sales agreements rather than lease agreements, the lessor reports cash receipts from a sales-type lease as cash inflows from operating activities.

At the beginning of the lease, the lessor reports the lease as a noncash investing activity (acquiring one asset and disposing of another) in the disclosure notes to the financial statements.

Lease Disclosures

Lease disclosure requirements are quite extensive for both the lessor and lessee. Virtually all aspects of the lease agreement must be disclosed. The guiding objective is that lessees and lessors provide disclosures that enable users of financial statements to assess the *amount, timing,* and *uncertainty* of cash flows arising from leases. Information disclosed is both qualitative and quantitative.

Qualitative Disclosures

A general description of the leasing arrangement is required, including information about variable lease payments, options, nonlease payments, and residual values. A reason for excluding variable lease payments from the measurement of the lessee's lease liability and the lessor's receivable is that they could result in unreliable measurements in the financial statements. Disclosing information in notes to financial statements about variable lease payments would be more useful than estimating and including the payments in assets and liabilities.

Quantitative Disclosures

LESSEE

- Finance lease costs, with separate disclosure of interest and amortization. (The total of these two is reported together in the income statement as lease expense.)
- Operating lease cost.
- Short-term lease cost.
- Variable lease cost.
- Weighted-average lease term of operating leases and finance leases.
- Weighted-average discount rate.
- A reconciliation of opening and closing balances of the right-of-use asset.
- Contractual obligations (and options that the lessee is "reasonably certain" to exercise) for each of the five succeeding fiscal years, plus a total for the remaining years.
- Table of future lease payments, segregated by type of lease, for each of the next five years, and total of payments for the remaining years, and (for finance leases) reconciled with the balance sheet liabilities.

LESSOR

- Information about lease contracts and significant assumptions and judgments.
- Table of lease revenues received.
- Lease sales disclosed separately from regular sales.
- Table of future lease payments, segregated by type of lease, for each of the next five years, and total of payments thereafter and (for sales-type leases) reconciled with the balance sheet receivables.
- Information about assets under operating lease.
- Information about risks associated with residual values.
- Information about significant changes in unguaranteed residual values.
- The gross investment and net investment in leases.

Financial Reporting Case Solution

©Pressmaster/
Shutterstock

1. **Are there advantages of leasing its equipment that HG should consider?** *(p. 835)* Leasing can offer many benefits. Leasing allows a company to acquire the use of an asset without having to pay the entire purchase price upfront. Leasing also can avoid a time-consuming purchase process. Also, if obsolescence is a concern, as it is for HG, leasing facilitates frequent replacement of assets.
2. **Would the effect on HG's income statement be the same if HG used finance leases instead of operating leases?** *(p. 846)* Finance leases involve the "front loading" of lease expense due to the fact that interest expense is higher initially than it is in the later years of a lease, while amortization expense for the right-of-use asset remains the same each period. Straight-line lease expense that occurs with operating leases defers expense recognition, making net income higher in the early years of the lease. Many lessees actively seek ways to structure leases as operating leases to take advantage of their income statement effects. ●

The Bottom Line

- **LO15–1** A variety of operational, financial, and tax incentives often make leasing an attractive alternative to purchasing. A lessee should classify a lease transaction as a finance lease if one or more of five classification criteria is met, and a lessor should record the lease as a sales-type lease. If none of the five criteria is met, both the lessee and the lessor classify the lease as an operating lease. The criteria are used to identify situations when the lease is economically similar to the purchase of an asset because the lessee obtains control of the underlying asset, meaning the ability to direct the use of the asset and obtain substantially all of its remaining benefits, in contrast to merely obtaining control over the use of the asset for a period of time. *(p. 835)*
- **LO15–2** When a lease is classified as a finance lease from the lessee's perspective and a sales-type lease from the lessor's perspective, (1) the lessee records a right-of-use asset and lease liability for the present value of the lease payments, and (2) the lessor records a lease receivable for the same amount and removes the asset from its books. The lessor recognizes interest revenue and the lessee recognizes interest expense at the effective rate times the outstanding balance. The lessee amortizes the right-of-use asset on a straight line basis. For an operating lease, the two components, interest and amortization, are shown as one lease expense in the income statement *(p. 841)*
- **LO15–3** The lessor records sales revenue and cost of goods sold for a sales-type lease with a selling profit. That happens when the present value of the lease payments exceeds the asset's carrying value. *(p. 846)*
- **LO15–4** In an operating lease, a sale is not recorded by the lessor. Instead, the periodic lease payments are accounted for as rent revenue by the lessor. The lessee records a right-of-use asset and lease liability at the present value of the lease payments. Interest expense is recognized at the effective rate times the outstanding balance. Amortization of the right-of-use asset is determined as the amount needed to cause the total lease expense (interest plus amortization) to be a straight-line amount equal to the lease payment. *(p. 847)*
- **LO15–5** In a lease of 12 months or less, the lessee can elect not to record a right-of-use asset and lease payable at the beginning of the lease term, but instead simply record lease payments as expense as they occur (short-cut method). *(p. 855)*
- **LO15–6** The lease term is the contractual lease term modified by any renewal or termination options that are reasonably certain to be exercised or not exercised. Options whose exercise is under the control of the lessor are automatically included. Lease payments include payments resulting from those options as well as excess guaranteed residual values. The calculation of the present value of lease payments at the beginning of the lease does not include any variable lease payments, unless those payments are "in-substance fixed payments" or if they are based solely on an index or rate. *(p. 855)*
- **LO15–7** For an arrangement to constitute a lease there must be an *identified asset,* and the lessee must have the right to *control its use.* Nonlease components that are included in periodic lease payments (like

maintenance) to be paid by the lessor are, in effect, indirectly paid by the lessee—and are expensed by the lessee. Insurance and property taxes included in periodic lease payments, however, do not transfer a good or service to the lessee and thus are considered part of lease payments. Initial direct costs incurred by the lessee are added to the right-of-use asset. If incurred by the lessor, they are recorded as a selling expense in a sales-type lease with a selling profit or expensed over the lease term in a sales-type lease with no selling profit or in an operating lease. The cost of a leasehold improvement is depreciated over its useful life to the lessee. (*p. 869*)

- **LO15–8** We report all cash flows as operating cash flows, except the principal portion of each payment in a finance lease, which is a cash flow from financing activities. Extensive disclosure requirements for lessees and lessors are designed to enable users of financial statements to assess the amount, timing, and uncertainty of cash flows arising from leases. (*p. 876*)
- **LO15–9** Accounting for leases is similar in most respects under U.S. GAAP or IFRS. The primary difference is that IFRS treats all leases as finance leases by the lessee. (*p. 852, 855 and 859*)

Sale-Leaseback Arrangements

APPENDIX 15

In a **sale-leaseback transaction**, the owner of an asset sells it and immediately leases it back from the new owner. Sound strange? Maybe, but this arrangement is common. In a sale-leaseback transaction two things happen:

1. The seller-lessee receives cash from the sale of the asset.
2. The seller-lessee pays periodic rent payments to the buyer-lessor to retain the use of the asset.

What motivates this kind of arrangement? The two most common reasons are (a) if the asset had been financed originally with debt and interest rates have fallen, the sale-leaseback transaction can be used to effectively refinance at a lower rate; or (b) the most likely motivation for a sale-leaseback transaction is to generate cash.

We account for this type of arrangement in one of two ways.

- **Sale-Leaseback Approach.** Record the sale of the asset (with any accompanying gain or loss) and then record a lease for the leaseback portion in accordance with the lease guidance described earlier in this chapter. As we see below, though, if the leaseback would qualify as a finance lease, no sale has occurred and this approach cannot be applied. Thus, the sale-leaseback approach is allowed only if the leaseback qualifies as an operating lease.
- **Financing Arrangement.** View the arrangement, not as a sale, but as a loan by the lessor to the lessee for the "sale" price. The asset remains on the lessee's books, and the leaseback is accounted for as debt. The "lease" payments are deemed to be repayment of the loan.

We apply the sale-leaseback approach only if the usual requirements for revenue recognition (Chapter 6) are met (so the sale can be recorded), and it wouldn't qualify as a finance lease, which would indicate the asset is effectively sold back to the lessee.[24]

Recording a sale-leaseback transaction follows the basic accounting concept of substance over form.

If the transaction does not qualify for sale-leaseback accounting, both the companies account for it as a financing arrangement. The reasoning for allowing sale-leaseback accounting only for situations in which the leaseback qualifies as an operating lease is quite logical when we remember the primary distinction between a finance lease and an operating lease. A finance lease is substantively a "sale," so if this is the nature of a leaseback, then we

[24] More specifically, *none* of the following conditions can be present if sale-leaseback accounting is to be used:
The seller-lessee has an option to buy back the asset at any price other than fair value (and other similar items must be available to buy).
The buyer-lessor has the option to sell the asset back to the seller-lessee that the buyer-lessor is "reasonably certain" to exercise.
The leaseback term is for 75% of the remaining economic life of the asset.
The present value of the lease payments accounts for 90% of the fair value of the asset.

have a sale by the lessee and then a sale back to the lessee, so we really have no sale at all. Essentially, the asset still belongs to the lessee, and cash comes to the lessee at the time of the transaction which is paid back in the form of periodic "lease payments." Of course, this is the nature of a loan, so we account for it that way.

Illustration 15–A1 demonstrates a sale-leaseback involving a leaseback that qualifies as an operating lease.

Illustration 15–A1
Sale-Leaseback

Teledyne Distribution Center was in need of cash. Its solution: sell its four warehouses for $1,000,000, their fair value, and then lease back the warehouses to continue using them. The warehouses had a carrying value on Teledyne's books of $900,000 (original cost $1,200,000). Other information:

1. The sale date is December 31, 2021.
2. The noncancelable lease term is 10 years and requires annual payments of $118,360 beginning December 31, 2021. The estimated remaining useful life of the warehouses is 15 years.
3. The annual lease payments (present value $800,000] provides the lessor with a 10% rate of return on the financing arrangement.*

The gain (or loss) on a sale-leaseback is recognized at the time of the sale.

December 31, 2021

Cash	1,000,000	
Accumulated depreciation ($1,200,000 – $900,000)	300,000	
Warehouses (cost)		1,200,000
Gain on sale-leaseback (difference)		100,000
Right-of-use asset (present value of lease payments)	800,000	
Lease payable (present value of lease payments)		800,000
Lease payable	118,360	
Cash		118,360

In an operating lease, the lessee records interest the normal way and then "plugs" the amortization at whatever amount is needed for interest plus amortization to equal the *straight-line* lease payment.

December 31, 2022

Interest expense [10% × ($800,000 – $118,360)]	68,164	
Lease payable (difference)	50,196	
Cash (lease payment)		118,360
Amortization expense ($118,360 – $68,164)	50,196	
Right-of-use asset		50,196

*$118,360 lease payments × 6.75902 (from Table 6) $n = 10, i = 10\%$ = $800,000 ($799,997.61 rounded) Present value

Since none of the criteria for a finance lease is met, the leaseback is recorded by the lessee as an operating lease.[25] The sale-leaseback, then, is deemed to be two distinct transactions: a legitimate sale creating a $100,000 gain and then an operating lease. On the flip side of the transaction, the lessor would record the purchase of a $1,000,000 asset and then record lease revenue of $118,360 each year over the term of the lease.

But now suppose the remaining useful life is 12 years instead of 15 and that the ten lease payments are $147,950, creating a present value of $1,000,000 instead of $800,000. According to the criteria in Illustration 15–2, either of those circumstances would require that the arrangement be accounted for as a finance lease, instead of an operating lease. As such, the transaction does not qualify for sale-leaseback accounting. We would view the arrangement, not as a sale, but as a loan by the lessor to the lessee for the $1,000,000 "sale" price. The

[25]The lease term is less than the major part of the expected useful life of the warehouses (67%). Also, the present value of lease payments ($800,000) is less than substantially all of the fair value of the warehouses ($1,000,000). None of the other three of the five criteria is present.

asset remains on the lessee's books. The "lease" payments would be considered to be repayment of the loan. It would be recorded as shown in Illustration 15–A2.

December 31, 2021		
Cash	1,000,000	
Notes payable		1,000,000
Notes payable	147,950	
Cash		147,950
December 31, 2022		
Interest expense [10% × ($1,000,000 – $147,950)]	85,205	
Notes payable (difference)	62,745	
Cash (lease payment)		147,950

Illustration 15–A2

Sale-Leaseback Deemed to Be a Loan

Though structured as a sale and then a lease, in substance, the lessee is actually borrowing $1,000,000.

The "lease" payments would be considered to be repayment of the loan.

Questions For Review of Key Topics

Q 15–1 One of the advantages of leasing rather than purchasing an asset is that leasing offers flexibility and a lower cost when disposing of the asset. Explain.

Q 15–2 The basic concept of "substance over form" influences lease accounting. Explain.

Q 15–3 How is interest expense determined in a finance lease transaction? How does the approach compare to other forms of debt (such as bonds payable or notes payable)?

Q 15–4 How are leases and installment notes the same? How do they differ?

Q 15–5 A lessee should classify a lease transaction as a finance lease if it is noncancelable and one or more of five classification criteria are met. Otherwise, it is an operating lease. What are these criteria?

Q 15–6 Lukawitz Industries leased non-specialized equipment to Seminole Corporation for a four-year period, at which time possession of the leased asset will revert back to Lukawitz. The equipment cost Lukawitz $4 million and has an expected useful life of six years. Its normal sales price is $5.6 million. The present value of the lease payments for both the lessor and lessee is $5.2 million. The first payment was made at the beginning of the lease. How should this lease be classified (a) by Lukawitz Industries (the lessor) and (b) by Seminole Corporation (the lessee)? Why?

Q 15–7 In accounting for a finance lease/sales-type lease, how are the lessee's and lessor's income statements affected?

Q 15–8 What is selling profit on a sales-type lease? How do we account for a sales-type lease with a selling profit?

Q 15–9 At the beginning of an operating lease, the lessee will record what asset and liability, if any?

Q 15–10 At the beginning of an operating lease, the lessor will record what asset or assets, if any?

Q 15–11 In accounting for an operating lease, how are the lessee's and lessor's income statements affected?

Q 15–12 Briefly describe the conceptual basis for asset and liability recognition under the right-of-use approach used by the lessee in a lease transaction.

Q 15–13 In a financing lease, "front loading" of lease expense and lease revenue occurs. What does this mean, and how is it avoided in an operating lease?

Q 15–14 The discount rate influences virtually every amount reported in connection with a lease by both the lessor and the lessee. What is the lessor's discount rate when determining the present value of lease payments? What is the lessee's discount rate?

Q 15–15 A lease that has a lease term (including any options to terminate or renew that are reasonably certain) of twelve months or less is considered a "short-term lease." How does a lessee record a lease using the shortcut approach available as an option for short-term leases?

Q 15–16 A lease might specify that lease payments may be increased (or decreased) at some future time during the lease term depending on whether or not some specified event occurs such as revenues or profits exceeding some designated level. Under what circumstances are contingent rentals included or excluded from lease payments? If excluded, how are they recognized in income determination?

Q 15–17 What is a purchase option? How does it affect accounting for a lease?

Q 15–18 A six-year lease can be renewed for two additional three-year periods, and it also can be terminated after only three years. How do the lessee and lessor decide the lease term to be used in accounting for the lease?

Q 15–19 Culinary Creations leased kitchen equipment under a five-year lease with an option to renew for three years at the end of five years and an option to renew for an additional three years at the end of eight years. The first three-year renewal option can be exercised for one-half the original and usual rate. What is the length of the lease term that Culinary Creations should assume in recording the transactions related to the lease?

Q 15–20 What situations cause us to remeasure a lease liability and right-of-use asset? How is that accomplished?

Q 15–21 Occasionally, a lease agreement includes a guarantee by the lessee that the lessor will recover a specified residual value when custody of the asset reverts back to the lessor at the end of the lease term. Under what circumstance can the guaranteed residual value influence the amounts recorded by the lessee and lessor? In that circumstance, how are the amounts affected?

Q 15–22 Compare the way a purchase option that is reasonably certain to be exercised and a lessee-guaranteed residual value are treated by the lessee and lessor when determining lease payments.

Q 15–23 What nonlease costs might be included as part of lease payments? How are they accounted for by the lessee in a finance lease when paid by the lessee? When paid by the lessor? Explain.

Q 15–24 The lessor's initial direct costs often are substantial. What are initial direct costs?

Q 15–25 When are initial direct costs recognized in an operating lease? In a sales-type lease with selling profit? In a sales-type lease with no selling profit? Why?

Q 15–26 What is the primary objective of the required lease disclosures for the lessor and lessee?

Q 15–27 Where can we find authoritative guidance for accounting for leases under IFRS?

Q 15–28 Could a finance lease under IFRS be classified as an operating lease under U.S. GAAP? Explain.

Q 15–29 When a company sells an asset and simultaneously leases it back, what criteria must be met to apply sale-leaseback accounting rather than accounting for the transaction as a loan ?

Q 15–30 Zimmern Machines sold equipment with a 10-year economic life to Bourdain Acres, while concurrently entering into an 8-year leaseback. Eight years is considered a major part of the economic life of the equipment. The sale agreement contains no option for Zimmern to repurchase the equipment or any other provision that would prevent its sale. Can Zimmern Machines account for the transaction as a sale-leaseback? Why?

Brief Exercises

connect

BE 15–1
Lease classification
● LO15–1

(Note: BE 15–1 and 15–2 are two variations of the same basic situation.)
Corinth Co. leased non-specialized equipment to Athens Corporation for an eight-year period, at which time possession of the equipment will revert back to Corinth. The equipment cost Corinth $16 million and has an expected useful life of 12 years. Its normal sales price is $22.4 million. The present value of the lease payments for both the lessor and lessee is $20.6 million. The first payment was made at the beginning of the lease. How should Corinth classify this lease?

BE 15–2
Lease classification
● LO15–1, LO15–2

Corinth Co. leased non-specialized equipment to Athens Corporation for an eight-year period, at which time possession of the equipment will revert back to Corinth. The equipment cost Corinth $16 million and has an expected useful life of 12 years. Its normal sales price is $22.4 million. The present value of the lease payments for both the lessor and lessee is $20.6 million. The first payment was made at the beginning of the lease. How should Athens classify this lease?

BE 15–3
Lessee and lessor; calculate interest; finance/sales-type lease
● LO15–2

A finance lease agreement calls for quarterly lease payments of $5,376 over a 10-year lease term, with the first payment on July 1, the beginning of the lease. The annual interest rate is 8%. Both the present value of the lease payments and the cost of the asset to the lessor are $150,000. What would be the amount of interest expense the lessee would record in conjunction with the second quarterly payment on October 1? What would be the amount of interest revenue the lessor would record in conjunction with the second quarterly payment on October 1?

BE 15–4
Finance lease; lessee; balance sheet effects
● LO15–2

(Note: BE 15–4, 15–5, and 15–6 are three variations of the same basic situation.)
A lease agreement that qualifies as a finance lease calls for annual lease payments of $26,269 over a six-year lease term (also the asset's useful life), with the first payment at January 1, the beginning of the lease. The interest rate is 5%. If the lessee's fiscal year is the calendar year, what would be the amount of the lease liability that the lessee would report in its balance sheet at the end of the first year? What would be the interest payable?

BE 15–5
Finance lease; lessee; income statement effects
● LO15–2

A lease agreement that qualifies as a finance lease calls for annual lease payments of $26,269 over a six-year lease term (also the asset's useful life), with the first payment at January 1, the beginning of the lease. The interest rate is 5%. If the lessee's fiscal year is the calendar year, what would be the amounts related to the lease that the lessee would report in its income statement for the year ended December 31(ignore taxes)?

BE 15–6
Sales-type lease; lessor; income statement effects
● LO15–3

A lease agreement that qualifies as a finance lease calls for annual lease payments of $26,269 over a six-year lease term (also the asset's useful life), with the first payment at January 1, the beginning of the lease. The interest rate is 5%. The lessor's fiscal year is the calendar year. The lessor manufactured this asset at a cost of $125,000. What would be the increase in earnings that the lessor would report in its income statement for the year ended December 31(ignore taxes)?

BE 15–7
Sales-type lease; lessor; calculate lease payments
● LO15–3

Manning Imports is contemplating an agreement to lease equipment to a customer for five years. Manning normally sells the asset for a cash price of $100,000. Assuming that 8% is a reasonable rate of interest, what must be the amount of quarterly lease payments (beginning at the commencement of the lease) in order for Manning to recover its normal selling price as well as be compensated for financing the asset over the lease term?

BE 15–8
Operating lease
● LO15–4

(Note: BE 15–8 and 15–9 are two variations of the same basic situation.)
At the beginning of its fiscal year, Lakeside Inc. leased office space to LTT Corporation under a seven-year operating lease agreement. The contract calls for quarterly rent payments of $25,000 each. The office building was acquired by Lakeside at a cost of $2 million and was expected to have a useful life of 25 years with no residual value. What will be the effect of the lease on LTT's earnings for the first year (ignore taxes)?

BE 15–9
Operating lease
● LO15–4

At the beginning of its fiscal year, Lakeside Inc. leased office space to LTT Corporation under a seven-year operating lease agreement. The contract calls for quarterly rent payments of $25,000 each. The office building was acquired by Lakeside at a cost of $2 million and was expected to have a useful life of 25 years with no residual value. What will be the effect of the lease on Lakeside's earnings for the first year (ignore taxes)?

BE 15–10
Short-term lease
● LO15–5

King Cones leased ice cream-making equipment from Ace Leasing. Ace earns interest under such arrangements at a 6% annual rate. The lease term is eight months with monthly payments of $10,000 due at the end of each month. King Cones elected the short-term lease option. What is the effect of the lease on King Cones' earnings during the eight-month term (ignore taxes)?

BE 15–11
Uncertain lease term
● LO15–6

Java Hut leased a specialty expresso machine for a 10-year noncancelable term. At the end of the 10-year term, Java Hut has four consecutive one-year renewal options. A replacement machine can be acquired, but due to an expensive installation process and Java Hut's lease term for its store, Java Hut expects to lease the machine for 12 years. What is the lease term?

BE 15–12
Uncertain lease payments
● LO15–6

On January 1, Espinoza Moving and Storage leased a truck for a four-year period, at which time possession of the truck will revert back to the lessor. Annual lease payments are $10,000 due on December 31 of each year, calculated by the lessor using a 5% discount rate. If Espinoza's revenues exceed a specified amount during the lease term, Espinoza will pay an additional $4,000 lease payment at the end of the lease. Espinoza estimates a 60% probability of meeting the target revenue amount. What amount should be added to the right-of-use asset and lease liability under the contingent rent agreement?

BE 15–13
Purchase option; lessor; sales-type lease
● LO15–2, LO15–3, LO15–6

Ace Leasing acquires equipment and leases it to customers under long-term sales-type leases. Ace earns interest under these arrangements at a 6% annual rate. Ace leased a machine it purchased for $600,000 under an arrangement that specified annual payments beginning at the commencement of the lease for five years. The lessee had the option to purchase the machine at the end of the lease term for $100,000 when it was expected to have a residual value of $160,000. Calculate the amount of the annual lease payments.

BE 15–14
Residual value; sales-type lease
● LO15–2, LO15–3, LO15–6

On January 1, James Industries leased equipment to a customer for a four-year period, at which time possession of the leased asset will revert back to James. The equipment cost James $700,000 and has an expected useful life of six years. Its normal sales price is $700,000. The residual value after four years is $100,000. Lease payments are due on December 31 of each year, beginning with the first payment at the end of the first year. The interest rate is 5%. Calculate the amount of the annual lease payments.

BE 15–15
Guaranteed residual value
● LO15–6

On January 1, Garcia Supply leased a truck for a four-year period, at which time possession of the truck will revert back to the lessor. Annual lease payments are $10,000 due on December 31 of each year, calculated by the lessor using a 5% discount rate. Negotiations led to Garcia guaranteeing a $36,000 residual value at the end of the lease term. Garcia estimates that the residual value after four years will be $35,000. What is the amount to be added to the right-of-use asset and lease liability under the residual value guarantee?

BE 15–16
Lessor's initial direct costs; sales-type lease
● LO15–3, LO15–7

Bryant leased equipment that had a retail cash selling price of $600,000 and a useful life of five years with no residual value. The lessor spent $530,000 to manufacture the equipment and used an implicit rate of 8% when calculating annual lease payments of $139,142 beginning January 1, the beginning of the lease. Lease payments will be made January 1 each year of the lease. Incremental costs of consummating the lease transaction incurred by the lessor were $15,000. What is the effect of the lease on the lessor's earnings during the first year, not including any effect of depreciation no longer required on the asset under lease (ignore taxes)?

BE 15–17
Is it a lease?
● LO15–7

Financial Machinery entered into an arrangement to provide Viable Bank with equipment for the bank's data center. The contract does not explicitly identify the equipment to be used to fulfill the contract. Financial Machinery has several data centers that are interchangeable and can service multiple customers at one time. Due to security processes in place for its customer data, Viable Bank insists on specific restrictions on the equipment to be utilized and the arrangement imposes restrictions on access and substitution by Financial Machinery for the term of the contract. Does the contract include an identifiable asset for purposes of determining whether it contains a lease?

BE 15–18
Is it a lease?
● LO15–7

Able Equipment entered into an arrangement to provide M. T. Bin Wholesale's data center with Able's newest server model. Due to security processes in place for its customer data, M. T. Bin requires access to the equipment and the ability to direct its use. Able can replace or reconfigure the equipment if Able finds it's financially advantageous to do so. Does the contract contain a lease?

BE 15–19
Nonlease payments
● LO15–2, LO15–7

On January 1, 2021, Jasperse Corporation leased equipment under a finance lease designed to earn the lessor a 12% rate of return for providing long-term financing. The lease agreement specified ten annual payments of $75,000 beginning January 1, and each December 31 thereafter through 2029. A 10-year service agreement was scheduled to provide maintenance of the equipment as required for a fee of $5,000 per year. Insurance premiums of $4,000 annually are related to the equipment. Both amounts were to be paid by the lessor and lease payments reflect both expenditures. At what amount will Jasperse record a right-of-use asset?

Exercises

connect

E 15–1
Lease classification
● LO15–1

Each of the four independent situations below describes a lease requiring annual lease payments of $10,000. For each situation, determine the appropriate lease classification by the lessee and indicate why.

	Situation			
	1	2	3	4
Lease term (years)	4	4	4	4
Asset's useful life (years)	6	5	6	6
Asset's fair value	$44,000	$45,000	$41,000	$38,000
Purchase option that is reasonably certain to be exercised?	No	Yes	No	No
Annual lease payments	Beg. of yr.	End of yr.	Beg. of yr.	End of yr.
Lessor's implicit rate (known by lessee)	5%	6%	5%	6%
Lessee's incremental borrowing rate	5%	5%	5%	5%

E 15–2
Finance lease; calculate lease payments
● LO15–2

American Food Services, Inc. leased a packaging machine from Barton and Barton Corporation. Barton and Barton completed construction of the machine on January 1, 2021. The lease agreement for the $4 million (fair value and present value of the lease payments) machine specified four equal payments at the end of each year. The useful life of the machine was expected to be four years with no residual value. Barton and Barton's implicit interest rate was 10%.

Required:

1. Prepare the journal entry for American Food Services at the beginning of the lease on January 1, 2021.
2. Prepare an amortization schedule for the four-year term of the lease.
3. Prepare the appropriate entries related to the lease on December 31, 2021.
4. Prepare the appropriate entries related to the lease on December 31, 2023.

(Note: You may wish to compare your solution to this exercise with that of E 14–20 which deals with a parallel situation in which the packaging machine was acquired with an installment note.)

E 15–3
Finance lease; lessee; balance sheet and income statement effects
● LO15–2

(Note: E 15–3, 15–4, and 15–5 are three variations of the same situation.)
On June 30, 2021, Georgia-Atlantic, Inc. leased warehouse equipment from IC Leasing Corporation. The lease agreement calls for Georgia-Atlantic to make semiannual lease payments of $562,907 over a three-year lease term, payable each June 30 and December 31, with the first payment at June 30, 2021. Georgia-Atlantic's incremental borrowing rate is 10%, the same rate IC uses to calculate lease payment amounts. Amortization is recorded on a straight-line basis at the end of each fiscal year. The fair value of the equipment is $3 million.

Required:
1. Determine the present value of the lease payments at June 30, 2021 (to the nearest $000) that Georgia-Atlantic uses to record the right-of-use asset and lease liability.
2. What amounts related to the lease would Georgia-Atlantic report in its balance sheet at December 31, 2021 (ignore taxes)?
3. What amounts related to the lease would Georgia-Atlantic report in its income statement for the year ended December 31, 2021 (ignore taxes)?

E 15–4
Sales-type lease; lessor; balance sheet and income statement effects
● LO15–2

On June 30, 2021, Georgia-Atlantic, Inc. leased warehouse equipment from IC Leasing Corporation. The lease agreement calls for Georgia-Atlantic to make semiannual lease payments of $562,907 over a three-year lease term (also the asset's useful life), payable each June 30 and December 31, with the first payment at June 30, 2021. Georgia-Atlantic's incremental borrowing rate is 10%, the same rate IC used to calculate lease payment amounts. IC purchased the equipment from Builders, Inc. at a cost of $3 million.

Required:
1. What amount related to the lease would IC report in its balance sheet at December 31, 2021 (ignore taxes)?
2. What amount related to the lease would IC report in its income statement for the year ended December 31, 2021 (ignore taxes)?

E 15–5
Sales-type lease; lessor; balance sheet and income statement effects
● LO15–3

On June 30, 2021, Georgia-Atlantic, Inc. leased warehouse equipment from Builders, Inc. The lease agreement calls for Georgia-Atlantic to make semiannual lease payments of $562,907 over a three-year lease term (also the asset's useful life), payable each June 30 and December 31, with the first payment at June 30, 2021. Georgia-Atlantic's incremental borrowing rate is 10%, the same rate Builders used to calculate lease payment amounts. Builders manufactured the equipment at a cost of $2.5 million.

Required:
1. Determine the price at which Builders is "selling" the equipment (present value of the lease payments) at June 30, 2021 (to the nearest $000).
2. What amount related to the lease would Builders report in its balance sheet at December 31, 2021 (ignore taxes)?
3. What line item amounts related to the lease would Builders report in its income statement for the year ended December 31, 2021 (ignore taxes)?

E 15–6
Finance lease; lessee
● LO15–2

(Note: E 15–6, 15–7, and 15–8 are three variations of the same basic situation.)
Manufacturers Southern leased high-tech electronic equipment from Edison Leasing on January 1, 2021. Edison purchased the equipment from International Machines at a cost of $112,080.

Related Information:	
Lease term	2 years (8 quarterly periods)
Quarterly rental payments	$15,000 at the beginning of each period
Economic life of asset	2 years
Fair value of asset	$112,080
Implicit interest rate	8%
(Also lessee's incremental borrowing rate)	

Required:
Prepare a lease amortization schedule and appropriate entries for Manufacturers Southern from the beginning of the lease through January 1, 2022. Amortization is recorded at the end of each fiscal year (December 31) on a straight-line basis.

E 15–7
Sales-type lease with no selling profit; lessor
● LO15–2

Edison Leasing leased high-tech electronic equipment to Manufacturers Southern on January 1, 2021. Edison purchased the equipment from International Machines at a cost of $112,080.

Related Information:	
Lease term	2 years (8 quarterly periods)
Quarterly rental payments	$15,000 at the beginning of each period
Economic life of asset	2 years
Fair value of asset	$112,080
Implicit interest rate	8%
(Also lessee's incremental borrowing rate)	

Required:
Prepare a lease amortization schedule and appropriate entries for Edison Leasing from the beginning of the lease through January 1, 2022. Edison's fiscal year ends December 31.

E 15–8
Sales-type lease with selling profit; lessor; calculate lease payments
● LO15–3

Manufacturers Southern leased high-tech electronic equipment from International Machines on January 1, 2021. International Machines manufactured the equipment at a cost of $85,000. Manufacturers Southern's fiscal year ends December 31.

Related Information:	
Lease term	2 years (8 quarterly periods)
Quarterly rental payments	$15,000 at the beginning of each period
Economic life of asset	2 years
Fair value of asset	$112,080
Implicit interest rate	8%

Required:
1. Show how International Machines determined the $15,000 quarterly lease payments.
2. Prepare appropriate entries for International Machines to record the lease at its beginning, January 1, 2021, and the second lease payment on April 1, 2021.

E 15–9
Lessor calculation of annual lease payments; lessee calculation of asset and liability
● LO15–2

Each of the three independent situations below describes a finance lease in which annual lease payments are payable at the beginning of each year. The lessee is aware of the lessor's implicit rate of return.

	Situation		
	1	2	3
Lease term (years)	10	20	4
Lessor's rate of return (known by lessee)	11%	9%	12%
Lessee's incremental borrowing rate	12%	10%	10%
Fair value of lease asset	$600,000	$980,000	$185,000

Required:
For each situation, determine:
a. The amount of the annual lease payments as calculated by the lessor.
b. The amount the lessee would record as a right-of-use asset and a lease liability.

E 15–10
Lessor calculation of annual lease payments; lessee calculation of asset and liability
● LO15–2

(Note: This is a variation of E 15–9 modified to assume lease payments are at the end of each period.)
Each of the three independent situations below describes a finance lease in which annual lease payments are payable at the *end* of each year. The lessee is aware of the lessor's implicit rate of return.

	Situation		
	1	2	3
Lease term (years)	10	20	4
Lessor's rate of return	11%	9%	12%
Lessee's incremental borrowing rate	12%	10%	10%
Fair value of lease asset	$600,000	$980,000	$185,000

Required:
For each situation, determine:
a. The amount of the annual lease payments as calculated by the lessor.
b. The amount the lessee would record as a right-of-use asset and a lease liability.

E 15–11
Lessee and lessor; sales-type lease with selling profit
● LO15–2, LO15–3

Eye Deal Optometry leased vision-testing equipment from Insight Machines on January 1, 2021. Insight Machines manufactured the equipment at a cost of $200,000 and lists a cash selling price of $250,177. Appropriate adjusting entries are made quarterly.

Related Information:	
Lease term	5 years (20 quarterly periods)
Quarterly lease payments	$15,000 at Jan. 1, 2021, and at Mar. 31, June 30, Sept. 30, and Dec. 31 thereafter
Economic life of asset	5 years
Interest rate charged by the lessor	8%

Required:

1. Prepare appropriate entries for Eye Deal to record the arrangement at its beginning, January 1, 2021, and on March 31, 2021.
2. Prepare appropriate entries for Insight Machines to record the arrangement at its beginning, January 1, 2021, and on March 31, 2021.

E 15–12
Lessee; finance lease; effect on financial statements
● LO15–2

(Note: E 15–12, 15–13, 15–14, and 15–25 are variations of the same situation.)
At January 1, 2021, Café Med leased restaurant equipment from Crescent Corporation under a nine-year lease agreement. The lease agreement specifies annual payments of $25,000 beginning January 1, 2021, the beginning of the lease, and at each December 31 thereafter through 2028. The equipment was acquired recently by Crescent at a cost of $180,000 (its fair value) and was expected to have a useful life of 12 years with no salvage value at the end of its life. (Because the lease term is only nine years, the asset does have an expected residual value at the end of the lease term of $50,995.) Crescent seeks a 10% return on its lease investments. By this arrangement, the lease is deemed to be a finance lease.

Required:

1. What will be the effect of the lease on Café Med's earnings for the first year (ignore taxes)?
2. What will be the balances in the balance sheet accounts related to the lease at the end of the first year of Café Med (ignore taxes)?

E 15–13
Lessee; operating lease; effect on financial statements
● LO15–4

At January 1, 2021, Café Med leased restaurant equipment from Crescent Corporation under a nine-year lease agreement. The lease agreement specifies annual payments of $25,000 beginning January 1, 2021, the beginning of the lease, and at each December 31 thereafter through 2028. The equipment was acquired recently by Crescent at a cost of $180,000 (its fair value) and was expected to have a useful life of 13 years with no salvage value at the end of its life. (Because the lease term is only nine years, the asset does have an expected residual value at the end of the lease term of $50,995.) Crescent seeks a 10% return on its lease investments. By this arrangement, the lease is deemed to be an operating lease.

Required:

1. What will be the effect of the lease on Café Med's earnings for the first year (ignore taxes)?
2. What will be the balances in the balance sheet accounts related to the lease at the end of the first year for Café Med (ignore taxes)?

E 15–14
Lessor; operating lease; effect on financial statements
● LO15–4

At January 1, 2021, Café Med leased restaurant equipment from Crescent Corporation under a nine-year lease agreement. The lease agreement specifies annual payments of $25,000 beginning January 1, 2021, the beginning of the lease, and at each December 31 thereafter through 2028. The equipment was acquired recently by Crescent at a cost of $180,000 (its fair value) and was expected to have a useful life of 13 years with no salvage value at the end of its life. (Because the lease term is only nine years, the asset does have an expected residual value at the end of the lease term of $50,995.) Crescent seeks a 10% return on its lease investments. By this arrangement, the lease is deemed to be an operating lease.

Required:

1. What will be the effects of the lease on Crescent's (lessor's) earnings for the first year (ignore taxes)?
2. What will be the balances in the balance sheet accounts related to the lease at the end of the first year for Crescent (ignore taxes)?

E 15–15
Sales-type lease; lessor; income statement effects
● LO15–3

King Company leased equipment from Mann Industries. The lease agreement qualifies as a finance lease and requires annual lease payments of $52,538 over a six-year lease term (also the asset's useful life), with the first payment at January 1, the beginning of the lease. The interest rate is 5%. The asset being leased cost Mann $250,000 to produce.

Required:

1. Determine the price at which the lessor is "selling" the asset (present value of the lease payments).
2. What would be the amounts related to the lease that the lessor would report in its income statement for the year ended December 31 (ignore taxes)?

E 15–16
Lessee; operating lease
● LO15–4

Baillie Power leased high-tech electronic equipment from Courtney Leasing on January 1, 2021. Courtney purchased the equipment from Doane Machines at a cost of $250,000, its fair value.

Related Information:	
Lease term	2 years (8 quarterly periods)
Quarterly lease payments	$15,000 at Jan. 1, 2021, and at Mar. 31, June 30, Sept. 30, and Dec. 31 thereafter
Economic life of asset	5 years
Interest rate charged by the lessor	8%

Required:

Prepare appropriate entries for Baillie Power from the beginning of the lease through December 31, 2021. December 31 is the fiscal year end for each company. Appropriate adjusting entries are recorded at the end of each quarter.

E 15–17
Lessee and lessor; operating lease
● LO15–4

On January 1, 2021, Nath-Langstrom Services, Inc., a computer software training firm, leased several computers under a two-year operating lease agreement from ComputerWorld Leasing, which routinely finances equipment for other firms at an annual interest rate of 4%. The contract calls for four rent payments of $10,000 each, payable semiannually on June 30 and December 31 each year. The computers were acquired by ComputerWorld at a cost of $90,000 and were expected to have a useful life of five years with no residual value. Both firms record amortization and depreciation semiannually.

Required:

Prepare the appropriate entries for both (a) the lessee and (b) the lessor from the beginning of the lease through the end of 2021.

E 15–18
Short-term lease
● LO15–5

Chance Enterprises leased equipment from Third Bank Leasing on January 1, 2021. Third Bank Leasing purchased the equipment at a cost of $1,000,000. Chance elected the short-term lease option.

Related Information:	
Lease term	1 year (12 monthly periods)
Monthly lease payments	$15,000 at Jan. 1, 2021, through Dec. 1, 2021
Economic life of asset	5 years
Interest rate charged by the lessor	8%

Required:

Prepare appropriate entries for Chance from the beginning of the lease through April 1, 2021.

E 15–19
Lessee; renewal option
● LO15–2, LO15–6

Natick Industries leased high-tech instruments from Framingham Leasing on January 1, 2021. Natick has the option to renew the lease at the end of two years for an additional three years. Natick is subject to a $45,000 penalty after two years if it fails to renew the lease. Framingham Leasing purchased the equipment from Waltham Machines at a cost of $250,177.

Related Information:	
Lease term	3 years (12 quarterly periods)
Lease renewal option for an additional	2 years (8 quarterly periods)
Quarterly lease payments	$15,000 at Jan. 1, 2021, and at Mar. 31, June 30, Sept. 30, and Dec. 31 thereafter
Economic life of asset	5 years
Interest rate charged by the lessor	8%

Required:

Prepare appropriate entries for Natick Industries from the beginning of the lease through March 31, 2021. Appropriate adjusting entries are made quarterly.

E 15–20
Variable lease payments
● LO15–2, LO15–6

On January 1, 2021, Wetick Optometrists leased diagnostic equipment from Southern Corp., which had purchased the equipment at a cost of $1,437,237. The lease agreement specifies six annual payments of $300,000 beginning January 1, 2021, the beginning of the lease, and at each December 31 thereafter through 2025. The six-year lease term ending December 31, 2026 (a year after the final payment), is equal to the estimated useful life of the equipment. The contract specifies that lease payments for each year will increase on the basis of the increase in the Consumer Price Index for the year just ended. Thus, the first payment will be $300,000, and the second and subsequent payments might be different. The CPI at the beginning of the lease is 120. Southern routinely acquires diagnostic equipment for lease to other firms. The interest rate in these financing arrangements is 10%.

Required:

1. Prepare the appropriate journal entries for Wetick and Southern to record the lease at its beginning.
2. Assuming the CPI is 124 at that time, prepare the appropriate journal entries related to the lease for Wetick at December 31, 2021.

E 15–21
Lessee; variable lease payments
● LO15–2, LO15–6

On January 1, 2021, QuickStream Communications leased telephone equipment from Digium, Inc. Digium's cash selling price for the equipment is $1,306,578. The lease agreement specifies six annual payments of $300,000 beginning December 31, 2021, and at each December 31 thereafter through 2026. The six-year lease is equal to the estimated useful life of the equipment. The contract specifies that lease payments for each year will increase by the higher of (a) the increase in the Consumer Price Index for the preceding year or (b) 3%. The CPI at the beginning of the lease is 120. Digium routinely leases equipment to other firms. The interest rate in these lease arrangements is 10%.

Required:

Prepare the appropriate journal entries for QuickStream to record the lease at its beginning date of January 1, 2021.

E 15–22
Lessee; variable lease payments
● LO15–2, LO15–6

On January 1, 2021, Taco King leased retail space from Fogelman Properties. The 10-year finance lease requires quarterly variable lease payments equal to 3% of Taco King's sales revenue, with a quarterly sales minimum of $400,000. Payments at the beginning of each quarter are based on previous quarter sales. During the previous 5-year period, Taco King has generated quarterly sales of over $650,000. Fogelman's interest rate, known by Taco King, was 4%.

Required:

1. Prepare the journal entries for Taco King at the beginning of the lease at January 1, 2021.
2. Prepare the journal entries for Taco King at April 1, 2021 First quarter sales were $660,000. Amortization is recorded quarterly.

E 15–23
Lessee; renewal options
● LO15–2, LO15–6

On January 1, 2021, Rick's Pawn Shop leased a truck from Corey Motors for a six-year period with an option to extend the lease for three years. Rick's had no significant economic incentive as of the beginning of the lease to exercise the 3-year extension option. Annual lease payments are $10,000 due on December 31 of each year, calculated by the lessor using a 5% interest rate. The agreement is considered an operating lease.

Required:

1. Prepare Rick's journal entry to record for the right-of-use asset and lease liability at January 1, 2021.
2. Prepare the journal entries to record interest and amortization at December 31, 2021.

E 15–24
Calculation of annual lease payments; residual value
● LO15–2, LO15–6

Each of the four independent situations below describes a finance lease in which annual lease payments are payable at the beginning of each year. The lessee is aware of the lessor's implicit rate of return.

	Situation			
	1	2	3	4
Lease term (years)	4	7	5	8
Lessor's rate of return	10%	11%	9%	12%
Fair value of lease asset	$50,000	$350,000	$75,000	$465,000
Lessor's cost of lease asset	$50,000	$350,000	$45,000	$465,000
Residual value:				
Estimated fair value	0	$ 50,000	$ 7,000	$ 45,000
Guaranteed fair value	0	0	$ 7,000	$ 50,000

Required:

For each situation, determine:

a. The amount of the annual lease payments as calculated by the lessor.
b. The amount the lessee would record as a right-of-use asset and a lease liability.

E 15–25
Lessor; sales-type lease; residual value effect on financial statements
● LO15–2, LO15–6

At January 1, 2021, Café Med leased restaurant equipment from Crescent Corporation under a nine-year lease agreement. The lease agreement specifies annual payments of $25,000 beginning January 1, 2021, the beginning of the lease, and at each December 31 thereafter through 2028. The equipment was acquired recently by Crescent at a cost of $180,000 (its fair value) and was expected to have a useful life of 12 years with no salvage value at the end of its life. (Because the lease term is only 9 years, the asset does have an expected residual value at the end of the lease term of $50,995.) Both (a) the present value of the lease payments and (b) the present value of the residual value (i.e., the residual asset) are included in the lease receivable because the two amounts combine to allow the lessor to recover its net investment. Crescent seeks a 10% return on its lease investments. By this arrangement, the lease is deemed to be a finance lease to the lessee.

Required:

1. What will be the effect of the lease on Crescent's earnings for the first year including (ignore taxes)?
2. What will be the balances in the balance sheet accounts related to the lease at the end of the first year for Crescent (ignore taxes)?

E 15–26
Lease concepts; finance/sales-type leases; guaranteed and unguaranteed residual value
● LO15–2, LO15–6

Each of the four independent situations below describes a sales-type lease in which annual lease payments of $100,000 are payable at the beginning of each year. Each is a finance lease for the lessee. Determine the following amounts at the beginning of the lease:

A. The lessor's
   1. Lease payments
   2. Gross investment in the lease
   3. Net investment in the lease

B. The lessee's
   4. Lease payments
   5. Right-of-use asset
   6. Lease liability

	Situation			
	1	2	3	4
Lease term (years)	7	7	8	8
Lessor's and lessee's interest rate	9%	11%	10%	12%
Residual value:				
Estimated fair value	0	$50,000	$8,000	$50,000
Guaranteed by lessee	0	0	$8,000	$60,000

E 15–27
Lessee; lessee guaranteed residual value
● LO15–2, LO15–6

On January 1, 2021, Maywood Hydraulics leased drilling equipment from Aqua Leasing for a four-year period ending December 31, 2024, at which time possession of the leased asset will revert back to Aqua. The equipment cost Aqua $412,184 and has an expected economic life of five years. Aqua expects the residual value at December 31, 2024, to be $50,000. Negotiations led to Maywood guaranteeing a $70,000 residual value.

Equal payments under the lease are $100,000 and are due on December 31 of each year with the first payment being made on December 31, 2021. Maywood is aware that Aqua used a 5% interest rate when calculating lease payments.

Required:

1. Prepare the appropriate entry for Maywood on January 1, 2021, to record the lease.
2. Prepare all appropriate entries for Maywood on December 31, 2021, related to the lease.

E 15–28
Calculation of annual lease payments; purchase option
● LO15–2, LO15–6

For each of the three independent situations below determine the amount of the annual lease payments. Each describes a finance lease in which annual lease payments are payable at the beginning of each year. Each lease agreement contains an option that permits the lessee to acquire the leased asset at an option price that is sufficiently lower than the expected fair value that the exercise of the option appears reasonably certain.

	Situation		
	1	2	3
Lease term (years)	5	5	4
Lessor's rate of return	12%	11%	9%
Fair value of leased asset	$60,000	$420,000	$185,000
Lessor's cost of leased asset	$50,000	$420,000	$145,000
Purchase option:			
Exercise price	$ 10,000	$ 50,000	$ 22,000
Exercisable at end of year:	5	5	3
Reasonably certain?	yes	no	yes

E 15–29
Finance lease; purchase options; lessee
● LO15–2, LO15–6

Federated Fabrications leased a tooling machine on January 1, 2021, for a three-year period ending December 31, 2023. The lease agreement specified annual payments of $36,000 beginning with the first payment at the beginning of the lease, and each December 31 through 2022. The company had the option to purchase the machine on December 30, 2023, for $45,000 when its fair value was expected to be $60,000, a sufficient difference that exercise seems reasonably certain. The machine's estimated useful life was six years with no salvage value. Federated was aware that the lessor's implicit rate of return was 12%.

Required:
1. Calculate the amount Federated should record as a right-of-use asset and lease liability for this finance lease.
2. Prepare an amortization schedule that describes the pattern of interest expense for Federated over the lease term.
3. Prepare the appropriate entries for Federated from the beginning of the lease through the end of the lease term.

E 15–30
Purchase option; lessor; sales-type lease; no selling profit
● LO15–2, LO15–6

Universal Leasing leases electronic equipment to a variety of businesses. The company's primary service is providing alternate financing by acquiring equipment and leasing it to customers under long-term sales-type leases. Universal earns interest under these arrangements at a 10% annual rate.

The company leased an electronic typesetting machine it purchased for $30,900 to a local publisher, Desktop Inc., on December 31, 2020. The lease contract specified annual payments of $8,000 beginning January 1, 2021, the beginning of the lease, and each December 31 through 2022 (three-year lease term). The publisher had the option to purchase the machine on December 30, 2023, the end of the lease term, for $12,000 when it was expected to have a residual value of $16,000, a sufficient difference that exercise seems reasonably certain.

Required:
1. Show how Universal calculated the $8,000 annual lease payments for this sales-type lease.
2. Prepare an amortization schedule that describes the pattern of interest revenue for Universal Leasing over the lease term.
3. Prepare the appropriate entries for Universal Leasing from the beginning of the lease through the end of the lease term.

E 15–31
Is it a lease?
● LO15–7

Warren Marina owns a large marina that contains numerous boat slips of various sizes. Warren contracts with boat owners to provide slips to house the customers' boats. Lucky Fisherman Fleet contracted with Warren to provide space for four of its fishing boats. The contract specifies that Lucky Fisherman's boats will be kept in identified slips in the marina. However, Warren has the right to shift the boats to other slips within its marina at its discretion, subject to the requirement to provide 45-foot slips per boat for a three-year period. Warren frequently rearranges its customers' boats to meet the needs of new contracts. Costs of reallocating space is low relative to the benefits of being able to accommodate more customers and their specific requests. Warren paid $16,000 on March 1, 2021, for the first year's accommodations. The market rate of interest is 5%.

Required:
Prepare the appropriate entry(s) for Lucky Fisherman Fleet at March 1, the commencement of the agreement.

E 15–32
Is it a lease?
● LO15–7

(This exercise is a variation of E 15–31.)

Warren Marina owns a large marina that contains numerous boat slips of various sizes. Warren contracts with boat owners to provide slips to house the customers' boats. Lucky Fisherman Fleet contracted with Warren to provide space for four of its fishing boats. The contract specifies that Lucky Fisherman's boats will be kept in identified slips in the marina. The agreement includes the requirement that Warren provide 45-foot slips per boat for a three-year period, space that Lucky Fisherman can modify with fenders, docklines, and equipment needed to conduct its fishing business. Warren cannot switch locations of the boats or modify the slips without Lucky Fisherman's consent. Warren paid $16,000 on March 1, 2021, for the first year's accommodations. The market rate of interest is 5%.

Required:
Prepare the appropriate entry(s) for Lucky Fisherman Fleet at March 1, the commencement of the agreement.

E 15–33
Nonlease payments; lessor and lessee
● LO15–2, LO15–7

On January 1, 2021, NRC Credit Corporation leased equipment to Brand Services under a finance/sales-type lease designed to earn NRC a 12% rate of return for providing long-term financing. The lease agreement specified the following:

a. Ten annual payments of $55,000 beginning January 1, 2021, the beginning of the lease and each December 31 thereafter through 2029.
b. The estimated useful life of the leased equipment is 10 years with no residual value. Its cost to NRC was $316,412.
c. The lease qualifies as a finance lease/sales-type lease.

d. A 10-year service agreement with Quality Maintenance Company was negotiated to provide maintenance of the equipment as required. Payments of $5,000 per year are specified, beginning January 1, 2021. NRC was to pay this cost as incurred, but lease payments reflect this expenditure.

e. A partial amortization schedule, appropriate for both the lessee and lessor, follows:

	Payments	Effective Interest	Decrease in Balance	Outstanding Balance
		(12% × Outstanding balance)		
				316,412
1/1/2021	50,000		50,000	266,412
12/31/2021	50,000	0.12 (266,412) = 31,969	18,031	248,381
12/31/2022	50,000	0.12 (248,381) = 29,806	20,194	228,187

Required:
Prepare the appropriate entries for both the lessee and lessor related to the lease on the following dates:
1. January 1, 2021
2. December 31, 2021

E 15–34
Lessor's initial direct costs; sales-type lease
● LO15–2, LO15–7

Terms of a lease agreement and related facts were as follows:
a. Incremental costs of commissions for brokering the lease and consummating the completed lease transaction incurred by the lessor were $4,242.
b. The retail cash selling price of the leased asset was $500,000. Its useful life was three years with no residual value.
c. The lease term was three years and the lessor paid $500,000 to acquire the asset.
d. Annual lease payments at the beginning of each year were $184,330.
e. Lessor's implicit rate when calculating annual rental payments was 11%.

Required:
1. Prepare the appropriate entries for the lessor to record the lease and related payments at its beginning, January 1, 2021.
2. Calculate the effective rate of interest revenue after adjusting the net investment by initial direct costs.
3. Record any entry(s) necessary at December 31, 2021, the fiscal year-end.

E 15–35
Lessor's initial direct costs; sales-type lease
● LO15–3, LO15–7

The lease agreement and related facts indicate the following:
a. Leased equipment had a retail cash selling price of $300,000. Its useful life was five years with no residual value.
b. The lease term was five years and the lessor paid $265,000 to acquire the equipment (thus, selling profit).
c. Lessor's implicit rate when calculating annual lease payments was 8%.
d. Annual lease payments beginning January 1, 2021, the beginning of the lease, were $69,571.
e. Incremental costs of commissions for brokering the lease and consummating the completed lease transaction incurred by the lessor were $7,500.

Required:
Prepare the appropriate entries for the lessor to record:
1. The lease and the initial payment at its commencement.
2. Any entry(s) necessary at December 31, 2021, the fiscal year-end.

E 15–36
Lessor's initial direct costs; operating lease
● LO15–4, LO15–7

The following relate to an operating lease agreement:
a. The lease term is 3 years, beginning January 1, 2021.
b. The leased asset cost the lessor $800,000 and had a useful life of eight years with no residual value. The lessor uses straight-line depreciation for its depreciable assets.
c. Annual lease payments at the beginning of each year were $137,000.
d. Incremental costs of negotiating and consummating the completed lease transaction incurred by the lessor were $2,400.

Required:
Prepare the appropriate entries for the lessor from the beginning of the lease through the end of the lease term.

E 15–37
Concepts; terminology
● LO15–2 through LO15–8

Listed below are several terms and phrases associated with leases. Pair each item from List A with the item from List B (by letter) that is most appropriately associated with it.

	List A	List B
_____	1. Effective rate times balance	a. PV of purchase option exercise price
_____	2. Revenue recognition issues	b. Lessor's net investment
_____	3. Lease payments plus residual value	c. Lessor's gross investment
_____	4. Periodic lease payments plus excess lessee-guaranteed residual value	d. Operating lease
		e. Depreciable assets
_____	5. PV of lease payments plus PV of residual value	f. Component of lease payments
_____	6. Initial direct costs	g. Nonlease payments
_____	7. Rent revenue	h. Amortization longer than lease term
_____	8. Purchase option	i. Disclosure only
_____	9. Leasehold improvements	j. Interest expense
_____	10. Cash expected to satisfy residual value guarantee	k. Control passed to lessee
_____	11. Payments expensed by the lessee	l. Lessee's lease payments
_____	12. Deducted in lessor's computation of lease payments	m. Might shorten lease term
		n. Sales-type lease selling expense
_____	13. Title transfers to lessee	
_____	14. Contingent rentals	

E 15–38
FASB codification research; reassessment of lease terms
● LO15–8

CODE

The lease term is the noncancelable period for which a lessee has the right to use an underlying asset, modified by any renewal or termination options that are "reasonably certain" to be exercised, or not exercised. Options whose exercise is under the control of the lessor are automatically included. The *FASB Accounting Standards Codification*® represents the single source of authoritative U.S. generally accepted accounting principles.

Required:

1. Access the FASB's Codification Research System at the FASB website (**www.fasb.org**). Determine the specific eight-digit Codification citation (XXX-XX-XX-X) that describes the guidelines for determining when the lessee should reassess the term of the lease.
2. List the disclosure requirements.

E 15–39
FASB codification research
● LO15–8

CODE

Access the FASB's Codification Research System® at the FASB website (**www.fasb.org**). Determine the specific seven- or eight-digit Codification citation (XXX-XX-XX-X) for accounting for each of the following items:

1. Definition of initial direct costs.
2. When a modification to a contract is reported as a separate contract (that is, separate from the original contract).
3. The disclosures required in the notes to the financial statements for a lessor.
4. The classification criteria for when a lessee classifies a lease as a finance lease and a lessor classifies a lease as a sales-type lease.

E 15–40
Sale-leaseback
● Appendix 15

To raise operating funds, Signal Aviation sold an airplane on January 1, 2021, to a finance company for $770,000. Signal immediately leased the plane back for a 13-year period, at which time ownership of the airplane will transfer to Signal. The airplane has a fair value of $800,000. Its cost and its book value were $600,000. Its useful life is estimated to be 15 years. The lease requires Signal to make payments of $102,771 to the finance company each January 1. Signal depreciates assets on a straight-line basis. The lease has an implicit rate of 11%.

Required:

Prepare the appropriate entries for Signal on

1. January 1, 2021, to record the transaction
2. December 31, 2021, to record necessary adjustments

E 15–41
Sale-leaseback; operating lease
● Appendix 15

To raise operating funds, National Distribution Center sold its office building to an insurance company on January 1, 2021, for $800,000 and immediately leased the building back. The operating lease is for the final 12 years of the building's estimated 20-year remaining useful life. The building has a fair value of $800,000 and a book value of $650,000 (its original cost was $1 million). The rental payments of $100,000 are payable to the insurance company each December 31. The lease has an implicit rate of 9%.

Required:

Prepare the appropriate entries for National Distribution Center on

1. January 1, 2021, to record the sale-leaseback
2. December 31, 2021, to record necessary adjustments

Problems

connect

P 15–1
Integrating problem; bonds; note; lease
● LO15–2

You are the new controller for Moonlight Bay Resorts. The company CFO has asked you to determine the company's interest expense for the year ended December 31, 2021. Your accounting group provided you the following information on the company's debt:

1. On July 1, 2021, Moonlight Bay issued bonds with a face amount of $2,000,000. The bonds mature in 20 years and interest of 9% is payable semiannually on June 30 and December 31. The bonds were issued at a price to yield investors 10%. Moonlight Bay records interest at the effective rate.
2. At December 31, 2020, Moonlight Bay had a 10% installment note payable to Third Mercantile Bank with a balance of $500,000. The annual payment is $60,000, payable each June 30.
3. On January 1, 2021, Moonlight Bay leased a building under a finance lease calling for four annual lease payments of $40,000 beginning January 1, 2021. Moonlight Bay's incremental borrowing rate on the date of the lease was 11% and the lessor's implicit rate, which was known by Moonlight Bay, was 10%.

Required:
Calculate interest expense for the year ended December 31, 2021.

P 15–2
Finance lease
● LO15–2

At the beginning of 2021, VHF Industries acquired a machine with a fair value of $6,074,700 by signing a four-year lease. The lease is payable in four annual payments of $2 million at the end of each year.

Required:
1. What is the effective rate of interest implicit in the agreement?
2. Prepare the lessee's journal entry at the beginning of the lease.
3. Prepare the journal entry to record the first lease payment at December 31, 2021.
4. Prepare the journal entry to record the second lease payment at December 31, 2022.
5. Suppose the fair value of the machine and the lessor's implicit rate were unknown at the time of the lease, but that the lessee's incremental borrowing rate of interest for notes of similar risk was 11%. Prepare the lessee's entry at the beginning of the lease.

(Note: You may wish to compare your solution to P 15–2 with that of P 14–12, which deals with a parallel situation in which the machine was acquired with an installment note.)

P 15–3
Lease amortization schedule
● LO15–2

On January 1, 2021, Majestic Mantles leased a lathe from Equipment Leasing under a finance lease. Lease payments are made annually. Title does not transfer to the lessee and there is no purchase option or guarantee of a residual value by Majestic. Portions of the Equipment Leasing's lease amortization schedule appear below:

Jan. 1	Payments	Effective Interest	Decrease in Balance	Outstanding Balance
				187,298
2021	20,000		20,000	167,298
2022	20,000	16,730	3,270	164,028
2023	20,000	16,403	3,597	160,431
2024	20,000	16,043	3,957	156,474
2025	20,000	15,647	4,353	152,122
2026	20,000	15,212	4,788	147,334
2027	20,000	14,733	5,267	142,067
—	—	—	—	—
—	—	—	—	—
—	—	—	—	—
2038	20,000	4,974	15,026	34,711
2039	20,000	3,471	16,529	18,182
2040	20,000	1,818	18,182	0

Required:
1. What is Majestic's lease liability at the beginning of the lease (after the first payment)?
2. What amount would Majestic record as a right-of-use asset?
3. What is the lease term in years?
4. What is the effective annual interest rate?
5. What is the total amount of lease payments?
6. What is the total effective interest expense recorded over the term of the lease?

P 15–4
Finance/sales-type lease; lessee and lessor
● LO15–1, LO15–2, LO15–3

Rand Medical manufactures lithotripters. Lithotripsy uses shock waves instead of surgery to eliminate kidney stones. Physicians' Leasing purchased a lithotripter from Rand for $2,000,000 and leased it to Mid-South Urologists Group, Inc., on January 1, 2021.

Lease Description:	
Quarterly lease payments	$130,516—beginning of each period
Lease term	5 years (20 quarters)
No residual value; no purchase option	
Economic life of lithotripter	5 years
Implicit interest rate and lessee's incremental borrowing rate	12%
Fair value of asset	$2,000,000

1. How should this lease be classified by Mid-South Urologists Group and by Physicians' Leasing?
2. Prepare appropriate entries for both Mid-South Urologists Group and Physicians' Leasing from the beginning of the lease through the second rental payment on April 1, 2021. Adjusting entries are recorded at the end of each fiscal year (December 31).
3. Assume Mid-South Urologists Group leased the lithotripter directly from the manufacturer, Rand Medical, which produced the machine at a cost of $1.7 million. Prepare appropriate entries for Rand Medical from the beginning of the lease through the second lease payment on April 1, 2021.

P 15–5
Lessee; operating lease; advance payment; leasehold improvement
● LO15–4

On January 1, 2021, Winn Heat Transfer leased office space under a three-year operating lease agreement. The arrangement specified three annual lease payments of $80,000 each, beginning December 31, 2021, and at each December 31 through 2023. The lessor, HVAC Leasing calculates lease payments based on an annual interest rate of 5%. Winn also paid a $100,000 advance payment at the beginning of the lease in addition to the first $80,000 lease payment. With permission of the owner, Winn made structural modifications to the building before occupying the space at a cost of $180,000. The useful life of the building and the structural modifications were estimated to be 30 years with no residual value.

Required:
Prepare the appropriate entries for Winn Heat Transfer from the beginning of the lease through the end of 2023. Winn's fiscal year is the calendar year.

P 15–6
Operating lease; scheduled payment increases
● LO15–4

On January 1, 2021, Sweetwater Furniture Company leased office space under a 21-year operating lease agreement. The contract calls for annual lease payments on December 31 of each year. The payments are $10,000 the first year and increase by $500 per year. Benefits expected from using the office space are expected to remain constant over the lease term.

Required:
Record Sweetwater's lease payment at December 31, 2025 (the fifth lease payment), and December 31, 2038 (the fifteenth lease payment).

P 15–7
Lease amortization schedule
● LO15–2, LO15–6

On January 1, 2021, National Insulation Corporation (NIC) leased equipment from United Leasing under a finance lease. Lease payments are made annually. Title does not transfer to the lessee and there is no purchase option or guarantee of a residual value by NIC. Portions of the United Leasing's lease amortization schedule appear below:

Jan. 1	Payments	Effective Interest	Decrease in Balance	Outstanding Balance
2021				192,501
2022	20,000		20,000	172,501
2023	20,000	17,250	2,750	169,751
2024	20,000	16,975	3,025	166,726
2025	20,000	16,673	3,327	163,399
2026	20,000	16,340	3,660	159,739
2027	20,000	15,974	4,026	155,713
—	—	—	—	—
—	—	—	—	—
—	—	—	—	—
2038	20,000	7,364	12,636	61,006
2039	20,000	6,101	13,899	47,107
2040	20,000	4,711	15,289	31,818
2041	35,000	3,182	31,818	0

Required:
1. What is the lease term in years?
2. What is the asset's residual value expected at the end of the lease term?

3. What is the effective annual interest rate?
4. What is the total amount of lease payments for United?
5. What is the total amount of lease payments for NIC?
6. What is United's net investment at the beginning of the lease (after the first payment)?
7. What is United's total effective interest revenue recorded over the term of the lease?
8. What amount would NIC record as a right-of-use asset at the beginning of the lease?

P 15–8
Reassessment of lease term
● LO15–2, LO15–4, LO15–6

On January 1, 2021, Rick's Pawn Shop leased a truck from Corey Motors for a six-year period with an option to extend the lease for three years. Rick's had no significant economic incentive as of the beginning of the lease to exercise the 3-year extension option. Annual lease payments are $10,000 due on December 31 of each year, calculated by the lessor using a 5% discount rate. Assume that at the beginning of the third year, January 1, 2023, Rick's had made significant improvements to the truck whose cost could be recovered only if it exercises the extension option, creating an expectation that extension of the lease was "reasonably certain." The relevant interest rate at that time was 6%.

Required:

1. Prepare the journal entry, if any, at the end of the second year for the lessee to account for the reassessment.
2. Prepare the journal entry, if any, at the end of the second year for the lessor to account for the reassessment.

P 15–9
Lease concepts; sales-type leases; guaranteed and unguaranteed residual value
● LO15–2, LO15–6

Each of the four independent situations below describes a sales-type lease in which annual lease payments of $10,000 are payable at the beginning of each year. Each is a finance lease for the lessee. Determine the following amounts at the beginning of the lease.

A. The lessor's
   1. Lease payments
   2. Gross investment in the lease
   3. Net investment in the lease
B. The lessee's
   4. Lease payments
   5. Right-of-use asset
   6. Lease liability

	Situation			
	1	2	3	4
Lease term (years)	4	4	4	4
Asset's useful life (years)	4	5	5	7
Lessor's implicit rate (known by lessee)	11%	11%	11%	11%
Residual value:				
Guaranteed by lessee	0	$4,000	$2,000	0
Unguaranteed	0	0	$2,000	$4,000
Purchase option:				
After (years)	none	3	4	3
Exercise price	n/a	$7,000	$1,000	$3,000
Reasonably certain?	n/a	no	no	yes

P 15–10
Sales-type lease; purchase option reasonably certain to be exercised before lease term ends; lessor and lessee
● LO15–3, LO15–6

Mid-South Auto Leasing leases vehicles to consumers. The attraction to customers is that the company can offer competitive prices due to volume buying and requires an interest rate implicit in the lease that is one percent below alternate methods of financing. On September 30, 2021, the company leased a delivery truck to a local florist, Anything Grows. The fiscal year for both companies ends December 31.

The lease agreement specified quarterly payments of $3,000 beginning September 30, 2021, the beginning of the lease, and each quarter (December 31, March 31, and June 30) through June 30, 2024 (three-year lease term). The florist had the option to purchase the truck on September 29, 2023, for $6,000 when it was expected to have a residual value of $10,000. The estimated useful life of the truck is four years. Mid-South Auto Leasing's quarterly interest rate for determining payments was 3% (approximately 12% annually). Mid-South paid $25,000 for the truck. Both companies use straight-line depreciation or amortization. Anything Grows' incremental interest rate is 12%.

Hint: A lease term ends for accounting purposes when an option becomes exercisable if it's expected to be exercised (i.e., a BPO).

Required:

1. Calculate the amount of selling profit that Mid-South would recognize in this sales-type lease. (Be careful to note that, although payments occur on the last calendar day of each quarter, since the first payment was at the beginning of the lease, payments represent an annuity due.)

2. Prepare the appropriate entries for Anything Grows and Mid-South on September 30, 2021.
3. Prepare an amortization schedule(s) describing the pattern of interest expense for Anything Grows and interest revenue for Mid-South Auto Leasing over the lease term
4. Prepare the appropriate entries for Anything Grows and Mid-South Auto Leasing on December 31, 2021.
5. Prepare the appropriate entries for Anything Grows and Mid-South on September 29, 2023, assuming the purchase option was exercised on that date.

P 15–11
Change in lease term; operating lease; lessor
● LO15–4, LO15–6

Universal Leasing leases electronic equipment to a variety of businesses. The company's primary service is providing alternate financing by acquiring equipment and leasing it to customers under long-term leases. Universal earns interest under these arrangements at a 10% annual rate.

Universal purchased an electronic typesetting machine on December 31, 2020, for $90,000 and then leased it to Desktop, Inc., a local publisher. The six-year operating lease term commenced January 1, 2021, and the lease contract specified annual payments of $8,000 beginning December 31, 2021, and on each December 31 through 2026. The machine's estimated useful life is 15 years with no estimated residual value.

The publisher had the option to terminate the lease after four years. At the beginning of the lease, there was no reason to believe the lease would be terminated.

Required:
1. Prepare the appropriate entries for Universal Leasing from the beginning of the lease through the end of 2021.
2. At the beginning of 2022, there was a significant indication that Desktop's economic incentive to terminate the lease had changed causing both companies to believe termination of the lease at the end of four years (three years remaining) is "reasonably certain." Prepare any appropriate entries for Universal Leasing at January 1, 2022, to reflect the change in the lease term.
3. Prepare the appropriate entries pertaining to the lease for Universal Leasing at December 31, 2022.

P 15–12
Lessee; renewal option
● LO15–2, LO15–6

High Time Tours leased rock-climbing equipment from Adventures Leasing on January 1, 2021. High Time has the option to renew the lease at the end of two years for an additional three years for $8,000 per quarter. Adventures purchased the equipment at a cost of $198,375.

Related Information:	
Lease term	2 years (8 quarterly periods)
Lease renewal option for an additional	3 years at $8,000 per quarter
Quarterly lease payments	$15,000 at Jan. 1, 2021, and at Mar. 31, June 30, Sept. 30, and Dec. 31 thereafter
Economic life of asset	5 years
Interest rate charged by the lessor	8%

Required:
1. Prepare appropriate entries for High Time Tours from the beginning of the lease through March 31, 2021. Appropriate adjusting entries are made quarterly.
2. Prepare an amortization schedule for the term of the lease.

P 15–13
Lessee and lessor; lessee guaranteed residual value
● LO15–2, LO15–6

On January 1, 2021, Allied Industries leased a high-performance conveyer to Karrier Company for a four-year period ending December 31, 2024, at which time possession of the leased asset will revert back to Allied. The equipment cost Allied $956,000 and has an expected useful life of five years. Allied expects the residual value at December 31, 2025, will be $300,000. Negotiations led to the lessee guaranteeing a $340,000 residual value.

Equal payments under the finance/sales-type lease are $200,000 and are due on December 31 of each year with the first payment being made on December 31, 2021. Karrier is aware that Allied used a 5% interest rate when calculating lease payments.

Required:
1. Prepare the appropriate entries for both Karrier and Allied on January 1, 2021, to record the lease.
2. Prepare all appropriate entries for both Karrier and Allied on December 31, 2021, related to the lease.

P 15–14
Lessee and lessor; lessor; sales-type lease with selling profit; residual value
● LO15–2, LO15–3

Newton Labs leased chronometers from Brookline Instruments on January 1, 2021. Brookline Instruments manufactured the chronometers at a cost of $200,000. The chronometers have a fair value of $260,000. Appropriate adjusting entries are made quarterly.

Related Information:

Lease term	5 years (20 quarterly periods)
Quarterly lease payments	\$14,547 at Jan. 1, 2021, and at Mar. 31, June 30, Sept. 30, and Dec. 31 thereafter
Economic life of asset	6 years
Estimated residual value of chronometers at end of lease term	\$25,823
Interest rate charged by the lessor	8%

Required:

1. Prepare appropriate entries for Newton Labs to record the arrangement at its commencement, January 1, 2021, and on March 31, 2021.
2. Prepare appropriate entries for Brookline Instruments to record the arrangement at its commencement, January 1, 2021, and on March 31, 2021.

P 15–15
Nonlease payments; lessor and lessee
● LO15–2, LO15–7

On January 1, 2021, Lesco Leasing leased equipment to Quality Services under a finance/sales-type lease designed to earn NRC a 12% rate of return for providing long-term financing. The lease agreement specified

a. Ten annual payments of \$56,000 beginning January 1, 2021, the beginning of the lease and each December 31 thereafter through 2029.
b. The estimated useful life of the leased equipment is 10 years with no residual value. Its cost to Lesco was \$322,741.
c. The lease qualifies as a finance lease/sales-type lease.
d. A 10-year service agreement with Quality Maintenance Company was negotiated to provide maintenance of the equipment as required. Payments of \$5,000 per year are specified, beginning January 1, 2021. Lesco was to pay this cost as incurred, but lease payments reflect this expenditure. Also included in the \$56,000 payments is an insurance premium of \$4,000 providing coverage for the equipment.

Required:

1. Prepare the appropriate entries for both the lessee and lessor related to the lease on January 1, 2021.
2. Prepare the appropriate entries for both the lessee and lessor related to the lease on December 31, 2021.

P 15–16
Lessor's initial direct costs; operating and sales-type leases
● LO15–2, LO15–3, LO15–4, LO15–7

Terms of a lease agreement and related facts were as follows:

a. The lease asset had a retail cash selling price of \$100,000. Its useful life was six years with no residual value (straight-line depreciation).
b. Annual lease payments at the beginning of each year were \$20,873, beginning January 1.
c. Lessor's implicit rate when calculating annual rental payments was 10%.
d. Costs of \$2,062 for legal fees for the lease execution were the responsibility of the lessor.

Required:

Prepare the appropriate entries for the lessor to record the lease, the initial payment at its beginning, and at the December 31 fiscal year-end under each of the following three independent assumptions:

1. The lease term is three years and the lessor paid \$100,000 to acquire the asset (operating lease).
2. The lease term is six years and the lessor paid \$100,000 to acquire the asset. Also assume that adjusting the lease receivable (net investment) by initial direct costs reduces the effective rate of interest to 9%.
3. The lease term is six years and the lessor paid \$85,000 to acquire the asset.

P 15–17
Nonlease costs; lessor and lessee
● LO15–2, LO15–7

Branif Leasing leases mechanical equipment to industrial consumers under sales-type leases that earn Branif a 10% rate of return for providing long-term financing. A lease agreement with Branson Construction specified 20 annual payments beginning December 31, 2021, the beginning of the lease. The estimated useful life of the leased equipment is 20 years with no residual value. Its cost to Branif was \$936,492. The lease qualifies as a finance lease to Branson. Maintenance of the equipment was contracted for through a 20-year service agreement with Midway Service Company requiring 20 annual payments of \$3,000 beginning December 31, 2021. Progressive Insurance Company charges Branif \$3,000 annually for hazard insurance coverage on the equipment. Both companies use straight-line depreciation or amortization.

Required:

Prepare the appropriate entries for both the lessee and lessor to record the second lease payment and depreciation on December 31, 2022, under each of three independent assumptions:

1. The lessee pays maintenance costs as incurred. The lessor pays insurance premiums as incurred. The lease agreement requires annual payments of \$100,000.
2. The contract specifies that the lessor pays maintenance costs as incurred. The lessee's lease payments were increased to \$103,000 to include an amount sufficient to reimburse these costs.
3. The lessee's lease payments of \$103,000 included \$3,000 for hazard insurance on the equipment rather than maintenance.

P 15–18
Lessee-guaranteed residual value; unguaranteed residual value; nonlease costs; different interest rates for lessor and lessee
● LO15–3, LO15–6, LO15–7

On December 31, 2021, Yard Art Landscaping leased a delivery truck from Branch Motors. Branch paid $40,000 for the truck. Its retail value is $45,114.

The lease agreement specified annual payments of $11,000 beginning December 31, 2021, the beginning of the lease, and at each December 31 through 2024. Branch Motors' interest rate for determining payments was 10%. At the end of the four-year lease term (December 31, 2025) the truck was expected to be worth $15,000. The estimated useful life of the truck is five years with no salvage value. Both companies use straight-line amortization or depreciation.

Yard Art guaranteed a residual value of $6,000. Yard Art's incremental borrowing rate is 9% and is unaware of Branch's implicit rate.

A $1,000 per year maintenance agreement was arranged for the truck with an outside service firm. As an expedient, Branch Motors agreed to pay this fee. It is, however, reflected in the $11,000 lease payments.

Required:

1. How should this lease be classified by Yard Art Landscaping (the lessee)? Why?
2. Calculate the amount Yard Art Landscaping would record as a right-of-use asset and a lease liability.
3. How should this lease be classified by Branch Motors (the lessor)? Why?
4. Show how Branch Motors calculated the $11,000 annual lease payments.
5. Calculate the amount Branch Motors would record as sales revenue.
6. Prepare the appropriate entries for both Yard Art and Branch Motors on December 31, 2021.
7. Prepare an amortization schedule that describes the pattern of interest expense over the lease term for Yard Art.
8. Prepare an amortization schedule that describes the pattern of interest revenue over the lease term for Branch Motors.
9. Prepare the appropriate entries for both Yard Art and Branch Motors on December 31, 2022.
10. Prepare the appropriate entries for both Yard Art and Branch Motors on December 31, 2024 (the final lease payment).
11. Prepare the appropriate entries for both Yard Art and Branch Motors on December 31, 2025 (the end of the lease term), assuming the truck is returned to the lessor and the actual residual value of the truck was $4,000 on that date.

P 15–19
Initial direct costs; sales-type lease
● LO15–2, LO15–7

Bidwell Leasing purchased a single-engine plane for its fair value of $645,526 and leased it to Red Baron Flying Club on January 1, 2021.

Terms of the lease agreement and related facts were

a. Eight annual payments of $110,000 beginning January 1, 2021, the beginning of the lease, and at each December 31 through 2027. Red Baron knows that Bidwell Leasing's implicit interest rate was 10%. The estimated useful life of the plane is eight years. Payments were calculated as follows:

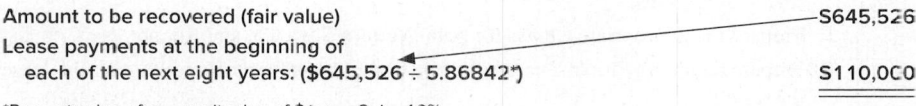

Amount to be recovered (fair value)	**$645,526**
Lease payments at the beginning of each of the next eight years: ($645,526 ÷ 5.86842*)	**$110,000**

*Present value of an annuity due of $1: $n = 8$, $i = 10\%$.

b. Red Baron's incremental borrowing rate is 11%.

c. Incremental costs of negotiating and consummating the completed lease transaction incurred by Bidwell Leasing were $18,099.

Required:

1. How should this lease be classified (a) by Bidwell Leasing (the lessor) and (b) by Red Baron (the lessee)?
2. Prepare the appropriate entries for both Red Baron Flying Club and Bidwell Leasing on January 1, 2021.
3. Prepare an amortization schedule that describes the pattern of interest expense over the lease term for Red Baron Flying Club.
4. Determine the effective rate of interest for Bidwell Leasing for the purpose of recognizing interest revenue over the lease term.
5. Prepare an amortization schedule that describes the pattern of interest revenue over the lease term for Bidwell Leasing.
6. Prepare the appropriate entries for both Red Baron and Bidwell Leasing on December 31, 2021 (the second lease payment). Both companies use straight-line depreciation or amortization.
7. Prepare the appropriate entries for both Red Baron and Bidwell Leasing on December 31, 2027 (the final lease payment).

P 15–20
Initial direct costs; sales-type lease with a selling profit
● LO15–3, LO15–7

(Note: This problem is a variation of P 15–19, modified to cause the lease to be a sales-type lease with a selling profit.)

Bidwell Leasing purchased a single-engine plane for $400,000 and leased it to Red Baron Flying Club for its fair value of $645,526 on January 1, 2021.

Terms of the lease agreement and related facts were

a. Eight annual payments of $110,000 beginning January 1, 2021, the beginning of the lease, and at each December 31 through 2027. Red Baron knows that Bidwell Leasing's implicit interest rate was 10%. The estimated useful life of the plane is eight years. Payments were calculated as follows:

Amount to be recovered (fair value)	$645,526
Lease payments at the beginning of each of the next eight years: ($645,526 ÷ 5.86842*)	$110,000

*Present value of an annuity due of $1: $n = 8$, $i = 10\%$.

b. Red Baron's incremental borrowing rate is 11%.

c. Incremental costs of consummating the completed lease transaction incurred by Bidwell Leasing were $18,099.

Required:

1. How should this lease be classified (a) by Bidwell Leasing (the lessor) and (b) by Red Baron (the lessee)?
2. Prepare the appropriate entries for both Red Baron Flying Club and Bidwell Leasing on January 1, 2021.
3. Prepare an amortization schedule that describes the pattern of interest expense over the lease term for Red Baron Flying Club.
4. Prepare the appropriate entries for both Red Baron and Bidwell Leasing on December 31, 2021 (the second lease payment). Both companies use straight-line depreciation or amortization.
5. Prepare the appropriate entries for both Red Baron and Bidwell Leasing on December 31, 2027 (the final lease payment).

P 15–21
Guaranteed residual value; sales-type lease
● LO15–2, LO15–5, LO15–6

(Note: P 15–21, 15–22, and 15–23 are three variations of the same basic situation.)

On December 31, 2021, Rhone-Metro Industries leased equipment to Western Soya Co. for a four-year period ending December 31, 2025, at which time possession of the leased asset will revert back to Rhone-Metro. The equipment cost Rhone-Metro $365,760 and has an expected useful life of six years. Its normal sales price is $365,760. The lessee-guaranteed residual value at December 31, 2025, is $25,000. Equal payments under the lease are $100,000 and are due on December 31 of each year. The first payment was made on December 31, 2021. Western Soya's incremental borrowing rate is 12%. Western Soya knows the interest rate implicit in the lease payments is 10%. Both companies use straight-line depreciation or amortization.

Required:

1. Show how Rhone-Metro calculated the $100,000 annual lease payments.
2. How should this lease be classified (a) by Western Soya Co. (the lessee) and (b) by Rhone-Metro Industries (the lessor)? Why?
3. Prepare the appropriate entries for both Western Soya Co. and Rhone-Metro on December 31, 2021.
4. Prepare an amortization schedule(s) describing the pattern of interest over the lease term for the lessee and the lessor.
5. Prepare all appropriate entries for both Western Soya and Rhone-Metro on December 31, 2022 (the second lease payment and amortization).
6. Prepare the appropriate entries for both Western Soya and Rhone-Metro on December 31, 2025, assuming the equipment is returned to Rhone-Metro and the actual residual value on that date is $1,500.

P 15–22
Unguaranteed residual value; nonlease payments; sales-type lease
● LO15–2, LO15–6, LO15–7

Rhone-Metro Industries manufactures equipment that is sold or leased. On December 31, 2021, Rhone-Metro leased equipment to Western Soya Co. for a four-year period ending December 31, 2025, at which time possession of the leased asset will revert back to Rhone-Metro. The equipment cost $300,000 to manufacture and has an expected useful life of six years. Its normal sales price is $365,760. The expected residual value of $25,000 at December 31, 2025, is not guaranteed. Equal payments under the lease are $104,000 (including $4,000 maintenance costs) and are due on December 31 of each year. The first payment was made on December 31, 2021. Western Soya's incremental borrowing rate is 12%. Western Soya knows the interest rate implicit in the lease payments is 10%. Both companies use straight-line depreciation or amortization.

Required:

1. Show how Rhone-Metro calculated the $104,000 annual lease payments.
2. How should this lease be classified (a) by Western Soya Co. (the lessee) and (b) by Rhone-Metro Industries (the lessor)? Why?
3. Prepare the appropriate entries for both Western Soya Co. and Rhone-Metro on December 31, 2021.

4. Prepare an amortization schedule(s) describing the pattern of interest over the lease term for the lessee and the lessor.
5. Prepare the appropriate entries for both Western Soya and Rhone-Metro on December 31, 2022 (the second lease payment and amortization).
6. Prepare the appropriate entries for both Western Soya and Rhone-Metro on December 31, 2025, assuming the equipment is returned to Rhone-Metro and the actual residual value on that date is $1,500.

P 15–23
Purchase option reasonably certain to be exercised before lease term ends; nonlease payments; sales-type lease
● LO15–3, LO15–6, LO15–7

Rhone-Metro Industries manufactures equipment that is sold or leased. On December 31, 2021, Rhone-Metro leased equipment to Western Soya Co. for a noncancelable stated lease term of four years ending December 31, 2025, at which time possession of the leased asset will revert back to Rhone-Metro. The equipment cost $300,000 to manufacture and has an expected useful life of six years. Its normal sales price is $365,760. The expected residual value of $25,000 at December 31, 2025, is not guaranteed. Western Soya Co. is reasonably certain to exercise a purchase option on December 30, 2024, at an option price of $10,000. Equal payments under the lease are $134,960 (including $4,000 annual maintenance costs) and are due on December 31 of each year. The first payment was made on December 31, 2021. Western Soya's incremental borrowing rate is 12%. Western Soya knows the interest rate implicit in the lease payments is 10%. Both companies use straight-line depreciation or amortization.

Hint: A lease term ends for accounting purposes when an option becomes exercisable if it's expected to be exercised (i.e., a BPO).

Required:

1. Show how Rhone-Metro calculated the $134,960 annual lease payments.
2. How should this lease be classified (a) by Western Soya Co. (the lessee) and (b) by Rhone-Metro Industries (the lessor)? Why?
3. Prepare the appropriate entries for both Western Soya Co. and Rhone-Metro on December 31, 2021.
4. Prepare an amortization schedule(s) describing the pattern of interest over the lease term for the lessee and the lessor.
5. Prepare the appropriate entries for both Western Soya and Rhone-Metro on December 31, 2022 (the second rent payment and amortization).
6. Prepare the appropriate entries for both Western Soya and Rhone-Metro on December 30, 2024, assuming the purchase option is exercised on that date.

P 15–24
Lessee and lessor; lessee guaranteed residual value
● LO15–2, LO15–7

On January 1, 2021, Nguyen Electronics leased equipment from Nevels Leasing for a four-year period ending December 31, 2024, at which time possession of the leased asset will revert back to Nevels. The equipment cost Nevels $824,368 and has an expected economic life of five years. Nevels expects the residual value at December 31, 2024, will be $100,000. Negotiations led to the lessee guaranteeing a $140,000 residual value.

Equal payments under the lease are $200,000 and are due on December 31 of each year with the first payment being made on December 31, 2021. Nguyen is aware that Nevels used a 5% interest rate when calculating lease payments.

Required:

1. Prepare the appropriate entries for both Nguyen and Nevels on January 1, 2021, to record the lease.
2. Prepare all appropriate entries for both Nguyen and Nevels on December 31, 2021, related to the lease.

P 15–25
Operating lease; uneven lease payments
● LO15–4, LO15–7

On January 1, 2021, Harlon Consulting entered into a three-year lease for new office space agreeing to lease payments of $5,000 in 2021, $6,000 in 2022, and $7,000 in 2023. Payments are due on December 31 of each year with the first payment being made on December 31, 2021. Harlon is aware that the lessor used a 5% interest rate when calculating lease payments.

Required:

1. Prepare the appropriate entries for Harlon Consulting on January 1, 2021, to record the lease.
2. Prepare all appropriate entries for Harlon Consulting on December 31, 2021, related to the lease.
3. Prepare all appropriate entries for Harlon Consulting on December 31, 2022, related to the lease.
4. Prepare all appropriate entries for Harlon Consulting on December 31, 2023, related to the lease.

P 15–26
Operating lease; uneven lease payments
● LO15–4, LO15–7

On January 1, 2021, Harlon Consulting entered into a three-year lease for new office space agreeing to lease payments of $7,000 in 2021, $6,000 in 2022, and $5,000 in 2023. Payments are due on December 31 of each year with the first payment being made on December 31, 2021. Harlon is aware that the lessor used a 5% interest rate when calculating lease payments.

Required:

1. Prepare the appropriate entries for Harlon Consulting on January 1, 2021, to record the lease.
2. Prepare all appropriate entries for Harlon Consulting on December 31, 2021, related to the lease.
3. Prepare all appropriate entries for Harlon Consulting on December 31, 2022, related to the lease.
4. Prepare all appropriate entries for Harlon Consulting on December 31, 2023, related to the lease.

P 15–27
Modification of a lease
● LO15–2, LO15–3, LO15–6

On January 1, 2021, Worcester Construction leased International Harvester equipment from Newton LeaseCorp. Newton LeaseCorp purchased the equipment from Wellesley Harvester at a cost of $958,158. Worcester's borrowing rate for similar transactions is 10%.

The lease agreement specified four annual payments of $200,000 beginning January 1, 2021, the beginning of the lease, and on each December 31 thereafter through 2023. The useful life of the equipment is estimated to be six years. The present value of those four payments at a discount rate of 10% is $697,370.

On January 1, 2023 (after two years and three payments), Worcester and Newton agreed to extend the lease term by two years. The market rate of interest at that time was 9%.

Required:

1. Prepare the appropriate entries for Worcester Construction on January 1, 2023, to adjust its lease liability for the lease modification.
2. Prepare all appropriate entries for Newton LeaseCorp on January 1, 2023, to record the lease modification.
3. Prepare all appropriate entries for Worcester Construction on December 31, 2023, related to the lease.
4. Prepare all appropriate entries for Newton LeaseCorp on December 31, 2023, related to the lease.

P 15–28
Finance lease; lessee; financial statement effects
● LO15–2, LO15–8

(Note: P 15–28, 15–29, and 15–30 are three variations of the same basic situation.) Werner Chemical, Inc., leased a protein analyzer on September 30, 2021. The five-year lease agreement calls for Werner to make quarterly lease payments of $391,548, payable each September 30, December 31, March 31, and June 30, with the first payment at September 30, 2021. Werner's incremental borrowing rate is 12%. Amortization is recorded on a straight-line basis at the end of each fiscal year. The useful life of the equipment is five years.

Required:

1. Determine the present value of the lease payments at September 30, 2021 (to the nearest $000).
2. What pretax amounts related to the lease would Werner report in its balance sheet at December 31, 2021?
3. What pretax amounts related to the lease would Werner report in its income statement for the year ended December 31, 2021?
4. What pretax amounts related to the lease would Werner report in its statement of cash flows for the year ended December 31, 2021?

P 15–29
Sales-type lease; lessor; financial statement effects
● LO15–2, LO15–8

Abbott Equipment leased a protein analyzer to Werner Chemical, Inc., on September 30, 2021. Abbott purchased the machine from NutraLabs, Inc., at a cost of $6 million. The five-year lease agreement calls for Werner to make quarterly lease payments of $391,548, payable each September 30, December 31, March 31, and June 30, with the first payment at September 30, 2021. Abbott's implicit interest rate is 12%. The useful life of the equipment is five years.

Required:

1. What pretax amounts related to the lease would Abbott report in its balance sheet at December 31, 2021?
2. What pretax amounts related to the lease would Abbott report in its income statement for the year ended December 31, 2021?
3. What pretax amounts related to the lease would Abbott report in its statement of cash flows for the year ended December 31, 2021?

P 15–30
Sales-type lease; lessor; financial statement effects
● LO15–3, LO15–8

NutraLabs, Inc., leased a protein analyzer to Werner Chemical, Inc., on September 30, 2021. NutraLabs manufactured the machine at a cost of $5 million. The five-year lease agreement calls for Werner to make quarterly lease payments of $391,548, payable each September 30, December 31, March 31, and June 30, with the first payment at September 30, 2021. NutraLabs' implicit interest rate is 12%. The useful life of the equipment is five years.

Required:

1. Determine the price at which NutraLabs is "selling" the equipment (present value of the lease payments) at September 30, 2021 (to the nearest $000).
2. What pretax amounts related to the lease would NutraLabs report in its balance sheet at December 31, 2021?
3. What pretax amounts related to the lease would NutraLabs report in its income statement for the year ended December 31, 2021?
4. What pretax amounts related to the lease would NutraLabs report in its statement of cash flows for the year ended December 31, 2021?

P 15–31
Sale-leaseback
● Appendix 15

To raise operating funds, North American Courier Corporation sold its building on January 1, 2021, to an insurance company for $500,000 and immediately leased the building back. The lease is for a 10-year period ending December 31, 2030, at which time ownership of the building will revert to North American Courier. The building has a carrying amount of $400,000 (original cost $1,000,000). The lease requires North American to make payments of $88,492 to the insurance company each December 31. The building had a total original useful life of 30 years with no residual value and is being depreciated on a straight-line basis. The lease has an implicit rate of 12%.

Required:

1. Prepare the appropriate entries for North American (a) on January 1, 2021, to record the transaction and (b) on December 31, 2021, to record necessary adjustments.
2. Show how North American's December 31, 2021, balance sheet and income statement would reflect the sale-leaseback.

Decision Makers' Perspective

©ImageGap/Getty Images

Apply your critical-thinking ability to the knowledge you've gained. These cases will provide you an opportunity to develop your research, analysis, judgment, and communication skills. You also will work with other students, integrate what you've learned, apply it in real world situations, and consider its global and ethical ramifications. This practice will broaden your knowledge and further develop your decision-making abilities.

Research Case 15–1
FASB codification; locate and extract relevant information and authoritative support for a financial reporting issue; finance lease; sublease of a leased asset
● LO15–2, LO15–8

CODE

"I don't see that in my intermediate accounting text I saved from college," you grumble to a colleague in the accounting division of Dowell Chemical Corporation. "This will take some research." Your comments pertain to the appropriate accounting treatment of a proposed sublease of warehouses Dowell has used for product storage.

Dowell leased the warehouses one year ago on December 31. The five-year lease agreement called for Dowell to make quarterly lease payments of $2,398,303, payable each December 31, March 31, June 30, and September 30, with the first payment at the lease's beginning. As a finance lease, Dowell had recorded the right-of-use asset and liability at $40 million, the present value of the lease payments at 8%. Dowell records amortization on a straight-line basis at the end of each fiscal year.

Today, Danielle Andries, Dowell's controller, explained a proposal to sublease the underused warehouses to American Tankers, Inc., for the remaining four years of the lease term. American Tankers would be substituted as lessee under the original lease agreement. As the new lessee, it would become the primary obligor under the agreement, and Dowell would not be secondarily liable for fulfilling the obligations under the lease agreement. "Check on how we would need to account for this and get back to me," she had said.

Required:

1. After the first full year under the warehouse lease, what is the balance in Dowell's lease liability? An amortization schedule will be helpful in determining this amount but is not required.
2. After the first full year under the warehouse lease, what is the carrying amount of Dowell's leased warehouses?
3. Obtain the relevant authoritative literature on accounting for derecognition of finance leases by lessees using the FASB's Codification Research System. You might gain access from the FASB website (**www.fasb.org**), from your school library, or some other source. To determine the appropriate accounting treatment for the proposed sublease, what is the specific seven-digit Codification citation (XXX-XX-XX) that Dowell would rely on to determine:
   a. if the proposal will qualify as a termination of a finance lease, and
   b. the appropriate accounting treatment for the sublease?
4. What, if any, journal entry would Dowell record in connection with the sublease?

Analysis Case 15–2
Classification issues; lessee accounting
● LO15–1, LO15–2, LO15–3, LO15–6, LO15–7

Interstate Automobiles Corporation leased 40 vans to VIP Transport under a four-year noncancelable lease on January 1, 2021. Information concerning the lease and the vans follows:

a. Equal annual lease payments of $300,000 are due on January 1, 2021, and thereafter on December 31 each year. The first payment was made January 1, 2021. Interstate's implicit interest rate is 10% and known by VIP.
b. VIP has the option to purchase all of the vans at the end of the lease for a total of $290,000. The vans' estimated residual value is $50,000 at the end of 7 years, the estimated life of each van.
c. VIP estimates the fair value of the vans to be $1,260,000. Interstate's cost was $1,050,000.
d. VIP's incremental borrowing rate is 9%.
e. VIP will pay the maintenance fees not included in the annual lease payments of $1,000 per year. The amortization method is straight-line.

Required:

1. If the vans' estimated residual value is $300,000 at the end of the lease term, how should the lease be classified by VIP? by Interstate?
2. If the vans' estimated residual value is $400,000 at the end of the lease term, how should the lease be classified by VIP? by Interstate?

3. Regardless of your response to previous requirements, suppose VIP recorded the lease on January 1, 2021, as a finance lease in the amount of $1,100,000 and that a bargain purchase option exists. What would be the appropriate journal entries related to the finance lease for the second lease payment on December 31, 2021?

Ethics Case 15–3
Leasehold improvements
● LO15–3

American Movieplex, a large movie theater chain, leases most of its theater facilities. In conjunction with recent operating leases, the company spent $28 million for seats and carpeting. The question being discussed over breakfast on Wednesday morning was the length of the depreciation period for these leasehold improvements. The company controller, Sarah Keene, was surprised by the suggestion of Larry Person, her new assistant.

Keene: Why 25 years? We've never depreciated leasehold improvements for such a long period.
Person: I noticed that in my review of back records. But during our expansion to the Midwest, we don't need expenses to be any higher than necessary.
Keene: But isn't that a pretty rosy estimate of these assets' actual life? Trade publications show an average depreciation period of 12 years.

Required:
1. How would increasing the depreciation period affect American Movieplex's earnings?
2. Does revising the estimate pose an ethical dilemma?
3. Who would be affected if Person's suggestion is followed?

Analysis Case 15–4
Lease concepts; Walmart
● LO15–1 through LO15–4

Real World Financials

Walmart Inc. is the world's largest retailer. A large portion of the premises that the company occupies are leased. Its financial statements and disclosure notes revealed the following information:

Balance Sheet
($ in millions)

	2017	2016
Assets		
Property:		
Property under finance lease obligations	$11,637	$11,096
Less: Accumulated amortization	(5,169)	(4,751)
Liabilities		
Current liabilities:		
Finance lease obligations due within one year	565	551
Long-term debt:		
Long-term finance lease obligations	6,003	5,816

Required:
1. The net asset "property under finance lease obligations" has a 2017 balance of $6,468 million ($11,637 – $5,169). Liabilities for these leases total $6,568 ($565 + $6,003). Why do the asset and liability amounts differ?
2. Prepare a 2017 summary entry to record Walmart's lease payments, which were $800 million.
3. What is the approximate average interest rate on Walmart's finance leases? (*Hint:* See Req. 2)
4. Discuss some possible reasons why Walmart leases rather than purchases most of its premises.

Communication Case 15–5
Where's the gain?
● Appendix 15

General Tools is seeking ways to maintain and improve cash balances. As company controller, you have proposed the sale and leaseback of much of the company's equipment. As seller-lessee, General Tools would retain the right to essentially all of the remaining use of the equipment. The term of the lease would be six years.

You previously convinced your CFO of the cash flow benefits of the arrangement, but now he doesn't understand the way you will account for the transaction. "I really had counted on that gain on the sale portion of the transaction to bolster this period's earnings. What gives?" he wondered. "Put it in a memo, will you? I'm having trouble following what you're saying to me."

Required:
Write a memo to your CFO. Include discussion of each of these points:
1. How the transaction should be accounted for.
2. Why General Tools will not get the gain the CFO had counted on.

IFRS Case 15–6
Lease classification; U.S. GAAP
● LO15–1, LO15–2, LO15–4, LO15–9
IFRS

Security Devices Inc. (SDI) needs additional office space to accommodate expansion. SDI wants to avoid income statement effects that would disrupt its attempts to "smooth" income over time.

Required:
1. Which lease classification would management prefer?
2. Would meeting its reporting objective be more difficult or less difficult under IFRS?

Data Analytics

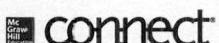

Data analytics is the process of examining data sets in order to draw conclusions about the information they contain. If you haven't completed any of the prior data analytics cases, follow the instructions listed in the Chapter 1 Data Analytics case to get set up. You will need to watch the videos referred to in the Chapters 1 - 3 Data Analytics cases. No additional videos are required for this case. All short training videos can be found here: **https://www.tableau.com/learn/training#getting-started**.

Data Analytics Case
Lease Liability and Debt Covenants
LO15–4

In the Data Analytics Cases in prior chapters, you used Tableau to examine a data set and create charts to compare two (hypothetical) publicly traded companies, Big Store and Discount Goods. In this chapter, you will examine GPS Corporation and Tru, Inc. on the basis of various financial metrics. In this case you examine the companies' pattern of leasing facilities, their transition to the new lease accounting standard in 2019, and the effect of that transition on debt covenants.

Since 2019 operating leases have been reported in the balance sheet of the lessee as a right-of-use asset and lease liability. Many companies at the time had debt covenants restricting the amount of debt during loan periods. GPS Corporation and True, Inc. had as part of their loan agreements a requirement that the debt to equity ratio could not exceed 2.0 without being in default of the loans.

Required:

For each of the two companies in the seven-year period, 2015-2021, use Tableau to calculate and display the trends for the long-term portion of the (a) capital lease liability, (b) the finance lease liability, and (c) the operating lease liability, as well as (d) the debt to equity ratio. Based upon what you find, answer the following questions:

1. Is GPS's pattern of leasing facilities as evidenced by its capital lease liability (a) increasing, (b) decreasing, or (c) remaining relatively the same over the period 2015-2018?
2. What is the capital lease liability for GPS in 2020?
3. Is Tru's operating lease liability (a) increasing, (b) decreasing, or (c) remaining relatively the same over the period 2019-2021?
4. Is GPS technically in violation of its debt covenant during the period 2019-2021?

Resources:

Download the "GPS_Tru_Financials.xlsx" Excel file available in Connect, or under Student Resources within the Library tab. Save it to the computer on which you will be using Tableau.

For this case, you will create a calculation to produce the debt to equity ratio to allow you to compare and contrast the two companies.

After you view the training videos, follow these steps to create the charts you'll use for this case:

- Open Tableau and connect to the Excel spreadsheet you downloaded.
- Starting on the Sheet 1 tab, drag "Company" and "Year" under Dimensions to the Columns shelf. Change "Year" to *discrete* by right-clicking and selecting "Discrete." Select "Show 'Filter'" and uncheck all the years except 2015 - 2021. Click "OK."
- Drag "Capital lease liability", "Lease liability – operating lease", "Lease liability – finance lease" under Measures into the Rows shelf.
- Add labels to the bars by clicking on "Label" under the "Marks" card and clicking the box "Show mark label." Format the labels to Times New Roman, bold, black and 10-point font. Edit the color of the years on the "Marks" card if desired by dragging "Year" on to the Color "Marks" card.
- Change the title of the sheet to be "Lease Liability" by right-clicking and selecting "Edit title." Format the title to Times New Roman, bold, black and 15-point font. Change the title of "Sheet 1" to match the sheet title by right-clicking, selecting "Rename" and typing in the new title.
- On the Sheet 2 tab, follow the procedure above for the company and year.
- Drag "Total liabilities" under Measures to the Rows shelf. Change to "Discrete" data following the same instructions and uncheck "Show Header" to eliminate the number from view.
- Drag "Total shareholders' equity" under Measures to the Rows shelf, change to discrete data, and uncheck "Show Header."
- Create a calculated field by clicking the "Analysis" tab at the top of the screen and selecting "Create Calculated field." Name the calculation "Debt to Equity Ratio." In the Calculation Editor window, drag "Total liabilities," type "/", then drag "Total shareholders' equity" from the Rows shelf. Make sure the window says that the calculation is valid and click OK.
- Drag the newly created "Debt to Equity Ratio" to the Rows shelf. Click on the "Show Me" and select "side-by-side bars." Add labels to the bars by clicking on "Label" under the "Marks" card and clicking the box "Show mark labels." Format the labels to Times New Roman, bold, black and 10-point font. Edit the color on the "Marks" card if desired.

- Change the title of the sheet to be "Debt to Equity Ratio" by right-clicking and selecting "Edit title." Format the title to Times New Roman, bold, black and 15-point font. Change the title of "Sheet 2" to match the sheet title by right-clicking, selecting "Rename" and typing in the new title.
- Format all other labels to be Times New Roman, bold, black and 12-point font.
- Once complete, save the file as "DA15_Your initials.twbx."

Continuing Cases

Target Case
● LO15–2, LO15–4
Real World Financials

Target Corporation prepares its financial statements according to U.S. GAAP. Target's financial statements and disclosure notes for the year ended February 3, 2018, are available in Connect. This material is also available under the Investor Relations link at the company's website (**www.target.com**).

Required:

In its Analysis of "Financial Condition: New Accounting Pronouncements," Target's financial statements for the year ended February 3, 2018, the company indicates that

> In February 2016, the FASB issued ASU No. 2016-02, Leases, to require organizations that lease assets to recognize the rights and obligations created by those leases on the balance sheet. The new standard is effective in 2019.

Refer to *Note 22: Leases.* New lease accounting guidance requires companies to record a right-of-use asset and a lease liability for all leases, with the exception of short-term leases, at present value. If Target had used the new lease accounting guidance in its fiscal 2017 (February 3, 2018) financial statements, what would be the amount reported as a liability for its leases, operating and capital (finance) combined [rounded to nearest $million]?

Hint: Assume the payments "after 2020" are to be paid evenly over a 16-year period and all payments are at the end of years indicated. Target indicates elsewhere in its financial statements that 6% is an appropriate discount rate for its leases.

Air France–KLM Case
● LO15–9
Real World Financials
IFRS

Air France–KLM (AF), a Franco-Dutch company, prepares its financial statements according to International Financial Reporting Standards. AF's financial statements and disclosure notes for the year ended December 31, 2017, are available in Connect. This material is also available under the Finance link at the company's website (**www.airfranceklm.com**).

Required:

1. In *Note 4:* Summary of accounting policies, part 4.14: Leases, AF states that "leases are classified as finance leases when the lease arrangement transfers substantially all the risks and rewards of ownership to the lessee. All other leases are classified as operating leases." Is this the policy companies using U.S. GAAP follow?
2. Is this the policy AF will follow when it begins applying the new lease guidance in the 2015 update to *IFRS 16*? Explain.

CPA Exam Questions and Simulations

ROGER CPA Review

Sample CPA Exam questions from Roger CPA Review are available in Connect as support for the topics in this chapter. These multiple-choice questions and task-based simulations include expert-written explanations and solutions, and provide a starting point for students to become familiar with the content and functionality of the actual CPA Exam.

CHAPTER

16 Accounting for Income Taxes

OVERVIEW

In this chapter we explore financial accounting and reporting for the effects of income taxes. The discussion defines and illustrates temporary differences, which are the basis for recognizing deferred tax assets and deferred tax liabilities, as well as permanent differences, which have no deferred tax consequences. You will learn how to adjust deferred tax assets and deferred tax liabilities when tax laws or rates change. We also discuss accounting for the tax effects of net operating losses as well as intraperiod tax allocation.

In December of 2017, the United States Congress passed a sweeping tax reform package. Congress reduced the corporate tax rate, accelerated the timing of the depreciation deduction for some assets, revised how and when a company can use net operating losses to reduce its tax bill, and made many other changes. This chapter is written under the new tax rules to make sure that instructors and their students have access to the most current content possible.

LEARNING OBJECTIVES

After studying this chapter, you should be able to:

- **LO16–1** Explain the conceptual underpinnings of accounting for temporary differences and the four-step method used to calculate income tax expense. *(p. 909)*
- **LO16–2** Describe the types of temporary differences that cause deferred tax liabilities and determine the amounts needed to record periodic income taxes. *(p. 912)*
- **LO16–3** Describe the types of temporary differences that cause deferred tax assets and determine the amounts needed to record periodic income taxes. *(p. 919)*
- **LO16–4** Describe when and how a valuation allowance is recorded for deferred tax assets. *(p. 925)*
- **LO16–5** Explain why permanent differences have no deferred tax consequences. *(p. 929)*
- **LO16–6** Explain how a change in tax rates affects the measurement of deferred tax amounts. *(p. 933)*
- **LO16–7** Describe when and how the tax effects of net operating losses are recognized in the financial statements. *(p. 937)*
- **LO16–8** Explain how deferred tax assets and deferred tax liabilities are reported in a classified balance sheet and describe related disclosures. *(p. 941)*
- **LO16–9** Demonstrate how to account for uncertainty in income tax decisions. *(p. 943)*
- **LO16–10** Explain intraperiod tax allocation. *(p. 946)*
- **LO16–11** Discuss the primary differences between U.S. GAAP and IFRS with respect to accounting for income taxes. *(p. 931)*

© wavebreakmediamicro/123RF

FINANCIAL REPORTING CASE

What's the Difference?

Laura Drake, the new CEO of Times-Lehrer Industries, has asked you, a senior member of the company's audit staff, to help her prepare for an upcoming board meeting. A brilliant engineer and operations specialist, Laura has limited knowledge of accounting and has been pouring over the past few years of the company's financial statements.

"Why isn't the amount we report as tax expense equal to the amount we pay in taxes?" Laura asked. "And I see we are blaming reduced profitability in 2017 on our net deferred tax assets and the fact that Congress passed rules reducing future tax rates. I thought assets and lower taxes were good things—why did they combine to hurt our earnings? And I also see we have something called a net operating loss carryforward—what's that?"

QUESTIONS

By the time you finish this chapter, you should be able to respond appropriately to the questions posed in this case. Compare your response to the solution provided at the end of the chapter.

1. Explain to Laura how differences between financial reporting standards and income tax rules might cause the income tax expense and the amount of income tax paid to differ. *(p. 909)*
2. How might a reduction in future tax rates affect deferred tax assets in a way that reduces current net income? *(p. 933, 935)*
3. What are net operating loss carryforwards and how can they provide cash savings? *(p. 923, 938)*

PART A

Temporary Differences

FINANCIAL Reporting Case

Q1, p. 909

● LO16–1

Let's say that JCorp's 2021 tax return indicates that the company is obligated to pay $25 million in income taxes. JCorp's CFO knows that another $5 million in income taxes is attributable to operations in 2021, but tax rules allow JCorp to delay paying that amount until subsequent tax years. The reason that JCorp is able to defer paying these taxes is that tax rules allow some revenues and expenses to be reported in its tax return in years other than when those amounts are reported in its income statement. These differences cause JCorp's pretax income in its tax return to differ from pretax income in its income statement.

Here are the financial reporting questions to consider:

1. What *tax liability* should JCorp report in its 2021 balance sheet? Tax rules obligate the company to pay $25 million now for the current tax year, but the company will eventually pay the remaining $5 million that was deferred to subsequent tax years.
2. What *tax expense* should JCorp report in its 2021 income statement? The company eventually will pay taxes of $30 million attributable to operations in 2021, but taxes to be paid for the current tax year are only $25 million.

For perspective on these questions, let's first consider the circumstances that give rise to deferred taxes and then think about how we should account for them.

Conceptual Underpinnings

The objectives of financial accounting and tax accounting are not the same.

In general, the revenues and expenses (and gains and losses) included in a company's tax return for a given year are the same as those reported in the company's income statement for the same year. However, in some instances, tax laws and financial accounting standards differ because the fundamental objectives of financial reporting and those of taxing authorities are not the same. Financial accounting standards are established to provide useful information to investors and creditors. Congress, on the other hand, establishes tax regulations to allow it to raise funds in a socially acceptable manner, as well as to influence the behavior of taxpayers. Congress uses tax laws to encourage activities it deems desirable, such as investment in productive assets, and to discourage activities it deems undesirable, such as violations of law.

Temporary differences **arise when tax rules and accounting rules recognize income in different periods.**

An income statement and a tax return both include revenues and expenses, but differences between financial accounting standards and tax rules create a **temporary difference** between pretax accounting income and taxable income. For temporary differences, the issue is not *whether* an amount is taxable or deductible, but *when.*

For example, assume that Watson Associates purchases $60 thousand of computer equipment in January of 2021. Watson estimates that the computer equipment will have an estimated useful life of three years. For financial reporting purposes, Watson records straight-line depreciation each year of $20 thousand.

	2021	2022	2023	Total
Income before tax and depreciation	$120	$120	$120	$360
Depreciation in the **income statement**	(20)	(20)	(20)	(60)
Pretax accounting income	$100	$100	$100	**$300**

However, tax rules allow Watson to deduct the entire $60 thousand cost of the equipment in 2021, essentially taking 100% of the depreciation for tax purposes in the year the asset is purchased. Assuming that Watson has a tax rate of 25%, deducting the entire depreciation in 2021 reduces taxable income that year by $60 thousand but leaves none of the depreciation to reduce taxable income in 2022 and 2023:

	2021	2022	2023	Total
Income before tax and depreciation	$120	$120	$120	$360
Depreciation on the tax return	(60)	(0)	(0)	(60)
Taxable income (tax return)	$ 60	$120	$120	**$300**
Tax rate	× 25%	× 25%	× 25%	
Tax payable	$ 15	$ 30	$ 30	

Let's first focus on pretax accounting income and taxable income. Notice that pretax accounting income and taxable income both are $300 thousand over the three-year depreciation period. But these two amounts differ from each other in each of the three years. In 2021, tax depreciation ($60 thousand) is greater than accounting depreciation ($20 thousand), causing taxable income to be less than pretax accounting income by $40 thousand. This difference is temporary and reverses in 2022 and 2023 when Watson recognizes depreciation expense in the income statement ($20 thousand in 2022 and $20 thousand in 2023) but can't deduct any more depreciation in the tax return.

Now let's consider tax expense and tax payable. We can see that Watson has *tax payable* of $15 thousand in 2021 but $30 thousand in 2022 and 2023. But what about *tax expense*? One simple approach would be to make tax expense in the income statement each period equal to whatever tax is payable in that period and ignore any future implications of the timing difference. However, that approach would not communicate to investors that Watson's current activities have resulted in future tax consequences. Investors might incorrectly think that Watson's low tax expense in 2021 and resulting high net income will persist in the future. Likewise, investors might not realize that Watson anticipates higher tax payable in the future ($30 thousand in 2022 and 2023 rather than only $15 thousand in 2021).

Instead, Watson should recognize income tax expense based on the "accrual concept," just like other expenses. That is, income tax expense reported each period should be the amount caused by that period's events and activities, regardless of the period in which the tax laws indicate a tax obligation exists. If tax laws allow a company to postpone paying taxes on activities reported in the current period's income statement, the company must report a **deferred tax liability** because the company anticipates those activities will lead to **future taxable amounts**. On the other hand, if tax laws require the company to pay more tax than is indicated by the activities reported in the current period's income statement, the company reports a **deferred tax asset** reflecting the benefit of **future deductible amounts**. Each year's tax expense reported in the income statement includes not only a current portion related to tax payable in the current year but also a deferred portion that includes any changes in deferred tax assets and liabilities.

Accounting for income taxes is consistent with the accrual concept of accounting.

The 4-Step Process

The discussion of differences between accounting income and taxable income leads us to the real key to understanding accounting for income taxes. Tax expense is not calculated directly but rather is the result of the combination of income tax payable and any changes in deferred tax assets and liabilities. We'll use the following color-coded 4-step process to maintain that perspective throughout the chapter.

Tax expense is a "plug" determined by changes in the tax payable, deferred tax assets, and deferred tax liability accounts.

1. **Calculate tax payable:** This is the amount of tax currently payable based on the current year's tax return.
2. **Calculate ending DTAs and DTLs:** In this step, we calculate the appropriate ending balances of the deferred tax assets (DTAs) and deferred tax liabilities (DTLs).
3. **Calculate change in DTAs and DTLs:** In this step, we determine the change (debit or credit) in each of the deferred tax assets and liabilities needed to move from their previous balances to the ending balances calculated in step 2.
4. **Plug tax expense:** In this final step, we combine the tax payable (step 1) and any changes in the deferred tax accounts (step 3) to determine income tax expense.

Types of Temporary Differences

You will see that we apply this 4-step process to four basic types of temporary differences. As shown in Illustration 16–1, temporary differences can be categorized by whether they are associated with a revenue (gain) or expense (loss), and whether that revenue or expense is recognized in the income statement before or after it is reported in the tax return.

Illustration 16–1
Types of Temporary Differences

	Revenues (or gains)	Expenses (or losses)
Reported in the income statement now, but on the tax return later	• Installment sales of property (installment method for taxes) • Unrealized gain from recording investments at fair value (taxable when asset is sold)	• Estimated expenses and losses (tax-deductible when paid) • Unrealized loss from recording investments at fair value or inventory at LCM (tax-deductible when asset is sold)
Reported on the tax return now, but in the income statement later	• Rent collected in advance • Subscriptions collected in advance • Other revenue collected in advance	• Accelerated depreciation on the tax return in excess of straight-line depreciation in the income statement • Prepaid expenses (tax-deductible when paid)

Note that:

- The temporary differences shown in the diagonal purple areas create *deferred tax liabilities* because they result in *taxable* amounts in some future year(s).
- The temporary differences in the opposite diagonal blue areas create *deferred tax assets* because they result in *deductible* amounts in some future year(s).

Next we'll walk through examples applying the 4-step process to each of these four types of temporary differences, starting with the Watson depreciation example we already began. You can refer back to this table throughout our discussion.

Additional Consideration

As shown in Illustration 16–1, temporary differences are primarily caused by revenues, expenses, gains, and losses being included in taxable income in a year other than the year in which they are recognized for financial reporting purposes. Other events that are beyond the scope of this textbook also can cause temporary differences and are briefly described in FASB ASC 740–10–25: Income Taxes–Overall–Recognition. Our discussions in this chapter focus on temporary differences caused by the timing of revenue and expense recognition, but it's important to realize that the concept of temporary differences embraces all differences that will result in taxable or deductible amounts in future years.

Deferred Tax Liabilities

Expense-Related Deferred Tax Liabilities

LO16–2

To determine taxable income, we add back to pretax accounting income any depreciation expense in the income statement and then subtract any tax deduction allowed on the tax return.

The lower right corner of Illustration 16–1 lists some common expense-related deferred tax liabilities. Perhaps the most common relates to depreciation. Tax laws typically permit the cost of a depreciable asset to be deducted in the tax return sooner than it is reported as depreciation expense in the income statement.[1] The difference in tax laws creates a temporary difference in taxable income and pretax accounting income. Taxable income will be lower in the initial years when the tax depreciation deduction is higher, but the situation reverses in later years when the pretax accounting income will be lower. Let's return to our Watson example to see how this works. Illustration 16–2 summarizes the relevant information.

Illustration 16–2 Expense Reported on the Tax Return *before* the Income Statement

Watson Associates purchased $60 thousand of equipment in early January of 2021. Watson estimates the equipment has a useful life of three years, so it depreciates the equipment straight line, with $20 thousand of depreciation expense 2021–2023. However, tax rules allow Watson to take the entire $60 thousand deduction for the cost of the equipment on its 2021 tax return. Watson has a 25% tax rate and pretax accounting income of $100 thousand in each of those years.

($ in thousands)		2021		2022		2023	Total
Pretax accounting income (income statement)		$100		$100		$100	**$300**
Depreciation expense in the income statement	$20		$20		$20		60
Depreciation deduction on the tax return	(60)		(0)		(0)		(60)
Temporary difference		(40)		20		20	0
Taxable income (tax return)		$ 60		$120		$120	**$300**

In 2021, taxable income is less than accounting income because depreciation deductions on the tax return are greater than tax expense on the income statement.

Now let's apply the 4-step process to account for the temporary difference related to depreciation expense that is shown in Illustration 16–2. The journal entry to record 2021 income taxes is shown in Illustration 16–2A.

[1]The accelerated depreciation method prescribed by the tax code is the modified accelerated cost recovery system (MACRS). However, the tax legislation passed in December 2017 permits companies to deduct 100% of the cost of many relatively short-lived assets in the year the asset is purchased. See Chapter 11's Appendix 11A for further discussion.

Illustration 16–2A Expense-Related Deferred Tax Liability—2021

($ in thousands)	Current Year 2021	Future Taxable Amounts 2022	2023	Total
Pretax accounting income	$100			
Temporary difference:				
Depreciation	(40)	$20	$20	$40
Taxable income (tax return)	$ 60			
Enacted tax rate	25%			25%
Tax payable	**$ 15**			**$10**

Deferred Tax Liability

	0
	10
	10

Journal Entry at the End of 2021

Income tax expense (to balance)	25	
Income tax payable (determined above)		15
Deferred tax liability (determined above)		10

Step 1: Tax payable: $15
Step 2: DTL end bal: $10
Step 3: DTL change: $10
Step 4: Tax exp plug: $25

Let's walk through the 4-step process ($ in thousands).

Step 1: Pretax accounting income for 2021 is $100, but taxable income is only $60 because of a $40 higher depreciation deduction for tax purposes. Tax payable for 2021 is recorded for $15 (= $60 taxable income × 25% tax rate).

Step 2: Watson needs to recognize a deferred tax liability of **$10** (= $40 taxable income × 25% tax rate) for the tax the company is allowed to defer to future years as a result of being able to depreciate the asset's entire cost the first period. We say that the temporary difference *originated* in 2021, because that is the year in which it gave rise to the deferred tax liability.

Step 3: The balance of the deferred tax liability account needs to be adjusted from its current balance (which is $0 in the origination year) to the amount that needs to be reported (the ending balance of **$10** from step 2). In this case, that adjustment is **$10.**

Step 4: Total tax expense equals $25. This amount includes a current portion that's payable now ($15 from step 1) plus the deferred portion that's represented by the increase in the deferred tax liability (**$10** from step 3).

Now, let's follow the determination of income taxes for this illustration all the way through the complete reversal of the temporary difference. Remember, we're assuming the pretax accounting income is $100 thousand each year and the only difference between that amount and taxable income is the difference in depreciation. We determine income tax expense for 2022 in Illustration 16–2B.

Let's look closer at Illustration 16-2B ($ in thousands). In 2022, Watson's tax payable is $30 (= $120 taxable income × 25% tax rate). There's no tax deduction for depreciation in 2022 because the tax rules enabled Watson to depreciate the asset's entire cost in 2021. The income statement, though, reports straight-line depreciation expense of $20, giving us a $20 difference between pretax income in the income statement and taxable income in the tax return. This difference reverses half of the $40 temporary difference that originated in 2021. So, at this point, the tax to be deferred to future periods is **$5**, and we need to reduce the balance of the deferred tax liability by **$5** (from **$10** in 2021 to **$5** in 2022). As a consequence, total tax expense is $25, equal to its current portion ($30 from step 1) combined with the *decrease* in the deferred tax liability (**$(5)** from step 3).

Illustration 16–2B Expense-Related Deferred Tax Liability—2022

($ in thousands)	Current Year 2022	Future Taxable Amounts 2023	Total
Pretax accounting income	$100		
Temporary difference:			
Depreciation	20	$20	$20
Taxable income (tax return)	$120		
Enacted tax rate	25%		25%
Tax payable	$ 30		$ 5

Step 1: Tax payable: $30
Step 2: DTL end bal: $5
Step 3: DTL change: $(5)
Step 4: Tax exp plug: $25

Deferred Tax Liability

	10
5	
	5

Journal Entry at the End of 2022

Income tax expense (to balance)	25	
Deferred tax liability (determined above)	5	
Income tax payable (determined above)		30

As you can see from Illustration 16–2C, Watson's accounting in 2023 is very similar to its accounting in 2022. Here's why, Watson's 2023 tax payable again is $30, and again there's no tax deduction even though the income statement reports straight-line depreciation expense of $20. This difference reverses the remaining $20 of the $40 temporary difference that originated in 2021. So, at this point, the tax to be deferred to future periods is **$0,** and we need to reduce the balance of the deferred tax liability by **$5** (from **$5** in 2022 to **$0** in 2023). As a consequence, total tax expense is $25, equal to its current portion ($30 from step 1) combined with the *decrease* in the deferred tax liability (**$(5)** from step 3).

Illustration 16–2C Expense-Related Deferred Tax Liability—2023

($ in thousands)	Current Year 2023	Future Taxable Amounts
Pretax accounting income	$100	
Temporary difference:		
Depreciation	20	$ 0
Taxable income (tax return)	$120	
Enacted tax rate	25%	25%
Tax payable	$ 30	$ 0

Step 1: Tax payable: $30
Step 2: DTL end bal: $0
Step 3: DTL change: $(5)
Step 4: Tax exp plug: $25

Deferred Tax Liability

	5
5	
	0

Journal Entry at the End of 2023

Income tax expense (to balance)	25	
Deferred tax liability (determined above)	5	
Income tax payable (determined above)		30

Notice that the deferred tax liability increased in 2021 when the temporary difference originated, and then decreased in 2022 and 2023 as the temporary difference reversed.

Deferred Tax Liability

($ in thousands)			
		10	2021 ($40 × 25%)
2022 ($20 × 25%)	5		
2023 ($20 × 25%)	5		
		0	Balance after 3 years

Balance Sheet and Income Statement Perspectives

Our perspective in this example so far has focused on the *income statement* effects of the depreciation. Another perspective starts with the *balance sheet.* An assumption underlying a balance sheet is that assets will be recovered (used or sold to produce cash), and liabilities will be settled (paid with cash). Those assets and liabilities typically create taxable or deductible amounts in the future when they are recovered or settled. Before that occurs, there is a temporary difference between the *book value* of assets and liabilities in the balance sheet and their equivalent **tax basis** (which is an asset or liability's original value for tax purposes reduced by any amounts included to date on tax returns). A difference between book value and tax basis, commonly called a *book-tax difference,* implies a future taxable or deductible amount, so we can use book-tax differences to calculate deferred tax assets and liabilities.

Deferred tax assets and liabilities can be computed from temporary book-tax differences.

The *tax basis* of an asset or liability is its original value for tax purposes reduced by any amounts included to date on tax returns.

To see how this works, let's look back at our depreciation example for Watson. At the end of 2021, the asset's book value is reported as $40 in the balance sheet, equal to the asset's original cost of $60 minus depreciation of $20 in 2021. However, for tax purposes, the asset was fully depreciated in 2021, so its tax basis at the end of 2021 was $0. This creates a book-tax difference of $40. That book-tax difference implies that, in the future, Watson will have $40 less tax deduction for depreciation, so Watson's *future taxable* income will be higher by $40. Therefore, Watson should recognize a deferred tax liability of $40 × 25% = $10.[2]

As shown in the following table, Watson can use the asset's book-tax difference each year to calculate the related deferred tax liability balance. That's just another way to accomplish **step 2** of our 4-step process. Notice that the deferred tax liability balances (**step 2**) and changes in those balances (**step 3**) shown in this table are the same as those shown in Illustration 16–2A, 16–2B, and 16–2C. The income statement and balance sheet perspectives are two complementary ways to accomplish the same objectives.

	Initial year 2021 12/31 balance	2022 Depr	2022 12/31 balance	2023 Depr	2023 12/31 balance
Depreciable asset:					
Accounting book value	$40*	$(20)	$20	$(20)	$0
Tax basis	0		0		0
Temporary difference	$40		$20		$0
Tax rate	25%		25%		25%
Deferred tax liability (DTL)	**$10**	→	**$5**	→	**$0**
DTL change needed	**$10**		**$(5)**		**$(5)**

* $60 (initial cost) − $20 (depreciation for 2021) = $40

[2]What if Watson sold the asset instead of using it? Watson's gain or loss on sale for tax purposes will be based on the asset's tax basis, so Watson will recognize $40 more gain (or less loss) for tax purposes. Watson once again will have a higher tax bill of $40 × 25% = $10, again implying a deferred tax liability of $10. Regardless of whether Watson keeps or sells the asset, Watson's $40 book-tax difference implies a deferred tax liability of $10.

Revenue-Related Deferred Tax Liabilities

Deferred tax liabilities also can be driven by temporary differences related to revenue recognition. The upper left corner of Illustration 16–1 lists some common revenue-related deferred tax liabilities. One example relates to installment sales. Income from selling properties on an installment basis is reported for financial reporting purposes in the year of the sale. But tax laws permit installment income to be reported in the tax return later, as cash is received. As a consequence, a temporary difference occurs because taxable income is less than accounting income in the year of an installment sale but higher than accounting income in later years when the installment receivable is collected. A numerical example is provided in Illustration 16–3.

Illustration 16–3 Revenue Reported on the Tax Return *after* the Income Statement

Kent Land Management reported pretax accounting income in 2021, 2022, and 2023 of $180 million, $100 million, and $100 million, respectively, which includes 2021 income of $80 million from installment sales of property. However, the installment sales are reported on the tax return when collected, in 2022 ($20 million) and 2023 ($60 million).* The enacted tax rate is 25% each year.

($ in millions)		2021		2022		2023	Total
Pretax accounting income (income statement)		$180		$100		$100	$380
Installment sale income in the income statement	$(80)		$ 0		$ 0		(80)
Installment sale income on the tax return	0		20		60		80
Temporary difference		(80)		20		60	0
Taxable income (tax return)		$100		$120		$160	$380

In 2021, taxable income is less than accounting income because income from installment sales is not reported on the tax return until 2022–2023.

Notice that pretax accounting income and taxable income total the same amount ($380) over the three-year period but are different in each individual year. In 2021, taxable income is $80 million *less* than accounting income because it does not include income from installment sales. That temporary difference reverses over the next two years. In 2022 and 2023, taxable income is *more* than accounting income, because income from the installment sales, reported in the income statement in 2021, becomes taxable during the next two years as installments are collected. Because tax laws permit the company to delay reporting this income as part of taxable income, the company is able to defer paying tax on that income. As shown in Illustration 16–3A, that tax is not avoided, just deferred.

We calculate income tax expense by following the four steps ($ in millions). Kent's 2021 tax payable is $25. With future taxable amounts of $80, taxable at 25%, a **$20** deferred tax liability should be recognized as of the end of 2021. Because no previous balance exists, we credit deferred tax liability for the entire **$20** change in the account. That amount combines with tax payable of $25 to give us income tax expense of $45.

In Illustration 16-3B we see that some of the initial $80 temporary difference reverses in 2022 as the company collects some of the installment receivable ($20) and includes that amount in 2022 taxable income.

Kent's 2022 tax payable is $30. Now that $20 of the installment income is taxed in 2022, the remaining $60 of the temporary difference remains to be taxed later, so future taxable amounts as of the end of 2022 are $60. This means that a deferred tax liability of **$15** should be shown in the balance sheet. Reducing the deferred tax liability from **$20** in 2021 to **$15** now requires us to debit the deferred tax liability for **$5**. As a consequence, total tax expense is $25 in 2022, equal to its current portion ($30 from step 1) combined with the decrease in the deferred tax liability (**(5)** from step 3).

Illustration 16-3C shows us that Kent's 2023 tax payable is $40. Now that the remaining $60 of the installment income is taxed in 2023, none of the temporary difference remains to be taxed later, so future taxable amounts as of the end of 2023 are $0. With no future taxable amounts remaining, the deferred tax liability should be reported as **$0** as of the end of 2023,

Illustration 16–3A Revenue-Related Deferred Tax Liability—2021

($ in millions)	Current Year 2021	Future Taxable Amounts 2022	2023	Total
Pretax accounting income	$180			
Temporary difference:				
Installment income	(80)	$20	$60	$80
Taxable income (tax return)	$100			
Enacted tax rate	25%			25%
Tax payable	$ 25			**$20**

Deferred Tax Liability

	0
	[20]
	10

Journal Entry at the End of 2021

Income tax expense (to balance)	[45]	
Income tax payable (determined above)		25
Deferred tax liability (determined above)		20

Step 1: Tax payable: $25
Step 2: DTL end bal: $20
Step 3: DTL change: $20
Step 4: Tax exp plug: $45

Illustration 16–3B Revenue-Related Deferred Tax Liability—2022

($ in millions)	Current Year 2022	Future Taxable Amounts 2023	Total
Pretax accounting income	$100		
Temporary difference:			
Installment income	20	$60	$60
Taxable income (tax return)	$120		
Enacted tax rate	25%		25%
Tax payable	$ 30		**$15**

Deferred Tax Liability

	20
[5]	
	15

Journal Entry at the End of 2022

Income tax expense (to balance)	[25]	
Deferred tax liability (determined above)	5	
Income tax payable (determined above)		30

Step 1: Tax payable: $30
Step 2: DTL end bal: $15
Step 3: DTL change: $(5)
Step 4: Tax exp plug: $25

requiring us to debit the deferred tax liability for **$15**. As a consequence, total tax expense is $25 in 2023, equal to its current portion ($40 from step 1) combined with the decrease in the deferred tax liability ((**15**) from step 3).

Now look at what happened to the deferred tax liability account over the life of the installment receivable. Do you see how the $80 temporary difference originates in 2021 and reverses in 2022 and 2023 as that $80 is taxed?

Illustration 16–3C Revenue-Related Deferred Tax Liability—2023

($ in millions)	Current Year 2023	Future Taxable Amounts
Pretax accounting income	$100	
Temporary difference:		
Installment income	60	$ 0
Taxable income (tax return)	$160	
Enacted tax rate	25%	25%
Tax payable	**$ 40**	**$ 0**

Step 1: Tax payable: $40
Step 2: DTL end bal: $0
Step 3: DTL change: $(15)
Step 4: Tax exp plug: $25

Deferred Tax Liability

	15
15	
	0

Journal Entry at the End of 2023

Income tax expense (to balance)	25	
Deferred tax liability (determined above)	**15**	
Income tax payable (determined above)		**40**

Deferred Tax Liability

($ in millions)			
		20	2021 ($80 × 25%)
2022 ($20 × 25%)	**5**		
2023 ($60 × 25%)	**15**		
		0	Balance after 3 years

Taking a balance sheet perspective, we can calculate the deferred tax liability each year by looking at the book-tax difference that exists for the installment receivable. The book value of the receivable starts in 2021 at $80, drops to $60 in 2022, and then drops to $0 in 2023 as cash is collected. There is no receivable from a tax perspective, because taxable income is only recognized as cash is collected, so the tax basis of the receivable is $0 each year. The book-tax difference in the receivable in a given year represents additional taxable income that Kent will pay tax on in the future. The balance of the deferred tax liability each year is calculated as the book-tax difference times the applicable tax rate. To make sure you understand this point, refer back to Illustration 16–3A, 16–3B, and 16–3C and compare the deferred tax liability balances (**step 2**) and changes in those balances (**step 3**) in those illustrations with those shown below.

	Initial year 2021	2022		2023	
	12/31 balance	**Cash received**	**12/31 balance**	**Cash received**	**12/31 balance**
Installment receivable:					
Accounting book value	$80	$(20)	$60	$(60)	$ 0
Tax basis	0		0		0
Temporary difference	$80		$60		$ 0
Tax rate	25%		25%		25%
Deferred tax liability (DTL)	**$20** →		**$15** →		**$ 0**
DTL change needed	**$20**		**$ (5)**		**$(15)**

Deferred Tax Assets

● LO16–3

The temporary differences illustrated to this point produce future *taxable* amounts when the temporary differences reverse. Sometimes, though, the future tax consequence of a temporary difference will be to *decrease* taxable income relative to accounting income. Such situations produce what's referred to as **future deductible amounts**. These have favorable future tax consequences that are recognized as **deferred tax assets**.

We report deferred tax assets for the future tax benefits of temporary differences that create future deductible amounts.

Expense-Related Deferred Tax Assets

As noted in the upper right corner of Illustration 16–1, one circumstance that requires recognition of a deferred tax asset is when an estimated expense is reported in the income statement when incurred but deducted on the tax return in later years when the expense is actually paid. Illustration 16–4 provides an example: a quality-assurance warranty (discussed extensively in Chapter 13).

Illustration 16–4 Expense Reported on the Tax Return after the Income Statement

RDP Networking reported pretax accounting income in 2021, 2022, and 2023 of $120 million, $100 million, and $100 million, respectively. The 2021 income statement includes an $80 million warranty expense that is deducted for tax purposes when paid in 2022 ($36 million) and 2023 ($44 million). The income tax rate is 25% each year.

($ in millions)		2021		2022		2023	Total
Pretax accounting income (income statement)		$120		$100		$100	$320
Warranty expense in the income statement	$80		$ 0		$ 0		(80)
Warranty expense on the tax return	0		(36)		(44)		80
Temporary difference		80		(36)		(44)	0
Taxable income (tax return)		$200		$ 64		$ 56	$320

At the end of 2021, the amounts needed to record income tax for 2021 would be determined as shown in Illustration 16–4A.

In 2021, taxable income is greater than accounting income because warranty expense is not reported on the tax return until 2022–2023.

Illustration 16–4A Expense-Related Deferred Tax Asset—2021

($ in millions)	Current Year 2021	Future Deductible Amounts 2022	2023	Total
Pretax accounting income	$120			
Temporary difference:				
Warranty expense	80	$(36)	$(44)	$(80)
Taxable income (tax return)	$200			
Enacted tax rate	25%			25%
Tax payable	**$ 50**			**$(20)**

Deferred Tax Asset	
0	
20	
20	

Journal Entry at the End of 2021

Income tax expense (to balance)	30	
Deferred tax asset (determined above)	20	
Income tax payable (determined above)		50

Step 1: Tax payable: $50
Step 2: DTA end bal: $20
Step 3: DTA change: $20
Step 4: Tax exp plug: $30

Let's review what happened ($ in millions). In 2021, RDP's tax payable is $50 (= $200 taxable income × 25% tax rate). The reason taxable income is $80 higher than pretax accounting income is that, while GAAP requires the $80 warranty expense to be subtracted from accounting income in 2021, the tax rules don't permit RDP to deduct the expense on the tax return until the cost of satisfying the warranty actually is paid, which in this case will occur over the next two years. So, when that total future deductible amount of $80 is deducted, taxable income will be reduced by that amount, saving a total of $20 at a 25% tax rate. To represent that future tax savings, RDP reports a **$20** deferred tax asset at the end of 2021. Because no previous balance exists, we debit the deferred tax asset for the entire **$20.** That amount combined with the tax payable credit of $50 gives us income tax expense of $30. RDP must pay $50 tax now, but as a result of something that happens in 2021 (selling goods under warranty), it will save $20 in future taxes, so the 2021 tax expense is actually only $30.

We follow the situation to 2022 in Illustration 16–4B.

Illustration 16–4B Expense-Related Deferred Tax Asset—2022

($ in millions)	Current Year 2022	Future Deductible Amounts 2023	Total
Pretax accounting income	$100		
Temporary difference:			
Warranty expense	(36)	$(44)	$(44)
Taxable income (tax return)	$ 64		
Enacted tax rate	25%		25%
Tax payable	$ 16		$(11)

Step 1: Tax payable: $16
Step 2: DTA end bal: $11
Step 3: DTA change: $(9)
Step 4: Tax exp plug: $25

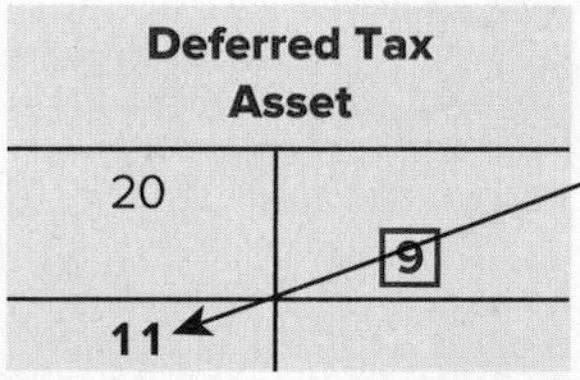

Journal Entry at the End of 2022

Income tax expense (to balance)	25	
Deferred tax asset (determined above)		9
Income tax payable (determined above)		16

Again, income tax expense is a combination of the tax payable now and the change in deferred tax ($ in millions). RDP's 2022 tax payable is $16. One reason it's not more is that taxable income is reduced by deducting $36 of last year's warranty expense. Because $36 of the original $80 temporary difference has reversed, only $44 remains to be deducted in the future. So, RDP should report a deferred tax asset of **$11** (= $44 × 25%) as of the end of 2022. So, RDP must credit the deferred tax asset for **$9** to reduce the $20 existing balance to that amount. This combines with the tax payable currently of $16 to give us income tax expense of $25.

We follow the example through the last year in Illustration 16–4C.

RDP's accounting in 2023 is very similar to its accounting in 2022. Tax payable is $14. As of the end of 2023, no future deductible amounts remain, so the balance in RDP's deferred tax asset should be **$0.** That requires RDP to credit the deferred tax asset for **$11,** which combines with the tax payable of $14 to give us income tax expense of $25.

At the end of 2021 and 2022, the company reports a deferred tax asset representing future income tax benefits. That deferred tax asset is reduced to zero by the end of 2023, after the tax savings the asset represented have been realized.

Illustration 16–4C Expense-Related Deferred Tax Asset—2023

($ in millions)	Current Year 2023	Future Deductible Amounts
Pretax accounting income	$100	
Temporary difference:		
Warranty expense	(44)	$ 0
Taxable income (tax return)	$ 56	
Enacted tax rate	25%	25%
Tax payable	**$ 14**	**$ 0**

Deferred Tax Asset

11	
	11
0	

Journal Entry at the End of 2023

Income tax expense (to balance)	25	
Deferred tax asset (determined above)		11
Income tax payable (determined above)		14

Step 1: Tax payable: $14
Step 2: DTA end bal: $0
Step 3: DTA change: $(11)
Step 4: Tax exp plug: $25

Deferred Tax Asset

($ in millions)			
2021 ($80 × 25%)	20		
		9	2022 ($36 × 25%)
		11	2023 ($44 × 25%)
Balance after 3 years	0		

We also can calculate this deferred tax asset each year using the book-tax difference that exists for the warranty liability in the balance sheet. The warranty liability starts in 2021 with a book value of $80, falls to $44 in 2022, and then to $0 in 2023 as the warranty liability is settled. The tax basis stays at $0 each year, because tax deductions are allowed only as cash is used to settle warranty claims, so there is no liability from a tax perspective. The book-tax difference in the warranty liability in a given year represents additional tax deductions RDP will receive in the future. This creates a deferred tax asset. The balance of the deferred tax asset each year is calculated as the book-tax difference times the applicable tax rate. You can verify that these deferred tax asset balances (**step 2**) and changes in those balances (**step 3**) are the same as those shown in Illustrations 16-4A, 16-4B, and 16-4C.

Income taxes payable in 2022 and 2023 are less because of the taxes prepaid in 2021.

	Initial year 2021	2022		2023	
	12/31 balance	**Warranty work**	**12/31 balance**	**Warranty work**	**12/31 balance**
Warranty liability:					
Accounting book value	$80	$(36)	$44	$(44)	$ 0
Tax basis	0		0		0
Temporary difference	$80		$44		$ 0
Tax rate	25%		25%		25%
Deferred tax asset (DTA)	**$20**	→	**$11**	→	**$ 0**
DTL change needed	**$20**		**$ (9)**		**$(11)**

Additional Consideration

Unlike most assets, management views deferred tax assets to be *less* desirable than deferred tax liabilities because deferred tax assets result from taxable income (and tax) being higher now than later. It's more desirable to delay paying taxes as long as possible. Therefore, all else equal, managers would prefer to recognize deferred tax liabilities, which result from having lower taxable income (and thus lower tax) now.

Revenue-Related Deferred Tax Assets

As shown in the lower left corner of Illustration 16–1, another type of temporary difference that gives rise to a deferred tax asset is a *revenue* that is taxed when collected but recognized in the income statement in later years when performance obligations are satisfied. Illustration 16–5 demonstrates this second type with a common example: deferred revenue.

Illustration 16–5 Revenue Reported on the Tax Return before the Income Statement

Tomorrow Publications reported pretax accounting income of $100 thousand in 2021, 2022 and 2023. The 2021 income statement does not include $80 thousand of magazine subscriptions received that year for one- and two-year subscriptions. Instead, that revenue will be recognized for financial reporting purposes in 2022 ($60 thousand) and 2023 ($20 thousand). The entire $80 thousand is included in taxable income in 2021. The income tax rate is 25% each year.

($ in thousands)		2021		2022		2023	Total
Pretax accounting income (income statement)		$100		$100		$100	$300
Subscription revenue in the income statement	$ 0		$ (60)		$ (20)		(80)
Subscription revenue on the tax return	80		0		0		80
Temporary difference		80		(60)		(20)	0
Taxable income (tax return)		$180		$ 40		$ 80	$300

In 2021, taxable income is greater than accounting income because subscription revenue is not reported in the income statement until 2022–2023.

Notice that this temporary difference produces *future deductible* amounts—amounts that are deducted from pretax accounting income to arrive at taxable income in future years. In 2021, taxable income is $80 thousand *more* than pretax accounting income because it includes the deferred subscription revenue not yet reported in the income statement. However, in 2022 and 2023 taxable income is *less* than accounting income because the subscription revenue is recognized and reported in the income statements but not on the tax returns of those two years.

In effect, tax laws require the company to prepay the income tax on this revenue, which is a sacrifice now but will benefit the company later when the revenue is recognized in the financial statements but not taxed. In the meantime, the company has an asset representing this future income tax benefit.

At the end of 2021, the amounts needed to record 2021 income tax expense would be determined as shown in Illustration 16–5A ($ in thousands). Tomorrow's 2021 tax payable is $45. Taxable income is greater than pretax accounting income in 2021, but the opposite will be true in 2022 and 2023 when the $60 and $20 of deferred revenue are included in the income statement but not taxable on the tax return. With future deductible amounts related to deferred revenue of $80, taxable at 25%, Tomorrow needs a **$20** deferred tax asset as of the end of 2021. Because no previous balance exists, Tomorrow debits the deferred tax asset for the entire **$20.** That amount reduces tax payable of $45 to give us income tax expense of $25.

Let's look at Illustration 16–5B to continue the example for 2022. Tomorrow's 2022 tax payable is $10. Because $60 of the original $80 temporary difference has reversed, only $20 remains. So, RDP should report a deferred tax asset of **$5** (= $20 × 25%) as of the end

Illustration 16–5A Revenue-Related Deferred Tax Asset—2021

($ in thousands)	Current Year 2021	Future Deductible Amounts 2022	2023	Total
Pretax accounting income	$100			
Temporary difference:				
Deferred revenue	80	$(60)	$(20)	$(80)
Taxable income (tax return)	$180			
Enacted tax rate	25%			25%
Tax payable	**$ 45**			**$(20)**

Deferred Tax Asset	
0	
20	
20	

Journal Entry at the End of 2021

Income tax expense (to balance)	25	
Deferred tax asset (determined above)	20	
Income tax payable (determined above)		45

Step 1: Tax payable: $45
Step 2 DTA end bal: $20
Step 3 DTA change: $20
Step 4 Tax exp plug: $25

Illustration 16–5B Revenue-Related Deferred Tax Asset—2022

($ in thousands)	Current Year 2022	Future Deductible Amounts 2023	Total
Pretax accounting income	$100		
Temporary difference:			
Deferred revenue	(60)	$(20)	$(20)
Taxable income (tax return)	$ 40		
Enacted tax rate	25%		25%
Tax payable	**$ 10**		**$ (5)**

Deferred Tax Asset	
20	
	15
5	

Journal Entry at the End of 2022

Income tax expense (to balance)	25	
Deferred tax asset (determined above)		15
Income tax payable (determined above)		10

Step 1: Tax payable: $10
Step 2 DTA end bal: $5
Step 3 DTA change: $(15)
Step 4 Tax exp plug: $25

of 2022, requiring that RDP credit the deferred tax asset for **$15** to reduce the $20 existing balance to that amount. This combines with the tax payable of **$10** to give us income tax expense of **$25**.

Here's another way to look at it. One reason taxable income is not a higher number in 2021 is that $60 of subscription revenue reported in the 2022 *income statement* was reported on the 2021 *tax return,* so the tax on that amount, $15 (= $60 × 25%) already has been paid. This $15 tax benefit in 2022 that was represented as part of the $20 deferred tax asset now has been realized, so $5 of the asset remains.

In Illustration 16–5C, we see what happens in the last year of our example.

Illustration 16–5C Revenue-Related Deferred Tax Asset—2023

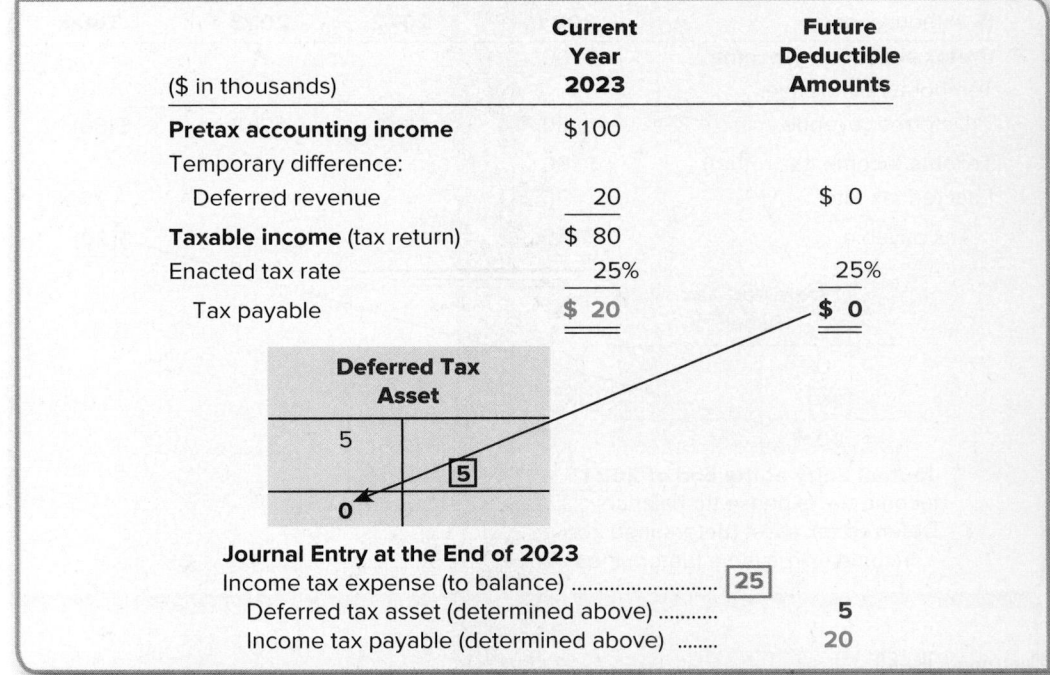

($ in thousands)	Current Year 2023	Future Deductible Amounts
Pretax accounting income	$100	
Temporary difference:		
Deferred revenue	20	$ 0
Taxable income (tax return)	$ 80	
Enacted tax rate	25%	25%
Tax payable	$ 20	$ 0

Step 1: Tax payable: $20
Step 2: DTA end bal: $0
Step 3: DTA change: $(5)
Step 4: Tax exp plug: $25

Deferred Tax Asset	
5	
	5
0	

Journal Entry at the End of 2023

Income tax expense (to balance)	25	
Deferred tax asset (determined above)		5
Income tax payable (determined above)		20

In 2023, the remaining deferred tax benefit is realized. Tomorrow's tax payable is $20. As of the end of 2023, no future deductible amount remains, so the balance in Tomorrow's deferred tax asset should be **$0**. So, Tomorrow should credit the deferred tax asset for **$5,** which combines with the tax payable of $20 to give us income tax expense of $25.

At the end of 2021 and 2022, the company reports a deferred tax asset for future income tax benefits. That deferred tax asset is reduced to zero by the end of 2023.

Deferred Tax Asset

($ in thousands)	Debit	Credit	
2021 ($80 × 25%)	20		
		15	2022 ($60 × 25%)
		5	2023 ($20 × 25%)
Balance after 3 years	0		

Of course, we instead could view this from a balance sheet perspective and calculate the deferred tax asset based on book-tax differences. The deferred revenue liability has a book value of $80 in 2021, and that liability reduces to $20 in 2022 and to $0 in 2023 when it is settled by providing the promised magazines. The tax basis of that liability remains at $0, because subscription receipts are included in taxable income in 2021 and no additional liability exists. The book-tax difference in deferred revenue represents less taxable income that the company will recognize in the future. This creates a deferred tax asset. The balance of the deferred tax asset each year is calculated as the book-tax difference times the applicable tax rate. Compare those balances (**step 2**) and changes in those balances (**step 3**) in the table below with those shown in Illustration 16–5A, 16–5B, and Illustration 16–5C to see that the balance sheet approach accomplishes the same steps in our 4-step process for calculating tax expense.

	Initial year 2021	2022		2023	
	12/31 balance	Goods provided	12/31 balance	Goods provided	12/31 balance
Deferred revenue:					
Accounting book value	$80	$(60)	$ 20	$(20)	$ 0
Tax basis	0		0		0
Temporary difference	$80		$ 20		$ 0
Tax rate	25%		25%		25%
Deferred tax asset (DTA)	**$20**	→	**$ 5**	→	**$ 0**
DTA change needed	**$20**		**$(15)**		**$(5)**

Valuation Allowance

● LO16–4

We recognize deferred tax assets for all temporary differences giving rise to future deductible amounts.[3] However, we then reduce a deferred tax asset by a valuation allowance if it is "more likely than not" that some portion or all of the deferred tax asset will not be realized.[4] Remember, a future deductible amount reduces taxable income in the future and saves taxes only if there is taxable income to be reduced when that deduction is available. So, a **valuation allowance** is needed if taxable income is anticipated to be insufficient to realize the tax benefit.

For example, let's say management has previously recorded a deferred tax asset of $8 million. However, due to declining income in some tax districts, management determines in 2021 that it's more likely than not that **$3** million of the deferred tax asset ultimately will not be realized in future years. The net deferred tax asset would be reduced by the creation of a valuation allowance as follows:

A valuation allowance is needed if it is more likely than not that some portion or all of a deferred tax asset will not be realized.

	($ in millions)	
Income tax expense (to balance)	3.0	
Valuation allowance		3.0

The effect is to increase income tax expense in the year the valuation allowance is established as a result of reduced expectations of future tax savings. In the 2021 balance sheet, the deferred tax asset would be reported at the net amount expected to reduce taxes in the future:

Deferred tax asset	$8
Less: Valuation allowance	(3)
	$5

This is not a new concept for you. You've reduced assets before using an allowance account. Suppose, for example, that you have accounts receivable of $8 million but expect that $3 million of that amount will not ultimately be collected from your customers. You would reduce the asset indirectly using an allowance:

Accounts receivable	$8
Less: Allowance for uncollectible accounts	(3)
	$5

[3] Unless the deductibility itself is uncertain. In that case, whether we recognize a deferred tax asset (and if so, its amount) is determined in accordance with FASB ASC 740–10–25: Income Taxes—Overall—Recognition, discussed in Part D of this chapter.

[4] "More likely than not" means a likelihood of more than 50%, FASB ASC 740–10–30: Income Taxes—Overall—Initial Measurement.

Additional Consideration

The decision as to whether a valuation allowance is needed should be based on the weight of all available evidence. The real question is whether or not there will be sufficient taxable income in future years for the anticipated tax benefit to be realized. After all, a deduction reduces taxes only if it reduces taxable income.

All evidence—both positive and negative—should be considered, and much managerial judgment is required. For instance, operating losses in recent years or anticipated circumstances that would adversely affect future operations would constitute negative evidence. On the other hand, a strong history of profitable operations or sizable, existing contracts would constitute positive evidence of sufficient taxable income to be able to realize the deferred tax asset.

We also must take into account any managerial actions that could be taken to recognize taxable income and thereby reduce or eliminate a valuation allowance. These tax-planning strategies include any prudent and feasible actions management might take to realize a tax benefit while it is available.

At the end of each reporting period, the valuation allowance is reevaluated. The appropriate balance is determined and the valuation allowance is adjusted—up or down—to create that balance. For instance, let's say that at the end of the following year, 2022, available evidence now indicates that only **$500,000** of the deferred tax asset ultimately will not be realized. We would adjust the valuation allowance to reflect the indicated amount:

	Valuation Allowance	
1/1/2022		3,000,000
	2,500,000	
12/31/2022		**500,000**

In the journal entry that adjusts the valuation allowance, income tax expense is the final plug. In this case, because the valuation allowance is reduced with a debit, income tax expense is reduced with a credit.

	($ in millions)	
Valuation allowance ($3 – $0.5)	**2.5**	
Income tax expense		2.5

Illustration 16–6 accompanied the January 28, 2017, annual report of **Sears Holdings Corporation**, which operates a variety of retailers including Sears, Kmart, and Lands' End. Sears disclosed that it had deferred tax assets of $5.6 billion, offset by a valuation allowance of $5.5 billion. Why such a large valuation allowance? Illustration 16–6 provides Sears's depressing explanation.

Illustration 16–6
Explanation for Valuation Allowance—Sears Holding Corporation

Real World Financials

Note 10 – Income Taxes (in part)

Management assesses the available positive and negative evidence to estimate if sufficient future taxable income will be generated to use the existing deferred tax assets. A significant piece of objective negative evidence evaluated was the cumulative loss incurred over the three-year periods ended January 28, 2017, January 30, 2016, and January 31, 2015. Such objective evidence limits the ability to consider other subjective evidence such as our projections for future income.

Source: Sears Holding Corporation

Given its large recent losses, Sears cannot argue convincingly that it's more likely than not that it will have sufficient future income to utilize its deferred tax assets, so it must record a large valuation allowance.

International Financial Reporting Standards

Valuation Allowances. Under U.S. GAAP, companies recognize deferred tax assets and then reduce those assets with an offsetting valuation allowance if it is not "more likely than not" that the asset will be realized. In contrast, under IFRS, deferred tax assets only are recognized to begin with if it is probable (defined as "more likely than not") that they will be realized. That means that we could see more deferred tax assets and offsetting valuation allowances under U.S. GAAP than how the same company would appear under IFRS.

Disclosures Linking Tax Expense with Changes in Deferred Tax Assets and Liabilities

We've now looked at four examples that illustrate revenue-related and expense-related deferred tax assets and deferred tax liabilities. In each example, the journal entry to recognize income tax expense included three items: income tax payable, the change in the deferred tax asset or liability, and (to balance) the income tax expense. We've also seen that changes in the valuation allowance sometimes needed to adjust deferred tax assets also can affect the tax expense. Companies provide disclosure in the notes that help investors see these relationships. As an example, let's look at the tax note that appeared in the January 28, 2017, annual report of **Shoe Carnival, Inc.**, a large footwear retailer.

Illustration 16–7A
Disclosure of Tax Expense—Shoe Carnival, Inc.

Real World Financials

Note 7 – Income Taxes (in part)
The provision for income taxes consisted of:

($ in thousands)	**2016**	**2015**	**2014**
Current:			
Federal	$ 13,366	$ 18,366	$14,575
State	1,997	2,267	1,800
Puerto Rico	250	249	350
Total current	15,613	20,882	16,725
Deferred:			
Federal	(153)	(3,000)	(1,229)
State	(1,228)	(145)	(115)
Puerto Rico	(1,456)	(318)	(1,149)
Total deferred	**(2,837)**	(3,463)	(2,493)
Valuation allowance	**1,456**	318	1,943
Total provision	$14,232	$17,737	$16,175

Source: Shoe Carnival, Inc.

Illustration 16–7A gives us enough information to reproduce a journal entry that summarizes Shoe Carnival's 2016 tax expense. We just need to understand that "total current" in the note refers to current tax expense, and therefore to tax payable, "total deferred" refers to the deferred portion of tax expense, which is equal to the net change in deferred tax assets and liabilities, "valuation allowance" refers to the change in the valuation allowance, and "total provision" refers to income tax expense. Here's the journal entry ($ in thousands):

	Debit	Credit	
Income tax expense (to balance)	14,232		
Deferred tax assets and liabilities	**2,837**		} $2,837 – 1,456 = **$1,381**
Valuation allowance		**1,456**	
Income tax payable		15,613	

We can see from this journal entry that Shoe Carnival's operations caused deferred tax assets and liabilities to change by a net debit of **$2,837** during 2016, and its valuation allowance changed by an offsetting credit of **$1,456**, so the total change in these accounts was $2,837 – $1,456 = **$1,381**. Let's verify that change by comparing the 2016 fiscal year end (January 28, 2017) and 2015 fiscal year end (January 30, 2016) balances of Shoe Carnival's deferred tax assets, valuation allowance and deferred tax liabilities, shown in Illustration 16–7B.

Illustration 16–7B

Disclosure of Deferred Tax Assets and Liabilities—Shoe Carnival, Inc.

Real World Financials

Note 7 – Income Taxes (in part)

Deferred income taxes are the result of temporary differences in the recognition of revenue and expense for tax and financial reporting purposes. The sources of these differences and the tax effect of each are as follows:

($ in thousands)	January 28, 2017		January 30, 2016	
Deferred tax assets:				
Accrued rent	$ 4,333		$ 4,321	
Accrued compensation	8,552		6,911	
Accrued employee benefits	555		532	
Inventory	1,125		737	
Self-insurance reserves	758		641	
Lease incentives	11,996		12,522	
Net operating loss carry forward	3,719		2,261	
Other	638		411	
Total deferred tax assets	31,676		28,336	
Valuation allowance	(3,717)		(2,261)	
Total deferred tax assets—net of valuation allowance	27,959		26,075	
Deferred tax liabilities:				
Depreciation	17,256		16,671	
Capitalized costs	1,103		1,153	
Other	0		32	
Total deferred tax liabilities	18,359		17,856	
Net deferred tax asset	**$ 9,600**	–	**$ 8,219**	**= $1,381**

Source: Shoe Carnival, Inc.

We can see from Illustration 16–7B that Shoe Carnival has a net deferred tax asset of **$9,600** as of January 28, 2017 (the end of its 2016 fiscal year), which is equal to its total deferred tax asset of $31,676 less its valuation allowance of $3,717 and its total deferred tax liability of $18,359. Compared to the net deferred tax asset of **$8,219** at January 30, 2016, we see that the net deferred tax asset has increased by **$9,600** – **$8,219** = **$1,381**. Now look back at our journal entry following Illustration 16–7A. Remember, that entry showed an increase in net deferred tax assets (**$2,837**) with an offsetting increase in the valuation allowance (**$1,456**), again totaling **$1,381**. See how the tax expense journal entry ties to the changes in deferred tax assets, liabilities, and valuation allowance?

Additional Consideration

The amount of change in deferred taxes reconciles perfectly between the tables shown in Illustrations 16–7A and 16–7B. That often won't be the case in practice because the table shown in Illustration 16–7A focuses on taxes from *continuing operations,* but the deferred tax assets and liabilities listed in Illustration 16–7B relate to all aspects of the company. For example, if Shoe Carnival had deferred taxes associated with discontinued operations or other comprehensive income items, or if it had added deferred tax assets and liabilities from an acquisition during the year, the change in deferred taxes from continuing operations wouldn't capture everything that affected deferred tax assets and liabilities.

Permanent Differences

PART B

● LO16–5

So far, we've dealt with temporary differences between the reported amount of an asset or liability in the financial statements and its tax basis. However, some differences aren't temporary. Rather, they are caused by transactions and events that under existing tax law will *never* affect taxable income or taxes payable. For example, interest received from investments in governmental bonds issued by state and municipal governments typically is permanently exempt from taxation. Interest revenue of this type is, of course, reported as revenue on the recipient's income statement, but not on its tax return—not now, not later. So, there is a **permanent difference** between pretax accounting income and taxable income. This situation will *not* reverse in a later year—the tax-free income will *never* be reported on the tax return.

FINANCIAL Reporting Case

Q3, p. 909

Illustration 16–8 provides examples of permanent differences that commonly occur in practice.

- Interest received from investments in bonds issued by state and municipal governments (generally not taxable).
- Investment expenses incurred to obtain tax-exempt income (not tax deductible).
- Life insurance proceeds on the death of an insured executive (not taxable).
- Premiums paid for life insurance policies when the payer is the beneficiary (not tax deductible).
- Compensation expense pertaining to some employee stock option plans (not tax deductible).
- Fines and penalties due to violations of the law (generally not tax deductible).
- Difference in tax paid on foreign income permanently reinvested in the foreign country and the amount that would have been paid if taxed at U.S. rates.[&]
- Portion of dividends received from U.S. corporations that is not taxable due to the dividends received deduction.[*]
- Tax deduction for depletion of natural resources (percentage depletion) that is allowed in excess of an already full-depleted asset's cost.[†]

[&]These differences were largely eliminated by tax reform instituted by the U.S. Congress in late 2017.
[*]When a corporation owns shares of another U.S. corporation, a percentage of the dividends from those shares is exempt from taxation due to the dividends received deduction. The percentage is 50% if the investor owns less than 20% of the investee's shares, 65% for over 20% ownership, and 100% for dividends from members of the same affiliated group.
[†]The cost of natural resources is reported as depletion expense over their extraction period for financial reporting purposes, but tax rules prescribe sometimes different percentages of cost to be deducted for tax purposes. There usually is a difference between the cost depletion and percentage depletion that doesn't eventually reverse.

Illustration 16–8
Differences without Deferred Tax Consequences

Provisions of the tax laws, in some instances, dictate that the amount of a revenue that is taxable or expense that is deductible permanently differs from the amount reported in the income statement.

Accounting for permanent differences is less complex than is accounting for temporary differences. We calculate taxes payable according to the tax law, and, since permanent differences are not temporary, we don't create a deferred tax asset or liability. Therefore, when we "plug" tax expense, tax expense is determined by tax payable. The term *permanent difference* doesn't refer to a difference between tax payable and tax expense—those are the *same* with respect to these. Rather, it refers to a difference between taxable income and pretax accounting income. That's why we adjust pretax accounting income in the illustrations that follow to eliminate permanent differences when calculating tax payable.

To compare temporary and permanent differences, we can modify Illustration 16–3 to include nontaxable income in Kent Land Management's 2022 pretax accounting income. We do this in Illustration 16–9.

Kent Land Management reported pretax accounting income in 2021 of $185 million, which included $80 million from installment sales of property **and $5 million interest from investments in municipal bonds in 2021.** The installment sales income is reported for tax purposes in 2022 ($20 million) and 2023 ($60 million). The enacted tax rate is 25% each year.

(Continued)

Illustration 16–9
Temporary and Permanent Differences

Illustration 16–9
(Concluded)

Step 1: Tax payable: $25
Step 2: DTL end bal: $20
Step 3: DTL change: $20
Step 4: Tax exp plug: $45

($ in millions)	Current Year 2021	Future Taxable Amounts 2022	2023	Total
Pretax accounting income	$185			
Permanent difference:				
Municipal bond interest	**(5)**			
Temporary difference:				
Installment income	(80)	$20	$60	$80
Taxable income (tax return)	$100			
Enacted tax rate	25%			25%
Tax payable	**$ 25**			**$20**

Deferred Tax Liability

	0
	20
	20

Journal Entry at the End of 2021

Income tax expense (to balance)	45	
Income tax payable (determined above)		25
Deferred tax liability (determined above)		20

The *effective tax rate* equals tax expense divided by pretax accounting income.

A company's **effective tax rate** is calculated by dividing the company's tax expense by its pretax accounting income. Permanent differences affect the effective tax rate because they affect pretax accounting income. Let's look at Illustration 16–9 to understand why that's the case. In Illustration 16–9, the effective tax rate is $45 million ÷ $185 million, or 24.3%. That's with **$5** million of municipal bond interest included in the $185 million of pretax accounting income. If, instead, pretax accounting income hadn't included the **$5** million of municipal bond interest, it would have been only $180 million, and the effective tax rate would have been $45 million ÷ $180 million, or 25%, equal to the statutory tax rate. Including the municipal bond interest in pretax accounting income increased the denominator of the effective tax rate, so it produced a lower effective tax rate.

Permanent differences affect a company's effective tax rate.

More generally, nontaxable revenues and gains like municipal bond interest are permanent differences that result in *higher* pretax accounting income, so they produce effective tax rates that are *lower* than the statutory tax rate. Likewise, nondeductible expenses and losses are permanent differences that result in *lower* pretax accounting income, so they produce effective tax rates that are *higher* than the statutory tax rate. Companies are required to report a reconciliation between their effective and statutory tax rates in disclosure notes, as shown in Illustration 16–10's example from **Shoe Carnival's** financial statements for the 2016 fiscal year (ended January 28, 2017).

Illustration 16–10
Effective Tax Rate—Shoe Carnival

Real World Financials

Note 9: Taxes (in part)
Effective Tax Rate Reconciliation

	2016	**2015**	**2014**
U.S. statutory tax rate	35.0%	35.0%	35.0%
State and local income taxes, net of federal tax benefit	2.1	2.7	3.1
Puerto Rico	0.2	0.3	0.2
Valuation allowance	4.0	0.7	4.7
Tax benefit of foreign losses	(3.6)	(0.6)	(4.3)
Other	0.0	0.0	0.1
Effective income tax rate	37.7%	38.1%	38.8%

Source: Shoe Carnival

Two points are important to note about Shoe Carnival's effective tax rate reconciliation. First, the federal statutory rate is 35% in this example—it did not drop to 21% until 2018. Second, the effect of Shoe Carnival increasing its valuation allowance during the year was to increase the effective tax rate. To better understand this, look back at Illustration 16–4A and 16–5A and note that establishing a deferred tax asset decreases tax expense so that the effective tax rate is closer to to the statutory rate (in those examples, 25%). Establishing a valuation allowance increases tax expense, thus increasing the effective tax rate. The valuation allowance essentially counteracts the beneficial effect of the deferred tax asset on tax expense, so when the valuation allowance increases, the effective tax rate increases.

Additional Consideration

Accounting for taxes on unrepatriated foreign earnings. Historically, the largest "permanent difference" in many companies' effective tax rate reconciliation related to taxes on foreign earnings. Companies often seek to minimize their tax bills by arranging their operations so their income is recognized outside the United States in jurisdictions that have low tax rates. Those lower tax bills are viewed as creating permanent differences so long as the company does not intend to "repatriate" the foreign earnings by transferring those earnings back to the United States. A company still must include the foreign income in pretax accounting income in the income statement, but that income is taxed at the lower foreign rate, so income tax payable is lower, tax expense is lower, and the company's effective tax rate is lower. As an example, **Coca-Cola**'s 2016 10-K indicated that the effect of such "earnings in jurisdictions taxed at rates different from the statutory federal rate" was to cut its effective tax rate in half, reducing the federal statutory rate of 35% to 17.5%. The 2017 "Tax Cuts and Jobs Act" passed by the U.S. Congress increased taxes on such foreign earnings, but some of these permanent differences still are likely to appear in effective tax rate reconciliations.

What if management changes its mind and decides to repatriate foreign earnings? In that case, the company must pay tax at a higher rate. For example, the 2017 tax act imposed a one-time "deemed repatriation tax" on accumulated, untaxed earnings of foreign corporations that was equal to 15.5% on earnings held as "cash and cash equivalents" and 8% on other earnings. These taxes can be significant—**Goldman Sachs Group, Inc.**, announced on December 27, 2017, that it would take a $5 billion charge associated with the tax act and attributed two-thirds of that amount to the repatriation tax. As a result, the company has a higher tax bill and higher tax expense. Pretax accounting income is unaffected (remember, that income appeared in the income statement in the period in which revenue was recognized), so the higher tax expense results in a higher effective tax rate. The company's original choice not to repatriate created a "permanent" difference that reduced the effective tax rate, and the company's later payment of this repatriation tax creates an offsetting "permanent" difference that goes in the other direction.

International Financial Reporting Standards

Non-Tax Differences Affect Taxes. Despite the similar approaches for accounting for taxation under *IAS No. 12*, "Income Tax," and U.S. GAAP, differences in reported amounts for deferred taxes are among the most frequent between the two reporting approaches.[5] The reason is that a great many of the nontax differences between IFRS and U.S. GAAP affect deferred taxes as well.

LO16–11

For example, we noted in Chapter 13 that we accrue a loss contingency under U.S. GAAP if it's both probable and can be reasonably estimated and that IFRS guidelines are similar, but the threshold is "more likely than not." This is a lower threshold than "probable." In this chapter, we noted that accruing a loss contingency (like warranty expense) in the income statement leads to a deferred tax asset if it can't be deducted on the tax return until a later period. As a result, under the lower threshold of IFRS, we might record a loss contingency and thus a deferred tax asset, but under U.S. GAAP we might record neither. So, even though accounting for deferred taxes is the same, accounting for loss contingencies is different, causing a difference in the reported amounts of deferred taxes under IFRS and U.S. GAAP.

[5] "Income Taxes," *International Accounting Standard No. 12* (IASCF), as amended effective January 1, 2014.

Concept Review Exercise

TEMPORARY AND PERMANENT DIFFERENCES

Mid-South Cellular Systems began operations in 2021. That year the company reported pretax accounting income of $150 million, which included the following amounts:

1. Compensation expense of $6 million related to employee stock option plans granted to organizers was reported in the 2021 income statement. This expense is not deductible for tax purposes.
2. An asset with a four-year useful life was acquired at the beginning of 2021. It is depreciated by the straight-line method in the income statement. For this asset, Mid-South uses MACRS on the tax return, causing deductions for depreciation to be more than straight-line depreciation the first two years but less than straight-line depreciation the next two years ($ in millions):

	Depreciation		
	Income Statement	**Tax Return**	**Difference**
2021	$150	$198	$ (48)
2022	150	264	(114)
2023	150	90	60
2024	150	48	102
	$600	$600	$ 0

The enacted tax rate is 25%.

Required:

Prepare the journal entry to record Mid-South Cellular's income taxes for 2021.

Solution

Because the compensation expense is not tax deductible, taxable income is not reduced by the $6 million deduction and is higher than accounting income by that amount.

($ in millions)	Current Year 2021	Future Taxable Amounts 2022	2023	2024	Total
Pretax accounting income	$150				
Permanent difference:					
Nondeductible compensation	**6**				
Temporary difference:					
Depreciation	(48)	$(114)	$60	$102	$48
Taxable income (tax return)	$108				
Enacted tax rate	25%				25%
Tax payable	$ 27				$12

Step 1: Tax payable: $27
Step 2: DTL end bal: $12
Step 3: DTL change: $12
Step 4: Tax exp plug: $39

Deferred Tax Liability

	0
	12
	12

Journal Entry at the End of 2021

Income tax expense (to balance)	39	
Income tax payable (determined above)		27
Deferred tax liability (determined above)		12

The necessary journal entry is:

Journal Entry at the End of 2021

Income tax expense (to balance)	39	
Income tax payable (determined above)		27
Deferred tax liability (determined above)		12

Note that tax expense of $39 equals 25% × $156. Because the compensation expense of $6 will never be deductible (its a permanent difference), tax expense is calculated as if the compensation expense had never occurred. On the other hand, the differences associated with depreciation are temporary, so a deferred tax liability is established to account for those effects.

PART C

Other Tax Accounting Issues

Tax Rate Considerations

LO16–6

To measure a deferred tax liability or asset, we multiply a temporary difference by the currently *enacted* tax rate that will be effective in the year(s) the temporary difference reverses.[6] A conceptual case can be made that these measurements should be based on the tax rates that are *expected* to apply in future periods, regardless of whether those rates have yet been enacted. However, this is one of many examples of the frequent trade-off between relevance and reliability. In this case, the FASB chose to favor reliability by waiting until an anticipated change actually is enacted into law before recognizing its tax consequences.

When Enacted Tax Rates Differ Between Years

A deferred tax liability (or asset) is calculated using enacted tax rates and laws.

Existing tax laws may call for enacted tax rates to differ in the future years in which a temporary difference is expected to reverse. When a phased-in change in rates is scheduled to occur, the specific tax rates of each future year are multiplied by the amounts reversing in each of those years. The total is the deferred tax liability or asset.

To illustrate, let's again modify our Kent Land Management illustration, this time to assume a scheduled change in tax rates. See Illustration 16–11.

Because the 2023 rate is higher (30% as opposed to 25%), the future taxable amount will generate a higher amount of tax ($18 million, rather than the $15 million that would be generated if the tax rate were 25% in 2023). That requires a larger increase in the deferred tax liability, and a higher corresponding tax expense. Be sure to note that, when the deferred tax liability of $23 million is established in 2021, the 2022 rate (25%) and the 2023 rate (30%) already have been enacted into law. In the next section, we discuss how to handle a change resulting from new legislation.

Changes in Enacted Tax Laws or Rates

FINANCIAL Reporting Case

Q2, p. 909

When an enacted tax rate changes, the deferred tax liability or asset should be adjusted and the effect shown in tax expense in the year the change is enacted.

Tax laws sometimes change. When such a change in a tax law or rate is enacted, any existing deferred tax liability or asset must be adjusted to reflect the effects of the change. Remember, a deferred tax liability or asset is meant to reflect the amount to be paid or recovered in the future. When legislation changes that amount, the deferred tax liability or asset also should change. The effect is reflected in operating income in the year the change in the tax law or rate is enacted.[7]

For clarification, let's consider what happened in late December of 2017. Congress revised the federal tax rate, dropping it from 35% to 21% starting in 2018. There was no change in tax *payable* for companies in 2017 because that amount was calculated using the

[6] The current U.S. federal corporate tax rate is 21% (having been revised downward from the 35% rate that applied to 2017). Most states tax corporate income at rates less than 10%. We use 25% as a combined rate in most of our illustrations to simplify calculations.
[7] FASB ASC 740–10–35: Income Taxes—Overall—Subsequent Measurement.

Illustration 16–11 Scheduled Change in Tax Rates

Kent Land Management reported pretax accounting income in 2021 of $185 million, which included $80 million from installment sales of property **and $5 million interest from investments in municipal bonds in 2021.** The installment sales income is reported for tax purposes in 2022 ($20 million) and 2023 ($60 million). The enacted tax rate is 25% in 2021 and 2022, but increases to 30% in 2023.

Step 1: Tax payable: $25
Step 2: DTL end bal: $23
Step 3: DTL change: $23
Step 4: Tax exp plug: $48

	Current Year	Future Taxable Amounts		
($ in millions)	**2021**	**2022**	**2023**	**Total**
Pretax accounting income	$185			
Permanent difference:				
Municipal bond interest	**(5)**			
Temporary difference:				
Installment income	(80)	$20	$60	
Taxable income (tax return)	$100			
Enacted tax rate	25%	25%	30%	
Tax payable	$ 25	$ 5 +	$18 =	**$23**
Deferred tax liability				

Deferred Tax Liability

	0
	23
	23

Journal Entry at the End of 2021

Income tax expense (to balance)	48	
Income tax payable (determined above)		25
Deferred tax liability (determined above)		23

old 35% rate. However, the change did affect the value of companies' deferred tax assets and liabilities, and adjusting those accounts in turn affected each company's 2017 tax expense *for the entire year* and, in some cases quite substantially, net income. To see this more clearly, let's walk through a couple of examples.

EFFECT OF A TAX RATE CHANGE ON A DEFERRED TAX LIABILITY Imagine that, late in 2017, GladCo has future taxable amounts of $1 million and a tax rate (federal plus state) of 39%, so has already recognized a deferred tax liability of $390,000 (= $1 million × 39%). GladCo then learns that Congress has enacted a law decreasing the tax rate for future years from 35% to 21%, meaning that GladCo's total enacted future tax rate has dropped from 39% to 25%. GladCo revises its deferred tax liability downwards on December 31, 2017, as follows:

Deferred Tax Liability

	Debit	Credit	
		390,000	($1 million × 39%)
Plug to revise DTL for decrease in tax rate	140,000		
		250,000	($1 million × 25%)

To recognize the effect of the change in future tax rates, GladCo debits its deferred tax liability for **$140,000**. There is no effect on 2017 tax payable, so the plug to tax expense is a credit of **$140,000**.

December 31, 2017

Deferred tax liability	140,000	
Income tax expense		140,000

GladCo is particularly glad about the decrease in future tax rates, because GladCo's future taxable amount of $1 million will cost it less in taxes under a 25% rate than under a 39% rate. In 2017, the period in which the future tax rate change is enacted, GladCo revises its estimate of future tax payments and consequently recognizes a reduction in tax expense (i.e., a tax benefit) that improves its 2017 net income. If, instead, future tax rates had been revised upward, GladCo would need to credit its deferred tax liability to revise it upward, and the offsetting debit would increase tax expense and reduce 2017 net income.

EFFECT OF A TAX RATE CHANGE ON A DEFERRED TAX ASSET Now let's modify our example to consider SadCo, which has future *deductible* amounts of $1 million and a tax rate (federal plus state) of 39%. SadCo already has recognized a deferred tax *asset* of $390,000 (= $1 million × 39%). Late in 2017, SadCo learns that Congress has enacted a law decreasing the future tax rate from 35% to 21%, such that SadCo's total enacted future tax rate has dropped from 39% to 25%. SadCo revises its deferred tax asset on December 31, 2017, as follows:

FINANCIAL Reporting Case

Q2, p. 909

Deferred Tax Liability

($1 million × 39%)	390,000		
		140,000	Plug to revise DTA for decrease in tax rate
($1 million × 25%)	**250,000**		

To recognize the effect of the change in future tax rates, SadCo credits the deferred tax asset for **$140,000** in the year the change is enacted. There is no effect on 2017 tax payable, so the plug to tax expense is a debit of **$140,000**.

December 31, 2017		
Income tax expense	140,000	
Deferred tax asset		140,000

While SadCo is generally happy that tax rates will decease in future years, it is sad with respect to the effect of that tax rate change on the value of its deferred tax assets, and therefore its 2017 net income. SadCo's future deductible amount of $1 million will reduce future tax payments by less with a 25% rate than with a 39% rate. So, that asset is less valuable, and SadCo must reduce the carrying value of its deferred tax asset. The offsetting increase in tax expense reduces SadCo's net income in 2017, the period in which it revises its deferred tax asset. If, instead, future tax rates had been revised upward, SadCo would debit its deferred tax asset to revise it upward, and the offsetting credit would decrease tax expense and increase 2017 net income.

The effect of a tax rate change on earnings can be very significant, given that the entire effect on the balances of deferred tax assets and deferred tax liabilities is included in earnings in the period the change in enacted. Whether the change increases or decreases earnings depends on the direction of the change as well as on whether a company has a net deferred tax asset or deferred tax liability. For example, the decrease in federal tax rates enacted in late 2017 required **Citigroup** to *decrease* earnings by $22.6 billion in the fourth quarter of 2017 as it reduced the carrying value of its large net deferred tax asset. On the other hand, the same decrease in tax rates *increased* **Verizon**'s earnings by $16.8 billion in the fourth quarter of 2017 as it reduced the carrying value of its large net deferred tax liability.

Multiple Temporary Differences

It would be unusual for any but a very small company to have only a single temporary difference in any given year. Having multiple temporary differences, though, doesn't change any of the principles you've learned so far in connection with single differences. We categorize all temporary differences according to whether they create (a) future taxable amounts or (b) future deductible amounts. The total of the future taxable amounts then is multiplied by the future tax rate to determine the appropriate balance for the deferred tax liability, and the total of the future deductible amounts is multiplied by the future tax rate to determine the appropriate balance for the deferred tax asset. This is demonstrated in Illustration 16–12.

Illustration 16–12
Multiple Temporary Differences

2021

During 2021, its first year of operations, Eli-Wallace Distributors reported pretax accounting income of $200 million, which included the following amounts:

1. Income (net) from installment sales of warehouses in 2021 of $8 million will be reported for tax purposes in 2022 ($6 million) and 2023 ($2 million).
2. Depreciation is reported by the straight-line method on an asset with a four-year useful life. On the tax return, deductions for depreciation will be more than straight-line depreciation the first two years but less than straight-line depreciation the next two years ($ in millions):

	Income Statement	Tax Return	Difference
2021	$ 50	$ 66	$(16)
2022	50	88	(38)
2023	50	30	20
2024	50	16	34
	$200	$200	$ 0

3. Estimated warranty expense will be deductible on the tax return when actually paid during the next two years. Estimated deductions are as follows ($ in millions):

	Income Statement	Tax Return	Difference
2021	$12		$12
2022		$ 4	(4)
2023		8	(8)
	$12	$12	$ 0

2022

During 2022, pretax accounting income of $200 million included an estimated loss of $2 million from having accrued a loss contingency. The loss is expected to be paid in 2024, at which time it will be tax deductible.

The enacted tax rate is 25% each year.

Look at Illustration 16–12A to see how Eli-Wallace determines the income tax amounts for 2021. Then look at Illustration 16–12B to see how those amounts are determined for 2022.

Illustration 16–12A: Deferred taxes with multiple differences—Initial Year

($ in millions)	Current Year 2021	Future Taxable (Deductible) Amounts 2022	2023	2024	Future Taxable Amounts (total)	Future Deductible Amounts (total)
Pretax accounting income	$200					
Temporary difference:						
Installment sales	(8)	$ 6	$ 2		$ 8	
Depreciation	(16)	(38)	20	$34	16	
Warranty expense	12	(4)	(8)			$ (12)
Taxable income (tax return)	$188				24	(12)
Enacted tax rate	25%				25%	25%
Tax payable	$ 47				$6	$ (3)

Step 1: Tax payable: $47
Step 2: DTA end bal: $3
DTL end bal: $6
Step 3: DTA change: $3
DTL change: $6
Step 4: Tax exp plug: $50

Deferred Tax Liability

	0
	6
	6

Deferred Tax Asset

0	
3	
3	

Journal Entry at the End of 2021

Income tax expense (to balance)	50	
Deferred tax asset (determined above)	3	
Income tax payable (determined above)		47
Deferred tax liability (determined above)		6

Illustration 16–12B Deferred taxes with multiple differences—Future year

($ in millions)	Current Year 2022	Future Taxable (Deductible) Amounts 2023	2024	Future Taxable Amounts (total)	Future Deductible Amounts (total)
Pretax accounting income	$200				
Temporary difference:					
Installment sales	$ 6	$ 2		$ 2	
Depreciation	(38)	20	$34	54	
Warranty expense	(4)	(8)			$ (8)
Estimated loss	2		(2)		(2)
Taxable income (tax return)	$166			$56	$(10)
Enacted tax rate	25%			25%	25%
Tax payable	**$ 41.5**			**$14**	**$ (2.5)**

Deferred Tax Liability	
	6
	8
	14

Deferred Tax Asset	
3	
	0.5
2.5	

Journal Entry at the End of 2021

Income tax expense (to balance)	50	
Deferred tax asset (determined above)		0.5
Deferred tax liability (determined above)		8.0
Income tax payable (determined above)		41.5

Step 1: Tax payable: $41.5
Step 2: DTA end bal: $2.5
DTL end bal: $14.0
Step 3: DTA change: $(0.5)
DTL change: $8.0
Step 4: Tax exp plug: $50.0

Of course, if a phased-in change in tax rates is scheduled to occur, it would be necessary to determine the total of the future taxable amounts and the total of the future deductible amounts for each future year, as outlined previously. Then the specific tax rates of each future year would be multiplied by the two totals in each of those years. Those annual tax effects then would be summed to find the balances for the deferred tax liability and the deferred tax asset.

Net Operating Losses

● LO16–7

A **net operating loss** (NOL) is negative taxable income: tax-deductible expenses exceed taxable revenues. Of course, there is no tax payable for the year a net operating loss occurs because there's no taxable income. In addition, tax laws permit a net operating loss to be used to reduce taxable income in subsequent profitable years. Why do the tax laws permit that offsetting? Well, let's consider two imaginary companies. Volatile Co. has negative income some years and positive income other years that averages out to $0 income over time. Stable Co. has $0 income each year. It wouldn't be fair to tax Volatile in the good years and provide no relief in the bad years, while not taxing Stable at all, because Volatile and Stable generate the same total amount of income over time. It's more fair to allow Volatile to offset the income in its good years with the losses in its bad years when determining how much tax it should pay.

The accounting question is: When should the tax benefit created by a net operating loss be recognized in the income statement? The answer is: In the year the loss occurs.

Net Operating Loss Carryforward

Current federal tax laws allow most companies to carry forward an NOL and offset it against taxable income in future years. Companies are limited to offsetting a maximum of 80% of the taxable income in any given year. If an NOL is big enough, it could be used in multiple future years to offset taxable income and reduce tax payable. NOLs don't expire. Rather, companies can carry forward an NOL indefinitely until it is used.

Additional Consideration

NOLs arising in tax years beginning before December 31, 2017 are treated differently. Those NOLs can offset 100% of the taxable income in each future year to which they are carried forward, but can only be carried forward 20 years before expiring.

Because NOL carryforwards offset future taxable income and therefore reduce tax payable, they produce cash savings for the company when future taxable income is generated. Large NOL carryforwards also can make an unprofitable company an attractive target for acquisition by a company that could use those NOL carryforwards to shelter its own future earnings from taxes. If the IRS determines that an acquisition is made solely to obtain the tax benefits of NOL carryforwards, the deductions will not be allowed. Still, the motivation for making an acquisition is difficult to determine, so it is not uncommon for companies to purchase other companies to obtain their NOL carryforwards.

FINANCIAL Reporting Case

Q3, p. 909

How do we account for an NOL prior to it being used to offset future taxable income? You have learned in this chapter that a deferred tax asset is recognized for the future tax benefit of temporary differences that create future deductible amounts. A **net operating loss carryforward** also creates future deductible amounts. Logically, then, a deferred tax asset is recognized for an NOL carryforward. Debiting the deferred tax asset associated with the NOL carryforward requires an offsetting credit to tax expense. This is demonstrated in Illustration 16–13A.

Illustration 16–13A Net Operating Loss Carryforward

During 2021, its first year of operations, American Laminating Corporation reported an operating loss of $120 million for financial reporting and tax purposes. The enacted tax rate is 25%.

($ in millions)	Current Year 2021	Future Deductible Amounts
Net Operating loss	$(120)	
NOL carryforward	120	$(120)
Taxable income (tax return)	$ 0	
Enacted tax rate	25%	25%
Tax payable	$ 0	**$30**

Step 1: Tax payable: $0
Step 2: DTA end bal: $30
Step 3: DTA change: $30
Step 4: Tax exp plug: $30

Deferred Tax Asset	
0	
30	
30	

Journal Entry at the End of 2021

Deferred tax asset (determined above)	30	
Income tax expense (to balance)		30

RECOGNIZING AN NOL CARRYFORWARD You can see that the reduction of tax expense associated with the NOL is recognized for accounting purposes in the year the net operating loss occurs. That reduction in tax expense is sometimes labeled a tax benefit in the income statement. Just as we reduce pretax *income* by tax *expense* to calculate net income, we reduce a pretax *loss* by its tax *benefit* to calculate a net loss. That way, the net after-tax operating loss shown on the income statement reflects the tax savings that the operating loss is expected to create.

Income Statement (partial)

	($ in millions)
Operating loss before income taxes	$(120)
Less: Income tax benefit	30
Net loss	$ (90)

UTILIZING AN NOL CARRYFORWARD In a future year, a company can reduce its tax payable by offsetting taxable income with an NOL carried forward from a prior year. Using an NOL is accounted for essentially the same way as using the deductible amounts associated with any deferred tax asset. Let's return to our example to see how this is done. American Laminating's loss in 2021 created a $120 million NOL carryforward. As demonstrated in Illustration 16–13B, American Laminating will be able to use $80 million of the NOL in 2022, and will carry forward the remaining $40 million to subsequent years.

Illustration 16–13B Utilizing an NOL Carryforward

During 2022, its second year of operations, American Laminating Corporation reported pretax income of $100 million for financial reporting and tax purposes. American is limited to offsetting 80% of income with its NOL carryforward in any tax year. The enacted tax rate is 25%.

($ in millions)	Current Year 2022	Future Deductible Amounts
Pretax accounting and taxable income	$(100)	
NOL carryforward (using 80% × $100)	80	$(40)
Taxable income (tax return)	$ 20	
Enacted tax rate	25%	25%
Tax payable	$ 5	$ 10

Deferred Tax Asset	
30	
	20
10	

Journal Entry at the End of 2022

Income tax expense (to balance)	25	
Deferred tax asset (determined above)		20
Income tax payable (determined above)		5

Step 1: Tax payable: $5
Step 2: DTA end bal: $10
Step 3: DTA change: $20
Step 4: Tax exp plug: $25

American Laminating's income statement would show $100 million of pretax accounting income reduced by $25 million of tax expense:

Income Statement (partial)

	($ in millions)
Pretax accounting income	$100
Less: Income tax expense	25
Net income	$ 75

Of the $25 million of tax expense, $5 million is payable currently, and the other **$20** million is a result of reducing (using up) the deferred tax asset associated with the NOL. An additional $40 million of NOL remains to reduce future taxable income, to be represented by a deferred tax asset of **$10** million (= $40 million × 25%) as of the end of 2022.

VALUATION ALLOWANCE Just as for all other deductible temporary differences, a deferred tax asset is recognized for an NOL without regard to the likelihood of having taxable income in future years sufficient to absorb future deductible amounts. However, the deferred tax asset is then reduced by a valuation allowance if it is "more likely than not" that some of the deferred tax asset will not be realized. Even though current tax law allows NOLs to be carried forward indefinitely, a valuation allowance still might be necessary because the same problems that give rise to current operating losses also make it less likely the company will be able to stay in business.

Additional Consideration

Net Operating Loss Carryback. For tax years beginning before January 1, 2018, companies could elect to carry an NOL *back* to the past two years and carry forward any amount of NOL that remained. Carrybacks produced an immediate refund of taxes paid in those prior years, so this option was very attractive. Carryback was limited to two years, and carryforward up to 20 years. The NOL could be offset against 100% of taxable income in those years.

For tax years beginning after December 31, 2017, NOL carrybacks are not allowed for most companies. Rather, NOLs can only be carried forward, and the use of an NOL carryforward in any individual tax year is limited to offsetting a maximum of 80% of taxable income in that year. However, the old rules still apply for property and casualty insurance companies, and some farm-related businesses are allowed to carry NOLs back two years and forward indefinitely. For these companies, there is no 80% limitation. Also, many states continue to allow NOL carrybacks with respect to calculating state income taxes. Therefore, it is useful to understand how to account for NOL carrybacks.

To illustrate accounting for NOL carrybacks, let's modify Illustration 16–13B to assume that we are dealing with a property and casualty insurance company, and that there was taxable income in the two years prior to the net operating loss. Note that the net operating loss must be applied to the earlier year first and then brought forward to the next year. If any of the NOL remains after reducing taxable income to zero in the two previous years, the remainder is carried forward to future years as a net operating loss carryforward.

During 2021, American Property and Casualty Insurance Corporation reported a net operating loss of $120 million for financial reporting and tax purposes. The enacted tax rate is 25% for 2021. Taxable income, tax rates, and income taxes paid in the two previous years were as follows:

	Taxable Income	Tax Rate	Income Taxes Paid
2019	$20 million	20%	$ 4 million
2020	$60 million	25%	$15 million

Here's how the income tax benefit of the net operating loss carryback and the net operating loss carryforward is determined:

	Prior Years		Current Year	Future Deductible
($ in millions)	**2019**	**2020**	**2021**	**Amounts (total)**
Net operating loss			$(120)	
NOL carryback	$(20)	$(60)	80	
NOL carryforward			40	$ 40
Taxable income (tax return)			$ 0	
Enacted tax rate	20%	25%	25%	25%
Tax refund	$ (4)	$ (15)	$ 0	
Deferred tax asset				**$ 10**

Step 1: Tax refund: $19
Step 2: DTA end bal: $10
Step 3: DTA change: $10
Step 4: Tax exp plug: $29

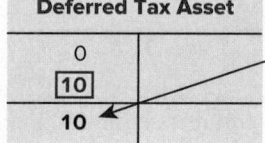

Deferred Tax Asset	
0	
10	
10	

Journal Entry at the End of 2021

Income tax refund receivable ($4 + $15, determined above)	19	
Deferred tax asset (determined above)	10	
Income tax expense (to balance)		29

(continued)

(concluded)

American's income statement would include the following:

		($ in millions)
Operating loss before income taxes		$(120)
Income tax benefit:		
Current: Tax refund from NOL carryback	$19	
Deferred: Tax savings from NOL carryforward	10	29
Net loss		$ (91)

The income tax benefit (reduction in tax expense) of both a net operating loss carryback and a net operating loss carryforward is recognized for accounting purposes in the year the NOL occurs. The net after-tax loss reflects the reduction of past taxes from the NOL carryback and future tax savings that the NOL carryforward is expected to create. In this example, the income tax benefit ($29 million) is less than it was when we assumed a carryforward only ($30 million; see Illustration 16–13A). This is because the tax rate in one of the carryback years (20% in 2019) was lower than the carryforward rate (25%).

Financial Statement Presentation

Balance Sheet Classification

All deferred tax liabilities, deferred tax assets, and any valuation allowance against deferred tax assets are classified as noncurrent in the balance sheet.[8] If these deferred tax accounts relate to the same tax-paying component of the company and the same tax jurisdiction, they are netted against each other and shown as a single net number in the balance sheet. For example, a company with a deferred tax liability of $10 million, a deferred tax asset of $5 million, and a valuation allowance of $2 million would show a net noncurrent deferred tax liability of $7 million [$10 million – ($5 million – $2 million)]. ● LO16–8

Sometimes components of a single company are viewed as different companies for tax purposes, or pay tax in different jurisdictions. If deferred tax liabilities and assets relate to components of a company that are separate for tax purposes, or relate to different tax jurisdictions, they should not be offset. So, in the prior illustration, if the deferred tax assets (and related valuation allowance) applied to a separate tax jurisdiction from the deferred tax liabilities, the company would show a noncurrent deferred tax liability of $10 million and a noncurrent deferred tax asset of $3 million ($5 million – $2 million).

Disclosure Notes

We've already seen many of the disclosures that companies have to present in the tax note, but there are a few we haven't yet covered.

INCOME TAX EXPENSE Illustration 16–7A shows **Shoe Carnival**'s disclosure of current tax payable, deferred tax, and tax expense. More generally, disclosure notes should indicate the following:

- Current portion of the tax expense (or tax benefit); that's new tax payable this period.
- Deferred portion of the tax expense (or tax benefit), with separate disclosure of amounts attributable to:
  - Portions that do not include the effect of separately disclosed amounts.
  - Operating loss carryforwards.

[8]Prior to 2017, companies were required to classify deferred tax assets and liabilities as current or noncurrent. To simplify presentation, the FASB required that all deferred tax items be classified as noncurrent when it issued "Balance Sheet Classification of Deferred Taxes," *Accounting Standards Update No. 2015–17: (Topic 740)* (Norwalk, Conn.: FASB, 2015).

- Adjustments due to changes in tax laws or rates.
- Adjustments to the beginning-of-the-year valuation allowance due to revised estimates.
- Tax credits.

Additional Consideration

Throughout this chapter, we have used a single journal entry to record entries for tax payable, deferred tax accounts, and tax expense. As an alternative, we could record two journal entries that separate (a) the current portion of the tax expense and (b) the deferred portion of the tax expense.

For example, in Illustration 16–12A, we recorded tax expense in a single journal entry:

December 31, 2021

Income tax expense	50	
Deferred tax asset	3	
Deferred tax liability		6
Income tax payable		47

Instead, we could have separately recorded the current and deferred tax expense:

December 31, 2021

Current income tax expense	47	
Income tax payable		47
Deferred income tax expense	3	
Deferred tax asset	3	
Deferred tax liability		6

Throughout the chapter, we demonstrated the simpler approach, but using two entries provides the same result.

DEFERRED TAX ASSETS AND DEFERRED TAX LIABILITIES Illustration 16–7B shows **Shoe Carnival**'s disclosure of its deferred tax assets and deferred tax liabilities. It shows a total net deferred tax asset of $9,600 thousand, which would appear in its January 28, 2017, balance sheet. More generally, companies must disclose the following:

- Total of all deferred tax liabilities.
- Total of all deferred tax assets.
- Total valuation allowance recognized for deferred tax assets.
- Net change in the valuation allowance.
- Approximate tax effect of each type of temporary difference (and carryforward).

EFFECTIVE TAX RATE RECONCILIATION Illustration 16–10 shows **Shoe Carnival**'s effective tax rate reconciliation. Companies are required to provide that reconciliation, indicating the amount and nature of each significant reconciling item.

NET OPERATING LOSS (NOL) CARRYFORWARDS Companies must disclose the amounts of any NOL carryforwards, as well as any applicable expiration dates. As indicated previously, NOLs arising in tax years beginning after December 31, 2017, can be carried forward indefinitely for federal taxes, but sometimes there are expiration dates that apply to older NOLs, to other unusual circumstances, or with respect to state or local taxes.

PART D

Coping with Uncertainty in Income Taxes

As you might imagine, most companies strive to legitimately reduce their overall tax burden and to reduce or delay cash outflows for taxes. However, even without these efforts to reduce taxes, most companies' tax returns will include many tax positions that are subject

LO16–9

to multiple interpretations. That is, the position management takes with respect to an element of tax expense might differ from the position the IRS or other taxing authorities might take on that same item. It can take many years to resolve uncertainty about whether management's tax positions will be challenged and, if challenged, whether those positions will be upheld. How should we account for such uncertain tax positions? Should we assume management will prevail, or assume that questioned positions will be disallowed and the company ultimately will owe more tax?

For example, assume that Derrick Company claims on its tax return a particular deduction it believes is legitimate, and that position saves the company $8 million in 2021 income taxes. Derrick knows that, historically, the IRS has challenged many deductions of this type. Since tax returns usually aren't examined for one, two, or more years, uncertainty exists.

TWO-STEP DECISION PROCESS To deal with that uncertainty, companies are only allowed to reduce tax expense (recognize a tax benefit) for a questionable position if it is "more likely than not" (defined as a greater than 50% chance) that the position will be sustained if challenged.[9] Guidance also prescribes how to *measure* the amount to be recognized if, in fact, it can be recognized. The decision, then, is a "two-step" process.

Step 1. A tax benefit may be reflected in the financial statements only if it is "more likely than not" that the company will be able to sustain the tax position, based on its technical merits.

Step 2. A tax benefit should be measured as the largest amount of benefit that is "cumulatively greater than 50 percent likely to be realized" (demonstrated later).

For the step 1 decision as to whether the position can be sustained, companies must assume that the position will be reviewed by the IRS or other taxing authority and litigated to the "highest court possible," and that the taxing authority has knowledge of all relevant facts.

NOT "MORE LIKELY THAN NOT" Let's say that in the step 1 decision, Derrick believes the more-likely-than-not criterion is *not* met. In that case, none of the tax benefit (reduction in tax expense) can be recorded in 2021, and income tax expense is recorded at the same amount as if the tax deduction was not taken.

If there's a 50% chance or less of the company's position being sustained on examination, tax expense can't be reduced to reflect the tax benefit.

Suppose, for instance, that Derrick's current income tax payable is $24 million after being reduced by the full $8 million tax benefit.[10] If it's *more likely than not* that the tax benefit isn't sustainable upon examination, the benefit can't be recognized as a reduction of tax expense. So, Derrick would record (a) current income tax payable that reflects the entire $8 million benefit of the deduction, (b) an additional tax liability that represents the obligation to pay an additional $8 million of taxes under the assumption that the deduction ultimately will not be upheld, and (c) tax expense as if the deduction had never been taken.

	($ in millions)	
Income tax expense (without **$8** tax benefit)	32	
Income tax payable (with **$8** tax benefit)		24
Liability—uncertain tax positions		8

The $8 million liability is recognized to account for the fact that it is probable that tax officials will disallow the tax treatment used to compute income tax payable.

The $8 million difference is the tax that Derrick didn't pay because Derrick took the deduction. That amount is potentially due if the deduction is not upheld later. Because the ultimate outcome probably won't be determined within the upcoming year, the *Liability—projected additional tax* is reported as a long-term liability unless it's known to be current.[11]

[9]FASB ASC 740: Income Taxes–Overall.

[10]For illustration, if pretax accounting income is $128 million, the tax rate is 25%, and the questionable deduction is $32 million, income tax payable would be ($128 million – $32 million) × 25% = $24 million.

[11]If a company has any deferred tax assets from net operating loss carryforwards, it generally nets the liability for uncertain tax positions against those deferred tax assets for presentation in the balance sheet, rather than presenting the liability separately. Why? The liability indicates that future tax will be *paid*, and the operating loss carryforward indicates that future tax will be *saved*, so the two are netted

MEASURING THE TAX BENEFIT Now, let's say that even though Derrick is aware of the IRS's tendency to challenge deductions of this sort, management believes it *is* more likely than not that the position will be upheld if later challenged. Since Derrick has determined in step 1 that yes, a tax benefit can be recognized, it now needs to decide how much. That's step 2.

Suppose the following table represents management's estimates of the likelihood of various amounts of tax benefit that would be upheld:

The largest amount that has a cumulative greater-than-50%-chance of being realized is **$6** million (10% + 20% + 25% = **55%**).

Likelihood Table ($ in millions)

Amount of the tax benefit that management expects to be sustained	$8	$7	**$6**	$5	$4
Percentage likelihood that the tax position will be sustained at this level	10%	20%	25%	25%	20%
Cumulative probability that at least that much tax position will be sustained	10%	30%	**55%**	80%	100%

The amount of tax benefit that Derrick can recognize in the financial statements (reduce tax expense) is **$6** million because it represents the largest amount of benefit that is more likely than not (greater than 50% probability) to be sustained. So, Derrick would record (a) current income tax payable that reflects the entire $8 million benefit of the deduction, (b) an additional tax liability that represents the obligation to pay an additional $2 million of taxes under the assumption that **$6** million of tax benefit ultimately will be upheld, and (c) tax expense as if there is a **$6** million tax benefit.

	($ in millions)	
Income tax expense (with **$6** tax benefit)	26	
Income tax payable (with $8 tax benefit)		24
Liability—uncertain tax positions ($8 – **$6**)		2

Only **$6** million of the tax benefit is recognized in income tax expense.

In summary, the highest amount Derrick might have to pay is the full $8 million of tax previously avoided because of the 2021 deduction; the least amount, of course, is zero. The most likely amount (the largest amount of benefit that is cumulatively greater than 50 percent likely to be realized) is **$6** million. Therefore, the most likely additional liability is $2 million ($8 – **$6** = **$2**). That's the amount we record as a *Liability—uncertain tax positions,* along with the current income tax payable of $24 million.

RESOLUTION OF THE UNCERTAINTY Now let's consider what happens in the future, when the uncertainty associated with the tax position has been resolved. We'll consider three scenarios. Note that, in each scenario, the "Liability—uncertain tax positions" gets reduced to zero in the period in which the uncertainty is resolved.

1. **Worst case scenario.** The entire position is disallowed, such that Derrick owes $8 million tax (plus any interest and penalties, which we are ignoring).

	($ in millions)	
Income tax expense	6	
Liability—uncertain tax positions	2	
Income tax payable (or cash)		8

2. **Best case scenario.** The entire position is upheld, so Derrick owes no additional tax.

	($ in millions)	
Liability—uncertain tax positions	2	
Income tax expense		2

3. **Expected scenario.** The $6 million position is allowed as expected, so Derrick owes the expected $2 million tax (plus any interest and penalties, which we are ignoring).

	($ in millions)	
Liability—uncertain tax positions	2	
Income tax payable (or cash)		2

Companies are required to include in the disclosure notes a clear reconciliation of the beginning and ending balance of their liability for unrecognized tax benefits. As an example, Illustration 16–14 includes an excerpt from the tax note of **Staples, Inc.** for its fiscal year ended January 28, 2017.

Illustration 16–14

Disclosure of Deferred Taxes—Staples, Inc.

Real World Financials

The following summarizes the activity related to the Company's unrecognized tax benefits, including those related to discontinued operations ($ in millions):

	2016	**2015**	**2014**
Balance at beginning of fiscal year	$136	$216	$281
Additions for tax positions related to current year	30	19	22
Additions for tax positions of prior years	8	5	36
Reductions for tax positions of prior years	(8)	(5)	(88)
Reduction for statute of limitations expiration	(22)	(69)	(17)
Settlements	(7)	(30)	(18)
Balance at end of fiscal year	$137	$136	$216

Source: Staples, Inc.

Additional Consideration

The Balance Sheet Focus of Income Tax Accounting. The way we account for income taxes is a useful example of the FASB's balance sheet emphasis. Rather than calculating tax expense directly, it always is calculated as whatever amount is implied by the combination of income tax payable and the changes that occurred during the period in deferred tax assets, deferred tax liabilities, the valuation allowance for deferred tax assets, and the liability for uncertain tax positions. In fact, we can summarize virtually everything in this chapter by visualizing the journal entry implied by the changes in those accounts and applying our 4-step process. Let's draw some T-accounts to see how that works:

	Deferred Tax Assets		**Val. Allowance—Deferred Tax Assets**		**Deferred Tax Liabilities**		**Liability for Uncertain Tax Positions**	
Beg. bal.	S			U		W		Y
	B	or B	C	or C	D	or D	E	or E
End. Bal.	T			V		X		Z

With these T-accounts in mind, think of a company's tax accounting as including the following steps:

1. Step 1: Determine income tax payable (tax rate times taxable income). We'll call that number "A."
2. **Step 2:** Determine **T**, **V**, **X**, and **Z**, the ending balances needed in the tax-related balance sheet accounts, as we do in the chapter.
3. **Step 3:** Determine whatever changes in those accounts, **B**, **C**, **D**, and **E**, are needed to reach their required ending balances. We already know the beginning balances in those accounts because they are the same as the ending balances reported in the prior period.
4. **Step 4:** Finally, calculate tax expense, F, as the amount necessary to balance the journal entry. For example:

Tax expense (to balance)	F	
Deferred tax asset (amount needed to achieve needed balance)	B	
Deferred tax liability (amount needed to achieve needed balance)		D
Valuation allowance—deferred tax asset (amount needed to achieve needed balance)		C
Liability—uncertain tax positions (amount needed to achieve needed balance)		E
Taxes payable (tax rate × taxable income)		A

(continued)

(concluded)

Tax expense (F) always is a "plug" figure to make the journal entry balance. Usually the plug is a debit, but sometimes its a credit to recognize a tax benefit. Depending on a company's particular circumstances, a debit or a credit could be required to reach the appropriate ending balance in each of the tax-related balance sheet accounts, **B, C, D,** and **E**. To the extent these accounts changed because of other transactions (for example, an acquisition), we would account for those effects first and then determine the amounts necessary to reach the appropriate ending balances.

Intraperiod Tax Allocation

LO16–10

You should recall that an income statement reports discontinued operations separately from income (or loss) from continuing operations to better allow the user of the statement to isolate irregular components of net income from those that represent recurring business operations.[12] This helps the user to more accurately project future operations. Because taxes are an important part of operations, the total income tax expense for a reporting period should be allocated between continuing and discontinued operations.

The related tax effect can be either a tax expense or a tax benefit. A gain on disposal of a discontinued operation increases taxable income, so it produces tax expense, while a loss on disposal reduces taxable income, so it produces a tax benefit.[13] For example, assume a company has $84 million of pretax income from continuing operations and $16 million of pretax income from discontinued operations. Assuming a 25% tax rate, the company would report the gain from discontinued operations net of tax in its income statement:

	($ in millions)
Income before tax and discontinued operation	$84
Less: Income tax expense ($84 × 25%)	(21)
Income before discontinued operations	63
Income from discontinued operations (net of $4 income tax expense)	12
Net income	$75

Similarly, if instead the company has $116 million of pretax income from continuing operations and $16 million of pretax *loss* from discontinued operations, the company would report the loss from discontinued operations net of tax in its income statement:

	($ in millions)
Income before tax and discontinued operation	$116
Less: Income tax expense ($116 × 25%)	(29)
Income before discontinued operations	87
Loss from discontinued operations (net of $4 income tax benefit)	(12)
Net income	$ 75

Allocating income taxes within a particular reporting period is intraperiod tax allocation.

Allocating income taxes among financial statement components in this way within a particular reporting period is referred to as *intraperiod tax allocation*. You should recognize the contrast with *inter*period tax allocation—terminology sometimes used to describe allocating income taxes between two or more reporting periods by recognizing deferred tax assets and liabilities. While interperiod tax allocation is challenging and controversial, intraperiod tax allocation is relatively straightforward and substantially free from controversy.

[12]Until recently, GAAP included another category, extraordinary items, for which income was shown separately net of tax. The FASB eliminated the concept of extraordinary items starting in 2016 when it issued "Income Statement—Extraordinary and Unusual Items," *Accounting Standards Update No. 2015-01:* (Subtopic 225-20) (Norwalk, Conn.: FASB, 2015).

[13]As discussed in Chapter 4, companies separately report (a) any gain or loss from running a discontinued operation prior to disposal, and (b) any gain or loss on disposal of a discontinued operation's assets. For simplicity we report a combined number.

OTHER COMPREHENSIVE INCOME Allocating income tax expense or benefit also applies to components of comprehensive income reported separately from net income. You should recall from our discussions in Chapters 4 and 12 that "comprehensive income" extends our view of income beyond conventional net income to include four types of gains and losses that traditionally haven't been included in income statements. The other comprehensive income (OCI) items relate to investments, postretirement benefit plans, derivatives, and foreign currency translation. When these OCI items are reported in a statement of comprehensive income and shown in accumulated other comprehensive income (AOCI) in shareholders' equity, they are reported net of their respective income tax effects.[14]

Additional Consideration

The huge federal tax rate reduction that occurred in 2017 created an interesting situation: "stranded tax effects." Here's what happened. Let's say a company had a large unrealized gain from an investment in a previous year. As you learned in Chapter 12, that gain would be reflected in AOCI, net of related tax expense. Because tax is not payable until an investment is sold, this gain would have created a deferred tax liability. So, when the tax rate was reduced by Congress, the company reduced its deferred tax liability to reflect the lower tax to be paid upon sale. But that fixed only one of two accounts affected. The gain sitting in AOCI was reduced by tax expense calculated under the old, high tax rate, and now, the extra tax above the newer, low tax rate was "stranded" in AOCI. Unless an adjustment was made, that inappropriately high tax would be reflected in earnings in some future period when the investment is sold. To solve this problem, the FASB issued ASU 2018-02,[15] which allowed companies to make a one-time adjustment that restated AOCI to reflect the new tax rate, with an offsetting entry to retained earnings.

Decision Makers' Perspective

Income taxes represent one of the largest expenditures that many firms incur. When state, local, and foreign taxes are considered along with federal taxes, the total bite can easily consume 25% to 30% of income. A key factor, then, in any decision that managers make should be the impact on taxes. Decision makers must constantly be alert to options that minimize or delay taxes. During the course of this chapter, we encountered situations that are not taxable (for example, interest on governmental bonds) and those that delay taxes (for example, using accelerated depreciation on the tax return). Astute managers make investment decisions that consider the tax effect of available alternatives. Similarly, outside analysts should consider how effectively management has managed its tax exposure and monitor the current and prospective impact of taxes on their interests in the company.

Consider an example. Large, capital-intensive companies with significant investments in buildings and equipment often have sizable deferred tax liabilities from temporary differences in depreciation. If new investments cause the level of depreciable assets to at least remain the same over time, the deferred tax liability can be effectively delayed indefinitely. Investors and creditors should be watchful for situations that might cause material paydowns of that deferred tax liability, such as impending plant closings or investment patterns that suggest declining levels of depreciable assets. Unexpected additional tax expenditures can severely diminish an otherwise attractive prospective rate of return.

You also learned in the chapter that deferred tax assets represent future tax benefits. One such deferred tax asset that often reflects sizable future tax deductions is a net operating loss (NOL) carryforward. When a company has a large net operating loss carryforward, a large amount of future income can be earned tax-free. This tax shelter can be a huge advantage, not to be overlooked by careful analysts.

Investment patterns and other disclosures can indicate potential tax expenditures.

[14] This can be accomplished by (a) presenting components of other comprehensive income net of related income tax effects or (b) presenting a single tax amount for all, and individual components shown before income tax effects with disclosure of the income taxes allocated to each component either in a disclosure note or parenthetically in the statement.

[15] "Income Statement—Reporting Comprehensive Income (Topic 220): Reclassification of Certain Tax Effects from Accumulated other Comprehensive Income" *Accounting Standards Update No. 2018-02:* (Norwalk, Conn.: FASB, 2018).

Net operating loss carryforwards can indicate significant future tax savings.

Managers and outsiders are aware that increasing debt increases risk. Deferred tax liabilities increase reported debt. As discussed and demonstrated in the previous chapter, financial risk often is measured by the debt to equity ratio, total liabilities divided by shareholders' equity. Other things being equal, the higher the debt to equity ratio, the higher the risk. Should the deferred tax liability be included in the computation of this ratio? Some analysts will argue that it should be excluded, observing that in many cases the deferred tax liability account remains the same or continually grows larger. Their contention is that no future tax payment will be required. Others, though, contend that is no different from the common situation in which long-term borrowings tend to remain the same or continually grow larger. Research supports the notion that deferred tax liabilities are, in fact, viewed by investors as real liabilities and investors appear to discount them according to the timing and likelihood of the liabilities' settlement.[16]

Deferred tax liabilities increase risk as measured by the debt to equity ratio.

Whenever managerial discretion can materially impact reported earnings, analysts should be wary of the implications for earnings quality assessment. We indicated earlier that the decision as to whether or not a valuation allowance is used, as well as the size of the allowance, is largely discretionary. Research indicates that an increase in a valuation allowance provides useful information, signaling that management is pessimistic about its ability to generate enough future income to benefit from the tax deductions provided by deferred tax assets.[17] However, research also indicates that some companies do use the deferred tax asset valuation allowance account to manage earnings upward to meet analyst forecasts.[18] More generally, a survey of nearly 600 corporate tax executives provides evidence that most top management care at least as much about tax expense and its effect on earnings per share as they do about the actual cash taxes that are paid by their companies, and that an important consideration in tax planning is increasing earnings per share.[19] Alert investors should not overlook the potential for companies using tax expense to manage their earnings.

In short, managers who make decisions based on estimated pretax cash flows and outside investors and creditors who make decisions based on pretax income numbers are perilously ignoring one of the most important aspects of those decisions. Taxes should be a primary consideration in any business decision. ●

Concept Review Exercise

MULTIPLE DIFFERENCES AND NET OPERATING LOSS

Mid-South Cellular Systems began operations in 2021. In 2022, its second year of operations, pretax accounting income was $88 million, which included the following amounts:

1. Insurance expense of $14 million, representing one-third of a $42 million, three-year casualty and liability insurance policy that is deducted for tax purposes entirely in 2022.
2. Insurance expense for a $2 million premium on a life insurance policy that guarantees a $50 million payment upon the death of the company president. The premium is not deductible for tax purposes.
3. An asset with a four-year useful life was acquired last year. It is depreciated by the straight-line method in the income statement. MACRS is used on the tax return, causing deductions for depreciation to be more than straight-line depreciation the first two years but less than straight-line depreciation the next two years ($ in millions):

[16]See Dan Givoly and Carla Hayn, "The Valuation of the Deferred Tax Liability: Evidence from the Stock Market," *The Accounting Review,* April 1992, pp. 394–410.

[17]See Greg Miller and Doug Skinner, "Determinants of the Valuation Allowance for Deferred Tax Assets under SFAS No. 109," *The Accounting Review,* April 1998, pp. 213–233.

[18]See Sonia O. Rego and Mary Margaret Frank, "Do Managers Use the Valuation Allowance Account to Manage Earnings Around Certain Earnings Targets?" *Journal of the American Taxation Association 28 (1),* 2006, pp. 43–65.

[19]See John Graham, Michelle Hanlon, Terry Shevlin, and Nemit Shroff, "Incentives for Tax Planning and Avoidance: Evidence from the Field," *The Accounting Review,* May 2014, pp. 999–1023.

	Income Statement	Tax Return	Difference
2021	$150	$198	$ (48)
2022	150	264	(114)
2023	150	90	60
2024	150	48	102
	$600	$600	0

4. Equipment rental revenue of $80 million is reported in the income statement, which does not include an additional $20 million of advance payment for 2023 rent. Because tax law requires that advance rent be taxed when it is received, $100 million of rental revenue is correctly reported on the 2022 income tax return.

The enacted tax rate is 25%.

Required:

1. Prepare the journal entry to record Mid-South Cellular's income taxes for 2022.
2. What is Mid-South Cellular's 2022 net income?

Solution

1. Prepare the journal entry to record Mid-South Cellular's income taxes for 2022.

($ in millions)	Current Year 2022	Future Taxable (Deductible) Amounts 2023	2024	Future Taxable Amounts (total)	Future Deductible Amounts (total)
Pretax accounting income	$ 88				
Permanent difference:					
Life insurance premium	2				
Temporary difference:					
Prepaid insurance	(28)	$14	$ 14	$ 28	
Depreciation	(114)	60	102	162	
Advance rent received	20	(20)			$ (20)
Net operating loss (tax return)	$(32)				
NOL carryforward	32				(32)
	$ 0			$190	$(52)
Enacted tax rate	25%			25%	25%
Tax payable	**$ 0**			**$47.5**	**$ (13)**

Deferred Tax Liability		Deferred Tax Asset	
	12*	0	
	35.5	13	
	47.5	13	

Step 1: Tax payable: $0
Step 2: DTA end bal: $13.0
DTL end bal: $47.5
Step 3: DTA change: $13.0
DTL change: $35.5
Step 4: Tax exp plug: $22.5

Journal Entry at the End of 2021

Income tax expense (to balance)	22.5	
Deferred tax asset (determined above)	13.0	
Deferred tax liability (determined above)		35.5

*The opening balance of the deferred tax liability relates to the temporary difference for depreciation in 2021, and is $48 x 25% = $12.

2. What is Mid-South Cellular's 2022 net income?

Pretax accounting income	$88.0
Income tax expense	(22.5)
Net income	$65.5

Financial Reporting Case Solution

©wavebreakmediamicro/123RF

1. **Explain to Laura how differences between financial reporting standards and income tax rules might cause the income tax expense and the amount of income tax paid to differ.** *(p. 909)* The differences in the rules for computing taxable income and those for financial reporting often cause amounts to be included in taxable income in a different year(s) from the year in which they are recognized for financial reporting purposes. Temporary differences result in future taxable or deductible amounts when the temporary differences reverse. As a result, tax payments frequently occur in years different from the years in which the revenues and expenses that cause the taxes are generated.
2. **How might a reduction in future tax rates affect deferred tax assets in a way that reduces current net income?** *(pp. 933 and 935)* Deferred tax assets capture anticipated reductions in future tax bills that result from the company being able to use tax deductions that have arisen from its operations but not yet been taken. For example, a company may have recognized $100 of warranty expense in the financial statements but won't report the expense on the tax return until future warranty costs are actually incurred. The company knows it has a $100 tax deduction coming when those costs are incurred, so it recognizes an asset for the anticipated tax savings arising from that deduction. Given a tax rate of 25%, the company would show a deferred tax asset of 25% × $100 = $25. Now imagine the company learns that future tax rates have been reduced to 20%. In that case, the future $100 tax deduction will save only $20 of taxes, so the value of its deferred tax asset has declined by $5. In the period the future tax rate change is enacted, the company is required to include in tax expense and therefore net income all of the effects of the tax rate change on existing deferred tax assets and liabilities. That's what happened in 2017 to companies that had big deferred tax assets when Congress enacted legislation that dramatically reduced future corporate tax rates.
3. **What are net operating loss carryforwards and how can they provide cash savings?** *(pp. 929 and 938)* When a company has negative taxable income on its tax return, it is permitted to carry that net operating loss forward and offset it against taxable income in future years and to not pay tax in those years. Such NOL carryforwards can produce valuable tax savings in those future years. ●

The Bottom Line

- **LO16–1** Temporary differences between pretax accounting income and taxable income produce future taxable or deductible amounts, which give rise to deferred tax liabilities and deferred tax assets. Consistent with the accrual concept, tax expense includes not only an amount associated with current tax payable, but also includes a deferred amount associated with changes in deferred tax assets and liabilities. To calculate tax expense, we follow a four-step process: (1) calculate tax payable, (2) calculate the ending balances of any deferred tax accounts or any liability for uncertain tax positions, (3) calculate changes in those accounts, and (4) plug for tax expense. (*p. 909*)
- **LO16–2** When the future tax consequence of a temporary difference will be to increase taxable income relative to pretax accounting income, future taxable amounts are created. The future tax consequences associated with those amounts are recognized as deferred tax liabilities. (*p. 912*)
- **LO16–3** When the future tax consequence of a temporary difference will be to decrease taxable income relative to pretax accounting income, future deductible amounts are created. The future tax consequences associated with those amounts are recognized as deferred tax assets. (*p. 919*)
- **LO16–4** Deferred tax assets are recognized for all deductible temporary differences. However, a deferred tax asset is then reduced by a valuation allowance if it is more likely than not that some portion or all of the deferred tax asset will not be realized. (*p. 925*)
- **LO16–5** Permanent differences between the reported amount of an asset or liability in the financial statements and its tax basis are those caused by transactions and events that under existing tax law will never affect taxable income or taxes payable. These are disregarded when determining both the tax payable currently, the deferred tax effect, and tax expense. (*p. 929*)

- **LO16–6** Deferred tax liabilities (and assets) are calculated by multiplying future taxable (and deductible) amounts by the currently enacted tax rates that will apply to them. If a change in a tax law or rate occurs, the deferred tax liability or asset is adjusted to reflect the change in the amount to be paid or recovered. That effect is reflected in tax expense in the year of the enactment of the change in the tax law or rate. (*p. 933*)
- **LO16–7** Tax laws permit a net operating loss (NOL) to be used to reduce taxable income in other, profitable years. For most companies, NOLs can be carried forward indefinitely and used to offset up to 80% of the taxable income reported in each future year. Some farm-related companies and insurance companies also are allowed to carry NOLs back two years to offset taxable income in those years and generate an immediate tax refund. The tax benefit of an NOL carryforward or carryback is recognized in the year of the loss. (*p. 937*)
- **LO16–8** Deferred tax assets and deferred tax liabilities are classified as noncurrent. If they relate to the same taxable component of the company and the same tax jurisdiction, they are netted against each other and shown as a single noncurrent net deferred tax asset or liability. Otherwise they are not offset against each other for purposes of balance sheet presentation. Disclosure notes should reveal additional relevant information pertaining to deferred tax amounts reported on the balance sheet, the components of income tax expense, and available operating loss carryforwards. (*p. 941*)
- **LO16–9** A tax benefit associated with an uncertain tax position may be reflected in the financial statements only if it is "more likely than not" that the company will be able to sustain the tax return position, based on its technical merits. It should be measured as the largest amount of benefit that is cumulatively greater than 50 percent likely to be realized. (*p. 943*)
- **LO16–10** Through intraperiod tax allocation, the total income tax expense for a reporting period is allocated among the financial statement items that gave rise to it: specifically, income (or loss) from continuing operations, discontinued operations, and prior period adjustments (to the beginning retained earnings balance). (*p. 946*)
- **LO16–11** Despite the similar approaches for accounting for taxation under IFRS and U.S. GAAP, differences in reported amounts for deferred taxes are among the most frequent between the two approaches because a great many of the *nontax* differences between IFRS and U.S. GAAP affect deferred taxes. (*p. 931*)

Questions For Review of Key Topics

Q 16–1 A member of the board of directors is concerned that the company's income statement reports income tax expense of $12.3 million, but the income tax obligation to the government for the year is only $7.9 million. How might the corporate controller explain this difference?

Q 16–2 A deferred tax liability (or asset) is described as the tax effect of the temporary difference between the financial statement carrying amount (book value) of an asset or liability and its tax basis. Explain this tax effect of the temporary difference. How might it produce a deferred tax liability? A deferred tax asset?

Q 16–3 Sometimes a temporary difference will produce future deductible amounts. Explain what is meant by future deductible amounts. Describe two general situations that have this effect. How are such situations recognized in the financial statements?

Q 16–4 The benefit of future deductible amounts can be achieved only if future income is sufficient to take advantage of the deferred deductions. For that reason, not all deferred tax assets will ultimately be realized. How is this possibility reflected in the way we recognize deferred tax assets?

Q 16–5 Temporary differences result in future taxable or deductible amounts when the related asset or liability is recovered or settled. Some differences, though, are not temporary. What events create permanent differences? What effect do these have on the determination of income taxes payable? Of deferred income taxes? Of tax expense?

Q 16–6 Identify three examples of differences with no deferred tax consequences.

Q 16–7 The income tax rate for Hudson Refinery has been 35% for each of its 12 years of operation. Company forecasters expect a much-debated tax reform bill to be passed by Congress early next year. The new tax measure would increase Hudson's tax rate to 42%. When measuring this year's deferred tax liability, which rate should Hudson use?

Q 16–8 In late 2017 the federal tax rate for subsequent years was decreased from 35% to 21%. How would this affect an existing deferred tax liability? How would the change be reflected in net income?

Q 16–9 A net operating loss occurs when tax-deductible expenses exceed taxable revenues. Tax laws permit the net operating loss to be used to reduce taxable income in future profitable years. How are loss carryforwards recognized for financial reporting purposes?

Q 16–10 How are deferred tax assets and deferred tax liabilities reported in a classified balance sheet?

Q 16–11 Additional disclosures are required pertaining to deferred tax amounts reported on the balance sheet. What are the needed disclosures?

Q 16–12 Additional disclosures are required pertaining to the income tax expense reported in the income statement. What are the needed disclosures?

Q 16–13 Accounting for uncertainty in tax positions is prescribed by GAAP in FASB ASC 740–10: Income Taxes–Overall (previously *FASB Interpretation No. 48 (FIN 48)*). Describe the two-step process required by GAAP.

Q 16–14 What is intraperiod tax allocation?

IFRS Q 16–15 IFRS and U.S. GAAP follow similar approaches to accounting for taxation. Nevertheless, differences in reported amounts for deferred taxes are among the most frequent between IFRS and U.S. GAAP. Why?

Brief Exercises

connect

BE 16–1
Temporary difference
● LO16–1, LO16–2

A company reports 2021 *pretax accounting income* of $10 million, but because of a single temporary difference, *taxable income* is only $7 million. No temporary differences existed at the beginning of the year, and the tax rate is 25%. Prepare the appropriate journal entry to record income taxes.

BE 16–2
Temporary difference; determine taxable income; determine deferred tax amount
● LO16–2

Kara Fashions uses straight-line depreciation for financial statement reporting and MACRS for income tax reporting. Three years after its purchase, one of Kara's buildings has a book value of $400,000 and a tax basis of $300,000. There were no other temporary differences and no permanent differences. Taxable income was $4 million and Kara's tax rate is 25%. What is the deferred tax liability to be reported in the balance sheet? Assuming that the deferred tax liability balance was $20,000 the previous year, prepare the appropriate journal entry to record income taxes this year.

BE 16–3
Temporary difference; determine taxable income; determine deferred tax amount for asset 100% depreciated in year of purchase
● LO16–2

Milo Manufacturing uses straight-line depreciation for financial statement reporting and is able to deduct 100% of the cost of equipment in the year the equipment is purchased for tax purposes. Four years after its purchase, one of Milo's manufacturing machines has a book value of $600,000. There were no other temporary differences and no permanent differences. Taxable income was $10 million and Milo's tax rate is 25%. What is the deferred tax liability to be reported in the balance sheet? Assuming that the deferred tax liability balance was $175,000 the previous year, prepare the appropriate journal entry to record income taxes this year.

BE 16–4
Temporary difference
● LO16–1, LO16–3

A company reports *pretax accounting income* of $10 million, but because of a single temporary difference, *taxable income* is $12 million. No temporary differences existed at the beginning of the year, and the tax rate is 25%. Prepare the appropriate journal entry to record income taxes.

BE 16–5
Temporary difference; income tax payable given
● LO16–3

In 2021, Ryan Management collected rent revenue for 2022 tenant occupancy. For financial reporting, the rent is recorded as deferred revenue and then recognized as revenue in the period tenants occupy rental property. For tax reporting, the rent is taxed when collected in 2021. The deferred portion of the rent collected in 2021 was $50 million. Taxable income is $180 million in 2021. No temporary differences existed at the beginning of the year, and the tax rate is 25%. Prepare the appropriate journal entry to record income taxes in 2021.

BE 16–6
Temporary difference; income tax payable given
● LO16–3

Refer to the situation described in BE 16–5. Suppose the deferred portion of the rent collected was $40 million at the end of 2022. Taxable income is $200 million. Prepare the appropriate journal entry to record income taxes in 2022.

BE 16–7
Valuation allowance
● LO16–3, LO16–4

At the end of the year, the deferred tax asset account had a balance of $4 million attributable to a temporary difference of $16 million in a liability for estimated expenses. Taxable income is $60 million. No temporary differences existed at the beginning of the year, and the tax rate is 25%. Prepare the journal entry(s) to record income taxes, assuming it is more likely than not that three-fourths of the deferred tax asset will not ultimately be realized.

BE 16–8
Valuation allowance
● LO16–3, LO16–4

VeriFone Systems is a provider of electronic card payment terminals, peripherals, network products, and software. In its 2015 annual report, the company reported deferred tax assets totaling about $398 million. The company also reported valuation allowances totaling about $332 million. What would motivate VeriFone to have a valuation allowance almost equal to its deferred tax assets?

BE 16–9
Temporary and permanent differences; determine deferred tax consequences
● LO16–2, LO16–3, LO16–5

Differences between pretax accounting income and taxable income were as follows during 2021:

	($ in millions)
Pretax accounting income	$300
Permanent difference	(24)
	276
Temporary difference	(16)
Taxable income	$260

The cumulative temporary difference as of the end of 2021 is $40 million (also the future taxable amount). The enacted tax rate is 25%. What is the deferred tax asset or liability to be reported in the balance sheet?

BE 16–10
Calculate taxable income
● LO16–2, LO16–5

Shannon Polymers uses straight-line depreciation for financial reporting purposes for equipment costing $800,000 and with an expected useful life of four years and no residual value. Assume that, for tax purposes, the deduction is 40%, 30%, 20%, and 10% in those years. Pretax accounting income the first year the equipment was used was $900,000, which includes interest revenue of $20,000 from municipal governmental bonds. Other than the two described, there are no differences between accounting income and taxable income. The enacted tax rate is 25%. Prepare the journal entry to record income taxes.

BE 16–11
Multiple tax rates
● LO16–6

J-Matt, Inc., had pretax accounting income of $291,000 and taxable income of $300,000 in 2021. The only difference between accounting and taxable income is estimated product warranty costs of $9,000 for sales in 2021. Warranty payments are expected to be in equal amounts over the next three years (2022–2024) and will be tax deductible at that time. Recent tax legislation will change the tax rate from the current 25% to 20% in 2023. Determine the amounts necessary to record J-Matt's income taxes for 2021 and prepare the appropriate journal entry

BE 16–12
Change in tax rate
● LO16–6

Superior Developers sells lots for residential development. When lots are sold, Superior recognizes income for financial reporting purposes in the year of the sale. For some lots, Superior recognizes income for tax purposes when the cash is collected. In 2020, Superior sold lots for $20 for which no cash was collected at the time of the sale. This cash will be collected equally over 2021 and 2022. The enacted tax rate was 40% at the time of the sale. In 2021, a new tax law was enacted, revising the tax rate from 40% to 25% beginning in 2022. Calculate the total amount by which Superior should change its deferred tax liability in 2021.

BE 16–13
Net operating loss carryforward
● LO16–7

During its first year of operations, **Nive.com** reported a net operating loss of $15 million for financial reporting and tax purposes. The enacted tax rate is 25%. Prepare the journal entry to recognize the income tax benefit of the net operating loss.

BE 16–14
Net operating loss carryback
● LO16–7

Insure Corporation reported a net operating loss of $25 million for financial reporting and tax purposes. Taxable income last year and the previous year, respectively, was $20 million and $15 million. The enacted tax rate each year is 25%. Assume that Insure qualifies as a type of company that is allowed to carry back an NOL to two prior taxable years, using the earliest year first. Prepare the journal entry to recognize the income tax benefit of the net operating loss.

BE 16–15
Tax uncertainty
● LO16–9

First Bank has some questions as to the tax-free nature of $5 million of governmental bonds held in its investment portfolio. This amount is excluded from First Bank's taxable income of $55 million. Management has determined that there is a 65% chance that the tax-free status of this entire amount of interest can't withstand scrutiny of taxing authorities. Assuming a 25% tax rate, what amount of income tax expense should the bank report?

BE 16–16
Intraperiod tax allocation
● LO16–10

Southeast Airlines had pretax earnings of $65 million. Included in this amount is income from discontinued operations of $10 million. The company's tax rate is 25%. What is the amount of income tax expense that Southeast would report in its income statement for continuing operations? How should the gain on disposal of a discontinued operation be reported?

Exercises

E 16–1
Temporary difference; taxable income given
● LO16–1, LO16–2, LO16–8

Alvis Corporation reports *pretax accounting income* of $400,000, but due to a single temporary difference, *taxable income* is only $250,000. At the beginning of the year, no temporary differences existed.

Required:

1. Assuming a tax rate of 25%, what will be Alvis's net income?
2. What will Alvis report in the balance sheet pertaining to income taxes?

E 16–2
Determine taxable income; determine prior year deferred tax amount
● LO16–2

On January 1, 2018, Ameen Company purchased major pieces of manufacturing equipment for a total of $36 million. Ameen uses straight-line depreciation for financial statement reporting and MACRS for income tax reporting. At December 31, 2020, the book value of the equipment was $30 million and its tax basis was $20 million. At December 31, 2021, the book value of the equipment was $28 million and its tax basis was $12 million. There were no other temporary differences and no permanent differences. Pretax accounting income for 2021 was $50 million.

Required:

1. Prepare the appropriate journal entry to record Ameen's 2021 income taxes. Assume an income tax rate of 25%.
2. What is Ameen's 2021 net income?

E 16–3
Determine taxable income; determine prior year deferred tax amount; 100% depreciation in year of purchase
● LO16–2

(This exercise is a variation of E 16–2, modified to have the asset fully depreciated in the year of purchase.) On January 1, 2018, Ameen Company purchased major pieces of manufacturing equipment for a total of $36 million. Ameen uses straight-line depreciation for financial statement reporting and deducted 100% of the equipment's cost for income tax reporting in 2018. At December 31, 2020, the book value of the equipment was $30 million. At December 31, 2021, the book value of the equipment was $28 million. There were no other temporary differences and no permanent differences. Pretax accounting income for 2021 was $50 million.

Required:

1. Prepare the appropriate journal entry to record Ameen's 2021 income taxes. Assume an income tax rate of 25%.
2. What is Ameen's 2021 net income?

E 16–4
Taxable income given; calculate deferred tax liability from book-tax difference
● LO16–2

Ayres Services acquired an asset for $80 million in 2021. The asset is depreciated for financial reporting purposes over four years on a straight-line basis (no residual value). For tax purposes the asset's cost is depreciated by MACRS. The enacted tax rate is 25%. Amounts for pretax accounting income, depreciation, and taxable income in 2021, 2022, 2023, and 2024 are as follows:

	($ in millions)			
	2021	**2022**	**2023**	**2024**
Pretax accounting income	$330	$350	$365	$400
Depreciation on the income statement	20	20	20	20
Depreciation on the tax return	(25)	(33)	(15)	(7)
Taxable income	$325	$337	$370	$413

Required:

For December 31 of each year, determine (a) the temporary book-tax difference for the depreciable asset and (b) the balance to be reported in the deferred tax liability account.

E 16–5
Taxable income given; calculate deferred tax liability from book-tax difference; 100% depreciation in the year of purchase
● LO16–2

(This exercise is a variation of E 16–4, modified to have the asset fully depreciated in the year of purchase.) Ayres Services acquired an asset for $80 million in 2021. The asset is depreciated for financial reporting purposes over four years on a straight-line basis (no residual value). Ayers deducted 100% of the asset's cost for income tax reporting in 2021. The enacted tax rate is 25%. Amounts for pretax accounting income, depreciation, and taxable income in 2021, 2022, 2023, and 2024 are as follows:

	($ in millions)			
	2021	**2022**	**2023**	**2024**
Pretax accounting income	$330	$350	$365	$400
Depreciation on the income statement	20	20	20	20
Depreciation on the tax return	(80)	(0)	(0)	(0)
Taxable income	$270	$370	$385	$420

Required:

For December 31 of each year, determine (a) the temporary book-tax difference for the depreciable asset and (b) the balance to be reported in the deferred tax liability account.

E 16–6
Temporary difference; income tax payable given
● LO16–3

In 2021, DFS Medical Supply collected rent revenue for 2022 tenant occupancy. For income tax reporting, the rent is taxed when collected. For financial statement reporting, the rent is recorded as deferred revenue and then recognized as revenue in the period tenants occupy the rental property. The deferred portion of the rent collected in 2021 amounted to $300,000 at December 31, 2021. DFS had no temporary differences at the beginning of the year.

Required:

Assuming an income tax rate of 25% and 2021 income tax payable of $950,000, prepare the journal entry to record income taxes for 2021.

E 16–7
Temporary difference; future deductible amounts; taxable income given
● LO16–3

Lance Lawn Services reports warranty expense by estimating the amount that eventually will be paid to satisfy warranties on its product sales. For tax purposes, the expense is deducted when the warranty work is completed. At December 31, 2021, Lance has a warranty liability of $2 million and taxable income of $75 million. At December 31, 2020, Lance reported a deferred tax asset of $435,000 related to this difference in reporting warranties, its only temporary difference. The enacted tax rate is 25% each year.

Required:

Prepare the appropriate journal entry to record Lance's income tax provision for 2021.

E 16–8
Identify future taxable amounts and future deductible amounts
● LO16–2, LO16–3

Listed below are 10 causes of temporary differences. For each temporary difference, indicate (by letter) whether it will create future deductible amounts (D) or future taxable amounts (T).

Temporary Difference

_____ 1. Accrual of loss contingency; tax-deductible when paid.
_____ 2. Newspaper subscriptions; taxable when cash is received, recognized for financial reporting when the performance obligation is satisfied.
_____ 3. Prepaid rent; tax-deductible when paid.
_____ 4. Accrued bond interest expense; tax-deductible when paid.
_____ 5. Prepaid insurance; tax-deductible when paid
_____ 6. Unrealized loss from recording investments at fair value; tax-deductible when investments are sold.
_____ 7. Warranty expense; estimated for financial reporting when products are sold; deducted for tax purposes when paid.
_____ 8. Advance rent receipts on an operating lease as the lessor; taxable when received.
_____ 9. Straight-line depreciation for financial reporting; accelerated depreciation for tax purposes.
_____ 10. Accrued expense for employee vacation days not yet taken; tax deductible when employee takes vacation in future.

E 16–9
Identify future taxable amounts and future deductible amounts
● LO16–2, LO16–3

(This is a variation of E 16–8, modified to focus on the balance sheet accounts related to the deferred tax amounts.)

Listed below are 10 causes of temporary differences. For each temporary difference indicate the balance sheet account for which the situation creates a temporary difference.

Temporary Difference

_____ 1. Accrual of loss contingency; tax-deductible when paid.
_____ 2. Newspaper subscriptions; taxable when cash is received, recognized for financial reporting when the performance obligation is satisfied.
_____ 3. Prepaid rent; tax-deductible when paid.
_____ 4. Accrued bond interest expense; tax-deductible when paid.
_____ 5. Prepaid insurance; tax-deductible when paid
_____ 6. Unrealized loss from recording available-for-sale investments at fair value; tax-deductible when investments are sold.
_____ 7. Warranty expense; estimated for financial reporting when products are sold; deducted for tax purposes when paid.
_____ 8. Advance rent receipts on an operating lease as the lessor; taxable when received.
_____ 9. Straight-line depreciation for financial reporting; accelerated depreciation for tax purposes.
_____ 10. Accrued expense for employee vacation days not yet taken; tax deductible when employee takes vacation in future.

E 16–10
Calculate income tax amounts under various circumstances
● LO16–2, LO16–3

Four independent situations are described below. Each involves future deductible amounts and/or future taxable amounts produced by temporary differences:

	($ in thousands) Situation			
	1	**2**	**3**	**4**
Taxable income	$84	$216	$196	$260
Future deductible amounts	16		20	20
Future taxable amounts		16	16	28
Balance(s) at beginning of the year:				
Deferred tax asset	2		9	4
Deferred tax liability		8	2	

The enacted tax rate is 25%.

Required:
For each situation, determine the:
a. Income tax payable currently.
b. Deferred tax asset—balance.
c. Deferred tax asset—change.
d. Deferred tax liability—balance.
e. Deferred tax liability—change.
f. Income tax expense.

E 16–11
Determine taxable income
● LO16–2, LO16–3

Eight independent situations are described below. Each involves future deductible amounts and/or future taxable amounts ($ in millions).

	Temporary Differences Reported First on: The Income Statement		The Tax Return	
	Revenue	**Expense**	**Revenue**	**Expense**
1.		$20		
2.	$20			
3.			$20	
4.				$20
5.	15	20		
6.		20	15	
7.	15	20		10
8.	15	20	5	10

Required:
For each situation, determine taxable income, assuming pretax accounting income is $100 million.

E 16–12
Deferred tax asset; taxable income given; valuation allowance
● LO16–4

At the end of 2020, Payne Industries had a deferred tax asset account with a balance of $25 million attributable to a temporary book-tax difference of $100 million in a liability for estimated expenses. At the end of 2021, the temporary difference is $64 million. Payne has no other temporary differences and no valuation allowance for the deferred tax asset. Taxable income for 2021 is $180 million and the tax rate is 25%.

Required:
1. Prepare the journal entry(s) to record Payne's income taxes for 2021, assuming it is more likely than not that the deferred tax asset will be realized.
2. Prepare the journal entry(s) to record Payne's income taxes for 2021, assuming it is more likely than not that only one-fourth of the deferred tax asset ultimately will be realized.

E 16–13
Deferred tax asset; income tax payable given; previous balance in valuation allowance
● LO16–4

(This is a variation of E 16–12, modified to assume a previous balance in the valuation allowance.)

At the end of 2020, Payne Industries had a deferred tax asset account with a balance of $25 million attributable to a temporary book-tax difference of $100 million in a liability for estimated expenses. At the end of 2021, the temporary difference is $64 million. Payne has no other temporary differences. Taxable income for 2021 is $180 million and the tax rate is 25%.

Payne has a valuation allowance of $10 million for the deferred tax asset at the beginning of 2021.

Required:
1. Prepare the journal entry(s) to record Payne's income taxes for 2021, assuming it is more likely than not that the deferred tax asset will be realized.
2. Prepare the journal entry(s) to record Payne's income taxes for 2021, assuming it is more likely than not that only one-fourth of the deferred tax asset ultimately will be realized.

E 16–14
FASB codification research; valuation allowance
● LO16–4

When a company records a deferred tax asset, it may need to also report a valuation allowance if it is "more likely than not" that some portion or all of the deferred tax asset will not be realized.

Required:

1. Access the FASB Accounting Standards Codification at the FASB website (**www.fasb.org**). What is the specific nine-digit Codification citation (XXX-XX-XX-XX) that describes the guidelines for determining the disclosure requirements pertaining to how a firm should determine whether a valuation allowance for deferred tax assets is needed?
2. What are the guidelines?

E 16–15
Multiple differences; calculate taxable income
● LO16–2, LO16–5

Southern Atlantic Distributors began operations in January 2021 and purchased a delivery truck for $40,000. Southern Atlantic plans to use straight-line depreciation over a four-year expected useful life for financial reporting purposes. For tax purposes, the deduction is 50% of cost in 2021, 30% in 2022, and 20% in 2023. Pretax accounting income for 2021 was $300,000, which includes interest revenue of $40,000 from municipal governmental bonds. The enacted tax rate is 25%.

Required:

Assuming no differences between accounting income and taxable income other than those described above:

1. Prepare the journal entry to record income taxes in 2021.
2. What is Southern Atlantic's 2021 net income?

E 16–16
Multiple differences
● LO16–2, LO16–3, LO16–5

For the year ended December 31, 2021, Fidelity Engineering reported pretax accounting income of $978,000. Selected information for 2021 from Fidelity's records follows:

Interest income on municipal governmental bonds	$32,000
Depreciation claimed on the 2021 tax return in excess of depreciation on the income statement	58,000
Carrying amount of depreciable assets in excess of their tax basis at year-end	88,000
Warranty expense reported on the income statement	26,000
Actual warranty expenditures in 2021	10,000

Fidelity's income tax rate is 25%. At January 1, 2021, Fidelity's records indicated balances of zero and $7,500 in its deferred tax asset and deferred tax liability accounts, respectively.

Required:

1. Determine the amounts necessary to record income taxes for 2021, and prepare the appropriate journal entry.
2. What is Fidelity's 2021 net income?

E 16–17
Multiple tax rates
● LO16–3, LO16–6

Allmond Corporation, organized on January 3, 2021, had pretax accounting income of $14 million and taxable income of $20 million for the year ended December 31, 2021. The 2021 tax rate is 25%. The only difference between accounting income and taxable income is estimated product warranty costs. Assume that expected payments and scheduled tax rates (based on recently enacted tax legislation) are as follows:

2022	$2 million	20%
2023	1 million	20%
2024	1 million	20%
2025	2 million	15%

Required:

1. Determine the amounts necessary to record Allmond's income taxes for 2021 and prepare the appropriate journal entry.
2. What is Allmond's 2021 net income?

E 16–18
Change in tax rates; calculate taxable income
● LO16–2, LO16–6

Arnold Industries has pretax accounting income of $32 million for the year ended December 31, 2021. The tax rate is 25%. The only difference between accounting income and taxable income relates to an operating lease in which Arnold is the lessee. The inception of the lease was December 28, 2021. An $8 million advance rent payment at the inception of the lease is tax-deductible in 2021 but, for financial reporting purposes, represents prepaid rent expense to be recognized equally over the four-year lease term.

Required:

1. Determine the amounts necessary to record Arnold's income taxes for 2021, and prepare the appropriate journal entry.
2. Determine the amounts necessary to record Arnold's income taxes for 2022, and prepare the appropriate journal entry. Pretax accounting income was $50 million for the year ended December 31, 2022.

3. Assume a new tax law is enacted in 2022 that causes the tax rate to change from 25% to 15% beginning in 2023. Determine the amounts necessary to record Arnold's income taxes for 2022, and prepare the appropriate journal entry.
4. Why is Arnold's 2022 income tax expense different when the tax rate change occurs from what it would be without the change?

E 16–19
Deferred taxes; change in tax rates
● LO16–2, LO16–6

Bronson Industries reported a deferred tax liability of $5 million for the year ended December 31, 2020, related to a temporary difference of $20 million. The tax rate was 25%. The temporary difference is expected to reverse in 2022, at which time the deferred tax liability will become payable. There are no other temporary differences in 2020–2022. Assume a new tax law is enacted in 2021 that causes the tax rate to change from 25% to 15% beginning in 2022. (The rate remains 25% for 2021 taxes.) Taxable income in 2021 is $30 million.

Required:
Determine the effect of the tax rate change and prepare the appropriate journal entry to record Bronson's income tax expense in 2021. What effect, if any, will enacting the change in the 2022 tax rate have on Bronson's 2021 net income?

E 16–20
Deferred taxes; change in tax rates
● LO16–3, LO16–6

Shwonson Industries reported a deferred tax asset of $5 million for the year ended December 31, 2020, related to a temporary difference of $20 million. The tax rate was 25%. The temporary difference is expected to reverse in 2022, at which time the deferred tax asset will reduce taxable income. There are no other temporary differences in 2020–2022. Assume a new tax law is enacted in 2021 that causes the tax rate to change from 25% to 15% beginning in 2022. (The rate remains 25% for 2021 taxes.) Taxable income in 2021 is $30 million.

Required:
Determine the effect of the tax rate change and prepare the appropriate journal entry to record Shwonson's income tax expense in 2021. What effect, if any, will enacting the change in the 2022 tax rate have on Shwonson's 2021 net income?

E 16–21
Multiple temporary differences; record income taxes
● LO16–2, LO16–3

The information that follows pertains to Esther Food Products:

a. At December 31, 2021, temporary differences were associated with the following future taxable (deductible) amounts:

Depreciation	$60,000
Prepaid expenses	17,000
Warranty expenses	(12,000)

b. No temporary differences existed at the beginning of 2021.
c. Pretax accounting income was $80,000 and taxable income was $15,000 for the year ended December 31, 2021.
d. The tax rate is 25%.

Required:
Determine the amounts necessary to record income taxes for 2021, and prepare the appropriate journal entry.

E 16–22
Multiple temporary differences; record income taxes
● LO16–2, LO16–3

The information that follows pertains to Richards Refrigeration, Inc.:

a. At December 31, 2021, temporary differences existed between the financial statement book values and the tax bases of the following ($ in millions):

	Book Value	Tax Basis	Future Taxable (Deductible) Amount
Buildings and equipment (net of accumulated depreciation)	$120	$90	$ 30
Prepaid insurance	50	0	50
Liability—loss contingency	25	0	(25)

b. No temporary differences existed at the beginning of 2021.
c. Pretax accounting income was $200 million and taxable income was $145 million for the year ended December 31, 2021. The tax rate is 25%.

Required:
1. Determine the amounts necessary to record income taxes for 2021, and prepare the appropriate journal entry.
2. What is the 2021 net income?

E 16–23
Net operating loss carryforward
● LO16–7

During 2021, its first year of operations, Baginski Steel Corporation reported a net operating loss of $360,000 for financial reporting and tax purposes. The enacted tax rate is 25%.

Required:

1. Prepare the journal entry to recognize the income tax benefit of the net operating loss. Assume the weight of available evidence suggests that future taxable income will be sufficient to benefit from future deductible amounts arising from the net operating loss carryforward.
2. Show the lower portion of the 2021 income statement that reports the income tax benefit of the net operating loss.

E 16–24
Net operating loss carryback
● LO16–7

Wynn Farms reported a net operating loss of $100,000 for financial reporting and tax purposes in 2021. The enacted tax rate is 25%. Taxable income, tax rates, and income taxes paid in Wynn's first four years of operation were as follows:

	Taxable Income	Tax Rates	Income Taxes Paid
2017	$60,000	15%	$ 9,000
2018	70,000	15	10,500
2019	80,000	25	20,000
2020	60,000	30	18,000

Required:

1. Prepare the journal entry to recognize the income tax benefit of the net operating loss. NOL carrybacks are not allowed for most companies, except for property and casualty insurance companies as well as some farm-related businesses. Assume Wynn is one of those businesses.
2. Show the lower portion of the 2021 income statement that reports the income tax benefit of the net operating loss.

E 16–25
Net operating loss carryback and carryforward
● LO16–7

(This exercise is based on the situation described in E 16–24, modified to include a carryforward in addition to a carryback.)

Wynn Farms reported a net operating loss of $160,000 for financial reporting and tax purposes in 2021. The enacted tax rate is 25%. Taxable income, tax rates, and income taxes paid in Wynn's first four years of operation were as follows:

	Taxable Income	Tax Rates	Income Taxes Paid
2017	$60,000	15%	$ 9,000
2018	70,000	15	10,500
2019	80,000	25	20,000
2020	60,000	30	18,000

Required:

1. NOL carrybacks are not allowed for most companies, except for property and casualty insurance companies as well as some farm-related businesses. Assume Wynn is one of those businesses. Prepare the journal entry to recognize the income tax benefit of the net operating loss.
2. Show the lower portion of the 2021 income statement that reports the income tax benefit of the net operating loss.

E 16–26
Identifying income tax deferrals
● LO16–2, LO16–3, LO16–5, LO16–7

Listed below are ten independent situations. For each situation indicate (by letter) whether it will create a deferred tax asset (A), a deferred tax liability (L), or neither (N).

Situation

_____ 1. Advance payments on insurance, deductible when paid.
_____ 2. Estimated warranty costs; tax deductible when paid.
_____ 3. Rent revenue collected in advance; cash basis for tax purposes.
_____ 4. Interest received from investments in municipal governmental bonds.
_____ 5. Prepaid expenses, tax deductible when paid.
_____ 6. Net operating loss carryforward.
_____ 7. Net operating loss carryback.
_____ 8. Straight-line depreciation for financial reporting; MACRS for tax purposes.
_____ 9. Organization costs expensed when incurred; tax deductible over 15 years.
_____ 10. Life insurance proceeds received upon the death of the company president.

E 16–27
Multiple temporary differences; balance sheet presentation
● LO16–2, LO16–3, LO16–5, LO16–6, LO16–8

At December 31, DePaul Corporation had the following cumulative temporary differences associated with its operations:

1. Estimated warranty expense, $16 million temporary difference: expense recorded in the year of the sale; tax-deductible when paid (one-year warranty).
2. Depreciation expense, $120 million temporary difference: straight-line in the income statement; MACRS on the tax return.
3. Income from installment sales of properties, $20 million temporary difference: income recorded in the year of the sale; taxable when received equally over the next five years.
4. Rent revenue collected in advance, $24 million temporary difference; taxable in the year collected; recorded as income when the performance obligation is satisfied in the following year.

Required:
Assuming DePaul will show a single noncurrent net amount in its December 31 balance sheet, indicate that amount and whether it is a net deferred tax asset or liability. The tax rate is 25%.

E 16–28
Multiple tax rates
● LO16–2, LO16–5, LO16–6

Case Development began operations in December 2021. When property is sold on an installment basis, Case recognizes installment income for financial reporting purposes in the year of the sale. For tax purposes, installment income is reported by the installment method. 2021 installment income was $600,000 and will be collected over the next three years. Scheduled collections and enacted tax rates for 2022–2024 are as follows:

2022	$140,000	20%
2023	260,000	25
2024	200,000	25

Pretax accounting income for 2021 was $810,000, which includes interest revenue of $10,000 from municipal governmental bonds. The enacted tax rate for 2021 is 20%.

Required:
1. Assuming no differences between accounting income and taxable income other than those described above, prepare the appropriate journal entry to record Case's 2021 income taxes.
2. What is Case's 2021 net income?

E 16–29
Multiple differences; multiple tax rates
● LO16–2, LO16–3, LO16–5, LO16–6

(This exercise is a variation of E 16–28, modified to include a second temporary difference.)

Case Development began operations in December 2021. When property is sold on an installment basis, Case recognizes installment income for financial reporting purposes in the year of the sale. For tax purposes, installment income is reported by the installment method. 2021 installment income was $600,000 and will be collected over the next three years. Scheduled collections and enacted tax rates for 2022–2024 are as follows:

2022	$140,000	20%
2023	260,000	25
2024	200,000	25

Case also had product warranty costs of $80,000 expensed for financial reporting purposes in 2021. For tax purposes, only the $20,000 of warranty costs actually paid in 2021 was deducted. The remaining $60,000 will be deducted for tax purposes when paid over the next three years as follows:

2022	$20,000
2023	24,000
2024	16,000

Pretax accounting income for 2021 was $810,000, which includes interest revenue of $10,000 from municipal bonds. The enacted tax rate for 2021 is 20%.

Required:
1. Assuming no differences between accounting income and taxable income other than those described above, prepare the appropriate journal entry to record Case's 2021 income taxes.
2. What is Case's 2021 net income?

E 16–30
Balance sheet classification
● LO16–8

As of December 31, 2021, Lange Company has the following deferred tax assets and liabilities:

Deferred tax assets	
Pension plans	$300,000
Inventory	200,000
Total deferred tax assets	$500,000

(continued)

(concluded)

Deferred tax liabilities	
Property, plant and equipment	$100,000
Gain on equity investments (unrealized)	350,000
Total deferred tax liabilities	$450,000

Required:

1. Assume that all of Lange's deferred tax assets and liabilities are in the same tax jurisdiction. How would deferred taxes be shown on Lange's balance sheet?
2. Assume that the deferred tax effects of Lange's pension plans and unrealized gains on investments occurred in a different tax jurisdiction from Lange's other deferred tax effects. How would deferred taxes be shown on Lange's balance sheet?

E 16–31
Concepts; terminology
● LO16–2 through LO16–8

Listed below are several terms and phrases associated with accounting for income taxes. Pair each item from List A with the item from List B (by letter) that is most appropriately associated with it.

List A	List B
____ 1. No tax consequences.	a. Deferred tax liability.
____ 2. Originates, then reverses.	b. Deferred tax asset.
____ 3. Revise deferred tax amounts.	c. 2 years.
____ 4. Operating loss.	d. Current and deferred tax consequence combined.
____ 5. Future tax effect of prepaid expenses; tax deductible when paid.	e. Temporary difference.
____ 6. Loss carryback.	f. Specific tax rates times amounts reversing each year.
____ 7. Future tax effect of estimated warranty expense.	g. Nontemporary differences.
____ 8. Valuation allowance.	h. When enacted tax rate changes.
____ 9. Phased-in change in rates.	i. Net deferred tax asset or liability.
____ 10. Balance sheet presentation.	j. "More likely than not" test.
____ 11. Individual tax consequences of financial statement components.	k. Intraperiod tax allocation.
____ 12. Income tax expense.	l. Negative taxable income.

E 16–32
Tax credit; uncertain tax position
● LO16–9

Delta Catfish Company has taken a position in its tax return to claim a tax credit of $10 million (direct reduction in taxes payable) and has determined that its sustainability is "more likely than not," based on its technical merits. Delta has developed the probability table shown below of all possible material outcomes ($ in millions):

Probability Table

Amount of the tax benefit that management expects to receive	$10	$ 8	$ 6	$ 4	$ 2
Percentage likelihood that the tax benefit will be sustained at this level	10%	20%	25%	20%	25%

Delta's taxable income is $84 million for the year. Its effective tax rate is 25%. The tax credit would be a direct reduction in current taxes payable.

Required:

1. At what amount would Delta measure the tax benefit in its income statement?
2. Prepare the appropriate journal entry for Delta to record its income taxes for the year.

E 16–33
Intraperiod tax allocation
● LO16–10

The following income statement does not reflect intraperiod tax allocation.

Required:

Recast the income statement to reflect intraperiod tax allocation.

INCOME STATEMENT
For the Fiscal Year Ended March 31, 2021
($ in millions)

Sales revenue	$830
Cost of goods sold	(350)
Gross profit	480
Operating expenses	(180)
Income tax expense	(54)
Income before discontinued operations	246
Loss from discontinued operations	(84)
Net income	$162

The company's tax rate is 25%.

E 16–34
FASB codification research
● LO16–6, LO16–8, LO16–10

Access the *FASB Accounting Standards Codification* at the FASB website (**www.fasb.org**). What is the specific eight-digit Codification citation (XXX-XX-XX-X) that applies to each of the following items:

1. The specific items to which income tax expense is allocated for intraperiod tax allocation.
2. The tax rate used to calculate deferred tax assets and liabilities.
3. The required disclosures in the notes to financial statements for the components of income tax expense.

CODE

Problems

P 16–1
Single temporary difference originates each year for four years
● LO16–2

Alsup Consulting sometimes performs services for which it receives payment at the conclusion of the engagement, up to six months after services commence. Alsup recognizes service revenue for financial reporting purposes when the services are performed. For tax purposes, revenue is reported when fees are collected. Service revenue, collections, and pretax accounting income for 2020–2023 are as follows:

	Service Revenue	Collections	Pretax Accounting Income
2020	$660,000	$620,000	$186,000
2021	750,000	778,000	260,000
2022	710,000	702,000	228,000
2023	716,000	720,000	200,000

There are no differences between accounting income and taxable income other than the temporary difference described above. The enacted tax rate for each year is 25%.

Required:

1. Prepare the appropriate journal entry to record Alsup's 2021 income taxes.
2. Prepare the appropriate journal entry to record Alsup's 2022 income taxes.
3. Prepare the appropriate journal entry to record Alsup's 2023 income taxes.

(*Hint:* You will find it helpful to prepare a schedule that shows the balances in service revenue receivable at December 31, 2020–2023.)

P 16–2
Temporary difference; determine deferred tax amount for three years
● LO16–3

Times-Roman Publishing Company reports the following amounts in its first three years of operation:

($ in thousands)	2021	2022	2023
Pretax accounting income	$250	$240	$230
Taxable income	290	220	260

The difference between pretax accounting income and taxable income is due to subscription revenue for one-year magazine subscriptions being reported for tax purposes in the year received, but reported in the income statement in later years when the performance obligation is satisfied. The income tax rate is 25% each year. Times-Roman anticipates profitable operations in the future.

Required:

1. What is the balance sheet account that gives rise to a temporary difference in this situation?
2. For each year, indicate the cumulative amount of the temporary difference at year-end.
3. Determine the balance in the related deferred tax account at the end of each year. Is it a deferred tax asset or a deferred tax liability?

P 16–3
Change in tax rate; single temporary difference
● LO16–2, LO16–6

Dixon Development began operations in December 2021. When lots for industrial development are sold, Dixon recognizes income for financial reporting purposes in the year of the sale. For some lots, Dixon recognizes income for tax purposes when collected. Income recognized for financial reporting purposes in 2021 for lots sold this way was $12 million, which will be collected over the next three years. Scheduled collections for 2022–2024 are as follows:

2022	$ 4 million
2023	5 million
2024	3 million
	$12 million

Pretax accounting income for 2021 was $15 million. The enacted tax rate is 25%.

Required:

1. Assuming no differences between accounting income and taxable income other than those described above, prepare the journal entry to record income taxes in 2021.
2. Suppose a new tax law, revising the tax rate from 25% to 20%, beginning in 2023, is enacted in 2022, when pretax accounting income was $20 million. No 2022 lot sales qualified for the special tax treatment. Prepare the appropriate journal entry to record income taxes in 2022.
3. If the new tax rate had not been enacted, what would have been the appropriate balance in the deferred tax liability account at the end of 2022? Why?

P 16–4
Change in tax rate; record taxes for four years
● LO16–2, LO16–6

Zekany Corporation would have had identical income before taxes on both its income tax returns and income statements for the years 2021 through 2024 except for differences in depreciation on an operational asset. The asset cost $120,000 and is depreciated for income tax purposes in the following amounts:

2021	$39,600
2022	52,800
2023	18,000
2024	9,600

The operational asset has a four-year life and no residual value. The straight-line method is used for financial reporting purposes.

Income amounts before depreciation expense and income taxes for each of the four years were as follows:

	2021	2022	2023	2024
Accounting income before taxes and depreciation	$60,000	$80,000	$70,000	$70,000

Assume the income tax rate for 2021 and 2022 was 30%; however, during 2022, tax legislation was passed to raise the tax rate to 40% beginning in 2023. The 40% rate remained in effect through the years 2023 and 2024. Both the accounting and income tax periods end December 31.

Required:

Prepare the journal entries to record income taxes for the years 2021 through 2024.

P 16–5
Change in tax rate; record taxes for four years
● LO16–2, LO16–5, LO16–6

The DeVille Company reported pretax accounting income on its income statement as follows:

2021	$350,000
2022	270,000
2023	340,000
2024	380,000

Included in the income of 2021 was an installment sale of property in the amount of $50,000. However, for tax purposes, DeVille reported the income in the year cash was collected. Cash collected on the installment sale was $20,000 in 2022, $25,000 in 2023, and $5,000 in 2024.

Included in the 2023 income was $15,000 interest from investments in municipal governmental bonds.

The enacted tax rate for 2021 and 2022 was 40%, but during 2022, new tax legislation was passed reducing the tax rate to 25% for the years 2023 and beyond.

Required:

Prepare the year-end journal entries to record income taxes for the years 2021–2024.

P 16–6
Multiple differences; temporary difference yet to originate; tax rates change
● LO16–2, LO16–3, LO16–6

You are the new accounting manager at the Barry Transport Company. Your CFO has asked you to provide input on the company's income tax position based on the following:

1. Pretax accounting income was $45 million and taxable income was $8 million for the year ended December 31, 2021.
2. The difference was due to three items:
   a. Tax depreciation exceeds book depreciation by $30 million in 2021 for the business complex acquired that year. This amount is scheduled to be $60 million in 2022 and to reverse as ($50 million) and ($40 million) in 2023 and 2024, respectively.
   b. Insurance of $12 million was paid in 2021 for 2022 coverage.
   c. A $5 million loss contingency was accrued in 2021, to be paid in 2023.
3. No temporary differences existed at the beginning of 2021.
4. The tax rate is 25%.

Required:

1. Determine the amounts necessary to record income taxes for 2021, and prepare the appropriate journal entry.
2. Assume the enacted federal income tax law specifies that the tax rate will change from 25% to 20% in 2023. When scheduling the reversal of the depreciation difference, you were uncertain as to how to deal with the fact that the difference will continue to originate in 2022 before reversing the next two years. Upon consulting **PricewaterhouseCoopers**' *Comperio* database, you found:

   .441 Depreciable and amortizable assets
   Only the reversals of the temporary difference at the balance sheet date would be scheduled. Future originations are not considered in determining the reversal pattern of temporary differences for depreciable assets. *FAS 109* [FASB ASC 740–Income Taxes] is silent as to how the balance sheet date temporary differences are deemed to reverse, but the FIFO pattern is intended.

You interpret that to mean, when future taxable amounts are being scheduled, and a portion of a temporary difference has yet to originate, only the reversals of the *temporary difference at the balance sheet date* can be scheduled and multiplied by the tax rate that will be in effect when the difference reverses. Future originations (like the depreciation difference the second year) are not considered when determining the timing of the reversal. For the existing temporary difference, it is assumed that the difference will reverse the first year the difference begins reversing.

Determine the amounts necessary to record income taxes for 2021, and prepare the appropriate journal entry.

P 16–7
Multiple differences; calculate taxable income; balance sheet classification
● LO16–2, LO16–3, LO16–5, LO16–8

Sherrod, Inc., reported pretax accounting income of $76 million for 2021. The following information relates to differences between pretax accounting income and taxable income:

a. Income from installment sales of properties included in pretax accounting income in 2021 exceeded that reported for tax purposes by $3 million. The installment receivable account at year-end 2021 had a balance of $7 million (representing portions of 2020 and 2021 installment sales), expected to be collected equally in 2022 and 2023.

b. Sherrod was assessed a penalty of $2 million by the Environmental Protection Agency for violation of a federal law in 2021. The fine is to be paid in equal amounts in 2021 and 2022.

c. Sherrod rents its operating facilities but owns one asset acquired in 2020 at a cost of $80 million. Depreciation is reported by the straight-line method, assuming a four-year useful life. On the tax return, deductions for depreciation will be more than straight-line depreciation the first two years but less than straight-line depreciation the next two years ($ in millions):

	Income Statement	Tax Return	Difference
2020	$20	$26	$ (6)
2021	20	35	(15)
2022	20	12	8
2023	20	7	13
	$80	$80	$ 0

d. For tax purposes, warranty expense is deducted when costs are incurred. The balance of the warranty liability was $2 million at the end of 2020. Warranty expense of $4 million is recognized in the income statement in 2021. $3 million of cost is incurred in 2021, and another $3 million of cost anticipated in 2022. At December 31, 2021, the warranty liability is $3 million (after adjusting entries).

e. In 2021, Sherrod accrued an expense and related liability for estimated paid future absences of $7 million relating to the company's new paid vacation program. Future compensation will be deductible on the tax return when actually paid during the next two years ($4 million in 2022; $3 million in 2023).

f. During 2020, accounting income included an estimated loss of $2 million from having accrued a loss contingency. The loss is paid in 2021, at which time it is tax deductible.

Balances in the deferred tax asset and deferred tax liability accounts at January 1, 2021, were $1 million and $2.5 million, respectively. The enacted tax rate is 25% each year.

Required:

1. Determine the amounts necessary to record income taxes for 2021, and prepare the appropriate journal entry.
2. What is the 2021 net income?
3. Show how any deferred tax amounts should be classified and reported in the 2021 balance sheet.

P 16–8
Multiple differences; taxable income given; two years; balance sheet classification; change in tax rate
● LO16–1, LO16–2, LO16–3, LO16–5, LO16–6, LO16–8

Arndt, Inc. reported the following for 2021 and 2022 ($ in millions):

	2021	2022
Revenues	$888	$980
Expenses	760	800
Pretax accounting income (income statement)	$128	$180
Taxable income (tax return)	$116	$200
Tax rate: 25%		

a. Expenses each year include $30 million from a two-year casualty insurance policy purchased in 2021 for $60 million. The cost is tax deductible in 2021.

b. Expenses include $2 million insurance premiums each year for life insurance on key executives.

c. Arndt sells one-year subscriptions to a weekly journal. Subscription sales collected and taxable in 2021 and 2022 were $33 million and $35 million, respectively. Subscriptions included in 2021 and 2022 financial reporting revenues were $25 million ($10 million collected in 2020 but not recognized as revenue until 2021) and $33 million, respectively. *Hint:* View this as two temporary differences—one reversing in 2021; one originating in 2021.

d. 2021 expenses included a $14 million unrealized loss from reducing investments (classified as trading securities) to fair value. The investments were sold and the loss realized in 2022.

e. During 2020, accounting income included an estimated loss of $6 million from having accrued a loss contingency. The loss was paid in 2021, at which time it is tax deductible.

f. At January 1, 2021, Arndt had a deferred tax asset of $4 million and no deferred tax liability.

Required:

1. Which of the five differences described in items a–e are temporary and which are permanent differences? Why?
2. Prepare a schedule that (a) reconciles the difference between pretax accounting income and taxable income and (b) determines the amounts necessary to record income taxes for 2021. Prepare the appropriate journal entry.
3. Show how any 2021 deferred tax amounts should be classified and reported on the 2021 balance sheet.
4. Prepare a schedule that (a) reconciles the difference between pretax accounting income and taxable income and (b) determines the amounts necessary to record income taxes for 2022. Prepare the appropriate journal entry.
5. Explain how any 2022 deferred tax amounts should be classified and reported on the 2022 balance sheet.
6. Suppose that during 2022, tax legislation was passed that will lower Arndt's effective tax rate to 15% beginning in 2023. Repeat requirement 4.

P 16–9
Determine deferred tax assets and liabilities from book-tax differences
● LO16–2, LO16–3

Corning-Howell reported taxable income in 2021 of $120 million. At December 31, 2021, the reported amount of some assets and liabilities in the financial statements differed from their tax bases as indicated below:

	Carrying Amount	Tax Basis
Assets		
Current		
Net accounts receivable	$ 10 million	$ 12 million
Prepaid insurance	20 million	0
Prepaid advertising	8 million	0
Noncurrent		
Investments in equity securities (fair value)*	4 million	0
Buildings and equipment (net)	360 million	280 million
Liabilities		
Current		
Deferred subscription revenue	14 million	0
Long-term		
Liability—compensated future absences	594 million	0

*Gains and losses taxable when investments are sold.

The total deferred tax asset and deferred tax liability amounts at January 1, 2021, were $156.25 million and $25 million, respectively. The enacted tax rate is 25% each year.

Required:

1. Determine the total deferred tax asset and deferred tax liability amounts at December 31, 2021.
2. Determine the increase (decrease) in the deferred tax asset and deferred tax liability accounts at December 31, 2021.
3. Determine the income tax payable currently for the year ended December 31, 2021.
4. Prepare the journal entry to record income taxes for 2021.

P 16–10
Net operating loss carryforward; multiple differences
● LO16–3, LO16–5, LO16–7

Fore Farms reported a pretax operating loss of $137 million for financial reporting purposes in 2021. Contributing to the loss were (a) a penalty of $5 million assessed by the Environmental Protection Agency for violation of a federal law and paid in 2021 and (b) an estimated loss of $12 million from accruing a loss contingency. The loss will be tax deductible when paid in 2022.

The enacted tax rate is 25%. There were no temporary differences at the beginning of the year and none originating in 2021 other than those described above.

Required:

1. Prepare the journal entry to recognize the income tax benefit of the net operating loss in 2021.
2. Show the lower portion of the 2021 income statement that reports the income tax benefit of the net operating loss.
3. Prepare the journal entry to record income taxes in 2022 assuming pretax accounting income is $160 million. No additional temporary differences originate in 2022.

P 16–11
Net operating loss carryback and carryforward; multiple differences
● LO16–3, LO16–5, LO16–7

(Note: this problem is a variation of P 16–10, modified to allow a net operating loss carryback.) Fore Farms reported a pretax operating loss of $137 million for financial reporting purposes in 2021. Contributing to the loss were (a) a penalty of $5 million assessed by the Environmental Protection Agency for violation of a federal law and paid in 2021 and (b) an estimated loss of $12 million from accruing a loss contingency. The loss will be tax deductible when paid in 2022.

The enacted tax rate is 25%. There were no temporary differences at the beginning of the year and none originating in 2021 other than those described above. Taxable income in Fores's two previous years of operation was as follows:

2019	$80 million
2020	$32 million

Required:

1. Prepare the journal entry to recognize the income tax benefit of the net operating loss in 2021. Assume Fore will carry back its NOL to prior years.
2. Show the lower portion of the 2021 income statement that reports the income tax benefit of the net operating loss.
3. Prepare the journal entry to record income taxes in 2022 assuming pretax accounting income is $160 million. No additional temporary differences originate in 2022.

P 16–12
Integrating problem—bonds, leases, taxes
● LO16–2, LO16–6, LO16–8

The long-term liabilities section of CPS Transportation's December 31, 2020, balance sheet included the following:

a. A lease liability with 15 remaining lease payments of $10,000 each, due annually on January 1:

Lease liability	$76,061
Less: Current portion	2,394
	$73,667

The incremental borrowing rate at the inception of the lease was 11% and the lessor's implicit rate, which was known by CPS Transportation, was 10%.

b. A deferred income tax liability due to a single temporary difference. The only difference between CPS Transportation's taxable income and pretax accounting income is depreciation on a machine acquired on January 1, 2020, for $500,000. The machine's estimated useful life is five years, with no salvage value. Depreciation is computed using the straight-line method for financial reporting purposes and the MACRS method for tax purposes. Depreciation expense for tax and financial reporting purposes for 2021 through 2024 is as follows:

Year	MACRS Depreciation	Straight-line Depreciation	Difference
2021	$160,000	$100,000	$60,000
2022	80,000	100,000	(20,000)
2023	70,000	100,000	(30,000)
2024	60,000	100,000	(40,000)

The enacted federal income tax rates are 20% for 2020 and 25% for 2021 through 2024. CPS had a deferred tax liability of $7,500 as of December 31, 2020. For the year ended December 31, 2021, CPS's income before income taxes was $900,000.

On July 1, 2021, CPS Transportation issued $800,000 of 9% bonds. The bonds mature in 20 years, and interest is payable each January 1 and July 1. The bonds were issued at a price to yield the investors 10%. CPS records interest at the effective interest rate.

Required:

1. Determine CPS Transportation's income tax expense and net income for the year ended December 31, 2021.
2. Determine CPS Transportation's interest expense for the year ended December 31, 2021.
3. Prepare the long-term liabilities section of CPS Transportation's December 31, 2021, balance sheet.

P 16–13
Multiple differences; uncertain tax position
● LO16–2, LO16–5, LO16–9

Tru Developers, Inc., sells plots of land for industrial development. Tru recognizes income for financial reporting purposes in the year it sells the plots. For some of the plots sold this year, Tru took the position that it could recognize the income for tax purposes when the installments are collected. Income that Tru recognized for financial reporting purposes in 2021 for plots in this category was $60 million. The company expected to collect 60% of each sale in 2022 and 40% in 2023. This amount over the next two years is as follows:

2022	$36 million
2023	24 million
	$60 million

Tru's pretax accounting income for 2021 was $88 million. In its income statement, Tru reported interest income of $16 million, unrelated to the land sales, for which the company's position is that the interest is not taxable. Accordingly, the interest was not reported on the tax return. There are no differences between accounting income and taxable income other than those described above. The enacted tax rate is 25%.

Management believes the tax position taken on the land sales has a greater than 50% chance of being upheld based on its technical merits, but the position taken on the interest has a less than 50% chance of being upheld. It is further believed that the following likelihood percentages apply to the tax treatment of the land sales ($ in millions):

Amount Qualifying for Installment Sales Treatment	Percentage Likelihood of Tax Treatment Being Sustained
$60	20%
50	20%
40	20%
30	20%
20	20%

Required:

1. What portion of the tax benefit of tax-free interest will Tru recognize on its 2021 tax return?
2. What portion of the tax benefit of tax-free interest will Tru recognize on its 2021 financial statements?
3. (a) What portion of the tax on the $60 million income from the plots sold on an installment basis will Tru defer on its 2021 tax return? (b) What portion of the tax on the $60 million income from the plots sold on an installment basis will Tru show as a deferred tax asset or liability in its 2021 financial statements? (c) How is the difference between these two amounts reported?
4. Prepare the journal entry to record income taxes in 2021, assuming full recognition of the tax benefits in the financial statements of both differences between pretax accounting income and taxable income.
5. Prepare the journal entry to record income taxes in 2021, assuming the recognition of the tax benefits in the financial statements you indicated in requirements 1–3.

Decision Maker's Perspective

©ImageGap/Getty Images

Apply your critical-thinking ability to the knowledge you've gained. These cases will provide you an opportunity to develop your research, analysis, judgment, and communication skills. You also will work with other students, integrate what you've learned, apply it in real-world situations, and consider its global and ethical ramifications. This practice will broaden your knowledge and further develop your decision-making abilities.

Analysis Case 16–1
Basic concepts
● LO16–1 through LO16–8

One of the longest debates in accounting history is the issue of deferred taxes. The controversy began in the 1940s and has continued, even after the FASB issued *Statement of Financial Accounting Standards No. 109* [FASB ASC 740: Income Taxes] in 1992. At issue is the appropriate treatment of tax consequences of economic events that occur in years other than that of the events themselves.

Required:

1. Distinguish between temporary differences and permanent differences. Provide an example of each.
2. Distinguish between *intraperiod* tax allocation and *interperiod* tax allocation (deferred tax accounting). Provide an example of each.
3. How are deferred tax assets and deferred tax liabilities classified and reported in the financial statements?

Integrating Case 16–2
Postretirement benefits
● LO16–3

FASB ASC 715–60: Compensation—Retirement Benefits—Defined Benefit Plans—Other Postretirement (previously *Statement of Financial Accounting Standards No. 106)* establishes accounting standards for postretirement benefits other than pensions, most notably postretirement health care benefits. Essentially, the standard requires companies to accrue compensation expense each year employees perform services, for the expected cost of providing future postretirement benefits that can be attributed to that service. Typically, companies do not prefund these costs for two reasons: (a) unlike pension liabilities, no federal law requires companies to fund nonpension postretirement benefits, and (b) funding contributions, again unlike for pension liabilities, are not tax deductible. (The costs aren't tax deductible until paid to, or on behalf of, employees.)

Required:

1. As a result of being required to record the periodic postretirement expense and related liability, most companies report lower earnings and higher liabilities. How might many companies also report higher deferred tax assets as a result of GAAP for postretirement plans?
2. One objection to current GAAP as cited in the chapter is the omission of requirements to discount deferred tax amounts to their present values. This objection is inappropriate in the context of deferred tax amounts necessitated by accounting for postretirement benefits. Why?

Integrating Case 16–3
Tax effects of accounting changes and error correction; six situations
● LO16–2, LO16–3, LO16–8

Williams-Santana Inc. is a manufacturer of high-tech industrial parts that was started in 2007 by two talented engineers with little business training. In 2021, the company was acquired by one of its major customers. As part of an internal audit, the following facts were discovered. The audit occurred during 2021 before any adjusting entries or closing entries were prepared. The income tax rate is 25% for all years.

a. A five-year casualty insurance policy was purchased at the beginning of 2019 for $35,000. The full amount was debited to insurance expense at the time.

b. On December 31, 2020, merchandise inventory was overstated by $25,000 due to a mistake in the physical inventory count using the periodic inventory system.

c. The company changed inventory cost methods to FIFO from LIFO at the end of 2021 for both financial statement and income tax purposes. The change will cause a $960,000 increase in the beginning inventory at January 1, 2020.

d. At the end of 2020, the company failed to accrue $15,500 of sales commissions earned by employees during 2020. The expense was recorded when the commissions were paid in early 2021.

e. At the beginning of 2019, the company purchased a machine at a cost of $720,000. Its useful life was estimated to be 10 years with no salvage value. The machine has been depreciated by the double declining-balance method. Its carrying amount on December 31, 2020, was $460,800. On January 1, 2021, the company changed to the straight-line method.

f. Additional industrial robots were acquired at the beginning of 2018 and added to the company's assembly process. The $1,000,000 cost of the equipment was inadvertently recorded as repair expense. The robots have 10-year useful lives and no material salvage value. This class of equipment is depreciated by the straight-line method for both financial reporting and income tax reporting.

Required:

For each situation:

1. Identify whether it represents an accounting change or an error. If an accounting change, identify the type of change.
2. Prepare any journal entry necessary as a direct result of the change or error correction, as well as any adjusting entry for 2021 related to the situation described. Any tax effects should be adjusted for through the deferred tax liability account.
3. Briefly describe any other steps that should be taken to appropriately report the situation.

Communication Case 16–4
Deferred taxes; changing rates; write a memo
● LO16–2, LO16–5, LO16–6

The date is November 15, 2017. You are the new controller for Engineered Solutions. The company treasurer, Randy Patey, believes that as a result of pending legislation, the currently enacted 40% income tax rate may be decreased for 2018 to 25% and is uncertain which tax rate to apply in determining deferred taxes for 2017. Patey also is uncertain which temporary differences should be included in that determination and has solicited your help. Your accounting group provided you the following information.

Two items are relevant to the decisions. One is the $50,000 insurance premium the company pays annually for the CEO's life insurance policy, for which the company is the beneficiary. The second is that Engineered Solutions purchased a building on January 1, 2016, for $6,000,000. The building's estimated useful life is 30 years from the date of purchase, with no salvage value. Depreciation is computed using the straight-line method for financial reporting purposes and the MACRS method for tax purposes. As a result, the building's tax basis is $5,200,000 at December 31, 2017.

Required:

Write a memo to Patey that:

a. Identifies the objectives of accounting for income taxes.

b. Differentiates temporary differences and permanent differences.

c. Explains which tax rate to use.

d. Calculates the deferred tax liability at December 31, 2017.

Real World Case 16–5
Disclosure issues; balance sheet classifications; Walmart
● LO16–2, LO16–3, LO16–4, LO16–8

Real World Financials

The income tax disclosure note accompanying the January 31, 2017, financial statements of **Walmart** is reproduced below:

	2017	2016	2015
Current:			
U.S. federal	$3,454	$5,562	$6,165
U.S. state and local	495	622	810
International	1,510	1,400	1,529
Total current tax provision	5,459	7,584	8,504
Deferred:			
U.S. federal	1,054	(704)	(387)
U.S. state and local	51	(106)	(55)
International	(360)	(216)	(77)
Total deferred tax expense (benefit)	745	(1,026)	(519)
Total provision for income taxes	$6,204	$6,558	$7,985

	2017	2016
Deferred tax assets:		
Loss and tax credit carryforwards	$3,633	$3,313
Accrued liabilities	3,437	3,763
Share-based compensation	309	192
Other	1,474	1,390
Total deferred tax assets	8,853	8,658
Valuation allowances	(1,494)	(1,456)
Deferred tax assets, net of valuation allowance	$7,359	$7,202
Deferred tax liabilities:		
Property and equipment	$6,435	$5,813
Inventories	1,808	1,790
Other	1,884	1,452
Total deferred tax liabilities	10,127	9,055
Net deferred tax liabilities	$2,768	$1,853

Required:

1. Focusing on only the first part of Note 9, relating current, deferred, and total provision for income taxes, prepare a summary journal entry that records Walmart's 2017 tax expense associated with income from continuing operations.
2. Calculate the actual change in Walmart's net deferred tax liability for fiscal 2017. Does that change reconcile with the change indicated in your summary journal entry? What besides continuing operations might affect deferred taxes?

Source: Walmart

C 16–6
Valuation allowance; Delta Air Lines
● LO16–4

Real World Financials

Delta Air Lines revealed in its 10-K filing that its valuation allowance for deferred tax assets at the end of 2013 was $177 million, dramatically lower than the over $10 billion recorded at the end of 2012. Here is an excerpt from a press report from Bloomberg in January 2014, regarding this allowance:

> Delta Air Lines Inc. (DAL) led shares of U.S. carriers higher after posting fourth-quarter profit that topped analysts' estimates and forecasting an operating margin of as much as 8 percent in this year's initial three months. . . . Airlines are benefiting from lower fuel prices, constraints on capacity growth, controls on operating costs and demand that's keeping planes full, said Ray Neidl of Nexa Capital Partners LLC, a Washington-based aerospace and transportation consulting firm. . . . Net income was $8.48 billion, including an $8 billion non-cash gain from the reversal of a tax valuation allowance.[20]

[20] Mary Schlangenstein, "Delta Leads Airline Stock Gains as Profit Beats Estimates," *Bloomberg*, January 21, 2014.

The following is an excerpt from a disclosure note to Delta's 2013 financial statements:

NOTE 12. INCOME TAXES *(In part)*

Deferred Taxes

The components of deferred tax assets and liabilities at December 31 were as follows ($ in millions):

	December 31,	
	2013	**2012**
Deferred tax assets:		
Net operating loss carryforwards	$ 6,024	$ 6,414
Pension, postretirement and other benefits	4,982	6,415
AMT credit carryforward	378	402
Deferred revenue	1,965	2,133
Other	698	881
Valuation allowance	(177)	(10,963)
Total deferred tax assets	$13,870	$ 5,282
Deferred tax liabilities:		
Depreciation	$ 4,799	$ 4,851
Intangible assets	1,704	1,730
Other	639	285
Total deferred tax liabilities	$ 7,142	$ 6,866

Required:

1. What is a valuation allowance against deferred tax assets? When must such an allowance be recorded? Use Delta's situation to help illustrate your response.
2. Is an amount recorded in a valuation allowance for a deferred tax asset permanent? Explain why Delta is able to reclaim its valuation allowance.
3. Consider the excerpt from Bloomberg's press release. Recalculate the effect on Delta's 2013 net income of the change in Delta's valuation allowance for its deferred tax assets.

Source: Delta Air Lines

Research Case 16–7
Researching the way tax deductions are reported on a corporation tax return; retrieving a tax form from the Internet
● LO16–2, LO16–3, LO16–4, LO16–8

The Internal Revenue Service (IRS) maintains an information site on the Internet that provides tax information and services. Among those services is a server for publications and forms which allows a visitor to download a variety of IRS forms and publications.

Required:

1. Access the IRS site at irs.gov. After exploring the information available there, navigate to the forms and instructions page.
2. Download the corporation tax return, Form 1120.
3. Note the specific deductions listed that are deductible from total income to arrive at taxable income. Are any deductions listed that might not also be included among expenses in the income statement? One of the deductions indicated is "net operating loss deduction." Will every company report this tax deduction in every year? If not, under what circumstances might a company report an amount for this item?

Analysis Case 16–8
Reporting deferred taxes; Ford Motor Company
● LO16–2, LO16–3, LO16–4, LO16–7

Real World Financials

Access the 2016 financial statements and related disclosure notes of **Ford Motor Company** from its website at **corporate.ford.com**.

Required:

1. In Note 21, find Ford's net deferred tax asset or liability. What is that number?
2. Does Ford show a valuation allowance against deferred tax assets? If so, what is the number, and what is Ford's explanation for it?
3. Does Ford have any NOL carryforwards? What is the amount of any carryforward, what deferred tax asset or liability is associated with it, and what effective tax rate does that imply was used to calculate its deferred tax effect?

Analysis Case 16–9
Reporting deferred taxes; Kroger Co.
● LO16–2, LO16–3, LO16–4, LO16–7, LO16–8

Real World Financials

Kroger Co. is one of the largest retail food companies in the United States as measured by total annual sales. The Kroger Co. operates supermarkets, convenience stores, and manufactures and processes food that its supermarkets sell.

Using EDGAR (**sec.gov**) or the company's website (**kroger.com**), check the company's annual report for the year ended January 28, 2017.

Required:

1. From the income statement, determine the income tax expense for the most recent year. Tie that number to the first table in disclosure Note 5: "Taxes Based on Income," and prepare a summary journal entry that records Kroger's tax expense from continuing operations in the most recent year.
2. In 2016 companies could classify their deferred tax assets and liabilities as current or noncurrent. From Kroger's Note 5, calculate the total (current + noncurrent) net deferred tax asset or liability as of January 28, 2017, and January 30, 2016. By how much did that amount change? To what extent did you account for that change in the journal entry you wrote for the first requirement of this case? Speculate as to the explanation of any difference.

Judgment Case 16–10
Analyzing the effect of deferred tax liabilities on firm risk; Macy's, Inc.
● LO16–8

Real World Financials

The following is a portion of the balance sheets of **Macy's, Inc.** for the years ended January 28, 2017 and January 30, 2016:

	January 28, 2017	January 30, 2016
LIABILITIES AND SHAREHOLDERS' EQUITY		
Current Liabilities:		
Short-term debt	$ 309	$ 642
Merchandise accounts payable	1,423	1,526
Accounts payable and accrued liabilities	3,563	3,333
Income taxes	352	227
Total Current Liabilities	5,647	5,728
Long-Term debt	6,562	6,995
Deferred Income Taxes	1,443	1,477
Other Liabilities	1,877	2,123
Shareholders' Equity		
Common stock (310.8 and 340.6 shares outstanding)	3	3
Additional paid-in capital	617	621
Accumulated equity	6,088	6,334
Treasury stock	(1,489)	(1,665)
Accumulated other comprehensive loss	(896)	(1,043)
Total Macy's, Inc. Shareholders' Equity	4,323	4,250
Noncontrolling interest	(1)	3
Total Shareholders' Equity	4,322	4,253
Total Liabilities and Shareholders' Equity	$19,851	$20,576

Required:

1. What is Macy's debt to equity ratio for the year ended January 28, 2017?
2. What would Macy's debt to equity ratio be if we excluded deferred tax liabilities from its calculation? What would be the percentage change?
3. What might be the rationale for not excluding long-term deferred tax liabilities from liabilities when computing the debt to equity ratio?

Source: Macy's, Inc.

Trueblood Accounting Case 16–11
Valuation allowances against deferred tax assets
● LO16–4

The following Trueblood case is recommended for use with this chapter. The case provides an excellent opportunity for class discussion, group projects, and writing assignments. The case, along with Professor's Discussion material, can be obtained from the Deloitte foundation at its website: **www.deloitte.com/us/truebloodcases.**

Case 13-10: LOL – Income Taxes

This case gives students an opportunity to better understand how valuation allowances against deferred tax assets are estimated and calculated. Students consider the sources of taxable income that can be used to determine whether a deferred tax asset is more likely than not to be realized in the future.

Trueblood Accounting Case 16–12
Uncertain tax positions
● LO16–9

The following Trueblood case is recommended for use with this chapter. The case provides an excellent opportunity for class discussion, group projects, and writing assignments. The case, along with Professor's Discussion Material, can be obtained from the Deloitte Foundation at its website: **www.deloitte.com/us/truebloodcases**.

Case 15-9: Settled or Not Settled

This case gives the students an opportunity to better understand accounting for uncertain tax positions. The case illustrates the judgments involved in applying FASB ASC 740 (formerly FIN 48).

Judgment Case 16–13
Intraperiod tax allocation
● LO16–10

Russell-James Corporation is a diversified consumer products company. During 2021, Russell-James discontinued its line of cosmetics, which constituted discontinued operations for financial reporting purposes. As vice president of the food products division, you are interested in the effect of the discontinuance on the company's profitability. One item of information you requested was an income statement. The income statement you received was labeled *preliminary* and *unaudited:*

RUSSELL-JAMES CORPORATION
Income Statement
For the Year Ended December 31, 2021
($ in millions, except per share amounts)

Sales revenue		$ 300
Cost of goods sold		(90)
Gross profit		210
Selling and administrative expenses		(50)
Income from continuing operations before income taxes		160
Income tax expense		(19)
Income from continuing operations		141
Discontinued operations:		
Loss from operations of cosmetics division	$(100)	
Gain from disposal of cosmetics division	16	(84)
Net income		$ 57
Per Share of Common Stock (100 million shares):		
Income from continuing operations		$ 1.41
Loss from operations of cosmetics division		(1.00)
Gain from disposal of cosmetics division		0.16
Net income		$ 0.57

You are somewhat surprised at the magnitude of the loss incurred by the cosmetics division prior to its disposal. Another item that draws your attention is the apparently low tax rate indicated by the statement ($19 ÷ 160 = 12%). Upon further investigation, you are told the company's tax rate is 25%.

Required:

1. Recast the income statement to reflect intraperiod tax allocation.
2. How would you reconcile the income tax expense shown on the statement above with the amount your recast statement reports?

Source: Russell-James Corporation

Data Analytics

connect

Data analytics is the process of examining data sets in order to draw conclusions about the information they contain. If you haven't completed any of the prior data analytics cases, follow the instructions listed in the Chapter 1 Data Analytics case to get set up. You will need to watch the videos referred to in the Chapters 1 - 3 Data Analytics cases. No additional videos are required for this case. All short training videos can be found here: **https://www.tableau.com/learn/training#getting-started**.

Data Analytics Case
Deferred Taxes and the Tax Cuts and Jobs Act of 2017
● LO16–4

In the Data Analytics Cases in the previous chapter, you used Tableau to examine a data set and create charts to examine two (hypothetical) publicly traded companies: GPS Corporation and Tru, Inc. as to their pattern of leasing facilities, their transition to the new lease accounting standard in 2019, and the effect of that transition on debt covenants. In this case, you examine the effect of the Tax Cuts and Jobs Act of 2017 on these companies' operations and financial position.

Required:

For each of the two companies in the ten-year period, 2012-2021, use Tableau to calculate and display the trends for (a) the provision for income taxes (b) the deferred tax liability, and (c) effective tax rate. Based upon what you find, answer the following questions:

1. Is Tru, Inc.'s provision for income tax (a) higher, (b) lower, or (c) roughly the same over the period 2018-2021 than in previous years?
2. Did Tru, Inc.'s deferred tax liability appear to benefit from the Tax Act?
3. What is the effective tax rate for GPS in 2017 (calculated as the provision for income tax divided by income from continuing operations plus provision for income tax)?
4. What is the effective tax rate for GPS in 2018 (calculated as the provision for income tax divided by income from continuing operations plus provision for income tax)?

Resources:

Download the "GPS_Tru_Financials.xlsx" Excel file available in Connect or under Student Resources within the Library tab. Save it to the computer on which you will be using Tableau.

For this case, you will create calculations to produce the effective tax rate to allow you to compare and contrast the effect of the 2017 Tax Act on the two companies.

After you view the training videos, follow these steps to create the charts you'll use for this case:

- Start Tableau and open the Excel spreadsheet you downloaded.
- Starting on the Sheet 1 tab at the bottom of the canvas to the right of the data source, drag "Company" and "Year" under Dimensions to the Columns shelf. Change "Year" to *discrete* by right-clicking and selecting "Discrete."
- Drag "Provision for income tax" and "Deferred income taxes" under Measures into the Rows shelf.
- Add labels to the bars by clicking on "Label" under the "Marks" card and clicking the box "Show mark labels." Format the labels to Times New Roman, bold, black and 10-point font. Edit the color of the years on the "Marks card" if desired by dragging "Year" on to the Color Marks card.
- Change the title of the sheet to be "Provision for Income Taxes and Deferred Tax Liability Trend 2012–2021" by right-clicking and selecting "Edit title." Format the title to Times New Roman, bold, black and 15-point font. Change the title of "Sheet 1" to match the sheet title by right-clicking, selecting "Rename," and typing in the new title.
- On the Sheet 2 tab, follow the procedure above for the company and year.
- Create a calculated field by clicking the "Analysis" tab at the top of the screen and selecting "Create Calculated Field." Name the calculation "Effective Tax Rate." In the Calculation Editor window, drag "Provision for income taxes" and type a division sign. Then type "(", drag "Income from continuing operations," type an addition sign, drag "Provision for income taxes," and type ")" from the Rows shelf. Make sure the window says that the calculation is valid and click OK.
- Drag the newly created "Effective Tax Rate" to the Rows shelf. Click on the "Show Me" and select "side-by-side bars." Add labels to the bars by clicking on "Label" under the "Marks" card and clicking the box "Show mark labels." Format the labels to Times New Roman, bold, black and 10-point font. Edit the color on the "Marks" card if desired.
- Change the title of the sheet to be "Effect of the Tax Cuts and Jobs Act of 2017" by right-clicking and selecting "Edit title." Format the title to Times New Roman, bold, black and 15-point font. Change the title of "Sheet 2" to match the sheet title by right-clicking, selecting "Rename" and typing in the new title.
- Format all other labels to be Times New Roman, bold, black and 12-point font.
- Once complete, save the file as "DA16_Your initials.twbx."

Continuing Cases

Target Case
● LO16-2, LO16-3, LO16-5, LO16-8, LO16-9

Real World Financials

Target Corporation prepares its financial statements according to U.S. GAAP. Target's financial statements and disclosure notes for the year ended February 3, 2018, are available in Connect. This material also is available under the Investor Relations link at the company's website (**www.target.com**).

Required:

1. From the income statement, determine the income tax expense for the year ended February 3, 2018. Tie that number to the second table in disclosure Note 23, "Provision for Income Taxes," and prepare a summary journal entry that records Target's tax expense from continuing operations for the year ended February 3, 2018.

2. Focusing on the third table in disclosure Note 23, "Net Deferred Tax Asset/(Liability)," calculate the change in net deferred tax assets or liability. By how much did that amount change? To what extent did you account for that change in the journal entry you wrote for the first requirement of this case? List possible causes of any difference.
3. Target's Note 23 indicates that "In December 2017, the U.S. government enacted the Tax Cuts and Jobs Act tax reform legislation (the Tax Act), which among other matters reduced the U.S. corporate income tax rate from 35 percent to 21 percent effective January 1, 2018. . . . We have recorded a provisional $352 million net tax benefit primarily related to the remeasurement of certain deferred tax assets and liabilities, including $381 million of benefit from the new lower rate, partially offset by $29 million of deferred income tax expense from our foreign operations." What's the effect on net income?
4. What is Target's liability for unrecognized tax benefits as of February 3, 2018? If Target were to prevail in court and realize $50 million more in tax savings than it thought more likely than not to occur, what would be the effect on the liability for unrecognized tax benefits and on net income?

Air France–KLM Case

● LO16–11

IFRS

AIRFRANCE KLM

Real World Financials

Air France–KLM (AF), a Franco-Dutch company, prepares its financial statements according to International Financial Reporting Standards. AF's financial statements and disclosure notes for the year ended December 31, 2017, are available in Connect. This material is also available under the Finance link at the company's website (**www.airfranceklm.com**).

Required:

1. What amounts are shown in AF''s December 31, 2017, balance sheet for deferred taxes?
2. Here's an excerpt from AF's notes to its financial statements:

> **Deferred taxes (in part)**
> The Group records deferred taxes using the balance sheet liability method, providing for any temporary differences between the carrying amounts of assets and liabilities for financial reporting purposes and the amounts used for taxation purposes, except for exceptions described in IAS 12 "Income taxes." The tax rates used are those enacted or substantively enacted at the balance sheet date.

Is this policy consistent with U.S. GAAP? Explain.

3. Here's an excerpt from one of AF's notes to its financial statements:

> **Deferred taxes (in part)**
> Deferred tax assets related to temporary differences and tax losses carried forward are recognized only to the extent it is probable that a future taxable profit will be available against which the asset can be utilized at the tax entity level.

Is this policy consistent with U.S. GAAP? Explain.

CPA Exam Questions and Simulations

ROGER CPA Review

Sample CPA Exam questions from Roger CPA Review are available in Connect as support for the topics in this chapter. These multiple-choice questions and task-based simulations include expert-written explanations and solutions, and provide a starting point for students to become familiar with the content and functionality of the actual CPA Exam.

CHAPTER

17

Pensions and Other Postretirement Benefits

OVERVIEW

Employee compensation comes in many forms. Salaries and wages, of course, provide direct and current payment for services provided. However, it's commonplace for compensation also to include benefits payable after retirement. We discuss pension benefits and other postretirement benefits in this chapter. Accounting for pension benefits recognizes that they represent deferred compensation for current service. Accordingly, the cost of these benefits is recognized on an accrual basis during the years that employees earn the benefits.

LEARNING OBJECTIVES

After studying this chapter, you should be able to:

- **LO17–1** Explain the fundamental differences between a defined contribution pension plan and a defined benefit pension plan. (*p. 978*)
- **LO17–2** Distinguish among the vested benefit obligation, the accumulated benefit obligation, and the projected benefit obligation (PBO). (*p. 983*)
- **LO17–3** Describe the five events that might change the balance of the PBO. (*p. 984*)
- **LO17–4** Explain how plan assets accumulate to provide retiree benefits and understand the role of the trustee in administering the fund. (*p. 990*)
- **LO17–5** Describe the funded status of pension plans and how that amount is reported. (*p. 992*)
- **LO17–6** Describe how pension expense is a composite of periodic changes that occur in both the pension obligation and the plan assets. (*p. 992*)
- **LO17–7** Record for pension plans the periodic expense and funding as well as new gains and losses and new prior service cost as they occur. (*p. 998*)
- **LO17–8** Understand the interrelationships among the elements that constitute a defined benefit pension plan. (*p. 1004*)
- **LO17–9** Describe the nature of postretirement benefit plans other than pensions and identify the similarities and differences in accounting for those plans and pensions. (*p. 1009*)
- **LO17–10** Explain how the obligation for postretirement benefits is measured and how the obligation changes. (*p. 1011*)
- **LO17–11** Determine the components of postretirement benefit expense. (*p. 1012*)
- **LO17–12** Discuss the primary differences between U.S. GAAP and IFRS with respect to accounting for postretirement benefit plans. (*pp. 998, 1001, 1005,* and *1007*)

©Peter Bernik/Shutterstock

FINANCIAL REPORTING CASE

United Dynamics

You read yesterday that many companies in the United States have pension plans that are severely underfunded. This caught your attention in part because you have your office interview tomorrow with United Dynamics. You hadn't really thought that much about the pension plan of your potential future employer, in part because your current employer has a defined contribution 401(k) plan, for which funding is not a concern. However, United Dynamics is an older firm with a defined benefit plan, for which funding is the employer's responsibility.

To prepare for your interview, you obtained a copy of United Dynamics' financial statements. Unfortunately, the financial statements themselves are of little help. You are unable to find any pension liability in the balance sheet, but the statement does report a relatively small "pension asset." The income statement reports pension expense for each of the years reported. For help, you search the disclosure notes. In part, the pension disclosure note reads as follows:

Note 7: Pension Plan

United Dynamics has a defined benefit pension plan covering substantially all of its employees. Plan benefits are based on years of service and the employee's compensation during the last three years of employment. The company's funding policy is consistent with the funding requirements of federal law and regulations. The net periodic pension expense for the company included the following components. The company's pension expense was as follows ($ in millions):

	2021	2020	2019
Current service costs	$ 43	$ 47	$ 42
Interest cost on projected benefit obligation	178	164	152
Return on assets	(213)	(194)	(187)
Amortization of prior service cost	43	43	43
Amortization of net gain	(2)	(1)	—
Net pension costs	$ 49	$ 59	$ 50

The following table describes the change in projected benefit obligation for the plan years ended December 31, 2021, and December 31, 2020 ($ in millions):

	2021	2020
Projected benefit obligation at beginning of year	$2,194	$2,121
Service cost	43	47
Interest cost	178	164
Actuarial (gain) loss	319	(40)
Benefits paid	(106)	(98)
Projected benefit obligation at end of year	$2,628	$2,194

The weighted-average discount rate and rate of increase in future compensation levels used in determining the actuarial present value of the projected benefit obligations in the above table were 8.1% and 4.3%, respectively, at December 31, 2021, and 7.73% and 4.7%, respectively, at December 31, 2020. The expected long-term rate of return on assets was 9.1% at December 31, 2021 and 2020.

(continued)

(continued)

The following table describes the change in the fair value of plan assets for the same two years ($ in millions):

	2021	2020
Fair value of plan assets at beginning of year	$2,340	$ 2,133
Actual return on plan assets	215	178
Employer contributions	358	127
Benefits paid	(106)	(98)
Fair value of plan assets at end of year	$2,807	$2,340

"Ouch! I can't believe how much of my accounting I forgot," you complain to yourself. "I'd better get out my old intermediate accounting book."

QUESTIONS

By the time you finish this chapter, you should be able to respond appropriately to the questions posed in this case. Compare your response to the solution provided at the end of the chapter.

1. Why is pension plan underfunding not a concern in your present employment? *(p. 981)*
2. Were you correct that the pension liability is not reported in the balance sheet? What is the liability? *(p. 983)*
3. What is the amount of the plan assets available to pay benefits? What are the factors that can cause that amount to change? *(p. 990)*
4. What does the "pension asset" represent? Are you interviewing with a company whose pension plan is severely underfunded? *(p. 992)*
5. How is the pension expense influenced by changes in the pension liability and plan assets? *(p. 993)*

PART A The Nature of Pension Plans

LO17–1

United States pension funds total more than $15 trillion, roughly 12 times the size of Japan's gross national product. This powerful investment base now controls a sizable portion of the stock market. At the company level, the enormous size of pension funds is reflected in a periodic pension cost that constitutes one of the largest expenses many companies report. The corporate liability for providing pension benefits is huge. Obviously, then, the financial reporting responsibility for pensions has important social and economic implications.

Pension plans are designed to provide income to individuals during their retirement years. This is accomplished by setting aside funds during an employee's working years so that at retirement the accumulated funds plus earnings from investing those funds are available to replace wages. Actually, an individual who periodically invests in stocks, bonds, certificates of deposit (CDs), or other investments for the purpose of saving for retirement is establishing a personal pension fund. Often, such individual plans take the form of individual retirement accounts (IRAs) to take advantage of tax breaks offered by that arrangement. In employer plans, some or all of the periodic contributions to the retirement fund often are provided by the employer.

Pension plans often enhance productivity, reduce turnover, satisfy union demands, and allow employers to compete in the labor market.

Corporations establish pension plans for a variety of reasons. Sponsorship of pension plans provides employees with a degree of retirement security and fulfills a moral obligation felt by many employers. This security also can induce a degree of job satisfaction and perhaps loyalty that might enhance productivity and reduce turnover. Motivation to sponsor a plan sometimes comes from union demands and often relates to being competitive in the labor market.

Additional Consideration

When established according to tight guidelines, a pension plan gains important tax advantages. Such arrangements are called *qualified plans* because they qualify for favorable tax treatment. In a qualified plan, the employer is permitted an immediate tax deduction for amounts paid into the pension fund (within specified limits). The employees, on the other hand, are not taxed at the time employer contributions are made—only when retirement benefits are received. Moreover, earnings on the funds set aside by the employer are not taxed while in the pension fund, so the earnings accumulate tax free. If you are familiar with the tax advantages of IRAs, you probably recognize the similarity between those individual plans and corporate pension arrangements.

For a pension plan to be qualified for special tax treatment it must meet these general requirements.

1. It must cover at least 70% of employees.
2. It cannot discriminate in favor of highly compensated employees.
3. It must be funded in advance of retirement through contributions to an irrevocable trust fund.
4. Benefits must vest after a specified period of service, commonly five years. (We discuss this in more detail later.)
5. It complies with specific restrictions on the timing and amount of contributions and benefits.

Qualified pension plans offer important tax benefits.

Sometimes, employers agree to annually contribute a specific (defined) amount to a pension fund on behalf of employees but make no commitment regarding benefit amounts at retirement. In other arrangements, employers don't specify the amount of annual contributions but promise to provide determinable (defined) amounts at retirement. These two arrangements describe defined contribution pension plans and defined benefit pension plans, respectively:

- **Defined contribution pension plans** promise fixed annual contributions to a pension fund (say, 5% of the employees' pay or, more often, to match contributions by workers). Employees choose (from designated options) where funds are invested—usually stocks or fixed-income securities. Retirement pay depends on the size of the fund at retirement.
- **Defined benefit pension plans** promise fixed retirement benefits defined by a designated formula. Typically, the pension formula bases retirement pay on the employees' (a) years of service, (b) annual compensation (often final pay or an average for the last few years), and sometimes (c) age. Employers are responsible for ensuring that sufficient funds are available to provide promised benefits.

Virtually all *new* pension plans are defined contribution plans.

Today, approximately three-fourths of workers covered by pension plans are covered by defined contribution plans, roughly one-fourth by defined benefit plans. This represents a radical shift from previous years when the traditional defined benefit plan was far more common. However, very few *new* pension plans are of the defined benefit variety. In fact, many companies are terminating long-standing defined benefit plans and substituting defined contribution plans. Why the shift? There are three main reasons:

1. Government regulations make defined benefit plans cumbersome and costly to administer.
2. Employers are increasingly unwilling to bear the risk of defined benefit plans; with defined contribution plans, the company's obligation ends when contributions are made.
3. There has been a shift among many employers from trying to "buy long-term loyalty" (with defined benefit plans) to trying to attract new talent (with more mobile defined contribution plans).

The two categories of pension plans are depicted in Illustration 17–1.

Illustration 17–1

Defined Contribution and Defined Benefit Pension Plans

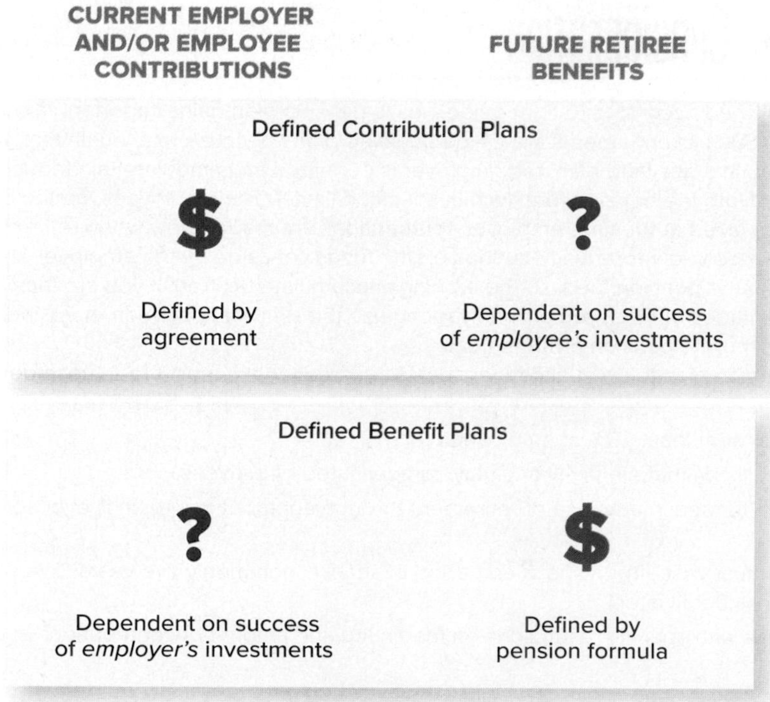

Both types of plans have a common goal: to provide income to employees during their retirement years. Still, the two types of plans differ regarding who bears the risk—the employer or the employees—for whether the retirement objectives are achieved. The two types of plans also have entirely different implications for accounting and financial reporting. Our discussion of defined contribution plans will be brief. Although these are now the most popular type of corporate pension plan, their relative simplicity permits a rather straightforward accounting treatment that requires little explanation. On the other hand, defined benefit plans require considerably more complex accounting treatment and constitute the primary focus of this chapter.

Defined Contribution Pension Plans

***Defined contribution plans* promise defined periodic contributions to a pension fund, without further commitment regarding benefit amounts at retirement.**

Defined contribution pension plans are becoming increasingly popular vehicles for employers to provide retirement income without the paperwork, cost, and risk generated by the more traditional defined benefit plans. Defined contribution plans promise fixed periodic contributions to a pension fund. Retirement income depends on the size of the fund at retirement. No further commitment is made by the employer regarding benefit amounts at retirement.

These plans have several variations. In money purchase plans, employers contribute a fixed percentage of employees' salaries. Thrift plans, savings plans, and 401(k) plans (named after the Tax Code section that specifies the conditions for the favorable tax treatment of these plans) permit voluntary contributions by employees. These contributions typically are matched to a specified extent by employers. Many employers match up to 50% of employee contributions up to the first 5% or 6% of salary.

When plans link the amount of contributions to company performance, labels include profit sharing plans, incentive savings plans, 401(k) profit sharing plans, and similar titles. When employees make contributions to the plan in addition to employer contributions, it's called a *contributory* plan. Sometimes the amount the employer contributes is tied to the amount of the employee contribution.[1] Variations are seemingly endless. An example from a recent annual report of **Microsoft Corporation** is shown in Illustration 17–2.

[1]One popular way for employer companies to provide contributions is with shares of its own common stock. If so, the arrangements usually are designed to comply with government requirements to be designated an employee stock ownership plan (ESOP).

Note 20 (in part)
Savings plan

We have a savings plan in the U.S. that qualifies under Section 401(k) of the Internal Revenue Code, and a number of savings plans in international locations. Participating U.S. employees may contribute a portion of their salary, subject to certain limitations. Beginning January 2016, we contribute fifty cents for each dollar a participant contributes in this plan, with a maximum employer contribution of 50% of the IRS contribution limit for the calendar year. Prior to January 2016, we contributed fifty cents for each dollar of the first 6% a participant contributed in this plan, with a maximum contribution of the lesser of 3% of a participant's earnings or 3% of the IRS compensation limit for the calendar year. Matching contributions for all plans were $734 million, $549 million, and $454 million in fiscal years 2017, 2016, and 2015, respectively, and were expensed as contributed.

Illustration 17–2
Defined Contribution Plan—Microsoft Corporation
Real World Financials

FINANCIAL Reporting Case

Q1, p. 978

Accounting for these plans is quite easy. Each year, the employer simply records pension expense equal to the amount of the annual contribution. Suppose a plan promises an annual contribution equal to 3% of an employee's salary. If an employee's salary is $110,000 in a particular year, the employer would simply recognize pension expense in the amount of the contribution:

For defined contribution plans, the employer simply records pension expense equal to the cash contribution.

Pension expense	3,300	
Cash ($110,000 × 3%)		3,300

The employee's retirement benefits are totally dependent upon how well investments perform. Who bears the risk (or reward) of that uncertainty? The employee would bear the risk of uncertain investment returns and, potentially, settle for far less at retirement than at first expected.[2] On the other hand, the employer would be free of any further obligation. Because the actual investments are held by an independent investment firm, the employer is free of that record-keeping responsibility as well.

Risk is reversed in a defined benefit plan. Because specific benefits are promised at retirement, the employer would be responsible for making up the difference when investment performance is less than expected. We look at defined benefit plans next.

Defined Benefit Pension Plans

When setting aside cash to fund a pension plan, the uncertainty surrounding the rate of return on plan assets is but one of several uncertainties inherent in a defined benefit plan. Employee turnover affects the number of employees who ultimately will become eligible for retirement benefits. The age at which employees will choose to retire as well as life expectancies will impact both the length of the retirement period and the amount of the benefits. Inflation, future compensation levels, and interest rates also have obvious influence on eventual benefits.

Defined benefit plans **promise fixed retirement benefits defined by a designated formula.**

Uncertainties complicate determining how much to set aside each year to ensure that sufficient funds are available to provide promised benefits.

This is particularly true when pension benefits are defined by a pension formula, as usually is the case. A typical formula might specify that a retiree will receive annual retirement benefits based on the employee's years of service and annual pay at retirement (say, pay level in the final year, highest pay achieved, or average pay in the last two or more years). For example, a pension formula might define annual retirement benefits as follows:

A pension formula typically defines retirement pay based on the employee's (a) years of service, (b) annual compensation, and sometimes (c) age.

$$1\tfrac{1}{2}\% \times \text{Years of service} \times \text{Final year's salary}$$

By this formula, the annual benefits to an employee who retires after 30 years of service, with a final salary of $100,000, would be

$$1\tfrac{1}{2}\% \times 30 \text{ years} \times \$100{,}000 = \$45{,}000$$

Typically, a firm will hire an **actuary**, a professional trained in a particular branch of statistics and mathematics, to assess the various uncertainties (employee turnover, salary

[2]Of course, this is not entirely unappealing to the employee. Defined contribution plans allow an employee to select investments in line with his or her own risk preferences and often provide greater retirement benefits and flexibility than defined benefit plans.

levels, mortality, etc.) and to estimate the company's obligation to employees in connection with its pension plan. Such estimates are inherently subjective, so regardless of the skill of the actuary, estimates invariably deviate from the actual outcome to one degree or another.[3] For instance, the return on assets can turn out to be more or less than expected. These deviations are referred to as *gains* and *losses* on pension assets. When it's necessary to revise estimates related to the pension obligation because it's determined to be more or less than previously thought, these revisions are referred to as *losses* and *gains,* respectively, on the pension liability. Later, we will discuss the accounting treatment of gains and losses from either source. The point here is that the risk of the pension obligation changing unexpectedly or the pension funds being inadequate to meet the obligation is borne by the employer with a defined benefit pension plan.

Pension gains and losses occur when the *pension obligation* is lower or higher than expected.

The key elements of a defined benefit pension plan are listed below:

1. The *employer's obligation* to pay retirement benefits in the future.
2. The *plan assets* set aside by the employer from which to pay the retirement benefits in the future.
3. The *periodic expense* of having a pension plan.

Neither the pension obligation nor the plan assets are reported individually in the balance sheet.

As you will learn in this chapter, the first two of these elements are not reported individually in the employer's financial statements. This may seem confusing at first because it is inconsistent with the way you're accustomed to treating assets and liabilities. Even though they are not separately reported, it's critical that you understand the composition of both the pension obligation and the plan assets because (a) they are reported as a net amount in the balance sheet, and (b) their balances are reported in disclosure notes. And, importantly, the pension expense reported in the income statement is a direct composite of periodic changes that occur in both the pension obligation and the plan assets.

The pension expense is a direct composite of periodic changes that occur in both the pension obligation and the plan assets.

For this reason, we will devote a considerable portion of our early discussion to understanding the composition of the pension obligation and the plan assets before focusing on the derivation of pension expense and required financial statement disclosures. We will begin with a quick overview of how periodic changes that occur in both the pension obligation and the plan assets affect pension expense. Next, we will explore how those changes occur, beginning with changes in the pension obligation followed by changes in plan assets. We'll then return to pension expense for a closer look at how those changes influence its calculation. After that, we will bring together the separate but related parts by using a simple spreadsheet to demonstrate how each element of the pension plan articulates with the other elements.

Pension Expense—An Overview

The annual pension expense reflects changes in both the pension obligation and the plan assets. Illustration 17–3 provides a brief overview of how these changes are included in pension expense. After the overview, we'll look closer at each of the components.

Illustration 17–3

Components of Pension Expense—Overview (More Later)

Components of Pension Expense

+	**Service cost** ascribed to employee service during the period
+	**Interest** accrued on the pension liability
−	**Return** on the plan assets*
	Amortized portion of:
+	**Prior service cost** attributed to employee service before an amendment to the pension plan
+ or (−)	**Losses or (gains)** from revisions in the pension liability or from investing plan assets
=	**Pension expense**

*The actual return is adjusted for any difference between actual and expected return, resulting in the expected return being reflected in pension expense. This loss or gain from investing plan assets is combined with losses and gains from revisions in the pension liability for deferred inclusion in pension expense. (See the last component of pension expense.)

Interest and investment return are financing aspects of the pension cost.

The recognition of some elements of the pension expense is delayed.

[3]We discuss changes in more detail in Chapter 20.

Next we explore each of these pension expense components in the context of its being a part of either (a) the pension obligation or (b) the plan assets. After you learn how the expense components relate to these elements of the pension plan, we'll return to explore further how they are included in the pension expense.

The Pension Obligation and Plan Assets

PART B

The Pension Obligation

FINANCIAL Reporting Case

Q2, p. 978

● LO17–2

Now we consider more precisely what is meant by the pension obligation. Unfortunately, there's not just one definition, nor is there uniformity concerning which definition is most appropriate for pension accounting. Actually, three different ways to measure the pension obligation have meaning in pension accounting, as shown in Illustration 17–4.

Illustration 17–4

Ways to Measure the Pension Obligation

1. **Accumulated benefit obligation (ABO)** The actuary's estimate of the total retirement benefits (at their discounted present value) earned so far by employees, applying the pension formula using **existing** compensation levels.
2. **Vested benefit obligation (VBO)** The portion of the accumulated benefit obligation that plan participants are entitled to receive regardless of their continued employment.
3. **Projected benefit obligation (PBO)** The actuary's estimate of the total retirement benefits (at their discounted present value) earned so far by employees, applying the pension formula using **estimated future** compensation levels. (If the pension formula does not include future compensation levels, the PBO and the ABO are the same.)

Later you will learn that the projected benefit obligation is the basis for some elements of the periodic pension expense. Remember, there is but one obligation; these are three ways to measure it. The relationship among the three is depicted in Illustration 17–5.

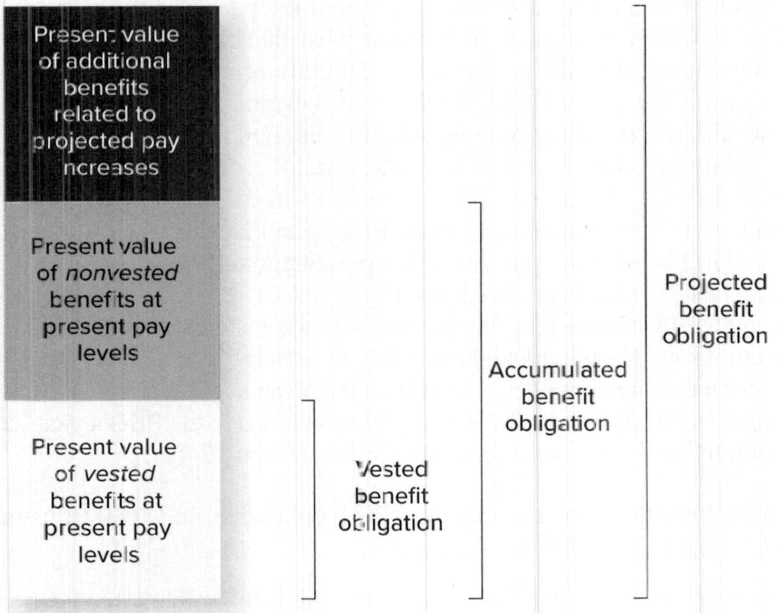

Illustration 17–5

Alternative Measures of the Pension Obligation

Now let's look closer at how the obligation is measured in each of these three ways. Keep in mind, though, that it's not the accountant's responsibility to actually derive the measurement; a professional actuary provides these numbers. However, for the accountant to effectively use the numbers provided, she or he must understand their derivation.

Accumulated Benefit Obligation

The *accumulated benefit obligation* ignores possible pay increases in the future.

The **accumulated benefit obligation (ABO)** is an estimate of the discounted present value of the retirement benefits earned so far by employees, applying the plan's pension formula using *existing compensation levels*. When we look at a detailed calculation of the projected benefit obligation later, keep in mind that simply substituting the employee's existing compensation in the pension formula for her projected salary at retirement would give us the accumulated benefit obligation.

Vested Benefit Obligation

Suppose an employee leaves the company to take another job. Will she still get earned benefits at retirement? The answer depends on whether the benefits are vested under the terms of this particular pension plan. If benefits are fully vested—yes. **Vested benefits** are those that employees have the right to receive even if their employment were to cease today.

The benefits of most pension plans vest after five years.

Pension plans typically require some minimum period of employment before benefits vest. Before the Employee Retirement Income Security Act (ERISA) was passed in 1974, horror stories relating to lost benefits were commonplace. It was possible, for example, for an employee to be dismissed a week before retirement and be left with no pension benefits. Vesting requirements were tightened drastically to protect employees. These requirements have been changed periodically since then. Today, benefits must vest (a) fully within five years or (b) 20% within three years with another 20% vesting each subsequent year until fully vested after seven years. Five-year vesting is most common. ERISA also established the Pension Benefit Guaranty Corporation (PBGC) to impose liens on corporate assets for unfunded pension liabilities in certain instances and to administer terminated pension plans. The PBGC provides a form of insurance for employees similar to the role of the FDIC for bank accounts and is financed by premiums from employers equal to specified amounts for each covered employee. It makes retirement payments for terminated plans and guarantees basic vested benefits when pension liabilities exceed assets. The vested benefit obligation is actually a subset of the ABO, the portion attributable to benefits that have vested.

Projected Benefit Obligation

● LO17–3

The *PBO* estimates retirement benefits by applying the pension formula using projected future compensation levels.

As described earlier, when the ABO is estimated, the most recent salary is included in the pension formula to estimate future benefits, even if the pension formula specifies the final year's salary. No attempt is made to forecast what that salary would be the year before retirement. Of course, the most recent salary certainly offers an objective number to measure the obligation, but is it realistic? Since it's unlikely that there will be no salary increases between now and retirement, a more meaningful measurement should include a projection of what the salary might be at retirement.[4] Measured this way, the liability is referred to as the **projected benefit obligation (PBO)**. Hereafter in the chapter, when we mention the "pension obligation," we are referring to the PBO. The PBO measurement may be less reliable than the ABO but is more relevant and representationally faithful.

To understand the concepts involved, it's helpful to look at a numerical example. We'll simplify the example (Illustration 17–6) by looking at how pension amounts would be determined for a single employee. Keep in mind though, that in actuality, calculations would be made (by the actuary) for the entire employee pool rather than on an individual-by-individual basis.

If the actuary's estimate of the final salary hasn't changed, the PBO a year later at the end of 2020 would be $139,715 as demonstrated in Illustration 17–6A.

CHANGES IN THE PBO Notice that the PBO increased during 2020 (Illustration 17–6A) from $119,822 to $139,715 for two reasons:

1. One more service year is included in the pension formula calculation (service cost).
2. The employee is one year closer to retirement, causing the present value of benefits to increase due to the time value of future benefits (interest cost).

[4]To project future salaries for a group of employees, actuaries usually assume some percentage rate of increase in compensation levels in upcoming years. Recent estimates of the rate of compensation increase have ranged from 0.5% to 5.75% with 3.8% being the most commonly reported expectation (Pension/OPEB 2017 Assumption and Disclosure Survey, PWC, 2017).

Illustration 17–6

Projected Benefit Obligation

Jessica Farrow was hired by Global Communications at the beginning of 2010. The company has a defined benefit pension plan that specifies annual retirement benefits equal to:

1.5% × Service years × Final year's salary

- Farrow is expected to retire at the end of 2049 after 40 years service.
- Her retirement period is expected to be 20 years.
- At the end of 2019, 10 years after being hired, her salary is $100,000.
- The interest rate is 6%.
- The company's actuary projects Farrow's salary to be $400,000 at retirement.*

What is the company's projected benefit obligation with respect to Jessica Farrow?

Steps to calculate the projected benefit obligation:

1. Use the pension formula (including a projection of future salary levels) to determine the retirement benefits earned to date.
2. Find the present value of the retirement benefits as of the retirement date (end of 2049).
3. Find the present value of retirement benefits as of the current date (end of 2019).

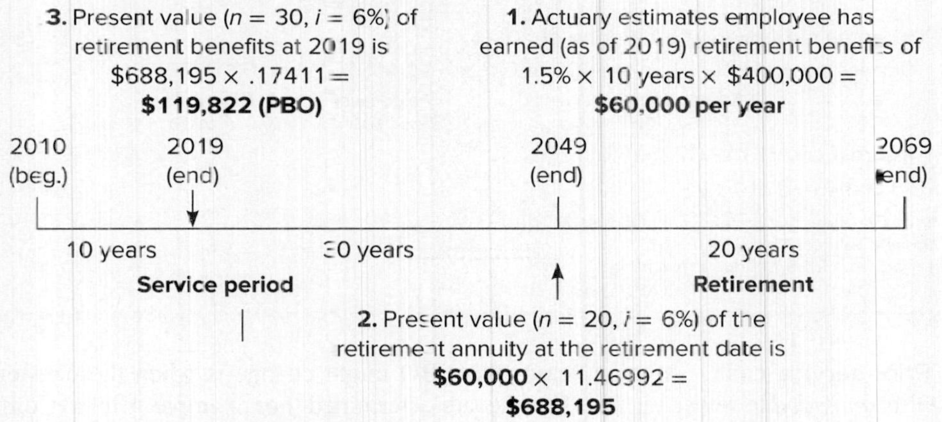
3. Present value (n = 30, i = 6%) of
retirement benefits at 2019 is
$688,195 × .17411 =
$119,822 (PBO)
1. Actuary estimates employee has
earned (as of 2019) retirement benefits of
1.5% × 10 years × $400,000 =
$60,000 per year
2010
(beg.)
2019
(end)
2049
(end)
2069
(end)
10 years
30 years
20 years
Service period
Retirement
2. Present value (n = 20, i = 6%) of the
retirement annuity at the retirement date is
$60,000 × 11.46992 =
$688,195

*This salary reflects an estimated compound rate of increase of about 5% and should take into account expectations concerning inflation, promotions, productivity gains, and other factors that might influence salary levels.

The actuary includes projected salaries in the pension formula. The projected benefit obligation is the present value of those benefits.

Using Excel, enter:
=PV(.06,20,– 60000)
Output: 688195

Using a calculator, enter: N 20 I 6 PMT – 60000 FV
Output: PV 688195

Illustration 17–6A

PBO in 2020

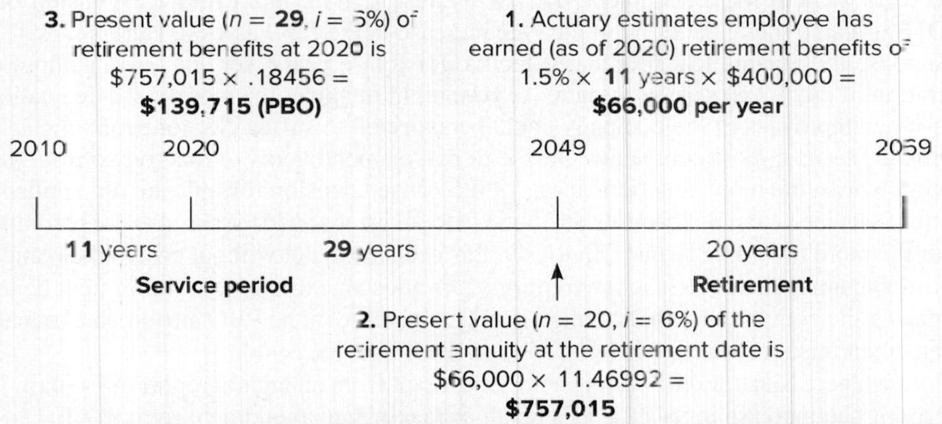
3. Present value (n = **29**, i = 6%) of
retirement benefits at 2020 is
$757,015 × .18456 =
$139,715 (PBO)
1. Actuary estimates employee has
earned (as of 2020) retirement benefits of
1.5% × **11** years × $400,000 =
$66,000 per year
2010
2020
2049
2069
11 years
29 years
20 years
Service period
Retirement
2. Present value (n = 20, i = 6%) of the
retirement annuity at the retirement date is
$66,000 × 11.46992 =
$757,015

In 2020 the pension formula includes one more service year.

Also, 2020 is one year closer to the retirement date for the purpose of calculating the present value.

These represent two of the events that might possibly cause the balance of the PBO to change. Let's elaborate on these and the three other events that might change the balance of the PBO. The five events are (1) service cost, (2) interest cost, (3) prior service cost, (4) gains and losses, and (5) payments to retired employees.

Each year's service adds to the obligation to pay benefits.

Interest accrues on the PBO each year.

1. Service cost. As we just witnessed in the illustration, the PBO increases each year by the amount of that year's **service cost**. This represents the increase in the projected benefit obligation attributable to employee service performed during the period. As we explain later, it also is the primary component of the annual pension expense.

2. Interest cost. The second reason the PBO increases is called the **interest cost**. Even though the projected benefit obligation is not formally recognized as a liability in the company's balance sheet, it is a liability nevertheless. And, as with other liabilities, interest accrues on its balance as time passes. The amount can be calculated directly as the assumed discount rate multiplied by the projected benefit obligation at the beginning of the year.[5]

Additional Consideration

We can verify the increase in the PBO as being caused by the service cost and interest cost as follows:

PBO at the beginning of 2020 (end of 2019)				$119,822
Service cost: (1.5% × 1 yr. × $400,000)	×	11.46992*	× 0.18456†	12,701
Annual retirement benefits from 2020 service		To discount to 2049	To discount to 2020	
Interest cost: $119,822 × 6%				7,189
PBO at the end of 2020				$139,712‡

*Present value of an ordinary annuity of $1; $n = 20$, $i = 6\%$.
†Present value of $1; $n = 29$, $i = 6\%$.
‡Differs from $139,715 due to rounding.

3. Prior service cost. Another reason the PBO might change is when the pension plan itself is *amended* to revise the way benefits are determined. For example, Global Communications in our illustration might choose to revise the pension formula by which benefits are calculated. Let's back up and assume the formula's salary percentage is increased in 2020 from 1.5% to **1.7%**:

1.7% × Service years × Final year's salary
(revised pension formula)

Obviously, the annual service cost from this date forward will be higher than it would have been without the amendment. This will cause a more rapid future expansion of the PBO. But it also might cause an immediate increase in the PBO as well. Here's why.

When a pension plan is amended, credit often is given for employee service rendered in prior years. The cost of doing so is called *prior service cost.*

Suppose the amendment becomes effective for future years' service only, without consideration of employee service to date. As you might imagine, the morale and dedication of long-time employees of the company could be expected to suffer. So, for economic as well as ethical reasons, most companies choose to make amendments retroactive to prior years. In other words, the more beneficial terms of the revised pension formula are not applied just to future service years, but benefits attributable to all prior service years also are recomputed under the more favorable terms. Obviously, this decision is not without cost to the company. Making the amendment retroactive to prior years adds an extra layer of retirement benefits, increasing the company's benefit obligation. The increase in the PBO attributable to making a plan amendment retroactive is referred to as **prior service cost**.[6]

For instance, Illustration 17–7 presents an excerpt from an annual report of **Ecolab, Inc.**, describing the increase in its PBO as a result of making an amendment retroactive.

[5]Discount rates recently reported have ranged from 2.1% to 4.5% with 4.1% being the most commonly assumed rate (Pension/OPEB 2017 Assumption and Disclosure Survey, PWC, 2017).

[6]Prior service cost also is created if a defined benefit pension plan is initially adopted by a company that previously did not have one, and the plan itself is made retroactive to give credit for prior years' service. Prior service cost is created by plan amendments far more often than by plan adoptions because most companies already have pension plans, and new pension plans in recent years have predominantly been defined contribution plans.

Note 1: Retirement Plans (in part)
. . . The Company amended its U.S. pension plan to change the formula for pension benefits and to provide a more rapid vesting schedule. The plan amendments resulted in a $6 million increase in the projected benefits obligation.

Source: Ecolab, Inc.

Illustration 17–7
Prior Service Cost—Ecolab, Inc.
Real World Financials

Let's put prior service cost in the context of our illustration.

At the end of 2019, and therefore the beginning of 2020, the PBO is $119,822. If the plan is amended on January 3, 2020, the PBO for that date could be recomputed as follows:

Retroactive benefits from an amendment add additional costs, increasing the company's PBO. This increase is the prior service cost.

	PBO without Amendment		PBO with Amendment	
1.	1.5% × 10 yrs. × $400,000	= $ 60,000	1.7% × 10 yrs. × $400,000	= $ 68,000
2.	$60,000 × 11.46992	= 688,195	$68,000 × 11.46992	= 779,955
3.	$688,195 × .17411	= 119,822	$779,955 × .17411	= 135,798

$15,976
Prior service cost

The **$15,976** increase in the PBO attributable to applying the more generous terms of the amendment to prior service years is the prior service cost. And, because we assumed the amendment occurred at the beginning of 2020, both the 2020 service cost and the 2020 interest cost would change as a result of the prior service cost. This is how:

PBO at the beginning of 2020 (end of 2019)	$119,822
Prior service cost (determined above)	15,976
PBO including prior service cost at the beginning of 2020	$135,798
Service cost: (1.7% × 1yr. × $400,000) × 11.46992* × 0.18456† (Annual retirement benefits from 2020 service; To discount to 2049; To discount to 2020)	14,395
Interest cost: $135,798‡ × 6%	8,148
PBO at the end of 2020	$158,341

*Present value of an ordinary annuity of $1; $n = 20$, $i = 6\%$.
†Present value of $1; $n = 29$, $i = 6\%$.
‡Includes the beginning balance plus the prior service cost because the amendment occurred at the beginning of the year.

Additional Consideration

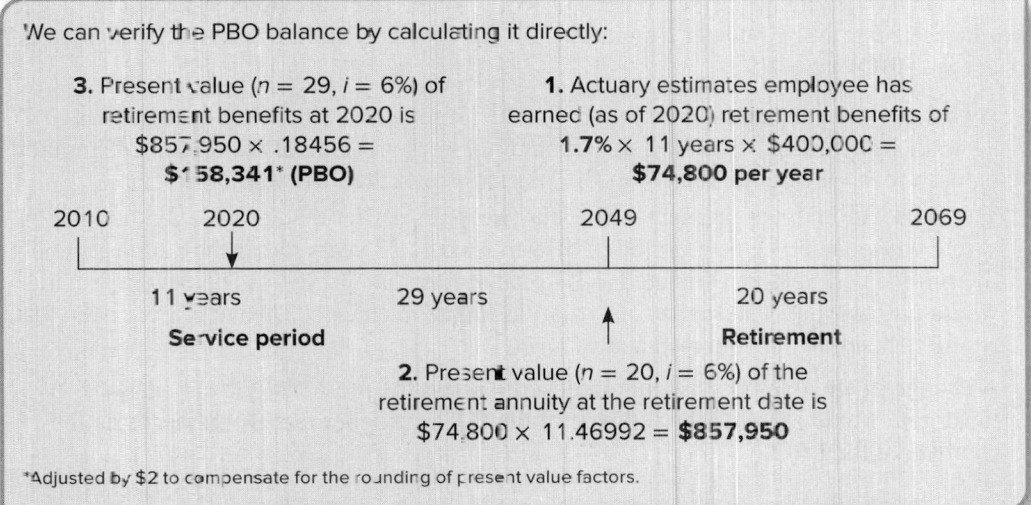

We can verify the PBO balance by calculating it directly:

3. Present value ($n = 29$, $i = 6\%$) of retirement benefits at 2020 is $857,950 × .18456 = **$158,341* (PBO)**

1. Actuary estimates employee has earned (as of 2020) retirement benefits of 1.7% × 11 years × $400,000 = **$74,800 per year**

2010 — 2020 — 2049 — 2069

11 years **Service period** (2010–2020); 29 years (2020–2049); 20 years **Retirement** (2049–2069)

2. Present value ($n = 20$, $i = 6\%$) of the retirement annuity at the retirement date is $74,800 × 11.46992 = **$857,950**

*Adjusted by $2 to compensate for the rounding of present value factors.

The pension formula reflects the plan amendment.

The plan amendment would affect not only the year in which it occurs, but also each subsequent year because the revised pension formula determines each year's service cost. Continuing our illustration to 2021 demonstrates this:

During 2021, the PBO increased as a result of service cost and interest cost.

PBO at the beginning of 2021 (end of 2020)	$158,341
Service cost: (1.7% × 1 yr. × $400,000) × 11,46992* × 0.19563' (Annual retirement benefits from 2021 service; To discount to 2049; To discount to 2021)	15,258
Interest cost : $158,341 × 6 %	9,500
PBO at the end of 2021	$183,099

*Present value of an ordinary annuity of $1; $n = 20$, $i = 6\%$.
'Present value of $1; $n = 28$, $i = 6\%$.

Decreases and increases in estimates of the PBO because of periodic re-evaluation of uncertainties are called gains and losses.

4. Gain or loss on the PBO. We mentioned earlier that a number of estimates are necessary to derive the PBO. When one or more of these estimates requires revision, the estimate of the PBO also will require revision. The resulting decrease or increase in the PBO is referred to as a *gain* or *loss,* respectively. Let's modify our illustration to imitate the effect of revising one of the several possible estimates involved. Suppose, for instance, that new information at the end of 2021 about inflation and compensation trends suggests that the estimate of Farrow's final salary should be increased by 5% to **$420,000**. This would affect the estimate of the PBO as follows:

	PBO *without* Revised Estimate		PBO *with* Revised Estimate	
1.	1.7% × 12 yrs. × $400,000	= $ 81,600	1.7% × 12 yrs. × **$420,000**	= $ 85,680
2.	$81,600 × 11.46992	= 935,945	$85,680 × 11.46992	= 982,743
3.	$935,945 × .19563	= 183,099	$982,743 × .19563	= 192,254

$9,155
Loss on PBO

Changing the final salary estimate changes the PBO.

The difference of **$9,155** represents a loss on the PBO because the obligation turned out to be higher than previously expected. Now there would be three elements of the increase in the PBO during 2021.[7]

The revised estimate caused the PBO to increase.

PBO at the beginning of 2021	$158,341
Service cost (calculated above)	15,258
Interest cost (calculated above)	9,500
Loss on PBO (calculated above)	**9,155**
PBO at the end of 2021	$192,254

If a revised estimate causes the PBO to be lower than previously expected, a gain would be indicated. Consider how a few of the other possible estimate changes would affect the PBO:

- A change in life expectancies might cause the retirement period to be estimated as 21 years rather than 20 years. Calculation of the present value of the retirement annuity would use $n = 21$, rather than $n = 20$. The estimate of the PBO would increase.
- The expectation that retirement will occur two years earlier than previously thought would cause the retirement period to be estimated as 22 years rather than 20 years and the service period to be estimated as 28 years rather than 30 years. The new expectation would probably also cause the final salary estimate to change. The net effect on the PBO would depend on the circumstances.
- A change in the assumed discount rate would affect the present value calculations. A lower rate would increase the estimate of the PBO. A higher rate would decrease the estimate of the PBO.

[7]The increase in the PBO due to amending the pension formula (prior service cost) occurred in 2020.

5. Payment of retirement benefits. We've seen how the PBO will change due to the accumulation of service cost from year to year, the accrual of interest as time passes, making plan amendments retroactive to prior years, and periodic adjustments when estimates change. Another change in the PBO occurs when the obligation is reduced as benefits actually are paid to retired employees.

Payment of retirement benefits reduces the PBO.

The payment of such benefits is not applicable in our present illustration because we've limited the situation to calculations concerning an individual employee who is several years from retirement. Remember, though, in reality the actuary would make these calculations for the entire pool of employees covered by the pension plan. But the concepts involved would be the same. Illustration 17–8 summarizes the five ways the PBO can change.

Illustration 17–8

Components of Change in the PBO

The Projected Benefits Obligation Changes as a Result of:

Cause	Effect	Frequency
Service cost	+	Each period
Interest cost	+	Each period (except the first period of the plan, when no obligation exists to accrue interest)
Prior service cost	+	Only if the plan is amended (or initiated) that period
Loss or gain on PBO	+ or –	Whenever revisions are made in the pension liability estimate
Retiree benefits paid	–	Each period (unless no employees have yet retired under the plan)

Illustration Expanded to Consider the Entire Employee Pool

For our single employee, the PBO at the end of 2021 is $192,254. Let's say now that Global Communications has 2,000 active employees covered by the pension plan and 100 retired employees receiving retirement benefits. Illustration 17–9 expands the numbers to represent all covered employees.

The PBO is not formally recognized in the balance sheet.

Illustration 17–9

The PBO Expanded to Include All Employees

The changes in the PBO for Global Communications during 2021 were as follows:*

	($ in millions)
PBO at the beginning of 2021† (amount assumed)	$400
Service cost, 2021 (amount assumed)	41
Interest cost: $400 × 6%	24
Loss (gain) on PBO (amount assumed)	23
Less: Retiree benefits paid (amount assumed)	(38)
PBO at the end of 2021	$450

*Of course these expanded amounts are not simply the amounts for Jessica Farrow multiplied by 2,000 employees because her years of service, expected retirement date, and salary are not necessarily representative of other employees. Also, the expanded amounts take into account expected employee turnover and current retirees.
†Includes the prior service cost that increased the PBO when the plan was amended in 2020.

Additional Consideration

We are witnessing an accelerating trend of companies changing the way they calculate pension costs. Traditionally the discount rate companies use to calculate the present value of their future pension obligations has been a *weighted average* of interest rates from a bond yield curve. A bond yield curve charts the difference (spread) between near-term and long-term rates. This average rate is used to calculate the present value of estimated future

(continued)

(concluded)

retirement payments of all employees regardless of their anticipated retirement dates. This is what we did in the illustrations in this section.

In a variation of this approach, called the "spot-rate" method, we use different rates for different employees depending on when they are expected to retire. Let's say a typical 30-year yield curve is "upward-sloping," reflecting an increase from 1% for very short-term bonds to 4% for 30-year maturities, averaging 3%. Using the traditional method, we would calculate the present value of everyone's expected retirement pay as well as the increase in that present value (remember, that's the service cost) using a 3% rate. We also use the same rate when determining interest cost each year. With the spot-rate approach, we instead would apply rates corresponding to expected retirement periods. For instance, we would use the 5-year "spot" rate, let's say 1.2%, for employees expected to retire in 5 years, the 10-year "spot" rate, let's say 2.9%, for employees expected to retire in 10 years, and the 15-year "spot" rate, let's say 3.4%, for employees expected to retire in 15 years.

After AT&T switched to the "spot-rate" method in 2014, several dozen companies followed suit in the next two years and the trend continues. Why the popularity? Idealistically, we could refer to the perception that the spot-rate method provides more precise measurements, but more likely it's because the method can improve a company's financial results, at least temporarily. How? It actually reduces both the *service cost* and *interest cost* components of pension expense and thus increases reported profit. Service cost is the present value of the year-to-year benefits earned by employees. The bulk of that cost is generated by employees farthest from retirement (far out on the yield curve). When the spot-rate approach is used with a typical upward sloping yield curve, that's when the highest interest rates are applied. Higher interest rates create lower present values, lower service cost in this case. Interest cost also is reduced. Lower spot interest rates apply to employees nearer retirement, and those are the ones with higher accumulated benefits. Lower rates times the higher balances, and higher rates times the lower balances, produce a lower interest cost.

These effects would be eliminated or reversed if the yield curve were to become flat or inverted. Although this has been historically infrequent, recent increases in short-term rates with little change in longer-term rates suggest a flattening of the curve and less motivation to use the spot-rate approach.

Pension Plan Assets

● LO17–4

So far our focus has been on the employer's obligation to provide retirement benefits in the future. We turn our attention now to the resources with which the company will satisfy that obligation—the **pension plan assets**. Like the PBO, the pension plan assets are not reported separately in the employer's balance sheet but are netted together with the PBO to report either a net pension asset (debit balance) or a net pension liability (credit balance). Its separate balance, too, must be reported in the disclosure notes to the financial statements (as does the separate PBO balance), and as explained below, the return on these assets is included in the calculation of the periodic pension expense.

FINANCIAL Reporting Case

Q3, p. 978

We assumed in the previous section that Global Communications' obligation is $450 million for service performed to date. When employees retire, will there be sufficient funds to provide the anticipated benefits? To ensure sufficient funding, Global will contribute cash each year to a pension fund.

A *trustee* manages pension plan assets.

The assets of a pension fund must be held by a **trustee**. A trustee accepts employer contributions, invests the contributions, accumulates the earnings on the investments, and pays benefits from the plan assets to retired employees or their beneficiaries. The trustee can be an individual, a bank, or a trust company. Plan assets are invested in stocks, bonds, and other income-producing assets. The accumulated balance of the annual employer contributions plus the return on the investments (dividends, interest, market price appreciation) must be sufficient to pay benefits as they come due.

When an employer estimates how much it must set aside each year to accumulate sufficient funds to pay retirement benefits as they come due, it's necessary to estimate the return those investments will produce. This is the **expected return on plan assets**. The higher the return, the less the employer must actually contribute. On the other hand, a relatively low return means the difference must be made up by higher contributions. In practice, recent estimates of the rate of return have ranged from 4.3% to 9%, with 7% being the most commonly reported expectation.[8] In Illustration 17–10, we shift the focus of our numerical illustration to emphasize Global's pension plan assets.

Illustration 17–10

How Plan Assets Change

Global Communications funds its defined benefit pension plan by contributing each year the year's service cost plus a portion of the prior service cost. Cash of $48 million was contributed to the pension fund at the end of 2021.

Plan assets at the beginning of 2021 were valued at $300 million. The expected rate of return on the investment of those assets was 9%, but the actual return in 2021 was 10%. Retirement benefits of $38 million were paid at the end of 2021 to retired employees.

What is the value of the company's pension plan assets at the end of 2021?

	($ in millions)
Plan assets at the beginning of 2021	$300
Actual return on plan assets (10% × $300)	30
Cash contributions	48
Less: Retiree benefits paid	(38)
Plan assets at the end of 2021	$340

A trustee accepts employer contributions, invests the contributions, accumulates the earnings on the investments, and pays benefits from the plan assets.

Recall that Global's PBO at the end of 2021 is $450 million. Because the plan assets are only $340 million, the pension plan is said to be *underfunded*. One reason is that we assumed Global incurred a $60 million prior service cost from amending the pension plan at the beginning of 2020, and that cost is being funded over several years. Another factor is the loss from increasing the PBO due to the estimate revision, since funding has been based on the previous estimate. Later, we'll assume earlier revisions also have increased the PBO. Of course, actual performance of the investments also impacts a plan's funded status.

An *underfunded* pension plan means the PBO exceeds plan assets.

It is not unusual for pension plans today to be underfunded. Historically, the funded status of pension plans has varied considerably. Prior to the Employee Retirement Income Security Act (ERISA) in 1974, many plans were grossly underfunded. The law established minimum funding standards among other matters designed to protect plan participants. The new standards brought most plans closer to full funding. Then the stock market boom of the 1980s caused the value of plan assets for many pension funds to swell to well in excess of their projected benefit obligations. More than 80% of pension plans were overfunded. As a result, managers explored ways to divert funds to other areas of operations. Today, a majority of plans again are underfunded. The economic crisis of 2008–2009 took its toll. Stock market declines reduced the funded status of pension plans from 108% at the end of 2007 to 79% at the end of 2008.[9] Despite recent positive stock market returns, at the beginning of 2018, about 80% of pension funds still were underfunded. One culprit has been low interest rates. Low interest rates hurt plans' funded status because the pension obligation is a present value calculation that increases with a lower discount rate. Even small interest rate changes have big effects on funded status. Many of the underfunded plans are with troubled companies, placing employees at risk. The PBGC guarantees are limited to about $5,000 per month, often less than promised pension benefits.

An *overfunded* pension plan means plan assets exceed the PBO.

[8] Pension/OPEB 2017 Assumption and Disclosure Survey, PWC, 2017.

[9] Goldman Sachs, "Global Markets Institute Accounting Policy Update: Pension review 2009–Fallout from funded status decline just beginning," Global Markets Institute, June 4, 2009.

Reporting the Funded Status of the Pension Plan

● LO17–5

FINANCIAL Reporting Case

Q4, p. 978

A company must report in its balance sheet a liability for the underfunded (or asset for the overfunded) status of its postretirement plans.

A company's PBO is not reported separately among liabilities in the balance sheet. Similarly, the plan assets a company sets aside to pay those benefits are not separately reported among assets in the balance sheet. However, firms do report the net difference between those two amounts, referred to as the "funded status" of the plan.[10] From our previous discussion, we see the funded status for Global to be the following at December 31, 2021, and December 31, 2020:

($ in millions)	2021	2020
Projected benefit obligation (PBO)	$450	$400
Fair value of plan assets	340	300
Underfunded status	$110	$100

Because the plan is underfunded, Global reports a net pension liability of $110 million in its 2021 balance sheet and $100 million in 2020. Be sure to note that the "net pension liability" is not an actual account balance. Instead, it's the PBO account balance and the Plan assets account balance simply reported in the balance sheet as a single net amount. If the plan becomes overfunded in the future, Global will report a net pension asset instead.

Now, let's look at all the ways that changes in the PBO and the pension plan assets affect pension expense.

PART C Determining Pension Expense

The Relationship between Pension Expense and Changes in the PBO and Plan Assets

● LO17–6

The costs are allocated to the periods the services are performed.

Like wages, salaries, commissions, and other forms of pay, pension expense is part of a company's compensation for employee services each year. Accordingly, the accounting objective is to achieve a matching of the costs of providing this form of compensation with the benefits of the services performed. However, the fact that this form of compensation actually is paid to employees many years after the service is performed means that other elements in addition to the annual service cost will affect the ultimate pension cost. These other elements are related to changes that occur over time in both the pension obligation and the pension plan assets. Illustration 17–11 provides a summary of how some of these changes influence pension expense.

We've examined each of the components of pension expense from the viewpoint of its effect on the PBO or on plan assets, using the Global Communications illustration to demonstrate that effect. Now, let's expand the same illustration to see how these changes affect *pension expense.* Illustration 17–12 provides this expanded example.

Components of Pension Expense

Illustration 17–12 demonstrates the relationship between some of the changes in the PBO and in plan assets and the components of pension expense: service cost, interest cost, the return on plan assets, prior service cost amortization, and net gain or loss amortization. Let's look at these five components of pension expense one at a time.

We report service cost as part of compensation costs, separate from the other components of pension expense.

1. SERVICE COST The $41 million service cost represents the increase in the projected benefit obligation attributable to employee service performed during 2021 (benefits earned by employees during the year). Each year, this is the first component of the pension expense. As we discuss later in Part D, we don't actually include the service cost along with the other components of pension expense we report in the income statement. Instead, we break out that component of the expense and include it in the income statement as part of the total compensation costs (salaries and payroll taxes, etc.).

[10]FASB ASC 715–30–25: Compensation—Retirement Benefits—Defined Benefit Plans—Pension—Recognition .

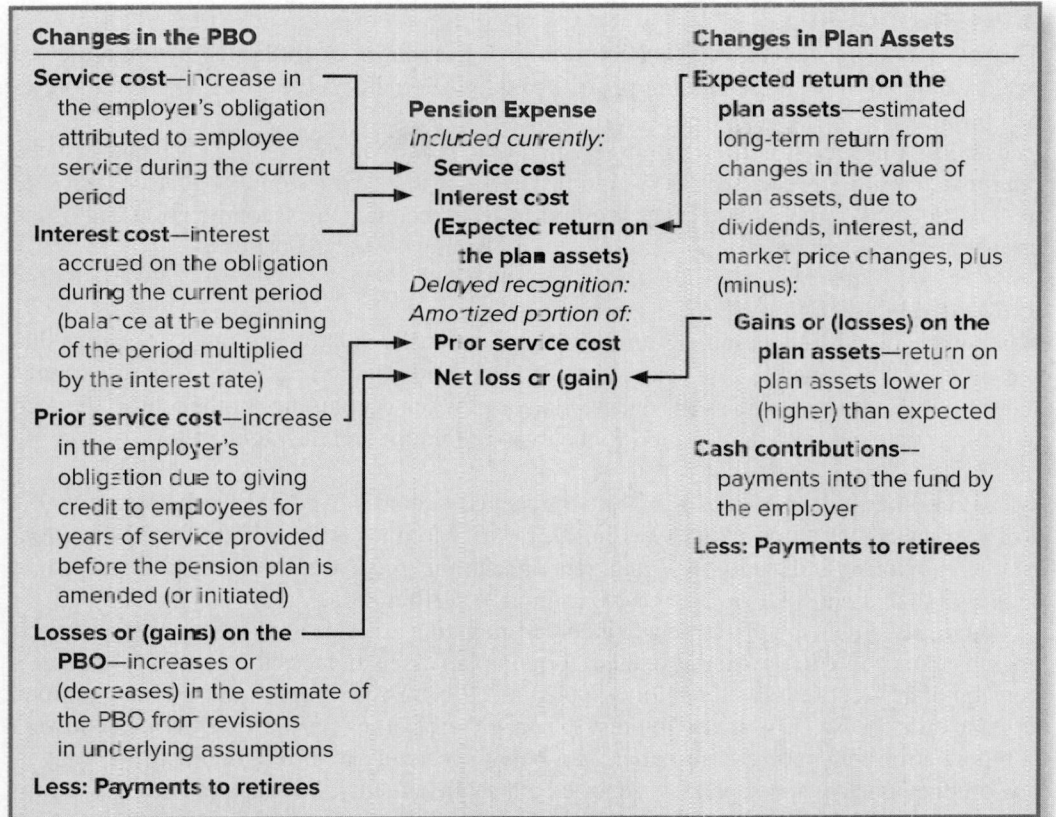

Illustration 17–11

Components of the Periodic Pension Expense

Changes in the PBO

Service cost—increase in the employer's obligation attributed to employee service during the current period

Interest cost—interest accrued on the obligation during the current period (balance at the beginning of the period multiplied by the interest rate)

Prior service cost—increase in the employer's obligation due to giving credit to employees for years of service provided before the pension plan is amended (or initiated)

Losses or (gains) on the PBO—increases or (decreases) in the estimate of the PBO from revisions in underlying assumptions

Less: Payments to retirees

Pension Expense

Included currently:

Service cost

Interest cost

(Expected return on the plan assets)

Delayed recognition:

Amortized portion of:

Prior service cost

Net loss or (gain)

Changes in Plan Assets

Expected return on the plan assets—estimated long-term return from changes in the value of plan assets, due to dividends, interest, and market price changes, plus (minus):

Gains or (losses) on the plan assets—return on plan assets lower or (higher) than expected

Cash contributions—payments into the fund by the employer

Less: Payments to retirees

FINANCIAL Reporting Case

Q5, p. 978

The pension expense is a composite of periodic changes that occur in both the pension obligation and the plan assets.

Illustration 17–12

Pension Expense

Reports from the actuary and the trustee of plan assets indicate the following changes during 2021 in the PBO and plan assets of Global Communications ($ in millions).

	PBO		Plan Assets
Beginning of 2021	$400	*Beginning* of 2021	$300
Service cost	41	Actual return on plan assets,	
Interest cost, 6%	24	10% (9% expected)*	30
Loss (gain) on PBO	23	Cash contributions	48
Less: Retiree benefits	(38)	Less: Retiree benefits	(38)
End of 2021	$450	*End* of 2021	$340

Assume a *prior service cost* of $60 million was incurred at the beginning of the previous year (2020) due to a plan amendment increasing the PBO. Also assume that at the beginning of 2021, Global had a *net loss* of $55 million (previous losses exceeded previous gains). The average remaining service life of employees is estimated at 15 years.

Global's 2021 Pension Expense Is Determined as Follows:	($ in millions)
Service cost	$41
Interest cost	24
Expected return on the plan assets ($30 actual, less $3 gain)	(27)
Amortization of prior service cost (calculated later)	4
Amortization of net loss (calculated later)	1
Pension expense	$43

*Expected rates of return anticipate the performance of various investments of plan assets. This is not necessarily the same as the discount rate used by the actuary to estimate the pension obligation. Assumed rates of return recently reported have ranged from 4.3% to 9%, with 7% being the most commonly assumed rate (Pension/OPEB 2017 Assumption and Disclosure Survey, PWC, 2017).

These are the changes in the PBO and in the plan assets we previously discussed (Illustration 17–9 and Illustration 17–10).

2. INTEREST COST

Interest cost is the discount rate times the PBO balance at the beginning of the year.

The interest cost is calculated as the interest rate (discount rate) multiplied by the projected benefit obligation at the beginning of the year. In 2021, this is 6% times $400 million, or $24 million.

The PBO balance is not separately reported as a liability in the company's balance sheet, but it is a liability nevertheless.[11] The interest expense that accrues on its balance is not separately reported in the income statement but instead becomes the second component of the annual pension expense.

3. RETURN ON PLAN ASSETS

The return earned on investment securities increases the plan asset balance.

Remember, plan assets comprise funds invested in stocks, bonds, and other securities that presumably will generate dividends, interest, and capital gains. Each year, these earnings represent the return on plan assets during that year. When accounting for the return, we need to differentiate between its two modes: the *expected* return and the *actual* return.

Actual versus expected return. We've assumed Global's expected rate of return is 9%, so its expected return on plan assets in 2021 was 9% times $300 million, or $27 million. But, as previously indicated, the actual rate of return in 2021 was 10%, producing an actual return on plan assets of 10% times $300 million, or $30 million.

Obviously, investing plan assets in income-producing assets lessens the amounts employers must contribute to the fund. So, the return on plan assets reduces the net cost of having a pension plan. Accordingly, the return on plan assets each year *reduces* the amount recorded as pension expense. Just as the interest expense that accrues on the PBO is included as a component of pension expense rather than being separately reported, the investment revenue on plan assets is not separately reported either. In actuality, both the interest and return-on-assets components of pension expense do not directly represent employee compensation. Instead, they are financial items created only because the future obligation to retirees must be funded currently.

The interest and return-on-assets components are financial items created rather than direct employee compensation.

Adjustment for loss or gain. A controversial question is *when* differences between the actual and expected return should be recognized in pension expense. It seems logical that since the net cost of having a pension plan is reduced by the actual return on plan assets, the charge to pension expense should be the actual return on plan assets. However, the FASB concluded that the actual return should first be adjusted by any difference between that return and the return amount that had been expected. So, it's actually the *expected* return that is included in the calculation of pension expense. In our illustration, Global's pension expense is reduced by the expected return of $27 million.

The return on plan assets reduces the net *cost of having a* pension plan.

The difference between the actual and expected return is considered a loss or gain on plan assets. Although we don't include these losses and gains as part of pension expense when they occur, it's possible they will affect pension expense at a later time. Soon, we will discuss how that might happen.

Any loss or gain is not included in pension expense right away.

4. AMORTIZATION OF PRIOR SERVICE COST Recall that the $60 million increase in Global's PBO due to recalculating benefits employees earned in prior years as a result of a plan amendment is referred to as the prior service cost. Obviously, prior service cost adds to the cost of having a pension plan. But when should this cost be recognized as pension expense? An argument can be made that the cost should be recognized as expense in the year of the amendment when the cost increases the company's pension obligation. In fact, some members of the FASB have advocated this approach. At present, though, we amortize the cost gradually to pension expense. Here's the rationalization.

Amending a pension plan, and especially choosing to make that amendment retroactive, typically is done with the idea that future operations will benefit from those choices. For that reason, the cost is not recognized as pension expense in the year the plan is amended. Instead, it is recognized as pension expense over the time that the employees who benefited

Prior service cost is recognized as pension expense over the future service period of the employees whose benefits are recalculated.

[11]As we discussed earlier and will revisit later, the PBO is combined with pension assets with the net difference reported in the balance sheet as either a net pension liability or a net pension asset.

from the retroactive amendment will work for the company in the future. Presumably, this future service period is when the company will receive the benefits of its actions.

In our illustration, the amendment occurred in 2020, increasing the PBO at that time. For the individual employee, Jessica Farrow, the prior service cost was calculated to be $15,976. Our illustration assumes that, for *all* plan participants, the prior service cost was $60 million at the beginning of 2020. The prior service cost at the beginning of 2021 is $56 million. The following section explains how this amount was computed.

One assumption in our illustration is that the average remaining service life of the active employee group is 15 years. To recognize the $60 million prior service cost in equal annual amounts over this period, the amount amortized as an increase in pension expense each year is $4 million:[12]

Amortization of Prior Service Cost	($ in millions)
Service cost	$41
Interest cost	24
Expected return on the plan assets	(27)
Amortization of prior service cost–AOCI	**4**
Amortization of net loss–AOCI	1
2021 pension expense	$43

By the straight-line method, prior service cost is recognized over the average remaining service life of the active employee group.

Be sure to note that, even though we're amortizing it, the prior service cost is not an asset, but instead a part of *accumulated other comprehensive income* (AOCI), a shareholders' equity account. This is a result of the FASB's disinclination to treat the cost as an expense as it is incurred. The Board, instead, prefers to consider it to be *other comprehensive income* (OCI) like the handful of losses and gains also categorized the same way and not reported among the gains and losses in the traditional income statement. You first learned about comprehensive income in Chapter 4 and again in Chapter 12. We'll revisit it again later in this chapter.

Prior service cost is not expensed as it is incurred. Instead, it is reported as a component of AOCI to be amortized over time.

The prior service cost balance in AOCI declines by $4 million each year:

Prior Service Cost–AOCI	($ in millions)
Prior service cost at the beginning of 2021	$56
Less: 2021 amortization	(4)
Prior service cost at the end of 2021	$52

5. AMORTIZATION OF A NET LOSS OR NET GAIN You learned previously that gains and losses can occur when expectations are revised concerning either the PBO or the return on plan assets. Illustration 17–13 summarizes the possibilities.

Illustration 17–13

Gains and Losses

	Projected Benefit Obligation	Return on Plan Assets
Higher than expected	Loss	Gain
Lower than expected	Gain	Loss

Gains and losses occur when either the PBO or the return on plan assets turns out to be different than expected.

[12]An alternative to this straight-line approach, called the *service method*, attempts to allocate the prior service cost to each year in proportion to the fraction of the total remaining service years worked in each of those years. This method is described in the chapter appendix.

We report gains and losses as OCI in the statement of comprehensive income.

Like the prior service cost we just discussed, we typically don't include these gains and losses as part of pension expense in the income statement, but instead report them as OCI in the statement of comprehensive income as they occur. We then report the gains and losses (net of subsequent amortization) on a cumulative basis as a net loss–AOCI or a net gain–AOCI, depending on whether we have greater losses or gains over time. We report this amount in the balance sheet as a part of *accumulated other comprehensive income* (AOCI), a shareholders' equity account.

There is no conceptual justification for not including losses and gains in earnings. After all, these increases and decreases in either the PBO or plan assets immediately impact the net cost of providing a pension plan and, conceptually, should be included in pension expense as they occur.

Nevertheless, the FASB allows companies to delay income statement recognition of gains and losses from either source. Why? For practical reasons.

Income Smoothing

The FASB acknowledged the conceptual shortcoming of delaying the recognition of a gain or a loss while opting for this more politically acceptable approach. Delayed recognition was favored by a dominant segment of corporate America that was concerned with the effect of allowing gains and losses to immediately impact reported earnings.

The practical justification for delayed recognition is that, over time, gains and losses might cancel one another out. Given this possibility, why create unnecessary fluctuations in reported income by letting temporary gains and losses decrease and increase (respectively) pension expense? Of course, as years pass there may be more gains than losses, or vice versa, preventing their offsetting one another completely. So, if a net gain or a net loss gets "too large," pension expense must be adjusted.

The FASB defines too large rather arbitrarily as being when a net gain or a net loss at the beginning of a year exceeds an amount equal to 10% of the PBO, or 10% of plan assets, whichever is higher.[13] This threshold amount is referred to as the "corridor." When the corridor is exceeded, the excess is not charged to pension expense all at once. Instead, as a further concession to income smoothing, only a portion of the excess is included in pension expense. The minimum amount that should be included is the excess divided by the average remaining service period of active employees expected to receive benefits under the plan.

Additional Consideration

Although the vast majority of companies choose to delay recognition of gains and losses as described above, companies are permitted to recognize them immediately (FASB ASC 715-30-35-25: Compensation—Retirement Benefits—Defined Benefit Plans—Pension—Subsequent Measurement—Gains and Losses). Recently, several companies have chosen that option. The option to recognize pension gains and losses immediately is called the *mark-to-market* (MTM) method. More than 50 companies, including **Verizon** and **FedEx**, have switched to MTM since 2010.

Why, you might ask, would a company voluntarily choose to report what often is a massive expense at the time of the adoption? While companies usually announce that the reason for the switch is to make operating performance easier to understand and provide a more current picture of pension plan performance (which it does), the more pragmatic reason is to get rid of large accumulated pension actuarial losses all at once, rather than amortizing them to pension expense over time and allowing them to negatively affect profits for years to come.

[13]For this purpose, the FASB specifies the market-related value of plan assets. This can be either the fair value or a weighted-average fair value over a period not to exceed five years. We will uniformly assume fair value in this chapter.

In our illustration, we're assuming a net loss–AOCI of $55 million at the beginning of 2021. Also recall that the PBO and plan assets are $400 million and $300 million, respectively, at that time. The amount amortized to 2021 pension expense is $1 million, calculated as follows:

A net gain or a net loss affects pension expense only if it exceeds an amount equal to 10% of the PBO, or 10% of plan assets, whichever is higher.

Determining Net Loss Amortization—2021	($ in millions)
Net loss (previous losses exceeded previous gains)	$55
10% of $400 ($400 is greater than $300): the "corridor"	(40)
Excess at the beginning of the year	$15
Average remaining service period	÷ 15 years
Amount amortized to 2021 pension expense	**$ 1**

Because the net loss exceeds an amount equal to the greater of 10% of the PBO or 10% of plan assets, part of the excess is amortized to pension expense.

The pension expense is increased because a net loss is being amortized. If a net *gain* were being amortized, the amount would be *deducted* from pension expense because a gain would indicate that balance of the net cost of providing the pension plan had decreased.

Amortization of a net gain would decrease pension expense.

Amortization of a net loss increases pension expense.

Amortization of the Net Loss–AOCI	($ in millions)
Service cost	$41
Interest cost	24
Expected return on the plan assets	(27)
Amortization of prior service cost–AOCI	4
Amortization of net loss–AOCI	**1**
2021 pension expense	$43

This amortization reduces the net loss–AOCI in 2021 by $1 million. Also recall that Global incurred (a) a $23 million loss in 2021 from revising estimates relating to the PBO and (b) a $3 million gain when the 2021 return on plan assets was higher than expected. These three changes affected the net loss–AOCI in 2021 as follows:

Net Loss–AOCI	($ in millions)
Net loss–AOCI at the beginning of 2021	$55
Less: 2021 amortization	**(1)**
Plus: 2021 loss on PBO	23
Less: 2021 gain on plan assets	(3)
Net loss–AOCI at the end of 2021	$74

New losses add to a net loss; new gains reduce a net loss.

Additional Consideration

The $74 million balance in net loss–AOCI at the end of 2021 would be the beginning balance in 2022. It would be compared with the 2022 beginning balances in the PBO and plan assets to determine whether amortization would be necessary in 2022. If you were to look back to our analyses of the changes in those two balances, you would see the 2022 beginning balances in the PBO and plan assets to be $450 million and $340 million, respectively. The amount amortized to 2022 pension expense will be $1.93 million, calculated as follows:

	($ in millions)
Net loss (previous losses exceeded previous gains)	$ 74
10% of $450 ($450 is greater than $340)	(45)
Excess at the beginning of the year	$ 29
Average remaining service period	÷ 15 years*
Amount amortized to 2022 pension expense	$1.93

*Assumes the average remaining service period of active employees is still 15 years in 2022, due to new employees joining the firm.

PART D

Reporting Issues

Recording Gains and Losses

● LO17–7

As we discussed earlier, gains and losses (either from changing assumptions regarding the PBO or from the return on assets being higher or lower than expected) are deferred and not immediately included in pension expense and net income. Instead, we report them as *other comprehensive income (OCI)* in the statement of comprehensive income. So Global records a *loss–OCI* for the **$23** million loss that occurs in 2021 when it revises its estimate of future salary levels causing its PBO estimate to increase. Global also records a **$3** million *gain–OCI* that occurred when the $30 million actual return on plan assets exceeded the $27 million expected return. Here's the entry:

Losses and gains are reported as OCI.

To Record New Gains and Losses	($ in millions)	
Loss–OCI (from change in assumption)	**23**	
PBO		**23**
Plan assets	**3**	
Gain–OCI ($30 actual return on assets minus $27 expected return)		**3**

The loss is an increase in the PBO due to a change in an assumption. In this entry, we are recording that increase in the PBO account balance. If the change in assumption had caused the PBO to be reduced instead, we would debit the PBO here and credit a gain–OCI.

Similarly, the gain due to the actual return on plan assets exceeding the expected return is an increase in plan assets. In the next section, we record a journal entry that increases plan assets for the expected return (as a component of pension expense), so the two adjustments together cause the plan assets account balance to reflect the *actual* return (expected increase plus the additional increase represented by the gain). Of course, if the actual return had been less than expected, we would debit a loss–OCI and credit plan assets here.

Additional Consideration

Just as we record new losses and gains as they occur, we also will record a change in the prior service cost account for any new prior service cost should it occur. For instance, if Global revised its pension formula again and recalculated its PBO using the more generous formula, causing a $40 million increase in the PBO, the company would record the new prior service cost this way:

To Record New Prior Service Cost	($ in millions)	
Prior service cost–OCI (increase in PBO due to plan amendment)	40	
PBO		40

If an amendment *reduces* rather than increases the PBO, the *negative prior service cost* would reduce both the prior service cost and pension liability.

International Financial Reporting Standards

● LO17–12

Accounting for Gains and Losses. Accounting for gains and losses in defined benefit plans (called "remeasurement" gains and losses under IFRS) under *IAS No. 19* is similar to U.S. GAAP, but there are two important differences.

The first difference relates to the make-up of the gain or loss on plan assets. As we know from the chapter, this amount under U.S. GAAP is the difference between the actual and expected returns, where the expected return is different from company to company and usually different from the interest rate used to determine the interest cost. Not so under IFRS, which requires that we use the same rate (the rate for "high grade corporate bonds") for both the interest cost on the defined benefit obligation (called projected benefit obligation or PBO under U.S. GAAP) and the interest revenue on the plan assets. In fact, under IFRS, we multiply that rate, say 6%, times the net difference between the defined benefit obligation (DBO) and plan assets and report the **net interest cost/income**:

	($ in millions)	
Net interest cost [6% × ($400 – $300)]	**6**	
Plan assets (6% × $300: interest income)	18	
DBO (6% × $400: interest cost)		24

Under IFRS, we have a *remeasurement* gain when the actual return on plan assets exceeds the "high grade corporate bond rate" we use to determine the net interest cost/income.

As a result, the remeasurement gain (or loss) under IFRS usually is an amount different from the gain (or loss) on plan assets under GAAP:

To Record Gains and Losses

U.S. GAAP

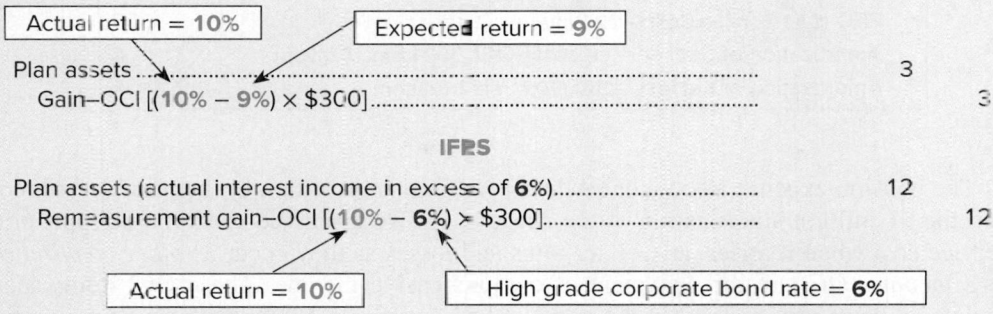

Actual return = 10% | Expected return = 9%

Plan assets	3	
Gain–OCI [(10% – 9%) × $300]		3

IFRS

Plan assets (actual interest income in excess of 6%)	12	
Remeasurement gain–OCI [(10% – 6%) × $300]		12

Actual return = 10% | High grade corporate bond rate = 6%

A second difference relates to the treatment of gains and losses *after* they are initially recorded in OCI. We've seen that U.S. GAAP requires that gains and losses (either from the actual return exceeding an assumed amount [*entries above*] or from changing assumptions regarding the pension obligation [*entries below*]) are to be (a) included among OCI items in the statement of comprehensive income when they first arise and then (b) gradually amortized or recycled out of OCI and into expense (when the accumulated net gain or net loss exceeds the 10% threshold). Similar to U.S. GAAP, under IFRS these gains and losses are included in OCI when they first arise, but unlike U.S. GAAP those amounts are *not subsequently amortized out of OCI and into expense.** Instead, under IFRS those amounts remain in the balance sheet as accumulated other comprehensive income. The initial entries, then, are the same:

Remeasurement gains and losses under IFRS are reported as OCI but, unlike under U.S. GAAP, are *not subsequently amortized to expense* and recycled into net income.

U.S. GAAP

Loss–OCI†	23	
PBO		23

IFRS

Remeasurement loss–OCI††	23	
DBO		23

*"Employee Benefits," *International Accounting Standard No. 19* (IASCF), as amended effective January 1, 2018.
†Subsequently amortized to expense and recycled from other comprehensive income to net income.
††Not subsequently amortized to expense; remains in accumulated other comprehensive income.

Recording the Pension Expense

Recall from Illustration 17–12 that Global's 2021 pension expense is $43 million. The expense includes the **$41** million service cost and the **$24** million interest cost, both of which, as we learned earlier, add to Global's PBO. Similarly, the expense includes a **$27** million expected return on plan assets, which adds to the plan assets.[14] These changes are reflected in the following entry:

Service cost	$41
Interest cost	24
Expected return on assets	(27)
Amortization of prior service cost	4
Amortization of net loss	1
Pension expense	$43

To Record Pension Expense	($ in millions)	
Pension expense (total)	43	
Plan assets (**$27** expected return on assets).	**27**	
PBO (**$41** Service Cost + **$24** Interest Cost). . . .		65
Amortization of prior service cost—OCI (2021 amortization)		4
Amortization of net loss—OCI (2021 amortization)		1

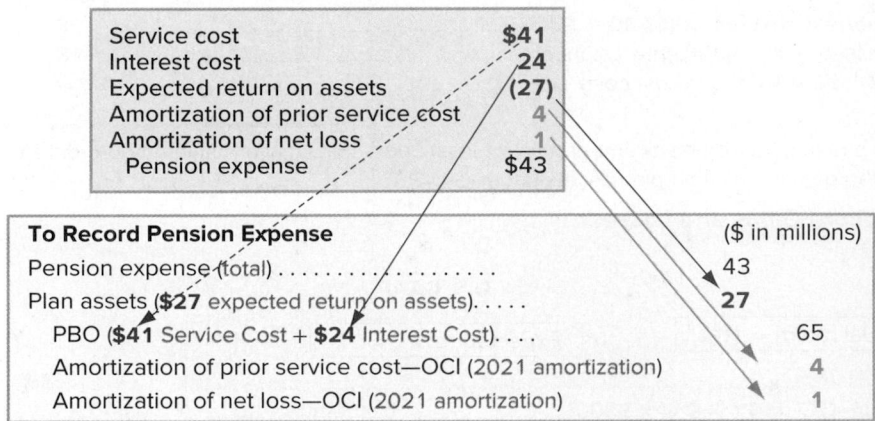

The pension expense also includes the $4 million amortization of the prior service cost and the $1 million amortization of the net loss. As we discussed earlier, we report prior service cost when it arises, as well as gains and losses as they occur as *other comprehensive income (OCI)* in the statement of comprehensive income. These OCI items accumulate as Prior service cost–AOCI and Net loss (or gain)–AOCI. So, when we amortize these AOCI accounts, we report the amortization amounts as OCI items in the statement of comprehensive income as well. These amortized amounts are being "reclassified" from OCI to net income.

New gains and losses and prior service cost are reported as OCI. So is the amortization of their accumulated balances.

Amortization reduces the Prior service cost–AOCI and the Net loss–AOCI. Since these accounts have debit balances, we credit the accounts for the amortization. If we were amortizing a net gain, we would *debit* the account because a net gain has a credit balance.

Remember, we report the funded status of the plan in the balance sheet. That's the difference between the PBO and plan assets. In this case, it's a net pension liability since the plan is underfunded; that is, the PBO exceeds plan assets.

Reporting Pension Expense in the Income Statement

The service cost component of pension expense is reported separate from the non-service cost components of pension expense in the income statement as part of the total compensation costs.

The service cost represents benefits earned by employees during the year as part of their compensation. Consistent with that characteristic, companies report the service cost component of pension expense in the income statement as part of the total *compensation costs* arising from services rendered by the employees during the period, separate from the other (non-service cost) components of pension expense. This presentation reflects the nature of service cost being different from that of the other elements of pension cost. The non-service cost components of pension expense are presented in the income statement also, but separate from the service cost component and outside the subtotal of income from operations.[15] In our illustration, then, Global will report the $41 million service cost as one of the

[14] The increase in plan assets is the $30 million *actual* return, but the $27 million *expected* return is the component of pension expense because the $3 million gain isn't included in expense. We saw in the previous section that the $3 million gain also increases plan assets.

[15] Accounting Standards Update No. 2017-07, *Compensation—Retirement Benefits (Topic 715): Improving the Presentation of Net Periodic Pension Cost and Net Periodic Postretirement Benefit Cost*, FASB: March, 2017).

elements of compensation expense and the $2 million net amount of the remaining components of pension expense on a separate line below income from operations:

		($ in millions)
Revenue		$xxx
Operating expenses		(xxx)*
Income from operations		$xxx
Other income (expenses), net	$xxx	
Non-service cost components of pension expense	2	(xxx)
Net income		$xxx

*The $41 million service cost component of pension expense is reported in the same line item or items of other employee compensation.

In addition, the components of pension expense are itemized in disclosure notes. For instance, **Kellogg Company** described the composition of its pension expense in the disclosure note in a recent annual report, shown in Illustration 17–14.

Illustration 17–14

Disclosure of Pension Expense—**Kellogg Company**

Real World Financials

Components of Pension Expense

($ in millions)	2017	2016	2015
Service cost	$ 96	$ 98	$114
Interest cost	164	174	206
Expected return on plan assets	(371)	(352)	(399)
Amortization of losses (gains)	(36)	323	303
Amortization of prior service costs (credits)	9	13	13
Curtailment and special termination benefits	(151)	1	(1)
Net expense	$(289)	$257	$236

Source: Kellogg Company

The components of pension expense are itemized in the disclosure note.

International Financial Reporting Standards

Prior Service Cost. Under *IAS No. 19*, prior service cost (called past service cost under IFRS) is combined with the current service cost and reported within the income statement rather than as a component of other comprehensive income as it is under U.S. GAAP.[16] For Global, the service cost would be recorded under IFRS in 2020 when the plan amendment occurred as follows ($ in millions):

● LO17–12

Service cost (service cost–2020 **plus $60**)	xxx	
DBO (service cost–2020)		xxx
DBO (past service cost)		**60***

*From Illustration 17–12.

Recording the Funding of Plan Assets

When Global adds its annual cash investment to its plan assets in 2021, the value of those plan assets increases by $48 million:

To Record Funding	($ in millions)	
Plan assets	**48**	
Cash **(contribution to plan assets)**		**48**

[16] "Employee Benefits," *International Accounting Standard No. 19* (IASCF), as amended effective January 1, 2016.

It's not unusual for the cash contribution to differ from that year's pension expense. After all, determining the periodic pension expense and the funding of the pension plan are two separate processes. Pension expense is an accounting decision. How much to contribute each year is a financing decision affected by cash flow and tax considerations, as well as minimum funding requirements. The Pension Protection Act of 2006 and the Employee Retirement Income Security Act of 1974 (ERISA) establish the pension funding requirements. Subject to these considerations, cash contributions are actuarially determined with the objective of accumulating (along with investment returns) sufficient funds to provide promised retirement benefits.

We saw earlier that when pension benefits are paid to retired employees, those payments reduce the plan assets established to pay the benefits and also reduce the obligation to pay the benefits, the PBO:

To Record Payment of Benefits	($ in millions)	
PBO	38	
Plan assets (payments to retired employees)		38

Now that we've recorded these entries for the funding of the pension plan and payment of benefits, we have recorded all the changes in the PBO and plan assets for 2021:

($ in millions)

PBO

Debit		Credit	
		400	Balance
		23	Loss
		41	Service cost
		24	Interest cost
Benefits paid	38		
		450	Balance

Plan Assets

Debit		Credit	
Balance	300		
Gain	3		
Expected return	27		
Funding	48		
		38	Benefits paid
Balance	340		

Remember, though, we don't report either of these balances separately in the balance sheet. Instead, we net the two and report a net pension liability of $450 – $340 = $110 million, the funded status of the pension plan.

Additional Consideration

Rather than recording each of these changes in the PBO and Plan Asset accounts, we could have recorded them in a single Net pension (liability) asset account. For example, we could have recorded the gains and losses, pension expense, and funding as follows:

	($ in millions)	
Loss—OCI (from change in assumption)	23	
Net pension (liability) asset		23
Net pension (liability) asset	3	
Gain—OCI ($30 actual return on assets – $27 expected return)		3
Pension expense	43	
Net pension (liability) asset		38
Amortization of prior service cost—OCI*		4
Amortization of net loss—OCI*		1
Net pension (liability) asset	48	
Cash		48

In fact many companies use this abbreviated approach. However, recording the changes in the accounts they affect (PBO and Plan assets) and then combining the two accounts for balance sheet reporting as we do in our illustrations has two advantages. First, we need the separate balances for the required disclosure in the pension disclosure note. Second, it is a more logical approach. It's easier to see that the return on plan assets is an increase in plan assets than it is to see it as a reduction in Net pension (liability) asset.

*Because Prior service cost—OCI and Net loss—OCI have debit balances, we amortize them with credits. We would amortize a Net gain—OCI (credit balance) with a debit.

Comprehensive Income

Comprehensive income, as you may recall from Chapter 4, is a more expansive view of income than traditional net income. In fact, it encompasses all changes in equity other than from transactions with owners.[17] So, in addition to net income, comprehensive income includes up to four other changes in equity. A statement of comprehensive income is demonstrated in Illustration 17–15, highlighting the presentation of the components of OCI pertaining to Global's pension plan.

Illustration 17–15
Statement of Comprehensive Income

		($ in millions)
Net income		$xxx
Other comprehensive income:		
Gain (loss) on AFS investments (unrealized)*	$ x	
Pension plan:		
Loss—due to revising a PBO estimate	**(23)**	
Gain—return on plan assets exceeds expected†	**3**	
Amortization of net loss	**1**	
Amortization of prior service cost	**4**	
Deferred gains (losses) from derivatives	x	
Gains (losses) from foreign currency transactions and exchange rate adjustments	x	xx
Comprehensive income		$xxx

Note: These amounts are shown without considering taxes. Actually each of the elements of comprehensive income should be reported net of tax. For instance, if the tax rate is 25%, the loss would be reported as $17.25 million: $23 million less a $5.75 million tax benefit.
*An unrealized loss also might occur from recording an impairment of a debt security investment as described in Chapter 12 and, as discussed in Chapter 14, applying the "fair value option" to a liability entails reporting in OCI any part of the change in the liability's fair value that's due to the liability's credit risk.
†From Illustration 17–12.

Other comprehensive income (OCI) items are reported both (a) as they occur and then (b) as an accumulated balance within shareholder's equity in the balance sheet as shown in Illustration 17–16.[18]

Illustration 17–16
Balance Sheet Presentation of Pension Amounts

Global Communication
Balance Sheets
For Years Ended December 31

	2021	2020
Assets		
Current assets	$xxx	$xxx
Property, plant, and equipment	xxx	xx
Liabilities		
Current liabilities	$xxx	$xxx
Net pension liability	**110**	**100**
Other long-term liabilities	xxx	xxx
Shareholders' Equity		
Common stock	$xxx	$xxx
Retained earnings	xxx	xxx
Accumulated other comprehensive		
Gain (loss) on AFS investments (unrealized)	xxx	xxx
Net loss*	**(74)**	**(55)**
Prior service cost*	**(52)**	**(56)**

*These are debit balances and therefore negative components of accumulated other comprehensive income; a net gain would have a credit balance and be a positive component of accumulated other comprehensive income.

[17]Transactions with owners primarily include dividends and the sale or purchase of shares of the company's stock.
[18]Companies can present net income and other comprehensive income either in a single continuous statement of comprehensive income or in two separate but consecutive statements.

Reporting OCI as it occurs and also as an accumulated balance is consistent with the way we report net income and its accumulated counterpart, retained earnings.

In addition to reporting the gains or losses (and other elements of comprehensive income) that occur in the current reporting period, we also report these amounts on a *cumulative* basis in the balance sheet. Comprehensive income includes (a) net income and (b) OCI. Remember that we report net income that occurs in the current reporting period in the income statement and also report accumulated net income (that hasn't been distributed as dividends) in the balance sheet as retained earnings. Similarly, we report OCI as it occurs in the current reporting period (see Illustration 17–15) and also report *accumulated other comprehensive income* in the balance sheet. In its 2021 balance sheet, Global will report the amounts as shown in Illustration 17–16.

Look back to the schedule shown earlier to see how the net loss–AOCI increased from **$55** million to **$74** million during 2021 and another schedule also shown earlier to see how the prior service cost–AOCI decreased from **$56** million to **$52** million. The pension liability represents the underfunded status of Global's pension plan on the two dates.

Income Tax Considerations

OCI items are reported net of tax, in both the (a) statement of comprehensive income and (b) AOCI.

We have ignored the income tax effects of the amounts in order to focus on the core issues. Note, though, that as gains and losses occur, they are reported along with their tax effects (tax expense for a gain, tax savings for a loss) in the statement of comprehensive income. This can be accomplished by presenting components of other comprehensive income either net of related income tax effects or *before* income tax effects with disclosure of the income taxes allocated to each component either in a disclosure note or parenthetically in the statement.[19] Likewise, AOCI in the balance sheet also is reported net of tax.

Putting the Pieces Together

● LO17–8

In preceding sections, we've discussed (1) the projected benefit obligation (including changes due to periodic service cost, accrued interest, revised estimates, plan amendments, and the payment of benefits); (2) the plan assets (including changes due to investment returns, employer contributions, and the payment of benefits); (3) prior service cost; (4) gains and losses; (5) the periodic pension expense (comprising components of each of these); and (6) the funded status of the plan. These elements of a pension plan are interrelated. It's helpful to see how each element relates to the others. One way is to bring each part together in a *pension spreadsheet.* We do this for Global Communications in Illustration 17–17.

You should spend several minutes studying this spreadsheet, focusing on the relationships among the elements that constitute a pension plan. Notice that the first numerical column simply repeats the actuary's report of how the PBO changed during the year, as explained previously (Illustration 17–9). Likewise, the second column reproduces the changes in plan assets we discussed earlier (Illustration 17–10). We've also previously noted the changes in the prior service cost—AOCI and the net loss—AOCI that are duplicated in the third and fourth columns. The fifth column repeats the calculation of the 2021 pension expense we determined earlier, and the cash contribution to the pension fund is the sole item in the next column.

Rather than report the PBO and plan asset balances separately, we combine those balances and report a single net pension liability or net pension asset in the balance sheet.

The last column shows the changes in the funded status of the plan. Be sure to notice that the funded status is the difference between the PBO (column 1) and the plan assets (column 2). That means that each of the changes we see in either of the first two columns also is reflected as a change in the funded status in the last column. The net pension liability (or net pension asset) balance is not carried in company records. Instead, we use this label to report the PBO and plan assets in the balance sheet as a single net amount.

Notice that each change in a formal account (light-shaded columns) is reflected in exactly two of those columns with one debit and one credit.

[19]Similarly, if any new prior service cost should arise due to a plan amendment, it too would be reported net of tax or with tax effects shown parenthetically.

Illustration 17–17

Pension Spreadsheet

	Recorded in Accounts						Reported Only
	PBO	**Plan Assets**	**Prior Service Cost —AOCI**	**Net Loss —AOCI**	**Pension Expense**	**Cash**	**Net Pension (Liability)/ Asset**
Balance, Jan. 1, 2021	**(400)**	**300**	56	55			**(100)**
Service cost	(41)				41		(41)
Interest cost	(24)				24		(24)
Expected return on assets		27			(27)		27
Adjust for: Gain on assets		3		(3)			3
Amortization of:							
Prior service cost—AOCI			(4)		4		
Net loss—AOCI				(1)	1		
Loss on PBO	(23)			23			(23)
Prior service cost (new)*	0		0				0
Contribution to fund		48				(48)	48
Retiree benefits paid	38	(38)					
Balance, Dec. 31, 2021	**(450)**	**340**	52	74	43		**(110)**

Note: ( )s represent credits to accounts, not necessarily decreases.
*This amount was $60 million in the 2020 pension spreadsheet.

When the PBO exceeds plan assets, we have a net pension liability. If plan assets exceed the PBO we have a net pension asset.

The net pension liability (or net pension asset) is not recorded in a journal entry but is the amount reported in the balance sheet as the PBO and plan assets combined.

International Financial Reporting Standards

Reporting Pension Expense. Under IFRS we separately report (a) the service cost component (including past service cost) and (b) the net interest cost/income component in the *income statement* and (c) remeasurement gains and losses as *other comprehensive income.*

● LO17–12

Decision Makers' Perspective

Although financial statement items are casualties of the political compromises of GAAP guidance, information provided in the disclosure notes fortunately makes up for some of the deficiencies.[20] Foremost among the useful disclosures are changes in the projected benefit obligation, changes in the fair value of plan assets, and a breakdown of the components of the annual pension expense. Other information also is made available to make it possible for interested analysts to reconstruct the financial statements with pension assets and liabilities included. We'll look at specific disclosures after we discuss postretirement benefits other than pensions because the two types of plans are reported together.

Investors and creditors must be cautious of the nontraditional treatment of pension information when developing financial ratios as part of an analysis of financial statements. The various elements of pensions that are not reported separately on the balance sheet and income statement (PBO, plan assets, gains and losses) can be included in ratios such as the debt to equity ratio or return on assets, but only by deliberately obtaining those numbers from the disclosure notes and adjusting the computation of the ratios. Similarly, without adjustment, profitability ratios and the times interest earned ratio will be distorted because pension expense includes the financial components of interest and return on assets.

Pension amounts reported in the disclosure notes fill a reporting gap left by the minimal disclosures in the primary financial statements.

[20]FASB ASC 715–20–50: Compensation–Retirement Benefits–Defined Benefit Plans–General–Disclosure.

Earnings quality (as defined in Chapter 4 and discussed in other chapters) also can be influenced by amounts reported in pension disclosures. Companies with relatively sizable unrecognized pension costs (prior service cost, net gain or loss) can be expected to exhibit a relatively high "temporary" earnings component. Recall that temporary earnings are expected to be less predictive of future earnings than the "permanent" earnings component. ●

Settlement or Curtailment of Pension Plans

Companies sometimes terminate defined benefit plans to reduce costs and lessen risk.

To cut down on cumbersome paperwork and lessen their exposure to the risk posed by defined benefit plans, many companies are providing defined contribution plans instead. When a plan is terminated, a change in earnings is reported at that time.[21] For instance, **Accenture** described the termination of its pension plan in the following disclosure note (see Illustration 17–18):

Illustration 17–18

Gain on the Termination of a Defined Benefit Plan—Accenture

Real World Financials

U.S. Defined Benefit Pension Plan Settlement Charges (in part)

In May 2017, the Company settled its U.S. pension plan obligations. Plan participants elected to receive either a lump-sum distribution or to transfer benefits to a third-party annuity provider. As a result of the settlement, the Company was relieved of any further obligation under its U.S. pension plan. During fiscal 2017, the Company recorded a pension settlement charge of $509,793, and related income tax benefits of $198,219.

Source: Accenture

Concept Review Exercise

PENSION PLANS

Allied Services, Inc., has a noncontributory, defined benefit pension plan. Pension plan assets had a fair value of $900 million at December 31, 2020.

On January 3, 2021, Allied amended the pension formula to increase benefits for each service year. By making the amendment retroactive to prior years, Allied incurred a prior service cost of $75 million, adding to the previous projected benefit obligation of $875 million. The prior service cost is to be amortized (expensed) over 15 years. The service cost is $31 million for 2021. Both the actuary's discount rate and the expected rate of return on plan assets were 8%. The actual rate of return on plan assets was 10%.

At December 31, 2021, $16 million was contributed to the pension fund and $22 million was paid to retired employees. Also, at that time, the actuary revised a previous assumption, increasing the PBO estimate by $10 million. The net loss AOCI at the beginning of the year was $13 million.

Required:

Determine each of the following amounts as of December 31, 2021, the fiscal year-end for Allied: (1) projected benefit obligation; (2) plan assets; and (3) pension expense.

Solution:

($ in millions)	Projected Benefit Obligation	Plan Assets	Pension Expense
Balances at Jan. 1	$ 875	$900	$ 0
Prior service cost	75		
Service cost	31		31
Interest cost [($875 + $75)* × 8%]	76		76
Return on plan assets:			
Actual ($900 × 10%)		90	
Expected ($900 × 8%)			(72)

(continued)

[21] FASB ASC 715–30–35: Compensation–Retirement Benefits–Defined Benefit Plans–Pension–Subsequent Measurement.

(concluded)

($ in millions)	Projected Benefit Obligation	Plan Assets	Pension Expense
Amortization of prior service cost ($75 ÷ 15 yrs)			5
Amortization of net loss			0†
Loss on PBO	10		
Cash contribution		16	
Retirement payments	(22)	(22)	
Balance at Dec. 31	$1,045	$984	$40

Note: The $18 million gain on plan assets ($90 – $72 million) is not recognized yet; it is carried forward to be combined with previous and future gains and losses, which will be recognized only if the net gain or net loss exceeds 10% of the higher of the PBO or plan assets.
*Since the plan was amended at the beginning of the year, the prior service cost increased the PBO at that time.
†Since the net loss ($13) does not exceed 10% of $900 (higher than $875), no amortization is required for 2021.

International Financial Reporting Standards

Accounting for pensions and other postretirement benefits under IFRS is specified in a recent IASB amendment to its postemployment benefit standard, *IAS No. 19*.

Under the update, the changes that constitute the components of the net pension cost are separated into (a) the cost of service (b) the cost of financing that cost, and (c) gains or losses from occasional remeasurement of the cost:

● LO17–12

Components	Include:	Recognized immediately in the:
Service cost	• Current service costs • Past service costs (if any)	• balance sheet • income statement
Net interest cost or Net interest income	• **Interest rate* × net pension liability (if DBO > plan assets)** **or** • **Interest rate* × net pension asset (if plan assets > DBO)**	• balance sheet • income statement
Remeasurement cost	• Remeasurements of service costs caused by changes in assumptions (i.e., assumptions like salary expectations length of service, length of retirement etc.) • Gains and losses arising from experience differing from what was assumed • Investment gains and losses on plan assets	• balance sheet • statement of comprehensive income as OCI

*High-quality corporate bond rate.

Classifying the Components of the Net Pension Cost

Using the amounts from our Global Communications illustration, the changes in 2021 would be classified as follows:

Profit & Loss (Income Statement)

	($ in millions)	
Service cost:		
Service cost—2021	$ 41	
Past service cost	0*	$41
Net interest cost/income:		
Net interest cost [6% × ($400 – $300)]		6
		$47

(continued)

We record (a) individual components of the net pension cost (in net income or OCI) and (b) the balance sheet accounts they affect.

Under IFRS, we report separately the three costs of having a defined benefit plan:

In net income:

- service cost
- net interest cost

In OCI:

- remeasurement cost

(concluded)

Other Comprehensive Income (OCI)

Remeasurement cost:		
Loss (gain) on DBO—change in salary estimate	$23	
Loss (gain) on plan assets: (10% − 6%) × $300	(12)	11
Net pension cost		$58

*Last year (2020), this amount was $60 million.

Notice that there is no separate interest cost and there is no "expected" return on plan assets. Instead, those two are essentially combined into a single measure, net interest cost/income. Because they're combined, the same rate (6% in this case) is used for both, rather than having one rate for interest cost and another for expected return on assets.

Recording the Components of the Net Pension Cost

We report the (a) service cost and (b) net interest cost/income in *net income* and (c) remeasurement costs in *OCI,* as each of those amounts occurs. By recording each component, we record the total pension expense. Remember, each component is a change in either plan assets (the return on assets) or the defined benefit obligation (all others), so as we record each component we also record its effect on plan assets and the defined benefit obligation, as follows:

To Record Net Pension Cost	($ in millions)	
Service cost	41	
DBO (service cost-2021)		41
DBO (past service cost: none in 2021)		0
Net interest cost [6% × ($400 − $300)]	6	
Plan assets (6% × $300: interest income)	18	
DBO (6% × $400: interest cost)		24
Plan assets (actual return in excess of 6%)	12	
Remeasurement gain on plan assets—OCI [(10% − 6%) × $300]		12
Remeasurement loss from change in assumption—OCI	23	
DBO (change in future salary estimate)		23

When Global adds its annual cash investment to its plan assets, the value of those plan assets increases by $48 million:

To Record Funding	($ in millions)	
Plan assets	48	
Cash (contribution to plan assets)		48

In our illustration, Global's retired employees were paid benefits of $38 million in 2021. Paying those benefits, of course, reduces the obligation to pay benefits (the DBO), and since the payments are made from the plan assets, that balance is reduced as well:

To Record Payment of Benefits	($ in millions)	
DBO	38	
Plan assets		38

Reporting the Components of the Net Pension Cost

We report the components of the total pension cost in the **statement of comprehensive income** as follows:

Revenue	$xxx
Operating expenses (including **$41 pension service cost**)	(xx)
Finance costs (including **$6 net interest cost on pensions**)	xx
Profit before tax	$xxx
Tax expense	(xx)
Net income	$xxx
Other comprehensive income	

(continued)

(concluded)

Remeasurement loss arising from change in pension assumption	**$ 23**
Remeasurement gain from the return on plan assets exceeding the interest rate	**(12)**
Comprehensive income	$xxx

Notice in the journal entries that we have no "pension expense" being recorded as we had using U.S. GAAP in the chapter. Instead, individual components of the net pension cost are recorded. The reason is that the amendment calls for components, not the net total, to be reported in the income statement and statement of comprehensive income.

Postretirement Benefits Other Than Pensions

PART E

LO17–9

As we just discussed, most companies have pension plans that provide for the future payments of retirement benefits to compensate employees for their current services. Many companies also furnish *other postretirement benefits* to their retired employees. These may include medical coverage, dental coverage, life insurance, group legal services, and other benefits. By far the most common is health care benefits. One of every three U.S. workers in medium- and large-size companies participates in health care plans that provide for coverage that continues into retirement. The aggregate impact is considerable; the total obligation for all U.S. corporations is measured in trillions of dollars.

Prior to 1993, employers accounted for postretirement benefit costs on a pay-as-you-go basis, meaning the expense each year was simply the amount of insurance premiums or medical claims paid, depending on the way the company provided health care benefits. The FASB revised GAAP to require a completely different approach. The expected future health care costs for retirees now must be recognized as an expense over the years necessary for employees to become entitled to the benefits.[22] This is the accrual basis that also is the basis for pension accounting.

In fact, accounting for postretirement benefits is similar in most respects to accounting for pension benefits. This is because the two forms of benefits are fundamentally similar. Each is a form of deferred compensation earned during the employee's service life and each can be estimated as the present value of the cost of providing the expected future benefits. **General Motors** described its plan as shown in Illustration 17–19.

Note 5: Other Postretirement Benefits (in part)

The Corporation and certain of its domestic subsidiaries maintain hourly and salaried benefit plans that provide postretirement medical, dental, vision, and life insurance to retirees and eligible dependents. . . . [GAAP] requires that the cost of such benefits be recognized in the financial statements during the period employees provide service to the Corporation.

Source: General Motors

Illustration 17–19

Disclosures—General Motors

Real World Financials

Despite the similarities, though, there are a few differences in the characteristics of the benefits that necessitate differences in accounting treatment. Because accounting for the two types of retiree benefits is so nearly the same, our discussion in this portion of the chapter will emphasize the differences. This will allow you to use what you learned earlier in the chapter regarding pension accounting as a foundation for learning how to account for other postretirement benefits, supplementing that common base only when necessary. Focusing on the differences also will reinforce your understanding of pension accounting.

[22]FASB ASC 715–60: Compensation—Retirement Benefits—Defined Benefit Plans—Postretirement.

What Is a Postretirement Benefit Plan?

Before addressing the accounting ramifications, let's look at a typical retiree health care plan.[23] First, it's important to distinguish retiree health care benefits from health care benefits provided during an employee's working years. The annual cost of providing *preretirement* benefits is simply part of the annual compensation expense. However, many companies offer coverage that continues into retirement. It is the deferred aspect of these *postretirement* benefits that creates an accounting issue.

Eligibility usually is based on age and/or years of service.

Usually a plan promises benefits in exchange for services performed over a designated number of years, or reaching a particular age, or both. For instance, a plan might specify that employees are eligible for postretirement benefits after both working 20 years and reaching age 62 while in service. Eligibility requirements and the nature of benefits usually are specified by a written plan, or sometimes only by company practice.

Postretirement Health Benefits and Pension Benefits Compared

Keep in mind that retiree health benefits differ fundamentally from pension benefits in some important respects:

1. The amount of *pension* benefits generally is based on the number of years an employee works for the company so that the longer the employee works, the higher are the benefits. On the other hand, the amount of *postretirement health care* benefits typically is unrelated to service. It's usually an all-or-nothing plan in which a certain level of coverage is promised upon retirement, independent of the length of service beyond that necessary for eligibility.
2. Although coverage might be identical, the cost of providing the coverage might vary significantly from retiree to retiree and from year to year because of differing medical needs.
3. Postretirement health care plans often require the retiree to share in the cost of coverage through monthly contribution payments. For instance, a company might pay 80% of insurance premiums, with the retiree paying 20%. The net cost of providing coverage is reduced by these contributions as well as by any portion of the cost paid by Medicare or other insurance.
4. Coverage often is provided to spouses and eligible dependents.

Determining the Net Cost of Benefits

To determine the postretirement benefit obligation and the postretirement benefit expense, the company's actuary first must make estimates of what the postretirement benefit costs will be for current employees. Then, as illustrated in Illustration 17–20, contributions to those costs by employees are deducted, as well as Medicare's share of the costs (for retirement years when the retiree will be 65 or older), to determine the estimated net cost of benefits to the employer:

Illustration 17–20
Estimating the Net Cost of Benefits

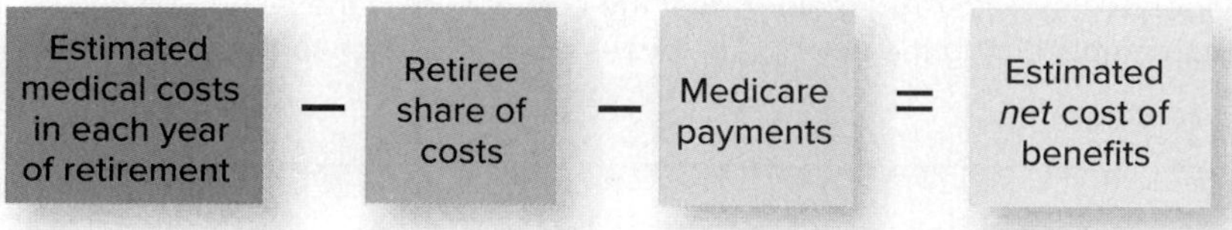

Remember, postretirement health care benefits are anticipated actual costs of providing the promised health care, rather than an amount estimated by a defined benefit formula. This makes these estimates inherently more intricate, particularly because health care costs in general are notoriously difficult to forecast. And, since postretirement health care benefits are partially paid by the retiree and by Medicare, these cost-sharing amounts must be estimated as well.

[23]For convenience, our discussion focuses on health care benefits because these are by far the most common type of postretirement benefits other than pensions. But the concepts we discuss apply equally to other forms of postretirement benefits.

On the other hand, estimating postretirement benefits costs is similar in many ways to estimating pension costs. Both estimates entail a variety of assumptions to be made by the company's actuary. Many of these assumptions are the same; for instance, both require estimates of

Many of the assumptions needed to estimate postretirement health care benefits are the same as those needed to estimate pension benefits.

1. A discount rate.
2. Expected return on plan assets (if the plan is funded).
3. Employee turnover.
4. Expected retirement age.
5. Expected compensation increases (if the plan is pay-related).
6. Expected age of death.
7. Number and ages of beneficiaries and dependents.

Of course, the relative importance of some estimates is different from that for pension plans. Dependency status, turnover, and retirement age, for example, take on much greater significance. Also, additional assumptions become necessary as a result of differences between pension plans and other postretirement benefit plans. Specifically, it's necessary to estimate

Some additional assumptions are needed to estimate postretirement health care benefits besides those needed to estimate pension benefits.

1. The current cost of providing health care benefits at each age that participants might receive benefits.
2. Demographic characteristics of plan participants that might affect the amount and timing of benefits.
3. Benefit coverage provided by Medicare, other insurance, or other sources that will reduce the net cost of employer-provided benefits.
4. The expected health care cost trend rate.[24]

Taking these assumptions into account, the company's actuary estimates what the net cost of postretirement benefits will be for current employees in each year of their expected retirement. The discounted present value of those costs is the expected postretirement benefit obligation.

The postretirement benefit obligation is the discounted present value of the benefits during retirement.

Postretirement Benefit Obligation

● LO17–10

There are two related obligation amounts. As indicated in Illustration 17–21, one measures the total obligation and the other refers to a specific portion of the total:

Illustration 17–21
Two Views of the Obligation for Postretirement Benefits Other Than Pensions

1. **Expected postretirement benefit obligation (EPBO):** The actuary's estimate of the total postretirement benefits (at their discounted present value) expected to be received by plan participants.
2. **Accumulated postretirement benefit obligation (APBO):** The portion of the EPBO attributed to employee service to date.

The accumulated postretirement benefit obligation (APBO) is analogous to the projected benefit obligation (PBO) for pensions. Like the PBO, the APBO is reported in the balance sheet only to the extent that it exceeds plan assets.

Measuring the Obligation

To illustrate, assume the actuary estimates that the net cost of providing health care benefits to Jessica Farrow (our illustration employee from earlier in the chapter) during her retirement years has a present value of $10,842 as of the end of 2019. This is the EPBO. If the

[24]Health care cost trend rates recently reported have ranged from –% to 3.9%, with 6.6% being the most commonly assumed rate. (Pension/OPEB 2017 Assumption and Disclosure Survey, PWC, 2017).

benefits (and therefore the costs) relate to an estimated 35 years of service and 10 of those years have been completed, the APBO would be[25]

$3,098 represents the portion of the EPBO related to the first 10 years of the 35-year service period.

$10,842	×	10/35	=	$3,098
EPBO		Fraction attributed to service to date		APBO

If the assumed discount rate is 6%, a year later the EPBO will have grown to **$11,493** simply because of a year's interest accruing at that rate ($10,842 × 1.06 = **$11,493**). Notice that there is no increase in the EPBO for service because, unlike the obligation in most pension plans, the total obligation is not increased by an additional year's service.

The APBO, however, is the portion of the EPBO related to service up to a particular date. Consequently, the APBO will have increased both because of interest and because the service fraction will be higher (service cost):

$3,612 represents the portion of the EPBO related to the first 11 years of the 35-year service period.

$11,493	×	11/35	=	$3,612
EPBO		Fraction attributed to service to date		APBO

The two elements of the increase in 2020 can be separated as follows:

The APBO increases each year due to (a) interest accrued on the APBO and (b) the portion of the EPBO attributed to that year.

APBO at the beginning of the year	$3,098
Interest cost: $3,098 × 6%	186
Service cost: (**$11,493** × 1/35) portion of EPBO attributed to the year	328
APBO at the end of the year	$3,612

Attribution

Attribution is the process of assigning the cost of benefits to the years during which those benefits are assumed to be earned by employees. We accomplish this by assigning an equal fraction of the EPBO to each year of service from the employee's date of hire to the employee's full eligibility date.[26] This is the date the employee has performed all the service necessary to have earned all the retiree benefits estimated to be received by the employee.[27] In our earlier example, we assumed the attribution period was 35 years and accordingly accrued 1/35 of the EPBO each year. The amount accrued each year increases both the APBO and the postretirement benefit expense. In Illustration 17–22 we see how the 35-year attribution (accrual) period was determined.

The attribution period does not include years of service beyond the full eligibility date even if the employee is expected to work after that date.

Some critics of this approach feel there is a fundamental inconsistency between the way we measure the benefits and the way we assign the benefits to specific service periods. The benefits (EPBO) are measured with the concession that the employee may work beyond the full eligibility date; however, the attribution period does not include years of service after that date. The counterargument is the fact that at the full eligibility date the employee will have earned the right to receive the full benefits expected under the plan and the amount of the benefits will not increase with service beyond that date.

Accounting for Postretirement Benefit Plans Other Than Pensions

● LO17–11

We account for pensions and for other postretirement benefits essentially the same way.

As we just discussed, it's necessary to attribute a portion of the accumulated postretirement benefit obligation to each year as the service cost for that year as opposed to measuring the actual benefits employees earn during the year as we did for pension plans. That's due to the fundamental nature of these other postretirement plans under which employees are

[25]Assigning the costs to particular service years is referred to as the *attribution* of the costs to the years the benefits are assumed earned. We discuss attribution in the next section.

[26]If the plan specifically grants credit only for service from a date after the employee's date of hire, the beginning of the attribution period is considered to be the beginning of that credited service period, rather than the employee's date of hire.

[27]Or any beneficiaries and covered dependents.

Illustration 17–22

Determining the Attribution Period

Jessica Farrow was hired by Global Communications at age 22 at the beginning of 2010 and is expected to retire at the end of 2049 at age 61. The retirement period is estimated to be 20 years.*

Global's employees are eligible for postretirement health care benefits after both reaching age 56 while in service and having worked 20 years.

Since Farrow becomes fully eligible at age 56 (the end of 2044), retiree benefits are attributed to the 35-year period from her date of hire through that date. Graphically, the situation can be described as follows:

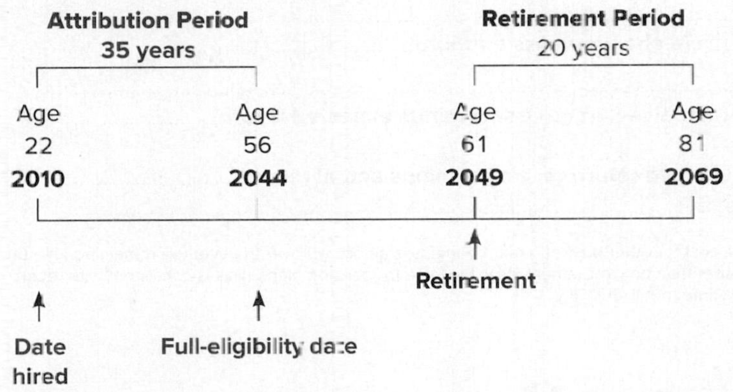

*You probably recognize this as the situation used earlier in the chapter to illustrate pension accounting.

The attribution period spans each year of service from the employee's date of hire to the employee's full eligibility date.

ineligible for benefits until specific eligibility criteria are met, at which time they become 100% eligible. This contrasts with pension plans under which employees earn additional benefits each year until they retire (see Illustration 17-23).

Illustration 17–23

Measuring Service Cost

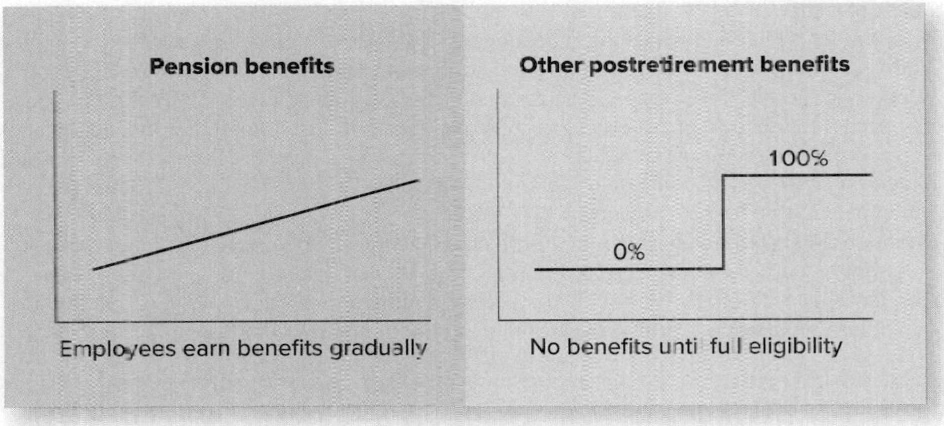

Measuring the service cost differs, though, due to a fundamental difference in the way employees acquire benefits under the two types of plans.

While we report service cost and the non-service components the same way as for pensions, the measurement of service cost is the primary difference between accounting for pensions and for other postretirement benefits. Otherwise, though, accounting for the two is virtually identical. For example, a company with an underfunded postretirement benefit plan with existing prior service cost and net loss–AOCI would record the following journal entries annually:

To Record Postretirement Benefit Expense

Postretirement benefit expense **(total)**	xx	
Plan assets **(expected return on assets)**	xx	
Amortization of net gain—OCI **(current amortization)**	xx	
APBO **(service cost + interest cost)**		xx
Amortization of net loss—OCI **(current amortization)**		xx
Amortization of prior service cost—OCI* **(current amortization)**		xx

(continued)

(concluded)

To Record Cash Funding of Plan Assets		
Plan assets	xx	
Cash **(contribution to plan assets)**		xx
To Record Gains and Losses		
Loss—OCI **(from change in assumption)**	xx	
APBO		xx
or		
APBO	xx	
Gain—OCI **(from change in assumption)**		xx
Plan assets	xx	
Gain—OCI **(actual return on assets minus expected return)**		xx
or		
Loss—OCI **(expected return on assets minus actual return)**	xx	
Plan assets		xx

*The prior service cost for other postretirement benefits is amortized over the average remaining time until "full eligibility" for employees rather than until retirement, as is the case for pension plans. This is consistent with recording "regular" service cost over the time to full eligibility.

Ethical Dilemma

Earlier this year, you were elected to the board of directors of Champion International, Inc., Champion has offered its employees postretirement health care benefits for 35 years. The practice of extending health care benefits to retirees began modestly. Most employees retired after age 65, when most benefits were covered by Medicare. Costs also were lower because life expectancies were shorter and medical care was less expensive. Because costs were so low, little attention was paid to accounting for these benefits. The company simply recorded an expense when benefits were provided to retirees. The FASB changed all that. Now, the obligation for these benefits must be anticipated and reported in the annual report. Worse yet, the magnitude of the obligation has grown enormously, almost unnoticed. Health care costs have soared in recent years. Medical technology and other factors have extended life expectancies. Of course, the value to employees of this benefit has grown parallel to the growth of the burden to the company.

Without being required to anticipate future costs, many within Champion's management were caught by surprise at the enormity of the company's obligation. Equally disconcerting was the fact that such a huge liability now must be exposed to public view. Now you find that several board members are urging the dismantling of the postretirement plan altogether.

A Comprehensive Illustration

We assumed earlier that the EPBO at the end of 2019 was determined by the actuary to be \$10,842. This was the present value on that date of all anticipated future benefits. Then we noted that the EPBO at the end of the next year would have grown by 6% to \$11,493. This amount, too, would represent the present value of the same anticipated future benefits, but as of a year later. The APBO, remember, is the portion of the EPBO attributed to service performed to a particular date. So, we determined the APBO at the end of 2020 to be \$11,493 × $\frac{11}{35}$, or \$3,612. We determined the \$328 service cost noted earlier for 2020 as the portion of the EPBO attributed to that year: \$11,493 × $\frac{1}{35}$.

Now, let's review our previous discussion of how the EPBO, the APBO, and the postretirement benefit expense are determined by calculating those amounts a year later, at the end of 2021. Before doing so, however, we can anticipate (a) the EPBO to be \$11,493 × 1.06, or \$12,182, (b) the APBO to be $\frac{12}{35}$ of that amount, or \$4,177, and (c) the 2021 service cost to be $\frac{1}{35}$ of that amount, or \$348. In Illustration 17–24 we see if our expectations are borne out by direct calculation.

Illustration 17–24

Determining the Postretirement Benefit Obligation

Assume the actuary has estimated the net cost of retiree benefits in each year of Jessica Farrow's 20-year expected retirement period to be the amounts shown in the calculation below. She is fully eligible for benefits at the end of 2044 and is expected to retire at the end of 2049.

Calculating the APBO and the postretirement benefit expense at the end of 2021, 12 years after being hired, begins with estimating the EPBO. Steps to calculate (a) the EPBO, (b) the APBO, and (c) the annual service cost at the end of 2021, 12 years after being hired, follow:

(a). 1. Estimate the cost of retiree benefits in each year of the expected retirement period and deduct anticipated Medicare reimbursements and retiree cost-sharing to derive the net cost to the employer in each year of the expected retirement period.
2. Find the present value of each year's net benefit cost as of the retirement date.
3. Find the present value of the total net benefit cost as of the current date. This is the **EPBO**.

(b). Multiply the EPBO by the attribution factor (service to date/total attribution period). This is the **APBO**. The **service cost** in any year is simply one year's worth of the EPBO.

(c). Multiply the EPBO by $\frac{1}{\text{total attribution period}}$.

The EPBO is the discounted present value of the total benefits expected to be earned.

The fraction of the EPBO considered to be earned this year is the service cost.

The fraction of the EPBO considered to be earned so far is the APBO.

The steps are demonstrated in Illustration 17–24A.

Illustration 17–24A

EPBO, APBO, and Service Cost in 2019

(a.1). Actuary estimates the net cost of benefits paid during retirement years:

(a.2). Present value [n = 1, 2, 3, 4, . . . 19, 20; i = 6%] of the net benefits as of the retirement date:

Year	Age	Net Benefit	Present Value at 2049
2050	62	5,000	$ 4,717
2051	63	5,600	4,984
2052	64	6,300	5,290
2053	65	3,000	2,376
~	~	~	~
2068	80	9,550	3,156
2069	81	10,300	3,212
			$62,269

The actuary estimates the net cost to the employer in each year the retiree is expected to receive benefits.

As of the retirement date, the lump-sum equivalent of the expected yearly costs is $62,269.

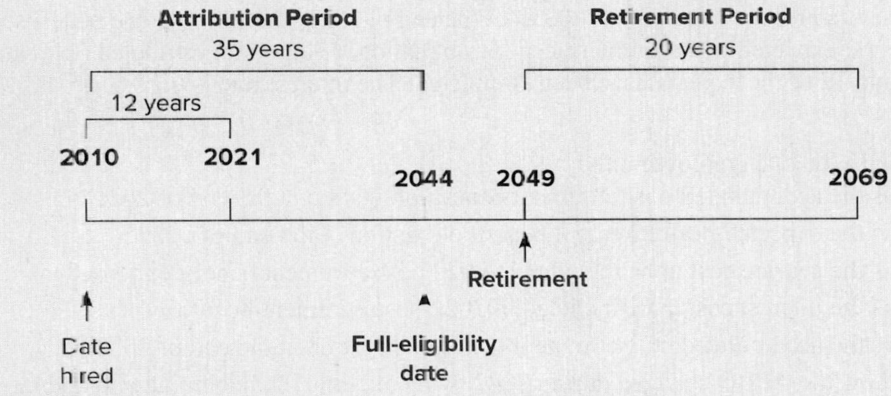

(a.3). Present value (n = 28, i = 6%) of postretirement benefits at 2021 is
$62,269 × 0.19563 = **$12,182** (EPBO)

(b). $12,182 × 12/35 = **$4,177** (APBO)

(c). $12,182 × 1/35 = **$348** (Service Cost)

The EPBO in 2021 is the present value of those benefits.

The APBO is the portion of the EPBO attributed to service to date.

The service cost is the portion of the EPBO attributed to a particular year's service.

Decision Makers' Perspective

Postretirement benefit amounts reported in the disclosure notes fill a reporting gap left by the minimal disclosures in the primary financial statements.

When they analyze financial statements, investors and creditors should be wary of the non-standard way companies report pension and other postretirement information. Recall that in the balance sheet, firms do not separately report the benefit obligation and the plan assets. Also, companies have considerable latitude in making the several assumptions needed to estimate the components of postretirement benefit plans. Fortunately, information provided in the disclosure notes makes up for some of the deficiency in balance sheet information and makes it possible for interested analysts to modify their analysis. As for pensions, the choices companies make for the discount rate, expected return on plan assets, and the compensation growth rate can greatly impact postretirement benefit expense and earnings quality. The disclosures required are very similar to pension disclosures. In fact, disclosures for the two types of retiree benefits typically are combined.[28] Disclosures include the following:

- Descriptions of the plans.
- Estimates of the obligations (PBO, ABO, vested benefit obligation, EPBO, and APBO).
- The percentage of total plan assets for each major category of assets (equity securities, debt securities, real estate, other) as well as a description of investment strategies, including any target asset allocations and risk management practices.
- A breakdown of the components of the annual pension and postretirement benefit expenses for the years reported.
- The discount rates, the assumed rate of compensation increases used to measure the PBO, the expected long-term rate of return on plan assets, and the expected rate of increase in future medical and dental benefit costs.
- Estimated benefit payments presented separately for the next five years and in the aggregate for years 6–10.
- Estimate of expected contributions to fund the plan for the next year.
- (a) Any changes to the net gain or net loss and prior service cost arising during the period, (b) the accumulated amounts of these components of accumulated other comprehensive income, and (c) the amounts of those balances expected to be amortized in the next year.
- Other information to make it possible for interested analysts to reconstruct the financial statements with plan assets and liabilities included. ●

Concept Review Exercise

OTHER POSTRETIREMENT BENEFITS

Technology Group, Inc., has an unfunded retiree health care plan. The actuary estimates the net cost of providing health care benefits to a particular employee during his retirement years to have a present value of $24,000 as of the end of 2021 (the EPBO). The benefits, and therefore the expected postretirement benefit obligation, relate to an estimated 36 years of service, and 12 of those years have been completed. The interest rate is 6%.

Required:

Pertaining to the one employee only:

1. What is the accumulated postretirement benefit obligation at the end of 2021?
2. What is the expected postretirement benefit obligation at the end of 2022?
3. What is the service cost to be included in 2022 postretirement benefit expense?
4. What is the interest cost to be included in 2022 postretirement benefit expense?
5. What is the accumulated postretirement benefit obligation at the end of 2022?
6. Show how the APBO changed during 2022 by reconciling the beginning and ending balances.
7. What is the 2022 postretirement benefit expense, assuming no net gains or losses and no prior service cost?

[28]FASB ASC 715-60-50: Compensation—Retirement Benefits—Defined Benefit Plans—Other Postretirement—Disclosure.

Solution:

1. What is the accumulated postretirement benefit obligation at the end of 2021?

$24,000	×	$^{12}/_{36}$	=	$8,000
EPBO 2021		Fraction earned		APBO 2021

2. What is the expected postretirement benefit obligation at the end of 2022?

$24,000	×	1.06	=	$25,440
EPBO 2021		To accrue interest		EPBO 2022

3. What is the service cost to be included in 2022 postretirement benefit expense?

$25,440	×	$^{1}/_{36}$	=	$707
EPBO 2022		Attributed to 2022		Service cost

4. What is the interest cost to be included in 2022 postretirement benefit expense?

$8,000 (beginning APBO) × 6% = $480

5. What is the accumulated postretirement benefit obligation at the end of 2022?

$25,440	×	$^{13}/_{36}$	=	$9,187
EPBO 2022		Fraction attributed to service to date		APBO 2022

6. Show how the APBO changed during 2022 by reconciling the beginning and ending balances.

APBO at the beginning of 2022 (from req. 1)	$8,000
Service cost (from req. 3)	707
Interest cost (from req. 4)	480
APBO at the end of 2022 (from req. 5)	$9,187

7. What is the 2022 postretirement benefit expense, assuming no net gains or losses and no prior service cost?

Service cost	$ 707
Interest cost	480
Actual return on the plan assets	(not funded)
Adjusted for: gain or loss on the plan assets	(not funded)
Amortization of prior service cost	none
Amortization of net gain or loss	none
Postretirement benefit expense	$ 1,187

Financial Reporting Case Solution

©Peter Bernik/Shutterstock

1. **Why is pension plan underfunding not a concern in your present employment?** *(p. 981)* In a defined contribution plan, the employer is not obliged to provide benefits beyond the annual contribution to the employees' plan. No liability is created. Unlike retirement benefits paid in a defined benefit plan, the employee's retirement benefits in a defined contribution plan are totally dependent on how well invested assets perform in the marketplace.
2. **Were you correct that the pension liability is not reported in the balance sheet? What is the liability?** *(p. 983)* Yes and no. The pension liability is not reported separately in the balance sheet. It is, however, combined with pension assets with the net difference reported in the balance sheet. The separate balance is disclosed, though, in the notes. For United Dynamics, the PBO at the end of 2021 is $2,628 million.
3. **What is the amount of the plan assets available to pay benefits? What are the factors that can cause that amount to change?** *(p. 990)* The plan assets at the end of 2021 total $2,807 million. A trustee accepts employer contributions, invests the contributions, accumulates the earnings on the investments, and pays benefits from the plan assets. So the amount is increased each year by employer cash contributions and (hopefully) a return on assets invested. It is decreased by amounts paid out to retired employees.
4. **What does the "pension asset" represent? Are you interviewing with a company whose pension plan is severely underfunded?** *(p. 992)* The pension asset is not the plan assets available to pay pension benefits. Instead, it's the net difference between those assets and the pension obligation. United Dynamics' plan assets exceed the pension obligation in each year presented, and therefore it is one of the few companies whose pension plan is overfunded.
5. **How is the pension expense influenced by changes in the pension liability and plan assets?** *(p. 993)* The pension expense reported on the income statement is a composite of periodic changes that occur in both the pension obligation and the plan assets. For United Dynamics in 2021, the pension expense included the service cost and interest cost, which are changes in the PBO, and the return on plan assets. It also included an amortized portion of prior service costs (a previous change in the PBO) and of net gains (gains and losses result from changes in both the PBO and plan assets). ●

The Bottom Line

- **LO17–1** Pension plans are arrangements designed to provide income to individuals during their retirement years. *Defined contribution* plans promise fixed annual contributions to a pension fund, without further commitment regarding benefit amounts at retirement. *Defined benefit* plans promise fixed retirement benefits defined by a designated formula. The employer sets aside cash each year to provide sufficient funds to pay promised benefits. (*p. 978*)
- **LO17–2** The *accumulated benefit obligation* is an estimate of the discounted present value of the retirement benefits earned so far by employees, applying the plan's pension formula to *existing* compensation levels. The vested benefit obligation is the portion of the accumulated benefit obligation that plan participants are entitled to receive regardless of their continued employment. The *projected benefit obligation* estimates retirement benefits by applying the pension formula to *projected future* compensation levels. (*p. 983*)
- **LO17–3** The PBO can change due to the accumulation of *service cost* from year to year, the accrual of *interest* as time passes, making plan amendments retroactive to prior years (prior service cost), and periodic adjustments when estimates change (gains and losses). The obligation is reduced as benefits actually are paid to retired employees. (*p. 984*)
- **LO17–4** The plan assets consist of the accumulated balance of the annual employer contributions plus the return on the investments less benefits paid to retirees. (*p. 990*)

- **LO17–5** The difference between an employer's obligation (PBO for pensions, APBO for other postretirement benefit plans) and the resources available to satisfy that obligation (plan assets) is the funded status of the pension plan. The employer must report the "funded status" of the plan in the balance sheet as a net pension liability if the obligation exceeds the plan assets or as a net pension asset if the plan assets exceed the obligation. (*p. 992*)
- **LO17–6** The pension expense is a composite of periodic changes in both the projected benefit obligation and the plan assets. Service cost is the increase in the PBO attributable to employee service and is the primary component of pension expense. The interest and return-on-assets components are financial items created only because the pension payment is delayed and the obligation is funded currently. Prior service cost is recognized over employees' future service period. Also, neither a loss (gain) on the PBO nor a loss (gain) on plan assets is immediately recognized in pension expense; they are recognized on a delayed basis to achieve income smoothing. (*p. 992*)
- **LO17–7** Recording pension expense causes the net pension liability/asset to change by the service cost, the interest cost, and the expected return on plan assets. Any amortization amounts included in the expense will reduce the *accumulated other comprehensive income* balances being amortized; for example, net loss—AOCI and prior service cost—AOCI. Similarly, the plan assets are increased by the annual cash investment. New losses and gains (as well as any new prior service cost should it occur) are recognized as other comprehensive income. The service cost is reported in the income statement separately from the other (non-service cost) components of the expense.(*p. 998*)
- **LO17–8** The various elements of a pension plan—projected benefit obligation, plan assets, prior service cost, gains and losses, pension expense, and the funded status of the plan—are interrelated. One way to see how each element relates to the other is to bring each part together in a *pension spreadsheet*. (*p. 1004*)
- **LO17–9** Accounting for postretirement benefits is similar in most respects to accounting for pension benefits. Like pensions, other postretirement benefits are a form of deferred compensation. Unlike pensions, their cost is attributed to the years from the employee's date of hire to the full eligibility date. (*p. 1009*)
- **LO17–10** The expected postretirement benefit obligation (EPBO) is the actuary's estimate of the total postretirement benefits (at their discounted present value) expected to be received by plan participants. The accumulated postretirement benefit obligation (APBO) is the portion of the EPBO attributed to employee service to date. (*p. 1011*)
- **LO17–11** The components of postretirement benefit expense are essentially the same as those for pension expense. (*p. 1012*)
- **LO17–12** Under both U.S. GAAP and IFRS we report all changes in the obligation and in the value of plan assets as they occur. The ways the changes are determined and reported are different, though, for some of the changes. The changes that constitute the components of the net pension cost are reported separately as (a) the service cost, (b) the net interest cost/income, and (c) remeasurement gains or losses. Under IFRS, *past service cost* is combined with current service cost and reported as service cost within the income statement rather than as a component of other comprehensive income as it is under U.S. GAAP. Under IFRS, gains and losses are not recycled from other comprehensive income as is required under U.S. GAAP (when the accumulated net gain or net loss exceeds the 10% threshold). Also, the interest cost and return on plan assets are replaced by net interest cost/income (interest rate times the difference between the DBO and plan assets). (*pp. 998, 1001, 1005,* and *1007*) ●

Service Method of Allocating Prior Service Cost

APPENDIX 17

When amortizing prior service cost, our objective is to match the cost with employee service. The straight-line method described in this chapter allocates an equal amount of the prior service cost to each year of the 15-year average service period of affected employees. But consider this: fewer of the affected employees will be working for the company toward the end of that period than at the beginning. Some probably will retire or quit in each year following the amendment.

An allocation approach that reflects the declining service pattern is called the **service method**. This method allocates the prior service cost to each year in proportion to the fraction of the total remaining service years worked in each of those years. To do this, it's necessary to estimate how many of the 2,000 employees working at the beginning of 2020 when the amendment is made will still be employed in each year after the amendment.

Let's suppose, for example, that the actuary estimates that a declining number of these employees still will be employed in each of the next 28 years as indicated in the abbreviated schedule below. The portion of the prior service cost amortized to pension expense each year is \$60 million times a declining fraction. Each year's fraction is that year's service divided by the 28-year total (30,000). This is demonstrated in Illustration 17A–1.

Illustration 17A–1

Service Method of Amortizing Prior Service Cost

Year	Number of Employees Still Employed (assumed for the illustration)	Fraction of Total Service Years		Prior Service Cost ($ in millions)		Amount Cost Amortized ($ in millions)
2020	2,000	2,000/30,000	×	\$60	=	\$ 4.0
2021	2,000	2,000/30,000	×	60	=	4.0
2022	1,850	1,850/30,000	×	60	=	3.7
2023	1,700	1,700/30,000	×	60	=	3.4
2024	1,550	1,550/30,000	×	60	=	3.1
—	—	—	×	—	=	—
2045	400	400/30,000	×	60	=	0.8
2046	250	250/30,000	×	60	=	0.5
2047	100	100/30,000	×	60	=	0.2
Totals	30,000	30,000/30,000				\$60.0
	Total number of service years					Total amount amortized

By the service method, prior service cost is recognized each year in proportion to the fraction of the total remaining service years worked that year.

The service method amortized an equal amount *per employee* each year.

Conceptually, the service method achieves a better matching of the cost and benefits. In fact, this is the FASB's recommended approach. However, GAAP permits the consistent use of any method that amortizes the prior service cost at least as quickly.[29] The straight-line method meets this condition and is the approach most often used in practice. In our illustration, the cost is completely amortized over 15 years rather than the 28 years required by the service method. The 15-year average service life is simply the total estimated service years divided by the total number of employees in the group:

30,000 years	÷	2,000	=	15 years
Total number of service years		Total number of employees		Average service years

Questions For Review of Key Topics

Q 17–1 What is a pension plan? What motivates a corporation to offer a pension plan for its employees?

Q 17–2 Qualified pension plans offer important tax benefits. What is the special tax treatment and what qualifies a pension plan for these benefits?

Q 17–3 Lamont Corporation has a pension plan in which the corporation makes all contributions and employees receive benefits at retirement based on the balance in their accumulated pension fund. What type of pension plan does Lamont have?

Q 17–4 What is the vested benefit obligation?

Q 17–5 Differentiate between the accumulated benefit obligation and the projected benefit obligation.

Q 17–6 Name five events that might change the balance of the PBO.

Q 17–7 Name three events that might change the balance of the plan assets.

[29] FASB ASC 715-30-35-13: Compensation—Retirement Benefits—Defined Benefit Plans—Pension—Subsequent Measurement—Prior Service Costs.

Q 17–8 What are the components that might be included in the calculation of net pension cost recognized for a period by an employer sponsoring a defined benefit pension plan?

Q 17–9 Define the service cost component of the periodic pension expense.

Q 17–10 Define the interest cost component of the periodic pension expense.

Q 17–11 The return on plan assets is the increase in plan assets (at fair value), adjusted for contributions to the plan and benefits paid during the period. How is the return included in the calculation of the periodic pension expense?

Q 17–12 Define prior service cost. How is it reported in the financial statements? How is it included in pension expense?

Q 17–13 How should gains or losses related to pension plan assets be recognized? How does this treatment compare to that for gains or losses related to the pension obligation?

Q 17–14 Is a company's PBO reported in the balance sheet? Its plan assets? Explain.

Q 17–15 What two components of pension expense may be negative (i.e., reduce pension expense)?

Q 17–16 Which are the components of pension expense that involve delayed recognition?

Q 17–17 Evaluate this statement: The excess of the actual return on plan assets over the expected return decreases the employer's pension cost.

Q 17–18 When accounting for pension costs, how should the payment into the pension fund be recorded? How does it affect the funded status of the plan?

Q 17–19 TFC Inc., revises its estimate of future salary levels, causing its PBO estimate to increase by $3 million. How is the $3 million reflected in TFC's financial statements?

Q 17–20 A pension plan is underfunded when the employer's obligation (PBO) exceeds the resources available to satisfy that obligation (plan assets) and overfunded when the opposite is the case. How is this funded status reported on the balance sheet if plan assets exceed the PBO? If the PBO exceeds plan assets?

Q 17–21 What are two ways to measure the obligation for postretirement benefits other than pensions? Define these measurement approaches.

Q 17–22 How are the costs of providing postretirement benefits other than pensions expensed?

Q 17–23 The components of postretirement benefit expense are similar to the components of pension expense. In what fundamental way does the service cost component differ between these two expenses?

Q 17–24 The EPBO for Branch Industries at the end of 2021 was determined by the actuary to be $20,000 as it relates to employee Will Lawson. Lawson was hired at the beginning of 2007. He will be fully eligible to retire with health care benefits after 15 more years but is expected to retire in 25 years. What is the APBO as it relates to Will Lawson?

IFRS **Q 17–25** The income statement of Mid-South Logistics includes $12 million for amortized prior service cost. Does Mid-South Logistics prepare its financial statements according to U.S. GAAP or IFRS? Explain.

IFRS **Q 17–26** How do U.S. GAAP and IFRS differ with regard to reporting gains and losses from changing assumptions used to measure the pension obligation?

Brief Exercises

connect

BE 17–1
Changes in the projected benefit obligation
● LO17–3

The projected benefit obligation was $80 million at the beginning of the year. Service cost for the year was $10 million. At the end of the year, pension benefits paid by the trustee were $6 million and there were no pension-related other comprehensive income accounts. The actuary's discount rate was 5%. What was the amount of the projected benefit obligation at year-end?

BE 17–2
Changes in the projected benefit obligation
● LO17–3

The projected benefit obligation was $80 million at the beginning of the year and $85 million at the end of the year. At the end of the year, pension benefits paid by the trustee were $6 million and there were no pension-related other comprehensive income accounts. The actuary's discount rate was 5%. What was the amount of the service cost for the year?

BE 17–3
Changes in the projected benefit obligation
● LO17–3

The projected benefit obligation was $80 million at the beginning of the year and $85 million at the end of the year. Service cost for the year was $10 million. At the end of the year, there were no pension-related other comprehensive income accounts. The actuary's discount rate was 5%. What was the amount of the retiree benefits paid by the trustee?

BE 17–4
Changes in the projected benefit obligation
● LO17–3

The projected benefit obligation was $80 million at the beginning of the year and $85 million at the end of the year. Service cost for the year was $10 million. At the end of the year, pension benefits paid by the trustee were $6 million. The actuary's discount rate was 5%. At the end of the year, the actuary revised the estimate of the percentage rate of increase in compensation levels in upcoming years. What was the amount of the gain or loss the estimate change caused?

BE 17–5
Changes in pension plan assets
● LO17–4

Pension plan assets were $80 million at the beginning of the year. The return on plan assets was 5%. At the end of the year, retiree benefits paid by the trustee were $6 million and cash invested in the pension fund was $7 million. What was the amount of the pension plan assets at year-end?

BE 17–6
Changes in pension plan assets
● LO17–4

Pension plan assets were $80 million at the beginning of the year and $83 million at the end of the year. The return on plan assets was 5%. At the end of the year, cash invested in the pension fund was $7 million. What was the amount of the retiree benefits paid by the trustee?

BE 17–7
Changes in pension plan assets
● LO17–4

Pension plan assets were $100 million at the beginning of the year and $104 million at the end of the year. At the end of the year, retiree benefits paid by the trustee were $6 million and cash invested in the pension fund was $7 million. What was the percentage rate of return on plan assets?

BE 17–8
Reporting the funded status of pension plans
● LO17–5

JDS Shipyard's projected benefit obligation, accumulated benefit obligation, and plan assets were $40 million, $30 million, and $25 million, respectively, at the end of the year. What, if any, pension liability must be reported in the balance sheet? What would JDS report if the plan assets were $45 million instead?

BE 17–9
Pension expense
● LO17–6

The projected benefit obligation was $80 million at the beginning of the year. Service cost for the year was $10 million. At the end of the year, pension benefits paid by the trustee were $6 million and there were no pension-related other comprehensive income (OCI) accounts requiring amortization. The actuary's discount rate was 5%. The actual return on plan assets was $5 million although it was expected to be only $4 million. What was the total pension expense for the year?

BE 17–10
Pension expense; prior service cost
● LO17–6

The pension plan was amended last year, creating a prior service cost of $20 million. Service cost and interest cost for the year were $10 million and $4 million, respectively. At the end of the year, there was a negligible balance in the net gain—pensions account. The actual return on plan assets was $4 million although it was expected to be $6 million. On average, employees' remaining service life with the company is 10 years. What was the total pension cost for the year?

BE 17–11
Net gain
● LO17–6

The projected benefit obligation and plan assets were $80 million and $100 million, respectively, at the beginning of the year. Due primarily to favorable stock market performance in recent years, there also was a net gain of $30 million. On average, employees' remaining service life with the company is 10 years. As a result of the net gain, what was the increase or decrease in pension expense for the year?

BE 17–12
Recording pension expense
● LO17–7

The Warren Group's pension cost is $67 million. This amount includes a $70 million service cost, a $50 million interest cost, a $55 million reduction for the expected return on plan assets, and a $2 million amortization of a prior service cost. How is the net pension liability affected when the pension cost is recorded?

BE 17–13
Recording pension expense
● LO17–7

Major Medical reported a net loss—AOCI in last year's balance sheet. This year, the company revised its estimate of future salary levels causing its PBO estimate to decline by $4 million. Also, the $8 million actual return on plan assets fell short of the $9 million expected return. How does this gain and loss affect Major's income statement, statement of comprehensive income, and balance sheet?

BE 17–14
Postretirement benefits; determine the APBO and service cost
● LO17–9, LO17–10

Prince Distribution Inc., has an unfunded postretirement benefit plan. Medical care and life insurance benefits are provided to employees who render 10 years service and attain age 55 while in service. At the end of 2021, Jim Lukawitz is 31. He was hired by Prince at age 25 (6 years ago) and is expected to retire at age 62. The expected postretirement benefit obligation for Lukawitz at the end of 2021 is $50,000 and $54,000 at the end of 2022. Calculate the accumulated postretirement benefit obligation at the end of 2021 and 2022 and the service cost for 2021 and 2022, as pertaining to Lukawitz.

BE 17–15
Postretirement benefits; changes in the APBO
● LO17–11

On January 1, 2021, Medical Transport Company's accumulated postretirement benefit obligation was $25 million. At the end of 2021, retiree benefits paid were $3 million. Service cost for 2021 is $7 million. Assumptions regarding the trend of future health care costs were revised at the end of 2021, causing the actuary to revise downward the estimate of the APBO by $1 million. The actuary's discount rate is 8%. Determine the amount of the accumulated postretirement benefit obligation at December 31, 2021.

Exercises

connect

E 17–1
Changes in the PBO
● LO17–3

Indicate by letter whether each of the events listed below increases (**I**), decreases (**D**), or has no effect (**N**) on an employer's projected benefit obligation.

Events

_____ 1. Interest cost.
_____ 2. Amortization of prior service cost.
_____ 3. A decrease in the average life expectancy of employees.
_____ 4. An increase in the average life expectancy of employees.
_____ 5. A plan amendment that increases benefits is made retroactive to prior years.
_____ 6. An increase in the actuary's assumed discount rate.
_____ 7. Cash contributions to the pension fund by the employer.
_____ 8. Benefits are paid to retired employees.
_____ 9. Service cost.
_____ 10. Return on plan assets during the year are lower than expected.
_____ 11. Return on plan assets during the year are higher than expected.

E 17–2
Determine the projected benefit obligation
● LO17–3

On January 1, 2021, Ravetch Corporation's projected benefit obligation was $30 million. During 2021, pension benefits paid by the trustee were $4 million. Service cost for 2021 is $12 million. Pension plan assets (at fair value) increased during 2021 by $6 million as expected. At the end of 2021, there were no pension-related other comprehensive income (OCI) accounts. The actuary's discount rate was 10%.

Required:
Determine the amount of the projected benefit obligation at December 31, 2021.

E 17–3
Components of pension expense
● LO17–6

Indicate by letter whether each of the events listed below increases (**I**), decreases (**D**), or has no effect (**N**) on an employer's periodic pension expense in the year the event occurs.

Events

_____ 1. Interest cost.
_____ 2. Amortization of prior service cost—AOCI.
_____ 3. Excess of the expected return on plan assets over the actual return.
_____ 4. Expected return on plan assets.
_____ 5. A plan amendment that increases benefits is made retroactive to prior years.
_____ 6. Actuary's estimate of the PBO is increased.
_____ 7. Cash contributions to the pension fund by the employer.
_____ 8. Benefits are paid to retired employees.
_____ 9. Service cost.
_____ 10. Excess of the actual return on plan assets over the expected return.
_____ 11. Amortization of net loss—AOCI.
_____ 12. Amortization of net gain—AOCI.

E 17–4
Recording pension expense
● LO17–6, LO17–7

Harrison Forklift's pension expense includes a service cost of $10 million. Harrison began the year with a pension liability of $28 million (underfunded pension plan).

Required:
Prepare the appropriate general journal entries to record Harrison's pension expense in each of the following independent situations regarding the other (non-service cost) components of pension expense ($ in millions):

1. Interest cost, $6; expected return on assets, $4; amortization of net loss, $2.
2. Interest cost, $6; expected return on assets, $4; amortization of net gain, $2.
3. Interest cost, $6; expected return on assets, $4; amortization of net loss, $2; amortization of prior service cost, $3 million.

E 17–5
Determine pension plan assets
● LO17–4

The following data relate to Ramesh Company's defined benefit pension plan:

	($ in millions)
Plan assets at fair value, January 1	$600
Expected return on plan assets	60
Actual return on plan assets	48
Contributions to the pension fund (end of year)	100
Amortization of net loss	10
Pension benefits paid (end of year)	11
Pension expense	72

Required:
Determine the amount of pension plan assets at fair value on December 31.

E 17–6
Changes in the pension obligation; determine service cost
● LO17–3, LO17–6

Pension data for David Emerson Enterprises include the following:

	($ in millions)
Discount rate, 10%	
Projected benefit obligation, January 1	$360
Projected benefit obligation, December 31	465
Accumulated benefit obligation, January 1	300
Accumulated benefit obligation, December 31	415
Cash contributions to pension fund, December 31	150
Benefit payments to retirees, December 31	54

Required:
Assuming no change in actuarial assumptions and estimates, determine the service cost component of pension expense for the year ended December 31.

E 17–7
Changes in plan assets; determine cash contributions
● LO17–4

Pension data for Fahy Transportation Inc., include the following:

	($ in millions)
Discount rate, 7%	
Expected return on plan assets, 10%	
Actual return on plan assets, 11%	
Projected benefit obligation, January 1	$730
Plan assets (fair value), January 1	700
Plan assets (fair value), December 31	750
Benefit payments to retirees, December 31	66

Required:
Assuming cash contributions were made at the end of the year, what was the amount of those contributions?

E 17–8
Components of pension expense
● LO17–6

Pension data for Sterling Properties include the following:

	($ in thousands)
Service cost, 2021	$112
Projected benefit obligation, January 1, 2021	850
Plan assets (fair value), January 1, 2021	900
Prior service cost—AOCI (2021 amortization, $8)	80
Net loss—AOCI (2021 amortization, $1)	101
Interest rate, 6%	
Expected return on plan assets, 10%	
Actual return on plan assets, 11%	

Required:
Determine pension expense for 2021.

E 17–9
Components of pension expense; IFRS
● LO17–6, LO17–12
IFRS

Refer to the situation described in E 17–8.

Required:
How might your answer differ if we assume Sterling Properties prepares its financial statements according to International Financial Reporting Standards (IFRS)? The interest rate on high-grade corporate bonds is 6%.

E 17–10
Determine pension expense
● LO17–6, LO17–7

Abbott and Abbott has a noncontributory, defined benefit pension plan. At December 31, 2021, Abbott and Abbott received the following information:

Projected Benefit Obligation	($ in millions)
Balance, January 1	$120
Service cost	20
Interest cost	12
Benefits paid	(9)
Balance, December 31	$143

(continued)

(concluded)

Plan Assets	
Balance, January 1	$ 80
Actual return on plan assets	9
Contributions 2021	20
Benefits paid	(9)
Balance, December 31	$100

The expected long-term rate of return on plan assets was 10%. There was no prior service cost and a negligible net loss—AOCI on January 1, 2021.

Required:

1. Determine Abbott and Abbott's pension expense for 2021.
2. Prepare the journal entries to record Abbott and Abbott's (a) pension expense, (b) funding, and (c) payment for 2021.

E 17–11
Components of pension expense; journal entries
● LO17–6, LO17–7

Pension data for Barry Financial Services Inc., include the following:

	($ in thousands)
Discount rate, 7%	
Expected return on plan assets, 10%	
Actual return on plan assets, 9%	
Service cost, 2021	$ 310
January 1, 2021:	
Projected benefit obligation	2,300
Accumulated benefit obligation	2,000
Plan assets (fair value)	2,400
Prior service cost—AOCI (2021 amortization, $25)	325
Net gain—AOCI (2021 amortization, $6)	330
There were no changes in actuarial assumptions.	
December 31, 2021:	
Cash contributions to pension fund, December 31, 2021	245
Benefit payments to retirees, December 31, 2021	270

Required:

1. Determine pension expense for 2021.
2. Prepare the journal entries to record (a) pension expense, (b) gains and losses (if any), (c) funding, and (d) retiree benefits for 2021.

E 17–12
PBO calculations; ABO calculations; present value concepts
● LO17–1, LO17–2, LO17–3

Clark Industries has a defined benefit pension plan that specifies annual, year-end retirement benefits equal to

$$1.2\% \times \text{Service years} \times \text{Final year's salary}$$

Stanley Mills was hired by Clark at the beginning of 2002. Mills is expected to retire at the end of 2046 after 45 years of service. His retirement is expected to span 15 years. At the end of 2021, 20 years after being hired, his salary is $80,000. The company's actuary projects Mills's salary to be $270,000 at retirement. The actuary's discount rate is 7%.

Required:

1. Estimate the amount of Stanley Mills's annual retirement payments for the 15 retirement years earned as of the end of 2021.
2. Suppose Clark's pension plan permits a lump-sum payment at retirement in lieu of annuity payments. Determine the lump-sum equivalent as the present value as of the retirement date of annuity payments during the retirement period.
3. What is the company's projected benefit obligation at the end of 2021 with respect to Stanley Mills?
4. Even though pension accounting centers on the PBO calculation, the ABO still must be disclosed in the pension disclosure note. What is the company's accumulated benefit obligation at the end of 2021 with respect to Stanley Mills?
5. If we assume no estimates change in the meantime, what is the company's projected benefit obligation at the end of 2022 with respect to Stanley Mills?
6. What portion of the 2022 increase in the PBO is attributable to 2022 service (the service cost component of pension expense) and to accrued interest (the interest cost component of pension expense)?

E 17–13
Determining the amortization of net loss or net gain
● LO17–6

Hicks Cable Company has a defined benefit pension plan. Three alternative possibilities for pension-related data at January 1, 2021, are shown below:

($ in thousands)	Case 1	Case 2	Case 3
Net loss (gain)—AOCI, Jan. 1	$ 320	$ (330)	$ 260
2021 loss (gain) on plan assets	(11)	(8)	2
2021 loss (gain) on PBO	(23)	16	(265)
Accumulated benefit obligation, Jan. 1	(2,950)	(2,550)	(1,450)
Projected benefit obligation, Jan. 1	(3,310)	(2,670)	(1,700)
Fair value of plan assets, Jan. 1	2,800	2,700	1,550
Average remaining service period of active employees (years)	12	15	10

Required:

1. For each independent case, calculate any amortization of the net loss or gain that should be included as a component of pension expense for 2021.
2. For each independent case, determine the net loss—AOCI or net gain—AOCI as of January 1, 2022.

E 17–14
Effect of pension expense components on balance sheet accounts
● LO17–7, LO17–8

Warrick Boards calculated pension expense for its underfunded pension plan as follows:

	($ in millions)
Service cost	$224
Interest cost	150
Expected return on the plan assets ($100 actual, less $10 gain)	(90)
Amortization of prior service cost	8
Amortization of net loss	2
Pension expense	$294

Required:

Which elements of Warrick's balance sheet are affected by the components of pension expense? What are the specific changes in these accounts?

E 17–15
Pension spreadsheet
● LO17–8

A partially completed pension spreadsheet showing the relationships among the elements that comprise the defined benefit pension plan of Universal Products is given below. The actuary's discount rate is 5%. At the end of 2019, the pension formula was amended, creating a prior service cost of $120,000. The expected rate of return on assets was 8%, and the average remaining service life of the active employee group is 20 years in the current year as well as the previous two years.

Required:

Fill in the missing amounts.

( )s indicate credits; debits otherwise ($ in thousands)	PBO	Plan Assets	Prior Service Cost / AOCI	Net Loss / AOCI	Pension Expense	Cash	Net Pension (Liability)/ Asset
Balance, Jan. 1, 2021	(800)	600	114	80			(200)
Service cost					84		
Interest cost, 5%	(40)						
Expected return on assets					(48)		
Adjust for:							
Loss on assets				6			
Amortization:							
Prior service cost							
Amortization:							
Net loss							
Gain on PBO							12
Prior service cost	0						
Cash funding						(68)	
Retiree benefits							
Balance, Dec. 31, 2021	(862)		108				

E 17–16
Determine and record pension expense and gains and losses; funding and retiree benefits
● LO17–6, LO17–7

Actuary and trustee reports indicate the following changes in the PBO and plan assets of Douglas-Roberts Industries during 2021:

Prior service cost at Jan. 1, 2021, from plan amendment at the beginning of 2018 (amortization: $4 million per year)	$28 million
Net loss—AOCI at Jan. 1, 2021 (previous losses exceeded previous gains)	$80 million
Average remaining service life of the active employee group	10 years
Actuary's discount rate	7%

($ in millions)

PBO		Plan Assets	
Beginning of 2021	$600	*Beginning* of 2021	$400
Service cost	80	Return on plan assets,	
Interest cost, 7%	42	8% (10% expected)	32
Loss (gain) on PBO	(14)	Cash contributions	90
Less: Retiree benefits	(38)	Less: Retiree benefits	(38)
End of 2021	$670	*End* of 2021	$484

Required:

1. Determine Douglas-Roberts's pension expense for 2021, and prepare the appropriate journal entry to record the expense.
2. Prepare the appropriate journal entry(s) to record any 2021 gains and losses.
3. Prepare the appropriate journal entry to record the cash contribution to plan assets.
4. Prepare the appropriate journal entry to record retiree benefits.

E 17–17
Concepts; terminology
● LO17–2 through LO17–8

Listed below are several terms and phrases associated with pensions. Pair each item from List A with the item from List B (by letter) that is most appropriately associated with it.

List A	List B
_____ 1. Future compensation levels estimated.	a. Actual return exceeds expected
_____ 2. All funding provided by the employer.	b. Net gain—AOCI
_____ 3. Credit to OCI and debit to plan assets.	c. Vested benefit obligation
_____ 4. Retirement benefits specified by formula.	d. Projected benefit obligation
_____ 5. Trade-off between being relevant and representationally faithful.	e. Choice between PBO and ABO
	f. Noncontributory pension plan
_____ 6. Cumulative gains in excess of losses.	g. Accumulated benefit obligation
_____ 7. Current pay levels implicitly assumed.	h. Plan assets
_____ 8. Created by the passage of time.	i. Interest cost
_____ 9. Not contingent on future employment.	j. Delayed recognition in earnings
_____ 10. Risk borne by employee.	k. Defined contribution plan
_____ 11. Increased by employer contributions.	l. Defined benefit plan
_____ 12. Caused by plan amendment.	m. Prior service cost
_____ 13. Loss on plan assets.	n. Amortize net loss—AOCI
_____ 14. Excess over 10% of plan assets or PBO.	

E 17–18
IFRS; actuarial gains and losses
● LO17–7, LO17–12
IFRS

Patel Industries has a noncontributory, defined benefit pension plan. Since the inception of the plan, the actuary has used as the discount rate the rate on high-quality corporate bonds, which recently has been 7%. During 2021, changing economic conditions caused the rate to change to 6%, and the actuary decided that 6% is the appropriate rate.

Required:

1. Does the change in discount rate create a gain, or does it create a loss for Patel under U.S. GAAP? Why?
2. Assume the magnitude of the change is $13 million. Prepare the appropriate journal entry to record any 2021 gain or loss under U.S. GAAP. If Patel prepares its financial statements according to U.S. GAAP, how will the company report the gain or loss?
3. Would your response to requirement 2 differ if Patel prepares its financial statements according to International Financial Reporting Standards (IFRS)?

E 17–19
Record pension expense, funding, and gains and losses; determine account balances
● LO17–6, LO17–7, LO17–8

Beale Management has a noncontributory, defined benefit pension plan. On December 31, 2021 (the end of Beale's fiscal year), the following pension-related data were available:

Projected Benefit Obligation	($ in millions)
Balance, January 1, 2021	$480
Service cost	82
Interest cost, discount rate, 5%	24
Gain due to changes in actuarial assumptions in 2021	(10)
Pension benefits paid	(40)
Balance, December 31, 2021	$536
Plan Assets	
Balance, January 1, 2021	$500
Actual return on plan assets	40
(Expected return on plan assets, $45)	
Cash contributions	70
Pension benefits paid	(40)
Balance, December 31, 2021	$570
January 1, 2021, balances:	
Pension asset	$ 20
Prior service cost—AOCI (amortization $8 per year)	48
Net gain—AOCI (any amortization over 15 years)	80

Required:

1. Prepare the 2021 journal entry to record pension expense.
2. Prepare the journal entry(s) to record any 2021 gains and losses.
3. Prepare the 2021 journal entries to record the contribution to plan assets and benefit payments to retirees.
4. Determine the balances at December 31, 2021, in (a) the PBO, (b) plan assets, (c) the net gain—AOCI, and (d) prior service cost—AOCI. Show how the balances changed during 2021. [*Hint:* You might find T-accounts useful.]
5. What amount will Beale report in its 2021 balance sheet as a net pension asset or net pension liability for the funded status of the plan?

E 17–20
Pension spreadsheet
● LO17–8

Refer to the data provided in E 17–19.

Required:

Prepare a pension spreadsheet to show the relationship among the PBO, plan assets, prior service cost, the net gain, pension expense, and the net pension asset.

E 17–21
Determine pension expense; prior service cost
● LO17–6, LO17–7

Lacy Construction has a noncontributory, defined benefit pension plan. At December 31, 2021, Lacy received the following information:

Projected Benefit Obligation	($ in millions)
Balance, January 1	$360
Service cost	60
Prior service cost	12
Interest cost (7.5%)	27
Benefits paid	(37)
Balance, December 31	$422
Plan Assets	
Balance, January 1	$240
Actual return on plan assets	27
Contributions, 2021	60
Benefits paid	(37)
Balance, December 31	$290

The expected long-term rate of return on plan assets was 10%. There were no AOCI balances related to pensions on January 1, 2021. At the end of 2021, Lacy amended the pension formula creating a prior service cost of $12 million.

Required:

1. Determine Lacy's pension expense for 2021.
2. Prepare the journal entry(s) to record Lacy's (a) pension expense, (b) gains or losses, (c) prior service cost, (d) funding, and (e) payment of retiree benefits for 2021.

E 17–22
IFRS; prior service cost
● LO17–7, LO17–12
IFRS

Refer to the situation described in E 17–21.

Required:

How might your solution differ if Lacy Construction prepares its financial statements according to International Financial Reporting Standards (IFRS)? Assume the actuary's discount rate is the rate on high-quality corporate bonds. Include any appropriate journal entries in your response.

E 17–23
Classifying accounting changes and errors
● LO17–8

Indicate with the appropriate letter the nature of each adjustment described below:

Type of Adjustment

A. Change in principle
B. Change in estimate
C. Correction of an error
D. Neither an accounting change nor an error

_____ 1. Change in actuarial assumptions for a defined benefit pension plan.

_____ 2. Determination that the projected benefit obligation under a pension plan exceeded the fair value of plan assets at the end of the previous year by $17,000. The only pension-related amount on the balance sheet was a net pension liability of $30,000.

_____ 3. Pension plan assets for a defined benefit pension plan achieving a rate of return in excess of the amount anticipated.

_____ 4. Instituting a pension plan for the first time and adopting GAAP for employers' accounting for defined benefit pension and other postretirement plans.

E 17–24
Postretirement benefits; determine APBO, EPBO
● LO17–10

Classified Electronics has an unfunded retiree health care plan. Each of the company's three employees has been with the firm since its inception at the beginning of 2020. As of the end of 2021, the actuary estimates the total net cost of providing health care benefits to employees during their retirement years to have a present value of $72,000. Each of the employees will become fully eligible for benefits after 28 more years of service but aren't expected to retire for 35 more years. The interest rate is 6%.

Required:

1. What is the expected postretirement benefit obligation at the end of 2021?
2. What is the accumulated postretirement benefit obligation at the end of 2021?
3. What is the expected postretirement benefit obligation at the end of 2022?
4. What is the accumulated postretirement benefit obligation at the end of 2022?

E 17–25
Postretirement benefits; determine APBO, service cost, interest cost; prepare journal entry
● LO17–10, LO17–11

The following data are available pertaining to Household Appliance Company's retiree health care plan for 2021:

Number of employees covered	2
Years employed as of January 1, 2021	3 (each)
Attribution period	25 years
Expected postretirement benefit obligation, Jan. 1	$50,000
Expected postretirement benefit obligation, Dec. 31	$53,000
Interest rate	6%
Funding	none

Required:

1. What is the accumulated postretirement benefit obligation at the beginning of 2021?
2. What is interest cost to be included in 2021 postretirement benefit expense?
3. What is service cost to be included in 2021 postretirement benefit expense?
4. Prepare the journal entry to record the postretirement benefit expense for 2021.

E 17–26
Postretirement benefits; determine EPBO; attribution period
● LO17–10, LO17–11

Lorin Management Services has an unfunded postretirement benefit plan. On December 31, 2021, the following data were available concerning changes in the plan's accumulated postretirement benefit obligation with respect to one of Lorin's employees:

APBO at the beginning of 2021	$16,364
Interest cost: ($16,364 × 10%)	1,636
Service cost: ($44,000 × 1/22)	2,000
Portion of EPBO attributed to 2021	
APBO at the end of 2021	$20,000

Required:

1. Over how many years is the expected postretirement benefit obligation being expensed (attribution period)?

2. What is the expected postretirement benefit obligation at the *end* of 2021?
3. When was the employee hired by Lorin?
4. What is the expected postretirement benefit obligation at the *beginning* of 2021?

E 17–27
Postretirement benefits; components of postretirement benefit expense
● LO17–11

Data pertaining to the postretirement health care benefit plan of Sterling Properties include the following for 2021:

	($ in thousands)
Service cost	$124
Accumulated postretirement benefit obligation, January 1	700
Plan assets (fair value), January 1	50
Prior service cost—AOCI	none
Net gain—AOCI (2021 amortization, $1)	91
Retiree benefits paid (end of year)	87
Contribution to health care benefit fund (end of year)	185
Discount rate, 7%	
Return on plan assets (actual and expected), 10%	

Required:
1. Determine the postretirement benefit expense for 2021.
2. Prepare the appropriate journal entries to record the (a) postretirement benefit expense, (b) funding, and (c) retiree benefits for 2021.

E 17–28
Postretirement benefits; amortization of net loss
● LO17–11

Cahal-Michael Company has a postretirement health care benefit plan. On January 1, 2021, the following plan-related data were available:

	($ in thousands)
Net loss—AOCI	$ 336
Accumulated postretirement benefit obligation	2,800
Fair value of plan assets	500
Average remaining service period to retirement	14 years (same in previous 10 years)

The rate of return on plan assets during 2021 was 10%, although it was expected to be 9%. The actuary revised assumptions regarding the APBO at the end of the year, resulting in a $39,000 increase in the estimate of that obligation.

Required:
1. Calculate any amortization of the net loss that should be included as a component of postretirement benefit expense for 2021.
2. Assume the postretirement benefit expense for 2021, not including the amortization of the net loss component, is $212,000. What is the expense for the year?
3. Determine the net loss or gain as of December 31, 2021.

E 17–29
Postretirement benefits; determine and record expense
● LO17–11

Gorky-Park Corporation provides postretirement health care benefits to employees who provide at least 12 years of service and reach age 62 while in service. On January 1, 2021, the following plan-related data were available:

	($ in millions)
Accumulated postretirement benefit obligation	$130
Fair value of plan assets	none
Average remaining service period to retirement	25 years (same in previous 10 years)
Average remaining service period to full eligibility	20 years (same in previous 10 years)

On January 1, 2021, Gorky-Park amends the plan to provide certain dental benefits in addition to previously provided medical benefits. The actuary determines that the cost of making the amendment retroactive increases the APBO by $20 million. Management chooses to amortize the prior service cost on a straight-line basis. The service cost for 2021 is $34 million. The interest rate is 8%.

Required:
1. Calculate the postretirement benefit expense for 2021.
2. Prepare the journal entry to record the expense.

E 17–30
Postretirement benefits; negative plan amendment
● LO17–11

Southeast Technology provides postretirement health care benefits to employees. On January 1, 2021, the following plan-related data were available:

	($ in thousands)
Prior service cost—originated in 2016	$ 50
Accumulated postretirement benefit obligation	530
Fair value of plan assets	none
Average remaining service period to retirement	20 years (same in previous 10 years)
Average remaining service period to full eligibility	15 years (same in previous 10 years)

On January 1, 2021, Southeast amends the plan in response to spiraling health care costs. The amendment establishes an annual maximum of $3,000 for medical benefits that the plan will provide. The actuary determines that the effect of this amendment is to decrease the APBO by $80,000. Management amortizes prior service cost on a straight-line basis. The interest rate is 8%. The service cost for 2021 is $114,000.

Required:

1. Calculate the prior service cost amortization for 2021.
2. Calculate the postretirement benefit expense for 2021.

E 17–31
Prior service cost; service method; straight-line method (Based on Appendix 17)

Frazier Refrigeration amended its defined benefit pension plan on December 31, 2021, to increase retirement benefits earned with each service year. The consulting actuary estimated the prior service cost incurred by making the amendment retroactive to prior years to be $110,000. Frazier's 100 present employees are expected to retire at the rate of approximately 10 each year at the end of each of the next 10 years.

Required:

1. Using the service method, calculate the amount of prior service cost to be amortized to pension expense in each of the next 10 years.
2. Using the straight-line method, calculate the amount of prior service cost to be amortized to pension expense in each of the next 10 years.

E 17–32
FASB codification research; postretirement benefit plan
● LO17–11

When a company sponsors a postretirement benefit plan other than a pension plan, benefits typically are not earned by employees on the basis of a formula, so assigning the service cost to specific periods is more difficult. The *FASB Accounting Standards Codification* represents the single source of authoritative U.S. generally accepted accounting principles.

Required:

1. Obtain the relevant authoritative literature on how a firm should attribute the expected postretirement benefit obligation to years of service using the *FASB Accounting Standards Codification* at the FASB website (**www.fasb.org**). Find the specific seven-digit Codification citation (XXX-XX-XX) that describes the guidelines for each of the following questions:
   a. What is the objective for attributing expected postretirement benefit obligations to years of service?
   b. When does the attribution period for expected postretirement benefits begin for an employee?
   c. When does the attribution period for expected postretirement benefits end for an employee?
2. What are the guidelines for each?

E 17–33
FASB codification research
● LO17–1, LO17–2, LO17–5

Access the *FASB Accounting Standards Codification* at the FASB website (**www.fasb.org**). Determine the specific eight-digit Codification citation (XXX-XX-XX-X) for accounting for each of the following items:

1. The disclosure required in the notes to the financial statements for pension plan assets.
2. Recognition of the net pension asset or net pension liability.
3. Disclosures required in the notes to the financial statements for pension cost for a defined contribution plan.

Problems

connect

(Note: Problems 1–5 are variations of the same situation, designed to focus on different elements of the pension plan.)

P 17–1
ABO calculations; present value concepts
● LO17–2, LO17–3

Sachs Brands's defined benefit pension plan specifies annual retirement benefits equal to 1.6% × service years × final year's salary, payable at the end of each year. Angela Davenport was hired by Sachs at the beginning of 2007 and is expected to retire at the end of 2041 after 35 years' service. Her retirement is expected to span

18 years. Davenport's salary is $90,000 at the end of 2021 and the company's actuary projects her salary to be $240,000 at retirement. The actuary's discount rate is 7%.

Required:

1. Draw a time line that depicts Davenport's expected service period, retirement period, and a 2021 measurement date for the pension obligation.
2. Estimate by the accumulated benefits approach the amount of Davenport's annual retirement payments earned as of the end of 2021.
3. What is the company's accumulated benefit obligation at the end of 2021 with respect to Davenport?
4. If no estimates are changed in the meantime, what will be the accumulated benefit obligation at the end of 2024 (three years later), when Davenport's salary is $100,000?

P 17–2
PBO calculations; present value concepts
● LO17–3

Sachs Brands's defined benefit pension plan specifies annual retirement benefits equal to 1.6% × service years × final year's salary, payable at the end of each year. Angela Davenport was hired by Sachs at the beginning of 2007 and is expected to retire at the end of 2041 after 35 years' service. Her retirement is expected to span 18 years. Davenport's salary is $90,000 at the end of 2021 and the company's actuary projects her salary to be $240,000 at retirement. The actuary's discount rate is 7%.

Required:

1. Draw a time line that depicts Davenport's expected service period, retirement period, and a 2021 measurement date for the pension obligation.
2. Estimate by the projected benefits approach the amount of Davenport's annual retirement payments earned as of the end of 2021.
3. What is the company's projected benefit obligation at the end of 2021 with respect to Davenport?
4. If no estimates are changed in the meantime, what will be the company's projected benefit obligation at the end of 2024 (three years later) with respect to Davenport?

P 17–3
Service cost, interest, and PBO calculations; present value concepts
● LO17–3

Sachs Brands's defined benefit pension plan specifies annual retirement benefits equal to 1.6% × service years × final year's salary, payable at the end of each year. Angela Davenport was hired by Sachs at the beginning of 2007 and is expected to retire at the end of 2041 after 35 years' service. Her retirement is expected to span 18 years. Davenport's salary is $90,000 at the end of 2021 and the company's actuary projects her salary to be $240,000 at retirement. The actuary's discount rate is 7%.

Required:

1. What is the company's projected benefit obligation at the beginning of 2021 (after 14 years' service) with respect to Davenport?
2. Estimate by the projected benefits approach the portion of Davenport's annual retirement payments attributable to 2021 service.
3. What is the company's service cost for 2021 with respect to Davenport?
4. What is the company's interest cost for 2021 with respect to Davenport?
5. Combine your answers to *requirements* 1, 3, and 4 to determine the company's projected benefit obligation at the end of 2021 (after 15 years' service) with respect to Davenport.

P 17–4
Prior service cost; components of pension expense; present value concepts
● LO17–3, LO17–6

Sachs Brands's defined benefit pension plan specifies annual retirement benefits equal to 1.6% × service years × final year's salary, payable at the end of each year. Angela Davenport was hired by Sachs at the beginning of 2007 and is expected to retire at the end of 2041 after 35 years' service. Her retirement is expected to span 18 years. Davenport's salary is $90,000 at the end of 2021 and the company's actuary projects her salary to be $240,000 at retirement. The actuary's discount rate is 7%.

At the beginning of 2022, the pension formula was amended to:

1.75% × Service years × Final year's salary

The amendment was made retroactive to apply the increased benefits to prior service years.

Required:

1. What is the company's prior service cost at the beginning of 2022 with respect to Davenport after the amendment described above?
2. Since the amendment occurred at the *beginning* of 2022, amortization of the prior service cost begins in 2022. What is the prior service cost amortization that would be included in pension expense?
3. What is the service cost for 2022 with respect to Davenport?
4. What is the interest cost for 2022 with respect to Davenport?

5. Calculate pension expense for 2022 with respect to Davenport, assuming plan assets attributable to her of $150,000 and a rate of return (actual and expected) of 10%.

P 17–5
Gain on PBO; present value concepts
● LO17–3, LO17–6

Sachs Brands's defined benefit pension plan specifies annual retirement benefits equal to 1.6% × service years × final year's salary, payable at the end of each year. Angela Davenport was hired by Sachs at the beginning of 2007 and is expected to retire at the end of 2041 after 35 years' service. Her retirement is expected to span 18 years. Davenport's salary is $90,000 at the end of 2021 and the company's actuary projects her salary to be $240,000 at retirement. The actuary's discount rate is 7%.

At the beginning of 2022, changing economic conditions caused the actuary to reassess the applicable discount rate. It was decided that 8% is the appropriate rate.

Required:
Calculate the effect of the change in the assumed discount rate on the PBO at the beginning of 2022 with respect to Davenport.

P 17–6
Determine the PBO; plan assets; pension expense; two years
● LO17–3, LO17–4, LO17–5

Stanley-Morgan Industries adopted a defined benefit pension plan on April 12, 2021. The provisions of the plan were not made retroactive to prior years. A local bank, engaged as trustee for the plan assets, expects plan assets to earn a 10% rate of return. A consulting firm, engaged as actuary, recommends 6% as the appropriate discount rate. The service cost is $150,000 for 2021 and $200,000 for 2022. Year-end funding is $160,000 for 2021 and $170,000 for 2022. No assumptions or estimates were revised during 2021.

Required:
Calculate each of the following amounts as of both December 31, 2021, and December 31, 2022:
1. Projected benefit obligation.
2. Plan assets.
3. Pension expense.
4. Net pension asset or net pension liability.

P 17–7
Determining the amortization of net gain
● LO17–6

Herring Wholesale Company has a defined benefit pension plan. On January 1, 2021, the following pension-related data were available:

	($ in thousands)
Net gain—AOCI	$ 170
Accumulated benefit obligation	1,170
Projected benefit obligation	1,400
Fair value of plan assets	1,100
Average remaining service period of active employees (expected to remain constant for the next several years)	15 years

The rate of return on plan assets during 2021 was 9%, although it was expected to be 10%. The actuary revised assumptions regarding the PBO at the end of the year, resulting in a $23,000 decrease in the estimate of that obligation.

Required:
1. Calculate any amortization of the net gain that should be included as a component of net pension expense for 2021.
2. Assume the net pension expense for 2021, not including the amortization of the net gain component, is $325,000. What is pension expense for the year?
3. Determine the net loss—AOCI or net gain—AOCI as of January 1, 2022.

P 17–8
Pension spreadsheet; record pension expense and funding; new gains and losses
● LO17–7, LO17–8

A partially completed pension spreadsheet showing the relationships among the elements that constitute Carney, Inc.'s defined benefit pension plan follows. Six years earlier, Carney revised its pension formula and recalculated benefits earned by employees in prior years using the more generous formula. The prior service cost created by the recalculation is being amortized at the rate of $5 million per year. At the end of 2021, the pension formula was amended again, creating an additional prior service cost of $40 million. The expected rate of return on assets and the actuary's discount rate were 10%, and the average remaining service life of the active employee group is 10 years.

( )s indicate credits; debits otherwise ($ in millions)	PBO	Plan Assets	Prior Service Cost	Net Loss	Pension Expense	Cash	Net Pension (Liability) / Asset
Balance, Jan. 1, 2021	(830)	580	20	93			(150)
Service cost	?				74		?

(continued)

(concluded)

( )s indicate credits; debits otherwise ($ in millions)	PBO	Plan Assets	Prior Service Cost	Net Loss	Pension Expense	Cash	Net Pension (Liability) / Asset
Interest cost	?				?		?
Expected return on asset		?					
Adjust for:							
Loss on assets		(7)		?			?
Amortization of:							
Prior service cost			?		?		
Net loss				?	?		
Loss on PBO	?			?			(13)
Prior service cost	?		?				?
Cash funding		?				?	84
Retiree benefits	?	?					
Balance, Dec. 31, 2021	?	775	?	?	?		?

Required:

1. Fill in the missing amounts.
2. Prepare the 2021 journal entry to record pension expense.
3. Prepare the journal entry(s) to record any 2021 gains and losses and new prior service cost in 2021.
4. Prepare the 2021 journal entries to record (a) the cash contribution to plan assets and (b) the payment of retiree benefits.

P 17–9
Determine pension expense; PBO; plan assets; net pension asset or liability; journal entries
● LO17–3 through LO17–8

U.S. Metallurgical Inc., reported the following balances in its financial statements and disclosure notes at December 31, 2020.

Plan assets	$400,000
Projected benefit obligation	320,000

U.S.M.'s actuary determined that 2021 service cost is $60,000. Both the expected and actual rate of return on plan assets are 9%. The interest (discount) rate is 5%. U.S.M. contributed $120,000 to the pension fund at the end of 2021, and retirees were paid $44,000 from plan assets.

Required:

1. What is the pension expense at the end of 2021?
2. What is the projected benefit obligation at the end of 2021?
3. What is the plan assets balance at the end of 2021?
4. What is the net pension asset or net pension liability at the end of 2021?
5. Prepare journal entries to record the (a) pension expense, (b) funding of plan assets, and (c) retiree benefit payments.

P 17–10
Prior service cost; calculate pension expense; journal entries; determine net pension asset or liability
● LO17–5 through LO17–7

Electronic Distribution has a defined benefit pension plan. Characteristics of the plan during 2021 are as follows:

	($ in millions)
PBO balance, January 1	$480
Plan assets balance, January 1	300
Service cost	75
Interest cost	45
Gain from change in actuarial assumption	22
Benefits paid	(36)
Actual return on plan assets	20
Contributions 2021	60

The expected long-term rate of return on plan assets was 8%. There were no AOCI balances related to pensions on January 1, 2021, but at the end of 2021, the company amended the pension formula, creating a prior service cost of $12 million.

Required:

1. Calculate the pension expense for 2021.
2. Prepare the journal entries to record (a) pension expense, (b) gains or losses, (c) prior service cost, (d) funding, and (e) payment of benefits for 2021.
3. What amount will Electronic Distribution report in its 2021 balance sheet as a net pension asset or net pension liability?

P 17–11
IFRS; calculate pension expense; journal entries; determine net pension asset or liability
● LO17–5 through LO17–7, LO17–12

IFRS

Refer to the situation described in P 17–10. Assume Electronic Distribution prepares its financial statements according to International Financial Reporting Standards (IFRS). Also assume that 10% is the current interest rate on high-quality corporate bonds.

Required:

1. Calculate the net pension cost for 2021, separating its components into appropriate categories for reporting.
2. Prepare the journal entries to record (a) the components of net pension cost, (b)gains or losses, (c) past service cost, (d) funding, and (e) payment of benefits for 2021.
3. What amount will Electronic Distribution report in its 2021 balance sheet as a net pension asset or net pension liability?

P 17–12
Determine pension expense; journal entries; two years
● LO17–3 through LO17–8

The Kollar Company has a defined benefit pension plan. Pension information concerning the fiscal years 2021 and 2022 are presented below ($ in millions):

Information Provided by Pension Plan Actuary:

a. Projected benefit obligation as of December 31, 2020 = $1,800.
b. Prior service cost from plan amendment on January 2, 2021 = $400 (straight-line amortization for 10-year average remaining service period).
c. Service cost for 2021 = $520.
d. Service cost for 2022 = $570.
e. Discount rate used by actuary on projected benefit obligation for 2021 and 2022 = 10%.
f. Payments to retirees in 2021 = $380.
g. Payments to retirees in 2022 = $450.
h. No changes in actuarial assumptions or estimates.
i. Net gain—AOCI on January 1, 2021 = $230.
j. Net gains and losses are amortized for 10 years in 2021 and 2022.

Information Provided by Pension Fund Trustee:

a. Plan asset balance at fair value on January 1, 2021 = $1,600.
b. 2021 contributions = $540.
c. 2022 contributions = $590.
d. Expected long-term rate of return on plan assets = 12%.
e. 2021 actual return on plan assets = $180.
f. 2022 actual return on plan assets = $210.

Required:

1. Calculate pension expense for 2021 and 2022.
2. Prepare the journal entries for 2021 and 2022 to record pension expense.
3. Prepare the journal entries for 2021 and 2022 to record any gains and losses and new prior service cost.
4. Prepare the journal entries for 2021 and 2022 to record (a) the cash contribution to plan assets and (b) the benefit payments to retirees.

P 17–13
Determine the PBO, plan assets, pension expense; prior service cost
● LO17–3, LO17–4, LO17–6

Lewis Industries adopted a defined benefit pension plan on January 1, 2021. By making the provisions of the plan retroactive to prior years, Lewis incurred a prior service cost of $2 million. The prior service cost was funded immediately by a $2 million cash payment to the fund trustee on January 2, 2021. However, the cost is to be amortized (expensed) over 10 years. The service cost—$250,000 for 2021—is fully funded at the end of each year. Both the actuary's discount rate and the expected rate of return on plan assets were 9%. The actual rate of return on plan assets was 11%. At December 31, the trustee paid $16,000 to an employee who retired during 2021.

Required:

Determine each of the following amounts as of December 31, 2021, the fiscal year-end for Lewis:

1. Projected benefit obligation.
2. Plan assets.
3. Pension expense.

P 17–14
Relationship among pension elements
● LO17–3 through LO17–8

The funded status of Hilton Paneling Inc.'s defined benefit pension plan and the balances in prior service cost and the net gain—pensions, are given below ($ in thousands):

	2021 Beginning Balances	2021 Ending Balances
Projected benefit obligation	$2,300	$2,501
Plan assets	2,400	2,591
Funded status	100	90
Prior service cost—AOCI	325	300
Net gain—AOCI	330	300

Retirees were paid $270,000, and the employer contribution to the pension fund was $245,000 at the end of 2021. The expected rate of return on plan assets was 10%, and the actuary's discount rate is 7%. There were no changes in actuarial estimates and assumptions regarding the PBO.

Required:

Determine the following amounts for 2021:

1. Actual return on plan assets.
2. Loss or gain on plan assets.
3. Service cost.
4. Pension expense.
5. Average remaining service life of active employees (used to determine amortization of the net gain).

P 17–15
Comprehensive—pension elements; spreadsheet
● LO17–8

The following pension-related data pertain to Metro Recreation's noncontributory, defined benefit pension plan for 2021 ($ in thousands):

	Jan. 1	Dec. 31
Projected benefit obligation	$ 4,100	$4,380
Accumulated benefit obligation	3,715	3,950
Plan assets (fair value)	4,530	4,975
Interest (discount) rate, 7%		
Expected return on plan assets, 10%		
Prior service cost—AOCI (from Dec. 31, 2020, amendment)	840	
Net loss—AOCI	477	
Average remaining service life: 12 years		
Gain due to changes in actuarial assumptions		44
Contributions to pension fund (end of year)		340
Pension benefits paid (end of year)		295

Required:

Prepare a pension spreadsheet that shows the relationships among the various pension balances, shows the changes in those balances, and computes pension expense for 2021.

P 17–16
Comprehensive—reporting a pension plan; pension spreadsheet; determine changes in balances; two years
● LO17–3 through LO17–8

Actuary and trustee reports indicate the following changes in the PBO and plan assets of Lakeside Cable during 2021:

Prior service cost at Jan. 1, 2021, from plan amendment at the beginning of 2019 (amortization: $4 million per year)	$32 million
Net loss-pensions at Jan.1, 2021 (previous losses exceeded previous gains)	$40 million
Average remaining service life of the active employee group	10 years
Actuary's discount rate	8%

($ in millions)

	PBO		Plan Assets
Beginning of 2021	$300	*Beginning* of 2021	$200
Service cost	48	Return on plan assets, 7.5% (10% expected)	15
Interest cost, 8%	24	Cash contributions	45
Loss (gain) on PBO	(2)	Less: Retiree benefits	(20)
Less: Retiree benefits	(20)		
End of 2021	$350	*End* of 2021	$240

Required:

1. Determine Lakeside's pension expense for 2021, and prepare the appropriate journal entries to record the expense as well as the cash contribution to plan assets and payment of benefits to retirees.
2. Determine the new gains and/or losses in 2021 and prepare the appropriate journal entry(s) to record them.
3. Prepare a pension spreadsheet to assist you in determining end of 2021 balances in the PBO, plan assets, prior service cost—AOCI, the net loss—AOCI, and the pension liability.
4. Assume the following actuary and trustee reports indicating changes in the PBO and plan assets of Lakeside Cable during 2022 ($ in millions):

	PBO		Plan Assets
Beginning of 2022	$350	*Beginning* of 2022	$240
Service cost	38	Return on plan assets,	
Interest cost at 8%	28	15% (10% expected)	36
Loss (gain) on PBO	5	Cash contributions	30
Less: Retiree benefits	(16)	Less: Retiree benefits	(16)
End of 2022	$405	*End* of 2022	$290

Determine Lakeside's pension expense for 2022, and prepare the appropriate journal entries to record the expense, the cash funding of plan assets, and payment of benefits to retirees.

5. Determine the new gains and/or losses in 2022, and prepare the appropriate journal entry(s) to record them.
6. Using T-accounts, determine the balances at December 31, 2022, in the net loss—AOCI and prior service cost—AOCI.
7. Confirm the balances determined in *requirement* 6 by preparing a pension spreadsheet.

P 17–17
Integrating Problem—Deferred tax effects of pension entries; integrate concepts learned in Chapter 16
● LO17–7

Reproduced below are the journal entries related to Illustration 17–12 in this chapter that Global Communications used to record its pension expense and funding in 2021 and the new gain and loss that occurred that year. To focus on the core issues, we ignored the income tax effects of the pension amounts.

	($ in millions)	
To Record Pension Expense		
Pension expense (total)	43	
Plan assets (expected return on plan assets)	27	
PBO ($41 service cost + $24 interest cost)		65
Amortization of prior service cost—OCI (2021 amortization)		4
Amortization of net loss—OCI (2021 amortization)		1
To Record Funding		
Plan assets	48	
Cash (contribution to plan assets)		48
To Record Payment of Benefits		
PBO	38	
Plan assets (retiree benefits)		38
To Record Gains and Losses		
Loss—OCI (from change in assumption)	23	
PBO		23
Plan assets	3	
Gain—OCI (from actual return exceeding expected return)		3

Required:

1. Recast these journal entries to include the income tax effects of the events being recorded. Assume that Global's tax rate is 25%. [*Hint:* Costs are incurred and recognized for financial reporting purposes now, but the tax impact comes much later—when these amounts are deducted for tax purposes as actual payments for retiree benefits occur in the future. As a result, the tax effects are deferred, creating the need to record deferred tax assets and deferred tax liabilities. So, you may want to refer back to Chapter 16 to refresh your memory on these concepts.]
2. Prepare a statement of comprehensive income for 2021, assuming Global's only other sources of comprehensive income were net income of $300 million and a $30 million unrealized holding gain on investments in securities available for sale.

P 17–18
Postretirement benefits; EPBO calculations; APBO calculations; components of postretirement benefit expense; present value concepts
● LO17–9, LO17–10

Century-Fox Corporation's employees are eligible for postretirement health care benefits after both being employed at the end of the year in which age 60 is attained and having worked 20 years. Jason Snyder was hired at the end of 1998 by Century-Fox at age 34 and is expected to retire at the end of 2026 (age 62). His retirement is expected to span five years (unrealistically short in order to simplify calculations). The company's actuary has estimated the net cost of retiree benefits in each retirement year as shown below. The discount rate is 6%. The plan is not prefunded. Assume costs are incurred at the end of each year.

Year	Expected Age	Net Cost
2027	63	$4,000
2028	64	4,400
2029	65	2,300
2030	66	2,500
2031	67	2,800

Required:
1. Draw a time line that depicts Snyder's attribution period for retiree benefits and expected retirement period.
2. Calculate the present value of the net benefits as of the expected retirement date.
3. With respect to Snyder, what is the company's expected postretirement benefit obligation at the end of 2021?
4. With respect to Snyder, what is the company's accumulated postretirement benefit obligation at the end of 2021?
5. With respect to Snyder, what is the company's accumulated postretirement benefit obligation at the end of 2022?
6. What is the service cost to be included in 2022 postretirement benefit expense?
7. What is the interest cost to be included in 2022 postretirement benefit expense?
8. Show how the APBO changed during 2022 by reconciling the beginning and ending balances.

P 17–19
Postretirement benefits; schedule of postretirement benefit costs
● LO17–9 through LO17–11

Stockton Labeling Company has a retiree health care plan. Employees become fully eligible for benefits after working for the company eight years. Stockton hired Misty Newburn on January 1, 2021. As of the end of 2021, the actuary estimates the total net cost of providing health care benefits to Newburn during her retirement years to have a present value of $18,000. The actuary's discount rate is 10%.

Required:
Prepare a schedule that shows the EPBO, the APBO, the service cost, the interest cost, and the postretirement benefit expense for each of the years 2021–2028.

P 17–20
Postretirement benefits; relationship among elements of postretirement benefit plan
● LO17–9 through LO17–11

The information below pertains to the retiree health care plan of Thompson Technologies:

Thompson began funding the plan in 2021 with a contribution of $127,000 to the benefit fund at the end of the year. Retirees were paid $52,000. The actuary's discount rate is 5%. There were no changes in actuarial estimates and assumptions.

($ in thousands)	2021 Beginning Balances	2021 Ending Balances
Accumulated postretirement benefit obligation	$460	$485
Plan assets	0	75
Funded status	(460)	(410)
Prior service cost—AOCI	120	110
Net gain—AOCI	(50)	(49)

Required:
Determine the following amounts for 2021:
1. Service cost.
2. Postretirement benefit expense.
3. Net benefit liability.

Office Depot, Inc., is a leading global provider of products, services, and solutions for workplaces. The following is an excerpt from a disclosure note in the company's annual report for the fiscal year ended December 31, 2017:

P 17–21
Pension disclosure; amortization of actuarial gain or loss; Office Depot, Inc.
● LO17–3 through LO17–7
Real World Financials

Required:

1. What amount did Office Depot report in its balance sheet related to the pension plan at December 31, 2017?

NOTE 13. EMPLOYEE BENEFIT PLANS (in part)

	Pension Benefits	
($ in millions)	2017	2016
Changes in projected benefit obligation:		
Obligation at beginning of period	$1,000	$1,094
Service cost	6	7
Interest cost	39	45
Actuarial (gain) loss	25	(8)
Benefits paid	(91)	(138)
Obligation at end of period	$ 979	$1,000
Change in plan assets:		
Fair value of plan assets at beginning of period	$ 870	$ 922
Actual return (loss) on plan assets*	114	84
Employer contribution	15	2
Benefits paid	(91)	(138)
Fair value of plan assets at end of period	908	870
Net liability recognized at end of period	$ (71)	$ (130)

*Expected return $48 and $55 in 2017 and 2016 respectively

2. When calculating pension expense at December 31, what amount, if any, did Office Depot include in its income statement as the amortization of unrecognized net actuarial loss (net loss—AOCI)? This AOCI account had a balance of $38 million at the beginning of the year and was the only AOCI account related to pensions. The average remaining service life of employees was 10 was years.
3. What was the pension expense?
4. What were the appropriate journal entries to record Office Depot's pension expense and to record gains and/or losses related to the pension plan?

Decision Makers' Perspective

©ImageGap/Getty Images

Apply your critical-thinking ability to the knowledge you've gained. These cases will provide you an opportunity to develop your research, analysis, judgment, and communication skills. You also will work with other students, integrate what you've learned, apply it in real-world situations, and consider its global and ethical ramifications. This practice will broaden your knowledge and further develop your decision-making abilities.

Judgment Case 17–1
Choose your retirement option
● LO17–1, LO17–3, LO17–4, LO17–5

"I only get one shot at this?" you wonder aloud. Mrs. Montgomery, human resources manager at Covington State University, has just explained that newly hired assistant professors must choose between two retirement plan options. "Yes, I'm afraid so," she concedes. "But you do have a week to decide."

Mrs. Montgomery's explanation was that your two alternatives are (1) the state's defined benefit plan and (2) a defined contribution plan under which the university will contribute each year an amount equal to 8% of your salary. The defined benefit plan will provide annual retirement benefits determined by the following formula: 1.5% × years of service × salary at retirement.

"It's a good thing I studied pensions in my accounting program," you tell her. "Now let's see. You say the state is currently assuming our salaries will rise about 3% a year, and the interest rate they use in their calculations is 6%? And, for someone my age, you say they assume I'll retire after 40 years and draw retirement pay for 20 years. I'll do some research and get back to you."

Required:

1. You were hired at the beginning of 2021 at a salary of $100,000. If you choose the state's defined benefit plan and projections hold true, what will be your annual retirement pay? What is the present value of your retirement annuity as of the anticipated retirement date (end of 2060)?
2. Suppose instead that you choose the defined contribution plan. Assuming that the rate of increase in salary is the same as the state assumes and that the rate of return on your retirement plan assets will be 6% compounded

annually, what will be the future value of your plan assets as of the anticipated retirement date (end of 2060)? What will be your annual retirement pay (assuming continuing investment of remaining assets at 6%)?

3. Based on this numerical comparison, which plan would you choose? What other factors must you also consider in making the choice?

Hint: The calculations are greatly simplified using an electronic spreadsheet such as Excel. There are many ways to set up the spreadsheet. One relatively easy way is to set up the first few rows with the formulas as shown below, then use the "fill down" function to fill in the remaining 38 rows, and use the Insert: Name: Define: function to name column B "n". Since contributions are assumed made at the end of each year, there are 39 years remaining to maturity at the end of 2021. Note that multiplying each contribution by $(1.06)^n$, where n equals the remaining number of years to retirement, calculates the future value of each contribution invested at 6% until retirement.

	A	B	C	D	E
1	End of	Years to			Future Value
2	Year	Retirement	Salary	Contribution	at Retirement
3	2021	39	100,000	= C3*0.08	= D3*1.06^n
4	=A3+1	= B3-1	= C3*1.03	= C4*0.08	= D4*1.06^n

Communication Case 17–2
Pension concepts
● LO17–2 through LO17–8

Stacy Persoff is the newly hired assistant controller of Kemp Industries, a regional supplier of hardwood derivative products. The company sponsors a defined benefit pension plan that covers its 420 employees. On reviewing last year's financial statements, Persoff was concerned about some items reported in the disclosure notes relating to the pension plan. Portions of the relevant note follow:

Note 8: Pensions

The company has a defined benefit pension plan covering substantially all of its employees. Pension benefits are based on employee service years and the employee's compensation during the last two years of employment. The company contributes annually the maximum amount permitted by the federal tax code. Plan contributions provide for benefits expected to be earned in the future as well as those earned to date. The following reconciles the plan's funded status and amount recognized in the balance sheet at December 31, 2021 ($ in thousands).

Actuarial Present Value Benefit Obligations:	
Accumulated benefit obligation (including vested benefits of $318)	$(1,305)
Projected benefit obligation	$(1,800)
Plan assets at fair value	1,575
Projected benefit obligation in excess of plan assets	$ (225)

Kemp's comparative income statements reported total pension expense of $108,000 in 2021 and $86,520 in 2020. Since employment has remained fairly constant in recent years, Persoff expressed concern over the increase in the pension expense. She expressed her concern to you, a three-year senior accountant at Kemp. "I'm also interested in the differences in these liability measurements," she mentioned.

Required:

Write a memo to Persoff. In the memo, do the following:

1. Explain to Persoff how the composition of the total pension expense can create the situation she sees. Briefly describe the components of pension expense. Include a description of how the service cost component is reported in the income statement.
2. Briefly explain how pension gains and losses are recognized in earnings.
3. Describe for her the differences and similarities between the accumulated benefit obligation and the projected benefit obligation.
4. Explain how the "Projected benefit obligation in excess of plan assets" is reported in the financial statements.

Judgment Case 17–3
Barlow's wife; relationship among pension elements
● LO17–8

LGD Consulting is a medium-sized provider of environmental engineering services. The corporation sponsors a noncontributory, defined benefit pension plan. Alan Barlow, a new employee and participant in the pension plan, obtained a copy of the 2021 financial statements, partly to obtain additional information about his new employer's obligation under the plan. In part, the pension disclosure note reads as follows:

Note 8: Retirement Benefits

The Company has a defined benefit pension plan covering substantially all of its employees. The benefits are based on years of service and the employee's compensation during the last two years of employment. The company's funding policy is consistent with the funding requirements of federal law and regulations. Generally, pension costs accrued are funded. Plan assets consist primarily of stocks, bonds, commingled trust funds, and cash.

The change in projected benefit obligation for the plan years ended December 31, 2021, and December 31, 2020

($ in thousands)	2021	2020
Projected benefit obligation at beginning of year	$3,786	$3,715
Service cost	103	94
Interest cost	287	284
Actuarial (gain) loss	302	(23)
Benefits paid	(324)	(284)
Projected benefit obligation at end of year	$4,154	$3,786

The weighted average discount rate and rate of increase in future compensation levels used in determining the actuarial present value of the projected benefit obligations in the above table were 7.0% and 4.3%, respectively, at December 31, 2021, and 7.75% and 4.7%, respectively, at December 31, 2020. The expected long-term rate of return on assets was 10.0% at December 31, 2021 and 2020.

The change in the fair value of plan assets for the plan years ended December 31, 2021 and 2020.

($ in thousands)	2021	2020
Fair value of plan assets at beginning of year	$3,756	$3,616
Actual return on plan assets	1,100	372
Employer contributions	27	52
Benefits paid	(324)	(284)
Fair value of plan assets at end of year	$4,559	$3,756

Included in the Consolidated Balance Sheets are the following components of accumulated other comprehensive income.

($ in thousands)	2021	2020
Net actuarial gain	$(620)	$(165)
Prior service cost	44	46

Net periodic defined benefit pension cost for fiscal 2021, 2020, and 2019 is included the following components.

($ in thousands)	2021	2020	2019
Service cost	$ 103	$ 94	$ 112
Interest cost	287	284	263
Expected return on plan assets	(342)	(326)	(296)
Amortization of prior service cost	2	2	1
Recognized net actuarial (gain) loss	(2)	2	4
Net periodic pension cost	$ 48	$ 56	$ 84

In attempting to reconcile amounts reported in the disclosure note with amounts reported in the income statement and balance sheet, Barlow became confused. He was able to find the pension expense on the income statement but was unable to make sense of the balance sheet amounts. Expressing his frustration to his wife, Barlow said, "It appears to me that the company has calculated pension expense as if they have the pension liability and pension assets they include in the note, but I can't seem to find those amounts in the balance sheet. In fact, there are several amounts here I can't seem to account for. They also say they've made some assumptions about interest rates, pay increases, and profits on invested assets. I wonder what difference it would make if they assumed other numbers."

Barlow's wife took accounting courses in college and remembers most of what she learned about pension accounting. She attempts to clear up her husband's confusion.

Required:

Assume the role of Barlow's wife. Answer the following questions for your husband.

1. Is Barlow's observation correct that the company has calculated pension expense on the basis of amounts not reported in the balance sheet?
2. What amount would the company report as a pension liability in the balance sheet?
3. What amount would the company report as a pension asset in the balance sheet?
4. Which of the other two amounts reported in the disclosure note would the company report in the balance sheet?
5. The disclosure note reports a net actuarial gain as well as an actuarial loss. Does the loss in 2021 indicate that the PBO is higher or is lower than previously expected due to some unspecified change in an actuarial assumption. How are these related? What do the amounts mean?
6. Losses and gains are reported in the statement of comprehensive income as they occur. These amounts accumulate as a net gain or net loss in the balance sheet as part of what account?

Communication Case 17–4
Barlow's wife; relationship among pension elements
● LO17–8

The focus of this case is question 1 in the previous case. Your instructor will divide the class into two to six groups, depending on the size of the class. The mission of your group is to assess the correctness of Barlow's observation and to suggest the appropriate treatment of the pension obligation. The suggested treatment need not be that required by GAAP.

Required:

1. Each group member should deliberate the situation independently and draft a tentative argument prior to the class session for which the case is assigned.
2. In class, each group will meet for 10 to 15 minutes in different areas of the classroom. During that meeting, group members will take turns sharing their suggestions for the purpose of arriving at a single group treatment.
3. After the allotted time, a spokesperson for each group (selected during the group meetings) will share the group's solution with the class. The goal of the class is to incorporate the views of each group into a consensus approach to the situation.

Real World Case 17–5
Types of pension plans; disclosures
● LO17–1

Real World Financials

Refer to the financial statements and related disclosure notes of **Microsoft Corporation** (**www.microsoft.com**).

Required:

1. What type of pension plan does Microsoft sponsor for its employees? Explain.
2. Who bears the "risk" of factors that might reduce retirement benefits in this type of plan? Explain.
3. Assuming that employee and employer contributions vest immediately, suppose a Microsoft employee contributes $1,000 to the pension plan during her first year of employment and directs investments to a bond mutual fund. If she leaves Microsoft early in her second year, after the mutual fund's value has increased by 2%, how much will she be entitled to roll over into an Individual Retirement Account (IRA)?
4. How did Microsoft account for its participation in the pension plan in fiscal 2017?

Ethics Case 17–6
401(k) plan contributions
● LO17–1

You are in your third year as internal auditor with VXI International, manufacturer of parts and supplies for jet aircraft. VXI began a defined contribution pension plan three years ago. The plan is a so-called 401(k) plan (named after the Tax Code section that specifies the conditions for the favorable tax treatment of these plans) that permits voluntary contributions by employees. Employees' contributions are matched with one dollar of employer contribution for every two dollars of employee contribution. Approximately $500,000 of contributions is deducted from employee paychecks each month for investment in one of three employer-sponsored mutual funds.

While performing some preliminary audit tests, you happen to notice that employee contributions to these plans usually do not show up on mutual fund statements for up to two months following the end of pay periods from which the deductions are drawn. On further investigation, you discover that when the plan was first begun, contributions were invested within one week of receipt of the funds. When you question the firm's investment manager about the apparent change in the timing of investments, you are told, "Last year Mr. Maxwell (the CFO) directed me to initially deposit the contributions in the corporate investment account. At the close of each quarter, we add the employer matching contribution and deposit the combined amount in specific employee mutual funds."

Required:

1. What is Mr. Maxwell's apparent motivation for the change in the way contributions are handled?
2. Do you perceive an ethical dilemma?

Real World Case 17–7
Types of pension plans; reporting postretirement plans; disclosures
● LO17–5, LO17–8

Real World Financials

Refer to the 2017 financial statements and related disclosure notes of **FedEx Corporation**. The financial statements can be found at the company's website (**www.fedex.com**).

Required:

1. Does FedEx sponsor defined benefit pension plans for its employees? Defined contribution plans? Postretirement healthcare plans?
2. What amount does FedEx report in its balance sheet in 2017 for its U.S. pension plans? For its postretirement healthcare plans?
3. FedEx reports three actuarial assumptions used in its calculations of the projected benefit obligation. What were those three amounts in 2017? Did reported changes in those assumptions from the previous year increase, decrease, or have no effect on the projected benefit obligation?

Analysis Case 17–8
Pension amendment
● LO17–5, LO17–8

Charles Rubin is a 30-year employee of Amalgamated Motors. Charles was pleased with recent negotiations between his employer and the United Auto Workers. Among other favorable provisions of the new agreement, the pact also includes a 14% increase in pension payments for workers under 64 with 30 years of service who retire during the agreement. Although the elimination of a cap on outside income earned by retirees has been generally viewed as an incentive for older workers to retire, Charles sees promise for his dream of becoming a part-time engineering consultant after retirement. What has caught Charles's attention is the following excerpt from an article in the financial press:

> Amalgamated Motors Corp. will record a $240 million charge due to increases in retirement benefits for hourly United Auto Workers employees.
>
> The charge stems from AM's new tentative labor contract with the UAW. According to a filing with the Securities and Exchange Commission, the charge amounts to 26 cents a share.
>
> The company warned that its "unfunded pension obligation and pension expense are expected to be unfavorably impacted as a result of the recently completed labor negotiations."

Taking advantage of an employee stock purchase plan, Charles has become an active AM stockholder as well as employee. His stockholder side is moderately concerned by the article's reference to the unfavorable impact of the recently completed labor negotiations.

Required:

1. When a company modifies its pension benefits the way Amalgamated Motors did, what name do we give the added cost? How is it accounted for?
2. What does AM mean when it says its "unfunded pension obligation and pension expense are expected to be unfavorably impacted as a result of the recently completed labor negotiations"?

Analysis Case 17–9
Effect of pensions on earnings
● LO17–7

While doing some online research concerning a possible investment you come across an article that mentions in passing that a representative of **Morgan Stanley** had indicated that a company's pension plan had benefited its reported earnings. Curiosity piqued, you seek your old Intermediate Accounting text.

Required:

1. Can the net periodic pension "cost" cause a company's reported earnings to increase? Explain.
2. Companies must report the actuarial assumptions used to make estimates concerning pension plans. Which estimate influences the earnings effect in *requirement* 1? Can any of the other estimates influence earnings? Explain.

Research Case 17–10
Researching the way employee benefits are tested on the CPA Exam; retrieving information from the Internet
● LO17–9, LO17–10, LO17–11

The board of examiners of the American Institute of Certified Public Accountants (AICPA) is responsible for preparing the CPA examination. The boards of accountancy of all 50 states, the District of Columbia, Guam, Puerto Rico, the U.S. Virgin Islands, and the Mariana Islands use the examination as the primary way to measure the technical competence of CPA candidates. The content for each examination section is specified by the AICPA and described in outline form in "blueprints".

Required:

1. Access the AICPA website on the Internet. The web address is **www.aicpa.org**.
2. Access the CPA exam section within the site. Locate the exam content portion of the section.
3. In which of the four separately graded sections of the exam are postretirement benefits tested?
4. From the AICPA site, access the Board of Accountancy for your state. What are the education requirements in your state to sit for the CPA exam?

Analysis Case 17–11
Pensions and other postretirement benefit plans; analysis of disclosure notes
● LO17–6, LO17–11

Real World Financials

Macy's, Inc., operates about 700 Macy's and Bloomingdale's department stores and furniture galleries in 45 states and U.S. territories as well as Bloomingdale's Outlet stores, macys.com, and bloomingdales.com. Refer to the financial statements for the year ended February 2, 2018, and related disclosure notes of the company. The financial statements can be found at the company's website (**www.macys.com**).

Required:

1. From the information provided in various portions of Note 9, reconcile the beginning balance ($1,232) and ending balances for Net loss—AOCI associated with Macy's pension plan. Clearly label each reconciling amount.
2. Macy's was required to amortize a portion of its Net loss—AOCI associated with its pension plan in the year ended February 2, 2018. Based on the calculation of that amount, determine the average remaining service life of the company's employees used in that calculation.
3. Suppose you believe that Macy's should have assumed a 1% lower health care cost trend during the year ended February 2, 2018. What would the accumulated postretirement benefit obligation have been on February 2, 2018, under that assumption?

Data Analytics

connect

Data analytics is the process of examining data sets in order to draw conclusions about the information they contain. If you haven't completed any of the prior data analytics cases, follow the instructions listed in the Chapter 1 Data Analytics case to get set up. You will need to watch the videos referred to in the Chapters 1 - 3 Data Analytics cases. No additional videos are required for this case. All short training videos can be found here: **https://www.tableau.com/learn/training#getting-started**.

Data Analytics Case
Reporting Pension Plans
● LO17–6

In the Data Analytics Cases in the previous chapter, you used Tableau to examine a data set and create charts to examine two (hypothetical) publicly traded companies: GPS Corporation and Tru, Inc., to examine the effect of the Tax Cuts and Jobs Act of 2017 on these companies' operations and financial position. Now, you examine the funded status of the two companies' pension plans and any changes in that funded status during the previous ten years. You will also observe the change in the way components of pension expense are reported in the income statement.

Required:

For each of the two companies in the ten-year period, 2012-2021, use Tableau to calculate and display the trends for (a) the amounts the companies report in their balance sheets pertaining to their defined benefit pension plans cost [projected benefit obligation minus pension plan assets] and (b) the pension service cost and pension non-service cost components of the net pension cost. Based upon what you find, answer the following questions:

1. In which years is GPS's pension plan underfunded during the period 2012-2021?
2. In which years is Tru, Inc.'s pension plan underfunded during the period 2012-2021?
3. In which year did the two companies begin reporting the service cost and non-service cost components of the net pension cost separately in their income statements?
4. What are the (a) service cost and (b) non-service cost components of the net pension cost for GPS in 2021?

Resources:

Download the "GPS_Tru_Financials.xlsx" Excel file available in Connect, or under Student Resources within the Library tab. Save it to the computer on which you will be using Tableau.

For this case, you will create calculations to produce the funded status of the companies' pension plans to allow you to compare and contrast the two companies.

After you view the training videos, follow these steps to create the charts you'll use for this case:

- Open Tableau and connect to the Excel spreadsheet you downloaded.
- Starting on the Sheet 1 tab, drag "($ in 000s) Company" and "Year" under Dimensions to the Columns shelf. Change "Year" to *discrete* by right-clicking and selecting "Discrete."
- Create a calculated field by clicking the "Analysis" tab at the top of the screen and selecting "Create Calculated field." Name the calculation "Pension liability/asset." In the Calculation Editor window, from Measures, drag "Projected Benefit Obligation", type a minus sign, and then drag "Pension Plan Assets". Make sure the window says that the calculation is valid and click OK.
- Drag the newly created "Pension liability/asset" to the Row shelf. Click on the "Show Me" and select "side-by-side bars." Add labels to the bars by clicking on "Label" under the "Marks card" and clicking the box "Show mark label." Format the labels to Times New Roman, bold, black and 10-point font. Edit the color on the "Color Mark" card if desired.
- Change the title of the sheet to be "Pension liability/asset" by right-clicking and selecting "Edit title." Format the title to Times New Roman, bold, black and 15-point font. Change the title of "Sheet 1" to match the sheet title by right-clicking, selecting "Rename" and typing in the new title.
- Format all other labels to be Times New Roman, bold, black and 12-point font.
- On the Sheet 2 tab, follow the procedure above for the company and year.
- Drag "Pension service cost" and "Pension non-service cost component" under Measures into the Rows shelf. Edit the axis by selecting the axis, right-clicking, and clicking on "Edit Axis. . .". Select "Fixed" and change the lower range to be -20 and the upper range to be 600 for both charts.
- Add labels to the bars by clicking on "Label" under the "Marks" card and clicking the box "Show mark labels." Format the labels to Times New Roman, bold, black and 10-point font. Edit the color on the "Marks" card if desired.
- Change the title of the sheet to be "Pension Service Cost and Non Service Cost" by right-clicking and selecting "Edit title." Format the title to Times New Roman, bold, black and 15-point font. Change the title of "Sheet 2" to match the sheet title by right-clicking, selecting "Rename" and typing in the new title.
- Format all other labels to be Times New Roman, bold, black and 12-point font.
- Once complete, save the file as "DA17_Your initials.twbx."

Continuing Cases

Target Case
● LO17–3
LO17–4
LO17–5
LO17–6

Target Corporation prepares its financial statements according to U.S. GAAP. Target's financial statements and disclosure notes for the year ended February 3, 2018, are available in Connect. This material also is available under the Investor Relations link at the company's website (**www.target.com**).

Target has both defined contribution and defined benefit pension plan. In Note 28 "Pension and Postretirement Health Care Plans," Target describes its defined benefit plans.

Required:

1. What were the changes in Target's Projected Benefits Obligation in the fiscal years ended February 3, 2018 (fiscal 2017), and January 28, 2017 (fiscal 2016), for its qualified pension plans?
2. What were the changes in Target's Pension Plan Assets in the fiscal years ended February 3, 2018, and January 28, 2017, for its qualified pension plans?
3. Were these pension plans overfunded or underfunded for the fiscal years ended February 3, 2018, and January 28, 2017?
4. What were the components of Target's Pension Expense in the fiscal years 2017, 2016, and 2015?

Air France–KLM Case
● LO17–12

IFRS

Air France–KLM (AF), a Franco-Dutch company, prepares its financial statements according to International Financial Reporting Standards. AF's financial statements and disclosure notes for the year ended December 31, 2017, are available in Connect. This material also is available under the Finance link at the company's website (**www.airfranceklm-finance.com**).

Required:

1. AF reported past service cost (called prior service cost under U.S. GAAP) in its income statement as part of net periodic pension cost. Is that reporting method the same or different from the way we report prior service cost under U.S. GAAP?
2. Look at Note 30.2, "Retirement Benefits." AF incorporates estimates regarding staff turnover, life expectancy, salary increase, retirement age, and discount rates. How did AF report changes in these assumptions? Is that reporting method the same or different from the way we report changes under U.S. GAAP?
3. AF does not report remeasurement gains and losses in its income statement. Where did AF report these amounts? Is that reporting method the same or different from the way we report pension expense under U.S. GAAP?
4. See Note 22. Did AF report (a) net interest cost or (b) net interest income in 2017?

CPA Exam Questions and Simulations

ROGER CPA Review

Sample CPA Exam questions from Roger CPA Review are available in Connect as support for the topics in this chapter. These multiple-choice questions and task-based simulations include expert-written explanations and solutions, and provide a starting point for students to become familiar with the content and functionality of the actual CPA Exam.

CHAPTER

18 Shareholders' Equity

OVERVIEW

We turn our attention from liabilities, which represent the creditors' interests in the assets of a corporation, to the shareholders' residual interest in those assets. The discussions distinguish between the two basic sources of shareholders' equity: (1) *invested capital* and (2) *earned capital.* We explore the expansion of corporate capital through the issuance of shares and the contraction caused by the retirement of shares or, equivalently, the purchase of treasury shares. In our discussions of retained earnings, we examine cash dividends, property dividends, stock dividends, and stock splits.

LEARNING OBJECTIVES

After studying this chapter, you should be able to:

- **LO18–1** Describe the components of shareholders' equity and explain how they are reported in a statement of shareholders' equity. (*p. 1047*)
- **LO18–2** Describe comprehensive income and its components. (*p. 1050*)
- **LO18–3** Understand the corporate form of organization and the nature of stock. (*p. 1051*)
- **LO18–4** Record the issuance of shares when sold for cash and for noncash consideration. (*p. 1058*)
- **LO18–5** Distinguish between accounting for retired shares and for treasury shares. (*p. 1062*)
- **LO18–6** Describe retained earnings and distinguish it from paid-in capital. (*p. 1068*)
- **LO18–7** Explain the basis of corporate dividends, including the similarities and differences between cash and property dividends. (*p. 1068*)
- **LO18–8** Explain stock dividends and stock splits and how we account for them. (*p. 1070*)
- **LO18–9** Discuss the primary differences between U.S. GAAP and IFRS with respect to accounting for shareholders' equity. (*pp. 1052* and *1058*)

©Peathegee Inc/Blend Images

FINANCIAL REPORTING CASE

Nike, Inc.

Finally, you have some uninterrupted time to get back on the net. Earlier today you noticed on the Internet that the market price of Nike's common stock was up almost 10%. You've been eager to look into why this happened, but have had one meeting after another all day.

You've been a stockholder of **Nike** since the beginning of the year when you bought your last pair of running shoes and saw how much they sell for these days. The dividends that you receive quarterly are nice, but that's not why you bought the stock; you were convinced at the time that the stock price was poised to rise rapidly. A few well-placed clicks of the mouse and you come across the following news article:

BEAVERTON, Ore.--(BUSINESS WIRE)—NIKE, Inc. (NYSE: NKE) announced today that its Board of Directors has approved a quarterly cash dividend of $0.20 per share on the company's outstanding Class A and Class B Common Stock. This represents an increase of 11% versus the prior quarterly dividend rate of $0.18 per share. The dividend declared today is payable on January 2, 2018, to shareholders of record at the close of business December 4, 2017.

"This marks NIKE's 16th consecutive year of increasing dividend payouts," said Mark Parker, Chairman, President, and CEO of NIKE, Inc. "Today's announcement, combined with the four-year $12 billion share repurchase program we announced in 2015, demonstrates our continued confidence in generating strong cash flow and returns for shareholders through our new Consumer Direct Offense as we continue to invest in fueling sustainable, long-term growth and profitability."

Nike also distributed a two-for-one split of both NIKE's Class A and Class B Common shares in the form of a 100% stock dividend on December 23, 2015.

QUESTIONS

By the time you finish this chapter, you should be able to respond appropriately to the questions posed in this case. Compare your response to the solution provided at the end of the chapter.

1. Do you think the stock price increase is related to Nike's share repurchase plan? (*p. 1062*)
2. What are Nike's choices in accounting for the share repurchases? (*p. 1063*)
3. What effect does the quarterly cash dividend of 20 cents a share have on Nike's assets? Its liabilities? Its shareholders' equity? (*p. 1069*)
4. What effect does the stock split have on Nike's assets? Its liabilities? Its shareholders' equity? (*p. 1072*)

The Nature of Shareholders' Equity

PART A

LO18–1

A corporation raises money to fund its business operations by some mix of debt and equity financing. In earlier chapters, we examined debt financing in the form of notes, bonds, leases, and other liabilities. Amounts representing those liabilities denote *creditors' interest* in the company's assets. Now we focus on various forms of *equity* financing. Specifically, in this chapter we consider transactions that affect shareholders' equity—those accounts that represent the *ownership interests* of shareholders.

In principle, shareholders' equity is a relatively straightforward concept. Shareholders' equity is a residual amount—what's left over after creditor claims have been subtracted from assets (in other words, net assets). You probably recall the residual nature of shareholders' equity from the basic accounting equation:

Net assets equal shareholders' equity.

$$\underbrace{\text{Assets} - \text{Liabilities}}_{\text{Net Assets}} = \text{Shareholders' Equity}$$

Shareholders' equity accounts denote the ownership interests of shareholders.

Ownership interests of shareholders arise primarily from two sources: (1) amounts *invested* by shareholders in the corporation and (2) amounts *earned* by the corporation on behalf of its shareholders. These two sources are reported as (1) paid-in capital and (2) retained earnings.

Despite being a seemingly clear-cut concept, shareholders' equity and its component accounts often are misunderstood and misinterpreted. As we explore the transactions that affect shareholders' equity and its component accounts, try not to allow yourself to be overwhelmed by unfamiliar terminology or to be overly concerned with precise account titles. Terminology pertaining to shareholders' equity accounts is notoriously diverse. Every shareholders' equity account has several aliases. Indeed, shareholders' equity itself is often referred to as *stockholders' equity* (**Walmart**), *shareowners' equity* (**General Electric**), *shareholders' investment* (**Target**), *stockholders' investment* (**FedEx**), *shareholders' equity* (**Apple**), *equity* (**Air France–KLM**), and many other similar titles.

Complicating matters, transactions that affect shareholders' equity are influenced by corporation laws of individual states in which companies are located. And, as we see later, generally accepted accounting principles provide companies with considerable latitude when choosing accounting methods in this area.

Legal requirements and disclosure objectives make it preferable to separate a corporation's capital into several separate shareholders' equity accounts.

Keeping this perspective in mind while you study the chapter should aid you in understanding the essential concepts. At a very basic level, each transaction we examine can be viewed simply as an increase or decrease in shareholders' equity, per se, without regard to specific shareholders' equity accounts. In fact, for a business organized as a single proprietorship, all capital changes are recorded in a single owner's equity account. The same concepts apply to a corporation. But for corporations, additional considerations make it desirable to separate owners' equity into several separate shareholders' equity accounts. These additional considerations—legal requirements and disclosure objectives—are discussed in later sections of this chapter. So, as you study the separate effects of transactions on retained earnings and specific paid-in capital accounts, you may find it helpful to ask yourself frequently, "What is the net effect of this transaction on shareholders' equity?" or, equivalently, "By how much are net assets (assets minus liabilities) affected by this transaction?"

Financial Reporting Overview

Before we examine the events that underlie specific shareholders' equity accounts, let's summarize how individual accounts relate to each other. The condensed balance sheet in Illustration 18–1 of Exposition Corporation, a hypothetical company, provides that perspective.

Illustration 18–1 depicts a rather comprehensive situation. It's unlikely that any one company would have shareholders' equity from all of these sources at any one time. Remember that, at this point, our objective is only to get a general perspective of the items constituting shareholders' equity. Although company records would include separate accounts for each of these components of shareholders' equity in the balance sheet, in practice Exposition would report a more condensed version similar to that in Illustration 18–1A and the one we will see later in the chapter when we look at the presentation by Abercrombie & Fitch (in Illustration 18–3).

The four classifications within shareholders' equity are paid-in capital, retained earnings, accumulated other comprehensive income, and treasury stock. We discuss these now in the context of Exposition Corporation.

Paid-in Capital

Paid-in capital consists primarily of amounts invested by shareholders when they purchase shares of stock from the corporation or arise from the company buying back some of those

Illustration 18–1

Detailed Shareholders' Equity Presentation

EXPOSITION CORPORATION
Balance Sheet
December 31, 2021
($ in millions)

Assets
$3,000
Liabilities
$1,000
Shareholders' Equity

Paid-in capital:		
Capital stock (par):		
Preferred stock, 10%, $10 par, cumulative, nonparticipating	$100	
Common stock, $1 par	55	
Common stock dividends distributable	5	
Additional paid-in capital:		
Paid-in capital—excess of par, preferred	50	
Paid-in capital—excess of par, common	260	
Paid-in capital—share repurchase	8	
Paid-in capital—conversion of bonds	7	
Paid-in capital—stock options	9	
Paid-in capital—restricted stock	5	
Paid-in capital—lapse of stock options	1	
Total paid-in capital		$ 500
Retained earnings		1,670
Accumulated other comprehensive income:		
Gain (loss) on AFS investments (unrealized)	(85)	
Net unrecognized gain (loss) on pensions	(75)	
Deferred gain (loss) on derivatives*	(4)	
Adjustments from foreign currency translation**	0	(164)
Treasury stock (at cost)		(6)
Total shareholders' equity		$2,000

*When a derivative designated as a cash flow hedge is adjusted to fair value, the gain or loss is deferred as a component of other comprehensive income and included in earnings later, at the same time as earnings are affected by the hedged transaction (described in Appendix A at the end of this textbook).
**Changes in foreign currency exchange rates are discussed elsewhere in your accounting curriculum. The amount could be an addition to or reduction in shareholders' equity.

Assets *minus* Liabilities equals Shareholders' Equity.

The primary source of paid-in capital is the investment made by shareholders when buying preferred and common stock.

Several other events also affect paid-in capital.

Retained earnings represents earned capital.

Illustration 18–1A

Typical Shareholders' Equity Presentation

Preferred stock, 10%, $10 par, cumulative, nonparticipating	$ 100
Common stock, $1 par	60
Additional paid-in capital	340
Retained earnings	1,670
Accumulated other comprehensive income:	
Gain (loss) on AFS investments (unrealized)	(85)
Net unrealized loss on pensions	(75)
Deferred loss on derivatives	(4)
Treasury stock	(6)
Total shareholders' equity	$2,000

shares or from share-based compensation activities. Later in this chapter and the next, we consider in more detail the events and transactions that affect paid-in capital.

Retained Earnings

Retained earnings is accumulated on behalf of shareholders and reported as a single amount. We discuss retained earnings in Part C of this chapter.

Treasury Stock

We discuss the final component of shareholders' equity—treasury stock—later in the chapter. It indicates that some of the shares previously sold were bought back by the corporation from shareholders.

● LO18–2

Accumulated Other Comprehensive Income

Comprehensive income includes net income as well as other gains, losses, and adjustments that change shareholders' equity but are not included in traditional net income.

Also notice that shareholders' equity of Exposition Corporation is adjusted for three events that are not included in net income and so don't affect retained earnings but are part of "other comprehensive income" and therefore are included as a separate component of shareholders' equity: accumulated comprehensive income.[1] **Comprehensive income** provides a more expansive view of the change in shareholders' equity than does traditional net income. It's the total *nonowner* change in equity for a reporting period. That is, comprehensive income encompasses all changes in equity other than those from transactions with owners. Transactions between the corporation and its shareholders primarily include dividends and the issuance or purchase of shares of the company's stock. Most nonowner changes are reported in the income statement. So, the nonowner changes other than those that are part of traditional net income are the ones reported as "other comprehensive income."

Comprehensive income extends our view of income beyond net income reported in an income statement to include four types of gains and losses not included in income statements.

1. Net holding gains (losses) on available-for-sale investments in debt securities.
2. Gains (losses) from and amendments to postretirement benefit plans.
3. Deferred gains (losses) on derivatives.
4. Adjustments from foreign currency translation.

The first of these are the gains and losses on some securities that occur when the fair values of these investments increase or decrease. As you learned in Chapters 12 and 14, gains and losses on available-for-sale investments in debt securities aren't included in earnings until they are realized through the sale of the securities but are considered components of *other comprehensive income (OCI)* in the meantime. Similarly, as we discussed in Chapter 17, net gains and losses as well as "prior service cost" on pensions sometimes affect other comprehensive income rather than net income as they occur. You have not yet studied the third and fourth potential components of other comprehensive income. As described in Appendix A, "Derivatives," at the back of this textbook, when a derivative designated as a "*cash flow hedge*" is adjusted to fair value, the gain or loss is deferred as a component of other comprehensive income and included in earnings later, at the same time as earnings are affected by the hedged transaction. Adjustments from changes in foreign currency exchange rates are discussed elsewhere in your accounting curriculum. These, too, are included in other comprehensive income (OCI) but not net income.

OCI is reported in the statement of comprehensive income.

AOCI is reported in the balance sheet.

OCI shares another trait with net income. Just as net income is reported periodically in the income statement and also on a *cumulative* basis as part of retained earnings, OCI too, is reported periodically in the statement of comprehensive income and also as **accumulated other comprehensive income (AOCI)** in the balance sheet along with retained earnings. In other words, we report two attributes of OCI: (1) components of comprehensive income *created during the reporting period* and (2) the comprehensive income *accumulated* over the current and prior periods.

The first attribute—components of comprehensive income *created during the reporting period*—can be reported either as (a) an expanded version of the income statement or (b) a separate statement immediately following the income statement. Regardless of the placement a company chooses, the presentation is similar. It will report net income, other components of comprehensive income, and total comprehensive income, similar to the presentation in Illustration 18–2. Note that each component is reported net of its related income tax expense or income tax benefit.[2]

[1]Comprehensive income was introduced in Chapter 4 and revisited in Chapters 12 and 17.

[2]This can be accomplished by presenting components of other comprehensive income, either net of related income tax effects (as in this presentation) or before income tax effects with disclosure of the income taxes allocated to each component either in a disclosure note or parenthetically in the statement.

Illustration 18–2
Statement of Comprehensive Income

		($ in millions)
Net income		$xxx
Other comprehensive income:		
Gains (loss) on AFS investments (unrealized), net of tax)*	$x	
Gain (loss) from amendments to postretirement benefit plans (net of tax)†	(x)	
Deferred gain (loss) on derivatives (net of tax)‡	(x)	
Adjustments from foreign currency translation (net of tax)§	x	xx
Comprehensive income		$xxx

*Changes in the fair value of some investments in debt securities (described in Chapter 12). An unrealized loss also might occur from recording an "other than temporary" impairment in excess of a credit loss for a debt investment. And, as discussed in Chapter 14, applying the "fair value option" to a liability entails reporting in OCI any part of the change in the liability's fair value that's due to the liability's credit risk.
†Gains and losses due to revising assumptions or market returns differing from expectations and prior service cost from amending the plan (described in Chapter 17).
‡When a derivative designated as a cash flow hedge is adjusted to fair value, the gain or loss is deferred as a component of comprehensive income and included in earnings later, at the same time as earnings are affected by the hedged transaction (described in the Derivatives Appendix to the text).
§Changes in foreign currency exchange rates when translating financial statements of foreign subsidiaries to U.S. dollars. The amount could be an addition to or reduction in shareholders' equity. (This item is discussed elsewhere in your accounting curriculum.)

The second measure—the comprehensive income *accumulated* over the current and prior periods—is reported as a separate component of shareholders' equity following retained earnings, similar to the presentation by Exposition Corporation in Illustration 18–1. Note that amounts reported here—accumulated other comprehensive income (AOCI)—represent the *cumulative* sum of the changes in each component of other comprehensive income created during each reporting period (Illustration 18–2) throughout all prior years.

Reporting Shareholders' Equity

You seldom, if ever, will see the numerous list of items shown in Illustration 18–1 reported in the presentation of paid-in capital. Instead, companies keep track of individual additional paid-in capital accounts in company records but ordinarily report these amounts as a single subtotal: additional paid-in capital. Pertinent rights and privileges of various securities outstanding, such as dividend and liquidation preferences, call and conversion information, and voting rights, are summarized in disclosure notes.[3] The shareholders' equity portion of the balance sheet of **Abercrombie & Fitch**, shown in Illustration 18–3, is a typical presentation format.

The balance sheet reports annual balances of shareholders' equity accounts. However, companies also should disclose the sources of the changes in those accounts.[4] This is the purpose of the **statement of shareholders' equity**. To illustrate, Illustration 18–4 shows how **Walmart** reported the changes in its shareholders' equity balances.

A statement of shareholders' equity reports the transactions that cause changes in its shareholders' equity account balances.

The current year changes that Walmart statements of shareholders' equity reveal are net income, other comprehensive income, the purchase of common stock, and dividends declared.

The Corporate Organization

● LO18–3

A company may be organized in any of three ways: (1) a sole proprietorship, (2) a partnership, or (3) a corporation. In your introductory accounting course, you studied each form. In this course, we focus exclusively on the corporate form of organization.

Corporations are the dominant form of business organization.

Most well-known companies, such as **Microsoft, Amazon,** and **General Electric,** are corporations. Also, many smaller companies—even one-owner businesses—are corporations.

[3]FASB ACS 505–10–50: Equity—Overall—Disclosure.
[4]FASB ASC 505–10–50–2: Equity—Overall—Disclosure.

Illustration 18–3

Typical Presentation Format—Abercrombie & Fitch

Real World Financials

Details of each class of stock are reported in the balance sheet or in disclosure notes.

ABERCROMBIE & FITCH CO.
Consolidated Balance Sheets
($ in thousands, except par value amounts)

	January 28, 2017	January 30, 2016
Stockholders' Equity:		
Class A common stock—$0.01 par value: 150,000 shares authorized and 103,300 shares issued at each of January 28, 2017, and January 30, 2016	1,033	1,033
Paid-in capital	396,590	407,029
Retained earnings	2,474,703	2,530,196
Accumulated other comprehensive (loss), net of tax	(121,302)	(114,619)
Treasury stock, at average cost—35,542 and 35,952 shares at January 28, 2017, and January 30, 2016, respectively	(1,507,589)	(1,532,576)
Total Stockholders' Equity	$1,243,435	$1,291,063

Source: Abercrombie & Fitch

Although fewer in number than sole proprietorships and partnerships, in terms of business volume, corporations are the predominant form of business organization.

Accounting for most transactions is the same regardless of the form of business organization.

In most respects, transactions are accounted for in the same way regardless of the form of business organization. Assets and liabilities are unaffected by the way a company is organized. The exception is the method of accounting for capital, the ownership interest in the company. Rather than recording all changes in ownership interests in a single capital account for each owner, as we do for sole proprietorships and partnerships, we use the several capital accounts described in the previous section to record those changes for a corporation. Before discussing how we account for specific ownership changes, let's look at the characteristics of a corporation that make this form of organization distinctive and require special accounting treatment.

International Financial Reporting Standards

● LO18–9

Use of the term "reserves" and other terminology differences. Shareholders' equity is classified under IFRS into two categories: share capital and "reserves." The term reserves is considered misleading and thus is discouraged under U.S. GAAP. Here are some other differences in equity terminology:

U.S. GAAP	IFRS
Capital stock:	Share capital:
Common stock	Ordinary shares
Preferred stock	Preference shares
Paid-in capital—excess of par, common	Share premium, ordinary shares
Paid-in capital—excess of par, preferred	Share premium, preference shares
Accumulated other comprehensive income:	Reserves:
Net gains (losses) on investments—AOCI	Investment revaluation reserve
Net gains (losses) foreign currency translation—AOCI	Translation reserve
(N/A: adjusting P, P, & E to fair value not permitted)	Revaluation reserve
Retained earnings	Retained earnings
Total shareholders' equity	**Total equity**
Presented after Liabilities	Often presented before Liabilities

Illustration 18–4 Statement of Shareholders' Equity—Walmart

Real World Financials

($ in millions)	Common Stock Shares	Common Stock Amount	Capital in Excess of Par Value	Retained Earnings	Accumulated Other Comprehensive Income (Loss)	Total Walmart Shareholders' Equity
Balances, January 31, 2014	3,233	$323	$2,362	$76,566	$(2,996)	$76,255
Consolidated net income	—	—	—	16,363	—	16,363
Other comprehensive loss, net of income taxes	—	—	—	—	(4,172)	(4,172)
Cash dividends declared ($1.92 per share)	—	—	—	(6,185)	—	(6,185)
Purchase of Company stock	(13)	(1)	(29)	(950)	—	(930)
Other	8	1	129	(17)	—	113
Balances, January 31, 2015	3,228	323	2,462	85,777	(7,168)	81,394
Consolidated net income	—	—	—	14,694	—	14,694
Other comprehensive income net of income taxes	—	—	—	—	(4,429)	(4,429)
Cash dividends declared ($1.96 per share)	—	—	—	(6,294)	—	(6,294)
Purchase of Company stock	(65)	(6)	(102)	(4,148)	—	(4,256)
Cash dividend declared to noncontrolling interest	—	—	—	—	—	—
Other	(1)	—	(555)	(8)	—	(563)
Balances, January 31, 2016	3,162	317	1,805	90,021	(11,597)	80,546
Consolidated net income	—	—	—	13,643	—	13,643
Other comprehensive income, net of income taxes	—	—	—	—	(2,635)	(2,653)
Cash dividends declared ($2.00 per share)	—	—	—	(6,216)	—	(6,216)
Purchase of Company stock	(120)	(12)	(174)	(8,090)	—	(8,276)
Other	6	—	740	(4)	—	736
Balances, January 31, 2017	3,048	$ 305	$ 2,371	$ 89,354	$ (14,232)	$ 77,798

Source: Walmart

Limited Liability

The owners are not personally liable for debts of a corporation. Unlike a proprietorship or a partnership, a corporation is a separate legal entity, responsible for its own debts. Shareholders' liability is limited to the amounts they invest in the company when they purchase shares (unless the shareholder also is an officer of the corporation). The limited liability of shareholders is perhaps the single most important advantage of corporate organization. In other forms of business, creditors may look to the personal assets of owners for satisfaction of business debt.

A corporation is a separate legal entity—separate and distinct from its owners.

Ease of Raising Capital

A corporation is better suited to raising capital than is a proprietorship or a partnership. All companies can raise funds by operating at a profit or by borrowing. However, attracting equity capital is easier for a corporation. Because corporations sell ownership interest in the form of shares of stock, ownership rights are easily transferred. An investor can sell his/her ownership interest at any time and without affecting the corporation or its operations.

Ownership interest in a corporation is easily transferred.

From the viewpoint of a potential investor, another favorable aspect of investing in a corporation is the lack of mutual agency. Individual partners in a partnership have the power to bind the business to a contract. Therefore, an investor in a partnership must be careful regarding the character and business savvy of fellow co-owners. On the other hand, shareholders' participation in the affairs of a corporation is limited to voting at shareholders'

Shareholders do not have a mutual agency relationship.

meetings (unless the shareholder also is a member of management). Consequently, a shareholder needn't exercise the same degree of care that partners must in selecting co-owners.

Obviously, then, a corporation offers advantages over the other forms of organization, particularly in its ability to raise investment capital. As you might guess, though, these benefits do not come without a price.

Disadvantages

Corporations are subject to extensive government regulation.

Paperwork! To protect the rights of those who buy a corporation's stock or who loan money to a corporation, the state in which the company is incorporated and the federal government impose extensive reporting requirements. Primarily, the required paperwork is intended to ensure adequate disclosure of information needed by investors and creditors.

Corporations create double taxation.

You read earlier that corporations are separate legal entities. As such, they also are separate taxable entities. Often this causes what is referred to as *double taxation.* Corporations first pay income taxes on their earnings. Then, when those earnings are distributed as cash dividends, shareholders pay personal income taxes on the previously taxed earnings. Proprietorships and partnerships are not taxed at the business level; each owner's share of profits is taxed only as personal income.

Types of Corporations

When referring to corporations in this text, we are referring to corporations formed by private individuals for the purpose of generating profits. These corporations raise capital by selling stock. There are, however, other types of corporations.

Not-for-profit corporations may be owned:
1. By the public sector.
2. By a governmental unit.

Some corporations, such as churches, hospitals, universities, and charities do not sell stock and are not organized for profit. Also, some not-for-profit corporations are government-owned—the **Federal Deposit Insurance Corporation (FDIC)**, for instance. Accounting for not-for-profit corporations is discussed elsewhere in the accounting curriculum.

Corporations organized for profit may be:
1. Publicly held and traded:
   a. On an exchange.
   b. Over-the-counter.
2. Privately held.

Corporations organized for profit may be publicly held or privately (or closely) held. The stock of publicly held corporations is available for purchase by the general public. You can buy shares of **Home Depot**, **Ford Motor Company**, or **Walmart** through a stockbroker. These shares are traded on the New York Stock Exchange. Other publicly held stock, like **Apple** and **Microsoft**, are available through Nasdaq (National Association of Securities Dealers Automated Quotations).

On the other hand, shares of privately held companies are owned by only a few individuals (perhaps a family) and are not available to the general public. Corporations whose stock is privately held do not need to register those shares with the Securities and Exchange Commission and are spared the voluminous, annual reporting requirements of the SEC. Of course, new sources of equity financing are limited when shares are privately held, as is the market for selling existing shares.

Privately held companies' shares are held by only a few individuals and are not available to the general public.

Frequently, companies begin as smaller, privately held corporations. Then as success broadens opportunities for expansion, the corporation goes public. For example, in 2012, the shareholders of **Facebook** decided to take public the privately held company. The result was the largest initial public offering since **Visa** in 2008. In 2014, **Alibaba's** IPO surpassed that of both companies to become the largest in history.

Hybrid Organizations

A corporation can elect to comply with a special set of tax rules and be designated an **S corporation**. S corporations have characteristics of both regular corporations and partnerships. Owners have the limited liability protection of a corporation, but income and expenses are passed through to the owners as in a partnership, avoiding double taxation.

Two particular business structures have evolved in response to liability issues and tax treatment—limited liability companies and limited liability partnerships.

A **limited liability company** offers several advantages. Owners are not liable for the debts of the business, except to the extent of their investment. Unlike a limited partnership, all members of a limited liability company can be involved with managing the business without losing liability protection. Like an S corporation, income and expenses are passed through to the owners as in a partnership, avoiding double taxation, but there are no limitations on the number of owners as in an S corporation.

A **limited liability partnership** is similar to a limited liability company, except it doesn't offer all the liability protection available in the limited liability company structure. Partners are liable for their own actions but not entirely liable for the actions of other partners. Professional firms, such as for law and CPAs, often are formed as LLPs.

The Model Business Corporation Act

Corporations are formed in accordance with the corporation laws of individual states. State laws are not uniform, but share many similarities, thanks to the widespread adoption of the **Model Business Corporation Act**.[5] This act is designed to serve as a guide to states in the development of their corporation statutes. It presently serves as the model for the majority of states.

The *Model Business Corporation Act* serves as the model for the corporation statutes of most states.

State laws regarding the nature of shares that can be authorized, the issuance and repurchase of those shares, and conditions for distributions to shareholders obviously influence actions of corporations. Naturally, differences among state laws affect how we account for many of the shareholders' equity transactions discussed in this chapter. For that reason, we will focus on the normal case, as described by the Model Business Corporation Act, and note situations where variations in state law might require different accounting. Your goal is not to learn diverse procedures caused by peculiarities of state laws, but to understand the broad concepts of accounting for shareholders' equity that can be applied to any specific circumstance.

Variations among state laws influence GAAP pertaining to shareholders' equity transactions.

The process of incorporating a business is similar in all states. The **articles of incorporation** (sometimes called the *corporate charter*) describe (a) the nature of the firm's business activities, (b) the shares to be issued, and (c) the composition of the initial **board of directors**. The board of directors establishes corporate policies and appoints officers who manage the corporation.

The number of shares authorized is the maximum number of shares that a corporation is legally permitted to issue, as specified in its articles of incorporation. The number of authorized shares is determined at the company's creation and can only be increased by a vote of the shareholders. At least some of the shares authorized by the articles of incorporation are sold (issued) at the inception of the corporation. Frequently, the initial shareholders include members of the board of directors or officers (who may be one and the same). Ultimately, it is the corporation's shareholders that control the company. Shareholders are the owners of the corporation. By voting their shares, they determine the makeup of the board of directors—who in turn appoint officers, who in turn manage the company.

Shareholders' investment in a corporation ordinarily is referred to as paid-in capital. In the next section, we examine the methods normally used to maintain records of shareholders' investment and to report such paid-in capital in financial statements.

Paid-in Capital

PART B

Fundamental Share Rights

In reading the previous paragraphs, you noted that corporations raise equity funds by selling shares of the corporation. Shareholders are the owners of a corporation. If a corporation has only one class of shares, no designation of the shares is necessary, but they typically are labeled *common* shares, or shares of *common* stock. Ownership rights held by common shareholders, unless specifically withheld by agreement with the shareholders, are as follows:

1. The right to vote on matters that come before the shareholders, including the election of corporate directors. Each share represents one vote.

[5] *Model Business Corporation Act*, the American Bar Association, 2016 Revision.

2. The right to share in profits when dividends are declared. The percentage of shares owned by a shareholder determines his/her share of dividends distributed.
3. The right to share in the distribution of assets if the company is liquidated. The percentage of shares owned by a shareholder determines his/her share of assets after creditors and preferred shareholders are paid.

Another right sometimes given to common shareholders is the right to maintain one's percentage share of ownership when new shares are issued. This is referred to as a *preemptive right.* Each shareholder is offered the opportunity to buy a percentage of any new shares issued equal to the percentage of shares he/she owns at the time. In most states, this right must be specifically granted; in others, it is presumed unless contractually excluded.

However, this right usually is not explicitly stated because of the inconvenience it causes corporations when they issue new shares. The exclusion of the preemptive right ordinarily is inconsequential because few shareholders own enough stock to be concerned about their ownership percentage.

Distinguishing Classes of Shares

It is not uncommon for a firm to have more than one, and perhaps several, classes of shares, each with different rights and limitations. To attract investors, companies have devised quite a variety of ownership securities.

If more than one class of shares is authorized by the articles of incorporation, the specific rights of each (for instance, the right to vote, residual interest in assets, and dividend rights) must be stated. Also, some designation must be given to distinguish each class.

Some of the distinguishing designations often used are these:

Terminology varies in the way companies differentiate among share types.

1. Class A, class B, and so on (**Tyson Foods**)
2. Preferred stock, common stock, and class B stock (**Hershey's**)
3. Common and preferred (**HP**)
4. Common stock and capital stock (**Alphabet, Inc.**)
5. Common and serial preferred (**Smucker's**)

It often is difficult to predict the rights and privileges of shares on the basis of whether they are labeled *common* or *preferred.*

In your introductory study of accounting, you probably became most familiar with the common stock–preferred stock distinction. That terminology has deep roots in tradition. Early English corporate charters provided for shares that were preferred over others as to dividends and liquidation rights. These provisions were reflected in early American corporation laws. But as our economy developed, corporations increasingly felt the need for innovative ways of attracting investment capital. The result has been a gradual development of a wide range of share classifications that cannot easily be identified by these historical designations.

To reflect the flexibility that now exists in the creation of equity shares, the Model Business Corporation Act, and thus many state statutes, no longer mention the words common and preferred. But the influence of tradition lingers. Most corporations still designate shares as common or preferred. For consistency with practice, the illustrations you study in this chapter use those designations. As you consider the examples, keep in mind that the same concepts apply regardless of the language used to distinguish shares.

Typical Rights of Preferred Shares

An issue of shares with certain preferences or features that distinguish it from the class of shares customarily called common shares may be assigned any of the several labels mentioned earlier. Very often the distinguishing designation is **preferred stock**. The special rights of preferred shareholders usually include one or both of the following:

1. Preferred shareholders typically have a preference to a specified amount of dividends (stated dollar amount per share or % of par value per share). That is, if the board of directors declares dividends, preferred shareholders will receive the designated dividend before any dividends are paid to common shareholders.

2. Preferred shareholders customarily have a preference (over common shareholders) as to the distribution of assets in the event the corporation is dissolved.

Preferred shareholders sometimes have the **right of conversion**, which allows them to exchange shares of preferred stock for common stock at a specified conversion ratio. For instance, in 2018, **Wells Fargo** had outstanding 12.4 million shares of convertible preferred stock. Alternatively, a **redemption privilege** might allow preferred shareholders the option, under specified conditions, to return their shares for a predetermined redemption price. Preferred shareholders have preference over common stockholders in dividends and liquidation rights. Each Wells Fargo preferred share is convertible into 32 common shares. Similarly, shares may be redeemable at the option of the issuing corporation (sometimes referred to as *callable*).

Shares might be:
1. *Convertible* into a specified number of another class of shares.
2. *Redeemable* at the option of:
   a. Shareholders.
   b. The corporation.

Preferred shares may be **cumulative** or **noncumulative**. Typically, preferred shares are cumulative, which means that if the specified dividend is not paid for a given year, the unpaid dividends (called *dividends in arrears*) accumulate and must be made up in a later dividend year before any dividends are paid on common shares. We see this illustrated in Part B of the chapter.

Preferred shares may be **participating** or **nonparticipating**. A participating feature allows preferred shareholders to receive additional dividends beyond the stated amount. If the preferred shares are fully participating, the distribution of dividends to common and preferred shareholders is a pro rata allocation based on the relative par amounts of common and preferred stock outstanding. Participating preferred stock, previously quite common, is rare today.

Remember that the designations of common and preferred imply no necessary rights, privileges, or limitations of the shares so designated. Such relative rights must be specified by the contract with shareholders. A corporation can create classes of preferred shares that are indistinguishable from common shares in voting rights and/or the right to participate in assets (distributed as dividends or distributed upon liquidation). Likewise, it is possible to devise classes of common shares that possess preferential rights, superior to those of preferred shares.

Is It Equity or Is It Debt?

You probably also can imagine an issue of preferred shares that is almost indistinguishable from a bond issue. Let's say, for instance, that preferred shares call for annual cash dividends of 10% of the par value, dividends are cumulative, and the shares must be redeemed for cash in 10 years. Although the declaration of dividends rests in the discretion of the board of directors, the contract with preferred shareholders can be worded in such a way that directors are compelled to declare dividends each year the company is profitable. For a profitable company, it would be difficult to draw the line between this issue of preferred shares and a 10%, 10-year bond issue. Even in a more typical situation, preferred shares are somewhat hybrid securities—a cross between equity and debt.

The line between debt and equity is hard to draw.

Sometimes the similarity to debt is even more obvious. Suppose shares are mandatorily redeemable—the company is obligated to buy back the shares at a specified future date. The fact that the company is obligated to pay cash (or other assets) at a fixed or determinable date in the future makes this financial instrument tantamount to debt. A mandatorily redeemable financial instrument must be reported in the balance sheet as a liability, not as shareholders' equity.[5] **Extended Stay America**, for instance, reported its mandatorily redeemable preferred shares as a liability in its 2018 balance sheet.

Mandatorily redeemable shares are classified as liabilities.

The Concept of Par Value

Another prevalent practice (besides labeling shares as common and preferred) that has little significance other than historical is assigning a par value to shares. The concept of par value dates back as far as the concept of owning shares of a business. Par value originally indicated the real value of shares. All shares were issued at that price.

We have inherited the archaic concept of par value from early corporate law.

During the late 19th and early 20th centuries, many cases of selling shares for less than par value—known as *watered shares*—received a great deal of attention and were the subject of

[5]FASB ASC 480–10–25–4: Distinguishing Liabilities from Equity—Overall—Recognition.

International Financial Reporting Standards

● LO18–9

Distinction between Debt and Equity for Preferred Stock. Differences in the definitions and requirements can result in the same instrument being classified differently between debt and equity under IFRS and U.S. GAAP. Under U.S. GAAP, preferred stock normally is reported as equity, but is reported as debt with the dividends reported in the income statement as interest expense if it is "mandatorily redeemable" preferred stock. Under IFRS, most non-mandatorily redeemable preferred stock (preference shares) also is reported as debt as well as some preference shares that aren't redeemable. Under IFRS (*IAS No. 32*), the critical feature that distinguishes a liability is if the issuer is or can be required to deliver cash (or another financial instrument) to the holder.[7] **Unilever** describes such a difference in a disclosure note:

> **Additional Information for U.S. Investors (in part)**
>
> **Preference Shares**
>
> Under IAS 32, Unilever recognizes preference shares that provide a fixed preference dividend as borrowings with preference dividends recognized in the income statement. Under U.S. GAAP such preference shares are classified in shareholders' equity with dividends treated as a deduction to shareholder's equity.

a number of lawsuits. Investors and creditors contended that they relied on the par value as the permanent investment in the corporation and therefore net assets must always be at least that amount. Not only was par value assumed to be the amount invested by shareholders, but it also was defined by early corporation laws as the amount of net assets not available for distribution to shareholders (as dividends or otherwise).

Shares with nominal par value became common to dodge elaborate statutory rules pertaining to par value shares.

Many companies began turning to par value shares with very low par values—often pennies—to escape the watered shares liability of issuing shares below an arbitrary par value and to limit the restrictions on distributions. This practice is common today.

Accountants and attorneys have been aware for decades that laws pertaining to par value and legal capital not only are bewildering but fail in their intent to safeguard creditors from payments to shareholders. Actually, to the extent that creditors are led to believe that they are afforded protection, they are misled. Like the designations of common and preferred shares, the concepts of par value and legal capital have been eliminated entirely from the Model Business Corporation Act.[8]

Many states already have adopted these provisions of the Model Act. But most established corporations issued shares prior to changes in the state statutes. Consequently, most companies have par value shares outstanding and continue to issue previously authorized par value shares. The evolution will be gradual to the simpler, more meaningful provisions of the Model Act.

Most shares continue to bear arbitrarily designated par values.

In the meantime, accountants must be familiar with the outdated concepts of par value and legal capital in order to properly record and report transactions related to par value shares. For that reason, most of the discussion in this chapter centers on par value shares. Largely, this means only that proceeds from shareholders' investment is allocated between stated capital and additional paid-in capital. Be aware, though, that in the absence of archaic laws that prompted the creation of par value shares, there is no theoretical reason to do so.

Accounting for the Issuance of Shares

Par Value Shares Issued For Cash

● LO18–4

When shares are sold for cash (see Illustration 18–5), the capital stock account (usually common or preferred) is credited for the amount representing stated capital. When shares have a designated par value, that amount denotes stated capital and is credited to the stock

[7]"Financial Instruments: Presentation," *International Accounting Standard No. 32* (IASCF), as amended effective January 1, 2016.

[8]American Bar Association, official comment to Section 6.21 of the *Model Business Corporation Act,* 2016 Revision.

account. Proceeds in excess of this amount are credited to paid-in capital—excess of par (also called additional paid-in capital).

Illustration 18–5

Shares Sold for Cash

When par value shares are issued, only the par amount is credited to the stock account; the remainder to paid-in capital in excess of par.

Dow Industrial sells 10 million of its common shares, $1 par per share, for $10 per share:

	($ in millions)	
Cash (10 million shares at $10 per share)	100	
Common stock (10 million shares at $1 par per share)		10
Paid-in capital—excess of par (remainder: 10 million shares at $10 − $1 = $9 per share)		90

No-Par Shares Issued for Cash

In states that allow no-par stock, when no-par shares are sold for cash (see Illustration 18–5A), the entire amount received is credited to the stock account.

Illustration 18–5A

No-Par Shares Sold for Cash

The total amount received from the sale of no-par shares is credited to the stock account.

Dow Industrial sells 10 million of its no-par common shares for $10 per share:

	($ in millions)	
Cash (10 million shares at $10 per share)	100	
Common stock		100

Shares Issued for Noncash Consideration

Occasionally, a company might issue its shares for consideration other than cash. It is not uncommon for a new company, yet to establish a reliable cash flow, to pay for promotional and legal services with shares rather than with cash. Similarly, shares might be given in payment for land, or for equipment, or for some other noncash asset.

Shares should be issued at fair value.

Even without a receipt of cash to establish the fair value of the shares at the time of the exchange, the transaction still should be recorded at the grant-date fair value of the shares to be issued. This treatment is consistent with the accounting requirement for employee share-based payment awards we discuss in the next chapter and with the general rule for accounting for noncash transactions.[9]

Illustration 18–6 demonstrates a situation where the quoted market price is the best evidence of fair value.

Illustration 18–6

Shares Sold for Noncash Consideration

The quoted market price for the shares issued might be the best evidence of fair value.

DuMont Chemicals issues 1 million of its common shares, $1 par per share, in exchange for a custom-built factory for which no cash price is available. Today's issue of *The Wall Street Journal* lists DuMont's stock at **$10 per share:**

	($ in millions)	
Property, plant, and equipment (1 million shares at **$10 per share**)	10	
Common stock (1 million shares at $1 par per share)		1
Paid-in capital—excess of par (remainder)		9

More Than One Security Issued for a Single Price

Although uncommon, a company might sell more than one security—perhaps common shares and preferred shares—for a single price. As you might expect, the cash received usually is the sum of the separate market values of the two securities. Of course, each is then recorded at its market value. However, if only one security's value is known, the second security's market value is inferred from the total selling price, as demonstrated in Illustration 18–7.

[9]FASB Accounting Standards Update (ASU) No. 2018-07, Compensation—Stock Compensation (Topic 718): Improvements to Nonemployee Share-Based Payment Accounting.

Illustration 18–7
More than One Security Sold for a Single Price

When only one security's value is known ($40 million), the second security's market value ($60 million) is assumed from the total selling price ($100 million).

AP&P issues 4 million of its common shares, $1 par per share, and 2 million of its preferred shares, $5 par, for $100 million. Today's issue of *The Wall Street Journal* lists AP&P's common at $10 per share. There is no established market for the preferred shares.

	($ in millions)	
Cash	100	
Common stock (4 million shares × $1 par)		4
Paid-in capital—excess of par, common		36
Preferred stock (2 million shares × $5 par)		10
Paid-in capital—excess of par, preferred		50

Because the shares sell for a total of $100 million, and the market value of the common shares is known to be **$40** million (4 million × $10), the preferred shares are inferred to have a market value of $60 million.

Additional Consideration

In the unlikely event that the total selling price is not equal to the sum of the two market prices (when both market values are known), the total selling price is allocated between the two securities, in proportion to their relative market values. You should note that this is the same approach we use (a) when more than one asset is purchased for a single purchase price to allocate the single price to the various assets acquired, (b) when detachable warrants and bonds are issued for a single price, and (c) in any other situation when more than one item is associated with a single purchase price or selling price.

Share Issue Costs

Share issue costs reduce the net cash proceeds from selling the shares and thus paid-in capital—excess of par.

When a company sells shares, it obtains the legal, promotional, and accounting services necessary to effect the sale. The cost of these services reduces the net proceeds from selling the shares. Since paid-in capital—excess of par is credited for the excess of the proceeds over the par value of the shares sold, the effect of share issue costs is to reduce the amount credited to that account. For example, in 2018, **AveXis, Inc.**, sold 4,509,840 shares of its $0.0001 par common stock at $102 per share. the company received net proceeds from the public offering of $431,857,000, after deducting underwriting discounts and commissions and other offering expenses. AveXis's entry to record the sale was as follows:

The cash proceeds is the net amount received after paying share issue costs.

Cash	431,857,000	
Common stock (4,509,840 shares at $0.0001 par per share) **The cash proceeds is the net amount received after paying share issue costs.**		451
Paid-in capital—excess of par (remainder)		431,856,549

Concept Review Exercise

EXPANSION OF CORPORATE CAPITAL

Situation: The shareholders' equity section of the balance sheet of National Foods, Inc., included the following accounts at December 31, 2019:

Shareholders' Equity	($ in millions)
Paid-in capital:	
Common stock, 120 million shares at $1 par	$ 120
Paid-in capital—excess of par	836
Retained earnings	2,449
Total shareholders' equity	$3,405

Required:

1. During 2020, several transactions affected the stock of National Foods. Prepare the appropriate entries for these events.
   a. On March 11, National Foods issued 10 million of its 9.2% preferred shares, $1 par per share, for $44 per share.
   b. On November 22, 1 million common shares, $1 par per share, were issued in exchange for eight labeling machines. Each machine was built to custom specifications so no cash price was available. National Food's stock was listed at $10 per share.
   c. On November 23, 1 million of the common shares and 1 million preferred shares were sold for $60 million. The preferred shares had not traded since March and their market value was uncertain.
2. Prepare the shareholders' equity section of the comparative balance sheets for National Foods at December 31, 2020 and 2019. Assume that net income for 2020 was $400 million and the only other transaction affecting shareholders' equity was the payment of the 9.2% dividend on the 11 million preferred shares ($1 million).

Solution:

1. During 2020 several transactions affected the stock of National Foods. Prepare the appropriate entries for these events.
   a. On March 11, National Foods issued 10 million of its preferred shares, $1 par per share, for $44 per share.

	($ in millions)	
Cash	440	
Preferred stock (10 million shares × $1 par per share)		10
Paid-in capital—excess of par, preferred		430

   b. On November 22, 1 million common shares, $1 par per share, were issued in exchange for eight labeling machines:

	($ in millions)	
Machinery (fair value of shares)	10	
Common stock (1 million shares × $1 par per share)		1
Paid-in capital—excess of par, common (1 million shares × $9)		9

The transaction was recorded at the fair value of the shares exchanged for the machinery.

   c. On November 23, 1 million of the common shares and 1 million preferred shares were sold for $60 million:

	($ in millions)	
Cash	60	
Common stock (1 million shares × $1 par per share)		1
Paid-in capital—excess of par, common		9
Preferred stock (1 million shares × $1 par per share)		1
Paid-in capital—excess of par, preferred (to balance)		49

Since the value of only the common stock was known, the preferred stock's market value ($50 per share) was inferred from the total selling price.

2. Prepare the shareholders' equity section of the comparative balance sheets for National Foods at December 31, 2020 and 2019.

NATIONAL FOODS, INC.
Balance Sheet
(Shareholders' Equity Section)

($ in millions)	2020	2019
Shareholders' Equity		
Preferred stock, 9.2%, $1 par (2020: $10 million + $1 million)	$ 11	$ —
Common stock, $1 par (2020: $120 million + $1 million + $1 million)	122	120
Paid-in capital—excess of par, preferred (2020: $430 million + $49 million)	479	—
Paid-in capital—excess of par, common (2020: $836 million + $9 million + $9 million)	854	836
Retained earnings (2019: $2,449 million + $400 million − $1 million)	2,848	2,449
Total shareholders' equity	$ 4,314	$ 3,405

Note: This situation is continued in the next Concept Review Exercise.

Share repurchases

● LO18–5

In the previous section we examined various ways stock might be issued. In this section, we look at situations in which companies reacquire shares previously sold. Most medium- and large-size companies buy back their own shares. Many have formal share repurchase plans to buy back stock over a series of years.

Decision Makers' Perspective

When a company's management feels the market price of its stock is undervalued, it may attempt to support the price by decreasing the supply of stock in the marketplace. A **Johnson & Johnson** announcement that it planned to buy back up to $5 billion of its outstanding shares triggered a buying spree that pushed the stock price up by more than 3%.

Decreasing the supply of shares in the marketplace supports the price of remaining shares.

When announcing plans to expand its stock buyback program to $90 billion, **Apple** chief executive Tim Cook said it "views its shares as undervalued."[10] Although clearly a company may attempt to increase net assets by buying its shares at a low price and selling them back later at a higher price, that investment is not viewed as an asset. Similarly, increases and decreases in net assets from that activity are not reported as gains and losses in the company's income statement. Instead, buying and selling its shares are transactions between the corporation and its owners, analogous to retiring shares and then selling previously unissued shares. You should note the contrast between a company's purchasing of its own shares and its purchasing of shares in another corporation as an investment.

FINANCIAL Reporting Case

Q1 , p. 1047

Unlike an investment in another firm's shares, the acquisition of a company's own shares does not create an asset.

Though not considered an investment, the repurchase of shares often is a judicious use of a company's cash. By increasing per share earnings and supporting share price, shareholders benefit. To the extent this strategy is effective, a share buyback can be viewed as a way to "distribute" company profits without paying dividends. Capital gains from any stock price increase are taxed at lower capital gains tax rates than ordinary income tax rates on dividends.

Buybacks zoomed to record highs when tax cuts freed up massive amounts of cash in 2017-18.

The Tax Cuts and Jobs Act of 2017 ignited an unprecedented volume of share buybacks, exceeding a trillion dollars in 2017 and 2018. The corporate tax rate was slashed from 35% to 21% and other provisions made it advantageous for companies to bring home from overseas huge sums of accumulated earnings from foreign operations. Suddenly finding themselves in the enviable position of having mountains of excess cash, share buybacks became the obvious choice for how to distribute that surplus to shareholders. **Apple**, alone, bought back over $23 billion of its own stock in March of 2018, a record amount for any U.S. company.

Companies buy back shares to offset the increase in shares issued to employees.

Perhaps the primary motivation for most stock repurchases is to offset the increase in shares that routinely are issued to employees under stock award and stock option

[10]Daisuke Wakabayashi, "Apple Boosts Buyback, Splits Stock to Reward Investors," Dow Jones Business Report, **www.nasdaq.com**, April 23, 2014.

compensation programs. **Microsoft** reported its stock buyback program designed to offset the effect of its stock option and stock purchase plans, as shown in Illustration 18–8.

> **Note 11: Stockholders' Equity (in part)**
> Our board of directors has approved a program to repurchase shares of our common stock to reduce the dilutive effect of our stock option and stock purchase plans.
> Source: Microsoft

Illustration 18–8
Disclosure of Share Repurchase Program—Microsoft
Real World Financials

Similarly, shares might be reacquired to distribute in a stock dividend, a proposed merger, or as a defense against a hostile takeover.[11]

Whatever the reason shares are repurchased, a company has a choice of how to account for the buyback.

1. The shares can be formally retired.
2. The shares can be called treasury stock.

Unfortunately, the choice is not dictated by the nature of the buyback, but by practical motivations of the company. ●

Shares Formally Retired or Viewed as Treasury Stock

FINANCIAL Reporting Case

Q2, p. 1047

When a corporation retires its own shares, those shares assume the same status as authorized but unissued shares, just the same as if they never had been issued. We saw earlier in the chapter that when shares are sold, both cash (usually) and shareholders' equity are increased, the company becomes larger. Conversely, when cash is paid to **retire stock**, the effect is to decrease both cash and shareholders' equity; the size of the company literally is reduced.

Reacquired shares are equivalent to authorized but unissued shares.

Out of tradition and for practical reasons, companies usually reacquire shares of previously issued stock without formally retiring them.[12] Shares repurchased and not retired are referred to as **treasury stock**. Because reacquired shares are essentially the same as shares that never were issued at all, treasury shares have no voting rights nor do they receive cash dividends. As demonstrated in Illustration 18–9, when shares are repurchased as treasury stock, we reduce shareholders' equity with a debit to a negative (or contra) shareholders' equity account labeled treasury stock. That entry is reversed later through a credit to treasury stock when the treasury stock is resold. Like the concepts of par value and legal capital, the concept of treasury shares no longer is recognized in most state statutes.[13] Some companies, in fact, are eliminating treasury shares from their financial statements as corporate statutes are modernized. **Microsoft** retires the shares it buys back rather than labeling them treasury stock. Still, you will see treasury shares reported in the balance sheets of many companies.

Accounting for Retired Shares

When shares are formally retired, we should reduce precisely the same accounts that previously were increased when the shares were sold, namely, common (or preferred) stock and paid-in capital—excess of par. The first column of Illustration 18–9 demonstrates this. The paid-in capital—excess of par account for American Semiconductor shows a balance of $900 million while the common stock account shows a balance of $100 million. Thus the 100 million outstanding shares were originally sold for an average of $9 per share above

[11] A corporate takeover occurs when an individual or group of individuals acquires a majority of a company's outstanding common stock from present shareholders. Corporations that are the object of a hostile takeover attempt—a public bid for control of a company's stock against the company's wishes—often take evasive action involving the reacquisition of shares.

[12] The concept of treasury shares originated long ago when new companies found they could sell shares at an unrealistically low price equal to par value to incorporators, who then donated those shares back to the company. Since these shares already had been issued (though not outstanding), they could be sold at whatever the real market price was without adjusting stated capital. Because treasury shares are already issued, different rules apply to their purchase and resale than to unissued shares. Companies can (1) issue shares without regard to preemptive rights of shareholders, or (2) distribute shares as a dividend to shareholders even without a balance in retained earnings.

[13] *The Revised Model Business Corporation Act* eliminated the concept of treasury shares in 1984 after 1980 revisions had eliminated the concepts of par value and legal capital. Most state laws have since followed suit.

Illustration 18–9

Comparison of Share Retirement and Treasury Stock Accounting—Share Buybacks

American Semiconductor's balance sheet included the following:

Shareholders' Equity	($ in millions)
Common stock, 100 million shares at $1 par	$ 100
Paid-in capital—excess of par	900
Paid-in capital—share repurchase	2
Retained earnings	2,000

Retirement			**Treasury Stock**	
Reacquired 1 million of its common shares				
Case 1: Shares repurchased at $7 per share				
Common stock ($1 par × 1 million shares)	1		Treasury stock (cost)	7
Paid-in capital—excess of par ($9 per share)	9			
Paid-in capital—share repurchase (difference)		3		
Cash		**7**	Cash	**7**
OR				
Case 2: Shares repurchased at $13 per share				
Common stock ($1 par × 1 million shares)	1		Treasury stock (cost)	13
Paid-in capital—excess of par ($9 per share)	9			
Paid-in capital—share repurchase	2*			
Retained earnings (difference)	1			
Cash		**13**	Cash	**13**

*Because there is a $2 million credit balance.

Formally retiring shares restores the balances in both the Common stock account and Paid-in capital—excess of par to what those balances would have been if the shares never had been issued.

When we view a buyback as treasury stock, the cost of acquiring the shares is debited to the treasury stock account.

par, or $10 per share. Consequently, when 1 million shares are retired (regardless of the retirement price), American Semiconductor should reduce its common stock account by $1 per share and its paid-in capital—excess of par by $9 per share. Another way to view the reduction is that because 1% of the shares are retired, both share account balances (common stock and paid-in capital—excess of par) are reduced by 1%.

How we treat the difference between the cash paid to buy the shares and the amount the shares originally sold for (amounts debited to common stock and paid-in capital—excess of par) depends on whether the cash paid is *less* than the original issue price (credit difference) or the cash paid is *more* than the original issue price (debit difference):

1. If a *credit* difference is created (as in Case 1 of Illustration 18–9), we credit paid-in capital—share repurchase.
2. If a *debit* difference is created (as in Case 2 of Illustration 18–9), we debit paid-in capital—share repurchase, but only if that account already has a credit balance. Otherwise, we debit retained earnings. (Reducing paid-in capital beyond its previous balance would create a negative balance, which can never happen.)

***Paid-in capital—share repurchase* is debited to the extent of its credit balance before debiting retained earnings.**

Payments made by a corporation to retire its own shares are viewed as a distribution of corporate assets to shareholders.

Why is paid-in capital credited in Case 1 and retained earnings debited in Case 2? The answer lies in the fact that the payments made by a corporation to repurchase its own shares are a distribution of corporate assets to shareholders.

In Case 1, only $7 million is distributed to shareholders to retire shares that originally provided $10 million of paid-in capital. Thus, some of the original investment ($3 million in this case) remains and is labeled *paid-in capital—share repurchase.*

In Case 2, more cash ($13 million) is distributed to shareholders to retire shares than originally was paid in. The amount paid in comprises the original investment of $10 million for the shares being retired, plus $2 million of paid-in capital created by previous repurchase transactions—$12 million total. Thirteen million is returned to shareholders. The additional $1 million paid is viewed as a dividend on the shareholders' investment, and thus a reduction of retained earnings.[14]

[14]In the next section of this chapter, you will be reminded that dividends reduce retained earnings. (You first learned this in your introductory accounting course.)

Additional Consideration

Some companies choose to debit retained earnings for the entire debit difference between the cash paid to repurchase shares and the par amount of those shares rather than allocate that difference in the prescribed way. **Target Corporation** is an example of a company that follows this approach. While this method lacks conceptual merit by itself, it's permitted by ASC 505-30-30-8, which states that "a corporation can always capitalize or allocate retained earnings for such purposes."

Accounting for Treasury Stock

We view the purchase of treasury stock as a temporary reduction of shareholders' equity, to be reversed later when the treasury stock is resold. The cost of acquiring the shares is "temporarily" debited to the treasury stock account (second column of Illustration 18–9). At this point, the shares are considered to be *issued, but not outstanding*.

The purchase of treasury stock and its subsequent resale are considered to be a "single transaction." The purchase of treasury stock is perceived as a temporary reduction of shareholders' equity, to be reversed later when the treasury stock is resold. The company "temporarily" debits the treasury stock account when acquiring the shares. The common stock account is not affected. Later, when the shares are resold, the treasury stock account will be credited, and any difference from the cash proceeds upon resale will be allocated to specific shareholders' equity accounts. Effectively, we consider the purchase of treasury stock and its subsequent resale to be a "single transaction."

Additional Consideration

The approach to accounting for treasury stock we discuss in this chapter is referred to as the "cost method." Another permissible approach is the "par value method." It is essentially identical to formally retiring shares, which is why it sometimes is referred to as the *retirement method of accounting for treasury stock.* In fact, if we substitute Treasury stock for Common stock in each of the journal entries we used to account for retirement of shares in Illustrations 18–9 and 18–11, we have the par value method. Because the method has virtually disappeared from practice, we do not discuss it further in this chapter.

BALANCE SHEET EFFECT Formally retiring shares restores the balances in both the Common stock account and Paid-in capital—excess of par to what those balances would have been if the shares never had been issued at all. If the amount paid in this subsequent repurchase is *less* than the amount received from the initial sale (credit difference), that net increase in assets (cash) is reflected as Paid-in capital—share repurchase. On the other hand, if the amount paid in this subsequent repurchase is *more* than the amount received from the initial sale plus any paid-in capital created by previous repurchases (debit difference), a net reduction in assets (cash) occurs and that reduction is reflected as a decrease in retained earnings.

In contrast, when a share repurchase is viewed as treasury stock, the cost of the treasury stock is simply reported as a reduction in total shareholders' equity. Reporting under the two approaches is compared in Illustration 18–10, using the situation described above for American Semiconductor after the purchase of treasury stock in Illustration 18–9 (Case 2). Notice that either way total shareholders' equity is the same.

Resale of Shares

After shares are formally retired, any subsequent sale of shares is simply the sale of new, unissued shares and is accounted for accordingly. This is demonstrated in the first column of Illustration 18–11.

Illustration 18–10

Reporting Share Buyback in the Balance Sheet

Retirement reduces common stock and associated shareholders' equity accounts.

($ in millions)	Shares Retired	Treasury Stock
Shareholders' Equity		
Paid-in capital:		
Common stock, 100 million shares at $1 par	$ 99	$ 100
Paid-in capital—excess of par	891	900
Paid-in capital—share repurchase		2
Retained earnings	1,999	2,000
Less: Treasury stock, 1 million shares (at cost)		(13)
Total shareholders' equity	$ 2,989	$ 2,989

Illustration 18–11

Comparison of Share Retirement and Treasury Stock Accounting—Subsequent Sale of Shares

After formally retiring shares, we record a subsequent sale of shares exactly like any sale of shares.

The resale of treasury shares is viewed as the consummation of the "single transaction" begun when the treasury shares were purchased.

American Semiconductor sold 1 million shares after reacquiring shares at $13 per share (Case 2 in Illustration 18–9).

Retirement			Treasury Stock		
Sold 1 million shares					
Case A: Shares sold at $14 per share					
Cash	**14**		Cash	**14**	
Common stock (par)		1	Treasury stock (cost)		13
Paid-in capital—excess of par		13	Paid-in capital—share repurchase		1
Or					
Case B: Shares sold at $10 per share					
Cash	**10**		Cash	**10**	
Common stock (par)		1	Retained earnings (to balance)	1	
Paid-in capital—excess of par		9	Paid-in capital—share repurchase	2*	
			Treasury stock (cost)		13

*Because there is a $2 million credit balance.

Allocating the cost of treasury shares occurs when the shares are resold.

The resale of treasury shares is viewed as the consummation of the single transaction begun when the treasury shares were repurchased. The effect of the single transaction of purchasing treasury stock and reselling it for more than cost (Case 2 of Illustration 18–9 and Case A of Illustration 18–11) is to *increase* both cash and shareholders' equity (by $1 million). The effect of the single transaction of purchasing treasury stock and reselling it for less than cost (Case 2 of Illustration 18–9 and Case B of Illustration 18–11) is to *decrease* both cash and shareholders' equity (by $3 million).

Note that retained earnings may be debited in a treasury stock transaction, but not credited. Also notice that transactions involving treasury stock have no impact on the income statement. Both of those statements—no credit (increase) to retained earnings and no impact on the income statement—follow the reasoning that a corporation's buying and selling of its own shares are transactions between the corporation and its owners and not part of the revenue recognition process.

Additional Consideration

Treasury Shares Acquired at Different Costs

Determining the cost of treasury stock sold is similar to determining the cost of goods sold.

Notice that the treasury stock account always is credited for the cost of the reissued shares ($13 million in Illustration 18–11). When shares are reissued, if treasury stock on hand has been purchased at different per share prices, the cost of the shares sold must be determined using a cost flow assumption—FIFO, LIFO, or weighted average—similar to determining the cost of goods sold when inventory items are acquired at different unit costs.

Concept Review Exercise

TREASURY STOCK

Situation: The shareholders' equity section of the balance sheet of National Foods, Inc. included the following accounts at December 31, 2020

Shareholders' Equity	**($ in millions)**
Paid-in capital:	
Preferred stock, 11 million shares at $1 par	$ 11
Common stock, 122 million shares at $1 par	122
Paid-in capital—excess of par, preferred	479
Paid-in capital—excess of par, common	854
Retained earnings	2,848
Total shareholders' equity	$4,314

Required:

1. National Foods reacquired common shares during 2021 and sold shares in two separate transactions later that year. Prepare the entries for both the purchase and subsequent sale of shares during 2021, assuming that the shares were (a) retired and (b) considered to be treasury stock.
   a. National Foods purchased 6 million shares at $10 per share.
   b. National Foods sold 2 million shares at $12 per share.
   c. National Foods sold 2 million shares at $7 per share.
2. Prepare the shareholders' equity section of National Foods' balance sheet at December 31, 2021, assuming the shares were both (a) retired and (b) viewed as treasury stock. Net income for 2021 was $400 million, and preferred shareholders were paid $1 million cash dividends.

Solution:

1. National Foods reacquired common shares during 2021 and sold shares in two separate transactions later that year. Prepare the entries for both the purchase and subsequent sale of shares during 2021, assuming that the shares were (a) retired and (b) considered to be treasury stock.
   a. National Foods **purchased** 6 million shares at $10 per share:

Retirement ($ in millions)			**Treasury Stock ($ in millions)**		
Common stock (6 million shares × $1)	6		Treasury stock (6 million shares × $10)	60	
Paid-in capital—excess of par (6 million shares × $7*)	42		Cash		60
Retained earnings (to balance)	12				
Cash		60			

*$854 million ÷ 122 million shares

   b. National Foods **sold** 2 million shares at $12 per share ($ in millions):

Cash	24		Cash	24	
Common stock (2 million shares × $1)		2	Treasury stock (2 million shares × $10)		20
Paid-in capital—excess of par		22	Paid-in capital—share repurchase		4

   c. National Foods **sold** 2 million shares at $7 per share ($ in millions):

Cash	14		Cash	14	
Common stock (2 million shares× $1 par)		2	Paid-in capital—share repurchase	4	
Paid-in capital—excess of par		12	Retained earnings (to balance)	2	
			Treasury stock (2 million shares × $10)		20

2. Prepare the shareholders' equity section of National Foods's balance sheet at December 31, 2021, assuming the shares were both (a) retired and (b) viewed as treasury stock.

NATIONAL FOODS, INC.
Balance Sheet
(Shareholders' Equity Section)
At December 31, 2021

($ in millions)	Shares Retired	Treasury Stock
Shareholders' Equity		
Preferred stock, 11 million shares at $1 par	$ 11	$ 11
Common stock, 122 million shares at $1 par	120	122
Paid-in capital—excess of par, preferred	479	479
Paid-in capital—excess of par, common	846*	854
Retained earnings	3,235†	3,245‡
Treasury stock, at cost; 2 million common shares	—	(20)
Total shareholders' equity	$4,691	$4,691

Note: This situation is continued in the next Concept Review Exercise.

*$854 – $42 + $22 + $12
†$2,848 – $12 + $400 – $1
‡$2,848 – $2 + $400 – $1

PART C

Retained Earnings

Characteristics of Retained Earnings

● LO18–6

In the previous section we examined *invested* capital. Now we consider *earned* capital, that is, retained earnings. In general, retained earnings represents a corporation's accumulated, undistributed net income (or net loss). A more descriptive title used by some companies is reinvested earnings. A credit balance in this account indicates a dollar amount of assets previously earned by the firm but not distributed as dividends to shareholders. We refer to a debit balance in retained earnings as a **deficit**. **Microsoft** reported a deficit for several years until retained earnings grew to a positive balance in 2013.

You saw in the previous section that the buyback of shares (as well as the resale of treasury shares in some cases) can decrease retained earnings. We examine in this section the effect on retained earnings of dividends and stock splits.[15]

Dividends

● LO18–7

Shareholders' initial investments in a corporation are represented by amounts reported as paid-in capital. One way a corporation provides a return to its shareholders on their investments is to pay them a **dividend**, typically cash.[16]

Dividends are distributions of assets the company has generated on behalf of its shareholders. If dividends are paid that exceed the amount of assets earned by the company, then management is, in effect, returning to shareholders a portion of their investments, rather than providing them a return on that investment. So most companies view retained earnings as the amount available for dividends.[17]

[15] Retained earnings also can be affected by some changes in accounting principles and for some corrections of errors, as we will see in Chapter 20. And, too, retained earnings can be affected by the repurchase of shares as we discussed in Part A of this chapter.

[16] Dividends are not the only return shareholders earn; when market prices of their shares rise, shareholders benefit also. Indeed, many companies have adopted policies of never paying dividends but reinvesting all assets they earn. The motivation is to accommodate more rapid expansion and thus, presumably, increases in the market price of the stock.

[17] Ordinarily, this is not the legal limitation. Most states permit a company to pay dividends so long as, after the dividend, its assets would not be "less than the sum of its total liabilities plus the amount that would be needed, if the corporation were to be dissolved at the time of the distribution, to satisfy the preferential rights upon dissolution of shareholders whose preferential rights are superior to those receiving the distribution." (Revised Model Business Corporation Act, American Bar Association, 2016.) Thus, legally, a corporation can distribute amounts equal to total shareholders' equity less dissolution preferences of senior equity securities (usually preferred stock).

Liquidating Dividend

In unusual instances in which a dividend exceeds the balance in retained earnings, the excess is referred to as a **liquidating dividend** because some of the invested capital is being liquidated. This might occur when a corporation is being dissolved and assets (not subject to a superior claim by creditors) are distributed to shareholders. Any portion of a dividend not representing a distribution of earnings should be debited to additional paid-in capital rather than retained earnings.

Any dividend not representing a distribution of earnings should be debited to paid-in capital.

Retained Earnings Restrictions

Sometimes the amount available for dividends purposely is reduced by management. A restriction of retained earnings designates a portion of the balance in retained earnings as being *unavailable for dividends*. A company might restrict retained earnings to indicate management's intention to withhold for some specific purpose the assets represented by that portion of the retained earnings balance. For example, management might anticipate the need for a specific amount of assets in upcoming years to repay a maturing debt, to cover a contingent loss, or to finance expansion of the facilities. Be sure to understand that the restriction itself does not set aside cash for the designated event but merely communicates management's intention not to distribute the stated amount as a dividend.

A restriction of retained earnings communicates management's intention to withhold assets represented by a specified portion of the retained earnings balance.

A restriction of retained earnings normally is indicated by a disclosure note to the financial statements. Although instances are rare, a formal journal entry may be used to reclassify a portion of retained earnings to an "appropriated" retained earnings account.

Normally a restriction of retained earnings is indicated by a disclosure note.

Cash Dividends

You learned in Chapter 14 that paying interest to creditors is a contractual obligation. No such legal obligation exists for paying dividends to shareholders. A liability is not recorded until a company's board of directors votes to declare a dividend. In practice, though, corporations ordinarily try to maintain a stable dividend pattern over time.

When directors declare a cash dividend, we reduce retained earnings and record a liability. Before the payment actually can be made, a listing must be assembled of shareholders entitled to receive the dividend. A specific date is stated as to when the determination will be made of the recipients of the dividend. This date is called the **date of record**. Registered owners of shares of stock on this date are entitled to receive the dividend—even if they sell those shares prior to the actual cash payment. To be a registered owner of shares on the date of record, an investor must purchase the shares before the **ex-dividend date**. This date usually is one business day before the date of record. Shares purchased on or after that date are purchased ex dividend—without the right to receive the declared dividend. As a result, the market price of a share typically will decline by the amount of the dividend, other things being equal, on the ex-dividend date. Consider Illustration 18–12.

FINANCIAL Reporting Case

Q3, p. 1047

Illustration 18–12

Cash Dividends

On June 1, the board of directors of Craft Industries declares a cash dividend of $2 per share on its 100 million shares, payable to shareholders of record June 15, to be paid July 1.

June 1—Declaration Date	($ in millions)	
Retained earnings	200	
Cash dividends payable (100 million shares at $2 per share)		200
June 14—Ex-Dividend Date		
No entry		
June 15—Date of Record		
No entry		
July 1—Payment Date		
Cash dividends payable	200	
Cash		200

At the declaration date, retained earnings is reduced and a liability is recorded.

Registered owners of shares on the date of record are entitled to receive the dividend.

A sufficient balance in retained earnings permits a dividend to be declared. Note that retained earnings is a shareholders' equity account representing a dollar claim on assets in general, but not on any specific asset in particular. Sufficient retained earnings does not ensure sufficient cash to make payment. These are two separate accounts having no necessary connection with one another.

DIVIDENDS ON PREFERRED SHARES As mentioned in Part A, preferred shares can be cumulative or noncumulative. Cumulative means that if the specified dividend is not paid for a given year, the unpaid dividends (called *dividends* in *arrears*) accumulate and must be made up in a later dividend year before any dividends are paid on common shares. Illustration 18–13 provides an example.

Illustration 18–13

Distribution of Dividends to Preferred Shareholders

The shareholders' equity section of Corbin Enterprises includes the items shown below. The board of directors declared cash dividends of $360,000, $500,000, and $700,000 in its first three years of operation—2020, 2021, and 2022, respectively.

Common stock	$3,000,000
Paid-in capital—excess of par, common	9,800,000
Preferred stock, 8%	6,000,000
Paid-in capital—excess of par, preferred	780,000

The preferred shareholders are entitled to dividends of $480,000 (8% × $6,000,000).

Determine the amount of dividends to be paid to preferred and common shareholders in each of the three years, assuming that the preferred stock is cumulative and nonparticipating.

	Preferred	Common
2020	$ 360,000*	$0
2021	500,000**	0
2022	580,000†	120,000 (remainder)

*Only $360,000 dividends are declared in 2020 so dividends in arrears are $120,000.
**$120,000 dividends in arrears plus $380,000 of the $480,000 current preference.
†$100,000 dividends in arrears ($480,000 – $380,000) plus the $480,000 current preference

Property Dividends

Because cash is the asset most easily divided and distributed to shareholders, most corporate dividends are cash dividends. In concept, though, any asset can be distributed to shareholders as a dividend. When a noncash asset is distributed, it is referred to as a **property dividend** (often called a *dividend in kind* or a *nonreciprocal transfer to owners*). **Gold Resource Corp.**, a Colorado gold mining company, announced in 2014 that it would make dividend payments in gold bullion instead of cash.

RFM Corporation conferred a property dividend to its shareholders when it distributed shares of **Philipine Townships, Inc.**, stock that RFM was holding as an investment. Securities held as investments are the assets most often distributed in a property dividend due to the relative ease of dividing these assets among shareholders and determining their fair values.

The *fair value* of the assets to be distributed is the amount recorded for a property dividend.

A property dividend should be recorded at the fair value of the assets to be distributed, measured at the date of declaration. This may require revaluing the asset to fair value prior to recording the dividend. If so, a gain or loss is recognized for the difference between book value and fair value. This is demonstrated in Illustration 18–14.

Stock Dividends and Splits

Stock Dividends

● LO18–8

A **stock dividend** is the distribution of additional shares of stock to current shareholders of the corporation. Be sure to note the contrast between a stock dividend and either a cash or property dividend. A stock dividend affects neither the assets nor the liabilities of the firm.

Illustration 18–14
Property Dividends

On October 1 the board of directors of Craft Industries declares a property dividend of 2 million shares of Beaman Corporation's preferred stock.

- Craft had purchased the Beaman shares in March as an investment (book value: $9 million).
- The investment shares have a fair value of $5 per share, **$10** million.
- The property dividend is payable to shareholders of record October 15, to be distributed November 1.

	($ in millions)	
October 1—Declaration Date		
Investment in equity securities *	1	
Gain on investments ($10 – $9)		1
Retained earnings (2 million shares at $5 per share)	**10**	
Property dividends payable		**10**
October 15—Date of Record		
No entry		
November 1—Payment Date		
Property dividends payable	**10**	
Investment in equity securities		**10**

*As an alternative to using a fair value adjustment account, we adjust for fair value directly to the investment account.

Before recording the property dividend, the asset first must be written up to fair value.

Also, because each shareholder receives the same percentage increase in shares, shareholders' proportional interest in (percentage ownership of) the firm remains unchanged. For example, if a 5% stock dividend is declared, every shareholder will receive additional shares of stock to equal 5% of the shares they currently own.

The prescribed accounting treatment of a stock dividend requires that shareholders' equity items be reclassified by reducing one or more shareholders' equity accounts and simultaneously increasing one or more paid-in capital accounts. The amount reclassified depends on the size of the stock dividend. For a "small" stock dividend, typically less than 25%, the fair value of the additional shares distributed is transferred from retained earnings to paid-in capital, as demonstrated in Illustration 18–15.[18]

Illustration 18–15
Stock Dividend

Craft declares and distributes a 10% common stock dividend (10 million shares) when the market value of the $1 par common stock is $12 per share.

	($ in millions)	
Retained earnings (10 million shares at $12 per share)	120	
Common stock (10 million shares at $1 par per share)		10
Paid-in capital—excess of par (remainder)		110

A "small" stock dividend requires reclassification to paid-in capital of retained earnings equal to the fair value of the additional shares distributed.

Additional Consideration

The entry above is recorded on the declaration date. Since the additional shares are not yet issued, some accountants would prefer to credit "common stock dividends distributable" at this point, instead of common stock. In that case, when the shares are issued, common stock dividends distributable is debited and common stock credited. The choice really is inconsequential; either way the $10 million amount would be reported as part of paid-in capital on a balance sheet prepared between the declaration and distribution of the shares.

STOCK MARKET REACTION TO STOCK DISTRIBUTIONS As a Craft shareholder owning 10 shares at the time of the 10% stock dividend, you would receive an 11th share. Since each is worth $12, would you benefit by $12 when you receive the additional

[18]FASB ASC 505–20: Equity—Stock Dividends and Stock Splits. In this pronouncement a small stock dividend is defined as one 20% to 25% or less. For filings with that agency, the SEC has refined the definition to comprise stock distributions of less than 25%.

share from Craft? Of course not. If the value of each share were to remain $12 when the 10 million new shares are distributed, the total market value of the company would grow by $120 million (10 million shares × $12 per share).

The market price per share will decline in proportion to the increase in the number of shares distributed in a stock dividend.

A corporation cannot increase its market value simply by distributing additional stock certificates. Because all shareholders receive the same percentage increase in their respective holdings, you, and all other shareholders, still would own the same percentage of the company as before the distribution. Accordingly, the per share value of your shares should decline from $12 to $10.91 so that your 11 shares would be worth $120— precisely what your 10 shares were worth prior to the stock dividend. You might compare Craft Industries to a pizza. Cutting the pizza into 16 slices instead of 12 doesn't create more to eat; you still have the same pizza, just a higher number of smaller slices. Any failure of the stock price to actually adjust in proportion to the additional shares issued probably would be due to information other than the distribution reaching shareholders at the same time.

Early rule-makers felt that per share market prices do not adjust in response to an increase in the number of shares.

Then, what justification is there for recording the additional shares at market value? In 1941 (and reaffirmed in 1953), accounting rule-makers felt that many shareholders are deceived by small stock dividends, believing they benefit by the market value of their additional shares. Furthermore they erroneously felt that these individual beliefs are collectively reflected in the stock market by per share prices that remain unchanged by stock dividends. Consequently, their prescribed accounting treatment is to reduce retained earnings by the same amount as if cash dividends were paid equal to the market value of the shares issued.[19]

This obsolete reasoning is inconsistent with our earlier conclusion that the market price per share *will* decline in approximate proportion to the increase in the number of shares distributed. Our intuitive conclusion is supported also by formal research.[20]

Capitalizing retained earnings for a stock dividend artificially reclassifies earned capital as invested capital.

Besides being based on fallacious reasoning, accounting for stock dividends by artificially reclassifying "earned" capital as "invested" capital conflicts with the reporting objective of reporting shareholders' equity by source. Despite these limitations, this outdated accounting standard still applies.

Companies sometimes declare a stock dividend in lieu of a real dividend.

Companies sometimes declare a stock dividend so they can capitalize retained earnings.

REASONS FOR STOCK DIVIDENDS Since neither the corporation nor its shareholders apparently benefits from stock dividends, why do companies declare them?[21] Occasionally, a company tries to give shareholders the illusion that they are receiving a real dividend.

Another reason is merely to enable the corporation to take advantage of the accepted accounting practice of capitalizing retained earnings. Specifically, a company might wish to reduce an existing balance in retained earnings—otherwise available for *cash* dividends—so it can reinvest the assets represented by that balance.

Stock Splits

FINANCIAL Reporting Case

Q4, p. 1047

A frequent reason for issuing a stock dividend is actually to induce the per share market price decline it causes. For instance, after a company declares a 100% stock dividend on 100 million shares of common stock, with a per share market price of $12, it then has 200 million shares, each with an approximate market value of $6. The motivation for reducing the per share market price is to increase the stock's *marketability* by making it attractive to a larger number of potential investors.

A large stock dividend is known as a stock *split*.

A stock distribution of 25% or higher can be accounted for in one of two ways: (1) as a "large" stock dividend or (2) as a **stock split**.[22] Thus, a 100% stock dividend could be labeled a 2-for-1 stock split and accounted for as such. Conceptually, the proper accounting treatment of a stock distribution is to make no journal entry. This, in fact, is the prescribed accounting treatment for a stock split.

[19]FASB ASC 505–20–30–3: Equity—Stock Dividends and Stock Splits—Initial Measurement.

[20]Taylor W. Foster III and Don Vickrey, "The Information Content of Stock Dividend Announcements," *Accounting Review* (April 1978); and J. David Spiceland and Alan J. Winters, "The Market Reaction to Stock Distributions: The Effect of Market Anticipation and Cash Returns," *Accounting and Business Research* (Summer 1986).

[21]After hitting a high in the 1940s, the number of stock dividends has declined significantly. Currently, about 3% of companies declare stock dividends in any given year.

[22]FASB ASC 505–20–25–2: Equity—Stock Dividends and Stock Splits—Recognition.

Additional Consideration

No cash dividends are paid on treasury shares. Usually stock dividends aren't paid on treasury shares either. Treasury shares are essentially equivalent to shares that never have been issued. In some circumstances, though, the intended use of the repurchased shares will give reason for the treasury shares to participate in a stock dividend. For instance, if the treasury shares have been specifically designated for issuance to executives in a stock option plan or stock award plan, it would be appropriate to adjust the number of shares by the stock distribution.

Since the same common stock account balance (total par) represents twice as many shares in a 2-for-1 stock split, the par value per share will be reduced by one-half. In the previous example, if the par were $1 per share before the stock distribution, then after the 2-for-1 stock split, the par would be $0.50 per share.

As you might expect, having the par value per share change in this way is cumbersome and expensive. All records, printed or electronic, that refer to the previous amount must be changed to reflect the new amount. The practical solution is to account for the large stock distribution as a stock dividend rather than a stock split.

Stock Splits Effected in the Form of Stock Dividends (Large Stock Dividends)

To avoid changing the par value per share, a company can refer to the stock distribution as a *stock split effected in the form of a stock dividend,* or simply a *stock dividend.* In that case, a journal entry increases the common stock account by the par value of the additional shares. Instead of reducing retained earnings in these instances, most companies reduce (debit) paid-in capital—excess of par to offset the credit to common stock (Illustration 18–16).

Craft declares and distributes a 2-for-1 stock split effected in the form of a 100% stock dividend (100 million shares) when the market value of the $1 par common stock is $12 per share.

	($ in millions)	
Paid-in capital—excess of par	100	
Common stock (100 million shares at $1 par per share)		100

Illustration 18–16

Stock Split Effected in the Form of a Stock Dividend

If a *stock split* is effected in the form of a stock dividend, a journal entry prevents the par per share from changing.

Notice that this entry does not reclassify earned capital as invested capital. Some companies, though, choose to debit retained earnings instead.

	($ in millions)	
Retained earnings	100	
Common stock (100 million shares × $1 par per share)		100

Some companies capitalize retained earnings when recording a stock split effected in the form of a stock dividend.

Netflix described a stock split in its disclosure notes as shown in Illustration 18–17.

Note 8—Stock Split

On June 23, 2015, the Company's Board of Directors declared a seven-for-one stock split in the form of a stock dividend that was paid on July 14, 2015, to all shareholders of record as of July 2, 2015. Outstanding share and per-share amounts disclosed for all periods presented have been retroactively adjusted to reflect the effects of the stock split.

Source: Netflix

Illustration 18–17

Stock Split Disclosure—Netflix

Real World Financials

In 2015, Netflix distributed a rare 7-for-1 stock split that dropped the price of Netflix shares from about $700 all the way down to about $100. Three years later, that stock price was $350 per share. If there had been no split in 2015, the price of one share would have been around $2,450.

Additional Consideration

A company choosing to capitalize retained earnings when recording a stock split effected in the form of a stock dividend may elect to capitalize an amount other than par value. Accounting guidelines are vague in this regard, stating only that legal amounts are minimum requirements and do not prevent the capitalization of a larger amount per share.

Source: FASB ASC 505–20–30–4: Equity–Stock Dividends and Stock Splits–Initial Measurement (previously "Restatement and Revision of Accounting Research Bulletins," *Accounting Research Bulletin* No. 43 (New York: AICPA, 1961), Chap. 7, sec. B, par. 14).

REVERSE STOCK SPLIT A reverse stock split occurs when a company decreases, rather than increases, its outstanding shares. After a 1-for-4 reverse stock split, for example, 100 million shares, $1 par per share, would become 25 million shares, $4 par per share. No journal entry is necessary. Of course the market price per share theoretically would quadruple, which usually is the motivation for declaring a reverse stock split. Companies that reverse split their shares frequently are struggling companies trying to accomplish with the split what the market has been unwilling to do—increase the stock price. Often this is to prevent the stock's price from becoming so low that it is delisted from a market exchange. Reverse splits are not unusual occurrences, particularly during stock market downturns. In 2017, **Xerox Corporation** declared a 1-for-4 reverse split.

FRACTIONAL SHARES Typically, a stock dividend or stock split results in some shareholders being entitled to fractions of whole shares. For example, if a company declares a 25% stock dividend, or equivalently a 5-for-4 stock split, a shareholder owning 10 shares would be entitled to 2 1/2 shares. Another shareholder with 15 shares would be entitled to 3 3/4 shares.

Cash payments usually are made when shareholders are entitled to fractions of whole shares.

Cash payments usually are made to shareholders for *fractional* shares and are called "cash in lieu of payments." In the situation described above, for instance, if the market price at declaration is $12 per share, the shareholder with 15 shares would receive 3 additional shares and $9 in cash ($12 × 3/4).

Decision Maker's Perspective

Profitability is the key to a company's long-run survival. A summary measure of profitability often used by investors and potential investors, particularly common shareholders, is the return on shareholders' equity, or simply return on equity. This ratio measures the ability of company management to generate net income from the resources that owners provide. The ratio is computed by dividing net income by average shareholders' equity. A variation of this ratio often is used when a company has both preferred and common stock outstanding. The return to common shareholders' equity is calculated by subtracting dividends to preferred shareholders from the numerator and using average common shareholders' equity as the denominator. The modified ratio focuses on the profits generated on the assets provided by common shareholders.

The return on equity is a popular measure of profitability.

Although the ratio is useful when evaluating the effectiveness of management in employing resources provided by owners, analysts must be careful not to view it in isolation or without considering how the ratio is derived. Keep in mind that shareholders' equity is a measure of the book value of equity, equivalent to the book value of net assets. Book value measures quickly become out of line with market values. An asset's book value usually equals its market value on the date it's purchased; the two aren't necessarily the same after that. Equivalently, the market value of a share of stock (or of total shareholders' equity) usually is different from its book value. As a result, to supplement the return on equity ratio,

Book value measures have limited use in financial analysis.

analysts often relate earnings to the market value of equity, calculating the earnings-price ratio. This ratio is simply the earnings per share divided by the market price per share.

A variation of this ratio often used by analysts is the price-earnings ratio, which is simply the inverse of the earnings-price ratio: the market price per share divided by the earnings per share.

Share retirement and treasury stock transactions can affect the return to owners.

To better understand the differences between the book value ratio and the market value ratio, let's consider the following condensed information reported by Sharp-Novell Industries for 2021 and 2020:

($ in thousands, except per share amounts)	2021	2020
Sales	$ 3,500	$ 3,100
Net income	125	114
Current assets	$ 750	$ 720
Property, plant, and equipment (net)	900	850
Total assets	$ 1,650	$ 1,570
Current liabilities	$ 550	$ 530
Long-term liabilities	540	520
Paid-in capital	210	210
Retained earnings	350	310
Liabilities and shareholders' equity	$ 1,650	$ 1,570
Shares outstanding	50,000	50,000
Stock price (average)	$ 42.50	$ 42.50

The 2021 return on shareholders' equity is computed by dividing net income by average shareholders' equity:

$$\$125 \div [(\$560 + \$520) \div 2] = \underline{\underline{23.1\%}}$$

The earnings-price ratio is the earnings per share divided by the market price per share:

$$\text{Earnings per share (2021)} = \$125 \div 50 = \$2.50$$
$$\text{Earnings-price ratio} = \$2.50 \div \$42.50 = \underline{\underline{5.9\%}}$$

A variation of this ratio often used by analysts is the price-earnings ratio, which is simply the inverse of the earnings-price ratio: the market price per share divided by the earnings per share.

Obviously, the return on the market value of equity is much lower than on the book value of equity. This points out the importance of looking at more than a single ratio when making decisions. While 23.1% may seem like a desirable return, 5.9% is not nearly so attractive. Companies often emphasize the return on shareholders' equity in their annual reports. Alert investors should not accept this measure of achievement at face value. For some companies this is a meaningful measure of performance; but for others, the market-based ratio means more, particularly for a mature firm whose book value and market value are more divergent.

Share retirement and treasury stock transactions can affect the return to owners.

Decisions managers make with regard to shareholders' equity transactions can significantly impact the return to shareholders For example, when a company buys back shares of its own stock, the return on equity goes up. Net income is divided by a smaller amount of shareholders' equity. On the other hand, the share buyback uses assets, reducing the resources available to earn net income in the future. So, managers as well as outside analysts must carefully consider the decision to reacquire shares in light of the current economic environment, the firm's investment opportunities, and cost of capital to decide whether such a transaction is in the long-term best interests of owners. Investors should be wary of buybacks during down times because the resulting decrease in shares and increase in earnings per share can be used to mask a slowdown in earnings growth.

The decision to pay dividends requires similar considerations. When earnings are high, are shareholders better off receiving substantial cash dividends or having management reinvest those funds to finance future growth (and future dividends)? The answer, of course, depends on the particular circumstances involved. Dividend decisions should reflect managerial strategy concerning the mix of internal versus external financing, alternative investment opportunities, and industry conditions. High dividends often are found in mature industries and low dividends in growth industries. ●

Dividend decisions should be evaluated in light of prevailing circumstances.

Ethical Dilemma

Interworld Distributors has paid quarterly cash dividends since 1985. The dividends have steadily increased from $.25 per share to the latest dividend declaration of $2.00 per share. The board of directors is eager to continue this trend despite the fact that revenues fell significantly during recent months as a result of worsening economic conditions and increased competition. The company founder and member of the board proposes a solution. He suggests a 5% stock dividend in lieu of a cash dividend to be accompanied by the following press announcement:

"In lieu of our regular $2.00 per share cash dividend, Interworld will distribute a 5% stock dividend on its common shares, currently trading at $40 per share. Changing the form of the dividend will permit the Company to direct available cash resources to the modernization of physical facilities in preparation for competing in the 21st century."

What do you think?

Concept Review Exercise

CHANGES IN RETAINED EARNINGS

Situation: The shareholders' equity section of the balance sheet of National Foods, Inc., included the following accounts at December 31, 2021:

Shareholders' Equity	($ in millions)
Paid-in capital	
Preferred stock, 9.09%, 11 million shares at $1 par	$ 11
Common stock, 122 million shares at $1 par	122
Paid-in capital—excess of par, preferred	479
Paid-in capital—excess of par, common	854
Retained earnings	3,245
Treasury stock, at cost, 2 million common shares	(20)
Total shareholders' equity	$4,691

Required:

1. During 2022, several events and transactions affected the retained earnings of National Foods. Prepare the appropriate entries for these events.
   a. On March 1, the board of directors declared a cash dividend of $1 per share on its 120 million outstanding shares (122 million – 2 million treasury shares), payable on April 3 to shareholders of record March 11.
   b. On March 5, the board of directors declared a property dividend of 120 million shares of **Kroger** common stock that National Foods had purchased in February as an investment (book value: $900 million). The investment shares had a fair value of $8 per share and were distributed March 30 to shareholders of record March 15.
   c. On April 13, a 3-for-2 stock split was declared and distributed. The stock split was effected in the form of a 50% stock dividend. The market value of the $1 par common stock was $20 per share.
   d. On October 13, a 10% common stock dividend was declared and distributed when the market value of the $1 par common stock was $12 per share. Cash in lieu of payments was distributed for fractional shares equivalent to 1 million whole shares.
   e. On December 1, the board of directors declared the 9.09% cash dividend on the 11 million preferred shares, payable on December 23 to shareholders of record December 11.
2. Prepare a statement of shareholders' equity for National Foods reporting the changes in shareholders' equity accounts for 2020, 2021, and 2022. Refer to the previous two Concept Reviews in this chapter for the 2020 and 2021 changes. For 2021, assume that shares were reacquired as treasury stock. Also, look back to the statement of shareholders' equity in Illustration 18–4 for the format of the statement. Assume that net income for 2022 is $225 million.

Solution:

1. During 2022, several events and transactions affected the retained earnings of National Foods. Prepare the appropriate entries for these events.

a. Cash dividend of $1 per share on its 120 million *outstanding* common shares (122 million – 2 million treasury shares), payable on April 3 to shareholders of record March 11. (Note: Dividends aren't paid on treasury shares.)

	($ in millions)	
March 1—Declaration Date		
Retained earnings	120	
Cash dividends payable (120 million shares at $1 per share)		120
March 11—Date of Record		
No entry		
April 3—Payment Date		
Cash dividends payable	120	
Cash		120

The declaration of a dividend reduces retained earnings and creates a liability.

b. Property dividend of 120 million shares of Kroger common stock:

	($ in millions)	
March 5—Declaration Date		
Investment in equity securities	60	
Gain on investments ($960 – $900)		60
Retained earnings (fair value of asset to be distributed)	960	
Property dividends payable		960
March 15—Date of Record		
No entry		
March 30—Payment Date		
Property dividends payable	960	
Investment in equity securities		960

The investment first must be written up to the $960 million fair value ($8 × 120 million shares).

The liability is satisfied when the Kroger shares are distributed to shareholders.

c. 3-for-2 stock split effected in the form of a 50% stock dividend:

	($ in millions)	
April 13		
Paid-in capital—excess of par*	60	
Common stock (60 million shares at $1 par per share)		60

*Alternatively, retained earnings may be debited.

120 million shares times 50% equals 60 million new shares—recorded at par.

d. 10% common stock dividend—with cash in lieu of payments for shares equivalent to 1 million whole shares:

	($ in millions)	
October 13		
Retained earnings (18 million shares* at $12 per share)	216	
Common stock (17 million shares at $1 par per share)		17
Paid-in capital—excess of par (17 million shares at $11 per share above par)		187
Cash (1 million equivalent shares at $12 market price per share)		12

*(120 million + 60 million) × 10% = 18 million shares

The stock dividend occurs after the 3-for-2 stock split; thus 18 million shares are distributed.

The $12 fair value of the additional shares is capitalized in this small stock dividend.

e. 9.09% cash dividend on the 11 million preferred shares, payable on December 23 to shareholders of record December 11:

	($ in millions)	
December 1—Declaration Date		
Retained earnings	1	
Cash dividends payable ($11 million par × 9.09%)		1
December 11—Date of Record		
No entry		
December 23—Payment Date		
Cash dividends payable	1	
Cash		1

f. Prepare a statement of shareholders' equity for National Foods reporting the changes in shareholders' equity accounts for 2020, 2021, and 2022.

NATIONAL FOODS
Statement of Shareholders' Equity
For the Years Ended December 31, 2022, 2021, and 2020
($ in millions)

These are the transactions from Concept Review Exercise—Expansion of Corporate Capital.

These are the transactions from Concept Review Exercise—Treasury Stock.

These are the transactions from Concept Review Exercise—Changes in Retained Earnings.

	Preferred Stock	Common Stock	Additional Paid-In Capital	Retained Earnings	Treasury Stock (at cost)	Total Share-holders' Equity
Balance at January 1, 2020		120	836	2,449		3,405
Sale of preferred shares	10		430			440
Issuance of common shares		1	9			10
Issuance of common and preferred shares	1	1	58			60
Net income				400		400
Cash dividends, preferred				(1)		(1)
Balance at December 31, 2020	11	122	1,333	2,848		4,314
Purchase of treasury shares					(60)	(60)
Sale of treasury shares			4		20	24
Sale of treasury shares			(4)	(2)	20	14
Net income				400		400
Cash dividends, preferred				(1)		(1)
Balance at December 31, 2021	11	122	1,333	3,245	(20)	4,691
Cash dividends, common				(120)		(120)
Property dividends, common				(960)		(960)
3-for-2 split effected in the form of a stock dividend		60	(60)			
10% stock dividend		17	187	(216)		(12)
Cash dividends, preferred				(1)		(1)
Net income				225		225
Balance at December 31, 2022	11	199	1,460	2,173	(20)	3,823

Financial Reporting Case Solution

©Peathegee Inc/Blend Images

1. **Do you think the stock price increase is related to Nike's share repurchase plan?** *(p. 1062)* The stock price increase probably is related to **Nike**'s buyback plan. The marketplace realizes that decreasing the supply of shares supports the price of remaining shares. However, the repurchase of shares is not necessarily the best use of a company's cash. Whether it is in the shareholders' best interests depends on what other opportunities the company has for the cash available.
2. **What are Nike's choices in accounting for the share repurchases?** *(p. 1063)* When a corporation reacquires its own shares, those shares assume the same status as authorized but unissued shares, just as if they never had been issued. However, for exactly the same transaction, companies can choose between two accounting alternatives: (a) formally retiring them or (b) accounting for the shares repurchased as treasury stock. In actuality, Nike uses alternative (a).
3. **What effect does the quarterly cash dividend of 20 cents per share have on Nike's assets? Its liabilities? Its shareholders' equity?** *(p. 1069)* Each quarter, when directors declare a cash dividend, retained earnings are reduced and a liability is recorded. The liability is paid with cash on the payment date. So, the net effect is a decrease in Nike's assets and its shareholders' equity. The effect on liabilities is temporary.

4. What effect does the stock split have on Nike's assets? Its liabilities? Its shareholders' equity? *(p. 1072)* Conceptually, the proper accounting treatment of a stock split is to make no journal entry. However, since Nike refers to the stock distribution as a stock split "in the form of a 100% stock dividend," a journal entry would increase the common stock account by the par value of the additional shares and would reduce paid-in capital—excess of par. This merely moves an amount from one part of shareholders' equity to another. Regardless of the accounting method, there is no change in Nike's assets, liabilities, or total shareholders' equity. ●

The Bottom Line

- **LO18–1** Shareholders' equity is the owners' residual interest in a corporation's assets. It arises primarily from (1) amounts invested by shareholders and (2) amounts earned by the corporation on behalf of its shareholders. These are reported as (1) paid-in capital and (2) retained earnings in a balance sheet. A statement of shareholders' equity reports the sources of the changes in individual shareholders' equity accounts. *(p. 1047)*
- **LO18–2** Comprehensive income encompasses all changes in equity except those caused by transactions with owners (like dividends and the sale or purchase of shares). It includes traditional net income as well as "other comprehensive income." *(p. 1050)*
- **LO18–3** A corporation is a separate legal entity—separate and distinct from its owners. A corporation is well suited to raising capital and has limited liability. However, it gives rise to "double taxation." Common shareholders usually have voting rights; preferred shareholders usually have a preference to a specified amount of dividends and to assets in the event of liquidation. *(p. 1051)*
- **LO18–4** Shares sold for consideration other than cash (maybe services or a noncash asset) should be recorded at the fair value of the shares or the noncash consideration, whichever seems more clearly evident. *(p. 1058)*
- **LO18–5** When a corporation retires previously issued shares, those shares assume the same status as authorized but unissued shares—just the same as if they had never been issued. Payments made to retire shares are viewed as a distribution of corporate assets to shareholders. When reacquired shares are viewed as treasury stock, the cost of acquiring the shares is temporarily debited to the treasury stock account. Recording the effects on specific shareholders' equity accounts is delayed until later when the shares are reissued. *(p. 1062)*
- **LO18–6** Retained earnings represents, in general, a corporation's accumulated, undistributed, or reinvested net income (or net loss). Distributions of assets are dividends. *(p. 1068)*
- **LO18–7** Most corporate dividends are paid in cash. When a noncash asset is distributed, it is referred to as a property dividend. The fair value of the assets to be distributed is the amount recorded for a property dividend. *(p. 1068)*
- **LO18–8** A stock dividend is the distribution of additional shares of stock to current shareholders. For a small stock dividend (25% or less), the fair value of the additional shares distributed is transferred from retained earnings to paid-in capital. For a stock distribution of 25% or higher, the par value of the additional shares is reclassified within shareholders' equity if referred to as a stock split effected in the form of a stock dividend, but if referred to merely as a stock split, no journal entry is recorded. *(p. 1070)*
- **LO18–9** U.S. GAAP and IFRS are generally compatible with respect to accounting for shareholders' equity. Some differences exist in presentation format and terminology. *(pp. 1052 and 1058)* ●

Quasi Reorganizations

APPENDIX 18

A firm undergoing financial difficulties, but with favorable future prospects, may use a **quasi reorganization** to write down inflated asset values and eliminate an accumulated deficit (debit balance in retained earnings). To effect the reorganization, these procedures are followed:

1. The firm's assets (and perhaps liabilities) are revalued (up or down) to reflect fair values, with corresponding credits or debits to retained earnings. This process typically increases the deficit.

2. The debit balance in retained earnings (deficit) is eliminated against additional paid-in capital. If additional paid-in capital is not sufficient to absorb the entire deficit, a reduction in stock may be necessary (with an appropriate restating of the par amount per share).
3. Retained earnings is dated. That is, disclosure is provided to indicate the date the deficit was eliminated and when the new accumulation of earnings began.

The procedure is demonstrated in Illustration 18A–1. The shareholders approved the quasi reorganization effective January 1, 2021. The plan was to be accomplished by a reduction of inventory by $75 million, a reduction in property, plant, and equipment (net) of $175 million, and appropriate adjustments to shareholders' equity.

Illustration 18A–1
Quasi Reorganization

The Emerson-Walsch Corporation has incurred operating losses for several years. A newly elected board of directors voted to implement a quasi reorganization, subject to shareholder approval. The balance sheet, on December 31, 2020, immediately prior to the restatement, includes the data shown below.

	($ in millions)
Cash	$ 75
Receivables	200
Inventory	375
Property, plant, and equipment (net)	400
	$1,050
Liabilities	$ 400
Common stock (800 million shares at $1 par)	800
Additional paid-in capital	150
Retained earnings (deficit)	**(300)**
	$1,050

When assets are revalued to reflect fair values, the process often increases the deficit.

To revalue assets:

Retained earnings	75	
Inventory		75
Retained earnings	175	
Property, plant, and equipment		175

The deficit, $550 ($300 + $75 + $175), can be only partially absorbed by the balance of additional paid-in capital.

To eliminate a portion of the deficit against available additional paid-In capital:

Additional paid-in capital	150	
Retained earnings		150

The remaining deficit, $400 ($300 + $75 + $175 − $150), must be absorbed by reducing the balance in common stock.

To eliminate the remainder of the deficit against common stock:

Common stock	400	
Retained earnings		400

The balance sheet immediately after the restatement would include the following:

Assets and liabilities reflect current values.

Because a reduced balance represents the same 800 million shares, the par amount per share must be reduced.

The deficit is eliminated.

	($ in millions)
Cash	$ 75
Receivables	200
Inventory	300
Property, plant, and equipment (net)	225
	$800
Liabilities	$400
Common stock (800 million shares at $0.50 par)	400
Additional paid-in capital	0
Retained earnings (deficit)	**0**
	$800

Note A: Upon the recommendation of the board of directors and approval by shareholders, a quasi reorganization was implemented January 1, 2021. The plan was accomplished by a reduction of inventory by $75 million, a reduction in property, plant, and equipment (net) of $175 million, and appropriate adjustments to shareholders' equity. The balance in retained earnings reflects the elimination of a $300 million deficit on that date.

Questions For Review of Key Topics

Q 18–1 Identify and briefly describe the two primary sources of shareholders' equity.

Q 18–2 Identify the three common forms of business organization and the primary difference between the way they are accounted for.

Q 18–3 Corporations offer the advantage of limited liability. Explain what is meant by that statement.

Q 18–4 Distinguish between not-for-profit and for-profit corporations.

Q 18–5 Distinguish between publicly held and privately (or closely) held corporations.

Q 18–6 How does the Model Business Corporation Act affect the way corporations operate?

Q 18–7 The owners of a corporation are its shareholders. If a corporation has only one class of shares, they typically are labeled common shares. Indicate the ownership rights held by common shareholders, unless specifically withheld by agreement.

Q 18–8 What is meant by a shareholder's preemptive right?

Q 18–9 Terminology varies in the way companies differentiate among share types. But many corporations designate shares as common or preferred. What are the two special rights usually given to preferred shareholders?

Q 18–10 Most preferred shares are cumulative. Explain what this means.

Q 18–11 The par value of shares historically indicated the real value of shares and all shares were issued at that price. The concept has changed with time. Describe the meaning of par value as it has evolved to today.

Q 18–12 What is comprehensive income? How does comprehensive income differ from net income? Where do companies report it in a balance sheet?

Q 18–13 How do we report components of comprehensive income *created during the reporting period?*

Q 18–14 The balance sheet reports the balances of shareholders' equity accounts. What additional information is provided by the statement of shareholders' equity?

Q 18–15 At times, companies issue their shares for consideration other than cash. What is the measurement objective in those cases?

Q 18–16 Companies occasionally sell more than one security for a single price. How is the issue price allocated among the separate securities?

Q 18–17 The costs of legal, promotional, and accounting services necessary to effect the sale of shares are referred to as share issue costs. How are these costs recorded? Compare this approach to the way debt issue costs are recorded.

Q 18–18 When a corporation acquires its own shares, those shares assume the same status as authorized but unissued shares, as if they never had been issued. Explain how this is reflected in the accounting records if the shares are formally retired.

Q 18–19 Discuss the conceptual basis for accounting for a share buyback as treasury stock.

Q 18–20 The prescribed accounting treatment for stock dividends implicitly assumes that shareholders are fooled by small stock dividends and benefit by the market value of their additional shares. Explain this statement. Is it logical?

Q 18–21 Brandon Components declares a 2-for-1 stock split. What will be the effects of the split, and how should it be recorded?

Q 18–22 What is a reverse stock split? What would be the effect of a reverse stock split on one million $1 par shares? On the accounting records?

Q 18–23 Suppose you own 80 shares of **Facebook** common stock when the company declares a 4% stock dividend. What will you receive as a result?

Q 18–24 (Based on Appendix 18) A quasi reorganization is sometimes employed by a firm undergoing financial difficulties but with favorable future prospects. What are two objectives of this procedure? Briefly describe the procedural steps.

Brief Exercises

connect

BE 18–1
Comprehensive income
● LO18–1

Schaeffer Corporation reports $50 million accumulated other comprehensive income in its balance sheet as a component of shareholders' equity. In a related statement reporting comprehensive income for the year, the company reveals net income of $400 million and other comprehensive income of $15 million. What was the balance in accumulated other comprehensive income in last year's balance sheet?

BE 18–2
Stock issued
● LO18–4

Penne Pharmaceuticals sold 8 million shares of its $1 par common stock to provide funds for research and development. If the issue price is $12 per share, what is the journal entry to record the sale of the shares?

BE 18–3
Stock issued
● LO18–4

Lewelling Company issued 100,000 shares of its $1 par common stock to the Michael Morgan law firm as compensation for 4,000 hours of legal services performed. Morgan's usual rate is $240 per hour. By what amount should Lewelling's paid-in capital—excess of par increase as a result of this transaction?

BE 18–4
Stock issued
● LO18–4

Hamilton Boats issued 175,000 shares of its no-par common stock to Sudoku Motors in exchange for 1,000 four-stroke outboard motors that normally sell in quantity for $3,500 each. By what amount should Hamilton's shareholders' equity increase as a result of this transaction?

BE 18–5
Retirement of shares
● LO18–5

Horton Industries' shareholders' equity included 100 million shares of $1 par common stock and a balance in paid-in capital—excess of par of $900 million. Assuming that Horton retires shares it reacquires (restores their status to that of authorized but unissued shares), by what amount will Horton's total paid-in capital decline if it reacquires 2 million shares at $8.50 per share?

BE 18–6
Retirement of shares
● LO18–5

Agee Storage issued 35 million shares of its $1 par common stock at $16 per share several years ago. Last year, for the first time, Agee reacquired 1 million shares at $14 per share. If Agee now *retires* 1 million shares at $19 per share. By what amount will Agee's total paid-in capital decline?

BE 18–7
Treasury stock
● LO18–5

The Jennings Group reacquired 2 million of its shares at $70 per share as treasury stock. Last year, for the first time, Jennings sold 1 million treasury shares at $71 per share. If Jennings now sells the remaining 1 million treasury shares at $67 per share, by what amount will retained earnings decline?

BE 18–8
Treasury stock
● LO18–5

In previous years, Cox Transport reacquired 2 million treasury shares at $20 per share and, later, 1 million treasury shares at $26 per share. If Cox now sells 1 million treasury shares at $29 per share and determines cost as the weighted-average cost of treasury shares, by what amount will Cox's paid-in capital—share repurchase increase?

BE 18–9
Treasury stock
● LO18–5

Refer to the situation described in BE 18–8. By what amount will Cox's paid-in capital—share repurchase increase if it determines the cost of treasury shares by the FIFO method?

BE 18–10
Cash dividend
● LO18–8
Real World Financials

Following is a recent **Microsoft** press release:

REDMOND, Wash. — Nov. 29, 2017 — Microsoft Corp. on Wednesday announced that its board of directors declared a quarterly dividend of $0.42 per share. The dividend is payable March 8, 2018, to shareholders of record on Feb. 15, 2018. The ex-dividend date will be Feb. 14, 2018.

Prepare the journal entries Microsoft used to record the declaration and payment of the cash dividend for its 7,720 million shares.

Source: Microsoft

BE 18–11
Effect of preferred stock on dividends
● LO18–7

The shareholders' equity of MLS Enterprises includes $200 million of no par common stock and $400 million of 6% cumulative preferred stock. The board of directors of MLS declared cash dividends of $50 million in 2021 after paying $20 million cash dividends in both 2020 and 2019. What is the amount of dividends common shareholders will receive in 2021?

BE 18–12
Property dividend
● LO18–7

Adams Moving and Storage, a family-owned corporation, declared a property dividend of 1,000 shares of **GE** common stock that Adams had purchased in February for $37,000 as an investment. GE's shares had a market value of $35 per share on the declaration date. Prepare the journal entries to record the property dividend on the declaration and payment dates.

BE 18–13
Stock dividend
● LO18–8

On June 13, the board of directors of Siewert Inc. declared a 5% stock dividend on its 60 million, $1 par, common shares, to be distributed on July 1. The market price of Siewert common stock was $25 on June 13. Prepare the journal entry to record the stock dividend.

BE 18–14
Stock split
● LO18–8

Refer to the situation described in BE 18–13, but assume a 2-for-1 stock split instead of the 5% stock dividend. Prepare the journal entry to record the stock split if it is *not* to be effected in the form of a stock dividend. What is the par per share after the split?

BE 18–15
Stock split
● LO18–8

Refer to the situation described in BE 18–13, but assume a 2-for-1 stock split instead of the 5% stock dividend. Prepare the journal entry to record the stock split if it is to be effected in the form of a 100% stock dividend. What is the par per share after the split?

BE 18–16
IFRS; reporting shareholders' equity
● LO18–9
IFRS

Nestlé S.A., the largest food and beverage company in the world, prepares its financial statements according to International Financial Reporting Standards. Its financial statements include ordinary share capital, translation reserve, and share premium. If Nestlé used U.S. GAAP, what would be the likely account titles for these accounts?

Exercises

connect

E 18–1
Comprehensive income
● LO18–2

The following is from the 2021 annual report of Kaufman Chemicals, Inc.:

Statements of Comprehensive Income	**Years Ended December 31**		
	2021	**2020**	**2019**
Net income	$856	$766	$594
Other comprehensive income:			
Change in net unrealized gains on AFS investments, net of tax of $22, ($14), and $15 in 2021, 2020, and 2019, respectively	34	(21)	23
Other	(2)	(1)	1
Total comprehensive income	$888	$744	$618

Kaufman reports accumulated other comprehensive income in its balance sheet as a component of shareholders' equity as follows:

($ in millions)	2021	2020
Shareholders' equity:		
Common stock	355	355
Additional paid-in capital	8,567	8,567
Retained earnings	6,544	5,988
Accumulated other comprehensive income	107	75
Total shareholders' equity	$15,573	$14,985

Required:
1. What is comprehensive income and how does it differ from net income?
2. How is comprehensive income reported in a balance sheet?
3. Why is Kaufman's 2021 balance sheet amount different from the 2021 amount reported in the disclosure note? Explain.
4. From the information provided, determine how Kaufman calculated the $107 million accumulated other comprehensive income in 2021.

E 18–2
FASB codification research; reporting other comprehensive income in shareholders' equity
● LO18–2
CODE

Companies are required to transfer "other comprehensive income" each period to shareholders' equity. The *FASB Accounting Standards Codification* represents the single source of authoritative U.S. generally accepted accounting principles. Obtain the relevant authoritative literature on reporting other comprehensive income in shareholders' equity using the *FASB Accounting Standards Codification* at the FASB website (**www.fasb.org**).

Required:
1. What is the specific nine-digit Codification citation (XXX-XX-XX-XX) that describes the guidelines for reporting that component of shareholders' equity?
2. What is the specific nine-digit Codification citation (XXX-XX-XX-XX) that describes the guidelines for presenting accumulated other comprehensive income on the statement of shareholders' equity?

E 18–3
Earnings or OCI?
● LO18–2

Indicate by letter whether each of the items listed below most likely is reported in the income statement as Net Income (**NI**) or in the statement of comprehensive income as Other Comprehensive Income (**OCI**).

	Items
_____	1. Increase in the fair value of available-for-sale (AFS) debt securities
_____	2. Gain on sale of land
_____	3. Loss on pension plan assets (actual return less than expected)
_____	4. Adjustment for foreign currency translation
_____	5. Increase in the fair value of investments in common stock securities
_____	6. Loss from revising an assumption related to a pension plan
_____	7. Loss on sale of patent
_____	8. Prior service cost in defined benefit pension plan
_____	9. Increase in the fair value of bonds outstanding due to change in general interest rates; fair value option
_____	10. Gain on postretirement plan assets (actual return more than expected)

E 18–4
Stock issued for cash; Wright Medical Group
● LO18–4
Real World Financials

The following is a news item reported by Reuters:

WASHINGTON, Jan 29 (Reuters)—**Wright Medical Group**, a maker of reconstructive implants for knees and hips, on Tuesday filed to sell 3 million shares of common stock.

In a filing with the U.S. Securities and Exchange Commission, it said it plans to use the proceeds from the offering for general corporate purposes, working capital, research and development, and acquisitions.

After the sale there will be about 31.5 million shares outstanding in the Arlington, Tennessee-based company, according to the SEC filing.

Wright shares closed at $17.15 on Nasdaq.

The common stock of Wright Medical Group has a par of $0.01 per share.

Required:
Prepare the journal entry to record the sale of the shares assuming the price existing when the announcement was made and ignoring share issue costs.

Source: Reuters

E 18–5
Issuance of shares; noncash consideration
● LO18–4

During its first year of operations, Eastern Data Links Corporation entered into the following transactions relating to shareholders' equity. The articles of incorporation authorized the issue of 8 million common shares, $1 par per share, and 1 million preferred shares, $50 par per share.

Required:
Prepare the appropriate journal entries to record each transaction.

Feb. 12	Sold 2 million common shares, for $9 per share.
13	Issued 40,000 common shares to attorneys in exchange for legal services.
13	Sold 80,000 of its common shares and 4,000 preferred shares for a total of $945,000.
Nov. 15	Issued 380,000 of its common shares in exchange for equipment for which the cash price was known to be $3,688,000.

E 18–6
Redeemable shares
● LO18–4

Williams Industries has outstanding 30 million common shares, 20 million Class A shares, and 20 million Class B shares. Williams has the right but not the obligation to repurchase the Class A shares if a change in ownership of the voting common shares causes J. P. Williams, founder and CEO, to have less than 50% ownership. Williams has the unconditional obligation to repurchase the Class B shares upon the death of J. P. Williams.

Required:
Which, if any, of the shares should be reported in Williams's balance sheet as liabilities? Explain.

E 18–7
Share issue costs; issuance
● LO18–4

ICOT Industries issued 15 million of its $1 par common shares for $424 million on April 11. Legal, promotional, and accounting services necessary to effect the sale cost $2 million.

Required:
1. Prepare the journal entry to record the issuance of the shares.
2. Explain how recording the share issue costs differs from the way debt issue costs are recorded.

E 18–8
Reporting preferred shares
● LO18–4, LO18–7

Ozark Distributing Company is primarily engaged in the wholesale distribution of consumer products in the Ozark Mountain regions. The following disclosure note appeared in the company's 2021 annual report:

Note 5. Convertible Preferred Stock (in part):
The Company has the following Convertible Preferred Stock outstanding as of September 2021:

Date of issuance:	June 17, 2018
Optionally redeemable beginning:	June 18, 2020
Par value (gross proceeds):	$2 500,000
Number of shares:	100,000
Liquidation preference per share:	$25.00
Conversion price per share	$30.31
Number of common shares in which to be converted:	82,481
Dividend rate:	6.785%

The Preferred Stock is convertible at any time by the holders into a number of shares of Ozark's common stock equal to the number of preferred shares being converted times a fraction equal to $25.00 divided by the conversion price. The conversion prices for the Preferred Stock are subject to customary adjustments in the event of stock splits, stock dividends and certain other distributions on the Common Stock. Cumulative dividends for the Preferred Stock are payable in arrears, when, as, and if declared by the Board of Directors, on March 31, June 30, September 30, and December 31 of each year.

The Preferred Stock is optionally redeemable by the Company beginning on various dates, as listed above, at redemption prices equal to 112% of the liquidation preference. The redemption prices decrease 1% annually thereafter until the redemption price equals the liquidation preference, after which date it remains the liquidation preference.

Required:
1. What amount of dividends is paid annually to a preferred shareholder owning 100 shares of the Series A preferred stock?
2. If dividends are not paid in 2022 and 2023, but are paid in 2024, what amount of dividends will the shareholder receive?
3. If the investor chooses to convert the shares in 2022, how many shares of common stock will the investor receive for his/her 100 shares?
4. If Ozark chooses to redeem the shares on June 18, 2022, what amount will the investor be paid for his/her 100 shares?

E 18–9
New equity issues; offerings announcements
● LO18–4

When companies offer new equity security issues, they publicize the offerings in the financial press and on Internet sites. Assume the following were among the equity offerings reported in December 2021:

New Securities Issues

Equity

American Materials Transfer Corporation (AMTC)—7.5 million common shares, $.001 par, priced at $13.546 each through underwriters led by Second Tennessee Bank N.A. and Morgan, Dunavant & Co., according to a syndicate official.

Proactive Solutions Inc. (PSI)—Offering of 9 million common shares, $0.01 par, was priced at $15.20 a share via lead manager Stanley Brothers, Inc., according to a syndicate official.

Required:
Prepare the appropriate journal entries to record the sale of both issues to underwriters. Ignore share issue costs.

E 18–10
Retirement of shares
● LO18–5

Borner Communications' articles of incorporation authorized the issuance of 130 million common shares. The transactions described below effected changes in Borner's outstanding shares. Prior to the transactions, Borner's shareholders' equity included the following:

Shareholders' Equity	($ in millions)
Common stock, 100 million shares at $1 par	$100
Paid-in capital—excess of par	300
Retained earnings	210

Required:

Assuming that Borner Communications retires shares it reacquires (restores their status to that of authorized but unissued shares), record the appropriate journal entry for each of the following transactions:

1. On January 7, 2021, Borner reacquired 2 million shares at $5 per share.
2. On August 23, 2021, Borner reacquired 4 million shares at $3.50 per share.
3. On July 25, 2022, Borner sold 3 million common shares at $6 per share.

E 18–11
Retirement of shares
● LO18–5

In 2021, Borland Semiconductors entered into the transactions described below. In 2018, Borland had issued 170 million shares of its $1 par common stock at $34 per share.

Required:

Assuming that Borland retires shares it reacquires, record the appropriate journal entry for each of the following transactions:

1. On January 2, 2021, Borland reacquired 10 million shares at $32.50 per share.
2. On March 3, 2021, Borland reacquired 10 million shares at $36 per share.
3. On August 13, 2021, Borland sold 1 million shares at $42 per share.
4. On December 15, 2021, Borland sold 2 million shares at $36 per share.

E 18–12
Treasury stock
● LO18–5

In 2021, Western Transport Company entered into the treasury stock transactions described below. In 2019, Western Transport had issued 140 million shares of its $1 par common stock at $17 per share.

Required:

Prepare the appropriate journal entry for each of the following transactions:

1. On January 23, 2021, Western Transport reacquired 10 million shares at $20 per share.
2. On September 3, 2021, Western Transport sold 1 million treasury shares at $21 per share.
3. On November 4, 2021, Western Transport sold 1 million treasury shares at $18 per share.

E 18–13
Treasury stock; weighted-average and FIFO cost
● LO18–5

At December 31, 2020, the balance sheet of Meca International included the following shareholders' equity accounts:

Shareholders' Equity	($ in millions)
Common stock, 60 million shares at $1 par	$ 60
Paid-in capital—excess of par	300
Retained earnings	410

Required:

Assuming that Meca International views its share buybacks as treasury stock, record the appropriate journal entry for each of the following transactions:

1. On February 12, 2021, Meca reacquired 1 million common shares at $13 per share.
2. On June 9, 2022, Meca reacquired 2 million common shares at $10 per share.
3. On May 25, 2023, Meca sold 2 million treasury shares at $15 per share. Determine cost as the weighted-average cost of treasury shares.
4. For the previous transaction, assume Meca determines the cost of treasury shares by the FIFO method.

E 18–14
Reporting shareholders' equity after share repurchase
● LO18–5

On two previous occasions, the management of Dennison and Company, Inc., repurchased some of its common shares. Between buyback transactions, the corporation issued common shares under its management incentive plan. Shown below is shareholders' equity following these share transactions, as reported by two different methods of accounting for reacquired shares.

($ in millions)	Method A	Method B
Shareholders' equity		
Paid-in capital:		
Preferred stock, $10 par	$ 150	$ 150
Common stock, $1 par	200	197
Additional paid-in capital	1,204	1,201
Retained earnings	2,994	2,979
Less: Treasury stock	(21)	
Total shareholders' equity	$4,527	$4,527

Required:

1. Infer from the presentation which method of accounting for reacquired shares is represented by each of the two columns.
2. Explain why presentation formats are different and why some account balances are different for the two methods.

E 18–15
Change from treasury stock to retired stock
LO18–5

In keeping with a modernization of corporate statutes in its home state, UMC Corporation decided in 2021 to discontinue accounting for reacquired shares as treasury stock. Instead, shares repurchased will be viewed as having been retired, reassuming the status of unissued shares. As part of the change, treasury shares held were reclassified as retired stock. At December 31, 2020, UMC's balance sheet reported the following shareholders' equity:

	($ in millions)
Common stock, $1 par	$ 200
Paid-in capital—excess of par	800
Retained earnings	956
Treasury stock (4 million shares at cost)	(25)
Total shareholders' equity	$1,931

Required:

Identify the type of accounting change this decision represents, and prepare the journal entry to effect the reclassification of treasury shares as retired shares.

E 18–16
Stock buyback; Ford press announcement
LO18–5
Real World Financials

The following excerpt is from an article reported in an online issue of **Bloomberg**.

(Bloomberg) **Ford Motor Co.** (F) said it will repurchase $1.8 billion of its shares to reduce dilution from recent stock grants to executives.

The par amount per share for Ford's common stock is $0.01. Paid-in capital—excess of par is $5.39 per share on average. The market price was $16.

Required:

1. Suppose Ford reacquires 112 million shares through repurchase on the open market at $16 per share. Prepare the appropriate journal entry to record the purchase. Ford considers the shares it buys back to be treasury stock.
2. Suppose Ford considers the shares it buys back to be retired rather than treated as treasury stock. Prepare the appropriate journal entry to record the purchase.
3. What does the company mean by saying that the buyback will serve "to offset dilution from executive compensation?"

Source: Bloomberg

E 18–17
Transactions affecting retained earnings
LO18–6, LO18–7

Shown below in T-account format are the changes affecting the retained earnings of Brenner-Jude Corporation during 2021. At January 1, 2021, the corporation had outstanding 105 million common shares, $1 par per share.

Retained Earnings ($ in millions)

		90	Beginning balance
Retirement of 5 million common shares for $22 million	2		
		88	Net income for the year
Declaration and payment of a $0.33 per share cash dividend	33		
Declaration and distribution of a 4% stock dividend	20		
		123	Ending balance

Required:

1. From the information provided by the account changes, you should be able to recreate the transactions that affected Brenner-Jude's retained earnings during 2021. Prepare the journal entries that Brenner-Jude must have recorded during the year for these transactions. (Hint: In lieu of revenues and expenses, use an account titled "Income summary" to close net income or net loss.)
2. Prepare a statement of retained earnings for Brenner-Jude for the year ended 2021.

E 18–18
Effect of cumulative, nonparticipating preferred stock on dividends—3 years
● LO18–7

The shareholders' equity of ILP Industries includes the items shown below. The board of directors of ILP declared cash dividends of $8 million, $20 million, and $150 million in its first three years of operation—2021, 2022, and 2023, respectively.

	($ in millions)
Common stock	$100
Paid-in capital—excess of par, common	980
Preferred stock, 8%	200
Paid-in capital—excess of par, preferred	555

Required:
Determine the amount of dividends to be paid to preferred and common shareholders in each of the three years, assuming that the preferred stock is cumulative and nonparticipating.

	Preferred	Common
2021		
2022		
2023		

E 18–19
Stock dividend
● LO18–8

The shareholders' equity of Core Technologies Company on June 30, 2020, included the following:

Common stock, $1 par; authorized, 8 million shares; issued and outstanding, 3 million shares	$ 3,000,000
Paid-in capital—excess of par	12,000,000
Retained earnings	14,000,000

On April 1, 2021, the board of directors of Core Technologies declared a 10% stock dividend on common shares, to be distributed on June 1. The market price of Core Technologies' common stock was $30 on April 1, 2021, and $40 on June 1, 2021.

Required:
Prepare the journal entry to record the declaration and distribution of the stock dividend on the declaration date.

E 18–20
Stock split; Hanmi Financial Corporation
● LO18–8
Real World Financials

Hanmi Financial Corporation is the parent company of Hanmi Bank. The company's stock split was announced in the following wire:

LOS ANGELES Jan. 20 *BUSINESS WIRE* —Hanmi Financial Corporation (Nasdaq), announced that the Board of Directors has approved a two-for-one stock split, to be effected in the form of a 100 percent common stock dividend. Hanmi Financial Corporation stockholders of record at the close of business on January 31 will receive one additional share of common stock for every share of common stock then held. Distribution of additional shares issued as a result of the split is expected to occur on or about February 15.

At the time of the stock split, 24.5 million shares of common stock, $.001 par per share, were outstanding.

Required:
1. Prepare the journal entry, if any, that Hanmi recorded at the time of the stock split.
2. What is the probable motivation for declaring the 2-for-1 stock split to be effected by a dividend payable in shares of common stock?
3. If Hanmi's stock price had been $36 at the time of the split, what would be its approximate value after the split (other things equal)?

Source: Business Wire

E 18–21
Cash in lieu of fractional share rights
● LO18–8

Douglas McDonald Company's balance sheet included the following shareholders' equity accounts at December 31, 2020:

	($ in millions)
Paid-in capital:	
Common stock, 900 million shares at $1 par	$ 900
Paid-in capital—excess of par	15,800
Retained earnings	14,888
Total shareholders' equity	$31,588

On March 16, 2021, a 4% common stock dividend was declared and distributed. The market value of the common stock was $21 per share. Fractional share rights represented 2 million equivalent whole shares. Cash was paid in lieu of the fractional share rights.

Required:
1. What is a fractional share right?
2. Prepare the appropriate entries for the declaration and distribution of the stock dividend.

E 18–22
FASB codification research
● LO18–1, LO18–5, LO18–8

CODE

Access the *FASB Accounting Standards Codification* at the FASB website (**www.fasb.org**). Determine the specific eight-digit Codification citation (XXX-XX-XX-X) for accounting for each of the following items:
1. Requirements to disclose within the financial statements the pertinent rights and privileges of the various securities outstanding.
2. Requirement to record a "small" stock dividend at the fair value of the shares issued.
3. Requirement to exclude from the determination of net income gains and losses on transactions in a company's own stock.

E 18–23
Transactions affecting retained earnings
● LO18–6 through LO18–8

The balance sheet of Consolidated Paper, Inc., included the following shareholders' equity accounts at December 31, 2020:

Paid-in capital	
Preferred stock, 3.8%, 90,000 shares at $1 par	$ 90,000
Common stock, 364,000 shares at $1 par	364,000
Paid-in capital—excess of par, preferred	1,437,000
Paid-in capital—excess of par, common	2,574,000
Retained earnings	9,735,000
Treasury stock, at cost; 4,000 common shares	(44,000)
Total shareholders' equity	$14,156,000

During 2021, several events and transactions affected the retained earnings of Consolidated Paper.

Required:
1. Prepare the appropriate entries for these events:
   a. On March 3, the board of directors declared a property dividend of 240,000 shares of Leasco International common stock that Consolidated Paper had purchased in January as an investment (book value: $700,000). The investment shares had a fair value of $3 per share and were distributed March 31 to shareholders of record March 15.
   b. On May 3, a 5-for-4 stock split was declared and distributed. The stock split was effected in the form of a 25% stock dividend. The market value of the $1 par common stock was $11 per share.
   c. On July 5, a 2% common stock dividend was declared and distributed. The market value of the common stock was $11 per share.
   d. On December 1, the board of directors declared the 8.8% cash dividend on the 90,000 preferred shares, payable on December 28 to shareholders of record December 20.
   e. On December 1, the board of directors declared a cash dividend of $0.50 per share on its common shares, payable on December 28 to shareholders of record December 20.
2. Prepare the shareholders' equity section of the balance sheet for Consolidated Paper, Inc., at December 31, 2021. Net income for the year was $810,000.

E 18–24
Profitability ratio
● LO18–1

Comparative balance sheets for Softech Canvas Goods for 2021 and 2020 are shown below. Softech pays no dividends and instead reinvests all earnings for future growth.

Comparative Balance Sheets
($ in thousands)

	December 31 2021	2020
Assets:		
Cash	$ 50	$ 40
Accounts receivable	100	120
Short-term investments	50	40
Inventory	200	140
Property, plant, and equipment (net)	600	550
	$1,000	$890

(continued)

(concluded)

Liabilities and Shareholders' Equity:		
Current liabilities	$ 240	$ 210
Bonds payable	160	160
Paid-in capital	400	400
Retained earnings	200	120
	$1,000	$ 890

Required:
1. Determine the return on shareholders' equity for 2021.
2. What does the ratio measure?

E 18–25
IFRS; equity terminology
● LO18–9

Indicate by letter whether each of the terms or phrases listed below is more associated with financial statements prepared in accordance with U.S. GAAP (**U**) or International Financial Reporting Standards (**I**).

Terms and phrases

_____ 1. Common stock
_____ 2. Preference shares
_____ 3. Liabilities often listed before equity in the balance sheet (statement of financial position)
_____ 4. Asset revaluation reserve
_____ 5. Accumulated other comprehensive income
_____ 6. Share premium
_____ 7. Equity often listed before liabilities in the balance sheet (statement of financial position)
_____ 8. Translation reserve
_____ 9. Ordinary shares
_____ 10. Paid-in capital—excess of par
_____ 11. Net gains (losses) on investments—AOCI
_____ 12. Investment revaluation reserve
_____ 13. Preferred stock

E 18–26
General ledger exercise; bonds; installment note, early extinguishment
● LO18–4, LO18–5, LO18–7

On January 1, 2021, the general ledger of Freedom Fireworks includes the following account balances:

Accounts	Debit	Credit
Cash	$ 42,700	
Accounts Receivable	44,500	
Supplies	7,500	
Equipment	64,000	
Accumulated Depreciation		$ 9,000
Accounts Payable		14,600
Common Stock, $1 par value		10,000
Paid-in Capital—Excess of Par		80,000
Retained Earnings		45,100
Totals	$158,700	$158,700

During January 2021, the following transactions occur:

January 2	Issue an additional 2,000 shares of $1 par value common stock for $40,000.
January 9	Provide services to customers on account, $14,300.
January 10	Purchase additional supplies on account, $4,900.
January 12	Repurchase 1,000 shares of treasury stock for $18 per share.
January 15	Pay cash on accounts payable, $16,500.
January 21	Provide services to customers for cash, $49,100.
January 22	Receive cash on accounts receivable, $16,600.
January 29	Declare a cash dividend of $0.30 per share to all shares outstanding on January 29. The dividend is payable on February 15. (*Hint:* Freedom Fireworks had 10,000 shares outstanding on January 1, 2021, and dividends are not paid on treasury stock.)
January 30	Reissue 600 shares of treasury stock for $20 per share.
January 31	Pay cash for salaries during January, $42,000.

The following information is available on January 31, 2021.
1. Unpaid utilities for the month of January are $6,200.
2. Supplies at the end of January total $5,100.

3. Depreciation on the equipment for the month of January is calculated using the straight-line method. At the time the equipment was purchased, the company estimated a service life of three years and a residual value of $10,000.
4. Accrued income taxes at the end of January are $2,000.

Required:

1. Record each of the transactions listed above in the "General Journal" tab (these are shown as items 1–10) assuming a FIFO perpetual inventory system. Review the "General Ledger" and the "Trial Balance" tabs to see the effect of the transactions on the account balances.
2. Record adjusting entries on January 31, in the "General Journal" tab (these are shown as items 11–14).
3. Review the adjusted "Trial Balance" as of January 31, 2021, in the "Trial Balance" tab.
4. Prepare a multiple-step income statement for the period ended January 31, 2021, in the "Income Statement" tab.
5. Prepare a classified balance sheet as of January 31, 2021, in the "Balance Sheet" tab.
6. Record closing entries in the "General Journal" tab (these are shown as items 15 and 16).
7. Using the information from the requirements above, complete the "Analysis" tab.
   a. Calculate the return on equity for the month of January. If the average return on equity for the industry for January is 2.5%, is the company more or less profitable than other companies in the same industry?
   b. How many shares of common stock are outstanding as of January 31, 2018?
   c. Calculate earnings per share for the month of January. (Hint: To calculate average shares of common stock outstanding take the beginning shares outstanding plus the ending shares outstanding and divide the total by 2.) If earnings per share was $3.60 last year (i.e., an average of $0.30 per month), is earnings per share for January 2018 better or worse than last year's average?

Problems

connect

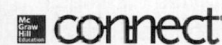

P 18–1
Various stock transactions; correction of journal entries
● LO18–4

Part A

During its first year of operations, the McCollum Corporation entered into the following transactions relating to shareholders' equity. The corporation was authorized to issue 100 million common shares, $1 par per share.

Required:

Prepare the appropriate journal entries to record each transaction.

Jan. 9	Issued 40 million common shares for $20 per share.
Mar. 11	Issued 5,000 shares in exchange for custom-made equipment. McCollum's shares have traded recently on the stock exchange at $20 per share.

Part B

A new staff accountant for the McCollum Corporation recorded the following journal entries during the second year of operations. McCollum retires shares that it reacquires (restores their status to that of authorized but unissued shares).

Required:

Prepare the journal entries that should have been recorded for each of the transactions.

		($ in millions)	
Sept. 1	Common stock	2	
	Retained earnings	48	
	Cash		50
Dec. 1	Cash	26	
	Common stock		1
	Gain on sale of previously issued shares		25

P 18–2
Share buyback—comparison of retirement and treasury stock treatment
● LO18–5

The shareholders' equity section of the balance sheet of TNL Systems Inc. included the following accounts at December 31, 2020:

Shareholders' Equity	($ in millions)
Common stock, 240 million shares at $1 par	$ 240
Paid-in capital—excess of par	1,680
Paid-in capital—share repurchase	1
Retained earnings	1,100

Required:

1. During 2021, TNL Systems reacquired shares of its common stock and later sold shares in two separate transactions. Prepare the entries for both the purchase and subsequent resale of the shares assuming the shares are (a) retired and (b) viewed as treasury stock.
   a. On February 5, 2021, TNL Systems purchased 6 million shares at $10 per share.

b. On July 9, 2021, the corporation sold 2 million shares at $12 per share.

c. On November 14, 2023, the corporation sold 2 million shares at $7 per share.

2. Prepare the shareholders' equity section of TNL Systems' balance sheet at December 31, 2023, comparing the two approaches. Assume all net income earned in 2021–2023 was distributed to shareholders as cash dividends.

P 18–3
Reacquired shares—comparison of retired shares and treasury shares
● LO18–5

National Supply's shareholders' equity included the following accounts at December 31, 2020:

Shareholders' Equity	($ in millions)
Common stock, 6 million shares at $1 par	$ 6,000,000
Paid-in capital—excess of par	30,000,000
Retained earnings	86,500,000

Required:

1. National Supply reacquired shares of its common stock in two separate transactions and later sold shares. Prepare the entries for each of the transactions under each of two separate assumptions: the shares are (a) retired and (b) accounted for as treasury stock.

February 15, 2021	Reacquired 300,000 shares at $8 per share.
February 17, 2022	Reacquired 300,000 shares at $5.50 per share.
November 9, 2023	Sold 200,000 shares at $7 per share (assume FIFO cost).

2. Prepare the shareholders' equity section of National Supply's balance sheet at December 31, 2023, assuming the shares are (a) retired and (b) accounted for as treasury stock. Net income was $14 million in 2021, $15 million in 2022, and $16 million in 2023. No dividends were paid during the three-year period.

P 18–4
Statement of retained earnings
● LO18–5, LO18–7

Comparative statements of retained earnings for Renn-Dever Corporation were reported in its 2021 annual report as follows.

RENN-DEVER CORPORATION
Statements of Retained Earnings
For the Years Ended December 31

	2021	2020	2019
Balance at beginning of year	$6,794,292	$5,464,052	$5,624,552
Net income (loss)	3,308,700	2,240,900	(160,500)
Deductions:			
Stock dividend (34,900 shares)	242,000		
Common shares retired (110,000 shares)		212,660	
Common stock cash dividends	889,950	698,000	0
Balance at end of year	$8,971,042	$6,794,292	$5,464,052

At December 31, 2018, common shares consisted of the following:

Common stock, 1,855,000 shares at $1 par	$1,855,000
Paid-in capital—excess of par	7,420,000

Required:

Infer from the reports the events and transactions that affected Renn-Dever Corporation's retained earnings during 2019, 2020, and 2021. Prepare the journal entries that reflect those events and transactions. (Hint: In lieu of revenues and expenses, use an account titled "Income summary" to close net income or net loss.)

P 18–5
Shareholders' equity transactions; statement of shareholders' equity
● LO18–6 through LO18–8

Listed below are the transactions that affected the shareholders' equity of Branch-Rickie Corporation during the period 2021–2023. At December 31, 2020, the corporation's accounts included:

	($ in thousands)
Common stock, 105 million shares at $1 par	$105,000
Paid-in capital—excess of par	630,000
Retained earnings	970,000

a. November 1, 2021, the board of directors declared a cash dividend of $0.80 per share on its common shares, payable to shareholders of record November 15, to be paid December 1.

b. On March 1, 2022, the board of directors declared a property dividend consisting of corporate bonds of Warner Corporation that Branch-Rickie was holding as an investment. The bonds had a fair value of $1.6 million, but were purchased two years previously for $1.3 million. Because they were intended to be held to maturity, the bonds had not been previously written up. The property dividend was payable to shareholders of record March 13, to be distributed April 5.

c. On July 12, 2022, the corporation declared and distributed a 5% common stock dividend (when the market value of the common stock was $21 per share). Cash was paid in lieu of fractional shares representing 250,000 equivalent whole shares.

d. On November 1, 2022, the board of directors declared a cash dividend of $0.80 per share on its common shares, payable to shareholders of record November 15, to be paid December 1.

e. On January 15, 2023, the board of directors declared and distributed a 3-for-2 stock split effected in the form of a 50% stock dividend when the market value of the common stock was $22 per share.

f. On November 1, 2023, the board of directors declared a cash dividend of $0.65 per share on its common shares, payable to shareholders of record November 15, to be paid December 1.

Required:

1. Prepare the journal entries that Branch-Rickie recorded during the three-year period for these transactions.
2. Prepare comparative statements of shareholders' equity for Branch-Rickie for the three-year period ($ in 000s). Net income was $330 million, $395 million, and $455 million for 2021, 2022, and 2023, respectively.

P 18–6
Statement of shareholders' equity
● LO18–1, LO18–3 through LO18–8

Comparative statements of shareholders' equity for Anaconda International Corporation were reported as follows for the fiscal years ending December 31, 2021, 2022, and 2023.

ANACONDA INTERNATIONAL CORPORATION
Statements of Shareholders' Equity
For the Years Ended Dec. 31, 2021, 2022, and 2023
($ in millions)

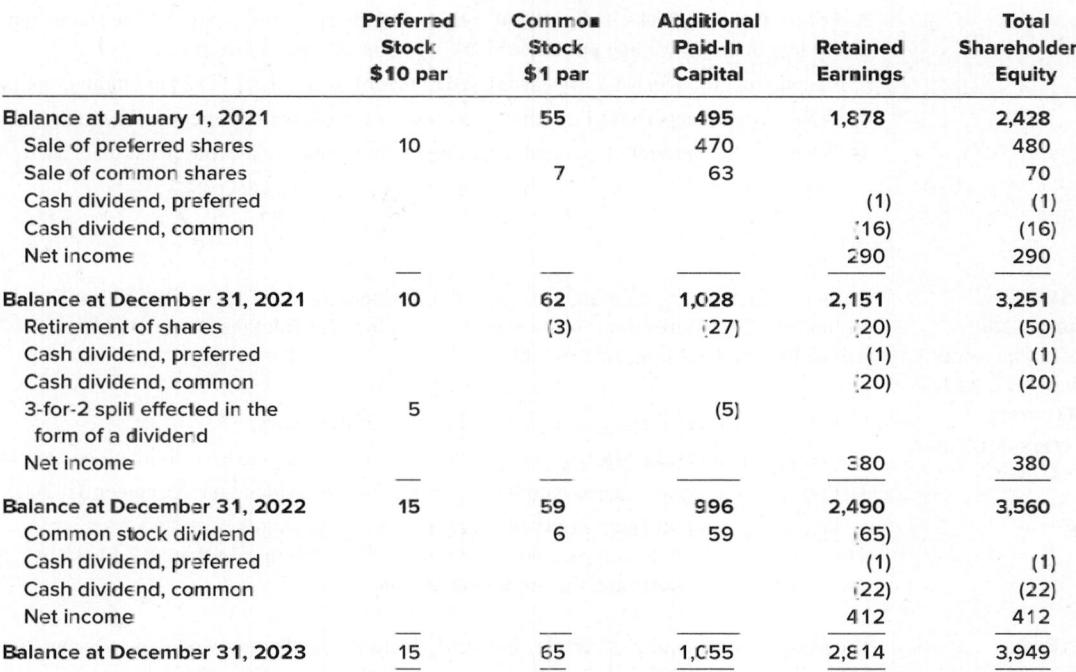

	Preferred Stock $10 par	Common Stock $1 par	Additional Paid-In Capital	Retained Earnings	Total Shareholders' Equity
Balance at January 1, 2021		55	495	1,878	2,428
Sale of preferred shares	10		470		480
Sale of common shares		7	63		70
Cash dividend, preferred				(1)	(1)
Cash dividend, common				(16)	(16)
Net income				290	290
Balance at December 31, 2021	10	62	1,028	2,151	3,251
Retirement of shares		(3)	(27)	(20)	(50)
Cash dividend, preferred				(1)	(1)
Cash dividend, common				(20)	(20)
3-for-2 split effected in the form of a dividend	5		(5)		
Net income				380	380
Balance at December 31, 2022	15	59	996	2,490	3,560
Common stock dividend		6	59	(65)	
Cash dividend, preferred				(1)	(1)
Cash dividend, common				(22)	(22)
Net income				412	412
Balance at December 31, 2023	15	65	1,055	2,814	3,949

Required:

1. Infer from the statements the events and transactions that affected Anaconda International Corporation's shareholders' equity during 2021, 2022, and 2023. Prepare the journal entries that reflect those events and transactions. (Hint: In lieu of revenues and expenses, use an account titled "Income summary" to close net income or net loss.)
2. Prepare the shareholders' equity section of Anaconda's comparative balance sheets at December 31, 2023 and 2022.

P 18–7
Reporting shareholders' equity; comprehensive income; Cisco Systems
● LO18–1 through LO18–4
Real World Financials

The following is a portion of the Statement of Shareholders' Equity from **Cisco Systems**' July 29, 2017 annual report.

CISCO SYSTEMS, INC.
Consolidated Statements of Equity (in part)

($ in millions)	Shares of Common Stock	Common Stock and Additional Paid-In Capital	Retained Earnings	Accumulated Other Comprehensive Income	Total Shareholder' Equity
Balance at July 25, 2016	5,029	$44,516	$19,396	$(326)	$63,586
Net income			9,609		9,609
Other comprehensive income (loss)				372	372

(continued)

(concluded)

Issuance of common stock	92	708			708
Repurchase of common stock	(118)	(1,050)	(2,656)		(3,706)
Shares repurchased for tax withholdings on vesting of restricted stock units	(20)	(619)			(619)
Cash dividends declared ($1.10 per common share)			(5,511)		(5,511)
Tax effects from employee stock incentive plans		(10)			(10)
Share-based compensation expense		1,540			1,540
Purchase acquisitions		168			168
Balance at July 29, 2017	4,983	$45,253	$20,838	$ 46	$66,137

Required:

1. How does Cisco account for its share buybacks? Treasury stock or retired shares?
2. For its share buybacks in the period shown, was the price Cisco paid for the shares repurchased (a) more or (b) less than the average price at which Cisco had sold the shares previously?
3. Reconstruct the journal entry Cisco used to record the buyback. The par amount of Cisco's shares is $0.001.
4. What two amounts caused the change in Cisco's comprehensive income for the period shown?
5. What was the amount of Accumulated other comprehensive income (loss) that Cisco reported in its July 29, 2017 balance sheet?

Source: Cisco Systems

P 18–8
Share issue costs; issuance; dividends; early retirement
● LO18–3, LO18–4, LO18–7

During its first year of operations, Cupola Fan Corporation issued 30,000 of $1 par Class B shares for $385,000 on June 30, 2021. Share issue costs were $1,500. One year from the issue date (July 1, 2022), the corporation retired 10% of the shares for $39,500.

Required:

1. Prepare the journal entry to record the issuance of the shares.
2. Prepare the journal entry to record the declaration of a $2 per share dividend on December 1, 2021.
3. Prepare the journal entry to record the payment of the dividend on December 31, 2021.
4. Prepare the journal entry to record the retirement of the shares.

(Note: You may wish to compare your solution to this problem with that of P 14–16, which deals with parallel issues of debt issue costs and the retirement of debt.)

P 18–9
Effect of preferred stock characteristics on dividends
● LO18–7

The shareholders' equity of Kramer Industries includes the data shown below. During 2022, cash dividends of $150 million were declared. Dividends were not declared in 2020 or 2021.

	($ in millions)
Common stock	$200
Paid-in capital—excess of par, common	800
Preferred stock, 10%, nonparticipating	100
Paid-in capital—excess of par, preferred	270

Required:

Determine the amount of dividends payable to preferred shareholders and to common shareholders under each of the following two assumptions regarding the characteristics of the preferred stock.

Assumption A —The preferred stock is noncumulative.
Assumption B —The preferred stock is cumulative.

P 18–10
Transactions affecting retained earnings
● LO18–4 through LO18–8

Indicate by letter whether each of the transactions listed below increases (**I**), decreases (**D**), or has no effect (**N**) on retained earnings. Assume the shareholders' equity of the transacting company includes only common stock, paid-in capital—excess of par, and retained earnings at the time of each transaction. (Some transactions have two possible answers. Indicate both.)

Transactions

N	1. Sale of common stock
_____	2. Purchase of treasury stock at a cost *less* than the original issue price
_____	3. Purchase of treasury stock at a cost *greater* than the original issue price
_____	4. Declaration of a property dividend
_____	5. Sale of treasury stock for *more* than cost
_____	6. Sale of treasury stock for *less* than cost
_____	7. Net income for the year
_____	8. Declaration of a cash dividend
_____	9. Payment of a previously declared cash dividend
_____	10. Issuance of convertible bonds for cash
_____	11. Declaration and distribution of a 5% stock dividend
_____	12. Retirement of common stock at a cost *less* than the original issue price
_____	13. Retirement of common stock at a cost *greater* than the original issue price
_____	14. A stock split effected in the form of a stock dividend
_____	15. A stock split in which the par value per share is reduced (not effected in the form of a stock dividend)
_____	16. A net loss for the year

P 18–11
Stock dividends received on investments; integrative problem
● LO18–8

Ellis Transport Company acquired 1.2 million shares of stock in L&K Corporation at $44 per share. They are reported by Ellis at "fair value through net income." Ellis sold 200,000 shares at $46, received a 10% stock dividend, and then later in the year sold another 100,000 shares at $43.

Hint: There is no entry for the stock dividend, but a new investment per share must be calculated for use later when the shares are sold.

Required:

Prepare journal entries to record these transactions.

P 18–12
Various shareholders' equity topics; comprehensive
● LO18–1, LO18–4 through LO18–8

Part A

In late 2020, the Nicklaus Corporation was formed. The corporate charter authorizes the issuance of 5,000,000 shares of common stock carrying a $1 par value, and 1,000,000 shares of $5 par value, noncumulative, nonparticipating preferred stock. On January 2, 2021, 3,000,000 shares of the common stock are issued in exchange for cash at an average price of $10 per share. Also on January 2, all 1,000,000 shares of preferred stock are issued at $20 per share.

Required:

1. Prepare journal entries to record these transactions.
2. Prepare the shareholders' equity section of the Nicklaus balance sheet as of March 31, 2021. (Assume net income for the first quarter 2021 was $1,000,000.)

Part B

During 2021, the Nicklaus Corporation participated in three treasury stock transactions:

a. On June 30, 2021, the corporation reacquires 200,000 shares for the treasury at a price of $12 per share.
b. On July 31, 2021, 50,000 treasury shares are reissued at $15 per share.
c. On September 30, 2021, 50,000 treasury shares are reissued at $10 per share.

Required:

1. Prepare journal entries to record these transactions.
2. Prepare the Nicklaus Corporation shareholders' equity section as it would appear in a balance sheet prepared at September 30, 2021. (Assume net income for the second and third quarter was $3,000,000.)

Part C

On October 1, 2021, Nicklaus Corporation receives permission to replace its $1 par value common stock (5,000,000 shares authorized, 3,000,000 shares issued, and 2,900,000 shares outstanding) with a new common stock issue having a $0.50 par value. Since the new par value is one-half the amount of the old, this represents a 2-for-1 stock split. That is, the shareholders will receive two shares of the $0.50 par stock in exchange for each share of the $1 par stock they own. The $1 par stock will be collected and destroyed by the issuing corporation.

On November 1, 2021, the Nicklaus Corporation declares a $0.05 per share cash dividend on common stock and a $0.25 per share cash dividend on preferred stock. Payment is scheduled for December 1, 2021, to shareholders of record on November 15, 2021.

On December 2, 2021, the Nicklaus Corporation declares a 1% stock dividend payable on December 28, 2021, to shareholders of record on December 14. At the date of declaration, the common stock was selling in the open market at $10 per share. The dividend will result in 58,000 (0.01 × 5,800,000) additional shares being issued to shareholders.

Required:

1. Prepare journal entries to record the declaration and payment of these stock and cash dividends.
2. Prepare the December 31, 2021, shareholders' equity section of the balance sheet for the Nicklaus Corporation. (Assume net income for the fourth quarter was $2,500,000.)
3. Prepare a statement of shareholders' equity for Nicklaus Corporation for 2021.

P 18–13
Quasi reorganization (based on Appendix 18)
● Appendix 18

A new CEO was hired to revive the floundering Champion Chemical Corporation. The company had endured operating losses for several years, but confidence was emerging that better times were ahead. The board of directors and shareholders approved a quasi reorganization for the corporation. The reorganization included devaluing inventory for obsolescence by $105 million and increasing land by $5 million. Immediately prior to the restatement, at December 31, 2021, Champion Chemical Corporation's balance sheet appeared as follows (in condensed form):

CHAMPION CHEMICAL CORPORATION
Balance Sheet
At December 31, 2021
($ in millions)

Cash	$ 20
Receivables	40
Inventory	230
Land	40
Buildings and equipment (net)	90
	$420
Liabilities	$240
Common stock (320 million shares at $1 par)	320
Additional paid-in capital	60
Retained earnings (deficit)	(200)
	$420

Required:

1. Prepare the journal entries appropriate to record the quasi reorganization on January 1, 2022.
2. Prepare a balance sheet as it would appear immediately after the restatement.

Decision Makers' Perspective

©ImageGap/ Getty Images

Apply your critical-thinking ability to the knowledge you've gained. These cases will provide you an opportunity to develop your research, analysis, judgment, and communication skills. You also will work with other students, integrate what you've learned, apply it in real-world situations, and consider its global and ethical ramifications. This practice will broaden your knowledge and further develop your decision-making abilities.

Real World Case 18–1
Initial public offering of common stock; Dolby Laboratories
● LO18–4

Real World Financials

Ray Dolby started **Dolby Laboratories** and since then has been a leader in the entertainment industry and consumer electronics. Closely held since its founding in 1965, Dolby decided to go public. Here's an AP news report:

DOLBY'S IPO EXPECTED TO PLAY SWEET MUSIC—The initial public offering market is hoping for a big bang this week from Dolby Laboratories Inc. The San Francisco company, whose sound systems and double-D logo are ubiquitous in the movie industry as well as in consumer electronics, plans to sell 27.5 million shares for $13.50 to $15.50 each. Founded by Cambridge-trained scientist Ray Dolby 39 years ago, the company started out manufacturing noise-reduction equipment for the music industry that eliminated the background "hiss" on recordings, and has since expanded to encompass everything from digital audio systems to Dolby Surround sound. The company's IPO, which is lead-managed by underwriters Morgan Stanley and Goldman Sachs Group Inc., is expected to do well not only because of its brand recognition, but also because of its strong financials. (*AP*)

Required:

1. Assuming the shares are issued at the midpoint of the price range indicated, how much capital did the IPO raise for Dolby Laboratories before any underwriting discount and offering expenses?
2. If the par amount is $0.01 per share, what journal entry did Dolby use to record the sale?

Source: Associated Press

Analysis Case 18–2
Statement of shareholders' equity
● LO18–1, LO18–3, LO18–6, LO18–7

The shareholders' equity portion of the balance sheet of Sessel's Department Stores, Inc., a large regional specialty retailer, is as follows:

SESSEL'S DEPARTMENT STORES, INC.
Comparative Balance Sheets
Shareholders' Equity Section

($ in thousands, except per share amounts)	Dec. 31, 2022	Dec. 31, 2021
Shareholders' Equity		
Preferred stock—$1 par value; 20,000 total shares authorized, Series A—600 shares authorized, issued, and outstanding, $50 per share liquidation preference	$ 57,700	$ —
Series B—33 shares authorized, no shares outstanding		
Common stock—$0.10 par; 200,000 shares authorized, 19,940 and 18,580 shares issued and outstanding at Dec. 31, 2022, and Dec. 31, 2021, respectively	1,994	1,858
Additional paid-in capital	227,992	201,430
Retained income	73,666	44,798
Total shareholders' equity	$361,352	$248,086

Disclosures elsewhere in Sessel's annual report revealed the following changes in shareholders' equity accounts for 2022, 2021, 2020:

2022:

1. The only changes in retained earnings during 2022 were preferred dividends on preferred stock of $3,388,000 and net income.
2. The preferred stock is convertible. During the year, 6,592 shares were issued. All shares were converted into 320,000 shares of common stock. No gain or loss was recorded on the conversion.
3. Common shares were issued in a public offering and upon the exercise of stock options. On the statement of shareholders' equity, Sessel's reports these two items on a single line entitled: "Issuance of shares."

2021:

1. Net income: $12,126,000.
2. Issuance of common stock: 5,580,000 shares at $112,706,000.

2020:

1. Net income: $13,494,000.
2. Issuance of common stock: 120,000 shares at $826,000.

Required:
From these disclosures, prepare comparative statements of shareholders' equity for 2022, 2021, and 2020.

Communication Case 18–3
IFRS; Is preferred stock debt or equity? Group interaction
● LO18–1, LO18–9
IFRS

An unsettled question in accounting for stock is this: Should preferred stock be recognized as a liability, or should it be considered equity? Under International Financial Reporting Standards, preferred stock (preference shares) often is reported as debt with the dividends reported in the income statement as interest expense. Under U.S. GAAP, that is the case only for "mandatorily redeemable" preferred stock.

Two opposing viewpoints are:

View 1: Preferred stock should be considered equity.
View 2: Preferred stock should be reported as a liability.

In considering this question, focus on conceptual issues regarding the practicable and theoretically appropriate treatment, unconstrained by GAAP.

Required:

1. Which view do you favor? Develop a list of arguments in support of your view prior to the class session for which the case is assigned.
2. In class, your instructor will pair you (and everyone else) with a classmate (who also has independently developed an argument).
   a. You will be given three minutes to argue your view to your partner. Your partner likewise will be given three minutes to argue his or her view to you. During these three-minute presentations, the listening partner is not permitted to speak.
   b. Then after each person has had a turn attempting to convince his or her partner, the two partners will have a three-minute discussion in which they will decide which view is more convincing and arguments will be merged into a single view for each pair.

3. After the allotted time, a spokesperson for each of the two views will be selected by the instructor. Each spokesperson will field arguments from the class in support of that view's position and list the arguments on the board. The class then will discuss the merits of the two lists of arguments and attempt to reach a consensus view, though a consensus is not necessary.

Research Case 18–4
FASB codification; comprehensive income; locate and extract relevant information and authoritative support for a financial reporting issue; integrative; Cisco Systems
● LO18–2

Real World Financials

Titan Networking became a public company through an IPO (initial public offering) two weeks ago. You are looking forward to the challenges of being assistant controller for a publicly owned corporation. One such challenge came in the form of a memo in this morning's in-box. "We need to start reporting comprehensive income in our financials," the message from your boss said. "Do some research on that, will you? That concept didn't exist when I went to school." In response, you sought out the financial statements of **Cisco Systems**, the networking industry leader. The following are excerpts from disclosure notes from Cisco's 2017 annual report:

	Years Ended		
	July 29, 2017	**July 30, 2016**	**July 25, 2015**
Net income	$9,609	$10,739	$8,981
Other comprehensive income (loss):			
Available-for-sale investments:			
Change in net unrealized gains, net of tax	(89)	92	(12)
Net (gains) losses reclassified into earnings, net of tax	50	1	(100)
	(39)	93	(112)
Cash flow hedging instruments:			
Change in unrealized gains and losses, net of tax	17	(59)	(140)
Net (gains) losses reclassified into earnings	74	16	136
	91	(43)	(4)
Net change in cumulative translation adjustment and actuarial gains and losses, net of tax	321	(447)	(498)
Other comprehensive income (loss)	373	(397)	(614)
Comprehensive income	$9,982	$10,342	$8,367

Required:

1. Locate the financial statements of Cisco in the Investor Relations section of Cisco's website. Search the 2017 annual report for information about how Cisco accounts for comprehensive income. What does Cisco report in its balance sheet for 2017 Accumulated other comprehensive income?
2. From the information Cisco's financial statements provide, show the calculation of the change in the accumulated other comprehensive income from the end of fiscal 2016 to the end of fiscal 2017. *Hint:* One component of the change is Comprehensive (income) loss attributable to noncontrolling interests.
3. Access the *FASB Accounting Standards Codification* at the FASB website (**www.fasb.org**). Identify the specific eight-digit Codification citation (XXX-XX-XX-X) from the authoritative literature that describes the two alternative formats for reporting comprehensive income.

Source: Cisco Systems

Real World Case 18–5
Share repurchase; stock split; cash dividends; Nike, Inc.
● LO18–5 through LO18–8

Real World Financials

Nike is the world's leading designer, marketer, and distributor of authentic athletic footwear, apparel, equipment, and accessories. The following is a press release from the company:

NIKE, INC. ANNOUNCES 11 PERCENT INCREASE IN QUARTERLY DIVIDEND

BEAVERTON, Ore.—(BUSINESS WIRE)—NIKE, Inc. (NYSE: NKE) announced today that its Board of Directors has approved a quarterly cash dividend of $0.20 per share. . . . This represents an increase of 11 percent versus the prior quarterly dividend rate of $0.18 per share. The dividend declared today is payable on January 2, 2018, to shareholders of record at the close of business December 4, 2017. "Today's announcement, combined with the four-year $12 billion share repurchase program we announced in 2015, demonstrates our continued confidence in generating strong cash flow and returns for shareholders through our new Consumer Direct Offense as we continue to invest in fueling sustainable, long-term growth and profitability."

When the share repurchase program was announced the company also declared a stock split distributed in the form of a 100% stock dividend. At that time Nike's 1,200 million shares were trading at $130 per share. Nike's shares have a stated value of $0.001 per share.

Required:

1. How did Nike account for the stock split? Prepare the appropriate journal entry.
2. Assume Nike repurchased 50 million shares after the stock split at an average price of $59 per share. The original issue price of the shares, after adjusting for the 6 stock splits since the shares were issued, was

$0.15 per share. What entry would Nike have recorded to account for the repurchase? Nike views the repurchase of stock as a formal retirement of shares.

3. Suppose Nike views the repurchase of stock as an acquisition of treasury stock. What entry would Nike have recorded to account for the repurchase?
4. How should Nike account for the cash dividend announced in the press release? Prepare the entries for the declaration and for the payment of the dividend.

Source: Nike, Inc.

Communication Case 18–6
Issuance of shares; share issue costs; prepare a report
● LO18–4

You are the newest member of the staff of Brinks & Company, a medium-size investment management firm. You are supervised by Les Kramer, an employee of two years. Les has a reputation as being technically sound but has a noticeable gap in his accounting education. Knowing you are knowledgeable about accounting issues, he requested you provide him with a synopsis of accounting for share issue costs.

"I thought the cost of issuing securities is recorded separately and expensed over time," he stated in a handwritten memo. "But I don't see that for IBR's underwriting expenses. What gives?"

He apparently was referring to a disclosure note on a page of a prospective investee's annual report, photocopied and attached to his memo. To raise funds for expansion, the company sold additional shares of its $0.10 par common stock. The following disclosure note appeared in the company's most recent annual report:

NOTES TO CONSOLIDATED FINANCIAL STATEMENTS

Note 10—Stock Transactions (in part)

In February and March, the Company sold 2,395,000 shares of Common Stock at $22.25 per share in a public offering. Net proceeds to the Company were approximately $50.2 million after the underwriting discount and offering expenses.

Required:

Write a formal memo to your supervisor. Briefly explain how share issue costs are accounted for and how that accounting differs from that of debt issue costs. To make sure your explanation is understood in context of the footnote, include in your memo the following:

a. At what total amount did the shares sell to the public? How is the difference between this amount and the $50.2 million net proceeds accounted for?
b. The appropriate journal entry to record the sale of the shares.

Analysis Case 18–7
Analyzing financial statements; price-earnings ratio; dividend payout ratio
● LO18–1

AGF Foods Company is a large, primarily domestic, consumer foods company involved in the manufacture, distribution, and sale of a variety of food products. Industry averages are derived from Troy's *The Almanac of Business and Industrial Financial Ratios* and Dun and Bradstreet's *Industry Norms and Key Business Ratios*. Following are the 2021 and 2020 comparative income statements and balance sheets for AGF. The market price of AGF's common stock is $47 at the end of 2021. (The financial data we use are from actual financial statements of a well-known corporation, but the company name used in our illustration is fictitious and the numbers and dates have been modified slightly to disguise the company's identity.)

Profitability is the key to a company's long-run survival. Profitability measures focus on a company's ability to provide an adequate return relative to resources devoted to company operations.

AGF FOODS COMPANY
Years Ended December 31, 2021 and 2020

($ in millions)	2021	2020
Comparative Income Statements		
Net sales	$ 6,440	$ 5,800
Cost of goods sold	(3,667)	(3,389)
Gross profit	2,773	2,411
Operating expenses	(1,916)	(1,629)
Operating income	857	782
Interest expense	(54)	(53)
Income from operations before tax	803	729
Income taxes	(316)	(287)
Net income	$ 487	$ 442
Average shares outstanding	181 million	181 million

(continued)

(concluded)

Comparative Balance Sheets

Assets		
Current assets:		
Cash	$ 48	$ 142
Accounts receivable	347	320
Marketable securities	358	—
Inventories	914	874
Prepaid expenses	212	154
Total current assets	1,879	1,490
Property, plant, and equipment (net)	2,592	2,291
Intangibles (net)	800	843
Other assets	74	60
Total assets	$ 5,345	$ 4,684
Liabilities and Shareholders' Equity		
Current liabilities:		
Accounts payable	$ 254	$ 276
Accrued liabilities	493	496
Notes payable	518	115
Current portion of long-term debt	208	54
Total current liabilities	1,473	941
Long-term debt	534	728
Deferred income taxes	407	344
Total liabilities	2,414	2,013
Shareholders' equity:		
Common stock, $1 par	180	180
Additional paid-in capital	21	63
Retained earnings	2,730	2,428
Total shareholders' equity	2,931	2,671
Total liabilities and shareholders' equity	$ 5,345	$ 4,684

Required:

1. Calculate the return on shareholders' equity for AGF. The average return for the stocks listed on the New York Stock Exchange in a comparable period was 18.8%. What information does your calculation provide an investor?
2. Calculate AGF's earnings per share and earnings-price ratio. The average return for the stocks listed on the New York Stock Exchange in a comparable time period was 5.4%. What does your calculation indicate about AGF's earnings?

Ethics Case 18–8
The Swiss label maker; value of shares issued for equipment
● LO18–4

Bricker Graphics is a privately held company specializing in package labels. Representatives of the firm have just returned from Switzerland, where a Swiss firm is manufacturing a custom-made high speed, color labeling machine. Confidence is high that the new machine will help rescue Bricker from sharply declining profitability. Bricker's chief operating officer, Don Benson, has been under fire for not reaching the company's performance goals of achieving a rate of return on assets of at least 12%.

The afternoon of his return from Switzerland, Benson called Susan Sharp into his office. Susan is Bricker's Controller.

Benson: I wish you had been able to go. We have some accounting issues to consider.
Sharp: I wish I'd been there, too. I understand the food was marvelous. What are the accounting issues?
Benson: They discussed accepting our notes at the going rate for a face amount of $12.5 million. We also discussed financing with stock.
Sharp: I thought we agreed; debt is the way to go for us now.
Benson: Yes, but I've been thinking. We can issue shares for a total of $10 million. The labeler is custom made and doesn't have a quoted selling price, but the domestic labelers we considered went for around $10 million. It sure would help our rate of return if we keep the asset base as low as possible.

Required:

1. How will Benson's plan affect the return measure? What accounting issue is involved?
2. Is the proposal ethical?
3. Who would be affected if the proposal is implemented?

Data Analytics

connect

Data analytics is the process of examining data sets in order to draw conclusions about the information they contain. If you haven't completed any of the prior data analytics cases, follow the instructions listed in the Chapter 1 Data Analytics case to get set up. You will need to watch the videos referred to in the Chapters 1 - 3 Data Analytics cases. No additional videos are required for this case. All short training videos can be found here: **https://www.tableau.com/learn/training#getting-started.**

Data Analytics Case
Rate of Return on Shareholders' Equity
● LO18–7

In the Data Analytics Cases in the previous chapter, you used Tableau to examine two (hypothetical) publicly traded companies: GPS Corporation and Tru, Inc. regarding changes in the funded status of their pension plans during the previous ten years as well as the way components of pension expense are reported in the income statement. This time, you investigate the ability of the two companies to provide a return to the investors who buy the companies' stock.

Required:

For each of the two companies in the most recent five-year period, 2017-2021, use Tableau to calculate and display the trends for (a) the debt to equity ratio, (b) the rate of return on shareholders' equity [Net income/Shareholders' equity], and (c) the rate of return on assets [Net income / Total assets]. Based upon what you find, answer the following questions:

1. Which of the two companies, GPS Corporation or Tru, Inc., finances a higher percentage of its assets by borrowed funds relative to funds invested by shareholders as measured by the debt to equity ratio in 2021?
2. Which of the two companies, GPS Corporation or Tru, Inc., indicates a higher profitability during the period 2017-2021 without regard to the sources of financing as measured by the return on assets ratio?
3. Which of the two companies, GPS Corporation or Tru, Inc., indicates a higher effectiveness of employing resources provided by owners during the period 2017-2021 as measured by the return on shareholders' equity ratio?
4. Management is using its borrowed funds to enhance the earnings for shareholders of (a) GPS Corporation, (b) Tru, Inc., (c) both firms, or (d) neither firm during the period 2017-2021?

Resources:

Download the "GPS_Tru_Financials.xlsx" Excel file available in Connect, or under Additional Student Resources within the Library tab. Save it to the computer on which you will be using Tableau.

For this case, you will create calculations to produce several long-term solvency ratios to allow you to compare and contrast the two companies.

After you view the training videos, follow these steps to create the charts you'll use for this case:

- Open Tableau and connect to the Excel spreadsheet you downloaded.
- Click on the "Sheet 1" tab at the bottom of the canvas, to the right of the Data Source at the bottom of the screen.
- Drag "($ in 000s) Company" and "Year" under Dimensions to the Columns shelf. Change "Year" to *discrete* by right-clicking and selecting "discrete." Select "Show Filter" and uncheck all the years except the last five years (2017 – 2021). Click OK.
- Create a calculated field by clicking the "Analysis" tab at the top of the screen and selecting "Create Calculated field." Name the calculation "Debt to equity ratio." In the Calculation Editor window, from Measures, drag "Total liabilities (Cur + noncur)", type a division sign, and drag "Total shareholders' equity" beside it. Make sure the window says that the calculation is valid and click OK.
- Drag the newly created "Debt to equity ratio" to the Rows shelf. Click on the "Show Me" and select "side-by-side bars." Add labels to the bars by clicking on "Label" under the "Marks" card and clicking the box "Show mark labels." Format the labels to Times New Roman, bold, black and 10-point font. Edit the color on the Color marks card if desired.
- Change the title of the sheet to be "Debt to Equity Ratio Trend" by right-clicking and selecting "Edit title." Format the title to Times New Roman, bold, black and 15-point font. Change the title of "Sheet 1" to match the sheet title by right-clicking, selecting "Rename" and typing in the new title.
- Click on the New Worksheet tab ("Sheet 2" should be open) to the right of the newly named "Debt to Equity Ratio Trend" sheet at the bottom of the screen.
- Follow the procedures above for the company and year.
- Create a calculated field by clicking the "Analysis" tab at the top of the screen and selecting "Create Calculated field." Name the calculation "Rate of return on shareholders' equity." In the Calculation Editor window, from Measures, drag "Net income/(loss)", type a division sign, and then drag "Total shareholders' equity". Make sure the window says that the calculation is valid and click OK.

- Create a calculated field by clicking the "Analysis" tab at the top of the screen and selecting "Create Calculated field." Name the calculation "Rate of return on assets." In the Calculation Editor window from Measures, drag "Net income/(loss)", type a division sign, and then drag "Total assets". Make sure the window says that the calculation is valid and click OK.
- Drag the newly created "Rate of return on shareholders' equity" and "Rate of return on assets" to the Rows shelf. Add labels to the bars by clicking on "Label" under the "Marks card" and clicking the box "Show mark label." Format the labels to Times New Roman, bold, black and 10-point font. Select Field in the upper right corner of the "Format Font" box and change to Percentage with one decimal place. Edit the color of the years on the Color marks card if desired.
- Edit the axis to utilize the same scale by right-clicking on the axis and selecting "Edit Axis. . .". Select Fixed and set from 0 to 0.35 for both charts.
- Change the title of the sheet to be "Rate of Return on Shareholders' Equity and Assets Trends" by right-clicking and selecting "Edit title." Format the title to Times New Roman, bold, black and 15-point font. Change the title of "Sheet 2" to match the sheet title by right-clicking, selecting "Rename" and typing in the new title.
- Format all other labels to be Times New Roman, bold, black and 12-point font.
- Once complete, save the file as "DA18_Your initials.twbx."

Continuing Cases

Target Case
● LO18–1, LO18–5

Target Corporation prepares its financial statements according to U.S. GAAP. Target's financial statements and disclosure notes for the year ended February 3, 2018, are available in Connect. This material also is available under the Investor Relations link at the company's website (**www.target.com**). Target refers to its Shareholders' Equity as Shareholders' Investment.

Required:

1. Refer to Target's Consolidated Statements of Shareholders' Investment. What are the six types of events and transactions that affected one or more of the Shareholders' Investment accounts in the year ended February 3, 2018?
2. Note 25, "Share Repurchase," provides the information we need to reconstruct the journal entry that summarizes Target's share repurchases in the year ended February 3, 2018. Provide that entry (dollars in millions, rounded to the nearest million).
3. Does Target account for share repurchases as (a) treasury stock or (b) retired shares?

Air France–KLM Case
● LO18–9
IFRS

Air France–KLM (AF), a Franco-Dutch company, prepares its financial statements according to International Financial Reporting Standards. AF's financial statements and disclosure notes for the year ended December 31, 2017, are available in Connect. This material is also available under the Finance link at the company's website (**www.airfranceklm.com**).

Required:

1. Locate Note 28.5 in AF's financial statements. What items comprise "Reserves and retained earnings" as reported in the balance sheet?
2. In its presentation of the components of the balance sheet, which is listed first, current assets or non-current assets? Does this approach differ from U.S. GAAP?
3. In its presentation of the components of the balance sheet, which is listed first, current liabilities or non-current liabilities? Does this approach differ from U.S. GAAP?
4. In its presentation of the components of the balance sheet, which is listed first, liabilities or shareholders' equity? Does this approach differ from U.S. GAAP?

CPA Exam Questions and Simulations

ROGER CPA Review

Sample CPA Exam questions from Roger CPA Review are available in Connect as support for the topics in this chapter. These multiple-choice questions and task-based simulations include expert-written explanations and solutions, and provide a starting point for students to become familiar with the content and functionality of the actual CPA Exam.

CHAPTER

19 Share-Based Compensation and Earnings Per Share

OVERVIEW

We've discussed a variety of employee compensation plans in prior chapters, including pension and other postretirement benefits in Chapter 17. In this chapter, we look at some common forms of compensation in which the amount of the compensation employees receive is tied to the market price of company stock. We will see that these *share-based* compensation plans—restricted stock awards, restricted stock units, stock options, and stock appreciation rights—create shareholders' equity, which was the topic of the previous chapter, and which also often affects the way we calculate earnings per share, the topic of the second part of the current chapter. Specifically, we view these as *potential common shares* along with convertible securities, and we calculate earnings per share as if the securities already had been exercised or converted into additional common shares.

LEARNING OBJECTIVES

After studying this chapter, you should be able to:

- **LO19–1** Explain and implement the accounting for restricted stock plans. (*p. 1106*)
- **LO19–2** Explain and implement the accounting for stock options. (*p. 1108*)
- **LO19–3** Explain and implement the accounting for employee share purchase plans. (*p. 1117*)
- **LO19–4** Distinguish between a simple and a complex capital structure. (*p. 1119*)
- **LO19–5** Describe what is meant by the weighted-average number of common shares. (*p. 1120*)
- **LO19–6** Differentiate the effect on EPS of the sale of new shares, a stock dividend or stock split, and the reacquisition of shares. (*p. 1120*)
- **LO19–7** Describe how preferred dividends affect the calculation of EPS. (*p. 1122*)
- **LO19–8** Describe how options, rights, and warrants are incorporated in the calculation of EPS. (*p. 1123*)
- **LO19–9** Describe how convertible securities are incorporated in the calculation of EPS. (*p. 1126*)
- **LO19–10** Determine whether potential common shares are antidilutive. (*p. 1128*)
- **LO19–11** Describe the two components of the proceeds used in the treasury stock method and how restricted stock is incorporated in the calculation of EPS. (*p. 1133*)
- **LO19–12** Explain the way contingently issuable shares are incorporated in the calculation of EPS. (*p. 1134*)
- **LO19–13** Describe the way EPS information should be reported in an income statement. (*p. 1136*)
- **LO19–14** Discuss the primary differences between U.S. GAAP and IFRS with respect to accounting for share-based compensation and earnings per share. (*pp. 1113, 1115,* and *1119*)

©Fabio Cardoso/Getty Images

FINANCIAL REPORTING CASE

Proper Motivation?

The coffee room discussion Thursday morning was particularly lively. Yesterday's press release describing National Electronic Ventures' choice of Sandra Veres as its president and chief operating officer was today's hot topic in all the company's departments. The press release noted that Ms. Veres's compensation package includes elements beyond salary that are intended to not only motivate her to accept the offer, but also to remain with the company and work to increase shareholder value. Excerpts from the release follow:

National Electronic Ventures, Inc., today announced it had attracted G. Sandra Veres, respected executive from the wireless communications industry, to succeed chairman Walter Kovac. Veres will assume the new role as CEO on Jan. 1, 2021. Ms. Veres will receive a compensation package at NEV of more than $1 million in salary, stock options to buy more than 800,000 shares of NEV stock, and a grant of restricted stock.

By the time you finish this chapter, you should be able to respond appropriately to the questions posed in this case. Compare your response to the solution provided at the end of the chapter.

QUESTIONS

1. How can a compensation package such as this serve as an incentive to Ms. Veres? *(p. 1105)*
2. Ms. Veres received a "grant of restricted stock." How should NEV account for the grant? *(p. 1106)*
3. Included were stock options to buy more than 800,000 shares of NEV stock. How will the options affect NEV's compensation expense? *(p. 1108)*
4. How will the presence of these and other similar stock options affect NEV's earnings per share? *(p. 1124)*

Share-Based Compensation

PART A

Employee compensation plans frequently include share-based awards. These awards are forms of payment whose value is dependent on the value of the company's stock. These may be outright awards of shares, stock options, or cash payments tied to the market price of shares. Sometimes only key executives participate in a stock benefit plan. In fact, CEOs typically get more than half their total compensation from share-based compensation plans. Typically, an executive compensation plan is tied to performance in a strategy that uses compensation to motivate its recipients. Some firms pay their directors entirely in shares. Actual compensation depends on the market value of the shares. Obviously, that's quite an incentive to act in the best interests of shareholders.

Although the variations of share-based compensation plans are seemingly endless, each shares common goals. Whether the plan is a stock award plan, a stock option plan, a stock appreciation rights (SARs) plan, or one of the several similar plans, the goals are to provide compensation to designated employees, while sometimes providing those employees with some sort of performance incentive. Likewise, our goals in accounting for each of these plans are the same for each: (1) to determine the fair value of the compensation and (2) to expense that compensation over the periods in which participants perform services. The issue is not trivial. Salary often is a minor portion of executive pay relative to stock awards and stock options.

FINANCIAL Reporting Case

Q1, p. 1105

The accounting objective is to record compensation expense over the periods in which related services are performed.

Restricted Stock Plans

● LO19–1

Executive compensation sometimes includes a grant of shares of stock or the right to receive shares. Usually, such shares are restricted in such a way as to provide some incentive to the recipient. Typically, restricted stock award plans are tied to continued employment. The two primary types of restricted stock plans are (a) restricted stock awards and (b) restricted stock units. Let's compare the two.

Restricted Stock Awards

Usually, restricted shares are subject to forfeiture if the employee doesn't remain with the company.

In a **restricted stock award**, shares actually are awarded in the name of the employee, although the company might retain physical possession of the shares. The employee has all rights of a shareholder, subject to certain restrictions or forfeiture. Ordinarily, the shares are subject to forfeiture by the employee if employment is terminated within some specified number of years from the date of grant. The employee usually is not free to sell the shares during the restriction period, and a statement to that effect often is inscribed on the stock certificates. These restrictions give the employee incentive to remain with the company until rights to the shares vest.

FINANCIAL Reporting Case

Q2, p. 1105

The compensation associated with a share of restricted stock is the market price at the grant date of an unrestricted share of the same stock. This amount is accrued as compensation expense with a credit to paid-in capital—restricted stock, over the service period for which participants receive the shares, usually from the date of grant to when restrictions are lifted (the vesting date).[1] Once the shares vest and the restrictions are lifted, paid-in capital—restricted stock is replaced by common stock and paid-in capital—excess of par. Any market price changes that might occur after the grant date that don't affect the total compensation. This is essentially the same as accounting for restricted stock units to be settled in shares, as demonstrated in the next section.

Restricted Stock Units

An increasingly popular variation of restricted stock awards is **restricted stock units (RSUs)**. In fact, RSUs have become a much more popular form of compensation than their restricted stock award cousins. A restricted stock unit is a right to receive a specified number of shares of company stock. It could be a performance bonus, a signing bonus, or regular compensation. The employee doesn't receive the stock right away. Instead, the shares are distributed as the recipient of RSUs satisfies the vesting requirement.[2] So, like restricted stock awards, the recipient benefits by the value of the shares at the end of the vesting period. Unlike restricted stock awards, though, the shares are not issued at the time of the grant. Delaying the increase in outstanding shares is more acceptable to other shareholders. As part of its share-based compensation plan, **Apple Inc.** provides compensation to executives in the form of RSUs as we see in Illustration 19–1.

Illustration 19–1

Restricted Stock Units (RSUs); Apple Inc.

Real World Financials

Share-Based Compensation (in part)

Share-based compensation cost for RSUs is measured based on the closing fair market value of the Company's common stock on the date of grant. . . . The Company recognizes share-based compensation expense over the award's requisite service period on a straight-line basis.

Source: Apple Inc.

Terms of RSUs vary.[3] Sometimes, the recipient is given the *cash equivalent* of the number of shares used to value the RSUs. Or, the terms might stipulate that either the recipient or the company is allowed to choose whether to settle in stock or cash.

Although RSUs delay the issuance of shares and avoid some administrative complexities of outright awards of restricted stock, accounting for RSUs to be settled in stock is

[1]Restricted stock plans usually are designed to comply with Tax Code Section 83 to allow employee compensation to be nontaxable to the employee until the date the shares become substantially vested, which is when the restrictions are lifted. Likewise, the employer gets no tax deduction until the compensation becomes taxable to the employee.

[2]Sometimes, the shares to be issued depend upon achieving specific performance targets.

[3]Typically, but not always, RSUs have no voting privileges and receive no dividends until the shares are issued at vesting. Sometimes, restricted stock does have voting privileges and receives dividends during the vesting period even though recipients are restricted from selling the shares until vesting.

essentially the same as for restricted stock awards. We see accounting for award plans or units of restricted stock demonstrated in Illustration 19–2.

Illustration 19–2

Restricted Stock Award Plans or Restricted Stock Units (RSUs)

Under its restricted stock unit (RSU) plan, Universal Communications grants RSUs representing 5 million of its $1 par common shares to certain key executives at January 1, 2021.

- The shares are subject to forfeiture if employment is terminated within four years.
- Shares have a current market price of $12 per share.

January 1, 2021
No entry

Calculate total compensation expense:

$12
× 5 million
= **$60 million**

The total compensation is the market value of the shares ($12) times 5 million shares.

The total compensation is to be allocated to expense over the four-year service (vesting) period: 2021–2024.

$60 million ÷ 4 years = $15 million per year

December 31, 2021, 2022, 2023, 2024	($ in millions)	
Compensation expense (**$60** million ÷ 4 years)	15	
Paid-in capital—restricted stock		15
December 31, 2024		
Paid-in capital—restricted stock (5 million shares at $12)	60	
Common stock (5 million shares at $1 par)		5
Paid-in capital—excess of par (difference)		55

The **$60** million is accrued to compensation expense over the four-year service period.

When restrictions are lifted, paid-in capital—restricted stock, is replaced by common stock and paid-in capital—excess of par.

On the other hand, *if the employee will receive cash* or can elect to receive cash, we consider the award to be a *liability* rather than equity, as is the case in Illustration 19–2. When an RSU is considered to be a liability, we determine its fair value at the grant date and recognize that amount as compensation expense over the requisite service period consistent with the way we account for restricted stock awards, RSUs, and other share-based compensation. However, because these plans are considered to be liabilities payable in cash, the credit portion of the entry as we recognize compensation expense each year is to Liability—restricted stock. And, it's necessary to periodically adjust the liability (and corresponding compensation) based on the change in the stock's fair value until the liability is paid. Note that this is consistent with the way we account for other liabilities. Accounting for share-based compensation that's considered to be a liability is demonstrated in Appendix 19B of the chapter.

If restricted stock shares or RSUs are forfeited because, say, the employee leaves the company, entries previously made related to that specific employee would simply be reversed. This would result in a decrease in compensation expense in the year of forfeiture. The total compensation, adjusted for the forfeited amount, is then allocated over the remaining service period.

Additional Consideration

An alternative way of accomplishing the same result is to debit deferred compensation for the full value of the RSUs ($60 million in the illustration) on the date they are granted.

	($ in millions)	
Deferred compensation (5 million shares at $12)	60	
Common stock (5 million shares at $1 par)		5
Paid-in capital—excess of par (difference)		55

If so, deferred compensation (a term used for delayed recognition of compensation expense) is reported as a reduction in shareholders' equity, resulting in a zero net effect on shareholders' equity. Then, deferred compensation is credited when compensation expense is debited over the service period. Just as in Illustration 19–2, the result is an increase in both compensation expense and shareholders' equity each year over the vesting period.

Stock Option Plans

Often, employees aren't actually awarded shares, but rather are given the option to buy shares in the future. In fact, **stock options** have become an integral part of the total compensation package for key officers of most medium and large companies. As with any compensation plan, the accounting objective is to report compensation expense during the period of service for which the compensation is given.

● LO19–2

Expense—The Great Debate

Stock option plans give employees the option to purchase (a) a specified number of shares of the firm's stock, (b) at a specified price, (c) during a specified period of time. One of the most heated controversies in standard-setting history has been the debate over the amount of compensation to be recognized as expense for stock options. At issue is how the value of stock options is measured, which for most options determines whether any expense at all is recognized.

Historically, options were measured at their **intrinsic values**—the simple difference between the market price of the shares and the option price at which they can be acquired. For instance, an option that permits an employee to buy a $25-per-share stock for $10 per share has an intrinsic value of $15. However, plans in which the exercise price equals the market value of the underlying stock at the date of grant (which describes most executive stock option plans) have no intrinsic value and therefore result in zero compensation when measured this way, even though the fair value of the options can be quite significant. **Facebook** CEO Mark Zuckerberg reaped a $3.3 billion gain in 2013 alone by exercising stock options.[4] To many, it seems counterintuitive to not recognize compensation expense for plans that routinely provide executives with a substantial part of their total compensation.

FINANCIAL Reporting Case

Q3, p. 1105

After lengthy debate, the FASB consented to encourage, rather than require, that the fair value of options be recognized as expense.

FAILED ATTEMPT TO REQUIRE EXPENSING This is where the controversy ensues. In 1993, the FASB issued an Exposure Draft of a new standard that would have required companies to measure options at their *fair values* at the time they are granted and to expense that amount over the appropriate service period. To jump straight to the punch line, the FASB bowed to public pressure and agreed to withdraw the requirement before it became a standard. The FASB consented to encourage, rather than require, that fair value compensation be recognized as expense. Companies were permitted to continue accounting under prior GAAP (the intrinsic value method referred to in the previous paragraph).[5] Before we discuss the details of accounting for stock options, it's helpful to look back at what led the FASB to first propose fair value accounting and later rescind that proposal.

As the 1990s began, the public was becoming increasingly aware of the enormity of executive compensation in general and compensation in the form of stock options in particular. The lack of accounting for this compensation was apparent, prompting the SEC to encourage the FASB to move forward on its stock option project. Even Congress got into the fray when, in 1992, a bill was introduced that would require firms to report compensation expense based on the fair value of options. Motivated by this encouragement, the FASB issued its exposure draft in 1993. The real disharmony began then. Opposition to the proposed standard was broad and vehement, and that perhaps is an understatement. Critics based their opposition on one or more of these three objections:

1. *Options with no intrinsic value at issue have zero fair value and should not give rise to expense recognition.* The FASB, and even some critics of the proposal, were adamant that options provide valuable compensation at the grant date to recipients.
2. *It is impossible to measure the fair value of the compensation on the grant date.* The FASB argued vigorously that value can be approximated using one of several **option pricing models**. These are statistical models that use computers to incorporate

[4] *San Jose Mercury News,* April 1, 2014.

[5] "Accounting for Stock Issued to Employees," *Opinions of the Accounting Principles Board No. 25* (New York: AICPA, 1972). This pronouncement is superseded and therefore is not codified in the *FASB Accounting Standards Codification*.

information about a company's stock and the terms of the stock option to estimate the options' fair value. We might say the FASB position is that it's better to be approximately right than precisely wrong.

3. *The proposed standard would have unacceptable economic consequences.* Essentially, this argument asserted that requiring this popular means of compensation to be expensed would cause companies to discontinue the use of options.

The opposition included corporate executives, auditors, members of Congress, and the SEC.[5] Ironically, the very groups that provided the most impetus for the rule change initially—the SEC and Congress—were among the most effective detractors in the end. The only group that offered much support at all was the academic community, and that was by-and-large nonvocal support. In reversing its decision, the FASB was not swayed by any of the specific arguments of any opposition group. Dennis Beresford, chair of the FASB at the time, indicated that it was fear of government control of the standard-setting process that prompted the Board to modify its position. The Board remained steadfast that the proposed change was appropriate.

There were consistent criticisms of the FASB's requirement to expense option compensation.

VOLUNTARY EXPENSING Prior to 2002, only two companies—**Boeing** and **Winn-Dixie**—reported stock option compensation expense at fair value. However, in 2002, public outrage mounted amid high-profile accounting scandals at **Enron**, **WorldCom**, **Tyco**, and others. Some degree of consensus emerged that greed on the part of some corporate executives contributed to the fraudulent and misleading financial reporting at the time. In fact, many in the media were pointing to the proliferation of stock options as a primary form of compensation as a culprit in fueling that greed. An episode of the PBS series *Frontline* argued that not expensing the value of stock options contributed to the collapse of Enron. For these reasons, renewed interest surfaced in requiring stock option compensation to be reported in income statements.

CURRENT REQUIREMENT TO EXPENSE Emerging from the rekindled debate was a GAAP revision that now requires fair value accounting for employee stock options, eliminating altogether the intrinsic value approach.[7] As you might expect, the proposal did not come without opposition. Many of the same groups that successfully blocked the FASB from enacting a similar requirement in 1995 led the opposition. Not surprisingly, at the forefront of the resistance were the high-tech companies that extensively use stock options as a primary form of compensating employees and thus are most susceptible to a reduction in reported earnings when that compensation is included in income statements. For example, consider **Apple Inc.**'s earnings for the 12 months ending March 27, 2004. Reported net income of $179 million would have been only $56 million, or 69% less, if the GAAP revision had been in effect then.[8]

Current GAAP now requires companies to record the value of options in their income statements.

It's important to note that the way we account for stock options has no effect whatsoever on cash flows, only on whether the value of stock options is included among expenses. This is not to say that companies haven't altered their compensation strategies. Already, we have seen a shift in the way some companies compensate their employees. Partly due to the negative connotation that has become associated with executive stock options, we've seen fewer options and more bonuses and restricted stock awards. Let's examine the way stock options are accounted for now.

We've witnessed a discernable shift in the way executives are compensated—fewer options, more stock awards and bonuses.

Recognizing the Fair Value of Options

Accounting for stock options parallels the accounting for restricted stock we discussed in the first part of this chapter. That is, we measure compensation as the fair value of the stock options at the grant date and then record that amount as compensation expense over the service period for which employees receive the options. Estimating the fair value requires the

[6]All of the then "Big Six" CPA firms lobbied against the proposal. Senator Lieberman of Connecticut introduced a bill in Congress that, if passed, would have forbidden the FASB from passing a requirement to expense stock option compensation.

[7]FASB ASC 718–Compensation.

[8]Alex Salkever, "What Could Crunch Apple Shares," *BusinessWeek*, July 12, 2004, p.11.

The fair value of a stock option can be determined by employing a recognized option pricing model.

use of one of several option pricing models. These mathematical models assimilate a variety of information about a company's stock and the terms of the stock option to estimate the option's fair value. The model should take into account the following:

An option pricing model takes into account several variables.

- Exercise price of the option.
- Expected term of the option.
- Current market price of the stock.
- Expected dividends.
- Expected risk-free rate of return during the term of the option.
- Expected volatility of the stock.

The techniques for estimating fair value have been among the most controversial issues in the debate.

The current GAAP for accounting for employee stock options modified the way companies actually measure fair value. It calls for using models that permit greater flexibility in modeling the ways employees are expected to exercise options and their expected employment termination patterns after options vest.[9] Option-pricing theory, on which the pricing models are based, is a topic explored in depth in finance courses and is subject to active empirical investigation and development. A simplified discussion is provided in Appendix 19A.[10]

The total compensation, as estimated by the options' fair value, is reported as compensation expense over the period of service for which the options are given. Recipients normally are not allowed to exercise their options for a specified number of years. This delay provides added incentive to remain with the company. The time between the date options are granted and the first date they can be exercised is the vesting period and usually is considered to be the service period over which the compensation expense is reported. The process is demonstrated in Illustration 19–3.

Illustration 19–3
Stock Options

At January 1, 2021, Universal Communications grants 10 million options to key executives.

- The options permit recipients to acquire 10 million of the company's $1 par common shares within the next eight years, but not before December 31, 2024 (the vesting date).
- The exercise price is the market price of the shares on the date of grant, $35 per share.
- The fair value of the options, estimated by an appropriate option pricing model, is $8 per option.

January 1, 2021

No entry

Calculate total compensation expense:

$ 8	Estimated fair value per option
× 10 million	Options granted
= **$ 80** million	Total compensation

Fair value is estimated at the date of grant.

The total compensation is allocated to expense over the four-year service (vesting) period: 2021–2024.

$80 million ÷ 4 years = $20 million per year

The value of the award is expensed over the service period for which the compensation is provided.

December 31, 2021, 2022, 2023, 2024	($ in millions)	
Compensation expense (**$80** million ÷ 4 years)	20	
Paid-in capital—stock options		20

Option compensation expense is based on the number of options expected to vest.

FORFEITURES If previous experience indicates that a material number of the options will be forfeited before they vest (due to employee turnover or violation of other terms of the options), we adjust the amount of compensation recorded (a) by estimating forfeitures or (b) as forfeitures occur.

[9]FASB ASC 718–10–55–30 to 32: Compensation—Stock Compensation—Overall—Implementation and Guidance Illustrations—Selecting or Estimating the Expected Term.

[10]An expanded discussion is provided in FASB ASC 718–10–55: Compensation—Stock Compensation—Overall—Implementation and Guidance Illustrations.

The default approach is to *estimate the percentage* of options that will be forfeited and adjust grant date calculation of the fair value of the options to reflect that expectation. For instance, if a forfeiture rate of 5% is expected, Universal's estimated total compensation would be 95% of $80 million, or $76 million. In that case, the annual compensation expense in Illustration 19–3 would have been **$19** million ($76 ÷ 4) instead of $20 million. We see the effect of that possibility in Illustration 19–3A.

Illustration 19–3A

Default Approach: Forfeiture Estimates with Revisions

Initial Expectation:	($ in millions)	
2021		
Compensation expense ([$80 × 95%] ÷ 4)	**19**	
Paid-in capital—stock options		**19**
2022		
Compensation expense ([$80 × 95%] ÷ 4)	**19**	
Paid-in capital—stock options		**19**
Estimate Revision after 2 Years:		
2023		
Compensation expense [($80 × 90% × 3/4) – (**$19** + **$19**)]	16	
Paid-in capital—stock options		16
2024		
Compensation expense [($80 × 90% × 4/4) – (**$19** + **$19** + $16)]	18	
Paid-in capital—stock options		18

The value of the compensation is originally estimated to be $76 million, or $19 million per year.

The expense each year is the current estimate of total compensation that should have been recorded to date less the amount already recorded.

What if that expectation changes later? Universal should adjust the cumulative amount of compensation expense recorded to date in the year the estimate changes. Suppose, for instance, that during the third year, Universal revises its estimate of forfeitures from 5% to 10%. The new estimate of total compensation would then be $80 million × 90%, or $72 million. For the first three years, the portion of the total compensation that should have been reported would be $72 million × ¾, or $54 million, and since $38 million (**$19** × 2) of that was recorded in 2021–2022 before the estimate changed, an additional $16 million would now be recorded in 2023. Then if the estimate isn't changed again, the remaining $18 million ($72 – $54) would be recorded in 2024.

When forfeiture estimates change, the cumulative effect on compensation is reflected in current earnings.

Additional Consideration

Notice that the $18 million in 2024 is the amount that would have been reported in each of the four years if Universal had assumed a 10% forfeiture rate from the beginning. Also be aware that this approach is contrary to the usual way companies account for changes in estimates. For instance, assume a company acquires a four-year depreciable asset having an estimated residual value of 5% of cost. The $76 million depreciable cost would be depreciated straight line at **$19** million over the four-year useful life. If the estimated residual value changes after two years to 10%, the new estimated depreciable cost of $72 would be reduced by the $38 million depreciation recorded the first two years, and the remaining $34 million would be depreciated equally, $17 million per year, over the remaining two years.

In an alternative approach, as a practical expedient, companies can elect to account for forfeitures of stock options (or restricted stock) when forfeitures actually occur, rather than estimate forfeitures that will occur during the vesting period. So rather than recording compensation expense and paid-in capital for the *net* amount of awards expected to vest, companies can choose to initially record compensation based on the total amount and then reduce compensation expense and paid-in capital only if and when forfeitures occur.[11] Once

Companies can elect to account for forfeitures when forfeitures actually occur rather than estimate them.

[11] ASC 718-10-35-3: Compensation—Stock Compensation-Overall-Subsequent Measurement Recognition of Compensation Costs over the Requisite Service Period

a forfeiture occurs, we adjust the cumulative amount of compensation expense recorded to date in the year the forfeiture occurs and thereafter in a manner similar to the way estimates are adjusted as demonstrated in Illustration 19–3A.

This election applies only to forfeitures related to employee turnover. A company must disclose its policy election for forfeitures (estimated or recorded as they occur). To demonstrate the alternative approach, let's say Universal Communications recorded compensation of **$20** million in both 2021 and 2022 based on the $80 million fair value of the stock options issued at the beginning of 2021. Then, in 2023, options with a fair value of $8 million when granted are forfeited due to executive turnover. Universal's journal entries would look like those in Illustration 19–3B.

Illustration 19–3B

Optional Approach: Revisions as Forfeitures Occur

December 31, 2021	($ in millions)	
Compensation expense ($80 million ÷ 4 years)	**20**	
Paid-in capital—stock options		**20**
December 31, 2022 ($ in millions)		
Compensation expense ($80 million ÷ 4 years)	**20**	
Paid-in capital—stock options		**20**
When Actual Forfeitures Occur in the Third Year:		
December 31, 2023 ($ in millions)		
Compensation expense ([($80 – $8) × ¾] – **$20** – **$20**)	14	
Paid-in capital—stock options		14
December 31, 2024 ($ in millions)		
Compensation expense ([($80 – $8) × 4/4] – **$20** – **$20** – $14)	18	
Paid-in capital—stock options		18

This election applies only to forfeitures related to employee *turnover*. For share-based plans with performance conditions (which we discuss later), companies must assess the probability that such conditions will be achieved. A company must disclose its policy election for forfeitures (estimated, or recorded as they occur).

When Options are Exercised

If half the options in Illustration 19–3 (for 5 million shares) are exercised on July 11, 2027, when the market price is $50 per share, the following journal entry is recorded:

Recording the exercise of options is not affected by the market price on the exercise date.

July 11, 2027	($ in millions)	
Cash ($35 exercise price × 5 million shares)	175	
Paid-in capital—stock options (½ account balance)	40	
Common stock (5 million shares at $1 par per share)		5
Paid-in capital—excess of par (to balance)		210

Notice that the market price at exercise is irrelevant. Changes in the market price of underlying shares do not influence the previously measured fair value of options.

When Unexercised Options Expire

If options that have vested expire without being exercised, the following journal entry is made (assuming the remaining 5 million options in our illustration are allowed to expire):

***Paid-in capital—stock options* becomes *paid-in capital—expiration of stock options* when options expire without being exercised.**

	($ in millions)	
Paid-in capital—stock options (account balance)	40	
Paid-in capital—expiration of stock options		40

In effect, we rename the paid-in capital attributable to the stock option plan. Compensation expense for the four years' service, as of the measurement date, is not affected.

Additional Consideration

Tax Consequences of Stock-Based Compensation Plans In Illustration 19–3 we ignored the tax effect. To illustrate the effect of taxes, let's assume Universal Communications' income tax rate is 25%.

For tax purposes, plans can either qualify as "incentive stock option plans" under the Tax Code or be "nonqualified plans." Among the requirements of a qualified option plan is that the exercise price be equal to the market price at the grant date. Under a qualified incentive plan, the recipient pays no income tax until any shares acquired from exercise of stock options are subsequently sold. On the other hand, the company gets no tax deduction at all. With a nonqualified plan the employee can't delay paying income tax, but the employer is permitted to deduct the difference between the exercise price and the market price at the exercise date. Let's consider both.

Tax treatment favors the employer in a nonqualified stock option plan.

Case 1. With an incentive plan, the employer receives no tax deduction at all. If Universal's plan qualifies as an incentive plan, the company will receive no tax deduction upon exercise of the options and thus no tax consequences.

Because an incentive plan provides no tax deduction, it has no deferred tax consequences.

Case 2. On the other hand, if we assume the plan does not qualify as an incentive plan, Universal will deduct from taxable income the difference between the exercise price and the market price at the exercise date. Recall from Chapter 16 that this creates a temporary difference between accounting income (for which compensation expense is recorded currently) and taxable income (for which the tax deduction is taken later upon the exercise of the options). We assume the temporary difference is the cumulative amount expensed for the options. The following entries would be recorded on the dates shown:

December 31, 2021, 2022, 2023, 2024	($ in millions)	
Compensation expense ($80 million ÷ 4 years)	20	
Paid-in capital—stock options		20
Deferred tax asset (25% × $20 million)	5	
Income tax expense		5

The difference between the market price at exercise and the eventual tax savings is recognized in income tax expense.

The after-tax effect on earnings is thus $15 million each year ($20 – $5).

If all of the options (for 10 million shares) are exercised on April 4, 2026, when the stock's market price is **$50** per share:

Cash ($35 exercise price × 10 million shares)	350	
Paid-in capital—stock options (account balance)	80	
Common stock (10 million shares at $1 par per share)		10
Paid-in capital—excess of par (to balance)		420
Income tax payable [(**$50** – $35) × 10 million shares × 25%]	37.5	
Deferred tax asset (4 years × $5 million)		20.0
Income tax expense (remainder)		17.5

A deferred tax asset is recognized now for the future tax savings from the tax deduction when the nonqualified stock options are exercised.

The tax consequences of all nonqualifying stock options as well as restricted stock plans also are accounted for in the manner demonstrated above.

International Financial Reporting Standards

● LO19–14

Recognition of Deferred Tax Asset for Stock Options. Under U.S. GAAP, a deferred tax asset (DTA) is created for the cumulative amount of the fair value of the options the company has recorded for compensation expense. The DTA is the tax rate times the amount of compensation.

Under IFRS, the deferred tax asset isn't created until the award is "in the money;" that is, it has intrinsic value. When it is in the money, the addition to the DTA is the portion of the intrinsic value earned to date times the tax rate.

Plans with Graded Vesting

The stock option plans we've discussed so far vest (become exercisable) on one single date (e.g., four years from date of grant). This is referred to as *cliff vesting*. More frequently, though, awards specify that recipients gradually become eligible to exercise their options rather than all at once. This is called *graded vesting*. For instance, a company might award stock options that vest 25% the first year, 25% the second year, and 50% the third year, or maybe 25% each year for four years.

In such a case, the company can choose to account for the options essentially the same as the cliff-vesting plans we've discussed to this point. It can estimate a single fair value for each of the options, even though they vest over different time periods, using a single weighted-average expected life of the options. The company then allocates that total compensation cost (fair value per option times number of options) over the entire vesting period.

Most companies, though, choose a slightly more complex method because it usually results in a lower expense.[12] In this approach, we view each vesting group (or tranche) separately, as if it were a separate award. See Illustration 19–4.

Illustration 19–4 Stock Options; Graded Vesting; Separate Valuation Approach

At January 1, 2021, Taggart Mobility issued 10 million executive stock options permitting executives to buy 10 million shares of stock for $25.

- The vesting schedule is 20% the first year, 30% the second year, and 50% the third year (graded-vesting).
- The value of the options that vest over the 3-year period is estimated at January 1, 2021, by separating the total award into three groups (or tranches) according to the year in which they vest (because the expected life for each group differs).

The fair value of the options as of January 1, 2021, is estimated as follows:

Vesting Date	Amount Vesting	Fair Value per Option
Dec. 31, 2021	20%	$3.50
Dec. 31, 2022	30%	$4.00
Dec. 31, 2023	50%	$6.00

We allocate the compensation cost for each of the three groups (tranches) evenly over its *individual* vesting (service) period:

	Compensation Expense in ($ in millions)				
Shares Vesting at	**2021**	**2022**	**2023**	**Total**	
Dec. 31, 2021	$ 7			$ 7	(10 M × 20% × $3.50)
Dec. 31, 2022	6	$ 6		12	(10 M × 30% × $4.00)
Dec. 31, 2023	10	10	$10	30	(10 M × 50% × $6.00)
Expensed	$23	$16	$10	= $49	

At any given date, a company must have recognized at least the amount vested by that date. The allocation in this instance meets that constraint:

- The $23 million recognized in 2021 exceeds the $7 million vested.
- The $39 million ($23 + $16) recognized by 2022 exceeds the $19 million ($7 + $12) vested by the same time.

Companies also can choose to use the straight-line method, which would allocate the $49 million total compensation cost equally to 2021, 2022, and 2023 at $16.333 million per year.

[12]The fair value of the options usually is higher when estimated using a single weighted-average expected life of the options rather than when estimated as the total of fair values of the multiple vesting groups.

International Financial Reporting Standards

LO19–14

When options have graded vesting, U.S. GAAP permits companies to account for each vesting amount separately, for instance, as if there were three separate awards as in the previous illustration, but also allows companies the option to account for the entire award on a straight-line basis over the entire vesting period. Either way, the company must recognize at least the amount of the award that has vested by that date.

Under IFRS, the straight-line choice is not permitted. Also, there's no requirement that the company must recognize at least the amount of the award that has vested by each reporting date.

Plans with Performance or Market Conditions

Stock option (and other share-based) plans often specify a performance condition or a market condition that must be satisfied before employees are allowed the benefits of the award. The objective is to provide employees with additional incentive for managerial achievement. For instance, an option might not be exercisable until a performance target is met. The target could be divisional revenue, earnings per share, sales growth, or rate of return on assets. The possibilities are limitless. On the other hand, the target might be market-related, perhaps a specified stock price or a stock price change exceeding a particular index. The way we account for such plans depends on whether the condition is performance-based or market-based.

The terms of performance options vary with some measure of performance that ties rewards to productivity.

PLANS WITH PERFORMANCE CONDITIONS Whether we recognize compensation expense for performance-based options depends (a) initially on whether it's probable[13] that the performance target will be met and (b) ultimately on whether the performance target actually is met. Accounting is as described earlier for other stock options. Initial estimates of compensation cost, as well as subsequent revisions of that estimate, take into account the likelihood of both forfeitures and achieving performance targets. For example, in Illustration 19–3, if the options described also had included a condition that the options would become exercisable only if sales increase by 10% after four years, we would estimate the likelihood of that occurring; specifically, is it probable? Let's say we initially estimate that it is probable that sales will increase by 10% after four years. Then, our initial estimate of the total compensation would have been unchanged at

If compensation from a stock option depends on meeting a performance target, compensation is recorded only if we feel it's probable the target will be met.

10 million	×	$8	=	$80 million
Options expected to vest		Fair value per option		Estimated total compensation

Suppose, though, that after two years, we estimate that it is *not* probable that sales will increase by 10% after four years. Then, our new estimate of the total compensation would change to

0	×	$8	=	$0
Options expected to vest		Fair value per option		Estimated total compensation

If it later becomes probable that a performance target will not be met, we reverse any compensation expense already recorded.

In that case, we would reverse the $40 million expensed in 2021–2022. No compensation can be recognized for options that don't vest due to performance targets not being met, and that's our expectation.

When we revise our estimate of total compensation because our expectation of probability changes, we record the effect of the change in the current period.

Conversely, assume that our initial expectation is that it is *not* probable that sales will increase by 10% after four years and so we record no annual compensation expense. But then, in the third year, we estimate that it is probable that sales will increase by 10% after four years. At that point, our revised estimate of the total compensation would change to

[13] "Probable" means the same as it did in Chapter 13, when we were estimating the likelihood that payment would be made for a loss contingency, and elsewhere when making accounting estimates. Probable is a matter of professional judgement (often 70–75%).

$80 million, and we would reflect the cumulative effect on compensation in 2023 earnings and record compensation thereafter:

If we had begun with our new estimate of total compensation, $20 million would have been expensed in each of the first three years.

2023		
Compensation expense [($80 × ¾) – $0]	60	
Paid-in capital—stock options		60
2024		
Compensation expense [($80 × 4/4) – $60]	20	
Paid-in capital—stock options		20

Additional **Consideration**

Suppose an executive who is granted some of the performance-based options in this example is eligible to retire on January 1, 2021. So, when the options are granted she would be eligible to receive the options even if the performance target (sales increase by 10%) is met after she retires. In that case, the service period needed to receive compensation effectively is one day. Should compensation for her be measured and expensed over the vesting period as in the example, or should that compensation be measured and recorded all at once (full amount if probable, zero otherwise) on the grant date?

It should be treated the same as the compensation for the other options. It should be treated as a performance condition consistent with those for which the requisite service period is not met prior to the performance target being met. So, we would record no compensation expense until meeting the performance target is probable, even if this is not until *after* the requisite vesting period is completed. An important secondary benefit of this treatment is that the employer doesn't need to assess at the grant date whether certain employees are retirement-eligible or will become retirement-eligible during the performance period in order to value the awards.[14]

This treatment differs from that of International Financial Reporting Standards, which specifies that if the performance target can be met *after* the requisite vesting period is completed, compensation should be measured and recorded all at once (full amount if probable, zero otherwise) on the grant date.

If the target is based on changes in the market rather than on performance, we record compensation as if there were no target.

PLANS WITH MARKET CONDITIONS If the award contains a market condition (e.g., a share option with an exercisability requirement based on the stock price reaching a specified level), then no special accounting is required. The fair value estimate of the share option already implicitly reflects market conditions due to the nature of share option pricing models. So, we recognize compensation expense regardless of when, if ever, the market condition is met.

DECLINE IN POPULARITY OF OPTIONS Recent years have witnessed a steady shift in the way companies compensate their top executives. At their peak in 1999, stock options represented about 78% of the average executive's incentive pay. In 2016, stock options accounted for 15% of S&P 500 executive pay while restricted stock accounted for 47.2%.[15] In the wake of past notorious accounting scandals, the image of stock options was tarnished in the view of many who believed that the potential to garner millions in stock option gains created incentives for executives to boost company stock prices through risky or fraudulent behavior. That image motivated many firms to move away from stock options in favor of other forms of share-based compensation, particularly restricted stock awards and, increasingly, restricted stock units. Also contributing to the rise of restricted stock is the feeling by many that it better aligns pay with performance. From the executive's perspective, restricted stock is a more certain, though potentially less lucrative, form of compensation.

[14]FASB Emerging Issues Task Force, *Accounting Standards Update No. 2014-12,* "Compensation—Stock Compensation (Topic 718): Accounting for Share-Based Payments When the Terms of an Award Provide That a Performance Target Could Be Achieved after the Requisite Service Period," June 2014.

[15]Source: Equilar. Copyright 2016.

Employee Share Purchase Plans

Employee share purchase plans often permit all employees to buy shares directly from their company at favorable terms. The primary intent of these plans is to encourage employee ownership of the company's shares. Presumably, loyalty is enhanced among employee-shareholders. The employee also benefits because, typically, these plans allow employees to buy shares from their employer without brokerage fees and, perhaps, at a slight discount. Some companies even encourage participation by matching or partially matching employee purchases.

● LO19–3

Share purchase plans permit employees to buy shares directly from the corporation.

When accounting for employee share purchase plans (ESPPs), we need to determine whether or not they provide compensation to employees. If not, we simply record the sale of new shares as employees buy them and do not record compensation expense. An ESPP is considered noncompensatory as long as

1. substantially all employees can participate,
2. employees have no longer than one month after the price is fixed to decide whether to participate, and
3. the discount is no greater than 5% (or can be justified as reasonable).

If these criteria for the plan being noncompensatory are *not* met, say the discount is 15%, accounting is similar to other share-based plans. The 15% discount to employees, then, is considered to be compensation, and that amount is recorded as expense.[16] Compensation expense replaces the cash debit for any employer-provided portion. Say an employee buys shares (no-par) under the plan for $850 rather than the current market price of $1,000. The $150 discount is recorded as compensation expense.

Cash (discounted price)	850	
Compensation expense ($1,000 × 15%)	150	
Common stock (market value)		1,000

Decision Makers' Perspective

In several previous chapters, we have revisited the concept of "earnings quality" (as first defined in Chapter 4). We also have noted that one rather common practice that negatively influences earnings quality is earnings management, which refers to companies' use of one or more of several techniques designed to artificially increase (or decrease) earnings. A frequent objective of earnings management is to meet analysts' expectations regarding projections of income. The share-based compensation plans we discuss in this chapter suggest another motive managers sometimes have to manipulate income. If a manager's personal compensation includes company stock, stock options, or other compensation based on the value of the firm's stock, it's not hard to imagine an increased desire to ensure that market expectations are met and that reported earnings have a positive effect on stock prices. In fact, as we discussed earlier, that is precisely the reaction these incentive compensation plans are designed to elicit. Investors and creditors, though, should be alert to indications of attempts to artificially manipulate income and realize that the likelihood of earnings management is probably higher for companies with generous share-based compensation plans.

Analysts should be aware of the possibility of earnings management as a way to increase managers' compensation.

One way managers might manipulate numbers is to low-ball the data that go into the option-pricing models. The models used to estimate fair value are built largely around subjective assumptions. That possibility emphasizes the need for investors to look closely at the assumptions reported in the stock option disclosure note, and particularly at how those assumptions change from year to year. ●

[16]FASB ASC 718–50–25–1 and 2: Compensation—Stock Compensation—Employee Share Purchase Plans—Recognition.

Concept Review Exercise

SHARE-BASED COMPENSATION PLANS

Listed below are transactions dealing with various stock benefit plans of Fortune-Time Corporation during the period 2021–2023. The market price of the stock is $45 at January 1, 2021.

a. On January 1, 2021, the company issued 10 million common shares to divisional managers under its restricted stock award plan. The shares are subject to forfeiture if employment is terminated within three years.
b. On January 1, 2021, the company granted incentive stock options to its senior management, exercisable for 1.5 million common shares. The options must be exercised within five years, but not before January 1, 2023. The exercise price of the stock options is equal to the fair value of the common stock on the date the options are granted. An option pricing model estimates the fair value of the options to be $4 per option. All recipients are expected to remain employed through the vesting date.
c. Recorded compensation expense on December 31, 2021.
d. A divisional manager holding 1 million of the restricted shares left the company to become CEO of a competitor on September 15, 2022, before the required service period ended.
e. Recorded compensation expense on December 31, 2022.

Required:

Prepare the journal entries that Fortune-Time recorded for each of these transactions. (Ignore any tax effects.)

Solution:

January 1, 2021

Restricted Stock Award Plan

No entry.

Total compensation is measured as 10 million shares × $45 = $450 million.

Stock Options

No entry.

Total compensation is measured as 1.5 million shares × $4 = $6 million.

December 31, 2021

Restricted Stock		*($ in millions)*
Compensation expense ($450 million ÷ 3 years)	150	
Paid-in capital—restricted stock		150
Stock Options		
Compensation expense ($6 million ÷ 2 years)	3	
Paid-in capital—stock options		3

September 15, 2022

Restricted Stock		
Paid-in capital—restricted stock (10% × $150)	15	
Compensation expense		15

December 31, 2022

Restricted Stock		
Compensation expense [$450 – (10% × $450) – $150 + $15] ÷ 2 years	135	
Paid-in capital—restricted stock		135
Stock Options		
Compensation expense ($6 million ÷ 2 years)	3	
Paid-in capital—stock options		3

Earnings Per Share

PART B

A typical corporate annual report contains four comparative financial statements, an extensive list of disclosure notes and schedules, and several pages of charts, tables, and textual descriptions. Of these myriad facts and figures, the single accounting number that is reported most frequently in the media and receives by far the most attention by investors and creditors is **earnings per share** (EPS). The reasons for the considerable attention paid to earnings per share certainly include the desire to find a way to summarize the performance of business enterprises into a single number.

Earnings per share is the single accounting number that receives the most media attention.

Summarizing performance in a way that permits comparisons is difficult because the companies that report the numbers are different from one another. And yet, the desire to condense performance to a single number has created a demand for EPS information. The profession has responded with rules designed to maximize the comparability of EPS numbers by minimizing the inconsistencies in their calculation from one company to the next.[17]

Comparability is an enhancing characteristic of relevant accounting information *(Conceptual Framework)*.

Keep in mind as you study the requirements that a primary goal is comparability. As a result, many of the rules devised to achieve consistency are unavoidably arbitrary, meaning that other choices the FASB might have made in many instances would be equally adequate.

International Financial Reporting Standards

Earnings per Share. The earnings per share requirements of U.S. GAAP, *FASB ASC 260: Earnings per Share*, are a result of the FASB's cooperation with the IASB to narrow the differences between IFRS and U.S. GAAP. A few differences remain. The differences that remain are the result of differences in the application of the treasury stock method, the treatment of contracts that may be settled in shares or cash, and contingently issuable shares.

● LO19–14

Basic Earnings Per Share

A firm is said to have a **simple capital structure** if it has no outstanding securities that could potentially dilute earnings per share. In this context, to dilute means to *reduce* earnings per share. For instance, if a firm has convertible bonds outstanding and those bonds are converted, the resulting increase in common shares could decrease (or dilute) earnings per share. That is, the new shares replaced by the converted bonds might participate in future earnings. So convertible bonds are referred to as **potential common shares**. Other potential common shares are convertible preferred stock, stock options, and contingently issuable shares. We will see how the potentially dilutive effects of these securities are included in the calculation of EPS later in this chapter. Now, though, our focus is on the calculation of EPS for a simple capital structure—when no potential common shares are present. In these cases, the calculation is referred to as **basic EPS**, and is simply earnings available to common shareholders divided by the weighted-average number of common shares outstanding.

A firm has a simple capital structure if it has no *potential common shares*.

● LO19–4

Basic EPS reflects no dilution, only shares now outstanding.

In the most elemental setting, earnings per share (or net loss per share) is merely a firm's net income (or net loss) divided by the number of shares of common stock outstanding throughout the year. The calculation becomes more demanding (a) when the number of shares has changed during the reporting period, (b) when the earnings available to common shareholders are diminished by dividends to preferred shareholders, or (c) when we attempt to take into account the impending effect of potential common shares (which we do in a later section of the chapter). To illustrate the calculation of EPS in each of its dimensions, we will use only one example in this chapter. We'll start with the most basic situation and then add one new element at a time until we have considered all the principal ways the calculation can be affected. In this way, you can see the effect of each component of earnings per share, not just in isolation, but in relation to the effects of other components as well. The basic calculation is shown in Illustration 19–5.

EPS expresses a firm's profitability on a per share basis.

[17]FASB ASC 260: Earnings per Share. The guidance is applicable only for public companies.

Illustration 19–5
Fundamental Calculation

Sovran Financial Corporation reported net income of $154 million in 2021 (tax rate 25%). Its capital structure consisted of the following:

Common Stock

Jan. 1 — 60 million common shares were outstanding
($ amounts in millions, except per share amount)

Basic EPS:

$$\frac{\overset{\text{Net income}}{\$154}}{\underset{\text{Shares outstanding}}{60}} = \$2.57$$

In the most elemental setting, earnings per share is simply a company's earnings divided by the number of shares outstanding.

Issuance of New Shares

Because the shares discussed in Illustration 19–5 remained unchanged throughout the year, the denominator of the EPS calculation is simply the number of shares outstanding. But if the number of shares has changed, it's necessary to find the *weighted average* of the shares outstanding during the period the earnings were generated. For instance, if an additional 12 million shares had been issued on March 1 of the year just ended, we calculate the weighted-average number of shares to be 70 million, as demonstrated in Illustration 19–6.

● LO19–5

Because the new shares were outstanding only 10 months, or $^{10}/_{12}$ of the year, we increase the 60 million shares already outstanding by the additional shares—weighted by the fraction of the year ($^{10}/_{12}$) they were outstanding. The weighted average is $60 + \mathbf{12}\ (^{10}/_{12}) = 60 + 10 = 70$ shares. The reason for time-weighting the shares issued is that the resources the stock sale provides the company are available for generating income only after the date the shares are sold. So, weighting is necessary to make the shares in the fraction's denominator consistent with the income in its numerator (see Illustration 19–6).

Illustration 19–6
Weighted Average

Sovran Financial Corporation reported net income of $154 million for 2021 (tax rate 25%). Its capital structure included the following:

Common Stock

Jan. 1 — 60 million common shares were outstanding
Mar. 1 — 12 million new shares were sold
($ amounts in millions, except per share amount)

Basic EPS:

$$\frac{\overset{\text{Net income}}{\$154}}{\underset{\text{Shares at Jan. 1}}{60} + \underset{\textbf{New shares}}{\mathbf{12\ (^{10}/_{12})}}} = \frac{\$154}{70} = \$2.20$$

Any new shares issued are time-weighted by the fraction of the period they were outstanding and then added to the number of shares outstanding for the entire period.

Stock Dividends and Stock Splits

● LO19–6

Recall that a stock dividend or a stock split is a distribution of additional shares to existing shareholders. But there's an important and fundamental difference between the increase in shares caused by a stock dividend and an increase from selling new shares. When new shares are sold, both assets and shareholders' equity are increased by an additional investment in the firm by shareholders. On the other hand, a stock dividend or stock split merely increases the number of shares without affecting the firm's assets. In effect, the same pie is divided into more pieces. The result is a larger number of less valuable shares. This fundamental change in the nature of the shares is reflected in a calculation of EPS by simply increasing the number of shares.

In Illustration 19–7, notice that the additional shares created by the stock dividend are *not* weighted for the time period they were outstanding. Instead, the increase is treated as if it occurred at the beginning of the year.

Illustration 19–7
Stock Dividends and Stock Splits

Sovran Financial Corporation reported net income of $154 million in 2021 (tax rate 25%). Its capital structure included the following:

Common Stock

Jan. 1	60 million common shares were outstanding
Mar. 1	12 million new shares were sold
June 17	**A 10% stock dividend was distributed**
	($ amounts in millions, except per share amount)

Basic EPS:

Net income

$$\frac{\$154}{60\ \mathbf{(1.10)} + 12\ (^{10}/_{12})\ \mathbf{(1.10)}} = \frac{\$154}{77} = \$2.00$$

Shares at Jan. 1 — New shares

Stock dividend adjustment

Shares outstanding prior to the stock dividend are retroactively restated to reflect the 10% increase in shares—that is, treated as if the distribution occurred at the beginning of the period.

The number of shares outstanding after a 10% stock dividend is **1.10** times higher than before. This multiple is applied to both the beginning shares and the new shares sold before the stock distribution. If this had been a 25% stock dividend, the multiple would have been 1.25; a 2-for-1 stock split means a multiple of 2, and so on.

Notice that EPS without the 10% stock dividend ($2.20) is 10% more than it is with the stock distribution ($2). This is caused by the increase in the number of shares. But, unlike a sale of new shares, this should not be interpreted as a "dilution" of earnings per share. Shareholders' interests in their company's earnings have not been diluted. Instead, each shareholder's interest is represented by more—though less valuable—shares.

A simplistic but convenient way to view the effect is to think of the predistribution shares as having been "blue." After the stock dividend, the more valuable "blue" shares are gone, replaced by a larger number of, let's say, "green" shares. From now on, we compute the earnings per "green" share, whereas we previously calculated earnings per "blue" share. We restate the number of shares retroactively to reflect the stock dividend, as if the shares always had been "green." After all, our intent is to let the calculation reflect the fundamental change in the nature of the shares.

Additional Consideration

When last year's EPS is reported in the current year's comparative income statements, it also should reflect the increased shares from the stock dividend. For instance, suppose EPS was $2.09 for 2020: $115 million net income divided by 55 million weighted-average shares. When reported again for comparison purposes in the 2021 comparative income statements, that figure would be restated to reflect the 10% stock dividend [$115 ÷ (55 × **1.10**) = $1.90], as shown:

Earnings per Share:	**2021**	**2020**
	$2.00	$1.90

The EPS numbers now are comparable—both reflect the stock dividend. Otherwise we would be comparing earnings per "green" share with earnings per "blue" share; this way both are earnings per "green" share.

Reacquired Shares

If shares were reacquired during the period (either retired or as treasury stock), the weighted-average number of shares is reduced. The number of reacquired shares is time-weighted for the *fraction of the year they were **not** outstanding,* prior to being *subtracted* from the number of shares outstanding during the period. Let's modify our continuing illustration to assume 8 million shares were reacquired on October 1 as treasury stock (Illustration 19–8).

Illustration 19–8

Reacquired Shares

Sovran Financial Corporation reported net income of $154 million in 2021 (tax rate 25%). Its capital structure included the following:

Common Stock

Jan. 1	60 million common shares outstanding
Mar. 1	12 million new shares were sold
June 17	A 10% stock dividend was distributed
Oct. 1	**8 million shares were reacquired as treasury stock**

($ amounts in millions, except per share amounts)

Basic EPS:

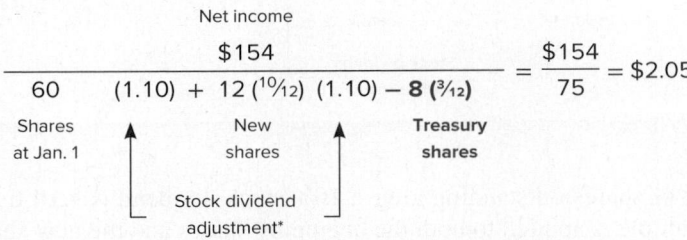

$$\frac{\overset{\text{Net income}}{\$154}}{60\,(1.10) + 12\,(^{10}/_{12})\,(1.10) - \mathbf{8}\,(^{3}/_{12})} = \frac{\$154}{75} = \$2.05$$

60 = Shares at Jan. 1; 12 (10/12) = New shares; 8 (3/12) = **Treasury shares**; (1.10) = Stock dividend adjustment*

*Not necessary for the treasury shares since they were reacquired after the stock dividend and thus already reflect the adjustment (that is, the shares repurchased are **8** million "new green" shares).

The **8** million shares reacquired as treasury stock are weighted by (3/12) to reflect the fact they were not outstanding the last three months of the year.

The adjustment for reacquired shares is the same as for new shares sold, except the shares are deducted rather than added.

Compare the adjustment for treasury shares with the adjustment for new shares sold. Each is time-weighted for the fraction of the year the shares were or were not outstanding. But also notice two differences. The new shares are added, while the reacquired shares are subtracted. The second difference is that the reacquired shares are not multiplied by 1.10 to adjust for the 10% stock dividend. The reason is the shares were repurchased after the June 17 stock dividend; the reacquired shares are 8 million of the new post-distribution shares. (To use our earlier representation, these are 8 million "green" shares.) To generalize, when a stock distribution occurs during the reporting period, any sales or purchases of shares that occur *before* the distribution are increased by the distribution. But the stock distribution does not increase the number of shares sold or purchased, if any, *after* the distribution.

Any sales or purchases of shares that occur before, but not after, a stock dividend or split are affected by the distribution.

Earnings Available to Common Shareholders

● LO19–7

The denominator in an EPS calculation is the weighted-average number of common shares outstanding. Logically, the numerator should similarly represent earnings available to common shareholders. This was automatic in our illustrations to this point because the only shares outstanding were common shares. But when a senior class of shareholders (like preferred shareholders) is entitled to a specified allocation of earnings (like preferred dividends), those amounts are subtracted from earnings before calculating earnings per share.[18] This is demonstrated in Illustration 19–9.

Preferred dividends reduce earnings available to common shareholders unless the preferred stock is noncumulative and no dividends were declared that year.

Suppose no dividends were declared for the year. Should we adjust for preferred dividends? Yes, if the preferred stock is cumulative—and most preferred stock is. This means that when dividends are not declared, the unpaid dividends accumulate to be paid in a future year when (if) dividends are subsequently declared. Obviously, the presumption is that although the year's dividend preference isn't distributed this year, it eventually will be paid.

[18]You learned in Chapter 18 that when dividends are declared, preferred shareholders have a preference (over common shareholders) to a specified amount.

Illustration 19–9
Preferred Dividends

Sovrar Financial Corporation reported net income of $154 million in 2021 (tax rate 25%). Its capital structure included the following:

Common Stock

January 1	60 million common shares were outstanding
March 1	12 million new shares were sold
June 17	A 10% stock dividend was distributed
October 1	8 million shares were reacquired as treasury stock

Preferred Stock, Nonconvertible

January 1–December 31 **5 million shares 8%, $10 par**

($ amounts in millions, except per share amount)

Basic EPS:

Net income — **Preferred dividends**

$$\frac{\$154 \quad -\$4^{*}}{60\ (1.10)\ +\ 12\ (^{10}\!/_{12})\ (1.10)\ -\ 8\ (^{3}\!/_{12})} = \frac{\$150}{75} = \$2.00$$

Shares at Jan. 1 — New shares — Treasury shares

Stock dividend adjustment

*8% × $10 par × 5 million shares.

Preferred dividends are subtracted from net income so that "earnings available to common shareholders" is divided by the weighted-average number of common shares.

We have encountered no potential common shares to this point in our continuing illustration. As a result, we have what is referred to as a simple capital structure. (Although, at this point, you may question this label.) For a simple capital structure, a single presentation of basic earnings per common share is appropriate. We turn our attention now to situations described as complex capital structures. In these situations, two separate presentations are required: basic EPS and diluted EPS.

Diluted Earnings Per Share

Potential Common Shares

Imagine a situation in which convertible bonds are outstanding that will significantly increase the number of common shares if bondholders exercise their options to exchange their bonds for shares of common stock. Should these potential shares be ignored when earnings per share is calculated? After all, they haven't been converted as yet, so to assume an increase in shares for a conversion that may never occur might mislead investors and creditors. On the other hand, if conversion is imminent, not taking into account the dilutive effect of the share increase might mislead investors and creditors. The profession's solution to the dilemma is to calculate earnings per share twice.

Securities such as these convertible bonds, while not being common stock, may become common stock through their exercise or conversion. Therefore they may dilute (reduce) earnings per share and are called **potential common shares**. A firm is said to have a **complex capital structure** if potential common shares are outstanding. Besides convertible bonds, other potential common shares are convertible preferred stock, stock options, and contingently issuable securities. (We'll discuss each of these shortly.) A firm with a complex capital structure reports two EPS calculations. **Basic EPS** ignores the dilutive effect of such securities; **diluted EPS** incorporates the dilutive effect of all potential common shares.

In a complex capital structure, a second EPS computation takes into account the assumed effect of *potential common shares*, essentially a "worst case scenario."

Options, Rights, and Warrants

● LO19–8

Stock options, stock rights, and stock warrants are similar. Each gives its holders the right to exercise their option to purchase common stock, usually at a specified exercise price. The dilution that would result from their exercise should be reflected in the calculation of diluted EPS, but not basic EPS.

Stock options are assumed to have been exercised when calculating diluted EPS.

To include the dilutive effect of a security means to calculate EPS *as if* the potential increase in shares already has occurred (even though it hasn't yet). So, for stock options (or rights, or warrants), we pretend the options have been exercised. In fact, we assume the options were exercised at the beginning of the reporting period, or when the options were issued if that's later. We then assume the cash proceeds from selling the new shares at the exercise price are used to buy back as many shares as possible at the shares' average market price during the year. This is demonstrated in Illustration 19–10.

Illustration 19–10
Stock Options

Sovran Financial Corporation reported net income of $154 million in 2021 (tax rate 25%). Its capital structure included the following:

Common Stock

January 1	60 million common shares outstanding
March 1	12 million new shares were sold
June 17	A 10% stock dividend was distributed
October 1	8 million shares were reacquired as treasury stock

(The average market price of the common shares during 2021 was $25 per share.)

Stock options give their holders (company executives in this case) the right to purchase common stock at a specified exercise price ($20 in this case).

Preferred Stock Nonconvertible

January 1–December 31 5 million shares 8%, $10 par

Incentive Stock Options

Executive stock options granted in 2016, exercisable after 2020 for 15 million common shares* at an exercise price of $20 per share

($ amounts in millions, except per share amounts)

Basic EPS (unchanged)

The stock options do not affect the calculation of *basic EPS.*

$$\frac{\overset{\text{Net income}}{\$154} \quad \overset{\text{Preferred dividends}}{-\$4}}{\underset{\text{Shares at Jan. 1}}{60}\,(1.10) + \underset{\text{New shares}}{12\,(^{10}\!/_{12})}\,(1.10) - \underset{\text{Treasury shares}}{8\,(^{3}\!/_{12})}} = \frac{\$150}{75} = \$2.00$$

(1.10) = Stock dividend adjustment

Diluted EPS

The calculation of diluted EPS assumes that the shares specified by stock options were issued at the exercise price and that the proceeds were used to buy back (as treasury stock) as many of those shares as can be purchased at the market price during the period.

$$\frac{\overset{\text{Net income}}{\$154} \quad \overset{\text{Preferred dividends}}{-\$4}}{\underset{\text{Shares at Jan. 1}}{60}\,(1.10) + \underset{\text{New shares}}{12\,(^{10}\!/_{12})}\,(1.10) - \underset{\text{Treasury shares}}{8\,(^{3}\!/_{12})} + \underset{\text{Exercise of options}}{\mathbf{(15^{*} - 12^{\dagger})}}} = \frac{\$150}{78} = \$1.92$$

(1.10) = Stock dividend adjustment

*Adjusted for the stock dividend. Prior to the stock dividend, the options were exercisable for 13 7/11 million of the "old" shares. Upon the stock dividend, the new equivalent of 13 7/11 became 15 million (13 7/11 × 1.10) of the "new" shares.

†**Shares Assumed Reacquired for Diluted EPS**

15	million shares
× $ 20	(exercise price)
$300	million
÷ $ 25	(average market price)
12	million shares reacquired

FINANCIAL Reporting Case

Q4, p. 1105

When we simulate the exercise of the stock options, we calculate EPS as if 15 million shares were sold at the beginning of the year. This obviously increases the number of shares in the denominator by 15 million shares. But it is insufficient to simply add the additional shares without considering the accompanying consequences. Remember, if this hypothetical scenario had occurred, the company would have had $300 million cash proceeds from the exercise of the options (15 million shares × $20 exercise price per share). What would have been the effect on earnings per share? This depends on what the company would have done with the $300 million cash proceeds. Would the proceeds have been used to buy more equipment? Increase the sales force? Expand facilities? Pay dividends?

Obviously, there are literally hundreds of choices, and it's unlikely that any two firms would spend the $300 million exactly the same way. But remember, our objective is to create some degree of uniformity in the way firms determine earnings per share so the resulting numbers are comparable. So, standard-setters decided on a single assumption for all firms to enhance comparability.

For diluted EPS, we assume the proceeds from exercise of the options were used to reacquire shares as treasury stock at the average market price of the common stock during the reporting period. Consequently, the weighted-average number of shares is increased by the difference between the shares assumed issued and those assumed reacquired. In our illustration, 15 million shares issued minus 12 million shares reacquired ($300 million ÷ $25 per share) equals 3 million net increase in shares.

The way we take into account the dilutive effect of stock options is called the *treasury stock method* because of our assumption that treasury shares are purchased with the cash proceeds of the exercise of the options. Besides providing comparability, this assumption actually is plausible because, if the options were exercised, more shares would be needed to issue to option-holders. And, as discussed in the previous chapter, many firms routinely buy back shares either to issue to option-holders or, equivalently, to offset the issuance of new shares.

Additional Consideration

Actual Exercise of Options. What if options are actually exercised during the reporting period? In that case, we include in the denominator of both basic and diluted EPS the actual shares issued upon the exercise of the options, time-weighted for the fraction of the year the new shares actually are outstanding. This is consistent with the way we include new shares sold under any circumstances.

In addition, we include in diluted EPS only the incremental shares that would have been issued prior to the actual exercise of the options if we pretend the options were exercised at the beginning of the period. We time-weight those shares by the fraction of the year the shares would be outstanding prior to the actual exercise of the options.

Let's say the options in our illustration were exercised on September 1 and that the average per-share price of the stock from the beginning of the year until September 1 was $24. In that case, we add 5 million shares to the denominator of both basic and diluted EPS: the 15 million shares actually issued × 4/12 of the year actually outstanding = 5 million shares. Then, for *diluted EPS only,* we add the incremental shares that would have been issued *prior to* the actual exercise of the options if we pretend the options were exercised at the beginning of the period. We assume the proceeds (15 million shares × $20 = $300 million) are used to buy shares back at $24.

15	million shares
× $ 20	(exercise price)
$300	million
÷ $ 24	(average market price prior to exercise)
12.5	million shares

The incremental number of shares to be included in the computation of diluted EPS, then, is weighted for the period the options were outstanding prior to exercise, the 8 months from January 1 to September 1, or 8/12 of a year.

No adjustment to the numerator

+ (15 − 12.5)(8/12)

Convertible Securities

● LO19–9

Sometimes corporations include a conversion feature as part of a bond offering, a note payable, or an issue of preferred stock. Convertible securities can be converted into (exchanged for) shares of stock at the option of the holder of the security. For that reason, convertible securities are potentially dilutive. EPS will be affected if and when such securities are converted and new shares of common stock are issued. In the previous section you learned that the potentially dilutive effect of stock options is reflected in diluted EPS calculations by assuming the options were exercised. Similarly, the potentially dilutive effect of convertible securities is reflected in diluted EPS calculations by assuming they were converted.

When we assume conversion, the denominator of the EPS fraction is increased by the additional common shares that would have been issued upon conversion.

By the *if converted method,* as it's called, we assume the conversion into common stock occurred at the beginning of the period (or at the time the convertible security is issued, if that's later). We increase the denominator of the EPS fraction by the additional common shares that would have been issued upon conversion. We increase the numerator by the interest (after-tax) on bonds or other debt or the preferred dividends that would have been avoided if the convertible securities had not been outstanding due to having been converted.

CONVERTIBLE BONDS Now, let's return to our continuing illustration and modify it to include the existence of convertible bonds (Illustration 19–11).

Illustration 19–11
Convertible Bonds

Sovran Financial Corporation reported net income of $154 million in 2021 (tax rate 25%). Its capital structure included the following:

Common Stock

Jan. 1	60 million common shares were outstanding
Mar. 1	12 million new shares were sold
June 17	A 10% stock dividend was distributed
Oct. 1	8 million shares were reacquired as treasury stock

(The average market price of the common shares during 2021 was $25 per share.)

Preferred Stock, Nonconvertible

January 1–December 31 5 million shares 8%, $10 par

Incentive Stock Options

Executive stock options granted in 2016, exercisable after 2020 for 15 million common shares* at an exercise price of $20 per share

Convertible Bonds

8%, $300 million face amount issued in 2020, convertible into 12 million common shares

($ amounts in millions, except per share amounts)

Basic EPS (unchanged)

The convertible bonds do not affect the calculation of *basic EPS*.

Net income: $154; Preferred dividends: −$4

$$\frac{\$154 - \$4}{60\,(1.10) + 12\,(^{10}/_{12})\,(1.10) - 8\,(^{3}/_{12})} = \frac{\$150}{75} = \$2.00$$

Shares at Jan. 1; New shares; Treasury shares; Stock dividend adjustment

Diluted EPS

The numerator is increased by the after-tax interest that would have been avoided.

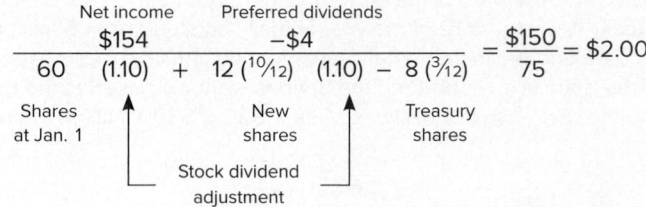

Net income: $154; Preferred dividends: −$4; **After-tax interest savings**

$$\frac{\$154 - \$4 + \mathbf{\$24 - 25\%\,(\$24)}}{60\,(1.10) + 12\,(^{10}/_{12})\,(1.10) - 8\,(^{3}/_{12}) + (15^{*} - 12) + \mathbf{12^{*}}} = \frac{\$168}{90} = \$1.87$$

Shares at Jan. 1; New shares; Treasury shares; Exercise of options; **Conversion of bonds**; Stock dividend adjustment

*Adjusted for the stock dividend. For example, prior to the stock dividend, the bonds were exercisable for $10^{10}/_{11}$ million of the "old" shares, which became 12 million ($10^{10}/_{11}$ × 1.10) of the "new" shares after the stock dividend.

We increase the denominator by the 12 million shares that would have been issued if the bonds had been converted. However, if that hypothetical conversion had occurred, the bonds would not have been outstanding during the year. What effect would the absence of the bonds have had on income? Obviously, the bond interest expense (8% × $300 million = $24 million) would have been saved, causing income to be higher. But saving the interest paid would also have meant losing a $24 million tax deduction on the income tax return. With a 25% tax rate that would mean paying $6 million more income taxes. So, to reflect in earnings the $18 million after-tax interest that would have been avoided in the event of conversion, we add back the $24 million of interest expense, but deduct 25% × $24 million for the higher tax expense.

Additional Consideration

The $300 million of convertible bonds in our illustration were issued at face value. Suppose the bonds had been issued for $282 million. In that case, the adjustment to earnings would be modified to include the amortization of the **$18** million bond discount. Assuming straight-line amortization and a **10-year** maturity, the adjustment to the diluted EPS calculation would have been

$$\frac{+[\$24 + (\$18 \div 10)] \times (1 - 25\%)^*}{12}$$

to reflect the fact that the interest expense would include the $24 million stated interest plus one-tenth of the bond discount.[19]

*This is an alternative way to represent the after-tax adjustment to interest since subtracting 25% of the interest expense is the same as multiplying interest expense by 75%.

Our illustration describes the treatment of convertible bonds. The same treatment pertains to other debt that is convertible into common shares such as convertible notes payable. Remember from our discussion of debt in earlier chapters that all debt is similar, whether in the form of bonds, notes, or other configurations.

Additional Consideration

Notice that we assumed the bonds were converted at the beginning of the reporting period since they were outstanding all year. However, if the convertible bonds had been issued during the reporting period, we would assume their conversion occurred on the date of issue. It would be illogical to assume they were converted before they were issued. If the convertible bonds in our illustration had been sold on **September 1**, for instance, the adjustment to the EPS calculation would have been

$$\frac{+[\$24 - 25\%(\$24)]\ (4/12)}{+12\ (4/12)}$$

to reflect the fact that the net increase in shares would have been effective for only four months of the year.

We assume convertible securities were converted (or options exercised) at the beginning of the reporting period or at the time the securities are issued, if later.

CONVERTIBLE PREFERRED STOCK The potentially dilutive effect of convertible preferred stock is reflected in EPS calculations in much the same way as convertible debt. That is, we calculate EPS as if conversion already had occurred. Specifically, we add shares to the denominator of the EPS fraction. We do not subtract the preferred dividends in the numerator because those dividends would have been avoided if the preferred stock had been converted. In Illustration 19–12, we assume our preferred stock is convertible into 3 million shares of common stock.

[19]See Chapter 14 if you need to refresh your memory about bond discount amortization.

Illustration 19–12

Convertible Preferred Stock

Sovran Financial Corporation reported net income of $154 million in 2021 (tax rate 25%). Its capital structure included the following:

Common Stock

Jan. 1	60 million common shares were outstanding
Mar. 1	12 million new shares were sold
June 17	A 10% stock dividend was distributed
Oct. 1	8 million shares were reacquired as treasury stock

(The average market price of the common shares during 2021 was $25 per share.)

Preferred Stock, Convertible into 3 million common shares*

January 1–December 31 5 million shares 8%, $10 par

Incentive Stock Options

Executive stock options granted in 2016, exercisable after 2020 for 15 million common shares* at an exercise price of $20 per share

Convertible Bonds

8%, $300 million face amount issued in 2020, convertible into 12 million common shares

($ amounts in millions, except per share amounts)

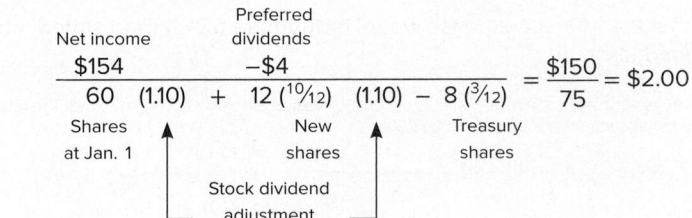

Basic EPS

$$\frac{\underset{\text{Net income}}{\$154}\ \underset{\text{Preferred dividends}}{-\$4}}{\underset{\text{Shares at Jan. 1}}{60\ (1.10)} + \underset{\text{New shares}}{12\ (^{10}/_{12})\ (1.10)} - \underset{\text{Treasury shares}}{8\ (^{3}/_{12})}} = \frac{\$150}{75} = \$2.00$$

(1.10 = Stock dividend adjustment)

Diluted EPS

$$\frac{\underset{\text{Net income}}{\$154}\ \underset{\text{Preferred dividends (crossed out)}}{\cancel{-\$4}}\ \underset{\text{After-tax interest savings}}{+\ \$24 - 25\%(\$24)}}{\underset{\text{Shares at Jan. 1}}{60}\ (1.10) + \underset{\text{New shares}}{12\ (^{10}/_{12})\ (1.10)} - \underset{\text{Treasury shares}}{8\ (^{3}/_{12})} + \underset{\text{Exercise of options}}{(15 - 12)}\ \underset{\text{Conversion of bonds}}{+12}\ \underset{\text{Conversion of preferred shares}}{\mathbf{+3^*}}} = \frac{\$172}{93} = \$1.85$$

(1.10 = Stock dividend adjustment)

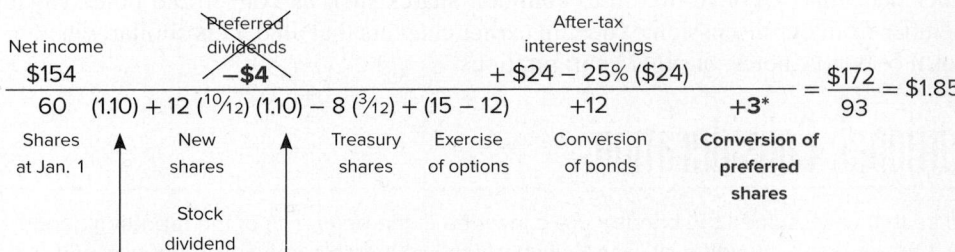

Since diluted EPS is calculated as if the preferred shares had been converted, there are no dividends.

*Adjusted for the stock dividend. For example, prior to the stock dividend, the preferred shares were convertible into $2\frac{8}{11}$ million of the "old" shares, which became 3 million ($2\frac{8}{11} \times 1.10$) of the "new" shares after the stock dividend.

The adjustment for the conversion of the preferred stock is applied only to diluted EPS computations. Basic EPS is unaffected.

However, when diluted EPS is calculated, we hypothetically assume the convertible preferred stock was *not* outstanding. Accordingly, no preferred dividends on these shares would have been paid.

Antidilutive Securities

At times, the effect of the conversion or exercise of potential common shares would be to increase, rather than decrease, EPS. These we refer to as **antidilutive securities**. Such securities are ignored when calculating both basic and diluted EPS.

Options, Warrants, Rights

● LO19–10

For illustration, recall the way we treated the stock options in our continuing illustration. In applying the treasury stock method, the number of shares assumed repurchased is fewer than the number of shares assumed sold. This is the case any time the buyback (average market) price is higher than the exercise price. Consequently, there will be a net increase in the number of shares, so earnings per share will decline.

On the other hand, when the exercise price is *higher* than the market price, to assume shares are sold at the exercise price and repurchased at the market price would mean buying back *more* shares than were sold. This would produce a net decrease in the number of shares. So EPS would increase, not decrease. These would have an antidilutive effect and would not be considered exercised. In fact, a rational investor would not exercise options at an exercise price higher than the current market price anyway. Let's look at the example provided by Illustration 19–13.

Antidilutive securities are ignored when calculating both basic and diluted EPS.

Illustration 19–13
Antidilutive Warrants

Sovran Financial Corporation reported net income of $154 million in 2021 (tax rate 25%). Its capital structure included the following:

Common Stock

Jan. 1	60 million common shares were outstanding
Mar. 1	12 million new shares were sold
June 17	A 10% stock dividend was distributed
Oct. 1	8 million shares were reacquired as treasury stock

(The average market price of the common shares during 2021 was **$25** per share.)

Preferred Stock, Convertible into 3 million common shares

January 1–December 31 5 million shares 8%, $10 par

Incentive Stock Options

Executive stock options granted in 2016, exercisable after 2020 for 15 million common shares at an exercise price of $20 per share

Convertible Bonds

8%, $300 million face amount issued in 2020, convertible into 12 million common shares

Stock Warrants

Warrants granted in 2020, exercisable for 4 million common shares* at an exercise price of $32.50 per share

Calculations

The calculations of both basic and diluted EPS are unaffected by the warrants because the effect of exercising the warrants would be antidilutive

*Adjusted for the stock dividend. For example, prior to the stock dividend, the warrants were exercisable for $3^7/_{11}$ million of the "old" shares, which became 4 million ($3^7/_{11}$ × 1.10) of the "new" shares after the stock dividend.

The $32.50 exercise price is higher than the market price, $25, so to assume shares are sold at the exercise price and repurchased at the market price would mean reacquiring more shares than were sold.

To assume 4 million shares were sold at the $32.50 exercise price and repurchased at the lower market price ($25) would mean reacquiring 5.2 million shares. That's *more* shares than were assumed sold. Because the effect would be antidilutive, we would simply ignore the warrants in the calculations.

In our continuing illustration, only the stock warrants were antidilutive. The other potential common shares caused EPS to decline when we considered them exercised or converted. In the case of the executive stock options, it was readily apparent that their effect would be dilutive because the exercise price was less than the market price, indicating that fewer shares could be repurchased (at the average market price) than were assumed issued (at the exercise price). As a result, the denominator increased. When only the denominator of a fraction increases, the fraction itself decreases. On the other hand, in the case of the warrants, it was apparent that their effect would be antidilutive because the exercise price was higher than the market price, which would have decreased the denominator and therefore increased the fraction.

When a company has a *net loss,* rather than net income, it reports a loss per share. In that situation, stock options that otherwise are dilutive will be antidilutive. Here's why. Suppose we have a loss per share of $2.00 calculated as ($150 million) ÷ 75 million shares = ($2.00).

Now suppose stock options are outstanding that, if exercised, will increase the number of shares by 5 million. If that increase is included in the calculation, the loss per share will be $1.88 calculated as ($150 million) ÷ 80 million shares = ($1.88). The *loss* per share *declines*. This represents an *increase* in performance—not a dilution of performance. The options would be considered antidilutive, then, and not included in the calculation of the net loss per share. Any potential common shares not included in dilutive EPS because they are antidilutive should be revealed in the disclosure notes.

Convertible Securities

For convertible securities, though, it's not immediately obvious whether the effect of their conversion would be dilutive or antidilutive because the assumed conversion would affect both the numerator and the denominator of the EPS fraction. We discovered each was dilutive only after including the effect in the calculation and observing the result—a decline in EPS. But there's an easier way.

To determine whether convertible securities are dilutive and should be included in a diluted EPS calculation, we can compare the "incremental effect" of the conversion (expressed as a fraction) with the EPS fraction before the effect of any convertible security is considered. This, of course is our basic EPS. Recall from Illustration 19–12 that basic EPS is $2.00.

For comparison, we determine the "earnings per incremental share" of the two convertible securities.

Conversion of bonds

The incremental effect (of conversion) of the bonds is the after-tax interest saved divided by the additional common shares from conversion.

After-tax interest savings

$$\frac{+\$24 - 25\%\ (\$24)}{+12} = \frac{\$18}{12} = \$1.50$$

Conversion of bonds

Conversion of preferred stock

The incremental effect (of conversion) of the preferred stock is the dividends that wouldn't be paid divided by the additional common shares from conversion.

Preferred dividends

$$\frac{+\$4}{+3} = \$1.33$$

Conversion of preferred shares

If the incremental effect of a security is *higher* than basic EPS, it is antidilutive. That's not the case in our illustration.

Order of Entry for Multiple Convertible Securities

A convertible security might seem to be dilutive when looked at individually but, in fact, may be antidilutive when included in combination with other convertible securities. This is because the *order of entry* for including their effects in the EPS calculation determines by how much, or even whether, EPS decreases as a result of their assumed conversion. Because our goal is to reveal the maximum potential dilution that might result, theoretically we should calculate diluted EPS using every possible combination of potential common shares to find the combination that yields the lowest EPS. But that's not necessary.

We can use the earnings per incremental share we calculated to determine the sequence of including securities' effects in the calculation. We include the securities in reverse order, beginning with the lowest incremental effect (that is, most dilutive), followed by the next lowest, and so on. This is, in fact, the order in which we included the securities in our continuing illustration.

Additional Consideration

Actually, the order of inclusion made no difference in our example, but would in many instances. For example, suppose the preferred stock had been convertible into 2.1 million shares, rather than 3 million shares. The incremental effect of its conversion would have been as follows:

Conversion of Preferred Stock

$$\frac{\overset{\text{Preferred dividends}}{+\$4}}{\underset{\text{Conversion of preferred shares}}{+2.1}} = \$1.90$$

On the surface, the effect would seem to be dilutive because $1.90 is less than $2.00, basic EPS. In fact, if this were the only convertible security, it would be dilutive. But, after the convertible bonds are assumed converted first, then the assumed conversion of the preferred stock would be *antidilutive.*

With Conversion of Bonds

$$\frac{\overset{\text{Net income}}{\$154} \quad \overset{\text{Preferred dividends}}{-\$4} \quad \overset{\text{After-tax interest savings}}{+\ \$24 - 25\%\ (\$24)}}{60\ (1.10) + 12\ (^{10}/_{12})\ (1.10) - 8\ (^{3}/_{12}) + (15 - 12) + 12} = \frac{\$168}{90} = \$1.867$$

Denominator: Shares at Jan. 1; New shares; Stock dividend adjustment; Treasury shares; Exercise of options; Conversion of bonds

With Conversion of Preferred Stock

$$\frac{\overset{\text{Net income}}{\$154} \quad \overset{\text{Preferred dividends}}{\cancel{-\$4}} \quad \overset{\text{After-tax interest savings}}{+\ \$24 - 25\%\ (\$24)}}{60\ (1.10) + 12\ (^{10}/_{12})\ (1.10) - 8\ (^{3}/_{12}) + (15 - 12) + 12 + \mathbf{2.1}} = \frac{\$172}{92.1} = \$1.868$$

Denominator: Shares at Jan. 1; New shares; Stock dividend adjustment; Treasury shares; Exercise of options; Conversion of bonds; **Conversion of preferred shares**

Although the incremental effect of the convertible preferred stock ($1.90) is lower than basic EPS ($2.00), when included in the calculation after the convertible bonds, the effect is antidilutive (EPS increases).

Because the incremental effect of the convertible bonds ($1.50) is lower than the incremental effect of the convertible preferred stock ($1.90), it is included first.

A convertible security might seem to be dilutive when looked at individually but may be antidilutive when included in combination with other convertible securities.

Concept Review Exercise

BASIC AND DILUTED EPS

At December 31, 2021, the financial statements of Clevenger Casting Corporation included the following:

Net income for 2021	$500 million
Common stock, $1 par:	
Shares outstanding on January 1	150 million shares
Shares retired for cash on February 1	24 million shares
Shares sold for cash on September 1	18 million shares
2-for-1 split on July 23	

(continued)

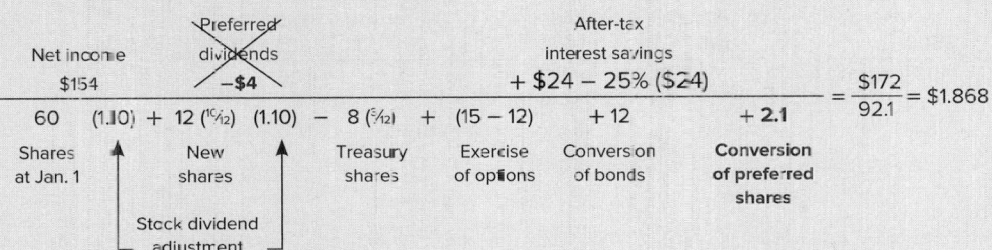

(concluded)

Preferred stock, 10%, $70 par, cumulative, nonconvertible	$ 70 million
Preferred stock, 8%, $50 par, cumulative, convertible into 4 million shares of common stock	$100 million
Incentive stock options outstanding, fully vested, for 4 million shares of common stock; the exercise price is $15	
Bonds payable, 10%, convertible into 20 million shares of common stock	$200 million

Additional data:

The market price of the common stock averaged $20 during 2021.

The convertible preferred stock and the bonds payable had been issued at par in 2019. The tax rate for the year was 25%.

Required:

Compute basic and diluted earnings per share for the year ended December 31, 2021.

($ amounts in millions, except per share amounts)

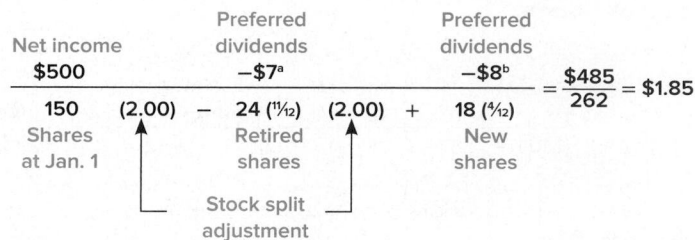

Solution:

Basic EPS

$$\frac{\underset{\text{Net income}}{\$500} \; \underset{\text{Preferred dividends}}{-\$7^{a}} \; \underset{\text{Preferred dividends}}{-\$8^{b}}}{\underset{\text{Shares at Jan. 1}}{150} \; \underset{\text{Stock split adjustment}}{(2.00)} - \underset{\text{Retired shares}}{24\,(^{11}/_{12})} \; \underset{\text{Stock split adjustment}}{(2.00)} + \underset{\text{New shares}}{18\,(^{4}/_{12})}} = \frac{\$485}{262} = \$1.85$$

Diluted EPS

$$\frac{\underset{\text{Net income}}{\$500} \; \underset{\text{Preferred dividends}}{-\$7^{a}} \; \underset{\text{Preferred dividends}}{-\$8^{b}} \; \underset{\text{After-tax interest savings}}{+\$20 - 25\%\,(\$20^{d})}}{\underset{\text{Shares at Jan. 1}}{150} \; \underset{\text{Stock split adjustment}}{(2.00)} - \underset{\text{Retired shares}}{24\,(^{11}/_{12})} \; \underset{\text{Stock split adjustment}}{(2.00)} + \underset{\text{New shares}}{18\,(^{4}/_{12})} + \underset{\text{Exercise of warrants}}{(4-3)^{c}} \; \underset{\text{Conversion of bonds}}{+20}} = \frac{\$500}{283} = \$1.77$$

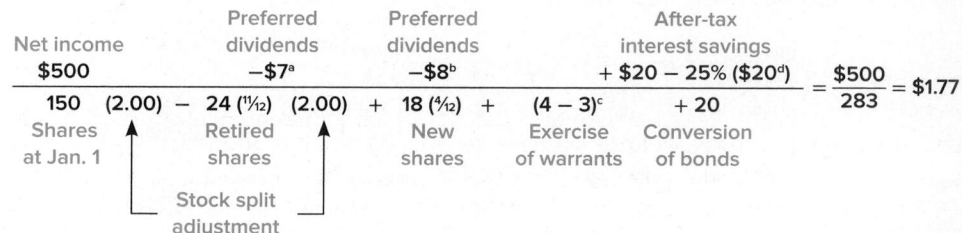

[a]10% × $70 million = $7 million
[b]8% × $100 million = $8 million

[c]Exercise of warrants:

	4 million shares
×	$ 15 (exercise price)
	$60 million
÷	$20 (average market price)
	3 million shares

[d]10% × $200 million = $20 million

Dilution:

Conversion of Bonds

$$\frac{\underset{\text{After-tax interest savings}}{+\$20 - 25\%\,(\$20)}}{\underset{\text{Conversion of bonds}}{+20}} = \$0.75$$

Conversion of 8% Preferred Stock

$$\frac{\underset{\text{Preferred dividends}}{+\$8}}{\underset{\text{Conversion of preferred shares}}{+4}} = \$2.00^{*}$$

*Because the incremental effect of conversion of the preferred stock ($2) is higher than EPS without the conversion of the preferred stock, the conversion would be *antidilutive* and is *not* considered in the calculation of diluted EPS.

Additional EPS Issues

Components of the "Proceeds" in the Treasury Stock Method

● LO19–11

In calculating diluted EPS when stock options are outstanding, we assume the options have been exercised. That is, we pretend the company sold the shares specified by the options at the exercise price and that the "proceeds" were used to buy back (as treasury stock) as many shares as can be purchased at the average market price of the stock during the year. The proceeds for the calculation should include the amount received from the hypothetical exercise of the options ($300 million in Illustration 19–10). But that's only the first of two possible components.

For the treasury stock method, "proceeds" include:

1. The amount, if any, received from the hypothetical exercise of options or vesting of restricted stock.
2. The total compensation from the award that's not yet expensed.

The second component of the proceeds is the total compensation from the award that's *not yet expensed.* If the fair value of an option had been $4 at the grant date, the total compensation would have been 15 million shares times $4, or $60 million. In our illustration, though, we assumed the options were fully vested before 2021, so all $60 million already had been expensed and this second component of the proceeds was zero. If the options had been only half vested, half the compensation would have been unexpensed and $30 million would have been added to the $300 million proceeds. This would have been the case, for instance, if our calculation was made after two years of a four-year vesting period.

Why do the proceeds include these two components? We might think of it like this. The "proceeds" include everything the firm will receive from the award: (1) cash, if any, at exercise and (2) services from the recipient (value of award given as compensation). The reason we *exclude the expensed portion* is that, when it's expensed, earnings are reduced, so that portion of the dilution already is reflected in EPS. Excluding that expensed portion from the proceeds avoids the additional dilution that would occur if we included those additional proceeds in our hypothetical buyback of shares. Hence, we avoid double-counting the dilutive effect of the compensation.

Restricted Stock Awards in EPS Calculations

As we discussed earlier, restricted stock awards and their cousins, restricted stock units (RSUs), have replaced stock options as the share-based compensation plan of choice. Like stock options, they represent potential common shares and their dilutive effect is included in diluted EPS. In fact, they too are included using the treasury stock method. That is, the shares are added to the denominator and then reduced by the number of shares that can be bought back with the "proceeds" at the average market price of the company's stock during the year. Unlike stock options, though, the first component of the proceeds usually is absent; employees don't pay to acquire their shares.

Also, only *unvested* restricted stock award shares and RSU shares are included in EPS calculations; fully vested shares are actually distributed and thus outstanding. The proceeds for the EPS calculation include the total compensation from the unvested restricted stock that's not yet expensed, the second component. For an example, refer back to the restricted stock in Illustration 19–2. The total compensation for the award is $60 million ($12 market price per share × 5 million shares). Because the restricted stock vests over four years, it is expensed as $15 million each year for four years. At the end of 2021, the first year, $45 million remains unexpensed, so $45 million would be the assumed proceeds in an EPS calculation. If we assume the average market price that year was, say, $15, the $45 million will buy back 3 million shares and we would add to the denominator of diluted EPS 2 million common shares:

$$\frac{\text{No adjustment to the numerator}}{5 \text{ million} - 3^{*} \text{ million} = 2 \textit{ million}}$$

*Assumed purchase of treasury shares

$45	million
÷ $15	average market price
3	million shares

*At the end of 2022, the *second* year, $30 million remains unexpensed, so assuming the average market price that year was, say, $12, we would add to the denominator of diluted EPS 2.5 million common shares:

$$\frac{\text{No adjustment to the numerator}}{5 \text{ million} - 2.5^{*} \text{ million} = 2.5 \textit{ million}}$$

*Assumed purchase of treasury shares

$30	million
÷ $12	average market price
2.5	million shares

Contingently Issuable Shares

● LO19–12 Sometimes an agreement specifies that additional shares of common stock will be issued, contingent on the occurrence of some future circumstance. For instance, in the disclosure note reproduced in Illustration 19–14, **Hunt Manufacturing Co.** reported contingent shares in connection with its acquisition of **Feeny Manufacturing Company**.

Illustration 19–14

Contingently Issuable Shares—Hunt Manufacturing Company

Real World Financials

Note 12: Acquisitions (in part)

The Company acquired Feeny Manufacturing Company of Muncie, Indiana, for 135,000 shares of restricted common stock with a value of $7.71 per share. Feeny Manufacturing Company is a manufacturer of kitchen storage products. The purchase agreement calls for the issuance of up to 135,000 additional shares of common stock in the next fiscal year based on the earnings of Feeny Manufacturing Company. . . .

Source: Hunt Manufacturing Company

At times, contingent shares are issuable to shareholders of an acquired company, certain key executives, or others in the event a certain level of performance is achieved. Contingent performance may be a desired level of income, a target stock price, or some other measurable activity level.

Contingently issuable shares are considered outstanding in the computation of diluted EPS.

When calculating EPS, **contingently issuable shares** are considered to be outstanding in the computation of diluted EPS if the target performance level already is being met (assumed to remain at existing levels until the end of the contingency period). For example, if shares will be issued at a future date if a certain level of income is achieved and that level of income or more was already reported this year, those additional shares are simply added to the denominator of the diluted EPS fraction.[20]

For clarification, refer to our continuing illustration of diluted EPS and assume 3 million additional shares will become issuable to certain executives in the following year (2022) if net income that year is $150 million or more. Recall that net income for Sovran Finanical in 2021 was $154 million, so the additional shares would be considered outstanding in the computation of diluted EPS by simply adding **3** million additional shares to the denominator of the EPS fraction. Obviously, the 2022 condition ($150 million net income or more) has not been met yet since it's only 2021. But because that level of income was achieved in 2021, the presumption is that it's likely to be achieved in 2022 as well.

Assumed Issuance of Contingently Issuable Shares (diluted EPS):

If a level of income must be attained before the shares will be issued, and income already is that amount or more, the additional shares are simply added to the denominator.

$$\frac{\text{No adjustment to the numerator}}{\begin{array}{c}\mathbf{+3} \\ \textbf{Additonal shares}\end{array}}$$

On the other hand, if the target income next year is $160 million, the contingent shares would simply be ignored in our calculation.

[20]The shares should be included in both basic and diluted EPS if all conditions have actually been met so that there is no *circumstance* under which those shares would not be issued. In essence, these *are no longer contingent shares.*

Summary of the Effect of Potential Common Shares on Earnings Per Share

You have seen that under certain circumstances, securities that have the potential of reducing earnings per share by becoming common stock are assumed already to have become common stock for the purpose of calculating EPS. The table in Illustration 19–15 summarizes the circumstances under which the dilutive effect of these securities is reflected in the calculation of basic and diluted EPS.

Illustration 19–15
When Potential Common Shares Are Reflected in EPS

	Is the Dilutive Effect Reflected in the Calculation of EPS?*	
Potential Common Shares	**Basic EPS**	**Diluted EPS**
• Stock options (or warrants, rights)	no	yes
• Restricted stock	no	yes
• Convertible securities (bonds, notes, preferred stock)	no	yes
• Contingently issuable shares	no	yes†

*The effect is not included for any security if its effect is antidilutive.
†Unless shares are contingent upon some level of performance not yet achieved.

Illustration 19–16 summarizes the specific effects on the diluted EPS fraction when the dilutive effect of a potentially dilutive security is reflected in the calculation.

Illustration 19–16
How Potential Common Shares Are Reflected in a Diluted EPS Calculation

	Modification to the Diluted EPS Fractions	
Potential Common Shares	**Numerator**	**Denominator**
• Stock options (or warrants, rights)	None	Add the shares that would be created by their exercise,* reduced by shares repurchased at the average share price.
• Restricted stock	None	Add shares that would be created by their vesting,* reduced by shares repurchased at the average share price.
• Convertible bonds (or notes)	Add the interest (after-tax) that would have been avoided if the debt had been converted.	Add shares that would be created by the conversion* of the bonds (or notes).
• Convertible preferred stock	Do not deduct the dividends that would have been avoided if the preferred stock had been converted.	Add shares that would have been created by the conversion* of the preferred stock.
• Contingently issuable shares: Issuable when specified conditions are met, and those conditions currently are being met	None	Add shares that are issuable.
• Contingently issuable shares: Issuable when specified conditions are met, and those conditions are not currently being met	None	None

*At the beginning of the year or when potential common shares were issued, whichever is later (time-weight the increase in shares if assumed exercised or converted in midyear).

Actual Conversions

When calculating EPS in our example, we "pretended" the convertible bonds had been converted at the beginning of the year. What if they actually had been converted, let's say on November 1? Interestingly, diluted EPS would be precisely the same. Here's why:

1. The actual conversion would cause an actual increase in shares of 12 million on November 1. These would be time-weighted so the denominator would increase by 12 ($^2/_{12}$). Also, the numerator would be higher because net income actually would be increased by the after-tax interest saved on the bonds for the last two months, [\$24 – 25% (\$24)] × ($^2/_{12}$). Be sure to note that this would not be an adjustment in the EPS calculation. Instead, net income would actually have been higher by [\$24 – 25% (\$24)] × ($^2/_{12}$) = \$3. That is, reported net income would have been \$157 rather than \$154.
2. We would assume conversion for the period before November 1 because they were potentially dilutive during that period. The 12 million shares assumed outstanding from January 1 to November 1 would be time-weighted for that 10-month period: 12 ($^{10}/_{12}$). Also, the numerator would be increased by the after-tax interest assumed saved on the bonds for the first 10 months, [\$24 – 25% (\$24)] × ($^{10}/_{12}$).

Notice that the incremental effect on diluted EPS is the same either way.

EPS would be precisely the same whether convertible securities were actually converted or not.

Not Actually Converted: **Converted on November 1:**

$$\frac{+\$24 - 25\%(\$24)}{+12} = \frac{+[\$24 - 25\%(\$24)] \times (^2/_{12}) + [\$24 - 25\%(\$24)] \times (^{10}/_{12})}{+12(^2/_{12}) \quad + 12(^{10}/_{12})}$$

Numerators: Assumed after-tax interest savings = Actual after-tax interest savings + Assumed after-tax interest savings

Denominators: Assumed conversion of bonds = Actual conversion of bonds + Assumed conversion of bonds

$$\frac{\$18}{12} = \frac{\$3 \quad + \quad \$15}{2 \quad + \quad 10}$$

Illustration 19–17 shows the disclosure note **Clorox Company** reported after the conversion of convertible notes during the year.

Illustration 19–17

Conversion of Notes—The Clorox Company

Real World Financials

Note 1: Significant Accounting Policies—Earnings Per Common Share (in part)

A \$9,000,000 note payable to Henkel Corporation was converted into 1,200,000 shares of common stock on August 1. . . . Earnings per common share and weighted-average shares outstanding reflect this conversion as if it were effective during all periods presented.

Source: The Clorox Company

● LO19–13

Financial Statement Presentation of Earnings Per Share Data

Recall from Chapter 4 that if a company disposes of a component of its operations, the company will report "discontinued operations" as a separate item within the income statement as follows:

Income from Continuing Operations
Discontinued operations
Net income

When the income statement includes discontinued operations, EPS data (both basic and diluted) must also be reported separately for income from continuing operations and net income. Per share amounts for discontinued operations would be disclosed either on the face of the income statement or in the notes to financial statements. Presentation on the face of the income statement is illustrated by the partial income statements of **H&R Block, Inc.**, from a recent annual report and exhibited in Illustration 19–18.

Basic and diluted EPS data should be reported on the face of the income statement for all reporting periods presented in the comparative statements. Businesses without potential

Illustration 19–18

EPS Disclosure—H&R Block

Real World Financials

Consolidated Income Statements (partial)
For the Years Ended April 30, 2017 and 2016

	2017	2016
Net income from continuing operations	$420,917	$383,553
Net loss from discontinued operations	(11,972)	(9,286)
Net income	$403,945	$374,267
Basic Earnings (Loss) Per Share:		
Continuing operations	$ 1.97	$ 1.54
Discontinued operations	(0.05)	(0.04)
Consolidated	$ 1.92	$ 1.50
Diluted Earnings (Loss) Per Share:		
Continuing operations	$ 1.96	$ 1.53
Discontinued operations	(0.05)	(0.04)
Consolidated	$ 1.91	$ 1.49

Source: H&R Block

common shares present basic EPS only. Disclosure notes should provide additional disclosures including the following:

1. A reconciliation of the numerator and denominator used in the basic EPS computations to the numerator and the denominator used in the diluted EPS computations. An example of this is presented in Illustration 19–19 using the situation described previously inIllustration 19–12.
2. Any adjustments to the numerator for preferred dividends.
3. Any potential common shares that weren't included because they were antidilutive.
4. Any transactions that occurred after the end of the most recent period that would materially affect earnings per share.

Illustration 19–19

Reconciliation of Basic EPS Computations to Diluted EPS Computations

Earnings per Share Reconciliation

($ in millions)	Income (Numerator)	Share (Denominator)	Per Share Amount
Net income	$154		
Preferred dividends	(4)		
Basic earnings per share	**150**	**75**	**$2.00**
Stock options	None	3*	
Convertible debt	18	12	
Convertible preferred stock	4	3	
Diluted earnings per share	**$172**	**93**	**$1.85**

Note: Stock warrants to purchase an additional 4 million shares at $32.50 per share were outstanding throughout the year but were not included in diluted EPS because the warrants' exercise price is greater than the average market price of the common shares.

*15 million − [(15 million × $20) ÷ $25] = 3 million net additional shares.

Additional Consideration

It's possible that potential common shares would have a dilutive effect on one component of net income but an antidilutive effect on another. When the inclusion of the potential common shares has a dilutive effect on "income from continuing operations," the effect should be included in all calculations of diluted EPS. In other words, the same number of potential common shares used in computing the diluted per-share amount for income from continuing operations is used in computing all other diluted per-share amounts, even when amounts are antidilutive to the individual per-share amounts.

Decision Makers' Perspective

We noted earlier in the chapter that investors and creditors pay a great deal of attention to earnings per share information. Because of the importance analysts attach to earnings announcements, companies are particularly eager to meet earnings expectations. As we first noted in Chapter 4, this desire has contributed to a relatively recent trend, especially among technology firms, to report **pro forma** earnings per share. What exactly are pro forma earnings? Unfortunately there is no answer to that question. Essentially, pro forma earnings are actual (GAAP) earnings increased due to the reduction of any expenses the reporting company feels are unusual and should be excluded. Always, though, the pro forma results of a company look better than the real results. **Broadcom Corporation**, a provider of broadband and network products, reported pro forma *earnings* of $0.49 per share. However on a GAAP basis, it actually had a *loss* of $3.29 per share. This is not an isolated example.

Make sure you pay lots of attention to the man behind the curtain. If any earnings figure says pro forma, you should immediately look for a footnote or explanation telling you just what is and is not included in the calculation.

When companies report pro forma results, they argue they are trying to help investors by giving them numbers that more accurately reflect their normal business activities, because they exclude unusual expenses. Analysts should be skeptical, though. Because of the purely discretionary nature of pro forma reporting and several noted instances of abuse, analysts should, at a minimum, find out precisely what expenses are excluded and what the actual GAAP numbers are.

Another way management might enhance the appearance of EPS numbers is by massaging the denominator of the calculation. Reducing the number of shares increases earnings *per share*. Some companies judiciously use share buyback programs to manipulate the number of shares and therefore EPS. There is nothing inherently wrong with share buybacks and, as we noted in Chapter 18, they can benefit shareholders. The motivation for buybacks, though, can sometimes be detected in the year-to-year pattern of net income and EPS. Companies whose growth rates in earnings per share routinely exceed their growth in net income may be using buybacks to artificially increase EPS.

One way analysts use EPS data is in connection with the price-earnings ratio. This ratio is simply the market price per share divided by the earnings per share. It measures the market's perception of the quality of a company's earnings by indicating the price multiple the capital market is willing to pay for the company's earnings. Presumably, this ratio reflects the information provided by all financial information in that the market price reflects analysts' perceptions of the company's growth potential, stability, and relative risk. The price-earnings ratio relates these performance measures with the external judgment of the marketplace concerning the value of the firm.

The price-earnings ratio measures the quality of a company's earnings.

The ratio measures the quality of earnings in the sense that it represents the market's expectation of future earnings as indicated by current earnings. Caution is called for in comparing price-earnings ratios. For instance, a ratio might be low, not because earnings expectations are low, but because of abnormally elevated current earnings. On the contrary, the ratio might be high, not because earnings expectations are high, but because the company's current earnings are temporarily depressed. Similarly, an analyst should be alert to differences among accounting methods used to measure earnings from company to company when making comparisons.

The dividend payout ratio indicates the percentage of earnings that is distributed to shareholders as dividends.

Another ratio frequently calculated by shareholders and potential shareholders is the dividend payout ratio. This ratio expresses the percentage of earnings that is distributed to shareholders as dividends. The ratio is calculated by dividing dividends per common share by the earnings per share.

This ratio provides an indication of a firm's reinvestment strategy. A low payout ratio suggests that a company is retaining a large portion of earnings for reinvestment for new facilities and other operating needs. Low payouts often are found in growth industries and high payouts in mature industries. Often, though, the ratio is merely a reflection of managerial strategy concerning the mix of internal versus external financing. The ratio also is considered by investors who, for tax or other reasons, prefer current income over market price appreciation, or vice versa. ●

Concept Review Exercise

ADDITIONAL EPS ISSUES

At December 31, 2021, the financial statements of Bahnson General, Inc. included the following:

Net income for 2021 (including a net-of-tax loss from discontinued operations of $10 million)	$180 million
Common stock, $1 par:	
Shares outstanding on January 1	44 million
The share price was $25 and $28 at the beginning and end of the year, respectively.	

Additional data:

- At January 1, 2021, $200 million of 8% convertible notes were outstanding. The notes were converted on April 1 into 16 million shares of common stock.
- An agreement with company executives calls for the issuance of up to 12 million additional shares of common stock in 2022 and 2023 based on Bahnson's net income in those years. Executives will receive 2 million shares at the end of each of those two years if the company's stock price is at least $26 and another 4 million shares each year if the stock price is at least $29.50.

The tax rate is 25%.

Required:

Compute basic and diluted earnings per share for the year ended December 31, 2021.

Solution:

($ amounts in millions, except per share amounts)

Basic EPS

$$\frac{\underset{\text{Net income}}{\$180}}{\underset{\text{Shares at Jan. 1}}{44} + \underset{\text{Actual conversion of notes}}{16\,(9/12)}} = \frac{\$180}{56} = \$3.21$$

Diluted EPS

$$\frac{\underset{\text{Net income}}{\$180} + [\$16 - 25\%(\$16)] \times (3/12)}{\underset{\text{Shares at Jan. 1}}{44} + \underset{\text{Actual conversion of notes}}{16\,(9/12)} + \underset{\text{Assumed conversion of notes}}{16\,(3/12)} + \underset{\text{Contingent shares}}{(2 + 2)}} = \frac{\$183}{64} = \$2.86$$

Convertible Notes: Notice that the effect on diluted EPS would be precisely the same whether the convertible notes were actually converted or not.

Converted on April 1:

$$\frac{\underset{\text{Net income including actual after-tax interest savings}}{\$180} + \underset{\text{Assumed after-tax interest savings}}{[\$16 - 25\%(\$16)] \times (3/12)}}{\underset{\text{Shares at Jan. 1}}{44} + \underset{\text{Actual conversion of notes}}{16\,(9/12)} + \underset{\text{Assumed conversion of notes}}{16\,(3/12)}} = \frac{\$183}{60}$$

Not Actually Converted:

Net income without actual after-tax interest savings		Assumed after-tax interest savings		
$171*		+ [$16 – 25% ($16)]	=	$183
44	+	16		60
Shares at Jan. 1		Assumed conversion of notes		

$$\frac{\$171^{*} + [\$16 - 25\%(\$16)]}{44 + 16} = \frac{\$183}{60}$$

*$180 – {[$16 – 25% ($16)] × (9/12)} = $171; after-tax interest from Apr. 1 to Dec. 31.

Contingently Issuable Shares:
Because the conditions are met for issuing 4 million shares (2 million for each of two years), those shares are simply added to the denominator of diluted EPS. The current share price ($28) is projected to remain the same throughout the contingency period, so the other 8 million shares (4 million for each of two years) are excluded.

Income Statement Presentation:
To determine the per share amounts for income before discontinued operations, we substitute that amount for net income in the numerator (in this case, that means adding back the $10 million loss from discontinued operations):

$$\text{Basic: } \frac{\$180 + \$10}{56} = \$3.39 \qquad \text{Diluted: } \frac{\$183 + \$10}{64} = \$3.02$$

Earnings per Share:	**Basic***	**Diluted**
Income before discontinued operations	$3.39	$ 3.02
Loss from discontinued operations	(0.18)	(0.16)
Net income	$ 3.21	$ 2.86

*Only diluted EPS is required on the face of the income statement. Basic EPS is reported in the EPS reconciliation shown in the disclosure note (below).

Disclosure Note:
Earnings per Share Reconciliation:

	Income (Numerator)	**Shares (Denominator)**	**Per Share Amount**
Basic Earnings per Share			
Income before discontinued operations	$190	56	$3.39
Loss from discontinued operations	(10)	56	(0.18)
Net income	$180	56	$ 3.21
Convertible debt	3	4	
Contingently issuable shares	—	4	
Diluted Earnings per Share			
Income before discontinued operations	$193	64	$3.02
Loss from discontinued operations	(10)	64	(0.16)
Net income	$183	64	$2.86

Financial Reporting Case Solution

©Fabio Cardoso/Getty Images

1. **How can a compensation package such as this serve as an incentive to Ms. Veres?** *(p. 1105)* Stock-based plans like the restricted stock and stock options that Ms. Veres is receiving are designed to motivate recipients. If the shares awarded are restricted so that Ms. Veres is not free to sell the shares during the restriction period, she has an incentive to remain with the company until rights to the shares vest. Likewise, stock options can be made exercisable only after a specified period of employment. An additional incentive of stock-based plans is that the recipient will be motivated to take actions that will maximize the value of the shares.

2. **Ms. Veres received a "grant of restricted stock." How should NEV account for the grant?** *(p. 1106)* The compensation associated with restricted stock is the market price of unrestricted shares of the same stock. NEV will accrue this amount as compensation expense over the service period from the date of grant to when restrictions are lifted.
3. **Included were stock options to buy more than 800,000 shares of NEV stock. How will the options affect NEV's compensation expense?** *(p. 1108)* Similar to the method used for restricted stock, the value of the options is recorded as compensation over the service period, usually the vesting period.
4. **How will the presence of these and other similar stock options affect NEV's earnings per share?** *(p. 1124)* If outstanding stock options were exercised, the resulting increase in shares would reduce or dilute EPS. If we don't take into account the dilutive effect of the share increase, we might mislead investors and creditors. So, in addition to basic EPS, we also calculate diluted EPS to include the dilutive effect of options and other potential common shares. This means calculating EPS as if the potential increase in shares already has occurred (even though it hasn't yet). ●

The Bottom Line

- **LO19–1** We measure the fair value of stock issued in a restricted stock plan and expense it over the service period, usually from the date of grant to the vesting date. (*p. 1106*)
- **LO19–2** Similarly, we estimate the fair value of stock options at the grant date and expense it over the service period, usually from the date of grant to the vesting date. Fair value is estimated at the grant date using an option-pricing model that considers the exercise price and expected term of the option, the current market price of the underlying stock and its expected volatility, expected dividends, and the expected risk-free rate of return. (*p. 1108*)
- **LO19–3** Employee share purchase plans allow employees to buy company stock under convenient or favorable terms. Most such plans are considered compensatory and require any discount to be recorded as compensation expense. (*p. 1117*)
- **LO19–4** A company has a simple capital structure if it has no outstanding securities that could potentially dilute earnings per share. For such a firm, EPS is simply earnings available to common shareholders divided by the weighted-average number of common shares outstanding. When potential common shares are outstanding, the company is said to have a complex capital structure. In that case, two EPS calculations are reported. Basic EPS assumes no dilution. Diluted EPS assumes maximum potential dilution. (*p. 1119*)
- **LO19–5** EPS calculations are based on the weighted-average number of shares outstanding during the period. Any new shares issued during the period are time-weighted by the fraction of the period they were outstanding and then added to the number of shares outstanding for the period. (*p. 1120*)
- **LO19–6** For a stock dividend or stock split, shares outstanding prior to the stock distribution are retroactively restated to reflect the increase in shares. When shares are reacquired, as treasury stock or to be retired, they are time-weighted for the fraction of the period they were not outstanding, prior to being subtracted from the number of shares outstanding during the reporting period. (*p. 1120*)
- **LO19–7** The numerator in the EPS calculation should reflect earnings available to common shareholders. So, any dividends on preferred stock outstanding should be subtracted from reported net income. This adjustment is made for cumulative preferred stock whether or not dividends are declared that period. (*p. 1122*)
- **LO19–8** For diluted EPS, it is assumed that stock options, rights, and warrants are exercised at the beginning of the period (or at the time the options are issued, if later) and the cash proceeds received are used to buy back (as treasury stock) as many of those shares as can be acquired at the average market price during the period. (*p. 1123*)
- **LO19–9** To incorporate convertible securities into the calculation of diluted EPS, the conversion is assumed to have occurred at the beginning of the period (or at the time the convertible security is issued, if later). The denominator of the EPS fraction is adjusted for the additional common shares assumed, and the numerator is increased by the interest (after-tax) on bonds and not reduced by the preferred dividends that would have been avoided in the event of conversion. (*p. 1126*)

- **LO19–10** If including potential common shares in the EPS calculation causes EPS to *increase* rather than decrease, then those potential common shares are considered *antidilutive* and are omitted from the calculation. (*p. 1128*)
- **LO19–11** For the treasury stock method, "proceeds" include (1) the amount, if any, received from the hypothetical exercise of options or vesting of restricted stock and (2) the total compensation from the award that's not yet expensed. (*p. 1133*)
- **LO19–12** Contingently issuable shares are considered outstanding in the computation of diluted EPS when they will later be issued upon the mere passage of time or because of conditions that currently are met. (*p. 1134*)
- **LO19–13** EPS data (both basic and diluted) must be reported for (a) income before any discontinued operations, (b) the discontinued operations, and (c) net income. Disclosures also should include a reconciliation of the numerator and denominator used in the computations. (*p. 1136*)
- **LO19–14** When options have graded vesting, unlike under U.S. GAAP, IFRS does not permit the straight-line method for allocating compensation or require that the company recognize at least the amount of the award that has vested by each reporting date. The earnings per share requirements of IFRS and U.S. GAAP are similar. The few differences that remain are the result of differences in the application of the treasury stock method, the treatment of contracts that may be settled in shares or cash, and contingently issuable shares. (*pp. 1113, 1115,* and *1119*)

APPENDIX 19A Option-Pricing Theory

Option values have two essential components: (1) intrinsic value and (2) time value.

Intrinsic Value

Intrinsic value is the benefit the holder of an option would realize by exercising the option rather than buying the underlying stock directly. An option that permits an employee to buy $25 stock for $10 has an intrinsic value of $15. An option that has an exercise price equal to or exceeding the market price of the underlying stock has zero intrinsic value.

Time Value

In addition to their intrinsic value, options also have a time value due to the fact that (a) the holder of an option does not have to pay the exercise price until the option is exercised and (b) the market price of the underlying stock may yet rise and create additional intrinsic value. All options have time value so long as time remains before expiration. The longer the time until expiration, other things being equal, the greater the time value. For instance, the option described above with an intrinsic value of $15, might have a fair value of, say, $22 if time still remains until the option expires. The $7 difference represents the time value of the option. Time value can be subdivided into two components: (1) the effects of time value of money and (2) volatility value.

TIME VALUE OF MONEY

An option's value is enhanced by the delay in paying cash for the shares.

The time value of money component arises because the holder of an option does not have to pay the exercise price until the option is exercised. Instead, the holder can invest funds elsewhere while waiting to exercise the option. For measurement purposes, the time value of money component is assumed to be the rate of return available on risk-free U.S. Treasury Securities. The higher the time value of money, the higher the value of being able to delay payment of the exercise price.

When the underlying stock pays no dividends, the time value of money component is the difference between the exercise price (a future amount) and its discounted present value. Let's say the exercise price is $30. If the present value (discounted at the risk-free rate) is $24, the time value of money component is $6. On the other hand, if the stock pays a dividend (or is expected to during the life of the option), the time value of money component is lower.

The value of being able to delay payment of the exercise price would be partially offset by the cost of forgoing the dividend in the meantime. For instance, if the stock underlying the options just described were expected to pay dividends and the discounted present value of the expected dividends were $2, the time value of money component in that example would be reduced from $6 to $4.

The time value of money component is the difference between the exercise price and its discounted present value minus the present value of expected dividends.

VOLATILITY VALUE

The volatility value represents the possibility that the option holder might profit from market price appreciation of the underlying stock while being exposed to the loss of only the value of the option, rather than the full market value of the stock. For example, fair value of an option to buy a share at an exercise price of $30 might be measured as $7. The potential profit from market price appreciation is conceptually unlimited. And yet, the potential loss from the stock's value failing to appreciate is only $7.

A stock's volatility is the amount by which its price has fluctuated previously or is expected to fluctuate in the future. The greater a stock's volatility, the greater the potential profit. It usually is measured as one standard deviation of a statistical distribution. Statistically, if the expected annualized volatility is 25%, the probability is approximately 67% that the stock's year-end price will fall within roughly plus or minus 25% of its beginning-of-year price. Stated differently, the probability is approximately 33% that the year-end stock price will fall outside that range.

Volatility enhances the likelihood of stock price appreciation.

Option-pricing models make assumptions about the likelihood of various future stock prices by making assumptions about the statistical distribution of future stock prices that take into account the expected volatility of the stock price. One popular option pricing model, the Black–Scholes model, for instance, assumes a log-normal distribution. This assumption posits that the stock price is as likely to fall by half as it is to double and that large price movements are less likely than small price movements. The higher a stock's volatility, the higher the probability of large increases or decreases in market price. Because the cost of large decreases is limited to the option's current value, but the profitability from large increases is unlimited, an option on a highly volatile stock has a higher probability of a large profit than does an option on a less volatile stock.

Summary

In summary, the fair value of an option is (a) its intrinsic value plus (b) its time value of money component plus (c) its volatility component. The variables that affect an option's fair value and the effect of each are indicated in Illustration 19A–1.

All Other Factors Being Equal, If the	The Option Value Will Be
Exercise price is higher	Lower
Term of the option is longer	Higher
Market price of the stock is higher	Higher
Dividends are higher	Lower
Risk-free rate of return is higher	Higher
Volatility of the stock is higher	Higher

Illustration 19A–1
Effect of Variables on an Option's Fair Value

APPENDIX 19B

Stock Appreciation Rights

Stock appreciation rights (SARs) overcome a major disadvantage of stock option plans that require employees to actually buy shares when the options are exercised. Even though the options' exercise price may be significantly lower than the market value of the shares, the employee still must come up with enough cash to take advantage of the bargain. This can be quite a burden if the award is sizable. In a nonqualified stock option plan, income taxes also would have to be paid when the options are exercised.[21]

In an SAR plan, the employer pays compensation equal to the increase in share price from a specified level.

SARs offer a solution. Unlike stock options, these awards enable an employee to benefit by the amount that the market price of the company's stock rises without having to buy shares. Instead, the employee is awarded the share appreciation, which is the amount by which the market price on the exercise date exceeds a prespecified price (usually the market price at the date of grant). For instance, if the share price rises from $35 to $50, the employee receives $15 cash for each SAR held. The share appreciation usually is payable in cash or the recipient has the choice between cash and shares. A plan of this type offered by **IBM** is described in Illustration 19B–1.

Illustration 19B–1

Stock Appreciation Rights—IBM Corporation

Real World Financials

Long-Term Performance Plan (in part)
SARs offer eligible optionees the alternative of electing not to exercise the related stock option, but to receive payment in cash and/or stock, equivalent to the difference between the option price and the average market price of IBM stock on the date of exercising the right.

Source: IBM Corporation

IS IT DEBT OR IS IT EQUITY?

In some plans, the employer chooses whether to issue shares or cash at exercise. In other plans, the choice belongs to the employee.[22] Who has the choice determines the way it's accounted for. More specifically, the accounting treatment depends on whether the award is considered an equity instrument or a liability. If the employer can elect to settle in shares of stock rather than cash, the award is considered to be equity. On the other hand, if the employee will receive cash or can elect to receive cash, the award is considered to be a liability.

If an employer can elect to settle in shares of stock rather than cash, the award is considered to be equity.

The distinction between share-based awards that are considered equity and those that are considered liabilities is based on whether the employer is obligated to transfer assets to the employee. A cash SAR requires the transfer of assets, and therefore is a liability. A stock option, on the other hand, is an equity instrument if it requires only the issuance of stock. This does not mean that a stock option whose issuer may later choose to settle in cash is not an equity instrument. Instead, cash settlement would be considered equivalent to repurchasing an equity instrument for cash.

If an employee can elect to receive cash, the award is considered to be a liability.

SARS PAYABLE IN SHARES (EQUITY)

When a SAR is considered to be equity (because the employer can elect to settle in shares of stock rather than cash), we estimate the fair value of the SARs at the grant date and accrue that compensation to expense over the service period. Normally, the fair value of a SAR is the same as the fair value of a stock option with the same terms. The fair value is determined at the grant date and accrued to compensation expense over the service period the same way as for other share-based compensation plans. The total compensation is not revised for subsequent changes in the price of the underlying stock. This is demonstrated in Case 1 of Illustration 19–B2.

The cash settlement of an equity award is considered the repurchase of an equity instrument

SARS PAYABLE IN CASH (LIABILITY)

When a SAR is considered to be a liability (because the employee can elect to receive cash upon settlement), we estimate the fair value of the SARs and recognize that amount as compensation expense over the requisite service period consistent with the way we account for options and other share-based compensation. However, because these plans are considered to be liabilities, it's necessary to periodically re-estimate the fair value in order to continually adjust the liability (and corresponding compensation) until it is paid. Be sure to note that this is consistent with the way we account for other liabilities. Recall from our discussions in Chapter 16, for instance, that when a tax rate change causes a change in the eventual liability for deferred income taxes, we adjust that liability.

Compensation expense reported to date is the estimated total compensation multiplied by the fraction of the service period that has expired.

[21]The tax treatment of share-based plans is discussed an earlier Additional Consideration box.

[22]Many such plans are called tandem plans and award an employee both a cash SAR and an SAR that calls for settlement in an equivalent amount of shares. The exercise of one cancels the other.

The periodic expense (and adjustment to the liability) is the fraction of the total compensation earned to date by recipients of the SARs (based on the elapsed fraction of the service period) reduced by any amounts expensed in prior periods. For example, if the fair value of SARs at the end of a period is $8, the total compensation would be **$80** million if 10 million SARs are expected to vest. Let's say two years of a four-year service period have elapsed, and $21 million was expensed the first year. Then, compensation expense the second year would be $19 million, calculated as (2/4 **of $80** million) minus $21. An example spanning several years is provided in Illustration 19B–2, Case 2.

We make up for incorrect previous estimates by adjusting expense in the period the estimate is revised.

Illustration 19B–2

Stock Appreciation Rights
Case 1: Equity

At January 1, 2021, Universal Communications issued 10 million SARs that, upon exercise, entitle key executives to receive compensation equal in value to the excess of the market price at exercise over the share price at the date of grant.

- The SARs vest at the end of 2024 (cannot be exercised until then) and expire at the end of 2028.
- The fair value of the SARs, estimated by an appropriate option pricing model, is $8 per SAR at January 1, 2021.
- The fair value re-estimated at December 31, 2021, 2022, 2023, 2024, and 2025, is $8.40, $8, $6, $4.30, and $5, respectively.

Case 1: SARs considered to be equity because Universal can elect to settle in shares of Universal stock at exercise

January 1, 2021
No entry
Calculate total compensation expense:

$ 8	Estimated fair value per SAR
× 10 million	SARs granted
= **$80 million**	Total compensation

Fair value is estimated at the date of grant.

The total compensation is allocated to expense over the four-year service (vesting) period: 2021–2024

$80 million ÷ 4 years = $20 million per year

December 31, 2021, 2022, 2023, 2024	($ in millions)	
Compensation expense (**$80** million ÷ 4 years)	20	
Paid-in capital—SAR plan		20

The value of the compensation is estimated each year at the fair value of the SARs.

Case 2: SARs considered to be a liability because employees can elect to receive cash at exercise

January 1, 2021
No entry

December 31, 2021	($ in millions)	
Compensation expense ($8.40 × 10 million × ¼)	21	
Liability—SAR plan		21
December 31, 2022		
Compensation expense [($8 × 10 million × 2/4) – $21]	19	
Liability—SAR plan		19
December 31, 2023		
Compensation expense [($6 × 10 million × 3/4) – $21 – $19]	5	
Liability—SAR plan		5
December 31, 2024		
Liability—SAR plan	2	
Compensation expense [($4.30 × 10 million × 4/4) – $21 – $19 – $5]		2

The expense each year is the current estimate of total compensation that should have been recorded to date less the amount already recorded.

If the fair value falls below the amount expensed to date, both the liability and expense are reduced.

Note that the way we treat changes in compensation estimates entails a catch-up adjustment in the period of change, *inconsistent* with the usual treatment of a change in estimate.

Remember that for most changes in estimate, revisions are allocated over remaining periods, rather than all at once in the period of change. The treatment is, however, consistent with the way we treat changes in forfeiture rate estimates, as we discussed earlier in the chapter.

The liability continues to be adjusted after the service period if the rights haven't been exercised yet.

Compensation expense and the liability continue to be adjusted until the SARs expire or are exercised.

December 31, 2025		($ in millions)
Compensation expense [($5 × 10 million × **all**) – $21 – $19 – $5 + $2]	7	
Liability—SAR plan		7

It's necessary to continue to adjust both compensation expense and the liability until the SARs ultimately either are exercised or lapse.[23] Assume, for example, that the SARs are exercised on October 11, 2026, when their fair value is $4.50, and executives choose to receive the market price appreciation in cash.

Adjustment continues after the service period if the SARs have not yet been exercised.

October 11, 2026		($ in millions)
Liability—SAR plan	5	
Compensation expense [($4.50 × 10 million × all) – $50]		5
Liability—SAR plan (balance)	45	
Cash		45

Let's look at the changes in the liability—SAR plan account during the 2021–2026 period:

The liability is adjusted each period as changes in the fair value estimates cause changes in the liability.

Liability—SAR Plan

			($ in millions)
		21	2021
		19	2022
		5	2023
2024	2		
		7	2025
2026	5		
2026	45		
		0	Balance after exercise

RESTRICTED STOCK UNITS PAYABLE IN CASH (LIABILITY)

Accounting for RSUs payable in cash is essentially the same as accounting for the SARs payable in cash.

Recall from our discussion in the chapter that restricted stock units (RSUs) give the recipient the right to receive a set number of shares of company stock after the vesting requirement is satisfied. But, sometimes the recipient is given the *cash equivalent* of those shares instead. *If the employee will receive cash* or can elect to receive cash, as in the case of an SAR, we consider the award to be a *liability.* We determine its fair value at the grant date and recognize that amount as compensation expense over the requisite service period and, like SARs payable in cash, we periodically adjust the liability (and corresponding compensation) based on the change in the stock's fair value until the liability is paid. Accounting for RSUs payable in cash (sometimes called phantom shares or phantom performance shares) is quite similar to accounting for the SARs payable in cash in case 2 of Illustration 19B–2. Of course, though, we would label the liability "Liability—restricted stock." And, we would determine the periodic value of the liability (and compensation) as the actual fair value of the shares, rather than an *estimated* fair value as is necessary when valuing SARs. ●

[23]Except that the cumulative compensation expense cannot be negative; that is, the liability cannot be reduced below zero.

Questions For Review of Key Topics

Q 19–1 What is restricted stock? How do restricted stock awards differ from restricted stock units (RSUs)? Describe how compensation expense is determined and recorded for a restricted stock award plan.

Q 19–2 Stock option plans provide employees the option to purchase (a) a specified number of shares of the firm's stock, (b) at a specified price, (c) during a specified period of time. One of the most controversial aspects of accounting for stock-based compensation is how the fair value of stock options should be measured. Describe the general approach to measuring fair value.

Q 19–3 The Tax Code differentiates between qualified option plans, including incentive plans, and nonqualified plans. What are the major differences in tax treatment between incentive plans and nonqualified plans?

Q 19–4 Stock option (and other share-based) plans often specify a performance condition or a market condition that must be satisfied before employees are allowed the benefits of the award. Describe the general approach we use to account for performance-based options and options with market-related conditions.

Q 19–5 What is a simple capital structure? How is EPS determined for a company with a simple capital structure?

Q 19–6 When calculating the weighted-average number of common shares, how are stock dividends and stock splits treated? Compare this treatment with that of additional shares sold for cash at midyear.

Q 19–7 Blake Distributors had 100,000 common shares outstanding at the beginning of the year, January 1. On May 13, Blake distributed a 5% stock dividend. Blake retired 1,200 shares on August 1. What is the weighted-average number of shares for calculating EPS?

Q 19–8 Why are preferred dividends deducted from net income when calculating EPS? Are there circumstances when this deduction is not made?

Q 19–9 Distinguish between basic and diluted EPS.

Q 19–10 The treasury stock method is used to incorporate the dilutive effect of stock options, stock warrants, and similar securities. Describe this method as it applies to diluted EPS.

Q 19–11 The potentially dilutive effect of convertible securities is reflected in EPS calculations by the if-converted method. Describe this method as it relates to convertible bonds.

Q 19–12 How is the potentially dilutive effect of convertible preferred stock reflected in EPS calculations by the if-converted method? How is this different from the way convertible bonds are considered?

Q 19–13 A convertible security may appear to be dilutive when looked at individually but might be antidilutive when included in combination with other convertible securities. How should the order be determined for inclusion of convertible securities in an EPS calculation to avoid including an antidilutive security?

Q 19–14 If stock options and restricted stock are outstanding when calculating diluted EPS, what are the components of the "proceeds" available for the repurchase of shares under the treasury stock method?

Q 19–15 Wiseman Electronics has an agreement with certain of its division managers that 50,000 contingently issuable shares will be issued next year in the event operating income exceeds $2.1 million that year. In what way, if any, is the calculation of EPS affected by these contingently issuable shares, assuming this year's operating income was $2.2 million? $2.0 million?

Q 19–16 Diluted EPS would be precisely the same whether convertible securities were actually converted or not. Why?

Q 19–17 When the income statement includes discontinued operations, which amounts require per share presentation?

Q 19–18 In addition to EPS numbers themselves, what additional disclosures should be provided concerning the EPS information?

Q 19–19 (Based on Appendix 19A) The fair value of stock options can be considered to comprise two main components. What are they?

Q 19–20 (Based on Appendix 19B) LTV Corporation grants SARs to key executives. Upon exercise, the SARs entitle executives to receive either cash or stock equal in value to the excess of the market price at exercise over the share price at the date of grant. How should LTV account for the awards?

Brief Exercises

connect

BE 19–1
Restricted stock award
● LO19–1

First Link Services granted 8 million of its $1 par common shares to executives, subject to forfeiture if employment is terminated within three years. The common shares have a market price of $6 per share on the grant date of the restricted stock award. Ignoring taxes, what is the total compensation cost pertaining to the restricted shares? What is the effect on earnings in the year after the shares are granted to executives?

BE 19–2
Restricted stock units
● LO19–1

Second Link Services granted restricted stock units (RSUs) representing 16 million of its $1 par common shares to executives, subject to forfeiture if employment is terminated within four years. After the recipients of the RSUs satisfy the vesting requirement, the company will distribute the shares. The common shares had a market price of $10 per share on the grant date. Ignoring taxes, what is the total compensation cost pertaining to the restricted stock units? What is the effect on earnings in the year after the shares are granted to executives?

BE 19–3
Stock options
● LO19–2

Under its executive stock option plan, National Corporation granted 12 million options on January 1, 2021, that permit executives to purchase 12 million of the company's $1 par common shares within the next six years, but not before December 31, 2023 (the vesting date). The exercise price is the market price of the shares on the date of grant, $17 per share. The fair value of the options, estimated by an appropriate option pricing model, is $5 per option. No forfeitures are anticipated. Ignoring taxes, what is the total compensation cost pertaining to the stock options? What is the effect on earnings in the year after the options are granted to executives?

BE 19–4
Stock options; forfeiture
● LO19–2

Refer to the situation described in BE 19–3. Suppose that unexpected turnover during 2022 caused the forfeiture of 5% of the stock options. What is the effect on earnings in 2022? In 2023?

BE 19–5
Stock options; forfeitures
● LO19–2

On January 1, 2021, Hugh Morris Comedy Club (HMCC) granted 1 million stock options to key executives exercisable for 1 million shares of the company's common stock at $20 per share. The stock options are intended as compensation for the next three years. The options are exercisable within a four-year period beginning January 1, 2024, by the executives still in the employ of the company. No options were terminated during 2021. The market price of the common stock was $25 per share at the date of the grant. HMCC estimated the fair value of the options at $9 each. One percent of the options are forfeited during 2022 due to executive turnover. What amount should HMCC record as compensation expense for the year ended December 31, 2022, assuming HMCC chooses the option to record forfeitures as they actually occur?

BE 19–6
Stock options; exercise
● LO19–2

Refer to the situation described in BE 19–3. Suppose that the options are exercised on April 3, 2024, when the market price is $19 per share. What journal entry will National record?

BE 19–7
Stock options; expiration
● LO19–2

Refer to the situation described in BE 19–3. Suppose that the options expire without being exercised. What journal entry will National record?

BE 19–8
Performance-based options
● LO19–2

On January 1, 2021, Farmer Fabrication issued stock options for 100,000 shares to a division manager. The options have an estimated fair value of $6 each. To provide additional incentive for managerial achievement, the options are not exercisable unless divisional revenue increases by 5% in three years. Farmer initially estimates that it is probable the goal will be achieved. How much compensation will be recorded in 2021, 2022, 2023?

BE 19–9
Performance-based options
● LO19–2

Refer to the situation described in BE 19–8. Suppose that after one year, Farmer estimates that it is *not* probable that divisional revenue will increase by 5% in three years. What journal entry will be needed to account for the options in 2022?

BE 19–10
Performance-based options
● LO19–2

Refer to the situation described in BE 19–8. Suppose that Farmer initially estimates that it is *not* probable the goal will be achieved, but then after one year, Farmer estimates that it *is* probable that divisional revenue will increase by 5% by the end of 2023. What journal entry(s) will Farmer record to account for the options in 2022 and thereafter?

BE 19–11
Options with market-based conditions
● LO19–2

On January 1, 2021, Farmer Fabrication issued stock options for 100,000 shares to a division manager. The options have an estimated fair value of $6 each. To provide additional incentive for managerial achievement, the options are not exercisable unless Farmer Fabrication's stock price increases by 5% in three years. Farmer initially estimates that it is not probable the goal will be achieved. How much compensation will be recorded in 2021, 2022, and 2023?

BE 19–12
EPS; shares issued; shares retired
● LO19–5, LO19–6

McDonnell-Myer Corporation reported net income of $741 million. The company had 544 million common shares outstanding at January 1 and sold 36 million shares on February 28. As part of an annual share repurchase plan, 6 million shares were retired on April 30 for $47 per share. Calculate McDonnell-Myer's earnings per share for the year.

BE 19–13
EPS; nonconvertible preferred shares
● LO19–7

At December 31, 2020 and 2021, Funk & Noble Corporation had outstanding 820 million shares of common stock and 2 million shares of 8%, $100 par value cumulative preferred stock. No dividends were declared on either the preferred or common stock in 2020 or 2021. Net income for 2021 was $426 million. The income tax rate is 25%. Calculate earnings per share for the year ended December 31, 2021.

BE 19–14
EPS; stock options
● LO19–8

Fully vested incentive stock options exercisable at $50 per share to obtain 24,000 shares of common stock were outstanding during a period when the average market price of the common stock was $60 and the ending market price was $60. What will be the net increase in the weighted-average number of shares outstanding due to the assumed exercise of these options when calculating diluted earnings per share?

BE 19–15
EPS; convertible preferred shares
● LO19–9

Ahnberg Corporation had 800,000 shares of common stock issued and outstanding at January 1. No common shares were issued during the year, but on January 1, Ahnberg issued 100,000 shares of convertible preferred stock. The preferred shares are convertible into 200,000 shares of common stock. During the year Ahnberg paid $60,000 cash dividends on the preferred stock. Net income was $1,500,000. What were Ahnberg's basic and diluted earnings per share for the year?

BE 19–16
EPS; restricted stock award
● LO19–11

Niles Company granted 9 million of its no par common shares to executives, subject to forfeiture if employment is terminated within three years. The common shares have a market price of $5 per share on January 1, 2020, the grant date of the restricted stock award. When calculating diluted EPS at December 31, 2021, what will be the net increase in the weighted-average number of shares outstanding if the market price of the common shares averaged $5 per share during 2021?

Exercises

connect

E 19–1
Restricted stock award plan
● LO19–1

Allied Paper Products, Inc., offers a restricted stock award plan to its vice presidents. On January 1, 2021, the company granted 16 million of its $1 par common shares, subject to forfeiture if employment is terminated within two years. The common shares have a market price of $5 per share on the grant date.

Required:

1. Determine the total compensation cost pertaining to the restricted shares.
2. Prepare the appropriate journal entries related to the restricted stock through December 31, 2022.

E 19–2
Restricted stock units
● LO19–1

On January 1, 2021, Tru Fashions Corporation awarded restricted stock units (RSUs) representing 12 million of its $1 par common shares to key personnel, subject to forfeiture if employment is terminated within three years. After the recipients of the RSUs satisfy the vesting requirement, the company will distribute the shares. On the grant date, the shares had a market price of $2.50 per share.

Required:

1. Determine the total compensation cost pertaining to the RSUs.
2. Prepare the appropriate journal entry to record the award of RSUs on January 1, 2021.
3. Prepare the appropriate journal entry to record compensation expense on December 31, 2021.
4. Prepare the appropriate journal entry to record compensation expense on December 31, 2022.
5. Prepare the appropriate journal entry to record compensation expense on December 31, 2023.
6. Prepare the appropriate journal entry to record the lifting of restrictions on the RSUs and issuing shares at December 31, 2023.

E 19–3
Restricted stock units; Facebook
● LO19–1

Real World Financials

Facebook Inc. included the following disclosure note in an annual report:

Share-Based Compensation (in part)
. . . compensation expense related to these grants is based on the grant date fair value of the RSUs and is recognized on a straight-line basis over the applicable service period.

The following table summarizes the activities for our unvested RSUs for the year ended December 31, 2017:

	Number of Shares	Weighted Average Grant Date Fair Value
Unvested at December 31, 2016	98,586	$ 82.99
Granted	36,741	147.28
Vested	(43,176)	83.74
Forfeited	(10,937)	91.76
Unvested at December 31, 2017	81,214	$110.49

Required:
1. Assuming a four-year vesting period, how much compensation expense did Facebook report in the year ended December 31, 2018, for the restricted stock units granted during the year ended December 31, 2017? (Round dollar amounts to the nearest million.)
2. Based on the information provided in the disclosure note, prepare the journal entry that summarizes the vesting of RSUs during the year ended December 31, 2017. (Facebook's common shares have a par amount per share of $0.000006. Round dollar amounts to the nearest million.)

Source: Facebook

E 19–4
Restricted stock units; forfeitures anticipated
● LO19–1

Magnetic-Optical Corporation offers a variety of share-based compensation plans to employees. Under its restricted stock unit plan, the company on January 1, 2021, granted restricted stock units (RSUs) representing 4 million of its $1 par common shares to various division managers. The shares are subject to forfeiture if employment is terminated within three years. The common shares have a market price of $22.50 per share on the grant date. Management's policy is to estimate forfeitures.

Required:
1. Determine the total compensation cost pertaining to the RSUs.
2. Prepare the appropriate journal entry to record the RSUs on January 1, 2021.
3. Prepare the appropriate journal entry to record compensation expense on December 31, 2021.
4. Suppose Magnetic-Optical expected a 10% forfeiture rate on the RSUs prior to vesting. Determine the total compensation cost.

E 19–5
Restricted stock units; forfeitures
● LO19–1

On January 1, 2021, David Mest Communications granted restricted stock units (RSUs) representing 30 million of its $1 par common shares to executives, subject to forfeiture if employment is terminated within three years. After the recipients of the RSUs satisfy the vesting requirement, the company will distribute the shares. The common shares had a market price of $12 per share on the grant date. At the date of grant, Mest anticipated that 6% of the recipients would leave the firm prior to vesting. On January 1, 2022, 5% of the RSUs are forfeited due to executive turnover. Mest chooses the option to account for forfeitures when they actually occur.

Required:
1. Prepare the appropriate journal entry to record compensation expense on December 31, 2021.
2. Prepare the appropriate journal entry to record compensation expense on December 31, 2022.
3. Prepare the appropriate journal entry to record compensation expense on December 31, 2023.

E 19–6
Stock options
● LO19–2

Heidi Software Corporation provides a variety of share-based compensation plans to its employees. Under its executive stock option plan, the company granted options on January 1, 2021, that permit executives to acquire 4 million of the company's $1 par common shares within the next five years, but not before December 31, 2022 (the vesting date). The exercise price is the market price of the shares on the date of grant, $14 per share. The fair value of the 4 million options, estimated by an appropriate option pricing model, is $3 per option. No forfeitures are anticipated. Ignore taxes.

Required:
1. Determine the total compensation cost pertaining to the options.
2. Prepare the appropriate journal entry to record the award of options on January 1, 2021.

3. Prepare the appropriate journal entry to record compensation expense on December 31, 2021.
4. Prepare the appropriate journal entry to record compensation expense on December 31, 2022.

E 19–7
Stock options: forfeiture of options
● LO19–2

On January 1, 2021, Adams-Meneke Corporation granted 25 million incentive stock options to division managers, each permitting holders to purchase one share of the company's $1 par common shares within the next six years, but not before December 31, 2023 (the vesting date). The exercise price is the market price of the shares on the date of grant, currently $10 per share. The fair value of the options, estimated by an appropriate option pricing model, is $3 per option. Management's policy is to estimate forfeitures. No forfeitures are anticipated. Ignore taxes.

Required:
1. Determine the total compensation cost pertaining to the options on January 1, 2021.
2. Prepare the appropriate journal entry to record compensation expense on December 31, 2021.
3. Unexpected turnover during 2022 caused an estimate of the forfeiture of 6% of the stock options. Prepare the appropriate journal entry(s) on December 31, 2022 and 2023 in response to the new estimate.

E 19–8
Stock options exercise; expirations
● LO19–2

Walters Audio Visual Inc. offers an incentive stock option plan to its regional managers. On January 1, 2021, options were granted for 40 million $1 par common shares. The exercise price is the market price on the grant date—$8 per share. Options cannot be exercised prior to January 1, 2023, and expire December 31, 2027. The fair value of the 40 million options, estimated by an appropriate option pricing model, is $1 per option.

Required:
1. Determine the total compensation cost pertaining to the incentive stock option plan.
2. Prepare the appropriate journal entry to record compensation expense on December 31, 2021.
3. Prepare the appropriate journal entry to record compensation expense on December 31, 2022.
4. Prepare the appropriate journal entry to record the exercise of 75% of the options on March 12, 2023, when the market price is $9 per share.
5. Prepare the appropriate journal entry on December 31, 2027, when the remaining options that have vested expire without being exercised.

E 19–9
Stock options; exercise
● LO19–2

SSG Cycles manufactures and distributes motorcycle parts and supplies. Employees are offered a variety of share-based compensation plans. Under its nonqualified stock option plan, SSG granted options to key officers on January 1, 2021. The options permit holders to acquire 12 million of the company's $1 par common shares for $11 within the next six years, but not before January 1, 2024 (the vesting date). The market price of the shares on the date of grant is $13 per share. The fair value of the 12 million options, estimated by an appropriate option pricing model, is $3 per option.

Required:
1. Determine the total compensation cost pertaining to the incentive stock option plan.
2. Prepare the appropriate journal entries to record compensation expense on December 31, 2021, 2022, and 2023.
3. Record the exercise of the options if all of the options are exercised on May 11, 2025, when the market price is $14 per share.

E 19–10
Employee share purchase plan
● LO19–3

In order to encourage employee ownership of the company's $1 par common shares, Washington Distribution permits any of its employees to buy shares directly from the company through payroll deduction. There are no brokerage fees and shares can be purchased at a 15% discount. During March, employees purchased 50,000 shares at a time when the market price of the shares on the New York Stock Exchange was $12 per share.

Required:
Prepare the appropriate journal entry to record the March purchases of shares under the employee share purchase plan.

E 19–11
Employee share purchase plan; Tesla Motors
● LO19–3
Real World Financials

Tesla Motors's disclosure notes for the year ending December 31, 2017, included the following regarding its $0.001 par common stock:

EMPLOYEE STOCK PURCHASE PLAN—Our employees are eligible to purchase our common stock through payroll deductions of up to 15% of their eligible compensation, subject to any plan limitations. The purchase price would be 85% of the lower of the fair market value on the first and last trading days of each six-month offering period. During the years ended December 31, 2017, 2016, and 2015, we issued 370,173, 321,788, and 220,571 shares under the ESPP for $71.0 million, $51.7 million, and $37.5 million, respectively. There were 1,423,978 shares available for issuance under the ESPP as of December 31, 2017.

Required:
Prepare the journal entry that summarizes Tesla's employee share purchases for the year ending December 31, 2017.

Source: Tesla Motors

E 19–12
EPS; shares issued; stock dividend
● LO19–5, LO19–6

For the year ended December 31, 2021, Norstar Industries reported net income of $655,000. At January 1, 2021, the company had 900,000 common shares outstanding. The following changes in the number of shares occurred during 2021:

Apr. 30	Sold 60,000 shares in a public offering
May 24	Declared and distributed a 5% stock dividend
June 1	Issued 72,000 shares as part of the consideration for the purchase of assets from a subsidiary

Required:
Compute Norstar's earnings per share for the year ended December 31, 2021.

E 19–13
EPS; treasury stock; new shares; stock dividends; two years
● LO19–5, LO19–6

The Alford Group had 202,000 shares of common stock outstanding at January 1, 2021. The following activities affected common shares during the year. There are no potential common shares outstanding.

2021	
Feb. 28	Purchased 6,000 shares of treasury stock.
Oct. 31	Sold the treasury shares purchased on February 28.
Nov. 30	Issued 24,000 new shares.
Dec. 31	Net income for 2021 is $400,000.
2022	
Jan. 15	Declared and issued a 2-for-1 stock split.
Dec. 31	Net income for 2022 is $400,000.

Required:
1. Determine the 2021 EPS.
2. Determine the 2022 EPS.
3. At what amount will the 2021 EPS be presented in the 2022 comparative financial statements?

E 19–14
EPS; stock dividend; nonconvertible preferred stock
● LO19–5, LO19–6, LO19–7

Hardaway Fixtures' balance sheet at December 31, 2020, included the following:

Shares issued and outstanding:	
Common stock, $1 par	$800,000
Nonconvertible preferred stock, $50 par	20,000

On July 21, 2021, Hardaway issued a 25% stock dividend on its common stock. On December 12, it paid $50,000 cash dividends on the preferred stock. Net income for the year ended December 31, 2021, was $2,000,000.

Required:
Compute Hardaway's earnings per share for the year ended December 31, 2021.

E 19–15
EPS; net loss; nonconvertible preferred stock; shares sold
● LO19–5, LO19–6, LO19–7

At December 31, 2020, Albrecht Corporation had outstanding 373,000 shares of common stock and 8,000 shares of 9.5%, $100 par value cumulative, nonconvertible preferred stock. On May 31, 2021, Albrecht sold for cash 12,000 shares of its common stock. No cash dividends were declared for 2021. For the year ended December 31, 2021, Albrecht reported a net loss of $114,000.

Required:
Calculate Albrecht's net loss per share for the year ended December 31, 2021.

E 19–16
EPS; stock dividend; nonconvertible preferred stock; treasury shares; shares sold
● LO19–5, LO19–6, LO19–7

On December 31, 2020, Berclair Inc. had 200 million shares of common stock and 3 million shares of 9%, $100 par value cumulative preferred stock issued and outstanding. On March 1, 2021, Berclair purchased 24 million shares of its common stock as treasury stock. Berclair issued a 5% common stock dividend on July 1, 2021. Four million treasury shares were sold on October 1. Net income for the year ended December 31, 2021, was $150 million.

Required:
Compute Berclair's earnings per share for the year ended December 31, 2021.

E 19–17
EPS; stock dividend; nonconvertible preferred stock; treasury shares; shares sold; stock options
● LO19–5 through LO19–8

(Note: This is a variation of E 19–16, modified to include stock options.) On December 31, 2020, Berclair Inc. had 200 million shares of common stock and 3 million shares of 9%, $100 par value cumulative preferred stock issued and outstanding. On March 1, 2021, Berclair purchased 24 million shares of its common stock as treasury stock. Berclair issued a 5% common stock dividend on July 1, 2021. Four million treasury shares were sold on October 1. Net income for the year ended December 31, 2021, was $150 million.

Also outstanding at December 31 were 30 million incentive stock options granted to key executives on September 13, 2016. The options were exercisable as of September 13, 2020, for 30 million common shares at an exercise price of $56 per share. During 2021, the market price of the common shares averaged $70 per share.

Required:
Compute Berclair's basic and diluted earnings per share for the year ended December 31, 2021.

E 19–18
EPS; stock dividend; nonconvertible preferred stock; treasury shares; shares sold; stock options exercised
● LO19–5 through LO19–8

(Note: This is a variation of E 19–16, modified to include the exercise of stock options.)

On December 31, 2020, Berclair Inc. had 200 million shares of common stock and 3 million shares of 9%, $100 par value cumulative preferred stock issued and outstanding. On March 1, 2021, Berclair purchased 24 million shares of its common stock as treasury stock. Berclair issued a 5% common stock dividend on July 1, 2021. Four million treasury shares were sold on October 1. Net income for the year ended December 31, 2021, was $150 million.

Also outstanding at December 31 were 30 million incentive stock options granted to key executives on September 13, 2016. The options were exercisable as of September 13, 2020, for 30 million common shares at an exercise price of $56 per share. During 2021, the market price of the common shares averaged $70 per share.

The options were exercised on September 1, 2021.

Required:
Compute Berclair's basic and diluted earnings per share for the year ended December 31, 2021.

E 19–19
EPS; stock dividend; nonconvertible preferred stock; treasury shares; shares sold; stock options; convertible bonds
● LO19–5 through LO19–9

(Note: This is a variation of E 19–17 modified to include convertible bonds).

On December 31, 2020, Berclair Inc. had 200 million shares of common stock and 3 million shares of 9%, $100 par value cumulative preferred stock issued and outstanding. On March 1, 2021, Berclair purchased 24 million shares of its common stock as treasury stock. Berclair issued a 5% common stock dividend on July 1, 2021. Four million treasury shares were sold on October 1. Net income for the year ended December 31, 2021, was $150 million. The income tax rate is 25%.

Also outstanding at December 31 were incentive stock options granted to key executives on September 13, 2016. The options are exercisable as of September 13, 2020, for 30 million common shares at an exercise price of $56 per share. During 2021, the market price of the common shares averaged $70 per share.

In 2017, $50 million of 8% bonds, convertible into 6 million common shares, were issued at face value.

Required:
Compute Berclair's basic and diluted earnings per share for the year ended December 31, 2021.

E 19–20
EPS; shares issued; stock options
● LO19–6 through LO19–9

Stanley Department Stores reported net income of $720,000 for the year ended December 31, 2021.

Additional Information:

Common shares outstanding at Jan. 1, 2021	80,000
Incentive stock options (vested in 2020) outstanding throughout 2021	24,000
(Each option is exercisable for one common share at an exercise price of $37.50.)	
During the year, the market price of Stanley's common stock averaged $45 per share.	
On Aug. 30, Stanley sold 15,000 common shares.	
Stanley's only debt consisted of $50,000 of 10% short-term bank notes.	
The company's income tax rate is 25%.	

Required:
Compute Stanley's basic and diluted earnings per share for the year ended December 31, 2021.

E 19–21
EPS; convertible preferred stock; convertible bonds; order of entry
● LO19–7, LO19–9

Information from the financial statements of Ames Fabricators, Inc., included the following:

	December 31	
	2021	2020
Common shares	100,000	100,000
Convertible preferred shares (convertible into 32,000 shares of common)	12,000	12,000
8% convertible bonds (convertible into 30,000 shares of common)	$1,000,000	$1,000,000

Ames's net income for the year ended December 31, 2021, is $500,000. The income tax rate is 25%. Ames paid dividends of $5 per share on its preferred stock during 2021.

Required:
Compute basic and diluted earnings per share for the year ended December 31, 2021.

E 19–22
EPS; restricted stock
● LO19–11

As part of its executive compensation plan, Vertovec Inc. granted 54,000 of its no-par common shares to executives, subject to forfeiture if employment is terminated within three years. Vertovec's common shares have a market price of $5 per share on January 1, 2020, the grant date of the restricted stock award, as well as on December 31, 2021. 800,000 shares were outstanding at January 1, 2021. Net income for 2021 was $120,000.

Required:
Compute Vertovec's basic and diluted earnings per share for the year ended December 31, 2021.

E 19–23
Record restricted stock; effect on EPS
● LO19–1, LO19–11

PHN Foods granted 18 million of its no-par common shares to executives, subject to forfeiture if employment is terminated within three years. The common shares have a market price of $5 per share on January 1, 2020, the grant date.

Required:
1. What journal entry will PHN Foods prepare to record executive compensation regarding these restricted shares at December 31, 2020 and December 31, 2021?
2. When calculating diluted EPS at December 31, 2021, what will be the net increase in the weighted average number of common shares with regard to the restricted stock shares if the market price of the common shares averages $5 per share during 2021?

E 19–24
New shares; contingently issuable shares
● LO19–6, LO19–12

During 2021, its first year of operations, McCollum Tool Works entered into the following transactions relating to shareholders' equity. The corporation was authorized to issue 100 million common shares, $1 par per share.

Jan.	2	Issued 35 million common shares for cash.
	3	Entered an agreement with the company president to issue up to 2 million additional shares of common stock in 2022 based on the earnings of McCollum in 2022. If net income exceeds $140 million, the president will receive 1 million shares; 2 million shares if net income exceeds $150 million.
Mar.	31	Issued 4 million shares in exchange for plant facilities.

Net income for 2021 was $148 million.

Required:
Compute basic and diluted earnings per share for the year ended December 31, 2021.

E 19–25
EPS; new shares; contingent agreements
● LO19–6, LO19–12

Anderson Steel Company began 2021 with 600,000 shares of common stock outstanding. On March 31, 2021, 100,000 new shares were sold at a price of $45 per share. The market price has risen steadily since that time to a high of $50 per share at December 31. No other changes in shares occurred during 2021, and no securities are outstanding that can become common stock. However, there are two agreements with officers of the company for future issuance of common stock. Both agreements relate to compensation arrangements reached in 2020. The first agreement grants to the company president a right to 10,000 shares of stock each year the closing market price is at least $48. The agreement begins in 2022 and expires in 2025. The second agreement grants to the controller a right to 15,000 shares of stock if she is still with the firm at the end of 2029. Net income for 2021 was $2,000,000.

Required:
Compute Anderson Steel Company's basic and diluted EPS for the year ended December 31, 2021.

E 19–26
EPS; concepts; terminology
● LO19–5 through LO19–13

Listed below are several terms and phrases associated with earnings per share. Pair each item from List A with the item from List B (by letter) that is most appropriately associated with it.

List A	List B
____ 1. Subtract preferred dividends.	a. Options exercised.
____ 2. Time-weighted by $^{5}/_{12}$.	b. Simple capital structure.
____ 3. Time-weighted shares assumed issued plus time-weighted actual shares.	c. Basic EPS.
	d. Convertible preferred stock.
____ 4. Midyear event treated as if it occurred at the beginning of the reporting period.	e. Earnings available to common shareholders.
	f. Antidilutive.
____ 5. Preferred dividends do not reduce earnings.	g. Increased marketability.
____ 6. Single EPS presentation.	h. Discontinued operations.
____ 7. Stock split.	i. Stock dividend.
____ 8. Potential common shares.	j. Add after-tax interest to numerator.
____ 9. Exercise price exceeds market price.	k. Diluted EPS.
____ 10. No dilution assumed.	l. Noncumulative, undeclared preferred dividends.

(continued)

(concluded)

List A	List B
_____ 11. Convertible bonds.	m. Common shares retired at the beginning of August.
_____ 12. Contingently issuable shares.	n. Include in diluted EPS when conditions for issuance are met.
_____ 13. Maximum potential dilution.	
_____ 14. Shown between per share amounts for net income and for income from continuing operations.	

E 19–27
FASB codification research
● LO19–2

The *FASB Accounting Standards Codification* represents the single source of authoritative U.S. generally accepted accounting principles. Obtain the relevant authoritative literature on stock compensation using the *FASB Accounting Standards Codification* at the FASB website (**www.fasb.org**).

Required:

1. What is the specific seven-digit Codification citation (XXX-XX-XX) that describes the information that companies must disclose about the exercise prices for their stock option plans?
2. List the disclosure requirements.

E 19–28
FASB codification research
● LO19–2, LO19–3, LO19–7, LO19–13

Access the *FASB Accounting Standards Codification* at the FASB website (**www.fasb.org**). Determine the specific seven-, eight-, or nine-digit Codification citation (XXX-XX-XX-X) for accounting for each of the following items:

1. Initial measurement of stock options.
2. The measurement date for share-based payments classified as liabilities.
3. The formula to calculate diluted earnings per share.
4. The way stock dividends or stock splits in the current year affect the presentation of EPS in the income statement.

E 19–29
Stock appreciation rights; settlement in shares
● Appendix 19B

As part of its stock-based compensation package, International Electronics (IE) granted 24 million stock appreciation rights (SARs) to top officers on January 1, 2021. At exercise, holders of the SARs are entitled to receive stock equal in value to the excess of the market price at exercise over the share price at the date of grant. The SARs cannot be exercised until the end of 2024 (vesting date) and expire at the end of 2026. The $1 par common shares have a market price of $46 per share on the grant date. The fair value of the SARs, estimated by an appropriate option pricing model, is $3 per SAR at January 1, 2021. The fair value re-estimated at December 31, 2021, 2022, 2023, 2024, and 2025, is $4, $3, $4, $2.50, and $3, respectively. All recipients are expected to remain employed through the vesting date.

Required:

1. Prepare any appropriate journal entry to record the award of SARs on January 1, 2021. Will the SARs be reported as debt or as equity?
2. Prepare the appropriate journal entries pertaining to the SARs on December 31, 2021–December 31, 2024.
3. The SARs remain unexercised on December 31, 2025. Prepare the appropriate journal entry on that date.
4. The SARs are exercised on June 6, 2026, when the share price is $50. Prepare the appropriate journal entry(s) on that date.

E 19–30
Stock appreciation rights; cash settlement
● Appendix 19B

(Note: This is a variation of E 19–29, modified to allow settlement in cash.)

As part of its stock-based compensation package, International Electronics granted 24 million stock appreciation rights (SARs) to top officers on January 1, 2021. At exercise, holders of the SARs are entitled to receive cash or stock equal in value to the excess of the market price at exercise over the share price at the date of grant. The SARs cannot be exercised until the end of 2024 (vesting date) and expire at the end of 2026. The $1 par common shares have a market price of $46 per share on the grant date. The fair value of the SARs, estimated by an appropriate option pricing model, is $3 per SAR at January 1, 2021. The fair value re-estimated at December 31, 2021, 2022, 2023, 2024, and 2025, is $4, $3, $4, $2.50, and $3, respectively. All recipients are expected to remain employed through the vesting date.

Required:

1. Prepare any appropriate journal entry to record the award of SARs on January 1, 2021.
2. Prepare the appropriate journal entries pertaining to the SARs on December 31, 2021–December 31, 2024.
3. The SARs remain unexercised on December 31, 2025. Prepare the appropriate journal entry on that date.
4. The SARs are exercised on June 6, 2026, when the share price is $50, and executives choose to receive the market price appreciation in cash. Prepare the appropriate journal entry(s) on that date.

E 19–31
Restricted stock units; cash settlement
● Appendix 19B

As part of its stock-based compensation package, on January 1, 2021, International Electronics granted restricted stock units (RSUs) representing 50 million $1 par common shares. At exercise, holders of the RSUs are entitled to receive cash or stock equal in value to the market price of those shares at exercise. The RSUs cannot be exercised until the end of 2024 (vesting date) and expire at the end of 2026. The $1 par common shares have a market price of $6 per share on the grant date. The fair value at December 31, 2021, 2022, 2023, 2024, and 2025, is $8, $6, $8, $5, and $6, respectively. All recipients are expected to remain employed through the vesting date. After the recipients of the RSUs satisfy the vesting requirement, the company will distribute the shares.

Required:
1. Prepare any appropriate journal entry to record the award of RSUs on January 1, 2021.
2. Prepare the appropriate journal entries pertaining to the RSUs on December 31, 2021–December 31, 2024.
3. The RSUs remain unexercised on December 31, 2025. Prepare the appropriate journal entry on that date.
4. The RSUs are exercised on June 6, 2026, when the share price is $6.50, and executives choose to receive cash. Prepare the appropriate journal entry(s) on that date.

Problems

connect

P 19–1
Stock options; forfeiture; exercise
● LO19–2

On October 15, 2020, the board of directors of Ensor Materials Corporation approved a stock option plan for key executives. On January 1, 2021, 20 million stock options were granted, exercisable for 20 million shares of Ensor's $1 par common stock. The options are exercisable between January 1, 2024, and December 31, 2026, at 80% of the quoted market price on January 1, 2021, which was $15. The fair value of the 20 million options, estimated by an appropriate option pricing model, is $6 per option. Ensor chooses the option to recognize forfeitures only when they occur.

Ten percent (2 million) of the options were forfeited when an executive resigned in 2022. All other options were exercised on July 12, 2025, when the stock's price jumped unexpectedly to $19 per share.

Required:
1. When is Ensor's stock option measurement date?
2. Determine the compensation expense for the stock option plan in 2021. (Ignore taxes.)
3. Prepare the journal entries to reflect the effect of forfeiture of the stock options on Ensor's financial statements for 2022 and 2023.
4. Is this effect consistent with the general approach for accounting for changes in estimates? Explain.
5. Prepare the journal entry to account for the exercise of the options in 2025.

P 19–2
Stock options; graded vesting
● LO19–2

Pastner Brands is a calendar-year firm with operations in several countries. As part of its executive compensation plan, at January 1, 2021, the company issued 400,000 executive stock options permitting executives to buy 400,000 shares of Pastner stock for $34 per share. One-fourth of the options vest in each of the next four years beginning at December 31, 2021 (graded vesting). Pastner elects to separate the total award into four groups (or tranches) according to the year in which they vest and measures the compensation cost for each vesting date as a separate award. The fair value of each tranche is estimated at January 1, 2021, as follows:

Vesting Date	Amount Vesting	Fair Value per Option
Dec. 31, 2021	25%	$3.50
Dec. 31, 2022	25%	$4.00
Dec. 31, 2023	25%	$4.50
Dec. 31, 2024	25%	$5.00

Required:
1. Determine the compensation expense related to the options to be recorded each year 2021–2024, assuming Pastner allocates the compensation cost for each of the four groups (tranches) separately.
2. Determine the compensation expense related to the options to be recorded each year 2021–2024, assuming Pastner uses the straight-line method to allocate the total compensation cost.

P 19–3
Stock options; graded vesting; measurement using a single fair value per option
● LO19–2

Refer to the situation described in P 19–2. Assume Pastner measures the fair value of all options on January 1, 2021, to be $4.50 per option using a single weighted-average expected life of the options assumption.

Required:

1. Determine the compensation expense related to the options to be recorded each year 2021–2024, assuming Pastner allocates the compensation cost for each of the four groups (tranches) separately.
2. Determine the compensation expense related to the options to be recorded each year 2021–2024, assuming Pastner uses the straight-line method to allocate the total compensation cost.

P 19–4
Stock options; graded vesting; IFRS
● LO19–2, LO19–14
IFRS

Refer to the situation described in P 19–2. Assume Pastner prepares its financial statements using International Financial Reporting Standards (IFRS).

Required:

Would your responses to *requirement* 1 and *requirement* 2 be the same using IFRS?

P 19–5
Stock option plan; deferred tax effect recognized
● LO19–2

Walters Audio Visual, Inc., offers a stock option plan to its regional managers. On January 1, 2021, 40 million options were granted for 40 million $1 par common shares. The exercise price is the market price on the grant date, $8 per share. Options cannot be exercised prior to January 1, 2023, and expire December 31, 2027. The fair value of the options, estimated by an appropriate option pricing model, is $2 per option. Because the plan does not qualify as an incentive plan, Walters will receive a tax deduction upon exercise of the options equal to the excess of the market price at exercise over the exercise price. The income tax rate is 25%.

Required:

1. Determine the total compensation cost pertaining to the stock option plan.
2. Prepare the appropriate journal entries to record compensation expense and its tax effect on December 31, 2021.
3. Prepare the appropriate journal entries to record compensation expense and its tax effect on December 31, 2022.
4. Record the exercise of the options and their tax effect if *all* of the options are exercised on March 20, 2026, when the market price is $12 per share.
5. Assume the option plan qualifies as an incentive plan. Prepare the appropriate journal entries to record compensation expense and its tax effect on December 31, 2021.
6. Assuming the option plan qualifies as an incentive plan, record the exercise of the options and their tax effect if *all* of the options are exercised on March 20, 2026, when the market price is $11 per share.

P 19–6
Stock option plan; deferred tax effect of a nonqualifying plan
● LO19–2

JBL Aircraft manufactures and distributes aircraft parts and supplies. Employees are offered a variety of share-based compensation plans. Under its nonqualified stock option plan, JBL granted options to key officers on January 1, 2021. The options permit holders to acquire 6 million of the company's $1 par common shares for $22 within the next six years, but not before January 1, 2024 (the vesting date). The market price of the shares on the date of grant is $26 per share. The fair value of the 6 million options, estimated by an appropriate option pricing model, is $6 per option. Because the plan does not qualify as an incentive plan, JBL will receive a tax deduction upon exercise of the options equal to the excess of the market price at exercise over the exercise price. The tax rate is 25%.

Required:

1. Determine the total compensation cost pertaining to the incentive stock option plan.
2. Prepare the appropriate journal entries to record compensation expense and its tax effect on December 31, 2021, 2022, and 2023.
3. Record the exercise of the options and their tax effect if *all* of the options are exercised on August 21, 2025, when the market price is $27 per share.

P 19–7
Performance option plan
● LO19–2

LCI Cable Company grants 1 million performance stock options to key executives at January 1, 2021. The options entitle executives to receive 1 million of LCI $1 par common shares, subject to the achievement of specific financial goals over the next four years. Attainment of these goals is considered probable initially and throughout the service period. The options have a current fair value of $12 per option.

Required:

1. Prepare the appropriate entry when the options are awarded on January 1, 2021.
2. Prepare the appropriate entries on December 31 of each year 2021–2024.
3. Suppose at the beginning of 2023, LCI decided it is not probable that the performance objectives will be met. Prepare the appropriate entries on December 31 of 2023 and 2024.

P 19–8
Net loss; stock dividend; nonconvertible preferred stock; treasury shares; shares sold; discontinued operations
● LO19–5 through LO19–7, LO19–13

On December 31, 2020, Ainsworth, Inc., had 600 million shares of common stock outstanding. Twenty million shares of 8%, $100 par value cumulative, nonconvertible preferred stock were sold on January 2, 2021. On April 30, 2021, Ainsworth purchased 30 million shares of its common stock as treasury stock. Twelve million treasury shares were sold on August 31. Ainsworth issued a 5% common stock dividend on June 12, 2021. No cash dividends were declared in 2021. For the year ended December 31, 2021, Ainsworth reported a net loss of $140 million, including an after-tax loss from discontinued operations of $400 million.

Required:
1. Determine Ainsworth's net loss per share for the year ended December 31, 2021.
2. Determine the per share amount of income or loss from continuing operations for the year ended December 31, 2021.
3. Prepare an EPS presentation that would be appropriate to appear on Ainsworth's 2021 and 2020 comparative income statements. Assume EPS was reported in 2020 as $0.75, based on net income (no discontinued operations) of $450 million and a weighted-average number of common shares of 600 million.

P 19–9
EPS from statement of retained earnings
● LO19–4 through LO19–6

(Note: This problem is based on the same situation described in P 18–4 in Chapter 18, modified to focus on EPS rather than recording the events that affected retained earnings.)

Comparative Statements of Retained Earnings for Renn-Dever Corporation were reported as follows for the fiscal years ending December 31, 2019, 2020, and 2021.

RENN-DEVER CORPORATION
Statements of Retained Earnings

	For the Years Ended December 31		
	2021	**2020**	**2019**
Balance at beginning of year	$6,794,292	$5,464,052	$5,624,552
Net income (loss)	3,308,700	2,240,900	(160,500)
Deductions:			
Stock dividend (34,900 shares)	242,000		
Common shares retired, September 30 (110,000 shares)		212,660	
Common stock cash dividends	889,950	698,000	0
Balance at end of year	$8,971,042	$6,794,292	$5,464,052

At December 31, 2018, paid-in capital consisted of the following:

Common stock, 1,855,000 shares at $1 par	$1,855,000
Paid in capital—excess of par	7,420,000

No preferred stock or potential common shares were outstanding during any of the periods shown.

Required:
Compute Renn-Dever's earnings per share as it would have appeared in income statements for the years ended December 31, 2019, 2020, and 2021.

P 19–10
EPS from statement of shareholders' equity
● LO19–4 through LO19–6

Comparative Statements of Shareholders' Equity for Locke Intertechnology Corporation were reported as follows for the fiscal years ending December 31, 2019, 2020, and 2021.

LOCKE INTERTECHNOLOGY CORPORATION
Statements of Shareholders' Equity
For the Years Ended Dec. 31, 2019, 2020, and 2021
($ in millions)

	Preferred Stock, $10 par	Common Stock, $1 par	Additional Paid-in Capital	Retained Earnings	Total Shareholders' Equity
Balance at January 1, 2019		$ 55	$ 495	$1,878	$2,428
Sale of preferred shares	10		470		480
Sale of common shares, 7/1		9	81		90
Cash dividend, preferred				(1)	(1)
Cash dividend, common				(16)	(16)
Net income				290	290
Balance at December 31, 2019	10	64	1,046	2,151	3,271
Retirement of common shares, 4/1		(4)	(36)	(20)	(60)

(continued)

(concluded)

LOCKE INTERTECHNOLOGY CORPORATION
Statements of Shareholders' Equity
For the Years Ended Dec. 31, 2019, 2020, and 2021
($ in millions)

	Preferred Stock, $10 par	Common Stock, $1 par	Additional Paid-in Capital	Retained Earnings	Total Shareholders' Equity
Cash dividend, preferred				(1)	(1)
Cash dividend, common				(20)	(20)
3-for-2 split effected in the form of a common stock dividend, 8/12		30	(30)		
Net income				380	380
Balance at December 31, 2020	10	90	980	2,490	3,570
10% common stock dividend, 5/1		9	90	(99)	
Sale of common shares, 9/1		3	31		34
Cash dividend, preferred				(2)	(2)
Cash dividend, common				(22)	(22)
Net income				412	412
Balance at December 31, 2021	$10	$102	$1,101	$2,779	$3,992

Required:
Infer from the statements the events and transactions that affected Locke Intertechnology Corporation's shareholders' equity and compute earnings per share as it would have appeared on the income statements for the years ended December 31, 2019, 2020, and 2021. No potential common shares were outstanding during any of the periods shown.

P 19–11
EPS; nonconvertible preferred stock; treasury shares; shares sold; stock dividend
● LO19–4 through LO19–7

On December 31, 2020, Dow Steel Corporation had 600,000 shares of common stock and 300,000 shares of 8%, noncumulative, nonconvertible preferred stock issued and outstanding. Dow issued a 4% common stock dividend on May 15 and paid cash dividends of $400,000 and $75,000 to common and preferred shareholders, respectively, on December 15, 2021.

On February 28, 2021, Dow sold 60,000 common shares. In keeping with its long-term share repurchase plan, 2,000 shares were retired on July 1. Dow's net income for the year ended December 31, 2021, was $2,100,000. The income tax rate is 25%.

Required:
Compute Dow's earnings per share for the year ended December 31, 2021.

P 19–12
EPS; nonconvertible preferred stock; treasury shares; shares sold; stock dividend; options
● LO19–4 through LO19–8, LO19–10

(Note: This is a variation of P 19–11, modified to include stock options.)

On December 31, 2020, Dow Steel Corporation had 600,000 shares of common stock and 300,000 shares of 8%, noncumulative, nonconvertible preferred stock issued and outstanding. Dow issued a 4% common stock dividend on May 15 and paid cash dividends of $400,000 and $75,000 to common and preferred shareholders, respectively, on December 15, 2021.

On February 28, 2021, Dow sold 60,000 common shares. In keeping with its long-term share repurchase plan, 2,000 shares were retired on July 1. Dow's net income for the year ended December 31, 2021, was $2,100,000. The income tax rate is 25%.

As part of an incentive compensation plan, Dow granted incentive stock options to division managers at December 31 of the current and each of the previous two years. Each option permits its holder to buy one share of common stock at an exercise price equal to market value at the date of grant and can be exercised one year from that date. Information concerning the number of options granted and common share prices follows:

Date Granted	Options Granted (adjusted for the stock dividend)	Share Price
December 31, 2019	8,000	$24
December 31, 2020	3,000	$33
December 31, 2021	6,500	$32

The market price of the common stock averaged $32 per share during 2021.

Required:
Compute Dow's earnings per share for the year ended December 31, 2021.

P 19–13
EPS; nonconvertible preferred stock; treasury shares; shares sold; stock dividend; options; convertible bonds; contingently issuable shares
● LO19–4 through LO19–11

(Note: This is a variation of P 19–12, modified to include convertible bonds and contingently issuable shares.) On December 31, 2020, Dow Steel Corporation had 600,000 shares of common stock and 300,000 shares of 8%, noncumulative, nonconvertible preferred stock issued and outstanding. Dow issued a 4% common stock dividend on May 15 and paid cash dividends of $400,000 and $75,000 to common and preferred shareholders, respectively, on December 15, 2021.

On February 28, 2021, Dow sold 60,000 common shares. In keeping with its long-term share repurchase plan, 2,000 shares were retired on July 1. Dow's net income for the year ended December 31, 2021, was $2,100,000. The income tax rate is 25%. Also, as a part of a 2020 agreement for the acquisition of Merrill Cable Company, another 23,000 shares (already adjusted for the stock dividend) are to be issued to former Merrill shareholders on December 31, 2022, if Merrill's 2022 net income is at least $500,000. In 2021, Merrill's net income was $630,000.

As part of an incentive compensation plan, Dow granted incentive stock options to division managers at December 31 of the current and each of the previous two years. Each option permits its holder to buy one share of common stock at an exercise price equal to market value at the date of grant and can be exercised one year from that date. Information concerning the number of options granted and common share prices follows:

Date Granted	Options Granted	Share Price
	(adjusted for the stock dividend)	
December 31, 2019	8,000	$24
December 31, 2020	3,000	$33
December 31, 2021	6,500	$32

The market price of the common stock averaged $32 per share during 2021.

On July 12, 2019, Dow issued $800,000 of convertible 8% bonds at face value. Each $1,000 bond is convertible into 30 common shares (adjusted for the stock dividend).

Required:
Compute Dow's basic and diluted earnings per share for the year ended December 31, 2021.

P 19–14
EPS; convertible preferred stock; convertible bonds; order of entry
● LO19–7, LO19–9, LO19–10

Information from the financial statements of Henderson-Niles Industries included the following at December 31, 2021:

Common shares outstanding throughout the year	100 million
Convertible preferred shares (convertible into 32 million shares of common)	60 million
Convertible 8% bonds (convertible into 13.5 million shares of common)	$900 million

Henderson-Niles's net income for the year ended December 31, 2021, is $520 million. The income tax rate is 25%. Henderson-Niles paid dividends of $2 per share on its preferred stock during 2021.

Required:
Compute basic and diluted earnings per share for the year ended December 31, 2021.

P 19–15
EPS; antidilution
● LO19–4 through LO19–10, LO19–13

Alciatore Company reported a net income of $150,000 in 2021. The weighted-average number of common shares outstanding for 2021 was 40,000. The average stock price for 2021 was $33. Assume an income tax rate of 25%.

Required:
For each of the following independent situations, indicate whether the effect of the security is antidilutive for diluted EPS.

1. 10,000 shares of 7.7% of $100 par convertible, cumulative preferred stock. Each share may be converted into two common shares.
2. 6.4% convertible 10-year, $500,000 of bonds, issued at face value. The bonds are convertible to 5,000 shares of common stock.
3. Stock options exercisable at $30 per share after January 1, 2023.
4. Warrants for 1,000 common shares with an exercise price of $35 per share.
5. A contingent agreement to issue 5,000 shares of stock to the company president if net income is at least $125,000 in 2022.

P 19–16
EPS; convertible bonds; treasury shares
● LO19–4 through LO19–6, LO19–9

At December 31, 2021, the financial statements of Hollingsworth Industries included the following:

Net income for 2021	$560 million
Bonds payable, 8%, convertible into 36 million shares of common stock	$300 million
Common stock:	
Shares outstanding on January 1	400 million
Treasury shares purchased for cash on September 1	30 million

Additional data:
The bonds payable were issued at par in 2019. The tax rate for 2021 was 25%.

Required:
Compute basic and diluted EPS for the year ended December 31, 2021.

P 19–17
EPS; options; convertible preferred; additional shares
● LO19–4 through LO19–9

On January 1, 2021, Tonge Industries had outstanding 440,000 common shares ($1 par) that originally sold for $20 per share, and 4,000 shares of 10% cumulative preferred stock ($100 par), convertible into 40,000 common shares.

On October 1, 2021, Tonge sold and issued an additional 16,000 shares of common stock at $33. At December 31, 2021, there were 20,000 incentive stock options outstanding, issued in 2020, and exercisable after one year for 20,000 shares of common stock at an exercise price of $30. The market price of the common stock at year-end was $48. During the year, the price of the common shares had averaged $40.

Net income was $650,000. The tax rate for the year was 25%.

Required:
Compute basic and diluted EPS for the year ended December 31, 2021.

P 19–18
EPS; stock options; nonconvertible preferred; convertible bonds; shares sold
● LO19–4 through LO19–9

At January 1, 2021, Canaday Corporation had outstanding the following securities:

600 million common shares
20 million 6% cumulative preferred shares, $50 par
6.4% convertible bonds, $2,000 million face amount, convertible into 80 million common shares

The following additional information is available:

- On September 1, 2021, Canaday sold 72 million additional shares of common stock.
- Incentive stock options to purchase 60 million shares of common stock after July 1, 2020, at $12 per share, were outstanding at the beginning and end of 2021. The average market price of Canaday's common stock was $18 per share during 2021.
- Canaday's net income for the year ended December 31, 2021, was $1,476 million. The effective income tax rate was 25%.

Required:
1. Calculate basic earnings per common share for the year ended December 31, 2021.
2. Calculate the diluted earnings per common share for the year ended December 31, 2021.

P 19–19
EPS; options; restricted stock; additional components for "proceeds" in treasury stock method
● LO19–1, LO19–2, LO19–4, LO19–8, LO19–11

Witter House is a calendar-year firm with 300 million common shares outstanding throughout 2021 and 2022. As part of its executive compensation plan, at January 1, 2020, the company had issued 30 million executive stock options permitting executives to buy 30 million shares of stock for $10 within the next eight years, but not prior to January 1, 2023. The fair value of the options was estimated on the grant date to be $3 per option.

In 2021, Witter House began granting employees stock awards rather than stock options as part of its equity compensation plans and granted 15 million restricted common shares to senior executives at January 1, 2021. The shares vest four years later. The fair value of the stock was $12 per share on the grant date. The average price of the common shares was $12 and $15 during 2021 and 2022, respectively.

The stock options qualify as an incentive plan. The restricted stock does not. The company's net income was $150 million and $160 million in 2021 and 2022, respectively.

Required:
1. Determine basic and diluted earnings per share for Witter House in 2021.
2. Determine basic and diluted earnings per share for Witter House in 2022.

Decision Makers' Perspective

©ImageGap/Getty Images

Apply your critical-thinking ability to the knowledge you've gained. These cases will provide you an opportunity to develop your research, analysis, judgment, and communication skills. You also will work with other students, integrate what you've learned, apply it in real-world situations, and consider its global and ethical ramifications. This practice will broaden your knowledge and further develop your decision-making abilities.

Real World Case 19–1
Restricted stock awards; Microsoft
● LO19–1

Real World Financials

Microsoft provides compensation to executives in the form of a variety of incentive compensation plans, including restricted stock award grants. The following is an excerpt from a disclosure note from Microsoft's 2017 annual report:

Note 20 Employee Stock and Savings Plans (in part)
Stock awards are grants that entitle the holder to shares of common stock as the award vests. Our stock awards generally vest over a five-year period. . . . During fiscal year 2017, the following activity occurred under our plans:

	Shares ($ in millions)	Weighted Average Grant-Date Fair Value
Stock awards:		
Nonvested balance, beginning of year	194	$36.92
Granted	84	55.64
Assumed in acquisitions	23	59.09
Vested	(80)	37.36
Forfeited	(20)	43.71
Nonvested balance, end of year	201	$46.32

Required:
If all awards are granted, acquired, vested, and forfeited evenly throughout the year, what is the compensation expense in fiscal 2017 pertaining to the previous and current stock awards? Assume forfeited shares were granted evenly throughout the three previous years, and perform all calculations to the nearest one-tenth of $1 million (e.g., $7,162,480 = $7,162.5).

Source: Microsoft

Communication Case 19–2
Stock options; basic concepts; prepare a memo
● LO19–2

You are assistant controller of Stamos & Company, a medium-size manufacturer of machine parts. On October 22, 2020, the board of directors approved a stock option plan for key executives. On January 1, 2021, a specific number of stock options were granted. The options were exercisable between January 1, 2026, and December 31, 2030, at 100% of the quoted market price at the grant date. The service period is for 2021 through 2026.

Your boss, the controller, is one of the executives to receive options. Neither he nor you have had occasion to deal with GAAP on accounting for stock options. He and you are aware of the traditional approach your company used years ago but do not know the newer method. Your boss understands how options might benefit him personally but wants to be aware also of how the options will be reported in the financial statements. He has asked you for a one-page synopsis of accounting for stock options under the fair value approach. He instructed you, "I don't care about the effect on taxes or earnings per share—just the basics, please."

Required:
Prepare such a report that includes the following:

1. At what point should the compensation cost be measured? How should it be measured?
2. How should compensation expense be measured for the stock option plan in 2021 and later?
3. If options are forfeited because an executive resigns before vesting, what is the effect of that forfeiture of the stock options on the financial statements?
4. If options are allowed to lapse after vesting, what is the effect on the financial statements?

Ethics Case 19–3
Stock options
● LO19–2

You are in your second year as an auditor with Dantly and Regis, a regional CPA firm. One of the firm's long-time clients is Mayberry-Cleaver Industries, a national company involved in the manufacturing, marketing, and sales of hydraulic devices used in specialized manufacturing applications. Early in this year's audit, you discover that Mayberry-Cleaver has changed its method of determining inventory from LIFO to FIFO. Your client's explanation is that FIFO is consistent with the method used by some other companies in the industry. Upon further investigation, you discover an executive stock option plan whose terms call for a significant increase in the shares available to executives if net income this year exceeds $44 million. Some quick calculations convince you that without the change in inventory methods, the target will not be reached; with the change, it will.

Required:

Do you perceive an ethical dilemma? What would be the likely impact of following the controller's suggestions? Who would benefit? Who would be injured?

Ethics Case 19–4
International Network Solutions
● LO19–5

International Network Solutions provides products and services related to remote access networking. The company has grown rapidly during its first 10 years of operations. As its segment of the industry has begun to mature, though, the fast growth of previous years has begun to slow. In fact, this year revenues and profits are roughly the same as last year.

One morning, nine weeks before the close of the fiscal year, Rob Mashburn, CFO, and Jessica Lane, controller, were sharing coffee and ideas in Lane's office.

Lane:	About the Board meeting Thursday. You may be right. This may be the time to suggest a share buyback program.
Mashburn:	To begin this year, you mean?
Lane:	Right! I know Barber will be lobbying to use the funds for our European expansion. She's probably right about the best use of our funds, but we can always issue more notes next year. Right now, we need a quick fix for our EPS numbers.
Mashburn:	Our shareholders are accustomed to increases every year.

Required:

1. How will a buyback of shares provide a "quick fix" for EPS?
2. Is the proposal ethical?
3. Who would be affected if the proposal is implemented?

Real World Case 19–5
Per share data; stock options; antidilutive securities; Best Buy Co.
● LO19–8

Real World Financials

The 2018 annual report of **Best Buy Co., Inc.**, reported profitable operations for the most recent six years. However, the company suffered a net loss in 2012. Best Buy reported the following for the twelve months ended March 3, 2012:

Basic (loss) earnings per share:	
Continuing operations	$ (2.89)
Discontinued operations	(0.47)
Basic (loss) earnings per share	$ (3.36)
Diluted (loss) earnings per share:	
Continuing operations	$ (2.89)
Discontinued operations	(0.47)
Diluted (loss) earnings per share	$ (3.36)
Dividends declared per Best Buy Co., Inc., common share	$ 0.62
Weighted average common shares outstanding (in millions)	
Basic	366.3
Diluted	366.3

Note: The calculation of diluted earnings per share assumes the conversion of the company's previously outstanding convertible debentures due in 2027 into 8.8 million shares common stock . . . and adds back the related after-tax interest expense of $1.5. . . . The calculation of diluted (loss) per share for the twelve months ended March 3, 2012, does not include potential dilutive shares of common stock because their inclusion would be antidilutive (i.e., reduce the net loss per share).

Required:

1. The note indicates that "diluted earnings per share assumes the conversion of the company's *previously outstanding* convertible debentures due in 2027 into 8.8 million shares common stock . . . and adds back the related after-tax interest expense of $1.5." Apparently, these bonds were actually converted during the year. By how much would diluted EPS have been different if the bonds had not been converted and were still outstanding?
2. Best Buy does not include potentially dilutive shares when calculating EPS for the twelve months ended March 3, 2012. Assume Best Buy had 40 million common equivalent shares and included them in the calculation, what would have been the amount of diluted loss per share for the twelve months ended March 3, 2012?

Source: Best Buy Co., Inc.

Analysis Case 19–6
EPS concepts
● LO19–4 through LO19–8

The shareholders' equity of Proactive Solutions, Inc., included the following at December 31, 2021:

Common stock, $1 par
Paid-in capital—excess of par on common stock
7% cumulative convertible preferred stock, $100 par value
Paid-in capital—excess of par on preferred stock
Retained earnings

Additional Information:

- Proactive had 7 million shares of preferred stock authorized of which 2 million were outstanding. All 2 million shares outstanding were issued in 2015 for $112 a share. The preferred stock is convertible into common stock on a two-for-one basis until December 31, 2023, after which the preferred stock no longer is convertible. None of the preferred stock has been converted into common stock at December 31, 2021. There were no dividends in arrears.
- Of the 13 million common shares authorized, there were 8 million shares outstanding at January 1, 2021. Proactive also sold 3 million shares at the beginning of September 2021 at a price of $52 a share.
- The company has an employee stock option plan in which certain key employees and officers may purchase shares of common stock at the market price at the date of the option grant. All options are exercisable beginning one year after the date of the grant and expire if not exercised within five years of the grant date. On January 1, 2021, options for 2 million shares were outstanding at prices ranging from $45 to $53 a share. Options for 1 million shares were exercised at $49 a share at the end of June 2021. No options expired during 2021. Additional options for 1.5 million shares were granted at $55 a share during the year. The 2.5 million options outstanding at December 31, 2021, were exercisable at $45 to $55 a share.

The only changes in the shareholders' equity for 2021 were those described above, 2021 net income, and cash dividends paid.

Required:

Explain how each of the following amounts should be determined when computing earnings per share for presentation in the income statements. For each, be specific as to the treatment of each item.

1. Numerator for basic EPS
2. Denominator for basic EPS
3. Numerator for diluted EPS
4. Denominator for diluted EPS

Analysis Case 19–7
EPS
● LO19–5 through LO19–8

"I guess I'll win that bet!" you announced to no one in particular.

"What bet?" Renee asked. Renee Patey was close enough to overhear you.

"When I bought my REC stock last year Randy insisted it was a mistake, that they were going to collapse. I bet him a Coke he was wrong. This press release says they have positive earnings," you bragged. Renee was looking over your shoulder now at the article you were pointing at:

CHICAGO (ACCOUNTING WIRE)—July 1, 2021—Republic Enterprise Companies, Inc. (REC), today reported net income attributable to REC of $3.6 billion for the quarter ended May 31, 2021. . . . Diluted earnings per share attributable to REC were $1.52 for the second quarter of 2021, compared with $1.21 for the second quarter of 2020.

Our Board of Directors . . . authorized the repurchase of shares of REC Common Stock, with an aggregate purchase price of up to $2.0 billion.

"A dollar fifty-two a share, huh?" Renee asked. "How many shares do you have? When do you get the check?"

Required:

1. Renee's questions imply that she thinks you will get cash dividends of $1.52 a share. What does earnings per share really tell you?
2. A previous press release indicated that "Share and per share amounts prior to the second quarter of 2019 have been restated to reflect the 1-for-20 reverse stock split effective June 30, 2019." What does that mean?
3. The press release indicates plans to repurchase shares of its own stock. Would the reduction in shares from a stock repurchase be taken into account when EPS is calculated? How?

Source: ACCOUNTING WIRE

Judgment Case 19–8
Where are the profits?
● LO19–4 through LO19–7, LO19–9

Del Conte Construction Company has experienced generally steady growth since its inception in 1973. Management is proud of its record of having maintained or increased its earnings per share in each year of its existence.

The economic downturn has led to disturbing dips in revenues the past two years. Despite concerted cost-cutting efforts, profits have declined in each of the two previous years. Net income in 2019, 2020, and 2021 was as follows:

2019	$145 million
2020	$134 million
2021	$ 95 million

A major shareholder has hired you to provide advice on whether to continue her present investment position or to curtail that position. Of particular concern is the declining profitability, despite the fact that earnings per share has continued a pattern of growth:

	Basic	Diluted
2019	$2.15	$1.91
2020	$2.44	$2.12
2021	$2.50	$2.50

She specifically asks you to explain this apparent paradox. During the course of your investigation you discover the following events:

- For the decade ending December 31, 2018, Del Conte had 60 million common shares and 20 million shares of 8%, $10 par nonconvertible preferred stock outstanding. Cash dividends have been paid quarterly on both.
- On July 1, 2020, half the preferred shares were retired in the open market. The remaining shares were retired on December 30, 2020.
- $55 million of 8% nonconvertible bonds were issued at the beginning of 2021, and a portion of the proceeds were used to call and retire $50 million of 8% debentures (outstanding since 2016) that were convertible into 9 million common shares.
- In 2019, management announced a share repurchase plan by which up to 24 million common shares would be retired. 12 million shares were retired on March 1 of both 2019 and 2020.
- Del Conte's income tax rate is 25% and has been for the last several years.

Required:
In preparation for your explanation of the apparent paradox to which your client refers, calculate both basic and diluted earnings per share for each of the three years.

Communication Case 19–9
Dilution
● LO19–9

"I thought I understood earnings per share," lamented Brad Dawson, "but you're telling me we need to pretend our convertible bonds have been converted! Or maybe not?"

Dawson, your boss, is the new manager of the Fabricating division of BVT Corporation. His background is engineering and he has only a basic understanding of earnings per share. Knowing you are an accounting graduate, he asks you to explain the questions he has about the calculation of the company's EPS. His reaction is to your explanation that the company's convertible bonds might be included in this year's calculation.

"Put it in a memo!" he grumbled as he left your office.

Required:
Write a memo to Dawson. Explain the effect on earnings per share of each of the following:
1. Convertible securities.
2. Antidilutive securities.

Analysis Case 19–10
Analyzing financial statements; price-earnings ratio; dividend payout ratio
● LO19–13

IGF Foods Company is a large, primarily domestic, consumer foods company involved in the manufacture, distribution, and sale of a variety of food products. Industry averages are derived from Troy's *The Almanac of Business and Industrial Financial Ratios* and Dun and Bradstreet's *Industry Norms and Key Business Ratios*. Following are the 2021 and 2020 comparative income statements and balance sheets for IGF. The market price of IGF's common stock is $47 during 2021. (The financial data we use are from actual financial statements of a well-known corporation, but the company name used in our illustration is fictitious and the numbers and dates have been modified slightly to disguise the company's identity.)

IGF FOODS COMPANY
Years Ended December 31, 2021 and 2020

($ in millions)	2021	2020
Comparative Income Statements		
Net sales	$6,440	$5,800
Cost of goods sold	(3,667)	(3,389)
Gross profit	2,773	2,411
Operating expenses	(1,916)	(1,629)
Operating income	857	782
Interest expense	(54)	(53)
Income from operations before tax	803	729
Income taxes	(316)	(287)

(continued)

(concluded)

IGF FOODS COMPANY
Years Ended December 31, 2021 and 2020

($ in millions)	2021	2020
Net income	$ 487	$ 442
Net income per share	$ 2.69	$ 2.44
Average shares outstanding	181 million	181 million
($ in millions)	2021	2020
Comparative Balance Sheets		
Assets		
Total current assets	$1,879	$1,490
Property, plant, and equipment (net)	2,592	2,291
Intangibles (net)	800	843
Other assets	74	60
Total assets	$5,345	$4,684
Liabilities and Shareholders' Equity		
Total current liabilities	$1,473	$ 941
Long term debt	534	728
Deferred income taxes	407	344
Total liabilities	2,414	2,013
Shareholders' equity:		
Common stock	180	180
Additional paid-in capital	21	63
Retained earnings	2,730	2,428
Total shareholders' equity	2,931	2,671
Total liabilities and shareholders' equity	$5,345	$4,684

Some ratios express income, dividends, and market prices on a per share basis. As such, these ratios appeal primarily to common shareholders, particularly when weighing investment possibilities. These ratios focus less on the fundamental soundness of a company and more on its investment characteristics.

Required:

1. Calculate 2021 earnings per share for IGF.
2. Calculate IGF's 2021 price-earnings ratio.
3. Calculate IGF's 2021 dividend payout ratio.

Analysis Case 19–11

Kellogg's EPS; PE ratio; dividend payout

● LO19–13

Real World Financials

While eating his **Kellogg**'s Frosted Flakes one January morning, Tony noticed the following article in his local paper:

Kellogg Company Reports Fourth-Quarter 2017 Results and Provides Guidance For 2018 (in part)

BATTLE CREEK, Mich., Feb. 8, 2018 /PRNewswire/—Kellogg Company (NYSE: K) Reported full-year 2017 net earnings $1,269 million, or $3.62 per share.

As a shareholder, Tony is well aware that Kellogg pays a regular cash dividend of $0.54 per share quarterly. A quick click on a price quote service indicated that Kellogg's shares closed at $67.98 on December 31.

Required:

1. Using the numbers provided, determine the price-earnings ratio for Kellogg Company for 2017.
2. What is the dividend payout ratio for Kellogg?

Source: Kellogg

Research Case 19–12
FASB codification; locate and extract relevant information and cite authoritative support for a financial reporting issue; change in classification of a share-based compensation instrument
● LO19–13

"Now what do I do?" moaned your colleague Matt. "This is a first for me," he confided. You and Matt are recent hires in the Accounting Division of National Paper. A top executive in the company has been given share-based incentive instruments that permit her to receive shares of National Paper equal in value to the amount the company shares rise above the shares' value two years ago when the instruments were issued to her as compensation. The instruments vest in three years. A clause was included in the compensation agreement that would permit her to receive cash rather than shares upon exercise if sales revenue in her division were to double by that time. Because that contingency was considered unlikely, the instruments have been accounted for as equity, with the grant date fair value being expensed over the five-year vesting period.

Now, though, surging sales of her division indicate that the contingent event has become probable, and the instruments should be accounted for as a liability rather than equity. The fair value of the award was estimated at $5 million on the grant date, but now is $8 million. Matt has asked your help in deciding what to recommend to your controller as the appropriate action to take at this point.

Required:

1. Obtain the relevant authoritative literature on accounting for a change in classification due to a change in probable settlement outcome using the *FASB Accounting Standards Codification* at the FASB website (**www.fasb.org**). To help explain to Matt the basic treatment of the situation described, determine the specific seven-digit Codification citation (XXX-XX-XX) you would rely on in applying that accounting treatment.
2. Prepare the journal entry to record the change in circumstance.

Data Analytics

Data analytics is the process of examining data sets in order to draw conclusions about the information they contain. If you haven't completed any of the prior data analytics cases, follow the instructions listed in the Chapter 1 Data Analytics case to get set up. You will need to watch the videos referred to in the Chapters 1 - 3 Data Analytics cases. No additional videos are required for this case. All short training videos can be found here: **https://www.tableau.com/learn/training#getting-started**.

Data Analytics Case
Market Performance
● LO19–13

In the Data Analytics Cases in the previous chapter, you used Tableau to examine two (hypothetical) publicly traded companies: GPS Corporation and Tru, Inc. regarding the return on stockholders' equity. This time, you investigate two aspects of the relative desirability of investing in stock of the two companies.

Required:

For each of the two companies in the most recent five-year period, 2017-2021, use Tableau to calculate and display the trends for (a) the dividend payout ratios [Dividends paid / Income from continuing operations] and (b) the price-earnings ratio [Market price per share / Diluted earnings per share]. The 2021 industry averages for the dividend payout and the price-earnings ratio are 20.2% and 27.9%, respectively. Based upon what you find, answer the following questions:

1. Which of the two companies, GPS Corporation or Tru, Inc., pays investors the greater percentage of its earnings during the period 2017-2021?
2. Is the dividend payout ratio for Tru, Inc. (a) increasing significantly, (b) decreasing significantly, or (c) remaining relatively the same during the period 2017-2021?
3. Is the price-earnings ratio for GPS Corporation (a) increasing significantly, (b) decreasing significantly, or (c) remaining relatively the same during the period 2017-2021?
4. Is the price-earnings ratio for Tru, Inc. (a) increasing significantly, (b) decreasing significantly, or (c) remaining relatively the same during the period 2017-2021?

Resources:

Download the "GPS_Tru_Financials.xlsx" Excel file available in Connect, or under Student Resources within the Library tab. Save it to the computer on which you will be using Tableau.

For this case, you will create several calculations to produce performance ratios to allow you to compare and contrast the two companies.

After you view the training videos, follow these steps to create the charts you'll use for this case:

- Open Tableau and connect to the Excel spreadsheet you downloaded.
- Click on the "Sheet 1" tab at the bottom of the canvas, to the right of the Data Source at the bottom of the screen.

- Drag "($ in 000s) Company" and "Year" under Dimensions to the Columns shelf. Change "Year" to *discrete* by right-clicking and selecting "Discrete." Select "Show Filter" and uncheck all the years except the last five years (2017 – 2021).
- Create a calculated field by clicking the "Analysis" tab at the top of the screen and selecting "Create Calculated Field." Name the calculation "Dividend payout ratio." In the Calculation Editor window, from Measures, drag "Dividends paid", type a division sign, drag "Income from continuing operations" beside it, and then type a multiplication sign and a "−1." Make sure the window says that the calculation is valid and click OK.
- Drag the newly created "Dividend payout ratio" to the Rows shelf. Click on the "Show Me" and select "side-by-side bars." Add labels to the bars by clicking on "Label" under the "Marks" card and clicking the box "Show mark labels." Format the labels to Times New Roman, bold, black and 9-point font. Edit the color on the Color marks card if desired.
- Change the title of the sheet to be "Dividends Payout Ratio Trend" by right-clicking and selecting "Edit title." Format the title to Times New Roman, bold, black and 15-point font. Change the title of "Sheet 1" to match the sheet title by right-clicking, selecting "Rename" and typing in the new title.
- Click on the New Worksheet tab ("Sheet 2" should open). to the right of the newly named "Dividend Payout Ratio Trend" sheet at the bottom of the screen.
- Follow the procedures above for the company and year.
- Create a calculated field by clicking the "Analysis" tab at the top of the screen and selecting "Create Calculated Field." Name the calculation "Price-earnings ratio." In the Calculation Editor window, from Measures, drag "Market price per share", type a division sign, and then drag "Diluted EPS". Make sure the window says that the calculation is valid and click OK.
- Drag the newly created "Price-earnings ratio" to the Rows shelf. Click on the "Show Me" and select "side-by-side bars." Add labels to the bars by clicking on "Label" under the "Marks" card and clicking the box "Show mark labels." Format the labels to Times New Roman, bold, black and 10-point font. Edit the color on the Color marks card if desired.
- Change the title of the sheet to be "Price-Earnings Ratio Trend" by right-clicking and selecting "Edit title." Format the title to Times New Roman, bold, black and 15-point font. Change the title of "Sheet 2" to match the sheet title by right-clicking, selecting "Rename" and typing in the new title.
- Format all other labels to be Times New Roman, bold, black and 12-point font.
- Once complete, save the file as "DA19_Your initials.twbx."

Continuing Cases

Target Case
- 19–1, LO19–2, LO19–8

Target Corporation prepares its financial statements according to U.S. GAAP. Target's financial statements and disclosure notes for the year ended February 3, 2018, are available in Connect. This material is also available under the Investor Relations link at the company's website (**www.target.com**). Target's share-based compensation includes several long-term incentive plans.

Required:

1. What are the three types of awards described in Note 26: Share-Based Compensation?
2. Based on the fair value of the awards granted, what was Target's primary form of share-based compensation for the year ended February 3, 2018?
3. Projections of future performance should be based primarily on continuing operations. What was diluted EPS for continuing operations in each of the most recent three years?
4. How many shares were included in diluted earnings per share but not basic earnings per share due to share-based compensation awards?

Air France–KLM Case
- LO19–9

IFRS

AIRFRANCE KLM

Air France–KLM (AF), a Franco-Dutch company, prepares its financial statements according to International Financial Reporting Standards. AF's financial statements and disclosure notes for the year ended December 31, 2017, are available in Connect. This material is also available under the Finance link at the company's website (**www.airfranceklm-finance.com**).

Required:

1. AF provides share-based compensation in the form of PPSs, Phantom Performance Shares. Recipients receive compensation in what form? Are such plans reported in the balance sheet as (a) assets, (b) liabilities, or (c) equity? [*Hint:* See Note 29: "Share-Based Compensation."]
2. Are AF's PPSs cliff vesting or graded vesting?

3. The PPS options are performance-based, which means that under either U.S. GAAP or IFRS the amount expensed depends on whether it's "probable" that the performance target will be met. Does this mean it's equally likely that performance-based PPSs will be expensed at the same amount under U.S. GAAP as under IFRS?
4. What amount(s) of earnings per share did AF report in its income statement for the year ended December 31, 2017? If AF used U.S. GAAP, would it have reported EPS using the same classification?

CPA Exam Questions and Simulations

ROGER CPA Review

Sample CPA Exam questions from Roger CPA Review are available in Connect as support for the topics in this chapter. These multiple-choice questions and task-based simulations include expert-written explanations and solutions, and provide a starting point for students to become familiar with the content and functionality of the actual CPA Exam.

CHAPTER

20 Accounting Changes and Error Corrections

OVERVIEW

Chapter 4 provided a brief overview of accounting changes and error correction. Later, we discussed changes encountered in connection with specific assets and liabilities as we dealt with those topics in subsequent chapters.

Here we revisit accounting changes and error correction to synthesize the way these are handled in a variety of situations that might be encountered in practice. We see that most changes in accounting principle are reported retrospectively. Changes in estimates are accounted for prospectively. A change in depreciation methods is considered a change in estimate resulting from a change in principle. Both changes in reporting entities and the correction of errors are reported retrospectively.

LEARNING OBJECTIVES

After studying this chapter, you should be able to:

- **LO20–1** Differentiate among the three types of accounting changes and distinguish among the retrospective, modified retrospective, and prospective approaches to accounting for and reporting accounting changes.(*p. 1172*)
- **LO20–2** Describe how changes in accounting principle typically are reported. (*p. 1174*)
- **LO20–3** Explain how and why some changes in accounting principle are reported prospectively. (*p. 1178*)
- **LO20–4** Explain how and why changes in estimates are reported prospectively. (*p. 1180*)
- **LO20–5** Describe the situations that constitute a change in reporting entity. (*p. 1182*)
- **LO20–6** Understand and apply the four-step process of correcting and reporting errors, regardless of the type of error or the timing of its discovery. (*p. 1186*)
- **LO20–7** Discuss the primary differences between U.S. GAAP and IFRS with respect to accounting changes and error corrections. (*pp. 1192*)

©TY Lim/Shutterstock

FINANCIAL REPORTING CASE

In a Jam

"What the heck!" Martin yelped as he handed you the annual report of **J.M. Smucker** he'd received in the mail today. "It looks like Smucker found a bunch of lost jelly. It says here that their inventory was $54 million last year. I distinctly remember them reporting that number last year as $52 million because my dad was born in '52, and I did a little wordplay in my mind about him 'taking inventory' of his life when he bought the red Mustang." He had circled the number in the comparative balance sheets. "When I bought Smucker shares last year, I promised myself I would monitor things pretty closely, but it's not as easy as I thought it would be."

As an accounting graduate, you can understand Martin's confusion. Flipping to the disclosure note on accounting changes, you proceed to clear things up for him.

QUESTIONS

By the time you finish this chapter, you should be able to respond appropriately to the questions posed in this case. Compare your response to the solution provided at the end of the chapter.

1. How can an accounting change cause a company to increase a previously reported inventory amount? (*p. 1175*)
2. Are all accounting changes reported this way? (*p. 118[illegible]*)

You learned early in your study of accounting that two of the qualitative characteristics of accounting information that contribute to its relevance and representational faithfulness are *consistency* and *comparability*. Though we strive to achieve and maintain these financial reporting attributes, we cannot ignore the forces of change. Ours is a dynamic business environment. The economy is increasingly a global one. Technological advances constantly transform both day-to-day operations and the flow of information about those operations. The accounting profession's response to the fluid environment often means issuing new standards that require companies to change accounting methods. Often, developments within an industry or the economy will prompt a company to voluntarily switch methods of accounting or to revise estimates or expectations. In short, change is inevitable. The question then becomes a matter of how best to address change when reporting financial information from year to year.

In the first part of this chapter, we differentiate among the various types of accounting changes that businesses face, with a focus on the most meaningful and least disruptive ways to report those changes. Then, in the second part of the chapter, we direct our attention to a closely related circumstance—the correction of errors.

Accounting Changes

PART A

Accounting changes fall into one of the three categories listed in Illustration 20–1.[1]

[1]FASB ASC 250: Accounting Changes and Error Corrections.

Illustration 20–1
Types of Accounting Changes

● LO20–1

Type of Change	Description	Examples
Change in accounting principle	Change from one generally accepted accounting principle to another.	• Adopt a new Accounting Standard. • Change methods of inventory costing. • Change from cost method to equity method, or vice versa.
Change in accounting estimate	Revise an estimate because of new information or new experience.	• Change depreciation methods.* • Change estimate of useful life of depreciable asset. • Change estimate of residual value of depreciable asset. • Change estimate of periods benefited by intangible assets. • Change actuarial estimates pertaining to a pension plan.
Change in reporting entity	Change from reporting as one type of entity to another type of entity.	• Consolidate a subsidiary not previously included in consolidated financial statements. • Report consolidated financial statements in place of individual statements.

*A change in depreciation methods is a change in estimate that is achieved by a change in accounting principle.

The correction of an error is another adjustment sometimes made to financial statements that is not actually an accounting change but is accounted for similarly. Errors occur when transactions are either recorded incorrectly or not recorded at all, as shown in Illustration 20–2.

Illustration 20–2
Correction of Errors

Type of Change	Description	Examples
Error correction	Correct an error caused by a transaction being recorded incorrectly or not at all.	• Mathematical mistakes. • Inaccurate physical count of inventory. • Change from the cash basis of accounting to the accrual basis. • Failure to record an adjusting entry. • Recording an asset as an expense, or vice versa. • Fraud or gross negligence.

Three approaches to reporting accounting changes and error corrections are used, depending on the situation.

The retrospective approach offers consistency and comparability.

1. Using the **retrospective approach**, financial statements issued prior to the change are adjusted to reflect the impact of the change whenever those statements are presented again for comparative purposes (comparative financial statements). For each year reported in the comparative statements, the balance of each account affected is adjusted to incorporate the change. So for the first period presented in the comparative statements, the adjusted balances include the cumulative effect of the change prior to that date. In other words, those statements are made to appear as if the newly adopted accounting method had been applied all along or that the error had never occurred. To achieve this result, a journal entry is created to adjust all account balances affected to what those amounts would have been. In addition, if retained earnings is one of the accounts that requires adjustment, that

adjustment is made to the beginning balance of retained earnings for the earliest period reported in the comparative statements of shareholders' equity. An advantage of this retrospective approach is that it achieves comparability among financial statements. All financial statements presented are prepared on the same basis. However, some argue that public confidence in the integrity of financial data suffers when numbers previously reported as correct are later superseded. On the other hand, proponents argue the opposite—that it's impossible to maintain public confidence unless the financial statements are comparable.

2. The **modified retrospective approach** requires application of the new standard only to the adoption period (that is, the current period) as well as adjustment of the balance of retained earnings at the beginning of the adoption period to capture the cumulative effects of prior periods without actually adjusting the numbers of the prior periods reported.
3. The **prospective approach** requires neither a modification of prior years' financial statements nor a journal entry to adjust account balances. Instead, the change is simply implemented in the period of the change, and its effects are reflected in the financial statements of the period of the change and future periods only.

The effects of a change are reflected in the financial statements of only the year of the change and future years under the *prospective approach*.

Now, let's look at each type of accounting change, one at a time, focusing on the selective application of these approaches.

Change in Accounting Principle

Accounting is not an exact science. Professional judgment is required to apply a set of principles, concepts, and objectives to specific sets of circumstances. This means choices must be made. In your study of accounting to date, you've encountered many areas where choices are necessary. For example, management must choose whether to use accelerated or straight-line depreciation. Is FIFO, LIFO, or average cost most appropriate to measure inventories? Should we adopt a new accounting standard early or wait until it's mandatory? These are but a few of the accounting choices management makes.

You also probably recall that comparability is an enhancing qualitative characteristic of financial reporting. To achieve this attribute of information, accounting choices, once made, should be consistently followed from year to year. This doesn't mean, though, that methods can never be changed. Changing circumstances might make a new method more appropriate. A change in economic conditions, for instance, might prompt a company to change accounting methods. The most extensive voluntary accounting change ever—a switch by hundreds of companies from FIFO to LIFO in the mid-1970s, for example—was a result of heightened inflation. Changes within a specific industry, too, can lead a company to switch methods, often to adapt to new technology or to be consistent with others in the industry. And, of course, a change might be mandated when the FASB codifies a new accounting standard. For these reasons, it's not uncommon for a company to switch from one accounting method to another. This is called a **change in accounting principle**.

Although consistency and comparability are desirable, changing to a new method sometimes is appropriate.

Decision Makers' Perspective—Motivation for Accounting Choices

It would be nice to think that all accounting choices are made by management in the best interest of fair and consistent financial reporting. Unfortunately, other motives influence the choices among accounting methods and whether to change methods. It has been suggested that the effect of choices on management compensation, on existing debt agreements, and on union negotiations each can affect management's selection of accounting methods.[2] For instance, research has suggested that managers of companies with bonus plans are more likely to choose accounting methods that maximize their bonuses (often those that increase net income).[3] Other research has indicated that the existence and nature of debt agreements

[2] R. L. Watts and J. L. Zimmerman, "Towards a Positive Theory of the Determination of Accounting Standards," *The Accounting Review*, January 1978, and "Positive Accounting Theory: A Ten Year Perspective," *The Accounting Review*, January 1990.

[3] For example, see P. M. Healy, "The Effect of Bonus Schemes on Accounting Decisions," *Journal of Accounting and Economics*, April 1985; and D. Dhaliwal, G. Salamon, and E. Smith, "The Effect of Owner versus Management Control on the Choice of Accounting Methods," *Journal of Accounting and Economics*, July 1992.

and other aspects of a firm's capital structure can influence accounting choices.[4] Whether a company is forbidden from paying dividends if retained earnings fall below a certain level, for example, can affect the choice of accounting methods.

A financial analyst must be aware that different accounting methods used by different firms, and by the same firm in different years, complicate comparisons. Financial ratios, for example, will differ when different accounting methods are used, even when there are no differences in attributes being compared.

Investors and creditors also should be alert to instances in which companies change accounting methods. They must consider not only the effect on comparability but also possible hidden motivations for making the changes. Are managers trying to compensate for a downturn in actual performance with a switch to methods that artificially inflate reported earnings? Is the firm in danger of violating debt covenants or other contractual agreements regarding financial position? Are executive compensation plans tied to reported performance measures? Fortunately, the nature and effect of changes are reported in the financial statements. Although a justification for a change is provided by management, analysts should be wary of accepting the reported justification at face value without considering a possible hidden agenda.

Choices are not always those that tend to increase income. As you learned in Chapter 8, many companies use the LIFO inventory method because it reduces income and therefore reduces the amount of income taxes that must be paid currently. Also, some very large and visible companies might be reluctant to report high income that might render them vulnerable to union demands, government regulations, or higher taxes.[5]

Another reason managers sometimes choose accounting methods that don't necessarily increase earnings was mentioned earlier. Most managers tend to prefer to report earnings that follow a regular, smooth trend from year to year. The desire to "smooth" earnings means that any attempt to manipulate earnings by choosing accounting methods is not always in the direction of higher income. Instead, the choice might be to avoid irregular earnings, particularly those with wide variations from year to year, a pattern that might be interpreted by analysts as denoting a risky situation.

Obviously, any time managers make accounting choices for any of the reasons discussed here, when the motivation is an objective other than to provide useful information, earnings quality suffers. As mentioned frequently throughout this text, earnings quality refers to the ability of reported earnings (income) to predict a company's future earnings.

A notable example of alleged "cooking the books" involved **General Electric**. The company was often suspected of using "cookie jar accounting," the practice of using unrealistic estimates and strategic choices of accounting methods to smooth out its earnings. In 2009, GE appeared to get its hand caught in the cookie jar for going beyond acceptable limits. GE paid a $50 million civil penalty to settle charges by the Securities and Exchange Commission accusing the company of violating U.S. securities laws four times between 2002 and 2003 to help it maintain a succession of earnings reports that beat Wall Street expectations each quarter from 1995 through 2004.

Let's turn our attention now to situations involving changes in methods and how we account for those changes. ●

The Retrospective Approach: Most Changes in Accounting Principle

● LO20–2

We report most voluntary changes in accounting principles retrospectively.[6] This means reporting all previous periods' financial statements as if the new method had been used in all prior periods. An example is provided in Illustration 20–3.

---

[4] R. M. Bowen, E. W. Noreen, and J. M. Lacy, "Determinants of the Corporate Decision to Capitalize Interest," *Journal of Accounting and Economics,* August 1981.

[5] This political cost motive is suggested by R. L. Watts and J. L. Zimmerman, "Positive Accounting Theory: A Ten Year Perspective," *The Accounting Review,* January 1990; and M. Zmijewski and R. Hagerman, "An Income Strategy Approach to the Positive Theory of Accounting Standard Setting/Choice," *Journal of Accounting and Economics,* August 1981.

[6] FASB ASC 250–10–45: Accounting Changes and Error Corrections—Overall—Other Presentation Matters.

Illustration 20–3

Change in Accounting Principle

Air Parts Corporation used the LIFO inventory costing method. At the beginning of 2021, Air Parts decided to change to the FIFO method. Income components for 2021 and prior years were as follows:

($ in millions)	Previous Years	2019	2020	2021
Cost of goods sold (LIFO)	$2,000	$400	$409	$418
Cost of goods sold (FIFO)	1,700	360	365	370
Difference	$ 300	$ 40	$ 44	$ 48
Revenues	$4,500	$875	$900	$950
Operating expenses	1,000	207	211	220

Air Parts has paid dividends of $40 million each year beginning in 2014. Its income tax rate is 25%. Retained earnings on January 1, 2019, was $700 million; inventory was $500 million.

LIFO usually produces higher cost of goods sold than does FIFO because more recently purchased goods (usually higher priced) are assumed sold first.

FINANCIAL Reporting Case

Q1, p. 1171

1. REVISE COMPARATIVE FINANCIAL STATEMENTS For each year reported in the comparative statements, Air Parts makes those statements appear as if the newly adopted accounting method (FIFO) had been applied all along. As you learned in Chapter 1, comparability is one of the important qualitative characteristics of accounting information.

When accounting changes occur, the usefulness of the comparative financial statements is enhanced with retrospective application of those changes.

Income Statements

($ in millions)	2019	2020	2021
Revenues	$875	$900	$950
Cost of goods sold (FIFO)	**(360)**	**(365)**	**(370)**
Operating expenses	(207)	(211)	(220)
Income before tax	308	324	360
Income tax expense (25%)	(77)	(81)	(90)
Net income	$231	$243	$270

The company recasts the comparative statements to appear as if the accounting method adopted in 2021 (FIFO) had been used in 2020 and 2019 as well.

Earnings per share each year, of course, also will be based on the revised net income numbers.

Balance Sheets

Inventory. In its comparative balance sheets, Air Parts will report 2021 **inventory** by its newly adopted method, FIFO, and revise the amounts it reported last year for its 2019 and 2020 inventory. Each year, inventory will be higher than it would have been by LIFO. Here's why:

FIFO usually produces *lower* cost of goods sold and thus *higher* inventory than does LIFO.

Since the cost of goods *available for sale* each period is the sum of the cost of goods *sold* and the cost of goods *unsold* (inventory), a difference in cost of goods sold resulting from having used LIFO rather than FIFO means there also is an opposite difference in inventory. Because cost of goods sold by the FIFO method is *less* than by LIFO, inventory by FIFO is *greater* than by LIFO. The amounts of the differences and also the cumulative differences over the years are calculated in Illustration 20–3A.

Retained earnings. Similarly, Air Parts will report **retained earnings** by FIFO each year as well. Retained earnings is different because the two inventory methods affect income differently. Because cost of goods sold by FIFO is *less* than by LIFO, income and therefore retained earnings by FIFO are *greater* than by LIFO.

When costs are rising, FIFO produces *lower* cost of goods sold than does LIFO and thus *higher* net income and retained earnings.

Comparative balance sheets, then, will report retained earnings for 2019, 2020, and 2021 at amounts $255, $288, and $324 million higher than would have been reported if the switch from LIFO had not occurred. These are the cumulative net income differences shown in Illustration 20–3A.

Retained earnings is revised each year to reflect FIFO.

Illustration 20–3A

Effects of Switch to FIFO

($ in millions)	Previous Years	2019	2020	2021
	Years Ending Dec. 31			
Cost of goods sold (LIFO)	$ 2,000	$400	$409	$418
Cost of goods sold (FIFO)	1,700	360	365	370
Differences	$ 300	$ 40	$ 44	$ 48
Cumulative differences (effect of not having used FIFO previously):				
Cost of goods sold (and inventory)	$ 300	**$340**	**$384**	**$432**
Income taxes (25%)	75	85	96	108
Net income and retained earnings	$ 225	$255	$288	$324

Comparative balance sheets, then, will report 2019 inventory **$340** million higher than it was reported in last year's statements. Likewise, 2020 inventory will be increased by **$384** million. Inventory for 2021, being reported for the first time, is **$432** million higher than it would have been if the switch from LIFO had not occurred.

Inventory, pretax income, income taxes, net income, and retained earnings all are higher.

By FIFO, cost of goods sold is lower.

The cumulative income effect increases each year by the annual after-tax difference in COGS.

Inventory, pretax income, income taxes, net income, and retained earnings all are higher.

Statements of shareholders' equity. Recall that a statement of shareholders' equity reports changes that occur in each shareholders' equity account starting with the beginning balances in the earliest year reported.

So, if retained earnings is one of the accounts whose balance requires adjustment due to a change in accounting principle (and it usually is), we must adjust the beginning balance of retained earnings for the earliest period reported in the comparative statements of shareholders' equity. The amount of the revision is the cumulative effect of the change on years prior to that date. Air Parts will revise its 2019 beginning retained earnings since that's the earliest year in its comparative statements. That balance had been reported in prior statements as $700 million. If FIFO had been used for inventory rather than LIFO, that amount would have been higher by **$225** million, as calculated in Illustration 20–3A. The disclosure note pertaining to the inventory change should point out the amount of the adjustment. The January 1, 2019, retained earnings balance reported in the comparative statements of shareholders' equity in Illustration 20–3B has been adjusted from $700 million to **$925** million.

Because it's the earliest year reported, 2019's beginning retained earnings is increased by the **$225** million cumulative income effect of the difference in inventory methods that occurred before 2019.

Illustration 20–3B

Comparative Statements of Shareholders' Equity

($ in millions)	Common Stock	Additional Paid-In Capital	Retained Earnings	Total Shareholders' Equity
Jan. 1, 2019			**$ 925**	
Net income (*revised* to FIFO)			231	
Dividends			(40)	
Dec. 31, 2019			$1,116	
Net income (*revised* to FIFO)			243	
Dividends			(40)	
Dec. 31, 2020			$1,319	
Net income (*using* FIFO)			270	
Dividends			(40)	
Dec. 31, 2021			$1,549	

A disclosure note should indicate that the beginning retained earnings balance has been increased by **$180** million, from **$700** million to **$880** million.

2. ADJUST ACCOUNTS FOR THE CHANGE Besides reporting revised amounts in the comparative financial statements, Air Parts must also adjust the book balances of affected accounts. It does so by creating a journal entry to change those balances from their current amounts (from using LIFO) to what those balances would have been using the newly adopted method (FIFO). As discussed in the previous section, differences in cost of goods sold and income are reflected in retained earnings, as are the income tax effects of changes in income. So, the journal entry updates inventory, retained earnings, and the income tax liability for the cumulative differences up to the year of the decision to change from the LIFO method to the FIFO method. Adjustments are made in the beginning of the year of change, so in our example, this would be for the differences generated up to January 1, 2021. Repeating a portion of the calculation we made in Illustration 20–3A, we determine the difference in cost of goods sold and therefore in inventory.

($ in millions)	Cumulative Difference pre-2019	2019	2020	Cumulative Difference pre-2021
Cost of goods sold (LIFO)	$2,000	$400	$409	
Cost of goods sold (FIFO)	1,700	360	365	
Difference	$ 300	$ 40	$ 44	**$384**

Cost of goods sold would have been $384 million *less* if FIFO rather than LIFO had been used in years before the change.

The $384 million cumulative difference in cost of goods sold also is the difference between the balance in inventory and what that balance would have been if the FIFO method, rather than LIFO, had been used before 2021. Inventory must be increased by that amount. Retained earnings must be increased also, but by only 75% of that amount because income taxes would have been higher by 25% of the change in pretax income.

Journal entry to record the change in principle

January 1, 2021		
Inventory (additional inventory if FIFO had been used)	384	
Retained earnings (additional net income if FIFO had been used)		288
Income tax payable (**$384** × 25%)		96

Inventory would have been $384 million more and cumulative prior earnings $288 more if FIFO rather than LIFO had been used.

Notice that the income tax effect is reflected in the income tax payable account. The reason is that, unlike for other accounting method changes, the Internal Revenue Code requires that the inventory costing method used for tax purposes must be the same as that used for financial reporting. For that reason, the tax code allows a retrospective change in an inventory method, but then requires that taxes saved previously ($96 million in this case) from having used another inventory method must now be repaid. However, taxpayers are given up to six years to pay the tax due. As a result, this liability has both a current portion (payable within one year) and a noncurrent portion (payable after one year), but is not a deferred tax liability.

Additional Consideration

What if the tax law did not require a recapture of the tax difference? Then there would be a credit to a deferred tax liability. That's because retrospectively increasing accounting income, but not taxable income, creates a temporary difference between the two that will reverse over time as the unsold inventory becomes cost of goods sold. Recall from Chapter 16 that in the meantime, there is a temporary difference reflected in the deferred tax liability.

If we were switching from FIFO to, say, the average method, we would record a deferred tax asset instead. For financial reporting purposes, but not for tax, we would be retrospectively decreasing accounting income, but not taxable income. This creates a temporary difference between the two that will reverse over time as the unsold inventory becomes cost of goods sold. When that happens, taxable income will be less than accounting income. When taxable income will be less than accounting income as a temporary difference reverses, we have a "future deductible amount" and record a deferred tax asset.

Note disclosure explains why the change was needed as well as its effects on items not reported on the face of the primary statements.

3. DISCLOSURE NOTES To achieve consistency and comparability, accounting choices once made should be consistently followed from year to year. Any change, then, requires that the new method be justified as clearly more appropriate. In the first set of financial statements after the change, a disclosure note is needed to provide that justification. The note also should point out that comparative information has been revised, or that retrospective revision has not been made because it is impracticable, and report any per share amounts affected for the current period and all prior periods presented. Disclosure of a recent change by **Abercrombie & Fitch** in a recent annual report provides us the example shown in Illustration 20–4.

Illustration 20–4
Disclosure of a Change in Inventory Method—Abercrombie & Fitch
Real World Financials

4. CHANGE IN ACCOUNTING PRINCIPLE

The Company elected to change its method of accounting for inventory from the lower of cost or market utilizing the retail method to the weighted average cost method. . . . In accordance with generally accepted accounting principles, all periods have been retroactively adjusted to reflect the period-specific effects of the change to the weighted average cost method. The Company believes that accounting under the weighted average cost method is preferable as it better aligns with the Company's focus on realized selling margin and improves the comparability of the Company's financial results with those of its competitors. Additionally, it will improve the matching of cost of goods sold with the related net sales and reflect the acquisition cost of inventory outstanding at each balance sheet date. The cumulative adjustment . . . was an increase in its inventory of $73.6 million and an increase in retained earnings of $47.3 million.

Source: Abercrombie & Fitch

The Modified Retrospective Approach

When a new FASB accounting standard update mandates a change in accounting principle, the Board typically specifies the way companies should implement the change and often allows companies to choose among multiple ways of accounting for the change. One approach usually allowed is the retrospective approach we discussed for voluntary changes in accounting principle, by which the new standard is applied retrospectively to all periods presented in the financial statements. And, as we discuss in the next section, sometimes the prospective approach is allowed or mandated for a new standard. A third approach the FASB sometimes allows is a *modified retrospective approach.* By this approach, we apply the new standard only to the adoption period (that is, the current period) and then adjust the balance of retained earnings at the beginning of the adoption period to capture the cumulative effects of prior periods without actually adjusting the numbers in the prior periods reported.

The Prospective Approach

● LO20–3

Although we usually report voluntary changes in accounting principles retrospectively, it's not always practicable or appropriate to do so.

Sometimes a lack of information makes it impracticable to report a change retrospectively so the new method is simply applied prospectively.

THE PROSPECTIVE APPROACH: WHEN RETROSPECTIVE APPLICATION IS IMPRACTICABLE For some changes in principle, insufficient information is available for retrospective application to be practicable. Revising balances in prior years means knowing what those balances should be. But suppose we're switching from the FIFO method of inventory costing to the LIFO method. Recall from your study of inventory costing methods that LIFO inventory consists of "layers" added in prior years at costs existing in those years. If another method has been used, though, the company likely hasn't kept track of those costs. So, accounting records of prior years usually are inadequate to report the change retrospectively. In that case, a company changing *to* LIFO usually reports the change prospectively, and the beginning inventory in the year the LIFO method is adopted becomes the base year inventory for all future LIFO calculations. A disclosure note should indicate reasons why retrospective application was impracticable. Prospective changes usually are accounted for as of the beginning of the year of change.

When **Books A Million, Inc.**, adopted the LIFO cost flow assumption for valuing its inventories, the change was reported in a disclosure note as shown in Illustration 20–5.

If it's impracticable to adjust each year reported, the change is applied retrospectively as of the earliest year practicable.

When it is impracticable to determine some period-specific effects. A company may have some, but not all, the information it needs to account for a change retrospectively.

Inventories (in part)

. . . the Company changed from the first-in, first-out (FIFO) method of accounting for inventories to the last-in, first-out (LIFO) method. Management believes this change was preferable in that it achieves a more appropriate matching of revenues and expenses. The impact of this accounting change was to increase "Costs of Products Sold" in the consolidated statements of operations by $0.7 million for the fiscal year. . . . The cumulative effect of a change in accounting principle from the FIFO method to LIFO method is not determinable. Accordingly, such change has been accounted for prospectively.

Source: Books A Million, Inc.

Illustration 20–5

Disclosure of a Change to LIFO—Books A Million, Inc.

Real World Financials

For instance, let's say a company changes to the LIFO inventory method effective as of the beginning of 2021. It has information that would allow it to revise all assets and liabilities on the basis of the newly adopted method for 2020 in its comparative statements, but not for 2019. In that case, the company should report 2020 statement amounts (revised) and 2021 statement amounts (reported without restatement for the first time) based on LIFO, but not revise 2019 numbers. Then, account balances should be retrospectively adjusted at the beginning of 2020 since that's the earliest date it's practicable to do so.

When it is impracticable to determine the cumulative effect of prior years. Another possibility is that the company doesn't have the information necessary to retrospectively adjust retained earnings, but does have information that would allow it to revise all assets and liabilities for one or more specific years. Let's say the records of inventory purchases and sales are not available for some previous years, which would have allowed it to determine the cumulative effect of applying this change to LIFO retrospectively. However, it does have all of the information necessary to apply the LIFO method on a prospective basis beginning in, say, 2019. In that case, the company should report numbers for years beginning in 2019 as if it had carried forward the 2018 ending balance in inventory (measured on the previous inventory costing basis) and then had begun applying LIFO as of January 1, 2019. Of course there would be no adjustment to retained earnings for the cumulative income effect of not using LIFO prior to that.

If full retrospective application isn't possible, the new method is applied prospectively beginning in the earliest year practicable.

THE PROSPECTIVE APPROACH: WHEN MANDATED BY AUTHORITATIVE ACCOUNTING LITERATURE Another exception to retrospective application of voluntary changes in accounting principle is when authoritative literature requires prospective application for specific changes in accounting methods. For instance, for a change from the equity method to another method of accounting for long-term investments, GAAP requires the prospective application of the new method.[7] Recall from Chapter 12 that when an investor's level of influence changes, it may be necessary to change from the equity method to another method. This could happen, for instance, if a sale of shares causes the investor's ownership interest to fall from, say, 25% to 15%, resulting in the equity method no longer being appropriate. When this situation happens, no adjustment is made to the remaining book value, sometimes called *carrying value* or *carrying amount* of the investment in equity affiliate account. Instead, the equity method is simply discontinued and the new method applied from then on in an investment in equity securities account. The balance in that investment account when the equity method is discontinued would serve as the new "cost" basis for writing the investment up or down to fair value on the next set of financial statements.

If a new accounting standards update specifically requires prospective accounting, that requirement is followed.

The prospective approach is also appropriate for a change *to* the equity method when, for instance, a purchase of additional stock causes the investor's share of ownership to increase from, say, 15% to 25%. The previous method is discontinued and the balance in the investment account at the date of the change (including any unrealized holding gains or losses that occurred prior to the date the investment qualifies for the equity method) is used as the starting balance for applying the equity method. Any cost of acquiring additional shares is added to that balance, and going forward that balance is adjusted for the investor's portion of investee earnings and dividends.

[7]FASB ASC 323–10–35–36: Investments—Equity Method and Joint Ventures—Overall—Subsequent Measurement—Decrease in Level of Ownership or Degree of Influence.

THE PROSPECTIVE APPROACH: CHANGING DEPRECIATION, AMORTIZATION, AND DEPLETION METHODS A change in depreciation methods is considered to be a change in accounting estimate that is achieved by a change in accounting principle. As a result, we account for such a change prospectively—precisely the way we account for changes in estimates. We discuss that approach in the next section.

Change in Accounting Estimate

LO20–4

You've encountered many instances during your study of accounting in which it's necessary to make estimates of uncertain future events. Depreciation, for example, entails estimates not only of the useful lives of depreciable assets, but their anticipated residual values as well. Anticipating uncollectible accounts receivable, predicting warranty expenses, amortizing intangible assets, and making actuarial assumptions for pension benefits are but a few of the accounting tasks that require estimates.

Revisions are viewed as a natural consequence of making estimates.

Estimates are an inherent and critical aspect of accounting. Unfortunately, though, estimates routinely differ from actual experience. IBM notes in the Management Discussion and Analysis section of its annual report that:

Real World Financials

> . . . For all of these estimates, it should be noted that future events rarely develop exactly as forecasted, and estimates require regular review and adjustment.". . . .
>
> Source: IBM Corporation

No matter how carefully known facts are considered and forecasts are prepared, new information and experience frequently force the revision of estimates. Of course, if the original estimate was based on erroneous information or calculations or was not made in good faith, the revision of that estimate constitutes the correction of an error.

A change in estimate is reflected in the financial statements of the current period and future periods.

Changes in accounting estimates are accounted for prospectively. When a company revises a previous estimate, prior financial statements are *not* revised. That is, they are not restated, and not retrospectively adjusted. Instead, the company merely incorporates the new estimate in any related accounting determinations from then on. So, it usually will affect some aspects of both the balance sheet and the income statement in the current period and future periods. If that effect is considered to be material or will materially affect future periods, a disclosure note should describe the effect of a change in estimate on income from continuing operations, net income, and related per share amounts for the current period.

When **Owens Corning** revised estimates of the useful lives of some of its depreciable assets, the change was disclosed in its annual report as shown in Illustration 20–6.

Illustration 20–6

Change in Estimate—Owens Corning

Real World Financials

> **Note 6: Depreciation of Plant and Equipment (in part)**
>
> . . . the Company completed a review of its fixed asset lives. The Company determined that as a result of actions taken to increase its preventative maintenance and programs initiated with its equipment suppliers to increase the quality of their products, actual lives for certain asset categories were generally longer than the useful lives for depreciation purposes. Therefore, the Company extended the estimated useful lives of certain categories of plant and equipment, effective. . . . The effect of this change in estimate reduced depreciation expense for the year ended . . . , by $14 million and increased income before cumulative effect of accounting change by $8 million ($.19 per share).
>
> Source: Owens-Corning Fiberglass Corporation

An example of another change in estimate is provided in Illustration 20–7.

The after-tax effect of the change in estimate is **$3** million [$400 million × (3% – 2%) = $4 million, less 25% of $4 million]. Assuming 100 million outstanding shares of common stock, the effect is described in a disclosure note to the financial statements as follows:

> **Note A: Warranties**
>
> In 2021, the company revised the percentage used to estimate warranty expense. The change provides a better indication of cost experience. The effect of the change was to decrease 2021 net income by **$3** million, or $0.03 per share.

Illustration 20–7
Change in Accounting Estimate

Universal Semiconductors estimates warranty expense as 2% of sales.

- After a review during 2021, Universal determined that 3% of sales is a more realistic estimate of its payment experience.
- Sales in 2021 are $400 million.
- The effective income tax rate is 25%.
- No account balances are adjusted.

Prior periods are not revised and no cumulative effect of the estimate change is reported. Rather, in 2021 and later years, the adjusting entry to record warranty expense simply will reflect the new percentage. In 2021, the entry would be as follows:

	($ in millions)	
Warranty expense (3% × $400 million)	12	
Warranty liability		12

Changing Depreciation, Amortization, and Depletion Methods

FINANCIAL Reporting Case

Q2, p. 1171

When a company acquires an asset that will provide benefits for several years, it allocates the cost of the asset over the asset's useful life. If the asset is a building, equipment, or other tangible asset, the allocation process is called *depreciation.* It's referred to as *amortization* if an intangible asset or *depletion* if a natural resource. In each case, estimates are essential to the allocation process. How long will benefits accrue? What will be the value of the asset when its use is discontinued? Will the benefits be realized evenly over the asset's life or will they be higher in some years than in others?

An exception to retrospective application of a change in accounting principle is a change in the method of depreciation (or amortization or depletion).

The choice of depreciation method and application reflects these estimates. Likewise, when a company changes the way it depreciates an asset in midstream, the change would be made to reflect a change in (a) estimated future benefits from the asset, (b) the pattern of receiving those benefits, or (c) the company's knowledge about those benefits. For instance, suppose Universal Semiconductors originally chose an accelerated depreciation method because it expected greater benefits in the earlier years of an asset's life. Then, two years later, when it became apparent that remaining benefits would be realized approximately evenly over the remaining useful life, Universal Semiconductor switched to straight-line depreciation. Even though the company is changing its depreciation method, it is doing so to reflect changes in its estimates of future benefits. As a result, *we report a change in depreciation method as a change in estimate, rather than as a change in accounting principle.*

Companies report a change in depreciation prospectively.

For this reason, a company reports a change in depreciation method (say to straight-line) *prospectively*; previous financial statements are not revised. Instead, the company simply employs the straight-line method from then on. The undepreciated cost remaining at the time of the change would be depreciated straight-line over the remaining useful life. Illustration 20–8 provides an example.

A company must justify any change in principle as preferable to the previous method.

Is a change in depreciation method a change in accounting principle, or is it a change in estimate? As we've seen, it's both. Even though it's considered to reflect a change in estimate and is accounted for as such, a change to a new depreciation method requires the company to justify the new method as being preferable to the previous method, just as for any other change in principle. A disclosure note should justify that the change is preferable and describe the effect of a change on any financial statement line items and per share amounts affected for all periods reported.

In practice, the situation arises infrequently. Most companies changing depreciation methods do not apply the change to existing assets, but instead to assets placed in service after that date. In those cases, of course, the new method is simply applied prospectively (see Illustration 20–9).

When it's not possible to distinguish between a change in principle and a change in estimate, the change should be treated as a change in estimate.

Sometimes, it's not easy to distinguish between a change in principle and a change in estimate. For example, if a company begins to capitalize rather than expense the cost of tools because their benefits beyond one year become apparent, the change could be construed as either a change in principle or a change in the estimated life of the asset. When the distinction is not possible, the change should be treated as a change in estimate. This treatment also is appropriate when both a change in principle and a change in estimate occur simultaneously.

Illustration 20–8
Change in Depreciation Methods

Universal Semiconductors switched from the sum-of-the-years'-digits depreciation method to straight-line depreciation in 2021.

- The change affects its precision equipment purchased at the beginning of 2019 at a cost of $63 million.
- The machinery has an expected useful life of five years and an estimated residual value of $3 million.

The depreciation prior to the change is as follows:

Sum-of-the-Years'-Digits Depreciation	($ in millions)
2019 depreciation	$20 ($60 × 5/15)
2020 depreciation	16 ($60 × 4/15)
Accumulated depreciation	$36

A change in depreciation method is considered a *change in accounting estimate resulting from a change in accounting principle.* So, Universal Semiconductors reports the change prospectively; previous financial statements are not revised. Instead, the company simply employs the straight-line method from 2021 on. The undepreciated cost remaining at the time of the change is depreciated straight-line over the remaining useful life.

The **$24** million depreciable cost not yet depreciated is spread over the asset's remaining three years.

Calculation of Straight-Line Depreciation	($ in millions)
Asset's cost	$63
Accumulated depreciation to date (calculated above)	(36)
Undepreciated cost, Jan. 1, 2021	$27
Estimated residual value	(3)
To be depreciated over remaining 3 years	**$24**
	3 years
Annual straight-line depreciation 2021–2024	$ 8

Adjusting entry (annually, "2021–2023")		
Depreciation expense (calculated above)	8	
Accumulated depreciation		8

Illustration 20–9
Change in Depreciation Method for Newly Acquired Assets—Rohm and Haas Company
Real World Financials

Note 12: Land, Buildings, and Equipment, Net (in part)

. . . the company changed its method of depreciation for newly acquired buildings and equipment to the straight-line method. The change had no cumulative effect on prior years' earnings but did increase [current year] net earnings by $9 million, or $0.14 per share. . . .

Source: Rohm and Haas Company

Change in Reporting Entity

● LO20–5

A reporting entity can be a single company, or it can be a group of companies that reports a single set of financial statements. For example, the consolidated financial statements of **PepsiCo Inc.** report the financial position and results of operations not only for the parent company, but also for its subsidiaries which include **Frito-Lay**, **Quaker**, **Tropicana**, and **Gatorade**. A **change in reporting entity** occurs as a result of (1) presenting consolidated financial statements in place of statements of individual companies or (2) changing specific companies that constitute the group for which consolidated or combined statements are prepared.[8]

Some changes in reporting entity are a result of changes in accounting rules. For example, companies like **Ford**, **General Motors**, and **IBM** must consolidate their manufacturing

[8]FASB ASC 810: Consolidation.

operations with their *financial* subsidiaries, creating a new entity that includes them both.[9] For those changes in entity, the prior-period financial statements that are presented for comparative purposes must be restated to appear as if the new entity existed in those periods.

However, the more frequent change in entity occurs when one company acquires another one. In those circumstances, the financial statements of the acquirer include the acquiree as of the date of acquisition, and the acquirer's prior-period financial statements that are presented for comparative purposes are not restated. This makes it difficult to make year-to-year comparisons for a company that frequently acquires other companies. Acquiring companies are required to provide a disclosure note that presents key financial statement information as if the acquisition had occurred before the beginning of the previous year. At a minimum, the supplemental pro forma information should display revenue, income from continuing operations, net income, and earnings per share.

A change in reporting entity is reported by recasting all previous periods' financial statements as if the new reporting entity existed in those periods.[10] In the first set of financial statements after the change, a disclosure note should describe the nature of the change and the reason it occurred. Also, the effect of the change on net income, income from continuing operations, and related per share amounts should be indicated for all periods presented. These disclosures aren't necessary in subsequent financial statements. **Hartford Life Insurance Company**, a financial services company, changed the composition of its reporting entity and described it this way (see Illustration 20–10):

A change in reporting entity requires that financial statements of prior periods be retrospectively revised to report the financial information for the new reporting entity in all periods.

Illustration 20–10

Change in Reporting Entity—Hartford Life Insurance

Real World Financials

1. Basis of Presentation and Accounting Policies (in part)

Hartford Life changed its reporting entity structure to contribute certain wholly owned subsidiaries, including Hartford Life's European insurance operations, several broker dealer entities and investment advisory and service entities from Hartford Life and Accident to the Company. The contribution of subsidiaries was effected to more closely align servicing entities with the writing company issuing the business they service as well as to more efficiently deploy capital across the organization. The change in reporting entity was retrospectively applied to the financial statements of the Company for all periods presented. The contributed subsidiaries resulted in an increase in stockholder's equity of approximately $1.3 billion.

Source: Hartford Life Insurance

Errors

The correction of an error is not actually an accounting change but is accounted for similarly. In fact, it's accounted for retrospectively, like a change in reporting entity and like most changes in accounting principle.

More specifically, previous years' financial statements that were incorrect as a result of the error are retrospectively restated to reflect the correction. And, of course, any account balances that are incorrect as a result of the error are corrected by a journal entry. If retained earnings is one of the incorrect accounts, the correction is reported as a prior period adjustment to the beginning balance in a statement of shareholders' equity (or statement of retained earnings if that's presented instead).[11] And, as for accounting changes, a disclosure note is needed to describe the nature of the error and the impact of its correction on operations. We discuss the correction of errors in more detail in Part B of this chapter. But first, let's compare the two approaches for reporting accounting changes and error corrections (see Illustration 20–11).

Previous years' financial statements are retrospectively restated to reflect the correction of an error.

[9]The issuance of *SFAS No. 94,* "Consolidation of All Majority-Owned Subsidiaries," [FASB ASC 810: Consolidation and 840: Leases] resulted in hundreds of entities consolidating previously unconsolidated finance subsidiaries.

[10]Any prior periods' statements are recast when those statements are presented again for comparative purposes.

[11]FASB ASC 250–10–45: Accounting Changes and Error Corrections—Overall—Other Presentation Matters.

Illustration 20–11

Approaches to Reporting Accounting Changes and Error Corrections

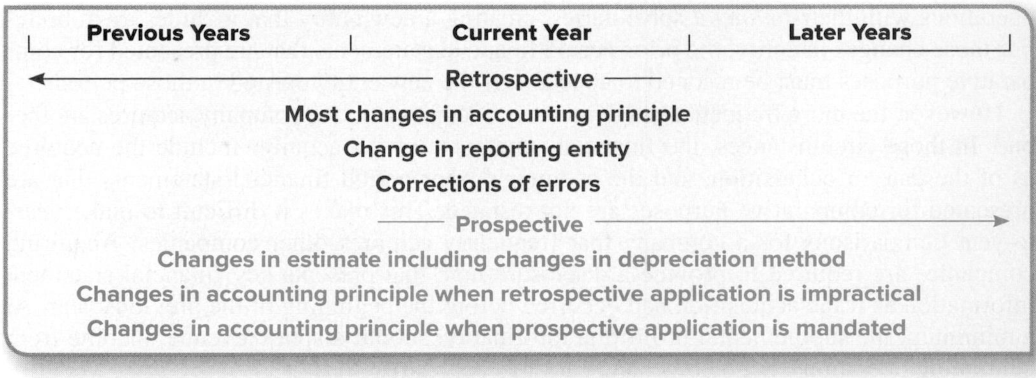

A comparison of accounting treatments is provided by Illustration 20–12.

Illustration 20–12 Accounting Changes and Errors: A Summary

	Change in Accounting Principle		Change in Estimate (including depreciation changes)	Change in Reporting Entity	Error
	Most Changes*	Exceptions†			
Method of accounting:	Retrospective	Prospective	Prospective	Retrospective	**Retrospective**
• Restate prior years' statements?	Yes	No	No	Yes	**Yes**
• Cumulative effect on prior years' income reported:	As adjustment to retained earnings of earliest year reported‡	Not reported	Not reported	As adjustment to retained earnings** of earliest year reported‡	**As adjustment to retained earnings of earliest year reported‡**
• Journal entry:	To adjust affected balances to new method	None, but subsequent accounting is affected by the change	None, but subsequent accounting is affected by the new estimate	Involves consolidated financial statements discussed in other courses	**To correct any balances that are incorrect as a result of the error**
• Disclosure note?	Yes	Yes	Yes	Yes	**Yes**

*Changes in depreciation, amortization, and depletion methods are considered changes in estimates.
†When retrospective application is impracticable, such as most changes to LIFO and certain mandated changes.
‡In the statement of shareholders' equity or statement of retained earnings.
**Except when one company acquires another one. Then the acquirer's prior-period financial statements that are presented for comparative purposes are not restated.

Concept Review Exercise

ACCOUNTING CHANGES

Modern Business Machines recently conducted an extensive review of its accounting and reporting policies. The following accounting changes are an outgrowth of that review:

1. MBM has a patent on a copier design. The patent has been amortized on a straight-line basis since it was acquired at a cost of $400,000 in 2018. During 2021, MBM decided that the benefits from the patent would be experienced over a total of 13 years rather than the 20-year legal life now being used to amortize its cost.

2. At the beginning of 2021, MBM changed its method of valuing inventory from the FIFO cost method to the average cost method. At December 31, 2020 and 2019, MBM's inventories were $560 and $540 million, respectively, on a FIFO cost basis but would have totaled $500 and $490 million, respectively, if determined on an average cost basis. MBM's income tax rate is 25%.

Required:

Prepare all journal entries needed in 2021 related to each change. Also, briefly describe any other measures MBM would take in connection with reporting the changes.

Solution:

1. **Change in estimate**

	($ in millions)	
Patent amortization expense (determined below)	34	
Patent		34

Calculation of Annual Amortization after the Estimate Change

	$400,000	Cost
$20,000		Old annual amortization ($400,000 ÷ 20 years)
× 3 years	(60,000)	Amortization to date (2018, 2019, 2020)
	340,000	Unamortized cost
	÷ 10	Estimated remaining life (13 years — 3 years)
	$ 34,000	New annual amortization

A disclosure note should describe the effect of a change in estimate on income from continuing operations, net income, and related per share amounts for the current period.

2. **Change in principle**

MBM creates a journal entry to bring up to date all account balances affected.

	($ in millions)	
Retained earnings (the difference in net income before 2021)	45	
Deferred tax asset ($60 million × 25%)	15	
Inventory ($560 million — $500 million)		60

For financial reporting purposes, but not for tax, MBM is retrospectively *decreasing* accounting income, but not taxable income. This creates a temporary difference between the two that will reverse over time as the unsold inventory becomes cost of goods sold. When that happens, taxable income will be less than accounting income. When taxable income will be less than accounting income as a temporary difference reverses, we have a "future deductible amount" and record a deferred tax asset.

Prior years' financial statements are revised to reflect the use of the new accounting method.

Also, MBM will revise all previous periods' financial statements (in this case 2020) as if the new method (average cost) were used in those periods. In other words, for each year in the comparative statements reported, the balance of each account affected will be revised to appear as if the average method had been applied all along.

Since it's the earliest year reported, 2020's beginning retained earnings is adjusted for the portion of the cumulative income effect of the change attributable to prior years.

Since retained earnings is one of the accounts whose balance requires adjustment (and it usually is), MBM makes an adjustment to the beginning balance of retained earnings for the earliest period (2020) reported in the comparative statements of shareholders' equity. Also, in the first set of financial statements after the change, a *disclosure note* describes the nature of the change, justifies management's decision to make the change, and indicates its effect on each item affected in the financial statements.

PART B

Correction of Accounting Errors

Nobody's perfect. People make mistakes, even accountants. When errors are discovered, they should be corrected.[12] Illustration 20–13 describes the steps to be taken to correct an error, if the effect of the error is material.[13]

● LO20–6

Illustration 20–13
Steps to Correct an Error

Step 1	A journal entry is made to correct any account balances that are incorrect as a result of the error.
Step 2	Previous years' financial statements that were incorrect as a result of the error are retrospectively restated to reflect the correction (for all years reported for comparative purposes).
Step 3	If retained earnings is one of the accounts incorrect as a result of the error, the correction is reported as a prior period adjustment to the beginning balance in a statement of shareholders' equity (or statement of retained earnings if that's presented instead).
Step 4	A disclosure note should describe the nature of the error and the impact of its correction on each financial statement line item and any per-share amounts affected for each prior period presented.

The retrospective approach is used for the correction of errors.

The correction of an error is treated as a prior period adjustment.

Prior Period Adjustments

Before we see these steps applied to the **correction of an error**, one of the steps requires elaboration. As discussed in Chapter 4, the correction of errors is the situation that creates prior period adjustments. A **prior period adjustment** refers to an addition to or reduction in the beginning retained earnings balance in a statement of shareholders' equity (or statement of retained earnings if that's presented instead).

In an earlier chapter, we saw that a statement of shareholders' equity is the most commonly used way to report the events that cause components of shareholders' equity to change during a particular reporting period. Some companies, though, choose to report the changes that occur in the balance of retained earnings separately in a statement of retained earnings. When it's discovered that the ending balance of retained earnings in the period prior to the discovery of an error was incorrect as a result of that error, the balance must be corrected when it appears as the beginning balance the following year. However, simply reporting a corrected amount might cause misunderstanding for someone familiar with the previously reported amount. Explicitly reporting a prior period adjustment on the statement itself avoids this confusion. The prior period adjustment is reported the same way whether the change in retained earnings is reported as a column in a statement of shareholders' equity or as a separate statement of retained earnings. For simplicity, our example assumes comparative statements of retained earnings:

A statement of retained earnings (or retained earnings column in a statement of shareholders' equity) reports the events that cause changes in retained earnings.

STATEMENTS OF RETAINED EARNINGS
For the Years Ended December 31, 2020 and 2019

	2020	**2019**
Balance at beginning of year	$600,000	$450,000
Net income	400,000	350,000
Less: Dividends	(200,000)	(200,000)
Balance at end of year	$800,000	$600,000

[12]Interestingly, it appears that not all accounting errors are unintentional. Research has shown that firms with errors that overstate income are more likely "to have diffuse ownership, lower growth in earnings and fewer income-increasing GAAP alternatives available, and are less likely to have audit committees," suggesting that "overstatement errors are the result of managers responding to economic incentives." M. L. DeFond and J. Jiambaolvo, "Incidence and Circumstances of Accounting Errors," *The Accounting Review,* July 1991.

[13]In practice, the vast majority of errors are not material with respect to their effect on the financial statements and are, therefore, simply corrected in the year discovered (step 1 only).

Now suppose that in 2021 it's discovered that an error in 2019 caused that year's net income to be overstated by **$20,000** (net income should have been $330,000). This means retained earnings in both prior years were overstated. Because comparative financial statements are presented and the current year is the year in which the error was discovered, the prior year would include a prior period adjustment as shown below:

STATEMENTS OF RETAINED EARNINGS
For the Years Ended December 31, 2021 and 2020

	2021	2020
Balance at beginning of year	$ 780,000	$600,000
Prior period adjustment		**(20,000)**
Corrected balance		$580,000
Net Income	500,000	400,000
Less: Dividends	(200,000)	(200,000)
Balance at end of year	$1,080,000	$780,000

The incorrect balance as previously reported is corrected by the prior period adjustment.

At least two years' (as in our example) and often three years' statements are reported in comparative financial statements. The prior period adjustment is applied to beginning retained earnings for the year following the error, or for the earliest year being reported in the comparative financial statements when the error occurs prior to the earliest year presented.[14]

Error Correction Illustrated

Now, let's discuss these procedures to correct errors in the context of a variety of the most common types of errors. Since there are literally thousands of possibilities, it's not practical to describe every error in every stage of its discovery. However, by applying the process to the situations described below, you should become sufficiently comfortable with the *process* that you could apply it to whatever situation you might encounter.

You shouldn't try to memorize how specific errors are corrected; you should learn the process needed to analyze whatever errors you might encounter.

As you study these examples, be sure to notice that it's significantly more complicated to deal with an error if (a) it affected net income in the reporting period in which it occurred and (b) it is not discovered until a later period.

Error Discovered in the Same Reporting Period That It Occurred

If an accounting error is made and discovered in the same accounting period, the original erroneous entry should simply be reversed and the appropriate entry recorded. The possibilities are limitless. Let's look at the one in Illustration 20–14.

Illustration 20–14
Error Discovered in the Same Reporting Period That It Occurred

G. H. Little, Inc., paid $3 million for replacement computers and recorded the expenditure as maintenance expense. The error was discovered a week later.

To Reverse the Erroneous Entry	($ in millions)	
Cash	3	
Maintenance expense		3
To Record the Correct Entry		
Equipment	3	
Cash		3

Note: These entries can, of course, be combined.

Error Affecting Previous Financial Statements, but Not Net Income

If an error did *not* affect net income in the year it occurred, it's relatively easy to correct. Examples are incorrectly recording salaries payable as accounts payable, recording a loss as an expense, or classifying a cash flow as an investing activity rather than a financing activity

[14] The retained earnings balances in years after the first year also are adjusted to what those balances would be if the error had not occurred, but a company may choose not to explicitly report those adjustments as separate line items

on the statement of cash flows. The 2017 financial statements of **Advanced Drainage Systems** announced that the company will restate its prior period financial statements and related financial information as filed with the Securities and Exchange Commission. The disclosure note reporting that restatement is reproduced in Illustration 20–15.

Illustration 20–15
Error Correction; Advanced Drainage Systems
Real World Financials

23. Restatement of Prior Period Financial Statements (in part)
Subsequent to the issuance of the Original Form 10-K, the Company identified errors in its historical consolidated financial statements related to the accounting for stock-based compensation for awards made to employees along with its accounting for certain executive stock repurchase agreements and executive termination payments, as described below.

As a result, the Company has restated its consolidated financial statements for the fiscal years ended March 31, 2016, 2015 and 2014. The restatement also affects periods prior to fiscal year 2014, with the cumulative effect of the errors reflected as an adjustment to the fiscal year 2014 opening stockholders' equity (deficit) balance.

Illustration 20–16 provides another example.

Illustration 20–16
Error Affecting Previous Financial Statements, but Not Net Income

MDS Transportation incorrectly recorded a $2 million note receivable as accounts receivable. The error was discovered a year later.

To Correct Incorrect Accounts	($ in millions)	
Notes receivable	2	
Accounts receivable		2

Step 1

Step 2 When reported for comparative purposes in the current year's annual report, last year's balance sheet would be restated to report the note as it should have been reported last year.

Step 3 Since last year's net income was not affected by the error, the balance in retained earnings was not incorrect. So no prior period adjustment to that account is necessary.

Step 4 A disclosure note would describe the nature of the error, but there would be no impact on net income, income from continuing operations, and earnings per share to report.

Error Affecting a Prior Year's Net Income

Most errors affect net income in some way. When they do, they affect the balance sheet as well. Both statements must be retrospectively restated; the statement of cash flows sometimes is affected, too. As with any error, all incorrect account balances must be corrected. Because these errors affect income, one of the balances that will require correction is retained earnings. Complicating matters, income taxes often are affected by income errors. In those cases, amended tax returns are prepared either to pay additional taxes or to claim a tax refund for taxes overpaid.

In Illustration 20–17, except as indicated, we ignore the tax effects of the errors and their correction to allow us to focus on the errors themselves rather than their tax aspects.

The effect of most errors is different, depending on *when* the error is discovered. For example, if the error in Illustration 20–17 is not discovered until 2022, rather than 2021, accumulated depreciation would be understated by another $1.4 million, or a total of **$4.2** million. If not discovered until 2024 or after, no correcting entry at all would be needed. By then, the sum of the omitted depreciation amounts ($1.4 million × 5 years) would equal the expense incorrectly recorded in 2019 ($7 million), so the retained earnings balance would be the same as if the error never had occurred. Also, the asset may have been disposed of—if the useful life estimate was correct—so neither the equipment nor accumulated depreciation would need to be recorded. Of course, any statements of prior years that were affected and are reported again in comparative statements still would be restated, and a disclosure note would describe the error and the impact of its correction on each financial statement line item and any per-share amounts affected for each prior period presented.

Most errors, in fact, eventually self-correct. An example of an uncommon instance in which an error never self-corrects would be an expense account debited for the cost of land. Because land doesn't depreciate, the error would continue until the land is sold.

Illustration 20–17

Error Affecting Net Income: Recording an Asset as an Expense

In 2021, internal auditors discovered that Seidman Distribution, Inc., had debited an expense account for the $7 million cost of sorting equipment purchased at the beginning of 2019. The equipment's useful life was expected to be five years with no residual value. Straight-line depreciation is used by Seidman.

Analysis

Correct (Should have been recorded)	($ in millions)		**Incorrect** (As recorded)	($ in millions)	
2019 Equipment	7.0		Expense	7.0	
Cash		7.0	Cash		7.0
2019 Expense	1.4		*Depreciation entry omitted*		
Accum. deprec.		1.4			
2020 Expense	1.4		*Depreciation entry omitted*		
Accum. deprec.		1.4			

During the two-year period, depreciation expense was understated by $2.8 million, but other expenses were overstated by $7 million, so net income during the period was understated by **$4.2** million. This means retained earnings is currently understated by that amount.

Accumulated depreciation is understated by $2.8 million.

To Correct Incorrect Accounts	($ in millions)	
Equipment	7.0	
Accumulated depreciation		2.8
Retained earnings		**4.2**

Step 1

The 2019 and 2020 financial statements that were incorrect as a result of the error are retrospectively restated to report the equipment acquired and to reflect the correct amount of depreciation expense and accumulated depreciation, assuming both statements are reported again for comparative purposes in the 2021 annual report.

Step 2

Because retained earnings is one of the accounts that is incorrect as a result of the error, a correction to that account of **$4.2** million is reported as a prior period adjustment to the 2021 beginning retained earnings balance in Seidman's comparative statements of shareholders' equity. A correction would be made also to the 2020 beginning retained earnings balance. That prior period adjustment, though, would be for the pre-2020 difference: $7 million – $1.4 million = $5.6 million. If 2019 statements also are included in the comparative report, no adjustment would be necessary for that period because the error didn't occur until after the beginning of 2019.

Step 3

Also, a disclosure note accompanying Seidman's 2021 financial statements should describe the nature of the error and the impact of its correction on each financial statement line item and any per-share amounts affected for each prior period presented (net income understated by $5.6 million in 2019 and overstated by $1.4 million in 2020).

Step 4

Additional Consideration

We ignored the tax impact of the error and its correction in Illustration 20–17. To consider taxes, we need to know whether depreciation was also omitted from the tax return and the depreciation methods used for tax reporting. Let's say that depreciation was omitted from the tax return also, and that straight-line depreciation is used by Seidman for both tax and financial reporting. The tax rate is 25%.

Total operating expenses (nontax) still would have been overstated by $4.2 million over the two-year period. But that would have caused taxable income to be understated and the tax liability and income tax expense to be understated by 25% of **$4.2** million, or $1.05 million. So net income and retained earnings would have been understated by only $3.15 million.

Operating expenses *overstated*	$ **4.20** million
Income tax expense *understated*	(1.05) million
Net income (and retained earnings) *understated*	$ 3.15 million

(continued)

(concluded)

To Correct Incorrect Accounts	($ in millions)	
Equipment	7.00	
Accumulated depreciation		2.80
Income tax payable (25% × **$4.2** million)		1.05
Retained earnings		3.15

If depreciation had been omitted from the income statement but not from the tax return, or if accelerated depreciation was used for tax reporting but straight-line depreciation for financial reporting, the credit to income tax payable in the correcting entry would be replaced by a credit to deferred tax liability.

Even errors that eventually correct themselves cause financial statements to be misstated in the meantime.

Some errors correct themselves the following year. For instance, if a company's ending inventory is incorrectly counted or otherwise misstated, the income statement would be in error for the year of the error and the following year, but the balance sheet would be incorrect only for the year the error occurs. After that, all account balances will be correct. This is demonstrated in Illustration 20–18.

Illustration 20–18

Error Affecting Net Income: Inventory Misstated

When analyzing inventory errors or other errors that affect cost of goods sold, you may find it helpful to visualize the determination of cost of goods sold, net income, and retained earnings.

In early 2021, Overseas Wholesale Supply discovered that $1 million of inventory had been inadvertently excluded from its 2019 ending inventory count.

Analysis:
U = Understated O = Overstated

2019		**2020**	
Beginning inventory		→ **Beginning inventory**	U
Plus: Net purchases		Plus: Net purchases	
Less: Ending inventory	U	Less: Ending inventory	
Cost of goods sold	O	Cost of goods sold	U
Revenues		Revenues	
Less: Cost of goods sold	O	Less: Cost of goods sold	U
Less: Other expenses		Less: Other expenses	
Net income	U	Net income	O
↓		↓	
Retained earnings	U	Retained earnings	*corrected*

If Error Is Discovered in 2020 (before closing)	($ in millions)	
Inventory	1	
Retained earnings		1

If Error Discovered in 2021 or Later
No correcting entry needed.

Step 1

Step 2

If the error is discovered in 2020, the 2019 financial statements that were incorrect as a result of the error are retrospectively restated to reflect the correct inventory amounts, cost of goods sold, and retained earnings when those statements are reported again for comparative purposes in the 2020 annual report. If the error is discovered in 2021, the 2020 financial statements also are retrospectively restated to reflect the correct inventory amounts and cost of goods sold (retained earnings would not require adjustment), even though no correcting entry would be needed at that point.

Step 3

Because retained earnings is one of the accounts incorrect if the error is discovered in 2020, the correction to that account is reported as a prior period adjustment to the 2020 beginning retained earnings balance in Overseas' statement of shareholders' equity. Of course, no prior period adjustment is needed if the error isn't discovered until 2021 or later.

Step 4

Also, a disclosure note in Overseas' annual report should describe the nature of the error and the impact of its correction on each year's net income (understated by $1 million in 2019, overstated by $1 million in 2020), income from continuing operations (same as net income), and earnings per share.

Other error corrections that benefit from a similar analysis are the overstatement of ending inventory, the overstatement or understatement of beginning inventory, and errors in recording merchandise purchases (or returns).

An error also would occur if a revenue or an expense is recorded in the wrong accounting period. Illustration 20–19 offers an example.

Illustration 20–19

Error Affecting Net Income: Failure to Record Sales Revenue

In 2021, General Paper Company discovered that $3,000 of merchandise (credit) sales the last week of 2020 were not recorded until the first week of 2021. The merchandise sold was appropriately excluded from 2020 ending inventory.

Analysis

	Correct (Should have been recorded)	($ in thousands)		**Incorrect** (As recorded)	($ in thousands)	
2020	Accounts receivable	3		No entry		
	Sales revenue		3			
2021	No entry			Accounts receivable	3	
				Sales revenue		3

2020 sales revenue was incorrectly recorded in 2021, so 2020 net income was understated. Retained earnings is currently understated in 2021. 2021 sales revenue is overstated.

Step 1

To Correct Incorrect Accounts	($ in thousands)	
Sales revenue	3	
Retained earnings		3

Note: If the sales revenue had not been recorded at all, the correcting entry would include a debit to accounts receivable rather than sales revenue.

Step 2

The 2020 financial statements that were incorrect as a result of the error are retroactively restated to reflect the correct amount of sales revenue and accounts receivable when those statements are reported again for comparative purposes in the 2021 annual report.

Step 3

Because retained earnings is one of the accounts incorrect as a result of the error, the correction to that account is reported as a prior period adjustment to the 2021 beginning retained earnings balance in General Paper's comparative statements of shareholders' equity.

Step 4

Also, a disclosure note in General Paper's 2021 annual report should describe the nature of the error and the impact of its correction on each financial statement line item (net income understated by $3,000 in 2020) and any per-share amounts affected for each prior period presented.

Ethical Dilemma

As a second-year accountant for McCormack Chemical Company, you were excited to be named assistant manager of the Agricultural Chemicals Division. After two weeks in your new position, you were supervising the year-end inventory count when the senior manager mentioned that two carloads of herbicides were omitted from the count and should be added. Upon checking, you confirm your understanding that the inventory in question had been deemed to be unsalable. "Yes," your manager agreed, "but we'll write that off next year when our bottom line won't be so critical to the continued existence of the Agricultural Chemicals Division. Jobs and families depend on our division showing well this year."

Hertz Global Holding, Inc. operates car rental businesses through the Hertz, Dollar, Thrifty, and Firefly brands. Illustration 20–20 reports information from a disclosure note in Hertz's annual report issued July 16, 2015. It shows how Hertz corrected its financial statements primarily because of depreciation errors on the sale of vehicles at retail locations.

Illustration 20–20
Error Correction; Hertz
Real World Financials

Note 2—Restatement (in part)

During the fourth quarter of 2013, the Company identified certain out of period misstatements totaling $46 million, of which $35 million ($21 million, net of tax) related to its previously issued consolidated financial statements for the years ended December 31, 2012, and prior. While these misstatements did not, individually or in the aggregate, result in a material misstatement of the Company's previously issued consolidated financial statements, correcting these misstatements in the fourth quarter of 2013 would have been material to that quarter . . . this Note 2 to the consolidated financial statements discloses the nature of the restatement matters and adjustments and shows the impact of the restatement matters on revenues, expenses, income, assets, liabilities, equity, and cash flows from operating activities, investing activities, and financing activities, and the cumulative effects of these adjustments on the consolidated statement of operations, balance sheet, and cash flows for 2012 and 2013. In addition, this Note shows the effects of the adjustment to opening retained earnings as of January 1, 2012, which adjustment reflects the impact of the restatement on periods prior to 2012. The cumulative impact of the out of period misstatements for all previously reported periods through December 31, 2013, including amounts associated with the revision previously reported in the 2013 Form 10-K/A, was approximately a $349 million reduction in pre-tax income and $231 million reduction in net income.

Source: Hertz

As mentioned at the outset, we've made no attempt to demonstrate the correction process for every kind of error in every stage of its discovery. However, after seeing the process applied to the few situations described, you should feel comfortable that the process is the same regardless of the specific situation you might encounter.

International Financial Reporting Standards

● LO20–7

Accounting Changes and Error Corrections. U.S. GAAP and International standards are largely converged regarding accounting changes and error corrections, but one difference concerns error corrections. When correcting errors in previously issued financial statements, IFRS (*IAS No. 8)* permits the effect of the error to be reported in the current period if it's not considered practicable to report it retrospectively, as is required by U.S. GAAP.[15]

Concept Review Exercise

CORRECTION OF ERRORS

In 2021, the following errors were discovered by the internal auditors of Development Technologies, Inc.

1. 2020 accrued salaries of $2 million were not recognized until they were paid in 2021.
2. A $3 million purchase of merchandise in 2021 was recorded in 2020 instead. The physical inventory count at the end of 2020 was correct.

Required:

Prepare the journal entries needed in 2021 to correct each error. Also, briefly describe any other measures Development Technologies would take in connection with correcting the errors. (Ignore income taxes.)

Solution:

Step 1:

1. To reduce 2021 salaries expense and reduce retained earnings to what it would have been if the expense had reduced net income in 2020.

[15]"Accounting Policies, Changes in Accounting Estimates and Errors," *International Accounting Standard No. 8* (IASCF), as amended effective January 1, 2014.

	($ in millions)	
Retained earnings	2	
Salaries expense		2

2. To include the $3 million in 2021 purchases and increase retained earnings to what it would have been if 2020 cost of goods sold had not included the $3 million purchases.

Analysis
U = Understated O = Overstated

2020		2021	
Beginning inventory		Beginning inventory	
Purchases	O	Purchases	U
Less: Ending inventory			
Cost of goods sold	O		
Revenues	‖		
Less: Cost of goods sold	O		
Less: Other expenses			
Net income	U		
Retained earnings	U		

	($ in millions)	
Purchases	3	
Retained earnings		3

Step 2:
The 2020 financial statements that were incorrect as a result of the errors would be *retrospectively restated* to reflect the correct salaries expense, cost of goods sold (income tax expense if taxes are considered), net income, and retained earnings when those statements are reported again for comparative purposes in the 2021 annual report.

Step 3:
Because retained earnings is one of the accounts that is incorrect, the correction to that account is reported as a *prior period adjustment* to the 2021 beginning retained earnings balance in the comparative statements of shareholders' equity.

Step 4:
Also, a *disclosure note* should describe the nature of the error and the impact of its correction on each financial statement line item and any per-share amounts affected for each prior period presented.

Financial Reporting Case Solution

©TY Lim/Shutterstock

1. **How can an accounting change cause a company to increase a previously reported inventory amount?** *(p. 1175)* Smucker didn't find any lost jelly. The company increased last year's inventory number by $2 million to reflect its change from *LIFO* to *FIFO* this year. If it had not revised the number, last year's inventory would be based on *LIFO* and this year's inventory on *FIFO*. Analysts would be comparing apples and oranges (or apple jelly and orange jelly). Retrospective application of an accounting change provides better comparability in accounting information.

2. **Are all accounting changes reported this way?** *(p. 1181)* Not all accounting changes are reported retrospectively. Besides most changes in accounting principle, changes in reporting entity and the correction of errors are reported that way, but some changes are reported prospectively instead. Changes in depreciation method, changes in accounting estimate, and some changes for which retrospective application is either impracticable or prohibited are reported prospectively in current and future periods only. ●

The Bottom Line

- **LO20–1** Accounting changes are categorized as (a) changes in *principle,* (b) changes in *estimates,* or (c) changes in *reporting entity.* Accounting changes can be accounted for retrospectively (prior years revised) or prospectively (only year of change and future years affected), or by the modified retrospective approach (apply a new standard only to the adoption period). (*p. 1172*)
- **LO20–2** Most voluntary changes in accounting principles are reported retrospectively. This means for each year reported in the comparative statements, we make those statements appear as if the newly adopted accounting method had been applied all along. A journal entry is created to adjust all account balances affected as of the date of the change. In the first set of financial statements after the change, a disclosure note describes the change and justifies the new method as preferable. It also describes the effects of the change on all items affected, including the fact that the retained earnings balance was revised in the statement of shareholders' equity. (*p. 1174*)
- **LO20–3** Some changes are reported prospectively. These include (a) changes in the method of depreciation, amortization, or depletion; (b) some changes in principle for which retrospective application is impracticable; and (c) a few changes for which an authoritative pronouncement requires prospective application. (*p. 1178*)
- **LO20–4** Changes in estimates are accounted for prospectively. When a company revises a previous estimate, prior financial statements are not revised. Instead, the company merely incorporates the new estimate in any related accounting determinations from then on. (*p. 1180*)
- **LO20–5** A change in reporting entity requires that financial statements of prior periods be retrospectively revised to report the financial information for the new reporting entity in all periods. (*p. 1182*)
- **LO20–6** When errors are discovered, they should be corrected and accounted for retrospectively. Previous years' financial statements that were incorrect as a result of an error are retrospectively restated, and any account balances that are incorrect are corrected by a journal entry. If retained earnings is one of the incorrect accounts, the correction is reported as a prior period adjustment to the beginning balance in a statement of shareholders' equity. And, a disclosure note should describe the nature of the error and the impact of its correction on operations. (*p. 1186*)
- **LO20–7** U.S. GAAP and international standards are largely converged with respect to accounting changes and error corrections. One remaining difference is that when correcting errors in previously issued financial statements, IFRS permits the effect of the error to be reported in the current period if it's not considered practicable to report it retrospectively, as is required by U.S. GAAP. (*p. 1192*) ●

Questions For Review of Key Topics

Q 20–1 For accounting purposes, we classify accounting changes into three categories. What are they? Provide a short description of each.

Q 20–2 There are three basic accounting approaches to reporting accounting changes. What are they?

Q 20–3 We report most changes in accounting principle retrospectively. Describe this general way of recording and reporting changes in accounting principle.

Q 20–4 Lynch Corporation changes from the sum-of-the-years'-digits method of depreciation for existing assets to the straight-line method. How should the change be reported? Explain.

Q 20–5 Sugarbaker Designs Inc. changed from the FIFO inventory costing method to the average cost method during 2021. Which items from the 2020 financial statements should be restated on the basis of the average cost method when reported in the 2021 comparative financial statements?

Q 20–6 Most changes in accounting principles are recorded and reported retrospectively. In a few situations, though, the changes should be reported prospectively. When is prospective application appropriate? Provide examples.

Q 20–7 Southeast Steel, Inc., changed from the FIFO inventory costing method to the LIFO method during 2020. How would this change likely be reported in the 2021 comparative financial statements?

Q 20–8 Direct Assurance Company revised the estimates of the useful life of a trademark it had acquired three years earlier. How should Direct account for the change?

Q 20–9 It's not easy sometimes to distinguish between a change in principle and a change in estimate. In these cases, how should the change be accounted for?

Q 20–10 For financial reporting, a reporting entity can be a single company, or it can be a group of companies that reports a single set of financial statements. When changes occur that cause the financial statements to be those of a different reporting entity, we account for the situation as a change in reporting entity. What are the situations deemed to constitute a change in reporting entity?

Q 20–11 The issuance of FASB guidance regarding consolidation of all majority-owned subsidiaries required **Ford Motors** to include a previously unconsolidated finance subsidiary as part of the reporting entity. How did Ford report the change?

Q 20–12 Describe the process of correcting an error when it's discovered in a subsequent reporting period.

Q 20–13 If merchandise inventory is understated at the end of 2020, and the error is not discovered, how will net income be affected in 2021?

Q 20–14 If it is discovered that an extraordinary repair in the previous year was incorrectly debited to repair expense, how will retained earnings be reported in the current year's statement of shareholders' equity?

Q 20–15 What action is required when it is discovered that a five-year insurance premium payment of $50,000 two years ago was debited to insurance expense? (Ignore taxes.)

Q 20–16 Suppose the error described in the previous question is not discovered until six years later. What action will the discovery of this error require?

IFRS **Q 20–17** With regard to the correction of accounting errors, what is the difference between U.S. GAAP and IFRS?

Brief Exercises

connect

BE 20–1
Change in inventory methods: FIFO method to the average cost method
● LO20–2

In 2021, the Barton and Barton Company changed its method of valuing inventory from the FIFO method to the average cost method. At December 31, 2020, B & B's inventories were $32 million (FIFO). B & B's records indicated that the inventories would have totaled $23.8 million at December 31, 2020, if determined on an average cost basis. Ignoring income taxes, what journal entry will B & B use to record the adjustment in 2021? Briefly describe other steps B & B should take to report the change.

BE 20–2
Change in inventory methods: average cost method to the FIFO method
● LO20–2

In 2021, Adonis Industries changed its method of valuing inventory from the average cost method to the FIFO method. At December 31, 2020, Adonis's inventories were $47.6 million (average cost). Adonis's records indicated that the inventories would have totaled $64 million at December 31, 2020, if determined on a FIFO basis. Ignoring income taxes, what journal entry will Adonis use to record the adjustment in 2021?

BE 20–3
Change in inventory methods: FIFO method to the LIFO method
● LO20–3

In 2021, J J Dishes changed its method of valuing inventory from the FIFO method to the LIFO method. At December 31, 2020, J J's inventories were $96 million (FIFO). J J's records were insufficient to determine what inventories would have totaled if determined on a LIFO cost basis. Briefly describe the steps J J should take to report the change.

BE 20–4
Change in depreciation methods
● LO20–3

Irwin, Inc. constructed a machine at a total cost of $35 million. Construction was completed at the end of 2017 and the machine was placed in service at the beginning of 2018. The machine was being depreciated over a 10-year life using the sum-of-the-years'-digits method. The residual value is expected to be $2 million. At the beginning of 2021, Irwin decided to change to the straight-line method. Ignoring income taxes, what journal entry(s) should Irwin record relating to the machine for 2021?

BE 20–5
Change in depreciation methods
● LO20–3

Refer to the situation described in BE 20–4. Suppose Irwin has been using the straight-line method and switches to the sum-of-the-years'-digits method. Ignoring income taxes, what journal entry(s) should Irwin record relating to the machine for 2021?

BE 20–6
Book royalties
● LO20–4

Three programmers at Feenix Computer Storage, Inc., write an operating systems control manual for Hill-McGraw Publishing, Inc., for which Feenix receives royalties equal to 12% of net sales. Royalties are payable annually on February 1 for sales the previous year. The editor indicated to Feenix on December 31, 2021, that book sales subject to royalties for the year just ended are expected to be $300,000. Accordingly, Feenix accrued royalty revenue of $36,000 at December 31 and received royalties of $36,500 on February 1, 2022. What adjustments, if any, should be made to retained earnings or to the 2021 financial statements?

BE 20–7
Warranty expense
● LO20–4

In 2020, Quapau Products introduced a new line of hot water heaters that carry a one-year warranty against manufacturer's defects. Based on industry experience, warranty costs were expected to approximate 5% of sales revenue. First-year sales of the heaters were $300,000. An evaluation of the company's claims experience in late 2021 indicated that actual claims were less than expected—4% of sales rather than 5%. Assuming sales of the heaters in 2021 were $350,000 and warranty expenditures in 2021 totaled $12,000, what is the 2021 warranty expense?

BE 20–8
Change in estimate; useful life of patent
● LO20–4

Van Frank Telecommunications has a patent on a cellular transmission process. The company has amortized the $18 million cost of the patent on a straight-line basis since it was acquired at the beginning of 2017. Due to rapid technological advances in the industry, management decided that the patent would benefit the company over a total of six years rather than the nine-year life being used to amortize its cost. The decision was made at the end of 2021 (before adjusting and closing entries). What is the appropriate adjusting entry for patent amortization in 2021 to reflect the revised estimate?

BE 20–9
Error correction
● LO20–6

When DeSoto Water Works purchased a machine at the end of 2020 at a cost of $65,000, the company debited Buildings and credited Cash $65,000. The error was discovered in 2021. What journal entry will DeSoto use to correct the error? What other step(s) would be taken in connection with the error?

BE 20–10
Error correction
● LO20–6

In 2021, internal auditors discovered that PKE Displays, Inc., had debited an expense account for the $350,000 cost of a machine purchased on January 1, 2018. The machine's useful life was expected to be five years with no residual value. Straight-line depreciation is used by PKE. Ignoring income taxes, what journal entry will PKE use to correct the error?

BE 20–11
Error correction
● LO20–6

Refer to the situation described in BE 20–10. Assume the error was discovered in 2023, after the 2022 financial statements are issued. Ignoring income taxes, what journal entry will PKE use to correct the error?

BE 20–12
Error correction
● LO20–6

In 2021, the internal auditors of Development Technologies, Inc., discovered that (a) 2020 accrued salaries of $2 million were not recognized until they were paid in 2021 and (b) a $3 million purchase of merchandise in 2021 was recorded as a debit to Purchases in 2020 instead. The physical inventory count at the end of 2020 was correct. Ignoring income taxes, what journal entries are needed in 2021 to correct each error? Also, briefly describe any other measures Development Technologies would take in connection with correcting the errors.

Exercises

connect

E 20–1
Change in principle; change in inventory methods
● LO20–2

During 2019 (its first year of operations) and 2020, Fieri Foods used the FIFO inventory costing method for both financial reporting and tax purposes. At the beginning of 2021, Fieri decided to change to the average method for both financial reporting and tax purposes.

Income components before income tax for 2019, 2020, and 2021 were as follows:

($ in millions)	2019	2020	2021
Revenues	$ 380	$ 390	$ 420
Cost of goods sold (FIFO)	(38)	(40)	(45)
Cost of goods sold (average)	(52)	(56)	(62)
Operating expenses	(242)	(250)	(254)

Dividends of $20 million were paid each year. Fieri's fiscal year ends December 31.

Required:

1. Prepare the journal entry at the beginning of 2021 to record the change in accounting principle. (Ignore income taxes.)
2. Prepare the 2021–2020 comparative income statements.
3. Determine the balance in retained earnings at January 1, 2020, as Fieri reported previously using the FIFO method.
4. Determine the adjustment to the January 1, 2020, balance in retained earnings that Fieri would include in the 2021–2020 comparative statements of retained earnings or retained earnings column of the statements of shareholders' equity to revise it to the amount it would have been if Fieri had used the average method.

E 20–2
Change in principle; change in inventory methods
● LO20–2

Aquatic Equipment Corporation decided to switch from the LIFO method of costing inventories to the FIFO method at the beginning of 2021. The inventory as reported at the end of 2020 using LIFO would have been $60,000 higher using FIFO. Retained earnings at the end of 2020 was reported as $780,000 (reflecting the LIFO method). The tax rate is 25%.

Required:

1. Calculate the balance in retained earnings at the time of the change (beginning of 2021) as it would have been reported if FIFO had been used in prior years.
2. Prepare the journal entry at the beginning of 2021 to record the change in accounting principle.

E 20–3
Change from the treasury stock method to retired stock
● LO20–2

In keeping with a modernization of corporate statutes in its home state, UMC Corporation decided in 2021 to discontinue accounting for reacquired shares as treasury stock. Instead, shares repurchased will be viewed as having been retired, reassuming the status of unissued shares. As part of the change, treasury shares held were reclassified as retired stock. At December 31, 2020, UMC's balance sheet reported the following shareholders' equity:

	($ in millions)
Common stock, $1 par	$ 200
Paid-in capital—excess of par	800
Retained earnings	956
Treasury stock (4 million shares at cost)	(25)
Total shareholders' equity	$1,931

Required:

Identify the type of accounting change this decision represents and prepare the journal entry to effect the reclassification of treasury shares as retired shares.

E 20–4
Change in principle: change to the equity method
● LO20–2

The Crump Companies, Inc. has ownership interests in several public companies. At the beginning of 2021, the company's ownership interest in the common stock of Silken Properties increased to the point that it became appropriate to begin using the equity method of accounting for the investment. The balance in the investment in equity securities account was $31 million at the time of the change. Accountants working with company records determined that the balance in an investment in equity affiliate account would have been $48 million if the equity method had been used previously.

Required:

1. Will Crump apply the new method retrospectively or apply the new method prospectively?
2. Suppose Crump is changing *from* the equity method rather than *to* the equity method. Will Crump apply the new method retrospectively or prospectively?

E 20–5
FASB codification research; change in accounting for investments
● LO20–2

Companies often invest in the common stock of other corporations. The way we report these investments depends on the nature of the investment and the investor's motivation for the investment. The *FASB Accounting Standards Codification* represents the single source of authoritative U.S. generally accepted accounting principles. Obtain the relevant authoritative literature on the disclosure of accounting policies using the *FASB Accounting Standards Codification* at the FASB website (**www.fasb.org**).

Required:

1. What is the specific nine-digit Codification citation (XXX-XX-XX-XX) that describes how to account for a change in the level of ownership to a percentage that will mandate use of the equity method for investments in common stock?
2. What are the specific requirements?

E 20–6
FASB codification research
● LO20–2

Access the *FASB Accounting Standards Codification* at the FASB website (**www.fasb.org**). Determine the specific eight-digit Codification citation (XXX-XX-XX-X) for accounting for each of the following items:

1. Reporting most changes in accounting principle.
2. Disclosure requirements for a change in accounting principle.
3. Illustration of the application of a retrospective change in the method of accounting for inventory.

E 20–7
Change in principle; change in inventory cost method
● LO20–2

Millington Materials is a leading supplier of building equipment, building products, materials, and timber for sale, with over 200 branches across the Mid-South. On January 1, 2021, management decided to change from the average inventory costing method to the FIFO inventory costing method at each of its outlets.

The following table presents information concerning the change. The income tax rate for all years is 25%.

	Income before Income Tax		
	FIFO	**Average Cost**	**Difference**
Before 2020	$30 million	$16 million	$14 million
2020	16 million	10 million	6 million
2021	20 million	18 million	2 million

Required:

1. Prepare the journal entry to record the change in accounting principle.
2. Determine the net income to be reported in the 2021–2020 comparative income statements.
3. Which other 2020 amounts would be reported differently in the 2021–2020 comparative income statements and 2021–2020 comparative balance sheets than they were reported the previous year?
4. How would the change be reflected in the 2021–2020 comparative statements of shareholders' equity assuming cash dividends were $2 million each year and that no dividends were paid prior to 2020?

E 20–8
Change in inventory methods; FIFO method to the LIFO method
● LO20–3

Flay Foods has always used the FIFO inventory costing method for both financial reporting and tax purposes. At the beginning of 2021, Flay decided to change to the LIFO method. As a result of the change, net income in 2021 was $80 million. If the company had used LIFO in 2020, its cost of goods sold would have been higher by $6 million that year. Flay's records of inventory purchases and sales are not available for 2019 and several previous years. Last year, Flay reported the following net income amounts in its comparative income statements:

($ in millions)	2018	2019	2020
Net income	$80	$82	$84

Required:

1. Prepare the journal entry at the beginning of 2021 to record the change in accounting principle. (Ignore income taxes.)
2. Will Flay apply the LIFO cost method retrospectively or apply the LIFO cost method prospectively?
3. What amounts will Flay report for net income in its 2019–2021 comparative income statements?

E 20–9
Change in inventory methods; FIFO method to the LIFO method
● LO20–3

Wolfgang Kitchens has always used the FIFO inventory costing method for both financial reporting and tax purposes. At the beginning of 2021, Wolfgang decided to change to the LIFO method. Net income in 2021 was correctly stated as $90 million. If the company had used LIFO in 2020, its cost of goods sold would have been higher by $7 million that year. Company accountants are able to determine that the cumulative net income for all years prior to 2020 would have been lower by $23 million if LIFO had been used all along, but have insufficient information to determine specific effects of using LIFO in 2019. Last year, Wolfgang reported the following net income amounts in its comparative income statements:

($ in millions)	2018	2019	2020
Net income	$90	$92	$94

Required:

1. Prepare the journal entry at the beginning of 2021 to record the change in accounting principle. (Ignore income taxes.)
2. Wolfgang should revise reported account balances retrospectively as of the beginning of what year?
3. What amounts will Wolfgang report for net income in its 2019–2021 comparative income statements?

E 20–10
Change in depreciation methods
● LO20–3

For financial reporting, Clinton Poultry Farms has used the declining-balance method of depreciation for conveyor equipment acquired at the beginning of 2018 for $2,560,000. Its useful life was estimated to be six years with a $160,000 residual value. At the beginning of 2021, Clinton decides to change to the straight-line method. The effect of this change on depreciation for each year is as follows:

($ in thousands)

Year	Straight-Line	Declining Balance	Difference
2018	$ 400	$ 853	$453
2019	400	569	169
2020	400	379	(21)
	$1,200	$1,801	$601

Required:

1. Will Clinton apply the straight-line method retrospectively or apply the straight-line method prospectively?
2. Prepare any 2021 journal entry related to the change.

E 20–11
Change in depreciation methods
● LO20–3

The Canliss Milling Company purchased machinery on January 2 2019, for $800,000. A five-year life was estimated and no residual value was anticipated. Canliss decided to use the straight-line depreciation method and recorded $160,000 in depreciation in 2019 and 2020. Early in 2021, the company changed its depreciation method to the sum-of-the-years'-digits (SYD) method.

Required:

1. Will Canliss apply the straight-line method retrospectively or apply the method prospectively?
2. Prepare any 2021 journal entry related to the change.

E 20–12
Book royalties
● LO20–4

Dreighton Engineering Group receives royalties on a technical manual written by two of its engineers and sold to William B. Irving Publishing, Inc. Royalties are 10% of net sales, receivable on October 1 for sales in January through June and on April 1 for sales in July through December of the prior year. Sales of the manual began in July 2020, and Dreighton accrued royalty revenue of $31,000 at December 31, 2020, as follows:

Receivable—royalty revenue	31,000	
Royalty revenue		31,000

Dreighton received royalties of $36,000 on April 1, 2021, and $40,000 on October 1, 2021. Irving indicated to Dreighton on December 31 that book sales subject to royalties for the second half of 2021 are expected to be $500,000.

Required:

1. Prepare any journal entries Dreighton should record during 2021 related to the royalty revenue.
2. What is the amount of the adjustment, if any, that should be made to retained earnings in the 2020 financial statements?

E 20–13
Loss contingency
● LO20–4

The Commonwealth of Virginia filed suit in October 2019, against Northern Timber Corporation seeking civil penalties and injunctive relief for violations of environmental laws regulating forest conservation. When the financial statements were issued in 2020, Northern had not reached a settlement with state authorities, but legal counsel advised Northern Timber that it was probable the ultimate settlement would be $1,000,000 in penalties. The following entry was recorded:

Loss—litigation	1,000,000	
Liability—litigation		1,000,000

Late in 2021, a settlement was reached with state authorities to pay a total of $600,000 to cover the cost of violations.

Required:

1. Prepare any journal entry(s) related to the change.
2. Is Northern required to revise prior years' financial statements as a result of the change?
3. Is Northern required to provide a disclosure note to report the change?

E 20–14
Warranty expense
● LO20–4

Woodmier Lawn Products introduced a new line of commercial sprinklers in 2020 that carry a one-year warranty against manufacturer's defects. Because this was the first product for which the company offered a warranty, trade publications were consulted to determine the experience of others in the industry. Based on that experience, warranty costs were expected to approximate 2% of sales. Sales of the sprinklers in 2020 were $2,500,000. Accordingly, the following entries relating to the contingency for warranty costs were recorded during the first year of selling the product:

Accrued liability and expense		
Warranty expense (2% × $2,500,000)	50,000	
Warranty liability		50,000
Actual expenditures (summary entry)		
Warranty liability	23,000	
Cash (or salaries payable, parts and supplies, etc.)		23,000

In late 2021, the company's claims experience was evaluated and it was determined that claims were far more than expected—3% of sales rather than 2%.

Required:

1. Assuming sales of the sprinklers in 2021 were $3,600,000 and warranty expenditures in 2021 totaled $88,000, prepare any journal entries related to the warranty.
2. Assuming sales of the sprinklers were discontinued after 2020, prepare any journal entries in 2021 related to the warranty.

E 20–15
Deferred taxes; change in tax rates
● LO20–4

Bronson Industries reported a deferred tax liability of $5 million for the year ended December 31, 2020, related to a temporary difference of $20 million. The tax rate was 25%. The temporary difference is expected to reverse in 2022, at which time the deferred tax liability will become payable. There are no other temporary differences in 2020–2022. Assume a new tax law is enacted in 2021 that causes the tax rate to change from 25% to 20% beginning in 2022. (The rate remains 25% for 2021 taxes.) Taxable income in 2021 is $60 million.

Required:

1. Determine the effect of the change and prepare the appropriate journal entry to record Bronson's income tax expense in 2021.
2. Is Bronson required to revise prior years' financial statements as a result of the change?
3. Is Bronson required to provide a disclosure note to report the change?

E 20–16
Accounting change
● LO20–4

The Peridot Company purchased machinery on January 2, 2019, for $800,000. A five-year life was estimated and no residual value was anticipated. Peridot decided to use the straight-line depreciation method and recorded $160,000 in depreciation in 2019 and 2020. Early in 2021, the company revised the total estimated life of the machinery to eight years.

Required:

1. What type of change is this?
2. Is Peridot required to revise prior years' financial statements as a result of the change?
3. Is Peridot required to provide a disclosure note to report the change?
4. Determine depreciation for 2021.

E 20–17
Change in estimate; useful life and residual value of equipment
● LO20–4

Wardell Company purchased a mini computer on January 1, 2019, at a cost of $40,000. The computer has been depreciated using the straight-line method over an estimated five-year useful life with an estimated residual value of $4,000. On January 1, 2021, the estimate of useful life was changed to a total of 10 years, and the estimate of residual value was changed to $900.

Required:

1. Prepare the appropriate adjusting entry for depreciation in 2021 to reflect the revised estimate.
2. Prepare the appropriate adjusting entry for depreciation in 2021 to reflect the revised estimate, assuming that the company uses the sum-of-the-years'-digits method instead of the straight-line method.

E 20–18
Classifying accounting changes
● LO20–1 through LO20–5

Indicate with the appropriate letter the nature of each situation described below:

Type of Change	
PR	Change in principle reported retrospectively
PP	Change in principle reported prospectively
E	Change in estimate
EP	Change in estimate resulting from a change in principle
R	Change in reporting entity
N	Not an accounting change

_____ 1. Change from declining balance depreciation to straight-line.
_____ 2. Change in the estimated useful life of office equipment.
_____ 3. Technological advance that renders worthless a patent with an unamortized cost of $45,000.
_____ 4. Change from determining lower of cost or net realizable value (LCNRV) for the inventories by the individual item approach to the aggregate approach.
_____ 5. Change from LIFO inventory costing to the weighted-average inventory costing.
_____ 6. Settling a lawsuit for less than the amount accrued previously as a loss contingency.
_____ 7. Including in the consolidated financial statements a subsidiary acquired several years earlier that was appropriately not included in previous years.
_____ 8. Change by a retail store from reporting warranty expense on a pay-as-you-go basis to estimating the expense in the period of sale.
_____ 9. A shift of certain manufacturing overhead costs to inventory that previously were expensed as incurred to more accurately measure cost of goods sold. (Either method is generally acceptable.)
_____ 10. Pension plan assets for a defined benefit pension plan achieving a rate of return in excess of the amount anticipated.

E 20–19
Error correction; inventory error
● LO20–6

During 2021, WMC Corporation discovered that its ending inventories reported on its financial statements were misstated by the following amounts:

2019	understated by	$120,000
2020	overstated by	150,000

WMC uses the periodic inventory system and the FIFO cost method.

Required:

1. Determine the effect of these errors on retained earnings at January 1, 2021, before any adjustments. (Ignore income taxes.)
2. Prepare a journal entry to correct the error in 2021.
3. Will WMC account for the error (a) retrospectively or (b) prospectively?

E 20–20
Error corrections; investment
● LO20–6

On December 12, 2021, an investment in equity securities costing $80,000 was sold for $100,000. The total of the sale proceeds was credited to the investment in equity securities account.

Required:

1. Prepare the journal entry to correct the error, assuming it is discovered before the books are adjusted or closed in 2021. (Ignore income taxes.)
2. Prepare the journal entry to correct the error assuming it is not discovered until early 2022. (Ignore income taxes.)

E 20–21
Error in amortization schedule
● LO20–6

Wilkins Food Products Inc. acquired a packaging machine from Lawrence Specialists Corporation. Lawrence completed construction of the machine on January 1, 2019. In payment for the machine Wilkins issued a three-year installment note to be paid in three equal payments at the end of each year. The payments include interest at the rate of 10%. Lawrence made a conceptual error in preparing the amortization schedule, which Wilkins failed to discover until 2021. As a result of the error, Wilkins understated interest expense by $45,000 in 2019 and $40,000 in 2020.

Required:

1. Determine which accounts are incorrect as a result of these errors at January 1, 2021, before any adjustments. Explain your answer. (Ignore income taxes.)
2. Prepare a journal entry to correct the error.
3. Will Wilkins account for the error (a) retrospectively or (b) prospectively?

E 20–22
Error correction; accrued interest on bonds
● LO20–6

At the end of 2020, Majors Furniture Company failed to accrue $61,000 of interest expense that accrued during the last five months of 2020 on bonds payable. The bonds mature in 2032. The discount on the bonds is amortized by the straight-line method. The following entry was recorded on February 1, 2021, when the semiannual interest was paid:

Interest expense	73,200	
Discount on bonds payable		1,200
Cash		72,000

Required:

Prepare any journal entry necessary to correct the error, as well as any adjusting entry for 2021 related to the situation described. (Ignore income taxes.)

E 20–23
Error correction; three errors
● LO20–6

Below are three independent and unrelated errors.

a. On December 31, 2020, Wolfe-Bache Corporation failed to accrue salaries expense of $1,800. In January 2021, when it paid employees for the December 27–January 2 workweek, Wolfe-Bache made the following entry:

Salaries expense	2,520	
Cash		2,520

b. On the last day of 2020, Midwest Importers received a $90,000 prepayment from a tenant for 2021 rent of a building. Midwest recorded the receipt as rent revenue. The error was discovered midway through 2021.

c. At the end of 2020, Dinkins-Lowery Corporation failed to accrue interest of $8,000 on a note receivable. At the beginning of 2021, when the company received the cash, it was recorded as interest revenue.

Required:

For each error:

1. What would be the effect of each error on the income statement and the balance sheet in the 2020 financial statements?
2. Prepare any journal entries each company should record in 2021 to correct the errors.

E 20–24
Inventory errors
● LO20–6

For each of the following inventory errors occurring in 2021, determine the effect of the error on 2021's cost of goods sold, net income, and retained earnings. Assume that the error is not discovered until 2022 and that a periodic inventory system is used. Ignore income taxes.

U = Understated O = Overstated NE = No effect

	Cost of Goods Sold	Net Income	Retained Earnings
1. Overstatement of ending inventory.	U	O	O
2. Overstatement of purchases.			
3. Understatement of beginning inventory.			
4. Freight-in charges are understated.			
5. Understatement of ending inventory.			
6. Understatement of purchases.			
7. Overstatement of beginning inventory.			
8. Understatement of purchases and understatement of ending inventory, by the same amount.			

E 20–25
Classifying accounting changes and errors
● LO20–1 through LO20–6

Indicate with the appropriate letter the nature of each adjustment described below:

Type of Adjustment

A. Change in accounting principle (reported retrospectively).
B. Change in accounting principle (exception reported prospectively).
C. Change in estimate.
D. Change in estimate resulting from a change in principle.
E. Change in reporting entity.
F. Correction of an error.

_____ 1. Change from expensing extraordinary repairs to capitalizing the expenditures.
_____ 2. Change in the residual value of machinery.
_____ 3. Change from FIFO inventory costing to LIFO inventory costing.
_____ 4. Change in the percentage used to determine warranty expense.
_____ 5. Change from LIFO inventory costing to FIFO inventory costing.
_____ 6. Change from reporting an investment by the equity method due to a reduction in the percentage of shares owned.
_____ 7. Change in the composition of a group of firms reporting on a consolidated basis.
_____ 8. Change from sum-of-the-years'-digits depreciation to straight-line.
_____ 9. Change from FIFO inventory costing to average inventory costing.
_____ 10. Change in actuarial assumptions for a defined benefit pension plan.

Problems

P 20–1
Change in inventory costing methods; comparative income statements
● LO20–2

The Cecil-Booker Vending Company changed its method of valuing inventory from the average cost method to the FIFO cost method at the beginning of 2021. At December 31, 2020, inventories were $120,000 (average cost basis) and were $124,000 a year earlier. Cecil-Booker's accountants determined that the inventories would have totaled $155,000 at December 31, 2020, and $160,000 at December 31, 2019, if determined on a FIFO basis. A tax rate of 25% is in effect for all years.

One hundred thousand common shares were outstanding each year. Income from continuing operations was $400,000 in 2020 and $525,000 in 2021. There were no discontinued operations either year.

Required:

1. Prepare the journal entry at January 1, 2021, to record the change in accounting principle. (All tax effects should be reflected in the deferred tax liability account.)
2. Prepare the 2021–2020 comparative income statements beginning with income from continuing operations (adjusted for any revisions). Include per share amounts.

P 20–2
Change in principle; change in method of accounting for long-term construction
● LO20–2

The Pyramid Company has used the LIFO method of accounting for inventory during its first two years of operation, 2019 and 2020. At the beginning of 2021, Pyramid decided to change to the average cost method for both tax and financial reporting purposes. The following table presents information concerning the change for 2019–2021. The income tax rate for all years is 25%.

	Income before Income Tax				
	Using Average Cost Method	Using LIFO Method	Difference	Income Tax Effect	Difference after Tax
2019	$ 90,000	$60,000	$30,000	$7,500	$22,500
2020	45,000	36,000	9,000	2,250	6,750
Total	$135,000	$96,000	$39,000	$9,750	$29,250
2021	$ 51,000	$46,000	$ 5,000	$1,250	$ 3,750

Pyramid issued 50,000 $1 par, common shares for $230,000 when the business began, and there have been no changes in paid-in capital since then. Dividends were not paid the first year, but $10,000 cash dividends were paid in both 2020 and 2021.

Required:

1. Prepare the journal entry at January 1, 2021, to record the change in accounting principle.
2. Prepare the 2021–2020 comparative income statements beginning with income before income taxes.
3. Prepare the 2021–2020 comparative statements of shareholders' equity. [Hint: The 2019 statements reported retained earnings of $45,000. This is $60,000 – ($60,000 × 25%).]

P 20–3
Change in inventory costing methods comparative income statements
● LO20–2, LO20–3

Shown below are net income amounts as they would be determined by Weihrich Steel Company by each of three different inventory costing methods ($ in thousands).

	FIFO	Average Cost	LIFO
Pre-2020	$2,800	$2,540	$2,280
2020	750	600	540
	$3,550	$3,140	$2,820

Required:

1. Assume that Weihrich used FIFO before 2021, and then in 2021 decided to switch to average cost. Prepare the journal entry to record the change in accounting principle and briefly describe any other steps Weihrich should take to appropriately report the situation. (Ignore income tax effects.)
2. Assume that Weihrich used FIFO before 2021, and then in 2021 decided to switch to LIFO. Assume accounting records are inadequate to determine LIFO information prior to 2021. Therefore, the 2020 ($540) and pre-2020 ($2,280) data are not available. Prepare the journal entry to record the change in accounting principle and briefly describe any other steps Weihrich should take to appropriately report the situation. (Ignore income tax effects.)

3. Assume that Weihrich used FIFO before 2021, and then in 2021 decided to switch to LIFO cost. Weihrich's records of inventory purchases and sales are not available for several previous years. Therefore, the pre-2020 LIFO information ($2,280) is not available. However, Weihrich does have the information needed to apply LIFO on a prospective basis beginning in 2020. Prepare the journal entry to record the change in accounting principle, and briefly describe any other steps Weihrich should take to appropriately report the situation. (Ignore income tax effects.)

P 20–4
Change in inventory methods
● LO20–2

The Rockwell Corporation uses a periodic inventory system and has used the FIFO cost method since inception of the company in 1982. In 2021, the company decided to change to the average cost method. Data for 2021 are as follows:

Beginning inventory, FIFO (5,000 units @ $30.00)		$150,000
Purchases:		
5,000 units @ $36.00	$180,000	
5,000 units @ $40.00	200,000	380,000
Cost of goods available for sale		$530,000
Sales for 2021 (8,000 units @ $70.00)		$560,000

Additional Information:

1. The company's effective income tax rate is 25% for all years.
2. If the company had used the average cost method prior to 2021, ending inventory for 2020 would have been $130,000.
3. 7,000 units remained in inventory at the end of 2021.

Required:

1. Prepare the journal entry at the beginning of 2021 to record the change in principle.
2. In the 2021–2019 comparative financial statements, what will be the amounts of cost of goods sold and inventory reported for 2021?

P 20–5
Change in inventory methods
● LO20–2

Fantasy Fashions had used the LIFO method of costing inventories, but at the beginning of 2021 decided to change to the FIFO method. The inventory as reported at the end of 2020 using LIFO would have been $20 million higher using FIFO.

Retained earnings reported at the end of 2019 and 2020 was $240 million and $260 million, respectively (reflecting the LIFO method). Those amounts reflecting the FIFO method would have been $250 million and $272 million, respectively. 2020 net income reported at the end of 2020 was $28 million (LIFO method) but would have been $30 million using FIFO. After changing to FIFO, 2021 net income was $36 million. Dividends of $8 million were paid each year. The tax rate is 25%.

Required:

1. Prepare the journal entry at the beginning of 2021 to record the change in accounting principle.
2. In the 2021–2020 comparative income statements, what will be the amounts of net income reported for 2020 and 2021?
3. Prepare the 2021–2020 retained earnings column of the comparative statements of shareholders' equity.

P 20–6
Change in principle; change in depreciation methods
● LO20–3

During 2019 and 2020, Faulkner Manufacturing used the sum-of-the-years'-digits (SYD) method of depreciation for its depreciable assets, for both financial reporting and tax purposes. At the beginning of 2021, Faulkner decided to change to the straight-line method for both financial reporting and tax purposes. A tax rate of 25% is in effect for all years.

For an asset that cost $21,000 with an estimated residual value of $1,000 and an estimated useful life of 10 years, the depreciation under different methods is as follows:

Year	Straight Line	SYD	Difference
2019	$2,000	$3,636	$1,636
2020	2,000	3,273	1,273
	$4,000	$6,909	$2,909

Required:

1. Describe the way Faulkner should account for the change described. Include in your answer any journal entry Faulkner will record in 2021 related to the change and any required note disclosures.
2. Suppose instead that Faulkner previously used straight-line depreciation and changed to sum-of-the-years'-digits in 2021. Prepare any journal entry Faulkner will record in 2021 related to the change and any required note disclosures.

P 20–7
Depletion; change in estimate
● LO20–4

In 2021, the Marion Company purchased land containing a mineral mine for $1,600,000. Additional costs of $600,000 were incurred to develop the mine. Geologists estimated that 400,000 tons of ore would be extracted. After the ore is removed, the land will have a resale value of $100,000.

To aid in the extraction, Marion built various structures and small storage buildings on the site at a cost of $150,000. These structures have a useful life of 10 years. The structures cannot be moved after the ore has been removed and will be left at the site. In addition, new equipment costing $80,000 was purchased and installed at the site. Marion does not plan to move the equipment to another site, but estimates that it can be sold at auction for $4,000 after the mining project is completed.

In 2021, 50,000 tons of ore were extracted and sold. In 2022, the estimate of total tons of ore in the mine was revised from 400,000 to 487,500. During 2022, 80,000 tons were extracted.

Required:

1. Compute depletion and depreciation of the mine and the mining facilities and equipment for 2021 and 2022. Marion uses the units-of-production method to determine depreciation on mining facilities and equipment.
2. Compute the book value of the mineral mine, structures, and equipment as of December 31, 2022.

P 20–8
Accounting changes; six situations
● LO20–1 LO20–3 LO20–4

Described below are six independent and unrelated situations involving accounting changes. Each change occurs during 2021 before any adjusting entries or closing entries were prepared. Assume the tax rate for each company is 25% in all years. Any tax effects should be adjusted through the deferred tax liability account.

a. Fleming Home Products introduced a new line of commercial awnings in 2020 that carry a one-year warranty against manufacturer's defects. Based on industry experience, warranty costs were expected to approximate 3% of sales. Sales of the awnings in 2020 were $3,500,000. Accordingly, warranty expense and a warranty liability of $105,000 were recorded in 2020. In late 2021, the company's claims experience was evaluated, and it was determined that claims were far fewer than expected: 2% of sales rather than 3%. Sales of the awnings in 2021 were $4,000,000, and warranty expenditures in 2021 totaled $91,000.

b. On December 30, 2017, Rival Industries acquired its office building at a cost of $1,000,000. It was depreciated on a straight-line basis assuming a useful life of 40 years and no salvage value. However, plans were finalized in 2021 to relocate the company headquarters at the end of 2025. The vacated office building will have a salvage value at that time of $700,000.

c. Hobbs-Barto Merchandising, Inc., changed inventory cost methods to LIFO from FIFO at the end of 2021 for both financial statement and income tax purposes. Under FIFO, the inventory at January 1, 2021, is $690,000.

d. At the beginning of 2018, the Hoffman Group purchased office equipment at a cost of $330,000. Its useful life was estimated to be 10 years with no salvage value. The equipment was depreciated by the sum-of-the-years'-digits method. On January 1, 2021, the company changed to the straight-line method.

e. In November 2019, the State of Minnesota filed suit against Huggins Manufacturing Company, seeking penalties for violations of clean air laws. When the financial statements were issued in 2020, Huggins had not reached a settlement with state authorities, but legal counsel advised Huggins that it was probable the company would have to pay $200,000 in penalties. Accordingly, the following entry was recorded:

Loss—litigation	200,000	
Liability—litigation		200,000

Late in 2021, a settlement was reached with state authorities to pay a total of $350,000 in penalties.

f. At the beginning of 2021, Jantzen Specialties, which uses the sum-of-the-years'-digits method, changed to the straight-line method for newly acquired buildings and equipment. The change increased current year net earnings by $445,000.

Required:

For each situation:

1. Identify the type of change.
2. Prepare any journal entry necessary as a direct result of the change, as well as any adjusting entry for 2021 related to the situation described.
3. Briefly describe any other steps that should be taken to appropriately report the situation.

P 20–9
Accounting changes; identify type and reporting approach

At the beginning of 2021, Wagner Implements undertook a variety of changes in accounting methods, corrected several errors, and instituted new accounting policies.

Required:

Indicate for each item 1 to 10 below the type of change and the reporting approach Wagner would use.

● LO20–1 through LO20–4

Type of Change (choose one)	Reporting Approach (choose one)
P. Change in accounting principle	R. Retrospective approach
E. Change in accounting estimate	P. Prospective approach
EP. Change in estimate resulting from a change in principle	
X. Correction of an error	
N. Neither an accounting change nor an accounting error	

Change:

1. By acquiring additional stock, Wagner increased its investment in Wise, Inc., from a 12% interest to 25% and changed its method of accounting for the investment to the equity method.
2. Wagner instituted a postretirement benefit plan for its employees in 2021. Wagner did not previously have such a plan.
3. Wagner changed its method of depreciating computer equipment from the SYD method to the straight-line method.
4. Wagner determined that a liability insurance premium it both paid and expensed in 2020 covered the 2020–2022 period.
5. By selling shares in Launch Corp, Wagner decreased its investment in the company from a 23% interest to 15% and changed its method of accounting for the investment from the equity method to the fair value through net income method.
6. Due to an unexpected relocation, Wagner determined that its office building, previously depreciated using a 45-year life, should be depreciated using an 18-year life.
7. Wagner offers a three-year warranty on the farming equipment it sells. Manufacturing efficiencies caused Wagner to reduce its expectation of warranty costs from 2% of sales to 1% of sales.
8. Wagner changed from LIFO to FIFO to account for its materials and work-in-process inventories.
9. Wagner changed from FIFO to average cost to account for its equipment inventory.
10. Wagner sells extended service contracts on some of its equipment sold. Wagner performs services related to these contracts over several years, so in 2021, Wagner changed from recognizing revenue from these service contracts on a cash basis to the accrual basis.

P 20–10
Inventory errors
● LO20–6

You have been hired as the new controller for the Ralston Company. Shortly after joining the company in 2021, you discover the following errors related to the 2019 and 2020 financial statements:

a. Inventory at 12/31/2019 was understated by $6,000.
b. Inventory at 12/31/2020 was overstated by $9,000.
c. On 12/31/2020, inventory was purchased for $3,000. The company did not record the purchase until the inventory was paid for early in 2021. At that time, the purchase was recorded by a debit to purchases and a credit to cash.

The company uses a periodic inventory system.

Required:

1. Assuming that the errors were discovered after the 2020 financial statements were issued, analyze the effect of the errors on 2020 and 2019 cost of goods sold, net income, and retained earnings. (Ignore income taxes.)
2. Prepare a journal entry to correct the errors.
3. Will Ralston account for the change (a) retrospectively or (b) prospectively?

P 20–11
Error correction; change in depreciation method
● LO20–6

The Collins Corporation purchased office equipment at the beginning of 2019 and capitalized a cost of $2,000,000. This cost included the following expenditures:

Purchase price	$1,850,000
Freight charges	30,000
Installation charges	20,000
Annual maintenance charge	100,000
Total	$2,000,000

The company estimated an eight-year useful life for the equipment. No residual value is anticipated. The double-declining-balance method was used to determine depreciation expense for 2019 and 2020.

In 2021, after the 2020 financial statements were issued, the company decided to switch to the straight-line depreciation method for this equipment. At that time, the company's controller discovered that the original cost of the equipment incorrectly included one year of annual maintenance charges for the equipment.

Required:

1. Ignoring income taxes, prepare the appropriate correcting entry for the equipment capitalization error discovered in 2021.
2. Ignoring income taxes, prepare any 2021 journal entries related to the change in depreciation methods.

P 20–12
Accounting changes and error correction seven situations; tax effects ignored
● LO20–1 through LO20–4, LO20–6

Williams-Santana Inc. is a manufacturer of high-tech industrial parts that was started in 2009 by two talented engineers with little business training. In 2021, the company was acquired by one of its major customers. As part of an internal audit, the following facts were discovered. The audit occurred during 2021 before any adjusting entries or closing entries were prepared.

a. A five-year casualty insurance policy was purchased at the beginning of 2019 for $35,000. The full amount was debited to insurance expense at the time.
b. Effective January 1, 2021, the company changed the salvage value used in calculating depreciation for its office building. The building cost $600,000 on December 29, 2010, and has been depreciated on a straight-line basis assuming a useful life of 40 years and a salvage value of $100,000. Declining real estate values in the area indicate that the salvage value will be no more than $25,000.
c. On December 31, 2020, merchandise inventory was overstated by $25,000 due to a mistake in the physical inventory count using the periodic inventory system
d. The company changed inventory cost methods to FIFO from LIFO at the end of 2021 for both financial statement and income tax purposes. The change will cause a $960,000 increase in the beginning inventory at January 1, 2022.
e. At the end of 2020, the company failed to accrue $15,500 of sales commissions earned by employees during 2020. The expense was recorded when the commissions were paid in early 2021.
f. At the beginning of 2019, the company purchased a machine at a cost of $720,000. Its useful life was estimated to be 10 years with no salvage value. The machine has been depreciated by the double-declining balance method. Its book value on December 31, 2020, was $460,800. On January 1, 2021, the company changed to the straight-line method.
g. Warranty expense is determined each year as 1% of sales. Actual payment experience of recent years indicates that 0.75% is a better indication of the actual cost. Management effects the change in 2021. Credit sales for 2021 are $4,000,000; in 2020 they were $3,700,000.

Required:

For each situation

1. Identify whether it represents an accounting change or an error. If an accounting change, identify the type of change.
2. Prepare any journal entry necessary as a direct result of the change or error correction, as well as any adjusting entry for 2021 related to the situation described. (Ignore tax effects.)

P 20–13
Accounting changes and error correction; seven situations; tax effects considered
● LO20–1 through LO20–4, LO20–6

(Note: This problem is a variation of P 20–12, modified to consider income tax effects.) Williams-Santana Inc. is a manufacturer of high-tech industrial parts that was started in 2009 by two talented engineers with little business training. In 2021, the company was acquired by one of its major customers. As part of an internal audit, the following facts were discovered. The audit occurred during 2021 before any adjusting entries or closing entries were prepared. The income tax rate is 25% for all years.

a. A five-year casualty insurance policy was purchased at the beginning of 2019 for $35,000. The full amount was debited to insurance expense at the time.
b. Effective January 1, 2021, the company changed the salvage values used in calculating depreciation for its office building. The building cost $600,000 on December 29, 2010, and has been depreciated on a straight-line basis assuming a useful life of 40 years and a salvage value of $100,000 Declining real estate values in the area indicate that the salvage value will be no more than $25,000.
c. On December 31, 2020, merchandise inventory was overstated by $25,000 due to a mistake in the physical inventory count using the periodic inventory system.
d. The company changed inventory cost methods to FIFO from LIFO at the end of 2021 for both financial statement and income tax purposes. The change will cause a $960,000 increase in the beginning inventory at January 1, 2022.
e. At the end of 2020, the company failed to accrue $16,400 of sales commissions earned by employees during 2020. The expense was recorded when the commissions were paid in early 2021.
f. At the beginning of 2019, the company purchased a machine at a cost of $720,000. Its useful life was estimated to be ten years with no salvage value. The machine has been depreciated by the double-declining balance method. Its book value on December 31, 2020, was $460,800. On January 1, 2021, the company changed to the straight-line method.
g. Warranty expense is determined each year as 1% of sales. Actual payment experience of recent years indicates that 0.75% is a better indication of the actual cost. Management effects the change in 2021. Credit sales for 2021 are $4,000,000; in 2020 they were $3,700,000.

Required:

For each situation

1. Identify whether it represents an accounting change or an error. If an accounting change, identify the type of change.
2. Prepare any journal entry necessary as a direct result of the change or error correction, as well as any adjusting entry for 2021 related to the situation described. Any tax effects should be adjusted for through Income tax payable or Refund—income tax.

P 20–14
Errors; change in estimate; change in principle; restatement of previous financial statements
● LO20–1, LO20–3, LO20–4, LO20–6

Whaley Distributors is a wholesale distributor of electronic components. Financial statements for the years ended December 31, 2019 and 2020, reported the following amounts and subtotals ($ in millions):

	Assets	Liabilities	Shareholders' Equity	Net Income	Expenses
2019	$740	$330	$410	$210	$150
2020	820	400	420	230	175

In 2021, the following situations occurred or came to light:

a. Internal auditors discovered that ending inventories reported on the financial statements the two previous years were misstated due to faulty internal controls. The errors were in the following amounts:

2019 inventory	Overstated by $12 million
2020 inventory	Understated by $10 million

b. A liability was accrued in 2019 for a probable payment of $7 million in connection with a lawsuit ultimately settled in December 2021 for $4 million.

c. A patent costing $18 million at the beginning of 2019, expected to benefit operations for a total of six years, has not been amortized since acquired.

d. Whaley's conveyer equipment was depreciated by the sum-of-the-years'-digits (SYD) basis since it was acquired at the beginning of 2019 at a cost of $30 million. It has an expected useful life of five years and no expected residual value. At the beginning of 2021, Whaley decided to switch to straight-line depreciation.

Required:

For each situation

1. Prepare any journal entry necessary as a direct result of the change or error correction, as well as any adjusting entry for 2021 related to the situation described. (Ignore tax effects.)
2. Determine the amounts to be reported for each of the five items shown above from the 2019 and 2020 financial statements when those amounts are reported again in the 2019–2021 comparative financial statements.

P 20–15
Correction of errors; six errors
● LO20–6

Conrad Playground Supply underwent a restructuring in 2021. The company conducted a thorough internal audit, during which the following facts were discovered. The audit occurred during 2021 before any adjusting entries or closing entries are prepared.

a. Additional computers were acquired at the beginning of 2019 and added to the company's office network. The $45,000 cost of the computers was inadvertently recorded as maintenance expense. Computers have five-year useful lives and no material salvage value. This class of equipment is depreciated by the straight-line method.

b. Two weeks prior to the audit, the company paid $17,000 for assembly tools and recorded the expenditure as office supplies. The error was discovered a week later.

c. On December 31, 2020, merchandise inventory was understated by $78,000 due to a mistake in the physical inventory count. The company uses the periodic inventory system.

d. Two years earlier, the company recorded a 4% stock dividend (2,000 common shares, $1 par) as follows:

Retained earnings	2,000	
Common stock		2,000

The shares had a market price at the time of $12 per share.

e. At the end of 2020, the company failed to accrue $104,000 of interest expense that accrued during the last four months of 2020 on bonds payable. The bonds, which were issued at face value, mature in 2025. The following entry was recorded on March 1, 2021, when the semiannual interest was paid, as well as on September 1 of each year:

Interest expense	156,000	
Cash		156,000

f. A three-year liability insurance policy was purchased at the beginning of 2020 for $72,000. The full premium was debited to insurance expense at the time.

Required:

For each error, prepare any journal entry necessary to correct the error, as well as any year-end adjusting entry for 2021 related to the situation described. (Ignore income taxes.)

P 20–16
Integrating problem; errors; deferred taxes; contingency; change in tax rates
● LO20–6

You are internal auditor for Shannon Supplies, Inc., and are reviewing the company's preliminary financial statements. The statements, prepared after making the adjusting entries, but before closing entries for the year ended December 31, 2021, are as follows:

SHANNON SUPPLIES, INC.
Balance Sheet
December 31, 2021

Assets	($ in thousands)
Cash	$2,400
Investment in equity securities	250
Accounts receivable, net	810
Inventory	1,060
Equipment	1,240
Less: Accumulated depreciation	(560)
Total assets	$5,200
Liabilities and Shareholders' Equity	
Accounts payable and accrued expenses	$3,320
Income tax payable	220
Common stock, $1 par	200
Additional paid-in capital	750
Retained earnings	710
Total liabilities and shareholders' equity	$5,200

SHANNON SUPPLIES, INC.
Income Statement
For the Year Ended December 31, 2021

Sales revenue		$3,400
Operating expenses:		
Cost of goods sold	$1,140	
Selling and administrative	896	
Depreciation	84	2,120
Income before income tax		1,280
Income tax expense		(320)
Net income		$ 960

Shannon's income tax rate was 25% in 2021 and previous years. During the course of the audit, the following additional information (not considered when the above statements were prepared) was obtained:

a. Shannon's investment portfolio consists of blue chip stocks held for long-term appreciation. To raise working capital, some of the shares with an original cost of $180,000 were sold in May 2021. Shannon accountants debited cash and credited investment in equity securities for the $220,000 proceeds of the sale.

b. At December 31, 2021, the fair value of the remaining equity securities in the investment portfolio was $274,000.

c. The state of Alabama filed suit against Shannon in October 2019, seeking civil penalties and injunctive relief for violations of environmental regulations regulating emissions. Shannon's legal counsel previously believed that an unfavorable outcome of this litigation was not probable, but based on negotiations with state attorneys in 2021, now believes eventual payment to the state of $130,000 is probable, most likely to be paid in 2024.

d. The $1,060,000 inventory total, which was based on a physical count at December 31, 2021, was priced at cost. Based on your conversations with company accountants, you determined that the inventory cost was overstated by $132,000.

e. Electronic counters costing $80,000 were added to the equipment on December 29, 2020. The cost was charged to repairs.

f. Shannon's equipment, on which the counters were installed, had a remaining useful life of four years on December 29, 2020, and is being depreciated by the straight-line method for both financial and tax reporting.

g. A new tax law was enacted in 2021 which will cause Shannon's income tax rate to change from 25% to 20% beginning in 2022.

Required:

Prepare journal entries to record the effects on Shannon's accounting records at December 31, 2021, for each of the items described above.

P 20–17
Integrating problem; error; depreciation; deferred taxes
● LO20–6

George Young Industries (GYI) acquired industrial robots at the beginning of 2018 and added them to the company's assembly process. During 2021, management became aware that the $1 million cost of the machinery was inadvertently recorded as repair expense on GYI's books and on its income tax return. The industrial robots have 10-year useful lives and no material salvage value. This class of equipment is depreciated by the straight-line method for financial reporting purposes and for tax purposes it is considered to be MACRS 7-year property. Cost deducted over 7 years by the modified accelerated recovery system as follows:

Year	MACRS Deductions
2018	$ 142,900
2019	244,900
2020	174,900
2021	124,900
2022	89,300
2023	89,200
2024	89,300
2025	44,600
Totals	$1,000,000

The tax rate is 25% for all years involved.

Required:

1. Prepare any journal entry necessary as a direct result of the error described.
2. Will GYI account for the change (a) retrospectively or (b) prospectively?
3. Prepare the adjusting entry for 2021 depreciation.

Decision Makers' Perspective

©ImageGap/Getty Images

Apply your critical-thinking ability to the knowledge you've gained. These cases will provide you an opportunity to develop your research, analysis, judgment, and communication skills. You also will work with other students, integrate what you've learned, apply it in real-world situations, and consider its global and ethical ramifications. This practice will broaden your knowledge and further develop your decision-making abilities.

Integrating Case 20–1
Change to dollar-value LIFO
● LO20–3

Webster Products, Inc., adopted the dollar-value LIFO method of determining inventory costs for financial and income tax reporting on January 1, 2021. Webster continues to use the FIFO method for internal decision-making purposes. Webster's FIFO inventories at December 31, 2021, 2022, and 2023, were $300,000, $412,500, and $585,000, respectively. Internally generated cost indexes are used to convert FIFO inventory amounts to dollar-value LIFO amounts. Webster estimated these indexes as follows:

2021	1.00
2022	1.25
2023	1.50

Required:

1. Determine Webster's dollar-value LIFO inventory at December 31, 2022 and 2023.
2. Will Webster account for the change (a) retrospectively or (b) prospectively?

Communication Case 20–2
Change in inventory method; disclosure note
● LO20–2

Mayfair Department Stores operates over 30 retail stores in the Pacific Northwest. Prior to 2021, the company used the FIFO method to value its inventory. In 2021, Mayfair decided to switch to the dollar-value LIFO retail inventory method. One of your responsibilities as assistant controller is to prepare the disclosure note describing the change in method that will be included in the company's 2021 financial statements. Kenneth Meier, the controller, provided the following information:

- Internally developed retail price indexes are used to adjust for the effects of changing prices.
- If the change had not been made, cost of goods sold for the year would have been $22 million lower. The company's income tax rate is 25% and there were 100 million shares of common stock outstanding during 2021.
- The cumulative effect of the change on prior years' income is not determinable.
- The reasons for the change were (a) to provide a more consistent matching of merchandise costs with sales revenue, and (b) the new method provides a more comparable basis of accounting with competitors that also use the LIFO method.

Required:

1. Prepare for Kenneth Meier the disclosure note that will be included in the 2021 financial statements.
2. Will Mayfair account for the change (a) retrospectively or (b) prospectively?

Ethics Case 20–3
Softening the blow
● LO20–1, LO20–2, LO20–3

Late one Thursday afternoon, Joy Martin, a veteran audit manager with a regional CPA firm, was reviewing documents for a long-time client of the firm, AMT Transport. The year-end audit was scheduled to begin Monday.

For three months, the economy had been in a down cycle and the transportation industry was particularly hard hit. As a result, Joy expected AMT's financial results would not be pleasant news to shareholders. However, what Joy saw in the preliminary statements made her sigh aloud. Results were much worse than she feared.

"Larry (the company president) already is in the doghouse with shareholders," Joy thought to herself. "When they see these numbers, they'll hang him out to dry."

"I wonder if he's considered some strategic accounting changes," she thought, after reflecting on the situation. "The bad news could be softened quite a bit by changing inventory methods from LIFO to FIFO or reconsidering some of the estimates used in other areas."

Required:

1. How would the actions contemplated contribute toward "softening" the bad news?
2. Do you perceive an ethical dilemma? What would be the likely impact of following up on Joy's thoughts? Who would benefit? Who would be injured?

Analysis Case 20–4
Change in inventory methods; concepts
● LO20–2, LO20–3

Generally accepted accounting principles should be applied consistently from period to period. However, changes within a company, as well as changes in the external economic environment, may force a company to change an accounting method. The specific reporting requirements when a company changes from one generally accepted inventory method to another depend on the methods involved.

Required:

Explain the accounting treatment for a change in inventory method (a) not involving LIFO, (b) from the LIFO method, and (c) to the LIFO method. Explain the logic underlying those treatments. Also, describe how disclosure requirements are designed to address the departure from consistency and comparability of changes in accounting principle.

Communication Case 20–5
Change in loss contingency; write a memo
● LO20–4

Late in 2021, you and two other officers of Curbo Fabrications Corporation just returned from a meeting with officials of the City of Jackson. The meeting was unexpectedly favorable even though it culminated in a settlement with city authorities that your company pay a total of $475,000 to cover the cost of violations of city construction codes. Jackson filed suit in November 2019 against Curbo Fabrications Corporation, seeking civil penalties and injunctive relief for violations of city construction codes regulating earthquake damage standards. Alleged violations involved several construction projects completed during the previous three years. When the financial statements were issued in 2020, Curbo had not reached a settlement with state authorities, but legal counsel had advised the company that it was probable the ultimate settlement would be $750,000 in penalties. The following entry was recorded:

Loss—litigation	750,000	
Liability—litigation		750,000

The final settlement, therefore, was a pleasant surprise. While returning from the meeting, conversation turned to reporting the settlement in the 2021 financial statements. You drew the short straw and were selected to write a memo to Janet Zenc, the financial vice president, advising the proper course of action.

Required:

Write the memo. Include descriptions of any journal entries related to the change in amounts. Briefly describe other steps Curbo should take to report the settlement.

Analysis Case 20–6
Two wrongs make a right?
● LO20–4

Early one Wednesday afternoon, Ken and Larry studied in the dormitory room they shared at Fogelman College. Ken, an accounting major, was advising Larry, a management major, regarding a project for Larry's Business Policy class. One aspect of the project involved analyzing the 2021 annual report of Craft Paper Company. Though not central to his business policy case, a footnote had caught Larry's attention.

Depreciation and Cost of Timber Harvested (in part)

($ in millions)	2021	2020	2019
Depreciation of buildings, machinery and equipment	$260.9	$329.8	$322.5
Cost of timber harvested and amortization of logging roads	4.9	4.9	4.9
	$265.8	$334.7	$327.4

Beginning in 2021, the Company revised the estimated average useful lives used to compute depreciation for most of its pulp and paper mill equipment from 16 years to 20 years and for most of its finishing and converting equipment from 12 years to 15 years. These revisions were made to more properly reflect the true economic lives of the assets and to better align the Company's depreciable lives with the predominant practice in the industry. The change had the effect of increasing net income by approximately $55 million.

"If I understand this right, Ken, the company is not going back and recalculating a lower depreciation for earlier years. Instead they seem to be leaving depreciation overstated in earlier years and making up for that by understating it in current and future years," Larry mused. "Is that the way it is in accounting? Two wrongs make a right?"

Required:

What are the two wrongs to which Larry refers? Is he right?

Research Case 20–7
FASB codification; researching the way changes in postretirement benefit estimates are reported; retrieving disclosures from the Internet
● LO20–4

It's financial statements preparation time at Center Industries, where you have been assistant controller for two months. Ben Huddler, the controller, seems to be pleasant but unpredictable. Today, although your schedule is filled with meetings with internal and outside auditors and two members of the board of directors, Ben made a request. "As you know, we're decreasing the rate at which we assume health care costs will rise when measuring our postretirement benefit obligation. I'd like to know how others have reported similar changes. Can you find me an example?" he asked. "I'd bet you could get one off the Internet."

Required:

Access a recent 10-K filing of a firm you think might have a postretirement healthcare plan. You may need to look up several companies before you find what you're looking for. Older, established companies are most likely to have such benefit plans. (Note: You may be able to focus your search by searching with key words and phrases in Google on the Internet.) Find the portion of the disclosures that reports the effect of a change in healthcare cost trends.

1. What information is provided about the effect of the change on the company's estimated benefit obligation?
2. Obtain the relevant authoritative literature on the disclosure of accounting policies using the *FASB Accounting Standards Codification* at the FASB website (**www.fasb.org**). What is the specific eight-digit Codification citation (XXX-XX-XX-X) companies rely on when disclosing the effect of a change in healthcare cost trends?

Analysis Case 20–8
Various changes
● LO20–1 through LO20–4

DRS Corporation changed the way it depreciates its computers from the sum-of-the-year's-digits method to the straight-line method beginning January 1, 2021. DRS also changed its estimated residual value used in computing depreciation for its office building. At the end of 2021, DRS changed the specific subsidiaries constituting the group of companies for which its consolidated financial statements are prepared.

Required:

1. For each accounting change DRS undertook, indicate the type of change and how DRS should report the change. Be specific.
2. Why should companies disclose changes in accounting principles?

Analysis Case 20–9
Various changes
● LO20–1 through LO20–4

Ray Solutions decided to make the following changes in its accounting policies on January 1, 2021:

a. Changed from the cash to the accrual basis of accounting for recognizing revenue on its service contracts.
b. Adopted straight-line depreciation for all future equipment purchases, but continued to use accelerated depreciation for all equipment acquired before 2021.
c. Changed from the LIFO inventory method to the FIFO inventory method.

Required:

For each accounting change Ray undertook, indicate (a) the type of change and (b) whether Ray should report the change retrospectively or prospectively.

Judgment Case 20–10
Accounting changes; independent situations
● LO20–1 through LO20–5

Sometimes a business entity will change its method of accounting for certain items. The change may be classified as a change in accounting principle, a change in accounting estimate, or a change in reporting entity. Listed below are three independent, unrelated sets of facts relating to accounting changes.

Situation I: A company determined that the depreciable lives of its fixed assets are presently too long to fairly match the cost of the fixed assets with the revenue produced. The company decided at the beginning of the current year to reduce the depreciable lives of all of its existing fixed assets by five years.

Situation II: On December 31, 2020, Gary Company owned 51% of Allen Company, at which time Gary reported its investment on a nonconsolidated basis due to political uncertainties in the country in which Allen was located. On January 2, 2021, the management of Gary Company was satisfied that the political uncertainties were resolved and the assets of the company were in no danger of nationalization. Accordingly, Gary will prepare consolidated financial statements for Gary and Allen for the year ended December 31, 2021.

Situation III: A company decides in January 2021 to adopt the straight-line method of depreciation for plant equipment. The straight-line method will be used for new acquisitions as well as for previously acquired plant equipment, for which depreciation had been provided on an accelerated basis.

Required:

For each of the situations described above, provide the information indicated below. Complete your discussion of each situation before going on to the next situation.

1. Type of accounting change
2. Manner of reporting the change under current generally accepted accounting principles: retrospectively or prospectively?
3. Should a disclosure note be provided in connection with the change?

Judgment Case 20–11
Inventory errors
● LO20–6

Some inventory errors are said to be "self-correcting," in that the error has the opposite financial statement effect in the period following the error, thereby "correcting," the original account balance errors.

Required:
Despite this self-correcting feature, discuss why these errors should not be ignored and describe the steps required to account for the error correction.

Ethics Case 20–12
Overstatement of ending inventory
● LO20–6

Danville Bottlers is a wholesale beverage company. Danville uses the FIFO inventory method to determine the cost of its ending inventory. Ending inventory quantities are determined by a physical count. For the fiscal year-end June 30, 2021, ending inventory was originally determined to be $3,265,000. However, on July 17, 2021, John Howard, the company's controller, discovered an error in the ending inventory count. He determined that the correct ending inventory amount should be $2,600,000.

Danville is a privately owned corporation with significant financing provided by a local bank. The bank requires annual audited financial statements as a condition of the loan. By July 17, the auditors had completed their review of the financial statements, which are scheduled to be issued on July 25. They did not discover the inventory error.

John's first reaction was to communicate his finding to the auditors and to revise the financial statements before they are issued. However, he knows that his and his fellow workers' profit-sharing plans are based on annual pretax earnings and that if he revises the statements, everyone's profit sharing bonus will be significantly reduced.

Required:
1. Why will bonuses be negatively affected? What is the effect on pretax earnings?
2. If the error is not corrected in the current year and is discovered by the auditors during the following year's audit, how will the error be reported in the company's financial statements?
3. Discuss the ethical dilemma Howard faces.

Continuing Cases

Target Case
● LO20–2,LO20–4

Target Corporation prepares its financial statements according to U.S. GAAP. Target's financial statements and disclosure notes for the year ended February 3, 2018, are available in Connect. This material is also available under the Investor Relations link at the company's website (**www.target.com**). Target's share-based compensation includes several long-term incentive plans.

Required:
1. Refer to Target's financial statements for the year ended February 3, 2018. Note 3 provides information on an accounting change Target made in fiscal 2017. Was this a change in estimate, a change in principle, a change in reporting entity, or an error correction?
2. Did Target account for and report the change retrospectively or prospectively?
3. What was the effect of the change on depreciation and amortization in fiscal 2016?
4. Was the effect of the change on earnings per share in fiscal 2017 an increase, a decrease, or no effect?

Air France–KLM Case
● LO20–2
IFRS

Air France–KLM (AF), a Franco-Dutch company, prepares its financial statements according to International Financial Reporting Standards. AF's financial statements and disclosure notes for the year ended December 31, 2017, are available in Connect. This material also is available under the Finance link at the company's website (**www.airfranceklm-finance.com**).

Required:
1. Refer to AF's disclosure notes, in particular Note 2: Restatement of Accounts 2016. Was the change described in the note a change in estimate, a change in principle, a change in reporting entity, or an error correction?
2. Is this the same approach AF would follow if using U.S. GAAP?

CPA Exam Questions and Simulations

ROGER CPA Review

Sample CPA Exam questions from Roger CPA Review are available in Connect as support for the topics in this chapter. These multiple-choice questions and task-based simulations include expert-written explanations and solutions, and provide a starting point for students to become familiar with the content and functionality of the actual CPA Exam.

CHAPTER

21 The Statement of Cash Flows Revisited

OVERVIEW

The objective of financial reporting is to provide investors and creditors with useful information, primarily in the form of financial statements. The balance sheet and the income statement—the focus of your study in earlier chapters—do not provide all the information needed by these decision makers. Here you will learn how the statement of cash flows fills the information gap left by the other financial statements.

The statement lists all cash inflows and cash outflows, and classifies them as cash flows from (a) operating, (b) investing, or (c) financing activities. Investing and financing activities that do not directly affect cash also are reported.

LEARNING OBJECTIVES

After studying this chapter, you should be able to:

- **LO21–1** Explain the usefulness of the statement of cash flows. *(p. 1215)*
- **LO21–2** Define cash equivalents. *(p. 1219)*
- **LO21–3** Determine cash flows from operating activities by the direct method. *(p. 1219)*
- **LO21–4** Determine cash flows from operating activities by the indirect method. *(p. 1221)*
- **LO21–5** Identify transactions that are classified as investing activities. *(p. 1222)*
- **LO21–6** Identify transactions that are classified as financing activities. *(p. 1223)*
- **LO21–7** Identify transactions that represent noncash investing and financing activities. *(p. 1224)*
- **LO21–8** Prepare a statement of cash flows with the aid of a spreadsheet or T-accounts. *(p. 1225)*
- **LO21–9** Discuss the primary differences between U.S. GAAP and IFRS with respect to the statement of cash flows. *(p. 1245)*

©Hero Images/Getty Images

FINANCIAL REPORTING CASE

Where's the Cash?

"What do you mean you can't afford a wage increase?" union negotiator Vince Barr insisted. "We've all seen your income statement. You had record earnings this year."

This is the first day of company negotiations with union representatives. As company controller, you know it's going to be up to you to explain the company's position on the financial aspects of the negotiations. In fact, you've known for some time that a critical point of contention would be the moderate increase in this year's profits after three years of level or slightly declining earnings. Not helping the situation is that the company has always used accelerated depreciation on its equipment, which it began replacing this year at considerably higher prices than it cost several years back.

QUESTIONS

By the time you finish this chapter, you should be able to respond appropriately to the questions posed in this case. Compare your response to the solution provided at the end of the chapter.

1. What are the cash flow aspects of the situation that Mr. Barr may be overlooking in making his case for a wage increase? How can a company's operations generate a healthy profit and yet produce meager or even negative cash flows? (*p. 1219*)
2. What information can a statement of cash flows provide about a company's investing activities that can be useful in decisions such as this? (*p. 1222*)
3. What information can a statement of cash flows provide about a company's financing activities that can be useful in decisions such as this? (*p. 1223*)

The Content and Value of the Statement of Cash Flows

PART A

Decision Makers' Perspective—Usefulness of Cash Flow Information

A fund manager of a major insurance company, considering investing $8,000,000 in the common stock of **The Coca-Cola Company**, asks herself, "What are the prospects of future dividends and market-price appreciation? Will we get a return commensurate with the cost and risk of our investment?" A bank officer, examining an application for a business loan, asks himself, "If I approve this loan, what is the likelihood of the borrower making interest payments on time and repaying the loan when due?" Investors and creditors continually face these and similar decisions that require projections of the relative ability of a business to generate future cash flows and of the risk associated with those forecasts.

● LO21–1

To make these projections, decision makers rely heavily on the information reported in periodic financial statements. In the final analysis, cash flows into and out of a business enterprise are the most fundamental events on which investors and creditors base their decisions. Naturally, these decisions focus on the prospects of the decision makers receiving cash returns from their dealings with the firm. However, it is the ability of the firm to

Investors and creditors require cash flows from the corporation.

Cash flows to investors and creditors depend on the corporation generating cash flows to itself.

generate cash flows to itself that ultimately determines the potential for cash flows from the firm to investors and creditors.

The financial statements that have been the focus of your study in earlier chapters—the income statement and the balance sheet—offer information helpful in forecasting future cash-generating ability. Some important questions, however, are not easily answered from the information these statements provide. For example, meaningful projections of a company's future profitability and risk depend on answers to such questions as the following:

- In what types of activities is the company investing?
- Are these activities being financed with debt? With equity? By cash generated from operations?
- Are facilities being acquired to accommodate future expansion?
- How does the amount of cash generated from operations compare with net income over time?
- Why isn't the increase in retained earnings reflected as an increase in dividends?
- What happens to the cash received from the sale of assets?
- By what means is debt being retired?

Many decisions benefit from information about the company's underlying cash flow process.

The information needed to answer these and similar questions is found in the continuous series of cash flows that the income statement and the balance sheet describe only indirectly. This underlying cash flow process is considered next. ●

Cash Inflows and Outflows

Cash continually flows in and out of an active business. Businesses disburse cash to acquire property and equipment to maintain or expand productive capacity. When no longer needed, these assets may be sold for cash. Cash is paid to produce or purchase inventory for resale, as well as to pay for the expenses of selling these goods. The ultimate outcome of these selling activities is an inflow of cash. Cash might be invested in securities of other firms. These investments provide cash inflows during the investment period in the form of dividends or interest and at the end of the investment period when the securities are sold. To raise cash to finance their operations, firms sell stock and/or acquire debt. Cash payments are made as dividends to shareholders and interest to creditors. When debt is repaid or stock repurchased, cash flows out of the firm. To help you visualize the continual process of cash receipts and cash payments, that process is diagrammed in Illustration 21–1. The diagram also previews the way we will later classify the cash flows in a statement of cash flows.

Illustration 21–1

Cash Inflows and Cash Outflows

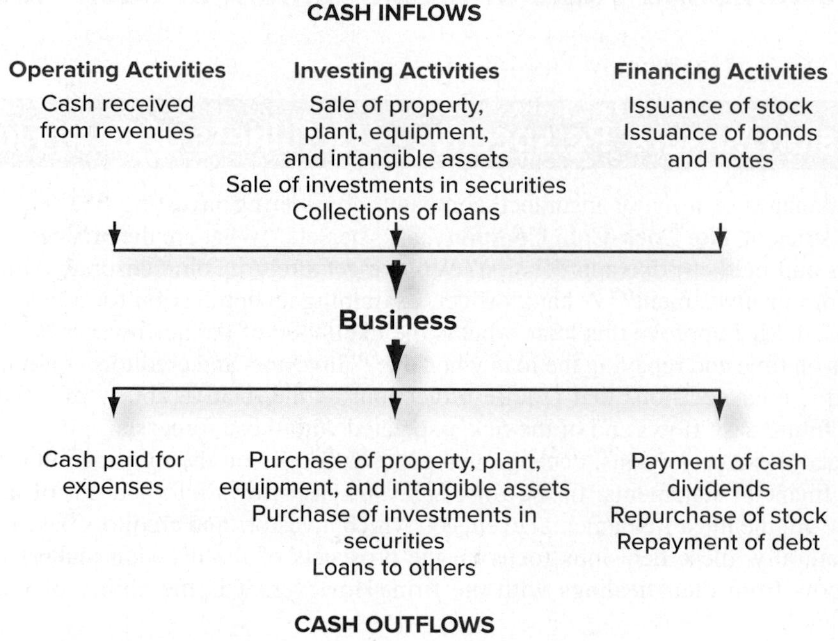

Embodied in this assortment of cash flows is a wealth of information that investors and creditors require to make educated decisions. Much of the value of the underlying information provided by the cash flows is lost when reported only indirectly by the balance sheet and the income statement. Each cash flow eventually impacts decision makers by affecting the balances of various accounts in the balance sheet. Also, many of the cash flows—those related to income-producing activities—are represented in the income statement. However, they are not necessarily reported in the period the cash flows occur because the income statement measures activities on an accrual basis. The statement of cash flows fills the information gap by reporting the cash flows directly.

The statement of cash flows provides information about cash flows that is lost when reported only indirectly by the balance sheet and the income statement.

Structure of the Statement of Cash Flows

A statement of cash flows is shown in Illustration 21–2. The statement lists all cash inflows and cash outflows during the reporting period. To enhance the informational value of the presentation, the cash flows are classified according to the nature of the activities that bring about the cash flows. The three primary categories of cash flows are (1) cash flows from operating activities, (2) cash flows from investing activities, and (3) cash flows from financing activities. Classifying each cash flow by source (operating, investing, or financing activities) is more informative than simply listing the various cash flows. Notice, too, that the noncash investing and financing activities—investing and financing activities that do not directly increase or decrease cash—also are reported. The GAAP requirement for the statement of cash flows was issued in direct response to *FASB Concept Statement 1,* which stated that the primary objective of financial reporting is to "provide information to help investors and creditors, and others assess the amounts, timing, and uncertainty of prospective net cash inflows to the related enterprise."[1]

Many companies suffered bankruptcy because they were unable to generate sufficient cash to satisfy their obligations. Doubtless, many investors in the stock of these firms would have been spared substantial losses if the financial statements had been designed to foresee the cash flow problems the companies were experiencing. A noted illustration is the dissolution of **W. T. Grant** during the 1970s. Grant, a general retailer in the days before malls, was a blue chip stock of its time. Grant's statement of changes in financial position (the predecessor of the statement of cash flows) reported working capital from operations of $46 million in 1972. Yet, if presented, a statement of cash flows would have reported cash flows from operating activities of negative $10 million. In fact, the unreported cash flow deficiency grew to $114 million in 1973, while working capital from operations was reported as having increased by $1 million. That year, without the benefit of cash flow information, investors were buying Grant's stock at prices that represented up to 20 times its earnings.[2]

More recently, even with cash flow information available, cash flow problems can go unheeded or unsuccessfully addressed. An example is the recent demise of **Toys R Us**. The relationship of cash inflows and cash outflows illustrates a major aspect of a leveraged buyout (LBO) which fashioned the latest ownership of the company. Typically, an LBO is followed by layoffs, spending cuts, and other cost-cutting initiatives designed to create enough cash inflow to meet the cash outflow needed to service the massive debt created by the LBO itself. In Toys R Us' final three years (2014—2016), by far the most notable aspect of its statement of cash flows was a positive cash flow from operations being dwarfed by a huge and accelerating cash outflow for debt repayments.

Realize, too, that an unprofitable company with good cash flow can survive. **Amazon.com** provides an excellent example. Founded as an online seller of books in 1995, Amazon didn't actually make a profit for a decade, but it did raise huge amounts of cash by selling stock, so much so that it was able to weather 10 years of sizable losses. The financing cash inflows funded expansion into many new lines of business and made up for the continual losses. Amazon now has enormously positive operating cash flows and is one of the largest companies in the world.

[1]FASB Concepts Statement No. 8, *Conceptual Framework for Financial Reporting* —Chapter 1, "The Objective of General Purpose Financial Reporting," and Chapter 3, "Qualitative Characteristics of Useful Financial Information" (a replacement of FASB Concepts Statements No. 1 and No. 2), September 2010.

[2]Cheryl A. Zega, "The New Statement of Cash Flows," *Management Accounting,* September 1988.

Illustration 21–2

Statement of Cash Flows

UNITED BRANDS CORPORATION
Statement of Cash Flows
For Year Ended December 31, 2021

Cash Flows from Operating Activities	($ in millions)	
Cash inflows:		
From customers	$98	
From investment revenue	3	
Cash outflows:		
To suppliers of goods	(50)	
To employees	(11)	
For interest	(3)	
For insurance	(4)	
For income tax	(11)	
Net cash flows from operating activities		$22
Cash Flows from Investing Activities		
Purchase of land	(30)	
Purchase of short-term investment	(12)	
Sale of land	18	
Sale of equipment	5	
Net cash flows from investing activities		(19)
Cash Flows from Financing Activities		
Sale of common shares	26	
Retirement of bonds payable	(15)	
Payment of cash dividends	(5)	
Net cash flows from financing activities		6
Net increase in cash		9
Cash balance, January 1		20
Cash balance, December 31		$29

Note X:
Noncash Investing and Financing Activities

Acquired $20 million of equipment by issuing a 12%, 5-year note	$20

Reconciliation of Net Income to Cash Flows from Operating Activities:

Net income	$12
Adjustments for noncash effects:	
Gain on sale of land	(8)
Depreciation expense	3
Loss on sale of equipment	2
Changes in operating assets and liabilities:	
Increase in accounts receivable	(2)
Decrease in inventory	4
Increase in accounts payable	6
Increase in salaries payable	2
Decrease in discount on bonds payable	2
Decrease in prepaid insurance	3
Decrease in income tax payable	(2)
Net cash flows from operating activities	$22

The statement of cash flows for United Brands Corporation (UBC), shown in Illustration 21–2, is intended at this point in the discussion to illustrate the basic structure and composition of the statement. Later, we will see how the statement of cash flows for UBC is prepared from the information typically available for this purpose. We will refer to UBC's statement of cash flows frequently throughout the chapter as the discussion becomes more specific regarding the criteria for classifying cash flows in the three primary categories and as we identify the specific cash flows to be reported on the statement. We will examine the content of the statement in more detail following a look at how this relatively recent financial statement has evolved to its present form over the course of the last several decades.

Cash, Cash Equivalents, and Restricted Cash

● LO21–2

Skilled cash managers will invest temporarily idle cash in short-term investments to earn interest on those funds, rather than maintain an unnecessarily large balance in a checking account. The FASB views short-term, highly liquid investments that can be readily converted to cash, with little risk of loss, as **cash equivalents**. Amounts held as investments of this type are essentially equivalent to cash because they are quickly available for use as cash. Therefore, on the statement of cash flows there is no differentiation between amounts held as cash (e.g., currency and checking accounts) and amounts held in cash equivalent investments. Similarly, sometimes companies might have contractual agreements that require specific amounts to be set aside for designated purposes, like debt repayment or workers' compensation claims. Those restricted cash amounts also are part of reported cash balances. So, when we refer in this chapter to cash, we are referring to the total of cash, cash equivalents, and restricted cash.

There is no differentiation between amounts held as cash and amounts held in cash equivalent investments or as restricted cash.

Examples of cash equivalents are money market funds, Treasury bills, and commercial paper. To be classified as cash equivalents, these investments must have a maturity date not longer than three months from the date of purchase. Flexibility is permitted in designating cash equivalents. Each company must establish a policy regarding which short-term, highly liquid investments it classifies as cash equivalents. The policy should be consistent with the company's customary motivation for acquiring various investments and should be disclosed in the notes to the statement.[3] A recent annual report of **ExxonMobil Corporation** provides this description of its cash equivalents (Illustration 21–3):

Each firm's policy regarding which short-term, highly liquid investments it classifies as cash equivalents should be disclosed in the notes to the financial statements.

Note 4: Cash Flow Information (in part)

The consolidated statement of cash flows provides information about changes in cash and cash equivalents. Highly liquid investments with maturities of three months or less when acquired are classified as cash equivalents.

Source: ExxonMobil Corporation

Illustration 21–3

Disclosure of Cash Equivalents—ExxonMobil Corporation

Real World Financials

Transactions that involve merely transfers from cash to cash equivalents (such as the purchase of a three-month Treasury bill), or from cash equivalents to cash (such as the sale of a Treasury bill), should not be reported on the statement of cash flows. The total of cash and cash equivalents is not altered by such transactions.[4] The cash balance reported in the balance sheet also represents the total of cash and cash equivalents, which allows us to compare the change in that balance with the net increase or decrease in the cash flows reported on the statement of cash flows.

Primary Elements of the Statement of Cash Flows

● LO21–3

This section describes the three primary activity classifications: (1) operating activities, (2) investing activities, and (3) financing activities. It also describes two other requirements of the statement of cash flows: (4) the reconciliation of the net increase or decrease in cash with the change in the balance of the cash account and (5) noncash investing and financing activities.

FINANCIAL Reporting Case

Q1, p. 1215

CASH FLOWS FROM OPERATING ACTIVITIES The income statement reports the success of a business in generating a profit from its operations. Net income (or loss) is the result

[3]A change in that policy is treated as a change in accounting principle.

[4]An exception is the sale of a cash equivalent at a gain or loss. This exception is described in more detail later in the chapter.

of netting together the revenues recognized during the reporting period, regardless of when cash is received, and the expenses incurred in generating those revenues, regardless of when cash is paid. This is the accrual concept of accounting that has been emphasized throughout your study of accounting. Information about net income and its components, measured by the accrual concept, generally provides a better indication of current operating performance than does information about current cash receipts and payments.[5] Nevertheless, as indicated earlier, the cash effects of earning activities also provide useful information that is not directly accessible from the income statement. The first cash flow classification in the statement of cash flows reports that information.

The cash effects of the elements of net income are reported as cash flows from operating activities.

Cash flows from operating activities are both inflows and outflows of cash that result from activities reported in the income statement. In other words, this classification of cash flows includes the elements of net income, but reported on a cash basis. The components of this section of the statement of cash flows, and their relationship with the elements of the income statement, are illustrated in Illustration 21–4.

Illustration 21–4

Relationship between the Income Statement and Cash Flows from Operating Activities (Direct Method)

Income Statement	**Cash Flows from Operating Activities**
Revenues:	**Cash inflows:**
Sales and service revenue	Cash received from customers
Investment revenue	Cash revenue received (e.g., dividends, interest)
Noncash revenues and gains (e.g., gain on sale of assets)	(Not reported)
Less: Expenses:	**Less: Cash outflows:**
Cost of goods sold	Cash paid to suppliers of inventory
Salaries expense	Cash paid to employees
Noncash expenses and losses (e.g., depreciation, amortization, loss on sale of assets)	(Not reported)
Interest expense	Cash paid to creditors
Other operating expenses	Cash paid to insurance companies and others
Income tax expense	Cash paid to the government
Net income	***Net cash flows from operating activities***

To see the concept applied, let's look again at the cash flows from operating activities reported by United Brands Corporation. That section of the statement of cash flows is extracted from Illustration 21–2 and reproduced in Illustration 21–5.

Illustration 21–5

Cash Flows from Operating Activities; Direct Method

Cash flows from operating activities are the elements of net income, but reported on a cash basis.

Cash Flows from Operating Activities:		
Cash inflows:		
From customers	$98	
From investment revenue	3	
Cash outflows:		
To suppliers of goods	(50)	
To employees	(11)	
For interest	(3)	
For insurance	(4)	
For income tax	(11)	
Net cash flows from operating activities		$22

[5]FASB Concepts Statement No. 8, *Conceptual Framework for Financial Reporting*—Chapter 1, "The Objective of General Purpose Financial Reporting," and Chapter 3, "Qualitative Characteristics of Useful Financial Information" (a replacement of FASB Concepts Statements No. 1 and No. 2), September 2010.

Cash inflows from operating activities exceeded cash outflows for expenses by $22 million. We'll see later (in Illustration 21–9) that UBC's net income from the same operating activities was only $12 million. Why did operating activities produce net cash inflows greater than net income? The reason will become apparent when we determine, in a later section, the specific amounts of these cash flows.

You also should be aware that the generalization stated earlier that cash flows from operating activities include the elements of net income reported on a cash basis is not strictly true for all elements of the income statement. Notice in Illustration 21–5 that no cash effects are reported for depreciation and amortization of property, plant, and equipment, and intangibles, nor for gains and losses from the sale of those assets. Cash outflows occur when these assets are acquired, and cash inflows occur when the assets are sold. However, as described later, the acquisition and subsequent resale of these assets are classified as investing activities rather than as operating activities.

Quite the opposite, the purchase and the sale of inventory are considered operating activities. The cash effects of these transactions—namely, (1) cash payments to suppliers and (2) cash receipts from customers—are included in the determination of cash flows from operating activities. Why are inventories treated differently from property, plant, equipment, and intangible assets when classifying their cash effects if they all are acquired for the purpose of producing revenues? The essential difference is that inventory typically is purchased for the purpose of being sold as part of the firm's current operations, while the other assets are purchased as investments to benefit the business over a relatively long period of time.

DIRECT METHOD OR INDIRECT METHOD OF REPORTING CASH FLOWS FROM OPERATING ACTIVITIES The presentation by UBC of cash flows from operating activities illustrated in Illustration 21–2 and reproduced in Illustration 21–5 is referred to as the **direct method**. The method is named for the fact that the cash effect of each operating activity (i.e., income statement item) is reported *directly* in the statement of cash flows. For instance, UBC reports "cash received from customers" as the cash effect of sales activities, "cash paid to suppliers" as the cash effect of cost of goods sold, and so on. Then, UBC simply omits from the presentation any income statement items that do not affect cash at all, such as depreciation expense.

● LO21–4

Another way UBC might have reported cash flows from operating activities is by the **indirect method**. By this approach, the net cash increase or decrease from operating activities ($22 million in our example) would be derived *indirectly* by starting with reported net income and working backwards to convert that amount to a cash basis. As we see later in the chapter, UBC's net income is **$12** million. Using the indirect method, UBC would replace the previous presentation of net cash flows from operating activities with the one shown in Illustration 21–6.

Be sure to note that the indirect method generates the same $22 million net cash flows from operating activities as the direct method. Rather than directly reporting only the

Illustration 21–6

Cash Flows from Operating Activities; Indirect Method

Cash Flows from Operating Activities		
Net income	**$12**	
Adjustments for noncash effects:		
Gain on sale of land	(8)	
Depreciation expense	3	
Loss on sale of equipment	2	
Changes in operating assets and liabilities:		
Increase in accounts receivable	(2)	
Decrease in inventory	4	
Increase in accounts payable	6	
Increase in salaries payable	2	
Decrease in discount on bonds payable	2	
Decrease in prepaid insurance	3	
Decrease in income tax payable	(2)	
Net cash flows from operating activities		$22

By the indirect method, UBC derives the net cash increase or decrease from operating activities indirectly, by starting with reported net income and working backwards to convert that amount to a cash basis.

components of the income statement that *do* represent increases or decreases in cash, by the indirect method we begin with net income—which includes both cash and noncash components—and back out all amounts that *don't* reflect increases or decreases in cash. Later in the chapter, we explore the specific adjustments made to net income to achieve this result. At this point, it is sufficient to realize that two alternative methods are permitted for reporting net cash flows from operating activities. Either way, we convert accrual-based income to cash flows produced by those same operating activities.

Notice also that the indirect method presentation is identical to what UBC reported earlier as the "Reconciliation of Net Income to Cash Flows from Operating Activities" in Note X of Illustration 21–2. Whether cash flows from operating activities are reported by the direct method or by the indirect method, the financial statements must reconcile the difference between net income and cash flows from operating activities. When a company uses the *direct method,* the company presents the reconciliation in a separate schedule, as UBC does. That presentation is precisely the same as the presentation of net cash flows from operating activities by the indirect method. On the other hand, a company choosing to use the indirect method is not required to provide a separate reconciliation schedule because the "cash flows from operating activities" section of the statement of cash flows serves that purpose. Most companies use the indirect method.[6]

It's important to understand, too, that regardless of which method a company chooses to report *operating* activities, that choice has no effect on the way it identifies and reports cash flows from *investing* and *financing* activities. We turn our attention now to those two sections of the statement of cash flows. Later, in Part C, we'll return for a more thorough discussion of the alternative methods of reporting the operating activities section.

FINANCIAL Reporting Case

Q2, p. 1215

CASH FLOWS FROM INVESTING ACTIVITIES Companies periodically invest cash to replace or expand productive facilities such as property, plant, and equipment. Investments might also be made in other assets, such as securities of other firms, with the expectation of a return on those investments. Information concerning these investing activities can provide valuable insight to decision makers regarding the nature and magnitude of assets being acquired for future use, as well as provide clues concerning the company's ambitions for the future.

● LO21–5

Cash outflows and cash inflows due to the acquisition and disposition of assets (other than inventory and assets classified as cash equivalents) are reported as cash flows from investing activities.

Cash flows from investing activities are both outflows and inflows of cash caused by the acquisition and disposition of assets. Included in this classification are cash payments to acquire (1) property, plant, and equipment and other productive assets (except inventories); (2) investments in securities (except cash equivalents and trading securities); and (3) nontrade receivables.[7,8] When these assets later are liquidated, any cash receipts from their disposition also are classified as investing activities. For instance, cash received from the sale of the assets or from the collection of a note receivable (principal amount only) represents cash inflows from investing activities. Be sure to realize that, unlike what the label might imply, any investment revenue like interest, dividends, or other cash return from these investments is not an investing activity. The reason, remember, is that investment revenue is an income statement item and therefore is an operating activity.

For illustration, notice the cash flows reported as investing activities by UBC. That section of the statement of cash flows is extracted from the complete statement in Illustration 21–2 and reproduced in Illustration 21–7.

UBC reports as investing activities the cash paid to purchase both land and a short-term investment. The other two investing activities reported are cash receipts for the sale of assets—equipment and land—that were acquired in earlier years. The specific transactions creating these cash flows are described in a later section of this chapter.

[6]According to the AICPA, *U.S. GAAP Financial Statements—Best Practices in Presentation and Disclosure,* a survey of 500 companies showed that 495 companies chose to use the indirect method, only 5 the direct method.

[7]Inflows and outflows of cash from buying and selling trading securities typically are considered operating activities because financial institutions that routinely transact in trading securities consider them an appropriate part of their normal operations.

[8]A nontrade receivable differs from a trade receivable in that it is not one associated with the company's normal trade; that is, it's not received from a customer. A trade receivable, or accounts receivable, is an *operating asset.* A nontrade receivable, on the other hand, might be a loan to an affiliate company or to an officer of the firm. To understand how the creation of a nontrade receivable is an *investing* activity, you might view such a loan as an investment in the receivable.

Cash Flows from Investing Activities:		
Purchase of land	$(30)	
Purchase of short-term investment	(12)	
Sale of land	18	
Sale of equipment	5	
Net cash flows from investing activities		$(19)

Illustration 21–7

Cash Flows from Investing Activities

Cash flows from investing activities include investments in assets and their subsequent sale.

The purchase and sale of inventories are not considered investing activities. Inventories are purchased for the purpose of being sold as part of the firm's primary operations, so their purchase and sale are classified as operating activities.

Also, the purchase and sale of assets classified as cash equivalents are not reported as investing activities. In fact, these activities usually are not reported on the statement of cash flows. For example, when temporarily idle cash is invested in a money market fund considered to be a cash equivalent, the total of cash and cash equivalents does not change. Likewise, when the cash is later withdrawn from the money market fund, the total remains unchanged. The exception is when cash equivalents are sold at a gain or a loss. In that case, the total of cash and cash equivalents actually increases or decreases in the process of transferring from one cash equivalent account to another cash equivalent account. As a result, the change in cash would be reported as a cash flow from operating activities. This is illustrated later in the chapter.

CASH FLOWS FROM FINANCING ACTIVITIES Not only is it important for investors and creditors to be informed about how a company is investing its funds, but also how its investing activities are being financed. Hopefully, the primary operations of the firm provide a source of internal financing. Information revealed in the cash flows from operating activities section of the statement of cash flows lets statement users know the extent of available internal financing. However, a major portion of financing for many companies is provided by external sources, specifically by shareholders and creditors.

FINANCIAL Reporting Case

Q3, p. 1215

● LO21–6

Cash flows from financing activities are both inflows and outflows of cash resulting from the external financing of a business. We include in this classification cash inflows from (a) the sale of common and preferred stock and (b) the issuance of bonds and other debt securities. Subsequent transactions related to these financing transactions, such as a buyback of stock (to retire the stock or as treasury stock), the repayment of debt, and the payment of cash dividends to shareholders, also are classified as financing activities.

Cash inflows and cash outflows due to the external financing of a business are reported as cash flows from financing activities.

For illustration, refer to Illustration 21–8 excerpted from the complete statement of cash flows of UBC in Illustration 21–2.

Cash Flows from Financing Activities:		
Sale of common shares	$ 26	
Retirement of bonds payable	(15)	
Payment of cash dividends	(5)	
Net cash flows from financing activities		$6

Illustration 21–8

Cash Flows from Financing Activities

Cash flows from financing activities include the sale or repurchase of shares, the issuance or repayment of debt securities, and the payment of cash dividends.

The cash received from the sale of common stock is reported as a financing activity. Since the sale of common stock is a financing activity, providing a cash return (dividend) to common shareholders also is a financing activity. Similarly, when the bonds being retired were issued in a prior year, that cash inflow was reported as a financing activity. In the current year, when the bonds are retired, that same amount of the resulting cash outflow is likewise classified as a financing activity.

At first glance, it may appear inconsistent to classify the payment of cash dividends to shareholders as a financing activity when, as stated earlier, paying interest to creditors is classified as an operating activity. But remember, cash flows from operating activities

Interest, unlike dividends, is a determinant of net income and therefore an operating activity.

should reflect the cash effects of items that enter into the determination of net income. Interest expense is a determinant of net income. A dividend, on the other hand, is a distribution of net income and not an expense.

Additional Consideration

Sometimes a cash receipt or a cash payment relates to more than one type of activity. For instance, proceeds from an insurance policy might be for a fire loss for a building that contained both equipment in use and inventory for sale. In that case, the cash proceeds pertaining to the building and the equipment is an investing activity and the part pertaining to inventory is an operating activity. It's not always clear how a cash flow should be allocated. When the "nature of the cash flows" cannot be determined by first looking for GAAP guidance, the classification should be based on the "predominance principle"—that is, judgment based on the activity that is likely to be the predominant source or use of cash flows.

RECONCILIATION WITH CHANGE IN CASH BALANCE One of the first items you may have noticed about UBC's statement of cash flows is that there is a net change in cash of $9 million. Is this a significant item of information provided by the statement? The primary objective of the statement of cash flows is not to tell us that cash increased by $9 million. We can readily see the increase or decrease in cash by comparing the beginning and ending balances in the cash account in comparative balance sheets. Instead, the purpose of the statement of cash flows is to explain *why* cash increased by $9 million.

To reinforce the fact that the net amount of cash inflows and outflows explains the change in the cash balance, the statement of cash flows includes a reconciliation of the net increase (or decrease) in cash with the company's beginning and ending cash balances. Notice, for instance, that on UBC's statement of cash flows, the reconciliation appears as follows:

The net amount of cash inflows and outflows reconciles the change in the company's beginning and ending cash balances.

Net increase in cash	$ 9
Cash balance, January 1	20
Cash balance, December 31	$29

● LO21–7

NONCASH INVESTING AND FINANCING ACTIVITIES Suppose UBC were to borrow $20 million cash from a bank, issuing a long-term note payable for that amount. This transaction would be reported on a statement of cash flows as a financing activity. Now suppose UBC used that $20 million cash to purchase new equipment. This second transaction would be reported as an investing activity.

Instead of two separate transactions, as indicated by Illustration 21–2, UBC acquired $20 million of new equipment by issuing a $20 million long-term note payable in a single transaction. Undertaking a significant investing activity and a significant financing activity as two parts of a single transaction does not diminish the value of reporting these activities. For that reason, transactions that do not increase or decrease cash, but that result in significant investing and financing activities, must be reported in related disclosures.

These **noncash investing and financing activities**, such as UBC's acquiring equipment (an investing activity) by issuing a long-term note payable (a financing activity), are reported in a separate disclosure schedule or note. UBC reported this transaction in the following manner:

Significant noncash investing and financing activities are reported also.

Noncash Investing and Financing Activities:
Acquired $20 million of equipment by issuing a 12%, 5-year note.

It's convenient to report noncash investing and financing activities on the same page as the statement of cash flows, as did UBC, only if there are few such transactions. Otherwise, precisely the same information would be reported in disclosure notes to the financial statements.[9]

[9] FASB ASC 230–10–50–6: Statement of Cash Flows—Overall—Disclosure—Noncash Investing and Financing Activities.

Examples of noncash transactions that would be reported in this manner are as follows:

1. Acquiring an asset by incurring a debt payable to the seller.
2. Acquiring use of an asset by entering into a lease agreement.
3. Converting debt into common stock or other equity securities.
4. Exchanging noncash assets or liabilities for other noncash assets or liabilities.

Noncash transactions that do not affect a company's assets or liabilities, such as the distribution of stock dividends, are not considered investing or financing activities and are not reported. Recall from Chapter 18 that stock dividends merely increase the number of shares of stock owned by existing shareholders. From an accounting standpoint, the stock dividend causes a dollar amount to be transferred from one part of shareholders' equity (retained earnings) to another part of shareholders' equity (paid-in capital). Neither assets nor liabilities are affected; therefore, no investing or financing activity has occurred.

Preparation of the Statement of Cash Flows

The objective in preparing the statement of cash flows is to identify all transactions and events that represent operating, investing, or financing activities and to list and classify those activities in proper statement format. A difficulty in preparing a statement of cash flows is that typical accounting systems are not designed to produce the specific information we need for the statement. At the end of a reporting cycle, balances exist in accounts reported in the income statement (sales revenue, cost of goods sold, etc.) and the balance sheet (accounts receivable, common stock, etc.). However, the ledger contains no balances for cash paid to acquire equipment, or cash received from sale of land, or any other cash flow needed for the statement. As a result, it's necessary to find a way of using available information to reconstruct the various cash flows that occurred during the reporting period. Typically, the information available to assist the statement preparer includes an income statement for the year and balance sheets for both the current and preceding years (comparative statements). The accounting records also can provide additional information about transactions that caused changes in account balances during the year. ● LO21–8

Additional Consideration

A transaction involving an investing and financing activity may be part cash and part noncash. For example, a company might pay cash for a part of the purchase price of new equipment and issue a long-term note for the remaining amount. In our previous illustration, UBC issued a note payable for the $20 million cost of the equipment it acquired. Suppose the equipment were purchased in the following manner:

Equipment	20	
Cash		6
Note payable		14

In that case, **$6** million would be reported under the caption "Cash flows from investing activities," and the noncash portion of the transaction—issuing a $14 million note payable for $14 million of equipment—would be reported as a "noncash investing and financing activity." UBC's statement of cash flows, if modified by the assumption of a part cash/part noncash transaction, would report these two elements of the transaction as follows:

Cash Flows from Investing Activities:		
Purchase of land	$(30)	
Purchase of short-term investments	(12)	
Sale of land	18	
Sale of equipment	5	
Purchase of equipment	**(6)**	
Net cash flows from investing activities		$(25)

(continued)

(concluded)

Noncash Investing and Financing Activities:
Acquired $20 million of equipment by paying cash and issuing a 12%, 5-year note as follows:

Cost of equipment	$20 million
Cash paid	(**6** million)
Note issued	$14 million

The typical year-end data is provided for UBC in Illustration 21–9. We have referred frequently to the statement of cash flows of UBC to illustrate the nature of the activities the statement reports. Now we'll see how that statement is developed from the data provided in that illustration.

Illustration 21–9

Comparative Balance Sheets and Income Statement

UNITED BRANDS CORPORATION
Comparative Balance Sheets
December 31, 2021 and 2020
($ in millions)

	2021		2020
Assets			
Cash	$ 29		$ 20
Accounts receivable	32		30
Short-term investments	12		0
Inventory	46		50
Prepaid insurance	3		6
Land	80		60
Buildings and equipment	81		75
Less: Accumulated depreciation	(16)		(20)
	$267		$221
Liabilities			
Accounts payable	$ 26		$ 20
Salaries payable	3		1
Income tax payable	6		8
Notes payable	20		0
Bonds payable	35		50
Less: Discount on bonds	(1)		(3)
Shareholders' Equity			
Common stock	130		100
Paid-in capital—excess of par	29		20
Retained earnings	19		25
	$267		$221

Income Statement

Revenues			
Sales revenue	$100		
Investment revenue	3		
Gain on sale of land	8	$111	
Expenses			
Cost of goods sold	60		
Salaries expense	13		

Illustration 21–9
(concluded)

Depreciation expense	3	
Bond interest expense	5	
Insurance expense	7	
Loss on sale of equipment	2	
Income tax expense	9	99
Net income		$ 12

Additional information from the accounting records:

a. Company land, purchased in a previous year for $10 million, was sold for $18 million.

b. Equipment that originally cost $14 million, and which was one-half depreciated, was sold for $5 million cash.

c. The common shares of Mazuma Corporation were purchased for $12 million as a short-term investment.

d. Property was purchased for $30 million cash for use as a parking lot.

e. On December 30, 2021, new equipment was acquired by issuing a 12%, five-year, $20 million note payable to the seller.

f. On January 1, 2021, $15 million of bonds (issued 20 years ago at their face amount) were retired at maturity.

g. The increase in the common stock account is attributable to the issuance of a 10% stock dividend (1 million shares) and the subsequent sale of 2 million shares of common stock. The market price of the $10 par value common stock was $13 per share on the dates of both transactions.

h. Cash dividends of $5 million were paid to shareholders.

In situations involving relatively few transactions, it's possible to prepare the statement of cash flows by merely inspecting the available data and logically determining the reportable activities. Few real-life situations are sufficiently simple to be solved this way. Usually, it is more practical to use some systematic method of analyzing the available data to ensure that all operating, investing, and financing activities are detected. A common approach is to use either a manual or electronic spreadsheet to organize and analyze the information used to prepare the statement.[10]

Reconstructing the events and transactions that occurred during the period helps identify the operating, investing, and financing activities to be reported.

Whether the statement of cash flows is prepared by an unaided inspection and analysis or with the aid of a systematic technique such as spreadsheet analysis, the analytical process is the same. To identify the activities to be reported in the statement, we use available data to reconstruct the events and transactions that involved operating, investing, and financing activities during the year. It is helpful to reproduce the journal entries that were recorded at the time of the transaction. Examining reconstructed journal entries makes it easier to visualize whether a reportable activity is involved and how that activity is to be classified.

Next, in Part B, we see how a spreadsheet simplifies the process of preparing a statement of cash flows. Even if you choose not to use a spreadsheet, the summary entries described can be used to help you find the cash inflows and outflows you need to prepare a statement of cash flows. For this demonstration, we assume the direct method is used to determine and report cash flows from operating activities. Appreciation of the direct method provides the backdrop for a thorough understanding of the indirect method that we explore in Part C.

[10]The T-account method is a second systematic approach to the preparation of the statement of cash flows. This method is identical in concept and similar in application to the spreadsheet method. The T-account method is used to prepare the statement of cash flows for UBC in Appendix 21B.

PART B

Preparing the SCF: Direct Method of Reporting Cash Flows from Operating Activities

Using a Spreadsheet

There can be no cash inflow or cash outflow without a corresponding change in a noncash account.

An important advantage gained by using a spreadsheet is that it ensures that no reportable activities are inadvertently overlooked. Spreadsheet analysis relies on the fact that, in order for cash to increase or decrease, there must be a corresponding change in an account other than cash. Therefore, if we can identify the events and transactions that caused the change in each noncash account during the year, we will have identified all the operating, investing, and financing activities that are to be included in the statement of cash flows.

Recording spreadsheet entries that explain account balance changes simultaneously identifies and classifies the activities to be reported on the statement of cash flows.

The beginning and ending balances of each account are entered on the spreadsheet. Then, as journal entries are reconstructed in our analysis of the data, those entries are recorded on the spreadsheet so that the debits and credits of the spreadsheet entries explain the changes in the account balances. Only after spreadsheet entries have explained the changes in all account balances can we feel confident that all operating, investing, and financing activities have been identified. The spreadsheet is designed in such a way that, as we record spreadsheet entries that explain account balance changes, we are simultaneously identifying and classifying the activities to be reported on the statement of cash flows.

We begin by transferring the comparative balance sheets and income statement to a blank spreadsheet. For illustration, refer to the 2021 and 2020 balances in the completed spreadsheet for UBC, shown in Illustration 21–9A. Notice that the amounts for elements of the income statement are ending balances resulting from accumulations during the year. Beginning balances in each of these accounts are always zero.

Following the balance sheets and income statement, we allocate space on the spreadsheet for the statement of cash flows. Although at this point we have not yet identified the specific cash flow activities shown in the completed spreadsheet, we can include headings for the major categories of activities: cash flows from operating activities, cash flows from investing activities, and cash flows from financing activities. Leaving several lines between headings allows adequate space to include the specific cash flows identified in subsequent analysis.

Spreadsheet entries duplicate the actual journal entries used to record the transactions as they occurred during the year.

The spreadsheet entries shown in the two changes columns, which separate the beginning and ending balances, explain the increase or decrease in each account balance. You will see in the next section how these entries were reconstructed. We number each entry and enter those numbers on the spreadsheet with the debit(s) and credit(s) of the entry as we go along. Although spreadsheet entries are in the form of debits and credits like journal entries, they are entered on the spreadsheet only. They are not recorded in the formal accounting records. In effect, these entries duplicate, frequently in summary form, the actual journal entries used to record the transactions as they occurred during the year.

To reconstruct the journal entries, we analyze each account, one at a time, deciding at each step what transaction or event caused the change in that account. Often, the reason for the change in an account balance is readily apparent from viewing the change in conjunction with that of a related account elsewhere in the financial statements. Sometimes it's necessary to consult the accounting records for additional information to help explain the transaction that resulted in the change.

You may find it helpful to diagram in T-account format the relationship between accounts to better visualize certain changes, particularly in your initial study of the chapter. The analysis that follows is occasionally supplemented with such diagrams to emphasize *why,* rather than merely *how,* specific cash flow amounts emerge from the analysis.

Although there is no mandatory order in which to analyze the accounts, it is convenient to begin with the income statement accounts, followed by the balance sheet accounts. We analyze the accounts of UBC in that order below. Although our analysis of each account culminates in a spreadsheet entry, keep in mind that the analysis described also is appropriate to identify reportable activities when a spreadsheet is not used.[11]

[11]The spreadsheet entries also are used to record the same transactions when the T-account method is used. We refer again to these entries when that method is described in Appendix 21B.

Illustration 21–9A

Spreadsheet—Direct Method

UNITED BRANDS CORPORATION
Spreadsheet for the Statement of Cash Flows

	Dec. 31, 2020	Changes				Dec. 31, 2021
		Debits		Credits		
Balance Sheet						
Assets						
Cash	20	(19)	9			29
Accounts receivable	30	(1)	2			32
Short-term investments	0	(12)	12			12
Inventory	50			(4)	4	46
Prepaid insurance	6			(8)	3	3
Land	60	(13)	30	(3)	10	80
Buildings and equipment	75	(14)	20X	(9)	14	81
Less: Accumulated depreciation	(20)	(9)	7	(6)	3	(16)
	221					267
Liabilities:						
Accounts payable	20			(4)	6	26
Salaries payable	1			(5)	2	3
Income tax payable	8	(10)	2			6
Notes payable	0			(14)	20X	20
Bonds payable	50	(15)	15			35
Less: Discount on bonds	(3)			(7)	2	(1)
Shareholders' Equity:						
Common stock	100			(16)	10	
				(17)	20	130
Paid-in capital—excess of par	20			(16)	3	
				(17)	6	29
Retained earnings	25	(16)	13			
		(18)	5	(11)	12	19
	221					267
Income Statement						
Revenues:						
Sales revenue				(1)	100	100
Investment revenue				(2)	3	3
Gain on sale of land				(3)	8	8
Expenses:						
Cost of goods sold		(4)	60			(60)
Salaries expense		(5)	13			(13)
Depreciation expense		(6)	3			(3)
Interest expense (bonds)		(7)	5			(5)
Insurance expense		(8)	7			(7)
Loss on sale of equipment		(9)	2			(2)
Income tax expense		(10)	9			(9)
Net income		(11)	**12**			**12**
Statement of Cash Flows						
Operating Activities:						
Cash inflows:						
From customers		(1)	98			
From investment revenue		(2)	3			
Cash outflows:						
To suppliers of goods				(4)	50	
To employees				(5)	11	
For interest				(7)	3	
For insurance				(8)	4	
For income tax				(10)	11	
Net cash flows						22

(continued)

Illustration 21–9A
(concluded)

	Dec. 31, 2020	Changes Debits		Changes Credits		Dec. 31, 2021
Investing Activities:						
Sale of land		(3)	18			
Sale of equipment		(9)	5			
Purchase of short-term investment				(12)	12	
Purchase of land				(13)	30	
Net cash flows						(19)
Financing Activities:						
Retirement of bonds payable				(15)	15	
Sale of common shares		(17)	26			
Payment of dividends				(18)	5	
Net cash flows						6
Net increase in cash				(19)	9	9
Totals			376		376	

X: As explained later, the **Xs** serve as a reminder to report this noncash transaction.

Income Statement Accounts

Amounts reported in the income statement usually are not the same as the cash effects of the items reported.

As described in an earlier section, cash flows from operating activities are inflows and outflows of cash that result from activities reported in the income statement. Thus, to identify those cash inflows and outflows, we begin by analyzing the components of the income statement. It is important to keep in mind that the amounts reported in the income statement usually do not represent the cash effects of the items reported. For example, UBC reports sales revenue of $100 million. This does not mean, however, that it collected $100 million cash from customers during the year. In fact, by referring to the beginning and ending balances in accounts receivable, we see that cash received from customers could not have been $100 million. Since accounts receivable increased during the year, some of the sales revenue recognized must not yet have been collected. This is explained further in the next section.

The cash effects of other income statement elements can be similarly discerned by referring to changes in the balances of the balance sheet accounts that are directly related to those elements. So, to identify cash flows from operating activities, we examine, one at a time, the elements of the income statement in conjunction with any balance sheet accounts affected by each element.

1. SALES REVENUE Accounts receivable is the balance sheet account that is affected by sales revenue. Specifically, accounts receivable is increased by credit sales and is decreased as cash is received from customers. We can compare sales and the change in accounts receivable during the year to determine the amount of cash received from customers. This relationship can be viewed in T-account format as follows:

Accounts Receivable

Beginning balance	30		
Credit sales *(increases A/R)*	100	?	Cash received *(decreases A/R)*
Ending balance	32		

We see from this analysis that cash received from customers must have been $98 million. Note that even if some of the year's sales were cash sales, say $40 million cash sales and $60 million credit sales, the result is the same:

Accounts Receivable

Beginning balance	30		Cash sales	$40
Credit sales	60	58 →	Received on account	58
Ending balance	32		Cash received	**$98**

Thus, cash flows from operating activities should include cash received from customers of **$98** million. The net effect of sales revenue activity during the year can be summarized in the following entry:

	($ in millions)	
Entry (1) Cash (received from customers)	**98**	
Accounts receivable (given)	2	
Sales revenue ($100 – $0)		100

Relating sales and the *change* in accounts receivable during the period helps determine the amount of cash received from customers.

The entry above appears as entry (1) in the completed spreadsheet for UBC, shown in Illustration 21–9A. The entry explains the changes in two account balances—accounts receivable and sales revenue. Since the entry affects cash, it also identifies a cash flow to be reported on the statement of cash flows. The **$98** million debit to cash is therefore entered in the statement of cash flows section of the spreadsheet under the heading of cash flows from operating activities.

Additional Consideration

The preceding discussion describes the most common situation—companies recognize revenue by selling goods and services, increase accounts receivable, and then collect the cash and decrease accounts receivable later. Some companies, though, often collect the cash in advance of satisfying performance obligations and thus record revenue and decrease deferred revenue. In those cases, we need to analyze any changes in the deferred revenue account for differences between revenue reported and cash collected. For instance, if UBC also had a $1 million increase in deferred revenue, the summary entry would be modified as follows:

	($ in millions)	
Entry (1) Cash (received from customers)	99	
Accounts receivable (given)	2	
Deferred revenue (given)		1
Sales revenue ($100 – $0)		100

Notice that we enter the cash portion of entry (1) as one of several cash flows on the statement of cash flows rather than as a debit to the cash account. Only after all cash inflows and outflows have been identified will the net change in cash be entered as a debit to the cash account. In fact, the entry to reconcile the $9 million increase in the cash account and the $9 million net increase in cash on the statement of cash flows will serve as a final check of the accuracy of our spreadsheet analysis.

Additional Consideration

Notice that bad debt expense does not appear in the income statement and allowance for uncollectible accounts does not appear in the balance sheet. We have assumed that bad debts are immaterial for UBC. When this is not the case, it's necessary to consider the write-off of bad debts as we determine cash received from customers. Here's why.

When using the allowance method to account for bad debts, a company estimates the dollar amount of customer accounts that will ultimately prove uncollectible and records both bad debt expense and allowance for uncollectible accounts for that estimate.

Bad debt expense	xxx	
Allowance for uncollectible accounts		xxx

(continued)

(concluded)

Then, when accounts actually prove uncollectible, accounts receivable and the allowance are reduced.

Allowance for uncollectible accounts	xxx	
Accounts receivable		xxx

In our illustration, we concluded that UBC received $2 million less cash ($98 million) than sales for the year ($100 million) because accounts receivable increased by that amount. However, if a portion of the change in accounts receivable had been due to write-offs of bad debts, that conclusion would be incorrect. Let's say, for instance, that UBC had bad debt expense of $2 million and its allowance for uncollectible accounts had increased by $1 million. Because the allowance for uncollectible accounts would be credited by $2 million in the adjusting entry for bad debts expense, necessarily there also would have been a $1 million debit to the account in order for there to have been a net increase (credit) in its balance of only $1 million. That debit would occur due to write-offs of bad debts totaling $1 million.

	($ in millions)	
Allowance for uncollectible accounts	1	
Accounts receivable		**1**

This would indicate that a portion ($1 million credit) of the total change in accounts receivable ($2 million debit) would have been due to write-offs of bad debts, and the remaining change ($3 million debit) would have been due to cash collections being less than sales revenue. Cash received from customers would have been only $97 million in that case. We can view this in the framework of our T-account analysis as follows:

Accounts Receivable

Beginning balance	30		
Credit sales	100	97	Cash received
		1	Bad debt write-offs
Ending balance	32		

The effect of write-offs of bad debts can be explicitly considered by combining all the accounts related to sales and collection activities into a single summary spreadsheet entry:

	($ in millions)	
Cash (received from customers)	97	
Accounts receivable ($32 – $30)	2	
Bad debt expense (from income statement)	2	
Allowance for uncollectible accounts ($3 – $2)		1
Sales revenue ($100 – $0)		100

This single entry summarizes all transactions related to sales, bad debt expense, write-offs of accounts receivable, and cash collections from sales.

The remaining summary entries are described in subsections 2 through 19. When including the entries on the spreadsheet, it is helpful to number the entries sequentially to provide a means of retracing the steps taken in the analysis if the need arises. You also may find it helpful to put a check mark (✓) to the right of the ending balance when the change in that balance has been explained. Then, once you have check marks next to every noncash account, you will know you are finished.

2. INVESTMENT REVENUE The income statement reports investment revenue of $3 million. Before concluding that this amount was received in cash, we first refer to the balance sheets to see whether a change in an account there indicates otherwise. A change in either of two balance sheet accounts, (a) investment revenue receivable or (b) long-term investments, might indicate that cash received from investment revenue differs from the amount reported in the income statement.

Changes in related accounts might indicate that investment revenue reported in the income statement is a different amount from cash received from the investment.

a. If we observe either an increase or a decrease in an *investment revenue receivable* account (e.g., interest receivable, dividends receivable), we would conclude that the amount of cash received during the year was less than (if an increase) or more than (if a decrease) the amount of revenue reported. The analysis would be identical to that of sales revenue and accounts receivable.

b. Also, an unexplained increase in a *long-term investment* account might indicate that a portion of investment revenue has not yet been received in cash. Recall from Chapter 12 that when using the equity method to account for investments in the stock of another corporation, investment revenue is recognized as the investor's percentage share of the investee's income, whether or not the revenue is received currently as cash dividends. For example, assume the investor owns 25% of the common stock of a corporation that reports net income of $12 million and pays dividends of $4 million. This situation would have produced a $2 million increase in long-term investments, which can be demonstrated by reconstructing the journal entries for the recognition of investment and the receipt of cash dividends:

	($ in millions)	
Long-term investments	3	
Investment revenue ($12 × 25%)		3
Cash ($4 × 25%)	1	
Long-term investments		1

A combined entry would produce the same results:

Long-term investments	2	
Cash ($4 × 25%)	1	
Investment revenue ($12 × 25%)		3

The $2 million net increase in long-term investments would represent the investment revenue not received in cash. This would also explain why there is a $3 million increase (credit) in investment revenue. If these events had occurred, we would prepare a spreadsheet entry identical to the combined entry above. The spreadsheet entry would (a) explain the $2 million increase in long-term investments, (b) explain the $3 million increase in investment revenue, and (c) identify a $1 million cash inflow from operating activities.

However, because neither an investment revenue receivable account nor a long-term investment account appears on the comparative balance sheets, we can conclude that $3 million of investment revenue was collected in cash. Entry (2) on the spreadsheet is as follows:

Because no other transactions are apparent that would have caused a change in investment revenue, we can conclude that $3 million of investment revenue was collected in cash.

	($ in millions)	
Entry (2) Cash (received from investment revenue)	3	
Investment revenue ($3 – $0)		3

3. GAIN ON SALE OF LAND The third item reported in the income statement is an $8 million gain on the sale of land. Recall that our objective in analyzing each element of the statement is to determine the cash effect of that element. To do so, we need additional information about the transaction that caused this gain. The accounting records—item (a) in Illustration 21–9—indicate that land that originally cost $10 million was sold for $18 million. The entry recorded in the journal when the land was sold also serves as our summary entry:

A gain (or loss) is simply the difference between cash received in the sale of an asset and the book value of the asset—not a cash flow.

	($ in millions)	
Entry (3) Cash (received from sale of land	18	
Land (given)		10
Gain on sale of land ($8 – $0)		8

The cash effect of this transaction is a cash increase of $18 million. We therefore include the debit as a cash inflow in the statement of cash flows section of the spreadsheet. However, unlike the cash effect of the previous two spreadsheet entries, it is not reported as an operating activity. The sale of land is an *investing* activity, so this cash inflow is listed under that heading of the spreadsheet. The entry also accounts for the $8 million gain on sale of land. The $10 million credit to land does not, by itself, explain the $20 million increase in that account. As we will later discover, another transaction also affected the land account.

It is important to understand that the gain is simply the difference between cash received in the sale of land (reported as an investing activity) and the book value of the land. To report the $8 million gain as a cash flow from operating activities, in addition to reporting $18 million as a cash flow from investing activities, would be to report the $8 million twice.

4. COST OF GOODS SOLD During the year, UBC sold goods that had cost $60 million. This does not necessarily indicate that $60 million cash was paid to suppliers of those goods. To determine the amount of cash paid to suppliers, we look to the two current accounts affected by merchandise purchases—inventory and accounts payable. The analysis can be viewed as a two-step process.

First, we compare cost of goods sold with the change in inventory to determine the cost of goods *purchased* (not necessarily cash paid) during the year. To facilitate our analysis, we can examine the relationship in T-account format:

Inventory

Beginning balance	50		
Cost of goods purchased *(increases inventory)*	**?**	60	Cost of goods sold *(decreases inventory)*
Ending balance	46		

Determining the amount of cash paid to suppliers means looking at not only the cost of goods sold, but also the changes in both inventory and accounts payable.

From this analysis, we see that **$56** million of goods were *purchased* during the year. It is not necessarily true, though, that **$56** million cash was paid to suppliers of these goods. By looking in accounts payable, we can determine the cash paid to suppliers:

Accounts Payable

		20	Beginning balance
Cash paid to suppliers *(decreases A/P)*	?	**56**	Cost of goods purchased *(increases A/P)*
		26	Ending Balance

We now see that cash paid to suppliers was $50 million. The entry that summarizes merchandise acquisitions is as follows:

Although $60 million of goods were sold during the year, only $50 million cash was paid to suppliers of these goods.

	($ in millions)	
Entry (4) Cost of goods sold ($60 – $0)	60	
Inventory ($46 – $50)		4
Accounts payable ($26 – $20)		6
Cash (paid to suppliers of goods)		**50**

5. SALARIES EXPENSE The balance sheet account affected by salaries expense is salaries payable. By analyzing salaries expense in relation to the change in salaries payable, we can determine the amount of cash paid to employees:

Salaries Payable

		1	Beginning balance
Cash paid to employees *(decreases salaries payable)*	**?**	13	Salaries expense *(increases salaries payable)*
		3	Ending Balance

This analysis indicates that only **$11** million cash was paid to employees; the remaining $2 million of salaries expense is reflected as an increase in salaries payable.

Viewing the relationship in journal entry format provides the same conclusion and also gives us the entry in our spreadsheet analysis:

	($ in millions)	
Entry (5) Salaries expense ($13 – $0)	13	
Salaries payable ($3 – $1)		2
Cash (paid to employees)		**11**

Although salaries expense was $13 million, only $11 million cash was paid to employees.

6. DEPRECIATION EXPENSE The income statement reports depreciation expense of $3 million. The entry used to record depreciation, which also serves as our summary entry, is

	($ in millions)	
Entry (6) Depreciation expense ($3 – $0)	3	
Accumulated depreciation		3

Depreciation expense does not require a current cash expenditure.

Depreciation is a noncash expense. It is merely an allocation in the current period of a prior cash expenditure (for the depreciable asset). Therefore, unlike the other entries to this point, the depreciation entry has no effect on the statement of cash flows. However, it does explain the change in the depreciation expense account and a portion of the change in accumulated depreciation.

7. INTEREST EXPENSE Recall from Chapter 14 that bond interest expense differs from the amount of cash paid to bondholders when bonds are issued at either a premium or a discount. The difference between the two amounts is the reduction of the premium or discount. By referring to the balance sheet, we see that UBC's bonds were issued at a discount. Since we know that interest expense is $5 million and that $2 million of the discount was reduced in 2021, we can determine that **$3** million cash was paid to bondholders by recreating the entry that summarizes the recording of bond interest expense.

	($ in millions)	
Entry (7) Interest expense ($5 – $0)	5	
Discount on bonds payable ($1 – $3)		2
Cash (paid for interest)		**3**

When bonds are issued at either a premium or a discount, bond interest expense is not the same as the amount of cash interest paid to bondholders.

Recording this entry on the spreadsheet explains the change in both the interest expense and discount on bonds payable accounts. It also provides us with another cash outflow from operating activities. Of course, if a premium were being reduced, rather than a discount, the outflow of cash would be *greater* than the expense.

Additional Consideration

If the balance sheet had revealed an increase or decrease in an accrued bond interest payable account, the entry calculating cash paid to bondholders would require modification. For example, if UBC had a bond interest payable account, and that account had increased (a credit) by $1 million, the entry would have been as follows:

	($ in millions)	
Entry (7) Interest expense	5	
(revised) Discount on bonds payable		2
Interest payable		1
Cash (paid to bondholders)		2

If the amount owed to bondholders increased by $1 million, they obviously were paid $1 million less cash than if there had been no change in the amount owed them. Similarly, if bond interest payable decreased by $1 million, the opposite would be true; that is, cash paid to them would have been $1 million more.

8. INSURANCE EXPENSE A decrease of $3 million in the prepaid insurance account indicates that cash paid for insurance coverage was $3 million less than the $7 million insurance expense for the year. Viewing prepaid insurance in T-account format clarifies this point.

Prepaid Insurance

Beginning balance	6		
Cash paid for insurance *(increases prepaid insurance)*	**?**	7	Insurance expense *(decreases prepaid insurance)*
Ending balance	3		

From this analysis, we can conclude that **$4** million was paid for insurance. We reach the same conclusion by preparing the following spreadsheet entry:

Since **$3** million of prepaid insurance was allocated to insurance expense, only **$4** million of the expense was paid in cash during the period.

	($ in millions)	
Entry (8) Insurance expense ($7 – $0)	7	
Prepaid insurance ($3 – $6)		3
Cash (paid for insurance)		**4**

The entry accounts for the change in both the insurance expense and prepaid insurance accounts and also identifies a cash outflow from operating activities.

9. LOSS ON SALE OF EQUIPMENT A $2 million loss on the sale of equipment is the next item reported in the income statement. To determine the cash effect of the sale of equipment, we need additional information about the transaction. The information we need is provided in item (b) of Illustration 21–9. Recreating the journal entry for the transaction described gives us the following entry:

Recreating the journal entry for the sale of equipment reveals a **$5** million cash inflow from investing activities.

	($ in millions)	
Entry (9) Cash (from the sale of equipment)	**5**	
Loss on sale of equipment ($2 – $0)	2	
Accumulated depreciation ($14 × 50%)	7	
Buildings and equipment (given)		14

The **$5** million cash inflow is entered in the statement of cash flows section of the spreadsheet as an investing activity. The $2 million debit to the loss on sale of equipment explains the change in that account balance. Referring to the spreadsheet, we see that a portion of the change in accumulated depreciation was accounted for in entry (6). The debit to accumulated depreciation in the entry above completes the explanation for the change in that account. However, the credit to buildings and equipment only partially justifies the change in that account. We must assume that the analysis of a subsequent transaction will account for the unexplained portion of the change.

Recognize too that the loss, like the gain in entry (3), has no cash effect in the current period. Therefore, it is not reported in the statement of cash flows when using the direct method.

10. INCOME TAX EXPENSE The final expense reported in the income statement is income tax expense. Since income tax payable is the balance sheet account affected by this expense, we look to the change in that account to help determine the cash paid for income tax. A T-account analysis can be used to find the cash effect as follows:

Income Tax Payable

		8	Beginning balance
Cash paid for income tax *(decreases the liability)*	?	9	Income tax expense
		6	Ending Balance

This analysis reveals that **$11** million cash was paid for income tax, $2 million more than the year's expense. The overpayment explains why the liability for income tax decreased by $2 million.

The same conclusion can be reached from the following summary entry, which represents the net effect of income tax on UBC's accounts.

	($ in millions)	
Entry (10) Income tax expense ($9 – $0)	9	
Income tax payable ($6 – $8)	2	
Cash (paid for income tax)		**11**

Additional Consideration

Entry (10) would require modification in the situation described below.

Note that UBC does not have a deferred tax account. Recall from Chapter 16 that temporary differences between taxable income and pretax accounting income give rise to deferred tax. If temporary differences had been present, which would be evidenced by a change in a deferred tax account, the calculation of cash paid for income tax would require modification. Assume, for example, that a deferred tax liability account had experienced a credit change of $1 million for the year. In that case, the previous spreadsheet entry would be revised as follows:

	($ in millions)	
Entry (10) Interest tax expense	9	
(revised) Income tax payable	2	
Deferred tax liability		1
Cash (paid for income tax)		10

As the revised entry indicates, only $10 million cash would have been paid in this situation, rather than $11 million. The $1 million difference represents the portion of the income tax expense whose payment is deferred to a later year.

11. NET INCOME The balance in the retained earnings account at the end of the year includes an increase due to net income. If we are to account for all changes in each of the accounts, we must include the following entry on the spreadsheet, which represents the closing of net income to retained earnings.

	($ in millions)	
Entry (11) Net income	12	
Retained earnings		12

This entry partially explains the change in the retained earnings account.

This entry does not affect amounts reported on the statement of cash flows. We include the entry on the spreadsheet analysis only to help explain account balance changes.

Balance Sheet Accounts

To identify all the operating, investing, and financing activities when using a spreadsheet, we must account for the changes in each account for both the income statement and the balance sheet. Thus far, we have explained the change in each income statement account. Since the transactions that gave rise to some of those changes involved balance sheet accounts as well, some changes in balance sheet accounts have already been explained. We now reconstruct the transactions that caused changes in the remaining balances.

With the exception of the cash account, the accounts are analyzed in the order of their presentation in the balance sheet. As noted earlier, we save the entry that reconciles the change in the cash account with the net change in cash from the statement of cash flows as a final check on the accuracy of the spreadsheet.

12. SHORT-TERM INVESTMENTS Since the change in accounts receivable was explained previously [in entry (1)], we proceed to the next asset in the balance sheet. The balance in short-term investments increased from zero to $12 million. In the absence of evidence to the contrary, we could assume that the increase is due to the purchase of short-term investments during the year. This assumption is confirmed by item (c) of Illustration 21–9.

The entry to record the investment and our summary entry is as follows:

The $12 million increase in the short-term investments account is due to the purchase of short-term investments during the year.

	($ in millions)	
Entry (12) Short-term investments ($12 – $0)	12	
Cash (purchase of short-term investment)		12

The $12 million cash outflow is entered in the statement of cash flows section of the spreadsheet as an investing activity. An exception is when an investment is classified as a "trading security," in which case the cash outflow is reported as an operating activity.

Additional Consideration

Recall that some highly liquid, short-term investments such as money market funds, Treasury bills, or commercial paper might be classified as cash equivalents. If the short-term investment above were classified as a cash equivalent, its purchase would have no effect on the total of cash and cash equivalents. In other words, since cash would include this investment, its purchase would constitute both a debit and a credit to cash. We would neither prepare a spreadsheet entry nor report the transaction on the statement of cash flows.

Likewise, a sale of a cash equivalent would not affect the total of cash and cash equivalents and would not be reported.

An exception would be if the cash equivalent investment were sold for either more or less than its acquisition cost. For example, assume a Treasury bill classified as a cash equivalent was sold for $1 million more than its $2 million cost. The sale would constitute both a $3 million increase and a $2 million decrease in cash. We see the effect more clearly if we reconstruct the transaction in journal entry format:

	($ in millions)	
Cash	3	
Gain on sale of cash equivalent		1
Cash (cash equivalent investment)		2

The spreadsheet entry to reflect the net increase in cash would be:

Entry (X) Cash (from sale of cash equivalents)	1	
Gain on sale of cash equivalent		1

The $1 million net increase in cash and cash equivalents would be reported as a cash inflow from *operating* activities.

When we discuss the "indirect method" of reporting operating activities later in the chapter, we will be adding noncash losses and subtracting noncash gains when adjusting net income for noncash income items. Unlike most losses and gains, though, we would not add a loss or deduct a gain on cash equivalents because they *do* affect "cash" as defined.

13. LAND The changes in the balances of both inventory and prepaid insurance were accounted for in previous summary entries (4) and (8). Land is the next account whose change has yet to be fully explained. We discovered in a previous transaction that a sale of land caused a $10 million reduction in the account. Yet, the account shows a net *increase* of $20 million. It would be logical to assume that the unexplained increase of $30 million was due to a purchase of land. The transaction described in item (d) of Illustration 21–9 supports that assumption and is portrayed in the following summary entry:

A $30 million purchase of land accounts for the portion of the $20 million increase in the account that was not previously explained by the sale of land.

	($ in millions)	
Entry (13) Land (given)	30	
Cash (purchase of land)		30

The $30 million payment is reported as a cash outflow from investing activities.

14. BUILDINGS AND EQUIPMENT When examining a previous transaction [entry (9)], we determined that the buildings and equipment account was reduced by $14 million from the sale of used equipment. And yet the account shows a net *increase* of $6 million for the year. The accounting records [item (e) of Illustration 21–9] reveal the remaining unexplained cause of the net increase. New equipment costing $20 million was purchased by issuing a $20 million note payable. Recall from the discussion in a previous section of this chapter that, although this is a noncash transaction, it represents both a significant investing activity (investing in new equipment) and a significant financing activity (financing the acquisition with long-term debt).

The journal entry used to record the transaction when the equipment was acquired also serves as our summary entry:

Investing in new equipment is a significant investing activity and financing the acquisition with long-term debt is a significant financing activity.

	($ in millions)	
Entry (14) Buildings and equipment (given)	20	
Notes payable (given)		20

Remember that the statement of cash flows section of the spreadsheet will serve as the basis for our preparation of the formal statement. But the noncash entry above will not affect the cash flows section of the spreadsheet. Because we want to report this noncash investing and financing activity when we prepare the statement of cash flows, it is helpful to "mark" the spreadsheet entry as a reminder not to overlook this transaction when the statement is prepared. Crosses (*X*) serve this purpose on the spreadsheet in Illustration 21–9A.

Additional Consideration

Leases

When a lessee acquires an asset and related liability as a result of a lease agreement, there is no inflow or outflow of cash. However, because a primary purpose of the statement of cash flows is to report significant operating, investing, and financing activities, the initial transaction is reported in the disclosure notes as a significant noncash investing activity (investing in the asset) and financing activity (financing it with debt). Then, the lessee's cash payments for leases are classified as financing activities in its statement of cash flows and presented separately from other financing cash flows. The lessor classifies its cash receipts from lease payments as operating activities in its statement of cash flows, after initially reporting its acquisition of a lease receivable and derecognition of the asset under lease as a significant noncash investing and financing activity in its cash flow disclosure note.

15. BONDS PAYABLE The balance in the bonds payable account decreased during the year by $15 million. Illustration 21–9, item (f), reveals the cause. Cash was paid to retire $15 million face value of bonds. The spreadsheet entry that duplicates the journal entry that was recorded when the bonds were retired is as follows:

	($ in millions)	
Entry (15) Bonds payable ($35 – $50)	15	
Cash (retirement of bonds payable)		15

The cash outflow is reported as a financing activity.

Additional Consideration

Statement of Cash Flows: Reporting a Bond Discount or Premium

Conceptually, the way we report cash flows for bonds depends on whether the bonds were issued at face amount, at a discount, or at a premium.[12] As we look at these three possibilities, note that in each case

- The same total amount that is borrowed (lent) and reported as a financing (investing) activity is the amount that should be reported later as a *financing* (*investing*) activity as the loan is repaid.
- All other amounts paid during or at the end of the term to maturity represent interest and should be reported among *operating* activities.

Though not codified in the *FASB Accounting Standards Codification,* this treatment is a preference of the SEC, though anecdotal evidence suggests that many firms do not follow this procedure in practice.

Bonds Issued at Face Amount

If bonds are issued at their face amount, the cash inflow at issuance ($15 million in the illustration) is a financing activity, as is the cash outflow at maturity to repay the amount borrowed ($15 million in the illustration). The periodic cash interest payments in the meantime are operating activities.

Bonds Issued at a Discount

If bonds are issued at a discount, the cash inflow at issuance is a financing activity, but that discounted price is less than the cash outflow for the face amount at maturity. The portion of the face amount paid at maturity to be reported as a financing activity, then, is limited to the amount borrowed (discounted price at issuance). The remaining portion of the payment (equal to the discount) represents interest and thus should be reported as an operating activity. This "extra" amount paid at maturity, together with the periodic cash interest payments during the term to maturity, comprise the total interest paid to borrow the cash received at issuance. This total amount of interest reported among operating activities is equal to the total interest expense reported during the term to maturity.

In Chapter 14's Illustration 14–3, for instance, $700,000 of bonds were issued at $666,633. That discounted price is the amount the issuer borrowed and the amount reported as a financing activity. When the $700,000 face amount is paid at maturity, $666,633 of that amount represents repayment of the borrowing and also is reported as a financing activity. The remaining $33,367 (the discount) represents interest (as do the periodic cash payments) and is reported among operating activities.

(continued)

[12]Lisa Koonce, "SCF Insights," Financial Accounting Standards Research Initiative, October 6, 2011.

(concluded)

We can better visualize this if we look again at the amortization schedule we saw earlier for the bonds issued at a discount

Date	Cash Interest	Effective Interest	Increase in Balance	Outstanding Balance
	(6% × Face amount)	(7% × Outstanding balance)	(Discount reduction)	
1/1/21				**666,633**
6/30/21	42,000	.07 (666,633) = 46,664	4,664	671,297
12/31/21	42,000	.07 (671,297) = 46,991	4,991	676,288
6/30/22	42,000	.07 (676,288) = 47,340	5,340	681,628
12/31/22	42,000	.07 (681,628) = 47,714	5,714	687,342
6/30/23	42,000	.07 (687,342) = 48,114	6,114	693,456
12/31/23	42,000	.07 (693,456) = 48,544	6,544	700,000
	252,000	**285,367**	33,367	

Operating

$700,000	Payment
− 33,367	Operating
$666,633	Financing

You may remember from Chapter 14 that the *total* interest, $285,367, is the $252,000 *cash* interest paid during the term to maturity plus the "extra" amount paid at maturity. The $33,367 extra amount is necessary for the bonds to pay an effective interest rate equal to what the market demanded when the bonds were purchased (which was higher than the rate at which the bonds made periodic interest payments).

To summarize, bonds issued at a discount are treated as follows on a statement of cash flows:

	Financing Activities	Operating Activities
INFLOWS:		
At issuance	**Face Amount minus Discount**	
OUTFLOWS:		
During term to maturity		**Cash Interest Payments**
At maturity	**Face Amount minus Discount**	**Discount**

Bonds Issued at a Premium

If the bonds are issued at a premium, the cash repaid at maturity is less than the cash received at issuance. So, when does the rest of the amount received at issuance get repaid? To see the answer, remember that when bonds are issued at a premium, it's because the stated interest rate at issuance is higher than the market interest rate. That means that the cash interest payments each period are higher than the interest expense amounts reported each period. The remainder of each payment is a repayment of the extra amount received at issuance. Similar to installment payments that include both (a) an amount for interest and (b) an amount that reduces the debt, we can think of these bond interest payments the same way. The cash payments cover the interest expense, and the remainder of each payment (the premium amortization) goes toward paying a portion of the amount borrowed. The interest expense portions of the cash payments are operating activities and the repayment portions are reported as financing activities each period. That means that the total amount of financing *outflows*, including payment of the face amount at maturity and the premium amortization over the life of the bonds, equals the total amount of financing *inflow* received when the bonds are issued.

(continued)

(concluded)

To see this more clearly, let's look again at the amortization schedule we saw in Chapter 14 for bonds issued at a premium:

Date	Cash Interest	Effective Interest	Decrease in Balance	Outstanding Balance
	(6% × Face amount)	(5% × Outstanding balance)	(Premium reduction)	
1/1/21				**735,533**
6/30/21	42,000	0.05 (735,533) = 36,777	**5,223**	730,310
12/31/21	42,000	0.05 (730,310) = 36,516	**5,484**	724,826
6/30/22	42,000	0.05 (724,826) = 36,241	**5,759**	719,067
12/31/22	42,000	0.05 (719,067) = 35,953	**6,047**	713,020
6/30/23	42,000	0.05 (713,020) = 35,651	**6,349**	706,671
12/31/23	42,000	0.05 (706,671) = 35,329	**6,671**	**700,000**
	252,000	**216,467**	**35,533**	

Operating

$700,000	Payment
+ **35,533**	**Financing**
$735,633	**Financing**

To summarize, bonds issued at a premium are treated as follows on a statement of cash flows:

	Financing Activities	Operating Activities
INFLOWS: *At issuance*	**Face Amount plus Premium**	
OUTFLOWS: *During term to maturity*	**Premium Amortization**	**Interest Expense**
At maturity	**Face Amount**	

Remember that when using the **indirect method**, we add any discount amortization back to net income each period. That's in order to adjust the interest expense that's subtracted within net income to cash interest paid, which is the amount we want to report as an operating activity. When we have a premium, though, the amount we want to report as an operating activity is interest expense, as we just saw. As a result, we do *not* adjust net income for premium amortization as we do for discount amortization.

16–17. COMMON STOCK The comparative balance sheets indicate that the common stock account balance increased by $30 million. We look to the accounting records—Illustration 21–9, item (g)—for an explanation. Two transactions, a stock dividend and a sale of new shares of common stock, combined to cause the increase. To create the summary entries for our analysis, we replicate the journal entries for the two transactions as described below.

Remember from Chapter 18 that to record a small stock dividend, we capitalize retained earnings for the market value of the shares distributed—in this case, 1 million shares times $13 per share, or $13 million. The entry is as follows:

Although this transaction does not identify a cash flow, nor does it represent an investing or financing activity, we include the summary entry to help explain changes in the three account balances affected.

		($ in millions)	
Entry (16)	Retained earnings (1 million shares × $13)	13	
	Common stock (1 million shares × $10 par)		10
	Paid-in capital—excess of par (difference)		3

Also recall from the discussion of noncash investing and financing activities earlier in the chapter that stock dividends do not represent a significant investing or financing activity. Therefore, this transaction is not reported in the statement of cash flows. We include the entry in our spreadsheet analysis only to help explain changes in the account balances affected.

The sale of 2 million shares of common stock at $13 per share is represented by the following spreadsheet entry:

The sale of common shares explains the remaining increase in the common stock account and the remaining increase in paid-in capital—excess of par.

	($ in millions)	
Entry (17) Cash (from sale of common stock)	26	
Common stock ([$130 – $100] – $10)		20
Paid-in capital—excess of par ([$29 – $20] – $3)		6

The cash inflow is reported in the statement of cash flows as a financing activity.

Additional Consideration

If cash is paid to retire outstanding shares of stock or to purchase those shares as treasury stock, the cash outflow would be reported in a statement of cash flows as a financing activity.

Together, the two entries above account for both the $30 million increase in the common stock account and the $9 million increase in paid-in capital—excess of par.

18. RETAINED EARNINGS The stock dividend in entry (16) above includes a $13 million reduction of retained earnings. Previously, we saw in entry (11) that net income increased retained earnings by $12 million. The net reduction of $1 million accounted for by these two entries leaves $5 million of the $6 million net decrease in the account unexplained.

Retained Earnings

		25	Beginning balance
(16) Stock dividend	13	12	Net income (11)
(18) ?	?		
		19	Ending balance

Without additional information about the $5 million decrease in retained earnings, we might assume it was due to a $5 million cash dividend. This assumption is unnecessary, though, because the cash dividend is described in Illustration 21–9, item (h).

Retained Earnings

		25	Beginning balance
(16) Stock dividend	13	12	Net income (11)
(18) **Cash dividend**	**5**		
		19	Ending balance

The summary entry for the spreadsheet is as follows:

The cash dividend accounts for the previously unexplained change in retained earnings.

	($ in millions)	
Entry (18) Retained earnings	5	
Cash (payment of cash dividends)		5

19. COMPLETING THE SPREADSHEET In preparing the spreadsheet to this point, we have analyzed each noncash account on both the income statement and the balance sheet. Our purpose was to identify the transactions that, during the year, had affected each account. By recreating each transaction in the form of a summary entry—in effect, duplicating the journal entry that had been used to record the transaction—we were able to explain the change in the balance of each account. That is, the debits and credits in the changes columns of the spreadsheet account for the increase or decrease in each noncash account. When a transaction being entered on the spreadsheet included an operating, investing, or financing activity, we entered that portion of the entry under the corresponding heading of the statement of cash flows section of the spreadsheet. Since, as noted earlier, there can be no operating, investing, or financing activity without a corresponding change in one or more of the noncash accounts, we should feel confident at this point that we have identified all of the activities that should be reported on the statement of cash flows.

The cash flows section of the spreadsheet provides the information to be reported in the statement of cash flows.

To check the accuracy of the analysis, we compare the change in the balance of the cash account with the net change in cash flows produced by the activities listed in the statement of cash flows section of the spreadsheet. The net increase or decrease in cash flows from each of the statement of cash flows categories is extended to the extreme right column of the spreadsheet. By reference to Illustration 21–9A, we see that net cash flows from operating, investing, and financing activities are: $22 million, ($19 million), and $6 million, respectively. Together these activities provide a net increase in cash of $9 million. This amount corresponds to the increase in the balance of the cash account from $20 million to $29 million. To complete the spreadsheet, we include the final summary entry:

	($ in millions)	
Entry (19) Cash	9	
Net increase in cash (from statement of cash flows activities)		9

As a final check of accuracy, we can confirm that the total of the debits is equal to the total of the credits in the changes columns of the spreadsheet.[13]

Ethical Dilemma

"We must get it," Courtney Lowell, president of Industrial Fasteners, roared. "Without it, we're in big trouble." The "it" Mr. Lowell referred to is the renewal of a $14 million loan with Community First Bank. The big trouble he fears is the lack of funds necessary to repay the existing debt and few, if any, prospects for raising the funds elsewhere.

Mr. Lowell had just hung up the phone after a conversation with a bank vice-president in which it was made clear that this year's statement of cash flows must look better than last year's. Mr. Lowell knows that improvements are not on course to happen. In fact, cash flow projections were dismal.

Later that day, Tim Cratchet, assistant controller, was summoned to Mr. Lowell's office. "Cratchet," Lowell barked, "I've looked at our accounts receivable. I think we can generate quite a bit of cash by selling or factoring most of those receivables. I know it will cost us more than if we collect them ourselves, but it sure will make our cash flow picture look better."

Is there an ethical question facing Cratchet?

[13]The mechanical and computational aspects of the spreadsheet analysis are simplified greatly when performed on an electronic spreadsheet such as Microsoft Excel.

International Financial Reporting Standards

Classification of Cash Flows. Like U.S. GAAP, international standards also require a statement of cash flows. Consistent with U.S. GAAP, cash flows are classified as operating, investing, or financing. However, the U.S. standard designates cash outflows for interest payments and cash inflows from interest and dividends received as operating cash flows. Dividends paid to shareholders are classified as financing cash flows.

● LO21–9

IAS No. 7, on the other hand, allows more flexibility. Companies can report interest and dividends paid as either operating or financing cash flows and interest and dividends received as either operating or investing cash flows. Interest and dividend payments usually are reported as financing activities. Interest and dividends received normally are classified as investing activities.

Typical Classification of Cash Flows from Interest and Dividend

U.S. GAAP	IFRS
Operating Activities	***Operating Activities***
Dividends received	
Interest received	
Interest paid	
Investing Activities	***Investing Activities***
	Dividends received
	Interest received
Financing Activities	***Financing Activities***
Dividends paid	Dividends paid
	Interest paid

The spreadsheet is now complete. The statement of cash flows can now be prepared directly from the spreadsheet simply by presenting the items included in the statement of cash flows section of the spreadsheet in the appropriate format of the statement. It will be the statement of cash flows we first saw in Illustration 21–2 early in this chapter.

The statements of cash flows from an annual report of **CVS Health Corp.** are shown in Illustration 21–10. Notice that the reconciliation schedule was reported by CVS in the statement of cash flows itself. Many companies report the schedule separately in the disclosure notes.

Illustration 21–10 Statement of Cash Flows—CVS Health Corp.

CVS HEALTH CORP.
Consolidated Statements of Cash Flows

	Year Ended December 31,		
($ in millions)	**2017**	**2016**	**2015**
Cash flows from operating activities:			
Cash receipts from customers	$176,594	$ 172,310	$ 148,954
Cash paid for inventory and prescriptions dispensed by retail network pharmacies	(149,279)	(142,511)	(122,498)
Cash paid to other suppliers and employees	(15,348)	(15,478)	(14,035)
Interest received	21	20	21
Interest paid	(1,072)	(1,140)	(629)
Income taxes paid	(2,909)	(3,060)	(3,274)
Net cash provided by operating activities	8,007	10,141	8,539

(continued)

Illustration 21–10 (concluded)

Real World Financials

($ in millions)	Year Ended December 31, 2017	2016	2015
Cash flows from investing activities:			
Purchases of property and equipment	(1,918)	(2,224)	(2,367)
Proceeds from sale-leaseback transactions	265	230	411
Proceeds from sale of property and equipment and other assets	33	37	35
Acquisitions (net of cash acquired) and other investments	(1,287)	(539)	(11,475)
Purchase of available-for-sale investments	(86)	(65)	(267)
Maturity of available-for-sale investments	61	91	243
Net cash used in investing activities	(2.932)	(2,470)	(13,420)
Cash flows from financing activities:			
Increase (decrease) in short-term debt	(598)	1,874	(685)
Proceeds from issuance of long-term debt	—	3,455	14,805
Repayments of long-term debt	—	(5,943)	(2,902)
Purchase of noncontrolling interest in subsidiary	—	(39)	—
Payment of contingent consideration	—	(26)	(58)
Dividends paid	(2,049)	(1,840)	(1,576)
Proceeds from exercise of stock options	329	296	362
Payments for taxes related to net share settlement of equity awards	(71)	(72)	(63)
Repurchase of common stock	(4,361)	(4,461)	(5,001)
Other	(1)	(5)	(3)
Net cash provided by (used in) financing activities	(6,751)	(6,761)	(4,879)
Effect of exchange rate changes on cash and cash equivalents	1	2	(20)
Net increase (decrease) in cash and cash equivalents:	(1,675)	912	(22)
Cash and cash equivalents at the beginning of the year	3,371	2,459	2,481
Cash and cash equivalents at the end of the year	$ 1,696	$ 3,371	$ 2,459
Reconciliation of net income to net cash provided by operating activities:			
Net income	$ 6,623	$ 5,319	$ 5,239
Adjustments required to reconcile net income to net cash provided by operating activities:			
Depreciation and amortization	2,479	2,475	2,092
Goodwill impairments	181	—	—
Losses on settlements of defined benefit pension plans	187	—	—
Stock-based compensation	234	222	230
Loss on early extinguishment of debt	—	643	—
Deferred income taxes and other noncash items	(1,334)	18	(252)
Change in operating assets and liabilities, net of effects from acquisitions:	53	135	(14)
Accounts receivable, net	(941)	(243)	(1,594)
Inventories	(514)	(742)	(1,141)
Other current assets	(341)	35	355
Other assets	3	(43)	2
Accounts payable and claims and discounts payable	1,710	2,189	2,834
Accrued expenses	(371)	131	892
Other long-term liabilities	38	2	(104)
Net cash provided by operating activities	$ 8,007	$ 10,141	$ 8,539

Source: CVS Health Corp.

Concept Review Exercise

COMPREHENSIVE REVIEW

The comparative balance sheets for 2021 and 2020 and the income statement for 2021 are given below for Beneficial Drill Company. Additional information from Beneficial Drill's accounting records is provided also.

Required:

Prepare the statement of cash flows of Beneficial Drill Company for the year ended December 31, 2021. Present cash flows from operating activities by the direct method and use a spreadsheet to assist in your analysis.

BENEFICIAL DRILL COMPANY
Comparative Balance Sheets
December 31, 2021 and 2020
($ in millions)

Assets	2021	2020
Cash	$ 24	$ 41
Accounts receivable	94	96
Investment revenue receivable	3	2
Inventory	115	110
Prepaid insurance	2	3
Long-term investments	77	60
Land	110	80
Buildings and equipment	220	240
Less: Accumulated depreciation	(35)	(60)
Patent	15	16
	$625	$588
Liabilities		
Accounts payable	$ 23	$ 30
Salaries payable	2	5
Interest payable	4	2
Income tax payable	6	8
Deferred tax liability	5	4
Notes payable	15	0
Bonds payable	150	130
Less: Discount on bonds	(9)	(10)
Shareholders' Equity		
Common stock	210	200
Paid-in capital—excess of par	44	40
Retained earnings	182	179
Less: Treasury stock (at cost)	(7)	0
	$625	$588

BENEFICIAL DRILL COMPANY
Income Statement
For Year Ended December 31, 2021
($ in millions)

Revenues		
Sales revenue	$200	
Investment revenue	6	
Gain on sale of treasury bills	1	$207
Expenses and losses		
Cost of goods sold	110	
Salaries expense	30	
Depreciation expense	5	
Amortization expense	1	
Insurance expense	3	
Interest expense	14	
Loss on sale of equipment	10	
Income tax expense	7	(180)
Net income		$ 27

Additional information from the accounting records:

a. Investment revenue includes Beneficial Drill Company's $3 million share of the net income of Hammer Company, an equity method investee.
b. Treasury bills were sold during 2021 at a gain of $1 million. Beneficial Drill Company classifies its investments in Treasury bills as cash equivalents.
c. Equipment that originally cost $60 million and was one-half depreciated was rendered unusable by a bolt of lightning. Most major components of the machine were unharmed and were sold for $20 million.
d. Temporary differences between pretax accounting income and taxable income caused the deferred tax liability to increase by $1 million.
e. The common stock of Wrench Corporation was purchased for $14 million as a long-term investment.
f. Land costing $30 million was acquired by paying $15 million cash and issuing a 13%, seven-year, $15 million note payable to the seller.
g. New equipment was purchased for $40 million cash.
h. $20 million of bonds were issued at face value.
i. On January 19, Beneficial issued a 5% stock dividend (1 million shares). The market price of the $10 par value common stock was $14 per share at that time.
j. Cash dividends of $10 million were paid to shareholders.
k. In November, 500,000 common shares were repurchased as treasury stock at a cost of $7 million.

Solution:

BENEFICIAL DRILL COMPANY
Spreadsheet for the Statement of Cash Flows

	Dec. 31 2020	Changes Debits		Credits		Dec. 31 2021
Balance Sheet						
Assets						
Cash	41			(20)	17	24
Accounts receivable	96			(1)	2	94
Investment revenue receivable	2	(2)	1			3
Inventory	110	(4)	5			115
Prepaid insurance	3			(8)	1	2
Long-term investments	60	(2)	3			
		(13)	14			77
Land	80	(14)	30X			110
Buildings and equipment	240	(15)	40	(10)	60	220
Less: Accumulated depreciation	(60)	(10)	30	(6)	5	(35)
Patent	16			(7)	1	15
	588					625
Liabilities						
Accounts payable	30	(4)	7			23
Salaries payable	5	(5)	3			2
Interest payable	2			(9)	2	4
Income tax payable	8	(11)	2			6
Deferred tax liability	4			(11)	1	5
Notes payable	0			(14)	15X	15
Bonds payable	130			(16)	20	150
Less: Discount on bonds	(10)			(9)	1	(9)
Shareholders' Equity						
Common stock	200			(17)	10	210

(continued)

(concluded)

BENEFICIAL DRILL COMPANY
Spreadsheet for the Statement of Cash Flows

	Dec. 31 2020	Changes Debits		Credits		Dec. 31 2021
Paid-in capital—excess of par	40			(17)	4	44
Retained earnings	179	(17)	14			
		(18)	10	(12)	27	182
Less: Treasury stock	0	(19)	7			(7)
	588					625
Income Statement						
Revenues:						
Sales revenue				(1)	200	200
Investment revenue				(2)	6	6
Gain on sale of Treasury bills				(3)	1	1
Expenses:						
Cost of goods sold		(4)	110			(110)
Salaries expense		(5)	30			(30)
Depreciation expense		(6)	5			(5)
Amortization expense		(7)	1			(1)
Insurance expense		(8)	3			(3)
Interest expense		(9)	14			(14)
Loss on sale of equipment		(10)	10			(10)
Income tax expense		(11)	7			(7)
Net income		(12)	27			27
Statement of Cash Flows						
Operating Activities:						
Cash inflows:						
From customers		(1)	202			
From investment revenue		(2)	2			
From sale of Treasury bills		(3)	1			
Cash outflows:						
To suppliers of goods				(4)	122	
To employees				(5)	33	
For insurance				(8)	2	
For bond interest				(9)	11	
For income tax				(11)	8	
Net cash flows						29
Investing Activities:						
Sale of equipment		(10)	20			
Purchase of LT investments				(13)	14	
Purchase of land				(14)	15	
Purchase of equipment				(15)	40	
Net cash flows						(49)
Financing Activities:						
Issuance of bonds		(16)	20			
Payment of dividends				(18)	10	
Purchase of treasury stock				(19)	7	
Net cash flows						3
Net decrease in cash		(20)	17			(17)
Totals			635		635	

BENEFICIAL DRILL COMPANY
Statement of Cash Flows
For Year Ended December 31, 2021
($ in millions)

Cash Flows from Operating Activities		
Cash inflows:		
From customers	$ 202	
From investment revenue	2	
From sale of Treasury bills	1	
Cash outflows:		
To suppliers of goods	(122)	
To employees	(33)	
For insurance	(2)	
For interest	(11)	
For income tax	(8)	
Net cash flows from operating activities		$ 29
Cash Flows from Investing Activities		
Sale of equipment	$ 20	
Purchase of long-term investments	(14)	
Purchase of land	(15)	
Purchase of equipment	(40)	
Net cash flows from investing activities		(49)
Cash Flows from Financing Activities		
Issuance of bonds	$ 20	
Payment of dividends	(10)	
Purchase of treasury stock	(7)	
Net cash flows from financing activities		3
Net decrease in cash		$(17)
Cash balance, January 1		41
Cash balance, December 31		$ 24
Noncash Investing and Financing Activities		
Acquired $30 million of land by paying cash and issuing a 13%, 7-year note as follows:		
Cost of land	$ 30	
Cash paid	15	
Note issued	$ 15	

PART C

Preparing the SCF: Indirect Method of Reporting Cash Flows from Operating Activities

Getting There through the Back Door

The presentation of cash flows from operating activities illustrated in Part B is referred to as the *direct method.* By this method, the cash effect of each operating activity (i.e., income statement item) is reported directly on the statement of cash flows. For instance, cash received from customers is reported as the cash effect of sales activities, and cash paid to suppliers is reported as the cash effect of cost of goods sold. Income statement items that have *no* cash effect, such as depreciation expense, gains, and losses, are simply not reported.

As we discussed previously, a permissible alternative is the *indirect method,* by which the net cash increase or decrease from operating activities is derived indirectly by starting with reported net income and working backwards to convert that amount to a cash basis. The derivation by the indirect method of net cash flows from operating activities for UBC is shown in Illustration 21–9B. For the adjustment amounts, you may wish to refer back to UBC's balance sheets and income statement presented in Illustration 21–9.

Illustration 21–9B

Indirect Method

The indirect method derives the net cash increase or decrease from operating activities indirectly, by starting with reported net income and "working backwards" to convert that amount to a cash basis.

Cash Flows from Operating Activities—Indirect Method
and
Reconciliation of Net Income to
Net Cash Flows from Operating Activities

Net Income	$12
Adjustments for noncash effects:	
Gain on sale of land	(8)
Depreciation expense	3
Loss on sale of equipment	2
Changes in operating assets and liabilities:	
Increase in accounts receivable	(2)
Decrease in inventory	4
Decrease in prepaid insurance	3
Increase in accounts payable	6
Increase in salaries payable	2
Decrease in income tax payable	(2)
Decrease in ciscount on bonds payable	2
Net cash flows from operating activities	$22

Notice that the indirect method yields the same $22 million net cash flows from operating activities as does the direct method. This is understandable when you consider that the indirect method simply reverses the differences between the accrual-based income statement and cash flows from operating activities. We accomplish this as described in the next two sections.

Components of Net Income that do not Increase or Decrease Cash

Amounts that were subtracted in determining net income but did not reduce cash are *added back* to net income to reverse the effect of their having been subtracted. For example depreciation expense and the loss on sale of equipment are added back to net income. Other things being equal, this restores net income to what it would have been had depreciation and the loss not been subtracted at all.

Similarly, amounts that were added in determining net income but did not increase cash are subtracted from net income to reverse the effect of their having been added. For example, UBC's gain on sale of land is deducted from net income. Here's why. UBC sold for $18 million land that originally cost $10 million. Recording the sale produced a gain of $8 million, which UBC appropriately included in its income statement. But did this gain increase UBC's cash? No. Certainly selling the land increased cash—by $18 million. We therefore include the $18 million as a cash inflow in the statement of cash flows. However, the sale of land is an investing activity. The gain itself, though, is simply the difference between cash received in the sale of land (reported as an investing activity) and the original cost of the land. If UBC also reported the $8 million gain as a cash flow from operating activities, in addition to reporting $18 million as a cash flow from investing activities, UBC would report the $8 million twice. So, because UBC added the gain in determining its net income but the gain had no effect on cash, the gain must now be subtracted from net income to reverse the effect of its having been added.

Components of Net Income That Do Increase or Decrease Cash

For components of net income that increase or decrease cash, but by an amount different from that reported in the income statement, net income is adjusted for changes in the balances of related balance sheet accounts to *convert the effects of those items to a cash basis.*

For example, sales of $100 million are included in the income statement as a component of net income, and yet, since accounts receivable increased by $2 million, only $98 million cash was collected from customers during the reporting period. Sales are converted to a cash basis by subtracting the $2 million increase in accounts receivable.

Here's another example. The income statement reports salaries expense as $13 million. Just because employees earned $13 million during the reporting period, though, doesn't necessarily mean UBC paid those employees $13 million in cash during the same period. In fact, we see in the comparative balance sheets that salaries payable increased from $1 million to $3 million; UBC owes its employees $2 million more than before the year started. The company must not have paid the entire $13 million expense. By analyzing salaries expense in relation to the change in salaries payable, we can determine the amount of cash paid to employees, as shown next:

Salaries Payable

		1	Beginning balance
Cash paid to employees *(decreases salaries payable)*	?	13	Salaries expense *(increases salaries payable)*
		3	Ending balance

This inspection indicates that UBC paid only $11 million cash to its employees; the remaining $2 million of salaries expense is reflected as an increase in salaries payable. From a cash perspective, then, by subtracting $13 million for salaries in the income statement, UBC has subtracted $2 million more than the reduction in cash. Adding back the $2 million leaves UBC in the same position as if it had deducted only the $11 million cash paid to employees.

Following a similar analysis of the cash effects of the remaining components of net income, those items are likewise converted to a cash basis by adjusting net income for increases and decreases in related accounts.

For components of net income that increase or decrease cash by an amount exactly the same as that reported in the income statement, no adjustment of net income is required. For example, investment revenue of $3 million is included in UBC's $12 million net income amount. Because $3 million also is the amount of cash received from that activity, this element of net income already represents its cash effect and needs no adjustment.[14]

Comparison with the Direct Method

The indirect method is compared with the direct method in Illustration 21–11, using the data of UBC. To better illustrate the relationship between the two methods, the adjustments to net income using the indirect method are presented parallel to the related cash inflows and cash outflows of the direct method. The income statement is included in the graphic to demonstrate that the indirect method also serves to reconcile differences between the elements of that statement and the cash flows reported by the direct method.

As a practical consideration, you might notice that the adjustments to net income using the indirect method follow a convenient pattern. *Increases* in related assets are deducted from net income (i.e., the increase in accounts receivable) when converting to cash from operating activities. Conversely, *decreases* in assets are added (inventory and prepaid insurance in this case). Changes in related liabilities are handled in just the opposite way. Increases in related liabilities are *added* to net income (i.e., the increases in accounts payable and salaries payable) while decreases in liabilities are subtracted (i.e., decrease in income tax payable).[15]

Of course, these are adjustments to net income that effectively convert components of income from reported accrual amounts to a cash basis. The other adjustments to net income (gain, depreciation, loss), as pointed out earlier, are to remove the three income statement components that have no effect at all on cash. This pattern is summarized in Illustration 21–12.

[14]We determined in Part B (subsection 2) that there is no evidence that cash received from investments differs from investment revenue.

[15]The adjustment for the decrease in bond discount is logically consistent with this pattern as well. Bond discount is a contra liability. It's logical, then, that an adjustment for a decrease in this account be added—the opposite of the way a decrease in a liability is treated.

Illustration 21–11 Comparison of the Indirect Method and the Direct Method of Determining Cash Flows from Operating Activities

Income Statement		Cash Flows from Operating Activities: Indirect Method		Direct Method	
		Net income	**$12**		
		Adjustments:			
Sales	$100	Increase in accounts receivable	(2)	Cash received from customers	$98
Investment revenue	3	*(No adjustment—no investment revenue receivable or long-term investments)*		Cash received from investments	3
Gain on sale of land	8	Gain on sale of land	(8)	*(Not reported—no cash effect)*	
Cost of goods sold	(60)	Decrease in inventory	4		
		Increase in accounts payable	6	Cash paid to suppliers	(50)
Salaries expense	(13)	Increase in salaries payable	2	Cash paid to employees	(11)
Depreciation expense	(3)	Depreciation expense	3	*(Not reported—no cash effect)*	
Interest expense	(5)	Decrease in bond discount	2	Cash paid for interest	(3)
Insurance expense	(7)	Decrease in prepaid insurance	3	Cash paid for insurance	(4)
Loss on sale of equipment	(2)	Loss on sale of equipment	2	*(Not reported—no cash effect)*	
Income tax expense	(9)	Decrease in income tax payable	(2)	Cash paid for income taxes	(11)
Net Income	**$ 12**	**Net cash flows from operating activities**	**$22**	**Net cash flows from operating activities**	**$22**

Illustration 21–12

Adjustments to Convert Net Income to a Cash Basis—Indirect Method

Type of Adjustment	To Adjust for Noncash Effect
Adjustments for Noncash Effects:	
Income statement components that have *no effect* at all on cash but are *additions* to income	Deduct from net income
Income statement components that have *no effect* at all on cash but are *deductions* from income	Add to net income
Changes in Operating Assets and Liabilities:	
Increases in assets related to an income statement component	Deduct from net income
Decreases in assets related to an income statement component	Add to net income
Increases in liabilities related to an income statement component	Add to net income
Decreases in liabilities related to an income statement component	Deduct from net income

Although either the direct method or the indirect method is permitted, the FASB strongly encourages companies to report cash flows from operating activities by the direct method. The obvious appeal of this approach is that it reports specific operating cash receipts and operating cash payments, which is consistent with the primary objective of the statement of cash flows. Investors and creditors gain additional insight into the specific sources of cash receipts and payments from operating activities revealed by this reporting method. Also, statement users can more readily interpret and understand the information presented because the direct method avoids the confusion caused by reporting noncash items and other reconciling adjustments under the caption *cash flows from operating activities.* Nonetheless, the vast majority of companies choose to use the indirect method. Reasons for this choice range from longstanding tradition to the desire to withhold as much information as possible from competitors.

Reconciliation of Net Income to Cash Flows from Operating Activities

As we discussed earlier, whether cash flows from operating activities are reported by the direct method or by the indirect method, the financial statements must report a reconciliation of net income to net cash flows from operating activities.[16] When the direct method is used, the reconciliation is presented in a separate schedule and is identical to the presentation of net cash flows from operating activities by the indirect method. In other words, Illustration 21–9B also serves as the reconciliation schedule to accompany a statement of cash flows using the direct method. Obviously, a separate reconciliation schedule is not required when using the indirect method because the cash flows from operating activities section of the statement of cash flows *is* a reconciliation of net income to net cash flows from operating activities.[17]

Remember that the direct and indirect methods are alternative approaches to deriving net cash flows from *operating* activities only. The choice of which method is used for that purpose does not affect the way cash flows from *investing* and *financing* activities are identified and reported.

The statements of cash flows from the annual report of **Amazon.com, Inc.**, which uses the indirect method, are shown in Illustration 21–13.

Illustration 21–13 Statement of Cash Flows—Indirect Method; **Amazon.com**, Inc. Real World Financials

AMAZON.COM, INC.
CONSOLIDATED STATEMENTS OF CASH FLOWS

($ in millions)	2017	2016	2015
CASH AND CASH EQUIVALENTS, BEGINNING OF PERIOD	**$ 19,334**	$15,890	$14,557
OPERATING ACTIVITIES:			
Net income	**3,033**	2,371	596
Adjustments to reconcile net income to net cash from operating activities:			
Depreciation of property and equipment, including internal-use software and website development, and other amortization, including capitalized content costs	**11,478**	8,116	6,281
Stock-based compensation	**4,215**	2,975	2,119
Other operating expense, net	**202**	160	155
Other expense (income), net	**(292)**	(20)	250
Deferred income taxes	**(29)**	(246)	81
Changes in operating assets and liabilities:			
Inventories	**(3,583)**	(1,426)	(2,187)
Accounts receivable, net and other	**(4,786)**	(3,367)	(1,755)
Accounts payable	**7,175**	5,030	4,294
Accrued expenses and other	**283**	1,724	913
Unearned revenue	**738**	1,955	1,292
Net cash provided by (used in) operating activities	**18,434**	17,272	12,039

(continued)

[16]Not-for-profit entities using the direct method don't have to provide a reconciliation, FASB ASC 230-10-40-29.

[17]It is permissible to present the reconciliation in a separate schedule and to report the net cash flows from operating activities as a single line item on the statement of cash flows.

Illustration 21–13 (concluded)

INVESTING ACTIVITIES:			
Purchases of property and equipment, including internal-use software and website development	**(11,955)**	(7,804)	(5,387)
Proceeds from property and equipment incentives	**1,897**	1,067	798
Acquisitions, net of cash acquired, and other	**(13,972)**	(116)	(795)
Sales and maturities of marketable securities	**9,988**	4,733	3,025
Purchases of marketable securities	**(13,777)**	(7,756)	(4,091)
Net cash provided by (used in) investing activities	**(27,819)**	(9,876)	(6,450)
FINANCING ACTIVITIES:			
Proceeds from long-term debt and other	**16,231**	621	353
Repayments of long-term debt and other	**(1,372)**	(354)	(1,652)
Principal repayments of capital lease obligations	**(4,799)**	(3,860)	(2,462)
Principal repayments of finance lease obligations	**(200)**	(147)	(121)
Net cash provided by (used in) financing activities	**9,860**	(3,740)	(3,882)
Foreign currency effect on cash and cash equivalents	**713**	(212)	(374)
Net increase (decrease) in cash and cash equivalents	**1,188**	3,444	1,333
CASH AND CASH EQUIVALENTS, END OF PERIOD	**$ 20,522**	$19,334	$15,890
SUPPLEMENTAL CASH FLOW INFORMATION:			
Cash paid for interest on long-term debt	**$ 328**	$ 290	$ 325
Cash paid for interest on capital and finance lease obligations	**319**	206	153
Cash paid for income taxes, net of refunds	**957**	412	273
Property and equipment acquired under capital leases	**9,637**	5,704	4,717
Property and equipment acquired under build-to-suit leases	**3,541**	1,209	544

Source: **Amazon.com**, Inc.

For most companies, expenditures for interest and for taxes are significant. Cash payments for interest and for taxes usually are specifically indicated when the direct method is employed, as is the case for **CVS Health Corp.** reported earlier in Illustration 21–10. When the indirect method is used, those amounts aren't readily apparent and must be *separately reported*, either on the face of the statement as Amazon.com does or in an accompanying disclosure note.

We use a spreadsheet to help prepare a statement of cash flows by the indirect method in Appendix 21A.

Decision Makers' Perspective—Cash Flow Ratios

We have emphasized the analysis of financial statements from a decision maker's perspective throughout this text. Often that analysis included the development and comparison of financial ratios. Ratios based on income statement and balance sheet amounts enjoy a long tradition of acceptance, from which several standard ratios, including those described in earlier chapters, have evolved. To gain another viewpoint, some analysts supplement their investigation with cash flow ratios. Some cash flow ratios are derived by simply substituting cash flow from operations (CFFO) from the statement of cash flows in place of net income in many ratios, not to replace those ratios, but to complement them. For example, the times interest earned ratio can be modified to reflect the number of times the cash outflow for interest is provided by cash inflow from operations, and any of the profitability ratios can be modified to determine the cash generated from assets, shareholders' equity, sales, and so on. Illustration 21–14 summarizes the calculation and usefulness of several representative cash flow ratios.

Illustration 21–14
Cash Flow Ratios

	Calculation	Measures
Performance Ratios		
Cash flow to sales	CFFO / Net sales	Cash generated by each sales dollar
Cash return on assets	CFFO / Average total assets	Cash generated from all resources
Cash return on shareholders' equity	CFFO / Average shareholders' equity	Cash generated from owner-provided resources
Cash to income	CFFO / Income from continuing operations	Cash-generating ability of continuing operations
Cash flow per share	(CFFO − preferred dividends) / Weighted-average shares	Operating cash flow on a per share basis
Sufficiency Ratios		
Debt coverage	Total liabilities / CFFO	Financial risk and financial leverage
Interest coverage	(CFFO + interest + taxes) / Interest	Ability to satisfy fixed obligations
Reinvestment	CFFO / Cash outflow for noncurrent assets	Ability to acquire assets with operating cash flows
Debt payment	CFFO / Cash outflow for LT debt repayment	Ability to pay debts with operating cash flows
Dividend payment	CFFO / Cash outflow for dividends	Ability to pay dividends with operating cash flows
Investing and financing activity	CFFO / Cash outflows for investing and financing activities	Ability to acquire assets, pay debts, and make distributions to owners

Cash flow ratios have received limited acceptance to date, due in large part to the long tradition of accrual-based ratios coupled with the relatively brief time that all companies have published statements of cash flows. A lack of consensus on cash flow ratios by which to make comparisons also has slowed their acceptance. In fact, companies are prohibited from reporting cash flow per share in the statement of cash flows. Nevertheless, cash flow ratios offer insight in the evaluation of a company's profitability and financial strength.[18] ●

[18]Proposals for informative sets of cash flow ratios are offered by Charles Carslaw and John Mills, "Developing Ratios for Effective Cash Flow Analysis," *Journal of Accountancy,* November 1991; Don Giacomino and David Mielke, "Cash Flows: Another Approach to Ratio Analysis," *Journal of Accountancy,* March 1993; and John Mills and Jeanne Yamamura, "The Power of Cash Flows Ratios," *Journal of Accountancy,* October 1998.

Financial Reporting Case Solution

©Hero Images/Getty Images

1. **What are the cash flow aspects of the situation that Mr. Barr may be overlooking in making his case for a wage increase? How can a company's operations generate a healthy profit and yet produce meager or even negative cash flows?** *(p. 1219)* Positive net income does not necessarily indicate a healthy cash position. A statement of cash flows provides information about cash flows not seen when looking only at the balance sheet and the income statement. Although cash flows from operating activities result from the same activities that are reported in the income statement, the income statement reports the activities on an accrual basis. That is, revenues reported are those earned during the reporting period, regardless of when cash is received, and the expenses incurred in generating those revenues, regardless of when cash is paid. Thus, the very same operations can generate a healthy profit and yet produce meager or even negative cash flows.
2. **What information can a statement of cash flows provide about a company's investing activities that can be useful in decisions such as this?** *(p. 1222)* Cash flows from investing activities result from the acquisition and disposition of assets. Information about investing activities is useful to decision makers regarding the nature and magnitude of productive assets being acquired for future use. In the union negotiations, for instance, Mr. Barr may not be aware of the substantial investments under way to replace and update equipment and the cash requirements of those investments. Relatedly, the relatively low depreciation charges accelerated depreciation provides in the later years of assets' lives may cause profits to seem artificially high given the necessity to replace those assets at higher prices.
3. **What information can a statement of cash flows provide about a company's financing activities that can be useful in decisions such as this?** *(p. 1223)* Information about financing activities provides insights into sources of a company's external financing. Recent debt issues, for instance, might indicate a need for higher cash flows to maintain higher interest charges. Similarly, recent external financing activity may suggest that a company might be near its practical limits from external sources and, therefore, may need a greater reliance on internal financing through operations. ●

The Bottom Line

- **LO21–1** Decision makers focus on the prospects of receiving a cash return from their dealings with a firm. But it is the ability of the firm to generate cash flows to itself that ultimately determines the potential for cash flows to investors and creditors. The statement of cash flows fills an information gap left by the balance sheet and the income statement by presenting information about cash flows that the other statements either do not provide or provide only indirectly. *(p. 1215)*
- **LO21–2** Cash includes cash equivalents. These are short-term, highly liquid investments that can readily be converted to cash with little risk of loss. *(p. 1219)*
- **LO21–3** Cash flows from operating activities are both inflows and outflows of cash that result from activities reported in the income statement. *(p. 1219)*
- **LO21–4** Unlike the direct method, which directly lists cash inflows and outflows, the indirect method derives cash flows indirectly, by starting with reported net income and working backwards to convert that amount to a cash basis. *(p. 1221)*
- **LO21–5** Cash flows from investing activities are related to the acquisition and disposition of assets, other than inventory and assets classified as cash equivalents. *(p. 1222)*
- **LO21–6** Cash flows from financing activities result from the external financing of a business. *(p. 1223)*
- **LO21–7** Noncash investing and financing activities, such as acquiring equipment (an investing activity) by issuing a long-term note payable (a financing activity), are reported in a related disclosure schedule or note. *(p. 1224)*

- LO21–8 A spreadsheet provides a systematic method of preparing a statement of cash flows by analyzing available data to ensure that all operating, investing, and financing activities are detected. Recording spreadsheet entries that explain account balance changes simultaneously identifies and classifies the activities to be reported on the statement of cash flows. (*p. 1225*)
- LO21–9 IFRS allows more flexibility than U.S. GAAP in the classification of cash flows. Companies can report interest and dividends paid as either operating or financing cash flows and interest and dividends received as either operating or investing cash flows. (*p. 1245*)

APPENDIX 21A Spreadsheet for the Indirect Method

A spreadsheet is equally useful in preparing a statement of cash flows whether we use the direct or the indirect method of determining cash flows from operating activities. The format of the spreadsheet differs only with respect to operating activities. The analysis of transactions for the purpose of identifying cash flows to be reported is the same. To illustrate, Illustration 21A–1 provides a spreadsheet analysis of the data for UBC.

Illustration 21A–1
Indirect Method

UNITED BRANDS CORPORATION
Spreadsheet for the Statement of Cash Flows

	Dec. 31, 2020	Changes: Debits		Changes: Credits		Dec. 31, 2021
Balance Sheet						
Assets						
Cash	20	(19)	9			29
Accounts receivable	30	(5)	2			32
Short-term investments	0	(12)	12			12
Inventory	50			(6)	4	46
Prepaid insurance	6			(8)	3	3
Land	60	(13)	30	(2)	10	80
Buildings and equipment	75	(14)	20*X*	(3)	14	81
Less: Accumulated depreciation	(20)	(3)	7	(4)	3	(16)
	221					267
Liabilities						
Accounts payable	20			(7)	6	26
Salaries payable	1			(9)	2	3
Income tax payable	8	(11)	2			6
Notes payable	0			(14)	20*X*	20
Bonds payable	50	(15)	15			35
Less: Discount on bonds	(3)			(10)	2	(1)
Shareholders' Equity						
Common stock	100			(16)	10	
				(17)	20	130
Paid-in capital—excess of par	20			(16)	3	
				(17)	6	29
Retained earnings	25	(16)	13			
		(18)	5	(1)	12	19
	221					267
Statement of Cash Flows						
Operating activities:						
Net income		(1)	12			
Adjustments for noncash effects:						
Gain on sale of land				(2)	8	
Depreciation expense		(4)	3			
Loss on sale of equipment		(3)	2			
Increase in accounts receivable				(5)	2	
Decrease in inventory		(6)	4			

(continued)

Illustration 21A–1

(concluded)

Decrease in prepaid insurance	(8)	3			
Increase in accounts payable	(7)	6			
Increase in salaries payable	(9)	2			
Decrease in income tax payable			(11)	2	
Amortization of bond discount	(10)	2			
Net cash flows					22
Investing activities:					
Purchase of land			(13)	30	
Purchase of short-term investment			(12)	12	
Sale of land	(2)	18			
Sale of equipment	(3)	5			
Net cash flows					(19)
Financing activities:					
Sale of common shares	(17)	26			
Retirement of bonds payable			(15)	15	
Payment of dividends			(18)	5	
Net cash flows					6
Net increase in cash			(19)	9	9
Totals		198		198	

Two differences should be noted between the spreadsheet in Illustration 21A–1 and the spreadsheet we used earlier for the direct method. First, in the statement of cash flows section of the spreadsheet, under the heading of "cash flows from operating activities," specific cash inflows and cash outflows are replaced by net income and the required adjustments for noncash effects. Second, we do not include an income statement section. This section is unnecessary because, using the indirect method, we are not interested in identifying specific operating activities that cause increases and decreases in cash. Instead, we need from the income statement only the amount of net income, which is converted to a cash basis by adjusting for any noncash amounts included in net income. The spreadsheet entries in journal entry form for the indirect method are illustrated in Illustration 21A–2.

Illustration 21A–2

Spreadsheet Entries for the Indirect Method

Entry (1)	Net income—CFOA	12	
	Retained earnings		12
	Establishes net income as the initial amount of cash flows from operating activities, to be adjusted to a cash basis by subsequent entries.		
Entry (2)	Cash (received from sale of land)	18	
	Land		10
	Gain on sale of land—CFOA		8
	Deducts the noncash gain added in determining net income, explains a portion of the change in land, and identifies a cash inflow from investing activities.		
Entry (3)	Cash (received from sale of equipment)	5	
	Loss on sale of equipment—CFOA	2	
	Accumulated depreciation	7	
	Buildings and equipment		14
	Adds back the noncash loss subtracted in determining net income, explains portions of the changes in accumulated depreciation and buildings and equipment, and identifies a cash inflow from investing activities.		
Entry (4)	Depreciation expense—CFOA	3	
	Accumulated depreciation		3
	Adds back the noncash expense subtracted in determining net income.		
Entry (5)	Accounts receivable	2	
	Increase in accounts receivable—CFOA		2
	Reduces net income to reflect $98 million cash received from customers rather than $100 million sales.		

(continued)

Illustration 21A–2
(concluded)

Entry (6)	Decrease in inventory—CFOA	4	
	Inventory		4
	Increases net income to reflect a deduction of $56 million cost of goods purchased rather than $60 million cost of goods sold.		
Entry (7)	Increase in accounts payable—CFOA	6	
	Accounts payable		6
	Increases net income to reflect a deduction of $50 million cash paid to suppliers rather than $56 million cost of goods purchased.		
Entry (8)	Decrease in prepaid insurance—CFOA	3	
	Prepaid insurance		3
	Increases net income to reflect a deduction of $4 million cash paid for insurance rather than $7 million insurance expense.		
Entry (9)	Increase in salaries payable—CFOA	2	
	Salaries payable		2
	Increases net income to reflect a deduction of $11 million cash paid to employees rather than $13 million salaries expense.		
Entry (10)	Amortization of discount on bonds—CFOA	2	
	Discount on bonds		2
	Increases net income to reflect a deduction of $3 million cash paid for bond interest rather than $5 million bond interest expense.		
Entry (11)	Income tax payable	2	
	Decrease in income tax payable—CFOA		2
	Reduces net income to reflect a deduction of $11 million cash paid for income taxes rather than $9 million income tax expense.		
Entry (12)	Short-term investment	12	
	Cash (purchase of short-term investment)		12
	Explains the increase in the short-term investment account and identifies a cash outflow from investing activities.		
Entry (13)	Land	30	
	Cash (purchase of land)		30
	Explains a portion of the change in the land account and identifies a cash outflow from investing activities.		
Entry (14)	Buildings and equipment	20	
	Notes payable		20
	Partially explains the changes in the buildings, equipment, and notes payable accounts and identifies a significant noncash investing and financing activity.		
Entry (15)	Bonds payable	15	
	Cash (retirement of bonds payable)		15
	Explains the decrease in the bonds payable account and identifies a cash outflow from financing activities.		
Entry (16)	Retained earnings	13	
	Common stock		10
	Paid-in capital—excess of par		3
	Partially explains the changes in the retained earnings, common stock, and paid-in capital—excess of par accounts.		
Entry (17)	Cash (from sale of common stock)	26	
	Common stock		20
	Paid-in capital—excess of par		6
	Partially explains the changes in the common stock and paid-in capital—excess of par accounts and identifies a cash inflow from financing activities.		
Entry (18)	Retained earnings	5	
	Cash (payment of cash dividends)		5
	Partially explains the change in the retained earnings account and identifies a cash outflow from financing activities.		
Entry (19)	Cash	9	
	Net increase in cash (from statement of cash flows activities)		9
	Reconciles the net increase in cash from operating, investing, and financing activities to the increase in the cash balance.		

Remember that there is no mandatory order in which the account changes must be analyzed. However, since we determine net cash flows from operating activities by working backwards from net income when using the indirect method, it is convenient to start with the spreadsheet entry that represents the credit to retained earnings due to net income. This entry corresponds to summary entry (11) using the direct method. By entering the debit portion of the entry as the first item under the cash flows from operating activities (CFOA), we establish net income as the initial amount of cash flows from operating activities, which is then adjusted to a cash basis by subsequent entries. Entries (2)–(4) duplicate the transactions that involve noncash components of net income. Changes in current assets and current liabilities that represent differences between revenues and expenses and the cash effects of those revenues and expenses are accounted for by entries (5)–(11). Summary entries (12)–(19) explain the changes in the balance sheet not already accounted for by previous entries, and are identical to entries (12)–(19) recorded using the direct method.

The statement of cash flows presenting net cash flows from operating activities by the indirect method is illustrated in Illustration 21A–3.

Illustration 21A–3

Statement of Cash flows—Indirect Method

All parts of the statement of cash flows except operating activities are precisely the same as in the direct method.

UNITED BRANDS CORPORATION
Statement of Cash Flows
For Year Ended December 31, 2021
($ in millions)

Cash Flows from Operating Activities		
Net income	$ 12	
Adjustments for noncash effects:		
Gain on sale of land	(8)	
Depreciation expense	3	
Loss on sale of equipment	2	
Changes in operating assets and liabilities:		
Increase in accounts receivable	(2)	
Decrease in inventory	4	
Decrease in prepaid insurance	3	
Increase in accounts payable	6	
Increase in salaries payable	2	
Decrease in income tax payable	(2)	
Decrease in discount on bonds payable	2	
Net cash flows from operating activities		$22
Cash Flows from Investing Activities		
Purchase of land	(30)	
Purchase of short-term investment	(12)	
Sale of land	18	
Sale of equipment	5	
Net cash from investing activities		(19)
Cash Flows from Financing Activities		
Sale of common shares	26	
Retirement of bonds payable	(15)	
Payment of cash dividends	(5)	
Net cash flows from financing activities		6
Net increase in cash		9
Cash balance, January 1		20
Cash balance, December 31		$29
Note X:		
Noncash Investing and Financing Activities		
Acquired $20 million of equipment by issuing a 12%, 5-year note		$20

APPENDIX 21B The T-Account Method of Preparing the Statement of Cash Flows

The T-account method serves the same purpose as a spreadsheet in assisting in the preparation of a statement of cash flows.

This chapter has demonstrated the use of a spreadsheet to prepare the statement of cash flows. A second systematic approach to the preparation of the statement is referred to as the T-account method. The two methods are identical in concept. Both approaches reconstruct the transactions that caused changes in each account balance during the year, simultaneously identifying the operating, investing, and financing activities to be reported on the statement of cash flows. The form of the two methods differs only by whether the entries for those transactions are recorded on a spreadsheet or in T-accounts. In both cases, entries are recorded until the net change in each account balance has been explained.

Some accountants feel that the T-account method is less time-consuming than preparing a spreadsheet but accomplishes precisely the same goal. Since both methods are simply analytical techniques to assist in statement preparation, the choice is a matter of personal preference. The following five steps outline the T-account method:

1. Draw T-accounts for each income statement and balance sheet account.
2. The T-account for cash should be drawn considerably larger than other T-accounts because more space is required to accommodate the numerous debits and credits to cash. Also, the cash T-account will serve the same purpose as the statement of cash flows section of the spreadsheet in that the formal statement of cash flows is developed from the cash flows reported there. Therefore, it is convenient to partition the cash T-account with headings for "Operating Activities," "Investing Activities," and "Financing Activities" before entries are recorded.
3. Enter each account's net change on the appropriate side (debit or credit) of the uppermost portion of each T-account. These changes will serve as individual check figures for determining whether the increase or decrease in each account balance has been explained. These first three steps establish the basic work form for the T-account method.
4. Reconstruct the transactions that caused changes in each account balance during the year and record the entries for those transactions directly in the T-accounts. Again using UBC as an example, the entries we record in the T-accounts are exactly the same as the spreadsheet entries we created in the chapter when using the spreadsheet method. The analysis we used in creating those spreadsheet entries is equally applicable to the T-account method. For that reason, that analysis is not repeated here. The complete T-account work form for UBC is presented below. Account balance changes are provided from balances given in Illustration 21–9.
5. After all account balances have been explained by T-account entries, prepare the statement of cash flows from the cash T-account, being careful also to report noncash investing and financing activities. The statement of cash flows for UBC appears in Illustration 21–2.

BALANCE SHEET ACCOUNTS
Cash (statement of cash flows)

		9			
Operating Activities:					
From customers	(1)	98	50	(4)	To suppliers of goods
From investment revenue	(2)	3	11	(5)	To employees
			3	(7)	For interest
			4	(8)	For insurance
			11	(10)	For income taxes
Investing Activities:					
Sale of land	(3)	18	12	(12)	Purchase of short-term investment
Sale of equipment	(9)	5	30	(13)	Purchase of land
Financing Activities:					
Sale of common stock	(17)	26	15	(15)	Retirement of bonds payable
			5	(18)	Payment of dividends

Accounts Receivable

	2		
(1)	2		

Short-Term Investments

	12		
(12)	12		

Inventory

		4	
		4	(4)

Prepaid Insurance

		3	
		3	(8)

Land

	20		
(13)	30	10	(3)

Buildings and Equipment

	6		
X(14)	20	14	(9)

Accumulated Depreciation

	4		
(9)	7	3	(6)

Accounts Payable

		6	
		6	(4)

Salaries Payable

		2	
		2	(5)

Income Tax Payable

	2		
(10)	2		

Notes Payable

		20	
		20	(14)X

Bonds Payable

	15		
(15)	15		

Discount on Bonds

		2	
		2	(7)

Common Stock

		30	
		10	(16)
		20	(17)

Paid-in Capital—excess of par

		9	
		3	(16)
		6	(17)

Retained Earnings

	6		
(16)	13		
(18)	5	12	(11)

X Noncash Investing and Financing Activity

INCOME STATEMENT ACCOUNTS

Sales Revenue

		100	
		100	(1)

Investment Revenue

		3	
		3	(2)

Gain on Sale of Land

		8	
		8	(3)

Cost of Goods Sold

	60		
(4)	60		

Salaries Expense

	13		
(5)	13		

Depreciation Expense

	3		
(6)	3		

Interest Expense

	5		
(7)	5		

Insurance Expense

	7		
(8)	7		

Loss on Sale of Equipment

	2		
(9)	2		

Income Tax Expense

	9		
(10)	9		

Net Income (Income Summary)

	12		
(11)	12		

Questions For Review of Key Topics

Q 21–1 Effects of all cash flows affect the balances of various accounts reported in the balance sheet. Also, the activities that cause some of these cash flows are reported in the income statement. What, then, is the need for an additional financial statement that reports cash flows?

Q 21–2 The statement of cash flows provides a list of all the cash inflows and cash outflows during the reporting period. To make the list more informative, the cash flows are classified according to the nature of the activities that create the cash flows. What are the three primary classifications?

Q 21–3 Is an investment in Treasury bills always classified as a cash equivalent? Explain.

Q 21–4 Transactions that involve merely purchases or sales of cash equivalents generally are not reported in a statement of cash flows. Describe an exception to this generalization. What is the essential characteristic of the transaction that qualifies as an exception?

Q 21–5 What are the differences between cash flows from operating activities and the elements of an income statement?

Q 21–6 Do cash flows from operating activities report all the elements of the income statement on a cash basis? Explain.

Q 21–7 Investing activities include the acquisition and disposition of assets. Provide three specific examples. Identify two exceptions.

Q 21–8 The issuance of stock and the issuance of bonds are reported as financing activities. Are payments of dividends to shareholders and payments of interest to bondholders also reported as financing activities? Explain.

Q 21–9 Does the statement of cash flows report only transactions that cause an increase or a decrease in cash? Explain.

Q 21–10 How would the acquisition of a building be reported on a statement of cash flows if purchased by issuing a mortgage note payable in addition to a significant cash down payment?

Q 21–11 Perhaps the most noteworthy item reported on an income statement is net income—the amount by which revenues exceed expenses. The most noteworthy item reported on a statement of cash flows is *not* the amount of net cash flows. Explain.

Q 21–12 What is the purpose of the "changes" columns of a spreadsheet to prepare a statement of cash flows?

Q 21–13 Given sales revenue of $200,000, how can it be determined whether or not $200,000 cash was received from customers?

Q 21–14 When an asset is sold at a gain, why is the gain not reported as a cash inflow from operating activities?

Q 21–15 When determining the amount of cash paid for income taxes, what would be indicated by an increase in the deferred income tax liability account?

Q 21–16 When using the indirect method of determining net cash flows from operating activities, how is warranty expense reported? Why? What other expenses are reported in a like manner?

Q 21–17 When using the indirect method of determining net cash flows from operating activities, how are revenues and expenses reported on the statement of cash flows if their cash effects are identical to the amounts reported in the income statement?

Q 21–18 Why does the FASB recommend the direct method over the indirect method?

Q 21–19 Compare the manner in which investing activities are reported on a statement of cash flows prepared by the direct method and by the indirect method.

IFRS **Q 21–20** Where can we find authoritative guidance for the statement of cash flows under IFRS?

IFRS **Q 21–21** U.S. GAAP designates cash outflows for interest payments and cash inflows from interest and dividends received as operating cash flows. Dividends paid to shareholders are classified as financing cash flows. How are these cash flows reported under IFRS?

Brief Exercises

connect

BE 21–1
Determine cash received from customers
● LO21–3

Horton Housewares' accounts receivable decreased during the year by $5 million. What is the amount of cash Horton received from customers during the reporting period if its sales were $33 million? Prepare a summary entry that represents the net effect of the selling and collection activities during the reporting period.

BE 21–2
Determine cash received from customers
● LO21–3

April Wood Products' accounts receivable increased during the year by $4 million. What is the amount of cash April Wood Products received from customers during the reporting period if its sales were $44 million? Prepare a summary entry that represents the net effect of the selling and collection activities during the reporting period.

BE 21–3
Determine cash paid to suppliers
● LO21–3

LaRoe Lawns' inventory increased during the year by $6 million. Its accounts payable increased by $5 million during the same period. What is the amount of cash LaRoe paid to suppliers of merchandise during the reporting period if its cost of goods sold was $25 million? Prepare a summary entry that represents the net effect of merchandise purchases during the reporting period.

BE 21–4
Determine cash paid to employees
● LO21–3

Sherriane Baby Products' salaries expense was $17 million. What is the amount of cash Sherriane paid to employees during the reporting period if its salaries payable increased by $3 million? Prepare a summary entry that represents the net effect of salaries expense incurred and paid during the reporting period.

BE 21–5
Bond interest and discount
● LO21–3, LO21–6

Agee Technology, Inc., issued 9% bonds, dated January 1, with a face amount of $400 million on July 1, 2021, at a price of $380 million. For bonds of similar risk and maturity, the market yield is 10%. Interest is paid semi-annually on June 30 and December 31. Prepare the journal entry to record interest at the effective interest rate at December 31. What would be the amount(s) related to the bonds that Agee would report in its statement of cash flows for the year ended December 31, 2021, if it uses the direct method?

BE 21–6
Bond interest and discount
● LO21–4, LO21–6

Refer to the situation described in BE 21–5. What would be the amount(s) related to the bonds that Agee would report in its statement of cash flows for the year ended December 31, 2021, if it uses the indirect method?

BE 21–7
Installment note
● LO21–3, LO21–6

On January 1, 2021, the Merit Group issued to its bank a $41 million, five-year installment note to be paid in five equal payments at the end of each year. Installment payments of $10 million annually include interest at the rate of 7%. What would be the amount(s) related to the note that Merit would report in its statement of cash flows for the year ended December 31, 2021?

BE 21–8
Sale of land
● LO21–3, LO21–4, LO21–5

On July 15, 2021, M.W. Morgan Distribution sold land for $35 million that it had purchased in 2016 for $22 million. What would be the amount(s) related to the sale that Morgan would report in its statement of cash flows for the year ended December 31, 2021, using the direct method? The indirect method?

BE 21–9
Investing activities
● LO21–5

Carter Containers sold marketable equity securities, land, and common stock for $30 million, $15 million, and $40 million, respectively. Carter also purchased treasury stock, equipment, and a patent for $21 million, $25 million, and $12 million, respectively. What amount should Carter report as net cash from investing activities?

BE 21–10
Financing activities
● LO21–6

Refer to the situation described in BE 21–9. What amount should Carter report as net cash from financing activities?

BE 21–11
Indirect method
● LO21–4

Sanders Awnings reported net income of $90 million. Included in that number were depreciation expense of $3 million and a loss on the sale of equipment of $2 million. Records reveal increases in accounts receivable, accounts payable, and inventory of $1 million, $4 million, and $3 million, respectively. What were Sanders' cash flows from operating activities?

BE 21–12
Indirect method
● LO21–4

Sunset Acres reported net income of $60 million. Included in that number were trademark amortization expense of $2 million and a gain on the sale of land of $1 million. Records reveal decreases in accounts receivable, accounts payable, and inventory of $2 million, $5 million, and $4 million, respectively. What were Sunset's cash flows from operating activities?

Exercises

E 21–1
Classification of cash flows
● LO21–3 through LO21–6

Listed below are several transactions that typically produce either an increase or a decrease in cash. Indicate by letter whether the cash effect of each transaction is reported on a statement of cash flows as an operating (**O**), investing (**I**), or financing (**F**) activity.

	Transactions
F	1. Sale of common stock.
____	2. Sale of land.
____	3. Purchase of treasury stock.
____	4. Merchandise sales.
____	5. Issuance of a long-term note payable.
____	6. Purchase of merchandise.
____	7. Repayment of a note payable.
____	8. Employee salaries.
____	9. Sale of equipment at a gain.
____	10. Issuance of bonds.
____	11. Acquisition of bonds of another corporation.
____	12. Payment of semiannual interest on bonds payable.
____	13. Payment of a cash dividend.
____	14. Purchase of a building.
____	15. Collection of nontrade note receivable (principal amount).
____	16. Loan to another company.
____	17. Retirement of common stock.
____	18. Income taxes.
____	19. Issuance of a short-term note payable.
____	20. Sale of a copyright.

E 21–2
Determine cash paid to suppliers of merchandise
● LO21–3, LO21–6

Shown below in T-account format are the beginning and ending balances ($ in millions) of both inventory and accounts payable.

Inventory

Beginning balance	90		
Ending balance	93		

Accounts Payable

		14	Beginning balance
		16	Ending balance

Required:

1. Use a T-account analysis to determine the amount of cash paid to suppliers of merchandise during the reporting period if cost of goods sold was $300 million.
2. Prepare a summary entry that represents the net effect of merchandise purchases during the reporting period.

E 21–3
Determine cash received from customers
● LO21–3

Determine the amount of cash received from customers for each of the three independent situations below. All dollars are in millions.

Situation	Sales Revenue	Accounts Receivable Increase (Decrease)	Cash Received from Customers
1	100	-0-	?
2	100	5	?
3	100	(5)	?

E 21–4
Summary entries for cash received from customers
● LO21–3

For each of the three independent situations below, prepare journal entries that summarize the selling and collection activities for the reporting period in order to determine the amount of cash received from customers and to explain the change in each account shown. All dollars are in millions.

Situation	Sales Revenue	Accounts Receivable Increase (Decrease)	Cash Received from Customers
1	200	-0-	?
2	200	10	?
3	200	(10)	?

E 21–5
Determine cash paid to suppliers of merchandise
● LO21–3

Determine the amount of cash paid to suppliers of merchandise for each of the nine independent situations below. All dollars are in millions.

Situation	Cost of Goods Sold	Inventory Increase (Decrease)	Accounts Payable Increase (Decrease)	Cash Paid to Suppliers
1	100	0	0	?
2	100	3	0	?
3	100	(3)	0	?
4	100	0	7	?
5	100	0	(7)	?
6	100	3	7	?
7	100	3	(7)	?
8	100	(3)	(7)	?
9	100	(3)	7	?

E 21–6
Summary entries for cash paid to suppliers of merchandise
● LO21–3

For each of the five independent situations below, prepare a journal entry that summarizes the purchases, sales, and payments related to inventories in order to determine the amount of cash paid to suppliers and explain the change in each account shown. All dollars are in millions.

Situation	Cost of Goods Sold	Inventory Increase (Decrease)	Accounts Payable Increase (Decrease)	Cash Paid to Suppliers
1	200	0	0	?
2	200	6	0	?
3	200	0	14	?
4	200	6	14	?
5	200	(6)	(14)	?

E 21–7
Determine cash paid for bond interest
● LO21–3

Determine the amount of cash paid to bondholders for bond interest for each of the six independent situations below. All dollars are in millions.

Situation	Interest Expense	Interest Payable Increase (Decrease)	Unamortized Discount Increase (Decrease)	Cash Paid for Interest
1	10	0	0	?
2	10	2	0	?
3	10	(2)	0	?
4	10	0	(3)	?
5	10	2	(3)	?
6	10	(2)	(3)	?

E 21–8
Determine cash paid for bond interest
● LO21–3

For each of the four independent situations below, prepare a single journal entry that summarizes the recording and payment of interest in order to determine the amount of cash paid for bond interest and explain the change (if any) in each of the accounts shown. All dollars are in millions.

Situation	Interest Expense	Interest Payable Increase (Decrease)	Unamortized Discount Increase (Decrease)	Cash Paid for Interest
1	20	0	0	?
2	20	4	0	?
3	20	0	(6)	?
4	20	(4)	(6)	?

E 21–9
Determine cash paid for income taxes
● LO21–3

Determine the amount of cash paid for income taxes in each of the nine independent situations below. All dollars are in millions.

Situation	Income Tax Expense	Income Tax Payable Increase (Decrease)	Deferred Tax Liability Increase (Decrease)	Cash Paid for Taxes
1	10	0	0	?
2	10	3	0	?
3	10	(3)	0	?
4	10	0	2	?
5	10	0	(2)	?
6	10	3	2	?
7	10	3	(2)	?
8	10	(3)	(2)	?
9	10	(3)	2	?

E 21–10
Summary entries for cash paid for income taxes
● LO21–3

For each of the five independent situations below, prepare a single journal entry that summarizes the recording and payment of income taxes in order to determine the amount of cash paid for income taxes and explain the change (if any) in each of the accounts shown. All dollars are in millions.

Situation	Income Tax Expense	Income Tax Payable Increase (Decrease)	Deferred Tax Liability Increase (Decrease)	Cash Paid for Taxes
1	10	0	0	?
2	10	3	0	?
3	10	0	(2)	?
4	10	3	2	?
5	10	(3)	(2)	?

E 21–11
Bonds; statement of cash flow effects
● LO21–3

Most Solutions, Inc., issued 10% bonds, dated January 1, with a face amount of $640 million on January 1, 2021. The bonds mature in 2031 (10 years). For bonds of similar risk and maturity the market yield is 12%. Interest expense is recorded at the effective interest rate. Interest is paid semiannually on June 30 and December 31. Most recorded the sale as follows:

January 1, 2021		
Cash (price)	566,589,440	
Discount on bonds (difference)	73,410,560	
Bonds payable (face amount)		640,000,000

Required:
What would be the amount(s) related to the bonds that Most would report in its statement of cash flows for the year ended December 31, 2021?

E 21–12
Installment note; statement of cash flow effects
● LO21–3, LO21–6

National Food Services, Inc., borrowed $4 million from its local bank on January 1, 2021, and issued a 4-year installment note to be paid in four equal payments at the end of each year. The payments include interest at the rate of 10%. Installment payments are $1,261,881 annually.

Required:
What would be the amount(s) related to the note that National would report in its statement of cash flows for the year ended December 31, 2021?

E 21–13
Identifying cash flows from investing activities and financing activities
● LO21–5, LO21–6

In preparation for developing its statement of cash flows for the year ended December 31, 2021, Rapid Pac, Inc., collected the following information:

	($ in millions)
Fair value of shares issued in a stock dividend	$ 65
Payment for the early extinguishment of long-term bonds (book value: $97 million)	102
Proceeds from the sale of treasury stock (cost: $17 million)	22
Gain on sale of land	4
Proceeds from sale of land	12
Purchase of Microsoft common stock	160
Declaration of cash dividends	44
Distribution of cash dividends declared in 2020	40

Required:
1. In Rapid Pac's statement of cash flows, what were net cash inflows (or outflows) from investing activities for 2021?
2. In Rapid Pac's statement of cash flows, what were net cash inflows (or outflows) from financing activities for 2021?

E 21–14
Identifying cash flows from investing activities and financing activities
● LO21–5, LO21–6

In preparation for developing its statement of cash flows for the year ended December 31, 2021, Millennium Solutions, Inc. collected the following information:

	($ in millions)
Payment for the early extinguishments of long-term notes (book value: $50 million)	$ 54
Sale of common shares	176
Retirement of common shares	122
Loss on sale of equipment	2
Proceeds from sale of equipment	8
Issuance of short-term note payable for cash	10
Acquisition of building for cash	7
Purchase of marketable securities (not a cash equivalent)	5
Purchase of marketable securities (considered a cash equivalent)	1
Cash payment for 3-year insurance policy	3
Collection of note receivable with interest (principal amount, $11)	13
Declaration of cash dividends	33
Distribution of cash dividends declared in 2020	30

Required:

1. In Millennium's statement of cash flows, what were net cash inflows (or outflows) from investing activities for 2021?
2. In Millennium's statement of cash flows, what were net cash inflows (or outflows) from financing activities for 2021?

E 21–15
Lease; lessee; statement of cash flows effects
● LO21–3, LO21–5, LO21–6

Wilson Foods Corporation leased a commercial food processor on September 30, 2021. The five-year finance lease agreement calls for Wilson to make quarterly lease payments of $195,774, payable each September 30, December 31, March 31, June 30, with the first payment at September 30, 2021. Wilson's incremental borrowing rate is 12%. Wilson records amortization on a straight-line basis at the end of each fiscal year. Wilson recorded the lease as follows:

September 30, 2021		
Right-of-use asset (calculated below)	3,000,000	
Lease payable (calculated below)		3,000,000
Lease payable	195,774	
Cash (first payment)		195,774

Calculation of the present value of lease payments

$195,774 × 15.32380* = $3,000,000
(rounded)

*Present value of an annuity due of $1: n = 20, i = 3% (from Table 6).

Required:

What would be the pretax amounts related to the lease that Wilson would report in its statement of cash flows for the year ended December 31, 2021?

E 21–16
Equity method investment; statement of cash flow effects
● LO21–3, LO21–5

On January 1, 2021, Beilich Enterprises bought 20% of the outstanding common stock of Wolfe Construction Company for $600 million cash. Wolfe's net income for the year ended December 31, 2021, was $300 million. During 2021, Wolfe declared and paid cash dividends of $60 million. Beilich recorded the investment as follows:

	($ in millions)	
Purchase		
Investment in Wolfe Construction shares	600	
Cash		600
Net income		
Investment in Wolfe Construction shares (20% × $300 million)	60	
Investment revenue		60
Dividends		
Cash (20% × $60 million)	12	
Investment in Wolfe Construction shares		12

Required:

What would be the pretax amounts related to the investment that Beilich would report in its statement of cash flows for the year ended December 31, 2021?

E 21–17
Indirect method reconciliation of net income to net cash flows from operating activities
● LO21–4

The accounting records of EZ Company provided the data below. Prepare a reconciliation of net income to net cash flows from operating activities.

Net income	$50,000
Depreciation expense	7,000
Increase in inventory	1,500
Decrease in salaries payable	800
Decrease in accounts receivable	2,000
Amortization of patent	500
Amortization of premium on bonds	1,000
Increase in accounts payable	4,000
Cash dividends	12,000

E 21–18
Spreadsheet entries from statement of retained earnings
● LO21–3 through LO21–8

The statement of retained earnings of Gary Larson Publishers is presented below.

GARY LARSON PUBLISHERS
Statement of Retained Earnings
For the Year Ended December 31, 2021
($ in millions)

Retained earnings, January 1		$200
Add:	Net income	75
Deduct:	Cash dividend	(25)
	Stock dividend (1 million shares of $1 par common stock)	(16)
	Property dividend (Garfield Company preferred stock held as a short-term investment)	(12)
	Sale of treasury stock (cost $53 million)	(10)
Retained earnings, December 31		$212

Required:
For the transactions that affected Larson's retained earnings, reconstruct the journal entries that can be used to determine cash flows to be reported in a statement of cash flows. Also indicate any investing and financing activities you identify from this analysis that should be reported on the statement of cash flows.

E 21–19
Relationship between the income statement and cash flows from operating activities (direct method and indirect method)
● LO21–3, LO21–4

The following schedule relates the income statement with cash flows from operating activities, derived by both the direct and indirect methods, in the format illustrated by Illustration 21–11 in the chapter. The amounts for income statement elements are missing.

Income Statement		Cash Flows from Operating Activities: Indirect Method		Direct Method	
		Net income	$?		
		Adjustments:			
Sales	$?	Decrease in accounts receivable	12	Cash received from customers	$ 612
Cost of goods sold	?	Increase in inventory	(24)		
		Decrease in accounts payable	(36)	Cash paid to suppliers	(420)
Salaries expense	?	Increase in salaries payable	12	Cash paid to employees	(66)
Depreciation expense	?	Depreciation expense	18	(Not reported—no cash effect)	
Insurance expense	?	Decrease in prepaid insurance	18	Cash paid for insurance	(24)
Loss on sale of land	?	Loss on sale of land	12	(Not reported—no cash effect)	
Income tax expense	?	Increase in income tax payable	12	Cash paid for income taxes	(42)
Net income	$?	**Net cash flows from operating activities**	$ 60	**Net cash flows from operating activities**	$ 60

Required:
Deduce the missing amounts and prepare the income statement.

E 21–20
Reconciliation of net cash flows from operating activities to net income
● LO21–3, LO21–4

The income statement and the cash flows from the operating activities section of the statement of cash flows are provided below for Syntric Company. The merchandise inventory account balance neither increased nor decreased during the reporting period. Syntric had no liability for insurance, deferred income taxes at any time during the period.

SYNTRIC COMPANY
Income Statement
For the Year Ended December 31, 2021
($ in thousands)

Sales		$ 312
Cost of goods sold		(188)
Gross margin		124
Salaries expense	$41	
Insurance expense	22	
Depreciation expense	11	
Depletion expense	5	
Interest expense	10	(89)
Gains and losses:		
Gain on sale of equipment		25
Loss on sale of land		(8)
Income before tax		52
Income tax expense		(26)
Net income		$ 26
Cash Flows from Operating Activities:		
Cash received from customers		$ 258
Cash paid to suppliers		(175)
Cash paid to employees		(37)
Cash paid for interest		(9)
Cash paid for insurance		(16)
Cash paid for income tax		(14)
Net cash flows from operating activities		$ 7

Required:
Prepare a schedule to reconcile net income to net cash flows from operating activities.

E 21–21
Cash flows from operating activities (direct method) derived from an income statement; cash flows from operating activities (indirect method)
● LO21–3, LO21–4

The income statement and a schedule reconciling cash flows from operating activities to net income are provided below ($ in thousands) for Peach Computers.

PEACH COMPUTERS
Income Statement
For the Year Ended December 31, 2021

Sales		$305
Cost of goods sold		185
Gross margin		120
Salaries expense	$41	
Insurance expense	19	
Depreciation expense	11	
Loss on sale of land	5	76
Income before tax		44
Income tax expense		22
Net income		$ 22

Reconciliation of Net Income
To Net Cash Flows from Operating Activities

Net income	$ 22
Adjustments for Noncash Effects	
Depreciation expense	11
Loss on sale of land	5
Changes in operating assets and liabilities:	
Decrease in accounts receivable	6
Increase in inventory	(13)
Decrease in accounts payable	(8)
Increase in salaries payable	5
Decrease in prepaid insurance	9
Increase in income tax payable	20
Net cash flows from operating activities	$ 57

Required:
1. Calculate each of the following amounts for Peach Computers:
   a. Cash received from customers during the reporting period
   b. Cash paid to suppliers of goods during the reporting period
   c. Cash paid to employees during the reporting period
   d. Cash paid for insurance during the reporting period
   e. Cash paid for income taxes during the reporting period
2. Prepare the cash flows from operating activities section of the statement of cash flows (direct method).

E 21–22
Indirect method; reconciliation of net income to net cash flows from operating activities
● LO21–4

The accounting records of Baddour Company provided the data below. Prepare a reconciliation of net income to net cash flows from operating activities.

Net loss	$5,000
Depreciation expense	6,000
Increase in salaries payable	500
Decrease in accounts receivable	2,000
Increase in inventory	2,300
Amortization of patent	300
Reduction in discount on bonds	200

E 21–23
Cash flows from operating activities (direct method)
● LO21–3

Portions of the financial statements for Myriad Products are provided below.

MYRIAD PRODUCTS COMPANY
Income Statement
For the Year Ended December 31, 2021
($ in millions)

Sales		$660
Cost of goods sold		250
Gross margin		410
Salaries expense	$110	
Depreciation expense	90	
Amortization expense	5	
Interest expense	20	
Loss on sale of land	3	228
Income before taxes		182
Income tax expense		91
Net Income		$ 91

MYRIAD PRODUCTS COMPANY
Selected Accounts from Comparative Balance Sheets
December 31, 2021 and 2020
($ in millions)

	Year		
	2021	**2020**	**Change**
Cash	$102	$100	$ 2
Accounts receivable	220	232	(12)
Inventory	440	450	(10)
Accounts payable	140	134	6
Salaries payable	80	86	(6)
Interest payable	25	20	5
Income tax payable	15	10	5

Required:
Prepare the cash flows from operating activities section of the statement of cash flows for Myriad Products Company using the *direct method.*

E 21–24
Cash flows from operating activities (indirect method)
● LO21–4

Refer to the data provided in E 21–23 for Myriad Products Company.

Required:
Prepare the cash flows from the operating activities section of the statement of cash flows for Myriad Products Company using the *indirect method.*

E 21–25
Cash flows from operating activities (direct method)—includes sale of cash equivalent
● LO21–3

Portions of the financial statements for Clear Transmissions Company are provided on the next page.

CLEAR TRANSMISSIONS COMPANY
Income Statement
For the Year Ended December 31, 2021
($ in thousands)

Sales		$1,320
Cost of goods sold		500
Gross margin		820
Salaries expense	$220	
Depreci		
Amortization expense	10	
Interest expense	40	
Loss on sale of cash equivalents	6	456
Income before taxes		364
Income tax expense		182
Net Income		$ 182

CLEAR TRANSMISSIONS COMPANY
Selected Accounts from Comparative Balance Sheets
December 31, 2021 and 2020
($ in thousands)

	Year		
	2021	**2020**	**Change**
Cash	$102	$100	$ 2
Accounts receivable	220	232	(12)
Inventory	440	450	(10)
Accounts payable	140	134	6
Salaries payable	80	86	(6)
Interest payable	25	20	5
Income tax payable	15	10	5

Required:
Prepare the cash flows from operating activities section of the statement of cash flows for Clear Transmissions Company using the *direct method.*

E 21–26
Cash flows from operating activities (indirect method)—includes sale of cash equivalent
● LO21–4

Refer to the data provided in E 21–25 for Clear Transmissions Company.

Required:
Prepare the cash flows from operating activities section of the statement of cash flows for Clear Transmissions Company using the *indirect method.*

E 21–27
Statement of cash flows; direct method
● LO21–3, LO21–5, LO21–6, LO21–8

Comparative balance sheets for 2021 and 2020, a statement of income for 2021, and additional information from the accounting records of Red, Inc., are provided below.

RED, INC.
Comparative Balance Sheets
December 31, 2021 and 2020
($ in millions)

	2021	**2020**
Assets		
Cash	$ 24	$ 110
Accounts receivable	178	132
Prepaid insurance	7	3
Inventory	285	175
Buildings and equipment	400	350
Less: Accumulated depreciation	(119)	(240)
	$ 775	$ 530
Liabilities		
Accounts payable	$ 87	$100
Accrued liabilities	6	11
Notes payable	50	0
Bonds payable	160	0
Shareholders' Equity		
Common stock	400	400
Retained earnings	72	19
	$ 775	$ 530

RED, INC.
Statement of Income
For Year Ended December 31, 2021
($ in millions)

Revenues		
Sales revenue		$2,000
Expenses		
Cost of goods sold	$1,400	
Depreciation expense	50	
Operating expenses	447	1,897
Net income		$ 103

Additional information from the accounting records:

a. During 2021, $230 million of equipment was purchased to replace $180 million of equipment (95% depreciated) sold at book value.

b. In order to maintain the usual policy of paying cash dividends of $50 million, it was necessary for Red to borrow $50 million from its bank.

Required:
Prepare the statement of cash flows of Red, Inc., for the year ended December 31, 2021. Present cash flows from operating activities by the direct method. (You may omit the schedule to reconcile net income with cash flows from operating activities.)

E 21–28
Pension plan funding
● LO21–3

Mayer Corporation has a defined benefit pension plan. Mayer's policy is to fund the plan annually, cash payments being made at the end of each year. Data relating to the pension plan for 2021 are as follows:

	December 31 ($ in millions)	
	2021	**2020**
Plan assets	$1,080	$900
Net pension expense for 2021:		
Service cost	$ 112	
Interest cost (6% × $850)	51	
Actual return on the plan assets (11% × $900 = $99)		
Adjusted for: $9 gain on the plan assets*	(90)	
Amortization of prior service cost	8	
Amortization of net loss	1	
	$ 82	

*(11% × $900) – (10% – $900)

Required:
Recreate the journal entries used to record Mayer's 2021 pension expense, gain on plan assets, and funding of plan assets in order to determine the cash paid to the pension trustee as reported in the statement of cash flows.

E 21–29
FASB codification research
● LO21–2

CODE

The statement of cash flows (as well as the balance sheet) includes within cash the notion of cash equivalents. The *FASB Accounting Standards Codification* represents the single source of authoritative U.S. generally accepted accounting principles.

Required:
Obtain the relevant authoritative literature on cash equivalents using the *FASB Accounting Standards Codification* at the FASB website (**www.fasb.org**). What is the specific seven-digit Codification citation (XXX-XX-XX) that describes the guidelines for determining what items should be deemed cash equivalents?

E 21–30
FASB codification research
● LO21–1, LO21–4, LO21–7

CODE

Access the *FASB Accounting Standards Codification* at the FASB website (**www.fasb.org**). Determine the specific eight-digit Codification citation (XXX-XX-XX-X) for accounting for each of the following items:

1. Disclosure of interest and income tax paid if the indirect method is used.
2. Primary objectives of a statement of cash flows.
3. Disclosure of noncash investing and financing activities.

E 21–31
Statement of cash flows; indirect method
● LO21–4, LO21–5, LO21–6, LO21–8, Appendix 21A

Refer to the data provided in E 21–27 for Red, Inc.

Required:
Prepare the statement of cash flows for Red, Inc., using the indirect method to report operating activities.

E 21–32
Statement of cash flows; T-account method
● LO21–3, Appendix 21B

Refer to the data provided in E 21–27 for Red, Inc.

Required:
Prepare the statement of cash flows (direct method) for Red, Inc. Use the T-account method to assist in your analysis.

Problems

connect

P 21–1
Classification of cash flows from investing and financing activities
● LO21–2, LO21–5 through LO21–7

Listed below are transactions that might be reported as investing and/or financing activities on a statement of cash flows. Possible reporting classifications of those transactions are provided also.

Required:
Indicate the reporting classification of each transaction by entering the appropriate classification code.

Classifications	
+I	Investing activity (cash inflow)
–I	Investing activity (cash outflow)
+F	Financing activity (cash inflow)
–F	Financing activity (cash outflow)
N	Noncash investing and financing activity
X	Not reported as an investing and/or a financing activity

Transactions

+I	1. Sale of land.
____	2. Issuance of common stock for cash.
____	3. Purchase of treasury stock.
____	4. Conversion of bonds payable to common stock.
____	5. Lease of equipment.
____	6. Sale of patent.
____	7. Acquisition of building for cash.
____	8. Issuance of common stock for land.
____	9. Collection of note receivable (principal amount).
____	10. Issuance of bonds.
____	11. Issuance of stock dividend.
____	12. Payment of property dividend.
____	13. Payment of cash dividends.
____	14. Issuance of short-term note payable for cash.
____	15. Issuance of long-term note payable for cash.
____	16. Purchase of marketable debt securities ("available for sale").
____	17. Payment of note payable.
____	18. Cash payment for five-year insurance policy.
____	19. Sale of equipment.
____	20. Issuance of note payable for equipment.
____	21. Acquisition of common stock of another corporation.
____	22. Repayment of long-term debt by issuing common stock.
____	23. Payment of semiannual interest on bonds payable.
____	24. Retirement of preferred stock.
____	25. Loan to another company.
____	26. Sale of inventory to customers.
____	27. Purchase of marketable securities (cash equivalents).

P 21–2
Statement of cash flows; direct method
● LO21–3, LO21–8

The comparative balance sheets for 2021 and 2020 and the statement of income for 2021 are given below for Wright Company. Additional information from Wright's accounting records is provided also.

WRIGHT COMPANY
Comparative Balance Sheets
December 31, 2021 and 2020
($ in thousands)

	2021	2020
Assets		
Cash	$ 42	$ 30
Accounts receivable	73	75
Short-term investment	40	15
Inventory	75	70
Land	50	60
Buildings and equipment	550	400
Less: Accumulated depreciation	(115)	(75)
	$ 715	$575
Liabilities		
Accounts payable	$ 28	$ 35
Salaries payable	2	5
Interest payable	5	3
Income tax payable	9	12
Notes payable	0	30
Bonds payable	160	100
Shareholders' Equity		
Common stock	250	200
Paid-in capital—excess of par	126	100
Retained earnings	135	90
	$ 715	$575

WRIGHT COMPANY
Income Statement
For Year Ended December 31, 2021
($ in thousands)

Revenues:		
Sales revenue		$380
Expenses:		
Cost of goods sold	$ 130	
Salaries expense	45	
Depreciation expense	40	
Interest expense	12	
Loss on sale of land	3	
Income tax expense	70	300
Net income		$ 80

Additional information from the accounting records:

a. Land that originally cost $10,000 was sold for $7,000.
b. The common stock of Microsoft Corporation was purchased for $25,000 as a short-term investment not classified as a cash equivalent.
c. New equipment was purchased for $150,000 cash.
d. A $30,000 note was paid at maturity on January 1.
e. On January 1, 2021, bonds were sold at their $60,000 face value.
f. Common stock ($50,000 par) was sold for $76,000.
g. Net income was $80,000 and cash dividends of $35,000 were paid to shareholders.

Required:

Prepare the statement of cash flows of Wright Company for the year ended December 31, 2021. Present cash flows from operating activities by the direct method. (You may omit the schedule to reconcile net income with cash flows from operating activities.)

P 21–3
Statement of cash flows; direct method
● LO21–3, LO21–8

The comparative balance sheets for 2021 and 2020 and the statement of income for 2021 are given below for National Intercable Company. Additional information from NIC's accounting records is provided also.

NATIONAL INTERCABLE COMPANY
Comparative Balance Sheets
December 31, 2021 and 2020
($ in millions)

	2021	2020
Assets		
Cash	$ 57	$ 55
Accounts receivable	181	170
Less: Allowance for uncollectible accounts	(8)	(6)
Prepaid insurance	7	12
Inventory	170	165
Long-term investment	66	90
Land	150	150
Buildings and equipment	290	270
Less: Accumulated depreciation	(85)	(75)
Trademark	24	25
	$852	$856
Liabilities		
Accounts payable	$ 30	$ 45
Salaries payable	3	8
Deferred tax liability	18	15
Lease liability	68	0
Bonds payable	145	275
Less: Discount on bonds	(22)	(25)
Shareholders' Equity		
Common stock	310	290
Paid-in capital—excess of par	95	85
Preferred stock	50	0
Retained earnings	155	163
	$852	$856

NATIONAL INTERCABLE COMPANY
Income Statement
For Year Ended December 31, 2021
($ in millions)

Revenues		
Sales revenue	$320	
Investment revenue	15	
Gain on sale of investments	5	$ 340
Expenses		
Cost of goods sold	125	
Salaries expense	55	
Depreciation expense	25	
Amortization expense	1	
Bad debt expense	7	
Insurance expense	13	
Interest expense	30	
Loss on sale of building	42	298
Income before tax		42
Income tax expense		20
Net income		$ 22

Additional information from the accounting records:

a. Investment revenue includes National Intercable Company's $5 million share of the net income of Central Fiber Optics Corporation, an equity method investee.

b. A long-term investment in bonds, originally purchased for $30 million, was sold for $35 million.

c. Pretax accounting income exceeded taxable income, causing the deferred income tax liability to increase by $3 million.

d. A building that originally cost $60 million, and which was one-fourth depreciated, was destroyed by fire. Some undamaged sections were sold for $3 million.
e. The right to use a building was acquired with a seven-year lease agreement; present value of lease payments, $80 million. Annual lease payments of $12 million are paid at Jan. 1 of each year starting in 2021.
f. $130 million of bonds were retired at maturity.
g. $20 million par value of common stock was sold for $30 million, and $50 million of preferred stock was sold at par.
h. Shareholders were paid cash dividends of $30 million.

Required:
1. Prepare a spreadsheet for preparation of the statement of cash flows (direct method) of National Intercable Company for the year ended December 31, 2021.
2. Prepare the statement of cash flows. (A reconciliation schedule is not required.)

P 21–4
Statement of cash flows; direct method
● LO21–3, LO21–8

The comparative balance sheets for 2021 and 2020 and the statement of income for 2021 are given below for Dux Company. Additional information from Dux's accounting records is provided also.

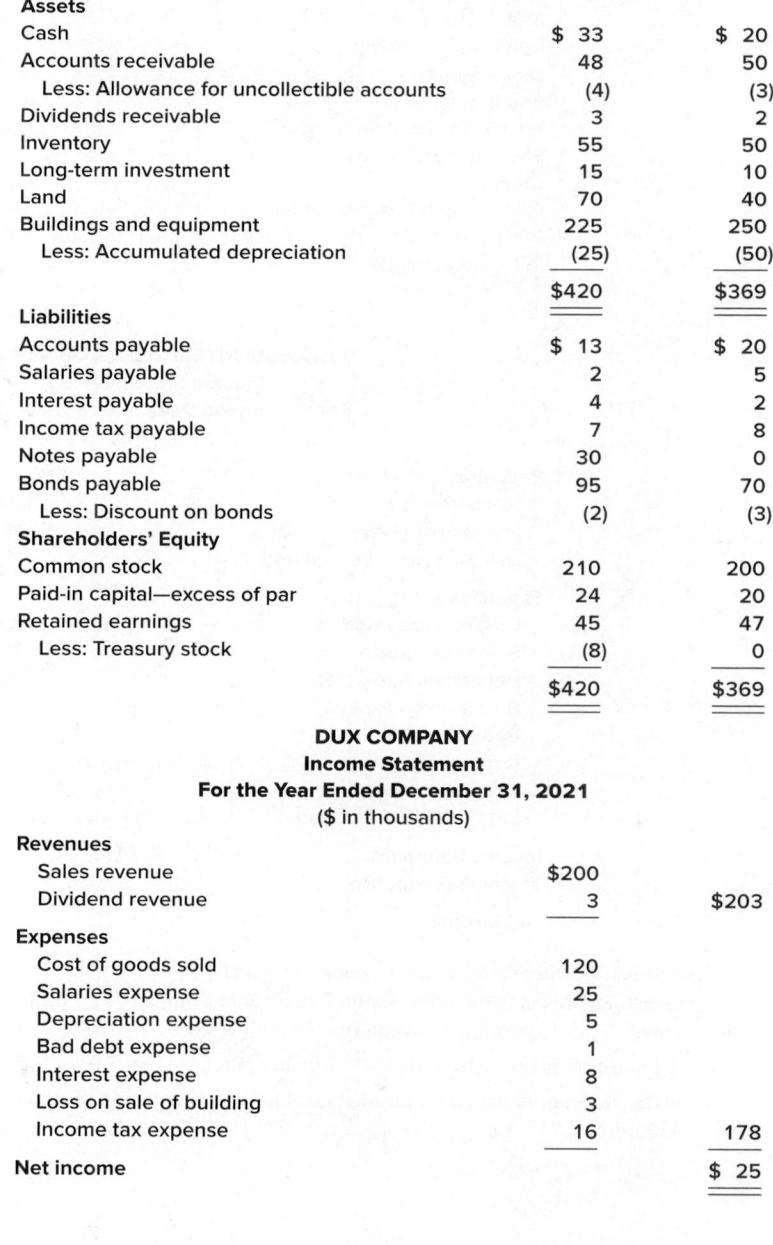

DUX COMPANY
Comparative Balance Sheets
December 31, 2021 and 2020
($ in thousands)

	2021	2020
Assets		
Cash	$ 33	$ 20
Accounts receivable	48	50
Less: Allowance for uncollectible accounts	(4)	(3)
Dividends receivable	3	2
Inventory	55	50
Long-term investment	15	10
Land	70	40
Buildings and equipment	225	250
Less: Accumulated depreciation	(25)	(50)
	$420	$369
Liabilities		
Accounts payable	$ 13	$ 20
Salaries payable	2	5
Interest payable	4	2
Income tax payable	7	8
Notes payable	30	0
Bonds payable	95	70
Less: Discount on bonds	(2)	(3)
Shareholders' Equity		
Common stock	210	200
Paid-in capital—excess of par	24	20
Retained earnings	45	47
Less: Treasury stock	(8)	0
	$420	$369

DUX COMPANY
Income Statement
For the Year Ended December 31, 2021
($ in thousands)

Revenues		
Sales revenue	$200	
Dividend revenue	3	$203
Expenses		
Cost of goods sold	120	
Salaries expense	25	
Depreciation expense	5	
Bad debt expense	1	
Interest expense	8	
Loss on sale of building	3	
Income tax expense	16	178
Net income		$ 25

Additional information from the accounting records:

a. A building that originally cost $40,000, and which was three-fourths depreciated, was sold for $7,000.

b. The common stock of Byrd Corporation was purchased for $5,000 as a long-term investment.

c. Property was acquired by issuing a 13%, seven-year, $30,000 note payable to the seller.

d. New equipment was purchased for $15,000 cash.

e. On January 1, 2021, bonds were sold at their $25,000 face value.

f. On January 19, Dux issued a 5% stock dividend (1,000 shares). The market price of the $10 par value common stock was $14 per share at that time.

g. Cash dividends of $13,000 were paid to shareholders.

h. On November 12, 500 shares of common stock were repurchased as treasury stock at a cost of $8,000.

Required:

Prepare the statement of cash flows of Dux Company for the year ended December 31, 2021. Present cash flows from operating activities by the direct method. (You may omit the schedule to reconcile net income to cash flows from operating activities.)

P 21–5
Statement of cash flows; direct method
● LO21–3
LO21–8

Comparative balance sheets for 2021 and 2020 and a statement of income for 2021 are given below for Metagrobolize Industries. Additional information from the accounting records of Metagrobolize also is provided.

METAGROBOLIZE INDUSTRIES
Comparative Balance Sheets
December 31, 2021 and 2020
($ in thousands)

	2021	2020
Assets		
Cash	$ 580	$ 375
Accounts receivable	600	450
Inventory	900	525
Land	675	600
Building	900	900
Less: Accumulated depreciation	(300)	(270)
Equipment	2,850	2,250
Less: Accumulated depreciation	(525)	(480)
Patent	1,200	1,500
	$6,880	$5,850
Liabilities		
Accounts payable	$ 750	$ 450
Accrued liabilities	300	225
Lease liability—land	130	0
Shareholders' Equity		
Common stock	3,150	3,000
Paid-in capital—excess of par	750	675
Retained earnings	1,800	1,500
	$6,880	$5,850

METAGROBOLIZE INDUSTRIES
Income Statement
For the Year Ended December 31, 2021
($ in thousands)

Revenues		
Sales revenue	$2,645	
Gain on sale of land	90	$2,735
Expenses		
Cost of goods sold	600	
Depreciation expense—building	30	
Depreciation expense—equipment	315	
Loss on sale of equipment	15	
Amortization of patent	300	
Operating expenses	500	1,760
Net income		$ 975

Additional information from the accounting records:

a. Annual payments of $20,000 on the finance lease liability are paid each January 1, beginning in 2021.

b. During 2021, equipment with a cost of $300,000 (90% depreciated) was sold.

c. The statement of shareholders' equity reveals reductions of $225,000 and $450,000 for stock dividends and cash dividends, respectively.

Required:

Prepare the statement of cash flows of Metagrobolize for the year ended December 31, 2021. Present cash flows from operating activities by the direct method. (You may omit the schedule to reconcile net income to cash flows from operating activities.)

P 21–6
Cash flows from operating activities (direct method) derived from an income statement and cash flows from operating activities (indirect method)
● LO21–3, LO21–4

The income statement and a schedule reconciling cash flows from operating activities to net income are provided below for Mike Roe Computers.

MIKE ROE COMPUTERS
Income Statement
For the Year Ended December 31, 2021
($ in millions)

Sales		$150
Cost of goods sold		90
Gross margin		60
Salaries expense	$20	
Insurance expense	12	
Depreciation expense	5	
Interest expense	6	(43)
Gains and losses:		
Gain on sale of equipment		12
Loss on sale of land		(3)
Income before tax		26
Income tax expense		(13)
Net income		$ 13

Reconciliation of Net Income to Net Cash Flows from Operating Activities

Net income	$ 13
Adjustments for noncash effects:	
Decrease in accounts receivable	5
Gain on sale of equipment	(12)
Increase in inventory	(6)
Increase in accounts payable	9
Increase in salaries payable	3
Depreciation expense	5
Decrease in bond discount	3
Decrease in prepaid insurance	2
Loss on sale of land	3
Increase in income tax payable	6
Net cash flows from operating activities	$ 31

Required:

1. Calculate each of the following amounts for Mike Roe Computers:
   a. Cash received from customers during the reporting period.
   b. Cash paid to suppliers of goods during the reporting period.
   c. Cash paid to employees during the reporting period.
   d. Cash paid for interest during the reporting period.
   e. Cash paid for insurance during the reporting period.
   f. Cash paid for income taxes during the reporting period.
2. Prepare the cash flows from operating activities section of the statement of cash flows (direct method).

P 21–7
Cash flows from operating activities (direct method) derived from an income statement and cash flows from operating activities (indirect method)
● LO21–3, LO21–4

The income statement and a schedule reconciling cash flows from operating activities to net income are provided below for Macrosoft Corporation.

MACROSOFT CORPORATION
Income Statement
For the Year Ended December 31, 2021
($ in millions)

Revenues and gains:		
Sales	$310	
Gain on sale of cash equivalents	2	
Gain on sale of investments	24	$336
Expenses and loss:		
Cost of goods sold	120	
Salaries	40	
Interest expense	12	
Insurance	20	
Depreciation	10	
Patent amortization	4	
Loss on sale of land	6	212
Income before tax		124
Income tax expense		62
Net income		$ 62

Reconciliation of Net Income to Net Cash Flows from Operating Activities

Net income	$ 62
Adjustments for noncash effects:	
Depreciation expense	10
Patent amortization expense	4
Loss on sale of land	6
Gain on sale of investment	(24)
Decrease in accounts receivable	6
Increase in inventory	(12)
Increase in accounts payable	18
Decrease in bond discount	1
Increase in salaries payable	6
Decrease in prepaid insurance	4
Increase in income tax payable	10
Net cash flows from operating activities	$ 91

Required:
Prepare the cash flows from operating activities section of the statement of cash flows (direct method).

P 21–8
Cash flows from operating activities (direct method and indirect method)—deferred income tax liability and amortization of bond discount
● LO21–3, LO21–4

Portions of the financial statements for Parnell Company are provided below.

PARNELL COMPANY
Income Statement
For the Year Ended December 31, 2021
($ in thousands)

Revenues and gains:		
Sales	$800	
Gain on sale of building	11	$811
Expenses and loss:		
Cost of goods sold	300	
Salaries	120	
Insurance	40	
Depreciation	123	
Interest expense	50	
Loss on sale of equipment	12	645
Income before tax		166
Income tax expense		78
Net income		$ 88

PARNELL COMPANY
Selected Accounts from Comparative Balance Sheets
December 31, 2021 and 2020
($ in thousands)

	Year		
	2021	2020	Change
Cash	$134	$100	$ 34
Accounts receivable	324	216	108
Inventory	321	425	(104)
Prepaid insurance	66	88	(22)
Accounts payable	210	117	93
Salaries payable	102	93	9
Deferred tax liability	60	52	8
Bond discount	190	200	(10)

Required:
1. Prepare the cash flows from operating activities section of the statement of cash flows for Parnell Company using the direct method.
2. Prepare the cash flows from operating activities section of the statement of cash flows for Parnell Company using the indirect method.

P 21–9
Cash flows from operating activities (direct method and indirect method)—cash equivalent
● LO21–3, LO21–4

Portions of the financial statements for Hawkeye Company are provided below.

HAWKEYE COMPANY
Income Statement
For the Year Ended December 31, 2021
($ in millions)

Sales		$900
Cost of goods sold		350
Gross margin		550
Operating expenses:		
Salaries	$232	
Depreciation	190	
Loss on sale of land	12	
Total operating expenses		434
Operating income		116
Other income (expense):		
Gain on sale of cash equivalents		4
Interest expense		(40)
Income before tax		80
Income tax expense		40
Net income		$ 40

HAWKEYE COMPANY
Selected Accounts from Comparative Balance Sheets
December 31, 2021 and 2020

	Year		
	2021	**2020**	**Change**
Cash	$212	$200	$ 12
Accounts receivable	395	421	(26)
Inventory	860	850	10
Accounts payable	210	234	(24)
Salaries payable	180	188	(8)
Interest payable	55	50	5
Income tax payable	90	104	(14)

Required:
1. Prepare the cash flows from operating activities section of the statement of cash flows for Hawkeye Company using the direct method.
2. Prepare the cash flows from operating activities section of the statement of cash flows for Hawkeye Company using the indirect method.

P 21–10
Relationship between the income statement and cash flows from operating activities (direct method and indirect method)
● LO21–3, LO21–4

The following schedule relates the income statement with cash flows from operating activities, derived by both the direct and indirect methods, in the format illustrated by Illustration 21–11 in the chapter. Some elements necessary to complete the schedule are missing.

Cash Flows from Operating Activities

Income Statement		**Indirect Method**		**Direct Method**	
		Net income	**$?**		
		Adjustments:			
Sales	$300	Decrease in accounts receivable	6	Cash received from customers	$?
Gain on sale of equipment	24	Gain on sale of equipment	(24)	(Not reported—no cash effect)	
		Increase in inventory	(12)		
Cost of goods sold	(?)	Increase in accounts payable	18	Cash paid to suppliers	(174)
Salaries expense	(42)	? in salaries payable	6	Cash paid to employees	(36)
Depreciation expense	(9)	Depreciation expense	9	Cash paid for depreciation	?
Interest expense	(?)	Decrease in bond discount	3	Cash paid for interest	(9)
Insurance expense	(21)	Decrease in prepaid insurance	9	Cash paid for insurance	(?)
Loss on sale of land	(6)	Loss on sale of land	6	(Not reported—no cash effect)	
Income tax expense	(27)	Increase in income tax payable	?	Cash paid for income taxes	(21)
Net Income	$?	**Net cash flows from operating activities**	**$ 54**	**Net cash flows from operating activities**	**$ 54**

Required:
Complete the schedule by determining each of the following missing elements:
1. Cash received from customers.
2. Cost of goods sold.
3. ? in salaries payable (Increase? Or decrease?).
4. Cash paid for depreciation.
5. Interest expense.
6. Cash paid for insurance.
7. Increase in income tax payable.
8. Net income.

P 21–11
Prepare a statement of cash flows; direct method
● LO21–3, LO21–8

The comparative balance sheets for 2021 and 2020 and the income statement for 2021 are given below for Arduous Company. Additional information from Arduous's accounting records is provided also.

ARDUOUS COMPANY
Comparative Balance Sheets
December 31, 2021 and 2020
($ in millions)

	2021	2020
Assets		
Cash	$ 109	$ 81
Accounts receivable	190	194
Investment revenue receivable	6	4
Inventory	205	200
Prepaid insurance	4	8
Long-term investment	156	125
Land	196	150
Buildings and equipment	412	400
Less: Accumulated depreciation	(97)	(120)
Patent	30	32
	$1,211	$1,074
Liabilities		
Accounts payable	$ 50	$ 65
Salaries payable	6	11
Interest payable (bonds)	8	4
Income tax payable	12	14
Deferred tax liability	11	8
Notes payable	23	0
Lease liability	75	0
Bonds payable	215	275
Less: Discount on bonds	(22)	(25)
Shareholders' Equity		
Common stock	430	410
Paid-in capital—excess of par	95	85
Preferred stock	75	0
Retained earnings	242	227
Less: Treasury stock	(9)	0
	$1,211	$1,074

ARDUOUS COMPANY
Income Statement For Year Ended
December 31, 2021
($ in millions)

Revenues and gain:		
Sales revenue	$410	
Investment revenue	11	
Gain on sale of Treasury bills	2	$423
Expenses and loss:		
Cost of goods sold	180	
Salaries expense	73	
Depreciation expense	12	
Amortization expense	2	
Insurance expense	7	
Interest expense	28	
Loss on sale of equipment	18	
Income tax expense	36	356
Net income		$ 67

Additional information from the accounting records:

a. Investment revenue includes Arduous Company's $6 million share of the net income of Demur Company, an equity method investee.

b. Treasury bills were sold during 2021 at a gain of $2 million. Arduous Company classifies its investments in Treasury bills as cash equivalents.

c. Equipment originally costing $70 million that was one-half depreciated was rendered unusable by a flood. Most major components of the equipment were unharmed and were sold for $17 million.

d. Temporary differences between pretax accounting income and taxable income caused the deferred tax liability to increase by $3 million.

e. The preferred stock of Tory Corporation was purchased for $25 million as a long-term investment.

f. Land costing $46 million was acquired by issuing $23 million cash and a 15%, four-year, $23 million note payable to the seller.

g. The right to use a building was acquired with a 15-year lease agreement; present value of lease payments, $82 million. Annual lease payments of $7 million are paid at the beginning of each year starting January 1, 2021.

h. $60 million of bonds were retired at maturity.

i. In February, Arduous issued a 5% stock dividend (4 million shares). The market price of the $5 par value common stock was $7.50 per share at that time.

j. In April, 1 million shares of common stock were repurchased as treasury stock at a cost of $9 million.

Required:

Prepare the statement of cash flows of Arduous Company for the year ended December 31, 2021. Present cash flows from operating activities by the direct method. (A reconciliation schedule is not required.)

P 21–12
Transactions affecting retained earnings
● LO21–5, LO21–6, LO21–8

Shown below in T-account format are the changes affecting the retained earnings of Brenner-Jude Corporation during 2021. At January 1, 2021, the corporation had outstanding 105 million common shares, $1 par per share.

Retained Earnings ($ in millions)

		90	Beginning balance
Retirement of 5 million common shares for $22 million	2		
		88	Net income for the year
Declaration and payment of a $0.33 per share cash dividend	33		
Declaration and distribution of a 4% stock dividend	20		
		123	Ending balance

Required:

1. From the information provided by the account changes you should be able to re-create the transactions that affected Brenner-Jude's retained earnings during 2021. Reconstruct the journal entries which can be used as spreadsheet entries in the preparation of a statement of cash flows. Also indicate any investing and financing activities you identify from this analysis that should be reported on the statement of cash flows.
2. Prepare a statement of retained earnings for Brenner-Jude for the year ended 2021. (You may wish to compare your solution to this problem with the parallel situation described in Exercise 18–17.)

P 21–13
Various cash flows
● LO21–3 through LO21–8

Following are selected balance sheet accounts of Del Conte Corp. at December 31, 2021 and 2020, and the increases or decreases in each account from 2020 to 2021. Also presented is selected income statement information for the year ended December 31, 2021, and additional information.

Selected Balance Sheet Accounts	2021	2020	Increase (Decrease)
Assets			
Accounts receivable	$ 34,000	$ 24,000	$10,000
Property, plant, and equipment	277,000	247,000	30,000
Accumulated depreciation	(178,000)	(167,000)	11,000
Liabilities and Stockholders' Equity			
Bonds payable	49,000	46,000	3,000
Dividends payable	8,000	5,000	3,000
Common stock, $1 par	22,000	19,000	3,000
Additional paid-in capital	9,000	3,000	6,000
Retained earnings	104,000	91,000	13,000

Selected Income Statement Information for the Year Ended December 31, 2021

Sales revenue	$155,000
Depreciation	33,000
Gain on sale of equipment	13,000
Net income	28,000

Additional Information:

a. Accounts receivable relate to sales of merchandise.

b. During 2021, equipment costing $40,000 was sold for cash.

c. During 2021, bonds payable with a face value of $20,000 were issued in exchange for property, plant, and equipment. There was no amortization of bond discount or premium.

Required:

Items 1 through 5 represent activities that will be reported in Del Conte's statement of cash flows for the year ended December 31, 2021. The following two responses are required for each item:

a. Determine the amount that should be reported in Del Conte's 2021 statement of cash flows.

b. Using the list below, determine the category in which the amount should be reported in the statement of cash flows.

O. Operating activity
I. Investing activity
F. Financing activity

	Amount	Category
1. Cash collections from customers (direct method).	____	____
2. Payments for purchase of property, plant, and equipment.	____	____
3. Proceeds from sale of equipment.	____	____
4. Cash dividends paid.	____	____
5. Redemption of bonds payable.	____	____

(AICPA adapted)

P 21–14
Statement of cash flows; indirect method; limited information
● LO21–4, LO21–8

The comparative balance sheets for 2021 and 2020 are given below for Surmise Company. Net income for 2021 was $50 million.

SURMISE COMPANY
Comparative Balance Sheets
December 31, 2021 and 2020
($ in millions)

	2021	2020
Assets		
Cash	$ 36	$ 40
Accounts receivable	92	96
Less: Allowance for uncollectible accounts	(12)	(4)
Prepaid expenses	8	5
Inventory	145	130
Long-term investment	80	40
Land	100	100
Buildings and equipment	420	300
Less: Accumulated depreciation	(142)	(120)
Patent	16	17
	$ 743	$ 604
Liabilities		
Accounts payable	$ 13	$ 32
Accrued liabilities	2	10
Notes payable	35	0
Lease liability	111	0
Bonds payable	65	125
Shareholders' Equity		
Common stock	60	50
Paid-in capital—excess of par	245	205
Retained earnings	212	182
	$ 743	$ 604

Required:

Prepare the statement of cash flows of Surmise Company for the year ended December 31, 2021. Use the indirect method to present cash flows from operating activities because you do not have sufficient information to use the direct method. You will need to make reasonable assumptions concerning the reasons for changes in some account balances. A spreadsheet or T-account analysis will be helpful. (*Hint:* The right to use a building was acquired with a seven-year lease agreement. Annual lease payments of $9 million are paid at January 1 of each year starting in 2021.)

P 21–15
Integrating problem; bonds; lease transactions; lessee and lessor; statement of cash flow effects
● LO21–3, LO21–5, LO21–6

Digital Telephony issued 10% bonds, dated January 1, with a face amount of $32 million on January 1, 2021. The bonds mature in 2031 (10 years). For bonds of similar risk and maturity the market yield is 12%. Interest is paid semiannually on June 30 and December 31. Digital recorded the issue as follows:

Cash	28,329,472	
Discount on bonds	3,670,528	
Bonds payable		32,000,000

Digital also leased switching equipment to Midsouth Communications, Inc. on September 30, 2021. Digital purchased the equipment from MDS Corp. at a cost of $6 million. The five-year lease agreement calls for Midsouth to make quarterly lease payments of $391,548, payable each September 30, December 31, March 31, and June 30, with the first payment on September 30, 2021. Digital's implicit interest rate is 12%.

Required:

1. What would be the amount(s) related to the bonds that Digital would report in its statement of cash flows for the year ended December 31, 2021, if Digital uses the direct method? The indirect method?
2. What would be the amounts related to the lease that *Midsouth* would report in its statement of cash flows for the year ended December 31, 2021?
3. What would be the amounts related to the lease that *Digital* would report in its statement of cash flows for the year ended December 31, 2021?
4. Assume *MDS* manufactured the equipment at a cost of $5 million and that *Midsouth* leased the equipment directly from *MDS.* What would be the amounts related to the lease that *MDS* would report in its statement of cash flows for the year ended December 31, 2021?

P 21–16
Statement of cash flows; indirect method
● LO21–4, LO21–8

Refer to the data provided in the P 21–4 for Dux Company.

Required:

Prepare the statement of cash flows for Dux Company using the *indirect method.*

P 21–17
Statement of cash flows; indirect method
● LO21–4, LO21–8

Refer to the data provided in the P 21–5 for Metagrobolize Industries.

Required:

Prepare the statement of cash flows for Metagrobolize Industries using the *indirect method.*

P 21–18
Statement of cash flows; indirect method
● LO21–4, LO21–8

Refer to the data provided in the P 21–11 for Arduous Company.

Required:

Prepare the statement of cash flows for Arduous Company using the *indirect method.*

(Note: The following problems use the technique learned in Appendix 21B.)

P 21–19
Statement of cash flows; T-account method
● LO21–3, LO21–8

Refer to the data provided in the P 21–4 for Dux Company.

Required:

Prepare the statement of cash flows for Dux Company. Use the T-account method to assist in your analysis.

P 21–20
Statement of cash flows; T-account method
● LO21–3, LO21–8

Refer to the data provided in the P 21–5 for Metagrobolize Industries.

Required:
Prepare the statement of cash flows for Metagrobolize Industries. Use the T-account method to assist in your analysis.

P 21–21
Statement of cash flows; T-account method
● LO21–3, LO21–8

Refer to the data provided in the P 21–11 for Arduous Company.

Required:
Prepare the statement of cash flows for Arduous Company. Use the T-account method to assist in your analysis.

Decision Makers' Perspective

©ImageGap/ Getty Images

Apply your critical-thinking ability to the knowledge you've gained. These cases will provide you an opportunity to develop your research, analysis, judgment, and communication skills. You also will work with other students, integrate what you've learned, apply it in real-world situations, and consider its global and ethical ramifications. This practice will broaden your knowledge and further develop your decision-making abilities.

Analysis Case 21–1
Distinguish income and cash flows
● LO21–1, LO21–3, LO21–4

"Why can't we pay our shareholders a dividend?" shouted your new boss. "This income statement you prepared for me says we earned $5 million in our first half year!"

You were hired last month as the chief accountant for Enigma Corporation, which was organized on July 1 of the year just ended. You recently prepared the financial statements below.

ENIGMA CORPORATION
Income Statement
For the Six Months Ended December 31, 2021
($ in millions)

Sales revenue	$75
Cost of goods sold	(30)
Depreciation expense	(5)
Remaining expenses	(35)
Net income	$ 5

ENIGMA CORPORATION
Balance Sheet
December 31, 2021
($ in millions)

Cash	$ 1
Accounts receivable (net)	20
Merchandise inventory	15
Equipment (net)	44
Total	$80
Accounts payable	$ 2
Accrued liabilities	7
Notes payable	36
Common stock	30
Retained earnings	5
Total	$80

You have just explained to your boss, Robert James, that although net income was $5 million, operating activities produced a net decrease in cash. Unable to understand your verbal explanation, he has asked you to prepare a written report explaining the apparent discrepancy between Enigma's profitability and its cash flows. To increase the chances of your boss's understanding the situation, you want to include in your report a determination of net cash flows from operating activities demonstrating how it is possible for operating activities to simultaneously produce a positive net income and negative net cash flows.

Required:
Determine net cash flows from operating activities by both the direct and indirect methods.

Judgment Case 21–2
Distinguish income and cash flows
● LO21–3, LO21–8

You are a loan officer for First Benevolent Bank. You have an uneasy feeling as you examine a loan application from Daring Corporation. The application included the following financial statements.

DARING CORPORATION
Income Statement
For the Year Ended December 31, 2021

Sales revenue	$100,000
Cost of goods sold	(50,000)
Depreciation expense	(5,000)
Remaining expenses	(25,000)
Net income	$ 20,000

DARING CORPORATION
Balance Sheet
December 31, 2021

Cash	$ 5,000
Accounts receivable	25,000
Inventory	20,000
Operational assets	55,000
Accumulated depreciation	(5,000)
Total	$100,000
Accounts payable	$ 10,000
Interest payable	5,000
Note payable	45,000
Common stock	20,000
Retained earnings	20,000
Total	$100,000

It is not Daring's profitability that worries you. The income statement submitted with the application shows net income of $20,000 in Daring's first year of operations. By referring to the balance sheet, you see that this net income represents a 20% rate of return on assets of $100,000. Your concern stems from the recollection that the note payable reported on Daring's balance sheet is a two-year loan you approved earlier in the year.

You also recall another promising new company that, just last year, defaulted on another of your bank's loans when it failed due to its inability to generate sufficient cash flows to meet its obligations. Before requesting additional information from Daring, you decide to test your memory of the intermediate accounting class you took in night school by attempting to prepare a statement of cash flows from the information available in the loan application.

Real World Case 21–3
Information from cash flow activities; FedEx
● LO21–3 through LO21–8

Real World Financials

Locate the financial statements and related disclosure notes of **FedEx Corporation** for the fiscal year ended May 31, 2017. You can locate the report online at **www.fedex.com**. Use the information provided in the statement of cash flows to respond to the questions below.

Required:
1. Is FedEx expanding its business or contracting its business, as evidenced by the investing activities.
2. Is FedEx raising as much cash through financing activities as the amount needed to fund its investing activities?
3. Determine the activities listed under financing activities for the most recent fiscal year. [*Hint:* FedEx's Statement of Changes in Common Stockholders' Investment (statement of shareholders' equity) will help you determine the nature of the stock activity.] What is the most notable financing activity reported over the most recent three years? What was the amount of cash received or paid for that activity each year?
4. What are the cash payments FedEx made for interest and for income taxes in the three years reported? (*Hint:* See the disclosure notes.)

5. Obtain the relevant authoritative literature on disclosure of interest and income taxes using the *FASB Accounting Standards Codification.* You might gain access at the FASB website (**www.fasb.org**). What is the specific eight-digit Codification citation (XXX-XX-XX-X) that specifies the way FedEx reports interest and income taxes separately that enabled you to answer *requirement* 3?

Source: FedEx Corporation

Analysis Case 21–4
Smudged ink; find missing amounts
● LO21–3, LO21–4

"Be careful with that coffee!" Your roommate is staring in disbelief at the papers in front of her. "This was my contribution to our team project," she moaned. "When you spilled your coffee, it splashed on this page. Now I can't recognize some of these numbers, and Craig has my source documents."

Knowing how important this afternoon's presentation is to your roommate, you're eager to see what can be done. "Let me see that," you offer. "I think we can figure this out." The statement of cash flows and income statement are intact. The reconciliation schedule and the comparative balance sheets are coffee casualties.

DISTINCTIVE INDUSTRIES
Statement of Cash Flows
For the Year Ended December 31, 2018
($ in millions)

Cash Flows from Operating Activities:		
Collections from customers	$213	
Payment to suppliers	(90)	
Payment of general & administrative expenses	(54)	
Payment of income taxes	(27)	
Net cash flows from operating activities		$ 42
Cash Flows from Investing Activities:		
Sale of equipment		120
Cash Flows from Financing Activities:		
Issuance of common stock	30	
Payment of dividends	(9)	
Net cash flows from financing activities		21
Net increase in cash		$183
Reconciliation of net income to cash flows from operating activities:		
Net income	$ 84	
Adjustments for noncash items:		
Depreciation expense	☐	
☐	☐	
☐	☐	
☐	☐	
☐	☐	
☐	☐	
Net cash flows from operating activities		☐

DISTINCTIVE INDUSTRIES
Income Statement
For the Year Ended December 31, 2021
($ in millions)

Sales revenue		$240
Cost of goods sold		96
Gross profit		144
Operating expenses:		
General and administrative	$54	
Depreciation	30	
Total operating expenses		84
Operating income		60
Other income:		
Gain on sale of equipment		45
Income before income taxes		105
Income tax expense		21
Net income		$ 84

DISTINCTIVE INDUSTRIES
Comparative Balance Sheets
At December 31
($ in millions)

	2021	2020
Assets:		
Cash	$ 360	☐
Accounts receivable (net)	☐	252
Inventory	180	☐
Property, plant, and equipment	450	600
Less: Accumulated depreciation	(120)	☐
Total assets	☐	☐
Liabilities and shareholders' equity:		
Accounts payable	$ 120	$ 90
General and administrative expenses payable	27	27
Income tax payable	66	☐
Common stock	720	690
Retained earnings	☐	141
Total liabilities and shareholders' equity	☐	☐

Required:

1. Determine the missing amounts, and reconstruct the comparative balance sheets.
2. Reconstruct the reconciliation of net income to cash flows from operating activities (operating cash flows using the indirect method).

Ethics Case 21–5
Where's the cash?
● LO21–1, LO21–3

After graduating near the top of his class, Ben Naegle was hired by the local office of a Big 4 CPA firm in his hometown. Two years later, impressed with his technical skills and experience, Park Electronics, a large regional consumer electronics chain, hired Ben as assistant controller. This was last week. Now Ben's initial excitement has turned to distress.

The cause of Ben's distress is the set of financial statements he's stared at for the last four hours. For some time prior to his recruitment, he had been aware of the long trend of moderate profitability of his new employer. The reports on his desk confirm the slight, but steady, improvements in net income in recent years. The trend he was just now becoming aware of, though, was the decline in cash flows from operations.

Ben had sketched out the following comparison ($ in millions):

	2021	2020	2019	2018
Income from operations	$140.0	$132.0	$127.5	$127.0
Net income	38.5	35.0	34.5	29.5
Cash flow from operations	1.6	17.0	12.0	15.5

Profits? Yes. Increasing profits? Yes. The cause of his distress? The ominous trend in cash flow which is consistently lower than net income.

Upon closer review, Ben noticed three events in the last two years that, unfortunately, seemed related:

a. Park's credit policy had been loosened; credit terms were relaxed and payment periods were lengthened.
b. Accounts receivable balances had increased dramatically.
c. Several of the company's compensation arrangements, including that of the controller and the company president, were based on reported net income.

Required:

1. What is so ominous about the combination of events Ben sees?
2. What course of action, if any, should Ben take?

Analysis Case 21–6
Information from cash flow activities; Kroger
● LO21–3 through LO21–8
Real World Financials

Refer to the financial statements and related disclosure notes of **The Kroger Company** for the fiscal year ending January 30, 2018. You can locate the report online from "investor relations" at **www.kroger.com**.

Notice that Kroger's net income has declined over the three years reported. To supplement their analysis of profitability, many analysts like to look at "free cash flow." A popular way to measure this metric is "structural free cash flow" (or as Warren Buffett calls it, "owner's earnings"), which is calculated as net income from operations, plus depreciation and amortization, minus capital expenditures.

Required:

Determine free cash flows for Kroger in each of the three years reported. Compare that amount with net income each year. What pattern do you detect?

Source: Kroger Company

Research Case 21–7
FASB codification; locate and extract relevant information and cite authoritative support for a financial reporting issue; cash flow classification
● LO21–4 through LO21–6

CODE

"This one's got me stumped," you say to no one in particular. "First day on the job; I'd better get it right." It's the classification of notes payable in the statement of cash flows that has you in doubt. Having received an "A" in Intermediate Accounting, you know that a note payable representing a bank loan is a financing activity, but this one is a note payable to your employer's primary merchandise supplier. "I wonder if the accounting is the same."

Required:

1. Obtain the relevant authoritative literature on cash flow classification using the *FASB Accounting Standards Codification.* You might gain access at the FASB website (**www.fasb.org**). What is the specific nine-digit Codification citation (XXX-XX-XX-X) that specifies the classification of notes payable to suppliers?
2. Is accounting the same for both short-term and long-term notes payable to suppliers?

IFRS Case 21–8
Statement of cash flows presentation, British Telecommunications
● LO21–3 through LO21–6, LO21–9

IFRS

Real World Financials

British Telecommunications Plc (BT), a U.K. company, is the world's oldest communications company. The company prepares its financial statements in accordance with International Financial Reporting Standards. Locate BT's statement of cash flows from it's website at **www.btplc.com**.

Required:

1. What are the primary classifications into which BT's cash inflows and cash outflows are separated? Is this classification the same as or different from cash flow statements prepared in accordance with U.S. GAAP?
2. How are cash inflows from dividends and interest and cash outflows for dividends and interest classified in BT's cash flow statements? Is this classification the same as or different from cash flow statements prepared in accordance with U.S. GAAP?

Source: British Telecommunications Plc (BT)

Data Analytics

connect

Data analytics is the process of examining data sets in order to draw conclusions about the information they contain. If you haven't completed any of the prior data analytics cases, follow the instructions listed in the Chapter 1 Data Analytics case to get set up. You will need to watch the videos referred to in the Chapters 1 - 3 Data Analytics cases. No additional videos are required for this case. All short training videos can be found here: **https://www.tableau.com/learn/training#getting-started**.

Data Analytics Case
Cash Performance Ratios
● LO21–3

In the Data Analytics Cases in previous chapters, we applied Tableau to examine two (hypothetical) publicly traded companies: GPS Corporation and Tru, Inc. regarding various aspects of their earnings performance and financial position. In this case, you focus on performance in terms of the ability of the two companies to generate a cash return on its assets.

Required:

While net income is usually considered the best measure of long-term profitability of a company, the cash return on assets ratio provides an indication of the actual cash flows generated by the company's assets during each individual reporting period, not influenced by any income measurements or income recognition.

For each of the two companies in the most recent five-year period, 2017-2021, use Tableau to calculate and display the trends for (a) the rate of return on assets [Net income / Total assets] and (b) the cash return on assets [Net cash flows from operating activities / Total assets].

Based upon what you find, answer the following questions:

1. What was (a) the return on assets in 2021 and (b) the cash return on assets for the most recent five years (rounded to the nearest one-half percent) for Tru, Inc.?
2. What was (a) the cash return on assets in 2021 and (b) the average cash return on assets (rounded to the nearest one-half percent) for the most recent five years for Tru, Inc.?
3. Which of the two companies, GPS Corporation or Tru, Inc., indicates the most erratic pattern of performance during the period 2017-2021 as measured by the cash return on assets ratio?

Resources:

Download the "GPS_Tru_Financials.xlsx" Excel file available in Connect, or under Student Resources within the Library tab. Save it to the computer on which you will be using Tableau.

For this case, you will create several calculations to produce two long-term profitability ratios to allow you to compare and contrast the two companies.

After you view the training videos, follow these steps to create the charts you'll use for this case:

- Open Tableau and connect to the Excel spreadsheet you downloaded.
- Click on the "Sheet 1" tab at the bottom of the canvas, to the right of the Data Source at the bottom of the screen.
- Drag "($ in 000s) Company" and "Year" under Dimensions to the Columns shelf. Change "Year" to *discrete* by right-clicking and selecting "discrete." Select "Show Filter" and uncheck all the years except the last five years (2017 – 2021).
- Create a calculated field by clicking the "Analysis" tab at the top of the screen and selecting "Create Calculated Field." Name the calculation "Return on assets ratio." In the Calculation Editor window, from Measures, drag "Net Income / (loss)", type a division sign, and drag "Total assets" beside it. Make sure the window says that the calculation is valid and click OK.
- Create a calculated field by clicking the "Analysis" tab at the top of the screen and selecting "Create Calculated Field." Name the calculation "Cash return on assets." In the Calculation Editor window, from Measures, drag "Net cash flows from operating activities", type a division sign, and then drag "Total assets". Make sure the window says that the calculation is valid and click OK.
- Drag the newly created "Return on assets ratio" and "Cash return on assets" to the Rows shelf. Add labels to the bars by clicking on "Label" under the "Marks" card and clicking the box "Show mark labels." Format the labels to Times New Roman, bold, black and 10-point font. Edit the color on the Color marks card if desired.
- Change the title of the sheet to be "Return on Assets vs. Cash Return on Assets" by right-clicking and selecting "Edit title." Format the title to Times New Roman, bold, black and 15-point font. Change the title of "Sheet 1" to match the sheet title by right-clicking, selecting "Rename" and typing in the new title.
- Click on the New Worksheet tab ("Sheet 2" should open) to the right of the newly named "Return on Assets vs. Cash Return on Assets" sheet at the bottom of the screen.
- Follow the procedures above for the "Company and "Year" but drag the company field to Columns shelf and the year field to Rows shelf.
- Double click on "Return on assets ratio" under Measures. Click on "Analysis," select "Totals" and then select "Show Column Grand Totals." Select "Totals" again, then select "Total All Using" and click on "Average."
- Right-click one of the number fields and select "format." Format the numbers to Times New Roman, bold, pink and 12-point font. Select Field and then SUM(Return on assets ratio) in the upper right corner and change to Percentage with one decimal place and change the Alignment to "Center."
- Change the title of the sheet to be "Average Return on Assets" by right-clicking and selecting "Edit title." Format the title to Times New Roman, bold, black and 15-point font. Change the title of "Sheet 2" to match the sheet title by right-clicking, selecting "Rename" and typing in the new title.
- Click on the "Average Return on Assets" tab at the bottom of the canvas, right-click and select "Duplicate." Select the duplicated sheet (now named "Average Return on Assets (2)).
- Drag the "SUM (Return on Assets)" pill from the "Marks" card back to Measures. Double click on "Cash Return on Assets" under Measures. Follow the procedure above to change the Totals to "Average" through the "Analysis" tab.
- Format the numbers to Times New Roman, bold, blue and 12-point font. Select Field and the SUM(Cash Return on Assets) in the upper right corner and change to Percentage with one decimal place and change the Alignment to "Center."
- Change the title of the sheet to be "Average Cash Return on Assets" by right-clicking and selecting "Edit title." Format the title to Times New Roman, bold, black and 15-point font. Change the title of "Sheet 2" to match the sheet title by right-clicking, selecting "Rename" and typing in the new title.
- Once complete, save the file as "DA21_Your initials.twbx."

Continuing Cases

Target Case

- LO21–2, LO21–3, LO21–4, LO21–5, LO21–6, LO21–7

Target Corporation prepares its financial statements according to U.S. GAAP. Target's financial statements and disclosure notes for the year ended February 3, 2018, are available in Connect. This material is also available under the Investor Relations link at the company's website (**www.target.com**).

Required:

1. Is Target's cash provided by operating activities increasing or decreasing?
2. Is Target's cash provided by operating activities more or less than net income in fiscal 2017?
3. Is Target increasing or decreasing its investment in property and equipment?
4. Is Target increasing or decreasing its long-term debt?
5. Some transactions that don't increase or decrease cash must be reported in conjunction with a statement of cash flows. What activity of this type did Target report during each of the three years presented?

Air France–KLM Case

- LO21–9

IFRS

AIRFRANCE KLM

Air France–KLM (AF), a Franco-Dutch company, prepares its financial statements according to International Financial Reporting Standards. AF's financial statements and disclosure notes for the year ended December 31, 2017, are provided in Connect. This material is also available under the Finance link at the company's website (**www.airfranceklm-finance.com**).

Required:

1. What are the three primary classifications into which AF's cash inflows and cash outflows are separated?
2. Is this classification different than cash flow statements prepared in accordance with U.S. GAAP?
3. In which classification are cash inflows from dividends included in AF's cash flow statements?
4. Is this classification different than cash flow statements prepared in accordance with U.S. GAAP?

CPA Exam Questions and Simulations

ROGER CPA Review

Sample CPA Exam questions from Roger CPA Review are available in Connect as support for the topics in this chapter. These multiple-choice questions and task-based simulations include expert-written explanations and solutions, and provide a starting point for students to become familiar with the content and functionality of the actual CPA Exam.

APPENDIX

A Derivatives

OVERVIEW

A complex class of financial instruments exists in financial markets in response to the desire of firms to manage risks. In fact, these financial instruments would not exist in their own right, but have been created solely to hedge against risks created by other financial instruments or by transactions that have yet to occur but are anticipated. These financial instruments often are called derivatives because they "derive" their values or contractually required cash flows from some other security or index. Accounting for derivatives is the focus of this Appendix.

LEARNING OBJECTIVES

After studying this Appendix, you should be able to:

- **LOA–1** Identify the major derivative securities used to hedge against risk. *(p. A-3)*
- **LOA–2** Understand the way we account for fair value hedges. *(p. A-8)*
- **LOA–3** Understand the way we account for cash flow hedges. *(p. A-12)*
- **LOA–4** Understand the way we account for foreign currency hedges. *(p. A-16)*
- **LOA–5** Describe disclosure requirements for derivatives and risk. *(p. A-18)*
- **LOA–6** Understand the extended method for interest rate swap accounting. *(p. A-18)*

". . . derivatives are financial weapons of mass destruction . . ."
— Warren Buffett, Berkshire Hathaway CEO

". . . the growing use of complex financial instruments known as derivatives does not pose a threat to the country's financial system . . ."
— Alan Greenspan, Federal Reserve Chairman

"Total world derivatives are $1000 trillion or 19 times the total world GDP of $54 trillion."
— Chuck Burr, *Culture Change*

In today's global economy and evolving financial markets, businesses are increasingly exposed to a variety of risks, which, unmanaged, can have major impacts on earnings or even threaten a company's very existence. Risk management, then, has become critical. Derivative financial instruments have become the key tools of risk management.[1]

[1]Almost all financial institutions and over half of all nonfinancial companies use derivatives.

Derivatives are financial instruments that "derive" their values or contractually required cash flows from some other security or index. For instance, a contract allowing a company to buy a particular asset (say steel, gold, or flour) at a designated future date at a *predetermined price* is a financial instrument that derives its value from expected and actual changes in the price of the underlying asset. Financial futures, forward contracts, options, and interest rate swaps are the most frequently used derivatives. We discuss each of these in the paragraphs that follow. Derivatives are valued as tools to manage or hedge companies' increasing exposures to risk, including interest rate risk, price risk, and foreign exchange risk. Companies may enter into derivatives to entirely or partially offset these risk exposures. The variety, complexity, and magnitude of derivatives have grown rapidly in recent years. Accounting standard-setters have scrambled to keep pace.

Derivatives are financial instruments that "derive" their values from some other security or *index*.

A persistent stream of headline stories has alerted us to multimillion-dollar losses by many companies and the financial collapse of Bear Stearns and AIG.[2] Focusing on these headlines, it would be tempting to conclude that derivatives are risky business indeed. Certainly they can be quite risky if misused, but the fact is, these financial instruments exist to lessen, not increase, risk. Properly used, they serve as a form of "insurance" against risk. In fact, if a company is exposed to a substantial risk and does not hedge that risk, it is taking a gamble. On the other hand, if a derivative is used improperly, it can be a huge gamble itself.

Derivatives serve as a form of "insurance" against risk.

The notional amount of the derivatives market vastly exceeds the total value of the assets they are intended to mimic or mirror. This implies that firms are using derivatives for purposes other than risk management. Many observers are fearful that the size of the derivatives market poses significant risk to the economy. Some caution that the vast derivatives market could even cause the entire global financial system to crash. Why? If interest rates rise, then many speculative interest rate swaps would incur losses. But it's the size of the interest rate swap market, and thus the size of the resultant losses, that prompts the anxiety. At the end of June 2017, the over-the-counter derivatives market was $542 trillion. Yes, that's 542 with twelve zeroes ($542,000,000,000,000). And, that's nearly 30 times the U.S. Gross Domestic Product. Seventy-seven percent of those derivatives ($416 trillion) are interest rate swaps.[3] Our focus here, though, is not on the risk posed by the speculative use of derivatives, but instead on the use of derivatives to reduce company risk.

Derivatives Used to Hedge Risk

Hedging means taking an action that is expected to produce exposure to a particular type of risk that is precisely the *opposite* of an actual risk to which the company already is exposed. For instance, the volatility of interest rates creates exposure to interest-rate risk for companies that issue debt—which, of course, includes most companies. So, a company that frequently arranges short-term loans from its bank under notes with floating (variable) interest rates is exposed to the risk that interest rates might increase and adversely affect borrowing costs. Similarly, a company that regularly reissues commercial paper (short-term notes) as it matures faces the possibility that new rates will be higher and cut into forecasted income. When borrowings are large, the potential cost can be substantial. So, the firm might choose to hedge its position by entering into a transaction that would produce a *gain* of roughly the same amount as the potential loss if interest rates do, in fact, increase. A way to do this is to enter into a contract that will exchange (swap) floating rate payments for fixed interest payments, or vice versa, without exchanging the underlying notes.

Hedging means taking a risk position that is opposite to an actual position that is exposed to risk.

● LOA–1

Hedging is used to deal with three areas of risk exposure: fair value risk, cash flow risk, and foreign currency risk. Hedging against *fair value risk* protects the existing financial instrument against changes in its fair value. Hedging against *cash flow risk* protects against market related fluctuations in the cash to be paid or cash to be received for items such as interest receivable and payable. Hedging against *foreign currency risk* protects against exchange rate fluctuations that would change the dollar equivalent that must be repaid or will be received on a contract denominated in a foreign currency. Let's look at some of the more common derivatives.

[2] Bear Sterns has since been sold at a bargain basement price to JP Morgan, and AIG has since recovered from its difficulties.
[3] Bank for International Settlements, *BIS Quarterly Review*, March 2018.

A *futures contract* allows a firm to sell (or buy) a commodity or a financial instrument at a designated future date, at today's price.

FUTURES A futures contract is an agreement between a seller and a buyer that requires the seller to deliver a particular *commodity* (a nonfinancial asset such as corn, gold, or pork bellies) at a designated future date, at a *predetermined* price. When the contract involves a *financial instrument,* such as a Treasury bond, Treasury bill, commercial paper, or a certificate of deposit, the agreement is referred to as a *financial futures contract.*[4] These contracts are actively traded on regulated futures exchanges.

To appreciate the way these hedges work, you need to remember that when interest rates *rise,* the market price of interest-bearing securities *goes down.* For instance, if you have an investment in a 10% bond and market interest rates go up to, say, 12%, your 10% bond is less valuable relative to other bonds paying the higher rate. Conversely, when interest rates *decline,* the market price of interest-bearing securities *goes up.* This risk that the investment's value might change is referred to as *fair value risk.* The company that issued the securities is faced with fair value risk also. If interest rates decline, the fair value of that company's debt would rise, a risk the borrower may want to hedge against. Later in this section, we'll look at an illustration of how the borrower would account for and report such a hedge.

Now let's look at the effect on a contract to sell or buy securities (or any asset for that matter) at preset prices. One who is contracted to *sell* securities at a *preset* price after their market price has fallen benefits from the rise in interest rates. Consequently, the value of the contract that gives one the right to sell securities at a preset price goes up as the market price declines. The seller in a futures contract derives a gain (loss) when interest rates rise (decline).[5] Conversely, the one obligated to buy securities at a preset price experiences a loss. This risk of having to pay more cash or receive less cash is referred to as *cash flow risk.*

The seller in a financial futures contract realizes a gain (loss) when interest rates rise (decline).

Another example of cash flow risk would be borrowing money by issuing a variable (floating) rate note. If market interest rates rise, the borrower would have to pay more interest. Similarly, the lender (investor) in the variable (floating) rate note transaction would face cash flow risk that interest rates would decline, resulting in lower cash interest receipts.

Let's look closer at how a financial futures contract can mitigate cash flow risk. Consider a company in April that will replace the $10 million of 5.5% notes it owes to its bank with an issuance of bonds to the public when the notes mature in June. The company is exposed to the risk that interest rates in June will have risen, increasing borrowing costs. To counteract that possibility, the firm might enter a contract in April to deliver (sell) bonds in June at their *current* price.

Here's what happens then. If interest rates rise, borrowing costs will go up for our example company because it will have to sell debt securities (like notes payable or bonds payable) at a higher interest cost (or lower price). But that loss will be offset (approximately) by the gain produced by being in the opposite position on Treasury bond futures. Take note, though, this works both ways. If interest rates go down causing debt security prices to rise, the potential benefit of being able to issue debt at that lower interest rate (higher price) will be offset by a loss on the futures position. Since there are no corporate bond futures contracts, the company buys Treasury bond futures, which will accomplish essentially the same purpose. In essence, the firm agrees to sell Treasury bonds in June at a price established now (April). Let's say it's April 6 and the price of Treasury bond futures on the Chicago Mercantile Exchange is quoted as 142.88.[6] Since the trading unit of Treasury bond futures is a 15-year, $100,000, 6% Treasury bond, the company might sell 70 Treasury bond futures to hedge the June issuance of debt. This would effectively provide a hedge of 70 x $100,000 x 142.88% = $10,001,600.[7]

A very important point about futures contracts is that the seller does not need to have actual possession of the commodity or financial instrument (the Treasury bonds, in this case), nor is the purchaser of the contract required to take possession of the commodity. In

[4]Note that a financial futures contract meets the definition of a financial instrument because it entails the exchange of financial instruments (cash for Treasury bonds, for instance). But, a futures contract for the sale or purchase of a nonfinancial commodity like corn or gold does not meet the definition because one of the items to be exchanged is not a financial instrument.

[5]The seller of a futures contract is obligated to sell the bonds at a future date. The buyer of a futures contract is obligated to buy the bonds at a future date. The company in our example, then, is the seller of the futures contract.

[6]Price quotes are expressed as a percentage of par.

[7]This is a simplification of the more sophisticated way financial managers determine the optimal number of futures.

fact, virtually all financial futures contracts are "netted out" before the actual transaction is to take place. This is simply a matter of reversing the original position. A seller closes out his transaction with a purchase. Likewise, a purchaser would close out her transaction with a sale. After all, the objective is not to actually buy or sell Treasury bonds (or whatever the commodity or financial instrument might be), but to incur the financial impact of movements in interest rates as reflected in changes in Treasury bond prices. Specifically, it will buy at the lower price (to reverse the original seller position) at the same time it's selling its new bond issue at that same lower price. The financial futures market is an "artificial" exchange in that its reason for existing is to provide a mechanism to transfer risk from those exposed to it to those willing to accept the risk, not to actually buy and sell the underlying financial instruments.

If the impending debt issue being hedged is a short-term issue, the company may attain a more effective hedge by selling Treasury *bill* futures since Treasury bills are 90-day securities, or maybe certificate of deposit (CD) futures that also are traded in futures markets. The object is to get the closest association between the financial effects of interest rate movements on the actual transaction and the effects on the financial instrument used as a hedge.

The effectiveness of a hedge is influenced by the closeness of the match between the designated risk being hedged and the financial instrument chosen as a hedge.

FORWARD CONTRACTS A **forward contract** is similar to a futures contract, but a forward contract calls for delivery on a designated date, whereas a futures contract permits the seller to decide later which specific day within the specified month will be the delivery date (if it gets as far as actual delivery before it is closed out). Also, unlike a futures contract, a forward contract usually is not traded on a market exchange. Instead, a forward contract usually is traded in an over-the-counter market using an "intermediary" that will find a seller for a buyer or a buyer for a seller (and who will customize the contract for the exact day, etc.)

To illustrate a nonfinancial forward contract, let's say a large restaurant chain uses 500,000 pounds of avocados each year during the week of Cinco de Mayo to satisfy demand for its famous guacamole. In recent years, the price of avocados has fluctuated significantly. To reduce the *cash flow risk* associated with price volatility, the restaurant chain enters into a forward contract with an intermediary on October 31 to buy 500,000 pounds of avocados at a set price of $2.80/lb, with delivery in six months (April 30). The *hedged transaction* is the forecasted purchase of avocados from suppliers. The *hedging instrument* is the forward contract arranged with an intermediary, and the *hedged risk* is the change in purchase price.

As we saw with the financial futures contract, the seller and purchaser do not need to exchange possession of the underlying commodity or financial instrument. Rather, the forward contract will settle by an exchange of cash between the restaurant chain and the intermediary based on the current market price for avocados on the date the contract is settled (April 30).

How does this impact the restaurant chain? If avocado prices rise and the restaurant chain pays more to suppliers of avocados, this increased cost will be offset by the gain from being in the opposite position in the forward contract. For example, assume the current market price on April 30 is $3.00/lb. The company would receive cash from the settlement of the forward contract of $100,000 ([$3.00 – $2.80] × 500,000 pounds). At the same time, if avocado prices fall and the restaurant pays less to purchase avocados, this decrease in cost will be offset by the loss on the settlement of the forward position. In either scenario, however, the restaurant chain achieves its goal of reducing the cash flow risk related to price volatility—when considering the purchase of avocados from suppliers and the settlement of the forward contract with the intermediary, the restaurant ends up paying $2.80/lb for avocados on April 30, regardless of the actual price movement.

OPTIONS **Options** frequently are purchased to hedge exposure to the effects of changing interest rates. Options serve the same purpose as futures in that respect but are fundamentally different. An option on a financial instrument—say a Treasury bill—gives its holder the right either to buy (call option) or to sell (put option) the Treasury bill at a specified price and within a given time period. Importantly, though, the option holder has no obligation to exercise the option. On the other hand, the holder of a futures contract must buy or sell within a specified period unless the contract is closed out before delivery comes due.

To illustrate an option contract, let's reconsider the example of the restaurant chain that wants to protect itself from rising avocado prices. Let's say the restaurant chain is able to enter into a contract that gives it the *option* to purchase 500,000 pounds of avocados in six months at the price of \$2.80/lb. Assume this option contract has an initial premium of \$0.02 per avocado, or \$10,000. That is, the market price of avocados at the inception of the contract is lower than the option price. If the price of avocados increases to \$3.00/lb, then the outcome of an option contract will be similar to what we saw with the forward contract: the restaurant chain will exercise its option and receive \$100,000 ([\$3.00 – \$2.80] × 500,000 pounds) upon settlement of the option. Because the restaurant chain paid an initial premium, the net cash effect of the option contract (\$90,000 = \$100,000 cash settlement – \$10,000 premium) would be less than in the forward contract example (\$100,000 cash settlement). Unlike the forward contract, however, if avocado prices fall, then the restaurant chain does not have to pay to settle the option contract. Rather, it will let the option expire.

FOREIGN CURRENCY FUTURES Foreign loans frequently are denominated in the currency of the lender (Japanese yen, Swiss franc, Euro, and so on). When loans must be repaid in foreign currencies, a new element of risk is introduced. This is because if exchange rates change, the dollar equivalent of the foreign currency that must be repaid differs from the dollar equivalent of the foreign currency borrowed.

Foreign exchange risk often is hedged in the same manner as interest rate risk.

To hedge against "foreign exchange risk" exposure, some firms buy or sell **foreign currency futures** contracts. These are similar to financial futures except specific foreign currencies are specified as the underlying in the futures contracts rather than specific debt instruments. They work the same way to protect against foreign exchange risk as financial futures protect against fair value or cash flow risk.

Interest rate swaps **exchange fixed interest payments for floating rate payments, or vice versa, without exchanging the underlying notional (principal) amounts.**

INTEREST RATE SWAPS Over 82% of derivatives are interest rate contracts, of which 75% are **interest rate swaps**. These contracts exchange (swap) fixed interest payments for floating rate payments, or vice versa, without exchanging the underlying debt instruments. For example, suppose you owe \$100,000 on a 10% fixed rate home loan. You envy your neighbor who also is paying 10% on her \$100,000 mortgage, but hers is a floating rate loan, so if market rates fall, so will her loan rate. To the contrary, she is envious of your fixed rate, fearful that rates will rise, increasing her payments. A solution would be for the two of you to effectively swap interest payments using an interest rate swap agreement. The way a swap works, you both would continue to actually make your own interest payments to your respective lenders, but you would exchange with each other the net cash difference between payments at specified intervals. So, in this case, if market rates (and thus floating payments) increase, you would pay your neighbor; if rates fall, she pays you. The net effect is to exchange the consequences of rate changes. In other words, you have effectively converted your fixed-rate debt to floating-rate debt; your neighbor has done the opposite.

Of course, this technique is not dependent on happening into such a fortuitous pairing of two borrowers with opposite philosophies on interest rate risk. Instead, banks or other intermediaries (such as hedge funds) offer, for a fee, one-sided swap agreements to companies desiring to be either fixed-rate payers or variable-rate payers. Intermediaries usually strive to maintain a balanced portfolio of matched, offsetting swap agreements.

Theoretically, the two parties to such a transaction exchange principal amounts, say the \$100,000 amount above, in addition to the interest on those amounts. It makes no practical sense, though, for the companies to send each other \$100,000. So, instead, the principal amount is not actually exchanged, but serves merely as the computational base for interest calculations and is called the *notional amount.* Similarly, the fixed-rate loan payer doesn't usually send the entire fixed interest amount (say 10% × \$100,000 = \$10,000) to the variable-rate payer and receive the entire variable interest amount (say 9% × \$100,000 = \$9,000). Generally, only the net amount (\$1,000 in this case) is exchanged. For example, Company A pays its lender 10% and Company B pays its lender a floating rate that is currently 9%, and the two companies have entered into an interest rate swap as illustrated in Illustration A–1.

Illustration A–1 Interest Rate Swap

Swap of annual payments on $100,000 notional amount; fixed interest rate: 10% ($10,000)

Floating rate: 9% at time of first payment ($9,000); Company B pays Company A $1,000.
The result of the swap:
Company A: $10,000 to lender – $1,000 cash settlement = $9,000 floating amount.
Company B: $9,000 to lender + $1,000 cash settlement = $10,000 fixed amount.

From an accounting standpoint, the central issue is not the operational differences among various hedge instruments, but their similarities in functioning as hedges against risk.

Accounting for Derivatives

A key to accounting for derivatives is knowing the purpose for which a company holds them and whether the company is effective in serving that purpose. Derivatives, for instance, may be held for risk management (hedging activities). The desired effect, and often the real effect, is a reduction in risk. On the other hand, derivatives sometimes are held for speculative position taking, hoping for large profits. The effect of this activity usually is to *increase* risk. Perhaps more important, derivatives acquired as hedges and intended to reduce risk may, in fact, unintentionally increase risk instead.

Derivatives not serving as hedges are extremely speculative due to the high leverage inherent in such investments.

It's important to understand that, serving as investments rather than as hedges, derivatives are extremely speculative. This is due to the high leverage inherent in derivatives. Here's why. The investment outlay usually is negligible, but the potential gain or loss on the investment usually is quite high. A small change in interest rates or another underlying event can trigger a large change in the fair value of the derivative. Because the initial investment was minimal, the change in value relative to the investment itself represents a huge percentage gain or loss. Their extraordinarily risky nature prompted Warren Buffett, one of the country's most celebrated financiers, to refer to derivatives as "financial weapons of mass destruction." Accounting for derivatives is designed to treat differently (a) derivatives designated as hedges and those not designated as hedges as well as (b) intended hedges that are highly effective and those that are not.[8]

Each derivative contract has a "fair value," which is an amount that one side owes the other at a particular moment.

The basic approach to accounting for derivatives is fairly straightforward, although implementation can be quite cumbersome. All derivatives, no exceptions, are carried on the balance sheet as either assets or liabilities at fair (or market) value.[9] The reasoning is that (a) derivatives create either rights or obligations that meet the definition of assets or liabilities, and (b) fair value is the most meaningful measurement.

Accounting for the gain or loss on a derivative depends on how it is used. Specifically, if the derivative is not designated as a hedging instrument or doesn't qualify as one, any gain or loss from fair value changes is recognized immediately in earnings. On the other hand, if a derivative is used to hedge against exposure to risk, any gain or loss from fair value changes is either (a) recognized immediately in earnings along with an offsetting loss or gain on the item being hedged or (b) deferred in comprehensive income until it can be recognized in earnings at the same time as earnings are affected by a hedged transaction. Which way depends on whether the derivative is designated as a (a) fair value hedge, (b) cash flow hedge, or (c) foreign currency hedge. Let's look now at each of the three hedge designations.

[8] Accounting Standards Update No. 2017-12, Derivatives and Hedging (Topic 815): Targeted Improvements to Accounting for Hedging Activities, (Norwalk, Conn.: FASB, January 2017).
[9] FASB ASC 815–10: Derivatives and Hedging—Overall.

● LOA–2

A gain or loss from a *fair value hedge* is recognized immediately in earnings, along with the loss or gain from the item being hedged.

FAIR VALUE HEDGES A company can be adversely affected when a change in either prices or interest rates causes a change in the fair value of one of its assets, its liabilities, or a commitment to buy or sell assets or liabilities. If a derivative is used to hedge against the exposure to changes in the fair value of an asset, liability, or a firm commitment, it can be designated as a **fair value hedge**. In that case, when the derivative is adjusted to reflect changes in fair value, the other side of the entry recognizes a gain or loss to be included *currently* in earnings. At the same time, though, the loss or gain from changes in the fair value (due to the risk being hedged)[10] of the item being hedged also is included currently in earnings. This means that, to the extent the hedge is effective in serving its purpose, the gain or loss on the derivative will be offset by the loss or gain on the item being hedged. In fact, this is precisely the concept behind the procedure, and offsetting effects of the derivative and the hedged item are reported in the same line item in the income statement.

The income effects of the hedge instrument and the income effects of the item being hedged should affect earnings at the same time and in the same income statement line item.

The reasoning is that as interest rates or other underlying events change, a hedge instrument will produce a gain approximately equal to a loss on the item being hedged (or vice versa). These income effects are interrelated and offsetting, so it would be improper to report the income effects in different periods. More critically, the intent and effect of having the hedge instrument is to *lessen* risk. And yet, recognizing gains in one period and counterbalancing losses in another period would tend to cause fluctuations in income that convey an *increase* in risk. However, to the extent that a hedge is not perfectly (but still highly) effective and produces gains or losses different from the losses or gains being hedged, the difference is recognized in earnings immediately.

Some of the more common uses of fair value hedges are:

- An interest rate swap to synthetically convert fixed-rate debt (for which interest rate changes could change the fair value of the debt) into floating-rate debt.
- A futures contract to hedge changes in the fair value (due to price changes) of aluminum, sugar, or some other type of inventory.
- A futures contract to hedge the fair value (due to price changes) of a firm commitment to sell natural gas or some other asset.

ILLUSTRATION OF FAIR VALUE HEDGE Because interest rate swaps comprise the majority of derivatives in use, we will use swaps to illustrate accounting for derivatives. Let's look at the example in Illustration A–2. The shortcut method can be used only for an interest-bearing asset or liability.

Illustration A–2
Interest Rate Swap—Shortcut Method

Wintel Semiconductors issued $1 million of 18-month, 10% notes payable to Third Bank on January 1, 2021. Wintel is exposed to the risk that general interest rates will decline, causing the fair value of its debt to rise. (If the fair value of Wintel's debt increases, its effective borrowing cost is higher relative to the market.)

- To hedge against this fair value risk, the firm entered into an 18-month interest rate swap agreement on January 1 and designated the swap as a hedge against changes in the fair value of the note.
- Although Wintel still will pay the 10% fixed rate to the bank, the swap calls for the company to *receive payment* from the intermediary based on a 10% fixed interest rate on a notional (principal) amount of $1 million and to *make payment* based on a floating interest rate tied to changes in general rates.* The actual cash settlement on the swap will be a net payment to the appropriate party.

As the Illustration will show, this effectively converts Wintel's fixed-rate debt to floating-rate debt. Cash settlement of the net interest amount is made semiannually at June 30 and December 31 of each year, with the net interest being the difference between the $50,000

(continued)

[10]The fair value of a hedged item might also change for reasons other than from effects of the risk being hedged. For instance, the hedged risk may be that a change in interest rates will cause the fair value of a bond to change. The bond price might also change, though, if the market perceives that the bond's default risk has changed.

Illustration A–2 (concluded)

fixed interest [$1 million × (10% × ½)] and the floating interest rate, times $1 million at those dates.

Floating (market) settlement rates were 9% at June 30, 2021, 8% at December 31, 2021, and 9% at June 30, 2022. Net interest receipts can be calculated as shown below. Fair values of both the derivative and the note resulting from those market rate changes are assumed to be quotes obtained from securities dealers.

	1/1/21	6/30/21	12/31/21	6/30/22
Fixed rate	10%	10%	10%	10%
Floating rate	10%	9%	8%	9%
Fixed payments [$1 million × (10% × ½)]		$ 50,000	$ 50,000	$ 50,000
Floating payments ($1 million × ½ floating rate)		45,000	40,000	45,000
Net interest receipts		$ 5,000	$ 10,000	$ 5,000
Fair value of interest rate swap	$0	$ 9,363	$ 9,615	$0
Fair value of note payable	$1,000,000	$1,009,363	$1,009,615	$1,000,000

*A common measure for benchmarking variable interest rates is LIBOR, the London Interbank Offered Rate, a base rate at which large international banks lend funds to each other.

When the floating rate declined from 10% to 9%, the fair values of both the derivative (swap) and the note increased. This created an offsetting gain on the derivative and a holding loss on the note. Both are recognized in earnings at the same time (at June 30, 2021). The changes in fair value (holding gains and losses) are presented in the same line item as the earnings effect of the underlying hedged item (interest expense). So, we record these holding gains and losses as decreases or increases in interest expense.

January 1, 2021		
Cash	1,000,000	
Notes payable		1,000,000
To record the issuance of the note.		
June 30, 2021		
Interest expense (10% × ½ × $1 million)	50,000	
Cash		50,000
To record interest on the note.		
Cash [**$50,000** – (9% × ½ × $1 million)]	5,000	
Interest expense		5,000
To record the net cash settlement on the swap.		
Interest rate swap* ($9,363 – $0)	9,363	
Interest expense		9,363
To record change in fair value of the derivative.		
Interest expense	9,363	
Note payable ($1,009,363 – $1,000,000)		9,363
To record change in fair value of the note due to interest rate changes.		

*This would be a liability rather than an investment (asset) if the fair value had declined.

The interest rate swap is designated as a fair value hedge on this note at issuance.

The swap settlement is the difference between the fixed interest (5%) and variable interest (4.5%).

The fair value of derivatives is recognized in the balance sheet.

The change in fair value of the derivative and of the hedged liability are reported in the same income statement line item as the earnings effect of the hedged item.

The net interest settlement on June 30, 2021, is $5,000 because the fixed rate is 5% (half of the 10% annual rate) and the floating rate is 4.5% (half of the 9% annual rate).

As with any debt, interest expense is the effective rate times the outstanding balance.

The settlement is the difference between the fixed interest (5%) and variable interest (4%).

The derivative is increased by the change in fair value. The note is increased by the change in fair value.

December 31, 2021		
Interest expense	50,000	
Cash (10% × ½ × $1,000,000)		50,000
To record interest on the note.		
Cash [**$50,000** – (8% × ½ × $1 million)]	10,000	
Interest expense		10,000
To record the net cash settlement on the swap.		
Interest rate swap ($9,615 – $9,363)	252	
Interest expense		252
To record the change in fair value of the derivative.		
Interest expense	252	
Note payable ($1,009,615 – $1,009,363)		252
To record the change in fair value of the note due to interest rate changes.		

The fair value of the swap increased by $252 (from $9,363 to $9,615). Similarly, we adjust the note's book value by the amount necessary to increase it to fair value. This produces a holding loss on the note that exactly offsets the gain on the swap. This result is the hedging effect that motivated Wintel to enter the fair value hedging arrangement in the first place.

At June 30, 2022, Wintel repeats the process of adjusting to fair value both the derivative investment and the note being hedged.

The net interest received is the difference between the fixed interest (5%) and floating interest (4.5%).

The swap's fair value now is zero.

June 30, 2022		
Interest expense	50,000	
Cash (10% × ½ × $1,000,000)		50,000
To record interest on the note.		
Cash [**$50,000** – (9% × ½ × $1 million)]	5,000	
Interest expense		5,000
To record the change in fair value of the derivative.		
Interest expense	9,615	
Interest rate swap ($0 – $9,615)		9,615
To record the change in fair value of the derivative.		
Note payable ($1,000,000 – $1,009,615)	9,615	
Interest expense		9,615
To record the change in fair value of the note due to interest rate changes.		
Note payable	1,000,000	
Cash		1,000,000
To repay the loan.		

The net interest received is the difference between the fixed rate (5%) and floating rate (4.5%) times $1 million. The fair value of the swap decreased by $9,615 (from $9,615 to zero).[11] That decline represents a holding *loss* that we recognize in earnings. Similarly, we record an offsetting holding gain on the note for the change in its fair value.

Now let's see how the book values changed for the swap account and the note:

	Swap		Note	
Jan. 1, 2021				1,000,000
June 30, 2021	9,363			9,363
Dec. 31, 2021	252			252
June 30, 2022		9,615	9,615 1,000,000	
	0			0

[11]Because there are no future cash receipts from the swap arrangement at this point, the fair value of the swap is zero.

The income statement is affected as follows:

	Income Statement + (−)	
June 30, 2021	(50,000)	Interest expense—fixed payment
	5,000	Interest expense—net cash settlement
	9,363	Interest expense—gain on interest rate swap
	(9,363)	Interest expense—loss on hedged note
	(45,000)	Net effect—same as floating interest payment
Dec. 31, 2021	(50,000)	Interest expense—fixed payment
	10,000	Interest expense—net cash settlement
	252	Interest expense—gain on interest rate swap
	(252)	Interest expense—loss on hedged note
	(40,000)	Net effect—same as floating interest payment
June 30, 2022	(50,000)	Interest expense—fixed payment
	5,000	Interest expense—net cash settlement
	9,615	Interest expense—gain on interest rate swap
	(9,615)	Interest expense—loss on hedged note
	(45,000)	Net effect—same as floating interest payment

As this demonstrates, the swap effectively converts fixed-interest debt to floating-interest debt.

Additional Consideration

Fair Value of the Swap

The fair value of a derivative typically is based on a quote obtained from a derivatives dealer. That fair value will approximate the present value of the expected net interest settlement receipts for the remaining term of the swap. In fact, we can actually calculate the fair value of the swap that we accepted as given in our illustration.

Since the June 30, 2021, floating rate of 9% caused the cash settlement on that date to be $5,000, it's reasonable to look at 9% as the best estimate of future floating rates and therefore assume the remaining two cash settlements also will be $5,000 each. We can then calculate at June 30, 2021, the present value of those expected net interest settlement receipts for the remaining term of the swap.

Fixed interest	10% × ½ × $1 million	$ 50,000
Expected floating interest	9% × ½ × $1 million	45,000
Expected cash receipts for both Dec. 31, 2021 and June 30, 2022		$ 5,000
		× 1.87267*
Present value of expected net interest settlement receipts for the remaining term		$ 9,363

*Present value of an ordinary annuity of $1: $n = 2$, $i = 4.5\%$ (½ of 9%) (from Table 4)

Fair Value of the Notes

The fair value of the note payable will be the present value of principal and remaining interest payments discounted at the market rate. The market rate will vary with the designated floating rate but might differ due to changes in default (credit) risk and the term structure of interest rates. Assuming it's 9% at June 30, 2021, we can calculate the fair value (present value) of the notes.

Interest	$50,000* × 1.87267† =	$ 93,633
Principal	$1,000,000 × 0.91573‡ =	915,730
Present value of remaining principal and interest payments		$1,009,363

*½ of 10% × $1,000,000
†Present value of an ordinary annuity of $1: $n = 2$, $i = 4.5\%$ (from Table 4)
‡Present value of $1: $n = 2$, $i = 4.5\%$ (from Table 2)

(continued)

(concluded)

Note: Often the cash settlement rate is "reset" as of each cash settlement date. Thus, the floating rate actually used at the end of each period to determine the payment is the floating market rate as of the beginning of the same period. In our illustration, for instance, there would have been no cash settlement at June 30, 2021, since we would use the beginning floating rate of 10% to determine payment. Similarly, we would have used the 9% floating rate at June 30, 2021, to determine the cash settlement six months later at December 31. In effect, each cash settlement would be delayed six months. Had this arrangement been in effect in the current illustration, there would have been one fewer cash settlement payment (two rather than three), but would not have affected the fair value calculations above because, either way, our expectation would be cash receipts of $5,000 for both December 31, 2021, and June 30, 2022.

● LOA–3

A gain or loss from a *cash flow hedge* is deferred as other comprehensive income until it can be recognized in earnings along with the earnings effect of the item being hedged.

CASH FLOW HEDGES The risk in some transactions or events is the risk of a change in cash flows, rather than a change in fair values. We noted earlier, for instance, that *fixed-rate* debt subjects a company to the risk that interest rate changes could change the fair value of the debt. On the other hand, if the obligation is *floating-rate* debt, the fair value of the debt will *not* change when interest rates do, but cash flows *will* change. If a derivative is used to hedge against the exposure to changes in cash inflows or cash outflows of an asset or liability or a forecasted transaction (like a future purchase or sale), it can be designated as a cash flow hedge. In that case, when the derivative is adjusted to reflect changes in fair value, the other side of the entry is a gain or loss. That gain or loss is *deferred* as a component of other comprehensive income. It's included in earnings later, at the same time as earnings are affected by the hedged transaction. When it ultimately is reported in the income statement, it's included in the same income statement line item as the effects of the hedged item. For instance, let's say that our inventory of carbon fibers is being hedged against the risk of rising carbon fiber prices. Any effect of such a price increase will affect cost of goods sold when we sell the item produced with the carbon fiber. So, we will defer (with a credit to Other comprehensive income) any increase in the value of the hedging contract. Then, when the item is sold, we *debit* Other comprehensive income and reduce (*credit*) cost of goods sold. Once again, the effect is to match the earnings effect of the derivative (reduce cost of goods sold, increasing earnings) with the earnings effect of the item being hedged (increase cost of goods sold, decreasing earnings), precisely the concept behind hedge accounting. Stated differently, when it's time for income statement recognition of the item being hedged, the accumulated change in the hedging contract that's been reflected in Accumulated other comprehensive income will be reclassified out of that account and into the income statement along with effect of the item being hedged.

To understand the deferral of the gain or loss, we need to revisit the concept of comprehensive income. Comprehensive income, as you may recall from Chapters 4, 12, 17, and 18, is a more expansive view of the change in shareholders' equity than traditional net income. In fact, it encompasses all changes in equity other than from transactions with owners.[12] So, in addition to net income itself, comprehensive income includes up to four other changes in equity that don't (yet) belong in net income, namely, net holding gains (losses) on investments in debt securities (Chapter 12), gains (losses) from, and amendments to, postretirement benefit plans (Chapter 17), gains (losses) from foreign currency translation, and deferred gains (losses) from derivatives designated as cash flow hedges and those designated as qualifying hedging relationships.[13]

[12]Transactions with owners primarily include dividends and the sale or purchase of shares of the company's stock.

[13]FASB ASC 220–10–55: Comprehensive Income—Overall—Implementation Guidance and Illustrations and Accounting Standards Update No. 2017-12, Derivatives and Hedging (Topic 815): Targeted Improvements to Accounting for Hedging Activities, (Norwalk, Conn.: FASB, January 2017).

Some of the more commonly used cash flow hedges are

- An interest rate swap to synthetically convert floating rate debt (for which interest rate changes could change the cash interest payments) into fixed rate debt.
- A futures contract to hedge a forecasted sale (for which price changes could change the cash receipts) of natural gas, crude oil, or some other asset
- A forward contract to hedge a forecasted purchase (for which price changes could change the cash payments) of a commodity.

A company can designate the hedged risk as the overall change in cash flows of a hedged item or the change in cash flows attributable to a contractually specified component of a forecasted purchase or sale or a contractually specified interest rate in floating rate debt.

ILLUSTRATION OF CASH FLOW HEDGE – INTEREST RATE SWAP Let's assume once again that we have a company that enters into an interest rate swap. However, in Illustration A–3, the company seeks to hedge exposure to cash flow risk from interest rate volatility.

Illustration A–3

Cash Flow Hedge; Interest Rate Swap (Floating Rate to Fixed Rate)—Shortcut Method

On January 1, 2021, Daley Company issues $2 million of floating-rate debt based on LIBOR, with interest paid annually.

- The debt has 3 years to maturity, and the LIBOR rate is 5% at issuance.
- Interest payments on the debt, based on beginning-of-year rates, are due on December 31 of each year and the variable rate is reset after the payment is made.
- Daley is exposed to the risk that interest rates will rise, which will increase its periodic interest payments.
- To hedge against this cash flow risk, the firm entered into a 3-year interest rate swap agreement on January 1 and designated the swap as a cash flow hedge because it protects Daley from having to make higher cash outflows for interest if interest rates do rise.
- The swap calls for the company to pay a fixed interest rate of 5% and receive LIBOR based on a notional amount of $2 million.
- Settlements on the interest rate swap will be made annually on December 31, with interest rate resets after settlement at beginning-of-year rates.
- Assume that LIBOR rates are reset to 5.5% at December 31, 2021, 6% at December 31, 2022, and 6.5% at December 31, 2023.
- The market value of the note in this case does not change as interest rate fluctuates, because the note carries a variable interest rate.
- Fair values of the interest rate swap at the end of each accounting period are assumed to be quotes by securities dealers.
- Net cash settlements for the swap can be calculated as shown below:

	1/1/21	12/31/21	12/31/22	12/31/23
Floating rate	5%	5.5%	4.75%	n/a
Fixed rate	5%	5%	5%	n/a
Receive floating ($2 million × floating rate)		$100,000	$110,000	$ 95,000
Fixed payments ($2 million × 6%)		100,000	100,000	100,000
Net cash settlement		$ 0	$ 10,000	$ (5,000)
Fair value of swap	$0	$ 18,463	$ (4,773)	$ 0

As the illustration shows, this swap effectively converts Daley's variable-rate debt to fixed-rate debt. Daley applies the shortcut method.

January 1, 2021		
Cash	2,000,000	
Notes payable		2,000,000
To record the issuance of the floating-rate debt.		

December 31, 2021

Interest expense (5% × $2 million)	100,000	
Cash		100,000
To record interest on the note.		
Interest rate swap	18,463	
Other comprehensive income (OCI)		18,463
To record the change in the fair value of the derivative.		

At December 31, 2021, there is no cash exchanged for settlement of the interest rate swap, because the cash settlement is based on beginning-of-year rates (when both the fixed and floating rates were 5%). Daley does recognize an increase in the fair value of the interest rate swap, related to the rising interest rates in the upcoming year. This increase in fair value is recognized in other comprehensive income.

December 31, 2022

Interest expense (5.5% × $2 million)	110,000	
Cash		110,000
To record interest on the note.		
Cash [(5.5% × $2 million) – $100,000]	10,000	
Interest expense		10,000
To record the net cash settlement on the swap.		
Other comprehensive income (OCI)	23,236	
Interest rate swap		23,236
To record the change in the fair value of the derivative from an asset of $18,463 to a liability of $4,773.		

The net interest received is the difference between the floating rate at the beginning of 2022 (5.5%) and the fixed rate (5%) times $2 million notional amount. The effect of the net cash settlement is to reduce total cash outflows related to interest at $100,000, which is equal to paying the fixed rate of 5%. This result is the cash flow hedging effect that motivated Daley to enter the derivative arrangement.

At December 31, 2023, Daley repeats the process of adjusting the derivative to fair value. Because interest rates have decreased and Daley anticipates a future net cash payment, the swap represents a liability to the company.

December 31, 2023

Interest expense (4.75% × $2 million)	95,000	
Cash		95,000
To record interest on the note.		
Interest expense	5,000	
Cash [(4.75% × $2 million) – $100,000]		5,000
To record the net cash settlement on the swap.		
Notes payable	2,000,000	
Cash		2,000,000
To repay the debt.		
Interest rate swap	4,773	
Other comprehensive income		4,773
To close out the swap contract account; the swap contract was a liability of $4,773 and now has a fair value of zero.		

The fair value of the interest rate swap has decreased to zero, as there are no additional cash settlements expected as the debt and the swap has now matured. All earnings effects of the hedged item and derivative instrument have been recorded in the income statement in interest expense.

Total cash flows are affected as follows

December 31, 2021	Payment on floating rate debt (5%)	($100,000)
	Net settlement on interest rate swap	0
	Total cash outflow	**($100,000)**
December 31, 2022	Payment on floating rate debt (5.5%)	($110,000)
	Net settlement on interest rate swap	10,000
	Total cash outflow	**($100,000)**
December 31, 2023	Payment on floating rate debt (4.75%)	($ 95,000)
	Net settlement on interest rate swap	(5,000)
	Total cash outflow	**($100,000)**

A company also can use a forward contract to hedge the risk of change in cash flows attributable to a component of a forecasted purchase of a commodity.

ILLUSTRATION OF CASH FLOW HEDGE – FORWARD CONTRACT Let's assume again that we have a company that seeks to hedge exposure to cash flow risk, this time from a forecasted purchase of a commodity. We see this in Illustration A–4, using the extended method. The shortcut method is not available because that can only be used for an interest-bearing asset or liability. A forecasted transaction of a nonfinancial commodity does not fit that category.

Illustration A–4

Cash Flow Hedge; Forward Contract

Cavalier Bicycles uses carbon fiber to produce its high-performance road and mountain bikes. Cavalier anticipates it will need to purchase 45,000 pounds of carbon fiber in July 2022 to make enough bikes to meet its fourth quarter sales demand. However, if the price of carbon fiber increases, the cost to produce the bikes will increase and in turn, lower Cavalier's profit margins.

To hedge against the risk of rising carbon fiber prices, on June 1, 2022, Cavalier enters into a forward contract with a third-party intermediary to buy 45,000 pounds of carbon fiber at the current spot price of $10 per pound. It designates the contract as a cash flow hedge of the anticipated carbon fiber purchase because the hedged risk is attributable to the volatility in carbon fiber prices.

The price in the contract expires on July 31, 2022, and the contract will settle by an exchange of cash with the intermediary based on the spot price on the settlement date. As the illustration will show, Cavalier uses this forward contract to lock in the cost of its inventory at the prevailing market price of $450,000 (45,000 pounds × $10 per pound).

June 1, 2022 No entry is required, because the current spot price equals the contract price. Thus, the forward contract has no value.

June 30, 2022		
Forward contract [($10.50 – $10.00) × 45,000]	22,500	
Other comprehensive income (OCI)		22,500
To record the change in the fair value of the derivative.		

At June 30, 2022, the spot prices of carbon fiber have increased. Therefore, the value of the forward contract to Cavalier Bicycles has increased, as Cavalier is able to buy carbon fiber at a lower price than current market prices. Thus, Cavalier reports the forward contract in the balance sheet as an asset, with the gain on the derivative as a component of other comprehensive income. Cavalier will recognize this gain in current earnings when the hedged item—in this case the cost of carbon fiber—is recognized in earnings (when the produced inventory is sold).

If Cavalier purchases 45,000 pounds of carbon fiber on July 2, 2022, for $10.50, it would make the following entries to record the purchase of inventory and the settlement of the forward contract:

July 2, 2022		
Inventory—Carbon Fiber	472,500	
Cash ($10.50 × 45,000 pounds)		472,500
To record purchase of inventory.		
Cash [($10.50 – $10.00) × 45,000]	22,500	
Forward contract		22,500
To record settlement of the forward contract.		

The $22,500 forward contract settlement offsets the amount paid to purchase the inventory at the prevailing market price of $472,500. The result is a net cash outflow of $10 per pound, and an effective hedge of the cash flow for the purchase of inventory.

Cavalier defers the income effects of the forward contract in other comprehensive income until the period in which it sells the inventory, affecting earnings through the same line item as the hedged item—cost of goods sold. That is, the effect of the inventory price is reflected in cost of goods sold as the inventory is sold, and effect of the forward contract ($22,500) previously deferred with a credit to Other comprehensive income now is reflected as a $22,500 decrease in cost of goods sold as Cavalier debits Other comprehensive income for that same amount previously deferred.

For example, let's say that Cavalier manufactures the carbon fiber into bicycles. Cavalier sells its bikes during the fourth quarter of 2022 for $1,200,000, and the total value of the inventory, which includes the cost of the carbon fiber purchase made on July 2, 2022, is $650,000. The sale is recorded as follows:

Fourth Quarter 2022		
Cash	1,200,000	
Sales revenue		1,200,000
To record sales.		
Cost of goods sold	650,000	
Inventory—Bikes		650,000
To record cost of goods sold.		
Other comprehensive income (OCI)	22,500	
Cost of goods sold		22,500
To match the earnings effect of the hedged item (cost of carbon fiber) and the forward contract.		

Because the hedged purchase of carbon fiber has now affected earnings, Cavalier also recognizes the previously deferred gain on the value of the forward contract into earnings.

● LOA–4

The possibility that foreign currency exchange rates might change exposes many companies to foreign currency risk.

FOREIGN CURRENCY HEDGES Today's economy is increasingly a global one. The majority of large "U.S." companies are, in truth, multinational companies that may receive only a fraction of their revenues from U.S. operations. Many operations of those companies are located abroad. Foreign operations often are denominated in the currency of the foreign country (the Euro, Japanese yen, Russian rubles, and so on). Even companies without foreign operations sometimes hold investments, issue debt, or conduct other transactions denominated in foreign currencies. As exchange rates change, the dollar equivalent of the foreign currency changes. The possibility of currency rate changes exposes these companies to the risk that some transactions require settlement in a currency other than the entities' functional currency or that foreign operations will require translation adjustments to reported amounts.

A **foreign currency hedge** can be a hedge of foreign currency exposure of the following:

- A firm commitment—treated as a fair value hedge.
- An available-for-sale security—treated as a fair value hedge.
- A forecasted transaction—treated as a cash flow hedge.
- A company's net investment in a foreign operation—the gain or loss is reported in *other comprehensive income* as part of unrealized gains and losses from foreign currency translation.[14]

HEDGE EFFECTIVENESS When a company elects to apply hedge accounting, it must establish at the inception of the hedge the method it will use to assess the effectiveness of the hedging derivative.[15] The key criterion for qualifying as a hedge is that the hedging relationship must be "highly effective" in achieving offsetting changes in fair values or cash flows based on the hedging company's specified risk management objective and strategy.

To qualify as a hedge, the hedging relationship must be highly effective in achieving offsetting changes in fair values or cash flows.

An assessment of this effectiveness must be made at least every three months and whenever financial statements are issued or earnings are reported. There are no precise guidelines for assessing effectiveness, but it generally means a high correlation between changes in the fair value or cash flows of the derivative and of the item being hedged, not necessarily a specific reduction in risk. Initial hedge effectiveness assessments are performed quantitatively and, at hedge inception, the company must elect whether it will perform subsequent assessments qualitatively or quantitatively. Hedge accounting must be terminated for hedging relationships that no longer are highly effective.

In Illustration A–2 (and in Illustration A–3), the loss on the hedged note exactly offset the gain on the swap. This is because the swap in this instance was highly effective in hedging the risk due to interest rate changes. However, the loss and gain would not have exactly offset each other if there were differences in critical terms of the swap and the hedged note. For instance, suppose the swap's term had been different from that of the note (say a three-year swap term compared with the 18-month term of the note) or if the notional amount of the swap differed from that of the note (say $500,000 rather than $1 million). In that case, changes in the fair value of the swap and changes in the fair value of the note would not be the same. The result would be a greater (or lesser) amount recognized in earnings for the swap than for the note. Because there would not be an exact offset, earnings would be affected. That is a desired effect of hedge accounting; to the extent that a hedge is effective, the earnings effect of a derivative cancels out the earnings effect of the item being hedged.

Imperfect hedges result in part of the derivative gain or loss being included in current earnings.

Hedge accounting also allows a company to exclude certain components of the change in the value of the derivative from the calculation of hedge effectiveness, including components of the change in time value, as well as differences in spot and forward or futures prices. The company amortizes the initial value of the excluded component and can either (a) elect to recognize the initial value of the excluded component in earnings immediately or (b) amortize it over the life of the hedging instrument. The difference between the initial value recognized in the income statement and the change in the fair value of the excluded component is reported as an item of Other comprehensive income.

FAIR VALUE CHANGES UNRELATED TO THE RISK BEING HEDGED In Illustration A–2, the fair value of the hedged note and the fair value of the swap changed by the same amounts each year because we assumed the fair values changed only due to interest rate changes. It's also possible, though, that the note's fair value would change by an amount different from that of the swap for reasons unrelated to interest rates. Remember from our earlier discussion that the market's perception of a company's creditworthiness, and thus its ability to pay interest and principal when due, also can affect the value of debt, whether interest rates change or not. In hedge accounting, we ignore those changes. We recognize only the fair value changes in the hedged item that we can attribute to the risk being hedged (interest rate risk in this case). For example, if a changing perception of default risk had caused the note's

Fair value changes unrelated to the risk being hedged are ignored.

---

[14]This is the same treatment previously prescribed for these translation adjustments by FASB ASC 830: Foreign Currency Matters.

[15]Remember, if a derivative is not designated as a hedge, any gains or losses from changes in its fair value are recognized immediately in earnings.

fair value to increase by an additional, say $5,000, our journal entries in Illustration A–2 would have been unaffected. Notice, then, that although we always mark a *derivative* to fair value, the reported amount of the *item being hedged* may not be its fair value. We mark a hedged item to fair value only to the extent that its fair value changed due to the risk being hedged.

● LOA–5

Disclosure of Derivatives and Risk

To be adequately informed about the effectiveness of a company's risk management, investors and creditors need information about strategies for holding derivatives and specific hedging activities. Toward that end, extensive disclosure requirements provide information that includes the following:

- Objectives and strategies for holding and issuing derivatives.
- A description of the items for which risks are being hedged.
- The location and fair value amounts of derivative instruments reported in the balance sheet.
- For forecasted transactions: a description, time before the transaction is expected to occur, the gains and losses accumulated in other comprehensive income, and the events that will trigger their recognition in earnings.
- Beginning balance of, changes in, and ending balance of the derivative component of other comprehensive income.
- The location and net amount of gain or loss reported in earnings.
- Qualitative and quantitative information about failed fair value or cash flow hedges: canceled commitments or previously hedged forecasted transactions no longer expected to occur.

The intent is to provide information about the company's success in reducing risks and, consequently, about risks not managed successfully. Ample disclosures about derivatives are essential to maintain awareness of potential opportunities and problems with risk management.

In addition, GAAP requires companies to provide enhanced disclosures indicating (a) how and why the company uses derivative instruments, (b) how the company accounts for derivative instruments and related hedged items, and (c) how derivative instruments and related hedged items affect the company's balance sheet, income statement, and cash flows.[16] The required disclosures include two tables, one that highlights the location and fair values of derivative instruments in the balance sheet, and another that indicates the location and amounts of gains and losses on derivative instruments in the income statement. The two tables distinguish between derivative instruments that are designated as hedging instruments and those that are not. The tables also categorize derivative instruments by each major type—interest rate contracts, foreign exchange contracts, equity contracts, commodity contracts, credit contracts, and other types of contracts and by income and expense line item (if applicable).

Even for some traditional liabilities, the amounts reported on the face of the financial statements provide inadequate disclosure about the degree to which a company is exposed to risk of loss. To provide adequate disclosure about a company's exposure to risk, additional information must be provided about (a) concentrations of credit risk and (b) the fair value of all financial instruments.[17]

● LOA–6

Extended Method for Interest Rate Swap Accounting

A shortcut method for accounting for an interest rate swap is permitted when a hedge meets certain criteria. In general, the criteria are designed to see if the hedge supports the assumption of "no ineffectiveness." Illustration A–2 of a fair value hedge met those criteria, in

[16]FASB ASC 815: Derivatives and Hedging.

[17]FASB ASC 825–10–50–1: Financial Instruments—Overall—Disclosure.

particular, (a) the swap's notional amount matches the note's principal amount, (b) the swap's expiration date matches the note's maturity date, (c) the fair value of the swap is zero at inception, and (d) the floating payment is at the market rate.[18] Because Wintel can conclude that the swap will be highly effective in offsetting changes in the fair value of the debt, it can use the changes in the fair value of the swap to measure the offsetting changes in the fair value of the debt. That's the essence of the shortcut method used in Illustration A–2. The extended method required when the criteria are *not* met for the shortcut method is described in this section (Illustration A–5 begins by describing the same scenario as in Illustration A–2). It produces the same effect on earnings and in the balance sheet as does the procedure shown in Illustration A–2. Note that the shortcut method is available only for interest rate swaps; the extended method applies to all other hedge instruments.

Illustration A–5

Interest Rate Swap (Fixed Rate to Variable Rate)—Extended Method

Wintel Semiconductors issued $1 million of 18-month, 10% notes payable to Third Bank on January 1, 2021. Wintel is exposed to the risk that general interest rates will decline, causing the fair value of its debt to rise. (If the fair value of Wintel's debt increases, its effective borrowing cost is higher relative to the market.)

- To hedge against this fair value risk, the firm entered into an 18-month interest rate swap agreement through an intermediary on January 1 and designated the swap as a hedge against changes in the fair value of the note.
- Although Wintel still will pay the 10% fixed rate to the bank, the swap calls for the company to *receive payment* from the intermediary based on a 10% fixed interest rate on a notional (principal) amount of $1 million and to *make payment* based on a floating interest rate tied to changes in general rates.
- Cash settlement of the net interest amount is made semiannually at June 30 and December 31 of each year, with the net interest being the difference between the $50,000 fixed interest [$1 million × (10% × ½)] and the floating interest rate, times $1 million at those dates. That is, if the floating rate is less than the fixed rate, Wintel will collect cash. If the floating rate is more than the fixed rate, Wintel will pay cash.
- Floating (market) settlement rates were 9% at June 30, 2021, 8% at December 31, 2021, and 9% at June 30, 2022.

Net interest receipts can be calculated as shown below. Fair values of both the derivative and the note resulting from those market rate changes are assumed to be quotes obtained from securities dealers.

	1/1/21	6/30/21	12/31/21	6/30/22
Fixed rate	10%	10%	10%	10%
Floating rate	10%	9%	8%	9%
Fixed payments				
[$1 million × (10% × ½)]		$ 50,000	$ 50,000	$ 50,000
Floating payments				
($1 million × ½ floating rate)		45,000	40,000	45,000
Net interest receipts		$ 5,000	$ 10,000	$ 5,000
Fair value of interest rate swap	$0	$ 9,363	$ 9,615	$0
Fair value of note payable	$1,000,000	$1,009,363	$1,009,615	$1,000,000

When the floating rate declined in Illustration A–5 from 10% to 9%, the fair values of both the derivative (swap) and the note increased. This created an offsetting gain on the derivative and holding loss on the note. Both are recognized in earnings the same period and in the same income statement line item (June 30, 2021). Because the changes in fair value (holding gains and losses) are presented in the same line item as the earnings effect of

[18]There is no precise minimum interval, though it generally is three to six months or less. Other criteria are specified by FASB ASC 815–20–25–104: Derivatives and Hedging—Hedging—General—Recognition—Shortcut Method, *SFAS No. 133* (para. 68), in addition to the key conditions listed here.

the underlying hedged item (interest expense), we record these holding gains and losses as decreases or increases in interest expense.

The interest rate swap is designated as a fair value hedge on this note at issuance.

The swap settlement is the difference between the fixed interest (5%) and variable interest (4.5%).

The fair value of derivatives is recognized in the balance sheet. All changes in fair value are presented in the same line item as the earnings effect of the hedged item.

The hedged liability (or asset) is adjusted to fair value as well.

January 1, 2021		
Cash	1,000,000	
Notes payable		1,000,000
To record the issuance of the note.		
June 30, 2021		
Interest expense (10% × ½ × $1 million)	50,000	
Cash		50,000
To record interest to Third Bank.		
Cash [**$50,000** – (**9%** × ½ × $1 million)]	5,000	
Interest rate swap ($9,363 – $0)	9,363	
Interest expense (to balance)		14,363
To record the net cash settlement, accrued interest on the swap, and change in the fair value of the derivative.		
Interest expense	9,363	
Notes payable ($1,009,363 – $1,000,000)		9,363
To record change in fair value of the note due to interest rate changes.		

The net interest settlement on June 30, 2021, is $5,000 because the fixed rate is 5% (half of the 10% annual rate) and the floating rate is 4.5% (half of the 9% annual rate). A holding gain ($14,363) is produced by holding the derivative security during a time when an interest rate decline caused an increase in the value of that asset. A portion ($5,000) of the gain was received in cash, and another portion ($9,363) is reflected as an increase in the value of the asset.

We also have a holding loss of the same amount. This is because we also held a liability during the same time period, and the interest rate change caused its fair value to increase as well. The changes in fair value (holding gains and losses) are presented in the same line item as the earnings effect of the underlying hedged item (interest expense). So, we record these holding gains and losses as decreases or increases in interest expense.

As with any debt, interest expense is the effective rate times the outstanding balance.

The cash settlement is the difference between the fixed interest (5%) and variable interest (4%).

Interest ($421) accrues on the asset.

The note is increased by the change in fair value.

December 31, 2021		
Interest expense (9% × ½ × $1,009,363)	45,421	
Notes payable (difference)*	4,579	
Cash (10% × ½ × $1,000,000)		50,000
To record interest to Third Bank.		
Cash [**$50,000** – (**8%** × ½ × $1 million)]	10,000	
Interest rate swap ($9,615 – $9,363)	252	
Interest revenue (9% × ½ × $9,363)		421
Interest expense (to balance)		9,831
To record the net cash settlement, accrued interest on the swap, and change in the fair value of the derivative.		
Interest expense	4,831	
Notes payable ($1,009,615 – $1,009,363 + $4,579)		4,831
To record the change in fair value of the note due to interest rate changes.		

*We could use a premium on the note to adjust its book value.

We determine interest on the note the same way we do for any liability, as you learned earlier—at the effective rate (9% × ½) times the outstanding balance ($1,009,363). This results in reducing the note's book value for the cash interest paid in excess of the interest expense.

The fair value of the swap increased due to the interest rate decline by $252 (from $9,363 to $9,615). The holding gain we recognize in earnings consists of that increase (a) plus the $10,000 cash settlement also created by the interest rate decline and (b) minus the $421 increase that results not from the interest rate decline, but from interest accruing on the asset.[19] Similarly, we adjust the note's book value by the amount necessary to increase it to fair value, allowing for the $4,579 reduction in the note in the earlier entry to record interest. Again, all amounts recognized in earnings are presented as interest expense.

At June 30, 2022, Wintel repeats the process of adjusting to fair value both the derivative investment and the note being hedged.

June 30, 2022		
Interest expense (8% × ½ × $1,009,615)	40,385	
Notes payable (difference)	9,615	
Cash (10% × ½ × $1,000,000)		50,000
To record interest to Third Bank.		
Cash [$50,000 – (9% × ½ × $1 million)]	5,000	
Interest expense (to balance)	5,000	
Interest rate swap ($0 – $9,615)		9,615
Interest revenue (8% × ½ × $9,615)		385
To record the net cash settlement, accrued interest on the swap, and change in the fair value of the derivative.		
Notes payable ($1,000,000 – $1,009,615 + $9,615)	0	
Interest expense		0
To record the change in fair value of the note due to interest rate changes		
Note payable	1,000,000	
Cash		1,000,000
To repay the loan from Third Bank.		

Interest expense is the effective rate times the outstanding balance.

The net interest received is the difference between the fixed interest (5%) and floating interest (4.5%).

The swap's fair value now is zero.

The net interest received is the difference between the fixed rate (5%) and floating rate (4.5%), times $1 million. The fair value of the swap decreased by $9,615 (from $9,615 to zero).[20] The holding loss we recognize in earnings consists of that decline (a) minus the $5,000 portion of the decline resulting from it being realized in cash settlement and (b) plus the $385 increase that results not from the interest rate change, but from interest accruing on the asset.

Now let's see how the book values changed for the swap account and the note.

	Swap (debit)	Swap (credit)	Note (debit)	Note (credit)
Jan. 1, 2021				1,000,000
June 30, 2021	9,363			9,363
Dec. 31, 2021	252		4,579	4,831
June 30, 2022		9,615	9,615	
			1,000,000	
Balance	0			0

[19] The investment in the interest rate swap represents the present value of expected future net interest receipts. As with other such assets, interest accrues at the effective rate times the outstanding balance. You also can think of the accrued interest mathematically as the increase in present value of the future cash flows as we get one period nearer to the dates when the cash will be received.

[20] Because there are no future cash receipts or payments from the swap arrangement at this point, the fair value of the swap is zero.

The income statement is affected as follows:

	Income Statement + (—)	
June 30, 2021	(50,000)	Interest expense
	0	Interest expense (no time has passed)
	14,363	Interest expense - interest rate swap
	(9,363)	Interest expense - hedged note
	(45,000)	Net effect—same as floating interest payment
Dec. 31, 2021	(45,421)	Interest expense
	421	Interest expense (passage of time)
	9,831	Interest expense - interest rate swap
	(4,831)	Interest expense - hedged note
	(40,000)	Net effect—same as floating interest payment
June 30, 2022	(40,385)	Interest expense
	385	Interest expense (passage of time)
	(5,000)	Interest expense - interest rate swap
	0	Interest expense - hedged note
	(45,000)	Net effect—same as floating interest payment

As this demonstrates, the swap effectively converts Wintel's fixed-interest debt to floating interest debt.

Additional Consideration

Private Company GAAP—Derivatives and Hedging. The Private Company Council (PCC) sought feedback from private company stakeholders and found that most users of private company financial statements find it difficult to obtain fixed-rate borrowing and often enter into an interest rate swap to economically convert their variable-rate borrowing into a fixed-rate borrowing. However, this arrangement caused significant variability in the income statements. The PCC concluded that the cost and complexity of hedge accounting outweighed the benefits for private companies.

In response to the PCC's conclusion, the FASB issued an Accounting Standards Update in 2014 that allows a simplified approach to make it easier for certain interest rate swaps to qualify for hedge accounting for private companies that is quite different from what is required for public companies.[21] This alternative allows a nonpublic company (that's not a financial institution) to apply hedge accounting to its interest rate swaps as long as the terms of the swap and the related debt are aligned. If the conditions are met, the company can assume the cash flow hedge is fully effective. Those applying the simplified hedge accounting approach will be able to recognize the swap at its settlement value, instead of at its fair value. As a result, the amount of interest expense recorded in the income statement approximates the amount that would have been recognized if the private company had borrowed at a fixed rate.

This alternative should significantly reduce the cost and complexity of accounting for derivatives and hedging transactions of private companies.

[21] *Accounting Standards Update No. 2014-03,* "Derivatives and Hedging (Topic 815): Accounting for Certain Receive-Variable, Pay-Fixed Interest Rate Swaps—Simplified Hedge Accounting Approach (a consensus of the Private Comapny Council)," (Norwalk, Conn.: FASB, January 2014).

The Bottom Line

- LOA–1 A variety of derivative securities are used to hedge against risk. These include futures contracts, forward contracts, and options.
- LOA–2 A fair value hedge is used to hedge against the exposure to changes in the fair value of an asset, liability, or a firm commitment.
- LOA–3 A cash flow hedge is used to hedge against the exposure to changes in cash inflows or cash outflows of an asset or liability or a forecasted transaction (like a future purchase or sale).
- LOA–4 A foreign currency hedge is used to hedge against the possibility that currency rate changes will cause some transactions that require settlement in a currency other than the entities' functional currency or that foreign operations will require translation adjustments to reported amounts.
- LOA–5 Ample disclosures about derivatives are essential to maintain awareness of potential opportunities and problems with risk management.
- LOA–6 The extended method produces the same results as the short-cut method and must be used unless (a) the swap's notional amount matches the note's principal amount, (b) the swap's expiration date matches the note's maturity date, (c) the fair value of the swap is zero at inception, and (d) the floating payment is at the market rate.

Questions For Review of Key Topics

Q A–1 Some financial instruments are called derivatives. Why?

Q A–2 Should gains and losses on a fair value hedge be recorded as they occur, or should they be recorded to coincide with losses and gains on the item being hedged?

Q A–3 Hines Moving Company held a fixed-rate debt of $2 million. The company wanted to hedge its fair value exposure with an interest rate swap. However, the only notional available at the time, on the type of swap it desired, was $2.5 million. What will be the effect of any gain or loss on the $500,000 notional difference?

Q A–4 What is a futures contract?

Q A–5 What is the effect on interest of an interest rate swap?

Q A–6 How are derivatives reported on the balance sheet? Why?

Q A–7 When is a gain or a loss from a cash flow hedge reported in earnings?

Exercises

connect

E A–1
Derivatives; hedge classification
● LOA–1

Indicate (by abbreviation) the type of hedge each activity described below would represent.

Hedge Type

FV	Fair value hedge
CF	Cash flow hedge
FC	Foreign currency hedge
N	Would not qualify as a hedge

Activity

_____1. An options contract to hedge possible future price changes of inventory.
_____2. A futures contract to hedge exposure to interest rate changes prior to replacing bank notes when they mature.
_____3. An interest rate swap to synthetically convert floating rate debt into fixed rate debt.
_____4. An interest rate swap to synthetically convert fixed rate debt into floating rate debt.
_____5. A futures contract to hedge possible future price changes of timber covered by a firm commitment to sell.
_____6. A futures contract to hedge possible future price changes of a forecasted sale of aluminum.
_____7. ExxonMobil's net investment in a Kuwait oil field.
_____8. An interest rate swap to synthetically convert floating rate interest on a stock investment into fixed rate interest.
_____9. An interest rate swap to synthetically convert fixed rate interest on a held-to-maturity debt investment into floating rate interest.

(continued)

(concluded) ______10. An interest rate swap to synthetically convert floating rate interest on a held-to-maturity debt investment into fixed rate interest.

______11. An interest rate swap to synthetically convert fixed rate interest on a stock investment into floating rate interest.

E A–2
Derivatives; interest rate swap; fixed rate debt
● LOA–2

On January 1, 2021, LLB Industries borrowed $200,000 from Trust Bank by issuing a two-year, 10% note, with interest payable quarterly. LLB entered into a two-year interest rate swap agreement on January 1, 2021, and designated the swap as a fair value hedge. Its intent was to hedge the risk that general interest rates will decline, causing the fair value of its debt to increase. The agreement called for the company to receive payment based on a 10% fixed interest rate on a notional amount of $200,000 and to pay interest based on a floating interest rate. The contract called for cash settlement of the net interest amount quarterly.

Floating (LIBOR) settlement rates were 10% at January 1, 8% at March 31, and 6% June 30, 2021. The fair values of the swap are quotes obtained from a derivatives dealer. Those quotes and the fair values of the note are as indicated below.

	January 1	March 31	June 30
Fair value of interest rate swap	0	$ 6,472	$ 11,394
Fair value of note payable	$200,000	$206,472	$211,394

Required:
1. Calculate the net cash settlement at March 31 and June 30, 2021.
2. Prepare the journal entries through June 30, 2021, to record the issuance of the note, interest, and necessary adjustments for changes in fair value.

E A–3
Derivatives; interest rate swap; fixed rate investment
● LOA–2

(This is a variation of E A–2, modified to consider an investment in debt securities.)

On January 1, 2021, S&S Corporation invested in LLB Industries' negotiable two-year, 10% notes, with interest receivable quarterly. The company classified the investment as available-for-sale. S&S entered into a two-year interest rate swap agreement on January 1, 2021, and designated the swap as a fair value hedge. Its intent was to hedge the risk that general interest rates will decline, causing the fair value of its investment to increase. The agreement called for the company to make payment based on a 10% fixed interest rate on a notional amount of $200,000 and to receive interest based on a floating interest rate. The contract called for cash settlement of the net interest amount quarterly.

Floating (LIBOR) settlement rates were 10% at January 1, 8% at March 31, and 6% June 30, 2021. The fair values of the swap are quotes obtained from a derivatives dealer. Those quotes and the fair values of the investment in notes are as follows:

	January 1	March 31	June 30
Fair value of interest rate swap	0	$ 6,472	$ 11,394
Fair value of the investment in notes	$200,000	$206,472	$211,394

Required:
1. Calculate the net cash settlement at March 31 and June 30, 2021.
2. Prepare the journal entries through June 30, 2021, to record the investment in notes, interest, and necessary adjustments for changes in fair value.

E A–4
Derivatives; interest rate swap; fixed rate debt; fair value change unrelated to hedged risk
● LOA–2

(This is a variation of E A–2, modified to consider fair value change unrelated to hedged risk.)

LLB Industries borrowed $200,000 from Trust Bank by issuing a two-year, 10% note, with interest payable quarterly. LLB entered into a two-year interest rate swap agreement on January 1, 2021, and designated the swap as a fair value hedge. Its intent was to hedge the risk that general interest rates will decline, causing the fair value of its debt to increase. The agreement called for the company to receive payment based on a 10% fixed interest rate on a notional amount of $200,000 and to pay interest based on a floating interest rate.

Floating (LIBOR) settlement rates were 10% at January 1, 8% at March 31, and 6% at June 30, 2021. The fair values of the swap are quotes obtained from a derivatives dealer. Those quotes and the fair values of the note are as indicated below. The additional rise in the fair value of the note (higher than that of the swap) on June 30 was due to investors' perceptions that the creditworthiness of LLB was improving.

	January 1	March 31	June 30
Fair value of interest rate swap	0	$ 6,472	$ 11,394
Fair value of note payable	$200,000	$206,472	$220,000

Required:

1. Calculate the net cash settlement at June 30, 2021.
2. Prepare the journal entries on June 30, 2021, to record the interest and necessary adjustments for changes in fair value.

E A–5
Derivatives; interest rate swap; fixed rate debt; extended method
● LOA–6

(This is a variation of Exercise A–2, modified to consider the extended method.)

On January 1, 2021, LLB Industries borrowed $200,000 from Trust Bank by issuing a two-year, 10% note, with interest payable quarterly. LLB entered into a two-year interest rate swap agreement on January 1, 2021, and designated the swap as a fair value hedge. Its intent was to hedge the risk that general interest rates will decline, causing the fair value of its debt to increase. The agreement called for the company to receive payment based on a 10% fixed interest rate on a notional amount of $200,000 and to pay interest based on a floating interest rate. The contract called for cash settlement of the net interest amount quarterly.

Floating (LIBOR) settlement rates were 10% at January 1, 8% at March 31, and 6% at June 30, 2021. The fair values of the swap are quotes obtained from a derivatives dealer. Those quotes and the fair values of the note are as follows:

	January 1	March 31	June 30
Fair value of interest rate swap	0	$ 6,472	$ 11,394
Fair value of note payable	$200,000	$206,472	$211,394

Required:

Prepare the journal entries through June 30, 2021, to record the issuance of the note, interest, and necessary adjustments for changes in fair value. Use the extended method demonstrated in Illustration A–5.

E A–6
Derivatives; interest rate swap; fixed-rate debt; fair value change unrelated to hedged
● LOA–6

(Note: This is a variation of Exercise A–5, modified to consider fair value change unrelated to hedged risk.)

On January 1, 2021, LLB Industries borrowed $200,000 from Trust Bank by issuing a two-year, 10% note, with interest payable quarterly. LLB entered into a two-year interest rate swap agreement on January 1, 2021, and designated the swap as a fair value hedge. Its intent was to hedge the risk that general interest rates will decline, causing the fair value of its debt to increase. The agreement called for the company to receive payment based on a 10% fixed interest rate on a notional amount of $200,000 and to pay interest based on a floating interest rate. The contract called for cash settlement of the net interest amount quarterly.

Floating (LIBOR) settlement rates were 10% at January 1, 8% at March 31, and 6% June 30, 2021. The fair values of the swap are quotes obtained from a derivatives dealer. Those quotes and the fair values of the note are as indicated below. The additional rise in the fair value of the note (higher than that of the swap) on June 30 was due to investors' perceptions that the creditworthiness of LLB was improving.

	January 1	March 31	June 30
Fair value of interest rate swap	0	$ 6,472	$ 11,394
Fair value of note payable	$200,000	206,472	220,000

Required:

1. Calculate the net cash settlement at June 30, 2021.
2. Prepare the journal entries on June 30, 2021, to record the interest and necessary adjustments for changes in fair value. Use the extended method demonstrated in Illustration A–5.

E A–7
Derivatives; fair value hedge—futures contract
● LOA–2

Arlington Steel Company is a producer of raw steel and steel-related products. On January 3, 2022, Arlington enters into a firm commitment to purchase 10,000 tons of iron ore pellets from a supplier to satisfy spring production demands. The purchase is to be at a fixed price of $63 per ton on April 30, 2022. To protect against the risk of changes in the fair value of the commitment contract, Arlington enters into a futures contract to sell 10,000 tons of iron ore on April 30 for $63/ton (the current price). The contract calls for net cash settlement, and the company must report changes in the fair values of its hedging instruments each quarter.

Required:

On March 31, the price of iron ore fell to $61/ton, and then to $60/ton on April 30.

1. Calculate the net cash settlement at April 30, 2022
2. Prepare the journal entries for the period January 1 to April 30, 2022, to record the firm commitment, necessary adjustments for changes in fair value, and settlement of the futures contract.

E A–8
Derivatives; foreign currency; cash flow hedge
● LOA–4

Cleveland Company is a U.S. firm with a U.S. dollar functional currency that manufactures copper-related products. It forecasts that it will sell 5,000 feet of copper tubing to one of its largest customers at a price of ¥50,000,000. Although this sale has not been firmly committed, Cleveland expects that the sale will occur in six months on June 30, 2022. Thus, Cleveland is exposed to changes in foreign currency exchange rates. To reduce this exposure, Cleveland enters into a six-month foreign currency exchange forward contract with a third-party dealer on January 1, 2022, to deliver ¥ and receive US$. The foreign exchange contract has the following terms:

Contract amount: ¥50,000,000
Maturity date: June 30, 2022
Forward contract rate: ¥105.00 = US $1.00

Yen / US$ Exchange rates:

Date	Spot rate	Forward rate for June 30
January 1	¥100.00/US $1.00	¥105.00/US $1.00
March 31	¥102.00/US $1.00	¥108.00/US $1.00
June 30	¥110.00/US $1.00	

Cleveland obtains the fair values of the forward exchange contract from the third-party dealer.

	January 1	March 31	June 30
Swap fair value	$0	$12,900	$21,645

Required:
1. Calculate the net settlement on June 30, 2022.
2. Prepare the journal entries for the period January 1 to June 30, 2022, to record the forward contract, necessary adjustments for changes in fair value, and settlement.

E A–9
Derivatives; cash flow hedge; interest rate swap; shortcut method
● LOA–3

On January 1, 2021, JPS Industries borrowed $300,000 from Austin Bank by issuing a three-year, floating rate note based on LIBOR, with interest payable semi-annually on June 30 and December of each year. JPS entered into a three-year interest rate swap agreement on January 1, 2021, and designated the swap as a cash flow hedge. The intent was to hedge the risk that interest rates will rise, increasing its semi-annual interest payments. The swap agreement called for the company to receive payment based on a floating interest rate on a notional amount of $300,000 and to pay a 6% fixed interest rate. The contract called for cash settlement of the net interest amount semi-annually, and the rate on each reset date (June 30 and December 31) determines the variable interest rate for the following six months.

LIBOR rates in 2021 were 6% at January 1, 5.5% at June 30, and 7% at December 31. The fair values of the swap on those dates, obtained by dealer quotes, were as follows:

	January 1	June 30	December 31
Swap fair value	$0	$(3,100)	$4,000

Required:
1. Calculate the net settlement on June 30, 2021.
2. Prepare journal entries for the period January 1 to December 31, 2021, to record the note payable and hedging instrument, necessary adjustments for changes in fair value, and settlement of the swap contract.

Problems

P A–1
Derivatives; interest rate swap
● LOA–2

On January 1, 2021, Labtech Circuits borrowed $100,000 from First Bank by issuing a three-year, 8% note, payable on December 31, 2023. Labtech wanted to hedge the risk that general interest rates will decline, causing the fair value of its debt to increase. Therefore, Labtech entered into a three-year interest rate swap agreement on January 1, 2021, and designated the swap as a fair value hedge. The agreement called for the company to receive payment based on an 8% fixed interest rate on a notional amount of $100,000 and to pay interest based on a floating interest rate tied to LIBOR. The contract called for cash settlement of the net interest amount on December 31 of each year.

Floating (LIBOR) settlement rates were 8% at inception and 9%, 7%, and 7% at the end of 2021, 2022, and 2023, respectively. The fair values of the swap are quotes obtained from a derivatives dealer. These quotes and the fair values of the note are as follows:

	January 1	December 31		
	2021	2021	2022	2023
Fair value of interest rate swap	0	$ (1,759)	$ 935	0
Fair value of note payable	$100,000	$98,241	$100,935	$100,000

Required:

1. Calculate the net cash settlement at the end of 2021, 2022, and 2023.
2. Prepare the journal entries during 2021 to record the issuance of the note, interest, and necessary adjustments for changes in fair value.
3. Prepare the journal entries during 2022 to record interest, net cash interest settlement for the interest rate swap, and necessary adjustments for changes in fair value.
4. Prepare the journal entries during 2023 to record interest, net cash interest settlement for the interest rate swap, necessary adjustments for changes in fair value, and repayment of the debt.
5. Calculate the book values of both the swap account and the note in each of the three years.
6. Calculate the net effect on earnings of the hedging arrangement in each of the three years. (Ignore income taxes.)
7. Suppose the fair value of the note at December 31, 2021, had been $97,000 rather than $98,241, with the additional decline in fair value due to investors' perceptions that the creditworthiness of Labtech was worsening. How would that affect your entries to record changes in the fair values?

P A–2
Derivatives; interest rate swap; comprehensive
● LOA–3

CMOS Chips is hedging a 20-year, $10 million, 7% bond payable with a 20-year interest rate swap and has designated the swap as a fair value hedge. The agreement called for CMOS to receive payment based on a 7% fixed interest rate on a notional amount of $10 million and to pay interest based on a floating interest rate tied to LIBOR. The contract calls for cash settlement of the net interest amount on December 31 of each year.

At December 31, 2021, the fair value of the derivative and of the hedged bonds has increased by $100,000 because interest rates declined during the reporting period.

Required:

1. Does CMOS have an unrealized gain or loss on the derivative for the period? On the bonds? Will earnings increase or decrease due to the hedging arrangement? Why?
2. Suppose interest rates increased, rather than decreased, causing the fair value of both the derivative and of the hedged bonds to decrease by $100,000. Would CMOS have an unrealized gain or loss on the derivative for the period? On the bonds? Would earnings increase or decrease due to the hedging arrangement? Why?
3. Suppose the fair value of the bonds at December 31, 2021, had increased by $110,000 rather than $100,000, with the additional increase in fair value due to investors' perceptions that the creditworthiness of CMOS was improving. Would CMOS have an unrealized gain or loss on the derivative for the period? On the bonds? Would earnings increase or decrease due to the hedging arrangement? Why?
4. Suppose the notional amount of the swap had been $12 million, rather than the $10 million principal amount of the bonds. As a result, at December 31, 2021, the swap's fair value had increased by $120,000 rather than $100,000. Would CMOS have an unrealized gain or loss on the derivative for the period? On the bonds? Would earnings increase or decrease due to the hedging arrangement? Why?
5. Suppose BIOS Corporation is an investor, having purchased all $10 million of the bonds issued by CMOS as described in the original situation above. BIOS is hedging its investment, classified as available-for-sale, with a 20-year interest rate swap and has designated the swap as a fair value hedge. The agreement called for BIOS to make *payment* based on a 7% fixed interest rate on a notional amount of $10 million and to *receive* interest based on a floating interest rate tied to LIBOR. Would BIOS have an unrealized gain or loss on the derivative for the period due to interest rates having declined? On the bonds? Would earnings increase or decrease due to the hedging arrangement? Why?

P A–3
Derivatives; interest rate swap; fixed rate debt; extended method
● LOA–6

(Note: This is a variation of P A–1, modified to consider the extended method demonstrated in Illustration A–5.)

On January 1, 2021, Labtech Circuits borrowed $100,000 from First Bank by issuing a three-year, 8% note, payable on December 31, 2023. Labtech wanted to hedge the risk that general interest rates will decline, causing the fair value of its debt to increase. Therefore, Labtech entered into a three-year interest rate swap agreement on January 1, 2021, and designated the swap as a fair value hedge. The agreement called for the company to receive payment based on an 8% fixed interest rate on a notional amount of $100,000 and to pay interest based on a floating interest rate tied to LIBOR. The contract called for cash settlement of the net interest amount on December 31 of each year.

Floating (LIBOR) settlement rates were 8% at inception and 9%, 7%, and 7% at the end of 2021, 2022, and 2023, respectively. The fair values of the swap are quotes obtained from a derivatives dealer. Those quotes and the fair values of the note are as follows:

	January 1	December 31		
	2021	2021	2022	2023
Fair value of interest rate swap	0	$ (1,759)	$ 935	0
Fair value of note payable	$100,000	$ 98,241	100,935	$100,000

Required:
Use the extended method demonstrated in Illustration A–5.

1. Calculate the net cash settlement at the end of 2021, 2022, and 2023.
2. Prepare the journal entries during 2021 to record the issuance of the note, interest, and necessary adjustments for changes in fair value.
3. Prepare the journal entries during 2022 to record interest, net cash interest settlement for the interest rate swap, and necessary adjustments for changes in fair value.
4. Prepare the journal entries during 2023 to record interest, net cash interest settlement for the interest rate swap, necessary adjustments for changes in fair value, and repayment of the debt.
5. Calculate the book values of both the swap account and the note in each of the three years.
6. Calculate the net effect on earnings of the hedging arrangement in each of the three years. (Ignore income taxes.)
7. Suppose the fair value of the note at December 31, 2021, had been $97,000 rather than $98,241, with the additional decline in fair value due to investors' perceptions that the creditworthiness of Labtech was worsening. How would that affect your entries to record changes in the fair values?

Broaden Your Perspective

©ImageGap/Getty Images

Apply your critical-thinking ability to the knowledge you've gained. These cases will provide you an opportunity to develop your research, analysis, judgment, and communication skills. You also will work with other students, integrate what you've learned, apply it in real-world situations, and consider its global and ethical ramifications. This practice will broaden your knowledge and further develop your decision-making abilities.

Real World Case A–1
Derivative losses; recognition in earnings
● LOA–4

The following is an excerpt from a disclosure note of **Johnson & Johnson:**

6. Fair Value Measurements (in part)
As of December 31, 2017, the balance of deferred net gains on derivatives included in accumulated other comprehensive income was $70 million after-tax. The Company expects that substantially all of the amounts related to forward foreign exchange contracts will be reclassified into earnings over the next 12 months as a result of transactions that are expected to occur over that period.

Required:

1. Johnson & Johnson indicates that it expects that substantially all of the balance of deferred net gains on derivatives will be reclassified into earnings over the next 12 months as a result of transactions that are expected to occur over that period. What is meant by "reclassified into earnings"?
2. What type(s) of hedging transaction might be accounted for in this way?

Communication Case A–2
Derivatives; hedge accounting
● LOA–3

A conceptual question in accounting for derivatives is this: Should gains and losses on a hedge instrument be recorded as they occur, or should they be recorded to coincide (match) with income effects of the item being hedged?

ABI Wholesalers plans to issue long-term notes in May that will replace its $20 million of 9.5% bonds when they mature in July. ABI is exposed to the risk that interest rates in July will have risen, increasing borrowing costs (reducing the selling price of its notes). To hedge that possibility, ABI entered a (Treasury bond) futures contract in May to deliver (sell) bonds in July at their *current* price.

As a result, if interest rates rise, borrowing costs will go up for ABI because it will sell notes at a higher interest cost (or lower price). But that loss will be offset (approximately) by the gain produced by being in the opposite position on Treasury bond futures.

Two opposing viewpoints are:

View 1: Gains and losses on instruments designed to hedge anticipated transactions should be recorded as they occur.

View 2: Gains and losses on instruments designed to hedge anticipated transactions should be recorded to coincide (match) with income effects of the item being hedged.

In considering this question, focus on conceptual issues regarding the practicable and theoretically appropriate treatment, unconstrained by GAAP. Your instructor will divide the class into two to six groups, depending on the size of the class. The mission of your group is to reach consensus on the appropriate accounting for the gains and losses on instruments designed to hedge anticipated transactions.

Required:

1. Each group member should deliberate the situation independently and draft a tentative argument prior to the class session for which the case is assigned.
2. In class, each group will meet for 10 to 15 minutes in different areas of the classroom. During that meeting, group members will take turns sharing their suggestions for the purpose of arriving at a single group treatment.
3. After the allotted time, a spokesperson for each group (selected during the group meetings) will share the group's solution with the class. The goal of the class is to incorporate the views of each group into a consensus approach to the situation.

APPENDIX

B GAAP Comprehensive Case

Target Corporation prepares its financial statements according to U.S. GAAP. Target's financial statements and disclosure notes for the year ended February 3, 2018, are available in Connect. This material is also available under the Investor Relations link at the company's website (**www.target.com**). This case addresses a variety of characteristics of financial statements prepared using U.S. GAAP. Questions are grouped in parts according to various sections of the textbook.

Part A: Financial Statements, Income Measurement, and Current Assets

A1. What amounts did Target report for the following items for the year ended February 3, 2018?
   a. Total revenues
   b. Income from current operations
   c. Net income or net loss
   d. Total assets
   e. Total equity

A2. What was Target's basic earnings per share for the year ended February 3, 2018?

A3. What is Target's fiscal year-end? Why do you think Target chose that year-end?

A4. Regarding Target's audit report:
   a. Who is Target's auditor?
   b. Did Target receive a "clean" (unmodified) audit opinion?

A5. Refer to Target's balance sheet for the years ended February 3, 2018, and January 28, 2017. Based on the amounts reported for accumulated depreciation, and assuming no depreciable assets were sold during the year, prepare an adjusting entry to record Target's depreciation for the year.

A6. Refer to Target's statement of cash flows for the year ended February 3, 2018. Assuming your answer to requirement 1 includes all depreciation expense recognized during the year, how much amortization expense was recognized during the year?

A7. Note 13 provides information on Target's current assets. Assume all prepaid expenses are for prepaid insurance and that insurance expense comprises $50 million of the $14,248 million of selling, general, and administrative expenses reported in the income statement for the year ended February 3, 2018. How much cash did Target pay for insurance coverage during the year? Prepare the adjusting entry Target would make to record all insurance expense for the year. What would be the effect on the income statement and balance sheet if Target didn't record an adjusting entry for prepaid expenses?

A8. By what name does Target label its balance sheet?

A9. What amounts did Target report for the following items on February 3, 2018?
   a. Current assets
   b. Long-term assets
   c. Total assets

d. Current liabilities
e. Long-term liabilities
f. Total liabilities
g. Total shareholders' equity

A10. What was Target's largest current asset? What was its largest current liability?

A11. Compute Target's current ratio and debt to equity ratio in 2018?

A12. Assuming Target's industry had an average current ratio of 1.0 and an average debt to equity ratio of 2.5, comment on Target's liquidity and long-term solvency.

A13. Why do you think Target has chosen to have its fiscal year end on January 30, as opposed to December 31?

A14. Regarding Target's audit report:

a. Who is Target's auditor?
b. Did Target receive a "clean" (unqualified) audit opinion?

A15. By what name does Target label its income statement?

A16. What amounts did Target report for the following items for the year ended February 3, 2018?

a. Sales
b. Gross margin
c. Earnings from continuing operations before income taxes
d. Net earnings from continuing operations
e. Net earnings

A17. What was Target's basic earnings per share for the year ended February 3, 2018?

A18. Does Target report any items as part of its comprehensive income? If so, what are they.

A19. Does Target prepare the statement of cash flows using the direct method or the indirect method?

A20. Which is higher, net earnings or operating cash flows? Which line item is the biggest reason for this difference? Explain why.

A21. What are the largest investing cash flow and the largest financing cash flow reported by the company for the year ended February 3, 2018?

A22. On what line of Target's income statement is revenue reported? What was the amount of revenue Target reported for the fiscal year ended February 3, 2018?

A23. Disclosure Note 2 indicates that Target generally records revenue in retail stores at the point of sale. Does that suggest that Target generally records revenue at a point in time or over a period of time? Explain.

A24. Disclosure Note 2 indicates that customers ("guests") can return some merchandise within 90 days of purchase and can return other merchandise within a year of purchase. How are Target's revenue and net income affected by returns, given that it does not know at the time a sale is made which items will be returned?

A25. Disclosure Note 2 indicates that "Commissions earned on sales generated by leased departments are included within sales and were $44 million . . . in 2017 . . ." Do you think it likely that Target is accounting for those sales as a principal or an agent? Explain.

A26. Disclosure Note 2 discusses Target's accounting for gift card sales. Does Target recognize revenue when it sells a gift card to a customer? If not, when does it recognize revenue? Explain.

A27. Disclosure Note 4 discussed how Target accounts for consideration received from vendors, which they call "vendor income." Does that consideration produce revenue for Target? Does that consideration produce revenue for Target's vendors? Explain.

A28. What is Target's policy for designating investments as cash equivalents?

A29. What is Target's balance of cash equivalents for the fiscal year ended February 3, 2018?

A30. What is Target's policy with respect to accounting for merchandise returns?

A31. Does Target have accounts receivable? Speculate as to why it has the balance that it has. (Hint, see Disclosure Notes 9, 11 and 13.)

A32. Does Target use average cost, FIFO, or LIFO as its inventory cost flow assumption?

A33. In addition to the purchase price, what additional expenditures does the company include in the initial cost of merchandise?

A34. Calculate the gross profit ratio and the inventory turnover ratio for the fiscal year ended February 3, 2018. Compare Target's ratios with the industry averages of 24.5% and 7.1 times. Determine whether Target's ratios indicate the company is more/less profitable and sells its inventory more/less frequently compared to the industry average.

A35. What retail indices (internally measured or externally measured) does Target use to measure the LIFO provision?

A36. Does Target adjust the retail value of inventory for permanent markups or permanent markdowns to effectively report inventory at the lower of cost or market?

A37. Target has agreements with certain vendors whereby Target does not purchase or pay for merchandise until the merchandise is ultimately sold to a customer. Are sales and cost of sales of this inventory included in Target's income statement? Is unsold inventory at the end of the year included as part of ending inventory in the balance sheet?

Part B: Property, Plant, and Equipment and Intangible Assets

B1. What amount ($ in millions) does Target report for net property and equipment for the year ended February 3, 2018? What is the largest category of property and equipment reported on the face of the balance sheet?

B2. What amount ($ in millions) of cash was used in the fiscal year ended February 3, 2018, to purchase property and equipment? Is this an increase or decrease compared to the previous year?

B3. Do you think a company like Target would have more research and development costs or more advertising costs? Explain.

B4. What is Target's fixed-asset turnover ratio for the fiscal year ended February 3, 2018? What is the ratio intended to measure?

B5. Does Target include any intangible assets in total assets (yes/no)? Hint: see Notes 15 and 16.

B6. Compare the property and equipment listed in the balance sheet with the list in Note 14. What are the estimated useful lives for recording depreciation? Is land listed in Note 14 (yes/no)?

B7. In Note 14, which depreciation method does Target use for property and equipment for financial reporting? Which depreciation method is used for tax purposes? Why might these methods be chosen?

B8. In Note 14, how does Target record repairs and maintenance expense?

B9. In Note 14, does Target report any impairment of property and equipment for the year ended February 3, 2018? If so, what was the amount and what were the reasons for the impairments?

B10. From Notes 15 and 16, were any impairments related to intangible assets recorded for the year ended February 3, 2018? If so, what was the amount and what were the reasons for the impairments?

Part C: Investments

Target does not have investments in stock or bonds. However, **CVS Health Corp.,** which purchased Target's pharmacy and clinical business during 2015, does have some investments. Access CVS's 2017 10K (issued on February 14, 2018) at **investors.cvshealth.com** to answer the following questions:

C1. CVS indicates in Note 1 that it has some short-term investments that consist of certificates of deposit (CDs).

a. How has CVS classified those CDs for accounting purposes?
b. Per CVS's balance sheet, what was the balance in CVS's short-term investments as of December 31, 2017 and December 31, 2016?
c. Per CVS's statement of cash flows, what cash transactions affected short-term investments during 2017?
d. Prepare a T-account that summarizes transactions affecting CVS's short-term investments during 2017. Speculate as to the explanation for any "plug" figure necessary to make the T-account balance.

C2. Per Note 1, CVS has equity-method investments in SureScripts, LLC and in Heartland Healthcare Services. CVS indicates that those investments are immaterial for the year ended December 31, 2017. Assuming that the Heartland investment is material:

a. How would Heartland's earnings affect CVS's income statement?
b. How would Heartland's earnings affect CVS's balance sheet?

Part D: Liabilities

D1. Target's Consolidated Statement of Financial Position (its balance sheet) discloses its current assets and current liabilities.

a. What are the four components of Target's current liabilities?
b. Are current assets sufficient to cover current liabilities? What is the current ratio for the year ended February 3, 2018? How does the ratio compare with the prior year?
c. Why might a company want to avoid having its current ratio be too low? Too high?

D2. Disclosure Note 2 discusses Target's accounting for gift card sales. Disclosure Note 18 indicates the amount of gift card liability that is recognized in Target's balance sheet.

a. By how much did Target's gift card liability change between February 3, 2018 and January 28, 2017?
b. How would the following affect Target's gift card liability (indicate "increase," "decrease," or "no change" for each):
   i. Sale of a gift card
   ii. Redemption of a gift card (the holder using it to acquire goods or services)
   iii. Increase in breakage estimated for gift cards already sold

D3. Disclosure Note 19 discusses Target's accounting for contingencies.

a. What is Target's approach for accruing losses for litigation claims associated with the data breach? Is their approach appropriate?
b. Target experienced a data breach in 2013, when "an intruder stole certain payment card and other guest information from our network." Target recorded a contingent liability for litigation associated with the data breach at that time, and continues to update that liability over time based on new information. Related new expenses for fiscal 2017 totaled $5 million. Prepare a journal entry to record Target's recognition of new expenses associated with the data breach litigation.

D4. Calculate the debt to equity ratio for Target at February 3, 2018. The average ratio for companies in the Discount Retailers industry sector in a comparable time period was 2.0.

D5. Calculate Target's times interest earned ratio for the year ended February 3, 2018. The coverage for companies in the Discount Retailers industry sector in a comparable time period was 6.9.

Part E: Leases, Income Taxes, and Pensions

E1. Refer to disclosure note 22 following Target's financial statements. What is the amount reported for "capital" leases (shown as the present value of minimum lease payments)? What is the total of those lease payments? What accounts for the difference between the two amounts?

E2. What is the total of the operating lease payments?

E3. New lease accounting guidance (discussed in Chapter 15) will require companies to report a liability for operating leases at present value as well as for capital leases (now called finance leases). If Target had used the new lease accounting guidance in its fiscal 2017 financial statements, what would be the amount reported as a liability for operating leases? Hint: Assume the payments "after 2020" are to be paid evenly over a 16 years period and all payments are at the end of years indicated. Target indicates elsewhere in its financial statements that 6% is an appropriate discount rate for its leases.

E4. Refer to *Note 22: Leases.* New lease accounting guidance requires companies to record a right-of-use asset and a lease liability for all leases, with the exception of short-term leases, at present value. If Target had used the new lease accounting guidance in its fiscal 2017 (February 3, 2018) financial statements, what would be the amount reported as a liability for its leases, operating and capital (finance) combined [rounded to nearest $ million]?

Hint: Assume the payments "after 2020" are to be paid evenly over a 16-year period and all payments are at the end of years indicated. Target indicates elsewhere in its financial statements that 6% is an appropriate discount rate for its leases.

E5. From the income statement, determine the income tax expense for the year ended February 3, 2018. Tie that number to the second table in Disclosure Note 23, "Provision for Income Taxes," and prepare a summary journal entry that records Target's tax expense from continuing operations for the year ended February 3, 2018.

E6. Focusing on the third table in Disclosure Note 23, "Net Deferred Tax Asset/(Liability)," calculate the change in net deferred tax assets or liability. By how much did that amount change? To what extent did you account for that change in the journal entry you wrote for the first requirement of this case? List possible causes of any difference.

E7. Target's Note 23 indicates that "In December 2017, the U.S. government enacted the Tax Cuts and Jobs Act tax reform legislation (the Tax Act), which among other matters reduced the U.S. corporate income tax rate from 35 percent to 21 percent effective January 1, 2018. . . . We have recorded a provisional $352 million net tax benefit primarily related to the remeasurement of certain deferred tax assets and liabilities, including $381 million of benefit from the new lower rate, partially offset by $29 million of deferred income tax expense from our foreign operations." What's the effect on net income?

E8. What is Target's liability for unrecognized tax benefits as of February 3, 2018? If Target were to prevail in court and realize $50 million more in tax savings than it thought more likely than not to occur, what would be the effect on the liability for unrecognized tax benefits and on net income?

E9. What were the changes in Target's Projected Benefits Obligation in the fiscal years ended February 3, 2018 (fiscal 2017), and January 28, 2017 (fiscal 2016), for its qualified pension plans?

E10. What were the changes in Target's Pension Plan Assets in the fiscal years ended February 3, 2018, and January 28, 2017, for its qualified pension plans?

E11. Were these pension plans overfunded or underfunded for the fiscal years ended February 3, 2018, and January 28, 2017?

E12. What were the components of Target's Pension Expense in the fiscal years 2017, 2016, and 2015?

Part F: Shareholders' Equity and Additional Financial Reporting Issues

F1. Refer to Target's Consolidated Statements of Shareholders' Investment. What are the six types of events and transactions that affected one or more of the Shareholders' Investment accounts in the year ended February 3, 2018?

F2. Note 25, "Share Repurchase," provides the information we need to reconstruct the journal entry that summarizes Target's share repurchases in the year ended February 3, 2018. Provide that entry (dollars in millions, rounded to the nearest million).

F3. Does Target account for share repurchases as (a) treasury stock or (b) retired shares

F4. What are the three types of awards described in Note 26:Share-Based Compensation?

F5. Based on the fair value of the awards granted, what was Target's primary form of share-based compensation for the year ended February 3, 2018?

F6. Projections of future performance should be based primarily on continuing operations. What was diluted EPS for continuing operations in each of the most recent three years?

F7. How many shares were included in diluted earnings per share but not basic earnings per share due to share-based compensation awards?

F8. Refer to Target's financial statements for the year ended February 3, 2018. Note 3 provides information on an accounting change Target made in fiscal 2017. Was this a change in estimate, a change in principle, a change in reporting entity, or an error correction?

F9. Did Target account for and report the change retrospectively or prospectively?

F10. What was the effect of the change on depreciation and amortization in fiscal 2016?

F11. Was the effect of the change on earnings per share in fiscal 2017 an increase, a decrease, or no effect?

F12. Is Target's cash provided by operating activities (a) increasing or (b) decreasing?

F13. Is Target's cash provided by operating activities more or less than net income in fiscal 2017?

F14. Is Target increasing or decreasing its investment in property and equipment?

F15. Is Target increasing or decreasing its long-term debt?

F16. Some transactions that don't increase or decrease cash must be reported in conjunction with a statement of cash flows. What activity of this type did Target report during each of the three years presented?

APPENDIX

C IFRS Comprehensive Case

Air France–KLM (AF), a Franco-Dutch company, prepares its financial statements according to International Financial Reporting Standards. AF's financial statements and disclosure notes for the year ended December 31, 2017, are available in Connect. This material is also available under the Finance link at the company's website (**www.airfranceklm.com**). This case addresses a variety of characteristics of financial statements prepared using IFRS, often comparing and contrasting those attributes of statements prepared under U.S. GAAP. Questions are grouped in parts according to various sections of the textbook.

Part A: Financial Statements, Income Measurement, and Current Assets

A1. What amounts did AF report for the following items for the fiscal year ended December 31, 2017?

a. Total revenues
b. Income from current operations
c. Net income or net loss (AF equity holders)
d. Total assets
e. Total equity

A2. What was AF's basic earnings or loss per share for the year ended December 31, 2017?

A3. Refer to AF's balance sheet and compare it to the balance sheet presentation in Illustration 2–14. What differences do you see in the format of the two balance sheets?

A4. What differences do you see in the terminology used in the two balance sheets?

A5. Describe the apparent differences in the order of presentation of the components of the balance sheet between IFRS as applied by Air France–KLM (AF) and a typical balance sheet prepared in accordance with U.S. GAAP.

A6. How does AF classify operating expenses in its income statement? How are these expenses typically classified in a U.S. company's income statement?

A7. How does AF classify interest paid, interest received, and dividends received in its statement of cash flows? What other alternatives, if any, does the company have for the classification of these items? How are these items classified under U.S. GAAP?

A8. Refer to AF's Note 30.2 "Description of the actuarial assumptions and related sensitivities."

a. What are the average discount rates used to measure AF's (a) 10–15 year and (b) 15 year and more pension obligations in the "euro" geographic zone in 2017?
b. If the rate used had been 1% (100 basis points) higher, what change would have occurred in the pension obligation in 2017? What if the rate had been 1% lower?

A9. In Note 4.6, AF indicates that "Sales related to air transportation are recognized when the transportation service is provided," so passenger and freight tickets are consequently recorded as 'Deferred revenue upon issuance.'"

a. Examine AF's balance sheet. What is the total amount of deferred revenue on ticket sales as of December 31, 2017?

b. When transportation services are provided with respect to the deferred revenue on ticket sales, what journal entry would AF make to reduce deferred revenue?
c. Does AF's treatment of deferred revenue under IFRS appear consistent with how these transactions would be handled under U.S. GAAP? Explain.

A10. AF has a frequent flyer program, "Flying Blue," which allows members to acquire "miles" as they fly on Air France or partner airlines that are redeemable for free flights or other benefits.
a. How does AF account for these miles?
b. Does AF report any liability associated with these miles as of December 31, 2017?
c. Although AF's 2017 annual report was issued prior to the effective date of ASU No. 2014-09, consider whether the manner in which AF accounts for its frequent flier program appears consistent with the revenue recognition guidelines included in the ASU.

A11. In Note 4.11, AF describes how it values trade receivables. How does the approach used by AF compare to U.S. GAAP?

A12. In Note 25, AF reconciles the beginning and ending balances of its valuation allowance for trade accounts receivable. Prepare a T-account for the valuation allowance and include entries for the beginning and ending balances and any reconciling items that affected the account during 2017.

A13. Examine Note 27. Does AF have any bank overdrafts? If so, are the overdrafts shown in the balance sheet the same way they would be shown under U.S. GAAP?

A14. What method does the company use to value its inventory? What other alternatives are available under IFRS? Under U.S. GAAP?

A15. AF's inventories are valued at the lower of cost or net realizable value. Does this approach differ from U.S. GAAP?

Part B: Property, Plant, and Equipment and Intangible Assets

B1. What method does Air France–KLM use to amortize the cost of computer software development costs? How does this approach differ from U.S. GAAP?

B2. AF does not report any research and development expenditures. If it did, its approach to accounting for research and development would be significantly different from U.S. GAAP. Describe the differences between IFRS and U.S. GAAP in accounting for research and development expenditures.

B3. AF does not report the receipt of any governments grants. If it did, its approach to accounting for government grants would be significantly different from U.S. GAAP. Describe the differences between IFRS and U.S. GAAP in accounting for government grants. If AF received a grant for the purchase of assets, what alternative accounting treatments are available under IFRS?

B4. AF's property, plant, and equipment is reported at cost. The company has a policy of not revaluing property, plant, and equipment. Suppose AF decided to revalue its flight equipment on December 31, 2017, and that the fair value of the equipment on that date was €12,000 million. Prepare the journal entry to record the revaluation, assuming that the journal entry to record annual depreciation had already been recorded. (Hint: you will need to locate the original cost and accumulated depreciation of the equipment at the end of the year in the appropriate disclosure note.)

B5. Under U.S. GAAP, what alternatives do companies have to value their property, plant, and equipment?

B6. AF calculates depreciation of plant and equipment on a straight-line basis, over the useful life of the asset. Describe any differences between IFRS and U.S. GAAP in the calculation of depreciation.

B7. When does AF test for the possible impairment of fixed assets? How does this approach differ from U.S. GAAP?

B8. Describe the approach AF uses to determine fixed asset impairment losses. (Hint: see Note 4.14.) How does this approach differ from U.S. GAAP?

B9. The following is included in AF's Disclosure Note 4.13: "Intangible assets are recorded at initial cost less accumulated amortization and any accumulated impairment losses." Assume that on December 31, 2017, AF decided to revalue its Other intangible assets (see Note 17), and that the fair value on that date was determined to be €500 million. Amortization expense for the year already has been recorded. Prepare the journal entry to record the revaluation.

Part C: Investments

C1. Read Notes 23 and 35.4. Focusing on investments accounted for at fair value through profit and loss (FVTPL):

a. As of December 31, 2017, what is the total balance of those investments in the balance sheet?

b. How much of that balance is classified as current and how much as noncurrent?

c. How much of the fair value of those investments is accounted for using level 1, level 2, and level 3 inputs of the fair value hierarchy? Given that information, assess the reliability (representational faithfulness) of this fair value estimate.

C2. Complete C1 again, but for investments accounted for as available for sale.

C3. Given your answer to C1c and C2c, which type of investment has more reliable (representational faithful) fair value estimates: FVTPL investments or available for sale investments?

C4. Read Notes 4.3. and 21.

a. When AF can exercise significant influence over an investee, what accounting approach does it use to account for the investment? How does AF determine if it can exercise significant influence?

b. If AF is involved in a joint venture, what accounting approach does it use to account for the investment?

c. What is the carrying value of AF's equity-method investments in its December 31, 2017, balance sheet?

d. How did AF's equity-method investments affect AF's 2017 net income from continuing operations?

Part D: Liabilities

D1. Read Notes 4.6 and 35. What do you think gave rise to total deferred income of €249 as of the end of fiscal 2017? Would transactions of this type be handled similarly under U.S. GAAP?

D2. Is the threshold for recognizing a provision under IFRS different than it is under U.S. GAAP? Explain.

D3. Note 32 lists "other provisions."

a. Do the beginning and ending balances of total provisions and retirement benefits shown in Note 32 for fiscal 2017 tie to the balance sheet? By how much has the total amount of AF's "other provisions" increased or decreased during fiscal 2017?

b. Write journal entries for the following changes in the litigation provision that occurred during fiscal 2017, assuming any amounts recorded on the income statement are recorded as "provision expense" and any use of provisions is paid for in cash. In each case, provide a brief explanation of the event your journal entry is capturing.

i. New provision.

ii. Use of provision.

c. Is AF's treatment of litigation provision under IFRS similar to how it would be treated under U.S. GAAP?

D4. Note 32.2 lists a number of contingent liabilities. Are amounts for those items recognized as a liability on AF's balance sheet? Explain.

D5. Examine the long-term borrowings in AF's balance sheet and the related note (33.2.2). Note that AF has convertible bonds outstanding that it issued in 2013. Prepare the journal entry AF would use to record the issue of convertible bonds. Prepare the journal entry AF would use to record the issue of the convertible bonds if AF used U.S. GAAP.

D6. AF does not elect the fair value option (FVO) to report its financial liabilities. Examine Note 35.3, "Market value of financial instruments." If the company had elected the FVO for all of its debt measured at amortized cost, what would be the balance at December 31, 2017, in the fair value adjustment account?

Part E: Leases, Income Taxes, and Pensions

E1. In Note 4, "Summary of accounting policies," part 4.14, "Leases," AF states that "leases are classified as finance leases when the lease arrangement transfers substantially all the risks and rewards of ownership to the lessee." Is this the policy companies using U.S. GAAP follow?

E2. Is this the policy AF will follow when it begins applying the new lease guidance in the 2015 update to *IFRS 17*? Explain.

E3. What amounts are shown in AF''s December 31, 2017, balance sheet for deferred taxes?

E4. Here's an excerpt from one of AF's notes to its financial statements:

Deferred taxes (in part)

The Group records deferred taxes using the balance sheet liability method, providing for any temporary differences between the carrying amounts of assets and liabilities for financial reporting purposes and the amounts used for taxation purposes, except for exceptions described in IAS 12 "Income taxes." The tax rates used are those enacted or substantively enacted at the balance sheet date.

Is this policy consistent with U.S. GAAP? Explain.

E5. Below is an excerpt from one of AF's notes to its financial statements:

Deferred taxes (in part)

Deferred tax assets related to temporary differences and tax losses carried forward are recognized only to the extent it is probable that a future taxable profit will be available against which the asset can be utilized at the tax entity level.

Is this policy consistent with U.S. GAAP? Explain.

E6. AF reported past service cost (called prior service cost under U.S. GAAP) in its income statement as part of net periodic pension cost. Is that reporting method the same or different from the way we report prior service cost under U.S. GAAP?

E7. Look at Note 30.2, "Retirement Benefits." AF incorporates estimates regarding staff turnover, life expectancy, salary increase, retirement age, and discount rates. How did AF report changes in these assumptions? Is that reporting method the same or different from the way we report changes under U.S. GAAP?

E8. AF does not report remeasurement gains and losses in its income statement. Where did AF report these amounts? Is that reporting method the same or different from the way we report pension expense under U.S. GAAP?

E9. See Note 22. Did AF report Net interest cost or Net interest income in 2017? How is that amount determined?

Part F: Shareholders' Equity and Additional Financial Reporting Issues

F1. Locate Note 28.5 in AF's financial statements. What items comprise "Reserves and retained earnings," as reported in the balance sheet? If Air France–KLM used U.S. GAAP, what would be different for the reporting of these items?

F2. In its presentation of the components of the balance sheet, which is listed first, current assets or non-current assets? Does this approach differ from U.S. GAAP?

F3. In its presentation of the components of the balance sheet, which is listed first, current liabilities or non-current liabilities? Does this approach differ from U.S. GAAP?

F4. In its presentation of the components of the balance sheet, which is listed first, liabilities or shareholders' equity? Does this approach differ from U.S. GAAP?

F5. AF provides share-based compensation in the form of PPSs, Phantom Performance Shares. Recipients receive compensation in what form? Are such plans reported in the balance sheet as (a) assets, (b) liabilities, or (c) equity? [*Hint:* See Note 29: "Share-Based Compensation."]

F6. Are AF's PPSs cliff vesting or graded vesting?

F7. The PPS options are performance-based, which means that under either U.S. GAAP or IFRS the amount expensed depends on whether it's "probable" that the performance target will be met. Does this mean it's equally likely that performance-based PPSs will be expensed at the same amount under U.S. GAAP as under IFRS?

F8. What amount(s) of earnings per share did AF report in its income statement for the year ended December 31, 2017? If AF used U.S. GAAP, would it have reported EPS using the same classification?

F9. Refer to AF's disclosure notes, in particular Note 2, "Restatement of Accounts 2016." Was the change described in the note a change in estimate, a change in principle, a change in reporting entity, or an error correction?

F10. Is this the same approach AF would follow if using U.S. GAAP?

F11. What are the three primary classifications into which AF's cash inflows and cash outflows are separated?

F12. Is this classification the same as or different from cash flow statements prepared in accordance with U.S. GAAP?

F13. In which classification are cash inflows from dividends included in AF's cash flow statements?

F14. Is this classification different from cash flow statements prepared in accordance with U.S. GAAP?

Glossary

Accounting equation Assets = Liabilities + Shareholders' Equity.

Accounting principles board (APB) the second private sector body delegated the task of setting accounting standards.

Accounts storage areas used to keep track of how transactions and events cause increases and decreases in the balances of financial elements.

Accounts payable obligations to suppliers of merchandise or of services purchased on account.

Accounts receivable receivables resulting from the sale of goods or services on account.

Accounts receivable aging schedule calculating the necessary allowance for uncollectible accounts by applying different percentages to accounts receivable balances depending on the length of time outstanding.

Accretion expense the increase in an asset retirement obligation that accrues as an operating expense.

Accrual accounting measures income according to the entity's accomplishments and resource sacrifices during the period from transactions related to providing goods and services to customers, regardless of when cash is received or paid.

Accruals when the cash flow comes after either expense or revenue recognition.

Accrued interest interest that has accrued since the last interest date.

Accrued liabilities expenses already incurred but not yet paid (accrued expenses).

Accrued receivables recognition of revenue earned before cash is received.

Accumulated benefit obligation (ABO) the discounted present value of estimated retirement benefits earned so far by employees, applying the plan's pension formula using existing compensation levels.

Accumulated other comprehensive income (AOCI) a component of stockholders' equity that reports the accumulated amount of other comprehensive income items in the current and prior periods.

Acid-test ratio current assets, excluding inventories and prepaid items, divided by current liabilities.

Acquisition costs the amounts paid to acquire the rights to explore for undiscovered natural resources or to extract proven natural resources.

Activity-based method allocation of an asset's cost base using a measure of the asset's input or output.

Actuary a professional trained in a particular branch of statistics and mathematics to assess the various uncertainties and to estimate the company's obligation to employees in connection with its pension plan.

Additions the adding of a new major component to an existing asset.

Adjusted trial balance trial balance after adjusting entries have been recorded.

Adjusting entries internal transactions recorded at the end of any period when financial statements are prepared.

Advance payment payment made at the beginning of the lease that represents prepaid rent

Agent facilitates transfers of goods and services between sellers and customers.

Allocation base cost of the asset expected to be consumed during its service life.

Allocation method the pattern in which the allocation base is expected to be consumed.

Allowance for uncollectible accounts contra account that reduces accounts receivable to the net amount expected to be collected. Also called the allowance for bad debts, the allowance for doubtful accounts, or the allowance for credit losses.

Allowance method recording bad debt expense and reducing accounts receivable indirectly by crediting the allowance for uncollectible accounts, a contra account to accounts receivable, for an estimate of the amount that eventually will prove uncollectible.

American institute of accountants (AIA)/american institute of certified public accountants (AICPA) national organization of professional public accountants.

Amortization cost allocation for intangibles.

Annual bonuses annual bonuses are one-time payments in addition to normal salary, typically tied to performance of the individual or company during a period

Annuity cash flows received or paid in the same amount each period.

Annuity due cash flows occurring at the beginning of each period.

Antidilutive securities the effect of the conversion or exercise of potential common shares would be to increase, rather than decrease, EPS.

Articles of incorporation statement of the nature of the firm's business activities, the shares to be issued, and the composition of the initial board of directors.

Asset retirement obligations (AROS) obligations associated with the disposition of an operational asset.

Asset turnover ratio measure of a company's efficiency in using assets to generate revenue.

Asset/liability approach recognition and measurement of assets and liabilities drives revenue and expense recognition.

Assigning using receivables as collateral for loans; nonpayment of a debt will require the proceeds from collecting the assigned receivables to go directly toward repayment of the debt.

Auditor's report report issued by CPAs who audit the financial statements that informs users of the audit findings.

Auditors independent professionals who render an opinion about whether the financial statements fairly present the company's financial position, performance, and cash flows in compliance with GAAP.

Average collection period indication of the average age of accounts receivable.

Average cost method assumes cost of goods sold and ending inventory consist of a mixture of all the goods available for sale.

Average days in inventory indicates the average number of days it normally takes to sell inventory.

Balance sheet a financial statement that presents an organized list of assets, liabilities, and equity at a particular point in time.

Balance sheet approach determining an income statement amount by estimating the appropriate carrying value of a balance sheet account and then adjusting the account as necessary to reach that carrying value.

Bank reconciliation comparison of the bank balance with the balance in the company's own records.

Basic EPS computed by dividing income available to common stockholders (net income less any preferred stock dividends) by the weighted-average number of common shares outstanding for the period.

Bill-and-hold occurs when a customer purchases goods but requests that the seller retain physical possession of the goods until a later date.

Billings on construction contract contra account to the asset *construction in progress* recognizing that a customer has been billed for work performed; subtracted from construction in progress to determine balance sheet presentation.

Board of directors establishes corporate policies and appoints officers who manage the corporation.

Bond indenture document that describes specific promises made to bondholders.

Bonds A form of debt consisting of separable units (bonds) that obligates the issuing corporation to repay a stated amount at a specified maturity date and to pay interest to bondholders between the issue date and maturity.

Book value assets minus liabilities as shown in the balance sheet.

Callable allows the issuing company to buy back, or call, outstanding bonds from the bondholders before their scheduled maturity date.

Capital budgeting the process of evaluating the purchase of operational assets.

Capital markets mechanisms that foster the allocation of resources efficiently.

Capital structure the mixture of liabilities and shareholders' equity in a company.

Cash currency and coins, balances in checking accounts, and items acceptable for deposit in these accounts, such as checks and money orders received from customers.

Cash disbursements journal record of cash disbursements.

Cash equivalents short-term investments that have a maturity date no longer than three months from the date of purchase.

Cash flows from financing activities both inflows and outflows of cash resulting from the external financing of a business.

Cash flows from investing activities both outflows and inflows of cash caused by the acquisition and disposition of assets.

Cash flows from operating activities both inflows and outflows of cash that result from activities reported on the income statement.

Cash receipts journal record of cash receipts.

Cash surrender value a determinable amount of money that can be received in exchange for surrendering a life insurance policy while the insured is still alive.

Cash-basis accounting measures income as the difference between cash receipts and cash disbursements during a reporting period from transactions related to providing goods and services to customers.

Certified Public Accountants (CPAS) licensed individuals who can represent that the financial statements have been audited in accordance with generally accepted auditing standards.

Change in accounting estimate a change in an estimate when new information comes to light.

Change in accounting principle switch by a company from one accounting method to another.

Change in reporting entity presentation of consolidated financial statements in place of statements of individual companies, or a change in the specific companies that constitute the group for which consolidated or combined statements are prepared.

Closing process the temporary accounts are reduced to zero balances, and these temporary account balances are closed (transferred) to retained earnings to reflect the changes that have occurred in that account during the period.

Commercial paper unsecured notes sold in minimum denominations of $25,000 with maturities ranging from 30 to 270 days.

Committee on Accounting Procedure (CAP) first private sector body that was delegated the task of setting accounting standards.

Comparability the ability to help users see similarities and differences among events and conditions.

Comparative financial statements corresponding financial statements from the previous years accompanying the issued financial statements.

Compensating balance specified balance (usually some percentage of the committee amount) a borrower of a loan is asked to maintain in a low-interest or noninterest-bearing account at the bank.

Complete depiction is complete if it includes all information necessary for faithful representation.

Complex capital structure potential common shares are outstanding.

Composite depreciation method physically dissimilar assets are aggregated to gain the convenience of group depreciation.

Compound interest interest computed not only on the initial investment but also on the accumulated interest in previous periods.

Comprehensive income change in shareholders' equity for the period from nonowner sources; equal to net income plus other comprehensive income. Traditional net income plus other nonowner changes in equity.

Conceptual framework deals with theoretical and conceptual issues and provides an underlying structure for current and future accounting and reporting standards.

Confirmatory value confirmation of investor expectations about future cash-generating ability.

Conservatism practice followed in an attempt to ensure that uncertainties and risks inherent in business situations are adequately considered.

Consignment a selling arrangement whereby the consignor physically transfers goods to another company ("consignee") to sell, while legal title and risk of ownership of those goods remain with the consignor during the consignment period.

Consistency permits valid comparisons between different periods.

Consolidated financial statements combination of the separate financial statements of the parent and subsidiary each period into a single aggregate set of financial statements, as if there were only one company.

Construction in progress asset account equivalent to the asset work-in-process inventory in a manufacturing company.

Contingently issuable shares additional shares of common stock to be issued, contingent on the occurrence of some future circumstance.

Contract an agreement that creates legally enforceable rights and obligations. Contracts can be explicit or implicit.

Contract asset asset recognizing that a seller has a conditional right to receive payment after satisfying a performance obligation.

Contract liability a label given to deferred revenue or unearned revenue accounts.

Control usually an investor can control the investee if it owns more than 50% of the investee's voting shares.

Conventional retail method application of the retail inventory method that excludes markdowns in the calculation of the cost-to-retail percentage, as a way to approximate the lower of average cost or market.

Convertible bonds bonds for which bondholders have the option to convert the bonds into shares of stock.

Copyright exclusive right of protection given to a creator of a published work, such as a song, painting, photograph, or book.

Corporation dominant form of business organization that acquires capital from investors in exchange for ownership interest and from creditors by borrowing.

Correction of an error an adjustment a company makes due to an error made.

Cost effectiveness the perceived benefit of increased decision usefulness exceeds the anticipated cost of providing that information.

Cost of goods sold cost of the inventory sold during the period.

Cost recovery method deferral of all gross profit recognition until the cost of the item sold has been recovered.

Cost-to-retail percentage ratio found by dividing goods available for sale at cost by goods available for sale at retail.

Coupons bonds name of the owner was not registered; the holder actually clipped an attached coupon and redeemed it in accordance with instructions on the indenture.

Credit losses losses due to failure by customers to pay amounts owed for purchase of goods or services; also called bad debts, impairments of receivables, and uncollectible accounts.

Credits represent the right side of the account.

Cumulative if the specified dividend is not paid in a given year, the unpaid dividends accumulate and must be made up in a later dividend year before any dividends are paid on common shares.

Current assets includes assets that are cash, will be converted into cash, or will be used up within one year from the balance sheet date (or operating cycle, if longer).

Current costs are the costs that would be incurred to purchase or reproduce an asset.

Current expected credit loss (CECL) model a model used to estimate credit losses (bad debts) for receivables as well as those debt investments that are accounted for as held to maturity or as available for sale.

Current liabilities expected to require the use of current assets for payment, and usually are payable within one year from the balance sheet date (or operating cycle, if longer).

Current maturities of long-term debt the portion of long-term notes, loans, mortgages, and bonds payable that is payable within the next year (or operating cycle, if longer), reported as a current liability.

Current ratio current assets divided by current liabilities.

Date of record specific date stated as to when the determination will be made of the recipient of the dividend.

Debenture bond backed only by the "full faith and credit" of the issuing corporation.

Debits represent the left side of the account.

Debt issue costs Costs of issuing debt securities are called *debt issue costs* and are accounted for the same way as bond discount.

Debt to equity ratio compares resources provided by creditors with resources provided by owners.

Decision usefulness the quality of being useful to decision making.

Default risk a company's ability to pay its obligations when they come due.

Deferred annuity first cash flow occurs more than one period after the date the agreement begins.

Deferred revenues cash received from a customer for goods or services to be provided in a future period.

Deferred tax asset taxes to be saved in the future when future deductible amounts reduce taxable income (when the temporary differences reverse).

Deferred tax liability taxes to be paid in the future when future taxable amounts become taxable (when the temporary differences reverse).

Deficit debit balance in retained earnings.

Defined benefit pension plans fixed retirement benefits defined by a designated formula, based on employees' years of service and annual compensation.

Defined contribution pension plans fixed annual contributions to a pension fund; employees choose where funds are invested—usually stocks or fixed-income securities.

Depletion allocation of the cost of natural resources.

Depreciation cost allocation for plant and equipment.

Derivatives financial instruments usually created to hedge against risks created by other financial instruments or by transactions that have yet to occur but are anticipated and that "derive" their values or contractually required cash flows from some other security or index.

Detachable stock purchase warrants the investor has the option to purchase a stated number of shares of common stock at a specified option price, within a given period of time.

Development costs for natural resources, costs incurred after the resource has been discovered but before production begins.

Diluted EPS incorporates the dilutive effect of all potential common shares in the calculation of EPS.

Direct financing lease lease in which the lessor finances the asset for the lessee and earns interest revenue over the lease term.

Direct method cash effect of each operating activity (i.e., income statement item) is reported directly on the statement of cash flows.

Direct write-off method an allowance for uncollectible accounts is not used; instead bad debts that do arise are written off as bad debt expense.

Disclosure including pertinent information in the financial statements and accompanying notes.

Disclosure notes additional insights about company operations, accounting principles, contractual agreements, and pending litigation written in notes that accompany the financial statements.

Discontinued operations the discontinuance of a component of an entity whose operations and cash flows can be clearly distinguished from the rest of the entity.

Discount arises when bonds are sold for less than face amount.

Discounting the transfer of a note receivable to a financial institution.

Distinct a good or service is *distinct* if it is both *capable of being distinct* and *separately identifiable from other goods or services in the contract.* It is capable of being distinct if the customer could use the good or service on its own or in combination with other goods and services it could obtain elsewhere. It is separately identifiable if the good or service is distinct in the context of the contract because it is not highly interrelated with other goods and services in the contract. Distinct goods and services are accounted for as separate performance obligations.

Dividend distribution to shareholders of a portion of assets earned.

Dollar-value LIFO (DVL) an inventory costing method comprising layers of dollar value from different periods and using cost indexes to adjust for changes in price levels over time.

Dollar-value LIFO retail method LIFO retail method combined with dollar-value LIFO.

Double-declining-balance (DDB) method 200% of the straight-line rate is multiplied by book value.

Double-entry system dual effect that each transaction has on the accounting equation when recorded.

DuPont framework depict return on equity as determined by profit margin (representing profitability), asset turnover (representing efficiency), and the equity multiplier (representing leverage).

Early extinguishment of debt debt is retired prior to its scheduled maturity date.

Earnings per share (EPS) the amount of income earned by a company expressed on a per share basis.

Earnings quality refers to the ability of reported earnings (income) to predict a company's future earnings.

Economic entity assumption presumes that economic events can be identified specifically with an economic entity.

Economic events events that directly affect the financial position of the company.

Effective interest method calculates interest revenue by multiplying the outstanding balance of the investment by the relevant interest rate.

Effective rate the actual rate at which money grows per year.

Effective tax rate equals tax expense divided by pretax accounting income.

Emerging Issues Task Force (EITF) responsible for providing more timely responses to emerging financial reporting issues.

Employee share purchase plans permit all employees to buy shares directly from their company, often at favorable terms.

Equity method used when an investor can't control, but can significantly influence, the investee. Under the equity method, the investor recognizes in its own income statement its proportionate share of the investee's income.

Estimates predictions of future events.

Ethics a code or moral system that provides criteria for evaluating right and wrong.

Ex-dividend date date. usually one business day before the date of record, and is the first day the stock trades without the right to receive the declared dividend.

Expected cash flow approach adjusts the cash flows, not the discount rate, for the uncertainty or risk of those cash flows.

Expected return on plan assets estimated long-term return on invested assets.

Expenses outflows or other using up of assets or incurrences of liabilities from delivering or producing goods, rendering services, or other activities that constitute the entity's ongoing operations.

Exploration costs for natural resources, expenditures such as drilling a well, or excavating a mine, or any other costs of searching for natural resources.

Extended warranties additional, extended service that covers new problems arising after the buyer takes control of the product.

External events exchanges between the company and separate economic entities.

F.O.B. (free on board) destination legal title to the goods does not pass from the seller to the buyer until the goods arrive at their destination (the customer's location); the seller is responsible for shipping costs and transit insurance.

F.O.B. (free on board) shipping point legal title to the goods passes from the seller to the buyer at the point of shipment (when the seller delivers the goods to the common carrier); the buyer is responsible for shipping costs and transit insurance .

Factor financial institution that buys receivables for cash, handles the billing and collection of the receivables, and charges a fee for this service.

Fair value bases measurements on the price that would be received to sell assets or transfer liabilities in an orderly market transaction.

Fair value hedge a derivative is used to hedge against the exposure to changes in the fair value of an asset or liability or a firm commitment.

Fair value option allows companies to report specified financial assets and liabilities at fair value.

Faithful representation exists when there is agreement between a measure or description and the phenomenon it purports to represent.

Finance leases lessee has, in substance, purchased the lease asset; assumed when one of five classification criteria is met.

Financial accounting provides relevant financial information to various external users.

Financial Accounting Foundation (FAF) responsible for selecting the members of the FASB and its Advisory Council, ensuring adequate funding of FASB activities, and exercising general oversight of the FASB's activities.

Financial Accounting Standards Board (FASB) the current private sector body that has been delegated the task of setting accounting standards.

Financial instrument cash; evidence of an ownership interest in an entity; a contract that imposes on one entity an obligation to deliver cash or another financial instrument, and conveys to the second entity a right to receive cash or another financial instrument; and a contract that imposes on one entity an obligation to exchange financial instruments on potentially unfavorable terms and conveys to a second entity a right to exchange other financial instruments on potentially favorable terms.

Financial leverage by earning a return on borrowed funds that exceeds the cost of borrowing the funds, a company can provide its shareholders with a total return higher than it could achieve by employing equity funds alone.

Financial reporting process of providing financial statement information to external users.

Financial statements primary means of communicating financial information to external parties.

Financing activities involve cash inflows and outflows from transactions with creditors (excluding trade creditors) and owners.

Finished goods products that have been completed in the manufacturing process but have not yet been sold.

First-in, first-out (FIFO) method assumes that items sold are those that were acquired first.

Fixed-asset turnover ratio the effectiveness of managers to use fixed assets to generate sales, measured as net sales divided by average fixed assets.

Foreign currency futures contract agreement that requires the seller to deliver a specific foreign currency at a designated future date at a specific price.

Foreign currency hedge if a derivative is used to hedge the risk that some transactions require settlement in a currency other than the entities' functional currency or that foreign operations will require translation adjustments to reported amounts.

Forward contract calls for delivery on a specific date; is not traded on a market exchange; does not call for a daily cash settlement for price changes in the underlying contract.

Franchise contractual arrangement under which the franchisor grants the franchisee the exclusive right to use the franchisor's trademark or tradename within a geographical area, usually for a specified period of time.

Franchisee individual or corporation given the right to operate a business involving the franchisor's products or services and use its name and other symbols for a specific period of time.

Franchisor grants to the franchisee the right to operate a business involving the franchisor's products or services and use its name and other symbols for a specific period of time.

Fraud an intentional act by one or more individuals among management, those charged with governance, employees, or third parties, involving the use of deception that results in a misstatement in the financial statements that are the subject of an audit.

Free from error information is free from error if it contains no errors or omissions.

Freight-in transportation-in; in a periodic system, freight costs generally are added to this temporary account, which is added to purchases in determining net purchases.

Full-cost method allows costs incurred in searching for oil and gas within a large geographical area to be capitalized as assets and expensed in the future as oil and gas from the successful wells are removed from that area.

Full-disclosure principle financial reports should include any information that could affect the decisions made by external users.

Functional intellectual property has significant stand-alone functionality, such that the intellectual property can perform a function or task, be played, or be aired over various types of media; the seller typically recognizes revenue at the point in time the customer can start using functional intellectual property.

Future deductible amounts the future tax consequence of a temporary difference will be to decrease taxable income relative to accounting income.

Future taxable amounts the future tax consequence of temporary difference will be to increase taxable income relative to accounting income.

Future value amount of money that a dollar will grow to at some point in the future.

Gains increases in equity from peripheral, or incidental, transactions of an entity.

General journal used to record any type of transaction.

General ledger collection of accounts that organizes the accounts and allows for keeping track of increases and decreases and resulting balances

Generally Accepted Accounting Principles (GAAP) set of both broad and specific guidelines that companies should follow when measuring and reporting the information in their financial statements and related notes.

Gift card transferable prepayments for a specified dollar value of goods or services to be delivered at a future date. Gift cards give rise to deferred revenue liabilities until they are redeemed or viewed as not going to be redeemed (broken).

Going concern assumption in the absence of information to the contrary, it is anticipated that a business entity will continue to operate indefinitely.

Governmental Accounting Standards Board (GASB) responsible for developing accounting standards for governmental units such as states and cities.

Gross method The buyer views a discount not taken as part of the cost of inventory; the seller views a discount not taken by the customer as part of sales of revenue.

Gross profit method (gross margin method) estimates cost of goods sold, which is then subtracted from cost of goods available for sale to estimate ending inventory.

Gross profit ratio gross profit (net sales minus cost of goods sold) divided by net sales.

Group depreciation method collection of assets defined as depreciable assets that share similar service lives and other attributes.

Half-year convention record one-half of a full year's depreciation in the year of acquisition and another half year in the year of disposal.

Hedging taking an action that is expected to produce exposure to a particular type of risk that is precisely the opposite of an actual risk to which the company already is exposed.

Historical cost original transaction value.

Horizontal analysis comparison by expressing each item as a percentage of that same item in the financial statements of another year (base amount) in order to more easily see year-to-year changes.

Illegal acts violations of the law, such as bribes, kickbacks, and illegal contributions to political candidates.

Implicit rate of interest rate implicit in the agreement.

Improvements replacement of a major component of an operational asset.

Income from continuing operations revenues, expenses (including income taxes), gains, and losses, excluding those related to discontinued operations and extraordinary items.

Income statement statement of operations or statement of earnings that is used to summarize the profit-generating activities that occurred during a particular reporting period.

Income statement approach estimating an income statement amount directly, rather than basing it on the change in a balance sheet account.

Indirect method the net cash increase or decrease from operating activities is derived indirectly by starting with reported net income and working backwards to convert that amount to a cash basis.

Initial direct costs costs incurred by the lessor that are associated directly with originating a lease and are essential to acquire the lease.

Installment sales method recognizes revenue and costs only when cash payments are received.

Institute of Internal Auditors national organization of accountants providing internal auditing services for their own organizations.

Institute of Management Accountants (IMA) primary national organization of accountants working in industry and government.

Intangible assets operational assets that lack physical substance; examples include patents, copyrights, franchises, and goodwill.

Interest "rent" paid for the use of money for some period of time.

Interest cost interest accrued on the projected benefit obligation calculated as the discount rate multiplied by the projected benefit obligation at the beginning of the year.

Interest rate swap agreement to exchange fixed interest payments for floating rate payments, or vice versa, without exchanging the underlying principal amounts.

Interest-bearing note receivable notes that state a principal and interest rate to be paid by a debtor to a creditor.

Internal control a company's plan to encourage adherence to company policies and procedures, promote operational efficiency, minimize errors and theft, and enhance the reliability and accuracy of accounting data.

Internal events events that directly affect the financial position of the company but don't involve an exchange transaction with another entity.

International Accounting Standards Board (IASB) objectives are to develop a single set of high-quality, understandable global accounting standards, to promote the use of those standards, and to bring about the convergence of national accounting standards and International Accounting Standards.

International Accounting Standards Committee (IASC) umbrella organization formed to develop global accounting standards.

International Financial Reporting Standards (IFRS) developed by the IASB and used by more than 100 countries.

Intrinsic value difference between the market price of the shares and the option price at which they can be acquired.

Inventory goods awaiting sale (finished goods), goods in the course of production (work in process), and goods to be consumed directly or indirectly in production (raw materials). Goods acquired, manufactured, or in the process of being manufactured for sale.

Inventory turnover ratio measures a company's efficiency in managing its investment in inventory.

Investing activities involve the acquisition and sale of long-term assets used in the business and nonoperating investment assets.

Journal a chronological record of all economic events affecting financial position.

Journal entry captures the effect of a transaction on financial position in debit/credit form.

Just-in-time (JIT) system a system used by a manufacturer to coordinate production with suppliers so that raw materials or components arrive just as they are needed in the production process.

Land improvements the cost of parking lots, driveways, and private roads and the costs of fences and lawn and garden sprinkler systems.

last-in, first-out (LIFO) method assumes that the last units purchased are the first ones sold.

Lease payments payments the lessee is required to make in connection with the lease.

Leasehold improvement account title when a lessee makes improvements to leased property that reverts back to the lessor at the end of the lease.

Lessee user of a leased asset.

Lessor owner of a leased asset.

Liabilities probable future sacrifices of economic benefits arising from present obligations of a particular entity to transfer assets or provide services to other entities in the future as a result of past transactions or events.

Licenses allow the customer to access the seller's intellectual property.

LIFO conformity rule if a company uses LIFO to measure taxable income, the company also must use LIFO for external financial reporting.

LIFO inventory pools groups of inventory units based on physical similarities of the individual units.

LIFO liquidation the decline in inventory quantity during the period.

LIFO reserve contra account to inventory used to record the difference between the internal method and LIFO.

Limited liability company owners are not liable for the debts of the business, except to the extent of their investment; all members can be involved with managing the business without losing liability protection; no limitations on the number of owners.

Limited liability partnership similar to a limited liability company, except it doesn't offer all the liability protection available in the limited liability company structure.

Line of credit allows a company to borrow cash without having to follow formal loan procedures and paperwork.

Liquidating dividend when a dividend exceeds the balance in retained earnings and returns invested capital to owners

Liquidity the ability of a company to convert its assets to cash to pay its current liabilities.

Long-term solvency an assessment of whether a company will be able to pay all its liabilities, which includes long-term liabilities.

Loss contingency existing, uncertain situation involving potential loss which will be resolved when some future event occurs.

Losses decreases in equity from peripheral, or incidental, transactions of the entity.

Lower of cost or market subsequent measurement of inventory applied by companies that use LIFO or the retail inventory method. This approach requires companies to report ending inventory at the lower of cost or market.

Lower of cost or net realizable value (NRV) subsequent measurement of inventory applied by companies that use FIFO, average cost, or any other method besides LIFO or the retail inventory method. This approach requires companies to report ending inventory at the lower of cost or net realizable value.

Management's discussion and analysis (MDA) provides a biased but informed perspective of a company's operations, liquidity, and capital resources.

Market (for inventory reporting) Current replacement cost of inventory, not to exceed net realizable value (NRV) or to be lower than NRV minus a normal profit margin.

Material has qualitative or quantitative characteristics that make it matter for decision making.

Measurement process of associating numerical amounts with the elements.

Model Business Corporation Act designed to serve as a guide to states in the development of their corporation statutes.

Modified accelerated cost recovery system (MACRS) federal income tax code allows taxpayers to compute depreciation for their tax returns using this method.

Modified retrospective approach accounting change is applied only to the adoption period with adjustment of the balance of retained earnings at the beginning of the adoption period to capture the cumulative effects of prior periods.

Monetary assets money and claims to receive money, the amount of which is fixed or determinable.

Monetary liabilities obligations to pay amounts of cash, the amount of which is fixed or determinable.

Monetary unit assumption states that financial statement elements should be measured in a particular monetary unit (in the United States, the U.S. dollar).

Mortgage bond backed by a lien on specified real estate owned by the issuer.

Multiple-step income statement format that includes a number of intermediate subtotals before arriving at income from continuing operations.

Natural resources oil and gas deposits, timber tracts, and mineral deposits.

Net income Net income is the difference between revenues and expenses.

Net markdown net effect of the change in selling price (increase, decrease, increase).

Net markup net effect of the change in selling price (increase, increase, decrease).

Net method The buyer considers the cost of inventory to include the net, after-discount amount, and any discounts not taken are reported as interest expense; the seller considers sales revenue to be the net amount, after discount, and any discounts not taken by the customer are included in sales revenue.

Net operating cash flow difference between cash receipts and cash disbursements from providing goods and services.

Net operating loss negative taxable income because tax-deductible expenses exceed taxable revenues.

Net operating loss carryforward offsets future taxable income with an NOL to provide a reduction of taxes payable in that future period; therefore, gives rise to a deferred tax asset because it is a future deductible amount.

Net realizable value estimated selling prices in the ordinary course of business, less reasonably predictable costs of completion, disposal, and transportation.

Neutral implies freedom from bias.

Non-gaap earnings actual (GAAP) earnings reduced by any expenses the reporting company feels are unusual and should be excluded.

Noncash investing and financing activities transactions that do not increase or decrease cash but that result in significant investing and financing activities.

Noncumulative if the specified dividend is not declared in any given year, it need never be paid.

Noninterest-bearing note notes for which the interest is deducted from the face amount of the note to determine the cash proceeds made available to the borrower at the outset.

Nonoperating income includes revenues, expenses, gains, and losses related to peripheral or incidental activities of the company.

Nonparticipating preferred shareholder dividends are limited to the stated amount.

Notes payable promissory notes (essentially an IOU) that obligate the issuing corporation to repay a stated amount at or by a specified maturity date and to pay interest to the lender between the issue date and maturity.

Notes receivable receivables supported by a formal agreement or note that specifies payment terms.

Objectives-oriented/principles-based accounting standards approach to standard setting stresses professional judgment, as opposed to following a list of rules.

Operating activities inflows and outflows of cash related to transactions entering into the determination of net income.

Operating cycle period of time necessary to convert cash to raw materials, raw materials to finished product, the finished product to receivables, and then finally receivables back to cash.

Operating income includes revenues, expenses, gains, and losses directly related to the principal revenue-generating activities of the company.

Operating leases fundamental rights and responsibilities of ownership are retained by the lessor and the lessee merely is using the asset temporarily.

Operating segment a component of an enterprise that engages in business activities from which it may earn revenues and incur expenses (including revenues and expenses relating to transactions with other companies of the same enterprise); whose operating results are regularly reviewed by the enterprise's chief operating decision maker to make decisions about resources to be allocated to the segment and assess its performance; for which discrete financial information is available.

Operational risk how adept a company is at withstanding various events and circumstances that might impair its ability to earn profits.

Option gives the holder the right either to buy or sell a financial instrument at a specified price.

Option pricing models statistical models that incorporate information about a company's stock and the terms of the stock option to estimate the option's fair value.

Ordinary annuity cash flows occur at the end of each period.

Organization costs costs related to organizing a new entity, such as legal fees and state filing fees to incorporate.

Other comprehensive income (OCI) changes in stockholders' equity other than transactions with owners and other than items that affect net income.

Paid-in capital invested capital consisting primarily of amounts invested by shareholders when they purchase shares of stock from the corporation.

Parenthetical comments/modifying comments supplemental information disclosed on the face of financial statements.

Participating preferred shareholders are allowed to receive additional dividends beyond the stated amount.

Patent exclusive right to manufacture a product or to use a process.

Pension plan assets employer contributions and accumulated earnings on the investment of those contributions to be used to pay retirement benefits to retired employees.

Performance obligations promises to transfer goods and services to a customer. They are satisfied when the seller transfers *control* of goods or services to the customer.

Periodic inventory system a system of accounting for inventory that involves an adjusting entry at the end of the period to update the balances of the inventory account and the cost of goods sold account for purchases, sales, and returns during the period.

Periodicity assumption allows the life of a company to be divided into artificial time periods to provide timely information.

Permanent accounts represent assets, liabilities, and shareholders' equity at a point in time.

Permanent difference difference between pretax accounting income and taxable income and, consequently, between the reported amount of an asset or liability in the financial statements and its tax basis that will not "reverse" resulting from transactions and events that under existing tax law will never affect taxable income or taxes payable.

Perpetual inventory system a system of accounting for inventory by continuously adjusting the balance of the inventory account for each purchase, sale, or return of inventory; the cost of goods sold account is adjusted for each sale or return of inventory by customers.

Pledging a promise to relinquish control of an asset (or assets) in payment (or partial payment) of an amount due.

Post-closing trial balance verifies that the closing entries were prepared and posted correctly and that the accounts are now ready for next year's transactions.

Posting transferring debits and credits recorded in individual journal entries to the specific accounts affected.

Potential common shares securities that, while not being common stock, may become common stock through their exercise, conversion, or issuance and therefore dilute (reduce) earnings per share.

Predictive value confirmation of investor expectations about future cash-generating ability.

Preferred stock typically has a preference (a) to a specified amount of dividends (stated dollar amount per share or percentage of par value per share) and (b) to distribution of assets in the event the corporation is dissolved.

Premium arises when bonds are sold for more than face amount.

Prepaid expenses costs of assets acquired in one period and expensed in a future period.

Prepayments/deferrals the cash flow precedes either expense or revenue recognition.

Present value Present value is today's equivalent of a particular amount in the future, after backing out the time value of money.

Principal controls goods or services and is responsible for providing them to the customer.

Prior period adjustment addition to or reduction in the beginning retained earnings balance in a statement of shareholders' equity due to a correction of an error.

Prior service cost the cost of credit given for an amendment to a pension plan to employee service rendered in prior years.

Product costs costs associated with products and expensed as cost of goods sold only when the related products are sold.

Profit margin on sales net income divided by net sales; measures the amount of net income achieved per sales dollar.

Projected benefit obligation (PBO) the discounted present value of estimated retirement benefits earned so far by employees, applying the plan's pension formula using projected future compensation levels.

Property dividend when a noncash asset is distributed.

Property, plant, and equipment tangible, long-lived assets used in the operations of the business, such as land, buildings, equipment, machinery, furniture, and vehicles, as well as natural resources, such as mineral mines, timber tracts, and oil wells .

Prospective approach effects of a change are reflected in the financial statements of only the year of the change and future years

Proxy statement contains disclosures on compensation to directors and executives; sent to all shareholders each year.

Purchase commitments contracts that obligate a company to purchase a specified amount of merchandise or raw materials at specified prices on or before specified dates.

Purchase discounts reductions in the amount to be paid if remittance is made within a designated period of time.

Purchase option a provision of some lease contracts that gives the lessee the option of purchasing the leased property during, or at the end of, the lease term at a specified price.

Purchase return a reduction in both inventory and accounts payable (if the account has not yet been paid) at the time of the return.

Purchases journal records the purchase of merchandise on account.

Quality-assurance warranty obligation by the seller to make repairs or replace products that are later demonstrated to be defective for some period of time after the sale.

Quasi reorganization a firm undergoing financial difficulties, but with favorable future prospects, may use a quasi reorganization to write down inflated asset values and eliminate an accumulated deficit.

Rate of return on stock investment

$$\frac{\text{Dividends} + \text{Share price appreciation}}{\text{Initial investment}}$$

Ratio analysis comparison of accounting numbers to evaluate the performance and risk of a firm.

Raw materials components purchased from suppliers that will become part of the finished product.

Realization principle realization principle bases revenue recognition on completion of the earnings process and reasonable certainty about collectibility.

Rearrangements expenditures made to restructure an asset without addition, replacement, or improvement.

Receivables a company's claims to the future collection of cash, other assets, or services.

Receivables turnover ratio indicates how quickly a company is able to collect its accounts receivable.

Recognition process of admitting information into the basic financial statements.

Redemption privilege might allow preferred shareholders the option, under specified conditions, to return their shares for a predetermined redemption price.

Refund Liability amount the seller estimates will be refunded to customers who make returns.

Related-party transactions transactions with owners, management, families of owners or management, affiliated companies, and other parties that can significantly influence or be influenced by the company.

Relevance one of the primary decision-specific qualities that make accounting information useful; made up of predictive value and/or feedback value and timeliness.

Replacement depreciation method depreciation is recorded when assets are replaced.

Residual asset carrying amount of a leased asset not transferred to the lessee.

Residual value or salvage value, the amount the company expects to receive for the asset at the end of its service life less any anticipated disposal costs.

Restoration costs costs to restore land or other property to its original condition after extraction of the natural resource ends.

Restricted stock shares issued in the name of the employee, subject to forfeiture by the employee if employment is terminated within some specified number of years from the date of grant.

Restricted stock units right to receive shares, subject to forfeiture by the employee if employment is terminated within some specified number of years from the date of grant.

Restructuring costs costs associated with plans by management to materially change either the scope or manner in which its company's operations are conducted.

Retained earnings amounts earned by the corporation on behalf of its shareholders and not (yet) distributed to them as dividends.

Retired stock shares repurchased and not designated as treasury stock, assuming the same status as authorized but unissued shares, as if never issued

Retirement depreciation method records depreciation when assets are disposed of and measures depreciation as the difference between the proceeds received and cost.

Retrospective approach financial statements issued in previous years are revised to reflect the impact of an accounting change whenever those statements are presented again for comparative purpose.

Return on assets (ROA) a company's profitability in relation to overall resources, measured as net income divided by average total assets.

Return on equity (ROE) amount of profit management can generate from shareholders' equity, measured as net income divided by average shareholders' equity.

Revenue/expense approach recognition and measurement of revenues and expenses are emphasized.

Revenues inflows of assets or settlements of liabilities (or a combination of both) from delivering or producing goods, rendering services, or other activities that constitute the entity's ongoing major or central operations.

Reverse stock split when a company decreases, rather than increases, its outstanding shares.

Reversing entries optional entries that remove the effects of some of the adjusting entries made at the end of the previous reporting period for the sole purpose of simplifying journal entries made during the new period.

Right of conversion shareholders' right to exchange shares of preferred stock for common stock at specified conversion ratio.

Right of return customers' right to return merchandise to retailers if they are not satisfied.

Rules-based accounting standards a list of rules for choosing the appropriate accounting treatment for a transaction.

S corporation has characteristics of both regular corporations and partnerships.

Sale-leaseback transaction the owner of an asset sells it and immediately leases it back from the new owner.

Sales discounts cash discounts; represent reductions not in the selling price of a good or service but in the amount to be received from a credit customer if the amount is paid within a specific period of time.

Sales journal records credit sales.

Sales return the return of merchandise for a refund or for credit to be applied to other purchases.

Sales-type lease lessor transfers control of lease asset to lessee, with or without a selling profit on the sale of the asset.

Securities and Exchange Commission (SEC) has the authority to set accounting standards for companies, but it relies on the private sector to do so.

Securities available-for-sale debt securities the investor acquires for purposes other than active trading or to be held to maturity.

Securities to be held-to-maturity debt securities for which the investor has the "positive intent and ability" to hold the securities to maturity.

Securitization the company creates a special purpose entity (SPE), usually a trust or a subsidiary; the SPE buys a pool of trade receivables, credit card receivables, or loans from the company and then sells related securities.

Selling profit when the fair value of the asset (usually the present value of the lease payments, or "selling price") exceeds the cost or carrying value of the asset sold.

Separation of duties an internal control technique in which various functions are distributed amongst employees to provide cross-checking that encourages accuracy and discourages fraud.

Serial bonds more structured (and less popular) way to retire bonds on a piecemeal basis.

Service cost increase in the projected benefit obligation attributable to employee service performed during the period.

Service life (useful life) the estimated use that the company expects to receive from the asset.

Service method allocation approach that reflects the declining service pattern of the prior service cost.

Short-term investments investments not classified as cash equivalents that the company has the ability and intent to sell within one year (or operating cycle if longer).

Significant influence effective control is absent but the investor is able to affect the operating and financial policies of the investee (usually is the case when investor holds between 20% and 50% of the investee's voting shares).

Simple capital structure a firm that has no potential common shares (outstanding securities that could potentially dilute earnings per share).

Simple interest computed by multiplying an initial investment times both the applicable interest rate and the period of time for which the money is used.

Single-step income statement format that groups all revenues and gains together and all expenses and losses together.

Sinking fund debentures bonds that must be redeemed on a prespecified year-by-year basis; administered by a trustee who repurchases bonds in the open market.

Source documents relay essential information about each transaction to the accountant, e.g., sales invoices, bills from suppliers, cash register tapes.

Special journal record of a repetitive type of transaction, e.g., a sales journal.

Specific identification method each unit sold during the period or each unit on hand at the end of the period is matched with its actual cost.

Specific interest method for interest capitalization, rates from specific construction loans to the extent of specific borrowings are used before using the average rate of other debt.

Stand-alone selling price the amount at which the good or service is sold separately under similar circumstances.

Start-up costs whenever a company introduces a new product or service, or commences business in a new territory or with a new customer, it incurs one-time costs that are expensed in the period incurred.

Statement of cash flows change statement summarizing the transactions that caused cash to change during the period.

Statement of shareholders' equity statement disclosing the source of changes in the shareholders' equity accounts.

Stock appreciation rights (SARS) awards that enable an employee to benefit by the amount that the market price of the company's stock rises above a specified amount, without having to buy shares.

Stock dividend distribution of additional shares of stock to current shareholders of the corporation.

Stock options employees aren't actually awarded shares, but rather are given the option to buy shares at a specified exercise price within some specified number of years from the date of grant.

Stock split stock distribution of 25% or higher, sometimes called a *large* stock dividend.

Straight-line method recording interest each period at the same dollar amount.

Straight-line method allocation of an equal amount to each year. For depreciation of plant and equipment or amortization of intangible assets, an equal amount of the allocation base is allocated to each year of the asset's service life. This method is sometimes used for allocation of other assets, and liabilities, over time.

Subordinated debenture the holder is not entitled to receive any liquidation payments until the claims of other specified debt issues are satisfied.

Subsequent event a significant development that takes place after the company's fiscal year-end but before the financial statements are issued

Subsidiary ledger record of a group of subsidiary accounts associated with a particular general ledger control account.

Successful efforts method requires that exploration costs that are known not to have resulted in the discovery of oil or gas be included as expense in the period the expenditures are made.

Sum-of-the-years'-digits (SYD) method systematic acceleration of depreciation by multiplying the depreciable base by a fraction that declines each year.

Supplemental schedules and tables reports containing more detailed information than is shown in the primary financial statements.

Symbolic intellectual property has usefulness to the customer that depends on the seller's ongoing activities, so it transfers a right of access; the seller recognizes revenue over the period of time the customer accesses the IP.

T-account account with space at the top for the account title and two sides for recording increases and decreases.

Tax basis of an asset or liability is its original value for tax purposes reduced by any amounts included to date on tax returns.

Technological feasibility established when the enterprise has completed all planning, designing, coding, and testing activities that are necessary to establish that the product can be produced to meet its design specifications including functions, features, and technical performance requirements.

Temporary accounts represent changes in the retained earnings component of shareholders' equity for a corporation caused by revenue, expense, gain, and loss transactions.

Temporary difference difference between pretax accounting income and taxable income and, consequently, between the reported amount of an asset or liability in the financial statements and its tax basis which will "reverse" in later years.

Time value of money money can be invested today to earn interest and grow to a larger dollar amount in the future.

Time-based methods allocates the cost base according to the passage of time.

Timeliness information that is available to users early enough to allow its use in the decision process.

Times interest earned ratio a way to gauge the ability of a company to satisfy its fixed debt obligations by comparing interest charges with the income available to pay those charges.

Trade discounts percentage reduction from the list price.

Trade notes payable formally recognized by a written promissory note.

Trademark (tradename) exclusive right to display a word, a slogan, a symbol, or an emblem that distinctively identifies a company, a product, or a service.

Trading securities debt securities the investor (usually a financial institution) acquires principally for the purpose of selling in the near term.

Transaction analysis process of reviewing the source documents to determine the dual effect on the accounting equation and the specific elements involved.

Transaction price the amount the seller expects to be entitled to receive from the customer in exchange for providing goods and services.

Transactions economic events.

Transportation-in freight-in; in a periodic system, freight costs generally are added to this temporary account, which is added to purchases in determining net purchases.

Treasury stock shares repurchased and not retired.

Troubled debt restructuring the original terms of a debt agreement are changed as a result of financial difficulties experienced by the debtor (borrower).

Trustee person who accepts employer contributions, invests the contributions, accumulates the earnings on the investments, and pays benefits from the plan assets to retired employees or their beneficiaries.

Unadjusted trial balance a list of the general ledger accounts and their balances at a particular date.

Understandability users must understand the information within the context of the decision being made.

Units-of-production method computes a depreciation rate per measure of activity and then multiplies this rate by actual activity to determine periodic depreciation.

Unrealized holding gains and losses gains and losses that arise from holding an investment during a period in which its fair value changes.

Valuation allowance indirect reduction (contra account) in a deferred tax asset when it is more likely than not that some portion or all of the deferred tax asset will not be realized.

Variable consideration transaction price is uncertain because it includes an amount that varies depending on the occurrence or nonoccurrence of a future event.

Verifiability implies a consensus among different measurers.

Vertical analysis expression of each item in the financial statements as a percentage of an appropriate corresponding total, or base amount, but within the same year.

Vested benefits benefits that employees have the right to receive even if their employment were to cease today.

Weighted-average interest method for interest capitalization, weighted-average rate on all interest-bearing debt, including all construction loans, is used.

With recourse the seller retains the risk of uncollectibility.

Without recourse the buyer assumes the risk of bad debts.

Work-in-process inventory products that are not yet complete in the manufacturing process.

Working capital differences between current assets and current liabilities.

Worksheet used to organize the accounting information needed to prepare adjusting and closing entries and the financial statements.

Subject Index

Notes: Page numbers followed by n indicate material in footnotes. General information about standards and standard-setting organizations may be found in this index. Specific standards and pronouncements are listed in the Accounting Standards Index.

A

Abandoned property, 729
Abbott Laboratories, Inc.
  business segment information disclosure, 139
  discontinued operations disclosure, 177, 181
  EPS disclosures on income statement, 184–185
  geographic area sales disclosure, 140
  statement of earnings, 167
AbbVie Inc., 546
Abercrombie & Fitch Co.
  change in inventory method, 507
  disclosure of change in inventory method, 1178
  shareholders' equity on balance sheet, 1051–1052
Abnormal shortages (inventory), 474
ABO. *See* Accumulated benefit obligation
Accelerated cost recovery system (ACRS), 616n23
Accelerated depreciation methods
  declining balance methods, 575–576
  prescribed by tax code, 929n1
  sum-of-the years'-digits method, 576–577
  switch to straight-line, 577
  use on tax return, 964
Accenture, 1006
Account relationships, 50–51
Accountants, ethical dilemmas for, 17
Accounting changes, 1171–1185
  in accounting estimate, 1180–1182
  in accounting principle, 1172–1173
  approaches to reporting, 1172–1173
  error correction. *See* Accounting errors, correction of
  IFRS *vs.* GAAP, 1192
  interim reporting of, 206
  in reporting entity, 1182–1183
  reporting on income statement, 183
  summary, 1172, 1184
Accounting equation, 48–50, 1048
  balance sheet and, 114
  debits and credits and, 51
  internal events and, 63
  transaction analysis, 53
Accounting errors, 1183–1184
  correction of, 1186–1192
    error affecting previous financial statements but not net income, 1187–1188
    error affecting prior year's net income, 1188–1192
    error discovered in same reporting period, 1187
  on income statement, 183–184
  prior period adjustments, 1186–1187
  description and examples, 1172
  IFRS *vs.* GAAP, 1192
  intention and, 1186n12
  inventory errors, 482–485
    error discovered the following year, 483–484
    error discovered two years later, 484
    visualizing the effect of, 483
  materiality of, 21, 125
  summary, 1184
Accounting estimate. *See* Change in accounting estimate; Estimates
Accounting principles
  change in. *See* Change in accounting principles
  full-disclosure principle, 30, 122
  GAAP. *See* Generally accepted accounting principles (GAAP)
  principles-based accounting standards, 16
  realization principle, 27, 278
Accounting Principles Board (APB), 9
Accounting Principles Board Interpretations, 9
Accounting Principles Board Opinions (APBOs), 9
Accounting Principles Board Statements, 9
Accounting process, 46–108. *See also* Accounting processing cycle
  basic model, 48–51
    accounting equation, 48–50
    account relationships, 50–51
  cash basis to accrual conversion, 80–82
Accounting processing cycle, 51–63
  adjusting entries, 63–70
    accruals, 67–69
    estimates, 69–70
    illustration, 69
    prepayments, 64–67
  closing process, 76–78
  financial statements, 72–76. *See also* Financial statements, preparation of
  illustration of, 55–61
  journal entries, 55–60
  ledger accounts, 60–61
  overview of, 52–55
  steps of, 52
  trial balance, 61
Accounting Research Bulletins (ARBs), 9
Accounting standards
  hierarchy of standard-setting authority, 10
  principles- *vs.* rules-based, 16–18
Accounting Standards Codification System (ASC), 11
Accounting standards development, 8–14
  historical perspective, 9–12
    codification, 10–11
    converging U.S. and international standards, 11–12
    early U.S. standard setting, 9
    FASB, 9–10
    international standards, 11–12
  standard-setting process
    due process, 12–13
    politics in, 13–14
Accounting Standards Updates (ASUs), 10, 14
Accounts, 50
Accounts payable
  defined, 118, 719
  recording payment of, 56
Accounts receivable
  as current assets, 115
  earnings quality, 371–372
  factoring, 363, 722
  initial valuation of, 344–349
    discounts, 344–346
      payment not within discount period, 345–346
      payment within discount period, 345
    sales returns, 346–349
      actual returns, 348
      estimated returns, 348–349
  measuring and reporting, summary, 354
  net *vs.* gross, in turnover calculation, 198n34
  performance obligations, revenue recognition and, 302
  receivables management, 370–372
  securitization, 364
  subsequent valuation of, 350–355
    allowance method (GAAP), 350–352
    direct write-off method, 350
    estimating allowance for uncollectible accts, 352–355
Accounts receivable aging schedule, 352
Accretion expense, 520
Accrual accounting model, 7–8
  adjusting entries, 64
  cash basis accounting compared, 7
  conversion from cash basis to, 80–82
  financial statement elements and, 23–24
  in GAAP, 81n7
  income tax reporting and, 911
  operating activities and, 190
  revenue recognition, 26
Accruals
  accrued liabilities, 67–68
  accrued receivables, 68–69
  of contingent liability for warranties, 735
  defined, 67
  reversing entries for, 86
Accrued interest, 805
Accrued interest payable, 724
Accrued liabilities, 723–726
  accrued interest payable, 724
  adjusting entries for, 67–68
  annual bonuses, 726
  balance sheet classifications, 118–119
  salaries, commissions, and bonuses, 724–726
Accrued receivables, 68–69
Accumulated benefit obligation (ABO), 983–984
Accumulated deficit account, 120
Accumulated depreciation, 65, 575, 612
Accumulated other comprehensive income (AOCI), 120, 187–188
  available-for-sale (AFS) securities, 657–658
  intraperiod tax allocation, 947
  pension plans and, 1004
    gain or loss on pension assets, 999
    prior service cost, 995–996, 1000
  reporting on balance sheet, 1004, 1050
  shareholders' equity, 1050–1051
Accumulated postretirement benefit obligation (APBO), 1011
Acid-test (quick) ratio, 131–132, 343, 745
ACL. *See* Allowance for credit losses, 361
Acquired in-process research and development, 546
Acquisition of companies
  asset retirement obligations (AROs), 520
  changes in reporting entity, 1182–1183

Acquisition costs
contract acquisition costs, 525
defined, 518
intangible assets by category, 515
natural resources, 515, 518
property, plant, and equipment by category, 515
Activity-based methods of depreciation, 574, 577–578
Activity ratios, 197–198
asset turnover, 197, 199–203
average collection period, 198, 201, 202, 370, 371
average days in inventory, 198, 428
fixed-asset turnover, 530
inventory turnover, 197, 198, 198n35, 201–203, 428
receivables turnover, 198, 201–203, 343, 370, 371
Actual interest limitation, 540
Actuary, 981
Additions, 612
Adelphia Communications, 15
Adidas, 298
Adjusted market assessment approach, 297
Adjusted trial balance, 52, 54, 69, 70, 76, 85
Adjusting entries, 63–70
accruals, 67–69
in cash-to-accrual conversion, 80–82
to correct book balance, 481
defined, 63
for depreciation expense in equity method, 677
end-of-period, 84, 88n8
estimates, 69–70
illustration, 69
prepayments, 64–67
deferred revenues, 65–66
prepaid expenses, 64–65
recording, 54
visualizing, with T-accounts, 81
worksheet used for, 84–85
Adobe Systems, 544–545
Advance collections, 726–729
Advance payments, leases, 873
Advanced Drainage Systems, 1188
Adverse opinion in auditor's report, 130
Advertising expenses, prepaid, 64
Aetna Inc., 778
Agent
defined, 294
examples, 294–295
seller as, 294–295
Agrium Inc., 598
AIA. *See* American Institute of Accountants
AICPA. *See* American Institute of Certified Public Accountants
AIG, A-3, A-3n2
Air France-KLM (AF), C-0–C-4
case, 12, 44, 108, 165, 236, 274, 337, 400, 457, 510, 569, 642, 715, 770, 832, 906, 974, 1045, 1102, 1168–1169, 1213, 1293
equity, 1048
ownership interest in WAM (Amadeus), 680
Alibaba, 1054
Allocation base, 573
Allocation method
cost allocation. *See* Cost allocation (assets)
defined, 573
effective interest method. *See* Effective interest method
intraperiod tax allocation. *See* Intraperiod tax allocation
Allowance method of accounts receivable, 350–352
accounts deemed uncollectible, 351
previously written-off accounts reinstated, 351–352
recognizing allowance for uncollectible accounts, 351
Allowance for credit losses (ACL), 361
Allowance for sales returns, 346n8
Allowance for uncollectible accounts
defined, 350
recognizing at end of first year, 351
recognizing at end of second year, 353
Almanac of Business and Industrial Financial Ratios, The (Troy), 1099
Alphabet, Inc., 24, 335
classes of shares, 1056
R&D costs, 541
Amazon.com
cash flow, 1217
as corporation, 1052
principal *vs.* agent, 295, 320, 334
revenue recognition, 279, 301
statement of cash flows, 1254–1255
Amazon Prime, 291
American Accounting Association, 17n26, 340n2
Pathways Commission, 4
American Eagle Outfitters, 469–470
American Institute of Accountants (AIA), 9
American Institute of Certified Public Accountants (AICPA), 9, 1043
Ameris Bancorp, 739
AmerisourceBergen, 442
Amgen, Inc.
disposition and depreciation, 639–640
patents, 522
Amiram, D., 781n3
Amortization
allocation method, 591
change in accounting estimate, 596–597, 1181–1182
changes in, 597–598
defined, 572, 591
of discount for debt investment, 648
error correction, 598–600
intangible assets, 117
assets not subject to, 593–595
assets subject to, 591–593
pension expense
net loss or net gain, 995–996
prior service cost, 994–995
right-of-use asset, 843–844, 849–850
sales-type lease with residual value, 861–862
Amortization base, 573
Amortization schedules
bonds
discount on bonds, 779, 786, 1241–1242
premium on bonds, 781
premium and discount compared, 782
straight-line method, 783
zero-coupon, 780
debt investment, 648, 649
installment notes, 789
leases, 840–841, 843, 849, 861–862
long-term notes payable
of right-of-use asset, 849–850
for troubled debt restructuring, 810
Anheuser-Busch InBev, 645
Ann Taylor, 874
Annual bonuses, 726
Annual financial statements. *See* Financial statements
Annual report, 122
Annual report disclosures, 122–130
auditor's report, 128–130
compensation of directors and executives, 126–128
disclosure notes, 122–125
management's discussion and analysis, 125
management's responsibilities, 125–126
Annuities
annuity due, 248–254
future value of annuity due, 250
future value of ordinary annuity, 248–249
ordinary annuity, 248–254
present value techniques
valuation of installment notes, 260–261
valuation of long-term bonds, 259–260
valuation of long-term leases, 260
valuation of pension obligation, 261–262
Annuity due
defined, 248
future value of, 250
present value of, 252–253
Antidilutive securities, 1128–1131
convertible securities, 1130
options, warrants, rights, 1128–1130
AOCI. *See* Accumulated other comprehensive income
APB. *See* Accounting Principles Board
APBO. *See* Accumulated postretirement benefit obligation
APBOs. See Accounting Principles Board Opinions
Apple Inc., 21, 38
audit report, 15
iTunes licensing, 298
principle *vs.* agent, 320
as publicly held corporation, 1054
restricted stock units, 1106
revenue recognition for gift cards, 301
shareholders' equity, 1048
stock buyback program, 1062
stock option expense, 1109
Applebee's International, Inc., 35
ARBs. See *Accounting Research Bulletins*
AROs. *See* Asset retirement obligations
Arthur Andersen, 15
Articles of incorporation, 1055
ASC. *See* Accounting Standards Codification System
Assessments, unasserted, 740–741
Asset(s)
in accounting equation, 48–49, 53
biological, 590–591
book value, 1074
change in depreciation method for newly acquired, 1182
costs to be capitalized, 516–525
intangible assets, 521–525
property, plant and equipment, 516–521
current, 74, 114–116 (*See also* Current assets)
accounts receivable, 115
cash and cash equivalents, 114–115
inventory, 115–116
prepaid expenses, 116
short-term investments, 115
defined, 24, 114
depreciable, gains and losses on sale of, 584–585
disposition of, 583–584
donated, 528–530
error in recording as expense, 1189
identified, for lease, 869
impairment of value (*See* Impairment of value)
leasing (*See* Leases)
long-lived, revenue-producing, 513
long-term, 116–118 (*See also* Long-term assets)
intangible, 117
investments, 116
other, 117–118
property, plant, and equipment, 116–117
lump-sum purchase of, 525–526
monetary *vs.* nonmonetary, 531n15
pension plan, 990–992
qualifying, for interest capitalization, 536
residual, 861
retirement of, 585
right-of-use (*See also* Leases; Right-of-use asset)
Asset exchanges, 531–534
fair value in, 531–532
lacking commercial substance, 532–534
gain situation, 533
loss situations, 534
Asset impairments, 175
Asset/liability approach, 32

Asset retirement obligations (AROs)
  capitalizing cost of, 518–521
  disclosure of, 521
Assets held for sale
  decision maker's perspective, 585
  impairment of value, 608–609
Asset turnover ratio, 197
Asset valuation
  accounts receivable (*See* Accounts receivable)
  bargain purchase situations, 524
  biological assets, 590–591
  book value (*See* Book value)
  costs to be capitalized, 516–525
   intangible assets, 521–525
   property, plant, and equipment, 516–521
  deferred payments, 526–528
  donated assets, 528–530
  equity securities issuances, 528
  fair value (*See* Fair value option)
  impairment of (*See* Impairment of value)
  initial valuation, 516, 525
  long-term leases, 260
  lump-sum purchases, 525
  noncash acquisitions, 526–534
  residual (*See* Residual value)
  revaluation in quasi reorganizations, 1079–1080
  software development costs, 544–545
  types of assets, 514–515
Assigning accounts receivable, 362
  assignee, 363
  assignor, 363
AstroNova Inc.
  shareholders' equity, 187
  statement of comprehensive income, 186
ASUs. See Accounting Standards Updates
AT&T, 680
AU Optronics, 768
Audit Committee, financial statement responsibility, 127
Auditor
  role in financial reporting, 14–15
  Section 404 requirements and, 15–16, 340
  user need for full disclosure and, 22, 125
Auditor's report, 128–130
Autodesk, 371
Available-for-sale (AFS) securities, 649, 656–661
  adjusting to fair value (2021), 657–658
  comprehensive income, 655–657
  credit losses for, 690–692
  financial statement presentation, 661
  impairment of, 660
  income taxes and, 663
  Intel, 684n30
  reporting, 661
  sale of, 658–660
  unrealized holding gains/losses and, 657
Average accumulated expenditures, 536–537
Average collection period, 198, 201, 202, 370, 371
Average cost method of inventory valuation, 414–416
  change in inventory method to, 480–481
  periodic average cost, 414–415
  perpetual average cost, 415–416
  weighted average unit cost, 415
Average days in inventory, 198, 428
AveXis, Inc., 1060
Avon Products, Inc., 391–392
AVX Corporation, 349

B

Bad debt(s). *See also* Accounts receivable, subsequent valuation of; Uncollectible accounts
  allowance method to account for, 1231–1232
  CECL model and, 355, 651, 688–689
  earnings quality and, 371–372
  estimating, 69, 183
  IFRS vs. GAAP, 361
  income smoothing and, 173
  increase in allowance for, 198n34
  notes receivable and, 360
  receivables management and, 370–371
  receivables turnover ratio and, 198
  reporting, 21
  sale without recourse and, 364, 368
  sale with recourse and, 364, 368
  trouble with, 339
Bad debt expense, 302, 352–354, 371, 378–379, 734
Balance sheet, 5. *See also* Statement of financial position
  accounts used in preparing SOCF
   bonds payable, 1240
   buildings and equipment, 1239
   common stock, 1242–1243
   completing the spreadsheet, 1244–1246
   land, 1239
   retained earnings, 1243
   short-term investments, 1238
  classification of elements, 113–121
   assets, 114–118
   liabilities, 118–119
   shareholders' equity, 119–120
  comparative, 1226
  deferred tax liabilities on, 915, 941
  defined, 73, 112
  disclosure notes for revenue recognition, 302
  held-to-maturity (HTM) securities on, 651
  illustration, 74
  limitations, 113
  preparation of, 73–75
  presentation, GAAP *vs.* IFRS, 120
  revenue recognition, at term *vs.* completion of contract, 310–311
  revising, for change in accounting principle, 1175–1176
  share repurchase effect on, 1065
  T-accounts, 1262–1263
  trading securities on, 656
  usefulness of, 112–113
Balance sheet accounts, 60
Balance sheet approach, 352
Bank of America, 363
Bank loans
  cash flow information on application for, 1215
  cash restrictions with, 342
  debits/credits involved in, 50
  fair value of, 801
  interest on, 721
  journal entry for, 54
  line of credit, 720–721
  short-term, 720–721
  transaction analysis of, 49
Bank reconciliation
  as cash control, 373–376
  reconciling items, 374
Bankruptcy remote, 364n19
Barcode scanning, 405–407
Bargain purchase, 524
Bargain purchase option (BPO), 838, 865
Basic accounting equation, 1048
Basic EPS (earnings per share), 184, 1119, 1123. *See also* Earnings per share
  earnings available to common shareholders, 1122–1123
  issuance of new shares, 1120
  reacquired shares, 1122
  stock dividends, 1120–1121
  stock splits, 1120–1121
Bay Isle Financial, 485
Bear Stearns, A-3, A-3n2, 719
Bearer bonds, 775
B Communications LTD, 768
Beneficial conversion feature, 796–797
Beresford, Dennis, 1109
Berkshire Hathaway, A-2
Best Buy Co., Inc.
  balance sheet, 148
  earnings per share, 1163
  income statement, 149
  inventories, 404
  recording lease transactions, 847
  Samsung's relationship with, 296
  summary compensation table, 127–128
  unredeemed gift card liability, 728
Bill-and-hold arrangements, 299–300
Bill and hold strategy, 372
Billings on construction contract, 305–306
Billings in excess of CIP, 311
Biological assets, 590–591
Black–Scholes model, 1143
Blair, Donald W., 126–127
Bloomberg, 1087
Board of directors, 1055
Boeing Corporation
  indefinite-lived intangible assets, 594
  lease payments, 269, 274
  voluntary stock option expensing, 1109
Bond discount, 1240–1241
Bond indenture, 774–775
Bond premium, 1240–1241
Bonds. *See also* Debt investments
  bond indenture, 774–775
  convertible, 795–798, 1126–1127
  defined, 774
  with detachable warrants, 798–800
  effective interest method, 778–784
  fair value of, 801–804
  issuance
   issued at discount, 777–778, 1240–1241
   issued at face amount, 1240
   issued at premium, 780–782, 1241
   issued between interest dates, 805–806
    bonds issued at discount, 806–807
    bonds issued at face amounts, 805–806
   recording at, 775–778
  long-term, valuation of, 259–260
  purchased at discount, 646
  sale of
   at discount, 647
   at premium, 647
   selling price determination, 775–777
  zero-coupon, 780
Bonds payable, balance sheet account, 1240
Bonus, annual, 725
Bonus depreciation, 616
Booking Holdings Inc., 334
Books A Million, Inc., 1178–1179
Book-tax difference, 915
Book (carrying) value
  accounts receivable, 378
  asset
   dispositions, 583–584, 586
   fair value higher than, 186, 587
   fair value less than, 587, 609
   fair value not determinable, 532, 534
   goodwill, 604n19, 605–606
   held for sale, 585, 608–609
   lack of commercial substance, 532–534
   recording asset at, 534
   recoverability of, 601, 603
   retirement, 585
  asset exchanges, 531–532
  bonds, 259
  company, 111, 119
  defined, 113
  depreciation and
   declining-balance method, 575–576
   partial year, 582–583
   straight-line, 575, 577
   sum-of-the years' digits, 577
   units-of-production, 578

Book (carrying) value—*Cont.*
derivatives, A-10, A-20–A-21
discontinued operations, 179–181
fixed-asset turnover ratio, 530
IFRS vs. GAAP, 587–588
impairment of value and, 600–606
intangible assets, 530n14, 594, 604
market value and, 113, 547
notes receivable, 359–360, 364
temporary book-tax difference, 915
Book value method in bond conversion, 798
Boston Scientific Corporation, 301
Bowen, R. M., 1174n4
BPO. *See* Bargain purchase option
Bribery, 125
Brigham, Eugene, 427n15
British Telecommunications Plc (BT), 1291
Broadcom Corporation, 1138
Brown, F. Donaldson, 200n36
Brown, K. L., 308n19
Brown, Robert Moren, 421n12
Buffett, Warren, A-2, A-7
Buildings. *See also* Property, plant, and equipment
capitalizing cost of, 518
typical acquisition costs, 515
Buildings and equipment balance sheet account, 1239
Burger King, 329
Burr, Chuck, A-2
BusinessCash.Com, 363
Buybacks. *See* Share buybacks

C

CA, Inc., 371, 544–545
Calendar year, 25
Callable bonds, 775, 798
Callable debt, 729–730
Callable preferred shares, 1057
Campbell, H., 173n4, 173n5
Campbell Soup Company, 25, 177
CAP. *See* Committee on Accounting Procedure
Capital, working (*See* Working capital)
Capital budgeting decisions, 530
Capital gains, 1062
Capitalization of costs
intangible assets, 521–525
copyrights, 522–523
franchises, 523
goodwill, 523–525
patents, 522
trademarks, 523
property, plant, and equipment, 516–525
asset retirement obligations, 518–521
building costs, 518
equipment costs, 516–517
land costs, 517
land improvements, 518
natural resources costs, 518
self-constructed assets, 535–540
interest capitalization, 536–540
overhead allocation, 536–537
software development costs, 543, 544–545
Capitalization period, 536
Capital leases, 123. *See also* Finance leases; Sales-type leases
Capital market efficiency, 421n12
Capital markets, 6
Capital structure, 133
complex, 1123
simple, 1119, 1123
Carrying value/amount, 1179. *See also* Book (carrying) value
Carslaw, Charles, 1256n18
Cash. *See also* Cash and cash equivalents
as current asset, 114–115
defined, 339
holding, risk *vs.* return, 343
petty, cash controls for, 376–377
restricted, 341–343
statement of cash flows and, 1219
Cash-basis accounting
accrual accounting compared, 7–8
conversion to accrual basis from, 80–82
defined, 7
net operating cash flow and, 7–8
operating activities and, 190
Cash and cash equivalents, 339–343. *See also* Cash equivalents
as current assets, 114–115
decision makers' perspective on, 343
internal control of, 340–341
cash disbursements, 341
cash receipts, 341
International Financial Reporting Standards, 342
restricted cash and compensating balances, 341–343
Cash controls
bank reconciliation, 373–376
petty cash, 376–377
Cash disbursements, 341
Cash disbursements journal, 88
Cash discounts, 345
Cash dividends, 1069–1070
alternative method of recording, 58
double taxation and, 1054
financing activities and, 194
investment revenue received as, 678
operating activities and, 679, 679n25
on preferred shares, 1057, 1070
Cash equivalents. *See also* Cash and cash equivalents
commercial paper as, 1238
as current assets, 114–115
defined, 340, 1219
investing activities and, 193
for liability assumption, 28
restricted stock used and, 1106, 1146
securities considered as, 123
statement of cash flows and, 190, 1219, 1223, 1254
Cash flow(s). *See also* Statement of cash flows (SCF)
bankruptcy and, 1217
cash inflows, 1216–1217
cash outflows, 1216–1217
categories of, 1217
classifying
financing activities, 194–195
IFRS *vs.* GAAP, 1245
investing activities, 193–194
noncash investing and financing activities, 195
operating activities, 190–193
from financing activities, 1223–1224
from investing activities, 1222–1223
negative, 7n6
net operating, 7, 80
from operating activities, 1219–1221
ratios, 1255–1256
statement of (*See* Statement of cash flows)
undiscounted *vs.* discounted estimates, impairment loss and, 602
Cash flow hedges, A-12–A-13
derivative designated as, 1050
forward contract, A-15–A-16
interest rate swap, A-13–A-15
Cash flow per share ratio, 1256
Cash flow risk
with forward contracts, A-5
with futures, A-4
hedging against, A-3
Cash flow to sales ratio, 1256
Cash flow statement
held-to-maturity (HTM) securities on, 651
trading securities on, 656
Cash to income ratio, 1256
Cash rebates, 737
Cash receipts, 341
Cash receipts journal, 88, 89–90
Cash return on assets (ROA), 1256
Cash return on shareholders' equity, 1256
Cash surrender value (life insurance), 688
Caterpillar, Inc.
self-constructed assets, 535
supplemental LIFO disclosures, 449
Caylor, Marcus, 745n31
CECL (current expected credit loss) model
allowance for uncollectible accounts, 352, 355
credit losses on HTM investments, 651
debt investments, 688–689
notes receivable, 360–361
troubled debt restructuring, 377
Certified Public Accountant (CPA), 9, 15, 17
CFA Institute, 10n12
Champions Oncology, Inc., 125
Change in accounting estimate, 183, 1180–1182
asset cost allocation, 596–597
depreciation, amortization, and depletion methods, 183, 1178, 1181–1182
loss projected for entire long-term contract, 312–314
periodic loss on profitable project, 311–312
summary, 1172
Change in accounting principle, 182, 1173–1179
change in estimate achieved by, 596
decision makers' perspective, 1173–1174
defined, 1173
income statement and, 182
income tax effects of, 1177
interim reporting of, 206
justification for, 182
lack of consistency due to, 129
mandated changes, 182
modified retrospective approach, 182, 1178, 1178
prospective approach, 182, 1178–1179
depreciation, amortization, and depletion methods, 1179
mandated by authoritative literature, 1179
retrospective application impractical, 1178–1179
retrospective approach, 182, 1172–1178
adjusting accounts for change, 1176–1177
disclosure notes, 1177–1178
revising comparative financial statements, 1175–1176
review, 1172
Change in inventory method
change to LIFO method, 481–482
earnings quality and, 485
IRS regulations, 482
most inventory changes, 480–481
Change in reporting entity, 1182–1183
Change in revaluation surplus, 186
Checks
checks outstanding, 374
payments made by, 339, 341
Chevron Corporation
depreciation and depletion, 640
oil and gas accounting disclosure, 549
Chicago Board of Trade, 685n34
Chicago Mercantile Exchange, A-4
Chipotle Mexican Grill, Inc., 547
CIP. *See* Construction in progress (CIP) account
Cisco Systems, Inc.
bad debts, 397
disclosure note excerpts, 1098
inventory write-offs by, 485
investments in securities
available-for-sale, 661–662
reporting shareholders' equity, 1093
Citigroup Inc., 803
Claims and assessments, 740–741
Cliff vesting, 1114
Closing entries, 54, 84–85
Closing process, 76–78, 84–85
Cloud computing arrangements, 545
Coca-Cola Company, The, 162
available-for-sale securities, 694, 695

cash flow information, 1215
debt to equity ratio, 793–794
estimated value of trademark, 521, 523
reporting investments, 645
subsequent events disclosure note, 740
taxes on unrepatriated foreign earnings, 931
trading securities, 694
zero-coupon bonds, 780
Codes of professional conduct, 17
Codification, 10–11
Cohan, William D., 719
Collateral
accounts receivable as, 362–363, 722
disclosure note for, 792
inventory as, 722
note receivable as, 365
for short-term loan, 722
Collins, Daniel W., 15n21
Comcast Corporation, 723
Comiskey, E. E., 685n31
Commercial paper
as cash equivalent, 115, 340, 1219, 1238
as current liability, 719
defined, 723
issuing, as temporary financing method, 723
special purpose entity and, 364
Commercial substance
requirement, 532–533
gain situation, 533
lack of commercial substance, 532
loss situation, 534
revenue recognition and, 287n9, 288, 303
Committed line of credit, 721
Committee on Accounting Procedure (CAP), 9
Committee of Sponsoring Organizations (COSO), 127, 129, 340
Commodity, A-4
Common shares, 1055–1056. *See also* Earnings per share (EPS)
accounting for issuance of, 1053–1060
for cash, 1058
issue costs, 1060
more than one security at single price, 1059–1060
for noncash consideration, 1059
no-par shares issued for cash, 1059
recording, 54
balance sheet account, 1242–1243
classes of, 1056
conversion of bonds to (*See* Convertible bonds)
fundamental rights, 1055–1056
par value concept, 1056–1058
potential,
as antidilutive securities, 1129, 1130
defined, 1119
as diluting EPS, 1123
effect on earnings per share, 1135
RSUs as representing, 1133
reporting in statement of cash flows, 1229–1230
return on equity and, 201
Common stock, 1055. *See also* Common shares
Community Health Systems
bad debt trouble, 339
goodwill impairment disclosure, 606
Compaq Computer Corporation, 523
Comparability, 22
Comparative financial statements
defined, 130
earnings per share reported in, 1119
prior period adjustment reported in, 483n14, 599n15, 1137
retrospective approach and, 1172–1176
revenue recognition and, 273
revising, for change in inventory method, 480–481
Compensating balances, 342, 721n6
Compensation of directors and executives, annual report disclosures, 126–128
Competitive disadvantage costs, 23
Completed contract method, 304
Complex capital structure, 1123
Composite depreciation method, 585–588
Compound interest, 240
Comprehensive income, 73
accumulated other [*See* Accumulated other comprehensive income (AOCI) or loss]
available-for-sale securities and, 656–657
defined, 24, 168, 185
flexibility in reporting, 185–186
GAAP *vs.* IFRS, 186
illustration, 186
other comprehensive income [*See* Other comprehensive income (OCI) or loss items]
pension plans and, 1003–1004
Computers in accounting, 48
Conceptual framework, 10, 18–31
asset/liability approach, 32
disclosure, 25, 30, 31
FASB project to improve, 31
financial statement elements, 23
GAAP *vs.* IFRS, 19, 32
measurement, 25, 27–30
objective of financial reporting, 20
qualitative characteristics of reporting information, 20–23
recognition, 25–26
expense, 25–26
general criteria, 25
revenue, 25
revenue/expense approach, 32
underlying assumptions, 23–25
economic entity assumption, 23–24
going concern assumption, 24–25
monetary unit assumption, 25
periodicity assumption, 25
Concession, creditor to debtor, 377n29
Confirmatory value, 21
ConocoPhilips, 590
Conservatism, 22
Consideration, variable, 290–293
Consignment
defined, 409
revenue recognition and, 301
Consistency, 22, 129, 1171
Consolidated financial statements, 674, 676–677
Construction costs. *See also* Self-constructed assets
allocation of revenue to each period, 309
revenue recognition for, 305–306
Construction in progress (CIP), 305–306, 311
Consulting services, related-party transactions, 124
Consumer Price Index (CPI), 431, 857, 858
Contingencies, 732–744
controversies in accounting for, 743
gain contingencies, 742
loss contingencies, 732–742
litigation claims, 738–739
product warranties and guarantees, 735–736
subsequent events, 739–740
unasserted claims and assessments, 740–741
Contingent loss, 741
Contingently issuable shares, 1134
Contract(s). *See also* Long-term contracts; Performance obligations
bond, 259, 775, 805–806
cloud-computing arrangements, 545
construction contracts, 281 (*See also* Long-term contracts)
contract acquisition costs, 525
current liabilities, 718
debt contracts, 688
deferred payment, 522
defined, 287
derivatives (*See* Derivatives)
disclosure of, 30
financial instrument as, 685
financing component of, 359
forward, A-5
franchise, 523–525
futures, A-4–A-5
lease (See Leases)
long-term liabilities, 119
loss contingencies in, 741–742
modifications, 288
multiple deliverables (See Multiple-deliverable arrangements)
partnerships and, 1053
performance obligations in, 278–280, 286, 344
customer options for additional goods or services, 289–290
distinct or separately identifiable, 283–284
prepayments, 289
sales contracts, 348n9
warranties, 289
purchase commitments, 487–488
contract period extending beyond fiscal year, 487–488
contract period within fiscal year, 488
R&D performance, 545
revenue recognition principle and, 278–280 (*See also* Revenue recognition)
shareholder, 1057
termination, 283
warranty contracts
extended, 737
quality-assurance, 735
Contract acquisition costs, capitalization of, 525
Contract asset, 302
Contract liability, 302
Contra revenue account, sales returns as, 294
Contributory pension plan, 980
Control
of good or service, revenue recognition and, 279, 280
surrender of, with receivables financing, 367–368
of use of identified asset, lease and, 870–871
of voting stock, 674
Conventional retail method, 472–473
Convergence efforts (FASB/IASB), 12
conceptual framework, 19, 30
convertible debt, 797
fair value option, 30
impediments to, 420
International Accounting Standards Board (IAS), 11
inventory cost flow assumptions and, 420
Convertible bonds, 795–798, 1126–1127
conversion option exercised, 797–798
defined, 775
IFRS *vs.* GAAP, 797
incremental effect of conversion, 1130–1131
induced conversion, 798
as potential common shares, 1119, 1123
Convertible preferred stock, 1127–1128
Convertible securities
convertible bonds, 1126–1127
convertible preferred stock, 1127–1128
EPS and, 1130
incremental effect of conversion, 1132, 1136
order of entry for multiple, 1130–1131
Cook, Tim, 1062
Cookie jar accounting, 371, 1174
Cooking the books, 1174
Copyright
capitalizing/amortizing, 525
as intangible asset, 117, 521–523
as Microsoft's revenue-producing assets, 513–515
Corporate charter, 1055

Corporate organization
disadvantages of, 1054
as dominant business organization form, 6
ease of raising capital, 1053–1054
hybrid organizations, 1054–1055
limited liability, 1053
Model Business Corporation Act, 1055, 1056, 1058
Revised Model Business Corporation Act, 1063n13, 1068n17
shareholders' equity and, 1052–1056
types of, 1054
Corporate takeover, 1063n11
Correction of an error. *See* Accounting errors, correction of
COSO. *See* Committee of Sponsoring Organizations (COSO)
Cost allocation (assets)
assets held for sale, 608–609
change in depreciation, amortization, or depletion method, 597–598
change in estimate, 596–597
error correction, 598–600
impairment of value (*See* Impairment of value)
measuring, 572–574
allocation base, 573–574
allocation method, 574
service life, 573
overview, 571–572
subsequent expenditures, 611–614
additions, 612
costs of defending intangible rights, 613–614
improvements, 612–613
rearrangements, 613
repairs and maintenance, 611
Cost approaches to fair value, 29
Cost-to-cost ratio, 308
Costco Wholesale Corporation
disclosure of inventory method, 479
inventory, 115
profitability analysis illustration, 201–203
Cost effectiveness, 22
Cost flow methods (of inventory estimation), 471–473
approximating average cost, 471–472
conventional retail method, 472–473
LIFO retail method, 472–473
Cost of goods sold
allocation of, 413–414
average cost method, 413, 471–472
change in accounting principle and, 1185
on comparative balance sheet, 1175–1177
defined, 404
depletion included in, 589
effect of inventory errors, 483–485
expense recognition, 27
freight-in on purchases and, 410
gross profit method, 467–468
gross profit ratio, 428
income statement account, 169, 1234
inventory method changes, 481–482
inventory write down as part of, 462
inventory turnover ratio, 198, 201
journal entries, 57, 59, 408, 847
LIFO retail method, 473–474
manufacturing company, 405
periodic inventory system and, 57, 407–408, 412, 415, 467
FIFO, 416
LIFO, 418, 421–424, 429
perpetual inventory system and, 405–406, 412, 415
FIFO, 417
LIFO, 418–419
purchase returns, 410
retail inventory method, 469
sales returns and, 347, 348, 349, 409–410
sales revenue and, 172
sales-type lease and, 862–863
statement of cash flows, 1221, 1225, 1226, 1229, 1234, 1249
voluntary changes in accounting principles and, 182
Cost indexes, dollar-value LIFO (DVL), 431
Cost method of accounting for treasury stock, 1065
Cost-to-retail percentage, 469
Coupon(s), 738
Coupon bonds, 775
Coupon rate, bond, 774
CPA. *See* Certified Public Accountant
CPI. *See* Consumer Price Index
Credit-adjusted risk-free rate, 519
Credit line, 721
Credit losses, 350. *See also* Bad debts; Uncollectible accounts
Creditors, 4, 5
cash flow perspective, 7
change in accounting methods, 1174
of corporations, 1053–1054
diluted EPS and, 1123
interest in company assets, 1047
paid-in capital and, 1056, 1058
pension information analysis, 1005, 1016
risk analysis by, 133–134
share-based compensation plans and, 1117
statement of cash flows, 1215–1216
Credit risk, bonds and, 802
Credits
accrued liabilities, 67–68
accrued receivables, 68
adjusting entries, 69–70
deferred revenues, 65–66
prepaid expenses, 64–67
on T-account, 50–51
Credit sales, 67, 343
Culture Change (Burr), A-2
Cumulative preferred shares, 1057
Current assets, 74, 114–116
accounts receivable, 115
cash and cash equivalents, 114–115
inventory, 115–116
liquidity ratios and, 131
prepaid expenses, 116
short-term investments, 115
Current cost, 27–28
Current expected credit loss model. *See* CECL (current expected credit loss) model
Current liabilities, 74, 112, 118–119. *See also* Liabilities
accrued liabilities, 67–68, 723–726
accrued interest payable, 724
salaries, commissions, and bonuses, 724–726
from advance collections, 726–729
collections for third parties, 728–729
deposits and advances from customer, 726–728
gift cards, 728
balance sheet classification, 74, 114
characteristics of, 718
decision makers' perspective, 744–745
defined, 719
discontinued operations, 181
noncurrent liabilities *vs.*, 729–731
current maturities of long-term debt, 729
obligations callable by creditor, 729–730
obligations expected to be refinanced, 730–731
open accounts and notes
accounts payable, 719
commercial paper, 723
short-term notes payable, 720–723
trade notes payable, 720
payroll-related liabilities, 747–748
working capital and, 131–132
Current market value, 28. *See also* Fair value; Fair value option (FVO)
Current maturities of long-term debt, 119, 729
Current ratio, 131–132, 343, 744–745
Current receivables, 343–373
accounts receivable, 343–373 (*See also* Accounts receivable)
subsequent valuation, 350–355
decision makers' perspective
earnings quality, 371–372
receivables management, 370–371
financing with, 362–369
disclosures, 367–368
illustration, 368
sale of receivables, 363–365
sale with recourse, 364–365
sale without recourse, 364
secured borrowing, 362–363
transfer as sale *vs.* secured borrowing, 366–367
transfers of notes receivable, 365–366
notes receivable, 356–361. See also Notes receivable
Current selling price, 469
Customer(s)
advance payments from, 727–728
major, reporting information by, 140
Customer options
for additional goods or services, 289–290
stock options compared with, 289n10
CVS Health Corporation
acquisition of Aetna Inc., 778
disclosure of change in inventory method, 481
purchase of Target pharmacy, 714
statement of cash flows, 1245–1246, 1255

D

Data analytics, 42
Data analytics cases
accumulated depreciation to fixed assets, 640–641
asset turnover, 568
balance sheet composition, 106–107
cash performance ratios, 1291–1292
deferred taxes and Tax Cuts and Jobs Act of 2017, 972–973
gross profit ratio, 456–457
inventory management, 509–510
investment in equity securities, 713–714
lease liability and debt covenants, 905
liquidity, 163–164
long-term solvency, 831–832
market performance, 1167–1168
net income and cash flows, 43
pension plan reporting, 1044
profitability ratios, 235–236
rate of return on shareholders' equity, 1101–1102
receivables management, 399–400
sales returns, 335–336
short-term debt, 768–769
Date of record, 1069
Dayton, George, 3–4
Deans, Sarah, 14n19
Debenture bond, 774
Debits
for adjusting entries, 64–66
on T-account, 50–51
Debt agreements
ethical dilemma involving, 133
existing violation of provision in, 730
modification of, 377, 807
reported net income and, 420
Debt coverage ratio, 1256
Debt to equity ratio, 133, 792
Debt investments
classifications
available-for-sale (AFS), 649, 656–661

comparing approaches, 661–663
held-to-maturity (HTM), 649–652, 650–652
overview, 649–650
trading (TS), 649
trading securities (TS), 652–656
transfers between reporting categories, 663–664
examples, 646–648
fair value option (FVO), 664–665
financial statement presentation and disclosure, 668–669
IFRS *vs.* GAAP, 663
impairment of, 688–693
credit losses for available-for-sale investments, 690–692
credit losses for held-to-maturity investments, 689–690
recording interest revenue, 647–648
recording purchase of, 647
Debt issue costs, 785–786
Debtor, financial difficulties for, 377n29
Debt payment ratio, 1256
Debt securities. *See also* Bonds; Debt investments; Notes receivable
held-to-maturity (HTM) investments
issuance of, 1223
Dechow, P. M., 8n8, 372n26
Decision makers perspective. *See* Management decision making
Decision usefulness of information, 23
Declaration date, 1069, 1071, 1077
Declining-balance depreciation methods, 575–576
Deere & Company, 289, 363
Default risk, 130, 792
Defense of intangible rights, 614
Deferrals, 64. *See also* Deferred *entries*; Prepayments
Deferred annuity, 253–254
Deferred charges, 118
Deferred income taxes, 678
Deferred payments, 525–528
Deferred revenue, 65–66
contract liability as, 302
customer advances as, 727
defined, 118
recognition of, 295, 310
Deferred tax asset (DTA), 911–912, 919–925
defined, 919
desirability of, *vs.* deferred tax liabilities, 922
disclosure notes, 942
effect of tax rate change on, 935
expense-related, 919–922
IFRS *vs.* GAAP, 1113
revenue-related, 922–925
Deferred tax liabilities, 911–918
balance sheet/income statement effects, 915
capital-intensive companies, 947–948
disclosure notes, 942
effect of tax rate change on, 934–935
expense-related, 912–915
revenue-related, 916–919
Defined benefit pension plans, 979–982, 1006
defined, 979
defined contribution plans compared, 979–980
IFRS *vs.* GAAP, 998–999
key elements of, 982
pension formula, 982
shift away from, 980, 1006
termination, 1006
Defined contribution pension plans, 979–981
defined, 979
defined benefits plans compared, 979–980
shift toward, 980
Definition criterion, 26
DeFond, M. L., 1186n12
Delayed recognition,
of compensation expense, 1106
justification for, 996
periodic pension accounting, 993
Dell Technologies Inc., 846
Delta Air Lines, 969
Depletion, 589–591
changes in accounting estimate, 1181–1182
changes in estimates, 596–597
changes in method, 597–598
cost allocation overview, 589
defined, 572, 589
error correction, 598–600
percentage depletion, 590
statutory, 590
Depletion base, 573
Deposits outstanding, 374
Deposits in transit, 374
Depreciable base, 573
Depreciation, defined, 572
Depreciation expense
adjustments for additional equity investments, 677
change in accounting estimate for, 183, 1181–1182
component-based approach, 580
cost allocation, 571–574
defined, 65, 575
disclosure note for error in, 1191–1192
income statement account, 1235
indirect method of reporting, 191–193
intangible assets, 117
leasehold improvements on operating leases, 873
partial-year, 582
recording, 65
in statement of cash flows, 1221–1227, 1229, 1235
Depreciation methods
activity-based methods, 574, 577–578
changes in estimates, 596–597
changes in method, 597–598
decision makers' perspective on selecting, 578–579
dispositions and, 583–584
assets held for sale, 585
retirements, 585
double-declining-balance (DDB) method, 615
error correction, 598–600
group and composite methods, 585–588
IFRS *vs.* GAAP, 580
MACRS (modified accelerated cost recovery system), 615–616
partial period depreciation, 581–583
replacement method 616–617
retirement methods, 616–617
time-based methods, 574–577
accelerated, 575–577
straight-line, 574–575
use of various, real-world financials, 579
Derivatives, A-2–A-29
accounting for, A-7–A-18
cash flow hedges, A-12–A-13
cash value hedges
decision makers' perspective on, 685–686
extended method for interest rate swap accounting, A-18–A-22
fair value changes unrelated to risk being hedged, A-17–A-18
fair value hedges, A-8–A-12
fair value of the notes, A-11–A-12
fair value of the swap, A-11
foreign currency hedges, A-16–A-17
forward contract, A-14–A-16
hedge effectiveness, A-17
interest rate swap A-13–A-16
defined, 685, A-3
disclosure of, A-18
Private Company Council (PCC) and, A-22
used to hedge risk, A-3–A-7
foreign currency futures, A-6
forward contracts, A-5
futures, A-4–A-5
interest rate swaps, A-6–A-7
options, A-5–A-6
Desir, R., 738n22
Detachable stock purchase warrants, 798–800
Deutsche Bank, 608
Development, GAAP definition, 541
Development costs
defined, 518
for intangible assets, 113, 522
for natural resources, 515, 518–519
Dhaliwal, D., 421n13, 1173n3
Dichev, I., 173n4
Diluted earnings per share, 184, 1123–1128
convertible securities, 1126–1128
convertible bonds, 1125–1127
convertible preferred stock, 1127–1128
options, rights, and warrants, 1123–1125
potential common shares, 1123
Dilutive effects
earnings per share and, 1118
of stock options, rights, or warrants, 1123–1125
in treasury stock method, 1125
Direct financing lease, 845n9, 864–865
defined, 864
requirements for, 865
Direct method
comparison with indirect method, 191–193, 1221–1222
defined, 191
FASB and, 1253
of reporting cash flows from operating activities, 1221–1222
spreadsheet used in, 1228–1246
balance sheet accounts, 1238–1246
comparison with indirect method, 1252–1253
completing the spreadsheet, 1244–1245
income statement accounts, 1230–1237
Direct write-off method of accounts receivable, 350
Directors, annual compensation disclosures, 126–128
Disclaimer in audit report, 130
Disclosure(s), 30–31. *See also* Disclosure notes
Disclosure notes
accounting changes
change in accounting principle, 1177–1178, 1185
change in estimate, 596, 1180
change in inventory, 482–484
accounting errors, 1183, 1186
error correction, 599–600
prior year's net income, 1188–1192
conveying financial information via, 5, 30, 122–125
contingent liability, 740–742
deferred assets and liabilities, 927–927
defined, 30
earnings per share, dilutive vs. antidilutive, 1130
examples of (*See* Disclosure notes, examples)
fair value disclosure, 682, 685
FASB project, 686
full-disclosure principle and, 30
gain contingencies, 742
impairment loss, 602
inventory
inventory change, 1176
inventory errors, 482–484
leases, 852, 853n14, 877
long-term debt, 791–795
loss contingencies, 735
materiality and, 183
pension plan, 982
income taxes and, 1004, 1005
management perspective, 1005
plan assets, 990
prior period adjustments, 184

Disclosure notes—*Cont.*
purchase commitments outstanding, 488
R&D expense, total, 543
shareholders' equity, 1051
restriction of retained earnings, 1069
sick pay accrual, 725–726
statement of cash flows, noncash investing and financing activities, 1224
substantial and unusual write-down, 462
summary of accounting policies and, 123
taxes
deferred tax assets and liabilities, 927, 928, 942
effective tax rate reconciliation, 942
net operating loss carryforwards, 942
income taxes, 941–942, 947, 1004, 1005
Disclosure notes, examples
accounting changes
accounting principle, 1177–1178
depreciation/amortization/depletion methods, 1181
estimate, 597, 1180
inventory method, 481, 482, 1178, 1179
reporting entity, 1183
accounting errors
depreciation errors, 1191–1192
net income
failure to record sales revenue, 1191
inventory misstated, 1190
recording asset as expense, 1189
previous financial statements, but not net income, 1188
restatement of prior period financial statements, 1188
amortization, 592
asset impairment, 602, 604
goodwill, 606, 608
asset retirement obligations, 521
debentures, 775
capitalized interest, 540
cash equivalents, 115, 340
contingent losses, 734, 740
accrual of, 734
litigation claims, 738, 739
potential, 732
unasserted claims, 741
credit lines, 721
current liabilities, 717, 720
commercial paper, 723
notes secured by inventory, 723
deferred tax assets and liabilities, 942
deferred taxes, 945
depletion, 590, 591
depreciation, 571, 579, 587
change in method, 598
discontinued operations, 178–181
earnings per share, 1137
contingently issuable share, 1134
conversion of notes, 1136
effective tax rate reconciliation, 930
examples, 122–123
financial statement presentation and, 668–669
financing with receivables, 367–368
government grants, 529
impairment loss, 602
income tax expense, 941–942
intangible assets, 117, 523, 592
indefinite life, 594, 604
interest capitalization, 540
inventories, 405
convert FIFO to LIFO, 482
convert LIFO to FIFO, 427, 481
inventory method, 470, 479
LIFO, 427, 403, 420, 422
LIFO liquidation, 424
LIFO reserve, 422–423
lower of cost or net realizable value, 463
types of inventory, 116
investments
change to equity method, 680
debt securities, 645
fair value disclosure, 668, 682, 685
long-term investments, 118
marketable securities, 649
transfer between categories, 664
leases
end of lease term, 852
initial lease transaction, 1239
liabilities
debt disclosures, 792
long-term liabilities, 119
net operating loss carryforwards, 942
in Nike's 2017 annual report, 122
noncash investing and financing activities, 195
notes receivable fair value, 360–361
noteworthy events and transactions, 124–125
oil and gas accounting, 549
pension plan, 977, 1005, 1016
defined benefit plan, 1006
other postretirement benefits, 1009
pension expense, 1001
property, plant, and equipment, 514
receivables as collateral, 362
restricted cash, 341–343
revenue recognition, 302
sale of receivables, 367–368
securitization program, 364
service life, 573
shareholders' equity
debt vs. equity for preferred stock, 1058
quasi reorganization, 1080
share repurpose program, 1063
stock split, 1073
software disclosure, 544–545
statement of cash flows
cash equivalents, 1219
noncash investing and financing activities, 1224
reconciliation schedule, 1293
subsequent events, 123–124, 740
Disclosure overload, 31
Discontinued operations, 177–181
defined, 177–178
income from, 170
reporting
component considered held for sale, 180–181
component has been sold, 178–180
interim reporting, 181, 206
Discount(s)
bond issued at, 775–778
in contracts with multiple performance obligations, 285
employee, retail inventory method and, 474
sales, 345, 474
trade, 344
Discount on bond investment, 647
Discount on notes payable, 527, 722n7
Discount on notes receivable, 296, 358, 359, 788
Discount (interest) rate
change in, as economic event, 48n1
estimating future cash flows, 28, 519
noninterest-bearing notes, 357, 360n15, 722, 722n7
notes receivable, 365–366
operating leases and, 853
Disposition of assets, 583–584
Distributions to owners, 24
Dividend(s), 1068–1070
cash dividends, 1069–1070
investment revenue and, 1233
payment of, 58, 1218, 1223
recording payment of, 58
cash *vs.* stock, 1076
defined, 1068
dividends in arrears, 1070
equity method and, 676
large stock dividends, 1073–1074
liquidating dividends, 1069
preferred shares, 1070, 1122–1123
property dividends, 1070
retained earnings restrictions, 1069
stock dividends, 1070–1072 (*See also* Stock dividends)
stock splits, 1072–1073
Dividend in kind, 1070
Dividend payment ratio, 1256
Dividend rights, 1056, 1068n17
Dividends account, 77
Dividends in arrears, 1057, 1070
Dolby Laboratories, 1096
Dollar General Corporation
inventories disclosures, 422
merchandise inventory, 459
Dollar-value LIFO (DVL), 431–433
advantages of, 433
cost indexes, 431
inventory estimation technique, 431–433
retail method, 477–480
Domino's Pizza, 874
Donated assets, 528–530
Double-declining-balance (DDB) method, 576, 582, 616
Double-entry system, 50
Double-extension method, 431
Double taxation, 1054
Dow Chemical Company, 521
DowDuPont Inc., 756
Dun and Bradstreet, *Industry Norms and Key Business Ratios*, 1099
DuPont framework
defined, 200
illustration, Costco and Walmart, 202
DVL. *See* Dollar-value LIFO

E

Early extinguishment of debt, 795
Earnings
non-GAAP, 176
permanent, 173–176, 199
transitory, 173, 199
Earnings per share (EPS), 1119–1138
actual conversions, 1136
antidilutive securities, 1128–1131
convertible securities, 1130
options, warrants, rights, 1128–1130
basic EPS, 184, 1119–1138
earnings available to common shareholders, 1122–1123
issuance of new shares, 1120
reacquired shares, 1122
stock dividends, 1120–1121
stock splits, 1120–1121
components of "proceeds" in treasury stock method, 1133
contingently issuable shares, 1134
decision makers' perspective, 1138
diluted EPS, 184, 1123–1128. *See also* Diluted earnings per share
convertible securities, 1126–1128
convertible bonds, 1126–1127
convertible preferred stock, 1127–1128
options, rights, and warrants, 1123–1125
potential common shares, 1123
earnings available to common shareholders, 1122–1123
financial statement presentation of, 1136–1137
IFRS *vs.* GAAP, 1119
on income statement, 184–185
interim reporting of, 206
issuance of new shares, 1120
potential common shares and, 1135
pro forma, 1138

reacquired shares, 1122
restricted stock awards in calculations of, 1133–1134
stock dividends, 1120–1121
stock splits, 1120–1121
Earnings quality, 173–176
accounts receivable and, 371–372
classification shifting, 173
deferred tax assets and, 948
defined, 173
earnings management and, 1117
impairment losses and, 609
income smoothing, 173
inventory management and, 429
inventory valuation changes and, 485
motivation for accounting choices and, 1174
non-GAAP earnings, 176
nonoperating income and, 175–176
operating income and, 173–175
other unusual items, 175
restructuring costs, 174–175
pensions and
pension disclosures, 1006
postretirement benefit expense and, 1016
predictive/confirmatory value and, 21
receivables management, 371–372
temporary *vs.* permanent earnings, 173, 199
Earnings-price ratio, 1075
Ecolab, Inc., 986–987
Economic entity assumption, 23–24
Economic environment of accounting, 3–5
cash *vs.* accrual accounting, 7–8
financial reporting and, 5–8
investment-credit decision, 6
Economic events, 48
Economist, The, 657
EDGAR (Electronic Data Gathering, Analysis, and Retrieval) system, 111, 161–163
Effective interest method
bonds, 647–648, 778–784
bonds sold at discount, 647
bonds sold at premium, 647, 780–782
defined, 648, 778
face amount of bonds, 778
fair value reporting, 775n1, 779
interperiod allocation of interest, 782–783
long-term notes receivable, 377
straight-line method, 783–784
zero-coupon bonds, 780
Effective (annual) interest rate
compensating balances and, 342, 721n6
compound interest calculation, 242, 242n1
defined, 240
formula, 68
HTM debt investments and, 689–690
noninterest-bearing notes and, 722
notes receivable and, 358n14, 359–360
simple interest calculation, 240
time value of money and (*See* Time value of money)
Effective tax rate, 930
EITF. *See* Emerging Issues Task Force
EITF Issue Consensuses, 10
El Paso Natural Gas Company (EPNG), 586–587
Electronic data processing systems, 48. *See also* Excel spreadsheets
Elliott, John A., 17n25
EMC Corporation, 390
Emerging Issues Task Force (EITF), 10
Employee compensation
bonuses (*See* Annual bonuses)
directors and executives. (*See* Executive compensation)
pension plans (*See* Pension plans)
postretirement benefits (*See* Postretirement benefit plans)
salaries, commissions, and bonuses
accrued liability for, 724
paid future absences, 724–726
share-based (*See* Share-based compensation)
Employee benefit plans, disclosure notes for, 123
Employee Retirement Income Security Act of 1974 (ERISA), 984, 991, 1002
Employee share purchase plan (ESPP), 1117
Employee stock ownership plan (ESOP), 980n1
Employees
discounts, 474
projecting future salaries for group of, 984n4
withholding taxes, 747
Employers' payroll taxes, 747
Enron Corporation, 15, 161
exotic derivatives, 686
voluntary stock option expensing, 1109
Enterprise Resource Planning (ERP), 48
EPBO. *See* Expected postretirement benefit obligation
EPS. *See* Earnings per share
Equipment. *See also* Property, plant, and equipment
acquisition costs, 515
cash flows involving, 1216, 1222–1223
costs to be capitalized, 516–517
depreciation on excavation, 589
loss on sale of, 1221, 1236
noncash purchase of, 1224
Equity. *See also* Shareholders' equity
defined, 24, 114
SARS payable in shares as, 1144
Equity financing. *See* Shareholders' equity
Equity investments, 669–686
characteristics of, 669
equity method
control and significant influence, 674–676
purchase of investment, 676
receiving dividends, 676
recording investment revenue, 675
fair value through net income, 669–674
adjust to fair value (2021), 670–671
adjust remaining inventory to fair value (2022), 672–673
financial statement presentation, 673–674
purchase investments, 670
recognize investment revenue, 670
sell investment, 671–672
further adjustments
additional depreciation, 677
fair value option (FVO), 682
financial instruments and investment derivatives, 685–686
investor has no significant influence (*See* Fair value through net income subentry)
investor has significant influence (*See* Equity method subentry)
reporting categories, 669
Equity method
changing conditions and, 680–682
change from another method, 680
change to another method, 680
comparison to fair value, 681
fair value option, 681
sale of investment, 681
consolidated financial statements and, 674
control and significant influence, 674–675
defined, 674
fair value through net income method *vs.,* 681–682
further adjustments
adjustments for other assets and liabilities, 678
changing conditions, 679–680
impairment of equity method investments, 679–680
investee reports net loss, 679
investment acquired mid-year, 679
no adjustments for land or goodwill, 678
reporting for, 678–679
IFRS *vs.* GAAP, 682
purchase of investment, 675
receiving dividends, 675
recording investment revenue, 676
strength of, 685
Equity multiplier, 200–203
Equity/net assets. *See* Shareholders' equity
Equity securities. *See also* Common shares; Preferred stock
issuance of, 528
ERISA. *See* Employee Retirement Income Security Act; Employee Retirement Income Security Act of 1974
Ernst & Young LLP, 15, 128
ERP. *See* Enterprise Resource Planning
Errors
accounting (*See* Accounting errors)
in bank reconciliation, 374
fraud *vs.,* 125
Escheatment laws, 729
ESPP. *See* Employee share purchase plan
Estee Lauder Companies Inc., 594
Estimates
actuarial, 981–982
adjusting entries for, 69–70
bad debts [*See* Bad debt(s)]
balance sheet and, 113
changes in (*See* Change in accounting estimate)
fair value estimates
credit loss calculation and, 690
distortion of, 372, 685
financial statements and, 668–669
share options, 1116
stock appreciation rights, 1146
inventories (*See* Inventory, estimation techniques)
Ethics in accounting
analytical model for ethical decisions, 17–18
bad debt expense, 354
capitalizing vs. expensing expenditures, 516
cash dividend *vs.* stock dividend, 1076
debt agreements, 133
donated assets, 528–529
ethical decision making, 17–18
ethics defined, 17
financial reporting and
financial statements and, 18
fraudulent reporting, 1109
intentional misstatement of revenue, 175
misreporting inventory, 1191
reform, 16–17
income smoothing and classification shifting, 173
lump-sum purchase price of building, 525
net realizable value of inventory, 462
non-GAAP earnings, 176
postretirement health benefits, 1014
professionalism and ethics, 17
purchase option with lease, 869
revenue recognition, 301
risk analysis, current ratio, 133
statement of cash flow, 1244
European Union (EU)
euro as monetary unit of, 25
International Financial Reporting Standards and, 11, 14

Ex-dividend date, 1069
Excel spreadsheets
amortization schedule creation with, 789n9
discount on bonds, 777
for interest calculations, 240, 241, 242
preparing statement of cash flows (*See* Spreadsheet preparation of statement of cash flows)
solving present and future value problems using, 255, 358, 359, 651, 842, 985
using a worksheet, 84–85
Exchanges, 531–534
fair value not determinable, 532
lack of commercial substance, 532–534
gain situation, 533
loss situation, 534
Executive compensation
disclosure of, 126–128
performance conditions for, 1112, 1115–1116
share-based (*See* Stock option plans)
Expected cash flow approach
defined, 519
in testing for impairment, 602
to warranty obligation estimation, 736
Expected cost plus margin approach, 297
Expected credit loss (ECL) model, 361. *See also* Current expected credit loss model
Expected postretirement benefit obligation (EPBO), 1011
attribution of costs, 1012
comprehensive illustration, 1014–1015
measuring obligation, 1011–1012
Expected return, 6
Expected return on plan assets, 991, 1016
Expedia, Inc., 295, 334
Expense(s). *See also entries for specific types of expenses*
in accrual accounting, 8
classification of, 169
defined, 24, 26, 169
interim reporting, 204
prepaid, 64–65, 116
revenue/expense approach, 32
Expense recognition, 26–27
Expense-related deferred tax liabilities, 912–915
Expensing in stock option plans
current requirement to expense, 1109
failed attempt to require expensing, 1108–1109
voluntary expensing, 1109
Explicit interest, 246–247
Exploration costs, 518
Extended Stay America, 1057
Extended warranties, 289, 737
External events, 48
External transactions
accounting processing cycle and, 51–63
identifying from source documents, 52
illustrations of, 55–61
posting to general ledger, 54
transaction analysis, 53
External users of financial information, 4–5
Extraordinary items, elimination of category, 174n7, 946n12
ExxonMobil Corporation, 1219

F

Face amount
bonds, 774
effective interest method, 778
issued between interest dates, 805
long-term notes, 787
reporting in statement of cash flows, 1240
debt investment, 646
Facebook
auditor's report, 128–129
initial public offering, 1054
R&D costs, 113
restricted stock units, 1150
Zuckerberg's stock options, 1108
Factoring receivables, 363–354, 722
FAF. *See* Financial Accounting Foundation
Fair value
adjusting trading securities to, 653–654
consideration less than, 524
GAAP measurement attribute, 28–30
as hard to estimate, 657, 673
hierarchy, 29
measuring stock options at, 1108
of notes receivable, disclosing, 361
recording asset at, 534
stock option, 143, 1109–1112
Fair value accounting standard, 13
Fair value estimates
credit loss calculation and, 690
distortion of, 372, 685
financial statements and, 668–669
share options, 1116
stock appreciation rights, 1146
Fair value hedges, A-8–A-12
defined, A-8
firm commitments, A-8
illustration, A-8–A-12
interest rate swaps
Fair value hierarchy, 29–30
Fair value option (FVO), 30
debt investments, 664–665
equity investments, 682
IFRS *vs.* GAAP, 663–665
Fair value risk
with futures, A-4
hedging against, A-3
Fair value through net income approach
adjust to fair value (2021), 670–671
adjust to fair value (2022), 672–673
equity method *vs.*, 681–682
financial statement presentation, 673–674
inaccurate estimates and, 685
purchase investments, 670
recognize investment revenue, 670
sell investment, 671–672
Fair value through other comprehensive income (FVOCI), 673
Fair value through profit or loss (FVPL), 673
Faithful representation, 20–22
FASAC. *See* Financial Accounting Standards Advisory Council
FASB. *See* Financial Accounting Standards Board
FASB Accounting Standards Codification, 10–11
FDIC. *See* Federal Deposit Insurance Corporation
Federal Deposit Insurance Corporation (FDIC), 1054
Federal Insurance Contributions Act (FICA), 747
FUTA. *See* Federal Unemployment Tax Act
FedEx Corporation
equity securities, 695
f.o.b. shipping point and, 409
information from cash flow activities, 1288
mark-to-market method of recognizing pension gains/losses, 996
pension plans, 1042
revenue source for, 168
stockholders' investment, 1048
Feeny Manufacturing Company, 1134
FIFO. *See* First-in, first-out method
Finance leases, 836–837. *See also* Finance/sales-type leases
defined, 836
direct, 845n9
expense recognition between operating and, 851–852
impact on lessee's statement of cash flow, 876
installment notes *vs.*, 839–841
operating lease compared, 836
Finance/sales-type leases, 841–845
defined, 836–837
recording amortization of right-of-use asset, 843–844
recording interest expense/revenue, 843
Financial accounting, 2–36
conceptual framework. See Conceptual framework
defined, 4
economic environment, 5–8
cash versus accrual accounting, 7–8
investment-credit decision, 6
ethics in accounting, 17–18
financial reporting
auditor's role, 14–15
financial reporting reform, 15
objectives-oriented approach and, 16
Pathways Commission visualization of, 4
primary objective of, 4, 6, 47
process for (*See* Accounting process)
standards development (*See* Accounting standards development)
Financial Accounting Foundation (FAF), 9–10, 12, 14
Financial Accounting Standards Advisory Council (FASAC), 10, 12
Financial Accounting Standards Board (FASB), 9–10, 12
Accounting Standards Codification System (ASC), 10, 1240
Accounting Standards Updates, 10
available-for-sale investments, 690
balance sheet approach to deferred tax liabilities, 945–946
bargain purchase, 524
cash equivalents, 1219
changes in accounting principle, 182, 1173, 1178
codification, 10–11
commercial substance requirement, 533
conceptual framework (*See* Conceptual framework)
contingent losses, 742
convergence efforts with IASB (*See* Convergence efforts)
convertible debt, 797
direct reporting method encouraged by, 193, 1253
earnings per share, 1119
Emerging Issues Task Force (EITF), 10, 11
employee stock options, 13
equity investment sales, 671
expected cash flow approach, 519, 736
exposure draft, 13, 31, 110
fair value accounting requirements, 14
financial instruments project, 686
framework for financial instrument accounting issues, 686
gains/losses as net income vs. OCI, 185
goodwill project, 606, 607
income smoothing, 996
investment disclosure, 685
lease classification, 837, 852
licenses, 298
OCI classification, 803
oil and gas exploration costs, 549
political pressure on, 13–14, 549
Private Company Council (PCC), 14, 607, A-22
receivables financing, 369
research and development costs, 541
pension/postretirement benefits, 994–996, 1009, 1014
prior service cost, 1020
revenue recognition, 278
revenues, 277
sales approach, 367
segment reporting, 138–140
shipping costs, 291
standard-setting process, 12–14

statement of cash flows, 1217, 1240
Statements of Financial Accounting Concepts (SFACs), 18–19
stranded tax effects, 947
Financial calculators
bond discount calculations, 777
finance lease calculations, 842
interest calculations, 240
present/future value functions, 241–242, 255
Financial crisis of 2008–2009
fair value estimates and, 13, 657
funded status of pension plans, 991
loss contingency accounting and, 355
troubled debt restructuring, 807
Financial disclosures. *See* Disclosure(s); Disclosure notes
Financial flexibility, 112
Financial futures contract, A-4
Financial information
adjusting entries for, 64
comparing
accrual accounting and, 8
change in accounting principle and, 182
conceptual framework and, 19
costs of providing, 23
decision making and, 4
filing, using EDGAR system, 111
Financial Accounting Foundation (FAF) and, 10n12
in financial statements (*See* Financial statements)
interim reporting of, 205
means of conveying, 5
monetary unit assumption and, 25
providers and users, 5
qualitative characteristics of, 20–23
requirements for providing, 25
Financial instrument. *See also entries for specific types*
defined, 685
disclosing fair value of, 792
investment derivatives and, 685–686
Financial intermediaries, 5, 845
Financial leverage, 134–135
capitalization mix and, 134
defined, 134
in DuPont framework, 200
favorable, 793
Financial position, ratios to evaluate, 131
Financial reporting
auditor's role in, 14–15
defined, 5
economic environment of accounting and, 5–8
cash *vs.* accrual accounting, 7–8
investment-credit decision, 6
interim reporting, 204–207
in accounting processing cycle, 76
discontinued operations, 181
earnings per share, 206
global perspective, 206
minimum disclosures, 206–207
required by SEC, 181
revenues and expenses, 205–206
unusual items, 206
objectives of, 23, 910
pension plans (*See* Pension plans)
reform, 15–16
qualitative characteristics of information, 20–23
segment information (*See* Operating segments)
use of term, 5
Financial Reporting Releases (FRRs), 9n11
Financial statement(s), 5. *See also entries for individual financial statements*
in accounting processing cycle, 54
cash basis to accrual conversion, 80–82
certification by executives, 126–127
comparative, 130
in annual reports, 1119
prior period adjustment, 483n14, 599n15, 1187
retrospective approach and, 1172, 1175–1176
revenue recognition and, 278
revising, 480–481
consolidated, 674, 676–677
defined, 72
disclosures for long-term notes, 792–793
elements of, 23–24
error affecting previous, 1187–1188
ethical issues with, 16–18, 125, 175, 1109, 1191
interim reporting of, 204–207
issued *vs.* available to be issued, 739n23
leasing's impact on, 852
overview, 5
preparation of
balance sheet, 73–75
between interest dates, 782–783
income statement, 72–73
statement of cash flows, 75
statement of comprehensive income, 73
statement of shareholders equity, 76
required management certification of, 125–126
revision of, 1175–1176
risk analysis using information in, 130–135
viewed as issued (GAAP), 124n7
Financial statement presentation
debt investments, 668–669
disclosure notes, 941–942
discontinued operations, 178–181
earnings per share (EPS), 1136–1137
equity securities, when investor lacks significant influence, 673
fair value through net income, 673–674
income tax expense, 941–942
net operating losses, 941
international standards on (*See* International Financial Reporting Standards (IFRS))
investments, 668
available-for-sale (AFS) securities, 661
held-to-maturity (HTM) securities, 651–652
trading securities (TS), 656
trading securities (TS), 656
Financing activities
cash flows from, 194, 1223–1224
noncash, 195
noncash investing and, 1224–1225
"off-balance-sheet," 839
on statement of cash flows, 75
Finished goods inventory, 115–116, 308n18, 404–405
Finite-life intangible asset, 546
First-in, first-out method (FIFO)
change from LIFO method to, 1175–1177
defined, 416
as inventory cost flow assumption, 414
periodic FIFO, 416
perpetual FIFO, 417
physical flow of inventory and, 420–421
Fiscal year
adoption of, 25
asset acquisition during, 581
closing process and, 76, 85, 87
expenditure capitalization and, 611
financial statements issued after close of, 731, 739–740
interest dates and, 782–783
interim period and. *See* Interim period
inventory
change in inventory method and, 482
physical inventory counts and, 406
lease obligations and, 877
purchase commitments and, 487–489
SEC reporting requirements and, 9n10, 23, 25n31
subsequent events and, 124, 739–740
unusual items and, 206
Fixed assets. *See* Property, plant, and equipment
Fixed-asset turnover ratio, 530
f.o.b. (free on board) destination, 409
f.o.b. (free on board) shipping point, 409
Forecasted transactions, A-18
Forced sale situations, 524, 807
Ford Motor Company
change in reporting entity, 1182
deferred taxes, 970
as publicly held corporation, 1054
stock buyback, 1087
Foreign Corrupt Practices Act of 1977, 125
Foreign currency futures contract, A-6
Foreign currency hedge, A-16–A-17
Foreign currency risk, A-3
Foreign currency translation, A-12, 947, 1050–1052
Forfeitures, stock options, 1110–1112
Form 10-K, 9n10, 122, 161, 291
Form 10-Q, 9n10
Forward contracts, A-5
cash flow hedge, A-15–A-16
Foster, Taylor W., III, 1072n20
401(k) plan, 980
Fractional shares, 1074
Franchise
acquisition costs, 515
amortization of, 592
balance sheet classifications, 117
bill-and-hold arrangement and, 299–300
costs to be capitalized, 521–522
defined, 523
as intangible asset, 511
revenue recognition and, 298–299
as symbolic intellectual property, 298
Franchisee, 299
Franchisor, 299
Frank, Mary Margaret, 948n18
Fraud, 124–125. *See also* Ethics in accounting
Fred's Inc., 508
Freedom from error, 22
Free-enterprise economy, 5
Free on board (f.o.b.), 409
Freight-in on purchases, 410
Freight-out, 410
Fringe benefits, 747
Frito-Lay, 1182
Frontline (television series), 1109
Full-cost method (oil and gas accounting), 548–549
Full-disclosure principle, 30, 122
Functional intellectual property, 298–299
Fundamental share rights, 1055–1056
FUTA (Federal Unemployment Tax Act), 747
Future cash flows
asset write-down and, 600
best estimate of, 519
capital budgeting and, 530
commercial substance and, 533
debt investment and, 646, 689, 777
estimating
expected cash flow approach and (*See* Expected cash flow approach)
financial reporting and, 20
long-term liabilities and, 119
predicting, 7, 21, 423
present value of, 28, 246, 519, 602
recoverability test and, 601, 603

Future deductible amount, 911, 919, 1177
Futures contracts, A-4–A-6
Future taxable amounts, 911–918, 930, 934–937
Future value (FV)
of annuity due, 250
compound interest and, 243
defined, 241
financial calculator and Excel to find, 255
of ordinary annuity, 248–249
of single amount, 241–242
solving for unknown values, 245
Future value (FV) tables, P1–P6
future value of $1, P1
future value of an annuity due of $1, P5
future value of an ordinary annuity of $1, P3
FVO. *See* Fair value option
FVOCI. *See* Fair value through other comprehensive income
FVPL. *See* Fair value through profit or loss

G

GAAP. *See* Generally accepted accounting principles (GAAP)
Gain(s)
on asset exchanges, 531
commercial substance requirement, 533
from continuing operations, 169
deferred, in cash flow hedges, A-12, A-16
defined, 24, 169
disposition of assets, 585–587
as element of financial statement, 24
holding gains (losses)
on interest rate swap, A-9, A-10, A-12, A-19–A-21
realized, 651–652, 659–660
unrealized (See Unrealized holding gains/losses)
pension
plan assets, 982, 998–999
projected benefit obligation (PBO), 985, 988–989, 991
in profit equation, 168
remeasurement gains on pension plans, 998–999, 1005, 1008–1009
on sale of land, 1226, 1229
unrealized, 185
Gain contingencies, 742
Gap, Inc., The, 42, 115
contingent lease payments disclosure, 857
fixed-asset turnover ratio, 530
Ross Stores, Inc., 530
Gatorade, 1182
General Electric, 14
"cookie jar accounting," 1174
as corporation, 1052
shareowners' equity, 1048
General journal, 52, 59, 70
General ledger, 50, 54, 60
Generally accepted accounting principles (GAAP)
accounts and notes receivable, 361
accrual basis requirement, 81n7
allowance method of accounting for bad debts, 350–352
asset/liability approach to, 32
assumptions underlying, 23–24
auditor role in, 14–15
balance sheet presentation, 120
bargain purchase situation, 524
cash and cash equivalents, 342
cash flow classification, 193
CECL (Current Expected Credit Loss) model, 352–355
classification of cash flows, 195
codification framework, 10–11
comprehensive income, 186
defined, 8–9
differences from IFRS [*See* Convergence efforts (FASB/IASB); International Financial Reporting Standards]
differences from tax rules, 166
disclosures of derivatives, A-18
earnings per share, 1119, 1138
equity method, 1293
evolution of, 31–32
extraordinary items, elimination of category, 174n7
fair value option, 361
FASB conceptual framework and, 10
gain contingencies, 742
geographic reporting, 139–140
going concern assumption, 24–25
goodwill, 525
government grants, 529
impairment of value, 604, 605
income statement presentation, 172
interim reporting, 206
International Financial Reporting Standards *vs.*, 12
inventory write-downs not permitted, 459
leases, 839, 845n9, 852, 855, 869
loss contingencies, 741–742
lower of cost or net realizable value (LCNRV), 463
participating interests, 368
Private Company Council (PCC), 14, 607, A-22
"probable" defined, 288
qualified opinion and, 130
recovery of impairments, 693
"Repo 105" transactions, 367
research and development (R&D), 543
revenue/expense approach, 32
revenue recognition (*See* Revenue recognition)
segment reporting, 138–139
software development costs, 544
statement of cash flow, 1217
stock option plans, 1108–1110, 1113, 1115
transfers of receivables, 369
underlying assumptions of, 23–25
unqualified options and, 128–129
use of fair value, 27–28
valuation allowances, 927
voluntary accounting changes, 182
General Mills, Inc., 371
current liabilities, 719–720
inventory cost flow assumptions, 420
uncollectible accounts, 385
General Motors Corporation
bonds, 801
change in reporting entity, 1182
investments disclosure, 649
long-lived, revenue-producing assets, 513
loss contingencies of, 732–733
other postretirement benefits disclosure, 1009
sale of receivables by, 363
Genuine Parts Company, 424
Geographic area, reporting by, 139–140
Giacomino, Don, 1256n18
Gibson Greetings, 686
Gift cards
breakage, 728
as current liability from advance collections, 728
revenue recognition and, 301
Gilead Sciences, 522
Givoly, Dan, 948n16
GlaxoSmithKline, 229–230, 640
Global perspective, *See* Convergence efforts; International Financial Reporting Standards (IFRS)
Glover, Jon, 835n1
Going concern
GAAP assumption, 24–25
indicators of, 129
Goldman Sachs Group, Inc.
debt issue costs, 785
funded status of pension plans, 991n9
taxes on unrepatriated foreign earnings, 931
Gold Resource Corp., 1070
Gold's Gym, 281
Goods on consignment, 409
Goods in transit, recording, 408–409
Goodwill, 523–525
acquisition costs, 515
annual impairment test, 677n24
consolidated financial statement reporting, 677–678
defined, 523,
equity investments and, 678
impairment of value, 174–175, 605–608
indefinite useful life of, 594
as intangible asset, 117, 514, 515, 521, 523–525
valuation of, 594
Google, 24, 27 *See also* Alphabet, Inc.
R&D costs, 541
sale of Motorola Mobility, 177
self-driving car expenditures, 541
Government grants, 529
Governmental Accounting Standards Advisory Council (GASAC), 11
Governmental Accounting Standards Board (GASB), 10n12, 11
Graded vesting, 1114, 1115
Graham, J., 173n4
Graham, John, 948n19
Green Mountain Coffee Roasters, Inc., 396
Greenspan, Alan, A-2
Gross margin method of estimating inventory, 467–468
Gross method
recording interest revenue, 647n1
recording purchase discounts, 410–411
recording sales discounts, 345
Gross profit
defined, 73, 846
income statement, 73, 171, 172
inventory management and, 427–428
operating income presented as, 170
as performance measure, 172
revenue recognition and, 306–313
sales-type leases and, 846–847
Gross profit method of estimating inventory, 467–469
Gross profit ratio, 428
Group depreciation method, 585–588

H

H&M, 120
H&R Block, Inc., 1136–1137
Hagerman, R., 1174n5
Hail, L., 12n17
Half-year convention, 583
Hanlon, Michelle, 948n19
Hanmi Financial Corporation, 1088
Harley-Davidson, 427–428
Hartford Life Insurance Company, 1183
Hayn, Carla, 948n16
Health care benefits, 1010, 1014
Healy, P. M., 421n13, 1173n3
Hedging
cash flow, 1049–1051, A-7, A-12–A-13, A-18
defined, A-3
fair value hedges, A-8–A-12
foreign currency hedge, A-17
forward contracts, A-15
FVO and, 665
hedge effectiveness/ineffectiveness,
interest rate swaps, A-13–A-15
Private Company Council (PCC) and, A-22
Heineken, 543
Held-to-maturity (HTM) securities, 649
credit losses for, 689–690
financial statement presentation, 651–652
impairment of, 651
sale of, 650–651
unrealized holding gains/losses not recognized, 650
Hershey Company, The, 371
classes of shares, 1056
partial income statement, 174
Hertz Global Holding, Inc., 1191–1192
Hewlett-Packard (HP), 38
acquisition of Compaq, 523

inventory of, 403–404
Historical cost
  defined, 28
  donated assets and, 529
  fair value vs., 30
  going concern assumption and, 25
  interest capitalization and, 536
  as measurement attribute, 27–28, 31, 113
  as verifiable, 22
Home Depot, Inc., The
  case, 231–232
  income statement, 170–171
  interim data in annual report, 205
  Macy's, 414
  partial income statement, 175–176
  performance obligations for, 283
  property, plant, and equipment, 567
  as publicly held corporation, 1054
  ratio analysis, 443
  specific identification of inventory, 414
Honeywell International Inc., 38
Hostile takeover attempt, 1063n11
Horizontal analysis, 130–131
Houston, Joel, 427n15
HP (Hewlett-Packard)
  classes of shares, 1056
  fair value disclosures of investment securities, 668
  zero-coupon debt, 827
Hunt Manufacturing Co., 1134
Hybrid organizations, 1054–1055
Hybrid securities, 796, 1057

I

IASB. *See* International Accounting Standards Board
IASC. *See* International Accounting Standards Committee
IBM Corporation, 38, 102, 157, 371
  change in reporting entity, 1182
  credit lines disclosure note, 721
  governmental incentives for, 529
  securities classification, 699
  service life disclosure, 573
  software development costs, 544–545
  stock appreciation rights, 1144
IFRS. *See* International Financial Reporting Standards
Illegal acts, 125
  cookie jar accounting, 371, 1174
  fraudulent financial reporting, 1109
Illegal contributions to political candidates, 125
IMA. *See* Institute of Management Accountants
Impairment of value
  assets held and used, 600–606
    finite-life intangible assets, 600–603
    property, plant, and equipment, 600–603
  assets held for sale, 608–609
  available-for-sale (AFS) securities, 660
  debt investments, 689–693
  equity method investments, 679–680
  held-to-maturity (HTM) securities, 651
  IFRS *vs.* GAAP, 603–604, 693
  impairment losses and earnings quality, 609
  indefinite-life intangible assets, 604–606
  summary of guidelines, 609
  when to test for, 601
Implicit rate of interest, 528, 787
Implied fair value, 604n19
Improvements, 612–613
Imputed rate of interest, 785
Incentive savings plan, 980
Incentive stock option plans, 1113. *See also* Employee share purchase plan; Employee stock ownership plan
Income
  AOCI (*See* Accumulated other comprehensive income)
  comprehensive (*See* Comprehensive income)
  from continuing operations (*See* Income from continuing operations)
  manipulation of (*See* Income manipulation; Income smoothing)
  net (*See* Net income)
  nonoperating income, 170, 175–176
  OCI (*See* Other comprehensive income)
  operating, 170, 174–175
  other income (expense), 170
Income approaches to fair value, 29
Income before taxes
  income statement, 172
  use of, 172
Income from continuing operations, 168–172
  defined, 169–170
  expenses, 169
  gains and losses, 169
  income statement formats, 171–172
  income tax expense, 170–171
  major components of, 170
  operating *vs.* nonoperating income, 170
  revenues, 168
Income from discontinued operations, 170
Income manipulation
  big bath accounting, 485, 609
  cookie jar accounting, 371, 1174
  earnings quality and, 173
  LIFO cost of goods sold, 429
  premature revenue recognition, 464
Income recognition. *See also* Revenue recognition
  over term *vs.* completion of contract, 310
Income smoothing
  classification shifting and, 173
  earnings quality and, 173
  pension expense and, 996–997
Income statement, 5
  accounting changes
    accounting estimate, 183
    accounting principle, 182
    amortization, 183
    depletion method, 183
    depreciation, 183
  accounts used in preparing SOCF
    cost of goods sold, 1234
    depreciation expense, 1235
    gain on sale of land, 1233–1234
    income tax expense, 1236–1237
    insurance expense, 1236
    interest expense, 1235
    investment revenue, 1232–1233
    loss on sale of equipment, 1236
    net income, 1237
    salaries expense, 1234–1235
    sales revenue, 1230–1232
  comprehensive income, 185–188
    accumulated other comprehensive income (AOCI), 187–188
    flexibility in reporting, 185–186
  correction of accounting errors, 183–184
    prior period adjustments, 183–184
  deferred tax liabilities on, 915
  defined, 72, 168
  disclosure notes for revenue recognition, 302
  discontinued operations reporting, 177–181
    when reported, 177–178
  earnings per share, 184–185
  earnings quality, 173–176
    classification shifting, 173
    income smoothing, 173
    non-GAAP earnings, 176
    nonoperating income, 175–176
    operating income and, 173–175
      other unusual items, 175
      restructuring costs, 174–175
  held-to-maturity (HTM) securities on, 651
  illustration, 1226–1227
  income from continuing operations, 168–172
    expenses, 169
    gains and losses, 169
    income statement formats, 171–172
    income tax expense, 170–171
    operating *vs.* nonoperating income, 170
    revenues, 168
  income statement formats, 171–172
  income tax expense, 170–171
  operating *vs.* nonoperating income, 170
  preparation of, 72–73
  presentation, GAAP *vs.* IFRS, 172
  recording pension expense in, 1000–1001
  revising, for change in accounting principle, 1175
  trading securities on, 656
Income statement approach, 354
Income summary account, 77
Income taxes
  available-for-sale (AFS) securities and, 663
  balance sheet focus, 945–946
  continuing operations, 171–172
  coping with uncertainty in, 942–947
    more-likely-than-not criterion, 943
    tax benefit measurement, 944
    two-step decision process, 943
    uncertainty resolution, 944–945
  decision makers' perspective, 947–948
  deferred, investment revenue and, 678
  disclosure notes for, 123
  double taxation, 1054
  financial statement presentation, 941–942
    balance sheet, 941–942
    disclosure notes, 941–942
      deferred tax assets and liabilities, 942
      effective tax rate reconciliation, 942
      income tax expense, 941–942
      net operating loss carryforward, 942
  IFRS *vs.* GAAP, 931
  intraperiod tax allocation, 946–947
  inventory costs and, 421–422
  net operating losses (NOL), 937–941
    carryforward, 937–939
    valuation allowance, 939
  pension plan and, 1004
  permanent differences, 929–931
  tax rate considerations, 933–935
    changes in tax laws or rates, 933–935
    tax rates differ between years, 933
  temporary differences
    conceptual underpinnings, 910–912
    deferred tax assets, 919–925. See also Deferred tax assets
    deferred tax liabilities, 912–918. See also Deferred tax liabilities
    defined, 910
    disclosures linking tax expense with changes in deferred tax assets/liabilities, 927–929
    multiple temporary differences, 935–937
    tax liability *vs.* tax expense, 909
    types of, 911–912
    valuation allowance, 925–927
  unrepatriated foreign earnings, 931

Income tax expense, 170–171, 1236–1237
Incremental borrowing rate, 853n15
Indefinite-life intangible assets, 604–606
Indefinite useful life, 594
Indirect manufacturing costs (overhead), 535
Indirect method
comparison with direct method, 1252–1253
cash increased/decreased, 1251–1252
cash not increased/decreased, 1251
reconciliation of net income to cash flows from operating acts, 1254–1255
of operating activities, 1221–1222
spreadsheet for, 1258–1261
Individual retirement accounts (IRAs), 778
Induced conversion, 796
Industry Norms and Key Business Ratios (Dun and Bradstreet), 1099
Infinite-life intangible asset, 546
Inflation, 435
Initial franchise fee, **257, 301–302**
Initial market transactions, 6
Initial markup, 476
Initial public offering (IPO), **6, 1042**
In-process research and development, 550
Input-based estimate of contract completion, 282
Installment notes, 789–790
finance leases *vs.*, 839–841
valuation of, 260–261
Installment payments, 836
Installment sales method, **297–299**
cost recovery method, **298–299**
for franchises, **301–302**
in tax reporting, **909–911**
Institute of Internal Auditors, 17
Institute of Management Accountants (IMA), 17, **352n**
In-substance fixed payments, 851
Insurance expense, 80, **861, 863, 1236**
Insurance policies, **685–686**
Intangible assets
acquisition of, 522
acquisition costs, 515
amortization of, 521–525
balance sheet classifications, 116, 117
biological assets, 590–591
cost allocation for
defense of intangible rights, 613
development costs, 113, 514, 518
costs to be capitalized, 521–525
copyrights, 522–523
franchises, 523
goodwill, 523–525
patents, 522
trademarks, 523
defense of intangible rights, 613
defined, 117, 514, 521
goodwill. See Goodwill
impairment of asset value
finite-life intangibles, 600–603
goodwill, 604–608
indefinite-life intangibles other than goodwill, 604–606
initial valuation of, 516, 522
revaluation option for, 594
as specifically identifiable, 523
valuation, IFRS *vs.* GAAP, 594–595
Intel Corporation, 116
intangible asset useful life disclosure, 591–592
inventories, 404
investment gains, 684n30
investments, 710–711
revenue recognition policy, 294
Intellectual property (IP), revenue recognition and, 298
Interbrand Sampson, 521
Interest
on advances from customers, 727
bond, 774
defined, 240
on short-term notes payable, 721
simple *vs.* compound, 240
Interest capitalization, 536–540
average accumulated expenditures, 536–537
disclosure, 540
interest rates, 537–540
period of capitalization, 536
qualifying assets, 536
Interest cost (pension), 986
changes in projected benefit obligation (PBO), 984–988
net interest cost/income, 999, 1005, 1007
as pension expense component, 982–983
Interest coverage ratio, 1256
Interest expense
capitalization of (*See* Interest capitalization)
as current liability, 721–722
effective interest method. See Effective interest method
explicit interest, 118, 246–247
income statement account, 1235
lease accounting
installment note vs. finance lease, 839–841
recording in finance lease, 843–844
no explicit interest, 247
on notes
notes receivable, 360
short-term notes payable, 721–722
operating leases, 849
used in preparation of statement of cash flows, 1235
Interest income/revenue
accounting for gains and losses, 999
accrued interest receivable, 68
as nonoperating income, 175
Interest method, effective. *See* Effective interest method
Interest rate
credit-adjusted risk-free rate, 519–520
debt security, 646
discount rate [See Discount (interest) rate]
effective [*See* Effective (annual) interest rate]
implicit, 787
imputing an, 787
in interest capitalization, 537–540
stated rate
annuities, 248
bonds, 259, 646–648, 774, 776, 780–783, 806
debt, 342
effective interest method and, 778
held-to-maturity investments, 689
noninterest-bearing note, 722
short-term notes payable, 721
unknown, solving for, 243–244
unrealistic, 360, 787–788
Interest rate futures, 685n34
Interest rate risk, A-3, A-6, A-17
Interest rate swaps, A-6–A-7
cash flow hedge, A-13–A-15
extended method for accounting for, A-18–A-22
Interest-bearing notes payable, 357–359, 721
Interim reporting, 181, 204–207
accounting changes, 206
earnings per share, 206
interim reports, 204
minimum disclosures, 206–207
revenues and expenses, 205–206
unusual items, 206
Internal control
bank reconciliation, 373–376
cash and cash equivalents, 340–341
cash controls (*See* Cash controls)
cash disbursements, 341, 373
cash receipts, 341
defined, 340
Section 404 of SOX, 15, 340
Internal events, 48, 63
Internal Revenue Service (IRS). *See also* Income taxes
challenges to tax positions, 943, 944
consent form to change accounting methods, 482
regulations
on change to LIFO method, 482
retail inventory method acceptable, 469n7
International Accounting Standards (IASs), 11
International Accounting Standards Board (IASB), 11–12
politics in standard setting, 14
convergence efforts with FASB. See Convergence efforts (FASB/IASB)
International Accounting Standards Committee (IASC), 11
International Financial Reporting Standards (IFRS), 11–12. *See also* Convergence efforts (FASB/IASB)
accounting changes and error corrections, 1192
accounts and notes receivable, 361
balance sheet presentation, 120
biological assets, 590–591
cash and cash equivalents, 342
cash flows classification, 195, 1245
classification of cash flows, 195
comprehensive income, 186
convertible bonds, 797
costs of defending intangible rights, 614
debt investments
accounting for, 663
transfers between categories, 664
depreciation, 580
earnings per share, 1119
equity investments, 673
equity method, 682
fair value option, 665
gain contingencies, 742
generally accepted accounting principles (GAAP) *vs.*, 12
government grants, 529
impairment of value, 693
finite-life intangible as, 603–604
goodwill, 607–608
indefinite-life int ass, 605
property, plant, and equip, 603–604
income statement presentation, 172
income taxes
non-tax difference affect taxes, 931
valuation allowances, 927
intangible assets valuation, 594–595
interim reporting, 206
inventory cost flow assumptions, 420
leases
operating leases, 852
reassessment of right-of-use asset and....., 852
shortcut method, 852
loss contingencies, 741
lower of cost or net realizable value (LCNRV), 463
pension plans
defined benefit plans, 998–999
postemployment benefit std, 1007–1009
prior service/past service cost, 1001
reporting pension expense, 1005
politics in setting, 14
"probable" defined, 288
property, plant, and equipment valuation, 587–588
research activities and development activities, 543
revenue recognition, 299
segment reporting, 139
shareholders' equity
debt *vs.* equity for p s..., 1058
terminology differences, 1052
software development costs, 593
stock options

- deferred tax asset, 1113
- graded vesting, 1115
- transfers of receivables, 369
- use of term "highly probable," 293

International Paper Company
- depreciation method disclosure, 579
- LIFO inventory valuation, 419
- long-lived, revenue-producing assets, 513

International standard setting, 11–12. *See also* International Financial Reporting Standards (IFRS)
Interperiod tax allocation, 946
Intraperiod tax allocation, 177, 946–947
Intrinsic values, 1108, 1142
Intuit, 544–545
Inventory
- activity ratios (*See* activity ratios)
- balance sheet classifications, 115–116, 1175
- change in accounting principle measuring, 1173
- change in inventory method
  - change to LIFO method, 481–482
  - most inventory changes, 480–481
- as current assets, 115–116
- defined, 403
- errors in, 482–485
  - error discovered the following year, 483–484
  - error discovered two years later, 484
  - visualizing the effect of, 483
- estimation techniques
  - gross profit method, 467–468
  - retail inventory method, 469–477 (*See also* Retail inventory method)
- expenditures included in, 409–412
  - freight-in on purchases, 410
  - purchase discounts, 410–411
  - purchase returns, 410
- inventory errors, 482–484
  - correction of, 1190–1191
  - discovered the following year, 483–484
  - discovered two years later, 484
- inventory management, 427–429
- misreporting, as ethical dilemma, 1191
- misstatement of, affecting net income, 1190
- periodic system (*See* Periodic inventory system)
- perpetual system (*See* Perpetual inventory system)
- physical counts of, 416
- purchase commitments, 487–488
  - contract beyond fiscal year, 488
  - contract within fiscal year, 487–488
- recording and measuring, 55, 403–429
  - inventory cost flow assumptions, 412–424
    - average cost method, 414–416
    - comparison of methods, 419–420
    - factors influencing method choice, 420–422
    - FIFO method, 416–417
    - IFRS, 420
    - LIFO method, 417–419
    - LIFO reserves and liquidation, 422–424
    - specific identification method, 414
  - inventory systems, 405–408
    - just-in-time (JIT) system, 427–429
    - periodic inventory system, 56–57, 407–408
    - perpetual inventory system, 56, 405–407
    - perpetual *vs.* periodic systems, 408
  - inventory types, 404–405
    - manufacturing inventory, 404–405
    - merchandising inventory, 404
  - physical units included in inventory, 407–408
    - goods in transit, 408–409
    - goods on consignment, 409
    - sales returns, 409
  - transactions affecting net purchases, 409–412
    - freight-in on purchase, 410
    - purchase discounts, 410–411
    - purchase returns, 410
    - recording transactions in perpetual and period system, 411–412
- shortages, retail inventory method and, 474
- subsequent measurement of, 459–466
  - lower of cost or market (LCM), 465–466
  - lower of cost or net realizable value (LCNRV), 460–463
- write-downs of (*See* Write-downs of assets)

Inventory cost flow assumptions, 412–424
- average cost method, 414–416
- comparison of methods, 419–420
- factors influencing method choice, 420–422
- FIFO method, 416–417
- IFRS, 420
- LIFO method, 417–419
- LIFO reserves and liquidation, 422–424
- specific identification method, 414

Inventory systems, 405–408
- just-in-time (JIT) system, 427–429
- periodic inventory system, 56–57, 407–408
- perpetual inventory system, 56, 405–407
- perpetual *vs.* periodic systems, 408

Inventory turnover ratio, 197, 198, 198n35, 201–203, 428
Inventory write-down, 459. *See also* Write-downs of assets
Investcorp, 588
Investing activities
- cash flows from, 193–194, 1222–1223
- noncash, 195, 1224–1225
- purchase and sale of securities. See Available-for-sale (AFS) securities; Held-to-maturity (HTM) securities; Trading securities (TS)
- sale of land as, 1222–1223, 1225, 1233–1234
- on statement of cash flows, 75

Investing and financing activity ratio, 1256
Investment(s)
- debt (*See* Debt investments)
- decision makers' perspective on accounting for, 684–685
- disclosure notes for, 122
- equity (*See* Equity investments)
- life insurance policies, 688
- as long-term assets, 116
- special purpose funds, 687–688

Investment-credit decision, 6
Investment derivatives, 685–686. *See also* Derivatives
Investment revenue
- income statement account, 1232–1233
- investment revenue receivable account, 1233

Investments by owners, 24
IP. *See* Intellectual property
Ittner, C. D., 726n11

J

Jackson, S. B., 372n25
JDS Uniphase, 485
Jiambalvo, J., 1186n12
J. C. Penney, 469
J. M. Smucker, 1171
Johnson & Johnson
- accounts receivable disclosure, 350
- case, 273–274
- fair value measurements disclosure, A-28
- stock buyback, 1062

Johnson, Sara, 657, 657n9
Journal, 52
- recording transactions in. *See* Journal entries
- sales journal, 88–89
- special journals, 52, 88
  - cash disbursements journal, 88
  - cash receipts journal, 88–90
  - purchases journal, 88

Journal entries, 52, 54–60
- accounting error correction, 1185, 1186
- accumulated depreciation, 586–587
- adjusting entries, 66–69
- retrospective approach, 1172–1173
- amortization expense, 592
- available-for-securities investments, 658–660, 692
- bad debt expense, 351, 378
- bank reconciliation, 374–376
- bonds
  - bonds payable, 1240
  - convertible, 796–798
  - issued at discount, 777–779
  - sold at premium, 781
  - straight-line method, 783
- cash equivalent sale, 1238
- change in accounting principle, 1184
- common stock, 55, 1242
- construction costs, 306, 310, 312–314
- cost of goods sold, 406
- credit loss expense, 690
- defined, 54
- depletion, 589
- depreciation expense, 1235
- detachable warrants, 800
- discounted note treated as sale, 366
- equipment sale, 1236, 1239
- equity investment, 528, 670–673, 681
- estimated returns, 348n10
- external transactions, 52, 55–56, 65–68
- fair value adjustment, 653–655, 658
- franchise revaluation, 595
- gift cards, 728–729
- goodwill, 524
- gross method vs. net method, 345
- installment note, 790
- interest as investment revenue, 648
- interest between interest dates, 807
- inventory transactions
  - change in inventory principle, 481, 1177
  - error correction, 483
  - summary entries, periodic system, 407
  - summary entries, perpetual system, 406
- leases
  - lease payments, 842, 849
  - lease liability, 260
  - lease term reassessment, 856
  - purchase option, 866
  - residual value equals estimated amount, 862
- natural resource costs, 520
- nonmonetary asset exchange, 531–533
- note exchanged for assets or services, 787–788
- oil and gas accounting, 548
- other comprehensive income (OCI), 998, 1008–1009
- payroll, 748
- pensions
  - net pension cost, 1007–1009
  - postretirement benefit expense, 1013
- petty cash system, 376
- prior period adjustment, 183, 599

Journal entries—*Cont.*
reclassification adjustment, 660
reconstructed, 1227, 1228, 1233
responsibility for recording, 88
revenue recognition, 279, 282, 285, 286
reversing entries, 85–87, 183
salaries expense, 1235
sales returns, 363, 365
spreadsheet entries for indirect method, 1258–1261
stock options, 1112
stock split, 1073
summary entry, 1244
supplies expense, 64
taxes
deferred taxes with multiple differences, 936–937
expense-related deferred tax asset, 919–921
expense-related deferred tax liability, 913–914
net operating loss carryforward, 938–940
one vs. two entries, 942
revenue-related deferred tax asset, 923–924
revenue-related deferred tax liability, 917–918
scheduled change in tax rate, 934
valuation allowance, 925–928
visualizing entries with T-accounts, 945–946
uncollectible accounts, 351–353
JP Morgan, A-3n2
JPMorgan Chase, 785
JP Morgan Chase & Company, 738, 803
Just-in-time (JIT) inventory system, 427–429

K

Kalay, A., 781n3
Kellogg Company
disclosure of pension expense, 1001
EPS and PE ratio, 1166
Keurig Green Mountain, Inc., 396
Kinney, William R., Jr., 15n21
Koonce, Lisa, 1240n12
Kroger Company
deferred taxes, 971
information from cash flow activities, 1290
inventory management, 419
inventory measurement, 403

L

Lacy, J. M., 1174n4
LA Fitness, 289
LaFond, Ryan, 15n21
Land. *See also* Property, plant, and equipment
acquisition costs, 515
balance sheet account, 1239
capitalizing cost of, 517
in consolidated financial statements, 675
defined, 515
purchase of, 1225, 1227, 1230, 1239
sale of, 1225, 1233–1234
Land improvements
acquisition costs, 515
capitalizing cost of, 518
defined, 515
Landsman, W. R., 372n27
Langenderfer, Harold Q., 17n26
Larcker, D. F., 726n11
Larson, R. K., 308n19
Last-in, first-out (LIFO) method
as inventory cost flow assumption, 414
change to FIFO method, 1175–1177
defined, 417
dollar-value LIFO retail method, 477–480
LIFO allowance, 422–423
LIFO conformity rule, 421, 579
LIFO inventory pools, 429–430
LIFO liquidations, 423–424
LIFO reserves, 422–423
LIFO retail method (of inventory estimation), 473
lower of cost or market approach (*See* Lower of cost or market)
methods of simplifying
dollar-value LIFO, 431–433
LIFO inventory pools, 429–430
periodic LIFO, 417–418
perpetual LIFO, 418–419
Laws
corporate, affecting shareholder equity transactions, 1048
illegal acts. See Illegal acts
legal life
intangible assets, 591
trademarks, 594
legal title (*See* Ownership)
litigation claims, 738–739
loss contingencies, class action lawsuits and, 733
par value and legal capital, 1058
tax laws
bonus depreciation, 616
changes in, 933–935, 942
congressional use of, 910
depreciation, 912
financial accounting standards vs., 910
installment income, 916
net operating losses, 937, 939
taxable income, 911, 922, 929
LBO. *See* Leveraged buyout
LCM. *See* Lower of cost or market (LCM) approach
LCNRV. *See* Lower of cost or net realizable value
Lease(s)
benefits of leasing, 835–836
classification of, 836–839
criteria for, 837–839
decision makers' perspective, 839
effect of residual value on, 864
direct financing, 864–865
disclosure notes for, 122
disclosures, 877
qualitative, 877
quantitative, 877
finance (*See* Finance leases; Finance/sales type leases)
impact of leasing on financial statements, 852
key criteria for, 869–871
long-term, 260
operating (*See* Operating leases)
sale-leaseback arrangements, 879–881
sales type (*See* Finance/sales-type leases; Sales-type leases)
short-term, 855
statement of cash flows and, 876–877, 1239
finance leases—lessee, 876
operating leases, 876
sales-type leases—lessor, 876–877
uncertainty in transactions, 855–869
lease payments, 857–858
dependent on index or rate, 858
in-substance fixed, 857
lease term, 856
modified terms, 857–858
purchase option, 865–867
remeasurement of lease liability, 868–869
residual value, 858–865 [See also Residual value (of leased property)]
summary of effects of uncertainties, 868
Leasehold improvements, 873
Lease payments, 838–839
nonlease components of payments
advance payments, 873
initial direct costs, 872–873
leasehold improvements, 873
residual value effect on size of, 860–061
uncertainty and, 857–858
variable
depending on index or rate, 858
in-substance fixed payments, 857
Lease terms
avoiding classification as finance leases, 839
modification of, 858
purchase option exercisable before end of, 837–838
reassessment of, 856
termination penalties, 867
uncertain, 856–857
Legal costs
franchises, 523
litigation claims, 738
patent filing, 542–543
Legal title to goods. *See* Ownership
Ledger accounts, 60–61
Lehman Brothers, 367
Lessee, 835
Lessor, 835
Leuz, C., 12n17
Lev, B., 429n16
Leverage. *See* Financial leverage
Leveraged buyout (LBO), 1217
Leverage ratio, equity multiplier, 200–203
Levitt, Arthur, Jr., 175n10
Liabilities
in accounting equation, 48–49, 53
accounting for, 774
accrued, 67–69
on balance sheet, 118–119
characteristics of, 718
corporate, for providing pension plans, 978
current, 74, 112 (*See also* Current liabilities)
defined, 24, 114, 718
long-term, 112 (*See also* Long-term liabilities)
non-debt, operating lease as, 848
payroll-related, 747–748
to be refinanced, classification of, 731
RSUs payable in cash as, 1146
SARS payable in cash as, 1144–1146
LIBOR rate, A-13
Licenses
defined, 298
revenue recognition and, 298–299
variable consideration and, 299
Life insurance policies, 688
LIFO. *See* Last-in, first-out method
Lime Energy, 175
Limited liability companies, 1053
Limited liability partnership, 1055
Lines of credit, 721
Link-chain method, 431
LinkedIn, 645
Liquidating dividend, 1069
Liquidity
decision makers' perspective on, 343, 744–745
defined, 112
ratio analysis for, 131
Liquidity ratios, 131–133
acid-test (quick) ratio, 131–132, 343, 745
current ratio, 131–132, 343, 744–745
decision makers' perspective, 343
List price, 284n8
Litigation claims, 738–739
Liu, X., 372n25
Loans
bank loans (*See* Bank loans)
cash inflow from financing activities, 194
cash outflow from investing activities, 193
cash restrictions with, 342
compensating balances, 342
credit sale as, 344
current liability, 119
installment notes, 248, 260–261, 789–790
interest capitalization example, 537–540
no explicit interest rate, 247
noninterest-bearing, 722
notes payable
long-term, 786–790
short-term, 720–723
notes receivable, 68, 115, 356–361, 365
origination fees, 835
present value of cash flows for, 527
sale-leaseback arrangement as, 879–881
secured, 362–363, 722–723

time value of money and, 295
troubled debt restructuring, 377–379, 807
Lockheed Martin Corporation, 140
Long-lived assets, 513. *See also* Property, plant, and equipment
depreciation expense, 65
reporting by geographic area, 137
Long-term assets, 116–118
intangible assets, 117
investments, 116
other, 117–118
property, plant, and equipment, 116–117
Long-term bonds, 259–260
Long-term contracts, 304–314. *See also* Contracts
accounting for profitable, 305–310
completed contract method, 304
losses, 311–314
periodic loss on profitable contract, 311–312
projected on entire project, 312–314
performance obligations in, 283–286
profitable, 305–310
accounting for accounts receivable, 305–305
accounting for cost of construction, 305–306
revenue recognition
based on percentage of completion, 308–309
comparison: over term vs. at completion, 310–311
upon completion, 307, 309
general approach, 306–307
Long-term debt
bonds (*See* Bonds; Bonds payable; Convertible bonds)
convertible bonds, 795–798
conversion option exercised, 797–798
induced conversion, 798
current maturities of, 119
debt issue costs, 785–786
disclosure notes for, 122
early extinguishment of, 795
effective interest method, 778–784
bonds issued at premium, 780–782
financial statements prepared between interest dates, 782–783
straight-line method, 783–784
zero-coupon bonds, 780
long-term notes, 786–795. *See also* Long-term notes
nature of, 773–774
reporting at fair value, 801–804
determining fair value, 801
mix and match, 803–804
reporting changes in fair value, 802–803
troubled debt restructuring, 807–811
debt continued with modified terms, 808–811
debt settled 808
Long-term investment account, 1233
Long-term leases, 260
Long-term liabilities
on balance sheet, 112 114
bonds payable (*See* Bonds payable)
classification as, 731
current vs. noncurrent liabilities, 719, 729–731
current maturities of long-term debt, 729
obligations callable by creditor, 729–730
refinancing short-term obligations, 730
defined, 119
notes payable (*See* Notes payable)
restructuring costs as 174
Long-term notes payable, 786–790
amortization schedules, 789
decision makers' perspective, 792–795
early extinguishment of debt, 795
exchanged for assets or services, 786–789
financial statement disclosures, 792–793
implicit rate of interest, 787
installment notes, 789–790
issued for cash, 786
net method of accounting for, 788n8
Long-term notes receivable, 359–360
Long-term solvency
defined, 112–113
ratio analysis for, 131
Loss contingencies, 732–742
defined, 733
IFRS *vs.* GAAP, 741
litigation claims, 738–739
product warranties and guarantees, 735–736
expected cash flow approach, 736
extended warranty contracts, 737
manufacturer's quality-assurance warranty, 735–736
subsequent events, 739–740
unasserted claims and assessments, 740–741
Losses
cash equivalent assets, 1223
from continuing operations, 169
defined, 24, 169
disposition of assets 584–585
as element of financial statement, 24
exchange lacks commercial substance, 534
on impairment of asset value. See Impairment of value
on long-term contracts, 309n20, 311–314
period loss for profitable project, 311–312
projected loss for entire project, 312–314
net operating losses 937–941
carrybacks, 940–941
carryforwards, 937–941
financial statement disclosures, 942
operating losses
as indicator of going concern, 129
valuation allowance and 926, 939
pensions
loss on pension plan assets, 982, 994
projected benefit obligation (PBO), 988, 989, 991
on purchase commitments, 487
contract beyond fiscal year, 488
contract within fiscal year, 487–488
in troubled debt restructuring, 361
Lower of cost or market (LCM) approach, 465–466
Lower of cost or net realizable value (LCNRV), 460–463
adjusting cost to net realizable value, 462–463
applying, 461–462
determining net realizable value, 460
IFRS vs. GAAP, 463
inventory write-downs, 460, 462, 463
lower of cost or market (LCM) rule compared with, 466
Lowe's Companies, Inc., 443
Lucent Technologies, 485
Lump-sum purchases, 525–526

M

Macy's, Inc.
asset impairment disclosure, 602–603
deferred tax liabilities, 971
pensions and postretirement benefit plans, 1043
reporting long-term debt, 829
retail inventory method use, 469
revenue recognition example, 279, 280
Maintenance
cost allocation of, 611
as nonlease components of lease payment, 871–872
Management
decision making. *See* Management decision making
income manipulation by. *See* Income manipulation
receivables, 364
responsibilities for financial disclosures, 125–126
Management approach to determining reportable operating segments, 133
Management decision making
accounting for investments, 684–685
asset grouping, 601
asset service life, 573
assets held for sale, 585
capital budgeting decisions, 530
cash flow ratios, 1255–1256
cash and receivables, 343
current liabilities, 744–745
debt and risk, 792–795
depreciation method selection, 578–579
earnings per share (EPS), 1138
earnings quality, 429
income taxes, 947–948
inventory management, 173, 427–429
IPOs, 1096–1100
leases
advantages of, 835–836
economic incentive for exercising options, 857
financial statement impact, 852
lease renewal options, 857
lease *vs.* buy, 839
motivation for accounting choices, 1173–1174
pension plan disclosures, 1005–1006
pension/postretirement information, 1016
receivables management, 370–372
return on equity, 1074–1075
share-based compensation plans, 1117
share buybacks, 1062
statement of cash flows, 1215–1216
Management intent, 118
Management's discussion and analysis (MD&A), 125
Management responsibilities, annual report disclosures, 125–126
Managerial accounting, 4n4
Mandatorily redeemable preferred stock, 1058
Manipulation of earnings. *See* Income manipulation
Manual accounting systems, 48
Manufacturer's quality-assurance warranty, 735–736
Manufacturing inventory, 404–405
Manufacturing operating cycle, 114
Manufacturing overhead, 404
Markdown, 470
cancellation, 470
net, 471–474
Market (cost of inventory), 465–466
Market approaches to fair value, 29
Market rate of interest
for debt investment, 646, 648, 779
for long-term notes payable, 787
Market participants, fair value and perspective of, 28
Market value
company's book value *vs.*, 113
current (*See* fair value)
bonds, 779
market value method of conversion, 798–800
relative, of bonds, 799
debt, 786n7
disclosure on financial statements, 792
equipment, 788
equity securities, 528, 1059

Market value method (bond conversion), 798
Mark-to-market (MTM) method, 996
Markup
additional, 470
cancellation, 470
on cost, 468
gross profit ratio and, 467–468
initial, 470
net, 470–474
Marriott International, 360
Mastercard, 363
Matching principle, 27n34
Materiality
accounting errors, 598–599
assessing, using quantitative benchmarks, 125
capitalization of expenditures and, 540, 611
cash payment vs. borrowing, 296
exposure draft, 13, 31, 32
illegal act impact, 125
income determination errors, 586
inventory error, 483
LIFO liquidation, 424
profit/loss disclosure and, 429
purchase discount lost, 411
rearrangements, 613
reported inventory differences, 428
straight-line method of interest determination, 784
threshold for, 21
U.S. Supreme Court's description of, 21
Material misstatement, 129
Maturity date
bond, 774, 775
cash equivalents, 114, 340, 1219
debt investment, 646, 647
installment notes, 790, 795
interest-bearing note receivable, 357
Maturity value
of bond, 774
of debt investment, 646
McAfee, 319
McCreevy, Charlie, 14, 14n20
McDonald's Corporation
franchise fee, 523
leasehold improvement accounting, 874
performance obligations, 332
revenue recognition, 279
McKesson Corporation, 422–423
McVay, S., 173n6
MD&A. *See* Management's discussion and analysis
Measurability criterion, 26
Measurement
cash flows, 18
description, 31
defined, 25
in conceptual framework, 19, 27–30
current cost, 28
fair value, 28–30
fair value option, 30
historical cost, 28
net realizable value, 28
present value, 18, 28
impairment of value, 601, 603–606
of inventories. See Inventory measurement
mixed attribute measurement model, 27
revenue recognition and, 26
subjective, 22
Measurement attributes, GAAP, 27–30
Medicaid, 1015
Medicare, 747, 748, 1010, 1011, 1014, 1015
Medtronic, 371
Merchandising inventory, 404
Merchandising operating cycle of, 114
Merck & Co., Inc., 15
accrual of a loss contingency, 734
depreciation method disclosure, 579
investments, 713
R&D costs, 541
Microsoft Corporation
acquisition of LinkedIn, 645
available-for-sale securities, 695
cash dividend press release, 1082
comprehensive income, 713
as corporation, 1052
defined contribution plan, 980–981
GAAP *vs.* non-GAAP earnings, 176
long-lived, revenue-producing assets, 513
pension plan disclosures, 1042
as publicly held corporation, 1054
R&D costs, 541
ratio analysis, 387
restricted stock awards, 1162
retained earnings deficit of, 1068
revenue recognition, 298
software development costs, 544–545
stock buyback program, 1063
Mielke, David, 1256n18
Miller, Greg, 948n17
Mills, John, 1256n18
Minimum disclosures for interim reporting, 206–207
Miscellaneous Cookie Jar Reserves, 371
Mitsubishi UFJ Financial Group, 680
Mixed attribute measurement model, 27
Mix and match in fair value option, 803–304
Model Business Corporation Act, 1055, 1056, 1058
Modified accelerated cost recovery system (MACRS), 615–616
expense-related deferred tax liabilities and, 912n1
Modified half-year convention, 583n6
Modified retrospective approach, 182, 1173, 1178
Modifying comments, 30
Mondi Limited, 591
Monetary assets, 246
Monetary liabilities, 246
Monetary unit assumption, 24–25
Money, time value of (*See* Time value of money)
Money purchase plans, 980
Moody's Investors Service, Inc., 776
Morgan Stanley, 680
lawsuit settlement, 766
pension earnings, 1043
Mortgage bond, 775
Mott, Dane, 14n19
Mueller Industries, 775
Mulford, C. W., 685n31
Multinational corporations, 11, 14, 420
Multiple-step income statement format, 172
Multiple temporary differences, 934–937
Murphy Oil Corporation, 600
Myers, L. A., 372n26

N

NASDAQ (National Association of Securities Dealers Automated Quotations), 29, 1054
Natural resources, 514
acquisition costs, 515
biological assets, 590–591
capitalizing costs of, 518
depletion of, 572, 589–591
Nelson, Mark W., 17n25
Nestlé S.A., 1083
Net assets, 119
Net cash flows
from financing activities, 194
from investing activities, 194
from operating activities, 190
Net income
accounting errors
failure to record sales revenue, 1191
inventory misstated, 1190
prior year, 1188–1192
in accrual accounting, 8, 8n7
as bottom line, 172, 173
components of
income tax expense, 170–171
operating/nonoperating income, 170
presentation of, 173
revenues/expenses/gains/losses, 168–170
that increase/decrease cash, 1251–1252
that don't increase/decrease cash, 1251
expense recognition, 27
income statement account, 171, 172, 1237
inventory costs and, 421–422
as performance measure, 172
reconciliation to cash flows, 1218, 1254–1255
Net markdown, 470
Net markup, 470
Net method
purchase discounts, 410–411
sales discounts, 345
Net operating cash flow, 7, 80
Net operating losses (NOL), 937–941
disclosure notes, 942
loss carryback, 940–941
loss carryforward, 937–939, 947
Net realizable value (NRV), 27–28, 350n11, 460–465
Netflix, 1073–1074
Neutrality, 20, 22
New share issuance, 1058–1060
earnings per share and, 1120
more than one security issued for single price, 1059–1060
preemptive rights and, 1056
share issue costs, 1060
shares issued for cash, 1058–1059
shares issued for noncash consideration, 1059
New York Stock Exchange, 1054
New York Times Company, 727
Nike, Inc.
acid-test ratio, 132
bad debts, 390–391
balance sheet, 111–120, 131
debt to equity ratio, 133
management report with financial statements, 126–127
shareholder's equity, 1047
share repurchase stock split, 1098–1099
times interest earned ratio, 134
working capital, 132
Nissan, 711–712
Nive.com, 953
No call provisions on bonds, 775
No ineffectiveness criterion, A-18
Nokia, 604
NOL. *See* Net operating losses (NOL)
Nominal rate of interest on bonds, 774
Noncancelable lease term, 836, 838, 867, 868, 873, 880
Noncash acquisitions, 525–535
deferred payments, 525–528
donated assets, 528–530
exchanges, 531–534
issuance of equity securities, 528
Noncash assets
exchanging, 1225
indirect method of reporting, 191–193
long-term note issued in exchange for, 786
property dividend, 1070
in troubled debt restructuring, 378
Noncash consideration, stock issued for, 1059
Noncash investing and financing activities
financial disclosure of, 195, 876
reporting in statement of cash flows, 1217–1218, 1224–1225
Noncommitted line of credit, 721
Noncumulative preferred shares, 1057, 1070
Noncurrent assets, 116. *See also* Long-term assets
Noncurrent liabilities. *See* Long-term liabilities
Non-debt liability, 848
Non-GAAP earnings, 176
Noncommitted line of credit, 721

- Noncumulative preferred shares, 1057
- Noninterest-bearing note, 357–359, 722
- Nonlease components of lease payment, 871–873
  - advance payments, 873
  - initial direct costs, 872–873
  - leasehold improvements, 873
- Nonmonetary asset exchange, 531
- Nonoperating income
  - defined, 170
  - earnings quality and, 175–176
  - operating income vs., 170
- Nonparticipating preferred shares, 1057
- Nonqualified stock option plan, 1113
- Nonreciprocal transfer to owners, 1070
- Nontrade receivable, 115, 343, 1222n8
- Nordstrom, 535
- Noreen, E. W., 1174n4
- Norfolk Southern Corporation, 565
- Normal shortages (inventory), 474
- Nortel Networks, 300, 326
- Notes payable, 118, 720, 786. *See also* Deferred payments; Long-term notes
  - balance sheet classification, 118
  - as current liability, 719
  - defined, 118
  - derivatives
    - cash flow hedge and, A-13–A-14
    - fair value hedge and, A-8–A9
    - interest rate swap and, A-19–A-21
  - discount on, 527–528
  - as noncash financing activity, 1226
  - recording, 54
  - short-term, 720–723
  - trade notes payable, 719–720
- Notes receivable, 356–361
  - CECL (current expected credit loss) model, 360, 361, 377
  - as current assets, 115
  - defined, 115, 343, 356
  - discounting, 296, 357–359
  - fair value disclosure, 361
  - fair value option, 361
  - IFRS vs. GAAP, 361
  - long-term, 359–360, 370–372
  - received solely for cash, 360
  - short-term interest-bearing notes, 357
  - short-term noninterest-bearing notes, 357–359
  - subsequent valuation of, 360–361
  - transfers of, 365–366
  - valuation of, 246–247
- Notes to financial statements. *See* Disclosure notes
- Noteworthy events and transactions, 124–125
- Not-for-profit corporations, 1054
- Notional value, A-6
- NRV. *See* Net realizable value
- NutraCea, 300
- Nvidia Corporation, 555
- NYSE (New York Stock Exchange), 29, 1047

O

- Objectives-oriented accounting standards, 16
- OCI. *See* Other comprehensive income; Other comprehensive income (OCI) or loss items
- Off-balance-sheet financing, 839
  - financial instruments, 686
  - leasing, 839, 857
  - notes payable, 794
- Off-balance-sheet risk, 122, 735
- Office Depot, Inc., 1038–1039
- Offsetting
  - net operating losses and, 937–940
  - in pension accounting, 996
  - permanent difference, 931
  - trial balance, 61
  - valuation allowances, 927–938
- Oil and gas accounting, 548–549
- One-line consolidation, 679
- Open accounts and notes
  - commercial paper, 723
  - short-term notes payable, 720–723
    - credit lines, 721
    - interest, 721–722
    - secured loans, 722
  - trade notes payable, 720
- Operating activities
  - cash flows from, 190–193, 1219–1221
    - direct method of reporting, 191, 1228–1246
    - indirect method of reporting, 191–192, 1250–1255
    - reconciliation of net income to, 1254
  - on statement of cash flows, 75
- Operating asset, 1222n8
- Operating cycle, 114
- Operating efficiencies, 611
- Operating income
  - defined, 170
  - earnings quality and
    - other unusual items, 175
    - restructuring costs, 174–175
  - income statement format, 172
  - nonoperating income vs., 170
- Operating leases, 847–855
  - advance payments, 873
  - classification of, 836–839
  - defined, 847
  - disclosure note, 862
  - disclosure requirements, 852, 853n14
  - discount rate and, 853
  - expense recognition between finance and, 851–852
  - finance lease compared, 847
  - GAAP vs. IFRS, 852
  - initial direct costs, 872–873
  - leasehold improvements, 873
  - recording lease expense/revenue, 848–850
    - amortization of right-of-use asset, 849–850
    - interest, 849
    - lessor and, 850
  - reporting lease expense/revenue, 850–852
    - finance vs. operating leases, 851–852
  - sale-leaseback arrangements, 879–881
  - statement of cash flow impact on, 875
- Operating segments
  - areas determined to be reportable, 138
  - defined, 138
  - reporting information by, 138–139
- Operational risk, 130
- Option pricing models, 1108
- Option-pricing theory, 1110, 1142–1143
  - intrinsic value, 1142
  - time value, 1142
    - time value of money, 1142
    - volatility value, 1143
- Options, A-5–A-6
  - bonds with detachable warrants, 798–800
  - conversion option, 795–798
  - fair value option (*See* Fair value option)
  - stock (*See* Stock option plans; Stock options)
- Oracle Corporation
  - revenue recognition, 277
  - software development costs, 544–545
- Orange County (California), exotic derivatives, 686
- Ordinary annuity, 248–254
  - future value of, 248–249
  - present value of, 250–252
  - solving for unknown values, 255–257
- Organization costs, 546–547
- Original transaction value, 28
- Other comprehensive income (OCI) or loss, 73
  - available-for-sale (AFS) securities, 657
  - credit risk and, 802
  - defined, 168, 185
  - intraperiod tax allocation and, 947
  - loss, 188
  - pension plans and, 995, 996
    - gain or loss on pension assets, 998, 999
    - prior service cost, 995, 1000
    - remeasurement gains and losses, 1005, 1008
  - reporting, 168
  - revaluation surplus, 186, 587–588
  - shareholders' equity and, 1050
  - valuation of intangible assets, 594
  - valuation of property, plant, and equipment, 587
- Other income (expense), 73, 170
- Output-based estimate of contract completion, 282
- Output measures
  - in activity-based depreciation, 577–578
  - leased assets, 843n8
  - of progress toward completion, 308–309
- Overdrafts, 342
- Overhead
  - allocation for self-constructed assets, 535–536
  - construction costs, 305
  - manufacturing, 404–405
- Overstock.com, 334
- Owens Corning, 1180–1181
- Owners' equity. *See also* Shareholders' equity
  - in accounting equation, 48–49, 53
- Ownership. *See also* Capital leases; Lessor; Shareholders' equity
  - distributions to owners, 24
  - of goods in transit, 408–409
  - title transfer at completion of contract, 281
  - title transfer upon delivery, 291
- Ozel, B., 781n3

P

- Paid future absences, 724–725
- Paid-in capital, 49, 119
  - on balance sheet, 75
  - classes of shares, 1056–1057
    - distinguishing designations, 1056
    - equity vs. debt, 1056–1057
    - preferred shares, 1056–1057
  - fundamental share rights, 1055–1056
  - issuance of shares
    - more than one security issued for single price, 1059–1060
    - no-par shares issued for cash, 1059
    - par value shares issued for cash, 1058–1059
    - resale of shares, 1065–1066
    - retired shares, accounting for, 1063–1064
    - share issue costs, 1060
    - share repurchases, 1062–1063
    - shares formally retired or viewed as treasury stock, 1063
    - shares issued for noncash consideration, 1059
    - treasury stock, accounting for, 1065
  - par value, 1057–1058
  - in shareholders' equity, 1048–1049
- PaineWebber Inc., 335, 371–372
- Parent company, 674
- Parenthetical comments, 30
- Parker, Mark G., 126–127
- Partial period depreciation, 581–583
- Participating interests, with receivables transfers, 368
- Participating preferred shares, 1057
- Par value
  - bond, 774
  - stock, 1057–1059
- Par value method, 1065
- Patents
  - acquisition costs, 515
  - amortization of, 592
  - costs to be capitalized, 521
  - defined, 522
- Pathways Commission of the American Accounting Association, accounting visualization, 4
- Paycheck, 747

Payments. *See also* Troubled debt restructuring
on account, 56
of cash dividends, 58
deferred, asset valuation and, 526–528
floating, A-6, A-9, A-19
prepayment. See Prepayments/deferrals
Payroll-related deductions, 729
Payroll-related liabilities, 747–748
PBGC. *See* Pension Benefit Guaranty Corporation
PBO. *See* Projected benefit obligation
PCC. *See* Private Company Council
PCM, Inc., 334
Peachtree Accounting Software, 48
Peasnell, K., 372n27
Penalties for lease termination, 867
Pension Benefit Guaranty Corporation (PBGC), 984
Pension expense
components of, 982–983, 992–996
amortization of net loss or net gain, 995–996
amortization of prior service cost, 994–995
illustration, 993
interest cost, 994
return on plan assets, 994
service cost, 992
income smoothing, 996–997
reporting, IFRS, 1005
reporting, in income statement, 1000–1001
Pension obligation, 983–990. *See also* Projected benefit obligation; Time value of money
accumulated benefit obligation (ABO), 983–984
alternative measures, 983
for employee pool, 989
projected benefit obligation (PBO), 983–989
revisions to, 982
valuation of, 261–262
vested benefit obligation (VBO), 983–984
Pension plan(s)
assets (*See* Pension plan assets)
bond yield curve to calculate costs for, 989–990
comprehensive income, 1003–1004
decision makers' perspective, 1005–1006
defined benefit, 979–982
defined contribution, 979–981
disclosure notes for, 122
income tax considerations, 1004
nature of, 978–983
pension obligation
accumulated benefit obligation (ABO), 983–984
projected benefit obligation (PBO), 983–989
vested benefit obligation (VBO), 983–984
pension spreadsheet, 1004–1005
plan assets, 990–992
projected benefit obligation (PBO) (*See* Projected benefit obligation)
qualified plan, 979
postretirement benefits other than (*See* Postretirement benefit plans)
recording funding of plan assets, 1001–1002
regulations, 979, 996, 1041
reporting funded status of, 992
reporting issues, 998–1006
comprehensive income, 1003–1004
decision makers' perspective, 1005–1006
funding of plan assets, 1001–1002
gains and losses, 998–999
income tax considerations, 1004
pension expense, 1000–1001
settlement or curtailment of plan, 1006
risk and, 970, 971, 972
settlement or curtailment of, 1006
shift from defined benefit to defined contribution, 986n6, 1006
tax advantages of, 979
underfunded, 991
Pension plan assets, 990–992
defined, 990
expected return on, 991, 993, 994
funded status of plan, 992, 1000
recording the funding of, 1002–1003
reporting funded status of, 1000
gains or losses on, 982, 993–996
Pension Protection Act of 2006, 1002
Pension spreadsheet, 1004–1005
Pep Boys, 874
PepsiCo, Inc.
change in reporting entity, 1182
debt to equity ratio, 793–794
Percentage depletion, 590
Percentage-of-completion method, 304–305
Performance obligations
defined, 279
identifying
customer options for additional goods or services, 289–290
prepayments, 289
warranties, 289
principals *vs.* agents, 294
revenue recognition
multiple obligations, 283–286
over period of time, 281–282
simple example, 279
transaction price allocated to, 284
transaction price determination, 284
Performance ratios, 131
Period costs, product costs *vs.*, 410n4
Periodic average cost (of inventory), 414–415
Periodic FIFO, 416
Periodic inventory system, 56–57, 407–408
cost flow assumptions (*See* Inventory cost flow assumptions)
defined, 407
freight costs in, 410
gross profit method and, 411–412, 467
perpetual system vs., 408
purchase returns in, 410
recording transactions under, 56, 411–412
retail inventory method, 469
Periodic LIFO, 417–418
Periodicity assumption, 24–25
Permanent accounts, 51
Permanent differences
accounting income *vs.* taxable income, 929–931
taxes on unrepatriated foreign earnings, 931
Perpetual average cost (of inventory), 415–416
Perpetual FIFO, 417
Perpetual inventory system, 56, 405–408
cost flow assumptions (*See* Inventory cost flow assumptions)
freight costs in, 410
gross profit method and, 411–412
periodic system vs., 408
purchase returns in, 410
recording transactions under, 56, 411–412
technology and inventory accounting, 406
Perpetual LIFO, 418–419
Petty cash fund, 376–377
Pfeiffer, R., 738n22
Pfizer, 117, 333
Pharmacyclics Inc., 546
Philippine Townships, Inc., 1070
Pier 1 Imports, Inc., 567
Pilgrim's Pride Corp., 398–399
Piper Jaffrey, 686
Pledging accounts receivable, 362, 722
Politics
in accounting standard setting, 13–14
illegal contributions, 125
Portfolio, securities, 654
Post-closing trial balance, 54, 78
Posting
adjusting entries, 63–65
defined, 54
reference
on general journal, 59
on ledger account, 60
Postretirement benefit plans, 1009–1015
accounting for, 1012–1015
APBO (*See* Accumulated postretirement benefit obligation)
comprehensive illustration, 1014–1015
decision makers' perspective, 1016
EPBO (*See* Expected postretirement benefit obligation)
health *vs.* pension benefits, 1010
net cost of benefits, 1010–1011
postretirement benefit obligation, 1015
attribution, 1012
measuring, 1011–1012
Potential common shares
antidilutive effect of, 1128–1130, 1137
convertible bonds as, 1119
diluted earnings per share and, 1123
dilutive effect of, 184
restrictive stock awards as, 1133
summary, effect on EPS, 1135
PPI. *See* Producer Price Index
Predictive value, 21
Predominance principle, 1224
Preemptive rights, 1056
Preferred dividends, 1122–1123
Preferred stock
convertible, 1057, 1127–1128
cumulative *vs.* noncumulative, 1057, 1070
dividends on, 1070
equity *vs.* debt, 1057, 1058
return on equity calculation and, 201
IFRS *vs.* GAAP, 1058
rights of, 1056–1057
Premium on bonds, 647
effective interest method, 778–784
reporting in statement of cash flows, 123
Premiums, promotional, 738
Prepaid advertising, 18, 64
Prepaid expenses, 64–65, 116
Prepaid insurance, 80
Prepayments, 64–67
contract performance obligations and, 289
deferred revenues, 65–66
prepaid expenses, 64–65
Present value (PV)
of an annuity
annuity due, 252–253
deferred annuity, 253–254
ordinary annuity, 250–252
valuation of installment notes, 260–261
valuation of long-term bonds, 259–260, 259–262
valuation of pension obligations, 261–262
compound interest and, 243
defined, 242
expected cash flow approach and, 519
financial calculator and Excel to find, 255
GAAP measurement attribute, 28
of single amount, 242–245
single cash amount, 246–247
solving for other values in formula, 243–245
solving for unknown values, 255–257
Present value (PV) tables
present value of $1, P2
present value of an annuity due of $1, P6
present value of an ordinary annuity of $1, P4

Price
list, 284n3
stand-alone selling, 284
transaction, 284
Price-earnings ratio, 1075, 1138
Priceline.com, 295
PricewaterhouseCoopers, 964
Principal
bond, 774
of debt investment, 646
seller as, performance obligation and, 294–295
Principles-based accounting standards, 16
Prior period adjustments, 183–184, 1186–1187
errors involving depreciation calculation, 599
for inventory error, 483–484
Prior service cost, 986
amortization of, pension expense, 994–995
service method of allocating, 1019
Private Company Council (PCC), 14
accounting for goodwill, 607
derivatives and hedging, A-22
Procter & Gamble, 686
Producer Price Index (PPI), 431
Product costs, 410
Product warranties and guarantees
expected cash flow approach, 736
extended warranty contracts, 737
manufacturer's quality-assurance warranty, 735–736
Production facilities, as assets, 513
Profit
gross. *See* Gross profit *entries*
selling, 846
Profitability
as liquidity indication, 343
risk and, 134–135
Profitability analysis, 197–201
activity ratios, 197–198
asset turnover, 197
inventory turnover, 198
receivables turnover, 198
DuPont framework, 200
illustration, 201–203
profitability ratios, 199–201
profit margin on sales, 199
return on assets, 199–200
return on shareholders' equity, 200–201
summary of ratios, 201
Profitability ratios, 199–201
profit margin on sales, 199
return on assets, 199–200
return on shareholders' equity, 200–201
Profit margin on sales, 199
Profit sharing plan, 980
Pro forma earnings per share, 1138
Projected benefit obligation (PBO), 983–989, 999
changes in, 984–988
gain or loss on PBO, 988
interest cost, 986
prior service cost, 986–988
retirement benefits payments, 989
service cost, 986
illustration considering entire employee pool, 989
Promissory note, 720
Property dividends, 1070
Property, plant, and equipment
costs to be capitalized, 516–525. *See also* Capitalization of costs
defined, 514
depreciation of (*See* Depreciation)
disclosure notes for, 123
impairment of value (*See* Impairment of value)
as long-term assets, 116–117
noncash acquisitions, 525–535
deferred payments, 525–528
donated assets, 528–530
exchanges, 531–534
issuance of equity securities, 528
research and development (R&D), 541–547
determining R&D costs, 541–544
R&D performed for others, 545
R&D purchased in business acquis, 546
start-up costs, 546–547
self-constructed assets, 535–540
interest capitalization, 536–540
overhead allocation, 535–536
software development costs, 544–545
typical acquisition costs, 515
valuation at acquisition
lump-sum purchases, 525–526
types of assets, 514–515
valuation of, IFRS *vs.* GAAP, 587–588
Prospective approach, 182
change in accounting principle, 1173
changing depreciation, amortization, depletion methods, 1179
when mandated by accounting literature, 1179
when retrospective approach not practical, 1178–1179
Protiviti, Inc., 15n23
Provision for income taxes, 170
Proxy statement, 127
Public Company Accounting Reform and Investor Protection Act of 2002, 10n12. *See also* Sarbanes-Oxley Act
Public Company Accounting Oversight Board AS 2201, 340
Purchase commitments, 487–489
contract period extends beyond fiscal year, 488
contract period within fiscal year, 487–488
defined, 487
Purchase discounts, 410–411
Purchase option, 837–838, 865–867
exercisable before the end of lease term, 867
termination penalties, 867
Purchase returns, 410
Purchases account, 57
Purchases journal, 88

Q

Quaker, 1182
Qualified institutional buyers, 785n4
Qualified opinion, 130
Qualified pension plans, 979
Qualified stock option plans, 1113
Qualitative characteristics of reporting information, 20–23
cost effectiveness, 22
enhancing qualitative characteristics, 22–23
fundamental qualitative characteristics, 20–22
Quality-assurance warranties, 289, 735
Quasi reorganizations, 1079–1080
Quick assets, 132, 132
Quick (acid-test) ratio, 131–132, 343, 745
QuickBooks, 48

R

Radio frequency identification (RFID) tags, 406
Rajan, M. V., 726n11
Rajgopal, S., 173n4, 173n5
Ralph Lauren Corporation, 232–234
Rate of return on stock investment, 6
Ratio analysis
activity ratios, 197–198
asset turnover, 197, 199–203
average collection period, 198, 201, 202, 370, 371
average days in inventory, 198, 428
fixed-asset turnover, 530
inventory turnover, 197, 198, 198n35, 201–203, 428
receivables turnover, 198, 201–203, 343, 370, 371
financial statements and, 131
cash flow ratios, 1255–1256
defined, 131
earnings-price ratio, 1075
gross profit, 428–429, 467–469
leverage ratio, equity multiplier, 200–203
liquidity ratios, 131–135
acid-test (quick) ratio, 131–132, 343, 745
current ratio, 131–132, 343, 744–745
performance ratios, 1256
price-earnings ratio, 1075, 1138
profitability ratios, 199–201
profit margin on sales, 199
return on assets (ROA), 135, 199–202, 366, 794, 839, 871, 982, 998, 1000, 1005, 1008, 1115
return on shareholders' equity (ROE), 131, 200–201, 1074–1075
solvency ratios, 133–135
debt to equity, 133, 792
times interest earned ratio, 134–135, 793–794
sufficiency ratios, 1256
Raw materials, 404
Real estate, revenue recognition, 295
Realignment costs, 174–175
Realization principle, 26, 278. *See also* Revenue recognition
Realized gains (losses), trading securities
AFS investments and, 657, 660–661, 668–669
fair value vs. equity method, 681
HTM investments and, 651–652
IFRS vs. GAAP, 673
Rearrangements, 613
"Reasonably assured," 856n16
Rebates, 737
Receivables
accounts receivable (*See* Accounts receivable)
accrued, 68–69
assignment of, 362–363
current (*See* Current receivables)
defined, 343
ethical issues, 354
factoring, 363–354, 722
financing with (*See* Current receivables, financing with)
nontrade receivables, 115, 194 343, 1222, 1222n8
notes receivable (*See* Notes receivable)
sale of, 363–365
trade receivables, 115, 343
used as collateral, 362–363, 365, 722
Receivables management, 382-383
Receivables turnover ratio, 198, 201–203, 343, 370, 371
Reclassification adjustment, AFS securities, 659–661
Recognition
defined, 25
description, 31
expense recognition, 26–27
general recognition criteria, 26
revenue recognition, 26 (*See also* Revenue recognition)
Reconstructed journal entries, 1227, 1228
Recording transactions. *See* Accounting processing cycle
Recourse obligation, 365
Recoverability test, 601, 603, 604
Redemption privilege, 1057
Refundable deposits, 726
Refund liability, 347
Rego, Sonia O., 948n18
Reinvestment ratio, 1256
Related-party transactions, 124
Relative market values (of bonds), 799
Relevance criterion, 20–21, 26
Reliability criterion, 26
Remeasurement cost, 1007–1008
Remeasurement gains and losses, 998
Renault, 711–712

Renegotiation of debt, 377–379, 807–811
Rent
 deferred rent revenue, 58–61, 66–67, 70
 prepaid rent, 56, 59–61, 65, 66
Rent expense
 journal entry for, 65–66, 70
 record lease payment as, 855, 857
Reorganization. *See also* Bankruptcy
 quasi reorganizations, 1079–1080
 restructuring costs, 174–175
Repairs, 611
Repatriation of foreign earnings, 931
Replacement cost (RC), 465–466
Replacement depreciation method, 617
Reportable operating segment, 138–139
Reporting entity, change in, 1183
Reporting unit, 605, 607
Representational faithfulness, 22, 27, 30, 31, 1171
Research, GAAP definition, 541
Research and development (R&D), 541–547
 acquired in-process, 546
 determining R&D costs, 541–544
 development, defined, 545
 expensing of costs, 522
 FASB requirements, 541
 for intangible assets, 117
 GAAP definitions, 541
 GAAP *vs.* IFRS, 543
 legal costs, 542–543
 purchased in acquisitions, 546
 R&D performed for others, 545
 R&D purchased in business acquisition, 546
 research, defined, 541
 software development costs, 544–545
 start-up costs, 546–547
Residual approach, 297
Residual asset, 861
Residual value
 asset, 573–576
  change in accounting estimate, 596–598
  impairment of value, 600
  intangible assets, 591, 594
  natural resources, 589, 590
 defined, 573
 leased property, 858–865
  amounts recorded by lessor and, 861–864
  guarantee of residual value, 859
  lease classification and, 864
  size of lease payment and, 860–861
 sales-type lease with, 861
Restoration costs, 518
Restricted cash
 compensating balances and, 341–343
 statement of cash flows and, 1219
Restricted stock, 126–127
Restricted stock awards
 defined, 1106–1107
 in EPS calculations, 1133–1134
 executive pay in, 127
 increase in popularity of, 1116
Restricted stock plans, 1106–1107
Restricted stock units (RSUs), 1106–1107
 in EPS calculations, 1133
 payable in cash, 1146
Restructuring costs, 174–175
Retail inventory method, 469–477
 cost flow methods, 471–473
  approximating average cost, 471–472
  conventional retail method, 472–473
  LIFO retail method, 472–473
 dollar-value LIFO retail method, 477–480
 LIFO retail method, 472–473
 other issues, 473–475
 retail terminology, 470–471
Retained earnings, 1049, 1068–1076
 in accounting equation, 49, 114
 balance sheet account, 1243
 capitalization of, 1074
 cash dividend payments and, 1069–1070
 characteristics of, 1068
 in shareholders' equity, 1049–1050
 decision makers' perspective, 1074–1075
 defined, 49, 120
 dividends, 1068–1070. See also Dividends
  cash dividends, 1069–1070
  liquidating dividend, 1069
  property dividends, 1070
  retained earnings restrictions, 1069
 error correction, 183
 ethical dilemma, 1076
 ownership interests reported as, 1048
 presentation of
  balance sheet, 74, 120, 1004
  income statement, 72, 75
  recording indirectly, 51
  revising, 1175
  statement of shareholders' equity, 76
  statement of cash flows, 1243
 prior period adjustments to, 183, 483–485, 1183, 1186–1187
 quasi reorganizations, 1080
 restrictions on, 1069
 share repurchase and, 1065
 stock dividends, 1070–1072
 stock splits, 1072–1073
 temporary accounts closed to, 76–77, 85, 87
Retired stock, 1063
Retirement benefits. *See also* Pension plans
 payment of, 989
Retirement of assets, 585
Retirement depreciation method, 617
Retirement method of accounting for treasury stock, 1065
Retrospective approach, 1174–1178
 change in accounting principle, 182, 1172–1173
  adjust accounts for change, 1176–1177
  disclosure notes, 1177–1178
  revise comparative financial statement, 1175–1176
Return. *See also* Rate of return
 defined, 4n2
 expected, 6
Return on assets (ROA), 135, 199–202, 366, 794, 839, 871, 982, 998, 1000, 1005, 1008, 1115
Return on pension plan assets, 991, 993, 994
Return on shareholders' equity (ROE), 131, 200–201, 1074–1075
Revaluation option
 global perspective, 186, 587–588, 594–595
 intangible assets, 594–595
Revaluation surplus, 587–588
Revenue(s)
 in accrual accounting, 8
 collected in advance, 726–729
 from continuing operations, 168
 deferred, 65–66, 727 (*See also* Deferred revenue)
 defined, 24, 26, 168, 277
 as element of financial statement, 24
 intentional misstatement of, 175
 interim reporting, 205–206
 from investments, 1222–1223
 recognition. See Revenue recognition
 relation to expenses, 27
 reported on tax return, 910
 revenue/expense approach, 32
 sales revenue. See Sales revenue
 service revenue. See Service revenue
Revenue recognition, 26, 276–337
 considerations when applying steps to, 280
 core principle, 278
 IASB *vs.* GAAP, 299
 introduction, 278–280
 long-term contracts, 304–314
  accounting for profitable, 305–310
   construction costs & a r, 305–306
   contract completion, 307, 309–310
   general approach, 306–307
   over time acc to % of c, 308–309
  losses, 311–314
   for entire project, 312–314
   periodic, for profitable project, 311–312
  over term *vs.* completion of contract, 310–311
 multiple performance obligations, 283–286
 over period of time, 281–82
 premature, 464
 principals *vs.* agents, 294–295
 at single point in time, 280–281
 special issues, 286
  contract identification, 287–288
  disclosures
   balance sheet, 302
   disclosure notes, 302
   income statement, 302
  performance obligation identification, 289–290
  revenue recognition when performance obl satisfied, 298–301
  transaction price allocated to performance obligations, 297
  transaction price determination, 290–297
 summary, 303
 summary of fundamental issues, 286
Revenue/expense approach, 32
Reversals
 IFRS vs. GAAP, 463, 603
 loss, 603
 revenue, 293, 294, 303, 349
 temporary difference, 913–914, 916
Reverse stock split, 1074
Reversing entries, 85–87
Revised Model Business Corporation Act, 1063n13, 1068n17
RFM Corporation, 1070
Ricks, William E., 421n12
Right of access, 298–298
Right of conversion, 1057
Right of return, 293–294
Right-of-use asset
 amortization of
  finance/sales-type lease, 843–844
  operating lease, 849–850
 asset recorded as, 835, 841
 modification of lease terms and, 858
Risk (uncertainty), 6
 cash flow risk, A-3
 bond issues, 776, 780
 credit risk, 519, 803, 836
 debt and, 792, 796
 debt issue costs, 785
 default risk, 130, 133–134, 792
 deferred tax liabilities and, 948
 defined, 4n2
 derivatives used to hedge, A-3–A-7
 disclosure of, for derivatives, A-18
 fair value risk, A-3
 foreign currency, A-3
 hedging against (*See* Derivatives)
 of holding cash, 343
 interest rate risk, A-3
 in lease transactions, 836, 838–839, 852–853, 857, 859, 871
 of liability classification, 719, 729
 management perspective, 792–795, 839, 852
 off-balance-sheet, 122
 operational risk, 130
 pension plans, 979–982, 1006
 predicting, 4
 price risk, A-3
 profitability and, 134–135
Risk analysis, 130–135, 744
 ethical issues, 133

financial statement information used in, 130–135
liquidity ratios, 131–133
solvency ratios, 133–135
ROA. *See* Return on assets
Rockness, Joanne W., 17n26
ROE. *See* Return on shareholders' equity
Rohm and Haas Company, 1182
Royalties, 299
RSUs. *See* Restricted stock units
Rules-based accounting standards, 16–18

S

SABMiller, 645
Salamon, G., 421n13, 1173n3
Salaries
accrued, 70, 86
as accrued liability, 118, 723–726
recording payment of, 58
salaries expense, 81, 86, 1234–1235
salaries payable, 118
Sales commission, 525
Sales discounts, 345
forfeited account, 346
methods of recording, 345
retail inventory method and, 474
Sales journal, 88–89
Sale-leaseback transaction, 879–881
Sales returns, 294, 346–349
defined, 347
recording, in inventory, 409
retail inventory method and, 474
Sales revenue
income statement account, 1230–1232
T-account and, 81
Sales-type leases, 835–837. *See also* Finance/sales-type leases
impact on lessor's statement of cash flow, 876–877
with residual value, 861
with selling profit, 846–847
Salkever, Alex, 1109n8
Salvage value, 573
Samsung, 296, 847
Sanofi-Aventis
depreciation, 580
financing with receivables, 398
Sarbanes-Oxley Act (SOX) of 2002, 10n12, 15–16
key provisions, 16
requirements of, 126
Section 401, non-GAAP earnings, 176
Section 404, 15, 340
Sarbanes-Oxley Compliance Survey (2011), 15n23
SARs. *See* Stock appreciation rights
Savings plan
Microsoft disclosure note about, 981
as money purchase plan, 980
SCF. *See also* Statement of cash flows
Schaff, William, 485
S corporation, 1054
Sears, 535
Sears Holdings Corporation
explanation for valuation allowance, 926
retail inventory method use, 469
SEC. *See* Securities and Exchange Commission
Secured borrowing, 362–363
Securities
antidilutive (*See* Antidilutive securities)
available-for-sale (*See* Available-for-sale (AFS) securities)
convertible (*See* Convertible securities)
equity method of reporting (*See* Equity method)
fair value of (*See* Fair value option)
held-to-maturity securities (*See* Held-to-maturity (HTM) securities)
hybrid (*See* Convertible securities)
trading securities (*See* Trading securities)
Securities Act (1933), 9
Securities Exchange Act of 1934, 9, 127
Securities and Exchange Commission (SEC)
accounting pronouncements, 10
annual reporting requirements, 122
on cookie jar accounting, 1174
creation of, 9
defined, 9
on disclosure of litigation, 738
EDGAR (Electronic Data Gathering, Analysis, and Retrieval), 111
expensing of stock options, 1109
Financial Reporting Releases (FRRs), 9n11
on fiscal year, 25
full-cost method (oil and gas accounting) and, 548–549
GAAP oversight by, 16
GE's "cookie jar accounting" and, 1174
on intentional misstatement of revenue, 175
interim reporting, 181
materiality assessment, 125
non-GAAP earning, 176
oil and gas accounting, 549
privately held corporations and, 1054
publications of, 9n11
registration of bond issues, 785
registration of commercial paper issues, 723
reporting requirements of responsibilities under SOX, 16
Staff Accounting Bulletins, 9n11
website, "Investor Relations" link, 111
Securitization, 364
Security (damage) deposits, 726
Segment information, 138–140
GAAP *vs.* IFRS, 139
by geographic area, 139–140
information about major customers, 139
by operating segment, 138–139
Segment reporting, 139
Self-constructed assets, 535–540
interest capitalization, 536–540
average accumulated expenditures, 536–537
disclosure, 537–540
interest rates, 537–540
period of capitalization, 536
qualifying assets, 536
overhead allocation, 535–536
Self-correcting errors, 484, 1188
Seller as principal or agent, 294–295
Selling price
bond, 775–777
current, 469
Selling profit, 846
Semtech Corporation, 514
Seneca Foods Corporation, 482
"Separately identifiable" criterion, 284
Separation of duties, 341
Serial bonds, 775
Service cost
changes in projected benefit obligation (PBO), 988–989
as component of net pension cost, 1007–1008
defined, 986
as pension expense component, 992–993
prior service cost, 987, 994–995, 1019
Service (useful) life
in asset cost allocation, 573
change in estimate of, 596–598
defined, 572–573
depreciation and, 574 (*See also* Depreciation entries)
employees, 995, 1009, 1020
extension of, 611
natural resource depletion and, 589
Service method approach, 995n12, 1019–1020
SFACs. See Statements of Financial Accounting Concepts
SFASs. See Statements of Financial Accounting Standards
Shakespeare, C., 372n26, 372n27
Share-based compensation, 1105–1118
decision makers' perspective, 1117
employee share purchase plans, 1117
restricted stock plans, 1106, 1108–1116
stock option plans, expensing debate, 1108–1109
Share buybacks (repurchases), 1062, 1064, 1138
Shareholders
earnings available to common, 1122–1123
ownership interests of, 1047. See also Shareholder's equity
risk-return trade-off, 793
significant influence and, 670
Shareholders' equity, 50, 114. *See also* Owners' equity
on balance sheet, 75, 119–120
corporate organization and, 1052–1056
financial reporting overview, 1048–1052
accumulated other comprehensive income, 1050–1051
paid-in capital, 1048–1049
retained earnings, 1049–1050
statement of shareholders' equity, 1051
treasury stock, 1050
IFRS *vs.* GAAP, 1052
nature of, 1047–1055
paid-in capital
classes of shares, 1056–1057
fundamental share rights, 1055–1056
issuance of shares, 1058–1066
par value, 1057–1058
reporting, 1051
retained earnings, 1068–1074 (*See also* Retained earnings)
return on, 1074–1075
return on equity calculation and, 201
statement of (*See* Statement of shareholders' equity)
Shareholders' rights
fundamental, 1055–1056
preferred shares, 1056–1057
protection of, 1054
summarized in disclosure notes, 1051
transference of, 1053
treasury shares, 1063
Shares. *See* Stocks
issuance of (*See* Paid-in capital, issuance of shares)
Shevlin, Terry, 948n19
Shipping costs, 291
Shoe Carnival, Inc.
disclosure of tax expense, 927–928
effective tax rate, 930–931
effective tax rate reconciliation, 942
Short-term interest-bearing notes, 357
Short-term investments, 115, 1238
Short-term leases, 855
Short-term noninterest-bearing notes, 357–359
Short-term notes payable, 720–723
credit lines, 721
secured loans, 722
Shroff, Nemit, 948n19
Sick days, 724
Siemens AG, 463
government grants to, 529
research and development, GAAP *vs.* IFRS, 567
statement of cash flows (partial), 195
Significant influence, 674
Silicon Valley, stock option compensation in, 13
Simple capital structure, 1119
Simple interest
compound interest vs., 240
defined, 240
Single entity concept, 679
Single statement approach, 73

Single-step income statement format, 171
Sinking fund debentures, 775
SiriusXM Radio Inc., 35
Six Flags Entertainment, 281, 808
Skaife, Hollis Ashbaugh, 15n21
Skinner, Doug, 948n17
SM Energy Company, 520
Smith, E., 421n13, 1173n3
Smithfield Foods, Inc., 722–723
Smucker's, 1056
Software development costs, 544–545
  amortization of, 593
  IFRS *vs.* GAAP, 593
Solvency, long-term, 112, 131
Solvency ratios, 133–135
  debt to equity ratio, 133
  times interest earned ratio, 134–135
Source documents, 52
Southwest Airlines, 274
SOX. *See* Sarbanes-Oxley Act
SPE. *See* Special purpose entity
Special journals, 52, 88
Special purpose entity (SPE), 364
Special purpose funds, 687–688
Specific goods LIFO, 429n17
Specific identification method, 414
Specific interest method, 540
Spiceland, J. David, 1072n20
Spot-rate method (to determine pension expense), 990
Spreadsheet preparation of statement of cash flows
  balance sheet accounts, 1238–1246
    bonds payable, 1240
    buildings and equipment, 1239
    common stock, 1242–1243
    land, 1239
    retained earnings, 1243
    short-term investments, 1238
  completing the spreadsheet, 1244–1246
  income statement accounts, 1230–1237
    cost of goods sold, 1234
    depreciation expense, 1235
    gain on sale of land, 1233–1234
    income tax expense, 1236–1237
    insurance expense, 1236
    interest expense, 1235
    investment revenue, 1232–1233
    loss on sale of equipment, 1236
    net income, 1237
    salaries expense, 1234–1235
    sales revenue, 1230–1232
  indirect method, 1258–1261
Sprint Corporation, 753
Staff Accounting Bulletins, 9n11
Stand-alone selling price, 284
Stand ready obligation, 743
Standard & Poor's Corporation, 776
Staples, Inc., 115
Starbucks Corporation, 123
  accrual of litigation contingencies, 739
  revenue recognition, 279
Start-up costs, 546–547
Stated rate of interest
  on bond, 774
  on debt investment, 646, 647
State Farm Insurance, 168
Statement of cash flows (SCF), 5
  cash, cash equivalents, and restricted cash, 1219
  cash inflows and outflows, 1216–1217
  classifying cash flows, 190–195
    financing activities, 194
    investing activities, 193–194
    noncash investing and financing acts, 195
    operating activities, 190–193
  decision makers' perspective, 1215–1216
  defined, 75, 168, 189
  illustration, 75, 1218
  impact on leases, 876–877
  indirect method of reporting operating activities, 680n25
  preparation of, 75, 1225–1227
    direct method, 1230–1237. See also Direct method
    indirect method, 1251–1255. See also Indirect method
    T-account method, 1261–1262
    using spreadsheet, 1228–1246.
  primary elements of, 1219–1225
    cash flows from financing activities, 1223–1224
    cash flows from investing activities, 1222–1223
    cash flows from operating activities, 1219–1222
    noncash investing and financing activities, 1224–1225
    reconciliation with change in cash balance, 1224
  reporting a bond discount or premium, 1240–1241
  structure of, 1216–1225
  usefulness of, 190
Statement of comprehensive income, 1051
  available-for-sale (AFS) securities on, 646, 661
  gain or loss on pension assets, 996
  held-to-maturity (HTM) securities on, 651
  OCI reported in, 947, 996, 998–1000, 1003
  preparation of, 73
  purpose of, 73
  trading securities reported on, 656
Statement of financial position, 5, 112. *See also* Balance sheet
Statement of operations, 5. *See also* Income statement
Statement of shareholders' equity, 5, 1051, 1053
  preparation of, 76
  prior period adjustment for, 183–184, 483–484, 599, 1183, 1186
  purpose of, 1051
  revising, for change in accounting principles, 1176
State employment taxes, 747, 748, 757
Statements of Financial Accounting Concepts (SFACs), 18–19
Statements of Financial Accounting Standards (SFASs), 10
Statutory depletion, 590
Stock(s)
  authorized shares, 1055
  common (*See* Common shares)
  compensation based on (*See* Share-based compensation)
  contingently issuable shares, 1134–1135
  EPS (*See* Earnings per share)
  issuance of (See New share issuance)
  preferred (*See* Preferred stock)
  retained earnings (*See* Retained earnings)
  retired shares (*See* Retired stock)
  sale of, as financing activity, 1216
  share buybacks (repurchases) [*See* Share buybacks (repurchases)]
  treasury stock (*See* Treasury stock)
Stock appreciation rights (SARs), 1143–1146
  debt *vs.* equity, 1144
  defined, 1143
  payable in cash (liability), 1144–1146
  payable in shares (equity), 1144
  restricted stock units payable in cash (liability), 1146
Stock award plans, 1106–1107
Stock buyback programs, 1062–1063
Stock dividends, 1070–1072. *See also* Dividends
  basic EPS and, 184, 1119
  large, 1073–1074
  reasons for, 1072
  statement of cash flows and, 1225
  stock market reaction to, 1071–1072
  stock splits as, 1072–1073
Stockholders' equity, 114, 119. *See also* Shareholders' equity
Stockholders' investment. *See* Shareholders' equity
Stock market
  crash of 1929, 9
  downturns, reverse stock splits and, 1074
  pension plans and, 978, 991
  reaction to stock dividends, 1071–1072
Stock option(s), 126. *See also* Options
  actual exercise of, EPS and, 1125
  antidilutive securities and, 1128–1130
  defined, 1108
  diluted EPS and, 1123–1125
  option-pricing theory, 1142–1143
Stock option plans
  decline in popularity of, 1116
  exercise of options, 1112
  expensing debate
    current requirement to expense, 1109
    failed attempt to require expensing, 1108–1109
    voluntary expensing, 1109
  fair value of options, 1109–1112
  forfeitures, 1110–1112
  graded vesting, 1114–1115
  market conditions, 1116
  performance conditions, 1115–1116
  qualified *vs.* nonqualified incentive plans, 1113
  stock appreciation rights (SARs), 1143–1146
  tax consequences of, 1113
  unexercised options expire, 1112
Stock rights
  antidilutive securities and, 1128–1130
  diluted EPS and, 1123–1125
Stock split, 1072–1074
  capitalizing retained earnings when recording, 1074
  defined, 1072
  effected as stock dividends, 1073–1074
  fractional shares, 1074
  reverse, 1074
Stock warrants
  antidilutive securities and, 1128–1130
  bonds with detachable stock purchase warrants, 798–800
  diluted EPS and, 1123–1125
  issue price allocation, 799
  more than one security issued for a single price, 1059–1060
Straight-line method (of depreciation), 783–784
  decision makers' perspective, 578
  frequency of use, 579
  overview of, 574–575
  stock option plans with graded vesting, 1114
  switch from accelerated method to, 574–575
Stranded tax effects, 947
Subordinated debenture, 775
Subscription revenues, 911
  Internet gaming, 282–286, 297, 302
  magazine, 64–66, 727, 922–924
Subsequent events
  disclosure notes, 123–124
  loss contingencies, 739–740
Subsidiary, investee as, 674
Subsidiary accounts, 51
Subsidiary ledger, 87–88
Substance over form concept, 837
Successful efforts method, 548–549
Sufficiency ratios, 1256
Summary of significant accounting policies, 123

Summary transactions, 57
Sum-of-the-years'-digits (SYD) method, 576–577
Sunbeam Corporation, 300, 335, 372
Supplemental schedules and tables, 30
Supplies,
prepayments, 64–65
recording purchase, 64
Swaps. *See* Interest rate swaps
Sycamore Networks, 485
Symantec Corp., 371
Symbolic intellectual property, 298

T

T-accounts
accounts receivable, 89
adjusting entries, 87
allowance for uncollectible accounts, 351, 353
converting from cash to accrual accounting, 81
debits and credits, 50
income tax accounting, 945–946
inventory accounts, 405
manufacturing cost flows, 405
revenue recognition
over contract term, 308
upon completion, 307
salaries expense, 81
sales revenue, 81, 89
statement of cash flows, 1227n10, 1261–1262
Tainted classification, 650n4
Take-Two Interactive Software, 293
Tandem plans, 1144n22
Target Corporation, 3–5, 124, 163
case, 44, 107–108, 165, 236, 274, 337, 400, 457, 510, 568–569, 642, 714, 769–770, 832, 906, 973, 1045, 1102, 1168, 1213, 1293
comprehensive case, B-0–B-5
financial statements for 2016, 5
leasehold improvement accounting, 874
repurchase of shares, 1065
retail inventory method use, 469
revenue recognition for gift cards, 301
shareholders' investment, 1048
Tarpley, Robin L., 17n25
Tax basis, 915
Tax benefits
of leasing, 836
LIFO inventory accounting, 420–422
net operating loss and, 937–941
Tax Cuts and Jobs Act of 2017, 616, 931, 973, 1062
Tax depreciation
ACRS, 616n23
MACRS, 616–617
Taxes
income (*See* Income taxes)
leasing and, 836
payroll, 747
property as nonlease component of lease payment, 871, 872
sales, as liability from advance collection, 729
Tax expense, tax liability *vs.*, 909
Tax laws
changes in laws or rates, 933–935, 942
deferred tax liability and, 911–912, 916, 922, 934–935, 1177
effect on retrospective reporting of changes, 1177
financial accounting standards vs., 910
net operating loss and, 939
permanent differences and (*See* Permanent differences)
temporary differences and (*See* Temporary differences)
Tax liability
deferred, 911
tax expense *vs.*, 909
Technological feasibility, 544
Technology
feasibility criterion, 544–545
inventory accounting and, 406–407
software development costs, 544–545
Temporary accounts
closing, 76
defined, 51
income statement, 75
in worksheet, 85
Temporary differences
defined, 910
in depreciation, 947
income tax effects, 909
deferred tax assets, 919–925
deferred tax liabilities, 912–918
multiple temporary differences
permanent differences compared, 929–931
valuation allowance, 925–927
disclosures, 927
multiple, 935–937
types of, 911–912
Tesla Motors, 1151
Thiagarajan, S. R., 429n16
Thrift pension plans, 980
Timberline Resources, 130
Time-based allocation method, 574
Time-based depreciation methods, 574
accelerated methods
declining-balance m., 575–576
sum-of-the years'-digits (SYD) m, 576–577
switch from accelerated to straight-line, 577
Timeliness, 22–23
Time lines
cause of loss contingency, 739–740
for cost allocation, 572
for R&D, 542, 544
Times interest earned ratio, 134–135, 793–794
Time value of money
basic annuities, 248–254
future value of an annuity, 248–249
present value of an annuity, 250–255
capital budgeting decisions and, 530
deferred payments, 526–528
defined, 239
future value of single amount, 241–242
option-pricing theory and, 1142
present value, 28 (*See also* Present value)
of an annuity, 250–254
single cash amount, 242–245
solving for unknown values, 255–257
revenue recognition and, 295–296
simple *vs.* compound interest, 240
summary of concepts, 262–263
Time-weighting
actual conversions, 1136
exercise of options, 1125
expenditures, 537–539
increase in shares, 1135
new shares, 1120
reacquired shares, 1122
Timing
bank reconciliation and, 374
cash flow
cash inflows, 1217
income taxes and accounts payable, 125
transferred assets, 369
current liabilities, 718
deferred tax liabilities, 948
expected cash flows, 736
expected return and, 6
expense recognition, 27, 170, 912
future benefits of intangible assets, 514, 521
HTM investment, sale of, 655
income smoothing and, 685
leasing, 872, 877
performance obligations, 283
pension plans, contributions and benefits, 979, 1011
postretirement benefits, 1011
revenue recognition, 26, 123, 170, 282, 284–285, 302, 305–307, 310
accounts payable, 745
temporary differences and, 912
Toll Brothers, 283
Toys R Us, 1217
Trade discounts, 344
Trademarks, 523
acquisition costs, 515, 593–594
balance sheet classifications, 117
costs to be capitalized, 521, 522, 525
defined, 523
as intangible assets, 514, 515, 592
legal life of, 594
symbolic IP as, 298
Tradename, 523
Trade notes payable, 719
Trade receivables, 115, 343, 1222n8
Trading securities (TS)
accounting for portfolios, 654
cash inflows from, 1194
cash outflows from, 1222
compared with HTM securities,
compared with AFS securities, 656, 657, 658
decision makers' perspective, 684
defined, 649
fair value adjustment, 670
fair value option (FVO) and, 664
fair value through net income and, 670
financial statement presentation, 656, 668
investments classified as, 193, 673, 1238
portfolio of, 654
reporting approach, 656
sale of, 654–655
adjusting to fair value (2021), 653–654
recording the transaction, 655
Tranches, 1114
Transaction(s)
defined, 49
related-party, 124
Transaction analysis
accounting equation and, 53
defined, 52
illustration, 49
Transaction price, 284
delivery component, 295
determination of, 290–297
payments by seller to customer, 296–297
right of return, 293–294
seller as principle *vs.* agent, 294–295
time value of money, 295–296
variable consideration, 290–293
financing component, 295–296
Transferee, 362
Transferor, 362, 366
Transportation-in account, 410
Treadway Commission, Committee of Sponsoring Organizations (COSO), 127, 129, 340
Treasury bills
as cash equivalent, 115, 340, 1219
futures, A-5
as liquid investment, 190, 1238
options, A-5
Treasury bond futures, A-4–A-5
Treasury stock, 120
accounting for, 1065
acquired at different costs, 1066
no dividends paid on, 1073
resale of, 1066
in shareholders' equity, 1050
shares viewed as, 1063
Treasury stock method, 1125
components of "proceeds" in, 1133
earnings per share (EPS) and, 1119, 1133
options, rights, and warrants as antidilutive securities, 1128–1130
restricted stock awards, 1133
Trial balance, 61
adjusted, 54, 69, 70, 72, 76, 85
post-closing, 78
unadjusted, 54, 61, 64–68, 84, 85
Tropicana, 1182

Troubled debt restructuring, 361
debt continued with modified terms, 808–811
debt settled, 808
defined, 377, 807
when receivable continued with modified terms, 377–378
when receivable settled outright, 378–379
Trustee, 990
2/10, n/30 terms, 345–346, 427–428
Two statement approach, 73
Two-step decision process, 943
Tyco International, 1109
Tyson Foods Inc.
classes of shares, 1056
receivables management, 398–388

U

Unadjusted trial balance, 54, 61
Uncertainty, 6. *See also* Risk
Uncollectible accounts. *See also* Bad debts
allowance for, 350
balance sheet approach to estimating, 350, 352
CECL (Current Expected Credit Loss) model, 352
for allowance for uncollectible accounts, 352, 355
for credit losses on HTM investments, 651
for debt investments, 688–689
notes receivable, 360–361
for troubled debt restructuring, 377
combined approaches to, 354
direct write-off of, 350
estimating allowance for, 352–355
income statement approach to estimating, 354
Underfunded pension plans, 991
Understandability, 23
Underwriters, 785
Unearned revenue, 118, 727. *See also* Deferred revenues
advances from customers, 727
balance sheet classification, 118
as contract liability, 302
defined, 118
Unexpired prepaid expense, 66
Unilever, 1058
Union Pacific, 741
Unit LIFO, 429
United States Steel Corporation, 734
Units-of-production method (of depreciation), 577–578
Unqualified auditor's report, 128, 129
Unrealized holding gains/losses
AFT debt investments and, 656–657
debt investments, 647
defined, 649
equity investments, 669–672
financial statement presentation, 668, 673
HTM debt investments and, 647, 650
reporting changes in fair value, 802
for trading securities, 652–655
transfer at fair value, 664
Unrepatriated foreign earnings, 931
U.S. Congress
depreciation incentives offered by, 421
foreign corrupt practices act (1977), 125
in hierarchy of standard-setting authority, 10
LIFO conformity rule and, 420
oil and gas accounting and, 549
on reporting of compensation expense, 1108–1109
Sarbanes-Oxley Act (SOX), 15, 16, 176n12
Securities Act (1933), 9
Securities Exchange Act (1934), 9
Tax Cuts and Job Act (2017), 616, 929, 931, 933, 947
use of tax laws, 910, 934, 935
U.S. Copyright Office, 522
Useful life, 573. *See also* Service (useful) life
indefinite, 594
of intangible asset, 591
U.S. Department of Energy, 548–549
U.S. Patent and Trademark Office, 522
U.S. Supreme Court, 21

V

Vacation days, 724
Vaccine stockpiles, 300
Valuation allowance
deferred tax asset reduced by, 925–927
IFRS *vs.* GAAP, 927
for net operating loss carryforward, 939
Variable consideration, 290–293
accounting for, 292
allocating, 297
constraint on recognizing, 292–293
estimating, 291–292
licenses and, 299
sales discounts as, 345
VBO. *See* Vested benefit obligation
Verifiability, 22
VeriFone Systems, 953
Verizon
mark-to-market method of recognizing pension gains/losses, 996
performance obligation, 279
Verizon Communications Inc., 597
Vertical analysis, 131
Vested benefit obligation (VBO), 973, 974, 983–984
Vesting
cliff vesting, 1114
graded vesting, 1114–1115
requirements for, 984
vesting date, 1106
Vickrey, Don, 1072n20
Virco Mfg. Corporation, 362
VISA
factoring by, 363
initial public offering, 1054
Vodafone Group, Plc., 157–158, 742
Volatility value, 1143
Volkswagen, 541
Voluntary conversion, 798
Voluntary deductions, 747
Voting rights of shareholders, 674n22, 1051, 1057, 1063

W

Walgreens Boots Alliance, Inc.
disclosure of cash equivalents, 340
disclosure of restricted cash, 342
restricted cash disclosure, 342
supplement LIFO reserve disclosures, 437
Walmart Stores, Inc., 160–161
interest capitalization disclosure, 540
main source of revenue for, 168
profitability analysis illustration, 201–203
property leases, 904
as publicly held corporation, 1054
retail inventory method use, 469
shareholders' equity on balance sheet, 1051–1052
statement of shareholders' equity, 1051, 1053
stockholders' equity, 1048
Walt Disney Company
disclosure note on capitalized interest, 513
other intangible assets, 514–515
self-constructed assets, 535
WAM (Amadeus), 680
Wancyzk, Mavis, 239
Warranties
contract performance obligations and, 289
extended warranty contracts, 737
manufacturer's quality-assurance, 735–736
Warrants
bonds with detachable, 798–800
stock (*See* Stock warrants)
Watered shares, 1057–1058
Watts, R. L., 1173n2, 1174n5
Weighted-average accumulated expenditures, 537
Weighted-average cost of inventory, 414
Weighted-average interest method, 537–540
Weighted-average number of shares outstanding, 1120
Wells Fargo & Company, 360–361, 1057
Weyerhaeuser Company
depreciation disclosure notes, 571
units-of-production method for depletion, 614
Winn-Dixie, 1109
Winters, Alan J., 1072n20
Work-in-process inventory, 404, 281–282
Working capital, 131–132
Worksheets, 84–85
WorldCom, Inc., 15, 516, 1109
Wright Medical Group, 1084
Write-downs of assets
assets held for sale, 608–609
assets held and used, 600, 604
inventory, 459–460
adjusting cost to net realizable value, 462–462
as big bath accounting, 485
lower of cost or net realizable value (LCNRV), 463–464
lower of cost or market (LCM), 465
obsolete inventory, 175, 427n14
larger-than-necessary, 13
property, plant, and equipment, 600, 605
recovery of, 691
Write-offs of uncollectible accounts, 362
W. T. Grant, 1217
Wysocki, P., 12n17

X

Xerox Corporation, 15, 1074

Y

Yamamura, Jeanne, 1256n18
Yelp, 176

Z

Zega, Charyl A., 1217n2
Zero balances of temporary accounts, 76
Zero-coupon bonds, 780
Zimmerman, J. L., 1173n2, 1174n5
Zmijewski, M., 1174n5
Zoo Doo Compost Company, 762
Zuckerberg, Mark, 1108

Accounting Standards Index

- Accounting Principles Board (APB)
  - *Interpretations*, 9
  - *Opinions*, 9
    - No. 21: "Interest on Receivables and Payables," 246n5
    - No. 22: "Disclosure of Accounting Policies," 123n6
    - No. 25: "Accounting for Stock Issued to Employees," 1108n5
    - No. 28: "Interim Financial Reporting," 206n40
  - *Statements*, 9
- Accounting Standards Codification System (ASC) of FASB, 11
  - No. 105-10: Generally Accepted Accounting Principles—Overall, 10n14
  - No. 205: Presentation of Financial Statements—Discontinued Operations
    - 20-15-2: A discontinued operation is also defined as business or nonprofit activity that is considered held for sale when acquired, 177n15
    - 20-20: Glossary, 178n16
    - 20-45-10: 22 Assets and liabilities held for sale are not offset and presented as a single net amount, but instead are listed separately, 181n22
    - 20-50: Disclosure, 181n21
    - 40: XXX, 129n13
  - No. 210-10-S99-2: SAB Topic 6.H—Balance Sheet—Overall—SEC Materials, 342n4
  - No. 220: Comprehensive Income
    - 10-45-1: Overall—Other Presentation Matters, 73n5, 185n27
    - 10-55: Overall—Implementation Guidance and Illustrations, A-12n13
    - 10-55 (para 24-27): Overall—Implementation Guidance and Illustrations—Case B, 658n11
    - 45: Other Presentation Matters, 5n5
  - No. 230: Statement of Cash Flows—Overall
    - 10-40-29: XXX, 1254n16
    - 10-45: Other Presentation Matters, 189n31
    - 10-50-6: Disclosure—Noncash Investing and Financing Activities, 1224n9
  - No. 235-10-50: Notes to Financial Statements—Overall—Disclosure, 123n6
  - No. 250: Accounting Changes and Error Corrections, 1171n1
    - 10-45: Overall—Other Presentation Matters, 182n23, 183n26, 1174n6, 1183n11
    - 10-S99-1: Overall—SEC Materials, SAB Topic 1.M: Assessing Materiality, 125n11
  - No. 260: Earnings per Share, 1119n17
  - No. 270-10-50: Interim Reporting—Overall—Disclosure, 206n40
  - No. 280: Segment Reporting, 23n29
    - 10-50-1: Overall—Disclosure, 138n18
    - 10-50-20 through 26: Overall—Disclosure, 138n19
    - 10-50-32 Overall—Disclosure, 138n19
    - 10-50-41: Overall—Disclosure, 139n21
  - No. 310: Receivables
    - 10-35-7: Overall—Subsequent Measurement—Losses from Uncollectible Receivables, 734n20
    - 40-15: Troubled Debt Restructurings by Creditors—Scope and Scope Exceptions, 377n29
  - No. 320: Investments—Debt and Equity Securities
    - 10-25: Overall—Recognition, 650n2, 650n4
    - 10-40: Overall—Derecognition, 655n6, 659n12
    - 10-45-13: Overall—Other Presentation Matters, 668n15
  - No. 321 Investments—Equity Securities
    - 10-35: Overall—General, 670n16, 674n20
    - 10-40: Overall—Derecognition, 671n17, 672n18
  - No. 323: Investments—Equity Method and Joint Ventures—Overall
    - 10-15-10: Scope and Scope Exceptions, 675n23
    - 10-35-31: Decrease in Investment Value, 580n26
    - 10-35-33: Increase in Level of Ownership or Degree of Influence, 580n27
    - 10-35-36: Subsequent Measurement—Decrease in Level of Ownership or Degree of Influence, 1179n7
  - No. 330: Inventory—Overall
    - 10-30: Initial Measurement, 410n3
    - 10-35-1A through 1C: Subsequent Measurement, 460n1
    - 10-35-8: Subsequent Measurement, 461n4
    - 10-S35-1: Subsequent Measurement, 462n5
  - No. 350: Intangibles—Goodwill and Other
    - 20-35: Goodwill—Subsequent Measurement, 522n7, 604, 677n24
    - 30-35: General Intangibles Other Than Goodwill—Subsequent Measurement—Determining the Useful Life of an Intangible Asset, 522n7, 593n13
    - 40-25: Internal-Use Software—Recognition, 545n27
  - No. 360: Property, Plant, and Equipment
    - 10-35-4 Overall—Subsequent Measurement—Depreciation, 574n1
    - 10-35-15 through 25: Overall—Subsequent Measurement—Impairment or Disposal of Long-Lived Assets, 600n16, 600n17
    - 10-45-5: Overall—Other Presentation Matters, 170n2
    - 10-45-9: Overall—Other Presentation Matters—Long-Lived Assets Classified as Held for Sale, 180n20
    - 10-45-9: Overall—Other Presentation Matters—Impairment or Disposal of Long-Lived Assets, 585n7
  - No. 410: Asset Retirement and Environmental Obligations
    - 20-25: Asset Retirement Obligations—Recognition, 519n4
    - 30-35-12: Environmental Obligations—Subsequent Measurement, 742n27
  - No. 420-10-20: Exit or Disposal Cost Obligations—Overall—Glossary, 174n8
  - No. 450: Contingencies
    - 10: Loss Contingencies, 733n18
    - 20-30: Loss Contingencies—Initial Measurement, 733n19
    - 20-50: Loss Contingencies—Disclosure, 740n24
    - 30-50: Gain Contingencies—Disclosure, 742n29
  - No. 460-10-25: Guarantees—Overall—Recognition, 743n30
  - No. 470: Debt
    - 10-45: Overall—Other Presentation Matters, 730n15, 730n16
    - 20-25: Debt with Conversion Options—Recognition, 796n13, 796n14
    - 20-40: Debt with Conversion Options—Derecognition, 798n16
  - No. 480-10-25-4: Distinguishing Liabilities from Equity—Overall—Recognition, 1057n6
  - No. 505: Equity
    - 10-50: Overall—Disclosure, 1051n3, 1051n4
    - 20: Stock Dividends and Stock Splits, 1071n18
    - 20-25-2: Recognition, 1072n22
    - 20-30-3: Initial Measurement, 1072n19
    - 30-30-8: "a corporation can always capitalize or allocate retained earnings for such purposes," 1065
    - 50-30-6: Equity-Based Payments to Non-Employees, 1059n9
  - No. 605: Revenue Recognition
    - 10-S99: Overall—SEC Materials, 277n2, 300n14
    - 45-50-2: Principal Agent Considerations—Disclosure—Shipping and Handling Fees and Costs, 410n5
  - No. 606: Revenue from Contracts with Customers, 864

Accounting Standards Codification System (ASC) of FASB—*Cont.*
10-05-4: Overall—Overview and Background—General, 278n*
10-25-18B: Overall–Identifying Performance Obligations, 291n12
10-32: Overall—Measurement—The Existence of a Significant Financing Component, 344n6, 348n9; 727n13
10-55-59-62: Overall–Implementation Guidance and Illustrations—Determining the Nature of the Entity's Promise, 298n13
No. 710-10-25: Compensation—General—Overall—Recognition, 725n10
No. 715: Compensation—Retirement Benefits
20-50: Defined Benefit Plans—General—Disclosure, 1005n20
30-25: Defined Benefit Plans—Pension—Recognition, 992n10
30-35-17: Defined Benefit Plans—Pension—Subsequent Measurement—Prior Service Costs, 1020n29
30-35-25: Defined Benefit Plans—Pension—Subsequent Measurement—Gains and Losses, 996
30–35: Defined Benefit Plans—Pension—Subsequent Measurement, 1006n21
60: Defined Benefit Plans—Postretirement, 968, 1009n27
60-50: Compensation–Retirement Benefits—Defined Benefit Plans—Other Postretirement—Disclosure, 1016n28
No. 718: Compensation—Stock Compensation, 1109n7
10-35-3: Overall—Subsequent Measurement—Recognition of Compensation Costs over the Requisite Service Period, 1111n11
10-55-30: Overall—Implementation and Guidance Illustrations, 1110n10
10-55-30 to 32: Overall—Implementation and Guidance Illustrations—Selecting or Estimating the Expected Term, 1110n9
50-25-1 and 2: Employee Share Purchase Plans—Recognition, 1117n16
No. 720-15-25-1: Other Expenses—Start-Up Costs—Recognition, 547n29
No. 730-10: Research and Development—Overall
20: Glossary, 541n22
25-1: Recognition, 541n21
55: Implementation Guidance, 542n23
No. 740: Income Taxes, 733, 943n9, 964, 967, 972
10: Overall, 952
10-25: Overall—Recognition, 733n18, 912, 925n3
30: Initial Measurement, 925n4
35: Subsequent Measurement, 933n7
No. 805: Business Combinations, 524n8, 546n28
20-25: Identifiable Assets and Liabilities, and Any Noncontrolling Interest—Recognition, 740n20
No. 810: Consolidation, 1182n8, 1183n9
No. 815: Derivatives and Hedging—Overall, A-7n8, A-18n16
20-25-104: Hedging—General—Recognition—Shortcut Method, A-19n18
No. 820: Fair Value Measurements and Disclosures, 28n37
No. 825-10: Financial Instruments—Overall
25: Recognition—Fair Value Option, 30n39, 361n16, 801n17
50-1: Disclosure, A-18n17
50-10: Disclosure—Fair Value of Financial Instruments, 361n16, 792n10
50-28: Disclosure—Fair Value Option, 682n28
No. 830: Foreign Currency Matters, A-17n14
No. 835: Interest
20-25: Capitalization of Interest—Recognition, 536n19
30: Imputation of Interest, 246n5, 784n5
30-15-3: Imputation of Interest—Scope and Scope Exceptions, 719n4
No. 840: Leases, 1183n9
No. 842: Leases, 838, 838n3, 848, 850, 852, 869, 833n, 835n
10-15-42A, Overall—Scope and Scope Exceptions—Lessor, 872n22
No. 845: Nonmonetary Transactions, 532n16
10-25-6: Overall—Recognition, 537n
No. 850-10-50: Related Party Disclosures—Overall—Disclosure, 124n8
No. 855: Subsequent Events, 124n7, 739n23
No. 860: Transfers and Servicing, 364n20, 367n*, 369, 369n24
10-40: Overall—Derecognition, 367n21, 368n22
10-55: Overall—Implementation Guidance and Illustration, 367n*
No. 932: Extractive Activities–Oil and Gas, 549n30
No. 958-605-15-2 and 25-2: Not-for-Profit Entities—Revenue Recognition—Scope and Scope Exceptions—Contributions Received, 529n12
No. 985: Software
20-25: Costs of Software to be Sold, Leased, or Marketed—Recognition, 544n25, 545n26
American Institute of Certified Public Accountants (AICPA)
Accounting Research Bulletins, 9
No. 43: "Restatement and Revision of Accounting Research Bulletins," 574n1, 1074n*
Professional Standards AU 240
"Consideration of Fraud in a Financial Statement Audit," 125n9
Professional Standards AU 250
"Consideration of Laws and Regulations in an Audit of Financial Statements," 125n10
Professional Standards AU 570
"The Auditor's Consideration of an Entity's Ability to Continue as a Going Concern," 129n13
Committee on Accounting Procedure (CAP)
Accounting Research Bulletins, 9
Financial Accounting Standards Board (FASB)
Accounting Standards Updates, 10,
No. 2011-03: "Transfers and Servicing (Topic 860): Reconsideration of Effective Control for Repurchase Agreements," 367n*
No. 2011-05: "Comprehensive Income (Topic 220): Presentation of Comprehensive Income," 5n5, 73n5
No. 2014-02: "Intangibles—Goodwill and Other (Topic 350): Accounting for Goodwill," 607n20
No. 2014-03: "Derivatives and Hedging (Topic 815): Accounting for Certain Receive-Variable, Pay-Fixed Interest Rate Swaps—Simplified Hedge Accounting Approach (a consensus of the Private Company Council)" A-22n21
No. 2014-09: "Revenue from Contracts with Customers (Topic 606)," 26, 26n33, 278, 278n*, 278n3, 287n9, 288, 305, 337, 344n6, 346n7
No. 2014-12: "Compensation—Stock Compensation (Topic 718): Accounting for Share-Based Payments When the Terms of an Award Provide That a Performance Target Could Be Achieved after the Requisite Service Period," 1116n14
No. 2015-01: "Income Statement—Extraordinary and Unusual Items," 174n7, 946n12
No. 2015-03: "Interest–Imputation of Interest (Subtopic 835-30) Simplifying the Presentation of Debt Issuance Costs," 785n5
No. 2015-11: "Inventory-Simplifying the Measurement of Inventory," 460n2
No. 2015-17: "Balance Sheet Classification of Deferred Taxes," 941n8
No. 2016-1: "Recognition and Measurement of Financial Assets and Financial Liabilities," 674n20
No. 2016-02: Leases, 845n9, 906
No. 2016-10: "Identifying Performance Obligations and Licensing," 291n12, 298, 298n13
No. 2016-13: "Financial Instruments—Credit Losses," 352n12, 355n13
No. 2017-07, "Compensation—Retirement Benefits (Topic 715): Improving the Presentation of Net Periodic Pension Cost and Net Periodic Postretirement Benefit Cost," 1000n15

No. 2017-12: "Derivatives and Hedging (Topic 815): Targeted Improvements to Accounting for Hedging Activities," A-7n8, A-12n13
No. 2017-14: XXX, 604n19
Emerging Issues Task Force, 10
Emerging Issues Task Force (EITF) Issue Consensuses, 10
Interpretations, 10
No. 14: "Reasonable Estimation of the Amount of the Loss," 733n19
No. 45: "Guarantor's Accounting and Disclosure Requirements for Guarantees, Including Indirect Guarantees of Indebtedness of Others," 743n30
No. 48: "Accounting for Uncertainty in Income Taxes," 733n18, 943n9, 952
Statements of Financial Accounting Concepts (SFAC) of FASB, 18–19
No. 1: "Objectives of Financial Reporting by Business Enterprises," 1217, 1217n1, 1220n5
No. 2: "Qualitative Characteristics of Accounting Information," 1217n2, 1220n5
No. 5: "Recognition and Measurement in Financial Statements of Business Enterprises," 19, 26, 25n32, 25–28,
No. 6: "Elements of Financial Statements," 18–19, 23, 23n30, 27n34, 27n35, 277n1, 288, 718n1, 726n12, 741n26, 763
No. 7: "Using Cash Flow Information and Present Value in Accounting Measurements," 18–19, 28, 28n36, 519n5, 736, 736n21, 760
No. 8: "Conceptual Framework for Financial Reporting—Chapter 1, The Objective of General Purpose Financial Reporting and Chapter 3, Qualitative Characteristics of Useful Financial Information," 18–19, 22, 26, 1217n1, 1220n5
Proposed Statement of Financial Accounting Concepts: Chapter 8: Notes to Financial Statements, 31, 31n41
Statements of Financial Accounting Standards (SFAS) of FASB, 10
No. 5: "Accounting for Contingencies," 733n18, 738n22
No. 16: "Prior Period Adjustments," 183n26
No. 19: "Financial Accounting and Reporting by Oil and Gas Producing Companies," 549n30
No. 57: "Related Party Disclosures," 124n8
No. 94: "Consolidation of All Majority-Owned Subsidiaries," 1183n9
No. 95: "Statement of Cash Flows," 189n31
No. 106: "Employers' Accounting for Postretirement Benefits Other Than Pensions," 968
No. 107: "Disclosures About Fair Value of Financial Instruments," 361n16, A-11n
No. 109: "Accounting for Income Taxes," 925n4, 948n17, 933n7, 964, 967
No. 115: "Accounting for Certain Investments in Debt and Equity Securities," 650n4, 668n15
No. 130: "Reporting Comprehensive Income," 185n27
No. 131: "Disclosures about Segments of an Enterprise and Related Information," 23n29, 138n18, 138n19, 139n21
No. 133: "Accounting for Derivative Instruments and Hedging Activities," A-19n18
No. 140: "Accounting for Transfers and Servicing of Financial Assets and Extinguishments of Liabilities," 367n21, 369n24
No. 142: "Goodwill and Other Intangible Assets," 522n7
No. 144: "Accounting for the Impairment of Long-Lived Assets and for Long-Lived Assets to Be Disposed Of," 170n2, 180n20
No. 146: "Accounting for Costs Associated with Exit or Disposal Activities," 174n8
No. 154: "Accounting Changes and Error Corrections—A Replacement of APB Opinion No. 20 and FASB Statement No. 3," 182n23
No. 157: "Fair Value Measurements," 28n37
No. 159: "The Fair Value Option for Financial Assets and Financial Liabilities," 30n39, 361n17
No. 165: "Subsequent Events," 124n7, 739n23
No. 166: "Accounting for Transfers of Financial Assets—An Amendment of FASB Statement No. 140," 364n30, 367n21, 368n22, 369n24
No. 168: "The FASB Accounting Standards Codification and the Hierarchy of Generally Accepted Accounting Principles—A Replacement of FASB Statement No. 162," 10n14
Technical Bulletins, 10
International Accounting Standards Board (IASB)
International Accounting Standards (IAS), 11
No. 1: "Financial Statement Presentation," 120, 120n4, 731n17
No. 2: "Inventories," 420, 420n9, 463, 463n6, 470n8,4
No. 7: "Statement of Cash Flows," 195, 195n33, 342, 342n5, 1245
No. 8: "Accounting Policies, Changes in Accounting Estimates and Errors," 1192, 1192n15,
No. 12: "Income Tax," 931, 931n5, 974
No. 16: "Property, Plant, and Equipment," 186, 186n29, 580, 580n4, 587, 587n9, 588n10
No. 19: "Employee Benefits," 998, 999n*, 1001, 1001n16, 1007
No. 20: "Government Grants," 529, 529n13, 548
No. 28: "Investments in Associates," 682, 682n29, 712
No. 32: "Financial Statements: Presentation," 797, 797n15, 1058, 1058n7
No. 34: "Interim Financial Reporting," 206, 206n39
No. 36: "Impairment of Assets," 603, 603n18, 607, 612
No. 37: "Provisions, Contingent Liabilities and Contingent Assets," 741, 742n28
No. 38: "Intangible Assets," 186n30, 543, 543n24, 548, 594, 594n14, 614n22
No. 39: "Financial Instruments: Recognition and Measurement," 664
No. 41: "Agriculture," 590n12
International Financial Reporting Standards (IFRS), 11–12
No. 5: "Noncurrent Assets Held for Sale and Discontinued Operations," 178n18
No. 8: "Operating Segments," 139, 139n20
No. 9: "Financial Instruments," 361, 361n18, 369, 369n23, 663–665, 663n14, 673, 673n19, 693, 695, 703, 709
No. 13: "Fair Value Measurement," 29n38
No. 15: XXX, 278n4
No. 16: XXX, 361
Accounting Trends and Techniques, 120n5, 580n5, 588n11
Public Company Accounting Oversight Board (PCAOB)
Auditing Standards
2201: "An Audit of Internal Control Over Financial Reporting That Is Integrated with An Audit of Financial Statements," 16, 128n12, 340, 340n1
Securities and Exchange Commission (SEC)
Accounting Series Release
No. 148: "Amendments to Regulations S-X and Related Interpretations and Guidelines Regarding the Disclosure of Compensating Balances and Short-Term Borrowing Arrangements," 342n4
Financial Reporting Releases, 9n10
Staff Accounting Bulletins, 9n10
No. 99: "Materiality," 125n11
No. 100: "Restructuring and Impairment Charges," 462n5
No. 101: "Revenue Recognition in Financial Statements," 277n2, 287n9, 300n14
No. 104: "Revenue Recognition," 277n2, 287n9, 300n14

Present and Future Value Tables

This table shows the future value of $1 at various interest rates (i) and time periods (n). It is used to calculate the future value of any single amount.

TABLE 1 Future Value of $1

$FV = \$1(1 + i)^n$

n/i	1.0%	1.5%	2.0%	2.5%	3.0%	3.5%	4.0%	4.5%	5.0%	5.5%	6.0%	7.0%	8.0%	9.0%	10.0%	11.0%	12.0%	20.0%
1	1.01000	1.01500	1.02000	1.02500	1.03000	1.03500	1.04000	1.04500	1.05000	1.05500	1.06000	1.07000	1.08000	1.09000	1.10000	1.11000	1.12000	1.20000
2	1.02010	1.03022	1.04040	1.05063	1.06090	1.07123	1.08160	1.09203	1.10250	1.11303	1.12360	1.14490	1.16640	1.18810	1.21000	1.23210	1.25440	1.44000
3	1.03030	1.04568	1.06121	1.07689	1.09273	1.10872	1.12486	1.14117	1.15763	1.17424	1.19102	1.22504	1.25971	1.29503	1.33100	1.36763	1.40493	1.72800
4	1.04060	1.06136	1.08243	1.10381	1.12551	1.14752	1.16986	1.19252	1.21551	1.23882	1.26248	1.31080	1.36049	1.41158	1.46410	1.51807	1.57352	2.07360
5	1.05101	1.07728	1.10408	1.13141	1.15927	1.18769	1.21665	1.24618	1.27628	1.30696	1.33823	1.40255	1.46933	1.53862	1.61051	1.68506	1.76234	2.48832
6	1.06152	1.09344	1.12616	1.15969	1.19405	1.22926	1.26532	1.30226	1.34010	1.37884	1.41852	1.50073	1.58687	1.67710	1.77156	1.87041	1.97382	2.98598
7	1.07214	1.10984	1.14869	1.18869	1.22987	1.27228	1.31593	1.36086	1.40710	1.45468	1.50363	1.60578	1.71382	1.82804	1.94872	2.07616	2.21068	3.58318
8	1.08286	1.12649	1.17166	1.21840	1.26677	1.31681	1.36857	1.42210	1.47746	1.53469	1.59385	1.71819	1.85093	1.99256	2.14359	2.30454	2.47596	4.29982
9	1.09369	1.14339	1.19509	1.24886	1.30477	1.36290	1.42331	1.48610	1.55133	1.61909	1.68948	1.83846	1.99900	2.17189	2.35795	2.55804	2.77308	5.15978
10	1.10462	1.16054	1.21899	1.28008	1.34392	1.41060	1.48024	1.55297	1.62889	1.70814	1.79085	1.96715	2.15892	2.36736	2.59374	2.83942	3.10585	6.19174
11	1.11567	1.17795	1.24337	1.31209	1.38423	1.45997	1.53945	1.62285	1.71034	1.80209	1.89830	2.10485	2.33164	2.58043	2.85312	3.15176	3.47855	7.43008
12	1.12683	1.19562	1.26824	1.34489	1.42576	1.51107	1.60103	1.69588	1.79586	1.90121	2.01220	2.25219	2.51817	2.81266	3.13843	3.49845	3.89598	8.91610
13	1.13809	1.21355	1.29361	1.37851	1.46853	1.56396	1.66507	1.77220	1.88565	2.00577	2.13293	2.40985	2.71962	3.06580	3.45227	3.88328	4.36349	10.69932
14	1.14947	1.23176	1.31948	1.41297	1.51259	1.61869	1.73168	1.85194	1.97993	2.11609	2.26090	2.57853	2.93719	3.34173	3.79750	4.31044	4.88711	12.83918
15	1.16097	1.25023	1.34587	1.44830	1.55797	1.67535	1.80094	1.93528	2.07893	2.23248	2.39656	2.75903	3.17217	3.64248	4.17725	4.78459	5.47357	15.40702
16	1.17258	1.26899	1.37279	1.48451	1.60471	1.73399	1.87298	2.02237	2.18287	2.35526	2.54035	2.95216	3.42594	3.97031	4.59497	5.31089	6.13039	18.48843
17	1.18430	1.28802	1.40024	1.52162	1.65285	1.79468	1.94790	2.11338	2.29202	2.48480	2.69277	3.15882	3.70002	4.32763	5.05447	5.89509	6.86604	22.18611
18	1.19615	1.30734	1.42825	1.55966	1.70243	1.85749	2.02582	2.20848	2.40662	2.62147	2.85434	3.37993	3.99602	4.71712	5.55992	6.54355	7.68997	26.62333
19	1.20811	1.32695	1.45681	1.59865	1.75351	1.92250	2.10685	2.30786	2.52695	2.76565	3.02560	3.61653	4.31570	5.14166	6.11591	7.26334	8.61276	31.94800
20	1.22019	1.34686	1.48595	1.63862	1.80611	1.98979	2.19112	2.41171	2.65330	2.91776	3.20714	3.86968	4.66096	5.60441	6.72750	8.06231	9.64629	38.33760
21	1.23239	1.36706	1.51567	1.67958	1.86029	2.05943	2.27877	2.52024	2.78596	3.07823	3.39956	4.14056	5.03383	6.10881	7.40025	8.94917	10.80385	46.00512
25	1.28243	1.45095	1.64061	1.85394	2.09378	2.36324	2.66584	3.00543	3.38635	3.81339	4.29187	5.42743	6.84848	8.62308	10.83471	13.58546	17.00006	95.39622
30	1.34785	1.56308	1.81136	2.09757	2.42726	2.80679	3.24340	3.74532	4.32194	4.98395	5.74349	7.61226	10.06266	13.26768	17.44940	22.89230	29.95992	237.37631
40	1.48886	1.81402	2.20804	2.68506	3.26204	3.95926	4.80102	5.81636	7.03999	8.51331	10.28572	14.97446	21.72452	31.40942	45.25926	65.00087	93.05097	1469.77160

This table shows the present value of $1 at various interest rates (i) and time periods (n). It is used to calculate the present value of any single amount.

TABLE 2 Present Value of $1

$$PV = \frac{\$1}{(1+i)^n}$$

n/i	1.0%	1.5%	2.0%	2.5%	3.0%	3.5%	4.0%	4.5%	5.0%	5.5%	6.0%	7.0%	8.0%	9.0%	10.0%	11.0%	12.0%	20.0%
1	0.99010	0.98522	0.98039	0.97561	0.97087	0.96618	0.96154	0.95694	0.95238	0.94787	0.94340	0.93458	0.92593	0.91743	0.90909	0.90090	0.89286	0.83333
2	0.98030	0.97066	0.96117	0.95181	0.94260	0.93351	0.92456	0.91573	0.90703	0.89845	0.89000	0.87344	0.85734	0.84168	0.82645	0.81162	0.79719	0.69444
3	0.97059	0.95632	0.94232	0.92860	0.91514	0.90194	0.88900	0.87630	0.86384	0.85161	0.83962	0.81630	0.79383	0.77218	0.75131	0.73119	0.71178	0.57870
4	0.96098	0.94218	0.92385	0.90595	0.88849	0.87144	0.85480	0.83856	0.82270	0.80722	0.79209	0.76290	0.73503	0.70843	0.68301	0.65873	0.63552	0.48225
5	0.95147	0.92826	0.90573	0.88385	0.86261	0.84197	0.82193	0.80245	0.78353	0.76513	0.74726	0.71299	0.68058	0.64993	0.62092	0.59345	0.56743	0.40188
6	**0.94205**	**0.91454**	**0.88797**	**0.86230**	**0.83748**	**0.81350**	**0.79031**	**0.76790**	**0.74622**	**0.72525**	**0.70496**	**0.66634**	**0.63017**	**0.59627**	**0.56447**	**0.53464**	**0.50663**	**0.33490**
7	0.93272	0.90103	0.87056	0.84127	0.81309	0.78599	0.75992	0.73483	0.71068	0.68744	0.66506	0.62275	0.58349	0.54703	0.51316	0.48166	0.45235	0.27908
8	0.92348	0.88771	0.85349	0.82075	0.78941	0.75941	0.73069	0.70319	0.67684	0.65160	0.62741	0.58201	0.54027	0.50187	0.46651	0.43393	0.40388	0.23257
9	0.91434	0.87459	0.83676	0.80073	0.76642	0.73373	0.70259	0.67290	0.64461	0.61763	0.59190	0.54393	0.50025	0.46043	0.42410	0.39092	0.36061	0.19381
10	0.90529	0.86167	0.82035	0.78120	0.74409	0.70892	0.67556	0.64393	0.61391	0.58543	0.55839	0.50835	0.46319	0.42241	0.38554	0.35218	0.32197	0.16151
11	**0.89632**	**0.84893**	**0.80426**	**0.76214**	**0.72242**	**0.68495**	**0.64958**	**0.61620**	**0.58468**	**0.55491**	**0.52679**	**0.47509**	**0.42888**	**0.38753**	**0.35049**	**0.31728**	**0.28748**	**0.13459**
12	0.88745	0.83639	0.78849	0.74356	0.70138	0.66178	0.62460	0.58966	0.55684	0.52598	0.49697	0.44401	0.39711	0.35553	0.31863	0.28584	0.25668	0.11216
13	0.87866	0.82403	0.77303	0.72542	0.68095	0.63940	0.60057	0.56427	0.53032	0.49856	0.46884	0.41496	0.36770	0.32618	0.28966	0.25751	0.22917	0.09346
14	0.86996	0.81185	0.75788	0.70773	0.66112	0.61778	0.57748	0.53997	0.50507	0.47257	0.44230	0.38782	0.34046	0.29925	0.26333	0.23199	0.20462	0.07789
15	0.86135	0.79985	0.74301	0.69047	0.64186	0.59689	0.55526	0.51672	0.48102	0.44793	0.41727	0.36245	0.31524	0.27454	0.23939	0.20900	0.18270	0.06491
16	**0.85282**	**0.78803**	**0.72845**	**0.67362**	**0.62317**	**0.57671**	**0.53391**	**0.49447**	**0.45811**	**0.42458**	**0.39365**	**0.33873**	**0.29189**	**0.25187**	**0.21763**	**0.18829**	**0.16312**	**0.05409**
17	0.84438	0.77639	0.71416	0.65720	0.60502	0.55720	0.51337	0.47318	0.43630	0.40245	0.37136	0.31657	0.27027	0.23107	0.19784	0.16963	0.14564	0.04507
18	0.83602	0.76491	0.70016	0.64117	0.58739	0.53836	0.49363	0.45280	0.41552	0.38147	0.35034	0.29586	0.25025	0.21199	0.17986	0.15282	0.13004	0.03756
19	0.82774	0.75361	0.68643	0.62553	0.57029	0.52016	0.47464	0.43330	0.39573	0.36158	0.33051	0.27651	0.23171	0.19449	0.16351	0.13768	0.11611	0.03130
20	0.81954	0.74247	0.67297	0.61027	0.55368	0.50257	0.45639	0.41464	0.37689	0.34273	0.31180	0.25842	0.21455	0.17843	0.14864	0.12403	0.10367	0.02608
21	**0.81143**	**0.73150**	**0.65978**	**0.59539**	**0.53755**	**0.48557**	**0.43883**	**0.39679**	**0.35894**	**0.32486**	**0.29416**	**0.24151**	**0.19866**	**0.16370**	**0.13513**	**0.11174**	**0.09256**	**0.02174**
24	0.78757	0.69954	0.62172	0.55288	0.49193	0.43796	0.39012	0.34770	0.31007	0.27666	0.24698	0.19715	0.15770	0.12640	0.10153	0.08170	0.06588	0.01258
25	0.77977	0.68921	0.60953	0.53939	0.47761	0.42315	0.37512	0.33273	0.29530	0.26223	0.23300	0.18425	0.14602	0.11597	0.09230	0.07361	0.05882	0.01048
28	0.75684	0.65910	0.57437	0.50088	0.43708	0.38165	0.33348	0.29157	0.25509	0.22332	0.19563	0.15040	0.11591	0.08955	0.06934	0.05382	0.04187	0.00607
29	0.74934	0.64936	0.56311	0.48866	0.42435	0.36875	0.32065	0.27902	0.24295	0.21168	0.18456	0.14056	0.10733	0.08215	0.06304	0.04849	0.03738	0.00506
30	**0.74192**	**0.63976**	**0.55207**	**0.47674**	**0.41199**	**0.35628**	**0.30832**	**0.26700**	**0.23138**	**0.20064**	**0.17411**	**0.13137**	**0.09938**	**0.07537**	**0.05731**	**0.04368**	**0.03338**	**0.00421**
31	0.73458	0.63031	0.54125	0.46511	0.39999	0.34423	0.29646	0.25550	0.22036	0.19018	0.16425	0.12277	0.09202	0.06915	0.05210	0.03935	0.02980	0.00351
40	0.67165	0.55126	0.45289	0.37243	0.30656	0.25257	0.20829	0.17193	0.14205	0.11746	0.09722	0.06678	0.04603	0.03184	0.02209	0.01538	0.01075	0.00068

This table shows the future value of an ordinary annuity of $1 at various interest rates (i) and time periods (n). It is used to calculate the future value of any series of equal payments made at the *end* of each compounding period.

TABLE 3 Future Value of an Ordinary Annuity of $1

$$FVA = \frac{(1+i)^n - 1}{i}$$

n/i	1.0%	1.5%	2.0%	2.5%	3.0%	3.5%	4.0%	4.5%	5.0%	5.5%	6.0%	7.0%	8.0%	9.0%	10.0%	11.0%	12.0%	20.0%
1	1.0000	1.0000	1.0000	1.0000	1.0000	1.0000	1.0000	1.0000	1.0000	1.0000	1.0000	1.0000	1.0000	1.0000	1.0000	1.0000	1.0000	1.0000
2	2.0100	2.0150	2.0200	2.0250	2.0300	2.0350	2.0400	2.0450	2.0500	2.0550	2.0600	2.0700	2.0800	2.0900	2.1000	2.1100	2.1200	2.2000
3	3.0301	3.0452	3.0604	3.0756	3.0909	3.1062	3.1216	3.1370	3.1525	3.1680	3.1836	3.2149	3.2464	3.2781	3.3100	3.3421	3.3744	3.6400
4	4.0604	4.0909	4.1216	4.1525	4.1836	4.2149	4.2465	4.2782	4.3101	4.3423	4.3746	4.4399	4.5061	4.5731	4.6410	4.7097	4.7793	5.3680
5	5.1010	5.1523	5.2040	5.2563	5.3091	5.3625	5.4163	5.4707	5.5256	5.5811	5.6371	5.7507	5.8666	5.9847	6.1051	6.2278	6.3528	7.4416
6	6.1520	6.2296	6.3081	6.3877	6.4684	6.5502	6.6330	6.7169	6.8019	6.8881	6.9753	7.1533	7.3359	7.5233	7.7156	7.9129	8.1152	9.9299
7	7.2135	7.3230	7.4343	7.5474	7.6625	7.7794	7.8983	8.0192	8.1420	8.2669	8.3938	8.6540	8.9228	9.2004	9.4872	9.7833	10.0890	12.9159
8	8.2857	8.4328	8.5830	8.7361	8.8923	9.0517	9.2142	9.3800	9.5491	9.7216	9.8975	10.2598	10.6366	11.0285	11.4359	11.8594	12.2997	16.4991
9	9.3685	9.5593	9.7546	9.9545	10.1591	10.3685	10.5828	10.8021	11.0266	11.2563	11.4913	11.9780	12.4876	13.0210	13.5795	14.1640	14.7757	20.7989
10	10.4622	10.7027	10.9497	11.2034	11.4639	11.7314	12.0061	12.2882	12.5779	12.8754	13.1808	13.8164	14.4866	15.1929	15.9374	16.7220	17.5487	25.9587
11	11.5668	11.8633	12.1687	12.4835	12.8078	13.1420	13.4864	13.8412	14.2068	14.5835	14.9716	15.7836	16.6455	17.5603	18.5312	19.5614	20.6546	32.1504
12	12.6825	13.0412	13.4121	13.7956	14.1920	14.6020	15.0258	15.4640	15.9171	16.3856	16.8699	17.8885	18.9771	20.1407	21.3843	22.7132	24.1331	39.5805
13	13.8093	14.2368	14.6803	15.1404	15.6178	16.1130	16.6268	17.1599	17.7130	18.2868	18.8821	20.1406	21.4953	22.9534	24.5227	26.2116	28.0291	48.4966
14	14.9474	15.4504	15.9739	16.5190	17.0863	17.6770	18.2919	18.9321	19.5986	20.2926	21.0151	22.5505	24.2149	26.0192	27.9750	30.0949	32.3926	59.1959
15	16.0969	16.6821	17.2934	17.9319	18.5989	19.2957	20.0236	20.7841	21.5786	22.4087	23.2760	25.1290	27.1521	29.3609	31.7725	34.4054	37.2797	72.0351
16	17.2579	17.9324	18.6393	19.3802	20.1569	20.9710	21.8245	22.7193	23.6575	24.6411	25.6725	27.8881	30.3243	33.0034	35.9497	39.1899	42.7533	87.4421
17	18.4304	19.2014	20.0121	20.8647	21.7616	22.7050	23.6975	24.7417	25.8404	26.9964	28.2129	30.8402	33.7502	36.9737	40.5447	44.5008	48.8837	105.9306
18	19.6147	20.4894	21.4123	22.3863	23.4144	24.4997	25.6454	26.8551	28.1324	29.4812	30.9057	33.9990	37.4502	41.3013	45.5992	50.3959	55.7497	128.1167
19	20.8109	21.7967	22.8406	23.9460	25.1169	26.3572	27.6712	29.0636	30.5390	32.1027	33.7600	37.3790	41.4463	46.0185	51.1591	56.9395	63.4397	154.7400
20	22.0190	23.1237	24.2974	25.5447	26.8704	28.2797	29.7781	31.3714	33.0660	34.8683	36.7856	40.9955	45.7620	51.1601	57.2750	64.2028	72.0524	186.6880
21	23.2392	24.4705	25.7833	27.1833	28.6765	30.2695	31.9692	33.7831	35.7193	37.7861	39.9927	44.8652	50.4229	56.7645	64.0025	72.2651	81.6987	225.0256
30	34.7849	37.5387	40.5681	43.9027	47.5754	51.6227	56.0849	61.0071	66.4388	72.4355	79.0582	94.4608	113.2832	136.3075	164.4940	199.0209	241.3327	1181.8816
40	48.8864	54.2679	60.4020	67.4026	75.4013	84.5503	95.0255	107.0303	120.7998	136.6056	154.7620	199.6351	259.0565	337.8824	442.5926	581.8261	767.0914	7343.8578

This table shows the present value of an oridinary annuity due of $1 at various interest rates (i) and time periods (n). It is used to calculate the present value of any series of equal payments made at the *end* of each compounding period.

TABLE 4 Present Value of an Ordinary Annuity of $1

$$PVA = \frac{1 - \frac{1}{(1+i)^n}}{i}$$

n/i	1.0%	1.5%	2.0%	2.5%	3.0%	3.5%	4.0%	4.5%	5.0%	5.5%	6.0%	7.0%	8.0%	9.0%	10.0%	11.0%	12.0%	20.0%
1	0.99010	0.98522	0.98039	0.97561	0.97087	0.96618	0.96154	0.95694	0.95238	0.94787	0.94340	0.93458	0.92593	0.91743	0.90909	0.90090	0.89286	0.83333
2	1.97040	1.95588	1.94156	1.92742	1.91347	1.89969	1.88609	1.87267	1.85941	1.84632	1.83339	1.80802	1.78326	1.75911	1.73554	1.71252	1.69005	1.52778
3	2.94099	2.91220	2.88388	2.85602	2.82861	2.80164	2.77509	2.74896	2.72325	2.69793	2.67301	2.62432	2.57710	2.53129	2.48685	2.44371	2.40183	2.10648
4	3.90197	3.85438	3.80773	3.76197	3.71710	3.67308	3.62990	3.58753	3.54595	3.50515	3.46511	3.38721	3.31213	3.23972	3.16987	3.10245	3.03735	2.58873
5	4.85343	4.78264	4.71346	4.64583	4.57971	4.51505	4.45182	4.38998	4.32948	4.27028	4.21236	4.10020	3.99271	3.88965	3.79079	3.69590	3.60478	2.99061
6	**5.79548**	**5.69719**	**5.60143**	**5.50813**	**5.41719**	**5.32855**	**5.24214**	**5.15787**	**5.07569**	**4.99553**	**4.91732**	**4.76654**	**4.62288**	**4.48592**	**4.35526**	**4.23054**	**4.11141**	**3.32551**
7	6.72819	6.59821	6.47199	6.34939	6.23028	6.11454	6.00205	5.89270	5.78637	5.68297	5.58238	5.38929	5.20637	5.03295	4.86842	4.71220	4.56376	3.60459
8	7.65168	7.48593	7.32548	7.17014	7.01969	6.87396	6.73274	6.59589	6.46321	6.33457	6.20979	5.97130	5.74664	5.53482	5.33493	5.14612	4.96764	3.83716
9	8.56602	8.36052	8.16224	7.97087	7.78611	7.60769	7.43533	7.26879	7.10782	6.95220	6.80169	6.51523	6.24689	5.99525	5.75902	5.53705	5.32825	4.03097
10	9.47130	9.22218	8.98259	8.75206	8.53020	8.31661	8.11090	7.91272	7.72173	7.53763	7.36009	7.02358	6.71008	6.41766	6.14457	5.88923	5.65022	4.19247
11	**10.36763**	**10.07112**	**9.78685**	**9.51421**	**9.25262**	**9.00155**	**8.76048**	**8.52892**	**8.30641**	**8.09254**	**7.88687**	**7.49867**	**7.13896**	**6.80519**	**6.49506**	**6.20652**	**5.93770**	**4.32706**
12	11.25508	10.90751	10.57534	10.25776	9.95400	9.66333	9.38507	9.11858	8.86325	8.61852	8.38384	7.94269	7.53608	7.16073	6.81369	6.49236	6.19437	4.43922
13	12.13374	11.73153	11.34837	10.98319	10.63496	10.30274	9.98565	9.68285	9.39357	9.11708	8.85268	8.35765	7.90378	7.48690	7.10336	6.74987	6.42355	4.53268
14	13.00370	12.54338	12.10625	11.69091	11.29607	10.92052	10.56312	10.22283	9.89864	9.58965	9.29498	8.74547	8.24424	7.78615	7.36669	6.98187	6.62817	4.61057
15	13.86505	13.34323	12.84926	12.38138	11.93794	11.51741	11.11839	10.73955	10.37966	10.03758	9.71225	9.10791	8.55948	8.06069	7.60608	7.19087	6.81086	4.67547
16	**14.71787**	**14.13126**	**13.57771**	**13.05500**	**12.56110**	**12.09412**	**11.65230**	**11.23402**	**10.83777**	**10.46216**	**10.10590**	**9.44665**	**8.85137**	**8.31256**	**7.82371**	**7.37916**	**6.97399**	**4.72956**
17	15.56225	14.90765	14.29187	13.71220	13.16612	12.65132	12.16567	11.70719	11.27407	10.86461	10.47726	9.76322	9.12164	8.54363	8.02155	7.54879	7.11963	4.77463
18	16.39827	15.67256	14.99203	14.35336	13.75351	13.18968	12.65930	12.15999	11.68959	11.24607	10.82760	10.05909	9.37189	8.75563	8.20141	7.70162	7.24967	4.81219
19	17.22601	16.42617	15.67846	14.97889	14.32380	13.70984	13.13394	12.59329	12.08532	11.60765	11.15812	10.33560	9.60360	8.95011	8.36492	7.83929	7.36578	4.84350
20	18.04555	17.16864	16.35143	15.58916	14.87747	14.21240	13.59033	13.00794	12.46221	11.95038	11.46992	10.59401	9.81815	9.12855	8.51356	7.96333	7.46944	4.86958
21	**18.85698**	**17.90014**	**17.01121**	**16.18455**	**15.41502**	**14.69797**	**14.02916**	**13.40472**	**12.82115**	**12.27524**	**11.76408**	**10.83553**	**10.01680**	**9.29224**	**8.64869**	**8.07507**	**7.56200**	**4.89132**
25	22.02316	20.71961	19.52346	18.42438	17.41315	16.48151	15.62208	14.82821	14.09394	13.41393	12.78336	11.65358	10.67478	9.82258	9.07704	8.42174	7.84314	4.94759
30	25.80771	24.01584	22.39646	20.93029	19.60044	18.39205	17.29203	16.28889	15.37245	14.53375	13.76483	12.40904	11.25778	10.27365	9.42691	8.69379	8.05518	4.97894
40	32.83469	29.91585	27.35548	25.10278	23.11477	21.35507	19.79277	18.40158	17.15909	16.04612	15.04630	13.33171	11.92461	10.75736	9.77905	8.95105	8.24378	4.99660

This table shows the future value of an annuity due of $1 at various interest rates (i) and time periods (n). It is used to calculate the future value of any series of equal payments made at the *beginning* of each compounding period.

TABLE 5 Future Value of an Annuity Due of $1

$$FVAD = \left[\frac{(1+i)^n - 1}{i}\right] \times (1+i)$$

n/i	1.0%	1.5%	2.0%	2.5%	3.0%	3.5%	4.0%	4.5%	5.0%	5.5%	6.0%	7.0%	8.0%	9.0%	10.0%	11.0%	12.0%	20.0%
1	1.0100	1.0150	1.0200	1.0250	1.0300	1.0350	1.0400	1.0450	1.0500	1.0550	1.0600	1.0700	1.0800	1.0900	1.1000	1.1100	1.1200	1.2000
2	2.0301	2.0452	2.0604	2.0756	2.0909	2.1062	2.1216	2.1370	2.1525	2.1680	2.1836	2.2149	2.2464	2.2781	2.3100	2.3421	2.3744	2.6400
3	3.0604	3.0909	3.1216	3.1525	3.1836	3.2149	3.2465	3.2782	3.3101	3.3423	3.3746	3.4399	3.5061	3.5731	3.6410	3.7097	3.7793	4.3680
4	4.1010	4.1523	4.2040	4.2563	4.3091	4.3625	4.4163	4.4707	4.5256	4.5811	4.6371	4.7507	4.8666	4.9847	5.1051	5.2278	5.3528	6.4416
5	5.1520	5.2296	5.3081	5.3877	5.4684	5.5502	5.6330	5.7169	5.8019	5.8881	5.9753	6.1533	6.3359	6.5233	6.7156	6.9129	7.1152	8.9299
6	6.2135	6.3230	6.4343	6.5474	6.6625	6.7794	6.8983	7.0192	7.1420	7.2669	7.3938	7.6540	7.9228	8.2004	8.4872	8.7833	9.0890	11.9159
7	7.2857	7.4328	7.5830	7.7361	7.8923	8.0517	8.2142	8.3800	8.5491	8.7216	8.8975	9.2598	9.6366	10.0285	10.4359	10.8594	11.2997	15.4991
8	8.3685	8.5593	8.7546	8.9545	9.1591	9.3685	9.5828	9.8021	10.0266	10.2563	10.4913	10.9780	11.4876	12.0210	12.5795	13.1640	13.7757	19.7989
9	9.4622	9.7027	9.9497	10.2034	10.4639	10.7314	11.0061	11.2882	11.5779	11.8754	12.1808	12.8164	13.4866	14.1929	14.9374	15.7220	16.5487	24.9587
10	10.5668	10.8633	11.1687	11.4835	11.8078	12.1420	12.4864	12.8412	13.2068	13.5835	13.9716	14.7836	15.6455	16.5603	17.5312	18.5614	19.6546	31.1504
11	11.6825	12.0412	12.4121	12.7956	13.1920	13.6020	14.0258	14.4640	14.9171	15.3856	15.8699	16.8885	17.9771	19.1407	20.3843	21.7132	23.1331	38.5805
12	12.8093	13.2368	13.6803	14.1404	14.6178	15.1130	15.6268	16.1599	16.7130	17.2868	17.8821	19.1406	20.4953	21.9534	23.5227	25.2116	27.0291	47.4966
13	13.9474	14.4504	14.9739	15.5190	16.0863	16.6770	17.2919	17.9321	18.5986	19.2926	20.0151	21.5505	23.2149	25.0192	26.9750	29.0949	31.3926	58.1959
14	15.0969	15.6821	16.2934	16.9319	17.5989	18.2957	19.0236	19.7841	20.5786	21.4087	22.2760	24.1290	26.1521	28.3609	30.7725	33.4054	36.2797	71.0351
15	16.2579	16.9324	17.6393	18.3802	19.1569	19.9710	20.8245	21.7193	22.6575	23.6411	24.6725	26.8881	29.3243	32.0034	34.9497	38.1899	41.7533	86.4421
16	17.4304	18.2014	19.0121	19.8647	20.7616	21.7050	22.6975	23.7417	24.8404	25.9964	27.2129	29.8402	32.7502	35.9737	39.5447	43.5008	47.8837	104.9306
17	18.6147	19.4894	20.4123	21.3863	22.4144	23.4997	24.6454	25.8551	27.1324	28.4812	29.9057	32.9990	36.4502	40.3013	44.5992	49.3959	54.7497	127.1167
18	19.8109	20.7967	21.8406	22.9460	24.1169	25.3572	26.6712	28.0636	29.5390	31.1027	32.7600	36.3790	40.4463	45.0185	50.1591	55.9395	62.4397	153.7400
19	21.0190	22.1237	23.2974	24.5447	25.8704	27.2797	28.7781	30.3714	32.0660	33.8683	35.7856	39.9955	44.7620	50.1601	56.2750	63.2028	71.0524	185.6880
20	22.2392	23.4705	24.7833	26.1833	27.6765	29.2695	30.9692	32.7831	34.7193	36.7861	38.9927	43.8652	49.4229	55.7645	63.0025	71.2651	80.6987	224.0256
21	23.4716	24.8376	26.2990	27.8629	29.5368	31.3289	33.2480	35.3034	37.5052	39.8643	42.3923	48.0057	54.4568	61.8733	70.4027	80.2143	91.5026	270.0307
25	28.5256	30.5140	32.6709	35.0117	37.5530	40.3131	43.3117	46.5706	50.1135	53.9660	58.1564	67.6765	78.9544	92.3240	108.1818	126.9988	149.3339	566.3773
30	35.1327	38.1018	41.3794	45.0003	49.0027	53.4295	58.3283	63.7524	69.7608	76.4194	83.8017	101.0730	122.3459	148.5752	180.9434	220.9132	270.2926	1418.2579
40	49.3752	55.0819	61.6100	69.0876	77.6633	87.5095	98.8265	111.8467	126.8398	144.1189	164.0477	213.6096	279.7810	368.2919	486.8518	645.8269	859.1424	8812.6294

This table shows the present value of an annuity due of $1 at various interest rates (i) and time periods (n). It is used to calculate the present value of any series of equal payments made at the *beginning* of each compounding period.

TABLE 6 Present Value of an Annuity Due of $1

$$PVAD = \left[\frac{1 - \frac{1}{(1+i)^n}}{i}\right] \times (1+i)$$

n/i	1.0%	1.5%	2.0%	2.5%	3.0%	3.5%	4.0%	4.5%	5.0%	5.5%	6.0%	7.0%	8.0%	9.0%	10.0%	11.0%	12.0%	20.0%
1	1.00000	1.00000	1.00000	1.00000	1.00000	1.00000	1.00000	1.00000	1.00000	1.00000	1.00000	1.00000	1.00000	1.00000	1.00000	1.00000	1.00000	1.00000
2	1.99010	1.98522	1.98039	1.97561	1.97087	1.96618	1.96154	1.95694	1.95238	1.94787	1.94340	1.93458	1.92593	1.91743	1.90909	1.90090	1.89286	1.83333
3	2.97040	2.95588	2.94156	2.92742	2.91347	2.89969	2.88609	2.87267	2.85941	2.84632	2.83339	2.80802	2.78326	2.75911	2.73554	2.71252	2.69005	2.52778
4	3.94099	3.91220	3.88388	3.85602	3.82861	3.80164	3.77509	3.74896	3.72325	3.69793	3.67301	3.62432	3.57710	3.53129	3.48685	3.44371	3.40183	3.10648
5	4.90197	4.85438	4.80773	4.76197	4.71710	4.67308	4.62990	4.58753	4.54595	4.50515	4.46511	4.38721	4.31213	4.23972	4.16987	4.10245	4.03735	3.58873
6	**5.85343**	**5.78264**	**5.71346**	**5.64583**	**5.57971**	**5.51505**	**5.45182**	**5.38998**	**5.32948**	**5.27028**	**5.21236**	**5.10020**	**4.99271**	**4.88965**	**4.79079**	**4.69590**	**4.60478**	**3.99061**
7	6.79548	6.69719	6.60143	6.50813	6.41719	6.32855	6.24214	6.15787	6.07569	5.99553	5.91732	5.76654	5.62288	5.48592	5.35526	5.23054	5.11141	4.32551
8	7.72819	7.59821	7.47199	7.34939	7.23028	7.11454	7.00205	6.89270	6.78637	6.68297	6.58238	6.38929	6.20637	6.03295	5.86842	5.71220	5.56376	4.60459
9	8.65168	8.48593	8.32548	8.17014	8.01969	7.87396	7.73274	7.59589	7.46321	7.33457	7.20979	6.97130	6.74664	6.53482	6.33493	6.14612	5.96764	4.83716
10	9.56602	9.36052	9.16224	8.97087	8.78611	8.60769	8.43533	8.26879	8.10782	7.95220	7.80169	7.51523	7.24689	6.99525	6.75902	6.53705	6.32825	5.03097
11	**10.47130**	**10.22218**	**9.98259**	**9.75206**	**9.53020**	**9.31661**	**9.11090**	**8.91272**	**8.72173**	**8.53763**	**8.36009**	**8.02358**	**7.71008**	**7.41766**	**7.14457**	**6.88923**	**6.65022**	**5.19247**
12	11.36763	11.07112	10.78685	10.51421	10.25262	10.00155	9.76048	9.52892	9.30641	9.09254	8.88687	8.49867	8.13896	7.80519	7.49506	7.20652	6.93770	5.32706
13	12.25508	11.90751	11.57534	11.25776	10.95400	10.66333	10.38507	10.11858	9.86325	9.61852	9.38384	8.94269	8.53608	8.16073	7.81369	7.49236	7.19437	5.43922
14	13.13374	12.73153	12.34837	11.98318	11.63496	11.30274	10.98565	10.68285	10.39357	10.11708	9.85268	9.35765	8.90378	8.48690	8.10336	7.74987	7.42355	5.53268
15	14.00370	13.54338	13.10625	12.69091	12.29607	11.92052	11.56312	11.22283	10.89864	10.58965	10.29498	9.74547	9.24424	8.78615	8.36669	7.98187	7.62817	5.61057
16	**14.86505**	**14.34323**	**13.84926**	**13.38138**	**12.93794**	**12.51741**	**12.11839**	**11.73955**	**11.37966**	**11.03758**	**10.71225**	**10.10791**	**9.55948**	**9.06069**	**8.60608**	**8.19087**	**7.81086**	**5.67547**
17	15.71787	15.13126	14.57771	14.05500	13.56110	13.09412	12.65230	12.23402	11.83777	11.46216	11.10590	10.44665	9.85137	9.31256	8.82371	8.37916	7.97399	5.72956
18	16.56225	15.90765	15.29187	14.71220	14.16612	13.65132	13.16567	12.70719	12.27407	11.86461	11.47726	10.76322	10.12164	9.54363	9.02155	8.54879	8.11963	5.77463
19	17.39827	16.67256	15.99203	15.35336	14.75351	14.18968	13.65930	13.15999	12.68959	12.24607	11.82760	11.05909	10.37189	9.75563	9.20141	8.70162	8.24967	5.81219
20	18.22601	17.42617	16.67846	15.97889	15.32380	14.70984	14.13394	13.59329	13.08532	12.60765	12.15812	11.33560	10.60360	9.95011	9.36492	8.83929	8.36578	5.84350
21	**19.04555**	**18.16864**	**17.35143**	**16.58916**	**15.87747**	**15.21240**	**14.59033**	**14.00794**	**13.46221**	**12.95038**	**12.46992**	**11.59401**	**10.81815**	**10.12855**	**9.51356**	**8.96333**	**8.46944**	**5.86958**
25	22.24339	21.03041	19.91393	18.88499	17.93554	17.05837	16.24696	15.49548	14.79864	14.15170	13.55036	12.46933	11.52876	10.70661	9.98474	9.34814	8.78432	5.93710
30	26.06579	24.37608	22.84438	21.45355	20.18845	19.03577	17.98371	17.02189	16.14107	15.33310	14.59072	13.27767	12.15841	11.19828	10.36961	9.65011	9.02181	5.97472
40	33.16303	30.36458	27.90259	25.73034	23.80822	22.10250	20.58448	19.22966	18.01704	16.92866	15.94907	14.26493	12.87858	11.72552	10.75696	9.93567	9.23303	5.99592